ISBN 978-1-5278-9803-5
PIBN 10923090

1 MONTH OF
FREE
READING

at
www.ForgottenBooks.com

By purchasing this book you are eligible for one month membership to ForgottenBooks.com, giving you unlimited access to our entire collection of over 1,000,000 titles via our web site and mobile apps.

To claim your free month visit:
www.forgottenbooks.com/free923090

M'Culloch's Universal Gazetteer.

A

DICTIONARY,

GEOGRAPHICAL, STATISTICAL, AND HISTORICAL,

OF THE VARIOUS

COUNTRIES, PLACES, AND PRINCIPAL

NATURAL OBJECTS IN THE WORLD.

BY J. R. M'CULLOCH, ESQ.

IN WHICH THE ARTICLES RELATING TO THE UNITED STATES HAVE BEEN GREATLY
MULTIPLIED AND EXTENDED, AND ADAPTED TO THE PRESENT CONDITION
OF THE COUNTRY, AND TO THE WANTS OF ITS CITIZENS.

BY DANIEL HASKEL, A.M.,

LATE PRESIDENT OF THE UNIVERSITY OF VERMONT.

" Nec omnia dicentur sed maxime insignia "

Illustrated with Seven large Maps.

IN TWO VOLUMES.

VOL. II.

NEW-YORK:

PUBLISHED BY HARPER & BROTHERS, 82 CLIFF-STREET.

1844.

UNIVERSAL GAZETTEER,

OR

DICTIONARY,

GEOGRAPHICAL, STATISTICAL, AND HISTORICAL.

IBARRA, a town of Ecuador, Columbia, in a delightful plain, on the Taguando, at the foot of the volcano Imbaru, 50 m. N.E. Quito, and on the high road between that city and Popayan. Lat. 0° 21' N., long. 78° 16' 34" W. Pop. unknown, but formerly estimated at 12,000. It was founded in 1597, is well built, and has a large and well built church, several convents, a college, formerly belonging to the Jesuits, a hospital, and many good private residences. Without the city are some suburbs, inhabited by the Indian population. It manufactures fine cotton and other fabrics. The district of which it is the capital produces sugar and wheat of the finest quality, and a good deal of cotton, the weaving of which and other materials into stockings, caps, gloves, flags, coverlets, &c., employs many of its inhabitants. (*Thompson's Alcedo, &c.*)

IBBERVILLE, an outlet of Mississippi r., which it leaves 14 m. below Baton Rouge, and 20 m. below it is joined by and lost in Amite r. It receives water from the Mississippi only at high flood, and is of no importance to navigation until its junction with Amite river.

IBERVILLE, parish, La. Centrally situated in the S. part of the state, and contains 350 sq. m. The borders of the streams only are sufficiently elevated for cultivation, where the soil is very fertile. It contained in 1840, 4201 neat cattle, 3088 sheep, 4088 swine; and produced 209,240 bushels of Indian corn, 30,994 of potatoes, 3,553,080 pounds of cotton, 3,789,800 of sugar. It had seventeen stores; one academy, 12 students, five schools, 133 scholars. Pop.: whites, 2523; slaves, 3887; free coloured, 85; total, 6495. Capital, Plaquemine.

ICELAND, a large island under the dominion of Denmark, in the N. Atlantic ocean, on the confines of the polar circle, generally considered as belonging to Europe, but which should rather, perhaps, be reckoned in America: between lat. 63° 30' and 66° 40' N., and long. 16° and 23° W. It is of a very irregular triangular shape, and is estimated to contain about 30,000 sq. m. Pop. (1834) 56,000, supposed to be spread over about two thirds of the island, the central portion being totally uninhabited, and imperfectly explored. Iceland appears to owe its existence to submarine volcanic agency, and to have been upheaved at intervals from the bottom of the sea. It is traversed in every direction by vast ranges of mountains: the principal ridges run chiefly E. and W., and, from these, inferior mountains branch off towards the coast, often terminating in rocky and bold headlands. All the coasts, but more especially the N. and W. are deeply indented with *fiords*, similar to those of Norway. The most extensive tract of level country is in the S.E. It is estimated that about a third part of the surface is covered with vegetation of some kind, while the other two thirds are occupied by snowy mountains or fields of lava. The general aspect of the country is the most desolate and dreary imaginable. The height of very few of the mountains has been correctly ascertained, and those said to attain an elevation of 7000 ft. are not the most lofty. The Yökuls, or enormous ice-mountains, are the greatest elevations: the most extensive of these is the Klofa Yökul in the E.; it lies behind the heights which line the S.E. coast, and forms, with little or no interruption, a vast chain of ice and snow mountains covering a surface of perhaps 3000 sq. m. The W. quarter contains, among other lofty heights, the Snafel Yökul, 4,580 ft. high. In the N. the mountains are not very high; but in the E. the Oroefa Yökul, 6350 ft. in elevation, is the most lofty of which any accurate measurement has been obtained. The celebrated volcano Hecla is in the S.W. quarter,

and about 20 m. inland. It is more remarkable for the frequency and violence of its eruptions than for its elevation, which is only about 5000 ft. (*See* HECLA.)

The bays and harbours along the coast are numerous and secure, but little known or frequented: the most so are those of Eyafiords on the N., Eyrarbacka on the S., and Reikiavik on the W. coast. The rivers, which are numerous and comparatively large, have mostly a N. or S. course. Although sufficiently wide, they are generally obstructed by rocks and shallows, and are too rapid to admit of navigation. There are several large lakes, of which My-vatn lake, in the N.E., is the most considerable: it is estimated at about 40 m. in circumference, and has upwards of 30 islands composed of lava. In no country have volcanic eruptions been so numerous as in Iceland, or spread over a larger surface. Besides more than 30 volcanic mountains, there exists an immense number of small cones and craters, from which streams of melted substances have been poured forth over the surrounding regions; nine volcanoes were active during the last century, four in the N., and the rest lying nearly in a direct line along the S. coast. Twenty-three eruptions of Hecla are recorded since the occupation of the island by Europeans: the first of these occurred in 1004. The most extensive and devasting eruption ever experienced in the island happened in 1783: it proceeded from the Skaptar Yökul, a volcano (or rather volcanic tract having several cones) near the centre of the country. This eruption did not entirely cease for about two years. It destroyed no fewer than 20 villages and 9000 human beings, or more than one fifth part of the then population of the island! On the S. and W. coasts, numerous islands have been from time to time thrown up; some of which still remain, while others have receded beneath the surface of the ocean, forming dangerous rocks and shoals. The Vestmanna islands, which lie about 15 m. from the E. coast, are a group consisting almost entirely of barren vitrified rocks: only one of them is inhabited.

Tracts of lava traverse the island in almost every direction. This substance chiefly occurs in isolated streams, having apparently flowed from the mountains; but in some parts there are continuous tracts, and along the S. coast, for 160 m. inland, the lava that spread over the country have been ejected from small cones rising immediately from the surface. The ground in this part is frequently broken by fissures and chasms, some of which are more than 3 m. in length, and upwards of 100 ft. in width. Besides the common lavas, Iceland abounds in other mineral masses indicative of an igneous origin; of these the most prevalent are tufa and submarine lava, obsidian, sulphur, &c. Whole mountains of tufa exist in every part. Sir G. Mackenzie observes, that the instance of tufa excepted, he saw no marks of stratification in any rock in the island, all the substances appearing to have been subjected to a degree of heat sufficient to reduce them to fusion; and that some, if not all, the Icelandic masses, which are not the produce of external eruptions, are really submarine lavas. The rocks not bearing external marks of heat, are mostly of trap, and contain all the varieties of zeolite, chalcedony, greenstone, porphyry, slate, &c.; the celebrated double refracting calcareous spar is found chiefly on the E. coast. Basaltic columns occur in many parts, especially on the W. coast, where they form several grottos; and that of Stappen bears a great resemblance to the cave of Fingal, in the island of Staffa.

Few metals are met with: iron and copper have been found; but the mines are not wrought. The supply of sulphur is inexhaustible; large mountains are incrusted with this substance, which, when removed, is again formed in crystals by the agency of the hot steam from below. Large quantities were formerly shipped; but latterly the supplies sent to foreign markets have been comparatively small.

By far the most remarkable phenomena of Iceland are the intermitting hot springs met with in several parts, and of all degrees of temperature. The water in some of these springs is at intervals violently thrown into the air to a great height. They have thence received the name of *geysers*, from the Icelandic verb *geysa*, to rage. The most celebrated of these springs are situated in a plain, about 16 m. N. from the village of Skalholt. The great geyser, or principal fountain of this kind, rises from a tube or funnel, 78 ft. in perpendicular depth, and from 8 to 10 ft. in diameter at the bottom, but gradually widening till it terminates in a capacious basin. After an emission, the basin and funnel are empty. The jets take place at intervals of about six hours; and when the water, in a violent state of ebullition, begins to rise in the pipe or funnel, and to fill the basin, subterraneous noises are heard, like the distant roar of cannon, the earth is slightly shaken, and the agitation increases till at length a column of water is suddenly thrown up, with vast force and loud explosions, to the height of 100 or 200 ft. And playing for a time like an artificial fountain, and giving off great clouds of vapour, the funnel is emptied, and a column of steam rushing up with great violence and a thundering noise, terminates the eruption. Such is the explosive force, that large stones thrown into the funnel are instantly ejected, and sometimes shivered into small fragments. (For an explanation of this phenomenon, see *Lyell's Geology*, ii., 309, 3d ed.) Some of the hot springs, near the inhabited parts of the island, are used for economical purposes; food is dressed over them; and in some places huts are built over small fountains, to form steam baths. In other parts of the island vast cauldrons of boiling mud are seen in a constant state of activity, sending up immense columns of dense vapour, which obscure the atmosphere a great way round.

That Iceland had formerly some extensive forests is apparent from authentic records, but they no longer exist: in fact, the climate seems to be now unsuitable for the growth of trees, those that are found at present being stunted and diminutive, and little better than underwood. Vast quantities of *surturbrand*, or fossil wood, are frequently found buried at a great depth beneath the surface.

Of the wild animals, foxes are the most numerous. Reindeer, which were introduced from Norway in 1770, in the intention of being domesticated, have increased very rapidly; but they are entirely wild, and are very difficult to kill. Bears are frequently brought down from the arctic regions on masses of floating ice; they sometimes commit great devastations, but are generally destroyed almost immediately after making the land. Nearly all kinds of seafowl inhabit the coasts and islands; and plovers, curlews, snipes, and a variety of game, are found in the interior. The eider duck is very plentiful; and the down taken from the nest is an important article of export. The birds are so familiar as to build their nests all round the roofs, and even inside the huts. A severe penalty is inflicted on those who kill them. The peasantry entertain a superstitious reverence, mingled with aversion, for the seal. The coasts, rivers, and lakes produce an abundance of fine fish; and it is from the sea that the Icelanders derive great part of their subsistence. Their fisheries are prosecuted with great activity; and at Niardivik, one of the fishing stations on the E. coast of the island, there are said to be 300 boats. Cod and haddock are plentiful on the coasts: of these, as well as of the other seafish, part is salted for exportation, but by far the greater part is dried for winter provision. The herring fishery is much neglected, as well as the inland fishery on the lakes and rivers.

The climate is more variable than that of the same latitudes on the continent. Great and sudden changes of temperature often occur; and it has frequently happened that after a night of frost, the thermometer during the day has risen to 70° Fahr. The intensity of the cold is much increased by the immense quantities of floating ice, which, being drifted from the polar regions, accumulate upon the coast. Fogs are frequent; but the air, on the whole, is reckoned wholesome. Thunder is seldom heard, but storms of wind and rain are frequent; and the *aurora borealis* and other meteors are much more common and brilliant here than in countries farther to the S. The sun is visible at midnight at the summer solstice, from the hills in the N. parts of the island. There is a prevalent opinion in Iceland, that the seasons in former ages were less unfavourable; but, there is probably no good foundation for this belief. The summers are necessarily short; but Dr. Henderson states that the cold is rarely more intense than in the S. of Scandinavia, and the winter he passed in the island was as mild as any he had experienced in Denmark or Sweden.

No grain is now cultivated, though traces exist of its having been formerly raised. Agriculture is limited to the rearing of various grasses for cattle, and haymaking is consequently the most important branch of rural industry. Potatoes have been introduced with some success; and several kinds of culinary vegetables are raised, but, with the exception of red cabbage, few attain perfection. The grasses are of the sorts common in other N. climates, and keep horses and other cattle in good condition during the summer. Many of the low mountains are covered with a coarse grass, which yields pretty good summer pasturage; and the meadows and valleys through which the rivers flow produce grass in tolerable abundance, which, when the weather allows of its being harvested, is made into hay. Seaweed and moss are eagerly devoured by the cattle in winter, when other good food fails, which is often the case. In 1834 it was estimated that there were about 500,000 head of sheep; from 35,000 to 40,000 head of black cattle; and from 50,000 to 60,000 horses in Iceland: goats are kept only in the N. The number of sheep appears to be increasing; they have remarkably fine fleeces, which are not shorn, but cast off entirely in the spring. The horses are hardy and small, seldom standing more than 14 hands high. There being no carriages of any description, they are principally used for carrying burdens; and the poorest peasant has generally four or five of these animals. Rents are paid mostly in produce; on the coasts in fish, in the interior in butter, sheep, &c. Tenants who are in easy circumstances generally employ one or more labourers, who, besides board and lodging, have from 10 to 19 specie dollars a year as wages. The whole population is employed either in fishing or feeding cattle, or both; those who breed cattle being, as compared with those who live by fishing, nearly as three to one. No manufactures, of any kind, are carried on for the purpose of trade. Every branch of industry is domestic, and confined chiefly to articles of clothing, such as coarse cloth, gloves, mittens, stockings, &c. The peasantry supply themselves with such furniture as their cottages require, and some manufacture silver trinkets, and snuff-boxes, and forge implements of iron. Every man can shoe his own horse; and in this land of primitive simplicity, even the bishop and chief justice are sometimes employed in this necessary occupation! The greater part of the trade is carried on by means of barter; the quantity of money in circulation is very small, few of the peasants possessing any. The merchants receive the articles for exportation at regulated prices, according to the state of the market, and pay for them in such foreign commodities as the inhabitants may require. The peasantry of the neighbourhood assemble annually at Reikiavik and the other principal settlements, and bring down with them wool, woollen manufactured goods, butter, skins, tallow, Iceland moss (*Lichen Icelandicus*), and sometimes a few cattle. In return for these they take back coffee, sugar, tobacco, snuff, a little brandy, rye, rye bread, wheaten flour, salt, soap, &c. The better class purchase linens and cotton goods, which have latterly come more into use. Those who live near the coasts bring to market dried cod and stock fish, dried salmon, whale, shark, and seal oils, seal skins, &c. The domestic produce has of late years, been considerable, and the export of wool amounts to from 3000 to 4000 skippunds annually.

The Icelanders are of Norwegian origin; they are tall, have a frank open countenance, a florid complexion, and flaxen hair. They seldom attain to an advanced age, but the females generally live longer than the men. They are hospitable; devotedly attached to their native land; remarkably grave and serious; and, indeed, apparently phlegmatic, but extremely animated on subjects which interest them. They have retained, with few innovations, the ancient modes of life and the costume of their race. Their principal articles of food are fish, fresh and dried, bread, made of imported corn, great quantities of rancid butter, game, and, in some parts, a porridge made of Icelandic moss. They sometimes use the flesh of the shark or sea-fish, when it has become tender from putrescence. Their huts, though larger, are not unlike those of the Irish: their dampness, with the darkness, filth, and stench of the fish, render them uninhabitable by strangers. The Icelandic, or original Scandinavian tongue, has been here preserved in all its ancient purity. The Icelanders are extremely attentive to their religious and domestic duties, and display in their dealings a scrupulous integrity. Perhaps there is no country in which the lower orders are so well informed. Domestic education is universal; and there are very few among them who cannot read and write, and many among the better class would be distinguished by their taste and learning in the most cultivated society of Europe. Even many of the peasantry are well versed in the classics; and the traveller is not unfrequently attended by guides who converse with him in Latin! In winter nights it is customary for a whole family to take their places in the principal apartment, where they proceed to their respective

4

tasks, while one, selected for the purpose, reads aloud some of their sagas (ancient tales), or such other historical narrative as can be found. Their stock of books is not large, but they lend to each other, and frequently copy what they borrow.

The island was formerly divided into four amts, or provinces, answering to the four cardinal points. The N. and E. are now merged into one, and the W. is presided over by the governor in person. This officer has the title of stif-tamtman; he is sometimes a native, but more frequently a Dane. Under him are the amtmen, or provincial governors, who possess a similar jurisdiction over their quarters. Each province is divided into sysels or shires, presided over by sysselmen, with authority similar to that of sheriffs; these collect taxes, hold petty courts, regulate assessments, &c. Under the sysselmen are repstfores, who are overseers of the poor, constables, &c. The talsrood, or chief justice, holds, with two assistants, a criminal court at Reikiavik, but very few cases are tried in the island, and all capital punishments are inflicted at Copenhagen. Crimes are rare, petty theft and drunkenness are the most common; the latter has been introduced chiefly by the crews of the Danish vessels that visit the coasts.

The island constitutes one bishopric; the bishop's salary does not exceed £300 per annum. There are about 184 pars.; but the clergy amount to upward of 300: their incomes are very small, and they are frequently among the poorest of the community. The only charitable institutions are, four hospitals, for the reception of those afflicted with leprosy, which, in the form of elephantiasis, was formerly very prevalent. Small-pox was formerly also very destructive. There are no workhouses, the sick and poor being almost universally supported by their own families. The principal school at Bessestedt, near the W. coast, has three masters, who teach classics, theology, and the Danish language; and several young men, after attending this school, go to Copenhagen to finish their studies. Reikiavik, the cap., on the S.W. coast, has little more than 500 resident inhab. chiefly Danes. Most of the villages are situated on the coasts, at convenient spots for the receipt and transport of merchandise.

The early and successful application of the Icelanders to the cultivation of literature is an anomaly in the history of learning. When most parts of continental Europe were in a state of rude ignorance, the inhab. of this remote island were well acquainted with poetry and history. The most flourishing period of Icelandic literature appears to have been from the 12th to the end of the 14th century. During the last three centuries, however, Iceland has produced many learned men, some of whom have risen to great eminence. The literature of the island in the present day may perhaps be said rather to have changed its character than declined from its ancient fame; the inhab. now attend more to solid branches of learning than to the poetical and historical romances of the ancient Icelandic sagas. Domestic education is carefully attended to; there is no want of modern books in Icelandic; and a printing press is actively employed in the island of Vidoe.

The discovery of Iceland by Europeans is attributed to a Norwegian pirate, about the year 860; but the earliest permanent settlement was effected by the Norwegians in 874. In little more than half a century, till the coasts were occupied by settlers; and about the year 928 the inhab. formed themselves into a republic, and established the Althing, or General Assembly of the nation, which was held annually at Thingvalla, in the S.W. and not abolished till 1800. The Icelanders maintained their independence for nearly 400 years; but during the 13th century became subject to Norway, and on the annexation of that kingdom to Denmark, Iceland was transferred along with it. (See Sir G. Mackenzie's Travels; Hooker's Tour in Iceland; Barrow's Visit to Iceland, 1834–5; Gaimard's Voyage en Island et Groenland, 1836; Henderson's Journal, &c.)

IDA, t., Monroe co., Mich. It has two schools, 50 scholars. Pop. 351.

IDRIA, a town of the Austrian empire, k. Illyria, duchy Carniola, circle Adelsberg, in a valley of the Carnic Alps, 22 m. W. by S. Laybach. Pop. (1836), 4185. The inhab. are principally engaged in mining; the quicksilver mines of Idria belonging to the Austrian government being, after those of Almaden in Spain, the richest and most celebrated in Europe. They yield annually from 3000 to 3500 cwt. of metal, about a sixth part of which is converted into the spol into vermilion, corrosive sublimate, and other preparations of mercury. The mine is rather more than 1000 ft. in depth. The formation in which it is situated is transition limestone, alternating with clay-slate, in which latter rock the quicksilver is found. It exists partly pure in globules among the slate; but it is mostly found in combination with sulphur, forming veins of cinnabar, &c., which vary greatly in thickness. The cinnabar ore is considered too poor to be wrought when it contains only from 15 to 18 per cent. of quicksilver, and is then usually abandoned in search of a better vein. The richest ore yields from 30 to 70 per cent. of metal. From 600 to 700 workmen are employed, of whom about 500 are miners. These are enrolled in a corps, and have a regular uniform. They are divided into three sections, which relieve each other, each working below for eight hours in the twenty-four, and the work incessantly going on. Within his eight hours, the labourer is required to perform a certain measurement of work, for which he receives 17 kreutzers (nearly 7d.). If he performs less or more than this measured extent, his pay is proportionally reduced or increased; but the number of those who gain less than the fixed sum is greater than of those who gain more. Besides their money pay, the miners get an allowance of corn sufficient for themselves and their families; and in illness, gratuitous medical aid. No lodging is found them; but they may purchase at a government store a number of articles of prime necessity, at fixed charges, generally below the ordinary market prices. The miners usually enter the service at fifteen years of age. After forty years' service, or earlier, if ill health overtake them, they are allowed to retire on full pay, and enjoy various privileges. The widows and orphans of miners are entitled to a pension, and about 25,000 florins are thus expended annually. The process of mining is said to be very unhealthy; the heat of the mine, varying from 60° up to 80° Fah., impregnates the atmosphere with volatilised mercury, which soon exerts all its characteristic effects on the constitutions of the miners. In some parts, the heat is so great, and the atmosphere so vitiated, that the workmen are obliged to relieve each other every two hours. The mine is very clean, and in its lower parts remarkably dry. In 1803, a violent conflagration broke out in the mine, destroying the whole of the works, with several of the workmen. Of the mercury produced at Idria, a small part goes to Trieste, whence it is exported chiefly to America; but by far the largest portion is sent to Vienna, partly for the plating of mirrors, but principally for the use of the gold and silver mines of Hungary and Transylvania.

Fifty years ago, Idria was notoriously a place of banishment for state prisoners and criminals, who were condemned to work in the mines. It is so no longer; no coercion is used, and no convicts are sent thither; the supply of labourers petitioning to be admitted is considerably greater than can be received into the service. The town and district of Idria is a mining intendency, with its own government; consisting of a director-general, an imperial comptroller of accounts, a secretary-general, and four councillors, who superintend all the departments of the public service, under the Council of Mines in Vienna. Idria has some German, primary, and other schools, and a small theatre. It had a school for instruction in mining, but it was abolished on the restoration of the Illyrian provinces to Austria. The aspect of the place is thus described by Turnbull: "We perceived the white church with its little steeple, perched on a small green knoll, and not far from it another insulated height, crowned with an antique-looking castle, erected by the Venetians during the time that they possessed Illyria, and which now serves as a residence for the bergrath, or director of the mines, and for the government offices connected therewith. Between these two heights, the town struggles along on very unequal ground; with a stream rushing through it, a second church in a sort of open market-place, some large buildings connected with the public administration, but scarcely any good shops or private houses." Mendicancy, or abject poverty, is, however, unknown. The mine was discovered by accident in 1497; it was afterward wrought by a company of Venetian merchants, and purchased by the house of Austria, who accorded the miners considerable privileges in 1575, since which the prosperity of Idria has been generally on the increase. (See the elaborate accounts of Frauche, in the Revue du Nord, vol. v., pt. ii.; Turnbull's Trav., i., 385–396; Berghaus, Oesterr. Nat. Encyc., &c.)

IGUALADA (an. Agua lata), a town of Spain, prov. Catalonia, N. m. N.W. Barcelona, and 296 m. E.N.E. Madrid; lat. 41° 40.' N, long. 1° 21' E. Pop. 7721. It stands on the Noya, a trib. of the Jour, in a rich plain, abounding with corn-fields and olive-grounds. It has some well-built streets, and a handsome suburb, the chief buildings being a par. church, two convents, a clerical college, hospital, and cavalry barracks. The inhab. are among the wealthiest and most industrious in Spain; and their manufactures, by which they are almost wholly supported, comprise cotton and woollen yarns and cloths, lace, and firearms, the last of which are highly esteemed. In the neighbourhood are several considerable paper-mills. Fairs, well attended, for manufactured produce, are held here in the beginning of Jan. and at the end of August. (Miñano.)

ILCHESTER, a bor., market town, and par. of England, co., Somerset, hund. Tintinhull, on the Yeo or Ivil (whence its name is derived), 18 m. E. Taunton, and 116 m. W.S.W.

ILDEFONSI (ST.).

London. Area of bor. and par., 690 acres: pop. in 1831, (including 120 prisoners in the jail), 1095. The town comprises four indifferently-built streets, and has but few public buildings. The church is remarkable for its octangular tower. A national school, and almshouses for sixteen women, are the only public charities. The county courthouse is handsome, and conveniently arranged. The jail, built on Howard's plan, is large, and well-regulated, and capable of accommodating upward of 900 prisoners, and was often quite full, when employed, as formerly, for a state prison and house of correction: it is now chiefly used for untried prisoners and debtors, the number of inmates averaging fifty. (Pris. Inspect. Rep.) The town, which has no manufactures, and little trade, derives its chief importance from the fact that a large portion of the county business is transacted here, the assizes being held at Ilchester alternately with Taunton, Wells, and Bridgewater. It is altogether, however, in a low, declining state, and pauperism is on the increase. Ilchester is a bor. by prescription, and sent two mems. to the H. of C. from the 26th of Edw. I. down to the passing of the Reform Act, when it was disfranchised: it was a mere nomination bor., in the patronage of the Duke of Cleveland. Markets on Wednesdays. Distinct traces of a Roman station, and the discovery of numerous Roman coins and antiquities, have led to the belief that this town occupies the site of the Ischalis of Ptolemy, the principal military station of the Romans in the West of England. It had 108 burgesses at the time of the Norman conquest. Still later, it was a place of considerable consequence, and was made, by patent of Edw. III., the assize town of Somerset.

ILDEFONSO (ST.), or LA GRANJA, a celebrated palace of the sovereigns of Spain, Old Castile, prov. Segovia, 42 m. N.N.W. Madrid, and 5 m. S.E. Segovia, on the N. declivity of the Sierra Guadarrama, built by Philip V. as a place of retirement during the hottest months of summer. "It is placed in a spot where the mountains fall back, leaving a recess sheltered from the hot air of the S. and from much of its sun, but exposed to whatever breeze may be wafted from the N.; the immediate activity towards the S. being occupied by the garden, which, though somewhat formal, is full of shade and coolness." (Inglis, i., 262.) The palace, which is of brick, plastered and painted, occupies three sides of a square, in the centre of which is the royal chapel. The principal front, looking towards the garden, is 530 ft. long, having two stories, with twelve rooms in a suite; the great entry, with its iron palisade, very much resembling that of Versailles. The interior is in everything regal; the ceilings of the apartments are painted in fresco, the walls decorated with noble mirrors, and the floors chequered with black and white marble, while the furniture, though somewhat antiquated, is highly enriched with jasper, verd-antique, and rare marbles. The upper rooms are adorned with the works of the first masters, chiefly of the Italian school, the lower apartments being used as a repository for sculpture. Many, however, of the best specimens once belonging to this palace, both in painting and sculpture, have been removed to the royal gallery of Madrid, which now possesses one of the richest collections in Europe. The gardens are laid out in the French style, with formal hedges and walks; and the trees, notwithstanding the labour with which the formation of these grounds was attended, are poor and starved; the chief feature, indeed, in these gardens, is the quantity of fine water, disposed in a variety of ways, and especially in the formation of fountains and works. "These," says Swinburne, "surpass all that I ever saw, not excepting the finest at Versailles. The jets d'eau send forth a clear crystal stream, which falls around like the finest dew: the most remarkable are eight fountains, dedicated to the different heathen deities, one of which, Fame, seated on a Pegasus, throws up from a trumpet a stream to the height of 129 ft. There are various other water-works, all adorned with statues of lead, varnished in imitation of brass; and the whole supply of water is procured from reservoirs on the hills above." (Swinburne, ii., 230.) The expense of constructing the garden alone, a large part of which was made by blasting out of the solid rock, must have been very great; and the entire expenditure on the palace gardens and water-works is stated by Townsend to have exceeded £6,000,000. In the town, which lies at a little distance below the palace, is a manufactory of mirrors, supported by the government, which, at the time when Townsend visited it, "proved a devouring monster, in a country where provisions were dear, fuel scarce, and carriage exceedingly expensive." Inglis says that the largest mirrors made there were 13½ ft. long, 8 ft. broad, and 6 in. deep. (Townsend, vol. ii.; Dillon, p. 85; Inglis, i., 261–265; Miñano.)

ILFRACOMBE, a seaport, market town, and par. of England, co. Devon, hund. Braunton, on the Bristol channel, 9 m. N. Barnstaple, 41 m. N.W. Exeter, and 172 m. W. by S. London. Area of par., 3690 acres. Pop. (1831), 3901.

ILLE-ET-VILAINE.

The town, consisting of one long street and a noble terrace facing the sea, extends W. from the harbour along the shore. The church, which stands at its upper end, is a large plain building containing some fine monuments: the living is attached to a prebend in Salisbury Cathedral. There are places of worship for Independents and Wesleyan Methodists, a large national school, and a girls' school of industry. The harbour is a natural basin formed by the curve of a very rocky shore, and a bold mass of rocks stretching nearly half way across the entrance of the recess shelters it from the northern storms. A battery and lighthouse stand on the top of this rocky mass, and the harbour is further defended by a pier 850 ft. in length, which has been lately put in excellent repair. There is safe anchorage for vessels of 230 tons, and ships can easily enter here when they cannot get up the Taw to Barnstaple; the consequence of which is, that Ilfracombe has taken away a great part of its coasting trade. The trade with Bristol, Swansea, and other ports in the Bristol channel is considerable; and many vessels are employed in the herring fishery. This port, in 1838, had 63 ships, of the burden of 3897 tons. Oats, barley, and fish are the chief articles of export. The town, however, depends in a great measure for its support on the numerous wealthy families that resort thither in summer since it has attained celebrity as a watering-place. The bathing is excellent, and the neighbourhood abounds with romantic scenery. Steam-packets run daily to and from Bristol, and at less frequent intervals, to and from Swansea, Tenby and Milford. The town is governed by a portreeve appointed by the lord of the manor. Markets, well supplied with fish, on Saturdays: fairs, April 14, and the first Saturday after Aug. 22.

ILLE-ET-VILAINE, a marit. dep. of France, in the N.W. part of the kingdom, formerly included in the prov. of Brittany; between lat. 47° 38' and 48° 42' 30'' N., and long. 1° and 2° 15' W., having W. Côtes-du-Nord and Morbihan, S. Loire Inférieure, E. Mayenne, and N. La Manche and the English channel. Length, N. to S., about 70 m. Area, 668,697 hectares. Pop. (1836), 547,250. The Menez mountains run through this dep. from E. to W.: but they rise to no great height, and the surface elsewhere is not hilly. The chief river is the Vilaine, which has mostly a S.W. course, and falls into the Atlantic in the dep. Morbihan; the Ille is one of its affluents. The Rance, which has its mouth in this dep., is connected with the Ille by a canal, extending from Dinan to Rennes, 52 m. in length, and wide and deep enough for vessels of 70 tons. Climate temperate, but very damp; fogs are frequent, and from 36 to 38 in. rain fall annually. Soil thin, and not generally fertile. In 1834, 397,496 hectares of land were arable, and 73,349 in pasture; forests, heaths, and waste lands occupying 146,078. Agriculture is in a backward state. Throughout the greater part of the dep. the land is parcelled out into small farms, one of 30 hectares being considered large. In 1835, of 143,550 properties subject to the contribution foncière, 60,990 were assessed at less than 5 fr., and 96,058 between 5 and 10 fr.; the number of considerable properties is much below the average of the deps. Principal crops, rye, oats, and barley; the dep. is not so suitable for wheat; and but little maize is grown; the annual quantity of grain produced is about 3,436,000 hectolitres, which is scarcely sufficient for home consumption; and the peasantry add to their corn chestnut flour, potatoes not being in general use: 13,900 hectares are in gardens and orchards; fruit is plentiful, and some very good cider is made; but the agricultural products of the greatest importance are flax and hemp, and the linen thread of the dep. is very highly valued. Both cattle and horses are of good breeds; many oxen from this dep. are fattened in Normandy for the Paris market. Dairy husbandry occupies a good deal of attention, and the beurre de Prévalaye, made in the neighbourhood of Rennes, is highly esteemed throughout France. The sheep are of an inferior kind. The sole, cod, mackerel, and other fisheries on the coast are extensive; and Cancale bay is celebrated for its oysters, with which Paris is in great part supplied. From 50 to 80 boats go annually from this dep. to the cod fishery of Newfoundland. Some copper, iron, argentiferous lead, and coal mines, and quarries of marble, granite, slate, limestone, &c., are wrought, but apparently not to any great extent. The manufactures consist chiefly of hemp and linen thread, packing and sail-cloth, cordage, flannels at Fougères, leather, &c. In the arrond. of Fougères there is a large government glass factory, partially wrought by steam, some of the products of which are equal to any made in Lyons. This dep. is divided into six arronds.; chief towns, Rennes, the cap., St. Malo, Fougères, Redon, Monfort, and Vitré. It sends seven mems. to the ch. of dep. Number of electors (1838–9), 2128. Total public revenue (1831), 11,116,307 fr. This dep. has produced many celebrated men, including M. de la Bourdonnaye, Maupertius, Savary, Vauban, Chateaubriand, and Broussais. (Hugo, art. Ille-et-Vilaine, &c.)

ILLINOIS, river, Ill., is formed by the junction of Kankakee and Des Plaines rivers. The Kankakee rises in the northern part of Indiana, and flows S.W. by W. into Illinois, where it receives the Iroquois river, which also rises in Indiana. They unite in the N. part of Iroquois co., Ill., whence the Kankakee flows N.W. to its junction with Des Plaines river, in the E. part of La Salle co. The Des Plaines river rises in Wisconsin ter., a few miles above the boundary of Illinois, and about 6 m. from lake Michigan. It flows S., generally over a bed of limestone rock, to its junction with the Kankakee, below which it is called Illinois river. It thence flows nearly W. to near Hennipen, in Putnam co., receiving Fox river from the N. at Ottawa, and Vermillion river from the S.E. near the foot of the rapids. It then flows to the S. and S.E., receiving Spoon river on the W. side and Sangamon river on the E., as far as Naples, in Morgan co. It then flows S. until it approaches within 5 m. the Mississippi, when it curves to the S.E., and finally of the E., to its junction with the Mississippi, Its length, exclusive of its windings, is about 260 m. It is navigable, at a moderate stage, the water, 210 m. to the foot of the rapids, and 9 m. farther to Ottawa, in high water, for steamboats. In extreme high water, the Mississippi backs up the Illinois about 70 m. From the head of navigation on the Illinois, a canal is in progress to Chicago, a distance, including 5½ m. in Chicago river, and a feeder of 4 m. from Fox river, of 106 m. It is 6 ft. deep, and 60 ft. wide at top; and is estimated to cost $8,654,337. This is one of the most important canals in the western country. It is not entirely completed, and an effort has been recently made to induce the creditors of Illinois to complete it; to whom the canal will be conveyed, until they shall have been completely paid.

ILLINOIS, one of the western United States, is bounded N. by Wisconsin ter., E. by lake Michigan and Indiana, S. by Ohio river, which separates it from Kentucky, and W. by Missouri and Iowa ter., from which it is separated by Mississippi river. It is between 37° and 42° 30′ N. lat., and between 87° 26′ and 91° 31′ W. long., and 10° 25′ and 14° 30′ W. long. from W. Its extreme length is 380 m., and its extreme width 230 m.; its mean width is 150 m.; containing, with the portion of lake Michigan within its limits, 59,300 sq. m., or 37,952,000 acres. The water area within its limits is 3730 sq. m. It is computed that 50,000 sq. m., or 32,000,000 acres, are arable land. The population in 1810 was 12,282; in 1820, 55,211; in 1830, 157,575; in 1840, 476,183; of whom 255,235 were white males; 217,019 white females; 1876 were coloured males; 1722, coloured females; Employed in agriculture, 105,337; in commerce, 2306; in manufactures and trades, 13,185; in mining, 782; in navigating the ocean, 63; do. lakes, rivers, and canals, 310; in the learned professions, 2031.

The state is divided into 87 counties, which, with their population in 1840, were as follows:

Counties	Pop. 1840	Counties	Pop. 1840
Adams	14,476	Lee	2,035
Alexander	3,313	Livingston	759
Bond	5,060	Logan	2,333
Boone	1,705	Macon	3,039
Brown	4,183	Macoupin	7,926
Bureau	3,067	Madison	14,433
Calhoun	1,741	Marion	4,742
Carroll	1,023	Marshall	1,848
Cass	2,981	McDonough	5,308
Champaign	1,475	McHenry	2,578
Christian	1,874	McLean	6,564
Clark	7,456	Menard	4,431
Clay	3,228	Mercer	2,352
Clinton	3,718	Monroe	4,481
Coles	9,616	Montgomery	4,490
Cook	10,201	Morgan	19,549
Crawford	4,479	Ogle	3,479
De Kalb	1,697	Peoria	6,153
De Witt	3,247	Perry	3,222
Du Page	3,535	Pike	11,728
Edgar	8,225	Pope	4,094
Edwards	3,070	Putnam	2,131
Effingham	1,675	Randolph	7,944
Fayette	6,328	Rock Island	2,610
Franklin	3,682	Sangamon	14,716
Fulton	13,142	Schuyler	6,972
Gallatin	10,760	Scott	6,315
Greene	11,951	Shelby	6,659
Hamilton	3,945	Stark	1,573
Hancock	9,946	Stephenson	2,800
Hardin	1,378	St. Clair	13,631
Henry	1,260	Tazewell	7,221
Iroquois	1,695	Union	5,524
Jackson	3,566	Vermillion	9,303
Jasper	1,478	Wabash	4,240
Jefferson	5,762	Warren	6,779
Jersey	4,885	Washington	4,810
Jo Daviess	6,180	Wayne	5,133
Johnson	3,626	White	7,919
Kane	6,501	Whiteside	2,514
Knox	7,060	Will	10,167
Lake	2,634	Williamson	4,457
La Salle	9,348	Winnebago	4,609
Lawrence	7,092		
		Total	476,183

Springfield, near the centre of the state, is the seat of government.

The general surface is level, or moderately undulating; the northern and southern portions are broken, and somewhat hilly, but no portion of the state is traversed by ranges of hills, and there is nothing in the state which can be denominated a mountain. That portion of the state which lies S. of a line from the mouth of Wabash river to the mouth of Kaskaskia river, is generally covered with timber, but N. of this the prairie country predominates. It is computed that two thirds of the surface of the state is covered with prairie. The eye sometimes wanders over immense plains covered with grass, and, in the season of them, adorned with flowers, with no other boundary of its vision but the distant horizon, though the view is often broken with occasional woodlands. Much of the prairie land is undulating and entirely dry. The dry prairies are generally from 30 to 100 feet higher than the bottom land on the rivers, and are often very fertile. They frequently extend from 6 to 12 m. in width. In many instances, there are copses or groves of timber, of from 100 to 2000 acres, in the midst of prairies, like islands in the ocean. This is a common feature of the country between Sangamon river and lake Michigan in the N. part of the state. There are extensive tracts called Barrens, which are not wanting in fertility, of a mixed character, uniting forest with prairie. The timber is generally scattering, and of a rough and stunted appearance. The surface is generally more uneven and rolling than the prairies. These tracts are commonly healthy, and abound with springs of pure water, and are better adapted for all descriptions of produce and all kinds of economic, wet and dry, than the richer and deeper mould of the river bottoms and the prairies. Illinois in general is abundantly supplied with timber, but it is unequally distributed, and on the prairies there is often a deficiency, which might be remedied by cultivation. The kinds of timber most abundant are oaks of various species, black and white walnut, ash of several kinds, elm, sugar maple, honey-locust, hackberry, linden, hickory, cotton-wood, pecans, mulberry, buckeye, sycamore, wild cherry, box, sassafras, and persimmon. In the S. and S.E. parts of the state are yellow poplar and beech; near the Ohio are cypress, and in several counties are clumps of yellow pine, and cedar. The undergrowth are redbud, papau, sumac, plum, crab-apple, grape vines, dogwood, spice bush, green brier, hazle, &c. The alluvial soil on the rivers produces cotton-wood and sycamore timber of amazing size. In some parts of the state are knobs or ridges of flint limestone, intermingled and covered with earth, elevated one or two hundred feet above the common surface. Back of the alluvions which border the streams there are bluffs, some in parallel ridges, and others of a conical form, formed of limestone rock, from fifty to one and two hundred feet high. Among these bluffs are ravines, which conduct the streams into the rivers. There are also in some parts sink-holes, or circular depressions like a basin, of various depths and extent, which discharge the water received by rains by evaporation or into the ground. There are few tracts of stony ground in the state; but quarries are to be found in the bluffs, and in the banks of the streams, and on the borders of the ravines throughout the state. The soil of the state is generally fertile. The vegetable productions are Indian corn, wheat, rye, oats, buckwheat, potatoes, turnips, cotton, hemp, flax, tobacco, castor bean. &c.

There were in the state, in 1840, 199,235 horses and mules, 698,274 neat cattle, 395,672 sheep, 1,495,254 swine, poultry valued at $309,984. There were produced 3,335,393 bushels of wheat, 88,251 of barley, 4,988,008 of oats, 88,197 of rye, 57,884 of buckwheat, 22,634,211 of Indian corn, 2,025,590 of potatoes, 658,007 pounds of wool, 17,742 of hops, 29,173 of wax, 200,947 of cotton, 564,296 of tobacco, 399,813 of sugar, 1150 of silk cocoons, 1976 tons of hemp and flax, 164,932 of hay. The products of the dairy were valued at $492,175; of the orchard, at $196,756; of lumber, at $903,666; of skins and furs, at $39,412. There were made 474 gallons of wine.

The most important mineral production of the state is lead, found in its N.W. part, and in Wisconsin, in inexhaustible quantities, of which 13,600,000 pounds have been smelted in one year. Galena is the centre of the lead trade. Salt springs are found in the E. and S. part, particularly near Shawneetown. The salt-works are here owned by the United States, and leased to the manufacturers. Coal abounds in the bluffs, and iron exists in various parts of the state. Bituminous coal abounds in the ravines and bluffs.

The climate is generally healthy, and the air, except in the neighbourhood of low and wet lands, is pure and serene. The average temperature through the year is from 50° to 53° of Fahrenheit. The winters are cold, and the summers, in the S. part, are quite warm.

Next to the great rivers Mississippi and Ohio on its borders, the Illinois is the largest river in this state, which it crosses diagonally, and, after a course of over 400 m. from its sources, it enters the Mississippi, 90 m. above the mouth of the Missouri. (See ILLINOIS RIVER.) Rock river rises in Wisconsin, crosses the N.W. part of the state, and, after a course of 300 m., mostly in this state, falls into the Mississippi. Kaskaskia river rises near the middle of the state, and, after a southwesterly course of 250 m., enters Mississippi river, 63 m. below the mouth of the Missouri. It is navigable for boats for 150 m. Sangamon is a large tributary of Illinois river. The Wabash runs chiefly in Indiana, but forms a part of the boundary between that state and Illinois. Little Wabash, after a course of 130 m., enters Wabash river, a little above the confluence of the latter with Ohio river. Peoria lake, which is an expansion of Illinois river, 200 m. from its mouth, is a beautiful sheet of water, 20 m. long and 2 m. broad.

Chicago, on lake Michigan, is the principal commercial depôt in the N. It has a tolerably good harbour, which has been improved by artificial works. Alton is the most commercial place on the Mississippi, 2¼ m. above the mouth of the Missouri. It has a good landing-place. The other principal places are Springfield, the capital, Quincy, Galena, Peoria, Vandalia, and Kaskaskia.

There were in this state in 1840, two commercial and 51 commission-houses engaged in foreign trade, with a capital of $338,803; 1348 retail stores, with a capital of $4,904,195; 405 persons employed in the lumber trade, with a capital of $93,350; 117 persons employed in internal transportation, who, with 268 butchers, packers, &c., employed a capital of $649,495. Home-made or family manufactures amounted to $693,567. Four fulling-mills and 16 woollen manufactories employed 34 persons, producing goods to the amount of 9540, with a capital of $96,306; 5 furnaces produced 158 tons of cast iron; 20 smelting-houses produced 8,755,000 pounds of lead, employing 72 persons, and a capital of $114,560; 22 persons produced 20,000 bushels of salt, with a capital of $10,000; 3 persons produced confectionary to the amount of $2840; 1 paper-mill produced $2000; 94 persons manufactured tobacco to the amount of $10,129; 68 persons manufactured hats and caps to the amount of $26,305, and straw bonnets to the amount of $1570, employing a capital of $12,918; 23 potteries employed 56 persons, produced articles to the amount of $95,746, with a capital of $10,995; 155 tanneries employed 306 persons, and a capital of $155,679; 695 other manufactories of leather, as saddleries, &c., produced articles to the amount of $847,217, with a capital of $98,503; 71 persons produced machinery to the amount of $37,730; 90 persons produced hardware and cutlery to the amount of $9730; 19 persons produced 90 cannon and 236 small arms; 7 persons manufactured the precious metals to the amount of $9400; 36 persons manufactured granite and marble to the amount $116,119; 995 persons produced bricks and lime to the amount of $963,396, with a capital of $104,648; 25 persons produced 519,673 pounds of soap, and 117,698 pounds of tallow candles, with a capital of $17,345; 150 distilleries produced 1,551,694 gallons, and eleven breweries 90,300 gallons, the whole employing 223 persons and a capital of $136,155; 307 persons produced carriages and wagons to the amount of $144,392, with a capital of $50,963; 98 flouring-mills produced 172,697 barrels of flour, and, with other mills, employed 2304 persons, and manufactured articles to the amount of $2,417,398, with a capital of $2,147,618; vessels were built to the amount of $39,200; 344 persons produced furniture to the amount of $64,410, with a capital of $52,392; 324 brick or stone houses, and 4123 wooden houses were built, employing 5737 persons, and cost $2,065,955; 45 printing-offices, 5 binderies, 3 daily, 2 semi-weekly, and 36 weekly newspapers, and 9 periodicals, employed 178 persons, and a capital of $71,300. The total amount of capital employed in manufactures was $3,136,512.

The Illinois College (New School Presbyterian), at Jacksonville, was founded in 1829; Shurtleff College (Baptist), in Upper Alton, in 1835; McKendree College (Methodist), in Lebanon, in 1834; McDonough College (Old School Presbyterian), at Macomb, in 1837. In these institutions there were, in 1840, 311 students. There were in the state 42 academies, with 1967 students; 1241 common and primary schools, with 34,876 scholars; and 27,502 white persons over 20 years of age who could neither read nor write. There is a state penitentiary at Springfield.

The Methodists have about 170 travelling preachers, and are the most numerous denomination; the Baptists have about 160 ministers; the Presbyterians of different descriptions, about 160 ministers; the Episcopalians have about 10 ministers; the Roman Catholics 10 priests: and there are some other denominations.

At the beginning of 1842, the State Bank of Illinois, with its branches, had an aggregate capital of $3,616,125; and a circulation of $2,861,598. The state debt, at the close of 1846, amounted to $13,465,662. There is a state penitentiary at Alton.

The governor is elected by the people for four years, but is eligible only four years in eight. A lieutenant-governor is elected at the same time, who is president of the senate, and, in case of the death, resignation, or absence of the governor, discharges his duties. The senators are elected for four years, and the representatives for two years. The number of senators shall never be less than one third, nor more than one half the number of representatives. The judges of the supreme court are appointed by the joint ballot of both houses of the legislature, and hold their offices during good behaviour. Every white male inhabitant over 21 years of age, who has resided in the state for six months next preceding an election, has the right of suffrage.

In 1836 this state adopted an extensive system of internal improvements, consisting of canals and railroads, most of which it must be left to another generation to complete. The Illinois and Michigan canal, the most important of them all, is in progress, and will probably be completed. It extends from Chicago river, about 5 m. from Chicago, to the head of steamboat navigation on the Illinois river, at Peru, 106 m., including a navigable feeder of 4 m., and a few miles of river navigation. It was commenced in 1836, and is estimated to cost, when completed, $8,654,337. It is 60 feet wide at top, and 6 feet deep. A large amount of money has already been expended upon it. A railroad extends from Springfield 53 m. to Merodosia on Illinois river. Coal Mine Bluffs railroad extends from Mississippi river 6 m. to the coal mine. Work has been done on several other railroads, but they are at present suspended.

In the latter part of the seventeenth century, Illinois was explored by the French from Canada, and some forts and trading posts were established. About 1720 several forts were built within the present limits of Illinois, of which fort Chartres was the most considerable. A chain of communication was formed from Canada to the mouth of the Mississippi river. The oldest document in the state is at Kaskaskia, which is a petition to Louis XV. for a grant of common fields, stating the great losses of the people the year before by an extraordinary flood. At the peace of 1763 this country, together with Canada, was ceded to the English. In 1765, Capt. Sterling, of the Royal Highland ers, took possession of Illinois, and was followed by several other commanders, who occupied fort Chartres. In the Revolutionary war, the Virginia militia, under General Clarke, subjugated fort Chartres, Kaskaskia, and conducted a successful expedition, in 1788, against Port Vincent, now Vincennes. In the same year the legislature of Virginia organised, in this remote region, the county of Illinois, which was afterward ceded to the United States. In 1800 the present territory of Illinois contained about 3000 inhabitants. In 1809 the territorial government was formed, and the population the next year amounted to 12,000. In 1812 a territorial government was formed, with a legislature and a delegate to Congress. In 1818 a state constitution was formed, and Illinois was received into the Union as the 22d state.

ILLINOIS, t., Washington co., Ark. It has 15 stores, two tanneries, three distilleries. Pop. 519.

ILLINOIS, CITY, p. v., Rock Island co., Ill., 172 m. N.N.W. Springfield, 901W. Situated between Rock river and the Mississippi.

ILLYRIA (KINGDOM OF), a portion of the Austrian empire, comprising the provs. of Carinthia, Carniola, and Istria, the islands of the Gulf of Quarnero, and the Illyrian Littorale. It lies between lat. 44° 25′ and 47° 7′ N., and long. 12° 14′ and 16° E., having N., Austria and Styria; E., the latter prov. and Croatia; W., the Tyrol and Italy; and S., the Adriatic sea. It is divided into the governments of Laybach and Trieste.

The divisions, with their extent and pop., are as follow:

Circles.	Area in sq m.	Towns.	Villages.	Pop.	Chief Towns.
Laybach	1,302	9	916	149,721	Laybach
Neustadt	1,536	13	1,936	193,423	Neustadt
Adelsberg	1,123	3	421	89,076	Idria
Klagenfurth	1,441	29	1,616	178,523	Klagenfurth
Villach	1,585	15	1,180	123,380	Villach
Trieste (Territory of.)					
Istria	2,178	29	479	78,813	Trieste
Goritz	1,837	10	441	211,020	Revigno
				176,570	Goritz
Troops in garrison				16,979	
	10,401	111	6,971	1,312,706	

Its N. part is covered by the central chain of the Alps and likewise by various offsets, constituting the southern limestone girdle of the Alpine system. The S. portion of the kingdom, comprising the government of Trieste, occupies the S. slope of this mountain-range towards the Adriatic. The main chain at the Gross Glockner (14,000 ft. high) takes the name of the Noric Alps, stretching its lofty

.—Illyria has
N. govern-
districts, which
mountainous
damp climate,
with difficulty,
crops; and the
part of the
The corn, in
up on poles
erections
, Slav.) are
. The most
want, and the
the valley of
vere as not
other hand,
circle of Idria,
. Excellent
," are grown
rotation of
succession of
ure) makes,
fivy; 3d year,
is broken up
quantity of
1837, in the
(*Official* re-

8,280 imp. qs.
7,648 —
3,030 —
,030 gallons.
,609 cub. feet.

in its pro-
on as the
charm-
fields are in
overed with
, around
aspect of
central Italy.
however, is
good hus-
has a
territory for
during
Olives and
landowner
for half
this part of
land and

770 imp. qs.
60 —
40 —
30 gallons.
10 —
0 cub. feet

grain pre-
: in the
ding must

ip chiefly
(marval)
The best
, grown
the oil of
the stones
are even
rely culti-
glia and

duce, and
act. The
ahed, in 1832,
ceed in Istria,

Save, the climate is much milder. At Laybach, the temperature of the year is 8·7 Réaum. The temperature of the government of Trieste presents a great contrast to that of the mountain districts. In the valley of the Isonzo, as during the same year amounted to 4500 pounds; but till 1837 this article was not included in the land-tax returns of the province.

The chief wild animals of the northern districts are the

crosses diagonally, and, after a course of over 400 m. from its sources, it enters the Mississippi, 20 m. above the mouth of the Missouri. (See ILLINOIS RIVER.) Rock river rises in Wisconsin, crosses the N.W. part of the state, and, after a course of 300 m., mostly in this state, falls into the Mississippi. Kaskaskia river rises near the middle of the state, and, after a south-westerly course of 250 m., enters Mississippi river, 63 m. below the mouth of the Missouri. It is navigable for boats for 150 m. Sangamon is a large tributary of Illinois river. The Wabash runs chiefly in Indiana, but forms a part of the boundary between that state and Illinois. Little Wabash, after a course of 130 m., enters Wabash river, a little above the confluence of the latter with Ohio river. Peoria lake, which is an expansion of Illinois river, 200 m. from its mouth, is a beautiful sheet of water, 20 m. long and 2 m. broad.

Chicago, on lake Michigan, is the principal commercial depôt in the N. It has a tolerably good harbour, which has been improved by artificial works. Alton is the most commercial place on the Mississippi, 2½ m. above the mouth of the Missouri. It has a good landing-place. The other principal places are Springfield, the capital, Quincy, Galena, Peoria, Vandalia, and Kaskaskia.

There were in this state in 1840, two commercial and 51 commission-houses engaged in foreign trade, with a capital of $228,803; 1348 retail stores, with a capital of $4,994,195; 605 persons employed in the lumber trade, with a capital of $93,350; 117 persons employed in internal transportation, who, with 968 butchers, packers, &c., employed a capital of $645,425. Home-made or family manufactures amounted to $393,567. Four fulling-mills and 14 woollen manufactories employed 34 persons, producing goods to the amount of $640, with a capital of $26,505; 5 furnaces produced 156 tons of cast iron; 20 smelting-houses produced 3,755,000 pounds of lead, employing 72 persons, and a capital of $114,560; 22 persons produced 20,000 bushels of salt, with a capital of $10,000; 3 persons produced confectionary to the amount of $2340; 1 paper-mill produced $9000; 24 persons manufactured tobacco the amount of $10,120; 68 persons manufactured hats and caps to the amount of $26,395, and straw bonnets to the amount of $1570, employing a capital of $13,918; 23 potteries employed 56 persons, produced articles to the amount of $26,740, with a capital of $10,925; 155 tanneries employed 306 persons, and a capital of $155,879; 695 other manufactories of leather, as saddleries, &c., produced articles to the amount of $347,317, with a capital of $98,503; 71 persons produced machinery to the amount of $37,730; 20 persons produced hardware and cutlery to the amount of $9750; 12 persons produced 20 cannon and 236 small arms; 7 persons manufactured the precious metals to the amount of $9460; 26 persons manufactured granite and marble to the amount $116,119; 995 persons produced bricks and lime to the amount of $263,295, with a capital of $104,645; 25 persons produced 519,673 pounds of soap, and 117,686 pounds of tallow candles, with a capital of $17,345; 150 distilleries produced 1,551,664 gallons, and eleven breweries 96,390 gallons, the whole employing 233 persons and a capital of $135,155; 307 persons produced carriages and wagons to the amount of $144,399, with a capital of $59,963; 98 flouring-mills produced 179,667 barrels of flour, and, with other mills, employed 2994 persons, and manufactured articles to the amount of $2,417,396, with a capital of $2,147,613; vessels were built to the amount of $39,909; 944 persons produced furniture to the amount of $84,410, with a capital of $63,392; 324 brick or stone houses, and 4123 wooden houses were built, employing 5737 persons, and cost $2,065,952; 45 printing-offices, 5 binderies, 3 daily, 2 semi-weekly, and 36 weekly newspapers, and 9 periodicals, employed 175 persons, and a capital of $71,300. The total amount of capital employed in manufactures was $3,136,512.

The Illinois College (New School Presbyterian), at Jacksonville, was founded in 1829; Shurtleff College (Baptist), in Upper Alton, in 1825; McKendree College (Methodist), in Lebanon, in 1834; McDonough College (Old School Presbyterian), at Macomb, in 1837. In these institutions there were, in 1840, 311 students. There were in the state 42 academies, with 1967 students; 1241 common and primary schools, with 34,876 scholars; and 27,502 white persons over 20 years of age who could neither read nor write. There is a state penitentiary at Springfield.

The Methodists have about 170 travelling preachers, and are the most numerous denomination; the Baptists have about 160 ministers; the Presbyterians of different descriptions, about 100 ministers; the Episcopalians have about 10 ministers; the Roman Catholics 10 priests: and there are some other denominations.

At the beginning of 1842, the State Bank of Illinois, with its branches, had an aggregate capital of $3,616,193; and a circulation of $2,861,886. The state debt, at the close of 1840, amounted to $13,465,662. There is a state penitentiary at Alton.

The governor is elected by the people for four years, but is eligible only four years in eight. A lieutenant-governor is elected at the same time, who is president of the senate, and, in case of the death, resignation, or absence of the governor, discharges his duties. The senators are elected for four years, and the representatives for two years. The number of senators shall never be less than one third, nor more than one half the number of representatives. The judges of the supreme court are appointed by the joint ballot of both houses of the legislature, and hold their offices during good behaviour. Every white male inhabitant over 21 years of age, who has resided in the state for six months next preceding an election, has the right of suffrage.

In 1836 this state adopted an extensive system of internal improvements, consisting of canals and railroads, most of which it must be left to another generation to complete. The Illinois and Michigan canal, the most important of them all, is in progress, and will probably be completed. It extends from Chicago river, about 5 m. from Chicago, to the head of steamboat navigation on the Illinois river, at Peru, 106 m., including a navigable feeder of 4 m., and a few miles of river navigation. It was commenced in 1836, and is estimated to cost, when completed, $6,654,337. It is 60 feet wide at top, and 6 feet deep. A large amount of money has already been expended upon it. A railroad extends from Springfield 53 m. to Merodosia on Illinois river. Coal Mine Bluffs railroad extends from Mississippi river 6 m. to the coal mine. Work has been done on several other railroads, but they are at present suspended.

In the latter part of the seventeenth century, Illinois was explored by the French from Canada, and some forts and trading posts were established. About 1720 several forts were built within the present limits of Illinois, of which fort Chartres was the most considerable. A chain of communication was formed from Canada to the mouth of the Mississippi river. The oldest document in the state is at Kaskaskia, which is a petition to Louis XV. for a grant of common fields, stating the great losses of the people the year before by an extraordinary flood. At the peace of 1763 this country, together with Canada, was ceded to the English. In 1765, Capt. Sterling, of the Royal Highlanders, took possession of Illinois, and was followed by several other commanders, who occupied fort Chartres. In the Revolutionary war, the Virginia militia, under General Clarke, subjugated fort Chartres, Kaskaskia, and conducted a successful expedition, in 1788, against Fort Vincent, now Vincennes. In the same year the legislature of Virginia organized, in this remote region, the county of Illinois, which was afterward ceded to the United States. In 1800 the present territory of Illinois contained about 3000 inhabitants. In 1809 the territorial government was formed, and the population the next year amounted to 12,000. In 1812 a territorial government was formed, with a legislature and a delegate to Congress. In 1818 a state constitution was formed, and Illinois was received into the Union as the 22d state.

ILLINOIS, t., Washington co., Ark. It has 15 stores, two tanneries, three distilleries. Pop. 519.

ILLINOIS, CITY, p. v., Rock Island co., Ill., 172 m. N.N.W. Springfield, 901W. Situated between Rock river and the Mississippi.

ILLYRIA (KINGDOM OF), a portion of the Austrian empire, comprising the provs. of Carinthia, Carniola, and Istria, the islands of the Gulf of Quarnero, and the Illyrian Littorale. It lies between lat. 44° 25′ and 47° 7′ N., and long. 13° 14′ and 16° E., having N., Austria and Styria; E., the latter prov. and Croatia; W., the Tyrol and Italy; and S., the Adriatic sea. It is divided into the governments of Laybach and Trieste.

The divisions, with their extent and pop., are as follow:

Circles	Area in sq m.	Towns	Villages	Pop.	Chief Towns
Laybach	1,302	9	915	149,721	Laybach
Neustadt	1,239	15	1,524	185,423	Neustadt
Adelsberg	1,120	8	421	98,076	Idria
Klagenfurth	1,441	33	1,516	179,523	Klagenfurth
Villach	1,382	15	1,130	129,560	Villach
Trieste (Territory of)	40	1	94	76,813	Trieste
Istria	3,175	38	679	$11,030	Rovigno
Gorits	1,537	10	441	176,570	Gorits
Troops in garrison	. .	. .	. .	16,972	
	10,491	111	4,971	1,312,788	

Its N. part is covered by the central chain of the Alps and likewise by various offsets, constituting the southern limestone girdle of the Alpine system. The S. portion of the kingdom, comprising the government of Trieste, occupies the S. slope of this mountain-range towards the Adriatic. The main chain at the Gross Glockner (14,000 ft. high) takes the name of the Noric Alps, stretching its lofty

other productions
used.

...ture.—Illyria has
...of the N. govern-
...8 districts, which
The mountainous
...and damp climate,
...with difficulty.
...al crops; and the
...1-3d part of the
...t. The corn, in
...strung up on poles
...and these erections
...(oove, Slav.) are
...house. The most
...to Kavant, and the
...is of the valley of
...so severe as not
...the other hand,
...circle of Idria,
...rsisted. Excellent
...tino," are grown
...tious rotation of
...al succession of
...manure) maize,
...barley; 3d year,
...and is broken up
The quantity of
...in 1837, in the
...ed. (*Official re-*

...ues.

61,288 imp. qu.	
187,648	—
252,088	—
720,009	—
73,950,000 gallons.	
5,122,609 cub. tziem.	

both in its pro-
...soon as the
...most charm-
...fields are in
...covered with
...poplars, around
...an aspect of
...central Italy.
...dry, however, is
...E., good hus-
...which has a
...e territory far
...overed during
...t. Olives and
...he landowner
...ming for half
...this part of
...used land and

..870 imp. qu.	
...040	—
...820	—
...900	—
...800 gallons.	
...390	—
...00 cub. tziem.	

grain pro-
...ion: in the
...ilding must

...mp chiefly
...s (marcni)
The best
...cco, grown
The oil of
The stones
...are even
...vely culti-
...eglia and

...duce, and
...product. The
...imbed, in 1838,
...11,291 pounds of raw silk. The silk produced in Istria,
during the same year amounted to 4500 pounds; but till
1837 this article was not included in the land-tax returns of
the province.

The chief wild animals of the northern districts are the

Save, the climate is much milder. At Laybach, the temperature of the year is 8·7 Réaum. The temperature of
the government of Trieste presents a great contrast to that
of the mountain districts. In the valley of the Isonzo, as

9

of the kingdom, comprising the government of Trieste,
occupies the S. slope of this mountain-range towards the
Adriatic. The main chain at the Gross Glockner (14,000 ft.
high) takes the name of the Noric Alps, stretching its lofty

... All this region contains extensive ice fields and glaciers. At the Ankogel the Noric Alps, taking a N.E. course, enter Styria; but a branch bounds the vale of the Drave on the N., and that of the Levant on the E., separating their waters from those of the Mur. The Carnic Alps form the S. boundary of the valley of the Drave, dividing it from that of the Save. Various summits in this chain are from 6000 to 8000 ft. high; and over one of them, the Loibel, the emperor Charles VI. constructed the road connecting the Drave and the Save valleys: its summit-level is 5477 ft. above the sea. At Mount Terglou, the Julian Alps break off, running S.E. towards the Adriatic and Dalmatia; E. of Idria they decline in height, forming an elevated plateau, remarkable for drought and sterility, owing to the porous nature of its constituent limestone. Besides the pass over the Loibel, various others connect the fruitful valleys of this rocksalic country, the most remarkable being the Kanscher, 5930 ft. high, between the Drave and the Lungau; the Wurzen, 3100 ft., and the Pass of Tarvis 2600 ft., leading from the valley of the Drave to that of the Tagliamento. The valleys of the Gail (an. *Vallis Julia*), the Levant, and Isonzo (*Vallis Juncosa*), in Carinthia, and of the Save and Wochein in Carniola, offer all the varieties of Alpine beauty, while in the S. those of the Isonzo and Wippach, especially the former, present a picture of the richest Italian cultivation. The only level tracts of any considerable extent lie S. of the Julian Alps towards the Adriatic, and in the Istrian peninsula.

The Carnic and Julian Alps are perforated by very numerous subterraneous cavities, which, by draining the surface of water, condemn whole districts to a melancholy sterility. Several of these caverns are celebrated for their great size and curious natural phenomena, as the Cave of Adelsberg in Carniola, the neighbouring Magdalen cavern, in which the "*Proteus Anguinus*" is found, &c. Through several of these the mountain torrents find subterraneous channels, to the great detriment of agricultural prosperity. (*See* ADELSBERG.)

The N. portion of Illyria is well watered. The Drau or Drave, rising in Tyrol, traverses Carinthia in all its length, and receives tributaries from both the N. and S. mountain barriers of that province. It is navigable from near Klagenfurt to its mouth in the Danube. The river second in importance is the Sau, or Save, which traverses Carniola with an E. course parallel to that of the Drave. The banks of the Upper Save are mostly level; but the mountains close in on the river near Reichenberg. It is navigable from near Laybach; and receives various affluents, both in Carniola and Croatia. The rivers falling on the S. side of the Alps to the Isonzo, Ausa, and Timavo. The Isonzo, traversing the beautiful vale of Frisul, and taking near its mouth the name of Sdobe, falls into the sea near Monfalcone. The Ausa falls into the sea near Buso; and the Timavo (*Timavus*), with a course of scarcely more than 1500 yards, is navigable up to its source. Istria is very scantily watered: the Quieto, its principal stream, falls into the sea near Cittanuova, and, as well as the Arsa on the E. side of the peninsula, is navigable for some miles of its course.

There are several lakes in the N., but none of any great extent. The Lake of Klagenfurth, 11 m. long, is united with the neighbouring city by a canal. At a short distance from it is the Ossiach Lake, 7 m. long, and connected with the Drave by the Laybach. Further N.W. lies the Mahlstadt Lake, 10 m. in length and 1 m. broad, with very picturesque banks. The Weissensee, the Feldensee, (an. *Lacus Peracius*), and, lastly, the remarkable Zirknitzer-see, are of smaller extent. The lake of Zirknitz has two islands, and receives its waters through subterraneous channels. During the spring, and the autumnal rains, it presents a sheet of water 4 m. long, and 1 m. broad; but in summer the waters recede, and leave a dry fertile surface, either used for hay, meadows, or raising summer corn. The openings by which the water rises and retires are then visible, and various names have been given them by the peasantry; such as *Ketter* (the kettle), *Betschek* (the cask), *Rethis* (the corn sieve), *Reschete* (the great sieve), *Sittenza* (the hair sieve), &c. When the lake is full, it has an abundance of fish, which disappear and return with the water. In Istria there is only one lake, that of Zeppitsch, near Chersano. The climate of Carinthia is most inclement. The mean temperature of the year at Klagenfurth is estimated by Blumenbach at 7° Réaum.; while, at Obervillach, the mean is 6°. The snow lies in the lower parts of the valley of the Drave till the middle or end of April; but in the valley of the Save, the climate is much milder. At Laybach, the temperature of the year is 8·7 Réaum. The temperature of the government of Trieste presents a great contrast to that of the mountain districts. In the valley of the Isonzo, as well as in Istria, the olive, vines, and other productions of a southern climate, are largely cultivated.

*Occupations of the People—Agriculture.—*Illyria has two distinct agricultural systems; that of the N. government, which is Alpine, and that of the S. districts, which are cultivated in the Italian fashion. The mountainous districts of Carinthia, situated in a cold and damp climate, and having a short summer, are tilled with difficulty. Rye and summer corn are the most usual crops; and the three-course system, according to which 1-3d part of the land is in fallow, is generally prevalent. The corn, in order to dry thoroughly, requires to be hung up on poles or railings of a peculiar construction; and these erections (called *Harfen*, Germ., and *Stog* or *Kosov*, Slav.) are often covered with a roof like that of a house. The most productive corn region is the valley of the Levant, and the district of Krappfeld. In the higher parts of the valley of the Drave, near Gottschee, the climate is so severe as not to allow of winter crops. Carniola, on the other hand, especially the valley of the Save, and the circle of Idria has a warm climate, and is highly cultivated. Excellent wheat and maize, especially the "*conquantino*," are grown to a great extent; and there is a judicious rotation of crops. Blumenbach states that the usual succession of crops on good farms is :—First year (with manure) maize, potatoes, flax, or millet; 2d year, wheat or barley; 3d year oats; 4th and 5th years, clover. The ground is broken up both with the plough, and by hacking. The quantity of land under cultivation, and its produce in 1837, in the government of Laybach, are thus stated. (*Official returns.*)

Distribution of Surface.	English Acres.	Produce.		
Arable land	685,490	Wheat	.	52,225 imp. qu.
		Maize	.	187,640
		Oats	.	502,160
		Barley	.	120,020
Vineyards	37,540	Wine	.	3,850,000 gallons
Forests	2,140,840	Timber	.	1,125,000 cub. tons
Meadows and gardens	779,...			
Commons	1,...			

The S. part of Illyria differs essentially both in its productions and cultivation from the N. As soon as the traveller enters the valley of the Isonzo, the most charming landscape is presented to his view. The fields are in the highest state of cultivation, and being covered with rows of mulberries, or with elms and poplars, around which the vines cluster, the country bears an aspect of profuse fertility, superior even to that of central Italy. The mode of irrigation pursued in Lombardy, however, is not practised in Friaul; and on advancing E. good husbandry is found to diminish. In Istria, which has a climate as well calculated as the Milanese territory for raising oranges and lemons, if they were covered during the winter, the farming system is execrable. Olives and sumach afford the principal crops both to the landowner and his *coloni*. The *métayer* system of farming for half the produce of the land, prevails likewise in this part of the empire. In this government the cultivated land and its produce were, in 1837, as follows :—

Distribution of Surface	English Acres.	Produce.		
Arable land	384,120	Wheat	.	73,870 imp. qu.
		Maize	.	127,040
		Barley and rye	.	91,430
		Oats	.	32,400
Vineyards	26,464	Wine	.	18,000,000 gallons
Olive grounds	11,548	Olive oil	.	301,900
Forests	644,144	Timber	.	344,000 cub. tons
Meadows	29,225			
Commons	739,212			

From these statements, it appears that the grain produced in Illyria is insufficient for its consumption: in the district of the "Litorale" wood for fuel and building must be procured from other districts.

Good flax is grown in all the valleys, and hemp chiefly in Friaul. Fruits of all kinds, especially chestnuts (*marones*) and figs, are abundant in the coast district. The best wines are those of Monfalcone and the Prosecco, grown near Trieste; but very little wine is exported. The oil of Istria is considered equal to that of Provence. The stones and refuse of the olive are used for fuel, and are even exported to Ancona. The olive is also extensively cultivated in the Quarnero islands, especially Veglia and Cherso.

Cheese is a considerable article of farming produce, and a good deal is exported. Silk is an increasing product. The two spinning establishments at Farra furnished, in 1836, 11,591 pounds of raw silk. The silk produced in Istria during the same year amounted to 4500 pounds; but in 1837 this article was not included in the land-tax returns of the province.

The chief wild animals of the northern districts are the

chamois, red deer, and roebuck, and less frequently the wolf, bear, and small lynx. In the S. provinces, the ortolan and the common partridge, quails, water-fowls, and birds of passage are common. The fishery in the gulf of Quarnero, and in the channels between the islands, furnishes an abundance of fish peculiar to those waters.

Mines.—The chief wealth of Illyria consists in the rich metallic veins found in its mountains. The N. mountain chain separating Carinthia from Styria, consists of transition formations, overlaying mica slate, which composes the great spine of the Noric Alps, and contains vast quantities of a very superior iron ore. This chain opens S. into several valleys, sending tributaries to the Drave; and in these secluded districts the various mining operations are carried on, favoured by the water-power afforded by the mountain torrents. In the valleys in the Lieser, Gurk, Olsa, Mettnitz, and Lavant, iron is the chief product. The mountains near Huttenberg are rivalled in productiveness only by the most prolific of the Swedish veins. The ore is chiefly the carbonate of iron. The usefulness of these mines to the country is much impeded by the interference of the government with the industrial occupations of its subjects. In fact, the limitations on the export of iron, and the vexatious hinderances to enterprise, are such as to cramp all speculation; and the quantity annually produced corresponds neither with the wealth of the mines nor with the wants of the empire. In Carniola the same description of ore is found, near Feistriz, in the valley of Wocheim; at Sava and Jauerburg, in the valley of the Save; and in Lower Carniola, near Hof. There are rich mines of lead at Bleiberg, in Carinthia, and of quicksilver in Idria. The latter are situated in the E. portion of the Julian Alps, on the right bank of the Isonzo. The ore is found in a schistose rock, breaking through the predominant limestone of that chain; and as the veins get deeper, they are said to become richer. Blasting is the usual method employed for obtaining the ore; and the workmen, on account of the depth and consequent heat of the mines, work by relays of eight hours each gang. The lowest point in the mine is 300 feet below the bed of the adjacent Idrizza. The following is the return of the produce of the mines of Illyria, for the year 1837:

Gold and silver .		Remarks
Lead . . .	56,487 cwt.	
Iron . . .	301,393 —	value £300,000.
Coal . . .	99,653 —	
Alum and graphite .	190 —	
Quicksilv. (from Istria)	3,296 —	value £75,000.

The other occupations of the people, though less important, exhibit an annual increase. The following table shows the increase in the number of registered manufacturers and traders between 1829 and 1837:

Districts.	Manufactories.		Comm. Establishments.		Trades.		Special Occupations.	
	1829.	1837.	1829.	1837.	1829.	1837.	1829.	1837.
Carinthia and Carniola	149	221	265	326	24,554	25,665	798,	707
Gov. of Triest. excl. of cap.	22	48	154	190	6,218	6,800	797	957
	171	269	419	516	30,772	32,465	1,595	1,660
Total, in 1829 . . . 33,957					Ditto, in 1837 34,910			

Most of the manufacturers in the above table are employed in converting the metals into hardware, &c. There is no return of the commercial establishments in Trieste, inasmuch as that city is not included in the tax registers of the kingdom, its taxes being commuted for a payment of 60,000 florins annually. Flax spinning and linen weaving are the common and supplementary employments of the peasantry during their leisure from tillage labour, and the quantity annually produced for home consumption and exportation is considerable. Common woollen fabrics are likewise manufactured for home consumption; and fine cloths are made at Klagenfurth. There are 21 glasshouses, but only two cotton factories in Illyria.

In the trade returns of Illyria, Trieste, being a free port, is usually excluded. Its exports are chiefly metal and timber. The inhabitants of the district of Gottschee are almost all pedlars, who travel through foreign countries with their wares. The trade of Carinthia, Carniola, and the Illyrian coast, exclusive of Trieste, according to the official report for 1837, was as follows:

Imports	7,304,257	florins.
Exports	7,197,505	do.

The amount of the trade of Trieste with the rest of the empire, during the same year, was:

Imports	31,281,533	florins.
Exports	12,712,889	do.

The exportation of metals is chiefly confined to the other provinces of the empire, Germany and Italy. Formerly a considerable export trade was carried on with England; but it has almost ceased since the interruption occasioned by the continental blockade, and the increased production of iron in England. The present customs regulations, which prohibit, by extravagant duties, the exportation of raw steel, have also been most prejudicial to the iron trade of Illyria. The shipping lists, in 1837, gave the following report:

Port.	Vessels.	No.	Tons.	Crews.
Trieste	Ships . . .	270	70,968	4,301
	Coasters .	172	6,836	914
	Barks . .	221	2,110	602
Rovigno	Coasters .	377	8,972	1,674
	Barks . .	191	1,540	670
	Total	1,307	90,451	8,358

Since 1837, great activity has prevailed in the shipping interest, chiefly owing to the exertions of a joint-stock company, named "The Austrian Lloyd's," which has 10 steamboats running between Trieste and the harbours of Dalmatia and the Levant. The Illyrian coast has many excellent harbours, few of which, however, are made available for purposes of commerce. Istria abounds with ports, many large enough to shelter whole fleets, the principal of which are Capo d'Istria, Pirano (Porto Rose), Quieto, Pola, Parenzo, Rovigno, &c., but these are now only frequented by the barks conveying salt, wine, oil, gall nuts, charcoal, bark, and other productions of the peninsula to Trieste and Venice. There are likewise some tolerable harbours in the Quarnero islands, among which the port of Lussin Piccolo is, perhaps, the most capacious.

The roads of Illyria are as good as in most parts of the Austrian empire. The valleys of the Drave and Save are used for communication between Tyrol and Salzburg, and Carinthia and Carniola. Two main lines of road lead from the capital to Trieste, one by Klagenfurth and Gorizz, the other by Laybach. From Gorizz the former has a branch to Venice and other parts of Italy, while the latter is connected by roads, following the vales of the Save and Drave, with Hungary and the military frontier provinces. The internal navigation is limited to rafts on the Save and Drave, by means of which rivers and their tributaries, much timber is floated down from the forests to the Danube.

Population.—The population of Illyria, in the course of 20 years, has increased in Carinthia and Carniola at the rate of 17·4 per cent., and in the Litorale at 30·3 per cent.

The inhabitants (with the exception of the German settlers and of the Italians who have immigrated into the southern circles) are of Slavonian origin, and the vernacular language of Carniola, which is used as a written dialect, is one of the purest of the Slavonic idioms. Carniola is divided into Upper and Lower, the seats of the *Gorenzi Krainzi* and the *Dolenzi Krainzi;* the former of which are the mountaineers of the Julian Alps, the latter the inhabitants of the valley of the Save. The *Vipesri,* in the valley of the Wippach; the *Erasckova,* on the Karst; the *Pixchens,* in the Poik valley; and the *Zeitske,* are perhaps only local names. The general denomination for the Illyrian Slavonians is "Windi or Wenden" (*Venedi*). The inhabitants of Friaul call themselves "*Furlani;*" the peninsula is occupied by the "*Istrians,*" and the Quarnero islands by "*Liburnii.*" Nearly one million of the inhabitants are Slavonians.

The condition of the Illyrian population, though certainly improving, is by no means prosperous. Like so many of the Slavonian inhabitants of the empire, they speak a language which has not for centuries been the vehicle of intellectual improvement, and from an early period they were governed by tyrants, who availed themselves of their feudal rights, to the injury of the people, without conferring on them any of the advantages incidental to that system. In fact, the Illyrians had no national existence till the time of Napoleon. The ephemeral kingdom of Illyria which he established infused a spirit into all classes, which awakened them from the lethargy of ages. Much still remains to be done towards ameliorating the condition of the peasant, yet the change in his condition for the better, within the present century, is very great. The mountaineers of Carinthia and Upper Carniola are the poorest and worst fed of the inhabitants. Among them "cretins," or idiots, are of frequent occurrence, and are recommended to their neighbours' charity by the superstitious notion that their presence in a family indicates good fortune. *Goître* is common among the mountaineers, and the mortality is so great as scarcely to admit of any increase in the population. The inhabitants of the valleys, especially those living near the

Save, are in a better condition, and in the district of Goritz enjoy a considerable degree of prosperity. Istria, with all its natural advantages, is worse cultivated, and less civilized than the rest of Illyria. The dress of the mountaineers resembles that of the peasant of Tyrol and Salzburg. The women wear peaked, broad-brimmed hats; and in Carinthia, instead of stays they wear a red girdle, sewn to the linen tunic or shift, which is seen between the upper part and skirts of the gown worn over it. Formerly the men of the Gail valley wore a gay dress of motley colours, from which the costume of harlequin in the Italian comedy is said to be derived; indeed, many of the figures in pantomimes are believed to have been originally caricatures of the Illyrian peasantry.

The institutions for education have greatly improved within the present century, and consisted, in 1837, of three lycea, or colleges, with 431 students; seven gymnasia, or grammar schools, with 1074 scholars; and 476 elementary schools for both sexes, attended by 38,354 children, or about one fourth of those who, from the statistical returns, were of a legal age to frequent the schools: 479 Sunday and repetition schools are attended by 19,688 young persons of both sexes. The criminal returns for the same year do not exhibit a greater proportion of crime to population than in the other provinces of the empire. Murder and crimes of violence, however, are frequent; for of 691 criminal investigations, 85 were cases of murder and manslaughter, 44 of stabbing, 12 of arson, and 34 of riot and outrageous conduct; making a total of 175 offences against the person. Illyria has three penitentiaries: one at Laybach, for Carinthia and Carniola; one at Capo d'Istria; and one at Gradiska; containing altogether 472 prisoners, of whom 20 were sentenced for less than one year, 397 for less than 10 years, 171 between 10 and 20 years, and four for life, and 33 were in the jails of the various criminal courts.

The prevailing religion is Roman Catholic in both governments; but in Carinthia there are 17,500 Lutherans, chiefly in the circle of Villach, and about 400 communicants of the Greek church. In the government of Trieste there are about 1650 persons of the united Greek confession, 600 Protestants, and 3000 Jews.

The Roman Catholics are under five bishops: those of Goritz, Labach, Trieste, Gurk, and Lavant; the last two of which are suffragans of the archbishop of Salzburg. There are 37 monasteries and convents in the kingdom, tenanted by 321 monks and 207 nuns; the number of the secular Catholic clergy is 9431, performing the pastoral duties of 967 parishes. The administration of this province is the same with that of the other German and Slavonic provinces of the empire. The cities of Laybach and Trieste are the seats of the respective governments; but the general court of appeals for civil and criminal cases throughout the kingdom is held at Klagenfurth, where also is the mining court of Illyria. The city of Trieste has, besides its praetor's court, a sanatory commission, with two lazarettos in the harbour, and numerous deputations at various places along the coast. (For further particulars, see CARINTHIA.)

ILMINSTER, a market town and parish of England, co. Somerset, hund. Abdick and Bulstone, on the Ivel, 10 m. S.E. Taunton, 4 m. S. by W. Bath, and 127 m. W. by S. London. Area of parish, 4300 acres; pop., in 1841, 3227. The town comprises two streets, intersecting each other at right angles, one of which is nearly a mile long: the houses are irregularly built, some being of stone or brick, and the greater part merely thatched. The church, formerly conventual, is cruciform, in the decorated Gothic style, and has a square embattled and pinnacled tower. There are also places of worship for Wesleyan Methodists and Independents, to which, as well as to the church, are attached well-frequented Sunday schools. A free grammar school was founded in 1550, and endowed with considerable estates; there is also a hospital for the maintenance of clergymen's widows. Ilminster was formerly an important woollen clothing town; but its industry is now confined to the weaving of narrow cloths, and is of little importance. Lacenet mills have been recently established, and give employment to several hands. Petty sessions are held in the market-house. Markets on Saturday: fairs for horses, live stock, cheese, &c. the last Wednesday in August.

IMOLA (an. *Forum Cornelii*), a town of N. Italy, Papal States, legat. Ravenna; on the Santerno and the Emilian way, 18 m. N.W. Forli, and 20 m. S.E. Bologna. Pop. about 9000. It is a town of some consideration; being a bishop's see, surrounded by ancient walls and ditches, and further defended by an old castle. It is tolerably well built, and has a cathedral and 15 other churches, numerous convents, a hospital, theatre, college, and a literary academy, of some celebrity, termed *de' Industriosi*, which has included among its members several distinguished individuals. It has manufactures of cream of tartar, called *tartaro de Bologna*, &c., and some trade in agricultural produce.

INDEPENDENCE, county, Ark. Situated towards the N.E. part of the state, and contains 1250 sq. m. Bounded E. by Big Black river. Watered by White river. It contained, in 1840, 6096 neat cattle, 1686 sheep, 19,399 swine; and produced 9151 bushels of wheat, 219,635 of Indian corn, 8708 of oats, 3879 of potatoes, 19,595 of tobacco, 18,939 of cotton. It had eight stores, eight grist-mills, three saw-mills, two tanneries, two distilleries, one printing-office, two weekly newspapers; one academy, 55 students, two schools, 45 scholars. Pop.: whites, 3146; slaves, 514; free coloured, 9; total, 3669. Capital, Batesville.

INDEPENDENCE, p. t., Allegany co., N.Y., 90 m. S.E. Angelica, 261 m. W. by S. Albany, 300 W. Organized in 1821. Watered by Cryder and Independence creeks, which flow into Genesee river. It contains four stores, one fulling-mill, one woollen factory, two grist-mills, three saw-mills, two tanneries; 13 schools, 371 scholars. Pop. 1440.

INDEPENDENCE, t., Warren co., N.J., 14 m. N.E. Belvidere. Drained by Pequest creek and its tributary, Bacon creek. Bounded S.E. by Musconetcong river. It contains one Friends' church; 11 stores, one furnace, six flouring-mills, three grist-mills, one oil-mill, two distilleries; 13 schools, 580 scholars. Pop. 2284.

INDEPENDENCE, p. t., Cuyahoga co., O., 10 m. S. Cleveland, 145 m. N.N.E. Columbus, 338 W. The Ohio canal and Cuyahoga river passes through it. It has eight schools, 178 scholars. Pop. 754.

INDEPENDENCE, p. v., capital of Jackson co., Mo., 146 m. N.N.W. Jefferson city, 1072 W. The Mormons attempted a settlement here, but were obliged by the inhabitants to remove. The Santa Fé traders take their departure from this place, and obtain here many of their supplies. It contains a courthouse, jail, several stores, and about 300 inhabitants.

INDIA (BRITISH), a very extensive empire, chiefly situated in the central portion of S. Asia, comprising the greater part of the peninsula of Hindostan, or India within the Ganges, with the island of Ceylon, the provinces of Assam, Cachar, Jynteah, Aracan, Martaban, Tavoy, Ye, and Mergui, in India beyond the Ganges, acquired from the Birmese in 1826; Prince of Wale's Island (Pulo Penang), Malacca, Singapore, &c., or the straits' settlements, situated on or adjacent to the Malay peninsula. These vast dominions lie between lat. 1° 30' and 31° 13' N., and long. 71° 45' and 140° E.; their principal boundaries being N.W. the Indian desert; N. the Himalaya, which, in the upper provinces of Agra and in Assam, separates them from the Chinese empire, Nepaul, and Bootan; E. the Birman empire and Siam, and S. and W. the Indian ocean, the bay of Bengal, and the Arabian sea. The area and pop. of the principal divisions of British India have been estimated as follows:

Division.	Area in sq. m.	Pop.
In Hindostan :		
Presidencies of Bengal and Agra	304,000	49,710,000
Presidency of Madras	139,900	14,045,000
Presidency of Bombay	65,100	6,240,000
Island of Ceylon	24,450	1,344,000
In India beyond the Ganges :		
(a) Provs. conquered from the Birmese (under the Bengal presid.)		
Assam	18,900	698,000
Jynteah	10,350	270,000
Cachar	}	70,000
Aracan	16,230	230,000
Tenasserim coast { Martaban, Tavoy and Ye, Mergui and Archipelago }	32,500	85,000
(b) Straits' settlements:		
Penang, or Prince of Wales's Island, and prov. Wellesley	}	
Malacca	1,570	154,500
Singapore		
Total	608,470	93,929,000

To the foregoing territories, under the immediate rule of the British, there may be added the tributary states of Berar, Oude, Mysore, Travancore, Cochin, Sattarah, the dominions of the Nizam, of the Rajpoot and Bundlecund chiefs, &c., which are substantially administered by British rulers, and are either entirely or in part surrounded by British territories, are estimated altogether to comprise about 433,000 sq. m., and a population of about 41,000.000.

The physical geography, products, inhabitants, industry, &c., of the several divisions, provinces, and districts of British India, will be found treated of under the head Hindostan, and in separate articles appropriated to each. The present article will, therefore, be principally occupied with those topics, such as the general government, the judicial and revenue systems, army, commerce, &c., of British India, that could not be conveniently introduced under any other head.

Government.—Previously to 1773, the government of that part of India that then belonged to the British was vested

11

in the East India Company. The body of proprietors of East India stock, assembled in general court, elected 24 directors, to whom the executive power was entrusted, the body of proprietors reserving exclusively to themselves all legislative authority. A vote in the court of proprietors was acquired by the holders of £500 of the company's stock; but to be a director, it was necessary to hold £2000 stock. The directors, with their chairman and deputy chairman, were chosen annually, and subsequently subdivided themselves, for despatch of business, into 10 separate committees. As early as 1707, the three principal presidencies into which British India is divided—those of Bombay, Madras, and Calcutta, were in existence. Each was governed by a president or governor, and a council of from nine to twelve members, appointed by commission of the company. All power was lodged in the president and council jointly, every question that came before them being decided by a majority of votes. In 1726, a charter was granted, by which the company were permitted to establish a mayor's court at each of the presidencies, consisting of a mayor and nine aldermen, empowered to decide in civil cases of all descriptions, with an appeal from their jurisdiction to the president and council. The latter were also vested with the power of holding courts of quarter sessions, for the exercise of penal judicature, in all cases excepting those of high treason, as well as a court of requests for the decision, by summary procedure, of pecuniary questions of inconsiderable amount. Added to this, the powers of justices of the peace were granted to the members of the council, and to them only, the president being at the same time commander in chief of all the military force stationed within his presidency. It will thus be readily seen that the officers of the company were recognised as the judges in their own cause in all cases; and that, notwithstanding the establishment of the mayors' courts, they still held all the judicial as well as the executive functions, both civil and military, in their own hands. An individual who became a member of the council was not debarred from subordinate functions; and from this circumstance especially it might have been expected that abuses would prevail; and to the abuses which thence arose, in fact, Mr. Mill attributes the embarrassments in which the affairs of the company afterwards became involved.

In 1773, the great increase in the territorial possessions of the company attracted the attention and excited the cupidity of the government at home; while the financial embarrassments of the company, and the abuses which had crept into the government of India, furnished ample grounds for interference. In consequence, the ministry introduced two bills into parliament, distinctly asserting the claim of the crown to the territorial acquisitions of the company, raising the qualification to vote in the court of proprietors from the possession of £500 to that of £1000 stock; giving to every proprietor possessed of £3000 two votes, of £6000 three votes, and of £10,000 four votes; limiting the annual election of the whole 24 directors to that of six only; vesting the government of Bengal, Bahar, and Orissa in a governor-general, with a salary of £25,000 a year, and four councillors, of £8000 each; rendering the other presidencies subordinate to that of Bengal; and establishing at Calcutta a supreme court of judicature, consisting of a chief justice, with £8000 a year, and three puisne judges, with £6000 a year each appointed by the crown. As subsidiary articles, it was proposed, that the first governor-general and councillors should be nominated by parliament in the act, and hold their office for five years, after which the patronage of those great offices should revert to the directors, but still subject to the approbation of the crown; that everything in the company's correspondence from India which related to civil or military affairs, to the government of the country, or the administration of the revenues, should be laid before ministers; that no person in the service either of the king or the company should be allowed to receive presents; and that the governor-general, councillors, and judges should be excluded from all commercial speculations and pursuits.

Mr. Pitt's famous India bill of 1784 established the board of control, consisting of six members of the privy council, appointed by the king, two of the principal secretaries of state being always members. The president of the board is, in fact, secretary of state for India, and is the officer responsible for its government, and for the proceedings of the board. The superintendence of the latter extends over the whole civil and military transactions carried on in India. It revises, cancels, or approves all despatches, letters, orders, or instructions proposed to be sent out by the court of directors to the government in India; it may also require the court to prepare and send out despatches on any given subject, couched in such terms as it may deem fit; it may transmit, in certain cases, orders to India, without the inspection of the directors, and has access to all the company's papers and records, and to all proceedings of the courts

of directors and proprietors. It is clear, therefore that, from 1784, when the board of control was established, the real sovereignty of British India was taken out of the hands of the company, and placed in those of ministers.

Under the act of 1833 (3 & 4 William IV., cap. 85), the company holds, under the superintendence of the board of control, the political government and patronage of British India, till the 30th of April, 1854; but its exclusive commercial privileges are no longer in existence. The supreme authority is vested in the governor-general, who is also governor of the presidency of Bengal. He is nominated by the court of directors, the nomination being subject to the approval of the sovereign, and is assisted by a council of five members, three of whom are appointed by the court of directors, from among persons who are or have been servants of the company; the fourth is also chosen in a similar manner, but from among persons unconnected with the company; and the fifth is the commander-in-chief, who takes rank and precedence immediately after the governor-general. The other presidencies have also their governors and councils, subordinate to the governor and council of the Bengal presidency; the presidency of Agra, however, comprising the upper provinces of Bengal, is at present administered by a lieutenant-governor only. The governor-general in council is competent to make laws for the whole of British India, which are binding upon all the courts of justice, unless annulled by higher authority. Parliament reserves to itself the right to supersede or suspend all proceedings and acts of the governor-general; and the court of directors has also power to disallow them. The foregoing remarks do not, however, in any way apply to Ceylon, which is quite independent of the jurisdiction of continental India, being placed directly under the colonial secretary of Great Britain. By the act of 1833, the salaries of the principal civic officers in India were fixed, that of the governor-general at £24,800 a year; the governors of the Bombay and Madras presidencies, £11,600; the ordinary members of the head council, £9600 each; and the members of the other councils, £6000 each yearly. (Parl. Acts, Reports, &c.; Mill's Hist., &c.)

Judicial System.—When, in 1793, the Marquis Cornwallis undertook his reform of the judicial and revenue systems of British India, that territory was in a most desplorable state. "The administration of justice through all its departments was most pernicious and depraved; the public revenue levied upon principles incompatible with the existence of private property; the people sunk in poverty and wretchedness; more than one third part of the country a desert, and the rest hastening to desolation." (Mill, v., 428.) Under the orders sent to India in 1786, the same individuals combined the business both of judicature and finance; being at once collectors of revenue, judges, and heads of the police. Lord Cornwallis endeavoured to separate these apparently incompatible offices, and distributed them among different individuals. He gave to native commissioners power to determine civil suits among natives to the value of 50 rupees, several of whom he established in each zillah or district, giving an appeal from their decisions to the zillah court, held in the principal town of the district, of which one of the company's servants was appointed the judge. The latter functionary was assisted by a registrar, and some other members from among the junior servants of the company, and natives duly qualified to expound the Hindoo or Mohammedan law. These courts had jurisdiction in cases to the amount of 1000 rupees. From them appeal might be made to four provincial courts established at Calcutta, Patna, Dacca, and Moorshedabad. These courts consisted of three judges, chosen from the civil department of the company's service, a registrar, one or more assistants from the junior European servants of the company, and three expounders of the native law—a cazee, mufti, and pundit. A higher tribunal, that of Sudder Dewannee Adawlut, was established at Calcutta, composed of the governor-general, his counsel, the head cazees, two mufties, two pundits, a registrar, and assistants. All suits of Europeans were exclusively tried in this court: appeal from it lay only to the king in council, in cases above the amount of 50,000 rupees. Four tribunals were erected in the four provincial courts, for criminal judicature, at which the judges, &c., of the civil tribunals officiated every month; the penal judicature was administered in most of the country districts only twice, but in that of Calcutta four times a year. The superior criminal tribunal was the Nizamut Adawlut, held at Calcutta, and constituted almost similarly to the Sudder Dewannee Adawlut.

But with all this machinery of legislation, nothing like a code of laws was promulgated. The Hindoo and Mohammedan population were governed by the rules laid down in their respective sacred books—the Shasters and the Koran —as interpreted by the ever-varying opinions of the pundits and cazees. The courts established on the European model were infected with all that multiplication of technical forms, which forms the worst feature of our own legal code, and all that delay and expensiveness of process, which tend to destroy the ends of justice, followed as a matter of course. The errors in the system adopted were great; but, considering the state in which Lord Cornwallis found affairs, it may be truly said that he effected a vast deal of good. He was actuated by the purest and most benevolent motives; and wisely endeavoured to respect, in as far as possible, the different legal codes of the various sects and natives comprised in the population of India.

Of late years, however, a disposition has grown up to unite again the judicial, magisterial, and revenue authorities which Lord Cornwallis had separated. A considerable change of this description was introduced by Sir T. Monro in the Madras territories, and more recently by Lord W. Bentinck in Bengal. (See *Revenue and Judicial Selections*; *Asiatic Journal, &c.*)

Within the cities of Calcutta, Madras, and Bombay, and also within the settlements of Penang, Singapore, and Malacca, English civil and criminal law is administered to both natives and Europeans, with the exception of their own laws of inheritance being preserved to the former. But beyond the limits of the above-mentioned cities, on the continent of India, the native laws have been made binding on Europeans as well as natives. The charter of 1833 provides that no one shall, by reason of his nation, colour, or faith, be disqualified from holding office under the company; and that, henceforth, there shall be no distinction of blood or nativity. "Upon this ground," says Mr. Crawford, "the legislative council of India, without vesting in the code of laws which, under direction of the same statute, was in course of preparation, passed a law subjecting Europeans to the same tribunals to which natives are subject, although these tribunals administer their own domestic laws respectively to Hindoos and Mohammedans, are cognisant of no others, repudiate expressly the laws of England, and are presided over by natives, or by unprofessional European servants of the East India Company, the first of whom rarely know a word of English; while the proceedings of the courts are both conducted in the native languages, to the express exclusion of the English tongue. This act, from its unpopular character, is commonly called by Europeans in India, the 'Black Act.'"

There can be no doubt that, under the act of 1833, Europeans gained great advantages by the abolition of the East India Company's monopoly and trade, the power to possess land, and the comparatively ample field which is thus opened to their enterprise. It is alleged, however, that, in so far as respects their rights, liberties, and laws, they are in a less favourable position than under the old system. Under the latter, British subjects, within the special jurisdiction of the king's courts, could only, like the native, be tried by their own laws, and the local government could enact no new law for their government not in accordance with the "laws of the realm." But by the new system, the governor-general in council may enact any laws whatsoever that shall be binding on British subjects, whether the same be consonant with the "laws of the realm," or otherwise. Under the old law, an appeal lay to the privy council, which any individual might institute. The privilege is now, however, cut off, and, under the modern system, nothing short of an act of parliament can repeal a law that has the sanction of the Indian authorities. Under the old system, British-born subjects in the provinces, that is, beyond the special jurisdictions of the king's courts, were, in civil cases, amenable only to courts presided over by their countrymen in the commission of the peace. Under the new system, they are amenable to the extent of £500 in value, to the pettiest native tribunal, presided over by a Mohammedan or Hindoo—judges equally ignorant of their manners, laws, and language, and, with few exceptions, viewing their religion with hatred or contempt. The appeal in the king's courts, which was a guarantee for their own laws, is taken away from them; and it lies to the chief native tribunal—of which the judges, indeed, are Englishmen, but is which the proceedings are in the native tongue—in which there is no one to advise the judges, and where an English barrister is not even permitted to plead. Under the old system, an appeal lay, even from the competent tribunals of the king's courts, to the king in council; under the new system no such appeal lies from the native tribunal, unless the value be above 25 times as much as it was before the innovation.

We believe, however, that, practically, little inconvenience has arisen, or is at all likely to arise, from most of these regulations. We may be quite sure that the power given to the governor-general and council of enacting laws will not be rashly or capriciously exercised. How exalted soever, these functionaries are not merely responsible to parliament, but to public opinion: the free press, now established in India, will not fail to advertise them of any error they may be likely to commit; while the growing at-

tention given to Indian affairs at home will tend to make them wary in their proceedings. We are less able to judge of the expediency of making British-born subjects responsible to the native tribunals; but even this is, we believe, less objectionable than it might appear to be.

Revenue system.—The land-tax constitutes the principal source of the revenue of British India, as it has always done of all eastern states. The governments of such countries may, in fact, be said to be the real proprietors of the land; but in India, as elsewhere, the cultivators have a perpetual, hereditary, and transferable right of occupancy, so long as they continue to pay the share of the produce of the land demanded by the government. The value of this right of occupancy to the rural population depends on the degree of resistance which they have been able to oppose to the exactions of arbitrary governments. In Bengal and the adjacent provinces of India, from the peculiarly timid character of the inhabitants, and the open and exposed nature of the country, this resistance has been trifling indeed, and, consequently, the value of the right of occupancy is the present, or *ryot* (an Arabic word, meaning subject), has been proportionally reduced. This, also, may be considered, though with some modifications, as being nearly the condition, in this respect, of the inhabitants of every part of the great plain of the Ganges, comprising more than half the population of Hindostan. But where the country is naturally difficult, the people have been able more effectually to resist the encroachments of the head landlord, or state, and to retain a valuable share in the property of the soil. This has been particularly the case along the ghauts, as in Bednore, Canara, Malabar, &c.; the inhabitants of which provinces not only lay claim to a right of private property in the soil, but have been generally ready to support their claim by force of arms. There can be no question, indeed, that the same modified right of property formerly existed everywhere; and it is indeed impossible that otherwise the land should ever have been reclaimed from the wilderness. But, in those parts of India which could be readily overrun by a military force, the right of property in the soil has long been little else than the right to cultivate one's paternal acres for behoof of others, the cultivators reserving only a bare subsistence for themselves.

Under the Mogul emperors, the practice in Bengal was to divide the gross produce of the soil, on the *mitayer* principle, into equal shares, whereof one was retained by the cultivator, the other going to government as rent or tax. The officers employed to collect this revenue were called *zemindars*; and in course of time their office seems to have become hereditary. It may be remarked that, in Persian, zemindar and landholder are synonymous; and this synonology, coupled with the hereditary nature of their office, which brought them exclusively into contact with the *ryot*, or occupier, as well as with the government, led many to believe that the zemindars were in reality the owners of the land, and that the ryots were their tenants. This, however, it is now admitted on all hands, was an incorrect opinion. The zemindars in reality were tax-gatherers, and were, in fact, obliged to pay to the government *nine tenths* of the produce collected from the ryots, retaining only one tenth as a compensation for their trouble; and, so long as the ryots paid their fixed contribution, they could not be ousted from their possessions, nor be in anywise interfered with.

But, notwithstanding what has now been stated, the perpetual or zemindary settlement, established by Lord Cornwallis in Bengal in 1793, was made on the assumption that the zemindars were the proprietors of the soil. His lordship, indeed, was far from being personally satisfied that such was really the case; but he was anxious to create a class of large proprietors, and to give them an interest in the improvement and prosperity of the country. It is clear, however, that this wish could not be realised without destroying the permanent rights of the ryots, for, unless this were accomplished, the zemindars could not interfere in the management of their estates. The interests of the zemindars, and the rights of the ryots, were plainly irreconcilable; and it was obvious that the former would endeavour to reduce the latter to the condition of tenants at will. But this necessary consequence was either overlooked or ineffectually provided against. The zemindars became, under condition of their paying the assessment, or quit-rent, due to government, proprietors or owners of the land. The amount of the assessment was fixed at the average of what it had been for a few years previously, and it was declared to be *perpetual* and *invariable* at that amount. When a zemindar fell into arrear with government, his estate might be either sold or resumed.

That the assessment was at the outset, and still is, too high, cannot well be doubted; and it must ever be matter of regret that the settlement was not made with the ryots, or cultivators, rather than with the zemindars; but, notwithstanding these and other defects, the measure was, on the whole, a great boon to India. Until the introduction of the perpetual system into Bengal, the revenue was raised in it, as it continues to be in the rest of India down to the present day, by a *variable* as well as a most oppressive land-tax. We all know what a pernicious influence tithe has had in this country; but suppose that, instead of amounting to 10, tithe had amounted to 50 per cent. of the gross produce of the soil, it would have been an effectual obstacle to all improvement; and the country would now have been in about the same state as in the days of Alfred, or of William the Conqueror.

In France, Italy, and other parts of Europe, where the *métayer* system is introduced, the landlord seldom or never gets half the produce, unless he also furnish the stock and farming capital, and, in most cases, the *seed*. But in India, neither the government nor the zemindars do anything of the sort: they merely supply the land, which is usually divided into very small portions, mostly about six, and rarely amounting to 24 acres. A demand on the occupiers of such patches for half the produce is quite extravagant, and hence the excessive poverty of the people, which is such as to stagger belief. Still, however, the perpetual system is vastly preferable in principle, and also in its practical influence, to any other revenue system hitherto established in India. It set limits to fiscal rapacity, and established, as it were, a rampart beyond which no tax-gatherer dared to intrude. The enormous amount of the assessment, and the rigour with which payment was at first enforced, ruined an immense number of zemindars. But their lands having come into new and more efficient hands, a better system of management was introduced, and the limitation of the government demand gave a stimulus to improvement unknown in any other part of Hindostan. This, in fact, was the grand *desideratum*. A land-tax that may be increased should the land be improved, is all but certain to prevent any such improvement being made. This has been its uniform operation in every country in the world that has had the bad fortune to be cursed with such a destructive impost. But a heavy land-tax, provided it be fixed and unsusceptible of increase, is no bar to improvements, unless in so far as it tends to deprive the proprietors and occupiers of land of the means of making them. There is, in such a case, no want of security, and the cultivator is not deterred from attempting improvements, or of bringing superior enterprise and industry to operate on his estate, by the fact that the tax will, in consequence, be increased.

The truth of what is now stated has been fully evinced in Bengal during the last 20 or 30 years; for both the population and the land-revenue of that part of our Indian empire has greatly increased. A great deal of waste land has been cultivated, and various works have been undertaken that would not be so much as dreamed of in any other part of our empire in the east. But, with all this, there has been but little, if any, improvement in the condition of the people of Bengal under our government. They, in fact, are practically excluded from, at least, all direct participation in the benefits resulting from the limitation of the assessment. They have merely exchanged one taskmaster for another. It is their landlords who have been the great gainers. The occupiers still, generally speaking, hold under the *métayer* principle, paying half, or even more, of their produce as rent; so that their poverty is often extreme, and their condition not unfrequently inferior even to that of the hired labourer, who receives the miserable pittance of two annas, or about 3d., a day as wages.

It seems, however, as if there were some strange fatality attending the government of India; and that the greatest talents and the best intentions should, when applied to legislate for that country, produce only the most pernicious projects. The perpetual settlement carried into effect by Lord Cornwallis in Bengal was keenly opposed by Lord Teignmouth, Colonel Wilkes, Mr. Thackeray, Sir T. Monro, and others, whose opinions on such subjects are certainly entitled to very great respect; and it would seem that the Board of Control became, at length, favourable to their views. In consequence of this change of opinion it was resolved to introduce a different system, under the superintendence of its zealous advocate, Sir Thomas Monro, into the presidency of Madras, or Fort St. George. This new system has received the name of the *ryotwar* settlement. It proceeds on the assumption that government possesses the entire property of the soil, and may dispose of it at pleasure; no middlemen or zemindars are interposed between the sovereign and the cultivators; the ryots being brought into immediate contact with the collectors appointed by government to receive their rents. It is impossible, however, to enter fully into the details of this system. They are in the last degree complicated, which of itself would be enough to show their inexpediency. The land is taxed, according to its quality, at rates varying from 6d. up to 70s. an acre. Thus, for example, if the land were mere *dry field*, without artificial irrigation, the land-tax would be about 3s.

an acre. If it had a supply of water capable of growing rice, the tax rises to 22s., or to nearly eight times the former rent; and if the irrigated land be a garden, or an orchard, the tax rises to 46s., or above 13 times the tax on dry land! In the first instance the natural and inherent fertility of the soil only is taxed; in the second, to that tax is added one on the capital and labour which the peasant or his ancestor laid out in reservoirs, canals, trenches, or wells. In the third, not only are all these taxed, but there is imposed besides an excise on fruits, garden stuffs, and potherbs. But the radical vice of the system is, that the lands are not let for a considerable number of years, or for ever. On the contrary, there is a constant tampering and interference with the concerns of the ryots. It is enacted, for example, that "at the end of each year the ryot shall be at liberty either to throw up a part of his land, or to occupy more, according to his circumstances." When, owing to bad crops, or other unforeseen accidents, a ryot becomes unable to pay his rent or assessment, it is declared that "*the village to which he belongs shall be liable for him to the extent of 10 per cent. on the rent of the remaining ryots, but no more.*" And, to crown the whole, the tehsildars, or native officers, employed in collecting the land-rents, or revenue, have been vested with powers to act as officers of police, to impose fines, and even to inflict corporal punishment almost at discretion!

It is really astonishing how acute and able men should have dreamed of establishing a system in an extensive and only half civilized country that every one must see would be destructive of the industry of the tenants, and would lead to the grossest abuses, were an attempt made to introduce it into the management even of a single estate in Great Britain. Mr. Tucker, a gentleman who resided long in India, and now occupies a place in the company's direction, has animadverted on this plan as follows: "My wish," says he, "is not to exaggerate; but when I find a system requiring a multiplicity of instruments, surveyors and inspectors, assessors, ordinary and extraordinary; potails, curnums, tehsildars, and cutcherry servants; and when I read the description given of these officers by the most zealous advocates of the system, their periodical visitations are pictured in my imagination as the passage of a flight of locusts, devouring in their course the fruits of the earth. For such complicated details, the most select agency would be required; whereas the agency we can command is of the most questionable character. We do not merely require experience and honesty to execute one great undertaking; the work is ever beginning and never ending, and calls for a perennial stream of intelligence and integrity. And can it be doubted that the people are oppressed and plundered by these multiform agents? The principle of the settlement is to take one third of the gross produce on account of government; and, in order to render the assessment moderate, Sir T. Monro proposed to grant a considerable deduction from the rates deducible from the survey reports. But, if it be moderate, how does it happen that the people continue in the same uniform condition of labouring peasants? Why do not the same changes take place here as in other communities? One man is industrious, economical, prudent, or fortunate; another is idle, wasteful, improvident, or unlucky. In the ordinary course of things, one should rise and the other fall: the former should, by degrees, absorb the possessions of the latter; should become rich, while his neighbour remained poor: gradations in society should take place; and, in the course of time, we might naturally expect to see the landlord, the yeoman, and the labourer. And what prevents this natural progression? I should answer, the *officers of government*. The fruits of industry are nipt in the bud. If one man produce more than his fellows, there is a public servant at hand ready to snatch the superfluity. And wherefore, then, should the husbandman toil, that a stranger may reap the produce?

"There are two other circumstances which tend to perpetuate this uniform condition. The ryots have no fixed possession; they are liable to be moved from field to field: this they sometimes do of their own accord, for the purpose of obtaining land supposed to be more lightly assessed; at other times the land is assigned by lot, with a view to a more equal and impartial distribution of the good and the bad, among the different cultivators. But these revolutions tend to destroy all local attachments, and are evidently calculated to take away one great incentive to exertion.

"The other levelling principle is to be found in the rule, which requires that the ryot shall make good the deficiencies of his neighbour to the extent of 10 per cent.; that is to the extent, probably, of his whole surplus earnings. Of what avail is it that the husbandman be diligent, skilful, and successful, if he is to be mulcted for his neighbour's negligence or misfortune? A. must pay the debt of B. If a village be prosperous it matters little, for the next village may have been exposed to some calamity; and from the abundance of the one we exact wherewithal to supply the deficiency of the other. Is it possible to fancy a system latter calculated to baffle the efforts of the individual, to repress industry, to extinguish hope, and to reduce all to one common state of universal pauperism?' (*Review of the Financial Policy of the E. I. Company*, p. 134.)

It will be afterward seen that, notwithstanding the long period of tranquillity that the Madras territories have enjoyed, the land revenue, instead of increasing, as it should have done under any reasonable system, has been progressively declining. The organization and maintenance of the existing ryotwar system is, in truth, the most discreditable fact connected with the history of British India. The assessment of the land revenue in Madras is, in every respect, quite as objectionable as the assessment established by Mehemet Ali, in Egypt (*See* Vol. I., p. 747); and it would seem, indeed, that the pacha had had the land revenue code of the Madras Presidency before him when he framed his code: if there be any substantial difference between the two, that of the pacha, arbitrary and oppressive though it be, is entitled to the preference.

We have already stated enough to show that a variable land-tax is, in all cases, most injurious to a country. It is understood to have been adopted by the authorities in India and England, in the expectation of enabling the government to participate in the advantages resulting from the improvement of the old lands, and from the bringing of new or waste land into cultivation. But it is clear, as well from the experience of Madras itself as of all other countries in which it has been tried, that a continually varying land-tax is an insuperable barrier to all improvement; and that it is, in fact, a powerful cause, not of advancement, but of poverty and barbarism. But the power of periodically revising the assessment might be retained without perpetually tampering with the occupiers. The only effect of this is to paralyse industry, to make those who are not poor counterfeit poverty, and to hinder any outlay of capital on the land. To obviate these disastrous consequences, the proper plan would be to assure the occupiers at a reasonable rate, and to make the assessment invariable for a period of at least 40 or 50 years. An arrangement of this kind would give the ryots that security of which they are now wholly destitute; and would, we are bold to say, do ten times more to improve the Presidency than all the other measures it is possible to adopt, save that of making the assessment perpetual. This plan is, in fact, beginning to be tried in some parts of India; and it has, we are assured, been attended with the best results.

The land revenue in most parts of British India is assessed under one or other of the systems now described; but in some parts of the Bengal provinces, in the ceded districts on the Nerbudda, and in the greater number of the native states, a different plan is adopted, which has received the name of the *village system*. This system, though defective in many respects, is incomparably superior to the ryotwar system, and, in some points, is even preferable to the perpetual system. It is a settlement made between the government and the cultivators, through the medium of the native village officers, who apportion the assessment without any direct interference on the part of the government functionaries. (*See* art. BOMBAY PRESIDENCY, for a short notice of this system of assessment.) It is difficult to state the proportion of the produce of a village paid to government. The authorities know little of the precise property of any of the proprietors: it is not the interest or the wish of the village that they should; and, if any member of the community fail to pay his share, that is a matter for the village at large to settle, and they usually come forward and pay it for him. These, however, are private arrangements; and the *mocuddim*, or headman, through whom the government settles with the cultivators, has no power from the government to enforce the assessment on the particular defaulter. The tax to be paid by each villager is settled by the villagers among themselves; the total assessment being calculated after inquiry into the property of the village—what it has paid and what it can pay—regular surveys of the village boundaries, and of its lands, having been previously made by government. The *mocuddim* or *potail* (headman) is elected by the villagers; and, if the latter become dissatisfied with him, they turn him out of office. This system may have, and doubtless has, its disadvantages; the potails may, from various motives, unequally assess the villagers; and the tendency to cultivate waste lands will not be so strong as under the perpetual settlement; but the latter effect is much more likely to be brought about under this than under the ryotwar system; nor does the village system involve the same inquisitorial acts on the part of government. If the amount of the tax charged on a village under this system were not too high, and if the amount, when once fixed, were made perpetual or invariable, for a period of at least 40 or 50 years, it would probably be as good a plan as could be devised for the assessment of the land-tax.

We may, in this place, compare the respective results which have followed under the different revenue systems,

but especially where the permanent and ryotwar systems of taxation have been established. In 1792-94, the total gross revenue of the four provinces of Bengal, Bahar, Orissa, and Benares, was £4,139,948, of which £3,549,589 consisted of land-tax, only £2,573,714 being, however, actually collected. In 1837-38, the total *gross* revenue amounted to £8,843,723, or to more than double its amount in 1792-94. The land-tax in 1837-38 amounted to £3,377,993, which was almost all collected. The produce of the other branches of revenue amounted, in 1837-38, to no less than £5,464,890, being nearly *five* times the produce in 1792-93, when the perpetual settlement was organised! It should also be observed that Bengal, which, but a short time previously to 1793, had been the theatre of a most frightful famine, has not since been afflicted with even a year of remarkable scarcity; while both famines and scarcities have been frequent in every other part of our dominions in Hindostan. In 1793, the highest estimate of the population of these provinces, exclusive of Benares, was 24,000,000; in 1835 it had risen to 37,500,000, or increased by more than a half in 33 years.

In the Madras Presidency the land-tax, in 1805-6, amounted to £3,469,977; in 1814-15 to £3,438,195; and in 1837-38 to only £3,149,781! being a decline of £380,000 a year; whereas the land-tax in Bengal during the same period had increased more than half a million! But how could it be otherwise? In Madras the tax, besides being assessed in the worst possible manner, is oppressively high; indeed, the land-tax paid by that Presidency is almost equal to that paid by the far richer and wealthier country of Bengal, Bahar, Orissa, and Benares, with more than double its population! The other taxes in Madras are also more onerous than in Bengal; and several, such as a monopoly of tobacco, a tax on fruit trees, on cow-dung used as fuel, and on arts and professions, are unknown in the latter. But notwithstanding, while in Bengal the land-tax amounts to little more than a third, it amounts in Madras to fully three fourths of the total revenue of the Presidency.

In the upper provinces of Bengal, now forming the government of Agra, where both the ryotwar and village systems prevail, and where the population is estimated to be about 15,000,000, or not quite half that of the four provinces of Bengal, Bahar, Orissa, and Benares, the land-tax, in the years 1806-7, was £2,103,410; in 1811-12 it was raised to £2,665,484; in 1819-20 to £3,061,932; and in 1829-30 to £3,766,566. In the short space of 23 years the tax had, therefore, been augmented by *the enormous sum of £1,663,156*. But this augmentation proved to have been a great deal too rapid; for in 1834-35 the land-tax realised in the upper provinces sank to £3,398,094, at the same time that the other branches of revenue amounted to only £796,867, making the land-tax 81 parts in 100 of the whole revenue. Two years afterward a dreadful famine broke out in the Agra provinces; and not only was little or no revenue collected, but the tax-receivers had to dole out relief to the tax-payers. In the Bombay Presidency, where fluctuating assessments prevail, the land-tax, in 1837-38, amounted to £1,727,717, collected at an expense of £364,717, or about one sixth part of its gross produce. The gross amount of all the other branches of the Bombay revenue amounted, during the same year, to only £389,119.

These statements conclusively demonstrate the vast superiority of the perpetual settlement, not merely as respects the prosperity of the country and the inhabitants, but also as a financial engine. Had the perpetual settlement been adopted in Madras when it was adopted in Bengal, we venture to say that the revenue of the former, instead of remaining stationary, or retrograding, would have advanced quite as rapidly as in the latter, while the population and wealth of the Presidency would have been proportionally increased.

Besides the lands subjected to the foregoing systems of assessment, a considerable extent of land in India is held rent-free. Throughout Hindostan and, indeed, we believe, throughout Asia, China perhaps excepted, a considerable portion of the land-tax is assigned to a great variety of parties, and for various purposes. Lands have been given to public officers as the reward of their services; to men of learning; to the favourites of sovereigns; for the maintenance of civil and military public establishments; and for the endowment of charitable, educational, and religious institutions. The grants, especially those for the use of temples, mosques, and shrines, were in perpetuity; and others became so through the usage of India. Inscriptions on stone and brass, found in most parts of India, attest the antiquity of these grants. One of them is supposed to be nearly coeval with the invasion of Britain by Julius Cæsar, and hundreds are of dates antecedent to the Norman invasion. (*Asiat. Researches*, I.; *Trans. of the Royal Asiat. Soc.*, passim.) The extent of these free tenure lands throughout India is very great. In the ceded territory under the Madras Presidency, comprising an area of 26,000 sq. m.,

15

they amount, as estimated by Sir T. Monro, to one fifth part of the entire surface. In the N.W. provinces of the Bengal Presidency (now Agra), embracing an area of 66,600 sq. m., the free tenure lands were ascertained by the British commissioners to amount to 44,861,776 begahs, the land-tax of which, if assessed in the usual manner, would have amounted to £1,236,000. From an inquiry made in 1777, it appeared that the rent-free lands in Bengal Proper amounted to 8,573,942 begahs, or 3,164,554 acres, which would have yielded a tax of £1,326,390 a year. It is deserving of notice, that the rent-free lands under the Agra Presidency were at the very threshold, as it were, of the Mohammedan power; and the territory in which they are included was in the possession of the Mohammedans *for six centuries*. But, notwithstanding their bigotry and despotism, they respected the free tenures. They also, much to their honour, respected them in a singular degree in Bengal, where most of them had originally consisted of tracts of waste or wild land, re-

claimed by the labour and capital of the grantees, or their heirs and successors. Lord Cornwallis, and the Indian council of his day, confirmed the possession of the rent-free lands to their holders, on the same perpetual tenure as the taxed lands; and it was enacted that those that held under a free tenure prior to 1765 should remain untaxed "for ever." It has been said that the present Indian government has manifested a strong disposition to seize upon the rent-free lands, or to subject them to a system of taxation; but, as a proceeding of this sort would be a flagrant violation of a solemn engagement, we do not believe that there is any real foundation for the statement.

The other principal sources of the public revenue are the sea and frontier customs, town and transit duties (the latter now abolished in Bengal, but still existing in Madras and Bombay), the salt and opium monopolies, &c.

We subjoin the following statement with respect to the revenue of the presidency of Bengal:—

ACCOUNT of the Revenue of the Bengal Presidency, in 1835-36, 1836-37, 1837-38, and 1838-39.

Revenue.	1835-36.	1836-37.	1837-38.	Estimate, 1838-39.
	Sica Rupees.	Co.'s Rupees.	Co.'s Rupees.	Co.'s Rupees.
Mint duties	3,90,313	2,11,174	5,04,363	6,21,390
Postoffice collections	5,66,911	6,05,803	5,59,464	6,03,504
Stamp duties	18,95,568	18,95,893	19,80,916	19,94,477
House-tax in Calcutta	2,62,448	2,69,876	2,38,069	2,15,850
Excise duties in do.	2,04,655	2,07,014	1,93,773	1,93,416
Judicial fees and fines	3,62,635	4,08,480	6,74,976	6,87,781
Miscellaneous civil receipts	3,62,809	30,39,587	32,06,999	32,69,677
Land revenue	2,97,96,194	3,39,37,571	3,37,79,031	3,46,96,368
Sayer and Abkarry, do.	21,07,195	23,05,962	21,86,946	21,96,571
Miscellaneous receipts in the revenue department	6,72,322	4,05,883	3,13,192	2,50,146
Receipts from the territory ceded by the Burmese	11,74,773	12,81,377	13,39,932	15,19,699
Customs	31,12,860	32,65,063	31,95,127	20,18,975
Sale of salt	1,65,56,690	1,50,03,860	1,73,03,830	2,15,90,370
Sale of opium	1,68,96,918	1,86,36,963	2,09,65,187	1,36,38,730
Marine and pilotage receipts	5,04,442	6,35,557	6,19,722	7,16,784
Revenues of Prince of Wales' Island, Singapore, and Malacca.	Sica Rs. Co.'s Rs.	Co.'s Rs. Co.'s Rs.		
Prince of Wales' Island	1,75,606 1,82,929 1,87,343	1,73,950		
Singapore	2,67,322 2,30,895 2,48,926	2,67,619		
Malacca	63,232 65,090 53,543	51,396	5,05,691 5,07,834 4,89,812 4,91,965
Subsidy received from the Nagpore government			7,70,306 6,80,000 16,00,000 6,45,330
Tributes from the Nizam, Rajpoot, and other states			4,36,484 25,09,060 12,14,868 9,47,580
Interest on arrears of revenue			3,51,418 8,36,797 7,16,045 6,80,451
Total gross revenue			7,68,42,922 8,80,62,584 9,06,11,545 8,78,18,704
Deduct allowances and assignments payable out of the revenues in accordance with treaties or other engagements			20,85,973 21,61,789 21,84,312 15,88,935
Charges of collecting the Revenue.	Sica Rs. Co.'s Rs. Co.'s Rs. Co.'s Rs.		7,47,56,949 8,59,00,802 8,84,27,233 8,62,39,769
Charges of collecting the stamp duties	1,36,786 1,34,617 1,35,189 1,16,699		
Charges of land, Sayer and Abkarry revenues	37,96,869 34,67,924 37,92,318 42,09,137		
Charges of customs	4,62,750 4,59,956 4,32,670 3,99,578		
Cost and charges of salt, including the quantity supplied to the French and Danish governments under convention	48,40,491 54,86,917 52,91,907 39,49,579		
Costs and charges of opium	45,90,495 54,95,995 65,97,949 65,67,759		1,37,82,305 1,50,38,709 1,62,00,033 1,52,95,661
Total nett revenue of Bengal presidency, after payment of allowances and assignments, and charges of collection			6,19,74,644 6,88,62,093 7,22,27,200 7,10,03,106

The total debt of India, in India, on the 30th of April, 1838, amounted to £38,949,393, bearing an interest of £1,427,368.

The army maintained in British India, consisted, in 1837 of the following effective force, in British, native, and contingent native troops:—

British.		Native.		Contingent Native.	
Staff	905	Staff (British)	312	Scindia	15,000
Horse artillery	1,927	On.'s officers of British birth.	3,416	Oude	10,000
Foot do.	4,354	Officers of Hindoo birth	3,416	Nizam { cavalry, 10,000 / infantry, 12,000 }	22,000
Engineers	77	Engineer corps	3,408		
Cavalry	2,585	Horse artillery	1,032	Baroda { cavalry, 3,000 / infantry, 4,000 }	7,000
Infantry	13,879	Foot do.	5,892		
Officers	735	Artillery train	1,302	Nagpore	1,000
		Cavalry	14,500	Holkar	3,000
		Infantry	124,961	Travancore, 3 battalions	3,000
	95,582		157,758	Cochin, 1 do.	1,000
				Mysore	4,000
				Cutch and Joudpoor	6,000
Total { British 95,582 / Native 157,758 / Native subsidiary 111,500 } 995,849				Rajpoot States { cavalry, 7,000 / infantry, 27,000 }	34,500
				Sattarah	5,000
					111,500

ABSTRACT VIEW of the Revenues (exclusive of commercial Assets realized in England) and Charges of British India, for the Years 1835-36, 1836-37, 1837-38, and 1838-39, including the Charges disbursed in Great Britain. *(Parl. Paper, No. 614, Sept. 1840.)*

Revenue.	1835-36.	1836-37.	1837-38.	1838-39, partly estimated.	Charges.	1835-36.	1836-37.	1837-38.	1838-39, partly estimated.
	Siccarupees.	*Co.'s rupees.*	*Co.'s rupees*	*Co.'s rupees*		*Siccarupees.*	*Co.'s rupees.*	*Co.'s rupees.*	*Co.'s rupees*
Bengal					Bengal				
Agra					Agra				
Madras					Madras				
Bombay					Bombay				
Total ordinary revenue.					Total ordinary charges of India.				
At 2s. per sicca rupee £					At 2s. per sicca rupee £				

The expense of the Anglo-Indian army, according to reports laid before parliament in 1830, was as follows:—

	£		£
Engineer Corps	68,874	Medical Staff	129,400
Artillery	698,463	Pioneers	74,511
Cavalry	1,609,534	Commissariat	614,397
Infantry	4,194,079	Sundries	2,176,787
Staff	431,400	Total	9,273,955

It may be observed, by the way, that this sum of £9,374,000 is more than double the sum annually expended on the Prussian army! Considerable additions have been made within the last half dozen years to the military force in India.

Each presidency has its separate army, commander-in-chief, staff, &c.; but the commander-in-chief of the supreme government has a general authority over the armies of all the presidencies. Among the native troops, called *Sepoys* (sepahies), there is a complete intermixture of tribes, castes, and creeds; but the infantry consists chiefly of Hindoos and the cavalry of Mohammedans. The Hindoo soldiers of the Bengal army are mostly of high caste, more than 90,000 being Brahmins. The soldiers of the Madras army are principally Rajpoots, and are reckoned the most persevering, hardy warriors; but they observe their religious customs so strictly, that the least deviation from them might have a dangerous effect on their discipline. The Bombay soldiers are the most easily disciplined, being generally of the lower castes. The troops are not raised by any forced levy or conscription; military service in India is quite voluntary, and is so popular that each regiment has a number of supernumeraries ready to take the place of such soldiers as die or leave. The men are well paid, clothed and fed. The corporal punishment of Hindoos is not allowed; imprisonment being in India, as in the French army, the principal engine by which discipline is kept up. In the former, however, the disgrace attending dismissal from the service, which is acutely felt by native soldiers, tends powerfully to preserve discipline and obedience. Each company has an English captain, lieutenant, and ensign, as well as a native captain, lieutenant, and ensign. The latter, however, are under the command of the British officers; so that, with the title and uniform of officers, they are, properly speaking, only subalterns or non-commissioned officers. The Indian army, when not in the field, is in camp the whole year through, a system which has contributed, in no small degree, to bring it to its present state of efficiency.

A good deal of conflicting evidence was given before the parliamentary committees, in 1828 and 1832, as to the real state of the Indian army, and the degree of dependence to be placed on it. On the whole, it would seem to be superior, in respect of discipline and organization, to any native army ever previously embodied in India; and so long as its discipline and efficiency are maintained unimpaired, it is

no doubt fully adequate to provide for the tranquillity of India, and its defence against Asiatic invaders. But the sepoys are decidedly inferior both in physical strength and mental energy to Europeans; and such being the case, we cannot help, how reluctantly soever, agreeing in opinion with those who think that the Indian army could not make any effectual opposition to anything like a corresponding force of French, Russian, or other European troops.

The Indian navy consists only of one frigate, four 18 gun brigs, six 10 gun corvettes and brigs, two armed steamers, and some other vessels: it is manned by about 300 European seamen, and from 500 to 700 natives, under about 140 British officers: it is attached to the Bombay presidency, which see.

Commerce.—Internal Trade.—Throughout the whole of the immense basin of the Ganges there is an extensive inland navigation; and this, also, is the case in the valleys of the larger rivers in the S.; but elsewhere the inland trade is greatly impeded by the want of roads, and the imperfect means of conveyance. With the exception of a few military roads, made by the English, none fit for carriages have been constructed in any part of the country; what are called high roads being, in fact, little better than broad and bad pathways, on which goods to a small extent are conveyed in carts, or, rather, very rude cars, drawn by a pair or more of oxen. Many kinds of goods are carried by pack-bullocks: on the N.W. frontiers of Hindostan, camels and horses are used; in the N., small horses, and even ponies and sheep are employed; but in most of the mountainous parts of Hindostan porters are the chief bearers of merchandise. The charge for conveying goods by land is estimated at an average of 100 m. at 56s. per ton, being about 28 times as much as the conveyance of the same weight of goods for 100 m. on the Ganges; and equal to more than half their freight by sea from Calcutta to London! It may hence be easily conceived, that the internal trade of the country is confined principally to the necessaries of life.

Corn, cotton, oleaginous plants, and sugar are the most important objects of inland commerce. The chief trade in rice takes place within the tract of the inundation of the Ganges: N. of lat. 23°, it is superseded by that of wheat and barley. Cotton is grown in every latitude in India; and is not, therefore, an article of very extensive internal commerce. Indian cotton is, speaking generally, coarse, dirty, and short in the staple; and is very inferior to most other kinds brought to the markets of Europe. But it is believed that this is not owing so much to any natural incapacity on the part of India to produce good cotton, as to the want of care in selecting the seed, and in the culture of the plant. In these respects, too, some very material improvements have been effected of late years; and a good deal of the cotton brought from India is now greatly superior to what it was a few years ago. But it is still susceptible of much improvement. It has been estimated that cotton fabrics, of the value of about £20,000,000 a year, are made by the population of British India, or of the value of £34,000,000 including the tributary states. Cotton goods from Great Britain are now imported to the value of about £3,000,000 a year, or about 1-17th part of the native manufacture. The real falling off in the amount of the Indian manufacture, in consequence of the import of British cottons, does not, however, amount to a million sterling a year; for we consume more than £300,000 worth of their cotton wool, and dispose of a million's worth of their fabrics in China. These statements sufficiently evince the fallacy of the often repeated complaints as to the destruction of the cotton manufacture of India by the importation of English goods; and it is necessary to add that, though the latter were imported to a much greater extent, the circumstance would be an advantage, not an injury, to India; for they would not be imported were they not cheaper, and consequently, more easily attainable than their own by the great bulk of the population. (For an account of the circumstances that led to the ruin of the manufacture of fine muslins in Dacca, see Vol. I., 723.

Sugar is a principal article of internal culture and trade. It is principally raised in the great plain of the Ganges. The average annual consumption of sugar in Hindostan has been estimated at between 1½ lbs. and 1¾ lbs. a head, which, for the British and tributary states, would amount to upwards of 650,000 tons; but we believe that this is a most exaggerated estimate, and that half the quantity would be much nearer the mark, though probably still in excess. The average consumption of salt is estimated at 15 lbs. per head, or upwards of 877,000 tons annually, which, at £8 a ton, including the tax, gives a total amount of between £7,000,000 and £8,000,000. This article is everywhere paid for chiefly in corn. The other staples of the inland trade are indigo, opium, silk, tobacco, nitre, oil-skins, drugs, hides, lime, timber, &c. The Malabar coast has some products peculiar to itself, as teak and sandal woods, black pepper, and cardamons. With these, and different metals, areca nuts, and spices obtained from other countries, woollen and cotton goods, and various manufactures and products of Europe and China, the corn, cotton, sugar, and other articles of the inland trade are paid for on that coast. But there is no extensive or well-organised system of inland trade in India. The different parts of the country are, in this respect, separate and unconnected. "The merchants of the upper provinces," says Mr. Trevelyan, "know nothing of the trade of the lower provinces; the merchants of the lower provinces know nothing of what is passing above Mirzapoor; and the maritime trade is a branch separate from both." This is a consequence, partly of the want of good roads and other easy modes of communication, but more, perhaps, of the internal duties laid on the transit of goods from one part of the country to another. These, however, have recently been abolished in Bengal; and there can be no doubt that this measure will be of signal advantage to that province, and is, in fact, one of the greatest boons conferred upon it by the English.

"In India, as under most uncivilised governments, the transit of goods within the country was made subject to duties; and upon all the roads and navigable rivers toll-houses or custom-houses (chokeys) were erected, which had power of stopping the goods till the duties were levied. By the rude and oppressive nature of the government, these custom-houses were exceedingly multiplied; and, in long carriage, the inconvenience of numerous stoppages and payments was very severe. As in all other departments of the government, so in this, there was nothing regular and fixed. The duties varied at different times and different places; and a wide avenue was always open to the extortion of the collectors. The internal trade of the country was, by these causes, subject to ruinous obstructions." (Mill, book iv., cap. 5.)

The pernicious consequences resulting from this state of things early engaged the company's attention; though, at first, their efforts were directed rather to obtain an exemption from the transit duties in favour of their own trade than to effect their abolition. In 1788, however, Lord Cornwallis, who was fully aware of their pernicious influence, adopted the judicious and decisive measure of abolishing the duties. But, unaccountable as it may seem, they were again restored in 1801; and were "frightfully increased" in 1810! Through the artificial impediments thus thrown in the way of internal commerce, the country was split, as it were, into a vast number of petty states, each surrounded by a line of custom-houses, and each jealous of the other. Metals, for example, passing from one town or district to another, were charged 10 per cent. ad valorem, and most other articles were charged from 5 to 10 per cent. "Hence, the power of carrying on business on a large scale, of using expensive machinery, and engaging numerous labourers, is contracted in an infinite degree; employments cannot be subdivided and improved; industry languishes;

and a general tendency exists towards that barbarous state of things in which every body is obliged to produce and manufacture everything he requires for his own consumption." (Trevelyan's Report, p. 5.)

Had the inland transit duties been productive of a large amount of revenue, that would have been some set-off against the enormous evils of which they were productive. But such was not the case. The expense of their collection, and the obstructions they threw in the way of communication, were such as to render their produce quite insignificant. At length, however, the pernicious influence of these duties in a commercial, and their inefficiency in a fiscal, point of view were clearly demonstrated by Mr. Trevelyan, then one of the secretaries to the Bengal government,* in the able report referred to above. In the course of the year (1836) following the publication of this report, the internal transit duties and town duties in Bengal were abolished; but their abolition in other parts of India, which it was expected would immediately follow, has not yet taken place.

External Trade.—In 1838-39, the values of the imports and exports of the four principal ports of British India were officially reported as under:—

	Imports.	Exports.
	£.	£.
Calcutta	3,082,406	3,844,771
Madras	381,255	445,296
Bombay	852,384	280,768
Singapore	1,961,130	1,204,069
Total	8,15×,275	5,734,419

Mr. Larpent furnished the following estimate of the trade of India and China with Great Britain in 1837-38:

Exports to England.	Estimat. value.	Mode in which Exports from India are paid for	Estimat. value.
	£.		£.
Indigo	2,000,000	British manufactures sent to India .	2,500,000
Sugar	600,000	Do. to China	500,000
Silk	750,000	Remittances of private fortunes	500,000
Silk piece-goods	350,000	E. I. Comp.'s home charges	3,000,000
Saltpetre	300,000	Opium sent to China . 2,400,000	
Rice	100,000	Cotton Do. . 1,000,000	
Sundry articles	1,000,000		
Bombay cotton, &c.	1,400,000		4,400,000
Tea from China	3,300,000	Less return of bullion from China to Cal. 1,800,000 cutta and Bombay }	1,800,000
Silk from do.	100,000	—say	3,000,000
Total .	9,800,000	Total .	9,600,000

Indigo grows luxuriantly from the equator to the 30th degree of lat.; but in India the best is produced in Bengal and Bahar, between lat. 23° and 27° N., and long. 84° and 90° E.; everywhere else the product is inferior. The annual produce of all the Bengal provs. has been estimated at about 9,000,000 lbs., produced on about 1,250,000 acres of cultivated land; the planters, at an average, farming about 2500 acres each. The prime cost of the article to the planters has been estimated at £1,680,000; the gross profit on which, including risk and charges to the port of exportation, amounts to 40 per cent. The production of silk in India is confined to Bengal, and the produce is inferior. (See art. HINDOSTAN.)

Opium was, for many years previously to the recent disturbances, and we believe still is, an article of great and rapidly increasing export to China, the Malay islands, and elsewhere. The poppy may be said to take the place in Indian agriculture that the vine and olive occupy in that of S. Europe. Its growth within the British territories has been confined to Bahar and the Benares districts; but in the prov. Malwah, most part of which is included in the dom. of Scindia, it is extensively cultivated, and pays an export duty on being shipped from ports under our dominion. Such has been the increased demand for opium in China and the Malay countries, that the exports were multiplied at least fourfold during the 20 years ending with 1838. It is produced under a monopoly, and yields a large revenue to government.

The principal export of cotton is to China; but the export to Great Britain is also pretty considerable, having amounted to about 40,000,000 lbs. in 1838. This, however, is not more than about 1-10th part of our whole annual consumption of cotton wool!

Notwithstanding the vast, and all but unlimited, capacities of British India for the production of sugar, its total export, in 1838, amounted to little more than two thirds of the export of sugar from the Mauritius! This miserable result was wholly, or almost wholly ascribable to the inferior quality of East Indian sugars, owing to the very rude and imperfect methods in which they are prepared; but it was

* Now Assistant-Secretary to the Treasury.

26

partly, also, ascribable to the circumstance of E. Indian sugars having been burdened, previously to 1836, with a duty of 8s. a cwt. over and above the duty charged on W. Indian sugars. But in the course of that year Bengal sugars were put on the same footing, in respect of duties, as those of the W. Indian colonies; and of late years a very great improvement has been effected in the manufacture of E. Indian sugars, the best of which are now about equal to the best of those from Jamaica and Demerara. In consequence of the circumstance now referred to, and of the recent high price of sugar in this country, the imports from India have increased with great rapidity.

We subjoin an account of the importation of sugar from British India, ex. Ceylon, during the three years before and the four years subsequent to the equalization of the duties:—

Years.	Imports.	Duty.	Years.	Imports.	Duty.
1833	111,721 cwts.	} 32s. per	1836	152,183 cwts.	
1834	76,513 —	{	1837	296,657 —	} 24s. per
1835	106,856 —	} cwt.	1838	498,854 —	} cwt.
			1839	564,794 —	

And we understand that this year (1840) the imports will amount to nearly 1,000,000 cwts., being considerably more than the importation from Jamaica!

The abolition of the discriminating duty, in favour of Bengal sugars, being founded on reason and equity, should certainly be extended to all parts of British India, and we think also to the sugar of the countries in India politically dependant on our government. Nothing, indeed, can be more unjust and inconsistent with sound principle, than to impose higher duties on the products of one portion or dependency of the empire than on those of another.

The corn of India, both rice and wheat, is inferior to that of most other countries, for the same reason that its cotton and sugar are inferior, both being the produce of a rude husbandry, and rude preparation. Rice is scalded instead of being kiln-dried; and wheat is never dried at all, except in the sun. It has been supposed that the latter might be largely imported under a free-corn trade into England; but we doubt whether there be any real grounds for such an opinion. Indian wheat is, speaking generally, very inferior to British wheat; and it could not be imported, in ordinary years, at less than 40s. to 44s. a quarter, supposing it to be exempt from all duty. Its price, free on board at Calcutta, may be taken at 15s. or 16s. a quarter; to which, if we add 18s. or 18s. for freight to England, and 8s. or 16s. for profits and landing charges here, it is abundantly plain that, except in high-priced years, it would not answer to import Indian corn.

STATEMENT of the Quantities of the Principal Articles imported into the United Kingdom from British India (the East India Company's Territories and Ceylon) in 1838:—

Articles.	Quantities.	Articles.	Quantities.
Coals, Logwood	ldes.	Cotton wool	lbs. 40,211,784
Cotton yarn		Saltpetre	cwts.
Coffee	7,756,963	Cubes nitre	cwts. 534,947
Cotton piece goods	pieces	Flax'd & lins'd husk	—
Embroidered muslin		Senna	lbs. 3,161,790
Ginger		Raw and waste	
Gum arabic		silk	1,151,309
Indigo	lbs.	Bandannas	
Snail-lac		Smooth, &c. pieces	
Borax, unrefined	cwts.	Sugar, unref.	cwts. 671,400
Hides, untanned		Tea	lbs.
Indigo	cwts. 6,574,184	Tin	cwts. 40,445
Cassia oil		Tobacco, un-	
Pepper		manufac.	lbs.
Rice	cwts. 383,385	Sheep's wool	1,697,396

STATEMENT of the Quantities and Declared Value of the Principal Articles of British and Irish Produce and Manufactures exported to the E. India Company's Territories and Ceylon in 1838:—

Articles.	Quantities.	Declared value.
		L.
Apparel, slops, &c.		81,845
Arms and ammunition		66,663
Beer and ale	tons.	75,844
Printed books		14,566
Brass and copper manuf.	cwts.	305,132
Cotton manufactures	yds.	1,791,226
Hosiery, lace, &c.		24,151
Cotton twist and yarn	yds.	649,405
Earthenware	pieces	55,241
Glass ware		54,300
Hardware and cutlery	cwts.	66,389
Iron and steel	tons.	157,767
Linen manufactures	yds.	26,230
Machinery		29,869
Plated wares and jewellery		42,579
Silk manufactures		14,854
Stationery		49,381
Woollen manufac. entered by		
the piece	pieces	184,529
Do. by the yd.	yds.	16,172
Other articles		249,379
Total declared value		**£3,476,155**

Previously to the recent discovery of nitrate of soda in S. America, Bengal and Bahar had a monopoly of the trade in saltpetre; and in 1828–29 the quantity exported from Calcutta was about 40,000 tons, valued at £160,000. The export of this article to England averages, at present, from 8000 to 10,000 tons a year. Dyes, shell-lac, linseed, safflower, sal-ammoniac, castor-oil, coffee (recently introduced with much success into Malabar, Mysore, Ceylon, &c.), tea from Assam, &c., tin, antimony, catechu, and pearl nuts, are other exports worthy of mention; and which owe their importance as much principally to the commercial enterprise and talent of Europeans. (*Lord's Report of 1840 on the Trade of India, &c.*)

For farther particulars as to British India the reader is referred to the article HINDOSTAN.

We subjoin a chronological statement of the principal territorial acquisitions made by the British in India.

Districts.	Date of Acquisition.	Districts.	Date of Acquisition.
Twenty-four Pergunnahs	1757	Goruckpore, Lower Doab, Bareilly, &c.	1801
Midnapoor, &c.	1760	Distr. in Bundlecund	1802
Burdwan, Midnapoor, and Chittagong	1760	Cuttack and Balasore	1803
Bengal, Bahar, &c.	1765	Upper Doab, Delhi territory, &c.	1803
Company's Jaghire in the vicinity of Madras—Chicacoya	1765	District in Gujrat	1805
Northern Circars	1766	Kumaon	1815
Zemindary of Benares	1775	Sugar and Scinde, Deoghur, &c.	1817
Island of Salsette	1775	Ahmednabad	1817
Nagore	1776	Candeish, &c.	1818
Guntoor Circars	1779	Ajmere	1818
Palaforza	1779	Poonah, Concan, S. Mahratta country	1818
Malabar, Dindigul, Seringapatam	1792	Distr. on the Nerbudda, Sumbhulpoor, &c.	1818
Coimbatore, Canara, Wynaad, &c.	1792	Lands in S. Concan	1820
Tanjore	1799	District in Saugor and Ahmednuggur	1820
Districts acquired by the Nizam—Tippoo Sultan in Tibet	1799	Singapore	1824
		Malacca	1824
		Assam, Aracan, Tavoy, Ye, Tenasserim	1826
Carnatic	1801	Coorg	1834

INDIA-BEYOND-THE-GANGES, sometimes called Indo-China, an extensive region of Asia, forming the eastern of its three great peninsulas, extending between the 7th and 26th degs. of N. lat., and the 92d and 109th of E. long., comprising the empires of Birmah, Siam, and Anam, the Malay peninsula, Laos, the Tenasserim provs., Aracan, Cathay, Cachar, Assam, and the Bengal districts of Sylhet, Tipperah, and Chittagong, which see.

INDIANA, one of the western United States, is bounded N. by Michigan state and lake; E. by Ohio; S. by Ohio river, which separates it from Kentucky; and W. by Illinois state, from which it is in part separated by Wabash river. It is between 37° 47' and 41° 50' N. lat., and between 84° 48' and 88° W. long., and between 7° 45' and 11° W. long. from W. It has a mean length of 260 miles, and a mean breadth of 140 miles, containing 36,000 square miles, or 23,040,000 acres. The Ohio river washes its southern border for 340 miles, and the Wabash runs on its western border for 150 miles. Lake Michigan washes its N.W. border. The population in 1800 was 5641; in 1810, 24,520; in 1820, 147,178; in 1830, 341,582; in 1840, 685,866. Of these there were 332,773 white males; 325,925 white females; 3731 coloured males; 3434 coloured females. Employed in agriculture, 148,806; in commerce, 3076; in manufactures and trades, 20,590; in mining, 233; in navigating the ocean, 89; ditto, canals, rivers, and lakes, 677; in the learned professions, 2237.

There are no mountains in Indiana, but the country bordering on Ohio river is broken and hilly. A range of hills runs parallel with Ohio river, from the mouth of the Great Miami to Blue river, sometimes approaching to within a few rods of the river, and at other times receding from it to the distance of 2 miles. Immediately below Blue river, the hills cease, and an immense tract of level land, covered with timber, is presented to the view. Strips of bottom and prairie land, covered with a heavy growth of timber, skirt all the principal rivers, excepting the Ohio, from 3 to 5 miles in width. With some few exceptions, the greater proportion of this state may be pronounced to be one vast level. The prairies and timber land alternate, and in general these kinds of land are more happily balanced than in other parts of the western country. Many prairies are long and narrow, so that the whole can be taken up, and timber be easily accessible by all the settlers. Even in the large prairies are those beautiful islands of timbered land, which form such a striking feature in the western prairies. The great extent of fertile land, and the happy distribution of rivers and springs, has been one cause of the very rapid increase of population in the state. For a wide extent on the north front of the state, between Wabash river and lake Michigan, the country is generally an extended plain, alternately prairie and timbered land, with a great proportion of swampy lands, and small lakes and ponds. The prairies bordering on Wabash river are particularly rich, having ordinarily a vegetable soil from 2 to 5 feet deep. Perhaps no part of the western world can show a greater extent of rich land in one body than that portion of the White river country, of which Indianapolis is the centre. The natural growth of the soil consists of oak of several kinds, ash, beech, buckeye, walnut, cherry, maple, elm, sassafras, linden, honey locust, cotton wood, sycamore, and mulberry. The principal productions are wheat, rye, Indian corn, oats, buckwheat, barley, potatoes; beef, pork, butter, cheese, &c.

The state is divided into 87 counties, which, with their population in 1840, were as follows. Newton county, formed in 1837, is not in the census, which makes 88 counties.

Counties.	Pop. 1840.	Counties.	Pop. 1840.
Adams	2,264	Lawrence	11,782
Allen	5,942	Madison	8,874
Blackford	1,226	Marshall	1,651
Bartholomew	10,046	Marion	16,080
Boone	8,121	Martin	3,875
Brown	2,364	Miami	3,048
Carroll	7,819	Monroe	10,143
Cass	6,600	Montgomery	14,438
Clark	14,595	Morgan	10,741
Clay	5,567	Noble	2,702
Clinton	7,504	Orange	9,602
Crawford	5,302	Owen	8,345
Daviess	6,720	Parke	13,499
Dearborn	19,327	Perry	4,655
Decatur	15,171	Pike	4,769
De Kalb	1,968	Porter	2,162
Delaware	8,843	Posey	6,883
Dubois	3,632	Pulaski	561
Elkhart	6,660	Putnam	16,843
Fayette	9,837	Randolph	10,684
Floyd	9,454	Ripley	10,292
Fountain	11,218	Rush	16,446
Franklin	13,349	Scott	4,242
Fulton	1,993	Shelby	12,005
Gibson	8,977	Spencer	6,305
Grant	4,875	St. Joseph	6,425
Greene	8,321	Starke	149
Hamilton	9,855	Steuben	2,578
Hancock	7,520	Sullivan	8,315
Harrison	12,459	Switzerland	9,920
Hendricks	11,264	Tippecanoe	13,734
Henry	15,128	Union	8,017
Huntington	1,570	Vanderburg	6,274
Jackson	8,961	Vermillion	8,274
Jasper	1,266	Vigo	12,076
Jay	3,863	Wabash	2,786
Jefferson	16,614	Warren	5,656
Jennings	8,829	Warwick	6,321
Johnson	9,352	Washington	15,565
Knox	10,657	Wayne	23,290
Kosciusko	4,170	Wells	1,822
La Grange	3,664	White	1,832
Lake	1,468	Whitely	1,237
La Porte	8,184	Total	685,806

Indianapolis, near the centre of the state, on White river, is the seat of government.

In 1840, there were in this state 241,036 horses and mules, 619,980 neat cattle. 675,982 sheep, 1,623,608 swine; poultry to the value of $357,594. There were produced 4,049,375 bushels of wheat, 129,621 of rye, 28,155,887 of Indian corn, 28,015 of barley, 5,981,605 of oats, 1,525,794 of potatoes, 1,237,919 pounds of wool, 1,820,306 of tobacco, 3,727,795 of sugar, 38,591 of hops. 30,647 of wax, 178,029 tons of hay, 8605 of hemp and flax; the products of the dairy were valued at $742,209; of the orchard, at $110,055; of lumber, at $430,971; of furs and skins, at $220,883. There were made 10,265 gallons of wine.

Iron and coal have been found in the state, and there are some salt springs, and Epsom salts are found in a cave near Corydon; but the mineral productions have no great interest.

The climate is generally pleasant and healthy, except in the vicinity of stagnant waters; the winters are mild in the southern part, and not very severe in the northern part, though the Wabash is frozen over so as to be passed upon the ice. In the central and southern parts snow seldom falls to a greater depth than 6 inches; but in the northern part it is sometimes a foot and a half deep. Peach trees blossom early in March, and the forest trees put forth leaves early in April. The winter is seldom longer than six weeks, though there are severe frosts in spring and autumn.

The Ohio river washes its whole S. border, affords great facilities for trade, and has some important places on its banks. The Wabash is the largest river, draining, with its branches, the greater part of the state. It is one of the finest tributaries of the Ohio, rises in the N.E. part of the state, crosses it N. of the middle, and flows S. near its W. line, and for 150 miles constitutes the boundary between this state and Illinois, and enters Ohio river, 30 miles above the mouth of Cumberland river. It is navigable in high water for 370 miles to La Fayette; but in low water it is obstructed by bars and ledges of rocks 15 miles below. White river, the largest tributary of the Wabash, consists of two main

branches, the East and West fork. The West fork rises near the border of Ohio, and traverses the whole breadth of the state. The East fork is nearly as great in extent, and in the volume of its waters. This river is about 200 miles long, and in its West fork is navigable in high water for 200 miles, to Indianapolis. It enters the Wabash about 100 miles from its mouth. The other principal tributaries of the Wabash are Salamanie and Mississinewa on the S., and Little river, Eel river, and Tippecanoe on the N. side. The Whitewater, in the S.E. part of the state, flows into Great Miami river, a little above its entrance into Ohio river. St. Joseph's river enters the N. part of the state, and flowing again into Michigan, it enters lake Michigan.

The largest place in the state is New Albany, on Ohio river, a little below the falls of the Ohio at Louisville; Indianapolis, the capital, Madison, and Evansville are flourishing places. Terre Haut, Lafayette, Logansport, and Fort Wayne, are growing centres of trade in the interior.

There were in 1840, 11 commercial and 96 commission houses engaged in foreign trade, with a capital of $1,207,400; 1901 retail stores, with a capital of $5,664,657; 767 persons employed in the lumber trade, with a capital of $90,374; 9705 persons engaged in internal transportation, who, with 237 butchers, packers, &c., employed a capital of $568,165.

The amount of home-made or family manufactures was $1,989,803. There were 94 fulling-mills and 37 woollen manufactories, employing 163 persons, producing to the amount of $66,867, and employing a capital of $77,954; 19 cotton factories, with 4062 spindles, employing 210 persons, producing articles to the amount of $133,400, with a capital of $148,500; seven furnaces produced 810 tons of cast iron, and one forge produced 20 tons of bar iron, employing 163 persons, and a capital of $57,720; 47 persons produced 949,040 bushels of bituminous coal, with a capital of $9,300; three paper-mills produced to the amount of $96,457, and other manufactures of paper produced to the amount of $54,008, the whole employing 100 persons, and a capital of $96,730; 981 persons manufactured flax or hemp to the amount of $6851; 89 persons manufactured tobacco to the amount of $65,653, with a capital of $94,706; hats and caps were manufactured to the amount of $192,844, and straw bonnets to the amount of $9045, the whole employing 163 persons, and a capital of $80,018; 496 tanneries, employed 978 persons, and a capital of $399,687; 579 other manufactories of leather, as saddleries, &c., produced articles to the amount of $730,031, and employed a capital of $247,549; 45, pottadies employed 79 persons, produced articles to the amount of $35,526, with a capital of $13,685; 26 persons produced drugs and paints to the amount of $47,798, with a capital of $17,964; 120 persons produced machinery to the amount of $133,802; 63 persons produced hardware and cutlery to the amount of $34,263; 47 persons manufactured 885 small arms; two persons manufactured the precious metals to the amount of $3500; 28 persons manufactured granite and marble to the amount of $6730; 1007 persons produced bricks and lime to the amount of $906,751, with a capital of $140,489; 30 persons made 1,135,560 pounds of soap, 298,929 pounds of tallow candles, and 111 pounds of wax or spermaceti candles, with a capital of $13,039; 393 distilleries produced 1,757,106 gallons, and 90 breweries produced 183,309 gallons, the whole employing 500 persons, and a capital of $299,316; five rope-walks, employing 11 persons, produced cordage to the amount of $6860, with a capital of $2870; 481 persons manufactured carriages and wagons to the amount of $163,136, with a capital of $78,116; 204 flouring-mills manufactured 294,694 barrels of flour; and with other mills, employed 2994 persons, produced articles to the amount of $4,339,134, and employed a capital of $2,077,018; vessels were built to the amount of $107,923; 564 persons produced furniture to the amount of $211,491, with a capital of $91,692; 346 brick or stone houses, and 4270 wooden houses were built, employing 5519 persons, and cost $1,941,312; 69 printing-offices, six binderies, four semi-weekly and 69 weekly newspapers, and three periodicals, employed 211 persons, and a capital of $58,500. The total amount of capital employed in manufactures was $4,139,043.

Indiana college at Bloomington, was founded in 1827; South Hanover College, at South Hanover, was founded in 1829; Wabash College, at Crawfordsville, was founded in 1833; the Indiana Asbury University, under the Methodists, was founded in 1839. In these institutions there were in 1840, 393 students. There were in the state 54 academies, with 2046 students, and 1521 common and primary schools, with 48,189 scholars. There were in the state 38,100 white persons over 20 years of age, who could neither read nor write.

In 1836, the Baptists had 334 churches and 218 ministers; the Presbyterians had 100 churches and 70 ministers; the Methodists about 70 circuit preachers; the Lutherans in 1840 had 30 congregations and eight ministers. Besides these there is a considerable number of Friends, some Episcapitans, Roman Catholics, and some Presbyterians, Methodists, and Baptists of different descriptions, not included in the above.

In the beginning of 1840, there was one bank, with 12 branches, in the state, with a capital of $2,395,231, and a circulation of $2,935,370. At the close of 1840, the state debt amounted to $13,667,433.

A governor is elected by the people for three years, and may be once re-elected. At every election of governor, a lieutenant-governor is elected, who is president of the senate, and in case of the death, resignation, or removal of the governor, discharges the duties of the office. The senators and representatives are apportioned among the counties, according to the number of male white inhabitants, over 21 years of age. There can be no fewer than 36 nor over 108 representatives. The representatives, and one third of the senators are elected annually by the people. The legislature meets in December, annually, at Indianapolis. The judges of the supreme and circuit courts are appointed for the term of seven years. The judges of the supreme court are appointed by the governor, with the consent of the senate; the chief justices of the circuit courts by the legislature; and the associate judges by the people. All male white inhabitants, over 21 years of age, who have resided in the state for one year next preceding the election, enjoy the right of suffrage.

The principal work of internal improvement undertaken by this state is the Wabash and Erie canal, which extends from Lafayette, at the head of steamboat navigation on the Wabash, 187 miles, to the navigable waters of lake Erie at Toledo, on Maumee bay. Eighty-seven and a half miles of this distance is in Ohio, and ninety-nine and a half are in Indiana. The completion of this great and important work was celebrated at Lafayette, July 4, 1843. The Whitewater canal extends from Lawrenceville, at the mouth of the river, 30 miles, to Brookville. It is completed thus far. It is designed to extend N. to Cambridge city, on the National road, to which something has been done. The whole length is at 76 miles. It is also to be extended by a branch to Cincinnati, which is in progress. The Madison and Indianapolis railroad, from Madison, on the Ohio river, 95 miles to Indianapolis, is in progress, and nearly completed. Other works of internal improvement have been projected and begun, but are at present suspended.

In 1702, Vincennes was settled by French soldiers of Louis XIV. from Canada. Separated from the rest of the world, they became assimilated to the savages by whom they were surrounded, and with whom they intermarried. At the peace between England and France in 1763, this country came into possession of the English. In the Revolutionary war the inhabitants took sides with the Americans, in consequence of which the general government ceded to them a tract of land about Vincennes. In 1787, the United States took possession of Vincennes, and erected a fort on the opposite side of the river, for a defence against the savages. The inhabitants at that period consisted of French, of Canadians, and of Indians. The victories and treaty of Wayne in 1795, put an end to Indian hostilities. In 1811, in consequence of depredations and murders, a military force was sent against the Indians; but the bloody battle of Tippecanoe, under General Harrison, compelled them to sue for peace. In 1816, Indiana was admitted to the Union as an independent state, having previously been under a territorial government. Since it became a state it has rapidly progressed in population and in improvement.

INDIANA, county, Pa. Situated centrally towards the W. part of the state, and contains 776 sq. m. Drained by head branches of the W. branch of Susquehanna r., and by branches of Conemaugh and Alleghany rivers. Coal and salt are abundant, and some iron ore is found. It contained in 1840, 16,190 neat cattle, 35,694 sheep, 24,377 swine; and produced 195,254 bushels of wheat, 78,021 of rye, 171,618 of Indian corn, 80,806 of buckwheat, 356,046 of oats, 163,507 of potatoes, 12,288 pounds of sugar, 483,980 bushels of bituminous coal, 70,890 of salt. It had three commission houses in foreign trade. 68 retail stores, one furnace, one forge, five fulling-mills, five woollen factories, three flouring-mills, 51 grist-mills, 74 saw-mills, 26 tanneries, seven distilleries, three potteries, three printing offices, three weekly newspapers; two academies, 55 students, 23 schools, 1413 scholars. Pop. 30,782. Capital, Indiana.

INDIANA, p. b., capital of Indiana co., Pa., 155 m. W.N.W. Harrisburg. 208 W. Incorporated in 1816. It contains a brick courthouse, jail of stone, four churches, two Presbyterian, a Methodist and Lutheran, an academy, a female seminary, 14 stores, one brewery, one pottery, two printing offices, two weekly newspapers; one academy, 23 students, one school, 70 scholars; 124 dwellings and 674 inhabitants.

INDIANA, t., Alleghany co., Pa., 10 m. N.E. Pittsburg. Bounded S. by Alleghany r. Drained by Deer and Pine creeks. It contains five stores, one fulling-mill, two wool-

len factories, one forge, one saw-mill; 12 schools, 480 scholars. Pop. 2697.

INDIANAPOLIS, p. v., or city, capital of Marion co., In., and of the state. Situated in Centre t., on the E. side of the W. fork of White r., which is navigable to this place for steamboats in high water. It was selected for the state capital, by commissioners appointed by the state in 1820, when it was covered by a dense forest. It was laid out in 1821. The national road passes through the place on Washington-street, the principal business street. This street is 120 feet wide, the other streets are 90 feet wide, with the exception of a circular street, which passes round the governor's house: this street is 80 feet wide. The streets cross each other at right angles, with the exception of four streets which diverge from the circular area around the governor's house, which cross the other streets diagonally. The place was originally laid out on a mile square, but additions have been made to it on different sides. The village or town city is laid out into squares of four acres, each divided into 12 lots. Through these squares are alleys of 30 feet from E. to W. and others of 15 feet from N. to S. The streets are named after different states of the union. On a rise of ground in the centre of the circular area, stands the governor's house, 60 feet square, and two stories high, with four elegant fronts. The courthouse is 60 by 53 feet and two stories high, with a lofty cupola. The state house is one of the most splendid buildings in the west. It is 180 feet long, 80 feet wide and 45 feet from the foundation to the top of the cornice, with an appropriate dome. It is on the model of the Parthenon at Athens, with a Doric portico on each front of ten Doric columns, with 13 pilastres on each side. It contains elegant halls for the two houses of the legislature, a court room, and rotunda. A bridge crosses White river, which cost $25,000. There is a steam flouring-mill 45 by 50 feet, and saw-mill 60 by 30. Besides the public buildings already mentioned, there are six churches, two Baptist, an Episcopal, Methodist, Presbyterian, and Lutheran. Centre town in which it is situated, contains 30 stores, one fulling-mill, one cotton factory with 500 spindles, one flouring-mill, four grist-mills, nine saw-mills, two oil-mills, two tanneries, one brewery, two printing offices, two binderies, two weekly, and one semi-weekly newspaper; three schools, 78 scholars. Pop. 2692. Fall creek enters White river, N.W. of the village, and Pogues run passes through its S. part.

INDIAN TERRITORY, is a tract of country W. of the settled parts of the United States, set apart for the residence of the Indian tribes which have been removed, chiefly from the southwestern states of the union, and of Florida. Here they are to be secured in governments of their own choice, subject to no other control of the United States than such as may be necessary to preserve the peace of the frontier, and between the several tribes. This territory is about 600 miles long from S. to N., and from 300 to 600 miles wide from E. to W. It is bounded N. by Platt r., E. by the states of Missouri and Arkansas, S. by Red river; and by a desert country on the W. The habitable part contains 190,000 sq. m. or about 76,800,000 acres. The number of the tribes occupying this territory, is about 70,000, exclusive of the wild tribes of the prairies. The whole number of Indians E. of the Rocky Mountains is nearly 300,000. The principal tribes in the Indian Territory are the Chickasaws, Choctaws, Creeks, Seminoles, and Cherokees, besides some indigenous tribes. These tribes are progressing in civilization, improvement, and the enjoyment of the comforts of settled life; and promise yet to redeem the Indian character from the opprobrium which has been cast upon it, as if they were incapable of civilization. Their condition, in the opinion of their best friends, and of themselves, has been improved by their removal; and it is to be hoped that they will never be disturbed in their present possessions. They receive considerable annuities from the United States, in compensation for the country which they left, and have ceded to the United States.

INDIES (EAST). Under this vague and ill-defined appellation are usually comprised Hindostan, India-beyond-the-Ganges, and the islands in the E. Archipelago.

INDIES (WEST). Under this term were formerly included not only the Caribbee and other islands in the Atlantic near the coast of America, but also all the countries included under the name of the Spanish Main. But at present the term is restricted so as to signify only the islands between lat. 10° and 27° N., and long. 60° and 85° W., comprising the larger and smaller Antilles; the former consisting of Cuba, Hayti, Jamaica, and Porto Rico; and the latter of the Virgin, Leeward and Windward groups, with the Bahamas, Trinidad, Tobago, and a few other islands. Of these, Hayti alone is independent. Cuba and Porto Rico belong to Spain; Jamaica, the Bahamas, Trinidad, Barbadoes, Antigua, Dominica, Grenada,

St. Lucia, &c., to Great Britain; Guadaloupe, Martinique, Marie Galante, &c., to France; St. Eustatius, Saba, and Curaçoa, to the Dutch; St. Croix, St. Thomas, and St. John, to the Danes; and St. Bartholomew to the Swedes. For further details, see in this Dictionary the several islands above named.

INDORE, a city of Hindostan, prov. Malwah, cap. of Holkar's dom., and the residence of that chief, a little N. of the Vindhyan Mountains, and 30 m. S. by E. Oojein: lat. 22° 42′ N., long. 75° 30′ E. Population very uncertain, it having fluctuated greatly at different periods. According to Malcolm (Central India, i. 496), it is now inconsiderable both in size and population, and, being but weakly fortified, is a place of small importance. It stands at nearly 2000 feet above the level of the sea, in a well wooded, pleasant, and healthy tract, and has been wholly built within the present century. Some of its streets are tolerably spacious, paved with granite slabs, and its houses often of two stories, and constructed partly of brick; but, speaking generally, it is mean and ill built, and contains no public edifice worthy of remark, except the palace, a massive quadrangular granite building, with decorations of carved wood.

The territories of Holkar comprise an area of 4250 sq. m., having N. and E. Scindia's dominion, and W. and S. territory of the Bombay presidency. By the treaty of 1818, Indore was placed on the footing of other subsidiary states, the British agreeing to maintain a force for its external and internal security; and Holkar to keep no useless troops, and to furnish us when required a contingent of 3000 horse. A British resident is accordingly stationed at this capital, and a British force at Mhow and Mahidpore. (Hamilton's E. I. Gaz.; Parl. Rep., &c.)

INDRE, an inland dep. of France, reg. centre, formerly included in the province Berri, between lat. 46° 22′ 30″ and 47° 15′ N., and long. 0° 51′ and 2° 13′ E.; having N. Loir-et-Cher, E. Cher, S. Creuse, and W. Vienne and Indre-et-Loire. Average length and breadth about 60 miles each. Area, 688,851 hectares. Population (1836) 257,350. Its surface is generally level, with a slope towards the N.W., in which direction nearly all its rivers run to join the Loire on the Cher. The Creuse bounds it W.; the other chief river is the Indre, whence it derives its name. The latter rises in the dep. Creuse, and has a course of about 94 miles through the centre of this and the succeeding dep. to its mouth in the Loire, below Tours. Châteauroux and Loches stand on its banks, but, like the other streams of this dep., it is innavigable. A pestiferous tract of pools and marshes, called the Brenne, extends throughout the centre and W. part of the dep., occupying about 1-10th part of the whole surface, and a more extensive tract towards the E. end, called the Pays de Champagne, is quite bare of wood, and infertile; but the remainder is mostly either under culture or covered with forests. In 1834, the arable land comprised 401,251 hectares, meadows 85,303 hectares, and forests and heaths 132,339 hectares. Agriculture is very backward; but more corn is grown than is required for home consumption, a result probably owing to the thinness of the population, as only about 1,480,000 hectolitres of all kinds are produced annually. The produce of wine amounts to about 450,000 hectolitres a year, which also is more than is consumed by the inhabitants. Fruits are good, and excellent hemp is raised. In 1830, there were 765,000 sheep in the dep., large flocks being fed on the Pays de Champagne. A good many oxen are fattened for the supply of Paris; and hogs for the markets of Auvergne and Limousin. Geese and other poultry are reared in large numbers, particularly in the Pays de Brenne. Fish are abundant; and leeches form an article of trade. Iron of good quality is found, and forges are numerous. Good gun-flints are obtained at Châteauroux. Next to iron goods and woollen cloths, the principal manufactures are those of cottons, woollen yarn, leather, tiles, earthenware, hats, paper, and parchment. The dep. exports corn, wine, cattle, wool, woollen cloths, iron and iron goods, &c., to double the value of its imports. In 1835, of 83,276 properties subject to the contribution foncière, 47,461 were assessed below 5 fr., and 13,002 between 5 and 10 fr.; the number of considerable properties is somewhat below the average of the deps. The peasantry are strongly attached to routine practices, and therefore little likely to better their condition. Education is little diffused; in 1836, 190 communes were without primary schools, and only 3250 persons, or 1-49th part of the population, were receiving public instruction. Indre is divided into 4 arronds.; chief towns Châteauroux, the capital, Le Blanc, Issoudun, and La Chatre. It sends four members to the cham. of dep. Number of electors (1838-39), 1632. Total public revenue (1831), 5,318,996 francs; expenditure in the same year, 2,772,904 francs.

INDRE-ET-LOIRE, a dep. of France, reg. of the W., formerly included in the province Touraine, comprising a

tract on both sides the Loire, between lat. 46° 46′ and 47° 43′ N., and long. 0° 2′ and 1° 21′ E., having N. Sarthe and Loir-et-Cher, E. the latter dep. and Indre, S. Indre and Vienne, and W. Maine-et-Loire. Area, 611,679 hectares. Population (1836), 304,270. Surface almost an entire plain, with a slope from both the N. and S. to the Loire, which runs through it, near its centre, from E. to W. The part of the dep. watered by the Loire is so productive and beautiful that it has been termed the garden of France; but the soil elsewhere is generally dry, thin, and poor, and in the N.W. there are some extensive pools and marshes. Heaths and wastes occupy nearly 1-6th part of the surface, and forests more than 1-10th. In 1834, 334,910 hectares were arable, 33,463 pasture-land, 35,094 vineyard, and 23,673 otherwise cultivated. Agriculture is tolerably well conducted, having been much improved of late years. The corn now produced is more than adequate to the supply of the dep.; in 1835, 2,790,780 hectolitres were harvested, 1,109,780 of which were wheat, and 845,630 oats. Beans, peas, &c., are of excellent quality. Wine is annually made of the value of 4,000,000 or 10,000,000 of francs, or about double what is required for home consumption; but it is generally inferior. About 140,000 quintals of hemp, worth 5,600,000 francs, are raised yearly; and liquorice, aniseed, coriander, angelica, truffles, &c., are cultivated. The culture of the mulberry-tree is increasing rapidly: in 1835, 42,000 kilog. cocoons were gathered. The chief exports of this dep. are its agricultural products: cattle are not reared in any great number, and most kinds of live stock are inferior. Manufacturing industry is in a rather active state. The woollen, leather, and silk manufactures of Tours have materially increased within the last few years. The file and rasp factory at Amboise employs 160 workmen, and consumes above 200,000 kilog. a year of fine steel. The manufactures of red lead and iron goods are important; and near Monthazon is the royal gunpowder factory and saltpetre refinery of Ripault, at which 250,000 kilog. of gunpowder are made annually. Indre-et-Loire is divided into three arronds., the chief towns of which are Tours, the capital, Chinon, and Loches. It sends four members, to the cham. of dep. No. of electors (1838-39), 2113. Total public revenue (1831), 7,765,125 francs. This is the native country of Descartes, who was born at La Haye on the 31st March, 1596: Indre-et-Loire has also produced Rabelais and Balzac. Agnes Sorel, Gabrielle d'Estrees, and the Duchess de la Vallière. (Hugo, art. Indre, and Indre-et-Loire; Official Tables.)

INDUS (Sindhu, Sansc.; Sub Sind, Pers.). A large river of S. Asia, forming during great part of its course the proper N.W. boundary of Hindostan, and lying between the 23d and 35th parallels of N. lat., and between the 67th and 81st degrees of E. longitude. The geography of this river, especially as regards its upper portion, is very imperfectly understood; but we shall endeavour to collect into a consistent account the information gained from the investigations of Major Rennell, Moorcroft, Burnes, Elphinstone, and other travellers. As the source of the river has not been visited by Europeans, its situation is at present only a matter of conjecture; but general consent seems to place it on the N. declivity of the Cailas branch of the Himalaya range, near the Chinese frontier town of Gorno, and not far from the lake Mansuroura and the sources of the Sutledje. The stream, called by the Chinese Singhe-tochu, takes a general W.N.W. course past Ladak, and receives the larger river Shyook, N.W. of Ladak, whence the united streams run through the country of little Thibet, and after cutting a passage through the great Himalaya range, in lat. 35° 30′ N., and long. 74° 20′ E., are joined, about 190 miles S. of the mountains, by the Aboo-Seen, and lower down at Attock, where it is 960 yards wide, and both deep and rapid, by the river of Cabul. The river is crossed here by a bridge of boats, constructed like that used by Alexander, and described by Arrian (lib. v. cap. 7). The bridge is only allowed to remain between November and April, when the river is low; and the construction of it is completed in the course of six days. S. of Attock, the Indus enters a plain, but soon afterwards winds among a group of mountains as far as Harrabah, whence it pursues a southward course to the sea, uninterrupted by hills, and expanding over the plain into various channels, which meet and separate again, but are rarely united into one body. The breadth of the river at Kaharee Ghat, in lat. 31° 22′ N., was found to be about 1000 yards, the deep part of the channel being only 100 yards across, and twelve feet deep. The banks in this vicinity are very low, and in summer are so much overflowed, that the stream expands in many places to a breadth of 13 miles. (Elphinstone, vol. II. p. 416.) In lat. 28° 55′, the Indus receives the Punjab rivers, and rolls past Mittun with a width of 2000 yards, and a depth near the left bank of four fathoms. "From this point to

Bukkur the main stream takes a S.W. course, with a direct channel, but frequently divided by sandbanks. Various narrow crooked branches also diverge from the parent stream, retaining a depth of from 8 to 15 feet of water; and these are navigated by boats ascending the Indus in preference to the great river itself. The country on both sides is of the richest nature, but particularly on the E. bank, where it is flooded from innumerable channels, cut for the purpose of throwing the water S.E. into the interior." (*Burnes' Bokhara*, vol. i. p. 260, 261.) About 17 miles S. of Bukkur, in lat. 27° 13', the Indus sends off a branch to the W. called the Larkhana river, after making a circuit, and expanding in one place into a large lake 12 miles broad, rejoins the main stream 30 miles below the point of separation. The insulated territory, called Chandokee, is one of the most fertile in the Sinde dominions. About 160 miles below Bukkur is Sehwan, in lat. 26° 23'; and between these points the river flows in a zig-zag course nearly S.W., the intervening country being richly watered and divided by its ramifications into numerous islets of the finest pasture. The distance between Sehwan and Hyderabad is 106 miles: the banks seldom exceed eight feet in height, and the neighbouring grounds are usually covered with tamarisks. The river throws off no branches in this part of its course, except the Fullalee (generally an unimportant stream), which leaves the Indus 12 miles above Hyderabad, and crossing the W. extremity of the Runn of Cutch, enters the Indian Ocean by the Khoree mouth. The main river opposite Hyderabad is 830 yards broad, and five fathoms deep; but the channel becomes narrower and deeper as it approaches Tatta, 65 miles below the capital. Shifting sandbanks also occur in many parts between these towns, to such an extent as to perplex the navigation. The course of the stream from Hyderabad is S.W. by S., with one decided turn below Jerruk, where it throws off the Piyaree leading to Mughribee, and entering the sea by the Seer mouth. The country N. of Tatta, which might be rendered one of the richest and most productive in the world, is devoted to sterility, presenting to the eye only dense thickets of tamarisk, saline shrubs, and other underwood. About five miles S. below Tatta is the commencement of the Delta of the Indus. The river here divides into two branches, that to the right being called Buggaur, while that to the left is known as the Sata. The latter is by far the larger of the two, and a little below the point of division has a breadth of 1000 yards: "it divides and subdivides itself into many channels, and precipitates its water into the sea by seven mouths, within the space of 36 miles; yet such is the violence of the stream, that it throws up sandbanks or bars; and only one mouth of this many mouthed sum is ever entered by vessels of 50 tons." (*Burnes' Bokhara*, vol. i. p. 207.) The Buggaur, on the other hand, flows in one stream as far as Darajee, within six miles of the sea, at which point it bifurcates, forming two arms, which fall into the ocean about 25 miles apart. A sand-bank, however, which crosses its upper part, close to the apex of the Delta, renders it unfit for navigation. The land embraced by the Buggaur and Sata extends at the junction of these rivers with the sea to about 70 miles; and so much, correctly speaking, is the *existing* Delta; but the river covers with its extreme a much wider space, and has two other mouths still farther E., viz. the Seer and Khoree, from which, however, the waters have been diverted by the rulers of Sinde into canals for the purpose of irrigation. If, therefore, these forsaken branches be included, the base of the Delta, measured in a straight line from the W. to the E. embouchure, extends 110 miles in a S.S.E. direction. Arrian estimates its extent at the base of Alexander's expedition at 1800 stadia, or nearly double that now assigned to it; but it seems doubtful whether we are to attribute this difference to any great changes in the bed of the river, or to the miscalculation of the Macedonian admiral, Nearchus. The inconstancy of the stream through the Delta makes the navigation both difficult and dangerous. The water is cast with such impetuosity from one bank to the other, that the soil is constantly falling in upon the river, and huge masses of clay hourly tumble into the stream, often with a tremendous crash. In some places, the water, when resisted by a firm bank, forms eddies and gulfs of great depth, in which the current is really terrific; and, in a high wind, the waves dash as in the ocean. It appears, indeed, from the *Report of the State and Navigation of the Indus*, by Lieuts. Carless, Wood, and Pottinger, notwithstanding the statement of Sir A. Burnes, of there being "an uninterrupted navigation from the sea to Lahore," that banks, bars, &c., offer such great obstructions, as effectually to prevent the river from ever becoming extensively available for the purposes of commerce. Vessels drawing eight feet water find themselves aground at the very entrance of the Seets mouth; the employment of ships is out of the question, and the

navigation of the *doondees*, or small native boats, is so tedious, that no communication of any importance could be kept up between Hyderabad and the sea, except by steamers, the use of which, in a country like Sinde, would be attended with extreme difficulty. There are also political obstacles to using the Indus as a channel of commerce. The people and princes are ignorant and barbarous: the former plunder the trader, and the latter overtax the merchant, so that goods are sent by land and by circuitous routes rather than by the Indus, their natural channel. The tides rise in the mouths of the Indus about nine feet at full moon, and both flow and ebb with great violence, particularly near the sea, where they flood and abandon the banks with equal and incredible velocity. This phenomenon was an object of great surprise to Alexander's fleet; and Arrian remarks (lib. vi. cap. 19), that "the ebbing and flowing of the waters was as in the great ocean, insomuch that the ships were left upon the dry ground: but what still more astonished Alexander and his friends was, that the tide, soon after returning, began to leave the ships, so that some were swept away by the fury of the tide and dashed to pieces, while others were driven on the banks and totally wrecked."

It is most probable, allowing for the exaggerations current on the subject, that the countries traversed by the Indus were less barbarous and uncivilized in the days of Alexander than at present: but admitting this, still we are disposed to reject the statement of Dr. Vincent, that there was then an extensive intercourse by means of the Indus between the Punjab and the coast of Malabar. He has even supposed that the vessels employed in that trade proceeded as far S. as Cape Comorin, and having doubled it, sailed N. along the coast of Coromandel! (*Voyage of Nearchus*, p. 11.) But there is no direct evidence of any such intercourse, and the presumption is all the other way. That there was a good deal of traffic on the river between its upper parts and Pattala (near the modern Tatta) is pretty certain; and the navigation of the mouths of the river must have been very different in antiquity from what it is at present, to admit of much intercourse taking place between Pattala and the ocean.

The tides are not perceptible more than 75 miles from the sea, or about 25 miles below Tatta. The quantity of water discharged by the Indus is stated by Sir A. Burnes to amount to 80,800 cubic feet per second, nearly as much as is discharged by the Mississippi, and *four times* as much as is discharged by the Ganges, the other great river of Hindostan. This discharge, provided the statement be accurate, must be attributed chiefly to the greater length of its course in high and snowy regions, to its numerous and large tributaries, and to the barren arid nature of the soil through which it passes; while the Ganges, on the other hand, expends its waters in irrigation, and blesses the inhabitants of its banks with rich and exuberant crops.

The Indus has numerous affluents, none of which, however, deserve any particular mention except the Sutledje, and the other rivers of the Punjab. Of these rivers, the Sutledje (the *Zaradrus* of Ptolemy), which is the most easterly of all, takes its rise near Garoo, on the great plain N. of the Himalaya mountains, enters the chain at Shipkee (where it is 10,484 feet above the sea), runs in a narrow mountain valley for upwards of 160 miles, and enters the S. plain at Ropur, whence its course is south-westward to its junction with the Indus. The other rivers of the Punjab, besides the Beas (the *Hyphasis* of Arrian), which is an affluent of the Sutledje, are, proceeding westward, the Ravee (the *Hydraotes* of Arrian), the Chenáb (*Acesines*), and the Jylum or *Hydaspes*. The last three, all of which rise on the S. slope of the great mountain range of N. India, join their waters with those of the Sutledje in lat. 29° 10' N., and long. 71° 19' E. The rivers of the Punjab are in general navigable up to the place where they issue from the mountains. (*Rennell's Hindostan*, p. 177, &c.; *Elphinstone's Caubul*, vol. ii., App.; *Burnes' Bokhara*, vol. i. passim; *Hamilton's Gazetteer*; *Ritter*, vols. v. and vi. passim.)

INDUSTRY, p. t., Franklin co., Me., 34 m. N.W. Augusta, 627 W. Incorporated in 1803. Bounded S.E. by Sandy river. It contains four stores, three grist-mills, three saw-mills, two tanneries; nine schools, 394 scholars. Pop. 1036.

INGHAM, county, Mich. Situated centrally toward the S. part of the state, and contains 580 sq. m. Organized in 1838. It contains Pine, Portage, and Swampy lakes. Watered by Red, Cedar, Willow, Mud, and Sycamore creeks. It contained in 1840, 2616 neat cattle, 172 sheep, 4358 swine; and produced 22,197 bushels of wheat, 18,923 of Indian corn, 10,947 of oats, 94,951 of potatoes, 37,782 pounds of sugar. It had four stores, one grist-mill, six saw-mills, one tannery; 13 schools, 367 scholars. Pop. 2498. There were but few settlements previous to the summer of 1837. Capital, Vevay.

INGRAM, p. t., Ingham co., Mich., 77 m. W. Detroit, 863 W. Pop. 873.

INGOLSTADT, a town of Bavaria, circ. Ratisbon, on the Danube, 23½ m. S.W. that city. "The pop. of this ancient and melancholy town is reduced to 9000 (1832), a number very disproportionate to its extent. It has recently been restored to the condition of a fortress, by the construction of very strong works on an improved plan. Its old fortifications had withstood sieges from the troops of the League of Schmalkald, from Gustavus Adolphus, and Duke Bernard of Saxe Weimar, and resisted Moreau for three months; but he, succeeding at length, caused them to be demolished. Ingolstadt lost its university, at which the celebrated Dr. Faustus studied, in 1800: it is now transferred to Munich." (*Murray's Handbook.*) It still possesses, however, a royal residence, nine churches, in one of which the Bavarian general, Tilly, was buried, and several hospitals and charitable institutions. It had formerly a considerable manufacture of woollen cloths; but this and its other branches of industry and trade has fallen into complete decay. (*Dist. Geog.; Stein, &c.*)

INNSBRUCK (Fr. *Inspruck*), a city of the Tyrol, of which it is the cap., on the Inn, 80 m. N. by E. Trent, and 248 m. W. by S. Vienna. Lat. 47° 16' 8" N., long. 11° 23' 43" E. Pop. (1838) 10,738. Its situation is highly picturesque. It stands in the middle of a valley, the sides of which are formed by mountains from 6000 to 8000 ft. high; and the Inn is crossed by a wooden bridge (whence the name of the city) from which a magnificent prospect is obtained. On and round this bridge one of the severest actions took place during the war of the Tyrolese, under Hofer, against the French. Innsbruck is divided into the old and new towns, and has five suburbs. The latter are larger and better built than the city itself, though badly paved. The houses of Innsbruck are mostly four or five stories high, built in the Italian style, with flat roofs, and are frequently ornamented with frescoes. Many have arcades below, occupied with shops. The object most attractive to strangers is the Franciscan, or Court church, an edifice containing numerous fine works of art. Among others, is the tomb dedicated to the emperor Maximilian[a], a splendid monument: it is ornamented with 24 bass-reliefs, representing the principal actions of his life, and is surrounded by 24 colossal bronze statues of persons celebrated in history, including Clovis, Theodoric, Arthur, Charles the Bold, Duke of Burgundy, Godfrey of Bouillon, Rodolph of Hapsburg, and many of the emperors of Austria, his descendants, &c. Here, also, is the mausoleum of the archduke Ferdinand of the Tyrol, and his wife, also adorned with bass-reliefs; the grave of Hofer; his statue in white marble, &c. There are numerous other churches, several of which are worth notice. The palace, an extensive building, has gardens extending along the Inn, which form a public promenade. In front of the Old Palace, the former residence of the archdukes of the Tyrol, and of some of the German emperors, is the "Golden Roof," a kind of oriel window, covered with a roof of gilt copper, and one of the curiosities of the place: this edifice is now used for the chancery-chamber. (*Kanzleigebäude.*) Innsbruck has a university of the 2d order, in which instruction is entirely gratuitous. It occupies an extensive and fine edifice, and has 18 professors, and exhibitions to the amount of 12,000 florins yearly. It is attended by about 350 students, and has attached to it a library, botanic garden, and cabinet of physical objects. The *Ferdinandeum*, founded in 1823 upon the model of the *Johanneum* of Grätz, is a museum devoted to the productions of the Tyrol in both art and natural history, and contains some interesting collections, particularly in the dep. of mineralogy. The seminary for noble ladies, founded by Maria Theresa in 1771, the gymnasium, ancient Jesuits' college, and various convents, provincial house of correction, council chamber, town-hall, theatre, and a handsome ball-room, are the other chief public buildings; a statue of Joseph II., and a triumphal arch raised by Maria Theresa, are among the most conspicuous ornaments of the city. Innsbruck is the seat of the state assembly, high judicial court, and other superior departments of the public service for the Tyrol and Vorarlberg: it has manufactures of silk, woollen, and cotton fabrics, leather, glass, and steel goods, and sealing-wax; and is the seat of a considerable trade between Italy and the other countries N. of the Alps. (*Oesterr. Nat. Encyc.; Berghaus; Turnbull's Austria, &c.*)

INVERARY a royal and parl. bor. and seaport of Scotland, co. Argyle, of which it is the cap., on a bay on the W. shore, and near the bottom of the arm of the sea called loch Fyne, 40 m. N.W. Glasgow. Pop. in 1831, 817, and not increasing. Inverary consists principally of two rows of houses, one of them fronting the bay, the other at right angles with it, running inward, and having a northern exposure. The houses, built on a uniform plan, are large and commodious; and the town is one of the neatest and cleanest, and its situation the most picturesque in Scotland. The public buildings are the par. church (in which public worship is alternately performed in Gaelic and English), and a handsome edifice by the water side, containing the court-house and other offices. In the immediate vicinity of the town, on the N., is Inverary castle, the chief residence of the ducal family of Argyle. It was built after a design by Adams, in 1749; but it is hardly worthy of the situation or of the family. It is an embattled structure, of two stories and a sunk floor, flanked with round overtopping towers, and surmounted with a square-winged pavilion. There is in the saloon a curious collection of old Highland arms, including some of those used by the Campbells in the battle of Culloden.

The family of Argyle was, for a lengthened period, one of the most illustrious in the annals of Scotland. Its chiefs were especially distinguished by their devotion to and support of the great principles of civil and religious freedom. Among other members of the family may be specified the Marquis of Argyle, beheaded in 1661; and his son and successor, who also fell a victim to arbitrary and unconstitutional power in 1685; Wodrow, and after him Mr. Fox in his Historical Fragment, have given singularly interesting accounts of the circumstances attending the trial and execution of the last-mentioned nobleman. The grandson of the first and the son of the last of these noble martyrs, created duke of Argyle and Greenwich, and commonly called the great Duke of Argyle, was celebrated both as a statesman and general. He was commander-in-chief in Scotland in 1715, and by his conduct on that and other occasions, was of signal service to the revolutionary establishment. Pope said of his Grace—

"Argyle, the state's whole thunder born to wield,
And shake-alike the senate and the field."

The staple commodity of Inverary is herrings, those of loch Fyne being celebrated for their superior excellence; but the fishing in the loch has latterly greatly declined, and in 1837 the quantity cured and packed in Inverary amounted to only 6334 barrels.

Inverary was erected into a bor. of barony in 1648. In a garden beside the church is a small obelisk, commemorative of the execution in this place, in 1685, of several gentlemen of the name of Campbell, on account of their adherence to presbyterianism. This bor. unites with Campbelton, Oban, and Irvine, in sending a mem. to the H. of C.; and in 1839-40 had 35 reg. voters. Edmund Stone, a self-taught mathematician, editor of *Euclid's Elements*, and author of a *Treatise on Fluxions*, and other works, was a native of Inverary. (*Beyond. Rep.; Wodrow's Hist. of Church of Scotland, passim.*)

INVERKEITHING, a royal and parl. bor., par., and seaport of Scotland, co. Fife, beautifully situated on rising ground on a bay on the N. bank of the frith of Forth, 12 m. N.W. Edinburgh. Pop. of town (1837), 2020. The town consists of a main street, and a smaller one branching off it, besides several wynds or lanes. Many of the houses are extremely old, and an air of antiquity generally marks the place. The only public buildings are the par. church, a dissenting chapel, the bor. school, and the town-house. About 12 in every 100 of the inhab. are, at an average, at school; a larger proportion than generally obtains elsewhere. There are three libraries in the bor. The par. abounds with coal, most of which is exported from St. David's, on Inverkeithing bay. There belong to the bor. 24 registered vessels, burden 1080 tons, employed chiefly in the coasting trade. A considerable number of English and foreign vessels resort to Inverkeithing for coal, bringing in exchange bark, timber, and bones for manure. There are in the immediate vicinity of the town a distillery, tan-work, ship-building yard, a salt work, a magnesia manufactory, and a brick work.

Inverkeithing was created a royal bor. by William the Lion in the 12th century. Its privileges included right of customs over a considerable district of country lying on the frith of Forth; but these have fallen into disuetude, with the exception of the duties at the markets held at Kinross and Tullybole, and the customs at North Queensferry. Even Edinburgh, at one time, paid an acknowledgment of superiority for some parts of the Calton Hill, but it was bought up, or relinquished. In the ridings of the Scottish parliament, the provost of Inverkeithing was entitled to precedence next to the provost of Edinburgh. Before the convention of royal burghs was appointed to be held at Edinburgh, Inverkeithing was the place of its meeting. This bor. unites with S. Queensferry, Dunfermline, Culross, and Stirling burghs, in sending a mem. to the H. of C., and in 1839-40, had 26 registered voters. (*Beyond. Rep.; New Stat. Acc. of Scot., § Fife, p. 229; Beauties of Scotland, iv., 118.*)

a Maximilian, who himself commenced this mausoleum, was after all not buried in it, but at Wienerische Newstadt, in Austria.

INVERLEITHEN, a par. and village of Scotland famous for its mineral well, co. Peebles, 22 m. S. by E. Edinburgh, and 5 m. E. by S. Peebles. It is situated in a romantic pastoral country, within ½ m. of the N. bank of the Tweed, and on both sides the Leithen, a tributary of that river. Pop. of the village, 447 ; not including summer visitors, the aggregate number of whom, in the season, may be about 1100. (*New Stat. Acc. of Scot., & Peeblesshire*, p. 31.) The mineral water has been analysed, and found to contain, per quart bottle, carbonate of magnesia, 10·2 grains ; muriate of lime, 19·4 ; and muriate of soda, 31. (*Ib.* 26.) The popularity of Inverleithen, as a watering-place, was greatly enhanced by the publication (in 1824) of Scott's novel, entitled " St. Ronan's Well," of which it was supposed to be the prototype. A yearly festival has been since instituted at Inverleithen, for the celebration of " the St. Ronan's Border Games ;" and the name of almost every street, or separate edifice, in the village, such as " Abbotsford Place," " Waverley Row," " Marmion Hotel," &c., refers to the illustrious novelist. Traquair-house, the seat of the noble family of that name, is in the immediate vicinity of Inverleithen. The first Earl of Traquair, lord treasurer of Scotland in the time of Charles I., was one of the most eminent statesmen of his day. Dr. Russell, author of the *History of Modern and Ancient Europe*, was born near the village, and was educated in it. The woollen manufacture has been introduced into Inverleithen. (*Crawfurd's Officers of State*, p. 405 ; *Factory Returns*, 1839.)

INVERNESS, a marit. co. of Scotland, and the most extensive in that part of the U. Kingdom : it stretches quite across the island from the E. to the W. sea, having N. the Moray frith and Ross-shire, W. the Atlantic ocean, S. Argyle and Perth, and E. Aberdeen, Banff, Moray, and Nairne. But it includes, exclusive of the mainland, the large island of Skye, with the smaller islands of Harris, N. and S. Uist, Benbecula, &c. Area, 2,716,880 acres, of which 1,943,280 belong to the mainland, and 773,700 to the islands ; the former having 54,400, and the latter 37,760 acres of water. Inverness-shire is, speaking generally, wild, mountainous, and rugged in the extreme. It is supposed that there is not more than 2½ per cent. of its surface not naturally covered with heath. Ben Nevis, the highest mountain in Great Britain, 4370 ft. above the level of the sea, is situated in Lochaber, near Fort William, in this co. Meallfourvonny, on the N. side of loch Ness, is 2730 ft. above the sea. The arable land, which is of very limited extent, is principally comprised in the low districts contiguous to the town of Inverness, in Strathspey (the low country on both sides the Spey), and in narrow glens along the other rivers and lakes. Climate very various ; but generally it may be said to be wet and stormy on the W. coast, severe in the interior, and comparatively mild and dry on the shore of the Moray frith. Principal rivers, Spey, Ness, and Beauly ; all which, but especially the first, have valuable salmon fisheries. The arable land of this co. was formerly divided into small patches, having usually a greater or less extent of hill pasture attached to them, and occupied by tenants at will. The latter lived in miserable huts ; and were at once excessively poor, idle, and disorderly. But the abolition of hereditary jurisdictions and clanship in 1748, and the carrying of good military and other roads into districts that were formerly quite impervious, by enabling the law to be everywhere brought into full operation, have completely repressed the feuds and disorders that formerly disgraced this and other Highland cos. The small holdings have also been very generally consolidated into sheep-farms, some of which are very extensive, and which are mostly stocked with cheviots. Arable farming has, also, been very much improved ; and, in consequence, there has been a very great increase in the quantity of disposable produce, and in the rent and value of the land. Good wheat is raised round the Moray frith ; but oats is the principal crop. The stock of black cattle is very large ; and cattle, sheep, and wool constitute the principal articles of export. In some districts there are extensive forests. There are no manufactures of any importance, nor any considerable town, except Inverness. Illicit distillation, that was once very prevalent, is now all but suppressed. Limestone, slate, and marble abound in most places ; but the want of coal renders the limestone of little value. Average rent of land, including islands, in 1810, 1s. 6d. per acre. Gaelic is the common language ; and the W. parts of the co., and some of the islands, it is the only one that is generally understood. Owing to the thinness of the pop., the co. is but ill supplied with schools, though in this respect, as in others, it is very much improved.

This co. is divided into two nearly equal portions, by a remarkable glen or valley, stretching N.E. and S.W. from the town of Inverness to loch Linnhe, opposite the island of Mull, on the W. coast. This glen, which is very narrow, consists principally of a chain of lakes, comprising loch Ness, loch Oich, and loch Lochy. Its surface being no

where more than 94 ft. above the level of the sea, advantage was taken of this circumstance, and of the continuous chain of lakes, to open a navigable communication between the E. and W. seas, avoiding, consequently, the lengthened and dangerous navigation by the Pentland frith. The entire length of this navigation, or of the Caledonian Canal, inclusive of the lakes, is rather more than 60 m. ; but the excavated part is little more than 23 m. It cost about £1,000,000, and is on a larger scale than any work of a similar class in any other part of the empire. It promises, however, to be a very unproductive undertaking ; and but for the invention of steamboats, which were unknown when it was commenced, it would have been nearly useless.

Inverness co. has 35 parishes : it sends one mem. to the H. of C. for the co. ; and the bor. of Inverness joins with Fortrose, Nairne, and Forres in sending a mem. Registered electors for the co. in 1839, 789. In 1831, Inverness-shire had 17,319 inhab. houses, 19,046 families, and 94,797 inhabs., of whom 44,510 were males, and 50,287 females. Valued rent, £73,166 Scotch. Annual value of real property in 1815, £125,568.

INVERNESS, the cap. of the above co., and of the Northern Highlands, a royal and parl. bor. and seaport of Scotland, on both sides the Ness, within a mile of its influx into the Moray frith, and at the N.E. extremity of the Great Glen of Scotland, forming the line of the Caledonian Canal, 112 m. N.W. by N. Edinburgh, and 81 m. N. by W. Aberdeen. The situation of Inverness is peculiarly striking and picturesque, standing, as it does, in the middle of a beautiful plain, of unequal extent in different directions, with the Moray frith on one side, and the back ground consisting of variously shaped hills, some of which are richly wooded, while others are bleak and rugged. " It is the boast of Inverness to unite two opposed qualities, and each in the greatest perfection ; the characters of a rich open lowland country with those of the wildest alpine scenery, both also being close at hand, and in many places intermixed ; while to all this is added a series of maritime landscape not often equalled." (*M'Culloch's Letters on the Highlands*, vol. i., p. 56.) The Ness, on whose banks the borough stands, is perhaps the shortest river in Scotland, flowing between loch Ness and the Moray frith, a distance of only 8 m. Pop. in 1831, 9888 ; of the town and par., in 1801, 8732 ; in 1831, 14,324 ; inhab. houses, 2195.

The most important portion of the town is on the right bank of the Ness. A handsome stone bridge of seven arches was erected across the river in 1685 : there is also a wooden bridge, built in 1808, at an expense of £4000. The principal streets lie E. or N., and consist generally of more elegant and substantial buildings than are to be found in any other town of the same size in Scotland. The streets, which are lighted with gas, are causewayed and flagged. The inhab. are supplied with water by pipes from the river. The shops, which supply the demand of an extensive district of country, are, in many instances, large and well stocked with goods. The villas in the suburbs are numerous and elegant, and the walks varied and commodious. The public buildings are the Exchange and Townhouse, near the centre of the town ; the jail, surmounted by a tower 130 ft. high ; the Assembly Rooms of the Northern Meeting ; Infirmary, Academy, the United Charity Institutions ; which last edifice occupies an elevated situation in the vicinity of the town. With regard to ecclesiastical buildings, the High Church is the most conspicuous ; the square tower attached to it was built by Oliver Cromwell, and the bell brought from the cathedral of Fortrose, on the N. banks of the Moray frith. There are two other parish churches, in one of which Gaelic alone is used ; and two chapels of ease. There are, also, dissenting chapels, belonging respectively to the Episcopalians, the United Associate Synod, the Independents, Baptists, Methodists, and Rom. Catholics.

Gaelic was formerly the only language spoken in Inverness and its neighbourhood ; and it is at this moment the ordinary speech of the lower orders, all of whom, however, understand and can speak English. It is admitted that the English language is spoken in greater purity by the middle and upper ranks in Inverness than in any other place in Scotland ; a distinction which is said, whether correctly or not, to have originated in the circumstance of Cromwell having stationed and long maintained an English garrison in the town. " The soldiers seem to have incorporated afterward with the inhab., and to have peopled the place with an English race ; for the language of the town has been long considered as peculiarly elegant." (*Johnson's Tour to the Hebrides*.) The Highland character, however, still predominates very considerably in the borough. In addition to the Gaelic language, the speech of the common people, their dress is more or less of Celtic fashion, and of home manufacture, such as the short coat, blue bonnet, plaid, rig and fur stockings, all of the coarsest materials.

The married women usually walk the streets and go to church without a bonnet; the maidens without either cap or bonnet; while the other parts of their dress are of the most simple and homely description.

Inverness enjoys eminent advantages as to education. The Academy, founded by subscriptions raised in the borough and elsewhere, is a chartered institution, and one of the most efficient seminaries in Scotland. The old grammar school, with its endowment, has merged into it. It is provided with a rector, and from four to six other teachers in the different departments of liberal study. There are various other excellent schools, some of which are endowed. Raining's Charity school, attended by about 250 scholars, is a useful institution. An infant school, which has been in operation for several years, is admirably conducted. The late Dr. Bell, of Madras, left £10,000 3 per cent. consols to the magistrates for the purposes of education. The number of female schools is very considerable; the better ranks in the northern counties generally sending their daughters thither to complete their education. Inverness has, besides, a mechanics' institute; various libraries, both subscription and circulating; two public reading-rooms, several printing presses, and three weekly newspapers. (New St. Acc. of Scot., § Inverness; Educat. Inquiry, Sect.; Parl. Paper, 1837, vol. 47.)

As to charitable institutions the most important are the infirmary and dispensary. There are no poor-rates in the town or parish, the poor being supported by the church collections, by special quarterly collections, and by the produce (£192) of certain sums bequeathed for the purpose. There are two funds (£3799 and £916) left for the support of decayed householders; also a fund (£87,000) for the education of boys of the name of Mackintosh; with sundry smaller sums in the hands of the magistrates.

Manufactures, in the proper sense of the term, may be said not to exist here, if we except those of linen, plaidings, and woollen stuffs, and a small hemp manufactory, principally for the making of bagging. Ship-building is carried on to some extent. There are breweries, distilleries, and tanworks. With regard to trade, Inverness is the centre of a custom-house district, which embraced, in 1835, 7597 tons of shipping, and 160 registered vessels; the port of Inverness possessing about half the vessels, and two thirds of the tonnage. While the town has regular traders, both steamers and sailing smacks, to Aberdeen, Leith, London, &c., on the E. coast; she has also a similar communication, by means of the Caledonian canal, with Glasgow, Liverpool, &c., on the W. coast; and also with Ireland. The canal passes within less than a mile of the borough; and Clachnaharry, (pop. 300), where it joins the Moray frith, is not more than a mile distant. There are three harbours, one of them for small craft, near the town, the others at the mouth of the river; while the canal wharves at Clachnaharry are also used for the loading and unloading of goods. Grain, at least oatmeal, used to be imported to Inverness; but oats are now exported to the amount of about 5000 bolls. Coal, almost the only kind of fuel used, is imported both from England and the frith of Forth. The imports consist generally of the various articles which the demand of a large district of country requires. The foreign imports consist of about 500 tons of hemp, and three or four cargoes of timber and Archangel tar. Inverness has several fairs; but the wool fair, in the month of July, attended by all the principal Highland sheep farmers, as well as by wool staplers and agents from England and the S. of Scotland, is the most eminent. Fully 100,000 stone of wool are annually sold at this market; while above the same number of sheep are also disposed of. The prices paid at this fair generally regulate those of all the other markets in the country. There are five banks in the borough, and a savings' bank.

Inverness is very ancient. In the 6th century it was the capital of the Pictish kingdom, when St. Columba of Iona went thither, ad ostium Nessa, with the view of converting the Pictish king to Christianity. An ancient castle stood on a rising ground E. of the town; but it was destroyed in the 11th century by Malcolm III., who built another on a commanding eminence near the river, which continued to be a royal fortress, till blown up, in 1746, by the troops of the Pretender. Inverness was erected into a royal borough by David I.; and various royal charters, confirming or extending its privileges, were subsequently conferred on it. The town was often an object of plunder to the lords of the isles and other Highland chiefs. A monastery, belonging to the Black Friars, existed in this place: but all traces of it have long since disappeared. The citadel referred to above, as constructed by Cromwell, was built in 1652–57, N. of the town, near the mouth of the river. Part of its ruins are still standing. Culloden Moor, the scene of the battle that decided the fate of the Pretender, Charles Stuart, is within 3 m. of the town. Since 1745, great improvements have been effected here. Previously to 1755, the post from Edin-

burgh to Inverness was conveyed by a man on foot! In 1745, the magistrates advertised for a saddler to settle in the borough; and in 1773, a cart, purchased by subscription, was first seen in the borough. No plan of regularly cleaning the streets was adopted till about the beginning of the present century. Inverness is now, however, superior perhaps to any town of its size in Scotland, as to all the necessaries, comforts, and luxuries of life. The borough united Forres, Fortrose, and Nairne, in sending a member to the House of Commons. Registered voters in 1839–40, 475. (Anderson's Highlands and Islands; Caledonia, vol. i.; and the works already referred to.)

INVERURY, a royal and parl. bor. and par. of Scotland, co. Aberdeen, in the angle formed by the confluence of the Don and Ury, 14 m. N.W. Aberdeen. Pop. of town (1831), 994; of town and par., 2052. The inhabitants are chiefly agriculturists, there being only about 36 weavers in the parish. The Aberdeenshire canal, begun in 1796, and completed in 1807, commences in the tide-way of the harbour of Aberdeen, and terminates at Port Elphinstone, near Inverury. The entire length is 18¼ m.; the surface width is 23 ft.; the depth 3¾ ft.; it has 17 locks; and its highest level is 168 ft. above low water mark. Keith Hall, the seat of the Earl of Kintore, who also holds the title of Lord of Inverury, is in the immediate vicinity of the borough. Arthur Johnston, editor of the Deliciæ Poetarum Scotorum, and who holds the next place to Buchanan among the Latin poets of Scotland, was born in the neighbourhood of Inverury in 1587. This borough unites with Banff, Cullen, Kintore, and Peterhead, in sending a member to the House of Commons. Registered voters in 1839–40, 94. (Boundary Reports; Beauties of Scotland, iv., 490.)

IONA. (See Hebrides.)

IONIA, county, Mich. Situated centrally towards the W. part of the state, and contains 576 sq. m. Organized in 1837. Drained by Grand river and its branches. It contained in 1840, 1886 neat cattle, 270 sheep, 3902 swine; and produced 32,362 bushels of wheat, 14,784 of Indian corn, 16,965 of oats, 93,500 of potatoes, 35,635 pounds of sugar. It had six stores, two grist-mills, nine saw-mills; eight schools, 290 scholars. Pop. 1923. Capital, Ionia.

IONIA, p. t., capital of Ionia co., Mich., 136 m. W.N.W. Detroit, 823 W. Watered by Grand river. It contains three stores, one grist-mill, three saw-mills; two schools, 140 scholars. Pop. 486. The village is situated on the N. side of Grand river, at the mouth of Prairie creek. Steamboats ply between this place and Grand Rapids, 35 m. below on the river, around which a steamboat canal is constructing, which will make the river navigable to this place. It contains a courthouse, jail, two churches, an Episcopal and Baptist; a U. States land office, several stores, 60 dwellings and about 350 inhabitants.

IONIAN ISLANDS, a collection of seven principal and many smaller islands on the W. and S. coasts of Greece, lying between the 36th and 40th parallels of N. lat., and between the 19th and 23d deg. of E. long., forming a republic under the protection of Great Britain. The principal islands, with their area, population, &c., are as follows:—

Islands.	Area in sq. m.	Total Pop. in 1836.		Aliens, &c.	Pop. to sq. m.
		Males.	Females.		
Corfu . .	227	35,361	29,836	9,306	327
Cephalonia .	348	34,264	24,233	936	162
Zante . .	156	18,676	15,673	1,217	226
Santa Maura .	190	9,597	8,098	180	95
Ithaca . .	44	4,942	4,702	106	213
Cerigo . .	116	4,198	4,561	37	73
Paxo . .	28	2,364	2,508	223	166
Total . .	1,097	110,416	98,745	12,027	196
		204,866			

These islands, a more minute description of which will be found under their several heads, have, generally speaking, rugged irregular coasts, and a very uneven surface; barren rocks and heath-covered hills forming nearly half their whole contents. Their geological formation is chiefly limestone, disposed in highly inclined strata, intermixed with grey foliated gypsum, and masses of sandstone; and there are few organic remains. The climate is mild, but subject to sudden changes. The sirocco, however, makes the heat occasionally oppressive, and the thermometer in summer frequently rises to 32° Réaum. Hurricanes (called here boracca) and earthquakes are frequent, especially in Zante. There fell, in 1838, 49·04 inches rain. Fine springs of fresh water are abundant on most of the islands. The soil is more favourable to grape cultivation than to the raising of corn; and hence more than three fourths of the surface available for tillage is laid out in currant-grounds, vineyards, and olive plantations, which are all managed with considerable skill. The land is chiefly in the hands of small proprietors, who let it out to tenants on the métayer system, receiving half the produce as rent.

The following table has been said to exhibit the employment of the surface, the nature and quantity of the produce, and the average market prices of the different articles in 1836. (*Off. Tables.*)

Kind of Produce.	Acres.	Amount of Produce.	Average Price.
			£. s.
Wheat	16,137	76,856 bush. *	3 11 per bush.
Indian corn, barley, &c.	37,497	188,000 —	8 4 —
Oats	8,945	28,915 —	1 8 —
Pulse	4,532	23,298 —	3 5½ —
Olive oil?	209,226	112,299 barr.	45 3 per barr.
Wine	115,182	316,147 —	17 4 —
Currants	15,749	17,945,149 lbs.	
Cotton	1,814	27,387 —	0 9½
Flax	1,310	74,986 —	0 5½
Total of land in crop	420,151	† Olives are most extensively cultivated in Corfu, grapes in Cephalonia, and currants in Zante. Olives are gathered in Dec. and currants in the middle of Sept	
Pasture land	64,960		
Total of available land	485,111		

* This should curiously be quarters; if not, the crop was worth little more than the seed.

The live stock on the islands, in 1836, consisted of 14,189 horses, 10,395 horned cattle, 95,950 sheep, and 66,896 goats, a number far too small to meet the demands of the population: and hence, large quantities are annually imported from Albania and Greece. The manufactures are not important. Soap is made at Corfu and Zante, and the value of the quantity exported, in 1836, is officially estimated at £12,000; earthenware, silk shawls, goat-hair carpets, coarse blankets, linen cloths, and sacking are also made to some extent. The islands, however, enjoy a considerable share of the commerce of the Mediterranean, owing to their convenient situation for the supply of the neighbouring continent. They import wheat and other grain; chiefly from Odessa, silks, cotton, and woollen fabrics, cured fish, British hardware, and colonial produce, the total value of which amounted, in 1836, to £242,525; and in the same year, they exported island produce and manufactures (olive-oil, currants, wine, valonia, cotton, salt, soap, and woven fabrics) to the amount of £689,588. The commerce of the islands is cramped by the high duties laid on exported articles. Their commercial relations will be best understood by the subjoined account of the entries and clearances of shipping in 1838.

Countries.	Inward.	Outward.
Ionian islands	122,220 tons	154,897 tons
English	32,114 —	20,480 —
Austrian	25,079 —	27,419 —
Russian	15,708 —	11,584 —
French	104 —	104 —
Neapolitan	3,203 —	3,150 —
Papal states	1,105 —	1,308 —
Sardinian	2,320 —	2,704 —
Greek	44,189 —	45,344 —
Turkish	3,426 —	3,647 —
All others	1,546 —	2 106 —
Total	**220,542 —**	**273,001 —**

These islands possess few manufactures properly so termed. The wives of the villani or peasants spin and weave a coarse kind of woollen cloth, sufficient in great part for the use of their families. A little soap is made at Corfu and Zante. The latter manufactures a considerable quantity of silk gros-de-Naples and handkerchiefs; the art of dyeing, however, is too little studied, and the establishments are on too small a scale. The peasantry, in general, are lazy, vain, delighting in display, and very superstitious. Those of Zante and Cephalonia are more industrious than the Corfiotes; in the first, particularly, their superior condition is probably to be ascribed, in part at least, to the fact that the nobles reside on their estates, and contribute by their example to stimulate industry. In Corfu, the taste for the city life, which prevailed in the time of the Venetian government, still operates to a great degree. The Corfiote proprietor resides but little in his villa; his land is neglected, while he continues in the practice of his forefathers, who preferred watching opportunities at the seat of a corrupt government, to improving their fortunes by the more legitimate means of honourable exertion and attention to their patrimony. In this respect, however, a material change for the better has taken place within the last thirty years.

The government of these islands, since 1817, has been vested in the high commissioner (who represents the sovereign of Great Britain), the senate, and the legislative assembly, which have jointly the title of the "parliament of the Ionian islands." The legislative assembly consists of 40 members, 29 of whom are *elected* by the synclete, or nobles of the different islands, Corfu, Cephalonia, and Zante returning seven each, Santa Maura four, and the three smaller islands sending four among them; the other 11 are styled *integral*, being officers appointed by the high commissioner. The assembly elects its own officers, fixes the amount of the supplies, and has the power of originating new laws. The senate, consisting of five members, is elect-

ed by the legislative assembly out of their own number; but the president is nominated by the commissioner, who likewise confirms the election of the rest. This body is legislative, so far that it has a veto on the proceedings of the assembly; but its chief business is to regulate affairs during the recess of the legislative body, to decide on matters submitted to it by the commissioner, and to nominate the officers under the general government, subject to the approbation of the commissioner. The assembly and senate are elected for five years; but may be dissolved at an earlier period, either by the sovereign or his representative, with whom ultimately rests the confirmation, or veto, of every measure, appointment, or proceeding of the general or local government. The separate islands, likewise, have each a council of five members, selected out of a list of ten, furnished by the *synclete*, besides whom five other active functionaries are nominated by the senate to act as an executive body. The judicial power is lodged in a supreme court at Corfu, comprising four ordinary and two extraordinary members, the latter being the commissioner and the president of the senate; of the former, two are native Ionians, nominated by the senate, and two are appointed by the commissioner, who may be either Ionian or British subjects. The ordinary members decide common causes, and, in case of difference of opinion, appeal to the extraordinary members. Subordinate to this court are four tribunals on each island, making 21 in the whole, and, under these again are justice-of-peace courts for minor offences and small civil suits; but the senate and high commissioner may reverse every decision whatever, if they think proper. The *sanità*, or health establishment, the police, and the army, are under the sole direction of the high commissioner. The public press is, likewise, under the immediate control of the commissioner and senate. The religious establishment consists of an archbishop and bishops, with the vicars or curates of the Greek church, which is the dominant sect. Full liberty, however, is given to the adherents of the Roman Catholic, Protestant, and other creeds.

The revenues of the Ionian islands are principally derived from the export duties on oil and currants of 19¼ per cent., on wine of 5 per cent., and on soap of 8 per cent. and *valorem*. The duties on imported merchandise are regulated by a tariff, and all articles not specifically included in it pay an *ad valorem* duty of 7 and 8 per cent. There are no direct taxes. The following is an official statement of the revenue and expenditure in these islands during 1837 :—

Revenue.		£.	Expenditure.	£.
Customs		34,885	General government	41,307
Transit duty		425	Local do. on the popular islands	88,649
Duty on oil		8,814	Contingencies of police and courts of justice	7,551
Do. currants		65,491		
Do. wine and spirits		2 891	Education	11,886
Do. tobacco		2,828	Brevets of public offices	9,944
Do. castje		8 046	Public works, roads, &c.	16,000
Stamp duties		13,726	Public buildings and mines	4,982
Port & est		2,044	Salaries	6,488
Monopoly of salt and gunpowder		3,316	Hospitals and other contingencies	8,515
Rents of public property		11,282	Collection of revenue	1,885
Rates for roads, &c.		9,221	Packet service	7,482
Assets and provedico surplus		4,805	Military protection	29,914
Miscellaneous		7,176		
		147,307		198,351

Owing to the want of an efficient police, and to the gross corruption of the Venetian government, under which impunity might be secured for the foulest crimes, the state of society in the Ionian islands, when they were placed under the ægis of England, was as bad as can be imagined. The inhabitants were at once lazy, ignorant, superstitious, cowardly, vindictive and blood-thirsty. Perjury was so common, as hardly to excite attention, and assassination was more frequent than in any other country, Corsica not excepted. But under the vigorous and equal government of Great Britain, a great change for the better has been effected; assassination is now comparatively rare, and the inhabitants are beginning to appreciate the advantages of honesty, fair dealing and industry.

Education in these islands has been progressively improving since the efforts of the Earl of Guildford first gave the impulse to the government and inhabitants. At Corfu there is a university in... ried by the government, in which instruction is furnished... y competent professors in classical and scientific subjects; and the same town contains an ecclesiastical seminary for young men intending to enter the priesthood of the Greek church. Each of the islands has a "secondary" school for instruction in the Greek, Italian, and English languages, writing, arithmetic, &c., and in the chief town of each island there is a central school for teaching reading, writing, and arithmetic. District schools are established in many parts of all the islands. The whole educational establishment is under the direction of a commission of public instruction, and the Ionian government make large annual grants for building school-rooms, providing books, slates, &c.

The only coins properly belonging to the Ionian islands, are a silver 3d. piece and a copper cent; but those mostly in circulation are Spanish doubloons and dollars, and Venetian dollars, received in payment for the produce exported to Spain and Italy. British silver coins are also occasionally met with. The chief standard of weight is the imperial troy pound of 5760 grains: 24 of these grains make a calco; 20 calci make an ounce, and 12 ounces compose a libbra sottile. The libbra grossa is equivalent to the pound avoirdupois, and 100 of these pounds make a talento. The English yard is the standard linear measure: 5½ yards make a camice, 220 yards a stadio, and 1760 yards a mile. The gallon (equivalent to the English gallon) contains eight bicoteli.

The Ionian islands are frequently mentioned in the ancient history of Greece, but only as detached governments, and not under their collective form. After having repeatedly changed masters during the middle ages, they at length became the possession of the Venetians early in the 15th century. They were thenceforward governed by an Italian proconsul; the Italian language was generally introduced into public acts and among the nobles; and Corfu was made the chief arsenal and port of the Venetian navy. In this state the islands continued till 1797, when they were seized by the French, who were confirmed in their possession by the treaty of Campo Formio. Two years afterwards they were taken by the Russians and Turks, and declared an independent republic, under their joint protection. The treaty of Tilsit, in 1807, restored them once more to the French, who retained them till 1814, when they were yielded to the English. By the arrangements of the congress of Vienna they were constituted a republic, and placed under the protection of Great Britain. The constitutional code was drawn up and ratified by the British government in July, 1817. (Tables of Rev.; Pop. Suppl., p. vii.; Parl. Rep., 1817–26; Priv. Off. Rep.; Turner's Levant, i., 99–106; and Burgess's Levant, vol. i.)

IOSCO, t., Livingston co., Mich. Watered by Grand river. It has one saw-mill; five schools, 103 scholars. Pop. 395.

IOWA, territory, is bounded N. by the British territory of the Hudson's bay company; E. by Wisconsin territory and the state of Illinois, from which it is separated by Mississippi river, and a line due N. from its source in Itasca lake to the British possessions; S. by the state of Missouri; and W. by Missouri river, to the entrance of White-earth river, and following this river N. to the British possessions. It is between 40° 30' and 49° N. lat. and between 90° and 108° W. long., and between 14° and 26° W. long. from W. It is over 600 miles long, and at a medium 250 miles broad, containing about 150,000 square miles, or 96,000,000 acres. The Indian title to a considerable portion of this territory has not been extinguished. The population, in 1840, was 43,111. Of these there were employed in agriculture, 10,469; in commerce, 355; in manufactures and trades 1629; in mining, 217; in navigating the ocean, rivers, and canals, 91; in the learned professions, 365.

The territory is divided into 18 counties, which, with their population, in 1840, were as follows:

Counties	Pop. 1840.	Counties	Pop. 1840.
Cedar	1,253	Jones	471
Clayton	1,101	Lee	6,093
Clinton	821	Linn	1,373
Delaware	168	Louisa	1,927
Demoines	5,575	Muscatine	1,942
Du Buque	3,059	Scott	2,140
Henry	3,772	Van Buren	6,146
Jackson	1,411	Washington	1,594
Jefferson	2,773		
Johnson	1,491	Total	43,111

Iowa city, on Iowa river, 33 m. W.N.W. Bloomington, was fixed on, in May 1839, as the permanent capital of Iowa.

The surface is moderately undulating, without mountains or high hills, excepting in the N. part, where are hills of considerable height. On the margins of the rivers are frequent ranges of bluffs, intersected by ravines. These bluffs are generally from 40 to 130 feet high, where a table-land usually commences, consisting of gently undulating timber land and prairie. The territory is well watered by rivers and creeks, the margins of which are skirted with a rich bottom land, and covered with timber. Probably nearly three fourths of the territory is destitute of trees, but the streams and timber are so happily diffused, that this circumstance occasions no great inconvenience. Some of the prairies have a level, others a rolling surface; some are covered with a rich coat of natural grass, affording excellent subsistence for cattle, and are frequently interspersed with hazel thickets and massafras shrubs, and, in the proper season, superbly decorated with flowers. The

soil, both on the bottom land and on the prairies, is generally good; the former consists of deep black mould, and in the latter it is intermingled with sandy loam, and sometimes with red clay and gravel. The soil on the high and rolling prairies will average from 18 to 24 inches deep, and on the bottom lands from 24 to 48 inches deep, and could not be exhausted by 100 years of successive cultivation. The productions of the soil are Indian corn, wheat, rye, oats, buckwheat, potatoes, pumpkins, melons, and all kinds of garden vegetables. The climate and soil are particularly favourable to the cultivation of fruit; and crab-apples, wild plums, strawberries, and grapes are indigenous and abundant. The ordinary yield of Indian corn is from 50 to 75 bushels to the acre; and of wheat from 30 to 35 bushels. Good wells are generally obtained at the depth of from 25 to 30 feet.

The agricultural statistics of 1840 give a favourable view of this country, which had very few white inhabitants before 1832. There were 10,794 horses or mules; 38,049 neat cattle, 15,354 sheep, 104,899 swine; poultry valued at $16,289. There were produced 154,693 bushels of wheat, 3793 of rye, 1,406,241 of Indian corn, 6212 of buckwheat, 216,385 of oats, 726 of barley, 234,063 of potatoes, 23,039 pounds of wool, 3132 of wax, 8706 of tobacco, 41,450 of sugar, 17,953 tons of hay, 313 of hemp or flax. The products of the dairy were valued at $23,809, of lumber at $50,980, of skins and furs at $33,504.

The bottom lands on the rivers, which are occasionally overflowed, are subject to bilious complaints, fevers, and agues; but, as the rivers are not sluggish, their borders are less unhealthy than in many portions of the west. With this exception, the climate is healthy. The winter commences in December and ends in March. The climate of the settled part is not colder than that of New-York and Philadelphia. Snow rarely falls to the depth of more than six or eight inches. As far N. as Prairie du Chien, the Mississippi is frozen over, so as to be passed on the ice, for from four to six weeks in the winter. The summers are warm without being oppressively hot, and are refreshed by frequent showers.

Mississippi river runs on its E. border for its whole length, and is navigable for steamboats to the mouth of St. Peter's river. St. Peter's river rises near the sources of Red river, and, after a course of 230 miles, enters the Mississippi nine miles below the falls of St. Anthony. Des Moine's river flows through its southern part, and enters the Mississippi at the foot of Des Moine's rapids, forming a part of its S.W. boundary. It is navigable for steamboats, in high water, for 100 miles, and for keelboats at all times. Iowa river is 300 m. long, and is navigable for steamboats to the junction of the Red Cedar, and for keelboats or light steamboats to Iowa city. Checcauque or Skunk river has a course of 150 miles, and affords much water-power. Red Cedar, the main branch of Iowa river, is navigable for keelboats, at certain seasons, for 100 miles from its mouth. The Wapsipinecon is 200 miles long, and enters the Mississippi 12 miles above Rock Island. It has a winding and rapid course, and affords much water-power. Maksquets river forms the S. boundary of the mineral region, and enters Mississippi river six miles below Bellevue. It has the finest water-power in the territory. It has a natural bridge of solid limestone of 40 feet span. Turkey river, after a course of 150 miles, enters the Mississippi, opposite to Cassville, passing through the N. part of the settled portion of the territory.

Burlington, on the Mississippi, 1429 miles above New Orleans, and 246 above St. Louis, is a place of extensive trade. Du Buque is the metropolis of the mineral region. Fort Madison, Bloomington, and Davenport on the Mississippi, are places of considerable business. Iowa city, the capital, of recent foundation, is a growing place.

There were, in 1840, 14 commission houses in foreign trade, with a capital of $93,300; 157 retail stores, capital $437,550; 29 persons employed in the lumber trade, with a capital of $16,250. Home-made, or family manufactures amounted to $85,966; three tanneries had a capital of $4400; two distilleries had a capital of $1300; six flouring-mills, 37 grist-mills, and 75 saw-mills, employed a capital of $166,650; 14 brick or stone, and 483 wooden houses, were built at an expense of $135,987; four printing-offices, and four weekly newspapers, employed a capital of $5700. The total capital employed in manufactures was $299,645.

The university of Iowa, at Mount Pleasant, in Henry co., has been chartered by the territorial legislature, under the direction of 21 trustees. Seven academies have been incorporated. In 1840 there was in operation one academy, with 25 students. There were 63 common and primary schools, with 1500 scholars.

The Methodists, Baptists, and Presbyterians are the principal religious denominations; but there are Episcopalians, Friends, Roman Catholics, and some others.

The chief Indian tribes of the territory are the Secs and

Foxes, the Chippewas, Ottawas, and Pottawatamies. The Sioux inhabit the N. part of the territory.

Iowa was separated from Wisconsin, and had a territorial government, which was organized July 4th, 1838. The governor and secretary of the territory, the judges, United States' attorney, and marshal, are appointed by the president of the United States, with the advice and consent of the senate. The people elect 13 members of the council, and 26 members of the house of representatives, who constitute the territorial legislature. The governor, who is also superintendant of Indian affairs, is appointed for three years. The council are elected once in two years, and the representatives annually. Every two years the people elect a delegate to the United States' congress. All white male citizens of the United States, over 21 years of age, and who have resided in the territory for six months next preceding an election, enjoy the right of suffrage.

In 1832 this territory was purchased of the Indians, and began to be settled by the whites in 1833. From November 1838 to June 1840, the amount of the sales of the public lands in the territory was $808,287.

Iowa, county, Wis. Situated in the S.W. part of the territory, and contains 1300 sq. m. Bounded N. by Wisconsin river, into which flow several small streams. Drained by E. and W. branches of Pekatonoskee river. Lead and copper ore are abundant. It contained, in 1840, 4867 neat cattle, 583 sheep, 7618 swine; and produced 12,945 bushels of wheat, 76,885 of Indian corn, 9424 of barley, 147,788 of oats, 49,326 of potatoes. It had 21 stores, 30 smelting houses, producing 8,900,000 pounds of lead, employing 126 persons, and a capital of $104,508, one flouring-mill, four grist-mills, seven saw-mills, one distillery, one printing-office, one weekly newspaper; 15 schools, 276 scholars. Pop. 3978. Capital, Mineral Point.

Iowa, city, p. v., capital of Johnson co., and of the territory of Iowa, 32 m. W.N.W. Bloomington, 35 m. S.S.W. Dubuque, 75 m. N. by W. Burlington. Situated on the E. side of Iowa river, which is navigable to this place for keelboats at all stages of the water. This place was the hunting-ground of the Indian until 1838. The first pirogue from the river, half a mile long and 180 yards wide, is devoted to a public promenade. The second elevation is about 19 feet, and the third about 30 feet above the promenade. Upon the brow of the last runs Capitol-street, 160 feet wide, and is met at right angles by Iowa avenue, of similar width. The capitol on Capitol-street, and fronting Iowa avenue, is of Grecian Doric architecture, 190 feet long and 60 feet wide, and two stories high above the basement. It is surmounted by a dome resting on 22 Corinthian columns. It is worthy of being the capital of the future state of Iowa. The village contains a church; 11 stores of different kinds, 125 dwellings, and about 800 inhabitants.

IPSAMBOUL. (See Nubia.)

IPSWICH, a parl. and mun. bor., river-port, and town of England, cap. co. Suffolk, on the Orwell, 69 m. S. Norwich, and 68 m. N.E. London; lat. 52° 4' N., long. 1° 9' E. Area of parliamentary borough, which includes 12 entire parishes and parts of six others, 7689 acres. Pop., in 1841, 24,940. The town occupies the foot of a range of hills gradually sloping to the river, which is navigable up to this point by vessels of 200 tons, and is crossed by a handsome iron bridge. The streets are irregularly built, and for the most part narrow; but some of them, which are new or have been recently widened, consist of neat and substantial buildings. On the whole, although combining a great many old-fashioned houses, the town presents a flourishing appearance, and is not only improving, but rapidly extending. It is lighted chiefly with gas, and the streets are either paved or macadamized. (Mun. Bound. Rep.) There are 12 parish churches, none very remarkable for architectural beauty, and several places of worship for dissenters. The other public buildings are the town-hall; the shire-hall; the custom-house, a respectable brick structure on the quay; a commodious market-house, erected in 1811; the corn exchange; the county jail, said to be well regulated; the borough jail; and the town library, kept as well as the grammar school, in an old building once a monastery of Black Friars. The grammar school, which was intended by its founder, Cardinal Wolsey, to form part of a college preparatory to Christ-church, Oxford, was chartered by Queen Elizabeth, in 1565. The master receives, for the instruction of 50 free boys (sons of freemen), a salary of £150 a year and a dwelling-house, and he is allowed to take boarders. A charity school; for maintaining, clothing, and educating 16 poor children, two national schools, and a Lancastrian school, furnished instruction, in 1834, to upwards of 600 children; and Sunday schools are attached to most of the churches and all the chapels. An institution for the support of widows and orphans of poor clergymen was established in 1704; and there are several almshouses. Two weekly papers are published in the town.

Ipswich formerly enjoyed a considerable share in the woollen and coarse linen trade; but manufacturing is now all but extinct, the trade in this district having been nearly annihilated by the cheaper goods made in the north. One manufacturer employs six looms in making cotton checks and cloths for sailors' shirts; and there is one sacking weaver. The town has no spinning factories; but small quantities of yarn are spun for the Norwich weavers. The present industrial establishments comprise two extensive iron foundries, a snuff-mill, a soap-work which produced, in 1838, 769,926 pounds hard soap, breweries, and two ship-building yards, with accommodations for constructing merchantmen of the largest size. The principal business of the town consists in the corn and coal trade. In 1838, 162,645 quarters of corn and grain (chiefly wheat and malt), exclusive of 57,805 cwts. of meal and flour, were exported hence to London and other parts of the kingdom; and upwards of 40,000 chaldrons of coal are annually imported and supplied by the Stourmarket canal to the W. parts of the county. A general foreign trade of some importance, especially in Norway timber, is carried on. In 1838, there belonged to the port 154 vessels, of 11,300 tons burden, and the gross customs' duties, in 1838, amounted to £41,897. Constant communication with London takes place by means of steam-packets, arriving and leaving on alternate days; and the intercourse will be still more facilitated, as soon as the Eastern Counties railway shall have been completed. (Handloom Weavers' and Municipal Boundaries Reports; Tables of Rev., Pop., &c.)

The corporation, the first charter of which was granted by King John, and confirmed by subsequent monarchs, appears to have been, previously to the passing of the Municipal Reform Act, one of the worst regulated and most corrupt in the kingdom, "every power intrusted to it, its property, its patronage, and its charities, having been used for election purposes." (Mun. Report.) The present municipal officers are 10 aldermen, one of whom is mayor, and 30 councillors; the borough being divided into five wards, and having a commission of the peace, under a recorder. Corporation revenue, in 1839, £4868. A court for the recovery of small debts is held every Tuesday. Ipswich has sent two members to the House of Commons since the reign of Edward I., the franchise, till the passing of the Reform Act, being vested in freemen (by birth, servitude, gift, or purchase) not receiving alms. The boundaries of the old borough have not been changed. Registered electors, in 1835-6, 1418.

The ancient name of the town was Gyppeswich, derived from its proximity to the confluence of the Gipping (now converted into a canal) with the Orwell. Its antiquity is proved by its destruction, in 991, by the Danes. In the reign of Edward the Confessor it comprised 300 burgesses. William the Conqueror erected a castle for its protection. Its ancient corporate privileges included admiralty jurisdiction over the river and port of Harwich, which was long subordinate to Ipswich, and an exemption from serving on county juries or holding county offices. During the 13th and 14th centuries, this town seems to have been a favourite resort of monks and clergymen, there being at that period not fewer than 21 churches and six religious houses. (Kirby's Suffolk Traveller; Acc. of Ipswich; Mun. Bound. Rep.; Mun. Corp. Rep.)

IPSWICH, p. t., port of entry, and one of the capitals of Essex co., Mass., 26 m. N.E. by N. Boston, 405 W. Incorporated in 1634. It contains seven stores, one lumber yard, one fulling-mill, two cotton factories with 3540 spindles, four tanneries; two academies, 90 students, nine schools, 966 scholars. Pop. 3600. The village is pleasantly situated on both sides of Ipswich river, 2 m. from its mouth. There is a stone bridge here with two arches, built in 1764. It contains a courthouse, jail, a bank; four churches, two Congregational, one Unitarian, and one Methodist; a female seminary, celebrated for the preparation of teachers. The falls, immediately above tide-water, afford good water-power. The river is navigable to this place. The tonnage in 1840 was 3739. The Eastern railroad passes through it, and three cars ply daily to Boston and Newburyport.

IRA, p. t., Rutland co., Vt., 70 m. S.S.W. Montpelier, 448 W. Organized in 1779. Watered by Castleton river. It contains a Baptist church; two saw-mills; five schools, 152 scholars. Pop. 431.

IRA, p. t., Cayuga co., N. Y., 169 m. W. Albany, 355 W. It contains a Presbyterian church; five stores, one fulling-mill, two grist-mills, four saw-mills; 16 schools, 795 scholars. Pop. 2683.

IRASBURG, p. t., capital of Orleans co., Vt., 45 m. N. by E. Montpelier, 561 W. Watered by Black river. Chartered in 1781, to Gen. Ira Allen, its principal proprietor, from whom it was named. It contains a courthouse, jail; two churches; two stores, one fulling-mill, one grist-mill, two saw-mills, one tannery; 10 schools, 371 scholars. Pop. 971.

IREDELL, county, N. C. Situated towards the W. part

of the state, and contains 800 sq. m. Bounded S.W. by Catawba river. Drained by South Yadkin river and its branches. It contained, in 1840, 15,770 neat cattle, 14,634 29,093 swine; and produced 109,444 bushels of wheat, 6419 of rye, 677,811 of Indian corn, 98,362 of oats, 22,839 of potatoes, 30,454 pounds of tobacco, 1,511,719 of cotton. It had 17 stores, 96 flouring-mills, 34 grist-mills, 19 saw-mills, three oil-mills, 14 tanneries, 150 distilleries; two academies, 130 students, one school, 25 scholars. Pop.: whites, 11,930;

Statesville.

IRELAND, a large and important island of Europe, in the N. Atlantic ocean. It is situated to the W. of Great Britain, being separated from the latter by St. George's channel on the S., the Irish sea in the middle, and the N. channel on the N.: the distance from St. David's Head, in S. Wales, across St. George's channel, to Carnsore point, in Ireland, is about 47 m.; the distance from Holyhead, in N. Wales, across the E. border of the Irish sea, to Dublin, about 55 m.; and the distance from the mull of Cantire, across the N. channel, to the opposite coast of Ireland, about 13½ m. And besides its proximity to England, Ireland has been long politically connected with that part of the empire; and since 1800, when its separate legislature was merged into the imperial parliament, it has formed a principal portion of the United Kingdom of Great Britain and Ireland.

Orbis Antiqui, i., 449.)

Ireland is situated between the parallels of 51° 25′ and N. lat., and of 6° and 11° W. long. It is of a rhomb figure; and though more compact than Great Britain, its coast is deeply indented, particularly on its S.W. and N. coast, with bays and arms of the sea. Its greatest length from Brow head, in Cork, to Fair head, in Antrim, is about 306 m.; and its greatest breadth from the W. coast of Mayo to the E. coast of Down, is about 182 m.; but in other places the breadth is much less, and there is no part of Ireland above 50 or 55 m. from the sea. Its area is estimated at 31,874 sq. m., of which 711½ sq. m. are water.

Face of the Country.—As contrasted with Scotland, or even the greater part of England, Ireland may be said to be a flat country. Still, however, the surface is in most parts much diversified; and even where it is quite flat, the prospect is generally bounded by hills or mountains in the distance. With the exception of the Devil's-bit and Sliebh-loom mountains, which run N.E. and S.W. for about 30 m., intersecting Tipperary, and dividing King's and Queen's counties, most of the other mountains in Ireland are parcelled out into groups, or form only short chains. The principal group is situated in the S.W. corner of the island, in the counties of Kerry and Cork, adjoining the celebrated lakes of Killarney. Gurrane Tual, in Macgillicuddy's Reeks, in this group, the highest mountain in Ireland, has an elevation of 3404 feet above the sea. The Wicklow mountains, in the county of Wicklow, on the E. coast of the island, cover a considerable area; Lugnaquilla, the highest, is about 3000 feet above the sea. Some of the glens in this mountain group are celebrated for their beauty. The Mourne mountains, in the S. part of the county of Down, are also of considerable extent; and some of their peaks attain to an elevation of above 2700 feet. The mountains of Donegal, and those in the N. parts of Leitrim and Sligo, and in the W. parts of Mayo and Galway, constitute a formidable barrier along the N.W. and the greater part of the W. coast, and serve at once to attract the moisture brought from the Atlantic, and to break the fury of the storms from that quarter. Some of the Irish mountains are rugged and precipitous; but the greater number are smooth and rounded, admitting of cultivation a considerable way up their sides, and sometimes to their very summits.

The central portion of Ireland consists of a vast tract of level land, broken in some places by a few undulating hill ranges; but for a great part of its extent nearly an uninterrupted flat, extending in some parts, as between Dublin and the bay of Galway, quite from sea to sea. This great level consists partly of rich cultivated land; but it also comprises a vast extent of bog, partly in Kildare, King's county, and Roscommon, and partly in Meath, Westmeath, and Queen's county. Though not continuous, these bogs differ but little in elevation; and being in many parts separated only by narrow ridges of dry land, they have received the common appellation of the bog of Allen. Several rivers have their sources in this bog, the highest part of which may be elevated about 280 feet above the level of the sea. There are several very extensive levels in other parts of the country; and some of them, particularly in Tipperary

and Limerick, are not inferior in fertility to any land in the empire.

Ireland is very well watered, having to boast of an unusual number of rivers and lakes. At the head of the former is the Shannon, which, as a channel of internal communication, is not inferior, if it be not superior, to any other river in the United Kingdom. Excepting the Shannon, and, perhaps, the Erne, there is no river of any consequence flowing westward. The Blackwater, Suir, Nore, and Barrow, all considerable streams; and the Lee and Bandon, which, though much smaller, have a good deal of commercial importance, pour their waters into the Atlantic on the S. coast; the Slaney, Liffey, Boyne, &c., discharge themselves into St. George's channel and the Irish sea; and the Bann and Foyle have their mouths on the N. coast. The Shannon, after rising at the base of the Culkeagh mountain, in Ulster, runs through the centre of the island, traversing, or, rather, expanding into the lakes Allen, Ree, Derg, &c.; and, after nearly insulating the province of Connaught and county of Clare, falls into the Atlantic, by an estuary of great length and width. This fine river is navigable for 214 miles, or throughout its entire course, except about six or seven miles above lough Allen. (See SHANNON.) The Blackwater or Broadwater is the chief river of Munster: it rises on the confines of Limerick and Kerry, and soon assumes an E. direction, which it generally preserves till about a dozen miles from its mouth, when it turns suddenly S., and falls into the ocean at Youghal harbour. Its course may be estimated at about 100 miles. The tide rises as high as Cappoquin, to which point it is navigable. Mallow, Fermoy, Lismore, and Youghal are on its banks. The Suir rises in the Sliebh loom mountains, and has generally a S. course till it approaches the Knock-me-le-down range of hills, which separates its basin from that of the Blackwater. It then turns E., and ultimately falls, together with the Barrow, into the estuary termed Waterford harbour. In a commercial point of view, this is one of the most valuable rivers of Ireland. Vessels of 500 tons come up it to Waterford; besides which city, Carrick, Clonmel, Cahir, &c. are situated on it. The Barrow is decidedly the most important of the Irish rivers, after the Shannon: it has been elsewhere described (Vol. I. p. 290.) The Nore, its chief tributary, holds its course between the Barrow and the Suir: it has a general direction S.S.E., running past Kilkenny, Thomastown, and Innistioge. It is navigable for considerable vessels to the latter place, and for barges to Thomastown. The Slaney, like the two foregoing rivers, has in general a S.E. course; it rises at the foot of mount Lugnaquilla, county of Wicklow, and falls into the arm of the sea termed Wexford haven. Wexford, Enniscorthy, Newtown Barry, Tullogh, &c. are built on it: it is navigable for barges as far as Enniscorthy. The Lee and the Bandon have both an E. course; the former on which Cork is situated, is navigable to that city for vessels of from 150 to 200 tons; the Bandon has its mouth in Kinsale harbour. The Boyne, celebrated in Irish history, has been elsewhere noticed. (Vol. I. p. 495.) The Liffey is remarkable only as the river on which the metropolis is situated. The Upper Bann rises near the Mourne mountains, and runs in lough Neagh, which receives several other large streams. The outlet of this lake is the lower Bann, which has a N. course to its mouth, five miles below Coleraine, to which point only it is navigable for boats, and that with difficulty, from the rapidity of its current. Its salmon and eel fisheries are highly important and valuable. The Foyle, formed by the confluence of several streams near Strabane, runs generally N.N.E., and discharges itself into lough Foyle. Strabane, Lifford, St. Johnstone, and Londonderry are on the Foyle, which is navigable to the latter city for the largest class of merchant-men, and to St. Johnstone for barges. The Erne, Arrow, Moy, Kenmare, &c., require no particular notice.

Ireland is more remarkable for the number and extent of her lakes, or, as they are there called, loughs, than either Scotland or England, though they must, perhaps, in general, yield to those of the sister island in point of picturesque beauty. Lough Neagh, in Ulster (which see), ranks pretty high among the secondary European lakes, inasmuch as it extends over 100,000 acres. Lough Erne, county of Fermanagh, consists of two considerable lakes, connected by a winding strait, on an island in which the town of Enniskillen is built. Both these lakes are full of islands, some large and thickly inhabited, many well wooded, and the whole so disposed, and accompanied by such a diversity of coast, as to form a vast number of rich and interesting prospects. Loughs Corrib, Mask, and the lakes of Killarney, so celebrated for their surrounding scenery, are the other principal lakes. (See KILLARNEY.) The total extent of the Irish lakes has been estimated, as already seen, at 455,399 acres; of which, 32,474 acres are included in Leinster, 44,656 in Munster, 183,796 in Ulster, and 194,477 in Connaught.

The term lough is also often applied in Ireland to arms of the sea nearly enclosed on all sides by the land, and frequently forming commodious harbours. Of these, the most celebrated are loughs Foyle and Swilly on the N., and Belfast and Strangford on the E. coast.

The Irish coast, particularly on the W. and S.W., is deeply indented with numerous bays, gulfs, and arms of the ocean, forming some noble havens. Ireland has been vaguely said to possess in all 14 harbours for the largest ships, 17 for frigates, and from 30 to 40 coasting vessels, independent of at least 94 good summer roadsteads. The principal inlets of the sea on the W. coast are Donegal, Sligo, Killala, Clew, Galway, Tralee, Brandon, Dingle, Bantry (a matchless bay) and Dunmanus bays, and the estuaries of the Shannon and Kenmare; and on the S. the harbours of Cork (one of the finest in Europe), Waterford, Dungarvan, Youghal, and the bays of Courtmacksherry, Clonakilty, &c. The E. coast has no good harbour; the principal inlets on that side being, exclusive of loughs Strangford and Belfast, the bays of Dublin, Dundalk, and Dundrum, and Wexford haven. The chief Irish headlands are Dunmore Head (which, exclusive of a few insignificant islands, is the most W. point of Europe) and Achil head, on the W. coast; cape Clear on the S., Carnsore point, on the S.E., and Fair and Malin heads, on the N. A great number of small islands and islets belong to Ireland, which lie chiefly along its W. coast. They are of little importance: the largest are Achil, Clare, N. and S. Arran, Valentia, and Rachlin (the *Ricina* of Ptolemy), on the N.E. coast.

The climate is more temperate and equable than that of other parts of Europe in the same latitude. The heat of summer is less oppressive, and the cold of winter less severe; and, when anything like immoderately hot or cold weather takes place, it lasts for a much shorter time. The great defect of the climate of Ireland is excess of humidity: not only is rain more frequent than in England, but the atmosphere, when there is no rain, is largely impregnated with moisture. This circumstance, the result of the insular position of Ireland, and of the prevalence of W. winds for three fourths of the year, accounts for the greater verdure of the country, and for the trees continuing in leaf much longer than in England. In the driest seasons, Ireland rarely suffers from drought, but the crops are often injured by too much wet. "It is a common saying in Ireland, that the very driest summers never hurt the land; for, although the corn and grass upon the high and dry grounds may get harm, nevertheless, the country in general gets more good than hurt by it: and when any dearths fall out to be in Ireland, they are not caused through immoderate heat and drought, as in most other countries, but through too much wet, and excessive rains." (*Boate's Nat. Hist. of Ireland,* 147, ed. 1652.) Hence, Ireland is naturally much better adapted for a grazing than for an agricultural country; a peculiarity noticed by Giraldus Cambrensis, in his *Topog. Hibernia,* who says that it is more fruitful of pasture than of fruit, and of straw than of grain. *Pascuis tamen quam frugibus, gramine quam grano fæcundior est insula.* It may also be worth mentioning, that the superiority of Ireland as a pastoral country was well known to the ancients. *Cati,* says Pomponius Mela, *ad maturanda semini inigri ; verbis adeo luxuriosa herbis, non lætis modo, sed etiam dulcibus, ut os exiguâ parte diei pecora impleant.* (*De Situ Orbis,* lib. iii., § 6.) It is alleged that the atmosphere is less humid now than formerly; and this is only what we should be inclined, *à priori,* to infer from the cutting down of the woods, and the great extension of cultivation. Were drainage as extensively practised in Ireland as in England, there can be little doubt that the climate would be still further improved; though, from the position of Ireland in respect of the Atlantic, it must necessarily be always distinguished for humidity. The average quantity of rain in a series of years was found to be 35 inches annually in Cork, and 34 in Derry. The changes of the seasons, and of the weather generally, are a good deal more uncertain even than in England; and the business of agriculture is proportionally hazardous. Thunder storms are less frequent and destructive in Ireland than in Britain. The mean temperature of the N. of Ireland is about 48°, of the middle 50°, and of the S. 52° Fahr. Its range at Dublin has been found to be between 14° 50' and 81° 50', the mean being about 49°. Peaches, grapes, and most other southern fruits do not ripen without much care and attention; but the broad-leaved myrtle grows luxuriantly in the S. counties, and the arbutus is not native to any other country so remote from the equator.

The *geology* of Ireland differs greatly from that of England, and in a general point of view rather resembles that of France; Ireland being, like the latter, a basin surrounded by mountains of a primary or transition character. The Mourne mountains, and others in the N.E., are composed chiefly of granite, mica-slate, gnwacké, porphyry, &c.,

similar to the mountain ranges on the opposite Scottish coast. Granite prevails in the Wicklow mountains, and is found, together with gneiss, mica-slate, horneblende, quartz, old red sandstone, &c., in Mayo, and other parts of the W. Clay-slate, felspar, primitive greenstone, and limestone, are the other chief primary and transition rocks. Limestone is a very prevalent formation, it being found over the whole country, except in a few of the N. and W. counties; in many places sandstone protrudes through it in the form of knolls. In the N., the trap-field of Antrim, the largest basaltic formation in Europe, extends over an area of 800 sq. m., and presents, in the Giant's Causeway, &c., the finest specimens of columnar basalt. No tertiary beds, containing shells, &c., like those of the London and Paris basins, have been discovered; but the limestone in most parts abounds with fossil remains. Coal, that most valuable of fossils, is found in the S. and E. The principal coalfield is that of Kilkenny, which rests, like the great coal formations of England, upon mountain limestone ; the other coal-fields are those of the counties Tipperary, Cork, Kerry, Limerick, lough Allen in Leitrim, Monaghan, and another in Ulster, N. of a line drawn between Dublin and Galway. Little coal is, however, raised; the produce of the Kilkenny coal-field, according to the return in the Railway Report, not exceeding 35,000 tons coal and 53,350 tons culm. The coal that is raised is also very inferior. In fact, Dublin, Belfast, Cork, and all the principal Irish towns, are supplied with coal from Great Britain. Iron is found in many parts of the country; and the great increase of iron-works in the earlier part of the 17th century is said by Boate to have been a principal cause of the destruction of forests in Ireland (p. 139). But these having been exhausted, and coal not having been found of such quality and in such quantity as to supply the deficiency, the Irish iron-works have been almost wholly abandoned. In Donegal and Galway, statuary marble, nearly equal to that of Italy, is found; and the black and gray marbles of Kilkenny are much prized, and exported to a considerable extent. There are copper and lead mines in Cork, Kerry, Wicklow, and other places. The copper mine of Allihies, in Cork, is said to employ from 1200 to 1500 men. Small quantities of gold and silver have been found in Wicklow. Indeed, some stream-works were wrought in the latter county, on account of government, previously to the rebellion in 1798; and it is said that as much gold was obtained as paid the expense. But some mining operations in Wicklow, commenced by government early in the present century, having failed, all attempts to obtain the precious metals have been since entirely abandoned. Copper is the only metal which at present appears to repay the labour and expense of raising it: the ore is mostly sent to Wales to be smelted. Antimony, manganese, serpentine of excellent quality, fullers' earth, gypsum, limestone, slate, with beryls, garnets, &c., are the other chief mineral products. (*Railway Report,* App. B., p. 66, &c.)

The deficiency of good coal in Ireland is less felt as regards domestic than manufacturing purposes. About 2,800,000 acres, or nearly 1-7th part of the entire surface, consists of bogs, which are capable of furnishing an almost inexhaustible supply of peat at very little more expense than that of the labour required in digging it. About 1,576,000 acres of this peat soil are estimated to consist of flat red bog; the remaining 1,255,000, called mountain bogs, lie on the surface of the uplands. The red peat bogs, which form a remarkable feature of this country, are chiefly comprised in the great central plain of Ireland; and the space bounded N. by a line drawn from Howth head to Sligo, and S., by another from Wicklow head to Galway, would include the greater portion of the Irish bogs. Unlike the English mosses, they are rarely level, but undulating; and in Donegal there is a bog completely diversified with hill and dale. These bogs consist of moist vegetable matter, containing a great deal of stagnant water; and, after heavy rains, fogs, &c., sometimes burst, and inundate or overwhelm the surrounding country. But they vary infinitely in wetness, as also in depth, composition, &c. The extensive bogs in the central part of the island, though separated from each other, have, as already stated, received the common name of the bog of Allen. The bogs in general rest upon a stratum of blue clay, based on limestone, and are invariably above the level of the sea ; their greatest elevation, however, not exceeding 486 ft. Many conflicting opinions have been entertained with respect to the origin of these bogs. It has been contended by some that they are of no great antiquity, and originated in the cutting down of the forests, after the invasion of Ireland by Henry II., or at a somewhat earlier period. It is alleged that the recumbent trees having intercepted and dammed up streams of water into the bogs, the rubbish carried along with them, the whole became gradually covered with a vegetation of moss, sedgy grass, rushes, and various aquatic plants. We believe, however, that there is but little foundation for this theory, and that the bogs owe their origin to natural causes, and not to a superstitious cutting down of the forests. The English did not, till long after the reign of Henry II., spread themselves over any considerable portion of the country, and could not, therefore, be the agents in any very remote and extensive destruction of its woods, which, in fact, were both numerous and extensive long after the bogs had attained to about their present extent. (See Beete's *Nat. Hist. of Ireland*, p. 118-122, ed. 1652.) The drainage and cultivation of these extensive portions of the surface of Ireland have long been regarded as objects of great national importance, and frequent attempts have been made to show that they might be effected at no very great expense. But there are but few examples in any part of the island, and those under very peculiar circumstances, of successful bog cultivation. The attempts to drain the bogs hitherto made in Ireland have not been very successful; and, even had they succeeded, it is doubtful whether the bogs would have produced any considerable return. It is, indeed, by no means clear, supposing them to be quite dried, that they would not, in most instances, be rendered still more worthless than at present. (*Wakefield*, L., 105.)

The bogs are not, however, without their value: they supply the inhabitants with fuel. In those parts, indeed, where bogs are scarce, they are the most valuable properties in the country. In not a few localities, they have been wholly cut out; and where this is the case, and other bogs are not easily accessible, the inhabitants have sustained great privations from the want of fuel.

The diversity of soils is not nearly so great in Ireland as in England. It has no stiff clay soils, such as those of Essex, Hants, Oxfordshire, &c., nor any chalk soils, as those of Hertford, Wilts, Sussex, &c. Sandy soils are also rare. Loam, resting on a substratum of limestone, predominates in Ireland; and, though often shallow, it is almost everywhere very fertile. A large part of Limerick, Tipperary, Roscommon, Meath, Longford, &c., consists of deep fine friable loam, and is, perhaps, not surpassed by any land in Europe; it is not permanently injured by the bad system of culture to which it is subjected, and, if kept clean, will yield an almost interminable series of corn crops; and how bad soever the order in which it is laid down to grass, it is in no long time covered with the finest pasture. The deep rich grazing lands on the banks of the Shannon and Fergus are not surpassed by the best in Lincolnshire. A good judge of such matters, Arthur Young, contends that, acre for acre, the soil of Ireland is superior to that of England, though, as the proportion of waste land in the former is much greater than in the latter country, we incline to think that this is an exaggerated statement; but, had Mr. Young confined his remark to the cultivable land in both countries, it would have been quite correct. In fact, if we deduct the bogs and mountains, we believe that Ireland is about the richest country, in respect of soil, in Europe. As a grazing country, she is probably superior to any other, and certainly is surpassed by none.

The botanist Schults not long ago complained that there were two great islands in Europe, Ireland and Sardinia, of which the *florae* were unknown; and he might, perhaps, have said the same of the *faunae* of the former. The arbutus and myrtle have been already mentioned, and besides these plants, most of those common to Britain are met with. The wild animals do not materially differ from those of England. Wolves formerly infested the country, but they were extirpated under Cromwell. The Irish greyhound, which was of use in clearing the country of these animals, is about 3 ft. in height, of a light colour, and of such strength and courage that it is said to be more than a match for the mastiff or bull-dog; it is now, however, nearly extinct. The numbers of deer have greatly declined with the clearance of the forests, and the progress of cultivation. The native Irish horse is seldom more than 13 hands high, very hardy, and sure footed: it is used for all kinds of labour. A large blood-horse is reared extensively in Meath, and is to be found in most of the rich grazing counties. The native Irish cattle, a breed with short legs, large bellies, and white faces, have been, to a considerable extent, superseded by the introduction of the Holderness, Staffordshire, and Devonshire breeds, either pure or crossed. As compared with England, but few sheep are raised in Ireland. The native Irish sheep is small, and covered with nearly as much hair as wool; but it is now uncommon in a pure state, having been crossed with various English breeds. Most of the Irish sheep are at present long-woolled, and are usually of large size. A breed of fine short-woolled sheep is peculiar to the mountains of Wicklow. Goats are very generally kept, and hogs are universal. The native Irish hogs are of the worst possible breed, being tall, lengthy, and narrow in the loins; but improved breeds are now common: they are fattened principally with potatoes, and one or more is to be found in every house. Everybody is acquainted with the story of

St. Patrick having extirpated reptiles from Ireland. And there is this much of foundation for the legend—that neither of the three species of snakes found in England is met with in the country, a circumstance which is most probably to be ascribed to the great humidity of the soil and climate. The toad is, however, said to be met with at the S.W. extremity of the kingdom; and frogs, having been introduced in the last century, are general denizens of the pools. Moles are unknown; and Irish oak timber is said not to harbour spiders, &c.

There is now, whatever may have been the case formerly, a great want of wood in most parts of Ireland. And, however rich the soil, the appearance of the country is, in most parts, indicative of the poverty and depressed condition of the bulk of the population. Generally speaking, what are called farm-houses and offices in England do not exist in Ireland; and the aspect of the cottages which, in the vast majority of instances, are of the most wretched description; the smallness of the fields, which, instead of hedges and ditches, or stone fences, are usually divided by turf dykes; and the badness of the house furniture, and of the agricultural implements, all impress the traveller with the most unfavourable convictions. But, how mortifying soever the contrast between the excellence of the soil and the state of the people, it is some satisfaction to know that it is less striking now than formerly. In many districts, a considerable advance has been made towards a better order of things; and the spirit of improvement has begun to send its seeds and spread its roots to most parts of the country. For table of provinces, &c., see next page.

Population.—The first authentic account of the population of Ireland is given by Sir William Petty, in his tract entitled the *Political Anatomy of Ireland*. Sir William was employed by government to superintend the survey and valuation of the forfeited estates, instituted during the protectorate; and so well did he execute his task that his survey continued, after the lapse of near two centuries, to be the standard of reference in the courts of law as to all points of property. He had, therefore, the best means of obtaining accurate information with respect to the numbers and condition of the people; and, as the result of his researches on these points are exceedingly curious, it is best to give them in his own words.

"The number of people in Ireland (1762) is about 1,100,000; viz., 200,000 English, Scotch, and Welsh Protestants and 800,000 Papists; whereof 1-4th are children unfit for labour, and 75,000 of the remainder are, by reason of their quality and estates, above the necessity of corporeal labour; so as there remains 750,000 labouring men and women, 500,000 whereof do perform the present work of the nation.

"The said 1,100,000 people do live in above 200,000 families or houses, whereof there are about 16,000 which have more than one chimney in each, and about 24,000 which have but one; all the other houses, being 160,000, are wretched, nasty cabins, without chimney, window, or door-shut, even worse than those of the savage Americans, and wholly unfit for the making merchantable butter, cheese, or the manufactures of woollen, linen, or leather.

"By comparing the extent of the territory with the number of people, it appears that Ireland is much under-peopled; forasmuch as there are above 10 acres of good land to every head in Ireland, whereas in England and France there are but four, and in Holland scarce one!" (*Polit. Anat. of Ireland*, p. 114, 118, ed. 1719.)

In 1721 an inquiry was instituted, by order of the House of Lords of Ireland, for ascertaining the pop., through the medium of the magistrates and established clergy, the result of which gives a pop. of 2,010,221. At this period, and for long after, Ireland was essentially a grazing country. To such an extent, indeed, was the pasturage system carried, that in 1727, during the administration of primate Boulter, a law was made to compel every occupier of 100 acres of land to cultivate at least five acres, under a penalty of 40s.!

According to the returns of the hearth-money collectors, which are believed to have been pretty accurate, the number of houses in Ireland in

Years.	Houses.		Pop.
1754 was .	395,439		2,372,634
1787 —	494,646	Which, allowing six	2,544,976
1777 —	448,426	inhabitants to each	2,690,556
1785 —	474,322	house, gives a pop-	2,845,932
1788 —	650,000	ulation of .	2,900,000
1791 —	701,102		4,206,612

In 1805, Mr. Newenham estimated the pop. at 5,375,456. An incomplete census was taken in 1812, from which the pop. was computed at 5,937,856. At length a complete census was taken in 1812, when Ireland was found to contain a pop. of 6,801,827. According to the last census, taken in 1831, the pop. amounted to 7,767,401: viz., Leinster, 1,909,713; Munster, 2,227,152; Ulster, 2,094,003; Connaught, 143,344; total, 7,767,401.

A TABLE of the Provinces and Counties of Ireland; specifying the Number of Baronies and Parishes, the Extent in square Miles and imperial Acres, with the Extent of the cultivable Land and the Space occupied by unimproved Mountains and Bogs, and by Lakes in each County in Ireland; and specifying, also, the Population of each County in 1831, and their Population per square Mile exclusive of Lakes; with Totals for the Provinces and Kingdom.

Provinces and Counties.	Number of Baronies.	Number of Parishes.	Extent in Square Miles.	Extent in English Statute Acres.	Whereof Cultivated Acres.	Whereof Unimproved Mountains and Bogs.	Whereof Lakes.	Pop. in 1831.	Pop. per sq. m. ex. Lakes.
LEINSTER.									
Carlow	5	30	344	219,863	196,833	22,630	. .	81,988	228·3
Dublin	6	107	356	248,651	237,819	10,812	. .	380,167	979·3
Kildare	10	113	613	392,435	325,988	66,447	. .	108,494	176·6
Kilkenny . . .	9	127	803	513,086	417,117	94,509	. .	193,686	241·0
King's	11	52	895	326,166	204,589	122,340	948	144,225	175·0
Longford . . .	6	23	419	263,645	192,506	55,347	15,592	112,558	280·6
Louth	4	61	322	208,961	191,545	14,916	. .	124,846	387·4
Meath	18	147	906	567,197	561,597	5,600	. .	176,826	190·4
Queen's . . .	8	50	620	304,810	238,836	66,972	. .	145,851	233·6
Westmeath . .	12	62	604	366,951	313,935	55,982	16,334	136,872	226·0
Wexford . . .	8	143	882	564,479	545,979	18,500	. .	182,713	208·0
Wicklow . . .	6	58	773	494,704	400,704	94,000	. .	121,557	157·2
Totals . . .	97	989	7,478	4,782,058	4,114,160	635,494	32,474	1,909,712	257·3
MUNSTER.									
Clare	9	79	1,254	802,328	534,113	250,504	18,495	258,322	211·0
Cork	16	249	2,765	1,769,363	1,698,803	700,760	. .	810,732	293·1
Kerry	8	83	1,795	1,146,790	581,180	552,860	14,680	263,126	148·4
Limerick . . .	9	125	1,054	674,783	582,802	91,981	. .	315,355	300·2
Tipperary . . .	10	196	1,653	1,013,173	819,608	182,147	11,392	402,563	257·2
Waterford . . .	7	74	736	471,981	353,947	118,034	. .	177,054	240·5
Totals . . .	59	816	9,187	5,879,873	3,930,838	1,905,368	44,658	2,227,152	244·2
ULSTER.									
Antrim	8	77	1,186	758,986	483,108	225,970	49,760	325,615	293·8
Armagh . . .	5	20	513	328,163	287,317	42,472	18,304	220,134	454·6
Cavan	7	20	746	473,448	421,402	30,008	21,957	227,923	382·2
Donegal . . .	6	42	1,820	1,165,107	520,736	644,371	. .	289,149	159·0
Down	8	60	955	611,404	502,677	108,569	158	352,012	369·6
Fermanagh . .	8	18	736	471,348	280,599	101,962	48,797	149,763	227·9
Londonderry . .	4	31	810	518,270	372,087	136,036	9,565	222,012	279·2
Monaghan . .	5	19	511	327,049	309,988	9,235	7,844	195,536	382·9
Tyrone	4	35	1,179	754,305	555,690	171,314	97,251	304,468	268·9
Totals . . .	54	322	8,456	5,408,070	3,754,368	1,469,922	193,796	2,986,692	280·1
CONNAUGHT.									
Galway . . .	16	116	2,360	1,510,509	955,713	476,957	77,929	414,684	185·9
Leitrim . . .	5	17	657	420,375	286,540	196,107	23,568	141,524	239·6
Mayo	9	66	2,117	1,355,048	871,904	425,194	57,940	366,328	187·0
Roscommon . .	6	56	938	600,405	453,555	131,063	94,787	249,613	273·4
Sligo	6	20	679	434,188	237,917	168,711	8,280	171,765	258·0
Totals . . .	48	295	6,765	4,320,608	2,805,160	1,330,092	194,477	1,343,914	206·0
General totals	259	2,436	31,874	90,300,608	14,603,473	5,340,736	455,399	7,767,401	249·2

The extraordinary increase of pop. since 1765 has arisen from a variety of causes. The bounty acts of the 19th Geo. III. (1785), on the exportation of corn from Ireland, appear to have given the first great stimulus to corn cultivation in that country; and consequently to the breaking up of the large grazing farms: and this stimulus was continued and increased by the high prices of corn during the late war. The impetus thus given to the subdivision of farms was vastly increased by the enfranchisement of the Catholics in 1793; for, the great majority of the landlords being eager to increase their political influence, by increasing the number of voters on their estates, did not hesitate to subdivide them into small patches for that object. The influence of these accidental causes was powerfully increased by the practice that has always prevailed in Ireland of dividing property, whether freehold or leasehold, equally among the children. This pernicious practice has, in fact, done more than anything else to stimulate early marriages, and the increase of the pop.; and to bring about that minute division of the land that is now the grand source of the misery of the country, and the most formidable, by far, of all the obstacles to its improvement. Wherever, indeed, such a practice prevails, the land is necessarily, in a few generations, split into minute portions, and is occupied by a redundant and beggarly pop.

Considering the extent of bogs in Ireland, the backward state of its agriculture, the deficiency of manufactures and trade, and the fewness of the great towns, its pop. is astonishingly dense. There was, in fact, in 1831, at an average of the entire kingdom, an individual for every 2·57 acres; whereas, in England, notwithstanding the number and magnitude of her great towns, and the vast amount of her manufacturing and commercial pop., there were 3·66 acres for every individual; and in Scotland there was only one individual to every eight acres! This wonderful density of pop. in Ireland is entirely ascribable to the interminable division and subdivision of the land, and the general dependance on the potato. But, however it may have originated, there can be no question that it is the immediate cause of the abject poverty and depressed condition of the great bulk of the people. It is not too much to say that there are at present (1840) more than double the number of persons in Ireland that it is, with its existing means of production, able either fully to employ, or to maintain in a moderate state of comfort.

Rural Economy.—Ireland, generally speaking, may be said to be a country of small farms and cottier cultivation. Few of the tillage farms extend to 40 acres; the great majority being about five acres, and varying from five to 10 and 15 acres. It is obvious that farms of this size cannot, except under peculiar circumstances, be well cultivated. They are too small to admit of a proper division of labour, or rotation of crops; at the same time that, owing to the poverty of the occupiers, the stock and implements are of the most inferior kind. Drainage, though the most essential of all improvements, is all but unknown in the greater number of Irish farms; and, in the smaller class of occupancies, the potato, owing to its supplying the greatest quantity of nutriment from a given space, is always a principal crop. Owing to the difficulty, and sometimes impossibility, of the occupiers of a small patch of land finding employment in the neighbourhood when they are not fully occupied at home, they acquire lazy, careless habits. In consequence, many of the most ordinary operations of husbandry, such as the cutting down of weeds, and even the harvesting of corn and potatoes at the proper time, are neglected; and, on the better class of farms, about twice the number of labourers are said to be required for their cultivation that would be necessary in England or Scotland.

There being few, and those only inconsiderable manufactures in Ireland, not less, perhaps, than 4-5ths of the pop. directly depend for employment and subsistence on the soil. The competition for small patches of land is consequently very keen, and the rents greater than the occupiers can afford, though not greater than might be paid for them, were they consolidated into properly sized farms, and cultivated on an improved system. In Ireland, in fact, the possession of a piece of ground has long been a condition all but indispensable to existence; and we need not, therefore, wonder that the occupiers should cling with desperate tenacity to their small patches. This has led in most parts to a sort of tacit but well understood agreement among the cottiers, or small farmers, to support each other against intruders; and, in the greater part of Ireland, it is as necessary to the quiet possession of the land to secure what is called the tenant's right, or the good-will of the occupier, as it is to make a bargain with the landlord. Any tenant who should neglect this indispensable precaution would run a great risk of being disturbed in, or violently ousted from, his possession. Indeed, most of the disturbances by which Ireland has been so long agitated and disgraced have been of an agrarian character, or have been directly or indirectly connected with the occupancy of the land.

It is not necessary to enter into any lengthened disquisitions as to the various circumstances which have led to that minute parcelling of the land that is the bane and curse of Ireland. The greatest influence is no doubt, as already stated, to be ascribed to the habit of providing for the sons, and sometimes, also, the daughters of the occupiers of land, by giving them shares of their father's holdings. Had there been a poor-law in Ireland as in England, compelling the landlords to provide for all residents on their estates, in the event of their becoming unable to provide for themselves, the landlords would, it is most probable, have long since devised means for checking the progress of subdivision. But, having no motive of this sort, and believing that the rents promised them by the small occupiers would be paid without any deduction, they either sanctioned or connived at the practice, till it had well nigh parcelled out the whole country into miserable patches, and filled it with a redundant and beggarly population. Latterly, however, they have become fully aware of the pernicious consequences of the practice, and have, in many instances, exerted themselves to check it; and to consolidate the small patches into considerable farms. Their efforts to bring about these desirable results have been facilitated by the operation of the act against sub-letting, passed in 1826; and the introduction of an assessment for the support of the poor will, no doubt, make them more anxious to hinder the too minute division of their estates, and their occupancy by parties likely to become paupers.

A good deal of what is peculiar in the mode of occupying land in Ireland has grown out of the circumstances under which it was originally acquired by the ancestors of its present owners. About 9-10ths of the land that was forfeited under Cromwell and William III.; and this vast amount of property was mostly either gratuitously bestowed upon, or was acquired at a very small sacrifice by, noblemen and gentlemen of fortune and influence in England. Such persons could not be expected to leave England to reside in Ireland; and, in point of fact, they very rarely visited their estates in the latter, but satisfied themselves with taking what rents they could get for them. There was no sympathy between them and their tenants: the religious and political principles of one party were opposed to those of the other. The landlords looked upon their tenants as a sort of unwilling bondsmen, who, if any favourable opportunity should present itself, would immediately shake off their dependance on them; and the tenants regarded the landlords as usurpers unjustly intruded on the estates of others, and as enemies to the religion and rights of the Irish people. Very few had, or could be expected to have, any confidence in the stability of such a state of things; and it could not be expected that landlords should care much about the permanent interests of such estates, or that they should lay out any considerable sum on their improvement. To build a farmhouse or offices was an outlay which, for a lengthened period, no Irish landlord ever incurred; and even to this day the old habit maintains an ascendancy, and the great majority of landlords lay out little or nothing on buildings. In consequence of this practice, and of the general smallness of the holdings, and the poverty of the occupiers, the farm-buildings, if we may so call them, of Ireland, are, as already stated, quite unworthy of the name; and, in most instances, are wretched in the extreme. Such a thing as a barn is hardly known among the smaller occupiers: and the corn is not unfrequently thrashed on the public roads, which serve as barn-doors.

The circumstances thus shortly stated, as to the acquisition of landed property in Ireland, account for the introduction of middlemen, and for the prevalence of absenteeism, and of partnership tenures. Nothing could be more natural than that an English gentleman, possessed of an Irish estate, should prefer letting it for a round sum to a middleman, allowing the latter to transact with the occupiers. This relieved him from an unpleasant duty, for which, most probably, he was wholly unfit, and gave him a certain income with little comparative trouble. It is true, however, that a middleman, having no permanent interest in the soil, or in the welfare of the estate or its occupiers, will look only to

temporary advantages, and will be more likely than a landlord to harass the cultivators, and to squeeze out of them all that they can possibly afford. Hence it is that, speaking generally, it is bad policy for proprietors to resort to the agency of middlemen; and they are rarely employed, except, as in the case of Ireland, property be supposed to be insecure, or it be very difficult or inconvenient for the landlord to deal directly with the occupiers. In Ireland, too, the injurious consequences naturally resulting from the employment of middlemen were materially aggravated by the state of the law; which authorized the landlord, in the event of the bankruptcy of a middleman to whom the occupiers had paid their rents, to come upon the latter, and to force them to pay their rents over again to him! The sub-letting act obviated this injustice; and no landlord, who has let an estate to a middleman, can now come upon a sub-tenant with a demand for rent bona fide paid to his immediate superior.

It should, however, be observed, that these remarks apply only to the employment of middlemen who have leases of a reasonable length. But many, perhaps we should say the majority, of Irish middlemen, hold under very long leases; and when such is the case, they are to be regarded as the real landlords, and have all their interests and feelings.

Much has been said about the injustice done to Ireland by the absence of many of her great proprietors. But, in point of fact, several of the absentee estates, as those of the Marquis of Hertford, the Duke of Devonshire, Earl Fitzwilliam, &c., are the best managed, the least subdivided, and most prosperous of any in the country. Besides, whatever might have been the case formerly, absenteeism is not at present more prevalent in Ireland than in Scotland or the N. of England, a pretty conclusive proof of the hollowness of the complaints with respect to it. The truth is, that it is not that many of the Irish landlords have been absentees, but that, whether absent or resident, they have too generally acted as if the improvement of their estates, and the prosperous condition of the occupiers, were matters with which they had little or no concern, that the depressed and backward condition of the country is to be ascribed.

The same circumstances that occasioned the employment of middlemen seem also to have led to the still more baneful practice of letting land in partnership to a number of tenants, sometimes, indeed, to a whole village, made conjointly and severally liable for the rent. It is needless to repeat what we have elsewhere stated (vol. i., 159), as to the pernicious influence of this mode of occupancy. Nothing can be imagined more destructive. Wherever it prevails it forms an insurmountable obstacle to all improvement. But though it have prevailed, and still prevails, in many very extensive districts, especially in the wilder and more mountainous parts of the country, we are glad to have to state that it is generally on the decline.

A system which has received the name of con-acre is very prevalent in most parts of Ireland, but especially in Connaught. By con-acre is meant a pernicious custom, prevalent among the landlords and occupiers of the larger class of farms, of letting to the peasantry, or cottiers, small slips of land, varying from a perch to half an acre, for a single season, to be planted with potatoes, or cropped. Old grass land is frequently let out on this system; and then it is usual to allow the surface to be pared and burned. An intelligent witness examined by the agricultural committee of 1833 (Thomas S. Linsay, Esq.) stated, that when he left Mayo, "the country appeared as if it were all on fire; I should say that 1-5th part of the surface of the co. is either burning, or is now covered with ashes." The rent got for land subjected to this abusive treatment is enormous, running from £7 to £12 or £13 an acre! Potatoes are invariably planted on con-acre land, when it is broken up from grass; and afterward it is usual to take from it successive crops of corn, till it be reduced to a caput mortuum, when it is left to be recovered by the vis medicatrix natura! Wherever this practice exists, there cannot, of course, be the least improvement; and nothing but the extraordinary fertility of the soil could enable it to produce anything under so destructive a system. (Agricul. Committee of 1833, Min. Evid., p. 333, &c.) In many parts the entire dependence of the peasantry being on these con-acre lands, when the crop fails they are reduced to the extreme of distress, and have rarely any choice between starvation and begging, unless it be to enlist under the banners of Captain Rock. Con-acre tenants dare not remove the crop from the ground without permission, which is seldom granted till the rent be paid. In most cases they are allowed to abandon the crop for the rent; but this is an alternative they make every sacrifice to avoid, as it involves the loss, not merely of their labour, but of their only means of supporting themselves during the ensuing season.

It has been said that the abject poverty of the Irish people is the principal cause of the depressed state of agriculture. But this very poverty has been principally occasioned by the circumstances now glanced at, and especially by the inveterate and most pernicious habit of splitting land. This has been at once the principal cause of the excessive increase and poverty of the pop., and of the wretched condition of agriculture; and no scheme for the improvement of Ireland that does not tend, directly or indirectly, to check or subvert this practice, is entitled to much consideration.

It must, however, be borne in mind that, though the previous remarks apply to the agriculture of Ireland taken as a whole, they do not apply to every estate, or even to every considerable district. Generally, it is most advanced in the eastern and northern counties, and is most backward in the S. and W., especially the latter. But in all the provinces some parts are much better cultivated than others; a few large estates still consist of pretty considerable farms; and the buildings upon and cultivation of some farms, occupied by landlords and principal tenants, would do no discredit to any part of the empire. These, however, are but exceptions, which it is to be hoped will every day become more numerous, to the ordinary state of things. The leading features of Irish agriculture are such as have been already described. No doubt, however, a spirit of improvement has insinuated itself into most quarters, which, notwithstanding the formidable difficulties in the way, can hardly fail to gather strength. The landlords, on whose conduct so much depends, are, as already stated, becoming more alive to their real interests. Improved implements of husbandry have been introduced into most parts of the country; while the ready communication with England by means of steam, and the boundless markets that have thence been opened for most articles of Irish produce, afford them ample means of improving their stock and culture. The course of tillage is still, however, the same in all its essential features; and hitherto the demand from England has led infinitely more to the extension of husbandry than to its amelioration. But it is difficult to suppose that it should not, also, have a ma-

torial influence on the latter; and, so soon as an effectual check is given to the practice, this, no doubt, will follow. It must, however, be kept in mind, that the introduction of a better system of agriculture is, by depriving useless hands, for a while injurious, rather than otherwise to the bulk of the labouring pop. "From north to south," say the midway commissioners, "indications of progressive improvement are everywhere visible; and most obviously accessible in the immediate influence of steam navigation. But all these signs of growing prosperity are, unhappily, so discernible in the condition of the labouring people, or the amount of the produce of their labour. The proportion of the latter reserved for their use is too small to be consistent with a healthy state of society. The pressure of a superabundant and excessive pop. is perpetually and fully acting to depress them." (Report 9.)

The great dependence of the Irish people is placed, everyone knows, on the potato; and so much so is this the case, that a large portion of the pop. but rarely tastes bread. Next to potatoes, oats, barley, and wheat, but especially oats, are the crops most commonly cultivated. Owing, however, to the humidity of the Irish climate, the climate is not well fitted for wheat and barley, which are, of course precarious and not of so good quality as in England; but it is admirably suited for the growth of oats, the crop of which has rapidly increased. Turnips are cultivated only in some of the best farmed districts, and are not looked upon in the light of a general crop.

Ireland, however, is much better adapted for a grazing than for an agricultural country; and, indeed, in this respect is the excellence of the soil, that in most parts it will yield, however fed and exhausted when laid down to grass, speedily to clothe itself with a rich and luxuriant crop of herbage. We have noticed the improvements made in stock in most parts of Ireland, by the introduction of improved breeds; and both the dairy and grazing system of husbandry have been materially extended.

ACCOUNT of the Quantities of Grain, including Flour and Meal, of the Growth of Ireland, imported into Great Britain from Ireland, in each Year, from 1800 to 1838 inclusive.

Years.	Wheat and Wheat Flour.	Barley, including Bere or Big.	Oats and Oatmeal.	Rye.	Pease.	Beans.	Malt.	Total.
	Quarters.	Quarters.	Quarters.	Quarters.	Quarters.	Quarters.	Quarters.	Quarters.
1800	749	75	2,411					
1801	160		675					
1802	166,751	7,126	341,151	306	113	1,665	3,305	461,...
1803	61,397	22,679	328,329	733	611	1,653	25	343,...
1804	70,071	2,561	340,682	386	1,670	2,560		316,...
1805	84,087	15,666	203,268	386	1,634	2,019		336,...
1806	102,376	3,857	357,077	431	1,320	2,361		468,...
1807	44,980	32,049	289,640	431	1,300	2,777		400,...
1808	42,677	30,355	370,374	572	75	3,605		508,...
1809	56,944	14,619	845,763	439	20	3,841		621,...
1810	758,383	3,231	575,737	21	235	4,667		897,...
1811	147,965	2,712	359,629	178	63	4,608		897,...
1812	138,283	42,123	691,400	446	77	4,636		977,...
1813	217,154	61,560	384,610	4	460	6,731		618,...
1814	185,478	18,770	397,537	307	438	6,371		661,...
1815	180,344	27,160	669,714	43	200	5,694		973,...
1816	191,631	68,654	411,117		15	2,273		680,...
1817	53,451	32,766	1,003,306	4	10	4,763		1,224,...
1818	145,179	36,257	739,613	2		2,904		977,...
1819	133,290	64,311	914,951	134		6,306		1,418,...
1820	423,407	37,695	1,162,940	180	2,474	4,680		1,682,...
1821	391,708	59,694	568,637	383	723	7,235		1,065,...
1822	453,064	32,522	1,108,487	198	325	6,540		1,880,...
1823	686,068	19,274	1,285,065	122	700	6,791	1,172	1,634,...
1824	338,384	44,600	1,490,864	290	1,421	11,355	10,269	2,902,...
1825	396,018	64,966	1,345,701	77	1,460	7,390	7,398	1,633,...
1826	314,851	97,791	1,343,597	566	1,326	15,037	579	1,656,...
1827	665,555	84,904	2,073,631	1,094	4,036	7,608	883	2,307,...
1828	628,184	97,140	1,673,688	454	4,035	34,445	3,011	2,307,...
1829	519,617	189,745	1,471,308	454	2,883	19,685	2,889	2,315,...
1830	580,777	185,400	1,855,701	235	4,146	15,008	10,869	2,690,...
1831	557,496	123,650	3,651,897	364	1,815	14,839	6,539	2,860,...
1832	799,202	101,767	1,782,660	165	2,646	16,114	7,917	2,737,...
1833	844,211	317,886	1,769,503	603	2,176	38,771	3,565	2,782,...
1834	779,505	134,468	1,288,767	614	3,447	94,556	10,357	2,670,...
1835	651,776	184,148	3,126,136	653	2,980	77,604	98,314	2,956,...
1836	306,757	187,473	2,874,875	1,046	60	95,629	4,174	3,030,...
1837	534,605	154,487	2,748,607	666	4,932	91,584	6,801	2,474,...
1838	542,563							

The rapid increase of the exports of corn and other raw produce from Ireland is very generally referred to as demonstrating the great improvement of agriculture; and, in so far as respects the increased exports of cattle, beef, and butter, the inference seems to be well founded. The breed of live stock has, as already stated, been very greatly improved; the system of stall-feeding has also been intro-

duced; and the increased exports of animal produce have been obtained, not only without any increase, but positive diminution, of the land in pasture. But it is otherwise with the extraordinary increase of the exports of corn and meal. The preceding table shows that the increased from less than a million of quarters previous to 1817 to nearly three and a half millions in 1836!

IRELAND.

one will venture to affirm that agriculture has improved in anything like a corresponding proportion; and as the condition of the bulk of the people has probably not varied very materially during this interval, there cannot be a doubt that the increased exports of corn are principally to be ascribed to the extension of tillage. The late Lord Clements says, in his tract on the Poverty of Ireland, that " the export of grain has increased most rapidly from those parts where no agricultural amendment whatever is visible." (P. 27.) It has there been occasioned partly and principally by the breaking up of grass land, and partly by the occupiers being tempted, by the facility and certainty of the market, to sell every bushel they can spare, subsisting themselves principally on potatoes, and retaining the worst corn for seed and their own use. We are afraid, too, that the increased exportation of pigs may, to a considerable extent, be accounted for in the same way; by the ready market afforded by the steamers, and the anxiety of the peasantry to procure the means of paying their rent, though at the expense of their comforts.

Rent of Land.—There are no means of forming any accurate estimates as to the amount of rent in Ireland. The property tax did not extend to it; and all that we have to trust to in determining its rental are estimates deduced from the rentals of particular estates, or from valuations made for the assessment of the local burdens and for the adjustment of the composition on account of tithe. The first, unless made with unusual care, is rarely much to be depended on, and leads almost always to exaggerated conclusions. In 1727, Mr. Brown computed the gross rental of Ireland, inclusive of quit-rents, tithes, &c., at £2,294,000; and, in 1778, Mr. Young estimated it at £6,000,000. (*Arthur Young's View of Ireland*, p. 232.) Mr. Wakefield, from minutes collected in his tour, estimated the average rental of Ireland at 27s. the Irish acre, or at 16s. 6½d. the imperial acre (vol. I., p. 305); and, notwithstanding the imperfect data on which it was founded, this estimate, though in excess, came pretty near the mark. No doubt, however, the elaborate estimate framed by Mr. Griffith, and contained in his evidence given in the Second Report of the Lords' Committee on Tithe (1832), is much more deserving of attention: it is principally bottomed on official valuations, and is probably, therefore, a little under the mark. According to Mr. Griffith, the average rent of Ireland, exclusive of the value of houses in Dublin, Cork, Belfast, Limerick, and other large towns, amounted to 12s. 5½d. per imp. acre. But the area assigned to the different counties and

to the kingdom in general, by Mr. Griffith, on which he estimates the rent, includes the fresh water lakes or loughs, comprising an area of 455,399 acres: these, however, should certainly be excluded. Supposing this to be done, the average rent per statute acre of the land of Ireland will, according to the data published by Mr. Griffith, be 12s. 9d. an acre; and allowing for the probable inferiority of the valuation for the local assessments to the actual rent, the latter may, perhaps, be taken at about 13s. 6d. an acre, at an average of the island, including mountains and bogs, but excluding lakes.

The annual value of the unimproved mountains and bogs has been variously estimated at from £500,000 to £700,000. If we suppose the latter to be the more correct sum, it will leave, on Mr. Griffith's hypothesis, a sum of about £13,000,000 for the *gross* rental of the cultivated land of Ireland, amounting to about 14,603,000 acres, equivalent, at an average, to about 16s. 5d. an acre; which, allowing for deficiencies in the valuation, may, perhaps, be increased to 17s. or 17s. 6d. an acre—a very high rent for a country occupied and farmed in the manner of Ireland. And to this, as already stated, has to be added the sum, frequently a large one, paid in most parts by new tenants to the previous tenants or their heirs, on account of *tenants' right.*[*]

Fisheries.—The seas round Ireland swarm with fish. Cod, ling, and hake are found in great abundance on the Nymph Bank to the S. of Waterford. Flat fish also abound in many parts. Large shoals of herrings visit the coast annually; and the bays and creeks furnish great quantities of the smaller and more delicate species, as pilchards, sprats, smelts, and sand-eels. The basking whale and sunfish are often seen off the W. coast. But the fishery has never been either largely or successfully carried on by the Irish; and, at this moment, the principal supply of salt fish is derived from Scotland. In 1764 a system of bounties was established to encourage the trade, but without any material success. It was revised in 1819 by a commission, which also gave loans for the purchase of boats and tackle. With such encouragement the number of fishermen and boats increased considerably during the ten years the system was in operation; but they declined proportionally on the withdrawal of the bounty.

The number, tonnage, and description of fishing vessels, and the number of men in 1829, when the system of bounties and loans commenced; in 1810, when it terminated; and in 1836, the latest year of which an official return has been made, were as follow:—

Years.	Decked.		Half-decked.		Open-sail.		Row-boats.		Total Men.
	No.	Men.	No.	Men.	No.	Men.	No.	Men.	
1829	394	1,908	421	2,948	2,051	10,581	4,899	21,422	36,159
1830	345	2,147	789	3,893	2,483	13,360	9,523	46,212	64,771
1836	215		870		1,812		7,864		54,119

There are salmon and eel fisheries in most of the great rivers. The salmon fisheries in the Bann, near Coleraine; the Foyle; the Billick, near Ballyshannon; the Boyne, above Drogheda; and in various other parts, are very productive. At an average of the nine years ending with 1836, the produce of the Foyle salmon fishery amounted to 2814 cwt. a year. Irish salmon, packed in ice, is principally exported to Liverpool, Bristol, and London.

Manufactures.—Ireland is not, and never has been, a manufacturing country. Its unsettled turbulent state, and the general dependence of the population on land, have hitherto formed insuperable obstacles to the formation of great manufacturing establishments in most parts of the country; while the want of coal, capital, and skilled workmen, and the great ascendancy of England and Scotland in all departments of manufacture, will, there is reason to think, hinder Ireland from ever attaining to eminence in this department. And it is needless to add, that while manufactured goods can be produced cheaper in Britain than in Ireland, so long will the interests of the latter be best promoted by their importation. It is, whatever the Irish demagogues may say to the contrary, a contradiction and an absurdity to suppose that either individuals or states can be enriched by producing at home what it would cost less to bring from abroad.

The woollen manufacture was carried on to some extent in Ireland previously to the revolution of 1688, soon after which, in compliance with the interested solicitations of the English manufacturers, the export of Irish woollens to foreign parts was prohibited, and oppressive duties laid on their importation into England. But, though it be impossible too severely to condemn this selfish insulting policy, there is no solid ground for supposing that it was productive of any real injury to Ireland. Though the acts complained of had never existed, the result would have been the same. It is quite nugatory to suppose that, under any circumstances, the woollens of Ireland should ever have been able to come into competition with those of England, either in the home market, or anywhere else.

The existing woollen manufacture of Ireland is carried on upon a small scale. At Dublin, and other parts in the vicinity, some cloth of a better description is made; and other branches are carried on to some extent in Kilkenny and other places, more especially at Mountmellick and Abbeyleix, in Queen's Co., and a few other places. It appears, from the returns of Mr. Stuart, inspector of factories, that there were in Ireland, in 1836, 31 woollen mills; but they were upon so small a scale as to employ, in all only 1231 persons. The railway commissioners estimate the value of the wool produced in Ireland at about £300,000 a year. It is not suitable for the manufacture of any cloths but those of a very low price; but it is well adapted for the manufacture of woollen stuffs, and hence the recent extension of that department at Mountmellick and Abbeyleix.

To compensate for the treatment of the woollen, the linen manufacture of Ireland was long the object of especial patronage. It was fostered and promoted by a number of statutes, and placed under the superintendence of a board, with an annual grant of public money for distribution in premiums and bounties. The board, however, has been discontinued for some years, and the grants withdrawn. The manufacture is chiefly confined to Ulster. In 1834, the last year for which there is an official return, the total value of unbleached linen sold in Ireland amounted to £2,580,697, of which that sold in Ulster produced, £2,109,305. The exports in 1835 amounted to 70,209,572 yards, of the estimated value of £3,725,654; and their value at present (1840) can hardly fall to exceed £4,000,000. The manufacture was at one time, and still is, very generally diffused over the country: the yarn being spun by the cottier's family, and woven by the cottier himself. But since the introduction of machinery for the spinning of

[*] The measurements of the different Irish counties given in this article are those of Mr. Griffith. They do not quite agree with those deduced from the Ordnance Survey, but the discrepancies are not material; and as the Survey will not be finished for some years, we thought it better to adopt Mr. Griffith's measurements throughout.

44

yarn, and of power-looms, the old system has been to a considerable extent abandoned, and the yarn is now principally spun by machinery. A good deal of cloth is also made by power-looms; but the greater part continues to be woven in the houses of the cottiers, who are supplied with yarn by the agents of the manufacturers. In fact, but for this change of system, the manufacture would have been wholly annihilated; as the manufacturers under the old domestic system could not have withstood the competition of Dundee, Leeds, &c. It is also of importance to observe that under the new plan the weavers, being regularly supplied with yarn, are kept constantly at work, and do not combine, at least to nearly the same extent as formerly, the incongruous occupations of weavers and farmers. This, as was to be expected, has tended to improve both businesses. In 1835 no fewer than 2,652,000 lbs. linen yarn were imported from Great Britain into Ireland; but we understand that, in consequence of the increased quantities of yarn produced by the mills in Belfast and the vicinity, the imports of yarn from Britain have since declined.

Distillation has been, for a lengthened period, an important business in Ireland. Previously to 1823, when the duty on spirits was 5s. 6d. a gallon, illicit distillation was extremely prevalent in Ireland; so much so, that the commissioners of revenue inquiry estimated the total annual consumption of spirits in Ireland at that period at 6,330,000 imp. galls., though, in 1822, no more than 2,396,367 paid duty! In 1823, the duty having been reduced to 2s. 10d. per imp. gall., 6,690,315 galls. paid duty in 1824, and 9,262,744 in 1825. In 1835 the duty was reduced to 2s. 4d. per gall.; and 1836 duty was paid on 12,296,342 galls. But it would seem as if the consumption had then attained to a maximum; for, in the course of 1839, and the present year (1840), great numbers of people in all parts of the kingdom were induced, principally by the exertions of a Roman Catholic priest, of the name of Matthew, to pledge themselves to abstain from spirituous liquors. It is difficult to learn the precise extent to which the temperance system has been carried, but there can be no doubt that it is in very extensive operation. In 1839 duty was only paid on 12,315,793 galls., being 1,980,623 galls. less than in 1838; and it is said that the falling off during the current year will be still more perceptible. It is to be hoped that the resolution evinced in maintaining and enforcing this system may equal the ardour displayed in its introduction. The habit of drinking spirits, formerly so prevalent among the Irish, particularly when sustained at fairs and other public meetings, was the source of innumerable outrages; nor can there be a doubt that the general diffusion of habits of temperance will be of signal service to the country.

The silk trade was introduced by French emigrants shortly after the Revolution: its chief seat was in Dublin; but since the repeal of the protecting duties it has declined, so as to be now nearly extinct, with the exception of tabbinet or Irish poplin, a mixed fabric of silk and worsted, for which there is a considerable demand. The first importation of cotton wool into Ireland, of which there is any authentic notice, took place in 1771. The manufacture was carried on with considerable success in several parts during the continuance of the protecting duties. On their withdrawal it declined for a while; but it has since revived, and is now prosecuted to a considerable extent, in the neighbourhood of Belfast, where there are several large mills, and at Portlaw, co. Waterford, where there is a mill, that employs above 1600 hands.

Account of the Factories in Ireland, and of the Number of Persons employed in them, in 1838.

	Mills.	Steam Engines	Water Wheels	Persons employed.
Cotton	24, having	19	52	4,693
Woollen	31	—	20	1,291
Silk	40	—	22	2,017
Total		38	50	11,670

The total power in steam engines and water wheels is said to be equivalent to 650 horse-power; and of the persons employed, 6592 are said to be under, and 5048 above 18 years of age. (Return by James Stuart, Esq., *Factory Report*, p. 235.)

Commerce.—The exportation of the raw produce of the soil has always formed the principal commercial business carried on in Ireland. During the late war, she supplied a large share of the provisions required for the army and navy serving-abroad; and she still sends large supplies to the colonial markets. Great Britain, however, is by far the best and most extensive market for all sorts of Irish produce; and her exports to this country, especially of corn and flour, and of butter, pigs, eggs, &c., have prodigiously increased. The conversion of grain into flour and meal has lately become an extensive business in Ireland; and many mills, erected for this purpose, are on an extensive scale, and are furnished with the best machinery.

By far the greater part of the trade of Ireland is carried on as a cross-channel trade with Great Britain, and especially with Liverpool, Bristol, and Glasgow. Its trade with foreign countries, and with British colonies, is comparatively inconsiderable. This is evident from the *first* and *third* of the following tables, the first of which gives an account of the estimated value of the total exports and imports of each port of Ireland in 1835; and the *third*, an account of the export and import trade of the island with foreign parts only, from 1830 to 1839, both inclusive.

I. Account of the estimated Value of the total Exports and the total Imports of each Port of Ireland in 1835.

Names of Ports.	Exports, 1835. Value.	Imports, 1835. Value.
Antrim and Killough	£ s. d.	£ s. d.
Arklow		
Ballycastle		
Belfast		
Ballytroan		
Ballycastle		
Ballyshannon		
Baltimore, &c.		
Bantry		
Bruckmoon		
Bandimilet		
Belfast		
Clare		
Coleraine and Portrush		
Cork		
Donaghadee		
Donegal		
Drogheda		
Dublin		
Dundalk		
Dungarvan		
Galway		
Kinsale		
Kilrush		
Kinsale		
Larne		
Limerick		
Londonderry		
Newcastle		
Newport		
Newry		
Ross		
Strangford		
Sligo		
Tralee		
Waterford		
Wexford		
Westport		
Wicklow		
Youghal		
Total	17,384,813 7 11	15,837,057 4 6

III. Account of the Trade of Ireland with Foreign Parts, during each of the Ten Years ending with 1839.

Years.	Imports. Official Value.	Exports. Official Value.	Exports. Real Value.
1830	£	£	£
1831			
1832			
1833			
1834			
1835			
1836			
1837			
1838			
1839			

The shipping of Ireland is but inconsiderable, compared with that of Great Britain; but it has increased considerably within the last ten years. We subjoin an account of the vessels built and registered, and of those belonging to Ireland, with their tonnage and men, in each of the 10 years, ending with 1839.

Years.	Built and Registered.		Belonging to Ireland.		
	Vessels	Tonnage	Vessels	Tonnage	Men.
1830	45	2,504	1,424	101,920	7,794
1831	39	2,435	1,447	105,774	8,041
1832	25	1,609	1,456	106,128	7,228
1833	31	2,219	1,492	110,945	8,368
1834	37	2,505	1,596	119,398	8,731
1835	39	2,521	1,627	124,739	9,292
1836	74	1,917	1,655	128,496	9,159
1837	38	3,204	1,664	139,509	9,465
1838	61	2,617	1,776	151,528	10,263
1839	69	4,014	1,889	149,249	11,364

Account of the Gross Customs' Revenue collected at the different Ports of Ireland in 1839.

Ports.	Customs.	Ports.	Customs.
	£		£
Ballimore	449	Limerick	149,780
Belfast	341,443	Londonderry	96,628
Coleraine	4,560	Newry	47,697
Cork	545,702	Sligo	29,450
Drogheda	9,236	Waterford	176,125
Dublin	928,607	Westport	9,181
Dundalk	18,816	Wexford	6,450
Galway	22,685		

IRELAND.

II. Account of the Quantities of the Principal Articles imported into and exported from Ireland in 1825 and 1835.

Imports.		1825.*	1835.	Exports.		1825.*	1835.
Coal, culm, and cinders	tons	738,483	1,061,378	Cows and oxen	num.	63,594	98,150
Corn, malt	qrs.	4,300		Horses	—	3,140	4,655
—— other sorts	—	5,861		Sheep	—	72,191	195,492
Hops	cwts.	13,944		Swine	—	65,919	376,191
Bark, oak, for tanners	tons	6,988		Grain, viz., wheat	qrs.	283,340	430,523
Beer and ale	tuns	333		—— barley	—	154,828	168,946
Fish, herrings	bbls.	111,328		—— oats	—	1,503,204	1,575,984
Salt	bush.	1,646,814		—— other grain	—	23,832	30,637
Sugar, British, refined	cwts.	66,398	43,987	Wheat, flour, and oatmeal	cwts.	599,194	1,934,480
Spirits, British and Irish	galls.		398,090	Potatoes	—		322,308
—— foreign, rum	—	33,995	37,240	Provisions, bacon and hams	—	382,978	379,111
—— brandy, copper	cwts.		4,550	—— beef and pork	—	604,253	370,172
Metals unwrought iron	tons	17,902	19,230	—— butter	—	474,151	827,009
—— lead	—		1,488	—— lard	—	35,961	70,987
—— tin	cwts.		1,236	Soap and candles	—	43	
Cast iron	tons		7,095		num.		32,944,800
Iron, foreign	—		3,290	Eggs	crates		2,975
Stones, slate	—		18,504		boxes		10,685
Wool, sheep's or lambs'	lbs.		811,560	Feathers	cwts.		6,432
Woollen yarn	—	579,051	65,118	Hides and calf-skins	num.		57,687
Linen yarn	—		2,632,000	Wool, sheep's or lambs'	bales		33
Cotton yarn	—	2,702,523	582,914		lbs.		784,184
Cotton manufactures	yds.	4,988,685	14,172,080	Flax and tow	cwts.	54,898	162,949
Woollen manufactures	—	3,384,918	7,884,000	Lead and copper ore	—		477,680
Wool, cotton, foreign	lbs.	4,065,920	2,646,256	Spirits	galls.	629,529	458,473
Wool, sheep, foreign	—		155,208	Beer	—		2,686,680
Linens, Irish	yds.		30,500		yds.	10,567,438	1,030,088
—— British	—		322,553	Cotton manufactures	pkgs.		4,583
Silks, raw and thrown, for'gn	lbs.	62,198	9,072	Cotton yarn	lbs.		12,498
Tinned plates	boxes		9,030		yds.	53,114,515	70,908,579
Leather	pkgs.		2,198	Linen	boxes		134
Wrought iron, hardwares	—		18,396		bales		7
Machinery and millwork	—		2,361	Silk manufactures	yds.		8,400
Glass and earthenware	—		14,109	Woollen manufactures	—		100,390
Haberdashery and apparel	—		4,177	Other articles	value	£486,300	£369,934
Ashes	cwts.	112,536	12,492				
Barilla	—		3,445	*Foreign and Colonial Merchandise.*			
Indigo	lbs.		90,183				
Hides	cwts.		74,897	Tea	lbs.		37,350
Tallow	—	196,147	87,867	Coffee	—	14,814	47,643
Sugar, foreign	—	980,634	189,080	Sugar	cwts.	26,239	2,992
Tea	lbs.	3,880,638	4,794,316	Molasses	—		7,597
Coffee	—	338,921	1,905,788	Tobacco	lbs.	1,118,296	92,751
Tobacco	—	3,904,934	4,487,746	Wine	galls.	64,995	73,063
Wines	galls.	965,949	304,031	Rum	—	25,807	1,195
Flax seed	bush.	535,331	946,458	Hemp	cwts.		1,720
Other articles	value	£2,091,973	£1,379,783	Other articles	value	£32,449	£19,775

* It is supposed that many of the articles which appear blank in the columns for 1825, and are detailed in those for 1835, are returned in the Aggregate Value of "other Articles."

The great preponderance of the customs' revenue of Dublin results from its being the principal port for the importation of wine, sugar, timber, and other taxed articles. Its trade is really inferior to that of Belfast, and but little superior to that of Cork.

Roads are generally well laid out, and kept in good order. They are made and maintained partly by turnpike trusts, but chiefly by county presentments, or assessments on the counties. The latter amounted, in 1838, to the sum of £427,093. The system of macadamizing was practised in Ireland for many years before it attracted public attention in Great Britain.

Canals.—The Grand canal, commenced in 1765, is carried from Dublin to Robertstown, 25 m. W., whence proceed two branches, that to the right to the Shannon harbour, on the Shannon, near Banagher, and thence on the W. of the river to Ballinasloe, 94 m. from Dublin, with a branch of 8½ m. to Kilbeggan; that to the left to Athy, 55 m. from Dublin, with a branch of 11 m. from Monasterevan to Portarlington and Mountmellick. The summit level is 200 feet above the sea. The Barrow is navigable from Athy, for small craft, to the Scars, 43 m., thence for larger vessels, by Ross to Waterford, 30 m.

The tonnage and tolls on the Grand canal, for the eight years ending with 1837, were,

Years.	Tonnage.	Toll.	Years.	Tonnage.	Toll.
1830	304,749	33,684	1834	225,473	38,192
1831	237,810	30,723	1835	215,308	38,030
1832	216,418	34,553	1836	228,770	38,953
1833	228,736	38,664	1837	215,910	37,357

The Royal canal, commenced in 1789, extends from Dublin to Tarmonbarry on the Shannon, 92 m., with a branch of 5 m. from Killashee to Longford. Its greatest height above sea level is 207 feet. The Shannon has been rendered navigable from Limerick almost to its source, and is traversed by steamboats both for passengers and goods. Its navigation is, however, exposed to considerable interrup-

tions; but works are now in progress by which these will be, in a great measure, obviated, and this grand channel of internal communication rendered much more available than at present. The Boyne navigation from Drogheda to Navan, and the Lagan from Belfast to lough Neagh, are partly river and partly still water. The Ulster canal, now in progress, is intended to connect loughs Neagh and Erne. The Suir Navigation Company was incorporated in 1837, for making a ship canal to Carrick-on-Suir. The railway, from Dublin to Kingstown, 6 m., is the only undertaking of the kind as yet completed. It is used chiefly for passengers. It was opened on December, 1834. The receipts for the three following years were, 1835, £39,089; 1836, £38,139; 1837, £33,290. The number of passengers during the same period was 3,608,000: annual average, 1,900,000; daily, 5,343.

Acts have been passed for railways from Dublin to Drogheda, from Belfast to Armagh, from Dundalk to Ballybay, and from Dublin to Kilkenny. The commissioners appointed to consider and report on a general system of railways for Ireland, drew up a most valuable and elaborate report, in which they recommend that government should construct some of the principal lines. This, however, is a question of great nicety and difficulty; and their recommendation has not hitherto been acted upon.

Banks.—Banking business in Ireland has long been, and perhaps still is, in a very unsatisfactory state. Till 1783, when the Bank of Ireland was incorporated, the business was wholly in the hands of individuals. The privileges given to the Bank of Ireland were similar to those of the Bank of England, as to the restriction of private banking establishments to six partners. Its capital, at its commencement, was £600,000, increased, in 1791, to £1,000,000; in 1797, to £1,500,000; in 1808, to £2,500,000; in 1821, to £3,000,000. Since 1824 there were 20 registered private banks; and since then several others were opened, but they have all closed or failed, except four in Dublin. The system of joint-stock banks came into operation, in Ireland, in 1825, excepting within a circle extending 65 m. round Dublin, of which the Bank of Ireland had a monopoly.

Account of the Joint Stock Banks existing in Ireland on the 5th day of January, 1838; specifying the Date of the Establishment of each Bank, the Number of its Branches, and the Number of Partners in each in the Years 1836, 1837, and 1838.

Banks.	Date of Establishment.	Number of Branches	Number of Partners		
			1836	1837	1838.
Hibernian Joint Stock Company	1825	By special Act 5 G. 4			
Provincial Bank of Ireland	1825	34	642	707	724
Northern Banking Company	1825	11	210	204	195
Belfast Company	1827	17	246	265	280
National Bank of Ireland	1835	16	216	230	462
Limerick National Bank of Ireland	1836	5	520	561	684
Agricultural and Commercial Bank of Ireland	1834	29	2,656	3,592	3,573
Clonmel National Bank	1836	3	165	467	446
Carrick-on-Suir National Bank	1836	1	391	418	571
Waterford National Bank	1836	4	434	461	613
Wexford and Enniscorthy National Bank	1836	2	389	417	599
Tipperary National Bank	1836	5	429	436	626
Tralee National Bank	1836	6	411	444	809
Ulster Banking Company	1836	9	656	789	679
Royal Bank of Ireland	1836	1	304	303	304
Southern Bank of Ireland	1837			92	
Cork National Bank	1837			415	530
Kilkenny National Bank	1837			379	546

The system of savings' banks, under an act passed in 1815, is gradually being extended. The total number of depositors and depositors was, in

	Depositors.		Deposits.
1835	£58,408		£1,804,653
1838	74,333		2,158,665

Revenue.—The revenue of Ireland is raised from the same sources as in England, except the assessed taxes, which have not been extended to Ireland. But, owing to the depressed situation of the great bulk of the Irish people, and their inability to consume taxed articles, the revenue of Ireland falls far short of what it might be expected to amount to from the magnitude of the population. In fact, the revenue of Ireland, with a population (in 1840) of 8,300,000, hardly equals that of Scotland with a population under 2,000,000.

We subjoin an account of the gross and nett receipt of the public revenue of Ireland in 1839, showing the rate per cent. which it cost to collect the gross revenue.

	Gross Receipt.	Nett Receipt, deducting Repayments.	Rate per cent. at which Gross Revenue was collected
	£ s. d.	£ s. d.	£ s. d.
Customs	2,021,389 1 3	2,016,584 9 7	11 1 0
Excise	1,419,629 1 8½	1,314,885 3 4	10 7 11½
Stamps	450,195 9 5	443,487 8 10	4 18 4
Post-office	285,369 2 7½	247,845 10 4	42 19 6½
Miscellaneous Fees	6,952 16 9	6,542 16 9	
Totals	4,376,699 11 8½	4,019,572 7 11½	11 14 8½

But, exclusive of the above, or of the public revenue, a considerable sum is annually raised by grand jury presentments, that is, by assessments on the counties made by the grand juries, for the constructing and keeping up of roads, prisons and bridewells, police and police establishments, and for charitable purposes, &c. In 1838 the presentments in question amounted to the gross sum of £1,131,046, distributed as follows:—

Account of the Sums raised by Grand Jury Presentments in Ireland, in 1838; specifying the Objects for which they were raised, and the Amount appropriated to each.

Descriptions of Works.	Total
	£ s. d.
1. New roads, bridges, pipes, gullets, quay walls, or cutting down hills, and filling up hollows and ditches	112,973 16 1¾
2. Repairs of roads, bridges, pipes, gullets, walls, &c	314,119 6 0
3. Court or sessions houses, erection or repairs	11,714 14 6
4. Jails, bridewells, houses of correction, building or repairing	8,728 7 0
5. All other prison and bridewell expenses, including salaries	98,615 0 8½
6. Police and police establishments, and pay mary to witnesses	161,763 3 10½
7. Salaries of all county officers not included above	110,513 7 2½
8. Public charities	117,810 19 0
9. Repayment of advances to government	129,051 13 5
10. Miscellaneous, and not included in the above	71,906 0 1
	1,131,951 3 1
Deduct re-presentments, &c.	7,910 0 0
Total for the whole of Ireland	1,131,046 3 0

This, however, is but a small sum, compared with what is assessed for similar purpose in England. But the weight of local taxation on the land will now be considerably increased, through the operation of the compulsory provision for the support of the poor; though that will be far more than counterbalanced by the good effects of which it cannot fail to be productive.

The constitution of Ireland is modelled on that of England; but, for a lengthened period, the native Irish, comprising the great bulk of the population, were effectually excluded from all participation in its benefits, and were, in fact, reduced to a state of *helotism.* This conduct, it is needless to add, was little less injurious to the conquerors than to the conquered. "As the English would neither in peace govern the Irish by the law, nor could in war root them out by the sword, they needs became pricks in their eyes and thorns in their sides." But nations are slow and reluctant learners; and that selfish short-sighted policy, whose effects were thus forcibly exposed by Sir John Davies (*Discoverie, &c.*, p. 195, ed. 1747) in the reign of James I., flourished in its full vigour down almost to our own times! The granting of the elective franchise to the Catholics, so late as 1793, was, in truth, the first great step in the progress to a better system, which was happily consummated by the repeal of the last remnant of the penal code in 1829. The odious distinctions by which society was formerly divided, have no longer any legal or statutory foundations. Adherence to the religion of their ancestors has ceased to entail upon the Catholics a denial of their political franchises; and all classes now participate equally in the rights and privileges granted by the constitution. The legislature consisted, previously to the union, of a chief governor, under the name of lord lieutenant, with power to appoint a deputy during absence, a house of lords and a house of commons. Under Henry VII., the prostration of the Irish parliament was effected, by transferring the right to hold parliaments, which had been vested in the lord lieutenant, and to originate bills, to the king and the English privy council. The first parliament, in which members were returned from all parts of Ireland, sat in the beginning of the reign of James I. The number of members varied at different times, but was ultimately fixed at 300, two for each county, two for Trinity College, and the remainder for cities and boroughs, the representatives for the latter being, in most instances, nominated by their proprietor or patron. Previously to 1768, the members held their seats for life, so that they could hardly be considered as representatives even of the Protestant part of the nation, and had but little sympathy with popular feelings. At this epoch, however, parliaments were made octennial. Since the union, Ireland has been represented in the imperial parliament by 28 temporal peers, elected for life by the whole body of Irish peers; four bishops, who sit according to annual rotation of sees; and, from the union till the passing of the Reform Act, it was represented in the House of Commons by 100 members, two for each county, two each for the cities of Dublin and Cork, one for Trinity College, and one each for the 31 boroughs of Armagh, Athlone, Bandon, Belfast, Carlow, Carrickfergus, Cashel, Clonmel, Coleraine, Downpatrick, Drogheda, Dundalk, Dungannon, Dungarvan, Ennis, Enniskillen, Galway, Kilkenny, Kinsale, Limerick, Lisburn, Londonderry, Mallow, New Ross, Newry, Portarlington, Sligo, Tralee, Waterford, Wexford, and Youghal. The Reform Act gave Ireland five additional members, which were assigned to Trinity College, Belfast, Galway, Limerick, and Waterford, which, consequently, have now two members each. The electoral franchise in counties is vested in the same classes as in England, with the substitution of £20 for £50, and 14 for 20 years; and in cities and boroughs in freemen resident within seven miles, and £10 freeholders. The electoral boundaries of the boroughs were fixed by a late statute. The executive government is vested in the lord lieutenant, or, in his absence, in the lords justices, generally the primate, lord chancellor, and commander of the forces, and a privy-council nominated by the crown, and consisting chiefly of the high judicial and ministerial functionaries. The lord lieutenant is assisted by a chief secretary, a member of the House of Commons, now usually also a cabinet minister; and who, being in effect secretary for Ireland, is especially responsible for its government. The salary of the lord lieutenant is £20,000 a year, with liberal allowances for residence and household.

The judicial establishment is vested, as in Great Britain, in the lord chancellor, removable at pleasure, assisted by the master of the rolls, and in 12 judges, four for each of the courts of Queen's Bench, Common Pleas and Exchequer, all of whom hold office during good behaviour. Two of the law judges go through each of the six circuits, into which the country is distributed, twice a year, to decide criminal and civil cases. The judges of the courts of Prerogative and Admiralty are generally practising barristers. A barrister also presides, along with the co. magistrates, at the Courts of Quarter Sessions. Petty sessions, at which at least two magistrates must be present, are held weekly, or once a fortnight, in every district. Each corporate town has a judge, or recorder, and local magistrates, elected by the corporation; and every manor has its courts under a seneschal or bailiff nominated by the proprietor. The lord chancellor has the power of appointing and removing the co. magistrates, for whose conduct he is responsible. An act passed in 1840 (3 & 4 Victoria, cap. 108,) for remodelling the municipal corporations in Irish towns. It gives the right of voting at municipal elections to all persons resident in boroughs, or within 7 m. of their boundaries, occupying houses, shops, or other premises within the same of the annual value of £10.

The conservation of the peace is committed, in the cos., to a lord lieutenant, aided by an indefinite number of deputy lord lieutenants, all nominated by the crown and by the high sheriff, selected, as in England, from lists prepared by the judges of assize. The police consists of a well-organized constabulary force, which consisted, on the 1st of January, 1839, of 8416 men, under an inspector general, two deputies, and four provincial inspectors, appointed by the crown, and removable at pleasure.

The larger towns have each a corporate police; and a military force, varying in numbers according to circumstances, is distributed throughout the country,

There are 45 county or town prisons, penitentiaries, and houses of correction, and 98 bridewells or places of temporary confinement. The superintendence of the prisons is committed to two inspectors-general, who make annual reports to parliament, and to a local inspector. The prisons are generally well constructed and regulated. The following table, extracted from the inspectors' reports, exhibits a view of the number of committals and convictions, and of the number of the latter visited with the highest and lowest grades of punishment, from 1826 to 1838 inclusive:

Year.	Committals.	Total.	Imprisonment. Six Months and under	Capital.	Executions.
1826	10,318	9,718	6,283	231	34
1827	18,031	10,367	6,546	346	57
1828	14,953	9,289	6,449	211	21
1829	15,271	9,449	6,526	224	38
1830	15,794	9,080	7,506	204	39
1831	16,192	9,695	9,460	307	37
1832	19,936	9,739	6,995	519	38
1833	17,819	11,444	8,536	237	39
1834	21,381	14,233	11,190	197	43
1835	21,205	15,210	12,787	179	57
1836	23,891	16,110	13,464	175	14
1837	14,804	9,536	6,106	184	10
1838	15,723	9,609	6,319	39	3

The regular troops stationed in Ireland at the undermentioned periods, during the last 10 years, have been as follows:

Years.	Artillery.	Cavalry.	Infantry.	Total.
1830	913	2,187	17,395	20,495
1833	934	2,064	21,061	24,059
1837	987	1,938	15,632	18,536
1839	985	1,643	14,076	16,344

The military department is under the control of the commander of the forces, whose head quarters is at Kilmainham. He has under him five general officers, who respectively command one of the five military districts into which the island is divided. The ordnance, which is a branch of that of Great Britain, has its chief station at the Pigeon House Fort: attached to it is the staff of the trigonometrical survey of Ireland. There is, at Kilmainham, a hospital for decayed and disabled soldiers, similar to that of Chelsea.

Religious Establishments.—The ecclesiastical arrangements that prevail in Ireland are at once anomalous and irrational. The Reformation never made any considerable progress in the country, the new doctrines being espoused only by the English settlers within the *pale*. But after protestantism had been adopted by the bulk of the English people, and had been made the established religion 'on this

---

side the water, it was determined to establish it as the state religion in Ireland. In pursuance of this resolution, the Catholic clergy were ejected from their livings, which were bestowed upon divines attached to the doctrines of the church of England. This change did not, however, occasion any corresponding change in the religious feelings of the people, who seemed, indeed, to become the more attached to their ancient faith, according as their clergy were treated with harshness and injustice. In every other country, the established religion, if there be one, is that of the great majority of the people; but in Ireland the established religion is and long has been that of a small minority—and that minority, be it observed, consists principally of the wealthy and best educated classes, who could, without difficulty, supply themselves with religious instruction! Such an arrangement is obviously inconsistent with every principle of sound policy and common sense. The grand object of an establishment should be the provision of religious instruction and consolation for the great bulk of the community, and especially for those who are too poor to be able to provide it for themselves. But, in Ireland, the reverse of all this obtains. The established religion is alien to and repudiated by nine tenths of the pop., who regard it as erroneous in principle, and as a usurpation upon the rights and property of their clergymen. These feelings are natural; and it is nugatory to suppose that they should be got rid of, so long as the existing arrangements are maintained. A Catholic establishment in England would not, in fact, be more irrational and absurd than a Protestant establishment in Ireland; and, so long as the latter is permitted exclusively to enjoy the revenues appropriated by the state for the support of religion, so long will it be an object of disgust and hostility to the Catholic people and clergy, that is, to the great majority of the nation, and be productive of the most implacable animosities.

Previously to the late regulations affecting the established church in Ireland, the country was divided into four archbishoprics, corresponding nearly with the four civil provinces, and these were further subdivided into 20 bishoprics, held by 18 bishops.

Under the new arrangements, the archbishoprics of Cashel and Tuam are reduced to bishoprics, thus dividing Ireland into the N. and S. provs., nearly according to a line drawn from the N. of Dublin co. to the S. of Galway Bay. The bishoprics are to be reduced to ten, and their annual incomes regulated as in the following table. The alteration takes place on the demise of the existing tenures. Those already altered are printed in italics:

	£
ARMAGH and Clogher	12,170
Meath and Clonmacnois	5,221
Derry and Raphoe	8,032
Down, Connor, and Dromore	5,896
Kilmore, Ardagh, and Elphin	7,478
Tuam, Killala, and Achonry	5,690
DUBLIN, Glandelagh, and Kildare	9,321
Ossory, Leighlin, and Ferns	6,530
Cashel, Emly, Waterford, and Lismore	7,354
Cloyne, Cork, and Ross	5,009
Killaloe, Kilfenora, Clonfert, and Kilmacduagh	4,530
Limerick, Ardfert, and Aghadoe	5,369

The other dignitaries are 33 deans, 26 precentors, 22 chancellors, 21 treasurers, 34 archdeacons, two provosts, and one sacristan. Besides these dignitaries, there are 178 prebendaries, and nine canons. Twelve of the cathedrals have subordinate corporations, consisting of five canons, 59 vicarschoral, and 15 choristers. The number of ecclesiastical pars. is 2348, consolidated into 1385 benefices. The parochial clergy is maintained by tithes and glebe lands, and in cities and large towns by minister's money.

The total amount of the income of the various members of the Protestant episcopal church, during the three years ending with 1831, was returned to parliament as follows:

	£
Archbishops and bishops	151,128
Deans and chapters	1,042
Economy estates of cathedrals	11,056
Other subordinate corporations	16,386
Dignities (not episcopal), and prebends	
Without cure of souls	34,482
Glebe lands	92,000
Tithes	555,000
Ministers' money	10,300
Total	865,535

There seem, however, to be good grounds for thinking that the income of the archbishops and bishops given above is below the truth. It principally consists of the rent of lands let on lease, or rather on leases renewable by fine. The total extent of land belonging to the different sees has

---

been returned to parliament at about 670,000 (669,374) acres. Now we have seen that the average rent of Ireland may be estimated at about 13s. 6d. an acre; and assuming the church estates to be only of a medium quality, which th y are believed to exceed, they should, on this hypothesis, be worth £452,250 a year. But supposing that, owing to the defective system under which they are occupied, they only produce 6s. an acre; still, even at that extremely low rate, their gross rental would amount to £201,000. At all events, it is sufficiently clear that the episcopal and glebe lands, if properly managed, would afford a revenue more than sufficient to provide for the religious instruction of the entire Protestant pop. of Ireland, without having recourse to tithes.

We also incline to think that the amount of tithe given in the above statement is considerably underrated. The following statement laid by Mr. Griffith, the engineer, before the lords' committee, on tithe, throws a good deal of light on this subject. It can of course, however, be regarded only as an approximation. "Ireland is divided into four archbishoprics, 28 bishoprics, 2450 parishes; and contains 20,490,000 statute acres. There are 1492 beneficed clergymen; 1539 pars. under the composition act, and 911 pars. not under the composition act. The

	£	s.	d.
Gross amount of the composition for those pars. which have compounded for tithe is	442,419	0	0
Average amount of the composition for those pars. which have compounded	287	9	0
Average proportion of the composition for tithe to £1 starling in the value of the land	0	1	3¼
According to the best data I have been able to procure, and from my own knowledge of the value of land in Ireland, I am of opinion that the gross annual value of the land, rated at a moderate rent, may be about . .	12,715,576	0	0
Rating the average amount of the tithe at 1s. 3¼d. in the pound sterling, of the value of land, it would appear that the gross amount of the tithe in Ireland would be	820,214	16	7
(If all the lands in Ireland were liable to tithe, which is not the case, consequently the total amount is less than that sum.)			
If we take the average amount of the compositions for those pars. which have compounded, and multiply that sum by 2450, the total number of pars. in Ireland, we shall have the sum of	704,313	15	0

This is probably the nearest approximation to the true amount of the tithe of Ireland."

In addition to the unpopularity attaching to the church of England in Ireland, from its being the church of a minority, the fact of its deriving the largest portion of its income from tithes, has tended materially to increase the odium under which it has long laboured. Tithe is everywhere a most vexatious and impolitic tax; but in Ireland it has been peculiarly noxious; for there the land being mostly split into small portions occupied by poor Catholic cottiers, the payment of tithe to Protestant clergymen is not only felt to be a most oppressive burden, but it is, at the same time, looked upon as a sacrifice imposed for the promotion and advantage of heresy and error. It has also been very unfairly assessed. By a resolution of the Irish H. of C. in 1735, grass lands obtained an exemption from tithe; so that while a tenth part of the produce of a potato garden or slip of land, on which, perhaps, a numerous family was dependent, went to the establishment, the herds of the opulent grazier contributed nothing to its support. Under such circumstances, we need not wonder that, for a lengthened period, the payment of tithes in Ireland has been made with extreme reluctance, and that their collection has, in innumerable instances, been productive of outrage and bloodshed. At last it became next to impossible, in many parts of the kingdom, to derive any revenue from this source; and in consequence it was attempted to substitute compositions or fixed payments for tithes in the room of tithes themselves. But, though productive of some advantage, this measure was comparatively useless, from its leaving the composition to be paid by the occupier and not by the landlord. To obviate this defect, an act was passed in 1838 (1 & 2 Victoria, cap. 109,) abolishing compositions for tithes, and substituting in their stead a fixed payment of three fourths of their amount, to be made by the landlords or others having a perpetual interest in the land. This act, by relieving the tithe-collector from the necessity of coming into contact with the great bulk of the occupiers,

has obviated a prolific source of predial disturbance, and been, in so far, advantageous. Still, however, it must not be supposed that either this or any other device should ever reconcile the Irish people to the appropriation of a large revenue to the exclusive use of the church of a small minority of their number. The effect of this preposterous arrangement is to insult and alienate the bulk of the population, who would be more or less than men if it ceased to encounter their rooted hostility.

The Roman Catholic church is arranged nearly in the same manner as the established church previously to the late changes. There are four archbishops, the same in name and provincial rank as those of the Protestant church, and 23 bishops. Eight of the bishops—Ardagh, Clogher, Derry, Down, and Connor, Dromore, Kilmore, Meath, and Raphoe—are suffragan to Armagh. Dublin has but three suffragans—Kildare and Leghlin united, Ferns, and Ossory. Six are suffragan to Cashel; namely, Ardfert and Aghadoe (usually called the Bishop of Kerry, Cloyne, and Ross) Cork, Killaloe, Limerick, Waterford, and Lismore. Tuam has four suffragans—Achonry, Clonfert, Killala, and Galway. The bishop of the united dioceses of Kilmacduagh and Kilfenora is alternately suffragan to the archbishops of Tuam and Cashel. The wardenship of Galway, formerly an exempt jurisdiction, subject only to the triennial visitation of the archbishop of Tuam, has been lately erected into a bishopric, under its former archie-piscopal jurisdiction. On the death of a bishop, the clergy of the diocese elect a vicar-capitular, who exercises spiritual jurisdiction during the vacancy. They also nominate one of their own body, or sometimes a stranger, as successor to the vacancy, in whose favour they postulate or petition the pope. The bishops of the province also present the names of two or three eligible persons to the pope. The new bishop is generally chosen from among this latter number; but the appointment virtually rests with the cardinals, who constitute the congregation de propaganda fide. Their nomination is submitted to the pope, by whom it is usually confirmed. In cases of old age or infirmity, the bishop nominates a coadjutor, to discharge the episcopal duties in his stead; and his recommendation is almost invariably attended to. The emoluments of a bishop arise from his parish, which is generally the best in the diocese, from licenses of marriage, &c., and from the cathedraticum. The last is an annual sum, varying from £2 to £10, according to the value of the parish, paid by the incumbent, in aid of the maintenance of the episcopal dignity. The parochial clergy are nominated exclusively by the bishop. The incomes of all descriptions of the Roman Catholic clergy of Ireland arise partly from fees on the celebration of births, marriages, and masses; and partly, and principally, perhaps, from Christmas and Easter dues, and other voluntary offerings. All places of worship are built by subscription. There are numerous monasteries and convents.

Exclusive of the injustice inflicted on the Roman Catholics of Ireland, by the seizure of the funds belonging to their church, and their appropriation to the support of the clergy of the church of England, they laboured for a lengthened period under the most degrading disabilities. The treaty of Limerick, in 1691, between the generals of William III. and those of James II., guaranteed to the Irish Roman Catholics the same religious privileges they had enjoyed during the reign of Charles II. But this treaty was most shamefully broken; and, during the reigns of Anne, George I., and George II., a series of acts were passed, constituting what has been called the Catholic penal code, which had for its object the extermination of the Roman Catholic religion in Ireland. It is unnecessary to recapitulate the provisions of these statutes. Their spirit was succinctly and truly described by Mr. Burke: "The laws made in this kingdom (Ireland) against papists were as bloody as any of those that had been enacted by the popish princes and states; and when these laws were not bloody, they were worse: they were slow, cruel, outrageous in their nature, and kept men alive only to insult in their persons every one of the rights and feelings of humanity." (Letter to Sir H. Langrishe.) Everybody knows that this atrocious code entirely failed of its object, and that, instead of being exterminated, the Roman Catholic religion gained new strength and vigour from the persecution to which it was exposed.

Per damna, per caedes, ab ipso
Duxit opes animumque ferro.

In the earlier part of the reign of George III., the leading statesmen of England became alive to the impolicy and mischievous operation of parts, at least, of the penal code; and its more offensive provisions were gradually repealed. In 1793, the elective franchise was conceded to the Roman Catholics; but they continued, down to a comparatively late period, to be excluded from the privilege of having seats in the legislature, of being members of corporations, and of holding numerous public offices of trust and emolument.

IRELAND.

At length, in 1829, the Roman Catholics were fully emancipated from all civil disabilities on account of religion, and were placed, as respects their political rights and franchises, nearly on the same footing as Protestants.

That this measure was a great boon to Ireland is most true; but, though it allayed, it was not enough to extinguish, religious feuds and animosities. Justice, and the most obvious dictates of policy, require, as already stated, either that the Roman Catholic should be made the established religion of Ireland, or, at all events, that the Roman Catholic clergy should participate, proportionally to the number of their flocks, in the emoluments now exclusively engrossed by the clergy of the church of England. It is a contradiction and an absurdity to suppose that a great and decisive majority should ever quietly submit to be deprived of privileges possessed by a minority. This, however, is the state of things in Ireland; and till it be radically and completely changed, the country will, no doubt, continue as heretofore, to be disgraced and distracted by religious dissensions.

The Protestant dissenters are found chiefly in Ulster. They are classed in congregations, an indefinite number of which forms a presbytery, and delegates, partly ministers and partly lay elders, form the general synod, which regulates the ecclesiastical concerns of the body, and is presided over by a moderator chosen annually. The Synod of Ulster is coexistent with the establishment of the Presbyterian doctrine and discipline in Ireland. The Southern Association, or Presbyterian Synod of Munster, was formed about 1660; the Presbytery of Antrim separated from the Synod of Ulster in 1727, and the Remonstrant Synod in 1829. The number of presbyteries and congregations in each body, and in the Seceding and Covenanters' synods are as follows:

	Presbyteries.	Congregations.
General Synod of Ulster	34	275
Presbyterian Synod of Munster	1	15
Presbytery of Antrim	1	13
Remonstrant Synod	4	27
Seceders	10	92
Covenanters	4	93

The Methodists are divided into two societies—the Wesleyan and the Primitive Wesleyan; the number in both societies is 26,944. The Independents, or Congregational Union, a separate body from Presbyterians or Methodists, have 26 congregations. Their classification in districts, stations, and missions, or missionary stations, is as follows:

	Districts.	Stations.	Ministers.
Wesleyans	11	46	16
Primitive Wesleyans	9	28	28

The Society of Friends, or Quakers, are most numerous in Dublin, the Queen's Co., and Armagh. The United Brethren, or Moravians, have establishments in Dublin and Antrim. The Jews have a synagogue in Dublin.

The numbers attached to each religious persuasion in Ireland, in 1834, were, according to the returns of the Commissioners of Public Instruction, as follow:

Denominations.	Number.	Centesimal Prop.
Established Church	852,064	10·726
Roman Catholics	6,427,712	80·212
Presbyterians	642,356	8·026
Other Dissenters	21,808	0·255

Or, in round numbers, out of every hundred souls, 11 are of the establishment, 81 Roman Catholics, and eight Protestant dissenters.

Education.—The principle of educating the great body of the people was fully recognised at the Reformation. An act of 28 Henry VIII. bound every beneficed clergyman by oath, on his incumbency, to keep or cause to be kept a school in his parish. A subsequent act of Elizabeth required the bishop and beneficed clergy of every diocese to maintain a grammar-school. But in nine cases out of ten, the oath and the act were alike disregarded; and the few schools that were organized were founded on sectarian principles, being intended for the exclusive use of the dominant sect. In 1733, a society was established by charter, for founding schools at the public expense, in which the children of the poor should be taught the elements of literature, and instructed in useful works. But though the avowed, this was not the real object of this society, which exerted itself to undermine the Catholic religion by educating Catholic children in the principles of the Protestant faith. But this attempt at proselytism was soon discovered; and the schools were, of course, deserted by all but Protestants, and have, in fact, served as so many foci for the dissemination of bigotry rather than of really useful instruction. But though thus thrown upon their own resources, the Catholic peasantry of Ireland are by no means uneducated, at least, if we understand by education, instruction in reading, writing, and the common rules of arithmetic. But we regret

to have to add, that the moral character of their education has too generally been of the most objectionable description; and that, instead of improving, it has not unfrequently tended to debase and pervert the mind, and to familiarize the young with immorality and disorder. In these respects, however, great improvements have been effected within these few years, and the character both of the country schoolmasters and of the school books (formerly of the worst possible description), have been greatly ameliorated.

In 1815, a society in Dublin, for the suppression of vice, received a large parliamentary grant for the instruction of the poor on the principles of the established church; and, in 1819, a society for the instruction of the poor, but professing to avoid any interference with the religious opinions of the pupils, received a much larger annual grant. The latter of these associations was called the Kildare-street Society, from the place of its meetings.

These societies failed, however, in producing a general effect. The grants of public money, by which the chartered schools were chiefly maintained, were withdrawn, from a conviction of their inefficacy, and of the abuses which had crept into their management. The grants to the society for the suppression of vice, and the Kildare-street Society, were also withdrawn, in consequence of their want of success, and of their real or supposed interference with the religious tenets of the pupils. In 1833, the public money hitherto parcelled out among these associations was vested in the lord lieutenant, to be expended in promoting the education of the children of every religious denomination, under the superintendence of commissioners forming a Board of National Education. Education in the national schools is strictly confined to the common and most useful branches of secular knowledge, the religious instruction of the pupils being, in every case, left to the care of their parents and the priests of the denominations to which they belong.

The commissioners comprise some of the highest dignitaries, both of the Protestant and Roman Catholic churches; and some distinguished Protestant and Catholic laymen. They seem to have discharged their important functions with great diligence and impartiality. The schools they assist in establishing, though opposed by the bigots of both factions, appear to be making the most satisfactory progress; and will, no doubt, be productive of great public benefit. We subjoin an account of the national schools in operation in Ireland on the 30th of September, 1836, specifying the number and sex of the children by which they were attended, and the number and sex of the teachers belonging to them.

Provinces.	No. of Schools.	No. of Children.			Teachers.	
		Males.	Females.	Total.	Male.	Female.
Ulster	503	35,615	21,466	58,161	497	162
Munster	364	30,434	28,392	58,726	349	246
Leinster	450	36,041	31,326	67,450	342	132
Connaught	142	11,097	7,084	18,191	106	44
	1521*	109,167	88,364	184,371	1244	496

The Sunday School Society, formed in 1809, for the moral and religious instruction of children unable to attend schools on week days, had in connection with it, in 1836, 2975 schools, attended by 20,585 teachers, and 214,164 pupils. It is maintained wholly by voluntary contributions.

There are three collegiate institutions for instruction in the higher departments of science and literature; Trinity College, Dublin, the only university entitled to confer degrees in all the faculties, the Roman Catholic College, at Maynooth, and the Academical Institution in Belfast; the details of each of these are given in the account of their respective localities. (*See* DUBLIN, MAYNOOTH, BELFAST.)

Poor.—Provision for the relief of disease and accidents, and for the preservation of health, is made by a board of health in Dublin, an infirmary in each county, fever hospitals in those districts most subject to that complaint, dispensaries, and lunatic asylums. These are wholly supported by assessments made by the grand juries of the counties in which they are established, except the dispensaries, which require a voluntary contribution, equal to the sum raised by county taxation. There are 10 district lunatic asylums, appropriated for the reception of patients from one or more counties, each of which contributes to its support in a fixed proportion, according to the number it is entitled to send thither. Besides the lunatics received into the asylums, there are 1586 which are distributed as follows: in houses of industry, and local asylums, 1234; in private asylums, 213; in infirmaries, 9; and in jails, 84. The state of disease, and that of sexes, as far as reported, are as follows:

	Mhota.	Ship-timp.	Cur.	Inner.	Total.	Mhota	Pem.
Total							

Notwithstanding the great natural advantages of the country, it has, as already seen, been overspread with a superabundant population, in such depressed circumstances as to be involved in the extreme of destitution on any failure of the potato crop; and there is also, at all times, much suffering, arising from the pressure of want. Down to a very recent period, there was no efficient provision for the relief of the poor, who, in consequence, had to depend wholly on private benevolence. Mendicity was practised to an extraordinary extent, and strangers in Ireland were shocked by the swarms and disgusted by the importunity of beggars of all ages and sexes, and in the most abject state of poverty, that infested the roads and public places. Such a state of things was a disgrace to a country pretending to be civilised; but discreditable as it was, it could not be materially improved without instituting a compulsory provision for the support of the poor, which was long successfully resisted, through the prevalence of false and unfounded theories with respect to its operation in this country. At length, however, sounder opinions gained an ascendancy; and parliament became impressed with the conviction that it was indispensable, in order to preserve the tranquillity of the country, to a season of scarcity, to make some more effectual provision for the support of the poor. This has been done by the act passed in 1838, which introduced the principle of compulsory assessment for the poor into Ireland; and which, while it will serve to protect the population from falling a sacrifice to the extremity of want, will be a new and powerful motive to the landlords to oppose the splitting of farms, and to take a greater interest than they had hitherto done in the condition of the cottiers and others inhabiting their estates. In both these respects, the compulsory assessments, when once brought fully into operation, will, we doubt not, be eminently useful. This new system is placed under the control of the Poor Law Commissioners for England, and several unions have been formed, and various workhouses have been, and others are in the course of being erected.

Races, Character, and Condition of the People.—It seems to be admitted on all hands that the first inhabitants of Ireland, of whom history has preserved any account, belonged to the great Celtic family. Much ingenious conjecture has been expended on the question whence Ireland derived her earliest colonists; and the claims of Britain, France, Spain, Scythia, and even Troy, to the honour of being the mother country of the Irish, have all been supported with some learning and much confidence. We shall not enter on this slippery arena; but shall content ourselves with observing that, owing to the greater proximity of Britain to the continent, it is most probable that she was peopled before Ireland; and the latter being nearer to Britain than to the continent, it is, for the same reason, most probable that she was either wholly peopled from Britain, or principally from her, but partly also from Gaul.

Though there be no direct evidence of the fact, it may, perhaps, be inferred that Ireland was visited at an early period by Phœnician, or rather Carthaginian ships; but, in those days, this must have been a long and perilous voyage; and there are no grounds whatever for thinking that it was of common occurrence, or that the Phœnicians ever made any settlement in the country.

The Irish belong to what is called the Gaelic division of the Celtic family; having, as is supposed, emigrated from Britain when the latter was invaded and settled by the Cimbri or Northern Celts. About the period when the Romans withdrew from Britain, a tribe called the *Scoti* began to acquire a preponderating influence in Ireland, which, from the 5th to about the 11th century, was thence called *Scotia*. But about the latter period this tribe, having effected a settlement on the W. coast of N. Britain, its name was transferred to that country, which still retains it, and Ireland again recovered its old name of Hibernia, Ierne, or Ireland. The greatest diversity of opinion exists, and as almost impenetrable obscurity hangs over every circumstance connected with the establishment of the Scoti in Ireland. Colonists from Belgium are known to have settled in it, and Pinkerton supposes that they were the progenitors of the Scoti; but this is disputed by Moore and others, who contend that the settlement of the Scoti in Ireland is comparatively recent; and that they were of Scandinavian origin.

But though these Belgian or Scandinavian immigrants succeeded in obtaining an ascendancy in parts of Ireland, they were not sufficiently numerous to make any considerable change in the language, character, or institutions of its Celtic inhabitants. "The conquering tribes themselves, one after another, became mingled with the general mass, leaving only in those few Teutonic words, which are found mixed up with the native Celtic, any vestige of their once separate existence." (*Moore's Ireland*, i., 56.)

The number of English settlers in Ireland was long inconsiderable. Till the plantation of Ulster, in the reign of James I., they were mostly, indeed, confined to the E. and S.E. counties; where, though they had partially changed the language, they had effected comparatively little change in the habits and manners of the people. The pop. of Connaught, and generally of all the western end of a large portion of the other parts of the island, may, even at this day, be considered as of nearly pure Celtic origin; and in several of the remoter districts Celtic is now the ordinary language of the common people. And, notwithstanding the differences that may easily be traced in different parts, from the intermixture of English and Scotch blood, the entire population has a peculiar and distinctive character, that is not to be mistaken. It may, in general, be said of the Irish, that they are ardent in their affection, credulous, vain, fond to excess of flattery, irascible, easily influenced by sudden impulses, and usually in extremes. They want not merely the foresight and prudence, but also the resolution and steady perseverance of the English and Scotch; and though their bravery be unquestionable, and they will undertake anything, they are apt, if they do not succeed at the first onset, to become dispirited, and to despond. They are eminently witty, hospitable and social; though often pusillanimous, prodigality is one of their distinguishing traits; as is their light-hearted, contented disposition: but this frequently degenerates into thoughtlessness; and, how advantageous soever in some respects, by disposing them to be satisfied with existing circumstances, tends to hinder their making any persevering and well-concerted efforts for their improvement.

Dr. Cramps[*], in his valuable essay on the employment of the people, has the following statements with respect to the character of the lower Irish: "Two leading and naturally allied features in the character of the lower Irish are idleness and inquisitiveness, especially when hired and employed to perform the work of others. The moment an overseer quits them, they inevitably drop their work, take snuff, and fall into chat as to the news of the day; no traveller can pass them without diverting their attention from the business in hand, and giving rise to numerous surmises as to his person, errand, and destination. The most trivial occurrence, especially in the sporting line, will hurry them, unless restrained, from their occupations. Even the sedentary manufacturer will, on such occasions, quit his employment. Nothing is more common than to see a weaver in the N. start from his loom on hearing a pack of hounds, and pursue them through a long and fatiguing chase. A tendency to pilfering and theft is very predominant among them, and connected with this vice is the prevalence of low cunning and lying; and, as their accompaniment, may be mentioned a fawning flattery. The blunt honesty, the bold independence of the English yeoman, are wanting; and in their stead too generally substituted the petty dishonesty of the vassal, the servility and artifice of the slave. Drunkenness is an evil of considerable magnitude is the mainspring of national vices. It is one to which the lower Irish are peculiarly addicted, and that from which the most serious obstructions arise to their industry and employment. That vile beverage, whiskey, so cheaply purchased, and so generally diffused, affords them an easy opportunity of gratifying this destructive passion. As one consequence of the general prevalence of ebriety, the lower Irish are remarkably riotous. I do not here so much allude to Whiteboyism, and other public disturbances, which owe their origin chiefly to other causes, as to their quarrels among themselves. Their fairs are frequently the scenes of confusion, riot, disturbance, and bloodshed. Combinations, too, risings, and outrage among tradesmen, are far from unusual, and on pretexts that are truly ridiculous. The Irish are, also, to a remarkable degree, lawlessly inclined. It is well known that instead of being anxious to apprehend offenders, or to assist the execution of the law, they are, in general, ready to give the former every assistance to escape; and to resist the latter, unless awed by superior force." (*Essay*, p. 170-175.)

We believe that this, though not a very flattering, is a perfectly fair statement; but the defects of national character, specified by Dr. Cramps, mostly originate in circumstances that either have been, or admit of being obviated. Drunkenness; happily, is now, one should think, in a fair way of being expunged from the list of Irish vices; and with it will disappear the riots and disturbances to which it gave birth. The idleness of the Irish is, as already stated, a consequence of the minute division of the land, and of the impossibility of its occupiers finding any regular or

[*] An intelligent physician at Limerick, whose work received a prize awarded by the Royal Irish Academy for the best essay on the subject of which it treats.

continuous employment. Irish labourers in England, when employed at piece-work, are remarkably industrious; and so, no doubt, they would be at home were they accustomed to constant work at fair wages. Their proneness to combination and outrage, and their readiness to obstruct the course of law, and to assist the escape of malefactors, are the natural consequences of centuries of oppression and misgovernment. Down to a very recent period the native Irish had not, and could not be expected to have, any confidence in the law. They were, in fact, a proscribed and enslaved race, among whom it would have been preposterous to look for "blunt honesty," and "bold independence." But though the "oppression and extortion" to which the Irish were formerly subject have wholly disappeared, their effects will, it is to be feared, be long visible. They can only, indeed, be removed by slow degrees; by government pursuing a consistent and impartial course; placing the Catholics on a level with the Protestants, in respect of religious endowments as well as of civil rights; diffusing sound instruction; discouraging agitation, and enforcing the empire of the law; and adopting every practicable method for preventing the further splitting of the land, and for promoting its consolidation into larger farms.

Wages in Ireland vary from about 1s. to about 6d. a day; but at neither rate is employment constant, and in parts of the country half the labourers are all but unoccupied for nearly half the year. Under such circumstances, it is needless to add that their food, clothes, &c., must, speaking generally, be of the most inferior description. In these respects, however, there are some material differences; and in the N.E. and Eastern counties, but especially the first, the condition of the peasantry is much superior to what it is in the S.W. and West.

We subjoin, from the Report on Railways*, the following statements with respect to the condition of the pop. in the N.E., S.E., E., and W. divisions of the country.

"In the first (N.E. division), they are better lodged, clothed, and fed than in the others: the wages of labour are higher, being, at an average, about 1s. per day; and their food consists chiefly of meal, potatoes, and milk. They are a frugal, industrious, and intelligent race; inhabiting a district for the most part inferior, in natural fertility, to the S. portion of Ireland, but cultivating it better, and paying higher rents in proportion to the quality of the land, notwithstanding the higher rate of wages.

"In the southern districts we find a population whose condition is, in every respect, inferior to that of the northern. Their habitations are worse; their food inferior, consisting at best of potatoes and milk, without meal: the wages of labour are found reduced from 1s. to 8d. per day; yet the peasantry are a robust, active, and athletic race, capable of great exertion, often exposed to great privations, ignorant, but eager for instruction; and readily trained, under judicious management, to habits of order and steady industry.

"The population of the midland (eastern) districts does not differ materially in condition from those of the south; but the inhabitants of the western district are decidedly inferior to both, in condition and appearance: their food consists of the potato alone, without meal, and in most cases without milk; their cabins are wretched hovels; their beds straw; the wages of labour are reduced to the lowest point, upon an average not more than 6d. per day. Poverty and misery have deprived them of all energy; labour brings no adequate return, and every motive to exertion is destroyed. Agriculture is in the rudest and lowest state. The substantial farmer, employing labourers, and cultivating his land according to the improved modes of modern husbandry, is rarely to be found among them. The country is covered with small occupiers, and swarms with an indigent and wretched population. It is true, that some landed proprietors have made great exertions to introduce a better system of agriculture, and to improve the condition of their immediate tenants; and a few of the lesser proprietors have made humble attempts to imitate them; but the great mass of the population exhibits a state of poverty bordering on destitution.

"The distinctions we have drawn as to the usual diet of agricultural labourers in the different parts of Ireland, are strictly applicable to those only who have regular employment. When they are out of work, which is the case in many places during three or four months of the year, the line is not so easily perceived. Then a reduction in the quantity as well as in the quality of their food takes place; but still, though on a diminished scale, their relative local degrees of comfort or of penury are maintained nearly according to the above classification. In no extremity of privation or distress, have the peasantry of the northern counties approached to a level with those of the W.; while Leinster and the greater part of the S., though sometimes reduced to the lowest condition, retain, generally, even in the most calamitous periods, a shade of superiority. There are districts, indeed, in every quarter of the land, where through peculiarities of situation, or other causes, distress falls with an equal pressure upon all; but such exceptions are rare, and so limited in extent, as scarcely to qualify the foregoing observations." (Report, p. 5.)

In another part of the same report, commissioners give the following information with respect to the deterioration in the condition of the lower orders: "Among the effects of the rapid increase of pop., without a corresponding increase of employment, the most alarming, though, perhaps, the most obviously to be expected, is a deterioration of the food of the peasantry. It could scarcely be thought, indeed, that their customary diet would admit of any reduction, save in quantity alone; yet it has been reduced as to quality also, in such a way as sensibly to diminish their comfort, if not to impair their health. Bread was never an article of common use among the labouring poor; but it is now less known by them than formerly. Milk is become almost a luxury to many of them; and the quality of their potato diet is generally much inferior to what it was at the commencement of the present century. A species of potato called the 'lumper,' has been brought into general cultivation, on account of its great productiveness, and the facility with which it can be raised from an inferior soil, and with a comparatively small portion of manure. This root, at its first introduction, was scarcely considered food good enough for swine; it neither possesses the farinaceous qualities of the better varieties of the plant, nor is it as palatable as any other, being wet and tasteless, and, in point of substantial nutriment, little better, as an article of human food, than a Swedish turnip. In many counties of Leinster, and throughout the provinces of Munster and Connaught, the 'lumper' now constitutes the principal food of the labouring peasantry; a fact which is the more striking, when we consider the great increase of produce, together with its manifest improvement in quality, which is annually raised in Ireland, for exportation and for consumption, by the superior classes." (p. 81.)

This certainly is a very unfavourable statement; but it is not possible that wealth should increase in the hands of the upper and middle classes, without the lower ultimately participating in its advantages.

History.—The early accounts of Ireland are singularly disfigured by fable. It was not invaded by the Romans, whose knowledge of it could, therefore, be derived only from the reports of the Britons, or of natives of Ireland in Britain. The fair presumption, however, is, that its inhabitants were then more barbarous than even those of this island.[†] In the 5th century Christianity was introduced into Ireland by St. Patrick, a native of N. Britain, who, in his youth, had been carried a captive into Ireland. Along with the gospel the British missionaries introduced the letters and learning of Rome; and a school founded at Armagh, not long after, became famous in most parts of Europe; but it would be as inconsequential to infer, from the fact of this and a few other schools existing in the country, that it was then distinguished by literature and civilisation, as it would be to allege that such was the case with the Western Islands, and the adjacent parts of the mainland of Scotland, in the 8th century, because there was then a celebrated monastery and school in Iona!

The accounts of the political state of Ireland, previously to the English invasion, are obscure and contradictory. This much, however, may be gleaned from them, that the island was parcelled out into a number of semi-independent states, which sometimes did, and sometimes did not, acknowledge their dependance on a chief prince or king of all Ireland. Incessant hostilities were waged by the petty sovereigns against each other, which were not even interrupted by the invasion of the Danes in the 9th century. The latter, in no very long space, became masters of the greater part of the coasts of the island; and occupied the ports of Dublin, Wexford, Waterford, and Cork, where they were taken by the English.

The successors of the petty sovereigns, or to the chiefs of clans or septs, were called *tanists*, and were generally elected from the family or kindred of the reigning prince or chieftain during his lifetime. Females were excluded from the succession, and minors were never chosen as tanists; the object being to have a prince of mature years always at the head of the seigniory or clan, who might be able to direct their operations, and to defend them from hostile attacks. The laws of the Irish were such as might

44

† Pomponius Mela, who has given an accurate an account of the soil of Ireland, and of the richness of its pastures, says, *Cultores ejus inconditi sunt et omnium virtutum ignari* (lib. iii., cap. 6). Strabo (lib. iv.) gives some extraordinary details respecting the Irish, which, however, he does not state of his own authority, but merely as having been reported to him.

be expected to prevail among a rude and barbarous people; and were administered in the open air by hereditary judges, denominated brehons. The most atrocious crimes might be compounded for by the payment of an eric, or fine; and, in all cases a considerable portion, and in some cases the whole of the fine "went to the lord, or chief of the sept, his interest obviously led him to encourage rather than to repress crime! The laws with respect to the succession to fixed property were such as would have alone served to extinguish all industry. "Through the whole country," says Leland, "the tenure of lands determined with the life of the possessor; and, as the crimes or misfortunes of men frequently forced them from one tribe to another, property was eternally fluctuating, and new partitions of lands made almost daily. Hence the cultivation of lands was only in proportion to the immediate demands of nature, and the tributes to be paid to superiors." (*Hist. of Ireland*, Introduct., p. 34.)

A people with such institutions could not be otherwise than barbarous; and such, in fact, they were. They had made little or no progress even in the most necessary arts; and were, with few exceptions, entire strangers to civilisation and refinement. "Neither was it possible to reform the evil customs that prevailed among the Irish, without altering their government; nor could that be accomplished by any other means than by their being subjected to some more civilised foreign power." (*Lyttleton's Henry II.*, v. 38.; where the reader will find an excellent account of the state of Ireland previously to the English invasion.)

Soon after the English conquest effected by Henry II., in 1171, the island was divided by John into 12 counties. But, though the king of England received the submission of the Irish chieftains, and was nominally lord of Ireland, his authority was, for a lengthened period, only partially recognised. The native families of O'Conor, O'Neil, O'Melaghlin, Byrne, and O'Toole, still asserted, and, to a certain degree, exercised sovereign authority in Connaught, Ulster, and part of the midland districts. Even in Leinster and Munster, where the English were principally settled, and which had partially adopted the laws and constitution of England, the sovereign authority was far from being generally or firmly established. The allegiance of several of the great feudal barons, who held extensive tracts of land, was frequently little better than nominal. The English families of De Burgh in the W., of Desmond in the S., and of Butler in the central parts, adopted the manners of the natives, and often became the declared and most dangerous enemies of their mother country. At one time there were nine counties palatine, with independent jurisdiction, in the part of the island subject to England, and distinguished by the name of the pale. The miseries resulting from the interminable disorders inseparable from such a state of things, were increased in 1315 by an invasion of the Scotch, under Edward, brother of Robert Bruce. He overran the greater part of the country, but was finally defeated and killed near Dundalk. The resources of the country were also wasted in subsidies, and its youth carried away to fight the battles of their masters on the continent, or in England, during the wars between the houses of York and Lancaster. After the death of Richard III., and the accession of Henry VII. had terminated this sanguinary struggle, Ireland was chosen by the defeated party of the Yorkists as a theatre on which to commence a system of operations for the dethronement of the new monarch. In consequence, Lambert Simnel was sent thither by the Duchess of Burgundy, as the descendant and representative of Edward IV. His title was acknowledged by the Anglo-Irish, and he was crowned in Dublin with all the ceremonies attendant on the inauguration of the ancient Irish sovereigns. A similar, though less vigorous effort was afterwards made in favour of Perkin Warbeck, whose title was also acknowledged in the S. of Ireland.

In 1495, a parliament assembled at Drogheda, under the presidency of Sir Edward Poynings, then lord-deputy, passed some very important statutes. By one of these, afterwards well known in Irish history by the name of "Poyning's Law," effectual provision was made for maintaining the ascendancy of the government of England over the legislature of Ireland. With this view it was enacted, that no parliament should in future be holden in Ireland without license from the king; and that no bill or draft of a law should be submitted to its consideration, without having been previously sent over to England by the Irish government for the approval, alteration, or rejection of the king; so that the power of the Irish parliament was thus, in fact, limited to the mere acceptance or rejection of bills approved or modified by the English government.

This act was much and justly complained of at a later period; but, when passed, it was a decidedly popular measure. Parliaments had previously been, for the most part, the mere instruments of the faction that happened to

be ascendant at the time; so that their enactments were often conflicting, and the administration wanted consistency. Poyning's law obviated, in some measure, these defects; and parliament henceforth became dependant rather on the government of England than on any particular faction or party in Ireland.

Early in the reign of Henry VIII. the spirit of insurrection broke out in a formidable shape. The chief authority had previously been exercised for a lengthened period by the rival families of the Fitzgeralds and Butlers, whose heads were the Earls of Kildare and Ormond. The former of these noblemen was at this period lord-lieutenant. On being summoned to England, to answer charges brought against his government, he appointed his son, Lord Thomas Fitzgerald, his deputy. The latter, on a false rumour of his father's execution in London, not only threw up the reins of government, but declared himself an open enemy to the English monarch, ravaged the pale, and laid siege to Dublin, where he was repulsed by the gallantry of the citizens. Having soon after surrendered to Lord Grey, the new lord-lieutenant, he was sent prisoner to England, where he explained his offences on the scaffold, along with several of his near relations, who, though unconnected with his acts, were unjustly implicated in their consequences. The introduction of the Reformed doctrines, which was effected with equal violence and contempt for the prejudices of those within and without the pale, brought a new element of discord into Ireland. The native Irish were devoted adherents of the church of Rome. Their hostility to the new doctrines did not, however, display itself openly during the reign of Henry, who, about this time, changed his title of lord to that of king of Ireland, nor in the reign of his protestant successor, Edward VI.; but it broke out with unrestrained fury in that of Elizabeth. O'Neil, who possessed nearly the whole of Ulster, instigated by the court of Spain, hoisted the standard of rebellion. He was supported by a Spanish armament, which took possession of Kinsale, without, however, being able to maintain itself in that position. After a lengthened contest O'Neil was forced, by the energetic and prudent measures of Lord Mountjoy, to an unconditional submission; and his subsequent flight from Ireland, on the imputed charge of another insurrection, terminated the war. Ulster was soon after divided into counties, and planted with numerous bodies of English and Scotch settlers, which laid the foundations of the improvement of that province, and gave it a distinctive character. The reign of James I., and the earlier part of that of Charles I., formed a period of undisturbed tranquillity. But the disputes between the latter and the English parliament afforded the Irish a flattering though fallacious prospect of regaining their independence and re-establishing their religion. To effect this object, an insurrection was secretly organized, on a very extensive scale, embracing, not only the native Irish, but many Roman Catholic families of English descent. This formidable conspiracy broke out in 1641. The treachery of one of the conspirators prevented Dublin from falling into their hands; but the insurrection broke out simultaneously in Ulster, and soon after spread into most other parts of the country. The most horrible excesses were committed by the conspirators, which were sometimes fearfully retaliated: and the country continued to be a prey to all the horrors of civil war till 1649, when Cromwell appeared in the field, at the head of a well-disciplined and powerful army. Having taken Drogheda by storm, he delivered it up to military execution; and such was the terror inspired by the fate of the city, that almost all the strongholds belonging to the party of the Catholics soon after fell into his hands, and the English supremacy was, for the first time established in every part of Ireland. The confiscations that followed Cromwell's success were upon so vast a scale that about *four fifths* of the soil was transferred to new proprietors, either parliamentary soldiers, or speculators, called adventurers, who had advanced money to carry on the war.

After this tremendous visitation Ireland continued tranquil, and began to advance considerably in prosperity, till the events connected with the revolution of 1688 again made it the theatre of fresh and sanguinary contests. After the flight of James II. from England, he landed, with a view to retrieve his fortunes, in Ireland, where he was received with open arms by the Catholics; and having brought with him from France a number of experienced troops and officers, partly Irish and partly French, he soon found himself at the head of a powerful army. Luckily, however, he was wholly without the talents necessary to ensure success in such an enterprise. The battle of the Boyne, on the 1st of July, 1690, gained by William III., turned the scale completely in favour of the latter; and the battle of Aughrim, on the 12th of July, 1691, when the British under Ginkell, afterwards earl of Athlone, obtained a decisive victory over the troops of James II., commanded by St. Ruth, who fell in the action, was the last great effort made by the Irish to achieve their independence. The remains of the Irish forces, having retreated to Limerick, capitulated under conditions embodied in the famous convention called the treaty of Limerick. We have already noticed the violation of this treaty. It is due, however, to the memory of our great deliverer, William III., to state that he was no willing party to its violation. This is entirely to be ascribed to the intolerance of the English and Irish protestants, who, flushed with victory, did not hesitate, despite the stipulations to the contrary in the treaty, to trample the Catholics under foot, and as far as possible to exterminate their religion. "By the total reduction," says Mr. Burke, "of the kingdom of Ireland, in 1691, the ruin of the native Irish, and in a great measure, too, of the first races of the English, was completely accomplished. The new interest was settled with as solid a stability as any thing in human affairs can look for. All the penal laws of that unparalleled code of oppression, which were made after the last event, were manifestly the effects of national hatred and scorn towards a conquered people whom the victors delighted to trample upon, and were not at all afraid to provoke. They were not the effects of their fears, but of their security. They who carried on this system looked to the irresistible force of Great Britain for their support in their acts of power." (*Letter to Sir H. Langrish*, p. 44.)

The violation of the treaty of Limerick being accompanied by the most extensive confiscations, and followed up by the enactment of the penal code, completed the prostration of Ireland. There being no longer any means of rising, nor even security at home, the aspiring Catholic youth sought employment and distinction in the service of France, which, for a lengthened period, drew large supplies of recruits from Ireland. Hence, by a singular contradiction, the same revolution that established freedom of conscience and a liberal system of government in England and Scotland, established an odious despotism and persecution in Ireland. In the words of Mr. Burke, "it established, in defiance of the principles of our revolution, the power of the smaller number, at the expense of the religious liberties of the far greater, and at the expense of the civil liberties of the whole."

But, as already stated, the penal code failed to effect its object; and, instead of being exterminated, the Catholics gradually acquired a still greater numerical superiority. At length, in the earlier part of the reign of George III., the rigour of the code began to be abated, and the Catholics ceased to be regarded as mere *feræ naturæ*.

One of the most curious chapters in Irish history is that connected with the embodying of the volunteers in 1782, and the revolution that was soon after effected in the constitution of Ireland. The difficulties in which Great Britain was then involved having occasioned the withdrawal of the greater number of the troops from Ireland, rumours were propagated of an expected invasion of the island by the French; and, to meet this contingency, the Protestants of Ulster and other parts took up arms, and formed themselves into volunteer corps. These bodies soon became sensible of their strength; and having appointed delegates and concerted measures, they proceeded to set about reforming the constitution. In this view they published declarations to the effect that Ireland was a free and independent kingdom, and that no power on earth, except that of the king, lords, and commons of Ireland, could legally enact laws to bind Irishmen. These declarations, which struck a direct blow at the superiority hitherto claimed and asserted by the British parliament, might, and most probably would, at another time, have been successfully resisted. But Great Britain, being then engaged in a desperate contest with her revolted colonies, and with almost all the great European powers, prudently made the concession demanded by the Irish volunteers; and the *Independence of Ireland* was proclaimed amid the most enthusiastic demonstrations of popular rejoicing.

In truth, however, this independence was apparent only. The wretched state of the elective franchise in Ireland was totally inconsistent with any thing like real independence; and so venal was the Irish parliament, that any minister, how unpopular soever, had no difficulty in securing a majority in that assembly. Hence the anticipations in which the more sanguine Irish patriots had indulged were destined soon to experience a most mortifying disappointment; and this, and the hopes inspired by the French revolution, terminated in the rebellion of 1798, which was not suppressed without a repetition of the former scenes of devastation and bloodshed.

The British government at length wisely determined to effect a legislative union between Great Britain and Ireland, and to suppress the separate legislature of the latter. This measure, notwithstanding a strenuous opposition, was happily carried, and took effect from the 1st of January,

1800. And, unless it were resolved or wished to put an end to all political connexion between the two countries, nothing could be more inexpedient and absurd than the existence of a separate independent legislature for Ireland. Perpetual jealousies could not have failed to arise between it and the legislature of Great Britain, which must necessarily in the end have led to estrangement, and probably separation. A legislative union was the only means of obviating these and other sources of mischief: its repeal would make Ireland a theatre for all sorts of projects and intrigues, and it would be sure to be followed, at no distant period, by the dismemberment of the empire. Its maintenance should, therefore, be regarded as a fundamental principle of policy: and, to give it permanence and stability, every effort should be made to remove all just grounds of complaint on the part of the Irish people, and to make the union one of national interest and affection, as well as of constitutional law.

IRONDEQUOIT, p. t., Monroe co., N. Y., 5 m. N. Rochester, 228 m. W. by N. Albany, 373 W. Bounded N. by lake Ontario, W. by Genesee river, E. by Irondequoit bay. It contains two grist-mills, three schools, 186 scholars. Pop. 1252.

IROQUOIS, county, Ill. Situated in the E. part of the state, and contains 1488 sq. m. Drained by Iroquois river, and its tributaries, sugar and spring creeks. Kankakee river, crosses its N. part. It contained in 1840, 2258 neat cattle, 9453 sheep, 7805 swine; and produced 15,864 bushels of wheat, 116,790 of Indian corn, 41,498 of oats, 8981 of potatoes, 17,464 pounds of sugar. It had seven stores, two grist-mills, three saw-mills, one tannery, one pottery; seven schools, 110 scholars. Pop. 1695. Capital, Montgomery.

IRKUTSK, GOVERNMENT OF. *See* SIBERIA.

IRKUTSK, a city of Asiatic Russia, cap. of Eastern Siberia, on the Angara, at its confluence with the Irkut, about 30 m. from the N.W. shore of Lake Baikal, 368 m. S.E. Krasnojarsk, and 1450 m. in nearly the same direction from Tobolsk; lat. 52° 16' 30" N., long. 104° 18' 45" E. (*Ermann, Reise um die Erde*, ii. 419, &c.) Pop., with its garrison, about 15,000. It is situated in a wide plain, 1940 ft. above the level of the sea; the mean temperature of the year being—0·3 R., or rather below the freezing point. The Angara, which is about 1600 ft. broad at Irkutsk, divides the city into two nearly equal parts. It is fortified and defended by a citadel, and has four suburbs. Of about 1900 private houses, only 50 are built of stone; the rest are chiefly of wood, or faced with painted planks. The streets are broad, but altogether unpaved; from the solidity of the ground, however, they are not dirty; and Erman says, that, in many respects, Irkutsk is much more agreeable than Tobolsk. It has 23 churches, 19 of which are constructed of stone; an exchange, also a stone edifice, and a good bazaar with numerous shops. The Baikal admiralty house and building docks on the Angara, medical college, gymnasium, and *gostinoi* of the Russo-American Company, are said to be worthy of a European city; the government-house, theatre, several convents and hospitals, and a prison, are among its other public edifices. It is the seat of an archbishop, and of a Russian governor, whose authority extends over the immense provinces of Irkutsk, Yakutsk, Okhotsk, Kamtschatka, and Russian America, including Bodega and the other settlements on the coast of California, distant nearly 1200 long.? It has numerous educational establishments, including, besides the gymnasium, with a library of 5060 volumes, an episcopal seminary, high school of navigation, with classes for instruction in the Tartar, Chinese, and Japanese languages; normal, secondary, Lancastrian, and other schools, and a cabinet of mineralogy. It has an imperial factory of woollen cloth for the supply of the troops in Siberia, manufactures of linen and other piece goods, glass, hats, soap, leather, &c.; and is the residence of numerous artisans in the different trades common in Europe. It is the great entrepôt for the commerce of N.E. Asia, importing tea, rhubarb, fruit, paper, silks, porcelain, and other manufactured goods from China by way of Kiachta, and furs, &c. from Kamtschatka, the Aleutian Islands, and Russian America; which articles are here exchanged for European goods sent from Petersburg and Moscow by way of Tobolsk. It has also some trade with Bokhara and Khokan. The total annual amount of its commerce is estimated at 4,000,000 paper roubles (or francs), one fourth of which has sometimes been transacted at its annual fair in June. (*Erman, reise um die Erde*, ii.; *Ritter, Asien Erdkunde*, ii. 129–134.; *Stein. Geog.*; *Dict. Geog.*, &c.)

IRRAWADI (*Erivati*, "the Great River"), an important Asiatic river, the principal in India beyond the Brahmaputra. It has its sources near the E. extremity of the Himmalaya range in Thibet, about lat. 28° N., and long. 97° 30' E., not far from the sources of the Lohit, a principal branch of the Brahmaputra. With the exception of who stretches to the W., at Bhamo and Ava, it flows generally S. through the centre of the Birman empire, which it traverses in its entire length, till it falls, by numerous mouths into the bay of Bengal (or rather the Eastern Ocean) between Cape Negrais and the Rangoon river, in about the 10th deg. of N. lat., and between 95° 25' and 97° E. long. Its course may be estimated at about 1200 m., during which it passes through 12 degs. of lat. It receives at Yandabo, lat. 21° 43' N., long. about 95° E., its principal tributary, the Ning-thee, or Kyen-dwen, from the N. Its delta commences about lat. 17° 45'. This is a vast alluvial plain, about 120 m. in length, N. and S., and wherewithal about as many miles across, intersected by a vast number of arms of the river, that frequently interlace each other. Of its numerous mouths, the Rangoon and Bassein rivers, forming respectively the E. and W. boundaries of the delta, are the principal. Most of its mouths are navigable for large craft; and those of Bassein and Rangoon for vessels drawing five fathoms' water. The harbour of Negrais, formed by the mouth of the river of same name, is said to be, without exception, the most secure in the bay of Bengal. The Bassein branch, which may be considered the proper continuation of the main stream of the Irrawadi, is about 700 yards in width at the point where the Rangoon river separates from it. From the apex of the delta to Yndaa above Ava, the breadth of the Irrawadi is seldom less than 1 m., and often 4 m. It may be ascended as far as Ava, at all seasons, by vessels of 200 tons; and in the rains they may proceed to the Mogaung river, a sailing distance of about 600 m. from the sea. Above Yndaa, the river suddenly contracts to 150 or 200 yards in breadth. It is navigable for canoes up to Bhamo; but in the dry season it is in many parts dangerous, from its passing over rocky ledges and through precipitous defiles. About 30 m. from its source, it has been observed with a width of 80 yards, during the dry season.

The current is not, in general, remarkably rapid; even above the Mogaung, the Irrawadi in the dry season, flows only at the rate of about 2 m. an hour. (Malcolm, i., 171.) But in the inundations, from June to Sept., it flows so rapidly that, in the delta, its current would be too powerful for boats to stem were it not for the assistance of the S.W. monsoon, which sets in the opposite direction. During its inundation it has a breadth of about 1 m. above Bhamo, and in some places below Ava of from 4 to 6 m. At the former place its rise is as much as 30 ft. at Ava about 30 ft., at Prome about 30 ft., and in its delta 10 ft. The latter region becomes at that period almost an uninterrupted expanse of water, it being at ordinary times little above the level of high tides. The quantity of water discharged by the Irrawadi, as compared with that discharged by the Ganges, is roughly estimated by Capt. Hannay, in the Asiat. Journ. of Bengal, as 1 to 1·52. In the plain of Pegu, and in the undulating country through which the Irrawadi flows in the middle part of its course, it incloses a great number of islands and sandbanks; though these, in various parts, would seem, from a comparison of the statements of Symes with those of Crawford to be less numerous than formerly. In the upper parts of its course, on its left or E. bank, the Irrawadi receives some large affluents, as the Shouaen Kha, Pho-loung or Bhamo river, Loung-tchaon, &c. Its chief affluents on the opposite side are the Mogaung and Ning-thee, which join it about the middle of its course. The last, as already stated, is its principal tributary; and after its junction, the Irrawadi receives no stream of any importance. Sabaing, the present metropolis, and Ava and Amerapura, former capitals of the Birman empire, Bhamo, the great mart for the Chinese trade with Birmah, Yandabo, Pagan, and Prome, are situated upon the main stream, and Rangoon and Bassein upon the branches bearing their names. Besides these cities, numerous towns and large villages are built on or near the banks of the river, the great mass of the Birman population being accumulated on the Irrawadi, leaving the rest of the country, in great part, an uninhabited desert.

The Irrawadi is to the Birman empire what the Nile is to Egypt, the source of life and abundance, and the main artery and great commercial highway of the country. "The number of trading boats on the river is astonishing. We pass scores every day, and sometimes hundreds; the largest of them carry 10,000 or 12,000 baskets of uncleaned rice, the smaller 200 or 300. Their chief lading seemed to be rice, salt, and gna-pee. In ascending they are for the most part drawn by the crew with a rope upon the bank, or propelled by sorting-poles; sailing only when the wind is fair, and neither too strong nor too weak. They are generally from three to four months in ascending from the delta to Ava.

"The boats on this river, though of all sizes up to 200 tons, are of but two general descriptions. All retain the canoe shape, sharp at each end. Large boats have one mast and a yard of long slender bamboo to which is sus-

pended a capacious sail. The sail is made in sections, the entire ones only being used in strong winds, and the others added at the sides when necessary. Sometimes a small sail is temporarily fastened above the yards to the rope, by which it is controlled. The deck extends from 5 to 16 ft. beyond the sides with large bamboos fastened beneath, making at once a platform for the men, when using their setting-poles, &c., and an outrigger to prevent their upsetting. The vessel itself is wholly covered with a regular Birman house, well thatched, which carries part of the cargo, and furnishes cabins to the family and boatmen. Over the roof is a platform, on which the men stand to work the sail. They are manned by from 15 to 25 or 30 men, and sometimes 40 or more." The smaller-sized vessels are of an elongated shape, like the foregoing, and do not merit a particular description.

"No one can ascend the river without being impressed with the hardihood, skill, energy, and good-humour of the Birman boatmen, and the happy adaptation of their boats to the navigation. In ascending, much of the way must be accomplished by setting-poles. For these they use straight bamboos, of a species which is almost solid and very strong. The end is applied not to the front of the shoulder, as with us, but above the collar-bone, or on the top of the shoulder. Bending forward till their heads touch the deck, they bring the resistance perpendicular to the spine, and their powers far greater power than is possible by our mode. When but slight exertion is required, the pole is applied as with us." (Malcolm's Trav. in S.E. Asia, i., 90, 91, 92, 97.)

Near the Irrawadi, in the prov. Sarawadi, are contained neat forests, covering the hill-ranges bounding the valley. Petrifactions of wood, bones, &c., are common along this river; and Mr. Crawford collected on its banks a great number of fossil remains, including those of two species of mastodon, the rhinoceros, hippopotamus, tapir, hog, ox, deer, antelope, gavial, alligator, emys, and trionix. (See Trans. of the Geolog. Soc., and Appendix to Crawford's Embassy, &c.) Coal (anthracite) has been discovered along its course, and about 40 m. S. Pagan are some rich petroleum wells on the E. bank, respecting which see Bennett, i., 373. (Pemberton, Rep. on the E. Frontier; Crawford, Geo. Symes, &c., in Asiat. Trans., &c.; Malcolm's Trav. in S.E. Asia; Asiat. Journ. of Bengal, &c., passim.)

IRVINE, a royal and parl. bor., seaport and market town of Scotland, co. Ayr, on a rising ground on the N. bank of the river of the same name, the estuary of which forms its harbour, 22 m. S.W. Glasgow, and 12 m. N. by W. Ayr. Pop. of the town and par., 1898. A suburb lies across the the S. of the river, which is connected with the town by a bridge, the widest and handsomest in the co. There are other suburbs, not in the royalty, but comprised, since 1898, within the parl. bor. The parish church, situated on a rising ground between the town and the river, with a handsome spire, is the most striking building in the bor. There is a chapel of ease, as also chapels belonging respectively to the Associated Synod and relief. On the N. of the town an academy was erected in 1814, at an expense of £3000, of which the burgh contributed £1633 4s. 6d.; the remainder being raised by public subscription. This seminary, which embraces all the branches of a learned and commercial education, has fully realised the object of its founders. There are various other schools; also several libraries, and a newsroom. Eglinton Castle, famous in the sporting world for the "tournament" held there in 1839, is in the immediate vicinity of the bor. There is a regular custom-house establishment in Irvine; coals are its chief article of export, of which from 40,000 to 60,000 tons a year are shipped, chiefly to Ireland. There are about 150 weavers, working in connexion with the Glasgow manufacturers, or for local consumption. There are three branch banks in the town. Irvine was created a royal bor. by Robert Bruce, in 1308. The Carmelites, or white friars, had a monastery here, founded in 1412, by Eanken of Fullarton. The bor. unites with Ayr, Campbelton, Oban, and Inverary, in returning a member to the House of Commons. Registered voters in 1839-40, 234. David Dickson, author of Therapeutica Sacra, and various other theological works, and subsequently professor of divinity in the University of Edinburgh, was long minister of Irvine. John Galt, author of Annals of the Parish, and other works, was a native of this place; and Burns was for a short time engaged in business here as a flax-dresser. (Boundary Reports; Pitcairn's History of the Family of Kennedy, p. 150-154; Keith's Scottish Dictionary, p. 456.)

IRVINE, p. v., capital of Estill co., Ky., 68 m. S.E. Frankfort, 534 W. Situated on the N.E. side of Kentucky river. It contains a courthouse, jail, several stores, and 280 inhabitants.

IRWIN, county, Ga. Situated in the S. part of the state, and contains 2070 sq. m. Bounded N.E. by Ocmulgee river. Drained by Alapaha and Little rivers, and their branches. It contained in 1840, 14,380 neat cattle, 1223 sheep, 9674 swine; and produced 2202 bushels of wheat, 33,546 of Indian corn, 11,179 of potatoes, 1782 pounds of rice, 21,100 of cotton, 23,770 of sugar. It had four stores, one flouring-mill, four grist-mills, one saw-mill; one school, 83 scholars. Pop., whites, 1772; slaves, 266; total, 2038. Capital, Irwinville.

Irwin, t., Venango co., Pa., 19 m. S.W. Franklin t. Drained by Scrub Grass creek. It contains one store, four grist-mills, four saw-mills, two tanneries, one pottery; four schools, 141 scholars. Pop. 1111.

IRWINTON, p. v., capital of Wilkinson co., Ga., 20 m. S. Milledgeville, 668 W. Situated 4 m. W. of Oconee river. It contains a courthouse, jail, an academy, a Methodist and Baptist church, several stores, 35 dwellings, and about 300 inhabitants.

IRWINVILLE, v., capital of Irwin co., Ga., 105 m. S. Milledgeville. Situated a little E. of Allapaha river. It contains a courthouse, jail, and a few stores and dwellings.

ISABELLA, county, Mich. Situated in the N. part of the settled portion of the peninsula, and contains 579 sq. m. Watered by Chippewa and Salt rivers. It is unorganized, and attached for judicial purposes to Ionia co.

ISCHIA (an. Ǣnaria, Inarime, and Pithecusa), an island of the Mediterranean, belonging to the k. of Naples, 8 m. S.W. from the promontory of Misenum, and 18 m. W.S.W. Naples. It is about 7 m. in length, and 20 in circ. having an area of 21 sq. m., and a pop. of about 24,000. Nearly in its centre is M. San Nicolo or Epomeo (an. Epomeus). This, though now an extinct, was formerly an active volcano, the eruptions of which are noticed by Strabo (lib. v.) and Pliny (lib. ii., § 88); and which burst forth with great fury in 1301, since which it has been quiescent. It is 2513 ft. above the level of the sea, and the whole island falls in a gentle slope from it to the sea, except on the N., where its sides are more abrupt. Ischia obviously, indeed, owes its origin to volcanic agency, and consists wholly of volcanic matters. Its bold and rocky shores present an imposing appearance from the sea; and the favourable impression it makes at a distance is not dispelled on landing, it being remarkable both for fertility of soil, and beauty of situation. Besides a great quantity of wine, it produces olives and a variety of fruits, with wheat, maize, pulse, and excellent herbage. It is well supplied with game, especially partridges. Sulphur and other useful mineral products are abundant, and there are numerous hot springs and natural vapour baths, especially at its N.W. extremity. The inhabitants are partly husbandmen, and partly sailors and fishermen. The manufacture of straw hats, baskets, and earthenware, are carried on to some extent.

Ischia is divided into two cantons; chief towns, Ischia and Foria; the former on the E. and the latter on the W. coast. Ischia, the cap. with 3000 inhabitants, is "a pretty town of white buildings, and the residence of a bishop. A round rock, as black as if just launched out of the bowels of a volcano, forms a kind of haven by means of a causeway communicating with the town; its summit and sides are covered with houses, old turrets, and ruinous fortifications, huddled together, and accessible only on one side by a steep winding road." (Swinburne, ii., 12.) On this rock stands an old fortress, in which the last princes of the house of Aragon took refuge, when Naples was conquered by the French. This building is now used as a state prison. Foria is ill built, and without a harbour.

The poets account for the volcanic phenomena of Ischia as for those of Vesuvius and Etna, by ascribing them to the violent efforts of Typhoeus and the other giants buried below them to escape from their prison:

> Apparet gravius barriss, quæ turbine nigro
> Fumentem premit Inprum, flammeoque rebelli
> One ejectantem.   Silius Italicus, xii., lin. 147.

See also Æneid, ix., lin., 714.

Ischia was, at a remote period, colonized by the Eretrians and Chalcidians, and afterward by Syracusans sent thither by Hiero, who, however, abandoned the island in consequence. It is said, of a violent eruption of Mount Epopeus, B.C., 470.

(Besides the authorities referred to above, see Cramer's Anc. Italy, ii., 183; Rampoldi, Corografia; Dict. Géog., &c.)

ISÈRE, a frontier dep. of France, in the E. part of the kingdom, formerly included in the prov. of Dauphiny; between lat. 44° 44' 30" and 45° 53' N., and long. 4° 46' and 6° 22' E., having E. Savoy, N. the dep. Ain, and W. Rhone, Loire, and Ardèche, from all which it is separated by the Rhone, S.W. Drôme, and S.E. Hautes Alpes. Length, N.W. to S.E., about 95 m.; average breadth about 40 m. Area, 899,031 hectares. Pop. (1836), 573,645. This dep. is very mountainous, especially its S.E. part, and its scenery is in general highly picturesque.* The Alpine chains that traverse it rise in the Col de Sayies to an elevation of 11,017 ft. (3356 métres), and in the Pic de Belladone to 10,302 ft. (3140 métres) above the level of the sea. Some of the valleys are spacious, and many very fertile; that of Graisi-

vaudan, through which the Isère flows, is one of the richest in France. There are a few plains in the N. and W., and numerous lakes and marshes, but some of the latter is of any considerable size. Next to the Rhone, the chief river is the Isère, which gives its name to the dep. It rises in the E. part of Savoy, runs with a tortuous course, generally S.W., and falls into the Rhone about 5 m. N.E.E. Valence, after a course of 168 m., 106 of which are navigable. Its chief affluents are the Romanche and Drac; considerable stands on its banks. West winds predominate in this dep., and the annual fall of rain is estimated at nearly 25 inches. In 1835, the arable lands were estimated at 316,267 hectares, meadows 66,718, vineyards 27,606, forests 106,620, and heaths, wastes, &c., 171,990 do. Agriculture is backward, but improving. In 1835, 2,739,190 hectolitres of corn, chiefly wheat and rye, were said to have been harvested, being a larger supply than was produced in any of the surrounding deps., besides 10,771,200 hectolitres of potatoes, or more than double the quantity grown in any other dep. of France. But it should be borne in mind that these returns, though given in the official tables, are but little to be depended on; and are, in fact, nothing but rough approximations. The vine is pretty generally cultivated, and the produce of wine amounts to about 450,000 hectolitres a year. Chestnuts, almonds, and other fruits abound, and large quantities of ratafia, and other liqueurs, are made. The number of mulberry trees had increased nearly a third between 1830 and 1834: in 1835, 430,256 kilog. cocoons were collected. Good cavalry horses and mules are bred. In 1830, the stock of black cattle amounted to about 127,000 head; the breed is generally small, but the cows are good milkers, and some superior cheese is made. The sheep yield excellent wool, and many flocks from the surrounding deps. are sent to pasture in summer in the mountains. Poultry are reared in great numbers. In 1835, of 295,450 properties subject to the contribution foncière, 80,639 were assessed at less than 5 fr., and 31,468 at from 5 to 10 fr.; the number of large properties is a good deal below the average of the departments.

Isère is one of the richest departments of France in respect of minerals, and mining is one of the chief occupations of its inhabitants. Gold and silver mines were wrought till the commencement of the present century. At present, iron, copper, zinc, and lead are the chief metallic products; but mercury, bismuth, antimony, and cobalt are likewise obtained; as are also coal, sulphur, alum, marble, granite, gypsum, &c. There are 10 large smelting furnaces, and numerous forges and steel factories. Paper, silk stuffs and yarn, coarse woollens, table linen, sail and packing cloth, gloves, especially at Grenoble, cotton and woollen yarn, crape, straw hats, mineral acids, &c., are the other chief manufactures. Lyons is the great entrepôt for the produce of Isère. The department is divided into four arrondissements, the chief towns of which are Grenoble, the capital, St. Marcellin, La Tour du Pin, and Vienne. It sends seven members to the chamber of deputies. Number of electors (1836–39), 1731. Total public revenue (1831), 19,351,872 fr. This department abounds with remarkable natural curiosities, and Roman and other antiquities. One of its most remarkable establishments is the GRANDE CHARTREUSE, which see. (Hugo, art. Isère; French Official Tables; Guide du Voyageur, &c.)

1818. See THAMES.

ISKARDO, a commercial town of Little Thibet, on the Upper Indus, about 120 m. N.W. Leh, but at present little known. It is reported to be a large fortress of irregular construction, and the capital of a district of same name. (Burns's Trav.)

ISLAMABAD, a town of India-beyond-the-Brahmaputra, belonging to the prov. of Bengal, district Chittagong, of which it is the cap., on the river Chittagong, 8 m. from the bay of Bengal, and 134 m. S.E. Dacca. Pop. 12,000 (Malcolm), about 2000 of whom are of Portuguese descent. "The streets are in good order, and the bazaar abundantly supplied with every sort of domestic and foreign produce. The mode of building, and the general aspect of everything, is decidedly Bengalee. About 300 vessels, chiefly brigs of from 40 to 100 tons, are owned in the place, and many vessels from other places resort thither. The chief exports are rice and salt. Large Maldive boats come annually, during the fine season, with cowries, tortoise-shell, cussels, cocoa-nuts, and coir for rope; and carry away rice and small manufactures." (Malcolm, i., 134.) This town is the emporium of a great extent of country, and the resort of numerous merchants. A kind of cotton canvass is made in its neighbourhood, and vessels of considerable burden are built. Islamabad has two Portuguese churches, and a large English school established in 1818. (Malcolm's Travels in S.E. Asia.)

ISLAND CREEK, p. t., Jefferson co., O., 145 m. E. by N. Columbus, 271 W. Pop. 1867.

ISLAY. See HEBRIDES.

ISLEBOROUGH, p. t., Waldo co., Me., 54 m. E. Augusta, 643 W. It consists of a large and several smaller islands in Penobscot bay, opposite to Northport and Lincolnville. Incorporated in 1789. It has good harbours and a fertile soil, and contains one store; five schools, 287 scholars. Pop. 777.

ISLE LA MOTTE, t., Grand Isle co., Vt., 13 m. W. St. Albans, 26 m. N.N.W. Burlington. It consists of an island in lake Champlain. Chartered in 1789. It has one store. Pop. 435.

ISLE OF WHITHORN. See WHITHORN.

ISLE OF WIGHT, co., Va. Situated in the S.E. part of the state, and contains 460 sq. m. Bounded N.E. by James river, W. by Blackwater river. It contained in 1840, 6351 neat cattle, 480 sheep, 22,990 swine; and produced 4136 bushels of wheat, 291,155 of Indian corn, 29,193 of oats, 77,082 of potatoes, 30,534 pounds of cotton. It had 31 stores, 19 grist-mills, nine saw-mills, one tannery, 295 distilleries; one academy, 21 students; 20 schools, 369 scholars. Pop.: whites, 4916; slaves, 3795; free coloured, 1266; total, 9972. Capital, Isle of Wight C.H.

ISLIP, p. t., Suffolk co., N. Y., 191 S. by E. Albany, 271 W. Bounded S. by the Atlantic. Great South bay, on its S. part, contains several islands belonging to it, and abounds with fish and wild game. It contains two churches, five stores, one grist-mill, one paper-mill, two tanneries; 10 schools, 577 scholars. Pop. 1889.

ISMAIL, a strongly-fortified town and harbour of Russia in Europe, in Bessarabia, on the N. side of the Kilian arm of the Danube, about 42 m. from the Black sea, lat. 45° 21' N., long. 28° 50' 15" E. Pop. 18,093. Ismail was stormed by the Russians, under Suwarrow, in 1790, by whom it was given up to an indiscriminating pillage and massacre. But latterly it has recovered, at least in part, from this barbarous devastation. It has a considerable trade, exporting corn, hides, tallow, &c. The customhouse and quarantine are of the first class. Owing to the shallowness of the water over the bar of the Kilian mouth, vessels bound for Ismail generally enter the Danube by the Sodineh or middle mouth. (See DANUBE, in this dictionary; and Hagemeister on the Black Sea, p. 94, Eng. trans.)

ISPAHAN (Aspadana), a celebrated city, formerly the cap. of Persia, and once so extensive and populous that the Persians said of it, in their inflated phraseology, " Sefaen niepe gihen"—" Ispahan is half the world." (Chardin, iii., 3.) It is situated in the prov. Irak Adjmi, of which it is the cap., as well as of a beglier-beglik, of the same name. 211 m. S. Teheran, and 283 m. S.S.W. Bushire. Lat. 32° 25' N., long. 51° 30' E. Pop. variously estimated at from 200,000 to 30,000; the latest estimate of Morier fixing it at 60,000. This, however, is unquestionably too low; and the population, most probably, exceeds 100,000. (Ritter, ix., 48.) This city, which was at the height of its glory during the reign of Shah-Abbas, in the 17th century, now presents to the traveller, in its buildings at least, little beyond the magnificent ruins of its former greatness. It stands in the midst of an extensive plain, abundantly watered by the Zenderood, a river about 600 ft. broad; and is surrounded by groves, avenues, and spreading orchards. "Among the first objects that struck our eyes," says Sir R. K. Porter, "were the numerous noble bridges, each carrying its long, level line of thickly-ranged arches to porch-like structures, some fallen into stately ruin, others nearly entire, but all exhibiting splendid memorials of the Sefi race. The S. avenue, through which we entered the town, terminated at the great bazaar of Shah-Abbas, the whole of which enormous pile is vaulted above to exclude heat, yet admit air and light. Hundreds of shops, without inhabitants, filled the sides of this once great emporium, the labyrinths of which we traversed for an extent of nearly 2 m., till we entered the Maidan Shah, another spacious theatre of departed grandeur." (Travels, ii., 37.) This vast oblong, formerly enriched with shops, in which every commodity of luxury and splendid manufacture was exposed, is of very large dimensions, being (according to Porter) 2000 feet long, and 700 feet broad, and in the centre of each of its sides stands some edifice remarkable for grandeur or character, while the remaining parts composing the square are occupied by uniform ranges of building, once used as apartments for the nobility and officers of the Persian court, the lower part being open, and forming a noble arched walk. On the N.W. side is the great painted gate of the bazaar, on which, in former times, stood the celebrated clock of Ispahan, and on the opposite side is the Meshed-Shah, a superb mosque built by Shah-Abbas, and dedicated to Mehedi, one of the 12 Imâms. The centre of the N.E. side is occupied by another mosque, called Looft Ullah, which faces the Ali-Kapi, a noble gate, surmounted by a dome, the marble ornaments of which still remain. Above the gate is a pavilion, pointed out as the place where Shah-Abbas was wont to sit and witness the games and exercises of his troops in the Maidan; but only a few wooden columns, pieces of glass, and decayed paintings remain to attest its former beauty, as described by Chardin. The summit of the tower commands a view of the city in its whole extent, presenting a succession of narrow, unpaved streets, ruinous houses, mosques, and shapeless structures, broken by groups of various tall trees, which once made part of the gardens attached to the houses now fallen to decay. In the S. part of the city is a large tract of pleasure-ground, called the Chahar-Bagh, which consists of a series of eight gardens, or paradises, watered by canals, basins, and fountains, adorned with numerous palaces or pavilions, and enclosed with four majestic walls. In the centre of the enclosure is the palace of the Chehel Sitoon, or forty pillars, the favourite residence of the later kings of the Sefi dynasty. Its front, which is entirely open to the garden, is sustained by a double range of columns, each shooting up from the united backs of four lions of white marble; and within are several large apartments, on which all the caprice and cost of Eastern magnificence have been lavished. The walls of the saloon, in particular, are embellished with large paintings, which, without exhibiting much taste, or correctness of design, are still useful as illustrations of the manners and habits of the Persians. The suburb of Julfa, which is situated S. of the Zenderood, and connected with the Chahar-Bagh by a bridge 1009 ft. long, having 34 arches, was originally founded for a body of Armenians, whom Shah-Abbas transplanted from their own country (Julfa on the Araxes), and stationed here, with full toleration of their religion, and many valuable mercantile privileges. They were known all over the East for their manufacturing industry; and their quarter, which was inhabited exclusively by Christians, formerly comprised 13 churches, and some of the handsomest private residences and gardens in the city, the population of this industrious quarter alone having exceeded 30,000 at the close of the 17th century. At present, however, it is little more than a mass of ruins, the few remaining houses being tenanted by a population whose moral condition, according to Sir R. K. Porter, has suffered a deterioration corresponding to the decline of their fortunes. The suburb of Abbasabad, which lay W. of the city, and that of the Guebers, or fire-worshippers, on the S. side, near Julfa, are entirely destroyed.

Ispahan has, within the last 30 years, begun to revive from its desolation; and the spontaneous efforts of the inhabitants, in trying to better their condition, were ably seconded by the exertions of Hadji Mohammed Hussein Khan, the Ameen-e-doulah, or second minister of the shah, who employed his immense wealth and influence in the improvement of his native city. A new palace, near the Shelal Sitoon, has been completed, and extensive repairs have been made in the bazaars, streets, and fountains; besides which, a large tract of land, close to the river, has been enclosed to form rice plantations, the produce of which now forms an important article of commerce. The manufacture of all kinds of woven fabrics, from the most costly gold brocade or figured velvet to the most ordinary calico or coarse cotton, is pursued on an extended scale; partly on raw materials raised in the surrounding district, and partly also on silk and cotton wool introduced from Ghilan and other provinces of Persia; many hands are also employed in making gold and silver trinkets, paper and paper boxes, pencases, ornamental book covers, firearms, sword-blades (of steel, from India), glass, and earthenware. These goods are sent to all parts of the East, Ispahan being the chief emporium in Persia, and on the great line of communication between India, Caubul, and China, to the E., and Turkey, Egypt, and the Mediterranean, on the W. Its trading prosperity, however, like that of Bushire, is much obstructed by the monopolies and injudicious taxes of the government. The inhabitants of Ispahan are considered the best manufacturers in Persia, and education seems to be very general. Every one above the lowest order can read and write; and artizans and shopkeepers are familiar with the works of their favourite poets. The merchants form a distinct class: frugal, and even penurious in their habits, they seldom make any display of wealth, and are extremely wary and circumspect in their commercial speculations, owing, no doubt, to the severity of their sufferings during national disturbances, when they have been usually selected as the first victims of plunder and oppression. Their houses are mean on the outside, with low, narrow entrances, but are often fitted up internally with great luxury. These merchants, with all their affectation of poverty, have capitals embarked in trade which vary from 50,000 to 150,000 tomans, and not only control in a great degree the whole trade of Persia, but are able also, it is said, to influence prices in the markets of W. Hindostan. Owing to insecurity and bad government, the interest of money in Ispahan varies from 12 to 36 per cent. a year; and the farming population are often compelled to pay 60 per cent. for the loans required to enable them to meet the exactions of the government. If trade exist at all under this wretched system, how great would it become under a government that should establish security, and give full scope to the enterprise and ingenuity of a people who are among the most industrious in W. Asia! (See *Hægemeister; Essai sur l'Aris Occidentale*, p. 266–275.)

The origin of Ispahan is uncertain; but its position seems to identify it with the *Aspadana* of Ptolemy. Under the caliphs of Bagdad, it became the capital of Irak, and rapidly increased in wealth, population, and trade. This rising prosperity, however, received a severe check during the invasion of Timour, who took the city in 1387, and gave it up to military execution. The troops massacred 70,000 of the inhabitants, whose heads, piled on the walls of Ispahan, long attested the merciless severity of the conqueror. From this desolation the city gradually revived under the Sefis; but it did not become the residence of royalty till Shah-Abbas the Great made it the metropolis of Persia, embellished it with stately mansions, and rendered it not only a luxurious capital, but filled it with merchants, artificers, and agriculturists from Europe as well as Asia, whose united industry soon made it the great emporium of the Asiatic world. The city was at this time 24 m. in circuit, and comprised, according to Chardin, 160 mosques, 48 colleges, 1800 caravanserais, 273 public baths, and 12 cemeteries; while the population amounted, according to the same author, to 600,000 persons; but other authors raise it to 1,100,000; and the inhabitants of 1400 villages are said to have derived their subsistence from its prosperity. There can, however, be no doubt that these statements are grossly exaggerated; and, in fact, merely show that Ispahan was then a rich, flourishing, and populous city. "Its bazaars," says Sir R. K. Porter, "were filled with merchandise from every quarter of the globe, mingled with rich bales of its own celebrated manufactures;" and the shah's court was the resort of ambassadors from the proudest kingdoms of the East, as well as of Europe. This prosperity, however, was but of short duration; for, in 1722, Persia was invaded by the Afghans, and Ispahan, after sustaining a siege of eight months, during which the adjacent country was laid waste by the barbarous policy of the enemy, was reduced to its present ruinous state: the walls were so completely destroyed that all traces of them are obliterated, the palaces dismantled and robbed of all their ornaments, and the people massacred without mercy. Nadir-Shah recaptured the city in 1727, but he took no steps to restore its ancient glory. The sovereigns have resided at Teheran during the last 70 years; and Ispahan has gradually fallen to a state of decay, from which even its commercial importance has not been able to preserve it." (*Chardin*, vol. iii., passim; *Porter's Travels*, ii., 37–68; *Ouseley's Travels*, iii., 62–68; *Ritter*, ix., 40–56.)

ISRAEL, t. Preble co., O. Organized in 1808. It contains the villages of Claysburg, Fairhaven, and Morning Sun. Pop. 1547.

ISRAEL'S RIVER, N. H., rises in the White Mountains, and enters Connecticut river at Lancaster.

ISSOIRE, a town of France, dep. Puy-de-Dôme, cap. arrond., on the Crouze, 19 m. S.S.E. Clermont. Pop. (1836) 3741. It is well built and clean; in its centre is a spacious market-place. It has manufactures of copper kettles and other copper wares, with some trade in walnut oil, hemp, and wine.

ISSOUDUN, a town of France, dep. Indre, of which it is the most important, though not nominally the chief town, cap. arrond., on the Theols, which is here crossed by three bridges, 16 m. N.E. Chateauroux, and on the high road between that city and Bourges. Pop. (1836) 9406. It stands partly on the declivity of a hill, and partly in the plain at its foot; it is said to be better laid out and built than any other town in the centre of France, and is remarkably clean. It owes its regularity and beauty principally to the numerous devastating fires it has undergone at different times; during one of which, in 1651, the citizens repulsed and put to flight the troops of Louis XIV., then investing the place. Issoudun was formerly a fortress of some strength, and possessed a large castle, a portion of which, now remaining, serves as a prison. The town has four churches, two hospitals, a new town-hall, barracks, a small theatre, and several public walks. It is the seat of a sub-prefecture, of a tribunal of original jurisdiction and commerce, and of a chamber of manufactures. It has linen and woollen cloth and parchment factories, and was formerly a place of considerable commercial activity; but it has not yet recovered the injury done to its industry by the revocation of the edict of Nantes. It is of great antiquity, having been one of the towns laid waste by the Bituriges to arrest the progress of Julius Cæsar. (*Hugo, &c.*)

ISTRIA. *See* ILLYRIA.

ITALY (Lat. *Italia*, Fr. *Italie*), one of the most celebrated and fertile countries of Europe, the seat of the greatest empire of antiquity, and of art, science, and civilisation, when the surrounding countries were immersed in barbarism. It is finely situated, comprising the whole of the central peninsula of S. Europe, with the extensive and rich country to the N. of the peninsula, and included between the Alps and the Mediterranean. It extends between lat. 37° 46′ and 46° 30′ N., and long. 6° 30′ and 18° 30′ E., having N.W. France and Savoy, N. Switzerland and the Tyrol, N.E. Carinthia, Carniola, and the Hungarian Littorale, E. the Adriatic, and on all other sides the Mediterranean. In antiquity, it was known by the names of *Hesperia, Ausonia, Saturnia, Œnotria, &c.*; but these names, though loosely applied to the whole country, were strictly applicable only to particular portions of its surface. Various derivations have been assigned to the term Italy. (See *Facciolati, Lexicon*, voce *Italia*.) In the first instance, it is said to have designated only its more S. portion, or what is now the peninsula of Calabria Ultra; but in the course of time it superseded every other name, and was gradually extended to the whole country from the Alps southward. In shape, Italy has been familiarly likened to a boot, the heel formed by the Terra d'Otranto, and the foot by Calabria, both in the Neapolitan dominion. The general direction of the Italian peninsula is S.E. and N.W.; its length, from mount St. Gothard to cape Spartivento, in Calabria, is nearly 730 English miles; its breadth varies from about 380 m. in N. Italy, to less than 80 m. near its centre; and in one part of Calabria it is no more than 18 m. from sea to sea. The area of the mainland may be estimated at about 100,000 sq. m.; but three large islands, Sicily, Sardinia, and Corsica, and many smaller, as Elba, Ischia, the Lipari group, &c., belong to Italy. It has long been divided into a number of independent states; the names, area, population, &c., of those at present existing in it may be seen in the following table:

States.	Area in Eng. sq. m.	Pop. by latest Census.		Pop. to sq. m.		Capitals.
Kingdom of Naples and Sicily:						
Naples	31,621	(1838) 6,021,284		190·4		Naples.
Sicily	10,510	(1836) 1,935,641		184·2		Palermo.
	42,131		7,956,925		188·	
Kingdom of Sardinia, &c.:						
Continental, &c. (excluding Savoy)	15,373	(1838) 3,561,998		231·7		Turin.
Insular	9,547	594,633		55·		Cagliari.
	24,920		4,966,631		164·	
Lombardo-Venetian kingdom	18,063	(1839) 4,707,530		260·6		Milan and Venice.
Illyrian government of Trieste (belonging to Austria)	4,055	471,470		116·2		Trieste.
	22,118		5,179,100		234·1	
Papal States	17,210	(1833) 2,732,436		158·7	Rome.
Grand Duchy of Tuscany	8,381	(1839) 1,481,079		176·6	Florence.
Duchy of Parma	2,966	(1833) 465,672		905·2	Parma.
" Modena	2,092	403,000		192·6	Modena.
" Lucca	413	(1836) 165,748		401·3	Lucca.
Republic of San Marino	22	7,600		345·	San Marino.
Total	119,555		22,478,199		188·	

(*Serristori, Statistica d'Italia: Bowring's Reports, &c.*)

Physical Geography.—The frontier of Italy is extremely well defined. She is defended on the N., N.E., and N.W. by the vast bulwark of the Alps, the passes of which might be easily guarded and made impervious to hostile attack. She has everywhere else a sea frontier; so that,

while she is protected by a natural rampart against attacks by land, she has every facility, by means of her extensive sea frontier and numerous ports, for internal and foreign commerce.

Though bounded by the Alps, only a comparatively

small portion of the surface of Italy is covered with alpine ramifications. The mountain system exclusively belonging to the peninsula is that of the Apennines. (See vol. i., 119, 120.) These mountains, which may be regarded as a continuation of the maritime Alps, at first run E. along the Mediterranean shores in the Sardinian territory; and then, turning gradually S., divide Tuscany from the Papal States, passing through the peninsula nearly in its centre, and sending off numerous branches on either side. At length, near lat. 40° 45', the main ridge divides into two separate chains, the principal of which continues S. to the extremity of Calabria, while the other runs E.S.E. through the Terra d'Otranto. The mean elevation of the Apennines is about 4000 ft.; Monte Corno, the summit of the Gran' Sasso d'Italia, in Abruzzo Ultra, is, however, 9521 ft. in height, and is capped with snow during the whole year; Monte Velino is 8183 ft.; and Monte Sibilla, in the Papal States, 7212 ft. high; and many other summits in central and extreme S. Italy approach the latter in elevation. The Apennines are much less rugged than the Alps, and abound with rich forests, and pasture land, on which numerous flocks of sheep are fed. They are of great service to the country, by the numerous rivers which have their sources in them, and by their influence in moderating the summer heats. Italy is also famous for its volcanoes; those of Etna, Vesuvius, and Stromboli, in the Lipari Islands, being, if not the greatest, by far the most celebrated and best known of any on the globe.

But, though for the most part mountainous, Italy has some plains of great extent and extraordinary fertility. Of these, the most extensive and richest is that of Lombardy, or of the Po. This whole plain extends from the foot of the Alps, near Susa, to the mouths of the Po, in the Adriatic, a distance of about 250 m., with a breadth varying from 50 to 120 m., including nearly the whole of the Lombardo-Venetian kingdom, the central portion of the Sardinian dominions, most part of the duchies of Parma and Modena, and the N. legations of the Papal States. This great plain is extremely well watered; the numerous rivers and streams that rise in the Alps, and pour down into the plain, afford a vast and inexhaustible supply of water; and from these an infinite number of canals have been cut, that diffuse the fertilizing element over the whole country, and give to its corn and rice fields, and its variegated meadows, extraordinary productiveness. The soil, though different in different parts, is for the most part loamy, and very fertile. The surface is generally divided into small farms of from 10 to 60 acres; and, if not scientifically, is at least carefully and economically cultivated. The fields are enclosed by lines of fruit-trees, mulberry-trees, poplars, and oaks; and their growth is so luxuriant that in many parts the country has the appearance of a vast forest. This plain has to boast of an immense number of cities, many of which are of great antiquity and considerable size, and all of them adorned with noble buildings and valuable works of art. Probably, on the whole, the plain of Lombardy may be called the garden of Europe; and, at all events, it is certainly the garden of Italy.

The next great plain stretches along the W. shore of Central Italy for about 200 miles, from Pisa, in Tuscany, to Terracina, between the Papal States and Naples. Within these limits are included the Tuscan maremma, great part of the campagna of Rome, and the Pontine marshes (anc. Pomptinæ paludes.) This plain is, in all respects, very different from the former. Though in antiquity, and to a certain extent, also, in the middle ages, it was celebrated for its fertility, and was highly cultivated and populous, it is now comparatively a desert. This is a consequence of the prevalence of malaria, which infects these districts to such an extent as to render them, at certain periods of the year, all but uninhabitable. They are necessarily, therefore, for the most part in pasture; and are occupied by a vagrant population, who reside in the country only in the healthy season. In the campagna of Rome the shepherds who have charge of the flocks are obliged, during the summer season, to repair every night to the city, or to some other town, as sleeping in the country would be fatal; it is then also extremely dangerous to travel by night through the Pontine marshes. The vagrant population of this extensive tract, and those who live on its borders, have all an emaciated, unhealthy, cadaverous aspect; and where the plain is cultivated, the labourers who come from other parts of the country to assist in the harvest frequently fall victims to the pernicious influence of the atmosphere, or have their constitutions injured for life. In the Tuscan maremme, the soil has in many places become, from neglect, sterile and unproductive; but in the campagna of Rome, and the Pontine marshes, the soil is, in most parts, extraordinarily fertile, is covered with a luxuriant vegetation, and, were it properly cultivated, would yield immense crops.

"There are no hills in the Campagna. Its undulations do not arise from elevations of the surface, but from depressions; it may be described as a plateau from 100 to 200 feet above the level of the sea, traversed by wide and shallow valleys, which occupy 1-4th or 1-5th part of its surface. Some of these valleys are dry, others have small sluggish streams, and they are from 50 to 150 feet deep. There is a strip of swamp along the sea-coast, probably two or three miles broad; but, with this exception, the Campagna di Roma seems to be generally dry; for the wet lands seen in some of its small valleys are such as we find in every country, and are not worth mentioning as an exception. Its present appearance is bleak and deserted in a remarkable degree. There are scattered clumps of brushwood; but the eye ranges over it for miles often without discovering a single timber tree, and I have seen nothing deserving the name of woodland or forest within its vast bounds. Fences are rare, except near Rome; a gentleman's country house, or villa, is not to be seen in it, nor a decent farm-house; and even the cottages are so few and far between, that in the 40 miles from Rome to the hills near Civita Vecchia, I am satisfied that I did not see 80 houses, of all descriptions. It is divided into immense estates, usually let in small lots, on the métayer system, and is kept mostly in pasture, not more than 1-5th or 1-10th part being under the plough, or rather hoe, for it is laboured with the latter.

"The Pontine marshes are 24 miles long, and probably 12 broad. The work of draining was commenced under the Roman republic, was continued by the emperors and popes, and is not yet entirely finished. The journey through them is monotonous beyond anything I have met with. A canal 30 feet broad, the grand trunk of the drainage, extends along the whole length, in a line mathematically straight. The soil thrown out of this canal forms a raised bank, five or six feet above the water, and 80 or 100 feet broad. An excellent road passes along this bank, with a double row of lofty trees on each side. You travel mile after mile on this road, on the same unvarying straight line, with the same endless vista before you, longing for some hill or valley, or turn of the road, to break the fatiguing uniformity of the scene. It was upon this canal that Horace travelled in a track-boat, on his journey to Brundusium. The marshes are not altogether uninhabited. A few houses are met with on the road, and others are seen in the distance. The surface is chiefly in pasture; but part is planted with tall reeds used for vine props, part covered with brushwood, probably raised for fuel, and some small patches are ploughed. Very little wet marsh is now visible till you come to the north or higher end, where there is a considerable tract still undrained. The general surface of the plain inclines eastward and southward, so that the inland part is actually lower than that towards the coast on the north; and, like the Neapolitan Campania, the level ground abuts sharply against the mountains." (Maclaren's Notes on Italy, p. 63.)

Various and very conflicting causes have been assigned for the increase of malaria, and the consequent depopulation of these extensive and once fertile territories. They were always, indeed, rather unhealthy; but their unhealthiness has been prodigiously aggravated in modern times. It is believed by many that its deterioration has been, in a considerable degree, owing to the wanton destruction of the woods and forests, by which the land was shaded in antiquity, and screened from the fiery beams of the summer sun. No doubt it is in part also a consequence of the obstructions that have been allowed to grow up in the courses and at the mouths of rivers, by which their waters have been formed into stagnant and noxious marshes. But the last-mentioned circumstance may itself be ascribed to what we believe has had by far the greatest influence, that is, to the decay of population and industry, occasioned by the irruptions of the barbarians, the ravages of war, and the influence of epidemics. The success that attended the efforts of the Grand Duke Leopold to reclaim some portions of the maremme, by establishing colonies in them, appears to have led many to believe that they were absolutely irreclaimable. Certainly, however, this is not the case. The great works, principally of a hydraulic character, that have of late years been undertaken and carried into effect in Tuscany, by which large tracts of the maremme have been converted into productive estates, show what may be done by judicious efforts on a large scale. Hitherto, indeed, the land that has been reclaimed, and made tolerably healthy, bears but a small proportion to what is still abandoned and pestiferous; but the example has been set, and what has been done and is doing in Tuscany, will probably lead to similar efforts being made by the Papal and Neapolitan governments. (See Tuscany.) Mr. Maclaren says of the Campagna di Roma, that, "having seen it I find it difficult to believe that, in the hands of an industrious people, enjoying the advantages of good government, it should not become fertile, populous, and healthy, and assume the cheerful aspect of Lombardy." (P. 67.)

51

The third great plain of Italy is that of Capitanata (Apulia), having Foggia in its centre. It comprises the greater portion of a tract of flat country, extending from the border of Samnium to Otranto, along the shore of the Adriatic, anciently included in Daunia, Japygia, Peucetia, and Messapia. The lower part of the Apulian plain is arid, the rivers decreasing both in size and frequency as we proceed farther S.; and in the provinces of Otranto and Bari the rain water is obliged to be carefully preserved in cisterns for the irrigation of the land. The upper portion of the plain, is more plentifully supplied with water, but it also has, in many parts, a sandy and thirsty soil. A great part of it is destitute of brush, house, or tree; it is farmed in large estates, and round about Lucera and elsewhere there is a good deal of arable land; but by far the greater portion of the surface consists of pastures, called *tavoliers*, into which immense flocks of sheep from the Abruzzi are driven to feed in the winter. (*See* APULIA, i., 193.) The sums paid for this privilege by the proprietors form a rich source of revenue to the crown of Naples, to which the *tapoliers* belong.

The level district round Naples is still well entitled to its ancient epithet of *Campania Felix*, being at once rich, well-cultivated (for Italy) and densely peopled. We borrow from Mr. Maclaren's Notes the following account of this famous plain. "Conceive a tract of *carse* land, 40 miles in length, by 15 or 20 in breadth, presenting a dead level like the surface of the ocean, and probably from 1 to 100 feet above it. In the midst of this vast area, there are two large islands; Vesuvius and its dependent hillocks constitute one of a round form, and about eight miles in diameter; a chain of hillocks, narrow ridges, and truncated cones, extending from Naples to Cape Misenum, covering a space of 12 miles in length, and three or four in breadth, constitutes the other. With the exception of these two elevated tracks, the whole district, as I have said, is a dead level. It is, in fact, a portion of the bottom of the ocean lifted up by subterranean agents, and converted into dry land. As might be expected, it does not rise by a series of small elevations to the outer hills of the Appennines; it abuts sharply against them, as the waters of the German ocean abut against the last level of the Lammermuir hills. I traversed this plain in three directions; and, excepting the mountains named, did not see a single hillock or eminence in it which would conceal a sheep. Though so level, it seems to be remarkably dry, and hence it is free of malaria. The vegetable soil, which is exposed in drains at some places, seems to be of great depth, and it is cultivated like a garden. It is put to what may be called a double use, first ploughed and sown with corn, and then at every interval of 50 or 100 feet there is a row of vines." (P. 57.)

Rivers and Lakes.—Few countries are better watered than Italy, whether in regard to springs, rivers, or lakes. The principal river is the Po, the *Eridanus* or *Padus* of the ancients; it issues from mount Viso in the Alps, on the confines of France, and receives, during its long course to the Adriatic, a vast number of tributary streams. It divides the great plain of Lombardy into two nearly equal parts, and is the grand receptacle for the streams flowing S. from the Alps, and for the lesser waters that flow N. from a part of the Appennine range:

> " Fired with a thousand raptures, I survey
> Eridanus through flowery meadows stray,
> The king of floods ! that, rolling o'er the plains,
> The towering Alps of half their moisture drains,
> And proudly swells with a whole winter's snows,
> Distributes wealth and plenty where he goes."

Of its numerous affluents, the most important are the Baltea, Sesa, Tessino, Adda, Chiesa, and Mincio, from the N.; and the Tanaro, Bormida, Trebia, famous for the great victory gained by Hannibal on its banks, and Parmo, on the S. The other large rivers of the N. of Italy, are the Adige, Brenta, Piave, and Tagliamento, all flowing S. from the Alps. In Central and Southern Italy no great river can be expected to arise, on account of the narrowness of the peninsula, and the central position of the Apennines, in which they have their sources. The Tiber is the other rivers of this part of Italy, it is interesting chiefly from its ancient renown, and the classical recollections associated with its name, than from its magnitude or intrinsic importance. Among others of this class are the Arno and Ombrone in Tuscany. Considerable differences of opinion have taken place as to the identity of the Rubicon, the S.E. boundary of Cisalpine Gaul, so famous in ancient history. It is generally, however, believed to be represented by the Fiumicino, which falls into the Adriatic 18 or 20 miles below Ravenna. An ancient law of the senate and people of Rome made it death to cross this river with arms in a hostile intention. Its passage by Cæsar, has been finely described by Lucan (lib. i., lin. 183—227); and his exclama-

tion on that occasion, "*jacta est alea,*" has passed into a proverb. In Naples, the only streams deserving the name of rivers are the Vulturno, the Garigliano, anciently the *Liris*, and the Ofanto, formerly the Aufidus, which flowing past Cannæ, is thence called *sanguineus* by Silius Italicus (lib. x. 390). The rivers which descend from the Appennines are apt, like other mountain currents, to swell suddenly, and to cause inundations in the level parts of the country, particularly towards the mouth of the Po.

> " Proluit insano contorquens vortice alvus
> Fluviorum rex Eridanus, camposque per omnes
> Cum stabulis armenta tulit." Georg. i., 481.

To restrain this, dykes or mounds have been erected in many places; and as the earthy substances brought down by the flood have, in many cases, raised the bed of the stream, and required fresh embankments, the mounds are often of considerable height, and have the appearance of aqueducts.

The most considerable of the Italian lakes are situated in the N.; including those of Garda, Maggiore, Como, Lugano, &c. In Central Italy are the lakes of Perugia (an. *Lacus Thrasimenus*), Bolsena, Bracciano, Celano or Fucino, Albano, &c. (see the names); and in the S. those of Averno and others, which, though insignificant in point of size, have acquired imperishable renown. Many considerable salt lagoons line the Mediterranean coast in various parts of Tuscany and the Papal States, and the shores of the Adriatic in the Venetian territories, and round the promontory of Gargano. Besides the Pontine marshes, at the S. extremity of the Pope's dominions, there are numerous marshy tracts of less extent in the Val di Chiana and other parts of Tuscany, in the plain of Salerno, and along the banks of the Po, especially in the region round its mouth. S. of the last mentioned tract, a considerable extent of bog-land, called the *Val di Commachio*, occupies a large portion of the papal legation of Ferrara. Italy has about 3000 miles of sea coast. Its chief capes and headlands are, Argentaro, Circello, Campanella, Spartivento, and Santa Maria di Leuca, on the Mediterranean, and the Testa di Gargano, and Cape Promontore (Istria), on the Adriatic. Of the gulfs or bays formed along its coasts, the principal are the Gulf of Taranto, on the S.E., between Apulia and Calabria; those of Genoa, Gaeta, Naples, Salerno, Policastro, Eufemia, and Gioja, on its W.; and those of Squillace, Manfredonia, and Trieste, on its E. shores.

Geology and Minerals.—Italy may be described as "a calcareous region enclosing a schistous band;" but volcanic action has been so prevalent, that the strata are often found extremely disarranged from their original position. North of Genoa, the primary formations in the Apennines include granite, gneiss, serpentine, quartz, clay slate, &c., often intermixed with transition limestone and grauwacké. Granite and gneiss are absent in the Apennine region of Central Italy, but they reappear in the S., where they predominate among the primary formations, from the Abruzzi to the farthest end of Calabria. They also exhibit themselves in the *Maremma*, near the surface ; the secondary formations in Tuscany being often intermixed with primary rocks, and in some instances overlain by them. The tertiary deposits of Italy are very extensive, and form the sub-Apennine region, or low hill ranges, extending along the flanks of the Apennines throughout the whole peninsula, consisting of sandstone, marl, coarse limestone, &c. These formations contain an abundance of marine shells, among which as many as 770 different species have been enumerated, half of them still inhabiting the adjacent seas. The alluvial plain of the Po abounds in fossil remains of mammalia, birds, and amphibia, and similar fossils have been discovered in the Neapolitan dominions. Several regions in the central and S. parts of Italy are almost wholly composed of volcanic products. Such are the Campagna di Roma, which abounds with a volcanic tufa, called *travertino*, of which great part of Rome is built; and the neighbourhood of Vesuvius, which is covered with lava and *scoria*. Numerous traces of extinct volcanoes exist, the craters of which have been converted into lakes. Such is the origin of the lakes of Bracciano, Vico, and Albano, in the neighbourhood of Rome.

Italy is less rich in metals than in most other things; it, however, is well supplied with iron ; it has also copper and lead ore, and the precious metals have been found, but in inconsiderable quantities. Tuscany is the chief seat of mining industry, and large quantities of iron are furnished by the island of Elba, belonging to that duchy. (*See* ELBA.) The most valuable mineral product of continental Italy is, however, the fine statuary marble of Carrara, in the Modenese territory. Marble of a similar kind, and nearly as good, is found at Seravezza, and other kinds are met with in almost every part of the peninsula. Great quantities of borax are found in Tuscany: sulphur, building stone, salt, nitre, alum, alabaster, crystal, &c., are the other chief mineral products; and the Apennines abound in basalt, dried lava, *pozzolana* sand, and other volcanic substances. Caverns of stalactites are met with in many parts, and mineral springs and vapours are of very frequent occurrence. (*Hoffmann, Europa; Lyell's Geology; Rampoldi; Dict. Geog.*)

The climate of Italy is delightful. Owing to its length from N. to S., and the great difference in the elevation of its surface, there is necessarily a considerable variation in the temperature of different parts; but, speaking generally, the air is throughout mild and genial; the excessive heats of summer are moderated by the influence of the mountains and the surrounding sea, and the cold of winter is hardly ever extreme. As respects temperature, it may be divided into four regions: the first, extending N. of the Apennines, and of lat. 43° 30′, and including the plain of Lombardy, has a climate somewhat similar to that of S. Germany, but warmer. In winter the lakes of Garda, Maggiore, &c., and the lagoons of Venice, are partially frozen; snow often falls, and the thermometer sometimes sinks to 14°; even in summer the N. wind is cold, and oranges, lemons, and other *agrumi* do not flourish in the open air. The second region, extending between lat. 43° 30′ and 41° 30′, includes the greater part of Tuscany and the Papal States, with the N. part of the kingdom of Naples. Within this band snow and ice are mostly confined to the mountain tops, and olives and *agrumi* of all kinds flourish luxuriantly without culture. The third region, from 41° 30′ to 39°, comprises the middle Neapolitan provinces. Snow is here very rare, and the finest fruits are found in the valleys throughout the winter. The fourth region embraces the S. part of Calabria, with Sicily, the Lipari islands, &c. Here the thermometer never falls to the freezing point, and the sugar-cane, Indian fig, papyrus palm, and other tropical plants are abundant on the low lands.

The following is a table of the median temperature of the year in different latitudes of Italy:

Places.	Lat. N.	Height above Sea.	Mean annual Temperature.
Milan	45° 28′	416 feet.	56° Fahr.
Bologna	44° 30′	225	57·44° —
Florence	43° 46′	200	59·4° —
Rome	41° 53′	167	60·4° —
Naples	40° 50′		61·9° —

Throughout most parts of Italy there are but three seasons in the year: a spring, which more than realises all that poets have said in its praise; a hot summer, and a short, and not severe, winter: most of the vegetable products, even in the N., flower by the end of March. Heavy rains prevail during October and November; W. and N.W. winds are the most prevalent; but the *libeccio* and *sirocco*, the *simoon* of the Arabs, also occasionally occur, and exert an oppressive, and in the S. an injurious, influence over the animal frame.

It is true, however, notwithstanding the mildness and general salubrity of the Italian climate, that large districts of the country are very unhealthy, and that the chances of longevity are less than in England and other countries under more inclement skies. But the unhealthiness complained of is not the effect of climate, but of circumstances connected with the physical geography of the country, and the want of industry: neither do we think that the inferior longevity of the Italians is to be ascribed to their climate, but to the depressed situation and poverty of the bulk of the people; the bad quality and scanty supply of food and clothes; the low state of medical science; and the want of cleanliness. The general climate may, indeed, be said to contribute indirectly to bring about these results, by encouraging slothful habits, and making the people less industrious than they would be were it more severe; and, no doubt, this is to some extent true. But it is to the want of enlightened institutions, a tolerant system of religion, and a free press, that the distressed state and heavy mortality of the people of Italy are mainly to be ascribed.

It has been supposed that the climate of Italy has undergone a considerable change, and that it is now less cold in winter than formerly. There seem to be good grounds for concurring in this opinion; and the change may be accounted for by the cutting down of the forests already alluded to, and by the changes that have taken place in the countries to the N. of Italy. (See *Hume's Essay on the Populousness of Ancient Nations*, and the authorities referred to in it.) We doubt, however, whether there be any foundation for the notion, that either the productiveness of Italy or its pop. has diminished. No doubt some extensive tracts, as the Tuscan *maremma*, the Campagna, and some parts of the Neapolitan territory, which in antiquity were occupied by a dense population, are now all but uninhabited; but, on the other hand, Lombardy has been signally improved, and is at this moment infinitely better cultivated and more populous than at any former period. On the whole, we incline to think, that whatever Italy may have lost in respect of population in certain districts, has been fully countervailed by a corresponding gain elsewhere; and that her decline from her ancient fame and influence has not been occasioned by any decline in the number of her sons.

General Aspect of Italy.—Speaking generally, nothing can surpass the beauty and diversity of the scenery of Italy. Its mountains have every variety of form and elevation: alternately smooth and rugged, they exhibit by turns gentle declivities and fine pastures, tremendous precipices and chasms, water-falls, deep and majestic forests, and summits, sometimes capped with snow, and sometimes emitting smoke and flames. Many of the valleys, as that of the Arno, are delightful beyond description; the plain of Lombardy is not less beautiful than rich, and even the half-desert tracts along the W. shore interest by their solitude and their vastness. The extent of the sea coast, and the number and magnitude of the bays, add also greatly to the beauty and variety of the landscape; while the clearness of the atmosphere gives to every object a brightness of colouring, and distinctness of outline, that can with difficulty be conceived by those accustomed to our cloudy and less brilliant skies. No wonder, then, that the beauty and richness, as well as the glory of the country, should have been a favourite theme of the ancient writers:

> "Sed neque Medorum silvæ, ditissima terra,
> Nec pulcher Ganges, atque auro turbidus Hermus,
> Laudibus Italiæ certent: non Bactra neque Indi,
> Totaque thuriferis Panchaia pinguis arenis.
> Hæc præcipue fruges et Bacchi Massicus humor
> Implevere; tenent oleamque, armentaque læta.
> Hic ver assiduum, atque alienis mensibus æstas:
> Adde tot egregias urbes, operumque laborem,
> Tot congesta manu præruptis oppida saxis,
> Fluminaque antiquos subter labentia muros.
> Salve, magna parens frugum, Saturnia tellus,
> Magna virum."—*Georg. ii.*, 136—174.

In respect of its *vegetable products*, Italy may be divided into six regions, according to its elevation. These are as follow:

Regions.	Elevation.	Products.
1. Of the plains	— to 1,200 ft.	Lentisk, myrtle, laurel, ilex, and cork trees, cotton, fig, olive, vine, pomegranate, &c.
2. Oak and chestnut	1,200 — 3,600	Oak, chestnut, &c., olive, vine, corn, &c.
3. Beech and fir	3,600 — 6,000	Beach, fir, larch, juniper, and wheat, barley, oats and maize, to some ft.
4. Subalpine region	6,000 — 6,600	Dwarf pine, arbutus, gentian, anemone.
5. Upper Alpine region	6,600 — 8,000	Androsace, saxifrage, and some other Alpine plants.
6. Region of snow	8,000 ft. and upwards.	Snowless moss, *Arenaria musciformis*, and a few other plants.

There is a much greater diversity of plants in the S. portion of the Apennine chain than in any other part of its extent: this diversity is the most marked in the second, or oak and chestnut region. The Italian or S. declivities of the Alps present a greater diversity of vegetation than those facing the N.; and more species of plants are found on them than on the Apennines. On the Alpine summits are seen the dwarf birch, juniper, and other plants of Lapland and Siberia, while at their feet flourish the fig, *Agave americana*, and *Cactus opuntia*. Mount Vesuvius has a Flora peculiar to itself. (*Hoffmann, Europa und Seine Bewohner*, iii., 64—68.)

Italy is much more an agricultural than a manufacturing country; but the indolence of a great part of the population, the remaining operation of the feudal system, and the backward state of agriculture, render the actual return far inferior to what the country is calculated to yield. Silk has become a most important product: its culture has increased very rapidly since the peace of 1815, and the total produce is now estimated at about 12,000,000 lbs. a year: wine and olives, particularly the latter, are also very important products; and there is a great abundance of the finest fruits. Corn is not so generally cultivated in Italy as in the more N. countries of Europe; but pulse and other vegetables are extensively raised. Particular parts of the country are appropriated to particular products. Lombardy is the chief corn country; in the Genoese and Tuscan territories, the culture of fruit, particularly of olives, predominates; while the unhealthy district of the Maremma and Campagna remains, as before stated, chiefly in a state of natural pasture. Skilful agriculture is principally confined to the N.; in the centre, with the exception of portions of Tuscany, and S. it is at a very low ebb; and throughout the kingdom of Naples the abundance of vegetable productions is owing more to the climate and soil than to the industry of the husbandman. The products of the N. parts of the peninsula are found there in abundance; and whole groves of olives are seen growing in the open country, interspersed with spices and other tropical products.

The pastures of Italy are stocked with large herds of black cattle, sheep, and goats: few horses are reared; and the breed is in little estimation, except in certain parts of the Neapolitan territory. Mules are more common, being found better adapted for the bad and mountainous roads. The operations of agriculture are performed by oxen. The buffalo is found in Italy, though hardly anywhere else in Europe. Hogs are fed in large herds in the forests, particularly in Calabria. The mountains and forests contain a number of wild animals; among others, the boar, stag, marmot, and badger. The lynx or tiger-cat is not uncommon in the mountains of Abruzzo; and the crested porcupine is supposed to be peculiar to the S. of Italy. Foxes, hares, and winged game, are sufficiently abundant. From the heat of the climate in the S. provinces, snakes and reptiles of different kinds are numerous. The rivers, lakes, and coasts abound with fish.

Manufactures and Trade.—Italy is not distinguished for manufactures: the chief are those of silk fabrics, silk thread, &c., which have their principal seat in Lombardy. Woollen and linen stuffs, straw plait, gauze, artificial flowers, straw hats, paper, parchment, leather, gloves, essences, and musical instruments, are among the other goods manufactured in Italy; but, generally speaking, the raw products of the country form its chief exports, and most manufactured articles, whether of necessity or luxury, are imported from foreign nations. Venice and Genoa engrossed a large proportion of the trade of Europe, till the discovery of the passage to the East by the Cape of Good Hope, and the enterprise of the Portuguese and Dutch, and after them the French and English, diverted European commerce into a new channel. From that period, the prosperity of those cities gradually decayed, and the first of them has sunk into comparative insignificance, while Italy at large has but a small portion only of her former commercial importance. In the Austrian, Papal, and Neapolitan territories, the exportation and importation of commodities is checked by impolitic duties and prohibitions; while, in the last two at least, little or nothing is done to promote trade or manufactures, by the improvement of roads, harbours, and such like public works. In Tuscany, a more liberal and enlightened policy is adopted, and Leghorn and Genoa still display a considerable degree of commercial wealth and activity.

Trieste, however, is at present the principal Italian port; but a good deal of its exports and imports are derived from and intended for Austria and Hungary. Italy is, next to Germany and Holland, the largest European importer of English goods. The exports to it, at an average of the six years ending with 1838, amounted to £2,736,151 a year. In 1838, the exports from the United Kingdom to Germany were £4,966,900; to Holland, £3,549,499; and to Italy, £3,076,931. Cotton stuffs and twist form about 2-3ds of our exports to Italy; the remainder being principally made up of sugar, coffee, and other colonial products; woollens, iron and steel, hardware, linens, fish, earthenware, coal, &c.

The exports to Great Britain are principally olive oil, brimstone, wine, kid and lamb skins, oak and cork bark, oranges and lemons, raw and thrown silk, partly imported direct and partly indirect through France; straw hats, wheat (a good deal at second hand from the Black sea), linseed, strumac, rags, &c. The trade with the Levant is very considerable; and a good deal is carried on with France, Austria, Greece, Switzerland, Germany, Russia, and America. Next to Trieste, Leghorn, and Genoa, the chief commercial ports are Civita-Vecchia, Naples, Gallipoli, Bari, Ancona, Venice, and Palermo: the principal inland commercial cities are Milan, Brescia, Verona, Bologna, Turin, Florence, Lucca, Rome, and Sinigaglia. Farther details respecting the trade of Italy will be found under these separate heads, and the different states into which the country is divided. We subjoin

A Statement of the Quantities of the principal Articles imported into the United Kingdom from Italy and the Italian Islands in 1838.

Articles.	Quantities.	Articles.	Quantities.
Brimstone (all kinds)	288,710 cwts.	Shumac	203,090 cwts.
Cheese	943 —	Raw and waste silk	379,294 lbs.
Cork (unmanufactured)	1,488 —	Lamb skins (undressed)	1,410,641 No.
Wheat	28,999 qrs.	Brandy	59,270 galls.
Gum Arabic	965 cwts.	Valionea	17,348 cwts.
Lemons and oranges	68,469 chests	Cotton wool	998,764 lbs.
Olive oil	1,385,754 galls.	Sheep's wool	1,758,894 —
Flax and linseed	209,171 bush.	Wine	435,909 galls.
Straw bonnets (or straw plait)	4,786 No.		

A Statement of the Quantities of the principal Articles exported from the United Kingdom to Italy and the Italian Islands, in 1838, specifying the declared Value of the Articles of British Produce and Manufacture.

British and Irish Produce and Manufactures.			Foreign and Colonial Produce.	
Articles.	Quantities.	Declared Value.	Articles.	Quantities.
		£		
Brass and Copper (manufactured)	7,609 cwts.	34,291	Cassia lignea	305,943 lbs.
Cotton (manufactured) and yarn		2,005,585	Cinnamon	50,179 —
Coals, culm, and cinders	95,709 tons	12,166	Cocoa-nuts	914,316 —
Earthenware	981,610 pieces	15,897	Coffee	2,306,822 —
Herrings, &c.	29,954 barrels	20,964	Cotton piece goods	92,980 pieces
Hardware and cutlery	8,227 cwts.	49,598	Ginger	2,334 cwts.
Iron and steel	20,593 tons	186,368	Shell-lac	311,737 lbs.
Linen (manufactured) and yarn		57,865	Indigo	540,846 —
Machinery, &c.		41,985	Pepper	297,905 —
Sugar (refined)	115,690 cwts.	226,373	Rum	107,164 galls.
Tin and pewter goods		38,345	Sugar (unrefined)	28,062 cwts.
Woollen manufactures, &c.		258,157	Tobacco (unmanufactured)	1,434,883 lbs.
Sundries		193,738	Cotton wool	2,930,756 —
Total British and Irish		£3,076,931		

The principal *roads* in Austrian Italy, Tuscany, &c., are pretty good, and some of them excellent; but in the Papal States and the Neapolitan dominions, they are in general very bad. In Central Italy, however, several ancient roads exist, in good preservation; as the Æmilian Way, and a part of the Appian, which constitute, in fact, the best routes in the territories of Parma, Modena, and the Pope. New and excellent roads have been opened from Genoa to Nice, Turin, and Leghorn, and from Leghorn to Grosetto. The road from Rome to Naples is extremely good; and a new road has been made from Naples to Brindisi, and the extremity of Calabria. Within the present century, also, magnificent roads have been carried over the Alps, by the passes of the Splügen, Simplon, St. Bernard, Mont Cenis, &c., and easy means of communication have thus been opened between Italy and transalpine Europe.

Canals are numerous, especially in the Austrian territories, the N. legations of the papal dominions, and the central part of Piedmont; they are chiefly, however, for irrigation only. But those of Pisa and Oento, and those from the Po to Ferrara and Reggio, are navigable.

Religion, Education, &c.—The population is entirely Roman Catholic, except a small portion inhabiting a few valleys of Piedmont, who profess the Protestant faith; some communicants of the Greek church, in the S. provinces

of Naples; and Jews, and strangers of various creeds, residing principally in the large cities, where they are allowed the free exercise of their different modes of worship. There are 38 Rom. Catholic archbishops, and an indefinite number of suffragan bishops, in Italy. The number of inferior ecclesiastics is surprisingly great; and the secular clergy are the principal teachers in their respective parishes. It has been said, with reference to public education in Italy, "It is quite a mistake to conceive that no advance has been made of late years in the department of education in Italy; so far is it otherwise, that the footing on which popular instruction has been placed is, on the whole, superior to what exists in France or England. Any one who will give himself the pains to enquire into the fact, will find that there are proportionally more Italians than Englishmen or Frenchmen who are able to read and write; and the children of the middling and lower classes in Lombardy and Tuscany have no reason to shrink from a comparison with their contemporaries in Protestant countries, as respects the quality of their acquirements. In every part of Italy the mind is perceptibly on the advance, more especially in the north." (*Journal of Education for April* 1833, p. 356.)

This, however, is a most inaccurate and unfair statement. It is true that elementary instruction is pretty generally diffused in N. Italy and Tuscany; but such is not the case in the Papal States and in Naples. And whatever may be the fact as to mere elementary instruction, most of the higher branches of education are very far behind in all parts of Italy. And what else could be expected in a country subject to irresponsible governments, and where the freedom of the press is almost unknown? There is not, and it would be folly to expect that these should be, any real instruction, in such a country, in either moral or political philosophy. People in N. Italy are taught to read and write; but there, as in most parts of the peninsula, this preliminary knowledge, instead of being turned to good account, is made a means of imbuing them with prejudices, and of enslaving their minds.

The most celebrated Italian universities are those of Pavia, Padua, Bologna, Pisa, Parma, Rome, Naples, &c.; but their ancient reputation has greatly fallen off; and it was originally owing, not so much to the superior instruction they afforded, as to the backwardness of the corresponding class of seminaries in other parts of Europe.

Of the societies instituted in the country with a view to the improvement of the language, the most celebrated is the *Academia della Crusca*, at Florence (see FLORENCE); others are established in Rome, Milan, Bologna, and other large towns. No part of modern Europe has surpassed Italy in the number of her sons eminent in literature, science, and the fine arts. This has been, in some degree, owing to her being the refuge of men of letters, when driven out of Greece by the invasion of the Turks; but far more to the early independence and wealth of the principal cities. Under the fostering influence of the latter, Dante, Petrarch, Boccaccio, Machiavelli, and a host of other great poets and prose writers, besides painters, sculptors, and musicians, flourished at a period when the literature and the arts of the rest of Europe were comparatively barbarous. The Italians still excel in works of imagination, and of pure science and antiquities; but the antipopular nature of their governments, the want of free institutions, and of a free press, drive them from the higher and more interesting walks of literature.

Italy is richer than any other country in monuments of antiquity and of the middle ages. Among the splendid relics of ancient grandeur are the Coliseum and Pantheon; the triumphal arches of Vespasian, Severus, and Constantine; the pillars of Trajan and Antonine; the obelisks, &c., at Rome; the amphitheatres of Verona and Pola; the catacombs of Naples; the ruins of the temples of Posidonia or Pæstum, simple, austere, massive, and of unknown antiquity; and, above all, the subterranean remains of Herculaneum and Pompeii. Almost every town possesses some memorial of antiquity, and there is scarcely a place or a stream of any size that is not imperishably associated with some circumstance of importance in ancient or modern history. Tivoli (*Tibur*), where Horace and Catullus had villas; the Alban Mount (*Mons Albanus*), surmounted by the temple of Jupiter Latialis; Frascati (*Tusculum*), the seat of Cicero's villa, whence the *Disputationes Tusculanæ*, the most beautiful of ethical disquisitions, derive their name; the lake Nemi (*Lacus Nemorensis*), sacred to Diana; the *Campi Phlegræi*, near Naples; the bay of Baiæ; the field of Cannæ; the lakes of Trasimene and Avernus, and a thousand other places, have all acquired an immortality of renown.

State of the People.—It is difficult to form any fair estimate of the real condition of the people of Italy. Having been long parcelled out into numerous small states, and subjected to different laws and customs, they are not a homogeneous nation; and it would be unjust, as well as inaccurate, to suppose them all alike. Nevertheless, they have many things in common; and the state of the peasantry is most parts of the country contrasts very disadvantageously with the fertility of the soil and the beauty of the climate.

> " How has kind Heav'n adorn'd the happy land,
> And scatter'd blessings with a wasteful hand!
> But what avail her unexhausted stores,
> Her blooming mountains and her sunny shores,
> With all the gifts that heav'n and earth impart,
> The smiles of nature and the charms of art,
> While proud Oppression in her valleys reigns,
> And Tyranny usurps her happy plains?
> The poor inhabitant beholds in vain
> The redd'ning orange and the swelling grain,
> Joyless he sees the growing oils and wines,
> And in the myrtle's fragrant shade repines;
> Starves in the midst of nature's bounty crost,
> And in the laden vineyard dies for thirst."

Perhaps, however, this is rather too unfavourable a picture. We agree with Dr. Moore in thinking that extreme indigence is accompanied with less wretchedness here than in most other European countries; a consequence partly and principally of the mildness of the climate, and partly of the temperance and contented disposition of the people. (*Moore's Italy*, ii., 240.) But, with all this, it is still true that there is in Italy a great deal, not merely of poverty, but of wretchedness and misery. With the exception of

the Neapolitan dominions, in which agriculture is at the lowest ebb, Italy may be said to be a country of small farms held on the *metayer* principle, or on condition of the occupier giving up half the produce to the proprietor; and where such a system of occupancy exists, there can be little or no improvement, nor any accumulation of wealth. In such a state of things, the occupiers live uniformly almost from hand to mouth, and are necessarily exposed to the most tremendous vicissitudes. Neither is there in Italy any regular state provision for the poor; and wherever this is the case, and especially in so densely peopled a country, there cannot fail to be innumerable instances of extreme suffering. The mortality that took place in Italy after the deficient harvest of 1817 was quite frightful, and mendicancy and misery prevail at all times to an extent unknown in better governed countries, though with fewer natural advantages. A superficial observer might suppose that the small farmers in the *Val d'Arno*, and other rich and beautiful districts of Tuscany and other parts of Italy, were in the enjoyment of most of the comforts of the Golden Age; but, in point of fact, they have to maintain a constant struggle with poverty. M. Lullin de Châteauvieux says of the occupiers in the Tuscan Arcadia, that, "on entering their houses, we find a total want of all the conveniences of life; a table more than frugal, and a general appearance of privation: they are unable to lay by anything as a reserve against unfavourable years." (*Lettres en Italie, &c.*, Eng. trans., p. 79.) And such is the case with the far greater portion of the country. Mr. Maclaren says, that the proportion of poor, tattered, and wretched persons at Naples is quite excessive. "On our way to Caserta, Baiæ, and other places, children eight or nine years old, absolutely naked, often kept running alongside our carriage for a quarter of a mile or more. In all the towns and villages near Naples, strangers are besieged with crowds of mendicants, whose importunities know no bounds." (*Letters*, p. 65.) Mr. Matthews says, that in "the Papal States all is slovenly and squalid; there seems to be no middle link in the chain of society between the cardinal and the beggar." (*Diary of an Invalid*, p. 283.) It is not so bad in the N. of Italy; but even there, the destitution and misery of the people are often such as almost to stagger belief. (See *Rose's Letters*, i., 138, &c.)

It were idle to expect that cleanliness should be a distinguishing characteristic of such a people. And, in fact, the towns, houses, and persons of the people, would, in most instances, be greatly improved by scrubbing, washing, and combing.

The reader may find in Moore, Matthews, and other writers, full details of the *cicisbeo* or *cavalier servente* system, peculiar to Italy. It is confined to the higher classes, and appears to be the natural result of a state of society in which marriages are adjusted on mere mercenary principles, the parties frequently meeting for the first time at the altar, and where there is little save affairs of gallantry to engage the attention of the men. The introduction of free institutions and of a free press into Italy would soon make an end of the race of cicisbeos. But till then, it is probable that this puerile display of apparent, if not real, disregard to the most important engagement of life, will maintain its ground.

Foundling hospitals abound in most parts of Italy; and are at once a powerful cause as well as an effect of the corruption of manners. They receive all classes of children, legitimate and illegitimate, rich and poor; and great numbers are annually sent to them. The mischievous consequences of such a state of things, and the disregard which it evinces for the most sacred obligations, are too obvious to require being pointed out. One of the greatest, and perhaps most indispensable reforms that could be effected in Italy would be the abolition of foundling hospitals. What is to be expected of those who do not scruple to send their children to die, or, if they escape death, to be brought up, independently of any care of theirs, in foundling hospitals? Such persons may have the cant of patriotism on their lips; but we may be quite sure that they will never incur any sacrifice to effect any object that they do not believe will directly conduce to the promotion of their own selfish ends and projects. We borrow from Mr. Matthews the following striking description of the ladies of Rome and of the Papal States.

"The women are in the grandest style of beauty. The general character of their figure is the majestic; they move about with the inceding tread of a Juno. The physiognomy of the Italian woman bears the stamp of the most lively sensibility, and explains her character at a glance. Voluptuousness is written in every feature; but it is that serious and enthusiastic expression of passion, the farthest removed from frivolity—which promises as much constancy as ardour, and to which love is not the capricious trifling gallantry of an hour of idleness, but the serious and sole occupation of life. There is an expression of energy and sublimity which bespeaks a firmness of soul and elevation of purpose equal to all trials; but this expression is too often mingled with a look of ferocity that is very repulsive. Black hair, and black sparkling eyes, with dark olive complexions, are the common characteristics of Italian physiognomy. A *blonde* is a rarity; the black eye, however, is not always bright and sparkling; it is sometimes set off with the soft melting languishment proper to its rival blue; and this, by removing all expression of fierceness, takes away everything that interferes with the bewitching fascination of an Italian beauty." (*Diary*, p. 115.)

Government.—There is nothing more than the shadow of popular representation in Italy. The little duchy of Lucca has, indeed, its senate of 36 representatives, of the classes of merchants, scholars, artisans, and cultivators, and some bodies in the city of Genoa and the island of Sardinia slightly trench on the power of the king of those dominions. The Lombardo-Venetian kingdom has also its two provincial assemblies; and in the kingdom of Naples there are nominally two legislative chambers, one of peers and the other of deputies. But the functions of the Sicilian chambers have, since 1815, been obsolete; and the provincial assemblies of the Lombardo-Venetian kingdom are divested of all legislative powers. Elsewhere the governments of Italy are absolute (San Marino being little more than a dependency of the pope), especially the popedom and the duchy of Modena; though in the former it is generally exercised with extreme mildness.

That a country so fertile and extensive as Italy, rich to the various products of a fruitful soil, enjoying an excellent climate, well situated for commerce, having a frontier covered by a range of almost impassable mountains, whose inhabitants are remarkably shrewd and intelligent, and which was civilized and powerful, possessed of great wealth, and studded over with innumerable free states, at a time when the rest of Europe was sunk in comparative barbarism, should have been thrown back in the career of improvement, and for centuries subjected to the sway of foreigners, and treated with all the ignominy of a conquered province, can hardly fail to excite the astonishment even of the most careless observer. To trace the various causes which have conspired to produce so striking an anomaly in the moral and political world, would require a lengthened essay, or rather a large volume. But if we mistake not, the same circumstance to which Italy principally owed her superiority in the 13th, 14th, and 15th centuries, has also been the principal cause of her subsequent degradation. The number of separate and independent communities into which Italy was then divided, by directly associating her inhabitants with the government of their respective cities, and making them feel that their own interests were identified with those of the community to which they belonged, powerfully excited their passions, and called forth all their energies. Those powers which had been dormant for centuries were again revived; Milan, Florence, Venice, Genoa, Pisa, &c., became the capitals of so many free states, distinguished by their wealth, and their progress in the arts: eloquence, poetry, history, architecture, painting, and every other pursuit that could either add to the comfort or the embellishment of society, were prosecuted with vigour and success. But this state of society, though it gave a powerful impulse to civilization, was also productive of the most implacable animosities. The disputes among the rival republics, from their limited territory, and their deeply affecting every individual, were prosecuted with all the eagerness of a personal and the rancour of a political quarrel. Sismondi's great work (*Républiques Italiennes du Moyen Age*) is chiefly filled with accounts of these conflicts. And such a state of society, how incompatible soever with the enjoyment of peace and tranquillity, unquestionably affords a fine field for the development of superior talent and mental energy.

Unfortunately, the contests between the different parties in Italy ended, as such contests almost always do, by making it an arena for the struggles, and subjecting it to the arms of foreigners. Germans, French, and Spanish troops, after being engaged in supporting the pretensions of one or other of the rival states, turned their arms against those they had supported, or who had invited them into their country, and, trampling on their liberties, imposed on them new and despotic masters.

Had Italy, when the republican governments were destroyed, had the good fortune to have been consolidated into one single and undivided monarchy, the people would have been fully compensated for the loss of political independence. According as local hatreds and party animosities subsided, the nation would have become animated with the same spirit, and would have been able to defend itself against foreign aggression; and the probability is, that in the course of time the people would have acquired power sufficient to soften the despotism of a government originating in conquest, and to recover possession of a portion at least of their former rights and privileges. But the subversion of the Italian republics was attended by no such result. Instead of being reduced under one, the country was divided among a hundred petty despots and despotical aristocracies. Nor was there any possibility of remedying these evils; for Austria, having obtained possession of the Milanese and Tuscany, was enabled to prevent any single state from acquiring a decided ascendancy, and to perpetuate and embitter those disastrous feuds and divisions which led to the ruin of the republics.

It would be an endless task to endeavour to describe the various effects of which this state of affairs has been productive. Ever since the subversion of the Florentine republic, in 1530, the Italians have ceased to exercise any perceptible influence over the deliberations of their multitudinous rulers. Parcelled out among foreign sovereigns, or sovereigns descended from foreigners, what interest could they feel in the contests of the Bourbons of Parma and Naples; the Austrians of Milan and Mantua; and the Loraine of Tuscany? They were not only deprived of their ancient liberties; but the constant state of vassalage in which their petty sovereigns were themselves held by the great transalpine powers prevented their acting in conformity either with the wishes or the real interests of their subjects. The national spirit was thus gradually destroyed; the Italians either ceased to have or to express an opinion on public affairs; they endeavoured to forget the stormy discussions in which they had been engaged, by plunging into the depths of sensuality; and from being the most active, intelligent, and industrious people in Europe, sunk into a state of sluggish indolence and apathy. "The victim by turns of selfish and sanguinary factions, of petty tyrants, and of foreign invaders, Italy has fallen like a star from its place in heaven; she has seen her harvests trodden down by the horses of the stranger, and the blood of her children wasted in quarrels not their own; *conquering* or *conquered*, in the indignant language of her poet (Filicaja), *still alike a slave*; a long retribution for the tyranny of Rome." (*Hallam's Middle Ages*, i., 358.)

In the latter part of the 18th century, Beccaria, Genovesi, Verri, Filangieri, and other eminent men, attempted to awaken their countrymen to a sense of their true interests; but their efforts were not attended by any corresponding success. At the epoch of the French revolution, the government of almost every state in Italy, with the exception of Tuscany, was a tissue of the grossest abuse. The use of torture was universal; civil and criminal processes were conducted secretly, and left to the decision of a single judge; a direct tax of 25 per cent. per annum was levied on all land and other tangible property; heavy transit duties were imposed on all commodities passing from one petty state to another; except in particular cases, the exportation of the raw produce of the soil was forbidden; the high roads were infested by robbers; morals were at the lowest ebb; and assassination was carried to an extent unknown anywhere else.

Whatever may have been the influence of French domination in other parts of the continent, there cannot, we apprehend, be a doubt, that, to Italy at least, it was most advantageous. Under Napoleon, who has a just title to be called its greatest benefactor, the countries now comprised in the Lombardo-Venetian kingdom formed the kingdom of Italy; Piedmont, Genoa, Parma, Tuscany, and Rome were united to France, and received her laws and institutions; and Naples was constituted into a subordinate kingdom, with improved and more liberal institutions. A vigorous and efficient police was everywhere organised; the oppressive shackles which the jealousy and short-sighted rapacity of the different petty states had imposed on the internal commerce of the country, were entirely removed, and full power was granted to export the various products of the soil; torture was abolished; a uniform code of laws was introduced; instead of the dark and mysterious proceedings of secret tribunals, justice was openly and impartially administered; science was protected and encouraged; the Italian soldiers emulated the discipline and bravery of their French allies; local prejudices and long-cherished antipathies wore on the wane; a *national* spirit was beginning to revive, and that energy which had for centuries been dissipated in frivolous and unimportant pursuits, was again exerted for the public benefit.

No doubt the government of the French in Italy was defective in many respects, and in some oppressive and arbitrary. But, notwithstanding these drawbacks, it was certainly far preferable either to that by which it was preceded, or to that by which it has been followed. Direct taxation was carried by the French to an unprecedented extent: and, latterly, the conscription was felt to be a severe hardship; but as the former was accompanied by the entire freedom of industry, and as the latter pressed indifferently on all classes, they were submitted to with little or no reluctance. "The Italians," says M. Sismondi, "partook of all the privileges of the conquerors: they became with them accustomed to the dominion of the law, to freedom of thought, and to military virtue; secure that, at no very distant period, when their political education should be accomplished, they would again be incorporated in that Italy, to the future liberty and glory of which they now directed their every thought." (*Progress and Fall of Italian Freedom, p. 36.*)

"Under the French," says Mr. Stewart Rose, "Italy enjoyed all the incalculable advantages of a code which allowed the cross-examination of witnesses, and gave publicity to all the proceedings of justice. This, indeed, was so under the ancient government of Venice; but a criminal code was given her by France, infinitely superior to what she possessed in the time of her republic. But the system of open pleadings and examinations, has given way to one which has abolished the oral examination of witnesses; and for these principles, perhaps yet more precious in Italy than elsewhere, has been substituted that of written depositions and secret applications to the judges. Under the government of France, the *annona* (laws regulating the trade in corn and other necessaries) laws slept, and justice, civil as well as criminal, was well and expeditiously distributed. At present, there is no one, uninfluenced by passion, who would not rather renounce a debt than endeavour to recover it by law." And in these respects, the laws continue at present nearly on the same footing as when Mr. Rose's work was published.

It is greatly to be regretted that, on the downfall of Napoleon, in 1814, provisions was not made for the consolidation of Italy into an independent state; but *diis aliter visum.* "The coalition destroyed all the good conferred on Italy by France." (*Sismondi.*) The old order of things was, to a considerable extent, restored; the republics of Venice and Genoa, indeed, disappeared, but the kingdom of Sardinia, the Papal States, and the dukedoms of Tuscany, Parma, Lucca, &c., were reintegrated nearly on the footing on which they stood before the revolution. Austria, however, had the lion's share in the new arrangements, having acquired the whole Milanese and *ci-devant* Venetian provinces; at the same time, that the dependent thrones of Tuscany, Modena, and Parma were filled by members of the house of Hapsburg, to which they look up for protection and support. Hence the influence of Austria is now all but omnipotent in N. and Central Italy, and it also predominates in the S., where the throne of Naples is again occupied by a Bourbon.

On the restoration of the old governments, a good many abuses which the French had rooted out were revived, and the nation was insulted and humiliated. With the division of the country into different states, an end was put to the equality of duties and the freedom of internal commerce; and those sectional prejudices and hatreds that had begun to be obliterated, again exhibited their odious characteristics. The open impartial justice, and the vigorous police, introduced by the French, were either wholly suppressed, or materially modified; and in the Papal States and Naples, especially the former, the priests again acquired a preponderating influence; and these are, once more, Turkey and Spain excepted, the worst governed of the European states.

The government of Austria in Italy cannot be justly said to be oppressive. But it is antipopular, jealous, and repulsive. This is evinced by the restraints laid on the press, and on the importation of books, and by its preventing the opening even of a school for elementary instruction without its express permission. The pressure of taxation and the conscription are less severely felt now than under the French, but this is more than countervailed by the defects in the administration of justice.

Under the French government, the prompt administration of justice, and the efficient police, almost wholly suppressed private assassination and public robbery: but they have again revived in the Papal States and Naples, though even there they are a good deal less frequent than formerly.

It is impossible to say how long the present order of things is destined to last; but at present, unless relief should come from without, the prospects of Italy are far from encouraging. The want of all sympathy with each other, and the jealousies that subsist among the different states, will, it is much to be feared, long oppose an insuperable obstacle to any united or persevering effort to throw off the yoke of their foreign masters; and even though such were not the case, there is a "softness of character, approaching to imbecility" (*Matthews*), that unnerves the Italians, and unfits them for sustaining the difficulties and perils that would have to be encountered in such a struggle.

ITALY (AUSTRIAN). Under this term are included all the Austrian territories within the limits of Italy, comprising the Lombardo-Venetian kingdom, and the government of Trieste, extending over a space of about 22,118 sq.

m., and having, in 1830, a population of 5,179,100. The government of Trieste is, however, included in the kingdom of Illyria; and the following statements will, therefore, apply only to the rest of the territory, being that, indeed, to which the name of Austrian Italy especially belongs.

The Lombardo-Venetian kingdom, one of the most valuable possessions under the Austrian sceptre, extends between lat. 44° 43' and 46° 41' N., and long. 8° 34' and 13° 30' E., having N. Carinthia, the Tyrol, and the Grisons, from which it is separated by the Alps; W. the Swiss cant. Ticino and Piedmont; S. the duchies of Parma and Modena, and the N. legations of the Papal States, from which it is chiefly divided by the Po; and E. the government of Trieste and the Adriatic. Area, population, subdivisions, &c., as follows:—

Government.	Delegations.	Area in sq. m.	Population.	Pop. to sq. m.
Lombardy	Bergamo	1,290	(1832) 355,942	270
	Brescia	1,365	384,742	287
	Como	1,385	386,703	714·4
	Cremona	470	197,786	270
	Crema-o-Lodi	728	163,384	211·6
	Milan	879	361,664	420·7
	Mella	1,045	470,903	471·6
	Pavia	517	132,595	95
	Sondrio	1,314	99,976	65
	Total	8,591	2,340,435	279
Venice	Padua	644	(1832) 350,514	544·5
	Rovigo	499	150,525	349
	Vicenza	574	387,547	340
	Verona	1,484	389,749	296
	Venice	1,055	389,137	347
	Treviso	735	351,728	546·6
	Friuli	2,709	373,971	136·5
	Belluno	514	137,440	44
	Total	9,245	2,099,284	217·4
	Total	14,053	4,448,473	246

But it would appear from the statements in the *Almanac de Gotha* for 1841, that the population in 1830 had increased to 4,577,483.

The N. part of this territory is mountainous; the S. flat, forming a portion of the plain of Lombardy. The Alpine chains on the N. frontier rise to an elevation of more than 13,000 ft. above the sea. By far the greater part of the surface, however, is flat; the flat lands, comprising the delegations of Pavia, Lodi, Crema, Cremona, Padua, Rovigo, Venice, and parts of Verona, Vicenza, Brescia, Milan, &c. The whole country abounds with rivers, all of which, except the Po, have more or less a S. course, and all contribute their waters to the Adriatic. The chief, after the Po, are the Ticino, between Lombardy and Piedmont, the Adda, Oglio, Chiese, Mincio, Adige, Brenta, Piave, Tagliamento, &c. At the foot of the Alpine chains, in the N. of Lombardy, are the lakes of Garda, Como, Maggiore, Lugano, Iseo, &c. The shores of the Adriatic are lined with extensive lagoons, in the midst of which is Venice. A succession of marshes extends along the banks of the Po, in the lower part of its course, and round its embouchure is a dreary tract of swampy ground scarcely enlivened by a single tree.

The central parts of the high mountain chain consist of granite and other primary formations; the lower hill ranges consist chiefly of secondary limestone. The country on the Po is a vast alluvial plain, containing numerous fossil remains. Traces of former volcanic action exist in the Euganean hills, an isolated group to the S.W. of Padua. Lava, basalt, iron, coal, turf, potter's clay, some copper, arsenic, marble, and alabaster, are the most important mineral products. The climate is generally healthy, except in the rice grounds along the Po, in the vicinity of Mantua, and near the Adriatic. The thermometer, though it keeps much higher in summer, generally sinks lower in winter in Lombardy than in England. The mean temperature of the year at Sondrio is 51 Fah., at Milan 53·6, and at Padua 56·6. More rain falls in this than in any other portion of the Austrian dominions; in the government of Venice the mean annual amount is estimated at 34 inches, and in Lombardy at 45 inches. The greatest fall is in autumn and winter.

The tops of the Alps are naked, covered with snow, and interspersed with glaciers; but their sides are for the most part covered with fir, larch, oak, pine, chestnut, and other trees, or natural pasturages. The plain country is continuously cultivated, and is one of the most productive portions of Europe. About four-fifths of the population of Lombardy depend directly or indirectly on agriculture; and nearly seven-tenths of the surface are under culture, the proportions in 100 parts being, arable lands 67, pasture 18, and wood 21. But, however adapted for cultivation on an extended scale, Lombardy is, as already stated, generally a country of small farms, cultivated on the *métayer* principle, and its agricultural inhabitants, though industrious, are mostly poor. Châteauvieux remarks, that over most of the country few of the farms exceed from 70 to 75 English acres, while few also have less than 10 or 12. The subdivision of the land is, however, much greater in the upland regions than in the plains: in the Milanese there are many farms of 120 acres. Most of the productive land in the mountains consists of pastures. Only the lower border of the mountain belt is arable; the land is there frequently cut into terraces, one above another, the divisions being occasionally supported by stone walls. The earth that fills these terrace-trenches is continually carried down to the lower levels by the action of rain, and other causes, and has to be brought up again every two or three years, often on peasants' backs, the routes being impracticable for vehicles. The vine, mulberry, walnut, and various other fruit trees, barley, rye, a little wheat, buckwheat, *panico*, millet, kitchen vegetables, hemp, and flax, are the chief agricultural products of this region. The land is here divided into the most minute portions; and being, as it were the one thing needful to existence, the greatest value is attached to its possession. The inheritance of an individual is often only a few square yards of land; and on the lake of Garda a similar extent of surface, cultivated with lemons or oranges, or the laurel (for its oil), serves to maintain a family. In the central region, or hill country, properties are less divided; though they are there split into small stewardships, worth from 15,000 to 20,000 francs. These farms are mostly the property of the higher classes, and of the inhabitants of cities. There is scarcely a single peasant proprietor, the peasantry being mere tenants, paying, in general, a rent of half the produce. A lease at a fixed rent, or a money rent, is extremely rare. Silk, wines, oranges, lemons, olives, and other fruits, corn, cheese, and cattle, are the chief products of this region; the culture of the silkworm is an important occupation of the peasants' families, and, with the money gained from this source, they provide themselves decently with the necessaries of life.

The aspect of this part of Lombardy is very pleasing. Flourishing villages, hamlets, and isolated houses are spread over it, connected by carriage roads made at the expense of the proprietors and communes, which latter possess a considerable portion of the soil in this and the next region. In the high flat country, or that part of the plain near the hills, small stewardships are not common. The system is that of *pigionanti*, or sharing-tenants; that is, tenants who pay a rent in money for their house, and a fixed rent in kind for their ground. In the low flat country, none of the property is communal; the farms let at from 10,000 to 60,000 fr. a year, and some as high as 96,000 fr.; and the farmers have considerable capital in stock, as cattle, implements, seed, and timber. In this region great numbers of cattle are fed. It has, like the high flat country, a siliceous bottom, with the difference, that here in every part water may be procured at a very little distance below the surface. In the delegation of Lodi, and its neighbourhood, the soil is so fertile and well watered, that the inhabitants have relinquished the growth of corn for that of the indigenous plants spontaneously produced. The meadows, constantly irrigated, are mowed, and spring again four times in the same year; and the value of the produce in grass is superior to that of the richest corn. (*Châteauvieux*, 275.)

The mode of irrigation deserves some notice. It is effected in the first place by *fontanili*, or excavations in the earth, in which are placed long tubes, from the bottom of which bubble up copious streams of water, analogous to Artesian wells. From the *fontanili* the water is conducted into a ditch, by which it is carried to irrigate the fields placed on a lower level. To these natural waters, derived from the subterraneous springs, replenished by a constant supply from the mountain region, are added a great mass of water drawn from the rivers by means of canals, some of which are navigable. The grounds thus inundated are let as high as 50 Milan livres (1*l*. 9*s*.) the *pertica* (about 3-15ths of an acre). The waters are diligently measured by rules, derived from hydrostatic laws, which have passed into a habitual practice. The canals are provided with graduated doors, which are raised or lowered according as the case may be: they are termed *maestri*. The measure is called *oncia*, and corresponds to the quantity of water which passes through a square hole, three Milanese inches high (an oncia of Milan equals two in. English), and 4 in. wide, open 1 in. below the surface of the water, which, with its pressure, determines a given velocity. Sometimes the same number of inches of water is given out by the day and the hour on different farms. The value of a property depends on the command and goodness of the water; if deprived of the fertilizing fluid, it would diminish rapidly in price. Hence the state of the waters is the object of local statutes, and of diligent care and attention. The absolute property of an inch of water is usually valued at from 10,000 to 15,000 fr.; but some

waters are valued as high as 30,000 fr. All proprietors are entitled to carry a new canal for the purpose of irrigation across the grounds of their neighbours, on paying the fair value of the ground occupied by the canal, and adding to it one quarter more. (*Bowring's Rep.*, p. 99.)

W. of the delegation of Lodi, between Milan and Pavia, a good deal of rice is grown. The district appropriated to its culture is divided by a great number of canals, lined with banks of turf, into squares of two or three acres each, within which the rice grows, in water, admitted by sluices, to the height of a few inches. The rice is sown after a single ploughing, and without any other preparation of the land. The sluices are opened to admit water when the plant is some inches in height, and is drained off again near the period at which the grain is ripe, to allow the land to dry before reaping. After having been reaped, the rice is tied into small sheaves, which lie heaped together some time before being thrashed. The soil remains dry till again ploughed. Rice is grown for three years successively on the same land, after which the ground is left fallow for two years, manured once, and produces in those two years a most abundant crop of hay. The produce of a crop of rice is estimated at double that of a crop of wheat. The rice grounds are let at fixed rents, of about 160 francs an acre; and even at this enormous rent, the farmers (who do not divide their profits with landlords) have often made large fortunes. The labour required is little, and not expensive; but it is very unhealthy. "Sickly labourers are seen passing along the banks, to superintend the distribution of the water. They are dressed like miners, in coarse cloth, and they wander about, pale as ghosts, in the reeds and near the sluices, which they have scarcely strength enough to open and shut. In crossing a canal, they are often obliged to plunge into the water, and they come out wet and covered with mud, carrying with them the germs of fever, which never fails to attack them. They are not the only victims, as the harvest men seldom get in their crop without being seized with rigors, the air in all the neighbouring places being deteriorated by the stagnant water. The avidity of the rice planters is, therefore, restrained by law, and they are prohibited extending this culture beyond certain limits." (*Châteauvieux*, p. 285.) The meadow lands in the irrigated country are, like the rice grounds, parcelled out into fields of two or three acres. After remaining about fifteen years in natural pasture (refreshed by a thick coat of manure every three years), they are ploughed and cleaned in autumn, and undergo five successive crops—hemp, followed by vegetables and oats, wheat followed by vegetables, maize and wheat. The land is then left to itself, and a crop of grass immediately springs up without any seed being sown. As soon as the sward becomes thick, the sluices are opened upon it as formerly. In the course of 20 years, 67 crops are obtained from the land: 61 for the food of cattle, five for that of man, and one for clothing. There are few countries in which cattle of every kind are more abundant. In 1836, there were in Lombardy 257,300 cows, 166,000 sheep, and 52,000 houses employed in agriculture, besides draught cattle, &c.

The famous Parmesan cheese is no longer made in Parma, but in a district 40 m. long by 20 wide, stretching from Abbiategrasso on the Tessino, to near the confluence of the Po and Adda. About 80,000 cows are set apart for its production. These come chiefly from Switzerland, the Tyrol, Bavaria, &c., being bought at from three to four years old; the supply of milk increases till they are six or seven years old, when it begins to fall off. About 11,500 are annually imported, at an average cost of from £14 to £15 each, but varying up to £20, (*Bowring*.); the sale of old cows, calves and whey, on which last many hogs are fed, is estimated to cover the cost of the young cows, and interest thereon; the profits of the butter and cheese remaining to the farmer. The cheese is made entirely of skimmed milk: the cows are only fed at stated times, and are stalled during a great part of the 24 hours to empty racks, a process which Arthur Young says he was assured was necessary to give the requisite richness to the milk! (ib. 188.) In the course of a year a cow produces at an average 900 *libbre grosse*, or 155 killogrammes of cheese. It is sold twice a year by the farmers, at about 1 franc per pound. The total quantity made is about 16,000,000 of *libbre grosse*, worth from 15 to 16,000,000 fr. There are about 12 lbs. butter for every 40 lbs. cheese. The value of the cheese and butter consumed and exported from Lombardy, is estimated at 23,360,000 fr. In the Milanese district, a fat cheese, called *strachino*, is made, especially at Gorgonzola, to which coagulated milk is carried from other parts. Its production is considered more profitable than that of Parmesan: it is sold at about 1 fr. 26 c. the killogr. Much of it is made from the wandering herds of cattle which descend in the autumn into the plains. The proprietors of these herds, called *bergamini*, belong to the mountain region: in summer they migrate in search of pasture to the N. side of the Alps, sometimes as far as the Grisons. The pasture lands of Lombardy are mostly in the mountains and low flat country; in its other regions, cattle, sheep, beasts of burden, and even goats, are scarce. The large farm-houses and offices throughout Lombardy are built on a uniform plan. They are of brick, and surround a square court-yard, on one side of which is the residence of the farmer, granary, and stables, all well arranged, while the other sides consist of covered porticoes, under which the fodder, carts, &c., are kept. Half the court is paved; the other half is an area on which to thrash the corn. A garden is attached to the building, the outer walls of which are covered with vines, producing a growth for ordinary use. Each of these farms has a métayer and his family, who usually hold it for generations. They consider it as a patrimony, and never think of renewing the lease, but go on from father to son, on the same terms, without writings. The stock of cattle, &c., belongs to the proprietor. In a farm near Marignano, Châteauvieux states that 100 cows and some animals for draught were kept on 85 acres of meadow land. The métayer estimated the average return of each cow at 226 fr., or of the whole at 20,960 fr. There were 35 acres of arable land, the produce reckoned at half that of the acres in grass, or 6990 fr. The gross produce of this farm of 120 acres was therefore 24,690 fr.; which sum was equally divided between the proprietor and métayer; the farmer out of his moiety paying the taxes and charges of irrigation, and the latter deducting from his share the whole expenses of cultivation.

The peasantry in the low flat country receive a part of their earnings in wages, and a part is produced from the share each has in the cultivation of the land. The ground is, as has been said, divided into portions of from 10 and 15 to 50 and 70 acres. Two acres are assigned to every man and his family, or three to families where there are two men. The farmer furnishes the oxen and horses to plough the ground, and advances the seed; the cultivator performs all the farm work till the crop be carried to the granary; 1-3d part of the buckwheat and beans, and 1-4th part of the rice, are then the share of the cultivator; out of which, however, he returns the farmer the seed formerly advanced, amounting to about 1-5th part of the rice, and 1-20th part of the buckwheat. In addition to their wages, the master allows most of his farm servants a small house and kitchen-garden rent free; and pays their capitation and other taxes, amounting to about 6 or 7 fr. each. The hire of a dairymaid, besides a certain quantity of provisions, varies from 115 to 220 fr. a year; that of a carter from 150 to 180 fr. Ordinary labourers get bread, rice soup, milk, &c., and from 62 to 80 fr. a year; drovers get only their food. Rice-reapers, wood-cutters, and mowers come mostly from the mountain districts of Parma and the Tyrol, and vine-dressers from Piedmont and the lake Maggiore: the wages of all vary from about 1 to 1½ fr. a day, with food.

In 1836, it was estimated that 770,000 Winchester bushels of rye, 3,795,000 maize, 5,392,700 maize, 893,000 rice. 10,110,000 cwt. hay, &c., 3,300,000 do. straw, 606,000 do. cheese, butter, and honey, 170,000 do. silk, and 1,916,000 diner wine, were produced in Lombardy.

We have fewer details respecting the agriculture of the Venetian provinces. The surface is estimated by Quadrio at 2,367,070 terrature, 1,102,198 of which are in the plain country, 747,989 being arable, or corn lands, 17,800 ricegrounds, and 189,000 meadows and pastures.

Maize is grown in considerable quantities near Verona, and the mulberry very extensively between that city and Mantua, and towards Vicenza. The mulberry-trees are frequently planted all round the corn-fields, and vines festooned from one tree to another, so that on the same ground three crops—silk, wine, and grain—are annually produced. From Verona to Vicenza the meadows are irrigated with great care as well as facility, by means of the numberless streams that flow into the Adige, the beds of which, being continually raised by the gravel they bring down, and artificially embanked, are, for the most part, above the general level of the plain. Notwithstanding the fertility of the soil, the inhabitants are generally poor: Several large farming establishments may be seen, but no comfortable cottages, or signs of wealth, among the peasantry, who bear a very indifferent character. The fields about Vicenza, however, are kept with great neatness, and cultivated with much industry, presenting a favourable contrast to those about Padua. On the road between these two cities all beauty of scenery disappears. "Willows in all their pollard ugliness, and long lank poplars trimmed to the top, afford a yearly crop of faggots, the only fuel of the country. The tops of the pollarded trees near Vicenza, may be seen cut almost in the shape of goblets, for the sake of holding the leaves of the maize placed there for drying. Potatoes are often cultivated

amidst the corn. On the road may be seen immense butts full of grapes, mounted upon clumsy wagons, to which they are secured by such iron rings and chains as would hold a frigate at her moorings, dragged along by four, six, or eight oxen, when a proper vehicle would not require more than a pair." (Condor's Italy, ii., 1, 11.) The grain produced in the Venetian provinces leaves a surplus over what is required to meet the home demand. Good husbandry diminishes as we proceed eastward, and Istria is a country which would scarcely repay it. That peninsula is a collection of barren limestone hills, interspersed with a few fertile valleys; it yields very little corn, and the expenses of cultivation nearly absorb the profits. Wood is scarce, and fuel has mostly to be brought from Carniola or elsewhere. The oils of Istria, however, are frequently as good as those of Tuscany, and form its chief export. Some of its wines, also, are good, but the inhabitants are more a commercial and sea-faring, than an agricultural or manufacturing people. (See ILLYRIA.)

The culture of silk, the most important staple of Northern Italy, is rapidly extending; and even in the delegs. of Pavia and Lodi, where the climate is unfavourable to the worm, the mulberry is gradually superseding the vine and olive. In the deleg. of Brescia alone, the oil crop has diminished within the last 35 years from 408,000 lbs. to half that quantity; while the produce of silk has risen within the same period from 1,300,000 to 3,600,000 lbs. The annual produce of silk in Lombardy, Venice, Tyrol, and Trentino, is estimated at about 7,000,000 lbs., or nearly 7-11ths of the total produce of Italy. The produce of the Lombardo-Venetian kingdom, in 1825, was 3,460,475 Milanese lbs., since which time it has consequently about doubled. The best silk in Lombardy is obtained in the district of Branza, between the lakes Maggiore and Como; and in the Venetian provs., from the delegs. of Treviso and Friuli. Milan and Bergamo are the great centres of the trade in silk; the former city and its neighbourhood being the chief seat of its manufacture. Next to the silk fabrics of Milan rank those of Como, Brescia, Cremona, and Mantua. Verona and its neighbourhood, with many Mantuan districts, produce the best sewing and twist silk. Vicenza and Bassano produce immense quantities of silk, chiefly double-threaded trains, and much also is obtained from Padua; but in all the latter named provs., quantity is more sought after than quality in the production of the article.

Next to silk, the chief manufactures are those of woollen and cotton fabrics, linen thread, paper, hats, iron goods, &c. In Lombardy there are several iron and copper works, with fabrics of earthenware, marble quarries, &c. (For details respecting the chief foreign trade of Austrian Italy, see MILAN, VENICE, TRIESTE, &c.) The mountain districts send wines into Switzerland and the Tyrol; and live stock, game, cheeses, butter, honey, firewood, charcoal, timber, granite, marble, slates, bricks, iron and steel, various implements, cloths, and a little hemp, into the flat country; out of which they receive, in exchange, wheat, rice, maize, and oil.

Taxation.—According to the new government survey, the value of the land in the Lombardo-Venetian kingdom has been estimated at 210,851,000 scudi (£38,656,000). The land-tax paid in Lombardy Proper amounts to from 21,080,000 to 22,000,000 lire, and in the Venetian provinces to about 12,000,000 lire, or together to about £1,040,000 a year, of nearly 2⅔ per cent on the assumed value of the capital. The valuation by which the land-tax is levied in the Milanese has not been altered since 1760: the tax has indeed been increased, but the increase has not been by any means equal to the increased value of the land. After 1760, the land-tax was as high as 48 centesimi per scudo, but since 1819 it has been reduced to 17·1 centesimi: in the Venetian provs. it has also been reduced, but it is still higher there than in the Milanese. It is supposed that the system, to be bottomed on the new survey, will equalize the taxation of the two portions of the kingdom. In the above survey, the valuation is guided by the amount of every kind of produce in ordinary years, and under an ordinary system of cultivation. The average prices, from 1823 to 1825, are taken as a guide, regard being had to local circumstances, distance from markets, &c. The expenses are deducted from the gross receipts, and calculated according to the system of farming. To allow for casualties, from 1-9th to 1-7th part is deducted from the nett proceeds for corn. 1-7th for flax, chestnuts, and olives, 1-15th for hay, 1-18th for wood. Churches, fortresses, and open spaces are free; but of all other buildings, the value is ascertained as nearly as possible, for the purpose of taxation. Machinery is free; not so mills or water-power. All buildings are assumed to be in an average state, and a reduction of from 20 to 40 per cent. is made for keeping them in repair.

The poll-tax is levied in places not subject to the tax on consumption. All individuals (except paupers), from the ages of 14 to 60, are liable; and it amounts to 3 lire 68 cen-

tushal for every inhab., without reference to his circumstances. In addition to this tax levied for the state, a sum not exceeding 3 lire may be imposed for the exigencies of the commune. The poll-tax, therefore, may reach, but can never exceed, the sum of 6 lire 66 cent. The injustice done to the humbler part of the pop., by imposing the same amount of poll-tax on them as on the higher classes, is in part compensated by the frequent practice of raising extraordinary impositions on the latter, in the shape of augmentations to the land-tax, and by the control of the communal property being almost entirely in the hands of the small proprietors. The collection of the taxes is farmed out on leases of three years; and the same person may be collector of several communes, or of several entire districts. The farmer of the taxes has power to proceed against defaulters, and in extreme cases to sell the land for arrears; but such proceedings are seldom or never heard of, and the collection is simple, and attended with very little expense.

The octrois, or taxes on consumption, which exist in walled towns differ in amount in different places. They do not everywhere comprise the same articles, but generally include wine, spirits, flour, bread, cattle, fish, oil, butter, cheese, hay, straw, wood, coals, and a few other articles. The mill-tax is levied at the mills, the others mostly at the town gates. At Milan, wine and vinegar are charged 1 lire 15 cent. the cwt., wheaten flour and bread about 1½ lire, hay and oats 86 cent., cheese 2 lire 90 cent., coals and sawed wood 57 cent., bricks and tiles 29 cent. per 100, oxen 7 lire 47 cent., pigs 3 lire 45 cent. each, &c. The taxes at Venice (which see) are higher. Taxes on trades have generally been abolished; in Milan, however, bakers and butchers are subject to an impost. Certain tradesmen in open towns are subject to taxes, which, like others, are farmed out to the best bidder, who usually compounds with the parties for a stipulated sum. The income and expenditure of the different cities of Lombardy amount annually to from 26,000,000 to 46,000,000 lire.

The import duties are heavy on most articles. Cotton, woollen, pewter, and tin manufactures, fine polished hardware, porcelain, and books allowed by the censorship, are admitted on payment of an ad valorem duty of 60 per cent. on their declared value. Silk fabrics- pay 20c. per lb. nett. The importation of salt is prohibited, salt, tobacco, saltpetre, and gunpowder being government monopolies. Here, as in England, the private cultivation of tobacco is disallowed, and the salt springs not made use of by the government, must be filled up. The revenue has risen less in consequence of an increase of duties than of increased production and consumption. The total public revenue of Lombardy amounts to upward of 50,000,000 lire a year. The public debt has been considerably reduced; and the interest, which is 5 per cent., is regularly paid.

Government, Education, &c.—The government of Austria in Italy is so liable to be disturbed, through the rooted dislike entertained by the Italians for the Germans, as to require the most vigilant attention on the part of the Austrian ministry. The policy of the latter has been to restrict the power and privileges of the nobles and large proprietors, who have generally been found at the head of any popular movement; and, at the same time, to conciliate the middle and lower classes. Accordingly, the representation in the provincial councils has been rendered apparently more popular than in the other parts of the empire. Each of the two provinces has its assembly, with attributes and powers similar to those of the German Stände; but their composition is wholly different. They have neither ecclesiastical members, nobles sitting in right of birth or property, nor deputies of close corporations; but with all this, the most effectual precautions are taken to hinder these assemblies having any popular bias. The members are appointed through the medium of a double, or rather a triple, stage of election. The two great classes of Contadini, the proprietors of land; and Cittadini, the inhab. of towns, are the primary electors, the suffrage depending on the payment of a certain amount of taxes. These primary electors return from their general body a council of electors, the members of which must possess a higher property qualification than is requisite for the primary electors. The council of election elect from the members of its own body a certain number of candidates, and from these candidates the crown selects those who are to act as members of the provincial assembly; and, as if all this were not enough to stifle anything like popular feeling, it reserves to itself, whenever it thinks fit, power to cancel all the proceedings, and to order a new selection! And even when elected, this assembly has no legislative powers; the will of the emperor being law. This is carried into effect by the viceroy, who is at present an Austrian archduke; and under whom there is a governor in each prov., assisted by a government council appointed in Vienna. Each deleg. has a delegate, or political superintendent, and a separate financial officer; each district a chancellor; and each commune a podestà. In

the chief town of each deleg. is a court of primary jurisdiction; in Milan and Venice are courts of appeal and of commerce; and a high court of revision sits in Verona. Trial by jury, and vivà voce pleadings and examinations, are unknown. And if we add to this, that a jealous censorship is established over the press, and that only certain foreign journals or books can be imported, we shall have a pretty good idea of the spirit of the government. Two foreign regiments are maintained for the police service, one in either government. Eight regiments of the line in the Austrian army are levied in these prov., but there is no militia. All males, whether noble or otherwise, are registered for military service at the age of 18, unless exempted from physical or other causes. From those thus registered the number required are taken by ballot; but are allowed to serve by approved substitutes, for whom, however, it is often necessary to pay large sums. The period of service is eight years, after which the soldier is entirely free. Whatever, therefore, may be said by such flimsy eulogists as Von Raumer, the government of Austria in Italy is undoubtedly a cold, repulsive, and jealous despotism. But it is not oppressive; and we agree with Mr. Turnbull, that in point of fiscal and military pressure, it is more lenient than that of the French; and all that may tend to the advancement of agriculture and commerce, and the material comforts of the people, is sedulously promoted.

It is, also, true that large sums are expended by the government in keeping up the roads and other public works, and in public education. A larger proportion of the pop. is educated in the Lombardo-Venetian kingdom than in any other prov. of the Austrian empire, except the Tyrol and Bohemia. By a law of 1822, every commune is obliged to maintain a primary school, either wholly or in part; and, in 1837, only 86 communes were without schools exclusively their own: 45 gymnasiums, 16 ecclesiastical seminaries, and 12 lyceums, exist in the chief towns, and there are two universities, those of Pavia and Padua, the former ranking as the first in Italy. But, notwithstanding all this apparatus, a really good education is unknown in Lombardy; and that which exists is better fitted to enslave and debase than to expand the mind. It is wholly under the direction of the clergy; and no school can be opened, or book used in a school, or other seminary, without the express sanction of the government. Even the Conversations-Lexicon has incurred the displeasure of this paternal government.

History.—The greater part of this portion of Italy, after the fall of the Western Empire, was successively possessed by the Heruli, Ostrogoths, Greeks, and Lombards; the latter held it from 568 till 774, when Charlemagne annexed it to the empire of the Franks, to which it remained attached till 888. From that period, except the territory of the Venetians, it generally belonged to the German emperors, till the establishment of the republic of Milan, in 1150. This republic was erected into a duchy in 1395, and, in 1535, came into the possession of the emperor Charles V. After the war of the Spanish succession, the duchies of Milan and Mantua were assigned to Austria, to which they have since belonged, with the exception of the short time they formed a part of the Cisalpine republic and French empire. Venice and its territory, which had existed as an aristocratic republic from the 7th century to 1797, was confirmed to Austria by the treaty of Vienna, in 1815. (Bowring's Reports on the Lomb.-Ven. States; Turnbull's Austria, &c., and the Journal of Education; Von Raumer's Italy, i., 194–203; Châteauvieux, Italy and its Agriculture, Rigby's Travels!, p. 14–26, 274–285, &c.)

ITALY, t., Yates co., N. Y., 15 m. W. Penn Yan, 196 m. W. by S. Albany. Bounded N.W. by Canandaigua lake. Drained by Flint creek. It has one store, one fulling-mill, two grist-mills, 10 saw-mills; nine schools, 496 scholars. Pop. 1634.

ITAWAMBA, county, Miss. Situated in the N.E. part of the state, and contains 900 sq. m. Drained by head branches of Tombigbee river. It contained in 1840, 12,391 neat cattle, 1909 sheep, 16,149 swine; and produced 5161 bushels of wheat, 903,259 of Indian corn, 5590 of oats, 10,316 of potatoes, 6719 pounds of tobacco, 270,970 of cotton. It had six stores, 10 grist-mills, seven saw-mills, one tannery, two distilleries; seven schools, 194 scholars. Pop. : whites, 4659; slaves, 720; free coloured three; total, 5375. Capital, Fulton.

ITHACA, p. t., capital of Tompkins co., N. Y., 168 m. W. by S. Albany, 40 m. S.E. Geneva, 295 W. Organised in 1821. Bounded N.E. by Cayuga lake. Watered by Fall, Cascadilla, and Six mile creeks, and Cayuga inlet, which afford good water-power. It contains 24 stores, two lumber yards, two woollen factories, one cotton factory, with 1572 spindles, two flouring-mills, one grist-mill, 10 saw-mills, one oil-mill, one paper-mill, three tanneries, one brewery, four printing-offices, two binderies, two weekly newspapers. Total capital in manufactures $670,250, one academy, 136 students, 15 schools, 1852 scholars. Pop. 5650. The village is situated on Cayuga inlet, 1½ m. from the head of the lake, and is navigable to this place for boats of 30 tons. The steamboat landing is on the lake. It is surrounded on three sides by an amphitheatre of hills, rising by a gentle ascent to the height of 500 feet. It contains 34 streets crossing each other at right angles, the back streets extending on to the hills, and commanding a beautiful prospect, including Cayuga lake. It contains a courthouse and jail in the same building, a county clerk's office, six churches, a Presbyterian, Dutch Reformed, Episcopal, Methodist, Baptist, and an African, two banks, a flourishing academy, a lyceum, most of the above named stores, and some of the manufacturing establishments, 700 dwellings, and about 4500 inhabitants. All the streams in the vicinity have falls. Fall creek descends 438 feet in the course of a mile, having successive falls of 70, 30, and 113 feet. The last is peculiarly grand. A tunnel is built, which conveys the water from the head of the fall to the machinery, which it made to move. Clinton hotel in this village, is one of the most spacious and splendid public houses in the state. It contains more than 150 rooms, and one of its halls is 190 feet long, and its dining room is 56 feet long. It has several porticos, the principal of which is supported by six Ionic columns.

The commercial facilities of Ithaca are as great as its manufacturing. It has a boatable communication to the city of New-York, by the Seneca and Erie canals. By the Ithaca and Owego railroad, 30 m. long, it communicates with Owego, on Susquehanna river. Its trade with Pennsylvania is considerable, receiving iron and coal in exchange for plaster, salt, lime, flour, and merchandise. It was founded in 1805, and chartered as a village in 1821. The scenery around the village is exceedingly picturesque and beautiful, and few places have more to interest the curious traveller.

ITHACA, one of the Ionian Islands, and celebrated in antiquity as the kingdom of Ulysses (*scopulos Ithaca, Laertia regna*, Virg., Æn. III., 275), 7 m. S. Santa Maura, 3 m. E. Cephalonia, and 17 m. W. the coast of Acarnania ; Point Marmaca, at its N. end, being in lat. 38° 30′ N., and long. 20° 39′ E. Length 14 m., breadth 4 m., area 44 sq. m. Pop., in 1836, 9644. It presents from the sea the appearance of a barren, rugged rock, deeply indented on its E. side by a gulf, at the bottom of which is Vathy, the port and cap. of the island, accurately described in the Odyssey :

> " A spacious port appears,
> Sacred to Phorcy's power, whose name it bears ;
> Two craggy rocks, projecting to the main,
> The roaring wind's tempestuous rage restrain ;
> Within, the waves in softer murmurs glide,
> And ships secure without their haulsers ride."—Pope.

About a third part of the surface is capable of cultivation, the greater part of which is laid out in vineyards. The chief produces are wine (esteemed in Greece as extremely delicious), olive oil, currants, barley, and a small quantity of wheat ; but the industry of the islanders is greatly impeded by the taxes levied on their exports by the Ionian government. After all, it appears probable that Ithaca has little to interest, beyond the associations connected with its ancient history. Many of the places mentioned by Homer can be traced, with great appearance of probability. The port *Phorcys* is clearly identical with Mole, and the inner harbour of Vathy seems to correspond with the *πεδλοχον 'Ρείθρον σχίνος* under Mount Neïon. In the S. part of the island, at no great distance from the shore, is a spring, rising at the foot of a rock still called *Korako*, and supposed to be the Arethusa of Homer. (See *Odys.*, v. 408.) Some ruins of Cyclopean walls, similar to those of Mycenæ and Tiryns, are considered by Dodwell to be the remains of the city of Ithaca, the residence of Ulysses. (See IONIAN ISLANDS.) (Dodwell, i., 66 ; Prie. Reports, &c.)

IVES (ST.), a parl. bor., seaport, and par. of Cornwall, at the W. extremity of the bay of same name, 18 m. W. Truro, and 250 m. W. by S. London. Area of par., 1850 acres ; pop. of parl. bor. (which includes the three pars. of St. Ives, Lelant, and Towednak), in 1831, 6378. It consists principally of one long street, branching S. into two smaller ; and the houses are generally of moderate size, and built in situations to suit the convenience of persons connected with the trade of the port. The church, a low but spacious building, erected in 1434, stands close to the sea : there are also two places of worship for dissenters, a national school, and two Sunday schools. A grammar school, founded by Charles I., has gone to decay. The town-hall and customhouse are the only other public edifices. The port has a pier, built by Smeaton, in 1770, at an expense of £10,000, within which small vessels lie aground at low water. Large ships may anchor in the bay, in 5 and 7 fathoms ; but, being quite exposed to the N. winds, it is not much frequented. Notwithstanding these disadvantages, however, this port had, in 1838, 117 ships, of 9019 tons burden, and

the customs revenue in 1838 amounted to £14,888. The pier dues are let for £830 a year, exclusive of the tax of 1d. per hhd. on the export of fish, which sometimes exceeds £600 in a single year. The principal employment of the inhabs. is the pilchard fishery, which of late has been carried on with more than ordinary success, and to a greater extent than in any other town of Devon or Cornwall. The season lasts from July to Sept., and in favourable years very large quantities are exported to the Mediterranean, a considerable supply being also furnished for the consumption of the town and neighbourhood. In 1831, the entire quantity exported amounted to 27,119 hhds., of which 18,141 hhds. were furnished by St. Ives alone. Several new mines have likewise been opened in the vicinity, affording additional employment to the people. The corporation, chartered in the reign of James II., was, down to the passing of the Municipal Reform Act, a close self-elected body of 11 members : it now comprises four aldermen, one of whom is mayor, and 12 councillors, and has a commission of the peace, under a recorder. Corporation revenue in 1839, was £298. The borough sent two members to the House of Commons from the 5th of Queen Mary down to the passing of the Reform Act, which deprived it of one member. Previously to the last-mentioned act, the franchise was vested in the inhabitants paying *scot and lot* ; the boundaries of the parliamentary borough were then also enlarged, by the addition of the two adjacent parishes of Lelant and Towednak. Registered electors, in 1839-40, 398. Markets on Wednesday and Friday ; cattle fair, Saturday before Advent. (*Mun. Rep. ; Parl. Bound. Rep. ; Com. Dict. &c.*)

IVIZA or IBIZA (an. *Ebusus*), an isl. in the Mediterranean, forming one of the Balearic group belonging to Spain, 50 m. E. by N. cape Nao in Valencia, and 42 m. S.W. Majorca ; the cap. on its S.W. side being in lat. 38° 53′ 16″ N., long. 1° 26′ 22″ E. It is of an irregular five-sided figure ; its length from N.E. to S.W. being 27 m., and its average breadth 15 m. The coast is irregular, indented by a great number of bays, the largest being thore of St. Antonio and Ivisa : the surface is hilly, and in many parts well wooded ; but there are several picturesque and fertile valleys having a soil well adapted for tillage. The climate is, in most respects, similar to that of Valencia and Catalonia : the winters are so mild that the thermometer seldom falls below 13° Réaum., and the heats of summer are tempered by the sea breezes. The chief products of the island are olives, wine, corn, flax, and hemp, different kinds of fruit, especially figs, for which it was celebrated even in the time of the elder Pliny. The salt-pans are so productive that salt is a chief article of exportation : large flocks of sheep are pastured on the hills, and the sea near the coast abounds with fish, the capture of which gives employment to many of the inhabitants. But, notwithstanding these advantages, the island is in great poverty, owing to the indolence of the inhabitants, and their slovenly mode of tillage. The Ivizans are of middle size, shrunk and sallow ; they speak a language similar to that spoken in Catalonia and Valencia, being a corrupt dialect of the ancient Romannce, once the common language of all S. Europe.

The capital Ivisa (which has a population of 5790 persons) is fortified, and has a good harbour. It is the residence of the governor, and a bishop's see. The chief buildings are a cathedral, six churches, two convents, two hospitals, and a public school.

Ivisa, the largest of two islands, called by Strabo *Pityusa*, or the pine-bearing islands, was early occupied by Phœnicians and Carthaginians, whence it has been called *Eberus Phœnissa* by Silius Italicus. (*Pun.*, lib. iii., i., 362.) It was taken from them by Q. Metellus, and remained subject to the Romans, and their successors the Vandals, till the conquest of Spain by the Moors in the 8th century. The Spaniards took the island in 1114, and attached it to the kingdom of Arragon, since which it has usually followed the fortunes of the larger islands, Majorca and Minorca. In 1708, during the war of the succession, it submitted to Sir John Leake with a British squadron, and was ceded to England, together with Minorca, at the peace of Utrecht. They continued in the possession of the British till the peace of 1814, when they were restored to Spain. (*Miñano, Dict. Gliog.*)

IVREA (an. *Eporedia*), a town of N. Italy, dom. of Sardinia, div. Turin, cap. prov. of the same name, on the Doire, 30 m. N.N.E. Turin. Pop. in 1838, loc. com., 8,473. It is an ill-built town defended by old fortifications, a citadel, and a small fortress upon an adjacent hill ; and has an ancient cathedral, supposed to occupy the site of a temple of Apollo, five other parish churches, several convents, a hospital, a seminary, and a large prison. Here are manufactures of silk fabrics and of organsined silk, and some recently established cotton works ; with markets for cheese, cattle, and other Alpine produce ; and for the iron obtained near Cogne, and other places in its vicinity. Eporedia is reported to have been colonised by the Romans in the time of Marius. It would appear from Tacitus (*Hist.*, i. 70) to have been a municipium as well as a colony. Strabo says that 35,000 Salassi, made prisoners by Terentius Varro, were sold here as slaves by public auction. Ivrea has been repeatedly taken by the French, and under the French empire was the capital of the department Doire. (*Dict. Géog.; Cramer's Italy, &c.*)

IZARD, county, Ark. Situated in the N. part of the state, and contains 1600 sq. m. Watered by White river, Little Red river, and branches of Spring river. It contained in 1840, 4115 neat cattle, 1138 sheep, 7704 swine; and produced 7764 bushels of wheat, 131,170 of Indian corn, 6771 of oats, 4307 of potatoes, 16,100 pounds of tobacco, 9648 of cotton. It had eight stores, six flouring-mills, two grist-mills, three distilleries; six schools, 151 scholars. Pop.: whites, 2099; slaves, 141; total, 2240. Capital, Athens.

IZARD, C. H. or Athens, p. v., capital of Izard co., Ark., 156 m. N. Little Rock, 1076 W. Situated on the N. side of White river, at the junction of the Big North fork. It contains a courthouse, jail, and several stores and dwellings.

J.

JACCA, a town of Spain, prov. Aragon, cap. of a partido of its own name, 56 m. N. by E. Saragossa, and 39 m. N.N. W. Huesca; lat. 42° 30′ N., long. 0° 24′ W. Pop., according to Miñano, 3012. It stands at the foot of one of the highest ridges of the Pyrenees, only 21 m. from the French frontier, in a wide and fertile valley, enclosed by the rivers Aragon and Gallego: it is surrounded by a strong wall, and entered by seven gates. The chief public buildings are a cathedral church, castle, military hospital, and five convents. The inhabitants are chiefly employed in agriculture and woollen weaving; but the difficulty of access to other places confines their industry to the supply of the town and immediate neighbourhood. The crops raised in the district comprise wheat, barley, pulse, &c., and fruits are abundant; but the severity of the climate during winter prevents it from producing many of the fruits of S. Europe.

Jacca was a place of some consideration in the time of the Romans, and was the *regio Jaccetania*. It was taken by M. P.-Cato, *anno* 195 A.C., and was made a station for the troops during the war with Spain. (*Miñano*.)

JACINTO, p. v., capital of Fishamingo co., Miss. 233 m. N.N.E. Jackson, 830 W. Situated on the head waters of Tuscumbia creek, and contains a courthouse, and also several stores and dwellings. Nett proceeds of the postoffice, $131.

JACKSON, county, Va. Situated toward the N.W. part of the state, and contains 420 sq. m. Bounded N.W. by Ohio river. Drained by Big Sand and Big Mill creeks, and branches of Little Kenawha river. It contained in 1840, 5059 neat cattle, 3179 sheep, 10,741 swine; and produced 28,290 bushels of wheat, 117,331 of Indian corn, 39,788 of oats, 4953 of potatoes, 5160 pounds of tobacco. It had eight stores, four flouring-mills, ten grist-mills, six saw-mills, four tanneries, two printing-offices, one weekly newspaper; eight schools, 153 scholars. Pop.: whites, 4803; slaves, 87; total, 4890. Capital, Ripley.

JACKSON, county, Ga. Situated centrally toward the N. part of the state, and contains 476 sq. m. Bounded S.W. by Appalachee r. Drained by branches of Oconee r. It contained in 1840, 5462 neat cattle, 3858 sheep, 11,339 swine; and produced 27,188 bushels of wheat, 212,435 of Indian corn, 99,396 of oats, 7199 of potatoes, 2230 pounds of tobacco, 903,670 of cotton. It had five stores, eight flouring-mills, 20 grist-mills, 15 saw-mills, three tanneries, 17 distilleries; 10 schools, 917 scholars. Pop.: whites, 5094; slaves, 2513; free coloured, 15; total, 8522. Capital, Jefferson.

JACKSON, county, Flor. Situated in the N. part of the territory, and contains 1500 sq. m. Bounded E. by Appalachicola river, W. by Choctawhatchee river. Drained by Chipola river and Holmes creek, and their branches. It contained in 1840, 15,907 neat cattle, 969 sheep, 14,584 swine; and produced 127,915 bushels of Indian corn, 38,105 of potatoes, 1,679,138 pounds of cotton. It had six stores, six grist-mills, six saw-mills; one academy, 95 students, two schools, 75 scholars. Pop.: whites, 2002; slaves, 2636; free coloured, 43; total, 4681. Capital, Marianna.

JACKSON, county, Ala. Situated in the N. part of the state, and contains 975 sq. m. Bounded S.E. and S. by Tennessee river. Drained by Paint Rock and Raccoon creeks, and their branches. It contained in 1840, 9098 neat cattle, 6551 sheep, 39,551 swine; and produced 20,956 bushels of wheat, 4557 of rye, 1,043,107 of Indian corn, 78,617 of oats, 35,920 of potatoes, 23,388 pounds of tobacco, 1,165,616 of cotton, 3404 of sugar. It had 12 stores, 24 grist-mills, seven saw-mills, four tanneries, 20 distilleries, one printing office, one weekly newspaper; 19 schools, 536 scholars.

Pop.: whites, 13,862; slaves, 1816; free coloured, 36; total, 15,715. Capital Bellefonte.

JACKSON, county, Miss. Situated in the S.E. part of the state, and contains 1175 sq. m. Bounded S. by the Gulf of Mexico. Watered by Pascagoula river and its branches. It contained in 1840, 8770 neat cattle, 795 sheep, 5229 swine; and produced 15,675 bushels of Indian corn, 21,080 of potatoes, 9800 pounds of cotton. It had three stores, three grist-mills, two saw-mills; three schools, 48 scholars. Pop.: whites, 1459; slaves, 494; free coloured, 82; total, 1995. Capital, Jackson C. H.

JACKSON, county, Tenn. Situated in the N. part of the state, and contains 625 sq. m. Watered by Cumberland r. and its branches. It contained in 1840, 8094 neat cattle, 13,782 sheep, 41,150 swine; and produced 37,085 bushels of wheat, 3632 of rye, 438,113 of Indian corn, 62,609 of oats, 20,301 of potatoes, 850,336 pounds of tobacco, 33,842 of cotton. It had 10 stores, 19 grist-mills, three saw-mills, 15 distilleries; four schools, 119 scholars. Pop.: whites, 11,536; slaves, 1236; free coloured, 110; total, 12,872. Capital, Gainesboro.'

JACKSON, county, O. Situated in the S. part of the state, and contains 400 sq. m. Millstone-grit is found in its N. and central parts, coal is abundant, and iron ore is found in its W. part. Watered by Little Scioto and Little Raccoon rivers, and by Salt and Symmes creeks. It contained in 1840, 8515 neat cattle, 12,062 sheep, 13,300 swine; and produced 56,749 bushels of wheat, 141,187 of Indian corn, 70,963 of oats, 15,606 of potatoes. It had 19 stores, nine grist-mills, 18 saw-mills, three tanneries, one distillery ; 11 schools, 233 scholars. Pop. 9744. Capital, Jackson.

JACKSON, county, Mich. Situated in the S. part of the state, and contains 790 sq. m. Organized in 1832. Watered by Grand river and its branches, and by branches of Kalamazoo river. Limestone is abundant. It contained in 1840, 12,565 neat cattle, 3090 sheep, 21,674 swine; and produced 186,649 bushels of wheat, 167,870 of Indian corn, 4364 of buckwheat, 11,806 of barley, 190,087 of oats, 147,668 of potatoes. It had 36 stores, four flouring-mills, three grist-mills, 25 saw-mills, four distilleries, one brewery, three printing-offices, three daily newspapers; three academies, 246 students, 37 schools, 1365 scholars. Pop. 13,130. Capital, Jackson.

JACKSON, county, Ia. Situated toward the S. part of the state, and contains 500 sq. m. Drained by Muscatatuck r., Driftwood, or E. fork of White river, and White and Salt creeks. It contained in 1840, 10,609 neat cattle, 8835 sheep, 28,005 swine; and produced 36,945 bushels of wheat, 2544 of rye, 386,498 of Indian corn, 73,212 of oats, 12,038 of potatoes, 22,986 pounds of tobacco, 26,678 of sugar. It had 30 stores, two flouring-mills, 12 grist-mills, nine saw-mills, seven tanneries, two distilleries, one pottery; 27 schools, 730 scholars. Pop. 8961. Capital, Brownstown.

JACKSON, county, Ill. Situated in the S. part of the state, and contains 576 sq. m. Organized in 1816. Bounded W. by Mississippi river. Watered by Muddy river and its tributaries. Salt is found on this river near Brownsville. It contained in 1840, 5963 neat cattle, 2264 sheep, 15,690 swine; and produced 18,530 bushels of wheat, 165,320 of Indian corn, 30,693 of oats, 19,644 of potatoes, 8520 pounds of tobacco, 11,160 of cotton, 2271 of sugar. It had six stores, six grist-mills, three saw-mills, one tannery, three distilleries. Pop. 3566. Capital Brownsville.

JACKSON, county, Iowa. Situated toward the N.E. part of the territory, and contains 698 sq. m. Bounded E. by Mississippi river. Drained by Macoquetais river and its branches. It contains iron ore, copper, tin, zinc, gypsum, and porcelain clay. The first is very abundant. It contained in 1840, 1619 neat cattle, 246 sheep, 4676 swine, and produced 6199 bushels of wheat, 48,519 of Indian corn, 8495 of oats, 20,098 of potatoes. It had six stores, two lumberyards, three grist-mills, four saw-mills. Pop. 1411. Capital, Bellevue.

JACKSON, county, Mo. Situated in the W. part of the state, and contains 525 sq. m. Bounded N. by Missouri r. Drained by Big and Little Blue rivers, and Fire Prairie cr. It contained in 1840, 10,896 neat cattle, 10,032 sheep, 37,963 swine; and produced 5336 bushels of wheat, 1449 of rye, 536,190 of Indian corn, 50,961 of oats, 22,410 of potatoes, 11,359 pounds of tobacco. It had 24 stores, three flouring-mills, 17 grist-mills, 19 saw-mills, two tanneries, 14 distilleries, one printing-office, one weekly newspaper; 14 schools, 613 scholars. Pop.: whites, 6345; slaves, 1261; free coloured, 6; total, 7612. Capital, Independence.

JACKSON, county, Ark. Situated toward the N.E. part of the state, and contains 800 sq. m. Bounded W. by Big Black river, N.E. by St. Francis river. Drained by Cache river. It contained in 1840, 2081 neat cattle, 400 sheep, 6312 swine; and produced 1754 bushels of wheat, 91,098 of Indian corn, 3011 of oats, 9600 of potatoes, 3664 pounds of tobacco, 11,700 of cotton. It had three stores, one cotton factory with six spindles, one flouring-mill, five grist-mills, three saw-mills, one tannery, one distillery. Pop.: whites, 1258; slaves 276; free coloured, 6; total, 1540. Capital, Elizabeth.

JACKSON, p. t., Waldo co., Me., 47 m. N.E. Augusta, 694 W. Drained by Marsh river. Incorporated in 1818. It had eight schools, 321 scholars. Assessors' valuation of real estate in 1842, $137,176. Pop. 853.

JACKSON, p. t., Coos co., N. H., 79 m. N. by E. Concord, 585 W. Incorporated in 1800. Drained by branches of Sawyer's river. It has one store, two grist-mills, two saw-mills, seven schools, 195 scholars. Pop. 584.

JACKSON, p. t., Washington co., N. Y., 40 m. N.E. Albany, 416 W. Organized in 1815. Bounded N. by Battenkill. It has one store, one academy, 18 students. Pop. 1730.

JACKSON, t., Northumberland co., Pa. It contains three stones, five grist-mills, three saw-mills, two tanneries, one pottery; four schools, 135 scholars. Pop. 1564.

JACKSON, p. t., Susquehanna co., Pa., 181 m. N.E. Harrisburg, 885 W. Drained by Lackawannock and Tunkhannock creeks. It has one store, two saw-mills; eight schools, 289 scholars. Pop. 754.

JACKSON, t., Dauphin co., Pa. Drained by Powell's and Armstrong's creeks. It contains two stores, five grist-mills, 11 saw-mills, one oil-mill, two distilleries; eight schools, 261 scholars. Pop. 1164.

JACKSON, t., Lycoming co., Pa., 14 m. N.W. Williamsport. Bounded E. by Lycoming creek, by branches of which it is drained. It has one grist-mill, two saw-mills; four schools, 125 scholars. Pop. 336.

JACKSON, p. t., Greene co., Pa. It has two stores, one tannery, one distillery; four schools, 76 scholars. Pop. 1090.

JACKSON, t., Lebanon co., Pa., 7 m. E. Lebanon. Drained by Tulpehocken creek. The Union canal crosses it. It contains four stores, two lumber-yards, three grist-mills, one saw-mill, two tanneries, two distilleries; one academy, 42 students, two schools, 111 scholars. Pop. 2308.

JACKSON, t., Tioga co., Pa., 20 m. N.E. Wellsborough. Drained by Seely's and Mill creeks. It contains four stores, two grist-mills, 36 saw-mills; six schools, 295 scholars. Pop. 1123.

JACKSON, t., Cambria co., Pa. It has three grist-mills, five saw-mills; five schools, 100 scholars. Pop. 693.

JACKSON, p. v., Lick t., capital of Jackson co., O., 26 m. S.E. Chillicothe, 63 m. S.E.E. Columbus, 377 W. It contains a brick courthouse, jail, six stores, two tanneries, a carding-machine, 50 dwellings, and about 300 inhabitants. The postoffice is called Jackson C. H.

JACKSON, t., Montgomery co., O. It has eight schools, 444 scholars. Pop. 1668.

JACKSON, t., Muskingum co., O. It has four schools, 200 scholars. Pop. 965.

JACKSON, t., Starke co., O. It has three schools, 30 scholars. Pop. 1547.

JACKSON, t., Trumbull co., O. It has nine schools, 400 scholars. Pop. 1194.

JACKSON, t., Allen co., O. It has one school, 20 scholars. Pop. 580.

JACKSON, t., Brown co., O. It has two schools, 110 scholars. Pop. 1253.

JACKSON, t., Champaign co., O. It has two stores, five saw-mills; eight schools, 98 scholars. Pop. 1431.

JACKSON, t., Clermont co., O. It has one school, 18 scholars. Pop. 883.

JACKSON, t., Coshocton co., O. Bounded E. by Muskingum r. It has one store; one school, 35 scholars. Pop. 1896.

JACKSON, t., Franklin co., O. Bounded E. by Scioto river. It has one school, 20 scholars. Pop. 784.

JACKSON, t., Guernsey co., O. Pop. 1153.

JACKSON, t., Hancock co., O. It has one grist-mill, one saw-mill; six schools, 198 scholars. Pop. 639.

JACKSON, t., Highland co., O. It has nine schools, 477 scholars. Pop. 2232.

JACKSON, t., Knox co., O. It has six schools, 284 scholars. Pop. 994.

JACKSON, t., Morgan co., O. It has one grist-mill, two saw-mills; seven schools, 200 scholars. Pop. 930.

JACKSON, t., Perry co., O. It has nine schools, 250 scholars. Pop. 1702.

JACKSON, t., Pickaway co., O. It has one flouring-mill, one grist-mill, three saw-mills; five schools, 135 scholars. Pop. 882.

JACKSON, t., Pike co., O. It has four stores, one flouring-mill, one saw-mill. Pop. 1094.

JACKSON, t., Clay co., Ia. It has three grist-mills, two distilleries; two schools, 82 scholars. Pop. 918.

JACKSON, t., Dearborn co., Ia. It has three stores, one woollen factory, two grist-mills, three saw-mills, one tannery, two distilleries; five schools, 150 scholars. Pop. 1007.

JACKSON, t., Fayette co., Ia. It has one school, 37 scholars. Pop. 1165.

JACKSON, t., Hancock co., Ia. It has four stores; three schools, 51 scholars. Pop. 1142.

JACKSON, t., Putnam co., Ia. It has one store, one flouring-mill, two grist-mills, two saw-mills, one oil-mill; two schools, 45 scholars. Pop. 923.

JACKSON, t., Ripley co., Ia. It has one school, 24 scholars. Pop. 4936.

JACKSON, t., Shelby co., Ia. It contains one store, one grist-mill, two saw-mills, one oil-mill, one tannery, two distilleries; one school, 20 scholars. Pop. 1511.

JACKSON, t., Washington co., Ia. It contains nine stores, one flouring-mill, six grist-mills, seven saw-mills, three tanneries; five schools, 111 scholars. Pop. 9463.

JACKSON, t., Wayne co., Ia. It has two academies, 110 students, three schools, 367 scholars. Pop. 3403.

JACKSON, p. t., capital of Jackson co., Mich., 79 m. W. Detroit, 549 W. Watered by Grand river. It contains 93 stores, one flouring-mill, five saw-mills, one distillery, one brewery, three printing-offices, three daily newspapers; one academy, 85 students, nine schools, 236 scholars. Pop. 2772. The village, situated on the E. bank of Grand river, which has here a fall of 6 feet, contains a courthouse, jail two churches, one Baptist and one Union; the state penitentiary, a branch of Michigan University, and a considerable portion of the above-named stores, mills, and manufactories.

JACKSON, p. v., capital of Butts co., Ga., 67 m. W. Milledgeville, 664 W. Situated 8 m. W. of Ocmulgee river. It contains a courthouse, jail, an academy, two churches, a Methodist and Presbyterian; nine stores, and about 20 dwellings. Nett proceeds of the postoffice, $222.

JACKSON, p. v., capital of Hinds co., Miss., and of the state, 1010 m. S.W. Washington city, D. C. Situated on the W. bank of Pearl river, which is navigable for boats to this place. It occupies a level spot half a mile square, and is regularly laid out a quarter of a mile from Pearl river. It contains an elegant state-house, which cost $600,000; a handsome governor's house, a state penitentiary, a large and fine building; a United States land office, two churches, a Methodist and Baptist; a bank, three printing-offices, issuing newspapers; 20 stores, a steam saw-mill, and 2100 inhabitants. A railroad, 45 miles long, connects it with Vicksburg on the Mississippi, which is continued 14 miles E. of Jackson, to Brandon.

JACKSON, p. v., East Feliciana par., La., 124 m. N.W. New-Orleans, 1142 W. It contains Louisiana College, founded in 1825, which has a president and eight professors, or other instructers, 18 alumni, 109 students, and 1850 vols. in its libraries. The commencement is on the first Wednesday in June. The village contains three academies, 262 students, two schools, 20 scholars. Pop. 932.

JACKSON, p. v., capital of Madison co., Tenn., 134 m. W.S.W. Nashville, 819 W. Situated on the N. side of the South fork of Forked Deer river. It contains a brick courthouse, 54 by 44 feet, a jail, market-house, a branch of the Union bank of Tennessee; two churches, two academies, 10 stores, and about 1200 inhabitants. It is about 60 miles in a direct line from Mississippi river.

JACKSON, p. v., capital of Cape Girardeau co., Mo., 198 m. S.E. Jefferson city, 866 W. Situated near a branch of Whitewater river, 10 m. W. of Mississippi river. It contains a courthouse, jail, a United States land-office, a printing-office, several stores, and about 600 inhabitants.

JACKSON, t., Johnson co., Mo. It has three schools, 105 scholars. Pop. 1596.

JACKSON, t., Monroe co., Mo. It has three academies, 125 students, five schools, 139 scholars. Pop. 2905.

JACKSON, t., Polk co., Mo. Pop. 1625.

JACKSON, C. H., p. v., capital of Jackson co., Va., 226 m. W.N.W. Richmond, 335 W. It contains a courthouse, store, and a few dwellings.

JACKSON, C. H., p. v., capital of Jackson co., Miss., 225 m. S.E. Jackson, the capital of the state, 1053 W. It contains a courthouse, store, and a few dwellings.

JACKSON, river, Va., rises by two branches in Pendleton co., and flowing southwardly about 50 miles, is joined by Cowpasture river to form James river. Falling Springs, one of its head branches, has a perpendicular fall of nearly 900 feet.

JACKSONBOROUGH, p. v., capital of Scriven co., Ga., 116 m. E.S.E. Milledgeville, 637 W. Situated on Beaverdam creek, near its entrance into Brier creek. It contains a courthouse, jail, Methodist church, and 15 or 20 dwellings.

JACKSONVILLE, p. v., capital of Telfair co., Ga., 115 m. S. Milledgeville, 768 W. Situated on the N. side of Ocmulgee river. It contains a courthouse, jail, several stores, and about 30 dwellings.

JACKSONVILLE, p. v., capital of Duval co., Flor., 252 m. E. Tallahassee, 801 W. Situated on the N. side of St. John's river, 30 miles from the bar. It contains a large courthouse, a jail, a church, an academy, and several dwellings.

JACKSONVILLE, p. v., capital of Benton co., Ala., 130 m. E.N.E. Tuscaloosa, 717 W. Situated 2 miles E. of Tallaschatchee creek, and contains a courthouse, and a few dwellings.

JACKSONVILLE, p. v., capital of Morgan co., Ill., 33 m. W. Springfield, 813 W. It is one of the largest inland towns of the state, on elevated ground, in the midst of a delightful prairie, which is fertile and well cultivated. It contains a spacious brick courthouse, a jail, a brick market house, a lyceum, a mechanics' association, a male and a female academy, four churches, a Presbyterian, Methodist, Episcopal, and Congregational; 24 stores, numerous mechanic shops, a steam flouring-mill and saw-mill, a cotton yarn factory, two oil mills, two carding-machines, one tannery, two printing-offices, two weekly newspapers, a bindery, and a periodical. Pop. 2500. It is the seat of Illinois college, founded in 1829, which has a president and four professors, or other instructers, 42 students, and 9000 volumes in its libraries. The commencement is on the third Wednesday in September.

JAEN, a prov. and kingdom of Spain, in Andalusia, between lat. 37° 30' and 38° 40' N., and long. 2° 50' and 4° 20' W. Its shape is that of an irregular four-sided figure; and it is bounded N. by the Sierra Morena and La Mancha, W. by Cordova, S. by Granada, and E. by Murcia. Greatest length, 85 miles; greatest breadth, 78 miles: area, 4430 sq. m. Pop. 277,000. This prov., situated in the upper part of the valley of the Guadalquivir, is encircled by lofty mountains, which make access difficult, and give to its borders a rude and mountainous character. The surface is chiefly an alternation of hills and valleys, formed by the Guadalimar, Herrumblar, and other affluents of the Guadalquivir. The climate, though damp in some parts, is, on the whole, healthy and favourable to vegetation. The soil on the hills, consisting of detritus from the primitive and transition rocks of the sierras Morena and Granada, is sandy and barren; but the valleys are extremely rich, and with moderate attention to tillage, might be made highly productive. Agriculture, however, is in the most degraded state; only a very small portion of the soil is tilled, and the produce is insufficient for the consumption of the province. Olives, wine, and other fruits of good quality, gall-nuts, woad, kermes, and shumac are abundant, and honey and silk are produced in small quantities. Cattle and horses, however, are pastured on a large scale, and a breed of the latter, peculiar to the neighbourhood of Ubeda, ranks as nearly equal to the Arabian. The mineral wealth of the province, which was celebrated even under the Romans, consists chiefly of iron, lead, and copper, with small quantities of silver; but lead and iron are the only ores now wrought. Veins of marble and jasper occur here as frequently as in Granada, but are not quarried, from want of spirit in the inhabitants. Manufacturing industry is quite insignificant: silk and woollen fabrics are made in some of the towns; but the chief branch of employment is in pottery, and particularly in making alcarrazas, a species of porous earthen jars, much used in Andalusia for keeping liquors cool in warm weather. (Bowles; Miñano; Dict. Géog.)

JAEN, a city of Spain, cap. of prov., and partido game name, and a bishop's see, on the Jaen, an affluent of the Guadalquivir, 37 m. N. Granada, and 123 m. E.N.E. Madrid. Pop., according to Miñano, 18,700. It is situated on the outskirts of the great Sierra de Susana, and is so surrounded by mountains, crossed by extremely bad roads, that few travellers have visited it. A recently made road, however, joining the high road between Cordova and Madrid, and passing through Baylen and the Puerto de Peñacerrado of the Sierra Morena, has made it more easy of access. The city, above which towers a Moorish castle commanding a fine view of the whole country, has extremely narrow streets, a cathedral, 12 parish churches, and 15 convents. The cathedral is of Corinthian architecture, 300 feet long by 190 feet in breadth, and built in a very pure style: the pavement is laid in chequered slabs of black and white marble, and the high altar is enriched with fine specimens of jasper and marbles. It also has some good pictures and sculptures. The city, which was celebrated, under the Moors, for its manufactures, still contains numerous fabrics of silk, linen and woollen cloths, and soaps, and has a thriving appearance. (Scott's Ronda and Gran., ii., 341.)

The remains of a Roman aqueduct, and various inscriptions, prove the antiquity of Jaen. Under the Moors, it rose to considerable importance, and successfully withstood the attacks of the kings of Castile. It was the theatre of war during the final struggles between the Moors and Spaniards in the 15th century, since which time it has never recovered its former consequence.

JAFFA, or YAFFA (an. Joppa), a town and port of Turkey in Asia, on the coast of Syria, pash. Damascus, sandjiak Gaza, 32 m. N.W. Jerusalem, and 60 m. S.S.W. Acre; lat. 32° 3' 25" N., long. 34° 46' 10" E. Population, according to Robinson, about 4000, one fourth of whom are Christians. It is fortified, and stands on a tongue of land projecting into the Mediterranean, and rising from the shore

in the form of an amphitheatre, at the top of which is a ruined castle. The port, defended by two batteries, is merely a long basin, enclosed by a ledge of rocks, extending from the S. side northward, directly in front of the town; but it is so choked up with sand as to be unapproachable by all except small coasting craft. The houses are chiefly of stone, and the streets are uneven, narrow, badly paved, and dirty: the principal public buildings, are three mosques, one Roman Catholic and two Greek churches, with three convents, and a good bazaar. The quarantine house, recently founded, is clean and well regulated: separate divisions, with a chapel attached to each, being allotted to the pilgrims of the several nations, chiefly Greek, who land here on their way to Jerusalem. A military establishment is kept up, comprising (according to Dr. Bowring) one regiment of infantry, with four battalions of 800 men, and three cavalry regiments, each having 700 men. A considerable traffic has recently been created by the disturbances in Syria for the supply of the Pacha's troops; but usually the town is dull, and little frequented by strangers, except at pilgrim time, when the population is often nearly doubled. Cotton is raised to some extent within the district; and in the neighbourhood are beautiful gardens of orange and lemon trees, tall waving cypresses, coral, and fragrant mimosas, intersected with enormous prickly pears. The fruit bears a high character, and forms a considerable article of export. Tradition assigns to Joppa an exceedingly ancient date. Joshua defined the possession of the tribe of Dan as including the "border before Joppa." (Josh. xix., 46.) In the time of Solomon it was, no doubt, a port of some consequence; for Hiram, king of Tyre, sent a letter to the former monarch, then engaged in building the temple at Jerusalem, saying, "We will cut wood out of Lebanon as much as thou shalt need; and we will bring it thee in floats by sea to Joppa, and thou shalt carry it up to Jerusalem:" and from this place Jonah took his passage in a ship going to Tarshish, when "he fled from the presence of the Lord." In the New Testament it is mentioned as the place where Peter had the vision which revealed to him the duty of preaching Christianity to the Gentiles as well as the Jews; and where he raised to life Dorcas, a faithful disciple, "full of good works and almsdeeds." Among the Greeks and Romans, also, Joppa had the reputation of being very ancient. It is stated by Pliny (Hist. Nat., lib. ix., § 5) to be the place where Andromeda was exposed to the sea monster, from which she was rescued by Perseus. Reland suspects that this fable may have its origin in, or be connected with, the history of Jonah." (Relandi Palestina, p. 864.) In A.D. 66, during the Jewish wars, it was repeatedly taken, and finally all but destroyed; and during the crusades it was so entirely ruined by Saladin, that it had scarcely any buildings left, except its two castles. It was soon afterward repaired by Louis IX. of France. The subsequent history of the place, till the close of the last century, is little known. In 1799 it was taken by Napoleon, after an obstinate and murderous siege. On this occasion Napoleon put to the sword about 1200 Turks that had formed part of the garrison of El Arisch, which, having previously capitulated, had been discharged, on their engaging not to serve against the French. But though their execution was, no doubt, justifiable, according to the laws of war, still it seems to have been an act of extreme and useless cruelty, and wholly at variance with the general conduct of Napoleon. (For farther particulars, see Bowring's Report on Syria; Wild's Narrative, vol. ii., p. 108–172; Robinson's Palestine, vol. i., p. 6–9.)

JAFFNA, a seaport town of Ceylon, near the N. extremity of the island, cap. of the distr. Jaffnapatam, 190 m. N. Colombo; lat. 9° 36' N., long. 79° 50' E. Pop. 8000? chiefly Mohammedan. The town stands on an inlet, navigable for boats, which communicates with the gulf of Manaar. It has near it a pentagonal fortress of some strength, which forms the head quarters of one of the principal garrisons in the island. As a commercial port, Jaffna is the third in Ceylon, ranking after Colombo and Point de Galle. Provisions are cheap; and from its salubrity the town is a favourite resort of the Dutch residents in Ceylon, who have named several small and verdant islands in the opposite roadstead after the principal cities of Holland.

JAFFREY, p. t., Cheshire co., N. H., 48 m. S.W. by S. Concord, 449 W. Incorporated in 1773. Grand Monadnock mountain lies in its N.W. part, more than 3000 feet high. It has several ponds, one of which is 400 rods long and 140 wide. Their outlets afford water-power. It contains a mineral spring, slightly impregnated with carbonate of iron and sulphate of soda. Near it is a deposite of yellow ochre, which has been exported. It contains three churches, a Congregational, Baptist, and Universalist; three stores, one woollen factory, one cotton factory, five grist-mills, three saw-mills; one academy, 40 students, 11 schools, 464 scholars. Pop. 1411.

JAGO (ST.), or SANTIAGO DE CUBA, a city of Cuba, cap. of its E. division, the second in population and magni-

trade, and the third in mercantile importance in the island, about 6 miles from the S. coast, on the river Santiago, the mouth of which forms its port, about 470 m. E.S.E. Havannah; lat. 19° 57' N., long. 76° 3' W. Pop. (1827) 26,738, of whom 828 were whites, 10,032 free coloured, and 7404 slaves. Santiago is well built, having wide streets and stone houses. It has a cathedral, several other churches, a college, hospital, and numerous convents and schools. The port is from N. to S. about 4 miles long, with an irregular breadth, and in some places rather narrow; but it has water sufficient for ships of the line, and is sheltered from wind on every side. Its entrance is narrow, and defended on the windward side by the Morro and Estrella castles. The city is very unhealthy; being hemmed in by mountains on three sides, the free circulation of air is greatly impeded, and the yellow fever commits great ravages in the rainy season. Santiago is the see of an archbishop, and the residence of a governor, who, in respect of civil and political affairs, is independent of the captain-general. It was the capital of Cuba till the beginning of the 18th century, when the Havannah was raised to that dignity; since which the importance of Santiago has diminished. Its trade has, however, of late years increased considerably. In 1827, the imports amounted in value to 1,441,943 dollars, and the exports to 1,570,536 dollars; ten-years afterward, in 1837, the imports were 2,299,399 dollars, and the exports 2,169,901 dollars. The gross customs revenue of the port amounted, in 1827, to 476,365 dollars, and in 1837, to 604,339 dollars. Santiago is the port where the copper ore of the Sierra de Cobre is shipped. It was founded by Diego Velasquez in 1514. (Humboldt; Turnbull's Cuba, p. 223, 224.)

JAGO (ST.), or SANTIAGO, a city of Chili, of which it is the capital and seat of government, in the province of the same name, on the Maypocho, at an elevation of 2800 ft. above the sea, 54 m. E.S.E. Valparaiso, and 270 m. N.N.E. Concepcion; lat. 33° 16' S., long. 80° 48' W. Pop., in 1830, estimated at 65,666. It is situated on the verge of the extensive and fertile plain of the Maypocho, and at a distance has a very imposing appearance, its domes and steeples rising among groves, vineyards, gardens, and maize fields. It is inferior to Lima and Buenos Ayres in its public buildings, but greatly surpasses them in cleanness and regularity, and is, upon the whole, one of the best cities in S. America as to appearance, convenience, and salubrity. Like other cities of Spanish origin, it is divided into quadras, that is, squares or compartments of buildings, 408 feet square, separated by streets about 13 yards across. The city-proper is on the S.W. bank of the Maypocho, and is connected with its suburb of La Chimba by a handsome stone bridge. On its S.E. side the city is separated from its suburb of Cañadilla by the Cañada, a handsome promenade 50 yards wide, planted with poplars; and at the S.W. extremity of Santiago is the suburb of Chuchunco. Fifteen years ago, the city-proper comprised 110 quadras, and the suburbs about the same number. It is built upon ground sloping greatly towards the W., of which circumstance advantage has been taken in supplying water for its consumption and under drainage, which latter is more perfect than in any other S. American city. The waters of the Maypocho are now frequently employed for the ornament as well as use of the city, there being numerous public fountains, reservoirs, &c. A solid brick rampart, 5 feet in breadth, and raised 10 feet above the ground, extends along the S. bank of the river, and protects the city against inundation from the river during the rains. Between it and the town is the Alameda, the favourite promenade of the inhabitants, planted with willows, and furnished with seats, reservoirs, &c. At the N.E. extremity of the city-proper is the hill of Santa Lucia, which destroyed Valparaiso and Concepcion in 1835. (Miers' Trav. in Chili, i. 426–429; Scarlett's S. Amer., &c.)

The houses of the city occupy a good deal of ground; most of them take up 1-4th part of a quadra. The rooms are ranged round the quadrangles or patios, the first being an outer paved courtyard, the second generally laid out as a parterre, and the third used for domestic purposes. The wide archway opening into the front patio is closed at night by a pair of large folding gates, but is always open during the day. The windows, looking into the two outer courtyards, are protected by iron gratings; but in Miers' time, 15 years ago, there were generally no windows in any of the other rooms, the door alone admitting light through a small grating. The front and sides of the houses facing the streets, where not blank walls, are divided into small rooms, and let out as shops. In the centre of the city is the Plaza, or great square, occupying an entire quadra. On its N.W. side are the directorial mansion, the palace of government, the prison, and the chamber of justice; on the S.W. side stand the cathedral and the old palace of the bishop, now occupied by the estado mayor; on the S.E. is a range of shops, &c., with an arcade in front; and the N.E. side is

composed of private residences. All these buildings, except the cathedral, are of brick, plastered and whitewashed. The palace is by far the best edifice as to its architecture: it consists of two stories, inclosing a large open quadrangle; the lower story comprises the armoury and treasury, and the upper story the great hall of audience and the ministers' offices. The cathedral is the only stone edifice in Santiago; it is constructed of limestone quarried in the Chimba suburb: its design is of the better order of Moorish architecture; but when seen by Miers, its front was only half finished. The bishop's palace is a heavy decayed building, and the arcade with the shops behind it is much dilapidated. In the centre of the square is an ornamental fountain, furnished with water by a subterraneous aqueduct. The city is mostly supplied hence with water for drinking, which is conveyed in barrels of 10 gallons each, two of which are a mule's load, and sold for 5d. a barrel. The largest public building, and that most admired by the natives, is the mint; but, according to Miers, it is an unsightly structure. It occupies an entire quadra, and, like the private houses, consists of a variety of offices arranged round three quadrangular courts. Its front, facing the shabby street in which it is situated, presents a series of heavy pilasters, supporting a rude cornice and a ponderous balustrade, and having in its centre a large arched portico. The entire edifice is of plain brick, and was, like the other public buildings, constructed by bricklayers sent out from Spain for the express purpose. The Consulado, a spacious plastered and whitewashed structure, in which the commercial tribunal, senate, and national congress meet; the customhouse and the handsome little theatre are worth notice. The city and suburbs are divided into five parishes. All the parish churches are mean; but not so those of the conventual establishments, which are somewhat numerous. One of the Jesuits' convents have been converted into a national college, and another is used for the public library and printing-office. The library contains several thousand printed volumes, and some curious MSS. relative to the Indian tribes.

Santiago has three markets: the principal is holden in the Bascuñal, a large open space at the foot of the bridge, and is tolerably well supplied with meat and vegetables. The other markets consist of mere moveable stands at either end of the Cañada; but meat, kitchen vegetables, fruits, and other requisites, are continually hawked about the streets on horses or mules, which precludes the necessity of sending to the markets. Fodder for horses is hawked about in a similar manner; and large quantities of lucerne, &c., are daily brought into the town, horses being kept by nearly every family. The horses of Santiago are generally well broken, and are more docile than those of Buenos Ayres. Most part of the adjacent country is devoted to the rearing of live stock; but, when cultivated, it produces good crops of wheat, the soil being excellent, and irrigated by many subterranean springs. The climate, were it not for the dreadful visitation of earthquakes, would be delightful; and, from its comparative coolness, European vegetables may be raised in great perfection. The vine is grown, and wine of good quality might be made if its manufacture were properly understood. In the outskirts of Santiago are numerous handsome quintas or villas, and the approaches to the city are mostly through lanes bounded by walls inclosing extensive vineyards and orchards, which yield a large revenue to their proprietors.

Santiago occupies the site of a previous Indian settlement: it was founded by Pedro de Valdivia in 1541. It has frequently suffered from earthquakes; but, with other towns of the interior of Chili, it escaped the catastrophe which destroyed Valparaiso and Concepcion in 1835. (Miers' Trav. in Chili, i. 426–429; Scarlett's S. Amer., &c.)

JAMAICA (Nat. Xaymaca), one of the Greater Antilles, and the largest and most valuable of the W. Indian islands, belonging to Great Britain. It lies in the Caribbean sea, between lat. 17° 44' and 18° 30' N., and long. 76° 19' and 78° 25' W., about 100 m. S. Cuba, and 190 m. W. Hayti, from which it is separated by the Windward channel. Shape nearly oval; greatest length, E. to W., 165 m.; average breadth, nearly 40 m. Area estimated at 6250 sq. m. Population not accurately ascertained, but it may probably amount to between 370,000 and 390,000. In 1835 there were 311,692 blacks.

The Blue mountains, a lofty range, run through the island, in its whole length, rising in some places to upwards of 7400 ft. in height. On the N. and S. sides of this range, the aspect of the country is extremely different. On the former, the surface rises gradually from the shore by undulating hills, separated by spacious valleys, watered by numerous rivulets, and clothed with pimento groves. The scenery on the S. side is much bolder. The shore is skirted by abrupt precipices and inaccessible cliffs; and the hill ranges towards the interior are more abrupt and less fertile. Between these ranges and the foot of the central chain are extensive savannahs, and wide plains cultivated with the sugar-cane, &c., the luxuriant beauty and verdure of which is set off by a boundless amphitheatre of forest

> "Insuperable bught of loftiest shade,
> Cedar, and branching palm."

The outline of the forest melts into the distant blue hills, and these again are lost in the clouds. The island is well watered. There are about 100 rivers, none of which, however, is navigable except for boats. Black river, which debouches on the S.W. coast, is the largest, but is only available for flat-bottomed boats and canoes for about 30 m. Like all the other streams, its current is very rapid.

From the geographical position of the island, so near the equator, the climate in the low grounds is necessarily very hot, with little variation throughout the year; the days and nights are, for the same reason, nearly of equal duration, there not being more than two hours difference between the longest day and the shortest. There is very little twilight; and we may add that, when it is noon in London, it is about seven o'clock in the morning in Jamaica. The medium temperature of the year near Kingston ranges between 70° and 80°, but little differences of elevation have a wonderful effect over the temperature and the salubrity of the climate. "At about 4200 ft. above the level of the sea, the temperature usually ranges between 55° and 65°; in the winter it falls even as low as 44°. There the vegetation of the tropics disappears, and is supplanted by that of temperate regions. Showers are common in the interior almost throughout the whole year, but they do not fall with the same violence as in the plains, and the quantity of rain appears to be less. The air is exceedingly humid, subject to dense fogs, and those rapid alternations of temperature peculiar to all mountain regions. While the pestilence of yellow fever rages in the low grounds, and along the coast of this island, cutting off its thousands annually, these elevated regions enjoy a complete immunity from its effects; for that bane of European life has never been known, in any climate, to extend beyond the height of 2500 ft. The inhabitants are said to enjoy a degree of longevity rarely attained in other countries, and to exhibit that ruddy glow of health which marks the countenance in northern climes, and forms a striking contrast to the pallid, sickly residents of the less elevated districts." (*Tulloch's Report on the Health of the Troops in the W. Indies*, p. 43.) The N. side of the island is said to be more healthy than the S.; but all insalubrity is supposed to cease at an elevation of 1400 ft. The midday heat is, during most part of the year, greatly modified by an invigorating sea breeze, called by Europeans the *doctor*, which sets in from eight to 10 o'clock in the morning, increases in force till about two, and declines with the sun, till, on the approach of evening, it is succeeded by the land wind from the mountains. When these winds become less regular, or altogether fail, as is sometimes the case before the rainy season, the atmosphere is exceedingly oppressive. The year is divided into a short wet season, which begins in April or May, and lasts about six weeks; a short dry season, from June to August; a long wet season, comprising September, October, and November; and a long dry season, which occupies the remaining four months, during which the weather is serene and pleasant, being comparatively cool. The annual fall of rain is nearly 50 in.; the amount has become less in proportion as the forests have been felled. More rain falls on the N. than the S. side of the island, and the average temperature is lower. The principal towns and military stations are on the S. side, and it is estimated that of the European troops employed in Jamaica, one seventh part died annually during the 20 years previously to 1837. Fevers, dysenteries, and diseases of the lungs or brain, are the most fatal. Fevers of a remittent character are more prevalent than in any of the other British stations in the W. Indies. Earthquakes are frequent, and sometimes dreadfully violent: in 1692 the town of Port Royal was submerged several fathoms beneath the ocean, by a catastrophe of this kind. Hurricanes mostly occur between July and October; and, though perhaps not so frequent as in the windward islands, they are sometimes most destructive. One of the most appalling of these visitations took place on the 3d of October, 1780. On this occasion the little seaport town of Savannah-la-Mar, on the S.W. coast of the island, was completely destroyed. During the tremendous conflict of the elements, the sea burst over it with irresistible fury, and in an instant swept into its abyss its inhabitants and their houses, leaving behind no vestige of either! Several hurricanes have occurred since, but, happily, none of them have had such frightful consequences. Jamaica contains no active volcano; but the traces of former volcanic action are sufficiently obvious. Micaceous schist, quartz, and rock spar are common; but, limestone, containing numerous shells, is the most prevalent geological formation. The island contains argentiferous lead, copper, iron, and antimony ores; and the Spaniards are

reported to have wrought both copper and silver mines. Mining industry is now, however, quite insignificant.

The turf-clad hills on the N. side of the island are chiefly composed of a chalky marl; elsewhere the soil is frequently of a deep chocolate colour, or a warm yellow or hazel. The latter, called the Jamaica *brick mould*, retains a good deal of moisture, and is among the best adapted for the sugar-cane throughout the W. Indies. But, though the soil be in some parts deep and fertile, Jamaica is not generally productive, and requires both skilful labour and manure to make it yield heavy crops. In 1789, only 1,907,589 acres were held under grants from the crown, and of this extent only 1,059,000 acres were under culture, leaving about 3,090,000 acres unproductive; from which circumstance it was hastily inferred by Edwards that "not more than one fourth part of the island is fit for any profitable cultivation, great part of the interior country being both impracticable and inaccessible." Of the above 1,059,000 acres, it was estimated that 639,000 (710 estates, averaging about 980 acres each), were occupied with sugar plantations; 280,000 acres taken up by 400 cattle-breeding farms, and 140,000 acres in cotton, indigo, coffee, pimento, ginger, &c. Indigo, cotton, and cocoa were formerly important staples, but these have mostly given way to other articles. Maize, Guinea corn, and rice are the principal grains cultivated; the latter, however, is not raised in great quantities. Maize yields two, and sometimes three crops a year, of from 15 to 40 bushels the acre. Calavances, a species of pea used by the negroes, the kinds of pulse and other garden vegetables common in Europe, thrive well in the mountains; and the markets of Spanish-town and Kingston are abundantly supplied with these, as well as native pot-herbs, &c., of excellent quality. The plantain, banana, yam, cassava, and sweet potato are indigenous; the first-named is the principal support of the blacks. Few countries offer so fine an assortment of tropical fruits. Among these is the bread-fruit tree, from Otaheite, originally introduced by Sir Joseph Banks. The orange, lemon, lime, vine, melon, fig, and pomegranate are met with, having probably been introduced by the Spaniards; and many other European fruits succeed in the cool mountain region. The sunflower is an article which has recently begun to be cultivated for its oil. It is said to be obtained from its seeds in greater quantity than from the castor nut; and the plant comes to maturity ten weeks after having been sown. Cinnamon has been naturalised in Jamaica; and the forests abound with dye-woods and guaiacum, iron-wood, braziletto, mahogany, green-heart, and other valuable kinds of timber, and woods fit for cabinet work. Various kinds of grasses are cultivated; the principal is Guinea grass, a product of so much importance, and growing so luxuriantly, that the grazing farms are for the most part covered with it. Horned cattle are excellent, and better or cheaper beef is not met with in any part of Europe. Oxen or mules are used for farm labour. Horses, an active and hardy breed, are reared for saddle and harness. Sheep, goats, and hogs are numerous: the latter are of a small breed, but their flesh is very good. Poultry, pigeons, &c., are kept in great numbers. The Europeans found many indigenous quadrupeds on the island, but none worthy of notice now exist, except the agouti, some monkeys, and rats, which last are in such immense numbers, and so destructive of the sugar-canes, that from 8 to 10 per cent. per annum of the sugar crop, while standing, is supposed to be destroyed by them. Great numbers of wild fowl are met with: and rice birds, esteemed great delicacies, visit the island in large flocks in October. Alligators inhabit some of the larger rivers, and many varieties of lizards and snakes are found; some of which are used as food by the natives. The mountain crab of Jamaica is highly prized. These singular animals come down by millions from the mountains to the sea, to deposit their spawn, from February to April, and return to their original habitations by the end of June. Copious accounts respecting them will be found in *Du Tertre, Brown, Edwards, &c.*

The European population consists of English, Irish, Scotch, French, German, and Portuguese settlers: the coloured races are divided, according to their share of negro blood, into *sambos, mulattos, quadroons,* and *mestises.* A few maroons, the descendants of Spanish slaves, inhabit parts of the interior. They formerly were a great annoyance to the colony, but being, at length, nearly exterminated, those that survived adopted a more peaceable mode of life. The total surface of Jamaica is generally estimated at about 4,000,000 acres; of which, according to a *Parl. Report* of 1839, 3,403,350 have been granted by the crown, on payment of a quit-rent to the government of the island, and of a land-tax of 2d. an acre; leaving 596,641 acres unaccounted for, and still vested in the crown. There are no means of ascertaining how much of the land assigned to individuals is actually under culture; but in 1838 only 2,588,056 acres paid quit-rent; leaving 815,303, probably less productive and valuable than the rest, but at any rate liable, if not to be resumed by the crown, to be sequestered by the governor of the island for non-payment of quit-rent. The attempt of Lord Sligo to resume such lands on behalf of the crown involved him in disputes with the house of assembly, which asserted a right to possess itself of them: and it would appear, in the absence of any express enactment on the subject, that the crown has no right to resume land once granted, except for the purpose of re-granting it to those who may pay up such quit-rents as are in arrear. (See *Rep. of the Colon. Land and Emigr. Commissioners in Jamaica, Report*, 1840, p. 11.) A large portion of the 396,541 acres unaccounted for is supposed to be held by individuals, owners of contiguous grants, and to be liable to a quit-rent. Nearly all the surface of Jamaica, therefore, appears to be occupied by private parties, and to be altogether under circumstances very different from what it was in Edwards's time, 50 years ago.

Few estates comprise more than 1200 acres; and the recent emancipation of the slaves has tended to split the land into more minute divisions. The large estates, especially those on which sugar is grown, have been latterly reduced to great difficulties for want of labourers. On some estates, on which 70 or 80 negro apprentices were formerly employed, more than 10 or a dozen can now be got to work ; and on other estates, previously wrought by hands, the ordinary number is said to have dwindled down to 20 or 25. The negroes are most anxious to become proprietors of land, which they accordingly purchase, in some parts, for £6 an acre, or thereby. They then devote the principal share of their attention to the culture of esculents and other necessaries on their own patch of ground, or raise arrow-root, ginger, &c., on speculation ; and work on the sugar and other estates only when it suits their inclination or convenience. Thus, in some districts, they will only work the four first days of the week ; and at critical periods of the crops it is necessary to offer high bribes to get them to leave their homes to assist on other days than Friday and Saturday. In some districts in the W. part of the island the wages of field labourers are 1s. 8d. a day, cane cutters get 2s. 8d., and the mill-yard people 3s. 4d. a day: in other parts the wages are higher, but the above may perhaps be considered a fair average for the island. Since their emancipation, the blacks who were formerly provided with lodgings and a piece of ground rent free have had to pay rent for them ; and a good deal of dissatisfaction has arisen from the manner in which this rent has been charged under the new system. It is said to have been estimated, in many cases, not according to the real worth of the premises, but according to the number of persons deriving subsistence from the land, so that the man with the largest family became liable to the heaviest rent! In order the better to command the services of the occupiers, the planters refused at first to give them leases, and stipulated that they might be ejected even at a week's notice. But this plan would seem to have defeated its own object; both by making the blacks inattentive to the culture of grounds held on so precarious a tenure, and by making them extremely anxious to acquire the property of a small piece of land. "Labour and rent," says Sir T. Metcalfe, "are the questions which agitate the island from one end to the other." Of late, however, the preferable plan of fixed rents, unconnected with labour, has been gaining ground ; and 5s. per week may be stated as about the average sum paid by the negro for a house and patch of land. It is due to the blacks to state that, since their emancipation, they have conducted themselves with the greatest propriety ; and, speaking generally, they are in comfortable circumstances.

The coffee and other estates on the N. side of the island have suffered much less, since the emancipation of the blacks, than the sugar estates in the S. In these places, coffee lands are cultivated by German and other European emigrants, and the climate, being there healthy, and well adapted to the constitution of Europeans, the latter have recently formed several flourishing settlements. Projects

are also on foot for increasing the labouring population, by carrying the liberated Africans thither rather than to Sierra Leone ; and by holding out inducements to European emigrants to settle in the central or higher regions of the island. But it is not very likely, seeing the facilities for emigration to other countries much more favourably situated, that the latter project, at least, should have any considerable success. The late great falling off in the exports from Jamaica is hardly, perhaps, greater than might nave been fairly anticipated. We need not here repeat the statements by which we have already endeavoured to show that it is nugatory to expect that the blacks, now that they are emancipated, should voluntarily undertake the hard labour they were compelled to undergo while in a state of slavery. (See GUIANA (BRITISH), HAYTI, &c.) A part, however, of the extraordinary falling off in the exports of sugar may be accounted for by the badness of the crops in the last year or two; and it is probable, as the negroes prefer living in the low plains, the only situations fitted for the growth of sugar, that they may, in some degree, resume its culture. Very great differences may be observed in the condition of different estates, owing, no doubt, to their various management, the degree of interest taken in the welfare of the labourers, &c. It is alleged that the smaller estates are best attended to, and that they have not been deteriorated by the change that has taken place. (See *Jamaica Report*, Part I., p. 8, 9.)

But besides the effects consequent to a transition from slave to free labour, Jamaica will, most probably, speedily have to sustain other and even more important changes. It is impossible that the sugar duties in this country should be permitted to continue, for any very lengthened period, on their present footing ; and even though they were, the rapid increase in the imports of sugar from Hindostan, which has all but boundless capacities for its production, will, it may be presumed, occasion, in the end, the abandonment of its culture in all the West Indian colonies that have not, like Cuba, and perhaps Demerara, unusual facilities for its production.

Commerce.—The history of the trade of Jamaica is not destitute of interest. For a long time after we obtained possession of the island, in 1655, the chief exports were cocoa, hides, and indigo. The wretched colonial policy of Spain had immersed Jamaica in sloth, poverty, and decay ; and even so late as 1772, the exports of sugar amounted to only 11,000 hhds. In 1774, they had increased to 78,000 hhds. of sugar, 26,000 puncheons of rum, and 6547 bags of coffee. The American war was very injurious to the W. Indian settlements, which may be said to be still suffering from its effects in the restrictions laid on the importation of food, lumber, &c., from the United States. The devastation of St. Domingo by the revolution of 1792, which in a few years annihilated a supply of 115,000 hhds. of sugar, previously exported by that island to Europe, gave a corresponding stimulus to its culture in Jamaica and elsewhere. The latter, which, at an average of six years preceding 1799, had produced only 83,000 hhds., exported in 1801 and 1802 upwards of 220,000 hhds., or 143,000 hhds. a year! The same cause gave a similar stimulus to the growth of coffee, which has been increased by the increasing demand for the article in Europe. In 1752 the exports of coffee from Jamaica amounted to only 60,000 lbs. ; in 1775, it amounted to 440,000 lbs. ; in 1797, it had increased to 7,931,621 lbs. ; and in 1839 the exports to England amounted to 19,311,000 lbs. The rise in the price of sugar, which so rapidly increased its production in Jamaica, occasioned a similar though less extensive increase in Cuba, Porto Rico, &c. ; and its subsequent cultivation in Brazil, Java, Louisiana, Guiana, the Mauritius, and other colonies, occasioned a heavy fall in its price, which involved the Jamaica planters in great distress. Notwithstanding a recent rally, there appears to be little prospect of prices attaining their old level, or of property in Jamaica being so valuable as formerly. In fact, as already stated, the presumption is all the other way ; and an entire revolution in the sugar trade may, at no distant period, be fairly anticipated.

ACCOUNT of the Quantities of the Principal Articles exported from Jamaica to the United Kingdom during each of the Six Years ending with 1839.

Articles.		1834.	1835.	1836.	1837.	1838.	1839.
Sugar (unrefined)	cwts.	1,256,253	1,148,760	1,054,042	903,933	1,053,181	765,078
Rum	galls.	2,994,067	2,450,272	2,116,094	2,044,075	2,303,790	1,654,232
Molasses	cwts.	2,800	982		83	294	59
Coffee	lbs.	18,988,883	11,154,307	14,834,896	9,950,679	12,593,746	9,423,197
Cotton	—	26,304	62,872		58,144	18,354	116,705
Ginger	cwts.	7,539	6,067		6,683	8,949	6,054
Pimento	lbs.	1,390,198	2,536,359		2,026,107	892,583	1,071,503
Arrow-root	—	230,087	218,710		92,567	90,153	89,970
Indigo	—	25,443	19,814		90,283	38,023	11,896
Saccades	—	34,379	33,663		27,438	33,418	28,403
Logwood	tons	8,195	5,996		4,439	4,816	5,908

JAMAICA.

Account of the Value of the different Kinds of Provisions imported into Jamaica during each of the Four Years, ending with 1838, and the first half of 1839.

Articles imported.	1835	1836	1837	1838	1839 first half of
	£.	£.	£.	£.	£.
Bacon, pork, and beef .	54,476	61,758	71,360	60,088	48,425
Flour, meal ale . . .	17,994	21,722	17,545	20,251	13,240
Bread, butter, & cheese	30,651	50,273	55,371	51,165	36,40
Corn, and meal . . .	113,011	114,798	149,832	131,419	70,228
Fish	118,047	117,529	68,074	120,742	59,094
Lard	5,398	8,333	9,159	7,461	8,012
Live stock	584	850	13,728	13,461	4,480
Rice, vegetables, &c. .	21,900	11,746	77,255	27,130	13,231
Total . . .	367,521	375,979	465,524	439,531	243,284

The total value of the imports from the United Kingdom amounted, in 1838, to £1,442,570. The value of the exports to Jamaica of late years has generally averaged £1,600,000 a year, being more than half the total amount of the exports to the British West India colonies. A considerable portion of the goods is, however, sent to Jamaica only as to an entrepot, being subsequently exported to the Spanish main. The principal ports (all of which are free), are Kingston, Ports Royal, and Morant, Black river, and Savanna-la-Mar on the S. coast; and Lucea and Montego bay, Falmouth, St. Ann, Ports Maria and Antonio, and Annotto bay, on the north.

Jamaica is divided into three counties; Middlesex in the centre, Surrey in the E., and Cornwall in the W. These are subdivided into 21 parishes, nine of which are comprised in the first, seven in the second, and five in the third named county. St. Jago de la Vega, or Spanish Town, is the seat of government; but Kingston is the largest town, and the real capital of the island. The executive power is vested in a governor, nominated by the crown, aided by a council of 12 members, appointed in like manner, of which the lieutenant-governor, chief justice, attorney general, and the bishop are members. The legislative power is vested in a house of assembly, of 45 members, two elected by every parish, and one by each of the chief towns, Kingston, Spanish Town, and Port Royal. All male inhabitants, of full age, and possessed of a freehold of £10 per annum, may vote for representatives. The latter ought to possess an estate of £300 a year, or personal property worth £1000. The assembly has all the privileges of the British House of Commons, and, like it, its utmost duration is seven years. Since 1728, the assembly and council have been the originators of all laws for the government of the island; the power of legislation having been then conferred upon the island, and a permanent revenue of £10,000 a year guaranteed by it to the crown. The salary of the governor is £5500 a year. Justice is administered in a supreme court, composed of the chief justice, and eight or 10 assistant judges, which sits three times a year for three weeks at Spanish Town. Courts of assize are holden three times a year in each county. Inferior courts of common pleas decide in causes to the value of £20 with costs, and ordinary justices of the peace in those not above 40s. There are admiralty and other special courts; and the governor presides at a court of chancery, from which appeal lies to the privy council. Since the emancipation of the slaves, courts of conciliation, similar to those established in Denmark (Vol. I., 750) and some other countries, have been instituted in numerous parishes; the blacks are frequently members of these tribunals, and are thus accustomed to the discharge of some of the most important social duties. Submission to their decisions is, of course, optional; but there, as elsewhere, they are usually acquiesced in. A new police force, of upwards of 400 constables, was established in 1840. There are prisons and houses of correction in each county and parish; but the system of prison discipline has been, until lately, very lax and inefficient. An act for its improvement has, however, recently passed the island legislature.

The military force usually amounts to about 3000 regular troops and a militia of from 16,000 to 18,000 men; in which latter body all the white males from 15 to 60 are bound to serve. The public expenditure on account of the religious establishments amounts to nearly £25,000 a year; ministers of other denominations besides the church being salaried by the government. Jamaica is under a bishop with a salary of £4000 a year, whose see extends over the Bahamas and Honduras. Nearly £15,000 are spent yearly on public instruction and charitable institutions. Education is pretty widely diffused, except in some parts towards the E. end of the island. But there and elsewhere numerous schools and churches have very recently been established; and it is said that the emancipated blacks have not been slow to avail themselves of the benefits resulting from the institution of savings' banks. The press is free, and several able publications are issued. The public revenue and expenditure amount, at an average, to nearly £500,000 a year.

JAMESTOWN.

each. The compensation money awarded to the proprietors for the liberation of the slaves amounted to £6,161,927, the average value of a slave, from 1822 to 1830, having been £44 15s. 2d. Within the present year the ordinary currency of the United Kingdom has been adopted in Jamaica.

Jamaica was discovered by Columbus in 1493, and was settled in 1503. It remained in the possession of Spain till 1655, when it was conquered by the English, to whom it has since belonged. (Parl. Papers; Edwards's Hist. of the West Indies, &c.)

JAMAICA, p. t., Windham co., Vt., 127 m. S. Montpelier, 449 W. Chartered in 1780. Watered by West river and its tributaries, which afford water-power. It contains dolomite or flexible marble, of a snow-white colour. It has two churches, a Congregational and Baptist, two stores, one fulling-mill, three grist-mills, eight saw-mills; 13 schools, 542 scholars. Pop. 1586.

JAMAICA, p. t., Queens co., N. Y., 12 m. E. New-York, 156 m. S. Albany, 238 W. In the N. is a range of hills which pass through Long Island; it is generally level, and in the S. are salt marshes. It contains the Union racecourse. It has nine stores, one lumber-yard, one woollen factory, five grist-mills, one saw-mill, two printing-offices, two weekly newspapers; one academy, 60 students, seven schools, 394 scholars. Pop. 3781. The village, situated on the Long Island railroad, was incorporated in 1814, and contains the county clerk's and surrogate's offices, five churches, a Presbyterian, Dutch Reformed, Episcopal, Methodist, and African, Union Hall academy, a female seminary, several stores, and manufacturing establishments, 200 dwellings, and about 1400 inhabitants.

JAMES, r., Va., is formed by the junction of Jackson's and Cowpasture rivers, in the Alleghany mountains. At the point where it begins to break through the Blue Ridge, it is joined by North river. It receives Rivanna river from the N., and Appomattox, at City Point, on its S. side, which last is its largest tributary, 130 m. long, and navigable through the greater part of its course. James river is 500 miles long, and enters Chesapeake bay near its mouth, through Hampton road. Lynchburg on the S. side, and Richmond on the N. side, are the principal places on it. A 40 gun ship may go up to Jamestown, and by lightening, to Harrison's bar, where there is 15 feet of water. Vessels of 200 tons go to Warwick, and of 120 tons to Rockets, just below Richmond, 110 m. from its mouth. There is a canal around the falls at Richmond, and above the river is navigable for batteaux, 220 miles. The country, of which this river is the outlet, abounds in tobacco, wheat, corn, hemp, coal, &c.

JAMES CITY, county, Va. Situated toward the E. part of the state, and contains 150 sq. m. Bounded S. by James river, N.E. by York river, W. by Chickahominy river. It contained in 1840, 2713 neat cattle, 914 sheep, 4712 swine; and produced 17,941 bushels of wheat, 86,500 of Indian corn, 34,765 of oats, 2904 of potatoes, 8000 pounds of tobacco, 6307 of cotton. It had five stores, two grist-mills; one college, 140 students, one academy, 36 students, four schools, 93 scholars. Pop.; whites, 1325; slaves, 1947; free coloured, 577; total, 3779. Capital, Williamsburg.

JAMESTOWN, t., Newport co., R. I., 3 m. W. Newport. It consists of Connanicut, a beautiful island in Narraganset bay, 8 miles long, and at a mean breadth 1 mile broad. The soil is a rich loam, adapted to grazing. The inhabitants are chiefly employed in agriculture. A ferry connects it with Newport, and another with South Kingston. On the S. end of the island is a lighthouse. It has one store, one grist-mill; one school, 59 scholars. Pop. 365.

JAMESTOWN, p. v., Ellicot t., Chautauque co., N. Y., 333 m. W. by S. Albany, 318 W. Situated chiefly on the N. side of the outlet of Chautauque lake, which affords a good water-power, used twice over at two separate dams, within the bounds of the village. Incorporated in 1826. It contains four churches, a Presbyterian, Congregational, Methodist, and Baptist, a bank, an academy, 15 stores, two woollen factories, a flouring-mill, four saw-mills, two furnaces, and plough factories, a sash factory, a pail and tub factory, a carding machine, and cloth dressing works, and other manufactories, 250 dwellings, and 2000 inhabitants. A steamboat plies daily through Chautauque lake, 22 miles to Mayville.

JAMESTOWN, James City co., Va., 8 m. S.W. Williamsburg, 65 m. E.S.E. Richmond. The first English settlement in the United States here, was founded in 1608. It was on a point of land projecting into James river, 32 miles from its mouth. A few ruins are all that remain of it.

JAMESTOWN, p. v., capital of Fentress co., Tenn., 134 m. E. by N. Nashville, 580 W. Situated on the side of Cumberland mountain. Laid out in 1827. It contains a courthouse, jail, a church, several stores, 30 dwellings, and about 150 inhabitants.

JAMESTOWN, p. v., Russell co., Ky., 90 m. S. Frankfort,

613 W. Situated 4 m. N. of Cumberland river. It contains a brick courthouse, a jail, several stores, and 160 inhabitants.

JANEIRO. See Rio de Janeiro.

JAPAN (EMPIRE OF), called *Niphon* by the Japanese, and *Yang-hoei* by the Chinese; an insular empire off the E. coast of Continental Asia, and opposite to the coast of Japan and the gulf of Tartary and Corea, from which it is separated by Manchooria. It comprises five large, and a great number of small islands, lying between the 30th and 50th parallels of N. lat., and between the 128th and 151st degrees of E. long.; bounded N. by the sea of Okotsk and the independent part of the island or peninsula of Tarakai, or Karafto (formerly known to English geographers as Saghalien); E. by the N. Pacific ocean; S. by the eastern sea of the Chinese; and W. by the sea of Japan, which communicates with the open ocean by the straits of La Perouse, Sangar, &c., running between the different islands. Our knowledge of Japan is very unsatisfactory: Dutch traders annually visit its only open port, Nangasaki; and the Russians have acquired some slight acquaintance with the country; but, though the talents of Kämpfer Thunberg, Krusenstern, Siebold, Meylan and Fischer, have been engaged in collecting materials for a good description of this very curious and interesting country, the cautious and jealous policy of the Japanese government with respect to the admission of foreigners, (caused, as in China, by the attempts of Jesuit missionaries to Christianise the country,) has hitherto, in a great measure, baffled the efforts of European inquirers into its internal arrangements and economy. The shores of Japan are, likewise, either so rocky or so extremely flat, and are so often enveloped in heavy and dangerous fogs, that exploring vessels cannot approach near enough to make an accurate survey of the coasts. It is necessary, therefore, to premise that the statements in this article, including the following table of the islands, &c., are to be regarded only as rude approximations to the truth:

Islands, and their Provinces.	Situation.	Extent in sq. m.	Chief Towns.
Japan Proper:			
Niphon, 53 provs.	Lat. 35° 30'—41° 20' N. Long. 131° 20'—149° E.	109,600	Jedo or Yedo, Miako, Osaka, Simonoseki.
Kiu-siu, 4 provs.	Lat. 31°—34° 10' N. Long. 129° 50'—133° E.	26,300	Kagosima, Sanga, Nangasaki.
Sikokf, 9 provs.	Lat. 29° 40'—34° 30' Long. 133°—135°	17,900	Tosa.
Iki / Insusima } 2 provs.	Lat. 34° 20' Long. 129° 20'	800	
Total of Japan Proper		155,200	
Japanese dependencies (called the gov. of Matsmai):			
Jesso	Lat. 41°—45° 30' N. Long. 140°—147° E.	68,500 ?	Matsmai, Khakodade.
Tarakai (or Karafto), S. part of	Lat. 46°—50° Long. 142°—145°	47,000 ?	
Kurile islands:			
Kunachir	Lat. 44°—46° Long. 145°—151°	1,800 ?	Ourbisch.
Iturup			
Urup			
Total of Japanese empire		266,600	

Physical Geography.—The three principal islands of Japan Proper, which alone have been explored by Europeans, have a very uneven surface, few plains being of any great extent, and the hilly country extensive and of a rocky character. Niphon, the largest, longest, and best known of these islands, contains a regular mountain chain, running N.N.E., the highest summit of which, called *Fusi*, is, according to Siebold, upward of 12,000 feet high, another also (*Sira-jema*) reaching an elevation of 8000 feet, and being covered with perpetual snow: the average height, however, is alleged to be so moderate, that the high ground generally admits of cultivation almost up to the dividing line of the watershed. The summits above named are active volcanoes, and many other hills emit either flames or smoke. Earthquakes are frequent, one in 1705 having destroyed nearly half of Yedo, and killed more than 100,000 of its inhabitants; thermal and mineral springs also are of very frequent occurrence, so that, on the whole, the islands of Japan may be considered the seat of great volcanic movements, connected, most probably, with those of Kamtschatka, and the islands of Formosa and the Asiatic Archipelago, all of which belong to a chain of heights almost as distinctly marked as the volcanic chain of America. The metallic riches of Japan are stated to be very great, comprising copper in sufficiently large quantities for an extensive exportation, a considerable quantity of sulphur, some lead, tin and iron, and a little gold and silver, the mines of the last two being under the exclusive superintendence of the government. The rivers of Japan, though numerous, are not long, on account of the peculiar narrowness of all the islands: few of them are navigable, and most might be characterized rather as torrents than rivers. The largest is the Yedogawa, in Niphon, rising in the large lake Oitz, or Bisa-no-oumi, and emptying itself, after a probable course of 60 or 80 miles, into the harbour of Osaka. The lake Fakonee, S.W. of Yedo, is treated with superstitious reverence by the natives. The climate in a country extending over so many parallels of lat. must, of course, vary extremely, the N. dependencies having a severely cold climate, while the S. parts of the empire are nearly as warm as the S. of France, though with a temperature considerably more variable, owing to their insular condition. In Kiu-siu and the S. parts of Niphon, as far N. as Yedo, the thermometer ranges between 104° and 29° Fahr., 89° being the average height in middle of summer, and 35° during the severest months ... ter. The winter cold, however, is much increased the ... prevalence of N. and N.E. winds; and the summer ... of July and August are moderated by cooling breezes from the S. and S.E. Rain is very frequent, falling more or less on two-thirds of all the days in the year, but more especially in June and July, which are the *satsuki*, or rainy months: hurricanes, also, and storms frequently occur, and are described as being very violent. (*Thunbury*, vol. iv., 68-90; *Siebold*, vol. i., p. 325.)

Agriculture.—Tillage is followed in Japan, not merely as a pursuit dictated by private interest, but also in obedience to a general and very peremptory law, which obliges all owners of land, under the penalty of confiscation, to keep their property in good productive condition, and, therefore, able (for this is the secret reason of the regulation) to pay a large land-tax to government or its officers. But, whatever may be the cause, the soil, though not naturally fertile, has been so much improved as to be rendered extremely productive. Few plants, except on the hills, are found in a natural state; and the face of the country, even on the mountain sides (which are formed into terraces, as in some parts of Italy and Persia), is so diligently cultivated, that, as Thunberg observes, " it would be difficult to find in the country a single nook of untilled land, even to the dry summits of the mountains;" and this is confirmed in all material respects by Siebold, one of the latest travellers in Japan. In the southern districts rice is raised in very large quantities, as it forms a principal article of food with the inhabitants; but wheat is little grown, and held in light estimation: barley, also buckwheat, a bean called *daidzen*, and another, the *soja dolichos* (from which the well known " soy" sauce is made), potatoes, melons, pumpkins, and cucumbers, are raised in great abundance; and the fruit trees of S. Europe, the orange, lemon, vine, peach and mulberry (the last of which is carefully reared for silk worms), are both plentiful and highly productive. Ginger and pepper are the chief spice plants. Cotton is cultivated in considerable quantities, and tobacco, introduced by the Jesuits, is very generally raised in the S. islands. The grand object of cultivation, however, next to rice, is the tea-plant, brought here from China in the 9th century. Not only are there large tea-plantations, with dyeing houses, &c.: but every hedge on every farm consists of the tea-plant, and furnishes the drink of the farmer's family and labourers. The finer sorts demand extreme care in their cultivation: the plants thrive best on well-watered hill-sides, and they are said to be manured with ried anchovies, and a liquor pressed out of mustard-seed. Among trees, the *Broussonetia papyrifera* is cultivated for 'ts bark, which is converted into cloth and paper; and the varnish tree (*Rhus vernix*, and called *scresine-ki* by the natives) for its gums, used in varnishing wooden furniture; the camphor laurel, also, the iron tree, the oak, fir, and cypress are common, and furnish products useful as well for home consumption as exportation. Of timber, however, there is an insufficiency, and supplies are obtained from the N. dependencies of Jesso and Saghalien. The plants, a great number of which are described by Siebold in his review of the climate and vegetables of Japan (vol. i., p. 280–292), are extremely beautiful, and many of them, as the *Clerodendron, Camelia*, and *Pyrus*, have been naturalized in England; and there is little doubt, the climate being so similar to ours, that numerous other specimens from the Japanese Flora may be introduced.

Cattle and other Animals.—Pasturage in a country inhabited by a people eating scarcely any animal food except fish, and so well supplied with cotton and silk that they feel no want of wool for the manufacture of clothes, must necessarily be very unimportant. Buffaloes and oxen are not numerous, and are used only for draught labour, and there are but few sheep, the progeny of a breed introduced by the Dutch soon after their settlement in Japan: the horses are of inferior size, and are only used by the nobility; there are neither mules nor asses, and pigs are found only in the neighbourhood of Nangasaki. Dogs are common, and a. e considered sacred animals, in consequence of the favour which they enjoyed from one of the *Mikados* or supreme emperors; and cats are even more esteemed, if possible, by the Japanese ladies than by the venerable spinsters of G eat Britain. Among the wild animals, may be enumerated bears, wild boars, foxes, wild dogs, deer, monkeys, hares, r ts, mice, and two small animals of the weasel kind peculiar) the country, and called the *ituts* and *tin*. Birds are numerous and of many varieties: falcons are highly valued, and pelicans, cranes, and herons are considered useful in destroying vermin and insects that are injurious to the interests of agriculture; the pheasants, ducks, and wild geese have splendid plumage; besides which there is a great variety of teal, storks, pigeons, ravens, larks, and other small birds. The common crow, however, and the parrot, have never yet been found in Japan. Among reptiles, snakes are not unfrequent, especially in the N. part of Niphon, and one variety, the *Oxrabami*, is of enormous size: tortoises also and lizards are of common occurrence; and the islands, particularly towards the S., abound with noxious insects, scorpions, centipedes, white ants, &c. An apterous phosphoric insect (*Lampyris japonica*) deserves notice as being similar in its habits to the fire-fly of America, but of an entirely different genus. The seas contain large quantities of fish, affording a main article of food to the inhabitants, and giving employment to "entire villages" of fishermen. The salmon, herring, cuttle-fish (*Sepia octopodia*), eel, perch (*Sciæna japonica* and *Callionymus japonicus*), with many others, are caught i great quantities: oysters, also, of a peculiar and delicio kind, are extremely abundant, and are used almost exc sively as food by many of the poor inhabitants about Ye o, where the fisheries lie. Whales and narwhals frequ atly visit the coast, and are caught by harpooning; the lesh is eaten, the whalebone serves various purposes, a 1 ambergris is extracted from the entrails. (*Thunberg*, ol. iii., 357, *ad finem; Siebold, passim.*)

Manufactures.—The industry of the Jape ese will bear to be compared with that of the Hindoos, or even Chinese. The artificers in copper, iron, and steel, hav a high character, and the swords of Japan rank second only to those made in Khorassan. Telescopes, thermom ers, watches and clocks, of good quality, are constructed t Nangasaki; and if the description of Meylan, in his exce lent work on Japan, of a very curious and complicated c ck, may be credited, some of the workmen possess a ve y high degree of mechanical ingenuity. Glass is made; ot the natives are not acquainted with the art of glass-blow ig. Printing was introduced in the 13th century, and is c aducted, as in China, by means of wooden blocks: eng vings are also made, but in a very clumsy manner. Silk ant otton fabrics, of good quality, are manufactured in quantitie almost sufficient for the consumption of the populati n. Porcelain, more highly esteemed even than that of Ch. a is formed from two peculiar kinds of earth, called *kaoli* i nd *petunsee*. The art of lacquering furniture with gold, silve , and various pigments, the secret of which was till lately almost exclusively confined to the Japanese, and hence called "japanning," is practised with great success; and the specimens that have reached Europe, and are now deposited, with many other curiosities, in the Royal Museum at the Hague, exceed in excellence every other sort of jai nned wares, though Meylan informs us that in the country they would only be esteem d second rate. The process is extremely tedious, and th gum requires long preparation for its conversion into va nish. Five coats, at least, are successively applied, and when dry, rubbed down and polished with stone; many of the more costly specimens, are inlaid with

70

mother-of-pearl. Good paper is made, from the maceration of the mulberry and other barks, the fibres of which are used in the manufacture of cordage. The art of building houses is little understood: they are almost universally constructed of timber, covered on the outside with plaster, and the insides consist usually of two stories, each of which, when divided, is parted off into close rooms by flimsy paper partitions, adorned, or rather disfigured, with garish and bold paintings. Of ship-building and navigation the Japanese have a very slight knowledge; and that is prevented from increasing by a law, which compels the people to build their ships in a particular fashion, somewhat similar to that of the Chinese junks. They are made of cedar, fir, or camphor-wood, and the merchant-vessels average about 70 feet in length, by 20 or 24 feet in breadth, their burden varying from 100 to 150 tons. Great numbers of ships are employed in trading with the different ports of the empire (the principal of which are Yedo, Soseki, Isinomaki, Saga, or Sakai, and Nangasaki), and many others besides are engaged in fishing; but it is manifest from their construction, (as seen in the plates accompanying Siebold's work,) that both hulk and rigging are wholly unfit for sea-navigation. (*Siebold*, i., 218–220.)

Trade and Commerce.—The internal trade of Japan is very extensive, and a variety of regulations are in force, the object of which is to protect and encourage home industry. The prices of goods are not enhanced by imposts of any kind; and communication between the great markets and all parts of the empire is facilitated by numerous coasting vessels and well maintained roads. The shops and markets, especially in Yedo, Miako, and Nangasaki, are well provided with almost every description of agricultural and manufactured produce, and the great fairs are crowded with people from the most distant parts of the country. Accounts also are published, from time to time, of the general state of trade and agriculture; and of the prices current for the chief articles of traffic at the trading towns of Yedo, Miako, Osaka, and Simoseki, on the island of Niphon, Sanga, Kokoura, and Nangasaki in Khu-siu, Tessa in Sikokf, and Matsmai in Jesso. Foreign commerce, however, so far from being encouraged, is vigorously opposed by the government, in consequence of the attempts of the Jesuit missionaries to Christianize the people. An edict, published in 1637, and still in force, makes it a capital offence for the natives to travel into other countries; and their seamen even, when accidentally cast on foreign shores, are, on their return, subjected to rigorous examination, and sometimes tedious imprisonment, to purify them from the supposed pollution contracted abroad. The Dutch, who were the first permitted to visit the empire after the expulsion of the Portuguese, had their earliest factory on the island of Firato; but they were removed, in 1641, by the emperor's orders, to Nangasaki, where, in common with the Coreans and Chinese, they are allowed to bring their goods for sale; but the number of vessels allowed to come each year, and the quantity of each description of wares to be sold, are strictly defined; and the residents in the factory are restricted to 11 only. The ships, immediately on their arrival, are minutely searched, and the crews are kept, during their stay in port, completely secluded from the natives, on the small island of Djesima, close to the harbour. All the business transactions are conducted by the Japanese, who also unload and reload the vessels. Besides these obstructions, the superintendent of the Dutch factory is obliged to send valuable presents, or rather tribute, to the siogun, and, once in four years, he makes an official visit to Yedo with great pomp, and gifts of more than usual value, costing with the journey about £3000. The imports comprise raw silk, woollen, cotton, and linen cloths of various kinds, sugar, dye-woods, seal-skins, pepper, and other spices, quick-silver, tin and iron, cinnabar, glass-wares, &c., from the Dutch, and silk, tea, sugar, dried fish, and whale oil from the Chinese: the exports consist chiefly of copper ingots (forming about 9-10ths of the whole), camphor, and, to a smaller extent, of silk fabrics, lacquered wares, porcelain, soja-dollchos, &c. (*Hagendorp, Coup d'Œil sur l'Isle de Java et l'Archipel. des Indes*, p. 385–400; *Siebold*, i., ch. 2 and 3; *Thunberg*, iii., 89–96.)

Accounts are kept in *thaits*, each of which is equivalent to 3½ Dutch florins, or 5s. 10d. English money, and the thail is composed of 10 *mas*, and the mas of 10 *condorins*. The gold coins are the *itzib*, worth 15 mas (or 8s. 9d.), and the *ko-bang*, equal to 64 mas (£1 7s. 4d.), and the *oban* valued at three kobangs. Large payments, however, are most commonly made in silver ingots of a fixed weight and value. The standard of weight is the Japanese *picoul*, equal to 130·9 English lbs. avordupois, and divided into 100 *cattys* and 1600 *tails*. The measure of length is the *settamy*, equivalent to 6 ft. 4 in. English; but road distance is reckoned by *ri*, or Japanese leagues, about 30 of which, according to Siebold (vol. i., p. 209), go to a degree of latitude (*Balbi*.)

Government and Laws.—The government of Japan is an hereditary, absolute monarchy. The supreme power was originally vested in an ecclesiastical emperor, called *Dairo-sama*, or *Mikado* ;[a] but in 1583 Joritomo, the emperor's *Sjôgun*, or military commander, usurped the chief civil power, and from that time to the present, notwithstanding its acknowledged illegality, the mikado, who is the only real emperor, has been a mere puppet-king, in a state of dependence on his sjôgún, his first officer, and the military chief of the empire. All enactments, however, must have the sanction of the emperor before they have legal force; he alone confers honorary distinctions on the sjôgún and the nobility, and he has the entire superintendence of religious affairs and education. Any farther connexion with sublunary affairs would, it is supposed, degrade the Son of heaven, and profane his holy character. His court is at Miako, where he lives secluded in a large palace, and surrounded by numerous officers, who treat him with almost divine honours. His person is considered too sacred to be exposed to the air, and the rays of the sun, and still less to the view of his subjects; and he is consequently confined within his palace: his hair, nails, and beard are not cleansed or cut by daylight, these operations being always performed when he is asleep; he never eats twice of the same plate, nor uses any vessel a second time; and they are invariably broken to prevent them from falling into unhallowed hands. The mikado's finances, however, are now restricted to the taxes collected from Miako and the surrounding territory, certain revenues from the treasury of the sjogún, and the fees paid on the admission to honourable dignities and offices. His income, indeed, is so small, and the number of his dependents so great, that he may truly be said to live in splendid poverty. The sjogún who has, as has been seen, usurped all the substantial power, holds his courts at Yedo, and exercises entire authority over the lives and property of the natives, controlled only by the laws enacted by former emperors, and which admit of little change. To him, also, directly belongs the local government of the five great towns, Yedo, Miako, Osaki, Sakai, and Naagasaki. The country is divided into 8 districts, which are subdivided into 68 provinces, and these again into 604 counties; the provinces are governed by princes called *daimio*, or high-named; and under them are governors of districts, called *sioma*, or well-named. The daimios are appointed by the sjogún, to whom they are accountable, with hostages for the proper exercise of their authority. They are entitled to the revenues of their respective provinces, which enable them, besides maintaining their state and dignity, to keep as armed force for the preservation of order, and to make outlays in repairing roads, and other public works. They reside usually in large towns, either maritime or situated on rivers, and their castles are defended by strong gates and lofty towers. Once a year, in token of subjection, they repair to the sjogún's court, at Yedo, attended by numerous and splendid retinues, and bearing valuable presents constituting a main portion of his yearly revenues. The executive department is confided to seven ministers, who undertake severally the departments of internal economy and finance, commerce and navigation, public works, police, civil and criminal legislation, war and religion. The supreme judicial council, called *gorodje*, is composed of five daimios, who assist the kubo in his decisions on political offences; and a senate of 15 daimois or nobles forms a subordinate court, that takes cognizance of civil and criminal cases.

The laws of Japan are severe, nay, even vindictive and sanguinary: fines are seldom imposed, and exile to the penal settlement of *Tsitso-su-sima* (inflicted on the nobles), banishment, imprisonment, torture, and death by decapitation, or impaling on a cross, are the ordinary penalties of crime, the shades of which are little distinguished. It frequently happens, also, that the courts visit with punishment not only the delinquents themselves, but their relatives and dependents, and even strangers, who have accidentally been spectators of their crimes; and hence the remarks of Montesquieu on the spirit of the Japanese laws are by no means incorrect: " *Ces lois, qui ne trouvent point d'innocens là où il peut y avoir un coupable, sont faites pour que tous les hommes se méfient les uns des autres, pour que chacun recherche la conduite de chacun, pour qu'il en soit l'inspecteur et le juge. Mais ces gens opiniâtres, capricieux, déterminés, bizarres, qui bravent tous les périls et tous les malheurs, sont-ils, corrigés ou arrêtés par la vue continuelle des supplices et ne s'y familiarisent pas ? Un législateur sage auroit cherché à ramener les esprits per un juste tempérament de peines et de récompenses, &c. ..... Mais le despotisme ne connoît pas ces ressorts, il ne mène pas par ces voies; il peut abuser de lui; mais c'est tout ce qu'il peut faire. Au Japon il a fait un effort, il est devenu plus cruel*

que lui-même." (*Esprit des Lois*, book vi., ch., 13.) The prisons are gloomy and horrid abodes, containing places for torture and private executions, besides numerous cells for solitary confinement. The police is extremely strict, and in the large towns each street has a chief officer, called the *ottoma*, who is responsible for the maintenance of order, the punishment of delinquents, and the registration of births, marriages, and deaths; besides these, four superintendents regulate the economy of the towns, and rigorously punish, often with death, the most trifling infraction of public order or peace, information of which is obtained by an established system of espionage.

Revenue.—The public revenues are derived from taxes on land and houses. The land is assumed to be the property of the state, and is rated according to the class of soil to which it belongs; the rate being said always to exceed one half and often three fourths of the produce; but it is difficult to believe that so heavy a tax can be collected. Tenants neglecting the proper cultivation of their land are punished by ejectment. Houses are rated according to the extent of street frontage, and the amount in which the holders are mulcted is greatly increased by forced presents to the civil officers, and dues for maintaining the temples and idols. The amount of the kubo's revenues cannot be ascertained; but it may be inferred that the land-tax, and the contributions from the daimios, who farm the taxes of their 68 provinces, must form a pretty large privy purse.

Armed Force.—The army in time of peace consists of 100,000 infantry, and 20,000 cavalry; the force during war being increased by levies from the different provinces, to 400,000 infantry, and 40,000 cavalry. The arms used by the infantry are the musket, pike, bow, sabre, and dagger; those of the mounted troops being the lance, sabre, and pistol. The artillery is confined to a few brass cannon and light pieces. The generals have no permanent office, but, in case of war or disturbance, are appointed by the sjogún and princes. Discipline and fortifications are little understood; and their batteries consist usually of a few odd-looking walls, raised without either order or apparent object. Japan, though an insular dominion, has no navy whatever; the ships, such as they are, being wholly used in trade.

Religion.—The form of religious worship in Japan especially the old form, has no resemblance whatever to any of the contemporary Chinese forms: the early inhabitants of Japan had a peculiar form, which, being respected as that of their ancestors, has maintained itself to this day, as well in the hut of the peasant as in the palace of the dairi. Being generally liked, it is not only tolerated, but even protected and venerated by government; even at the present time, it might have been the positive religion of the Japanese, if political causes had not obliged the subjects openly to acknowledge one of the sects of Buddh. The doctrines, views, and interpretations of the ancient rites of the Japanese worship are in no essential points similar to those of Buddh; and though, by contact of 1000 years, they appear to have more or less amalgamated, yet they are kept rigorously distinct by Japanese theologians. The old religion is the *Sin-sin* (lit. *faith in Gods*), or, according to Siebold, the *Kamiso-mitzi*, or way to the *kami*, or gods, the other being a modern Chinese term for it. This sect regards the founders of the empire to be sprung from *Ten-syoo-dai-zin*,* the supreme deity, and to have descended from heaven upon the Japanese land; and their title *Ten-zi* is a recognition of their divine origin. The race is never extinct; for in case of a failure in the succession, a descendant is supposed to be sent from heaven (though, in fact, privately selected by the emperor from the families of the nobles) to the childless ten-zi. The spirit of their ruler is immortal, and this also confirms the faith of the people in the immortality of the soul, in connexion with which they also believe in a future retribution of their good and evil deeds during life on earth. Their paradise is called *Takama-hahara*, and their hell *Ne-no-kuni*. The supreme Deity is too great to be addressed in prayer, save through the mediation of the Mikado, the Son of Heaven, or of inferior spirits called *kami*, of which 492 were born spirits, and 2640 are canonized mortals. For these *kami*, who seem to be regarded somewhat like the saints of the Romish calendar, as intercessors with God, temples are specially erected; and in every Japanese dwelling is a kind of oratory, in which the natives, morning and evening, offer their prayers to the supreme Deity. Large gates and triumphal arches lead to the temples, which, with the dwellings of the priests and other buildings, frequently form extensive and stately edifices. Various eatables are offered as sacrifices to the *kami*, and anciently even human victims were immolated to reconcile the hostility of evil spirits. The priests of this sect are allowed to marry.

* The name *Dairi* is by some writers applied to the emperor; but this is incorrect, for he *dairi* is simply the term for the court of the Mikado, who is here called *Dairi-sama* (lord of the dairi).

* This deity, however, though practically considered as supreme, is only the descendant of more ancient gods the most remote of whom, was according to the Japanese mythology, self-created out of infinite and eternal chaos.

JAPAN (EMPIRE OF).

The Buddhist form of worship is supposed to have been introduced from China, through Corea, in the 6th century of the Christian era; and the dogmas of that religion are divided into a higher and lower doctrine of faith. According to the first, man derives his origin from *nothing*, and therefore has no evil in himself; the impressions of the material world bringing out the evil in him, and fostering its growth. This evil is to be counteracted by following the bent of the soul within, which is neither more nor less than the Deity guiding our actions. The human body having sprung from nothing, will, after death, return to nothing: but the soul survives, that of the wicked floating eternally in the void of space, while that of the good will repose in the palace of the Deity, whence, if the denizens of this lower world should ever need the aid of a virtuous man, it will be sent from heaven to occupy another body. From this curious view of the Esoteric doctrine of the priests, let us to the more popular and practical tenets of the people.

the other side (i. e., in the outer world), before the judge *Emma*, stands a large mirror, in which the acts of all mankind are imaged forth. Near this mirror two spirits, who observe and report the deeds of every person, and a third records them in a book, by which souls of the dead will ultimately be judged, and, according to their sentence, sent to their places of rewards and punishments. *Amida*, the saving deity, is the god of paradise; and the way to ensure a journey on the *Gokurak*, or road to paradise (one only out of six to which departed spirits may be sent), is an obedience to five commandments —viz., not to lie, not to commit adultery, not to kill any living creature, not to get drunk, and not to steal. One of the roads for the dead is *Trikuryo*, the road to the hell of animals; and hence the Buddhists of Japan believe in the transmigration of souls into animals as well as men.

Of the religion of Buddh, as now professed, there are many ramifications, and much superstition prevails. Jew-*mabos*, or monks of the mountain, live a secluded and ascetic life; and blind monks, who deprive themselves of sight that they may not behold the vice around them, are common throughout Japan. Occasionally, in pursuance of vows, men are met running about the streets entirely naked, on a round of visits to different temples; multitudes of religious beggars also are to be seen with shaven heads; and singing girls, in the assumed habit of nuns, procure from the rich considerable sums. The sect of *Syuntoo*, which professes the morality of Confucius, is quite separate from any of the creeds above described, and has existed in Japan since A.D. 50. Here, as in China, its only object is the inculcation of a virtuous life in this world, without reference to an after state of existence. (*The above account of the religion of the Japanese is chiefly taken from Dr. Burger's paper, in the Chinese Repository for Nov., 1833; but the statements are confirmed in every important point by a private English authority.*)

Population, Manners, &c.—The population of Japan has been variously stated: but no estimate yet put forth has the slightest pretension to accuracy. The most moderate estimate, however, fixes it at rather more than fifty millions, exclusive of the inhabitants of the Japanese dependencies. They are divided into eight classes, the princes, nobles, priests, soldiers, civil officers, merchants, artisans, and labourers either agricultural or otherwise: the caste system is strictly pursued, and each follows the employment of his fathers, whatever his talents may be for a different pursuit. The people, physically considered, appear to be a mixed breed of Mongolian and Malay blood, they regard themselves as aborigines. They are, in general, well made, active, and supple, having yellow complexions, small deeply set eyes, short flattish noses, broad heads, and thick black hair, which, however, is not allowed to be worn except on the crown, the sides of the head being kept constantly shaved. The dress of the Japanese consists of several loose silken or cotton robes, worn over each other, the family arms being usually worked into the back and breast of that which covers the rest. To these is added, on state occasions, a robe of ceremony, and the higher classes wear with it a sort of trousers called hakkama (resembling a full-plaited petticoat drawn up between the legs). with one or more swords, according to the rank of the parties. The lower orders are prohibited from wearing swords. The men shave the front and crown of the head, the rest being gathered and formed into a tuft, covering the bald part: the women, on the contrary, wear their hair long, and arranged in the form of a turban, stuck full of pieces of highly polished tortoiseshell; and they paint their faces red and white, and stain their lips purple, and their teeth black. Hats are worn only in rainy weather; but the fan is an indispensable appendage to all classes of the Japanese. Their gait is awkward, owing partly to their clumsy shoes: but that of the women is the worst, in consequence of their practice of so tightly bandaging the hips, as to turn their feet inward. On the other hand, they do not deform them-

72

JAROSLAVL.

selves by confining their feet in tight shoes, like the Chinese. Polygamy is not practised even by the nobles, and far more freedom is permitted to the female sex than in China: many are well educated, and almost all play on musical instruments. Concubines are kept in numbers, varying according to the means of the owner; but they hold a rank much inferior to that of wives: prostitutes are found in every town in greater numbers than in any country of Asia, except Hindostan; and so little discredit is attached to their profession, that they are visited by married females; and received back without remark into respectable society.

Respecting the moral condition of a people so little known, it would be rash to venture any remarks. They are alleged, by Siebold and others to be intelligent, and desirous of increasing their knowledge by inquiries; they study medicine and astronomy, and their observations are as correctly made as their rude instruments will allow. Almanacks are compiled at Miako, the great centre of the national science and literature. The history of Japan has been written with great care by some of its learned writers; and their works on botany and zoology contain good descriptions and tolerable engravings of the plants and animals indigenous to their islands. Poetry, also, is cultivated, and there is a prevalent taste for music. The Japanese language has no relation to the Chinese, nor, indeed, to any known Asiatic language, except that of the Ainos, who inhabit Jesso and Tarakai. Klaproth, in his *Recherches Asiatiques*, Siebold in his *Travels*, A. de Remusat in his *Illustrations of the Japanese Language*, and Meylan in his able work on Japan, present some curious details, the exhibition of which does not fall within the province of this Dictionary. It is a polysyllabic language, has an alphabet of 47 letters, and is written in four different sets of characters, one of which (the *katakana*) is used exclusively by the males, while another (the *hiragana*) is appropriated to the females. The Chinese character also is in use among the learned.

History.—Marco Polo was the first to make known to Europeans the existence of a country called by him *Zipangu*, but since proved to be identical with Japan. In 1542, Mendez-Pinto, a Portuguese, was cast by storm on these shores, and a Portuguese settlement from Malacca was soon after made at Nagasaki, the commercial relations of which, with the inhabitants, were very considerable and highly lucrative to the settlers, till the interference, in 1585, of Jesuit missionaries with the religious profession of the inhabitants, led to the persecution and final expulsion of the traders. The Dutch soon afterwards (in 1600), with great difficulty, prevailed on the Japanese to allow them to trade on condition of not interfering with the national religion; but the vexatious and harassing regulations by which the trade is obstructed, and the very limited extent allowed to it, make it a matter of question how far the factory should be kept up by the Dutch government. The Russians tried, some years ago, to establish commercial relations with Japan; but their proposals were declined, and the envoys were ordered not to return on pain of death. The internal history of Japan is almost unknown; and the statements that have reached us through Kämpfer, Thunberg, and others, are too loose to be admitted as authentic history. (*Voyages de Thunberg au Japon*, vols. iii., and iv., *passim*; Siebold's *Voyage au Japon* (French edit.), vols. i., ii., and v.; *Meylan's Illust. of Japan; Kämpfer's Hist. of Japan*; and several excellent papers in the *Asiatic Journal for 1839–40*.)

JAROSLAVL, or YAROSLAV, a gov. of Russia in Europe, chiefly between the 57th and 59th degs. of lat. and the 38th and 42d of long., having N. Novogorod and Vologda, E. Kostroma, S. Vladimir, and W. Tver. Length, N. to S., about 160 m.; greatest breadth nearly the same. Area estimated at about 19,800 sq. m. (*Schnitzler*). Pop. (1838), 916,000. Surface almost wholly flat, in some parts marshy, and in general only moderately fertile. The Wolga traverses this government in its centre; the other chief rivers are its tributaries the Mologa, Schekma, &c., all which have, more or less, an E. direction. The lake of Rostof, in the S., is eight miles long by six broad, and there are nearly 40 other lakes of less size. Rye, barley, wheat, oats, peas, &c., are grown, and Schnitzler estimates the annual produce of corn at about three millions of chetwerts: a quantity insufficient for the inhabitants who are partly supplied from the adjacent provinces by means of the Wolga. Its hemp and flax are excellent, and cherry and apple orchards are numerous. The gardeners of Jaroslavl and Rostof are famed throughout Russia, and many are met with at Petersburg. Timber is rather scarce. The rearing of live-stock, excepting horses, is little pursued; but the fisheries in the Wolga are important. This government is, however, more noted for its manufacturing than its rural industry. In 1830, there were 105 factories, employing 7970 hands, chiefly in the towns of Jaroslavl, Rostof, and Ouglitch. Linen, cotton, and woollen stuffs, leath-

w, silk, paper, hardware, and tobacco are the principal manufactures: but, independently of the heads above mentioned, the peasants are almost everywhere partially occupied with weaving stockings and other fabrics, and making gloves, hats, harness, wooden shoes, and various rural implements. Commerce is facilitated by several navigable rivers and good roads.

Jaroslavi is subdivided into ten districts; chief towns Jaroslavi, Rostof, and Ougitch. Its population is Russian; and the women are proverbial (among Russians) for their beauty. Only about 1-17th part of the inhabitants reside in towns. In respect of education, the government is comprised under the division of Moscow; and, in 1832, had 19 public schools, and 1141 scholars, besides nine ecclesiastical seminaries, with 1607 students.

JAROSLAVL, a city of European Russia, cap. of the above gov., and of a circ. of same name, on the Wolga, at the mouth of the Kotorosth, 212 m. N.E. Moscow; lat. 57° 37′ 30″, long. 40° 16′. Pop. (1838), 23,856. It is well built, though mostly of wood; and is defended by a fort at the confluence of the two rivers. In its broad main street, which is ornamented with trees, are many handsome stone houses; and three convents and numerous churches contribute to give Jaroslavi an imposing appearance. The Demidof lyceum in this city, founded in 1803, has a good library, a cabinet of natural history, a chemical laboratory, and printing-press, and ranks immediately after the Russian universities. It was originally endowed with lands, to which 3578 serfs were attached, and with a capital of 368,600 silver roubles; since which it has received other valuable benefactions. The same educational course is pursued as in the universities, and lasts three years. The establishment is placed under a lay-director and an ecclesiastic, and has eight professors, two readers, and 40 pensionary students. Jaroslavi has also an ecclesiastical seminary, with 500 students. A large stone exchange (Gostinoi dvor), a hospital, foundling asylum, house of correction, and two workhouses, are the other chief public edifices. This city is the residence of a governor, and the see of an archbishop. It has about 40 different factories, including three of cotton, four of linen, and two of silk fabrics, eight tanneries, and several tobacco, hardware, and paper-making establishments. Its leather and table linen are much esteemed. The position of Jaroslavi on the Wolga contributes to promote its commerce, which is very considerable. Its manufactures are sent to Moscow and Petersburg, and a great many are sold at the fair of Markarief. Two annual fairs are held in Jaroslavi.

This is a city of considerable antiquity, being founded in 1025 by the famous Jaroslav, son of Vladimir the Great, who annexed it to the principality of Rostov. It fell under the dukes of Moscow, in 1495. Peter the Great was the first to give it commercial importance, by establishing its linen manufactures, since which its prosperity has been progressive. (Schnitzler, Dict. Géog.)

JAROSLAW or JAROSLAU, a town of the Austrian empire, Galicia, circ. Przemisl, on the San, a tributary of the Vistula, 16 m. N.N.W. the town of Przemisl. Pop. (1838) 7964, among whom are many Jews. It has a castle belonging to Prince Czartorinsky, a cathedral, and several other churches, a high school and girl's school, and manufactures of woollen and linen cloths, rosoglio, and wax candles. It has an extensive trade in those goods, and in wooden wares, honey, bleached wax, flax, and Hungarian wines, considerable quantities of all which are sent to Dantzic, though less than formerly. It has some rather large fairs; the principal is that holden on the 15th of Aug. (Berghaus; Stein; Oesterr. Nat. Encyc.)

JASPER, county, Ga. Situated near the centre of the state, and contains 460 sq. m. Bounded W. by Ocmulgee river, by branches of which, and by Murder creek, and other branches of Oconee river, it is watered. It contained in 1840, 4956 neat cattle, 3173 sheep, 18,298 swine; and produced 35,426 bushels of wheat, 443,903 of Indian corn, 20,362 of oats, 5270 of potatoes, 5,058,680 pounds of cotton. It had 13 stores, one flouring-mill, four grist-mills, two saw-mills, two tanneries, 12 distilleries, 19 breweries; five academies. 257 students, eight schools, 238 scholars. Pop.: whites, 4921; slaves, 6155; free coloured, 35; total, 11,111. Capital, Monticello.

JASPER, county, Miss. Situated S.E. of the centre of the state, and contains 650 sq. m. Watered by branches of Leaf, and Chickasawha rivers. It contained in 1840, 16,984 neat cattle, 1848 sheep, 14,029 swine; and produced 2350 bushels of wheat, 160,245 of Indian corn, 1514 of oats, 2377 of potatoes, 5612 pounds of rice, 1,180,185 of cotton. It had seven stores, four flouring-mills, four grist-mills, four saw-mills, one tannery, two distilleries; one academy, four students. three schools, 83 scholars. Pop.: whites, 2701; slaves, 255; free coloured, 9; total, 2965. Capital, Paulding.

JASPER, county, Ia. Situated towards the N.W. part of the state, and contains 980 sq. m. Watered by Iroquois

river, which is navigable in high water, and by Pipe and Sugar creeks. It contained in 1840, 2711 neat cattle, 1541 sheep, 4947 swine; and produced 5078 bushels of wheat, 47,070 of Indian corn, 2305 of oats, 3278 of potatoes, 1440 pounds of sugar. It had one flouring-mill, two grist-mills, one saw-mill, two oil-mills. Pop. 1987.

JASPER, county, Ill. Situated near the E. part of the state; and contains 475 sq. m. Watered by Embarras river and branches of Little Wabash river. It contained in 1840, 1549 neat cattle, 693 sheep, 4350 swine; and produced 2880 bushels of wheat, 39,487 of Indian corn, 3490 of oats, 3079 of potatoes, 1700 pounds of tobacco, 5470 of sugar. It had two grist-mills, two saw-mills. Pop. 1472.

JASPER, county, Mo. Situated in the S.W. part of the state, and contains 980 sq. m. Watered by branches of Neosho and Osage rivers. Capital, Jasper. Formed since the census of 1840.

JASPER, p. t., Steuben co., N. Y., 18 m. S.W. Bath, 230 m. W. by S. Albany, 969 W. Drained by Bennet's and Tuscarora creeks. It has two stores, one grist-mill, 15 saw-mills; 12 schools, 369 scholars. Pop. 1187.

JASPER, p. v., capital of Marion co., Tenn., 114 m. S.E. Nashville, 634 W. Situated 3 m. N. of Tennessee river, and 1 m. N.W. of Sequatchee river, Cumberland Mountain, 1 m. distant, presents a handsome view. It contains a courthouse, jail, eight stores, 40 dwellings, and about 250 inhabitants.

JASPER, p. v., capital of Hamilton co., Flor., 90 m. E. Tallahassee. Situated 8 m. N. of Suwannee river. It contains a courthouse, and a few stores and dwellings.

JASPER, p. v., capital of Dubois co., Ia., 194 m. S.S.W. Indianapolis, 655 W. Situated on the N.W. side of Patoka creek. It contains a courthouse, three stores, and about 80 inhabitants.

JASPER, p. v., capital of Jasper co., Mo., 163 m. S.W. Jefferson city, 1109 W. Situated on a branch of Neosho river.

JASSY (an. Jassiorium Municipium), a town of Moldavia, of which it is the cap. on the Bagial, a tributary of the Pruth, about 190 m. N.N.W. Galacz, and 160 m. W.N.W. Odessa, lat. 47° 8′ 30″ N., long. 270 30′ 15′ E. The pop., which is vaguely said to have once amounted to 80,000, and during the present century to 30,000, has been reduced, by war, pestilence, and fire, to less than 20,000. It is situated in a fertile country, partly on a hill, and partly in the valley beneath, and covers a large surface, the houses being interspersed with gardens and plantations. Its fortifications were demolished in 1788, and its only defence is now a small fortress on an eminence, opposite the residence of the hospodar. About 4700 houses, including all its handsomest residences, were destroyed by fire in 1822; since which, Jassy has presented a miserable appearance. Of the 6000 houses it is now said to contain, about 200 only are of stone or brick, and not more than 50 have a second floor. The principal street is wide, and lined with low shops; the other streets are narrow and crooked: they are paved only with logs, and in wet weather are impassable from the mud, while in dry weather they are enveloped in clouds of dust. There is a total want of cleanliness; and this, with the proximity of marshes, and the exhalations which rise from the imperfectly covered sewers, render the town, especially its lower part, very unhealthy. Jassy is the see of a Greek archbishop, whose residence is perhaps the most remarkable public edifice. It has many Greek churches and chapels, a Roman Catholic, and a Lutheran church, numerous convents, a hospital, three public baths, a large building appropriated to a Wallachian printing establishment, the only one in the province, a gymnasium, established in 1644, a Lancastrian school, and a school of handicrafts for females, founded in 1834. It has few manufactures; some canvass is, however, made in the town for export to Constantinople, and the trade in wine, flax, corn, hides, wool, wax, honey, and tallow, is considerable, especially at the fairs. The town has so often suffered from fire, that, to be secure, some of the merchants deposit their most valuable wares in chests in the high church of St. Nicholas. The boyars, or principal inhabitants, have a great passion for pageantry and gaming, and are illiterate in the extreme. Their costume is a mixture of Oriental and European, and the showy dresses of the upper classes strikingly contrast with the general wretched appearance of the population. Like the rest of Moldavia, Jassy swarms with beggars (Dict. Géog.; Macmichael's Journey from Moscow to Constantinople, p. 83; Stein.)

JASZ-BERENY, a town of Hungary, distr. Jagyzia, of which it is the cap., on both sides the Zagyva, here crossed by a stone bridge, 40 m. E. Pesth. Pop. 15,530. It has a large and handsome Roman Catholic parish church, several other churches, a Franciscan convent, Roman Catholic gymnasium, high school, and a town-hall, in which are kept the archives of Jagyzia and Great and Little Cumania. In the centre of the town stands a marble obelisk, erected in 1797 in honour of the archduke John; and within the precincts of the convent, on an island in the Zagyva, the traveller is shown a tomb, reported to be that of Attila! The town has a large trade in corn, horses, and cattle, which latter are reared in great numbers in its vicinity. (*Oesterr. Nat. Encyc.; Berghaus.*)

JAUER, a town of the Prussian dom., prov. Silesia cap. circ. of same name, on the Neisse (which, by its inundations, often does much damage), 19 m. S. by E. Leignitz. Pop. (1638) 5847. It is the seat of the judicial courts for the circle, &c.; has a house of correction, a Lutheran, and five Roman Catholic churches, a free school, and fabrics of linen and woollen cloths.

JAVA, a large and fine island of the eastern archipelago, 1st division, belonging principally to the Dutch, and the centre, as well as the most valuable, of their possessions in the East. It lies between the 6th and 9th degs. S. lat., and the 105th and 115th E. long.; separated from the Sumatra on the W. by the Straits of Sunda, E. by those of Bali from the island of that name; having N. the Sea of Java between it and Borneo, and S. the Indian ocean. Its general configuration is not unlike that of Cuba, except that it is not curved, and it also resembles Cuba in its extent, fertility, products, and commercial value, while it supports five times its amount of population. Its length, W. to E., is about 660 m.; breadth varying from 40 to 130 m. Area, inclusive of the neighbouring island of Madura, estimated at about 45,700 sq. m. Population between five and six millions, among whom are nearly 100,000 Chinese, with Malays, natives of Bali and other islands of the Archipelago, a few Arabs, Moors, and Bengalese, and several thousand Europeans, mostly Dutch, but about 900 English, including those employed in the mercantile navy.

Physical Geography, &c.—Most part of the surface is mountainous. A mountain chain, obviously of volcanic origin, runs W. and E. entirely through the centre of the island, its peaks varying in elevation from 5000 to probably 12,000 feet. All these peaks are of a conical form, and, with few exceptions, each appears to have originated in a distinct convulsion of nature. All have been at some period active volcanoes; in most of them, however, volcanic agency is now apparently extinct, though, from some, eruptions occasionally take place, and sulphureous vapours are emitted, especially after rain. The S. coast is usually bold and rocky, and being exposed to all the violence of the ocean, is unsafe for shipping; the N. shore is, on the contrary, low and marshy, and has many tolerable harbours and roadsteads, affording sufficient shelter to trading vessels, the sea being generally smooth. Rivers numerous; but very few of any size. The largest is the Solo, which runs through nearly the centre of the island, and disembogues on the N. coast, opposite Madura. Its length may be estimated at 400 m., 7-8th of which are navigable for vessels of 900 tons. Surakarta, the capital of the chief native prince, is on its banks; five or six other rivers are at all times navigable for a few miles from the coast, and probably 50 more are in the wet season used for the conveyance of rafts and rough produce downwards. There are many extensive swamps, and in the mountains many small lakes occupy the craters of extinct volcanoes.

Basalt, hornblende, and other volcanic formations are abundantly intermixed among the primary rocks of the mountain region. On either side of the mountain chain coarse limestone and argillaceous iron-stone are very prevalent formations, and are covered, especially in the lower parts of the country, with a volcanic soil of great richness, in some places 12 feet in depth. The N. coast rests entirely upon coral. Metals are few. Mineral springs of various kinds are met with, besides naphtha and petroleum wells, and in one district is a cluster of hills which eject a mixture of mud and salt water*, like the mud-volcano of Maccaluba, in Sicily. (*See* ARAGONA, vol. i., p. 145.)

The seasons are divided into the wet and dry. The former accompanies the monsoon from October to March or April; the latter, the E. monsoon, which lasts during the rest of the year. On the N. coast, where the thermometer sometimes rises to 90° Fah., the climate is very unfavourable to Europeans; but in the interior, at an elevation of 4000 feet, where the temperature ranges between 50° and 60°, no deleterious influence is to be apprehended from the atmosphere. Thunder storms and earthquakes are frequent, but hurricanes are unknown.

Java has a most luxuriant vegetation. It is distinguished by the number and excellence of its fruits and other vegetable products, which comprise many of the most valuable common to tropical climates. Dense forests of teak and other trees, useful for shipbuilding, cover a great part of the interior, especially towards the E. end of the island. The teak of Java is inferior in hardness and solidity to that of Malabar, but it is superior in those respects to that of

* From the salt water thus ejected 250 tons of salt are said to be made annually. (*Smith's Hist.* i., 27.)

Birmah; and is said to excel every other variety in durability. The sago, and many other palms, the very curious pitcher-plant (*Nepenthes distillatoria*), and two virulently poisonous plants, the *anchar* and the *chetik*, are natives of the island. The latter, which is peculiar to Java, is a large creeping shrub, and identical with the celebrated *upas*, formerly supposed, but on no good foundation, to be, like Avernus, destructive of birds flying over it. The aggregate number of mammalia has been estimated at 50, including the royal and black tigers, rhinoceros, several kinds of deer, the wild hog, wild Javan ox, buffalo, &c. Crocodiles and other large reptiles infest the mouths of the rivers and the marshes; and upwards of 20 venomous serpents are enumerated, including some of enormous size. Birds are in immense variety; the bird of paradise visits Java, from Gilolo, Papua, and the other islands to the E.; and the edible nests of the sea swallow (*Hirundo esculenta*) form an important and valuable article of trade for the Chinese markets. This singular product is obtained in the greatest perfection from deep, damp, and all but inaccessible caves along the rugged parts of the sea coast. These are the property of government; and, when they can be easily guarded, produce a considerable revenue. Mr. Crawfurd, who for several years superintended the collection of the valuable caverns of Karang-bolang, on the S. coast of the island, estimates the value of the nests obtained from them at about 140,000 dollars a year, collected at an expense of about 11 per cent. The nests are taken twice a year; and if no unnecessary violence be done, the operation seems to be but little injurious: at all events, the quantity is but little increased by the caves being left untouched for a year or two. The nests are assorted into three qualities, the best being the whitest, or those taken away before they have been soiled by the food or faeces of the young bird. The supply of nests being limited and unsusceptible of increase, and being, at the same time, highly prized by the rich and luxurious Chinese, on account of their real or supposed invigorating powers, they bring enormous prices; the finest sorts selling for £5 or £8 per lb.! and the inferior for 24s. or 25s. per do.! They are collected, but in smaller quantities, in other parts of the Archipelago: their total export to China is estimated at from 240,000 to 245,000 lbs. (*Crawfurd*, iii., 4384-37; *Commerc. Dict.*)

Industry, &c.—"The Javanese are a nation of husbandmen. To the crop the mechanic looks immediately for his wages, the soldier for his pay, the magistrate for his salary, the priest for his stipend, and the government for its tribute. The wealth of a province or village is measured by the extent and fertility of its land, its facilities for rice irrigation, and the number of its buffaloes. The proportion, at an average, of the inhab. engaged in agriculture to the rest of the pop., may be stated at 3½ or 4 to 1; and it is probable that if the whole island were under cultivation, no area of land of the same extent in any other quarter of the globe could surpass it, either in the quantity, quality, or value of its vegetable productions." (*Raffles*, i., 117-420.) At present, only about 1-3d part of the surface is supposed to be under culture; and yet Java not only produces enough of corn for its own consumption, as is a usual but most injurious practice in India. The peasant and his family bestow their labour exclusively on their own possessions, and consider their culture rather as an enjoyment than a task. It is here only that their industry assumes an active and systematic character: the women take a large share of the labour. The work of the plough, the harrow, and mattock, with all that concerns the important operations of irrigation, are performed by the men, but the lighter labours of sowing, transplanting, reaping, and housing, belong almost exclusively to the women.

The implements of agriculture are few and simple; but as well as the agricultural processes, they are more perfect, and imply a greater degree of intelligence than those of the Hindoos, and perhaps indeed, than those of any Asiatic people, the Chinese excepted. The Javanese plough, like the Hindoo, has no share. The sock is tipped with a few ounces of iron, and the earth board is carved out of the body of the plough; the wood is teak, the yoke of bamboo cane. One man conducts the plough, and with a long whip guides the cattle, which never exceed two in number. The Javanese harrow is a large rake, with a single row of teeth. The same yoke and cattle are used for it as for the plough, and over its beam a bamboo cane is placed, on which the person who guides it sits to give a necessary weight to the implement. The hoe is very indifferent; its edge only tipped with a little iron, and its handle about 2½ feet long. The Javanese sickle is a very peculiar instrument. Its object is to nip off separately each ear of rice with a few inches of the straw; for which purpose it is grasped in the right hand, and the operation effected with a dexterity acquired by habit. The whole farming stock of a villager may be purchased for little more than one third part of the yearly produce of his land; or for about 15 or 16 dollars, including a pair of buffaloes. These animals usually serve all agricultural and other purposes in place of horses. Cattle of every description are plentiful throughout Java; but the cows are inferior, and yield little milk. Sheep, goats, and hogs are numerous.

Rice is the principal food of all classes: it is grown not only along the whole of the sea coast, but in all the low grounds and ravines where water is to be had. Wherever rice is cultivated by immersion, the land is divided into small chequers of about 200 or 300 sq. yards, surrounded by dykes not exceeding 1½ feet high, to retain the water for irrigation. When this culture depends on the periodical rains, the charge of these dykes constitutes, as far as irrigation is concerned, the only care of the husbandman; but the greater quantity of the grain of Java is raised by the help of artificial irrigation. The principal care of the husbandman is to dam the brooks and mountain streams as they descend from the hills, and before the difficulty has occurred which would be presented by their passing through deep ravines. From this circumstance, the crests of the mountains, and the valleys at their feet, are best supplied with water, and there, consequently, is the finest and richest husbandry. The slopes of the mountains are formed into terraces highly cultivated, and the valleys are rendered almost impassable from the frequency of the water courses. The art of forcing rice by artificial irrigation is found only to prevail in the most improved parts of the eastern archipelago, and in the best lands. This mode does not depend upon the seasons; and hence we see in the best parts of Java, where it chiefly obtains, rice in every state of progress, at any given season, and in the same district, within, indeed, the compass of a few acres. In one little field, or, rather, compartment, the husbandman is ploughing or harrowing; in a second, he is sowing; in a third, transplanting; in a fourth, the grain is beginning to flower; in a fifth, it is yellow; and in the sixth, the women, children, and old men are busy reaping. Lands which may be inundated at pleasure almost always yield a white and a green crop within the year; and to take two white crops from them, whether a judicious practice or otherwise, is very common. Mr. Crawfurd states that he has seen lands which have produced, time immemorial, two yearly crops of rice. Two varieties of rice are raised in Java, one a large, productive, but delicate kind, requiring about seven months to ripen, and the other small, hardy, and less fruitful, which ripens in little more than five months. The first is always cultivated in rich lands, where one annual crop only is taken; but where two crops are raised, the other variety is grown. The rapid growth of the latter has, indeed, enabled the husbandman, in a few happy situations, to reap six crops in two years and a half.

Rice, of whatever description, is reaped and stored in the same way. The whole field is not reaped at once, but each portion of the grain is taken successively as it ripens; so that, in the desultory manner in which the operation is performed, a very small field, with many reapers, may occupy a period of 10 or 12 days in reaping. With the singular sickle before mentioned, the ears are nipped off, and forthwith transported to the village by the manual labour of the reapers, for cattle or carriages are very rarely used. At the village, the corn is sufficiently dried by a day or two's exposure in a powerful sun, when it is tied in sheaves or bundles, and deposited in little granaries of wicker work, one of which is attached to every cottage. Grain is never threshed by treading it out by means of cattle. It sometimes, chiefly in the case of mountain rice, becomes necessary to separate the seed from the straw, which is done by treading, or rather rubbing, the sheaf between the feet, an operation effected with considerable dexterity. Commonly the grain is stored for use, and transported to market in the straw. The operation of husking is performed by the women in large wooden mortars, with pestles of the same material. (*Crawfurd's Indian Archipelago*, i., 348-365.) Rice is most y grown in the E. part of the island, whence it is sent in large quantities to Batavia for exportation, or to Samarang, from which port a good deal is shipped for China, and the islands of the archipelago.

Coffee, which has now become the great commercial of Java, is grown in the uplands, the best situations being the valleys from 3000 to 4000 ft. above the level staple sea. The coffee plant grows from 12 to 16 ft. in height; it attains to maturity in about five years, and con-

tinues to bear well for the succeeding 10 or 12 years, each tree yielding, at an average, 1½ lb. coffee. The chief peculiarity of the coffee culture in Java is the planting of the *dadap* tree (*Erythrina indica*), in rows alternately wi the coffee plants, for the purpose of affording shelter to the latter. Coffee is raised principally in the W part of the island, where the residency of Preangers fu.nishes at least 1-3d part of the total produce.

Sugar is, also, an important staple. That best known in European markets is called Jaccatra sugar: it is grown near Batavia, where numerous sugar mills have been erected of late years. Formerly the sugar-mills and grounds were almost wholly in the hands of the Chinese, but this is no longer the case: the Europeans share the culture of sugar with the Chinese, and having the advantage of machinery, surpass the latter both in the quantity and quality of their produce. The Chinese, however, by their frugality and business-like habits, are supposed to reap the greatest profit from the sugar culture. A species of sugar, obtained by fermenting the juice of a tree, is much used by the natives.

The increase in the production of sugar in Java since 1825 has been most extraordinary: the quantity exported in 1837 having been about *twenty-five* times greater than in 1825! At present, the export of sugar may be estimated at about 48.000 tons. The increase in the growth of indigo and of coffee has been even greater than that of sugar; and Java is now become one of the principal sources for the supply of these important products. (See *post*.)

In 1830, the government officially announced that the cultivation of spices, previously prohibited in Java, would for the future be free to all parties desirous of engaging in it; and, farther, that every facility would be given to such persons, by supplying them with whatever information, and even the seed, they might require. "This measure (says the Batavian correspondent of the *Singapore Free Press*, June 13, 1839), is no doubt preparatory to the abandonment of the Spice islands, which have always been more of a dead weight on the government than anything else." The government have had already considerable success with cinnamon, the produce of which, in 1830, was estimated at 40,000 lbs. Pepper was formerly grown in great quantities, and its culture under the new system has again revived. Long pepper is indigenous, but has been greatly neglected. Tobacco is a considerable staple; and cotton is grown in almost every part of Java, except its W. extremity. Maize is grown in the plains, and wheat, rye, oats, and barley in the hilly tracts, but the latter only in small quantities. A great variety of pulses and vegetable oils, the sweet potato, cocoa, betel-leaf, cannella bark, and pistachio nuts, are among the other chief articles of culture. In 1828, the Dutch colonial government made some useless attempts to force the cultivation of the vine, and since then no attempt has been made to grow tea, but this also appears to have failed. The introduction of the silk worm has been more successful, and the growth of silk has made considerable progress.

Java is one of the finest and most flourishing colonies in the world. Labour is very cheap: in the European districts, labourers get only 4d. a day, and in the native districts only from 3d. to 4½d. a day; but they are, notwithstanding, in a much better condition than the inhab. of Bengal, being generally well fed and clothed, and, for the climate, well housed. Their food is principally rice or maize, with a little sugar; their clothing is chiefly of cotton, and in the centre of the island it is mostly the manufacture of the country; but they consume a greater quantity of manufactured articles of good quality than the Bengalese. Each peasant has his hut of bamboo, &c., which costs only from about 5s. to 10s. in the first instance, and is usually surrounded by a small garden. The proprietary right to the land, except in a few districts, belongs everywhere to the sovereign. No law nor usage give to the oldest occupant the land he has reclaimed from waste, or the farm he has enriched by his industry. As a matter of convenience, the same cultivator may continue to occupy the same field for life, and his family may afterward succeed, but none can retain possession against the will of the sovereign, or even of his own immediate superior. Half the produce of wet land, and a third part of that of dry, was formerly exacted by the government, but at present the government takes only one fifth part of the produce; nor has any proprietor purchasing land of the government the right to demand more of the native occupant, except for lands which the proprietor himself may have brought into cultivation, for which he may demand one third part, or less, according to the productiveness of the land. It is not uninstructive to compare this moderate assessment with the exorbitant amount taken from the occupiers in Hindostan, and to mark the results exhibited in the impoverishment of the inhab. of British India and the stationary state of the country, and the comfort of the Japanese labourer, and the great and rapidly increasing prosperity of Java.

No permission is necessary from the Dutch government

for Europeans wishing to go to Java, but a licence from the colonial governor is necessary to remaining there. Europeans are permitted to buy and sell lands in the W. provs., and to hold leases in the N. The principal conditions are the payment of a tax of 1 per cent. on the estimated value of the property ; that the proprietor shall not exact more than the before-mentioned proportion of produce as rent; and that he shall keep the roads and bridges in repair. The European proprietors receive their rents in kind; and are obliged to take their produce to Batavia to be shipped. The free cultivation of every article of produce is allowed, except the poppy. The extent of estates held in property in 1830, was about 5000 sq. miles, divided between 20 or 30 European, and 10 or 12 Chinese proprietors ; of which, about 1800 sq. m. were held by 8 or 10 British-born subjects; but in the interval the quantity of land held by Europeans has been materially increased. British and other foreign proprietors are treated precisely in the same way as the Dutch. The Chinese possess a somewhat less extent. Large capitals have been expended on the lands held by Europeans in irrigation, the construction of sugar-mills and mills for husking rice, and the introduction of machinery from Europe. The introduction of European capitalists and residents has greatly improved the condition of the natives, who are always ready to enter their service. Theft and robbery, though common elsewhere, are seldom heard of on estates belonging to Europeans, and there are no instances of personal violence done the latter. About 500 sq. m. in the territories of the native princes are leased by Europeans. (*Maclaine's Evidence, in Parl. E. I. Reports ; Earl, Eastern Seas.*) The natives cultivate the rest of the land according to their ancient customs and usages, paying a rent to the government, partly in money, and partly in kind. A village system is very prevalent, by which every commune has its own lands, the culture of which it has a right to direct, and which is conducted for the benefit of the inhabs. in common. This is particularly the case in its E.: the produce is afterward divided (after deducting the rent) into equal parts, according to the number of hands engaged in its production. The land belonging to a commune varies generally from about 40 to 100 acres ; and the extent allotted to each individual from one half to two acres.

Manufactures are few, and principally domestic : the peasant's family fabricates almost every article required for its own use. Cotton goods are woven ; and a cubit's length of cotton cloth, five spans in breadth, is considered a sufficient day's work by the Javanese weaver.

The Javanese and Indian islanders, in general, are wholly unacquainted with the art of manufacturing fine clothes of any kind ; all their fabrics are of a coarse, though durable texture ; and all the labours of the loom are performed by women only. Of calico-printing the Javanese are entirely ignorant ; but they have a singular substitute for it. The part not intended to be coloured, they daub over with melted wax. The cloth, thus treated, is thrown into the dyeing-vat, and the interstices take the colour of the pattern. If a second or third colour have to be added, the operation is repeated on the ground made by the first application of wax ; more wax is applied, and the cloth is once, or oftener, consigned to the vat. The greater refinement that is attempted, the more certain seems to be the failure. This awkward substitute for printing adds 100 per cent., at least, to the price of the cloth. "The latter," Mr. Crawfurd says, "is 450 per cent. on the price of the raw material ;" and he adds, "This is a picture of the rude condition of manufacturing industry, of the waste of labour and of time, which results, in an uncivilized stage of society, from imperfect machinery, indolence, unskilfulness, and the want of the division of labour." (*Ind. Archip.*, i., 179, 180.) And yet, unskilful as the manufacturing industry of the Javanese is, it generally excels that of the other islanders of the archipelago. Leather and saddlery are made at Solo, boots and shoes at Samarang, mats, and hats of bamboo, &c., coir, fishing-nets, paper from the bark of the *Morus papyrifera*, bricks, cabinet-work, carved wooden articles, boats and ships, in the construction of which the natives are tolerably versed, and krisses, matchlocks, and other arms, &c., are, exclusive of cottons, the chief manufacture has. Copper and brass pans are made, but their manufacture has

very much declined. Almost all the manufactured goods used by Europeans are imported. Java is the only island of the E. archipelago in which salt is made to any extent: along the N. coast there are numerous salt pans, from which a great deal more of the article is obtained than is required for home consumption—a quantity estimated at 32,000 tons annually. The salt-marshes and other inlets of the sea, are often embanked for the rearing of fish in large numbers.

In architecture, the Javanese surpass the other natives of the E. archipelago; and many structures of stone and brick, some in a style of superior magnificence (as, for instance, the temple of Boro Budor), exist in different parts of the island. But the art of building has declined since the middle of the 13th century, and the modern Japanese do not even understand the art of turning an arch, though arches are seen in every ancient structure remaining in Java. The *karatons*, or palaces of the native princes, are walled inclosures, laid out on a uniform plan, and comprising numerous buildings. They were formerly constructed of hewn stone, but at present consist only of ill-burnt bricks and ill-concocted mortar. After these, the better sort of residences are called *pendapas*, a word derived from the Sanscrit; and the edifice is, therefore, probably of Indian origin. In most of these a thatched roof is supported by four wooden pillars, round which is an awning of light materials, supported by moveable props of bamboo; and the whole is closed in by a temporary paling, and divided into apartments by light partitions. The chief materials of the houses of the Javanese are the bamboo, rattan, palmetto leaf, and wild grass. The house of a peasant in a populous part of Java, where materials are not the most abundant, will not exceed the value of 60 days' labour. In the dwellings of the chiefs there is generally, in a conspicuous part of the house, a kind of state bed, rather for display than use ; but an ordinary bed is usually only the bamboo floor of the cottage; or, at best, a bench of the same flimsy material, on which a mat and small pillow are laid, and the peasant retires to rest without undressing. Food is served up on salvers or trays of wood or brass. A few Chinese porcelain dishes are used occasionally, but neither spoons, knives, nor forks.

The *commerce* between Java and Holland, which amounts to five sevenths of the whole external trade, is chiefly carried on by the *Nederlandsch Handel Maatschappy*, or Dutch Commercial Society, which includes some of the most wealthy persons in the mother country. We subjoin a table, showing the value of the export and import trade of Java and Madura in 1837, specifying the principal countries with which these islands traded.

Countries.	Imports from. £	Exports to. £
Netherlands	875,392	2,501,534
Great Britain	337,533	89,134
France	50,613	90,968
Hamburg	14,019	15,154
America	18,458	77,330
Isle of France	4,075	3,541
Persian Gulf		8,734
British India	37,119	5,975
Siam	10,981	3,668
China	74,488	187,857
Japan	80,635	25,208
Australia	4,915	13,348
E. Archipelago	687,361	617,334
Other countries	8,494	14,572
Total	1,815,608	3,884,151

The principal articles of import are linens and cotton manufactures, chintzes, muslins, &c. (to the value of £595,226 in 1837), provisions, wipes and spirits, iron and iron goods, woollen goods, haberdashery, glass, copper wares, &c., from Europe and America ; opium from the Levant and Bengal ; sacking, linens, wheat, &c., from India ; porcelain, tea, tobacco, silk and silk goods, from China ; copper and camphor from Japan ; gambier, coffee, tin, cotton, gold dust, benzoin, sandal-wood, &c., for exportation, from the rest of the archipelago. The following account of the principal produce exported from Java during each of the 10 years ending with 1837, shows, better than anything else can do, the wonderful progress recently made by this noble island :

Years.	Coffee.	Pepper.	Indigo.	Hides.	Cloves.	Nutmegs.	Sugar.	Tin.	Rice.	Rattans.	Mace.	Areca.
	Piculs	Piculs	Lbs.	No.	Piculs	Piculs	Piculs	Piculs	Piculs	Piculs	Piculs	Leaguers
1828	216,182	8,226	23,010	49,400	1,852	1,648	25,970	19,505	15,534	31,301	600	534
1829	221,612	6,104	46,368	44,321	2,434	1,159	73,780	23,034	15,192	30,466	190	1,397
1830	286,740	5,061	22,063	30,249	803	1,304	108,640	21,426	13,321	5,000	177	1,827
1831	209,086	7,836	42,854	83,271	1,581	2,550	120,208	30,032	18,037	5,188	143	1,497
1832	314,173	7,075	168,211	82,385	5,114	3,849	345,872	47,801	23,072	14,323	947	2,000
1833	280,166	5,497	217,480	75,421	1,042	1,171	210,947	44,304	30,344	16,731	603	1,644
1834	455,008	7,704	250,421	53,000	4,040	4,209	367,131	39,163	23,379	14,000	1,192	1,432
1835	405,971	11,868	535,753	139,995	4,506	5,022	438,543	46,836	25,577	4,065	1,606	2,075
1836	498,077	7,006	407,798	120,000	2,185	3,846	509,513	47,739	36,430	49,968	991	1,477
1837	684,947	12,487	422,492	93,071	2,925	3,778	676,085	44,417		23,530	1,213	1,063

The value of the principal articles exported from Java in 1836 was—

	Florins.		Florins.
Arrack	115,295	Nutmegs	1,711,610
Hides	217,715	Sugar	9,632,141
Indigo	1,139,382	Tobacco	760,860
Coffee	15,090,382	Tin	2,718,810
Pepper, round	125,035	All other articles,	
Rice	3,369,615	and treasure	7,367,833
Mace	306,988		
Cloves	153,636	Total value	42,961,642

The total amount of import and export duties received in 1837, was £435,669. In the same year, 1648 ships, of the aggregate burden of 102,416 lasts (under the Dutch flag, 1423 of 79,902 lasts), entered; and 1891 ships, burden 111,299 lasts (1636 of 85,571 lasts, under Dutch colours), cleared out of the different ports of Java and Madura.

The internal traffic is comparatively small, though few countries have better means of communication. A carriage road, extending from one extremity of Java to the other, 800 m. in length, was made by General Daendels; but it is alleged that its construction cost the lives of 12,000 natives.

The Chinese weights are invariably used in commercial transactions at Batavia, and throughout Java and the other Dutch possessions in India. These are the picul, and the cattie, which is its hundredth part. The picul is commonly estimated at 125 Dutch or 133¼ lbs. avoirdupois, but at Batavia it has been long reckoned equal to 136 lbs. avoird. The bahar is three, and the timbang five piculs. The coyang of rice is equivalent to 3300 lbs. Dutch. The coins in use are similar to those current in the Netherlands. Spanish dollars are received at the custom-house in Batavia, at the rate of 109 for 260 florins.

Government, &c.—Java is divided into 19 provinces or residencies, or, including Madura, into 20, each governed by an European resident, assisted by a secretary, and as many sub-residents as may be deemed necessary. The residencies are subdivided into arrondissements or regencies, the administration of which, especially in respect to the police, is confined to native chiefs, termed regents. The colonial government at Batavia exerts a full and complete power over all the Dutch colonies in the E. seas. The governor-general in the capital is the representative of the king of Holland, and commander-in-chief of the forces by land and sea. He is assisted by a secretary-general, and a colonial council of four members, who must be of Dutch extraction, born in Holland, or one of its dependencies, and 30 years of age; and who can exercise no other functions while they remain councillors. Justice is administered in the last resort in a supreme court at Batavia, which has jurisdiction in all cases above the value of 500 florins. Three subordinate civil and criminal tribunals, and three courts martial, subordinate to a central court in the capital, are established in Batavia, Samarang, and Sourabaya. A member from each of these courts makes a circuit at least every three months into the residencies under its control, to preside at a court of assize, composed besides of four native chiefs chosen annually by the government, on the recommendation of the natives. The permanent tribunals of the residencies are the *land-raaden*, composed of the resident, four members selected from among the regents, a secretary, &c. In each arrondissement and commune there are justices of the peace, with authority in petty cases. The Chinese are governed by their own laws, under functionaries chosen by them, who are responsible to the Dutch for the behaviour of the rest. There are few slaves belonging to Europeans in Java. The greatest religious toleration exists, and ministers of all Christian sects are equally remunerated by the government. Superior schools are established in the chief towns, and primary schools in most of the residencies. The squadron stationed in Java sometimes comprises several ships of the line, but in time of peace usually consists only of a few frigates and corvettes. There is, besides, a colonial navy of light vessels (schooners, gun-boats, &c.), which forms a separate branch of service, though both are generally placed under the command of the admiral of the royal squadron, who has the title of Director of the Dutch East India Navy. Besides the foregoing force, a flotilla of cruisers, manned by native Javanese, is supported by the different marine residencies. The land forces consist of 11 battalions of infantry, a corps of pioneers, two battalions of artillery, a regiment of hussars, and a portion of a squadron of lancers. In all, there are about 8000 Europeans in the Dutch Javanese army, being about equal to 1·3d part of those serving in British India. But, notwithstanding the heavy expense incurred in its government, Java is one of the few colonial dependencies that in ordinary years remit a considerable revenue to the mother country.

The territories of the native princes comprise about one fourth part of Java and its inhabitants, in the centre, S., and S.E., of the island, the capital of the Susuhunan, or empire of Java, being at Surakarta, on the Solo, and that of the sultan, at Djocjokarta. The religion of both these dynasties is the Mohammedan, which prevails over almost the whole of the country. The *Javanese*, as a nation, are the most advanced of any in the E. archipelago. They only, of those inhabiting that region, have a native calendar, and have made considerable progress in the arts and sciences of civilized life. They appear to have received these originally from Hindostan, together with the Hindoo religion, which is supposed to have prevailed over Java, till its conquest by the Mohammedans in 1478. Copious details respecting the manners, customs, &c., of this curious and interesting people, the antiquities of the island, &c., which would take up too much space in the present work, may be found in Raffles, Crawfurd, and other writers on the subject.

The history of Java cannot be traced, with any degree of confidence, farther than the latter portion of the 12th century. From that time down to the establishment of Mohammedanism, at the close of the 15th century, the religion of the people was a modified Hindooism; and a number of independent states existed in Java. The ruins of Mojopahit, one of the principal capitals of these several states, are among the most extensive in the East. This city had between two opposite gates, the remains of which still exist, a breadth of about 2 m., which would give a circuit of 12 m., if the enclosure had been a square. The Hindoo kingdom of Mojopahit was overturned by the Arabs in 1479.

The Portuguese reached Java in 1511, and the Dutch in 1595. The latter founded Batavia in 1619, and gradually consolidated their power on the island, though for a long period engaged in continual wars with the native sovereigns. In 1811, Java was taken by a British force from Hindostan, and held till 1816, when, in pursuance of the treaty of Paris, it was restored to the Dutch. (*Hogendorp, Coup d'Œil sur l'Isle de Java; Raffles's Hist. of Java; Crawfurd's Indian Archipelago; Parl. Reports; Hamilton, E. I. Gaz., &c.*)

JAVA, co., N. Y., Wyoming co., N. Y., 287 m. W. Albany, 332 W. Organized in 1832. Drained by Cattaraugus, Buffalo, and Seneca creeks, the two latter flowing from Little Cattaraugus lake or pond, of 300 acres. It contains one church, three stores, two fulling-mills, one grist-mill, one saw-mill; 23 schools, 898 scholars. Pop. 2231.

JAY, co., Ia. Situated in the E. part of the state, and contains 370 sq. m. Watered by Salamanie river. It contained in 1840, 2839 neat cattle, 646 sheep, 7491 swine; and produced 9703 bushels of wheat, 69,842 of Indian corn, 5773 of oats, 9135 of potatoes, 16,018 pounds of sugar. It had eight stores, four grist-mills, two saw-mills; one school, 30 scholars. Pop. 3863.

JAY, p. t., Franklin co., Me., 99 m. W.N.W. Augusta, 607 W. Incorporated in 1795. Bounded S. by Androscoggin river. It contains three stores, two saw-mills; 18 schools, 760 scholars. Assessors' valuation of real estate in 1842, $230,077. Pop. 1750.

JAY, p. t., Orleans co., Vt., 58 m. N. Montpelier, 574 W. Chartered in 1792. The western range of the Green mountains runs through its W. part; on the E. it is level and fertile. Watered by branches of Missisque river. It has two saw-mills; three schools, 122 scholars. Pop. 308.

JAY, p. t., Essex co., N. Y., 18 m. N. Elizabeth, 148 m. N. Albany, 523 W. It contains two churches, a Methodist and Baptist; five stores, two fulling-mills, five forges, two saw-mills; 13 schools, 582 scholars. Pop. 2236.

JAY, C.H., p. v., cap. of Jay co., Ia., 100 m. N.E. Indianapolis, 523 W. Situated on the N. side of Salamanie river, and contains a courthouse, and a few dwellings.

JAXARTES, a celebrated river of antiquity, now very generally acknowledged to be identical with the Sir-Daria, the chief stream of the Kirghis-steppe. It rises in the Kachkar-Davan, a W. branch of the Tiang-khang range, in lat. 42° 30′ N., and long. 73° 50′ E. Its course to Kokan is W.S.W. about 180 m.; but at that point it takes a N.N.W. direction for about 300 m., as far as Akmetschet, in lat. 45° N., long. 66° 5′ E., where the channel divides, the N. and larger branch retaining the name Sir, while that to the S. is called Kouvan-Daria: their mouths in the Caspian sea lie about 40 m. apart, but are both in long. 61°. The entire length of the Sir, including its windings, cannot be much less than 900 m.; and it is both broad and deep, which may be attributed to its being the sole recipient of the waters on the N. side of the great chain separating the khanate of Kokan from Chinese Turkestan. It has no affluent of any great size; its banks, which are low and sandy, are usually flooded in summer and at the beginning of winter; and the water is described as being loaded with a whitish brown deposit. The ruins of temples and habitations in the Karakoum sands at the lower part of its course clearly prove that its banks were once peopled by a race far more civilized than the brigand Kirghis, who now wander over the Steppe. (*Lévchine*, p. 1, ch. v.)

Herodotus gives the name *Araxes* to a large river full of fish, and studded with islands, situated in a vast immeasurable plain. (*See* i., 201–216.) Some geographers have conjectured that he meant the Amoo (*Oxus*), others the Wolga; but D'Anville, Heeren, and Mannert, clearly show, from the position of the Massagetæ relatively to the Issedones, that no other river but the Sir could have been meant by the father of history. Ancient geographers agree in stating that the Jaxartes flowed into the Caspian sea, an assertion, perhaps, not quite so erroneous as modern critics have supposed, if any credit be attached to the investigations of Mouravief and Berg on the level of the country between the Caspian and Aral seas, which lead to the supposition that these great salt-lakes were once united. This conjecture, also, if it be correct, at once accounts for the great breadth (E. and W.) given to the Caspian by all the ancient writers. With respect to the term *Araxes*, which was used by the old authors as applicable to at least *five* distinct rivers, it is now regarded as generic, meaning simply any *rapid* stream, like the modern *Aras*. (See D'Anville's valuable paper, *Des Fleuves du Nord d'Araxes*, in vol. xxxvi. of the *Histoire de l'Acad. des Inscriptions*.) Herodotus, whose geography is in general so very accurate, was probably led into what Rennell calls his "prodigious mistake" respecting the direction of the Araxes, by not knowing that this name was held in common by several Eastern rivers. (Comp. *Rennell; Geog. of Herod.*, i., p. 270–272, and 296–293, with *Heeren's Reflections; Asia*, ii., p. 340, &c., and *Mannert, Geographie der Alten Griechen und Römer.* Th. ii., b. 2.)

JEAN D'ANGELY (ST.), a town of France, dep. Charente Inférieure, cap. arrond., on the Boutonne, which here begins to be navigable for vessels of from 30 to 40 tons, 33 m. S.E. by E. La Rochelle. Pop. (1836) 5342. It is ill built, but clean and cheerful. It has an ancient abbey, a handsome public hall, some baths, a theatre, and other places of entertainment, and a brisk trade in wine, brandy, and timber.

JEDBURGH, a royal and parl. bor. and market town of Scotland, co. Roxburgh, of which it is the cap., in a narrow valley on the Jed, about 2 m. above its junction with the Teviot, near the termination of the Cheviots, 40 m. S.E. Edinburgh, and 43 m. N.E. by N. Carlisle. Pop. of bor. and par. in 1811, 4454; in 1831, 5647; of which the bor. had 3709. It consists of four leading streets, which cross each other at right angles, and are wide and well built. Around the town are several beautiful villas. The Townhead, a street parallel with the river, consists of old houses which, with their inhabitants, are said for generations to have undergone little or no change. The public buildings are the Castle (built on the site of the ancient castle of Jedburgh, once a royal residence), containing a bridewell and prison, the county hall, the town-house, and churches belonging respectively to the Associate Synod and Relief. The parish church consists of the western portion of the abbey, founded by David I., in the 12th century, and will be noticed below. A majority of the people are dissenters. The denomination of dissenters, termed *Reliof*, had its origin here in 1754. The grammar school of Jedburgh, an endowed seminary, has long been eminent. It has the honour to reckon among its pupils Thomson, the illustrious author of the "Seasons," and of the "Castle of Indolence," born in the parish of Ednam, in this county, on the 11th of September, 1700. It may also be mentioned that Dr. Thos. Somerville, author of "The History of Great Britain, during the Life of Queen Anne," was minister of Jedburgh. The only charitable institution is a dispensary, founded in 1807; and open to the population of the adjoining district.

The woollen manufacture has been introduced into Jedburgh. The fabrics made are blankets, carpets, flannels, hosiery, &c.: there are three mills, driven by water, which employ 104 hands, exclusive of the stocking-weavers, who carry on their business in their own houses. (*Factory Returns, Parl. Papers, Jan.* 1839.) There is an establishment for the manufacture of printing presses, under a patent, conducted by the inventor, Mr. Hope, an iron founder in the borough. There are two branch banks and a savings' bank.

The abbey of Jedburgh, belonging to the canons regular of St. Augustine, must, when entire, have been one of the most magnificent ecclesiastical structures in Scotland. It exhibits different styles of architecture, according to the taste prevailing at the different periods when it was built. The walls of the nave, central tower, and choir, remain; and, though (the two last especially) much dilapidated, they sufficiently attest its ancient grandeur. The N. transept, which has a beautiful traceried window, is entire. There are two magnificent Norman doors in this edifice, one at the W. end, and the other in the S. wall of the nave, close to the transept. Indeed, the ruin generally affords

fine examples of the Saxon, Norman, and early English styles, the latter being admirably exemplified in the long range of narrow pointed windows above the arches of the middle part of the nave, and in the blank arches of the W. end. The altar, or E. end of the choir, the cloisters, and the chapter-house, have disappeared. We regret to have to add, that this noble ruin has been disfigured, and its character, in fact, destroyed, by "fitting up the W. end of the nave in a most barbarous style as the parish church." (*Morton.*) Luckily, this piece of miserable patchwork is as uncomfortable as it is unseemly; so that it is to be hoped it may be abandoned, and the ruins restored, in as far as possible, to their former state. (For farther information with respect to this fine ruin, see the learned and valuable work, entitled *Monastic Annals of Teviotdale*, by the Rev. James Morton.)

A monastery for gray friars was founded in this town by the citizens, in 1513; but of it all traces have disappeared. Here may still be seen the house in which Queen Mary lodged after her visit to the Earl of Bothwell, at Hermitage. Mary continued in it several days, owing to a sickness she had contracted in her unfortunate journey. The apartment which she occupied was on the third story, and is in tolerable preservation.

Jedburgh was erected into a royal borough in the 12th century, but the castle, the site of which is now occupied by the jail and bridewell, is supposed to have been of earlier date. After having been for some time in the possession of the English, the castle was taken by the Scotch, in 1409, and demolished. Like other borderers, the citizens of Jedburgh were anciently more celebrated for their martial than for their peaceful virtues. Their favourite weapon was a partizan or halbert, known by the name of the "Jethart (Jedburgh) staff." Their war-cry, or slogan, was "Jethart's here." The term "Jethart Justice," which implies execution before trial, is supposed to have originated in the many instances of lynch law, executed here on border marauders. (*Scott's Border Minstrelsy*, i., 50.) The eldest son of the Marquis of Lothian, descended from the ancient border family of the Kers of Fernihirst, for centuries the feudal superiors of the borough, has the title of Lord Jedburgh.

Jedburgh unites with N. Berwick, Haddington, Lauder, and Dunbar in sending a member to the House of Commons. Registered voters in 1839–40, 201. Corporation revenue, £644 1s. 4d. (*Keith's Scot. Bishops*, p. 302–452; *Redpath's Border Hist.; Chalmers's Caledonia.*)

JEDDO. *See* Yeddo.

JEFFERSON, co., N. Y. Situated in the N. part of the state, and contains 1125 sq. m. Bounded W. by lake Ontario, N.W. by St. Lawrence river. Watered by Black river, Indian, Chaumont, and Perch rivers, and Stony and Sandy creeks. It contained in 1840, 78,604 neat cattle, 165,300 sheep, 60,518 swine; and produced 406,721 bushels of wheat, 18,396 of rye, 445,973 of Indian corn, 36,641 of buckwheat, 74,540 of barley, 447,936 of oats, 1,345,818 of potatoes, 3905 pounds of hops, 519,254 of sugar. It had two commission houses in foreign trade, 186 retail stores, 15 lumber-yards, six furnaces, one forge, one smelting-house producing 300,000 pounds of lead, 22 fulling-mills. 11 woollen factories, one cotton factory with 1000 spindles, eight flouring-mills, 34 grist-mills, 109 saw-mills, four oil-mills, one paper-mill, 31 tanneries, nine distilleries, eight breweries, one rope-walk, four printing-offices, five weekly newspapers; one college, 200 students; two academies, 123 students; 292 schools, 11,348 scholars. Pop. 60,984. Capital, Watertown.

JEFFERSON, co., Pa. Situated towards the N.W. part of the state, and contains 1900 sq. m. Watered by Clarion or Toby's river and Redbank river, and their branches. Iron ore and coal are abundant. It contained in 1840, 5773 neat cattle, 7342 sheep, 8042 swine; and produced 43,509 bushels of wheat, 24,467 of rye, 23,369 of Indian corn, 14,501 of buckwheat, 77,077 of oats, 64,110 of potatoes, 27,067 pounds of sugar. It had 19 stores, two fulling-mills, one woollen factory, 14 grist-mills, 68 saw-mills, six tanneries, two distilleries, one printing-office, one weekly newspaper; one academy, 25 students; eight schools, 336 scholars. Pop. 7253. Capital, Brookville.

JEFFERSON, co., Va. Situated in the N.E. part of the state, and contains 225 sq. m. Watered by Shenandoah river. Bounded W. by Opequan creek. It contained in 1840, 11,915 neat cattle, 67,289 sheep, 72,467 swine; and produced 516,969 bushels of wheat, 41,975 of rye, 988,532 of Indian corn, 4230 of barley, 71,999 of oats, 151,443 of potatoes. It had 57 stores, two furnaces, three fulling-mills, three woollen factories, 18 flouring-mills, 17 grist-mills, 40 saw-mills, six tanneries, four distilleries, one brewery, two potteries, two printing-offices, two weekly newspapers; eight academies, 262 students; 19 schools, 475 scholars. Pop.: whites, 9383; slaves, 4157; free coloured 602; total, 14,082. Capital, Charleston.

JEFFERSON.

JEFFERSON, co., Ga. Situated towards the E. part of the state, and contains 680 sq. m. Watered by Ogeechee river and its branches, and by Briar creek, which last flows into Savannah river. It contained in 1840, 8863 neat cattle, 2369 sheep, 16,537 swine; and produced 16,301 bushels of wheat, 1128 of rye, 273,416 of Indian corn, 4391 of oats, 8770 of potatoes, 5,448,120 pounds of cotton. It had 13 stores, nine grist-mills, nine saw-mills; one school, 12 scholars. Pop.: whites, 2,877; slaves, 4342; free coloured, 35; total, 7254. Capital, Louisville.

JEFFERSON, co., Flor. Situated in the central part of the territory, and contains 702 sq. m. It extends from Georgia to the gulf of Mexico; bounded E. by Ocilla river. Mickasuky lake, in its W. part, has an outlet, which, uniting with several other streams, forms a considerable mill-stream, and suddenly sinks into the earth and disappears. It contained in 1840, 8112 neat cattle, 732 sheep, 9680 swine; and produced 125,540 bushels of Indian corn, 2015 of oats, 25,904 of potatoes, 7200 pounds of rice, 5500 of tobacco, 1,854,715 of cotton, 46,850 of sugar. It had nine stores, four grist-mills, two saw-mills, one tannery; one academy. 90 students; five schools, 94 scholars. Pop.: whites, 2162; slaves, 3549; free coloured, 2; total, 5713. Capital, Monticello.

JEFFERSON, co., Ala. Situated a little N. of the centre of the state, and contains 1040 sq. m. Watered by Locust fork of Black Warrior river and its branches. It contained in 1840, 8028 neat cattle, 2331 sheep, 22,163 swine; and produced 34,157 bushels of wheat, 353,751 of Indian corn, 52,790 of oats, 9314 of potatoes, 4003 pounds of tobacco, 796,867 of cotton. It had five stores, one flouring-mill, 95 grist-mills, 11 saw-mills, one oil-mill, four tanneries, five distilleries; 33 schools, 464 scholars. Pop.: whites. 5486; slaves, 1636; free coloured, 9; total, 7131. Capital, Elyton.

JEFFERSON, co., Miss. Situated in the S.W. part of the state, and contains 630 sq. m. Bounded N. by Mississippi river. Watered by Coles's creek and branches of Bayou Pierre and Homochitto rivers. It contained in 1840, 15,156 neat cattle, 7365 sheep, 21,823 swine; and produced 364,979 bushels of Indian corn, 30,261 of potatoes, 6918 pounds of rice, 14,038,479 of cotton. It had 11 stores, 153 grist-mills, eight saw-mills, two tanneries, two printing-offices, two weekly newspapers; one academy, 40 students; 15 schools, 318 scholars. Pop.: whites, 2369; slaves, 9176; free coloured, 85; total, 11,630. Capital, Fayette.

JEFFERSON, par., La. Situated in the S.E. part of the state, and contains 720 sq. m. Bounded N. by Barrataria bay of the gulf of Mexico. It contained in 1840 7891 neat cattle, 6781 sheep, 2778 swine; and produced 193,813 bushels of Indian corn, 82,250 of potatoes, 104,260 pounds of rice, 5,134,500 of sugar. It had 143 stores, 51 lumber-yards, six saw-mills, one tannery, one distillery, one pottery, two printing-offices, two daily and one weekly newspapers, and three periodicals: one academy, 55 students; two schools, 29 scholars. Pop.: whites, 4866; slaves, 4095; free coloured, 618; total, 10,470. Capital, La Fayette.

JEFFERSON, co., Tenn. Situated toward the E. part of the state, and contains 356 sq. m. Bounded N.W. by Holston river. Watered by French broad river and its branches. It contained in 1840, 7782 neat cattle, 9998 sheep, 27,717 swine; and produced 80,449 bushels of wheat, 1619 of rye, 396,644 of Indian corn, 77,967 of oats, 12,681 of potatoes, 10,107 pounds of tobacco, 4863 of sugar. It had 17 stores, two fulling-mills, three woollen factories, one cotton factory with 500 spindles. 12 flouring-mills, 38 grist mills, 30 saw-mills, three oil-mills, one powder-mill, 49 tanneries, 18 distilleries, one pottery; one college, 80 students; one academy, 45 students; 13 schools, 390 scholars. Pop.: whites, 10,662; slaves, 1282; free coloured, 132; total, 12,076. Capital, Dandridge.

JEFFERSON, co., Ky. Situated toward the N. part of the state, and contains 304 sq. m. Bounded N.W. by Ohio river. Watered by Floyd's fork of Salt river and Pond creek. The falls of the Ohio at Louisville, around which is a canal, are opposite to this county. It contained in 1840, 12,716 neat cattle, 14,971 sheep, 43,966 swine; and produced 115,175 bushels of wheat, 16,969 of rye, 665,890 of Indian corn, 1739 of barley, 156,093 of oats, 60,604 of potatoes, 75,280 pounds of tobacco, 3944 of sugar. It had one commercial and 11 commission houses in foreign trade, 270 retail stores, three lumber-yards, 11 flouring-mills, 17 grist-mills, 17 saw-mills, one oil-mill, one paper-mill, one glass factory, nine tanneries, nine distilleries, two breweries, two potteries, seven printing-offices, two binderies; five daily, seven weekly, and three semi-weekly newspapers, and one periodical; one college, 80 students; 15 academies, 452 students; 37 schools, 814 scholars. Pop.: whites, 26,987; slaves, 8596; free coloured, 763; total, 36,346. Capital, Louisville.

JEFFERSON, co., O. Situated in the E. part of the state, and contains 396 sq. m. Bounded E. by Ohio river. Drain-

ed by Cross, Short, and Yellow creeks. It contained in 1840, 8566 neat cattle, 18,814 sheep, 11,879 swine; and produced 287,486 bushels of wheat, 2001 of rye, 177,450 of Indian corn, 2963 of barley, 215,378 of oats, 50,189 of potatoes, 1000 pounds of sugar. It had three commission houses in foreign trade, 62 retail stores, five lumber-yards, one furnace, one fulling-mill, seven woollen factories, one cotton factory with 10,294 spindles, 24 flouring-mills, three grist-mills, 22 saw-mills, one oil-mill, one paper-mill, ten tanneries, four distilleries, two breweries, two printing-offices, two weekly newspapers: one college, 208 students; two academies, 170 students: 44 schools, 2906 scholars. Pop. 25,030. Capital, Steubenville.

JEFFERSON, co., Ia. Situated in the S.E. part of the state, and contains 360 sq. m. Bounded S.E. by Ohio river. Drained by Muscatatuck, Indian, Kentucky, Big, and Lewis creeks. Organized in 1809. It contained in 1840, 12,000 neat cattle, 13,477 sheep, 15,135 swine; and produced 85,483 bushels of wheat, 2195 of rye, 250,007 of Indian corn, 64,264 of oats, 14,597 of potatoes, 13,180 pounds of tobacco, 8503 of sugar. It had 35 stores, one furnace, one cotton factory with 800 spindles, 10 flouring-mills, 21 grist-mills, 31 saw-mills, one oil-mill, one paper-mill, 12 tanneries, one brewery, two printing-offices, four weekly newspapers; one college, 72 students; one academy, 22 students; 16 schools, 1050 scholars. Pop. 16,614. Capital, Madison.

JEFFERSON, county, Ill. Situated centrally in the S. part of the state, and contains 576 sq. m. Drained by branches of Big Muddy and Little Wabash rivers. It contained, in 1840, 19,370 neat cattle. 6001 sheep, 17,955 swine; and produced 11,503 bushels of wheat, 271,800 of Indian corn, 27,797 of oats, 8602 of potatoes, 29,372 pounds of tobacco, 13,862 of cotton. It had 10 stores, 14 grist-mills, three saw-mills, two tanneries, two distilleries; one academy, 125 students, eight schools, 375 scholars. Pop. 5762. Capital, Mount Vernon.

JEFFERSON, county, Mo. Situated in the E. part of the state, and contains 500 sq. m. It abounds with minerals and mineral springs. It contained, in 1840, 7753 neat cattle, 4909 sheep, 13,533 swine; and produced 13,350 bushels of wheat, 182,195 of Indian corn. 20,195 of oats, 4836 of potatoes, 3060 pounds of tobacco. It had six stores, one smelting-house producing 775,000 pounds of lead, three flouring-mills, four grist-mills, four saw-mills, two tanneries, one distillery; five schools, 88 scholars. Pop.: whites, 3980; slaves, 394; free coloured, 12; total, 4296. Capital, Hillsboro'.

JEFFERSON, county, Ark. Situated a little S.E. of the centre of the state, and contains 1180 sq. m. Watered by Arkansas river. It contained, in 1840, 7443 neat cattle, 639 sheep, 8671 swine; and produced 157,560 bushels of Indian corn, 6730 of potatoes, 639,750 pounds of cotton. It had seven stores, 17 grist-mills, 11 saw-mills. Pop.: whites, 1551; slaves, 1010; free coloured, five; total, 2566. Capital, Pine Bluffs.

JEFFERSON, county, Wis. Situated towards the S. part of the territory, and contains 576 sq. m. Watered by Rock river. It contained, in 1840, 1045 neat cattle, 32 sheep, 1783 swine; and produced 6647 bushels of wheat, 186 of rye, 15,192 of Indian corn, 4465 of oats, 14,410 of potatoes, 13,050 pounds of sugar. It had one store, four saw-mills; six schools, 138 scholars. Pop. 914. Capital, Jefferson.

JEFFERSON, county, Iowa. Situated in the S.W. part of the territory, and contains 360 sq. m. Drained by Checauque or Skunk river. Limestone and coal are found. It had, in 1840, 3075 neat cattle, 2118 sheep, 7172 swine; and produced 4273 bushels of wheat, 80,675 of Indian corn, 10,795 of oats. It had seven stores, five saw-mills; seven schools, 151 scholars. Pop. 2773. Capital, Fairfield.

JEFFERSON, p. t., Lincoln co., Me., 21 m. S.E. Augusta, 613 W. It lies around Damariscotta pond, the outlet of which forms Damariscotta river. Incorporated in 1807. It has 11 stores, two fulling-mills, four grist-mills, nine saw-mills, four tanneries; 15 schools, 345 scholars. Pop. 2214.

JEFFERSON, p. t., Coos co., N. H., 108 m. N. Concord, 569 W. Watered by Israel's river, and Pondichery pond, the source of St. John's r. It has one store, one grist-mill, two saw-mills; five schools, 200 scholars. Pop. 575.

JEFFERSON, p. t., Schoharie co., N. Y., 56 m. W. Albany, 375 W. Drained by branches of Delaware r. and of Schoharie cr. It contains three stores, three fulling-mills, two grist-mills, six saw-mills, three tanneries; two academies, 157 students; 16 schools, 609 scholars. Pop. 2033.

JEFFERSON, t., Morris co., N. J., 15 m. N.W. Morristown. Drained by Rockaway river. Hopatcong lake, in the S.W. part, on Hamburg mt., is from three to four miles long, and one mile broad; it covers 3000 acres, and forms the feeder of the summit-level of the Morris canal. Iron is extensively found in the mountains. It has two stores, 11 blommeries, two grist-mills, three saw-mills, one distillery; five schools, 175 scholars. Pop. 1412.

JEFFERSON, p. t., Greene co., Pa., 204 m. W. by S. Harrisburg, 220 W. Bounded E. by Monongahela r. Watered

79

by Ten Mile er. It contains seven stores, three grist-mills, four saw-mills, two tanneries, two distilleries ; five schools, 185 scholars. Pop. 1296.

JEFFERSON, t., Alleghany co., Pa. It has three stores, two flouring-mills, five grist-mills, eight saw-mills, seven distilleries ; 10 schools, 450 scholars. Pop. 1779.

JEFFERSON, t., Fayette co., Pa. It has two stores, one woollen factory, one glass factory, one flouring-mill, three grist-mills, five saw-mills ; three schools, 75 scholars. Pop. 1316.

JEFFERSON, p. t., capital of Ashtabula co., O., 904 m. N.E. Columbus, 338 W. Watered by Grand river. It contains a brick courthouse, 40 by 50 feet, an academy, three stores a printing-office, issuing a weekly newspaper ; five schools, 110 scholars. Pop. 710.

JEFFERSON, t.,' Adams co., O. It has five stores, one flouring-mill ; one school, 50 scholars. Pop. 937.

JEFFERSON, t., Franklin co., O., 12 m. E. by N. Columbus. It has 12 schools, 300 scholars. Pop. 1040.

JEFFERSON, t., Guernsey co., O. It has three schools, 90 scholars. Pop. 941.

JEFFERSON, t., Jackson co., O. It has one school, 26 scholars. Pop. 732.

JEFFERSON, t., Fayette co., O. It has two stores, three grist-mills, six saw-mills ; 12 schools, 300 scholars. Pop. 1949.

JEFFERSON, t., Montgomery co., O. It has three grist-mills, three saw-mills, eight distilleries ; five schools, 260 scholars. Pop. 1895.

JEFFERSON, t., Madison co., O., 14 m. W. Columbus. The National road passes through it. It has five stores, one grist-mill, one saw-mill, two tanneries ; 12 schools, 370 scholars. Pop. 607.

JEFFERSON, t., Muskingum co., O. It has two flouring-mills, three saw-mills ; six schools, 376 scholars. Pop. 1369.

JEFFERSON, t., Preble co., O. The National road passes through it. Pop. 2164.

JEFFERSON, t., Richland co., O. It has nine schools, 950 scholars. Pop. 2295.

JEFFERSON, t., Tuscarawas co., O. It has two saw-mills ; five schools, 336 scholars. Pop. 992.

JEFFERSON, t., Miami co., Ia. It has two stores, two flouring-mills, two saw-mills ; one school, 25 scholars. Pop. 481.

JEFFERSON, t., Putnam co., Ia. It has two stores, three grist-mills, four saw-mills ; eight schools, 489 scholars. Pop. 1129.

JEFFERSON, p. v., capital of Ashe co., N. C. 202 m. W.N.W. Raleigh, 366 W. Situated on a branch of New r. It contains a courthouse, jail, and several stores and dwellings.

JEFFERSON, p. v., capital of Jackson co., Ga., 95 m. N. Milledgeville, 630 W. It contains a courthouse, jail, an academy, several stores, and about 25 dwellings.

JEFFERSON, p. v., capital of Cherokee co., Ala., 152 m. N.E. Tuscaloosa, 677 W. Situated on the N. side of Coosa r. It has a courthouse, and several stores and dwellings.

JEFFERSON, p. t., capital of Jefferson co., Wis., 36 m. E. Madison, 834 W. It has one school, 49 scholars. Pop. 250.

JEFFERSON, t., Cole co., Mo. It has one academy, 42 students, two schools, 30 scholars. Pop. 9042.

JEFFERSON CITY, p. v., cap. of Cole co., Mo., and of the state, 936 W. Situated on the S. bank of Missouri river, on elevated and uneven ground, and contains a state-house, a large and elegant house for the governor, a state penitentiary, an academy, several large mercantile houses, a fine steam saw-mill, 200 dwellings, and 1174 inhabitants.

JEFFERSONTON, p. v., capital of Camden co., Ga. Situated at the head of navigation on Santilla r. It contains a courthouse, jail, several stores, and about 30 dwellings.

JEFFERSONVILLE, p. v., Clark co., Ia., 117 m. S. by E. Indianapolis, 597 W. Situated on a high bank of Ohio r., just above Louisville, and the falls in the r. It has a good landing-place, a ship-yard, iron foundry, six stores, a grist and saw mill, a printing-office, and about 800 inhabitants. It has a fine view of the surrounding country, including the falls.

JELLALABAD, or JULALABAD, a town of Afghanistan, in a fertile plain, and on the high road between Caubul and Peshawur, 80 m. E. by N. the former, and 60 m. W.N.W. the latter ; lat. 34° 30′ N., long. 70° 23′ E., Sir A. Burnes says, " It is one of the filthiest places I have seen in the E. It is a small town, with a bazaar of 50 shops, and a population of 2000 people ; but its number increases tenfold in the cold season, as the people flock to it from the surrounding hills. Julalabad is the residence of a chief of the Barukzye family, who has a revenue of about seven lacs of rupees a year. The Caubul river passes a quarter of a mile N. of the town, and is about 150 yards wide : it is not fordable." (Bokhara, &c., ii., 105.)

JEMME (EL). See Tysdrus.

JENA, a town of central Germany, grand duchy of Saxe-

Weimar, circ. Weimar-Jena, cap. district, on the Saale, 12 m. E. Weimar, and 41 m. N.E. Halle : lat. 50° 56′ 28″ N, long. 11° 37′ 15″ E. Pop., in 1838, 5817. (Berghaus.) The town, which is walled, and has handsome suburbs outside its four gates, lies in a valley, between two abrupt eminences, on the left bank of the river, which is here crossed by a handsome stone bridge. The streets are wide, and some of the houses are large and well built, many being highly ornamented with rude and grotesque sculpture. The ducal palace, containing a library and museum, with a good collection of minerals and animals, one Roman Catholic and three Protestant churches, three hospitals, a lunatic asylum, and the university-house, are the chief public buildings. It is a place of considerable eminence for literature, and the seat of a university, founded in the 17th century by the sovereign princes of the Ernestine branch of the house of Saxony, in whom the patronage and appointment of the professors is still vested. The constitution is similar to that of other German universities ; it has faculties of divinity, law, medicine, and philosophy, with 26 ordinary professors, composing a senatus academicus, for examining students and conferring degrees : there are also 17 extraordinary professors, and a few privat-docenten, or private tutors. The salaries of the ordinary professors range between £80 and £180, those of the " extraordinary" varying from £30 to £90, which are increased by fees from pupils, each of whom pays at the rate of about five rix dollars, or 15s. 6d., for the course. The remuneration of the tutors depends wholly on the number of their pupils. The annual expenditure of the university, including the expense of theological and other seminaries, the library (comprising 100,000 volumes), veterinary school, collections, botanical garden, prizes, officers, &c., amounts to about 38,000 dollars, or nearly £6000, a year. A fund, also, similar to that in Göttingen, with a capital of £4600, is employed in pensioning the widows of professors ; and an academical refectory fund (Speise-anstalt), supported by endowments and yearly grants from the grand dukes of Saxe Weimar, Coburg, and Meiningen, furnishes daily meals at several ordinaries for 129 indigent students. The number of students has averaged 500 during the last 10 or 12 years ; an attendance far more limited (owing to a great measure to the murder of Kotzebue, perpetrated by a student in Jena) than in the middle of the last century, when 2000 were in actual residence at the same time. (Journ. of Educat., vols. v. and ix.) Living in Jena is considered cheaper than at almost any other university of Germany ; and a student may live respectably, and enjoy for half the money the same education he could command in Scotland : but, notwithstanding this inducement, very few visit it, either from other German states or foreign countries. The industry of the town, which is considerable, comprises the manufacture of coarse linen fabrics, hats, tobacco, &c. ; and three annual fairs are very numerously attended.

Jena is famous in modern history, from its vicinity having been the scene of the great battle of the 14th of October, 1806, between the grand French army under Napoleon, and the Prussians, commanded by the king and the duke of Brunswick, the latter of whom was mortally wounded in the action. The French gained a complete and decisive victory. The Prussians lost above 20,000 men, killed and taken in the course of the day, with all their cannon, baggage, &c. In fact, their army may be said to have been totally destroyed ; as most of the troops who escaped from the field were soon after compelled to surrender.

JENNER, t., Somerset co., Pa., 12 m. N.W. Somerville bor. Drained by Beaver-dam run, on which coal is found. It contains three stores, one fulling-mill, one woollen factory, six grist-mills, 10 saw-mills, three tanneries, four distilleries ; one school, 37 scholars. Pop. 1480.

JENNINGS, county, Ia. Situated towards the S.E. part of the state, and contains 380 sq. m. Organized in 1816. Watered by Graham's fork and North fork of Muscastsack r., and Sand cr., which afford good water-power. It contained, in 1840, 7383 neat cattle, 7087 sheep, 14,318 swine ; and produced 56,691 bushels of wheat, 1971 of rye, 170,115 of Indian corn. 81,953 of oats, 9515 of potatoes, 21,425 pounds of tobacco. It had 18 stores, two flouring-mills, 10 grist-mills, 16 saw-mills, nine tanneries, four distilleries, one printing-office, one weekly newspaper ; two academies, 133 students, 17 schools, 555 scholars. Pop. 8899. Capital, Vernon.

JERICHO, p. t., Chittenden co., Vt., 52 m. N.W. Montpelier, 528 W. Bounded S.W. by Onion or Winooski river. Watered by Brown's r. which affords water-power. Chartered in 1763. It contains two churches, a Congregational and Baptist, and has some Methodists and Universalists, three stores, one grist-mill, six saw-mills, two tanneries ; one academy, 50 students, nine schools, 455 scholars. Pop. 1685.

JERSEY, an island of the English channel, belonging to Great Britain, and the principal of that group known as the Channel Islands, in St. Michael's bay, 12 m. W. the coast of France, and 65 m. S. Portland Bill, its N.W. point being in lat. 49° 14' N., long. 2° 28' W. Shape somewhat oblong; greatest length, E. to W., 12 m.; average breadth, 5 m.; area, 39,980 acres. Pop., in 1831, 36,582. The entire N. side of the island, and portions of the N.E. and S.W. sides, are defended by bold, precipitous rocks, rising to upward of 250 feet above the sea, and all around it are almost innumerable rocky islets, separated from the cliffs by the operation of the tides, which set with great force and rapidly round the Channel islands. The surface has a general inclination from N. to S., on which side the coast approaches the level of the sea. There is little table land; but elevated hill ranges run southward, bounding deep and narrow vales, watered by small rivers. Jersey, geologically considered, is, like the other islands in the same group, composed of secondary rocks, resting on granitic formations. True granite is not observed; but syenite, which is largely quarried, and exported as granite, is very prevalent, passing in some parts into porphyry and greenstone: it is covered in the S. and more level tracts by schistus and clay-slate, intermingled here and there with a clay conglomerate. Iron and manganese, the only metals that occur, are not wrought. The climate, though damp, owing to frequent rains and intense sea fogs, is remarkably mild. "The island," says Dr. Hooper, "enjoys an early spring and a lengthened autumn. Vegetation being usually active and forward in March, and the landscape far from naked at the end of December. Spring is marked by unsteadiness of temperature and harsh variable weather, with a prevalence of E. winds; and this disadvantage is felt particularly in May, which often fails to bring with it the expected enjoyments. March is mild, and October yet milder." (*Observ. on the Top. Clim. and Diseases of Jersey.*) The soil in the higher parts is gritty, being composed of detritus from the rocks and sea-sand, mixed with vegetable mould; but in the valleys there is a great depth of alluvial matter, washed down by violent rains from the higher lands; and these tracts, where not swampy, are extremely fertile. The S.W. corner of the island is a mere assemblage of sandy and barren hillocks. Agriculture is backward, owing partly to the minute division of property, occasioned by the law of gavelkind, and partly to the insufficiency of rural labourers. The value of land ranges between £120 and £160 per acre, and rents vary from £4 10s. to £8 15s., according to the distance from St. Helier's. Farms average about four acres, few exceeding 10: the occupiers are, consequently, for the most part poor; and even if they were possessed of adequate capital, the limited size of the farms is an invincible obstacle to the introduction of an improved system of farming. Some tendency towards improvement has, however, recently manifested itself. The ponderous Jersey plough, known as the *grande querue*, though not wholly discarded, is likely soon to be supplanted by the Norfolk plough.

The rotation of crops, as applicable to the soil and climate, is pretty well understood, and absolute fallows are rarely, if ever, seen. Wheat crops, cut partly in August, produce, according to the official returns, nearly five quarters per acre, and the gross yearly produce is said to amount to 13,800 qrs. of wheat and 2900 qrs. of barley. But the culture of neither wheat, barley, nor oats is found to be profitable, and they are, therefore, chiefly imported. Potatoes are raised in large quantities, the common sometimes exceeding 80,000 lbs. per acre; but the sea-weed used as manure gives them an unpleasant flavour. Parsnips and mangel-würzel are largely cultivated. Lucerne is highly in favour with the farmers, as it will grow on soils unfit for other purposes: four crops in a year are not unusual, and the land is afterward fed off. A large portion of the cultivable land (one quarter, according to Quayle) is occupied by apple trees, and the exports of apples and cider have been steadily increasing for some years, the export of the latter amounting to 2000 hhds. a year, according to the latest returns. The annual yield of apples averages 20 hhds. per acre. The *pear-main* is a good eating apple; but the pride of the island is the *chaumontelle* pear, often a pound in weight, and sold occasionally at the rate of £5 per hundred. The *calmar* pear is also well esteemed, and peach-apricots, melons, and strawberries are abundant, and noted for size and flavour. Timber trees grow in the hedge-rows, and unite with the fruit trees in giving to the scenery softness and richness rarely equalled. "In fact," says Mr. Inglis, "Jersey appears like an extensive pleasure-ground, one immense park, thickly sodded with trees, beautifully undulating, and dotted with cottages." (I., p. 35.) The manure universally used in dressing the land is sea-weed or *vraic*, the gathering of which is restricted by the island legislature to two seasons, the middle of March and the end of July, times of great interest to the natives. On grass lands the *vraic* is used in its natural state; but for other purposes it is burnt. Cattle breeding is a favourite and highly profitable pursuit here, and in the other Channel islands; and the treasure highest in a Jerseyman's estimation is his cow. (*Quayles' Agric. Jersey.*)

The Jersey cow (usually called the Alderney cow in England) materially differs from that of Guernsey, which is larger, and resembles the short-horned Devonshire breed. It has a fine, curved, tapering horn, slender nose, fine skin, and deer-like form. Its purity is maintained by breeding in and in; and, in order to preserve the breed intact, the legislature has prohibited the importation of other breeds under heavy penalties. Milch cows produce daily, at an average, 10 quarts of milk, and one pound of butter (eight quarts of the former producing one pound of the latter), the yearly produce of a cow being estimated at £10. The price of a good cow varies from £10 to £15. The butter is chiefly sent for sale to St. Helier's market, or exported to England; the quantity sent thither in 1830 amounted to 25,000 lbs. Sheep are little reared. The Jersey horse is a cross of the Cossack, procured through the residence of some Russian cavalry on the island in 1800: it is a strong, hard working animal; but no attention is paid to the improvement of the breed. The oyster fishery employs many of the natives; but lately it has been on the decline, owing to the competition of the French fishermen of Granville. The fishery is most active from February to May, and the exports of oysters in 1835 amounted to 150,000 bushels. In the same year 1470 dozen lobsters were sent to London from Jersey. The conger-eel and herring fishery, formerly highly productive, has been almost superseded by the deep-sea cod fishery, which employs nearly 80 vessels of 8000 tons, and gives employment during the summer to 1300 Jerseymen. The fish are chiefly sent to Brazil, 16,000 barrels, of 126 lbs. each, being sent thither in 1835.

The trade of Jersey has increased rapidly during the last 50 years, and its commercial relations, formerly confined to England and France, now extend to the chief countries of Europe, the W. Indies, and S. America. This increasing prosperity is proved by the returns of ships belonging to St. Helier's. That port, in 1817, had only 79 vessels of 8167 tons; while in 1837 (after a gradual increase) it had 964 ships of 23,826 tons, exclusive of about 500 fishing smacks, chiefly used in the oyster fisheries. Indeed, so great has been the increase of business of late years, that the erection of a new and larger pier at St. Helier's is in contemplation. The trade with England is subject to certain regulations intended to prevent contraband traffic; but every article of the growth, produce, or manufacture of Jersey is admitted into the mother country on payment of the duties imposed on similar commodities grown, produced, or manufactured here. The island receives from England, its general merchant, cotton and woollen fabrics, and hosiery, hardware and cutlery, earthenware and glass, soap and candles, and about 20,000 tons of coals yearly; in exchange for which it sends apples and cider, cattle, potatoes and potato-spirit, oysters, and granite. The imports from France consist of wine and brandy (70,000 gallons of the former and 50,000 gallons of the latter), skins, fruit, and poultry; for which coals, bricks, and potatoes are sent in exchange. The island is supplied with fir and oak timber (1400 loads of fir and 500 ditto of oak yearly) from Sweden and Norway, with hemp, linen fabrics, and tallow from Russia, with wheat and barley (about 22,000 qrs. annually) from Prussia and Denmark, and with cheese, geneva, and tiles from Holland; the exports to these countries chiefly consisting of cider and sugar from Brazil, with which this island has extensive dealings, employing 20 ships of 4000 tons, and importing thence about 600 tons of sugar and 4700 cwts. of coffee. The imports from Spain, Portugal, and Sicily average yearly 70,000 gallons of wine and 100,000 gallons of brandy. The Jersey merchants also trade with Honduras for mahogany, sent chiefly to England. The manufacturing industry of the island is almost confined to ship-building, shoe making, and hosiery. Ship-building is carried on to a considerable extent, in consequence of the timber imported into the island being exempted from all duty; though, if it be proper to lay a duty on the timber employed in ship-building in Britain, it is not easy to see why timber employed for the same purpose in Jersey should be exempted from the duty. Shoemaking is pretty extensively carried on, and about 13,000 pairs of boots and shoes, chiefly of French leather, are sent annually to British N. America. The hosiery business has greatly declined, owing to the use of machine-made stockings; and the persons now employed in it depend almost entirely on the demand of the island. The communication with England is kept up by means of steamers to and from Southampton four times a week, and by mail-packets twice a week to and from Weymouth. On the arrival of the steamers from Southampton, packets leave for St. Malo and Granville, returning on the alternate days. Traders are constantly sailing to and from London, Bristol, and other English ports.

The vernacular language of the island is French, which is used in the churches and courts of law: the upper ranks speak it in its purity, but the lower classes speak Jersey-French, a *patois* compounded of old Norman French with

Gallicised English. English, however, is becoming daily
more prevalent, and most of the country people understand
and speak it. "The Jerseymen, especially the lower orders,
are characterized by blunt independence, often amounting
to brusquerie, excessive love of gain, and unceasing industry. The minute division of property prevents them from
acquiring an independence, while at the same time the
actual ownership of land protected by legal privileges, gives
them a freedom of sentiment which no tenant at will can
enjoy. Their parsimony, however, is not only prejudicial
to themselves, as leading them to begrudge provender to
their most valuable cows, but is also injurious to others,
whom they overreach in bargaining." (Inglis.) Their fare
is simple and inexpensive, consisting principally of soupe-à-
choux, a compound of lard, cabbage, and potatoes: conger-
eel soup and pickled pork are rarities reserved for festive
occasions. The chaumontelle pear is commonly eaten with
tea; cider is the general substitute for beer. The higher
classes seldom give entertainments or exchange civilities,
and are, much divided by party spirit. The old parties of
Magot and Chariot have given way to the liberal Rose and
the exclusive high church and state Laurel. Literature is
forgotten amid island politics; and even the press, so powerful an engine in England, has scarcely any influence in
Jersey. The English residents must be considered as a
class quite distinct from the natives, with whom they have
little intercourse; they amount to about 4000, being chiefly
half-pay officers with their families, attracted by the cheapness of living and the mildness of the climate.

The revenues of Jersey have greatly increased of late
years, for at an average of the three years ending with 1819,
they only amounted to £4600 a year, whereas, in 1836, they
exceeded £14,600, arising from licenses to tavern-keepers,
market-tolls, harbour-dues, duties on wine and spirits, &c.
These revenues, after the current expenses of the government and the interest on the public debt (amounting in 1840
to £61,376) have been paid, are applied to the public works
and general improvement of the island. The expense of
the militia and English troops (exceeding £30,000 yearly)
is defrayed by the British government, and the salaries of
the governor and his officers are provided for from the great
tithes of the 12 parishes. French and Spanish coins were
until lately current in Jersey; but in 1832 the French government called in its old silver coins, since which time English sovereigns and silver have been commonly circulated.
The exchange varies from eight to nine per cent. in favour
of England, so that an English shilling passes for 13d., and
a sovereign for £1 1s. 6d. Jersey currency.

Jersey and Guernsey have long enjoyed peculiar privileges granted by John and succeeding monarchs. No process in either of the islands, commenced before an island
magistrate, can be carried out of it, and no person convicted
of felony out of the said islands is to forfeit his inheritance
in them, so as to deprive his heirs of their lawful possessions. They are exempted from the jurisdiction of the
British courts, except that of the admiralty, and have an
immunity from all taxes except what are voted by the island
legislature.

Jersey is governed by a local legislature, and a distinct
judicature under the ultimate control of the sovereign in
council. The legislative assembly, called the states, consists
of 36 members, viz., 12 jurats elected for life by the rate-
payers of the island, the 12 rectors of the 12 parishes into
which Jersey is divided, and the 12 constables of parishes,
chosen triennially by the parishioners. It is convened by
the bailiff, who always presides, either in person or by deputy; and its chief business is to raise money for the public
service, and to pass laws for the government of the island;
which, however, continue in force only three years, unless
ratified by the sovereign in council. The governor, as the
king's representative, has a veto on all the proceedings of
the states, but never uses it, except in cases which concern
"the special interest of the crown." The Jersey court of
judicature, called the "royal court," is composed of the
bailiff, who here represents the sovereign, and of the same
12 jurats who sit in the states. The officers are, the attorney-general, solicitor-general, high-sheriff or viscount clerk,
or greffier, and six pleaders appointed by the bailiff, and
styled avocats du barreau. This court has cognizance of all
pleas, suits, and actions, whether real, personal, or criminal,
arising within the island, except cases of treason and coining, which are referred to the sovereign in council. A code
of laws, compiled in 1771, and sanctioned by the king, is the
fundamental statute law; but it is extremely defective, and
is continually changed by the enactment of new laws. The
custom of gavelkind obtains, with respect to the disposal of
real property: the eldest son, however, by common usage,
takes half the estate, and the rest is equally divided. Personal property may be devised, but when left intestate is
divided among the children, two thirds going to sons, and
one third to daughters. Debts are recoverable by legal
process in the royal court. Insolvents may be compelled to

give up (renoncer) their property, for the benefit of creditors,
and either the vicomte may sequestrate it, to pay demands
entered against it, or the court may grant the debtor a respite
of a year and a day for payment of his debts. Persons not
possessing land or houses may be arrested for debt; but
property is attached before the person; and landed proprietors cannot be imprisoned till after a judgment. Debts contracted in England can be sued for in Jersey, if not of more
than six years' standing: debts contracted in Jersey are
recoverable within 10 years.

The military government of the island is conducted by a
lieutenant-governor, who has the custody of the fortresses,
and the command of both the regular troops and militia.
The chief fortresses are fort Regent, Elizabeth castle, and
Mt. Orgueil castle, all on the S. coast. The island is farther defended by a chain of martello towers, redoubts, and
batteries, which encircle it. The militia, in which all male
natives, from the age of 17 to 65, are liable to serve, comprises six regiments and 2500 men, exclusive of an artillery
battalion of 600 men. The regular troops in time of peace
seldom exceed 300 men; but 7600 men were quartered in
the island during last war. Since the reign of James I., the
church of England has been the established religion of Jersey, which is under the ecclesiastical direction of the bishop
of Winchester. Every parish has a church, and the service
is usually performed in French, except at St. Heiler's, where
English is the language of the congregation. The Independents, Wesleyans, and Baptists have chapels in which service is conducted both in French and English; and there
are two places of worship for Roman Catholics. Two free
grammar-schools were established in Jersey in the 15th
century; but the endowments are small. Two public
schools on the national system were established by subscription some years since, and are now in successful operation.

The remains of Roman fortifications and the discovery
of coins belonging to the emperors, prove Jersey to have
been a military station, and under the Franks it formed a
part of the region called Neustria. The Normans invaded
the Channel islands in the 9th century; and when the
duchy of Normandy was annexed to the crown of England
at the conquest, they came under the British dominion.
The French have repeatedly tried to wrest from us these
islands, which, by their proximity to the coast of France,
seem to be their natural property; but they have uniformly
failed. The last attempt was made in 1781 by a detachment of 700 soldiers, under the Baron de Rullecourt, who
surprised and captured the garrison, but were finally
compelled to escape to their vessels after a desperate
encounter with the native militia under Major Pierson, in
the streets of St. Heiler's. (Cæsarea; Inglis's Channel
Islands, vol. i. passim; Quayle's Agricultural Survey of
Jersey; Geol. Transl. vol. i.)

JERSEY, county, Ill. Situated in the W. part of the
state, and contains 300 sq. m. Bounded S. Mississippi
river, W. by Illinois river. Watered by Macoupin creek,
a tributary of Illinois river. It contained in 1840, 7130 neat
cattle, 4274 sheep, 14,807 swine; and produced 39,430
bushels of wheat, 258,924 of Indian corn, 48,077 of oats,
16,971 of potatoes, 15,096 pounds of tobacco, 1806 of cotton.
It had five stores, five flouring-mills, eight grist-mills, six
saw-mills, four tanneries, one pottery, one printing-office,
one weekly newspaper. Pop. 4535. Capital, Jerseyville.

JERSEY, p. t., Licking co., O., 35 m. N.E. Columbus,
384 W. It contains one store, four schools, 207 scholars.
Pop. 939.

JERSEY CITY, city, Hudson co., N. J., 589 N.E. Trenton, 8' E. Newark, 234 W. Situated on the W. side of
Hudson river, opposite to New York, a little over a mile
W. from it. It is connected with the city of New-York by
a steam ferry, with three boats, constantly plying. The
ground on which it is built projects into Hudson river,
having bays N. and S. of it. It is laid out into lots, 25 feet
front by 100 feet deep, distributed into 45 blocks of two
acres each, with broad streets crossing each other at right
angles. It contains four churches, an Episcopal, Dutch
Reformed, Methodist and Roman Catholic, a lyceum,
with a handsome brick edifice, a female academy, a high
school for boys, a bank, 23 stores, capital $27,000, an
extensive pottery, where beautiful delfware is produced to
the amount of $200,500 annually, a flint glass factory,
employing 100 hands, and producing plain and cut glass to
the amount of $200,080, annually, three lumber yards,
capital 3000, two iron foundries, two printing offices, one
bindery, two weekly newspapers, 11 schools, 330 scholars.
Pop. 3072. Directly W. of Jersey city is a settlement
called Harsimus, which contains a Baptist church, an iron
foundry, a ropewalk, a starch factory and about 30 dwellings. N. of this is another settlement called Pavonia,
which contains three carpet factories, and about 50 dwellings. The whole number of dwellings in the city and
suburbs is about 400. The city assessor in 1843 reported

the number of inhabitants as 3730, which is an increase of about 760 over the census of 1840. The adjoining suburbs contain about 1500, making the whole, 5250. The city is neatly built, and many of its inhabitants do business in the city of New-York. The New-Jersey railroad, which is continued to Philadelphia, and the Patterson and Hudson railroads commence here, and have a fine depot. The Morris canal, 101 m. long, connecting the Hudson and Delaware rivers, terminates here, and has a large basin. The Thatched Cottage Garden, in the S. part of the city, is a beautiful place of summer resort. Jersey city received a city incorporation in 1690, and is governed by a mayor and board of aldermen. Though it is in a different state, yet its contiguity and relations to the city of New-York, render it in fact, a suburb of that city, with the prosperity of which is own growth and prosperity are intimately connected.

JERSEY SHORE, p. b., Mifflin t., Lycoming co., Pa., 89 m. N.N.W. Harrisburg, 209 W. Situated on the N. side of the West branch of Susquehanna river. It contains a Methodist church, 11 stores, two tanneries: two schools, 58 scholars, 160 dwellings and 325 inhabitants. Incorporated in 1826. The western division of the Pennsylvania canal passes through the place.

JERSEYVILLE, p. v., capital of Jersey co., Ill., 71 m. S.W. Springfield, 821 W. Beautifully situated in Jersey prairie, and contains a courthouse, and about 30 dwellings. Nett proceeds of the post office, $390.

JERUSALEM (Heb. Kadushâ ; Gr. Kαδύτις by Herodotus, and 'Iερσσóλυμα by Strabo and later writers; mod. Arab. El-Koddes), a famous city of Palestine, interesting from its high antiquity, but far more from its intimate connection with the history of the Jews, and the eventful life of the great founder of Christianity ; 198 m. S.S.W. Damascus, 33 m. E. Jaffa, and 76 m. S. by E. Acre; lat. 31° 46' 34" N., long. 35° 31' 34" E. Population according to the official report of Mr. Consul Moore, 16,000, of whom about two thirds are Mohammedans. The population has been estimated by some travellers at 20,000 ; but it has not had so many inhabitants for some years, except at Easter, when the Moslem and Christian pilgrims swell the population to nearly a half more than its ordinary amount. The city stands on a hill, between two small valleys, in one of which, on the W., the brook Gihon runs with a S.E. course, to join the brook Kedron, in the narrow valley of Jehoshaphat, E. of Jerusalem. The first view of the city from the W. is thus described by Robinson:—"As we approach Jerusalem, the road becomes more and more rugged, and all appearance of vegetation ceases : the rocks are scantily covered with soil, the verdure is burnt up, and there is an entire absence of animal life. A line of embattled walls, above which rose a few cupolas and minarets, suddenly presented itself to my view. I was disappointed in its general appearance; but this feeling originated not so much from the aspect of the town as from the singularity of its position, surrounded by mountains, without any cultivated land to be seen, and not on any high road." (Pal. and Syr. i. p. 36.) The opposite view, however, from the Mount of Olives, is much more attractive, for it commands the whole of the city, and nearly every particular building, including the church of the Holy Sepulchre, the Armenian convent, the mosque of Omar, St. Stephen's gate, the round-topped houses, and the barren vacancies within its circle. (Henniker's Trav. p. 174.) The modern city, built about 309 years ago, is entirely surrounded by walls, barely 2½ m. in circle, flanked here and there with square towers. The four principal gates are those of Damascus and Jaffa on the W., that of Zion on the S., and St. Stephen's on the E. The interior is divided by two valleys, intersecting each other at right angles into four hills, on which history, sacred and profane, has stamped the imperishable names of Zion, Acra, Bezetha, and Moriah. Zion is now the Armenian and Jewish quarter ; Acra is better known as the lower city and Christian quarter ; while the mosque of Omar, with its sacred inclosure (called by the Turks el Haram Scheref), occupies the hill of Moriah. The streets are narrow, like those of all Syrian towns ; the houses, except those belonging to the Turks, are shabby, and the shops poorly supplied. Dr. E. Robinson, of the United States, however, remarks, "that he was agreeably disappointed, and found the houses better built, and the streets cleaner, than those of Alexandria, Smyrna, or Constantinople." (Geog. Journ. ix. p. 299.) The public buildings are not numerous, and excepting those consecrated to religious worship, there are none worthy of notice. The baths also and bazaars are mostly inferior to similar establishments in other parts of the E.

The boundaries of the old city, said by Pliny to be longe clarissima urbium Orientalis non Judaeæ modo (Hist. Nat. lib. v. § 13), are so imperfectly marked, that no fact can be deduced respecting them from the elaborate researches of D'Anville, Clarke, Niebuhr, and others, save only that

they varied at different periods; and that, when most extensive, at the æra of its destruction, its treble row of walls embraced a circuit of 33 stadia, including Mount Moriah, Mount Zion, Acra, Bezetha, &c. (Rolandi Palestina, p. 835.) But the walls having been wholly destroyed, it is impossible to trace their exact situation.[*] It is impossible also to describe in detail the many spots within the modern city which blind superstition or minute criticism has fixed on as the scenes of events connected with the history of the patriarchs, and the sufferings of Christ; but some places are ascertained beyond a doubt, which all travellers visit with interest, and which command universal respect. There can, for example, be no question, that the mount (Moriah) on which the mosque of Omar now stands was once crowned with the House of the Lord, built by Solomon, at a cost and with a magnificence of which we can form no adequate idea (1 Kings, caps. vi. and vii.). This great glory of Judea, after standing for above 400 years, was first rifled, and soon after destroyed by Nebuchadnezzar, king of Babylon. A second temple, built on the site of the first, by the Jews, after their return from the Babylonish captivity, was so much enlarged and improved by Herod the Great, as to be little inferior to that of Solomon. Tacitus calls it, immensæ opulentiæ templum; and he truly adds, nulla intus Deum effigie, vacuum sedes, et inanis arcana. (Hist. lib. v. § 8, 9.) Notwithstanding the efforts of Titus for its preservation, this structure, the palladium of the Jewish nation, was totally destroyed during the siege of Jerusalem, A.D. 70. The mosque of Omar, which occupies this sacred site, stands on an elevated four-sided plateau, about 1500 ft. long, and 1000 ft. broad, supported on all sides by massive walls, built up from the lower ground. The lowest portion of these walls is supposed by Dr. Robinson to belong to the ancient temple, and to be referable to the time of Herod at least, if not of Nehemiah and Solomon. The mosque, el Sakhara, the erection of which was begun by the caliph Omar, in 637, is of an octagonal shape, surrounded by a lead-covered dome, above which is a glittering crescent. It has four entrances, one of which, towards the N., is adorned by a fine portico, supported by eight Corinthian pillars of marble. Its 46 windows are of stained glass, and the walls are faced below with blue and white marble, and above with glazed tiles of various colours, forming a beautiful mosaic of texts from the Koran. It is altogether a fine specimen of light and elegant Oriental architecture ; and the building contrasts singularly with the severity of the surrounding scenery. The interior is not allowed to be entered except by the followers of the prophet; and Dr. Richardson, an English physician, is one of only four Christians who have been admitted within its walls. (See Robinson's Pal., vol. i. Append. p. 290.) "The arrangements," he remarks, "are so managed as to keep up the external octagonal shape. The inside of the wall is white, and without ornament; and the floor is of gray marble. A little within the W. door, is a flat polished slab of green marble, forming part of the floor, and regarded with peculiar respect by the Mohammedans; a little beyond is a series of 24 blue marble pillars supporting the roof, and inside these are four large square columns, forming the support of the dome, which rises about 100 ft. above the floor. The central part is railed round, a single door admitting the devotee to the sacred stone, called the Hadjr el Sakhara, on which is shown the print of Mahomet's foot when he was translated to heaven. The whole interior is extremely beautiful, and the effect is much heightened by the blending of colours in the pillars that run round the mosque." (Richardson's Travels in the Med., &c., vol. ii. p. 286.) Within the same enclosure, near its S. wall, is another mosque, of square shape, called El-Aksa. The cupola is spherical, and ornamented with arabesque paintings and gildings of great beauty. Between the mosques is a handsome marble fountain for ablutions. On the opposite side of the city, in the Latin quarter, called Harat el Nassara, is the church of the Holy Sepulchre, a building in the Byzantine style, erected by Helena, mother of Constantine the Great, in the centre of a court or enclosure, filled at pilgrim-time with pedlars of every description, especially venders of relics and rosaries. The building resembles Roman Catholic churches in general, but is greatly inferior, notwithstanding its valuable marbles, to many of the sacred edifices in Rome. Immediately in front of the entrance, which is guarded by Moslem sol-

* Josephus most distinctly says that the Romans left only the W. wall standing, with the towers Phaselus, Hippicus, and Mariamne, and that the remainder was razed to the ground. Τὸ δ᾽ ἄλλον ἅπαντα τῆς πόλεως περίβολον οὕτως ἐξωμάλισαν οἱ κατασκάπτοντες, ὡς μηδὲ πώποτ᾽ οἰκηθῆναι πίστιν ἄν ἔτι παρασχεῖν τοῖς προσελθοῦσι. (Jud. Bell. lib. vii. c. 1) This assurance of an eye-witness, and the knowledge that two subsequent and very destructive sieges left scarcely any remains even of a more recent city, suffices to show how little credit is due to any of those antiquarian speculations, however ingenious.

JERUSALEM.

diers (who receive a tax from all the pilgrims), is a slightly elevated marble slab, called the "stone of unction," on which, according to the monks, our Lord's body was laid, to be anointed by Joseph of Arithmathea; and near it are 17 steps, conducting to the supposed mount Calvary, now a handsome dome-covered apartment several feet above the floor of the church, floored and lined with the richest Italian marbles: in the crypt beneath is a circular silver plate with an aperture in the centre, through which the arm reaches the identical hole in which the cross was fixed! The great object of interest, however, is the Holy Sepulchre itself, an oblong structure 15 ft. long by 10 ft. in breadth, roofed in with a handsome ceiling corresponding to the richness of the silver, gold, and marble decorating its interior: it stands directly under the great dome of the church, and is divided into two chambers, the first containing the stone on which the angel sat when he addressed the affrighted women, "Why seek ye the living among the dead? he is not here, but is risen," and the other being the sepulchre to which he pointed, saying, "Behold the place where they laid him." The inner compartment, lined with verd antique, is only large enough to allow four persons to stand by the side of a plain white marble sarcophagus of the ordinary dimensions, over which hang seven large and 44 smaller lamps, always kept burning. Around the large circular hall, which is surrounded by a gallery supported on pillars, and roofed by a vast dome, are oratories for the Syrians, Copts, Maronites, and other sects who have not, like the Greeks, Armenians, and Roman Catholics, chapels in the body of the church. The Greek chapel at the E. end of the hall is parted off by a curtain, and is incomparably the most elegant and highly decorated: the Latin chapel closely resembles those seen in Italy, and has a gallery with a fine organ: that belonging to the the gallery. Various parts of the church Armenians is by monks and pilgrims, as the scenes of connected with the last sufferings of Christ; and to such an extent is superstition carried, that a stone is exhibited and gravely declared to be that on which our Saviour was placed, when put in the stocks! The faith, indeed, of intelligent men is most severely tested during a visit to this church: there cannot, however, be a doubt that it stands on the hill of Calvary, and it probably includes the site of the crucifixion; but there seems to be little ground for the assumption, that the tomb and site of the cross were so near to each other as to be inclosed by the iding. In an antechamber near the entry are same bal, the most authentic probably of which-are spurs of Godfrey of Bouillon. The tombs of Godfrey and his brother Baldwin were destroyed during a fire which took place in 1808, and have not been restored, owing to the ill-will felt by the Greek Christians towards the Romish church, to which these monarchs belonged. Westward of the church just described in the Hardt-el-Nassara, or Christians' street, is the Franciscan convent of St. Salvador, called by way of distinction Il Convento delle Terra Santa, a large stone building, having several courts and gardens enclosed within a strong wall. The funds are supplied by contributions sent from Rome and other Catholic countries, and the inmates comprise from 60 to 80 monks, chiefly Italian and Spanish, by whom European strangers visiting the Holy City are hospitably entertained. The church attached to the convent is gaudily furnished with candlesticks, images, &c., and has a good organ. E. of the above stands the Greek monastery, a well supported establishment with a small subterranean church. The city castle, close to the gate of Jaffa, is supposed to have been built on the ruins of the Turris Psephina of old Jerusalem: it comprises a few towers connected by curtains, and has a few old guns mounted on broken carriages. Close by it, on the ascent to the hill of Zion, is the Armenian convent, in the best-looking district of the city, comprising within its precincts rooms sufficient to accommodate a thousand pilgrims, and a large garden: the conventual church is spacious, and most elaborately ornamented; the floor is paved in the most delicate mosaic. E. of the convent is a small Armenian chapel, marking the site of the house of Annas, the high-priest; and just outside the gate of Zion is another chapel, supposed to occupy the site of the house of Caiaphas: these positions seem to be far from improbable. (Compare Joseph. Antiq. lib. xviii. c. 3, with St. John xviii. 94.) Not a vestige remains of the ancient buildings on Mount Zion. where David built a palace, his own residence, and that of his successors, whence it was emphatically called the "City of David." Its limits are, however, well defined by the aqueduct which conveyed water from Jerusalem to Bethlehem. The hill-side is now used as a Christian burial-ground. N. of the city, in the district called Acra, are the ruins of Herod's palace, and about 300 yards to the S.E., near the reputed pool of Bethesda, is the residence of the mutsellim, or Turkish governor, supposed, though with

little show of reason, to occupy the site of the prætorium of Pontius Pilate. It is a large straggling building, having a flat roof, which commands a complete view of the mosque of Omar: it stands in the principal street of the modern city, called by the Turks Hardt-el-Allam, and by the Christians Via Dolorosa, the monks having fixed on it as the line of route along which our Saviour was led from the hall of judgment to Calvary. The Jewish quarter (Hardt-el-Yahoud) occupies the hollow between the hills of Zion and Moriah: it contains seven mean and small synagogues; and the numerous private dwellings, how comfortable soever inside, have uniformly mean and ill-built exteriors, owing, it is said, to the fear of exciting among the Mohammedans any suspicion of the wealth of the despised nation. The poorer Jews are supported by charitable contributions obtained from their fellow-countrymen in Europe, especially in Germany and Spain. (Turner, ii. 264.) The Turks reside on the E. side of the city all round the great enclosure of Mount Moriah. The suburbs of Jerusalem abound with interesting remains of less questionable antiquity and authenticity than most of those within the modern walls. Close to the gate of Jaffa is the pool of Gihon, near which, in a village of the same name. "Zadok the priest, and Nathan the prophet, anointed Solomon king over Israel" (1 Kings, i. 34.), and, at a later period, Hezekiah "stopped the upper watercourse of Gihon, and brought it straight down to the W. side of the city of David." (2 Chron. xxxii. 30.) S. of Mount Zion is the valley of Hinnom, in which are numerous tombs hollowed out of the rock, and a building, once used by the Armenians as a charnel-house. The E. boundary of Jerusalem is formed by the valley of Jehosaphat, which divides it from the Mount of Olives. Proceeding up this valley, the traveller soon arrives at

"Siloa's brook, that flow'd
Fast by the oracle of God."

The source of these celebrated waters, which now, at least, are brackish and sulphureous, lies close under the walls of Hardt-el-Schereef on Mount Moriah; but the pool is rather more than 1-4 m. below it. "The stream," says Mr. Robinson, "issues by an underground passage from a rock, and falls into a small basin of no great depth. It was once covered with a chapel, erected to commemorate the miraculous cure of the man born blind." (St. John, ix., 1-7.) The descent to the lower pool, which is remarkable for its daily ebbing and flowing, is by a flight of 30 steps, whence it has acquired the name of the "fountain of stairs." On the E. side of the brook Kedron, now a mere rivulet, running in a valley so closely pent up as to deserve the name of a mountain-gorge, especially at its N. extremity, are four sepulchres, constructed, unlike most in Judea, above ground, and designated the tombs of the patriarchs; one of them is alleged to be the burial-place of Zacharias, the son of Barachias. (See Matt. xxiii., 29, 35.) S. of these tombs, and under the shadow of the temple of Solomon, is the favourite burial-ground of the Jews, among all of whom the dearest wish is, that they may lay their bones near those of their long-buried ancestors, and be ready for the summons of Jehovah, when He "shall come up to the valley of Jehosaphat, and there judge all the heathen round about." (Joel, iii., 12.) Further, N.E. are the gardens of Gethsemane, enclosed by a wall, and still in a sort of ruined cultivation, and the Mount of Olives, a hillock covered with stunted herbage, and with patches here and there of the tree with which it was once abundantly clothed. Here every spot has its grotto and legend, and on the hill the precise place is pointed out whence the Saviour ascended to heaven. The Empress Helena built on it a monastery, which the Turks have converted into a mosque; somewhat to the N. is the Church of the Ascension, now in the hands of the Greek Christians. N. of the bridge, over the brook Kedron, and about 250 yds. from St. Stephen's Gate, is the reputed tomb of the Virgin Mary, comprising, besides several cenotaphs, a subterraneous chapel, in which lamps are kept constantly burning, and services daily celebrated according to the rites of the Greek Church. Passing thence up the bank of the Kedron, and crossing the hill Bezetha, the stranger is conducted to the excavations called "the Tombs of the Kings." The road down to them is cut in the rock, and a stone doorway leads to a kind of antechamber, now at least, open at the top, and measuring 50 feet in length by 40 feet in breadth. It is ornamented by a beautifully carved cornice, and in the S.W. corner a door, formed of a single stone slab, admirably adapted to its framework, and easily working on its hinges, leads into a series of chambers, round which are niches in the rock for the reception of the dead. It is very probable that these are the "royal caves" described by Josephus, as situated close to the N. boundary of the ancient city (see Bell. Jud., lib. v., c. 4); but whether they contained the bones of the sons of David (2 Chron. xxxii., 33) or those of Helena, queen of Adiabene (as Drs. Clarke and Pococke have supposed), is a matter as to which no certain conclusions can be drawn.

Jerusalem, considered as a modern town, is of very slight importance. Superstition and fanaticism constitute the principal bond by which the population Christian, Jew, and Moslem, are held together. The Jew despises the Christian, and the follower of the prophet looks down with contempt on both; but pilgrims of each of the three creeds resort thither in such numbers as to increase the population a half; and heavy taxes are levied on all for the benefit of the pacha. The convents are supported by wealth sent from foreign countries, and a great influx of property takes place from the thousands of annual visiters, rich and poor, so that Jerusalem draws largely on Jaffa, Damascus, Nablous, and other places; but it has no industry whatever—nothing to give it commercial importance, unless, indeed, we may mention a trade, now almost wholly engrossed by the monks of the *Terra Santa* convent, in shells, beads, and relics, whole cargoes of which are shipped from Jaffa for Italy, Spain, and Portugal. The shells are of mother-of-pearl sculptured, and the beads are manufactured either from date-stones or a hard kind of wood called *Mecca fruit*. Rosaries and amulets are also made of the black fœtid limestone, and are highly valued in the East as charms against the plague. (See *Bowring's Report on Syria*, p. 21.) The retail trade seems to be equally insignificant. "The bazaar, or street of shops," says Mr. Robinson, "is arched over, dark, and gloomy, the shops are paltry, and the merchandise exposed for sale of an inferior description. This is the only part of Jerusalem where any signs of life are shown; and even here the pulsations of the expiring city are faint and almost imperceptible, its extremities being already cold and lifeless. In the other quarters of the town you may walk about a whole day without meeting with a human creature." Well, then, may the Jews, who still indulge the hope of restoring their metropolis to its pristine greatness, lament, with the prophet Jeremiah, "From the daughter of Zion all her beauty is departed. Jerusalem hath grievously sinned; therefore she is removed. The adversary hath spread out his hand, and the heathen hath entered into her sanctuary. All her people sigh and seek bread: see, O Lord, and consider, for I am become vile." (Lam., i., 6–11.) Nothing, indeed, can well be conceived so vile, so degrading, as the mummeries enacted in the Holy City, especially during the Easter festival. The monks, who are the servants of Mammon rather than of Christ, act on these occasions as showmen, and masters of the ceremonies; and even the pilgrims, who crowd to the Sepulchre in such numbers as to make order impossible, too frequently exhibit the greatest levity and unconcern. "What a scene was before me," says Mr. Turner. "The whole church was absolutely crammed with pilgrims, men and women halloing, shouting, singing, and violently struggling in to near the Sepulchre. One man in the contention had his right ear literally torn off." (ii., 198.) A few years ago, during the representation of the blasphemous pantomime, entitled "the Holy Fire" (intended to represent the descent of the Holy Spirit), the pressure was so intense, 6000 persons being assembled on the ground-floor, that great numbers fainted, a general confusion ensued, and upwards of 300 were either suffocated or crushed to death. (*Wilde's Narr.*, vol. ii., p. 212.) In fact, the whole scene is revolting to every rational and really devout Christian. Such, however, is the strength of superstition, that a pilgrimage to Jerusalem is still regarded, in many parts, as an act of the highest merit, and as bringing with it the assurance of eternal felicity.

The local government of Jerusalem is conducted by the *mutsellim*, or military governor; the *sania-khadi*, or chief of the police; the *mufti*, or chief judge; the *cape-serds* or superintendent of the mosque of Omar; and the *shbaoki*, or town-major; all of whom, except the *mufti*, hold their appointment under the pacha of Damascus.

Jerusalem has been usually supposed to be identical with the Salem of which Melchizedek was king in the time of Abraham, anno 1913 A.C., according to Archbishop Usher. When the Israelites entered the Holy Land 500 years afterward, it was in the possession of the Jebusites, descendants of Canaan. Joshua, soon after his entrance into Canaan, "fought against Jerusalem, and took it, and smote it with the edge of the sword, and set the city on fire" (Judges, i., 8); but the citadel on Mount Zion was held by the Jebusites till they were dislodged by David, who made Jerusalem the metropolis of his kingdom, and his dwelling in "the strong-hold of Zion." (2 Sam., v. 7.) He enlarged the city and built a beautiful palace: it was further embellished by his son Solomon, who, in the years 1012–1004 B.C., erected the temple already referred to. Palestine was afterward successively invaded by the Egyptians, Assyrians, and Babylonians, the last of whom, under Nebuchadnezzar (B.C. 588), took and destroyed the city,

burned the temple, and carried the people captive to Babylon. After a bondage of nearly 70 years the Jews were restored to their city, by Cyrus the Persian, and about anno 515 B.C. they rebuilt the temple, under the superintendence of Zerubbabel and Nehemiah. Alexander the Great is said, by Josephus, to have visited Jerusalem in peace, and to have respected the religion of the Jews; but the best critics reject this statement as inconsistent with the ascertained events in the life of Alexander, and unworthy of credit. (*Ancient Universal History*, viii., 536, 8vo; *Mitford's Greece*, vii., 532.) Ptolemy Soter, one of Alexander's generals, seized upon Syria and Palestine, sacked the Holy City, and carried off a large portion of its inhabitants to Alexandria. Later monarchs of the Macedonian empire, who attempted to introduce the pagan worship, were successfully opposed by the Maccabees, and the liberty of Judæa was at length restored, anno 165 B.C. The all-absorbing power of Rome finally put a period to Jewish independence, the whole of Syria being reduced by Pompey, and made a proconsular province. This great general, who took Jerusalem after a stout resistance, entered the temple, and explored its inmost recesses; and it is mentioned to his honour, that he touched none of the precious relics, or of the vast wealth accumulated in the sanctuary. *Victor ex illo fano nihil attigit.* (Cicero pro L. Flacco, § 28.) Jerusalem, however, was merely tributary, and had not lost its nominal sovereignty (in other and prophetic words, the *sceptro had not departed from Judah, nor a lawgiver from between his feet until Shiloh had come*, Gen. xlix., 10) till after the birth of Christ, when it became the residence of a procurator. The repeated rebellions of the Jews at length roused the vengeance of the Romans; and, A.D. 70, the city was taken by Titus, after one of the most memorable and destructive sieges of which history has preserved any account. The Jews, though rent by intestine factions, defended themselves with invincible obstinacy; they contemptuously rejected every proposal for a surrender, and braved alike the attacks of the Romans, and the still more dreadful attacks of famine. But their resistance was unavailing, except for their own destruction; and the city, being taken, was completely destroyed, along with the temple, three towers only being left as memorials of its existence and destruction. According to Josephus, no fewer than 1,100,000 persons fell in the siege, exclusive of above 100,000 taken prisoners. But, notwithstanding what has been alleged in defence of this statement by Brotier (Note ad lib. v., § 13, *Hist. Tacit*) and others, there can be no reasonable doubt that it is grossly exaggerated. The statement of Tacitus would seem to be infinitely more probable, though we incline to think that even it is, perhaps, beyond the mark. "Pervicacissimus quisque illuc perfugerat; eoque seditiosius agebant. Tres duces, totidem exercitus: prœlia, dolus, incendia inter ipsos, et magna vis frumenti ambusta. *Multitudinem obsessorum, omnis œtatis, virilis ac muliebris sexus,* SEXCENTA MILLIA *fuisse accepimus.* Arma cunctis qui ferre possent; et plures, quam pro numero, audebant. Obstinatio viris feminisque par; ac si transferre sedes cogerentur, major vitæ metus quam mortis."[*] (*Hist.*, lib. v., c. 12 & 13.) Adrian razed the city to the ground, ploughed up a great part of the surface, and built on its site the Roman town of Ælia Capitolina. The condition of Jerusalem at this period is well described by Milman:

> " Her tale of splendour now is told and done;
>   Her wine-cup of festivity is spilt,
>   And all is o'er, her grandeur and her guilt.
>   Her gold is dim, and even her music's voice;
>   The Heathen o'er her perish'd pomp rejoices;
>   Her streets are razed, her maidens sold for slaves,
>   Her gates thrown down, her elders in their graves:
>   Her feasts are holden mid the Gentiles' scorn,
>   By stealth her priesthood's holy garments worn;
>   Oh! long foretold, though long accomplish'd this,
>   Her home is left unto her desolate."
>   *Fall of Jerusalem.*

When Christianity, in the reign of Constantine, became the established religion of the Roman empire, Jerusalem, in name at least, was restored by the zealous Helena. The idol temples were destroyed, and several churches and other buildings were erected on sites supposed to be connected with the events of Christ's history; in short, no efforts and expense were spared in the attempt to raise the Holy City to its rank as the metropolis of Christendom. The period of prosperity thus commenced terminated in 636, by the conquest of Omar, who made the city tributary, heavily taxed the pilgrims, and desecrated the site of the temple, by erecting on it a mosque in honour of Mohammed.

After being more than 400 years subject to the Arabian

[*] It should be acknowledged, however, that the errors of Josephus, like those of Herodotus, Diodorus, Arrian, and others, in mere numbers, may, perhaps, be attributed less to the author's inaccuracy than to the old-fashioned writing in MSS., in which the numeration is affected by single letters, and mistakes, though easily occurring, are detected with extreme difficulty. In general points of history and topography, Josephus's works should be considered the vade-mecum of the traveller in Palestine.

JESI.   JESSORE.

caliphs, Jerusalem fell into the hands of the Turks, who proved still more oppressive masters than any of their predecessors. The resentment and sympathy of the princes and people of Christendom were awakened by Peter the hermit, and the crusades were undertaken to rescue the natives and pilgrims of Palestine, and, above all, the holy sepulchre, from the dominion of infidels. The Christian army reached Jerusalem in the summer of 1099. "Godfrey of Bouillon erected his standard on Mount Calvary: the time of the siege was fulfilled in forty days of calamity and anguish, during which the soldiers suffered intensely from hunger and thirst. At length, on a Friday, the day and hour of the Passion, Godfrey stood victorious on the walls of Jerusalem; his example was followed on every side by the emulation of valour; and about 450 years after the conquest of Omar, the Holy City was rescued from the Mohammedan yoke. A bloody sacrifice was offered to the God of the Christians: resistance might provoke, but neither age nor sex could mollify their implacable rage; they indulged themselves three days in a promiscuous massacre, and the infection of the dead bodies produced an epidemic disease." (Gibbon, xi., 84.) Saladin, 88 years afterward, appeared in arms before Jerusalem; some feeble and hasty efforts were made for its defence, but within 14 days the banners of the prophet were erected on its walls. Saphadin, the brother of Saladin, destroyed, in 1218, all that remained of the fortifications of this devoted city, and reduced the population to a servile subjection to the Mohammedans. A series of changes subsequently occurred: but Jerusalem came finally into the hands of Selim in 1519, since which the Turkish flag has always floated over its sacred places. For more than three centuries its fortunes have been stationary: crowds of pilgrims fill its streets at one season of the year, creating a temporary activity, and increasing the revenues of the Turkish officers; but at all other times its condition recalls forcibly the complaint of Jeremiah: "The city sits solitary that was full of people: she is become as a widow: she that was great among the provinces is become tributary. Her gates are desolate. . . . All her beauty is departed. . . . Filthiness is in her skirts."

Among the principal authorities for this article are Richardson's Travels along the Med., ii., 231, &c.; Hennikor's Travels, p. 173-198; Clarke, iv., 288-394; Elliot's Travels in Turkey, ii., 416-499; Robinson's Trav. in Pal. and Syr., ch. 5-9; Châteaubriand, Voyage, &c., ii., 116-180; Wilde's Narrative, ii., 180-250; Taciti Hist., libs. v. and vi.; Josephus, passim, but particularly Jud. Bell., l. vi. and vii.; and the Bible.

JERUSALEM, p. t., Yates co., N. Y., 5 m. W. Penn Yan, 190 m. W. Albany, 325 W. Watered by the W. branch of Crooked lake, and by a creek which enters its N. part. Organized in 1789. Jemima Wilkinson, the founder of the Shakers, resided and died here in 1819. It contains two churches, a Presbyterian and Baptist, three stores, one gristmill, 12 saw-mills, two tanneries; 16 schools, 836 scholars. Pop. 2635.

JERUSALEM, p. v., capital of Southampton co., Va., 70 m. E.S.E. Richmond, 189 W. It is situated on Nottaway river, and contains a courthouse, jail, and several stores and dwellings. Nett proceeds of the postoffice $135.

JERUSALEM, p. v., Hempstead t., Queens co., N. Y., 176 m. S. by E. Albany, 256 W. It contains a Friends' church, three stores, and about 25 dwellings, not very compact. Inhabited chiefly by Friends. The postoffice is called Jerusalem South.

JESI (an. Æsium), a town of Central Italy, papal states, deleg. Ancona, on the Flumedino (an. Æsis), 16 m. W.S.W. Ancona. Pop. about 6000. It is walled, and has a handsome main street, three large squares, a cathedral, and six other churches, many convents, a theatre, &c. It is a bishop's see. It has manufactures of silk and worsted stockings. Æsium anciently bore the rank of a Roman colony. Numerous antiquities exist on the banks of the river in its neighbourhood. (Rampoldi, Corografia, &c.)

JESSAMINE, county, Ky. Situated centrally towards the E. part of the state, and contains 256 sq. m. Bounded S. by Kentucky river. Watered by Hickman's creek. It contained in 1840, 5636 neat cattle, 15,685 sheep, 23,650 swine; and produced 61,896 bushels of wheat, 59,121 of rye, 477,942 of Indian corn, 76,160 of oats, 19,041 of potatoes, 73,283 pounds of tobacco. It had 23 stores, one woollen factory, one cotton factory, with 392 spindles, 11 flouring-mills, 30 grist-mills, 26 saw mills, one paper mill, five tanneries, 12 distilleries; three academies, 66 students, 21 schools, 521 scholars. Pop.: whites, 5799; slaves, 3473; free coloured, 144; total, 9396. Capital Nicholasville.

JESSELMERE, or JAYSULMEER, a state of N.W. Hindostan, prov. Rajpootana, subsidiary to the British, and one of the five principal Rajpoot principalities, between the 25th and 28th parallels of N. lat. and the 69th and 73d of E. long. Area estimated at 10,000 sq. m. Pop. perhaps, near 300,000. (Burnes in Geog. Journ., iv., 3.) Surface

uneven, and intersected with rocky hills: it is not watered by any considerable stream, has little arable land, and is hardly, in truth, more productive than the sandy desert that encompasses it. Cultivation is, consequently, very limited; and the parts which are cultivated yield only the coarser grains which form the food of the inhabitants. Irrigation is effected with great labour, chiefly by means of very deep wells and tanks; but large and spacious tanks occur every 2 or 3 m., and rain water is carefully preserved, the periodical rains being scanty and uncertain. The heat of summer is oppressive, but the cold of winter is sufficiently great for the tanks to be covered with ice every morning during a part of January. Mineral products few; the chief are primary limestone and lithographic stone: no metals appear to be found. Wood is scarce. The better kind of houses are of stone; the others mere conical grass huts. The open nature of the country frees it from the most formidable wild animals: foxes, wolves, hyenas, and jackalls are indeed met with, as are several kinds of antelopes, game of various kinds, wild ducks, &c.; but the uncertainty of water hinders both the animal and vegetable kingdom from thriving. Jesselmere is better suited for grazing than agriculture; but neither herds nor flocks are numerous. The horned cattle are of medium size, and indifferent quality: the sheep, though small, have excellent wool. The mass of the population consists of Bhattee Rajpoots. The commerce of Jesselmere is perfectly insignificant; what little wealth it does possess arises from its being on the chief road between Central India and the Indus; and the duties imposed on the transit of merchandise passing by it constitutes the chief resources of its ruler. It has no exports of its own; and is only manufacture is that of woollen cloth of a very fine texture, but in no demand elsewhere. Indigo and cotton cloths are imported from Malwah, sugar from Jeypoor and Delhi, iron and brass from Nagore. From twenty to twenty-five thousand maunds of opium pass annually through Jesselmere to Sinde; the return articles of transit thence being sulphur, assafetida, rice, and tobacco. The revenues of the rajah do not amount to two lacs of rupees yearly, more than half of which is derived from transit duties. The remainder is made up of fines, levies, salt taxes, and the land revenue, which latter is about 1-10th or 1-11th part of the nett produce. About 2000 rupees are derived yearly from the salt monopoly, some portion of which article is obtained in the principality; but most of it, as well as of grain, is imported from the neighbouring states. Jesselmere contains two towns and 84 villages; but, except in its capital, it everywhere betrays the strongest marks of poverty.

JESSELMERE, a town of N.W. Hindostan, prov. Rajpootana, cap. of the above rajahship, 190 m. W.N.W. Joudpore; lat. 26° 56' N., long. 70° 54' E. Population probably 20,000. (Burnes.) It is of an oval shape, about 2 m. in circuit, and surrounded by a rampart of loose stones. At its S.W. angle is a fort built on a scarped rock about 80 or 100 ft. higher than the city; and it presents a commanding appearance externally, and is in reality a place of considerable strength. It is of a triangular shape, its two longer sides, about 300 yards in length each, facing the W. and N. The only entrance is on the N. side, leading through several narrow and strong gates. The whole of the works are of firm substantial masonry, and comprise a vast number of towers (the natives say 175). These stud the brow of the hill on all sides, and give it a very remarkable appearance; some are as much as 40 ft. in height. This fortress is the residence of the rajah, and is supplied with water from wells 80 fathoms deep. The town is regularly laid out, and, for an eastern city, its streets are wide. Its houses are lofty, spacious, terrace-roofed, and built entirely of a hard yellow limestone, sometimes elegantly carved. Some opulent merchants reside at Jesselmere, it being on the great commercial route from Malwah to the port of Kurachee (Burnes in Geog. Journ., iv., 105-115.)

JESSORE, a distr. of British India, presid. and prov. Bengal, chiefly between the 22d and 24th degs. of N. lat., and the 89th and 90th of E. long.; having N. the main stream of the Ganges, separating it from the distr. Rajeshaye; E. Dacca and Backergunge; W. Nuddea and the 24 Pergunnahs, and S. the bay of Bengal. Length, N. to S., about 160 m.; average breadth, 32 m. Area, 5180 sq. m. Pop. (1822) 1,183,590. It is a flat country, intersected by numerous interlacing branches of the Ganges; its S. part comprises a portion of the region called the Sunderbunds; and on the shore are many extensive marshes, in which salt is largely made on government account. The soil is very fertile, and a good deal of rice is grown. Indigo, tobacco, mulberry, betel nut, and long pepper, are also raised; but a great proportion of the land is uncultivated, and covered with jungle. Chief towns, Jessore or Moorley, the residence of the Zillah authorities, Culna, and Manudpoor. Land revenue (1829-30), £190,935. (Hamilton's E. I. Gaz.; Parl. Reports.)

JEWETT'S CITY, p. v., Griswold t., New-London co., Ct., 8 m. N.E., Norwich, 47 m. E. by S. Hartford, 365 W. Situated on the E. side of Quinnebaug river, at the mouth of Patchaug river. It contains a Congregational church, a bank, five stores, three cotton factories, and about 1600 inhabitants. The water-power is very extensive. The Norwich and Worcester railroad passes through the place.

JEYPOOR, or JYEPORE, a city of N.W. Hindostan, prov. Rajpootana, cap. of a subsidiary state of the same name, in a barren valley, 130 m. S.W. Delhi; lat. 26° 53' N., long. 75° 37' E. Pop. estimated at 60,000. This is altogether the handsomest and most regularly built city of Hindostan. It is surrounded by a battlemented wall of gray stone, flanked with towers, and defended or commanded by a citadel and a line of forts on the adjacent heights, a few hundred feet in elevation. Jeypore is laid out, like most modern European and American cities, in regularly square blocks of houses. A main street, 2 m. long, and about 40 yards broad, traverses it from W. to E., and is crossed at right angles by four others of equal width, though much shorter. At the points of intersection are spacious market-places; and there are two good squares, which, like the principal streets, are crowded with shops. The great thoroughfares are, however, disfigured by hovels, platforms, and stalls, erected along the centre of them, which detract greatly from their appearance. The houses are generally two stories high, but some are three or four stories, with ornamented windows and balconies, and are often adorned with frescoes and sculptures. The chief public edifice of Jeypoor is a magnificent palace, constructed, it is said, by an Italian architect, in the 15th century, for the rajah Jey Singh, under whose reign this city was one of the principal seats of Hindoo learning. This palace, with its fine gardens, occupies about 1-8th part of the city; a sketch and account of it may be found in the Mod. Trav., i., 68. Jeypoor has numerous temples, in the purest Hindoo style, and some are of larger dimensions than are to be found in any other city of Upper Hindostan. (Heber; Boileau, in Asiat. Journ., 1836, ii., 206; Hamilton's E. I. Gaz., &c.)

JHYLUM, JELUM, or BEHUT (an. Hydaspes), a river of the Punjab, Hindostan, which rises in the S.E. extremity of Cashmere, and, after a course of about 450 m., at first N.W. or W., and afterwards S.W., joins the Chenab (Acesines), about lat. 31° 10' N. During most part of its source it is not fordable; and at Jelalpoor, in lat. 32° 40', it has been found, even when not at its highest point, 1800 yards broad, and 14 feet deep. It is correctly described by Arrian as "muddy and rapid," having a current of three or four miles an hour. Its banks are interesting as the scenes of several of the exploits of Alexander, but it is impossible to indicate their localities. Not far from the Jhylum is the famous tope of Manykiala. (See Punjab, Indus, &c.)

JO-DAVIESS, county, Ill. Situated in the N.W. part of the state, and contains 724 sq. m. Watered by Fever river, and Apple and Plum creeks. It abounds with lead and copper ore. It was named in honour of Gen. Joseph H. Daviess of Kentucky, who was killed in the battle of Tippecanoe, in 1811. It contained in 1840, 4107 neat cattle, 519 sheep, 8405 swine; and produced 18,560 bushels of wheat, 163,934 of Indian corn, 32,955 of oats, 59,940 of potatoes. It had 17 stores, 20 smelting-houses, producing 8,755,000 pounds of lead, and one smelting-house producing gold to the amount of $200, nine flouring-mills, 10 saw-mills; one academy, 70 students, nine schools, 297 scholars. Population 6180. Capital, Galena.

JIDDA. See Djidda.

JOANNINA. See Yannina.

JOHANNISBERG, or MOUNT ST. JOHN, a hill famous for its vineyards, with a castle, in the duchy of Nassau, near the E. bank of the Rhine, on the N. confines of the district called the Rhingau, 16 m. W. by N. Mentz. This hill formerly belonged to an abbey, the monks of which planted the vineyard towards the end of the 11th century. It comprises, excluding the portion which produces only ordinary wine, about 53 arpents and its produce in average years is estimated at about 25 tons of 1260 bottles each. The soil is composed of the débris of various coloured stratified marl. The grapes are gathered as late as possible, or when they are dead ripe. Its choicest produce, called Schloss-Johannisberger, is admitted to be the very finest of all the Rhenish wines, being distinguished by its high flavour and perfume, by an almost total want of acidity, and by its being improved the longer it is kept. The finest growths in the best years fetch enormous prices, sometimes as much as 12 florins the bottle! The vintages of 1779, 1783, 1801, 1811, and 1822, enjoy a high reputation. Schreiber says that the vineyard is worth from 75,000 to 80,000 florins a year nett revenue. After the secularization of the abbey of Fulda, this vineyard became successively the property of the late king of the Netherlands, Marshal Kellerman, and Prince Metternich, its present owner, to whom it was presented by the late emperor of Austria.

The prince has repaired the castle, which he occasionally occupies. (Schreiber, Guide du Rhin, p. 176; Henderson on Wines, p. 223, &c.)

JOHNSBURG, p. t., Warren co., N.Y., 88 m. N. Albany, 457 W. Bounded E. by Hudson river, and watered by its tributaries. It contains a Baptist church, three stores, one fulling-mill, two grist-mills, nine saw-mills, two tanneries; 10 schools, 302 scholars. Pop. 1139.

JOHNSON, county, N.C. Situated a little E. of the centre of the state, and contains 660 sq. m. Bounded N.E. by Contenay creek. Watered by Neuse river, and its branches, and by Little river. It contained in 1840, 11,152 neat cattle, 9011 sheep, 32,503 swine; and produced 9142 bushels of wheat, 2806 of rye, 337,797 of Indian corn, 23,459 of oats, 63,561 of potatoes, 461,169 pounds of cotton. It had four stores, 20 grist-mills, 15 saw-mills; two academies, 49 students, five schools, 77 scholars. Pop.: whites, 6896; slaves, 3476; free coloured, 127; total, 10,599. Capital, Smithfield.

JOHNSON, county, Tenn. Situated in the N.E. part of the state, and contains 200 sq. m. Watered by Watauga river, and its branches, which flow into the S. branch of Holston river. It contained in 1840, 2878 neat cattle, 3789 sheep, 4510 swine; and produced 5690 bushels of wheat, 4530 of rye, 93,463 of Indian corn, 9859 of buckwheat, 53,710 of oats, 28,118 of potatoes. It had three stores, one furnace, 19 forges, 21 grist-mills, 13 saw-mills, one tannery, 21 distilleries. Pop.: whites, 9493; slaves, 161; free coloured, 4; total, 9658. Capital, Taylorsville.

JOHNSON, county, Ia. Situated a little S. of the centre of the state, and contains 390 sq. m. Watered by White river and Sugar creek, and their branches. It contained in 1840, 9800 neat cattle, 11,797 sheep, 30,797 swine; and produced 46,118 bushels of wheat, 497,095 of Indian corn, 81,747 of oats, 14,121 of potatoes, 49,848 pounds of tobacco, 29,294 of sugar. It had 25 stores, three flouring-mills, 10 grist-mills, 12 saw-mills, nine tanneries, seven distilleries; one academy, 40 students, six schools, 160 scholars. Pop. 9352. Capital, Franklin.

JOHNSON, county, Ill. Situated in the S. part of the state, and contains 486 sq. m. Bounded S. by the Ohio river. Watered by Cash river and Big Bay creek. Towards its S. part are ponds and marshes. It contained in 1840, 2250 neat cattle, 1807 sheep, 12,786 swine; and produced 7514 bushels of wheat, 150,541 of Indian corn, 13,583 of oats, 3430 of potatoes, 18,319 pounds of tobacco, 34,787 of cotton, 6846 of sugar. It had 12 stores, 11 grist-mills, two saw-mills, one tannery, three distilleries; 12 schools, 380 scholars. Pop. 3626. Capital, Vienna.

JOHNSON, county, Iowa. Situated in the W. part of the territory, and contains 610 sq. m. Watered by Iowa river, and its branches. It contained in 1840, 987 neat cattle, 44 sheep, 1075 swine; and produced 10,700 bushels of Indian corn, 3400 of potatoes. It had four stores, two saw-mills. Pop. 1491. Capital, Iowa city.

JOHNSON, county, Mo. Situated in the W. part of the state, and contains 785 sq. m. Drained by Blackwater river, and its tributaries, and by tributaries of Osage river. It contained in 1840, 7617 neat cattle, 5578 sheep, 33,553 swine; and produced 4135 bushels of wheat, 230,375 of Indian corn, 69,958 of oats, 10,325 of potatoes, 24,557 pounds of tobacco, 1126 of cotton. It had 10 stores, 16 grist mills, six saw-mills, one tannery, six distilleries; 11 schools, 266 scholars. Pop.: whites, 3611; slaves, 556; free coloured, 4; total, 4471. Capital, Warrensburg.

JOHNSON, county, Ark. Situated towards the W. part of the state, and contains 900 sq. m. Watered by Arkansas river, and its tributaries. It contained in 1840, 5344 neat cattle, 823 sheep, 13,702 swine; and produced 7082 bushels of wheat, 197,935 of Indian corn, 6635 of oats, 15,466 of potatoes, 14,755 pounds of tobacco, 296,468 of cotton. It had 15 stores, nine grist-mills, five saw-mills, one powder-mill, three tanneries; six schools, 138 scholars. Pop.: whites, 3830; slaves, 591; free coloured, 3; total, 3433. Capital, Clarksville.

JOHNSON, p. t., Lamoille co., Vt., 36 m. N.N.W. Montpelier, 550 W. Chartered in 1792, first settled in 1784. Watered by Lamoille river and some of its branches. Mc-Connel's falls in Lamoille river, has a perpendicular descent of 15 feet, and a little below a natural bridge of solid rock, 80 feet wide. Soapstone and other potter's clay are found. It contains two churches, four stores, one woollen factory, one grist-mill, six saw-mills, two tanneries, one printing-office, one weekly newspaper; one academy, 160 students, 12 schools, 492 scholars. Pop. 1410.

JOHNSON, t., Champaign co., O. It has two stores, two grist-mills, three saw-mills; six schools, 190 scholars. Pop. 1913.

JOHNSON, t. Licking co., O. It has two schools, 65 scholars. Pop. 216.

JOHNSON, t., Barry co., Mich. It has two schools, 40 scholars. Pop. 227.

JOHNSON, t., Lagrange co., Ia. It has one store, one saw-mill, two schools, 29 scholars. Pop. 275.

JOHSON t., Crawford co., Mo. Pop. 743.

JOHN'S (ST.), a city and sea-port of New-Brunswick, on its S. coast, and the largest and most important town, though not the cap. of that colony. It is built on rocky and very irregular ground on a small peninsula, on the N. side of the St. John river, near its mouth, in the bay of Fundy, 130 m. W.S.W. Halifax, and 190 m. E.N.E. Augusta, in the State of Maine. Lat. 45° 20' N.; long. 66° 3' W. Pop. (1834) 12,885. Its harbour is commodious and spacious; and though a bar across its entrance dries at low water, the rise of the tides is such (from 25 to 30 feet) that large vessels enter the port at high water. The entrance to the harbour is between a bold headland bounding the river to the E., and Partridge Island, about two m. S. of the town, which has a lighthouse and a fort. Another fortress guards the harbour, at Carleton, opposite St. John's, and on a commanding height immediately above the town itself is Fort Howe, now in ruins. At ebb tide, a mud flat extends for some distance in front of St. John's; but at high water it is covered, and the aspect of the place is most imposing. A projecting rock separates the town into the upper and lower coves. The former, containing the wharfs and warehouses, is the principal division; but the lower has been much improved by the erection of a line of barracks. Several of the streets are inconveniently steep, and in winter even dangerous; though much labour has been employed to level and adapt them for carriages. The houses, principally of brick, are regularly arranged, and on the whole handsome; but ornament has not been much studied. The chief public buildings are, a handsome stone courthouse, recently erected on high ground above the middle of the town, the marine hospital, poor-house, jail, two Episcopal churches, a Scotch church, and Roman Catholic, Methodist, and Baptist chapels. The grammar school, which has an endowment of £135 a year, had, in 1836, 495 scholars; and there are other schools, and several religious and charitable associations. The provincial bank has a capital of £50,000, and within these few years' another bank, with a capital of £150,000, has been established. St. John's has a chamber of commerce; a savings' bank, and a marine insurance company; two public libraries, and a good news-room; and four or five well conducted weekly papers.

On the opposite bank of the river is the little town of Carleton, under the municipal government of St. John's, comprising a good many new buildings, a church, with some saw-mills, and building docks. St. John's is a corporate city, which, including Carleton, is divided into six wards, and governed by a mayor, recorder, six aldermen, and six assistants. The mayor, recorder, and other chief officers, are appointed by the governor; the aldermen being elected annually by the freemen.

St. John's is a free port, and the great commercial emporium of New-Brunswick. In 1836, 2549 ships, of the aggregate burden of 289,610 tons, entered; and 2389 ships, burden 298,127 tons, cleared out from the port and its outports. In the same year 81 vessels of the burden of 24,679 tons were built at St. John's. Several ships, averaging 400 tons, belonging to this port, are employed in the Pacific and eastern oceans in the seal and whale fishery. The herring fishery in the harbour affords from 10,000 to 15,000 barrels a year, besides salmon and shad.

"Fifty years ago, the site of this thriving city, with the exception of a few straggling huts, was covered with trees. This was its condition at the peace of 1783; and when we now (1833) view it with its population, its stately houses, its public buildings, its warehouses, its wharfs, and the majestic ships which crowd its port, we are more than lost in forming a conjecture of what it may become in less than a century. Its position will ever command the trade of the vast and fertile country, watered by the lakes and streams of the river St. John, and it will flourish, as all towns have flourished, through which the bulk of the exports and imports necessarily pass into the countries in which they are situated." (M'Gregor's British America, ii., 13–20; Bouchette; Wedderburns; Parl. Papers, &c.).

JOHN (ST.) or the ST. JOHN'S RIVER, called by the Indians Looshtook, "the long river;" the principal river of New-Brunswick, and next to the St. Lawrence, the finest in British America. The area of its basin is estimated by Darby at 19,200 sq. m. The St. John rises from two principal sources, about lat. 46° 10' N., and long. 70° W. in the territory N. of the State of Maine, disputed between Great Britain and the U. States. It flows through this territory, at first N.E., for about 100 m.; and then takes a bold curved sweep to the E., as far as long. 67° 50', where it leaves the disputed country, and enters the province of New-Brunswick. It then flows, first in a S. direction for about one fourth part of its course; then E. for perhaps 80 m.; and lastly S. for at least 50 more; when it discharges

itself into the bay of Fundy, a little below the city of St. John, about lat. 45° 20', and long. 66°, after an entire course, estimated by Darby at 380 miles.

Independent of any artificial improvement, the St. John is, in the greatest part of its course, one of the most navigable of the Atlantic rivers, being much less impeded by rapids, shoals, or falls, than any other stream between it and the Hudson. (Darby, p. 195.) At its mouth, which forms St. John's harbour, it is 5 m. wide; and at Fredericton, 85 m. up, it is half a mile wide. Vessels of 50 tons come up to Fredericton; and barks of 90 tons ascend to the Great Falls, about 200 m. from its mouth; above which it is fitted only for boats. It is unfortunate, however, that almost at the very entrance of this river, about a mile above St. John's, its bed contracts to about 400 feet in width, and is crossed by a formidable rocky bar; on which there is seldom more than 17 feet of water which only admits of the passage of vessels at certain times of the tide. The waters of the river at low ebb are in this place about 12 feet higher than the sea, and at high water about 5 feet lower; so that in every tide there are two falls—one outwards, and one inwards. The only time of passing with safety is when the waters on either side of the bar are about level, which happens twice in a tide, continuing nearly 20 minutes at a time. The tide is not perceptible much above Fredericton; where it rises to from 6 to 16 inches. The Great Falls, near lat. 47°, consist of one principal cataract, perhaps 50 feet high, and some smaller ones of several feet each, extending altogether for ⅓ mile along the stream, and having a total fall of about 75 feet. Though very inferior in respect of magnitude compared with that of Niagara, the falls of the St. John are said to be more picturesque. Its entire descent, from its mouth to its source, is estimated by Darby at probably 1000 feet. (View, &c., p. 197.)

Besides St. John's, Carleton, Gage-town and Fredericton, there is no place of any consequence on the banks of the St. John; but the country through which it flows is well cleared and settled, and is said to be greatly superior in fertility to the river basins of New-England. (See MAINE.) (Darby's View of the U. States; M'Gregor's Brit. Amer., &c.).

JOHN'S (ST.), a town of the island of Newfoundland, of which it is the capital, on its S.E. coast. Lat. 47° 33' N., long. 52° 29' W. Population of the town and its electoral district, comprising 13,412 acres (1836), 18,985, being about one fourth part of the total population of the colony. It stands at the inner end of an excellent harbour, the narrow entrance to which has 12 fathoms water in the centre of the channel. It is protected by several strong batteries and forts, and a lighthouse is constructed on a rock at the N. side of its entrance. The town extends along nearly the whole N. side of the port. It consists mostly of one main street, about one mile in length, and from 40 to 50 feet broad, from which, at almost every step, stages, called fish-flakes, project into the sea. There are some good stone and brick houses, and other handsome buildings, erected principally since the great fires that devastated St. John's in 1816 and 1817; but most part of the town is built of wood, and, with all its improvements, it still bears the aspect of a mere fishing station.

"In the time of war, St. John's is a place of great importance. There are a great number of shops, and a still greater number of public houses, in proportion to its size, in this than in most towns. Commodities were formerly dear; at present, shop goods are as low as in any town in America; and fresh meat, poultry, and vegetables, though not so cheap as on the continent, are not unreasonably dear.

"The population of St. John's fluctuates so frequently, that it is very difficult to state its numbers, even at any one period. Sometimes, during the fishing season, the town appears full of inhabitants; at others it seems half deserted. At one time, they depart for the seal fishery: at another to different cod-fishing stations. In the fall of the year, the fishermen arrive from all quarters to settle their accounts with the merchants, and procure supplies for the winter. At this period St. John's is crowded with people; swarms of whom depart for Prince Edward island, Nova Scotia, and cape Breton to procure a livelihood in those places, among the farmers, during winter. Many of them never return again to the fisheries, but remain in those colonies; or often, if they have relations in the U. States, and sometimes when they have not, find their way thither.

"Society in St. John's, particularly when we consider its great want of permanency, is in a much more respectable condition than might be expected; and the morals and social habits of the inhabitants are very different from the description of Lieutenant Chappell, who represents the principal inhabitants as having risen from the lowest fishermen, and the rest composed of turbulent Irishmen, both alike destitute of literature. The fishermen, who are principally Irishmen, are by no means altogether destitute of education: there are few of them but can read and write; and they are, in general, neither turbulent nor immoral. That they soon become in Newfoundland, as well as in all the other colonies, very different people to what they were before they left Ireland, is very certain. The cause is obvious; they are more comfortable, and they work cheerfully. When, after a fishing season of almost incredible fatigue and hardship, they return to St. John's, and meet their friends and acquaintances, they indulge, it is true, in idleness for a short time, and occasionally in drinking; but when the hazardous life they follow is considered, we need not be surprised that they do so, especially in a place where rum is as cheap as beer in England." (*M'Gregor's British America.*)

Fort Townshend, on a steep height above the town, was formerly the residence of the governor; but a new edifice has been recently built for this purpose, on so extravagant a scale, that it is said to have cost £50,000. The customhouse, church, and other public buildings present nothing remarkable. A Lancastrian school, with a government endowment of £100 a year, was attended, in 1836, by 216 boys; a girls' school with 600 pupils, chiefly supported by voluntary subscriptions to the amount of about £360 a year; and various other schools and benevolent associations. The inhabitants are generally possessed of the rudiments of education; and many of them pretty well informed. Several weekly newspapers are published in the town. Most of the population are Roman Catholics, and this is the see of a Roman Catholic bishop. It is a good deal agitated by party contentions.

In 1826, 116 vessels, of the aggregate burden, of 9208 tons, were fitted out, at St. John's, for the seal fishery. Agriculture is scarcely pursued at all in the neighbourhood; the ground being rugged and stony. Potatoes form the chief crop. Provisions and other commodities, are dearer than on the American continent, from which, indeed, they are mostly imported. (For the import and export trade of St. John's, &c., see NEWFOUNDLAND.) (*M'Gregor's Brit. America.* i., 165–173; *Parl. Papers.*)

JOHNSTON, a manufacturing town of Scotland, par. of Paisley, co. Renfrew, on the Black Cart, 3 m. W. Paisley, and 10 m. W. by S. Glasgow. The rise of this town is remarkable, being more rapid than that of any other place in Scotland: the ground on which it now stands began, for the first time, to be feued or let on building leases in 1781, when it contained only *ten* persons. In Oct. 1782, nine houses were built, and two more were being erected. In 1792, the inhabitants amounted to 1434; in 1811 to 3647; in 1831 to 5817; and now (1840) they are estimated at upwards of 7000. The place was formerly called "the Brig o'Johnsons," from a bridge over the river in the immediate vicinity. It is built on a regular plan, and lighted with gas. There are two squares, besides numerous streets, and public works. The houses are, for the most part, of good mason-work, two and three stories in height. To each house is attached an adequate extent of garden ground. It has an established church, and various dissenting places of worship. In the immediate neighbourhood is Johnston castle, the residence of Mr. Houston, lord of the manor. There are excellent grammar and English schools; two reading-rooms, three public libraries, several printing presses; various booksellers, lawyers, medical practitioners, &c. The civil polity of the town is managed by a committee elected annually by the feuars. A monthly justice of peace court is held in the assembly rooms.

Johnston is chiefly distinguished for its manufactures. It had, in 1839, 15 cotton-mills, employing in all 1456 persons, of whom 684 were between 9 and 18 years old. This is exclusive of mills at Elderslie, Linwood, and other places in the immediate vicinity. With two slight exceptions, the mills are all propelled by water. There are, besides, in Johnston, two brass and two iron foundries, on an extensive scale; with five machine manufactories, employing 120 individuals, as well as various minor branches of industry. The Glasgow and Ardrossan Canal, projected in 1806, has been completed only from Glasgow to Johnston. It was on this canal that light iron boats, or gig-boats, for the rapid conveyance of passengers, were first (1831) tried and established. The Glasgow, Paisley, Kilmarnock, and Ayr railway passes Johnston; so that this village enjoys every advantage in the way of internal intercourse. Near Johnston are four collieries. (*Factory Reports; Fowler's Com. Direct. for Renfrewshire; New Stat. Acc. of Scotland,* § *Renfrewshire,* p. 201–203.)

JOHNSTON, t., Trumbull co., O., 10 m. N.E. Warren. The postoffice is called Johnsonville, 154 m. N.W. Columbus, 208 W. The town has eight schools, 970 scholars. Pop. 869.

JOHNSTON, t., Providence co., R. I., 5 m. W. Providence. Watered by Wanmequatucket, and Powchasset rivers, and

Cedar brook. Incorporated in 1759. It contains three churches, two Baptist and one Friends, six stores, 16 cotton factories with 86,800 spindles, five dyeing works, five gristmills, four saw-mills; 12 schools, 625 scholars. Pop. 3477.

JOHNSTOWN, p. t., capital of Fulton co., N. Y., 46 m. N.W. Albany, 410 W. Drained by Garoga and Cayudutta creeks. It contains 96 stores, two lumber-yards, two fulling-mills, six grist-mills, 25 saw-mills, one paper-mill, two printing-offices, two weekly newspapers; two academies, 235 students, 94 schools, 109 scholars. Population 5408. The village situated 4 m. N. of Mohawk river, was incorporated in 1808, and contains a courthouse, jail, county clerks office, six churches, two Presbyterian, a Lutheran, Methodist, Dutch Reformed and Episcopal, a bank, various mills and manufactories, 150 dwellings, many of them neat, and about 1000 inhabitants. The courthouse, jail and Episcopal church were erected by Sir William Johnson, about the year 1771.

JOHNSTOWN, p. b., Conemaugh t., Cambria co., Pa., 150 m. W. Harrisburg, 176 W. Situated at the junction of Little Stoney creek, with Conemaugh river. It is regularly laid out on 200 acres of ground, with streets crossing each other at right angles. It contains six stores, two grist-mills, three saw-mills, six schools, 190 scholars. Population 1213. The railroad from Hollidaysburg over the Alleghany mountains, connecting the parts of the Pennsylvania canal, here terminates. The western division of the canal has a large basin, in the centre of the village.

JOIGNY (an. *Joviniacum*), a town of France, dep. Yonne, cap. arrond., on the Yonne, 15 m. N.W. by N. Auxerre. Pop. (1836), 4760. A handsome quay runs along the bank of the river, above which the town rises on a steep declivity, crowned with the remains of an ancient castle. Joigny is surrounded with old walls, and entered by six gates; it has two suburbs, with one of which it is connected by a handsome stone bridge of six arches across the Yonne. The streets are narrow, steep, and inconvenient; but some of the houses are good. It has a cathedral built in the 15th century, two other Gothic churches cavalry barracks, &c., with vinegar and other factories. (*Hugo,* art. *Yonne.*)

JONESBURG, p. v., capital of Camden co., N. C. (See CAMDEN C. H.)

JONES, county, N. C. Situated in the S.E. part of the state, and contains 380 sq. m. Watered by Trent river and its branches, flowing into Neuse river. It contained in 1840, 4730 neat cattle, 4069 sheep, 15,951 swine; and produced 5204 bushels of wheat, 1184 of rye, 323,580 of Indian corn, 8705 of oats, 38,529 of potatoes, 30,490 pounds of rice, 1,150,308 of cotton. It had two stores, 15 grist-mills, eight saw-mills, two distilleries; four schools, 81 scholars. Pop.: whites, 1947; slaves, 2818; free coloured, 180; total, 4945. Capital, Trenton.

JONES, county, Ga. Situated in the central part of the state, and contains 380 sq. m. Bounded W. by Ocmulgee river. Drained by its branches, and by branches of Oconee river. It contained in 1840, 8948 neat cattle, 4219 sheep, 19,853 swine; and produced 23,778 bushels of wheat, 274,938 of Indian corn, 18,755 of oats, 10,581 of potatoes, 2,754,565 pounds of cotton. It had seven stores, five furnaces, 29 forges, three flouring-mills, eight grist-mills, nine saw-mills, three oil-mills, two tanneries, 11 distilleries; four academies. 155 students, seven schools. 211 scholars. Pop.: whites, 4417; slaves, 5619; free coloured, 29; total, 10,065. Capital, Clinton.

JONES, county, Miss. Situated in the S.E. part of the state, and contains 679 sq. m. Drained by branches of Leaf river. It contained in 1840, 7329 neat cattle, 693 sheep, 9984 swine; and produced 24,880 bushels, of Indian corn, 3080 of potatoes, 9878 pounds of rice, 23,199 of cotton. Pop.: whites, 1194; slaves, 164; total, 1358. Capital, Ellisville.

JONES, county, Iowa. Situated a little N. of the centre of the territory, and contains 576 sq. m. Watered by Wapsipinecon and Macoquetais rivers, and their branches. It contained in 1840, 609 neat cattle, 65 sheep, 1502 swine; and produced 3537 bushels of wheat, 14,856 of Indian corn, 2286 of oats, 7333 of potatoes. It had two grist-mills, two saw-mills. Pop. 471. Capital, Edinburg.

JONES, t., Hancock co., Ia. It has one store, one grist-mill. Pop. 460.

JONESBOROUGH, p. t., Washington co., Me., 143 m. E.N.E. Augusta, 725 W. Bounded S. by Englishman's bay. Incorporated in 1809. Watered by Chandler's river. It contains three stores, one grist-mill, three saw-mills; four schools, 176 scholars. Pop. 392.

JONESBOROUGH, p. v., capital of Washington co., Tenn., 263 m. E. Nashville, 412 W. Situated on Little Limestone creek, near its source, and 10 m. S. of Holston river. It contains a courthouse, jail, two churches, a Methodist and Presbyterian, two academies, several stores, a printing-office, various mechanic shops, 150 dwellings, and about 900 inhabitants.

JONESBOROUGH, p. v., capital of Union co., Ill., 197 m. S. by E. Springfield, 831 W. It contains a courthouse, jail, two churches, seven stores, a carding machine, and about 30 dwellings and 200 inhabitants.

JONESPORT, p. t., Washington co., Me., 147 m. E. by N. Augusta, 730 W. Bounded E. by Englishman's bay, W. by Addison bay, and a small stream flowing into it. It has a good harbour, and contains four stores, one saw-mill; four schools, 250 scholars. Pop. 576.

JONESTOWN, p. v., Lebanon co. Pa., 29 m. E. Harrisburg, 139 W. Situated near the confluence of Great and Little Swatara creeks, and contains three churches, a Presbyterian, a Lutheran and a German Reformed, several stores, and about 100 dwellings.

JONESVILLE, p. v., capital of Lee co., Va., 384 m. W. by S. Richmond, 459 W. Situated on a branch of Powell's river. It contains a courthouse, jail, a church, free to all denominations, four stores, 50 dwellings, and 250 inhabitants. It is surrounded by mountains.

JONESVILLE, p. v., capital of Hillsdale co., Mich., 92 m. W.S.W. Detroit, 540 W. Situated on the E. bank of St. Joseph's river, and contains a courthouse, jail, a Presbyterian church, eight stores, two flouring-mills. It is handsomely located, and is flourishing.

JORDAN (Arab. Sheriat-el-Kebir), a river of Palestine famous in sacred history; it rises in lat. 39° 35′ N. long. 33° 26′ E., a few miles N. of Banias (the an. Caesarea Philippi), in a small pool formerly called Phiala, on the W. slope of Djebel-es-Sheikh, the Antilibanus or mount Hermon of antiquity. After a S. course of about 40 m., during which it crosses the fenny Bahr-el-Hool (an. L. Merom), it opens into the lake Tabariah or Gennesareth, close to the ancient town of Bethsaida. At the S. end of this fine sheet of water (15 m. long, and about seven broad), on and near which occurred so many striking scenes in the history of Christ, the Jordan enters a narrow pent-up valley called el-Ghor, and after running through it with a tortuous southerly course of about 90 m., empties its waters into the Dead sea, its entire length being about 150 m. The discoveries of Burckhardt in the Wady-el-Araby, which he traced completely up from the Red sea to the lake Asphaltites, have led to the supposition that before the volcanic movement which so altered the surface, this river had a continuous course down this valley to the gulf of Akabah (see DEAD SEA). Its tributaries on the W. side are mere torrents, one of the largest of which is the brook Kedron, rising in the suburbs of Jerusalem: the E. affluents comprise the Sheriat-el-Mandhur (an. Jarmak, Gr. 'Iepdpad), and the Wady Zerka, which is the scriptural Jabbok (see Deut. iii., 16). The breadth and rapidity of the stream vary in different parts and at different seasons. The floods occur in Feb. and March, and at that season, when filled with the melted snow of Mount Lebanon, it is from 30 to 70 yards wide, and about 17 feet deep, with a current so rapid that it is not safe even for an expert swimmer to bathe in it. In the dry seasons it is low, and has a comparatively languid current; and to this circumstance, probably, may be attributed the discrepancies in the statements respecting the nature and magnitude of the river. The channel, however, as appears from Maundrell and Burckhardt, having cut its way through a loose sandy soil, is much deeper now than formerly, and the waters, even in floods, run within narrower limits. A second and higher bank now skirts the actual bank at about a furlong's distance on either side, and the intervening space is so filled up with bushes and trees (tamarisks, willows, oleanders, myrtles, &c.) that the stream is completely hidden from view till its upper and dry channel has been passed. Lord Lindsay says: " The river is concealed till you are close upon it, by dense thickets of trees, reeds, and bushes, ' the pride of Jordan' (Zech. xi., 3), growing luxuriantly to the very water's edge. The lions, hippopotami, &c. (Jer. xlix., 19), that formerly haunted these thickets are extinct; but wild boars are still found there." " The nightingales," says the same writer, " sing in the cool, starlight night from the trees ; and the scene altogether was most delightful." (Travels in Egypt, and the Holy Land, vol. ii., p. 65.)

The water is described by Robinson as being rather warm than cold, of a white sulphureous colour, but free from any taste or smell. On analysis, however, it proves to be strikingly dissimilar to that of the Dead sea ; for while the latter contains one fourth part of its weight of salts, the former has only 1-300th part of the proportion of solid matter, contained in the water of the lake. (See Dr. Marcet, Phil. Trans. for 1807.)

The Jordan has been the scene of many events in which biblical scholars must be deeply interested. This river valley was the dwelling of Lot, who " pitched his tents towards Sodom," the men whereof " were wicked, and sinners before the Lord exceedingly." Here the four kings, persecuted by the five powerful princes close to the Salt (or Dead) sea, fought, and regained their liberty ; and the pow-

er of the latter was afterward destroyed by divine interference. (Comp. Gen. xiv., 1-12., with xix., 24-26.) At a later, but still very early historical period, when the chase of Israel were returning, after an absence of four centuries, to the possessions of Abraham, the great sheikh of a nation that was yet only in the nomad state, the ark, by command of Jehovah, was carried by the priests before the people-into the stream, and " the waters which came down from above, stood and rose up upon a heap; and those that came down toward the sea of the plain, even the Salt sea, failed, and were cut off; and the people passed over right against Jericho." (Josh. iii., 14-16.) It is said that the prophets Elijah and Elisha afterward divided its waters to prove their divine mission, and the special fact that " the spirit of Elijah doth rest on Elisha." (2 Kings, ii.) In Christian times, it has been celebrated as the stream in which Jesus Christ received from John the baptism which prepared him for the ministrations destined to exercise so important an influence over mankind. By modern devotees in Palestine, as elsewhere, indeed, throughout Christendom, the spirit of this institution has been forgotten, and a superstitious attention to the form substituted in its stead ; hence, every year pilgrims, at the great Easter season (about April), are found rushing, young and old, rich and poor, sick and sound, men, women, and children, into the stream. " All," says Mr. Elliott (an English clergyman), " carried with them the piece of cloth with which they wished to be enveloped after death." The Moslems ridicule these vain ablutions, and their violation of decorum ; and the Protestant cannot but lament the degradation they exhibit. (Burckhardt's Syria, p. 360-412.; Elliott's Travels, vol. ii. p. 476; Stephens's Inc. of Travel, vol. ii., p. 261; G. Robinson's Palestine, vol.·i., p. 69-75 ; Mod. Trav., &c.)

JORULLO, JURUYO, or XURULLO, an active volcano of Mexico, state of Valladolid, in an extensive plain 70 m. S.S.W. the city of that name, and 80 m. from the Pacific; remarkable not only for its extent, but as being the only volcano of any consequence that has originated in New Spain since its conquest by Europeans. Its origin was, perhaps, one of the most tremendous and extraordinary phenomena that has ever been witnessed ; for in one night there issued from the earth a volcano 1,600 ft. high, surrounded by more than 2000 apertures, which still continue to emit smoke. Humboldt, who visited Jorullo, describes its appearance and formation nearly as follows:—" A vast plain extends from the hills of Aguasarco, to near the villages of Teipa and Petatlan, from 2460 to 2624 ft. above the level of the sea. In the midst of a tract of ground, in which porphyry, with a base of greenstone, predominates, basaltic cones appear, the summits of which are crowned with evergreen oaks, small palm trees, &c., their beautiful vegetation forming a singular contrast with the aridity of the plain, laid waste by volcanic fire. Till the middle of the 18th century, fields cultivated with sugar-cane and indigo occupied the extent of ground between the rivers Cuitamba and San Pedro. From June 1759, hollow subterranean noises, accompanied by frequent earthquakes, succeeded one another for from 50 to 60 days. At length, in the night between the 28th and 29th of Sept., a tract of ground from three to four sq. m. in extent, which goes by the name of Malpays, rose up in the shape of a bladder. The bounds of this convulsion are still distinguishable in the fractured strata. The Malpays, near its edges, is only 39 ft. above the old level of the plain called the Playas de Jorullo ; but the convexity of the ground thus thrown up increases progressively toward the centre to an elevation of 524 ft. Flames were now seen to issue forth, it is said, for an extent of more than half a sq. league ; fragments of burning rocks were thrown up to prodigious heights; and, through a thick cloud of ashes, illumined by volcanic fire, the softened surface of the earth was seen to swell up like an agitated sea. The rivers of Cuitamba and San Pedro precipitated themselves into the burning chasms. Thousands of small cones, from six to nine ft. in height, called by the natives hornitos (ovens), issued forth from the Malpays, from each of which a thick vapour ascends to the height of from 30 to 50 ft. In many of them a subterraneous noise is heard, which appears to announce the proximity of a fluid in ebullition. In the midst of the ovens, six large masses, elevated from 1319 to 1640 ft. each above the old level of the plains, spring up from a chasm, the direction of which is from N.N.W. to S.S.E. The most elevated of these enormous masses, the great volcano of Jorullo, bears some resemblance in shape to the Puys of Auvergne, in France. It is continually burning, and has thrown up from the N. side an immense quantity of scorified and basaltic lavas, containing fragments of primitive rocks. These great eruptions of the central volcano continued till Feb. 1760. In the following years they became gradually less frequent; but the plains of Jorullo even at a great distance from the scene of the explosion were long uninhabitable, from the excessive heat which prevailed in them."

The Cuitamba and San Pedro totally disappeared on the occasion above mentioned; but two new streams are now seen bursting through the argillaceous vault of the *hornitos*, having the appearance of mineral waters, in which the thermometer rises to 126° Fahr. The Indians give these streams the names of the former rivers, because, in several parts of the Malpays, great masses of water, with which they are supposed to be continuous, are heard to run in a direction from E. to W., as the Cuitamba and San Pedro did originally. Jorullo is situated in the great volcanic band of Mexico, which runs E. and W., nearly at right angles, to the Cordillera, including the peaks of Orizaba, Puebla, Toluca, Tancitaro, Colima, &c.; and of which Humboldt conjectures the Revillagigedo islands, in the Pacific, may mark the continuation. (*Humboldt*; *Ward's Mexico*; *Encyp. Americana*; *Mod. Trav.*, xxvi., 191–197.)

JOUDPOOR, or MARWAR, a state of N.W. Hindostan, subsidiary to the British, and the most extensive and powerful in Rajpootana; between the 24th and 28th deg. of N. lat., and the 70th and 75th deg. of E. long.; having E. the states of Odeypoor and Jeypoor, and the British territory of Ajmere, from which it is separated by a mountain range; N. Bicanere; W. Jesselmere, and Sinde; and S. some smaller rajahships on the borders of Gujrat. Length, E. to W. about 360 m., by about the same breadth. Area estimated at 70,000 sq. m. Pop. uncertain.

Joudpoor and Jesselmere (which see), may be taken as pretty fair types of the several Rajpoot states of N.W. India; the former being, however, the most extensive and valuable of any, and the latter the least so. The wealth of Joudpoor has been much undervalued; and it has been erroneously considered as a portion of the sandy desert. Its exports in wheat are considerable; the soil is favourable to many other kinds of grain; and its central parts are highly productive. The country consists generally of open plains; the hills being almost confined to the S. The soil is not arid (as in Jesselmere, Bicanere, &c.); but is almost everywhere watered by torrents, and affluents of the Loonee or Salt river. This river rises in Ajmere, and flows through the centre of Joudpoor to enter the Rann of Cutch. Its waters are distributed over the adjacent wheat lands, which extend along its banks from Ajmere to the Runn, by means of earth aqueducts, sometimes a mile in length. The fields are surrounded with dykes to prevent the egress of the water; and being thus irrigated, Joudpoor produces heavy crops of barley, *bejree*, *jowaree*, and other kinds of grain. Neither the climate nor soil is favourable to the poppy; but an inferior kind of opium is grown in the E., where it is an article of large consumption and export. Tobacco is produced in some parts; but not in a sufficient quantity to supersede the necessity of importing it from Gujrat. Cotton is an important article of produce. Marwar is celebrated for its camels, which may be purchased in every village, at from 30 to 60 rupees each, and which have contributed greatly to the commercial importance of the state, by facilitating the conveyance of almost every kind of goods. Goats, sheep, and hogs are numerous; mutton is good, but the wool is not so much prized as that of the poorer countries. (*Jesselmere, &c.*) Salt is a very important article of produce. Large tracts are impregnated with it, especially about Punchpadder, on the Loonee towards Cutch. It is got by digging pits of about 120 ft. by 40 and about 10 ft. deep in the saline soil. A jungle shrub is then thrown in upon the water which exudes; this assists the crystallisation, and in the course of two years, the moisture having evaporated, a mass of salt, sometimes from four to five feet deep, is left. The commerce of Joudpoor is extensive; its great emporium being Pallee, about 40 m. S.E. of the capital. This town is the entrepôt between the W. coast and Upper India, and the channel by which the Malwah opium is exported to China and W. Asia. The chief trade of Pallee is in opium, and for six years, preceding 1834, the exports were never less than 1500 camel loads, and oftener 3000. A camel carries 10 maunds of 40 seers, and the Pallee maund exceeds that of Bombay, which would give an annual export of from 20 to 34 thousand maunds. The opium is sent by land to Kurachee in Sinde, a distance of 500 m., whence it is shipped to Damaun. The expenses of this journey are very great, as exorbitant transit duties are claimed by the rajahs of Joudpoor, Jesselmere, and other states, through which the opium passes. The Joudpoor government alone demands 50 rupees per camel load: It is customary with the Pallee merchants to consign their opium to contractors, who agree to deliver it safe at Damaun, uninjured by weather, plunder, or otherwise; on the receipt of 380 rupees for each camel load. Marwar exports wheat of superior quality to Ajmere, Bicanere, &c.; and has most extensive dealings in salt, with which it supplies the upper provs. of Bengal, and, indeed, all parts of Upper India. It imports from Sinde, by its return camels, rice, assafœtida, sulphur, &c.; from Lahore, Cashmere shawls; from Delhi and Jeypore, metals, woollen and cotton cloths, and sugar. From Cutch

it receives spices, cocoa-nuts, coffee, dates, &c.; ivory from Africa, and European goods from Bombay. Its commercial importance has risen wholly within the last 70 years. The inhab. are chiefly Rhatore Rajpoots, a handsome and brave race of men of the purest castes; the rajah of this tribe, and being considered its legitimate head, has a paramount influence far beyond the limits of his own territory. Bhats, Chunars, and Juats, the last of whom are the cultivators, comprise most of the remaining inhab. The government is a kind of feudal monarchy, the chiefs, who are both numerous and powerful, holding their lands on the tenure of military service. It is said, that the rajah can bring into the field, at ordinary times, a force of 60,000 men, besides mercenaries. The subsidy he furnishes to the British government is 1500 men. The revenues of the state are considerable. The royal lands yield 37 lacs of rupees annually, and the town of Pallee alone yields half a lac monthly. The assessment on the land is always paid in kind, varying from one third to one eighth part of the produce. Within the limits of Marwar, there are said to be 5000 towns and villages, many consisting of from 500 to 1000 houses. Chief towns, Joudpoor, the cap., Pallee, Nagore, and Meerta.

JOUDPOOR, a town of Hindostan, prov. Rajpootana, cap. of the above rajahship; in a hollow surrounded by rocky eminences, and on a soil destitute of water; 100 m. W. of Ajmere. Pop. probably 60,000 (*Burnes*). Notwithstanding the magnitude of this city, we have no recent accounts respecting it. Near it is the residence of the rajah, a fort about three quarters of a mile in circuit, placed on a low mountain, and said to have some resemblance to Windsor Castle. (*Burnes*, in *Geog. Journ.*, vol. iv., &c.)

JUAN-DEL-RIO (ST.), a town of Mexico, state of Queretaro, and cap. dist., of its own name. 31 m. N. W. Mexico and 164 m. S. W. Tampico; lat. 20° 25′ N., and long. 99° 30′ W. Pop. 16,000 ? It is described by Poinsett as a neat and tolerably well-built town, in an extensive plain, 6,400 ft. above the sea, and on the S. bank of a stream, crossed here by a fine bridge of five stone arches: S. of it rises a hill of basaltic rock, the summit of which is crowned with a pretty chapel and spire. The private residences are of stone, and are large, roomy, and well furnished: there is an excellent inn; but the prices of provisions are extravagantly high. The town is surrounded by gardens and orchards; and nothing can exceed the beauty and fertility of the neighbouring country. Indian corn is the chief article of culture; but the ear is much smaller than that of the corn grown in the United States. (*Poinsett's Notes on Mexico*, p. 176–179.)

JUAN-DE-LA-FRONTERA (SAN), a town of Buenos Ayres, near the Chilian frontier, cap. prov., and on the river of same name, 195 m. N. Mendoza; lat. 31° 4′ S., long. 68° 33′ W. Pop. estimated at from 16,000 to 20,000. Though much inferior in extent and pop. to Mendoza, Miers says it possesses much greater capabilities for becoming a flourishing city. Its climate is delightful, though the temp. rises sometimes to 108° F.; but, owing to the latter circumstance, the grape ripens exceedingly well, and very good wine is made. The territory round San Juan, besides being highly productive, has the advantage of being free from the incursions of the Indians, and some years ago a British agricultural colony was about to be established there. The prov. San Juan produces wheat, barley, maize, olives, figs, pasturage, garden vegetables, and all the fruits of the temperate zone in great luxuriance; and in times of scarcity, corn has been sent from San Juan to Buenos Ayres, a distance of above 1000 m. This, however, can never answer under ordinary circumstances, from the great expense attending the land carriage; but it is different with its wines and brandies, which, after all charges, may be sold in most of the provs. of the interior, and even at Buenos Ayres, at a fair profit. They are in general demand among the lower classes, and the quantity exported to other parts of the Republic is little short of that sent from Mendoza. The mountain ranges in the neighbourhood of San Juan yield fine statuary marble, gypsum, sulphur, alum rock, and copperas, and the earth in its vicinity is strongly impregnated with sulphate of soda, which is extracted by washing for medical purposes. (*Miers, Chile at La Plata*, i., 239–243; *Barish. Buenos Ayres*, 315–317.)

JUAN-FERNANDEZ, a group comprising two chief and several smaller islands in the S. Pacific ocean, about 400 m. W. of the coast of Chili: lat. 33° 40′ S., long. 79° W. The largest of these islands, and the only one inhabited, is called *Mas-a-tierra*, to distinguish it from *Mas-a-fuera*, a lofty volcanic rock, about 90 m. W. It is from 10 to 12 m. long, and about 6 m. broad, its area being nearly 70 sq. m. The coast line is very irregular, with frequent bays and headlands; and the chief harbours are Port English, on the S. side, visited by Anson in 1741; Port Juan, on the W.; and Cumberland bay on the N. side of the island. Its northern half is a lofty basaltic formation, intersected with narrow, but fruitful and well-wooded valleys; while to the

S. the land, though less elevated, is rocky and barren. The fig and vine flourish on the hill sides, and among the larger trees are the sandal, cork, and a species of palm called chats, bearing a rich fruit. Goats are found in a wild state, and on the rocky shores are seals and walruses: fish are plentiful, especially cod. The island is very subject to earthquakes, two of which (viz. in 1751 and 1835) are described as having done great damage. In the earthquake of 1835, an eruption burst through the sea about a mile from the land, where the depth is from 50 to 80 fathoms: smoke and water were ejected during the day, and flames were seen at night. (*Geog. Journ.*, vi., 1.)

Juan-Fernandez (which is popularly applied only to the island of *Mas-a-tierra*) was discovered by a Spanish navigator, who gave it to his own name, and formed an establishment, which was afterward abandoned. The buccaneers of the 17th century made it a place of resort during their cruises on the coast of Peru; and more recently it was the solitary dwelling, during four years, of a Scotchman, called Alexander Selkirk, whose adventures are supposed to have given rise to De Foe's inimitable novel of *Robinson Crusoe*. In 1750, the Spanish government formed a settlement, and built a fort; which, however, with the town, was all but destroyed by an earthquake in the following year. They were rebuilt somewhat farther from the shore; and were still inhabited, and in good order, when Cateret visited the island in 1767, but they were soon after abandoned. (*Ibid.* iv., 2.) The Chilian government established a penal colony here in 1819; but this has been discontinued on account of its expense. The island has lately been taken on lease from the Chilian government by an enterprising American, who has brought thither about 150 families of Tahitians, with the intention of cultivating the land, rearing cattle, and so improving the port of Cumberland bay, that it may become the resort of whalers, and other vessels navigating the Pacific ocean. (*Ibid.*; *Dict. Geog., &c.*)

JUANPORE, a distr. of British India, prov. Allahabad, chiefly between the 25th and 26th degs. of N. lat., and the 82d and 83d of E. long; having N. Oude, and the distr. Azimghur; E. Benares; S. the Ganges, separating it from Mirzapoor; and W. Allahabad. Area, 1890 sq. m. Pop. uncertain. Its surface is slightly undulating. The river Goomty runs through it in a S.E. direction. The soil is sandy, but generally well cultivated, and irrigated with care, except towards the Oude frontier, where there is much waste land covered with jungle. This district has improved greatly since it has been brought under the British government, and it is now the principal seat of the sugar cultivation in the central provs. of the Bengal presidency. Some sugar lands in Juanpore let as high as 10 rupees the begah; from 6 to 8 rupees being the average rent of sugar lands in the adjacent districts. (*Syn. in Parl. Rep.*, 1840.) The buildings and villages, though still very indifferent, have been latterly much improved. The land is generally divided into such small portions, that a few years since the incomes of very few landholders exceeded £50. Education appears to be at an extremely low ebb; and the people have always been rather celebrated for turbulence. The remains of many mud forts are to be seen, but none of stone exists, except that of the cap., Juanpore. Land revenue (1829–30), £139,350.

JUANPORE, a town of British India, presid. Bengal, cap. of the above distr., on the Goomty, 36 m. N.W. Benares. Pop. doubtful. Though now decayed and comparatively insignificant, it was previously to the middle of the 15th century a place of importance, and the cap. of an indep. sovereignty. It was annexed to the Mogul empire by Akbar, under whom was built its magnificent bridge over the Goomty, which is now in perfect preservation, and is one of the finest works of the kind in India. A stone fort, a mosque of great beauty, and a number of ruined edifices and monuments, attest the former greatness of Juanpore. The modern town is wholly built of mud; it is, however, the residence of the collector, judge, and other chief British authorities of the district. (*Hamilton; Mod. Trav., &c.*)

JUGGERNAUT (*Jaggannat'ha*, "the lord of the world"), a town and celebrated temple of Hindostan, the latter being one of the chief places of Hindoo pilgrimage, and, according to Hamilton, the most sacred of all the religious establishments of the natives of India. The town stands on the sea coast of the distr. of Cuttack, presid. Bengal, prov. Orissa, beside a branch of the Mahanudda, 45 m. S. Cuttack, and 260 m. S.W. Calcutta; lat. 19° 49' N., long. 85° 54' E. It contains nearly 5000 houses, with 30,000 inhab. It is for the most part mean and dirty, consisting of low brick buildings, with here and there large *serais* and some handsome residences. The chief street is wholly composed of religious edifices, interspersed with plantations; and at its S. end stands the great temple of the divinity or idol. This structure is imposing only from its immensity; its execution is rude and inelegant, and its form unpleasing

to the eye. It is built of coarse red granite, and was completed in 1198, at a cost of from 40 to 50 lacs of rupees (£400,000 to £500,000). The establishment of which it forms a part comprises about 50 temples dedicated to various deities, within a nearly square area inclosed by a stone wall 24 ft. high, and measuring 676 ft. in length on two of its sides, and 670 ft. on the two others. The principal gate of entrance to this area is on the E. side, from which a broad flight of 22 steps leads to a terrace raised about 25 ft., and inclosed by a second wall 445 ft. square. On this terrace is the first apartment, called the Bhog Mandap, a building 80 ft. square, in which the great idol is worshipped during the bathing festival; and in a line, and connected with it by a low portico, is the antechamber, opening into the great tower or sanctuary. This tower rises to 180 ft. above the area on which it is raised, or rather more than 200 ft. above the ground, and forms a valuable landmark to mariners on this dangerous coast. Its ground plan is 28 ft. square within the building; its shape is conical, its walls are externally covered with stone statues in relief, and its roof is ornamented with representations of monsters of various kinds. Little pains, however, appear to have been taken in the sculpture of these decorations, and of late the temple has had an outer coating of chunam or mortar, while its figures have been daubed with red paint: within this sanctuary, seated on their thrones, are the rude statues of three of the most revered deities of Hindoo faith—Juggernaut or Vishnu, his brother Balarama or Mahadeo, and his sister Subhadra or Kali; the temple being devoted to all three, though particularly to the first. Adjacent to this edifice are two other temples, much smaller, and of a pyramidal form. The E. gate of entrance to the outer inclosure is flanked by colossal figures of lions or griffins in a sitting posture, and porters of Hindoo mythology. In front of it is a column, remarkable for its light and elegant appearance, composed of a single block of dark basalt, 40 ft. high and 8 in diameter, supporting a sitting figure of the god Huniman. This pillar was brought thither from the half-ruined black pagoda of Kanaruk (which see), less than a century since. On the N.E. side of the temple is the collection of bungalows forming the European station.

All the land within a distance of 20 m. from the pagoda is accounted holy by the Hindoos, and is held rent free by the cultivators and others, on condition of their performing certain services in and about the temple. The priests and other persons deriving their subsistence from the establishment are said to amount to 3000 families! exclusive of 400 families of cooks, to prepare the holy food so much sought after by pilgrims.

"The provisions, &c., furnished daily for the *idol* and his attendants consist of 230 *seers* of rice, 97 *seers* of *kallsi*, 24 of *mung*, 188 of clarified buffalo's butter, 98 of molasses, 35 of vegetables, 100 of milk, 13 of spices, 20 of salt, and 23 of lamp oil. The holy food is presented to the idol three times a day, and the gates are cautiously shut during this presentation, and none but a few personal servants of the idol are allowed to be present. This meal lasts for about an hour, during which period the dancing girls attached to the temple (consisting of 120) dance and sing in the room with many pillars. On the ringing of a large bell, the doors are thrown open, the food is removed, and the rajah of *Khurda*, as high priest of the temple, divides it with the priests. (*Trans. As. Soc.*, iii., 253.)

The images of Juggernaut, Balarama, and Subhadra are nothing more than wooden busts, about 6 ft. high, fashioned into a rude resemblance of a human head, resting on a sort of pedestal. They are painted white, black, and yellow respectively, with grim, distorted features, and decorated with different coloured head-dresses. The two brothers have arms projecting forward, horizontally, from the ears; the sister is without arms. These monstrous figures may, in general, be seen daily, and are publicly exposed twice a year; when Juggernaut and his brother, after undergoing certain ablutions, assume the form of Ganesa, the elephant-headed god, a transformation effected by means of a mask. Thus dressed, they are placed on the high terrace, overlooking the outer wall of the temple, surrounded by crowds of priests, who fan them to drive away the flies, while the multitude below gaze in stupid admiration. But the grand festival, or *rat'h latra*, takes place in March, when the sun has entered Aries. This has been described as follows by a British eye-witness, for some years resident at Poori, Juggernaut. "Three large *rat'hs*, or cars of wood, are prepared for the occasion, of which the first (intended for Jaggannat'ha) has 16 wheels, each 6 ft. in diameter; the platform, to receive the idol, is 26 ft. square, and the whole car is fully 45 ft. from the ground. The wood-work is ornamented with images of different idols, and painted, and the car has a lofty dome covered with English woollens of the most gaudy colours, bought at the import warehouse in Calcutta; a large wooden image is placed on one side as a charioteer, and several wooden horses are susjoaded in front of the car with their legs in the air. (An exact model of the car of Juggernaut, about 3 ft. square and 4 ft. in height, exists in the museum of the Royal Asiatic Society.) Six strong cables are fastened to the rat'h, by which it is dragged on its journey. The concourse of pilgrims is always very great, and a loud shout from the multitude announces the approach of Jaggannat'ha, who is carried from the temple by a number of priests, appointed for the purpose. A short time after, the rajah of Khurda, as hereditary high priest, makes his appearance in a state palanquin of a strange construction, followed by large state elephants, and generally alights near the rat'h of the idol Balabhadra. The latter, and Subhadra, are placed upon two separate rat'hs, like that of Jaggannat'ha, except being a little smaller, the one having only 14 wheels, and the other 12. The rajah is surrounded by a large train of priests, and immediately prostrates himself before the idol Jaggannat'ha, amid the shouts of pilgrims and the piercing noise of the shrill silver trumpets; he then, with a broom, sweeps the floor of the car, and is presented by the priests with a silver vessel, containing essence of sandal-wood, with which the floor is sprinkled all around the idol. The rajah receives from Jaggannat'ha, as a mark of hohour, a garland of flowers, which the priests take from the image, and put round the rajah's neck. The rajah then descends from the principal car, and proceeds barefooted to the car of each of the other idols, and endeavours to propel them forward, without which ceremonies it is supposed they could not afterward be moved. On a signal being given, a most active scene commences, and several thousand men, each holding a small green branch in his hand, come running to the rat'hs, clearing their way through the crowd for a considerable distance in regular files. They immediately lay hold upon the cables, each man having first touched the car with his branch; and then, aided by the pilgrims (men and women), pull the rat'hs to their destination, taking care to keep their faces towards the idol, who is driven to his garden-house, where he is worshipped for four days, and then returns in the same way to the temple." (Mansbach in Trans. Asiat. Soc., iii., 256-259.) Besides that described, 12 other principal, and many minor festivals are celebrated during the year. The worship of Juggernaut is attended by every sect and class of Hindoos, who meet on equal terms, *all casts being abolished within the precincts of the temple.*

That excess of fanaticism, which is said to have prompted the pilgrims to court death, by throwing themselves, in crowds, under the wheels of Juggernaut, either never existed, or has long ceased to actuate the worshippers of the idol. During four years that Mr. Mansbach witnessed the festivals, only three cases of self-immolation occurred; one of these was probably accidental, and the two others were suicides, committed by sufferers to rid themselves of painful diseases. The greatest misrepresentations were formerly circulated in Europe respecting the number of widow burnings, pilgrims, loss of life, &c., at Juggernaut. It is true, that, for many miles round the temple, the sides of the roads are literally whitened with the bones of devotees, who have perished by the way-side. But this is not the result of any violent modes of destruction, voluntary or otherwise. If a Hindoo have reason to believe dissolution at hand, he forthwith collects his remaining strength, and, should he fortunately succeed in dragging his diseased body within sight of the sacred edifice, he will lie down in peace, and die with a perfect confidence of future happiness; besides which, thousands set out on a pilgrimage thither in health, and in the full intention of returning, whose subsistence failing by the way, devote themselves (in fact they can do nothing else) to death by starvation. An unfounded clamour was long raised in England against the government of British India for promoting idolatry, as it was said, by continuing to exact taxes on the pilgrims to Juggernaut, Gaya, and other places, as had previously been done by the native sovereigns. But the levy of taxes on pilgrims seems rather an odd way of promoting idolatry! However, in deference to the well-intentioned, though absurd misrepresentations propagated in England on the subject, these taxes have been repealed, to the great satisfaction of the "idolaters." The number of pilgrims to this and other shrines has since greatly increased; and the natives are extremely well pleased by this act of liberality on the part of the government. It may be right to mention, that no part of the pilgrim-tax ever came into the general funds of the government, but was wholly laid out on the repair of roads, and the maintenance of a proper police at the different places of pilgrimage. (See *Asiatic Researches,* vols. viii., x., xv.; *Trans. of the Royal Asiat. Soc.,* vol. iii.; *Hamilton's Hindostan and E. I. Gaz.; Asiatic Journal, &c.*)

JULIERS (Germ. *Jülich*), a town of Rhenish Prussia, gov. circle, on the Roër, a tributary of the Maese, 23½ m. W. Cologne, and 16½ m. N.E. Aix-la-Chapelle. Pop. 3130. It has a strong citadel, three churches, a fine old town-

hall, circle court of justice, police court, high school, &c., and manufactures of woollen cloth, leather, and vinegar.

Juliers is believed to be identical with *Juliacum,* in Antonine's Itinerary. After the extinction of the Roman dominion, it became the property of independent counts of the Germanic empire, who were created dukes by the emperor Charles IV., in 1356. The family of the dukes of Juliers becoming extinct in 1609, the town was taken by Prince Maurice of Nassau in the following year; in 1622 it was taken by the Spaniards, who held it till 1659. In 1794, it was taken by the French, who afterward made it the capital of the department Roër. The former duchy of Juliers is the most W. portion of the Prussian dominion, and is remarkable for its fertility, and its linen manufacture. (*Dict. Géog.; Schreiber; Berghaus.*)

JULIET, p. v., cap. of Will co., Ill., 165 m. N.E. by E. Springfield, 742 W. Situated on both sides of Des Plaines river, which affords water-power. The Illinois and Michigan canal here crosses the river. It contains two churches, a Methodist and an Episcopal; one saw-mill, one grist-mill, and about 700 inhabitants. In the precinct are two academies, 50 students; seven schools, 210 scholars. Pop. 2558.

JUMBOSEER, a town of British India, presid. Bombay, distr. Baroach, on a river of the same name. 25 m. N.N.W. Baroach. In 1820, it had upwards of 10,000 inhab. It carries on a considerable trade with Bombay, to which it sends cotton, grain, oil, and piece goods. (*Hamilton's E. I. Gaz.*)

JUMILLA, a town of Spain, prov. Murcia, 36 m. N. by W. Murcia, and 75 m. S.S.W. Valencia. Pop, according to Miñano, 6967. It is situated on the S. slope of a hill, at the summit of which is a castle commanding the town; streets straight and of moderate width, but not paved; the public buildings comprise two churches, two convents, a public granary, and a hospital. The town contains about 30 oil and corn mills, two soap manufactories, and an establishment for making firearms: the salt-pans being under the direction of government, cannot be in a prosperous condition. A considerable fair is held here, December 2. The climate, though not so genial as in neighbouring towns situated at a less elevation, is, on the whole, salubrious; and corn and fruit are abundant. Grazing, however, is the principal pursuit of the people in and near the town; and Miñano states that, at an average, 35,000 head of sheep and goats are pastured on the surrounding hills.

Jumilla was taken from the Moors, who, having founded or rebuilt it, gave it its present name, by a king of Arragon: it was again taken from Arragon by Henry of Trastamare, who made it subject to the crown of Castile. (*Miñano, Dict. Géog.;* see MURCIA.)

JUMNA (Sanscr. *Yamuna,* the *Jomanes* of Pliny), a river of Hindostan, and the chief tributary of the Ganges. It rises on the S.W. side of the great Himalaya range, about lat. 30° 55' N., and long. 78° 24' E., and has been traced to an elevation of about 11,900 ft. above the sea, at the foot of an abrupt mountain nearly 4000 ft. higher. Over the wall of this mountain falls a streamlet, probably caused by the melting of the snows on the summit, and which appears to be the true source of the river. For some miles, the Jumna proceeds through a glen not more than about 40 yards in width at its bottom, and bounded by mural precipices of granite many thousand feet in height. The stream is here concealed by a thick bed of frozen snow, which arches over the course of the river beneath, supported by the shelving walls of the ravine. About half a mile below the point to which the Jumna has been traced, is Jumnotri, a celebrated place of pilgrimage and ablution with the Hindoos. At this spot are numerous hot ferruginous springs, some of which rise in the rocky wall 10 or 12 ft. above the bed of the river; and, having melted the snow for 20 or 30 yards round, mix with the waters of the Jumna, rendering them sensibly warm. Some of the springs are hot enough to boil rice, their temperature having been found as high as 194·7° Fahr., or near the point at which water is converted into steam at that elevation, about 10,840 ft. above the sea. Before arriving at them, the Jumna is only about 3 ft. in width and a few inches deep; but these, causing a continual melting of the snow, contribute greatly to augment its supply of water. About 30 m. below its source, the Tonse unites with the Jumna; and though double the size of the latter, takes its name. From this point to Delhi the river flows generally in a S. direction; it thenceforward gradually declines to the S.E. Throughout its whole course it usually runs parallel to the Ganges, the tract between the two rivers, called the *Doab,* varying from 20 to 80 m. in width. At its emerging from the hilly region, about lat. 30° 15', the bed of the Jumna, which is 1000 yards broad, is full in the rains, though in the dry season the river is not more than 100 yards across. It is not usually very deep, being fordable in several places above Agra; in its progress through the province of Delhi

It divides into various branches, inclosing large islands. It joins the Ganges at Allahabad, where its breadth is fully equal to that of the latter river. Its entire length is estimated at 780 m. It receives no tributaries of any consequence in the upper part of its course, but in the lower, the Chumbul, Sind, Betwah, and Cane, join it from the S., and the Sinde from the N. Delhi, Agra, Allahabad, Etawah, and Kalpee are on its banks. From its shallowness, the Jumna is little serviceable to commerce, and its waters in the great plain of the upper provinces are so impregnated with natron that vegetation is rather hindered than promoted by its inundations. The country to the W. of the Delhi is, however, fertilized by the canal of Ali Mordan Khan, cut from it immediately after its leaving the hills; and the upper portion of the Doab is irrigated in a similar manner by the Zabeta Khan's canal, 200 m. in length, which also commences at the foot of the hilly region, and proceeds to Delhi. (*Geog. Journ.*, iv.; *Fraser*; *Hodgson*; *Asiat. Researches*, xii.; *Hamilton, E. I. Gaz.*, &c.)

JUNGEYPOOR, a town of Hindostan, prov. Bengal, distr. Moorshedabad, on an arm of the Ganges, 25 m. N.N.W. Moorshedabad. It is one of the principal stations in the British territories for the culture of the silk-worm. The mulberry is cultivated to a great extent from annual shoots, and large quantities of indigo are also grown in the neighbourhood.

JUNGLE MEHALS, a distr. of British India, presid. and prov. Bengal, between lat. 22° 30′ and 24° N., and long. 86° and 88° E.; having N. the distr. Beerbhoom, E. Burdwan, S. Hooghly and Midnapore, and W. Ramghur. Area, 6990 sq. m. Pop. (1829-30) 1,304,740. "The name of this district implies a waste territory and backward stage of civilization; yet it appears, from the report of the circuit judge in 1815, that no instance of gang-robbery or arson had occurred during the previous six months, and in India, where a country furnishes few materials for history, it may be presumed to be going on tolerably well." (*Hamilton's E. I. Gaz.*) Total land revenue (1829-30) £44,942.

JUNIATA, co., Pa. Situated a little S. of the centre of the state, and contains 360 sq. m. Organized in 1831. Watered by Juniata river and Licking and Tuscarora creeks. It contained in 1840, 11,089 neat cattle, 12,023 sheep, 18,604 swine; and produced 219,859 bushels of wheat, 62,919 of rye, 162,659 of Indian corn. 17,728 of buckwheat, 3035 of barley, 156,079 of oats, 56,290 of potatoes. It had 33 stores, nine fulling-mills, 11 flouring-mills, 17 grist-mills, 59 saw-mills, 21 tanneries, five distilleries, three printing-offices, three weekly newspapers; two academies, 53 students; 63 schools, 2375 scholars. Pop. 11,080. Capital, Mifflin.

JUNIATA, river, Pa., is formed by Raystown and Frankstown branches, which rise at the foot of the Alleghany mountains. Pursuing a winding course after the junction, it unites with Susquehanna river, on the E. side of Perry county, 15 m. above Harrisburg. The Pennsylvania canal passes along Juniata river, and the Frankstown branch, to Hollidaysburg.

JUNIATA, p. t., Perry co., Pa., 30 m. W.N.W. Harrisburg. 121 W. Watered by Raccoon, Buffalo, and Little Buffalo creeks. It contains Bloomfield village, the capital of the county, and has three stores, one furnace, five flouring-mills, 15 saw-mills, two tanneries, one pottery; 11 schools, 440 scholars. Pop. 1450.

JUNIUS, p. t., Seneca co., N. Y., 8 m. N. Waterloo, 189 m. W. Albany, 339 W. It has one store, one tannery; six schools, 213 scholars. Pop. 1394.

JURA, a frontier dep. of France, region of the E., formerly included in Franche Comté, between lat. 46° 16′ and 47° 18′ N., and long. 5° 19′ and 6° 12′ E., having N. Haute Saône, E. Doubs and a part of Switzerland, S. Ain, and W. Saône-et-Loire and Côte d'Or. Length, N.W. to S.E., 70 m. Area, 496,930 hectares. Pop. (1836) 315,355. More than two thirds of the surface, principally in the S. and E., is covered with mountain ranges belonging to the Jura system, the principal summit of which, the Récolet, 5633 ft. high, is in this department. Rivers, numerous: the chief are the Doubs and Ain. There are several small lakes, and in the N.W. some large marshes. In the plains the atmosphere is moist and heavy, while in the mountains it is dry, and the winters long and severe. In 1834, the arable lands were estimated at 183,113 hectares; meadows at 50,547; vineyards at 21,027; forests at 115,614; heaths, wastes, &c., at 79,000 do. According to Hugo, *l'agriculture du département paroît aussi perfectionnée qu'elle peut être.* Sufficient corn is grown for home consumption, chiefly wheat, barley, maize, and oats. In 1835, the crop of potatoes was estimated at 719,000 hectol. Upwards of 400,000 hectol. of wine are produced annually, some of which is very good. The mountains afford excellent pasture, on which many black cattle are fed; and *chàlets* are established on them, as in Switzerland. The butter and cheese of the department are much esteemed. In 1830, the horn-

ed cattle amounted to nearly 155,000 head: sheep are much less numerous. Horses and mules are extensively bred; and hogs, poultry, and bees are also very plentiful. In 1835, of 122,941 properties subject to the *contributions foncière*, 61,337 were assessed at less than 5 fr., and 19,865 at from 5 to 10 fr. The number of large properties is much below the average of the departments. There are several iron mines, and quarries of marble, alabaster, and gypsum. The department has also ores of lead, coal, copper, and even gold, but no mines of those metals are at present wrought. Iron forges and paper factories are numerous; cotton and linen fabrics, chamois and other leather, glue, mineral acids, and marble ornaments are among the other chief manufactures. Watches and trinkets are made of Morez, and ivory, bone, horn, marble, and wooden articles are sent all over Europe from the turning establishments of St. Claude. Jura is divided into four arrondissements: chief towns, Lons-le-Saulnier, the capital, Dôle, Poligny, and St. Claude. It sends four members to the chamber of deputies. Number of electors (1836-30), 1156. Total public revenue (1830), 7,532,947 fr. (*Hugo*, art. *Jura*; *French Official Tables*, &c.)

JURA MOUNTAINS, a chain of central Europe, usually classed with the Alpine system, and including the mountains of W. Switzerland, and those between the lake of Geneva, the Rhone, the Soane, and the Doubs. The range commonly thus designated has a length of about 160 m., with an average breadth of 30 m., commencing S. on the banks of the Rhone, and running N.E. to the junction of the Rhine and Aar; but connected mountains of analogous composition run N. through Suabia and Franconia, and S.W. along the right bank of the Rhone to the vicinity of Narbonne, so that the Jura range, in its most extended sense, has a length of about 600 m. The Swiss Jura, of which alone any notice will here be given, consists of several long parallel chains, enclosing narrow longitudinal valleys, such as the Val de Joux (in which is the mountain-lake of the same name, 3960 feet above the sea), the Val Travers, the Val de Ruz, and the valleys of the Valserine, Doubs, Birs, and other rivers. Transverse valleys, similar to those in the main Alpine system, are of rare occurrence, and the range throws off only one lateral spur, viz. the chain of mount Jorat, passing between the lakes of Geneva and Neuchâtel, and joining the Bernese Alps. The slope is rapid on the Swiss side, but more gentle towards France; and the ridge, as seen from a distance, presents a regular undulating line with rounded dome-like summits, contrasting strongly with the abrupt crags and towering peaks of the Alps. The chain sinks, as it advances N.: the culminating point, *le Reculet*, is 5633 feet high, and eight others rise above 5000 feet: the roads across the ridge have an elevation varying from 3600 to 2500 feet above the sea. Snow lies on the highest ground about seven months in the year, and there are no glaciers. The geological constitution of the Jura mountains, which has been described at length by Von Buch, Boué, and also by different writers in the *Geological Transactions* (London), is limestone of the oolitic series. The strata comprise most of the varieties lying between the lias and the compact limestone, answering to the Portland stone of English geologists; and the beds are thrown up at high elevations, thus causing the formation of those longitudinal valleys which are a characteristic feature of the Jura. On the S.E. slopes, and, as Lyell observes, exactly opposite the principal openings by which great rivers descend from the Alps, lie numerous "erratic" blocks of extraordinary magnitude. How these granite fragments came to their present situation is wholly matter of conjecture; but if it be true, as Lyell supposes, that the limestone layers of the Jura were upraised by some internal commotion, it is not improbable that these boulders were detached from the Alpine summits, and transported to lower platforms, which have been subsequently elevated. (*Princ. of Geol.*, vol. iii., p. 424.) The vegetation of the Jura nearly resembles that of the Alps: box-trees are very abundant on the N.W. side, and the hills near Poligny are covered with firs, the timber of which furnishes materials for the industry of the population during the winter months, and is also a considerable article of trade with the surrounding districts. Many of the villagers, also, on the mountain sides, and in the valleys, are employed in making watch-movements, which find a ready market at Geneva and other towns in which the watch trade is extensively pursued. (*Bruguière*; *Orographie de l'Europe*; *Ebel, Manuel du Voy. en Suisse*, &c.)

JURA, one of the Hebrides, which *see*.

JUTLAND, a large prov. of Denmark, formerly comprising the whole continental portion of the Danish dominions, but which is now restricted to the part of the peninsula belonging to Denmark to the N. of Sleswick, extending from about 55½° to nearly 58° N. lat., being about 170 m. in length, and from 60 to 80 in breadth, comprising an area of 9530 sq. m. Pop. in 1834, 395,959, having increased at the

rate of about one per cent. per annum during the present
century. It is of an oblong form, with the addition of a
triangle towards the N. Surface generally flat. It has few
rivers, and none of any considerable magnitude; but it is
deeply indented, and in part traversed by inlets or arms
(fiords) of the sea. Soil very various. In the middle it is
dry, sandy, and occupied by extensive heaths; on both
shores it is more fertile; and on the W. coast, particularly
towards the S., there are large tracts of very rich marsh-
land, defended by dykes from being overflowed by the sea.
Agriculture, though still backward, has made great progress
during the present century. Rye, oats, and buckwheat are
the crops most generally raised; and they, along with cattle of excellent quality, horses, and butter, form the principal articles of export. Hogs are so very plentiful, that Jutland has been called "the land of bacon and rye bread!"
Fish very abundant in the fiords or inlets of the sea. Minerals and manufactures unimportant. Principal towns,
Aalborg, Aarhuus, Wyborg, &c.

K.

KAFFA, or THEODOSIA, a seaport town of European
Russia, on the S.E. coast of the Crimea, lat. 45° 1' 37'' N.,
long. 35° 23' 37'' E. Pop. 7250. It is believed to stand on
the site of the ancient Theodosia, founded by Milesian
colonists in remote antiquity. The Athenians carried on a
great trade with this city, importing from it vast quantities
of corn, with slaves, lumber, and naval stores, hides, and
honey.* After undergoing many revolutions, it fell, in the
13th century, into the possession of the Genoese, who rebuilt it, and made it the chief seat of their power during
the lengthened period of their ascendancy in the Black sea.
In 1475 it was taken by the Turks; but it continued, down
to its conquest by the Russians, to be a large, populous
town. It, however, suffered severely from this event, partly
in consequence of the devastations committed by the Russian soldiery, and partly through the emigration of its Tartar inhabitants. Latterly, however, it has begun to revive;
though, owing to the superior advantages enjoyed by Kertsch
as an *entrepôt* for the trade of the sea off Azoff, it does not
seem very probable that Kaffa will ever recover her former
importance. The road, or bay of Kaffa is very extensive,
and capable of accommodating a great number of vessels.
It has deep water throughout; the holding ground is good;
and, with the exception of the E., it is sheltered from all
winds. (*Hagemeister on the Trade of the Black Sea*, p. 64,
&c., Eng. Trans.; *Purdy's Sailing Directions for the Black
Sea, &c.*, p. 208, &c.).

KAIRA, a distr. of British India, presid. Bombay; between lat. 22° 15' and 23° 35' N., and long. 72° 40' and 73°
30' E.; area, 1827 sq. m. Pop. (1832) 484,735. It consists
principally of territory ceded at different times by the Pe-
ishwa and Guicowar, is well watered, and contains a great
deal of good soil; but having been inhabited by a turbulent
population, it was greatly neglected before it came into our
hands; and it still contains much waste land; while education appears to be in a lower state than in most British
districts. The land is assessed under the village system,
and the assessment is realized without difficulty. Total
amount of land revenue (1818–19), £175,875.

KAIRA, a town of Hindostan, cap. of the above collectorate,
113 m. N.N.W. Surat; lat. 22° 47' N., long. 72° 48' E. It
is a neat town, surrounded by bastioned ramparts and walls
in good repair. Its streets are narrow, but tolerably clean,
and its houses are solid, lofty, and adorned with a great
deal of carving. Its chief public buildings are the district
courthouse, a handsome Grecian edifice, a large and secure
prison, a church, an English school, and, near the centre
of the town, a large Jain temple. The cantonment of
Kaira, about 1½ m. distant, is, unfortunately (like many of
the settlements in India founded by the British), in a very
unhealthy situation; but it is extensive and well laid out,
with good barracks, a hospital, a regimental school, and a
tolerable English library.

KAIRWAN, or KEERWAN, a large city of N. Africa,
at present the chief source of Mohammedan bigotry in that
country, regency Tunis, 85 m. S. from the city of Tunis;
lat. 35° 37' N., long. 9° 57' E. Pop. estimated, but on no
good authority, at 50,000. It is situated in a barren sandy
plain, and is surrounded by a low wall: the public buildings comprise a large citadel and several mosques, two of
which are extremely magnificent, supported, as Shaw affirms, "by an almost incredible number of pillars." (*Travels*, p. 116.) The houses are clean and respectable; and
the streets wide, and ornamented with columns, capitals,
and highly raised Cufic inscriptions. It is regarded as the
second town in the regency; and its Kadee, or governor,

---

may be said to be almost independent of the Bey of Tunis.
He fixes the price of provisions, which are said to be, though
certainly not on account of his interference, a half cheaper
than at Tunis; but, with all his influence, he cannot so far
overrule the bigotry of the inhabitants as to ensure a good
reception to the Christian traveller, who, if he ventures
within the walls, must take on himself all the risks of his
enterprise. Kairwan is famous for its yellow Morocco boots
and slippers, the delicate dye of which it has hitherto been
found impossible to equal. It was formerly a place of great
literary eminence, possessing well-endowed institutions and
good libraries, from which Europeans have derived a large
portion of their knowledge of Arabic literature; but of its
present claims to such distinction we have no information, as
the extreme jealousy of the people shuts out all local inquiry.
Kairwan is supposed by Shaw to occupy the site of the
Vicus Augusti in Antonine's Itinerary; but, notwithstanding the deference due to so great an authority, this supposition is contested by Temple and others, apparently on
pretty good grounds. The present city was built about
A.C. 670, about the time when Africa was invaded by the
generals of the Omaiyade, khalif Moawyad I. In 800, the
governor of W. Africa threw off his allegiance to the kha-
liph, declared himself independent, and established his
capital at Kairwan. In 969, the seat of government was
transferred to Cairo, since which its importance, though
still considerable, has materially declined. (*Shaw's Travels; Temple's Excursions*, vol. ii., p. 93–96.)

KAISARIAH, (an. *Mazaca*, and afterward *Cæsarea*), a
town of Asiatic Turkey, prov. Karamania, sandjiak of its
own name, on the Karsu (an. *Melas*), a tributary of the
Euraphrates, 140 m. E.N.E. Konieh, and 136 m. S.E. Angora, lat. 38° 43' N., long. 35° 29' 20'' E. Pop., according
to Kinneir, about 25,000, of whom 2000 are Armenians,
Greeks, and Jews. It is situated on the E. side of a fertile
plain of great length, and in a recess formed between two
spurs projecting from the lofty, snow-covered mount Erd-
jish, the *Argæus* of antiquity. The houses, though built
of stone and brick, have a mean appearance, and the streets
are said to be the filthiest in Turkey. It is surrounded by
a wall now in ruins, and in the suburb are some interesting
remains of a Roman city. Several mosques, one Greek
and two Armenian churches, a convent, and some mausoleums, are the chief public buildings. Kaisariah is the emporium of an extensive trade, and the resort of merchants
from all parts of Asia Minor and Syria, who come to purchase cotton cultivated in the vicinity in great quantities,
and sold either in a raw state or when manufactured into
cloth. Cotton thread and cloth constitute the chief articles
of industry; and there are some tanneries of yellow Morocco leather. The land in the neighbourhood is fertilized
by the inundations of the Karsu, and produces an abundance of large and delicious-flavoured fruits and vegetables.
The climate is very healthy, except within the town, where
epidemics prevail, owing to the offal, &c., left in the streets
to decay and infest the air.

Mazaca, the ancient capital of Cappadocia, took the name
of Cæsarea, in honour of Tiberius. Its antiquity is attested
by Strabo, who also gives an excellent description of the
neighbouring mountain. It was the residence of the kings
of Cappadocia previously to its being annexed to the Roman
empire, after which it continued to increase in size and
beauty. An amphitheatre and many temples were erected;
and in the reign of Valerian, when Shapoor I., king of Persia pillaged the city, and massacred its inhabitants, it is
said to have had a population of 400,000 persons, though
this is most probably far beyond the mark. (*Gibbon*, i.,
430.) Its dimensions were contracted by Justinian, who
rebuilt the walls: it was raised to the dignity of an apostolic see, and gave birth to St. Basil. Having been destroyed by an earthquake, it was afterward rebuilt, and by turns
became subject to the sultans of Iconium, the princes of
Karaman, and the grand seignior. (*Kinneir's Asia Minor*,
p. 96–106; *Geog. Journ.*, vol. viii.; *Dict. Geog.*, &c.)

KAISARIAH, a ruined town and seaport of Palestine.
(See CÆSAREA.)

KALAMAZOO, r., Mich., rises in Hillsdale co., and after
a course, with many windings, generally W.N.W. of 200
m., it enters lake Michigan, 41 m. N. of the mouth of St.
Joseph river, and 99 m. S. of the mouth of Grand river. Its
average depth for 8 m. from its mouth, is 19 feet. A bar at
its entrance into the lake has six or seven feet at low water.
Near its mouth it is from 300 to 490 feet broad. It is navigable at all seasons to Allegan, a distance of 38 miles, for
boats of 50 tons. The banks of this river are generally
low, but in the county of Allegan, they rise from 25 to 30
feet high. This river, and its tributaries, afford much
water-power.

KALAMAZOO, county, Mich. Situated toward the S.W.
part of the state, and contains 576 sq. m. It contained in
1840, 7861 neat cattle, 2604 sheep, 12,665 swine; and produced 161,168 bushels of wheat, 125,683 of Indian corn, 1415

* For an account of the trade of the Athenians with this emporium, see
Clarke's Connection of Russian, English, and Swiss Coins, p. 91.

93

of buckwheat, 5979 of barley, 187,866 of oats, 71,285 of potatoes, 44,439 pounds of sugar. It had 20 stores, one fulling-mill, six grist-mills, 22 saw-mills, two tanneries, two distilleries, two printing-offices, two weekly newspapers; one academy, 30 students, 50 schools, 1739 scholars. Pop. 7380. Capital, Kalamazoo.

KALAMAZOO, p. t., capital of Kalamazoo co., Mich., 141 m. W. Detroit, 605 W. Watered by Kalamazoo river. It contains 13 stores, one fulling-mill, two grist-mills, five saw-mills, one tannery, two distilleries, two printing-offices, two weekly newspapers; two schools, 110 scholars. Pop. 1290. The village is situated on the W. bank of Kalamazoo river, and contains a courthouse, jail, a branch of the Michigan bank, a branch of the University of Michigan, the Huron Literary Institute, a Presbyterian church, 10 stores, and a flouring-mill. The central railroad is located through it, but not completed so far.

KALIDA, p. v., capital of Putnam co., O., 114 m. N.W. Columbus, 479 W. Laid out in 1834 on the N.E. bank of Ottawa river, 1½ m. from its junction with Auglaize river. It contains a courthouse, jail, three stores, 25 dwellings, and 150 inhabitants. Nett proceeds of the postoffice, $107.

KALISZ, a city of Poland, and the most westerly in the Russian dominions, cap. palat. of the same name, on an island in the Prosna, immediately within the Russian frontier, 198 m. W.S.W. Warsaw, and 70 m. S.E. Posen. Pop. estimated at 15,000, of whom 2509 are Jews. This is one of the finest cities in the kingdom. It is surrounded by old walls flanked with towers, and entered by five gates; and has a citadel founded by Casimir the Great. Its streets are broad and well paved, and several are planted with trees; its houses are generally good. The most remarkable public edifices are the former palace of the voivodes, now occupied by the courts of law, the cathedral, church of St. Nicholas, and the Lutheran church. Besides the cathedral, there are five Roman Catholic churches and six convents, a synagogue, a Roman Catholic gymnasium or lyceum, with a fine library, and large scientific collections, a military school with 200 students, several superior female seminaries, elementary schools, attended by about 300 children of both sexes, a school of midwifery, &c. (Horschelmann's Stein. i., 701.) It has also a theatre, a house of charity, and three hospitals. Kalisz is a town of some industry, having manufactures of linen and woollen cloths, and leather. A fine road leads to Opatowek, a village about 6 m. distant E.S.E., celebrated for its large manufacture of woollens, and its gardens, which form the favourite resort of the inhabitants of Kalisz. This city was founded about 655, and was long the residence of the dukes of Great Poland. Near it, in 1706, the Poles totally defeated the Swedes; and in September 1835, a grand military muster and review took place at Kalisz, attended by the sovereigns of Russia, Austria, and Prussia. (Malte-Brun; Tableau de la Pologne; Balbi; Stein.)

KALPEE, or CALPEE, a large and populous town of British India, presid. and prov. Agra, on the S. bank of the Jumna, 45 m. S.W. Cawnpore. It is a place of considerable trade, being an entrepôt for the transport of cotton from the S.W. of India to the Gangetic prova.; and has also manufactures of sugar-candy, paper, &c.

KALUGA, a government of Russia in Europe, near its centre; chiefly between lat. 53° 30', and 55° 30' N., and long. 33° 40' and 37° E., having W. the gov. of Smolensk, N. the latter and Moscow, E. Tula, and S. Orel. Area, according to Köppen, 11,470 sq. m. Pop., in 1838, 915,000. Surface an almost uninterrupted plain, watered by numerous rivers, of which the Oka and its tributaries are the principal. Climate tolerably mild for the latitude. Soil mostly either sandy or hard clay, and not fertile. Forests occupy more than half the surface. Arable lands rather more than 2-5ths; but a good deal of manure is required to render the latter even moderately productive; and the agricultural produce is not adequate to the consumption of the inhabitants. Rye is principally grown; but oats, wheat, and barley, are also cultivated; as are hemp and flax. Cattle not numerous, and but little valued; but there are in the government two extensive studs for the breeding of superior horses. The fisheries are insignificant; little game is met with. Bog-iron is found, but in no great quantity, and a good deal has to be imported to supply the various iron works. This government being so little suitable for agriculture, the attention of its inhabitants has been naturally turned towards manufacturing industry; in this respect, Kaluga ranks immediately after the governments of Moscow and Vladimir. In 1830, 18,600 workmen were employed in distilleries and manufactures of sailcloth, linen and cotton goods, leather, soap, candles, and hardware. The manufacture of beet-root sugar has been lately introduced. Nearly all the peasants' families employ a considerable portion of their time in weaving. Many of the merchants in this government are opulent, and some have commercial transactions with foreign countries, through

Archangel. The chief exports are oils, spirits, potash, honey, linen, sailcloth, and other manufactured goods. The chief commercial towns are Kaluga and Borofsk. The inhabitants are nearly all of the Russian stock. Kaluga is divided into 11 districts, and is under the same military governor with Tula. Its scholastic institutions are under the university of Moscow, but they are extremely deficient; and it has only one printing press, which is the property of the crown!

KALUGA, a town of Russia in Europe, cap. of the above gov., on the Oka, near where it suddenly turns eastward, 105 m. S.E. Moscow. Lat. 54° 30' 27" N.; long. 36° 17' 12" E. Pop. 25,080. (Schmitzler.) Though comprising no more than about 3800 houses, it is said to occupy a space of 10 versts, or little short of 7 m. in circumference, and is divided into three quarters by the Oka and its tributary the Kaloujka. It is an ill built town, with narrow, crooked, and ill-paved streets, and wooden houses. There are, however, some good public edifices, as the high church, government-house, town-hall, and theatre. Of the 24 churches, 23 are of stone; a convent, also a stone building, gymnasium, seminary for poor children of noble birth, foundling asylum, several workhouses and hospitals, and a house of correction, are the other chief public establishments. Kaluga is one of the most important manufacturing and commercial towns in the empire: it has five sailcloth factories employing 400 weavers, and 1000 spinners, between 30 and 40 oil factories, numerous tan-yards, some sugar-refineries, and manufactures of woollen cloth, cotton fabrics, hats, paper-hangings, earthenware, soap, vitriol, &c. Besides carrying on an extensive internal trade, its merchants make large exports of lamb-skins, Russia leather, and wax, to Dantzic, Breslau, Berlin, and Leipsic. (Schnitzler, La Russia, p. 133-138; Possart, Russland, p. 517-520.)

KAMINIETZ (Polish, Kaminiec Podolski), a town of Russian Poland, gov. Podolia, of which it is the cap., on the Smotrycz, about 12 m. from its junction with the Dniester, 215 m. S.E. Kief, and 300 m. N.W. Odessa; lat. 48° 40' 30" N., long. 27° 1' 30" E. Pop., in 1830, according to an official document, 15,599; but this is probably much exaggerated: in 1822 it had only 600 houses and 8000 inhabitants, many of whom were Jews. It is irregularly laid out, with narrow streets, and wooden houses. It has, however, some conspicuous edifices of stone and other solid materials; including the cathedral, dedicated to St. Peter and St. Paul, a Gothic building containing 15 altars and a nave, supported by 150 columns. Near it is a column supporting a statue of the Saviour. The church of the Dominicans, originally constructed of wood, in 1360, was rebuilt in stone after the expulsion of the Turks in the 18th century. There are in all five Roman Catholic, and four Greek churches, and one Armenian church, a fine edifice, completed in 1767. The Roman Catholics have several convents. The other chief public buildings are the government library, circle school, and new gymnasium, commenced in 1837.

The town was formerly walled, but its works were levelled, by order of the Russian government, in 1812.[*] It is, however, still defended by a citadel and another fortress. The former, situated on a steep isolated rock overlooking the town, might be made impregnable, but it is commanded by some more lofty adjacent heights. Kaminiec was however, for a lengthened period, the principal bulwark of Poland on the side of Turkey. It was founded by the sons of Olgherd, in 1331, after that prince had wrested Podolia from the Tartars. It was soon after fortified, and in 1374 attained the rank of a city. It remained attached to Poland till its final capture by the Russians in the year 1793, except from 1672 to 1699, during which it was in the possession of the Turks. (Schnitzler, La Russie, p. 500, 501.; Possart, Das Kaiserth, Russl., p. 873.)

KAMTSCHÁTKA, a large peninsula at the N.E. extremity of Asia, forming a part of the Russian government of Irkutsk, and bounded N. by the country of the Tschuktchi, E. by the Aleutian archipelago, and W. by the sea of Okhotsk. It lies between the 51st and 62d parallels of N. lat., and the 156th and 167th deg. of E. long.; has a length of about 800 m., and a breadth varying from 100 to 250 m., the area being very loosely estimated at 80,000 sq. m. Supposed population only 5000, of whom about 1500 are Russians. The coast line on the W. side is tolerably regular, the gulf of Penginsky, at its N. end, forming the only considerable exception; but on the E. side are several extensive bays, enclosed respectively between the capes Chipunsky, Kronotsky, Kamtschatka, Ozernoy, and Olutorsky, the last of which is near the N.E. end of the peninsula: C. Lopatka (lat. 51° 0' 15" N., long. 156° 2' 13" E.) is the S. extremity of Kamtschatka. The coast generally speaking, is abrupt and rocky, especially on the E. side, and the peninsula, when viewed from the sea

[*] Balbi (1857) says they have been since restored.

presents the appearance of a barren and desolate rock; but in the interior there are plains of considerable extent, having a soil well adapted for tillage. The high lands, which cover about two thirds of the entire surface, consist of a chain of volcanic mountains, running in a S.S.W. direction. Many volcanoes in this chain have been ascertained by Erman and Lutké to be in a high state of action; and it seems very probable that, geologically considered, they form only one extremity of a great volcanic belt, continued through the Kurile and Japanese islands, Formosa, and the islands of the E. Indian archipelago.

The following statements are drawn up from the observations of the naturalists in Commodore Lutké's expedition, in 1827-30, and of Professor Erman, who visited Kamtschatka in 1829. In the main range running N. from C. Lopatka, 13 summits, with craters and hot springs, have been observed within the 51st and 56th parallels, one other height being isolated, and lying W. of the principal chain. The elevation of nine summits has been accurately measured, and appears to be as follows:—

Asatschinsky	8340 ft.	Krenotsky	10,610 ft
Vilutchinsky	6846	Klutchevsky	16,500
Avatcha	8760	Tolbachka	8960
Koriata	11,120	Chevelatch	10,500
Japonov	8080		

The most active are Asatschinsky, Avatcha, and Klutchevsky. The scoriæ and ashes thrown from the first, in 1828, were carried as far as Petropaulovsky, 120 versts distant; and it appears to be more or less in continual activity. In 1827 there was a violent eruption of Mount Avatcha, during which, besides lava and stones, a very large quantity of water was ejected; a phenomena remarked also by Humboldt in the volcano of Karkuariso, a little N. of Chimborazo, in the Colombian Andes, and known to have occurred, though in a less degree, during the eruptions of Etna and Vesuvius. At the summit is a crater several hundred yards in circumference, formed by a wall 20 feet high, composed of porphyry, felspar, and trachyte; and on the E. side, at an elevation of about 5800 feet, is another crater, now extinct, and similar both in origin and appearance to the Somma of Mount Vesuvius. Klutchevsky, which, in common with six others, continually emits smoke, was during the last century in very violent action, sometimes for a year or two at a time, sending forth vitrified stones, lava, pumice, and water: after having been comparatively quiet for about 46 years, it broke out again during Erman's visit in 1829. It presents a large base, swelling in an elliptic curve, and crowned by four cones: its geological components are trachyte, Labrador felspar, obsidian, and lava, and on its sides are numerous thermal springs of high temperature. Indeed, the general formation of Kamtschatka is of igneous origin, comprising porphyry, jasper, felspar, schist, trachyte, dolomite, &c.; the W. side, however, is composed of Neptunian, secondary, and tertiary rocks, among which may be distinguished various beds of lignites, sandstone, iron-sand, and chalk, in the last of which are found large quantities of yellow amber: fossil shells in great variety have been discovered in all the secondary and tertiary formations of this interesting peninsula. The shape of Kamtschatka precludes the possibility of there being any extensive rivers; and, accordingly, those met with resemble torrents more than rivers, being either nearly dry, or flooded and rapid: the Kamtschatka river, however, is alleged to be capable of admitting vessels of 100 tons about 150 m. up the stream.

The severity of the climate, though considerable, has been greatly exaggerated. The average temperature in the middle of winter is about 10° Réaum.; that of summer is about 7°; but the difference seems greater, owing to the prevalence of raw piercing winds, and thick fogs. Still, if any judgment may be formed from the health of the inhabitants, it cannot be unwholesome; for they are robust and long-lived, and there are few diseases, except small-pox, syphilis, &c., introduced by the Russians, who also corrupted the population by familiarizing them with the use of ardent spirits. (See Dobell's Travels, vol. i. p. 67.) The vegetation is generally considered to be very limited; but the limits are prescribed by man rather than by nature. Rye, barley, potatoes, cabbages, turnips, hemp and flax, with several other plants peculiar to the country, may be raised successfully, with moderate attention; but the people are, with few exceptions, devoted to hunting, able to live on game and dried fish, and extremely loath to engage in the more civilizing, though less exciting pursuit of agriculture, the first attempts at which date no further back than 1810. Among the fruits may be mentioned the raspberry, red currant, whortle-berry, cranberry, a delicious species of strawberry called knajnika, a wild cherry called cheremka, and a kind of apricot or plum. The forest trees comprise the birch, fir, larch, poplar, cedar, willow, and juniper. Pasturage has hitherto been little followed; but the abundance of grass shows that if there was an inclination towards it, the pursuit would be profitable. The animals usually hunted comprise bears, lynxes, sea and river otters, rein-deer, foxes of different colours, sables, beavers, &c.; and the number of skins exported is supposed to average about 38,000 a year, chiefly of foxes and sables. Among the birds, the principal are moor-game of different kinds, and many varieties of water-fowl, the eggs of which, saturated with oil, constitute the chief food of the inhabitants. The fish caught in the rivers comprise many varieties of salmon, some of which are peculiar to the country, all serving most essentially to supply winter food: the sea also abounds with cod, herrings, &c., and seals; walruses and whales furnish oil, exclusively employed for domestic purposes.

The trade of Kamtschatka, owing to the exactions of the Russian governors, who, in consequence of their great distance from Petersburg, or even Tobolsk, have few checks on their own cupidity, is of course extremely limited. Taxes are taken in skins; and the people complain bitterly, that no equitable system of taxation has been authorised by the imperial government. Hence, wholly left to the mercy of individual officers, they justly apprehend the insecurity of property, and want the chief motive for improving the natural resources of the country: labour is confined to the supply of merely temporary necessities, domestic comforts are little known or cared for, and affluence is scarcely ever attained even by the most provident and laborious. Furs and dried fish are exported from Petropaulowsky, chiefly by the Russians and Dutch, who bring in exchange rice, flour, coffee, sugar, brandy, and whiskey.

The natives, comprising the two tribes of the Kamtschatdales and Koriaks, who differ more in mode of life than in physical conformation, are of low stature, but stout and broad in the shoulders, with large heads, flat and broad faces, prominent cheek-bones, thin lips, lank black hair, and eyes deeply sunk in the head. Their features seem to identify them with the Mongolian race to which they are certainly more closely allied than to the Esquimaux, with whom Cochrane and Langsdorff have erroneously classed them. The Kamtschatdales are described by Dobell as being shy and averse to strangers, but at the same time intelligent, and fully capable of improvement, if endeavours were made to instruct them in the arts of civilised life. Honesty, openness of character, and extreme hospitality, are prevailing features among them; yet it has been remarked by more than one traveller, that their morals have been much debased by the introduction of felons from Siberia, and the quartering of Russian troops at Petropaulowsky: drunkenness has since that period been an increasing evil, and now threatens to be as destructive to the Kamtschatdales as to the Indian tribes of N. America. Their employment, when not agricultural, is hunting and fishing. They live in fixed habitations; but their dwellings are low, comfortless, and extremely filthy, sunk in the ground in the winter mouths, and raised on posts during summer, to facilitate the curing of fish, which is hung up on lines to dry. In travelling they use dogs instead of horses. These animals somewhat resemble the English shepherd-dog, are extremely intelligent, and endure an almost incredible degree of labour and privation. They are fed during the winter, when they are principally used, on offal and decayed fish, and in the summer are allowed to roam abroad, and shift for themselves. Few Kamtschatdales have less than six, and some upwards of twenty, the whole number of dogs being estimated at 3000. When used for draught they are harnessed, two and two to a sledge, one particularly well trained being placed in front as leader. The sledge is in the shape of an oblong basket about three feet long, and raised three feet from the ground: the driver usually sits sideways, like a lady or horseback, and urges the dogs by throwing at them a stick, which he afterwards catches with great dexterity. Occasionally parties travel in company; "and then," says Dobell, "the eagerness and impatience of the dogs, and the rivalry of the kyowrskits, or drivers, are worthy to be compared with the exertions of the high-blooded coursers of Newmarket; nor does the management and driving of the dogs require much less skill and attention than are needed in the latter case, to arrive at perfection, and gain the palm of victory." The Koriaks, who inhabit the N. part of the peninsula, a wandering tribe, subsist on the produce of their herds of rein deer, which they also use to draw their sledges. The number of Koriaks is unknown, and they are not included in the estimates of the population.

Kamtschatka was first known to the Russians in 1696, when Vladimir Atlasov invaded the peninsula, and made great part of it tributary to Peter the Great. The conquest was completed in 1706, since which, regular tribute has been paid, in furs, to the governor of Irkutsk. There are four districts, each of which is governed by a toion, or lieutenant, whose business is to preserve peace, enforce the orders of government, and collect the tribute, the quantity of which varies according to the character of the governor, and the favour which particular persons happen to edjoy. The commander of the troops resides at Petropaulowsky, which for some years has been the principal place. Its population, however, does not exceed 700, while that of Nishni-Kamtschatk, the former capital, has scarcely 150 persons. Boicheresk, a small harbour on the W. side of Kamtschatka, has a population of about 200. (*Cochrane's Travels in Siberia*, ii., 27-56; *Lutke's Voyages*, iii., 64-98; *Erman; Reise um die Erde*, i., 415-420; *Dobell's Kamtschatka, &c.*, i., 1-188.)

KANDAHAR. *See* CANDAHAR.

KANNAGHERRY (*Khanagiri*), a town of Hindostan, prov. Bejapoor, formerly the capital of a Hindoo principality, 19 m. N.W. Bijnagur. It is beautifully situated in a valley, enclosed by wooded declivities, and partially encircled by a rivulet. The principal street is very spacious, and at one extremity is a fine pagoda to Krishna, the interior of which is elaborately ornamented with stucco bass-reliefs. Various other temples have been converted into dwelling houses or stables by the Mussulman population; and the vicinity abounds with fragments of Hindoo monuments. (*Hamilton's E. I. Gaz.*)

KANAWHA, river, Va., rises in Ashé co., N. C., and is usually called New river, until its union with Gauley river. It runs N. and N.W., and enters Ohio river 232 miles below Pittsburg, at Point Pleasant. About 100 miles from its mouth are the Great Falls, where the river descends perpendicularly 50 feet. On its bank, 66 miles from its mouth, are Kanawha Salt Works. The river is here 150 yards wide. The salt region extends 15 miles on the river, and salt is manufactured to the amount of 1,500,000 bushels annually, and is capable of being indefinitely increased. The salt water is procured by boring through a formation of rock, from 300 to 500 feet deep, and the water rises in copper or tin tubes, to secure it from mixing with fresh water. After rising to the level of the surface of the river along its margin, it is raised 40 feet to the top of the bank by forcing pumps, moved by steam engines. Bituminous coal is abundant in the vicinity, and is employed to evaporate the water. These works employ in various ways about 1000 men, and the salt is afforded from 30 to 35 cents a bushel. Green Briar river, from the N.E., unites with New river, just before it passes the Iron mountain. Gauley river enters it from the E., 80 miles from its mouth, below which it is called Great Kanawha. Below this, Elk river enters it from the N.E., and Coal river from the S. The whole extent of this river, to its remote sources, is not far from 400 miles by the course of the river.

KANAWHA, county, Va. Situated in the W. part of the state, and contains 9000 square miles. Watered by Great Kanawha river and its tributaries, Elk and Cole rivers. It abounds with coal and salt springs. It contained in 1840, 2690 neat cattle, 3810 sheep, 7944 swine; and produced 34,539 bushels of wheat, 203,075 of Indian corn, 52,657 of oats, 7686 of potatoes, 7490 pounds of sugar, 1,000,000 bushels of salt, 6,225,000 of bituminous coal. It had 29 stores, two flouring-mills, 13 grist-mills, 27 saw-mills, nine oil-mills; 19 schools, 406 scholars. Pop., whites, 10,910; slaves, 2500; free coloured, 97; total, 13,567. Capital, Charleston.

KANAWHA, C. H., called also Charleston, p. v., capital of Kanawha co., Va., 313 m. W.N.W. Richmond, 350 W. Situated on the N. bank of Great Kanawha river, 60 miles from its entrance into Ohio river, at the junction of Elk river. Its main street extends a mile on the Kanawha, and reaches to Elk river. It contains a courthouse, jail, a branch of the bank of Virginia, a masonic hall; a female academy, two churches, a Presbyterian and Methodist; 13 stores, two steam saw-mills, one steam flouring-mill, one tannery, and about 130 dwellings. The Kanawha is here 300 yards wide, and 90 feet deep at low water, and is navigable by steamboats.

KANE, county, Ill. Situated on the N.E. part of the state, and contains 1296 square miles. Watered by Fox river and its branches, which afford water-power. Organized in 1836. It contained in 1840, 7838 neat cattle, 1982 sheep, 17,279 swine; and produced 150,110 bushels of wheat, 1263 of rye, 151,310 of Indian corn, 5080 of barley, 167,468 of oats, 77,373 of potatoes, 1400 pounds of sugar. It had 16 stores, six grist-mills, 22 saw-mills, two distilleries; 39 schools, 1153 scholars. Pop. 6501. Capital, Geneva.

KANOJE (*Kanyacubja*), a town of Hindostan, prov. Agra, and, according to Rennell, possibly the an. *Calinipaxa* mentioned by Pliny, about 3 miles from the Ganges, 118 m. E. by S. Agra, and 67 m. W.N.W. Lucknow; lat. 270 4' N., long. 790 47' E. It is now degraded into a mere second-rate town of the district of Etaweh; but it is mentioned by Ferishta as having been once the capital of the principal kingdom along the Ganges, comprising the modern provinces of Delhi, Agra, Oude, and Serinagur. The Indian histories

are full of accounts of its grandeur and extent, and for a distance of 6 miles the traveller now wanders over a tract covered with scattered ruins of brick and other buildings. The most perfect vestige of the ancient Hindoo city is a portion of a small and rude pagoda, its interior adorned with figures of Lakshmi and Rama, surrounded by the Hindoo pantheon in miniature. There are several handsome tombs, mosques, and other Mohammedan edifices in the town, Kanoje having been taken by the Mohammedans under Mahmood of Ghizni, in 1018. Under the Moguls it gave its name to a circar; but it soon lost its importance, and, to complete its ruin, it was sacked by the Mahrattas in 1761. The modern Kanoje consists of only a single street, and presents nothing worthy of note, except a citadel, close to which is the termination of a canal communicating with the Ganges. (*Rennell, Memoir, &c.; Lord Valentia; Hamilton's E. I. Gaz.*)

KANSAB, river, Indian territory, rises near the Rocky mountains, crosses the Indian territory, and enters Missouri river on the W. line of Mo. Its whole length is about 1200 m. 900 of which is navigable. It is 340 yds. wide at its mouth.

KARA-HISSAR. *See* AFIUM KARA-HISSAR.

KARAK, or KHARRACK (the *Icarus* of Arrian), an island of the Persian gulf, now belonging to the British, lat. 290 13' N., long. 500 21' E., 35 m. N.W. Bushire. It has an area of 12 or 13 square miles, with a population of about 300 or 400. "It affords a safe anchorage at all seasons, but more particularly during the severe gales which blow from the N.W., and are the prevailing winds in this sea. The greater part of the island is so rocky, that little use can be made of it; but the E. side, being somewhat lower than the other parts, is capable of being cultivated. It has abundance of water. The inhabitants gain a livelihood by gardening and fishing, and manufacture a small quantity of common cloth for their own consumption. The island of Corgo, lying about 1½ miles or 2 miles N. Karak, contains about 2 square miles, and is of a light sandy soil. It has also plenty of water, but not of so good a quality as that of Karak; and although not inhabited at present, it is capable of being cultivated, and will produce both wheat and barley during the rainy seasons." (*Kinneir's Pers. Empire*, p. 18, 19.) Pearls of a superior colour and description are fished around the coasts of both islands. The Dutch, after having been obliged to abandon their factory at Bussorah, founded an establishment at Karak in 1748. They were, however, driven from it by the Arabs, about 1765. Karak was subsequently occupied by the Persians; and in 1807, for a short period, by the French. During our recent disagreement with the shah of Persia, the British resident, previously stationed at Bushire, removed thither; and the island was taken possession of by an English force in 1839. Its acquisition will give us the complete command of the Persian gulf, and will be also serviceable from its affording a secure anchorage for our ships, and a station where they may water and refit. (*Kinneir's Pers. Empire; Asiat. Journal.*)

KARAMAN, a town of Asiatic Turkey in Karamania, 58 m. S.S.E. Konieh; lat. 370 10' N., long. 330 5' E. Pop. 15,000? It stands at the E. extremity of a large plain, and at the foot of the lofty range of Bediorin-dagh, a branch of mount Taurus: it covers with its squares and gardens a large area; the houses are of mud and sun-dried bricks, and have a mean, wretched appearance; but the climate is salubrious, and water abundant. The public buildings, comprise four mosques, with the ruins of others, numerous khans and hummums, and a castle on a height, now mouldering to decay. Karaman trades with Kaisariah, Smyrna, and Tarsus, in cotton fabrics, hides, and nut-galls; and it has a pretty extensive manufacture of blue cotton cloth, worn by the lower classes.

Karaman, which occupies the site of the ancient *Laranda*, is said to have been founded by Karaman Ogba, a powerful prince living in the 14th century. It was the capital of a Turkish kingdom, which lasted from the time of the partition of the Seljuck dominions of Iconium till 1486, when Karamania was subjected by the Ottoman emperor Bajazet II. Konieh then became the seat of the pachalic, and from that period Karaman has been gradually falling into decay. (*Kinneir's Asia M.*, p. 211; *Leake's Tour*, p. 99.)

KARAMANIA. *See* TURKEY IN ASIA.

KARASUBASAR, a town of European Russia, Crimea, 15 m. E. Simpheropol, inhabited by Tartars, Greeks, Russians, Jews, Armenians, &c. Pop., according to the official returns, nearly 11,000, which, if they may be depended upon, show a great increase within the last dozen years. Streets narrow, winding, and dirty. There are several graceful looking mosques, a new Roman Catholic church, a large building, or khan, occupied by shops, &c. It is celebrated for the manufacture of a very superior sort of red and yellow morocco leather, and it contains several tanneries, candle and soap works, potteries, and tile-works. It is also the great mart of the Crimea for fruit, wine, and cattle. There is a weekly market, and a great annual fair. (*Schnitzler, La Russie, &c.*, p. 726; *Lyell*, i., 356.)

98

KARLSBURG. See CARLSBURG.

KARS, a town of Turkish Armenia, formerly capital of a pachalic of the same name, on the Arpah-Chai, a tributary of the Aras or Araxes, 85 m. N.E. Erzeroum, and 100 m. E. by S. Trebizond; lat. 40° 25′ N., long. 41° 10′ E. Pop., in 1835, not exceeding 2000 families. It is situated on the N. side of a plain, which, though about 4000 feet high, is extremely fertile; a part of it is walled, and there is a citadel, which, however, is commanded by heights within musket-shot on the other side the river. Two stone bridges unite the two portions of the city divided by the river, which encircles the walled portion on three sides. The houses of the citadel are tolerably large and well built, but those in the town below are of the underground architecture usual in the Armenian villages. The public buildings comprise several mosques, and one Armenian church outside the walls: the Armenian convent is uninhabited and in ruins. Kars being the centre of a fine corn-growing district, had formerly a considerable trade in farming produce; but it was nearly destroyed during the Russian invasion, and is only slowly recovering.

Kars, the origin of which is doubtful (*Tournefort*, ii., p. 295), was formerly a large town, with a population of nearly 6000 families; but it is now little better than a heap of ruins. During the Russian occupation, a large part of the Turkish population abandoned it, while at the same time the Armenians emigrated with the retreating army of the Russians, leaving many deserted villages and much unoccupied land. The present inhabitants are about half Turkish and half Armenian; the former being described as turbulent and impatient of subordination to the pacha of Erzeroum, under whom it is now a sandjiak, and the residence of a mutzellim. (*Smith and Dwight's Miss. Researches*, p. 21; *Geog. Journ.*, vi., 190.)

KASAN, one of the eastern governments of Russia in Europe, having N. Viatka, E. Orenburg, S. Simbirsk, and W. Nijegorod. Area, 24,000 sq. m. Pop. 1,200,000, partly Russians and partly Tchouvaches, of Finnish origin, Tartars, &c. It is traversed for a considerable distance by the Wolga, the Kama, one of the principal affluents of the latter, and by some lesser streams, and is interspersed with numerous lakes. Surface generally flat, but in parts undulating and hilly; soil almost everywhere fertile, producing with very imperfect culture, abundant crops of rye, wheat, hemp and flax, &c. Forests very extensive, covering nearly half the surface. Climate in winter very severe; but the summer, though short, is generally fine. Grazing is not well understood, and but little attention is given to the rearing of cattle. The fishery in the Kama is very productive. There are numerous distilleries, saw-mills, and potashworks, with tanneries, &c. More than half the government belongs to the crown, which, in 1816, had 336,661 peasants. The public revenue in 1827 amounted to 5,905,314 roubles, of which 4,699,342 consisted of the obrok or rent paid by the peasants belonging to the crown. (*Schnitzler*, *La Russie*, &c., p. 665, &c.)

KASAN, a city of European Russia, cap. of the above government, on the Kasanka, about 4 m. above where it falls into the Wolga; lat. 55° 47′ 36″ N., long. 49° 21′ 9″ E. Pop., in 1833, 57,000. After being burnt down by Pougatcheff in 1774, Kasan was rebuilt, by order of Catherine II., on a more regular plan. It was again the prey of an accidental conflagration in September, 1815, by which it was more than half destroyed; but, like Moscow, it has risen from its ashes larger and better built than ever. It stands on very uneven ground, interspersed with lakes, and consists, like most other Russian cities, of three parts: the kremlin or citadel, on a considerable eminence; the town, property so called; and the slobodes, or suburbs. The town is well built, and has broad and spacious squares and market-places. In the suburbs, which are principally occupied by the Tartar population, the houses are of wood, and the streets disgustingly filthy. Principal buildings, the grand cathedral, founded in 1552; the cathedrals of St. Peter and St. Paul, with several other cathedrals and churches, some of them built in the course of the present century. The convent of Bogoroditskoi Kasanskoi, rebuilt by the emperor Alexander; the hotel of the general governor; the archiepiscopal palace; the hotel of the nobles; the bazaar; the military hospital; the arsenal, &c. Kasan is one of the most literary towns in Russia. It has a university, founded in 1804, but which was not opened till 1814. It had, in 1835, 79 principal and subordinate professors, 236 pupils, and a library of above 26,500 volumes. Its principal object is to supply instruction in the eastern languages, or in Arabic, Persian, Turkish, Tartar, and Mongul; within the last three or four years a professorship has been established for giving instruction in the Chinese language and literature. The city has also one of the four great theological academies, with a gymnasium, an observatory, a grammar-school, a Tartar school, a school for the instruction of school-masters, &c.; and several journals and publications issue from its press, among which are comprised some works in the

Turkish language. A great cloth manufactory, established by Peter the Great, is now the property of individuals, and employs about 1000 work-people; and there are, besides, manufactories of cottons, hardware, earthenware, and tiles, with tanneries, soap-works, distilleries, &c. Kasan is the seat of an admiralty; and vessels are constructed for the navigation of the Wolga and the Caspian. It also carries on an extensive trade, for which its situation adjoining the Wolga gives it peculiar facilities. About 15,000 of the population are Mohammedans. The rest, with the exception of a few Protestants, belong to the established Greek church. (*Schnitzler*, *La Russie*, &c., p. 671, &c.)

KASCHAU, a royal free city of Hungary, in the circ. on this side the Theiss, co. Abaujvar, on the Hernad, 123 m. N.E. Pesth. Pop. 13,600. It is well built, with fine squares, and regularly laid out streets; and has 13 Roman Catholic and two Lutheran churches, besides a theatre, and several other handsome public buildings. The chief public establishments are a royal academy, with a library of 10,000 volumes, and a fine collection of natural history; a gymnasium, an episcopal seminary, a school for nobles (adeliges Konvikt), and a military asylum: it is the seat of a county-assembly and court of justice; and has manufactories of tobacco, cutlery, earthenware, paper, &c.; and a large transit trade with Poland. (*Oesterr. Nat. Encyc.*; *Bergkaus*.)

KASKASKIA, r., Ill., rises in Champaign co., and, after a course, by its windings, of 400 m., enters Mississippi river, 120 m. above the mouth of Ohio r. It has been navigated by a steamboat to Carlyle.

KASKASKIA, p. v., capital of Randolph co., Ill., 142 m. S. Springfield, 834 W. Situated on the W. side of Kaskaskia r., 7 m. above its entrance into Mississippi r. Settled by the French in 1683, and, when ceded to Great Britain in 1763, contained 100 families. It contains a brick courthouse, a jail, a U. States land-office, a Roman Catholic church, a nunnery, a female boarding-school, four stores, 100 dwellings, and about 500 inhabitants, chiefly of French descent.

KATADIN, mt., Me., between the E. and W. branches of Penobscot r., has an isolated summit of 5300 feet above tide water. The view from its summit is extensive and grand.

KATRINE (LOCH), a lake of Scotland, in the district of Monteith, in the S.W. part of Perthshire, on the confines of Stirlingshire, 8 m. W. Callander, and 5 m. E. from loch Lomond. This, which is the most westerly and largest of a chain of lochs, consisting of lochs Venacher, Achray, and Katrine, the principal feeders of the Teith, is about 10½ m. in length, and from 1½ to 2 m. in width, of a serpentine form, and very deep. It is imbosomed among lofty mountains, divided by deep ravines, whose sides, in parts clothed with wood down to the water's edge, and in parts consisting of bold, rugged precipices, give it every variety of wild, picturesque scenery. Still, however, it was but seldom visited, and little known, till Scott made it the scene of his fine poem of *The Lady of the Lake*, when it at once attained the maximum of celebrity, and has since been annually resorted to by crowds of visiters. At the E. end of the loch, between it and loch Achray, is the celebrated pass of the Trosachs, so beautifully described in stanzas 11–13 of the first canto of *The Lady of the Lake*.

KAZAMEEN, a town of Asiatic Turkey, prov. Irak-Arabi, on the W. bank of the Tigris, 3 m. N. Bagdad. Pop. 8000, chiefly Persians, who have been induced to settle here on account of its being the burying-place of two celebrated imams, to whose memory a noble mosque has been erected. It is ornamented with two gilded cupolas, and, like those of Meshed Ali and Kerbela, is supported by the contributions of pilgrims. The town has a decent bazaar, many coffee-houses, three hummums, and a caravanserai; and on the opposite side of the river is the tomb of Imam Abu Hanafi, another Mohammedan saint. (*Kinneir*.)

KEATING, t., M'Kean co., Pa. Watered by Alleghany r. and its tributaries. It has one academy, 30 students, four schools, 135 scholars. Pop. 893.

KEDGEREE, a town of British India, prov. Bengal, on the W. side of the Hooghly river, near its mouth, lat. 21° 55′ N., long. 88° 16′ E. It stands in a low, swampy situation; but, according to Hamilton, it is, notwithstanding, much healthier than Diamond Harbour, and ships of war, unless compelled by strong reasons, should never go higher up the river. A lighthouse has been erected a few miles farther down, and of late years one has been established at Kedgeree, the charge for which, on ships sailing under British or American flags, is 3d. per ton per annum. The charge for pilotage to Kedgeree is half the full pilotage from the sea to Calcutta. A government marine officer is stationed at this town, who makes daily reports of the ships which arrive and sail. (*Hamilton's E. I. Gaz.*; *Commercial Dict.*)

KEDJE, a town of Beloochistan, prov. Mukran, of which it is the cap., on a rivulet, by which the surrounding district is well irrigated, 274 m. S.W. Khelat; lat. 29° 34′ N., long. 69° 28′ E. Population unknown, but it is said to have once contained 3000 houses. It stands clustered around the base of a precipice, on which is a fortress; and was formerly a place of considerable trade, which having declined, the town has fallen into decay. (*Pottinger's Baloochistan*, p. 304.)

KEENE, p. t, with Charlestown, the capital of Cheshire co., N. H., 80 m. W.N.W. Boston, 48 m. W.S.W. Concord; 494 W. Watered by Ashuelot r. It contains five churches (a Congregational, two Methodist, a Baptist, and a Universalist), 25 stores, one furnace, one woollen factory, two glass factories, three grist-mills, seven saw-mills, one oil-mill, two tanneries; two academies, 261 students, 13 schools, 695 scholars. Pop. 2616. The village is beautifully situated on a plain, a little E. of Ashuelot river. The principal street is a mile long, of ample width, and ornamented with trees. It contains a courthouse, a jail, a bank, and a printing-office, issuing a weekly newspaper. A canal from Ashuelot river affords water-power. It is regarded by travellers as one of the pleasantest villages in New England.

KEENE, p. t., Essex co., N. Y., 138 m. N. Albany, 513 W. It contains Mt. Marcy, the highest summit of the Adirondack mountains, 5467 feet above tide water in Hudson river. Drained by branches of Ausable r., which afford extensive water-power. It contains one store, one fulling-mill, one forge, three grist-mills, two saw-mills, one oil-mill, two tanneries; four schools, 185 scholars. Pop. 730.

KEENE, p. t., Coshocton co., O., 99 m. N.E. by E. Columbus, 344 W. It contains two churches, three stores, one grist mill, two saw-mills; one school, 56 scholars. Pop. 1043.

KEESVILLE, p. v., Ausable and Chesterfield ts., Clinton and Essex cos., N. Y. Situated on both sides of Ausable r., which divides the counties. The river affords extensive water-power. It contains four churches (a Congregational, Methodist, Baptist, and Roman Catholic), an academy, a bank, 12 stores, one woollen factory, two flouring-mills, four saw-mills, one forge, one rolling-mill, one extensive nail factory, producing 1000 tons annually, a printing-office, 250 dwellings, and about 1800 inhabitants. Its growth has been very rapid. Its trade in iron, nails, and lumber is very extensive.

KEHL, a town of Baden, circ. Middle Rhine, on the Rhine, immediately opposite Strasbourg, and 10 m. N.W. Offenburg. Population about 1000, or, with its immediate environs, nearly double that number. It was formerly a fortress, and was esteemed an important bulwark of Germany. It was fortified by Vauban in 1688, ceded by France to Baden in 1697, taken by the French in 1703, 1733, 1793, and 1796; by the Austrians, also, in the latter year; and retaken by the French in the succeeding. After the peace, its works were dismantled, Germersheim being fortified by the German Confederation in its stead. The town is connected by a bridge of boats with the opposite bank of the Rhine, near Strasbourg. Its inhabitants employ themselves chiefly in transit trade. (*Berghaus; Schreiber; Dict. Géog.*)

KEIGHLEY, or KIGHLEY, a market-town and parish of England, in the W. riding of co. York, wap. Staincliff and Ewcross, on an affluent of the Aire, 16 m. W.N.W. Leeds, and 178 m. N.N.W. London; area of parish, 10,160 acres. Pop., in 1831, 11,176, being an increase of 92 per cent. on that of 1811. The town is beautifully situated in a valley close to the range called the Blackstone Edge : and, though irregularly built, comprises many handsome stone houses : it is well paved, sufficiently supplied with water, and lighted with gas. A neat and commodious courthouse, and a spacious market-place, were erected in 1833, and more recently a Mechanics' Institute has been built on ground given by the Earl of Burlington, who has large possessions in and near the town. The church was built in 1805, on the site of one erected in the reign of Henry I., and is a large and handsome structure, with a lofty steeple, containing a fine peal of bells : the living is a rectory in the gift of the Duke of Devonshire. There are also places of worship for Independents, Baptists, Wesleyan new connexion, and Primitive Methodists, Swedenborgians, and the Society of Friends; and to all of these, as well as to the church, large Sunday schools are attached, furnishing religious instruction to about 1600 children of both sexes. A free grammar-school, founded and well endowed in 1713, a girls' national school, and an infant school, are the chief educational establishments; and a mechanics' Institute, founded in 1825, and now in union with that at Leeds, has conferred many benefits on the working classes. The worsted manufacture, especially of coarse stuffs, merinos, and worsted yarns, is carried on to a considerable extent; and the produce is sold in the piece-halls of Halifax and Bradford : 38 worsted-mills gave, in 1836, employment to 2125 hands, and five cotton-mills to 198 hands; about 1800 looms are at work within the parish. The Leeds and Liverpool canal, which passes near the

town, affords a cheap conveyance for manufactures, &c., and establishes a communication with Hull on the one hand, and Liverpool on the other. A court of requests is held here for the recovery of debts under 40s.; and under the Boundary Act Keighley is a polling-place for the W. riding. It is also the head of a union, comprising six parishes; the expense of maintaining the poor of this parish having amounted to £1538 in 1839. Markets, well supplied, on Wednesday : fairs, May 8th and 9th, and November 7th, 8th, and 9th.

Keighley is known in the history of the great civil war as having been the scene of an encounter, in 1645, between the king's troops, and a division of the parliamentary army, under Colonel Lambert. Its name is derived from an old family called Keighley, one of whose members married a Lord Cavendish, from whom the present Duke of Devonshire and the Earl of Burlington are descended. (*Baines's Gazetteer of Yorkshire; Parl. Rep.*)

KELLS, a town of Ireland, prov. Leinster, co. Meath, adjacent to the Blackwater, on the top and sides of a gentle hill, 35 m. N.W. Dublin, and 21 m. W. Drogheda. Population, in 1831, 4396, since which it has not increased. It consists of three principal and some smaller streets, and has some good houses ; but, generally speaking, it is a poor, mean place, and is neither lighted nor watched. Here is a fine old church, contiguous to which is a pillar or round tower 99 feet in height. It has also a Roman Catholic chapel, a courthouse, market-house, bridewell, fever hospital, a national school, and a school supported by Lady Headfort. A lace manufactory is said, in the *Municipal Boundary Report*, to employ 100 hands; and there is also an extensive brewery.

This is a very old town ; a synod having been held here in 1152, and a castle erected on the site of the present market-place in 1178. Here also was a monastery, some remains of which still exist, and are called St. Columb Kill's House, from the name of its reputed founder. In one of the streets is a fine stone cross. The borough returned two members to the Irish H. of C., but was disfranchised at the union. The magnificent seat of the Headfort family is in its vicinity. Postoffice revenue, in 1830, £878; in 1838, £715.

KEITH, a bor. of barony and market-town of Scotland, co. Banff, on both sides the Isla, a tributary of the Deveron, 41½ m. N.W. Aberdeen. Pop., in 1801, inc. the par., 2894; in 1831, 4464. Keith is, properly speaking, composed of three towns, namely, Old Keith and New Keith, on the S. of the river, and Fife Keith, on the N., the whole lying in the centre of an amphitheatre of hills. The first, which is very old, is but of mean appearance and irregular shape ; the second, begun to be erected in 1750, stands on a gentle eminence to the S.E. of the former, and consists of one principal street divided into several portions ; the third, or Fife Keith, which had its origin in 1816, is connected with the two former towns by two bridges over the Isla. New Keith is the largest and best built of the three divisions in question, and in it most part of the public buildings are situated, such as the parish church, a Gothic building, with a tower 104 feet high, and a Roman Catholic chapel, after the plan of St. Maria de Vittoria at Rome. It has also an Episcopal chapel, and two meeting-houses, belonging to the Associate Synod. The means of education are very ample. There are four subscription libraries. A considerable number of persons are employed in weaving woollen and linen cloth for the Aberdeen manufacturers; and it has also about 35 "customer weavers" employed by private persons for articles of local consumption. There are three branch banks in the borough. In addition to weekly markets, Keith has four annual fairs, all of considerable importance : Summer-eve Fair, held in September, is by far the greatest fair in the north for cattle and horses.

A skirmish took place in 1745 at Old Keith, between the forces of the Pretender and those in the royal service, in which the former had the advantage, and carried off 15 prisoners. James Ferguson, the celebrated self-taught astronomer, was born in the vicinity of Keith : the only school he ever attended was one at Keith, and that for only three months. He died in 1776. His "Autobiography" is well known. (*Beauties of Scot.*, vol. iv., §*Banffshire; Chambers's Gaz.*)

KELLY, t., Union co., Pa., 9 m. N. New Berlin. Drained by Buffalo creek and its branches. It has one store, one flouring-mill, one saw-mill ; five schools, 200 scholars. Pop. 788.

KELLY, t., Ottawa co., O. It consists of Cunningham's island in lake Erie. Pop. 66.

KELSO, an inland market-town of Scotland, co. Roxburgh, beautifully situated on the left bank of the Tweed, near the point where it is joined by the Teviot, 38 m. S.E. Edinburgh, and 20¼ m. S.W. Berwick-upon-Tweed. Pop. in 1801, 3298 ; in 1831, 4200 : of the town and parish at the latter date, 5114. The town, which is peculiarly neat and handsome, consists of four principal and some smaller streets. The former meet in a square or market-place in the centre of the town, consisting of well-built houses, which, like those in other parts, are mostly of freestone and slated. On the E. side of this square is the townhouse, an edifice of two stories, with a pediment in front supported by four Ionic columns, surmounted by a handsome balustrade, and dome springing from the centre of the roof. The old parish church being a "misshapen pile," a new or second parish church was built here in 1837 in the Elizabethan style, with a quadrangular tower 70 feet high. The bridge across the Tweed, from a plan of the late Mr. Rennie, is said to have been the prototype of Waterloo bridge over the Thames by the same architect. It has five elliptical arches: its total length is 494 feet; the breadth of the roadway is 25 feet; and the greatest height from the bed of the river, 43 feet. It was finished in 1803, at an expense of £18,000. In the immediate vicinity of the town, on the W., is Fleurs, the seat of the ducal family of Roxburgh, the feudal superiors of the borough; a mansion erected in 1718, but recently repaired and modernized, combining, as Sir W. Scott has observed, "the ideas of ancient baronial grandeur with those of modern taste." But the most prominent object is or around Kelso is its venerable abbey, founded in 1128 by David I. for Tyronensian monks, and endowed with immense possessions and privileges. Its form is that of a Latin cross, and it affords a fine specimen of the Saxon or early Norman style of architecture. It has long been in a state of dilapidation; but the Scotch reformers are guiltless of the demolition of this noble fabric: for, having been occupied as a place of security by the townspeople in 1545, it was then battered down by the English under the Earl of Hertford. The parts now remaining are the N. and S. aisles, each having two round towers, with two sides of the central tower, now only 91 feet high. The thickness of the lower walls is 5½ feet. The pillars are clustered; the arches circular. Part of the ruin served as the parish church from 1649 till 1771, when it was deserted, from the idea of insecurity, for another place of worship. The Roxburgh family have of late laudably exerted themselves to repair and perpetuate this fine ruin. Kelso has been characterized by Scott, in his "Autobiography" (p. 39), as "the most beautiful, if not the most romantic, village in Scotland." "It presents objects," he says, "not only grand in themselves, but venerable from their associations." The best view of the town and environs is from the bridge.

In addition to the old and new parish churches previously noticed, there are five other places of worship in the town, belonging respectively to the Episcopalians, Cameronians, Original Seceders, Relief and Associate Synod.

There are 10 schools in the borough and parish, attended by about 780 scholars: so that about a seventh part of the people are, at the same time, being educated; and this without including Sunday schools, of which there are six. Kelso has six subscription libraries; the oldest, containing about 5000 volumes, having been instituted in 1750. The "Kelso Physical and Antiquarian Society" would do credit to a much larger town. There are two reading-rooms; two newspapers, one published weekly, the other twice a week. Kelso was the first provincial town in Scotland that introduced the printing-press. (Irving's Scot. Poets, i., 75.) The first edition of Scott's Minstrelsy of the Scottish Border was printed in Kelso by James Ballantyne, who afterward brought the typographical art to high perfection in Edinburgh, where he carried on the printing business in partnership with Scott.

A dispensary was founded here in 1777. Poor-rates were introduced in 1795; and yield, with other sources of income, a sum of about £1300 a year. About 40 children are educated at the expense of the parish.

The carrying of leather, and the manufacture of woollen cloths, linen, stockings, and hats, which are the chief branches of industry, do not together employ more than 150 hands, and some of these branches are disappearing. The town, which is chiefly dependant on its retail trade, is remarkable for its numerous handsome shops. It has a weekly corn-market, at which a great deal of business is transacted; and several annual fairs for cattle and sheep. There are four banks in the town, exclusive of savings' bank.

Kelso was originally a species of suburb to the borough of Roxburgh, on the opposite bank of the Tweed. But the foundation of the abbey gave Kelso a more important character: and on the final destruction of Roxburgh, in the 15th century, its inhabitants transferred themselves thither. No traces now remain of the borough of Roxburgh, and but few of its castle; though the latter was for centuries one of the most important border fortresses. In 1460 James II., having taken the town of Roxburgh and demolished it, laid siege to the castle, during which he was killed by the bursting of a cannon. The queen, attended by her infant son, James III., encouraged the besiegers, and, in a few days, the fortress was compelled to surrender. It was then de-

stroyed; since which time it has remained in ruins, though partially repaired by the English, under Somerset, in 1547. Soon after the Reformation the lands and possessions of the abbey were conferred on the ancient family of Kerr, of Cessford, in the hands of whose descendants, the family of Roxburgh, they still remain. Kelso has repeatedly suffered from conflagrations, not in warlike times merely, but in pacific, as in 1686 and 1738. (Morton's Monastic Annals of Teviotdale; Haig's Hist. of Kelso; Tennant's Tour; New Stat. Acc. of Scotland; Roxburghshire, p. 298.)

KELSO, p. t., Dearborn co., Ia., 84 m. S.E. Indianapolis, 531 W. Pop. 1458.

KEMPER, county, Miss. Situated in the E. part of the state, and containing 750 sq. m. Drained by head branches of Sucarnochee river. It contained in 1840, 12,251 neat cattle, 1999 sheep, 20,814 swine; and produced 7885 bushels of wheat, 238,017 of Indian corn, 10,441 of oats, 22,907 of potatoes, 1646 pounds of tobacco, 3,996,565 of cotton. It had 16 stores, eight cotton factories, with 48 spindles, three flouring-mills, 13 grist-mills, 13 saw-mills; four academies, 133 students, 11 schools, 285 scholars. Pop.: whites, 4612; slaves, 3040; free coloured, 11; total, 7663. Capital, De Kalb.

KEMPSVILLE, p. v., Princess Ann co., Va., 10 m. S.E. by E. Norfolk, 116 m. S.E. by E. Richmond, 240 W. Situated at the head of tidewater on the E. branch of Elizabeth river. It contains a Baptist church, several stores, 30 dwellings, and about 200 inhabitants.

KEMPTEN (an. Campodunum), a town of Bavaria, circ. Swabia and Neuberg, cap. distr. of same name, on the Iller 50 m. S.S.W. Augsburg. Pop. about 8000. It consists of two parts, an old town surrounded with walls nearly encircled by the new town. The former is the commercial portion of Kempten; the latter, seated on higher ground, comprises the abbey, where was formerly held the court of the abbot of Kempten, an ecclesiastic possessing, besides the town, an independent territory of 340 sq. m., ceded to Bavaria in 1802. Kempten has a fine collegiate church, aqueduct, and theatre, a hospital, foundling asylum, public library, &c.; and is the seat of the council for the circle, courts of law for the circle and town, a board of tolls, a gymnasium, and high-school. It has manufactures of linen and cotton fabrics, and a brisk trade in these goods, and in wool, cattle, and Italian produce. The Iller becomes navigable near Kempten. Adjacent to the town is the eminence of Hilarmont, on which are the ruins of a fortress supposed to be Roman, and where various Roman coins have been found. (Stein; Berghaus, &c.)

KENANSVILLE, capital of Duplin co., N.C. Situated on the S. side of a branch of N.E. Cape Fear river. It contains a courthouse, jail, and several stores and dwellings. Nett proceeds of the postoffice, $112.

KENDAL (KIRBY), a market town, parl. bor. and par. of England, co. Westmoreland, ward of same name, 40 m. S. C.7lsle, and 219 m. N.N.W. London. Pop. of parl. bor. (which comprises the townships of Kendal, Kirkland, and Nethergraveship), in 1831, 11,577. The town, on the side of a hill, at the bottom of which the river Kent (crossed here by three bridges), runs nearly N. and S., consists principally of one long street on the line of the Carlisle road, and a lateral street leading down to the river on the Appleby road. The houses are well built of stone, and being whitened, and roofed with blue slates, have a remarkably clean and neat appearance. The town-hall is an elegant building, and the market for butchers' meat is neat and commodious. At the N.W. end of the town is a large and well arranged workhouse, and near it is a house of correction. The other principal buildings are a handsome hall, belonging to the society of "odd fellows," the assembly and news rooms, theatre, and several extensive factories. The church near the S. entrance of the town, is a large Gothic structure, with a square tower: the living is a vicarage, in the gift of Trinity College, Cambridge. There are two other episcopal places of worship, and 16 belonging to different denominations of dissenters. Among the educational establishments, are a well endowed grammar-school, with university exhibitions, a blue-coat charity, a green-coat school, a large national school, supported both by endowment and subscription, a school of industry, an infant school, and several Sunday schools. There is also a thriving mechanics' institute. The charitable institutions comprise a hospital for old unmarried women, endowed with £100 a year, a dispensary, and a lying-in charity; and the corporation have the trust of charitable funds to a considerable amount.

Kendal has long been noted for its weaving industry; and in the reigns of Richard II. and Henry IV., special laws were enacted for the protection of its manufactures. The present manufactures comprise linseys, serges, baizes, the coarser kinds of kerseymere, and carpets. There are 12 woollen-mills, which employed, in 1838, 942 hands; and about 3000 persons are employed in weaving, and otherwise preparing cloth. The wages of adult weavers (good hands) average 10s. a week, when fully employed; but the trade has been lately much depressed, and great distress has consequently prevailed among the working classes, who, in 1836-39, were half their time without employment. The marble works, for cutting and polishing marble, quarried at Kendal Fell, employ several hands; and the machinery is very ingenious. There is water communication by a canal with Lancaster, and a railway has been projected to connect Kendal with Carlisle northward, and with the N. Union, and the other great lines of England. A joint-stock bank, and two private banking establishments, furnish ample accommodation to the manufacturers, and a savings' bank has a large number of depositors. Two newspapers of opposite politics, the "Westmoreland Gazette," and "Westmoreland Advertiser," are published every Saturday, and are said to be well conducted, and pretty extensively circulated.

Kendal was first incorporated by Queen Elizabeth, and a second charter was granted by Charles I. Under the Municipal Reform Act, it is divided into three wards, the municipal officers being a recorder, a mayor, and five other aldermen, with 18 councillors. Corporation revenue, in 1839, £1448. The Reform Act conferred on Kendal the important privilege of sending one member to the House of Commons: the electoral boundaries include the townships of Kendal and Kirkland, with those parts of Nethergraveship which adjoin Kendal. Registered electors, in 1839-40, 351. Markets well attended, on Saturday: cattle fairs, March 22d, April 29th, and November 8th.

Near Kendal, on the opposite side of the river, are the ruins of a castle, commandingly situated on a rocky eminence, and celebrated as the birthplace of Catherine Parr, one of the queens of Henry VIII. A large portion of the outer wall and two towers still remain to mark out its former extent.

KENDALL, county Ill. Situated in the N.E. part of the state, and contains 394 sq. m. Watered by Fox river. Organized since the census of 1840. Capital, Yorkville.

KENDALL, p. t., Orleans co., N.Y., 249 m. W. by N. Albany, 306 W. Bounded N. by lake Ontario. It has two stores, one grist-mill, three saw-mills; 14 schools, 664 scholars. Pop. 1692.

KENILWORTH, a market town and par. of England, co. Warwick, hundred Knighthon, 5 m. N. Warwick, 19 m. S.E. Birmingham, and 96 m. N.N.W. London. Area of par. 6460 acres: pop., in 1831, 3097. It is delightfully situated on an affluent of the Avon, and consists chiefly of one long street, about 1 m. in length, part of the road from Warwick to Coventry. In the lower part of the town is the church, a Gothic building of different periods, having a handsome tower and spire: and near it are the ruins of an abbey, valued at the dissolution of the monasteries at £644. On the higher ground are several handsome houses; and at the top of the hill on which the town stands are the ruins of a castle, the ancient fame of which has been made familiar to all Europe by the Magician of the North. There are several places of worship for dissenters, to each of which, as well as to the church, are attached well attended Sunday-schools. A free-school was founded in 1794, and there is a large national school. Among other charities are almshouses for 16 widows, and an apprentice fund. Ribands, gauzes, and combs are made here; and there are chemical works for the preparation of Glauber salts, salammoniac, and Prussian blue; but they are not important. Markets on Wednesday; horse and cattle fairs, April 30, and September 30.

Kenilworth Castle, whose extensive ruins bear ample testimony to its ancient splendour and magnificence, was erected in 1120 by Geoffry de Clinton, treasurer and chamberlain to Henry I.: and in the reign of Edward I. the Earl of Leicester held a tournament here, which was attended by 100 knights with their ladies. The estate afterward reverted to the crown, and was given by Queen Elizabeth to her unworthy favourite, Dudley, Earl of Leicester, who is said to have expended on its improvement £60,000, a vast sum for those days. "The outer wall," says Sir W. Scott, "inclosed seven acres, a part of which was occupied by extensive stables and by a pleasure-garden, with its trim arbours and parterres; and the rest formed the large base-court or outer yard of the noble castle, which was itself composed of a huge pile of castellated buildings surrounding an inner court. A large and massive keep, called Cæsar's Tower, was of uncertain though great antiquity: and that noble and massive pile, which yet bears the name of Lancaster's Buildings, was erected by John of Gaunt, 'time honoured Lancaster.' The external wall was on the S. and W. sides adorned and defended by a lake partly artificial, across which was a stately bridge, and on the N. side was a barbican, which, even in its present ruinous state, is equal in extent and superior in architecture to the baronial castle of many a northern chief. Beyond the lake lay an extensive

chase, full of deer and game, and abounding with lofty trees. Queen Elizabeth twice visited this noble palace; and here, in 1575, she was entertained, with her whole court, with princely magnificence during 17 days, at the enormous expense of £1000 per diem. The castle was plundered and ultimately left in a state of ruin by Cromwell's soldiers, who appropriated to themselves the adjacent lands. After various changes, the estate came into the possession of Hyde, Earl of Clarendon, and is still held by that noble family." (Sir W. Scott's "Kenilworth," vol. ii., with notes; Bingley's Beauties of England and Wales, &c.)

KENNEBEC, river, Me., next to the Penobscot, the most important river in the state. It has its principal source in the outlet of Moosehead lake; but 20 m. below it receives Dead river, the longer branch, which rises within 5 m. of the Chaudiere, the latter flowing into the St. Lawrence. Its length is about 200 m. Its largest tributary, Androscoggin, enters it from the W., 18 m. from the ocean. It is navigable for large ships 12 m. to Bath, for sloops of 150 tons, 40 m. to Hallowell, and for small sloops 2 m. farther, to Augusta, the head of tidewater, and for boats to Waterville, 18 m., above Augusta. It has falls of about 10 ft. high at Norridgewock, and rapids for 2 m. below, and falls at three other places, affording extensive water-power. Bridges cross it at Augusta, at Canaan, and at Norridgewock. The navigation is closed for four months in the year at Hallowell; but below Bath it is open at all seasons. The most important towns on the river are Bath, Gardiner, Hallowell, Augusta, Waterville, and Norridgewock. The country on its borders is fertile, and it is the medium of an extensive trade.

KENNEBEC, county, Me. Situated in the central part of the state, and contains 1050 sq. m. Kennebec river runs centrally through it, from N. to S., and, with its tributaries, affords extensive water-power. It contained in 1840, 35,396 neat cattle, 29,759 sheep, 11,825 swine; and produced 86,514 bushels of wheat, 9830 of rye, 154,049 of Indian corn, 3510 of buckwheat, 57,057 of barley, 139,563 of oats, 1,165,389 of potatoes, 18,915 pounds of sugar. It had 235 stores, three lumber-yards, one furnace, 19 fulling-mills, three woollen factories, 55 grist-mills, 131 saw-mills, four oil-mills, three paper-mills, 43 tanneries, one distillery, one pottery, seven printing-offices, four binderies, six weekly newspapers, and one periodical; 2 colleges, 95 students, 19 academies, 1311 students, 334 schools, 17,163 scholars. Pop 50,923. Capital, Augusta.

KENNEBEC, p. t., port of entry, York co., Me., 75 m. S.S.W. Augusta, 517 W. Situated on the S.W. side of Kennebunk river, at its entrance into the Atlantic, and has a good harbour. It has some ship-building, and considerable shipping employed in the coasting trade and the fisheries. It contains seventeen stores, one cotton factory, with 1064 spindles, one grist-mill, three saw-mills, one printing-office, one weekly newspaper; two academies, 175 students, 11 schools, 885 scholars. Pop. 2323.

KENNEBUNKPORT, p. t., York co., Me., 78 m. S.S.W. Augusta, 590 W. Situated on the N.E. side of Kennebunk river, opposite to Kennebunk, with which its trade is connected. It contains 11 stores, two furnaces, one fulling-mill, four grist-mills, four saw-mills, three tanneries; one academy, 30 students, 13 schools, 1150 scholars. Pop. 2763.

KENNERY (CAVE-TEMPLES OF). (See SALSETTE.)

KENNET, t., Chester co., Pa., 28 m. S.W. Philadelphia. Drained by Red Clay creek. It contains a Friends' church, four stores, two grist-mills, six saw-mills; six schools, 236 scholars. Pop. 1290.

KENSINGTON, a town and par. of England, co. Middlesex, hund. Ossulston, in the suburbs of London, 1½ m. W. Hyde Park corner, comprising (with the hamlets of Bayswater, Earl's Court, Brompton, and Little Chelsea) an area of 2690 acres. Population, in 1831, 20,900. It consists of a main street forming a part of the London road, and of several subordinate streets running from it N. and S., one of which leads into a handsome square. The houses are well built, and many good detached residences are scattered in the outskirts. The parish church is a plain but spacious building, erected in 1690; and the living is a vicarage in the gift of the Bishop of London. There are also two district churches, and a proprietary episcopal chapel, with several places of worship for dissenters (the largest of which, built in 1794, belongs to the Independents). A large charity school, national and Lancastrian schools, and several private boarding schools, furnish instruction to all classes; and there are numerous charities for the relief of the aged and sick poor. The trade of the town chiefly depends on the many families of rank and wealth resident in and round it.

Kensington is the chief locality of a poor-law union, comprising, besides itself, the parishes of Chelsea, Fulham, Hammersmith, and Paddington. The expense of maintaining the poor of this parish amounted to £6900 in 1839 (Poor Law Comm. 4th Report.)

The palace, which, with its gardens, forms the chief object of attraction, is an irregular brick building, purchased by William III. of the Earl of Nottingham. Among other additions made by that monarch, the whole S. front was rebuilt under the direction of Sir C. Wren, and the interior received great improvements and embellishments: the W. front was rebuilt by Kent, in the reign of George II. The state rooms comprise 12 handsome chambers, well adapted for occasions of ceremony; but few of them, except the galleries, are of commanding proportions. The staircase, painted by Kent, is intended to represent a number of spectators on a court day; and the artist has introduced several portraits of characters connected with the court of George I.: the style, however, is bizarre, and in very bad taste. The presence chamber is now hung with pictures, many of which were highly valued by the late president West. This palace was the residence of William and Mary, Anne, George I., and George II., all of whom (except George I.) died within its walls. George III. removed the town residence of the court to St. James's; and Kensington palace has since been allotted to junior members of the royal family. The childhood of Queen Victoria was spent in it; and it has been for many years the town residence of the Duke of Sussex. His royal highness's library is very valuable, especially the collection of biblical works, including about 300 rare MSS. The gardens, planted with fine trees, occupy an area of about 350 acres, and have been for many years an attractive public promenade. Holland House, a brick structure, in the Elizabethan style, at the W. end of Kensington, was built in 1607, and descended in the reign of Charles I. to the Earl of Holland. Addison occupied it after his marriage with the dowager Countess of Warwick. In 1765 it was purchased by Henry Fox, Lord Holland, in whose family it still remains. The library is 112 ft. in length, and contains a valuable collection of books, especially in Spanish and Portuguese literature. There are many good pictures, and in the hall is a sitting statue of C. J. Fox. About 200 acres of land are attached to the house, which is one of the finest residences in the vicinity of London.

KENSINGTON, p. t., Rockingham co., N. H., 41 m. S.E. Concord, 478 W. Incorporated in 1737. It contains two Christian churches, two stores, two grist-mills, one saw-mill; three schools, 160 scholars. Pop. 665.

KENSINGTON, p. t., Philadelphia co., Pa., 100 m. E. by S. Harrisburg, 140 W. It is a suburb of Philadelphia, lying in the N.E. part along Delaware river. It is governed by 15 commissioners, but is generally regarded as a part of the city. It had in 1840, one consolidation-house in foreign trade, 12 retail stores, capital $107,980, seven lumber-yards, capital 116,580, nine woollen factories, 15 cotton factories, with 9700 spindles, one glass factory, four rope-walks, one brewery, three tanneries; six academies, 676 students, five schools, 874 scholars. Pop. 22,314. (See PHILADELPHIA.)

KENT, a marit. co. in the S.E. part of England, being the nearest of any in the kingdom to the continent, having N. the Thames and German ocean, E. and S.E. the German ocean and the straits of Dover, S. Sussex, and W. Surrey. Its greatest length, from Deptford to the N. Foreland, is about 64, and its greatest breadth about 30 m. Area, 996,680 acres, of which above 900,000 are said to be arable, meadow, and pasture. This is a finely diversified and beautiful county. Two parallel ridges of hills traverse its whole extent from E. to W. The upper, or most northerly of these ranges, extending from Westerham, to the confines of Surrey, to Dover, being composed chiefly of chalk, and thence called the chalk ridge; while the lower, or most southerly range, about 8 m. from the former, is usually called the ragstone range, from its consisting principally of ragstone and ironstone. The country to the N. of the upper range, including the isles of Sheppey, Grain, and Thanet (see THANET), is generally very fertile, and contains a good deal of marshy and of rich loamy land, producing the finest wheat. Romney Marsh, a celebrated grazing district (see ROMNEY MARSH), and the Weald, lie to the S. of the lower or ragstone range. The latter, which extends into Sussex and Surrey, is a very singular tract. Its soil is generally stiff and clayey, but in parts sand predominates. For a lengthened period it formed an immense forest; but was gradually, though slowly, brought into tillage. Its soil continues to be particularly well adapted to the growth of timber, especially oak, which here attains to the greatest luxuriance. At this moment most inclosures in the Weald are surrounded with oaks, and every wood and coppice is full of them. "When viewed from the adjoining hills, which command a prospect over the whole of it, the Weald exhibits the most delightful scene that can be imagined. It appears to the eye an extensive level country (the few hills in it being so small and inferior to those whence it is viewed), covered with all the richness of both art and nature; the variety of small inclosures of corn and meadow, and the houses, seats, and villages, promiscuously interspersed among the large and towering oaks, which grow over the whole face of it, have the most pleasing effect and represent to us, even at this time, something, though a great improvement of its original state, in the idea of an inhabited and well cultivated forest." (Hasted's Kent, i., 222, 223, 8vo. ed.) From its proximity to the continent the climate of Kent is colder in winter, and the E. winds in spring are said to be more piercing than in other counties in the same parallel more to the W.; but, on the other hand, the summers are warmer, and its autumns less liable to wet, which renders it especially fitted for the production of corn and fruit. Agriculture is in a very advanced state in Kent, and it has a greater variety of products than any other county in the kingdom. Its wheat, barley, beans, and pease are all excellent. With the exception of the Isle of Thanet, turnips are extensively raised on the light soils. Hops are produced in large quantities, especially in the district between Maidstone and Canterbury. Most part of the cherries, filberts, plums, and other fruits brought to the London markets, are supplied by the orchards between Maidstone and Tunbridge, &c.; while the Isle of Thanet and other places furnish supplies of spinach and of various seeds. Though Kent feeds large numbers of cattle, it cannot be called a grazing county; the stock of sheep is, however, very large. Romney Marsh has a peculiar breed that furnishes long, combing wool. There is a great deal of timber in other parts of the county, exclusive of the weald. Property much divided, and there are no great estates. Size of farms very various; but, owing to the sort of garden culture carried on in many parts, they are mostly, perhaps, rather small; many varying in extent from 10 to 30 acres, while there are but few above 200 or 250 acres. Average rent of land, in 1810, 17s. 5d. an acre. The yeomanry of Kent are a very superior class; and, besides their own, some of them occupy extensive hired farms. All lands in Kent, unless specially exempted by an act of the legislature, are held by the tenure of gavelkind; descending, in the event of the father dying intestate, not to the oldest son, but to all the sons alike in equal portions; and if there be no sons, they divide equally among the daughters. This is supposed to have been the common tenure in England before the conquest; but, exclusive of Kent, it now obtains in but a few places. Some estates have been disgavelled, or excepted by a special act of parliament, from this tenure; and partition is now, in most instances, prevented by testament. But such lands as are not disgavelled or entitled by testament, are invariably disposed of in the way stated above. (Hasted's Kent, i., 311–361, 8vo ed.) The customs that prevail with respect to the entry to farms operate injuriously on agriculture; and owing, as is said, to the prevalence of smuggling on the coasts, and the abuse of the poor laws, the peasantry were lately supposed to be a good deal demoralized; but both of these sources of disorder are now in the way of being obviated. Ironstone is abundant in many parts; and previously to the employment of coal in the making of iron, the Weald, from the abundance of its timber, was a principal seat of the iron trade; but this has been long abandoned. With the exception of ship-building carried on at Deptford, Woolwich, Chatham, and other places, manufactures are unimportant; they consist of paper, made at Maidstone and Dover, gunpowder at Dartford and Faversham; and toys at Tunbridge. Exclusive of the Thames, the principal rivers are the Medway (which see), Stour, Rother, Darent, and Ravensbourne. Kent is divided into the two nearly equal divisions of E. and W. Kent, each having its own court of sessions. Principal towns, Greenwich, Deptford, Chatham, Rochester, Canterbury, and Dover. It is divided into 5 lathes, 63 hundreds, and 15 liberties, and 411 parishes. It sends 16 members to the House of Commons, viz., two for each of the two divisions of the county; two for each of the boroughs of Canterbury, Rochester, Dover, Greenwich, Maidstone, and Sandwich, and one each for Chatham and Rye. Registered electors for the county, in 1839–40, 16,005, being 7344 for the E., and 8661 for the W. division. In 1831, Kent had 82,144 inhabited houses; 97,142 families; and 479,155 inhabitants, of whom 234,572 were males, and 244,583 females. Sum expended on the relief of the poor, in 1838–39, £300,043. Annual value of real property, in 1815, £1,687,443. Profits of trades and professions in do. £1,696,229.

KENT, cn., R. I. Situated in the centre of the state, and contains 196 sq. m. Drained by Pawtuxet and Flat rivers, which afford water-power. It contained in 1840, 4696 neat cattle, 10,406 sheep, 4221 swine; and produced 494 bushels of wheat, 6006 of rye, 64,119 of Indian corn, 3805 of barley, 11,915 of oats, 136,344 of potatoes. It had 163 stores, capital $179,610; five lumber-yards, nine fulling-mills, five woollen factories, 45 cotton factories 200,010 spindles, 27 grist-mills, 25 saw-mills, two rope-walks, two tanneries; six academies, 250 students; 64 schools, 1974 scholars. Pop. 13,083. Capital, East Greenwich.

Pawtuxet river rises in Providence county, but flows chiefly in Kent county. As this river is famous for its manufacturing establishments, the following summary of the villages on it is here given, from information received in 1843. *Pawtuxet*, near its mouth, has one cotton factory with 1500 spindles, with a head and fall in the river of 4 ft. It has two churches and six stores. *Clarksville* has one cotton-factory and bleaching works. The river has 6 ft. head and fall. It has one store, and about 450 inhabitants. *Natick* has six cotton factories with 30,000 spindles, and the river has 20 ft. head and fall. It has one Union church, six stores, and 1600 inhabitants. *Greenville*, at the forks of the river, has two cotton factories with 6000 spindles, with a head and fall in the river of 15 ft. It has three stores and 400 inhabitants. On the S. branch of Pawtuxet river, *Tuft's Mills* has one cotton factory with 4000 spindles, printing works, and a head and fall in the river of 9 ft. It has one Baptist church and two stores. *Centreville* has one woollen factory, three cotton factories with 6000 spindles, 15 ft. head and fall in the river, a Methodist church, two stores, and 350 inhabitants. *Compton Mills* has four cotton factories with 12,000 spindles, printing and bleaching works, and 15 ft. head and fall in the river. It has a Baptist church, six stores, and 1600 inhabitants. *Washington* has four cotton factories with 8000 spindles, and 16 ft. head and fall. It has one Union church, five stores, and 600 inhabitants. On the N. branch of the river, *Lippit* village has one cotton factory with 5000 spindles, bleaching works, and 15 ft. head and fall in the river. *Phœnix* has two cotton factories with 6000 spindles, and 15 ft. head and fall in the river. It has one church, four stores, and 400 inhabitants. *Centreville* has one cotton factory with 3000 spindles, with 9½ ft. head and fall in the river, and 300 inhabitants. *Arkwright* has two cotton factories with 6000 spindles, with 15 ft. head and fall in the river. It has a Baptist church, two stores, and 500 inhabitants. *Fiskville* has one cotton factory with 2000 spindles, a head and fall of 12 ft. in the river, and 200 inhabitants. *Jacksonville* has one cotton factory with 3000 spindles, 12 ft. head and fall in the river, one store, and 300 inhabitants. *Hope Mills* has one cotton factory with 3000 spindles, one store, and 300 inhabitants. *Scituate Mills* has two cotton factories with 5000 spindles, 15 ft. head and fall in the river, and 400 inhabitants. Such are the manufacturing villages and establishments on the single river Pawtuxet, of moderate extent, and its two main branches. A similar view is given of the villages and establishments on Blackstone river, under the article CUMBERLAND (which see). Together, they will give something of a picture of Rhode Island, which had the honour of commencing the cotton manufacture in the United States, and which, in proportion to its population, is still in advance of all the other states in prosecuting it. The above establishments are in Warwick and Coventry, Kent county, and Cranston and Scituate, in Providence county. For the purpose for which they are introduced, it is better to give them in connexion, than to distribute them to the several towns.

KENT, co., Del. Situated in the centre of the state, and contains 640 sq. m. Bounded E. by Delaware bay. Drained by Jones, Little Duck, and Motherkill creeks. It contained in 1840, 17,477 neat cattle, 13,780 sheep, 27,080 swine; and produced 85,342 bushels of wheat, 21,745 of rye, 396,231 of Indian corn, 3680 of buckwheat, 894,231 of oats, 68,375 of potatoes. It had 66 stores, nine lumberyards, one fulling-mill, one woollen factory, 20 grist-mills, 16 saw-mills, four tanneries, one printing-office, one periodical; two academies, 65 students; 46 schools, 2997 scholars. Pop.: whites, 13,518; slaves, 497; free coloured, 5627; total, 19,872. Capital, Dover.

KENT, co., Md. Situated towards the N.E. part of the state, and contains 340 sq. m. Bounded W. by Chesapeake bay. Watered by Chester river. It contained in 1840, 8498 neat cattle, 9109 sheep, 14,991 swine; and produced 133,147 bushels of wheat, 3250 of rye, 502,439 of Indian corn, 1118 of buckwheat, 277,393 of oats, 33,363 of potatoes. It had 270 stores, one fulling mill, one woollen factory, seven flouring-mills, 13 grist-mills, five saw-mills; three colleges, 80 students; one academy, 18 students; 14 schools, 474 scholars. Pop.: whites, 5616; slaves, 2735; free coloured, 2491; total, 10,842. Capital, Chester.

KENT, co., Mich. Situated in the W. part of the state, and contains 576 sq. m. Watered by Grand river and its tributaries, which afford good water-power. Limestone, salt springs, and gypsum are found. It contained in 1840, 1971 neat cattle, 922 sheep, 3460 swine; and produced 18,750 bushels of wheat, 13,390 of Indian corn, 17,220 of oats, 16,700 of potatoes, 3080 pounds of sugar. It had two commission houses in foreign trade, 12 retail stores, one furnace, two flouring-mills, one grist-mill, 18 saw-mills, one tannery, one printing-office, one weekly newspaper; nine schools, 363 scholars. Pop. 2587. Capital, Kent, or Grand Rapids.

KENT, p. t., Litchfield co., Ct., 51 m. W. Hartford, 293 W. Watered by Housatonic river, which affords water-

power. It contains an extensive bed of iron ore. Incorporated in 1739. It contains three churches, a Congregational, Methodist, and an Episcopal; six stores, three furnaces, two forges, three grist-mills, four saw-mills, one tannery, two distilleries; 12 schools, 475 scholars. Pop. 1739.

KENT, p. t., Putnam co., N. Y., 99 m. S. Albany, 292 W. Drained by Crotus river. It contains six stores, two fulling-mills, one woollen factory, one flouring-mill, five grist-mills, four saw-mills; 10 schools, 490 scholars. Pop. 1630.

KENTON, co., Ky. Situated in the N. part of the state, and contains 150 sq. m. Bounded N. by Ohio river, E. by Licking river. It contained in 1840, 4529 neat cattle, 7523 sheep, 13,774 swine; and produced 53,978 bushels of wheat, 4738 of rye, 294,653 of Indian corn, 28,989 of oats, 10,823 of potatoes, 601,774 pounds of tobacco, 5589 of sugar. It had 33 stores, two lumber-yards, one furnace, one forge, one cotton factory with 2232 spindles, two flouring-mills, 14 grist-mills, four saw-mills, two tanneries, two printing-offices, two weekly newspapers; 15 schools, 354 scholars. Pop.: whites, 7031; slaves, 751, free coloured 34; total, 7816. Capital, Independence.

KENTON, p. t., cap. of Hardin co., O., 71 m. N.W. Columbus, 442 W. Drained by a head branch of Blanchard's fork of Auglaize river. Bounded S.W. by Scioto river, on the N. side of which the village is situated. It contains six churches, a Lutheran, Methodist, Seceders, Associate Reformed, and Disciples; four stores; three schools, 189 scholars; 75 dwellings, and about 400 inhabitants.

KENTUCKY, one of the western United States, is bounded N. by Ohio, Indiana, and Illinois, from which it is separated by Ohio river; on the E. by Virginia; S. by Tennessee; and W. by Missouri, from which it is separated by Mississippi river. It is between 36° 30' and 39° 10' N. lat., and between 82° and 89° 30' W. long., and between 5° and 10° W. long. from W. Its greatest length is about 400 m., and its breadth varies from 5 to 170 m., containing 40,500 sq. m., or 25,920,000 acres. The population in 1790 was 73,677; in 1800, 220,959; in 1810, 406,511; in 1820, 564,317; in 1830, 688,844; in 1840, 779,828; of whom 182,258 were slaves. Of the free population, 305,393 were white males; 284,430 do. females; 3761 were coloured males, 3556 do. females. Employed in agriculture, 197,738; in commerce, 3448; in manufactures and trades, 23,217; in navigating the ocean, 44; do. canals, rivers, and lakes, 969; in mining, 331; in the learned professions, 2487.

There are 90 counties in this state, which, with their population in 1840, were as follows:

Counties	Total Pop.	Counties	Total Pop.
Adair	8,466	Hopkins	9,171
Allen	7,329	Jefferson	36,346
Anderson	5,452	Jessamine	9,396
Barren	17,234	Kenton	7,816
Bath	9,763	Knox	5,722
Boone	10,034	Laurel	3,079
Bourbon	14,478	Lawrence	4,730
Breathitt	2,195	Lewis	6,306
Breckin	7,055	Lincoln	10,187
Breckenridge	8,944	Livingston	9,025
Bullitt	6,334	Logan	13,615
Butler	3,898	Madison	16,355
Caldwell	10,365	Marion	11,032
Calloway	9,791	Mason	15,719
Campbell	5,214	M'Cracken	4,745
Carroll	3,969	Meade	5,780
Carter	2,905	Mercer	18,720
Casey	4,939	Monroe	6,526
Christian	15,587	Montgomery	9,332
Clark	10,802	Morgan	4,603
Clay	4,607	Muhlenburg	6,964
Clinton	3,863	Nelson	13,637
Cumberland	6,090	Nicholas	8,745
Daviess	8,331	Ohio	6,592
Edmonson	2,914	Oldham	7,380
Estill	5,535	Owen	8,232
Fayette	22,194	Pendleton	4,455
Fleming	13,268	Perry	3,089
Floyd	6,845	Pike	3,567
Franklin	9,420	Pulaski	9,620
Gallatin	4,003	Rockcastle	3,409
Garrard	10,480	Russell	4,238
Grant	4,192	Scott	13,668
Graves	7,465	Shelby	17,768
Grayson	4,461	Simpson	6,537
Greene	14,212	Spencer	6,581
Greenup	6,297	Todd	9,991
Hancock	1,515	Trigg	7,716
Hardin	16,357	Trimble	4,479
Harlan	3,015	Union	6,673
Harrison	12,472	Warren	15,446
Hart	7,031	Washington	10,596
Henderson	9,548	Wayne	7,399
Henry	10,015	Whitley	4,673
Hickman	3,864	Woodford	11,740
		Total	**779,924**

Frankfort, on Kentucky river, 60 m. from its entrance into the Ohio river, is the seat of government.

Cumberland mountains run on the S.E. border of the state, and send off spurs which extend into its eastern part, rendering it mountainous. The Cumberland range divides this state from Virginia. A tract along the Ohio river, from 5 to 30 m. wide, is broken and hilly, extending through the whole length of the state. But the hills are gently rounded, and are fertile to their tops, with narrow valleys between them of great fertility. Along the margin of the Ohio, with an average width of one mile, are bottom lands subject to periodical inundations. Between the hilly tract on Ohio river, the mountainous country in the eastern section, and Green river, is a tract of 100 m. long, and 50 m. wide, beautifully undulating, with a black and fertile soil, which has been denominated the garden of Kentucky. The forest growth of this region is black walnut, cherry, honey locust, buckeye, papaw, sugar maple, elm, ash, hawthorn, coffee-tree, yellow poplar, with an abundance of grape vines of a large size. The country in the S.W. part of the state, between Green and Cumberland rivers, has been improperly denominated barrens, as the soil is far from being poor. It is thinly wooded with short oak timber, and is covered, in summer, with a high grass. The whole state, below the mountains, rests on an immense bed of limestone, generally about 8 ft. below the surface, in which are frequent apertures, in which the waters of the rivers sink into the earth, causing the large rivers to be greatly diminished in the summer season, and some of the smaller ones entirely to disappear. In no part of the country do the rivers suffer so great a diminution in the dry season, as in Kentucky. The rivers have generally worn deep channels in the calcareous rocks over which they flow. Stupendous precipices are formed on Kentucky river, where the banks in many places are 300 feet high, of solid limestone, with a steep and elevated ascent above them. In the S.W. part of the state, between Green and Cumberland rivers, are several wonderful caves. The Mammoth cave, in Edmonson county, 130 m. from Lexington on the road to Nashville, is one of the most remarkable caves in the world. It has been explored to a great distance, and is, with good reason, supposed to extend for 6 or 10 m. The earth at the bottom is strongly impregnated with nitre, which has been, to a considerable extent, manufactured from it.

Wheat, tobacco, and hemp, are the staple productions of the state, but Indian corn, rye, oats, barley, buckwheat, flax and potatoes, and cotton for domestic use, are extensively cultivated. Apples, pears, peaches, and plums, are the most common fruit. Horses, horned cattle, pork, bacon, and lard, are extensively exported.

There were in the state in 1840, 395,853 horses or mules; 787,898 neat cattle; 1,008,240 sheep; 2,310,533 swine; poultry to the value of $536,439; there were produced 4,803,152 bushels of wheat; 17,491 of barley; 39,847,190 of Indian corn; 1,321,373 of rye, 7,155,974 of oats, 8169 of buckwheat, 1,055,895 of potatoes, 1,786,247 pounds of wool, 38,445 of wax, 53,436,909 of tobacco, 16,376 of rice, 691,456 of cotton, 1,377,535 of sugar, 86,306 tons of hay, 9996 of hemp and flax. The products of the dairy amounted to $931,363; of the orchard, to $434,935; of lumber, to $130,329. There were made 2299 gallons of wine. These productions bear testimony to the fertility of the soil.

Among the mineral productions, iron ore, coal, salt, and lime are abundant. Salt was produced in 1840, to the amount of 566,167 bushels, and is extensively exported. The greater parts of the exports of this state pass down the Mississippi to New-Orleans, and the chief imports are brought in steamboats, through the same river and the Ohio, and its various tributaries in this state.

The climate of this state is generally salubrious. The waters are mild, being only of two or three months' continuance, but the atmosphere is moist. The spring and autumn are delightful. The extremes of heat and cold during the year are less than in some other parts of the country.

Ohio river winds along the N. border of this state for 637 m. Cumberland river rises in the E. part of the state, passes into Tennessee and, with Tennessee river, crosses the W. part of the state, and both enter Ohio river. They are the largest rivers of the state, and among the largest tributaries of the Ohio. Both are extensively navigable. Cumberland river enters the Ohio 59 m. above the junction of the latter with the Mississippi. Tennessee river enters Ohio river 11½ m. below the mouth of Cumberland river. Big Sandy river, for a considerable distance, forms the boundary between this state and Virginia. Kentucky river rises in Cumberland mountains; and through a deep, rocky bed, enters the Ohio 77 m. above Louisville. It is navigable for steamboats 60 m. to Frankfort. Licking, Salt and Green rivers are extensively navigable, and fall into Ohio river. Mississippi river runs on the W. border of the state.

Louisville is much the largest, and most commercial place in the state. Lexington is distinguished for its beauty, for the refinement of its inhabitants, and for its extensive business. Maysville on the Ohio, and Frankfort on Kentucky river, are important places.

There were in the state, in 1840, five commercial and 50 commission houses in foreign trade, with a capital of

$868,760; 1685 retail stores, with a capital of $9,411,886; 571 persons employed in the lumber trade, with a capital of $105,995; 101 persons employed in internal transportation, who, with 183 butchers, packers, &c., employed a capital of $153,650. Home-made, or family manufactures, amounted to $2,522,462. Forty woollen manufactories employed 980 persons, manufactured to the amount of $151,946, with a capital of $138,800; 58 cotton manufactories, with 12,389 spindles, employed 593 persons, producing articles to the amount of $389,380, with a capital of $316,113; 17 furnaces, producing 39,395 tons of cast iron, and 13 forges, &c., producing 3537 tons of bar iron, employed 1108 persons, and a capital of $449,800; 27 persons produced 2195 tons of anthracite coal, with a capital of $14,150; 213 persons produced 588,187 tons of bituminous coal, with a capital of $76,667; 991 persons produced 219,695 bushels of salt, with a capital of $163,565; 160 persons produced granite and marble to the amount of $49,362, with a capital of $6212; seven paper-mills employed 47 persons, and produced articles to the amount of $44,000, employing a capital of $47,500; hats and caps were produced to the amount of $901,310, and straw bonnets to the amount of $4483, the whole employing 194 persons, and a capital of $118,838; 587 persons manufactured tobacco to the amount of $413,399, with a capital of $230,400; 397 tanneries employed 979 persons, and a capital of $567,954; 584 other manufactories of leather, as saddleries, &c., produced articles to the amount of $738,646, with a capital of $389,833; one glasshouse produced to the amount of $3088, with a capital of $500; 16 potteries employed 51 persons, and produced articles to the amount of $24,609, with a capital of $9870; 11 powder-mills employed 58 persons, and produced 292,500 pounds of gunpowder, with a capital of $49,000; 25 persons produced paints and drugs to the amount of $85,994, and turpentine and varnish to the amount of $9905, with a capital of $16,630; 38 persons produced confectionary to the amount of $26,050 with a capital of $14,250; 111 rope-walks employed 1888 persons, and produced cordage to the amount of $1,593,276, with a capital of $1,023,130; six persons produced musical instruments to the amount of $4500, with a capital of $5900; 149 persons produced machinery to the amount of $46,074; 30 persons produced hardware and cutlery to the amount of $33,350; 369 persons produced 2341 small arms, with a capital of $19,060; 21 persons manufactured the precious metals to the amount of $19,060; 657 persons produced bricks and lime to the amount of $949,919; 516 persons manufactured 3,989,496 pounds of soap, 563,635 do. of tallow candles, and 315 pounds of spermaceti or wax candles, with a capital of $28,785; 889 distilleries produced 1,763,685 gallons, and 59 breweries produced 214,589 gallons, the whole employing 1692 persons, and a capital of $315,308; 533 persons produced carriages and wagons to the amount of $168,734, with a capital of $79,378; 256 flouring-mills produced 273,688 barrels of flour, and, with other mills, employed 2067 persons, and produced to the amount of $3,437,937, with a capital of $1,680,689; 452 persons manufactured furniture to the amount of $273,350, with a capital of $120,295; 465 stone or brick houses, and 1757 wooden houses employed 2983 persons, and cost $1,039,173; 34 printing-offices, three binderies, five daily, seven semi-weekly, and 26 weekly newspapers, and eight periodicals, employed 936 persons, and a capital of $86,525. The whole amount of capital employed in manufactures was $5,945,980. The manufactures of this state are less than those of a number of others; but the idea of such progress in the year 1840 would have astonished Daniel Boone.

Transylvania University, at Lexington, was founded in 1798, and is a flourishing institution. Centre College, at Danville, was founded in 1822; St. Joseph's College, at Bardstown (Roman Catholic), was founded in 1819; Augusta College, at Augusta (Methodist), was founded in 1825; Cumberland College, at Princeton (Cumberland Presbyterian), was founded in 1825; Georgetown College, at Georgetown (Baptist), was founded in 1829; Bacon College, at Harrodsburg, was founded in 1836; St. Mary's College, Marion county (Roman Catholic), was founded in 1837. Transylvania University has a flourishing medical department, and there is a medical institution at Louisville. In these institutions there were, in 1840, 1419 students. There were 116 academies and grammar schools, with 4996 students; 932 common and primary schools, with 24,641 scholars. There were in the state 40,018 white persons over 20 years of age, who could neither read nor write.

The Baptists, the most numerous denomination, had in 1836, 500 churches, about 380 ministers, and 35,860 communicants. The Methodists had 100 travelling preachers, and 31,369 communicants. The Presbyterians had 190 churches, and about 10,089 communicants. The Episcopalians had a bishop, and 13 ministers. The Roman Catholics had a bishop, and 24 ministers. There were a considerable number of Cumberland Presbyterians and Reformed Baptists, two societies of Shakers, and one of Unitarians.

Near the commencement of 1840, there were three banks with 11 branches, with an aggregate capital of $7,789,002, and a circulation of $3,476,307. In January, 1840, five of these banks had suspended specie payments. In June 25th, 1842, the state debt amounted to $3,085,500, for internal improvements.

The first constitution was formed in 1790, and in 1799 the present constitution was formed. A governor is elected for four years by the people, and is ineligible for the next seven years. A lieutenant governor is chosen at the same time, who is president of the senate, and who, in case of the death or absence of the governor, discharges the duties of his office. The senators are elected for four years, one quarter of them being chosen annually. Their number cannot be over 38, the present number, nor less than 24. The representatives are elected annually, and apportioned every four years among the different counties, according to the number of electors. The present number, 100, is the highest which the constitution allows, and there can never be less than 58. The general assembly meets annually at Frankfort, on the first Monday of November. Every free white male citizen, who is 21 years of age, and who has resided two years in the state or county in which he offers his vote, is entitled to the right of suffrage. Votes are given openly, or viva voce, and not by ballot. The judges of the different courts hold their offices during good behaviour.

The most important work of internal improvement is the Louisville and Portland canal, 2¼ m. long, around the rapids in Ohio river. It admits steamboats of the largest class, is 50 ft. wide at the surface, is excavated 10 ft. deep in a compact limestone, and has an entire lockage of 23 ft. It cost $730,000. The navigation of Kentucky, Green, and Licking rivers has been extensively improved by dams and locks. A railroad extends from Lexington to Frankfort. It is designed to be continued to Louisville, but is for the present suspended. Several other railroads have been projected.

This state once belonged to Virginia. It was first explored in 1769–70, by Daniel Boone, an enterprising hunter. The first permanent settlement was made in 1774, at Harrodsburg. Until Wayne's treaty in 1795, it was continually exposed to incursions from the Indians. The first newspaper was issued at Lexington, Aug. 28th, 1787. Kentucky became a state, and was admitted into the Union, in 1792. It was separated from Virginia in 1786, after which it had a territorial government until 1792.

KENTUCKY, river, Ky., rises by three principal branches in the Cumberland mountains, denominated the North, Middle and South forks, which unite in Estill co. It pursues a circuitous course, in its lower parts generally N.N.W., and enters Ohio river at Carrollton, 534 miles below Pittsburg. The navigation from the mouth of the river to its forks, has been improved by the construction of 17 dams, creating as many pools, and 17 locks, connecting them. The dams are from 350 to 500 feet long, and from 20 to 25 feet and the locks are 178 feet long and 38 feet wide, and dams pth of the water for the whole course is 6 feet, so that there is a less draft than this, can navigate it all times. The difference of level overcome by the locks is 216 feet, and the cost of the whole work is estimated at $2,297,409. The distance from the mouth to the forks is 260 miles, but, in a direct line, only 112 miles.

KEOKUK, p. v., Lee co., Iowa. Situated on the W. side of Mississippi river, at the foot of the lower rapids. It derives its chief importance from the change of freight and the storage of merchandise at a low stage of the water in Mississippi river. Government are about removing the obstructions in the river. It is laid out on a mile square, and contains 150 or 200 inhabitants.

KERBELA, or MESHED HOSSEIN, a town of Asiatic Turkey, prov. Irak-Arabi, 50 m. S.W. Bagdad. Pop. uncertain, but very considerable. It stands on a plain about 6 miles W. of the Euphrates, with which it is connected by a canal said to be more ancient even than the one of Alexander. It has five gates, a well supplied bazaar, and seven caravanserais; but the chief ornaments of the city are the tomb of Hossein, adorned with a gilded cupola, and a noble mosque. Its chief lustre has been derived from Hossein, son of Ali by Fatima, the daughter of the prophet, who was slain near it, and to whose tomb numerous pilgrims of the sect of Ali flock from all quarters, but especially from Persia, to pay their devotions. It is subject to the Turks, but still the majority of the inhabitants are Persians; and it has always been a favourite object of their king to obtain possession of this place, as well as of Meshed Ali and Kazameen, both of which are, like Kerbela, the resort of pilgrims. The environs of the town and borders of the canal are shaded by extensive plantations of palm-trees, and the walls, which are upward of 2 miles in circuit, are kept in good repair, to secure the riches of the holy city against the predatory excursions of the Wahabees, by whom it was plundered some years ago.

Kerbela occupies the site of Volagesia, a small town built

by Vologese, one of the Parthian kings, contemporary with Nero and Vespasian. (Kinneir.)

KERESOUN (an. Cerasus), a town and seaport of Asiatic Turkey, on the S. shore of the Black sea, pach. Trebizond, from the town of which mace it is distant 94 m. W. by S.; lat. 40° 57' 10'' N., and long. 38° 24' E. Pop. about 300, half being Armenian and Greek. It stands on an elevated rocky promontory bounding an extensive bay to the E., and appears to have been formerly a place of great strength. A considerable part of the ancient wall still exists; but the present town is in a ruinous condition, and the people bear the appearance of being in abject poverty. There is some little trade in corn with the Crimea; and trading vessels are built in the bay under the city walls.

Cerasus was visited by Xenophon on his return with the ten thousand; and he calls it a "Hellenic colony, situated in the country of the Colchi." (Anab., v., 3.) It is also said to be the native country of the cherry, which hence received its name. It was here that Mithridates ordered his wives and sisters to be poisoned after the battle of Cabira, when it fell into the hands of Lucullus; but that it was, as Arrian states, identical with the Pharnacia which was the residence of the kings of Pontus, is, to say the least, extremely doubtful. (Cramer's Asia Minor, i., 281.) Kerasoun was conquered by Mahmoud II., and has since been attached to the Turkish empire. (Kinneir's Asia Minor, p. 339; Rennell's Geog. of W. Asia, vol. ii.)

KERKOUK (Demetrias, Strab.; Corcura, Ptol.), a large town of Asiatic Turkey, in Lower Kurdistan, cap. sandjiak, 109 m. S.E. Mosul, and 130 m. N. Bagdad. Pop. according to Kinneir, 13,000. It is situated on a commanding eminence nearly perpendicular on all sides, below which is an extensive suburb: it is surrounded by a mud wall, but beyond this are extensive suburbs. Besides numerous mosques, it has three Roman Catholic churches and one Armenian do. The streets are narrow and filthy; and the meanness of the houses leaves no doubt with respect to the poverty and wretchedness of the inhabitants. The surrounding district is uneven and hilly; and on the N. side a low range of barren and rocky mountains separates it from the fine plain of Altun-Kupri. In the passes through these mountains are numerous naphtha pits, yielding an inexhaustible supply of that useful commodity, which is sent in earthen jars all over the neighbouring country. (Kinneir's Persia, p. 298; Olivier, iv., 300.)

KERMAN (an. Caramania), a prov. of Persia, between lat. 25° 30' and 31° 30' N., and long. 54° 30' and 60° 20' E., having N. Khorassan, E. Afghanistan and Belochistan, S. the Persian gulf, and W. the provs. Fars and Laristan. Shape triangular; extreme length, 380 m.; breadth, 250 m.; supposed area, 65,000 sq. m. Pop. alleged to be under 600,000, having greatly decreased of late years through the wars of extermination waged by the Persians on the Guebres or Parsees. Kerman, generally speaking, is mountainous; but the elevation of the high ground varies considerably from mere hills to lofty ridges, scarcely lower than those of the great mass in which they originate. The principal range divides Nurmansheer from Laristan, and thence runs W. with many ramifications. The interior of the province is not irrigated by a single river, and the natives could not possibly exist, but for a few mountain springs, and the diligence used in cutting kanzas or subterraneous reservoirs, for watering the land. The Rud Shuir, which runs through the S. part of Kerman into the Persian gulf, is at present very imperfectly known. The climate is accounted the least healthy of any part of Persia; the hills, which are clad with snow nearly all the year, being extremely cold, and the long narrow valleys between them oppressively hot. The winds from the mountains are deliciously cool; but, as they bring with them agues, and epidemic fevers, the natives prefer sultry weather. The N. portion of the province, and that close on the coast, are arid sterile deserts; but in Nurmansheer and a few other central districts where irrigation has been properly followed up, layers of alluvial soil and rich vegetable mould are found to be exceedingly productive. Wheat, maize, and barley; cotton, tobacco, saffron, and madder, are raised with facility, and in the greatest perfection. Dates, oranges, lemons, grapes, almonds, and pistachios, with other fruits of S. Europe, are of common occurrence; and mulberry-trees are largely cultivated for the silk-worms, in breeding which the inhabitants have attained considerable celebrity. The gum-plants, the produce of which is not less esteemed than that from Arabia, comprise the asafœtida, mastic galbanum, sanderic ammoniac, sarcocolla, and tragacanth. Much attention is likewise given to the cultivation of the white rose, from which is distilled an attar or essence highly valued in Asia. Pasturage, however, is a more favourite pursuit than tillage. The breed of sheep peculiar to this province called dumbder, is small and short-legged, with a long bushy tail; its wool fetches a higher price in the market than that of any other variety in Persia. Camels also, and goats, are bred in great numbers, as their hair is thought to make a fibre at once stronger and more delicate than that of animals reared elsewhere. Oxen and horses are little attended to. The forests are infested with wild beasts of the cat and bear tribe; and these are many species of serpents, some being highly venomous. On the S. coast sea-fish is abundant; but the pearl fishery, once very profitable, has been abandoned in consequence of the too great depth of the oysterbeds. The mineral riches might be made a source of considerable wealth, for most metals are abundant; but iron, copper, and sulphur are the only products hitherto obtained. The manufactures comprise fine woollen fabrics, carpets, goats' and camels' hair shawls, coarse linens, and a peculiar kind of matchlock, much esteemed in the east. These articles, with chenna, a yellow dye, fruits, gums, &c., are either sent N. by caravans, or exported from the port of Gombroon.

The inhabitants were formerly almost exclusively Guebers; but the number of these is now less than 40,000. The Persians constitute the chief mass of the population; but there are also many Belooches and Arabs of different tribes. The government is vested in a beglerbeg; and the province is divided into nine districts, each of which is under a hakim or Beutenant. The taxes on land, and imposts on manufactured goods, are very oppressive, and operate as a great hinderance to industry. The S. part of Kerman, called Moghostan, is not subject to Persia, but to the imam of Muscat, who receives from it a yearly tribute of 7000 tomans. The Arabs of various tribes are governed by their respective sheiks. (*Kinneir's Persia*, p. 194–201; *Pottinger's Travels*, p. 220–227; *Hagemeister sur l'Asie Occidentale*, &c.)

KERMAN, or BERJAN (an. Carmana), a city of Persia, and cap. of the above prov., 230 m. E. Shiraz, and 340 m. S.E. Ispahaan; lat. 29° 58' N., long. 56° E. Pop. estimated by Pottinger at 30,000. This city, which was once more prosperous and extensive than at present, stands on the W. side of an extensive plain, so close to the mountains as to be completely commanded by two of them. The walls, pierced by four gates, are high and built of mud, flanked outside by a dry ditch, 20 yards wide, and 10 yards deep. On the S. side of the town is a citadel, in which the governor resides. The bazaar, well supplied with every article of necessity and luxury, is covered in with very elegant domes, built of a beautiful blue stone procured in the adjoining mountains. There are nine good caravanserais within the walls, several mosques, baths, &c.; but most of these are in a ruinous condition. The trade of Kerman, however, is still very considerable; and it is celebrated for its manufactures of shawls, carpets, and matchlocks, which are exported to Khorassan, Balk, and Khiva, Arabia, Sinde, and all parts of India. The shawls of Kerman are of coarser quality, but approaching nearly to that of Cashmere in colour and general appearance to the interior Cashmeres. Immense quantities of the commoner kinds are sent to all parts of Turkey; they are about two yards square, very low in price, and are generally worn by the lower classes in W. Asia.

Kerman, formerly one of the proudest cities of the Persian empire, owed much of its former opulence to its situation on the road from Bokhara to Gombroon, a port which has been almost superseded by Bushire. Domestic and foreign wars, however, with repeated pillages, have all but ruined it. In 1794 it was besieged and taken by Aga Mahommed Khan; the walls and public buildings were then levelled to the ground, a licentious soldiery were allowed to pillage it during three months, vast numbers of the inhabitants were put to death, and 30,000 are said to have been sent into exile. From these calamities Kerman is only very slowly recovering, nor does the present state of its trade warrant the conclusion that it will ever attain its former importance. (*Kinneir's Persia*, p. 198; *Pottinger, &c.*)

KERMANSHAW, or KERMANSHAH, a city of Persia, the cap. of Persian Kudistan and of a district bearing its own name, 82 m. W.S.W. Hamadan, and 290 m. S.W. Ispahaan; lat. 34° 26' N., long. 47° 13' 15" E. Pop. 30,000? It stands a short distance from the right bank of the Kerkah or Karaza, in a beautiful plain open to the S., but inclosed on every other side by lofty mountains. It is surrounded by a substantial brick wall, having round towers at its four angles and a deep ditch in front. The citadel, strongly fortified, is the residence of the beglerbeg, who belongs to the royal family of Persia. The streets are narrow, crooked, and unpaved; but the town is adorned with many gardens, has 14 hammams or public baths, four mosques, several bazaars, and a spacious caravanserai kept in tolerable repair. Its manufactures consist chiefly of woollen carpets and swords, mostly sent to Bagdad, with cotton, very delicious grapes, and other products of the rich soil belonging to the district. Considerable advantages accrue to the town in consequence of its situation on the great caravan road between Persia, Bagdad, &c., and Asiatic Turkey. Great improvements have been made by the existing dynasty in its fortifications and public buildings, and it has become the residence of one of the members of the reigning family; so that its population and general importance have been steadily increasing during the present century.

About 6 miles E. of Kermanshaw, on the road to Hamadan and in the N. range of mountains, are the excavations and sculptures of Taki Bostan. The most considerable of these is an arch cut in the rock, 60 feet high, 20 feet deep, and 24 feet wide; on the top is an emblematic figure, finished by two angels, the sculpture of which is tolerably perfect and in good taste. At the extremity of the arch is the figure of a mounted warrior clothed in chain armour, with a shield on his left arm, a lance in his right hand, a quiver at his side, and a tiara on his head. The horse is well proportioned and tolerably carved. The representation of a boar-hunt occupies the entire left side of the arch; it is remarkably well executed, "some parts being so exquisitely finished (according to Kinneir) that they would not have disgraced the finest artists of Greece and Rome." At the upper end of another cave, similar in shape and size to that just described, is a basso-relievo of two kings in the costume of Persepolis, and wearing globular crowns identifying them with members of the Shapour dynasty. Near the entrance of this cave, also, are three figures, two of which are treading on the third, who is prostrate. The origin of these sculptures is a matter of doubtful conjecture: some attribute them to Semiramis, while by others they are ascribed to the successor of Alexander; but, if Silvestre de Sacy's translations of the Pehlvi Inscriptions be correct, we cannot greatly err in attributing them to the monarchs of the Sassanian dynasty. (For farther particulars, see *Ker Porter's Travels*, ii., 163–204; *Kinneir's Persia*, 129–136; and, above all, *Ritter's* very full and satisfactory description, *Erdkunde von Asien*, part ix., p. 367–386.)

The date of the foundation of Kermanshaw is not accurately known, but it is generally attributed to Bahram (Varanes IV.), the son of Shapour II., about 400 years after Christ. Kobad improved it, and built a citadel, which, after having been almost destroyed by the Turks, was re-established by Koali-khan, when he restored its independence in 1723.

KERRY, a marit. co. in the S.W. part of Ireland, prov. Munster, having N. the estuary of the Shannon, E. and S. the cos. of Limerick and Cork, and W. the Atlantic ocean. Area, 1,148,730 acres, of which 529,969 are unimproved mountain and bog, and 14,669 water, including the lakes of Killarney, so famous for their scenery (see KILLARNEY). This county is particularly wild, rugged, and mountainous. Macgillicuddy's Reeks, the highest mountains in Ireland, lie to the W. of Killarney; and several other mountain ridges rise to above 2000 feet in height. The coast is deeply indented by Tralee and Dingle bays, and the estuary of the Kenmare: Dunmore Head, between the bays now named, in lat. 52° 7' 30" N., long. 10° 28' W., is the most westerly land in Ireland, and consequently in the United Kingdom. The climate is particularly mild, but also extremely moist. The soil on the low grounds mostly rest on a limestone bottom; it is very fertile, and produces fine herbage, which the mildness and moisture of the climate maintains in a constant state of verdure throughout the year. The arbutus flourishes in the greatest vigour round Killarney, and in other places in this county. Large flocks of goats are fed on the mountains, which also depasture great numbers of the pure Irish breed of middle-horned cattle. There are some rather extensive dairy farms; but, speaking generally, agriculture is at a very low ebb. Tillage farms are, for the most part, very small, and the occupiers miserably poor. The potato is the only article they reserve to themselves; cattle, corn, butter, pigs, eggs, &c., all go to market to make up the rent. Still, however, improvements are taking place; good roads now lead into districts that were formerly next to impervious; and some landlords, among whom Lords Headly and Lansdowne deserve to be especially noticed, have laboured, with considerable success, to introduce an improved system of management on their estates, and to meliorate the condition of the occupiers. In some parishes the greater part of the tillage is performed by means of the loy or spade; but Scotch and other improved ploughs are beginning to be introduced. The seaweed, which abounds along the seashore, furnishes an ample supply of manure; but it is in most parts neglected or injudiciously applied. Houses and cabins as bad as possible. Property mostly in very large estates; but some of them are leased for ever. Average rent of land 5s. 1d. per acre, being, Donegal excepted, the lowest in the kingdom. The Irish language is in many parts used to the total exclusion of the English; and, in consequence, old customs and habits maintain their ground in a remarkable degree. Minerals, though in a great measure unexplored, are, no doubt, of considerable value and importance. Copper mines have been wrought near Killarney, and one is now wrought on a small scale at Cahirciveen. Valentia Island produces good slate for roofing and flagging. Manufactures can hardly be said to exist. Principal rivers,

Feale, Lane, the outlet of the lakes of Killarney, Roughan, and Maug. Principal towns, Tralee, Killarney, and Dingle. Kerry is divided into eight baronies and 83 parishes; and sends three members to the House of Commons, viz., two for the county, and one for the borough of Tralee. Registered electors for the county, in 1836-30, 1218. In 1831, Kerry had 41,994 inhabited houses; 45,094 families; 263,196 inhabitants, of whom 131,696 were males, and 131,430 females.

KERSHAW, district, S. C. Situated a little N.E. of the centre of the state, and contains 792 sq. m. Bounded N.E. by Lynch's creek. Drained by Wateree river, and its tributaries, and by Little Lynch's creek. First settled by Irish Quakers in 1750. It contained in 1840, 11,494 neat cattle, 15,984 sheep, 2004 swine; and produced 4744 bushels of wheat, 160,300 of Indian corn, 11,525 of oats, 10,080 of potatoes, 14,475 pounds of cotton. It had 29 stores, one cotton factory, with 190 spindles, eight flouring-mills, 34 grist-mills, 10 saw-mills, four tanneries, one printing-office, one weekly newspaper; three academies, 187 students, 13 schools, 304 scholars. Pop.: whites, 3988; slaves, 8043; free coloured 250; total, 12,181. Capital, Camden.

KERTSCH, a seaport town of European Russia, in the Crimea, on a spacious bay on the W. side of the straits of Yenicale, lat. 45° 21' 30'' N., long. 36° 25' 90'' E. It has recently been a good deal improved; and contains some handsome edifices, and from 2000 to 3000 inhabitants. This town occupies the site of the ancient Panticapæum, the seat of the Bosphorian kings, and once the residence of Mithridates. A mound in its vicinity is said to be the tomb of that formidable and inveterate enemy of Rome; but this is contradicted by the most authentic accounts, which represent Mithridates as having been buried, by order of Pompey, in the sepulchre of his ancestors at Sinope. (Biographis Universalis, art. Mithridates.) The quarantine for the sea of Azoff has been established here; and it seems probable that it will at no distant period supersede Taganrog as the emporium of that sea. Corn, salt, and hides, are the principal articles of export. In the outer road, five or six miles from the town, there are 19 feet water; in the inner bay there are 14 feet, and close in shore it shoals to from 9 to 11 feet. (Hagemeister on the Trade of the Black Sea, p. 63, Eng. trans.)

KESMARK (Germ. Kaisersmarkt), a royal free town of Hungary, co. Zips, on the Poprad, a tributary of the Vistula, at the foot of the Tatra mountains, 139 m. N.E. Pesth. Pop. 4330, of whom about 2500 are Protestants. It is surrounded with old and decayed double walls, and entered by three gates, near one of which the Emperor Sigismund, in 1433, erected a large tower, to protect the town against the attacks of the Hussites. Paget says, " In Kesmark there is nothing remarkable, except the ruins of an old castle which formerly belonged to the family of Tököly, by whose restless ambition, and warlike talents, Hungary was involved in a series of civil wars, which, but for Sobieski's timely aid, would probably have ended in delivering the whole country into the power of the Turks." (Vol. i., p. 443.) Kesmark has, however, several handsome public buildings, as the town-hall, with an elegant tower, and a large Roman Catholic church; besides a Roman Catholic high school, Protestant lyceum, girls' school, &c. Many of its inhabitants are linen weavers and dyers; others carry on a brisk trade with Galicia in wine and garden produce. (Oest. Nat. Encyc.; Berghaus; Paget's Hungary.)

KESWICK, a market town of England, co. Cumberland, ward of Allerdale, par. of Crosthwaite, on the Greta, in a well cultivated valley at the foot of Skiddaw, and contiguous to the N. end of Derwent water, or Keswick lake, 22 m. S. by W. Carlisle, and 18 m. E. by N. Whitehaven. Population in 1831, 2159. This neat and finely situated town, which may be regarded as the capital of the English lakes, consists principally of one long street of well-built houses. It has manufactures of linsey-woolsey stuffs, and fancy waistcoatings; black lead pencils are also made in the town, of lead from the famous mine in Borrowdale; and the potting of char taken in the lake is a considerable business. Copper mines were formerly wrought in the vicinity, but they have been long abandoned. The principal dependence of the place is on the crowds of visiters to the adjacent lakes and mountains, who are here supplied with lodgings, guides, conveyances, &c. It has a free school, a national school, a workhouse, and two museums, containing many fine specimens of natural history peculiar to the county. Property, which at present produces above £300 a year, was bequeathed in 1842, by Sir John Banks, Chief Justice of the Common Pleas, for behoof of the poor of this, his native town.

Keswick lake, or Derwent-water, is about three miles in length, by rather more than one mile in breadth, extending over an area of 1989 acres. It has numerous small islands, is embosomed among lofty mountains, and, from its picturesque scenery, is deservedly called the " gem" of the lakes. (Tattersall's Guide to the Lakes, p. 37, &c.)

KESZDI-VASARTHELY (Germ. Neumarkt), a town of Transylvania, in the Szekler-Land, 45 m. N.E. Cronstadt. Population about 5008. It has a Protestant gymnasium, several breweries and distilleries, and manufactures of hats, paper, and cloth; but it is chiefly noted for its military establishments. It is the head quarters of the second regiment of Szekler infantry, in the Transylvanian military frontier, and has a celebrated military school. " This institution was founded by the late emperor, and is supported partly by a royal grant and partly by the Szeklers themselves. The regulation of it is entirely in the hands of the government. On the foundation there are 100 boys, from six to eighteen years of age, who are fed, clothed, and taught, free of all expense. A few additional scholars are admitted on the payment of about 16s. per month. The children, when they have finished their education, are drafted into the infantry, and often rise to the rank of officers. The course of education, besides drilling, exercising, &c., includes writing, reading, arithmetic, geography, mathematics, military drawing, and the German language. In fact, all the lessons are given in German, all the books are German, and the children are even obliged to speak German to each other. The national language is never heard within the walls of the school." Hence the Szeklers affirm, that the grand object of the school is to denationalise their children, and make them renounce their native language; so that its institution, whatever may be its ultimate influence, tends, in the mean time, to keep alive the distrust which these borderers entertain of the Austrian government. (Paget's Hungary, i., 417, 418; Oest. Nat. Encyc.)

KESZTHELY, a market town of Hungary, in the circ. on the other side of the Danube, co. Szalad, near the W. end of lake Balaton, 36 m. S.W. Veszprim, and 96 m. S. Presburg. Population 7000. " It is," says Mr. Paget, " a thriving little town, and of considerable importance from the great school of agriculture founded here by Count George Festetits, and known as the Georgicon, which, though no longer in so flourishing a state as formerly, has still several professors and practical teachers maintained at the count's expense." The object of this establishment is to form useful and well-instructed officers and accountants for the management of estates, to give instruction in particular branches of husbandry to the peasantry, and to furnish opportunities for farmers to improve their knowledge of agriculture. Eight or ten scholars are pensioned by the count, the rest being independent students; and the school is divided into six sections, 1. for scientific agriculture, and its auxiliary sciences; 2. for the law of property, as affecting landlords and tenants; 3. for practical husbandry, as taught to the peasantry; 4. for forest-planting and the chase; 5. for horse-breeding, training, &c.; and 6. for teaching girls the branches of knowledge connected with housekeeping. The complete course appointed for the pensioners lasts three years; but others may select their pursuits, and limit themselves to one or two years, as they think proper, the theoretical course lasting from the beginning of November to the end of August. In the Georgicon, large apartments are fitted up as lecture-rooms, depositories for philosophical instruments, museums, &c.; chambers are set apart for the pensioners; and the lower floors are occupied by the farming servants and their families, and by a spacious workshop for carpenters and coopers. The outbuildings comprise stalls for fattening cattle, a shed for sheep, a granary, brewhouse, and a house for silkworms and the winding of silk; gardens and orchards of different kinds are laid out for the purpose of teaching horticulture in all its branches, and on a farm set apart from the rest of the count's estates, practical instruction is given in the rotation of crops after the Norfolk system. In fact, the Georgicon, as described at length by Dr. Bright, (to whose valuable work we beg to refer the reader, who desires further information,) is a most complete establishment, and if conducted at present with the same vigour as at the time of the Doctor's visit, it cannot fail of being highly serviceable. The other educational institutions are a Catholic gymnasium, a high and normal school. The public buildings comprise besides the Georgicon, a fine castle, in which Count Festetits resides, and which contains a library of 15,000 volumes, two Catholic churches, a convent, and a hospital. Wine, from the extensive vineyards in the neighbourhood, is a considerable article of trade, and several hands are employed in weaving woollen fabrics. (Paget's Hungary, vol. ii., 577; Bright's Travels in Lower Hungary, p. 380-389.)

KETSKEMET, or KU7KEMET, a market town of Hungary, circ. on this side of the Danube, co. Pesth, 50 m. S.E. the cap., lat. 46° 54' 22'' N., and long. 19° 43' E. Population 36,000 ? The houses are generally low, the streets long, narrow, and crooked, and the surrounding districts of a dull monotonous character: there are five churches, (two Roman Catholic, and one each belonging to Greeks,

Lutherans, and Calvinists,) a Franciscan convent, a reformed college and gymnasium, a Piarist college, a normal school and a school of design, an orphan asylum, and a military hospital. The breeding of horses, cattle, and sheep is the chief employment of the population; and there are some tanneries and soap factories. (Oest. Encyc.; Dict. Géog.)

KETTERING, a market town and par. of England, co. Northampton, Huxloe hund., on an affluent of the Nen, 14 m. N.E. Northampton, and 65 m. N.N.W. London. Area of par., 2840 acres. Population in 1831, 4099. The centre of the town comprises a spacious area, surrounded by well-built houses and shops, with a commodious sessions-house; but in the suburbs are many low thatched tenements that have a mean and wretched appearance. The church, considered a fine specimen of ecclesiastical architecture, has an elegant embattled tower at its W. end, surmounted by a light crocketed spire: the living is a rectory. There are places of worship also for Wesleyan Methodists, Baptists, Independents, and the Society of Friends. Sunday schools are attached to all, except the last; and there is a small free school; this, and an almshouse for six poor widows, are the only endowed charities of the town. Several hundred weavers are engaged at Kettering, and the neighbouring villages of Rothwell and Desborough, in making silk plush for hats: a great number of hands were formerly employed in woollen and worsted weaving, but this branch of industry appears to have declined of late years. Petty sessions are held here on alternate weeks. Markets on Saturday, but for cattle and sheep on alternate Fridays. Fairs, Thursday before Easter, Friday before Whitsuntide, Thursday before Oct. 11, for horses, cattle, and farming stock.

KEW, a village of England, co. Surrey, hund. Kingston, on the S. bank of the Thames, about 6 m. W. London, forming with Petersham a united parish, having an area of 580 acres, and a population, in 1831, of 1440 persons. This village, which is connected with Brentford on the opposite side of the river by a handsome stone bridge of seven arches, consists principally of the houses on and near a large and neatly kept green. The parish church is a small brick structure with a turret at the W. end: it was intended only as a chapel of ease to Kingston, of which Kew was an appendage, till it was made a separate parish in 1769. Many handsome residences are scattered over the village, but none deserves particular mention except Kew House, for many years the favourite residence of George III. and his queen. It was taken on lease from S. Molyneux, Esq. by Frederick, Prince of Wales, and was greatly improved in its interior fittings by Kent. George III. continued to hold it on lease, and it is still occupied by members of the royal family or persons belonging to their households. Near this house, but close to the river-bank, a new palace was commenced by George III. under the directions of Wyatt; but the situation and plan of the building proved to be very ill chosen: it was never completed, and was ultimately taken down in 1827. The gardens, comprising about 120 acres, were laid out by Sir W. Chambers, for Frederick, Prince of Wales. The botanic garden contains a fine collection of native and exotic plants. In the pleasure-gardens are different grotesque, if not very elegant, buildings; the largest and most celebrated being an octagonal Chinese pagoda of 10 stories and 163 feet high, from the top of which is an extensive view of the surrounding country. The pleasure-grounds are kept in good order, and are accessible to the public on Sundays and Thursdays, from 12 till sunset, from midsummer to Michaelmas. The botanic garden and arboretum are open daily during all seasons from one to three in the afternoon.

KEYNSHAM, a market town and par. of England, co. Somerset, hund. of its own name, at the confluence of the Chew with the Avon, 5 m. E.S.E. Bristol, and 108 m. W. London. Area of parish, 3230 acres. Population, in 1831, 2142. The town is built on a rock, and consists of a single street, about a mile long. The church, which stands in the centre of the town, is a large and handsome edifice, with a fine lofty tower at its W. end, and some curious monuments: the living is a vicarage, in the gift of the Duke of Buckingham. The Wesleyan Methodists and Baptists have places of worship, to each of which, as well as to the church, Sunday schools are attached. A well conducted charity school also furnishes a plain education to poor children of both sexes. The river Chew runs through the E. end of Keynsham, and falls into the Avon at the bridge, which is of stone, and consists of 15 arches: another bridge crosses the Chew on the Bath road. The tides of the Avon ascend up to this town. The clothing trade, formerly considerable, has now almost wholly fallen to decay, though a few people are still employed in spinning and winding for the clothiers of Bradford and Shepton-mallet. Coarse linen-weaving has been introduced within the last 15 years, with little success; but a good deal of

business is carried on in malting. Petty sessions are held here on the market day, Thursday. Fairs, March 24, and August 15 for cattle and cheese.

KEYTESVILLE, p. v., capital of Chariton co., Mo., 91 m. N.W. Jefferson city, 985 W. Situated on a branch of Grand Chariton river, and 15 m. N. from its entrance into Missouri river. It contains a courthouse, jail, four stores, a saw and grist-mill, and about 200 inhabitants.

KEY WEST, p. v., capital of Monroe co., Flor. Situated on the N.W. end of Thompson's island, which is 4 m. long and 1 m. wide. It is one of the Florida keys, has a fine harbour, admitting vessels requiring 27 feet of water, and capable of being well fortified. It may become the key to the gulf of Mexico, as the passage here is safer, and 90 miles nearer, than round the Tortugas, and has nine feet of water. It was incorporated in 1829, and contains a courthouse, jail, an Episcopal church, 32 stores and groceries, two large warehouses, 110 dwellings, about 500 inhabitants. It has a lighthouse. About 17,000 bushels of salt are manufactured annually, by solar evaporation. The thermometer ranges from 50 to 90 degrees of Farenheit. The whole island lies on a bed of limestone, about a foot beneath its surface, and wells are dug into the rock, to the level of the sea, where fresh water is obtained; but rain water is chiefly used. Most of the inhabitants are employed as wreckers, and receive on an average $77,000 annually. About 15 vessels are annually wrecked on the Florida reef, and the inhabitants of Key West are employed in saving the property.

KHARAN, a town of Beloochistan, prov. Sarawan, 100 m. S.W. Khelat. Population uncertain, but probably about 3000. It is situated in an extremely mountainous district bearing the same name, and is surrounded by a mud-built wall with bastions. It is the residence of a sirdar, who can send into the field about 800 excellent and hardy soldiers. The camels of Kharan are the strongest and most active in these regions, a circumstance that gives their masters a decided superiority over their neighbours in their predatory pursuits. (Pottinger's Travels, p. 130.)

KHARKOFF, a government of European Russia, having on the N. Tchernigoff and Koursk, on the E. Voronesz, on the S. Ekaterinoslaff, and on the W. Poltava. Area, 17,450 sq. m. Population, in 1828, 1,234,000. This, like the other governments of Little Russia, has a flat, monotonous surface, and a very fertile soil. It has nearly 470,000 deciatines of forests. Principal rivers, the Donetz, Orkol, and Vorskla; but none of them are navigable, at least, for any considerable distance. All sorts of corn are raised, the produce in ordinary years amounting to above 5,000,000 chetwerts, of which, about 1,600,000 are exported. Flax and hemp, tobacco, hops, &c., are also raised, and the potato is extensively grown. Cattle excellent: there are few peasants without bees. With the exception of distilleries, which are numerous, and some tanneries, and establishments for the preparation of tallow and saltpetre, manufacturing industry can hardly be said to exist. The population consists of Little Russians, Great Russians, and Cossacks. Some regiments of cavalry are colonised in this government. (Schnitzler, La Russie, &c., p. 471.)

KHARKOFF, the cap. of the above government, on the Lopanh, lat. 49° 59' 27" N., long. 36° 26' 39' E. Population 13,000. It is built of wood; has narrow, crooked, and dirty streets: the ramparts by which it was formerly surrounded have been converted into gardens and public walks. It is the residence of the provincial authorities, and has a cathedral, a gymnasium, an ecclesiastical seminary, &c. Kharkoff is the seat of a university, founded in 1804, which, in 1835, had 55 professors and masters, and 349 pupils. It possesses a pretty good library, and a valuable collection of medals. This town is the seat of a considerable commerce. Four fairs are held each year, of which that called Krechtchenski (Jan. 3-15), and that of the Trinity, are the most extensive. If we may depend upon the official accounts, merchandise to the amount of 31,544,774 roubles was brought, in 1823, to the first of these fairs, above two thirds of which was disposed of. One of the other fairs is exclusively, or principally, for wool. (Schnitzler, ut supra.)

KHELAT, or KELAT, a city of Beloochistan, of which it is the cap, and a fortress of considerable strength, now in possession of the British; on an elevated site, on the W. side of a highly cultivated plain about 250 m. N. the Indian Ocean, and 240 m. S. by W. Candahar; lat. 29° 7' N., long. 65° 45' E. Population estimated by Pottinger at 20,000, chiefly Beloochee, Brahooes, Hindoos, and Afghans. The town, of an oblong form, is described by Pottinger, in 1810, as encompassed on three sides by a mud wall, 18 or 20 feet high, flanked at intervals of 250 paces, by bastions pierced, as well as the wall itself, with numberless loopholes for matchlocks. The defence of the fourth side is formed by the W. face of the hill, on which the town is partly built, being cut away perpendicularly. On the summit of this eminence stands the palace of the khan, enclosed by a mud wall, with bastions, kept in better repair than any other portion of the fortifications. In 1839, Major Willshire said, " The defences of the fort, as in the case of Ghizpee, far exceeded in strength what I had been led to suppose from previous report; and the towering height of the inner citadel was most formidable both in appearance and reality." (Parl. Report on Khelat.) It is, however, commanded by heights to the N. and W.; it has three gates, and above 2500 houses within the walls; and, in 1810, about half as many more were comprised in the suburbs. The houses are of half-burnt brick, on wooden frames, and plastered over with mud or chunam. The streets are generally broader than is common in the E., and have a raised footway on either side; but their centre is a receptacle for all sorts of filth; and they are dark and gloomy, from the upper stories of the houses nearly meeting. The markets are well furnished with flesh, vegetables, and other necessaries, at a cheap rate; and the town is supplied with excellent water by a spring which, according to Pottinger, is tepid during the night, but after sunrise becomes cold, and remains so the whole day. Some water-mills are turned by the stream from this source. Khelat has some trade and manufactures, respecting which *see* BELOOCHISTAN (Vol. i, p. 341.)

The many outrages alleged to have been committed on the followers of the army of the Indus, at the instigation of the khan of Khelat, compelled our interference; and Khelat was taken by storm by the British, Nov. 13, 1839, after a siege of a few hours.

KHERSON, a gov. in the S. part of Russia, in Europe, on the N. shore of the Black sea, between the rivers Dniestr, on the W., and Dniepr, on the E. Area, variously estimated at from about 25,000 to 30,000 sq. m. Pop., in 1838, including the military colonies, 765,000. Besides the great boundary rivers, already specified, it is divided into two not very unequal portions by the Bug. In the N. part of the government, the surface is undulating and covered with immense forests; but elsewhere it consists mostly of an immense steppe, without trees, and covered with grass the height of a man. Generally, the portion on the W. side of the Bug is decidedly more fertile than that on the E. side. Climate is extreme, the rivers being mostly frozen over for a short time during winter, while in the summer the thermometer rises sometimes to above 25° of Réaumer. Agriculture has made little progress, and is but a secondary pursuit, the rearing of cattle and sheep forming the chief employment of the inhabitants. The breed of sheep has been much improved, and is now the best in the empire. Among the horned cattle, buffaloes are common. Flax and hemp, tobacco, saffron, liquorice, &c., are all cultivated; and a good deal of an inferior acid wine is made. There are establishments for the cleaning and sorting of wool, tanneries, tallow and candle works, with manufactories of cloth, &c. The commerce of the government centres entirely at Odessa and Kherson, and is very extensive.

KHERSON, the cap. of the above government, on an eminence on the right bank of the Dniepr, about 60 m. above Kinbourn fort, at the entrance of the aestuary to that river, lat. 46° 37' 46" N., long. 32° 36' 33" E. It was founded in 1778, was fortified in 1780, and soon after became a large and flourishing town. Owing, however, to the difficulty of navigating the river, which, for 15 m. below Kherson, is shallow, and encumbered with shifting sand banks. Odessa, founded in 1792, soon took precedence of it as a commercial emporium, and it began to decline. Its population amounted in 1834, according to the official accounts quoted by M. Schnitzler, to 24,508. It is divided into four distinct parts: the citadel, the admiralty, and the Greek and military suburbs. Within the first are the government buildings, arsenal, prison, barracks, and the cathedral. The latter is the burial place of the celebrated Prince Potemkin, the powerful favourite of Catherine II., who died near Yassy, in 1791.* In the admiralty are the docks, for constructing ships of war, cut out of the limestone rock. They are sent down the river on machines, called camels, but only when there is a large flood. The Greek suburb is inhabited by the burgesses, and the military suburb by sailors and artizans.

Within these few years a part of the mast trade that used formerly to be confined to Riga, has been transferred to Kherson; and, besides masts, staves, planks, flax and hemp, corn, cordage, tallow, wool, of which it is a principal market, &c., are sent down the Dniepr to Kherson. But, owing to the cataracts and other obstructions to the navigation of the river between Ekaterinoslaff and Alexandrofsk, these shipments can only be made in spring and autumn; and, when the commodities have reached Kherson, it is found most convenient to ship them coastwise in small vessels for Odessa, where they are put on board the ships in quarantine. In fact, owing to the difficulties now mentioned, the greater part of the corn from the ci-devant Polish provinces, shipped at Odessa, is not brought down the Dniepr, but is conveyed direct to its destination, in wagons drawn by oxen.

John Howard, the celebrated English philanthropist, expired at Kherson, on the 20th of January, 1790, and is interred about 3 m. N. from the town, where an obelisk has been erected to his memory. (For farther particulars as to Kherson, see the article DNIEPR, in this dictionary, the works of Clarke and Lyall, as referred to in the note below; Schnitzler, La Russie, &c., p. 722; Hagemeister's Report on the Commerce of the Black Sea, English trans., p. 70, &c.)

KHIVA, KHAREZM, or ORGUNJE (an. Chorasmia), an indep. khanat of Turkestan, in Central Asia, properly comprising only a narrow strip of fertile land along the Oxus, in the lower portion of its course. Of late years, however, it has established a supremacy over the wandering Turkman hordes to the S. and W., and holds Merv (Mard), with its territory, on the road between Khorassan and Bokhara. Its dominion is believed at present to extend between the 36th and 44th degrees of N. lat., and 53d and 64th of E. long., having E. the Karakalpack territories and Bokhara, S. Affghanistan and the Persian province of Khorassan, W. the Caspian, and N. the Kirghiz steppe and the sea of Aral. The population of this extensive territory is estimated by Sir A. Burnes at only 300,000 at most, nearly the whole surface consisting of unproductive sandy wastes. The Oxus is the great fertiliser of the tract it passes through; many canals communicating with it have been cut for the purpose of irrigation, some of which are 30 m. in length; and the cultivated lands in the neighbourhood of the capital are surrounded with wet ditches. The climate and products are much the same as in Bokhara; the summer is warm, the air dry, and evaporation rapid; the winter is short, and ice lasts only a few days at a time. Agriculture is better attended to in the small extent of productive land comprised in this khanat than in some of the neighbouring countries. The lands, after being irrigated, are manured; but animal manure is scarce, from the faeces of the cattle being used as fuel, and their being seldom stalled. Wheat, barley, *djugari* (*Holcus saccharatus*), millet, sesamum, oleaginous plants, lentils, fruits, linseed, cotton, hemp, flax, and some rice, are grown. The vine thrives well; but the inhabitants, being chiefly Mohammedans, little or no wine is made. The distillation of brandy from raisins has, however, been introduced by the Persians; and, out of the capital, the inhabitants indulge pretty freely in its use. An intoxicating liquor, as well as a narcotic product for smoking, is obtained from hemp. Little tobacco is grown. Many of the fruits are good, and the melons are excellent; but the culture of fruit-trees is nearly abandoned for that of grain or fodder. Wood is sufficiently abundant in the N., and is not dear in the capital; but over all the desert the only vegetation is a few stunted bushes. Horned cattle are few; sheep and goats are much more numerous, their flesh, with that of the horse, forming the chief animal food of the inhabitants. Camels are the principal beasts of burden, and almost every khivan possesses one. Agriculture and cattle rearing occupy most of the settled population; but some cotton and silk stuffs, shawls, &c., are made by the women, and exported to the neighbouring countries. The dominant race in Khiva, as in Bokhara, is the Uzbek, to which the khan belongs; the rest of the population consists of Oulgours, Turkmans, Karakalpacks, Tadjiks, about 3000 families, chiefly prisoners of war from Bokhara, and a few Affghans, Jews, Armenians, Persians, Eimauks, Kirghiz, &c. The Uzbeks enjoy no privileges over the rest, but they compose the chief portion of the khan's army. The Turkmans are altogether nomadic, and live principally by plunder, especially the capture and sale of slaves. They seize upon the subjects of Russia on the Caspian, and make many inroads into Khorassan: Bokhara and the whole of Turkestan is supplied by them with Persian captives. In 1835, according to Sir A. Burnes, there were as many as 3000 Russian slaves in Khiva; and the capture of her subjects was one of the principal causes (and a substantial one certainly) of the late hostile attempts of Russia against Khiva. It is estimated that from 30,000 to 40,000 of the population of the khanat are slaves. They have frequently a piece of land given them to cultivate, or are permitted to exercise some handicraft, paying an annual rent to their masters for the privilege, from the produce of which they are afterward frequently able to ransom themselves. No foreign slave, however, even after the purchase of his liberty, is permitted to leave the country. Meyendorf, in comparing this khanat with that of Bokhara, observes: "Though the inhabitants of the two countries are of the same race, and profess the same religion, the schools of Khiva have never

* Dr. Clarke says that the body was taken up by order of the Emperor Paul, and deposited in " the first hole that could be found." (Vol. ii., p. 298, 8vo ed.) But Dr. Lyall assures us that this humane order, though given, was not obeyed. (i., p. 214.)

enjoyed the same reputation as those of Bokhara; the Khivans are more barbarous than the Bokharese, as is attested by an inferior agriculture, worse habitations, a more limited commerce, less wealth, and a more savage mode of life.'" (*Voyage à Boukhara*, p. 111.) According to Burnes, the Khivans are at best but an organized banditti, protected by the natural strength of their country. The trade of such a country may be dismissed in a few words. Four routes exist for communication with Russia: one through the Kirghiz steppe, W. of the Aral sea, to Orenburg; a second by way of Sarachak, or Sarachik, on the Oural, also to Orenburg; a third through Sarachak to Astrakhan; and a fourth from Khiva to Karaghaan, on the E. shore of the Caspian, whence goods are sent by sea to Astrakhan (*Heeren, Researches on Asiatic Nations, &c.*, transl. ii., 268.) About 2000 camels go annually to Orenburg, Astrakhan, and some towns of Caubul and Persia, with wheat, barley, silk and cotton fabrics, and yarn; and about a dozen large boats come annually from Astrakhan to Karaghaan and the gulf of Manghislak, with the products of Russia and the west, to be exchanged for those brought by the caravans from Khiva. The chief imports are slaves, coin, iron, and copper, wrought and unwrought; handkerchiefs, wax, honey, sugar, tea, which, as in Bokhara, is a favourite article; cochineal, spices, hardware, &c. The commerce with Persia is insignificant. The merchandise which goes to Astrabad is conveyed on camels, at a charge averaging from 3½ to 4 roubles per *pood*, under the conduct of Turkman guides. The trade of Khiva is solely in the hands of Turkmans, Khivans, and Persians; none but Mohammedan merchants being suffered to transact business within the khanat. No foreign merchants pass through or into the country with ease or safety; when not openly robbed of a large portion of their goods, the caravans are delayed by the khan's officers, the bales of merchandise are opened, and much property has been at times extorted. The khan demands duties at the port of Manghislak, on the Caspian, which lies opposite Astrakhan, and sometimes (it is said) on the Jaxartes, E. of the Aral sea. In order to reach Bokhara by a route avoiding Khiva altogether, the Russians attempted, in 1820, to send caravans by way of the latter river; but the khan took umbrage at a measure which turned the traffic from his own territories, and sent an army to the Jaxartes, which intercepted a caravan, and occasioned the destruction of its merchandise. Since then, no attempt has been made by the Russians to follow any route other than that through this khanat; but no intercourse of a really friendly nature has taken place between the two countries.

The commercial duties realized by the khan amount to, perhaps, half his total revenue, which latter is roughly estimated, by Helmersen (*China, &c.*, p. 45), at 2,000,000 roubles, or francs: the remainder of this sum being made up of one fifth of the produce of every predatory excursion of his subjects, a family tax of three ducats a year, taxes on war-horses, on land cultivated by slaves, &c. A regular transit duty of 2½ per cent. *ad valorem* is levied on all kinds of merchandise passing through the country. The government is despotic: for judicial affairs, each town has its *atalyk*, or judge; and in the capital is a central court of justice in the last resort, composed of the *cadi* or chief priest, the four ministers, and other members nominated by the khan. The khan may sometimes raise a force of 10,000 men, and has a park of nine pieces of ordnance. His troops, which are mostly cavalry, are entirely composed of Uzbeks and Turkmans, and armed like those of Bokhara: some of the Turkmans carry bows and arrows. There are in the khanat, besides Merv, only two towns worth notice, Khiva, the capital and seat of government, and Orgunje, the chief commercial town, and largest of the two.[*] Khiva was tributary to Bokhara till the late khan rendered it independent, early in the present century. The present khan, in 1832, led a hostile army into Merv, which he subdued, but he has always maintained friendly relations with Bokhara. Political relations have long existed between Russia and Khiva, envoys having been sent from the one to the other as early as the time of Peter the Great. Latterly, the Russians have determined to put an end to the robberies committed by this horde; and, though the impracticable nature of the country has hitherto hindered them from reaching Khiva, there is little doubt of their ultimate success.

KHIVA, a town of Central Asia, cap. of the above khanat, and residence of the khan; in an irrigated and fertile plain near the Oxus, 290 m. W.N.W. Bokhara, and 790 m. S.S.E. Openburg, on the high road between those two cities: lat. 41° 40' N., long. 59° 23' E. Pop. doubtful, but probably from 10,000 to 12,000. The town is surrounded by a mud wall and wet ditch, and contains about 700 houses, the suburbs comprising 1200 more. Khiva has a palace, which,

[*] This is the statement of Burnes. Hagemeister (1839) says that all the commerce of the khanat is now centred in the capital.

like nearly all the rest of the dwellings in the town, and in the khanat generally, is of mud, though placed upon an eminence composed of stone. The only stone buildings in the town are three mosques, one having a handsome minaret; a school, and a caravanserai. Khiva is externally picturesque, being surrounded with gardens; but its streets are so narrow as scarcely to admit a laden camel. Its population is very mixed; its chief trade is in slaves, for which it is the largest mart in Independent Turkestan. (*Helmersen, China, Bokhara, &c.; Meyendorf; Mouraviof; Burnes, &c.; Heeren; Chinese Repository; Zimmermann's Memoir on Khiva*, 1841.)

KHOI, a town of Persia, prov. Azerbijan, and cap. of a distr. 70 m. N.W. Tabriz. Pop., according to Smith, about 5000 families, or 30,000 inhab. It is situated on a tributary of the Kur, about 25 m. N. from the lake of Uruiah, and is a handsome, well-built town, in much better repair than most others in Persia. It has few mosques or large public buildings, but the regular streets, shaded with avenues of trees, give the town, on the whole, an appearance of respectability, and even grandeur. A large and handsome bazaar, with a caravanserai, furnishes ample accommodation to the merchants, who carry on a considerable trade with Turkey and E. Persia. The suburbs were formerly inhabited by about 600 Armenians, but their number has greatly decreased since the war with Russia, when most of them migrated N. of the Araxes. The plain of Khoi is celebrated as the scene of a great battle fought in 1514 between Shah Ismael and Selim I., in which the Turks, though the most numerous, were signally defeated. (*Smith and Dwight's Miss. Researches*, p. 315; *Joubert, Voyage en Perse*, p. 148; *Ritter, Asia*, ix.; *Kinneir's Persian Empire*, p. 154.)

KHOJEND, a town of Indep. Turkestan, in Central Asia, khanat of Khokan, near its W. extremity, cap. distr. of same name, on the Jaxartes, 90 m. W. Khokan, and said to be as populous as that city, or Samarcand, from which it is 150 m. N.E. It is built on rising ground, and protected by walls, which, however, are much decayed on the S. and W. sides. It is surrounded by wet ditches, and intersected by canals. It is of high antiquity, and near it, Nazarov says, are some remarkable ruins. Khojend has manufactures of coarse cotton goods, and a brisk trade in these, and in Russian merchandise. It is the station at which the caravans entering the khanat from Bokhara pay toll; as the town of Uech is for those entering from the Chinese dominion. (*Nazarov; Helmersen; Ritter, Asien Erdkunde*.)

KHOKAN, KOKAN, or FERGHANA, an indep. khanat of Turkestan, in Central Asia, between lat. 40° and 45° N., and long. 67° and 75° E.; having N. the Kirghiz steppe, E. and S.E. Chinese Turkestan, S. the table-land of Pamere and Bokhara, and W. the desert territory of the Karakalpacks. It is, for the most part, mountainous, comprising a portion of the region which forms the W. wall of the great table-land of E. Asia. (*See* ASIA, vol. i., p. 162.) The Jaxartes (Sir or Sihoon), which rises not far beyond the E. boundary, traverses it E. to W., about its centre, watering many fertile tracts. Khokan is divided into eight provinces or districts. Great extremes of climate are experienced at different seasons. The produce are very similar to those of the countries to the S. and W. This khanat has a greater extent of cultivable and pasture land than Bokhara. In the S., corn and fruits, especially grapes and melons, grow in great perfection; and a proverb of Central Asia praises the "pomegranates of Khojend with the apples of Samarcand." This was the patrimonial kingdom of the Emperor Baber, who celebrates in lively terms its beauty and fertility. Cotton and the mulberry are articles of constant culture, silk being the chief staple, and one for which Khokan is famous. The pastures on the Jaxartes are excellent: sheep are the principal live stock, and wool is an important product. The camel, horse, and ass are extensively used; and horse-flesh is a common article of food. Game is very plentiful. Coal, iron, copper, jasper, lapis lazuli, &c., are the chief mineral products. The use of coal has been long known in Kokhâm, since Abulfeda speaks of "stones that flame and burn" being found there; and this important mineral may, at no very distant period, become a powerful auxiliary in civilizing this, at present, semi-barbarous region. The inhabitants are mostly Uzbeks; to which race, as in Bokhara and Khiva, the khan belongs. They are Mohammedans, and equally bigoted and strict in their religious customs with the Bokharese. The dialect they use is the Jagatai-Turkish. The rest of the population are chiefly Tadjiks (*see* BOKHARA, vol. i., p. 394) and Kirghis, who inhabit the N. and E. The Tadjiks are deprived of the rights of property, which they enjoy in Khiva and Bokhara, and are only suffered to cultivate the soil under the Uzbeks. After agriculture, and the rearing of sheep and silkworms, the chief occupation of the people is the manufacture of embroidered

silks and cotton goods. The former are much worn by the
Kirghiz hordes; the latter are sent in large quantities to
Bokhara, the returns being made in Russian goods, as iron,
steel, woollen cloths, otter skins, cochineal, vitriol, sandal-
wood, &c. Shawls and other Indian manufactures come
from Cashmere and the Punjab, by Caubul and Balkh. The
rest of the trade is chiefly with Budukshan; the intercourse
with Chinese Turkestan is very limited, owing to the ill-
feeling that exists between the khan of Khokan and the
Chinese authorities. The same cause renders the intercourse between Yarkund and Bokhara less frequent; the
nearest and best route between those cities leading through
the valley of the Jaxartes. This route, though passing
over mountains on which travellers experience difficulty
of breathing, is passable except in the three summer
months, when it is flooded by the melting of the snow. It
may be travelled by a caravan in 45 days; and merchandise may be conveyed from Bokhara as far as Khokan in
carts, the route between those two cities being the best in
all Independent Turkestan. Some Russian caravans from
Semipalatinsk, Petropawlawsk, &c., go by the route through
Khokan to the Chinese frontier: and three times the quantity of Russian piece goods are sold in this country than go
to Bokhara. Of late, indeed, this khanat has begun to
have a very active trade with Russia, the caravans engaged in this trade passing through a region much less interrupted by marauding parties than those between Russia
and Bokhara, through the territories of Khiva and the Little Kirghiz horde. According to Burnes, a commercial
intercourse is also kept up between Khokan and Constantinople. A duty of 2½ per cent., ad valorem, is laid on all
merchandise imported by Sooncite Mussulmans, and 5 per
cent. on the goods of all other individuals passing the frontier; but these duties are levied with little regularity. Internal commerce is entirely free, as in Bokhara, and the
trade is second only to that of the last-named country.
Each town possesses at least one caravanserai; and has
stated fairs, at which a good deal of business is transacted.
Besides the capital, the chief towns are Andejan, Khojend,
Turkestan, and the others which give names to the several
provinces. The government is despotic; the khan maintains an army of about 10,000 cavalry, which he can, on
an emergency, increase to 30,000. According to some Chinese records, it would appear that this country was formerly subject to China; it has, however, for many ages thrown
off its allegiance. In the early part of the present century,
many of the adjacent Kirghiz tribes were reduced to subjection; but about 1830, the khan having supported the
Mohammedans of Cashgar against their Chinese masters,
was totally defeated in a great battle, and his territories
invaded by the latter; since which the power of Khokan
has been on the decline. This and the neighbouring countries are interesting, from having been the seats of nations
whose armies have frequently changed the political face of
Asia, and even, in some degree, of Europe. Besides giving
birth to Baber, the conqueror of Hindostan, who succeeded
the throne of Ferghana in 1494, Khokan, and its vicinity,
abounds with localities intimately connected with the history of Jenghis Khan and Timour.

It is probable that this country will, at no distant period,
be united to Russia. The boundary between Russia and
Khokan, as determined about 1828, was fixed at the Kuk-su,
or "Blue river," long. 67° 30' E.?; but, according to the
Asiatic Journal (August, 1834, p. 374), the Russians had
then crossed that river, and erected forts on the Khokan side.

KHOKAN, a city of Central Asia, cap. of the above khanat,
and seat of its government, on the Jaxartes, 230 m. N.E.
Samarcand, and about the same N.W. Cashgar. It is reported to be about half the size of Bokhara, which is supposed to contain 150,000 inhabitants. Khokan is an open
town, but contains a palace fortified with a wall of mud, of
which material most of the houses in the town are constructed. The only exceptions are three bazaars, built of stone,
open twice a week for the purposes of trade; some ancient
monuments in different parts of the city, and some large
stables constructed of brick, and belonging to the khan.
There are a great many mosques and public schools, and
several caravanserais. Among the population are many
Cashmerians, and some Hindoos, Jews, Nogai-Tartars, and
Russians. The streets are narrow, unpaved, and unpleasant; but its vicinity is very productive, and sprinkled with
numerous gardens, cultivated fields, meadows, and villages.
(*Wanthen.* in *Bengal Jour.*; *Nazarov*; *Meyendorf*; *Burnes*;
Ritter, Asien Erdkunde, v. 730-784, &c.)

KHONSAR, a town of Persia, prov. Irak-Adjimi, 82 m.
W.N.W. Ispahan; lat. 33° 7' N., long. 50° 95' E. It is said,
by Kinneir, to contain 2500 families, or from 12,000 to 18,000
people. Its situation is singularly interesting and romantic,
at the base of two ranges of mountains, running parallel to
each other, and so near to each other, that the houses occupy the
bottom, and, at the same time, the face of the hills to some
height. The town is about 6 m. long, but only ½ m. broad,

and each house is separated and surrounded by its own
garden. The hills afford an ample supply of water; and
the appearance of the black and barren rocks, without a
particle of vegetation hanging over the gardens, forms a
striking contrast with the luxuriant and variegated foliage
of the plantation. No corn of any kind is grown in the valley; but the fruit is so abundant, that it alone enables the
inhabitants to procure in return every article either of necessity or convenience. According to Kinneir, it yields an
annual revenue of 5000 tomâns, exclusive of a payment
usually made in dried fruits and cotton chintz. (*Kinneir's
Persia,* p. 128.)

KHOOLOOM, KHULM, or TASH-KURGHAN, a town
of the khanat of Koondooz,[a] in Central Asia, on the Khulm
river, a tributary of the Oxus, and on the high road between
Balkh and Koondooz, 40 m. E. by S. the former, and 66 m.
W. by S. the latter city. Lat. 36° 29' N., long. about 68 E.
Population estimated by Burnes in 1832 at 10,000; and if
this number may be depended upon, either the population
had greatly declined during the preceding ten years, or
(which is most probable) the number of houses had been
much exaggerated by Moorcroft, by whom they were estimated at 20,000. According to the latter, "The houses are
built of clay and sun-dried bricks, of one story, with domes,
in the usual fashion of the country, and each stands by itself
in a walled enclosure, often containing fruit trees. The
streets are straight, of a moderate breadth, intersecting each
other at right angles, and have commonly a stream of water
running through them. The town is surrounded by a wall
of earth, with wooden gates; a sufficient protection against
sudden incursions of horsemen, but none against artillery.
It is also guarded by two forts, one on an eminence, on the
right bank of the river to the S.E.; the other on the left
bank, and on the plain: both are of earth, and of no strength.
There are four tolerably good serais for travellers. The inhabitants are chiefly Tadjiks and Caubulees, with a sprink-
ling of Usbeks. The shops for dyes and drugs are usually
kept by Hindoos, who also act, in a small way, as bankers.
The venders of dried fruits are mostly from Caubul. The
Usbeks engage little in traffic. They are all, rich and poor,
dressed much alike, in long gowns of striped cotton ginghams. Bazaars are held every Monday and Thursday,
when horses, asses, mules, camels, cows, sheep, and goats
are brought to their respective markets. A sheep sells at
from two to four rupees; they are of the large tail variety,
and the fat of the tail, and along the back, is commonly one
third of the weight of the sheep, including the bones. Cotton cloths, cotton in the pod, tanned leather, raw hides,
fuel, grapes, raisins, pistachio nuts, pomegranates, dried
plums, rock salt, brown leather boots with iron-shod heels,
dyes, as the pomegranate bark, madder (indigenous), and
indigo, from Hindostan, are exposed for sale, along with
blankets of fine wool from Chitral, and raw wool from
thence and Budukhshan. Printed chintzes, quilts, and turbans are also brought from India. Coarse saddlery is much
in request. There is one market entirely for melons, which
are raised in the neighbourhood in great quantities. The
sheep and furs of Koondooz are exchanged at Yarkund for
tea, disposed of in Turkestan, at an advance of 600 per cent.
The following were the prices of different articles at the
time of our visit: Mutton, four to five *pyssa*[b] per *charak*
(2½ lbs.); beef, three pyssa ditto; sheep-tail fat, eight ditto;
sheep butter, 28 ditto; cow butter, 20 ditto; oil, 16 ditto;
rice, four ditto; wheat flour, seven *pyssa* for four *charaks*;
barley about 1½ maund for a rupee, &c. The workmen in
wood, leather, and metals were very indifferent, but demanded high wages, half to three quarters a rupee per day.
Most of them, in fact, had lands, and were in some degree
independent of labour." (id., 449-452.) Khooloom has
long been the station for receiving the custom duties on all
merchandise coming from the W. into Koondooz; which
duties amount to 2½ per cent. ad val.

"Old Khulm (now entirely destroyed) is situated about 4
m. from Tash-Kurghan. It was a place of importance in
the time of Khilich Ali (a former chief of Balkh); but its
situation on the plain exposed it to predatory incursions;
and the Hazaurehs dammed up or diverted the course of
the river, upon which the fertilization of its soil depended.
The chief, therefore, removed his capital to Tash-Kurghan,
much to the regret of the people of Khulm, whose orchards
had been celebrated throughout the E. for the quantity and
quality of their produce." (*Moorcroft's Trav.,* ii., 453, &c.;
Burnes's Bokhara; *Madras Journal of Literature, &c.,*
passim.)

KHORASSAN (country of the sun), a prov. of Persia,
lying between the 31st and 36th parallels of N. lat., and the
33d and 60th degrees of E. long., being bounded N.E. and N.
by the Oxus and country of Balkh; S. by Caubul and Beh-

---

[a] A paper in the *Madras Journal of Science,* &c., for January, 1840, notes
that Khooloom is an independent town; adding, however, that "nothing is
more variable than the limits of a khanat in Asia."

[b] A pyssa is the fifteenth part of a Mahomed-Shahie rupee.

sea; and W. by Irak, Asterabad, and Dughestan. Its boundaries, however, have been very different at different times; and its present area, which is small comparatively with the great extent of country that it comprised prior to the invasion of the Afghans, is roughly estimated at about 80,000 sq. m. Population uncertain. Its surface is much diversified by plains and mountains; a large portion consists of arid rocks, destitute of vegetation or fresh water, and of salt and sandy deserts, among which may be found a few fertile oases. The Elburz range of mountains crosses the N. part of the province eastward; and between this lofty ridge and the Caspian sea is an immense uninterrupted plain, which includes the steppe of Khiva, and forms a part of that extensive flat called by the natives Dusht-el-Kipchaak. That portion of the plain which belongs to Khorassan is without a single cultivated spot or permanent habitation; and its scanty population comprises only a few tribes of wandering Turkmans. At the foot of the mountains, however, there are many rich valleys, watered by numerous rivulets, and formerly well peopled, and cultivated. This district, known in Persia as the Attack, once comprised several large towns, all of which are now in ruins, and totally deserted, in consequence of the incessant attacks of the Turkmans, who have obtained full possession of the whole tract. The Elburz mountains and ramifications southward, which penetrate from 60 to 100 m. into the plain. This range contains considerable quantities of iron, which, however, is not wrought: the turquoise mines of Nishapoor are rich, and, if managed with skill, would yield large revenues; but the combined demands of the Persian government on the tenants of the land have led to the closing of many of the most productive mines. (See Nishapoor.) In this portion of the country are many fertile tracts, which, were there any security for property, would no doubt be cultivated and well peopled. The valley of Mushed is of great length, commencing about 10 m. N.W. of Sheerwan, and extending in a S.W. direction for upward of 50 m. beyond Mushed. Its breadth varies from 15 to 30 m., and it comprises, besides Mushed (which has a population of 50,000), the towns of Chinaraun, Radkan, and Koochan, with a great extent of good land, cultivated by Koordish settlers. The W. limit of Khorassan is nearly that of the great saline desert, which forms its predominating feature. This tract, which, though considerably more lofty, is considered by Fraser (p. 251), to be connected with the desert N. of the Elburz ridge, skirts the districts of Toheran, Kashan, and Ispahan, insulates that of Yezd, and extends from Tourabees southward to the confines of Fars, Kerman, and Seistan, including hardly any habitable country except that near Boojdoon and Ghayn. Its E. limit is pretty correctly indicated by a line connecting the towns of Herat, Subzawar, Furrah, and Doozhak. The nature of this desert varies much in different parts. In some places it produces a few of those plants that thrive in a salt soil, while in others it consists of a crackling crust of dry earth, covered with salt efflorescence: a considerable portion is marshy, and in the lower parts water accumulates during winter, which is evaporated in the hot weather, leaving lakes of salt on a bed of mud. Again, in certain districts, sand abounds in plains, interspersed with waving hillocks, easily moved by the wind, and sometimes so light and impalpable as to prove not only disagreeable, but extremely dangerous to travellers, who not unfrequently are buried in its heaps. Of the rivers of Khorassan, the Tedzen (an. Oxus) is next in size to the Oxus: it appears to rise near Seraks, and, after receiving the Meshed and other streams, falls into the Caspian sea in lat. 39° 41′ N. The rivers of the interior are few and inconsiderable, and, for the most part, are lost in the sand, like the Zenderoon of Ispahan. The climate of Khorassan varies according to the nature and elevation of the districts into which it is divided. In some parts it is temperate, in others extremely cold. The deserts are infested by the simoom, which is as fatal here as in Arabia. (See vol. i., p. 194.) The cultivated districts produce the grains and fruits of S. Europe, with assafœtida, frankincense, and other gums; but timber is rare. Cattle-breeding is the chief employment of the nomad race that wanders over the desert; and the camels and goats of Khorassan are celebrated for their fine, soft hair, which is an article of trade in the markets of Mushed and Nishapoor, the two largest towns of the province. The inhabitants of the settled districts are Tadjiks or Persians, properly so called, and their number has been estimated at 1,800,000. The Ilyats, or nomads, comprise Turkmans, Djeizers, and other Turkish tribes, and there are about 20,000 Kurds in the N. part of the province. The religion of all the inhabitants is Mohammedan, and most of them belong to the sect of Ali. The province is divided into several little governments; but the authority of the king of Persia extends only over the cities of Mushed, Nishapoor, Turshish, and Tabas, with their dependencies. The S. parts belong to the Afghans, and the Usbek Tartars and Turkmans wander over the N. and E., acknowledging only their native khans.

These wild tribes carry on incessant hostilities, invading each other's territories with bodies of irregular horse, who, after ravaging the country and burning the villages, carry off the inhabitants into slavery. (Fraser's Khorassan, p. 240, and Appendix; Kinneir's Persia, p. 181, &c.)

KHOTAN, or ILITSI, a town of Chinese Turkestan, prov. Yarkund, on the high road between that city and Lassa, 990 m. E.S.E. the former: lat. 37° 10′ N., long. about 79° E. It is principally occupied by Usbeks; and is said to be celebrated for "its musk, and the beauty of its inhabitants." Khotan, according to Abulfeda and other Mohammedan geographers, was formerly a town of great consequence: it is still a place of considerable size, enclosed by ramparts of earth, and, though ill built, has broad streets. It is the station of a Chinese governor and garrison; has manufactures of silk fabrics, leather, paper, &c., and a brisk trade in these and various other articles, including ya, the jasper of the ancients. (Malcomson, Ritter, Klaproth, &c.)

KIACHTA, or KIAKHTA, a town of Asiatic Russia, gov. and prov. Irkutsk, being the centre of the trade and political intercourse between the Russian and Chinese empires. It stands immediately within the Siberian frontier, on a rivulet of the same name, a tributary of the Selenga, and upon a plateau elevated about 2290 feet above the sea, 55 m. S. by E. Selenginsk, and 190 m. S.E. Irkutsk: lat. 50° 21′ 5″ N., long. 106° 26′ 15″ E. (Erman.) Population between 4000 and 5000. It is divided into an upper and lower town: the former, or the fortress of Troitzkoi Sawsk, was founded when the first commercial treaty took place between Russia and China, in 1728. The town within is regularly laid out, in the form of a square; in the centre of which is the bazaar, or market-place, a wooden building. Except a chapel of stone, and some of the public offices, built partly of brick, Kaichta is constructed wholly of wood. The church, government-house, barracks, and watch tower are the chief public edifices within the town: the various courts and government-offices, imperial rhubarb depôt, custom-house, &c., are in one of the suburbs. The lower town, a few versts distant, consists of only about 50 houses, inhabited by merchants, who conduct the trade with the Chinese, and some of whom are said to be very rich.

On the Chinese side of the boundary is the Mongolian village of Mia-mia-tchin (the place of trade), which, like the Russian town, is laid out in a square form, and surrounded by a palisade. It is ill built, and has only from 1200 to 1500 inhabitants, all males, no women being allowed to reside in it. All the mercantile transactions are conducted between this village and Lower Kiachta; and the merchants of the two places visit each other without let or hindrance. The goods bought by the Russians are immediately sent to Upper Kiachta, to be examined by the custom-house authorities. The Russians exchange furs, sheep and lamb skins, Russian and Silesian broadcloths, Russian and morocco leather, coarse linens, cattle, and especially bullion, for tea, raw and manufactured silks, nankeens, porcelain, sugar candy, rhubarb, tobacco, musk, &c. At the Kiachta December fair the tea bought by the Russians is, at an average, said to amount to 69,000 chests, or 4,898,000 lbs. of this pektoe; besides a large quantity of an inferior kind, much of which is consumed by the Siberians and nomadic Tartars. But, according to Klaproth, the accounts of the Russian commerce with China have been much exaggerated; the total amount of the trade seldom reaching 24,000,000 francs a year, and frequently not a fourth part of that sum. In 1831 the Russian exports by way of Kiachta amounted to 4,655,536 francs, and the imports to 6,756,836 francs. Goods may be conveyed from Kiachta to European Russia either by land or water (by the lake of Baikal, the Angara, Yenisei, &c.); in the former mode the journey occupies a year, and in the latter three years, or rather three very short summers, the rivers being for a great part of the year frozen over. (Erman, Reise um die Erde; Klaproth, Mémoires, &c., i. 37—86; Ritter, Official Papers, and Priv. Inform.)

KIDDERMINSTER, an important manufacturing and market town, parl. bor. and par. of England, co. Worcester, hund. Halfshire, on the Stour, an affluent of the Severn, 13 m. N. Worcester, 16 m. W.S.W. Birmingham, and 118 m. N.W. London. Area of the entire par., 19,800 acres: pop. of parl. bor. (which includes the old borough, and a small portion of the "foreign" district in 1831), 16,609. (Bound. Comn.'s Estimate.) The town, divided by the river, which is here crossed by a stone bridge, into two unequal parts, is irregularly built, but has several good streets, well paved, lighted with gas, and kept clean by an underground sewerage. In the centre of the market-place is the town hall, a capacious brick structure, comprising, besides several other rooms, a large council-chamber for corporation meetings, quarter sessions, &c.; the lower part of the building is used as a butchers' market, and underneath is a cellar that has sometimes been used as a jail. The church, which stands in a fine open space, on the brow of a hill, and close to the river, is a large Gothic edifice, richly adorned, and surmounted by a lofty pinnacled tower, the whole being in excellent repair. The interior has accommodation for 3000 persons, and contains several fine old monuments. Connected with the church, at its E. end, is a Gothic chapel or chantry, now appropriated to the use of the grammar-school. On the E. side of the town is the fine district chapel of St. George, erected in 1823, at an expense of £18,131; the altar-piece is embellished with a representation of the descent from the cross, in carpet-work, executed with much taste and brilliancy of colouring: there are also places of worship for Independents, Baptists, Wesleyan Methodists, and Unitarians. The grammar-school, chartered by Charles I., has estates attached to it worth about £300 a year; but though all the inhabitants are entitled to send their sons here to be educated, free of expense, it is of little practical utility, and is attended by only about 20 boys. A free school, founded in 1795, provides instruction for about 25 boys, chiefly dissenters. There are numerous Sunday schools, attended by about 3000 children; and four national schools, and three Lancastrian schools, furnish instruction to above 1100 children. The charitable institutions comprise several almshouses and a dispensary, with some clothing and benefit societies. Near the town is a chalybeate spring, the road to which is an agreeable and fashionable promenade, and in the suburbs are some elegant villas, inhabited by the wealthy manufacturers.

Kidderminster has been noted for its weaving industry since the time of Henry VIII., in whose reign it had a considerable trade in broadcloth. Linsey-woolseys were afterwards introduced, and were superseded, in their turn, by poplins, bombazeens, and carpets. The fabrics now made are carpets, finger-rugs, bombazeens, coverings for buttons, and waistcoat pieces. The carpet manufacture, introduced in 1735, has long been the staple business of the town: there are no power-looms, the carpets being all woven by the hand: and carpet-weaving is the principal trade; bombazeens are not extensively made, and button-coverings have only been lately introduced. The carpet fabrics comprise Brussels or pile carpets, Kidderminster or in-grain carpets, and Venetian carpets: the proportion of each, in 1838, is seen from the following table. (*Hand-loom Weavers' Report*, part v., 530.)

Description of Fabric.	No. of Manufactures	Looms.	Hands employed.
Brussels carpets	24	1785	1905 men
Kidderminster or Scotch	11	210	351 women
Venetian	7	45	1760 children
Total	24	2040	4016

Skilful and industrious carpet-weavers earn 27s. a week; but the average wages (1839) did not exceed 14s., the quantity woven averaging 24 yards a week. Button-makers and waistcoat piece makers earn about 12s. or 13s. a week: bombazeen-weavers (of whom there are about 70) earn only 7s. a week, but the work is light, and is principally performed by women and aged persons. There are six worsted mills, which employ 622 hands, and factory wages average from 3s. to 6s. a week for children and women, and from 12s. to 25s. for men. The moral condition of the weavers is said to have deteriorated of late years, chiefly in consequence of an obstinate strike in 1828, when wages were lowered 17 per cent., and when also the manufacture took root in other places. Since that period, the weavers are said to have been discontented and improvident, and, in fact, to have in a great measure changed their character. Rents are also said to have been considerably reduced. The manufactures and trade of the town are greatly facilitated by the Staffordshire and Worcestershire canal, which passes close to the town; and its communications have been further improved by the opening of the Birmingham and Gloucester railway.

Kidderminster is a borough by prescription, and received its charter of incorporation in 12 Charles I. Since the Municipal Reform Act it has been divided into three wards, the government being vested in a recorder, six aldermen, and 18 councillors. Quarter and petty sessions, and a court of requests for the recovery of small debts, are held in the town-hall. Corporation revenues, in 1839, £1910. In the reign of Edward I., Kidderminster sent two members to the House of Commons, but the privilege being either lost or disused, it ceased to be represented; and this populous and industrious town had no voice in the legislature till the Reform Act conferred on it the privilege of sending one member to the House of Commons. The electoral limits comprise the old municipal borough, and a small portion of the "foreign" district. There were 300 qualifying tenements in 1831, and 469 regular electors in 1839—40. Market on Thursday: fairs, Holy Thursday, June 20, September 4, and November 28, for horses, cattle, linen and woollen cloth. (*Parl. Reports.*)

KIDWELLY, or CYDWELI, a bor. market-town, and par. of S. Wales, co. Caermarthen, and hund. of its own

114

name, on the Gwendraeth-Vechan, 9 m. S. Caermarthen, and 179 m. W. London. Population, in 1831, 1631, there being a decrease of 98 persons during the preceding 10 years. It is divided by the river into two townships, Old Kidwelly being on the W., and New Kidwelly on the E. or left bank. The former was once surrounded by walls, with three gates, one of which is yet standing; but the houses have fallen to decay, and consist at present of little more than hovels. New Kidwelly, which is joined to the other by a stone bridge, has several respectable houses and numerous cottages. On a rocky eminence overlooking the Old Town, stands the castle, said to have been built soon after the Norman conquest, and now a large and imposing ruin in tolerable preservation, with many of its apartments and staircases still entire: the W. gateway is a noble specimen of architecture, and some of the towers at the angles retain their arched roofs of stone. The battlements command magnificent views of Caermarthen bay and the country on both sides the Towy. The church, which is in the New Town, is an old cruciform structure, with a tower and spire 170 feet high: the transepts are now in ruins, and the centre aisle is the only part used for service. The ruins of a priory of black monks adjoin the church. The living is a vicarage in the gift of the crown; and connected with it is a rural deanery in the diocese of St. David's. There are places of worship also for Calvinist and Wesleyan Methodists, Presbyterians, and other dissenters. A free-school is supported by funds in the hands of the corporation, and one other school is maintained by subscription. The industry of Kidwelly is chiefly employed in working coal, smelting iron, and making tin plates. It is not a place of much trade, however, owing to the choking up of the river, which is almost useless, notwithstanding the construction of wharfs, staiths, and other improvements, by Lord Cawdor. There is canal communication with Pembrey, where there is a commodious quay; and a canal and tram-road connect it also with Llanelly, which has a flourishing and increasing trade. Kidwelly forms a part of the duchy of Lancaster, but is governed by its own mayor and 12 aldermen, whose privileges were not affected by the Municipal Reform Act. Markets on Friday: cattle fairs, May 24, July 22, and Oct. 29.

KIEF, a government of Russia in Europe, lying lengthwise along the right bank of the Dniepr, having N. the government of Minsk, W. Volhynia and Podolia, and S. Kherson. Area estimated at about 20,000 sq. m. Population, in 1838, 460,000. Principal rivers, Dniepr, by which it is bounded all along the E., Pripet, which traverses its N. division, Teterff, and other affluents of the Dniepr. Surface flat; soil very fertile, so much so, that, though agriculture is very indifferent, the return to most sorts of grain is said to be as six to one. Cattle numerous, large, and of a fine breed. Horses small. Forest extensive. In its N. parts there are considerable marshes. Manufactures, exclusive of those carried on in the houses of the peasantry, can hardly be said to exist. Commerce trifling, and mostly in the hands of the Jews. Principal town Kief.

KIEF, the cap. of the above government, and the former residence of the grand dukes of Russia, on the Dniepr, a little below the confluence of the Desna with that river; lat. 50° 26' 53" N., long. 30° 27'. Population 26,000. This is a very ancient city. It was the earliest seat of the Christian religion in Russia, and was for a considerable period the capital of the empire. But it subsequently underwent many vicissitudes; being sometimes subject to the Lithuanians, the Tartars, and the Poles. In 1686, however, it was finally ceded to Russia, and has ever since continued in her possession. The town consists of three parts—the old town, on an eminence elevated considerably above the river; Pitcherak, or the citadel, more to the S., and on a still higher eminence; and the lower town, or Podohk, on a plain along the river. The first or old town contains the cathedral of St. Sophia, founded in 1037, and, an object of the greatest veneration on the part of the Russians. It is stated by Dr. Pinkerton, that the sum appropriated by the Russian government for the support of this the most ancient cathedral in the empire, with its priests, deacons, singers, &c., amounts to only £98 a year! (*Russia, &c.*, p. 217.) The citadel is surrounded by a rampart. Within it is the arsenal, erected by Catherine II., a large handsome building, containing an extensive supply of arms. But the principal object of curiosity in the citadel is the famous monastery of Pitcherak, with its cathedral. It derives its name from *pitchora*, a cavern, because in the vaults beneath are preserved the bodies of several Russian saints. The tower or belfry of the cathedral, deemed by the Russians a master-piece of architecture, rises to the height of 304½ feet. The theological academy of Kief, founded in 1661, in the Podolsk, is one of the most celebrated in Russia. In 1836, it was attended, according to Schnitzler, by 1500 pupils; but this, we suspect, must be an exaggeration, and we should think that 350, or 400, would be nearer the mark. In 1833 a university was founded at Kiel, intended to replace that of Wilna, suppressed after the late Polish revolution. It had, in 1837, 88 professors and assistants, about 200 pupils, and a library with more than 45,000 vols. The university buildings are said to be at once large and handsome. One of the most remarkable edifices in the lower town is the exchange, a very large building, the great hall of which can accommodate 3000 persons. The houses are, for the most part, of wood, and the streets narrow, crooked, and mostly unpaved. The town is principally dependent on the pilgrimages to the cathedral and the monastery, and on the academy. In 1799 a fair, formerly held at Dubno, was transferred thither. It takes place during January, and is attended by all the surrounding nobles, as well as by great numbers of merchants and other descriptions of people. They rendezvous in the exchange. Provisions of all kinds are abundant and exceedingly cheap. (See *Schnitzler, La Russie, &c.*, p. 449–458; *Pinkerton; Lyall's Travels*, i., &c.)

KIEL, a town of Denmark, cap. Holstein, on the N. shore of the prov., at the bottom of a beautiful bay, lat. 54° 19' 43" N. long. 10° 8' 18" E. Population 11,060. It is handsome, well built, and thriving. The university, founded in 1665, has had many distinguished men among its professors: it has a valuable library, comprising 70,000 volumes, and is attended-at present by about 300 students. There is also an excellent grammar school, with an orphan-house, a workhouse, &c. The church of St. Nicholas is a fine old building; a handsome royal palace stands on a hill adjoining the town, and the public baths deserve notice.

Kiel is the seat of government, and, since 1834, of the supreme court of appeal for the duchies of Sleswick, Holstein, and Lauenburg. There is an extensive hat manufacture; and starch, tobacco, refined sugar, &c., are also produced. The harbour is safe, and has water sufficient for large ships. A good deal of trade and ship-building is carried on. Packet boats sail regularly for Copenhagen; and the road from Hamburg to Kiel being equal to any in England, this route is much frequented by travellers visiting the Danish metropolis. The Holstein canal, forming a navigable communication between the Eyder and the Baltic, unites with the latter two miles from the two. There is a great annual fair in January.

KILDA (ST.), or HIRT, a small island belonging to Scotland, the most remote of the Hebrides or Western Islands, in the Atlantic ocean, 60 m. W. from Uist; lat. 57° 50' N., long. 8° 36' 30" W. It is about 3 m. in length by 2 m. in breadth, and contains about 4000 acres: having attached to it a few dependent and inferior islets. Except at the landing place on its S. side, and at a rocky bay on the N., the island is wholly fenced round with lofty inaccessible precipices. The landing place, now noticed, affords, except during southerly winds, good anchorage. St. Kilda is principally occupied by four hills; and though the soil be but thin and poor, it is, owing to the moisture and mildness of the climate, covered with luxuriant verdure, and affords pasture for some hundreds of sheep and a few cows. A small portion of the surface is in tillage, and produces the variety of barley called bear or big, and oats; but owing to the frequent and tremendous storms by which the island is visited, the crops are exceedingly precarious, and are not unfrequently destroyed. The inhabitants consist of 22 families, of five or six individuals each, who live together in one poor hamlet. The island being resorted to by vast numbers of sea fowl, the inhabitants are principally engaged in fowling, and are mainly dependent on the eggs, flesh, and feathers of the birds. Fowling is here, as in all similar localities, an extremely perilous occupation, and one requiring great nerve and dexterity. Fishing is, also, a considerable resource. The people are filthy in their habits, destitute of most of the comforts of life, and apparently unhealthy and short-lived. The island belongs to a single proprietor, who lets it to a middleman, by whom it is let to the inhabitants. The latter pay their rents in feathers and bear. The population has long been stationary.

Recently a considerable improvement has been effected in the condition of these remote islanders by the visits paid them in the course of summer by steamers from various parts of Great Britain. A few years ago they were extremely ignorant; but they are said to be, in this respect, a good deal improved through the residence among them of a worthy and most attentive clergyman. (*Martin's Voyage to St. Kilda*, 4th ed. Lond. 1753; *Pullarton and Baird on the Highlands and Islands of Scotland*, p. 56, &c.)

KILDARE, an inland co. of Ireland, prov. Leinster, having N. Meath, E. Dublin and Wicklow, S. Carlow and W. King's and Queen's cos. It contains 392,435 acres, of which 66,447 are unimproved bog and waste, consisting principally of portions of the bog of Allen (which see).

Surface mostly flat, or but slightly undulating; and, with the exception of the bog, the soil is mostly clayey and fertile. The famous common, called the *curragh* of Kildare, in the centre of the county contains about 5000 acres; and is said to be unmatched for the softness of its turf, and the richness of its verdure. When Mr. Wakefield visited Ireland, agriculture in this county was in the worst possible state (i., 419); but, though still very far behind, it has been materially improved in the interval; and better implements, better stock, and improved processes have been pretty generally introduced. There are some very large estates; but property is, notwithstanding, a good deal divided. Farms vary in size from 5 up to 260 and even 300 acres; and have, indeed, been less subdivided in this than in most Irish counties. Average rent of land, 13s. an acre. Minerals and manufactures unimportant. Principal rivers Barrow, Liffey, and Boyne; the last-mentioned river having its principal source in this county, near Carbury. It is also intersected by the Grand canal, and by its branch leading to Monasterevan and Athy. It has no considerable town, Athy being the most populous. Kildare contains 10 baronies, and 113 parishes: it returns two members to the House of Commons, both for the county. Registered electors in 1837–38, 1944. In 1831, it had 17,135 inhabited houses, 18,771 families, and 108,494 inhabitants, of whom 54,472 were males, and 53,922 females.

KILIA, a small town of European Russia, in Bessarabia, on the N. bank of the Kilia, an arm of the Danube, about eight miles from its mouth. It has some trade; but owing to the shallowness of the water over the bar at the mouth of this arm of the river, it is not likely ever to become a place of any importance.

KILKENNY, an inland co. of Ireland, prov. Leinster, having N. Queen's county; E. Carlow and Wexford, from which it is separated by the Barrow; S. Waterford, from which it is separated by the Suir; and W. the latter and Tipperary. Area, 513,786 acres; of which 96,306 are unimproved mountain and bog. Surface in parts hilly, the surface is mostly either flat, or but slightly undulating. Soil of various qualities; but, for the most part, it rests on a limestone bottom, and is light, loamy, and, in the valleys, particularly fertile. The remarks made under the head of Kildare as to the improvements that have been made in agriculture, since 1813, apply equally to this county. In some districts the dairy husbandry is extensively carried on. Property mostly in very large estates. Farms of various sizes, but generally small. Partnership tenures are not uncommon; and farm-houses and cottages are in general very inferior. Average rent of land, 17s. an acre, being among the highest rented counties in Ireland. There are extensive beds of coal in this county, and collieries have been wrought at Castlecomer for more than a century; but, owing to the excess of sulphur, the coal is but little used for domestic purposes, and is principally employed in malting, lime-burning, &c. The woollen manufactures formerly established in this county are now nearly extinct (see next article); and, if we except the grinding of corn into meal and flour, and some breweries, distilleries, and tanneries, the manufactures now carried on in the county are quite inconsiderable. Kilkenny is intersected by the Nore, and bounded on the E. by the Barrow, and on the S. by the Suir; so that it has peculiar facilities for the shipping of its produce, which may be conveyed either to Waterford by the Barrow and the Suir, or to Dublin by the Barrow and the Grand canal. It contains nine baronies and 127 parishes; and sends three members to the House of Commons, being two for the county, and one for the borough of Kilkenny. Registered electors for the county in 1838–39, 1314. In 1831, Kilkenny had 31,007 inhabited houses, 33,509 families, and 193,695 inhabitants; of whom 93,977 were males, and 99,709 females.

KILKENNY, an inland city and parl. bor. of Ireland, prov. Leinster, cap. of the above co., on the Nore, which is here crossed by two handsome stone bridges, 78 m. S.W. Dublin, and 27 m. N. by W. Waterford.

Kilkenny, with its liberties, which are comprised in the parl. bor., extends over a space of 16,400 imp. acres, and forms a county of itself; the portion on the E. side the river, called St. Cannice, or Irishtown, being on the estate of the bishop of Ossory. The city and liberties had, in 1821, 23,230; and in 1831, 23,741 inhabitants, of whom above 21,000 are Roman Catholics. Mr. Inglis says, that Kilkenny is well built (excepting the suburbs), beautifully situated, and a very interesting town. The principal streets are parallel to the river, but there are many cross streets. Of 3942 houses belonging to the town in 1831, 1857 were thatched, and 985 slated. The principal structure are the castle and cathedral. The former, which is of great antiquity, having been built by Strongbow, has been long the property and residence of the Ormonde family. It has recently been almost entirely rebuilt, and has been rendered a very commodious as well as a magnificent residence.

115

The church of St. Cannice, the cathedral of the see of Ossory, is a large venerable pile of Gothic architecture: it has several monuments, and near it is a round or pillar tower 108 feet high: the bishop's palace and the deanery are also close by. The county of the city comprises the parishes of St. Mary, St. Patrick, St. John, and St. Cannice. The church of St. Mary is an elegant modern building; that of St. John, which was the chapel of the monastery of the same name, has been restored, so as to preserve the character of its former singular style of architecture, in which the windows are replicated in such close succession that the intervals are merely mullions, whence it is called the Lantern of Kilkenny. There is a Roman Catholic chapel in each parish, that of St. Mary's being looked upon as the bishop's cathedral. Chapels are also attached to the Presentation convent, and to the Dominican and Capuchin friaries. The ruins of the Franciscan and of the Dominican, or Black Abbey, add greatly to the interest of the place.

A public grammar school, endowed by one of the earls of Ormonde, and elevated to the rank of a royal college by James II., has accommodation for 80 resident pupils: the house, which stands in a retired situation, on the banks of the Nore, was rebuilt, at the public expense, towards the close of the last century: the children of the inhabitants of Kilkenny are admitted at half price. Here is also a charter school in which 24 boys are instructed in weaving, a seminary for candidates for the R. Catholic priesthood at Birchfield, a large female school, conducted in the best possible manner by the nuns of the Presentation convent, a parochial school, and a female orphan house. There are about 1200 pupils in the public, and 1500 in the private schools.

The principal charitable institutions are the infirmary for the county, the fever hospital, and the house of industry, which has attached to it a lunatic asylum, wholly independent of the county district asylum: there are at present (1839) in it 16 male and 16 female patients, but the want of adequate accommodation obliges some of them to be lodged at night in the house of correction. There are several almshouses, and two loan funds. The charitable society affords relief to sick tradesmen and to their widows: the benevolent society to the bedridden poor.

A public walk, called the Mall, extends upwards of a mile along the bank of an unfinished canal and of the Nore. Here, also, is a small library, a news-room, a mechanics' friends' society, and a horticultural society. Races are held in September.

A charter, granted to the city by William Earl Marshal, was repeatedly confirmed by successive sovereigns. Elizabeth combined the two boroughs into a single corporation. The ruling charter is that of James I. By it the governing body consists of a mayor, two sheriffs, 18 aldermen, 36 common councilmen, and an unlimited number of freemen. Previously to the union, Kilkenny and Irishtown sent four members to the Irish House of Commons; and, since then, they have sent one member to the Imperial House of Commons. The right of voting was formerly in the freemen and freeholders of the co. of the city, the freedom of the city being obtained by birth, servitude, or by gift of the corporation. Registered electors in 1838–39, 663. The corporation, which is very wealthy, defrays all charges for lighting, paving, &c.

The mayor, and aldermen who have served as mayors, are justices of the peace for the city. A criminal court is held under their jurisdiction quarterly, and a court of record, on Tuesdays and Fridays, for sums above £36. The portreeve of Irishtown also holds a weekly court for the recovery of debts under 40s. in that borough. The assizes for the county and city are held here; as are the general sessions of the peace, in a quarterly rotation with three other places. The courthouse, built on the site of Grace's Old Castle, is a spacious and elegant building, with sufficient accommodation for the public business of the county and city. The county prison is at a small distance from the town; that of the city is ill constructed, and limited in its means of accommodation.

The Ormonde family have laudably exerted themselves at different periods to introduce manufactures into Kilkenny. In this view, Pierce, the third earl, brought over a colony of Flemings skilled in the making of tapestry and carpets, but without success. The first marquis expended large sums in attempts to establish the linen manufacture. That of frieze, after being carried on for a considerable period, was eventually transferred to the neighbouring town of Carrick-on-Suir. Wool-combing was also introduced, and the manufacture of blankets was extensively carried on; but this also has all but entirely failed. Mr. Inglis represents the woollen manufacturers of Kilkenny as being, at the period of his visit, without employment, and in the greatest distress; and the *Railway Commissioners* state that the entire value of the woollen goods produced within the

districts of Cork, Kilkenny, Meath, and Carrick-on-Suir, did not (in 1838) amount to £30,000 a year! Several flour and corn-mills have recently been erected in or near the city, and there are several distilleries, breweries, and tanneries, and a starch manufactory: but the principal dependence of the town is on its retail trade, of which it is an exclusive centre. Within about one mile from the city are some celebrated marble quarries, and a sawing and polishing mill. The marble is extremely beautiful; it has a black ground variegated with madrepore, bivalve shells, and other organic matter; it takes a fine polish, and makes beautiful chimney-pieces, and such like articles. Kilkenny coal neither emits flame nor smoke; but its sulphureous exhalations unfit it for domestic purposes. Two weekly newspapers are published in Kilkenny; and it has also branches of the bank of Ireland, and of the provincial and national banks. Postoffice revenue, in 1830, £1523; in 1836, £2113. Markets on Wednesdays and Saturdays, in the covered area of the Tholsel or town-house. Fairs on the 28th of March, and Corpus Christi day, for cattle and wool, are frequented by purchasers from all parts of the country.

Kilkenny derived its name from a church or cell dedicated to St. Cannice, or Kenny. It appears to have been a place of some importance before the arrival of the English; for Strongbow built a fortress here, which was enlarged and strengthened by William Earl Marshal, and subsequently by the earls of Ormonde, in whose possession it has continued for centuries. Parliaments were frequently held in this city; and a famous statute, passed in 1371, for regulating the intercourse between the English and the native Irish, is still quoted by the title of the Statute of Kilkenny. In the wars of 1641, the assembly of the confederated Catholics held its meetings here, in a building which is still, on that account, an object of curiosity to strangers. In 1650, it surrendered to Cromwell.

Kilkenny enjoys many advantages, independently of its locality as a central point of communication to all parts of the S. of Ireland. Its situation, equally picturesque and salubrious, the circumstance of its being the ancient and continued residence of one of the principal Anglo-Irish families, and of the bishop and dignitaries of the diocese, as well as of many highly respectable inhabitants, and the vicinity of numerous resident landholders of large property, have all contributed to increase its rank and respectability. The higher classes here may vie with those of the capital; but we regret to say, that the situation of the labouring classes is as bad as possible. Mr. Inglis says, " I found the working population in a miserable condition; hundreds subsisting on the chance contributions levied on the farmers round the country, and hundreds more subsisting at the very lowest point at which life can be sustained." And we understand that this paragraph may, with little modification, be applied to the state of the lower classes at this moment (1840). (*Inglis's Ireland*, i., 88; *Boundary Municipal and Railway Reports, &c.*)

KILLARNEY (TOWN AND LAKE OF). The town of Killarney in Ireland, co. Kerry, so celebrated for the fine scenery in its vicinity, is situated about 1½ m. from the E. margin of the lake, 162 m. S.W. Dublin, and 44 m. E. by N. Cork. Pop. in 1821, 7014: in 1831, 7910. The town took its rise from iron and copper works in its neighbourhood, now discontinued from want of fuel; but, for a lengthened period, it has been principally indebted for its support and celebrity to the attractions of the surrounding scenery. I has three pretty good streets, with many bad alleys, and close, filthy lanes and yards. Mr. Inglis says, that it has a large pauper population, and a vast number of idle persons; which, indeed, is a common characteristic of all places much resorted to by strangers. The great drawback, on a visit to Killarney, has hitherto been the number and importunity of the beggars by whom its streets and environs have been infested. It is to be hoped that something effectual may be done, through the agency of the Poor-law Commissioners, or otherwise, to rid the town of this nuisance.

The principal buildings are the parish church, built in 1802; a large, low, heavy Roman Catholic chapel, a Methodist meeting-house, a national school, a fever hospital with a dispensary, an almshouse for aged females, founded and endowed by Lady Kenmare, a market-house, theatre, courthouse, and bridewell. In New-street is a convent for nuns of the order of the Presentation. Attached to their convent is a school, in which these benevolent and excellent ladies give gratis and very superior instruction to about 400 girls: Lord Kenmare contributes £100 a year towards defraying the expense of this school, and also clothes 30 of the girls. General sessions are held four times a year; petty sessions on Tuesdays, and a manor court monthly; a party of the constabulary has a station here. It has several good inns, which, in the visiting season, are much frequented. The only manufactures, if we may so call them, carried on in the town, are those of toys and fancy articles, made of the wood of the arbutus, which is here very abundant. It has a considerable trade in corn, groceries, woollens, coarse linens, &c.; and it has some tanneries, two breweries, and a large flour-mill. Markets on Saturdays; fairs on 4th July, 8th August, 7th October, 11th and 28th November, and 28th December. Post-office revenue in 1836, £394; in 1838, £694. Branches of the agricultural and national banks were opened in 1835. The town is built on the estate of the Earl of Kenmare, whose house and grounds lie between it and the lakes.

The lake of Killarney, or lough Lane, consists properly of three lakes, connected by a winding channel, through which vessels pass from one to the other. It lies at the E. extremity of the extensive range of mountains called MacGillicuddy's Reeks, and has in its immediate vicinity, or rather, indeed, rising from its banks, the highest summits in Ireland. The larger division of the lake, or that portion called the lower lake, occupies an area of about 3000 Irish acres; its W. shore is formed by the mountains of Tomies and Glenna, respectively 2150 and 2600 ft. above the level of the sea, having their precipitous sides well clothed with finest trees; on the opposite shore is the striking contrast of flat land in a high state of cultivation, ornamented by the fine demesne of Lord Kenmare. There are said to be no fewer than 30 islands, many of which are extremely picturesque, in the lower lake. One of these islands, Innisfallen, has been admired by every traveller. Arthur Young says that it is the most beautiful spot in the United Kingdom, and perhaps in Europe. It contains about 20 acres, is extremely well wooded, and has every variety of tranquil beauty and sylvan scenery. On the S. shore of this lake is the fine ruin of Muckross Abbey, the property of Mr. Herbert. This lake is, in some parts, very deep. Between Glenna mountain and Ross island, the largest in the lake, the soundings give 42 fathoms; and as the surface of the lake is about 50 ft. above the level of the sea, it follows that its bottom is 208 ft. below that level.

The middle lake occupies about 640 Irish acres; it lies immediately under the Foss or Tusk mountain, elevated about 1800 ft. above the level of the sea. The strait which joins the middle and upper lake is about 3 m. in length, having, in many places, the appearance of a beautiful river. The upper lake contains about 720 Irish acres. It lies in a hollow, formed by some stupendous mountains, among which are Gurran Tual, the highest in Ireland, rising 3404 ft. above the level of the sea; so that its scenery is in the highest degree magnificent and sublime. "Here," says Mr. Wakefield, "nature assumes her roughest and most terrific attire to astonish the gazing spectator, who, lost amid wonder and surprise, thinks he treads enchanted ground; and while he scarcely knows to which side he shall direct his attention, can hardly believe that the scenes he sees around him are not the effects of delusion, or the airy phantoms of the brain, called into momentary existence by the creative powers of a fervid imagination. Here rocks piled upon rocks rise to a towering height; there one mountain rears its head in succession above another, and sometimes a gigantic range seems to overhang you, forming a scene that may be more easily conceived than described. Such sublime scenes cannot be beheld but with a mixed sensation of pleasure and awe, and on a contemplative mind they must make a deep and lasting impression," (vol. i., p. 66). In other places, however, especially on the E. shores of the lower and middle lakes, the scenery is of the softest and most agreeable kind, consisting of finely wooded promontories, ornamented with rivers and seats, verdant islands, &c.; and it is in the contrast between these and whatever is most wild and rugged, that we find the great charm of Killarney.

The lakes of Killarney receive the Flesk and several other streams, their redundant waters being carried off by the Lane. The latter issues from the N.W. extremity of the lower lake, and after pursuing a W.N.W. course for about 10 m., falls into Castlemaine harbour, at the bottom of Dingle bay. It is well stocked with salmon and white trout, and also with pearl oysters, whence pearls have been repeatedly taken. Were it desirable, it is said that the Lane might, at a small expense, be made navigable from the sea to the lake. (Inglis's Ireland; Young's Ireland, 4to ed.; Windele's Cork, &c., p. 294, &c.)

KILLBUCK, p. t., Holmes co., O., 69 m. N.E. Columbus, 326 W. Watered by Killbuck creek. It contains two grist-mills, two saw-mills, two tanneries, one distillery. Pop. 916.

KILLIECRANKIE, a celebrated pass through the Grampian mountains in Scotland, co. Perth, about 15 m. above Dunkeld. It is about ½ m. in length. The road is cut out of the side of one of the contiguous mountains; and below it at the foot of a high precipice, in the bottom of the ravine, the river Garry dashes along over rugged rocks, but so shaded with trees as hardly to be seen. At the N. extremity of this pass, the revolutionary army under Mackay was defeated in 1689, by the troops of James II. under the fa-

mous Graham of Claverhouse, Viscount Dundee, who fell in the moment of victory.

KILLINGLY, p. t., Windham co., Ct., 47 m. E. Hartford, 378 W. Bounded W. by Quinebaug r., and watered by its tributary Five-mile r. It has three fine villages, all in sight of each other, and is one of the largest cotton-manufacturing towns in the state. Chartered in 1708. It contains six churches, three Congregational and three Baptist; 22 stores, one furnace, one woollen factory, 16 cotton factories, with 21,998 spindles, eight grist-mills, 11 saw-mills; one academy, 100 students, 30 schools, 998 scholars. Population, 3685. The Norwich and Worcester railroad passes through it.

KILLINGWORTH, p. t., Middlesex co., Ct., 49 m. S. by E. Hartford, 326 W. Bounded W. by Hammonasset river. It contains three stores, one fulling-mill, four grist-mills, six saw-mills, two tanneries; seven schools, 171 scholars. Pop. 1123.

KILMARNOCK, an eminent manufacturing town, parl. bor., bor. of regality, and par. of Scotland, district of Cunningham, co. Ayr, on level ground on the N. bank of the Irvine, and on the small stream Kilmarnock or Fenwick, a tributary of the former; 20 m. S.W. by S. Glasgow, and 12 m. N.N.E. Ayr. Pop. of the parish of Kilmarnock, in 1801, 8079, in 1831, 18,093; but the population of the parl. bor., which includes the suburb of Riccarton, on the E. bank of the river, is at present, 1848, estimated at above 26,000.

The main street, forming part of the high road between Ayr and Glasgow, is upwards of 1 m. in length, and is regularly built. The houses, generally of freestone (which is found in great abundance in the immediate vicinity), are erected in a handsome substantial style. Similar remarks are applicable to all the modern portions of the town. Kilmarnock has recently been extended greatly towards the S. and E., and in these directions there are many handsome buildings, particularly Glencairn-street and Glencairn square. The older streets are narrow and irregular; but the magistrates having obtained an act for improving the town, about the beginning of the present century, judicious measures were adopted to carry its objects into effect; and Kilmarnock is now, on the whole, one of the neatest and best-built manufacturing towns in Scotland. The streets were lighted with gas in 1823.

Among the public buildings are the two parish churches, one of which, the High church, after the plan of St. Martin's in London, is surmounted by a tower 80 ft. in height; a third parish church, named above, erected in 1836, in the Gothic style of architecture; the academy; and the town-hall, a neat modern building in the centre of the town, on an arch over the water of Kilmarnock. The merchants' society have built a spacious inn, which, in point of architecture, is an ornament to the town. There are five bridges over the Kilmarnock within the town, and two over the Irvine between Kilmarnock and Riccarton, all substantial structures.

In addition to the three-parish churches, one of which is collegiate, there are two chapels belonging to the United Associate Synod; and the Relief, Cameronians, Independents, and original Seceders have each a chapel. The Roman Catholics, though they amounted, in 1838, to between 600 and 700, have no local priest, but one from Ayr visits them periodically. The Dissenters, including Roman Catholics, who are almost all Irish, comprise rather more than a third part of the whole community.

Kilmarnock is well furnished with the means of education. There are, in the country part of the parish, three schools, and three teachers; in the town 20 schools and 22 teachers. Of these seminaries, only three are endowed; the remainder being private or voluntary. The academy, erected in 1808 at the joint expense of the landowners and private contributors, and which is at once a parish school and a joint-stock establishment, is an efficient and useful institution, amply realizing the objects of its founders. The total number of pupils at all the schools in 1838, was upwards of 3000 (New Stat. Acc., ut supra), or about a seventh part of the population. This, too, is exclusive of 21 Sunday schools, attended by 1929 scholars. Two of the above schools are free: one for boys, and one for girls. There are several public libraries, and scientific and literary associations, three printing presses, and a weekly newspaper. And it may be worth mentioning that the first edition of Burns's Poems was printed here in 1786.

Poor-rates have been introduced. The number of paupers, including those who receive either occasional or permanent relief, and pauper lunatics (4), is 563; and the gross amount of funds for their relief only £1659. Kilmarnock has a dispensary, a workhouse, and a few benevolent associations for charitable purposes.

But this town is chiefly eminent as a place of trade and manufacture. It seems originally to have been distinguished for its manufacture of the broad flat woollen bonnets formerly worn by all but the entire Scotch peasantry; and of striped night-caps. These articles, called "Kilmarnock" bonnets and caps, are still manufactured to a very considerable extent, as are forage caps for the army. The incorporation of bonnet-makers (exclusive of other parties), manufactured, in 1831, 18,790 dozen of bonnets, of the value of £12,000; and the quantity is estimated to have increased a sixth since. The carpet manufacture was introduced nearly a century ago; but the value produced, in 1791, was only £24,400; whereas, in 1839, including Brussels, Venetian, Turkey, and Scotch carpets and rugs, its gross value was estimated at £150,000. Five woollen-mills in the town and neighbourhood engaged in spinning worsted or woollen yarns for the carpet factories and bonnet-makers, employed, in 1836, about 200 hands. (*Factory Returns*, 1839.) The manufacture of worsted printed shawls is at present the most extensive business carried on in Kilmarnock. It was first begun in Scotland in 1824, at Greenholm, in this neighbourhood, by Mr. William Hall, an ingenious and enterprising calico-printer; and so rapidly did it extend, that, during the year ending 1st June, 1831, there were manufactured 1,196,814 shawls, the value of which might be about £900,000. (*Bound. Rep. ut supra*.) The business affords employment at this moment (1840) to about 1400 persons, including weavers and printers; and the value of the shawls annually produced is estimated at about £340,000. There are extensive tanneries, and the boot and shoe trade is very considerable. About 2400 pairs are made weekly, of which ¾ are exported. Machinery is also produced to a considerable extent, and there are a number of inferior manufactures. In the *Statistical Account* of the parish in 1791, the gross annual value of its different manufactures was estimated at £86,850; whereas, in the *New Statistical Account* of the parish, published in 1840, the annual value of its leading manufactures is estimated at above £476,000; and, including inferior articles, the whole may probably amount to about £550,000! The weaving of cotton (by hand-loom), in connexion with the Glasgow market, is carried on to a great extent in Kilmarnock, as in all the towns in the W. of Scotland; chiefly, in this instance, by the Irish residents. There are four branch banks in the town, and a savings' bank. Coal is abundant in the neighbourhood, about 150,000 tons a year being exported. The market days are Tuesday and Friday.

The port of Kilmarnock is at Troon, on the Ayrshire coast, with which it is connected by a railroad 9½ miles in length. This was the first public railway constructed in Scotland, the act for its construction having passed in 1808, though it was not finished till 1812. Horse power is used. A branch of the Glasgow, Paisley, Kilmarnock, and Ayr railway, now in the course of being constructed, is to communicate with this town; so that it will soon possess the readiest means of intercourse with different seaports, and with all the most important towns in the W. of Scotland.

Kilmarnock was originally a mere appendage of the baronial manor of the Boyds, lords of Kilmarnock, attainted in 1745, who had their seat in the neighbourhood. Its first charter as a free borough of barony was granted by James VI. in 1591; a second was granted in 1672. The Reform Bill erected Kilmarnock into a parliamentary borough, conferring on it, along with Renfrew, Port-Glasgow, Dumbarton, and Rutherglen, the privilege of sending a member to the House of Commons. Constituency of the burgh in 1839–40, 630, being equal to the aggregate constituency of the other burghs. The municipal property of Kilmarnock is valued at £7899; its debts at £3675. Under the municipal Reform Act it is governed by a provost, four bailies, a treasurer, and 19 councillors.

KILMARNOCK, p. t., Piscataquis co., Me., 104 m. N.E. Augusta, 701 W. Watered by Piscataquis river and its branches. It has one saw-mill, one tannery; five schools, 122 scholars. Pop. 319.

KILMARNOCK, p. v., Lancaster co., Va., 90 m. E. Richmond, 150 W. Situated on a small creek, which enters Chesapeake bay. It contains four churches, two Methodist, a Baptist, and an Episcopal; three stores, and about 150 inhabitants.

KILLINGTON PEAK, Vt., a summit of the Green mountains in the S. part of Sherburne township, 3924 feet above tide-water.

KILRENNY, a royal and parl. bor., seaport, and par. of Scotland, co. Fife, on the N.E. shore of the firth of Forth, near the mouth of that great estuary, 90 m. N.E. Edinburgh, and 9½ m. S. by W. St. Andrews. Its burghal privileges embrace Cellardykes, sometimes called Nether Kilrenny, distant ½ m. S.E. Pop. 1705. Kilrenny is a place of no importance; but Cellardykes engages extensively in the herring and whale fishery, and is a thriving village. Kilrenny, which was created a royal borough in 1707, unites with Cupar, St. Andrews, and three small adjacent boroughs, in sending a member to the House of Commons. Registered electors in 1839–40, 49. Municipal revenue £44.

KILRUSH, a seaport town of Ireland, S.W. part of the

co. Clare, on the innermost extremity of a creek on the N° side of the estuary of the Shannon, 37 m. W. Limerick, and 20 m. E. by N. from Loophead, at the mouth of the Shannon. Pop. in 1831, 4000. It exports considerable quantities of corn, meal, and flour: the herring fishery is also carried on to some extent: and it has a pier, and a patent slip for the repair of vessels. It is a creek belonging to the port of Limerick. Its chief buildings are the parish church, Roman Catholic chapel, Methodist meeting house, market-house, custom-house, courthouse, and bridewell. It has a school on the foundation of Erasmus Smith, and some other schools. A manor court is held monthly; general sessions at Easter and Michaelmas, and petty sessions on Tuesdays. It is a coast-guard and constabulary station. Markets on Saturdays: fairs, May 10, and October 12. Postoffice revenue, in 1830, £253; in 1836, £402. Branches of the Agricultural and National Banks were opened in 1835. A mail car plies daily to Ennis. Turf for fuel is brought coastways by boats in large quantities.

KILSYTH, a bor. of barony, market and manufacturing town of Scotland, co. Stirling, in a valley 10½ m. N. by E. Glasgow, and 16 m. S.W. by S. Stirling. Pop. 2250. The town is irregularly built. The only public buildings are the parish church, with a lofty spire, and a chapel belonging to the Relief. The Independents have a small congregation, but no separate meetinghouse. The inhabitants are chiefly employed as cotton weavers in connexion with the manufacturers of Glasgow. In 1831, it had 494 weavers: it has now upwards of 500. About 20 persons are employed as sickle-makers. Iron stone and coals abound in the neighbourhood. The Forth and Clyde canal passes within ¼ mile to the S., and contributes greatly to the prosperity of the district. Of the *præsidia*, or forts, erected by Agricola in his fourth campaign, several mouldering remains may yet be traced. (*Taciti Agricola*, cap. 23.) They were generally about 2 m. apart, and built nearly in the direction afterward occupied by the wall of Antoninus. This wall, or *Graham's Dyke*, as it is vulgarly termed, built by the Emperor Antoninus Pius, about the year 140, as a protection against the Caledonians on the N., ran across the isthmus between the Forth and Clyde, and passed within five furlongs of Kilsyth on the S. Kilsyth gives its name to a great victory gained in its vicinity (15th August, 1645), by the Marquis of Montrose over the Covenanters, commanded by General Baillie. Sir James Livingston (a branch of the noble house of Linlithgow), was created Viscount Kilsyth (1661), on account of his loyalty during the civil wars: but the title was attainted, and the estates forfeited in the person of the third viscount, who joined the rebellion in 1715. "Religious revivals," as certain fanatical displays recently (1839) got up in various places throughout Scotland have been termed, originated at Kilsyth, but now (January 1841) they seem to have entirely subsided, not merely here, but everywhere else. (*Nimmo's Hist. of Stirlingshire*, edit. 1817; *Chalmer's Caledonia*.)

KILWINNING, a market and manufacturing town and bor. of barony, Scotland, in the district of Cunningham, co. Ayr, on a rising ground on the right bank of the Garnock, 3 m. N.N.W. Irvine, and 21 m. S.W. Glasgow. Pop. including the contiguous village of Byres, in 1801, 1843; in 1839, 2350. The town consists chiefly of one street, but there are various narrow lanes. The modern additions to the town are substantial and elegant. The only public buildings are the parish church (with a spire), and two dissenting chapels. Eglinton Castle, famous for the tournament held there in 1839, is in the immediate vicinity. The inhabitants are chiefly employed in the weaving of cottons, gauzes, &c., for the Paisley and Glasgow manufacturers. Number so employed about 500. Lime and coal abound in the district around. The Glasgow and Ayr railway passes close to Kilwinning.

Kilwinning is celebrated for its abbey, founded by Hugh de Moreville, constable of Scotland, in 1140, and dedicated to St. Winning. It was, at the Reformation, one of the richest in the kingdom. It is said that the foreign architect who built the abbey was the first to introduce the craft of Free Masonry into Scotland. The lodge of Kilwinning, as the mother lodge of the kingdom, was in the habit of granting charters to other lodges, all of which append the word Kilwinning to their name: but the institution of the Grand Lodge of Scotland at Edinburgh has nearly superseded the dignity of Kilwinning as a mother lodge. Kilwinning is the seat of a body of archers, which existed at least as early as 1488, and is still in a flourishing condition. (*Old Stat. Acc. of Scotland*, § *Kilwinning*; *Keith's Scot. Bishops*, p. 407; *Chamber's Gazetteer*.)

KINCARDINESHIRE, or THE MEARNS, a marit. co. of Scotland, having N. the co. of Aberdeen, from which it is for the most part separated by the Dee and Avon, E. the German ocean, by which it is bordered for above 30 miles, and S. and W. Forfar. It is of a triangular shape. Area, 244,480 acres, of which 1290 are water. The Grampian mountains occupy the western, central, and most of the northern parts of the county, extending from Battock-hill, 2811 feet high, on its W. confines, to Stonehaven on the E. coast. The arable land consists principally of the district denominated the How of the Mearns, being a portion of Strathmore, or a continuation of the How of Angus, extending from Stonehaven and Marykirk to within a few miles of Stonehaven. It comprises about 50,000 acres of comparatively low, fertile, and well cultivated land, with many thriving plantations. On the E., the How is divided by a range of low hills which separate it from what is called the Coast district, containing about 68,000 acres, about a half of which is in a high state of cultivation. There is also a narrow glen or district of arable land along the Dee. Property is in a few hands. Arable farms of all sizes, many small; some from 400 to 500 acres, and the proportion of small farms decreasing. Hill pastures let in immense tracts. Improvements began in this county about the middle of last century, and have been carried on since the close of the American war, and especially during the last dozen years, with great spirit and extraordinary success. Commodious farm-houses have been erected, and new and level roads constructed in districts where formerly there were only wretched footpaths. The following statements, extracted from the notice of the parish of Fettercairn in this county, in the New Statistical Account of Scotland, comparing its present state with its state when the old statistical account was published about half a century ago, may, with little modification, be applied to the entire county. "The improvements in this parish, either begun or completed, since the last statistical account was published, have been of great extent. Much waste ground has been reclaimed, and converted into productive arable land. Extensive plantations have been formed, which are now, generally, in a thriving state, and adding to the shelter of the fields, the beauty of the landscape, the resources of the proprietors, and the benefit of the neighbourhood. Better accommodations in the dwelling-houses, farm-steadings, and enclosures, have been provided. By means of extensive and judicious draining, the salubrity of the atmosphere has been improved, the state of disease has been altered, and the health of the people promoted. By the introduction of thrashing-mills, and other useful inventions, agricultural labour has been greatly diminished. By the formation of so many new roads, internal communication and access to markets have been very much facilitated. Enlarged means of intellectual, moral, and religious improvement, have been called into operation; and it is to be hoped that the habits, manners, and enjoyment of the people have, in some measure, kept pace with the increase of these advantages." (P. 157.) Average rent of land in 1810, 6s. 8d. an acre. Lime is the only mineral of any importance. The manufacture of the beautifully jointed and painted wooden snuff-boxes, now in very general demand, originated at Laurencekirk, in this county, about 1790; but Cannock and Mauchline, in Ayrshire, have become the principal seats of the manufacture. Principal rivers, Dee, N. Esk, Bervie, Dye, &c., on some of which are considerable salmon fisheries. It contains 19 parishes, and one royal borough, Inverbervie, which is quite inconsiderable. It sends one member to the House of Commons for the county, and Inverbervie joins with Montrose, Arbroath, and other boroughs, in returning a member. Registered electors for county, in 1839–40, 914. In 1831 Kincardine had 6979 inhabited houses, 7136 families, and 31,431 individuals, of whom 15,016 were males, and 16,415 females. Valued real property in 1815, £94,816.

KINCARDINE, a seaport town of Scotland, in a detached part of the co. Perth, par. of Tullialian, on the N. side of the frith of Forth, 21 m. W.N.W. Edinburgh. Pop. of par. in 1831, 3556; of town, about 3000. The streets are mostly narrow, irregular, and dirty; but the houses are good, especially those in the newer parts of the town. It has a good quay and harbour, and a good roadstead, affording convenient anchorage for vessels of large burden. Ship-building is carried on to a considerable extent, and the town has an extensive coasting trade. The parish church is at Tullialian, but there is a dissenting meeting-house in the town. The different parties in the town to whom vessels belong have formed themselves into a Kincardine Mutual Assurance Company, the value of the property so insured being at present (1840) estimated at £70,000. There are two branch banks in the town; and a regular ferry is established with the opposite side of the river.

KINDERHOOK, county, Mo. Situated a little S.W. of the centre of the state, and contains 390 square miles. Bounded N. by Osage river, and drained by its branches. Formed since the census of 1840. Capital, Oregon.

KINDERHOOK, p. t. Columbia co. N.Y., 19 m. S. Albany, 251 W. Watered by Kinderhook creek. Organized in 1788. It contains 23 stores, two fulling-mills, one woollen factory, three grist-mills, three saw-mills, one printing-office, one weekly newspaper; two academies, 314 students, 11 schools,

378 scholars. Pop. 3512. The village contains two churches, a Dutch Reformed and Baptist; a bank, an academy, 209 dwellings, and about 1900 inhabitants.

KINGHORN, a royal and parl. bor., seaport, and par. of Scotland, co. Fife, on an eminence, overhanging a small bay on the N. banks of the frith of Forth, 9 m. N. by E. Edinburgh, and 3 m. S.W. Kirkcaldy. Pop. of bor. in 1831, 1610; of bor. and par., 2579. The town was, not long since, one of the most irregularly built in Scotland; but it has of late undergone many improvements in this respect, and most of the older houses (which had two flats or stories, with outside stairs facing the street) have been superseded by more modern and better buildings. The only public edifices are the parish church, a dissenting chapel, a town-hall, jail, and a handsome schoolhouse, recently erected by subscription. Fifty poor children are educated gratuitously on the bequest of the late Mr. Philip of Kirkcaldy, and are clothed, and provided with books and other school utensils. There are several subscription libraries. The chief branch of industry is flax-spinning; three flax-mills, driven by steam, employing 250 hands. In addition to these, about 150 individuals are employed in the weaving of different linen fabrics. A few persons engaged in fishing. The harbour is bad, and scarcely any shipping is ever seen in it. Pettycur, about a mile W., is a better harbour; but its chief business is derived from its being one of the seats of the ferry across the frith of Forth to Leith and Newhaven.

Kinghorn lays claim to great antiquity: it is certain that it was created a royal bor. as early as the 13th century. It was originally a royal residence, but lost that dignity on the death of Alex. III., who was killed (1285) by falling over a rugged and lofty eminence about a mile W. of the town. Kinghorn unites with Burntisland, Dysart, and Kirkcaldy, in sending one member to the House of Commons. Registered voters in 1839–40, 40. (Bound. Returns; Factory Returns; Beauties of Scotland; Sibbald's Hist. of Fife.)

KING'S COUNTY, an inland co. of Ireland, prov. Leinster, having N. Westmeath, E. Kildare, S. Tipperary and Queen's County, and W. Roscommon, Galway, and Tipperary. Area, 598,166 acres. A portion of the bog of Allen covers a very considerable tract in the more northerly parts of this county, while on the S. it is partially encumbered with ramifications of the Devil's Bit and Sliebhbloom mountains. On the whole, the unimproved bog and mountain are supposed to occupy 123,349 acres, of which, however, the far greater portion belongs to the bog. Soil of an average degree of fertility. Estates mostly very large. Tillage farms small, but some of those devoted to grazing are very extensive. Subtenancy is less common here than in most parts of Ireland; but its rural economy is, notwithstanding, but little different from that of the surrounding counties. (See KILDARE, MEATH, &c.) Average rent of land 12s. an acre. Silver has been found at Edenderry, but if we except limestone, it has no minerals of any real importance; manufactures can hardly be said to exist. Its chief town is Birr or Parsonstown. It is bounded on the W. by the Shannon, and on the S. by the Little Brosna, while it is intersected by the Greater Brosna and the Grand canal. It is divided into 11 baronies and 58 parishes, and returns two members to the House of Commons, both for the county. Registered electors, in 1838–39, 1440. In 1831, King's County had 24,256 inhabited houses; 26,072 families; and 144,225 inhabitants, of whom 71,287 were males, and 72,908 females.

KING'S, county, N.Y. Situated on the W. end of Long Island, and contains 76 sq. m. Organized in 1683. The soil is a sandy loam, highly cultivated, and affording an abundance of vegetables for the New-York market. Coney Island lies in the Atlantic, on its S. shore, and is much resorted to for sea-bathing. It contained in 1840, 5078 neat cattle, 48 sheep, 8360 swine; and produced 24,964 bushels of wheat, 8537 of rye, 81,894 of Indian corn, 3033 of buck-wheat, 72,460 of oats, 95,805 of potatoes. It had five commercial and commission houses in foreign trade, capital $109,500; 209 retail stores, capital $515,500; six lumber-yards, capital $77,000; capital invested in drugs, paints, &c., $38,000; one tannery, nine distilleries, one brewery, 14 rope-walks, five printing-offices, two binderies, one daily, four weekly, one semi-weekly newspapers, and one periodical. Total capital in manufactures, $1,806,936; 23 academies, 1396 students; 53 schools, 3980 scholars. Pop. 47,613. Capital, Brooklyn.

KING AND QUEEN'S, county, Va. Situated in the E. part of the state, and contains 338 sq. m. Bounded N.E. by Piankatank river, S.W. by Mattapony river. It contained in 1840, 8609 neat cattle, 8571 sheep, 14,806 swine; and produced 40,368 bushels of wheat, 9063 of rye, 343,384 of Indian corn, 36,919 of oats, 13,925 of potatoes, 8190 pounds of tobacco, 63,975 of cotton. It had six stores, four flouring-mills, 95 grist-mills, eight saw-mills; 14 academies, 362 students, 13 schools, 296 scholars. Pop. whites, 4405; slaves, 5037; free coloured, 460; total, 10,882. Capital, King and Queen C. H.

KING AND QUEEN, C. H., p. v., capital of King and Queen co., Va., 53 m. E.N.E. Richmond, 141 W. Located on a plain, three fourths of a mile from Mataponay river. It contains a courthouse, jail, a flouring-mill and grist-mill in the vicinity, and about 70 inhabitants.

KINGFIELD, p. t., Franklin co., Me., 50 m. N.N.W. Augusta, 646 W. Incorporated in 1808. Watered by Seven Mile brook, which flows into Kennebec river. It contains three stores, one grist-mill, two saw-mills; 9 schools, 270 scholars. Pop. 671.

KING GEORGE, county, Va. Situated toward the E. part of the state, and contains 234 sq. m. Bounded N.E. by Potomac river, S.W. by Rappahannock river. Organized in 1720. It contained in 1840, 4348 neat cattle, 4623 sheep, 6689 swine; and produced 37,556 bushels of wheat, 3029 of rye, 234,270 of Indian corn, 36,097 of oats, 5082 of potatoes, 22,918 pounds of tobacco, 4165 of cotton. It had six stores, 11 grist-mills, three saw-mills; five academies, 37 students; 10 schools, 192 scholars. Pop., whites, 2393; slaves, 3283; free coloured, 276; total, 5972. Capital, King George C. H.

KING GEORGE, C. H., p. v., capital of King George co., Va., 80 m. N.N.E. Richmond, 76 W. It contains a courthouse, jail, one store, 15 dwellings, and about 90 inhabitants.

KINGSBURY, p. t., Washington co., N. Y., 56 m. N. Albany, 498 W. Watered by a branch of Wood creek. It contains the village of Sandy Hill, the half shire county seat, and has three churches, 14 stores, one fulling-mill, one woollen factory, one grist-mill, one saw-mill, one printing-office, one bindery, one weekly newspaper; three academies, 164 students; 16 schools, 850 scholars. Pop. 2772. The Champlain canal passes through it.

KINGSESSING, p. t., Philadelphia co., Pa., 7 m. S.W. Philadelphia, 96 m. E. by S. Harrisburg, 134 W. Bounded S.E. by Delaware river, which contains several islands belonging to it. Bounded E. by Schuylkill river, W. by Darby creek. On the banks of the Schuylkill is a fine botanic garden, founded by Bartram, the celebrated naturalist. It has six stores, two lumber-yards, two saw-mills. Pop. 1389.

KINGSTON-ON-THAMES, a bor., market town, and par. of England, co. Surrey, loc. dt. in hund. of its own name, but with separate jurisdiction, on the S. of the right bank of the Thames, 12 m. S.W. London. Pop. of the par. in 1831 (exc. Ham and Hook), 5680; and with the hamlet of Hampton-wick (which is included within the new boundary of the mun. bor.), 7459. "The town extends from N. to S. about ½ m. along the Thames, crossed here by an elegant stone bridge of five arches, opened in 1828, and rather more than a ¼ m. from E. to W. Nearly continuous lines of houses, however, diverge from the body of the town along the two principal high roads towards London, almost to the bottom of Kingston Hill, and on the road to Portsmouth as far as the parish boundary, 1½ m. from the town. On the opposite side of the river is Hampton-wick, which may be fairly considered to constitute a part of Kingston, though the communication has recently been much lessened by a toll levied on all passengers crossing the new bridge." (Mun. Bound. Rep.) The town is well paved and lighted with gas. The streets are narrow and irregular; but there is a spacious market place, in which is the town-hall, erected in the reign of James I., containing some curious pictures and carvings of high antiquity. The Lent assizes for the county, which were formerly held in it, have been for some years transferred to a neighbouring brick edifice built for the purpose; and attached to it is a small jail, used for the temporary accommodation of prisoners. The church is large, but plain, with a low square tower, and appears to have been erected at different periods, commencing with the reign of Richard II.: the living is a vicarage, in patronage of King's College, Cambridge. There are places of worship for several denominations of Dissenters. It has a grammar school, founded in 1560, furnishing instruction to between 30 and 40 boys; a boys' and girls' national school, supported by subscriptions; an almshouse for six aged men and as many women; and a dispensary. Kingston is not a place of much trade. Considerable business is done in malting, there being 15 malting-houses in or near the town; and there are also some flax and oil-mills; but most of the townspeople are dependant on their retail dealings with the neighbouring gentry. A large and well attended corn market is held every Saturday; and the fairs are on Thursday, Friday, and Saturday in Whitsun week, Aug. 2, 3, and 4, and Nov. 13, for horses, toys, pedlary, &c.

Kingston, first incorporated by King John in 1199, and chartered by many subsequent monarchs, has been governed since the passing of the Municipal Reform Act by a recorder, six aldermen, and 18 councillors; the borough being divided into three wards. Corporation revenue in 1839, about £3600; but of this £3466 arose from the sale of property. Rental of houses rated within the borough, £13,571. Members were sent by it to the House of Commons in the reigns of

Edward I. and II.; but the burgesses were relieved from the burden on petition, and the franchise has not since been renewed. Roman coins, urns, and other antiquities, that have been dug up in considerable quantities, prove Kingston to have been inhabited by those early conquerors of Britain. It received its name, King's-town (its more ancient appellation being More-ford) from its having been the residence of our Saxon monarchs, 8 of whom were crowned here, some in the market-place, and others in a very ancient chapel once attached to the church, but now destroyed. A general council was held here by Egbert in 838, and attended by the chief prelates and nobility of the realm. The town continued during several centuries, to be a place of high consideration, and in royal favour. (Lyson's Environs of London, art. Kingston; Mun. Bound. Report, &c.)

KINGSTON, the largest and most commercial city of Jamaica, though not the capital of the island; on its S. coast, on the N. side of a fine harbour, on the verge of an alluvial plain surrounded by an amphitheatre of mountains. Lat. 17° 56' 6" N., long. 76° 53' 15" W. Pop. loosely estimated at 35,000; but no accurate census has ever been taken, and this estimate is probably exaggerated. It is built on ground gently shelving to the verge of the sea, and was originally comprised in an oblong space, 1 m. in length, by ½ m. in breadth, but it has of late years extended considerably beyond these limits.

The streets in Lower Kingston are long and straight, crossing each other at right angles, like those of the new town of Edinburgh; the houses in general are two stories high, with verandahs above and below. There are two churches, an English and a Presbyterian, both handsome structures, especially the former, which is built on an elevated spot overlooking the city. Kingston has several dissenting chapels, two synagogues, a hospital founded in 1776, numerous other charitable institutions, a free school established in 1729, with an endowment of £1500 a year, a workhouse and house of correction, commercial subscription rooms, an athenæum, a society of agriculture, arts, and sciences, a savings' bank, and a theatre.

The mountain chain forming the boundary of the plain on which Kingston stands, terminates to the E. in a narrow ridge, whence a long narrow tongue of land extends to Port Royal, forming the S. boundary of Kingston harbour, a land-locked basin, in which ships of the largest burden may anchor in perfect security. It is strongly fortified. Its entrance, between Port Royal on the E., at the extremity of the tongue of land already noticed, and the opposite coast, is defended by Fort Charles, near Fort Royal, on the one hand, and by the Apostles' Battery, Fort Henderson, and Fort Augustine, on the other. The depth of water in the centre of the channel leading to the harbour is, where shallowest, 4 fathoms, and in the harbour itself it varies from 6 to 10 fathoms. About 2 m. N. of Kingston is Up-Park Camp, the only government barracks in the island, consisting of two long and parallel lines of buildings, two stories high, occupying, together with the parade ground, &c., between 200 and 300 acres. Not far from this station is the "Admiral's Pen," the former residence of the naval commander-in-chief; but which has been abandoned for several years. Stony-hill garrison is about 7 m. N. Kingston, at an elevation of about 2000 feet above the sea.

Kingston engrosses by far the largest portion of the trade of the island. In 1839, the value of the imports amounted to £439,394, of which products of the value of £168,589 were from Great Britain; £116,000 from British colonies; £137,307 from the United States; and £22,999 from other foreign countries. The value of the exports during the same year amounted to £747,419; of which, products worth £397,215 were shipped for Great Britain; £8909 for British colonies; £31,363 for the United States; and £308,391 for foreign countries. The staple article of export is, of course, sugar, of which the shipments amounted, in the course of the above year, to 3686 hhds., 83,686 cwts.; and those of rum to 9454 pun., 292,296 galls. During the same year, 435 ships, of the burden of 67,936 tons, entered; and 495 ships, of the burden of 51,814 tons, cleared out.

The corporation of Kingston consists of a mayor, 12 aldermen, and 12 common-councilmen. Corporation revenue, about £33,400 a year. The town was founded in 1693, in consequence of the destruction of Port Royal by an earthquake; but it was not incorporated till 1802. (Jamaica Almanack; Encycl. Americana; and Parliamentary Papers.)

KINGSTON, in Upper Canada. See TORONTO.

KINGSTON, p. t., Rockingham co., N. H., 33 m. S.S.E. Concord, 471 W. Chartered in 1694. Two ponds, one of 300 acres, and another of 980 acres, by their outlets give rise to Powow river, which flows into Merrimac river. It contains a Congregational church, six stores, four grist-mills, three saw-mills, two tanneries; six schools, 452 scholars. Pop. 1258.

KINGSTON, p. t., Plymouth co., Mass., 34 m. S.S.E. Bos

ton, 443 W. It has a good harbour, a branch of Plymouth harbour, into which a small stream, Jones river, empties. It has a considerable number of vessels engaged in the cod and mackerel fisheries. It contains three churches, a Congregational, Unitarian, and Baptist, four commercial houses in foreign trade, seven stores, one cotton factory, with 900 spindles, three grist-mills, three saw-mills; seven schools, 298 scholars. Pop. 1440.

KINGSTON, p. v., South Kingston t., capital of Washington co., R. I., 33 m. S. by W. Providence, 391 W. It contains a courthouse, jail, a Congregational church, a bank, and about 30 dwellings.

KINGSTON, p. t., capital of Ulster co., N. Y., 57 m. S. by W. Albany, 316 W. Bounded E. by Hudson river, S.E. by Rondout creek. Organized in 1788. Watered by Esopus creek. It contains three commission houses in foreign trade, 40 stores, four lumber-yards, one woollen factory, four grist-mills, three saw-mills, three tanneries, three printing-offices, one bindery, three weekly newspapers; 10 schools, 941 scholars. Pop. 5904. The village on Esopus creek, 3 m. from its entrance into Hudson river, contains 10 streets, crossing each other at right angles, and has a courthouse, jail, and county offices in a large stone edifice which cost $45,000, four churches, a Dutch Reformed, Methodist, Episcopal, and Baptist, a bank, an academy and female seminary, 275 dwellings, and about 2000 inhabitants. The village, then called Esopus, was burned by the British in 1777.

KINGSTON, p. v. Middlesex co., N. J., 14 m. N.E. Trenton, 193 W. Situated partly in Somerset co., on Millstone river, midway between New-York and Philadelphia. It contains a Presbyterian church, an academy, four stores, a grist-mill, saw-mill, and woollen factory, about 50 dwellings, and 300 inhabitants. The Delaware and Raritan canal passes through it, with locks at this place.

KINGSTON, p. t., Luzerne co., Pa., 127 m. N.E. Harrisburg, 228 W. It has eight stores, one furnace, three fulling-mills, one woollen factory, six grist-mills, four saw-mills, one paper-mill, one powder-mill, four tanneries; 8 schools, 411 scholars. Pop. 2604. The village formerly called Wyoming, is in the S. part of the t., and contains a church, several stores and about 50 dwellings. It is celebrated for the massacre of the Americans by the British and tories under Col. Butler, in violation of a treaty of surrender, in 1778.

KINGSTON, p. v., capital of Roane co., Tenn., 112 m. E. by S. Nashville, 536 W. Situated on a point formed by the junction of Holston and Clinch rivers, 60 miles below Knoxville by the course of the river. It contains a courthouse, jail, eight stores, and about 500 inhabitants.

KINGSTON, p. v., capital of Autauga co., Ala., 92 m. S.E. Tuscaloosa, 836 W. Situated on Autauga creek, which flows into Alabama river. It contains a courthouse, and several stores and dwellings. Nett proceeds of the post-office, $184.

KINGSTON, p. v., capital of Lenoir co., N.C., 80 m. S.E. by E. Raleigh, 296 W. Situated on the N. side of Neuse river. It contains a courthouse, jail, and several stores and dwellings. Nett proceeds of the postoffice, $646.

KINGSVILLE, p. t., Ashtabula co., O., 213 m. N.E. Columbus, 349 W. Bounded N. by lake Erie. Watered by Ashtabula and Conneaut creeks. It has eight schools, 396 scholars. Pop. 1412.

KING WILLIAM, county, Va. Situated toward the E. part of the state, and contains 270 sq. m. Bounded N.E. by Mattapony river, S.W. by Pamunkey river. It contained in 1840, 6441 neat cattle, 4977 sheep, 12,727 swine; and produced 26,534 bushels of wheat, 5876 of rye, 350,370 of Indian corn, 45,140 of oats, 17,436 of potatoes, 11,171 pounds of tobacco, 36,632 of cotton. It had nine stores, six flouring-mills, 17 grist-mills, four saw-mills; 12 academies, 221 students; five schools, 65 scholars. Pop.: whites, 3150; slaves, 5799; free coloured, 298; total, 9256. Capital, King William, C. H.

KING WILLIAM, C. H., p. v., capital of King William co., Va., 36 m. N.E. by E. Richmond, 136 W. It has a courthouse, jail, clerk's office, all of brick, neatly enclosed with an iron railing, and shaded with a grove of locust trees, one store, 16 dwellings, and about 75 inhabitants.

KINGWOOD, p. t., Hunterdon co., N. J., 29 m. N.W. Trenton, 166 W. Bounded W. by Delaware river. Drained by Lockatong creek. It contains eight stores, six grist-mills, four saw-mills, three distilleries; six schools, 157 scholars. Pop. 2047.

KINSWOOD, p. v., capital of Preston co., Va., 234 m. N.W. Richmond, 210 W. Situated 2 m. W. of Cheat river, and contains a courthouse, jail, three stores, 30 dwellings, and about 150 inhabitants.

KINROSS, a small inland co. of Scotland, on the W. confines of Fife, being entirely surrounded by the latter co. and that of Perth. Area, 50,560 acres, of which 4480 are water, consisting principally of Lochleven. (See next article.) Surface varied: in the lower district, to the N. and

W. of the lake, the soil is clayey, sandy, and moderately fertile; but in the upper districts it is mostly moorish, mossy, and unproductive. Agriculture a good deal improved; but it labours under great disadvantages from the backwardness of the climate. Property much subdivided, being mostly occupied by resident proprietors holding of the estate of Kinross under payment of a fee or quit rent. The manufactures are of little importance; and though it has limestone and freestone quarries, it has no coal. Average rent of land, in 1810, 9s. 10d. an acre. Kinross and Milnathort are the only towns. It is divided into seven parishes, and is united with Clackmannan and certain parishes in the S.W. part of Perth in returning a member to the House of Commons. Registered electors in this county in 1839-40, 463. In 1831, Kinross had 1594 inhabited houses, 2019 families, and 9072 inhabitants, of whom 4519 were males, and 4553 females. Valued rent, £30,193 Scotch; annual value of real property in 1815, £3,805.

KINROSS, a market town of Scotland, co. Kinross, of which it is the cap., in an open vale on the W. shore of Lochleven, and on the high road between Edinburgh and Perth, 21 m. N.W. by N. Edinburgh, and 134 m. S. by E. Perth. Population of town and parish in 1831, 2927; of which the town had 2890. The town formerly consisted of a series of narrow tortuous lanes, but the main street, along the public road, of comparatively recent erection, is wide and substantially built, though not entirely straight. The other portions of the town are irregular, narrow, and of an inferior description. The public buildings are the parish church, built in 1832, in the Gothic style, an elegant structure; the county hall, which also contains the public jail, erected in 1826; and two places of worship in connexion with the Associate Synod. On the margin of the lake, in the immediate vicinity of the town, is Kinross House, built on the site of an ancient castle, long the residence of the earls of Morton, by Sir William Bruce, architect to Charles II., and now the seat of the feudal superior of the burgh (Sir Graham Montgomery). This mansion was originally intended for the residence of the Duke of York, afterward James VII. of Scotland, in the event of his being prevented by the Exclusion Bill from succeeding to his brother. (*New Stat. Acc. of Scotland, Kinross*, p. 9.) Gas was introduced into the burgh in 1835; the gas works being placed about a mile N. of the town, at an equal distance from it and the neighbouring village of Milnathort. Of the whole inhabitants of the parish, 1940, or less than a half, belong to the Established church; the rest are dissenters. (*Ib.*, p. 22.)

Kinross, including the town and parish, has no fewer than twelve schools, of which one only is endowed, and four are taught by females: in 1838, rather more than the seventh part of the population were at school. (*Ib.*, p. 24.) There are two public libraries, three juvenile libraries of a religious kind, and a public reading room.

Kinross was famous of old for its cutlery; afterward for the manufacture of Silesia linen. But both those have ceased. Cotton weaving, in connexion with Glasgow, and more recently the manufacture of tartan shawls, plaiding, and such like articles, are now the principal employments. Damask weaving, for the Dunfermline manufacturers, has also been introduced. There were, in 1839, 334 cotton weavers, 48 weavers of tartan shawls, &c., and 14 of damask; total, 396. A mill for carding and spinning wool, in connexion with the tartan manufacture, was opened, in 1836, at West Tillyochie, 3 m. from Kinross. There are four annual fairs, chiefly for cattle, held at Kinross; and it has a branch bank of the British Linen Company.

Lochleven, on the banks of which the town is built, has of late been subjected to a considerable drainage. Its circuit is 12 m., being three less than formerly; and its mean depth has been reduced from 18½ to 14 feet. Its fishery, which opens on 1st January, and closes on 1st September, yields a yearly rent of £804. Notwithstanding its diminished size, Lochleven is still a very fine sheet of water. It contains three islands, of which two are important; St. Serf's, on the E., on which are the ruins of a priory belonging to the canon regulars of St. Augustine; and the Castle Isle, on the W., so named from its castle, once a royal residence, and in which, as everybody knows, Queen Mary was confined from 16th June, 1567 to 2d May, 1568. During her imprisonment here she was forced to sign an instrument resigning the crown to her infant son. The battle of Langside, which decided her fate in Scotland, took place on the 13th May, only eleven days after her escape from Lochleven. Andrew Winton, author of the *Crowykil of Scotland*, was prior of the monastery of St. Serf. Michael Bruce, the poet, who died in 1767, in the 21st year of his age, was born in Kinnesswood, on the N.E. shore of Lochleven, and received the principal part of his education in Kinross.

The borough of Kinross has no public property; but the inhabitants have, since 1742, had recourse to a voluntary assessment, varying from £90 to £25 a year, for municipal purposes; this fund is placed under the management of a board or committee, consisting of a dozen members chosen annually at a public meeting held for the purpose; thus constituting a species of municipal government.

KINSALE, a parl. bor. and seaport town of Ireland, co. Cork, on the E. side of the Bandon, a little way above its mouth in St. George's channel, 14 m. S. Cork, and 7 m. N. from the lighthouse on the Old Head of Kinsale, which last is in lat. 51° 36' 45" N., long. 8° 32' 16" W. Pop. of parl. bor., in 1831, 6897. It is mostly built along the water's edge, but extends in parts up a pretty steep hill, so that many of its streets are of inconvenient access; they are generally also narrow and dirty; the houses have for the most part an antiquated appearance, and some of them are said to be built in the Spanish fashion. The harbour is excellent. There are 12 feet water over the bar at the river's mouth at low ebb; and at the anchorage within the bar, off Cove, there are 4 or 5 fathoms water within half a cable's length of the shore, and large vessels may lie close to the town. It was formerly strongly fortified; and Fort Charles, on the E. side the river, is now converted into a barrack. It has an ancient parish church, a modern and handsome Roman Catholic chapel, another Roman Catholic chapel attached to a convent, and two Methodist meeting-houses; with a suite of assembly rooms, a town-hall, prison, fever hospital, and dispensary. Exclusive of Fort Charles, there is another extensive barrack adjoining the town. Here is an endowed school, founded in 1767; it has also charity schools for Roman Catholics and Protestants, Sunday schools, &c. The corporation, which claims to be such by prescription, confirmed by several charters, consisted of a sovereign and an unlimited number of burgesses and freemen; but no person was entitled to his freedom de jure, this being a matter of grace and favour in the hands of the council, consisting of the sovereign, burgesses, and common speaker. Previously to the union, Kinsale returned two members to the Irish House of Commons; and is has since returned one member to the imperial House of Commons, who, down to the passing of the Reform Act, was elected by the sovereign, burgesses, and freemen. The village of Scilly, contiguous to the town, is comprised within the limits of that present parliamentary borough, which includes a space of 273 acres, and had, in 1838–39, 274 registered electors. Corporation revenue about £550 a year.

Notwithstanding the excellence of its port, and its fine river, which is navigable for a considerable way above the town, the trade of Kinsale is but trifling, the value of its exports in 1835 not having exceeded £13,479. It is, consequently, in a depressed condition; and is one of the few Irish towns in which the postoffice revenue declined between 1830 and 1836. There is a brewery in the town, and some flour mills in the vicinity.

The principal dependence of the town is on its fisheries, which supply Cork and the surrounding country. Every kind of fish is taken; and the sales of fresh fish are said occasionally to average £500 per week. The fishermen are esteemed the most skilful of any in Ireland; and, being well acquainted with the coasts, they are good pilots, which obtained for them an exemption from impressment during the late war. They generally fish in good sea boats of from 15 to 20 tons, called hookers, and earn from 9s. to 12s. per week; several boats are also employed in the lobster fishery. Oysters of a large size were formerly abundant, but are said to be decreasing, from the want of a judicious and properly enforced code of fishery regulations. The number of boats, including those of the neighbouring inlet of Courtmacsherry, were, in 1836—

Vessels.	No.	Tonnage.	Men.
Half decked	11	196	74
Open sail	8		56
Row boats	115		760
Total	134	196	890

In the summer season Kinsale is resorted to by sea-bathers.

Kinsale is a place of some note in Irish history. It was taken in 1601 by a Spanish armament, but was retaken during the same year. James II. landed here in March, 1689; but it was taken by the troops of William III., under the Earl, afterward Duke of Marlborough, in the following year. It had formerly a royal dock-yard; and during the late war the harbour was a good deal resorted to by king's ships. (*Railway, Fishery, Boundary, &c., Reports.*)

KINTORE, a royal and parl. bor. and market town of Scotland, co. Aberdeen, on the line of road from Aberdeen to Inverness, 11 m. N.W. Aberdeen, and 3 m. S.E. by S. Inverury. Pop., in 1831, 409. It is an unimportant place. The Aberdeenshire canal passes it on the W. The borough lays claim to great antiquity: its earliest extant charter is dated 1506, confirming others of older date.

Kintore gives the title of earl to a branch of the ancient

family of Keith, descended, in the 17th century, from a younger son of the sixth earl Marischal. It unites with Elgin, Banff, Cullen, Inverury, and Peterhead, in sending one member to the House of Commons. Registered voters in 1839–40, 36. (*Boundary Returns; Beauties of Scotland*).

KIRBY-MOORSIDE, a market town and par. of England, N. riding co. York, wap. Ryedale, on the Dove, an affluent of the Derwent, 22 m. N. by E. York, and 192 m. N. by W. London. Area of par., comprising five townships, 13,700 acres. Pop. of township in 1831, 1892. The town, which is very small, stands on the S. side of the N. York moors, and is nearly encompassed by steep hills. The parish church, in a romantic situation, is about 1 m. distant, and the living is a vicarage, in the gift of the Lord Chancellor. There are places of worship also for Calvinistic and Wesleyan Methodists, and for the Society of Friends. The river turns several corn-mills; limestone is dug in the neighbourhood; and the malting trade is carried on, the surrounding district being very productive of grain. Its only historical celebrity is owing to the fact that George Villiers, Duke of Buckingham, the profligate favourite of Charles II. (a part of whose estates lay here), retired thither after his disgrace at court, and ended his days in seclusion and poverty. Markets on Wednesday; cattle and horse fairs, Whit-Wednesday and September 18.

KIRGHIS (STEPPE OF THE), a country of W. Asia, in the N. part of Independent Turkestan, between the 44th and 55th parallels of N. lat., and 53° and 89° E. long.; bounded N. by the Oui, a tributary of the Tobol, and a line of forts connecting Zverenogolovsk, Petropavlovsk, and Omsk; E. by the Irtish and the Chinese stations, extending S. as far as the 49d parallel; S. by the khanates of Khokan, Bokhara, and Khiva; and W. by the Oural and the Caspian sea. Length, about 1400 m.; breadth, 1100; probable area, 1,533,000 sq. m. Pop. of the three hordes composing the Kirghis nation, 2,390,000. According to M. Alexis de Lévchine, from whose able work, *La Description des Hordes et des Steppes des Kirghis-Kazaks,* lately published, we are enabled to give many interesting details respecting this nomadic people, who, till now, have been comparatively unknown. The Kirghis steppe is not, as hitherto generally supposed, a mere flat and unvaried plain, but is intersected by numerous mountain ridges, and even in its more level parts is covered with round hillocks, causing considerable undulations on the surface. Offsets of the Oural range occupy a large amount of surface in the W. and N.W. parts of the steppe. The W. continuations of the Altai range run in very irregular ridges close to the Chinese frontier, and finally connect themselves about the 49d parallel with the W. part of the Mus-tagh, or Thian-chan range. It would be difficult to reduce the ridges in the centre of the steppe to any system; but the principal are all N. of the 48th deg. of N. lat. The Kara-taou mountains separate the Kirghis steppe southward from the khanate of Khokan. The geological constituents and mineral riches of these mountains are little understood; the central masses appear to consist of granite, gneiss, serpentine quartz, &c., on which are superimposed silicious and clay-slate, blue limestone, coal slate, with various secondary and other rocks. Lead, copper, and iron, with a small quantity of silver, are found in these mountains, but the present state of the country makes mining wholly impossible.

The waters of the Kirghis steppe comprise, besides the two land-locked seas, the Caspian and the Aral (which see), a considerable number of lakes and rivers. Among the former (most of which are salt), the largest are the Balkat (115 m. long), Isik (90 m. by 30 m.) in the S.E. angle of the steppe, the Kourdaigtne, Teali, Tchagti, Ouba-gan Denghis, and Aksakal-Barbi lakes, with many others of smaller extent. The chief rivers are, 1. the Sur-Deria (anc. *Jaxartes ?*) rising in the Muz-tagh, about lat. 41° N., and long. 76° E., having a course S.W. to Kokan, and thence N.W. through the sandy plains of Kisti-houm and Kara-koum into the sea of Aral, its entire length somewhat exceeding 800 m.; and, 2. the Irtish, rising in Chinese Turkestan on the W. side of the Great Altai, entering the steppe in the 49th par., forming its E. boundary up to 53° N., and receiving on its W. banks the Ichim, the Tobol, and other tributaries, which intersect with their streams the entire N. half of the steppe. Numerous smaller rivers fall into the different lakes, and many others are almost unknown to Europeans.

The climate is remarkable for its extremes of heat and cold. In the middle and little hordes, that is, in the N. and N.W. parts of the steppe, the thermometer often falls to 20°, and sometimes 30° below freezing point (Réaum.). The rivers and plains are covered with ice, and the hills with a thick coating of snow; while strong winds from the N.E. increase the intensity of the cold, and hurricanes, called *bourans,* often uproot forest trees, and carry away both man and beast, causing dreadful, and often irremediable, destruction. (*Lévchine,* p. 5.) In summer, on the sun-

trary, the temperature often rises to 30° Réaum. (100° Fahr.) in the shade: the oppressiveness of the heat is much increased also by the sandy nature of the soil, and the paucity of rivers and forests over so vast an extent of country. This great variability of temperature, however, and the rapid transition from one extreme to the other, are said not to be so prejudicial to the health either of natives or travellers as might have been expected: agues, indeed, and fevers, are common in the marshy districts; but, generally speaking, the people are robust and long-lived. Rain is very rare, even on the mountain sides: dews refresh the soil in some parts, but by far the largest portion of the surface is dried up and rendered useless by the entire absence of atmospheric moisture. Trees and shrubs are only found on the banks of rivers, and at the foot of the mountains near the Russian frontier, where the soil is the most capable of cultivation: the principal are elms, poplars, willows, wild plum, juniper, and liquorice trees, (the latter very abundant, and their produce forming a principal article of trade,) wormwood, alkanet, tragacanth, various kinds of euphorbia, anemonies, camomile, asparagus, garlic and onions, horse-radish, wild oats, rye, &c. Short coarse grass generally covers the plains, on which also the sal-soda plant grows in great perfection. Agriculture, as a branch of industry, cannot be said to exist. Some land about the rivers is roughly tilled, for the purpose of raising millet, rye, and barley; but the pursuit, except by the Karakalpaks, S. of the Sir-Daria, is generally despised, being only followed by the poorest classes, and then chiefly by women. The wild animals of this region comprise the wolf, wild boar, fox, Cossack dog, wild goat, and hare, all of which roam in great numbers over every part of the steppe: the boar, buffalo, antelope, wild horse, beaver, and water-rat, are plentiful in some districts; and there are likewise tigers, lynxes, and other varieties of the cat tribe, in the S. Among the birds may be mentioned the eagle, falcon, cormorant, pelican, stork, heron, goose, and pheasant, with many smaller birds. The lakes and rivers abound with seals, and with several kinds of fish, sturgeon, pike, perch, and carp, being the most common. The domestic animals of the Kirghis are the sheep, goat, horse, and camel, the rearing of which constitutes the chief employment of this nomad race. Larger flocks of sheep are, perhaps, nowhere to be found, some of the richer inhabitants possessing upward of 20,000 head. The animals are strong and large, weighing from 100 to 150 lbs., and they have long coarse wool, and enormous tails, sometimes 30 lbs. in weight. They endure with astonishing patience the long privations of food and drink to which they are subject, soon recovering in spring their plump and healthy appearance.

The advantages derived by the people from these animals are immense: their flesh and milk supply them with food, and the wool furnishes felt for covering the tents and other purposes, while at the same time they serve as a standard of value, and form a chief article of export: for, according to Haxthausen, 1,600,000 sheep are sent off every year and sold in Russia, Bucharia, and China. Goats, very similar to those of Thibet, are chiefly used as guides in leading the sheep from pasture to pasture, as the latter will not move without them; their flesh is eaten, and the down concealed under their red shaggy hair is an extremely valuable article of trade. The camels (most of which have two humps, the single-humped variety being too delicate for the climate) are here, as elsewhere in Asia, the chief beasts of burden. They are indispensable to a nomad people, like the Kirghis, for transporting their women and children, their property and trading stock; nor is it unusual for the rich to possess 200 or even 400 of these animals. Their hair is spun and made into garments, the milk and flesh are used as food, and the skins of the younger animals make warm pelisses. The camels are extremely docile, and carry burdens varying from 14 to 18 poods (from 5 to 6 cwts.), travelling during long journeys at the rate of 25 or 35 m. a day. Camel breeding is pursued to a considerable extent in the great (or S.) horde, and great numbers are sent to Persia and India. Horned cattle are very little bred, except in the middle horde; and they were not introduced into the country till about a hundred years ago. Horses are reared in immense numbers, particularly in the N. part of the steppe, where there is a grass called kovil admirably suited for horse pasture. A kirghis's wealth is usually reckoned by the number of his horses; and the richest among them have as many as 8000 or 9000. They are small, but strong, and extremely rapid in their movements; they can travel from 30 to 50 m. without stopping for days together, and, like the other domestic animals, are inured to great privations and long abstinence from food and water. Various expedients are adopted to procure pasturage for the cattle during winter, by making enclosures, raking away the snow, &c.; but still they feel most severely the absence of nourishing food, and great numbers, especially of sheep, are lost every year. (Lévchine, p. 406-415.)

The population of the Kirghis steppe, however different the origin of its several sections, has long become amalgamated; and they are now but one people, inhabiting the same kind of country, speaking the same language, professing the same religion, and characterized by nearly uniform habits and customs. The following table of the races and tribes of the three great hordes gives also some insight into the distribution of the population:

Hordes and Races.	No. of Tribes.	Tents.	Individuals.
1. *Little Horde :*			
Alimouly	6		
Baiouly	12	100,000	500,000
Djetic-ourouq	7		
2. *Middle Horde:*	17		
Arghins	5		
Naimans	3	165,000	820,000
Kiptchak	2		
Ouvak-Ghirin	3		
3. *Great Horde :*			
Ouisun			
Toulati	10	75,000	400,000
Sergus			
Dondt	9		
Total	88	400,000	2,000,000

Hence it appears that the terms "great" and "little" are wholly misapplied. The little horde was still greater in the 14th century than at present; the great horde, however, is generally respected, as being the most ancient. The Kirghis, physically considered, are closely allied to the Mongol Turkmans. Their faces are not so flat and broad as those of the Kalmuks; but their small black eyes, small mouths, prominent cheek-bones, and almost beardless chins, prove their similarity to the Mongols, which has been strengthened also in recent years by frequent marriages with Kalmuk and Mongol women, whom they often bring away by force into their own country. The hair of the men is usually dark brown; but the women have black hair, fresh complexions, and brilliant, animated eyes, which, however, are ill-contrasted with lean cheek-bones, coarse skins, and a shapeless slovenly person. Both sexes are strong and healthy, long-lived, and capable of enduring, to an extraordinary extent, both cold and hunger; in fact, if they were not thoroughly inured to every kind of privation, they could not live in this country. The men take the most violent exercise, being often almost wholly on horseback for days together; but, in the height of summer, and during the winter, they spend their time in listless indolence, sleeping, drinking koumiss, their favourite beverage, and listening either to stories or the rude music of their national instruments, a reed pipe and a rude kind of violin. Household labour and tillage are undertaken wholly by the women, who, as in other parts of Asia, are treated almost like slaves. The Kirghis language is a very corrupt dialect of the Turkish, so interlarded with local words, that it is almost unintelligible by the Turks of Kazan and Khiva. Few can read, still fewer are able to write, and he that knows enough of Arabic to read the Koran is reckoned a paragon of erudition. Their poetry, however, clearly shows them to be an imaginative people. The dwellings of the Kirghis, who are distinctly pastoral, having no fixed station except in winter, consist of rude tents composed of wooden trellis-work covered with felt, having an opening at top serving at once for window and chimney: their dimensions average about 20 ft. in diameter, and 12 ft. in height; the ground (bare earth) is covered with felt, or carpeting; the inside is hung with straw mats, or red cloth; and the furniture consists only of a few boxes and warlike implements. The food of the people is very simple, consisting almost altogether of the flesh and milk of their flocks and herds. Bread is not known; but balamik, or porridge made of millet, rye, or wheat, is in common use. Rice, being an object of import, is very dear, and is used only by the rich. Smoked horse-hams, colts' haunches, and camels' humps, are esteemed great delicacies. Eremetchik, a rich cheese made from mares' milk, is likewise highly valued; a thinner and inferior kind, called brouts, is much used by the lower orders, and constitutes almost the only article of food on those marauding expeditions, which give such zest to the life of a Kirghis. Fish are eaten only by the lowest orders, chiefly by those living on the banks of rivers; and game is little valued. The favourite drinks are the everlasting koumiss, a whey made from mares' milk, and a spirit distilled from koumiss, alleged to be both strong and palatable. Arak (made by distilling rice), and tea, are luxuries enjoyed only by the wealthy classes. The dress of this nomad people is long and full, little suited, according to our notions, for the horse exercise, in which they are chiefly engaged: two or more tchapanes, or loose gowns of velvet, silk, or cotton, according to rank; a leathern belt fastening the robe and securing a knife and tobacco-bag; a round cap surmounted by another, when abroad, of felt or other warm materials, conically shaped, and with broad flaps; very full and highly ornamented trowsers are worn, by the men at least, over the gown, which is tucked underneath; and large pointed, high heeled boots, complete the costume. The heads of the men are usually kept shaved, with the exception of a forelock; but those of the women are adorned with long plaits running down the back. The female costume, in other respects, differs little from that of the males, except that the robe is close in front, and the bonnets are high, shaped like truncated cones, and surrounded by veils, which serve both for shade and warmth.

The employments of the men consist in an attendance on their flocks and herds, and in hunting antelopes, boars, and wild horses. Very generally, however, they join with the life of the huntsman that of the robber and man of violence, attacking and plundering caravans crossing their steppe, or seeking vengeance for some real or imagined insult from a neighbouring tribe. They are cowards in regular warfare, soon discouraged, and, when unhorsed in close conflict, wholly vanquished. Respecting their *barantas* or feuds, M. Lévchine says, " Rien de plus affreux, de plus funeste, que l'esprit de vengeance dans les Kirghis, et les suites de ce préjugé, de cette disposition cruelle, qui leur tenant lieu de la satisfaction que les lois seules doivent accorder, pervertissent la véritable bravoure. Leurs vengeances sont toutes dirigées par la passion effrénée du pillage, qui les ruine, les perd, les démoralise complètement; ces *barantas* consistent dans des vols ou des rapts mutuels de bestiaux, d'où résultent souvent entre eux des combats sanglants. Aujourd'hui, tout homme offensé volé, ou simplement mécontent, rassemble une bande de cavaliers, arrive chez son ennemi, attaque ses habitations, et lui enlève ses haras et ses bestiaux. Voilà l'héroïsme ; voilà en quoi consiste la grandeur chez les Kirghis!" (*Desc. des Kirghis-Kazaks*, p. 351.) These barantas had become so frequent and extensive in 1819–1820, that the population of the hordes, especially the little horde, was much thinned; the trade in cattle was all but destroyed; and thousands of families, unable to support life in their own country, emigrated to the government of Orenburg, and other parts of Russia. The arms of the warrior Kirghis are the lance, sabre, bow and arrows, a long-handled axe, called *tabalaas*, and a clumsy kind of gun; their defensive armour being a coat of mail, and sometimes a helmet. Among a people so disorderly, it is impossible that internal industry should flourish. Weaving is carried on for the supply of family wants; cordage is manufactured from horses' and goats' hair, a coarse soap is made of grease and vegetable ashes, and the skins of sheep and goats are converted into a rude kind of leather. Blacksmiths, and other workers in metal, make the ornaments attached to horse-furniture, belts, sword-blades, spears, &c.; but every article is of the coarsest quality and worst make. The trade now carried on by these people with other nations is much less considerable than it was half a century ago, in consequence of the loss of cattle and horses by the barantas. The Russians and Chinese have pretty large dealings with them, and a brisk trade is also carried on with Khiva, Khokan, and Little Bucharia. The trading-posts of the Russians are at Orenburg (the most important of them all) Troïsk, Petropavlovsk, Omsk, Semipalatinsk and Ouralsk, those of the Chinese being at Tchuguchak (Chin. *Talbaratai*) and Kuldsha (Chin. *Ili*).

The business, usually carried on in summer and autumn, is conducted wholly by barter, the Kirghis furnishing sheep, horses, horned cattle, camels, goats, goats' hair and wool, the skins of horses, sheep, and other animals, wild as well as domestic, and antelopes' horns; in return for which they receive from the Russians iron and copper implements, thimbles, needles, cutlery, padlocks, hatchets, velvets, brocades, silk-stuffs, linens, ribands, looking-glasses, snuff, &c.; from the Chinese, silver, silk goods, porcelain, japanned wares, and tea ; and from the Khurians and Bokharians, cotton goods, quilted dresses, rice, swords, fire-arms, and powder. Independently of the trade they carry on at the outposts, considerable traffic takes place with the caravans crossing the steppe between Khiva, Khokan, and the Russian frontier. The Kirghis are usually employed as protectors and guides in the journey over these wilds; great delays often occur owing to the caprice of the guides, and the travellers, if they be not entirely plundered of their property, are in general heavily mulcted by the khans, through whose pastures they are obliged to pass. In fact, says Lévchine, the experience of more than a century shows the impossibility of success in overland trade in W. Asia, so long as these tribes maintain their instinctive love of war and brigandage. The chief caravan routes are, 1, from Kabultzof to Khiva, across the Oust-ourt plateau, between the Aral and Caspian seas; 2, from Orenburg to Bokhara (64 days), over the Moguldjar mountains and across the Karakoum and Kisil-koum deserts; 3, from Semipalatinsk to Khokan (40 days). These roads, however, are so vaguely

laid down, and so often varied, that distances cannot be computed with any accuracy.

The government to which these people are subject cannot be properly compared with any form common to civilised countries. Geographers have termed it patriarchal and despotic ; but, in fact, there is no system of government; for even where a khan, or sovereign, is chosen, he is usually elected only by a few of the tribes, who obey only as long as they are pleased with their ruler, while the rest refuse all obedience, and probably take up arms against him. He may issue orders, but he cannot enforce compliance; and even where there is an absolute infraction of the laws of the Koran, by which they profess to be guided, the delinquent's punishment is more frequently inflicted by private revenge than by the decision of a public judge. The punishments are founded on the *lex talionis*, and consist commonly of the bastinado, maiming, and strangling ; but, if the offending party consent, almost all crimes may be atoned for by payments of sheep and horses. The khan must be elected from the highest class, known as the *white best*, those, in fact, who can lay claim to noble descent : the rest of the people belong to the *black best*,* and these are the only remaining distinctions among a people who, a century ago, were, of all others, the nicest in establishing family pre-eminence.

With respect to religion, it is difficult to say whether the Kirghis have any particular form. They acknowledge a supreme creative intelligence ; but some worship according to the dogmas of the Koran, and others mingle Islamism with an old kind of idolatry, while a third section of the population believe in the existence not only of a good deity, called *Kondaï* but also of a wicked spirit, *Chaïtane*, the author of all evil. In the existence of inferior spirits, and in witchcraft and sorcery, the people have universally the most implicit faith ; and the *badjis* travelling through the steppe reap great pecuniary advantage by imposing on their credulity.. The divinations, however, of the *Jaourountchis*, and other self-styled prophets, are not, according to Lévchine's account, a whit more absurd than the impositions anciently practised by the priests of Delphi (p. 336-338). The exercises of religion meet with little attention: long and frequent prayers do not suit the Kirghis ; they fast too often by compulsion to do so by choice ; and they are not so friendly to cleanliness as to relish the ablutions enjoined by the Mohammedan religion. In fact, with the exception of extreme credulity, these is hardly a trace of religious sentiment among them.

The history of the *Kirghis-Kazaks* cannot be traced with much probability beyond the 16th century. Earlier historians, commencing even with Herodotus, inform us, that the steppe was inhabited by a people living in felt tents, and otherwise assimilated to the great Mongolian family ; but these were *Nogais*, not Kirghis, being more civilised, and in all probability the builders of those temples and houses the ruins of which are still visible. (See *Herodotus*, iv. 24, 46 ; *Heeren's Researches, Asia,* ii. 265-268 ; and *Lévchine*, p. 117-135.) The name of the Kirghis first appears in Russian history about the middle of the 16th century ; but Ferdusi, in the 11th century, speaks of Kazaks characterised by the same habits as the Kirghis; though it does not appear that they then lived on the great steppe E. of the Aral. They first became nominally subject to Russia in 1740; but the rule of that country has never been felt but by the tribes adjoining the frontier. As to the native khans, so also to the Russian government, obedience is paid only when it is convenient, a rapid journey into the interior soon carrying them out of reach, when it suits their purpose to plunder rather than trade. It remains to be proved whether the efforts now in progress at Orenburg, to introduce civilisation into the steppe by educating young Kirghis, will accomplish the professed object of making them, instead of a burden and nuisance, useful and obedient subjects of Russia. (*Lévchine, Desc. des Hordes des Kirghis-Kazaks, traduite du Russe par Perry de Pigny; Hagemeister, sur l'Asie Occidentale ; Klaproth's Asia Polyglotta, &c.*)

KIRKCALDY, a royal and parl. bor., seaport, and manufacturing town of Scotland, co. Fife, on the N. shore of the frith of Forth, 10 m. N. Leith. Population of parliamentary borough, which includes the greater part of the parish, and the Link-town of Abbotshall, about 5659. The town consists principally of a single street, parallel to the shore, which, being measured from Bridgetown on the W. to East born on the E., is nearly 2 m. in length; and being united on the E. with Pathhead, Sinclairtown, &c., included in the borough of Dysart (which see), the mass of the town is about 4 m. in length. Having been originally laid out and built with no attention to any general plan, but according to the taste, convenience, and means of the parties,

_____
* These huts are, in all probability, synonymous with the Khedjes among the Mongolians, mentioned by Timkofsky.

this street was formerly of the most irregular description being narrow, crooked, and the houses frequently mean and poor. In 1821, however, an act was obtained for widening, paving, and lighting the streets, and otherwise improving the town; and since that period several new lines of houses have been constructed, and many important improvements effected. The houses in the main street are now mostly of a very respectable class, and the shops are good, and handsomely fitted up. There are several cross streets, some of which lead up the ascent of the hill to the N.; and there are several between the high street and the shore. The town is well supplied with water, paved, and lighted with gas. Among the public buildings are, the parish church, rebuilt in 1807, Abbotshall church, within the parliamentary borough; two quoad sacra churches belonging to the establishment; various dissenting chapels; and a handsome town-house, including a jail, courthouse, &c., with a spire erected in 1826. Besides the parochial schools, and a variety of private seminaries, some of which are said to be extremely well conducted, two charity schools have been opened in this town for the education of 260 children of both sexes, on an endowment by Mr. Philp, late merchant here, who bequeathed about £70,000 for the foundation and maintenance of these and similar schools in Pathhead and Kinghorn. The education given in these schools is of the more common and ordinary description; and the children are supplied with books, are clothed, and receive a small sum on leaving school. Kirkcaldy has also a mechanic's library, and two or three subscription coffee-rooms and libraries.

The inhabitants of Kirkcaldy are honourably distinguished by their enterprise, both as manufacturers and traders. The staple manufacture of the town is that of coarse linen, including sheetings, ticks, dowlas, canvas, &c. It appears, from the Factory Returns, that in 1838 there were in Kirkcaldy 10 flax-mills driven by steam, employing in all about 500 workpeople. It also appears that in 1839 there were 954 looms at work in the parishes of Kirkcaldy and Abbotshall, exclusive of a few in private cotton factories. The average neat weekly earnings of the weavers amount to at least 8s. 6d. a week; but superior hands make 10s. 6d., and some even as much as 15s. The wages of machine makers, masons, and other artisans, are higher, averaging from 16s. to 18s. a week. The moral and intellectual condition of the weavers is said to be very good; and there is very little apparent distress among them. There are no poor rates; the poor being supported by voluntary contributions.

Kirkcaldy has also a rope-walk, bleach-fields, two iron foundries, a distillery, a tile and brick work, a pottery, and two or three breweries. Branches of the bank of Scotland, the Commercial and National banks, and the Glasgow bank, are established here; and there is also a savings' bank. A chamber of commerce has been established for several years.

The harbour, near the E. end of the town, consists of an inner and outer basin. It is wholly artificial, being formed of three piers, and unfortunately dries, at low water; but, notwithstanding this drawback, the town possesses a good deal of shipping, and carries on a pretty extensive trade. There belonged to the port and its dependent creeks, in 1836, 178 ships, of the aggregate burden of 13,496 tons: vessels from Kirkcaldy have been for a lengthened period engaged in the N. sea whale-fishery; but here, as elsewhere, this business has been recently on the decline. Two shipping companies carry on a regular intercourse, by means of smacks and steamers with London and Leith; and there is a good deal of trade with the N. of Europe, whence hemp, flax, timber, tar, &c., are imported, and of which manufactured goods, coal, &c., are exported. Gross customs' revenue, in 1836, £6695. There is a weekly corn market, which is extremely well attended; and the town's markets for butchers' meat, fish, &c. are well supplied.

Kirkcaldy was made a royal bor. by Charles I. in 1644. It had attained about this period to considerable wealth and distinction; but it subsequently encountered severe losses, and about the middle of last century it had only two ferry-boats and one coasting vessel! But since 1783, and especially since the close of the American war, its manufactures, commerce, and population have steadily increased. It is now governed by a provost, two bailies, and 17 councillors. Corporation revenue, in 1820-46, £2950, principally the produce of the ferry between the town and Leith.

Kirkcaldy unites with Dunfermline, Dysart, and Kinghorn, in sending one member to the House of Commons. Registered electors in this bor., in 1839-40, 437.

Kirkcaldy has to boast of being the birth-place of Adam Smith, the author of the "Wealth of Nations," born here on the 5th of June, 1723. His father being comptroller of customs at this port, Smith received the rudiments of his education in the parish-school; and he afterwards resided here, with little interruption, from 1766 to 1776, occupied in the elaboration of his great work, which appeared in the last mentioned year. Raith, the magnificent seat of the Ferguson family, is in the immediate vicinity of the town. (Parl. Papers; Private Information.)

KIRKCUDBRIGHT, a marit. co. of Scotland, or, as it is more frequently termed, a stewartry, in the most southerly portion of that kingdom, comprising the E. half of the district known by the name of Galloway. It is bounded on the E., N., and W. by the counties of Dumfries, Ayr, and Wigtown, and on the S. by the Irish sea and the Solway Frith. Area, 523,760 acres, of which from 1-4th to 1-3d part may be arable. Surface much diversified, but in general hilly, and in extensive districts mountainous. The highest part of the Kell's range has an elevation of 2650 feet; and Cairnsmuir of Fleet, on the bay of Wigtown, rises to the height of 2329 feet. The greater number of the hills are bleak and barren; but in parts, particularly on the confines of Ayrshire, they afford good sheep pasture. The arable lands lie principally to the S. of a line drawn from the middle of the par. of Irongray to Gatehouse; but Criffel, 1831 feet high, on the Solway frith, and some other considerable hills, lie within this tract. Climate in the lower districts mild but moist; in the upper districts it is sometimes severe. Except along the Solway frith, the soil even of the arable land of the stewartry has seldom a smooth, continuous surface: it is very often broken with gravelly knolls, but the hollows between these consist principally of a gravelly or basely loam, and are often extremely productive, and particularly well adapted for the turnip husbandry; and in wet summers the arable knolls are covered with luxuriant crops, while many of those that do not admit of cultivation yield excellent pasture. Principal crops barley and oats; but wheat is also raised. Within the last few years the turnip culture has made great progress. Arable husbandry has been greatly improved since the peace: furrow draining is now extensively practised, and latterly bone-dust has been successfully employed in the raising of turnips. But the soil and climate are better suited for grazing than cropping, and the principal attention of the farmer is given to the former. The breed of polled cattle, peculiar to this county and that of Wigtown, is well known to be one of the best in the empire: they are principally sent up by land when half fat to the Norfolk fairs; but they are now, with sheep, sometimes fattened off on turnip and sent by steam to Liverpool. Farm buildings have been vastly improved, and the roads, which were formerly execrable, are now nowise inferior to those of any other county in the empire. There are some very large estates; but property is, notwithstanding, more subdivided in this than in most other Scotch counties. Farms of medium size, and all let on 19 year leases. This county and Wigtown are mostly subdivided by the dry stone walls known, from this district, by the name of "Galloway dykes," and which, when well built, make an excellent fence. Average rent of land, in 1836, 7s. 3d. an acre. Manufactures and minerals unimportant: lime, coal, and freestone are all imported, principally from Whitehaven, on the opposite side of the Solway frith. The granite used in the construction of the Liverpool docks is mostly obtained from near Creetown, in this county. Principal rivers, Dee, Fleet, and Urr: the salmon fisheries on the fleet are valuable. Principal town, Kirkcudbright. This county has 28 parishes, and sends one member to the House of Commons, for the county, while the borough of Kirkcudbright joins with Dumfries, Annan, and other boroughs in returning a member. Registered electors for the county, in 1839-40, 1288. In 1831 Kirkcudbright had 6954 inhabited houses; 8823 families; and 40,590 inhabitants, of whom 18,569 were males, and 21,691 females. Valued rent £114,127 Scotch. Annual value of real property, in 1815, £313,200.

KIRKCUDBRIGHT, a royal and parl. bor., and seaport of Scotland, cap. of the above county on the Dee, about 6 m. above its confluence with the Solway frith, 24 m. S.W. Dumfries, and 83 m. S.S.W. Edinburgh. Population of borough, in 1831, 2690. It is a finely situated, well built, respectable country town. The streets intersect each other at right angles, and the houses, which are mostly all two stories high, have a respectable appearance. A large and handsome Gothic church, with a spire, (which, however, is not very well proportioned,) was erected in 1838, at an expense of £6785; it has also a jail erected in 1816; an excellent academy, with a room for the public subscription library; the remains of an old castle, once the property of the lords Kirkcudbright. Exclusive of the parish church, there are places of worship belonging to the United Secession, Rowites, and Roman Catholics; but the congregations are small, dissent having made but little progress in this vicinity. The harbour of Kirkcudbright is the best in the S. of Scotland. At low ebb in neap tides there is about 10 feet water in the river; and as the tide then rises about 18 feet, there is at all times water to float the largest ships. The access to the Dee will be much facilitated by the lighthouse which is about being erected on the little Ross at its mouth. But, despite its fine harbour, Kirkcudbright, owing to its limited communication with the interior, and the thinness of the population in the vicinity, has very little trade. In 1840, 55 vessels of the burden of 2110 tons belonged to the port. A branch of the Bank of Scotland is established in the town. Ship-building is carried on to some extent: but it has no other manufacture worth notice. A regular steam communication is established between this town and Liverpool; and were the road to Ayr, distant about 55 m., improved, it might become an advantageous port for the landing and embarkation of such travellers between Liverpool and Glasgow as might be inclined to shorten the passage by sea to the narrowest limits. There is no bridge over the Dee nearer than Tongland, about 2 m. farther up the river, and to which it is navigable; but passengers, with horses and carriages, are ferried over in a flat-bottomed boat, with but little inconvenience. The town is lighted with gas; and is supplied with water brought from a distance by pipes.

It was made a royal borough by James II. in 1455. Under the municipal reform act, it is governed by a provost, two bailies, and 14 councillors. It unites with Dumfries, Annan, Sanquhar, and Lochmaben, in sending one member to the House of Commons. Registered electors, in 1839-40, 100. Corporation revenue in 1839-40, £1130 10s.: its pecuniary affairs have been exceedingly well managed; and it has at this moment the whole property contained in the charter of James II. The town's revenues are employed to defray the expenses of the academy, and the charges on account of lighting the town, supplying it with water, &c., for which no assessment is imposed on the inhabitants; and till within these few months, the whole expense of supporting the jail, the principal one in the county, was paid out of the revenues of the borough.

The environs of the town are extremely beautiful. The rising grounds on each side the river, from Tongland to the sea, are embellished with plantations. St. Mary's Isle, the residence of the earls of Selkirk, adjoins the town on the S. Kirkcudbright is a very desirable place of residence for people of small fortune; provisions of all sorts are abundant and cheap; house-rent is very low: a good education may be had for a mere trifle; the society is superior to that in most small towns; and there is a ready means of communicating with Edinburgh on the one hand, and with Liverpool and London on the other.

KIRKHAM, a manufacturing and market town, and par. of England, co. Lancaster, hund. Amounderness, in the low district, called the Fylde, 7 m. W. by N. Preston, 27 m. N. Liverpool, and 220 m. N.N.W. London. Area of par., which contains 18 townships and chapelries, 41,850 acres: pop., in 1831, 11,630: do. of Kirkham township, 3470. The town, though small, is handsome and well built. The church, a large modern structure, was erected, in 1822, at an expense of £5000: the tower, however, is ancient, and its interior, which accommodates nearly 2000 persons, is ornamented with several fine old monuments, carefully replaced in the new building. The living is a vicarage, in the patrimony of the dean and canons of Christchurch, Oxford, the chapelries in the out-townships being in the gift of the incumbent. Within the town, are places of worship for Wesleyan Methodists, Independents, Swedenborgians and Roman Catholics, with attached Sunday schools, attended 'y about 500 children; and connected with the church, a national school for boys and girls. A grammar-school, founded in 1670, enjoys a good reputation, and is attended by 80 or 100 boys: it is managed by a principal and two under-masters; the instruction given is purely classical: a charity school, established in 1760, for clothing and educating 40 girls, is alleged to be respectably conducted; and the Roman Catholics have two large schools for the children of that religion, which has numerous adherents in and round the town. The industry of Kirkham, 20 years ago, was confined to the manufacture of sail-cloth, cordage, and coarse linens, of materials brought from the Baltic; but now, the cotton manufacture is extensively carried on, and gives employment, within the par., to several hundred hands. In 1838, there were two flax-mills within the par., which employed 542 hands. The Lancaster canal, the Lancaster and Preston railway (opened in 1840), and the Preston and Wyre railway afford cheap and expeditious conveyance both for passengers and goods. Petty sessions are held once a fortnight, and a court of requests for debt under 40s., six monthly. A savings' bank is established here. Markets on Thursday; fairs, Feb. 4, and 5, April 20, and Oct. 18. (Baines's Gaz. of Lancaster, vol. ii., art. Kirkham.)

KIRKLAND, p. t., Oneida co., N. Y., 100 m. W.N.W.

134

Albany, 386 W. Watered by Oriskany creek. It contains Clinton v., the seat of Hamilton college. (See Clinton.) It has 12 stores, four fulling-mills, one woollen-factory, two cotton-factories, with 3000 spindles, one flouring-mill, three grist-mills, eight saw-mills, three tanneries, one distillery; two colleges, 190 students; one academy, 38 students; 18 schools, 718 scholars. Pop. 2384.

KIRKLESI, KIRK-EKLESI, or KIRK-KILISSA (meaning the town of forty churches), a town of European Turkey, prov. Roumelia, cap. circ. of its own name, 30 m. E. Adrianople, and 106 m. W.N.W. Constantinople; lat. 41° 50' N., long. 16° 55' E. Pop., according to Dr. Walsh, about 4500 families, or 22,000 individuals (2-5ths being Greeks). It is a large, dirty, ruinous town, surrounded with old walls defended by a citadel, and has a bazaar, several mosques and hummums, but no Greek church, the public celebration of the rites belonging to that religion being here extended with heavy penalties. The neighbourhood produces an abundance of grapes, melons, and other fruits; and a good deal of wine is made. The Turkish inhabitants are described as a rude, brutal, and ignorant rabble, treating all with contempt who speak any language in addition to their own. The Greeks, on the contrary, are described by Dr. Walsh as "a large and thriving community, who have established two good schools on the monitorial system for the instruction of their children, a degree of refinement to be met with in only one other town of Turkey." (Walsh's Journey from Constantinople, p. 137.)

KIRKWALL, a royal and parl. bor. market town, and seaport of Scotland, in Mainland, or Pomona, the largest of the Orkney islands, of which it is the cap., on the N.E. side of the island, at the head of an open bay exposed to the N.; lat. 58° 59' 31" N. long. 3° 23' 8" W., 95 m. N. by E. John O'Groats, and 208 m. N. Edinburgh. Pop., in 1831, 3065. The town consists chiefly of one narrow and inconvenient street, about 1 m. in length, parl　the bay. The houses have generally their gables to the street, and most of them bear the marks of antiquity. But new and handsome houses are gradually being erected, both in the town and neighbourhood. Here most of the country gentry reside, at least during winter, and the society of this remote place is esteemed equal, if not superior, to that of any provincial town of its size in Scotland. The only public building of a modern date is the town hall, with piazzas in front, containing a jail, assembly-rooms, court-rooms, &c. The principal building in Kirkwall is the cathedral, erected in the 12th century, and dedicated to Magnus, one of the Scandinavian earls of Orkney, who, having been assassinated in 1110, was canonized after his death. This venerable Gothic structure, which has been enlarged at different times, is, after the cathedral of Glasgow, the most entire in Scotland: it is in the form of a cross, its extreme length being 236 ft., its greatest width 56 ft., the height of the roof 71 ft., and that of the spire 140 ft. But the original spire having been destroyed by lightning in 1671, the present spire is modern, and it is, also, unworthy of the building. About 160 yds. S. from the cathedral are the ruins of two ancient edifices, viz. the Earl's Palace, built by Patrick Stewart, Earl of Orkney, and the Bishop's Palace. In the latter, Haco, king of Norway, died on his return to Orkney, after the unsuccessful battle of Largs, in 1263, and James V. occupied it on his visit to the island in 1540. The remains of Kirkwall castle, on the W., are still visible. The cathedral formed the cemetery of many Scandinavian kings, nobles, and warriors. The par. church, consisting of the choir of the cathedral, is collegiate. There are also chapels belonging respectively to the Associate Synod, Original Seceders and Independents. There are 19 schools, attended by 422 scholars. The town has several libraries, a museum, and a printing-press. There are no poor-rates, the poor being supported by the church collections, by the usual parochial dues, and by the produce (£16 4s. 5d.) of a bequest for the purpose. Malcolm Laing, the historian of Scotland, was born in the vicinity of Kirkwall, and educated at the grammar-school of the bor.; and at his death, in 1818, his remains were interred in St. Magnus' cathedral.

Rye straw raised in Orkney having been found to be peculiarly suitable for the manufacture of straw plait for ladies' bonnets, the business is carried on in Kirkwall to a considerable extent, though not so much so as formerly. It has also two distilleries, and some weaving is carried on for domestic use.

The herring, cod, and lobster fishery is prosecuted to considerable extent. The town is the seat of the courts of law for the whole of Orkney. Kirkwall has a customhouse, which comprises all the harbours in the Orkneys: total number of vessels, 77; tonnage, 4313: gross customs' revenue, in 1838, £1121. A steamboat plies between Leith and Kirkwall weekly, touching at Aberdeen, Wick, and other intermediate ports: in summer it goes as far as

Lerwick, in Shetland. Kirkwall has an annual fair in the month of August, which lasts about two weeks; and the greater part of all the mercantile business of the Orkney islands is negotiated at this fair. The town has two branch banks.

Kirkwall was made a royal bor. by James III. in 1486. It unites with Cromarty, Wick, Dingwall, Dornoch, and Tain, in sending a member to the House of Commons. Registered electors, in 1839–40, 143. Municipal revenue, £59 5s. 5d. (*Messrs. Anderson's Highlands and Islands of Scotland*, sect. ix.; *Barry's Hist. of Orkney; Keith's Scottish Bishops*, Russell's ed., 1824, 219–220.)

KIRKWOOD, t., Belmont co., O. The National road passes through its S.E. corner. It has two schools, 38 scholars. Pop. 227a.

KIRRIEMUIR, a bor. of barony, market and manufacturing town of Scotland, co. Forfar, in a pleasant situation, partly on a flat and partly on an inclined plain, along the N. brow of a picturesque glen, through which the streamlet Gerry runs, 15 m. N. by W. Dundee, and 5 m. N.W. Forfar. The Grampians are within 3 m. of the town, on he N. The view from its upper part, about 400 ft. above the level of the sea, is most extensive and striking, having the Grampian range on the N., and the whole extent of the splendid valley of Strathmore on the S. Pop. in 1801, 3321; in 1831, 4814; and in 1837, 4965.

The form of the town has some analogy to an anchor. The streets are lighted with gas. The only public buildings are, the Trades' Hall, the property of 12 friendly societies of the town and parish; the parish church; and chapels belonging respectively to the Associate Synod, the Relief, the Original Seceders, and the Episcopalians. There are 16 schools in the parish, of which three are endowed, one supported by subscriptions, and 12 unendowed. There are two bequests for education, the one educating about 36 boys, the other 20 boys and 20 girls. There are two libraries, and a news-room. About 50 years ago, only one newspaper came to the parish; the number is now about 200 a week. (*New Stat. Acc. of Scotland, § Forfarshire*, p. 189.) Dr. M'Crie, the distinguished biographer of John Knox, and Dr. Jamieson, the Scotch lexicographer, were once dissenting clergymen in Kirriemuir.

Though inland, and devoid of ready communication with the sea, Kirriemuir has attained to considerable eminence in the manufacture of the coarser kinds of linen fabrics, such as Osnaburgs, sail-cloth, bagging, imitation Russia sheeting, &c. This branch of business, which is carried on chiefly in connexion with the Dundee manufacturers, was introduced soon after the rebellion of 1745. During the year ending November, 1799, 1,814,874 yards were stamped. In 1833, 52,000 webs were woven, containing no fewer than 6,769,000 yards, or nearly *quadruple* the quantity produced at the beginning of the century. The quantity produced is now (1841) estimated at about 7,000,000 yards: the number of persons, including apprentices, employed in the town and vicinity, is about 3000. (*Stat. Acc., ut supra*.) "Although the yarns must be carried from the shore in carts, and along roads constructed on the common principles, and although the cloths, when manufactured, must be carried back by the same rude conveyance, such is the ingenuity of the weavers, and such their industry, that we are not only able to come into competition with our rivals in the more favoured towns on the coast, but even to bear away from them the palm of victory. Hence it is, that several mill-spinners in Melrose and Dundee, towns possessing many natural advantages, to which we can lay no claim, have been accustomed to send their yarns to be woven in this distant quarter." (*Ib.* p. 199.) But the communication has recently been much improved, at least with Dundee, inasmuch as the Dundee and Newtyle railroad extends to Glammis, 5 m. distant from Kirriemuir. There is also a railway between Arbroath and Forfar, the last place being also distant 5 m. The Kirriemuir weekly market is one of the best in the country. There are, besides, four annual fairs for cattle, horses, and sheep. There is a branch bank here: also a savings' bank. About 50 years ago, the revenue from the postoffice amounted, in one year, after paying the necessary expenses, to 8d.: it now (1833) amounts to £380 per annum. (*Ib.* p. 185.)

Kirriemuir is governed by a bailie, nominated by the feudal superior (Lord Douglas). The peace is preserved by a body of constables, chosen annually.

KIRTLAND, p. t., Lake co., O., 167 m. N.E. by N. Columbus, 353 W. Here the Mormons first settled, and built a great temple, which has, since their departure, been partially occupied by the Western Reserve Teachers' Seminary, a highly respectable institution, for males and females. It contains three stores, two grist-mills, six saw-mills; one academy, 165 students; 18 schools, 607 scholars. Pop. 1778. The Teachers' Seminary has a principal, and three male and three female assistants, and had, in its

various departments, in 1842, 368 students. It exerts a happy influence on the cause of schools in the vicinity.

KISHM (the *Oaracta* of an. Greek authors), the largest island in the Persian gulf, and the chief of a group situated near its mouth, extending between lat. 35° and 36° 30' N., and long. 26° and 27° E., comprising Ormus, Kenn-Anjar, Larak, and many smaller islands. Kishm is of an elongated shape, nearly 60 m. in length E. to W., and 12 m. in its greatest breadth. Population estimated at 5000. It is separated from the main land by Clarence straits, a narrow and intricate channel, navigable, however, for large ships, the soundings varying from 4 to 12 fathoms. A ridge of hills extends from one extremity to the other of the island on its S. side; the rest of the surface is mostly plain. Sandstone is the predominant formation. The surface is generally arid and barren, and is in parts extensively incrusted with saline efflorescence; but a few portions are remarkably productive. The N. part of the island is the most fertile and populous: the soil there consists of a black loam, on which wheat, barley, vegetables, melons, grapes, and dates in large quantities are produced. The island at present yields corn enough for home consumption. Boats from all parts of the gulf come to Kishm for wood. Cattle and poultry are reared; the former are scarce, but goats are bred in considerable numbers, and thrive well. The greatest exercise of the goats are jackals, with which the island is much infested; antelopes of a superior breed, partridges, and rock pigeons, abound, and wild fowl in winter. The inhabitants are chiefly Arabs; they employ themselves in fishing, agriculture, and the manufacture of cloth, and reside chiefly in villages and hamlets scattered along the coasts. Kishm is said to have once contained upwards of 300 towns and villages, but at present it has nothing like half that number. The chief towns are Kishm at its E., and Basidoh at its W. extremity, and Loft on its N. side. Kishm, with about 2000 inhabitants, seems to have been formerly of considerable commercial importance. It is surrounded by a high mud wall, flanked with towers, on which a few old guns are mounted. Streets narrow and dirty, houses flat roofed, and some of them large and neatly fitted up. The bazaar is plentifully supplied with many kinds of vegetables and fruits from Persia; and good wines, dried fruits, silk and cotton cloths, and carpets of the richest patterns, may be obtained. The town has a brisk trade and a bustling appearance, many native vessels calling for food and water, or to take pilots for the Kishm channel. It is the residence of the sheikh. A few coasting vessels are built here with timber from the Malabar coast. Basidoh, or Bassadore, once belonged to the Portuguese, and the ruins of their town and fort may still be traced. It is admirably situated in most respects, and healthy, but ill supplied with water. Being the principal station in the gulf for ships of the Indian navy, it has several European houses and public buildings, including a hospital, store and guard houses, a billiard-room, fives-court, &c.; and it is the residence of the commander of the Indian squadron. Its port is difficult to enter, but vessels have good anchorage in six or seven fathoms quarter of a mile from the shore. Loft is at present a town of only 600 inhabitants: it was bombarded by the English in 1809. Vessels may lie before it in four and a half fathoms water completely landlocked.

The island of Anjar, 3 m. S. of Kishm, is of volcanic origin, five or six miles in circuit, and uninhabited, though the remains of a town and reservoir are still visible on its N. side. It is covered with pits of salt and metallic ores; and between it and Kishm is an excellent anchorage. Larak, to the S.E., is also of volcanic origin, and inhabited only by a few fishermen. The Great and Little Tombs, about 25 m. S.W. Kishm, are low and uninhabited. The small islets between Kishm and the main land are verdant and covered with wood, a circumstance rare in the adjacent parts of Persia. Nearchus visited and described this island group; and Arrian affirms that in his time was to be seen in Kishm the sepulchre of its first king Erythras, from whom the gulf was named *Mare Erythræum*. These islands are now governed by a sheikh, tributary to the imâm of Muscat. (*Kempthorne and Whitelock*, in *Geog. Journal*, v. 277–280, viii., 176–182; *Kinneir's Pers. Emp.*, p. 14–16.)

KISKIMINETAS, p. t., Armstrong co., Pa., 187 m. W.N.W. Harrisburg, 211 W. Bounded S W. by Conemaugh or Kiskiminetas river. Salt is found. The Pennsylvania canal passes through a part of the town. It contains nine stores, one fulling-mill, two grist-mills, eight saw-mills, three tanneries, two distilleries; 13 schools, 370 scholars. Pop. 2287.

KITTANING, p. b., capital of Armstrong co., Pa., 182 m. W.N.W. Harrisburg, 226 W. Situated on Alleghany river. It contains a courthouse, jail of stone, county offices of brick, two churches, a Presbyterian, and one common to German Reformed and Lutherans, an academy,

10 stores, three tanneries, 100 dwellings and about 600 inhabitants.

KITTATINNY, mountains, a branch of the Alleghany, extend through the N. part of N. J., cross the Delaware river at the Delaware Water Gap, and pass through Pennsylvania and the W. part of Maryland into Virginia.

KITTERY, p. t., York co., Me., 100 m. S.W. Augusta, 497 W. Bounded S.W. by Piscataqua river, S.E. by the Atlantic. A bridge connects it with Portsmouth, N. H. Another bridge connects it with Badger's island, on which is a United States navy-yard. It has a good harbour for small vessels, of which it has a number, employed chiefly in the fisheries. It has some ship-building. It contains seven stores, two grist-mills, one saw-mill; 11 schools, 644 scholars. Pop. 2435.

KLATTAU, a town of Bohemia; cap. circ. of same name, on the Bradlenka, 70 m. S.W. Prague. Population 5700. It is well built, and has a castle, some handsome churches, a council-house, with a tower 150 feet in height, containing a bell weighing 90 *centners*, a gymnasium, high school, two hospitals, and manufactures of woollen cloth and stockings. It is said to have been founded in the 8th century.

KNARESBOROUGH, a parl. bor. market town, and par. of England, co. York, W. riding, wap. Claro, 16¼ m. W. by N. York, the same dist. N. Leeds, and 189 m. N. by W. London. Pop. of parl. bor. (which includes, besides the old bor., parts of Scriven and Knaresborough townships), in 1831, 6923. The town is beautifully situated on a slope, N.E. of the Nidd, "the stream of which is rapid, deep, and very serviceable for turning the wheels of mills and machinery connected with the linen trade:" (*Baund. Rep.*) Two stone bridges cross it, one above, and the other below the town; and on a beetling crag, close over the turret stands a ruined castle, opposite to which, on the other side the river, is a curious dropping well, the water of which runs from a source 50 feet above, and trickles through a porous limestone rock, with sufficient rapidity to deliver about 20 gallons per minute. At no great distance is an oratory, carved out of the rock, and a mile lower down the stream are the ruins of a priory, founded by Richard, brother of Henry III., and a cavern known as St. Robert's Cave, where Eugene Aram, now so well known through Sir Lytton Bulwer's novel, committed the murder in 1745, of which he was convicted 15 years after. The streets of Knaresborough are broad, regularly laid out, well paved, and lighted with gas: the houses are almost wholly of stone, and many of them large and handsome. The market-place is extensive, and there is a good market-house. The courthouse occupies the centre of the old castle, and another part of it is used as a prison for the liberty of the forest of Knaresborough. The parish church is of considerable antiquity, but little beauty. The Independents, Wesleyan Methodists, and Roman Catholics, have also places of worship, and the Sunday schools of the church and chapels are attended by upwards of 800 children. A charity school, two national schools, an infant school, a school of industry, and two other schools, furnish instruction to about 900 children, chiefly belonging to the working classes. The public institutions of the town comprise a public library, dispensary, lying-in charity, savings' bank, and Bible society.

The industry of Knaresborough is chiefly confined to linen-weaving. The trade has been long established, and a large amount of capital is vested in mills, warehouses, and machinery. Two flax-mills employed, in 1838, 142 hands, and about 400 looms are at work on various descriptions of linen. Business, however, has greatly declined within the last 12 years, before which nearly 700 looms were employed. Wages in 1839 were very low (averaging 9s. 6d. a week), and the weavers were often out of work, and is extreme poverty. The diminution of the trade is attributed to the powerful competition of lower-priced Scotch and Irish goods, and to the want of coal and the increasing use of cotton and union cloths. Other workmen's wages were as follows: carpenters 4s., and masons 3s. 6d. a day, blacksmiths 24s. a week. Tailors, shoemakers, &c., have little to do on account of the poverty of the place, and form a very small proportion of the inhabitants. There being no carriage in the neighbourhood, coals are brought 18 m. by land carriage; and the want of this mineral is a great drawback upon its manufactures. Knaresborough has a great corn market, and from this place and Ripon the manufacturing districts of the W. riding are principally supplied. (*Handloom and Bound. Reports.*)

Knaresborough is a borough by prescription, governed by a bailiff. The parliamentary franchise was granted in 1st Mary (1553), since which the borough has sent two members to the House of Commons. The right of voting till 1832 was vested in the owners of 84 burgage tenures, all of which, excepting four, being the property of the Duke of Devonshire, the members were his nominees. The

Boundary Act enlarged the limits of the borough by including in its parts of the townships of Scriven and Knaresborough: and, in 1831, there were 369 qualifying tenements. Registered electors, in 1839–40, 277. Markets on Wednesday, and on alternate Wednesdays fair cattle. Sheep fairs, Wednesdays and Thursdays after January 13. Wednesday after March 12, May 5 and 7, August 12. Statute fairs, Tuesday and Wednesday after October 10, and Wednesday after November 22.

KNIGHTON, a market town, parl. bor. and par. of N. Wales, co. Radnor, on the Teme, 28 m. S.S.W. Shrewsbury, and 138 m. W.N.W. London. Area of parish, 228 acres: pop., in 1831, 1939. Pop. of parl. bor. 1976. The town comprises two chief streets, intersecting each other at right-angles, and the gentle acclivity on which it stands not only gives it a picturesque appearance but greatly contributes to its cleanliness. A small modern-built church, subordinate to that of Stowe, in Shropshire, and a chapel for Calvinist-Methodists, are the only places of worship; and the charities comprise a free-school and an almshouse.

Knighton is principally occupied by tradesmen, mechanics, maltsters, &c.: it has no manufactures, the wool-dyeing and spinning business having ceased to exist. Wool-sampling is carried on to some extent, though much less than before 1811, when a large establishment failed. The market is very large and important: it is attended by dealers even from Birmingham and its neighbourhood, who come for meat, poultry, eggs, butter, cheese, &c.; and butchers' meat is sometimes sent to London. Petty sessions are held here monthly. The officers of the borough are a bailiff, burgesses and constables; but they have little or no authority, and the bailiff's only business is to collect the chief rents of the manor, which belongs to the Earl of Oxford. The boundaries of the parliamentary borough were not changed by the Boundary Act of 1832, and in 1839–40 there were 500 registered electors in the borough of New Radnor, to which Knighton is contributory.

Knighton is called by the Welsh *Tref-y-cloudd*, or "the town on the dyke," from the circumstance of its position close to Offa's dyke, which enters the parish on the N., and running due S. about two miles, may be traced through several parishes into the county of Hereford. (*Nicholson's Guide; Parl. Rep.*)

KNOWLTON, p. t., Warren co., N. J., 66 m. N. by W. Trenton, 224 W. Bounded W. by Delaware river, which passes Blue mountain in the N.W. corner of the town, at the celebrated Water Gap. Drained by Paulinskill and its branches. It contains two churches, a Presbyterian and Episcopal, several stores, seven grist-mills; 16 schools, 768 scholars. Pop. 2207.

KNOX, county, Tenn. Situated centrally in the E. part of the state, and contains 864 sq. m. Watered by Holston and French Broad rivers, which unite 4 m. above Knoxville. The Holston thence flows S.W. through the lower part of the county. It contained in 1840, 13,875 neat cattle, 11,604 sheep, 36,959 swine; and produced 161,491 bushels of wheat, 2940 of rye, 716,166 of Indian corn, 83,289 of oats, 18,050 of potatoes, 18,679 pounds of tobacco, 19,784 of cotton, 8490 of sugar. It had three commission houses in foreign trade, 39 retail stores, two forges, one fulling-mill, 19 flouring-mills, 41 grist-mills, 35 saw-mills, two oil-mills, one paper-mill, 11 tanneries, 41 distilleries, one pottery, three printing-offices, four weekly, and one semi-weekly newspapers, and one periodical; one college, 100 students; two academies, 112 students; three schools, 69 scholars. Pop.: whites, 13,378; slaves, 1934; free coloured, 173; total, 15,485. Capital, Knoxville.

KNOX, county, Ky. Situated in the S.E. part of the state, and contains 495 sq. m. Watered by Cumberland river, and its tributaries. It contained in 1840, 6384 neat cattle, 5718 sheep, 17,911 swine; and produced 7873 bushels of wheat, 2530 of rye, 263,141 of Indian corn, 54,231 of oats, 11,176 of potatoes, 9355 pounds of tobacco, 3618 of cotton, 10,056 of sugar. It had nine stores, one flouring-mill, seven grist-mills, two saw-mills, one tannery, 26 distilleries; three schools, 76 scholars. Pop.: whites, 5082; slaves, 526; free coloured, 164; total, 5772. Capital, Barbourville.

KNOX, county, O. Situated near the centre of the state, and contains 615 sq. m. Vernon river, and its tributaries afford good water-power. It contained in 1840, 27,773 neat cattle, 54,799 sheep, 44,421 swine; and produced 391,217 bushels of wheat, 14,895 of rye, 518,205 of Indian corn, 93,576 of buckwheat, 1052 of barley, 439,069 of oats, 80,587 of potatoes, 119,459 pounds of tobacco, 233,284 of sugar. It had over 36 bushels of edible grains exclusive of potatoes, to every individual of its population. It contained 36 stores, three fulling-mills, four flouring-mills, 23 grist mills, 57 saw-mills, 29 oil-mills, 20 tanneries, six distilleries, one brewery, one pottery, three printing-offices, three weekly newspapers; one college, 130 students; two academies, 57 students; 169 schools, 6697 scholars. Pop. 29,579. Capital, Mount Vernon.

Knox, county Ia. Situated towards the S.W. part of the state, and contains 540 sq. m. Organized in 1882. Bounded W. by Wabash river, S. by White river, E. by the W. fork of White river. It contained in 1840, 11,996 neat cattle, 10,106 sheep, 25,896 swine; and produced 51,670 bushels of wheat, 861 of rye, 605,553 of Indian corn. 194,946 of oats, 16,664 of potatoes, 39,113 pounds of tobacco, 37,691 of sugar. It had 49 stores, one cotton factory, with 1056 spindles, two flouring-mills, 16 grist-mills, 16 saw-mills, one oil-mill, six tanneries, two distilleries, one brewery, two potteries, two printing-offices, two weekly newspapers; three academies, 168 students; 86 schools, 880 scholars. Pop. 10,657. Capital, Vincennes.

Knox, county, Ill. Situated towards the N.W. part of the state, and contains 782 sq. m. Formed in 1825, but organized in 1830. Watered by Henderson and Spoon rivers, and their tributaries. It contained in 1840, 7596 neat cattle, 6887 sheep, 22,854 swine; and produced 65,676 bushels of wheat, 396,797 of Indian corn, 182,910 of oats, 39,763 of potatoes. It had 16 stores, two flouring-mills, four grist-mills, 15 saw-mills, four distilleries; one college, one academy, 75 students; 16 schools, 469 scholars. Pop. 7089. Capital, Knoxville.

Knox, p. t., Waldo co., Me., 25 m. N.E. by E. Augusta, 629 W. Incorporated in 1819. Pop. 597.

Knox, p. t., Albany co., N. Y., 21 m. W. Albany, 391 W. Drained by Norman's hill and branches, and a branch of Fones' creek. It contains a Presbyterian and Methodist church, six stores, one grist-mill, 15 saw-mills, four tanneries; one academy, 21 students; 13 schools, 686 scholars. Pop. 2243.

Knox, t., Holmes co., O. It has nine stores, two fulling-mills, two saw-mills, one oil-mill; one school, 30 scholars. Pop. 1129.

KNOXVILLE, p. v., capital of Crawford co., Ga., 15 m. S.W. by W. Milledgeville, 711 W. Situated 4 m. E. of Flint river, and contains a courthouse, jail, an academy, seven stores, 25 dwellings, and about 160 inhabitants. Net proceeds of the postoffice $241.

KNOXVILLE, p. v., with a city incorporation, capital of Knox co., Tenn., 183 m. E. by S. Nashville, 496 W. Situated at the head of steamboat navigation on the N. side of Holston river, a branch of Tennessee river. It is regularly laid out, and contains a fine stone courthouse, a brick jail, three churches, two Presbyterian and a Methodist, an academy, a female seminary, five wholesale and nine retail stores, four grist-mills, three saw-mills, and other manufacturing establishments, three printing-offices, two weekly newspapers, 200 dwellings and 1500 inhabitants. It is the largest town in E. Tennessee. Much of the Hiwassee railroad, extending from this place through Athens, Ga., to Augusta, is graded. It is the seat of East Tennessee college, founded in 1807, which has a president and four professors or other instructors, 56 students, and 3307 volumes in its libraries. The commencement is on the first Wednesday in August. It has a fine college edifice, which occupies a commanding position in the western suburbs of the town, and has a fund of $25,000, besides 15,000 acres of land

KNOXVILLE, p. v., capital of Knox co., Ill., 160 m. W.N.W. Springfield, 820 W. Situated on an elevated and fertile prairie, near Haw creek. Laid out in 1832, and contains a courthouse, 50 dwellings and about 300 inhabitants.

KNUTSFORD (corrupted from C'nute's Ford, so called because the Danish Canute crossed here with his army,) a market town and par. of England, co. Chester, hund. Bucklow, 11½ m. W. by N. Macclesfield, 12½ m. S. by W. Manchester, and 154 m. N.N.W. London. Area of par., 4300 acres. Population, in 1831, 3600: do. of Nether and Over Knutsford townships. 2040. The town is divided into two parts, called Over and Nether Knutsford, by the brook Birken, an affluent of the Bodlin, which rises about half a mile S. In Nether Knutsford are the market-place, sessions-house, and county jail, the last of which is said to be spacious and well conducted. The church, a modern structure of brick and stone, with a square tower, was built in 1741, when this parish was separated from that of Rostherne: the living is a vicarage in private patronage. Another church, at Over Knutsford, is in the patronage of Lord de Tabley. The other places of worship are for Wesleyan Methodists, Independents, and Unitarians; and Sunday-schools are connected with the two former, as well as the churches. The charities of the town comprise a free school, founded in the reign of Edward VI., and well endowed by an ancestor of the Leigh family, a school of industry for 160 girls, supported by the Egerton family (who support also another school at Rostherne), and a parochial school for 70 boys. The manufactures of shag, cotton velvet, sewing thread, worsted, and leather, employ many hands; but the supply of the wants of the opulent gentry in the neighbourhood is a chief source of support to the tradespeople. Races are held annually in July, and

are well attended. Knutsford is the election town for the N. division of Cheshire, and sessions are held in July and October. Markets on Saturday. Cloth and taxile fairs, Whit Tuesday, July 10 and November 8.

KOMORN. See COMORN.

KONIEH (an. Iconium,) a town of Asiatic Turkey, prov. Karamania, cap. of a pash. and sandjiak of its own name, 37° m. E. by S. Smyrna, and 132 m. S. Angora; lat. 37° 54' N., long. 32° 30' E. Population about 30,000, chiefly Turks. It extends over the plain E. and S. far beyond the walls, which are about two miles in circumference. Snow-covered mountains surround the level country. on every side except the E., where a dreary plain extends to the horizon. The walls were built by the Seljuk sultans, of materials taken from more ancient edifices; and the figures in alto relievo which ornament the gates are alleged by Kinneir to be among the finest in Turkey. In the middle of the town a small eminence is covered with the remains of a fortified palace, once inhabited by the Seljuk princes. The present public buildings comprise 12 large and numerous smaller mosques (that of Sultan Selim having been built on the model of St. Sophia at Constantinople), several medresses or colleges (only one of which, the Capan madrasa, is now inhabited), two Armenian churches, four public baths, and seven khans for the accommodation of merchants. The importance of Konieh belongs to the past; for it has now dwindled into insignificance, and exhibits every mark of desolation and decay. A few carpets and some morocco leather are manufactured here: but trade is in a very languishing state, and far the greater portion of the adjacent territory is permitted to lie waste.

Iconium, the capital of Lycaonia, mentioned by Herodotus and Xenophon as being on the great post road between Sardis and Susa, is reported by Strabo to have been a well-built town, situated in a fine country, and is celebrated in gospel history as having been the scene of St. Paul's persecution by the unbelieving inhabitants. (See Acts, xiv., 1-7.) After the taking of Nicea by the crusaders in 1099, it became the residence of the Seljuk sultans of Roum, by whom it was much embellished and enlarged. Frederick Barbarossa expelled them in 1189; but after his death, they re-entered their capital, and lived in splendour till the irruption of Jenghis-Khan, and his son Holukow, who broke the power of the Seljuks. Konieh has been included in the dominions of the Grand Seignior since the time of Bajazet, who finally extirpated the Ameers of Karamania. (Kinneir's Asia Minor, p. 217-231; Leake's Tour, p. 48.)

KONIGGRATZ (Boh. Kralowy-Hradec), a town of Bohemia, cap. circ. of same name. on the Elbe, 63 m. E.N.E. Prague. Population (1838) 6094. It is fortified; and has three suburbs, some large barracks, a fine cathedral, Jesuits' college, episcopal seminary, gymnasium, high school, and a celebrated orphan asylum. Woollen cloth weaving is the chief employment of the inhabitants. It was taken several times by the Prussians during the last century. (Oesterr. Nat. Encyc.; Berghaus, &c.)

KONIGSBERG, a large city of the Prussian states, now the cap. of the prov. of Prussia Proper, and of a reg. and circ. of the same name, as it formerly was of the monarchy, on the Pregel, near where it falls into the Frische Haff. lat. 54° 42' 11'' N., long. 20° 29' 15'' E. Population (1838) 68,000. A bar at the mouth of the Pregel prevents vessels drawing more than five or six feet water ascending the river to Königsberg; so that its port is properly at Pillau, at the junction of the Frische Haff with the Baltic. A part of Königsberg is built on an island formed by the Pregel, the houses being founded on piles, as at Venice and Amsterdam. Opposite to this island, and on the N. bank of the river, stands the rest of the city, consisting of the old town, and a quarter to the E. called Lobenicht. The circumference of these three quarters, which properly form the city, hardly exceeds two miles; but the suburbs are very widely spread, and the wall that encloses the whole is no less than nine miles in circumference; but a large portion of the included space consists of gardens and open fields. According to Dr. Granville, the "streets are long, narrow, dirty, ill paved, and very offensive, lined by lofty old-fashioned houses, the basement stories of which project far out in the shape of terraces, with their flights of steps guarded by antiquated brass railings, which are not only very inconvenient for the passage of carriages, but render that of pedestrians a work of real danger. Everywhere, in fact, houses and inhabitants are as old fashioned as if the court of the old dukes of Brandenburg were still held here." (Travels to Petersburg, &c., i., 247.)

The old town contains the town-house, rebuilt in 1774; an anatomical theatre, a hospital for the widows and orphans of citizens, and many large warehouses. The quarter to the E. of the old town contains a large hospital on the river side, a mint, theatre, and orphan-house. Here also is the old royal palace or castle, now the government-house. The insulated part of the town contains the council-house, exchange, and university buildings. Its orphan-house is also a conspicuous edifice; but none of these rival the cathedral, which, besides its architecture and ornaments, is remarkable for its organ, erected in 1721, containing 5000 pipes, and for several monuments of the old dukes of Brandenburg, the founders of the monarchy. There are seven bridges over the Pregel.

Königsberg is the seat of the government of the province, and of a court of appeal and a tribunal of commerce. Its university, founded in 1544, had Kant, who died here in 1804, for one of its professors, and is attended by about 450 students. The city has besides three gymnasiums, two seminaries for preachers, with numerous schools, a royal literary society, a celebrated observatory, and various other literary establishments, a blind asylum, &c. There are manufactures of woollens, cottons, leather, gloves, lace, wax, soap, refined sugar, &c., with breweries and distilleries on a large scale. The great articles of export consist of wheat, rye, barley, oats, pease, tares, flax and hemp, timber, linseed, ashes, bristles, &c.; the imports being colonial products, cotton, and cotton-twist, wine, dye stuffs, spices, oil, coals, &c. For an account of the shipping entered and cleared out at Königsberg; see PILLAU.

KOOM, a city of Persia, prov. Irak-Adjemi, district of its own name, 166 m. N. by W. Ispahan, and 80 m. S. by W. Teheran; lat. 34° 45′ N., long. 50° 59′ E. Population, according to Ouseley, about 8000; but Morier regards this statement as exaggerated. It stands in an extensive plain, and on the banks of a small river rising at no great distance, and lost eastward in the great salt desert. On approaching the city, the remains of habitations, gardens, and tombs become so numerous as to evince that this district was formerly very populous. Among the sepulchral ruins are upwards of 100 tombs of *imámzadahs* (descendants of imáms), distinguished by their tiled cupolas; and there is a very beautiful college, with a celebrated mosque and mausoleum dedicated to the memory of Fatima, the daughter of imám Reza, and containing the tombs also of Sefi I. and shah-Abbas II. The dome is lofty, and, with the interior, was, a few years ago, covered with gilt plates, by the mother of the late shah. Futtee Ali. Koom, though formerly a place of some trade in fruit, silk, soap, sword-blades, and white earthenware, has sunk into utter insignificance: the bazaars hardly contain 40 shops, and the only employment of the inhabitants is the cultivation of a little corn and rice. In fact, the place is little more than a mass of ruins, and at least two-thirds of the buildings have been untenanted for half a century. Its sanctity, however, as a place of refuge and pilgrimage is generally celebrated throughout Persia, and devotees still order their bones to be brought here for sepulture.

Koom is conjectured by D'Anville to stand on the site of the ancient *Cheana*, visited by Alexander. In the Shah Nameh it is named as an ancient city, and its foundation assigned to Kai-Kobad. More dependence, however, may be placed on the statement of D'Herbelot, that it was either founded or rebuilt by the Saracens, about the beginning of the ninth century. Timur-Leng destroyed it, but it regained its importance under the Sefi dynasty. In Chardin's time there were 15,000 houses, 20 large mosques, extensive bazaars, and a handsome bridge over the river; but in 1722, when the Afghans invaded Persia, they pillaged and all but destroyed the city: repeated earthquakes have also much damaged the remaining buildings, and Koom is now only a melancholy ruin. (*Kinneir's Persia*, 116; *Ouseley*, iii., 99-106; *Ritter's Asia*, vol. vi., part 2, p. 30-33.)

KOONDOOZ, an indep. khanat of Central Asia, between the 35th and 39th deg. N. lat., and the 66th and 73d E. long., at present comprising, as tributary states, Budukshan and many other small chiefships N. of the Hindoo Koosh. It has N. the territ. of Hissar, Durwaz, &c.; S., the Bolor-Tagh mountains, separating it from the Chinese dom.; S., Cauristan, and the Hindoo Koosh, which divides it from Caubul; and W. a part of Afghanistan, and the territ. of Balkh. The central part of this dominion, or Koondooz Proper, seems to be situated on a lower level than the surrounding provinces. It is of limited dimensions, is enclosed by ranges of low hills, and watered by two of the principal tributaries of the Oxus, in the upper part of its course. It is in many parts so marshy that the roads are obliged to be constructed on piles of wood, fixed among noxious and rank vegetation. The climate is most pestiferous: snow lies for three months in the winter, but the heat in summer is often excessive. The soil is, however, very fertile, and produces abundant crops of grain. In the marshy grounds rice is the chief product, and in the drier grounds wheat and barley. The revenues of the chief are derived, as in the other E. states, from the land: they are paid principally in kind, and are said to amount to a third part of the produce of the soil. Apricots, plums, cherries, &c., are plentiful, as are most necessaries of life; a good deal of silk, also, is produced on the banks of the Oxus. Since the conquest of Budukshan, that fine province has been in a great measure depopulated, its inhabitants being carried off to cultivate the lands of Koondooz, where they die rapidly, from the effects of the climate. The surrounding provinces have mostly both a rich soil and a good climate. The inhabitants of Koondooz are mostly Tadjiks. (*See* BOKHARA, i., 303.) The khan or *meer* is, however, an Uzbek, Koondooz appearing to be the most southerly region into which the Uzbeks ever penetrated, and afterward succeeded in establishing their dominion. The army, comprising about 20,000 cavalry, with six pieces of artillery, consists chiefly of Uzbeks; but most of the civil employments under the state remain in the hands of the native population. By adopting this line of policy, and governing his subjects in general with justice, and (for an eastern despot) considerable mildness, the present sovereign of Koondooz has consolidated his power: he is supposed to be rich, and his army, at least, is a match for any one likely to oppose it. The khan frequently makes "*chupaivals*" (chupans), or predatory incursions, into the neighbouring territory of Balkh, the Hazareh country, &c., for prisoners, whom he sells for slaves; and the chief of Chitral pays his tribute in human beings, who, being also sold by the khan, form a principal article of export from Koondooz. His encouragement of this nefarious traffic, and his behaviour to the unfortunate population of Budukshan, are the principal blemishes in the character of the khan: he has the merit of protecting commerce, and traders in his dominion are secure from exaction or molestation. There is a considerable trade between Koondooz and the Chinese province of Yarkund, and sometimes an exchange of presents. Tea is an important article of consumption. European and other foreign luxuries are derived from Bokhara, in exchange for slaves and cattle sent to its markets. At present, of all the Uzbek states, Koondooz is the most adverse to British influence.

Koondooz, the nominal cap., is in a wide valley, near the confluence of two rivers, about lat. 36° 50′ N., and long. 69° 10′ E. It has formerly been a large town, but its pop. does not now exceed 1500. It has a mud fort, surrounded by a ditch, and is the winter residence of the chief. The largest town in the khan's dominion is Khooloom (which see). (*Burnes's Trav.*, ii., 179-198; vol. iii., 175, 176, 275-281; *Ritter, Erdkunde, von Asien*, v., 810-815.)

KOROTCHA, or KAROTCHA, a town of Russia in Europe, gov. Koursk, cap. circ. on the river of the same name, and on the road from Voronezh to Kharkoff, 100 m. S.W. the former city. Pop. nearly 10,000. It is well built, and surrounded by numerous gardens; and has several churches, nearly all, however, constructed of timber, a circle high school, hospital, and a saltpetre manufactory; with an extensive trade in apples, for which its vicinity is famous. Korotcha was founded by Michael Fedorovitch, in 1658, as a barrier against the incursions of the Crim-Tartars. (*Schnitzler, La Russie; Possart, Das Russland.*)

KORTRIGHT, p. t., Delaware co., N. Y., 15 m. N. Delhi, 69 m. W.S.W. Albany, 365 W. Bounded S.E. by Delaware river, and drained by its tributaries. It contains three stores, two fulling-mills, three grist-mills, four saw-mills, three tanneries; 17 schools, 695 scholars. Pop. 2441.

KOSCIUSKO, co., Ia. Situated towards the N. part of the state, and contains 567 sq. m. Watered by Tippecanoe river and other streams, which afford water-power. It contained in 1840, 5037 neat cattle, 2923 sheep, 11,880 swine; and produced 20,600 bushels of wheat, 146,161 of Indian corn, 38,445 of oats, 21,955 of potatoes, 1569 pounds of tobacco, 68,945 of sugar. It had 13 stores, four grist-mills, 10 saw-mills, one tannery, one distillery; six schools, 216 scholars. Pop. 4170. Capital, Warsaw.

KOSCIUSKO, p. v., cap. of Attala co., Miss., 67 m. N.E. Jackson, 958 W. Situated on a branch of Yocannockany river, a head branch of Pearl river, and contains a court-house, Baptist church, an academy, and several dwellings. Nett proceeds of the postoffice, $142.

KOSTENDIL, or GHIUSTENDIL (*Justiniana secunda*), a town of European Turkey, prov. Roumelia, and cap. sandjak of its own name, 107 m. N. Salonica, and 192 m. W.N.W. Adrianople. Pop., according to Stein, 6000. It stands on the N. declivity of the Karasu mountains, at a short distance from the right bank of the Strouma (the ancient *Strymon*), and is defended by a crenelated wall flanked with square towers. A bazaar, governor's palace, and several sulphur baths, are the only public establishments. Employment is given to a portion of the population by the silver and iron mines of the neighbouring mountains.

KOSTROMA, a gov. of Russia in Europe, between 59° 45′ and 59° 12′ N. lat., and 40° 27′ and 48° E. long., having N. the gov. of Vologda, W. Jaroslavl, S. Wladimir and Nijegorod, and E. Viatka. Area. 30,400 sq. m. Pop. in 1838, 956,000. Surface flat, with some undulations. It is indifferently fertile, being marshy in the N., while in the S. it is sandy and clayey. Climate severe, but healthy. It is watered by the Wolga, and by its important tributaries, the Onja and Vetlouga. Principal corn crop, rye; but the quantity grown is insufficient for the consumption. Flax and hemp are largely produced. Cattle few, and but little attended to. This, however, is not the case with the forests, which are extensive, valuable, and better taken care of than those of most other governments. The rivers and lakes furnish abundance of fish. The inhabitants particularly excel in the preparation of Russia leather, and there are various fabrics of cloth and linen. Many of the peasants are masons, carpenters, &c., who seek for employment in the summer season in the contiguous governments; and many are employed at home in the making of charcoal, pitch and tar, moss, of which there is an immense consumption, boats, rafts, &c. (*Schnitzler, La Russie, &c.*, p. 132, &c.)

KOSTROMA, the cap. of the above gov., on the Wolga, at the confluence of the Kostroma with that river. Pop. nearly 10,000. Situation elevated and agreeable; houses mostly of stone; the rampart of earth by which it was formerly surrounded has been converted into a promenade. It has a handsome cathedral, two large convents, a great number of churches, and a large stone building, or bazaar, for the security, exhibition, and sale of merchandise. There are several tanneries, with manufactures of linen, Prussian blue, soap, and tallow; a bell-foundry. Various fairs, and a considerable commerce.

KOTAH, a town of Hindostan, prov. Rajpootana, the cap. of a subsidiary state of the same name, with an area of about 4480 sq. m. (*Sutherland.*) The town, on the Chumbul, 126 m. S.W. Agra, is large and populous, with some good and well-stocked bazaars, and a great number of temples and substantial private houses. The entrances to Kotah are through double gateways, and its walls are surrounded by a fosse hewn in the solid rock. Its chief public edifice is the palace of the rajah, rendered conspicuous by its lofty white turrets, and enclosed by a separate line of works. Kotah has manufactures of cloth and other articles of native consumption. Its territory is among the most flourishing of India, and about 28 years ago its gross revenue was estimated at 47 lacs of rupees, and its armed force at 25,000 men. (*Hamilton's E. I. Gaz.; Parl. Reports.*)

KOTOCH, a rajahship of N.W. Hindostan, subject to the maharajah of the Punjab; about lat. 32° N., and between long. 76° and 77° E., having W. and S. the territory of the Punjab, and N. and E. the rajahships Chamba, Kulu, and Mundi, separating it from the great range of the Himalayas. It comprises a portion of the upper valley of the Beas (*Hyphasis*), and is about 60 m. in length N. to S. by from 25 to 40 m. in breadth. Its natural products are few, but they might be much improved under an enlightened government. Opium is largely cultivated on the Kulu frontier, and cotton is raised, and furnishes the material from which the finer cloths of Hoshyarpur are manufactured. When this place was visited by Moorcroft, these cloths were sent in large quantities to W. Asia; agents from very remote places attended at Hoshyarpur, made advances to the weavers, and took the cloth in the rough from the loom, bleaching, washing, and packing it each in his own fashion to suit the market of his country. We have not learned whether any change has been effected in the interval in this trade. Superior wheat and rice are raised. Firs of large size grow in some tracts along the Beas; and in one part is an extensive bamboo forest. Rhubarb and the mulberry-tree are abundant; and iron is found, but the ore is not wrought. Shujaspoor, about lat. 31° 48', and long. 76° 38' E., is the capital. (*Moorcroft's Travels*, i., 130-162.)

KOURSK, a government in the S. part of European Russia, having that of Orloff on the N., Voronetz on the E., Kharkoff on the W., and Tchernigoff on the W. Area, 16,006 sq. m. (?) The estimates of the pop. differ very widely, but Schnitzler concludes that it cannot fall short of 1,680,000. Surface flat, or slightly undulating; soil very fertile; forests not very extensive, and in some parts there is a scarcity of wood. There are no navigable rivers, the want of which is one of the greatest drawbacks on the government. The climate is mild and healthy. Notwithstanding the backward state of agriculture, Hassel says that wheat and rye yield nine for one; but this is most probably exaggerated. Corn is kept in silos, or caves, sometimes for six or 10 years together, and there is always a large surplus for exportation. Hemp and flax, tobacco, hops, &c., are also produced. The pastures, which are excellent, afford ample provision for large herds of oxen, with horses, sheep, &c. There are in this government above 300,000 free peasants. Manufactures considerable and improving, consisting of coarse cloth for the army and the peasantry, leather, soap, saltpetre, spirits, earthenware, &c. Public instruction has made no considerable progress. There were, in 1831, in the government, 25 educational establishments, attended

by 4095 pupils, being only one pupil for every 390 individuals!

KOURSK, a town of European Russia, cap. of the above gov., lat. 51° 43' 41" N., long. 36° 20' 15" E. Pop. (in 1830) 22,447. It had a citadel and ramparts, but the former is in ruins, and the latter have been converted into public walks. Situation elevated; houses principally of wood, but many of stone; streets narrow, crooked, and ill paved. There are two convents, numerous churches, with a gymnasium, a normal school, a hospital, a foundling hospital, &c. It is a thriving, industrious town, having numerous tanneries, tile and earthenware works, wax and tallow foundries, &c. It carries on an extensive commerce with Petersburg, Moscow, and Odessa, sending to them cattle and horses, tallow, leather, wax and honey, hemp, and furs. Koreenaia Poustyn, a convent in the vicinity of Koursk, is celebrated for a miraculous image of the Virgin, and for a great fair held annually on the ninth Friday after Easter, resorted to equally by merchants and pilgrims. The value of the horses, cattle, and other articles exposed to sale at this fair in 1833, amounted, according to the official accounts, to about 30,000,000 roubles. But if this be not exaggerated, it is, at all events, greatly above the average. (*Schnitzler, La Russie, &c.*, p. 239; *Malte Brun, &c.*)

KRABNOJARSK, or KRASNOYERSK, a town of Asiatic Russia, gov. Yeniseisk, of which it is the cap., in a plain of great beauty and fertility, on the Yenisei, and on the high road between Tobolsk and Irkutsk, 280 m. E. by S. Tomsk, lat. 56° 1' N., long. 92° 57' 10" E. Pop. about 4000. Though small, this is a town of some importance, being the emporium of a wide extent of country. It is pretty well built; its two principal streets are broad, and its houses, which are mostly faced with planks, are painted in bright colours. Its chief public buildings are, several churches; an edifice partly of stone, occupied by the government offices; and a large factory, devoted to several branches of industry, especially coach-building, and the manufacture of Russia leather. The district subordinate to this town is the most productive in the whole provinces of Irkutsk of grain, cattle, horses, &c. "Flour is sold at Krasnojarsk generally at from 15 to 30 kopecks the pood, according to the goodness of the crops; excellent beef for 1¼ to 2 roubles, and other things in proportion. Fish and game are also in abundance; and the neighbourhood is famous for wild goats, the flesh of which is not inferior to venison." (*Dobell's Travels in Siberia*, ii., 96.) Within the last 20 years this town has been rising considerably in importance; and it has now a brisk traffic in Chinese and agricultural produce. Erman, who visited it in 1829, found it supplied with Madeira and other European wines, and speaks in high terms of a literary club, which he says may rank with those of a superior order in Europe. (*Erman, Reise um die Erde*, i., 26-30; *Dobell's Travels, &c.*)

KREMNITZ (Hung. *Kermnez-Banya*), a royal town of Hungary, co. Bars, and one of the principal mining and coining towns in the kingdom; in a deep valley, 10 m. W. Neusohl, and 88 m. E.N.E. Presburg. Pop. about 4680. The walled town comprises a castle and about 40 houses, one of which is the mint, ranged round an open space in which the market is held. In the suburbs are nearly 686 dwelling houses, and many mining offices; and about ¼ m. distant are the smelting furnaces. Kremnitz is ill paved, dirty, and disagreeable. It has five churches, one with a lofty gilt and coppered steeple and very gaudy internal ornaments, three chapels, a Protestant meeting-house, two hospitals, a royal infirmary for miners, a gymnasium, normal and girls' schools, a Lutheran grammar school, &c.; and it is the seat of municipal and mining tribunals, and of a mint, and councils of mines and forests.

The Kremnitz mines have 11 or 12 principal shafts, attached to which are 18 or 20 washing works (*pochwerken*). The best mines belong to private companies, but the richest veins of Kremnitz are now for the most part exhausted, and a considerable portion of the former workings is under water. The mines at present yield about 15,000 marcs of silver, and 250 do. of gold a year. These metals, however, are rarely found pure, but much intermixed with copper, lead, arsenic, &c. Quartz is the matrix of the ore, which is first reduced by the hammer to small pieces about the size of the stones used for Macadamizing roads: the ore is next exposed to the stamping-mill, by which it is pulverized; it is then washed over slanting frames; sometimes roasted to drive off the sulphur, arsenic, &c.; and is finally smelted. "The object of this process, which lasts four and twenty hours, is to separate the noble from the ignoble metals, which is effected by the oxydation of the latter. At the moment the oxydation is complete, a bright bluish-white metallic lustre spreads itself over the whole surface of the liquid metal. The impure metals are then allowed to run off, a stream of warm water is passed over the gold and silver to cool them; the solid mass is taken out, cut up into bars, weighed, and sent off to the mint, where the gold and silver are separated and coined. The smelting-houses of Kremnitz are the best in Hungary: instead of the common bellows, they have the double-cylinder bellows worked by water, which maintains a constant blast; and the loss of lead, instead of being 20 lbs. to the mare, is reduced to 12." (Paget's Hungary, i., 396, 397.)

All the gold and silver produced in Hungary, whether by private individuals or by the government, should be coined at Kremnitz. Delius stated, in 1772, that between 1740 and that year nearly 100,000,000 guldens of gold and silver had been obtained from the mines of Schemnitz and Kremnitz, and coined at the latter town; and that 3,000,000 guldens a year still came from Schemnitz, Kremnitz, and Transylvania. (See Bright's Travels in Lower Hungary, p. 177, 178.) The amount of gold and silver coined at Kremnitz is now about £250,000 a year (2,500,000 florins s. m.); but it is probable that this is considerably less than the amount produced, as it is known that a good deal finds its way to Vienna in bars. The silver is mostly coined into pieces of 20 kreutzers (zwanzigers), and the gold into ducats and half ducats." (Paget, i., 304, 305.) Kremnitz has also a royal vitriol factory, two paper-mills, and manufactures of earthenware and vermilion. It is abundantly supplied with excellent water by a water-course carried by a former archbishop of Gran, at his own expense, from the Thurocz to Kremnitz, a distance of 50 m. (Paget's Hungary, i., 390-397; Bright's Travels, 167-182; Oxster., Nat. Encyc.; Berghaus.)

KRISHNA, or KISTNAH (the name of a supposed incarnation of Vishnu), a river of Hindostan, and one which bounds an important division of that country; the S. part of the peninsula being entitled "India S. of the Krishna." It rises in the W. ghauts, near lat. 18° N., and long. 74°, not far from Sattarah; and runs, with a very tortuous course, E. for about 700 m., through the provs. Bejapoor, Beeder, and Hyderabad, and between the British distr. of Masulipatam and Guntoor, falling into the ocean on the Coromandel coast by several mouths near lat. 16°, long. 81°. Its course lies mostly through a mountainous country, greatly elevated above the sea; its channel is of very irregular depth, much broken by rocks and rapids, and it is altogether ill adapted for navigation, except in the N. Circars, where it is available for large boats. In the highlands, the craft usually employed on it are round inamboo wicker baskets, covered with half-tanned hides, and directed with paddles. Its chief tributaries are the Joongabonda, Gutpurba, and Malpurba, from the S.; and the Seema, Mussy, &c., from the N. Sattarah is the principal city on its banks. It is said to be much more productive of gems, &c., than the Godavery, diamonds, chalcedonies, onyx, and other precious stones, and some gold, being found in its sediment in the dry season. (Hamilton's E. I. Gaz.)

KULDSHA, GULDSCHA, or ILI, a city of Chinese Turkestan, cap. prov. Ili or Ele, in lat. 49° 46' N., long. 82° 43' 19", about 450 m. N.E. Yarkand. It is said to be 18 Chinese li (about 5 m.) in circuit, surrounded by mud walls and wet ditches, and entered by six gates. The town, according to Helmersen, is much better built than either Kotan or Bokhara: the houses are either of stone or wood, seldom of earth, and the streets are traversed by running streams. The inhabitants are mostly Chinese; there are, however, about 1500 Toorkee families, who profess Mohammedanism, but whose dress, customs, &c., resemble those of the rest of the population. The inhabitants of Kuldsha are very industrious, and devoted to commerce. Almost every house has a shop, frequently filled with expensive merchandise; besides which, the streets abound with moveable stalls, and hawkers going about to sell their wares, Kuldsha being the entrepôt of an extensive region, peopled by nomadic Kalmuck tribes. It is the residence of a Chinese governor. (Helmersen, Ritter, Asien, Erdkunde, i., 402-404.)

KULU, a rajahship of N.W. Hindostan, tributary to the maharajah of the Punjab, about lat. 30° N., and long. 77° E., having E. the great range of the Himalaya, N. Lahool, W. Chamba and Kotoch, and S. Mundi. Length, N. to S., about 40 m. Area and pop. uncertain. Wheat, barley, and buckwheat, but only a little rice, are raised. No kitchen vegetables are grown. Tobacco, though cultivated in most of the gardens, does not thrive very well, and a narcotic preparation from hemp is used as a substitute. The climate being much colder than in the adjacent rajahships of Kotoch and Mundi, woollen instead of cotton fabrics are used for clothing. Kulu, or Sultanpoor, the cap., about lat. 31° 57' N., and long. 77° 10' E., is of no great extent or pop.: it stands at the confluence of the Beas and Serberi. The part next the river, forming the southern or lower town, is occupied by the buildings in which the rajah resides with his family and attendants. The upper part of the town consists of the houses of shop keepers and artificers, and is separated from the lower division by a small bazaar. A

few coarse chintzes, blankets, and cotton cloths, with opium and musk, are the chief articles of traffic; the three former are brought from the plains, and paid for with the two latter. The principal traders are Hindoo pilgrims, of whom a vast number assemble here on their way to holy places in the mountains. (Moorcroft i. 171., &c.)

KUMAON, or KEMAON, a prov. of N. Hindostan, under the British presid. of Bengal, comprising with Kumaon Proper, that portion of Gherwal S.E. of the Alcananda river; chiefly between lat. 29° and 31° N., and long. 78° and 81° E., having N.W. Independent Gherwal, N.E. the great range of the Himalaya, separating it from Thibet, S.E. Nepaul, and S.W. the prov. Delhi. Area estimated at nearly 11,000 sq. m. The whole country is overspread by mountains successively ascending from S.W. to N.E., till they reach the height of 25,000 ft. The Ganges in the upper part of its course, the Kalee, and a few of their tributaries, are the chief rivers: there are no lakes of any consequence. The lower portion of the prov. is covered with jungle interspersed with groups of saul, mango, and other timber trees, and tracts of high reedy grass. The central hilly region is an almost uninterrupted forest: above the elevation of 2500 feet the vegetation of the tropics gives place to the pine, oak, rhododendron, &c. The fruits and vegetables of Europe are common, and thrive well. Wheat, mendua, and other dry grains are those principally grown, but rice, also is cultivated alternately with the foregoing, a regular rotation of crops being pursued. Hemp is raised in large quantities, and grows luxuriantly to the height of 12 or 14 ft.; little cotton is raised, though it is of excellent quality. The sovereign has the entire property of the soil; and the great functionaries under the native gov. were always paid by grants of land, or by perquisites arising from the soil. The operations of tillage, except ploughing and harrowing, are chiefly performed by women. The implements and operations of husbandry are similar to those in the plains of Hindostan. Irrigation is frequently effected by aqueducts carried a considerable distance, and water-mills, scarce in Hindostan, are here common. The breeds of horned cattle are small, but yield very good milk; there are domestic camels, but they are small, and goats and sheep are principally used for the transport of goods. Elephants, tigers, leopards, and various kinds of deer abound. Copper, iron, and lead mines are wrought to some extent; and garnets, rock crystal, and bitumen are met with. Manufactures very few; they include blankets, coarse canisters, hempen cloths, coarse cottons, bamboo mats and baskets, wooden vessels, &c. Artisans are sufficiently numerous in the towns, but their work exhibits little neatness. It is singular that, though the saw, plane, and turning-lathe, are unknown to joiners, the goldsmiths are acquainted with the use of the spirit blow-pipe. The inhabs. at large are more inclined to commerce than agriculture. They carry iron, copper, ginger, turmeric, and other hill drugs and roots into the plains of N. Hindostan, where they exchange them for coarse chintz, cotton cloths, sugar, tobacco, coloured glass, beads, and hard-ware; and they frequently travel to execute mercantile commissions as far as Furruckabad and Lucknow. The traders of more capital send the products of India across the mountains into Thibet, where they are exchanged for hawks, musk, coarse camlets, wax, incense, and other drugs and roots, the produce of that country; and borax, salt, and gold dust from Tartary. In the marts of Kumaon, the chief of which are Mandi, Kastpoor, Chilkia, Alseighur, and Najibabad, sugar-candy, spices, European broadcloths, coral, &c., from the S., are exchanged for shawl wool, coarse shawls, China silk, saffron, hides, &c. Large periodical fairs are held at the above places, whence necessaries are procured, there being no village markets in Kumaon. The country is thinly peopled; the inhabitants are of two distinct races, the dominant being the Hindoo; and the supposed aborigines a race apparently of Tartar origin, many of whom, called doms, appear to have been reduced to a state of slavery by their Hindoo conquerors. The native government was despotic in an oppressive degree till the British took possession of the country in 1815; since which, the condition of Kumaon and its inhabitants has been progressively ameliorated. Total public revenue (1822-23), 186,126 rup., of which the land-tax furnished 176,664: public expenditure, 21,984 rup. Kumaon, like many other parts of N. Hindostan, contains numerous places of Hindoo pilgrimage, and many Hindoo temples. Almora is the cap., which see. (Asiatic Researches, xvi.; Hamilton's E. I. Gaz.)

KUR, (an Cyrus), a river of western Asia, in Georgia, having its rise within the Turkish dominions, not far from Kars, on a S. offset of the Caucasian range, dividing the tributaries of the Caspian from those of the Black sea, in lat. 41° N., and long. 42° 30' E. It assumes its name near the town of Akiskar, whence it flows about 80 m. E.N.E. to Gori. Its course thenceforward is S.E., by Tiflis, through the plain of Kara, and afterward through a lower plain, bounding with salt marshes, and in which are several mud volcanoes and petroleum springs. The latter of these plains is frequently overflowed by the river. The total length of the Kur, as measured along its windings, somewhat exceeds 530 m.: its chief affluents are, h. the Alazan, from the main Caucasian ridge, joining the main stream in lat. 40° 56′ N., and long. 45° 51′ E.; and, 2d, the Aras (an. *Araxes*), which rises near Erzeroum, curves northward round mount Amsat, and thence runs S.E., and afterwards N.E., to its junction with the Kur, at Djwat. The Kur at this point is 140 yards broad, and may be navigated by large boats to its mouth on the W. side of the Caspian sea, a distance of about 180 m. Fishing villages are established on its lower banks, and great wealth is accumulated from the proceeds of these fisheries. A delta at the mouth projects considerably into the Caspian sea.

KURACHEE, or KARACHEE, the principal seaport of Sinde, N.W. Hindostan, on the E. side of an inlet of the Indian ocean, 80 m. S.W. Hyderabad, and about 16 m. from the W. arm of the Indus; lat. 24° 56′ N., long. 67° 19′ E. It is built on a low, barren, sandy shore, and is walled. In 1812 there were 2830 houses, within the walls, but the pop. did not reach 13,660. The town is irregularly laid out, and the streets are so narrow that two people can scarcely walk abreast. The houses are chiefly of mud and are mean, obtained in great abundance from the coast.

Karachee has a considerable trade with Cutch, Bombay, and the principal ports on the Malabar coast. Its harbour is commodious, perfectly safe in all winds, and, though not deep, is capable of sheltering vessels of 200 or 300 tons; so that it is of greater commercial importance than any of the ports on the Indus, which can only be reached from the sea by flat-bottomed boats. Nearly all the Malwa opium exported seaward is shipped at Karachee. In 1837, about 80 fishing boats, of from 10 to 15 tons, belonged to this port; and most of the men engaged in the fisheries of Sinde are from Karachee, and are superior in intelligence and appearance to the other inhabitants of the coast. Karachee was bombarded and taken in a few hours by a small British force, on the 3d of Feb., 1839. (*Geog. Journal*, v., 263; *Asiatic Journal*, 1839.)

KURDISTAN, an extensive country of W. Asia, comprised chiefly within the basin of the Tigris, and belonging partly to Turkey and partly also to Persia; being bounded N. by Armenia, E. by Azerbijan and Irak-Adjmi, S. by Khuzistan and the pach. of Bagdad, and W. by Diarbekir and Algezira. Area roughly estimated at 380,000 sq. m. Pop. 560,000, Kurds only, not including other races. Its surface generally is very unequal; but the mountains are much loftier and more frequent in its N. part, the plains in the latter being also considerably more elevated than in S. Kurdistan; and hence there is a great difference of climate in the two sections into which the country is divided. The principal ranges are the Djebel-tagh and Nimrod; the culminating summit being the snow-covered mount Elwund, rising 7500 ft. above the surrounding plain, and about 12,000 ft. above the sea. The geological constitution of these mountains consist of serpentine hornblende, and other primary rocks, covered, except in the highest parts, by transition limestone, old red sandstone, and various calcareous formations with other rocks, ascending even, in some parts, to the London clay. (*Ainsworth's Assyria, &c.*, p. 237–288.) The principal rivers are the Tigris, Diala, Great and Little Zab, Korah, and Kabur. Extensive and rich pasture grounds support great numbers of sheep and goats, the rearing of which constitutes the chief employment of the population, and their produce almost the whole wealth of the country. Hence in the Kurd dialect (which is a patois, composed chiefly, though not entirely, of Arabic and Persian,) the word *mdhl*, which means wealth generally, applies in a primary and more particular sense to flocks of sheep. Jaubert says that 1,500,000 sheep and goats are annually supplied to Constantinople from Kurdistan. Each flock comprises from 1500 to 2000 animals, and the time required to take them to their destination is somewhat more than 77 months: we believe, however, that the number is not half so great as M. Jaubert has represented. The consumption of London is under 1,500,000 sheep a year; and that of Constantinople, we venture to say, is not a third part so great. As respects the produce of the soil, the N. part produces the grains and fruits of middle Europe, while in the S. the plains and valleys produce, in addition, rice, cotton, tobacco, with a great variety of fruits: excellent timber is found in the forests, and nut-galls form a large article of export at Iskanderoon and Smyrna. Good cultivation, indeed, prevails in the vicinity of the towns, and more especially between Mosul and Bagdad, where the country, at the time of Kinneir's visit, seemed to be in a much more improved state than any other district he had visited in this part of the world. (*Persian Empire*, p. 205.) The agriculture of Kurdistan may elsewhere, however, be regarded as in the most primitive condition; and the implements of husbandry are less effective, even, than those of the neighbouring provinces, which owe almost everything to nature and very little to industry.

The Kurds, who inhabit this country, and give to it its distinctive appellation, are commonly considered as a mixed breed of Mongols and Usbek Tartars; though this is doubtful. They are Mohammedans, of the sect of Omar; their dress much resembles that of the Turks, but it is lighter, and they do not wear the turbans or the long beard. A red bonnet is their usual head-dress, and the outer garment is a cloak of black goat-skin. They are excellent horsemen, and the exercise of the lance, with other military amusements, are points in which they particularly excel. Improvisation is commonly, and, on the whole, not unsuccessfully practised; and their music, though rude, proves that they have a tolerable acquaintance with the art. There are two castes of Kurds, characterised by very different habits. Those of Turkish Kurdistan have fixed habitations, are acquainted with the working of metals, weaving, and other arts, and live subject to their native princes, and governed by their own laws. The nomad Kurds are chiefly found in Persian Kurdistan and in the pachaliks of Diarbekr and Mosul; often roaming over the desert in search of plunder to the neighbourhood even of Damascus and Aleppo. The love of theft and brigandage is a marked feature in the whole race, without exception; and this accounts for their usual carelessness and improvidence about property, for which there is no security. At the same time, all writers agree, that when visited by travellers they exercise the most generous hospitality, and often force handsome presents on their departing guests. (*Joubert*, p. 86.) The tents of the wandering tribes are low, hastily put together, constructed of coarse black cloth, and generally divided into two parts for the men and women. A defence of reed hurdles surrounds the encampment in which the tents are pitched, and the horses ready saddled are tied to stakes close to the encampment. Females meet with better treatment among them than in the rest of Asia; neither sex can marry without the permission of relatives, and the constancy of the contracting parties is commonly tried during a long engagement previously to marriage, which with them is considered a sacred and indissoluble tie. Hence the women are considered more as companions than slaves; they are treated with respect, and there is a freedom and openness in their character not to be found in other women of Turkey or Persia.

Turkish Kurdistan comprises the pachaliks of Mosul and Chehresour, with small parts of the pachaliks of Van and Bagdad. Persian Kurdistan is divided into four districts, Ardelan, Kermanshah, and Kinghiavor; Kermanshah being the cap. and the residence of a beglerbeg. Neither the sultan, however, nor the king of Persia, has any substantial power, their utmost authority being limited to the exaction of tribute, the payment of which they cannot always enforce. (*Kinneir's Persia*, p. 296–316; *Joubert, Voyage en Perse*, p. 75–89; *Ritter, Asien*, iv., 595, &c.)

KURILE ISLANDS, a chain of small islands connecting the peninsula of Kamtschatka with the large islands forming the empire of Japan: they are chiefly dependent on Russia, but the three farthest S. belong to Japan. They extend between lat. 43° 46′ and 51° N., and long. 143° 30′ and 156° 29′ E., and occupy a length of more than 780 m. Pop. unknown, but very small. The surface is very irregular, some of the heights rising nearly 6000 ft. above the ocean, while in other parts deep and narrow valleys are almost on a level with the sea. Volcanic eruptions and earthquakes are of common occurrence; and the geological constituents of the islands, examined by Lothé and others, being wholly of igneous origin, indubitably show their connexion with the great volcanic band passing S.S.W. from Kamtschatka to the island of Formosa, through more than 30 deg. of lat. The shores are abrupt and difficult of approach; the coast currents are very violent, especially on the E. or ocean side; and continual fogs hovering over the islands, render access extremely difficult. The animals and plants differ little from those found in Kamtschatka; and the minerals consist chiefly of iron, copper, and sulphur. The inhabitants mostly engage in hunting and fishing; the former supplying them not only with meat, but also with furs, which serve as money for the Russian Americans, Japanese, and Dutch; while the latter furnishes oil, whalebone, and spermaceti. Agriculture is confined to the islands belonging to Japan. The inhabitants of the N. islands resemble the Kamtschatdales in honesty, openness of character, hospitality, and shyness, to strangers. Those in the S. islands are Ainos, a race similar to the Japanese. These islands were discovered before the year 1713 and 1730; but it must be acknowledged that they are very little known, even after the lapse of more than a century, and the labours of Broughton, Krusenstern, and other travellers. (*Lothé's Voyages*, tome iii.; *Dict. Géog.*)

KURNOUL, a town of British India, presid. Madras, cap. of a subdivision of the Balaghaut ceded district, which formerly composed an independent Patan principality. It stands on the Toombuddra, 90 m. N.E. Bellary, defended on two sides by that river and its tributary the Hundry, and on the W. stongly fortified, those of its bastions being 50 ft. high, and covered to the parapets of the curtain by a steep glacis. S. of the fort, is the pettah, or open town, of considerable extent and population. Kurnoul was considered impregnable by the natives, and neither Hyder nor Tippoo ever attempted its capture; but it was taken by the British, in 1815, after a siege and bombardment of a single day. (Hamilton's E. I. Gas.)

KUTAIAH, (an. Cotyæum), a town of Asiatic Turkey, cap. of the prov. Anatoli and of a Sanjiak, 180 m. E.N.E. Smyrna, and 134 m. W. by S. Angora; lat. 39° 25' N., long. 30° 15' 15'' E. Population, according to Kinneir, about 55,000 (of whom 10,000 are Armenians, and 5000 Greeks).

The city stands at the foot of a cluster of mountains called the Pursak-Dagh, in which rises the Pursak (an. Thymbrius), flowing N. to the Black sea. The streets, though steep and narrow, contain many handsome and well supplied fountains, and many of the private houses are large and well furnished. Besides 50 mosques, 29 of which have stone minarets, one Greek, and four Armenian churches, there are 30 hummums or public baths, and 20 khans. The house architecture is very similar to that of Constantinople; and good gardens attached to many of the private residences take off much of the sombre appearance common to Turkish towns. The surrounding country is well watered, and extremely productive: grain, cotton, nut-galls, and different fruits, are raised in large quantities for exportation; and goats and sheep are pastured for their hair and wool, which fetch high prices in the markets of Smyrna and Constantinople. (Kinneir's Asia Minor, p. 226; Olivier, Voyage, &c., tom. vi., 409.)

KUTCH, or CUTCH, a small state of N.W. Hindostan, subsidiary to the British, between lat. 22° 45' and 23° 45' N. and long. 69° 35' and 71° 5' E., having N. and E. the Runn, separating it from Sinde Rajpootana and Gujrat, S. the gulf of Kutch, and W. the ocean, and an arm of the Indus, which divides it from Sinde. Its shape is elongated; greatest length, E. to W., 160 m.; average breadth 45 m. Area, nearly 7400 sq. m. Population uncertain. It is in general arid and barren; but its scenery is bold, forming a great contrast to that of the adjacent provinces on the W. and N. A chain of rocky hills runs through it in its whole length, dividing it into two nearly equal parts. This chain is of no great height, but its peaks rise in wild and volcanic cones of primary formations. It unites at its W. end with another mountain chain, running nearly parallel to it on the N. side: and from both many ramifications are given off. The streams of the province are mere torrents, dry when the rains have ceased; there is no navigable river. The scarcity of water is, in fact, one of the greatest drawbacks on the country; and the streams flowing N. of the mountains are all so brackish that in the hot season they are not drunk even by the cattle. Good water is, however, usually found 20 ft. below ground. The surface is mostly sandy, the sand resting on strata of clay; but near the hills the country is covered with volcanic matters, which in India are of rare occurrence. Coal and iron of good quality, bituminous and ligneous petrifactions, and fossil animals of a late geological period, are found; and there are some mineral springs yielding alum and other salts in large quantities. The country is generally bare of wood; date trees are pretty common. and the nem, peepul, and babool, are met with round the villages, but the tamarind, banyan, and mango, are rare, and the cocoa nut is reared with difficulty even on the sea coast. The arable land is chiefly in the narrow valleys between the mountain ranges towards the S. shore, which latter, is the best watered portion of Kutch. Less corn is grown than is necessary for home consumption; and it is imported from Gujrat, Malabar, and Sinde, in return for cotton, &c. The Kutch horse is of a good breed; but other domestic animals, except goats, are generally very inferior.

That singular tract, the Runn of Kutch, is thus described by Burnes:—"It extends from the Indus to the W. confines of Gujrat, a distance of about 200 Eng. m. In breadth it is about 35 m.; but there are, besides, various belts and ramifications, which give it an extent of about 7000 sq. m. It has no herbage, and vegetable life is only discernible in the shape of a stunted tamarisk bush, which thrives by the suction of the rain water that falls near it. It differs as widely from the sandy desert as it does from the cultivated plain; neither does it resemble the steppes of Russia; but may justly be considered of a nature peculiar to itself. It has none of the characteristics of a marsh; it is not covered or saturated with water, but at certain periods; it has neither weeds nor grass in its bed, which, instead of being slimy, is hard, dry, and sandy, of such a consistency as never to become clayey, unless from a long continuance of water on an individual spot; nor is it otherwise fenny or swampy. It is a vast expanse of flat, hardened sand, encrusted with salt sometimes an inch deep (the water having been evaporated
134

by the sun), and at others, beautifully crystallised in large lumps. So much is the whole surrounding country imbued with this mineral, that all the wells dug on a level with the Runn become salt." (Burnes, i., 317, 318.) Fresh water is, in fact, obtained only on what may be called the peninsulas and islands of the Runn, tracts of land elevated above the rest of that region, covered with verdure, and moderately peopled by a pastoral race. The Runn has every appearance of having been an inland sea; and indeed the natives of Kutch have a tradition that it was such about three centuries ago, and that Nerona, Bitaro, and other places on its limits, were formerly seaports. This is apparently confirmed by ship nails, and stones shaped like those still used as anchors, being frequently met with; and in one instance the hull of a vessel of some size was found imbedded in the soil. (See Burnes; Macmurdo, in Bomb. Trans. ii., &c.) During the S.W. monsoon the sea overflows a large part of the Runn; and it is also sometimes partially inundated by the Loonee, Bunas, Suadruwatee, and other rivers which lose their waters in it.

The mirage is here continually presented in wonderful perfection; and the wild ass, the only inhabitant of this desolate region, appears often to the traveller at a distance as large as an elephant.

Kutch has undergone many political vicissitudes which have been singularly connected with natural phenomena. In 1762, the ruler of Sinde, unable to conquer this province, threw a bund or dam across the Phurraun, the E. arm of the Indus, and converted the N.W. portion of Kutch from a fruitful rice district into a sandy waste. In 1819, a violent earthquake shook every fortress throughout Kutch; destroyed Bhooj and Anjar; submerged Sindree; and upheaved the Ullah bund (mound of God) across the former course of the Phurraun, a tract of soft clay and shells, 50 m. long, perhaps 16 broad, and many feet in height. In 1826, the Indus burst through the Ullah bund, and, after an interval of 65 years, resumed its former channel, with a depth at Sindree of three fathoms; a circumstance which may perhaps restore to Kutch a portion of its former commercial importance. (See Burnes; Lyell's Geology, &c.)

The chief towns of this province are Bhooj, the capital, Mandavee the principal port, Luckput, Moondra, Anjar, and Kotara. The exports are chiefly cotton, ghee, and oil, which are transported in coasting vessels of from 25 to 300 tons. The natives excel in naval architecture, and are noted for their skill and daring as seamen and pilots. "Among the timid navigators of the east," says Burnes, "the mariners of Kutch is truly adventurous: he voyages to Arabia, the Red sea, and Zanguebar, bravely stretching out on the ocean after quitting his native shore. For a trifling reward he will put to sea in the rainy season, and his adventurous spirit is encouraged by the Hindoo merchants of Mandavee, an enterprising and speculating body of men." (i., 6, 7.) The government is analogous to that which prevailed in many countries of Europe, in the middle ages. The rao is the head of a kind of feudal aristocracy, each member of which is absolute within his own domains. The rao can summon them all to his standard, with their followers, but he must pay them; the number of chieftains is about 200, their annual revenue varying from 100 to 30,000 rupees each. The state revenue of Kutch does not exceed 16 lacks of rupees a year, of which rather more than a half belongs to the rao, and the rest to members of his family. The Jharejahs to which sect the rao and his chieftains belong, are of Sindian origin, and are a degraded, ignorant, and sensual race, who pass their lives in indolence and drunkenness. They uniformly marry Rajpoot women; and their pride is so great, that, lest their daughters should disgrace them by marrying into inferior ranks, they are said sometimes to destroy them in infancy. The abolition of female infanticide has formed the subject of an express stipulation between the British government and the rao; but there is reason to believe that it still prevails. The religion of the population is a mixture of the Hindoo and Mohammedan, and it is difficult to say which predominates. Our first subsidiary connexion with Kutch took place in 1819. The rao furnishes one battalion of infantry to our subsidiary force. The British resident is stationed at Bhooj. (Burnes; Lyell's Geology; Med. Trav.; Hamilton's E. I. Gas.; Asiat. Journ.)

KUTTENBERG (Boh. Kuttnahora), a town of Bohemia, and, after Prague, Reichenberg, and Eger, the most populous in the kingdom, circle Czaslau, 38 m. E.S.E. Prague. Pop. (1838), with its four suburbs, 8545; and it had double that number of inhabitants before the failure of the veins of silver in the mines near it. The latter, however, still furnish copper, lead, arsenic, and zinc; and mining industry is the principal dependance of the inhabitants. The town has several public edifices, the principal being the church of St. Barbara, a fine Gothic building. It has also a high school, a military school, an Ursuline convent, a hospital, and factories for printing cottons and spinning cotton yarn. A good deal of starch is made for exportation to Silesia.

The first silver groschens were struck here in 1296. (*Berghaus ; Oesterr. Nat. Encyc., &c.*)

KUZISTAN, (an. *Susiana*), a prov. of Persia, situated between lat. 29° and 33° N., and long. 47° and 51° 30′ E., being bounded N.W. by the pachalik of Bagdad, N. by Louristan, E. by Faristan, and S. by the Persian gulf. Length about 340 m., breadth 120 m. : supposed area, 9000 sq. m. The country is divided, according to Kinneir, between the territories of the Chab-Sheikh, and those forming the government of Shuster. The Chab territories extend from the Chab to the confluence of the Karoen (an. *Choaspes ?*) and Absal, and from the shore of the Caspian sea to the range of hills skirting the valley of Ram-Hormuz ; this part of the country consists principally of sandy plains and morasses, wholly destitute of vegetation. Eastward, also, intersected by the river Tub, on the banks of which are a few cultivated spots, is a desert, about 30 farsangs long, and varying in breadth from 10 to 16 farsangs. The most fertile spots in this part of Kuzistan are near Dorak, the capital of the Chab territories, and in the delta of the Euphrates : in the latter, dates and rice are produced in great abundance on well-irrigated lands, the rice harvest taking place in August and September. The grain harvest is in April and May ; but the produce is insufficient for the consumption of the district. The N. and W. parts of the country afford tolerable pasturage ; and here the wandering tribes, comprising the greater part of the population, pitch their tents. The chief towns of the Chab territory are Dorak (the capital, with a population of 8000, and a manufacture of Arabian cloaks, largely exported), Ahwaz, Eadian, and Mashoor. The territories attached to the government of Shuster comprise the fairest part of Kuzistan : four noble rivers, with their tributaries, irrigate the plain in every direction. Its riches in Strabo's time consisted of cotton, rice, sugar, and grain, yielding a hundred-fold ; but it is at present, owing to the rapacity of the government, little better than a forsaken waste, the only signs of cultivation, being near Benduteed and Hawezen. Indeed nothing can be more lamentable, than the misrule, robbery, and utter absence of industry, which characterizes this part of Persia. (*See* SHUSTER, and *Kinneir's Persia*, p. 85–87.)

L.

LABRADOR, an immense peninsula of British N. America, opposite the island of Newfoundland, from which it is separated by the strait of Belleisle, extending between the 50th and 64th parallels of N. lat., and between long. 56° and 79° W. ; being bounded S. by Canada and the gulf of St. Lawrence, E. by the Atlantic ocean, N. by Hudson's straits, and W. by Hudson's bay. Fixed population estimated at only 4000. It is generally described as one of the most dreary and naked regions of the globe, exhibiting scarcely anything except rocks destitute of vegetation. But, though this be in appearance when seen from off the coast, on penetrating a little into its interior, the surface is found to be thickly clothed with pines, birches, and poplars ; and with a profusion of delicate berries. It is everywhere most copiously irrigated by brooks, streams, ponds, and lakes. A chain of high mountains appears inland ; but their height is not correctly known. Mount Thoresby, near the coast, is 2720 ft. high. The well-known Labrador felspar is found chiefly in the vicinity of Nain. The prevailing rock is gneiss, overlaid by a bed of sandstone, alternately red and white, and strongly marked with iron near the surface : above this again are varieties of secondary limestone, arranged in parallel strata, and full of shells. A few miles from the shore, the secondary formations disappear, leaving gneiss and mica-slate on the surface. (*Geog. Journ.*, vol. iv., p. 208.) The climate is extremely severe, the thermometer occasionally falling below zero of Fahr. ; the summers are of short duration, with an average day temperature of 56°. The prevailing winds, on the E. coast, are from W.S.W. to N.W. ; there is less fog than on the neighbouring island of Newfoundland, and the straits of Belleisle are never frozen over. Corn will not ripen ; but potatoes, cabbages, spinach, and turnips answer pretty well. The wealth of the country, however, consists chiefly in the abundance of fish on its coasts. Whales, cod, salmon, and herrings, are extremely plentiful. The Labrador fishery is nearly confined to the S.E. tract, opposite Newfoundland ; within a few years it has increased six-fold, and it now rivals that of Newfoundland. During the fishing season, about 300 schooners come from the latter to the fishing stations of Labrador, and about half the produce is sent to St. John's, the remainder being exported to England, Lisbon, and the Mediterranean, by English and Jersey houses unconnected with Newfoundland. The American fishing vessels average about 400, principally sloops and schooners, manned by crews varying from nine to 13 hands, making a

total of about 6000 men. Each man catches, at an average, about 100 quintals of fish during the season ; and the oil is in the proportion of one ton to 300 quintals. They frequent chiefly the N. part of the coast, clean their fish on board, and leave Labrador early in September. About 10 ships from Quebec, and 120 from Nova Scotia and New Brunswick, carry away fish and furs to the value of about £60,000 a year ; the French, also, send a few vessels, but they are not successful fishermen. From 16,000 to 18,000 seals are taken in the spring and autumn, producing about 350 tons oil ; and the export of furs of wolves, bears, foxes, and beavers caught in the interior, was valued, in 1832, at £3130. The total value of the produce of Labrador, during the same year, amounted to £302,050, exclusive of the articles sent to London by the Moravians settled at Nain.

The native population of Labrador are Esquimaux ; and of all the tribes settled on the shores of America, these are the most filthy, disgusting, and miserable. They form an exception to all others in their appearance, stature, and manner of living ; and are at once hated and despised by the other Indian tribes. They are of small stature, and in their language, persons, and manners, bear a near resemblance to the Greenlanders. Their food consists chiefly of the flesh of seal, reindeer, and fish ; and their dress is entirely of skins. Their houses in winter resemble caverns sunk in the earth, and consist only of one apartment, which, though not very large, generally contains several brothers, or other relatives, with their wives and children. In summer, they dwell in tents of a circular form, constructed of poles, and covered with skins sewed together, which they are continually moving from place to place. They have always a great number of dogs about their camp ; which, besides serving to guard the habitation, and to draw the sledges, are occasionally used as food, and their skins made into clothing. The European residents are English, Irish, or Jersey servants, left in charge of the property in the fishing-season, and who also employ themselves in catching seals. Their principal settlements are at Bradore bay, l'Ance-le-blanc, and Forteau bay, the last being by far the most considerable. The Moravians formed their first settlement in 1752. Their habits, and quiet unobtrusive life, render them comparatively unknown. They trade with the Esquimaux, bartering coarse cloths, powder, shot, guns, and edged-tools, for furs, oils, &c. Their influence is alleged to have been very beneficial to the natives, not only in changing their religious belief, but in improving both their moral and physical condition. Murder and acts of violence, are much less frequent than formerly ; and mutual enmities have been removed. Their boats, houses, and fishing implements are better constructed, and many of them have begun to exercise foresight and economy. The Moravian settlements are at Nain, Okkak, Hopedale, and Hebron, all on the E. coast.

The coast of Labrador was first discovered by Sebastian Cabot, in 1498 ; but it was not visited till 1501, when Corte Real called it *Terra Labrador* (cultivable land), to distinguish it from Greenland, which he named *Terra verde*. The name is now applied not only to the E. coast, but to the whole peninsula, including that part on Hudson's bay called the E. Main. (*M'Gregor's British America*, vol. i., p. 153 ; *Geog. Journ.*, vol. iv., &c.)

LACCADIVE ISLANDS, (*Laksha-Dwipa*, " a lac of isles,") a group in the Indian ocean, lying chiefly between lat. 10° and 15° N., and long. 79° and 74° E., about 75 m. from the Malabar coast. There are 19 principal isles, but the largest is not more than 6 sq. m. in extent. Most of them are surrounded by rocks and coral reefs : the water near them, however, is deep, and they are separated by several wide channels, frequented by ships passing from India to Persia and Arabia. They are inhabited by a race of Mohammedans called Moplays. They do not yield grain, but produce an infinite quantity of cocoa-nuts, from the husks of which the inhabitants form *coir* cables, which are more elastic and durable than hemp, as the sea-water, instead of rotting, preserves them. These islands are well supplied with fish, and export the small shells called cowries, which pass as coin all over India. Jagery, a little betel nut, plantains, a few eggs and poultry, and coral for conversion into lime, are their remaining exports ; but they are of little importance, and the inhabitants are wretchedly poor. Vasco da Gama discovered these islands in 1499 ; they were dependent on Cannanore till ceded by Tippoo, in 1799 ; and came into our possession with the rest of that sovereign's dominions. (*Hamilton's E. I. Gaz.*)

LACKAWANNOCK, t., Mercer co. Pa., 6 m. S.W. Mercer. Drained by Little Neshaanock cr., which flows into Shenango cr. It contains three stores, one flouring-mill, four grist-mills, five saw-mills, two tanneries, one distillery ; 14 schools, 500 scholars. Pop. 2139.

LACKAWANNOCK, t., Pa., rises in Wayne and Susquehanna counties, and flowing S. and S.W. for about 30 m., it enters Susquehanna river at Pittstown. It affords extensive water-power, and has large quantities of anthracite coal on its borders.

LACKAWANNA, p. t., Luzerne co., Pa. It has one furnace, four schools, 140 scholars. Pop. 363.

LACKAWAXEN, r., Pa., rises chiefly in Wayne county, flows through a deep valley, receives Dyberry, Middle, and Wallenpaupack creeks, and enters Delaware river in Pike county. A dam across Delaware river at the mouth of the Lackawaxen supplies the Delaware and Hudson canal with water, and enables boats from the canal to cross the Delaware, and they thence follow Lackawaxen river 25 m. to Honesdale, at the junction of Dyberry creek, whence there is a railroad across Moosuck mountain, 16 miles, to Carbondale.

LACKAWAXEN, p. t., Pike co., Pa., 182 m. N.E. Harrisburg, 279 W. Bounded N.E. by Delaware r. Watered by Lackawaxen cr. It has three schools, 59 scholars. Pop. 780.

LACON, p. v., capital of Marshall co., Ill., 99 m. N. Springfield, 816 W. Situated on the E. side of Illinois river. It has three stores, and about 20 dwellings. It is surrounded by a fertile country. Nett proceeds of the post-office, $203.

LADAKH, an independent country of W. Thibet, between the 32d and 36th degs. of N. lat., and the 76th and 79th degs. of E. long.; bounded on the N. and N.E. by the Karakoram mountains, which divide it from the Chinese provinces of Yarkund and Khoten, E. by Chasthan, Rodokh, and Gurdokh; S. and S.E. by Himalaya, separating it from Cashmere, and the territories of Bissahar, Kulu, and Chamba; and W. by Baltes, or Little Thibet. Length, N. to S., rather more than 290 m.; average breadth, 150 m. Area, according to Moorcroft, about 30,000 sq. m., who also estimates the population at from 150,000 to 180,000, chiefly of the Thibetan race. The country is divided into four districts; Ladakh Proper in the centre, Nobra to the N., Zanskar S.W., and Pití S.E. It is an inhospitable land, its surface being, for the most part, a succession of lateral mountain ranges belonging to the Himalaya, the lowest range rising nearly to the limit or perpetual snow. Lé, the capital, is more than 11,500 ft. above the level of the sea, and some parts of the province Nobra are 3000 ft. higher. The passes that lead into Ladakh from the S., are above 16,000 ft. high, and many summits in the central part of the country are much more lofty. Narrow and deep valleys of great length, watered by considerable rivers, intervene between the mountain ranges, and comprise nearly all the cultivable land of the country. The chief of these valleys is that of the Upper Indus, here called the Singh-kha-bab. This extends S.E. and N.W. through the greatest part of the country. The Indus, while within Ladakh, receives the Zanskar, Zakat, and Dras rivers; the Shakut, its chief affluent N. of the Himalaya, flows principally through Ladakh, but does not join the main stream till it has passed into Little Thibet. Nearly all the rivers of Ladakh are tributary to the Indus; in the S., however, are several which join the Sutlege, of which the Pití is the chief. There are some considerable lakes. The country is subject to extremes of temperature. Frost, snow, and sleet commence early in September, and continue, with little intermission, till the beginning of May. From the middle of December to the beginning of February, Moorcroft found the thermometer out of doors at night seldom above 15° Fahr. But during the summer the sun shines with great power; and at Lé, in July, the thermometer has been found, at noon, to stand, in the sun, at 134° Fahr., and between Lé and Pití, at 160° higher. The atmosphere is in general dry and clear: what little rain occurs falls chiefly during the summer months. The mountains being of primitive formation, the soil consists almost entirely of disintegrated rocks washed into the bottoms by the action of thaws and torrents. The decomposed granite and felspar clothes these portions of the surface with a coating of clay, sand, gravel, and pebbles, which skill and industry can only render productive. Both climate and soil being thus hostile to vegetable life, the general aspect of the country, where not cultivated, is of extreme sterility: a few willows and poplars are the only timber trees, and the chief verdure consists of the Tartaric furze, with a few tufts of wormwood, hyssop, dog-rose, and other plants of the desert.

Notwithstanding these unpromising circumstances, the harvests of Ladakh are by no means niggard; and year after year equally abundant crops are raised from the same land; without its ever being suffered to lie fallow, and without any attempt at an alteration of produce. The mountain sides are formed into a succession of terraces, supported by stone breast-works, down which stone channels conduct a plentiful supply of water, and the detritus from the rock. The stone dykes are not only disposed to form terraces near the towns and villages, but in spots remote from human habitations, where they are constructed by the peasantry, and suffered to remain undisturbed for several years,

perhaps for some generations, till a quantity of earth is collected.

The field thus gained from the mountain has next to be supplied with manure. As wood is very scarce, the faeces of cattle are mostly used as fuel. But Moorcroft says that the floors of the houses are strewed with a coating of gravel, three or four inches thick, which is removed from time to time, and this, with the ashes of the burnt fuel, forms almost the only manure that sustains the nutritive properties of the soil. Wheat, barley, and buckwheat are the chief grains cultivated. The wheat is of three, the barley of two varieties: one of the latter, the skerekh, or naked barley, is a superior kind, especially for malting, but it degenerates in a lower level, as in the adjacent plains of Hindostan. Wheat and barley are usually sown in May, and reaped in September, the great heat of the sun in summer fully compensating for the shortness of that season. At Pituk, near Lé, more than 10,000 feet above the sea, barley is said by Moorcroft to be ready for the sickle in two months from the time of sowing. The plough is entirely of wood, generally willow, except the point, which is formed of a small piece of iron. The furrow is not more than four or five inches deep; but the earth is well broken, and the seed is afterward carefully covered over. Ploughing is performed by a pair of shos (a hybrid male between the yak, bos grunniens, and common cow), or nubus, driven without reins, but, with the utmost precision, by the voice, or by a wand. The ground is ploughed twice; the grain is sown broad cast in the furrow, or planted by the dibble. Corn is frequently reaped while green, and laid on the ground in flat bundles to ripen more completely. In very dry soils the grain is pulled up by the roots, the straw being valuable for fodder; in early soils it is cut close to the ground by a curved, short-bladed sickle, which is, perhaps, quite as well adapted for the purpose as that of Europe. There is no great variety of kitchen vegetables; but, onions, carrots, turnips, and cabbages are raised in some places, and carraway, mustard, and tobacco are grown in a few gardens. Plenty of apricots and apples are raised everywhere; but few other kinds of fruits. Pears, grapes, and melons are imported from the neighbouring countries. Lucerne grows with great luxuriance in some parts, and a species of saintfoin is met with in the mountains; but the most valuable source of fodder is, perhaps, the prangos (Prang. patularia, Lindley). This plant, which is a perennial delighting in a poor, sterile soil, but growing in every variety of site, except actual swamp, is common in the W. of Ladakh, and varies in size, from a cluster of leaves and flowers, to from 12 to 18 feet in circumference. In August or September the plants are cut to within two or three inches from the ground, on which they are laid in bundles to dry, and afterward piled like other kinds of fodder, on the house-tops. The prangos require no shelter. In the winter, about one cwt. for 24 hours is considered sufficient for 90 sheep, or 30 lambs. Healthy sheep fed upon it are said to become fat in 90 days: it is also excellent food for cattle and horses, though perhaps less so than for sheep. Rhubarb is an abundant indigenous product; and Moorcroft supposes the facilities for obtaining a supply of it to be so great, that our E. Indian dealers in the article might easily undersell the merchants of Russia or Turkey in the European market.

The yalk-male or zho is principally used for the transport of burdens; horses are few and small, though active and hardy. The native breeds of sheep, though mostly larger than those of India, are much smaller than the sheep of Chan-than. One species, the Purik sheep, is very diminutive; but it gives two lambs in twelve months, about three pounds of wool a year, at two shearings, and its mutton is excellent. Being domesticated like the dog, it is maintained at a very small cost. The shawl-wool goat is the common breed in this and the neighbouring countries; the fleece is finer in Ladakh than elsewhere. The latter is cut once a year; the wool, picked out, is sent to Cashmere, and the hair made into ropes, coarse sacking, and blankets, for home consumption. The wild animals are not numerous: the ibex, wild sheep, wild asses, and a kind of wild horse, are the principal. The leopard, jaguar, ounce, bear, and lynx are rare. Fish are very plentiful, but the prevailing religion prevents their being used as food.

Sulphur is found in some places, and soda in great plenty on the Indus, and in the N. lead, iron, and copper are said to exist; and gold in the sands of the Shayuk, but the government, from politic or superstitious motives, has prohibited the search for this metal.

The native trade of Ladakh is of no great amount; but its transit trade is important, from the country being the great thoroughfare for the commercial intercourse between Thibet, Cashmere, Turkestan, China, and even Russia, on one hand, and Cashmere, the Punjab, and the plains of Hindostan on the other. Ladakh is the entrepôt for the goats' wool, of which the Cashmere shawls are made, and which is partly supplied from this country, but chiefly from Rodokh and

Chan-than. About 800 camel loads are annually exported to Cashmere, to which country, by ancient custom and consequence, the export is exclusively confined, all attempts to convey it elsewhere being punished by confiscation. In like manner, it is considered illegal in Rodokh and Chan-than to allow a trade in shawl wool, except through Ladakh; and, in the latter, impediments are opposed to any import from Yarkund, though the wool of that province be of superior quality and cheaper. The fleece of the wild goat is exported in smaller quantities to Cashmere, and wrought into shawls, soft cloth, and linings for shawl-wool stockings; this material is softer and warmer than the ordinary shawl wool, but is much less used for shawls. Sheeps' wool is wrought into cloths exported to Kotook, Kulu, &c.; and many Chan-than sheep are exported to the mountain states, where they are extensively used as beasts of burden, carrying from 25 lbs. to 30 lbs. weight. Tea comes from China through Lassa and Yarkund, and is exported in considerable quantities to Cashmere and the Punjab: inferior kinds of the same shrub are imported from the British territories of Bhusahar, and used by the lower classes in Ladakh. According to Moorcroft, 100 maunds of black tea, from Bhusahar, are imported annually into Lé, where it sells at 15 Mohammed Shahi rupees a maund. Borax and salt from Thibet; silks, silver ingots, and various manufactured articles from China; felts, camlets, dried sheep skins, steel, boots, Russia leather, brocades, velvets, and broadcloths, horses, and drugs from Yarkund; cooking vessels, water-pots, and about 300 maunds of dried apricots, &c., yearly from Baltee; shawls, chintzes, copper-tinned vessels, and other domestic utensils, and grain, from Cashmere and the Punjab; ghee, honey, raisins, and grain from Hindostan; and iron and hardware manufactures, wooden tea-cups in large numbers, &c., from Bhusahar, are the remaining principal imports into Ladakh. The imports from Yarkund, of Russian goods, &c., are mostly intended for the Punjab. The dried fruits from Baltee are exchanged for foreign wool, and the goods from Cashmere and the Punjab are partly re-exported into the Thibet provinces.

The government, as regards the people, is despotic; but the rajah has very little real power, being controlled by the lamas, or priesthood, by whom he is occasionally deposed. The business of the state is carried on by the khalun, or prime minister, the deputy khalun, the lom-pa, or chief military officer, the treasurer, who is a lama, and the master of the horse. The towns and districts are governed by inferior khaluns; and the magistracy is discharged by officers called nar-pon, and by the head men of villages. Most of these are paid by assignments of land, and by claims on the people for contributions, or articles of daily use. The rajah, khaluns, and lom-pa divide among them the produce of the imports on merchandise in transit, and carry on a trade in shawl wool and tea, from which most of their income is derived.

The revenue of the state is roughly estimated at about five lacs of rupees a year.

In spiritual affairs Ladakh is subordinate to the authority of the supreme pontiff of the Buddhists, the grand lama of Thibet, who appoints the chief lamas of this country. The lamas are very numerous, every family in which there is more than one son furnishing one, who is a family priest, attached to a monastic institution or college, though living ordinarily among the people, and conducting the rites of their daily worship. All profess poverty and celibacy, though a man who has been married is admissible into their order. The lamas do not confine themselves to strictly religious duties, but cultivate the land, rear sheep and goats, and take an active share in the fiscal and political administration. There are many conventual establishments for females.

Mohammedanism has of late made great progress in the S. and W., but the mass of the population are still Buddhists. "Their religious belief and practice seems to be a strange mixture of metaphysics, mysticism, morality, fortunetelling, juggling, and idolatry. The doctrine of the metempsychosis is curiously blended with tenets and precepts very similar to those of Christianity, and with the worship of grotesque divinities. The lamas recognise a sort of trinity, or a triad consisting of a paramouns deity, a prophet, and a book; and the people are exhorted to truth, charity, resignation, mutual forbearance, and good will. The religious service performed daily at the temples attached to monasteries consist chiefly of prayers and chanting, in which the mystic sentence, ' Oom mane pace me sen,' is frequently repeated, and the whole is accompanied with the music of wind instruments, chiefly harmonizing with tabrets and drums." (Moorcroft's Trav., i., 340, 341, 344.)

The military force consists of a peasant militia, very ill equipped and insufficient; and there is little to prevent Ladakh falling permanently under the dominion of some one of its more powerful neighbours.

There is little wealth in the country, but what exists is equally diffused, and the great body of the people are in easy circumstances. They pay no money taxes to the state; but are bound, to suit and service, both domestic and military, and furnish contributions in kind for the support of the rajah and the governors of districts. The people are in general mild and timid, frank, honest, and moral, when not corrupted by communication with the dissolute Cashmerians; but they are indolent, exceedingly dirty, and addicted to intoxication. Their food is nourishing, and consists chiefly of rice, meal porridge, bread, vegetables, tea, wheaten cakes, and once a day the flesh of sheep, goats, or yaks. The wealthy drink grape juice and water or sherbet, the poorer classes a kind of beer, called chang, made of fermented barley. All orders and both sexes dress chiefly in woollens; to which the men add mantles of flowered chintz, and broad ends or velvet caps, and the women cloaks of cotton, China satin, or Benares brocade lined with sheep skin, the wool inward, and numerous ornaments. Both sexes wear leather boots, in which they take great pride. Some curious domestic customs prevail: among others, polyandry is common, the younger sons of a family being subordinate husbands to the wife of the elder brother; and when the latter dies, his property, authority, and widow devolve upon the next brother.

History.—Ladakh originally formed one of the provinces of the kingdom of Thibet; but when the Chinese conquered that country, they did not extend their sway to Ladakh, which seems to have retained its own princes. About 170 years ago the Kalmuck Tartars invaded Ladakh, and the rajah fled to the governor of Cashmere, who, with the permission of Aurungzebe, reconquered the country for the rajah. From that time a small annual present was made to the emperor of Delhi through the governor of Cashmere. Ranjeet Singh took possession of Ladakh, and exacted a tribute; but, since his death, there is reason to believe that the country has recovered its former independence. A small annual tribute or present is, however, said to be sent to the authorities of Gardokh, on behalf of the government of Lassa. (Moorcroft's and Trevett's Travels, i., 282-360; Lloyd's and Gerard's Tour in the Himalaya; Trans. of the Asiat. Soc., i., 49-58; Asiatic Journal, vol. xviii., &c.)

LADAKH, or Lé, the cap. of the above country. (See Lé.)

LADOGA (LAKE), a lake of Russia in Europe, surrounded by the governments of Petersburg, Olonetz, and Wyborg in Finland, and extending from lat. 59° 57' to 61° 46', and from long. 29° 50' to 33° 30' E. Though the largest collection of fresh water in Europe, there is but little accessible information respecting it. Its length, N.W. to S.E., is about 125 m.; greatest breadth about 70 m. Area estimated at from 6200 to 6300 sq. m. Its depth is very unequal. It receives about 60 rivers, the chief of which are the Voxa, connecting it with the Saima lake in Finland; the Svir, by which the surplus waters of the lake Onega are poured into it; the Volkhov, by which it communicates with lake Ilmen; and the Siass, like the latter, from the S. It discharges its surplus waters by the Neva into the gulf of Finland. Its shores are generally low; on its N.W. and S. banks are situated Serdobol, Kronsberg, Kexholm, Schlusselburg, and New Ladoga. It has several islands, chiefly toward its N. extremity; and is so full of rocks and quicksands, and subject to storms, that, to avoid it, Peter the Great began, in 1718, the Ladoga canal, from New Ladoga, on the Volkhov, to Schlusselburg, on the Neva, along the S. shore of the lake, a distance of about 70 m. This work was finished under the empress Anne, in 1732: it is 74 feet broad, and, according to the season, from 4 to 8 or 9 feet deep, and has 20 large, besides many smaller sluices. It is annually navigated by an immense number of boats, chiefly with merchandise for Petersburg. The canals of Siass and Svir form, with that of Ladoga, a continuous chain of communication round the S. and S.E. shores of the lake; and the canal of Tikhvine (Novgorod) places it in direct connexion with the Wolga. (Schnitzler, La Russie; Possart, &c., passim; Dist. Geog.)

LADRONES, or MARIANNE ISLANDS, a group of islands in the N. Pacific ocean, belonging to Spain, between the 13th and 21st deg. of N. lat., and the 144th and 146th of E. long. There are about twenty of them; but five only are inhabited, and those lie near the S. extremity of the cluster. They are so close together, and are also so broken, as well in irregular in their form and position, as to appear like fragments, disjoined from each other, at remote periods, by some sudden convulsion of nature. Those fragments have now a very barren and unpromising aspect. In particular spots, indeed, there are scattered patches of verdure; but, in general, little better than naked rocks appear; and scarcely a tree or shrub is visible among them. The coast of the islands consists mostly of black or dark brown rocks, honeycombed in many parts by the action of the waves. Their geological constitution is almost wholly volcanic, and some volcanoes have been in action in modern times. The climate is generally serene and temperate, the tropical heats being much diminished by the regular sea-breezes. During the months of July and August, however, the weather is intolerably hot; and at the season of the W. monsoons, between June and Oct., the most tremendous hurricanes are experienced at the full and change of the moon. The surface of the interior is much broken, and rises into high hills, and even mountains; but the soil in the valleys is of great fertility, and, if properly cultivated, would produce abundantly most of the intertropical plants. Anson visited the Ladrones in 1742, and describes Tinian as abounding with everything necessary to human subsistence and comfort; and being withal of a most pleasant and delightful appearance, diversified by a happy intermixture of valleys and gently rising hills; the woods consisting of tall and well-spread trees, with fine lawns interspersed. The same island being, however, visited by subsequent navigators, among others by Byron, was found to have become an uninhabitable wilderness, overgrown with impenetrable thickets. The reason of this change was, that the Spaniards, by whom these islands had been conquered, had, for what reason it seems difficult to conjecture, removed the inhabitants from Tinian to another island, and after their departure it soon degenerated into a state of nature, and, when last visited, was nothing better than a wild and savage wilderness. This statement, however, does not apply to the whole group: for Kotzebue informs us that cotton, indigo, rice, Indian corn, sugar, and the plantain thrive in other islands, and produce abundant supplies for the population. Cattle, horses, mules, and asses are numerous, and the lama has been introduced with success from Peru. Wild hogs also are found in great numbers, many of them of a large size, weighing 200 lbs., particularly on the island of Saypan. They are very fierce, and, when hunted by dogs, make a formidable resistance. The fish that are found on the coast are very unwholesome, and produced pernicious effects on the crews of the ships both of Anson and Byron. The tripang, or holothuria, is caught by the natives, and sold to the Chinese. The country is infested with musquitoes, and with endless varieties of loathsome insects. The natives are tall, robust, and active; the men wear scarcely any covering, and the women only a petticoat of mat. Both sexes stain their teeth black, and many tattoo their bodies. Their huts are formed of wood from the palm tree, and divided by mats into several apartments devoted to distinct uses. They are good swimmers, and extremely clever in managing their canoes, in which, with a good wind, they will sail at the rate of 20 m. an hour. Their number, in the middle of the 17th century, is supposed to have amounted to 150,000; though this is probably far beyond the mark; but the race has been so much thinned by the cruelties practised on them by the Spaniards, that the present Indian population scarcely exceeds 4000. Guajan, the largest island, contained, in 1816, only one Indian family, its inhabitants (5390) consisting of settlers from Mexico and the Philippine islands. The capital is San Ygnacio de Agana, which, in 1816, had 3190 inhabitants, and was the seat of the Spanish governor. The number of Spaniards is very small.

The Ladrone islands were originally discovered by Magellan, who called them Las Islas de las Ladrones, or The islands of Thieves; because the Indians stole everything made of iron within their reach. At the latter end of the 17th century they obtained the name of the Marianas, or Marianne islands, from the queen of Spain, Mary Anne of Austria, mother of Charles II., at whose expense missionaries were sent thither to propagate the Christian faith.

LA FAYETTE, parish, La. Situated in the S. part of the state, and contains 1800 sq. m. Bounded S. by the gulf of Mexico, W. by lake Mermentau and river. Watered by Vermilion river. Much of the surface is too wet for cultivation. It contained, in 1840, 30,088 neat cattle, 9638 sheep, 5730 swine; and produced 83,812 bushels of Indian corn, 2,913,534 pounds of cotton, 1,016,534 of sugar. It had two tanneries; one academy, five students; six schools, 97 scholars. Pop.: whites, 4474; slaves, 3233; free coloured, 134; total, 7841. Capital, Vermilionville.

LA FAYETTE, county, Miss. Situated in the N. part of the state, and contains 790 sq. m. Watered by Tallahatchee river and its branches. It contained, in 1840, 9609 neat cattle, 3570 sheep, 20,136 swine; and produced 9495 bushels of wheat, 270,848 of Indian corn, 19,705 of oats, 27,190 of potatoes, 7517 pounds of tobacco, 1,399,008 of cotton. It had 19 stores, 19 grist-mills, 13 saw-mills, five tanneries; one college, 33 students; two academies, 84 students, seven schools, 158 scholars. Pop.: whites, 3676; slaves, 2849; free coloured, 13; total, 6531. Capital, Oxford.

LA FAYETTE, county, Mo. Situated towards the W. part of the state, and contains 450 sq. m. Bounded N. by Missouri river, by branches of which, and of La Mine river, it is drained, and which afford water-power. It contained, in 1840, 1900 neat cattle, 7280 sheep, 37,582 swine; and produced 9316 bushels of wheat, 339,833 of Indian corn, 56,896 of oats, 16,801 of potatoes, 7492 pounds of tobacco, 617 of

cotton. It had 13 stores, four grist-mills, seven saw-mills; eight schools, 254 scholars. Pop.: whites, 4780; slaves, 1990; free coloured, 95; total, 6815. Capital, Lexington.

LA FAYETTE, county, Ark. Situated in the S.W. part of the state, and contains 1260 sq. m. Watered by Red river and its branches. It contained, in 1840, 2331 neat cattle, 215 sheep, 6695 swine; and produced 136,560 bushels of Indian corn, 13,900 of potatoes, 1,312,590 pounds of cotton. Pop.: whites, 555; slaves, 1644; free coloured, one; total, 2200. Capital, Lewisville.

LA FAYETTE, p. t., Onondaga co., N. Y., 130 m. W. by N. Albany, 300 W. Drained by Onondaga cr. It contains two churches, a Presbyterian and Methodist, four stores, three flouring-mills, three grist-mills, eight saw-mills, three tanneries; 17 schools, 881 scholars. Pop. 2600.

LA FAYETTE, t., Coshocton co., O. It has one school, 69 scholars. Pop. 848.

LA FAYETTE, t., Medina co., O. It has one flouring-mill, three grist-mills, five saw-mills, two tanneries, one distillery; four schools, 197 scholars. Pop. 937.

LA FAYETTE, p. v., capital of Walker co., Ga., 190 m. N.W. of Milledgeville, 634 W. Situated near the head waters of Chattooga river, and contains a courthouse, jail, an academy, two churches, a Methodist and Baptist, seven stores, about 100 dwellings, and 600 inhabitants. Nett proceeds of the postoffice, $1969.

LA FAYETTE, p. v., capital of Tippecanoe co., Ia., 70 m. N.W. Indianapolis, 639 W. Situated on the E. side of Wabash r., at the head of steamboat navigation, 10 m. below the mouth of Tippecanoe r. It contains a courthouse, jail, market-house, a bank, an academy, seven churches, two Presbyterian (one Old and one New school), a Baptist, Methodist, Universalist, and an African, 21 stores, two flouring mills, two saw-mills, one paper-mill, one carding and fulling mill, 350 dwellings, and about 2000 inhabitants. The Wabash and Erie canal is in operation, and connects this place with lake Erie.

LA FAYETTE, city, capital of Jefferson par., La., 2 m. W. by S. New-Orleans, 1174 W. Situated on the N. side of Mississippi r. Though capital of a different parish, it is virtually a continuation of New-Orleans, which joins it on the N.E. It contains a courthouse, jail, and county officers, a market-house, and three churches. A steam-ferry here crosses the Mississippi, and the New-Orleans and Carrolton railroad passes in Naides-street through the city. It has one academy, 35 students, two schools, 153 scholars. Pop. 3807.

LA FAYETTE, C. H., p. v., capital of La Fayette co., Ark., 162 m. S.W. Little Rock, 1237 W.

LA FOURCHE, river, La., an outlet of the Mississippi, which it leaves above Donaldsonville, and flows S.E., 98 m., to the gulf of Mexico.

LA FOURCHE INTERIOR, parish, La. Situated in the S.E. part of the state, and contains 1100 sq. m. The surface is level, and much of it too wet for cultivation. Watered by Bayou La Fourche. It contained, in 1840, 4461 neat cattle, 1253 sheep, 9812 swine; and produced 188,747 bushels of Indian corn, 30,383 of potatoes, 1,509,000 pounds of cotton, 9,945,000 of sugar. It had three saw-mills, four schools, 56 scholars. Pop.: whites, 3098; slaves, 3046; free coloured, 71; total, 7303. Capital, Thibodeauxville.

LAGO-NEGRO, or LAGONERO, a town of the kingdom of Naples, prov. Basilicata, on the high road from Naples to Calabria, 12 m. N.E. Policastro. Population about 5800. (Rampoldi.) It is well built, and has an old castle, a hospital, and several other charitable institutions, with manufactures of woollen cloth and caps, and a large weekly market.

LAGOS, a fortified seaport town of Portugal, prov. Algarve, cap. of a comarca of its own name, 18 m. E. by N. cape St. Vincent, and 114 m. S. by E. Lisbon; lat. 37° 6' N., long. 8° 40' W. Population, 6793. It is built on the shore of a large bay sheltered N. and W. by hills covered with vines and fruit trees. The streets are narrow, and the houses generally small; but there are several handsome and regularly-built public edifices, among which are two parish churches, a military asylum, town hospital, and three convents, two of which are in the suburbs. The neighbourhood abounds in wine, figs, and other fruits, with pulse of different kinds; but, as in the rest of Algarve, there is a great scarcity of corn, which is imported from Alemtejo and the ports of Spain. The fishery of tunnies, anchovies, &c., is very considerable, and the produce, after being salted, is sent by sea to other parts of the kingdom. (Miñano.)

LA GRANGE, county, Ia. Situated in the N.E. part of the state, and contains 360 sq. m. Organized in 1832. Watered by Fawn, Pigeon, and Little Elkhart rivers. It contained in 1840, 4406 neat cattle, 3057 sheep, 7944 swine; and produced 58,163 bushels of wheat, 97,953 of Indian corn, 1619 of barley, 79,107 of oats, 27,278 of potatoes, 26,084 pounds of sugar. It had 12 stores, four grist-mills, 15 saw-mills, two tanneries, six distilleries; one college,

69 students, two academies, 130 students, 21 schools, 494 scholars. Pop. 3084. Capital, Lima.

LA GRANGE, p. t., Penobscot co., Me., 97 m. N.N.E. Augusta, 695 W. Watered by small tributaries of Penobscot river. It has one grist-mill; four schools, 149 scholars. Pop. 236.

LA GRANGE, t., Duchess co., N. Y. Bounded W. by Wappinger's creek. Drained by Sprout's creek, which flow into Hudson river. It has five stores, one woollen factory, two grist-mills, two saw-mills, one tannery; five schools, 125 scholars. Pop. 1851.

LA GRANGE, p. t., Loraine co., O., 8 m. S. Elyria, 111 m. N.N.E. Columbus, 370 W. Watered by the E. and W. branches of Black river. It has one store, one grist-mill, two saw-mills; eight schools, 249 scholars. Pop. 1001.

LA GRANGE, p. v., capital of Oldham co., Ky., 35 m. N.W. by W. Frankfort, 577 W. Situated about 6 m. from Ohio river. It contains a courthouse, jail, an academy, several stores, and 933 inhabitants.

LA GRANGE, p. v., capital of Troup co., Ga., 121 m. W. Milledgeville, 729 W Situated 5 m. E. of Chattahoochee river. It contains a courthouse, jail, three churches, a Baptist, Methodist, and Presbyterian, an academy, and about 1000 inhabitants.

LA GRANGE, p. v., Franklin co., Ala., 136 m. N. by Tuscaloosa, 776 W. Situated on elevated ground 16 m. S. of Tennessee river. It contains La Grange college, under the direction of the Methodists, founded in 1831, which has a president and five professors or other instructers, 50 alumni, 160 students, and 1980 vols. in its libraries.

LA GRANGE, p. t., Cass co., Mich., 171 m. W. by S. Detroit, 619 W. Watered by a branch of Dowagiake river and Putnam's creek. It contains two stores, one grist-mill, two saw-mills, one tannery, one distillery. Pop. 760.

LAGUNA. See TENERIFFE.

LAHORE, an independent kingdom of Hindostan. (See PUNJAB.)

LAHORE, a city of the Punjab, Hindostan, and in Runjeet Singh's time, the cap. of his dominions, on the Ravee (Hydraotes). 220 m. N.N.E. Delhi; lat. 31° 33′ N., long. 74° 98′ E. Lahore is surrounded by a brick wall about 30 ft. high, which extends for about 7 m., and is continuous with the fort. The latter, in which the rajah resides, is surrounded by a wall of no great strength, with loop-holes for musketry; a branch of the Ravee washes the foot of its N. face, but it has no moat on either of the remaining sides. The palace within this enclosure is of many stories, and entirely faced with a kind of porcelain enamel, on which processions and combats of men and animals are depicted. Many of these are so perfect as when first placed in the wall. Several of the old buildings are in ruins; others are entire, and throw into shade the meaner structures of more recent date. Runjeet Singh cleared away some of the rubbish, and repaired or refitted some of the ruined buildings of Jehangire and Shah Jehan; but his alterations were not always made with good feeling or taste. The great square and buildings of the principal mosques were converted into a place of exercise for his Sipahi infantry, and he stripped the dome of the mausoleum of Anof Jah of its white marbles to apply them to the erection of some insignificant apartments in the garden-court of the mosque. The divan-am, or general hall of audience, is a large apartment supported by many pillars. The divan-khas, or private audience-hall, is a suite of small chambers, offering nothing remarkable." (Moorcroft, i. 104, 105.)

Lahore is said to have been formerly 12 cos (about 19 m.) in circ. Burnes says that the ancient cap. extended E. to W. for 5 m., and had an average breadth of 3 m., as may be learned by the ruins. Whatever, indeed, may have been its actual extent, it is clear, from the remains of buildings beyond the walls, that it was once much more extensive than at present. The modern city occupies the W. angle of the ancient cap., and the portion of it within the walls is apparently very populous. Moorcroft, who visited Lahore in 1819, says: "The streets were crowded to an extent beyond any thing I ever witnessed in an Indian city. The houses were in general of brick, and five stories high, but many were in a very crazy condition. The chief bazaar follows the direction of the city wall, and is not far distant from it. The street is narrow, and this inconvenience is aggravated by platforms in front of the shops, on which the goods are displayed under projecting pent-houses of straw to protect them from the sun and rain. Through the centre of the remaining contracted space runs a deep and dirty drain, the smell from which was very offensive. The population consists of Mohammedans, Hindoos, and Sikhs, the former in the greatest number." (i. 105, 106.) Moorcroft states that he saw only one mosque of any size or magnificence; but Burnes, a later traveller, says there are two or three : the principal, or king's mosque, a large building of red sandstone brought

by Aurungzebe from near Delhi, had, however, been desecrated into a powder magazine.

Across the Ravee, about 4 m. N. Lahore, is the "Shah Dura," or mausoleum of the emperor Jehangire, a monument of great beauty. "It is a quadrangular building, with a minaret at each corner rising to the height of 70 ft. It is built chiefly of marble and red stone, which are alternately interlaid in all parts of the building. The sepulchre is of most chaste workmanship, with its inscriptions and ornaments arranged in beautiful mosaic; the shading of some roses and other flowers is even preserved by the different colours of the stone. Two lines of black letters, on a ground of white marble, announce the name and title of the "Conqueror of the World," Jehangire; and about a hundred different words in Arabic and Persian, with the single signification of God, are distributed on different parts of the sepulchre. The floor of the building is also mosaic. It is probable that this beautiful monument will soon be washed into the Ravee, which is capricious in its course near Lahore, and has lately overwhelmed a portion of the garden wall that environs the tomb." (Burnes' Bokhara, &c., i. 137.) The Shalimar, or garden of Shah Johan, is another magnificent remnant of Mogul grandeur. It is about ½ m. in length, and has three terraces, each rising above the other. A canal, brought from a great distance, intersects it, and throws up numerous fountains to cool the atmosphere. Runjeet Singh removed some of its marble houses, and replaced them by others of stone.

The bazaars of Lahore do not exhibit much appearance of wealth : the commerce of the Punjab is centred at Umritzir. Lahore was captured by Sultan Baber in 1590, and was for some time the seat of the Mogul government in India. It was for awhile in the possession of the Afghans, and was repeatedly sacked by Shah Zemaun, ex-king of Caubul. (Hamilton's E. I. Gaz.; Moorcroft; Burnes, &c.)

LAKE, county, O. Situated in the N.E. part of the state, and contains 220 sq. m. Bounded N. by Lake Erie. Watered by Grand and Chagrin rivers. It contained in 1840, 11,566 neat cattle, 29,394 sheep, 10,293 swine; and produced 82,570 bushels of wheat, 2979 of rye, 121,136 of Indian corn, 4717 of buckwheat, 7236 of barley, 62,255 of oats, 81,462 of potatoes, 64,931 pounds of sugar. It had 72 stores, five fulling-mills, 11 grist-mills, 41 saw-mills, 10 tanneries, two distilleries, two printing-offices, one bindery, three weekly newspapers; one college, 49 students; two academies, 285 students; 105 schools, 3815 scholars. Pop. 13,719. Capital, Painesville

LAKE, county, Ia. Situated in the N.W. corner of the state, and contains 405 sq. m. Bounded by Lake Michigan, S. by Kankakee river. Watered by Calumic and Deep rivers, and Cedar and Eagle creeks. It contained in 1840, 9085 neat cattle, 453 sheep, 4631 swine; and produced 15,636 bushels of wheat, 97,675 of Indian corn, 29,176 of oats, 16,583 of potatoes, 3090 pounds of tobacco. It had two stores, one grist-mill, four saw-mills, one printing-office; seven schools, 116 scholars. Pop. 1468. Capital, Crown Point.

LAKE, county, Ill. Situated in the N. part of the state, and contains 425 sq. m. Bounded E. by lake Michigan. Watered by Des Plaines river. It contained in 1840, 5349 neat cattle, 169 sheep, 7714 swine; and produced 92,566 bushels of wheat, 34,069 of Indian corn, 3755 of buckwheat, 2994 of barley, 46,406 of oats, 71,532 of potatoes. It had three stores, one grist-mill, six saw-mills, one tannery; 16 schools, 361 scholars. Pop. 2634. Capital, Little Fort.

LAKE, p. t., Starke co., O., 124 m. N.E. Columbus, 385 W. It has three schools, 87 scholars. Pop. 2139.

LAKE, t., Logan co., O. It contains the v. of Bellefonte, the capital of the co., and has seven schools, 283 scholars. Pop. 1175.

LAKE, t., Wayne co., O. Watered by the E. fork of Mohiccan creek. Pop. 1144.

LAKE LANDING, p. v., capital of Hyde co., N.C., 215 m. E. Raleigh, 400 W. Situated on the S. side of Matimuskeet lake. It contains a courthouse, and several stores and dwellings.

LAKE PLEASANT, p. t., capital of Hamilton co., N. Y., 81 m. N.W. by N. Albany, 450 W. It is 30 m. long, and mostly a wilderness, with mountains, valleys and lakes. Drained by Sacandaga and Indian rivers. The v. is on the W. side of Pleasant lake, and contains a court-house, county clerk's office, a church, a store, and 10 or 12 dwellings. The t. has one saw-mill, two schools, 107 scholars. Pop. 296.

LAKE PROVIDENCE, p. v., capital of Carroll par., La., 366 m. N.N.W. New Orleans, 1154 W. Situated on the W. bank of Mississippi river, between it and lake Providence. The v. contains a courthouse, jail, a hospital, and several stores and dwellings. The lake consists of an ancient bed of Mississippi river, which has taken a different course. The N. and S. entrance to the river has become completely closed, and it has its outlet into Tenáza river.

LALAND or LAALAND, an island of the Danish archipelago, in the Baltic, between lat. 54° 38' and 54° 58' N., and long. 11° 53' E.; forming, with Falster, from which it is separated by the narrow but now navigable channel of Guldborg, a prov. of the kingdom. Length, E. to W., 35 m.; average breadth about 13 m. Area, 450 sq. m. Pop. 47,000. It is low, and is in parts liable to inundations; its shores are much indented by the sea, and it has some considerable bays. In its centre is the town of Mariebôe, 5 m. in length by 2 in breadth. The climate is said to be unhealthy; but the soil is very fertile, and it is looked upon as the most productive of the Danish islands. Principal crops, wheat, rye, barley, and oats. Hemp and hops are also produced, and great quantities of apples. Oak, and other kinds of timber abound. Mineral products and manufactures few and insignificant. Laland has an active trade in agricultural produce; the chief seat of which is Nakshow, the capital, a town of 2800 inhabitants on the W. coast.

LALITA-PATAN, a considerable town of Nepaul, N. Hindostan, about 1½ m. S. Catmandoo, stated to have had, in 1803, a population of 24,000. It is said to be a handsomer town than Catmandoo, and to possess some fine public edifices.

LAMBALLE, a town of France, dep. Côtes-du-Nord cap. cant., on the declivity of a hill, beneath which runs the high road from Paris to Brest. 12 m. E.S.E. St. Brieuc. Pop. (1836), 4396. It is well built, has an industrious and thriving population, is surrounded by old walls, and has two suburbs, a communal college, public library, with manufactures of woollens, linens, parchment, leather, &c.; and a considerable trade in agricultural produce. (*Dict. Géog. &c.*)

LAMAR, p. t., Clinton co., Pa., 16 m. N.E. Bellefonte. Drained by Big Fishing and Cedar creeks. It contains four stores, one furnace, one forge, three flouring-mills, 10 saw-mills, two tanneries; one school, 25 scholars. Pop. 1863.

LAMOILLE, r., Vt., one of the four largest rivers on the W. side of the Green mountains, rises in Greensboro', Orleans co., and flowing S.W. for some distance, it turns to the W.N.W., and enters lake Champlain, in the N. part of Colchester. Its falls afford extensive water-power.

LAMOILLE, county, Vt., Situated toward the N.W. part of the state, and contains 490 sq. m. Watered by Lamoille river, and its branches. The western range of the Green mountains, containing Mansfield mountain, its highest peak, 4279 feet above tide water, is in its S. part. It contained in 1840, 16,555 neat cattle, 40,991 sheep, 7987 swine; and produced 21,070 bushels of wheat, 1604 of rye, 98,463 of Indian corn, 70,797 of oats, 472,583 of potatoes, 295,476 pounds of sugar. It had 26 stores, 11 fulling-mills, two woollen factories, 12 grist-mills, 54 saw-mills, one oil-mill, nine tanneries, one printing-office. one weekly newspaper; one academy, 100 students; 105 schools, 3816 scholars. Pop. 10,475. Capital, Hydepark.

LAMEGO, a city of Portugal. prov. Beira, and cap. of a comarca of its own name, near the left bank of the Douro, 44 m. E. Oporto, and 192 m. N.N.E. Lisbon; lat. 41° 4' N., long. 79 40' W. Pop. 9000. It stands at the foot of the Sierra de Penide (an offset of the Sierra Estrella), on the little river Balsamone, just before its junction with the Douro, and is divided into three quarters, two of which are occupied by the cathedral and bishop's palace, gardens, &c., while the third comprises the square, and a long street crossed by others of smaller size. A cathedral of Gothic architecture, built by order of Don Henrique, the father of the first king of Portugal, four convents, and a hospital, are the chief public establishments. The marshy lands, near the town, are very rich, producing an abundance of fine wines and delicious fruits; but these advantages are more than countervailed by the badness of the roads, which makes communication with Oporto, and other places all but impossible. (*Milsans.*)

LAMPETER, t., Lancaster co., Pa., 43 m. S.E. Harrisburg, 117 W. Bounded N.W. by Conestoga river, S.E. by Pequea creek. Watered by Mill creek. It contains three commission houses in foreign trade, nine stores, two fulling-mills, one woollen-factory, seven saw-mills, three tanneries, eight distilleries; 13 schools, 669 scholars. Pop. 3260.

LAMPEDUSA. LAMPION, and LINOSA: three islands in the Mediterranean, collectively called the Pelagian Isles, belonging to the kingdom of Naples, and lying between lat. 35° 27' and 38° N., and long. 12° and 13° E., about midway between Malta and the shore of Tunis. Lampedusa, the sm. *Lopadusa*, by far the largest, is about 13½ m. in circuit. Its shores are precipitous, but it has a tolerable harbour on its S. side. Its surface is level; the E. extremity has been cultivated by an English speculator; the W. end of the island is covered with dwarf olive

trees and other wood, much of which is cut for fuel, and sent to Malta and Tripoli. Both Lampion and Linosa are uninhabited, except by rabbits and goats; the former island has, however, some interesting traces of ancient buildings; the latter presents distinct marks of volcanic origin. (*Smyth's Sicily*, p. 264—269, &c.)

LANARKSHIRE, or CLYDESDALE, an inland co. of Scotland, having N. the cos. of Dumbarton and Stirling; E. West Lothian, Mid Lothian, and Peebles; S. Dumfries; and W. Ayr and Renfrew. It extends from Queensbury Hill, on the borders of Dumfries-shire, to near Renfrew, a distance of 55 m., comprising nearly the whole country drained by the Clyde (which see) and its tributaries, the Douglass, Avon, N. and S. Calder, &c. Area, 604,800 acres, of which from a third to a half are supposed to be arable. It is divided into three wards, each of which is characterized by peculiarities of surface, soil, and climate. The *upper ward*, of which Lanark is the principal town, includes nearly two thirds of the co., comprising the district bounded by Peebles on the E., Dumfries on the S., and Ayr on the W. This district consists for the most part of mountains, hills, and wide dreary moors; the only cultivable land lying along the banks of the Clyde and Douglass. Some of the mountains in this ward have an elevation of above 2300 ft. The *middle ward*, having Hamilton in its centre, has a comparatively level surface, the low grounds along the Clyde extending to a much greater distance, and the hills by which they are bounded on either side being of very inferior altitude. The *lower ward*, though of small dimensions as compared with either of the others, is the most fertile and best cultivated; and, having the city of Glasgow within its limits, it is by far the most populous, important, and wealthy of the three. The climate in the upper ward is often very severe; in the middle and lower ward it is comparatively mild and humid, especially in the latter. The soil of the middle and lower wards is principally a retentive clay, but in parts it is loamy, sandy, gravelly, &c. Agriculture, though formerly backward, has of late been greatly improved: drainage, which is here quite essential, is now prosecuted with the greatest vigour; and bone dust is extensively employed in the raising of turnips. The draught horses of this county have long enjoyed the highest reputation of any in Scotland. Ayrshire cows are generally introduced; and a good deal of cheese is made in imitation of Dunlop. There are several valuable orchards in what is called the *trough* of the Clyde, between the mouth of the S. Calder and the lowest waterfall. Farm houses and offices rank with those in the best improved districts. Property mostly in very large estates; farms of all sizes, and let generally on leases for 19 years. Average rent of land in 1810, 9s. 10d. an acre. The minerals of this county, particularly its iron and coal, are of the highest importance. The command of cheap and abundant supplies of the latter has been the principal cause of the extraordinary progress made by Glasgow in manufacturing industry; and more recently, the command of coal, added to the discovery of the peculiarly valuable carboniferous iron-stone (provincially *black-band*), have made Lanarkshire one of the principal seats of the British iron trade. In 1834, about 48,000 tons of iron were produced by the different iron works in this county; and so astonishing has been the subsequent progress of the trade, that in 1840 about 210,000 tons of iron were produced in this county, and various new furnaces were then, also, in the course of being erected! The principal iron works are those of Dundyvan, Gartsherrie, Summerlee, and Calder. (See vol. i. 995.) Lead is also rather extensively produced at Leadhills in this county. With respect to manufactures and commerce, it is sufficient to say that they are of the highest importance; and to refer for details to the article Glasgow, where they are principally concentrated. Each of the three wards into which this co. is divided has a sheriff substitute to superintend its judicial affairs. The Forth and Clyde canal is partly, and the Monkland canal wholly, in the co., and it has also several railways. It is divided into 47 parishes, and sends three members to the House of Commons, one being for the co. and two for the city of Glasgow; the boroughs of Lanark, Airdrie, and Hamilton, unite with Linlithgow and Falkirk in returning a member. Registered electors for the county, in 1839-40, 3964. In 1831 Lanarkshire had 58,745 inhabited houses: 64,876 families; and 316,819 inhabitants, of whom 150,299 were males, and 166,590 females. At present (1840) the population is probably not under 450,000. Valued rent, £169,139 Scotch: annual value of real property, in 1815, £686,531.

LANARK, a royal and parl. bor. and market town of Scotland co., Lanark, of which it is the cap., on an elevated plateau, 1½ m. from the Clyde, 30 m. S.W. Edinburgh, and 23 m. S.E. by E. Glasgow. Population of the town and parish, in 1821, 6087; in 1831, 7872: of the town only, in 1831, 4966. It consists of one leading street in the direction of E. and W., with several subsidiary streets and 'mes. The streets are well paved; but many of the houses are mean, being thatched with broom, heath, or straw, and exhibiting strong marks of poverty or decay: but the older buildings are gradually being superseded by new and better edifices. The only public buildings are the County Hall, including a jail, the parish church, two chapels belonging to the Kelief, and one to the Associate Synod. Several handsome baronial seats are in the near vicinity.

Various sums have been bequeathed, at different times, for the promotion of education. Twenty-eight boys are supported at the grammar-school; and, in addition to the school fees being paid, each gets an annual sum, varying from £2 to £3. There is, besides, a charity school for 50 children. The total number of schools in the parish is eight; total average attendance, 450: there is a subscription library and a reading room in the town. (New Stat. Account of Scotland, § Lanarkshire, p. 95, 27.) William Lithgow, the traveller, and Gavin Hamilton, the historical painter, were natives of the borough; and General Roy, the celebrated engineer, and author of "The Military Antiquities of the Romans in Britain," was educated at the grammar school.

Hand-loom weaving, in connection with the Glasgow manufacturers, is extensively carried on, there being, in the borough, 702 weavers. They work, at an average, above 12 hours a day. In order to eke out the slender pittance of the family, the wives of the married men engage in winding the weft on pirns; and, as if to perpetuate this poverty-stricken business, the children, both males and females, are usually employed in the work from an early age. About 120 females are employed in embroidering lace. Boots and shoes are made in a small extent for export. There are two branch banks in the town.

Lanark and its vicinity have many remains of antiquities. The Castle Hill, on the S. of the town, was once the site of a royal residence; but every trace of it has disappeared. The old church, the date of which is unknown, and St. Nicholas's chapel, have been allowed to go to ruins. There are, in the neighbourhood, distinct vestiges of two Roman camps, supposed by General Roy to have been the work of Agricola: one of them measures 600 yards in length, and 490 in breadth.

This borough seems to have been more important in ancient than in modern times. In 978, Kenneth II. held in it an assembly of the states of the realm. It was a royal borough as early as the 12th century. Lanark was the scene of the first military exploit of Sir William Wallace (who resided here for some time after his marriage with the co-heiress of Lamington), who killed (1296) Hazelrigg, the English sheriff, and drove his soldiers from the town. This borough formerly had the custody of the standard weights of Scotland: they are still preserved; but the act of 1824, introducing the imperial standard, has superseded their use.

Lanark unites with Falkirk, Linlithgow, Airdrie, and Hamilton, in sending a member to the House of Commons. Registered voters, in 1839-40, 280. The Falls of Clyde are in the near vicinity of the town; Bonnington Linn, 30 ft.; Corra Linn. 120 ft.; and Stonebyres, 84 ft.; the two former are to the E.; the latter to the W. of the town. Another remarkable object is the Cartland Crags, a deep chasm, formed by the Mouss, a small tributary of the Clyde, over which a bridge of three arches has been thrown (1825) whose two piers are each 146 ft. in height, about equal to the length of the bridge. (Chalmer's Caledonia; Boundary Reports, &c.)

LANARK (NEW), a manufacturing village of Scotland, co. Lanark, on the bank of the Clyde, close to the river, and bounded on the N. by steep and beautifully wooded hills, 1 m. S. of the borough of Lanark. Population 1901. The village consists of a series of cotton-mills, and of two streets, in which the work-people live; and so little space intervenes between the river and the hills, that there is room for only two lines of edifices. No person is allowed to reside here, unless he be connected with the factories.

The mills were founded, in 1784, by Mr. David Dale; and Arkwright, the father of the cotton manufacture, was for a while a partner in them. (Baines's Hist. of the Cotton Manufacture, p. 193.) Mr. Dale was afterwards succeeded by his son-in-law, Robert Owen, whose attempts (first made at New Lanark) to reduce to practice his absurd projects for the renovation of society, are well known: but Owen ceased, in 1827, to have any interest in the business. The mills at present give employment to 1669 individuals, of whom 361 are under 18 years of age. (Factory Reports, 1839, p. 304.) The hours of labour are limited to 11½ hours a day throughout the year; and the people are peculiarly decent and respectable. (New Stat. Account of Scotland, § Lanarkshire, p. 22.) A school is established in the works, for the education of the children, and is attended by about 300 pupils. (Ib. 27.) It may be

mentioned, that teaching by objects, and what is called (how justly we shall not stop to inquire) the intellectual, system of education, was originally practised at the mills of New Lanark, about the beginning of the century. There are two funeral societies, from which, on the death of a member or his wife, the family receive £4, on that of a child £2. The sum is collected as occasion requires, the society not accumulating any funds.

LANCASHIRE, or LANCASTER, a marit. co. of England, on its W. coast, having N. Cumberland and Westmoreland, E. Yorkshire, S. Derbyshire and Cheshire, and W. the Irish sea; by which it is in various parts deeply indented. Its most northerly portion, consisting of the hundred of Furness, is separated from the main body of the county by the intervention of Morecambe bay and a small portion of Westmoreland. Area, 1,130,940 acres, of which about 880,000 are supposed to be arable, meadow, and pasture. The hundred of Furness is generally rugged and mountainous; and the E. parts of the co. along the Yorkshire border are occupied by portions of, or offsets from, the great central or inner range of English mountains; but with these exceptions, the country is generally flat; and in the S. part of the co. an extensive plain stretches from Formby Point and Liverpool on the W., to Oldham on the E. Sandy loam and sand are the prevailing soils in the lower districts, in which, however, there are several extensive mosses; peat soil prevails in the moors. The climate is mild and salubrious; but more humid, perhaps, than any other in England. This co. is wholly indebted to manufactures and commerce for its vast population, wealth, and importance; for, as respects agriculture, it is, though considerably improved, one of the most backward in the empire. There is a great want of drainage. Few thrashing machines have been introduced; and agricultural implements are generally very imperfect. Potatoes are more extensively cultivated in this than in any other English county; and this, no doubt, is one cause why few turnips are raised. Grazing is more attended to than tillage husbandry; large quantities of hay are produced, and there is a good deal of dairying. Lancashire is believed to be the original seat of the long-horned breed of cattle; but they are now so crossed and intermixed with others, as to be seldom found pure. There are some large estates; but property is, notwithstanding, a good deal subdivided. Tillage farms for the most part rather small, and usually held on leases, a tenure too short to admit of the occupiers undertaking any very expensive improvements. Farm buildings generally good. Average rent of land. 22s. 5½d. an acre. Exclusive of the minerals, this county has vast beds of coal, and to that, more perhaps than anything else, its extraordinary progress in manufactures is to be ascribed. It is the grand seat of the cotton manufacture, which has grown up with a rapidity wholly unexampled in the history of industry. Manchester, Preston, Bolton, Oldham, Blackburn, Ashton, Bury, Chorley, Wigan, and other towns, where the manufacture is principally carried on, and Liverpool, the grand emporium of the trade of the co., have increased with equal rapidity. Manchester is now, beyond all dispute, the first manufacturing town in the world; and the trade and navigation of Liverpool are inferior only to those of London. Besides that of cotton, the woollen manufacture is extensively carried on at Rochdale and other places in this county, as is that of silk, flax, paper, hats, &c. The extension of manufactures and trade has been at once a cause and a consequence of the extension of the facilities for conveyance, by means of canals, railways, &c., which traverse this county in every direction, and bring it, as it were, into immediate communication with almost every other part of the empire. Lancashire was, indeed, the first county to construct a navigable canal (the Duke of Bridgewater's); and the opening of the Manchester and Liverpool railway, with locomotive engines, in 1830, forms a new and most important era in the history of internal communication. It is to be regretted that, notwithstanding the great extent of seacoast belonging to this county, it does not possess a single good harbour. Liverpool is the best, but the access to it is much embarrassed by sand-banks. Lancashire is a county palatine, and contains seven hundreds, four boroughs, and 70 parishes, many of which are very extensive. It sends 26 members to the House of Commons, being four for the county, two each for the boroughs of Manchester, Liverpool, Oldham, Bolton, Preston, Lancaster, Wigan, and Blackburn, and one each for Rochdale, Bury, Clitheroe, Ashton, Salford, and Warrington. Registered electors for co. in 1839-40, 97,796, being 9648 for the N., and 18,148 for the S. division. In 1831, Lancashire had 298,130 inhabited houses, and 1,336,854 inhabitants, of whom 650,389 were males, and 686,465 females. Sum paid for the relief of the poor in 1838-39, £318,049. Annual value of real property in 1815, £3,130,043. Profits of trade and professions in ditto, £2,929,078.

LANCASTER, a mun. and parl. bor., and seaport t. and

par. of England, cap. of the above co., locally situated in hunds. Amounderness and Lonsdale, but with separate jurisdiction, on the S. bank of the Lune, 46 m. N. by E. Liverpool, and 205 m. N. by W. London; lat. 54° 4′ N., long. 2° 48′ W. Area of parish (comprising 17 townships), 66,160 acres; pop. of ditto, in 1831, 23,817. Estimated pop. of parl. borough (which includes parts of Skerton and Bulk townships), in 1839, 15,600. The town stands on a gentle slope facing the Lune, which is crossed here by a handsome stone bridge of five arches: and the summit of the hill is crowned by the bastions of its fine old castle, and the lofty tower of the parish church. Nearly the whole town is built of freestone, from quarries in the neighbourhood: the houses generally are well constructed, and many are large and handsome. The streets, however, with one or two exceptions, are inconveniently narrow, and badly paved. Lancaster is lighted with gas, under an act passed in 1824, and is well supplied with water from springs and wells. The principal public building is the castle, once a magnificent structure, originally built in the 11th century, but renovated by John of Gaunt, Duke of Lancaster, during the reign of Edward III.: it was repaired at the end of the 16th century, and much enlarged in 1788, when it was converted, at an expense of £140,000, into assize and county courts, jail, female penitentiary, &c. The walls enclose an area of 10,925 sq. yards. The prison is conducted on the system of classification, and silent labour: above 160 debtors and 200 criminals have been confined in it at an average of the last few years. Among the other public buildings, exclusive of the churches, are the town-hall, erected in 1781, the custom-house, on St. George's Quay, having a portico and pediment supported by four Ionic columns, the assize-house, the assembly-room, the theatre, the public baths, and the market-house. The county lunatic asylum, on Lancaster Moor, is a quadrangular building, with a handsome Doric front, occupying, with its grounds, about 5 acres: it accommodates 300 patients, and is said to be humanely and judiciously conducted. The parish church, which stands on the "green and shapely knoll" of Castle-hill, is of the same date as the castle, and consists of a central and two side aisles of equal length, terminated by a well proportioned and lofty tower at its W. end: it was all but rebuilt in 1759. Its richly-carved stalls, and other curious carvings in the chancel, and its fine monuments, are universally admired. The living is a vicarage, of the clear annual value of £1700; and the incumbent nominates the ministers of St. John's and St. Ann's, two district churches, as well as those of all the chapelries within the parish. There are also places of worship for Rom. Catholics, Presbyterians, Independents, Wesleyan and Association Methodists, to each of which, as well as to the churches, Sunday schools are attached, furnishing religious instruction to about 1900 children. There is a meeting-house for the Society of Friends. The school-charities comprise an ancient grammar school under two masters, greatly modified in 1834, and now furnishing a good classical and general education to about 60 boys; a boys' national school, united with an old blue-coat charity, attended by 360 boys (30 of whom are clothed), a girls' national school established in 1830, and attended by 130 girls, a charity school for clothing and instructing 60 girls, a Catholic charity school, attended by 90 children of both sexes, and a Lancastrian school with 200 children. Among the other public charities may be mentioned Penny's Hospital, endowed with land worth £340 a year, and affording a residence, clothing, and small stipend, to 12 poor men, Gillison's Hospital, for the reception of eight unmarried women, each of whom has a stipend of £4 a year; Gardyner's Almshouses for four old men, a dispensary, and house of recovery; a lying-in charity, and a benevolent society. (*Charity Comm.*, 13th *Rep.*, p. 269–273.) Bible, church missionary, and tract societies, are also respectably supported. The chief literary establishments are "the Amicable" book society, the mechanics' library, and three newsrooms. A newspaper, called the "Lancaster Gazette," is published every Saturday. A savings' bank was established in 1823, and there is a joint-stock bank, entitled the Lancaster Banking Company, with 81 partners.

The port of Lancaster, which includes Wyre and Preston, had formerly a considerable share in the trade with the W. Indies; inasmuch as it appears that in 1799 (during which year 345 ships, of the burden of 28,540 tons, entered the port) 57 vessels of the burden of 12,800 tons came from the W. Indies only, to which there sailed, during the same year, 52 ships with cargoes of the estimated value of £3,500,000 sterling. In consequence of the superior facilities enjoyed by Liverpool, this branch of commerce is now all but extinct; a few vessels are engaged in the trade with N. America and the Baltic, but the great bulk of the shipping consists of coasters. In 1836, there belonged to the port and its dependant creeks, 131 ships, of the burden of 9633 tons. Vessels of above 200 tons load and unload in Glasson dock, constructed in 1787, about 5 m. below the town, to and from

which their cargoes are conveyed by means of lighters. Gross customs' revenue, in 1839, £41,296. The manufactures of Lancaster comprise cotton fabrics, silk thread, linen thread, and sail-cloth. The cotton trade, introduced in 1806, is in a thriving condition; and there were, within the parish, in 1839, eight cotton-mills, with 1373 hands: there are also three silk-throwing mills, with 300 hands, and a small flax-mill. The sail-cloth business has declined, and not more than 30 pieces a week are now made; whereas, at an average of the five years ending with 1801, the produce of the sail-cloth looms amounted to about 400 pieces a week, or upwards of 20,000 in the year. Cabinet-work and upholstery are made in considerable quantities for exportation; and there are candle and soap establishments, and two extensive ship-yards. On the whole, Lancaster, notwithstanding its distance from the great coal-field of S. Lancashire, may be said to be in a thriving condition; and the accelerated increase of the population since 1821, may be probably attributed to the increasing demand for factory labour. The Lancaster canal skirts the town, and about ½ m. to the N.E. it crosses the Lune by a noble aqueduct bridge of five arches, erected by Rennie at a cost of £48,000. The Lancaster and Preston junction railway, which completes the communication between Lancaster and London by a line of road about 230 m., is itself 20½ m. long, and was constructed at a cost of £380,000. Its success down to the present time has not realised the hopes of the projectors: but it will most probably be extended to Carlisle, and perhaps, in the end, to Glasgow, when it will be sure to pay.

Lancaster is one of the most ancient of the English boroughs; its first charter having been granted by King John, and confirmed by subsequent monarchs. The present municipal borough is divided into three wards, and governed by six aldermen (one of whom is mayor) and 18 councillors: it has a commission of the peace under a recorder. Corporation revenue, in 1839, about £1600. Assizes are held in Lent and summer; and the quarter sessions on Jan. 4, April 5, June 28, and Oct. 19. The wapentake court of Lonsdale, for debts under 40s., is held on the first Wednesday in each month; and the borough court sits every fourth Thursday for the recovery of debts to any amount incurred within the borough. The right to send representatives to parliament was first exercised in 1293 (23 Edward I.), but it ceased in 1359, and was not resumed till 1547, since which Lancaster has regularly sent two members to the House of Commons. Previously to the Reform Act, the right of election was vested in the freemen and inhabitants. The limits of the old parliamentary borough were extended by the Boundary Act, so as to include parts of the townships of Skerton and Bulk. Registered electors, in 1839–40, 1311. Lancaster has two weekly markets on Wednesday and Saturday, but chiefly on the latter; and fairs are held 1st May, 5th July, and 10th October, for cattle and cheese.

Lancaster is supposed to have been a Roman station. Urns, altars, and other antiquities have been discovered, and the *ad* *castor* given by the Saxons, serves to confirm the fact. The Normans found the town in a state of decay: the ancient city reduced to a village, and the Roman castrum little better than a ruin. It was given by William the Conqueror to Roger de Poleton, who built a castle on the site of the ruined castrum: a flourishing town soon gathered round; the burgesses of Lancaster acquired extensive privileges from their lords, and it continued to increase in importance. King John conferred "the honour of Lancaster" on his favourite Gilbert Fitz-Reinfrede, and gave it a charter. The first earl of Lancaster was created in 1266; and in 1351, Henry Earl of Derby was advanced, by special charter, to the title and dignity of Duke of Lancaster, with power to have a chancery in the county, and "to enjoy all other liberties and regalities belonging to a count palatine." John of Gaunt, fourth son of Edward III., married Blanch, the duke's daughter, and, by virtue of this alliance, succeeded to the title. His son, Henry of Bolingbroke, first Earl of Derby, and afterward Duke of Hereford, became Duke of Lancaster on his father's death in 1398, and finally King of England in 1399, from which time to the present this duchy has been associated with the regal dignity. Lancaster espoused the royalist cause during the parliamentary war, and was visited by the Jacobite troops in the rebellions of 1715 and 1745. (*Baines's Lancashire*, vol. ii.; *Parl. and Off. Reports*; *Private Inform.*)

LANCASTER, county, Pa. Situated towards the S.E. part of the state, and contains 926 sq. m. It was the fourth county established in the state; and, in natural advantages, is one of the richest. Susquehanna river washes its S.W. border for more than 40 m. Watered by Conestoga, Pequea, Conewango, and other creeks, tributary to Susquehanna river, which afford good water-power. Iron ore, and sulphate of magnesia, or Epsom salts, are extensively found. It contained (in 1840, 51,164 neat cattle, 41,957 sheep, 75,696 swine; and produced 1,199,377 bushels of wheat, 448,710 of rye, 1,307,000 of Indian corn, 19,073 of buckwheat, 1,375,673 of oats, 255,500 of potatoes, 46,860 lbs. of tobacco. It had over 34 bushels of edible grains, exclusive of potatoes, to every individual of its population. It had 66 stores, capital $1,445,005, two lumber-yards, capital $207,000, 12 fulling-mills, 10 woollen factories, one cotton factory, with 2000 spindles, 120 flouring-mills, 135 grist-mills, 104 saw-mills, two oil-mills, 57 tanneries, 102 distilleries, eight breweries, eight potteries, nine printing-offices, nine weekly news-papers. Total capital in manufactures, $1,213,484; four academies. 918 students; 206 schools, 6745 scholars. Pop. 94,953. Capital, Lancaster.

LANCASTER, county, Va. Situated in the E. part of the state, and contains 161 sq. m. Bounded E. by Chesapeake bay, S.W. by Rappahannock river. It contained in 1840, 3684 neat cattle, 2995 sheep, 7304 swine; and produced 25,750 bushels of wheat, 105,445 of Indian corn, 43,866 of oats, 7172 of potatoes, 10,272 pounds of cotton. It had 10 stores, seven grist-mills, one saw-mill; seven schools, 140 scholars. Pop., whites, 1903; slaves, 2478; free coloured 967; total, 4625. Capital, Heathville.

LANCASTER, district, S. C. Situated in the N. part of the and contains 594 sq. m. Bounded S.W. by Catawba state, N.E. by Lynches creek. Watered by Five Mile, Wax-haw, Cane, Sugar, Little Lynches, and other creeks. It contained in 1840, 10,943 neat cattle, 6165 sheep, 15,915 swine; and produced 23,910 bushels of wheat, 308,832 of Indian corn, 96,371 of oats, 13,793 of potatoes. 2,647,676 pounds of cotton. It had 18 stores, two flouring-mills, nine grist-mills, eight saw-mills, 19 oil-mills; one academy, six students; 17 schools, 446 scholars. Pop., whites, 5565; slaves, 6035; free coloured, 167; total, 9967. Capital, Lancaster.

LANCASTER, p. t., capital of Coos co., N. H. 101 m. N. by W. Concord, 582 W. Bounded N.W. by Connecticut river. Watered by Israel's river and its branches, which afford water-power. It contains two churches, a Congregational and Methodist; six stores, one fulling-mill. one grist-mill, four saw-mills. one printing-office, one weekly newspaper; one academy, 129 students; 11 schools, 408 scholars. Pop. 1316. The village is situated on Israel's river, 1 m. from its entrance into Connecticut river, and contains a courthouse, jail, a Congregational church, and an academy. A bridge here crosses Israel's river, and another Connecticut river, leading to Guildhall, Vt.

LANCASTER, p. t. Worcester co., Mass., 36 m. W. by N. Boston. 416 W. Watered by Nashua river. Incorporated in 1653. It contains four stores. two fulling-mills, one wool-len factory, two cotton factories. with 256 spindles, four grist-mills, five saw-mills, one printing-office, one bindery; one academy, six students; 12 schools, 365 scholars. Pop. 2019. The village, at the centre, contains a Unitarian church, a bank, an academy, and about 75 dwellings.

LANCASTER, p. t., Erie co., N. Y., 280 m. W. Albany, 387 W. Organized in 1833 from Clarence. Drained by Elli-cott's, Cayuga, and Seneca creeks. It contains four churches, a Presbyterian, Methodist, Lutheran, and a Roman Catho-lic: four stores, two fulling-mills, three grist-mills, 12 saw-mills; nine schools, 390 scholars. Pop. 3083.

LANCASTER, city, capital of Lancaster co., Pa., 37 m. E.S.E. Harrisburg, 62 m. W. Philadelphia, 111 W. It is in 42° 9' 38" N. lat., and 76° 90' 33" W. long. Pop. in 1830, 6063; in 1830, 7704; in 1840, 9417. It was for many years the seat of government of the state, which was removed to Harris-burg in 1812. Situated 1½ m. W. of Conestoga creek, which has been improved by a slack-water navigation to Susque-hanna river. a distance of 18 m. There are nine dams and locks, with a depth of water never less than 4 feet. The pools are from 1 to 3 m. long, and from 250 to 300 feet wide. The locks are 100 feet by 22, and the lifts vary from 7 to 9 feet. These dams create an extensive water-power. The city is regularly laid out, with broad streets, crossing each other at right angles, and are chiefly paved and curbed, and are neatly kept. The ancient buildings, erected by the early German settlers, are principally one story high; but the houses more recently built are lofty, commodious, and many of them elegant, and are equal in convenience and beauty to any in the state. The town plat contains a square of 2 m., containing 2500 acres, which are crossed by Conestoga creek in the E. part. Franklin college was established here in 1787, designed especially for the improvement of the German population, and respectably endowed; but after continuing in operation for two years, it declined to a respectable gram-mar school. The Lancaster county academy was establish-ed in 1827, and endowed with a fund of $3000. There are in the city a Lancasterian school, two public libraries, a reading room, and several other institutions of a literary character. It contains a brick courthouse at the intersec-tion of King and Queen streets, the two principal streets; a jail, 12 churches, two Lutheran, a German Reformed, Episcopal, Moravian, Presbyterian, Methodist, New Jerusa-lem, Roman Catholic, and two African; a theatre, and a mechanics' institute. The surrounding country is fertile, populous, and highly cultivated. The great western turn-

pike from Philadelphia to Pittsburg, and the Philadelphia and Columbia railroad, pass through it. It contained in 1840, two commission houses in foreign trade, capital $38,600, 32 retail stores, capital $842,750, three lumber-yards, capital $16,000, three furnaces, machinery manufactured to the amount of $12,500, two tanneries, 13 distilleries, four brew-eries, four potteries, two rope-walks, five printing-offices, three binderies, six weekly newspapers, two of which are in the German language; two academies, 84 students; 13 schools, 1715 scholars. In addition, there are in the town-ship one cotton factory, with 2000 spindles, three flouring-mills, three grist-mills, two saw-mills, five distilleries; three schools, 190 scholars. Pop., exclusive of the city, 800. Lancaster was first chartered as a borough in 1754, when it had 500 houses and 2000 inhabitants. This charter was confirmed in 1777. It was incorporated as a city in 1818, and has a mayor, recorder, and aldermen, a select and com-mon council. The mayor is elected by the councils, and all of the above form the mayor's court.

LANCASTER, p. v., Hocking t., capital of Fairfield co., O., 30 m. S.E. Columbus, 375 W. Situated on a head branch of Hockhocking river. It was laid out in 1800, with broad streets and convenient alleys, crossing each other at right angles. It contains a courthouse, jail, a bank, a town-hall, a market-house. seven churches, 16 stores, one woollen fac-tory, one iron foundry, two flouring-mills, two printing-offices, numerous mechanics, 400 dwellings, many of them elegant, and 2872 inhabitants. A lateral canal, 9 m. long, connects this place with the Ohio canal.

LANCASTER, p. v., capital of Garrard co., Ky., 57 m. S. by E. Frankfort, 555 W. Situated 5 m. E. of Dick's river. It contains a courthouse, jail, several stores, and 480 inhabi-tants.

LANCASTER, p. t., O., Jefferson co., Ia., 74 m. S.S.E. In-dianapolis. 574 W. Pop. 1787.

LANCASTER, C. H. p. v., capital of Lancaster co., Va., 83 m. E. by N. Richmond, 144 W. It contains a courthouse, jail, three stores, 30 dwellings, and about 180 inhabitants.

LANCASTER, C. H., capital of Lancaster dist., S. C. 72 m. N.N.E. Columbia, 434 W. It contains a fine courthouse and jail of stone, Franklin academy, several stores, and about 450 inhabitants.

LANCIANO, a town of the Neapolitan dominions, prov. Abruzzo Citra, cap. dist. and cant., or circondario, 6 m. from the Adriatic, and 18 m. S. Pescara. Pop. (ex. circ.) in 1833, 11,953. It is built on the summit of three hills, in a healthy and pleasant situation; and has a cathedral, several churches and convents, an archbishop's palace, a diocesan seminary, and other schools, a tribunal of primary jurisdiction, &c. This is a very ancient city; and in the middle ages it was distinguished by its proficiency in manufactures, and by the extent of the commerce carried on at its fairs; but these have both greatly declined. (Del Re, Descrizioni della Due Sicilie, ii., 291, &c.)

LANDAFF, or LLANDAFF (Llan-tâf, church of the Tâf), a town and par. of S. Wales, co. Glamorgan, hund. Kibbor, on the W. bank of the Tâf, 2 m. N.W. Cardiff, and 27 m. W. Bristol. Area of par., 9366 acres. Pop., in 1831, including the hamlets of Canton, Eiay, Fairwater and Ga-baith, 1759. The town is at present little more than an in-considerable village, with about a dozen respectable resi-dences, and several cottages; nor would it be worth notice, except from its being a bishop's see, and containing a hand-some cathedral. This sacred edifice was built early in the 12th century on the site of one still more ancient; but its W. end, with its fine front, and rich Norman doorways, and elegant pinnacled towers, have been allowed to fall into de-cay. The cathedral now comprises a choir short nave and transepts: its total length, from E. to W., including the Ladye-chapel behind the altar, is 263 feet, breadth of the body 65 feet, and height, from the floor to the centre of the roof, 119 feet. Very extensive repairs, but in very bad taste, were effected in 1751, at an expense of £7000. The new front, built about 80 feet within the original Norman W. end, has a Venetian window, Ionic pilasters, and flower-pot jars on the parapet; and till lately the fine Gothic altar was enclosed within a Grecian portico. The chapter-house, S. of the church, is in the decorated English style, with a central pillar; but it is fast falling into the same ruinous condition as the monuments and the episcopal palace, which were defaced and all but destroyed by Owen Glendwr. The choral services have been disused for some years, and the building is now employed as a parish church, the service being occasionally in the Welsh language. The see of Llandaff (created in the 6th century), comprises all the county in which it is situated, and Monmouthshire, ex-cept seven parishes. It is the poorest of all the English bishoprics, the annual income, including preferments, at an average of the three years ending with 1831, being only £924, and it has, consequently, been hitherto held in com-mendam with the deanery of St. Paul's, London, and the rectory of Bedwas. On the first avoidance of the see, how-over, the sum of £3150 is to be paid out of the episcopal augmentation fund, to raise the income to £4908, and a further allowance of £300 is to be made till the residence be restored. The patronage of the see comprises the cathedral appointments with eight livings, and the chapter comprises 11 dignitaries, besides the bishop; there are also two vicars-choral. Llandaff has no market, and is wholly dependant for its supplies on Cardiff, except for vegetables, which it sends in considerable quantities to that market. Cattle fairs, Feb. 9, and Whit-Monday, well attended. (*Nicholson's Cambrian Guide; Parl. Rep., &c.*)

LANDAFF, p. t., Grafton co., N. H., 64 m. N.W. Concord, 539 W. Chartered in 1764. Watered by Great Ammonoo-suck and Wild Ammonoooosuck rivers. It has three saw-mills; nine schools, 440 scholars. Pop. 957.

LANDAU, a strongly fortified town belonging to the German confederation, in Rhenish Bavaria, on the Queich, a tributary of the Rhine, 54 m. S. by W. Mayence, and 46 m. N.N.E. Strasburg. Pop., according to Berghaus, 6100, exclusive of the Bavarian garrison of 6000 men. This fortress is considered a *chef-d'œuvre* of Vauban, who commenced the construction of its works in 1680. It is an octagon, with seven bastions, as many demi-lunettes, and several out-works: its ditches are filled from the Queich. The barracks and magazines are bomb-proof. The town was almost entirely consumed by fire in 1688, since which, it has been regularly laid out, and has some good public edifices, including the principal church with a lofty tower, two convents, the town-hall, court of justice, and a civil and military hospital. In the centre of the town is a spacious parade ground. Some extensive vinegar factories have been established here within the last few years. The gates are closed at an early hour, after which, neither ingress nor egress is permitted.

The history of Landau is little else than that of a succession of sieges, blockades, captures, and other military events. It was founded by the Emperor Rodolph, of Hapsburg, and made a free town of the empire in the 14th century. During the 30 years' war, it was repeatedly taken and re-taken by the Swedes, Imperialists, French, &c., and in the 18th century it was many times taken or besieged by the French and Germans. It was generally held by the French from the peace of Nimeguen, in 1680 to 1815, when it was restored to Germany by the second treaty of Paris. (*Schreiber, Guide du Rhin, 71, 72; Berghaus; Stein; &c.*)

LANDERNEAU, a town and river port of France, dep. Finistère, on the Elorn, 12 m. E.N.E. Brest. Pop. (1836) 4035. It is ill built, and badly paved; but its quays are good, and its port admits vessels of from 300 to 400 tons. It has a large and fine marine hospital, formerly an Ursuline convent, and considerable manufactures of linen cloth and leather.

LANDES, a dep. of France, and one of the largest, though the poorest, in the kingdom, reg. S.W., chiefly between lat. 43° 30' and 44° 30' N., and long. 0° 7' and 1° 32' W., having N. Gironde, E. Lot-et-Garonne and Gers, S. Basses Pyrenées, and W. the bay of Biscay. Length and greatest breadth about 70 m. each. Area, 915,139 hect. Pop. (1836) 284,918. This dep. derives its name from an extensive tract of heath, marsh, and other waste land, with a loose sandy soil, about 300 feet above the level of the sea, termed the "Landes," which occupies 731,142 hect., or nearly 4-5ths of its total surface, besides a considerable portion of the adjacent dep. of the Gironde. This extensive and almost desert plain is for the most part a dead flat, interspersed with patches of pasture or cultivated land, clumps of pines, scattered habitations of a miserable kind, and a few wretched hamlets; and bounded towards the sea by a chain of *dunes* or sandy downs, inside which is a succession of lagoons frequently communicating with each other, and occasionally with the sea by openings between the *dunes*. The *dunes* extend along the shore nearly from the mouth of the Gironde to the Pyrenees, forming a chain from 140 to 150 m. in length, by about 5 m. in width, and from 100 to 150 feet in height. They consist of loose shifting sand thrown up by the sea. They are continually changing in form and position, according to the prevalent winds; but have a general tendency to move easterly, in which direction they are said to advance about 25 yards a year; and in process of time they would infallibly overspread the whole country, unless arrested and fixed by planting them with pine or other trees, as is done in Holland. Occasionally immense masses of sand have shifted their position through the agency of tempests, as in the African and Arabian deserts. The church and a considerable part of the village of Mimizan was overwhelmed by an inundation of this sort. The increase of the *dunes* having prevented the egress into the sea of many small rivulets, the lagoons have been formed, the largest of which is 7 m. in length and about as many in width. These also continue to extend, since the shifting sands have been gradually shallowing the channels by which they communicate with the sea. The surface of the "Landes" is usu-

ally parched and arid, except for about four months of the year, when the rains form extensive pools in its depressed portions, varying to the depth of several feet. These are often covered with sand carried over them by the wind, when they are called *blouses*, and are exceedingly dangerous to strangers. To avoid such dangers, and to travel more speedily through the loose soil, the inhabitants use long staffs having notches for the feet, 1, 2, or 3 feet above their lower extremity; so that a person of ordinary stature, when in walking order, has at a distance the appearance of a giant 8 feet high. The inhabitants are very expert at the use of these singular helps to locomotion. The Adour, and its tributary the Midouze, bound the "Landes" to the S.E. and form the N. limit of the fertile portion of this dep. The soil is there light, but productive. Maize, millet, wheat, rye, saffron, hemp, flax, &c., are grown: in the arrond. St. Sever, about 250,000 kilog. of linseed-oil are produced annually, and about 390,000 hectol. of wine, certain kinds of which, termed the *vins de sables*, rival some of the growths of the Gironde. The culture of the mulberry is on the increase. Agriculture is exceedingly backward. The "Landes" are mostly appropriated to the rearing of sheep, of which, in 1830, the dep. had 400,000. The lower classes in the "Landes" appear to be very wretched. Shut out from communication with the more civilized parts of the kingdom by the absence of great roads, they live in a half-savage state, clothed chiefly in sheepskins, on which or on straw they usually lie at night. Their food is principally maise or rye bread, with pilchards, which are prized in proportion to their rancidity; maize or millet porridge, and pea-soup with sour lard and some spice, to which they occasionally add garlic or a little fried bacon. They are grossly ignorant, and degraded both physically and mentally. In 1835, of 40,446 properties subject to the *contribution foncière*, 14,870 were assessed at less than 5 fr., 5978 at from 5 to 10 fr., and 6166 at from 10 to 20 fr. Goats, hogs, and poultry are frequently kept by the peasantry, and bees are numerous. The pine forests furnish abundance of deals, pitch, tar, rosin, &c.; and coal, iron ore, potters' clay, &c., are met with. Manufactures unimportant; some smelting furnaces and forges, employing about 500 hands, and some tanneries, oil-mills, and glass and earthenware factories, comprise almost all the manufacturing establishments. The trade of the dep. is chiefly in cattle, wines, timber, and agricultural produce. Landes is divided into three arronds., and sends three members to the chamber of deputies. Number of electors (1836-39) 1145. Chief towns, Mont-de-Marsan, the capital, St. Sever, and Dax. Total public revenue (1831) 3,585,809 fr.; expenditure, 2,576,700 fr.: leaving a surplus of more than 1,000,400 fr.: a large sum, considering the poverty of the dep. (*Hugo, art. Landes; French Official Tables, &c.*)

LANDGROVE, p. t., Bennington, Vt. co., 98 m. S. by W. Montpelier, 442 W. Chartered in 1780. Watered by branches of West river. Situated on the Green mountains. A stage road across them, from Manchester to Chester, passes through it. It contains a Methodist church, one store, two saw-mills; three schools, 115 scholars. Pop. 344.

LANDISBURG, p. v., Perry co., Pa., 99 m. W. Harrisburg, 117 W. Situated on the N. side of Sherman's creek, and contains a Presbyterian church, four stores, and about 400 inhabitants.

LANDSBERG, a town of the Prussian dom., prov. Brandenburg, gov. Frankfort, cap. circ., on the Warta, a tributary of the Netz, here crossed by an excellent bridge, 36 m. N.E. Frankfort on the Oder. Population, in 1836, according to Berghaus, 9970; but it is stated by Von Zedlitz (*Der Preussische Staat*, ii. 218) to be nearly 12,000, among whom are many Jews. Landsberg is divided into the Old and New Town, and has several suburbs. It is walled, and is one of the best built towns in the province. It has several churches, a house of correction, the inmates of which are made to support themselves by the manufacture of woollen cloths, a hospital, an orphan asylum, a high school, &c. It is a principal mart for corn and wool, the greater part of the produce of Pomerania, the Neumark, and W. Prussia, being brought thither for export by the Oder. The town has also brisk manufactures of woollen goods, leather, and paper, and numerous breweries and distilleries. Landsberg is the seat of a circle assembly, a circle and town tribunal of the first class, boards of taxation, forest economy, and agriculture, and the superintendency of the drainage of the vale of the Warta. (*Deichhauptmannschaft für das Warte-brüch.*) The town was repeatedly taken and retaken by the Swedes and the Imperialists in the 30 years war. (*Von Zedlitz; Berghaus, &c.*)

LANDSCRONA, a fortified seaport town of Sweden, prov. Malmœ, on a tongue of land projecting into the Sound. 16 m. N.E. Copenhagen; lat. 55° 51' 58" N.; long. 12° 49' 47" E. Pop. 3970. It has strong walls, a citadel, and other works; is well laid out, and has a safe and well sheltered harbour, with 20 ft. water.

LAND'S END, a headland at the W. extremity of the

in Cornwall, celebrated as being the most westerly land in England; lat. 50° 4' 8" N., long. 5° 41' 31" W. It is formed of granite cliffs, which rise about 60 ft. above the level of the sea. These assume, in some places, the appearance of shafts, and are as regular as if they had been cut by the chisel. About 1 m. W. from the Land's End, are the rocks called the Longships, on the largest of which is a light-house, with a fixed light, having the lantern elevated 66 ft. above high water mark.

LANDSHUT, a town of Bavaria, circ. Lower Bavaria, on the Isar, 36 m. N.E. Munich. Pop. 8000. It is divided into an old and a new town, has a suburb on an island in the Isar, with which it is united by two bridges, and is partly surrounded by old walls and ditches. It consists of two principal and many smaller streets; the houses, which are of brick, are mostly environed by gardens. The town has a very picturesque appearance, from the antique archi-tecture of its buildings, and the number of its towers and spires; that of St. Martin's church being one of the loftiest in Germany. It has an old castle, the residence of the dukes of Bavaria in the 13th century; a Cistercian abbey, in which they were buried; a royal palace, an old town-hall, a hospital for decayed citizens, two other hospitals, three convents, a lyceum, gymnasium, chirurgical and ec-clesiastical seminaries, and various other schools. In 1808, the university of Ingolstadt was removed thither; but in 1826 it was transferred to Munich, since which Landshut has declined. It still, however, has manufactures of wool-len cloths, stockings, tobacco, paper, cards, &c., with nu-merous distilleries and breweries, and some trade in corn, cattle, and wool. (Berghaus; Stein, &c.)

LANE-END. See POTTERIES.

LANESBOROUGH, p. t., Berkshire co., Mass., 135 m. W., by N. Boston, 388 W. Incorporated in 1765. Drained by head branches of Hoosatonic and Hoosick rivers. White marble is extensively found. It contains three churches, a Congregational, Episcopal, and Baptist, two stores, one grist-mill, four saw-mills; two academies, 45 students, six schools, 160 scholars. Pop. 1140.

LANGDON, p. t., Sullivan co., N. H., 50 m. W. by S. Concord, 456 W. Incorporated in 1787. Watered by Cold river, a branch of which crosses the t. It contains a Con-gregational church, a store, one grist-mill, four saw-mills, 12 schools. 215 scholars. Pop. 615.

LANGELAND, an island of the Danish archipelago, in the Baltic, between Laland and Funen, extending from lat. 54° 43' to 55° 30' N., and between long. 10° 46' and 11° E. Length N.N.E. to S.S.W. 33 m.; average breadth 2½ m. Area about 80 sq. m. Pop. about 17,000. Its shores are generally uniform, except on the W., where they are bro-ken by numerous inlets. Its surface is more elevated than that of the adjacent islands, but it is generally quite flat. Climate healthy. Chief products, corn, potatoes, fruits, and flax. A good many cattle are reared, and the fisheries are productive. Rudkiobing, on the W. coast, with 1580 in-habitants, is the chief town, and centre of the trade, which is tolerably active. (Dict. Géog., &c.)

LANGENSALZA, a town of Prussian Saxony, gov. Er-furt, cap circ. of its own name, on the Salza, 19½ m. N.W. Erfurt. Pop. (1836) 7142. It is well built, walled, and fur-ther defended by a castle; and has four churches, four hos-pitals, a lazaretto, an orphan asylum, a high school, a pub-lic library, and a theatre. It is the seat of a district coun-cil, a board of taxation, judicial courts for the town and circle, the Thuringian Agronomical Society, &c. It has manufactures of various descriptions of woollen, linen, and cotton fabrics, a saltpetre factory, with dyeing houses, breweries, distilleries, and paper-mills. (Von Zedlitz, Der Preussische Staat; Berghaus; Hörschelmann, Stein, &c.)

LANGHOLM, a bor. of barony and market town of Scot-land, co. Dumfries, in the bosom of a wooded valley on the Esk, and on the line of the road between Edinburgh and Carlisle, 21½ m. N. by W. the latter, and 39 m. S. by E. the former. Pop. of town, in 1831, 2384; of town and parish 3576. It is intersected by the Esk, New Langholm (found-ed in 1778) being on the W. side of the river. The latter is regularly built, of a triangular form. The old town con-sists chiefly of one street on the line of the road. In it are the town-hall and jail, ornamented with a spire, and the parish church. There are, also, chapels belonging respect-ively to the Associate Synod and Relief. The communica-tion between the different parts of the borough is main-tained by a fine bridge.

There are nine schools in the parish, of which two are endowed; total average attendance, 275. These are two subscription libraries, to one of which the late Thomas Tel-ford, the celebrated engineer, a native of the district, be-queathed £1000. William Julius Mickle, the translator of the Lusiad, was a native of the borough; and Sir John and Sir Pultney Malcolm were born in the neighbourhood.

The poor are supported partly by church collections and partly by assessment.

13

A cotton-mill driven by water, erected in 1788, gave, in 1835, employment to about 160 persons; but at present (1840) it is suspended. There are in the town above 300 weavers, of whom 70 are employed in the stocking trade; there are also two small woollen mills, with a distillery brewery, and two branch banks.

Langholm was created a burgh of barony in 1610. Gil-nockie Tower, the residence of "Johnie Armstrong," the famous border freebooter in the time of James V., is in the neighbourhood, but has long been in ruins. (New Stat Acc. of Scotland, § Dumfries-shire, p. 416.)

LANGRES (an. Andomatunum and Civitas Lingonum), a town of France, dep. Haute-Marne, being the largest and most populous town in the dep., though not its capital: it is, however, the capital of an arrondissement, and occupies, next to Briançon, the highest elevation of any town in the kingdom, 16 m. S.S.E. Chaumont, and 39 m. N.N.E. Dijon. Pop. in 1836 (ex. com.), 6191. It is surrounded with walls, flanked by towers, and is well built, its streets being regu-lar, wide, and clean. The principal public edifice of Lan-gres, its ancient cathedral, has a choir, the peristyle of which, of the Corinthian order, is supposed to have formed part of a Roman temple; the edifice itself, though of uncer-tain date, is very ancient, excepting the grand entrance, constructed in the 18th century. The bishopric of Langres was founded as early as the 3d century. Langres has a handsome town-hall, a theatre, a public library with 3000 volumes, a school of drawing, several hospitals, and a fine public promenade. It is distinguished by its cutlery, which is its chief branch of industry.

The Lingones are noticed by Cæsar as being attached to the Romans (De Bello Gallico, lib. 1, § 26, 40); they after-ward became foederati, or allies of the Romans; and their city is characterized by Frontinus as opulentissima. (Lib. iv., cap. 3.) Among the remains of antiquity, of which it has still to boast, are several triumphal arches; one of which, now included in the town walls, supposed to have been erected in honour of the two Gordians, circa anno 240, has a frieze on its entablature, indicating a high state of the arts. It suffered numerous disasters in the dark ages, being taken and burned by Attila, and again destroyed by the Vandals in 407. Louis VII. annexed it to the French crown. Diderot was a native of Langres, where he first saw the light in 1713. (Hugo, art. Haute Marne, &c.; D'Anville, Notice de l'Anciens Gauls, p. 417.)

LANGUEDOC, one of the old provinces of France, in the S. part of the kingdom, now distributed among the de-partments of Ardèche, Aude, Gard, Haute Garonne, He-rault, Haute-Loire, Lozère, and Tarn.

LANNION, a town and river port of France, dep. Côtes-du-Nord, cap. arrond., on the Guer, 35 m. W.N.W. St. Brieuc. Pop. (1836) 5296. It is ill built and triste; its port on the river is bordered by a spacious quay, but within the last 40 years vessels of 250 tons have been unable to come up to the latter. It has a church erected in the 12th centu-ry, two hospitals, barracks, and a communal college; it is the seat of a sub-prefecture, and a court of primary juris-diction, and has manufactures of linen fabrics, and an ac-tive trade in agricultural produce.

LANSING, t., Tompkins co. N. Y., 175 m. W. Albany; 300 W. Bounded S.W. by Cayuga lake. It is drained by Salmon creek, flowing into the lake, which affords water-power. Organized in 1817. It contains four churches, two Presbyterian and two Methodist, eight stores, one furnace, three fulling-mills, three flouring-mills, three grist-mills, eleven saw-mills, one oil-mill, three tanneries, one distille-ry; 34 schools, 1307 scholars. Pop. 3673.

LANSINGBURG, p. t., Rensellaer co., N. Y., 10 m. N.E. Albany, 154 m. N. New-York, 380 W. Bounded W. by Hudson river. on which the soil is a gravelly and clay loam, with elevated ground on the E. and N.E. The village is on the bank of Hudson river, which, by a dam between it and Troy, 1100 ft. long and 9 ft. high with a lock, is naviga-ble to this place for sloops at all times. It is regularly laid out in blocks or oblong squares, 400 ft. by 260 ft., with spa-cious streets and convenient alleys, crossing each other at right angles. It is two miles long, and half a mile wide and was incorporated in 1787. It contains six churches, two Presbyterian, an Episcopal, Methodist, Baptist, and Universalist, an academy, and several public libra-ries, two extensive manufactories of oil floor-cloths, one large brush and bellows manufactory, 450 dwellings, and about 3000 inhabitants. The t. has (mostly in the village) 47 stores, capital $240,100, one grist-mill, one tannery, one brewery, two printing-offices, one weekly newspaper. To-tal capital in manufactures, $240,700; one academy, 100 students; nine schools, 547 scholars. Pop. 3330. Several sloops and towboats are employed in the river trade. A bridge across Hudson river connects it with Waterford. Mohawk river enters the Hudson river, by several mouths, opposite to the place.

LANZEROTTA, one of the Canary islands, which see.

K   153

LAODICEA AD LYCUM, an ancient city of Phrygia, in Asia Minor, chiefly interesting as being the site of one of the seven primitive Christian churches, on the Lycus, a tributary of the Meander, 190 m. E.S.E. Smyrna, lat. 37° 38' N., long. 29° 15' N. The site of this town, once ranking as the second in Phrygia, is marked only by the deserted ruins of public buildings; and hence the neighbouring hamlet, inhabited only by a few squalid Turks, has received the name of *Eski-hissar*, "old castle." (*Elliot's Travels*, ii. 97.) The remains are very extensive; and the whole surface within the walls is strewed with pedestals and fragments, indicating by their size and workmanship the former luxury and magnificence of the city. The largest ruin is that of an oblong amphitheatre, having an area of 1000 sq. ft. Many of the seats are still in tolerable preservation, and at the W. end is a vaulted passage about 140 ft. long and designed for the horses and chariots entering the arena. A Greek inscription on the mouldings informs us that it was completed in the reign of the emperor Vespasian A.D. 68, after having occupied twelve years in building. There are remains also of an odeum, two theatres, and a fabric which Chandler supposed had been a senate-house and exchange. The soil in and about the city is hard, dry, and porous, bearing many indications of an igneous origin; and Laodicea has at many different times suffered greatly from earthquakes.

Laodicea, so called from the wife of its founder, Antiochus II., was long an inconsiderable place, notwithstanding the beneficence of Hiero, Zeno the philosopher, and his son Polemo. After its sufferings, however, in a siege by Mithridates, the Romans strengthened and enlarged it, so that at length, about the Christian era, it became, next to Apamea Cibotos, the largest city of Phrygia, and vied in importance with the cities on the coast. There can be little doubt that it was visited by St. Paul in the course of his missionary tour through Asia Minor, and perhaps the Christian converts of Laodicea, as well as those of Colossæ and Hierapolis (*Pambouk*), both neighbouring towns, were the results of the apostle's preaching. In the epistle to the Colossians (iv., 16) mention is made of an epistle to the Laodiceans; and though some critics have maintained that it is identical with that to the Ephesians, the more probable conjecture is that it has not come down to us. The persecution which raged in Asia Minor during the latter part of the first century tended somewhat to abate the zeal of the Laodicean Christians, and hence the rebuke in the Revelations. Of the subsequent history of this city for several centuries we know little: it was generally in a prosperous condition under the Roman emperors, and was flourishing even in 1196, when Frederic Barbarossa visited it on his way to the third crusade. Soon afterward, however, it was repeatedly attacked and ravaged by the Turks, and finally came into their hands in the beginning of the 14th century, since which it has been a mere ruin, "wretched and miserable, and poor and naked." (Rev., iii., 14–22.)

Laodicea *ad Lycum* must not be confounded with Laodicea *combusta* (now *Ladik*) 19 m. N.W. Konieh, also a considerable city, of which there are extensive ruins. (*Chandler*, i., 259; *Elliot*, ii., 97.)

LAODICEA AD MARE, in Syria. (*See* LATAKIA.)

LAON (Lat. *Laudunum*), a town of France, dep. Aisne, of which it is the cap., on the summit of a steep hill, 32 m. W.S.W. Mezieres, and 74 m. N.E. Paris. Lat. 49° 33' 54" N., long. 3° 37' 27" E. Pop. in 1836, ex. com., 7896. The town is about 1 m. in length, narrow in the centre, expanded at either extremity, and surrounded by old walls, flanked with numerous small towers. Except its main street, it is ill built and *triste*; but it has pleasant promenades, a healthy situation, and fertile neighbourhood. It has a large Gothic cathedral, with four towers, rebuilt in 1114; a large old abbey, now occupied by the prefecture; a public library, comprising 17,000 volumes; extensive barracks, a remarkable leaning tower, two hospitals, a town hall, communal college, theatre, *dépôt de mendicité*, &c. It is the seat of a tribunal of original jurisdiction; and has manufacture of nails, leather, copperas, earthenware, &c. Coarse woollens, and some other articles, are made in the *dépôt de mendicité*.

Laon has been sometimes supposed, but on no good grounds, to occupy the site of the *Bibrax* mentioned by Cæsar. (*See* AUTUN, vol. i., 255.) In the middle ages it was distinguished by its industry and wealth; its bishopric was one of the most lucrative in the kingdom; and the position and importance of the town made it be regarded as a kind of second capital. It was, however, far more distinguished by the spirit which animated its inhabitants, and by their persevering efforts to emancipate themselves from the feudal tyranny of their bishops, and to establish a municipal government and the regular administration of justice under magistrates of their own selection. They succeeded in establishing a government of this sort so early as the year 1110; and maintained it, at the cost of many great

sacrifices, for above two centuries, or till 1331, when it was finally abolished by a royal ordonnance. (For an account of the *commune* of Laon, see the interesting and instructive work of M. Thierry, *Lettres sur l'Histoire de France*, Nos. 16–18.)

Laon, was, in 1814, the scene of some severe fighting between the French and the allies. The Prussians under Blucher having occupied the town, their position was unsuccessfully attacked, on the 9th of March, by the French, under Napoleon; and the Prussians having cut to pieces and dispersed the corps of Marmont during the night, Napoleon was obliged to withdraw from before the town on the 11th. (*Hugo*, art. *Aisne*.)

LAOS, or the SHAN COUNTRY, a country of India beyond the Brahmaputra, which may be roughly estimated to extend between lat. 15° and 24° N., and long. 98° and 106° E.; having N. the Chinese province Yun-nan; W. the Birmese empire, from which it is separated by the Thanlweng river; S. the Tenasserim provinces, Siam and Camboja, and E. Tonquin and Cochin China, from which a lofty mountain chain divides it. Our knowledge of this extensive region is extremely limited; and what little we do know relates almost exclusively to N. Laos, or the portion bordering on the Birmese and Chinese empires. The country appears to be comprised in the basins of two large rivers, the Menam, which afterward waters Siam, and the Mekamkong, or river of Camboja, in the middle portion of its course. The Laos territories formerly comprised eight or nine larger and several smaller distinct states; but of late the Siamese have conquered most of these, and the rest are principally tributary to the surrounding nations, especially the Birmese and Chinese. The Laos population in the Siamese dominion is estimated by Mr. Crawfurd at 840,000; to which we may perhaps add nearly 200,000 for the population of N. Laos, &c.; making a total of somewhat more than a million. The country is fertile, but all accounts agree that is in general very poorly cultivated and thinly inhabited. The smaller villages are mere collections of huts; and a great part of the population consists of small migratory hordes, who have no permanent habitation. The labour of cultivation is thrown principally on the women. The fields are ploughed about the beginning of the rains in August, and the crop is reaped in February. The *Oryza glutinosa* is the only variety of rice that is raised; and, as there it no market for surplus grain, it sells in plentiful years at an extremely low price. The implements of husbandry are, rude ploughs, drawn by two oxen or buffaloes, harrows, spades, and hoes. The hire of a labourer averages a quarter of a rupee a day; but hired labourers are few, and the cultivators assist each other by turns in their various operations. The grain is cut with the common sickle, and threshed by treading out with oxen. Tobacco, with sugar-canes and mulberries, are generally raised; and the country yields pepper, cardamoms, different sorts of indigo, benzoin, stick lac, and other gums, betel, numerous fruits, an abundance of teak and sapan wood, a species of sandal-wood, &c. It abounds with elephants, which are exported in considerable numbers; and with buffaloes, oxen, and other animals found in the adjacent countries. There are, however, no sheep. Asses are used as beasts of burden; but wagons are frequently employed in the conveyance of goods. Gold is found in parts of N. Laos; but in such trifling quantities as hardly to afford the ordinary low rate of wages of the country to those engaged in sifting and washing the sand in which it is found. Tin ore is abundant; and iron, lead, copper, antimony, and silver are met with. Some of these metals are smelted and wrought; but the ores are principally sent in a rough state to Birmah. Silk and cotton fabrics, paper made from the bark of a creeping plant, leather, date-sugar, and gunpowder, are the chief manufactures. There are, however, gold, silver, and iron smiths, mat-makers, potters, embroiderers, and a variety of petty artisans. Spinning and weaving are usually performed by women, who, as in Birmah, conduct a good deal of the retail trade. Some commerce is carried on with the immediately adjacent countries. The inhabitants exchange their lac, sapan-wood, and other dyes, parroquet skins, ivory, rhinoceros' horns, wax, tin, lead, &c., with the Tonquinese for sulphur, cinnabar, gamboge, orpiment, borax, musk, silks, gold thread, embroidery, steel cutlery, paper, crockery, &c. About 50 merchants come annually from Tonquin, each with 20 or 30 horse-loads of merchandise. Large quantities of salt, with spices, woollen cloths, &c., are imported from Rangoon, to which the Laos merchants take jaghery, drugs, dyes, silks, cottons, lacquered wares, gold, silver, copper, and other metals, partly native produce, and partly obtained from China. The intercourse with the Tenasserim provinces is increasing; and some British cotton and woollen goods, salt, &c., are bought by the Shans at Martaban. In N. Laos, however, the people are not dependent on the coast for salt, a good deal, though of inferior quality, being there collected in the plains. A caravan occasionally comes from Siam.

The government is what is commonly, though incorrectly, called a hereditary despotism. The king is assisted by four councillors. The laws, derived from the institutes of Menu, are administered by the councillors, under whom are eight inferior judges. Their general tenor is the same as that of the Siamese laws, but they are not generally enforced with so much rigour. Unlike most E. countries, the people have a right of property in the soil, and may dispose of it at pleasure: waste land may be occupied by any one, and if he cultivate it, he establishes a right to its exclusive possession. In N. Laos, a small military force is kept up. The Shans are said by Kæmpfer to resemble the southern Chinese: but Captain Low thinks them more analogous to the Birmans; to whose dress, habits, customs, &c., their own are very similar. Various books have been written in the Shan language, which is little different from the Pali: it is written in a character similar to the Birmese.

Some of the most striking and venerated Buddhist temples are said to exist in this country. The most noted is that of Nang-rung, N.W. of Zimmel, the cap. of N. Laos. The chief city of S. Laos, Lanchang, is reported to be both populous and comparatively well built. The inhabitants assert that they are the stock whence the Siamese sprung, and this the latter do not hesitate to acknowledge. The emigration of the Siamese southward from Laos is conjectured by Captain Low to have been about the year 638. (Low's Hist. of Tenasserim, in Journ. of Royal Asiatic Soc. v., 245–253; Crawfurd's Embassy to Siam; Hamilton's E. I. Gaz. &c.)

LAPEER, county, Mich. Situated toward the N.E. part of the settled portion of the state, and contains 826 sq. m. Drained by Flint river, and branches, Belle river, and Kearsley creek. It contained in 1840, 3813 neat cattle, 1197 sheep, 6397 swine; and produced 35,479 bushels of wheat, 19,521 of Indian corn, 2055 of buckwheat, 3293 of barley, 96,009 of oats, 23,351 of potatoes, 65,533 pounds of sugar. It had six stores, four grist-mills, 14 saw-mills, four tanneries, two printing-offices, two weekly newspapers: one academy, 20 students, 23 schools, 570 scholars. Pop. 4265. Capital, Lapeer.

LAPEER, p. t. capital of Lapeer co., Mich., 61 m. N. Detroit, 564 W. The village is situated on Flint river, at the confluence of Farmer's creek. It contains a courthouse, jail, six stores, two flouring-mills, one saw-mill. There is much good water in the vicinity. The township contains three saw-mills, two tanneries, two printing-offices, two weekly newspapers; five schools, 198 scholars. Pop. 746.

LAPLAND, the most northerly country of Europe; belonging partly to Russia, and partly to Sweden, between lat. 64° and 71° N., and long. 10° and 49° E.; bounded N. by the Arctic ocean, E. by the White sea, S. by Sweden and Finland, and W. by the Atlantic ocean. Area 150,000 sq. m., about 2-3ds of which belong to Russia. Population vaguely estimated at 69,000, of whom only 8000 are Laplanders, the rest being Swedes, Norwegians, and Russians. That part of Lapland which lies along the N. shore of the gulf of Bothnia, is an extensive plain, abounding in immense forests of spruce and Scotch fir; but at the distance of 80 m. from the sea, the ground becomes gradually elevated, and is at last full of lofty mountains, composed chiefly of primitive and transition rocks, very rich in copper, and other metallic ores. These, between the lat. of 67° and 69° 30′, rise to a height of from 5500 to 6300 feet, which, in this hyperborean region, is 2700 feet above the line of perpetual congelation. These central mountains are the highest in Lapland. The ranges continue all the way to the N. caps, but decline gradually in height. The principal rivers of Lapland are the Torneo, which, taking its rise in the highest mountains, near lat. 68° 30′, holds a course first S.E., and afterward nearly S., receiving tributary streams from the right and left, till it reaches the N. extremity of the gulf of Bothnia, at the town of Torneo. The Kemi, a river almost equally large, rises in the N.E., flows S., and falls into the gulf of Bothnia, not far from the Torneo. The Lulea and Pitea both rise in the mountains of the N.W., in about lat. 68°, and flow S.E., nearly parallel to each other, till they also reach the gulf of Bothnia. In N. Lapland, above lat. 68° 30′, the slope of the ground is N. The Tana, which is the principal river in the N.E., and the Alten, the largest in the N.W., both run into the Arctic ocean. All these, like the rivers of Switzerland, are comparatively small in winter, and become mighty streams in summer, on the melting of the snows. Lapland abounds in lakes: that, called Enara, or Inджаге, in Russian Lapland, in lat. 69°, is of great size. Several of the others are likewise extensive, and are traversed by considerable rivers. The climate of Lapland is celebrated for extreme coldness; but, in fact, it is milder than that of any other region under the same parallel. The coasts of Norwegian Lapland and Finmark are free from ice early in May, whereas

the sea of Siberia is never open till the end of July. The climate of one part of the country, also, differs very much from that of another. In the maritime districts, the temperature is pretty uniform: the winters are not severe, but the summers are raw and foggy; while, in the interior, the winter is intensely cold, but the heat of summer is steady and freezifying. The mean annual temperature at the N. caps (lat. 71° 11′ 30″) is 6° higher than at Enontekis in the interior (in lat. 68° 30′). Yet, at the latter, the thermometer rises in July to 64°, while at the cape it seldom reaches 59°. In truth, the summer begins in May and ends in September; but in the valleys, among the mountains, corn ripens in the short space of three months. The sun being so many hours above the horizon, the heat is then intense, and the clouds of insects are exceedingly troublesome. The cold of winter, on the contrary, is frequently so intense as to freeze brandy, or spirits of wine; and the rivers in the interior are covered with ice to the depth of several feet. Towards the N., the sun remains for many weeks below the horizon in winter, and in summer is as long without setting. During the long night of winter, however, the darkness is relieved by the brightness of the moon and stars, and the vivid coruscations of the aurora borealis. The twilight is also such, that, during several hours each day, it is possible to read without a lamp or candle.

The vegetable productions of the maritime and mountainous districts differ as widely as the climate. In the low country, particularly near the shores of the gulf of Bothnia, are large forests of spruce, Scotch fir, and other resinous trees; potatoes, turnips, and other vegetables, are cultivated; and roses, carnations, &c., deck the gardens during the brief months of summer. In a colder region the spruce disappears, the Scotch fir being the only tree of that class that braves its severity. It, in its turn, declines in vigour, till it totally disappears; and its place is supplied by the birch, which again yields to the Salix glauca, a plant unknown in Britain, and peculiar to cold climates. The Rubus Chamæmorus, Rubus arcticus, and other berry-bearing plants, are here numerous, and support even an additional degree of cold; but we arrive soon after at a climate where nothing is to be seen but a few of the hardiest plants, such as the dwarf birch, with the Salix lapponica, Orobis lapporterra, and other trees and shrubs peculiar to the country. A few mosses still keep their ground; but before reaching the point of perpetual congelation, there is here, as in other countries quite destitute of every species of vegetation, neither plant nor animal to be seen. The reindeer's lichen is of a bright yellow colour, which, as the plant withers, becomes snow-white; it thrives better near the fir forests than in the loftier regions of birches, and a plain covered with this moss forms a Lapland meadow. It is the winter food of the cattle, and, when ground, is used as flour by the inhabitants. Rich pastures also are furnished by the bear's moss (Muscus polytricha), which, on account of its softness and elasticity, is made into beds and mattresses, alleged by travellers to be superior to any in Europe. The root of the Angelica, and the stem of the Fraxinus, are used as food, and of all the grains barley is that which thrives best: but the potato yields a surer harvest, and, if generally cultivated, might afford sufficient sustenance for the inhabitants. The turnip and cabbage, introduced by the Russians, succeed well on the low lands. The best agriculturists are the Flemish colonists, who have raised corn at Alten, in lat. 70°, which may safely be pronounced the N. limit of husbandry; but tillage, generally, is in a very backward state.

Among the animals of Lapland, the reindeer is the most valuable. It serves as the principal beast of burden; its milk is highly valued; its flesh supplies the chief nourishment of the people during a part of the year; its sinews are made into thread; its horns into spoons, and other domestic utensils; and its skin furnishes a great part of their dress. The reindeer bears a great resemblance to the stag, but is much smaller, being in general only four feet in height from the foot to the top of the back, and but two feet long in the body. It is remarkable equally for the elegance of its shape, the beauty of its palmated horns, and the ease with which it supports itself during a long winter of nine months. In summer it feeds on grass, and is extremely fond of the herb called the great water-horse tail; but in winter it refuses hay, and obtains its whole nourishment from the reindeer moss. It thrives best in the cold dry regions of central Lapland, where numerous herds roam at large the whole year round, under the care of shepherds assisted by dogs. The reindeer, indeed, form the chief wealth of the natives. The poorer classes have from 50 to 200; the middle classes from 300 to 700; and the affluent often above 1000 head. The females are driven home morning and evening to be milked, and yield about as much milk as the goat. Horses, oxen, goats, and sheep are common; and in the forests are bears, gluttons, wolves, elks, hares, martens, squirrels, and lemming-rats. Birds of passage arrive in flocks every summer; capercailles, grouse, partridges, and aquatic fowl, are very plentiful near the coast, and summergeyers and eagles soar nearly to the line of perpetual snow. The rivers are stored with salmon, herring, and other fish; and in July and August insects abound in such enormous quantities, that Wahlenberg has supposed that their dead bodies serve as an excellent manure for the soil.

The Laplanders, who call themselves *Some*, are most probably a tribe of Tschoude or Finns, though difference of situation has, in the course of ages, produced a fundamental difference of character. The Finns, an industrious though unpolished race, were encouraged to form colonies in Lapland about a century ago; and their number has since increased rapidly, while that of the Laplanders has been stationary, perhaps on the decline. Of the 27,000 inhabitants of Norwegian Lapland, there are not, it is thought, above 6000 Laplanders. They have swarthy complexions, black short hair, wide mouths, hollow cheeks, and long and pointed chins. They are strong, active, and hardy; but they suffer much from disease, and few live beyond fifty. Dishonesty is general among them, and dram-drinking is often carried to a fatal excess. They were not converted to Christianity till the 17th century. Those of them, Russian province are professedly of the Greek church, while those subject to Sweden are Lutherans. But notwithstanding the efforts of the missionaries, they are still very ignorant both of the doctrines and duties of Christianity, and retain many heathen superstitions.

The reindeer Laplanders live either wholly or principally on the produce of their herds, building their rude huts during summer in the moss pastures of the elevated country, and in winter on the level tracts inhabited by other nations; but the fishing Laplanders confine themselves to the banks of lakes and rivers, and catch fish and beavers, which, as well as skins and venison, they exchange with the Russians and Swedes for spirituous liquors, meal, salt, and tobacco.

The clothing of these half-civilised tribes is abundantly coarse, consisting of a woollen cap, a coat commonly of sheep skin, with the wool inward, and a great coat, either of kersey, or of reindeer skin, with the hair outward. They have no stockings, but a kind of pantaloons of coarse cloth, or tanned leather, fitted close to the legs; their shoes are made of reindeer's skin, the sole being taken from the forehead, and the upper leather from the legs. The women dress nearly in the same manner, but with the addition of some rude ornaments; and, in the case of the more affluent, of mantles and aprons of Russia linen or cotton. These, and leather for the boots of the men, are obtained in the petty traffic of the Laplanders with the Swedes. When travelling, and exposed to the winter blast, it is customary for the natives to cast a hood over the head, neck, and shoulders, leaving only a small opening, through which they see and breathe.

The language of the Laplanders is a Finnish dialect; but it contains so many obsolete and foreign words, that they are not intelligible by the inhabitants of Finland, nor indeed can the tribes in one part understand the language spoken by those of another. The Laponic has been mixed more than the other Finnish tongues with the German and Scandinavian, and hence its principal roots and derivations bear much less affinity with those in the languages of Upper Asia. (*Malte-Brun, Geog. Univ.; Schnitzler, La Russie,* p. 608, &c.)

LAPORTE, county, Ia. Situated toward the N.W. part of the state, and contains 460 sq. m. Bounded N.W. by lake Michigan, on which Michigan city is situated. Drained by Kankakee and Galllon rivers, and Trail creek. It contained in 1840, 9065 neat cattle, 4849 sheep, 19,104 swine; and produced 221,461 bushels of wheat, 270,742 of Indian corn, 6430 of buckwheat, 7545 of barley, 166,994 of oats, 77,594 of potatoes, 56,964 pounds of sugar. It had six commission houses in foreign trade, 34 stores, two flouring-mills, 11 grist-mills, 25 saw-mills, two tanneries, four distilleries, one pottery, three printing-offices, three weekly newspapers; two academies, 195 students; 36 schools, 642 scholars. Pop. 8184. Capital, La Porte.

LA PORTE, p. v., capital of Laporte co., Ia., 145 m. N. by W. Indianapolis, 660 W. It contains a courthouse and a number of stores and dwellings. Nett proceeds of the postoffice, $662.

LAR, a town of Persia, cap. of the prov. of Laristan, 130 m. W.N.W. Gombroon, and 192 m. S.E. Shiraz; lat. 27° 30' N., long. 59° 45' E. Pop. 12,000 ? It stands at the foot of a range of hills in an extensive plain, covered with palm trees. The houses generally are commodious and neatly furnished, and there are several handsome public buildings. The governor's house, in the middle of the city, is surrounded by a strong wall, flanked with towers. The bazaar, which is in good repair, is alleged to be the best structure of the kind in Persia: it is very ancient, and built

on a similar plan to that of Shiraz, but on a much greater scale, with loftier arches, greater length and breadth, and superior workmanship. The castle, on the top of a hill, overlooking the town, is now in ruins. Rain-water being the only water to be found in this parched and arid country, is collected during the wet season in large cisterns, similar to those in the island of Ormus.

Lar was formerly the capital of an Arabic dynasty, destroyed by Shah Abbas II. It is at present in a state of decay; but it still manufactures fire-arms, gunpowder, and cotton fabrics, exchanged at Shiraz and Gombroon for coffee, sugar, Indian silks, and European manufactures. (*Kinneir,* p. 83.)

LARGS, a bor. of barony, and seaport of Scotland, co. Ayr, beautifully situated on a bay of the same name, and overhung on the land side by richly-wooded hills, 22 m., dir. dist. by land, W. by S. Glasgow, and 40 m. by water. Stationary pop. 1900; but in summer, there are sometimes 1000 visiters at sea-bathing. It has an elegant suite of public baths, with a reading-room and library, and various circulating libraries. Though not built on any regular plan, it contains many excellent and substantial houses. The parish church, with its spire and clock, is eminently conspicuous. Many gentlemen's seats are in the neighbourhood.

Largs is celebrated in history as the scene of a great battle, fought in 1263, between Haco, king of Norway, and the troops of Alexander II., in which the former was signally defeated. The cairns and tumuli, erected by permission of the conquerors, by the Norwegians over their slain, are still visible on the S. side of the village.

LARISSA (Turk. *Yenischer*), a town of European Turkey, prov. Trikala, 25 m. N.W. Volo, and 70 m. E.S.E. Yanina; lat. 39° 32' N., long. 22° 40' 15" E. Pop. according to Holland, 90,000, though but little stress can be laid on this statement. It is a walled town, and is situated on the Salembria (an. *Penéus*), crossed here by a bridge of 10 arches. This river approaches it through a tract of woodland, almost concealing it from view, and then flows close at the foot of a convent of dervishes, two large Turkish mosques, and several groups of lofty buildings, soon after disappearing among the woods. The winter floods, which come down from the mountains with great force, frequently occasion great damage to the clay-built houses in the lower part of the town. Internally, Larissa is mean and irregular; near its centre is an open space, having some good bazaars; but the streets are generally ill built, narrow, and filthy; and both houses and people seem to be in the most abject condition. Besides the mosques, there is a Greek metropolitan church; and these, with some baths and a khan, constitute all the public buildings of the place. There is very little trade, and the bazaars are ill supplied with manufactured goods. The plains surrounding Larissa consist of a fine alluvial soil, and are extremely fertile. They produce large crops of Indian corn, wheat, and tobacco; and northward are rich sheep pastures. In fact, there wants nothing but good government and good laws to render Larissa and its vicinity industrious, rich, and populous; but no improvement of any kind need be expected under the worn out, imbecile despotism of Turkey.

If Holland and Dodwell be correct in their opinion, that the modern Larissa occupies the site of the ancient city of the same name, it is of very high antiquity, claiming, in competition with Phthia, the honour of being the birthplace of Achilles, hence called *Larissaus*, and being probably identical with the Πελάσγικον Ἄργος, mentioned by Homer in his catalogue of the Greek forces. (*Il. b.,* 681.) At a subsequent period it acquired some celebrity from its adoption of the democratical form of government, and from its zealous support of the Athenian cause during the Peloponnesian war. (Comp. *Aristot. Pol.* v., 6, with *Thuc.* ii., c. 22.) It afterward fell into the hands of Philip of Macedon and his successors, under whom it remained till the subversion of their empire by the Romans. It appears to have declined under the early Roman emperors from its ancient importance. Lucan says of it:

" Atque olim Larissa potens—— ."

(*L.b. vi., lin. 355.*)

The town and its neighbourhood were subject in ancient times to the same violent and sudden inundations which now cause such extensive mischief. (*Holland's Travels,* p. 388-396; *Dodwell's Tour,* i., 100; *Cramer's Greece,* i., 396.)

LARISTAN, a small prov. of Persia, part of the ancient *Caramania,* extending along the N. shore of the gulf of that name, between 26° and 29° N. lat., and 53° and 56° E. long., bounded N.W. by Fars, and N.E. by Kerman. Area, 16,000 sq. m. Pop. uncertain. It is the poorest and least productive province of Persia, diversified indeed with plains and mountains, extending to the sea, but so arid and so destitute of wholesome water that, were it not for the peritifical rains, which fill the cisterns of the natives, and enable them to cultivate the date-tree, with small quantities of wheat and barley, it would be quite uninhabitable. The coast is in the possession of different Arab tribes, who, under their respective sheikhs, maintain their independence, paying only a trifling tribute to the king. They are chiefly pirates by profession, and reside in small towns or road ports scattered along the shore of the gulf: the chief of these are Congoon, having about 3600 inhabitants; Nakhile, opposite the island of Shitwar; and Mogoo, which has one of the most secure roadsteads in the gulf. The interior of the country has not been visited by Europeans. Lar is the capital, which see. (*Kinneir's Persia*, p. 51.)

LARNE, a seaport town of Ireland, co. Antrim, on a creek of the inlet of the sea called Larne lough, 16 m. N. by E. Belfast. Pop. in 1831, 2615. It consists of an old and a new town, and has, besides, the parish church, a Roman Catholic chapel, three Presbyterian and one Methodist meeting houses, and a national school. A manor-court is held every six weeks, and petty sessions every fortnight. It is a constabulary and coast-guard station. It formerly carried on a brisk trade in salt, but its traffic is now chiefly confined to the export of linen, grain, and provisions. These amounted in 1835 to the value of £98,208, of which linen was estimated to make £40,000. Considerable quantities of lime are also exported. Coal is the principal article of importation. The harbour is land-locked, and is an admirable one for the smaller class of vessels, which cannot find depart at all times of the tide. Markets on Tuesdays; fairs on July 31, Dec. 1, and on the first Monday of every month. Post-office revenue in 1830, £453; in 1836, £516. A branch of the Belfast bank was opened in 1836. Fish is abundant, particularly mackerel, hake, cod, and mullet; salmon is taken near the entrance of the bay. The fishermen do not restrict themselves to the fishing, but are also agriculturists, and go to sea only when there is a prospect of a large take.

LARNICA (an. *Citium?*), a seaport town of the island of Cyprus, on its S.E. shore, at the bottom of the bay of Salines, 23 m. S.E. Nicosia, lat. 34° 54′ 30″ N., long. 33° 49′ 45″ E. Pop. estimated by Kinneir at 5000. It consists of an upper and a lower town: the latter, called the Marina, is built along the sea shore; the other is a little more inland, and on higher ground. The houses, with the exception of a few belonging to the Frank merchants, are built of mud bricks dried in the sun, and are mean in the extreme; they have mostly, however, very fine gardens, but these being inclosed by high walls, contribute little or nothing to the beauty of the town, as seen from the streets. It is the seat of a Greek bishopric, and in the upper town is the cathedral and convent of St. Saviour, and the lower has a mosque, a convent, the chapel of St. Lazarus, and the remains of a castle constructed by the princes of the house of Lusignan. Being situated on the verge of a marshy plain, screened by high mountains from the cooling influence of the N. winds, and having near it extensive lagoons, which in summer produce huge quantities of salt, Larnica is hot, and, at certain seasons, unhealthy. It has no good water, except what is brought to it by an aqueduct constructed, in 1747, by a Turkish emir. There is no harbour, but the bay, which opens to the S.E., and derives its name from the salt lagoons, affords good anchorage in deep water at no great distance off shore.

Such has been the influence of that rapacious and intolerant despotism under which this noble island has long groaned (*see* CYPRUS) that Larnica, though so poor and decayed, is now its second city, the emporium of its commerce, and the principal residence of the foreign consuls. The exports consist of wheat, several cargoes of which are exported to Spain and Portugal, with barley, cotton, silk, wine, and drugs; the imports are rice and sugar from Egypt, and cloth, hardware, and colonial produce from Malta and Smyrna. When Kinneir visited the island, this traffic was carried on in Levantine ships, under English colours, and such, probably, is still the case. (*Kinneir's Asia Minor, &c.*, 183; *Drummond's Travels*, 140.)

Drummond, Pococke, and the Abbé Mariti, concur in opinion that Larnica occupies the site of the ancient *Citium*; while Kinneir and others suppose the latter to have been near a cape, still called Chitti, a few miles S.W. from Larnica, where there are numerous tumuli and hillocks of rubbish. The probability, however, seems to be in favour of the supposition that the site of Larnica and Citium are really identical. (*Drummond*, p. 250; *Clarke*, iv., 29, 8vo edition.)

Citium was founded by the Phoenicians at a very remote period, and will be forever memorable as the birthplace of Zeno, the founder of the stoical system of philosophy. Perhaps we may be excused for saying that, notwithstanding his error in supposing pleasure and pain to be absolutely indifferent, no system of philosophy has ever been proposed so well fitted as that of Zeno to imbue its votaries with

the purest principles of benevolence, and the most heroic magnanimity.*

> " Secta fuit, servare modum, finemque tenere,
> Naturamque sequi, patriaeque impendere vitam ;
> Nec sibi, sed toti genitum se credere mundo."
>
> Lucan, ii., v., 380.*

Cimon, the great Athenian commander, either died at the siege of Citium or immediately after he had taken it. The epoch of the destruction of the city is unknown.

LA SALLE, co., Ill. Situated towards the N.E. part of the state, and contains 1364 sq. m. Watered by Illinois, Fox, and Vermillion rivers and their branches, which afford extensive water-power. The Illinois and Michigan canal passes through the county. It contained in 1840, 7306 neat cattle, 2216 sheep, 11,269 swine; and produced 118,843 bushels of wheat, 131,165 of Indian corn, 141,565 of oats, 49,524 of potatoes. It had 43 stores, 6½ grist-mills, 24 saw-mills, one tannery, one pottery, two printing-offices, two weekly newspapers; 27 schools, 115 scholars. Pop. 9348. Capital, Ottawa.

LA SALLE, p. t. Monroe co., Mich., 62 m. S.S.W. Detroit, 461 W. It has three schools, 197 scholars. Pop. 905.

LASSA, or H'LASSA (*Land of the Divine Intelligence*), a city of the Chinese empire and the cap. of Thibet, prov. Oui, 300 m. E. by N. Katmandoo, the cap. of Nepaul; lat. 29° 30′ N., long. 91° 49′ E. Pop. uncertain, but conjectured to be about 34,000. It is situated on the Galdjao, a tributary of the Sanpo, about 26 m. from its confluence with that river, in an extensive and fertile plain about 60 m. long and 35 m. broad, surrounded by lofty mountains. It is not walled, but its streets and houses, towers, bazaars, and handsome detached residences, indicate its importance, comparatively at least, with other towns of the kingdom. The houses are built of a brown stone, are two or three stories high, with tolerably lofty rooms, and, though somewhat grotesque, give the idea of wealth and respectability. The great temple of Buddha, which is likewise the residence of the Dalai Lama, the pontifical sovereign of Thibet, stands on the hill Botá-la, in the W. part of the city, and consists of an extensive range of square-shaped buildings, crowned in the centre with a gilded dome, and occupying altogether an area of about 40 begahs. It comprises, according to the Chinese geographers, 10,000 apartments, varying in size and grandeur according to the supposed dignity of the idols which they respectively contain. Contiguous to the temple, on its four sides, are the four celebrated monasteries of Brepkung, Sera, Ghalden, and Samyitl, all alleged to be inhabited by upwards of 4000 monks, and much resorted to by the Chinese and Mongols as schools of philosophy and Buddhism. In and near the city are five other temples, built on the same general plan, but very inferior in size and splendour to that just described. Lassa, besides being the resort of zealous Buddhists from all parts of China, Turkestan, Nepaul, &c., is a place of considerable trade in silk, wool, and goats' hair, woollen cloths and cashmeres, velvets, linens, sandwich, bezoar, various kinds of fruit, silver bullion, gold dust, and precious stones, chiefly with N. Hindostan, Nepaul, Bootan, Great Bucharia, and China; and in the markets, where the goods are exposed for sale on mats, regularly-appointed market inspectors fix the prices, from which no deviation is allowed. Handicraft is much followed, and with great success; and the lapidaries, workers in metal, and engravers are not inferior to the Chinese. (*Ritter, Erdkunde von Asien*, iii., p. 237-251; *Rennell's Hindostan*, p. 306; *Hamilton's Gaz.*)

LATAKIA, or LADIKIEH (an. *Laodicea ad mare*), a town of Syria, in the pash. of Aleppo, 90 m. S.W. Aleppo, and 74 m. S. by E. Iskenderoon; lat. 35° 36′ 30″ N., long. 35° 43′ E. Pop., according to Mr. Consul Moore, 3000; but, according to Mr. Barker, 10,000. (*Bowring's Report on Syria*, p. 114.) The town comprises an upper and a lower part, separated by gardens and plantations. The lower portion, called the *Scala*, consists of a double street, running parallel to the shore, and another leading down to it from the upper town, having coffee-houses, and places of resort for seafaring people. The port is a small, shallow basin, with a narrow entrance, and well sheltered, except westward: on its N. side is a ruined castle, standing on a rock connected by arches with the mainland; and at the E. end are the custom-house, landing-place, and several large warehouses. The upper town, which is in a very dilapidated state, in consequence of the damage occasioned by frequent earthquakes, consists of several narrow and irregular streets: the houses are constructed of cut stone, flat-roofed, usually two stories high, with an inner court. The greatest ornament of the place is a triumphal gate, between 30 ft. and 40 ft. in height, encircled near its summit by a handsome entablature: its four arches are in the Roman style of architecture, and, as the general appearance of the

---

* The reader who wishes for further information with respect to the Stoics will do well to refer to Smith's *Moral Sentiments*, part vii., sect. ii., cap. 1, and to the *Esprit des Loix of Montesquieu*, liv. xxiv., cap. 10.

157

building denotes great antiquity, it was probably erected in honour of Julius Cæsar, or, perhaps, Germanicus. (*Kinneir*.) The corners are adorned with handsome Corinthian pilasters, and one of its fronts exhibits a basso-relievo, with arms and martial instruments.

At no great distance is a mosque, built from the ruins of another ancient edifice, with Corinthian columns; and amid the rocks and crags N. of the town is a large necropolis, containing numerous square sarcophagi, similar to those seen in the island of Milo. There are three other mosques and two Greek churches. The bazaars are poor and insignificant, and the only considerable article of trade is tobacco, raised near the town in large quantities, and highly prized all over the Levant, and at Constantinople. It pays a duty on being reaped of 34 piastres per cantar, and of 3 piastres per cantar on exportation.

The produce of cotton in the Latakia district is not usually more than sufficient for the consumption of the country; but, when it exceeds it, the excess is exported to the French and Italian ports, the average price being 1900 piastres per cantar. Bees'-wax, scammony, and sponge, are the other chief articles of export. The imports comprise sugar, coffee, spices, cotton twist, and printed goods, woollen cloths, shawls, and tin. The trade of Latakia, however, is much restricted by the badness of its port, which is so choked up with mud and sand as to be inaccessible to vessels of more than 100 tons burden. Subjoined is an account of the number and tonnage of the vessels, and the value of their cargoes, that entered and left the port of Latakia in 1835, 1836, and 1837:

Years	Arrived.			Departed.		
	Ships.	Tonnage.	Values.	Ships.	Tonnage.	Values.
			£.			£.
1835	102	11,157	104,590	86	10,440	85,275
1836	106	12,382	131,347	102	11,647	99,713
1837	93	10,216	92,925	93	10,447	90,736

Latakia is the representative of the ancient Laodicea, so named by its founder, Seleucus Nicator, in honour of his mother, and was a town of considerable importance before the conquest of Syria by the Romans. It was visited by Julius Cæsar when on his way from Egypt to Pontus, and is styled Juliopolis on some of its medals. During the civil wars, Dolabella, with his fleet and army, was shut up in it by Cassius, and obliged to surrender. It became a bishop's see early in the Christian era, and was held by the Christians when the crusaders invaded Syria. It was afterward included in the empire of Saladin, and was finally added to the Turkish dominions by Selim I., in 1517. The ruins of the ancient city fully attest its size and grandeur, and offer ready building materials to the modern inhabitants. The acropolis stood on a tabular summit S.E. of the town, but nothing remains of it but a few wells and cisterns. *Kinneir's Asia Minor,* p. 163–169; *Olivier, Voyage en Syrie, &c.,* iv., 129; *Bowring's Report.*)

LATIMORE, t., Adams co., Pa., 15 m. N.E. Gettysburg. Watered by Bermudian creek and its tributaries. York sulphur springs lie 3 m. S. of the village. It contains one store, one fulling-mill, two woollen factories, four gristmills, six saw-mills, two oil-mills, one distillery, one pottery; six schools, 200 scholars. Pop. 1013.

LAUBEN, or LUBEN, a town of Prussian Silesia, gov. Liegnitz, cap. circ. of its own name, on the Queis, 40 m. W.S.W. Liegnitz. Pop. 3640. It is surrounded with old walls, and garrisoned by invalids. It is the seat of judicial courts for the town and circle; has a Roman Catholic and three Protestant churches, a gymnasium, an orphan asylum, two hospitals, a school for teaching the art of spinning woollen yarn, and some trade in linen and woollen fabrics. (*Von Zedlitz; Berghaus.*)

LAUDER, a royal and parl. borough and market-town of Scotland, co. Berwick, distr. of Lauderdale, of which it is the cap., on the Leader, a tributary of the Tweed, on the line of road between Edinburgh and Coldstream, 24 m. S.E. of the former, and 22 m. N.W. by W. of the latter. Pop. (which has long been stationary) 1075. The only public buildings are the parish church, a dissenting chapel, the town-house, and jail. Thirlstane castle, the ancient residence of the noble family of Lauderdale, is within ½ m. of the town. There are four schools, of which one is endowed, average attendance 300; and four subscription libraries. A common, comprising 1695 acres, is divided among the burgesses. In 1482, Cochrane and other minions of James III. were hanged by order of the Earl of Arran, and other noblemen, over the parapet of a bridge in the vicinity of this town. Lauder unites with Haddington, Dunbar, Jedburgh, and N. Berwick in sending one member to the House of Commons. Registered electors in 1839–40, 52.

LAUDERDALE, co., Ala. Situated in the N.W. part of the state, and contains 672 sq. m. Bounded S. by Tennessee river, which receives from it Shoal, Cypress, Blackwater, and other creeks. A canal extends around the

Muscle shoals in this county. This work is not completed to the head of the shoals. The part in use is 35½ m. long, and cost $571,895. The whole canal 30 Brown's ferry, at the head of the shoals, is estimated to cost $1,281,057. It contained in 1840, 13,515 neat cattle, 10,466 sheep, 44,281 swine; and produced 46,363 bushels of wheat, 2613 of rye, 741,673 of Indian corn, 107,345 of oats, 18,394 of potatoes, 16,467 pounds of tobacco, 3,691,199 of cotton, 1367 of sugar. It had one commercial and two commission houses in foreign trade, 25 retail stores, one cotton factory with 285 spindles, five flouring-mills, 43 grist-mills, 21 saw-mills, six tanneries, 15 distilleries, two printing-offices, two weekly newspapers; two academies, 85 students; 17 schools, 439 scholars. Pop.: whites, 9447; slaves, 4969; free coloured, 69; total, 14,485. Capital, Florence.

LAUDERDALE, co., Miss. Situated in the E. part of the state, and contains 700 sq. m. There were in 1840, 20,500 pounds of cotton produced. It had 10 grist-mills, 19 sawmills. Pop.: whites, 3992; slaves, 1383; free coloured, 13; total, 5358. Capital, Marion.

LAUDERDALE, co., Tenn. Situated in the W. part of the state, and contains 375 sq. m. Bounded W. by Mississippi river, N. by Forked Deer river, S. by Big Hatchee river. Watered by Coal creek. It contained in 1840, 4687 neat cattle, 802 sheep, 13,670 swine; and produced 3756 bushels of wheat, 179,985 of Indian corn, 8779 of oats, 15,545 of potatoes, 2127 pounds of rice, 43,353 of tobacco, 233,595 of cotton. It had three stores, 14 grist-mills, four saw-mills, one tannery, three distilleries; seven schools, 194 scholars. Pop.: whites, 9492; slaves, 1910; free coloured, 2; total, 3435. Capital, Ripley.

LAUENBURG, a town of the Danish dom., cap. of the duchy of same name, on the Elbe, 28 m. S.E. Hamburg. Pop. about 3400. It has the ruins of a castle formerly occupied by the dukes of Saxe Lauenburg, a church, a hospital, and a large market-place. A brisk transit trade is carried on between the Elbe and Lubeck through this town, which is also a station for collecting tolls on the Elbe, amounting to between 40,000 and 50,000 rix-dollars a year. (*Horschelmann's Stein,* i., 594; *Berghaus.*)

LAUGHERY, t., Dearborn co., Ia. It contains nine stores, one flouring-mill, one saw-mill; one academy, 30 students; four schools, 125 scholars. Pop. 1501.

LAUNCESTON, a parl. and mun. bor., market town, and par. of England, co. Cornwall, in the N. division of hund. East. on the Attery, a tributary of the Tamar, 19 m. E.N.E. Bodmin, 30 m. N.N.W. Plymouth, and 209 m. W. by S. London. Pop. of parl. bor. (which comprises, besides the old bor., the parishes of St. Stephen, St. Thomas, Lawhitton, and S. Petherwin), in 1831, 5394. The town consists of two chief avenues on the London and Tavistock roads, intersecting each other almost at right angles, crossed by several narrow and mean-looking streets. It was formerly surrounded by walls, parts of which, with two gates, are yet standing. The ruins of an ancient castle cover a large extent of ground, and attest its former strength and importance. A part of it to keep was once used as a county jail, but the prisoners are now sent to Bodmin, which has been the assize town since 1838. A small guildhall is the only public building devoted to civil purposes. The church, a handsome Gothic structure built of granite blocks, enriched with curiously-carved ornaments, has a lofty tower at its W. end: the living is a curacy of the yearly value of £116. There are places of worship also for Wesleyans and Baptists, with attached Sunday schools. A grammar school, founded by Queen Elizabeth, has, according to the charity commissioners, fallen into a state of decay, there having been no master since 1821. Bacon's charity school is in nearly as useless a condition, and the only place of instruction for the poor in the national school, attended by about 260 children. Numerous money charities are chiefly distributed by the corporation. This is neither a manufacturing nor a commercial town. Serge-weaving and woolspinning formerly employed a considerable number of hands, but the trade has wholly disappeared. The removal of the assizes and quarter sessions has, also, deprived the town of much of its activity, and it now depends chiefly on its retail trade and on its markets, which are large and well attended. Market-day, Saturday. Cattle fairs, first Thursday in March, third ditto in April, Whit-Monday, July 6, Nov. and Dec. 6.

Launceston, otherwise called Dunheved, received its first charter from Richard, Earl of Cornwall, in the 13th century, and its privileges were confirmed by Richard II., and many subsequent sovereigns. It is governed under the Municipal Reform Act by four aldermen and 12 councillors, but it has no commission of the peace. Corporation revenue in 1838, £430. Launceston returned two members to the House of Commons from the 23d Edward I. down to the passing of the Reform Act, which deprived it of one member. Previously to this act, the members, though formally elected by the corporation, were, in fact, mere nominees of the propri-

else, the Duke of Northumberland. Besides depriving it of one member, the Reform Act enlarged the limits of the borough, as stated above. Registered electors in 1839–40, 362.

LAUREL, co., Ky. Situated towards the S.E. part of the state, and contains 400 sq. m. Bounded W. by Rockcastle river, and drained by its branches. It contained in 1840, 3081 neat cattle, 3622 sheep, 11,174 swine; and produced 1664 bushels of wheat, 29,960 of Indian corn, 92,335 of oats, 5680 of potatoes, 4166 pounds of tobacco. It had seven stores, 13 grist-mills, six saw-mills, two tanneries; seven schools, 158 scholars. Pop.: whites, 2964; slaves, 149; free coloured, 6; total, 3079. Capital, London.

LAUREL MOUNTAINS lie W. of the main Alleghany range, extend from Pennsylvania across Virginia to Kentucky; and, under the name of Cumberland mountains, divide Virginia from Kentucky, and cross Tennessee, terminating near its S. border.

LAURENCE, or LAWRENCE (ST.), the principal river of N. America, and, when considered, as it should be, in connexion with the chain of lakes or inland seas of which it is the outlet, is one of the largest rivers in the world, extending, from W. to E., through about 27° of long., and about 8° of lat. Regarding, then, the St. Laurence in this point of view, or as a general name for the connecting line of that great river or water system that unites with the Atlantic in the gulf of St. Laurence, its remotest source will probably be found to be the St. Louis, an affluent of lake Superior, rising in the table-land of the Huron country, near the sources of the Mississippi, flowing S., and of the Red river, flowing N. It receives different names in different parts of its course, being, as already seen, at first the St. Louis; between lake Superior and lake Huron, the St. Mary; between lakes Huron and Erie, the St. Clair and Detroit; between lakes Erie and Ontario, the Niagara; and from Ontario to Montreal it is sometimes called the Cataraqui or Iroquois, its course from Montreal to the sea being the St. Laurence, properly so called, but it is now usually called the St. Laurence from lake Ontario to the sea. Considered in this point of view, its entire course, from its source to its mouth in the gulf of St. Laurence, in about long. 64° 30′ W., may be estimated at upwards of 2200 m. Besides traversing lakes Superior, Huron, Erie, and Ontario, the lake St. Clair, and some similar sheets of water, are mere enlargements of its bed. Lake Michigan, also, is included in its basin, which is roughly estimated by Darby to comprise an area of upwards of 500,000 sq. m., including the largest collection of fresh water to be found on the surface of the globe. (Darby's Geog. View, &c., 200, 201, 216, 231.) For considerably more than half its extent, the St. Laurence forms the boundary line between the British N. American territories and those of the United States.

The source of the St. Louis is estimated at about 1192 ft. above the sea level. (Darby, 201.) The elevation of the river to succeeding portions of its course, with the estimated area of the great inland seas and smaller lakes, of which it is the connecting link and outlet, are exhibited in the following table :—

	Elevation above the level.	Mean depth.	Mean length.	Mean breadth.	Area.
	Feet.	Feet.	Miles.	Miles.	Sq. miles.
Lake Superior	641	900	340	90	30,000
— Huron	596	900	250	95	16,000
— Michigan	600	900	300	50	15,000
— Erie	565	120	230	35	5,000
— Ontario	231	492	130	50	5,400
River St. Laurence and smaller lakes		50			1,200
Total water surface					72,600

The St. Laurence varies very considerably in breadth, in the middle part of its course inclosing a great many islands, and forming numerous rapids. In those parts of St. Mary, St. Clair, Detroit, and Niagara rivers, where no large islands are met with, the breadth of the stream is usually from ½ m. to 2 or 3 m. At the Sault of St. Louis, 5 m. above Montreal, the river narrows to 5 furlongs; and at Quebec, it is not more than 1214 yards across; but between those cities its average width is 2 m. From Quebec, the breadth of the St. Laurence begins to increase rapidly. Immediately beyond the island of Orleans it is 11 m. broad; where the Saguenay joins it, 18 m.; at Point Peïes, upwards of 29 m.; at the bay of Seven Islands, 70 m.; and at the island of Anticosti, about 220 m. from Quebec, it rolls a flood into the ocean nearly 100 m. across.

The basin of the St. Laurence is supposed by Darby to contain "more than the half of all the fresh water on this planet!" Taking the area, mean depth of the lakes, &c., as given above, their solid contents will amount to 1,547,911,782,360,000 cubic ft. of water, being sufficient to envelope the entire earth with a watery covering 3 in. in depth. (Darby, p. 232.)

The annual discharge, however, though prodigiously great, does not, from the nature of the basin, bear so considerable a proportion to the contained body of fluid as might, perhaps, have been expected. Darby, from observations made at three different places, estimated the hourly discharge at the enormous amount of 1,572,704,000 cubic ft. This estimate, continues Darby, "exceeds by more than a half the quantity which, on another occasion, I estimated for the Mississippi; and though contrary to my own opinion when I first arrived on the banks of the St. Laurence, I am convinced it falls below reality." (Geogr. View, 223.)

The source of the St. Laurence (St. Louis) being 1192 ft. above the level of the sea, the average fall of the river will, perhaps, be somewhat more than 6 inches a mile. But this fall is very unequally distributed, on account of the many, and in one instance stupendous, cataracts, rapids, &c., interspersed along the river's course. The Niagara, between lakes Erie and Ontario, has within the short distance of 35 m. a descent of at least 334 ft., 164 of which are contributed by the Great Falls. The St. Mary, between lakes Superior and Huron, has a fall of 22 ft. in 960 yards; and the rapids are so numerous and dangerous between Kingston and Montreal, that an extensive line of canal navigation has been cut, at a vast expense, to connect lake Ontario with the Ottawa, and enable ships to avoid this portion of the river. (For a more particular description of the Great Falls, the chief lakes through which the St. Laurence passes, and other parts of the basin, see arts. NIAGARA, and lakes SUPERIOR, HURON, ERIE, &c.)

The great Canadian lakes, especially the three upper lakes, receive few tributaries of any consequence; but the St. Laurence, in the middle and lower part of its course, is augmented by several considerable rivers, of which the Ottawa, from the N., uniting with it near Montreal, and the Saguenay, also from the N., uniting with it 120 m. below Quebec, are the most important.

The St. Laurence is said by Darby to be as remarkable for its uniformity throughout the year in the diurnal and monthly expenditure of its waters as the Mississippi is for its continual change. A rise of 3 ft. is a more remarkable phenomenon in the former than a rise of 30 would be in the latter. The two rivers differ widely also in numerous other particulars. The waters of the Mississippi are turbid; those of the St. Laurence and its lakes are highly transparent. In the course of the Mississippi few lakes or enlargements occur, its banks are low, much of the surface within its basin consists of open grassy plains, and before it disembogues it divides into numerous channels; the St. Laurence, on the contrary, consists, in great part, of a chain of vast lakes; as its bed enlarges, it has shelving or precipitous banks, generally covered with primeval forests; and, instead of a delta, it forms at its mouth a large estuary.

The St. Laurence is the great commercial thoroughfare of our Canadian provinces, and the northern states of the American union. Its banks, and those of its lower lakes, are studded with flourishing cities and towns, as Quebec, Montreal, St. Francis (Three Rivers), Cape Vincent (Kingston), Toronto, Buffalo, Oswego, &c., and others are daily springing into existence. The rise of the tide is perceptible as high as Three Rivers, 426 m. up the St. Laurence, and nearly midway between Quebec and Montreal. The river is navigable for ships of the line to Quebec, and for ships of 600 tons to Montreal, 580 m. from the sea, though the navigation be in some places obstructed by rocks and shoals. Beyond the latter point, however, a succession of rapids, especially between Cornwall and Johnston, unfits it for the navigation of other than flat-bottomed boats of from 10 to 15 tons. Further up, Ontario and Erie are navigable for ships of the largest size, as is the Niagara river, both above and below the falls. The falls of Niagara are avoided by the Welland canal, a work undertaken by a company incorporated in 1825. This canal, into the formation of which the Ouse, Welland, and Chippeway rivers enter, is 43½ m. in length, 56 ft. in breadth at its surface, and 26 ft. at its base, 8½ ft. deep; and has 37 wooden locks, 100 ft. long, 22 ft. wide, and capable of admitting ships of 125 tons. Detroit river is no more than 7 or 8 ft. in depth, and the lake and river of St. Clair are navigable only for steamboats and schooners; but beyond this, a wide navigation for ships of any magnitude extends nearly to the falls of St. Mary. Boats of 6 ft. draught may reach the foot of these falls, but they cannot ascend them, though canoes, at great risk, sometimes venture to shoot downwards. The falls of St. Mary are generally avoided by a portage of 2 m.

It is thus seen that there is a continued navigation for vessels of medium burden from the head of lake Huron to Kingston on lake Ontario, and from Montreal to the mouth of the St. Laurence. The water communication between Kingston and Montreal is effected chiefly by a chain of canals, the principal being the Rideau canal, constructed by the Canadian, or rather the English government, connecting lake Ontario with the Ottawa. Rideau river and lake, the Indian lake, and the Little Cataraqui, form parts of its course. It admits vessels of 195 tons. The Grenville and La Chine canals, with the Ottawa, continue the communication to Montreal; the Grenville canal is, however, only adapted for vessels not exceeding 20 ft. in width. On the side of the United States, the Grand Erie, Oswego, and Champlain canals (see NEW-YORK, ERIE, LAKE, &c.) unite the basin of the St. Laurence with the basins of the Hudson and Susquehanna; as the Ohio and Pennsylvania canals (see OHIO, PENNSYLVANIA) do with the basin of the Mississippi. A canal line has been proposed in Upper Canada between lakes Huron and Ontario, by which an easy and direct navigation of less than 300 m. might be substituted for one that is roundabout and often difficult, of between 600 and 700 m. Few improvements of the kind would, looking at the map merely, appear to be more advantageous and easily effected; but, in point of fact, a height of 387 ft. would have to be surmounted in a short distance by the canal; and we therefore incline to think that the project is at least premature, and should be postponed till the province be richer and better able to bear the expense.

Strong tides prevent the St. Laurence being covered with compact ice below Quebec; but the enormous masses driven in every direction by the winds and current render that portion of the river unnavigable for nearly half the year. Between Quebec and Montreal the water communication is totally suspended by the frost from the beginning of December to the middle of April. The navigation of Ontario closes in October. During the winter the N.E. part of that lake, from the bay of Quinto to Sackett's Harbour, is frozen across, and the rest of its surface is usually frozen to a considerable distance from the shore. Lake Erie is not so much encumbered with ice as lake Ontario, while lakes Huron and Michigan are more encumbered. On lake Superior the ice is said to extend to 70 m. from its shores. The frost, however, by no means stops commercial intercourse, but forms the rivers and lakes into excellent roads, on which vehicles of all descriptions are used. Among these are ice-boats, built like other vessels with a rudder, mast, sail, &c., and resting on iron skates attached at either end to cross-bars under stem and stern. One of these ice-boats has, it is said, sailed before the wind from Toronto to Fort George on Niagara, a distance of 40 m., in little more than three quarters of an hour! (Darby, Geog. View, &c., &c. Laurence Basin, p. 200-251; New-York Gaz., 11-36, &c.)

LAURENCE, or LAWRENCE (ST.), GULF OF, a bay of the Atlantic, chiefly between the 46th and 51st deg. of N. lat., and the 57th and 65th of W. long., bounded N. by Lower Canada and Labrador, E. by Newfoundland, S. by Nova Scotia and cape Breton, and W. by New Brunswick and the peninsula of Gaspé (Lower Canada). At its N.W. extremity it receives the river St. Laurence; and it communicates with the ocean on the N.E. by the strait of Belle-isle, between Labrador and Newfoundland, on the S.E. by its principal outlet, the channel called St. Paul's, between Newfoundland and cape Breton, and on the S. by the Gut of Canso, between cape Breton and Nova Scotia. It contains the large islands of Anticosti and Prince Edward; and the Magdalen islands, a group about lat. 47° 30', and between long. 61° 27' and 62° W., inhabited by perhaps 1000 Canadian, French, English, and Irish settlers, who carry on a profitable fishery. The shores of the gulf are generally precipitous, barren, and inhospitable; and its coasts are very prevalent. A powerful current sets continually from Hudson's strait into the gulf, through the strait of Belle-isle, and meeting the stream from the estuary of the St. Laurence forms a dangerous race off the S. coast of Newfoundland. (Purdy's Memoir of the Atlantic, p. 105, 144; Encyc. Amer.)

LAURENS, district, S. C. Situated toward the N.W. part of the state, and contains 920 sq. m. Bounded N.E. by Ennoree r., S.W. by Saluda r. Watered by Reedy r. and its branches. The Saluda is navigable for boats carrying 70 bales of cotton. It contained in 1840, 20,475 neat cattle, 13,260 sheep, 47,222 swine; and produced 186,296 bushels of wheat, 685,473 of Indian corn, 175,436 of oats, 30,676 of potatoes, 19,895 pounds of tobacco, 5,910,368 of cotton. It had 96 stores, two woollen factories, 20 flouring-mills, 41 grist-mills, 34 saw-mills, 10 tanneries, 64 distilleries; six academies, 906 students, 44 schools, 905 scholars. Pop.: whites, 12,579; slaves, 8011; free coloured, 101; total, 21,584. Capital, Laurensville.

LAURENS, county, Ga. Situated a little S.E. of the centre of the state, and contains 780 sq. m. Watered by Oconee r. and its tributaries. It contained in 1840, 17,045 neat cattle, 2947 sheep, 17,355 swine; and produced 9048 bushels of wheat, 146,990 of Indian corn, 4086 of oats, 10,745 of potatoes, 615,892 pounds of cotton. It had 11 stores, 16 grist-mills, six saw-mills, two distilleries; three academies, 90 students, two schools, 36 scholars. Pop.: whites, 3078;
slaves, 3692, free coloured, 5; total, 5386. Capital, Dublin.

LAURENS, p. t., Otsego co., N. Y., 15 m. S.W. Cooperstown, 83 m. W. Albany, 358 W. Organized in 1810. Drained by Otsego cr. It contains a Presbyterian church, six stores, one furnace, three fulling-mills, three grist-mills, nine saw-mills, two tanneries, one distillery; 16 schools, 540 scholars. Pop. 2172.

LAURENSVILLE, p. v., capital of Laurens dist., S. C., 70 m. N.W. by W. Columbia, 496 W. Pleasantly situated at the head of Little r., and contains a neat courthouse, a jail, a church, a public library, several stores, 40 dwellings, and about 300 inhabitants.

LAUSANNE, a city of Switzerland, cap. canton of Vaud, at the termination of a spur from the chain of the Jura, being, according to Ebel, 480 ft. above the level of the lake of Geneva, from the N. shore of which it is about 1 m. distant, and 30 m. N.E. Geneva: lat. 46° 31' 5" N., long. 6° 47' 32" E. Pop. 14,196. It is finely situated on three eminences, and their intervening valleys; but, from being on uneven ground, its streets are steep and irregular; they are also generally narrow and ill paved, and the interior of Lausanne by no means corresponds with its exterior appearance. It is divided into six quarters, the city and five suburbs, and is now an open town, but on its S. side are some remains of ancient walls. At the highest point of the city is the castle, a massive square building of stone, flanked at its angles by four brick towers. It was originally the residence of the bishops of Lausanne, but is now the council-house of the canton: its terrace, and that of the cathedral, commands magnificent views of the vicinity, the lake, and, far beyond, the stupendous mountains of Savoy. The church, formerly the cathedral, a vast Gothic building, founded about 1000, but not finished till the 13th century, is certainly the finest religious edifice in Switzerland. It has two large towers, one supporting an elegant spire, the summit of which is 240 ft. above the ground, and a fine round window of stained glass, 30 ft. in diameter: in the interior are some singular specimens of architecture; and among others the tomb of Amadeus VIII., duke of Savoy. This personage, after abdicating the dukedom, which he had greatly enlarged, and governed with singular ability, was elected pope, by the title of Felix V., under which name he is best known in history. But another pope having been elected, about the same time, by a different party in the church, Felix, to terminate the schism, resigned the tiara in 1449. He died within two years of this event. (See Biographie Universelle, art. Savoie, Amé VIII.) The church of St. Francis; the cantonal college, with a library and museum, comprising collections of antiquities and minerals found in the neighbourhood; the bishop's palace, now appropriated to a school of mutual instruction and the district prison; the cantonal hospital, a fine edifice in the Tuscan order; the lunatic asylum of Champ d'Air; the new penitentiary, established in 1832, and organized like that of Philadelphia; the barracks, theatre, charity schools, and post-office, are the other chief public buildings.

Lausanne will be ever famous in literary history, from its having been the residence of Haller, Tissot, Voltaire, and Gibbon. The house occupied by the latter, and which he wrote the last half of his immortal work, is in good preservation, and is the grand object of attraction to all travellers to Lausanne. "It was here," to borrow the beautiful passage in which Gibbon has perpetuated the memory of the event. "it was here, on the day or rather night of the 27th of June, 1787, between the hours of 11 and 12, that I wrote the last lines of the last page, in a summer-house in my garden. After laying down my pen, I took several turns in a berceau, or covered walk of acacias, which commands a prospect of the country, the lake, and the mountains. The air was temperate, the sky was serene, the silver orb of the moon was reflected from the waters, and all nature was silent. I will not dissemble the first emotions of joy on recovery of my freedom, and, perhaps, the establishment of my fame. But my pride was soon humbled, and a sober melancholy was spread over my mind, by the idea that I had taken an everlasting leave of an old and agreeable companion, and that whatsoever might be the future date of my history, the life of the historian must be short and precarious." When Inglis visited Lausanne a few years ago, the library of the historian is said to have been complete, but it has, we believe, been dispersed in the interval. Voltaire, previously to his settling at Ferney (which see), lived at Monrepos, a little distance from Lausanne, on the Bern road; and Byron wrote his "Prisoner of Chillon" at Ouchy, the port of Lausanne, on the lake.

Lausanne is now, as in the days of Gibbon, distinguished by its good society, and is in all respects a desirable place of residence for those who are neither anxious to be rich, nor ambitious of political distinction. "I noticed," says Inglis, "many new houses erecting, and very few old houses to let. The inhabitants, too, are steadily on the increase;

and the number of resident strangers is also greater every year. There are some cheaper places of residence than Lausanne; but no one, perhaps, where education is cheaper or better. Houses rent is decidedly lower in the neighbourhood of most English provincial towns than here; but the prices of provisions are lower than in England. The inns are extremely good, and not excessively expensive; and at the principal *tables d'hôte* the traveller will find an excellent and even elegant repast. There are, besides, several good coffee-rooms, where the best French and Swiss papers are regularly received." (*Switzerland, &c.*, p. 170.) There are, also, several public baths and libraries, and a chapel, in which the English as well as the Lutheran and Roman Catholic service is performed. A steamer sails daily from Ouchy to Geneva, and the E. end of the lake.

Lausanne is the seat of the superior courts of justice, and authorities of the canton of Vaud, of the councils of health and public instruction, the inspector of militia, and military commandant of the canton, &c. It has an academy, with 14 professors, founded in 1537, a college for the French language, with schools of military science, horsemanship, and drawing, and numerous literary societies. Its manufactures are of little importance. Woollen cloths, paper, leather, and a few other articles, are made, but in small quantities. The celebrated actor, John Kemble, is buried in the cemetery of St. Pierre, about two miles from Lausanne, where a monument is erected to his memory.

Lausanne derived its name from the ancient *Lausanium*, which stood a little to the W., in the plain of Vidy. Various Roman remains have been discovered there and elsewhere in the vicinity. Before the Reformation, Lausanne was a rich bishopric. It was taken in 1536, by the Bernese, and governed by an officer from Bern till 1798, when it fell into the hands of the French, who made it the capital of the department of the Leman. (*Ebel, Manuel Suisse*, p. 229-245; *Inglis's Switzerland; Coxe's Switzerland*, ii., 54.)

LAURENS, p. t., Northampton co., Pa., 100 m. E.N.E. Harrisburg, 295 W. Watered by Lehigh r., and Laurel run. The mountains contain coal, and the "Beaver Meadow" mine is in Spring mountain in this town. This mine is extensive and celebrated. It has five stores, two lumberyards, five saw-mills; five schools, 137 scholars. Pop. 1590.

LAVAL, a town of France, dep. Mayenne, of which it is the cap. on the Mayenne, and on the high road from Paris to Brest, 159 m. W.S.W. the former city, and 42 m. E. Rennes. Pop. (1826) 15,590. The town-proper is on a steep declivity on the W. bank of the river, across which it communicates with a suburb of about half its own size by two stone bridges, one built within the last 16 years. *Laval n'offre en général qu'un amas confus de vieilles maisons, séparées par des rues noires, escarpées, étroites, et tortueuses.* (*Dict. Géog.*) But though ill built, Hugo says it is well paved. It is surrounded with old walls, parts of which are in good repair; and contains many antiquated buildings, among which is the *château*, formerly the residence of the dukes of Laval, with a ponderous round tower, now serving as a prison. Many of the private houses have stood for centuries, and are curious specimens of Gothic architecture, though chiefly built of timber. The church of the Trinity, on the site of a former temple of Jupiter, those of *des Cordeliers* and St. Vénérand, and the new linen hall, are handsome edifices; but the prefecture, town-hall, theatre, and most of the other public buildings, are of a very ordinary description. It is the seat of tribunals of original jurisdiction and commerce, and has two hospitals, a communal college, a public library with 16,000 volumes, and a Trappist convent. It has considerable manufactures of linen stuffs and thread, with fabrics of cotton handkerchiefs, callicoes, flannel, &c., numerous bleaching grounds, tanneries, and marble works. It is also the entrepôt for the linen fabrics and yarn made in the adjacent cantons; markets being held in it every Saturday for such goods, and for wines, brandy, timber, iron, wool, &c., in which it has a considerable traffic. Laval was founded by Charles-le-Chauve, in the 9th century, to arrest the incursions of the Bretons. It was taken by Earl Talbot in 1465, but retaken by the French in the succeeding year. It suffered greatly in the Vendean war at the close of the last century. (*Hugo, art. Mayenne; Guide du Voyageur, &c.*)

LAVAUR, a town of France, dep. Tarn. cap. arrond. on the Agout, here crossed by a stone bridge, 20 m. S.W. Alby. Pop. in 1836 (ex. com.), 4692. It is divided into an old and a new town, both of which are ill built. It has a communal college, a public library, with 3500 volumes, with manufactures of silk stuffs, chiefly for furniture; and is the entrepôt for the silk goods of Upper Languedoc. In the 13th century, it was a stronghold of the Albigenses; but, after a lengthened and vigorous resistance, it was taken in 1211 by Simon de Montfort, by whom it was treated with the utmost barbarity. (*Hugo, art. Tarn.*)

LAVENHAM, or LANHAM, a market town and par. of England, co. Suffolk, hund. Babergh 15 m. W.N.W. Ips-

wich, and 57 m. N.E. London. Area of par., 2530 acres. Population, in 1831, 2107. The town on a branch of the river Bret, in a valley encompassed by hills on all sides except the S., comprises several small streets, with a spacious market-place, having a stone cross in its centre. The church, which has a steeple 142 feet high, is a handsome structure, partly of freestone, but partly, also, of curious inlaid flint-work: the porch is of highly ornamental architecture, and the timber ceiling and several pews in the interior are exquisitely carved, somewhat in the style of Henry VII.'s chapel in Westminster Abbey: the living is a rectory in the patronage of Caius college, Cambridge. There are also places of worship for Wesleyan Methodists and Independents, with attached Sunday schools. The charities comprise a free school, founded in 1647, and endowed with about £71 a year, some almshouses, and minor bequests for the poor. Lavenham had formerly a considerable business in the weaving of blue cloths, serges, and other woollen stuffs; but this has fallen to decay, and has been replaced of late years by the manufacture of hempen cloth, which here, as well as at Haverhill, employs a considerable number of hands. Lavenham is a borough by prescription; and land within the manor descends to the youngest son, according to the custom of borough-English. It is one of the polling places for the county. Markets small and ill-attended, on Tuesday. Fairs for butter and cheese, Shrove-Tuesday and Oct. 10.

LAVORO (TERRA DI), a prov. of S. Italy, in the kingdom of NAPLES, which see.

LAWRENCE, county, Ala. Situated in the N. part of the state, and contains 725 sq. m. Bounded N. by Tennessee river. Watered by Big Nance's and Town creeks, and their branches. It contained in 1840, 14,895 neat cattle, 7874 sheep, 40,781 swine; and produced 30,978 bushels of wheat, 1,319,590 of Indian corn, 38,414 of oats, 16,913 of potatoes, 3755 pounds of tobacco, 5,187,960 of cotton. It had seven stores, four flouring-mills, 14 grist-mills, nine saw-mills, 17 tanneries, six distilleries; one academy, 35 students, 20 schools, 600 scholars. Pop.: whites, 7143; slaves, 6145; free coloured, 25; total, 13,313. Capital, Moulton.

LAWRENCE, county, Miss. Situated towards the S. part of the state, and contains 790 sq. m. Watered by Pearl river, and its branches. It contained in 1840, 11,347 neat cattle, 3046 sheep, 22,286 swine; and produced 2399 bushels of wheat, 916,554 of Indian corn, 6736 of oats, 35,657 of potatoes, 83,394 pounds of rice, 1,658,523 of cotton. It had seven stores, 10 grist-mills, 18 saw-mills, one tannery, one distillery; one academy, 30 students, five schools, 116 scholars. Pop.: whites, 3648; slaves, 2273; total, 5920. Capital, Monticello.

LAWRENCE, county, Tenn. Situated in the S. part of the state, and contains 680 sq. m. Drained by Shoal creek and its branches, and by head branches of Buffalo river. It contained in 1840, 7344 neat cattle, 6103 sheep, 28,025 swine; and produced 25,976 bushels of wheat, 316,305 of Indian corn, 57,355 of oats, 6916 of potatoes, 9447 pounds of tobacco, 12,443 of cotton. It had nine stores, seven furnaces, four cotton factories, with 1180 spindles, four flouring-mills, 16 grist-mills, seven saw-mills, four tanneries, 17 distilleries; 10 schools, 312 scholars. Pop.: whites, 6370; slaves, 735; free coloured, 16; total, 7121. Capital, Lawrenceboro'.

LAWRENCE, county, Ky. Situated in the E. part of the state, and contains 650 sq. m. Bounded E. by Big Sandy river. Drained by Little Sandy river and Tygerts creek, and their branches, and Blaine's creek, which afford extensive water-power. It contained in 1840, 13,636 neat cattle, 6691 sheep, 15,851 swine; and produced 5496 bushels of wheat, 9009 of rye, 141,596 of Indian corn, 30,459 of oats, 6005 of potatoes. It had six stores, one flouring-mill, one grist-mill, two saw-mills, one tannery, two distilleries; eight schools, 176 scholars. Pop.: whites, 4639; slaves, 77; free coloured, 1; total, 4736. Capital Louisa.

LAWRENCE, county, O. Situated in the extreme S. part of the state, and contains 430 sq. m. Bounded S. by Ohio river, which runs on its border for 48 m. Drained by Indian, Guyandot, and Symmes creeks. Iron ore and mineral coal are abundant, and an excellent clay for stone ware is found. It contained in 1840, 13,456 neat cattle, 18,367 sheep, 12,853 swine; and produced 31,938 bushels of wheat, 283,300 of Indian corn, 74,404 of oats, 8334 of potatoes, 5385 pounds of sugar. It had 24 stores, 10 furnaces, one forge, one flouring-mill, 19 grist-mills, 14 saw-mills, six tanneries, two potteries; 56 schools, 1616 scholars. Pop. 9738. Capital, Burlington.

LAWRENCE, county, Ia. Centrally situated towards the S. part of the state, and contains 426 sq. m. Watered by the E. fork of White river, and its branches, which afford good water-power. It contained in 1840, 12,643 neat cattle, 16,991 sheep, 31,380 swine; and produced 75,610 bushels of wheat, 561,705 of Indian corn, 195,453 of oats, 12,500 of potatoes, 11,523 pounds of tobacco, 43,937 of sugar. It had 27 stores, one fulling-mill, three woollen factories, one cotton factory, with 140 spindles, six flouring-mills, 22 grist-mills, 14 saw-mills, eight tanneries, 11 distilleries, one pottery; three academies, 51 students, 36 schools, 1222 scholars. Pop. 11,782. Capital, Bedford.

LAWRENCE, county, Ill. Situated towards the S.E. part of the state, and contains 560 sq. m. Organized in 1821. Bounded E. by Wabash river, W. by Fox river. Watered by Embarrass river and Raccoon creek. It contained in 1840, 11,691 neat cattle, 5794 sheep, 18,180 swine; and produced 32,837 bushels of wheat, 422,747 of Indian corn, 73,948 of oats, 10,177 of potatoes. It had 16 stores, one flouring-mill, seven grist-mills, six saw-mills, one pottery; 16 schools, 533 scholars. Pop. 7092. Capital, Lawrenceville.

LAWRENCE, county, Ark. Situated towards the N.E. part of the state, and contains 1300 sq. m. Bounded S.E. by Cache river. Watered by Big Black river and its branches. It contained in 1840, 5792 neat cattle, 2273 sheep, 17,908 swine; and produced 11,636 bushels of wheat, 161,355 of Indian corn, 2948 of oats, 7058 of potatoes, 9409 pounds of tobacco, 13,974 of cotton. It had six stores, eight grist-mills, two saw-mills, two tanneries, eight distilleries; five schools, 115 scholars. Pop.: whites, 2569; slaves, 267; free coloured, 6; total, 2835. Capital, Jackson.

LAWRENCE, t., St. Lawrence co., N. Y., 85 m. N.E. Canton, 230 m. N.N.W. Albany. Bounded N.W. by St. Regis river. Watered by Deer river. It contains seven stores, two fulling-mills, one woollen factory, two grist-mills, four saw-mills; 13 schools, 564 scholars. Pop. 1845.

LAWRENCE, t., Mercer co., N. J., 6 m. N.E. Trenton. Drained by Assanpink creek. It contains two stores, three grist-mills, one saw-mill, two tanneries; one academy, 49 students; two schools, 92 scholars. Pop. 1556.

LAWRENCE, t., Tioga co., Pa., 90 m. N.E. Wellsborough. Watered by Tioga river. It contains five stores, one flouring-mill, three grist-mills, 10 saw-mills; five schools, 209 scholars. Pop. 888.

LAWRENCE, p. t., Washington co., O., 9 m. N.E. Marietta, 114 m. E.S.E. Columbus, 305 W. Watered by Little Muskingum river, which affords water-power. Iron ore, coal, salt springs and Seneca oil are found. Pop. 571.

LAWRENCE, t., Stark co., O., 9 m. N. Burlington. The Ohio canal passes through it. It has six schools, 150 scholars. Pop. 2043.

LAWRENCE, t., Tuscarawas co., O. Watered by Tuscarawas river. The Ohio canal passes through it. It contains the villages of Bolivar and Zoar. It has six commission houses in foreign trade, six retail stores, two furnaces, one flouring-mill, one grist-mill, two saw-mills, one oil-mill, three tanneries, one brewery, one pottery; seven schools, 530 scholars. Pop. 1893.

LAWRENCE, t., Marion co., Ia. It has nine stores, one grist-mill, three saw-mills, two tanneries; two schools, 36 scholars. Pop. 1437.

LAWRENCEBURG, p. v., capital of Lawrence county, Tenn., 75 m. S.S.W. Nashville, 729 W. Situated on the E. side of a branch of Shoal creek. It contains a courthouse, jail, a Methodist church, two academies, and 250 inhabitants.

LAWRENCEBURG, p. v., capital of Anderson co., Ky., 12 m. S. Frankfort, 554 W. Situated 2 m. W. of Kentucky river, and contains a courthouse, jail, and several stores and dwellings. Nett proceeds of the postoffice, $349.

LAWRENCEBURG, p. v., capital of Dearborn co., Ia., 86 m. S.E. Indianapolis, 514 W. Situated on the N. bank of Ohio river, two miles below the mouth of Great Miami river, at the entrance of Whitewater canal into Ohio river. It contains a courthouse, jail, several manufactures, a printing-office, a number of stores; and 1900 inhabitants. The proceeds of the postoffice, $609. The lower part of the village is liable to be overflowed at high water, so as to oblige the inhabitants to remove to the upper part of the houses, and to visit each other in skiffs. The new town is built in a more elevated position, on the second bank, and contains some elegant dwellings.

LAWRENCEVILLE, p. v., Lawrence t., Mercer county, N. J., 6 m. N. by E. Trenton, 172 W. Situated on a fertile plain, and contains a Presbyterian church, a store, and a flourishing academy.

LAWRENCEVILLE, borough, Pitt t., Alleghany county, Pa. Beautifully situated on the E. side of Alleghany river, 2½ m. above Pittsburg. It contains many elegant villas, and has three churches, a Presbyterian, Methodist, and Episcopal, the United States Alleghany arsenal, from which arms and military equipments are sent to the S. and W., an academy, 10 stores, and 867 inhabitants. It may be regarded as a suburb of Pittsburg.

LAWRENCEVILLE, p. v., capital of Brunswick co., Va., 73 m. S.S.W. Richmond, 169 W. Situated on a branch of Meherrin river, and contains a handsome courthouse, jail,
154

county clerk's office, an elegant Masonic hall, an Episcopal church, four stores, two tanneries, and about 300 inhabitants.

LAWRENCEVILLE, p. v., capital of Montgomery co., N. C., 115 m. W.S.W. Raleigh, 369 W. Situated on the E. side of Yadkin river, and contains a courthouse, jail, and a number of stores and dwellings. Nett proceeds of the postoffice, $62.

LAWRENCEVILLE, p. v., capital of Gwinnett co., Ga., 84 m. N.W. Milledgeville, 666 W. Situated 8 m. S.E. of Chattahoochee river, near the head branches of Yellow river. It contains a courthouse, jail, a Methodist church, an academy, 10 stores, 30 dwellings and about 300 inhabitants. There are a Presbyterian and Baptist church in the vicinity.

LAWRENCEVILLE, p. v., capital of Lawrence co., Ill., 159 m. S.E. by E. Springfield, 697 W. Situated on the W. bank of Embarrass river. It contains a brick courthouse, five stores, a grist and saw-mill, 70 dwellings and about 420 inhabitants.

LAWRENCEVILLE, p. v., capital of Monroe co., Ark., 75 m. E. Little Rock, 1015 W. Situated on a small stream, 5 m. E. of White river, and contains a courthouse, and a few dwellings.

LAYBACH (Germ. Laibach, Illyr. Lublena, an. Æmona), a city of the Austrian dominion, capital Illyria, and of the circle of the same name, comprising the duchies of Carinthia and Carniola, in which latter Laybach is situated, on the navigable river of the same name, a tributary of the Save, 80 m. S.W. Gratz, 72 m. E.S.E. Agram, and 54 m. N.E. Trieste. Lat. 46° 1′ 45″ N.; long. 14° 48′ 40″ E. Population, in 1838, 13,079 (Berghaus), comprising Germans, Italians, Illyrians, and Greeks. Laybach consists of the town-proper, five suburbs, and three adjacent hamlets. The town is situated on uneven ground, and has narrow and irregular streets, several of which, however, are well paved, and have foot-paths, while most of them are kept clean by running streams. Though ill laid out, Laybach is tolerably well built; and has several handsome public edifices, among which are the cathedral, St. James's church, and that of the Ursuline nuns, the Gothic town hall, the lyceum, to which an agricultural garden is attached; the theatre, masquerade-hall, barracks, military school, Auersperg palace, &c. The town is grouped round the castle hill: the castle being now converted into a house of correction, and state prison. Laybach has, in all, 10 churches, two hospitals, two convents, a gymnasium, a female school, a normal school, an ecclesiastical seminary, and orphan, lunatic, and other asylums. It is the see of a bishop, and the seat of the government of the circle, and of criminal, commercial, and mining tribunals for the province, town, and district judicial courts, the board of tolls, salt duties, and customs for the kingdom of Illyria, the agricultural society of Carniola, the museum for the duchy, a philharmonic society, &c. It has two large sugar-refineries, and fabrics of linen stuffs, porcelain, paper, leather, &c.: its silk and woollen manufactures have fallen into decay. A considerable transit trade is carried on between Laybach and Trieste, Croatia and S. Germany. Within the last dozen years some extensive marshes in its vicinity have been in a great measure drained, which has rendered the town much more healthy. Æmona was destroyed by Attila in 452, but having been restored, is said to have been enlarged and fortified by Narses. It is celebrated in diplomatic history for the congress held here in 1821. (Oesterr. Nat. Encyc.; Berghaus; Stein; Turnbull's Austria.)

LE', or LEH, improperly called LADAKH, a city of Central Asia, and capital of the principality of Ladakh, in Thibet; is the valley of the Upper Indus, at the foot and on the slope of some low hills on the N. side of the river, from which it is separated by a sandy plain about 2 m. broad, 150 m. S.E. Iskardo, and 930 m. N.W. by W. Lassa. Lat. 34° 10′ N.; long. 77° 45′ E. It is enclosed by a wall, furnished at intervals with conical and square towers, and extending on either side to the summit of the hills. The streets are disposed without any order, and form a most intricate labyrinth; and the houses are built contiguously, and run into each other so strangely, that from without it is difficult to determine the extent of each. The number of houses is said, by the natives, to be about 1000; but Moorcroft supposed they could scarcely exceed 500. They usually vary from one to two or three stories in height, and are furnished with light wooden balconies. The walls are sometimes wholly or in part of stone, but in general of large unburnt bricks, whitened outside with lime. The roofs are flat, and like the ceilings, formed of small trunks of poplar trees, above which a layer of willow shoots is laid, covered by a coating of straw, and that again by a bed of earth. They constitute a very insufficient defence against the weather, as during rain the water soon softens the earth, and pours down into the apartment. The rooms though frequently of good size, are rarely above seven or eight feet high, and unprovided with chimneys, though in the kitchen there is sometimes a square hole, which acts as an imperfect ventilator. The doors are made of planks of poplar mortised together: iron nails are rarely used, as they are too costly, the iron ore of the country being little wrought for want of fuel. A few fells and sheepskins, and a bench or two with a large box, constitute the principal articles of furniture. The temples are built of the same materials as the houses, and pillars of timber, like those in private dwellings, support the beams, being little more. In fact, than the stems of the poplar or willow, stripped of their bark and painted. The most considerable building in Lé is the palace of the rajah, which has a front of 250 feet, and is several stories in height. The population is chiefly of the Tibetan stock, but numerous Cashmerians have settled in Lé, and intermixed with the natives. Lé is the seat of an active commerce in shawl-wool, brought thither from the surrounding territory, from Lassa, Chinese Turkestan, &c., to be transported to Cashmere; and a silver coin is struck at this city, from bars of silver imported from China, which is in general circulation throughout the whole of Western Thibet. (*Moorcroft's Trav.*, i., 315-339; *Hamilton's E. I. Gaz.*)

LEACOCK, p. t., Lancaster co., Pa., 44 m. E.S.E. Harrisburg, 118 W. Bounded S.E. by Pecquea creek, W. by Conestoga river. Drained by Mill creek. It contains six stores, one woollen factory, three flouring-mills, three saw-mills, three distilleries, two potteries; nine schools, 287 scholars. Pop. 3337.

LEADHILLS, a mining village of Scotland, co. Lanark, in an alpine region, in an irregular valley 1300 feet above the level of the sea, and surrounded by wild heathy hills rising to the height of 2450 feet. Population 1150. The mining village of Wanlockhead, though only one mile distant, is in the county of Dumfries. Both villages are inhabited solely by persons connected with the mines; which, however, belong to different proprietors, and are wrought by different companies. With regard to Leadhills, the mineral district comprises a space about three miles in length by two and a half in breadth, and is principally composed of graiwacke, and graiwacke slate, which range from S.W. to N.E. These strata are associated with transition clay-slate, called edge matter, from its vertical position, through which the metalliferous veins pass. The principal lead veins run S.E. and N.W., with a dip to the E. of one foot in three. The common and compact galena, or lead glance, are the principal ores, and furnish all the lead used in the arts. The vein stones are quartz, calcareous spar, brown spar, sparry ironstone, heavy spar, &c. Silver is contained in the lead, but in too small quantity to repay its extraction. The Leadhills mines are rented by the Scotch mining company from the proprietor, who receives every 6th bar of lead, as seigneorage. The produce varies much in different years. It has lately been above 700 tuns a year; but it has sometimes been more than double that amount. The mines have been wrought from a very remote period. Gold is found in all the neighbouring streams, disseminated in minute particles among the clay more immediately covering the rocks and also occasionally intersperced in quartz. The search for this precious metal was formerly conducted on a pretty large scale, under royal authority; but never with much success: and all attempts of the kind, except by the curious, have long since ceased.

Leadhills has a chapel belonging to the established church, a school, and an excellent library founded in 1741. Allan Ramsay, the Scotch poet, was a native of this place. The miners of Leadhills are regarded as more than usually intelligent, moral, and respectable. (*Jameson's Mineralogy of the County of Dumfries; New Stat.' Acc. of Scotland, § Lanarkshire, p. 335-337; and § Dumfries-shire, p. 359-365.)

LEAKE, county, Miss. Situated near the centre of the state, and contains 576 sq. m. Watered by Pearl river, and its branches. It contained in 1840, 4576 neat cattle, 251 sheep, 10,951 swine; and produced 1077 bushels of wheat, 67,670 of Indian corn, 3163 of oats, 21,873 of potatoes, 2069 pounds of tobacco, 375,987 of cotton. It had six stores, five grist mills, four saw-mills; four schools, 69 scholars. Pop.: whites, 1614; slaves, 545; free coloured, 5; total, 2165. Capital, Carthage.

LEAKESVILLE, p. v., capital of Green co., Miss., 190 m. S.E. Jackson, 1092 W. Situated on the W. side of Chickasawhae river, and contains a courthouse, and several dwellings.

LEAMINGTON PRIORS, a town, par., and watering place of England, co. Warwick, in Kenilworth, div. of hund. Knightslow, on the Leam, a trib. of the Avon, 9 m. E. Warwick, and 97½ m. N.W. London. Area of par., 1720 acres. Population, in 1811, 543; in 1821, 2183; in 1831, 6209; and in 1841, supposed to be upwards of 14,000!

an unparalleled increase, occasioned by the growing celebrity of its mineral waters, and its many attractions as a place of fashionable resort. This town, which, 30 years ago, was an inconsiderable village, has now many noble and opulent residents; and the elegance of its squares, streets, crescents, and terraces, and of its numerous public and private edifices, justly entitle it to a place among the handsomest and best built towns in the kingdom. It formerly stood only on the S. of the river; but within the last few years it has been extended to the opposite side, with which it is connected by two handsome stone bridges: one of these, widened and beautified in 1840, has received the name of "Victoria Bridge;" the other, about a quarter of a mile lower down the Leam, on the estate of M. Wise, Esq., at whose expense it has been built, was also opened in 1840. The numerous hotels are nowise inferior to those of Bath, Cheltenham, and other fashionable watering places. It has, also, many suburban villas and detached residences.

The waters, to which Leamington owes its celebrity, embrace 11 different streams, uniting, in a single spot, saline, sulphureous, and chalybeate waters. That which most abounds, and which is known as " the Leamington waters," has been analysed by Drs. Lambe and Loudon*: it consists chiefly of the sulphate of magnesia and soda, in combination with muriate of soda, or common salt: the waters are used internally by dyspeptic and chronic patients; and have been found very useful when applied externally in cutaneous diseases and rheumatism.

The following Table, drawn up from the Analysis of Dr. Lambe, shows the number of Grains of Mineral Salts contained in a Gallon of Water from Two of the principal Springs.

Description of Salts.				Old Bath.	New Bath.
Carbonate of iron	.	.	.	1/6	
Muriate of magnesia	.	.	.	11½	30
" " soda	.	.	.	630	229
Sulphate of lime	.	.	.	11½	62·7
" " soda	.	.	.	120	149

The pump-rooms and baths are fitted up with every degree of elegance, combined with comfort and utility. They are constantly supplied with water from the springs; and these, with the assembly rooms, public libraries, music hall, and numerous promenades and pleasure gardens, form the principal attractions, and contribute chiefly to the amusement of the visiters. The church is a good specimen of Gothic architecture ; there is also a district church, episcopal chapel, and places of worship for Roman Catholics, Wesleyan Methodists, and other sects. A national school, an infant school, and several Sunday schools, are well supported ; and there are also several very excellent charitable institutions, particularly the " Warneford Hospital," endowed by Dr. Warneford, where the poor have the benefit of gratuitous baths, and of the best medical advice.

Being in the centre of a fine sporting county, Leamington has of late years become the head quarters of many of the leading Nimrods of the day. Three packs of hounds are hunted regularly during the season ; and its spring races (held on the Warwick course), its annual steeple chases, hunting club, and other similar attractions, have rendered it in this respect no mean competitor even of Melton Mowbray.

It could not, of course, be expected that manufactures should grow up in a place where pleasure forms the principal pursuit. The business of the town is, consequently, confined to the supply and retailing of articles required by the resident gentry and visiters ; and the latter are generally so numerous, as to make it a scene of bustle and activity during the greater part of the year. It has two weekly newspapers.

The Warwick and Northampton canal passes close to the town, and, by its union with other lines of canal communication, gives it all the advantage of an extensive inland navigation. It is only one stage of 10 m. from the Coventry station of the London and Birmingham railway.

The surrounding country, which is highly picturesque, furnishes an almost endless variety of pleasing rides and excursions, diversified by the fine residences of the Warwick, Clarendon, Leigh, Willoughby, and other families ; the ruins of Kenilworth castle, Guy's cliff, and other spots equally interesting to the tourist and the antiquary. (*Private Information.*)

LEBANON, an extensive and very celebrated range of mountains in W. Asia, connected northward with the tableland of Anatolia, thence running S.S.W. in two nearly parallel chains through Syria and Palestine, and finally connecting itself with mounts Horeb and Sinai near the gulf of Suez. The W. chain, called Djebel-Libaa, the Li-

* See "A Practical Dissertation on the Waters of Leamington Spa," &c., by Dr. Chas. Loudon.

LEBANON.

Serus proper of antiquity, detaches itself from the mountains of Asia Minor at the gulf of Iskenderoon; it is cut through by the deep channel of the Orontes, in lat. 37° 7', and as it proceeds southward, at an average distance of 24 m. from the Mediterranean, it increases in height, till, in lat. 34° 12', the culminating point of the chain, Djebel Makmel attains an elevation of 12,000 feet above the sea. Many summits in this part rise considerably above the limits of perpetual snow; and even in lat. 39° 50' the ancient Carmel and the twin summits of Ebal and Gerizim, so famous in the history of the Israelites (Deut. xi., 29), are conspicuous from their towering height; but more southward the mountains sink much lower, and are traced with some difficulty S. of Gaza. The E. chain, now called Djebel-es-Sheikh, and identical with the Anti-Libanus of Strabo (lib. xvi.), detaches itself from the range of Taurus, about 60 m. E. of that last mentioned; it attains the extreme altitude of about 5000 feet in lat. 33° 20', under the ancient name of mount Hermon, and after maintaining a considerable elevation as far S. as the 32d parallel, becomes lower and less regular as it skirts the Dead sea on its E. side, and finally is connected with the sandy hills of Arabia; this chain, indeed, is much less defined throughout its course, and altogether inferior in proportion to the chain running along the coast. The valley of Bakaah (an. Cœlo-Syria), which separates these chains, is about 100 m. long, and varies from 10 to 20 m. in breadth, having an elevation near the sources of the Orontes exceeding 2000 feet above the sea; and southward is the valley of the Jordan, which Burckhardt has traced through Arabia to the gulf of Akabah (see JORDAN). Besides the Orontes and Jordan, which are the two great rivers of this mountain system, a smaller stream, called the Leitanie, rises near Baalbec, and flows S.W. into the Mediterranean, a few miles N. of Tyre. The geology of mount Lebanon seems to have been little investigated by travellers; but from scattered hints collected from Richardson, Burckhardt, Robinson, and Elliott, it may be inferred that the general formation is carboniferous and mountain-limestone, with grauwacke and slate rising to the surface in the higher parts. The limestone in many parts is very porous, easily acted on by air and water, and rapidly worn into hollows of various shapes and sizes, which have been formed into sepulchres and caves, formerly the hiding-places of the persecuted Jews and Christians, (Elliott, ii., 237.) Basalt, and other igneous rocks, appear E. and S. of lake Tiberias, and the heights skirting the Dead sea, present granite, gneiss, dolomite, &c. Iron and coal are abundant in some parts of the range. The former is wrought in two districts; but, owing to the distance from which the fuel has to be brought for smelting the ore, the produce of the mines is scarcely sufficient for the consumption of the pachalik. The coal-mines which, during several years past, have been wrought by Mehemet Ali, are situated about eight hours' distance from Beyrout, at an elevation of about 2500 feet above the sea. The seams vary from three feet to four and a half feet in thickness; but the coal, though abundant, is rather sulphureous. In 1828, they employed 114 work-people in alternate gangs, day and night, at daily wages of three piastres (7¼d.) each. The quantity of coal dug up in 1837 amounted to all to only about 4000 tons. Iron-pyrites are found mixed with the coal, and smelting furnaces have been erected near the pits; but the returns are quite insignificant.

The principal animals found on mount Lebanon are, the roe-deer, the antelope, the goat, the mountain-sheep, the jerboa, &c.; with eagles, hawks, ravens, herons, and pelicans. The general aspect of the mountain scenery is thus described by an intelligent English clergyman: "Our route lay directly across mount Lebanon, the chief part of which is nearly barren. Almost the only tree which it nourishes is the fir, and consequently the view is not of a character to interest a lover of scenery. From the sea and the plains the range forms a noble object for the eye to rest on; but when once the ascent is begun, few of the component elements of a beautiful prospect are discernible. Deep ravines, indeed, and rugged beetling precipices meet one at every turn, and render travelling both painful and hazardous; but there are neither glaciers nor waterfalls, neither lakes nor rivers, no verdant fields nor smiling valleys, no extensive forests, no floral richness, and no rural villages: even the cedars, once "the glory of Lebanon" (Isa. ix., 13) have deserted it, and are replaced by the umbrella-topped fir. In one spot only called Bisharri, nearly opposite Tripoli, eight gigantic cedars, and a few of inferior size, attest the splendour of their by-gone race. The large trees measure about 36 feet round the trunk, and more than 100 feet between the extreme points of the opposite branches; while at the base, or a little above, they send out five limbs, each measuring 12 or 15 feet in circumference. At another spot W. of Bisharri, little known, and seldom visited, this same interesting tree is found in much

greater numbers, but of inferior growth. The mountaineers cut down the cedars for their charcoal and tar, which latter article is used medicinally to heal the wounds and diseases of the camel and the other animals." (Elliott's Travels, vol. ii., p. 255.)

"In fact," says another traveller, "it is impossible to view these patriarchs of the vegetable world, the remains of vast forests that once supplied Jerusalem with its finest timber and its choicest incense, without feeling the truth, aptness, and precision of the prophecies concerning them: 'The rest of the trees of his forest shall be few, that a child may write them.' 'Lebanon is ashamed and hewn down. The high ones of stature shall be hewn down: Lebanon shall fall by the mighty one.'" (Isaiah, x., 19, 33, 34; and xxxiii., 9.) It must not be supposed, however, from Mr. Elliott's description, that the whole mountain region is barren and uninteresting; for there are many fertile and well-peopled valleys, inhabited by an industrious people (about 35,000), chiefly Maronite Christians, occupied in the silk and dyeing trades, and in raising wine, corn, tobacco, and cotton. Dr. Bowring describes them as "an active and laborious race, who turn to good account such parts of the soil as are suited to tillage; and in no part of Syria," says he, "is there no obvious an activity, in none are the inhabitants so prosperous or so happy. The agricultural implements are rude: the plough is occasionally seen; but spade husbandry is much more used; and the steepness of the hill sides requires a succession of terraces for cultivation. Almost every male inhabitant is a small proprietor of land; and some of the emirs are large owners, either cultivating their estates themselves, or letting them out to tenants. Some of the convents produce a wine called Vino d'Oro, of good quality, both red and white; but it is often spoiled by the practice of boiling, and the use of skins. The tobacco of mount Lebanon rank also as the best in Syria. The quantity of raw silk produced in the district, exclusive of Tripoli, amounts annually to 240,000 okes, the price, in 1836, being from 120 to 125 piastres per oke: of this quantity 2-3rds are exported, and the rest consumed in the country. The weaving industry of mount Lebanon, however, is perhaps superior to its agriculture; for Mr. Consul Moore reports, that of about 1900 looms employed in this district, 300 were engaged in producing silk and cotton stuffs of the better qualities, 300 in weaving the abba, or coarse woollen garment of the peasantry, and 600 in making coarse cotton shirting. The manufacture and weaving of silk thread is likewise pursued to a considerable extent; and the annual consumption of gold for this trade averages about 50,000 drachms. Exorbitant taxes are, however, a great hindrance to industry; and it is only matter for surprise that, notwithstanding they are mulcted of nearly half their earnings, these people maintain their "proud bearing and independent character." (Bowring's Report on Syria; Elliott's Travels, vol. ii.; Robinson's Syria, vol. ii.; Rolandi, Pal. cap. xlviii., p. 311, &c.)

LEBANON, county, Pa. Situated toward the S.E. part of the state, and contains 348 sq. m. Drained by Swatara creek, and its branches. The Union canal passes through it. Large quantities of cast and of wrought iron are produced. It contained in 1840, 14,781 neat cattle, 10,977 sheep, 13,780 swine; and produced 215,429 bushels of wheat, 147,254 of rye, 239,631 of Indian corn, 232,691 of oats, 101,639 of potatoes. It had 56 stores, seven lumberyards, four fulling-mills, two woollen factories, three furnaces, three forges, 37 grist-mills, 24 saw-mills, one oil-mill, one paper-mill, 23 tanneries, 12 distilleries, two breweries, five potteries, four printing-offices, three weekly newspapers, and one periodical; four academies, 134 students, 36 schools, 397 scholars. Pop. 21,872. Capital, Lebanon.

LEBANON, p. t., York co., Me., 86 m. S.W. Augusta, 526 W. Bounded W. by Salmon Falls river, and drained by its branches. It has six stores, four grist-mills, five saw-mills, one tannery, one printing-office, one weekly newspaper; 20 schools, 981 scholars. Pop. 2272.

LEBANON, p. t., Grafton co., N. H., 4 m. S. Hanover, 29 m. N.W. Concord, 484 W. Bounded W. by Connecticut river. Chartered in 1761. Watered by Mascomy river, which has considerable falls, affording good water-power. There are falls here in Connecticut river, around which is a canal with locks. Lymane's bridge across Connecticut river, connects this township with Hartford, Vt. It contains four churches, a Congregational, Methodist, and two Universalist, nine stores, one fulling-mill, one woollen factory, three grist-mills, 10 saw-mills, one oil-mill; one academy, 100 students, 16 schools, 601 scholars. Pop. 1734.

LEBANON, p. t., New London co., Ct., 31 m. S.S.E. Hartford, 352 W. Watered by Yantic river, and its branches. It contains four churches, three Congregational, and a Baptist, three stores, two fulling-mills, two woollen factories, four grist-mills, seven saw-mills; one academy, 25 students; 16 schools, 346 scholars. Pop. 2194. The principal village is chiefly on one street, over a mile long, and 30 rods wide, so that families living on opposite sides of the street are not very near neighbours. The houses are not very compact, but some of them are neat.

LEBANON, p. t., Madison co., N. Y., 167 m. W. by N. Albany, 333 W. Watered by Chenango river, and its branches. It has one store, two fulling-mills, four grist-mills, nine saw-mills, two tanneries, one distillery; 16 schools, 641 scholars. Pop. 1794.

LEBANON, p. t., Hunterdon co., N. J., 40 m. N. by W. Trenton, 200 W. Watered by the S. branch of Raritan river. It contains 12 stores, 12 grist-mills, 17 saw-mills, three oil-mills, five tanneries, 11 distilleries; 12 schools, 285 scholars. Pop. 2949.

LEBANON, p. b., Lebanon t., capital of Lebanon co., Pa., 94 m. E. Harrisburg, 134 W. Situated on the Union canal. It is regularly laid out, and contains a large brick court-house, county offices, a stone jail, five churches, a German Reformed, Lutheran, Methodist, Mennonist, and Roman Catholic, four warehouses on the canal, nine stores, one grist-mill, one clover-mill, 300 dwellings, and about 2000 inhabitants. The township contains 17 stores, three lumber-yards, eight grist-mills, five tanneries, three distilleries, two breweries, three potteries, four printing-offices, three weekly newspapers, and one periodical; one academy, 50 students; 11 schools, 639 scholars. Pop. 6197.

LEBANON, p. v., capital of Warren co., O., 35 m. W.S.W. Columbus, 489 W. Situated between two branches of Turtle creek, and contains a courthouse, jail, a public library, three churches, a Presbyterian, Methodist, and Baptist, several stores, two iron founderies, two woollen factories, one grist-mill, one saw-mill, two printing-offices, two weekly newspapers, and about 1500 inhabitants.

LEBANON, p. v., capital of Russell co., Va., 285 m. W. by S. Richmond, 364 W. Situated on a branch of Clinch river, and contains a stone courthouse, jail, and several stores and dwellings, chiefly of wood. Nett proceeds of the postoffice, $182.

LEBANON, p. v., capital of Marion co., Ky., 39 m. S.W. by S. Frankfort, 581 W. Situated on a small branch of Salt river. It contains a courthouse, jail, several stores, and 546 inhabitants.

LEBANON, p. v., capital of Wilson co., Tenn., 32 m. E. Nashville, 692 W. Situated on a branch of Cumberland river, and contains a courthouse, jail, an academy, a female seminary, three churches, a Cumberland Presbyterian, Methodist, and Baptist, 12 stores, 100 dwellings, and about 1200 inhabitants, a portion of whom are coloured. There is a large steam cotton factory in the vicinity, and a college has been chartered to be denominated "The Southern University," which is not yet in operation. A spring rises near the courthouse, forming a creek, 15 or 20 feet wide. It was laid out in 1801, in the midst of a cedar grove, and incorporated in 1807.

LEBANON, p. v., capital of Boone co., Ia., 25 m. N.W. Indianapolis, 560 W. It contains a courthouse, jail, several stores, and about 150 inhabitants.

LEBANON, p. v., St. Clair co., Ill., 71 m. S. Springfield, 797 W. It contains three churches, two Methodist, and an Episcopal, eight stores, one steam flouring-mill, and about 70 dwellings. It contains M'Kendree College, under the direction of the Methodists, founded in 1834, which has a president and three professors, or other instructors, and 47 students. The commencement is on the second Wednesday in October.

LEBANON, p. v., capital of De Kalb co., Ala., 119 m. N.E. Tuscaloosa, 675 W. Situated on the N. side of Big Willa creek, and contains a courthouse, and several dwellings. Nett proceeds of the postoffice, $58.

LEBRIJA (an. Nabrissa), a town of Spain, prov. Cadiz, in the flat of the Guadalquivir, 29 m. S. by W. Seville, and 3 m. N. Cadiz. Pop., according to Miñano, 6716. A par. church, built of the materials of an old mosque, four convents, a hospital, and a well-endowed classical college, are the chief public buildings; and there is also a ruined castle of considerable extent. Being situated in the midst of an extensive and marshy flat, Lebrija is extremely unhealthy, especially during the heats of summer; but the circumjacent alluvial soil is highly productive. The town has fabrics of glass, earthenware, blankets and sacking, soap, bricks, tiles, and mortar. (Miñano.)

LECCE (an. Sybaris and Lupia), a city of the Neapolitan dom., prov. Otranto, cap. distr. and canton, on the road from Brindisi to Otranto, about 22 m. S.S.E. the former city, and the same distance N.W. the latter; lat. 40° 21' 16" N., long. 18° 16' 7" E. Pop. 14000. "The circ. of the present city at least equals that of Foggia; its houses are infinitely larger; and it is even supposed that it would conjointly admit a population of 20,000. The city is fortified by a wall and towers, in bad condition, above a deep ditch; and possesses, moreover, a castle or citadel. It comprises the usual appendages of a provincial capital, a luminary, tribunal, and theatre; and adds to these a large

manufactory of tobacco, the produce of which, as snuff, is highly esteemed throughout the kingdom. The principal gate of entrance to Lecce is very magnificent, though in a strange overloaded style of architecture. The facility with which the stone of the country is wrought,* has proved of great advantage to the architectural embellishments of Lecce; but it has also afforded a fatal facility of propagating the extravagant and almost incredibly bad taste exemplified in every building of consequence. Their magnitude alone is imposing to the spectator, while their innumerable absurdities disgust him. Among these edifices the churches are pre-eminent: they exhibit all the grotesque barbarity of the Gothic, without any of its spiry lightness; and their interior decorations, though by no means in the same style, are not likely to make up for these defects. The inside of the cathedral, dedicated to St. Oronzio, the patron of the city, is simple and unoffending. A few inlaid marbles, and some indifferent paintings, constitute its only ornaments. The ceiling is of brown carved wood, richly gilt; and it has, though perhaps not strictly adapted to a place of public worship, a handsome effect. In the principal square is an antique column brought from Brindisi: it supports the statue of the protecting saint, and near its base is a fountain without water, adorned by a small equestrian statue of one of the sovereigns of Naples. The inhabitants of Lecce are mostly in easy circumstances, and renowned for their courteous polished manners. The climate is oppressively hot during the summer, and the porous quality of the material of which the town is built is supposed to absorb the damp in the morning, only to emit it again at sunset, to the great prejudice of the health of the natives. The common disorder of the country, to which strangers are more particularly subject, is an intense catarrh, known by the name of ceride, which is frequently attended by serious fever, and often turns to internal inflammation." (Craven's Tour, &c., p. 134-137.) The surrounding district is one of great fertility; but bears all the marks of bad government, and backward cultivation. It supplies, however, silk, wool, flax, cotton, oil, wine, &c., of good quality, in which the city is said to have an active trade. Lecce has also manufactures of lace, linen thread, woollen cloth, and cotton and silk fabrics. The centre of the city is somewhat busy, being occupied by the artisans; but the other parts of the town are so very deserted, that Craven says he has frequently found himself the only person walking in one of the most considerable streets. The produce of Lecce is mostly exported from Otranto, or from San Cataldo on the Adriatic.

Sybaris, or Lupia, on the site of Lecce, was very ancient. Augustus remained in it for some days after his return to Italy on hearing of the death of Cæsar. In the middle ages it was called Lycium. It was made the fief of an earl soon after the establishment of the Normans; and Tancred, one of its earls, succeeded to the crown of Naples in 1189. The novelist Ammirato, and the anatomist Baglivi, were natives of Lecce. (Craven's Naples, i., 134-138; Swinburne's Two Sicilies, i., 379, 380, &c.; Craven's Ancient Italy, ii., 307.)

LECHLADE, a market town and par. of England, co. Gloucester, hund. Brightwell's Barrow, at the confluence of the Lech with the Isis, 13 m. E. Cirencester, and 68 m. W. by N. London. Area of par., 3080 acres. Pop., in 1831, 1944. The town consists chiefly of a single, long, wide, and well-built street; and the river (which is navigable up to this place, a distance, by water, of 146½ m. from London,) is crossed by a good stone bridge. The church, a handsome stone structure, built in the reign of Henry VII., has a tower and spire at its W. end. There are also places of worship for Wesleyan Methodists and Baptists: and two Sunday-schools. The principal importance of Lechlade is derived from its site at the junction of the Thames navigation with the Thames and Severn canal, which makes it the seat of an extensive transit trade in butter, cheese, corn, malt, &c.; but this will probably suffer a material diminution when the Great Western railway is completed. Markets on Tuesday; fairs, Aug. 5 and 12, and Sept. 9, for cattle and cheese.

LECTOURE (an. Lactora, and Civitas Lactoratensis), a town of France, dep. Gers, cap. arrond., on the summit of a steep isolated rock, 19 m. N. Auch. Pop., in 1836 (ex. com.), 2999. It was formerly surrounded with a triple range of strong walls, the remains of which still exist. It is traversed by a wide, regularly built, and clean street, at one end of which is a hospital occupying the site of an ancient castle, and at the other a handsome Gothic church, built by the English. Nearly all the other streets are fort ièsées et fort irrigulières. Near the above church is

___

* "The stone of the country is of a fine white, or soft when taken out of the quarry that it may be moulded like wax, and will receive any form the slightest strokes of the chisel impress it with; yet, by remaining exposed to the air, it very soon acquires a proper degree of consistency." (Swinburne's French, i., 269.)

old episcopal palace, now the town-hall, sub-prefecture, and court of primary jurisdiction. In the town-hall are portraits of Marshal Lannes, Duc de Montebello, and other distinguished individuals, natives of the town; a marble statue of the marshal is also erected on the public esplanade. Lectoure has manufactures of serge and coarse woollen cloths, and a considerable trade in cattle, wines, brandy, and agricultural produce.

Lectoure, though not mentioned by the ancient geographers, has several Roman antiquities; the chief is a votive altar, in good preservation, which dates from the time of Gratian. At the foot of the hill on which the town is built is a fountain of excellent water; its modern name is Hondelie, derived, as is said, from its ancient name, Fons Deliæ; having been consecrated to Diana, who had a temple in the vicinity.

Lectoure belonged, for a lengthened period, to the counts of Armagnac. The last of that family having been besieged in it in 1473, by the troops of Louis XI., commanded by the Cardinal of Alby, surrendered on terms which the cardinal offered and swore to observe. No sooner, however, had the blood-thirsty perfidious ecclesiastic got the count into his power, than he ordered him to be assassinated, and gave up the town to military execution. (L'Art de vérifier les Dates, part ii., tom. ix., 390, 8vo. ed.; Huge, art. Gers, &c.)

LEDBURY, a market town and par. of England, co. Hereford, hund. Radlow, near the Leden, a trib. of the Severn, 13 m. E. Hereford, 14 m. S.W. Worcester, and 105 m. W.N.W. London. Area of par., 8630 acres. Pop., in 1831, 3632. The town, situated on the slope of a hill at the E. angle of the co., at the extremity of the Malvern hills, comprises two principal intersecting streets, with others of inferior character. Many of the houses are handsome, and built of stone quarried in the neighbourhood. The church, which is of Norman architecture, with more recent alterations and additions, comprises a nave, aisles, and chancel, with a chapel; and the tower, which is detached from the rest of the building, is surmounted by a fine spire 60 ft. high: the living is a vicarage, in private patronage. There are, also, places of worship for Independents, Baptists, Wesleyan Methodists, with well-attended Sunday schools attached to each; a national school for both sexes, partly supported by the produce of two or three old charities; and a school of industry for girls. The free school, founded in the 16th century, formerly had the reputation of being a good classical school; but the endowment is very trifling, and the instruction is now confined to reading, writing and arithmetic. St. Catherine's hospital, for poor men and women, founded by Hugh Folyot, bishop of Hereford, in 1232, comprises a master, chaplain, seven brethren, and three sisters, each of whom, in addition to a comfortable dwelling and some yearly allowances, receives a stipend of 6s. a week. The hospital, lately rebuilt, is a handsome structure, with two wings, and a chapel and hall in the centre; and it is proposed, as the estates are increasing in value, to raise the number of inmates to 94. The parish is unusually rich in money charities, distributed chiefly by the clergy and churchwardens. (See Char. Comm. 2d Rep., p. 2.) Ledbury was celebrated during the reigns of Elizabeth and James I., for its extensive manufactures of broadcloth and silk; but they are now quite extinct. Malting, tanning, and the weaving of sacking, employ a considerable number of hands; but the present importance of the town is derived from its being the chief market of a district producing large quantities of hops, cider, and perry. Stone and marble are quarried in the neighbourhood, and the latter is sent to various parts for chimney-pieces, slabs, &c. The conveyance of heavy goods is much facilitated by the Hereford and Gloucester canal, which passes close to the town. Among the country seats in the vicinity of Ledbury, the largest and finest is Eastnor Castle, erected by the late Lord Somers. Markets on Tuesday; first Monday after Feb. 1., Monday before Easter, and May 12; June 22; first Tuesday in Aug., Oct. 2, and Monday before Dec. 21, for cattle, cheese, hops, &c.

Ledbury was anciently a parliamentary borough, and in the reign of Edward I., twice returned members to the House of Commons; but the privilege was not preserved.

LEDYARD, p. t., New London co., Ct., 47 m. S.E. Hartford, 365 W. Organized in 1836, from the N. part of Groton. Bounded W. by Thames river. Watered by Poquetannuck river on the N., and by head branches of Mystic river on the S.E. It contains three stores, four grist-mills, three saw-mills; 14 schools, 523 scholars. Pop.1871. The village on the E. side of Thames river, at Gales ferry, contains about 35 dwellings. The township was named from the celebrated traveller John Ledyard, who was a native of Groton.

LEDYARD, p. t., Cayuga co., N. Y., 171 m. W. Albany, 317 W. Organized in 1823. Bounded W. by Cayuga lake, into which its streams flow. It contains three churches, a

Presbyterian, Methodist, and an Episcopal, four stores, one fulling-mill, one flouring-mill, two grist-mills, four saw-mills; two academies, 91 students, 17 schools, 570 scholars. Pop. 2143. The steamboat from Ithaca to Cayuga bridge touches daily at Aurora village, on the lake, in this town ship.

LEE, county, Va. Situated in the S.W. corner of the state, and contains 512 sq. m. It lies between Cumberland and Powell's mountains, and is drained by Powell's river and its branches. It contained in 1840, 10,408 neat cattle, 10,493 sheep, 34,205 swine; and produced 26,766 bushels of wheat, 6498 of rye, 446,111 of Indian corn, 102,512 of oats, 23,349 of potatoes, 23,438 pounds of tobacco, 38,848 of sugar. It had 12 stores, five bloomeries, 24 flouring-mills, 31 grist-mills, 14 saw-mills, one oil-mill, seven tanneries, 18 distilleries; seven schools, 136 scholars. Pop.: whites, 7899; slaves, 580; free coloured, 32; total, 8441. Capital, Jonesville.

LEE, county, Ga. Situated toward the S.W. part of the state, and contains 600 sq. m. Watered by Flint river and its tributaries. It contained in 1840, 16,467 neat cattle, 738 sheep, 12,072 swine; and produced 2153 bushels of wheat, 105,805 of Indian corn, 2072 of oats, 7326 of potatoes, 1,067,140 pounds of cotton. It had seven stores, eight grist-mills, five saw-mills; one academy, 23 students, seven schools, 146 scholars. Pop.: whites, 2469; slaves, 3946; free coloured, 3; total, 4590. Capital, Starkville.

LEE, county, Ill. Situated in the N. part of the state, and contains 720 sq. m. Bounded S. by Des Plaines river. Watered by Fox river, and a head branch of Green river, flowing through Winnebago swamp. It contained in 1840, 1900 neat cattle, 318 sheep, 4469 swine; and produced 27,415 bushels of wheat, 45,500 of Indian corn, 22,085 of oats, 18,715 of potatoes. It had eight stores, one grist-mill, five saw-mills; seven schools, 192 scholars. Pop. 2035. Capital, Dixon.

LEE, county, Iowa. Situated in the S. part of the ter., and contains 600 sq. m. Organized in 1837. Bounded N.E. by Skunk river, S.W. by Des Moines river. Watered by Sugar, Halfbreed, and Lost creeks. The latter loses itself in a low prairie, 4 m. from Mississippi river. It contained in 1840, 5654 neat cattle, 2124 sheep, 13,898 swine; and produced 21,395 bushels of wheat, 312,714 of Indian corn, 30,473 of oats, 2641 of potatoes, 4363 pounds of sugar. It had seven stores, 14 schools, 346 scholars. Pop. 6093. Capital, Fort Madison.

LEE, p. t. Penobscot co., Me., about 125 m. N.E. Augusta. Drained by a tributary of Mattawamkeag river. Organized in 1832. It has eight stores; eight schools, 264 scholars. Pop. 794.

LEE, p. t., Strafford co., N. H., 36 m. E. by S. Concord, 485 W. Watered by Lamprey and Oyster rivers, the latter issuing from Wheelwrights pond. In its N. part, containing 165 acres. It contains two churches, a Baptist and a Christian, three stores, one grist-mill, five saw-mills; seven schools, 314 scholars. Pop. 995.

LEE, p. t., Berkshire co., Mass., 128 m. W. Boston, 6 m. S.E. Lenox, 366 W. Incorporated in 1777. Marble, iron ore, and limestone are abundantly found. Watered by Housatonic river, which affords good water-power. It is particularly distinguished for its extensive paper manufactories, which in 1840, employed 202 persons, producing to the amount of $250,000, employing a capital of $229,000. It has three churches, a Congregational, Methodist, and Baptist, 11 stores, two forges, one cotton factory, with five spindles, one grist-mill, seven saw-mills, 13 paper-mills, one printing-office; one academy, 42 students, 10 schools, 603 scholars. Pop. 2498.

LEE, p. t., Oneida co., N. Y., 114 m. W.N.W. Albany, 328 W. Drained by Canada and Fish creeks. Organized in 1811. It contains a church, five stores, two furnaces, one fulling-mill, one woollen-factory, one flouring-mill, two grist-mills, 14 saw-mills, four tanneries, one distillery; 25 schools, 898 scholars. Pop. 2936.

LEEDS, a parl. and mun. bor., par., and celebrated manufacturing town of England, being the great centre of the woollen cloth trade, co. York, W. riding, locally situated in wap. Skyrack, on both sides the navigable river Aire, 23 m. W.S.W. York, 29 m. N. Sheffield, and 170 m. N. by W. London; lat. 53° 47' 30" N., and long. 1° 33' W. Area of par. and of parl. and mun. bor. (which are all co-extensive), 21,450 acres: pop. of parl. bor. in 1831, 123,393; pop. of town, at same period, 71,602.

The principal and best part of Leeds stands on the slope of a hill N. of the Aire, and the buildings cover a space of about 1000 acres. The town, speaking generally, is irregularly built, with narrow and crooked streets; but the centre and W. end comprise several handsome streets lined with fine houses. Briggate, in the centre of the town, is the largest, and is as wide as Oxford-street, London. Three stone bridges, and two of cast iron, on the bow and string principle, cross the river Aire, on the S. side of which are the extensive suburbs of Holbeck and Hunslet, containing some large factories. The town is well paved, flagged, and lighted with gas. Hitherto the supply of water has been rather deficient; but extensive works are now (1840) on the point of being completed, by which an abundant supply of excellent water will be conveyed into the town from the Harewood hills, 5 or 6 m. distant, at an estimated expense of about £90,000. Among the public buildings, the cloth-halls deserve particular notice: the Mixed-cloth Hall, at the corner of Wellington-street, built in 1758, is a quadrangular building, 380 feet long, and 200 feet broad, enclosing an open area, and having about 1800 stands. The White-cloth Hall, for the sale of undyed goods, on the plan of the former, was erected in 1775: it has about 1200 stands. A third building of the same description, in Albion-street, but smaller, intended to accommodate traders not licensed to sell in the other halls, has been long abandoned. Close to the Mixed-cloth Hall, is a handsome edifice, called the "Commercial Buildings," which would do credit to the metropolis, appropriated chiefly to news and concert rooms, but partly also to trading purposes. The courthouse, in which the petty and quarter sessions of the borough and the Michaelmas sessions of the W. riding are held, is a well-arranged building for police purposes; but the jail attached to it is too small for the wants of the town and borough. The corn exchange faces Briggate: its front is of the Ionic order, and has a niche in the centre, with a statue of Queen Anne. The central market, erected at an expense of £35,000, is large, handsome, and commodious: there are also two other markets. The cavalry barracks, erected in 1820, on the N. side of the town, are well built and very extensive, occupying, with the parade grounds, nearly 12 acres. The workhouse is not of a size corresponding with the wants of so large a parish, but it is well conducted by a board of 20 overseers. The cost of maintaining the poor of the town amounted in 1830 to £33,870, the expenditure for the entire parish in the same year being £32,196. The hall of the Philosophical and Literary Society, a theatre on the S. side of the town, and two commodious bath establishments, are the only other public buildings besides the churches: of the latter, many are handsome. The parish church, now in course of being rebuilt, on the site of an old Gothic edifice, is in the perpendicular English style, and from its appearance promises to be one of the largest and handsomest churches in England: the living is a vicarage, worth £1900 a year, and having the patronage of nine ecclesiastical benefices. There are six other churches within the town, and 32 places of worship for dissenters, of which six belong to Wesleyan Methodists, seven to seceders from Methodism, seven to Independents, two to Unitarians, two to Baptists, and two to Roman Catholics, one of the latter being an elegant building, with a spire 150 feet high. An Independent chapel, recently erected in the E. Parade, at a cost of £12,000, has a handsome Doric portico, and is the finest Grecian edifice in the town. Each en-township likewise has its episcopal chapel, and one or more places of worship for dissenters. A spacious cemetery, on Woodhouse Moor, occupying 10 acres of ground, was opened in 1835, for the use of persons of all religious denominations: in the centre is a chapel, beneath which are large vaults. The Sunday schools give religious instruction to about 11,400 children, 4000 of whom are connected with the church, and the rest with the various denominations of dissenters. The establishments for general education comprise, 1. A well endowed grammar school, founded in 1552, which gives free instruction in classics and the elements of mathematics to the sons of all residents in Leeds, and enjoys the reputation of being ably and successfully conducted. 2. A national school, attended by upward of 400 children. 3. An extremely well conducted Lancastrian school, established in 1811, and giving instruction to 360 boys. 4. A model infant school, where a considerable number of persons have been trained for the teaching of infant schools, and where about 120 children receive regular instruction. 5. Marshall's schools, comprising a boys', girls', and infants' school. Several other schools are supported by the Wesleyans, or by voluntary subscription. St. John's Charity, founded in 1705, has for its object the training of girls to become household servants: its income is about £500 a year, and the management is vested in a committee of subscribers. The chief societies for the promotion of literature and science are the Philosophical and Literary Society, which has a handsome hall and library in Park Row, the Literary Institution, having an extensive library, and the Mechanics' Institute, which has a hall in E. Parade. Leeds has also a school of medicine, a society for the promotion of the fine arts, and five subscription libraries. The charitable institutions comprise, besides the schools already mentioned, an infirmary, founded in 1767, supported by subscriptions, amounting to £2800 annually, and accommodating 150 in-patients; a fever hospital, called the "House of Recovery;" a lying-in hospital, an eye, and

ear infirmary, and a public dispensary. There are likewise several endowed charities for the aged poor, and other benevolent institutions, the gross revenues of which exceed £4000 a year. Party politics run pretty high in Leeds. It has five weekly newspapers; one of which, the "Leeds Mercury," is one of the ablest and most widely circulated of the provincial papers.

We subjoin an account of the area, population, &c., of the different parts of the borough, drawn up from the census and other official returns:

Townships.	Area in Acres.	Pop. in 1831.	Rental in 1827.
			£.
Town of Leeds (comprising 9 townships)	3,393	71,602	242,677
Armley	1,040	5,150	10,340
Beeston	1,770	2,125	4,206
Bramley	2,410	7,468	22,660
Chapel Allerton	3,040	1,934	9,052
Farnley	2,470	1,591	4,609
Headingley with Burley	2,400	2,049	16,025
Holbeck	768	11,210	26,466
Hunslet	1,180	12,074	27,162
Potter-Newton	1,240	888	6,229
Wortley	900	5,944	15,690
Totals	21,454	123,393	379,345

Leeds owes its great and long-continued eminence as a manufacturing town, partly to its advantageous situation, and partly to the industry and ingenuity of its inhabitants. It stands in a fertile country, intersected with rivers, and possessing rich and all but inexhaustible beds of coal. The natural facilities afforded by its position for procuring raw materials, and for disposing of its manufactured produce, have been vastly extended by artificial means. On the one hand, it communicates with the Humber, and, consequently, with the German ocean, by means of the Aire and Calder navigation, which allows vessels of 120 tons to come up to the town; and on the other hand, it communicates with the Mersey and Liverpool by the Leeds and Liverpool canal. Railways have recently, also, been opened to York, Hull, and by Derby to London, and a new line is just completed between Leeds and Manchester. The weekly receipts of the N. Midland railway, which connects Leeds with York and Derby, average about £2500, and those of the Leeds and Selby railway (which has been leased to the York and North Midland railway) are estimated at £800. The staple manufacture is the weaving of woollen cloths; but the spinning of flax and worsted is also an important branch of industry, as will be seen from the following official returns of the factories in operation within the borough in 1839:

Description of Mills.	Nos.	Horse-power.	Engines and wheels.	Persons employed.
Woollen Mills	106	2,408	107	9,736
Flax	44	1,486	11	6,430
Worsted	13	414	15	3,146
Silk	3	64	1-135	115
Total	166	4,117	144-135	19,482

The woollen manufacture of Leeds and its neighbourhood is carried on in two ways—on the domestic system, and by means of factories. According to the former plan, the business is conducted by a number of small masters, generally possessed of very limited capitals, who have in their houses from two to four looms, and employ, besides themselves and their families, from three to seven journeymen. Formerly they used to carry the wool by hand labour through all the stages of its manufacture, till it was made into undressed cloth; but for several years past they have availed themselves, in the performance of various processes, of the public mills that have been erected, mostly on a joint-stock principle, in all the villages within the district where this system prevails. By this means the domestic cloths are produced as good and cheap as those made in factories. The wages of hand-loom weavers in and about Leeds varied (in 1840) from 12s. to 16s. a week. The factory system owes its existence to the improvements of machinery subsequent to 1790; and, though strongly opposed by the domestic clothiers, has greatly improved the manufacture, and raised Leeds to its present eminence as a mart for superfine broadcloths. The master manufacturers, who necessarily either possess or have the command of large capitals, employ a greater or less number of workmen, in one or more large factories, under their own inspection, or that of their superintendents. In these factories the whole processes are carried forward, from the breaking of the wool to the finishing of the cloth for the consumer. Power-looms, however, have hitherto been but little employed in the weaving of fine cloths, not one fourth part of those produced being now (1840) woven by their agency. The woollen fabrics manufactured at Leeds comprise broadcloths, ladies' cloths, kerseys, swansdowns, and beavers. The goods sold to the merchants in a rough or undressed state are finished in dyehouses and dressing-shops, which of themselves give employment to upward of 3000 persons. The sale of cloths is partly effected in the different cloth-halls, on the mornings of Tuesday and Saturday, between 11 and 12 ; but of late years, or since the manufacturers began wholly to finish their goods, the cloth-halls have lost a good deal of their importance, and a full half of the business that used to be entirely carried on in them is now transacted in private counting-houses. Shaloons, stuffs, and camlets are made to some extent ; and immense quantities of unfinished stuffs are brought here to be finished from Bradford and Halifax. Some of the flax mills are superb establishments ; large quantities of linen yarn are sent to Barnsley to be manufactured into linens, and large quantities are also sent to Ireland and France : canvass, sacking, and linens are also made to some extent in the town. The wages of linen weavers (of whom there are about 700) have been gradually falling during the last 18 years, and are now from 7½ to 15 per cent. under those given in 1823, owing, it is said, to the influx of Irish weavers, and the keen competition of the Scotch manufacturers. The present average wages per week are 8s. 4d. nett, when in full work, which is seldom the case. The weavers are characterised by the hand-loom commissioners as intelligent, sober, and steady, but in extreme poverty. The manufacture of machinery employs a great number of hands ; and there are likewise extensive chemical works, large glass-houses, potteries, making goods almost exclusively for exportation, extensive tobacco-mills, and soap-works, which produced, in 1839, 770,968 lbs. hard soap. The greatest portion of the manufacturing operations of the town are carried on by means of steam-engines, of which between 300 and 400 are at work within the parish. A branch of the Bank of England is established at Leeds, and there are five joint-stock banking companies, besides two private banking establishments, and a savings' bank, with a large number of depositors.

Leeds was first incorporated as a municipal borough in the reign of Charles I., and received its charter in the 13th of Charles II. Under the Municipal Reform Act it is divided into 12 wards, and the government is vested in a recorder, mayor, 15 aldermen, and 48 councillors: corporation revenue, in 1839, £13,623. Recently a body of police has been organised, similar to that of the metropolis. Notwithstanding its importance, as the first clothing town of the British empire, Leeds was not represented in parliament till 1832, when the Reform Act conferred on it the important privilege of sending two members to the H. of C. Registered electors, in 1839-40, 6182. Markets on Tuesday and Saturday ; cattle fairs on alternate Wednesdays; and for horses and hardware, July 10th, 11th, October 8th, and November 9th.

Leeds is mentioned by Bede, and in the Domesday survey. Leland, early in the 16th century, describes it as a market town, subsisting chiefly by clothing, reasonably well built, and as large as Bradford, but considerably less than Wakefield. The clothing trade had been introduced about 60 years before Leland's time, and the town has since gradually risen, by the persevering industry of its inhabitants, till it has become the third manufacturing town of the first manufacturing nation of the world. (Baines's Gazetteer of Yorkshire ; Parsons' Annals of Leeds, passim ; Parl. Rep. ; and valuable Priv. Inform.)

LEEDS, p. t., Kennebec co., Me., 21 m. W. by S. Augusta, 587 W. Bounded W. by Androscoggin r. Incorporated in 1802. It has a large pond on its N.E. border. It contains four stores, one fulling-mill, one grist-mill, five saw-mills, one tannery ; one academy, nine students, 12 schools, 604 scholars. Pop. 1736.

LEEK, a manufacturing market town, and par. of England, co. Stafford, hund. Totmonslow, on the Churnet, a tributary of the Trent, 19 m. S. Macclesfield, and 134 m. N. by W. London. Area of parish (comprising 10 townships), 34,370 acres. Population, in 1831, 10780. Population of Leek-and-Lowe township, 6374. It is situated in the mountainous part of the county called the Moorlands, on the road between London and Manchester, and consists of a principal street lined with some good modern houses, and crossed by several narrow and irregular avenues. The church is an old Gothic structure, with a square tower : in the churchyard are the remains of a Danish cross, 10 feet high : the living is a vicarage in the gift of Earl Mansfield. There are places of worship for Independents, Wesleyan and Newconnexion Methodists, and the Society of Friends. Except a small endowed school, there is no public day-school, and nearly all the instruction which the children receive is given in the Sunday-schools attached to the different places of worship. In the Wesleyan school there are upward of 1000 children. A mechanics' institute confers important benefits on the manufacturing population. An almshouse for eight widows, and some other charities, have endowments amounting to £180 a year. (Charity Commercial Reports.)

Leek has long been the seat of a manufacture of broad

silks and plain ribands, many of the latter being woven by power-looms, of which there are about 100 in the town. There are about 150 broad-looms and 180 engine-looms ; but only half the former and one third the latter were at work in January, 1839. About 30 of the broad-looms have jacquard machinery attached. The riband-looms are chiefly employed in making plain black sarsnets: galloons and ferrets are made exclusively by power-looms. The silk-mills, of which there were seven in 1838, employing 784 hands, embrace not only the weaving of ribands by power-looms, but the throwing and spinning of silk, and its twisting into sewing silk, braid, &c. The hand-loom weavers are chiefly employed on checked or figured silk neckerchiefs, and a few gros-de-naples and figured gown-pieces, the best black ribands, and silk serges of superior quality. These goods are prepared chiefly for the London market ; but the sewing silks, twist, and ribands are mostly for exportation. The hand-loom work is given out warped and wound to undertakers, who possess a varying number of looms, and employ journeymen and apprentices, to the former of whom they pay the warehouse price, after deducting for loom-rent, &c. There are about 40 undertakers in the broad trade, and they appear to be superior both in habits and condition to the same class in most other places, many of them possessing convenient and substantial dwelling-houses, the highest stories of which are used as workshops. The journeymen are an inferior class, living in £5 cottages ; but, though their houses be poor and mean, they are clean, as are also the persons and dress of the weavers and their families : the wives are commonly piecers and doublers, or overlookers in the factories, or else, if at home, wind silk ; the children easily get employment in the factories, at wages varying from 1s. to 2s. 6d. a week. The weekly earnings of the broad-loom weavers vary from 7s. 6d. to 9s. nett. The weekly nett wages of the power-loom weavers average 10s., and the women working in the mills usually earn from 5s. to 5s. 6d. From these rates there has been little variation ; and the trade of Leek generally appears to be of a steadier character than that of other towns engaged in the same manufacture. The cotton trade, which has extended itself thither from Lancashire, is not extensive: one small mill employs 60 hands. Coal is procured from the neighbouring Blue hills, in quantities amply sufficient for the wants both of the manufacturers and the population generally. (Hand loom Weavers' Report, part iv., p. 344.)

Leek is one of the polling places for the N. division of Staffordshire. Courts leet and baron are held annually by the lord of the manor (Earl Mansfield), who elects a constable for the civil government of the town. Markets on Wednesday : fairs for cattle and pedlary, February 7th, Easter Wednesday, May 18th, Whit Wednesday, July 3d and 28th, October 10th. and November 13th.

LEELANAN, county Mich. Situated on the E. shore of lake Michigan. Bounded N.E. by Grand Traverse bay. It has been recently formed, and is unorganized.

LEESBURG, p. v., capital of Loudon co., Va., 153 m. N. Richmond, 34 W. Situated on an elevated plain, at the foot of Kitoctin mountain. It is regularly laid out, and the streets are paved. It is supplied with water in pipes from a spring at the foot of the mountain, and contains a courthouse in a square, enclosed by a brick wall, a jail, county offices, three churches, an Episcopal, Methodist, and Presbyterian, a bank, an academy, female seminary, 24 stores, two printing-offices, two weekly newspapers, 500 dwellings, and about 9000 inhabitants. It is one of the most active and busy places in the state, and is governed by a mayor and 12 councillors.

LEETOWN, p. v., Jefferson co., Va., 177 m. N. by W. Richmond, 70 W. Situated on a small branch of Opequan cr. Named from Gen. Lee of the revolutionary army, who resided here. It contains a store, a flouring-mill, several dwellings, and about 80 inhabitants.

LEEUWARDEN, a town of Holland, prov. Friesland, of which it is the cap., on the Ee, 31 m. N. W. Groningen, lat. 53° 12' 14" N., long. 5° 47' 33" E. Population about 17,698. It is surrounded by an earth rampart and ditch, and intersected by numerous canals, the banks of which, like the ramparts, are planted with trees. It is well built ; its streets are wide and regular ; and it has several handsome public edifices, including the palace of the prince of Orange, the town-hall, arsenal, exchange, and house of correction. It has 12 churches, in one of which the princes of Orange are buried, a synagogue, a Latin school, a branch of the Society of Public Good, a printing establishment, and considerable manufactures of linen fabrics, paper, Friesland-green, &c., and a large general trade. It is the seat of tribunals of primary jurisdiction and commerce, and the residence of a provincial commandant, a military governor, a provincial head of police, and a receiver of taxes for the province. It sends four members to the provincial assembly. (De Cloet, &c.)

LEGHORN (Ital. Livorno, Fr. Livourne), a city and seaport, being the principal emporium of Italy, in the grand duchy of Tuscany, prov. Pisa, on the Mediterranean, 62 m. W.S.W. Florence, lat. 43° 33′ 5″ N., long. 10° 16′ 45″ E. Population, with its suburbs, in 1836, 76,397; of whom about 4000 are supposed to be Jews and Greeks, and 3000 other foreigners. It is of a square form, and about 2½ m. in circ., surrounded with new walls, and entered by five gates. It is a neat, clean, and well-built city, and its general air of animation, activity, and business is singularly opposed to the listless idleness of the inland towns of Italy. Its streets are in general wide and well paved, especially that which runs in a direct line from the gate of Pisa to the harbour, enlarging near its centre into a spacious square. The N. part of the city, called *Venezia Nova*, is intersected by canals, and comprises numerous wharfs, warehouses, and other buildings adapted to commerce. Leghorn has an outer and inner harbour, and a good roadstead. The outer harbour is protected by a fine mole, built by Cosmo II., which runs N.N.W. upward of half a mile into the sea. The port is apt to become encumbered with mud, and the water within is rather shallow, varying from 8 feet in the inner basin to 18 or 19 feet at the end of the mole. The outer harbour is, therefore, unfit for ships of more than 400 tons; and the inner harbour, called the *Darsena dei navicelli*, is only used for repairing ships, and for the reception of galleys and other small craft. A lighthouse, the lantern of which is 170 feet above the sea, is built on a rock a little S.W. from the mole. The roadstead lies W.N.W. from the harbour, between it and the Melora bank. The latter is a sand, 4 m. in length by two in breadth, lying N. and S. about 4 m. from shore. It has mostly from three to a half fathoms water over it; but towards its S. extremity, on some rocky points which project above the water, the Melora tower has been constructed to serve as a sea-mark. During S. winds there is sometimes a heavy sea in the roads; but the holding ground is good, and, with sufficient anchors and cables, and ordinary precaution, there is no danger. The lazaretto, said to be one of the best in Europe, lies on a little island to the S., about 1 m. from the tower.

The public and private buildings of Leghorn do not require any very particular notice. They are generally well adapted to their purposes, without being magnificent. The chief public edifices are the ducal palace, the arsenal, the *duomo* or cathedral, a Gothic building, designed by Vasari, six other parish churches, two Greek churches, chapels belonging to the English factory, and the Dutch and German Protestants, an Armenian, and a Maronite Arab church, a synagogue, the largest and finest in Europe after that of Amsterdam, a mosque, three hospitals, the female charity school of St. Peter and St. Paul, the theatre, and the public baths. Leghorn has also a citadel, an old castle, constructed in 1596 by Ferdinand I., two lazarets, besides that before mentioned, two *monti-di-pietà*, a workhouse, a house of refuge, a savings' bank, a large public school, established in 1746, and which has about 350 pupils, schools of navigation and artillery, architecture, painting, mutual instruction, &c., and an academy of sciences, letters, and arts, with a library of 6000 volumes open to the public. The city possesses few works of art, except a fine marble statue of the grand duke Ferdinand I., supported by four kneeling figures in bronze:

it stands on the quay of the inner harbour, and is said to have been executed by John of Bologna. In the cemeteries beyond the walls, however, are some good specimens of sculpture. The English burial-ground, or *Campo Inglese*, contains the remains of Smollett, and of several other distinguished Englishmen.

From being in a marshy situation, Leghorn is not quite salubrious, though, during the reign of the present grand duke, great improvements in draining, &c., have been effected, by which the public health has been much benefited. There are no good wells in the city; and water is brought by an aqueduct from Colognole, 12 m. distant; one of the most remarkable monuments in the town is the *Cisterna*, belonging to this aqueduct, whence water is distributed through the town.

Leghorn has a considerable coral fishery, but the greater portion of its inhabitants are engaged in manufactures and commerce: it produces woollen caps, straw hats, glass, paper, soap, starch, cream of tartar, &c.; and it has numerous coral and alabaster factories; and rope-walks, building-docks for merchant vessels, tanneries, &c. It was made a free port by Cosmo I., about the middle of the 16th century; and the comparative security and freedom which foreigners have long enjoyed in Tuscany, still more than its advantageous situation, have rendered Leghorn one of the first commercial cities of Italy. Its exports are similar to those from the other Italian ports, consisting principally of raw and manufactured silks, straw hats, straw plaiting, and straw for plaiting, all excellent; oil, fruits, wines, wool, boracic acid (recently become a very important article), rags, cheese, marble, argol, paper, anchovies, coral, manna, hemp, lamb skins, timber, with wheat and other species of corn from the Black sea, Egypt, and Barbary; cotton from Egypt, brimstone from Sicily, &c. The export at second hand of produce from the Euxine and the Levant has, however, rather declined of late years; the English, Americans, and other nations now generally importing such produce direct from Odessa, Smyrna, Alexandria, &c. The imports comprise most sorts of commodities, with the exception of some of those produced in Italy, as sugar, coffee, and all sorts of colonial produce; raw cotton, cotton and woollen stuffs, cotton twist, and other manufactured goods; salted fish, indigo, and other dye-stuffs, rice, hardware, earthenware, and metals; hides, &c. Ships, with corn on board, may unload within the limits of the lazaretto, without being detained to perform quarantine, a circumstance which has contributed to make Leghorn one of the principal *dépôts* for the wheat of the Black sea.

The old complicated system of currency has been abolished, and accounts are now kept in new *lire*, which are in all respects equal to French francs. Accounts may be converted from old into new lire at the rate of six old to five new lire.

We regret that there are no detailed statements of the export trade of Leghorn; but it is believed to amount in all to from £1,500,000 to £2,000,000 a year. We subjoin an account of the quantities of some of the principal articles of foreign produce imported into Leghorn during each of the six years ending with 1840, with the stocks on hand on the 1st of January, 1841.

Articles.			1835.	1836.	1837.	1838.	1839.	1840.	Stock, 1841.
			Lbs.	Lbs.	Lbs.	Lbs.	Lbs.	Lbs.	Lbs.
Sugar, Havanna	.	. . .	903,908	3,694,800	1,844,900	1,471,700	2,437,390	3,002,400	255,000
Crushed	.	. . .	6,047,000	4,187,000	8,919,000	7,822,400	6,853,000	9,515,000	1,804,040
Loaf	.	. . .	224,300	130,500	103,430	112,500	163,400	163,000	44,000
Brazil	.	. . .	1,321,500	946,300	14,600	132,000	556,100	443,000	30,000
East India and Santos	.	.	1,271,000	3,111,600	3,696,300	558,790	955,300	2,244,000	200,000
Coffee	.	. . .	1,757,300	2,670,000	2,205,400	1,458,300	1,799,000	4,025,000	730,000
Cotton	.	. . .	1,397,900	1,781,700	2,316,000	908,700	1,834,400	478,000	738,000
Pepper	.	. . .	1,048,500	1,319,500	1,334,400	965,000	1,086,000	756,000	115,000
Indigo	.	. Cases	75	254	183	109	92	200	64
Do.	.	Seroons	195	46	76	107	71	131	22

The quintal (100 lbs.) of Leghorn is about equal to 77 lbs. avoird. The cantara varies from 88 to 160 lbs.; the rotolo = 3 lbs. Port charges are the same on native and foreign ships. The anchorage dues on a vessel of 300 tons amount to 112 old lire, or to £3 14s. sterling; besides which, she must have a bill of health, which costs 7s. 2d. sterling. These, if she clear out in ballast, are the only charges to which she is subject; but if she clear out loaded, the bill of health will cost about 9s. sterling; and there is besides a charge of about 3d. for each bill of lading.

There are no other port charges whatever. Good water may be had at about 11d. a tun; and beef, bread, and fuel are all reasonably cheap. There are companies for the insurance of ships, though not of lives or houses. Leghorn is the residence of consuls from all the principal states of Europe. This port is supposed to be the ancient *Portus Herculis* or *Labronis*; but it has no remains of antiquity. In the 15th century, it was a mere village surrounded by

swamps, and it owes all its eminence and prosperity to the munificence of the Medici family, and the liberality of the present rulers of Tuscany. (*Repetti, Dizionario Geog Della Toscano; Rampoldi Corografia; Condor's Italy*, 51-55; *Commerc. Dict., &c.*)

LEGNAGO, a fortified town of Austrian Italy, deleg. of Verona, 22 m. S.E. the city of that name, on the Adige here crossed by a wooden bridge, with two draw-bridges, and on the high road from Mantua to Padua. Pop. 5790. (*Oesterr. Nat. Encyc.*) Its situation is said to be unhealthy: but it has several churches, convents and barracks: a powder magazine, a theatre, a hospital, a royal gymnasium, a manufacture of dyed leather, and some trade in corn, rice, and silk. Legnago is supposed to have been founded towards the latter period of the Lombard monarchy. It was fortified in the 16th century; and taken by the French, in 1796, after a three days' siege. (*Dict. Géog., &c.*)

LEHIGH, river, Pa., is a wild, picturesque, and rapid

stream, rising in Luzerne co. Its general course is south-eastward to Allentown in Lehigh co., where it turns suddenly eastward, and flows along the N. side of South mountain, and enters Delaware river at Easton. Its length is about 80 m. Along this river, by means of locks, dams, and canals, a navigation is opened from Easton to White-haven in Northampton co., 84½ m., of which 30½ consist of pools, 30½ of canals, and 2½ of locks, and the remainder of sluices. This forms a most important opening to the coal region, to which railroads are continued. Much lumber descends this stream.

LEHIGH, county, Pa. Situated in the E. part of the state, and contains 359 sq. m. It mostly lies in a valley, between South and Blue or Kittatiny mountains. Bounded N.E. by Lehigh river, and watered by its tributaries. The German language is generally spoken, though the English is understood. It contained in 1840, 18,198 neat cattle, 13,448 sheep, 32,695 swine; and produced 176,468 bushels of wheat, 391,830 of rye, 207,098 of Indian corn, 3018 of barley, 202,015 of oats, 116,061 of potatoes, 4000 pounds of tobacco. It had 76 stores, six lumber-yards, one furnace, one forge, two fulling-mills, six woollen factories, four flouring-mills, 62 grist-mills, 46 saw-mills, seven oil-mills, one paper-mill, three powder-mills, 29 tanneries, 119 distilleries, one pottery, four printing-offices, four weekly newspapers; one academy, 30 students; 56 schools, 2303 scholars. Pop. 25,787. Capital, Allentown.

LEHIGH t., Northampton co., Pa. Bounded W. by Lehigh river. Watered by Indian creek. In the N.W. corner of the t. is the Lehigh Water Gap, where the river passes Blue mountain. At the Gap is the village of Lehigh Gap, 104 m. N.E. Harrisburg, 198 W., which contains a mill for preparing hydraulic cement from water limestone found in the vicinity, and about 25 dwellings. Here is a toll bridge over Lehigh river. The town contains eight stores, one fulling-mill, six grist-mills, three saw-mills; nine schools, 510 scholars. Pop. 2049.

LEICESTER, an inland co., of England, nearly in the centre, having N. the counties of Derby and Nottingham, E. Lincoln and Rutland, S. Northampton and Warwick, and W. Stafford and Derby. It is of an oblong form; greatest length about 46 m.; greatest breadth, about 28 m. Area, 515,840 acres, of which about 480,000 are supposed to be arable, meadow, and pasture. Surface, varied and uneven; but except in the district of Charnwood Forest, to the S. of Loughborough, the hills do not attain to any considerable elevation, and are susceptible of the highest cultivation. The soil consists mostly of clayey and sandy loams; and in some parts, especially along the Soar, there are very rich and extensive meadows. The pastures are generally excellent; and this is much more a grazing than an agricultural county. It is famous for its breeds of cattle, sheep, and horses; all of which were much improved by the skill and long-continued exertions of the celebrated Mr. Robert Bakewell, of Dishley in this county. It is, however, true, that the Dishley breed of long-horned-cattle, so famous a few years ago, are everywhere losing ground; and that even in this their native county they are now very generally superseded by the short-horns, and other breeds. The fine rich cheese called Stilton is princially made in this county, in the farms round Melton Mowbray. The Leicester sheep, though of different varieties, all yield long combing wool. Horses are reared in considerable numbers. Barley is the principal corn crop; but wheat and oats are also extensively cultivated. Property mostly in large estates; farms of all sizes, and mostly held at will. Average rent of land, in 1810, 27s. 2¾d. an acre. Coal is wrought at Ashby-de-la-Zouch, and other places; and iron and lead, with lime, and slates, are also products of this county. Leicester is the grand seat of the manufacture of woollen stockings, caps, mitts, &c.; the business being principally carried on in the towns of Leicester, Loughborough, and Hinckley; but it is also widely diffused throughout the county. Hats are made at Loughborough and other places; and this is one of the principal malting counties. Owing to the openness of the country, the number of resident gentry, and other recommendations, Leicester has been long famous as a hunting county. Melton-Mowbray, in the centre of the sporting district, has accommodations for a vast number of horses; and during the season, is crowded with visiters, foreign as well as domestic. Principal river Soar; and the county is intersected by several canals and railways. Leicester is divided into six hundreds and 216 parishes. It sends six members to the House of Commons, being four for the county and two for the city of Leicester. Registered electors for the county, in 1839–40, 9033, being 4179 for the N., and 4854 for the S. division of the county. In 1841, there were in this county 44,549 inhabited houses, and 215,855 inhabitants, of whom 97,556 were males, and 99,447 females. Sum paid for the relief of the poor in 1838–39, £63,115. Annual value of real property in 1815, £951 908. Profits of trades and professions in do. £319,608.

168

LEICESTER, a paril. and mun. bor., and an important manufacturing town of England, cap. co. same name, hund. W. Gescote, on the E. bank of the Soar (crossed here by two stone bridges, one of which is about to be taken down, and replaced by one of iron on the same site), 25 m. S.S.E. Derby, 34 m. E. by N. Birmingham, and 87 m. N. by W. London. Area of mun. and parl. bor. (which are coextensive, and include the old borough and its liberties, with the part called the Castle-view), 3960 acres. Population, in 1831, 40,512, and, in 1840, 57,986. Assessed rental, in 1835, £78,733. The town, though irregularly built, has a respectable appearance, the streets being clean, and the houses in the principal thoroughfares substantial and neat. The main street forms a part of the great N. road, and is joined near its centre by several other handsome streets: at the corner of one of these stands the news-room, a square building of Ionic architecture, lately erected, and forming one of the chief ornaments of the place; and in it also is the New Hall, built in 1831, having apartments for concerts, the Mechanics' Institute, and the Museum of the Philosophical Society. The paving, lighting, and general economy are well conducted, and have been greatly improved within the last few years: water is plentifully supplied from pumps and wells. The public buildings devoted to civil purposes comprise the Guildhall, an old and unpretending building; the Assembly-rooms, with a commodious adjoining theatre; the Exchange, a plain structure, in the market-place: the borough jail, too small for the proper classification of the prisoners; the county jail, a massive structure, enclosing an area of four acres, built in 1824 at a cost of £75,000; and the county lunatic asylum, built in 1791. Among the ecclesiastical edifices, are eight parish or district churches, and 24 places of worship for dissenters. St. Nicholas, the oldest church, is a structure of Norman architecture, supposed to have been built of the materials of the adjoining Roman wall; it has a square W. tower between the nave and chancel. St. Mary's, in the Saxon and early English style, has a lofty tower and steeple. St. Martin's, the largest church in the town, is a cruciform structure, erected at different periods between the 13th and 16th centuries, and surmounted by a plain spire. The other churches, two of which have been very recently opened, are commodious. The Baptist chapel deserves notice, as being the scene of the pastoral labours of Robert Hall, one of the most able and eloquent divines of his day. Connected with the various places of worship are 24 Sunday schools, furnishing religious instruction to nearly 4000 children; besides which, one national, two Lancastrian, and three infant schools are attended by about 1300 scholars, and two parochial schools by 220 boys, who are clothed as well as educated. A collegiate school, lately established for members of the Church of England, and a proprietary school, open to all religious denominations, have each about 100 pupils. The Female Asylum clothes, maintains, and educates 16 girls, between the ages of 13 and 16, and afterward provides them with situations of domestic service. Four weekly newspapers are published in the town.

Leicester possesses many valuable charities, some of which are in the trust of the corporation, others connected with particular parishes. The grammar-school was founded by Queen Elizabeth in 1564, and endowed with lands belonging to the duchy of Lancaster, and several subsequent benefactions: the master's income from the endowment amounts to £123 a year; but is to be reduced to about £43 on the decease of the present master. This school is now nearly useless, being attended by only three or four boys, whereas, a few years ago, there were 80 scholars, 30 of whom boarded with the master. Newton's Charity possesses funds amounting to £750 a year, and supports, either wholly or in part, 12 schools in Leicester and other towns mentioned by the testator. The school at Leicester is a substantial brick building, near St. Nicholas's church, with a house adjoining, in which the master lives rent-free. He has a salary of £100 a year, besides coal and candle, and the use of a large garden. There are 100 boys in the school, sons of poor inhabitants belonging to the established church, who are clothed as well as instructed. Trinity Hospital is an extensive establishment, comprising a chapel, and range of apartments for 80 old men and women, who receive each 3s. a week, with other advantages. The revenue of this charity amounted, in 1835, to £830. Wigston's Hospital, is a structure of perpendicular architecture in St. Martin's Churchyard, erected in 1521, and endowed with estates, the rental of which exceeds £500 a year, but which, it is affirmed, if let like the estates of private individuals, would produce upward of £5000 a year. Each of the 24 inmates has an apartment and garden, with 4s. a week, and the chaplain or confrater has a stipend of £37 a year, with a house and garden. It may be worth mentioning that Chillingworth, and the learned Dr. Samuel Clarke, both filled this situation. Some smaller alms-houses, loan funds, and bequests to a considerable amount, exist in relieving the distress of the poor of the neighbourhood.

Leicester is the principal seat of the manufacture of woollen hosiery, including mits and caps, and of Berlin gloves and Lille thread; and is supposed at present (1840) to have 19800 frames, and 24,000 individuals engaged in these departments, exclusive of those engaged in the subordinate departments of machine making, wool combing, dyeing, &c. There were in the borough, in 1838, according to the returns of the Factory Inspector, 23 worsted-mills, employing 1418 hands, and three cotton winding-mills. The business of the former is to produce the yarn used by the stocking manufacturers. The wages of stocking weavers vary from 8s. to 16s., of glove-makers from 12s. to 15s., those of wool combers and dyers from 15s. to 90s., and those of machine makers from 25s. to 32s. a week. Manufacturing operations are greatly facilitated by a plentiful supply of coal from the Whitwich, Salbeon, and Derbyshire coal-fields. A canal joining the Trent, a railway to Swannington, and the recently opened Midland Counties' Railway, connecting Leicester with the London and Birmingham line at Rugby, and with the North Midland line at Derby, furnish abundant means of transport for manufactured produce, and have been, and no doubt will continue to be, of essential service to the town. Buildings are rapidly increasing in every direction, and in the neighbourhood are many elegant villas, occupied chiefly by manufacturers. The Leicestershire Banking Co., established in 1829, has its chief office in the town.

Leicester is a borough by prescription, incorporated by King John, and governed till 1835 by a charter of the 41st Elizabeth. The old corporation was a self-elected, close, and irresponsible body; and was long distinguished by its political exclusiveness and intolerance. The municipal officers under the Municipal Reform Act are a recorder, mayor, 14 aldermen, and 42 councillors, the borough being divided into seven wards. Corporation revenue in 1839, £21,069; but of this £19,387 was derived from the sale of property; and we believe that its ordinary revenue may amount to about £3000 or £4000. Assizes and quarter sessions are held here, and there is a court for the recovery of small debts. The borough has sent two members to the House of Commons since the reign of Edward I.; the franchise, till the Reform Act, being vested in the freemen (by birth, servitude, or gift), and the inhabitants paying scot and lot. The boundaries of the present parliament borough include, as already stated, besides the old borough, the liberties (which comprise part of the parishes of St. Mary and St. Margaret, together with the Newarke) and the extra-parochial part, called the Castle-view. Registered electors, in 1839–40, 3687, of whom 3650 were freemen. Markets on Wednesday and Saturday; horse and cattle fairs, March, Saturday before and after Easter, May 12, 13, 14, July 5, and Oct. 10.

Leicester occupies the site of Ratæ, an important Roman station, mentioned in Antonine's Itinerary. Near the Jewry wall five Roman pavements have lately been discovered, one of which is remarkable for its extent and beauty. Its Saxon name, Legeoceaster, is derived from its position on the Soar, anciently called the Leire. After the Norman conquest, a castle was built here, which, with the town, was nearly destroyed during the civil wars between Henry II. and his sons; but both were subsequently rebuilt by the earls of Lancaster; and during the reigns of the Lancastrian princes, the castle was often a royal residence, and the seat of parliament. Ultimately, however, it was pulled down in the reign of Charles I. During the great civil war, the town was successively occupied by the king and the parliamentary troops: the latter, by way of "purifying" the large church of St. Martin's, converted it into cavalry barracks. In a meadow near the town are some remains of a monastery of Black Canons, founded in 1143, the ruins of which amounted, at its dissolution, to £1062. Cardinal Wolsey expired in this abbey on the 29th Nov., 1530, having been compelled, by sickness, to take refuge here when on his way to London to be tried for high treason. The stocking-frame was introduced into Leicester about the close of the 17th century, since which time it has been steadily rising in manufacturing importance. (Char. and Mun. Commission Report; Thornby's Hist. of Leicester; Private Information.)

LEICESTER, p. t., Addison co., Vt., 73 m. S.S.W. Montpelier, 477 W. Chartered in 1763. Watered by Otter creek and Leicester river, the latter issuing from lake Dunmore, which is 4 m. long and three fourths of a mile wide, and lies between this town and Salisbury. It contains a Baptist church, and some Congregationalists and Universalists, and has three schools, 66 scholars. Pop. 603.

LEICESTER, p. t., Worcester co., Mass., 48 m. W. Boston, 408 W. Incorporated in 1713. Watered by French river, and branches of Blackstone river. The site is elevated. It contains five churches, a Congregational, Unitarian, Epis-

copal, Baptist, and Friends', an old and well endowed academy, founded in 1784, seven stores, three grist-mills, seven saw-mills, six tanneries; one academy, 219 students, 10 schools, 544 scholars. Pop. 1707.

LEICESTER, t., Livingston co., N. Y., 5 m. W. Geneseo, 232 m. W. Albany. Bounded E. by Genesee river. It contains three stores, one woollen factory, two grist-mills, three saw-mills, one tannery, two distilleries; one academy, 29 students; 16 schools, 758 scholars. Pop. 2415.

LEIGH, a manufacturing market town, and par. of England, co. Lancaster, hund. W. Derby, 12 m. W. Manchester, and 21 m. E.N.E. Liverpool. Area of parish (comprising the township of West Leigh, Astley, Atherton, Bedford, Pennington, and Tildesley), 11,969 acres; pop. in 1841, 22,230. Pop. of West Leigh and Pennington townships, comprising the town of Leigh, 3945. The town, consisting of two chief, and other subordinate streets, has a few well built houses, mixed with others of an inferior character. The church is a lofty stone structure, but low and decayed on the N. side: the living is a vicarage in the gift of Lord Lilford. Chapels of ease have also been erected in the townships of Astley, Chowbent, and Tildesley; the patronage of which is vested in the incumbent of Leigh. There are places of worship for Roman Catholics, Independents, Wesleyan and new connexion Methodists, and Swedenborgians, exclusive of others in the out townships; and upward of 4000 children are taught in the Sunday schools connected with the churches and chapels. The charities of the parish comprise the grammar-school, founded in 1655, but poorly endowed, and attended in 1836 by about 30 boys, 40 of whom were pay-scholars; and the free schools of Pennington and Astley, with some apprentice-funds, and minor bequests. (Char. Comm. 19th Rep.)

Leigh occupies a very respectable station among the cotton manufacturing towns of Lancashire. The business, which was formerly almost confined to weaving fustians, now embraces all the processes and branches of the cotton and mixed goods manufacture; and, according to Mr. Baines, upward of 8000 hands were employed, in 1834, chiefly in the townships of West Leigh, Tildesley, Atherton, and Bedford, in spinning and weaving cotton and silk, both by hand and power. In 1838, there were in the parish 19 cotton-mills, and one silk-mill, employing 2458 hands; eight of these mills are situated in Tildesley township, which has also two large factories for machinery. These branches of industry are greatly promoted by the abundance of coal and lime in the neighbourhood, and by the easy canal and railway communication with Liverpool and Manchester. A branch of the Duke of Bridgwater's canal unites here with the Leeds and Liverpool canal, and the Leigh and Kenyon tram-road connects the town with the Liverpool and Manchester railway, the communication being continued N. by the Bolton and Leigh railway: the latter, 7½ m. in length, was constructed at an expense of £10,000 per mile. The grass-lands of the parish are particularly rich, and the dairies round the town yield a cheese held in deserved estimation. Market on Saturday; and fairs, well attended for cattle, cheese, &c., April 24, and Dec. 7. (Baines's History of Lancashire, 4to edit.; Parl. Rep.)

LEIGH READ, county (formerly Mosquito), Flor. Situated on the E. side of the peninsula, with a great extent of territory. Watered by St. John's river and its tributaries, and other smaller streams. Capital, New-Smyrna. Pop. in 1840, 73.

LEIGHTON-BUZZARD (or, more properly, Leighton-Beau-desert), a market town and parish of England, co. Bedford, hund. Manshead, on the Ouzel, a tributary of the Ouse, 5 m. S. by W. Woburn, and 36 m. N.N.W. London. Area of parish, including five townships, 8990 acres. Pop. of par., 1831, 5194; do. of town, 1841, 6053. The streets are irregularly laid out, ill paved, and not lighted with gas; the supply of water is chiefly derived from wells. It has a fine pentagonal cross in an open area near the market-house, supposed to have been erected at the beginning of the 14th century: it consists of two stories, and is 36 ft. high. The church, formerly collegiate, is a large cruciform Gothic structure, with a tower and steeple rising from the intersection of its nave and transepts: the living is a vicarage, attached to a prebend in Lincoln cathedral. There are places of worship for Baptists and Wesleyan Methodists, and the Society of Friends (here a numerous body) have a large meeting-house. Besides Sunday-schools, there is a well-endowed charity-school for the gratuitous instruction of poor children; and a large Lancastrian school, for both sexes, supported by voluntary contributions. Wilkes's almshouses, founded in 1630, have an average yearly income of £200, and furnish lodging and stipend to eight poor widows: there are several other charitable foundations. (See Char. Comm. 12th Rep.) Lace-making, formerly a considerable branch of industry in Leighton-Buzzard, has been all but extinguished by the frame-lace trade of Nottingham. Straw-plaiting here, as in other towns of Bedfordshire, employs many females; but the principal trade is in corn and timber, the conveyance of which to the London market is greatly facilitated by the Grand Junction canal and the London and Birmingham railway, which has a station at this place. Markets large and well attended, on Tuesday; fairs for cattle, horses, and grain, on the 3d Tuesday in April, July 26, Oct. 24, and Nov. 7. One of the largest horse fairs in the S. of England is held on Whit-Tuesday.

LEINSTER, one of the four large provinces into which Ireland is divided on the E. side of the island, comprising the counties of Dublin, Kildare, Carlow, Kilkenny, Kings and Queens, Longford, Louth, Meath, Westmeath, Wicklow, and Wexford. For an account of its extent and population, see *anti*, p. 32.

LEIPSIC (Germ. *Leipzig*), a celebrated commercial city of the kingdom of Saxony, being, next to Hamburg, the chief trading city of Germany, and the first book emporium in the world. It stands on the White Elster (a tributary of the Saale), where it is joined by the Pleisse and Parde, 60 m. W.N.W. Dresden, and 90 m. S.E. Halle; lat. 51° 20′ 16″ N., long. 12° 21′ 45″ E. Pop. (1837) 47,514, nearly all Protestants. Its appearance, at a distance, is not imposing: it stands in a wide plain, which, though fertile, is unvaried by a single eminence to relieve its sameness. It occupies but a small extent of ground compared with its population, the total number of houses in it and its suburbs being only about 1460. These, however, are very lofty; many of them being six stories high, independent of three or four additional in the pyramidal roof; and each story, like the houses in the old town of Edinburgh, is usually occupied by a separate family. Few towns exhibit so much of the carved masonry which characterised the old German style of building, joined with so much stateliness. The streets are narrow, but the various markets and squares are large, clean, and neat. Leipsic is far inferior in elegance and beauty to Dresden; but it is better built than Frankfort, and has a decided air of comfort and substantiality. The suburbs are well laid out, and separated from the town by a succession of pleasant gardens, occupying the glacis, and other parts of the ancient fortifications. The great market-place, in the centre of the town, is rendered one of the most striking squares in Europe, by the quaint architecture of its surrounding buildings. In one of these, the *Rathhaus*, the allied sovereigns met to congratulate each other after the battle of Leipsic (see *post*). The *Konigshaus*, formerly a residence of the electors and kings of Saxony, was occupied by Napoleon in 1813. The Auerbach cellar, at no great distance, is noted as that in which Göethe has laid the celebrated carousel scene in Faust; and tradition says that Faust himself used to frequent it. At the S.E. part of the town is the castle of Pleissenburg, which withstood the attacks of Tilly in the 30 years' war, long after the town had surrendered. Its lower part is now a wool magazine, and its upper part an observatory 228 ft. high, from the summit of which a commanding view is obtained of Leipsic and its plain. The ramparts of the town have been laid out as public walks; and its gates have been recently removed. The church of St. Nicholas is a handsome square edifice, and of a species of Corinthian architecture; its interior is ornamented with numerous paintings by Œser, a celebrated Saxon artist of the last century. The other most remarkable public buildings are the *Augusteum Paulinum*, &c., belonging to the university, the cloth hall, booksellers' exchange, and new postoffice. There are many good hotels.

The university is the only one in Saxony; and ranks as one of the first, as well as most ancient, in Germany. It was founded in 1409 by some professors and students from the university of Prague. It is divided into four nations, the Saxon, Misnian, Franconian, and Silesian; and has faculties of theology, law, medicine, and philosophy. It had, in 1834, 34 ordinary professors; six in the faculty of theology, five in that of law, 10 in that of medicine, and 13 in that of philosophy. The total amount of their salaries was $32,410 (about £7735), independently of certain small fees and minor emoluments. There are, besides, many extraordinary professors teaching modern languages, &c., who do not belong to the Senatus Academicus. Leipsic university, though still well attended, has at present fewer students than usual: at the beginning of 1840 the total number was 925.

The greater number of lecture-rooms are here, as in Heidelberg, within the university buildings. Most of the students live within the walls of the Old Paulinum, without reference to their particular department of study: the only qualification necessary to entitle them to the bursary enjoyed there being an examination as to their proficiency in learning. Some students are allowed both board and lodging in the Paulinum; others are only entitled to a seat at the public table. "The university is rich in endowments for stipends to scholars; but with respect to such funds as are applicable to its maintenance and to scientific purposes, it is one of the poorest in Germany. An inventory of its

164

property, which has been lately made public, states its means toward these latter objects to amount to 3699 dollars per annum only, not more than £800. It appears, from a statement of its yearly disbursements, that Saxony does not expend as much on this, its sole university, as the Prussian treasury expends upon the least of its provincial universities. The disbursements in question amount to $56,315 about (£8000), not including scholars' stipends, nor support of the poor (*armenwesen*); and the proportion of that sum which is derived from the national revenue is but $23,486, or about £3690. The property of the university is valued at $1,100,000 (about £156,000); and out of this capital, which consists chiefly of house property, besides a small portion of meadow and arable land, some wood, and a few shares of mines, the yearly interest on $650,000 is applicable to benevolent purposes; the interest on the remainder, about $450,000, is therefore all that is available for the current expenses of the university." (*Journ. of Education*, No. xv. 151, 152.)

The *Augusteum* contains a library of 100,000 vols. (*Horschelmann*), and the university has also a museum of natural history, a botanic garden, anatomical theatre, laboratory, clinical and lying-in establishments, &c. Leipsic has, besides, a civic school, and attached to it a school of general knowledge, opened in 1834, several other superior and free schools, primary schools, &c., numerous learned associations, a public library, with 50,000 printed vols., and 2000 MS., and various scientific collections. Several hospitals; orphan, foundling, deaf and dumb, and lunatic asylums, and a house of correction nearly complete the public establishments. There are some private galleries of paintings, and other works of art, but none deserve particular notice.

Leipsic is a manufacturing town of considerable importance. Among its chief manufactures are silken and half-silken goods, stockings, leather, hats, playing and other cards, paper hangings, oil-cloth, wax, lights, starch, soap, sealing-wax, parchment, tobacco, gold and silver articles, liqueurs, chocolate, &c. Artisans of almost every kind reside in the town. Berghaus says, that of 153 book and map sellers belonging to the kingdom of Saxony, 122 live in Leipsic; and that of 50 printing-offices in the kingdom, Leipsic has 22, with 210 presses, including seven printing machines (*schnell pressen*).[*] There are also various silk-dyeing and woollen spinning factories; and a large wool market is held annually in May.

But the distinguishing characteristic of the commerce of Leipsic, is its book trade. Leipsic is, in fact, the grand emporium of the literature of Germany; a distinction of great importance, seeing that the number of readers and writers is greater in that than in any other country of Europe. The literary deluge which commenced in Germany immediately after the peace of 1814, continues to increase. Instead of 2000 works, which were then about the annual complement, there are now 8000. In 1837, the catalogue of the Easter fair announced 4353 new works, and that of the Michaelmas fair 3538; making a total of 7391 in the year, or an increase of 362 over the number in 1836. Of this number Prussia contributed 2169, and Saxony (kingd. of), 1342 publications. In the German book-trade, it is the custom for almost every house, either in the country or abroad, which publishes or sells German books, to have its agent at Leipsic, who receives and distributes its publications in the same way that the London booksellers receive and distribute English publications. The great sale of new works takes place at the Easter fair, when 600 booksellers sometimes assemble to settle their annual accounts. "The German author will submit to any degree of exertion, that his work may be ready for publication by that important season, when the whole brotherhood is in labour from the Rhine to the Vistula. Whatever the period of gestation may be, the time when he shall come to the birth is fixed by the almanack. If the auspicious moment pass away, he willingly bears his burden 12 months longer, or till the next advent of the Bibliopolical Lucina." (*Russell's Tour in Germany*, i. 231, 232.)

The fairs of Leipsic are the most celebrated in Germany. They are held at the new year, at Easter, and at Michaelmas. The last two are the most important. Above 20,000 dealers are said to have been present at the Easter fair in 1832, and above 13,000 at that of Michaelmas. They should close in eight days, but they generally last three weeks; and while they continue, Leipsic is the great mart of central Europe for all kinds of merchandise. According to the author of *Germany and the Germans*, in 1835–36, who visited Leipsic at one of its fairs;—"The whole appearance of the

* *Blg. Lander, &c.* (1836), iv. 178. According to the statement in the *Handbuch fur Reisende, art. Leipzig*, there were, in 1835, 104 booksellers; 20 printing-offices, with 250 presses, including three printing machines impelled by steam; five type and two stereotype foundries; and several lithographic and copperplate engraving establishments; and 10,000,000 sheets of letter-press, &c., were estimated to be annually printed at Leipsic.

town was unique; the streets, markets, and promenades were crowded, not only with the natives of every part of Europe, but even with those of Asia, Africa, and America; every house, yard, and porch, was converted into a bazaar for the display of merchandise, cottons, woollens, and silks of all shades; and, from every loom in Europe, were streaming like flags from the windows of the lofty houses; and although the Prussian tariff was in full force, yet I was informed by a merchant that the market was inundated with smuggled English manufactures." The statement respecting the attendance of Asiatics and Americans is, at least, no hyperbole: exclusive of Turks, Greeks and Wallachians, Georgians, Armenians, and even Persians are present; and from 380 to 400 guests sit down daily at the *tables d' hôte* of some of the principal inns. It is estimated that the produce of the sale of books at the Easter fair amounts to 3,000,000 dolls.

The establishment of the Prussian Customs' Union (commercial league) led many intelligent persons to apprehend that, however advantageous the new system might be for the commercial interests of Germany at large, it would exercise a most prejudicial influence over the trade of Leipsic, by materially injuring, if not annihilating, its fairs. The result, however, seems to show that the customs' union has had a directly contrary effect; and that, so far from being ruinous to the Leipsic fairs, it is the very thing which is most likely to arrest, or rather, perhaps, to retard their fall; as by giving an immense impulse to the internal communications within the German states, it has in some degree compensated Leipsic for the gradual falling off in its commerce with foreign nations. The Leipsic fairs were long the great marts whence Russia, even to the borders of China, Poland, the provinces of the Danube, and many of the Turkish provinces, were supplied with manufactures. At the Michaelmas fair of 1839, however, (the latest of which we have any detailed report), the falling off in the numbers of foreign purchasers was particularly remarkable. Of these, Greeks from Wallachia and Moldavia were the most numerous, their principal purchases being German broadcloths, ordinary English and German cotton goods, and French silks; about 7000 centners of manufactured goods of all kinds being then entered for those principalities. The number of Russians was small; but a great many Jews, from Galicia, Prussian Poland, and Cracow, as usual, attended: whose chief object seemed to be that of introducing, in spite of every obstacle, manufactured goods of all descriptions into Russia. Few buyers went from Austria, Switzerland, or Italy, Trieste and Frankfort being much more convenient marts for them: France and England, also, sent fewer than usual. Still, at the Michaelmas fair of 1839, the number of buyers and sellers, as well as the amount of business done, was as great, if not greater, than on former occasions.

Broadcloths, made within a circle of from 10 to 40 German miles round Leipsic, are among the most important articles brought to the fairs, and from 90,000 to 100,000 pieces were sold on the above occasion, mostly for consumption in the states of the Union. Other woollen goods, both English and German, were sold in considerable quantities, but at very low prices. British printed calicoes form an important item; but the heavier and coarser descriptions of cotton goods are said to be in a great measure driven out of the Leipsic market by German manufactures. French and Swiss silks are rapidly increasing in demand, their use having greatly increased of late among the middle classes in Germany. Hides and leather, lace and embroidery, linens, hosiery, hardware and cutlery, clocks, jewelry, French china, quills, furs, isinglass, &c., are among the other goods that are most extensively met with at the Leipsic fairs. (*Bowring's Report on the Prussian Commercial Union; Append.* xxv., 259–265.)

The Leipsic and other German booksellers have, since 1834, erected an exchange for their exclusive use. The building as originally planned was to be three stories high, 112 ft. in length, and 48 ft. in depth; and the estimated cost of its erection was from £4500 to £5000. Among the other establishments in aid of commerce, are a fire and a life assurance office, each with a capital of $1,000,000; a company for insurance against *hail*; a discount bank, with a capital of $250,000, &c. The Elbe American Company, established in 1825, has its seat in this town, which has also a savings' bank, and a *mont-de-piété*. A railroad has been completed between Leipsic and Dresden; and another to Magdeburg would, it was expected, be completed by the end of the present year (1840). These works have not yet (as was anticipated) diminished the price of necessaries in Leipsic, provisions, fuel, house-rent, &c., being higher in it than in most capitals. The dollar of Leipsic is equal to 3*s.* 4½*d.* sterling; 100 ells = 61·83 Eng. yards; 100 Leipsic lbs. = 103 avoird.

Leipsic is the cap. of the prov. and district of same name, and the seat of the judicial courts, &c. At the end of the 10th century it was only a little Slavonian village; but during the 12th it was fortified, and its two principal fairs established. Its new year's fair commenced in 1458, and its book trade originated in 1545.

The vicinity of Leipsic, and, indeed, the town itself, was, in October 1813, the scene of a most tremendous conflict. Napoleon having concentrated at this point such of his forces as he had been able to collect from the different parts of Germany, to the amount of about 135,000 men, was attacked on the 16th by the allied army, under prince Schwartzenberg, Blucher, and other generals, accompanied by the emperors of Russia and Austria, the king of Prussia, &c. The allied forces amounted to at least 250,000 men. The struggle, which was fierce, obstinate, and bloody in the extreme, terminated at nightfall without any decided advantage to either party. It was renewed on the 18th, when a Saxon brigade went over, during the heat of the action, from the French to the allies, which, combined with their superior force, gave the latter an advantage that all the genius of Napoleon, seconded by the valour and devotion of the French, could not counteract. Though the French maintained their ground during the day, a retreat became indispensable; and owing to the accidental blowing up of a bridge, a part of the French army was cut off; so that Napoleon lost 25,000 men, who fell into the hands of the allies as prisoners, exclusive of the far greater number who fell in the previous battles. Prince Poniatowski, who may emphatically be said to have been the last of the Poles, after displaying prodigies of valour, lost his life in the retreat on the 19th, having been drowned in attempting to cross the Elster. This great victory completely emancipated Germany from the yoke of the French, and opened the road to Paris to the allies. (*Berghaus, Herschelmann, Stein; Handb. für Kaufleute,* and *Comm. Dict.: Conversations' Lexicon; Strang, Russell; Bowring's Rep. on the Prussian Com. Union, &c.*)

LEIRIA, a city of Portugal, prov. Estremadura, on the small river Lis, 42 m. S.S.W. Coimbra, and 72 m. N.N.E. Lisbon, lat. 39° 30′ N., long. 8° 52′ W. Pop. 2500. The town, which, though in a fine country, is small, sombre, and wretched-looking, has no fewer than 19 parish churches (one of which is likewise a cathedral), and three convents! A considerable fair is held here on the 25th of March.

LEITH, a seaport and parl. bor. of Scotland, co. Mid Lothian, on both sides of a small river of the same name, at its confluence with the frith of Forth, on a flat sandy shore, 2 m. N. by E. of the centre of Edinburgh, of which city it is the port. Pop., in 1755, 7200 (*Maitland's History of Edinburgh*, p. 500); in 1811, 20,363; in 1831, 25,855; inhabited houses, 1641; number of persons to a house, 15·755. The great proportion of persons to houses results from the fact that Leith, like Edinburgh, has houses of considerable height, and that several families (in some instances six or eight) live under the same roof in different *flats* or stories, each having access by a "common stair," which communicates with every story of the tenement.

The river divides the town into two portions, called N. and S. Leith, of which the latter (the original borough) contains 18,437, and the former 7416 inhabitants. They are connected by two drawbridges, and by an elegant stone bridge at the W. extremity of the town. Leith is united to Edinburgh by a splendid road (partly paved, and partly macadamized), called Leith Walk, but it is so filled up with buildings that it may be regarded rather as a street than a road. Part of the town of Edinburgh on the W. stretches into the parish of S. Leith. The buildings in the older parts of Leith are huddled together without order or regularity, and the streets and lanes are, for the most part, narrow, crooked, and filthy. The new streets to the S. and E., however, form striking exceptions, being not much inferior to the best in Edinburgh. In S. Leith, the only two leading streets (Constitution-street and the Kirkgate) branch off from the N. termination of Leith Walk in the form of an acute angle. The street called the Shore, fronting the harbour on the S., is lofty and substantial. On the S.E. of S. Leith are Leith Links, or downs, a common belonging to the borough, ¾ m. in length by nearly ½ m. in breadth. The best buildings are erected on its skirts, chiefly on its N. and W. sides. A project for a new town, as an extension of N. Leith on the W., on a regular plan, was fixed on upwards of twenty years ago, and has been partially carried into effect: but building, both in this quarter and throughout the town generally, has been nearly suspended for about a dozen years. Both N. and S. Leith are lighted with gas, and supplied with water by the Edinburgh Water Company. The length of the borough, from E. to W., is 1¼ m., the mean breadth ½ m. The mean and dirty village of Newhaven, ¾ m. to the W., is inhabited almost exclusively by fishermen, who chiefly supply Leith and Edinburgh with fish. A low-water stone pier has been erected in Newhaven by the Mid Lothian and Fife ferry trustees for the use of the passage boats.

The public buildings in Leith are numerous. The Trinity-house of Grecian architecture, on the W. side of the Kirk-gate, was founded in 1555, and rebuilt in 1817. The parish church of S. Leith, opposite the Trinity-house, is a plain uninteresting edifice, built in the 16th century, and lately divested of its spire and other ornaments. The parish church was at Restalrig, 1 mile E. of the borough, till the Reformation, when it fell a sacrifice to the destructive zeal of the Presbyterians; since which the present building, originally a chapel dedicated to St. Mary, has served that purpose. The *quoad sacra* church of St. John, in Constitution-street, recently erected, is a spacious Gothic edifice; it has a lofty massive octagonal spire, with two schools attached to it, and forms altogether one of the most imposing objects in the borough. The present parish church of N. Leith is a modern structure of Grecian architecture, on an elevated situation with a spire 158 ft. high. There are two *quoad sacra* churches in this parish, both neat buildings, one of them (erected in 1840, for the special use of mariners) in the centre of N. Leith, the other at Newhaven. A handsome place of worship, in connexion with the establishment, has been erected (1841) in S. Leith, and endowed by John Gladstone, Esq. of Fasque, a native of the borough, it has attached to it a school and a hospital. The buildings, which are in the Gothic style, form three sides of a square; the estimated expense is £10,000; exclusive of the endowment, the amount of which has not yet transpired. Among the other public buildings may be mentioned the jail, a new edifice of Saxon architecture; the town-hall, in Constitution-street, erected in 1828, perhaps the most chaste and elegant specimen of modern architecture in the town; the Exchange Buildings, a large and spacious Grecian structure, extending to 180 ft. in front, and comprising a hotel, assembly rooms, and a reading room; the Leith bank; the Custom-house, close to the harbour on the N.; the High-school, at the S. corner of Leith Links; Dr. Bell's school; various dissenting chapels, particularly an episcopal one; and the Seafield baths, erected by a joint-stock company; in 1713, at the E. extremity of the town, at an expense of £8,000.

In regard to religious instruction, there are, in addition to the two parish and four *quoad sacra* churches (including Mr. Gladstone's), three chapels belonging to the Associate Synod, and one respectively to the Relief, Independents, and Episcopalians. The living of N. Leith is, next in Greenock, the highest in the Scottish church, being about £800 a year, arising principally from the tithe of fish land ed at Newhaven, and from the rent of the glebe, which is *fened*, or let on building leases. The church of S. Leith is collegiate, or is served by two ministers.

Literature and education cannot be said to have received, at least till of late, much encouragement in Leith. With the exception of the High school, an efficient seminary, built by subscription in 1805, little else has been done in furtherance of either. In 1833–34, the proportion of young persons at school was said to amount to one tenth of the population; but since then a school, founded by the late Dr. Bell of Madras, has come into operation; and an infant school, and two other seminaries for the children of the humbler classes, have also been opened in the interval. Dr. Bell's school was founded, and is supported, by a bequest of £10,000, left by that great patron of education. There are several subscription libraries; and a philharmonic society for the cultivation of music, both vocal and instrumental. But there are no literary or scientific associations; and no newspaper is published in Leith. The near vicinity of Edinburgh may account for these and similar facts. Dr. Henry Hunter, translator of Lavater's "Physiognomy," &c., and John Logan, author of "Sermons and Poems," were successively ministers of S. Leith; and John Home, author of "Douglas," and Hugo Arnot, the historian of Edinburgh, were natives of the borough.

With the exception of the Trinity house, Bell's bequest, and Gladstone's hospital and school, Leith has no important charities. The Trinity house, the funds of which are devoted to the relief of decayed sailors or their widows, supports, by monthly or quarterly payments, from 170 to 180 pensioners of various classes (but the number is necessarily variable); besides assisting sailors who have been shipwrecked, or are otherwise in distress. The recipients of the charity formerly lived in the house, but now they are all out-pensioners. There was a charity called King James's Hospital, for the support of old women; but the building has disappeared, though the funds, which are trifling, are still devoted to their original object. The other charitable institutions are the humane society, dispensary, and casualty hospital; the society for the relief of the destitute sick, and Leith boys' charity school. There are various friendly societies.

Poor-rates were introduced into S. Leith in 1817; into N. Leith in 1822: aggregate numbers of permanent and occasional paupers, at an average of three years, 1458, or nearly the 18th part of the whole pop., exclusive of 22 lunatics.

supported out of the poor's funds. Average assessment, £2829 17s. 2d.; average church collections and other parish dues, £303 4s. 6d.; total, £3666 1s. 8d.

Leith labours under all but incurable disadvantages in respect to its port. At low water, the tide recedes above a mile from the shore; and the stream of the rivulet by which the town is bisected is so tiny, that it is even insufficient to clear away the mud from the harbour. Various efforts have been made to overcome these natural difficulties; but hitherto with no very marked success. In 1720, a dock was formed on the E. side of the river, and that portion of the present pier which is of stone was erected; and in 1777 a small quay, called the custom-house quay, was built. But the increasing commerce of Leith soon rendered these trifling improvements quite inadequate to the demands of the port; and accordingly, in 1799, the magistrates of Edinburgh, who, till recently, had the uncontrolled management of all public matters connected with the town and port of Leith, obtained an act of parliament, authorizing them to borrow £160,000, for the construction of wet-docks. In consequence, two docks were constructed on the N. side of the harbour, between 1800 and 1817, each measuring 250 yards in length by 100 in breadth, and comprising together about 10½ imperial acres. Attached to them are three graving-docks, each 136 ft. long by 45 ft. wide at bottom; and 150 ft. long by 73 ft. wide at the top; with an entrance 36 ft. wide. At average spring tides, the depth of water in the docks is 16 ft. 9 in.; and at neap tides, 4 ft. less. The total expense was £985,000, of which £265,000 was borrowed by the city from government, at 5 per cent.; of which 3 per cent. was to be paid annually, and 2 per cent. to be accumulated as a sinking fund for the liquidation of the debt. The city gave as security a mortgage over all their Leith property, and a concurrent claim, with other creditors, over the entire municipal property of Edinburgh, besides ceding certain effects to the admiralty.

In addition to these great works, others were undertaken in 1831–32; viz., an elongation of the pier to the extent of 2550 ft., making it altogether upward of half a mile in length; and the formation of a covering bulwark on the opposite side, 1500 ft. in extent. The expense of the former (£98,000) was borne by the city of Edinburgh; the latter (£12,000) by government. The object of these works was to deepen the water in the channel: which has been effected to the extent of about 2 ft. But after all that has been done, the harbour is all but dry for low water, and there are only 17 ft. water over the bar, at its mouth, at high water spring tides, and only 14 ft. at neap tides. In fact, no vessel of above 400 tons burden can approach the harbour at the highest tides; and sometimes not even vessels of that burden: and though the act of parliament, called the Edinburgh and Leith Agreement Bill (July, 1838) has placed the docks under parliamentary commissioners, and allowed a certain large sum, viz., £125,000, to be conditionally expended in improving the harbour, it is all but certain that the physical disadvantages under which it labours can never be successfully overcome; that Leith can never be anything better than a very indifferent tide harbour; and that the expenditure of further sums upon it would be a very questionable proceeding.

The harbour of Edinburgh should, in fact, have been constructed more to the W., at Trinity or Granton. Indeed, after much opposition on the part of Leith and Edinburgh, a bill was carried through parliament in 1837, for the construction of a low-water pier at Trinity, half a mile W. of the harbour of Leith; but various circumstances, which it is unnecessary to notice, make it pretty certain that no effort will be made to carry that measure into effect. Luckily, however, the Duke of Buccleuch has been, for some years, engaged in the construction of a low-water pier on his estate of Granton, 1 m. W. of Leith; an undertaking of great national importance, and worthy an individual of great wealth and public spirit. This splendid work, by far the greatest of its kind attempted in Scotland, will, when completed, secure for Edinburgh all the advantages of a deep-water harbour, accessible at all times. The pier, which is constructed in the most approved manner, is to project into the sea about 1700 ft., shaped like a T, with its head to the N., having harbours and landing-places on both sides. It has been partially open for upward of two years, but its business has hitherto been principally confined to the accommodation of steamers. The duke has erected a large edifice for a hotel, warehouses, and other buildings; and there can be little doubt that, in no very lengthened period, the principal part of the shipping business carried on at Leith will be transferred to Granton. The latter communicates, by an excellent road, with the New Town of Edinburgh.

But notwithstanding these unfavourable circumstances, the commerce of Leith, from its being the port of Edinburgh, is very considerable, and has been slowly but steadily improving. It carries on a limited trade with Australia, the E. and W. Indies, China, the Mediterranean, Canada, and the United States; but its chief foreign trade is with Holland and the N. of Europe. With regard to its domestic trade, there are three companies who have altogether 18 vessels trading with London; and there are other companies, which have vessels trading with Hull, Newcastle, Liverpool, Greenock, Glasgow, Aberdeen. Wick, Helmsdale, Orkney, Shetland, Dundee, Stirling, &c.; the Greenland fishery once employed seven vessels, but now only five. The communication by steam with London is three times a week during winter, and at least five times during summer; with Newcastle twice, and Hull once a week; Hamburg once a fortnight; Stirling twice a day; the opposite coast of Fife three times a day; and a regular communication with every important place on the E. of Scotland from Lerwick in Shetland, and Kirkwall in Scotland, to Berwick-upon-Tweed. The steamers sail from Leith, Newhaven, the chain pier at Trinity or Granton, but now mostly from the latter.

Account of the Vessels, with their Tonnage, that entered the Port of Leith in the under-mentioned Years:

Years.	British Vessels.	Tons.	Foreign Vessels.	Tons.
1841	125	17,846	211	65,726
1842	229	24,898	47	7,112
1843	263	43,216	176	47,198
1845	284	33,345	164	47,157
1846	349	42,699	202	47,314
1849	222	30,433	303	2,436
1850	234	48,524	416	35,330

The Gross Amount of Customs' Dues collected at Leith during the following Years has been—

Years.	£.	s.	d.	Years.	£.	s.	d.
1833	386,910	9	11	1837	595,402	19	3
1834	386,863	19	6	1838	511,970	6	8
1836	514,974	3	5	1839	573,685	13	7

There belonged to Leith, in 1840, 176 vessels, of the aggregate burden of 19,954 tons; the amount of shipping, coastways and foreign, that entered and left the port during the same year, was 400,913 tons. Leith is consequently superior, as a port, to either Greenock or Glasgow, taken separately.

An attempt was recently made by a joint stock-company to introduce flax-spinning, and the manufacture of the coarser linen fabrics, into the town; but it was found impossible, as in most similar cases, artificially to raise up a manufacturing population, or successfully to come into competition with places, such as Kirkcaldy and Dundee, where the business had been gradually formed, and long established. The company, organised in Leith for this purpose, erected extensive premises, and employed 140 persons in mill-spinning in 1839; but their operations have entirely ceased, and their buildings are unemployed. A glass and bottle work has existed for a century, but one of seven furnaces only one is now at work. A pottery has just been commenced, and promises to be successful. Ship-building has long been ca..'d on to a great extent, and employs more capital than any other business in Leith. There are various extensive rope and sail works, distilleries, breweries, and iron foundries; a sugar refinery; a soap manufactory, which produced, in 1839, 2,130,250 lbs. of hard, and 51,573 lbs. of soft soap; a small linen manufactory, &c. There are nine incorporated trades; an incorporation of maltmen; a merchant company; a chamber of commerce; six banks, and a savings' bank. A branch of the Edinburgh and Dalkeith railway, brought into the town, terminates at the quay, opening an easy communication with the extensive collieries in the E. of Mid Lothian. The intercourse with Edinburgh by stage coaches is every quarter of an hour.

Leith existed as a town as early as the 13th century. The old church of N. Leith, dedicated to St. Ninian, close by the river, and long disused as a place of worship, was founded in 1493. It is now in ruins; but its cemetery is still used as a burial ground. A bridge over the river, built, in 1493, by Robert Ballenden, abbot of Holyrood house, was used till 1788, when the first drawbridge was erected. Leith is of no small note in the history of Scotland, having been the scene of more military service than perhaps any other town in the kingdom. It was often burnt and plundered. It was once walled on the land side, but all traces of such fortifications have disappeared. Leith was taken possession of by Cromwell, who laid a heavy assessment on the inhabitants, and erected a citadel, of which some portions still remain. It was long famous for its races, which took place at ebb-tide, on a tract of sand E. of the town; but they were transferred to Musselburgh Links in 1816. A martello tower on the sands, at some distance from the pier-head, was built, at an expense of £17,088, for the defence of the port, during the late war: the port is farther

defended, by a battery of nine guns, a little to the W. of Cromwell's fort.

Leith was long dependent on Edinburgh. So early as the 14th century, the latter obtained a grant from King Robert Bruce, of its harbour and mill; a right which was confirmed or extended by subsequent grants either from the crown, or Logan of Restalrig, the baronial superior of the place. The municipal government of the burgh was, as already stated, substantially vested in the town-council of Edinburgh, who had the entire management of the port. But the Scottish Borough Reform Bill, which came into operation in November, 1833, totally changed this state of things, and conferred on the inhabitants of Leith what they had long struggled to obtain, the uncontrolled exercise of their own municipal affairs; vesting them in a provost, four bailies, and 23 councillors, chosen by popular election. This act did not, however, extend to the rights of Edinburgh over the harbour and docks, nor to the revenue arising therefrom. But the Edinburgh and Leith Agreement Bill (July, 1838) made each town, in every respect, independent of the other. By this statute, the affairs of the harbour and docks are vested in 11 commissioners (of whom five are nominated by Her Majesty's Treasury, and three by the town-councils of Edinburgh and Leith respectively), whose proceedings, however, require the sanction of the treasury. The debt on the docks due to government, at the date of the passing of the act, was £228,374; and the commissioners are empowered to borrow a sum not exceeding £125,000 (the government postponing the security granted to it pro tanto), for the improvement of the port, provided the whole particulars and estimates receive the authority of the treasury. Government is also ready to postpone its claims to such annual sums as may be required for maintaining or extending the efficiency of the port. Certain sums are also directed by the act to be paid out of the harbour revenue to the city of Edinburgh for certain specified purposes. The income of the harbour and docks, in 1840, was £28,012 4s. 7½d.; the expenditure, embracing every item under the act, was £16,490 8s. 6d.; leaving a surplus of £9321 16s. 7½d.; which, with former savings, makes a total surplus of £16,444 4s. 1½d.

Previously to the passing of the Reform Bill in 1832, Leith had no parliamentary representative. But that act confirmed on it, with Portobello and Musselburgh, the right to send one member to the House of Commons. Registered voters in 1840-41, 1301; being more than two thirds of the entire constituency. (Revenue Tables; Boundary Reports; Campbell's Hist. of Leith; Chalmers's Caledonia; and Private Information.)

LEITMERITZ, a fortified town of Bohemia, cap. circ. of its own name, on the Elbe, here crossed by a bridge about 600 ft. in length, built partly of stone and partly of wood, 33½ m. N.N.E. Prague. Pop. (1837) 3687. It is well built, and has a handsome cathedral and other churches, a gymnasium, a theological seminary, a high school and girls' school, with manufactories of straw hats and chicory, and a considerable traffic in agricultural produce, and fish caught in the Elbe. The wines of its circle are the best of any in Bohemia, which, however, is no very high recommendation. It is a bishop's see, and the seat of a circle council. (Oesterr. Nat. Encyc., &c.)

LEITRIM, a marit. co. of Ireland, prov. Connaught, having N. Donegal bay, E. Fermanagh and Cavan, S. Longford, and W. Roscommon and Sligo. It is long and narrow, stretching N.N.W. and S.S.E. nearly 50 m. Area 420,375 acres, of which 126,167 are mountain and bog, and 25,369 water, including lough Allen, near the source of the Shannon, which is also in this county. Leitrim is wild, and generally mountainous; but in the valleys and low grounds the soil, which is incumbent on limestone, is mostly very fertile. Property in very large estates. Tillage farms small, and not unfrequently let on partnership leases. Agriculture perhaps improving, but in an excessively backward and depressed state. There is no rotation: corn follows corn as long as the soil will bear anything; or if the series be interrupted, it is only to make way for potatoes or flax; and when the land is exhausted, its recovery is left to the vis medicatrix naturæ; clover and turnips are nearly unknown; and here, as in most other districts of Ireland, the potato is the all but sole dependence of the bulk of the population. The habitations of the occupiers are mostly miserable huts; and, except in a few instances, office-houses, in the proper sense of the term, can hardly be said to exist. Average rent of land, 10s. 6d. an acre. Some coarse linen is made for home consumption. Leitrim contains five baronies, and 17 parishes. It sends two members to the House of Commons, both for the county. Registered electors, 1839-40, 2692. In 1831, Leitrim had 94,200 inhabited houses, and 141,524 inhabitants, of whom 69,451 were males, and 72,073 females.

LEMBERG (Polish, Lwow, Latin, Leopolis), a city of the Austrian dominions, cap. Galicia, on the Peltew, a tributary of the Bug, 185 m. E. by S. Cracow, and nearly 370 m. N.E Vienna. Lat. 49° 51' 40'' N.; long. 24° 2' 45'' E. Pop., exclusive of the garrison and strangers, in 1837, 54,965, of whom above 20,000 were Jews. Lemberg was formerly an important fortress; but the demolition of its fortifications was begun early in the last century, and completed under Joseph II., when its ramparts were planted with trees, and laid out in public walks. It has still, however, two castles, one within the town, and the other, the ruined castle of Lowenberg, on an adjacent eminence to the N. The city proper is small, but it has four suburbs, each equalling it in extent; and comprising the handsomest buildings. The lofty towers and cupolas of the cathedral and other churches, and the massiveness of its public structures, give Lemberg an air of grandeur, particularly when viewed from a distance. The city has narrow, dirty streets, and old houses; but the suburbs are generally well built, and have several wide, straight, and tolerably well paved streets, and some spacious squares. The houses are mostly of free-stone, two or three stories high, but roofed only with shingles. The chief public edifices are the palace of the Armenian archbishop in the Cracow suburb; the Dominican church, which has a fine monument by Thorwaldsen; one of the two synagogues, the old Jesuits' college, the new council-house, the governor's residence, the general hospital, and the large barracks in the Zolkiew suburb. Lemberg has upward of 20 churches, including a Greek and an Armenian cathedral, nine Roman Catholic parish churches, and Lutheran and Calvinist meeting-houses; a Greek, an Armenian, and six Roman Catholic convents, five hospitals, and a theatre. Its university, established in 1784, and remodelled in 1817, had, in 1833, 1311, and in 1837, 1321 students; of which, in the former year, 485 studied divinity, 399 philosophy, 242 law, and 185 medicine. It has also an imperial academy, a Roman Catholic, and a Greek ecclesiastical seminary, two gymnasia, two high schools, a school of arts and sciences, a normal school, a Jewish female, and many elementary schools; a provincial museum, chiefly for the natural and other products of Galicia, and a valuable public library, said to be rich in works in Polish literature. Lemberg has also an asylum for the children of soldiers, a military swimming school, a workhouse, and a prison for political offenders. It is the seat of the provincial government; of the courts for the city and circle, a court of appeal for the province, &c.: and has Roman Catholic, united Greek, and Armenian archbishops, and Lutheran and Calvinist superintendents.

Lemberg has manufactures of cotton and woollen stuffs, with dye works, distilleries, tanneries, and a few printing establishments; but it is much more a commercial than a manufacturing city. Next to Brody, with which it has a constant intercourse, it is the chief trading town of Galicia. Its position on the high road from Odessa to Silesia and Warsaw, renders it an emporium for much of the produce of S. Russia, Moldavia, and Wallachia, in its transit to central Europe. Large fairs are held at Lemberg; the most important is that called Drei Konigs Messe (Three Kings' Fair), which lasts six weeks from January 14, and attracts a vast concourse of Jewish, Christian, and even Mohammedan merchants. The Russians bring to the fairs large quantities of peltry from Siberia and Tartary, which they exchange for the woollen and cotton goods and hardware of Austria. Large herds of cattle arrive at Lemberg from Moldavia and Bessarabia, being thence distributed to different parts of Austria and Silesia. Lemberg is also one of the principal corn-markets of the Austrian empire. Corn is sent from it to Przemysl, on the San, where it is shipped for Dantzic; and it is, also, though more rarely, sent from it to some of the nearest stations on the Dniestr, for shipment for Odessa. But, owing to the length and difficulty of the navigation to either of these great emporiums, there is usually a very wide difference between the prices in them and in Lemberg. Thus, on the 26th of November, 1838, wheat sold at Lemberg for 15s. a quarter, whereas its price at Dantzic on the 20th of the same month, was no less than 41s. 6d.; the difference amounting to 26s. 4d. a quarter, being the measure of the cost of conveyance from Lemberg to Dantzic! We may remark, by the way, that this fact sets in a very striking point of view the absurdity of the statements so frequently put forth in our newspapers, contrasting the prices in this country with those in foreign markets, and ascribing their excess in England wholly to the influence of our corn-laws. Lemberg was founded in the 13th century. It was taken by Casimir I. of Poland, in 1340. It was besieged in 1648 by the famous Cossack chief, Bogdan Khmielnicki, who threatened its extermination, but withdrew on receiving a large ransom. In 1672 it was taken by the Turks; and in 1705 it was taken and sacked by Charles XII. of Sweden, when it ceased to be of much consequence as a fortress. It came into the possession of Austria in 1772, since which it has progressively advanced in wealth and population. (Oesterr.

Nat. Encyc.; Malte-Brun, Tableau de la Pologne, ed. 1833, i., 419; *Private Information*.)

LEMGO, a town of Germany. (See LIPPE-DETMOLD.)

LEMINGTON, p. t., Essex co., Vt., 94 m. N.E. Montpelier, 591 W. Bounded S.E. by Connecticut r., and watered by small tributaries of it, one of which has a cascade of 50 ft. Chartered in 1762. It has two saw-mills; two schools, 40 scholars. Pop. 124.

LEMNOS (Turk. *Stalimene*), an island of the Grecian Archipelago, belonging to the dom. of the Porte, 43 m. S.E the promontory of mount Athos, and about the same distance W. from the mouth of the Hellespont, mount Ther ma being in lat. 39° 53' 46'' N., long. 25° 8' 32'' E. Area about 150 sq. m. Pop. said to amount to 12,000 chiefly Greeks. It is of an irregular quadrilateral shape, being nearly divided into two peninsulas, by two deep bays or indentations of the sea, Port Paradise on its N. and Port St. Antonio on its S. side. The latter, which is spacious and land-locked, has good anchorage for large ships. The E. side presents to the sea a bold rock, *Monte Santo*, called by Æs chylus the 'Ερμαιον λέπας Λήμνου, in his brilliant description of the watch-fires between mount Ida and Mycenæ: a rocky bank projects from it upwards of 8 m. into the sea. The appearance of Lemnos is far from picturesque: barren, rocky, though not very high, mountains cover about two thirds of its surface, and scarcely a tree is to be seen, except in some of its narrow valleys, which are verdant and fertile, especially on its W. side. The whole island bears the strongest marks of volcanic action: the two highest mountains have craters; there are several thermal springs, and the rocks in many parts resemble the burnt and vitrified scoriæ of furnaces. One of its mountains, indeed, appears, from a fragment of a Greek poet preserved by Nicander, to have been constantly emitting flame and smoke; and hence, we may account for the fact of this island being sacred to Vulcan, who, when precipitated from heaven, is said to have fallen on its hospitable shores:—

" Lemnos earn deo : nec fama notior Æthua
Aut Liparæ domus."   *Val. Flaccus*, lib. ii., lin. 95.

This island has been long famous for its furnishing a peculiar siliceous earth or bole, celebrated for its detergent and medical qualities, called *Terra Lemnia* and *Terra Sigillata*, from its being impressed with a peculiar seal or mark. Galen visited the island in the second century, for the express purpose of making himself acquainted with this earth; and he states that it was then dug up with many religious ceremonies. (*De Simpl. Medic.*, lib. ix.) This practice has been continued down to our own times, or, at all events, to a very late period. The earth is dug up on the 6th of August, in the presence of the chief men of the island: when a sufficient quantity is extracted, the hole is filled up; the bags or parcels are then sealed, and, a few being sent to the grand seignior, the governor is accountable for the value of the others. But the reputation of the Lemnian earth is now much fallen off, and the demand for it has proportionally declined. (*Ancient Universal History*, viii., 346, 8vo. ed.)

At present the high grounds of the island are grazed by sheep; but the W. and S. valleys prod ce corn, good grapes and figs, cotton and mulberry trees. The climate, however, is too cold to ripen oranges and lemons; and the island frequently suffers from the locust.

The wine of Lemnos is of two sorts, both red; the best fetches about 8 paras per oka, or 2s. 3d. per bottle. It produces more than sufficient grain for its own consumption, the rest, with some wine, being sent to Mytilene; but its chief exports are ewe-milk cheese, silk, cotton, and wool. Wheat sells for 4 piastres (6s.) the bushel, barley for 8 paras the oke, and cheese for the same. The inhabitants are divided between agriculture and fishing, and the women (celebrated for their beauty) are employed in weaving cotton cloths. The Turks resemble those of the other islands, both in dress and manners; but the costume of the Greek women is remarkable as well as picturesque. It consists of a short scarlet jacket, with long sleeves, loose in front, and reaching only a few inches down the back, very short petticoats, wide calico trousers gathered at the ankles, yellow Turkish slippers, and a white handkerchief tied like a turban round the head. The town, Castro (the ancient *Myrina*), on the W. side, contains three Greek churches; and its port, or rather cove, is defended by a little pier, and commanded by a citadel on the overhanging rocks. Ships are built here; and the natives are excellent seamen. Pop. 9000. The other port is St. Antonio on its S. side, at the bottom of the bay already noticed. (*Walpole's Memoirs*, ii., 54, &c.)

Lemnos, according to Pliny, had a labyrinth more remarkable than that of Crete or of Egypt. It was supported by 140 columns, and its gates were so admirably adjusted, as to be turned by a child. (*Quorum in officinis turbines ita libratos pependerunt, ut puero circumeunte torneerentur.*) It was the work of three architects, one of whom, Theodosea, was a native of the island. Its remains are said to have been extant in Pliny's time. (*Hist. Nat.*, lib. xxxvi., cap. 13.) No certain traces of this famous edifice have been discovered in modern times; but this is probably a consequence of the island having been seldom visited by scientific travellers, or of the changes occasioned by the action of volcanoes, or other natural convulsions.

The first inhabitants of the island are said to have been Thracians. In the reign of Thoas, the only Lemnian king mentioned in history, the Lemnian women are said, in imitation of the Amazons, to have treacherously killed all the males (*Herodot.*, lib. vi., cap. 138); and hence any premeditated and detestable murder or other crime was long after called a "Lemnian action." Miltiades reduced the Lemnians under the sway of Athens.

LEMON, t., Butler co., O. It contains the villages of Middletown, Monroe, and Amanda. Watered by Dick's cr., flowing into Great Miami r., which bounds it on the N.W. It has five schools, 110 scholars. The Miami canal passes through it, from which there is a side cut to Lebanon. Pop. 2052.

LEMPSTER, p. t., Sullivan co., N. H., 40 m. W. Concord, 405 W. Drained by branches of Sugar and Cold rivers, which afford water-power. Chartered in 1761; first settled in 1764. It contains three churches, a Congregational, Methodist, and Baptist; two stores, one fulling-mill, one woollen factory, three grist-mills, six saw-mills; 21 schools, 319 scholars. Pop. 941.

LENA, a large river of N. Asia, the principal in E. Siberia, extending through 19° N. lat., and falling into the Arctic ocean. It rises in lat. 52° 30' N., and long. 105° E. on the W. slope of the lofty granite range, skirting the N.W. shore of the lake Baikal; and from the source as far as Ust Kulsk, a distance of 350 m., it pursues a N. course; but at that point it is turned E. by a chain of hills, and runs in a very tortuous channel E.N.E. for about 1000 m. to Yakutsk, the metropolis of E. Siberia, where it is a wide and noble river. Its general course from Yakutsk is N. down to the apex of the extensive delta formed at its mouth, the distance between those two points being about 700 m. If the distances along the stream, carefully measured on J. Arrowsmith's map, be correct, the entire length of this gigantic river is probably somewhat more than 2100 m. The basin of the Lena, according to Ritter, covers an area of about 840,560 sq. m., the principal tributaries above Yakutsk being the Kirenga, Vitim, and Olekma, on its E. side, while below that city, the main stream is joined E. by the Aldan, rising by several sources in the Stanovoi range, and W. by the Bilui, which rises on the E. side of the hills dividing the Lena basin from that of the Yenisei. The Lena has an extremely tortuous course with a sluggish stream, and encloses numerous islands. Mr. Dobell, who travelled up the stream from Yakutsk to Irkutsk, describes it as "one of the safest navigable rivers, of its size, in the whole world, its course being only very rapid in the spring, at the breaking up of the frost, when numerous tributary rivers and torrents, bursting their icy fetters, rush with impetuosity into the maternal bosom of the Lena. The river, at these times, is a truly sublime spectacle, particularly where it passes through what are called the *gates*, which confine it in a narrow channel between rugged cliffs rising perpendicularly nearly 300 ft. above the stream. The dashing and eddying of the stream in its course from one side to the other is terribly grand; and yet the native boatmen manage to descend the river without injury, even at this season. The forests on the bank are principally of spruce and the yellow pine, both of a large growth; and the soil on the mountains appears rich and good, and capable of producing grain of all sorts. Most of the farming settlements, however, are either on the level spots along the edge of the river, or on the declivities of the mountains. Below Yakutsk, the face of the country is very different; the river rolls thence through vast and almost uninhabited plains, covered with snow and ice, which never wholly melts, and beneath which have been found the carcasses of mammoths, rhinoceroses, and other fossil animals." (*Dobell's Siberia*, ii., 62–69; *Lyell's Geology*, i., 140–144.)

LENAWEE, county, Mich. Situated in the S.E. part of the state, and contains 735 sq. m. Watered by Raisin and Ottawa rivers. It contains iron ore. It had in 1840, 14,917 neat cattle, 6034 sheep, 22,972 swine; and produced 167,891 bushels of wheat, 3468 of rye, 190,538 of Indian corn, 8188 buckwheat, 5989 of barley, 151,111 of oats, 112,534 of potatoes, 23,945 pounds of sugar. It had six commission houses in foreign trade, 43 stores, three flouring-mills, nine grist-mills, 44 saw-mills, one tannery, three distilleries, three printing-offices, three weekly newspapers; one college, 20 students; one academy, 27 students; 114 schools, 6204 scholars. Pop. 17,889. Capital, Adrian.

LENHAM, a decayed market town and par. of England, co. Kent, lathe of Aylesford, hund. Eyhorne, near the source of the Len, a tributary of the Medway, 13 m. W.

Canterbury, and 40 m. E.S.E. London. Area of par., 6849 acres. Pop. in 1841, 2214. The town consists of a principal street, on the high road between Maidstone and Canterbury, intersected by another of smaller size. The church has a square tower, and 16 curiously carved stalls in its interior, which are supposed to have belonged to the abbot and monks of St. Augustine, at Canterbury, who had large estates within the parish. The market has been long disused; and the inhabitants are almost entirely engaged in agriculture.

LENOIR, county, N. C. Situated towards the S.E. part of the state, and contains 360 sq. m. Watered by Neuse r. and its branches. It contained in 1840, 7060 neat cattle, 4530 sheep, 17,490 swine; and produced 9060 bushels of wheat, 4200 of rye, 1,734,000 of Indian corn, 8570 of oats, 61,950 of potatoes, 1860 pounds of rice, 944,300 of cotton. It had 11 stores, one cotton factory, 11 flouring-mills, 17 grist-mills, 14 saw-mills, 12 distilleries; two academies, 30 students; six schools, 167 scholars. Pop.: whites, 3687; slaves, 3083; free coloured, 235; total, 7005. Capital, Kingston.

LENOIR, p. v., capital of Caldwell co., N. C. Situated on Lower cr., a branch of Catawba r., and contains a court-house, jail, a store, and several dwellings.

LENOX, p. t., capital of Berkshire co., Mass., 132 m. W. Boston, 370 W. Organized in 1767. Watered by Housatonic r. It has one furnace, one fulling-mill, one grist-mill, six saw-mills, one oil-mill, one tannery, one printing-office, one weekly newspaper; one academy, 80 students; seven schools, 341 scholars. Pop. 1213. The village is pleasantly situated on elevated ground, and contains a courthouse, jail, three churches, a Congregational, Methodist, and Episcopal; several stores, an academy, founded in 1803, and endowed with half a township of land in Maine, then belonging to Massachusetts, which has produced a handsome fund, and about 40 dwellings, many of them neat. Its pure air and agreeable society render it a pleasant place of summer resort.

LENOX, p. t., Madison co., N. Y., 121 m. W.N.W. Albany, 256 W. Bounded N.E. by Oneida cr., and Oneida lake, into which it flows. Drained by Cowasslon cr., a branch of Chittenango cr. The Erie canal passes through it. It contains seven churches, two Presbyterian, two Episcopal, two Baptist, and a Methodist; 15 stores, three fulling-mills, three grist-mills, 16 saw-mills, four tanneries; one academy, 36 students; 14 schools, 1463 scholars. Pop. 5440.

LENOX, p. t., Susquehanna co., Pa., 170 m. N.N.W. Harrisburg, 274 W. Watered by Tunkhannock cr. It contains two stores, two grist-mills, six saw-mills; eight schools, 940 scholars. Pop. 800.

LENOX, p. t., Ashtabula co., O., 15 m. S. Ashtabula, 291 N.E. Columbus, 335 W. It has one store, one grist-mill, four saw-mills; five schools, 166 scholars. Pop. 550.

LENTINI (an. *Leontium*), a town of Sicily, prov. Syracuse, on a hill washed by the river Porcari (an. *Lissus*), near the lake of Lentini, or Biveri, 14 m. S.S.W. Catania, and 20 m. N.W. Syracuse. Pop. in 1831, 7376. The country round is now, as of old, extremely fertile; and the inhabitants are chiefly employed in its culture, in the fishery on the lake, and the sale of the produce so obtained. In the winter season the lake, which is the largest in Sicily, is about 19 m. in circumference, but in summer its circumference is reduced to 6 or 9 m.; the exhalations from the mud that is thus left dry rendering the town and district very unhealthy: the fishery yields its proprietor, the prince of Butera, a considerable sum.

The ancient city of Leontium, founded by a colony of Chalcidians in the first year of the 13th Olympiad (*Thucydides*, lib. vi.), most probably occupied the exact site of the modern town; but the ground has been so much shaken and changed by natural convulsions, such as that of the great earthquake of 1693, that few vestiges of the ancient city can now be traced. When it was taken by the Romans under Marcellus, it was one of the principal cities of Sicily, as is sufficiently evinced by the notices of it in various writers, and especially by the detailed description which Polybius has left of its state at that period. "The city of Leontium," says he, "considered in its general position, faces the N. Through the middle of it runs a level valley, which contains the public buildings allotted to the administration of the government and justice, and, in a word, the whole that is called the Forum. The two sides of the valley are enclosed by two hills, which are rough and broken along their whole extent. But the summit of these hills is flat and plain, and is covered with temples and houses. There are two gates to the city: one of them is in the southern extremity of the city, and conducts to Syracuse; the other is on the opposite side, and leads to those lands so famed for their fertility, called the Leontine fields. Below the hill that stands on the W. side of the valley, flows the river Lissus; and on the same side, likewise, there is a row of houses built under the very precipice, and in a line parallel to the river. Between these houses and the river lies the road that has been mentioned." (*Hampton's Polybius*, iii., 105.)

In his third oration against Verres, Cicero repeatedly refers to Leontium, and celebrates the extraordinary fertility of its territory, *Ager Leontinus caput est rei frumentariae.* (*In Verrem*, lib. iii., cap. 23.) The famous orator, Gorgias, whose eloquence was instrumental in persuading the Athenians to undertake their fatal expedition against Sicily, was a native of Leontium.

LEOMINSTER, a parl. bor., market town, and par. of England, co. Hereford, hund. Wolphy on the Lugg, an affluent of the Wye, 11 m. N. Hereford, and 121 m. W.N.W. London. Area of par. and parl. bor., which are co-extensive, 9290 acres. Pop. in 1841, 4916. The town consists of a good principal street, about ⅔ m. long, intersected by several narrow and inconvenient lanes. There are several handsome inns and private residences, and being well paved and well lighted, it has, on the whole, a respectable appearance. The town-hall, called the butter-cross, in consequence of the butter-market being held in the lower part, is an odd-looking structure of timber and plaster, standing on oak pillars, with Ionic capitals. A market-house was erected in 1803, near which is a small jail. The church, which exhibits the architecture of several periods, has a tower 100 ft. high: the vicarage is in the gift of the lord chancellor. The Baptist, Wesleyan Methodists, Moravians, and the Society of Friends, have each places of worship; and well-attended Sunday schools are attached to the church and to various chapels. A free grammar school, founded and endowed by Queen Mary, "has entirely ceased to furnish gratuitous education, and has become a private school: the corporation appoints the master; but beyond paying him an annual stipend of £20, they have no concern in the management of the school." (*Mun. Corp. Rep.*) An almshouse, dispensary, and house of industry, are the only other public establishments.

Leominster was formerly one of the principal seats of the glove manufacture; but latterly the business has been on the decline. Hats are made, and coarse woollens, but the latter only to a small extent. Tanning is extensively carried on. The principal dependence of the town is, however, on its retail trade with the adjacent county. Coal is brought from Shropshire, partly by canal and partly by wagons, from the Clee Hills.

"The land in the borough and in the out-parish is in a great degree held, often in small portions, by the residents in the town. The country round produces, besides the common agricultural produce, apples and hops in great abundance. In the immediate neighbourhood of the town the meadow lands are let for £4 or £5 per acre. Further off the average rent is about £3; in the out parish the rent was represented to us much lower, seldom exceeding 36s., and sometimes falling as low as 12s. The actual farms vary in size from 80 to 400 acres. There are few tenures for lives. Some lands and houses belonging to the corporation are let for long terms, subject to three joint lives, but renewable, as the lives fall in, for fines certain. There are some leases for terms of years; but the greater number of holdings are from year to year, the leases for years expiring, and the tenant holding on. A considerable quantity of the land is occupied by the proprietors. The wages of labouring men, finding their own food, are, on an average of the whole year, about 10s. per week." (*Bound. Rep.*)

Leominster is a borough by prescription, and received several charters between 1554 and 1706, the governing charter till 1835 having been 36 Charles II. The municipal officers are, a recorder, four aldermen, and 12 councillors. Corporation revenue, in 1839, £596, of which, however, £66 12s. was derived from the sale of property. Quarter and petty sessions are held in the town-hall, and there is a court for the recovery of debts under £100. The parliamentary franchise was granted in 23 Edward I., since which time the borough has sent two members to the House of Commons, the voters, down to the passing of the Reform Act, being resident burgesses and inhabitants paying scot and lot. The Boundary Act made the parliamentary borough co-extensive with the parish. Registered electors in 1839-40, 694. Markets on Friday; large fairs for cattle, farming produce, &c., February 13, May 13, September 4, and November 8. (*Mun. and Bound. Rep.*, &c.)

LEOMINSTER, p. t., Worcester co., Mass., 44 m. W.N.W. Boston, 417 W. Incorporated in 1740. A main branch of Nashua r. affords water-power. Combs are extensively manufactured. It contains five churches, a Congregational, Unitarian, Methodist, Baptist, and Universalist; 12 stores, one fulling-mill, one flouring-mill, four grist-mills, 18 saw-mills, two tanneries; one academy, 25 students, 12 schools, 632 scholars. Pop. 2069.

LEON, an ancient kingdom of Spain, between lat. 40° 10' and 43° N., and long. 4° and 7° W.; bounded N. by Asturias, E. by Old Castile, S. by Estremadura, and W. by

Galicia: greatest length, 209 m.; breadth, 138 m.: area, 10,573 sq. m. Pop., according to Miñano, 1,236,988. The whole of this region is included in the basin of the Douro, and is intersected by several large tributaries of that river, the principal being the Pisuerga, Esla, and Tormes. The N. and S. districts are mountainous, the former comprising various offsets from the Asturian chain, and the latter being skirted by the central chain of the Peninsula, two of the highest summits of which are the Sierra de Gredos, 10,532 ft., and the Peña de Francia, 5089 ft. This hilly country produces the loftiest and best oaks in Spain, and is rich in iron ore, some portion of which is smelted, and made into hardware goods. The inhabitants of the Asturian mountains are a distinct race, robust, and simple in their manners, engaged during summer in pasturing cattle, mules, and the migratory flocks of sheep that pass at that season through their country, and at other times employed in tillage and in collecting Iceland moss, which is here very abundant, madder, and medicinal plants, which they sell in the markets of Leon and Madrid.

The less elevated parts of Leon contain many tracts, which afford excellent pasture, and dairy-farming might be pursued with great profit, were it not for the want of enterprise, security, and even tolerable roads. Maize, olives, wheat, and flax are cultivated in some parts; but there is a great want of irrigation. The wine of Salamanca is said to be of good quality; but that raised on the borders of Galicia is execrably bad. Leon has no public manufactures worth notice, except that of hardware; but there is a good deal of domestic manufacturing of woollen and linen stuffs for home consumption. The canal of Castile, constructed about 35 years ago, passes northward up the valley of the Pisuerga; but it was never finished, and contributes very little, if at all, to the advantage of the districts through which it passes. Townsend, who visited Leon while the works were in progress, has given a good account of the undertaking. (Vol. i., p. 367.)

The kingdom of Leon is divided into six provinces, Leon and Palencia in the N.; and Zamora, Toro, Valladolid, and Salamanca in the S.; the principal cities and towns being Leon, Valladolid, Ciudad Rodrigo, and Salamanca. This country was anciently inhabited by the *Vettones* and *Callaici*, and formed a part of the Roman *Terraconensis*. Don Pelayo and his successors during the 8th century, formed this district into a kingdom, called after its capital, and connected with that of Asturias. It was first added to Castile, in 1037, but continued as a unsettled state till 1230, when it was finally united to the dominions of Ferdinand III., king of Castile. (*Miñano; Mod. Trav.*)

LEON, a city of Spain, cap. kingd. and prov. of same name 50 m. S. Oviedo, and 176 m. N.W. Madrid; lat. 42° 43' N., long. 5° 17' 43" W. Pop., according to Miñano, 5300. This ancient city, once the capital of an independent kingdom and the residence of its sovereigns, stands on a kind of peninsula formed by the Bernesga and the Torio. It is surrounded by decayed walls, and bears in its narrow, dirty, unpaved streets, and almost ruinous houses, the indications of poverty and wretchedness. Among the public buildings, the largest is the cathedral, a Gothic structure, with a lofty spire deservedly admired for its lightness and elegance: the ecclesiastical establishment comprises a bishop and 40 canons. There are no fewer than 13 parish churches in the town and suburbs, and two canonical houses for Augustinian monks, with seven other monasteries. There are also four hospitals, one of which is for foundlings. The inhabitants are employed in linen weaving, in knitting stockings and caps, and making leather gloves; there are, also, some tanneries and soap factories. The surrounding country is bold and beautiful; but agriculture is in the most degraded state. Hay-making, however, though common here, is not usual in other parts of Spain.

Leon was founded prior to the reign of the Roman emperor, Galba: it was called by the Romans *Legio septima Germanica*, from the circumstance of that legion being stationed here: it was the first large town recovered from the Moors, after whose expulsion in 722, it was the residence of Christian kings during more than three centuries. (*Miñano; Townsend*, vol. i., p. 375, &c.)

LEON, a town of Mexico, state of Guanaxuato, in a fertile plain, and on the road from Guanaxuato to Lagos, 36 m. W.N.W. the former city. Pop. estimated at about 6500. It has three convents, a college, and a hospital, and carries on some trade in corn, &c.

LEON, county, Flor. Situated in the centre of the N. part of the territory, and contains 1894 sq. m. Bounded S. by Appalachee bay of the gulf of Mexico, W. by Ocklockney river. Watered by Wakully river. The whole county abounds with lakes, ponds, subterranean rivers, sink holes, and large springs. A railroad runs from Tallahassee, 20 m. to St. Mark's, and crosses the lower part of the county. It contained in 1840, 11,890 neat cattle, 1796 sheep, 18,339 swine; and produced 239,961 bushels of Indian corn, 3315 of oats,

43,789 of potatoes, 3898 pounds of tobacco, 5,530,844 of cotton, 29,159 of sugar. It had two commercial and 10 commission houses in foreign trade, 44 retail stores, six grist-mills, nine saw-mills, two printing-offices, one bindery, two weekly newspapers; one academy, 41 students; four schools, 180 scholars. Pop., whites, 3481; slaves, 7231; free coloured, 21; total, 10,713. Capital, Tallahassee.

LEAS, p. t., Cataraugus co., N. Y., 311 m. W. by S. Albany, 343 W. Watered by Conewango creek and its branches. It has 12 saw-mills; 11 schools, 473 scholars. Pop. 1338.

LEON (ISLA DE), a long and narrow island close to the S. coast of Spain, prov. Cadiz, and separated from the mainland only by the narrow but deep channel of Santi Petri, crossed by the bridge of Zuarzo, which being the only point of approach to the important city of Cadiz, is defended by strong redoubts. It is about 8 m. long by about 2 in breadth, and consists almost entirely of a dreary sandy waste, abounding with salt-water marshes. Cadiz occupies a small peninsula at the extremity of a long sandy isthmus, separated from the rest of the island by a line of fortifications called the Cortadura. (See CADIZ.)

There are two other towns, the chief of which is San Fernando, otherwise called Isla, and sometimes Leon, 6½ m. S.E. Cadiz. Pop., in 1830 (as estimated by Inglis), 33,000; but it has decreased 9000 since 1816, and is probably at present, 1846, under 25,000. "Isla," says Inglis, "is certainly one of the prettiest towns in Spain, and I never saw a cleaner and handsomer avenue than its principal street, which is about 1½ m. long. Every house is of the purest white, and every range of windows on every house has its green verandah." The principal buildings are the hôtel-de-ville, in the great square, and the great church, which is not only remarkable for architectural beauty, but also for a curious mausoleum, called the Pantheon, intended for the interment of the clergy. In 1808, when Mr. Jacob visited this town (which is quite of modern growth, having been built in the middle of the last century), it was inhabited by a numerous population, more active and industrious than in any other part of Spain; most of whom either belonged to the navy, or were engaged in the then busy dockyard of Caracca.

But its present condition is the very reverse of prosperous; for "Isla," says Mr. Inglis (1830), "is a sadly fallen town: the great naval school, and extensive docks of Caracca once gave employment, and life and prosperity to Isla; but now there is not a ship on the stocks, nor a pupil in the college." (Miñano; Inglis, vol. ii., p. 59; Med. Trav., &c.)

LEON DE NICARAGUA, a city of Central America, and the former cap. of the state of Nicaragua; in a savannah near a volcano, by whose eruptions it has occasionally suffered; about 90 m. N.W. Grenada, and 5 m. from the N.W. shore of the lake of Leon; lat. 12° 20′ N., long. 90° 54′ W. Pop., estimated by Thompson in 1829, at 35,000 (Official Visit, &c., 451); but it has since been greatly reduced by the revolutions that have taken place within its walls, and the decay consequent on the removal of the seat of government to Grenada. It is surrounded by old walls; and has several suburbs, a cathedral, and three other churches, several convents, a hospital, and a college. It is a bishop's see; and was originally founded, in 1523, on the spot now called Old Leon, but was removed to its present site in 1532.

LEONARD (ST.), a town of France, dep. Haute Vienne, cap. cant., on a hill near the Vienne, here crossed by a handsome bridge, 19 m. E. Limoges. Pop. in 1836, ex. com., 3504. It was fortified in the 15th century; and has manufactures of coarse woollens, paper, earthenware, &c.

LEONARDSTOWN, p. v., capital of St. Mary's co., Md., 87 m. S. Annapolis, 64 W. Situated on Britton's river, a short branch of Potomac river, 5 m. from its mouth. It contains a courthouse, jail, and several stores and dwellings. Nett proceeds of the postoffice, $350.

LEONESSA, a town of central Italy, in the Neapolitan dominions, prov. Abruzzo Ultra, 36 m. W.S.W. Teramo, and 14 m. N.N.E. Rieti. Pop., in 1830, 7000. It has several churches and convents, and some large annual fairs. It is situated in a wild rugged country, in an amphitheatre, surrounded by mountains which, in winter, intercept the sun's rays for half the day, and render the climate very severe. (Del Re, ii., 336.)

LEONFORTE, a town of Sicily, intend. Catania, dist. Nicosia, cap. cant., in a hollow of M. Tavi, near the Giaretta, and 37 m. W.N.W. Catania. Pop., in 1831, 10,678. Smythe says it is a fine town, in a healthy situation. It is surrounded with walls, and has a large square, from which two long and well built streets diverge. Its trade in corn, wine, oil, and silk is considerable, and it has a large annual fair. A good deal of asphaltum is found in its vicinity. (Smythe's Sicily; Ortolani Dizionario di Sicilia, &c.)

LEONI, p. t., Jackson co., Mich., 71 m. W. Detroit, 546 W. Watered by Grass lake, the source of Grand river. It

contains four stores, two grist-mills, two saw-mills, two distilleries; one school, 70 scholars. Pop. 1087.

LEPANTO (TOWN AND GULF OF), Lepanto (an. Naupactus), a seaport town of W. Greece, on the N. shore of the gulf of Lepanto, about 3½ m. E.N.E. from the castle of Roumelia, at its entrance, and 1 m. W. from the mouth of the Morino, lat. 38° 21′ 30″ N., long. 21° 46′ E. Pop. 3689? It is built on the side of a hill surmounted by a castle of little strength, whence two walls come down to the sea, enclosing the town on either side. The harbour, within the town, is shallow, and fit only for small craft, and the place has very little trade. In antiquity, Naupactus was a place of considerable importance. It was occupied by the Athenians during the Peloponnesian war; and after many vicissitudes was nearly destroyed by an earthquake during the reign of Justinian. Its present walls are built on the foundations of those by which it was surrounded in antiquity. (Cramer's Ancient Greece, ii., 107, &c.)

But, how unimportant soever, Lepanto has given its name to the extensive gulf on which it is situated, anciently the Corinthiacus Sinus, or bay of Corinth. The entrance to the gulf, between the ruined castles of the Morea and Roumelia, at the bottom of the gulf of Patras, is only about 1 m. across. Within, it expands into a magnificent basin, stretching E., with a little inclination to the S., to Maxi, a distance of about 78 m., being, where widest, about 20 m. across. Corinth, whence it formerly derived its name, is situated near its S. extremity. It has many fine bays and harbours; and in antiquity, there were several considerable towns on its banks. Between the castles, at its entrance, there are from 30 to 35 fathoms water; and within the gulf, the water is generally very deep, there being no soundings in the centre at 300 fathoms.

Lepanto has, also, given its name to one of the greatest conflicts of modern times. Philip II., king of Spain, the Pope, and the Venetians, entered, in 1570, into a league against the Turkish sultan, Selim, who, having conquered Cyprus, and become very powerful at sea, threatened to invade Italy. The Turks, being apprised of the intentions of the confederates, assembled a powerful fleet in the gulf of Lepanto, having a large land force on board. The allies, commanded by Don John of Austria, having made equally great preparations, the two armaments encountered each other on the 7th of October, near the mouth of the gulf of Lepanto. The contest was long, bloody, and destructive; and was maintained, on both sides, with invincible courage and resolution. In the end, however, the allies gained a complete victory. The Turks lost above 25,000 men, killed, and 10,000 taken prisoners, and with the exception of 30 or 40 galleys, that effected their escape, their whole fleet was either taken or destroyed. The Christians lost about 10,000 men, killed in the engagement, or who died of their wounds. Estimating it by the number of men engaged, this was certainly the greatest sea-fight that has taken place in modern times. It was, also, the first signal victory achieved over the Turks, and diffused the greatest joy throughout Christendom. Owing, however, to the contentions among the allied admirals, the results were not such as might have been expected. (Modern Universal History, XXVII., 416-423, 8vo ed.; Watson's Philip II., book 5.)

LE RAY, t., Jefferson co., N. Y., 8 m. N.E. Watertown, 158 m. N.W. Albany. Bounded S. by Black river. Watered by Indian river. It contains 10 stores, one flouring-mill, two grist-mills, four saw-mills, two tanneries, one distillery; one academy, 95 students; 25 schools, 1194 scholars. Pop. 3721.

LERIDA (an. Ilerda), a fortified town of Spain, Catalonia, 85 m. W. Barcelona, and 72 m. E. by S. Saragossa, lat. 41° 36′ N., long. 0° 46′ E. Pop., according to Miñano, in 1826, 12,600; but, according to Hörschelmann in 1822, 15,820. It is situated on the Segre (crossed here by a handsome bridge), under the protection of a hill on which are seen the ruins of a castle now going to decay, but formerly of considerable strength. Owing to the excess of stagnant water in the vicinity, Lerida is unhealthy, and fevers prevail in spring and summer. A good quay, however, has been lately constructed, which not only keeps out the river, but forms a fine promenade. Its principal street is nearly 1 m. long; but the rest of the town is confined, and the houses are generally ill built. A cathedral, three parish churches, a military hospital, and a priests' college (formerly celebrated as a university, but suppressed by Philip V.), are the chief public buildings; but none requires notice except the cathedral. A double flight of steps leads to the terrace on which the church gates open; the principal front is embellished with six fluted Corinthian pilasters, between which are three doors with finely-wrought iron gates, and the building is surmounted by two handsome square towers. The surrounding country, being thoroughly irrigated, is extremely productive, particularly in wine, for which its gravelly siliceous soil is well salted. Silkworms, also, are reared in considerable quantities. It has some silk and other fabrics, tanning, &c., but they are not very important. (*Townsend*, i., 194–197 ; *Miñano*.)

Lerida derives its chief celebrity from its connexion with Roman history. In the plain below *Ilerda*, Scipio (*anno* 216 A.C.) gained a signal victory over the Carthaginian Hanno; and about 150 years afterward it was rendered famous by the difficulties under which Julius Cæsar was placed when encamped in its neighbourhood. He had taken possession of a plain shut in between the rivers *Singa* and *Sicoris*, and defended by a deep intrenchment, while at the same time Petreius and Afranius, Pompey's generals, were encamped on a hill between him and Ilerda. In the intermediate space is a small plain, in the centre of which rises an eminence which, if seized and fortified, would enable its occupier to cut off all communication with the city. For this, during five hours, the opposing armies maintained a doubtful conflict; but in the end, fortune declared in favour of Afranius, and Cæsar retreated to his camp. At the same time, also, the disastrous intelligence was brought to him that, by the melting of the snow, his bridges had been broken down, the country laid under water, and all communication cut off with those districts by which his army was provisioned. Famine was the immediate consequence; and Cæsar himself says, "*Militum vires inopia frumenti diminuerat, atque incommoda in dies augebantur; at tam paucis diebus magna erat rerum facta commutatio, ac se fortuna inclinaverat, ut nostri magnâ inopiâ rerum conflictarentur; illi omnibus abundarent rebus, superioresque haberentur.*" Cæsar, however, without loss of time, set his men to work, and having made a sufficient number of light and portable canoes, set a party up the river during the night, who, with these boats, effected a landing, and fortified a camp. *Huc legionem postea transducit; atque ex utrâque parte pontem instituturm perficit biduo. Ita comitatus, et qui frumenti causâ processerant, tuto ad se recipit.* (*Cæs. de Bell., Civ. I., c.* 48–54.) Lerida has sustained many sieges; it was taken by storm in 1707, during the war of the succession; and the French again besieged it in 1810.

LEROY, p. t., Genesee co., N. Y., 236 m. W. by N. Albany, 370 W. Watered by Allen's creek and its branches. It contains 17 stores, two fulling-mill, two flouring-mills, two grist-mills, three saw-mills ; one academy, 116 students ; 12 schools, 486 scholars. Pop. 4393.

LEROY, p. t., Bradford co., Pa., 143 m. N. by E. Harrisburg, 253 W. It contains two stores, two fulling-mills, two woollen factories, two grist-mills, four saw-mills ; four schools, 178 scholars. Pop. 679.

LEROY, t., Lake co., O., 4 m. N.E. Chardon, 185 m. N.E. Columbus. Watered by a branch of Grand river. It has one saw-mill ; 10 schools, 400 scholars. Pop. 898.

LERWICK, an eminent fishing station and bor. of barony, on Mainland, the largest of the Shetland or Zetland islands, of which it is the capital, on the W. margin of the sound of Bressay, opposite Bressay island. Pop. in 1801, 1706 ; in 1831, 2750. The town (¼ m. in length) is built along the curvature of the bay, and consists of a number of white houses, of from two to three stories in height, with their gables in the Norwegian style, turned to the street, but disposed with the utmost irregularity, and as utter disregard of every convenience, except that of being as near as possible to the water. The town-hall, parish church, and two dissenting chapels (Independents and Methodists), are the only public buildings. The harbour, which is entirely land-locked by Bressay island, is so ample, that it might contain nearly the whole British navy. Bressay sound is a rendezvous for Davis straits and Greenland whale ships, which here take on board supplies of provisions, and complete their crews with seamen belonging to the islands, whom they part with on their return. This has always been one of the principal stations of the Dutch herring fishery ; but the fishery is now chiefly in the hands, not merely of the inhabitants of Lerwick, but of the islanders generally, who resort thither for the purpose. The Lerwick station (exclusive of those of Unst and Walls) in 1834, had 7540 boats, decked and undecked, employed in the fishery, manned by 3364 persons. The produce of the fishery during the same year amounted to 98,500 barrels of herrings, gutted and ungutted. Cod and other species of white fish are caught in the bay and neighbouring sea, and are also extensively exported. There is a manufactory of straw-plaiting for gentlemen's hats and ladies' bonnets ; a branch of business carried on both in the Orkneys and Zetland islands. Woollen stockings, underclothing, and gloves, all wrought with the hand, and sometimes of extraordinary fineness, are exported from Lerwick. It has a custom-house, the gross revenue of which, in 1839, was £754 5s. 4d. ; and a branch bank. The shopkeepers are in the habit of shutting their shops during breakfast and dinner. Provisions are abundant, and about a half cheaper than on the main land of Scotland. There are several sailing smacks to Leith, and a steamer weekly in summer.

Lerwick was built above 200 years ago, principally for the accommodation of the Dutch fishermen, 2000 of whose

busses were then said to have been often collected in Bressay sound. It has, however, been more prosperous during the last 30 years than at any previous period. Port Charlotte, for the protection of the town from attacks by sea, stands a little to the S. The inhabitants are of Scandinavian descent. (*Messrs. Anderson's Highlands and Islands; Hibbert's Description of the Shetland Islands; Parl. Reports.*)

LESINA, and LISSA, two islands of the Adriatic, belonging to the circle of Spalatro, in Dalmatia, the first 25 m. S., and the second 33 m. S.W., Spalatro. United area, 260 sq. m. Pop. about 14,000. Both islands are in great part mountainous, but they have, notwithstanding, a considerable extent of lower and productive land. Lesina (an. *Pharos* or *Pharia*) is said to be one of the most fertile islands in the Adriatic, with a great variety of valuable products. Corn is raised on the low grounds, but the quantity is insufficient to supply the consumption of the inhabitants; among its other products are wine, oil, figs, almonds, saffron, oranges, aloes, and honey. It has considerable numbers of sheep, and these, with wool and cheese, are among the articles of export. The products of Lissa (the an. *Issa*), are similar to the above, and in it, also, the supply of corn is insufficient for the consumption. The wine of Lissa, which was commended by Athenæus, is now sadly degenerated. The inhabitants of these islands are chiefly employed in fishing, and great quantities of fish are taken round their shores. They both furnish good marble, and prepare rosemary oil, liqueurs, &c. The town of Lesina, near the W. extremity of the island of same name, has about 1600 inhabitants, and is the see of a bishop, whose diocess comprises the islands Lesina, Lissa, and Brazza. In Lissa, which in antiquity had several flourishing towns, are San Giorgio, with one of the best harbours in Dalmatia, and Comissa, with 2100 inhabitants. (*Fortis's Trav. in Dalmatia*, p. 319–340 ; *Berghaus, &c.*)

LESLIE, a bor. of barony, and manufacturing town of Scotland, co. Fife, on an eminence on the left bank of the Leven, 1½ m. N. of the public road between Kirkcaldy and Cupar-Fife, 7½ m. N. by W. the former, and 9½ S.W. by S. the latter. Pop., in 1831, 1821. It consists chiefly of one street, and contains a parish church and three dissenting chapels. Leslie House, the seat of the noble family of Rothes, is in the immediate vicinity. Leslie has five mills for flax-spinning, employing above 450 hands. Weaving of cotton, in connexion with the Glasgow manufacturers, and of the coarse species of linen fabrics, prevails to a considerable extent, and gives employment to nearly 300 individuals. There are also three rather extensive bleach-fields. The nearest market town is Kirkcaldy. Leslie has existed as a village for upwards of 300 years. Dr. Pitcairn, the celebrated physician and Latin poet, was born at Pitcairn, the family seat, in the neighbourhood of the town. At Strathhenry, near this place, the seat of his maternal grandfather, Adam Smith, author of the "Wealth of Nations," when only three years of age, was carried away by a party of gypsies. The inhabitants of the village are remarkable for their rage for religious and political discussions. The first "Political Union" formed in Scotland was at Leslie, in 1831.

LETART, t., Meigs co., O. Situated in a bend of Ohio river, which bounds it on the E., S., and W., opposite to Letart's rapids, caused by the projection of a hill into Ohio river. It contains three stores, one grist-mill, four saw-mills ; four schools, 198 scholars. Pop. 642.

LETCHER, county, Ky. Situated in the S.E. part of the state, and contains 200 sq. m. Organized in 1842, chiefly from Perry county. Drained by a head branch of Kentucky river. Capital, Letcher C. H.

LETTERKENNY, an inland town and river port of Ireland, co. Donegal, prov. Ulster, on the Swilly, 4 m. from the S.W. extremity of the lough of the same name, and 18 m. W.S.W. Londonderry. Pop., in 1831, 2160. It consists of a square and a single street; and has a parish church, a Roman Catholic chapel, three Presbyterian meeting-houses, a national school, a fever hospital, with a dispensary, courthouse, and bridewell. General sessions are held in April and October, petty sessions every Wednesday, and it is a constabulary station. Markets on Fridays; fairs on the 1st Friday in Jan., 19th May, 10th July, 3d Friday in Aug., and 8th Nov. Some trade is carried on in the export of corn and other raw produce, the river admitting vessels of 130 tons to come up from the lough to near the town. Post-office revenue, in 1830, £389 ; in 1836, £490. A branch of the Belfast bank was opened here in 1835.

LETTERKENNY, t., Franklin co., Pa. Blue mountain covers its W. part. It contains six stores, one fulling-mill, one flouring-mill, three grist-mills, six saw-mills, six tanneries ; 11 schools, 598 scholars. Pop. 1918.

LEUCTRA, an ancient village of Greece, in the Theban territories, now Leftra or Lefka, 9 or 10 m. W.S.W. Thebes. It is at present only a heap of ruins, but is famous in ancient history for the victory gained in its vicinity, on the 8th of July, since 371 B.C., by the Thebans, under Epaminondas and Pelopidas, over the Spartans. The latter were superior in number and, perhaps, also in discipline and military skill, to their adversaries; but the ability of their generals enabled the Thebans to achieve, despite every disadvantage, the greatest triumph ever won by one Greek army over another. Cleombrotos, the Spartan king, was left dead on the field, with many of his principal officers, and the flower of his troops. Sparta lost with this battle the ascendancy she had so long enjoyed among the Grecian states. (Xenophon, Hellen., lib. vi., cap. 4; Diodorus Siculus, lib. xv.; Mitford's Greece, v., 90, 8vo ed.)

LEUTSCHAU (Hungar. Löcse), a royal free town of Hungary, co. Zips, of which it is the cap., on a hill 120 m. N.E. Pesth. Pop. (1837) 5175, of whom Berghaus says three eighths are Protestants. It is old and ill built, but has a large and handsome square, a Gothic church with the largest organ in Hungary, a large old town-hall, a new council-house, and several other edifices, the oldest Lutheran gymnasium in Hungary, a Roman Catholic gymnasium, a high school, a noble female seminary, and an asylum for soldiers' children. It produces linen fabrics and mead, of which last a good deal is sent into Poland. (Austrian Encyc.; Berghaus.)

LEVANT, a term applied to designate the coasts of Europe, Asia, and Africa, along the Mediterranean, from cape Matapan round the Ægean sea, Asia Minor, and Syria, to the western confines of Egypt. In the middle ages, the trade with these countries was almost exclusively in the hands of the Venetians, Genoese, and other Italians, who gave to them the general designation of Levante, or eastern countries. But the term Levant being no longer vernacular in the languages of the nations now principally engaged in the trade with the countries referred to, it seems to be falling into disuse.

LEVANT, p. t., Penobscot co., Me., 80 m. N.N.E. Augusta, 673 W. Watered by the Kandúskeag stream and its branches. It has three stores, one grist-mill, six saw-mills, one tannery; 494 scholars in schools. Pop. 1061.

LEVEN, a bor. of barony, seaport, and manufacturing town of Scotland, co. Fife, on a level at the mouth of the river of the same name, on the N. shore of the frith of Forth, 19½ m. N. by E. Edinburgh: on the W. of the river is its suburb of Dubbieside, or Inverleven. Pop. of both, 2083. Leven consists of two principal, and not very regular streets, running parallel to each other E. and W., with a variety of bye-lanes and detached houses. The communication between Leven and its suburb has hitherto been maintained by a suspension-bridge over the river, but a stone bridge is now (1840) being built. The only public buildings are the parish church, and chapels belonging to the Relief and the Associate Synod. There is, also, a small congregation of Independents. There are two libraries, a mechanics' institute, and a great variety of friendly societies.

Leven is chiefly remarkable for its manufactures. There are, either in the town or its immediate vicinity, six mills for spinning flax, driven partly by water and partly by steam, employing about 250 hands. There are, besides, 170 hand-loom weavers of coarse linens, of whom 22 are females. It has also a foundry for cast-iron, a saw-mill and wood yard, a mill for bruising bones, a brick and tile work, and an ochre mill.

The harbour is formed by a creek at the mouth of the river. At spring-tides it admits vessels of about 300 tons, but it dries at low water, and is, owing to sand-banks, extremely difficult of access. It has a small quay, quite insufficient for the growing trade of the place. Two brigs belonging to the port are employed chiefly in the American trade, and five sloops are engaged as coasters. In 1835 the value of the imports amounted to £43,190, and that of the exports to £68,483. A steamer sails twice a day to Leith in summer, and once in winter. The game of golf is much played on the links or downs of Dubbieside. (New Stat. Acc. of Scotland, § Fifeshire, p. 264–277.)

LEVERETT, p. t., Franklin co., Mass., 83 m. W. by N. Boston, 363 W. Incorporated in 1774. Drained by two small streams which flow into Connecticut river. It contains three churches, two Congregational and a Baptist; three stores, one fulling-mill, one woollen factory, two grist-mills, eight saw-mills, two tanneries; five schools, 337 scholars. Pop. 873.

LEWES, a parl. bor., market town, and par. of England, co. Sussex, rape and hund. of its own name, on the Ouse (crossed here by a stone bridge), 7 m. N.E. Brighton, and 49 m. S. London. Pop. of parl. bor., which comprises, with the old bor., parts of seven out-parishes, in 1841, 18,263. It is principally situated on a steep declivity W. of the Ouse, which here cuts through the chalk hills; but it partly, also, stands on the level ground on the E. side, sheltered by the South Downs, that rise abruptly almost close to the river banks. The streets are broad, well built, paved, and

lighted with gas; and the town generally has an appearance of wealth and respectability. The chief public building is the Assize-hall, in High-street, erected in 1812, at an expense of £15,000, comprising two courts, a council chamber, and other apartments. The house of correction, built on the plan of Howard, in 1794, was greatly enlarged in 1817, and now contains about 70 capacious rooms for prisoners, with 15 cells for solitary confinement. The silent system, with hard labour, is rigidly enforced, and the jail is, on the whole, well conducted; but "the building is incurably bad, in an unhealthy situation, and much too confined for the wants of so large a county." (Prison Inspectors' 4th Rep., part I.) There are six churches, and the ecclesiastical livings comprise four rectories, two of which are in the patronage of the crown. There are, likewise, seven places of worship for Wesleyan and Calvinist Methodists, Baptists, Independents, Unitarians, &c., to which, as well as to the churches, are attached well-attended Sunday schools. The free grammar school, supposed to have been founded in 1512, provides gratuitous instruction in classics, &c., to 12 boys, the sons of burgesses; and there is a university exhibition for the scholars, tenable for four years, of the annual value of £35. National, Lancastrian, and infant schools furnish elementary instruction for the children of the poor; and there are several endowed charities and benevolent institutions for the relief of the aged, sick, and indigent. Lewes had formerly an extensive trade in wool, but this has greatly declined; and the present traffic of the place, independently of a very considerable retail trade with the resident gentry of the district, is chiefly in grain, malt, sheep, and cattle; it is estimated that upwards of 80,000 sheep are sold annually at the September and October fairs. The Ouse is navigable up to the town, and there is a considerable trade with London, through Newhaven, its port. (See NEWHAVEN.) Lewes is a borough by prescription, and is governed by two headboroughs and two constables, elected by the burgesses; but these officers are subject to the jurisdiction of the county magistrates. The Lent and summer assizes are held here, and the quarter sessions for the E. division of Sussex are held in January, April, June, and October. This borough has sent two members to the House of Commons since the reign of Edward I., the franchise, down to the passing of the Reform Act, being vested in the scot and lot payers within the borough. The Boundary Act enlarged the limits of the borough so as to include, with the old borough, parts of the parishes of Southover, St. Anne's, St. Thomas-in-the-Cliffs, and S. Malling Registered electors in 1839–40, 863. Lewes is the place of election for the members for the E. division of Sussex, and the head of a poor-law union comprising seven parishes. Markets on Tuesday; cattle fairs, May 6 and Whit-Tuesday; large sheep fairs, Sept. 21 and Oct 2.

The fact of Lewes being a Roman station seems extremely doubtful, but it had acquired its present name (said to be derived from læswe, the Anglo-Saxon word for pastures) at least two centuries before the Norman conquest. William the Conqueror fixed on Lewes as the site of one of those fortresses by which he kept in awe his Saxon subjects; and considerable remains of it still exist on a commanding height N.W. of the town. One gateway is nearly entire; and the keep, which is in tolerable preservation, has recently been tastefully repaired. E. of the town also are the ruins of a very ancient and wealthy priory, the walls of which enclosed an area of about 33 acres: at the dissolution of the monasteries, its revenues amounted to £1090. (Horsfield's Antiquities of Lewes, vol. i.; Parl. Reports, &c.)

LEWIS. See HEBRIDES.

LEWIS, co., N. Y. Situated in the N. part of the state, and contains 1122 sq. m. Watered by Black river and its tributaries, and by Beaver, Moose, Indian, and Oswegatchie rivers. It contained in 1840, 31,136 neat cattle, 36,865 sheep, 18,076 swine; and produced 85,191 bushels of wheat, 2473 of rye, 46,984 of Indian corn, 8496 of buckwheat, 90,971 of barley, 144,860 of oats, 634,316 of potatoes, 5460 pounds of hops, 257,476 of sugar. It had 39 stores, two furnaces, five fulling-mills, two woollen factories, three flouring-mills, 11 grist-mills, 80 saw-mills, two oil-mills, 16 tanneries, one distillery, two printing-offices, two weekly newspapers; one academy, 120 students; 151 schools, 5939 scholars. Pop. 17,830. Capital, Martinsburg.

LEWIS, co., Virginia. Situated in the N.W. part of the state, and contains 1600 sq. m. Drained by Little Kanawha river, and its branches, and by the W. fork of Monongahela river and its branches. It contained in 1840, 19,257 neat cattle, 14,971 sheep, 19,692 swine; and produced 47,091 bushels of wheat, 4867 of rye, 253,110 of Indian corn, 80,161 of oats, 94,376 of potatoes, 11,698 pounds of tobacco, 93,784 of sugar. It had 16 stores, one fulling-mill, 24 grist-mills, 17 saw-mills, one oil-mill, five tanneries, three distilleries; nine schools, 219 scholars. Pop.: whites, 7989; slaves, 124; free coloured 38; total, 8151. Capital, Weston.

LEWIS AND REHOBOTH.   LEXINGTON.

LEWIS, co., Ky. Situated in the N.E. part of the state, and contains 375 sq. m. Bounded N. by Ohio river. Drained by Salt Lick and Kinniconick creeks, which flow into Ohio river. It contained in 1840, 3424 neat cattle, 4762 sheep, 6957 swine; and produced 16,752 bushels of wheat, 150,156 of Indian corn, 40,493 of oats, 3792 of potatoes, 19,976 pounds of sugar. It had 12 stores, six saw-mills, four tanneries; 10 schools, 964 scholars. Pop.: whites, 5873; slaves, 406; free coloured, 27; total, 6303. Capital, Clarksburg.

LEWIS, co., Mo. Situated in the N.E. part of the state, and contains 500 sq. m. Bounded E. by Mississippi river. Drained by Fabius and Wyaconda rivers. It contained in 1840, 4479 neat cattle, 4934 sheep, 26,248 swine; and produced 45,583 bushels of wheat, 9718 of rye, 376,452 of Indian corn, 27,081 of oats, 16,199 of potatoes, 17,886 pounds of tobacco, 3762 of sugar. It had 22 stores, two flouring-mills, nine grist-mills, nine saw-mills, one tannery, three distilleries; 17 schools, 368 scholars. Pop.: whites, 4906; slaves, 1065; free coloured, 9; total, 6040. Capital, Waterloo.

LEWIS, p. t., Essex co., N. Y., 181 m. N. Albany, 506 W. Drained by Boquet river and its tributaries. It contains two stores, five forges, two fulling-mills, one grist-mill, 33 saw-mills; 11 schools, 389 scholars. Pop. 1505.

LEWIS, t., Lycoming co., Pa. It has three stores, one grist-mill, seven saw-mills; two schools, 60 scholars. Pop. 542.

LEWIS, p. t., Brown co., O., 121 S.W. by S. Columbus, 483 W. Bounded S. by Ohio river, E. by Whiteoak creek. Drained by Bullskin creek. It has two schools, 56 scholars. Pop. 2437.

LEWIS AND REHOBOTH hund., Sussex co., Del. It has eight stores; one academy, 40 students; three schools, 36 scholars. Pop. 1834.

LEWISBOROUGH, p. t., Westchester co., N. Y., 130 m. S. by E. Albany, 296 W. Drained by Croton river. It has four schools, 85 scholars. Pop. 1619.

LEWISBURG, p. b., Buffalo t., Union co., Pa., 67 m. N. Harrisburg, 177 W. Situated on the W. bank of Susquehanna river, a little below the mouth of Buffalo creek. A lateral canal, half a mile long, connects it with the W. branch canal. It contains two churches, 13 stores, one furnace, one grist-mill, one saw-mill, two tanneries, one distillery, one printing-office, one weekly newspaper; one academy, 20 students; four schools, 230 scholars. Pop. 1930.

LEWISBURG, p. v., cap. of Greenbrier co., Va., 214 m. W. Richmond, 251 W. It contains a courthouse, jail, three churches, a Presbyterian, Methodist, and Baptist; a lyceum, an academy, six stores, a printing-office, 200 dwellings, and about 1000 inhabitants. It is 9 m. W. of the celebrated White Sulphur springs.

LEWISBURG, p. v., cap. of Marshall co., Tenn., 54 m. S. Nashville, 703 W. It contains a courthouse and several stores and dwellings. Nett proceeds of the post-office $133.

LEWISBURG, p. v., cap. of Conway co., Ark., 45 m. N.W. Little Rock, 1110 W. Situated on the N. bank of Arkansas river, and contains a courthouse and several stores and dwellings.

LEWISHAM, a populous village and par. of England, co. Kent, lathe Sutton-at-Hone, and half-hund. Blackheath, on the Ravensbourne, a trib. of the Thames, 4½ m. S.E. London. Area of par., which includes the hamlet of Sydenham, 5220 acres. Pop. in 1841, 12,376. The village consists chiefly of a long street, lined with good houses, and extending about 2 m. along the Hastings road. The lanes leading in different directions abound with handsome villas and detached residences, inhabited by opulent merchants and retired citizens, attracted thither by the beauty of the scenery and superior salubrity of the air. The church, which stands near the centre of the village, is a handsome structure, erected in 1832 on the site of an older but still modern building, accidentally destroyed by fire. There are, also, places of worship for Wesleyan Methodists and Independents; and in Sydenham, besides a district church and Episcopal chapel, there are three dissenters' chapels. A grammar school, founded in 1647, and now under the trusteeship of the Leather-sellers' Company of London, is conducted by an upper and under master, and is alleged to be well attended. A charity school, three subscription day schools, and several Sunday schools, have been established for teaching poor children; and there are almshouses for six poor women, and minor charitable bequests. The trade of the village is almost confined to the supply of the families resident within the parish, but at Loampit hill some marl and chalk pits furnish considerable quantities of lime, and there are some large brick and tile fields.

LEWISTON, p. t., Niagara co., N. Y., 297 m. W. by N. Albany, 410 W. Bounded W. by Niagara river. It contains four commission houses, 18 stores, one flouring-mill, 174

two grist mills, seven saw-mills; one academy, 122 students; 12 schools, 793 scholars. Pop. 2533. The village is situated on the E. side of Niagara river, 7 m. from lake Ontario, at the head of steamboat navigation. A railroad connects it with Buffalo by Niagara falls, which connects with the Lockport and Niagara falls railroad. It has a good steamboat landing, and steamboats regularly ply to Oswego, and other places on lake Ontario. There is a ferry across the river to Queenstown. A grant of lands by the state for the support of schools has yielded a fund of $6000. This place was burned during the last war, and deserted from December, 1813, to April, 1815. It was incorporated in 1822, and has become a flourishing village, with 800 or 900 inhabitants.

LEWISTOWN, p. t., Lincoln co., Me., 31 m. S.W. Augusta, 577 W. Bounded S.W. by Androscoggin river, which has here a fall of 47 feet in a distance of 12 or 15 rods, creating an extensive water-power. A bridge, 1600 feet long, crosses the river at the foot of the falls. It contains six stores, four fulling-mills, two woollen factories, one grist-mill, two saw-mills, two tanneries; 14 schools, 742 scholars. Pop. 1801.

LEWISTOWN, p. b., cap. of Mifflin co., Pa., 57 m. N.W. Harrisburg, 146 W. Situated on the N. side of Juniata river, at the junction of Kishcoquillas creek. The Pennsylvania canal passes through it. It has a number of streets crossing each other at right angles, and contains a large courthouse and the county offices, on a public square at the centre of the place; four churches, an academy, and about 400 dwellings. It has five commission houses in foreign trade, 17 retail stores, one furnace, one grist-mill, one saw-mill, three tanneries, one brewery, one pottery, one printing office, one weekly newspaper; one academy, 27 students; seven schools, 231 scholars. Pop. 2038.

LEWISTOWN, p. v., cap. of Fulton co., Ill., 56 m. N.W. Springfield, 833 W. It is 4 m. E. of Spoon river, and 12 m. W. of Illinois river. It contains a neat courthouse, jail, three stores, about 50 dwellings, and 398 inhabitants.

LEWISVILLE, p. v., cap. of Lafayette co., Ark., 168 m. S.W. Little Rock, 1237 W.

LEXINGTON, distr., S. C. Situated a little W. of the centre of the state, and contains 900 sq. m. Bounded N. W. by Broad river, which, being joined by Saluda river, becomes Congaree river, and S.W. by N. Edisto river. Watered by Saluda river and its tributaries. It contained in 1840, 15,619 neat cattle, 6144 sheep, 27,198 swine; and produced 96,549 bushels of wheat, 24,084 of oats, 27,733 of potatoes, 24,000 pounds of rice, 454,191 of cotton. It had 13 stores, one cotton factory, 46 grist-mills, 18 saw-mills, three distilleries; three academies, 105 students; nine schools, 146 scholars. Pop.: whites, 7401; slaves, 4683; free coloured, 25; total, 12,111. Capital, Lexington C.H.

LEXINGTON, p. t., Somerset co., Me., 63 m. N. by W. Augusta, 657 W. Incorporated in 1833. Watered by a small tributary of Kennebec river. It has one saw-mill; eight schools, 215 scholars. Assessors' valuation of real estate in 1842, $37,507. Pop. 564.

LEXINGTON, p. t., Middlesex co., Mass., 11 m. N.W. Boston, 451 W. Watered by branches of Shawsheen river. The village contains a Unitarian and Baptist church, and about 40 dwellings. This place is celebrated as the spot where the first blood was shed in the Revolutionary war, April 19th, 1775. On the spot where eight men of the militia were killed by the British, the legislature of Massachusetts has caused a monument to be erected, to perpetuate the event, and the memory of the brave men who offered up their lives as a first sacrifice to the independence of their country. The country around flew to arms, and severely chastised the British in their return to Boston, and Major Pitcairn, who led the British in this expedition, was soon after killed in the battle of Bunker hill. The township has seven stores, two grist-mills; one academy, 25 students; six schools, 411 scholars. Pop. 1642.

LEXINGTON, p. t., Greene co., N. Y., 28 m. W. Catskill, 55 m. S.W. Albany, 365 W. Drained by Schoharie creek. Organized in 1813. It contains two churches, a Presbyterian and a Baptist; six stores, one fulling-mill, two grist-mills, 15 saw-mills, three tanneries; 13 schools, 363 scholars. Pop. 2813.

LEXINGTON, p. v., Rockbridge co., Va., 146 m. W. Richmond, 188 W. Situated on an elevated bank, on the W. side of North river, a branch of James river. It contains a neat courthouse, a jail, four churches, two Presbyterian, a Methodist, and a Baptist; a military institute; Andrew Smith's female seminary, named from its founder; 11 stores, 170 dwellings, and 1900 inhabitants. It is the seat of Washington college, originally endowed by the father of his country "whose name it bears, with 100 shares of stock of the James River Company, which yield annually about $23500; and additionally by others. It was founded in 1812, has a president and five professors or other instructers, 126 alumni, 136 students, and 2700 volumes in its libraries.

The commencement is on the last Thursday in June. The Virginia Military Institute, conducted on the plan of the United States military academy at West Point, located here, has three professors and 60 cadets.

LEXINGTON, p. v., cap. of Davidson co., N. C., 117 m. W. Raleigh, 238 W. Situated a little W. of Abbott's creek, a branch of Yadkin river, and contains a courthouse, jail, and several stores and dwellings. Nett proceeds of the post-office, $256.

LEXINGTON, p. v., cap. of Oglethorpe co., Ga., 70 m. N. Milledgeville, 584 W. It contains a courthouse, jail, two churches, a Presbyterian and Methodist; a public library, several stores and mechanic shops, 40 dwellings, and about 250 inhabitants. "Meson Academy," which has the name of its founder, who gave to it $5000, is located here. The building, a substantial brick edifice, two stories high, cost $4000, and the other $4000 have been invested in bank stock. Nett proceeds of the postoffice, $508 annually.

LEXINGTON, p. v., cap. of Holmes co., Miss., 65 m. N. Jackson, 997 W. Situated on a branch of Yazoo river, and contains a courthouse, jail, and several stores and dwellings. Nett proceeds of the postoffice, $345.

LEXINGTON, p. v., cap. of Henderson co., Tenn., 117 m. W.S.W. Nashville, 801 W. Situated on Beech river, a branch of Tennessee river. It contains a courthouse, jail, two churches, a Cumberland Presbyterian and a Methodist; seven stores, one flouring-mill, one saw-mill, two tanneries, 32 dwellings, and 220 inhabitants.

LEXINGTON, city, cap. of Fayette co., Ky., is situated on Town fork of Elkhorn river, in 38° 6′ N. lat., and 84° 18′ W. long., 24 m. E.S.E. Frankfort, 85 m. S. Cincinnati, 74 m. E. Louisville, 562 W. Pop. in 1820, 5279; in 1830, 6404; in 1840, 6997; in 1843, 7500. In 1797, it contained only 50 dwellings, partly frame and partly hewn logs, with the chimneys on the outside. It was named by a party of hunters who were encamped on the spot when they received the intelligence of the battle of Lexington, Massachusetts, at the commencement of the Revolutionary war. It is one of the oldest towns in the state, and was formerly its seat of government. The city is laid out on 2 m. square, with broad streets from 69 to 75 feet wide, crossing each other at right angles, man_ of them paved and beautifully built, mostly with_ y number of the out lots of this large area are not built on, but are occupied for pasture. It contains a brick courthouse, jail, two banking houses, a market house, the Transylvania University and Medical Hall, city school-house, a lunatic asylum, orphan asylum, city poor-house, nine churches, a Baptist, Reformed Baptist, African Baptist, a Methodist, an African Methodist, two Presbyterian, an Episcopal, and a Roman Catholic; 30 dry goods and 50 groceries and clothing stores, one flouring-mill, 12 bagging and rope factories, three iron-foundries for stoves, &c., and 2000 dwellings, many of them neat, and some of them elegant. The churches are built of brick, of which the Reformed Baptist has the largest congregation, and the Methodist the next largest. The city school has from 250 to 500 scholars, with several competent teachers of both sexes. There are also several private schools, of a high order, for males and females. There is a public square near the centre of the place, surrounded by fine brick edifices, on which the market is situated. In the vicinity of the city are numerous elegant country seats. Seven Macadamized roads lead from the city in different directions, and a railroad extends 94 m. to Frankfort, and is designed to be continued to Louisville. The city is more distinguished for its intelligent and polished society, and as an elegant place of residence, than for the bustle of business. Transylvania University is one of the oldest institutions of the kind beyond the Alleghany mountains, and is highly respectable. It was founded in 1798, and newly organized in 1818; has classical, medical, and law departments, a president and 14 professors or other instructors, 105 students in the classical department, 271 in the medical, and 73 in the law department, making 451 in the whole, and 12,943 volumes in its libraries. The commencement is on the second Thursday in September. The graduates of the medical department, amounting to 1112, are more numerous than those of any similar institution in the United States, excepting the medical department of the university of Pennsylvania at Philadelphia. The lectures commence on the first Monday in November.

According to the census of 1840, the city had two commission houses in foreign trade, capital $35,500; 72 retail stores, capital $692,285; value of machinery produced, $12,800; hardware, cutlery, &c., $10,000; one woollen factory; nine rope-walks, capital 186,860; three tanneries, one brewery, four printing-offices, one bindery, three weekly and two semi-weekly newspapers, and seven periodicals. Total capital in manufactures, $498,340; one college, 389 students; two academies, 65 students; 12 schools, 686 scholars.

Lexington was incorporated as a city in 1831. It is in

the midst of one of the most fertile districts of country in the United States.

LEXINGTON, t., Stark co., O. It has one school, 27 scholars. Pop. 1627.

LEXINGTON, p. v., cap. of Scott co., Ia., 59 m. S. by E. Indianapolis, 577 W. Situated 11 m. W. of Ohio river, and contains a courthouse, jail, three stores, and about 200 inhabitants.

LEXINGTON C.H., p. v., cap. of Lexington distr., S. C., 12 m. W. Columbia, 518 W. It contains a courthouse, jail, several stores, 20 dwellings, and about 120 inhabitants.

LEYDEN (Lat. *Lugdunum Batavorum*), a celebrated city of Holland, being the fourth in the kingdom of the Netherlands, on the Old Rhine, 21 m. S.W. Amsterdam, and 10 m. N.E. the Hague; lat. 52° 9′ 23″ N., long. 4° 29′ 36″ E. Pop. in 1837, 36,110. It is surrounded by ramparts and a wet ditch, and is entered by ancient gateways. "On the outer side of the *singel*, or ditch, which everywhere encompasses the town, except where it is cut by the Rhine, is planted a beautiful double avenue of trees, forming agreeable walks for the citizens; and on the inner side rise the low green mounds, which serve the purpose of walls to this venerable city." (*Chambers.*) Like other Dutch towns, Leyden is traversed by canals, crossed by numerous bridges; though, as its trade is but trifling, the canals are of little use. The streets are usually long, broad, and well built; and there are some striking public edifices, and the town has an antique, venerable appearance. Barrow compares the *Breede Straat* (Broad street) of Leyden to the High street of Oxford, reckoned among the finest in Europe. He says, "In the first place, it is much wider, and at least three times the length; and, contrary to the usual practice of laying out streets by the Dutch, it has the same gently-winding turn, but wants the gradual ascent, which constitutes so much to the beauty of the High street of Oxford. The houses in that of Leyden are generally superior and more picturesque; and, though the number of colleges of ancient architecture, with their turrets, towers, and spires in Oxford exceed the number of public ba in the Broad street of Leyden, there is one, at least, which will bear comparison with the most picturesque collegiate High-street. This is the old Hôtel de Ville, built, as appears by an inscription in front, in the year 1574. It has a tall spire, somewhat remarkable in its architecture, and not inelegant. It is built of a dark blue stone, which has the appearance of black marble, and its prominent parts are tipped with gilding. The body of the building has nearly 30 windows on a line in front, three pediments, or gables, highly ornamented, a handsome balustrade, surmounted by a ridge of stone globes, and the whole front of this remarkable piece of architecture may be said to be

'With glist'ning spires, and pinnacles adorn'd.'

The ground-floor of the town-house is appropriated as a market for butchers' meat, but this is not seen from the street. Nothing can exceed the cleanliness of Leyden in all its streets, whether those with or those without canals. The former, with their quays, are particularly neat; and the bridges are mostly of stone, of which, they pretend to say, there are not fewer than 150." (*Tour in Holland*, 76, 77.) In the council and audience chambers, on the first floor of the town-hall, are several valuable paintings, as the Last Judgment, by Lucas Van Leyden; a large picture, representing the state of the city and its inhabitants during its siege by the Spaniards, including a portrait of the heroic burgomaster Vanderwerf. The church of St. Peter, founded in 1321, one of the finest Gothic edifices in Holland, contains the tombs of Boerhaave, the Meermans, Scaliger, Camper. &c. Near this church is a large open square, ornamented with trees, and having a canal in its centre; it was formerly covered with houses, accidentally destroyed by the blowing up of a boat laden with gunpowder in the canal, in 1807. About 150 persons lost their lives on this occasion. The church of St. Pancras has also a most imposing front, and the tomb of Vanderwerf. In the centre of the city is a ruined tower, called the burg, of uncertain but ancient date, erected on the only elevated spot of ground for many miles round.

Leyden is a very dull, inanimate town, without manufactures, trade, or bustle of any kind. But it is, notwithstanding, a most desirable residence for men of learning and research. Its university, which, for a lengthened period, was one of the most celebrated in Europe, was founded by the Prince of Orange, in 1575, to reward the inhabitants for their bravery, and as some compensation for the sufferings they sustained during the siege of the city by the Spaniards. It soon attained to the highest estimation, being deservedly esteemed one of the very best of the continental schools for the study of classics, law, medicine, and divinity. Among its professors, are the illustrious names of Douza, Joseph Scaliger, Daniel Heinsius, Gomarus, Arminius, Boerhaave, Van Swieten, Leeuenhoeck, Sgravewade, Burman, Ruhnken, &c. Grotius and Descartes were of the number of its pupils, as were Evelyn, Fielding, and Goldsmith. And though no longer so celebrated as formerly, it is still extremely well conducted, has valuable libraries and scientific collections, and able and learned professors. In 1835, it had in all 647 students; of whom, 250 studied law, 212 divinity, 131 medicine, and 54 philosophy. The college buildings are detached, and, in fact, are placed at considerable distances from each other, in different streets: they are all plain stone and brick, and sufficiently evince, by their appearance, that they have been intended for use and not for ornament. The principal of these buildings, which is very old, and was formerly a religious house, stands on the W. side of the city; its hall, in which the *senatus academicus* meets, is adorned with a fine portrait of William prince of Orange, founder of the university, and upwards of 100 portraits of professors in historical succession. The senatus consists of 33 professors; and as this university requires no test of religious faith, either from its professors or scholars, it comprises all sects and denominations, both Christian and Jewish. Most of the lectures are delivered in Latin, and the public announcement of the course is in that language. The students, who wear no particular dress, reside in lodgings in the town; and the greater number subscribe to a club-house and reading-room, supplied with German and French publications. The students of Leyden bear a high character for diligence; but, among other drawbacks, junior students have to act for six weeks as fags to those of older standing; and duelling is said not to be laid under any efficient restraint.

The museum of natural history, attached to the university, surpasses most others in Europe, being mainly indebted for its excellence to the public spirit of the Dutch naval officers and foreign *employees*, who take every opportunity of forwarding natural curiosities to their native country; but it also owes much to the acquisition of the valuable collection of birds by Temminck, and to the labours of travellers and collectors sent by the *senatus* to Africa, S. America, and other parts of the globe. The museum, which is open *gratis* to all classes, consists of an upper and under story, occupying four sides of a large court. The classification of the animal kingdom is according to the system of Cuvier; and such is the zeal manifested in perfecting the collections that Mr. Chambers mentions that 2500 guilders, or £208 sterling, had recently been paid by the university for one shell of a *nautilus*, to complete the series of such specimens!

The museum of Egyptian antiquities is particularly rich in *papyri*, jewellery, and gold ornaments; and comprises monuments from the ruins of Carthage, and the largest collection of Etruscan bronzes N. of the Alps. Siebold's extensive and valuable Japanese museum is also in Leyden. The library of the university has 60,000 printed volumes, and 14,000 MSS., more than 2000 of which are Arabic. The botanic garden, which comprises several acres, and is extremely well laid out, has an extensive series of specimens, arranged according to the systems of Linnæus and Jussieu, with extensive conservatories for rearing and preserving tropical plants, &c.

Leyden has a good observatory, 17 churches (one of which is Roman Catholic), two hospitals: an *hôtel des invalides*, a fine new edifice; an arsenal, custom-house, chamber of commerce; societies of Dutch literature, science, and poetry; branches of the Society of Public Good, the National Economical Society, and the Dutch Society of the Fine Arts; an academy of design, &c. It is said that all the children belonging to the city are being instructed: a small fee is exacted in the poor schools, which is not usual in Holland.

In the 17th century the manufacture of fine woollen cloth was extensively carried on at Leyden; and it is said to have had, in 1659, about 3000 houses, and 100,000 inhabitants. (*De Gloet*, p. 266.) Perhaps, however, this statement is exaggerated; but it is said by Busching to have had in 1733, 10,891 houses, which, at an average of seven individuals to a house, would give a population of about 77,000. (*Busching's Geography*, &c., 490, English edition.) Latterly, however, it has increased very considerably; its population, which in 1837 exceeded 36,000, having been under 28,000 in 1817. Its cloth manufacture has been for a lengthened period comparatively unimportant; but it is still carried on to some extent, particularly the manufacture of coarse cloths, and of counterpanes, rugs, &c. It also carries on some other branches of industry, and has a considerable traffic in wool, butter, and other articles of agricultural produce. It is connected by canals with Haarlem, Delft, and the Hague.

During the latter part of the 17th, and the greater part of the 18th century, the most interesting as well as the most celebrated branch of industry carried on at Leyden was that of printing and publishing. Many of the best and most beautiful of the Dutch editions of the classics, in

176

12 mo., 8vo. and 4to., including most of those by the Elsevirs, issued from the presses of this city, and would alone have conferred on it imperishable renown. A good deal of printing and publishing is still carried on; but we shall look in vain among the works now published here, or, we may add, anywhere else, for any that will bear a comparison with the *chefs-d'œuvre* alluded to above.

The siege of Leyden by the Spaniards in 1574 is one of the most memorable events in the history of the great struggle made by the United Provinces to emancipate themselves from the blind and brutal despotism of Spain. The inhabitants displayed the most invincible courage and resolution. Valdez, the Spanish general, despairing of being able to carry the town by storm, endeavoured to cut off all communication between it and the surrounding country, and to effect its reduction by famine. He completed his lines of circumvallation, and so far succeeded in his object, as to entail the most tremendous suffering on the inhabitants, without, however, shaking their determination to die rather than give up their city to the enemy. At length the country round the town having been laid under water, a squadron of flat-bottomed boats laden with provisions and stores made its way through the Spanish lines to the city. This was decisive of the fate of the siege; the Spaniards being obliged immediately to raise it, after having incurred a very heavy loss. (*Watson's Philip, II.*, b. 13.)

Leyden has given birth to some highly distinguished individuals. Rembrandt was born (in 1606) in its immediate vicinity; and it is the native place of Gerard Douw, Vandervelde, Mieris, Jan Steen, and other distinguished painters; and of Vossius, Heinsius, Muschenbrock, Van Swieten, John Boerhoff, better known as John of Leyden, founder of the Anabaptists, &c. The learned and laborious geographer Philip Cluvier, or *Cleverius*, though a native of Dantzic, resided principally in Leyden, where his learned and excellent works on the geography of ancient Germany, Sicily, and Italy, and his valuable *Introductio in Universam Geographiam* were published. He died here, in depressed circumstances, in 1623, at the early age of 43. (*Martinière*, art. *Leyden*. *Biographie Universelle*, &c.)

LEYDEN, p. t., Franklin co., Mass., 102 m. W.N.W. Boston, 412 W. Watered by Green river a branch of Deerfield river. Incorporated in 1809. It has one store, two grist-mills, three saw-mills; five schools, 250 scholars. Pop. 632.

LEYDEN, p. t., Lewis co., N. Y., 15 m. S. Martinsburg, 120 m. N.W. Albany, 425 W. Drained by Moose and Black rivers. It has five stores, two grist-mills, nine saw-mills, one printing office, one weekly newspaper; 10 schools, 656 scholars. Pop. 2438.

LEYTON (LOW), a village and par. of England, co. Essex, hund. Becontree, on the river Lea, 5 m. N.E. London. Area of par., 2200 acres. Population, in 1841, 3374. Low Leyton is situated on the low grounds near the E. bank of the river; but further E., connected by a long straggling street, is Leytonstone, on an eminence, comprising several handsome villas, chiefly tenanted by London merchants and traders. The church, a brick building with a low tower, is remarkable only as having been the scene of the pastoral labours of Strype the antiquary, who held the vicarage 68 years, and was buried here in 1737. A Roman Catholic chapel and chapel for Wesleyan Methodists, are the other places of worship; and the parish has, besides Sunday schools, a boys' free school, a school of industry for girls, and several minor charities.

LIBAU (Lettish, *Lepeta*), a seaport town of Russia, gov. Courland, on the Baltic, beside the lake Libau, 105 m. W. by S. Mittau. Population probably 5000. It is walled, and entered by a gate from the N. Its streets are narrow, and mostly unpaved; and its market-place, though large, is irregular. The houses are of timber, and only one story high. It has Lutheran, Roman Catholic, and Calvinistic churches, a hospital, and an orphan asylum. The port, though commodious, has only from eight to 12 feet water, and cannot, therefore, be entered by vessels of much burden. It has, however, a considerable trade: most part of the produce of Courland, as cattle, linseed, corn, hides, tallow, &c., being exported from it. Its imports are chiefly colonial products, manufactured goods, wine, oil, fruits, &c. In 1833, the value of the exports from Libau amounted to 5,011,899 roubles, and that of the imports to 617,754 r. (*Schnitzler, La Russie*, 385; *Possart, Das Kaiserth, Russl.* 415, &c.)

LIBERIA, a republican state of W. Africa, founded in 1821, by free blacks from the U. States of N. America, under the auspices of the American Colonization Society. Its territory extends along the Guinea coast for about 225 m., with a breadth inland of 20 or 30 m., chiefly between lat. 4° and 7° N., and long. 9° and 12° W. Population estimated at 4000 colonists, besides natives. The coast is generally low, but the country gradually rises towards the interior, and at about 20 or 30 m. from the sea, the hills are of considerable elevation. Several rivers fall into the Atlantic within the colony,—as the St. John, St. Paul, Mesurado, &c.; but they are navigable only by small vessels for short distances. The want, indeed, of any great navigable river that might have opened an intercourse with the interior, is a heavy drawback on the prosperity of this colony; and will always hinder it from becoming a place of much commercial importance. The soil is said to be fruitful, and the climate better, or rather less destructive, than in most other parts of the coast. Rice, cotton, coffee, sugar, indigo, bananas, cassava, and yams are raised; and camwood, palm-oil, ivory, hides, wax, and pepper, are among the exports. The settlement is visited by traders from the interior, and some trade is carried on with Europe and America, partly in colonial shipping. The care of the local interests and subordinate affairs of the colony is confided to native colonists, and it has two legislative chambers; but the powers of government are, notwithstanding, substantially vested in the agent of the American Colonization Society. Its object, in fact, was to serve as an outlet for the blacks, who might there enjoy that independence and consideration which long-cherished prejudices hinder them from enjoying in the United States; and it was supposed that the being able to dispose of manumitted slaves by sending them to this colony would promote the practice of manumission; but we believe that in this respect it has had very little influence, only about 2000 liberated slaves having been sent to it. A good many blacks rescued from slave-ships on the African coast and elsewhere have been landed here. Primary schools have been opened for the instruction of the blacks, and it has several churches, and a printing press. But we understand, that, on the whole, the condition and prospects of the colonists are far from satisfactory; though they are not worse than might have been reasonably anticipated.

The chief town, Monrovia, on Cape Mesurado, lat. 6° 25′ N., long. 10° 35′ W., has about 1200 inhabitants. (*Encyc. Americana; Murray's Encyc. of Geog.*, Amer. edit., lib. 45; *Macqueen's Survey of Africa.*)

LIBERTY, county, Ga. Situated in the S.E. part of the state, and contains 650 sq. m. Bounded S.E. by the Atlantic, S.W. by Alatamaha river. Watered by Little Cannoochee river, and Taylors creek. It contained in 1840, 14,381 neat cattle, 3196 sheep, 7539 swine; and produced 90,847 bushels of Indian corn, 6182 of oats, 88,474 of potatoes, 223,297 pounds of rice, 1,347,421 of cotton, 6450 of sugar. It had nine stores, seven grist-mills, six saw-mills: six academies, 166 students; one school, seven scholars. Pop.: whites, 1645, slaves, 5561, free-coloured, 35: total. 7241. Capital, Riceborough.

LIBERTY, p. t., Waldo co., Me., 20 m. E. Augusta, 616 W. Incorporated in 1827. Watered by branches of Muscongus and St. George rivers. It has two stores, five saw-mills; six schools. 321 scholars. Pop. 895.

LIBERTY, p. t., Sullivan co., N. Y., 119 m. S.W. Albany, 304 W. Bounded S.W. by Delaware river, and drained by small streams flowing into it. It contains two churches, a Presbyterian and an Episcopal, five stores, one fulling-mill, three grist-mills, 17 saw-mills, two tanneries; 11 schools. 588 scholars. Pop. 1569.

LIBERTY, p. t., Tioga co., Pa., 123 m. N. by W. Harrisburg, 233 W. It has three stores, one woollen-factory, four grist-mills, five saw-mills; one school, 50 scholars. Pop. 1128.

LIBERTY, t., Columbia co., Pa. Drained by Chelisquaque and Mahoning creeks. It has two stores, five grist-mills, five saw-mills; four schools, 170 scholars. Pop. 1328.

LIBERTY, p. v., capital of Bedford co., Va., 142 m. W. by S. Richmond, 217 W. It contains a courthouse, jail, three churches, two Baptist, and one Free, a masonic hall, five stores, one tobacco factory, two tanneries, 70 dwellings, and about 400 inhabitants.

LIBERTY, p. v., capital of Casey co., Ky., 60 m. S. Frankfort, 583 W. Situated on the N. side of Green river. It contains a courthouse, jail, and 125 inhabitants.

LIBERTY, t., Adams co., O. It has three stores, one flouring-mill, three grist-mills, three saw-mills; one school, 50 scholars. Pop. 1498.

LIBERTY, t., Butler co., O. It has four schools, 106 scholars. Pop. 1479.

LIBERTY, t., Clinton co., O. It has three stores, two flouring-mills, two grist-mills, three saw-mills; seven schools, 421 scholars. Pop. 1049.

LIBERTY, t., Crawford co., O. It has eight schools, 590 scholars. Pop. 1469.

LIBERTY, t., Fairfield co., O. The Ohio canal passes through it. Pop. 2778.

LIBERTY, t., Highland co., O. It contains Hillsborough village the capital of the county, and has 36 stores, seven

fulling-mills, seven flouring-mills. 26 grist-mills, 24 saw-mills, three oil-mills, 26 tanneries, one distillery, three potteries, two printing offices, two weekly newspapers; two academies, 150 students; 12 schools, 800 scholars. Pop. 3521.

LIBERTY, t., Licking co., O. It has 10 schools, 430 scholars. Pop. 1118.

LIBERTY, t., Seneca co., O. It has one store, one grist-mill, three saw-mills; seven schools, 281 scholars. Pop. 1084.

LIBERTY, p. v., capital of Union co., Ia., 73 m. E. by S. Indianapolis, 515 W. It contains a handsome brick courthouse, a jail, an academy, a church, four stores, a steam-saw-mill, 80 dwellings, and about 500 inhabitants.

LIBERTY, p. v., capital of Clay co., Mo., 159 m. W.N.W. Jefferson city, 1072 W. It contains a courthouse, jail, five churches, a Presbyterian, Dutch Reformed, Baptist, Methodist, and Roman Catholic, two academies, 14 stores, and 1192 inhabitants.

LIBERTY, p. v., capital of Amite co., Miss., 101 m. S.S.W. Jackson, 1181 W. Situated on the W. fork of Amite river.

LIBOURNE, a town and river-port of France, dep. Gironde, cap. arrond. on the Dordogne, at its juncture with the Isle, 22 m. E.N.E. Bordeaux. Population in 1836, ex. com., 9084. Few towns in France are so regularly and well built. Its streets are wide and clean, its houses elegant, and it is surrounded with good walls and agreeable promenades. Among the chief public edifices are extensive cavalry barracks, a theatre, a public library, with 3000 vols., and a handsome brick and stone bridge of nine arches across the Dordogne. The port, at high water, has from 10 to 16 feet water, admitting vessels of 300 tons burden. Libourne is the seat of a sub-prefecture, of tribunals of primary jurisdiction and commerce, and a sub-commissariat of marine. It has manufactures of woollen stuffs, articles of military equipment, glass and cordage, and docks for ship-building. It is an entrepôt for salt and agricultural produce destined for Bordeaux. It was founded by Edward I. of England, in 1286. (*Hugo*, art. *Gironde, &c.*)

LICHFIELD, a city, parl. bor., and co. of itself, locally situated in county Stafford, hund. Offlow, 15 m. N. Birmingham, 29 m. W. Leicester, and 108 m. N.W. London. Area of the county of city (which is co-extensive with the parl. bor.) 3180 acres. Population, in 1841, 14,754. The city, which stands in a fine valley, on a small affluent of the Trent, is irregularly built, with narrow streets; but it is well paved and lighted, many of the houses are handsome, and its general appearance is respectable. The chief public buildings, besides the churches, are the guildhall, a neat stone edifice, on the top of which are carved the city arms; the market-house, occupying the site of an old market-cross; the bishop's palace in the Close, and a small theatre. Lichfield is an episcopal see, and has a noble cathedral on the N. side of the town, close to a fine sheet of water. It is built chiefly in the decorated Gothic style peculiar to the 12th and 13th centuries, and comprises a nave, choir, and transepts, with a ladye-chapel. It measures, from E. to W., 410 feet, and is 153 feet wide, measured along the transepts. There are three towers, the central one of which, rising from the intersection, is surmounted by a light steeple, and has a total height of 280 feet: the towers at the W. end are each 180 feet high. The body of the church is spacious and lofty, supported by pillars formed of clustered columns with neat foliated capitals: the roof is beautifully groined, the choir is elegantly furnished, and there are several fine monuments, one of which is to Dr. Johnson, the great lexicographer, a native of this city, where he first saw the light on the 18th of September, 1709. The exterior parts of the building are highly ornamented with sculpture and tracery-work: the W. front displays a multitude of figures in *alto relievo*, illustrative of passages in Bible history, and on the roof is a statue of Charles II., erected by Bishop Hacket, who exerted himself during many years to repair the damages inflicted on the cathedral by the parliamentary troops in the great civil war. The whole building was thoroughly repaired in 1787–90, at an expense of £6000. The chapter comprises a dean, six residentiary canons, 14 prebendaries, and five priest vicars. At an average of the three years ending with 1831, the nett revenues of the bishopric of Lichfield amounted to £3923 a year; and at an average of the seven years ending with 1834, the revenues of the cathedral amounted to £1673 a year. In the city is one parish church and two chapelries, in the patronage of the dean and chapter; besides which, there are places of worship for Independents, Wesleyan Methodists, Roman Catholics, and other bodies of dissenters. Among the educational establishments are several Sunday schools, three national schools, an English charity school, and a free grammar-school, founded by Edward VI., and stated to

M   177

be in a flourishing condition, with 21 free boys and several stipendiary pupils boarding with the masters: among the former pupils of this school are the illustrious names of Ashmole, Addison, Garrick, Johnson, and Woollaston. The charitable institutions comprise two almshouses, a hospital for clergymen's widows and orphans, a mendicity society, and a dispensary. Lichfield has no trade or manufactures of importance: a carpet factory gives employment to about 150 persons, and 23 others are returned as belonging to a worsted-mill. Its trade is chiefly local, arising out of the wants of the town and neighbourhood, and there is little show of activity among those engaged in business. The Grand Junction canal passes near the city, and the Birmingham and Derby railway at a distance of about six miles. The city was anciently governed by a guild, dissolved by Edward VI., who gave it a charter of incorporation, subsequently confirmed by Charles II. Under the Municipal Reform Act, the borough is divided into two wards, and the municipal officers are, a recorder, mayor and five other aldermen, and 18 councillors. Corp. rev., in 1839, £3160; but of that £921 was derived from the sale of property. Quarter and petty sessions are held in the guildhall, and there is a court for the recovery of debts under 40s. Since the 33d Edward I., Lichfield has, with some intermissions, sent two members to the House of Commons. Previously to the Reform Act, the franchise was vested in the freeholders of the county of the city of Lichfield, and in the freemen and burgage-holders of the city. The boundaries of the present parliamentary borough include the county of the city, and the place called "the Close," belonging to the cathedral. Registered electors, in 1839–40, 876. Markets on Tuesday and Friday; fairs, January 10, Shrove-Tuesday, and Ash-Wednesday, for cattle, sheep, bacon, and cheese; May 12, for sheep and cattle; and first Tuesday in November for cheese.

LICK, t., Jackson co., O. Named from a salt lick within its limits. Watered by a branch of Scioto river, on which Jackson village the capital of the county is situated. It contains 10 stores, two tanneries; two schools, 53 scholars. Pop. 882.

LICKING, r., Ky., rises in Floyd co., and after a course of 180 miles enters Ohio river at Newport, opposite to Cincinnatti. In dry summers it has little water; but in winter and spring, flat boats descend it for 70 or 80 miles. By means of dams and locks, it is proposed to improve the navigation of this river, but the work is at present suspended.

LICKING, r., O., a considerable W. branch of Muskingum river, with which it unites opposite to Tanesville.

LICKING, county, O. Situated a little S.E. of the centre of the state, and contains 666 sq. m. Watered by Licking river, and its branches, which afford extensive water-power. Iron ore is abundant. The Ohio canal passes through it. It contained in 1840, 31,354 neat cattle, 53,419 sheep, 45,504 swine; and produced 546,471 bushels of wheat, 9184 of rye, 831,794 of Indian corn, 15,735 of buckwheat, 465,309 of oats, 107,941 of potatoes, 121,093 pounds of tobacco, 109,383 of sugar. It had 11 commission houses in foreign trade, 103 retail stores, one furnace, six fulling-mills, three woollen factories, 13 flouring-mills, 23 grist-mills, 72 saw-mills, two oil-mills, 21 tanneries, five distilleries, one brewery, three printing-offices, four weekly newspapers; five academies, 555 students, 129 schools, 6917 scholars. Pop. 35,096. Capital. Newark.

LICKING, t., Muskingum co., O. Watered by Licking river. The Ohio canal passes through it. It contains two churches, two flouring-mills, four saw-mills, and two large ancient mounds. Pop. 1051.

LICKING CREEK, t., Bedford co., Pa. It has three stores, four grist-mills, four saw-mills, one printing-office, one weekly newspaper. Pop. 894.

LIECHTENSTEIN (PRINCIPALTY OF), an independent state of S. Germany, and, according to most authorities, the least in extent and population throughout Europe; between lat. 47° 5' and 47° 18' N., and long. 9° 28' and 9° 38' E.; having S. the Swiss canton of the Grisons; W. the canton of St. Gall, from which it is separated by the Rhine; and E. the Austrian duchy of Vorarlberg. Area, 53 sq. m. Population 6300. The surface is mostly mountainous; a range of the Grison Alps traverses it, separating the Rhine from the Samina, a tributary of the Ill. Cattle-breeding, agriculture, timber cutting, and cotton spinning, especially the first, are the chief occupations of the inhabitants. Corn, wine, fruit, and flax, are the principal articles of culture. The government is vested in the prince, and in an assembly of deputies of the clergy and rural proprietors. Appeal from the court of original jurisdiction in Vaduz, lies to the court of chancery in Vienna, in which the prince has a seat; and thence to the court of appeal at Innsprück. Vaduz, or Liechtenstein, the capital, is a town of less than 1000 inhabitants. The principality furnishes a contingent of 55 men to the army

of the German confederation; it has one vote in the full council of that body, and together with other small states (see GERMANY, vol. i. p. 979), a vote in the committee, and the 16th place in the German diet. The prince of Liechtenstein is one of the richest proprietors of Europe: his estates in other parts of Germany, but especially in Moravia, extend over nearly 2200 sq. m.; and his annual revenue is estimated at 1,200,000 florins; of which amount, however, his independent sovereignty yields only about 20,000 florins. (Berghaus; Almanach de Gotha, &c.)

LIEGE (Dutch, Luyk, Germ. Lüttich), an important commercial and manufacturing city and river port of Belgium, cap. prov. of same name; on the Maese, 134 m. S. by W. Maestricht, and 54 m. E. by S. Brussels; lat. 50° 39' 22' N., long. 5° 31' 42'' E. Population, in 1836, 58,600; but the city is surrounded by a neighbourhood with a dense population employed in branches of industry similar to its own. Its population in the middle of the 15th century is said to have amounted to 190,000; latterly, however, it has been increasing. It is situated on the declivity and at the foot of a hill, and is consequently divided into an upper and lower town. The latter stands at the confluence of the Ourthe with the Maese, and is intersected by many branches of the last named river, which are enclosed by stone walls, and crossed by numerous bridges. The chief bridge, the Pont de l'Arche, thrown across the main stream of the Maese, is 153 yards in length, 49 feet in breadth, and has six arches, varying in diameter from 50 to 55 feet. A convenient quay for commercial purposes extends both above and below this bridge, for the whole length of the town along the Maese, which is navigable for small vessels as far as this city. Liege was formerly fortified, but its fortifications have been almost entirely destroyed. It is defended on the N.W. by a large citadel, lately rebuilt, and on the E.S.E. by Fort Chartreuse; besides which there are only a few outworks. There are 10 suburbs. Liege is, generally speaking, ill built. In both the lower and upper town the streets are narrow, and in the latter they have the additional disadvantage of being so steep as to be ascended in many places by flights of steps. Among the 11 squares, are two tolerably spacious; in one of which stands the town hall, and in the other the theatre. The townhall, comprising the provincial courthouse and prison, is a dark stone building, of great extent and magnificence, with two open courts, surrounded with a colonnade resembling that of the ducal palace at Venice. It was formerly the residence of the prince bishops of Liege. The cathedral of St. Lambert stood in this square; but it was destroyed by the French revolutionary forces in 1794, and no traces of it exist. The church of St. Jacques, in the decorated Gothic, is the architectural glory of the city. It was completed in 1552. Its interior is astonishingly magnificent, and displays some of the finest specimens of tracery and fret-work that is anywhere to be met with. It has a noble organ, but its statues and paintings are inferior. St. Croix, and some of the other churches, of which there are 21 Roman Catholic and one Protestant, are also fine structures. The theatre is a handsome modern building, surrounded by an arcade: from the square in which it stands a piece of water runs to encircle the town on its W. side, bordered by a promenade planted with trees. The buildings of the university stand beside the Maese, on the ruins of a church of the Jesuits. This institution, founded by the late king of Holland, in 1816, has faculties of theology, law, and physic; 46 professors, and usually from 400 to 500 students. It possesses a cabinet of mineralogy, with upwards of 5600 specimens, a cabinet of 3000 fossils, found in the vicinity, and other scientific collections, and a library comprising many curious MSS.

According to Mr. Chambers, "The sight of Liege at once reminds us of an English manufacturing town. We hail its engine chimneys and smoke as emblems both of wealth and advancement in the mechanical arts; and as we drive into its busy streets, and pass along its open quays thronged with commerce, we are apt to inquire of ourselves, can all this be on the continent, and not in one of the manufacturing districts of England?" (Tour, &c., in 1838.) In fact, Liege may be regarded as the Birmingham of the continent. It owes this distinction to its situation in a district abounding with coal and iron, and which also affords zinc, lead, copper, sulphur, alum, marble, and slate. The coal-field of Liege is the most extensive in the province of the same name, being five leagues in length, with a breadth varying up to two leagues. Coal is, however, rather expensive, the cost of raising it having been estimated, in 1836, at about 10 francs per ton.

The manufacture of cannons and fire-arms is that for which Liege and its environs are most celebrated. The royal cannon-foundry in this city, instituted in 1802, produces at an average nine pieces of ordnance weekly, partly brass and partly iron. There are numerous manufactories of fowling-pieces, muskets, pistols, &c. The guns of Liege are said to be cheaper than those of England; but there is not, we believe, any real ground for such an assertion; they may perhaps cost less money, but then they are not nearly so well finished, nor so good, as English guns. Had they been really cheaper, the manufacture would not have declined so rapidly as it did during the four years ending with 1838.

Account of the quantities of fire-arms manufactured at Liege in each of the four years ending with 1838.

Years.	Fowling Pieces.			Pistols.		Muskets and Military Fire Arms
	Single Barrels.	Double Barrels.	Severe Barrels.	Horse.	Pocket.	
1835	189,690	64,390	7,129	15,727	49,290	14,864
1836	145,844	54,548	5,639	22,088	7e,314	11,764
1837	113,062	55,841	16,816	15,464	42,784	28,200
1838	96,125	51,238	15,966	10,354	32,537	24,542
Total	412,386	98,4,14	44,749	63,686	416,108	202,291

(Briavoienne, Industrie en Belg. ii. 308.)

In 1835, the most flourishing year of the manufacture, the value of the fire-arms issued from the different factories of Liege was estimated at 7,000,000 francs.

Steam engines and machinery are largely produced in Liege and in the adjacent busy and populous village of Seraing, about 2 m. to the S.W., on the opposite bank of the Meuse. The palace of the former prince-bishops at that place having been bought in 1817 by the Messieurs Cockerill, Englishmen, they established in it the largest hardware manufacturing establishment in Belgium, or indeed on the continent. It is devoted to the construction of steam-engines and other descriptions of machinery, and to forging and manufacturing iron and iron goods.

It is said that 60 steam-engines of the aggregate power of 695 horses, are constantly employed in this factory, with from 2000 to 2500 workmen, 500 of whom are miners. Most of the locomotive engines upon the Belgian railways, the engines for steam vessels, &c., used in Belgium, have been made here, and many have also been sent to other parts. But we regret to have to add, that from some cause or other, Messrs. Cockerill have recently been involved in serious difficulties; and if their works be now carried on, it must, we apprehend, be through the advances that have been made them.

Liege has also manufactories of files, nails, stoves, and hardware of all kinds; watches, jewellery, bronze, and other ornaments; woollen and cotton fabrics, hats, glue, tobacco, paper, chemical products, &c.; with numerous dyeing houses, tanneries, and distilleries. It has an exchange, a chamber of commerce, a bank, with the privilege of coining money, a savings' bank, a mont-de-piété, numerous hospitals, and benevolent institutions, superior and elementary schools, and various learned societies. A railway connects Liege with Louvain and Brussels.

In the 7th century, a village named Legia occupied a part of the site of the present city. In 712, the ancient cathedral was founded, and Liege was erected into a bishopric. In the 10th century its bishops were raised to the rank of independent sovereign princes. In the succeeding ages continual wars and disturbances prevailed between the burghers, who were ardently attached to popular institutions, and the prince-bishops. It was taken on the 30th of October, 1468, by Charles the Bold, duke of Burgundy, and barbarously delivered up to military execution. During the French ascendancy, it became the capital of the department of Ourthe. (*Vandermaelen, Dict. of the Prov. Liege; Briavoienne, de l'Industrie en Belgique; Hoeschling, Chambers, &c.*)

LIEGNITZ, a town of Prussia, prov. Silesia, cap. gov. and circ. of Liegnitz, on the Katzbach, at its confluence with the Schwarzwasser, 46 m. W. by N. Breslau; lat. 51° 12' 48" N., long. 16° 9' 47" E. Pop. (1837), 11.607. It was merely a fortress of some strength, but now has only gates without walls; and its ramparts being planted with trees and laid out in gardens, serve only for public walks. It is an old, but a handsome, well built town: it has several suburbs, two Lutheran, and three Roman Catholic churches; a fine chapel—the Ferstenscapelle—in which are buried the princes of the line of Piast, a dynasty which gave 24 kings to Poland, and 123 dukes to Liegnitz, from 775 to 1675, when the family became extinct; the old castellated palace of those princes in the centre of the town, surrounded by a wet ditch, an ancient council-house, a gymnasium, an academy, established in 1810 for the sons of Silesian gentlemen, whether Roman Catholics or Protestants, an orphan asylum, a workhouse, a hospital, &c. Outside the town is a good cemetery. Liegnitz is the seat of the superior judicial courts, boards of taxation, and weights and measures, &c., for its government, and the head-quarters of several battalions of fusileers, of a landwehr or militia bat-

talion, and of a commandant of police. It manufactures woollen, cotton, and linen stuffs, stockings, lace, Prussian blue, and starch, and has several breweries and bleaching grounds, and an active trade in its own produce, and in madder and other products raised in the adjacent country. The gardeners in the vicinity are said to be the most expert of any in Silesia. On the 15th of August, 1760, Frederick the Great totally defeated the Austrian general Laudohn in the neighbourhood of this town; Frederick made his dispositions with so much skill as to render it impossible for Marshal Daun, who commanded another Austrian army, to come to Laudohn's assistance. (*Von Zedlitz, Das Pruss. Staat, iii., 148; Berghaus, Allg. Länder, &c.*)

LIERRE, a town of Belgium, prov. Antwerp, cap. canton, at the confluence of the Great and Little Nethe, 10 m. S.E. Antwerp. Population in 1836, 13,800 (*Houschling*). It is well built, and has several churches, a convent, a hospital, manufactures of cotton and woollen stuffs, with cotton-printing establishments, distilleries, breweries, and a number of oil-mills, rape seed being largely cultivated in its vicinity. It sends three deputies to the provincial assembly. (*Dict. Geog., &c.*)

LIFFORD, an inland town of Ireland, prov. Ulster, co. Donegal, of which it is the cap. It is situated on the extreme E. verge of the county, on the Foyle, immediately below the confluence of the Finn and Mourne rivers, 14 m. S.S.W. Londonderry. Population, in 1831, 1066. It is connected by a fine bridge over the Foyle with the town of Strabane in Tyrone, of which it is now merely a dependency. It consists of two small streets, and has a parish church, a Roman Catholic chapel, a Presbyterian meeting-house, a barrack, and a courthouse and prison for the county of Donegal. It sent two members to the Irish parliament till the Union, when it was disfranchised.

LIGOR, a town of S.E. Asia, cap. of a Malay principality, dependent on Siam, on the Ta-yung near its mouth, in the gulf of Siam, about lat. 8° 17' N., long. 100° 12' E. Population, estimated at 5000; chiefly Siamese, Malays, and Chinese. It appears to have been formerly more populous; but it was captured by the Burmese, and its inhabitants carried off, in 1760, and again in 1785. It has brick ramparts, and a wet ditch; and in 1805, 14 cannon were mounted on its walls. Within the town are many brick temples and pyramids, one having a gilt spire, a conspicuous object at sea; but all the dwelling-houses are of less solid materials. Two or three Chinese junks trade with Ligor, bringing cotton, and taking back tin, black pepper, rattans, &c. The rajah of Ligor has extensive authority, with the power of capital punishment over all the Malay states, tributary to Siam. (*Crawfurd's Siam, &c., ii., 211; Hamilton's E. I. Gazetteer.*)

LIGONIER, p. t., Westmoreland co., Pa., 149 m. W. Harrisburg, 175 W. Drained by Loyalhanna creek, and its branches. It contains four flouring-mills, one saw-mill, two tanneries, one distillery; nine schools, 394 scholars Pop. 2904. The village contains two churches, a Presbyterian and Methodist, three stores, 40 dwellings and 294 inhabitants.

LIMA, the cap. city of Peru, and, next to Mexico, the most magnificent in the countries formerly comprised in Spanish America, on the Rimac (whence, by corruption, the name of the city), in a delightful valley, from 500 to 600 feet above the level of the ocean, 6 m. from its port of Callao, on the Pacific, and about 300 m. S.S.E. Truxillo: lat. 19° 2' 45" S., long. 77° 17' 15" W. Population variously estimated, but it may probably amount to between 50,000 and 60,000. The great chain of the Andes passes within 30 leagues of the city; but its spurs approach to within three fourths of a league from its gates, and form an amphitheatre, within which Lima is built. The Rimac, which separates the city from its suburb, San Lazaro, is crossed by an excellent stone bridge of six arches, which, being furnished with recesses and seats, is a favourite promenade. The city, about two miles in length E. to W., by one and a quarter mile in its greatest breadth, is of a triangular, or rather, semicircular, shape, the base or long diameter, being formed by the river. Elsewhere, Lima is surrounded by a parapet wall, about seven miles in circuit, from 18 to 25 feet high, and about nine feet thick: it is pierced by six gates, open from 4 A.M. to 11 P.M., and is defended by 35 bastions. Except at some of the bastions, the wall is too narrow for the mounting of artillery; and it is merely sufficient to protect the town against any sudden attack by an Indian force, for which purpose it was constructed, in 1685. At the S.E. extremity of the city is a small citadel, in which are the artillery barracks, and a military dépôt. When seen from Callao roads, or even from a less distance, Lima has an imposing appearance, its numerous domes and spires giving it quite an oriental aspect. Like the other Spanish cities of America, it is laid out in quadras, or squares of houses, 480 feet each way, and divided by streets 33½ feet wide, intersecting each

179

other at right angles. The courses of the streets do not follow the cardinal points, but vary from E. to S.E., that the walls may cast a shade both in the morning and afternoon." In 1791, the city, with its suburb, El Cercado, contained 209 *quadras*, and 355 streets. Since then little or no improvement has been made; not a single new dwelling having been built within the walls during the last 30 years. (*Three Years in the Pacific*, i., 279.) Through the centre of nearly all the streets directed E. to W. runs a stream of water, three feet wide, used as a receptacle for all the filth thrown out from private dwellings. Most of the refuse is, however, got rid of by the Turkey buzzards, which swarm in Lima, and, like dogs in Lisbon, are the most efficient, or rather, the only scavengers. The streets are paved with round pebbles, and the narrow foot-paths with flat stones, in very bad repair. The same plan extends to the suburb of San Lazaro. The city is divided into four quarters, and each of these into 35 *barrios*. For each barrio an *alcalde*, or district magistrate, is selected from among the inhabitants. For religious purposes, it is divided into eight parishes. On account of the frequency of earthquakes, few houses are more than one story high, or if there be two stories, the walls of the upper consist of only cane, or wattled reeds, plastered over with clay, and whitewashed or painted. This kind of architecture is applied to even the churches and other public edifices, their upper parts being of wood-work, covered with stucco. The lower parts of the houses are mostly constructed of *adobes*, or sun-dried bricks, made of clay and chopped straw. The roofs are uniformly flat. Some of the better sort of houses have a terrace on the top, formed of large thin baked bricks; the common dwellings are usually roofed only with thin rafters, cane, and mats, covered with a layer of earth an inch or more thick; but as it rarely or never rains with any violence in Lima, these light roofs sufficiently answer their purpose, at the same time that they are not so easily thrown down by an earthquake, and when thrown down are incomparably less dangerous than if they were constructed of more solid materials. Most of the houses have a *patio*, or court yard, in front, with a large arched gateway opening to the street, over which is a heavy balcony. The walls of the *patios* are painted without and within with various devices, in fresco. Till of late years, few of the windows had either glass or sashes. Almost every house has a stream of water running through its precincts, which is used for domestic purposes. Gardens are rare.

In the centre of the city is the *Plaza Mayor*, or *de la Independencia*, the principal square and market-place. It is a fine open space, the size of a quadra. On its E. side are the cathedral, the *sagrario* or principal parish church, and the archbishop's palace; the last, a large superior edifice, is now partly occupied by the Peruvian senate. On the N. is what was once the viceroy's residence, an old unsightly structure, now appropriated to the courts of justice and other government offices. On the W. side are the *cabildo* or town-hall, a Chinese looking edifice, the city jail, and other offices; and on the fourth side is a colonnade before a row of private houses. The above public buildings have all ranges of mean looking shops in their lower story. The booths of small traders cover nearly a third part of the area of the square. In the centre is a fine bronze public fountain, 40 feet high, raised upon a level table of masonry 40 feet on each side, ornamented with eight lions supporting a statue of Fame, and supplied with excellent water from the Rimac.

A considerable portion of the area of the city is occupied by convents and churches. Besides a great many convents and nunneries, with churches attached, Lima has 57 churches, and 25 chapels belonging to hospitals, colleges, &c. (For an account of the churches, convents, &c. when in their splendour, see *Ulloa, Voyage de l'Amérique*, i., 498, &c.) The cathedral, founded by Pizarro, and in which he is buried, is a large fine edifice, 186 feet in front by 320 deep; but its effect is injured by gaudy colouring and grotesque ornaments. At either corner of the front is an octagonal tower, rising nearly 200 feet from its base, which is 46 feet high. These towers, having been thrown down by the earthquake of 1746, were rebuilt in 1800. In the belfries are several fine-toned bells, the largest of which weighs 310 quintals. The interior of the cathedral is magnificent. It is divided into three naves, and paved with large earthen tiles. The roof, which is beautifully panelled and carved, is supported by arches springing from a double row of square stone pillars. The high altar is in the Corinthian order, and its columns, cornices, and mouldings, are either cased with pure silver or are richly gilt. The seats and pulpit in the choir are exquisitely carved, and there are two large and fine-toned organs. "The riches which have been lavished at various times upon the interior of this edifice, are scarcely to be credited anywhere but in a city which once paved a street with ingots of silver to do

honour to a new viceroy. The balustrades surrounding the great altar, and the pipes of the organ, were of silver. It may be mentioned, as a proof of the abundance of silver ornaments, that in 1821, one and a half ton of silver was taken from the various churches in Lima without being missed, to meet the exigencies of the state." ' (*Caldcleugh's Travels in S. America*, ii., 56.) The *sagrario* has a fine façade, and its interior is very splendid and richly adorned. The roof is lofty and beautifully panelled, and in the centre is a cupola resting on the four corners formed by the intersection of the cross aisle. Several of the other parish churches are worthy a visit. Some of the conventual churches are remarkably rich. That of the Dominicans, 300 feet long by 80 broad, has a steeple 180 feet high, being the loftiest in Lima.

The revolution secularized a good deal of church property; but previously to that event, the Dominican convent is said to have had a rental of 50,000 dollars a year, and a large library, some good paintings, and numerous reliques, &c., including a statue of the Madonna studded with gems, said to be of enormous value. Some of the cells belonging to it were richly furnished. The Franciscan convent is among the oldest and largest in Lima. Its buildings cover two quadras, and its church, which is next in size to the cathedral, is gorgeously adorned. Its monks derive a considerable revenue from manufacture of shrouds, of which they have, or at least had, the monopoly. In addition to the convents, there are *casas de exercicio*, into which females retire during Lent, to perform acts of penance; and in the convent of Recoleto are similar cells for men. The number of monks and nuns here and in other parts of Peru was formerly very great; but it is now quite otherwise, and the influence of the revolution in turning out these lazy, dissolute drones, and in lessening the deference the inhabitants had been accustomed to pay to all priests, how undeserving soever, has been most beneficial. There are two founding asylums, and 11 public hospitals, one of the latter, St. Andres, having 600 beds. Attached to it, is an different botanic garden; and adjoining it, is the medical college of San Fernando, established in 1809. Lima has a university, founded in 1571: it occupies a handsome building, and is partly supported by congress, and partly by the produce of an annual *bull-bait!* The students, of whom there are only from 30 to 50, generally reside within the walls of the institution. The Peruvian house of representatives holds its sittings in an apartment in the university. The former palace of the Inquisition is now appropriated to a jail, and to the national museum, which, though in extremely bad order, possesses valuable collections of minerals, and Peruvian antiquities. Lima has several ecclesiastical colleges, and seminaries, and a nautical academy. The colleges, however, are now in anything but a flourishing state; but, on the other hand, numerous Lancasterian, and other primary schools, have sprung up, and it is alleged that all the white children are taught to read and write. Notwithstanding the low state of the university, it is affirmed that education has made a great advance in the Peruvian capital since the revolution, and its emancipation from the control of the priests is, at all events, an immense step in advance: a considerable number of modern scientific and other works are said to be annually imported from Europe.

There is a good theatre, but of rather a singular form, it being a long oval, with the stage occupying the greater part of one of its sides. Bull-fights were formerly celebrated at Lima with an *éclat* that rivalled those of Seville: and though abolished by San Martin in 1822, they appear to have revived. The amphitheatre, *Plaza del Acho*, in the suburb of San Lazaro, where they are held, has an area 400 feet in diameter, surrounded by a barrier seven feet high, and three tiers of boxes raised on brick pillars, with accommodations for from 10,000 to 12,000 spectators. Cock-fighting is a favourite public diversion; the cock-pit, or *coliseum*, is an area 50 feet in diameter, surrounded by steep benches and a tier of boxes, which, on Sundays and holydays, are usually crowded by visiters. Outside the walls, is the pantheon, a general cemetery established early in the present century. It is a square inclosure, laid out in walks and gardens, the surrounding wall being full of niches for the reception of corpses. These are generally deposited without coffins, their decay being accelerated by the application of unslaked lime. Before the establishment of the pantheon, the dead were always buried in churches; but this is now prohibited, and hearses belonging to the pantheon are provided for the performance of funerals, which are not allowed to traverse the streets after noon. Immediately without the suburb of San Lazaro, are some excellent public baths. The road from Callao to Lima is quite straight, and for nearly the last two miles is fenced on either side by a brick wall and parapet, shaded with trees, and irrigated by running streams. At intervals of 100 yards are ornamental stone seats; but the whole work, together with the fine gateway at its upper end, by which the city is entered, has been suffered to fall into decay.

The vicinity of Lima, where not covered with villas and pleasure-grounds, is very productive of maize, barley, various other grains, beans, kitchen vegetables, fruits, sugar, rice, tobacco, yams, potatoes, &c.; grapes are abundant, and yield some pretty good wine; olives thrive well; and water-melons are important articles of culture, being largely consumed in the city during the hot months. But agriculture and horticulture, like every other branch of industry, is much neglected. As very little rain falls at Lima, artificial irrigation is indispensable. The Incas had cut numerous trenches and canals in the neighbourhood, which the Spaniards finding ready to their hands, took some care to keep in order; but at present it is said that the drains for conveying the water from the city are so bad that the water is either suffered to run to waste, or to stagnate and generate noxious effluvia. The refuse of the city might be made a valuable manure for the soil; but such is the carelessness and indolence of the inhabitants, that it is either thrown into the canals, or conveyed without the walls, or to the river's brink, where it is suffered to accumulate in fermenting mounds of immense size. (Peru as it is, i., 20, 21.) Live stock are fed in great numbers near Lima, large quantities of animal food being consumed in the city. The demand for poultry is immense, especially for geese and turkeys. The slaughter of pigs is supposed to exceed 20,000 a year: the trade of the pork-butcher is one of the most lucrative, after that of the baker and lottery-man. Cook-stands for fish (which are good and cheap) and fried pork, are to be found at the corner of every square. From 40 to 50 head of oxen, and from 300 to 400 sheep, are killed daily for the Lima market; the beef is very good; the mutton of inferior quality. Pastry and sweetmeat criers are seen everywhere in the streets; and mazamorras or pap-shops are very common. Pap boiled with or without fruit or vegetable acid, and sweetened with sugar or molasses, constitutes the Limeaian dish "mazamorra," which is as great a favourite in Lima as roast beef in London. Few of the dishes, however, suit the taste of strangers, from their being, with the exception of poultry, either steeped in lard, or highly seasoned with pepper. Most families in inferior circumstances provide themselves with ready cooked food from the streets. Water for drinking, which is almost wholly supplied from the large fountain in the Plaça Mayor, is carried round the city by asses and other beasts of burden, carriages of most kinds being rare. The climate of Lima has been much praised: the extremes of heat and cold are never experienced; within the city the thermometer, in the shade, never falls in winter under 66° Fahr., nor rises in summer above 89°, its usual station being about 85° to 80°-aired apartments. The ordinary daily range of temperature is only 3° or 4°. The year is divided between the dry and the moist season; the former begins in November, the latter in May; and throughout the winter (May to October) a drizzly mist often prevails in the morning and evening, and light breezes are abandoned for woollens, &c. Cool breezes from the S.W. blow for three fourths of the year; and the hot rays of the sun are generally intercepted by a layer of clouds. Earthquakes occur every year, particularly after the mists disperse, and have usually been very destructive at intervals of 50 or 60 years; but Lima is free from storms. Epidemics are few. The climate, however, or rather, perhaps, the neglect of sanatory regulations and of cleanliness, seems to have an enervating tendency, as shown in the degeneracy of most of the inhabitants, especially the whites. The rate of mortality is also very high. The average number of deaths may, it is said, be estimated, from a tolerably correct return, from 1532 to 1835 inclusive, at 2800 a year, which, in 20 years, would amount to more than its total population. (See Peru as it is, i., 23-28.) Hence the population increases little, if at all. In 1793 it was stated to be 52,627; and in 1818 it amounted, according to a census, to 64,608; 27,545 males and 26,553 females. Of this number nearly 28,000 were whites and creoles, 6000 mestizos, from 5000 to 6000 Indians, and about 20,000 negroes or blacks. The negroes are chiefly employed as domestics and mechanics; the mestizos in trade and agriculture. The physical and moral character of the white inhabitants of Lima is Andalusian. The ladies are celebrated for beauty and fineness of figure, but want freshness of complexion. They wear a very remarkable walking-dress, peculiar to this city and Truxillo. "This dress consists of two parts, one called the saya, the other the manta. The first is a petticoat made to fit so tightly that, being at the same time quite elastic, the form of the limbs is rendered distinctly visible. The manta, or cloak, is also a petticoat, but, instead of hanging about the heels, as all honest petticoats ought to do, it is drawn over the head, breast, and face; and is kept so close by the hands, which it also conceals, that no part of the body, except one eye, and sometimes only a small portion of one eye, is percepti-

ble. A rich coloured handkerchief, or a silk band and tassel, are frequently tied round the waist, and hang nearly to the ground in front." (Hall's Travels, i., 108, 109.) Within doors the ladies adopt the English or French costume, with a profusion of jewellery. The morals of both sexes have been represented as lax in a high degree, but they are probably not worse (which, however, is not saying much) than in most other large cities of S. America. The author of Three Years in the Pacific says: "Intrigues are carried on to a great extent in fashionable circles; but I think there is more virtue and morality to be met with in the second ranks." (Ib., 166.) Extravagance in living, dress, and gambling are carried on to a great extent; and smoking is universal among both men and women.

Lima was made an archbishop's see in the 16th century; and was long the grand entrepôt for the trade of all the W. coast of S. America; but a considerable part of the foreign trade of Peru is now carried on through Buenos Ayres; and the former is also in the habit of importing European goods at second hand from Valparaiso and other parts in Chili. It is still, however, the great emporium of Peru. Its exports consist principally of bullion and specie, vicuñas, and sheeps' wool, bark, chinchilla skins, salt-petre, copper, tin, sugar, &c. The imports are principally woollen and cotton stuffs, cutlery, and hardware from England; silks, brandy, and wines from Spain and France; stock fish from the U. States; snuff, indigo, tar, naphtha, &c., from Mexico; tobacco from Colombia, with timber for the construction of ships and houses from Guayaquil; wheat, flour, dried fruits, and bullion from Chili; Paraguay tea from Paraguay, spices, quick-silver, perfumery, &c. (For full details as to these matters, see Peru.) The manufacturing industry of Lima is but insignificant. It has some smelting-houses, which, in 1834, produced 15,891 marcs of silver, but for several previous years double that quantity had been reduced to bars. It has also a mint, at which, from 1786 to 1820, 2,557,914 marcs of silver were coined. (Moyen, Reise, &c.; Rev. &c., Tables, 1839; Com. Dict.)

About three leagues to the S.E. is the favourite watering-place Chorrillos, resorted to by people of rank and fashion for several months in the summer, and by invalids during the winter. "It is only a small fishing village, constructed of cane and mud. The Indian owners of the houses let them to the bathers at a high rate during the bathing season; and some persons either take them for a term of years, or construct other light summer-houses for themselves. Chorrillos is sheltered from the S.W. blast by an elevated promontory, called the Moro-Solar." Numerous Peruvian antiquities lie scattered over the rich, but now partly waste and desolate, plain between this town and Lima.

Lima was founded by Pizarro in 1535, under the title of Ciudad de los Reyes, "City of Kings." It suffered severely from the earthquakes of 1678 and 1746, the latter leaving only 20 houses standing out of 3080; and again by those of 1784, 1806, and 1828. San Martin entered it on the 19th July, 1821, and proclaimed the independence of Peru at Lima on the 28th of the same month. (Ulloa, Voyage de l'Amérique, i., 429-466, gives by far the most complete and authentic account of Lima previously to the great earthquake of 1746. See also Three Years in the Pacific, 1835, i., 344-403; i., 1-309; Mayen Reise um die Erde, ii., 56-64; Poeppig; Scarlet's Trav.; Stevenson's S. America, i., 143-236; Caldcleugh, &c., in Mod. Trav., vol. xxviii.; Hall's Trav. in S. America; Encyc. Americana, &c.)

LIMA, p. t., Livingston co., N. Y., 213 m. W. Albany, 309 W. Bounded E. by the outlet of Hemlock and Honeoye lakes, and drained by a tributary of R. It contains a Presbyterian and a Methodist church, the Genesee Wesleyan seminary, well endowed and flourishing. seven stores, one furnace, one fulling mill, two flouring-mills, two grist-mills, three saw-mills; one academy, 350 students; nine schools, 250 scholars. Pop. 2176.

LIMA, p. v., capital of La Grange co., Ia., 175 m. N. by E. Indianapolis, 578 W. Situated on the N. bank of Pigeon r., and contains a courthouse, jail, two churches, a Methodist and Presbyterian, six stores, and about 300 inhabitants. The town has nine stores, one grist-mill, three saw-mills; one college, 90 students, one academy, 80 students, three schools, 75 scholars. Pop. 584.

LIMEHOUSE. See London.

LIMERICK, an inland co. of Ireland, prov. Munster, having N. the estuary of the Shannon, by which it is separated from Clare, E. Tipperary, S. Cork, and W. Kerry. Area, 674,783 acres, of which 91,981 are unimproved mountain and bog. Except on the S. W., and N.E. extremities, the surface is generally flat. Climate mild, but very humid. Soil for the most part excellent, and applicable to every purpose of husbandry. Mr. Young describes a large tract, called the "Golden Vale," as the best land he had ever seen. Tillage has of late years been much extended in this county, but is, speaking generally, very backward: still, however, there is a good deal of grass land, and grazing husbandry and the dairy are both extensively pursued. Some fine long-horned cattle are bred and fattened in this county, particularly in the low grounds along the Shannon. The pernicious system of *con acre* (see *anté*, p. 40) has, unfortunately, been widely spread in Limerick; and, though there has been a great increase in the exportation of wheat, wheat flour, oats, butter, and most other articles from the county, and a material improvement in stock, and in the implements of husbandry, it is believed that the condition of the cottiers, and the smaller class of occupiers, has been but little, if at all, improved. The latter, in fact, are in the most abject state; and it would seem that in Ireland, as in Italy and some other countries, the peasantry are frequently most wretched where the land is finest. Property in very large estates: tillage farms mostly very small, but some of the grazing farms are extensive. Average rent of land, 18s. 8d. an acre. Minerals and manufactures, excepting some departments of the latter, carried on in the city of Limerick, of no importance. Principal rivers, exclusive of the Shannon, Maig, Deale, and Mulkern. Limerick is divided into nine baronies and 125 parishes, and sends four members to the H. of C., viz., two for the county and two for the city of Limerick. Registered electors for the county, in 1839-40, 2708. In 1831 this county had 44,801 inhabited houses, 52,847 families, and 315,355 inhabitants, of whom 153,625 were males and 161,730 females.

LIMERICK, a city, parl. bor., river-port, and co. of a city in Ireland, prov. Munster, on the Shannon, 107 m. S.W. Dublin, and 55 m. E. Loophead at the mouth of the Shannon; lat. 52° 40' N., long. 8° 35' W. It is principally situated on the S.E. side of the river, within the county of Limerick, but partly also on its N. side, within the county Clare. The county of the city, which is identical with the parliamentary borough, includes an area of about 27,000 imperial acres; and had, in 1831, a population of 66,775, whereof the city and suburbs had 44,100, and the rural districts, or liberties, 22,675. Limerick is the fourth city of Ireland in respect of size and importance. It owes this distinction to its situation at the head of the estuary of the Shannon, which has made it the emporium of the extensive and fertile districts watered by that great river. It is divided into—1. the English town, now the oldest and most decayed portion, on King's island, formed by a detached arm of the Shannon; 2. Irishtown, immediately S. of the above; and, 3. the New town, to the W. of the latter, and called Newtown-Pery, from Pery, the family name of the earl of Limerick, on whose estate it is built. Popularly the first two divisions are called the Old, and the latter the New town. The country part of the city of the county, including Thomond bridge on the W. side of the river, and many other extensive lines of cottages, is called the Liberties. The contrast between the different parts of the city is very striking. The Old town is said, in the Municipal Boundary Report, to be "one vast mass of filth, dilapidation, and misery, which nothing but the general employment of the people throughout the country can correct, because the unemployed poor are attached to the large' crumbling city, where they can find, at a cheap rate, something like a roof to cover them." The New town, on the other hand, which has been wholly constructed within the last half century, is but little inferior to the best parts of Liverpool. It is well built, and the streets, which are broad and straight, cross each other at right angles. It has a handsome square, in which is a column surmounted by a statue of Mr. Spring Rice, now Lord Monteagle, to whom the city is much indebted. The houses in the liberties are mostly mere cabins, occupied by a very poor agricultural population. The main arm of the Shannon is crossed by two bridges, one of which, Thomond bridge, originally constructed in the early part of the 13th century, is now being rebuilt; the other, or Wellesley bridge, of five arches, each 70 feet in span, a very handsome structure, was completed in 1827. There are three bridges over the smaller arm of the Shannon, between English-town and Irish-town.

The county of the city has 13 parishes, and eight parts of parishes, besides an extra parochial district. Six of the parishes being within the city properly so called, which is also the seat of the see of Limerick. The cathedral, a large Gothic pile, has a lofty tower, a handsome interior, and many monuments, among which is that of Donogh O'Brien, king of Thomond. The embattled tower of this cathedral, 120 feet in height, commands a fine view of the city and adjacent country, including the course of the Shannon. None of the parochial churches seem to be worth notice, except St. Minichin's church, formerly the most ancient in the kingdom, but lately rebuilt, which, though small, is for situation and architecture by far the handsomest sacred edifice in the city. According to the Roman Catholic divisions, the city consists of five parishes, that of St. John being the bishop's mensal, and its church is considered the cathedral. The other places of worship are large, and some of elegant structure. There are friaries of the

Augustine. Dominican, and Franciscan orders, all of which have large chapels attached to them. Nearly 9-10ths of the inhabitants are Catholics. The Presbyterians, Quakers, Wesleyan and primitive Methodists, and Independents, have each a place of worship. The handsomest public building in Limerick is the new savings' bank, built this year (1840). It is a Doric structure, of cut limestone. The savings' bank was established Jan. 1820, and had, Nov. 20, 1840, £115,708 6s. 7d. deposits. The diocesan school for the dioceses of Limerick, Killaloe, and Kilfenora, is kept in the head-master's house. The literary and scientific institutions are the Limerick Institution and the Mechanics' Institute, and the Literary and Scientific Society. Those for charitable purposes connected with education are the Blue-coat school, founded in 1717; and free schools attached to the parishes and friaries, in which great numbers of children are instructed. No fewer than 1900 children are educated in the schools of the "Christian Brothers," to whom the city is much indebted. There are also schools founded on bequests of Mrs. Villiers and Dr. Hall. In the Old town is a school for females, conducted by the "Sisters of Mercy," assisted by a small grant from the Education Board: a Sunday school is also established in it, for the females employed during the week at the lace factories. In the New Town is a large female school, conducted by the nuns of the Presentation Convent. These three schools are attended by about 800 children, who, according to the statement of Mr. Inglis, are "able to write well, are perfectly instructed in reading, and exhibit in their appearance and behaviour the utmost order and neatness." In another school for females, 120 poor children are educated and clothed. The other charitable institutions are the County Hospital; Barrington's Hospital and City of Limerick Infirmary, a large building, containing 125 beds, built and munificently presented to the city by Sir Joseph Barrington and his four sons; the Lunatic Asylum for Limerick, Clare, and Kerry, opened in 1822, cost about £30,000; it has accommodation for about 200 patients. Its inmates in 1837 were 144 males and 143 females, maintained at a total expense of £4685, or at £15 11s. each; it is said to be very well conducted. Here is also a Fever and Lock Hospital; a Lying-in Hospital; a House of Industry, which accommodates 450 inmates, who contribute to their maintenance by their work; several endowed almshouses; a Magdalen Asylum; an Asylum for the Blind; a mendicity institution; a charitable loan fund; and a charitable pawn-office, founded by Matt. Barrington, Esq., on the plan of the *Mont-de-Piété* at Paris. Places of public amusement are not much encouraged. The theatre, a handsome building, was sold to the Augustine monks, and has been fitted up by them for a chapel, and a suite of assembly-rooms is applied to other purposes; but there is a small "circus" temporarily erected, occasionally used as a theatre. The Hanging Gardens, built by Mr. Roche, are formed of tiers of terraces, raised upon arches, on the uppermost of which is a range of green-houses, commanding a fine view of the city, river, and adjacent country. Limerick is the head-quarters of the S.W. military district, and has barracks for 1450 men. Three newspapers are published in the town, each twice a week, so that a paper issues daily from the press.

The corporation, which lays claim to prescriptive privileges, confirmed by a series of charters from the time of King John, consists of a mayor, two sheriffs, and an indefinite number of aldermen, burgesses, and freemen. The mayor has a sword of state and cap of maintenance carried before him on public occasions; he is also admiral of the river, with jurisdiction extending from 3 m. above the city to the open sea. The aldermen are elected for life; the freedom of the corporation is obtained by birth, marriage, or apprenticeship. The municipal affairs are chiefly transacted in an assembly of the freemen, called the Court of d'Oyer Hundred, revived by an act passed in 1823. Under the Irish Municipal Reform Act, the city will be governed by a mayor and eight aldermen. The city sent two members to the Irish House of Commons; and from the Union to the passing of the Reform Act, it sent one member to the Imperial House of Commons. The last-mentioned act conferred on it the privilege of sending two members to the Imperial House of Commons. Registered electors, in 1836-39, 2052. The system of local taxation is said, in the Municipal Boundary Report, to be exceedingly unfair; and to press with unjust and extreme severity on the agriculturists in the liberty.

Assizes are held twice a year for the county of the city, by the judges of assize; courts of general sessions every quarter, and petty sessions every week, at which the mayor and city magistrates preside. A court of civil jurisdiction, which is empowered to hold pleas to any amount, is held on Wednesdays; and a Court of Conscience for suits under 40s. every Thursday. Civil bill cases are tried before the assistant barrister of the county, who holds a court for this purpose within the city twice a year. The income of the corporation amounts to between £4000 and £5000 per annum. The city courthouse is a plain building : the county courthouse, a handsome structure, was erected in 1810, at an expense of £12,000. The prisons for the county and city are within the municipal limits. The former, erected in 1821, at an expense of £25,000, has a Doric portico, and is, perhaps, the finest building in the city : it is constructed on the radiating plan, and is said to be extremely well managed. The city jail, a gloomy quadrangular edifice, is comparatively ill adapted for its purpose.

There are three institutions for the promotion of manufactures and trade ; the Chamber of Commerce, a society of merchants, incorporated by royal charter in 1815, for the promotion and protection of trade, and who have expended considerable sums of money for these objects ; the Agricultural Association ; and the trustees for the Promotion of Industry, in whom a fund of £7000 was vested by the London Distress Committee. Limerick, however, can hardly be said to have any manufactures. The linen manufacture, which had attained to some magnitude, and that of cotton, which had been introduced, are all but extinct. The manufacture of leather gloves, for which the city was once famous, has not entirely disappeared ; though gloves, sold under the name of "Limerick gloves," are now not unfrequently made in Cork. There is a great deal of embroidering in lace ; and three lace factories now at work give employment to from 1200 to 1400 females. Several large flour mills have also been erected ; and near the town is a large distillery, and several breweries ; but, owing to the influence of " Father Matthew," most of the breweries have ceased working, and the distillery is comparatively idle ; one paper-mill still exists, and that is all ; and two iron foundries. Limerick has for many years been famed for its fishing-hooks, sent to all parts of the United Kingdom and America.

The great support of Limerick is her trade, which is very extensive. She is, as already stated, the great mart for the country traversed by the Shannon, and that immediately connected with it. Her exports, like those of most Irish towns, consist mostly of corn and provisions, including beef, pork, butter, &c. ; the value of these articles having amounted, in 1835, when prices were very low, to £717,600 out of a total export of £726,430 ! At present (1840) the value of the exports from Limerick is probably not under £1,900,000. The imports consist principally of manufactured goods, coal and iron, tea, sugar, tobacco, wine, salt fish, timber, &c. The Bank of Ireland, and the provincial and other banks, have branches here. The gross customs' duties amounted, in 1830, to £84,782, in 1835, to £146,843, and in 1839, to £148,782. The postoffice revenue, in 1830, was £5653 ; in 1838, £7903.

This city, and, indeed, the whole kingdom, has derived great advantages from the improvements made in the navigation of the Shannon, and the steamers introduced on that river ; and it will derive still farther advantages from the improvements that are projected in respect to it. The estuary of the Shannon forms one of the finest bays in the world : vessels of very large burden approach within a few miles of the city ; and ships of 400 or 500 tons unload at its quays. But as the strand along the latter dries at low water, and as its bottom is hard, vessels of considerable burden have been seriously injured on their grounding. To obviate this inconvenience, it has been proposed to construct a weir or dam across the river, a little below the town, which would give a constant depth of from 16 to 18 feet water for a considerable distance upward. A loan of above £55,000, guaranteed by the harbour dues, has been advanced by government for this purpose, and for the construction of quays ; and a farther loan is about to be given to complete these necessary improvements. In connection with the trade of the port may be mentioned the commercial buildings, erected in 1836, by a company of shareholders, with apartments for the chamber of commerce, a library, &c. ; the custom-house, and the exchange. There belong at present (1840) to Limerick 65 vessels, of the aggregate burden of 13,000 tons, besides 42 smaller craft, 1300 tons register.

There can be no doubt that the trade, wealth, and population of Limerick, are rapidly increasing, but at the same time there is a vast deal of misery in it ; and we regret to have to state that a large proportion of the lower classes, especially in the old town, are all but wholly destitute, and are exposed to extreme and almost incredible privations. That so much squalid poverty and abject misery should exist along with so much wealth and comfort, is a painful and a mortifying anomaly. We should think it well worth public consideration to inquire whether some national effort should not be made to relieve this and some other Irish cities of a portion of their pauper inhabitants, by sending them to the colonies ; and whether measures should not be taken to prevent a recurrence of the evil, by preventing the building of any very inferior houses. The present state of the poor in Limerick is disgraceful to the country, and discreditable to civilization.

Limerick was formerly fortified, and, from its commanding the first bridge above the embouchure of the Shannon, was an important military station. It was occupied, after the battle of Aghrim, by the troops of James II. : it capitulated to the English army under Ginkell, afterward Earl of Athlone, on the 3d of October, 1691. The capitulation, or, as it has been usually called, the treaty of Limerick, was very favourable to the besieged, and, indeed, to the Irish nation, or, at all events, to the Catholics. But it was afterward most shamefully violated by the conquering party, and its most important stipulations were openly set aside and trampled upon. The remains of its fortifications add considerable beauty and interest to this ancient city. "King John's Castle, from which the city arms are taken, forms part of the castle barracks, and the stone upon which the capitulation was signed is still in existence, and is regarded with watchful care by the citizens." (*Irish Boundary and Municipal Reports ; Railway Report ; Inglis's Ireland*, i., 294–312 ; and *Private Information*.)

LIMERICK. p. t., York co., Me., 76 m. S.W. Augusta, 535 W. Chartered in 1787. Watered by Little Ossipee river. It contains nine stores, three grist-mills, four saw-mills, one printing-office, one weekly newspaper, one academy, 139 students ; 613 scholars in schools. Pop. 1508.

LIMERICK, p. t., Montgomery co., Pa., 78 m. E. Harrisburg, 167 W. Bounded S.E. by Schuylkill river. It has two stores, one lumber-yard, one tannery. Pop. 1786.

LIMESTONE, co., Ala., situated in the N. part of the state, and contains 575 sq. m. Bounded S. by Tennessee river. Watered by Elk river and its branches, and Swan, Piney, and Limestone creeks. It contained in 1840, 10,746 neat cattle, 9050 sheep, 43,891 swine ; and produced 39,588 bushels of wheat, 2373 of rye, 980,240 of Indian corn, 113,615 of oats, 39,560 of potatoes, 13,512 pounds of tobacco, 5,772,946 of cotton. It had 11 stores, 20 grist-mills, 10 saw-mills, three tanneries, nine distilleries, one printing-office, one weekly newspaper, five academies, 155 students, 23 schools, 683 scholars. Pop. : whites, 7498 ; slaves, 6840 ; free coloured, 36 ; total, 14,374. Capital, Athens.

LIMESTONE, t., Lycoming co., Pa. It has two stores, one woollen-factory, one grist-mill, six saw-mills ; four schools, 125 scholars. Pop. 809.

LIMINGTON, p. t., York co., Me., 70 m. S.W. Augusta, 541 W. Bounded N.E. by Saco river, S.E. by Little Ossipee river. Incorporated in 1792. It contains seven stores, one fulling-mill, one grist-mill, one saw-mill ; 18 schools, 347 scholars. Pop. 2210.

LIMOGES (an. *Lemovicum*), a city of France, dep. Haute Vienne, of which it is the capital, on the declivity of a hill, at the foot of which is the Vienne, which is here crossed by an old stone bridge of six arches, 110 m. N.E. Bordeaux, and 215 m. S.S.W. Paris : lat. 45° 49' 53" N., long. 1° 15' 23" W. Pop. in 1836, ar. can., 23,963. It is divided into the "city" and the "town." The former occupies the site of the ancient Celtic and Roman city near the river, and is ill-built ; its streets being narrow and ill-paved, and its houses built of wood above the ground floor. The latter division, which is of modern date, on the upper part of the hill, is open, well built, surrounded with pleasant promenades, and particularly healthy. The cathedral, built in the 13th century, is of granite, and in the Gothic style. It has an imposing appearance at the first glance ; but when examined in detail, it presents many incongruities : among others, one end of the choir has base-reliefs, representing the labours of Hercules! Another church, which stands in the highest part of the town, has an elegant steeple, 296 ft. in height, a conspicuous object at a great distance. The bishop's palace is a very handsome modern granite building, surrounded with gardens. Limoges has a good town-hall, several hospitals, an exchange, mint, theatre, prison, cavalry barracks, public baths, and many public fountains. One of the latter, the *Fontaine d'Aigouleno*, has a basin 38 ft. in circumference, supposed to be hewn out of a single piece of granite, and supplies the upper part of the town at every season with abundance of good water, derived from an ancient subterraneous aqueduct. Besides this aqueduct, few Roman antiquities are found in Limoges : the site of an amphitheatre, the traces of which existed in 1823, is now occupied by the *Place d'Orsay*. Limoges is the seat of a prefecture and royal court, tribunals of primary jurisdiction and commerce, and a chamber of manufactures ; it has a royal college, university academy, diocesan seminary, royal societies of agriculture, arts, and sciences, schools of drawing and commerce, a public library with 12,000 volumes, museums of natural history, antiquities, the fine arts, and machinery, a *mont-de-piété*, and many charities. Its manufactures, which are important, include glass and earthenware, broadcloths, cassimeres, druggets, and other woollen fabrics, callicoes, cotton, linen, and hempen yarn, hats, wax candles, cards, paper, glue, &c. It has numerous tanneries, cotton and woollen dyeing-houses, and iron forges: its wax-bleaching factories rival those of Mans, and its brandy and liqueurs are in high repute. Being situated at the junction of several great roads, it is an entrepôt for the trade of several departments, with Thoulouse and the S. of France, and deals extensively, not only in its own manufactured goods, but in agricultural produce, salt, iron, copper, and brass wares, &c. Marshal Jourdan, and Dupuytren, the famous surgeon, were natives of Limoges. (*Hugo,* art. *Haute Vienne; Guide du Voyageur, &c.*)

LIMOUSIN, one of the old provs. of France, in the central part of the kingdom, now distributed among the departments of Corrèze, Creuse, Haute Vienne, and Dordogne.

LIMOUX, a town of France, dep. Aude, cap. arrond., on the Aude, 13 m. S.S.E. Carcassonne. Pop., in 1836, *ex. com.,* 6666. It is generally well built, paved, and lighted, and has a large parish church, a hospital, two public halls, a theatre, and a small picture gallery. It is the seat of tribunals of primary jurisdiction and commerce, a chamber of manufactures, &c. Its woollen manufactures produce annually from 11,600 to 12,000 pieces of broadcloth, worth from 6 to 17 francs an ell; it has also several woollen yarn factories, tanneries, and oil-mills; and is an entrepôt for lyon goods, in which, and in wines, oil, soap, and leather, it has an active trade. Its environs are highly picturesque and fertile. (*Hugo,* art. *Aude, &c.*)

LINCOLN, a marit. co. of England, on the E. coast, having N. the Humber, E. the German ocean, S. the counties of Cambridge, Northampton, and Rutland, and W. Leicester, Nottingham, and York. This is a very extensive county, comprising 1,671,040 acres, of which about 1,465,000 are said to be arable, meadow, and pasture. Though but little diversified in respect of surface, Lincoln is naturally divided into the districts of the *Wolds,* the *Moors,* and the *Fens.* The Wolds, a ridge from 8 to 10 m. in breadth, extend from Spilsby N. to Barton on the Humber; the soil is principally sandy loam on a chalk bottom, of very various degrees of fertility, but now much improved, and generally producing very excellent crops. The Moors stretch N. and S., from the Humber to Grantham; the heath by which they were formerly covered has now mostly disappeared, and they are now very productive of oats. The Fens comprise all the flat parts of the county, from Wainfleet on the Wash round by the mouth of the Nen to the borders of Rutland; most part of this district is usually included within the great level of the Fens. (*See* BEDFORD LEVEL, vol. i., 309.) Lincoln is one of the most productive counties in the empire; and improvements of all sorts have been prosecuted in it for many years past with extraordinary spirit and success. Large tracts in the Wolds and Moors, that 30 or 40 years ago were all but unproductive, now yield heavy crops of barley, oats, and turnips. This great improvement has been chiefly brought about by the liberal use of bone manure, which has been applied for a longer period, and on a more extensive scale in this than in any other county.

The excellence of the pasture in the Fens is too well known to require any especial notice: immense sums have been expended on their drainage, and in the recovering of land from the sea. Formerly the Fens were frequented by vast numbers of aquatic fowl; but since their drainage these have much fallen off, and the *decoys* for their capture are now of much less importance: geese, however, are still bred in considerable numbers, and are regularly plucked four or five times a year for their feathers. Previously to the improvement of the Wolds, rabbit warrens were very common, but they are now comparatively scarce. The native sheep of the Fens were remarkable for their size and the extraordinary length of their wool; they have, however, been so much crossed with New Leicesters, that it is now difficult to find one of the genuine breed. Some of the finest of the London dray-horses are bred in the Fens. The cattle depastured in the Fens are principally short-horns, and attain to a great size.

Property very variously divided, there being estates of all sizes, from £25,000 a year down to £5, but the great majority small. In the district called the Isle of Axcholme, in the N.W. part of the county, between the rivers Trent and Ancholme, the inhabitants live together in hamlets or villages as in France, and almost every householder is a proprietor, their properties varying from 1 to 50 acres. They are very industrious, and live very meanly. Size of farms various: in the Wolds and Moorish district they are mostly very large, but elsewhere they are rather small. They are generally held under leases of 7 and 14 years. Mr. Young has passed a high but well-merited eulogy on the enterprise and intelligence of the higher class of farmers in this county. (*Survey,* p. 48.) Average rent of land in 1810, 18*s.* 11*d.* an acre; but now (1840) probably 21*s.* Manufactures and minerals of no importance. The custom of *Borough English,* by which, if the father die intestate, the younger son succeeds to the paternal property, to the seclusion of his

184

elder brothers, prevails at Stamford, in this county. Principal rivers, Trent, Witham, Welland, and Ancholme. The Witham has been made navigable from Boston to Lincoln; and the Fosdyke canal extends from the latter to the Trent, near Torksey, completing an internal navigation between the Wash and the Humber. The county is popularly divided into the parts of *Lindsey* on the N., *Kesteven* on the S.W., and *Holland* on the S.E., and contains 33 hundreds, wapentakes, and sokes, with the city of Lincoln and the boroughs of Stamford, Boston, and Grantham. It is divided into 699 parishes, and sends 13 members to the House of Commons, viz., four for the county, two for the city of Lincoln, two each for the boroughs of Boston, Grantham, and Stamford, and one for Great Grimsby. Registered electors for the county, in 1839–40, 18,876, whereof 8729 are for the S., or the Holland, and 10,147 for the N., or the Lindsey, division. In 1841 it had 73,038 inhabited houses, and 362,717 inhabitants, of whom 181,802 were males, and 180,915 females. Sum paid for the relief of the poor, 1838–39, £63,115. Annual value of real property, in 1815, £2,096,611; profits of trades and professions in do., £373,672.

LINCOLN, a city, parl. and mun. bor., and market town of England, capital of the above county, on the Witham, 36 m. S. Hull, and 131 m. N. London; lat. 53° 94′ N., long. 0° 30′ W. The parliamentary borough (which the Boundary Act left unchanged externally, the only additions being the bail and close in its centre) is divided into 16 parishes, and had, in 1841, a population of 99,775. It is situated on the S. slope and at the foot of a hill, on the top of which is the cathedral, a striking object for many miles round. The streets are irregularly laid out; the largest and handsomest runs N. and S. up the hill on which the cathedral stands. A small part of the town, comprising two parishes, is on the S. side of the river, and is connected with the main body by one principal and two smaller bridges. The streets are well paved, lighted with gas, and supplied with water from public conduits. The principal and most interesting public building is the cathedral, erected at different times, from the 12th to the 15th century, and consequently exhibiting several varieties of architecture: the prevailing style, however, is early English, of a particularly rich and beautiful character. The closeness of the surrounding buildings is a great disadvantage to the display of architectural effect; but there is a tolerably open space towards the E. The church consists of a nave with its aisles, four transepts, a choir, chancel, and lady-chapel: three towers rise above the building, two at the W. end, 180 feet high, and one at the intersection of the nave and transepts, rising 203 feet above the floor: they are all gorgeously decorated with varied tracery, pillars, pilasters, windows, &c. The W. and principal front, in which are three fine doors, is distinguished by its beauty and magnificence; and, from the variety of its styles, is certainly the workmanship of three distinct and distant eras. According to Britton, the interior dimensions are as follow: Entire length, 482 feet, of which 223 feet belong to the nave, and the rest to the choir and lady-chapel; width of choir and nave, 80 feet; height of ditto, 80 feet; and width of W. front 174 feet; length of principal transepts, 233 feet; breadth of ditto, 66 feet. The great transepts, at the ends of which are circular windows, exhibit a good specimen of the English style; and the aisles on the E. side are divided into chapels and vestries: the choir, which is parted off from the nave by a stone screen, is of most elaborate composition; and the great E. window ranks as the second in England, in point of size and fine proportions. Attached to the E. side of the cathedral is the chapter-house, a structure differing from most others of the same nature in being *ten-sided,* and not octagonal: its groined roof is supported by an umbilical pillar, consisting of a circular shaft, cased by ten small fluted columns. The cloisters are on the N. side, and over them is the library, built by Dean Honeywood, at the end of the 17th century, containing a large collection of books, with some curious specimens of Roman antiquities. The cathedral bell, or "Great Tom of Lincoln," originally cast in 1610, having been cracked, was recast, with six other bells, into the present large bell and two quarter bells. The diameter of the great bell is 6 feet 16½ inches; and its weight 5⅓ tons, or about a ton heavier than the old one. At the time of the dissolution, Lincoln cathedral was one of the finest and most sumptuously adorned in the kingdom. There were then taken from it 2621 oz. gold, and 4285 oz. silver plate, besides precious stones of great value. It had formerly many costly sepulchres and monumental records; but the zealots at the Reformation either pulled them down or defaced them, so that, in 1549, scarcely a perfect tomb remained; and the little they left undestroyed was demolished by Cromwell's soldiers, by whom the cathedral was converted into barracks. The ruins of the bishop's palace, which was destroyed at the last mentioned epoch, stand S. of the church, and comprise a fine hall, an old gateway, and part of the kitchen. Adjoining these ruins, a modern house has been erected, which is oc-

LINCOLN.

copied by the bishop during his stay in Lincoln. (See *Britton's Account of Lincoln Cathedral.*)

Among the tombs yet in a tolerably perfect state are those of Catherine Swinford, wife of John Gaunt, duke of Lancaster, of their daughter Joan, and of several bishops and deans of the cathedral. The present establishment comprises a bishop, dean, precentor, subdean, chancellor, and 36 prebendaries, with four vicars-choral, and 20 choristers. The nett revenue of the see amounted, at an average of the three years ending with 1831, to £4549; but, on the next avoidance of the see, the income will be raised to £5000, with a farther allowance of £500, till a suitable residence be built. the limits of the diocese are hereafter to comprise only the counties of Lincoln and Nottingham. (*Orders in Council, Aug. 1838.*) The cathedral revenues, which nett £6926 a year, are equally divided between the dean, precentor, subdean, and chancellor; and the vicars-choral divide £115 yearly. Besides monasteries, nunneries, and other buildings devoted to pious uses, Lincoln had formerly upwards of 50 churches; but of these only 11 remain, exclusive of the cathedral, most of them being small and much mutilated. St. Peter at Gowts, evidently an old conventual church, and three other churches S. of the Witham, have lofty square Norman towers. An additional church is in course of being built by subscription. There are several places of worship for Roman Catholics, Wesleyan Methodists, and other Dissenters; and attached to them are Sunday schools, which, according to the parliamentary returns, were attended, in 1833, by 700 children. The national school (also a Sunday school) had 474 scholars in the same year; and there were two infant schools with 330 children. The grammar school, still held in the Grey Friars' chapel, was founded in 1583: it is well endowed, and the instruction, not confined to classics, is given by an upper and under master, who, in addition to their salaries from the corporation, receive fees from the boys, and take boarders. About 40 boys are stated to have been in attendance in 1837. The Bluecoat school, established in 1602, is endowed with landed property worth nearly £2000 a year, and furnishes clothing and instruction (with apprentice-premiums on leaving) to 56 boys. The master has £94 a year, with a house, coal, and candle, and the establishment is said to be well conducted. Wilkinson's school, which is very slenderly endowed, furnishes instruction to 16 boys. Lincoln is very rich in endowed charities, among which, as one of the principal and most useful, may be mentioned Sir Thomas White's loan-fund, for deserving and needy tradesmen, the assets of which are estimated at £850. (*Char. Comm. Report,* part iv.) A general dispensary, lunatic asylum, county hospital, and lying-in hospital, are the chief modern charities; and a flourishing mechanics' institute, several libraries, two news-rooms, and some book societies, are well supported. It has also three weekly newspapers. Among the buildings devoted to civil purposes are the county jail and courthouse, lately rebuilt from Smirke's designs, on the site of the old castle, a few remains of which are still standing on the hill W. of the cathedral. The county jail, constructed on Howard's plan, is well conducted. The Guildhall is an ancient Gothic building; but the borough courthouse is modern, and the jail is stated to be too small to admit of the classification of prisoners. The market house, a small theatre, and an assembly-room, are the only other public buildings; but there are several markets. W. of the town is a good race-course, near which is a large building, now dismantled, but used during the late war as a military dépôt. As respects ancient remains, few towns in England exhibit so many and so interesting as Lincoln. Saxon, Norman, and pointed arches, doorways with turrets, walls, mullions of windows, and other fragments, appear in every direction. Leland says there were "five gates in the wauls of the citie;" and of these the Chequer Gate in the Close, and the Stonebow crossing the High-street, are still in good preservation. John of Gaunt's palace and stables present some good examples of the Norman and early English style.

The trade of Lincoln consists chiefly in the exchange of the raw produce of the surrounding district for manufactured and other commodities. Large quantities of flour are sent to Manchester and London. There are some tanyards, malthouses, and tobacco manufactories, and extensive breweries produce excellent ale. It communicates by the Fossdyke canal with the Trent and its tributaries and canals; and the Witham navigation, running S.E. past Boston, connects it with the North sea. The Lincoln and Lindsey banking company and a private bank are established here; and there is a savings' bank.

Lincoln received its first charter from Henry II., which was confirmed by several subsequent monarchs, its governing charter till 1837 being that of Charles I. Under the Municipal Reform Act the city is divided into two wards, and is governed by six aldermen (one of whom is mayor) and 18 councillors. It has also a commission of the peace under a recorder. The assizes and quarter sessions are held for the city and county, and there is a court for the recovery of small debts. Corporation revenue about £3000.

Lincoln has regularly sent two members to the House of Commons since the reign of Henry III., the electors previously to the Reform Act being the freemen of the city. The Boundary Act includes the insulated part called the bail and close in the parliamentary borough; and those districts round the city called the liberties, which previously had not been represented, were added to the county. Registered electors in 1839-40, 1041. Lincoln is likewise the election town of the N. division of the county.

Lincoln stands on the line of the great Roman road called *Ermine-street*; and derives its name from its occupying the site of the Roman military station called *Lindum*. It was fortified by the Saxons; and at the time of the Domesday survey was one of the richest and most populous cities of England, having 1070 houses, and an extensive trade. The castle was built by William the Conqueror, in 1086; and the prosperity of the town was farther advanced by Henry I., who cleared out the foss-dyke, and made it navigable. The town was annexed to the duchy of Lancaster at the end of the 13th century; and about the middle of the 14th century it was inhabited by the celebrated John of Gaunt, duke of Lancaster, who not only improved the castle, but procured for the town many valuable privileges. In the civil wars of Charles I. the king came to Lincoln, and convened the nobility and freeholders of the county. The inhabitants promised to support the royal cause; but in 1643 the city was in the hands of the parliamentarians. The royalists recaptured it; but were again dispossessed, both of the town and castle, on the 5th of May, 1644.

LINCOLN, county, Me. Situated in the S. part of the state, and contains 950 sq. m. Bounded S. by the Atlantic. Watered by Kennebec, Sheepscot, Damariscotta, Muscongus, and St. George's rivers. It has many fine harbours. It contained in 1840, 30,904 neat cattle, 71,473 sheep, 15,005 swine; and produced 97,813 bushels of wheat, 16,534 of rye, 104,366 of Indian corn, 62,965 of barley, 62,290 of oats, 1.14 ,407 of potatoes. It had 14 commercial houses in foreign trade, 385 stores, 13 lumber-yards, capital invested in the fisheries, $187,906, three furnaces, 27 fulling-mills, two flouring-mills, 66 grist-mills, 178 saw-mills, one paper-mill, 47 tanneries, two potteries, four printing-offices, one bindery, four weekly newspapers; one college, 17 students; 12 academies, 396 students; 396 schools, 18,986 scholars. Pop. 63,517. Capitals, Wiscasset, Topsham, and Warren.

LINCOLN, county, N. C. Situated toward the W. part of the state, and contains 1200 sq. m. Bounded E. by Catawba r. Watered by Little or South Catawba r., and its branches, and by Buffalo cr., which flows into Broad r. in S. C. It contained in 1840, 26,372 neat cattle, 18,699 sheep, 45,636 swine; and produced 185,373 bushels of wheat, 8810 of rye, 787,225 of Indian corn, 84,394 of oats, 42,860 of potatoes, 25,000 pounds of tobacco, 1,479,306 of cotton. It had 30 stores, one fulling-mill, one cotton factory with 1264 spindles, 36 flouring-mills, 90 grist-mills, 75 saw-mills, seven oil-mills, one paper-mill, 20 tanneries, 356 distilleries, two printing-offices, two weekly newspapers; five academies, 160 students; 80 schools, 1925 scholars. Pop.; whites, 19,636; slaves, 5386; free coloured, 116; total, 25,160. Capital, Lincolnton.

LINCOLN, county, Ga. Situated toward the N.E. part of the state, and contains 290 sq. m. Bounded N.E. by Savannah r., S. by Little r., N. by Broad r. Watered by Fishing and Soap creeks. It contained in 1840, 4089 neat cattle, 2130 sheep, 19,951 swine; and produced 14,148 bushels of wheat, 97,167 of Indian corn, 22,484 of oats, 3633 of potatoes, 3,906,799 pounds of cotton. It had seven stores; three academies, 125 students, 12 schools, 362 scholars. Pop.; whites, 2527; slaves, 3329; free coloured, 20; total, 5886. Capital, Lincolnton.

LINCOLN, county, Tenn. Situated in the S. part of the state, and contains 650 sq. m. Watered by Elk r. and its tributaries. It contained in 1840, 20,718 neat cattle, 32,039 sheep, 98,665 swine; and produced 104,557 bushels of wheat, 8805 of rye, 1,436,575 of Indian corn, 239,808 of oats, 29,029 of potatoes, 24,162 pounds of tobacco, 464,518 of cotton, 19,361 of sugar. It had 21 stores, one cotton factory with 196 spindles, 12 flouring-mills, 40 grist-mills, 52 saw-mills, 14 tanneries, 87 distilleries, one printing-office, one weekly newspaper; four academies, 135 students; 41 schools, 1944 scholars. Pop.: whites, 17,217; slaves, 4231; free coloured, 55; total, 21,493. Capital, Fayetteville.

LINCOLN, county, Ky. Situated a little E. of the centre of the state, and contains 438 sq. m. Drained by Dick's r. and its branches, and by head branches of Green r. It contained in 1840, 14,445 neat cattle, 17,544 sheep, 23,595 swine; and produced 39,888 bushels of wheat, 81,981 of rye, 455,260 of Indian corn, 93,957 of oats, 11,017 of potatoes, 9634 pounds of sugar. It had eight stores, five flouring-mills, five grist-mills, three saw-mills, three oil-mills, seven tanneries, four distilleries; three academies, 150 students; five schools, 153

M* 193

scholars. Pop.: whites, 6582; slaves, 3430; free coloured, 155; total, 10,187. Capital, Stanford.

LINCOLN, county, Mo. Situated in the E. part of the state, and contains 576 sq. m. Drained by Cuivre r. and its branches. It contained in 1840, 10,065 neat cattle, 8196 sheep, 22,800 swine; and produced 97,321 bushels of wheat, 347,925 of Indian corn, 65,382 of oats, 13,566 of potatoes, 549,659 pounds of tobacco, 18,063 of sugar. It had 23 stores, five flouring-mills. 22 grist-mills, 13 saw-mills, one oil-mill, five tanneries, 10 distilleries; one academy, 72 students; 11 schools, 277 scholars. Pop.: whites, 5673; slaves, 1572; free coloured, 4; total, 7449. Capital, Troy.

LINCOLN. p. t., Penobscot co., Me., 117 m. N.E. Augusta, 712 W. Bounded N.W. by Penobscot r. Watered by Mataunscook r. It has six stores, one grist-mill, five saw-mills; 11 schools, 401 scholars. Assessors' valuation of real estate, in 1849, $97,371. Pop. 1121.

LINCOLN. p. t., Grafton co., N. H., 62 m. N. by W. Concord, 543 W. Drained by Pemigewasset r. It has three schools, 34 scholars. Pop. 76.

LINCOLN, p. t., Addison co., Vt., 35 m. S.W. Montpelier, 495 W. Watered by New Haven r. It contains a Friends' church; one store, three forges, seven saw-mills; eight schools, 310 scholars. Pop. 789.

LINCOLN, p. t., Middlesex co., Mass., 16 m. W. by N. Boston, 434 W. Bounded W. by Concord r. Beaver pond gives rise to a branch of Charles r. It contains a Congregational church; two stores, two grist-mills, three saw-mills; four schools, 200 scholars. Pop. 696.

LINCOLNTON, p. v., capital of Lincoln co., N. C., 172 W. by S. Raleigh, 409 W. Situated on the E. side of South or Little Catawba r. It contains three churches, a Methodist, Lutheran, and Presbyterian; two academies, nine stores, 85 dwellings, and 836 inhabitants. In the vicinity are a cotton factory, a paper-mill, and a rolling-mill, being, and nail factory.

LINCOLNTON, p. v., capital of Lincoln co., Ga., 96 m. N.E. Milledgeville, 566 W. It contains a courthouse, jail, a Baptist church, an academy, several stores, and 15 or 20 dwellings.

LINCOLNVILLE, p. t., Waldo co., Me., 10 m. S. Belfast 49 m. E. by S. Augusta, 636 W. Bounded E. by Penobsco r. It has a fine harbour in the N. part, called Duck Trap and has considerable coasting trade. It contains 10 stores, one fulling mill, three grist-mills, seven saw-mills; one academy, 30 students; 15 schools, 722 scholars. Pop. 3084.

LINDEN, p. v., capital of Marengo co., Ala., 80 m. S. by W. Tuscaloosa, 869 W. Situated on the S. side of Chickasaw r., a tributary of Tombigbee r., and 8 m. E. of it. It contains a courthouse, jail, and several stores and dwellings. Nett proceeds of the postoffice, $431.

LINDFIELD, a town and par. of England, co. Sussex, hund. Burley-Arches, rape of Pevensey, 14 m. N. by E. Brighton, and 33 m. S. London. Area, 3350 acres. Pop., in 1831, 1485. This town deserves notice for its useful institutions for instructing poor children of both sexes not only in reading, writing, &c., but also in the processes of agriculture, and various manual operations, as spinning and netting, printing, shoe-making, &c., and also for a benevolent society, giving pecuniary and other relief to poor persons not willing to receive parochial aid. The church is an old structure of plain exterior, with a low square tower; and the living is a curacy in the gift of the archbishop of Canterbury. There are two places of worship for Dissenters. Fairs for cattle and horses, May 12; and for sheep and lambs (the largest in the county), August 5.

LINGEN, an island of the E. Archipelago, off the N.E. coast of Sumatra, lying on the equator, and near long. 105°. It is about 50 m. in length, by 30 in its greatest breadth; having a healthy climate, and producing abundant supplies of fruit and poultry. Its geological formation indicates the presence of tin; and it furnishes some little gold. Its inhabitant may be considered as presenting the type of the Malay race in its greatest purity. (*Hamilton's E. I. Gaz.*)

LINDLEY, t., Steuben co., N. Y., 25 m. S.E. Bath, 221 m. E.S.E. Albany. Watered by Tioga r. The Corning and Blossburg railroad passes through it. It has two stores, one grist-mill, five saw-mills; three schools, 135 scholars. Pop. 636.

LINKLAEN, p. t., Chenango co., N. Y., 122 m. W. Albany. 336 W. Drained by branches of Ostelic r. It has one store, one fulling-mill, four saw-mills, two tanneries; nine schools. 354 scholars. Pop. 1949.

LINLITHGOW, a proyal and parl. bor. of Scotland, co. Linlithgow, of which it is the cap. in a valley on the S. bank of a lake of the same name, 15 m. W. by S. Edinburgh. Pop., in 1831, 3187. The town consists of one main street, along the line of road between Edinburgh and Falkirk, with several lanes branching off on both sides. The houses, with few exceptions, have an ancient and decayed appearance; the streets are lighted with gas, and macadamized. In addition to the town-hall, and jail, the

most prominent public building is the parish church, erected in the 12th century, but afterwards much enlarged and repaired. This, which is one of the best specimens of Gothic architecture in Scotland, is 168 ft. in length, 100 in breadth, including the aisles, and 90 in height, exclusive of the steeple: the latter, rising about 140 ft. above the ground, terminates in an imperial crown. The exterior had formerly a row of statues, of which one only remains, that of St. Michael, the tutelary saint of the borough: it is divided by a partition wall; the E. half only being used as the parish church; the other is unemployed.

The royal palace of Linlithgow is finely situated on an eminence projecting into the lake. This magnificent ruin is of a quadrangular form. It was begun so early as the 15th century; and was greatly enlarged and improved by James V., but was not finished till the reign of James VI. (James I. of England), who built the N. side of the quadrangle, after his visit to Scotland in 1617. The W. side of the palace is the most ancient; and here the apartment is still pointed out where the unfortunate Queen Mary first saw the light on the 7th of December, 1542. The palace was entire and habitable till 1746, when it was burnt, either intentionally, or through accident, by the troops under General Hawley. It covers an acre of ground; and though roofless, ruined, and desolate, its appearance sufficiently justifies the eulogium of Scott:—

    " Of all the palaces so fair,
     Built for the royal dwelling,
    In Scotland, far beyond compare,
     Linlithgow is excelling."
               Marmion, canto 4, stanza 15.

The hexagonal Cross well, in front of the town-house, about 20 ft. in height, is surmounted by a lion rampant supporting the arms of Scotland. The sculpture, by which it is adorned, is very complex; and the water is made to pour in great profusion from the mouths of 13 grotesque figures. This well, constructed in 1805, is said to be a fac-simile of the one previously existing, constructed in 1620.

There are two chapels belonging to the Associate Synod, and one to the Independents. Poor-rates have not been introduced; the poor being supported by the interest of certain funds left in mortmain, for the purpose, and by the church collections, and other parish dues.

There is a borough school endowed by the town, but no parish school. There are nine schools in the parish, all unendowed except the one referred to: total scholars, 547, or rather above a ninth part of the entire population. There are various reading-rooms, and a mechanics' library.

Linlithgow has little or no trade, but depends chiefly on its advantages as a provincial capital. Tanning and preparing leather, said to have been introduced by the soldiers of Cromwell, is the oldest and the staple branch of business, giving employment to nearly 100 hands. There are two extensive distilleries, a brewery, and a small glue manufactory. There are about 80 hand-loom weavers (cotton and linen). There is only one branch bank. The Union canal, between Edinburgh and Glasgow, and the Edinburgh and Glasgow railway, pass close along side the town. Blackness, on the frith of Forth, 5 m. distant, is its port.

Linlithgow was made a royal borough in the 12th century. In 1513, in an aisle of the parish church, the apparition is said to have appeared to James IV., that warned him against the expedition into England which terminated in the fatal battle of Flodden. (*Pitscottie's Hist. of Scotland*, i. 264, 265.) When passing through this town, on the 23d Jan., 1570, the Regent Murray (illegitimate brother of Queen Mary) was shot by Hamilton, of Bothwellhaugh, partly in revenge for a private injury, and partly from political motives. The house whence the shot was fired has been taken down, and replaced by a modern edifice. The White or Carmelite friars had a monastery here, founded in 1290; but all traces of it have disappeared. In addition to certain town dues, the municipal property consists chiefly of land; and the ancient custom of annually riding the marches, though disused in almost every other borough in Scotland, is here regularly observed. Corporation revenue, £716; number of councillors, 27. Linlithgow unites with Falkirk, Airdrie, Lanark, and Hamilton, in sending a member to the House of Commons. Registered voters in 1839-40, 119. (*Parl. Papers; Private Information.*)

LINLITHGOW. *See* LOTHIAN (WEST).

LINN, county, Iowa. Situated in the W. part of the ter., and contains 720 sq. m. Organized in 1837. Watered by Wabesipinica r Buffalo cr., and Cedar fork of Iowa r. It contained in 1840, 1491 neat cattle, 259 sheep, 2323 swine; and produced 5068 bushels of wheat, 94,654 of Indian corn, 9993 of oats, 5609 of potatoes, 4604 pounds of sugar. It had six stores, four saw-mills; one school, 12 scholars. Pop. 1373. Capital, Marion.

LINN, county, Mo. Situated toward the N. part of the state, and contains 588 sq. m. Watered by Locust and Yellow creeks. It contained in 1840, 9434 neat cattle 1819 sheep, 6976 swine; and produced 2634 bushels of wheat, 41,175 of Indian corn, 4584 of oats, 1600 of potatoes, 7012 pounds of tobacco. It had four stores. Pop. : whites, 2102; slaves, 143; total, 2245. Capital, Linneus.

LINTON, a market town and par. of England. co. Cambridge, hund. Chilford, 10 m. S.S.E. Cambridge, 48 m. N. by E. London. Area of par., 3683 acres. Pop. in 1831, 1676. The town, which stands on the line of a Roman road, and at the foot of the chalk downs communicating with the Chiltern range, comprises several irregular streets and lanes, lined in part with good brick houses, but with a much greater number of low thatched cottages. The church is a low structure in the pointed style, with a high embattled tower and handsome interior, the living being a vicarage in the gift of the bishop of Ely. There are places of worship, also, for Wesleyan Methodists and Baptists, and two Sunday schools, one of which belongs to the church. The market-house is a small square building. Tanning is the chief trade of the place, and in the neighbourhood are extensive nursery grounds, occupied by gardeners, florists, and seedsmen, who take their produce to the Cambridge market. Markets on Thursday: fairs for horses and lambs, Holy Thursday and July 26.

LINTON, t., Coshocton co., O. Organized in 1811. It contains Plainfield v., and has one fulling-mill, one sawmill. Pop. 1198.

LINTZ, or LINZ, a city of the Austrian dom., cap. of Upper Austria, on the Danube, which is here crossed by a wooden bridge 980 yards in length, 41 m. S. E. Passau, and 96 m. W. by N. Vienna; lat. 48° 18' 54" N., long. 14° 16' 45" E. Pop., in 1834, 22.318, ex. garrison. It consists of the city proper and three suburbs, which, as in Vienna, are more extensive than the city itself. All travellers speak favourably of Linz. "This beautiful city has nothing of Germany in it, except its language. The houses, all handsome and lofty, are stuccoed and painted, chiefly white, but many yellow or light brown. Almost all have architectural decorations and columns, friezes over the windows, and Venetian blinds outside. Balconies with flowers salute the eye at every turn; and not only on the broad, spacious "Place," but in the back streets also, the houses are lofty and elegant, and all look as clean, and white, and fresh, as if newly decorated and painted. We are sensible here of a decided change in the atmosphere. The sky is cloudless; the heat not oppressive : and there is a peculiar soft balminess in the air. The people, too, are handsome and well clothed, and look happy. Linz is celebrated for the beauty of its women, and, as far as I can judge, justly." (Turnbull's Austria, i., 131, 138.) The principal streets are wide and regular, though, according to the Austrian Encyc., most of them are badly paved, and the houses shingle-roofed. Linz has few remarkable public buildings. The churches are generally handsome; several have glittering cupolas, and many are richly gilded, and adorned with good paintings. The Landhaus, formerly a Franciscan convent, is the place of assembly for the states of the province, and accommodates the principal government offices. The schloss, or castle, on an eminence overlooking the Danube, was once the residence of the dukes of Austria, but is now the prison and penitentiary for the province. In the principal square is a marble column, erected in 1723, between statues of Jupiter and Neptune, to commemorate the escape of the city from the double attack of the plague and the Turks.

Linz is among the few German towns not encircled with continuous walls. Under the superintendence of the archduke Maximilian, it has recently been surrounded with a chain of 32 isolated forts, 23 being on the right, and 9 on the left bank of the Danube, at the distance of 1, 2, or 3 m. from the town. They communicate with each other by a covered way, and are placed at regular intervals in the plain or along the slopes and tops of the hills, in a circuit of 9 m. The highest eminence near the city, the Pöstlinberg, on the opposite side of the Danube, is surrounded by a circlet of 5 towers forming a citadel. Each tower is 30 ft. high, and 80 ft. in diameter, built within a hill of sand, and sunk into the earth, so that the roof alone projects ; and each has a glacis on the side farthest from the town. Each consists of three stories; the lower serving as a storehouse and a powder-magazine, the middle as a lodging for troops, the third being the platform on the summit, which, when not used, is covered by a temporary roof. The platform is mounted with 10 guns, so arranged that they can be brought to bear upon any point with the greatest facility, and command the glacis by a cross-fire in every direction. There are also guns on the lower story bearing upon the ditch, to frustrate any attempt to cross it. In this mode of fortification each fort must be made the object of a separate attack; and the expense is trifling compared with the common method. But it is very doubtful whether it will oppose so effectual a resistance to an invading army as a single fortress, or an adequate scale, constructed according to the approved principles of the art. Owing to the demolition of

the fortifications at Ulm by the French during the late war, there was not previously to the erection of these works, any fortress to defend the valley of the Danube between the frontier of France and Vienna. (Murray's Handb. for the S. of Germany, &c.)

Linz is the seat of the provincial government, and the assembly of nobles for Upper Austria, and of tribunals and councils for the whole state and the city ; and is the see of a bishop.' It has a lyceum, where courses of lectures are given in theology, philosophy, and medicine : the library belonging to this institution comprises about 40,000 vols. ; but as they consist, for the most part, of works on Roman Catholic theology, and such like subjects, they are of little or no use: if it were really a good collection, it would be of material service, for the reading-room is open to all the town, and under certain regulations, books may be taken home. Drawing-schools, and collections of mathematical and philosophical instruments are attached to the lyceum. It has also a gymnasium, an ecclesiastical seminary, a provincial academy of arts, an imperial collection of economical models, a normal high school, and school of arts, with three subordinate schools, two military schools, a school of engineering, a female school attached to the convent of the Ursuline nuns, and other seminaries ; a military and another large hospital, various charitable institutions, a private deaf-and-dumb asylum, a musical society, &c., with large barracks, a custom-house, a bank, and a small but fine theatre. The public gardens in the vicinity are favourite places of resort.

Linz has a large imperial factory of broadcloth, carpets, and other woollen stuffs, which occupies seven contiguous houses, and is said, at one period, to have employed directly and indirectly 23,000 individuals: but this was most probably very far beyond the mark ; and the introduction of machinery has since occasioned a material diminution of the numbers employed. Considerable quantities of the red woollen caps made here are sent to Turkey. Linz has other woollen factories, with manufactures of cotton and silk goods, leather, gold lace, cards, tobacco, &c. Two fairs are held annually, one at Easter, and the other at the Assumption ; and the transit trade by the Danube, especially since Linz became a station for the steamers on the river, is very considerable, and employs several of the inhabitants. Two railways meet at Linz: one goes N. to Budweis in Bohemia 67 m., and was the first constructed in Germany ; and the other to Gmunden on the Traun. It is intended to carry on the latter to Grätz, by way of Leoben and Brück : the mineral products of Styria, &c., will then be brought by it to the Danube, as the salt which supplies Upper Austria is at present from the Salzkammergut. (Turnbull's Austria, ii., 372.)

Linz is supposed to have been known to the Romans, and it is said to possess some Roman antiquities. It was purchased by Leopold II., margrave of Austria, in 1036. In 1626, during the civil war of Upper Austria, it opposed a long and successful resistance to Fadinger, the peasant leader, who was mortally wounded before its walls. The suburbs were then, however, destroyed by fire, and the castle and a part of the city suffered severely from the same cause in 1800. (Oesterr. Nat. Encyc. ; Berghaus ; Turnbull's Austria, &c.)

LIPARI ISLANDS, a group in that part of the Mediterranean called the Tyrrhenian sea: they are a dependency of Sicily, from the N. coast, of which they are from 10 to 40 m. distant, forming a part of the intend. of Messina, between lat. 38° 20' and 38° 55' N., and long. 14° 15' and 15° 15' E. Aggregate population of the group estimated at about 22,000, of whom about 12.500 belong to the town of Lipari. (See post.) There are seven principal islands, Lipari, Vulcano, Stromboli, Salini, Panaria, Felicudi and Alicudi ; and a number of adjacent islets and rocks. They are all mountainous, rising abruptly on their W. side, and shelving down gradually towards the E. ; and in addition to its uniformity, each island, with scarcely an exception, has a high isolated rock off its N. shore, a peculiarity extending even to the distant isle of Ustica. They are evidently of volcanic origin, being composed chiefly of hornstone and granite, covered with lava, scoriæ, pumice stone, and other volcanic products. Stromboli (which see), the most northerly of the islands, has the only volcano in Europe which is in constant activity. Lipari and Vulcano have also craters in which volcanic phenomena are occasionally manifest. Dolomieu, and others, suppose, with some show of reason, that Panaria and the adjacent islets of Dattolo, Basiluzza, Lisca. &c., which are circularly disposed, once formed parts of the rim of the crater of an immense volcano, which has now disappeared under the waves, but which may have been the Euonymos of the ancients. (Dolomieu, p. 105–108.)

The climate is highly salubrious, and the air pure and refreshing ; storms and earthquakes are, however, frequent. Where the volcanic substances have been decomposed so as to form soil, it is very fertile ; but it absorbs moisture so rapidly, that the inhabitants are obliged to construct capacious cisterns, in which rain-water is carefully preserved for irrigation, and other purposes. Lipari, the centre and largest of these islands, is about 18 m. in circuit. It was peopled by a colony of Cnidians, and is described by Strabo as having a fleet, and commanding the other islands. (Strabo, lib. vi.) Its interior is rugged and broken, presenting hills of vitrified volcanic substances, which, though at least 3000 years old, present no symptoms of decomposition; but it has, notwithstanding, two considerable plains, and some deep valleys, which are well cultivated, and productive. Exclusive of about a three months' supply of corn, it produces large quantities of fruit, especially grapes, with figs, prickly pears, olives, &c.: it also produces cotton, beans, and peas. Some wine is made; that called Malvasia being highly esteemed in Naples. Most of the grapes are, however, converted into raisins: they are prepared by placing the ripened grapes in an alkaline ley of ashes, more or less impregnated with salt, and afterward exposing them to the meridian sun. By this means, an extremely luscious raisin is produced. The agricultural products of the other islands are much the same as those of Lipari: in some, a few oxen are reared, but cattle are generally scarce and lean, the pasture being fit only for goats. Lipari was celebrated in antiquity for its hot springs and sudatories; they are now, however, but little used. The only spring in the island is hot. (Russell's Sicily, p. 274.) Lipari is the great magazine whence Europe is supplied with pumice-stone, its surface being almost wholly composed of that singular substance. Though so abundant in that island and Vulcano, pumice-stone is not found either in the neighbourhood of Etna or in the regions of extinct volcanoes on continental Europe, and only in small quantities in Vesuvius. It is of various kinds and degrees of specific gravity, one variety being so light as to float on water. It is used to polish marbles, metal, pasteboard, &c., and fetches from £8 to £10 a ton in the London market. Other volcanic products, as sulphur, nitra, sal ammoniac, pozzolana, bitumen, &c., are among the chief exports from the Lipari islands, and in these an active trade is carried on. The principal crater in Vulcano, the most S. of the islands, is covered with efflorescences and incrustations of the above products. Alum, however, which was formerly a great staple, and from which the Romans anciently derived a considerable revenue, and the Lipariot merchants great profits, now scarcely exists as an article of commerce: the failure of its production is supposed to be owing to a diminution in the intensity of the subterranean fires.

Sulphur is still exported, but not to the extent that it might be, from the notion that the vapour arising from its purification infects the air and injures vegetation. Salina is so called from the salt-pans on its S.E. shore, which produce enough of that article for the supply of all the islands. The pinna marina, from whose silky filaments the Romans made imperial robes, abounds on the shores of Salina. Next to pumice-stone, wine, raisins, currants, olives, salt, and sulphur, soda, capers, coral, and fish are the chief articles of export. The natives are generally poor, though few are in the extreme of poverty. They are industrious, hardy, and make good seamen; but they are immoral, filthy in their habits, and infested with scabies. (Dolomieu, Voyage aux Iles de Lipari, 1–140; Smyth's Sicily, 248–279.)

These islands were called Hephæstiades by the Greeks, and Vulcania by the Romans, from their emitting smoke and flames; such places being supposed to be either inhabited by, or under the immediate protection of Vulcan. Vulcano, however, was more especially sacred to the god of fire, and is said by Virgil to be

"Vulcani domus, et Vulcania nomine tellus."

They were also frequently called Æolian isles, from Æolus, one of their sovereigns. This prince having learned, according to Pliny, to foretel, from observations made on the smoke of the volcanoes, the coming changes of the winds, was said by the poets to have the latter under his command. (Hist. Nat., lib. iii., cap. 9.) Virgil has described the power and functions of Æolus as ruler of the winds, in one of the finest passages of the "Æneid:"

"Hic vasto rex Æolus antro
Luctantes ventos, tempestatesque sonoras
Imperio premit, ac vinclis et carcere frenat.
Illi indignantes magno cum murmure montis
Circum claustra fremunt. Celsa sedet Æolus arce,
Sceptra tenens; mollitque animos, et temperat iras.
Ni faciat, maria ac terras cælumque profundum
Quippe ferant rapidi secum, verrantque per auras."
Æneid, i., lin. 56–59.

LIPARI, the cap. town of the above group of islands, and of a canton under the intend. of Messina in Sicily, on a steep declivity on the E. side of Lipari island; lat. 38° 27′ 50″, long. 14° 57′ 50″ E. Pop. 12,500. It is healthy, but crowded, irregular, and dirty, with narrow streets and ruinous public edifices; of which last, the finest are the Capa-
188

chin convent, a hospital, a nunnery, and the bishop's palace. The castle, which encloses the cathedral and some other edifices, is erected in a commanding situation, on the summit of a huge volcanic rock. From fragments of a Cyclopian wall and other remains, Smyth conjectures that this was the identical Acropolis which the Romans, about anno 250 B.C., attempted to carry by escalade, but were repulsed and driven back with great loss by the Carthaginians under Hamilcar. (Sicily, &c., 264.) The greater part of the present fortress was built by Charles V., after Barbarossa had plundered the town in 1544. The cathedral is a neat edifice, but has been much injured by lightning. A college is established here, under which are eight schools in different parts of the islands. Several Greek and other antiquities exist in and round the town: an excellent statue on the Marina, supposed to have been erected in honour of Timasitheus, has, "by the addition of a copper nimbus, been converted into a saint." Lipari has an active trade in the produce of the islands with Messina, Palermo, Naples, &c. Its bay or harbour, nearly 2 m. in circuit, has deep water and tolerably good holding ground, but, from want of a mole, it is not at all times secure. (Smyth's Sicily, 263, 264, and Appendix; Russell; Brydone; Rampoldi, &c.)

LIPETZK, a town of Russian Europe, gov. Tambof, on the Voronêje, an affluent of the Don, 80 m. W. by S. Tambof. Pop. nearly 6000. It has several churches, most of them of wood. It had at the end of the last century an imperial cannon foundry, employing nearly 1500 hands, but it appears to be no longer in activity. A mineral spring, frequented by numerous visiters, was converted into a spa, under the auspices of Peter the Great, a statue of whom was erected in the town by one of its citizens, in 1839. (Possart; Das Kaiserth; Russland, 570.)

LIPPE-DETMOLD, a principality of N.W. Germany, between lat. 51° 47′ 30″ and 52° 11′ N., and long. 8° 36′ and 9° 20′ E., having N.E. and E. territories belonging to Hesse-Cassel, Hanover, Waldeck, and Prussia, and being elsewhere surrounded by Prussian Westphalia. Area, 445 sq. m. Pop. in 1838, 82,970, the great bulk of whom, with the reigning family, are Calvinists. The country is in general hilly, especially its S.W. part, where the Teutoburgerwald separates the basins of the Rhine and the Weser. The latter river forms a part of the N. boundary: the Werra, one of its tributaries, is the other principal stream of Lippe-Detmold. The climate is one of the mildest and most agreeable in N. Germany. The mean temperature of the year, in the valleys and plains, is about 49° Fahr.; and that of the winter no lower than 35° Fahr. Agriculture is the chief occupation of the inhabitants. Corn, of various descriptions, beans and peas, rape seed, flax and hemp, are the principal articles of culture. The country is well wooded, particularly with oak and beech; and timber is one of its most important products. A good many sheep and hogs, and excellent horses, are bred; and the rearing of bees is extensively pursued. About 29,000 centners of salt are annually produced from salt springs; and marble, lime, and iron are obtained in small quantities. The weaving of linen fabrics, and the spinning of linen yarn from the flax produced in the territory, partially occupy the rural population. Berghaus says there are 9000 looms in the principality; and that linen goods to the value of 1,080,000 rix dollars a year are made. There are some woollen cloth and glass factories, tanneries, distilleries, and papermills; and Lemgo has a manufacture of meerschaum pipebowls. These articles, after timber, linen stuffs, and yarn, and cattle, are the chief articles of export. The government is a constitutional and hereditary monarchy, remodelled in 1819; and vested in the prince, and a representative body or diet of 21 members; seven elected by the nobility and knights, seven by the inhabitants of towns, and seven by those of the rural district. The diet is convoked every two years, and no new tax can be imposed without its consent. All questions relative to taxation are decided by the states in one assembly, by a majority of votes; on other questions, though the states deliberate together, they vote in two separate assemblies, the deputies of the nobles and knights forming one by themselves. Appeal lies from the civil and criminal tribunals of the principality to the high court at Wolfenbüttel. The people are better informed than in many parts of Germany, the princes of Lippe-Detmold having, for a lengthened period, been liberal patrons of public instruction. The gymnasium at Lemgo, and the high school and teachers' seminary at Detmold, are in high estimation, many of the most celebrated scholars of Germany having received the early portion of their education at one or other of those establishments. Lippe-Detmold furnishes 691 men to the army of the German Confederation: it has one vote in the full council of that body; and, along with other small states (see vol. i., p. 979), a vote in the committee.

Detmold, the chief town and residence of the prince, and seat of government, and of the superior judicial courts, &c.

on the Werra, 54 m. E. Munster, lat. 51° 56′ N., and long. 8° 53′ 15″ E., has 2560 inhabitants. Lemgo, on the Bega, the largest town in the principality, has 3600 inhabitants, with a seminary for noble females, and an orphan asylum. (*Berghaus, Allg. Länder, &c., iv., 477–479; Cannabich; Alm. de Gotha.*)

LIPPE SCHAUMBURG. (See SCHAUMBURG LIPPE.)

LIRIA, a city of Spain, prov. Valencia, 18 m. N.W. Valencia. Pop., according to Miñano, 10,356. It partly occupies a space between two hills, but it is partly, also, on an acclivity, the summit of which is crowned by the ancient parish church *de la Sangre*. Another parish church, a chapel, and two monasteries, are the only other public buildings; and the town generally has a mean and neglected appearance. Its inhabitants are principally employed in the distillation of brandy, soap-boiling, tanning, tile-making, and the weaving of linen fabrics. The neighbourhood is well irrigated, and extremely productive. The marble quarried near Liria is also celebrated for its whiteness and fineness of grain. Its existence is traced by the Spaniards up to the time of the Carthaginians, who founded here the town of *Edera*. Under the Romans, it was called *Edeta*, and was the capital of the country of the *Edetani*. Having passed successively into the hands of the Goths and Moors, it was finally added to the crown of Arragon in 1239.

LISBON (Port. *Lisboa*; an. *Olisipo*, and afterward *Felicitas Julia*), the cap. city and principal seaport of Portugal, in the commerce of its own name, admirably situated for commerce on the right bank, and near the mouth of the Tagus, 172 m. S. Oporto, 290 m. N.W. Cadiz, and 280 m. W.S.W. Madrid. Lat. 38° 42′ 24″ N. Long. 9° 5′ 29″ W. Pop. 250,000, ? among whom are many negroes and mulattoes. The city (as seen from the river), covering several hills with palaces, churches, convents, and dazzling white houses, that rise more or less abruptly from the quays, has a magnificent and imposing appearance; but when the traveller has once landed, the delusion vanishes, for nothing can be more literally correct than the poetic description of Byron:—

> " Whoso entereth within this town,
> That sheening far, celestial seems to be,
> Disconsolate will wander up and down
> 'Mid many things unsightly to strange ee;
> For hut and palace show like filthily;
> The dingy denizens are rear'd in dirt;
> Ne personage of high or mean degree
> Doth care for cleanness of surtout or shirt,
> Though shent with Egypt's plague, unkempt, unwashed, unhurt."
> *Childe Harold, cant. i.*

The streets are badly paved, and generally narrow, and the houses, with here and there a latticed window, have a melancholy appearance; while in filthiness and impurity of every description Lisbon may vie with Constantinople. Mrs. Baillie, who resided here for more than two years, describes the streets as bending forth " the most pestilential effluvia. Dogs of every mongrel breed, lank, lean, and voracious, lie about the streets in alarming numbers. Indeed, Lisbon maintains no other scavengers." The police, however, has been improved, and scavengers now cleanse the wider streets; but the greater part of the city is still worthy its ancient notoriety for the want of cleanliness, and even decency. The E. quarter of the town not having been destroyed by the earthquake of 1755, is the oldest, and has very narrow, irregular lanes, skirted by high, old-fashioned, and half-ruinous houses; but lower down in the plain to which the ravages of that calamity were confined, the town has been rebuilt in a regular manner, and exquisitely paved, and there are a few squares and open spaces, which contrast strikingly with the mean appearance of other parts. Lisbon is an open town; and its suburbs are so nearly connected with it that it is difficult to define its limits. Measuring, however, from the small river Alcantara eastward, to the termination of the continuous buildings, we find it to be about 3½ m. in length; the breadth varies from 1 m. to 1½ m., so that its total area comprises about 3000 acres. The whole of this space, however, is not covered with buildings; for in many parts there are extensive plantations and gardens, public squares, and a vast extent of ground unoccupied, except by ruins and rubbish, the monuments of the catastrophe of 1755. Some houses, also, have been thrown down, and others greatly injured by subsequent shocks; and there is, perhaps, no great presumption in anticipating, owing to the frequency of these phenomena, that Lisbon may, one day or other, again become the subject of a similar calamity to that by which it was so nearly destroyed. One of the largest squares is the *Praça do Commercio*, in the centre of which is the equestrian statue, in bronze, of Joseph I. The front, towards the river, is open, and flights of stone steps descend from it to the water: on the E. side are the custom-house, E. India-house, and exchange; the public library is on the W. side, and facing the river stands a fine building called the *Junta do Commercio*. The *Rocio* is another fine square, communicating with that last mentioned by several well-built and uniform streets; and in it stand the scorched and blackened ruins of a large mansion recently burned down, once the palace of the Inquisition, but afterward converted into government offices. In this square were celebrated the *Autos da fé* which once disgraced Portugal, even more than the rest of Catholic Europe. Of these streets, one, the *Rua d' Ouro*, is chiefly occupied by goldsmiths and jewellers: the silversmiths live in the *Rua da Prata* (Silver-street), and a third is filled with the shops of cloth merchants and embroiderers. The haberdashers and other tradesmen have likewise their streets called after the trade to which they are appropriated. Among the other squares and remarkable places of Lisbon may be mentioned the *Praça da Figueira*, used as a market for fowls and vegetables; the *Praça das Amoreiras*, in the centre of which is a large reservoir whence water is supplied to the various fountains of the city; and the *Praça de Alegria*, now as formerly celebrated as the Rag-fair or Monmouth-street of Lisbon. The *Salitre*, leading from the *Passeio Publico* in the N. quarter of Lisbon, forms a cool and shady promenade; the quarter of Buenos Ayres, on the slopes W. of the town, is airy and pleasant, comprising many handsome residences, and a line of good streets runs westward over the bridge of Alcantara, connecting Lisbon with the suburb of Belem. The houses above the shops, and many others also, are let in separate floors, as at Madrid and Paris; and a common passage, not remarkable for cleanliness, leads to the door and staircase of each. The police, so inefficient during the peninsular war, that Byron might correctly have pictured Lisbon as

> " That purple land, where law secures not life,"

was much improved after the peace by the establishment of Novion's police; but this useful body was broken up at the time of Don Miguel's expulsion, and property as well as life are almost as unsafe as ever. The streets, with the exception of a few great thoroughfares, are scarcely lighted at all : there are plenty of fountains; but water is not used to clean the streets, and there is no attempt at sewerage. Fires are frequent; but they are not destructive, owing, in part, to the solid construction of the buildings, and in part to the unfrequent use of domestic fires, and the formation of the *aguadores* or Galician water-carriers into corps, stationed at different parts, to convey water from the fountains on the first alarm. These, of whom there are about 7000, are generally employed in Lisbon to carry burdens and perform the more severe labour. The work of porters, however, at the custom-house and India-house is done by the Portuguese, to the entire exclusion of the Gallegos.

Few cities of Europe are so scantily supplied with fine public buildings. The custom-house, exchange, and India-house are large and handsome ; but besides these there are scarcely any except the churches and convents, which crown the hills and look like palaces and fortresses. Some of the former, rebuilt since the earthquake, are very spacious, and profusely decorated in the worst taste. The principal of these are, 1. the cathedral, a large Moorish building restored in 1770, and situated on the slope of the hill on which is the *castello*, or citadel : 2. the church *De Coração de Jesus*, the largest and most sumptuous sacred edifice built since 1755, surmounted by a finely-proportioned dome, and remarkable as containing a mausoleum dedicated to the foundress, the queen Maria ; 3. the ancient church of the Martyrs, erected on the spot where Alphonso I. mounted the walls of Lisbon, and took it from the Moors ; and, lastly, the elegant but still unfinished church of Santa Engracia, which not having been touched for the last thirty years, a proverb has come into use, entitling all incomplete undertakings *as obras de Santa Engracia*. Here, also, we may mention the church of San Geronimo, at Belem, built by King Emanuel in 1499, and exhibiting a fine specimen of the ornamental Gothic and Arabic styles : in the interior is a royal mausoleum. The convents, which are of large size, form a principal feature in the town ; but, since the suppression of the monasteries in 1835, Lisbon has lost much of its monkish appearance, the buildings have been converted to public uses, or sold to private individuals, and the wealth has been thrown into the national treasury. The English built a Protestant chapel in 1823 contiguous to a cemetery, in which, among other remains of our countrymen, lie those of the celebrated Henry Fielding, who died here on the 8th of October, 1754, at the early age of 48. Among the other public structures, the fine aqueduct of *Aguas livres*, deserves notice as one of the greatest works of modern Europe, and to which will bear comparison even with the grand specimens of ancient bridge-architecture. It brings water from several springs about three leagues N.W. of the city : its course is partly under ground ; but as it approaches Lisbon, and crosses the deep valley of the Alcantara, it is carried over 35 bold marble arches for a length of about 2400 feet. The water enters the town at the *Praça das Amoreiras*, where, as before mentioned, is distributed to the various fountains, and whence the Gallician aguadeiros draw the supplies, which they sell from house to house, and hawk about the streets. The palace of Necessidades, in which the present queen has resided and held her courts ever since the death of the late king-consort at Ajuda, is small and mean-looking, and the palace of Bemposta is equally unworthy so imposing a name; but the palace of Ajuda, near Belem, lately completed, is a large building, and, notwithstanding its faults of architecture, may rank as one of the finest in Europe. The arsenal, postoffice, mint, corn-exchange, two public hospitals (one of which, called the hospital of San Joseph, is extremely well conducted, and has accommodation for 1500 patients, with an attached school of medicine), the nobles' college, and the palace of Calhariz, are the only other national buildings of any importance, except the theatres. The opera-house of San Carlos is a large building of good proportions, with a noble box in the centre for the royal family; and Dr. Wilde, a very recent traveller in Portugal, says that the opera enjoys a well deserved popularity, the singing being very good, and the ballet really admirable. (*Narrative*, vol. L., p. 49.) There is a theatre for the performances of the national drama; but it is small and mean, and the plays as well as the performers are of a very inferior character. Lisbon, also, like Madrid and Seville, has a bull-ring, the size of which, however, will bear no comparison with the latter, nor are the performances so splendidly appointed or well managed. It must be observed, however, that the people of Lisbon visit the opera rather in obedience to fashion than from any desire for amusement: the bull-fights are not attended, as in Spain, by the *élite* of society, and the national drama is chiefly supported by the *bourgeoisie*. Out-of-door amusements are seldom sought, except water-excursions, in which the people generally take great pleasure; the shores of the Tagus are indeed most beautiful; the country on the opposite side offers many interesting objects, as Almada, Barreiro Seixal, Setubal the convent of Arrabida, &c.

The literary and educational institutions of the Portuguese capital comprise, 1. a Royal Academy of Sciences, founded in 1778, having a good library and museum, and publishing memoirs and scientific works; 2. a patriotic literary society established in 1822, and sending forth a journal of its proceedings, a society for promoting national industry, and the following establishments, either wholly or in part supported by the government, viz. a school of commerce attended by about 150 pupils, a royal naval academy, a royal academy of engineering, a school of surgery, a music school, 12 schools of logic and rhetoric, 19 others for classical instruction, and 18 primary schools; but they are ill supported, inefficiently conducted, and have been, till very lately, remarkable rather for their antiquated style of instruction than for positive and general usefulness. The national public library of Lisbon in the Praça do Commercio has been much enriched by the addition of books formerly belonging to the monasteries, and probably now contains about 150,000 printed vols. besides MSS. The library of the Cortes in the *Hospicio real de nossa Senhora das Necessidades*, where that body holds its sittings, comprises about 30,000 vols.: and that belonging to the theological seminary of San Vicente de Fora has about 18,000 vols.: but the books in these collections are chiefly ecclesiastical and old, while the departments of science, modern literature, and modern history, are almost entirely neglected. In fact, Portuguese literature, down to a very recent period, had been, for many years, in a state of stagnation, and the institutions of Lisbon are now only slowly recovering from the lethargy in which they had been buried upward of thirty years. Besides the establishments already described may be mentioned the royal printing-office, and the cabinet of natural history and botanic garden at Ajuda. Several newspapers are published in the capital; but they are without exception badly conducted, and exercise very little influence either on society or government.

The harbour, or road of Lisbon, is one of the finest in the world, and the quays, which extend nearly 2½ m. along the banks, are at once convenient and beautiful. Fort St. Julian, built on a steep projecting rock, marks the N. entrance of the Tagus; and on it is a lighthouse, rising 120 feet above the sea level. Two large banks called the N. and S. Cachops, obstruct the river mouth, and on the middle of the latter stands the Bugio fort and lighthouse, the latter being 86 feet high. The least depth of water on the bar in the N. channel is four fathoms, and in the S. six fathoms; and there is little danger in entering the port, except during ebb tides, which run out at the rate of 7 m. an hour. Inside the harbour the water from nearly 20 fathoms in mid-channel shoals gradually to the edge; but in some parts vessels may come within 200 yards of the

shore. Lisbon, however, with all the advantages of its position and the excellence of its port, which commands the navigation of the Tagus, holds but a low rank in respect of commerce and industry. The despotism, intolerance, and imbecility of the government have weighed down the national energies, and the insecurity both of life and property, in consequence of bad laws and an inefficient police, have paralyzed industry of every description.

The foreign trade of Lisbon, formerly of considerable importance, but, perhaps, at all times, much overrated, has rapidly declined since the emancipation of Brazil. She had formerly about 400 ships, varying in burden from 300 to 600 tons, engaged in the South American trade; but at present only 50 vessels are employed in that trade, the average burden of which does not exceed 150 tons. Indeed, the produce of Portugal now sent to foreign countries is almost entirely conveyed to its destination in foreign ships. A small number of sea-going ships belonging to the port, probably about 60, of the aggregate burden of 9000 tons, are engaged in foreign trade, partly with the E. Indies and China, but chiefly between Setubal (or St. Ubes), and Cork, exporting salt in return for butter; and about 300 small craft are employed in the coasting trade. The following table shows the number and flags of different ships that entered and left Lisbon in 1837–38:—

Countries.	Arrived.		Departed.	
	1837.	1838.	1837.	1838.
British	250	244	226	245
Portuguese	368	371	339	371
Swedish	88	79	43	79
Dutch	86	50	65	49
Russian	18	21	16	24
Spanish	14	27	14	27
Sardinian	19	21	16	21
American	11	19	18	18
French	34	29	23	26
Hanoverian	9	17	9	17
Hamburgh	12	15	11	13
All others	89	39	94	96
Total of ships	1,173	1,092	1,100	1,049

The exports comprise wine, oil, fruit, and salt; among the imports are, hemp, flax, and linen cloths from Russia; iron, steel, salt fish, timber, pitch, and tar from the Baltic; linens, corn, &c., from Holland and Germany; silks from France; and cotton and woollen goods, cod-fish, hardware, ale and porter, linen, coals, and earthenware from England, which engrosses fully 7-8ths of the trade in foreign bottoms. The following account of the articles imported into Lisbon from the British dominions, in 1837, though imperfect, will give a tolerable idea of the present state of the trade between the two countries.

Description.	Quantities.	Value.
		£.
Cotton goods	10,992 cases	226,344
Woollen do.	1,240 bales	158,448
Linen do.	618 cases	17,625
Butter	12,535 firkins	30,739
Cod-fish	80,887 quintals	67,791
Wrought metals	13,055 cases	22,988
Drugs	2,836 —	12,469
Earthenware	837 casks	6,132
Tea	814 chests	16,629
Coals	678,866 arrobas	11,514
Iron	12,509 quintals	8,885
All other articles		45,458
Total value of Imports from Great Britain		776,388

The port of Lisbon is open to all nations, all articles, except corn and flour, being allowed to be warehoused. Goods so admitted, with the exception of vinegar, wine, and spirits, pay 15 per cent. on the new tariff valuation of 1837, and, where there is no tariff, *ad valorem*. All exported articles pay a duty of four per cent. *ad valorem*. There is no regular warehousing and bonding system at Lisbon: all imported dry goods are allowed to lie in the custom-house stores two years, and liquids six months, without charge, provided they are intended for consumption, and pay duty accordingly; otherwise, if re-exported, they pay two per cent. The port charges on a foreign ship of 300 tons, entering with a general or mixed cargo, and clearing out with the same, average 56,960 reis, or £11 4s., nearly four fifths of which are tonnage and light dues, the former being 100 reis, and the latter 50 reis, per ton. There are three respectable insurance companies (the Fidelidade, Restauração, and Bonança), in any of which insurances may be effected.

Lisbon has some fabrics of silk, paper, and soap; there are sugar refineries, tan-yards, and potteries; and its goldsmiths and jewellers are among the most expert in Europe; but in every pursuit it is to be perceived a want of energy and industry, to be traced perhaps to the character of the people as well as to political causes. With respect to the character of other artisans, Mrs. Baillie observes: "It is surprising how ignorant, or at least superficially acquainted, the Portuguese are with the commonest branches of handicraft: a carpenter is awkward and clumsy, spoiling every work he attempts, and the way in which the doors and wood-work, even of good houses, are finished, would have suited the rudest ages. Their carriages of all kinds, from the fidalgo's family coach to the peasant's market-cart, their agricultural implements, cutlery, locks and keys, &c., are ludicrously bad. They seem to disdain improvement, and are so infinitely below par, so strikingly inferior to the rest of Europe, as to form a sort of disgraceful wonder in the midst of the 19th century !" (*Lisbon*, i., 74.)

The Bank of Lisbon, founded in 1822, had a capital, in 1833, of about £691,100, divided into £100 shares. The bank discounts bills not having more than three months to run, at five per cent.; and it enjoys the singular, but valuable, privilege of having its claims on all estates paid in full, provided the estate amounts to so much, other creditors being obliged to content themselves with a division of the residue, if there be any. For weights and measures, see PORTUGAL; and for commercial details, see *Commercial Dictionary*, art. LISBON.

The climate of Lisbon is variable, but, on the whole, healthy and genial, notwithstanding the cold, piercing winds from the sea, which are disagreeable even during the scorching heats of summer, with the thermometer at 98° and 100°: heavy rains prevail in November and December, but it seldom snows: cold, clear weather is usual in January, and spring commences about the middle of February.

The population of Lisbon is of an extremely varied character : nearly a third of the lower orders are Gallegos, blacks, or mulattos, who, though the worst used and least considered, have a just claim to rank as the most hardy and industrious people of the capital. Genoese, Spaniards, and a few French also are employed as gardeners or as innkeepers, cooks, and stewards. The lower orders of the Portuguese, who are seen, perhaps, to more disadvantage in Lisbon than in any other part of the kingdom, are remarkable for their indolence and disregard of the comforts of life; but we believe that these evils are owing, in a great measure, to the total want of education, the influence of a debasing superstition, and the badness of the government. Garlick, rancid oil, dried fish, and goat cheese, which constitute their favourite food, are easily procurable; and so unconquerable is the predilection for the *dolce far niente*, occasioned in part, no doubt, by their climate, that they very seldom work, except for a bare subsistence. That contempt of cleanliness which is more or less evinced by all but the very highest classes, is most striking and revolting in the lower orders, whom Mr. Semple has well described as a "swarthy, meagre race, generally clothed in rags, and filthy beyond endurance." Irascibility and revengefulness are features of character common to all the inhabitants of the peninsula; but to these the Portuguese add cowardice, and hence assassinations and night attacks are far more common than in Spain. There can be no question, however, that the statements of travellers on this subject are greatly overcharged, or at least do not apply to the present period. Honesty and veracity are virtues seldom met with, unless among the merchants and better class of tradesmen; but there are some exceptions, particularly among domestic servants, who are usually respectful, attentive, and attached to the families by whom they are employed. The merchants are an important body, not inactive in business, and tolerably wealthy, considering the great diminution of their resources since the separation of Brazil from the crown of Portugal; but their habits are modelled on those of foreign countries, or from intercourse with the English and French, many of whom, especially the former, have considerable commercial establishments in Lisbon, and constitute, in fact, its best society. The Portuguese of the aristocratic classes are more grave, reserved, and proud than the Spaniards, against whom all orders of the people entertain a deep-rooted, national antipathy. Their neighbours, however, are not far behind them in the violence of their prejudices, if we may judge from the Spanish proverb : "Strip a Spaniard of all his virtues, and you make him a good Portuguese." Lisbon, as a place of residence, is somewhat dull, especially after Madrid. There are no public walks or lounges, like the Prado and the Puerta de Sol, unless, indeed, the "Tapada," a kind of paddock, on the road to Belem, may be reckoned among them; and even if there were, they would probably be little frequented; nor are the evenings rendered less tedious by the nightly tertulia, a pleasing feature in the society of Madrid. Families live much among themselves, seldom seeing company; neither sex is disposed to much exercise; and their chief pleasure is during summer, when they live in the retirement of their beautiful quintas, a great number of which are situated where

" Cintra's glorious Eden intervenes
In variegated maze of mount and glen."

The dress of the middle and higher classes of men differs little from that in England, except a cloak or loose great coat is commonly worn over the dress both in winter and summer. The ladies spend absurdly large sums on their

wardrobe; but their dress is tawdry and showy. Jewellery and gay-coloured shawls and mantillas are highly fashionable, coloured shoes being worn by walkers even in the filthiest streets of the city; indeed, ostentation and glare are prevailing features in the costume of Lisbon females, which forms a striking but unfavourable contrast with the sombre but luxurious dress of the ladies of Madrid. The diet of the people of Lisbon differs exceedingly from that of the French or Spaniards. Oil and garlic, the former usually thick and rancid, are unvarying ingredients at breakfast and supper, which are the principal meals : Indian corn and barley often supply the place of wheat; tea is little used, but chocolate is indispensable at breakfast, the accompaniments being hot beefsteaks, fish, &c. Rice is the invariable accompaniment of dinner, served up with boiled beef, ham, and fried sausages, all which are eaten promiscuously. The *cuisine*, however, of the higher classes is somewhat better conducted; but want of taste in these matters is universal. Fish is excellent, and cheap; and its market, in quality and variety at least, might vie with that of London. Abstemiousness in eating is little practised, even by the tender sex; but temperance in the use of wine is almost universal. Domestic habits are much more common among the middle classes in Lisbon than in Madrid; but both men and women marry at a very early age, and the parties are generally indifferent, often even unknown, to each other, the parents being the only negotiators of these unions, which may justly be called *marriages de convenance*. This accounts for, and perhaps in some measure excuses, the prevalence of conjugal infidelity, which is quite as common here, though scarcely so obtrusive, as in Spain.

The vicinity of Lisbon, ugly and uninteresting as is the city itself, presents most striking and delightful scenery. Orange and olive trees, cypresses, and judas trees, grow not only in the gardens, but in the open country. To the E. and N. of Lisbon are numerous quintas or country-houses, with rich plantations and vineyards; and about 6 m. N.W. of the capital is Cintra (which *see*), a place that holds the same relation to Lisbon as a resort of Sunday visiters, that Richmond does to London, and the romantic beauties of which have been celebrated by Byron, in language full of poetic beauty, and admirably descriptive of the scenery :

" The horrid crags by toppling convent crown'd,
The cork-trees hoar, that clothe the shaggy steep,
The mountain moss by scorching skies imbrown'd,
The sunken glen, whose sunless shrubs must weep,
The tender azure of the unruffled deep,
The orange tint that gild the greenest bough,
The torrents that from cliff to valley leap,
The vine on high, the willow-branch below,
Mix'd in one mighty scene, with varied beauty glow."
Childe Harold, c. i.

The valley of Collares, extending W. from Cintra, is one of the best cultivated, as well as richest, spots in the kingdom, and may very correctly be termed the nursery-garden of Lisbon, since the markets of that city are chiefly supplied from this quarter with fruit and vegetables ; neither must it be forgotten that the genuine Carcavella wine is made from a peculiar grape raised in this district. About 8 m. from Cintra is the palace and convent of Mafra, called, though very improperly, the Escurial of Portugal. (See MAFRA.) W. and S.W. of Lisbon the country is not so well cultivated, the hills (formed of basalt, covered with limestone) being more rocky and naked, and extending W. several miles beyond Belem. This suburb (for though Belem is 2¼ m. from the Praça do Commercio, it is connected by a nearly continuous line of streets), inhabited by a population of about 8000 persons, chiefly belonging to the nobility and wealthy citizens, may justly be called the west end of Lisbon. The church of San Geronimo has already been mentioned. The tower of Belem, another striking object, is the great custome-station of the port, whence the officers board all vessels entering the Tagus; close to it is a good quay, and without the village are the castle of Ajuda, and the *quinta de Rahuta*, with gardens, menageries, &c., open to strangers.

Lisbon was anciently called *Olisipo*, a name derived, as some say, from a legend that it was founded by Ulysses! The Romans changed its name to *Felicitas Julia*, giving it the privileges of a *municipium*, and the ruins of an ancient theatre near the cathedral warrant the inference that it was then a place of some note. The Moors captured the city in A.D. 716, and, with some slight exceptions, it remained under their power till, in 1145, Alphonso I. made it one of the capitals of Christendom. In the 14th century, Ferdinand I. surrounded the city with walls; but it attained no great importance till the reign of Emanuel the Great (1495—1521), who made it the principal port of the kingdom at a time when the Portuguese were distinguishing themselves above the other nations of Europe in maritime discovery, and wealth was fast pouring in from the recently explored regions of the east. Its subsequent history is so intimately connected with that of Portugal, that we beg to refer the reader to that article. But we cannot conclude this brief account of Lisbon without noticing the earthquake of 1755, by far the most tremendous, and most extensively felt, that has occurred in modern times. On the first of November, in the above year (a festival-day, on which all the churches were lighted up, and crowded with devotees), a sound like that of thunder was heard under ground, and immediately afterwards a violent shock threw down the greater part of the city, destroying about 60,000 human beings in six minutes! The sea first retired and laid the bar dry; it then immediately rolled in, rising 50 feet, or more, above its ordinary level. The neighbouring mountains, among the highest in Portugal, were impetuously shaken, and some of them opened at their summits, which were split and rent, huge masses of rock being thrown down into the subjacent vallies. But the most extraordinary circumstance was the subsidence of an extensive marble quay, on which great crowds had collected for safety. It suddenly sunk with all the people on it, and not one of their bodies ever floated to the surface; nor were those in boats and vessels, on the Tagus, much more fortunate, great numbers being destroyed in the whirlpool occasioned by this catastrophe. This earthquake destroyed also the seaport of Setubal, and a village about 20 m. from Morocco, with nearly all their inhabitants; violent shocks being, at the same time, felt all over W. Europe, in N. Africa, and even in the W. Indies and S. America. (*Lyell's Geology*, ii., 239.) From this disaster, Lisbon has never entirely recovered. The celebrated Marquis de Pombal, the chief minister of Portugal at the time, exerted himself to have it rebuilt on a regular plan, and to him it owes the few good streets in the neighbourhood of Rocio, the rest of the city presenting either ancient and crazy buildings crowded together in the greatest disorder, or heaps of ruins and rubbish allowed to lie where they fell 85 years ago, monuments, at once, of the indolence of the Portuguese, and of a calamity which all but annihilated one of the capitals of Europe. In 1807, the French army, under Junot, occupied Lisbon for a short time after their defeat at Vimiera; but they were soon driven from it by the combined Anglo-Portuguese army. Lord Wellington, in the same year, fortified the famous lines of Torres Vedras, which, in 1809, proved a sufficient defence against a fresh invasion of the French under Massena. (*Baillie's Lisbon*, 2 vols. passim; *Matthew's Diary of an Invalid*, p. 10-98; *Sir C. Broke's Travels in Spain and Morocco*, i., p. 8-11; *Wilde's Narr.*, i., p. 40-62; *Mod. Trav.* and *Priv. Information*.)

LISBON, p. t., Lincoln co., Me., 33 m. S.S.W. Augusta; 572 W. Bounded S. by Androscoggin river, which has here considerable falls, and across which is a bridge to Durham. It has two fulling-mills, one woollen-factory, two grist-mills, six saw-mills, two tanneries; 11 schools, 439 scholars. Pop. 1539.

LISBON, p. t., Grafton co., N. H., 89 m. N.N.W. Concord, 534 W. Chartered in 1763. Watered by great Ammonoosuc river and its branches. Limestone and iron ore are found. It was called Concord until 1824. It contains four churches, two Freewill Baptist and two Methodist; four stores, one fulling-mill, one woollen factory, two grist-mills, four saw-mills; 13 schools, 987 scholars. Pop. 1689.

LISBON, p. t., New-London co., Ct., 7 m. N. Norwich, 43 m. S.E. Hartford, 361 W. Watered by Quinnebaug and Shetucket rivers, which unite in the S. part of the township, and take the name of the latter. It contains two Congregational churches, three stores, one fulling-mill, one woollen factory, one cotton factory, with 1036 spindles, two grist-mills, six saw-mills; nine schools, 343 scholars. Pop. 1059.

LISBON, p. t., St. Lawrence co., N. Y., 10 m. W. Canton, 218 m. N.N.W. Albany, 485 W. Bounded N.W. by St. Lawrence river. Watered by Grass river. It has five saw-mills; 23 schools, 796 scholars. Pop. 3508.

LISBON, v., capital of Calcasieu par., La. Situated on the E. side of Calcasieu river, and contains a courthouse, a store, and several dwellings.

LISBURN, an inland town and parl. bor. of Ireland, co. Antrim, prov. Ulster, on the Lagan, and on the high road from Belfast to Dublin, 8 m. S.S.W. the former, and 80 m. N. by E. the latter. Pop. of old bor., in 1821, 4664; do., in 1831, 5918.

This is one of the handsomest, best built, and cleanest towns in the N. of Ireland. It consists principally of a main street along the great road. Its church has been constituted the cathedral of the united dioceses of Down and Connor. It has also a Roman Catholic chapel, a Presbyterian, two Methodist, and one Quaker meeting-house; a subscription school, Sunday schools, and a school for the education of Quaker children, supported by the voluntary subscriptions of its members, several almshouses, the infirmary for the county, a market-house, which contains a suite of assembly-rooms, and a courthouse, formerly a place of worship, for the Huguenot settlers. (See *post*.)

200

"The Lagan river, on which the town is situated, runs into the head of Belfast lough, and divides the county of Antrim from that of Down: it also separates a small suburb from Lisburn, no part of which is in the borough, though in the same parish (now in parl. bor.) Great improvements have been effected of late years in Lisburn by the Marquis of Hertford, who is the owner of the fee of the whole town, and of a considerable part of the surrounding country. A canal runs from lough Neagh into the river Lagan near the town, by which agricultural produce is conveyed to Belfast. Damask of the most beautiful description is manufactured in the town, as well as muslin and linen, though the two latter branches have fallen off considerably. There are also extensive vitriol works on an island formed by the canal and river." (*Parl. Boundary Report.*)

Under a patent from Charles II., Lisburn sent two members to the Irish House of Commons. The franchise was originally vested in the potwallopers; but was restricted by the 35 Geo. 3, cap. 29, to the £5 householders. Since the Union it has sent one member to the Imperial House of Commons. The present parl. bor. comprises 1265 acres; and had, in 1831, a pop. of 6961. Registered electors, 1839-40, 319.

Court-leets are held twice a year; a manorial court of record, with jurisdiction to the amount of £30; and another for debts to 40s., every third Wednesday. Petty sessions are held every Tuesday. It is a constabulary station. Markets on Tuesday: fairs, July 21, and Oct. 5. Post-office revenue, in 1830, £546; in 1836, £512. A branch of the Northern Bank was opened here in 1835.

This, which was long an obscure place, owed its first rise to the erection of a castle by Lord Conway, about 1627. It repulsed the Irish forces under Sir Phelim O'Neal with great slaughter, at the breaking out of the rebellion of 1641, and in 1644 baffled a similar attempt by General Monroe, who, a few years afterward, was defeated near the town by the parliamentary forces. Shortly after the Revolution, a body of Huguenots who emigrated from France on the revocation of the edict of Nantes, settled here, and introduced the finer branches of the linen manufacture, to which the town is mainly indebted for its prosperity. The castle, which was burned down with a part of the town in 1707, has never been rebuilt.

LISIEUX (an. *Lexovium*), a town of France, dep. Calvados, cap. arrond., on the Touques, 27 m. E. by S. Caen. Pop. in 1836, 11,473. It has but one good street, which forms part of the road between Caen and Evreux; all the others are narrow and crooked, and the houses built of wood, antiquated, and triste. It is, however, very well situated, and has environs of much beauty. Its cathedral, a Gothic edifice of the 12th century, has attached to it a fine chapel, dedicated to the Virgin, of a later date than the rest of the building. The bishop's palace, surrounded by noble gardens, the hospital, and the theatre, are all good buildings. Lisieux is the seat of courts of primary jurisdiction and commerce, and of a communal college, and has manufactures of woollen and cotton cloths, flannel, cotton and linen yarn, leather and brandy, and many bleaching factories and dyeing houses. It was formerly a fortress of some consequence, and was frequently besieged and captured during the middle ages. (*Hugo*, art. *Calvados, &c.*)

LISKEARD, or LESKARET, a parl. and mun. bor., market town, and par. of England, co. Cornwall, hund. West, 11 m. E. Bodmin, and 205 m. W. by S. London. Area of par., which is very nearly co-extensive with the parl. bor., 7740 acres. Pop. of parl. bor. in 1841, 4287. The town, which is meanly built with narrow streets, appears still more irregular in consequence of its site, partly in a hollow and partly on rocky heights, the foundations of some of the houses being on a level with the chimneys of others. Of late years, however, it has been considerably improved by the erection of large and handsome modern houses in the immediate environs. The town-hall, built at the beginning of the 18th century, is a large and somewhat elegant building supported by granite pillars. The church is a spacious Gothic structure, with a tower of more recent erection: the living is a vicarage, of the nett annual value of £302. An ancient free grammar school has been allowed, with other charities, to go to decay (*Comm. 22d Report*, part I.); but a national school for both sexes, and a school of industry for girls, are efficiently conducted and well attended.

Liskeard, once a town of some consequence in the duchy as the principal place for the coining and stamping of tin, has at present neither manufactures nor commerce, but it is the market of an extensive agricultural district. It has some trade in the metals of the adjacent mining districts, and there are likewise tanneries, rope-walks, &c. Markets on Saturday; fairs, Shrove-Monday, Monday before Palm-Sunday, Holy-Thursday, Aug. 15, Oct. 2, and the Monday after Dec. 6, for horses, cattle, sheep, and corn.

Liskeard (an. *Lis-kerrat*, meaning a fortified post) received its charter of incorporation in 1946, from Richard, earl of Cornwall, which was subsequently conferred by several sovereigns, and among others by Queen Elizabeth. According to the Municipal Reform Act, it is governed by a mayor, these other aldermen, and 12 councillors; it has a commission of the peace under a recorder. Corporation revenue, in 1888, £337. From the reign of Edward III. down to the passing of the Reform Act this borough enjoyed the privilege of sending two members to the House of Commons, who, for many years previously, though formally elected by the freemen, were, in fact, nominees of the Earl of St. Germaine. The Reform Act deprived Liskeard of one of its members, and at the same time enlarged the borough so as to comprise the entire parish with such parts of the old borough as are without the parish. Registered electors in 1839–40, 296.

LISLE, or LILLE (Flem. Ryssel), a strongly fortified city of France, dep. du Nord, of which it is the capital, on the canal connecting the Scarpe and Lys, in a plain 9 m. from the Belgian frontier, and 194 m. N.N.E. Paris. Lat. 50° 37′ 59″ N.; long. 3° 4′ 31″ E. Pop. in 1836, 73,000. The shape of the city is oval; length, N.W. to S.E. 1½ m., and greatest breadth about half as much. It is surrounded by a line of walls and bastions; beyond which, on its N.W. side, is the citadel, a regular pentagon, with a double ditch and extensive outworks, containing excellent barracks, officers' quarters, and magazines. This fortress has been considered the chef-d'œuvre of Vauban; and, in fact, Lille is one of the strongest cities of Europe. It is entered by seven gates, the most southerly of which, or the Porte de Paris, is a handsome Doric arch, built in 1682, to commemorate the military conquests of Louis XIV., who is represented as crowned by Victory in a group over the centre, the sides of the arch being flanked by colossal statues of Hercules and Minerva (or Mars, according to Hugo). Few French towns are generally so well laid out as this, though some parts, principally inhabited by the manufacturing population, are of the most wretched description. These are nearly 200 streets, the principal of which are straight and wide; and 27 squares and market places, the largest, the Grande Place, being 170 yards in length by nearly 85 in breadth. The houses are mostly modern, and in a solid, plain style, built chiefly of brick, but in part of stone from the neighbouring quarries. Few have more than two or three stories. Of late years, many have been built with sewns in front; and foot-pavements are becoming pretty general in the principal thoroughfares. Lille has many large and conspicuous public edifices. The Hôtel de Ville is the most remarkable, though without any great beauty. It is a heterogeneous assemblage of buildings of different epochs, appropriated to various purposes, as the tribunal of commerce, council of prud'hommes, dépôt for the octrois, police office, the residences of the chief civil and military authorities, society of sciences and arts, museum of natural history, cabinet of physical objects, library of archives, &c. It was originally a palace, constructed by the dukes of Burgundy in the 13th century; and was inhabited in 1584 by the Emperor Charles V. In 1700 it was partly destroyed by fire; but its ancient hall of conclave still exists, ornamented with some fine wainscoting, and several good paintings by Arnold de Vuez. In its chapel are preserved the likenesses of all the counts and countesses of Flanders, of the House of Burgundy. The church of St. Catherine, of simple and elegant architecture, is unfortunately hidden by mean buildings: it still possesses the magnificent painting by Rubens, of the martyrdom of St. Catherine, which luckily escaped destruction during the phrensy of the Revolution; though the elaborately ornamented door of the choir was then carried off. A telegraph is erected on the tower of this edifice. The church of St. Maurice, built in the 12th century, is the largest and oldest in the city, but its tower, which had become unsafe, has been removed. Previously to the Revolution, it possessed numerous excellent paintings; and it has still a St. Nicholas by Vanderburgh, and a martyrdom of St. Maurice by L. Jan. St. Madeline, with a handsome cupola, is the only other church worthy of remark. There are five hospitals. The Hôpital Général, founded in 1723, is a fine but unfinished pile of building, of great extent, and usually accommodating 1500 patients. The Hôpital Comtesse, founded in the 13th century, by the daughter of Baldwin, count of Flanders, and emperor of Constantinople, though it suffered greatly from fire in 1467, preserves all the characteristics of its original style of architecture. Its chapel has some good paintings by Vuez. The military hospital is large, well aired, and altogether one of the best in France. In 1814, a school of military surgery was established in it. Several large barracks are situated in different parts of the city. Lille has had a mint since a very early period; and of late, steam has been used in its machinery. The Protestant church, synagogue, abattoir or public slaughter-house, exchange, the prisons, the theatre, constructed in 1785, concert-hall, and prefecture, the last

three being handsome buildings, are the remaining principal public edifices. Lille has numerous benevolent institutions, a communal college, a public library with 21,000 volumes, well arranged, and comprising some valuable MSS., and a gallery of paintings comprising some admirable works of Vandyke, Rubens, Vuez, and other masters of the Flemish, French, and other schools. In 1834, this gallery was enriched by a collection of designs from the Italian school. A royal academy of music, academies of drawing, architecture, botany, &c.; a botanic garden, and various learned societies, nearly complete the list of public establishments. The canal on which Lille is built has several branches navigable for small trading vessels, which pervade the city. In its progress by and through Lille, different parts of this canal are called the upper, middle, and lower Doule; along the middle Doule, or portion between the town and the citadel, is a fine esplanade, the favourite resort of the upper classes. The middle Doule is here crossed by a handsome bridge, the Pont Royal, or Pont Napoléon; the other bridges are in no wise remarkable. Lille has few public fountains, and, in fact, the want of good water is the greatest inconvenience suffered by the city: nearly the whole supply has to be drawn from the canal, and is of very indifferent quality. Beyond the walls are several suburbs, chiefly inhabited by the manufacturing population.

Lille is one of the chief seats of the French cotton manufactures. Calicoes, cotton handkerchiefs, indiennes, stockings, and cotton yarn, are the goods principally produced. The manufacture of table-linen; linen thread, and lace, is also considerable; and fine woollen cloths, velvets, serges, hats, leather, paper, beet-root sugar, geneva, soap, and mineral acids, are made, some to a greater, and some to a less extent. Government has here a tobacco manufactory and a saltpetre refinery; and the neighbourhood is studded with bleaching grounds and oil-mills; and it is in the centre of some very extensive beet-root plantations. Steam power is extensively employed in the different manufactures. We regret to have to add, that the condition of the work-people employed in the different factories appears to be, speaking generally, exceedingly bad, and decidedly worse, indeed, than in any other town of France. M. Villermé says, that four out of every 13 persons are in a state of absolute indigence. Between 3000 and 4000 cotton weavers and twisters live in small, damp, ill-ventilated, under-ground cellars, crowded to excess, and in the most deplorable state of poverty. After describing these wretched hovels, and the still more wretched furniture, M. Villermé adds: "Je voudrais ne rien ajouter à ce détail des choses hideuses qui révèlent un premier coup d'œil, la profonde misère des malheureux habitans; mais je dois dire que, dans plusieurs des lits dont je viens de parler, j'ai vu reposer ensemble des individus des deux sexes et d'ages très différens, la plupart sans chemise et d'un saleté repoussante. Père, mère, vieillards, enfans, adultes, s'y pressent, s'y entassent." (Tableau Physique, &c., des Ouvriers, i., 82, 83.)[*] After all, however, the cellars do not appear to be the worst lodgings, for the garrets, which are quite as ill furnished, are much more exposed to the inclemencies of the weather, and are inhabited by those who cannot afford to buy fuel. The linen thread spinners, the most numerous class next to the cotton spinners and weavers, are said to be favourably distinguished from the latter by their greater cleanliness, sobriety, and better general conduct, though their wages be even lower. Villermé roughly estimates that a workman's family, consisting of a father, wife, and child of from 10 to 12 years of age, who are all employed, might make, at an average, about 915 fr. (£36) a year; of which the rent would absorb from 40 to 80 fr. (say 60 fr.), and food nearly 540; leaving less than 150 fr. for the expenses of furniture, clothing, washing, fire, candle, and tools; so that any slight indulgence, want of employment, or illness, could not fail to plunge the family into the utmost want. It is not easy, however, to see why, with such average wages, the labouring population of Lille should be so much depressed. A family of three persons, who should receive £30 a year in England, would be reckoned anything but badly off; and if it were true, as is commonly affirmed, that the cost of living is a third less in France than in England, a Lille working man with £36 a year should be as well off as a Manchester workman with £50. In truth and reality, however, there is no such difference between the cost of living in the two countries; and both it and the rates of wages in each approach much more nearly to an equality than is generally supposed. The Lille workmen, unlike those of Lyons, are not prone to engage in insurrections; and no French manufacturing town has, in proportion, so many mutual benefit societies. These, however, are so badly organized and conducted as to be next to useless; their meetings are always

[*] The details by M. le V100. de V. Burgemont, formerly préfet of the dep., are still more revolting. (See his Rapport, &c., centre in Chal. March, 1834.)

held in a *beer shop*; and at the end of the year, all the money in the chest above a small amount is divided among the contributors, to be spent on the spot, "and the new year commences with the formation of a new fund, the ultimate destination of which is the same." (*De Villeneuve-Bargement*.) Drinking is, in fact, the prevailing vice and sole amusement of the workmen of Lille. Though most prevalent among the cotton weavers, &c., it is not a consequence of the introduction of the cotton manufacture; for long previously to that epoch many of the work-people were accustomed to work only three days in the seven, and to spend the other four in pot-houses.

Lille is the seat of courts of primary jurisdiction and commerce, a *conseil des prud'hommes*, forest inspection, &c., and is the head quarters of the 16th military division of France.

It is supposed to have been founded in 640; and successively belonged to the counts of Flanders, the kings of France, and the dukes of Burgundy. In 1667 it was taken by Louis XIV.; and being improved and fortified by Vauban, was definitely annexed to the crown of France. It has sustained several sieges, of which the most celebrated was that by the allies, under the Duke of Marlborough and Prince Eugene in 1708. It was bravely defended by Marshal Boufflers; but notwithstanding the gallantry of the garrison, and the fact that the French had a powerful army in the field, it was ultimately obliged to surrender. In 1792 it was bombarded by the Austrians. (*Hugo*, art. *Nord; Fillermá, Tableau Physique, &c., des Ouvriers*, i., 78-107; *Guide du Voyageur; Dict. Géog., &c.*)

LISLE, p. t., Broome co., N. Y., 123 m., W.S.W. Albany, 310 W. Watered by Toughnioga river and its tributaries. It contains six stores, two fulling-mills, three grist-mills, 10 saw-mills, two tanneries; 11 schools, 542 scholars. Pop. 1560.

L'ISLE, or L'ILE, a town of France, dep. Vaucluse, cap. cant., on an island in the Sorgues, a tributary of the Rhone, 12 m. E. by S. Avignon. Pop., in 1836, 4818. It manufactures woollen fabrics and yarn, tram and organzine silk, and leather, and has some trade in silk, madder, oil, and wine.

LISMORE, an inland town of Ireland, cos. Waterford and Cork, on the Blackwater, 93 m. E.N.E. Cork. Pop., in 1831, 2894. The town has been much improved of late years, principally through the exertions, and at the expense, of the Devonshire family, who have large possessions in this part of Ireland. It stands on an eminence overlooking the river, across which is a fine bridge, built at the expense of the Duke of Devonshire. Lismore was formerly the seat of a bishopric, now united with Cashel and Waterford. The cathedral is in good preservation, and handsomely fitted up; it has also a large Roman Catholic chapel, a Presbyterian and a Methodist meeting house, an excellent grammar-school, built and endowed by the Devonshire family, some alms-houses, a courthouse, a fever hospital, and a dispensary. But the great attraction of Lismore is its magnificent old castle, founded by King John, in 1195. It is nobly situated on a rock rising perpendicularly from the river. This large and venerable pile was once the property of the famous Sir Walter Raleigh; and, after numerous vicissitudes, came through the Boyles, into the possession of the Devonshire family, by whom it has been greatly improved and embellished. It is now in complete repair, and is occasionally visited by its noble owner. It has withstood several sieges. In 1785, the Duke of Rutland, then Lord Lieutenant of Ireland, held a court here, and issued some proclamations, dated from the castle.

Lismore returned two members to the Irish House of Commons till the Union, when it was disfranchised. A manor court holds pleas to the extent of £10 every third week. Petty sessions are held on alternate Wednesdays. It is a constabulary station. The trade of the town is inconsiderable, though a canal has been constructed, by the Duke of Devonshire, from it to near Cappoquin, where the river becomes navigable. There is a salmon fishery close to the town, the produce of which is mostly sent to London. Postoffice revenue, in 1830, £273; in 1835, £371.

LISSA (Polish *Leszno*), a town of the Prussian dom., prov. Posen, circ. Fraustadt, near the border of Silesia, 12 m. E.N.E. Fraustadt. Pop. about 8600, of whom 4000 are Jews. It is walled, and has three Lutheran churches, a Roman Catholic church, a synagogue and Jewish school, a gymnasium, two hospitals, a fine council-house, and a theatre. Its streets are mostly narrow and dirty; and the greater number of the houses are of wood. The neighbouring castle was formerly the property of the Leczinsky family, of which Stanislaus, the last king of Poland, was a member; but it is now the residence of the princes Sulkowski, to whom the town belongs. Lissa is the seat of a police court, and a board of taxation: a considerable manufacture of woollen cloth is carried on in it and its neighbourhood; and, besides woollen stuffs, it trades in furs, wines, and hardware. Lissa was an unimportant village, when a num-

ber of Protestants, driven from Silesia, Bohemia, and Moravia, by the persecutions of the 17th century, settled in it; and to these immigrants it owes its present consequence. (*Von Zedlitz, Das Preussische Staat*, iii., 156, &c.)

LISSA. *See* LESINA.

LITCHFIELD, county, Ct. Situated in the N.W. part of the state, and contains 885 sq. m. Watered by Housatonic river and its branches, and by branches of Naugatuck and Farmington rivers, which afford extensive water power. Iron ore is found in Salisbury and Kent, and is extensively manufactured. The surface is elevated, and in some parts mountainous. It contained in 1840, 58,749 neat cattle, 119,990 sheep, 30,431 swine; and produced 33,574 bushels of wheat, 198,143 of rye, 919,378 of Indian corn, 39,665 of buckwheat, 316,687 of oats, 568,680 of potatoes, 735 pounds of silk cocoons, 42,354 of sugar. It had 131 stores, 14 furnaces, and 36 forges, employing a capital of $413,580, 34 fulling mills, 18 woollen factories, five cotton factories, with 6334 spindles, 49 grist-mills, 130 saw-mills, one oil-mill, two paper-mills, 49 tanneries, six distilleries, two printing-offices, two weekly newspapers; total capital in manufactures $994,947; 12 academies, 480 students; 233 schools, 9667 scholars. Pop. 40,448. Capital, Litchfield.

LITCHFIELD, p. t., Kennebec co., Me., 11 m. S.S.W. Augusta, 584 W. Incorporated in 1795. Watered by Cobbosseconte river and its branches. It contains six stores, three grist mills, two saw-mills; 17 schools, 732 scholars. Pop. 2293.

LITCHFIELD, t., Hillsborough co., N. H., 30 m. S. by E. Concord. Bounded W. by Merrimac river, across which are two ferries. It contains a Christian church, two stores, one grist-mill, three saw-mills; five schools, 161 scholars. Pop. 480.

LITCHFIELD, p. t., capital of Litchfield co., Ct. 100 m. N.N.E. New-York, 32 m. W. Hartford, 396 W. Incorporated in 1724, when it was a frontier town. It contained four ecclesiastical societies, Litchfield proper, south Farms, Northfield and Milton. The village, incorporated in 1818, is pleasantly situated on the level summit of a hill, and is chiefly built on two streets, crossing each other at right angles, at the intersection of which is a handsome public square. It contains a neat courthouse, a jail, two churches, a Congregational and Episcopal, a female academy, one of the oldest, and long the most respectable in the state, a branch of the Phœnix bank of Hartford, two printing-offices, 80 dwellings, many of them neat and handsome, and about 500 inhabitants. A law school was established here in 1784 by Hon. Tapping Reeve, with whom Hon. James Gould was associated in 1798. This was long the most celebrated institution of the kind in the United States, and here many of its principal civilians have been educated. It has been discontinued since the death of its distinguished conductors. In the S. part of the town is the largest pond in the state, a beautiful sheet of water, containing 900 acres. It is watered by head branches of Naugatuck and Shepaug rivers, which afford water-power. South Farms, in the S. part of the town, contains a congregational church, an academy, and a considerable number of dwellings. The town contains 21 stores, seven fulling-mills, five woollen factories, one forge, one paper-mill, six grist-mills, four saw-mills, one oil-mill, six tanneries, two printing-offices, two weekly newspapers; three academies, 55 students; 31 schools, 946 scholars. Pop. 4038.

LITCHFIELD, p. t., Herkimer co., N. Y., 83 m. N.W. Albany, 387 W. Organized in 1796. Drained by head branches of Unadilla river, and by Steele's creek. It contains four stores, two grist-mills, 12 saw-mills, three tanneries, one distillery; 11 shools, 455 scholars. Pop. 1672.

LITCHFIELD, p. t., Bradford co., Pa., 156 m. N. Harrisburg, 265 W. Drained by branches of Wepassening creek. It has one grist-mill, 11 saw-mills; five schools, 945 scholars. Pop. 817.

LITCHFIELD, p. t., Medina co., O., 113 m. N.E. by N. Columbus, 360 W. It has two stores, one grist-mill, two saw-mills; six schools, 945 scholars. Pop. 787.

LITCHFIELD, p. t., Hillsdale co., Mich. Watered by St. Joseph's river and its branches. It has one saw-mill, three schools, 112 scholars. Pop. 691.

LITCHFIELD, p. v., capital of Grayson co., Ky., 100 m. S.W. by W. Frankfort, 651 W. It contains a courthouse, jail, and several stores and dwellings.

LITTLE BEAVER. t., Beaver co., Pa. Drained by a branch of Little Beaver creek. It contains six stores, one fulling-mill, one woollen factory, four grist-mills, five saw-mills, four tanneries, one pottery; seven schools, 244 scholars. Pop. 1254.

LITTLE BRITAIN, t., Lancaster co., Pa., 90 m. S.E. Lancaster. Watered by Octarara creek and Conewingo creek. It contains 10 stores, one fulling-mill, one woollen factory, eight flouring-mills, nine grist-mills, eight saw-mills, five tanneries; nine schools, 300 scholars. Pop. 3048.

LITTLE COMPTON, p. t., Newport co., R. I., 32 m.

R.R. Providence, 433 W. Bounded S. by the Atlantic, W. by the E. entrance of Narragansett bay. First settled in 1674. It contains two churches, a Congregational and Friends, six stores; seven schools, 144 scholars. Pop. 1397.

LITTLE CREEK, hund., Kent county, Del. It has six stores; three schools, 294 scholars. Pop. 2050.

LITTLE CREEK, hund., Sussex co., Del. It has 13 stores, one forge, 11 grist-mills, 22 saw-mills, four tanneries; one academy, 53 students; 16 schools, 274 scholars. Pop. 2973.

LITTLE FALLS, p. t., Herkimer co., N. Y., 91 m. W.N.W. Albany, 390 W. Organized in 1889. Watered by Mohawk river, on the S. side of which runs the Erie canal, and on the N. side, the Utica and Schenectady railroad. The Mohawk here passes a mountain barrier, and the river descends 42 feet in three fourths of a mile, chiefly by two long rapids. The mountain abounds with rock crystals. The village, is a most romantic situation, contains five churches, a Presbyterian, Episcopal, Methodist, Baptist, and Roman Catholic, an academy, and various mills, and manufactories. A fine marble aqueduct crosses the river, and forms a feeder of the Erie canal. It is 214 feet long, 16 feet wide, with walls 14 feet high and 4 broad at top, with three arches, the central one of 70 feet span, and the side ones of 50 feet each. The canal originally passed the brow of the mountain here with difficulty, by expensive digging and embankment. In widening the canal, more ample room is obtained by occupying a part of the bed of the river, between an island and the S. bank. There are in the town, mostly in the village, 34 stores, three furnaces, two forges, one fulling-mill, one flouring-mill, two grist-mills, four saw-mills, three paper-mills, four tanneries, one distillery, one brewery; eight schools, 810 scholars. Pop. 2651.

LITTLE FALLS, p. v., Passaic co., N. J., 79 m. N.N.E. Trenton, 4 m. W.S.W. Paterson, 245 W. Situated chiefly on the S. bank of Passaic river, which here has a fall which may be used under a head of 33 feet, affording a great water-power. It contains four stores, two cotton factories, with 9498 spindles, one grist-mill, three saw-mills, one woollen carpet factory, a schoolhouse, used also as a church, 50 dwellings and about 356 inhabitants. The Morris canal passes the river by an aqueduct, below the falls. It is five and a half miles from Acquackanock landing.

LITTLE ROCK, city, capital of Pulaski co., Ark., and of the state, is 1065 m. from Washington city. It is in 34° 40' N. lat., and 92° 13' W. lon. on the S. bank of Arkansas river, 300 m. from its mouth by the course of the river. It is on a high bluff, elevated from 150 to 200 feet above the river, and is the first place in which rocks occur, above its mouth. Steamboats navigate to this place at all stages of the water, and in high water, above it to Fort Gibson, in the Indian territory, about 1400 m. from its mouth, which renders Little Rock extensively accessible by water from the interior. It is regularly laid out, and contains a statehouse, courthouse, jail, five churches, a Presbyterian, Episcopal, Methodist, Baptist, and Roman Catholic, two banking houses, a theatre, an academy, a United States arsenal, a United States land office, a penitentiary, 21 stores, two steam saw and grist mills, two printing-offices, 560 dwellings, and about 2600 inhabitants. It was laid out and first settled in 1820.

LITTLETON, p. v., Grafton co., N. H., 83 m. N.N.W. Concord. 544 W. Bounded N.W. by Connecticut river. The rapids, called Fifteen-mile falls, are opposite to this town, and the river runs in foaming waves, which cannot be passed by boats. Three bridges cross the river in this town. Watered by Ammonoosuck river. Incorporated in 1784, though previously granted by another name. It contains two churches, a Congregational and Baptist, four stores, one grist-mill, six saw-mills. Pop. 1778.

LITTLETON, p. t., Middlesex co., Mass., 26 m. W.N.W. Boston, 496 W. Incorporated in 1715. It contains two churches, a Congregational and Baptist, four stores, two grist-mills, two saw-mills; six schools, 303 scholars. Pop. 997.

LITTLE VALLEY, p. t., Cattaraugus co., N. Y., 300 m. W. by S. Albany, 342 W. Watered by Alleghany river, and its branches. It has three stores, one grist-mill, seven saw-mills; four schools, 265 scholars. Pop. 700.

LIVERMORE, p. t., Oxford co., Me., 30 m. W. Augusta, 596 W. Watered by Androscoggin river, which affords water-power. Incorporated in 1715. It contains nine stores, two fulling-mills, three grist mills, five saw-mills, two tanneries; 22 schools, 1146 scholars. Pop. 2745.

LITHUANIA, a country comprising a considerable portion of the ancient kingdom of Poland, at present parcelled among the Russian governments of Wilna, Grodno, and Minsk (which see).

LITTORALE (AUSTRIAN). See ILLYRIA, TRIESTE, &c.

LITTORALE (HUNGARIAN). See HUNGARY, CROTIA, FIUME, &c.

LIVADIA (an. Lebadia or Lebadeia), a city of indep. Greece, which, under the Turks, gave its name to the prov. comprising E. and W. Hellas, in Boeotia, on the Hercyna, about 6 m. W. the lake Copais, 25 m. W.N.W. Thebes, and 50 m. N.W. Athens; lat. 38° 26' N.; long. 22° 57' E. Pop. uncertain; but before the Greek revolution it was estimated at 10,000. At that period it had 1500 houses, many of which were good, though its streets were dirty, narrow, and inconvenient. Its site is very striking, occupying several fantastic knolls and crags at the entrance of a deep defile in a branch of the Heliconian chain. The river Hercyna, which rises in a full stream and with great force from beneath a rock close to the town, rolls in foaming torrents over masses of rock: it is augmented near its source, by a tributary stream from the cavern of Trophonius. The ancient city, called Mideia by Homer, is supposed to have been built, in part at least, upon the lofty heights which overhang the modern town, and upon which the remnants of a citadel are still visible; with additional buildings constructed by the Catalons, when they were in possession of this country.

Previously to the revolution, Livadia was the seat of a voivode or governor, and a cadi and had six mosques, and as many Greek churches. It had also a considerable trade in the produce of the surrounding territory, and of Attica, with Constantinople, and foreign countries. Madder, corn, oil, kermes, cotton, and honey, were among its principal exports, which it formerly sent to Trieste, Venice, Leghorn, Genoa, and even London. The port at Aspropiti, the ancient Anticyra, on the Corinthian gulf, is 18 m. W.S.W. from the town. Livadia was burned by the Turks in 1821, and subsequently in part destroyed by the Greeks in an attack upon the Turkish garrison. There can be little doubt, however, that since the revolution it has recovered some portion of its former trade and prosperity. Its greatest drawback is the unhealthiness of its situation. It suffers from great extremes of temperature; the air is frequently loaded with dense fogs, and in summer is vitiated by pestilential effluvia from the neighbouring lake of Copais. In 1785-86, the plague carried off 6000 of the inhabitants. (Clarke's Travels, vii., 146-170; Hughes, i., 237-349; Holhouse, p. 260-264.)

The entire celebrity and, perhaps, even existence of Livadia, was owing to its being the seat of the famous oracle or cavern of Trophonius. Dr. Clarke has sufficiently identified the site of this celebrated cavern; but the reasons he has alleged in favour of the hypothesis, that the Hercyna is identical with the fountain of Lethe, or waters of oblivion, are far from conclusive. (Clarke, vii., 161, 8vo. ed.) Clarke supposes the fountain that now issues from below the cavern to be that which anciently received the name of Mnemosyne, or waters of memory; but this fountain may formerly have been divided into two, or one of the ancient fountains may have disappeared through some convulsion of nature: at all events, it would seem to be pretty clear from the statement of Pausanias, that there were within the sacred precinct the two fountains of Lethe and Mnemosyne, exclusive of the source of the Hercyna. (Pausanias, lib. ix., cap. 39.)

This was one of the most formidable of the Greek oracles. The Hieron, or sacred cavern, was surrounded by bare, rugged, and high precipitous rocks. Thither those anxious to consult the oracle were, after long preparation, conducted at night through a grove; and could not fail to be deeply impressed by the solemnity of the place, and by the roaring of the waters of the Hercyna bursting forth from their subterranean caverns. Having arrived at the Hieron, the votary, after addressing a prayer to the statue of Trophonius by Dædalus, descended into the adytum, a narrow and deep aperture excavated in the rock, and, so doubt, leading to some great natural cave or chasm. Those who ventured down into this hidden recess seem, generally, to have experienced rather rough treatment. Trophonius was not to be questioned with impunity. The votaries, when they came forth from the abyss, were usually much exhausted, and had no distinct recollection of what they either heard or saw. Generally, however, the mysteries of this dread cavern made a deep impression on their minds, and entailed upon them a settled melancholy the remainder of their lives; so that it was a proverbial expression in Greece to say of a gloomy or melancholy individual, that "he had come from the cave of Trophonius." No doubt, however, the priests took care to modify their treatment of the votaries, as well as their responses, according to their rank and their bounty to the temple. Pausanias, who descended into the adytum, and describes what occurred to himself, states that when he came out he was so confused as to have lost his senses. But this visit must have taken place no less as the middle of the second century, after the oracle had been long on the decline; and when, probably, it had been stripped of half the horrors by which it had formerly struck terror into those who attempted
195

LIVERPOOL.

tempted to penetrate by its means through the veil that conceals futurity from mortal eyes.

The accounts of Trophonius, the reputed founder of the oracle, vary extremely. This much, however, seems to be agreed upon; that he was a mortal to whom, after his death, divine honours were paid; and that he was supposed to be endowed, like Apollo, with the power of foreseeing and predicting future events. It is probable that the gloomy grandeur of the place, and the discovery of some hidden cavern, where all sorts of impostures might be easily practised, first suggested the idea of making it the seat of an oracle; and there seems little doubt that it was indebted to the same circumstances for its celebrity and its votaries.

According to Dr. Clarke, the present town of Livadia occupies that part of the consecrated ground formerly covered by the grove of Trophonius; but this is merely one of those conjectures in which that ingenious person is, on all occasions, too prone to indulge. Pausanias says, that Lebadea was as much ornamented by temples, statues, and other splendid works of art, as any city of Greece. A statue of Trophonius by Praxiteles was deservedly reckoned among its principal treasures. (For further particulars as to Lebadéa, see *Pausanias*, lib. ix., cap. 39; *Voyage D'Anacharsis*, cap. 34; *Potter's Grecian Antiquities*, book ii., cap. 10, &c.)

LIVERPOOL, a parl. and mun. bor. and seaport of England, being, next to London, the greatest emporium of the British empire, and, in fact, of the world, in the county Lancaster, hund. W. Derby, on the E. or right bank of the Mersey, 32 m. W. by S. Manchester, 67 m. W.S.W. Leeds, and 196 m. N.N.W. London; lat. 53° 22' 30" N., long. 2° 57' W. The pop. of the parl. and mun. bor., at the undermentioned periods, has been:

Townships.	1801.	1811.	1821.	1831.
Liverpool	77,653	90,376	118,972	165,175
Kirkdale				2,591
Everton	499	913	2,109	4,518
W. Derby	2,666	3,886	6,384	9,612
Toxteth Park	2,049	5,864	12,829	94,067
Total of parl. bor.	82,957	105,851	140,811	205,954

The present (1841) population of the parliamentary borough is 286,487.

Liverpool stands partly on flat ground, along the edge of the river, and partly and principally on a gently rising declivity. Besides quintupling its population during the last half century, it has been more improved, during that period, than any other town in England, not excepting Manchester. Before that time, narrow, inconvenient, and ill-paved streets, lined with dull, heavy looking houses, were its characteristic features; but so great is the alteration effected chiefly through the exertions of the corporation and the public spirit of the citizens, that at present no town or city in the three kingdoms, except their capitals, has wider or handsomer streets, more sumptuous public buildings, or better constructed and more substantial private dwellings. The corporation is alleged to have expended on improvements, between 1786 and the present time, no less than £1,700,000! The present limits of the town comprise about two-thirds of the parliamentary borough: its length from N. to S. (measured from Brunswick dock to the gasworks in Vauxhall road) is about two and a half miles, its breadth from the river to the church at Edgehill, one and three quarter miles, and its area somewhat exceeds 2500 acres. The central point, from which many of the principal avenues diverge, is the open space partly occupied by St. John's church, and the railway station: the diverging roads are, 1. Dale-street, a fine broad avenue running S.W. to the townhall and exchange buildings, and continued under the name of Water-street, to George's dock; 2. Whitechapel, and Paradise-street, leading to the custom-house; 3. Lime-street, Renshaw-street, Berry-street, and Great George-street, running nearly S. towards Toxteth park; 4. the London road, taking an E. direction towards the Zoological gardens; and, lastly, Byrom-street, and Scotland road, leading to the house of correction in Kirkdale. The principal streets, independent of those above mentioned, are Castle-street, opposite the townhall, Lord-street, Church-street, Hanover-street, Bold-street, Rodney-street, Mount Pleasant, St. Anne's-street, and the Vauxhall road; and among the principal squares may be mentioned Great George's, Queen's, Abercrombie, Clayton, and Cleveland. Liverpool is abundantly supplied with good water by two companies, from 10 large reservoirs in various quarters of the town; and there are two gas companies, which supply gas not only to the streets, shops, and factories, but likewise to a large proportion of the most respectable private houses.

Corporation and Government Buildings.—The townhall, which stands at the N. end of Castle-street, was commenced in 1749; its interior having been destroyed by fire

in 1795, it has been since rebuilt at a cost of more than £110,000. It has a rustic basement, supporting a range of Corinthian columns and pilasters: in the S. front is a handsome portico, and the building is surmounted by a light and elegant cupola, above which is a colossal figure of Britannia. The interior comprises, besides a handsome suite of apartments for the mayor, a noble ball-room, 90 feet long by 40 feet both in breadth and height; a council-room, committee-rooms, town-clerk's, treasurer's, and surveyor's offices, &c.; the grand stair-case, under the cupola, is a magnificent specimen of modern architecture: on the landing is a colossal statue of Canning. The exchange buildings, which form three sides of the square in which the townhall stands, were completed in 1808, at a cost of £110,848, raised by £100 shares. The principal front is 197 feet in length: and the area, enclosed by the entire building, somewhat above 11,000 square yards: in the N. and S. fronts are two magnificent porticos, each supported by eight Corinthian columns and surmounted by a carved entablature with stone figures: a very handsome balustrade runs round the entire building. Piazzas extend round the basement, for the convenience of the merchants, in hot or rainy weather. In the interior is a magnificent news-room, originally 94 but now 125 feet in length, by 51 feet 9 inches in width, having an arched roof supported by two rows of columns: above this is a splendid room for the underwriters, while, in other parts of the building, are numerous counting-houses and offices, warehouses, &c. The Liverpool exchange is, in fact, one of the best specimens of Grecian architecture in England; and, perhaps, the noblest structure erected in modern times for purely commercial purposes. In the centre of the square is a monument in honour of Nelson, executed in bronze, on a marble basement: it represents the dying hero his foot on a prostrate enemy, receiving a naval crown from Victory. The principal group is surrounded by emblematical figures; but the whole are stiff, affected, and unnatural; and it neither redounds to the credit of the town nor of the artists. W. of the exchange stands the newsroom-house, a low, plain, stone building, with two principal entrances; in the interior are two spacious rooms, used as news price and crown courts, with various other apartments for the judges, magistrates, jurors, &c.; the whole was erected, in 1828, at a cost of £19,312, exclusive of some recent alterations occasioned by the removal of the assizes of the W. Derby and Salford hundred from Lancaster. The buildings for the custom-house, excise-office, &c., at the S. end of Castle-street, recently erected on the site of the old dock, which was filled up for the purpose, form a very extensive pile, remarkable for simplicity and massiveness; it covers an area of 6769 square yards, and has an extreme length of 467 feet measured from E. to W., with a total height of 67 feet, the length of its wings being 235 feet, and their breadth 94 feet. Grand porticos, supported by Ionic columns adorn the centre, and E. and W. fronts; and above the centre is a large dome, lighted by 16 windows, and embellished with pilasters. The basement, through which there is a public passage connecting Castle-street with Park-lane, is used for storing bonded goods: the centre is occupied by the great stair-cases and the long room, 146 feet in length, 70 feet wide, and 45 feet high, lighted from the great dome: the W. or river wing is occupied by the various offices of the custom-house; and the E. wing contains the excise-office, dock-offices, post-office, and stamp-office. This splendid and useful building was erected in consequence of an arrangement between the corporation of Liverpool and the government, brought about through the mediation of Messrs. Canning and Huskisson. The corporation gave the land, valued at £90,000, and erected the building, which, at the cost of 90 years, is to be ceded to government, on the latter paying for it the sum of £150,000, by annual instalments of £25,000. A splendid building, called St. George's hall, on the site of the old barracks, near St. John's church, is about to be erected by the corporation, at a cost of £80,000, for the accommodation of the assize courts, and which is also to comprise a room for public meetings, a music hall, &c.

Literary Establishments, &c.—Though so extensively busied in trade and commerce, the merchants of Liverpool are honourably distinguished by their attention to, and patronage of, science and literature; and the town has several valuable institutions for their promotion. The principal among these is the Royal Institution in Colquitt-street, formed, in 1814, at the suggestion of the late Mr. Roscoe, by the subscription of £100 shares, and chartered in 1822: it comprises academical schools, public lectures on various subjects, laboratories and philosophical apparatus, a collection of books, and a museum. The building, with a portico and two wings, contains suites of rooms well adapted to the purposes of the institution. The lecture-room is capable of accommodating 500 persons; and the natural history department of the museum, occupying two

204

floors of the building, is, perhaps, the largest and most valuable in the kingdom, after the British museum, and and that of the Zoological Society in London. The institution has also a fine collection of casts from the Elgin, Æginetan, and Phigalian marbles, and from some of the most celebrated statues of antiquity. A school of medicine, with an anatomical theatre and dissecting-rooms, is attached to the establishment, which has also very recently opened a grammar school, in a neat-looking Doric building, in Soul-street. The mechanics' institute, in Mount-court, near St. James's cemetery, intended not only to meet the wants of the working classes, but also to bring them in contact with those in higher spheres of life, was opened in 1837: its buildings, which, with courts, &c., cover nearly an acre of land, given by the corporation, were erected at a cost of above £11,000. The front, in the Ionic style, has a heavy appearance: the grand theatre is capable of accommodating 1900 persons: it has a museum, and a library, with 7300 volumes. It has attached to it three schools; one for the children of tradesmen and mechanics, furnishing a plain and useful education at a cost of £3 3s. a year; another for the children of those in more affluent circumstances, or who choose to pay the fees, which amount to £10 10s. a year; and a third school for adults, held only in the evening. Lectures are delivered twice a week: and prizes are given twice a year, for proficiency in the various branches of study. The institute has upwards of 3300 members and 1900 pupils, at the different schools, and is, unquestionably, a most important educational establishment. The Literary, Scientific, and Commercial Institution, in St. Anne's-street, established in 1835, has a library of 2200 volumes, a news-room, and a theatre for lectures; classes are formed for languages, &c., lectures are delivered on different branches of science and literature, and meetings are held for discussion. The annual subscription is 20s.; and it is well attended, chiefly by commercial and law clerks, medical students, &c. The Medical institution, in Mount Pleasant, lately built at a cost of about £6000 (one third of which, with the land, was contributed by the corporation, and the rest by the medical practitioners of the town), has a circular-shaped front, of the Ionic order, 105 feet in length, and 35 feet in height; and in the interior are various large apartments, used as libraries, museums, lecture-rooms, &c. Apothecaries' hall, belonging to a joint-stock company, may be mentioned here, not as a place of medical instruction, but as conferring important benefits on the profession and the public, by importing and manufacturing medicines of the purest quality: the building is handsome, and aft the arrangements most complete. The Collegiate institution, intended to furnish a superior education on conservative principles, is now (1841) being erected in Shaw-street, at an expense of £30,000.

Closely connected with the above are the news-rooms, among which the Athenæum holds the highest station. The building, opened in 1799, is large, but plain; 360 proprietors subscribe to form a yearly income of £1390; the library comprises 17,500 volumes; and the news-room is spacious, and well provided with the publications of the day. The Lyceum, a much handsomer building, erected at an expense of more than £11,000, supported by about 930 proprietors, paying guinea-subscriptions, has a library of about 30,000 volumes, in an elegant circular room lighted from a cupola. The Union news-room in Duke-street is also a very respectable building; and there is an important news-room, already noticed, in the Exchange.

The celebrity and example of Mr. Roscoe, the most distinguished of all the citizens of Liverpool, had a wonderful influence in creating and diffusing a taste for literature among all classes of his townsmen. The first editions of his valuable and standard works, the lives of Lorenzo de Medici and of Leo X., were published here. They were printed by Mr Copsey, the author of the poem entitled the "Press," and are enduring monuments of his skill and excellence as a typographer. The life of Poggio Bracciolini, by the Rev. William Shepherd, a distinguished Unitarian minister of the town, is also one of the products of the Liverpool press; as is the edition of Burns' works, with his life, in four volumes, by the justly celebrated Dr. Currie.

A newspaper, entitled the Courant, was published at Liverpool in 1712; but it did not succeed; and the earliest of the existing Liverpool papers appeared on the 28th of May, 1756, having been continued, under different names, to the present time. At this moment, Liverpool has the following newspapers, viz: The Albion, Times, Mercury, Chronicle, and Journal, which, more or less, strongly advocate reform principles; and the Courier, which is Conservative, are published once a week. The Mail, a Tory journal, appears on alternate days; and The Standard, of similar politics, twice a week. Gore's Advertiser and Myers' Commercial Advertiser, are devoted wholly to commercial matters.

Places of Amusement, &c.—Liverpool has three theatres and an assembly-room. The Theatre Royal, built in 1817, has a plain, uninteresting exterior; but the inside is large, elegantly fitted up, and well adapted for hearing. The company of comedians ranks as the best out of the metropolis. The Amphitheatre, in Great Charlotte-street, is used for equestrian exercises; and the Liver theatre, open during the winter months, exhibits burlettas and melodramas, equally gawdy and noisy with those presented in the minor theatres of London. The Wellington Rooms in Mount Pleasant, erected by subscription in 1815, comprise a fine suite of saloons, the largest of which, the ball-room, is 90 feet in length. Among other places of amusement, may be mentioned the Zoological and Botanic Gardens, and the race-course at Aintree, where meetings are held in May and July. The first of these, now an object of considerable attraction, in consequence of its rich and rapidly increasing collection, is situated in the E. suburbs. The grounds, comprising about 10 acres, are tastefully laid out; and the disposition of the buildings is well adapted for the accommodation of the animals. The Botanic Garden is chiefly interesting to the student; but admission is easy to all who are disposed to visit the grounds, situated in Edge lane, close to the E. boundary of the borough. The baths, on the W. side of George's Dock, forming a handsome feature in the town, are extremely well constructed, sufficiently large, and admirably provided with accommodation for visitors. There is, also, a floating bath for more adventurous bathers. Liverpool, however, like London, Manchester, and other densely peopled towns, is very imperfectly supplied with means for the healthy amusement of the great bulk of its inhabitants. St. James's Walk, near the Cemetery, and the Prince's Parade, on the river-bank, are fashionable and well-kept promenades; but there are no exercising-grounds, no open parks, no breathing-places for the use of the labouring population, the majority of whom, we are sorry to say, have for their homes, at most, two small rooms, more frequently only one, and often only a damp, miserable cellar. Such an addition to Liverpool would be a great boon; and a corporation which, unlike many others, has honourably distinguished itself by its attention to improvements, will, it is to be hoped, lend its powerful aid to the bringing about so desirable a result.

Markets, &c.—The markets of Liverpool are better supplied, perhaps, than any town in the British empire. Ireland and Scotland, particularly the former, furnish grain, live stock, bacon and butter, in immense quantities; and the Isle of Man, Anglesea, North Wales, and Cheshire, send excellent poultry and eggs, with butter and other farm produce; neither can any town in England, the metropolis not excepted, boast of market accommodation equal to Liverpool. The largest market building in St. John's, completed in 1822, at a cost of £35,613, covering a space of nearly two acres, being 183 yards in length, by 45 in breadth. It is a light and lofty structure, having its roof supported by 116 cast iron pillars; the walls are lined with 36 shops, and upwards of 400 stalls and standings run in four ranges up and down the interior. It is brilliantly lighted with gas; and on the whole, the regulations are so good, that it may be said to be unrivalled both for size and convenience. St. James's market, at the end of Great George-street, though only one-third the size of that last mentioned, is still a large and well constructed building, regulated by the corporation; as is St. Martin's market in Scotland-road. There are six other markets, besides a fish hall. The total expenditure by the corporation for these buildings has amounted to about £83,000. The Corn Exchange, in Brunswick-street, erected by a subscription of £16,000 in £100 shares, has a plain but handsome frontage, and affords every convenience for business.

Churches, Chapels, and Cemeteries.—Liverpool, which, till 1699, was a chapelry attached to Walton-on-the-hill, was constituted by act 10 & 11 Will. 3. c. 36, a distinct parish divided into two medieties: the parish churches are St. Nicholas and St. Peter's, and the livings are rectories, each valued at £615 a year, and in the gift of the corporation. There are, however, 33 other churches, and their number is increasing every year. Large sums of money have been expended in their erection and internal decorations; but with the exception of the most recent buildings, their exterior architecture is heavy, bizarre, and inappropriate. The following table exhibits the gradual increase of these buildings, their means of accommodation, the value of the incumbencies, and the nature of the patronage, the letter C meaning corporation patronage, and R that of the rectors of Liverpool, the rest marked P, being in the hands of trustees or private individuals. It is necessary to premise, however, that by the provisions of the Municipal Reform Act, the corporation to which so many of the churches now belong, is to dispose of the advowsons, and to apply the produce to the improvement of the town, and, in point of fact, several of them have been already sold.

Patron.	Names of Churches.	Com-pletion.	Church Room.	Annual Value.
				£.
P	St. Nicholas'	1698	1,528	615
P	St. Peter's	1704	1,357	615
R	St. Matthew's	1707	529	107
R	St. Stephen's	1792	850	180
C	St. George's	1734	890	130
C	St. Thomas'	1748	1,470	138
C	St. Paul's	1769	2,000	165
P	St. Anne's	1772	700	99
P	St. James's, Toxteth Park	1774	900	148
P	St. John's	1786	1,600	226
P	All Saints'	1789	1,400	
P	Trinity Church	1792	1,300	250
P	Christ Church	1797	2,300	105
P	St. Mark's	1805	1,714	270
P	St. George's, Everton	1814	950	200
P	St. Andrew's	1815	1,800	294
P	St. Philip's	1815	1,200	250
P	St. Mary's, Edgehill	1822	1,300	170
P	St. Michael's, Toxteth Park	1823	850	810
P	St. David's (Welsh church)	1827	1,600	400
P	St. Michael's	1827	1,300	490
C	St. Martin's	1829	2,000	530
P	St. Bride's	1830	1,400	295
P	St. Catherine's	1831	1,200	250
P	St. Luke's	1831	1,850	405
P	St. Augustine's, Everton	1831	1,500	200
R	St. Matthias'	1834	1,800	
P	St. Jude's, Edge-hill	1834	1,500	
P	St. John Baptist, Toxteth Park	1834		
P	St. Mary's, Kirkdale	1838		
	St. Saviour's	1840		
	St. Bartholomew's	1841	1,200	
	St. Silas'	1841	1,300	
	St. Clement's, Windsor	1841	1,300	
	St. Barnabas	1841	1,000	
	Another building			

Besides these churches there are three or four episcopal chapels not recognised by the authorities. The dissenters, likewise, are highly important and respectable, whether considered in respect of station, numbers, or character. Several of the churches and chapels have contiguous grave-yards; but the noxious practice of burying the dead within the town will probably be soon discontinued, in consequence of the formation of three large and well laid out cemeteries. St. James's cemetery, formed out of a stone quarry behind St. James's Walk, is planned with great elegance; another at Low Hill, is called the Necropolis; the third, or St. Mary's cemetery, is at Kirkdale.

The first Presbyterian congregation was established in 1672, and a second about thirty years afterward: the Baptists settled themselves here in 1714, and the Independents in 1777. The first Wesleyan Methodist chapel was opened in Mount Pleasant in 1791; and the New-connexion Methodists (or Kilhamites) built a chapel in 1798. Many of these places of worship are large and commodious, though few exhibit much exterior elegance, except perhaps those in Brunswick-street and Paradise-street, and the Scotch Kirk in Rodney-street. The dissenting places of worship existing in 1839 may be thus classified:

Wesleyan Methodists	8	Roman Catholics	5	
Baptists	7	New-connexion Methodists	3	
Welsh dissenters	7			
Scotch Kirk and dissenters	6	Society of Friends	1	
Unitarians	3	Other dissenters	6	
Independents	5	Total	51	

There are also two ships in the river set apart for religious service, the Marine church belonging to the establishment, and the Bethel ship for various denominations of dissenters.

Schools.—The Manchester Statistical Society was engaged during nearly a year collecting statistics on the state of education in Liverpool; and from their report of 1836 we learn that 39,700 children, of both sexes and all ages, were then being instructed, being rather more than half the entire population between the ages of 5 and 15: of this number, 4000 belonging to the upper classes were in private schools, 12,000 of the lowest order were in dames' schools, and the remainder in schools either endowed or supported by subscriptions. Among the endowed schools, the principal are the two corporation schools, formed in 1825, on the foundation of an old grammar school, that had been extinct since 1803: both are conducted on the national system, and infant schools are attached to each, so that, in all, above 1900 children are taught in them. There are two Church of England schools recently opened, which instruct 669 boys, 465 girls, and 595 infants. The Blue-coat Hospital, instituted in 1709, provides clothing, food, diet, and instruction for 250 boys and 100 girls. The building, which is of brick, has a handsome appearance, and the instruction, on the Madras system, is said to be as perfect as that of any school in England conducted on the same plan. The school of industry, established in 1810, is intended for training girls for domestic service. The number is limited to 100, and a few of the more deserving scholars have board and lodging, as well as a good plain education. Christ-church National schools educate 210 boys, 250 girls, and 290 infants; and Everton National school has 66 boys and 60 girls. Among the other schools, the best conducted are, Waterworth's school, in Hunter-street; St. Patrick's charity-school; the Duncan-street schools, supported by the Society of Friends; the Renshaw-street school (perhaps the best of all), maintained by the Unitarians, and the Caledonian school in Oldham-street. There is likewise a blind school, with 105 pupils, and a school for the deaf and dumb, attended by 50 children.

These statements are highly creditable to the liberality of all classes in Liverpool. The principal schools, too, are, speaking generally, well conducted; their buildings are airy and suitable, and the means of instruction, slates, books, &c., are abundantly provided. Still, however, when it is considered that 12,000 children belong to dames' schools, kept in small, ill-ventilated, and almost unfurnished rooms, and often in damp cellars, by persons wholly incapable, from age and ignorance, of giving instruction, it must be acknowledged that there is a great deficiency in the means of education, for so large a town: so important a matter should not be left to the avarice and caprice of parents, nor to the voluntary charity of individuals and societies, but should be taken up by the authorities, who should establish schools, under competent masters, in every parochial district, at the expense of the town at large. (*Corp. Rep., App.* p. iv., p. 2738: *Manch. Stat. Soc. Report*, 1836.)

Charitable Institutions, &c.—Liverpool has many handsome and respectable edifices devoted to charitable purposes, among which may be mentioned, the Charitable Institution-house in Slater-street, intended, though on a much smaller scale than the Exeter Hall of London, to give similar accommodation (*without charge*) to all religious and charitable institutions established in Liverpool, for committees, public meetings, &c. The Infirmary in Brownlow-street (removed from Shaw's-brow in 1824), was erected at a cost of £37,200; it is a large, chaste, and elegant building, with an extent of masonry, and a number of front windows, that give it, when seen from the street, an appearance of grandeur exceeded by few other buildings in the town. There are 20 wards, comprising excellent accommodation for 234 patients, and the medical staff attached to the institution equals in ability and attention that of any hospital out of the metropolis. The fever hospital, with 110 beds, supported out of the poor-rate, is a valuable institution. The lunatic asylum, lately erected at a cost of £11,000, to supply the place of an older establishment, has a handsome exterior, and comprises accommodation, with spacious airing-grounds for 60 patients, many of whom, as at York, belong to the higher classes. The Lock hospital, connected with the infirmary, was opened in 1834, with accommodation for 60 patients; and the same number of persons suffering from accidents and acute diseases, are received at the Northern hospital in Leeds-street, opened in 1834, which has 60 beds. Three dispensaries (one of which, in Vauxhall-road, is a large and elegant building, comprising accommodation for in-patients and medical students,) furnish gratuitous advice and medicine for the sick-poor, who are likewise attended by the resident officers at their own habitations; and there is an ophthalmic infirmary and dispensary, with which is connected an institution for diseases of the ear. The Ladies' charity affords relief to about 1200 lying-in women every year, with supplies of linen, &c.; the other principal charities are the Stranger's Friend Society, relieving about 1000 persons yearly, with a similar institution called the Charitable Society, the Penitentiary and Refuge for the Destitute, both intended for the reformation of degraded females, the Marine Humane Society, and the District Provident Society.

There are likewise many religious societies, the chief of which are the Bible Society (by far the largest in point of income), the Society for Promoting Christian Knowledge, and the Mariner's Church Society.

Port and Docks.—The rapid rise of the port of Liverpool to its present consequence, though, no doubt, principally owing, like that of the town itself, to the astonishing increase of manufactures and population in the extensive district of which it is the grand emporium, is also, in part, owing to the facilities that have been given to navigation, and commerce by the construction of wet and dry docks. The entrance to the estuary of the Mersey is a good deal encumbered with sand-banks, and is crossed by a bar, which, however, has at low water spring tides, where deepest, 11 feet water; and as the tides rise 21 feet at neap and 31 feet at spring tides, there is water for the largest ships; and as the channels are indicated by light vessels, and well marked with buoys, there is no difficulty in making the port. In fact, since the opening of the Victoria channel (by dredging) in Oct. 1839, vessels of the largest size cross the bar at 1st quarter flood: 14,000 vessels passed this channel in 12 months from its opening.

But the land around being low, the ships in the river are exposed to risk from gales of wind; and to obviate this inconvenience, and to facilitate their loading and unloading, the docks have been constructed, which constitute the great glory of the town. The first wet dock in the British empire was opened here in 1708, a second about half a century afterward, and since that period many more have been constructed on a very magnificent scale, and furnished with all sorts of conveniences, so that the aggregate area of those now in use amounts to above 110 acres, and the quay-space is nearly 7¼ m. in length. The following table shows the water-area and length of quay-space, with the date of construction of the different docks.

Docks.	Date of Completion.	Water-area.	Length of Quay-space.
		Sq. Yards.	Yards.
I. Dry Basins.			
Prince's Basin	1821	39,309	600
Coburg do.		1,925	135
George's do.	1795	16,373	455
George's Ferry do.		1,344	162
Old Dock God	1715	7,737	447
Queen's Basin	1796	20,391	601
S. Ferry do.		5,307	205
Total of dry Basins		70,495	2,565
II. Wet Docks.			
Clarence Dock & Lock	1830	99,313	914
Half-tide Basin		17,875	595
Waterloo Dock & Lock	1834	52,356	1,075
Trafalgar do.	1846	39,446	1,350
Victoria do.	1849	39,036	480
Prince's do.	1821	87,129	1,618
George's do.	1795	59,764	1,091
Canning do.		19,136	800
Salthouse do.	1795	55,695	760
King's do.	1796	57,776	878
Queen's do.	1796	51,566	1,286
Half-tide do.	1796	13,185	427
Brunswick do.	1832	60,821	1,662
Union Dock		9,945	480
Coburg Dock	1832	29,623	573
Total of dock areas		482,305	13,050

Among these the King's Dock, being contiguous to the King's Tobacco Warehouse, receives all the vessels from Virginia and other parts laden with tobacco; the Queen's and Brunswick Docks are occupied by ships laden with timber from Honduras, Canada, and the Baltic; the Canning Dock receives coasting vessels which exchange corn and provisions for colonial produce. Salthouse Dock accommodates ships from the Levant, Irish traders, and English coasting vessels; and the Clarence, Trafalgar, and Cobourg Docks, are exclusively for the accommodation of steamers, the latter being appropriated to the Transatlantic and Mediterranean steamers. All these works are defended on the side next the river by a strong sea-wall upward of 2½ m. in length. All precautions are taken to prevent the accumulation of mud in the docks by the use of steam-dredging machines, and strict rules, enforced by a vigilant police force, are established to maintain good order and prevent both fire and depredations.

The docks are all constructed on the estate of the corporation, and are managed by commissioners appointed by act of parliament. The bonding and other warehouses do not, however, belong to the dock estate, but are private property. Most of them are in the immediate vicinity of the docks, but some are at a considerable distance; and there is not, in this respect, the same accommodation in the Liverpool as in the London docks, where the warehouses being built along the dock-quays, goods are loaded and unloaded with the greatest possible facility. This difference in the situation of the warehouses in the two ports leads to a difference in the mode of discharging and loading ships in each: in London this is done by the servants of the different dock companies; whereas in Liverpool it is effected by gangs of private labourers, called lumpers, who contract, for a specific sum, to load or unload a vessel. A great reduction was effected in the Liverpool dock dues in 1836, and they are now extremely moderate.

Commerce.—Though now of such paramount importance,

little more than two centuries have elapsed since this great emporium was correctly described as "the little creek of Liverpool," being then merely a dependency of Chester! And so late as 1700, it had only about 5000 inhabitants, and 84 ships, of the burden of 5789 tons! The progress of the town in the interval in commerce, and in the accumulation of wealth and population, has been quite unprecedented in the history of industry. It is not, however, difficult to discover the causes of the all but apparently miraculous progress of Liverpool. A good deal must be ascribed to the enterprise, sagacity, and persevering industry of the merchants; but she is, no doubt, mainly indebted for her rise and the vast magnitude of her commerce, to her fortunate position, and, above all, to the increase of manufactures in Manchester and the surrounding district. The situation of Liverpool necessarily renders her a principal seat of the trade between Ireland and Great Britain; and as the population and trade of the former increased, it could not fail proportionally to increase the trade of this port. The gradual filling up of the Dee, and the consequent decline of Chester as a harbour, has also proved of no little advantage to Liverpool, by rendering her the great mart for the salt of Nantwich, and other places in Cheshire, the exportation of which to foreign parts, employs a great amount of shipping. Unquestionably, however, Liverpool would never have attained to half her present size, or importance, but for the cotton manufacture. But being the port through which Manchester, Oldham, Bury, Bolton, Ashton, and other great seats of that manufacture, could most conveniently obtain supplies of the raw material, and export their manufactured products, she has increased with every increase in this great department of industry; and it is no exaggeration to affirm, that the creative influence of the wonderful inventions and discoveries of Hargreaves, Arkwright, Crompton, Cartwright, and the other founders and improvers of the cotton manufacture, has been, though not so direct, quite as powerful, in the docks and warehouses of Liverpool as in the mills of Manchester.

The congenerous businesses of the slave trade and privateering appear to be the only departments of an exotic character, and not bottomed on any natural facility, that have ever been carried on to any great extent from Liverpool. The slave trade began in 1728; and was prosecuted vigorously and successfully down to the abolition of the trade in 1806, when it employed 111 ships, of the burden of 25,949 tons. It was apprehended by many that the abolition of this nefarious, though lucrative, traffic, would be a severe blow to the prosperity of the port. But so rapid was the increase of the legitimate and more natural branches of her trade, that it was but little felt at the time, and was very soon forgotten.

It is probable that the acquaintance with the slave trade may have given a stimulus to privateering; but, at all events, it was carried on to a great extent from Liverpool, both in the American and last French wars, especially in the former. In 1779, no fewer than 120 privateers belonged to the port, carrying each from 10 to 20 guns!

It is extremely difficult, or rather we should say quite impossible, to form any correct estimate of the total amount of the trade of Liverpool. Probably, however, the aggregate annual value of the imports and exports do not fall much short of the amazing sum of £40,000,000, if they do not rather exceed that amount! In 1834 it was estimated as follows by Mr. Myers, an extensive and intelligent merchant of the town:—

Exports.	£.	Imports.	£.
Woollens	4,000,000	Irish trade	4,000,000
Cotton stuffs and yarn	12,000,000	Raw cotton	5,500,000
Linens	1,000,000	Other articles	5,000,000
Hardware	1,300,000		
Earthenware	250,000		
Silk	150,000		
Salt and other articles	1,000,000		
Total	19,700,000		14,500,000

The vast preponderance of Liverpool in the Cotton Trade is obvious from the following

Account of the Quantities (in Packages) of the different Species of Cotton imported into Great Britain from 1830 to 1840, both inclusive; specifying the whole Quantity imported into Liverpool:—

Descriptions of Cotton.	1830.	1831.	1832.	1833.	1834.	1835.	1836.	1837.	1838.	1839.	1840.
	Packages.	Packages.	Packages.	Packages.	Packages.	Packages.	Packages.	Packages.	Packages.	Packages.	Packages.
American	618,287	608,287	888,786	654,768	733,328	763,199	164,707	844,812	1,194,800	814,300	1,237,300
Brazil	191,465	169,908	114,565	162,193	103,646	143,572	148,715	177,005	127,500	90,300	80,300
Egyptian	14,732	26,194	41,163	3,893	7,377	43,721	34,053	41,193	29,700	23,500	38,000
East India	35,019	78,754	100,896	94,698	86,008	117,865	219,493	145,174	107,300	132,800	216,300
W. India, &c.	11,791	11,304	6,490	12,646	17,485	22,796	23,506	27,791	29,400	56,000	92,300
Total imports into G. Britain	871,487	903,267	908,229	930,216	951,024	1,091,253	1,201,374	1,175,975	1,496,600	1,116,300	1,590,500
Total imports into Liverpool	783,695	791,588	779,071	840,953	841,474	970,717	1,023,578	1,036,005	1,395,415	1,019,220	1,415,941

LIVERPOOL.

Four fifths of the trade between the United Kingdom and the United States now centres in Liverpool; and she also has a large share of the trade with S. America and the W. Indies. She also carries on a considerable trade with the E. Indies and China, though in this department she is far surpassed by London. Indeed, the ships and products of Liverpool are to be found in every port, in every part of the world, accessible to merchantmen.

Account of the Principal Articles of East and West Indian and other Produce imported into Liverpool during each of the five Years ending the 31st December, 1840; and the Stocks on Hand on the 31st December, each Year. (From the Tables published by the Brokers' Association) :—

	Packages and Quantities.	Imports.					Stocks.				
		1836.	1837.	1838.	1839.	1840.	1836.	1837.	1838.	1839.	1840.
Ashes, Canadian	barrels	17,300	14,600	15,700	12,600	9,500	Pot. 8,700 Prt. 2,100	4,500 2,500	7,380 2,300	7,590 2,390	3,640 1,790
Brimstone	tons	13,130	14,400	16,500	6,730	18,000	7,500	7,860	11,600	4,600	16,300
Cassia Lignea	cases	3,980	3,105	1,440	1,550	1,100	3,500	4,100	5,300	900	900
Cocoa	brls. and bags	6,980	5,508	2,980	750	5,120	6,500	8,600	1,800	900	1,300
Coffee, West India, Br. Pl.	casks	7,900	5,370	8,108	6,240	6,443	2,400	1,359	3,206	1,900	2,390
Do. and Ceylon	brls and bags	4,350	5,408	6,520	3,350	4,550					3,500
East India and Cape	casks and bags	5,605	4,650	1,430	6,200	3,178	9,000	6,000	5,308	12,600	28,300
Foreign	ditto	1,050	9,080	7,350	14,610	5,780					1,640
Dyewoods, Logwood	tons	6,900	9,000	9,800	14,000	11,550	2,360	1,000	360	1,800	4,350
Fustic	ditto	3,700	3,100	3,200	7,300	7,000	3,300	1,900	200	1,900	1,200
Nicaragua Wood	ditto	3,400	1,700	750	900	1,720	4,600	4,000	4,300	3,490	3,400
Ginger, West India	brls. and bags	2,400	2,150	3,000	1,000	1,018	600	1,500	2,700	2,000	1,000
East India	bags and pockets	27,000	22,000	25,000	36,500	7,150	250	50.	467	850	500
Gum, Arabic	cases	1,250	1,740	1,250	2,650	1,720	450	600	610	750	1,200
Hides, Ox and Cow	number	260,000	330,000	650,000	963,000	315,500	71,000	66,800	44,400	12,000	10,000
East India	ditto	360,000	275,000	171,000	413,000	343,400	90,000	24,000	21,400	39,000	20,000
Horns, Buenos Ayres	ditto	28,500	36,000	62,000	77,000	29,000	11,000	8,900	3,800	300	Scarce
Indigo, East India	chests	2,000	900	1,300	1,650	1,800	400	250	350	600	1,150
Spanish	serons	1,400	2,700	1,170	2,800	1,175	400	70	45	80	450
Lac Dye	chests	950	1,850	2,000	1,540	1,610	500	1,150	1,750	1,800	3,500
Shell	ditto	2,040	2,550	3,150	3,500	4,025	500	1,850	1,750	4,300	3,400
Linseed	quarters	22,000	34,000	27,950	15,054	18,100	2,500	1,300	none		800
Madder, Dutch	casks	3,350	3,800	3,290	3,150	1,516	290	460	100	280	600
French	ditto					3,075	550	500	40	160	400
Madder Roots	bales, &c.	10,800	10,950	6,700	11,300	12,400	2,100	4,200	1,900	2,350	3,000
Molasses	casks	12,000	11,200	10,100	8,280	7,500	1,500	900	2,800	1,400	2,400
Olive Oil	tuns	3,800	3,310	4,010	2,400	8,100	1,200	1,500	1,100	600	300
Palm Oil	tons	11,000	8,000	10,320	14,300	12,270	1,200	1,800	2,100	3,300	4,200
Pepper, East India	bags and pockets	29,000	21,900	19,000	24,000	11,550	23,000	20,000	20,000	28,000	15,000
Pimento	brls. and bags	4,300	3,300	1,350	1,000	1,900	6,300	8,000	6,900	5,500	3,000
Rice, East India	bags	21,500	103,000	66,000	96,000	77,900	1,500	30,000	15,000	20,000	4,500
Rum, West India	puncheons	11,500	6,700	7,500	8,000	7,300	9,000	5,800	5,300	4,300	4,400
East India	ditto	none	300			900					100
Foreign	ditto	none		620	4,000	30	none	200	none	2,800	100
Saltpetre, East India	bags	33,800	43,000	30,800	58,000	29,300	13,000	13,500	7,400	22,000	14,000
Nitrate of Soda	ditto	31,800	32,000	31,400	25,000	53,100	18,000	24,500	27,000	17,000	4,000
Sugar, British Plantation	hhds. and tons.	56,500	47,200	46,500	33,000	29,000	17,600	7,200	14,000	10,000	4,200
East India	bags	43,000	91,000	90,000	92,000	107,300	20,000	32,000	20,000	22,000	11,300
Mauritius	ditto	52,500	51,000	62,500	82,000	50,300	16,000	11,000	12,500	17,000	16,500
Manilla, Java, &c.	cases, &c.	20,000	30,000	19,000	8,300	9,040	12,000	10,000	6,500	6,500	14,000
Havana	boxes	none	850	620	none	4,800	none	550	none	none	3,085
Brazil	chests	4,600	2,500	3,400	6,500	7,180	1,600	1,900	1,300	2,000	4,000
Do. and other Foreign	casks, brls., &c.	7,800	6,500	10,500	11,400	18,515	1,000	3,300	7,000	7,000	9,500
Sumac	bags	53,500	33,300	28,000	51,200	50,900	6,500	5,500	11,900	8,000	8,700
Tar	barrels	36,500	25,000	55,000	45,000	42,100	7,000	3,800	19,000	18,700	14,000
Tallow, European	casks	21,800	19,800	25,000	39,000	23,130	3,300	1,800	3,500	8,500	7,200
Timber	casks, boxes, &c.	450	440	300	740	1,070	60	250	85	450	900
Tobacco	hhds.	10,300	6,600	8,100	10,780	10,750	9,800	5,100	5,200	7,900	3,800
Turpentine	barrels	105,000	102,000	120,000	70,300	89,000	23,600	21,000	52,000	35,000	60,000

The following Table shows the distribution of the foreign and domestic trade of Liverpool :—

Account of the Number and Tonnage of the Vessels that entered and cleared from the Port of Liverpool, in 1839, specifying the Countries from whence they came, or for which they cleared, or the Departments in which they were engaged.

Countries.	Inward.				Outward.			
	British.		Foreign.		British.		Foreign.	
	Ships.	Tons.	Ships.	Tons.	Ships.	Tons.	Ships.	Tons.
Europe generally	850	180,027	658	113,658	738	108,104	723	126,938
Africa	78	21,435	13	1,397	116	29,898	14	3,088
Asia	191	46,379	1	166	150	56,683	5	1,730
America, viz.—								
British Northern Colonies	371	168,518	.	.	308	145,130		
British West Indies	171	44,781	.	.	237	57,045		
Foreign	23	4,015	9	1,401	49	9,638	38	8,869
United States	132	73,580	474	244,504	141	76,326	400	243,231
South American States	238	54,987	6	1,157	248	58,004	16	3,842
Total	1,904	543,839	1,161	362,985	2,038	548,607	1,254	388,064
Isles of Guernsey, Jersey, &c.	17	1,509	1	50	27	2,874		
Isle of Man	398	14,519	. 3	281	254	12,949		
Irish trade	2,009	481,434	.	.	2,705	417,730		
Other coasters	5,190	505,456	.	.	5,676	502,195		
Total	10,401	1,546,810	1,165	362,605	10,700	1,496,055	1,254	388,051

Subjoined is—

A Statement of the total Number of Vessels, of all Descriptions, with their total Tonnage, that entered the Liverpool Docks in the Years ending the 24th of June, 1840 and 1839, with the Amount of the Tonnage and other Duties payable by the same.

Year.	No. of Vessels.	Tonnage.	Duties on Tonnage.			Duties on Goods.			Lighthouse Duties.			Floating Light Duties.			Other Duties.			Total.		
			L.	s.	d.	L.	s.	d.	L.	s.	d.	L.	s.	d.	L.	s.	d.	L.	s.	d.
1840	15,998	2,445,708	99,921	2	3	85,975	11	2	6,937	0	11	3,118	19	4	9,394	13	3	197,477	18	6
1839	15,445	2,158,691	91,680	8	5	74,874	13	1	6,158	3	4	2,739	2	4	8,771	8	11	174,222	16	1

There belonged to Liverpool, on the first of Jan., 1840, 1123 ships, of the registered burden of 203,176 tons, manned by 13,958 seamen.

The gross customs revenue of Liverpool, in 1840, amounted to £4,604,144, while that of London, in the same year, amounted to £11,068,053. But it would be a great error to suppose that the trade of the metropolis exceeded that of Liverpool in this proportion. Cotton wool, and other raw materials for manufactures, on which low duties are paid, form the principal part of the foreign imports at Liverpool;

Whereas London imports comparatively few of these articles, her trade being principally in articles of direct consumption, as sugar, tea, coffee, wines, corn, &c., on which high duties are paid. This circumstance accounts, in part at least, for the comparatively large amount of the revenue received in the latter; and, allowing for doubt whether the foreign trade of London very much exceeds that of Liverpool. We subjoin an

Account of the Number of Ships, and their Tonnage, that entered the Ports of London and Liverpool, from parts, during each of the Three Years ending the 5th of January, 1841.

Years ended 5th January.	London				Liverpool			
	British Ships.		Foreign Ships.		British Ships.		Foreign Ships.	
	No.	Tons.	No.	Tons.	No.	Tons.	No.	Tons.
1839	4,396	883,985	1,737	277,969	2,048	567,791	1,198	402
1840	4,880	988,887	2,375	357,165	2,979	559,920	1,165	390
1841	4,547	934,880	2,381	354,456	2,187	573,339	1,305	468

But the extraordinary progress of the town, in population and commerce, will be best exhibited by the subjoined

Statement of the Christenings, Burials, Deaths, Marriages, and Population of the Parish of Liverpool, from with an Account of the Vessels and their Tonnage entering the Docks, and of the Amount of the Dock and Duties on such Ships in different Years since 1700.

Year.	Christenings.	Burials.	Marriages.	Houses.	Population.	Vessels.	Tonnage.	Dock Duties.
								£.
1700	132	194	35	1,142	5,714			
1705	243	209	41		8,168	84	5,780	
1710	238	211	40	1,634				
1714	346	247	57					
1716	234	262	73			113	8,286	
1720	410	393	58	2,387	11,833			
1721	378	482	63					
1722	357	261	56			131	8,700	870 12
1728	350	429	79					847 11
1730	397	307	149	2,430	12,074	412	18,679	
1735	451	578	122					
1737	496	479	131			171	19,016	
1740	501	513	182	3,600	18,080			
1744	658	587	198			182	12,775	
1749	880	779	377					
1750	978	1,075	300					
1751	923	617	258	3,700	18,500	220	29,176	
1754	918	875	394	5,156	25,787			2,197 14
1760	948	599	408			1,245		2,339 6
1763	1,037		529			1,782		2,141 1
1770	1,317	1,562	433	6,800	35,600	2,073		4,142 17
1773	1,397	1,100	500	6,940	34,604	2,914		4,795 1
1777	1,579	1,146	461		34,107	2,361		4,610 4
1784	2,043	1,635	816	7,110		3,098		6,387 11
1790	2,244	1,763	805	8,865	55,732	4,223		10,637 11
1791	2,491	2,166	854			4,045		11,645 11
1795	2,677	2,164	1,101			4,478		12,987 16
1801	2,797	2,768	1,234	11,784	77,708	5,080	450,719	28,365 6
1805	3,482	2,841	1,239			4,618	463,488	33,364 15
1811	4,183	3,079	1,596	16,102	94,376	5,616	611,190	34,738 18
1812	3,890	2,546	1,116			4,599	446,788	44,403 7
1813	3,535	2,534	1,220			5,341	547,486	50,177 13
1815	4,062	3,208	1,725			6,440	709,849	76,915 8
1818	4,315	3,372	1,519			6,779	754,690	98,538 8
1819	4,548	3,728	1,664			7,840	867,318	110,127 1
1820	4,718	3,157	1,633			7,276	805,653	94,419 11
1821	4,629	3,497	1,632	20,339	118,972	7,810	839,846	94,586 17
1822	4,734	3,379	1,402			8,136	898,908	102,403 17
1823	4,734	3,538	1,796			8,916	1,010,619	115,783 1
1824	5,029	4,132	1,806			10,001	1,180,914	130,911 11
1825	5,305	4,143	2,066			10,837	1,293,790	198,691 19
1826	5,527	5,268	2,021			9,601	1,228,318	121,000 19
1827	5,910	3,944	2,109			9,503	1,295,313	134,472 14
1828	5,387	3,722	2,120			10,703	1,311,111	141,389 15
1829	5,857	3,747	2,220			11,363	1,357,957	147,397 4
1830	5,813	3,845	2,340	27,361	165,921	11,214	1,411,964	151,390 17
1831	7,258	5,042	2,474			12,537	1,592,436	152,455 4
1832	7,867	5,883	2,535			12,998	1,540,057	170,047 6
1833	7,767	5,225	2,875			12,984	1,390,461	188,980 16
1834	7,759	5,881	2,805			13,444	1,692,870	191,799 17
1835	8,154	4,749	2,969			13,941	1,766,498	198,637 18
1836	8,536	5,296	2,781			14,959	1,947,613	221,994 10
1837	8,750	6,875	2,893			15,038	1,958,964	173,852 10
1838	9,398	4,867	3,189			14,889	2,096,906	146,290 5
1839	9,851	5,265				15,445	2,136,691	156,355 1
1840	9,542					15,998	2,445,708	175,190 14

Next to the consequence of Liverpool as a trading port, is its high importance as a packet station, second probably to none in the world, except London. The packet-ships, or those to New York and other parts of the United States, which, for size, excellent accommodation, and speed, are the objects of general admiration, leave the port daily; and five large steamships of unexampled magnitude are engaged in the same service. Packets are also regularly sent to the E. Indies, Rio de Janeiro, Laguayra, Buenos Ayres, Lima, Lisbon, Oporto, and the Mediterranean; and a steamer trades regularly between Liverpool and Havre. Of the numerous steam-vessels engaged in the service, no fewer than 94 sail to and from Dublin; ... are on the Waterford, and three on the Belfast station; seven ply to and from Glasgow; and several connect Liverpool with the Isle of Man, Drogheda, ford, Cork, Bristol, Dumfries, Carlisle, Whitehaven. A host of river-steamers, also, are constantly plying passengers at the various ferries of the Mersey, or running up and down the stream. In short, nothing can be more striking, or better convince the stranger of the great scale on which the entire business of Liverpool is conducted, than the view from the Cheshire shore of the forest of masts, extending upwards of 2 m. along the quays and banks; the activity ever visible in all the docks; the towering warehouses along the quays, instinct with life and labour; the ships constantly entering and leaving the bay, and the almost innumerable steamers of every size

N°

quality, packets, ferry-boats, and tugs, rapidly coursing up and down the river to their several destinations.

A commercial town like Liverpool must necessarily have many joint-stock banking companies and private banks. Among the former are the Branch bank of England, the Royal bank of Liverpool, the Bank of Liverpool, the Borough bank, the Liverpool Union bank, the Commercial bank of England, the Liverpool Commercial bank, the Manchester and Liverpool district banking company, the Liverpool banking company, the Albion bank, the N. and S. Wales bank, and the Central bank of Liverpool; there are, also, four private banks and a savings' bank.

Manufactures. — Liverpool is not, properly speaking, a manufacturing town; but the vast magnitude of its foreign commerce necessarily demands the practice of a great number of domestic trades, some connected with the shipping, and others dependent on the peculiar nature of the traffic of the port. There are several large sugar refineries, an extensive pottery conducted by the Herculaneum company, iron and brass foundries, public breweries, roperies, glass-staining works, and alkali works. The manufacture of soap is more extensively carried on here than in any town of the kingdom, the quantity produced in 1839 being 43,546,119 lbs. of hard, and 6,388,930 lbs. of soft, soap. Ship-building is also carried on to a great extent; and not only capital merchant ships and steamers have been launched from the slips in different places of the town, but many large ships of war have been built for the government.

There are several wind-mills, and steam-engines for grinding corn, colours, dye-woods, &c., and numerous and large manufactories of chain cables, anchors, &c. The making of watches and watch-movements employs a great number of hands, and large quantities of these articles are exported, with files and tools, produced on a large scale in and near the town. Steam-engines of the best and most powerful kind are made in four establishments, from which have proceeded many of the engines employed on board the largest steamships; and this business is every year increasing in importance.

Canals and Railroads. — The commerce of Liverpool has been greatly promoted by the facilities which it enjoys for inland transport, greater perhaps than those belonging to any other town of Great Britain, except Manchester. The Irwell and Mersey navigation (for which an act was obtained in 1790), was the first effort to improve on the resources of nature, almost contemporary with which was the Weaver navigation. By means of the former, raw cotton and cotton goods were conveyed by water to and from Manchester, while, by the latter, the salt of Cheshire was furnished with equal facilities for its transit to Liverpool. The Sankey-brook navigation, completed in 1768, the duke of Bridgewater's canals, the Trent and Mersey or Grand Trunk canals, and the Leeds and Liverpool canal, were finished in rapid succession, so that, in 1816, the port of Liverpool had a complete water communication, directly or indirectly, not only with the great manufacturing towns of Lancashire, Cheshire, and Yorkshire, from which it derives its chief articles of export, but likewise with the S. counties, and, in fact, with nearly every part of England. The following table, drawn up from "Priestley's History of Canals," exhibits some particulars respecting the size, levels, &c., of the above-mentioned undertakings.

Canals.	[⚓] Miles.	[⚓⚓] Feet.	[⚓] Feet.	Rise and Fall. Feet.	Estimated cost. L.
Mersey and Irwell navigation	20	. .	. .	R. 79	.
Weaver do.	24	. .	. .	F. 49	.
Sankey-brook do.	12	48	8½	F. 78	.
Duke of Bridgewater's canal	38½	. .	. .	F. 88½	220,000 ?
Trent and Mersey	93	. .	. .	R. 256 / F. 154	296,000
Leeds and Liverpool	134	48	5	R. 416 / F. 436	1,200,000
Total length of canal communication	301½				

* The cost of the above seems to have somewhat exceeded 300,000*l*.

Very large fortunes have been realised by the above undertakings; and, notwithstanding the successful competition of railways, they still bring in large incomes to the proprietors. The facility of transit, however, both for passengers and goods, has been vastly increased since the opening of the railways, by which Liverpool is brought within an hour's distance of Manchester, and both are brought within four hours of Birmingham, and nine hours of the metropolis. The act for the Liverpool and Manchester railway was obtained in 1826; the works were completed in 1830, at a cost of £876,000, or more than double the estimate laid before parliament, and the line was finally opened on the

15th September of that year, a day that will be long remembered, from its connexion with the melancholy death of Mr. Huskisson, one of our most enlightened commercial statesmen. This railway, which is 32 m. in length, has inclined planes at Sutton, Rainhill, and Liverpool; the last run through two tunnels, both commencing at Edgehill, where are stationary engines for drawing the trains up the inclined plane. The tunnel, for the conveyance of goods down to Wapping, close by the King's dock, is 2250 yds. long, with a rise of 1 in 53, and the interior, which is lighted with gas, has a height of 16 ft., with a breadth of 22 ft. Another tunnel, for passengers, comes out in the open space at Lime-street: it was opened in 1836, and has a length of 2250 yds. with a rise of 1 in 62, a width of 25 ft., and a height of 17 ft. This station has a Corinthian façade of 330 ft. in length, built at a cost of nearly £77000. The whole is in very elegant style, and will bear to be compared with the terminus of the London and Birmingham railway in the metropolis. This undertaking has been of immense advantage to the trade of Liverpool and Manchester, and the profits to the shareholders have been such as to allow, for some years back, dividends of 10 per cent., exclusive of the accumulation of a surplus fund. The traffic on this railway during the six years ending with 1836, was as follows:

Years.	Merchandise. Tons.	Coals. Tons.	Number of Passengers.
1831	108,589	11,245	445,047
1832	139,443	49,335	356,945
1833	195,704	91,309	386,492
1834	210,732	117,008	432,327
1835	230,629	116,348	473,847
1836	265,617	135,698	484,560
Total of 6 years	1,140,687	515,685	2,579,368

This return, which does not include immense quantities of cattle, sheep, and hogs, conveyed from Liverpool into the interior of the country, authorizes the supposition that the present yearly average may amount to about 500,000 passengers, and 250,000 tons, of merchandise, besides about 150,000 tons of coals; and these numbers will no doubt be very much higher, as soon as the completion of the Manchester and Leeds, and the Sheffield and Manchester railways shall have established a connexion of this nature between Liverpool, the clothing district of Yorkshire, and the important hardware manufacturing town of Sheffield. The Grand Junction railway, running S. to Birmingham, leaves the line between Liverpool and Manchester at Newtown bridge: it is 97½ m. long, and was constructed at a cost of £1,800,000: the expenses have been found to average 35 per cent. on the gross returns, and the last dividends were £7 10s. per £100 share. A railway is opened to Birkenhead and Chester. The North Union railway connects Liverpool northward with Bolton, Blackburn, Preston, and Lancaster: nor, it is probable, will many years elapse before there be a railway communication with Glasgow and Edinburgh through Carlisle, thus bringing the great commercial port of W. England within 10 hours' distance, both of the English and Scotch metropolis, and within a much smaller distance of all the great manufacturing towns S. of the Tweed.

Corporate establishment, &c. — Liverpool received its first charter of incorporation in 1203, with others from subsequent monarchs. William III. granted it a new charter in 1695, which was confirmed, with a few alterations, by George II. and III.; and by the provisions of this charter the town was governed down to the passing of the Municipal Reform Act in 1835. The borough is now divided into 16 wards, the corporate officers comprising a mayor, with 15 other aldermen, and 48 councillors. Corporation revenues, in 1839 (exclusive of £43,101, the amount of interest and money borrowed, and £13,695 accruing from the sale of property), £307,537. The corporation has the right, under an act passed in 1825, of nominating persons to fill subordinate corporate offices, and is empowered to make "laws for regulating the police of the town, the docks and the port generally, for lighting and watching the town, and for the suppression of disorderly and immoral practices." Quarter and petty sessions are held by the recorder, who is appointed by the crown, and a few years ago the assizes for W. Derby and Salford were removed here from Lancaster. The police, conducted under a commissioner, is formed, like that of the metropolis, into divisions, with superintendents, inspectors, sergeants, &c., and is said to be extremely efficient in suppressing crime, and maintaining order both in the town and port. This, indeed, is fully proved by two reports (1836-40), of the Watch Committee of Liverpool, to which the reader is referred for full information. The force at present comprises about 600 men, including inspectors, &c., the expense of maintaining which somewhat exceeds £50,000 a year. Liverpool has three prisons: the borough jail, erected on the plan of Howard, and formerly for some years a dépôt for French prisoners, is airy, well constructed, and most efficiently managed; the Bridewell is also well spoken of by the inspectors, who, however, give a very unfavourable account of the county House of Correction in Kirkdale, which, though well built and excellently arranged for the accommodation of prisoners, is alleged to be most inefficiently conducted. (See *Inspector's Reports,* 1837-38-39.)

The provision for the poor, in so populous a town as Liverpool, is, of course, on a large scale. The total poor-rates of the borough in 1839 amounted to £41,294, of which, £34,564 were expended solely for the relief and maintenance of the poor. The poorhouse, which, from its extent, might well be called a little town, is one of the largest in the kingdom; and the building arrangements admit at the same time of perfect classification, according to the principles of the Poor Law Amendment Act, while at the same time they allow of considerable indulgence to the sick and aged: most of the common trades are pursued within the building, and the pauper mechanics instruct the children in different branches of handicraft. In fact, it would be difficult to find a more complete and more efficiently conducted workhouse than that of Liverpool. The borough has enjoyed the privilege of sending two members to the House of Commons, since the 25th of Edward I. Down to the passing of the Reform Act, the elective franchise was vested in the freemen and free burgesses. The Boundary Act enlarged the borough so as to include the out-township of Kirkdale, Everton, W. Derbey, and Toxteth park. Registered electors, in 1839-40, 14,970.

Condition of the People, &c.—Owing to the rapid rise of Liverpool, its population consists, in a great degree, of adventurers, not merely from all parts of the United Kingdom, but of the world, attracted to it in the expectation, which, in many instances, has not been disappointed, of making a fortune. In such a society there is necessarily less prejudice, and fewer conventional and established observances, than in other and differently constituted towns. There is here, in fact, the greatest toleration for all sorts of individuals, and all sorts of opinions. Exclusiveness in Liverpool is, speaking generally, entirely out of the question; and you meet everywhere with people of all grades, all occupations, and all countries. It would be idle in such a place to look for that polish, and careful avoidance of debateable or irritating topics, that distinguish more aristocratical societies; but, on the other hand, it is free from the sameness and insipidity which characterizes the latter. The ostentatious display of wealth made by a lucky hit or successful speculation, is sometimes, no doubt, offensive enough; but, on the whole, society in Liverpool is, from its variety and ease, superior to that in most other purely mercantile towns. The higher class of merchants, having connexions and correspondents in most parts of the world, are generally very well informed; and some of them are honourably distinguished by their taste in literature, science, and the fine arts. All classes are eminently enterprising (sometimes, perhaps, to excess), vigilant, and industrious; and possess, in a high degree, the qualities that go to form successful merchants and traders.

The situation of the lower or labouring classes in Liverpool, is, in many respects, less satisfactory than could be wished. Owing to the intimate intercourse it has with Ireland, and the small expense at which an individual may be brought over from Dublin, vast crowds of Irish labourers land at Liverpool, where they constitute a large proportion of the labouring population. Many of these persons are often, especially on their landing from Ireland, and before they find employment, reduced to a state bordering on destitution; and even after they obtain employment, they are frequently in a very depressed condition. In the departments of skilled labour, such as those of carpenters, smiths, bricklayers, &c., which are not interfered with by the Irish, wages are high; and employment being pretty constant, such workmen as are sober and industrious are in comfortable circumstances.

In consequence partly of the unfavourable condition of so many of the lower classes, but partly, also, of the vast amount of property that is here, always, as it were, in a state of transition, passing from the warehouses to the shops, and from the warehouses to the conveyances by which it is to be carried to its ultimate destination, there is at once a great incentive to indulge in dishonest practices, and great opportunities for depredation. It is estimated in the *Report on the Constabulary Force* (p. 18), that the cost of those who live on the public by other than honest practices in Liverpool, amounts to not less than £708,000 a year![*] But though, no doubt, the amount is very large, we are strongly inclined to think that this statement is grossly ex-

[*] It should be observed, indeed, that the 708,800*l*. includes the supposed earnings of a certain class of females. It is, however, to say the least, inaccurate and absurd to class them with thieves and pickpockets. They make no influence, but not necessarily a dishonest livelihood. But with all this, we believe the statement to be unworthy of credit.

aggerated: supposing each of these thimble-riggers to make, at an average, £30 a year, which, probably, is beyond the mark, this sum would give an aggregate number of above £23,600, which must obviously be very far above the real number.

In Liverpool, as in Manchester and Glasgow, a very large proportion of the labouring classes are miserably lodged. It is certain that there are in Liverpool no fewer than about 8080 cellars, all of which are ill ventilated, and most of them, at the same time, dark, damp, and filthy. It is supposed that from 38,000 to 40,000 of the lower classes are lodged in these wretched abodes; and when such is the case, the wonder is not that fever and epidemics occasionally prevail in Liverpool, but that they are not incomparably more frequent and fatal.

Exclusive of cellars, there are supposed to be in Liverpool about 2400 courts. These, which are of very contracted dimensions, have a narrow entrance from the street, and are built all round, except at the entrance: the number of houses in courts varies very much, but at an average they may be taken at about six, though, as some of the families occasionally receive lodgers, the population is often much greater; but taking it at only 30 persons to a court (six families of five individuals each), it follows that about 72,000 persons must be lodged in these dark and noisome recesses! Courts are found in a great variety of places, and many have recently been constructed in Toxteth Park. All the filth of the different families is collected in the court, whence it is not usually removed, though often producing an intolerable stench, above once a year, if so frequently; and the contiguous houses being neither sufficiently supplied with water, nor sufficiently ventilated, nor kept clean, they are but little better, and, in many cases, even worse than the cellars. Indeed, not a few of the courts are seldom or never free from fever, and other pestilential diseases! (*See* the valuable *Evidence of Dr. Duncan before the Committee on the Health of Towns.*)

Such a state of things calls for the immediate interference not merely of the local authorities, but of the legislature; and the most effective measures should be adopted for improving, and, in certain cases, shutting up and pulling down such residences.

The general healthiness of Liverpool is evinced by the fact that, notwithstanding the powerful countervailing influence of the circumstances now alluded to, the mortality is not supposed to exceed 1 in 32: in 1837, it was as high as 1 in 28, but that was an unusually unhealthy year. No doubt, with proper sanatory regulations, properly enforced, the average mortality might be reduced to 1 in 37 or 38, or perhaps less.

The northern part of the town, as far S. as Whitechapel and Dale-street, is seated on a coarse, red, diluvial clay; the remainder is the new red sandstone; and, having a gentle declivity, it has every facility for drainage. It is also protected by hills from the cold, withering, N.N.E. winds; the climate, in fact, though humid, is more equable than in most other places; the sea breezes temper the heat of summer, and the cold of winter is usually from six to eight degrees below that under the same latitude on the E. coast of the island. Its situation is, therefore, one of the healthiest that can be imagined, and to this must be attributed the lightness of the mortality, despite the powerful countervailing influences noticed above.

The suburbs of Liverpool are of great extent, and increasing rapidly in population and importance. Many of these and Edgehill are pretty thickly covered with neat rows of houses and handsome villas; Bootle is becoming a large place, and the spa of Waterloo is rising in favour as a suburban residence. At the same time the W. shores of the Mersey are becoming densely peopled: Birkenhead; Woodside; Royal Rock Ferry, and Park; New Ferry, opposite which the lazarettoes are moored: Seacombe, and Egremont, have risen in a few years, from the meanest hamlets, to towns of considerable size and regular construction. Eastham Hotel is delightfully situated, and has long been a great thoroughfare for coaches to all parts of Cheshire and Wales. Close to the river's mouth, a watering place, called New Brighton, has been formed, the streets of which are regular and handsome. The rise of suburbs so extensive and beautiful, almost wholly during the last 30 years, furnishes another indication of the vast resources of Liverpool, derived from its great and constantly increasing commercial consequence. (*Enfield's History of Liverpool; Baines's Lancashire,* 4to ed.; *Stranger in Liverpool;* and valuable *private information.*)

LIVERPOOL, p. b. and t., Perry co., Pa., 29 m. N. by W. Harrisburg, 139 W. Bounded E. by Susquehanna river. The borough is situated at the mouth of a creek, on the W. bank of Susquehanna river, where are extensive iron works. It contains several stores, two schools, 109 scholars, and 451 inhabitants. The township contains three stores, one fulling-mill, one woollen factory, six flouring-mills, 10 saw-mills, one tannery, two distilleries; five schools, 198 scholars. Pop. 763.

LIVERPOOL, p. t., Medina co., O., 125 m. N.E. by N. Columbus, 356 W. Watered by Rocky river, and its branches. It has a sulphur spring, a rich petroleum spring, supposed to indicate the existence of bituminous coal, salt springs, and iron ore. It contains two stores, one fulling-mill, two grist-mills, three saw-mills, two tanneries, one distillery; three schools, 195 scholars. Pop. 1500.

LIVINGSTON, county, N. Y., centrally situated towards the W. part of the state, and contains 509 sq. m. Watered by Genesee river and Canaseraga creek. Gypsum and bog iron ore, and salt and sulphur springs, are found. The latter, at Avon, are much frequented. It contained in 1840, 29,849 neat cattle, 163,395 sheep, 37,858 swine; and produced 893,050 bushels of wheat, 3694 of rye, 184,730 of Indian corn, 96,488 of buckwheat, 84,275 of barley, 305,519 of oats, 348,369 of potatoes, 119,430 pounds of sugar. It had 121 stores, 12 furnaces, 14 fulling-mills, four woollen factories, six flouring-mills, 96 grist-mills, 66 saw-mills, four paper-mills, one oil-mill, two rope-walks, 13 tanneries, eight distilleries, two breweries, one pottery, four printing-offices, two binderies, four weekly newspapers; six academies, 732 students; 176 schools, 8766 scholars. Pop. 35,140. Capital, Geneseo.

LIVINGSTON, parish, La. Situated towards the S.E. part of the state, and contains 730 sq. m. Bounded W. and S. by Amite river. Watered by Tickfaw river, and its branches. It contained in 1840, 4361 neat cattle, 898 sheep, 10,373 swine; and produced 26,712 bushels of Indian corn, 1500 of oats, 30,497 of potatoes, 82,830 pounds of rice, 250,445 of cotton. It had four stores, 12 grist-mills, two saw-mills, one tannery; one academy, 60 students; one school, 71 scholars. Pop.: whites, 1533; slaves, 739; free coloured, 43; total, 2315. Capital, Springfield.

LIVINGSTON, county, Ky. Situated in the N. part of the state, and contains 330 sq. m. Bounded N. and W. by Ohio river, S.W. by Tennessee river, N.E. by Tradewater creek. Watered by Cumberland river. It contained in 1840, 9984 neat cattle, 7693 sheep, 99,036 swine; and produced 27,746 bushels of wheat, 421,415 of Indian corn, 36,140 of oats, 10,605 of potatoes, 1,222,500 pounds of tobacco. It had 19 stores, one forge, nine grist-mills, four saw-mills, three tanneries, four distilleries; two academies, 63 students; 10 schools, 361 scholars. Pop.: whites, 7326; slaves, 1568; free coloured, 99; total, 9025. Capital, Smithland.

LIVINGSTON, county, Mich. Situated a little E. of the central part of the settled portion of the peninsula, and contains 576 sq. m. Drained by head branches of Shiawasse, Cedar, and Huron rivers. Iron ore is found in the S.E. part, and Saline springs near the centre. It contained in 1840, 7931 neat cattle, 1903 sheep, 10,959 swine; and produced 84,943 bushels of wheat, 89,081 of Indian corn, 77,943 of oats, 93,647 of potatoes. It had 19 stores, three flouring-mills, five grist-mills, 14 saw-mills; 59 schools, 1777 scholars. Pop. 7430. Capital, Howell.

LIVINGSTON, county, Ill. Situated in the N.E. part of the state, and contains 1086 sq. m. Drained by Vermillion river, and its branches. It contained in 1840, 1015 neat cattle, 1113 sheep, 4266 swine; and produced 13,014 bushels of wheat, 55,845 of Indian corn, 14,999 of oats, 3576 of potatoes, 10,591 pounds of sugar. It had one store, one woollen factory, one grist-mill, three saw-mills; three schools, 44 scholars. Pop. 759. Capital, Pontiac.

LIVINGSTON, county, Mo. Situated towards the N.W. part of the state, and contains 510 sq. m. Watered by Grand river and its branches. It contained in 1840, 5639 neat cattle, 1863 sheep, 17,925 swine; and produced 1768 bushels of wheat, 135,598 of Indian corn, 4000 of oats, 3587 of potatoes. It had 13 stores; five schools, 90 scholars. Pop.: whites, 4068; slaves, 241; free coloured, 2; total, 4265. Capital, Chillicothe.

LIVINGSTON, p. t., Columbia co., N. Y., 37 m. S. Albany, 334 W. Bounded W. by Hudson river. Watered by Roeliff-Jansen's or Ancram creek. It belongs to Livingston's Manor and contains a Dutch Reformed Church, six stores, two fulling-mills, one woollen factory, two furnaces, four flouring-mills, four grist-mills; 10 schools, 447 scholars. Pop. 2190.

LIVINGSTON, p. t., Essex co., N. J., 9 m. N.W. Newark, 58 m. N. E. Trenton, 224 W. Bounded W. by Passaic river, by small tributaries of which it is drained. It contains three stores, one saw-mill; five schools, 158 scholars. Pop. 1637.

LIVINGSTON, p. v., capital of Sumpter co., Ala., 68 m. S.W. Tuscaloosa, 836 W. Situated on the E. side of Sucharnochee river, a branch of Tombigbee river. It contains a courthouse, jail, and several stores and dwellings. Nett proceeds of the postoffice, $1029.

LIVONIA (Russ. Liflandia, Germ. Livland, or Lief-land), a mark. gov. of European Russia, on the Baltic, having N. the gov. of Revel, E. the lake Peipus, separating it from

the gov. of Petersburg, and the govs. of Pskov and Vitebsk, S. the latter and Courland, and W. the gulf of Livonia. Length N. to S. about 150 m.; average breadth, 117 m. Area, including the island Œsel, in the Baltic, 17,388 sq. m. Pop. in 1836, 740,160. The coast and the greater part of the surface are flat and marshy; but in the districts of Venden and Dorpat are some hills of considerable elevation; Harberg, one of these, being nearly 1100 ft. in height. There are several extensive lakes: the principal, Virtserf, 24 m. in length, by from 9 to 6 m. in breadth, communicates with the lake Peipus by the Embach. Besides the last named, the chief rivers are the Dwina, which forms the S. boundary, the Evst, and the Bolder-Aa. The soil, though in some parts loamy, is in general sandy, but, being abundantly watered, it is, by proper measuring, rendered very productive. Rye and barley are the principal crops, and more of both is grown than is required for home consumption. Wheat and oats are less cultivated; buckwheat is raised on sandy soils; flax, hops, and pulse are also produced; and the potato culture is on the increase: fruits are of very indifferent quality. In some districts, agriculture is tolerably well conducted. The forests are an important source of wealth, and supply excellent timber; they abound, not only with game, but also with wolves, which are sometimes very destructive to the cattle. The rearing of live stock, though not altogether neglected, does not receive adequate attention; the breed of black cattle is, however, in the course of being improved. Horses and sheep are very inferior. The fisheries, both on the coast and in the fresh waters, are important. Chalk, alabaster, and other calcareous materials are abundant.

Rural industry and the distillation of spirits are by far the most important occupations. The manufactures of this government are, however, more extensive than those in its vicinity. The peasantry spin linen yarn, and weave their own cloths; and in the towns, especially Riga, there are sugar refineries, and tobacco, woollen cloth, cotton, linen, glass, and other factories, which employed, in 1837, about 5800 hands, and produce goods to the amount of 11,080,000 roubles a year. (Possart.) The N. part of Livonia formerly constituted a portion of Esthonia, and the S. a part of Lithuania. The population consists of Esthonians, Lithuanians, Russians, Germans, and (along a portion of the coast) Lives, the most ancient inhabitants of the country, and from whom it has derived its name. About 58,000 of the inhabitants reside in the towns, and these, as well as the nobles, clergy, &c., are chiefly of German descent. Until 1804, the Esthonians and Lithuanians were in a state of predial slavery; now, however, they are free, but without the right to hold real property. The prevailing religion is the Lutheran; there are only about 12,000 individuals of the Greek Church, and other professions of faith. Education is tolerably advanced in the towns, and the university of Dorpat, in this government, is the first in the empire. But, after all, only 1 in 143 of the inhabitants is said to be receiving public instruction. Livonia has a governor-general, whose authority extends over the governments of Pskov, and the other Baltic provinces; but it has its own provincial assembly, magistracy, &c., and has preserved many peculiar privileges, among which is that of exemption from the state monopoly of ardent spirits. It was divided into nine districts by Catherine II.: Riga is the capital and centre of its commerce; the other chief towns are Dorpat, Pernau, Pellin, and Arensburg in the island Œsel. Livonia was conquered by the Danes in the 12th century, and held by the Teutonic knights from 1346 to 1561. It afterward belonged to Poland, and next to Sweden; but was definitively annexed to Russia, by the treaty of Nystadt, in 1721. (Schnitzler's La Russie, 580-565; Possart; Das Kaiserth; Russland.)

LIVONIA, p. t., Livingston co., N. Y., 8 m. E. Geneseo, 291 m. W. Albany, 363 W. Organized in 1808. Bounded S.W. by Conesus lake. Watered by the outlet of Hemlock lake which latter enters its S. border. It contains four churches, two Presbyterian, a Methodist, and Baptist; also stores, four fulling-mills, one woollen factory, two flouring-mills, one grist-mill, three saw-mills, one tannery, two distilleries; one academy, 40 students; 15 schools, 881 scholars. Pop. 2719.

LIVONIA, p. t., Wayne co., Mich., 16 m. S.W. Detroit, 540 W. Watered by branches of Rouge river. It has one store, four saw-mills; nine schools, 366 scholars. Pop. 1168.

LIXURI. See CEPHALONIA.

LIZARD'S POINT, a bold headland, on the British channel, being the most southerly promontory of England, on the S. coast of Cornwall, 22 m. S.S.E. the Land's End, lat. of highest lighthouse. 49° 57' 41" N., long. 30 11' 5" W. The Lizard is famous in navigation, from its being the point whence ships usually take their departure from the channel, and, being, also, the best place for a land-fall when homeward bound. It is surmounted by two lighthouses with fixed lights, at a short distance from each other, the lanters of the one being 255 ft. and of the other 291 ft. above the level of the sea. Some steep rocks, called the Stags, lie to the S. of the Lizard.

LLAMPETER, or LLAN-BEDR, a parl. bor., market town, and par. of S. Wales, co. Cardigan, hund. Moyddyn, 25 m. E. by N. Cardigan, and 190 m. W. by N. London. Pop. of parl. bor., which is contributory to Cardigan, about 1860, that of the entire parish being, in 1831, 1197. The town, which stands on a slope about 1-2 m. N. of the Teify (crossed here by a stone bridge), appears to have been larger formerly than at present, when a score of tolerably built houses, and about 100 cottages comprise the whole of its private dwellings. The church, which stands on an eminence at the N. end of the town, is very ancient, and, being shaded with venerable yews, has a very picturesque appearance: there are also two chapels for Calvinistic Methodists and Presbyterians. The chief ornament of the place is the College of St. David's, a handsome Gothic structure, erected in 1825. This institution, founded by George IV. in 1822, at the suggestion of the late Dr. Burgess, then bishop of St. David's, and endowed with six livings, is intended to furnish clerical instruction for the clergy of the S. part of the principality, and has already done much to raise a profession, which, owing partly to the misconduct and partly to the poverty and ignorance of its members, had fallen into merited disrepute. The students reside within the college, the business of which is conducted by the principal, who gives theological instruction, and is assisted by Greek, Hebrew, Welsh, and other professors. The course of instruction lasts during two years, and is attended, at an average, by about 60 students, whose necessary expenses do not exceed £55 a year. The bishop of the diocese, who is the visiter, ordains none except graduates of the English universities, or certificated students of Llampeter College. The town is of little trading importance. Markets on Saturday. Fairs, well attended, for horses, cattle, and hogs, Jan. 11, Wednesday in Whitsun-week, July 16, first Saturdays in Aug. and Sept., Oct. 19, and first Saturday in Nov. The town is incorporate, governed by a portreeve, and sessions are held annually by the county magistrate on the second Wednesday in October.

LLANDEILO-FAWR, a market town and parish of S. Wales, co. Caermarthen, hunds. Caro and Perfedd, on the Towy, 13 m. E. by N. Caermarthen, and 199 m. W. by N. London. Pop. of parish (including 10 hamlets), in 1841, 5471: do. of township, in 1831, 1983. The town, situated in the beautiful and interesting vale of the Towy, is small and ill-built, the only public buildings being an old church, and four places of worship for dissenters. Newton Park, the residence of Lord Dynevor, and Golden Grove, belonging to Earl Cawdor, are the principal country-seats of the neighbourhood, which is very productive, and has some rich mines of coal and iron. A railway connects this coalfield with the port of Llanelly. Quarter sessions are held here; and Llandeilo-fawr is one of the polling places for the county. Markets, well supplied with corn, &c., on Saturday; fairs, Feb. 20, May 5, and 12, June 21, Aug. 23, and Nov. 12.

LLANDOVERY, a mun. bor. and market town of S. Wales, par. Llandingad, co. Caermarthen, hund. Perfedd, 22 m. E.N.E. Caermarthen, and 192 m. W. by N. London. Pop. of bor. in 1841, 1709. The town, agreeably situated in the upper part of the vale of the Towy, at a short distance from that river, has one principal avenue, and eight other streets lined with respectable houses. The keep of an old castle, destroyed by Cromwell, occupies the summit of an insulated rock, and forms a chief feature of the place. The parish church stands a little S. of the town, and there are likewise four places of worship for dissenters, with attached Sunday schools. National and Lancastrian schools are established, and there are almshouses and other charities for the aged poor. There is little trade or traffic of any kind in Llandovery; but it is a municipal borough, governed, since the Municipal Reform Act, by a mayor and three other aldermen, with 12 councillors. The petty sessions for the hundred of Perfedd are held here, and Llandovery is one of the polling places at the elections for the county. Markets on Wednesday and Saturday: cattle fairs Wednesday after Jan. 17, the 2d Wednesday after Easter, Whit-Tuesday, July 31, and Nov. 26.

LLANELLY, a parl. bor., seaport, market town, and par. of S. Wales, co. Caermarthen, hund. Caerwallon, 12 m. S.E. Caermarthen, 10½ m. W.N.W. Swansea, and 174 m. W. by N. London. Pop. of parl. bor. in 1841, 11,153. The town is irregularly built, on a creek near the sea shore; but some of the houses are good, and the place, on the whole, appears to be thriving. The church is an old, irregular structure, remarkable as having two towers, one embattled, and the other surmounted by a steeple: the living is a vicarage, and within the parish are two chapels-of-ease. Dissenters also of different denominations have several places of worship. A free school and two other schools, chiefly supported by subscription, furnish instruction to the children of the poor; and there are four charities for the relief of the sick and aged. Llanelly is situated in the midst of the rich mineral basin of S. Wales. Four large collieries at Llangennech employ upwards of 500 persons; and the abundance of excellent coal, a part of which is exported to France, Spain, and the Mediterranean, for the use of steamboats, has caused the establishment of the Llanelly and Cambrian copper-works. The ore is imported chiefly from Cornwall; and the copper-cakes and sheathing are sent to Liverpool, and other ports of the kingdom. There are also two iron-foundries, but both use air furnaces, and of no great importance. The town has four docks, two of which are floating basins, the largest being capable of accommodating no less than 50 vessels of 500 tons register. This port had, in 1836, 79 ships, of the aggregate burden of 3637 tons, and the register tonnage cleared out averaged, for the five years preceding 1839, 54,000 tons. The gross customs' revenue in 1829, amounted to £3650.

The interests of the town have been recently much promoted by the construction of a railway, with branches into different parts of the fine coal-field near Llanelly; and it is probable that Llanelly will, at no distant period, become one of the principal trading ports of the principality. The parliamentary borough, which is contributory to that of Caermarthen, includes the borough hamlet, with some additions. Registered electors in both boroughs, in 1839-40, 997. The borough is governed by a portreeve and burgesses, and had formerly both civil and criminal jurisdiction. Markets on Thursday and Saturday: fairs on Ascension-day and September 30.

LLANGADOG-FAWR, a market town and par. of S. Wales, co. Caermarthen, hund. Perfedd, on the Towy, here crossed by a handsome stone bridge, 19 m. E. by N. Caermarthen, and 197 m. W. by N. London. Pop. of parish, in 1841, 2004. The town has two pretty wide streets, with a few well-built houses and numerous cottages, an old church, and three dissenting places of worship, being the only public buildings, besides a ruinous old castle. Woollen stockings, and coarse woollen cloths, are made here; but the chief business is the sale of farm produce at the fairs and markets, which are very considerable. Markets on Thursday; fairs, March 12, July 9, Thursday after September 11, and 2d Thursday after October 10.

LLANGOLLEN, a town of N. Wales, co. Denbigh, hund. Chirk, on the Dee, 20 m. S.W. Chester, and 195 m. N.W. London. Pop. of par., in 1841, 4006. The town, beautifully situate in a deep, narrow vale, enclosed by lofty mountains, and watered by the Dee, which is crossed here by a good stone bridge, consists of one principal and a few smaller streets, lined with old and mean houses, interspersed with a few handsome modern dwellings, among which are three large and commodious inns. The church, in the early English style has service performed in it both in English and Welsh: the living is a chapelof-ease at a hamlet within the parish, and the dissenters have three places of worship. The inhabitants derive their chief support from summer visiters, who, in making the tour of N. Wales, usually make some stay here, in order to enjoy the fine scenery of this vale, which in some respects excels that of the vales of Clwyd and Festiniog. Many families, also, reside here during summer, so that Llangollen may be considered as a sort of watering place. The Reform Act made it a polling-place for the co. Markets on Saturdays. Fairs, March 17, May 31, and Aug. 31.

About 1 m. from Llangollen, situated on a high and steep conical hill, are the ruins of the castle of Dinas Bran, once a fortress of considerable strength; and about 1 m. beyond, nearly in the same direction, are the majestic remains of Vale-crucis abbey, still in tolerable preservation: the name of this abbey is derived from a pillar or cross, situated in an adjoining field, supposed to be of high antiquity. Four miles from the town, and in another direction is the Cydylltan aqueduct, by which the Ellesmere canal is conveyed across the Dee, a noble structure of 19 arches, raised 198 ft. above the river, at a cost of £47,000.

LLANIDLOES, a parl. bor., market town and par. of N. Wales, co. of Montgomery, hund. Llanidloes, at the confluence of the Clevedon with the Severn, 37 m. W.S.W. Shrewsbury, and 158 W.N.W. London. Pop. of town in 1841, 4961. It is situated in a valley on the E. bank of the Severn (crossed here by a handsome stone bridge of three arches), and is surrounded on all sides by lofty hills: the buildings have increased rapidly, and several respectable houses have been substituted for others composed of wood and plaster, which formerly gave the place a mean appearance. A new town-hall stands nearly in the centre of the town. The church, built in 1549 on the site of an older structure, and very recently repaired, is chiefly remarkable for a ceiling of delicately carved oak, and for a square tower of great antiquity. There are also places of worship for

Independents, Wesleyans, Calvinistic Methodist, Baptists, and the Society of Friends. Instruction is furnished in day-schools to about 70 children; but education is little valued, and the mass of the population are described, we hope too strongly, as being "cradled in ignorance, and inured to vice both by habit and example."

Flannel and other woollens are the principal articles man-ufactured in Llanidloes, and the present improved condition of the town is wholly attributable, to its trade in these arti-cles. The spinning of wool is conducted in six mills, em-ploying 180 hands; but the cloth is wholly made by hand-looms. The quantity of flannel annually manufactured averages 4300 pieces: there were 815 looms at work in 1838, which employed 590 men, 176 women, and 106 children. The wages of the best weavers are 10s. a week; but the average is about 7s. Spinners earn about 12s. The weavers are stated to be drunken, improvident, dishonest, and in-subordinate: " in fact, between poverty on the one hand, and want of education on the other, the condition of Llan-idloes presents a picture darker, by many shades, than any town of its size in the principality, except Merthyr-Tydvil." (Hand-loom Rep.) Within the parish is the lofty mount-ain of Plinlimon, or more properly, Pumlumon, " the five-peaked hill," on which are the sources of the Severn, Wey, and Rheidiol; and at the foot of the range there are slate quarries and lead mines, the produce of which contributes to the support of the place.

Llanidloes is a corporate town having a mayor, coroner, and other officers, elected at a court-leet; it was not con-sidered sufficiently important to be included in the provis-ions of the Municipal Reform Act. The Reform Act made it a parl. borough, contributory to Montgomery, which sends one member to the House of Commons, and the electoral limits comprise, besides the town, a considerable extent of surface on both sides the Severn. Registered electors, in 1839-40, 97. Markets on Saturday: fairs April 3, May 11, June 21, July 17, September 13, October 3, and 25.

LLANRWST, a town of N. Wales, partly in co. Caernar-von and partly also in co. Denbigh, on the Conway, 37 m. W. Chester, and 183 m. N.W. London. Pop. of par., in 1831, 3601. The town, in a spacious vale, surrounded by lofty and well-wooded hills, stands chiefly on the E., but partly also on the W. bank of the Conway, which is cross-ed here by an elegant bridge, constructed in 1636, from the designs of Inigo Jones. Three considerable streets, lined with tolerably built houses, branch from a spacious market-place, in the middle of which is the town-hall, a substan-tial brick structure. The church, an old and small building, has adjoining to it the Gwydir chapel, a square castellated edifice, originally erected as a family mausoleum by the Wynne family, and now used as a place of worship. It has many monuments; but its chief celebrity is owing to its con-taining the remains of the great Llewellyn, removed thither from the abbey of Aberconway, in which they were origin-ally interred. There are also 11 chapels for dissenters with-in the parish, and some good Sunday schools. Llanrwst formerly noted for its harp manufacture, depends at present almost entirely on its retail trade; for the spinning and knit-ting of wool is become quite insignificant. It derives con-siderable advantages from its position on the Conway, which brings up vessels of 60 tons burden to Trefriew with coal, lime, timber, &c., in return for slate and iron. Gwy-dir castle, a rather large and very elegant modern structure, is situated about ½ m. from the town. Markets on Tues-days and Saturdays. Fairs on March 8, April 25, June 10, Aug. 10, Sept. 17, Oct. 25, and Dec. 11.

LLANTRISSENT, a parl. bor., market town, and par. of S. Wales, co. Glamorgan, hund. Miskin, 10 m. N.W. Cardiff, and 140 m. W. London. Pop. of parl. bor. in 1831, 956: ditto of par., 2789. The town, which stands on a commanding eminence overlooking the vale of Glamorgan, consists only of three or four narrow and irregular streets, lined with old and ill-built houses. The town-hall and market-house were erected by the Bute family, who are lords of the manor, and the principal landowners in the par-ish. The church is a large structure in the Norman style, the living being a vicarage in the gift of the dean and chap-ter of Gloucester cathedral. There are also two chapels of ease in the out-townships, and several places of worship for dissenters, with attached Sunday schools. The ruins of an old castle with a high tower stands close to the town; and, at a short distance, are some interesting remains of an old monastery. Llantrissent has a very little trade; but coal, lead, and iron are found in considerable quantities, in the hamlet of Pentyrch, and sent to Cardiff for exportation. The charter of the borough was granted by Edward III., and the government is vested in a portreeve, constable, and 12 aldermen, whose privileges were left untouched by the Municipal Reform Act. Llantrissent is a parliamentary borough, contributory with Cowbridge to Cardiff, which sends one member to the House of Commons. Registered electors in 1839-40, 785, of whom 183 belonged to Llantris-

sent. Markets on Friday: fairs, Feb. 13, May 12, Aug. 12 and Oct. 30.

LLERENA, a town of Spain prov. Estremadura, 39 m. N. Seville, and 200 m. S.W. Madrid. Pop., according to Miñano, 6495. It stands on a plain at the foot of the Sierra San Bernardo, which separates Estremadura from Seville, and has two parish churches, four convents, and a hos-pital. The inhabitants are chiefly engaged in grazing sheep and cattle on the rich pastures of the vicinity, and in col-lecting oak-bark, galls, and timber from the neighbouring forests.

LO (ST.), (an. Briovera), a town of France, dep. La Manche, of which it is the cap., on the Vire, and on the high road between Paris and Cherbourg, 156 m. W. by N. the former. Pop. (1836) ex. com., 8890. It is ill laid out; streets steep and irregular; they mostly lead from a square in the highest and central part of the town, which has sev-eral of the principal public buildings: among these the few that deserve notice are, the church of Nôtre-dame with two lofty spires; that of St. Croix, built in 685, and considered the best specimen of Saxon architecture in France; the prefecture, a handsome new edifice; the town-hall, judicial court, prison, hospital, theatre, and a bridge of six arches over the Vire. The environs are picturesque and agreeable. St. Lo is the seat of tribunals of original jurisdiction and commerce, a chamber of manufactures, and a communal college. It has a public library with 2500 vols., a philharmonic society, societies of agriculture and com-merce, manufactures of fine woollen cloths, druggets, can-vass, serges, calicoes, lace, cutlery, &c., and considerable trade in thread, iron, salt, butter, cider, honey, cattle, &c. It derives its present name from a bishop of Costance in the sixth century. (Hugo, art. Manche.)

LOANGO, a kingdom of W. Africa, on the Atlantic ocean, bounded N. by Mayomba, and S. by Congo, from which it is separated by the Zaire. The coast is high and abrupt, but the hills are covered with earth and luxuriant vegetation. The soil is generally a stiff loam, and very productive; but near the coast is an extremely fine sand, that is carried about by the lightest breeze. The lakes and rivers, of which there is a considerable number, abound with fish, and in the forests are found tiger-cats, onzes, hyenas, monkeys, antelopes, hares, and other game. The climate is excessively hot: it sometimes rains, but the dews are sufficient for vegetation. Almost the only grains are maninc, maize, and a species of pulse called mazegre, rude-ly cultivated by women, who merely stir the ground to the depth of an inch, and cover up the grain, to prevent its be-ing devoured by birds; and even this slender culture is con-fined to small patches round the villages. The rest of the country is covered with luxuriant herbage, rising to the height of eight feet, allowed by the people to grow, ripen, and wither, without being applied to any use. Sometimes, however, they set fire to it, producing a wide extended con-flagration over the whole country, the coast appearing from the sea to be on fire. The finest fruits grow wild, and the sugar-cane attains an extraordinary size. The tree called the mapou is distinguished, like the baobab, by the enor-mous dimensions of its trunk. Palm trees are very plenti-ful, particularly that species from which the natives extract their favourite liquor. The potato and yam are also abun-dant. The Chinese hog is the only animal reared for do-mestic use, the natives having altogether neglected the breeding of sheep, cattle, and horses, formerly introduced by the Portuguese, and still abundant at their settlement of Paul de Loanda. The inhabitants usually reside in villages or clusters of straw huts in the midst of palm groves. They seem to be in the lowest state of degradation, being incorrigibly indolent, debauched, filthy, cowardly, and su-perstitious in the extreme. The country is divided among several chiefs, who, though often at war with each other, acknowledge the supreme authority of the king of Loango, the capital. The latter is elective and absolute; but the judicial power is vested in the cabals or assemblies of the different villages. Loango, called Boral by the natives, about 2 m. from the coast, in lat. 4° 36' S., long. 11° 37' E., has been said to have a population of 15,000 persons. It is nothing more than a collection of huts. This and the ports of Kabenda and Majumba, also, in Loango, were for-merly among the principal slave marts on the coast of Guin-ea; and notwithstanding the efforts that have been made for the suppression of the traffic, we doubt whether it be materially diminished. (For farther and ample information as to this country, see Voyages à la Côte Occidentale de l'Af-rique, by Degrandpré, passim; and Prevost, Histoire Gén-érale des Voyages, vol. iv. 579-610.)

LOCHES, a town of France, dep. Indre-et-Loire, cap. ar-rond. on a hill beside the Indre, 224 m. S.E. Tours. Pop. (1836), ex. com., 3600. It is irregularly laid out, and its streets are narrow; but it is clean, and has many good hou-ses. Its castle, on a plateau, at the summit of the hill on which the town is situated, has gained considerable notorigity in French history. It appears to have been built in the last ages of the western empire, and is one of the most remarkable remains of that period now existing in France. Charles VII. defended it successfully against the English; Louis XI. made it a state prison; and here, Cardinal Balue, of infamous memory, was confined in an iron cage for 11 years. It is now mostly destroyed, what remains being occupied by the sub-prefecture, prison, &c. The palace of Charles VII., now the municipality, is a large oblong building on the bank of the Indre; it was long the residence of Agnes Sorel, whose remains are deposited in a chapel in a tower of her erection. The church of Loches, originally founded *circa anno* 430, is a singular piece of architecture, with four steeples, two of which are about 160 ft. high. Loches communicates with the little town of Beaulieu by several bridges over the Indre. It is the seat of a tribunal of primary jurisdiction, and a communal college; and has manufactures of linens and coarse woollen cloths, paper, leather, &c. (*Hugo*, art. *Indre-et-Loire*.)

LOCHMABEN, a royal and parl. bor. and market town of Scotland, co. Dumfries, in a level country, surrounded by several lochs, or lakes, 10 m. N.E. Dumfries, and 26 m. N.W. Carlisle. Pop. 1012. The town consists of one extremely wide street, more or less overgrown with grass. The public buildings are a town-house, parish church, and a dissenting chapel. It has no manufactures. The schools are good; and there is a subscription library; but it is also neither about the poorest royal bor. in Scotland. Lochmaben owes its origin to the protection afforded by a castle of the same name, built in the 12th century by the Bruces, lords of Annandale, from whence King Robert Bruce was descended. The site of this fortress, surrounded by a deep fosse and moat, is still called the Castle-hill. Robert Bruce built another strong castle on a peninsula on the S.E. side of the Castle loch, which, with its outworks, covered nearly 16 acres. The walls, in the few places where they are still entire, are 12 ft. thick. It was preserved as a border fortress till the union of the crowns; since which it has gradually gone to decay. Bruce parcelled out the barony of Lochmaben, called the "Four Towns," as it contains four villages, among his retainers, in small patches, on the condition that the occupants should furnish a certain amount of provisions for the use of the royal fortress. These persons, who are called the "king's kindly tenants," had no written title to the lands; and at present, in case of a sale, a simple deed of conveyance is sufficient; and the succession is taken up without any feudal service. Owing to a misunderstanding between these tenants and the keeper of Lochmaben palace, Charles II., in 1664, guaranteed to them the perpetuity of their leases, and relieved them from every burden, except the rents and services paid by their ancestors in 1602, which are nominal merely. The tenants are a poor but contented class, having little intercourse with the rest of the community. Johnstone, of Annandale, is the hereditary keeper of the royal palace, and, as such, receives the nominal rents in question. Many of the inhabitants of the borough, like the "king's kindly tenants," are owners of small patches of land, there being within the borough no fewer than 141 small proprietors! Lochmaben unites with Annan, Sanquhar, Dumfries, and Kirkcudbright in sending a member to the House of Commons. Registered voters in 1840-41, 41. (*New Stat. Acc. of Scotland*, § *Dumfries*, p. 376-397; *Chalmers's Caledonia*, § *Dumfries-shire*; *Forsyth's Picture of Scotland*.)

LOCHWINNOCH, a manufacturing town of Scotland, co. Renfrew, on the Calder, a stream which terminates in castle Semple loch. 15½ m. S.W. Glasgow, and 25½ m. N. Ayr. Pop. in 1831, 9645. The town, which is sheltered in every direction, except the S.E., either by rising grounds, or thick plantations, has a main street (½ m. long), with several streets crossing it at right angles. There are, in the town, the parish church, a chapel belonging to the Associate synod, several public libraries, and various friendly societies. Manufactures were early introduced into Lochwinnoch, but those of linen and silk have disappeared. Thread-making was introduced in 1792: at one time there were about 90 thread-mills in the place, but the business is now nearly discontinued. Cotton is the staple manufacture. Three cotton-mills employ about 600 persons, and there are above 218 weavers employed by the manufactures of Glasgow and Paisley. There is a small power-loom factory, a small mill for carding and spinning wool, and one of the best corn-mills in Scotland. The line of the Glasgow and Ayr railroad, opened in 1840, passes close to Lochwinnoch. (*Ibid*; and *Factory Reports*, 1839, p. 306, 307.)

LOCKE, p. t., Cayuga co., N. Y., 20 m. S. Auburn, 155 m. W. Albany, 319 W. Organized in 1802. Watered by the inlet of Owasco lake. It contains two churches, a Presbyterian and Baptist; four stores, two fulling-mills, one woollen factory, three grist-mills, seven saw-mills; 12 schools, 516 scholars. Pop. 1654.

LOCK HAVEN, p. v., cap. of Clinton co., Pa., 107 m.

N.W. by N. Harrisburg, 190 W. Situated on the S. side of the W. branch of Susquehanna river. It contains a court-house, jail, four stores, and about 150 inhabitants.

LOCKERBIE, a market town of Scotland, co. Dumfries, in the centre of a rich and fertile country, on the road between Carlisle and Glasgow, 37 m. N.W. the former, and 66 m. S.E. the latter. Pop. in 1831, 1414. It is neat and regularly built: its only public buildings are the parish church, and a chapel belonging to the Associate synod. Lockerbie has long been distinguished for its excellent schools. There are two public libraries and a reading room. These are two fairs and 10 markets annually. The fairs are exclusively, or at least principally for lambs and wool. When the border feuds had so far ceased (after the union of the crowns) as to allow a slight intercourse between the English and Scotch, the sheep farmers of the S. of Scotland assembled here to meet the English dealers. This was the origin of these fairs, which have been long very important. The Lammas fair (second Monday in August) is the largest lamb fair in Scotland. The 10 markets have each a somewhat different object: one of them being for hiring servants, another for black cattle and horses; while those in winter are principally for pork, which is largely produced in the vicinity. There are two branch banks in the town. (*New Stat. Acc. of Scotland*, § *Dumfries*, p. 451.)

LOCKPORT, p. t., Niagara co., N. Y., 277 m. W. by N. Albany, 408 W. Organized in 1824. Bounded S. by Tonawanda creek, and watered by streams flowing into lake Ontario. It contains one commission house in foreign trade, 65 stores, five flouring-mills, nine grist-mills, 45 saw-mills, one oil-mill, one paper-mill, three tanneries, one distillery one brewery, one pottery, four printing-offices, one bindery; three academies, 503 students; 30 schools, 2670 scholars Pop. 5125. The village is situated on the Erie canal, and was founded in 1821, became the capital of the county in 1822, was incorporated in 1829, and is included in a parcel telegram, 12 m. long and 1½ m. wide, containing 1690 acres It is laid out with considerable regularity, on both sides of the canal, which here rises or descends by five double locks, of 12 feet lift each. W. of the village, the canal is excavated through the mountain ridge for the distance of 3 m., at an average depth of 30 feet, in limestone rock. This gives the waters of lake Erie an uninterrupted flow of 30 m. to this place, creating, by the fall at the village, a great water-power. It contains a stone courthouse, including a jail, and a fire-proof county clerk's office; an academy, a female seminary, a lyceum, 11 churches, two Presbyterian, two Episcopal, a Methodist, Baptist, Lutheran, two Friends', and a Roman Catholic, mostly of stone or brick, and an African; most of the above stores and manufacturing establishments, with many mechanic shops, 750 dwellings, and about 6000 inhabitants.

LOCKPORT, p. v., Will co., Ill., 170 m. N.E. by N. Springfield, 747 W. Situated on the Illinois and Michigan canal, at the termination of the lake level, 34½ m. from Chicago. The canal has here a fall of 90 feet, by two locks, producing, by its surplus water, great water-power. Des Plaines river has also, in the vicinity, a fall of 15 feet, affording to the place great facilities for manufacturing.

LODEVE, a town of France, dep. Herault, cap. arrond., on the Ergue, at the foot of the Cevennes, 27 m. W.N.W. Montpelier. Pop. in 1836, *ex. sem.*, 11,071. It is ill built; is surrounded by old fortifications; and has an old cathedral, formerly a bishop's see. In Lodève and its neighbourhood, from 7000 to 8000 work-people are employed in the manufacture of woollen cloth for the army, and nearly all the inhabitants of the town are in some manner connected with this business, at least three fourths of the population belonging to weavers' families. About 60,000 pieces of broadcloth are made annually. The government demand for this cloth being pretty constant, the people engaged in its manufacture have also pretty constant employment, and their condition is consequently better than that of most of those in the ordinary departments of industry. Wages are high; the men getting from 1 fr. 75 c. to 3 fr. a day; women from 75 c. to 1½ fr.; and children from 50 to 80 c. The workpeople are said to be active, industrious, and particularly sober; and the proportion of illegitimate to the total births in Lodève is said not to be more than one in 30, while in the department generally it is one in 19. This singular statement, the accuracy of which seems at first sight very questionable, is partly, at least, accounted for by the fact that the weavers usually marry early, that there is no garrison in the town, that some of the females lie in at Montpelier. Besides, and elsewhere, and that there is no foundling asylum in the whole arrondissement. (*Hugo*, art. *Herault*; *Villermé*; *Tableau Physique*, &c., *des Ouvriers*, t. 319-333.)

LODI, a city of Lombardy, cap. deleg. Lodi and Crema, on the Adda, here crossed by a wooden bridge, and on the road from Milan to Piacenza, 18 m. S.E. Milan. Lat. 45° 18′ 31″ N., long. 9° 20′ 22″ E. Pop. (1838) 18,962. It is sitnated on slightly rising ground, and surrounded by old walls, and entered by four gates. It is generally well built, and has broad and regular streets, an old citadel, now dismantled, and converted into barracks by the emperor Joseph II., numerous churches, a large hospital, a theatre, several handsome palaces, and a large market-place, surrounded with arcades. The church *della Incoronata* is said to have been designed by Bramante; it has a fine rotunda, and is ornamented with frescoes and paintings by Calisto, a pupil of Titian. In the cathedral is the "Murder of the Innocents," by the same artist. Lodi is a bishop's see; and the seat of the governor, assembly, and superior judicial courts for the delegates; it has a royal and ecclesiastical gymnasium, a public library, a normal school, founded by Joseph II., orphan and foundling asylums, a workhouse, a *mont de piété*, a large porcelain factory; and manufactures of linen fabrics, &c. It is the centre of the trade in Parmesan cheese. (See ITALY, AUSTRIAN, in this vol., p. 58.)

Lodi is famous in modern history for the victory achieved here on the 10th of May, 1796, by Napoleon, in his first Italian campaign. The cannon of the Austrians swept the bridge behind which they were drawn up; but it was, notwithstanding, forced by the French at the point of the bayonet, and the Austrian army totally defeated. On this occasion, the intrepidity and gallantry of Napoleon shone as conspicuously as his skill as a tactician. (*Oesterr., Nat. Encycl.*; *Borghaus*; *Conder's Italy, &c.*)

LODI, p. t., Seneca co., N. Y., 254 m. W. Albany, 318 W. Bounded W. by Seneca lake. Watered by streams flowing into it, and by others flowing E. into Cayuga lake. It contains a Dutch Reformed church, six stores, three fulling-mills, one woollen factory, one flouring-mill, five grist-mills, eight saw-mills, one oil-mill, two tanneries; 15 schools, 670 scholars. Pop. 3238.

LODI, t., Bergen co., N. J., 5 m. S.W. Hackensack. Bounded E. and S.E. by Hackensack river, W. and S.W. by Passaic river. Along the latter are many fine country seats. It has one store, one dyeing and printing works, three grist-mills, three saw-mills; two schools, 52 scholars. Pop. 687.

LODI, p. t., Washtenaw co., Mich., 43 m. W. Detroit, 590 W. Pop. 1677.

LOFFODEN ISLES, a group of islands on the coast of Norway, between lat. 67° 40' and 69° 30' N., and long. 11° 46' and 19° 20' E. There are five larger and several smaller islands, having in all from 3000 to 4000 inhabitants. The principal are (taking a S.W. direction) Andöen, Langöen, and Hindöen, which is the largest of the whole group, and, with six others, forms, on the side of the Norwegian continent, the great gulf of West Fiord. The coasts of these islands are extremely irregular, and they rise into lofty and rugged mountains, covered with perpetual snow, and in some places with glaciers. There are no trees, but only a few stunted shrubs, grass, and cryptogamous plants; nor are these islands of any importance, except on account of the fisheries, which are very extensive and valuable. "In the beginning of February the cod-fish set in from the ocean, and occupy the banks in West fiord. These banks are from 3 to 10 m. out in the fiord, at a depth of from 60 to 80 fathoms; and the fish crowd so much together while depositing their spawn that it is said a deep sea lead is often interrupted in its descent to the bottom through these *fish-hills*. The fishermen assemble in the month of January at the different stations, and the fish are caught by nets and long lines, set at night and taken up in the morning. An outfit or company consists of two boats, each having five men, and provided with six or eight nets; and every 20 or 30 of these companies have a large tender to bring out their provisions, nets and lines, and to take the produce to market. This fish are cured as round or stock fish till April, after which they are split, salted, and carried to Drontheim, or other places, to be dried on the rocks, like the Scotch dried cod. The stock-fish are merely gutted and hung up, two together, across poles, and are dried, without salt, in the wind. In a medium year (1827) there were 2916 boats fishing in 83 different stations, accompanied by 194 tenders, the number of men in all being 15,394. The produce amounted to 16,436,690 fish, which, when dried, would weigh 8800 tons; there were, also, 21,530 barrels of cod-oil, and 8000 barrels of cod-roe. This important winter-fishery ends in the middle of April. The herring fishery on these shores is of much less consequence. (*Laing's Norway, p. 399-403.*)

LOGAN, co., Va. Situated in the W. part of the state, and contains 2930 sq. m. Drained by Guyandotte and Coal rivers. It contained in 1840, 5431 neat cattle, 9454 sheep, 10,300 swine; and produced 7136 bushels of wheat, 870,930 of Indian corn, 96,404 of oats, 10,852 of potatoes, 0983 pounds of tobacco, 5946 of sugar. It had five stores, five flouring-mills, 52 grist-mills, four saw-mills, seven tanneries, two distilleries; 19 schools, 370 scholars. Pop.: whites, 4136; slaves, 150; total, 4309. Capital, Logan.

LOGAN, co., Ky. Situated in the S., towards the W. part of the state, and contains 600 sq. m. Drained by tributaries of Green and Cumberland rivers. It contained in 1840, 7992 neat cattle, 45894 sheep, 11,878 swine; and produced 49,758 bushels of wheat, 1197 of rye, 205,440 of Indian corn, 63,580 of oats, 4438 of potatoes, 383,597 pounds of tobacco, 96,397 of cotton. It had eight stores, two woollen factories, 13 cotton factories with 180 spindles, four flouring-mills, 11 grist-mills, seven saw-mills, one oil-mill, one tannery, six schools, 389 scholars. Pop.: whites, 8479; slaves, 4586; free coloured, 310; total, 12,615. Capital, Russellville.

LOGAN, co., O. Situated a little N.W. of the centre of the state, and contains 426 sq. m. Organized in 1818. Watered by Miami river, and its branches. It contained in 1840, 6370 neat cattle, 7638 sheep, 9343 swine; and produced 24,154 bushels of wheat, 2406 of rye, 528,235 of Indian corn, 1544 of buckwheat, 70,135 of oats, 9983 of potatoes, 24,306 pounds of sugar. It had 17 stores, one fulling-mill, two flouring-mills, 71 grist-mills, 13 saw-mills, two tanneries, one pottery, one printing-office, one weekly newspaper; 39 schools, 1806 scholars. Pop. 14,015. Capital, Bellefontaine.

LOGAN, co., Ill. Situated near the centre of the state, and contains 890 sq. m. Drained by Sugar creek, and other branches of Sangamon river. It contained in 1840, 3672 neat cattle, 3661 sheep, 16,604 swine; and produced 19,499 bushels of wheat, 234,490 of Indian corn, 22,230 of oats, 6998 of potatoes. It had five stores, six grist-mills, six saw-mills; seven schools, 180 scholars. Pop. 2333. Capital, Postville.

LOGAN, t., Clinton co., Pa., 90 m. N.E. Bellefonte. Drained by Big Fishing creek. It has two stores, four saw-mills, one tannery; one school, 30 scholars. Pop. 1187.

LOGAN, p. t., Dearborn co., Ia., 87 m. S.E. Indianapolis, 356 W. It has six stores, one fulling-mill, two flouring-mills, two grist-mills, six saw-mills, two tanneries, one distillery, five schools, 140 scholars. Pop. 1366.

LOGAN C.H., p. v., cap. of Logan co., Va., 351 m. W. Richmond, 366 W. Situated on the E. side of Guyandotte river. It contains a courthouse, jail, and several stores and dwellings.

LOGANSPORT, city and p. v., cap. of Cass co., Ia., 72 m. N. Indianapolis, 616 W. Situated on Wabash river, at the junction of Eel river, and has a fine water-power. It contains a courthouse, jail, an academy, six churches, two Presbyterian, a Methodist, an Episcopal, a Baptist, and a Roman Catholic; 12 stores, several flouring and saw mills, 400 dwellings, and about 2000 inhabitants. A fine bridge crosses the Wabash river, and another crosses Eel river. The Wabash and Erie canal passes through it, and its location for business is highly promising.

LOGHUR (*Lohogur*, "the iron fort"), a strong hill fort of Hindostan, prov. Aurungabad, in the British territories, 30 m. N.W. Poonah. From the perpendicular height of the rock on which it is built, this fortress could not, if properly defended, be taken by storm. It is supplied with water by numerous tanks and springs, and has extensive excavated magazines. It came into the possession of the British in 1818.

LOGRONO (an. *Juliobriga*), a town of Spain, in Old Castile, prov. Soria, on a spacious plain on the Ebro, which is here crossed by a handsome bridge, 57 m. W.S.W. Pampeluña, and 156 m. N.E. Madrid. Pop., according to Miñano, 8210. It comprises, beside several good streets, two fine squares, with a collegiate church, five parish churches, eight convents, and two hospitals. It has tanneries, distilleries, and fabrics of saddles, hats, and candles.

LOHEIA, a seaport town of Arabia, the most northerly in the territ. of Yemen, on the Red sea, 175 m. N.N.W. Mocha, lat. 15° 41' 20", long. 42° 46' 14". It stands on low ground, sometimes inundated by the sea. Its port is an shallow that vessels of even small burden are obliged to anchor at a considerable distance off shore. The environs are arid and sterile, and the town is ill supplied with water. It is not walled, but is defended by several towers at equal distances round it, though only one of these is defencible by cannon. A few houses are of stone, but the greater part are mere mud huts, thatched with grass, with a straw mat for the door, and rarely any windows. The chief edifices are, a mosque, with the tomb of the Mohammedan saint who founded the town; the governor's residence; the custom-house, and some coffee warehouses. The coffee shipped at Loheia is inferior to that of Mocha; but it, notwithstanding, carries on a considerable trade in it with Cairo, through Djidda. Lime is prepared in the neighbourhood by the calcination of coral, and near the town is a salt mine. (*Niebuhr, Voyage de l'Arabie, i., 243; Geographical Journal, &c.*)

LOIR-ET-CHER, a dep. of France, reg. centre, between lat. 47° 15' and 46° 10' N., and long. 0° 30' and 90 15' E., having N. Eure-et-Loire, E. Loiret and Cher, S. Indre and

Indre-et-Loire, and W. the latter and Sarthe. Length. N.W. to S.E., 80 m.; breadth varying from 20 to 45 m. Area, 685,971 hectares. Pop. (1836) 944,043. Surface mostly plain, with a general inclination towards the W. The Loire intersects the department nearly in its centre in a direction from E. to W.; the other chief rivers are, in the N. the Loir, a tributary of the Sarthe; and in the S. the Cher, Boneboeuf, and Cosson, affluents of the Loire. In the S. of the department are numerous pools and marshes, which in the arrondissement of Romorantin cover nearly 3400 hectares. In 1834, it was estimated that 369,627 hectares of the surface were arable, 31,634 occupied with pastures, 98,591 with vineyards, 70,210 with woods, and 80,896 with heaths, wastes, &c. More corn is grown than is required for home consumption; in 1835, 1,261,598 hectolitres were harvested, chiefly oats and wheat. The annual produce of wine is estimated at above 900,000 hectolitres, some of which is of a pretty tolerable quality: the wines are principally made into brandy and vinegar; but a peculiar variety, of a very deep, dark hue, is extensively employed to deepen the colour of other red wines, and to give a reddish tint to white wines. (Jullien, Topographic, p. 84.) Beans and peas, fruit, hemp, liquorice, and beet-root, are raised in considerable quantities. In 1830, about 277,000 sheep were kept in the department, the annual produce of their wool being estimated at 795,000 kilog. A good many poultry and bees also are reared. The rural population is, however, in a very depressed condition; the labouring class occupy miserable huts, and in one village the habitations are said to be mere caves dug in the rock. In 1835, of 95,051 properties subject to the contribution foncière, 49,780 were assessed at less than 5 fr., and 14,364 at from 5 to 10 fr. Iron, turf, and alabaster are met with, but the most valuable mineral product is flint: the most extensive beds of which in France are in the S. part of this department, which has furnished the greater part of the gun flints used in France. A good workman produces in a day 400 flints of the first, and 600 of the second quality; and it is accordingly estimated that 100 workmen, assisted by their families, will probably send to market in a year 30,000,000 flints of all descriptions. The employment is very injurious to the health, and it is alleged that most workmen die of chest diseases before they attain to 30 years of age. The department has several iron forges, tile and glass factories, potteries, &c., with manufactures, though on a small scale, of serge, woollen cloth, and other woollen fabrics, cotton and hempen cloths, paper, leather, chemical products, &c. It is divided into three arrondissements; chief towns, Blois, the capital, Romorantin, and Vendôme. It sends three members to the chamber of deputies. Number of electors in 1839-39, 1570. Total public revenue in 1831, 5,968,150 fr. (Hugo, art. Loir-et-Cher; Official Tables, &c.)

LOIRE (an. Liger), the principal river of France, through the central part of which it flows in a W. direction to its embouchure in the Atlantic. Its basin, which comprises nearly one fourth part of the kingdom, has the basin of the Seine on the N.E., that of the Garonne on the S.W., and that of the Rhone on the E. It rises in mount Gerbier de Jonc, on the W. declivity of the Cevennes, in the department Ardèche, about lat. 44° 38' N., long. 4° 30' E., at an elevation of 4563 ft. above the sea. (Bruguiers.) Its general direction is N.N.W. to near Orleans, after which it flows mostly W.S.W. to its mouth near Palmboeuf, in about lat. 47° 15' N., and long 2° 15' W. Its entire course is estimated at 670 m., of which 512 are navigable. Before losing itself in the ocean, it spreads out into a considerable estuary; below Nantes it is between 2 and 3 m. in width; but its navigation in the lower part of its course is rendered difficult by shallows and numerous islands. Ships of 900 tons, though built at Nantes, are loaded at Palmboeuf or St. Nazaire; and all ships of considerable burden unload nearly 30 m. below Nantes, their cargoes being conveyed to that city by lighters. During the first 40 m. of its course, the Loire has an average descent of more than 50 ft. a mile; its rate of descent afterward averages 4 ft. a mile. Its current is everywhere rapid, and its inundations are frequently productive of much damage; to prevent which, extensive embankments have been erected along its banks below Orleans.

The tide rises to about 5 m. below Nantes. Its chief tributaries are the Maine, Eadre, and Brive from the N.; and the Allier, Loiret, Cher, Indre, Vienne, and Sevre-Nantaise from the S. It is connected with the Seine by means of the Orleans, Briare, and Nivernais canals; with the Rhone by the canal du Centre; and with Brest harbour and the English channel by the Nantes and Brest canal. To obviate the impediments to navigation from sandbanks, &c., above Orleans, a lateral canal, commenced in 1822, has been constructed along the river: it begins opposite the mouth of the Briare canal, in the department of Loiret, and runs along its S.W. bank till it terminates opposite the canal du Centre, in the department of Allier. The entire

length of this canal is 123 m. The scenery along the Loire, though in parts very fine, is generally surpassed by that of the Rhone. Some very important cities stand on its banks, among which, reckoning from its source, may be specified, Roanne, Nevers, Orleans, Blois, Tours, Saumur, Ancenis, and Nantes. (Dict. Géog., &c.)

LOIRE-HAUTE, an inland dep. of France, between lat. 44° 45' and 45° 24' N., and long. 3° and 4° 30' E., having N. Puy-de-Dôme and Loire, S.E. Ardèche, and S.W. Lozère and Cantal. Area, 498,560 hectares. Pop. 295,331. It is generally mountainous, with a slope to the N. The Cevennes mountains run along its S.W. border, and a range, passing off laterally from them, intersects the department about its centre, and afterward bounds the department of Loire on the W. But most of its mountains belong to the volcanic system of France. The Loire and Allier are the principal rivers, and receive numerous small streams within the department: there are many small lakes and pools around Le Puy, and elsewhere. The bottoms of the valleys are fertile, but not the other parts of the department; by far the greater portion of the surface being stony or sandy. In 1834, it was estimated that there were 226,072 hectares of arable land, 79,432 ditto meadow, 74,030 ditto woods and forests, and 90,239 ditto heaths, &c. Agriculture is extremely backward; half the arable land is constantly in fallow, and the occupiers are miserably poor. Sufficient corn, chiefly rye, with some wheat, is, however, grown for home consumption; but about 50,000 hectol. of wine are annually imported. The natural pastures are good, and their irrigation is pretty well conducted. In 1830, there were about 168,000 head of cattle, and 278,000 sheep in the department; the latter yielding about 350,000 kilog. a year of wool. The rural population is, in general, very poor; and about 3000 individuals annually leave the department in search of employment in the provinces around, as reapers, road-makers, day labourers, &c.; and usually return, after about six months' absence, with sums supposed to average about 70 fr. each. The land is very much subdivided. In 1835, of 93,838 properties subject to the contribution foncière, 45,587 were assessed at less than 5 fr., and there are fewer large properties in this than in any other department of France, Corrèze only excepted. Haute-Loire yields annually about 200,000 metrical quintals of coal, worth as many francs, and a little iron, zinc, and antimony. Manufactures either do not exist at all, or are confined to common linen fabrics, lace, tiles, bricks, earthenware, silk riband, and organzine in small quantities. Le Puy has a small lace manufacture, and is the great entrepôt for the small bells (grelots) used by the muleteers and waggoners of the S. of France. Haute-Loire is divided into three arrondissements; chief towns, Le Puy, the capital, Bronde, and Yssengeaux. It sends three members to the chamber of deputies. Number of electors (1838-39) 1219. This department is said to be taxed in a disproportionately high degree as compared with its means: in 1831, the public revenue derived from it was 4.319,731 francs; expenditure, 2,324,390 francs. (Hugo, art. Haute-Loire; French Official Tables.)

LOIRE-INFERIEURE, a maritime dep. of France, formerly included in the prov. of Brittany, between lat. 46° 50' and 47° 50' N., and long. 1° and 2° 30' W., having N. Morbihan and Ille et-Vilaine, E. Maine-et-Loire, S. Vendée, and W. the Atlantic. Area, 681,704 hectares. Pop. in 1836, 470,768. The Loire has its mouth in this department, which it intersects from E. to W. near its centre. The Erdre, Sevre-Nantaise, Maine, Moine, &c., affluents of the Loire, are the other chief rivers, all of them being navigable for some distance. The Vilaine skirts the N.W. extremity of the department, and communicates with the Loire by the canal between Nantes and Brest. Lakes and pools are estimated to cover 7200 hectares; the chief of these is the Grand Lieu, in the S., 4 m. in length by about the same in breadth. There are only a few hills of insignificant elevation in the N.E.; but along a part of the coast is a succession of sandy downs (dunes), which, not having been fixed by any artificial method, are gradually extending themselves, and have quite buried the old village of Escoublac. On various parts of the shore, as at Guérandé, &c., the sea has receded to a considerable extent. The isles of Noir-Moutiers and Bacin belong to this department. In 1834, about 321,600 hectares were arable, and 105.069 in pasture; vineyards occupied 26,346 hectares, orchards 10,984 hect., woods 33,075 hect., and heaths, wastes, &c., 199,352 hect., or nearly one fifth part of the entire surface. The country, on the S. bank of the Loire, is much superior in fertility to that on the N., and it is nearly all under culture; but agriculture is everywhere in the most backward state. There are a great number of little proprietors, many of whom engage themselves as labourers on the larger farms, who hold from 1 to 10 acres of land, farmed by their families. Very few properties yield a rental of 8000 francs (£1300) a year. The largest farms seldom extend beyond

200 acres; the greater number varying from 160 to 200. In 1835, of 198,059 properties subject to the *contribution fonciére*, 67,796 were assessed at less than 5 fr., and 19,814 at from 5 to 10 fr. In the vicinity of Nantes there are lands which let as high as £3 10s. an acre; while in the vicinity of Châteaubriant the average rent is under 10s., and in other parts of the department it is still less. Leases generally run from three to five and seven years; seldom beyond the latter term. Few farms are let for a money rent. Some farmers pay a stipulated quantity of grain for the arable, and money for the pasture land; but the far greater number hold on the métayer principle, paying half the gross produce to the proprietor. The reader will not, consequently, be surprised to learn that the farmers are without capital or intelligence, that their implements and cattle are very inferior, and that the land, which is not half tilled, is usually left fallow every other year, and frequently for several years together. The usual wages of farm labourers vary from 7½d. to 8½d. a day; women get from 4d. to 7¼d. During harvest, wages are about half as much higher. Little butchers' meat is consumed by the agricultural population. Their food consists principally of bread, butter, or fat, cabbage soup, buckwheat, pancakes and potatoes. Paupers are very numerous in winter, and in the rural districts there is no adequate provision for their support. The occupiers are in general miserably lodged, frequently sleeping in the same apartment with their cattle. They are not in debt, but have no money; are strongly attached to routine practices, and move on without an effort to improve their condition. (*Consular Report.*)

The produce of corn is estimated at about 1,400,000 hectolitres a year, principally wheat, buckwheat, and rye; a good many turnips are raised as food for cattle and sheep. The produce of wine is estimated by Jullien at 900,000 hectolitres, but the quality is inferior; about 300,000 hectolitres are consumed in the department, the rest being principally converted into brandy. The annual produce of cider may be about 130,000 hectolitres. The pastures on the banks of the Loire are excellent, and feed great numbers of cattle. The cows are good milkers, and the vicinity of Nantes is famous for its butter. In 1830, the stock of sheep amounted to 239,000 head, producing 250,000 kilog. of wool. The horses, though not large, are strong and handsome. The forests, which abound with oaks, feed a good many hogs. Bees are numerous, and the honey and wax of the department have a high reputation. The pilchard and herring fisheries are important: the former employs 2500 fishermen on the water, and a great many women in salting and barrelling the pilchards on shore. The manufacture of salt, from the extensive salt-pans at Noirmoutiers, Guerande, Croisic, &c., employs about 7000 hands, and furnishes produce worth above 900,000 fr. a year. Bog iron is plentiful, and is smelted in the arrondissements of Ancenis and Châteaubriant. A tin mine is wrought at Piriac. Granite, coal, turf, porcelain, clay, &c., are the other chief mineral products. There are two royal cannon foundries and several building docks in the department, and manufactures of sail-cloth, rope, glass, porcelain, tiles, paper, leather, &c. The trade of this department centres almost entirely in Nantes (which see). It is divided into five arrondissements; chief towns, Nantes, the capital, Châteaubriant, Ancenis, Paimbœuf, and Savenay. It sends seven members to the chamber of deputies. Registered electors (1838–39) 2208. Total public revenue in 1831, 27,040,954 fr.; expenditure, 10,880,664 fr. (*Hugo*, art. *Loire Inférieure*; *French Official Tables*; *Parl. Report.*)

LOIRET, a dep. of France, region centre, between lat. 45° 13' and 48° 16' N., and long. 3° 42' and 4° 45' E., having N. Eure-et-Loir, Seine-et-Oise, and Seine-et-Marne, E. Yonne, S. Nièvre, Cher, and Loir-et-Cher, W. the last-named department. Area, 657,679 hectares. Pop. (1836) 319,189. Surface for the most part level, but in the N. is a chain of hills separating the basins of the Loire and the Seine. The Loire traverses the S. half of the department, generally in a W. direction. It receives the Loiret, which rises within the department, and joins the Loire after a short course; being, however, navigable for boats nearly to its source. S. of the Loire, the country is marshy, uncultivated, and infertile; but, in other parts, it is very productive, particularly in the W. districts. In 1833, the arable lands were said to comprise 394,590 hectares, meadows 24,464 ditto, vineyards 39,892 ditto, and forests 99,474 ditto. Agriculture is in a comparatively forward state. The corn grown, which is chiefly oats and wheat, exceeds the quantity required for home consumption. The annual produce of wine is estimated at 1,900,000 hectolitres; two thirds of which is exported, under the name of *vins d'Orléans*, and the rest consumed at home, or converted into brandy or vinegar. None of the wine is of a superior quality, but the better sorts are esteemed as *vins ordinaires*. Cider is made in the arrondissement of Montargis. Various fruits, with flax, hemp, saffron, &c., are grown; and of late the culture

---

of beet-root for sugar has gained ground. The different branches of rural industry are all pursued by the same individuals who simultaneously grow corn, garden produce, and wine; and rear cattle, sheep, poultry, &c. In 1830, it was estimated that there were in the department nearly 100,000 head of black cattle, and 400,000 sheep. The latter have been improved by crossing with English breeds. In 1835, of 118,143 properties subject to the *contribution fonciére*, 56,061 were assessed at less than 5 fr., and 13,457 at between 5 and 10 fr.; but there were, at the same time, a considerable number of large properties. The manufacture of coarse broadcloths and other woollen fabrics are said to employ a large number of hands; and Orleans has manufactures of fine cloth, flannels, woollen yarn, &c.; but the commerce of that city has latterly declined. Cotton yarn, vinegar, white lead, paper, parchment, earthenware, &c., are also produced; and there are numerous distilleries. Meung is celebrated for the leather; Montargis and Pithiviers are the chief seats of the French saffron trade; and the latter town is celebrated for its *pâtisons d'amandes*, and *pâtés d'alouettes*. The department is divided into four arrondissements: chief towns, Orleans, Gien, Montargis, and Pithiviers. It sends five members to the chamber of deputies. Number of electors (1838–39) 2693. Total public revenue (1831), 14,001,984 fr.; expenditure, 5,270,185 fr. (*Hugo*, art. *Loiret*; *Official Tables*, &c.)

LOKEREN, a town of Belgium, prov. E. Flanders, cap. canton, on the Deurne, and on the road from Ghent to Antwerp, 19 m. E.N.E. Ghent. Population, in 1836, 16,153. Its appearance is that of a "large, quiet, Flemish village." It is celebrated for its linen fabrics; and has also manufactures of cotton goods, flannels, lace, hats, and soap, with cotton printing establishments, bleaching grounds, breweries, distilleries, oil-mills, &c. It has large weekly markets, and a considerable trade in its native products, and those of the adjacent country. (*De Cloet*; *Henschling*, &c.)

LOMBARDY. See ITALY (AUSTRIAN).

LOMBOK, an island of the eastern archipelago, between lat. 8° and 9° N. and long. 116° and 117° E., separated on the W. from Bali by the strait of Lombok, and on the E. from Sumbawa by the strait of Aliss, the last being the most commodious passage through the Sunda chain of islands.

Lombok is of a rhomboidal shape; its length may be estimated at 53 m.; average breadth, 45 m. Area, probably 2400 sq. m. A mountain chain covered with forests, runs W. to E. through the S. portion of the island, and an isolated height, the peak of Lombok, rises in the N. to 8000 feet above the sea. Several rivers disembogue on the N., E., and W. coasts. The country is populous, fertile, and well cultivated. Rice is raised by artificial irrigation, as in the Carnatic; and abundant supplies of bullocks, hogs, poultry, vegetables, &c. may be obtained at the commodious port of Ampanam, on the W. coast. The inhabitants are Mohammedans, and more civilised than the E. islanders in general. They carry on a considerable trade with Java, Borneo, and other Malay islands. Lombok and Mataram are the chief towns; the last is the residence of the rajah, who is tributary to the sultan of Bali. (*Crawfurd's Indian Archipelago*; *Hamilton's E. I. Gazetteer*.)

LOMOND (BEN), this mountain attains to an elevation of 3195 feet above the level of the sea. From its vicinity to Glasgow, from which it is distant N.W. 27 m., and its position between lochs Lomond and Katrine, it is by far the best known and most frequently visited of any of the highland mountains. Its summit, which is composed of micaceous slate, mixed with quartz, commands a great extent of view. "The whole extent of loch Lomond, with its wooded isles, appears just beneath. Loch Long, loch Katrine, loch Earn, and the river Clyde, form the principal waters. The mountains of Arran appear very distinct; and to the N. alps upon alps fill up the amazing view." (*Pennant's Tour in Scotland*, ii. 176, ed. 1790.)

LOMOND (LOCH), a lake of Scotland, between the cos. of Stirling and Dumbarton. Its most southerly extremity being 6½ m. N from the town of Dumbarton. This, which is the largest of the Scotch, and, indeed, of the British lakes, is a noble sheet of water, of a triangular shape, being 24 m. in length N.N.W. and S.S.E., and where broadest, along its S. shore, it is from 7 to 8 m. across; but its upper portion, from Rowerdiannan inn, N. to Ardleesh, is comparatively narrow, being only about 1 m. in breadth. Its area is estimated at about 25,000 acres; its most usual depth is about 90 fathoms; but in some places it has a depth of 80, and even of 120 fathoms. It is studded with numerous islands, some of which are of considerable size, and finely wooded. The scenery of this lake is varied and magnificent. Its N. extremity stretches into a wild, rugged, and dreary country. On the E. side Ben Lomond, one of the most stupendous of the Grampian mountains, rises from its margin; but on descending the lake, the character of the scenery changes; the mountains become less precipitous; the glens between them are well wooded, and filled with gentlemen's seats; and on the S. it is bounded by a low, rich, fertile, and well cultivated country. Its surface level is from three to five feet higher in winter than in summer: and it is generally about 22 feet above the sea level. It receives several streams, of which the Endrick, which flows into its S.E. corner is the most considerable. Its surplus waters are conveyed away by the river Leven, which, issuing from its S. extremity, falls into the frith of Clyde, close to Dumbarton. In summer it is much resorted to by tourists; and a steamer is established on the lake for their accommodation. It may be worth while to state that the waters of this lake were violently agitated at the period of the great earthquake at Lisbon in 1755.

LONATO, a town of Lombardy, deleg. Brescia, cap. distr. on the summit of a hill, 12 m. E.S.E. Brescia. Population 3630. It is walled and defended by a castle, has four churches, a hospital, a *monté de piété*, cavalry barracks, and manufactures of silk twist and saltpetre.

LONDON (Lat. *Londinium*, Fr. *Londres*), the metropolis of the U. Kingdom of Great Britain and Ireland, and the most populous, wealthy, and commercial city, of which we have any accounts, is situated partly and principally on the N. bank of the Thames, in the county of Middlesex, and partly on its S. bank, in the county of Surrey, about 45 m. above the river's mouth at the Nore, and 15 below the highest tideway. The site on the N. side is high and dry, but on the S. it is so low as to be under the level of the highest tides; though by a well constructed system of drainage it is kept perfectly free from wet. The subsoil is a hard clay, known to geologists by the name of London clay, lying in the middle of the great chalk basin, extending from Berkshire to the E. coast. In several places the clay is covered by thick beds of gravel. Lat. of St. Paul's Cathedral, 51° 30' 49" N. long. 5' 48" W. Greenwich. Exclusive of the city of London, properly so called, the metropolis comprises the city of Westminster, the boroughs of Southwark, Lambeth, and Marylebone, and other contiguous districts, which, though formerly distinct, are now combined into one vast mass of houses.

The population of the metropolis (including the cities of London and Westminster, and the five parliamentary boroughs as determined by the Boundary Act, with the parish of Chelsea,) has been as follows, according to the last four decennial returns:—

Divisions.	1801.	1811.	1821.	1831.	Rate of increase in 30 Years
London, city of	128,129	120,984	125,434	122,983	27·4
Westminster, city and lib.	132,310	162,085	182,035	201,312	27·6
Marylebone borough	97,542	128,668	174,254	240,254	146·0
Finsbury bor.	134,576	167,130	204,721	289,1,23	99·4
Tower Hamlets bor.	194,560	237,487	291,650	-357,245	82·5
Chelsea, par. of	11,604	18,262	26,880	32,371	197·6
Southwark, parl. bor.	94,719	108,760	125,662	134,117	41·5
Lambeth bor.	48,526	74,106	101,589	140,813	201·5
Metropolis	885,196	1,013,051	1,224,388	1,506,429	69·0

The population of London, including the whole metropolis, according to the census of 1841, was 2,568,321; viz., within the walls, 54,626; without the walls, 70,382; London and suburbs, 1,873,676; London and Westminster, 347,861; Southwark, 93,648; Lambeth, 116,886.

London is of great antiquity. It is said by Tacitus to have been in the days of Nero, *copia negotiatorum et commeatuum maximè celebre*. (*Annal.* lib. xiv. § 33.) It suffered severely in the revolt of Boadicea; but it speedily recovered from that disaster, and has always been the largest and most important of the British towns. It is indebted for its early and long-continued prosperity to its admirable situation. Though 45 m. from the sea, it enjoys, owing to its position on a great navigable river, all the advantages of an excellent seaport, vessels of 800 tons burden coming up to London bridge. Had it been built lower down, it would have been less healthy and more exposed to hostile attacks; and had it been higher up, it would have been deprived of the inestimable advantage of a deep-water harbour.

The Romans surrounded London with walls. It is probable that its limits were then commensurate with the part of the city said to be "within the walls," reaching from the end of Lendenhall-street to the top of Ludgate hill, and from the Thames to London Wall and Little Britain. The wall appears to have inclosed it along the water as well as on the land sides. The great Roman roads called Watling-street and Ermin-street, as well as the *viæ vicinales*, centred in London.

The continued and rapid increase of buildings renders it difficult to ascertain the extent of the metropolis at any particular period. If we include in it those parts only that present a solid mass of houses, its length, from E. to W., may be taken at 5½ m., and its breadth, from N. to S., at about 3½ m. There is, however, a nearly continuous line of houses from Blackwall to Chelsea, a distance of nearly 7 m., and from Walworth to Holloway of 4½ m. The extent of surface covered by buildings is estimated at about 15 sq. m., or nearly 10,800 acres, so that M. Say, the celebrated French economist, did not really indulge in hyperbole when he said, *Londres n'est plus une ville: c'est une province couverte de maisons!*

Notwithstanding its immense size, it is not difficult for strangers to make their way in London. The Thames runs through it lengthwise from W. to E., and most of the great lines of the streets are parallel to the river, being intersected at variable distances by lines of cross streets, or of streets running N. and S. Of the former, or of the longitudinal streets parallel to the river, there are two principal lines: the most northerly of these enters London on the W. by the Bayswater-road, passing in front of the fine terraces facing the N. side of Hyde Park; it then passes through Oxford street, about 1½ m. in length, to St. Giles's, where it bends through a mass of inferior buildings; leaving this, it is prolonged by Holborn, a wide and handsome street about 1 m. in length; whence it proceeds through Skinner-street, Snowhill, and Newgate-street, till it reaches Cheapside, one of the greatest thoroughfares in the city. It next passes through the Poultry, having the Bank and the Exchange on the one hand, and the Mansion-House on the other, along Cornhill, to Leadenhall-street; and is thence continued, by Whitechapel and the Mile-end-road, into the country. Its entire length, from Hyde park to the Regent's canal, Mile-end, is above six miles.

The other great longitudinal street to the S. of that now traced, enters London on the W. at Hyde park corner. This is by far the most splendid of the entrances into the metropolis. Kensington gardens appear like an ornamental forest; Hyde park gradually rises to the splendid terraces on the N., and is bordered on the E. by magnificent houses, or rather palaces: on the left is the handsome entrance to Hyde park and the W. front of Apsley House, the town residence of the Duke of Wellington, *decus et tutamen patriæ*; and on the right the bold arch and gate leading to the Queen's palace; the Green park, apparently stretching to the towers of Westminster Abbey; and a long line of splendid buildings, with the Norwood hills in the distance. The promise of a magnificent city is not belied by an advance through Piccadilly. This, which is the first street of London traversed by the traveller from the W., is 1 m. in length, and is principally built only on one side, being open on the other to the Green park. It contains many magnificent private residences, and shops. On reaching the E. end of Piccadilly, the continuous line of street deflects to the right through the Haymarket, whence it proceeds to the E. along the splendid line of E. Pall-mall, through Trafalgar square, and past St. Martin's Church, till it unites with the Strand: this, though formerly in many places narrow and encumbered, is now a truly magnificent street: it follows pretty closely the line of the river, from which it is not far distant; and, besides two churches in its centre, has Exeter hall on its N., and Somerset house on its S. side. Contiguous to the latter is Wellington-street, leading to Waterloo bridge. The Strand terminates at the ancient gate, called Temple Bar, the boundary of the city on the W.; the great line of street being thence prolonged through Fleet-street, at the E. end of which, on the right, is a fine street, leading to Blackfriars bridge; and on the left Farringdon-street, one of the widest in the city, which it is intended to prolong to Islington: from Fleet-street the grand line continues up Ludgate hill, till it reaches St. Paul's, the noblest edifice in the kingdom.

At the E. end of St. Paul's Churchyard, the wider channel of communication joins in Cheapside the grand northern line already traced, coming from Oxford-street, Holborn, &c.; but another branch of the former line runs nearer the river, through Watling-street, Eastcheap, and Tower-street, to the wide and airy area of Tower hill, whence it may be traced either in a straight line through Ratcliffe-highway, N. of the London docks, or close by the river along Wapping and Shadwell, where the lines again form a single street leading to the W. India docks. The streets E. of the Tower are narrow, and lined with mean houses mostly occupied by persons connected with shipping. This line is altogether about 6 m. in length.

Another line of street which unites with that last described, may be considered as beginning at Vauxhall bridge, close to which is an open quay, ½ m. long, exhibiting a view of the river and of the archiepiscopal palace of Lambeth. The line of road is, however, soon separated from the river by ranges of buildings, along which it passes, till

219

It reaches Abingdon-street. At the termination of the latter it runs on, having Westminster Abbey on the left, and the Houses of Parliament and Westminster Hall on the right: after leaving these and Westminster bridge on the right, it connects with Parliament-street, and then with the spacious street called Whitehall, in which are the Treasury, Horse Guards, and Admiralty, separating it from St. James's park on the left, and the Banqueting Hall, with other handsome mansions shutting out the view of the river. The magnificence of the buildings in this short line of street is unequalled, except by those at the W. entrance of Piccadilly, and by the splendid terraces of the Regent's Park. Beyond Whitehall is Charing Cross, with the National Gallery, College of Physicians, &c., forming three sides of Trafalgar square, surrounding the site of the intended Nelson monument. Here the line, bending E. with the river, unites with the Strand, already noticed.

Among the principal streets running from N. to S., the first and most westerly is the Edgeware-road, with its continuations, Park-lane, Grosvenor-place, and Vauxhall-bridge road, which, for the most part, bound the metropolis westward. The second, proceeding eastward, is the line formed of Portland-place, Regent-street, and Waterloo-place, extending between the Regent's and St. James's parks, and forming by far the noblest public thoroughfare in London, as well from the width of road as from the beauty of the houses and shops on either side. At its S. termination is a granite column, surmounted by a bronze statue of the duke of York, brother to George IV. A little N. of Piccadilly the line curves through the Quadrant, a handsome range of buildings, bordered on each side by colonnades, of fluted Doric pillars forming an arcade over the footways: from this point it continues northward to Oxford-street, where it expands into a circus, and then, resuming its former dimensions, proceeds to the church in Langham-place: here, by a slight curve westward, it opens into Portland-place, which, from its containing the residences of the principal ambassadors, may be called the diplomatic quarter of London: the architecture of its houses is less showy than that seen in Regent-street; but the magnificent scale on which they are built sufficiently indicates the rank of their occupiers. Park-crescent and Park-square, opening into the Regent's park, form a noble finish to the whole. The third great N. and S. line is a continuation southward of the road from Hampstead; it passes through Tottenham-court-road to the E. end of Oxford-street, from which point its course may be traced through narrow streets down St. Martin's-lane to Charing-cross; but, though a busy, it is an intricate thoroughfare, and is devoid of architectural interest. The other principal N. and S. lines consist of Grey's Inn-lane and Chancery-lane; Goswell-street and Aldersgate-street; and the line of street commencing at the Regent's canal on the N., successively called Kingsland-road, Shoreditch, Norton-Folgate, Bishopsgate-street, and Gracechurch-street: at the S. termination of the latter this line passes over London bridge, and is thence prolonged across the borough as far as Kennington church in Surrey: its length is about 4 m., which may be considered the breadth of London in this quarter. The portion of this line at and near London bridge affords some of the finest points for viewing London and the animated scenery on the river. Exclusive of the above, there are an infinite number of cross streets, some of which are of great importance. Among others, a splendid line has recently been opened from Finsbury-square through Moorgate-street, Princes-street, and King William-street, to London bridge.

In addition to the various routes intersecting each other in different directions, a grand line of road embraces the greater part of London on the N., in a manner not unlike that in which the Boulevards encircle Paris. It commences in the Uxbridge-road, and has a N.E. course as far as King's cross, St. Pancras, where, turning eastward, it ascends Pentonville hill, and, entering the City-road, terminates in Finsbury-square.

In Southwark the great roads from the different bridges unite at the well-known posting-house called the Elephant and Castle. They are generally wide and well-built streets, though, with the exception of Blackfriars'-road, inferior to the principal thoroughfares N. of the river. A line of street, extending from Westminster-road to the borough, connects these several roads with each other.

Unlike Edinburgh and many other great towns, the houses in London are not, with the exception of those in the Temple and Inns of Court, divided into stories (*Scotticè* "flats"); but in the vast majority of instances belong to or are hired by one individual, by whom, however, portions of them are frequently let to lodgers. They have usually a story sunk below the level of the street, comprising the kitchen and other offices, above which are usually four stories. The smaller, and by far the most numerous, class of houses have narrow fronts, containing one room or shop in the front of the street floor and that immediately above it, the stair and 212

a smaller apartment occupying the back part: the two upper floors are frequently divided into smaller apartments. Every house has the inestimable advantage of having an abundant supply of water; and in all the better class of houses it is supplied to the top as well as to the under story. Except in the very worst parts of the town, all the refuse water and drainage of the house is conveyed by a covered drain to the sewer, or grand receptacle in the centre of the street, sunk below the line of the lateral drains. Most houses have cellars opposite to them under the street, for the stowage of coal and such like articles. No filth is ever laid down upon the streets, which have universally flagged footpaths along each side; and, notwithstanding the concourse of horses, and the grinding of the pavement by carriages, the streets are, speaking generally, extremely well kept.

But until a comparatively late period the architecture of the streets and houses of London was but little in harmony with the wealth of the inhabitants and the richness of the interior of the houses. Internal comfort was long the only, as it still is (and it is to be hoped will long continue to be) the grand object of the Londoner. Provided his house were clean, commodious, and well and handsomely furnished, he cared little about its external appearance. Hence it was that the interminable rows of dull-looking brick houses, erected with little or no regard to uniformity, led strangers to remark that the best streets resembled long walls pierced with holes for doors and windows. Even Bond-street was said, in 1810, by an intelligent foreigner, to be "an ugly, inconvenient street, the attractions of which it is difficult to understand." But the same author (*Simond*) adds: "You cannot pass the threshold without being struck with the look of order and neatness of the interior. Instead of the abominable filth of the common entrance and common stairs of a French house, here you step from the very street on a neat floorcloth or carpet, the wall painted or papered, a lamp in its glass ball hanging from the ceiling, and every apartment in the same style. All is neat, compact, and independent."

With the exception, indeed, of St. Paul's, Westminster Abbey, Somerset house, and a few other churches and public buildings, London displayed, till within the last few years, little architectural elegance. In our own times, however, the erection of immense and splendid ranges of buildings in every direction has made our metropolis so superior to most other capitals in appearance as it has long been in wealth, cleanliness, and comfort. The line of Regent-street has been already mentioned, to which may be added the Regent's park, "affording an enchanting landscape bounded by hills, and more than half surrounded by a large circuit of magnificent buildings, worthy the capital of the world." Belgrave and Eaton squares, and the adjoining streets and squares on the estate of the marquis of Westminster, with the terraces in Carlton gardens, have all been raised within the last 20 years, and are probably unequalled for symmetry and magnificence. Within a still shorter space a splendid city has been built on the elevated ground on the N. side of Hyde park; and these, with the magnificent new buildings in Pall-mall and St. James's-street, Trafalgar-square, &c., render the W. end of the city of London a residence worthy the wealthiest aristocracy in the world. But the improvements effected of late years in the city, or oldest part of the town, have been equally great and striking. The new streets that lead from the bank to London bridge on the one hand, and to Moorfields on the other, are on a grand scale; and when it is borne in mind that the ground which they traverse was previously occupied by a dense mass of houses, which had to be purchased at a high price, it will be seen that they do as much credit to the public spirit as to the taste of the citizens. Four new and noble bridges over the Thames form no small addition to the improvements of the last 30 years. On the whole, therefore, though it cannot be said of George IV., that, like Augustus, he found a capital of brick, and left one of marble, it is certain that during his reign an extraordinary impulse was given to architectural improvement, which, so far from being exhausted, promises to give the metropolis still greater splendour and convenience.

With extremely few exceptions, almost all the houses in London are built of brick. But within the last few years those in the principal streets have been mostly plastered or stuccoed over, and their fronts made so exactly to imitate the finest freestone, that it is sometimes no easy matter to distinguish between them. This method of dressing up houses has contributed most materially to the improved appearance of the town. Those, indeed, who have been accustomed to stone structures, are apt to associate ideas of insecurity and of rapid decay with stuccoed fabrics; but, provided the walls be well built, and the plaster be kept in repair by occasional painting, stuccoed houses are, in fact, all but imperishable. The cheapness of stucco, too, allows it to be applied to the inferior class of houses; while, from the facility with which it may be moulded, it permits an elaborateness of ornament that could not be executed in stone at many times the cost. Belgrave-square, and the magnificent terraces of Carlton gardens, Hyde park, &c., owe most part of their elegance to the judicious application of stucco.

It is much to be regretted that the Thames, which, from its breadth and depth, might be the greatest ornament of the city, as well as the principal source of its wealth and prosperity, is so closely pent up by wharfs, warehouses, and other buildings, that its banks are almost shut out from the view, except where it is crossed. It is, however, fronted by the custom-house and Somerset house, the Adelphi terrace, and by the Temple gardens, and some private houses in Whitehall. But the most magnificent views of the river, and, indeed, in some respects, of the city, are obtained from the bridges.

Divisions.—The most popular division of London is into three parts—the city, the west end, and the borough ; Temple Bar dividing the city from the west end, and the river separating both these portions from the borough. This division is necessarily vague, and, for specific purposes, different divisions are made. The city of London, strictly considered, is situated nearly in the centre of the metropolis, and is the seat of commerce on the largest scale. The city of Westminster is W. of the city of London : it contains the royal palaces, the houses of parliament, the law courts, most of the public offices, and the town residences of nearly all the nobility and aristocracy. The cities of London and Westminster, however, do not comprise above an eighth part of the area, or a fourth part of the population of the whole of what may be considered the metropolis. For the purpose of parliamentary elections, the metropolis is divided into seven districts : the cities of London and Westminster, as above stated ; the borough of Finsbury, N. of the city of London ; the Tower hamlets, E. of Finsbury and the city ; Marylebone, N. of the city of Westminster ; and two districts S. of the river, Southwark on the E. and Lambeth on the W. side.

The area of the city of London, which comprises only a small portion of the metropolis, is roughly estimated at about 600 acres. Its boundary line leaving the Thames at Temple-lane, passes northward, crossing Fleet-street at Temple Bar, and Holborn at "Holborn Bars." Turning eastward, it thence takes an undulating course, enclosing Smithfield, Finsbury circus, and Bishopsgate-street, S. of Spittal-square. It thence passes S.E. through Petticoat-lane, to Aldgate, from which point the boundary, pursuing a S.S.W. course, reaches the Thames by a very irregular line, excluding the tower. The city is divided into 108 parishes, of which 97 are said to be "within," and 11 "without," the walls. This division is now merely nominal, the ancient city boundary having long disappeared, although the city gates, where the walls passed the great thoroughfares, were standing in the last half of the 18th century.

The E. boundary of the city of Westminster coincides with the W. boundary of London at the Thames and Temple : it thence runs N.W. to the junction of Tottenham-court-road and Oxford-street. The latter street constitutes the whole N. boundary as far as the W. extremity at Kensington gardens. From this point a very irregular line, running to Chelsea hospital, forms the W. boundary. It then turns to the S.W. along the Serpentine, on leaving which it goes S. until it reaches the Thames near Chelsea hospital.

The five metropolitan boroughs, being parliamentary only, and not municipal, need not be minutely described. Marylebone includes the three parishes of Marylebone, Paddington, and St. Pancras ; Finsbury comprises nine parishes and the Rolls' liberty ; and the Tower hamlets include fifteen ; Southwark embraces not only the municipal borough, but the parishes of Bermondsey and Rotherhithe ; and Lambeth comprises Camberwell and Newington, as well as the parish of its own name.

Parks, Squares, &c.—The W. end of the town is beautified and rendered healthy by four extensive parks, appropriately called the lungs of London. They are open to the public ; and, though each has a different character, they all afford ample scope for healthy amusement and exercise to those resident in their vicinity. Hyde park (once the manor of Hyde, and belonging to the abbey of Westminster), lying W. of the road leading from Piccadilly and Oxford-street, contains about 400 acres, and has a large and deep artificial lake, crossed by a handsome bridge of five arches ; this lake, which is nearly straight, is, by an absurd misnomer, called the Serpentine river. The whole of this park is an open field of much beauty, dotted with trees, and traversed by carriage-ways, which, in fine weather, during the season, are covered with gay and fashionable equipages. Kensington gardens, lying W. of the park, and separated from it by a breech and wall, are open to the public, and constitute a fine shady promenade. St. James's park, extending from Whitehall to Buckingham palace, is less than one fifth part of Hyde park, and not so open ; its site being low, damp, and

marshy. Within these few years, however, the central part has been tastefully laid out, and what was a dirty, straight canal, running through a marsh, has become a handsome, varied sheet of water, dotted with islands forming the abode of numerous aquatic birds, and surrounded by lawns, shrubberies, and lofty trees. The avenues on the N. side of this park are open to all pedestrians, but only to the horses and carriages of some privileged members of the aristocracy. The S. drive is open to all private and hackney carriages. The Green park is a triangular piece of ground, about as large as St. James's, from which it gradually rises to Piccadilly : it is an open and pleasant promenade, and forms a sort of miniature Hyde park, but at present (1841) it is undergoing extensive alterations, for it remains to be seen whether they will be improvements. The Regent's park, which is as large as Hyde park, is perhaps the handsomest of all : it was formed during the regency in the last years of the reign of George III. It is situated to the N. of Portland place, on high ground, surrounded by splendid buildings, and is tastefully laid out. This park is not, however, what it professes to be, a place for the accommodation and recreation of the public : on the contrary, the public is shut out from three fourths of its extent ; and some even of its finest portions have been let to individuals, who have built villas upon them ! This is a scandalous abuse of the public property ; and it is really astonishing that it should have been allowed to be perpetrated, almost without notice. The gardens of the Zoological Society are situated on the N. side of this park ; and it is said that the central portion is about to be laid out as a garden for the Botanic Society. The probability, indeed, seems to be that the public will, at no distant period, be shut entirely out from this park, and left to admire its beauties from the dusty drive by which it is surrounded.

On the E. side of the Regent's park, near Park-square, is the large building most inaptly styled the Colosseum. It is a 16 sided polygonal structure, with a magnificent portico and cupola. It is principally occupied by an immense panoramic view of the metropolis, taken from the ball on the top of St. Paul's cathedral. But though the patience and elaborate execution of the artist (Mr. Hornor) be entitled to every praise, little else can be said in favour of his undertaking.

The squares of London are pretty numerous in all parts, though the largest and handsomest are in the W. end. In many the houses are in the first style of architecture, and the central gardens beautifully laid out. Grosvenor, Berkeley, and Hanover squares lie between Oxford-street and Piccadilly, and are, on the whole, the most fashionable ; though the newly-formed Belgrave-square, in Pimlico, bids fair to rival, and even surpass, them as a favourite residence of the aristocracy. St. James's-square, S. of Piccadilly, with Portman, Manchester, and Cavendish squares, N. of Oxford-street, are mostly occupied by the nobility and gentry. Farther E. are Russel and Bedford squares, and a cluster of squares to the N. of these, chiefly the residences of wealthy merchants. Lincoln's-inn-fields, S. of Holborn, is one of the largest and best-built squares, and its enclosure is more beautifully laid out than any other in the metropolis. Finsbury-square lies N. of the city, and near it is Finsbury circus, a round enclosure with a pretty garden. Many other squares, formed of good houses, respectably inhabited, are to be found in all parts of the town and neighbourhood.

Several of the best squares are decorated with statues ; among which may be remarked that of Pitt, by Chantrey, in Hanover-square ; of Fox, by Westmacott, in Bloomsbury-square ; of the duke of Bedford, by the same sculptor, in Russel-square ; and those of William III., Anne, and George I. in St. James's, Queen's, and Leicester squares. Other statues are placed in different parts of the metropolis, among which are the equestrian statue of Charles I., by Le Sueur, at Charing-cross ; of James II., by Gibbons, behind Whitehall ; of George III., by Wyat, Pall-mall ; of the late duke of Kent, in Park crescent ; and of Canning, in Palace-yard, adjoining the House of Commons. Near the E. entrance to Hyde park is a statue, copied from a figure at Rome, said, but without any authority, to be that of Achilles. It is of brass and was formed out of cannon captured by the duke of Wellington, in whose honour it was erected, and to whom it is inscribed, by the ladies of England ! But with all due deference, it is not easy to imagine anything more absurd. What has the duke of Wellington, by far the most illustrious Englishman of his age, in common with a colossal gladiator, that a statue of the latter should be erected in his honour ?

The monument on Fish-street hill, built in 1671-77, to commemorate the burning of London, is a fluted Doric column. 202 feet in height, designed by Sir Christopher Wren. The pedestal is decorated by a representation, in relief, of the destruction of the city, sculptured by Cibber : at the top of the column is a gallery affording a view of the E. part of the metropolis, and on the summit is a blazing urn, recently regilt. It is a noble column, and, had it been better situated, would have been one of the greatest ornaments of the city. A short English inscription on the pedestal ascribed, without the slightest foundation, the conflagration it is designed to commemorate to the treachery and malice of a popish faction. Pope alluded to this when he says,

" Where London's column pointing to the skies,
Like a tall bully, lifts the head and lies."

But, in 1830, this offensive inscription was obliterated, in pursuance of a resolution of the Court of Common Council.

The York column is a plain Doric pillar of granite, surmounted by a bronze colossal statue of the duke of York. The height of the column is 194 feet, and above the capital is an iron gallery, from which a good view is obtained of the W. end of the town. This column, erected in 1833, is situated on the N. side of St. James's park, at the lower end of Waterloo-place.

Bridges.—The Thames, averaging 1000 feet in width, is crossed by six bridges, built at an aggregate expense of more than £5,000,000. A wooden structure had been thrown across the river early in the 11th century; but the frequent and costly repairs indispensable for its maintenance led to the construction of one of more durable materials. A stone bridge, of pointed architecture, was completed in 1209, which, by means of occasional renovations, was kept standing till 1834. Down to the middle of last century, this was the only bridge between London and Southwark. The great inconvenience of a circuitous journey from the west end of the town to the city before the river could be crossed by carriages, induced parliament, in 1736, to make a grant for the erection of Westminster bridge at the court end of the metropolis. Blackfriars bridge (intended by its projectors to have been called Pitt bridge) was built about 20 years after, the expense of its construction being defrayed by a toll exacted during 19 years. Westminster and Blackfriars bridges were built of Portland stone, which being too soft to resist the constant attrition of the water, and of the ice in winter, their piers were so much worn as to threaten their entire destruction: latterly, however, their piers have been cased with granite, and they have been otherwise thoroughly repaired at a very heavy expense. The three bridges erected within the present century have completed the connexion of all the important districts N. and S. of the Thames. Two of these, Vauxhall and Southwark bridges, have iron arches, the centre arch of the latter being 240 feet in width. Waterloo bridge, which Canova said was itself "worth a visit from the remotest corner of the earth," is of granite, and has nine elliptical arches, each 127 feet wide. This bridge was built by a joint-stock company; but, owing to the want of any great thoroughfare leading to or from it through the city, and to the influence of the toll on passengers and carriages crossing the river by its means, it is very little frequented, and has turned out a most unprofitable undertaking. The demolition of the old London bridge was owing less to its decayed state than to the defects of its original construction. The piers and starlings between its numerous arches (21 at the period of its removal) occupied so large a portion of the water-way as to obstruct the course of the water both during the flow and ebb of the tide, especially the latter. At low water, indeed, there was a difference of nearly 5 feet between the level of the water on the upper and lower sides of the bridge. This, by occasioning a dangerous fall and eddy in the water for a considerable time both before and after low water, interrupted the navigation, and occasioned every now and then fatal accidents. At length it was determined to abate the nuisance, by pulling down the old bridge, and erecting in its stead a new structure, whose arches should be of such a size as not sensibly to affect the flow of the water. The new London bridge, like the Southwark and Waterloo bridges, was planned by the late John Rennie. It is built of granite, the span of the centre arch being 152 feet; and, whether we consider its magnitude, or the beauty and simplicity of its structure, it is certainly one of the finest specimens of bridge architecture in the world. The heavy expense of this fabric has been partly defrayed by a duty on all coal brought into the pool, and partly from the revenues of property appropriated for the support of " London bridges."

The following table comprises a statement of the principal particulars connected with the different bridges belonging to the city:

Name.	Date of Completion.	Cost.	No. of Arches.	Length.	Breadth.	Span of Central Arch.
London bridge	1831	£.1,150,000	5	920 ft.	56 ft.	152 ft.
Southwark —	1819	800,000	3	700	44	240
Blackfriars —	1770	300,000	9	995	42	100
Waterloo —	1817	1,150,000	9	1,240	42	127
Westminster —	1788	389,000	15	1,220	44	76
Vauxhall —	1814	300,000	9	800	40	78

214

The tunnel, which, unlike the bridges, passes *under*, and not over, the stream, effects a connexion between its banks nearly 2 m. below London bridge. The erection of a bridge in the centre of the port of London was of course impracticable, and the mode of uniting the two shores, without injury to the shipping interest, was long a difficult problem for engineers. It was at length solved by Mr. Brunel, who designed the tunnel, and has so nearly brought it to a close, that its completion may be looked upon as certain. It consists of a hollow brick cylinder, or pipe, subdivided into two roadways, each 15 feet high and 12 feet broad. Notwithstanding the danger attending the performance of the work, owing to the perpetual oozing through and occasional bursting in of the river, the loss of life during the 15 years it has occupied has been very inconsiderable. But how curious soever in other respects, we incline to think that the tunnel never will be of much practical utility. The difficulty of the descent will always be a considerable obstacle to its extensive use. It was begun by a private company, but has been mostly carried on by grants of public money.

Palaces and Houses of Parl.—St. James's at the W. end of Pall-Mall, is an irregular, mean-looking brick building, totally unworthy the name of palace: it was erected by Henry VIII., on the site of a hospital for female lepers, which existed in the 11th century. The interior, however, is handsomely fitted up and it is well adapted for court levées and drawing-rooms, which are mostly held in it. The chapel attached to this edifice is substantially the same that was used for the ancient hospital.

Buckingham Palace, at the W. end of St. James's Park is remarkable only for its extravagant cost, amounting nearly to £1,000,000, though perhaps the poorness of its effect is in some measure attributable to its depressed situation. It occupies the site of Arlington house, pulled down by John Sheffield duke of Buckingham, who erected in its stead a plain, but handsome residence. Having been purchased by George III. in 1762, it became the favourite abode of Queen Charlotte. Under George IV., whose rage for building was as decided as his taste was equivocal, Buckingham House was entirely rebuilt; and became, in 1837, the town-residence of the Queen. The principal front, to the east, forms three sides of a square, its narrow, projecting wings inclosing a space about 80 yards in width, in front of which is a marble arch (a miniature imitation of that of Constantine at Rome), with a bronzed iron gate. This arch does not, however, harmonise with the rest of the building; and the front is not only destitute of all grandeur, but is mean and paltry. The garden façade, an elevation of the Corinthian order, resting on a rustic basement, is in better taste. The interior is magnificently fitted up. A staircase of white marble leads to the picture gallery, drawing-rooms, and throne room, the latter of which is enriched with well executed bass reliefs. The picture-gallery, above 180 ft. in length, is filled with works of the best masters. The library and private rooms of the queen are in the basement story; but owing to the singular taste of George IV. for low rooms, the stories are not sufficiently high, and many of the apartments are badly lighted.

This, perhaps, is the only palace surrounded by what is frequently a puddle. Such, however, is really the case: for the ground in front of the palace not being paved, but merely covered with a compound, half gravel and half clay, it becomes, in wet weather, a most offensive puddle.

The old houses of Parliament stood upon ground formerly occupied by the palace of Westminster. Their appearance was far from imposing; but a certain degree of antiquated splendour, the associations connected with their history, and the importance of the purposes to which they were appropriated, made them respectable in the eyes of Englishmen. They were, however, wholly destroyed on the 16th October, 1834, by a fire, which for some time endangered the contiguous edifices of Westminster Hall and Abbey. A building, expected to be one of the finest ornaments of the metropolis, is now in the course of erection on the same spot, from the designs of Mr. Barry. It is to have a splendid river front, nearly 700 ft. in length, with a terrace and stairs leading down to the water; and at the S.W. angle, over the royal entrance, is to be erected a lofty tower, in the perpendicular English style, which, judging by the design, will be a fine specimen of modern architecture. It is to be hoped that the opportunity now offered, and so unlikely to recur, of raising a building worthy a great nation, may not be neglected, and that the new houses of Parliament may, when completed not only form a striking feature in the metropolis, but supply those accommodations for the despatch of the business of the legislature, the want of which was so much felt in the old buildings.

The government offices are, generally speaking, handsome edifices. The Council-office, on the western side of Whitehall, a modern structure of Palladian architecture, is generally admired. The Treasury, which joins that last mentioned, is an old brick edifice, which once formed part of

LONDON.

Cardinal Wolsey's palace. The Board of Control has a fine Ionic portico, but is, otherwise, a plain building. The Ordnance and Admiralty offices make no pretensions to display; and the "Horse Guards," which does pretend to it, is in very bad taste. Many of the public offices are in Somerset House, once a palace, occupied by Edward VI. and Elizabeth. The old building was taken down in 1775; and the present quadrangular structure, designed by Sir William Chambers, was completed in 1786, and divided into government offices. The street front is only 200 ft. in length but that facing the river is 800 ft. in length and is one of the noblest elevations in London. An eastern wing was added by King's college, in 1838, in completion of the architect's design.

On the river bank, in the E. part of the city, is the Tower,

  "With many a foul and midnight murder fed."

This rude fortress, about quarter of a mile below London bridge, was begun by William the Conqueror in 1078. The original building, now called the White tower, was completed in 1098. Additions were made by Henry III. in 1240, by Edward IV. in 1465, and the whole was substantially repaired by Charles II. in 1663. The Grand storehouse, a large building N. of the White tower, begun by James II., was completed by William III.; and numerous houses have been erected in it for the residence of officers connected with the establishment. The tower was a royal palace during more than five centuries. It was long, also, and still in fact is, a state prison; and several royal personages, and some of our highest nobles, and most distinguished commoners, have perished in this edifice, some by the hand of the public executioner, and some by the dagger and bowl of the assassin. It anciently contained several detached masses of building, most of which have now disappeared. The original tower, now called the White tower, still remains the principal edifice. The Martin tower is now called the Jewel tower. The Lantern tower, the Royal palace and the Mint, have been pulled down. Of the remainder of the old building vestiges may be traced under silver names. The present edifices consist of the church of St. Peter ad vincula, the Ordnance office, the record offices, the jewel office, armories, and barracks. In the small-arm armory complete stands of arms for 150,000 men are kept in constant readiness. The whole is surrounded by a moat, filled with water from the Thames, and the outer bank has been recently turned into pleasure grounds. The tower is open to visitors, who pay 6d. to see the armories and a similar sum to inspect the regalia. The ménagerie, formerly the best in England, having been superseded by that belonging to the Zoological Society in the Regent's park, was broken up a few years ago.

The Mint, formerly in the Tower, but now on Tower-hill, is a stone building of Greek architecture, consisting of a centre and wings. The workshops and officers occupy about 6000 square yards, and the machinery for coining is complete and efficient. The salaries of the officers and workmen amount to £15,000 a year, and the money coined in 1838 consisted of £2,855,365 in gold, £701,170 in silver, and £1565 in copper. The gold is computed at the Mint price of £3 17s. 10½d. per oz. troy, or 46·7 sovs. to the lb. troy; the silver at 5s. 6d. per oz., or 66s. to the lb. troy; and the copper at £234 per ton, or 24 pence to the lb. avoird.

Postoffice.—The Postoffice, in the centre of the metropolis, near St. Paul's, a large, handsome building, completed in 1829, of Portland stone, is 390 ft. long, 130 ft. wide, and 64 ft. high. The façade is adorned with three Ionic porticoes, over the central and largest of which is a plain pediment. Within this portico is the great hall, 80 ft. by 60 ft., divided into three compartments by rows of Ionic columns on granite pedestals; passages lead from it to the principal offices. This establishment consists of three branches, to each of which separate clerks are attached: viz. the general or inland, the foreign, and the district post, until lately termed the twopenny post. The business of the district post extends to a distance of 12 m. from the general postoffice.

There are at least two deliveries each day at every place within this line, and seven to all places within 2 m. of the chief office: intermediate places have three, four, or five deliveries. The general postoffice extends its operations all over the world, and letters are despatched by it every day. Twenty-five mails leave London every evening at 8 P.M., several of which are forwarded by railway, and there are 5 morning mails. No letters leave London on Sundays. Most of the evening mails reach town on their return between 6 and 7 o'clock in the morning. Early enough for the delivery of the letters before 10 A.M. in the city, and 11 A.M. in the other parts of London. Within 3 m. of the Postoffice are 228 receiving-houses, of which three are exclusively for the district post letters; the other houses receive all letters without distinction. Till 1839 there were two classes of receiving-houses, one for general and the other for twopenny post letters; but that distinction has

now ceased. There are about 100 country receiving-houses belonging to the district postoffice. In 1838, the last year of the old rates of postage, the gross revenue of the London postoffice amounted to £706,964.

Religious Establishments and Buildings.—London is a bishop's see, the highest in rank in the kingdom under the archbishops. The diocese till very lately comprehended 199 parishes in Middlesex, 396 in Essex, 56 in Hertfordshire, and four in Buckinghamshire, in all 650, containing a population of 1,728,665 persons; but, by the new ecclesiastical arrangements of 1836, it will in future comprise all the parishes of Middlesex, 23 in Surrey, 10 only in Essex, and nine in Kent, making a total of 241 parishes. The nett revenue of the diocese, at an average of the three years ending with 1831, is £13,929 per annum, and owing to the building now going on upon the bishop's estate, it will, at no distant period, amount to three or four times that sum; but, on the death of the present incumbent, the income of the see is to be fixed at £10,000 a year nett; the number of benefices returned to the commissioners (640), averaging a nett income of nearly £490 per annum. The number of curates was 332, whose income averaged something less than £100. Of the benefices, 75 were crown livings, 66 were in the gift of archbishops and bishops, and 277 belonged to private persons; the remainder were in the possession of corporate or ecclesiastical bodies. The number of parishes in the city of London is 113, of which 97 are within the walls, and 16 in the liberties: the 97 parishes are very small, and only 37 of them have churches; those belonging to the others either having been burned down at the great fire of 1666, and not rebuilt, or having been since removed to make room for improvements. Three additional churches have been built in the liberties, making the whole number now in the city, 76. Westminster contains 10 parishes, four of which were formed early in the last century, in consequence of the great increase of population at the W. end of the town, and one very recently; two only of these parishes, St. Margaret's and St. John's, are considered to form the city of Westminster, the other eight being descendants the liberties. Westminster was erected into a bishopric by Henry VIII. in 1541; and the whole of Middlesex, exclusive of the city of London and the parish of Fulham, was assigned as its diocese; but this bishopric existed only nine years, at the expiration of which the ecclesiastical government reverted to its former channel. Within the present century 12 district churches have been built, making the whole number 22. The other parishes of the metropolis amount to 46: the districts, which are partially, but not wholly, divided from the larger parishes, being those in which the incumbent is not endowed with a portion of the tithes, amount to 45, making the number of churches 91. This will make the whole of the churches in the metropolis amount to 180, without including chapels of ease. It is right, however, to state that those numbers may not exactly agree with other accounts; for so many parishes extend into two districts, so many are partly in the city and partly in the suburbs, that no two estimates of their number are found to agree. Besides the churches and chapels belonging to the establishment, there are nine chapels of the kirk of Scotland, 14 Roman Catholic chapels, 18 foreign Protestant churches and chapels, seven synagogues, and above 250 places of worship for dissenters and separatists of all denominations.

St. Paul's, the cathedral church of London, is not only the great architectural glory of the metropolis, but of the empire. This noble structure stands in an elevated situation at the top of Ludgate Hill, on the site of the former cathedral, destroyed during the great fire of 1666. Its foundations were laid on the 21st of June, 1675; and Sir Christopher Wren, by whom it was designed, and under whose direction the work was carried on, lived to complete the stupendous edifice, the last stone of which was laid by his son in 1710. It is built in the form of a Latin cross, with an additional arm or transept at the W. end to give breadth to its front, and has a semicircular projection at the E. end for the altar, and semicircular porticoes at either end of the transept. It is 510 feet in length, E. to W., the length of the cross, exclusive of the circular porticoes, is 250 feet, the breadth of the W. façade with the turrets, 180 feet, and the height of the walls 110 feet. An immense dome, or cupola, rising over the centre, is surmounted by a lantern, ball, and cross, the latter being elevated 360 feet above the level of the floor, and 370 feet above the pavement of the church-yard. The two turrets, or belfries, in the W. front, are each 222 feet in height. The walls are decorated by two stories of coupled pilasters arranged at regular distances, those below being of the Corinthian and those above of the Composite order. The whole building is of Portland stone; and the excellence of its foundations, and the massive solidity of its walls and piers, warrant the inference that it will be as lasting as it is magnificent.

St. Paul's, it is frequently said, is copied, or at least closely imitated from St. Peter's at Rome; and to some extenthis is true. But it is a copy that bears the impress of transcendent genius; and may be said to be to St. Peter's what the Æneid is to the Iliad and Odyssey. The fronts of both cathedrals are the parts, perhaps, in which they are most deficient; but in neither instance was the architect allowed to follow out his own conceptions. Bramante and Michael Angelo wished to have the portico of St. Peter's formed on the plan of the Pantheon, and Wren was obliged to modify his masterly designs so as to make them acceptable to those to whom he was obliged to defer. The belfries of St. Paul's gives it a character very different from that of St. Peter's. Neither is the dome of the latter so spherical as that of the British cathedral, nor is it so striking a feature of the building, being placed so far behind the lofty façade as to be almost invisible to a person standing near the edifice. But in the vastness of its proportions St. Peter's as far exceeds St. Paul's as the latter does the largest of the English churches. Perhaps, also, it is superior to St. Paul's in the harmony of its parts; the dome, though so grand a feature in the latter, being, it is very generally admitted, too large for the other parts of the building. But the English cathedral is, though longo intervallo, second only to St. Peter's; and is unquestionably the noblest of transalpine and of Protestant temples.

The interior of St. Paul's is chaste and imposing; but, owing to the want of ornament, it has rather a naked and austere appearance. Latterly it has been attempted to obviate this defect by placing within the cathedral monuments erected at the public expense to eminent individuals, among whom may be specified Lord Nelson, Abercrombie, Dr. Johnson, Howard the philanthropist, Sir William Jones, Sir Joshua Reynolds, &c. But these, with very few exceptions, do no credit either to the artists or the country, and are totally unworthy the temple they do nothing but encumber.

The remains of Sir Christopher Wren are deposited in one of the vaults of the cathedral; and before the entrance to the choir is the following appropriate inscription to his memory:

SUBTUS. CONDITUR. HUJUS. ECCLESIÆ. ET. URBIS.
CONDITOR. CHRISTOPHORUS WREN. QUI VIXIT.
ANNOS. ULTRA. NONAGINTA. NON. SIBI. SED.
BONO. PUBLICO. LECTOR. SI. MONUMENTUM. REQUIRIS.
CIRCUMSPICE.
OBIIT. XXV. FEB. ANNO. MDCCXXIII.
ÆTAT. 91.

Individuals ascend by an inside stair to the stone gallery which surrounds the exterior gallery above the colonnade; and by a more difficult ascent they reach the Golden Gallery, which crowns the apex of the dome, at the base of the lantern. The view from this latter point, on a clear day, is certainly unrivalled. The entire metropolis, vast as it is, appears to be spread out at the spectator's feet. The broad and silvery line of the river, crossed by numerous bridges, and bearing on its bosom thousands of vessels, gives infinite grandeur and variety to the scene. At this height, the people, horses, and carriages in the streets, and everything else on the surface, appear so greatly diminished, that the bustle of the crowd has been, not inaptly, compared to that of a swarm of emmets. Owing to the usual density of the smoke, this splendid view is seldom seen in perfection. It appears to the greatest advantage early in a clear summer morning, before the fires are lighted.

The more adventurous visiters not only ascend to the top of the cupola, but enter the lantern, and thence make their way into the copper ball by which it is crowned. The diameter of the latter is 6 feet 2 inches.

The whole cost of this noble structure amounted to £747,944, being less than the sum that has been thrown away on Buckingham Palace! It was, as has been often remarked, finished in 35 years, under the superintendence of one architect, by one master mason (Mr. Strong), and during the incumbency of one Bishop of London (Dr. Henry Compton). St. Peter's, on the contrary, was 145 years in building, during which time no fewer than 12 architects were employed upon it, and 19 popes sat in the papal chair! (See Brayley's Account of St. Paul's, in the Survey of London and Middlesex, ii., 249–310; Aikin's Essay on St. Paul's; Britton's Account of St. Paul's; Elmes's Life of Sir Christopher Wren, &c.)

It is greatly to be regretted that St. Paul's is so much hemmed in by the surrounding buildings. The view of the grand façade, with the dome rising above it, from the E. end of Ludgate-street, is, however, uncommonly fine; and far a good view of a portion of the building is now obtained from the opening made at the S. end of the new Postoffice. The dome appears to great advantage from the bridges and the river; and is seen at a great distance from all parts of the surrounding country, towering above the smoke by which the city is generally enveloped.

The effect of the smoke on the structure is not a little curious. In the parts protected from the weather it adheres, and the building has, in consequence, a black and sooty appearance; while, on the other hand, the parts exposed to the weather seem bleached or whitened. But this sort of pie-bald aspect has not the bad effect that might à priori have been expected.

Westminster Abbey, which, next to St. Paul's, is the grand ecclesiastical edifice of London, dates from the 13th century, though portions of the edifice, erected by Edward the Confessor, may still form part of the building. Great additions were made to it by Henry VII., who built the splendid chapel that still bears his name; and at the beginning of the last century the two towers at the W. front were added, from designs furnished by Sir Christopher Wren. In 1803, a considerable part of the building was destroyed by fire; but it has since been completely repaired, and Henry VIIth's chapel renovated in its original style. It is 360 feet long, and 195 wide, within the walls. Though built at many different times between the reigns of Henry III. and Henry VII., and never quite completed, it offers one of the finest specimens of the pointed style in England. It is in the form of a cross, the shape of which, externally at least, at the E. end, is almost obliterated by 12 small chapels, of which that of Henry VII. is the most magnificent and beautiful. The great variety of the abbey renders anything like a general description impossible. The N. side, with its beautiful gate, may be considered the principal front; though the view is much injured by the interference of St. Margaret's church. It presents a line of ornamental turreted buttresses and pointed windows, with a fanciful sculptured porch, decorated with immense flying buttresses, lofty pinnacles, and a large wheel window 32 feet in diameter. The most striking view of the interior is from the W. entrance, where the lofty pointed aisles, clustered columns, rich tracery work, and monumental decorations, judiciously lighted by painted windows, present a harmonious effect well calculated to arrest the attention of the most incurious. Many of the most illustrious statesmen, orators, warriors, philosophers, divines, poets, and distinguished individuals of all sorts, celebrated in the annals of the empire, are buried within its precincts; and their monuments, which are distributed all over the Abbey, give it the highest interest, and deeply impress the mind with feelings of awe and veneration. Since its restoration, in 1820, Henry VII.'s chapel has formed one of the most beautiful adjuncts to the Abbey; it is universally considered a gem, and is, undoubtedly, a most beautiful specimen of its style.

The other churches of London have no pretensions to be compared with those last mentioned. Of those which escaped the great fire of 1666, St. Saviour's in the borough, and the Temple church, deserve special mention. The former, recently restored to much of its ancient freshness, is a good specimen of the architecture of the 14th century; the latter is still more ancient, the greater portion being of the 13th century, and some parts very probably of the 12th: it is remarkable for its peculiar architecture, and particularly for the beautiful Roman arch forming the entrance to the building. It is now under repair. After the fire, several churches were built by Sir Christopher Wren, but the fame of St. Paul's has obscured the lustre of his other works. Bow church, in Cheapside, St. Bride's, Fleet-street, and St. Stephen's, Walbrook, are the most admired of Sir Christopher's churches. The latter is entitled to the highest praise. "He has not omitted a single beauty of which the design was capable; but has supplied them all with infinite grace." (Dallaway's Anecdotes, p. 142.) In the early part of last century several churches were erected, of which St. Martin's, St. George's, Hanover-square, and St. George's, Bloomsbury, have very fine porticoes, especially St. Martin's, which is a noble structure. Within the last 30 years, however, a complete change, and, which is worse, a great deterioration, has taken place in our ecclesiastical architecture. St. Pancras church, and some others, may, perhaps, be excepted from this censure; but an extreme poverty of architectural talent has been shown in designing the new churches of London, which are quite unworthy of these formerly erected, and of the city. The places of worship for dissenters are, with three or four exceptions, plain brick buildings, well arranged for the accommodation of large congregations, but constructed with little attention to ornament or taste.

Lambeth Palace.—One of the most extensive and handsome buildings S. of the Thames is Lambeth palace, on the river bank, nearly opposite the new houses of parliament. The original building, erected in 1191, was first intended for a college of canons; but as the pope refused his consent to its establishment, it was converted into an archiepiscopal palace, and has ever since been the town residence of the primate of all England. Great additions were made to it by Archbishop Boniface, and in the 15th century Archbishop Chichele built a square stone tower towards the river, called the Lollard's Tower, from the fact of some of those early reformers having been confined in it. Subsequent additions were

216

made by Cranmer, Pole, Parker, Juxon, Sancroft, and Tillotson; but the whole, as seen from the outside, is a heavy, dull-looking, brick structure, little interesting except from its antiquity. The late additions, however, completed in 1833, at a cost, including internal fittings, of nearly £80,000, are executed in much better taste. The new buildings, of Bath stone, stand in the gardens, E. of the old palace: the principal edifice is a beautiful and imposing structure, the ornamental portions, which are particularly rich, being copied from Westminster and St. Alban's abbeys. The entrance front, flanked with square towers, is 160 feet in length, the opposite or garden front being 30 feet longer. The principal rooms are of fine proportions, and richly though chastely embellished, the wood-work being almost wholly of oak. The library is perhaps one of the finest features of the interior; and though remarkably plain in its decorations and furniture, produces, from its great size, a very imposing effect. It contains upwards of 25,000 volumes, among which are many rare works in classics, divinity, &c.; and the MSS., some connected with the history of the see, and others of a miscellaneous character, are extremely valuable. In the older parts of the building the chief rooms are the long gallery, containing a curious collection of paintings, chiefly portraits of former prelates, the great hall, with an open roof of oak, presenting on the whole one of the finest specimens in the country of internal Gothic decorations; and the chapel, a small but extremely elegant apartment, fitted up with oak stalls, pews, and an exquisitely carved pulpit and screens. The altar-piece, however, ill accords with the rest of the fittings, being of the Corinthian order, painted and gilt. The park and gardens belonging to the palace occupy about 18 acres: they are completely walled round; nearly four acres are appropriated to the kitchen garden, the rest being tastefully planted, and laid out in shrubberies.

Cemeteries.—The crowded state of most of the metropolitan churchyards, and the growing conviction of their injurious influence on the health of the neighbourhoods in which they are placed, have, within these few years, suggested the establishment of public cemeteries at some distance from town. The first of these, at Kensal green, occupying a piece of ground of 46 acres in extent, tastefully planted and laid out, was opened in 1832. It is situated about 2 m. N.W. London: and has chapels for the performance of the funeral service according to the rites of the Church of England, or of those whose friends profess a different creed. The success of this undertaking, which was long opposed by ignorant prejudice, has since led to the construction of other cemeteries. That at Highgate, consecrated in 1838, and occupying about 20 acres, is in a beautiful situation N. of London, offering a splendid view, of great extent; but the decorations, though handsome, are too showy to be in keeping with the destination of the spot. The Norwood cemetery is 6 m. S. of the city, and is double the size of that last mentioned. The buildings in this cemetery are of a superior character; and an ingenious machinery is used to lower the coffins slowly and silently into the vaults beneath. Two cemeteries have been recently completed, at Abbey park, Stoke Newington, and Earl's court, Brompton; and others are contemplated. Hitherto, however, the new cemeteries have been too far from town, and too expensive, to be used by the poorer classes; and the formation of more convenient cemeteries, for the accommodation of the latter, is a matter eminently deserving the attention of the authorities.

Commerce.—London is not only the capital of a great empire, but is the first commercial city of the world. Her intercourse extends to the remotest countries, and her merchants are pre-eminent for wealth, enterprise, and integrity. The establishments connected with commerce are on a scale commensurate with the vast amount of business to be transacted. The public buildings for commercial purposes consist chiefly of the Bank of England, East India House, Royal Exchange, Excise Office, Custom-house, and Cornmarket.

The Bank of England, from its first incorporation in 1694 to the year 1734, transacted its affairs at Grocers' hall, in the Poultry. The first stone of the present building was laid in 1732; 40 years afterward the E. and W. wings were added, and in 1781 the church of St. Christopher was taken down to make room for farther additions. Until 1825 this edifice exhibited a great variety of incongruous styles; but endeavours have since been made, and with some success, to produce uniformity. The building is insulated, and covers eight acres: its shape is an irregular parallelogram. The longest side measuring 440 ft. Many of the rooms in the interior, such as the court-room, pay-hall, and dividend-office, are spacious and well proportioned; but the largest and loftiest of all is the rotunda, a circular hall, 57 ft. in diameter, and crowned by a handsome cupola and lantern. The chief transactions connected with the funds take place in this apartment. The affairs of the bank of England are

managed by a governor, deputy-governor, and 24 directors, elected annually. The business is conducted by about 900 clerks, whose salaries amount to about £190,000.

In 1833 the charter of the bank was continued till 1845, its capital being then also fixed at about £11,000,000 lent to government at 3 per cent. Branch banks in connexion with the bank of England have, since 1826, been established in most large towns, the chief business of which is to discount bills, issue notes, and transmit money to and from London. The profits of the bank accrue from interest on exchequer bills, discounts of commercial bills, interest on the capital lent to government, an allowance of about £130,000 a year for managing the public debt, and some other sources. The dividend received by the proprietors is 8 per cent.

STATEMENT of the Liabilities and Assets of the Bank of England, on the 31st of March, 1840, that is, of the Bank Notes in circulation, and the public and private Deposits held by the Bank, on the one hand, and of the Securities and Bullion in her possession, on the other.

Liabilities.	£		Assets.	£
Circulation.			Public Securities .	. 14,003,000
London 12,441,000			Private ditto	. 6,721,000
Country 3,954,000			Advances.	
	16,395,000		Bills of Exchange . 3,387,000	
Deposits, public .	2,404,000		Exchequer	
Ditto, private .	4,124,000		Bills.	
			Stock, &c. 265,000	
	20,923,000			3,652,000
				21,326,000
			Bullion .	4,416,500
				25,772,000

Of about 80 private banking-houses at present in London, three were in existence before the bank of England, viz., those of Messrs. Child, Temple Bar, Messrs. Hoare, Fleet-street, and Messrs. Snow, in the Strand. Within the last few years various joint stock banking companies have been established in the city, on the model of the Scotch banks; and the fair presumption seems to be that they will, at no distant period, entirely supersede the private bankers.

The Royal Exchange, originally erected by Sir T. Gresham, in 1566, was burnt down in the great fire; having been rebuilt within three years, and extensively repaired between 1820 and 1826, it was again destroyed by fire on the 10th of January, 1838. Latterly it formed a spacious quadrangle, surrounded by lofty stone buildings, with a covered colonnade running round the whole interior, and the N. and S. sides of the exterior. In the centre was a statue of Charles II. by Spiller, and in niches of the inner wall were statues of most of our monarchs, from the time of Edward I. A new exchange, on a grand scale, from a design by Mr. Tite, is now (1841) in the course of being constructed, which, when completed, will be one of the greatest ornaments of the city. At present the business usually transacted in the Exchange is carried on in the large court of the Excise-office in Broad-street. At a rough estimate, 3000 merchants and brokers have their places of business within half a mile of the Exchange, and meet there to carry on operations by which the commercial affairs of the world are powerfully influenced. Several houses, together with two churches, have been removed, to make room for the new edifice, the chief front of which is to face the Poultry and Mansion house.

The East India House, in Leadenhall-street, is the place where the East India Company's business is chiefly transacted: it was first built in 1726, but has been subsequently so much altered and enlarged, that scarcely any part of the old edifice now remains. It had a stone front, with a portico supported by six fluted Ionic columns, above which are a frieze and pediment ornamented with sculpture. The interior comprises numerous and handsome apartments, of which the largest are the court-room, the committee-room, and the two sale-rooms; in the E. wing are the library and museum, the former rich in Asiatic literature and rare Oriental MSS., the latter abounding with Indian curiosities, the spoils of successful wars waged with the native monarchs. The museum is open every Saturday. The East India Company has now, however, become exclusively a political institution; the act 3 & 4 Will. 4, prolonging the charter till 1854, having debarred the company from the privilege of trading.

River and Port.—What is legally termed the port of London extends 6½ m. below London bridge to Bugsby's hole, beyond Blackwall; though the actual port, consisting of the upper, middle, and lower pools, does not reach beyond Limehouse. The whole of the latter space is generally covered with vessels; a channel, only 300 ft. wide, being left clear for craft passing up and down the river. The port having been long insufficient for the proper accommodation of the shipping resorting to London, and being often blocked up by fleets of merchantmen, the quays also being heaped with bales, boxes, bags, and barrels, in such confusion that the most barefaced robberies were committed with impunity, the necessity of farther protection for merchandise became evident. Accordingly, at the close of last century, it was determined to excavate wet docks, capable of accommodating a large number of ships, with contiguous warehouses, the whole being enclosed by high walls. The West India docks, the first of these establishments, and the largest belonging to the port, were opened in 1803. They are situated about 4 m. down the river: including the City canal, a work intended for another object, but now a part of this establishment, they comprise about 295 acres, one fourth part of which is covered with water, the rest being occupied with quays and warehouses, the latter of great magnitude, and furnished with every convenience. They have an import and an export dock, with sufficient accommodation for 500 large merchantmen. The London docks, about 1½ m. from London bridge, were opened in 1805. They cover about 100 acres of ground, of which nearly a third is water. The vaults beneath the warehouses contain cellarage for 65,000 pipes of wine, and one of them has an area of seven acres. The tobacco warehouses are very extensive. The East India docks, smaller than those above described, and farther down the river, were opened in 1809. Their water-area is 30 acres, and their great depth (23 ft.) enables them to accommodate vessels of very large size. About the same time the Commercial docks were constructed on the S. side of the river, or rather the old docks for the Greenland ships were enlarged and provided with warehouses for bonding foreign corn. This dock covers 49 acres, 40 of which are water: they are used for vessels engaged in the Baltic and E. country commerce. The St. Katharine's docks, opened in 1828, are the nearest to London bridge, being just below the Tower. They enclose 24 acres, of which 11½ are water. The warehouses, which are on a very extensive scale, are close to the quays, having the lower or basement story open for the purpose of receiving or delivering goods from and to vessels that are being laden or unladen: the arcades are supported by iron columns of great strength. These docks have all been constructed, at a vast expense, by joint stock companies; and have on the whole been profitable concerns, though they have redounded infinitely more to the advantage of the port than to that of their projectors.

The number of colliers frequenting the port has often suggested the idea of excavating docks for their accommodation in the Isle of Dogs, opposite Greenwich; but nothing has yet been effected towards the execution of this plan. According to the present system, that part of the port below the lower pool serves as a place of anchorage for the colliers, only a certain number of which are allowed to be in the pool at once, and a flag is hoisted to notify when it is full. On the flag being hauled down, the first collier in rank enters the pool, and the others follow until the number is completed, when the flag is again hoisted; the rest wait their turn. The following statement of the quantities of coal and culm brought into the port at different periods, from 1820 to 1840, both inclusive, shows the consumption of coal in London. The great increase within the last dozen years is chiefly owing to the introduction of steam navigation and gas lighting.

	Tons.		Tons.
1820	1,692,335	1834	2,080,547
1825	1,820,975	1835	2,299,820
1830	2,005,364	1836	2,399,550
1831	2,053,673	1837	2,629,390
1832	2,147,290	1838	2,581,085
1833	2,014,805	1839	2,625,323

ACCOUNT of the Coal imported into London in 1840, specifying the Ports whence the Coal was shipped, and the Number of Cargoes and Tons imported from each.

Ports whence shipped	Cargoes.	Tons.
Newcastle	5,144	1,161,286
Sunderland	2,474	730,146
Stockton	1,535	435,369
Blyth	317	83,397
Scotland, ports of	250	31,800
South Wales, do.	234	60,069
Yorkshire, do., &c.	563	80,481
Total of Coal	9,166	2,561,590
Culm	8	1,915
Cinders	10	3,434
Total	9,188	2,566,889

The new Custom-house, a handsome building by the river-side, between London bridge and the Tower, was opened for business in 1817. The old one was burnt down in 1814, though not before the present building was begun, the former having been found inconveniently small. The river-front, 480 ft. in length, is built of Portland stone, and though rather plain, is decorated by three porticoes, each

supported by six Ionic columns. The long room, where the public business is transacted, is 186 ft. in length, and 69 ft. in width and height.

The vast extent of the trade of London will be apparent from the subjoined statement of the gross customs revenue from 1834 to 1840, both inclusive:

1834	£10,897,283	1838	£11,254,734
1835	11,773,616	1839	11,431,245
1836	12,156,279	1840	11,083,053
1837	11,188,036		

Now, as the total gross customs revenue of the United Kingdom amounted, in 1839, to £23,498,486, it would seem, from this statement, that the trade of London only equalled that of all the rest of the kingdom! This, however, would be a fallacious inference. The imports into several of the other great trading ports, including those of Liverpool, Hull, Dundee, &c., consist principally of cotton, wool, flax, and other raw materials of our manufactures, which are mostly admitted at low duties; whereas the imports into London consist principally of articles of consumption, as sugar, tea, coffee, wine, corn, &c., on which high duties are paid. But after making every allowance for the circumstances now stated, still the foreign trade of London is of prodigious and unparalleled extent. She may truly be said to be universi orbis terrarum emporium; and owing to her being the grand mart of all the rich, extensive, and densely peopled districts included within the basin of the Thames, we do not think, provided the country continue to prosper, that there is any ground for apprehending any falling off in the commerce of London. It is impossible to form any accurate estimate of the total value of the produce conveyed into and from London; but including the home and foreign markets, we believe it will not be overrated at the prodigious sum of eighty millions sterling.

Some idea, however imperfect, may be formed of the extent and distribution of the trade of London from the following statements.

ACCOUNT of the Ships entering the Port of London from 1823 to 1840, both inclusive, distinguishing between British and Foreign Ships from Foreign Parts, and Coasters.

Years.	Foreign Parts.				Coasters.	
	British.		Foreign.			
	Vessels.	Tonnage.	Vessels.	Tonnage.	Vessels.	Tonnage.
1823	3,031	611,751	965	161,795	19,079	2,191,262
1824	3,132	607,106	1,513	244,984	19,483	2,374,662
1825	3,309	753,363	1,743	302,122	19,327	2,300,308
1826	3,490	679,918	1,560	215,564	20,493	2,441,548
1827	4,018	760,162	1,531	221,688	17,977	2,226,640
1828	4,004	767,212	1,193	193,929		
1829	4,108	794,050	1,504	231,606	No. of coasters not stated during these five years.	
1830	3,910	714,229	1,264	207,506		
1831	4,140	769,989	1,547	209,159		
1832	3,774	640,057	86	144,211		
1833	3,461	671,289	1,201	179,848	19,336	2,247,231
1834	3,796	737,093	1,730	215,093	20,005	2,305,567
1835	3,740	740,245	1,067	184,493	20,471	2,761,862
1836	3,415	772,046	1,203	215,475	20,765	2,618,299
1837	1,079	82,738	1,547	210,133	21,382	2,818,780
1838	4,268	803,998	1,727	277,901	21,802	3,008,790
1839	4,503	949,467	2,276	557,163	21,112	3,426,705
1840	4,347	911,760	2,221	554,436		

An ACCOUNT of the Number and Tonnage of those Ships that entered the Port of London with Cargoes from Foreign Parts, in 1839, distinguishing the Countries whence they came.

Countries.	British.		Foreign.	
	Ships.	Tons.	Ships.	Tons.
Russia	608	136,407	38	16,941
Sweden	14	2,577	34	27,302
Norway			121	34,805
Denmark	17	1,915	252	22,085
Prussia	234	29,394	343	72,949
German States	245	60,009	281	18,497
Netherlands	390	34,954	284	41,858
France	435	70,422	395	49,529
Portugal, Azores, and Madeira	316	39,925	3	362
Spain and Canaries	280	34,134	44	4,880
Italian States	161	23,306	87	14,062
Ionian Islands	21	2,953		
Turkey and Continental Greece	54	8,851		
Morea and Greek Islands	21	3,032	1	289
Egypt	2	301		
Tripoli, Barbary, and Morocco	36	5,194		
Foreign Possessions in Africa	3	432	1	690
Ditto in Asia	36	16,760	2	645
China	36	29,051		
United States of America	15	7,278	68	32,558
British West Indies	43	8,190	27	6,368
Foreign Continental Colonies in America	57	17,081	3	873
Total	3,166	586,051	2,355	371,453

An Account of the Number and Tonnage of Coasting Vessels that have entered the Port of London, in each Year from 1833 to 1839, both inclusive.

Years.	General Coasters, excluding Colliers.		Irish Traders.		Total.	
	Vessels.	Tonnage.	Vessels.	Tonnage.	Vessels.	Tonnage.
1833	13,416	2,394,640	1,964	14,520	15,331	2,510,741
1834	15,981	2,441,495	1,463	147,962	20,669	2,403,457
1835	19,528	2,404,299	1,165	164,172	24,471	2,764,184
1836	19,267	2,456,499	1,045	134,279	22,765	2,610,974
1837	20,362	2,242,994	1,471	167,492	21,322	2,411,596
1838	20,288	2,267,741	1,259	160,139	21,502	2,907,158
1839	20,205	2,995,643	907	142,259	21,112	2,745,501

An Account of the Number and Tonnage of Ships that entered the Port of London in 1839 with Cargoes from the Colonies and Dependencies of England:

Colonies.	Ships.	Tons.
Gibraltar	90	£2,707
Malta	12	2,250
British Possessions in Africa	851	35,110
" " Asia	230	109,138
America, viz :		
British Northern Colonies	228	94,441
" West Indies	845	89,548
The Whale Fisheries	22	9,310
Isles of Guernsey, Jersey, and Man	546	47,129
Total	**1,503**	**417,149**

There belonged to the port of London, in 1840, 2950 ships, of the total burden of 581,000 tons, manned by 32,000 seamen! This, which is by far the greatest amount of shipping that ever belonged to any single port, will appear the more extraordinary when it is recollected that the colliers almost all belong to Newcastle, Sunderland, and other ports in the N. An immense number of barges are employed in the loading and unloading of colliers and other vessels in the river. The out-of-doors establishment of the customs, which is mostly all employed in the business of the port, comprises about 1250 individuals.

Insurance of houses, ships, lives, &c., is carried on to a far greater extent in London than anywhere else in the world. Marine insurances are mostly effected by private parties; but other insurances are generally made by joint stock companies. Some of these have been most successful, and have accumulated vast sums. It is believed, however, that not a few of these companies are of a very questionable description; and the conviction seems to be gradually gaining ground, that some public regulations should be laid down for the formation and guidance of insurance companies, so as to protect the insured against the extravagance, mismanagement, or bad faith of the directors.

Regent's Canal.—The port of London is connected with the Irish sea by a chain of canals of which only the Regent's canal, begun in 1812 and opened in 1820, comes within the scope of this account. The Regent's canal, which is 45 feet wide at the surface, 5 feet in depth, and 8½ m. long, begins on the N.W. side of London, at the Paddington branch of the Grand Junction canal, and proceeds in an irregular semi-circle to Limehouse, where is a large basin connected with the Thames. This canal has two tunnels, one of which, passing under Islington, is 5 furlongs in length. The descent, amounting to 90 feet, is effected by means of 13 locks; and it is crossed by 37 bridges. Five branches, terminating in basins, extend into different parts of the town.

Manufactures, retail trade, and markets.—London presents itself under too many points of view to be called a manufacturing city; yet it is the seat of many, and of some very extensive manufactures, several of which have their distinct quarters.

The silk manufacture is conducted on a very large scale in Spitalfields, Bethnal Green, and Mile-end town, and employs about 10,580 looms, belonging to 4300 families, the total population supported by the business being about 22,008. The trade fluctuates extremely, owing chiefly to the caprices of fashion, and great numbers of workmen are often thrown out of employment; but the distress, so often said to prevail in this densely-peopled district, is owing at least as much to the drunken and improvident habits of many of the weavers as to any falling off in the demand for labour. The net wages of plain silk weavers, when fully employed, range from 9s. to 11s. 6d., and those of velvet weavers from 15s. to 23s. a week. With respect to physical condition, this numerous body are a diminutive, impoverished, and feeble race, unable to withstand disease, and not long lived, circumstances attributable to close in-door employment, bad air, bad lodging, and bad food. We have elsewhere noticed the tendency to epidemic fevers in close and ill-drained neighbourhoods, and in no part of London are the fatal effects of lodging in close courts and cellars more visible than in Spitalfields and Bethnal Green. (See *Dr. S. Smith's Evidence before the Committee on Health of Towns*, p. 1–7.)

Porter is the favourite beverage of the lower and also of a considerable proportion of the middle classes of London. The breweries in which this favourite liquor is prepared are mostly on a very great scale; and are, indeed, by far the most gigantic manufacturing establishments in the metropolis, greatly exceeding anything of the kind to be found anywhere else. The capital vested in a first-rate brewery, and its dependent public houses in different parts of the town, is usually quite immense. The principal establishments produce from 160,000 to 260,000 barrels a year, principally porter, but partly also ale. It has been estimated, that nearly 2,000,000 barrels, or 72,000,000 gallons of porter and ale are brewed for consumption in London only, besides which great quantities are sent to different parts of the United Kingdom, and exported to the E. and W. Indies, the United States, and continental Europe. The splendid teams of horses in the drays belonging to the chief breweries of London are among the objects most worthy of admiration in the metropolis. There are several very extensive distilleries, vinegar factories, chemical works, soap-boiling houses, most of which are situated on the S. side of the river. The quantity of soap made in London in 1839 amounted to 38,065,175 lbs. of hard and 820,863 lbs. of soft soap. About 20 large engineering establishments employ several hundred workmen in making steam-engines and other machinery, chiefly in Lambeth and Southwark.

The principal sugar refineries are in Whitechapel, E. of the city. Watchmakers, who are numerous, reside principally about Clerkenwell. The finest cutlery and hardware is made in London, and the manufacture of metals of all kinds is carried on to a great extent. Coach-building is an important business; and the carriages of London are not only the handsomest, but the best built and most durable of any in the empire. Great numbers are made for exportation. Upward of 1500 hands are employed in constructing musical instruments, and in engraving music. Ship-building, and many trades connected with shipping, are extensively carried on. E. of London bridge. Owing to the extent to which the division of labour is carried, the tradesmen and artisans of London have attained to the greatest proficiency in their respective callings; and these cannot be a question that the cabinet-makers, printers, tailors, shoemakers, &c., of the metropolis are quite unrivalled.

The following list of professional men and tradesmen is compiled from Pigot's Directory for 1840:

Physicians	980	Brokers	760
Surgeons and apothec.	1,880	Tailors and clothiers	3,940
Druggists	475	Drapers and hosiers	1,220
Barristers	1,050	Hat and bonnet mak.	1,280
Attorneys	3,090	Dress and stay makers	1,900
Architects	230	Shoemakers	2,920
Artists	350	Hair-dressers	920
Engravers	300	Dyers	360
Schoolmasters, &c.	1,230	Coal merchants	680
Merchants	1,340	Hotel and tav. keepers	640
Stock-brokers	230	Publicans, &c.	5,309
Insurance and general brokers	230	Coffee and eating-houses	970
Watch and clock mak.	580	Wine merchants	1,050
Jewellers, &c.	490	Livery-stable keepers	660
Booksellers	780	Butchers	1,680
Stationers	570	Fishmongers	500
Printers and type-found.	620	Poulterers	150
Musical instrument mak.	380	Corn dealers and bakers	2,500
Coachmakers	380		
Builders	360	Confectioners	450
Bricklayers and masons	900	Grocers	1,390
Carpenters	1,560	Greengrocers, &c.	1,730
Cabinet-makers and upholsterers	1,100	Cheesemongers	1,050
		Milkmen	960
Painters and paperhangers	1,750	Small shopkeepers	2,700
		Tobacconists	1,060
Ironmongers, &c.	830	Pawnbrokers	330

The extent of the retail trade of London can only be conjectured. By a rough estimate, made by counting the pages of one of our most copious directories, the number of houses employed in business cannot be much under one hundred thousand, to one half of which shops are attached. The trades, generally speaking, are mixed indiscriminately, though some remains are still traceable of the ancient custom of particular trades congregating in particular places. Thus we still find coach-makers in Long-Acre, stay-makers in Holywell-street, booksellers in Paternoster-row, and bankers in Lombard-street. A good deal of business used to be transacted by itinerant venders, who made the streets resound with their cries. Recently, however, these have been diminished by the abolition of the dustmen, and muffin-boys' bells, the newsmen's horns, and the early cries of the chimney-sweepers. Fashionable shops attract attention by a magnificent and gorgeous display of wares: their windows, which in most cases comprise almost their entire front, are, in many instances, made of the finest plate glass, set in brass frames, and their interior is frequently lined with mirrors. Every sort of device is, as may be expected, used by the shopkeepers to attract customers.

Markets, &c.—London has about 50 markets for provisions, in nearly all of which goods are sold by retail as well as wholesale, though the majority of the inhabitants purchase at shops distinct from the markets. Smithfield is the great mart for live stock, which is sold on Mondays and Fridays. No fewer than 1,403,490 sheep were sold here in 1836, and 183,362 head of cattle. We may remark, by the way, that Smithfield market is situated in the very centre of the city; and this circumstance, by obliging the stock to be driven to and from it through crowded streets, makes it a very great nuisance. Frequent attempts have been made to have it removed to the suburbs; but hitherto without effect. London is also totally unprovided with proper slaughter-houses, or *abattoirs*. Exclusive of the stock brought to Smithfield market, a good many cattle and sheep are now imported in steamers, and privately sold; and in the colder months slaughtered cattle and sheep are extensively imported, particularly from the E. coast. Newgate and Leadenhall markets, with the Whitechapel carcass butchers, supply most of the butchers of the town and neighbourhood; and these, as well as all the other markets, also supply retail customers, and have a good supply of vegetables, poultry, game, eggs, and butter. Covent Garden market is the principal vegetable mart in London, and the immense supply of the finest fruits and vegetables, and the beauty of the plants on sale, make it well worth a visit. The Borough and Spitalfields markets are also chiefly supplied with vegetables. Billingsgate is the great fish-market, whence fish of all sorts are distributed to the shops and markets in different parts of the town. The supply of salmon, brought in ice from all parts of the kingdom, and of turbot, cod, lobsters, and oysters, is quite immense. Hungerford market is also a well-supplied fish *depôt*; but at this and Farringdon markets (both of them new and handsome establishments), butchers' meat, fruit, and vegetables are also sold. The corn market is held in Mark-lane, and is attended almost exclusively by wholesale dealers.

Different statements have, from time to time, been put forth of the consumption of the principal products brought into London; but, with the exception of coal, and one or two other articles, there are no means by which to arrive at anything like a correct conclusion in such matters. Allowing for the carcasses imported by steam and otherwise, the annual consumption of butchers' meat may at present be estimated at about 190,000 bullocks, 1,500,000 sheep, 25,000 calves, and 25,000 pigs, exclusive of great quantities of bacon and hams. The consumption of poultry, game, and eggs is also immense; but there are no means by which to estimate its amount. It may, however, be mentioned, that from 70 to 75 millions of eggs are annually imported into London from France and other foreign countries, exclusive of those brought from the different parts of Great Britain! About 12,000 cows are kept in the city and its environs for the supply of milk and cream; and if we add to their value that of the cheese and butter brought into the city, the expenditure on dairy produce will appear to be enormous. The consumption of wheat may, perhaps, be estimated at about 1,900,000 quarters a year; and the vast number of horses in London, and their high keep, must occasion an immense consumption of oats. The value of the fish, vegetables, &c., consumed in the city, has been set down by some intrepid calculators; but the data on which they formed their estimates were of too loose and unsatisfactory a character, to entitle them to any credit.

External and internal Communication.—The recent introduction of railways has already effected a great alteration in the intercourse of the provincial districts with the metropolis; but so short a time has elapsed since they have come into operation, that it is difficult to estimate their ultimate results; which, however, there can be little doubt, will be advantageous alike to the metropolis and provinces.

The Birmingham railway, opened through its whole extent in Sept. 1838, commences near Euston-square, where it has a splendid terminus. Trains leave both London and Birmingham eight times in the 24 hours, and the time occupied in the transit is five hours. The line is continued N. in two directions, by the Grand Junction and North Union railways, to Liverpool and Manchester, Preston and Lancaster; and, by the Derby and N. Midland railways, to Derby and Nottingham, Sheffield, Leeds, and York. The whole distance to the last place is now accomplished in 10 hours. The Southampton railway, which cost £1,780,000, was opened in 1838; it has its London station S. of the river, close to Vauxhall bridge, whence trains set out frequently during the day, and reach Southampton in three hours. The Great Western railway, constructed at a cost of £4,500,000, very lately opened as far as Bath, has its London terminus at Paddington, where are ample warehouses and accommodations both for passengers and goods.

The Eastern Counties railway, opened as far as Brentwood, has its London terminus in Shoreditch, and is intended to run to Great Yarmouth, through Romford, Ingatestone, Chelmsford, Colchester, Ipswich, and Norwich, a total distance of 126 m. Trains leave London seven times a day. The Greenwich railway, the earliest in operation of all the lines connected with the metropolis, commences in the borough, opposite to St. Saviour's church, and ends in London-street, Greenwich: its entire length is 3½ m. Trains run from both termini every quarter of an hour, from 8 A.M. to 10 P.M. The London and Croydon railway, opened through its entire length in June, 1839, branches from the Greenwich line about 2 m. from London bridge, and thence turning S., passes by New-Cross and Sydenham to Croydon, the whole distance being 10½ m. The cost of this undertaking was £750,000, and the passengers have hitherto averaged about 1800 a day. Trains leave both ends 12 times a day. The traffic will, of course, be vastly increased, when the two extension lines to Brighton and to Dover shall have been opened. The Blackwall railway, recently opened, connects the E. and W. India docks with the metropolis: trains run from each terminus every quarter of an hour. The North Eastern railway has very lately been opened as far as Broxbourne.

TABLE showing the Length of the Metropolitan Railways, and the amount of Traffic during the week ending Aug. 30, 1840. (*Railway Times*, Sept. 5, 1840.)

Name of Railway	Total Length	Length opened	Passengers	Total Receipts (£)
Birmingham Railway	112½	112½	28,600	16,340
Blackwall do.	3½	3½	53,760	576
Croydon do.	10½	10½	11,110	782
Eastern Counties' do.	126	17½	6,325	543
Great Western do.	118	46	24,314	6,363
Greenwich do.	3½	3½	38,440	1,505
South-western do.	77	77	19,380	7,111
	451½	267½	169,736	32,716

In 1837, before any very extensive railway was opened, the number of stage-coaches licensed to run between London, and places above 20 m. distant, was about 600, conveying nearly 4000 passengers each day. In 1838 the average daily number of persons travelling by three railways alone, the Great Western, Birmingham, and South-western, was about 3506, and the average daily number by the Eastern Counties line was nearly 1000 in 1839. As the railways in 1840 accommodate about 25,000 passengers a day, and many stages still continue to run, the present number of passengers is at least six-fold that of 1837.

The steamboats constantly plying on the river, and making daily excursions to Margate, Ramsgate, Gravesend, and other places below London, as well as to Richmond and various intermediate places up the river, have been roughly estimated to take 10,000 passengers a day. The weather, however, very much influences the number, as the majority of steamboat passengers proceed on excursions of pleasure rather than business. Saturdays and Sundays are the grand days.

The means of internal communication and of communication with places in the immediate neighbourhood of town, are also considerable. The number of short stages and omnibuses is altogether about 900, which, reckoning to each six journeys a day, and 12 passengers to each journey, convey every day about 54,000 persons. A new mode of internal communication between the E. and W. parts of London, has lately been effected by small steamers continually traversing the river, and taking a great number of passengers. Two or three of these convenient vessels leave London bridge every quarter of an hour during the summer, for Westminster and Chelsea.

Hackney-coaches were introduced more than 200 years ago; and previously to the introduction of cabriolets, in 1820, were very numerous; but it is a singular and not easily explained fact that, with very few exceptions, the hackney-coaches of London are the dirtiest and most uncomfortable carriages imaginable. Cabriolets being cheaper and more rapidly driven than the old lumbering hackney-coaches, speedily deprived the latter of a great part of their employment. Various improvements have been made from time to time in these one-horse carriages, and many now in use are clean, neat, and commodious.

Parcels Delivery Company.—A few years ago, a private company started under this name, which, while it has met with considerable success, has certainly been very serviceable to the public. The company have established a considerable number of vans, which traverse London in all directions with great regularity, three or four times a day, delivering parcels at more reasonable charges than had been imposed by the old private carriers. Another company, of the same kind, and conducted with equal spirit and ability, has been more recently established.

Literature.—London ranks almost as high as a literary

as a commercial city. Notwithstanding the factitious encouragement given to learning and science in Oxford and Cambridge, London is the favourite resort of literary and scientific men. Its immense population, the wealth and intelligence of its inhabitants, and the circumstance of its being the seat of government, attract aspiring individuals from all parts of the empire, especially those ambitious to distinguish themselves in literature or politics. The practical, common-sense character of the philosophy and literature of England is probably, indeed, in no small degree owing to its being principally cultivated in London, where the writers, by mixing with the world, learn to avoid those over-refined theories and fanciful distinctions, in which reclus speculators are so apt to indulge. With the exception of the provincial newspapers, the whole periodical literature of England centres in London. The number of persons engaged in this department, as authors, publishers, printers, &c., is quite immense. London has no fewer than 13 daily newspapers, and 67 that appear at other intervals. Many of these journals display great, and some consummate talent; and, considering the extreme rapidity with which articles for the daily journals must be written, and the want of time for revision, they are certainly extraordinary performances. So far as respects its newspaper press, London is infinitely superior to every other city; and however one-sided, prejudiced, and little to be depended on in party matters, it is not easy to imagine that it is likely to gain much in ability, variety, and interest.

It appears, from the Stamp-office Returns, that of 58,516,869 stamps issued to the different newspapers published in the United Kingdom during the year ended the 15th Sept., 1830, no fewer than 29,197,363, or the half of the whole, were issued to those published in London! And when the superior ability and information of the London press is taken into account, its preponderance will appear still more striking. During the same year, the total amount of the duty on advertisements paid by the newspapers of the United Kingdom amounted to £123,600, of which £46,221 was derived from the metropolitan journals.

A prodigious number of weekly, monthly, and quarterly magazines, reviews, and other publications, issue from the London press; and though many of these be of a very trashy and worthless description, a considerable number are of a widely different character, and are well fitted to amuse, instruct, and improve the reader. By far the greater number of these publications appear on the last day of every month, known among booksellers as "Magazine day;" when the great publishing houses make up and forward innumerable parcels, containing every variety of works, to their correspondents in all parts of the kingdom.

The magnitude and importance of the periodical press of the metropolis will be best seen from the following statement, drawn up from the catalogue for 1841 published by Messrs. Longmans and Co.

Description of Periodical.	Number.	Price.
Weekly magazines, &c.	24	1d. to 9d.
— parts of entire works	2	3d. — 1s.
Monthly magazines, &c.	147	1d. — 6s.
— parts of entire works	85	3d. — 25s.
Quarterly reviews, &c.	22	1s. —1s. 6d.
— parts of entire works	8	2s. 6d.—21s.
Transactions of learned societies	72	5s. — 52s.
Law Reports	20	
Total of periodicals (not newspapers)	383	
Newspapers, daily morning	8	
— daily evening	6	
— thrice a week	4	average 6d.
— twice a week	4	
— once a week (not Sunday)	29	
— Sunday	19	4d. — 10d.
Total of newspapers	80	

The greater number of the works written in Scotland are now published in Edinburgh; but nearly the whole of the works written in England and Ireland are published in London. The latter, in fact, is to the literature of Britain what Leipsic is to that of Germany, or Paris to that of France. The London publishers have agents all over the country, to whom they send all new publications; so that in the few instances in which books are printed at Oxford or Cambridge, or other provincial towns, it is usual to send them to London to be published.

Education.—London, unlike most other European capitals, had no university empowered to grant degrees till 1836, when one was established by royal charter (renewed in 1837) for "the advancement of religion and morality, and the promotion of useful knowledge, without distinction of rank, sect, or party. This institution differs (and, as we think, advantageously) from all other universities, in its having nothing to do with the business of education, being constituted for the sole purpose of ascertaining the proficiency of candidates for academical distinctions. It is, in fact, a Board of Examiners, empowered to grant degrees in science and literature to such candidates as are found,

on examination, to have attained the required proficiency. The senate, or board, consists of a chancellor, vice-chancellor, and 33 other members. The faculties are those of arts, law, and medicine, in each of which are several examiners, amounting in the whole to 29, of whom 10 are members of the senate. The sittings are held in Somerset-house, and the examinations are half-yearly. The greatest number of candidates for degrees has hitherto been furnished by the University and King's colleges. The former of these, opened in 1828, is governed by a council and senate of professors: the course of education embraces classics, pure and mixed science, history, jurisprudence, and medicine, religion being wholly excluded. The success of the medical school, which has for some years been the largest in London, has led to the erection of a good hospital close to the college. The general classes have not been so well attended as the sanguine friends of the establishment at first expected; but the attendance is likely to be increased from the addition to it of an excellent junior school, the instruction in which forms a good preparation for higher studies. This school, which for some years has averaged 350 scholars, is conducted by the classical professors, who pay a rent for the use of apartments within the college. King's college is a similar establishment to that last mentioned, and is similarly conducted, except that religion is taught in accordance with the principles of the Church of England. The general classes are well attended, and the junior school has upwards of 400 boys. The medical school is small. The buildings of these establishments are handsome and commodious; and the portico of University college is one of the finest in London.

Among the literary and scientific establishments of the metropolis, one of the most important and best supported is the Royal Institution in Albemarle-street. The building (recently new fronted in good taste, with fourteen Corinthian columns,) is exceedingly well arranged, and comprises a good library and reading room, a theatre for lectures, capable of accommodating 900 persons, and a chemical laboratory supposed to be one of the largest and best supplied with apparatus in Europe. Lectures on various subjects are delivered by the professors and other gentlemen temporarily engaged; and the important investigations made here by the late Sir Humphry Davy, Mr. Faraday, and others, have conferred on the institution a well-merited celebrity. Next in importance to that just mentioned is the London Institution, in Finsbury Circus, Moorfields, the objects of which are very similar, though not so fully and scientifically carried out. Lectures are given on literature, the fine arts, &c., once or twice a week from November to May: the library is both large and well selected, and the reading rooms are supplied with all the English and foreign literary journals. The Russell Institution, in Great Coram street, is similar in most respects to those just described; but, owing to the recent falling-off in its funds, its usefulness is at present very much circumscribed.

The welfare and improvement, also, of the working classes, and of young men generally, has been greatly promoted within the last 15 years by the establishment of mechanics' institutions in different parts of London. The earliest of these, called, par excellence, "The Mechanics' Institute," in Southampton Buildings, Holborn (opened in 1824), has about 1100 members; and the attendance on the lectures delivered in the theatre of this establishment, shows that the inhabitants of the metropolis in humble life are quite as anxious for improvement as their more wealthy neighbours. Classes are established for languages, and the library, which comprises 7000 volumes, is said to be well selected. The Western Literary Institution, the City Institution, in Aldersgate-street, and other establishments of the same kind in various districts, have since been founded, and have uniformly contributed to improve the intellect and morals of the working classes.

Among the many endowed schools in the metropolis, the most celebrated are: 1. Westminster School, founded by Queen Elizabeth in 1560, for the free instruction, clothing, board, and lodgment of 40 boys, called king's scholars, and for the gratuitous education of four others called bishop's boys; this school formerly enjoyed a high reputation; but its numbers have within the last dozen years fallen from 300 to 100. The king's scholars are elected for merit, and four years after their election are sent as students to Christ-church, Oxford, or as scholars to Trinity college, Cambridge. The school forms part of the collegiate establishment of the abbey. 2. The Charterhouse (corrupted from Chartreux), founded in 1611, and endowed with property, the gross rental of which in 1815 was £20,000 a year, is intended for the liberal education of 73 youths, 29 of whom are supported at the universities by exhibitions varying from £80 to £100 a year, and tenable for eight years. Besides the 44 foundation boys, the school is attended by others, whose number fluctuates according to the reputation of the masters, &c. A few years ago, after the im-

LONDON.

provements introduced by Dr. Russell, this school had a very high character, and the pupils were very successful in the competition for honours at Oxford and Cambridge; but since his retirement the school has been in a languishing condition. 3. Merchant Tailors' school, founded in 1561, in Suffolk-lane, Thames-street. The statutes provide that a classical education be furnished gratis for 100 boys, and for 150 others at rates varying from 5s. to 2s. 6d. a quarter. The scholars are examined once a year, and the most advanced are sent to St. John's College, Oxford, where 37 valuable fellowships were founded by Sir Thomas White for the encouragement of boys brought up in this school. 4. St. Paul's school, established in 1515 by Dean Colet, and placed by him under the direction of the Mercers' Company, provides a free education for 153 boys, the most advanced of whom are sent to Oxford and Cambridge, with exhibitions varying from £30 to £100 in value. Lord Camden has been a liberal benefactor to this school. The present building was erected in 1824; the gross income of the school is upwards of £6000. 5. Christ's hospital, more commonly known as the Blue-coat school, is of the noblest institutions in the city. It was incorporated by Edward VI. in 1553, and owes its origin to the active benevolence of some distinguished citizens. It was intended to maintain, clothe, and educate the young and helpless; and 360 boys and girls were admitted soon after its foundation. A second charter from Charles II., in 1673, provided for the education of 40 boys in mathematics and other learning calculated to qualify them for the sea-service. The management of the institution is vested in a body of governors (472 in 1840), who have each contributed, at least £400, to the funds of the institution; but very recently the qualification for a governor has been raised to £500. An individual, on becoming a governor, is entitled to present one boy; and he has usually a presentation once every succeeding three years. The present (1840) revenue of the hospital, arising from rents, and all other sources, amounts to about £70,000 a year, and its expenditure to nearly as much. Its establishment is London, on the site of the Old Grey Friars' monastery, accommodates, at present, 787 boys; and it has attached to it a subsidiary establishment at Hertford, for the younger children, where there are now 494 boys and 76 girls; making in all 1367 children, maintained, clothed, and well educated by the establishment. There are schools for grammar, mathematics, writing, and drawing. The Grecians, or those most advanced in the grammar-school, are sent with valuable exhibitions to Oxford and Cambridge, and those in the mathematical school are placed with commanders of ships, and equipped with clothing and nautical instruments, at the hospital's expense. Others are apprenticed to different trades. A magnificent building, called the Great Hall, erected by public subscription, and finished in 1829, opens towards Newgate-street, and is one of the finest ornaments of the city. The hall, in which the children breakfast, dine, and sup, is 187 feet in length, 51 in width, and 46½ feet high. Occasionally they sup in public, and on these occasions there is a great concourse of strangers to witness the spectacle, which is of the most interesting description. The whole interior arrangements deserve the highest praise; and every attention is paid to the health and comfort of the children. The well-known dress of the boys, which has not been changed since the formation of the institution, is, however, not merely antiquated, but inconvenient and uncomfortable; and it is certainly high time that it were modified. Presentations can only be obtained from the governors; who, speaking generally, exercise their patronage with the greatest disinterestedness. 6. The city of London school, established in 1835, may be said to have resulted from the inquiries of the charity Commissioners. A Mr. Carpenter had left an estate for a school, and the value of the property had greatly increased without any proper application of the funds. Repeated inquiries and remonstrances at length induced the corporation to establish a school on the site of Honey-lane market, Cheapside. The system of instruction is good, and the school is attended by upwards of 400 boys. The buildings, occupying a space of 180 feet long and 80 feet broad, are commodiously contrived, and have externally some pretensions to architectural elegance. Independently of the endowed schools, almost every parish supports a free school by voluntary contributions, and thus about 19,000 children of both sexes are clothed and educated. The number of private and Sunday schools is considerable, but cannot be accurately estimated. The National society, which has its model school in the Sanctuary at Westminster, gives instruction in various schools to nearly 9000 children, and upwards of 3000 are taught in the Lancastrian method by the British and Foreign School society, which has a good normal and model school in the Borough road. Much, however, still remains to be done towards giving a sound elementary education to the children of the industrious classes; though,

at the same time, it must be admitted that more has been effected in this respect during the last 20 years than our ancestors had done during entire centuries. (*Carlisle; Educ. Rep.*)

The charges on account of education at most of the public schools in London are oppressively high, far higher, indeed, than they ought to be: and this circumstance, combined with the want of schools in many districts, and the wish to improve their health, has led to the practice, so general in London, of sending children to the countries of the town to be boarded and educated. But the education in very many of these boarding establishments is of a very worthless description; and it is really surprising that no effort should have been made, by subjecting the masters to examination, establishing proprietary boarding-schools, or otherwise, to improve the quality of these suburban seminaries.

British Museum.—This truly national institution, established in 1753, is a grand repository of books, MSS., statues, coins, and other antiquities, specimens of animals and minerals, &c., and is, in most respects, one of the richest in Europe. It is principally deposited in Montague-house, formerly the residence of the duke of Montague. Great Russell-street, Bloomsbury. The nucleus of the collection was purchased by government of Sir Hans Sloane's executors for £90,000, and the museum was first opened to the public in January, 1759. But Montague-house, though spacious as a private residence, has long been found inadequate to the proper accommodation of the vast and continually increasing collections that belong to the museum; and in consequence a new quadrangular building has been designed by Sir R. Smirke, a part of which is already completed, and open to the public. In 1755, the Harleian MSS. were purchased, and the Cottonian library was removed from Dean's Yard, Westminster: in 1757 the royal library, founded by Henry VIII. out of the libraries of the suppressed monasteries, and enlarged by his different successors, was presented by George II. George III., in 1763, gave a valuable collection of pamphlets on the civil wars, and between 1806 and 1818 the Lansdowne, Hargrave, and Burney MSS. were purchased at an expense of £36,408. Various presents have been made from time to time, but the most valuable addition of late years has been the library of George III., collected at an expense of £200,000, and presented to the museum by his successor.[*] Modern English publications are added, free of expense, in consequence of a privilege which this establishment enjoys in common with the two universities, and some other bodies, of receiving gratis a copy of every book entered at Stationer's Hall; and about £3000 a year are expended in the purchase of old and foreign works, in the latter of which, however, the library is still extremely deficient. There are about 330,000 printed books, and 27,000 MSS., exclusive of charters. The want of a catalogue raisonné is much felt by the great majority of persons who resort to the library for study or research. The reading-rooms are open from nine till four in the winter, and till seven in the evening during four summer months. The average number of readers is about 280 a day. Admission is procured by a recommendatory letter either to one of the trustees, or to the chief librarian; and every facility is given by the numerous attendants for the most extensive research. No books are allowed to be taken out, it being supposed that such permission would lead to frequent and heavy losses; but, provided the value of the books were previously deposited, we incline to think that certain descriptions of works might be lent out with great advantage. In the department of antiquities may be mentioned the collection of Egyptian monuments, including the famous Rosetta stone (see vol. i. p. 821), acquired at the capitulation of Alexandria, in 1801; the Townley marbles, purchased at £28,000; the Phigalian and the Elgin marbles, the cost of which was £35,000; the latter include the statues of Theseus and Ilissus, and the sculptures in alto-relievo, from the friezes of the Parthenon. The collection of minerals was, for many years, deficient in various important particulars; but the recent additions purchased from Messrs. Hawkins and Mantell are extremely valuable; and now, both for size and classification, this department will bear to be compared with any mineralogical collection in Europe. The department of zoology is rich in birds and insects, but poor in other respects, especially in mammalia. The collection of medals, which has been accumulating since the foundation of the museum, consists of about 20,000 coins, above 6000 being purchased with the Hamilton collection of Herculanean antiquities, in 1772. The coins can only be seen by an order from a trustee, or a private introduction to the officer to whose charge they are entrusted. The public days at the museum are Mondays, Wednesdays, and Fridays, when all persons

---
[*] It is much to be regretted that this library had not been placed in an accessible situation in the W. end of the town.

have free admission from 10 to four, and in the summer months from 10 to seven. The building is closed during the first weeks of January, May, and September. The establishment is governed by 48 trustees, 23 of whom are official; and to these the officers are responsible. The chief acting trustees, with whom the appointment of the officers rests, are the archbishop of Canterbury, the lord chancellor, and the speaker of the House of Commons. (*Parl. Rep. on British Museum*, 1835.)

Literary and Scientific Societies.—Before the present century the learned societies of London were few in number, and very comprehensive in their objects. The great advancement of the physical sciences, in recent times, and the increased ardour with which every branch of knowledge has been cultivated, have produced a corresponding increase in the number of learned associations, and in all recent instances each body has confined its operations within a limited sphere. The following list comprises the principal societies, with the dates of their formation, the objects contemplated by them, and the publications made at their expense:

The Royal Society; physical and mathematical sciences. Instituted early in the 17th century; incorporated 1663. "Philosophical Transactions," from the year 1665.

The Society of Antiquaries. Instituted 1717; incorporated 1751. "Archæologia," from the year 1770.

Medical Society. Established 1773. "Vetusta Monumenta," from 1747.

Society of Arts. Established 1754, for the encouragement of the arts, commerce, and manufactures of Great Britain, by granting rewards. "Transactions," from the year 1783.

Linnean Society; natural history. Established 1788; incorporated 1802. "Transactions," from the year 1791.

Royal Institution. Established 1799, for the application of science to the ordinary purposes of life. "Journal," from 1810.

Horticultural Society. Established 1804; incorporated 1809. "Transactions," from 1812.

Royal Medico-Chirurgical Society. Established 1805. Chartered 1831. "Transactions," from the year 1808.

Geological Society. Established 1807; incorporated 1826. "Transactions," from 1811.

Society of Civil Engineers. Established 1817; incorporated 1828. "Transactions," from 1834.

Royal Astronomical Society. Established 1820; incorporated 1831. "Memoirs," from 1822.

Medico-Botanical Society. Established 1821. "Transactions," from 1834.

Royal Asiatic Society. Established 1823; incorporated 1824. "Transactions," from 1827 to 1835; "Journal," from 1834.

Royal Society of Literature. Founded 1821; incorporated 1825. "Transactions," from 1827.

Zoological Society. Instituted 1825; incorporated 1829. "Transactions," from 1833.

Royal Geographical Society. Chartered 1830. "Journal," from 1831.

Entomological Society. Established 1833 or 1834.

Statistical Society. Established 1834. "Journal," from 1837.

Architectural Society.

Royal Institute of British Architects. Established 1835; incorporated 1838. "Transactions," from 1836.

Royal Botanic Society. Chartered 1639.

Nearly all these societies hold meetings twice a month, from November to June inclusive; at which papers are read illustrative of matters connected with the objects of each association.

The following Table, taken from Gilbert's "Clerical Almanack" for 1842, supplies several details with respect to some of the more important of these societies, on which, it is believed, considerable reliance may be placed: [See top of next column.]

Picture Galleries.—The present national collection of pictures is of recent foundation, and should only be looked upon as the nucleus of one that may hereafter be worthy of the country. It occupies the W. wing of the National gallery, erected 1834-37, at the public expense, on the N.W. side of Trafalgar-square, facing Whitehall and Parliament-street, unquestionably the finest situation in the metropolis. The building has a front of 500 feet, with a portico and dome in its centre, supported by Corinthian columns. But whether it were owing to the limited means placed at the disposal of the architect, or to some incapacity on his part, the fabric is neither worthy of its site, its object, nor of the country. Unfortunately, too, the defects of its exterior are not countervailed by any superiority of internal economy, the apartments for the exhibition of the pictures being miserably deficient in point of size, and ill-arranged. The pictures, which consist of the Angerstein

Name of Society.	Date of Institution or Incorporation.	No. of Ordinary Members.	No. of Foreign and Corresponding Members.	Total No. of Members.
*Royal . .	1660	2,501	764	64	828
*Antiquarian .	1751	1,217	701	39	740
Society of Arts .	1783	1,500	88	39	573
*Linnean . .	1788	500	673	38	576
*Royal Institution	1800	3,774	816	30	941
*Horticultural .	1804	6,590	1,360	362	1,722
*Medical and Chirurg.	1805	770	460	51	451
*London Institution	1807	2,500	900	1	901
*Geological .	1807	1,500	745	36	381
Royal Institution	1810	1,125	700	2	708
*Civil Engineers .	1818	1,111	186	105	351
*Astronomical .	1820	574	202	43	345
*Royal Society of Lit.	1823	736	164	36	160
*Asiatic . .	1823	597	667	752	879
*Zoological .	1826	14,504	3,011	164	5,175
*Geographical .	1830	1,462	651	60	711
Architectural .	1831	225	85	19	148
*Institute of British Architects	1834	600	126	70	215
Statistical .	1834	760	300	30	418
Entomological .	1834	276	200		
Camden .	1863	579	200	None.	160

* Those marked with an asterisk have charters.

collection, purchased in 1824, of Sir G. Beaumont's collection, given by him in 1826, and of others, partly presented and partly purchased, amounting in all to about 170, are arranged in five rooms, of such diminutive size, that they will contain only a few more pictures, and none of large size. About half the pictures belong to the Italian school; and of these the *Ecce Homo*, and the Mercury, Venus, and Cupid, of Correggio; the Raising of Lazarus, by Sebastian del Piombo; the Bacchus and Ariadne, of Titian; and the Holy Family, by Murillo, are reckoned the most valuable. The works of the two Caracci, N. and G. Poussin, and Claude, may be here seen in their highest perfection; and there are some fine specimens of the English school, by Reynolds, Hogarth, Gainsborough, Wilson, and Lawrence. The gallery is open to the public on the first four days of the week: on Friday and Saturday students are permitted to copy the pictures. The Royal academy, which at present (by permission of government) occupies the remainder of this edifice, was established in 1768, for the instruction of young artists: lectures are delivered in anatomy, painting, sculpture, and architecture, and daily instructions are given to the students by the keeper, and other academicians. The annual exhibition of this corporate society usually comprises about 1200 specimens of art, and is one of the favourite lounges during the summer months. The profits of the exhibition, besides paying all the expenses of the schools, contribute to form incomes for the most deserving artists, while studying at Rome. (See *Comm. Report on the Arts, &c., of* 1836.) The society of British artists exhibits annually a good collection of pictures; but, as a whole, they are very inferior to those exhibited by the academy. The British institution, and society of painters in water colours, have also exhibitions, and their rooms are crowded during the fashionable season.

Theatres and Music.—The great theatres of modern London present a curious contrast to the rude and confined buildings called the Globe, Blackfriars, and Old Drury, in the time of Shakspeare, in which neither scenery, decorations, nor the comfort of the audience were at all considered.

The two patent theatres, Drury Lane and Covent Garden, contiguous to each other, have handsome exteriors, and very extensive and highly decorated insides. They enjoy or rather are supposed to enjoy, the exclusive privilege of representing tragedy and comedy, the legitimate drama; but the declining taste of those who visit the theatres, and the caprice of managers, have led to the frequent introduction of spectacles, and other pieces, to which music and scenery contribute more than the actors on the stage. Late dinner hours and other circumstances have of late years occasioned a great falling off in the taste for theatrical exhibitions, which are now visited more by strangers than by residents in London. At present few theatres are profitable. The Haymarket theatre, which has recently enjoyed more than ordinary prosperity, is of smaller size, and therefore better adapted for hearing, than the immense houses above mentioned: it is open during about eight months of the year, including the recesses of the two patent theatres. Besides these, there are several minor theatres, the names, localities and objects of which are given in the table at the top of next page.

Among these, Astley's deserves particular notice, for the excellent horsemanship displayed by M. Ducrow and his *corps dramatique*: it is certainly superior to the Franconi theatre at Paris.

LONDON.

Names.	Localities.	Objects.
The Prince's Theatre	King-st., St. James's	Operas and farces
English Opera House	Strand	Operas and farces
Adelphi	Ditto	Spectacles and burlettas
Strand	Ditto	Burlettas
Olympic	Wych-street	Ditto
Queen's	Tottenham-crt. Rd.	Ditto
City of London	Norton Falgate	Melodrama
Garrick	Goodman's Fields	Ditto
Sadler's Wells	Clerkenwell	Ditto
Astley's	Lambeth	Melodr. and horsemans.
Surrey	Blackfriars'-road	Melodrama
Victoria	Waterloo-road	Ditto

The Italian Opera house, in the Haymarket is the largest theatre in London. It scarcely, however, deserves the name of a national theatre, inasmuch as the singers, dancers, and musicians are chiefly foreigners, and as it depends for its support chiefly on the patronage of the court, nobility, and higher classes, many of whom hold private boxes, at rents averaging from £120 to £400 a year. All the patronage of rank and wealth, however, cannot, owing to the enormous cost of the performances, make it a good speculation for the manager, who at the end of an anxious season has frequently to lament heavy losses. The established London concerts consist of the ancient, phil-harmonic, and sacred-harmonic concerts, all of which are well and fashionably attended: many others are given by professional persons, for their own benefit, in the different public rooms at the W. end. The promenade concerts lately introduced, in imitation of those at Paris, promise, by the high favour which they enjoy, to improve the musical taste of the people, which has undoubtedly been on the increase during the last few years.

Benevolent Institutions.—There are above 70 establishments in London for the cure of disease; of which, 27 are properly hospitals; 25 dispensaries, where medicine and advice are gratuitously administered; nine are infirmaries for special diseases; and 11 lying-in charities. There are also 18 asylums for orphans and otherwise destitute persons, and various other benevolent establishments. The principal are the following:

1. St. Bartholomew's hospital, in West Smithfield, was first founded in the 12th century, and refounded by Henry VIII. in 1546. The building, a spacious quadrangular structure, is principally modern, having been finished in 1770. It makes up 600 beds, and receives annually about 5000 in-patients, and 6000 out-patients. Necessity is the only recommendation to this institution; and patients are received without limitation. The medical staff is equal to any in the metropolis. The staircase was gratuitously painted by Hogarth. 2. Guy's hospital. St. Thomas's-street, Southwark, founded in 1721, contains accommodation for 500 in-patients, and has an excellent museum and theatre of anatomy. This magnificent hospital, which consists of two quadrangles and two wings, was founded and endowed by Thomas Guy, a bookseller, who expended £18,793 upon the building, and left £219,419 for its endowment—the largest sum, perhaps, that has ever been expended by any individual on similar purposes. Recently, however, Guy's hospital has met with another benefactor, but little inferior, in point of liberality, to its founder; a citizen, of the name of Thomas Hunt, having bequeathed to it, in 1829, the princely sum of £200,000! The medical school attached to this hospital, while under the superintendence of the late Sir Astley Cooper, was one of the most extensive, and probably, also, the best in the empire. 3. St. Thomas's hospital, in High-street, Borough, was formed out of two other charities by Edward VI., and re-built in 1693. Additions were made in 1732, and a large part was rebuilt in 1836. It contains 18 wards, and 485 beds. The annual expenditure is about £10,000. 4. St. George's hospital, near Hyde park corner, lately rebuilt, has a fine front, 200 feet in length, facing the Green park. It accommodates 460 in-patients. 5. The Middlesex hospital, near Oxford-street, founded in 1745, receives 300 in-patients, and relieves numerous out-patients. 6. London hospital, in Whitechapel, was founded in 1749. Its wards accommodate about 250 patients. 7. Westminster hospital, rebuilt in 1833, near the Abbey, receives 230 in-patients. The University College and King's College hospital, and Charing Cross hospital, are smaller establishments of the same nature, each accommodating about one hundred and twenty patients.

Medical schools are connected with the above hospitals, in which lectures are delivered by the officers, and which are attended, altogether, by about 1900 students.

Bethlehem hospital, or Bedlam, is appropriated exclusively to the insane poor; it was founded in 1675, in Moorfields, whence it was removed, in 1814, to St. George's fields. The present building, received some extensive additions in 1838, and is now 697 feet in length, being at once extensive and magnificent. The rooms are large and airy, well warmed and ventilated, and are sufficient for the accommodation of an immense number of patients. St. Luke's, City-road, established for a similar purpose in 1751, accommodates 300 persons.

The Foundling hospital, Brunswick-square, was founded by Capt. Coram, in 1739, but the building was not commenced till 1742. It was established for the indiscriminate admission of deserted children; but the numbers were found to increase so rapidly, that the funds failed, and in 1760 the mode of admission was so much altered, that it is now nominally only a Foundling hospital. The number of children averages about 450, and they are maintained till the age of 15, when they are either apprenticed or otherwise provided for. The revenue is about £13,000 per annum.

The Magdalen hospital, Blackfriars-road, was established in 1748, for the reformation of erring females: the object is said to be attained in the majority of cases, more than two thirds of the females admitted being either restored to their friends, or provided with the means of procuring an honest livelihood.

The Philanthropic institution, St. George's fields, was founded, in 1788, for the reception and reform of young criminals discharged from prison. It provides them with immediate means of subsistence, and instructs them in some trade, so as to prevent the otherwise almost inevitable necessity of their returning to their former habits.

Hotels and Taverns.—The hotels, taverns, and coffee-shops of all classes, may be reckoned at something more than 6000. There are about 30 great hotels, situated chiefly at the W. end of the town, in the neighbourhood of Piccadilly. "In these establishments," says Prince Puckler Muskau, "everything is infinitely richer and more abundant than on the continent." The commercial and other inns, amounting to nearly 400, are scattered throughout the metropolis. They are generally respectable establishments, some of them being quite as commodious, if not so elegant, as the fashionable hotels. The establishments of licensed victuallers, under which denomination are included all places for the retail sale of spirits, amount to about 5000. Many of these are respectably conducted, though some are of an opposite character. The publicans furnish their guests not only with beer and spirits, but also with dining accommodation, &c. The gin or dram shops have been very much embellished of late years; and many of them are so handsomely, and even splendidly, fitted up, that they have acquired and are entitled to the name of "gin palaces." But notwithstanding the number and magnificence of these establishments, there is no real room or ground for the prevalent opinion, as to the increase of intemperance. No doubt it is much too widely diffused; but it nevertheless admits of demonstration, that, as compared with the population, the consumption of spirits in the metropolis is now very decidedly less than in the reign of George II., and the greater part of that of George III.; and that there has been a corresponding improvement in the habits of the lower classes.

The eating-houses and coffee-rooms, where spirits are not sold, amount to about 700; and are more numerous in the city than in Westminster. There are about 600 beer-shops. Numerous private houses are let out in lodgings, and many families receive boarders. The expense of living in these establishments varies, of course, with the quality of the house and the means of the guest. A lodger at an inn can hardly be accommodated, on a decent scale, much below 10s. a day, including all expenses of board, food, and servants; the maximum of the scale will, of course, depend on the habits or caprice of the guest. Board and lodging in private houses may be obtained at a somewhat lower rate than at hotels; but a single man in lodgings usually dines at an eating-house, and families generally prefer boarding at their own cost. A dinner (without wine) at an ordinary eating-house costs from 1s. 6d. to 2s.; and seldom exceeds 5s. at the more elegant establishments. In most cases the guest may depend on every attention; and at the superior houses he will find all the luxuries of the season.

Clubs.—There are about 40 clubs in the metropolis. A few of these establishments, such as Brookes's, Boodle's, and White's are of ancient date; but their present arrangements and constitution are of recent introduction. The accommodation they afford to gentlemen only occasionally visiting town, and to others desirous of enjoying the luxuries of a splendid establishment, at a moderate expense, and of meeting with a great variety of society, has made them popular among the upper classes. The club-houses are mostly edifices of a very superior character; and add much to the magnificence of the squares and streets in which they are situated. Each club consists of a limited number of members, varying from 1000 to 1500; they are admitted by ballot, pay a certain sum at entrance, from 10 to 25 guineas, and an annual subscription, varying from 5 to 10 guineas. The club-houses are fitted up with every luxury of a fashionable hotel, have excellent libraries, take in the best periodical publications, and provide dinners, coffee, wines, &c., at reasonable prices. A large building has recently been opened, in Regent-street, in the view of supplying strangers, frequenting the clubs, with beds. Some of the clubs are avowedly of a political character, and others are devoted exclusively to certain classes. Among these may be specified the Curlton, Reform, City Conservative, United Service, Oxford and Cambridge, Traveller's, Oriental, and West India; but most clubs are open, on election, to all gentlemen without reference to party or profession. Most of the club-houses are at the W. end of the town, particularly in Pall-mall and St. James's-street. The building erected for the Reform club, by Mr. Barry, is the finest structure belonging to this class of edifices; and is fitted up with equal taste and magnificence. The city of London has two club-houses, which, in point of elegance and luxury, may vie with those of the W. end. The number of members in the different clubs may be about 30,000.

Courts of Law.—The Courts of chancery, Queen's bench, Common pleas, and exchequer (the respective provinces of which are described in the article ENGLAND AND WALES), occupy apartments on the W. side of Westminster hall. This hall, built by William Rufus, was long supposed to be the largest room in Europe unsupported by pillars. It measures 270 feet in length by 74 in breadth, and is 90 feet high; but these dimensions have been much surpassed by the great plate-glass hall of Ravenhead, which is 339 feet long, and 155 feet broad, with a proportional height. Westminster hall has been used for coronation banquets, the last of which was given when George IV. was crowned. Parliaments have often met in it, and it is occasionally appropriated to important trials; among which may be specified, that of Charles I., and more recently those of Warren Hastings and Lord Melville. Ordinarily, however, it is a mere promenade for lawyers during the sitting of the courts. The lord chancellor sits out of term-time in the hall of Lincoln's-inn, near which is another court, occupied by the Vice-Chancellor.

The Central criminal court, the jurisdiction of which extends to all places within 18 m. of St. Paul's, was established in 1834. Its sittings are held at the Old Bailey, a stone building close to Newgate, once a month, and generally last five or six days at a time. There are two halls, of confined dimensions, in both of which the judges are engaged in trying prisoners during the sessions.

The court of Bankruptcy is in Basinghall-street, within the city of London; the court for the relief of Insolvent Debtors, in Portugal-street, Lincoln's Inn Fields; the Maritime and Palace courts are in Scotland-yard, Charing Cross; the Ecclesiastical and Admiralty courts, in Doctors' Commons, near St. Paul's; courts of Requests, which give summary judgment in case of small debts, are numerous in different quarters. The jurisdiction of the several courts is explained in the article ENGLAND AND WALES, vol. i., p. 854.

Inns of Court.—The inns of court, originally colleges for legal study, are now little more than residences for lawyers, or indeed for any one who chooses to hire chambers in them. They are not incorporated, and, cannot, consequently, make bye-laws; but, by prescription, their customs have obtained the force of laws. A law student, before being called to the bar, has now only to be entered as member of one of these inns, and to dine a certain number of times in the common hall, in order to qualify himself for the exercise of his profession. This is termed "eating" his way to the bar.

The chief inns are the Inner and Middle Temple, originally built by the knights templars in Fleet-street, in the reign of Henry II.; Lincoln's inn, in Chancery-lane, Gray's inn. Subordinate to the Temple, are Clifford's, Lyon's, Clement's, and New inns. Furnival's inn belongs to Lincoln's inn; Staple's inn and Barnard's inn are attached to Gray's inn. These are called Inns of Chancery. Thavies inn, and some others, have become mere private residences.

Prisons.—There are 10 criminal prisons, of which three are in the city of London. 1. Newgate, under the control of the corporation, is a building, the architecture of which is singularly characteristic of its destination. Newgate was a prison early in the 13th century; but the present edifice was erected in 1778, and again repaired after the riots of 1780. This, which may be called the great metropolitan jail, contains accommodation for about 400; but occasionally upward of 1000 are crowded within its walls. In front of this prison all the criminals of London and Middlesex, capitally convicted, suffer the last penalty of the law. It is said to be one of the worst regulated jails in the kingdom, "a fruitful source of demoralization to its unhappy inmates, and a reproach to the character of the city corporation." (*Police Inspectors' 4th Report*, p. 251.) 2. The Bridewell,

near Blackfriars Bridge (once a royal palace), is a house of correction for vagrants, pilferers, or disorderly persons, summarily convicted before the lord mayor and aldermen. The number confined averages 100: the prisoners are classified, the silent system adopted, and the tread-wheel generally used. 3. Giltspur-street compter, opposite St. Sepulchre's church, a plain edifice with a stone front, is used as a place of confinement for untried prisoners, and as a house of correction for offences less grave than those of the inmates of Newgate. Its use is restricted to persons convicted by the city magistrates. It holds about 160, and there is no classification. 4. Clerkenwell prison, belonging to the county of Middlesex, is one of a similar character with the last. It serves, also, as an auxiliary to Newgate, receiving prisoners remanded from the police courts, or committed for trial at the general sessions. Its inmates average 180, and some attempt has lately been made at classification. 5. Cold-Bath-Field prison, a very extensive brick building, near Gray's-inn-lane, is a house of correction for Middlesex; and contains felons, misdemeanants, and rogues and vagabonds. It is an insulated brick building, containing spacious courts and airy grounds. The classification is good, and the silent system is followed, connected with hard labour. A large tread-mill employs 390 prisoners at a time. This prison accommodates upward of 1900. 6. The Westminster bridewell, begun in 1831 and finished in 1834, is surrounded by a lofty wall, with a complete roadway outside: it is built on the Panopticon principle, and has a courtyard in the centre 250 ft. in diameter, with prisons round it for 600 persons; but the average number confined is 350. The arrangement of the building is said to be excellent; and the window of the governor's house commands a complete view of all the day-rooms and yards, and of the two tread-wheels. Instruction is given to juvenile offenders. The silent system is adopted, and a good classification maintained. 7. The penitentiary, at Millbank, Westminster, built on the Panopticon principle, has no peculiar connexion with the metropolis, but is intended for the confinement and reformation of criminals whose sentence of transportation or death has been commuted. It contains accommodation for 1100 prisoners; but the number of inmates averages about 600. The building is insulated, and is surrounded by a wall enclosing 18 acres of ground. The penitentiary is managed by a committee, nominated by the secretary of state, and the chaplain is the governor. 8. The Surrey county jail is in Horsemonger-lane, Newington Causeway. It contains about 250 prisoners, debtors as well as criminals; and there is little classification. The top of the building is used as a place of execution. 9. The borough compter, in Tooley-street, is also a prison for debtors and criminals. The management and discipline of this prison are stated to be exceedingly defective: the average number of inmates is about 40. 10. The Brixton House of Correction is exclusively confined to prisoners sentenced to hard labour at the assizes and sessions, or by magistrates, under summary convictions. Hard labour and the silent system are rigorously enforced. The average number of prisoners is 280. (*Inspec. of Pris. Rep.*, 1836–37.)

The following Table, drawn up from the Jail Returns and Prison Inspectors' Reports for 1839, furnishes some details respecting the economy of the Metropolitan Jails:

Prisons	Power of Accommodation	Prisoners Mich. 1839.	Ann. Cost of each Prisoner	Total Annual Exp.
	Prisoners		£. s.	£.
City of London :				
Newgate	500	183	6 11	6,789
Giltspur-street	540	172	10 1	5,880
Bridewell	126		9 6	5,106
Borough compter	40	37	4 6	965
City of Westminster :				
Tothill Fields	1,120	840	4 16	7,904
Gen. Penitentiary	1,120	484	10 8	13,594
County of Middlesex :				
Clerkenwell	460	370	5 7	1,130
Cold Bath-Fields	1,500	1,142	7 6	17,042
County of Surrey :				
Horsemonger-lane	600	230	9 1	4,454
Brixton	590	284	12 11	4,291

The principal prisons for debtors are, 1. The Queen's Bench, in the borough, chiefly used for debtors on process from the court of Queen's Bench, but also for persons committed for libels, contempts, &c. It is a spacious healthy prison, containing 234 rooms, in which 500 persons have occasionally been confined at once. The rules of the bench, a district in which prisoners may purchase the liberty of residing out of prison, include a space of nearly 1 sq. m. 2. The Fleet prison, in Farringdon-street, is chiefly for debtors under process from the court of Common Pleas, and those committed under Exchequer process, and for contempts of the court of Chancery. The rules of the Fleet are not so extensive as those of the Queen's Bench. 3. White Cross-street prison, in the street of that name, in the city, is inconveniently built and wretchedly managed.

LONDON.

Its confined extent, when compared with the average number of the inmates, and the filth and disorder prevalent in every part of it, are not a little discreditable to the corporation of London. 4. The Marshalsea prison, in the borough of Southwark, is a small prison for persons committed by the Marshalsea court, and for debtors arrested under process in the Palace court. It is a quadrangular building, containing about 60 rooms, and accommodation for about 100 prisoners. The debtors' prisons have been comparatively deserted since the new act respecting imprisonment for debt came into active operation; but before that time they were often inconveniently and even unwholesomely crowded:

Crimes.—The crimes committed in London are both grave and numerous; but the degree of demoralization, as compared with the population, is not greater than in other places offering the same facility for successful depredation, and having an equal amount of poverty. The Report of the Constab. Comm. gives the following statement of depredators known to the metropolitan police in 1838:

Burglars and housebreakers .	217
Highway-robbers .	38
Pickpockets and common thieves	4,430
Coiners and utterers of base coin	345
Forgers .	3
Swindlers, &c.	390
Horse and dog stealers	152
Begging-letter impostors	136
Disorderlies, habitual	2,796
Vagrants .	2,295
Street prostitutes .	6,371

The annual average of convictions during several late years within the metropolis amounts to 3300 (or about 1 in 415 of the entire population), more than half of which are for slight crimes demanding six or three months' imprisonment. Capital offences, except murder, are now generally punished by transportation for life to the new penal settlement, Norfolk island. The executions in London have averaged two annually during the last seven years. The various crimes of the metropolis are undoubtedly on the decrease; and the frequent notification of them at present is more owing to the vigilance of the police, who detect and prosecute offenders, than to any actual increase of crime. (*Constab. F. Com. Report,* i., 13.)

The following Table exhibits the Average Annual Amount of Crimes in and about London during six years (1834-1839), and the Proportion which the Convictions in Middlesex and the four Metropolitan Counties respectively, as compared with their estimated Population in 1837, bear to those in England and Wales, as compared with its Population: the Convictions in the latter being considered as 1:

Classes of Crimes.	Average Annual Convictions.				
	Middlesex.	Ratio.	Metr. Co.	Ratio.	Eng. & Wales.
I. Offences against the person :					
Murder	1	·55	4·6	1·23	18
Manslaughter . . .	8.6	·97	18	·95	101
Attempts to maim, &c. .	2.6	1·46	18	1·37	70
Assaults	187	2·13	279	1·94	909
Child-stealing . . .	1.6	8·03	2	3·55	3
Sexual crimes . . .	20	1·64	47	1·37	187
Total of Class I.	218	1·92	328	1·47	1,188
II. Offences against property, with violence:					
Housebreaking and burglary	77	1·19	105	1·24	642
Robbery and attempts at do.	18	·90	87	·98	224
Total of Class II.	95	1·66	192	1·17	866
III. Malicious offences against property	·6	·10	4	·44	48
IV. Offences against prop. without violence :					
Horse and cattle stealing	25	·43	60	·60	248
Larceny, &c. . .	1,882	1·56	2,586	1·26	12,041
Embezzlement and fraud	104	2·04	155	1·88	506
Total of Class IV.	2,011	1·55	2,711	1·55	12,295
V. Forgery and coining	105	2·12	147	2·34	235
VI. Other offences					
Riots, &c. . . .	27	·67	45	·92	667
Smuggling . . .	nil.	1	1·34	4	
Poaching	nil.	1	7	·22	113
Keeping bad houses .	26·6	3·77	50	2·26	70
Various	14·6	1·40	30	1·40	76
Total of Class VI.	68	·99	105	·76	730
Average yearly amount of all convictions .	2,498	1·54	4,570	1·42	16,092

The increase of crime in and about London has been chiefly in attempts to maim and kill (82 per cent.), in sexual offences (33 per cent.), in frauds (73 per cent.), and in larcenies (24 per cent.): there has been a decrease in robberies of 22 per cent., in riots of 33 per cent., and in poaching, &c., of 117 per cent. Independently of the offences tried by juries, many others are summarily punished by the police magistrates, amounting annually to about 17,000, exclusive of fines for disorder and drunkenness. (*Off. Criminal Tables.*)

Police.—Till 1830, the police of London had the reputation of being the most defective establishment of the kind in Europe. A great reformation, however, has been effected within the last few years, and the metropolis is now, perhaps, superior in this respect to any other in Europe. There are 11 police offices, two of which are in the city, and one in Southwark. These are :

The Guildhall in the City.	Great Marlborough - street,
The Mansion House, do.	Oxford-street.
Bow-street, near Covent Garden.	Worship-street, Finsbury-square.
Hatton Garden, near Holborn.	Lambeth - street, White-chapel.
Queen-square, Westminster.	Union Office, Southwark.
High-street, Marylebone.	Thames Police, Wapping.

The first two of these offices are regulated by the city authorities; the rest are under the control of the secretary of state. Magistrates sit every day at each office, to hear and determine cases of misdemeanor and breach of the peace, as well as to examine and commit for trial all persons accused of felonies, to administer oaths, swear in constables, and to perform other magisterial functions. A number of officers is appropriated to each establishment, and to the Thames office a river police is attached.

The chief instrument of preserving the peace of the metropolis, is the metropolitan police, established in 1829. This body is dispersed over the whole of London, excepting the city, which is protected by a distinct body, of similar character, but less effective and worse disciplined. The city police is under the control of the corporation : the other force is governed by two commissioners, who communicate directly with the secretary of state for the home department. The whole body is divided into 17 companies, to each of which is attached a conveniently situated station-house : each company is placed under a superintendent, who has under him 4 inspectors, 16 sergeants, and 144 constables. Their duties reach beyond the metropolis, extending from Brentford on the W. to the borders of Essex on the E., and from Norwood on the S. to Highgate N. The expense is defrayed by an assessment limited to 8d. in the pound on the parish rates, the deficiency being made up by the treasury. The city, as before said, is not under the charge of the metropolitan police, but is protected by a body of men organized on the plan, and in imitation of the arrangements of that body, but placed under the city authorities. The city police, consisting of 500 men, is divided into six companies, to each of which belong inspectors, sergeants, and constables, and the whole is immediately under the control of a superintendent. All the constables, both of the city and metropolitan police, wear a blue uniform, with the number of each man, and a letter designating the division to which he belongs on the collar of his coat. They are constantly on duty, day and night; but the force is increased at night. The services conferred on the community by the metropolitan police may be in some measure estimated from some details furnished in the report of the constabulary force commissioners. It is there stated, that in the years 1836-38 they saved 200 lives, and prevented 152 suicides, rendered assistance at 2430 accidents and 775 fires, relieved or conducted to a place of safety, 1870 sick, insane, or otherwise helpless persons, and restored to their homes 2830 lost or strayed children : during the same period, they recovered from thieves, &c., £14,490, and from careless exposure and drunkards, £54,219. (*Constab. Force Com. Report,* i., 292.)

Pauperism and Mendicity.—London, with all its wealth, contains much misery and indigence, a large proportion of which, however, is attributable more to demoralization than mere misfortune. Since the Poor Law Amendment Act, most of the metropolitan parishes have placed themselves under its regulations, only 21 parishes still adhering to the old system of maintaining their poor. The money expended in maintaining the poor within the metropolis in 1839 amounted, according to the last Report of the Poor Law Commissioners, to £374,744. The mendicants, a class almost wholly separate from the paupers, pursue their vocation almost as regularly and with as much success as tradesmen. The Mendicity Society have laboured usefully to expose the impositions of mendicants; but neither their agents nor the new police have been able to suppress them. " Of the London beggars, nine out of ten are gross impostors and convicted vagrants; and of these the very worst are the blind and cripples. The records of the above society afford surprising proofs of the profligacy of the regular street-beggars, and the inveteracy of their idle and dishonest habits. The metropolitan police, in 1837, apprehended 4300 mendicants." (*Metr. Police Off. Rep.,* 1838) The really indigent are relieved by an excellent institution, entitled the Refuge for the Destitute, which provides a meal and a bed

234

for those who give satisfactory proof of requiring such assistance. The private lodgings of mendicants are crowded, unwholesome, and literally sinks of iniquity.

Water.—The supply of London with water was anciently procured from brooks running through the city. The increase of inhabitants made these sources insufficient; while, at the same time, they became less accessible, owing to the encroachment of buildings. To remedy this inconvenience, water was brought by leaden pipes in the 13th century from Tyburn, then a mere country village, into the city, where it flowed into conduits from which the inhabitants drew it at pleasure. In the beginning of the 17th century Sir Hugh Middleton projected, and, in despite of the greatest difficulties, carried into effect, in 1613, his plan for bringing the water of two copious springs in Hertfordshire to London, by an aqueduct, called the New River, 40 m. in length, including windings. The Thames has long been one of the great sources of supply; and, as early as 1581, waterwheels and other hydraulic machinery were established at London bridge. These wheels, which at one time raised 45,000 hogsheads per day, were wholly removed when the old bridge was pulled down. The greater number, however, of the existing water companies derive their supply from the Thames, the water being filtered in immense reservoirs. In 1834 the following account of the houses supplied with water, and of the quantity furnished to each, with the different rates of charge, was laid before the House of Commons:

Water Companies.	Houses supplied.	Total yearly supply.	Daily average to each.	Charge per house.
		Galls.	*galls.*	*£. d.*
New River Company -	70,145	114,625,000	441	17 3
Chelsea do.	13,609	13,755,000	408	27 0
Grand Junction do.	8,780	21,700,000	367	33 6
W. Middlesex do.	16,000	20,000,000	125	45 6
E. London do.	46,451	37,015,504	121	22 0
S. London do.	22,940	8,400,000	100	44 4
Lambeth do.	16,604	11,300,000	174	17 0
Southwark do.	7,000	7,000,000	100	21 0
Total water supply	99,006	237,017,761		

In every street in London there are fire-plugs or cocks, at any of which a copious supply of water should be obtained in a few minutes in case of fire; though it must be admitted that the supply has sometimes, through neglect, been very long delayed, to the great injury of property. Much of the water is also used in watering the streets and improving the drainage; indeed, scarcely a third part of the supply is used for purposes strictly domestic. Abundant springs of the finest water may be procured in all parts of London, by boring below the clay strata; but no public measures have yet been taken to ensure a supply from this source. (See *Metr. Water Rep.*, 1840.)

Sewers.—The sewers of London constitute a system of drainage unknown to any other modern city; and, though out of sight and hardly appreciable by others than engineers, they have excited the astonishment of all who have investigated the subject. Their depth is, in most cases, sufficient to drain the deepest cellars in each neighbourhood, and the size of the main branches rival that of the celebrated Roman Cloaca Maxima. We have no means of ascertaining their length; but that portion under the commissioners of the city is 15 m. in length, and an extent of 100,000 ft., or nearly 20 m., was constructed under the Westminster board between 1807 and 1834. The sewerage, however, is still imperfect in some low neighbourhoods, especially about Wapping, Stepney, Bethnall-green, Westminster, &c., and wherever this is the case, the health of the poor is greatly deteriorated. The malignant fevers that occasionally make fearful ravages in poor districts are, indeed, mainly attributable to the absence or bad state of the drains. (See *Rep. of Commissioners on Health of Towns, and Dr. Arnot's Evidence.*) The sewers are under the authority of seven boards of commissioners, whose jurisdiction extends to a circle of 10 m. radius, measured from the post-office. The first sewers were constructed in 1498.

Paving.—The streets of London are not only well paved for carriages, but they have also on both sides, for the accommodation of pedestrians, smooth and usually wide flagged footways, raised some inches above the carriage way. This advantage it enjoys in common with most English towns; but few cities on the continent are provided with a similar convenience, though Paris has in some measure followed the example. In streets wide enough to admit of it. The paving is under the control of numerous boards, each of which has its particular district. It is conjectured that the amount expended in paving the streets of London exceeds £200,000 per annum. Pavement was first laid down in the metropolis in 1417, in Holborn. In 1615, the plan of having footways of broad stone was begun, but it was not universal until the middle of last century. For some time past the principal streets have been paved with granite,

mostly brought from Aberdeen. Very recently, however, portions of them have been paved with wood; and how singular soever it may appear, it is believed that this pavement will be more durable than granite, at the same time that it will be incomparably more advantageous, by lessening the wear and tear of carriages, the quantity of dust, and the noise.

Lighting.—The metropolis is excellently lighted with gas, even in its most remote and secluded parts. Without going back to the year 1416, when lanterns were first hung out before citizens' houses, or even three centuries later, when an act passed to compel housekeepers to light up a lamp for five hours during the dark nights; many may remember the old oil lamps, which were said by a foreigner to "edge the streets with two long lines of brightish little dots indicative of light, but yielding very little." M. Simond somewhat exaggerated the deficiency; but still the difference between the old and present plan of lighting is so great as to make it difficult to believe that the inhabitants could have been satisfied with the oil lamps, "few and far between," that are still to be seen in the less populous parts of the suburbs. Gas was first tried in London in 1807, but with little success, as no means had then been discovered for removing its impurities. Pall-mall had been for some years the only street thus lighted, when, in 1816, a charter was obtained by a gas company, which slowly but certainly extended its operations. The profit of this speculation led to the formation of other companies; but it was not till 1820 that any considerable portion of the metropolis adopted gas. From that period, however, public bodies and private traders began rapidly to introduce it into their establishments, and parochial boards adopted the luminous gas-jets in lieu of the sickly, glimmering oil-lanterns. There are now 19 gas companies, who may probably produce, at an average, 10,000,000 cubic feet of gas every 24 hours. The number of lights is variously estimated: in 1823 it was stated before a committee of the House of Commons, that the number was nearly 60,000. Several companies have since been formed, and some of the old ones have doubled and trebled their produce, so that the whole number at present may, perhaps, be reckoned at 100,000.

Fires.—London has suffered from fire oftener, perhaps, than any other capital, except Constantinople; but the precautions taken in rebuilding the city, after the great fire of 1666, were calculated to prevent the recurrence of such a calamity. The streets were made much wider, bricks and stones were substituted for wood, and party walls were built between adjacent buildings. At a subsequent period the Building Act (14 Geo. 3, c. 78) compelled the erection of thick party walls between the separate tenements, and obliged each parish to keep one or more fire-engines always ready for service. The various insurance offices also began to maintain fire-engines at their own expense, attended by bodies of well-disciplined firemen; and in 1825 some of the largest of these establishments entered into an arrangement, by which all their force was put under one superintendent. The fire-brigade association was gradually joined by the other offices, and at present all the London insurance offices contribute to support this most efficient establishment. One superintendent now guides the whole, aided by about 100 foremen and engineers, who are placed at 18 different stations in all parts of the town and suburbs. The firemen, who are all numbered, wear a uniform of dark gray and a strong leathern helmet; a third part of their body is always on duty, and they are provided with the best means of extinguishing fires, and rescuing persons in danger. The average number of fires for the five years ending with 1837, was 495 per annum; of which number 340 were slight, and 29 only extremely destructive.

Health.—"The metropolis has in itself all the elements of a healthy city. If the tides leave the banks of the Thames exposed, that great river sweeps through the city from W. to E., and the winds rush fresh over its waters. The land rises in undulations to Hampstead heath, and the Surrey hills; pure water is abundant, and would flow under almost every street; the artificial heat and gas, noisome as it sometimes is, ascends in a vast column to the sky, and is replaced by under-currents from the surrounding country." (*App. to Regist. Gen. 2d. Rep.*)

But notwithstanding these favourable circumstances, London was long exceedingly unhealthy, and down to 1666 was hardly ever free from the plague. This excess of mortality was, no doubt, occasioned by the wretched state of the town. The streets were then narrow, crooked many of them unpaved, and generally filthy: the houses, built of wood and lofty, were dark, irregular, and ill-contrived: each story projected over the one below, so that they almost met at the top, thereby precluding, as much as possible, the escape of foul and the access of pure air: the shops were also furnished with enormous signs, which being suspended crosswise in the middle of the street, tended still further to prevent ventilation: the sewers were, at the same time, in a very imperfect state, the drains which conveyed away the filth not being arched over, but running above ground; and if we add to this the deficiency of water, and the prevalence of sluttishness in-doors, which then existed to an extent not easily to be imagined,[*] we need not wonder at the ravages made by the plague and other diseases. (See *Maitland's Hist. of London*, passim; and *Heberden's Tract on Diseases*, p. 71.)

In 1593, the deaths by the plague within the bills of mortality amounted to 11,503: in 1603 to 30,561; in 1623 to 35,403; in 1636, to 10,400; and in the dreadful pestilence of 1665 they rose to 68,596! And it is impossible to say how soon it might have again burst forth, had it not been for the severe but providential visitation of the great fire by which it was immediately followed, and which, by destroying the most crowded and ill-built parts of the city, afforded an opportunity, which was luckily embraced, of rebuilding them on a better and more commodious plan. Very severe regulations were then also laid down for the enforcement of cleanliness; and the supply of water being at the same time augmented and better distributed, and the drains greatly improved and arched over, London has not since been visited by any very destructive epidemics. Still, however, the mortality during the first half of last century was very great; and Short, Corbyn Morris, Price, and other well-informed writers of the period, indulge in bitter complaints of the severe drain on the country, occasioned by the waste of life in London. The population appears, indeed, to have declined between 1740 and 1750; and during the 10 years ending with 1768, the deaths appear to have amounted at an average to 22,596 a year, while the births did not exceed 15,710. (*Price*, vol. ii., p. 26.) Probably, however, some portion of this enormous discrepancy is apparent only, and may be accounted for by omissions in the registers of births. But it is, notwithstanding, abundantly certain that the deaths very materially exceeded the births at the period referred to; and that they preserved this ascendancy down to a much later period. The mortality in 1765–1775 was estimated at about 1 in 20, or 5 per cent. of the existing population; but from this period a very material change for the better began to take place. In 1790 the births, for the first time, exceeded the burials; and during the ten years ended with 1800, there was an excess in the total number of births of 51,000 over the total number of deaths. This excess has since continued to increase, so that it is plain, supposing no unfavourable change to take place, that London might go on adding indefinitely to her population, without drawing a single recruit from the country.

The following table represents the mean annual mortality *per cent.* in the metropolis, and in England and Wales, from 12 classes of disease. (*2d Rep. App.*, p. 13.)

Classes of Disease.	Metropolis.	England and Wales.
1. Epidemic Endemic Contagious	·747	·652
2. Nervous system	·497	·382
3. Respiratory Organs	·770	·398
4. Circulating do.	·045	·024
5. Digestive do.	·160	·129
6. Urinary do.	·013	·011
7. Generative do.	·008	·081
8. Locomotive do.	·021	·014
9. Integumentary system	·004	·003
10. Uncertain	·285	·296
11. Old age	·219	·237
12. Violent deaths	·075	·081
All causes of death	2·500	2·103
Population to 1 sq. mile	90,803	369

At present the average rate of mortality in London is estimated at about 2·6 per cent.; but the degree of mortality varies widely in different districts, increasing with the poverty and bad accommodations, and diminishing with the wealth and improved accommodations of the inhabitants. Thus, while the annual rate of mortality is 2·9 per cent. in Whitechapel, and 3·2 in Southwark and St. Giles's, it is less than 2·0 in St. George's, Hanover-square, and St. Pancras. "It is found, indeed, from a comparison of the several metropolitan districts, that, *ceteris paribus*, the mortality increases as the density of the population increases, and that where the density and the population are the same, the rate of mortality depends on the efficiency of the ventilation, and of the means employed for the removal of impurities." (*App. to Regist. Gen. 1st Rep.*) Epidemic diseases in crowded parts of London are attended with nearly

double the mortality that belongs to them in more airy districts; and diseases of the respiratory system are increased a half in close neighbourhoods. Mr. Farr's statements in the Report of the Registrar-general, as to the importance of ventilation and drainage, are fully corroborated by Dr Southwood Smith, Dr. Arnot, and other authorities.

Summer is the healthiest, winter the most fatal season, and this rule has prevailed since the beginning of the last century. The deaths out of 100 living (1838) averaged in Jan., Feb., March, ·85; in Apr., May, June, ·70; in July, Aug., Sep., ·60; in Oct., Nov., Dec., ·66.

Increase of Population.—It is much to be regretted that there are no accurate accounts of the population of London previously to the census of 1801. The population of the city was, however, estimated by Graunt, the well-informed author of the famous *Treatise on Bills of Mortality*, at 384,000 in 1661, and adding 1-5th to this for the population of Westminster, Lambeth, Stepney, and other outlying parishes, he estimated the entire population at about 460,000. (*Observations*, &c., 5th ed., p. 82, and p. 105.) In all large towns, except (as in Petersburg) there be a great excess of military, the number of females is, in modern times, found invariably to exceed that of males; but, if we may depend on Graunt's estimate, the reverse was the case in the city of London at the epoch referred to, for he makes the number of males 199,112, and of females only 184,896 (p. 83). In 1696 the population of the city and the out parishes was carefully estimated, by the celebrated Gregory King, at 527,560; and considering the great additions that had been made to the metropolis between the Restoration and the Revolution, this increase does not seem to be greater than we should have been led to infer from Graunt's estimate. The population advanced slowly during the first half of last century, and, indeed, as already stated, it fell off between 1740 and 1750. In his tract on the population of England, published in 1782, Dr. Price estimated the population of London, in 1777, at only 543,420 (p. 5). But there can be no doubt that this estimate, like that which he gave of the population of the kingdom, was very decidedly under the mark; and the probability seems to be that, in 1777, London had from 640,000 to 650,000 inhabitants.[*] Its population amounted, including Chelsea, as has been already seen, to 898,198 in 1801, and to 1,506,469 in 1831; and in 1841 it exceeded the prodigious sum of 2,000,000 (see the commencement of this article)—the greatest number of human beings ever, we believe, congregated within the same space in any age or country.

"———— Opulent, enlarged, and still
Increasing London! Babylon of old,
Not more the glory of the earth than she,
A more accomplished world's chief glory now."

London is, no doubt, principally indebted for her extraordinary rise and unexampled magnitude, to her admirable situation, on a great navigable river within a short distance of the sea, and in the centre of a rich and fertile country, of which she is naturally the emporium. Her river enables her to obtain abundant supplies of all the bulkier descriptions of products, not only from all parts of the United Kingdom, but also of the world, at the lowest possible cost. The advantages thence resulting have been great and obvious. A city in an inland situation never could have attained to anything like the colossal magnitude of London. Indeed, almost all great cities, in all ages of the world, have been built either on the seashore or on the banks of some great navigable river. Paris is probably the largest city that ever existed without any very great command of water carriage. But her advance has been slow compared with that of London; and notwithstanding the advantage she has long enjoyed, from being the capital of a powerful monarchy, and the residence of a polished and luxurious court, her population is not, at this moment, half that of London.

The extraordinary growth of the latter during the present century seems to be mainly attributable to the same causes that have increased wealth and population in other parts of the empire, that is, to the progress of arts, manufactures, and commerce. Though not in the manufacturing districts, London is now, by means of canals, railways, and other improved means of communication, intimately connected with them; and the many advantages she enjoys as a trading and commercial port, will always secure for her a large share of the shipments of manufactured products. London has also derived a vast accession of influence from her being the place where the dividends on the public debt are paid, where all transfers of stock are effected, and where all the important pecuniary transactions of the empire are ultimately adjusted. And how paradoxical soever it may at first sight appear, it is certainly true that the very magnitude of London is an efficient cause of her continued increase. The greater a city becomes, the greater is the

---

[*] Erasmus, who visited England in the reign of Henry VIII., and was well acquainted with the country, ascribes the prevalence of the sweating sickness (a species of plague) and the plague to the incommodious form and bad exposition of the houses, the filthiness of the streets, and the dirtiness within doors. In a letter to Cardinal Wolsey's physician, he says, speaking of London, "*Conclavia sola fere strata sunt argilla, tum scirpis palustribus, qui subinde sic renovantur, ut fundamentum maneat aliquoties annos viginti sub se fovens sputa, vomitus, mictum tum canum et hominum, projectam cervisiam, et piscium reliquias, aliasque sordes non nominandas.*"

[*] See the Tracts of the Rev. Mr. Howlett and of Mr. Wales, in answer to Dr. Price.

scope she affords for the exercise of every talent and acquirement, and for the gratification of every taste and desire; and the more powerful, consequently, are the motives by which she attracts all sorts of individuals, whether aspiring or careless, industrious or idle, grave or gay, virtuous or profligate.*

Vast as London is, the chances are, should the country continue to prosper, that she will continue to increase in magnitude for centuries to come; and the progress she has already made, unprecedented as it has been, may, not improbably, be surpassed by that which she is yet destined to make.

Habits.—The peculiarities of character belonging to the inhabitants of London must be learned from studying the manners of the middle and lower classes, as the higher classes, who reside here only during four or five months in the spring and summer, and leave whole districts almost uninhabited during the rest of the year, can scarcely be entitled Londoners. Great activity and unwearied diligence in business, a shrewd perception of character, and an ever-watchful regard of self-interest, not unmingled, however, with generosity, are the chief characteristics of the native population. Owing to the extreme subdivision of employment, and the undivided attention which most individuals give to their own pursuit, the citizens are, for the most part, singularly expert in it, and proportionally ignorant of everything else. This, however, is less so now than formerly; the extensive circulation of cheap publications having diffused information as to many topics of which the bulk of the population had formerly the most imperfect ideas. The leading merchants and tradesmen have generally houses in the outskirts of the town, and in the country, to which they retire after business hours during summer; and many, indeed, occupy these houses during the whole year. But to the inferior shopkeepers and tradesmen, summer and winter make little difference. Some, indeed, rusticate at a hot and dusty watering-place for a week or fortnight, trying, often in vain, to rid themselves of the turmoil and anxiety of business; but of a far larger number it may be said, as of John Gilpin, that

"For twice ten tedious years still they
No holyday had seen."

The London tradesman, unlike the Parisian, is essentially domestic in his habits, and his visits to his friends or club are only the exception to his ordinary regularity. Most classes of skilled workmen receive high wages, which having been little, if at all reduced since 1815, they are at present much better off than during the war. Their circumstances vary, of course, according to their prudence: few save money, but all live well, using butchers' meat to an extent unknown anywhere else, and dressing, on holydays at least, in a style equal to that of the classes above them. Many, not contented with one holyday in the week, keep a second, known as "St. Monday," sometimes spent in drunken revelry, but more frequently in country excursions with their families: others, however, work unremittingly from one year's end to the other, content with an annual Easter or Whit Monday's trip to Shoeness in a steamer, or with a pic-nic excursion to Hampstead or Blackheath. In the bright mornings of summer large social parties may be seen leaving town in vans, provided with good fare, and not unfrequently with a minstrel ready to furnish music for a dance on the green. The lowest class of all, whose means of existence are precarious, disreputable, or dishonest, have pleasures peculiar to themselves; but cleanliness and respectability of appearance are little studied by them, and they are almost unknown to the rest of the people, except when their wants or delinquencies intrude them on the public notice.

Environs.—The metropolis is surrounded by a country of varied surface and great productiveness. The ground on the E., W., and S. extends in a flat along the river, which is prevented from inundating it at high water by extensive embankments, probably constructed by the Romans. But on the N. the ground rises gradually to an elevation of 300 or 400 ft., and the flat on the S. is also bounded by grounds which attain to a like elevation. The picturesque hills of Surrey, near Dulwich and Norwood, are studded with the villas of wealthy merchants and others, who retire here from the bustle of town; and Blackheath, more to the E., and nearer the river, though not so fashionable as in the days when Greenwich had a palace and a court, is still a

favourite resort in summer, and the buildings have increased since the railway has furnished fresh facilities for communication with London. N. of the metropolis lie Hampstead and Highgate, both of which, owing to their height, command extensive views of Hertfordshire, Surrey, and other counties: these villages, as well as Hornsey, Stamford hill, and Walthamstow, are filled with respectable residences, chiefly occupied by persons who daily visit London in pursuit of business. This prevalent fashion among the wealthy Londoners of fixing their abode in the suburbs has been greatly encouraged by the easy communication afforded by the numerous omnibuses and coaches which run to and fro at all hours of the day, and till late at night. Owing to this circumstance, the population of the city proper has decreased considerably since the commencement of the present century; indeed, it may now be called a collection of shops and warehouses rather than of residences for families. The suburban villas vary in size and grandeur, according to the means of their proprietors, but comfort and neatness are their universal characteristics.

Corporation.—The city of London is under the government of the lord mayor, two sheriffs, 25 aldermen, 206 common councilmen, a recorder, and other officers, and is divided, for municipal purposes, into 26 wards, each of which is under the government of an alderman. The Saxon denomination for the governor of London was *portgref* or *portreve*, which, about a century after the Conquest, was changed to mayor. This officer was appointed by the crown till 1215, when the citizens obtained the right of electing their own mayor. The mode of election now followed was fixed in 1475 by an act of the common council.

The lord mayor is annually chosen from the body of aldermen, at a court held at Guildhall on Michaelmas day, and is sworn in to the duties of his office on the 9th of November following. A grand pageant takes place on the occasion, followed by a sumptuous dinner and bell held at the Mansion-house. In most instances, the alderman next in seniority to the lord mayor is elected his successor. He is always free of one of the great city companies, and must have served the office of sheriff. The lord mayor is second only to the sovereign within the city, and at the sovereign's death he takes his seat at the privy council, and signs before every other subject. His powers are similar to those of a lord-lieutenant of a county, and his authority extends over the whole city and a portion of the suburbs.

The division of the city into wards appears to have been made very early in the 13th century; there were 24 wards, which became 25 in the year 1393, by a division of the ward of Farringdon. In 1550 a great part of the borough of Southwark was formed into a ward, and called Bridge Ward Without; but it is now merely a nominal ward, giving a name to the senior alderman, who, on the occasion of a vacancy, is removed to it from his own ward, and is then called "the father of the city."

The following is an alphabetical list of the names of the wards, with an indication of their situation, the number of common councilmen, and the number of houses in each :*

1. Aldersgate, on both sides of Aldersgate-street, including the Post-office. Com. coun. 8; houses, 726.
2. Aldgate, at the E. end of the city, includes the E. ends of Leadenhall-street, and Fenchurch-street, and Crutched-friars, called Alegate in the old list of 1285, given by Maitland. Com. coun. 8 (6); houses, 770.
3. Bassishaw (corrupted from Basinge's-haugh) includes little more than Basinghall-street. Com. coun. 4; houses 130.
4. Billingsgate, from Billingsgate market to near Fenchurch-street. Com. coun. 10 (8); houses 343.
5. Bishopsgate, both sides of Bishopsgate-street. Com. coun. 14; houses 1460.
6. Bread-street, E. of St. Paul's, and S.W. of Cheapside. Com. coun. 12 (8); houses 290.
7. Bridge Within, London bridge and Fish-street hill, includes the Monument. Com. coun. 15 (8); houses, 198.
8. Bridge Without, part of the borough of Southwark.
9. Broad-street, between Bishopsgate ward and Coleman-street, includes the Bank; this is, apparently, the Lothberry of the ancient list. Com. coun. 10 (8); houses 560.
10. Candlewick, between Lombard-street and London bridge, named from Cannon-street, which was formerly called Candlewick-street. Com. coun. 8 (6); houses 210.
11. Castle Baynard, from St. Paul's to the Thames. Com. coun. 10 (8); houses 542.
12. Cheap, both sides of the E. end of Cheapside and the Poultry, including Guildhall. This is probably Ward Fori in the ancient list. Com. coun. 12 (8); houses 360.
13. Coleman-street, includes Lothbury, part of London wall, and Finsbury circus. Com. coun. 6 (8); houses 761

* The clear and comprehensive account given by Seneca of the motives which drew so great a concourse of people to imperial Rome, applies without the alteration of a syllable to London: "Aspice agedum hanc frequentiam, cui vix urbis immensae tecta sufficiunt. Ex magnâ cipite et coloniis suis, ex toto denique orbe terrarum confluxerunt. Alios adduxit ambitio, alios necessitas officii publici, alios imposita legatio, alios luxuria, opulentum et opportunum vitiis locum quaerens; alios liberalium studiorum cupiditas, alios spectacula: quosdam traxit amicitia, quosdam industria, latam ostendendâ virtuti nacta materiam: quidam venalem formam attolerunt, quidam venalem eloquentiam. Nullum non hominum genus concurrit in urbem, et virtutibus et vitiis magna praemia ponentem."—*Consolat. ad Helviam*, cap. 6.

* At first each ward sent two councillors, but the number has been gradually increased, till it reached 240 in the whole: but a regulation very recently made will reduce the number to 206; the alteration will take place at the next election (1840), and the changes are indicated by being placed between brackets.

14. Cordwainers, S.E. of Cheapside; includes Bow church. Com. coun. 8 (6); houses 315.

15. Cornhill, a small ward on both sides of Cornhill, includes the Exchange. Com. coun. 6; houses, 167.

16. Cripplegate, reaches from Wood-street, Cheapside, to the boundary of the city on the N.; it includes Fore-street and the Barbican. Com. coun. 16; houses 2072.

17. Dowgate, between Southwark bridge and London bridge, includes Merchant Tailor's school. Com. coun. 8 (6); houses, 203.

18. Farringdon Within, includes St. Paul's cathedral, part of Cheapside, Newgate-street, and Ludgate-street, and reaches the river near Blackfriar's bridge; this and the following are the "Lodgate and Newgate" of the old list. Com. coun. 17 (14); houses 1008.

19. Farringdon Without, includes Smithfield, the Old Bailey, the Fleet, part of Holborn, and the whole of Fleet-street. Com. coun. 16; houses 3030.

20. Langbourne, includes Fenchurch-street, a part of Lombard-street. Com. coun. 10 (8); houses 500.

21. Lime-street, includes the East India House, and a small space around it. Com. coun. 4; houses 190.

22. Portsoken, eastward of Houndsditch and the Minories. Com. coun. 5 (8); houses 1916.

23. Queenhithe on the river, W. of Southwark bridge. Com. coun. 6; houses 350.

24. Tower, from Tower hill to Billingsgate, includes the Custom house. Com. coun. 12 (8); houses 530.

25. Vintry, on the Thames, and both sides of Southwark bridge. Com. coun. 9 (6); houses 380.

26. Walbrook, S. of the Mansion house, includes the Mansion house, and the church of St. Stephen's, Walbrook. Com. coun. 8 (6); houses 366.

Houses in the whole, 16,466.

The aldermen are chosen by such householders as are freemen, and pay an annual rent of £10. Each alderman is elected for life, and has the direction of the business of his ward, under the superintendence of the lord mayor. They are all justices of the peace within the city. The sheriffs are elected every year, on Midsummer day, by the corporation and freemen, and are sheriffs of the county of Middlesex, as well as of the city of London; they enter on their duties, and are sworn in at Westminster on Michaelmas day. The common councilmen are chosen by the householders in all the several wards, except Bridge Without. The common councilmen are the representatives of the inhabitants in the "Court of Common Council," which is composed of the lord mayor, aldermen, and common councilmen. This court disposes of the corporation funds, makes laws for the regulation of the city, and nominates certain of the city officers. Its sittings are usually public, and its title is "honourable."

The Livery consists of freemen of the city, who are also free of one or other of the city companies. Each of these companies was, at its formation, intended to comprise the different individuals within the city, properly so called, engaged in the peculiar department of industry called by its name; and had powers to enact bye-laws, and to lay down regulations for the government of the trade. Thus, for example, no one could commence business within the city of London as grocer, mercer, or goldsmith, without being free of the grocers', mercers', or goldsmiths' companies. And this freedom could only be acquired by inheritance, serving an apprenticeship to a freeman, or paying a fine, or otherwise, as the company might choose to order; and, after admission, all individuals had to conform in the conduct of their business to the rules and regulations laid down by the company. But the inconveniences of this system gradually became obvious; and it has, in consequence, been so much modified that the privileges of the different incorporated companies no longer oppose any obstacle to individuals from distant parts of the country establishing themselves in business within the city, nor interfere in any degree with the management of their concerns. In fact, any one who pleases may now purchase at Guildhall a license entitling him to trade within the city for £5, without being free of, or having anything to do with any company. The city companies have, in truth, become charitable rather than political, or even municipal institutions. Some of them have a great deal of property. The principal companies obtained very large grants of land in Ulster during the reign of James I.; and most of them are trustees for sums of money and other property bequeathed by benevolent individuals. They expend their revenues partly in festivities, but principally in pensions to widows and decayed brethren, the support of schools, &c. There are in all 81 companies, of which 40 have halls, where they transact business, keep their records, and hold festivals. Some of these halls are very fine fabrics; that of the goldsmiths, in Foster-lane, rebuilt since 1831, is a magnificent structure; and, were it in a situation where it could be seen, would be one of the principal ornaments of the city.

238

The following 12 are called the *Great Companies*, and from one or other of them the lord mayor must be elected:

Mercers.	Merchant Tailors.
Grocers.	Haberdashers.
Drapers.	Salters.
Fishmongers.	Ironmongers.
Goldsmiths.	Vintners.
Skinners.	Clothworkers.

There are about 12,000 liverymen, in whom, previously to the passing of the Reform Act in 1832, the right of returning the four members of the House of Commons for the city was exclusively vested. A Common Hall is an assembly of the liverymen, called together at the requisition of a considerable number of their body: the lord mayor is the president by right of office.

The Guildhall, where the corporation meetings, festivals, and common halls are held, stands at the N. end of King-street, Cheapside. Having been much damaged in the great fire of 1666, it was replaced by the present edifice, constructed of the materials of the old building. The front, added in 1789, is in a heterogeneous style. The great hall, 153 ft. in length by 48 in breadth, and 53 in height, built and paved of stone, is capable of accommodating 6000 persons; at least that number were present at the grand entertainment given by the corporation to the allied sovereigns in 1814. At each end of the hall is a magnificent painted glass window, in the pointed style; but the roof is flat, panelled, and inappropriate; and the whole requires to be renovated and made consistent with the original character of the building. In the hall are statues erected by the corporation in honour of Lord Chatham and his son, the Right Honourable William Pitt, Nelson, and Alderman Beckford. On the pedestal of the latter is inscribed the famous reply made, in 1770, by Beckford, who was then lord mayor, and one of the members for the city, to the answer of his majesty (George III.) to an address and remonstrance of the common council. At the W. end of the hall are the two wooden giants called Gog and Magog, the subject of so many popular tales. In the council-chamber, where the lord mayor, aldermen, and common council hold their courts, is a statue of George III. by Chantrey; it has also a library containing books of reference, relative chiefly to the history of London, and the affairs of the city, and various other rooms for the use of the corporation.

The city has its peculiar courts of law, most of which are held in the Guildhall. The lord mayor's court, for actions of debts and trespass, and for appeals, is composed of the lord mayor, recorder of the city, and aldermen. The sheriffs hold courts of record four days every week. The Chamberlain's court, held daily, decides disputes between masters and apprentices, and admits qualified persons to the freedom of the city. Courts of Petty Session for small offences are held daily at the Mansion house, by the lord mayor and an alderman, and at the Guildhall by two aldermen. There are also several minor courts.

The revenues of the corporation of London amounted, according to the commissioners' report, to £132,035 in 1831, and to £160,194 in 1832. These large funds are derived from rents of houses and land, market-tolls, bequests, interests on government securities, and a few other sources. The expenditure in the year 1831 was £149,411, and in 1832 £169,256: the chief items consist of salaries to municipal officers, maintenance of police and prisons, corporation entertainments, purchase of securities, and payment of debts. The lord mayor has £8000 a year allowed him to support the dignity of his office, and a splendid official residence. This, which is called the Mansion-house, stands at the E. of the Poultry, nearly opposite the Bank. It is a large structure, begun in 1739, and finished in 1753, in the Palladian style, with a fine Corinthian portico, on a lofty rustic basement. The grand or Egyptian hall (in which, however, there is nothing Egyptian!), the ball-room, and the saloon, are magnificent apartments; but some of the private apartments, occupied by the lord mayor, are but indifferently lighted. The plate used at civic entertainments belongs to the corporation, and is very valuable.

The city of Westminster was anciently governed by the abbot; but since the Reformation it has been under the authority of civil officers nominated by the dean. The chief magistrate is the high steward, generally a nobleman, who holds the office for life; the next is the high bailiff, chosen by the high steward, who also holds the office for life. There are 16 burgesses, whose offices are similar to those of aldermen, each having jurisdiction in a separate ward; out of these are elected two head burgesses, one for the city and the other for the liberties, who take rank after the high bailiff; each burgess has an assistant: there is also a high constable, who has authority over the other constables. The court of Quarter Sessions is held at the Westminster town-hall four times a year. The court of St. Martin's-le-Grand is held for the trial of personal actions relating to

LONDON.

that part of the liberties. The court-leet is held under the authority of the dean, for choosing officers, removing nuisances, and similar matters.

Southwark was anciently governed by its own officers, but since the year 1327, it has been for many purposes subject to the lord mayor, who governs by a steward and bailiff, the former of whom holds a court of record every Monday at the Town-hall in the Borough High-street. Another court is held at Bankside for the Clink liberty, a mean, densely peopled district, to the westward of London bridge.

Parliamentary Representation.—Down to the passing of the Reform Act, in 1832, the metropolis sent eight members to the House of Commons, viz., four for the city, elected by the liverymen; two for the city of Westminster, elected by scot and lot voters; and two for the borough of Southwark, also elected by scot and lot voters. In addition to the above, the Reform Act created four new boroughs, out of parts of the metropolis not included in the former boroughs, viz., those of Marylebone, Finsbury, the Tower Hamlets, and Lambeth, giving to each two members. Hence the metropolis now returns 16 members to the House of Commons, elected by the £10 householders, and those previously in possession of the franchise. Subjoined is an

ACCOUNT of the Parliamentary Boroughs in the Metropolis, with the Number of their Representatives, and the Electors registered in each in the Year 1839-40:

Places.	Members.	Electors. 1839-40.
London (City of)	4	19,064
Westminster	2	14,354
Marylebone	2	11,823
Finsbury	2	13,974
Tower Hamlets	2	13,551
Southwark	2	5,047
Lambeth	2	6,547
Total	**16**	**83,922**

Historical Notice.—Nothing is known of London previously to the invasion of the Romans; and it may be doubted, from the silence of Julius Cæsar, whether it then existed, or, at all events, whether it had attained to any considerable magnitude. But, however this may be, it is clear, from the statement of Tacitus (*Annal.*, lib. XXXIII., cap. 14), already referred to, that so early as the reign of Nero it was an important emporium, though not distinguished by the title of a colony; and it is doubtful whether it ever attained to this honour.

After the Romans had left Britain, and the Saxons had divided the country among themselves, London is supposed to have become the capital of the E. Saxon kingdom. On the introduction of Christianity into England, it was one of the first places to embrace the new faith, and early became a bishop's see. St. Paul's, and St. Peter's, in Westminster, were first founded about this time. In the paucity of intelligence concerning the period of the heptarchy, all we hear of London is, that it suffered severely from fire in 764, 798, and 891, on each of which occasions it is said to have been nearly destroyed. As soon as England had been united under one monarch, it appears to have become the metropolis of the empire; and, in 833, a wittenagemot, or parliament was held in it to consult on the best means of repelling the Danes, who were ravaging the eastern counties. It was, however, sacked by the Danes in 839: in 982 it was nearly destroyed by fire; and in 994, the inhabitants purchased a temporary remission from the attacks of the Danes, by paying them a high ransom.

At the Conquest, London submitted to William, and soon after received a charter in the English language, the original of which is still preserved. Within the 60 years following the Norman conquest it suffered severely by fire on five different occasions; but being then built principally of wood, it was easily repaired from the timber furnished by the extensive forests of Islington and Hornsey, which still existed when Fitzstephen wrote in the succeeding century. London was then unpaved, and, if we may believe the statement of contemporary historians, the rafters of the roof of Bow church, which were blown off by a hurricane in 1091, struck into the ground to a depth of 20 feet. The same hurricane caused so high a tide in the Thames, that the wooden bridge, which had stood 200 years, was carried away by the stream. On the accession of Henry I. in 1100 a new charter was granted to the city, which restored its ancient privileges, as they existed before the Norman conquest, relieved the inhabitants from many oppressive services, such as compulsory entertainment of the king's household, and abolished several barbarous customs of the Saxon period. The citizens acquired by this charter the privilege of choosing their own magistrates. The Norman monarchs, it is true, seldom respected corporate privileges, even when conceded by themselves; but still this charter was valuable as furnishing a standard to which to refer in future disputes with the prerogative, and it is said to have served as the model

from which Magna Charta was taken. About the middle of the 12th century, it was determined to build a stone bridge over the Thames. The first wooden bridge having, as already stated, been carried away in 1091, was replaced by another, which was burned down in 1136. The bridge erected instead of the latter became so ruinous in less than 30 years, that it was thought a stone bridge would be less costly in the end than the continual repairs required to keep up these unsubstantial, though cheaper structures. The new bridge, begun in 1176, and finished in 1209, was a noble work for the time, and may be said to have been the very bridge taken down in 1832, though frequent alterations, additions, and repairs, had considerably impaired its identity. Three years after its erection a dreadful loss of human life was occasioned by a fire on the bridge, described in Stowe's Chronicle: "The tenth of July at night the city of London upon the S. side of the river of Thames, with the Church of our Ladie of the Canons in Southwarke, being on fire, and an exceeding great multitude of people passing the bridge, sodainely the N. parte, by blowing of the S. winde, was also set on fire, and the people which were even now passing the bridge, perceiving the same, would have returned, but were stopped with fire, and it came to passe, that as they protracted time, the S. ende was fired, so that people thronging themselves betwixt the two fires, there came to side them many ships and vessels, into the which the multitude so undiscreetly pressed, that the ships being drowned, it was saide, there were destroyed about three thousand persons." About this time an order was made by the court of aldermen that no house should in future be built without party walls 3 feet in thickness, and 16 feet in height. This order, dated in 1181, was intended to obviate the frequent fires by which London had so often been partially destroyed; but it appears to have been little, if at all, attended to, and is interesting principally from its being the first document in which the chief magistrate of London is designated lord mayor. He had hitherto been called the chief bailiff.

In the year 1211 the citizens began to form a deep ditch, 200 feet wide, without the city wall on all sides, as a means of defence against King John. In 1218 the forest of Middlesex was cleared, and the citizens of London were permitted to purchase land, and build there. Thus was begun that part of the metropolis which stands N. of the city, and is now so populous. In 1221, Henry III. laid the first stone of the present Westminster Abbey. In 1236, the first water pipes were laid down in the city, which had previously been supplied with water from wells and rivulets running through it into the Thames; but these had gradually been either obliterated or converted to common sewers. The principal was the River of Wells, which, though hidden, still runs under what was lately Battlebridge, in the New Road, passes down through Clerkenwell and Turnmill-street, and falls into the Thames close by Blackfriars bridge. This river may be yet seen, or at least heard, through the iron gratings by which it is covered, in some of the obscure streets in Clerkenwell. This stream at low tide pours into the Thames a mass of black mud on the N. side of Blackfriars bridge. Holbourne, now a common sewer running through Holborn Hill, fell into the river of Wells. Walbrook entered the city by Winchester-street, and fell into the Thames at Dowgate: its name still remains. Langbourn ran with a swift stream from Aldgate to a spot near the Mansion house, where it fell into Walbrook. Several of the wells remain to this day: Holywell, in Shoreditch; Clerkenwell, in Clerkenwell, Agnes le Clear, Perilous Pond, &c. The city, however, had gradually so much increased, that the supply was inadequate to the demand; in many cases, too, the new houses had encroached upon the little streams, and made it inconvenient to approach them. The water pipes now laid down brought a copious stream from six springs at Tyburn, a village on the site of the present Oxford-street, near its W. end, and conveyed it to one main pipe, 6 inches diameter, to the city of London. Afterward various leaden cisterns, named conduits, were constructed for the reception of water: the first of these was built in Westcheap (now Cheapside) in 1285; these were taken down, not only because increasing traffic rendered ground valuable, but because they had become comparatively unnecessary, from the introduction of a mode of supplying water to every tenement.

In 1930 several buildings were added to the Tower, one of which was appropriated to three leopards, presented in 1235 by the Emperor Frederic to Henry III., who, a few years after, assessed the city at the rate of 6d. per day for the maintenance of these animals, and 1½d. per day for that of their keeper. In 1282, during a great frost, such masses of ice were brought down the Thames, that five arches of London bridge were destroyed. In 1304 the first recorder was appointed. In 1326, in consequence of the facility with which felons made their escape from London across the bridge into the adjoining village of Southwark, which, until then, was beyond the mayor's jurisdiction, Edward III.

LONDON.

granted a charter assigning this village to the city forever, and empowering the city magistrates to act in Southwark as in London. This jurisdiction still continues.

As we advance in time records of events crowd so thick upon us, that we can only notice such as have produced permanent effects; and in so doing, we must proceed rather by centuries than years. Of fires, pestilences, famines, and riots, it may be enough to say that they were extremely numerous; but except the great fire of 1666 none had any lasting consequences.

In 1354 the office of alderman, which had hitherto been annual only, was rendered more important by a law which made the aldermen irremovable for life, unless on some especial cause. In 1381 the rebellion of Wat Tyler, and his death by the hands of the lord mayor, occasioned the addition of the drawn dagger to the coat of arms of the city, where it still appears. During this century many improvements were effected in cleaning and paving the streets, and clearing out the water-courses and great city ditch; but an effectual bar was raised to these measures in 1392, when, in consequence of the refusal of a loan of £10,000 to King Richard by the corporation, the mayor, sheriffs, aldermen, and principal citizens were imprisoned, heavy penalties exacted, the city franchises abrogated, and the courts removed to York. Heavy bribes effected a removal of several of these grievances; but the city did not recover its proper influence till the accession of Henry IV.

In the following century the progress of improvement was still more rapid. Lamps were lighted at night in the streets so early as 1416. Holborn, a part of the Strand, and other principal thoroughfares were paved; additional conduits and water-pipes were laid down; the old wooden houses began to disappear, and were replaced by respectable edifices of brick: and the city wall was repaired. The bricks used for these purposes were made in Moorfields. The slaughtering of cattle within the walls was forbidden in consequence of the bad effects produced by it in the absence of sewers. An indication of some attention to the police of the city appears in the erection of stocks in every ward, for the punishment of disorderly persons. Guildhall, Leadenhall, and Crosby-house, in Bishopsgate-street (a portion of which has recently been restored), were built in this century.

In the 16th century the advance was much greater; an unusually long exemption from those civil wars which had so much injured London under the Plantagenets, gave leisure to introduce those improvements which distinguish a modern town from a city of the middle ages. The city watch was now improved, nuisances were removed, street paving became more general, and regulations were made for supplying the town with provisions, so as to prevent the frequent famines which had before arisen, more from defective arrangements than from real want. The removal of monasteries had also a great effect in improving London: 54 large and many smaller establishments made way for factories, schools, charitable asylums, and hospitals. St. James's palace was built, the park was laid out, and many fine buildings were erected in Westminster. The two cities were now first joined by a number of handsome mansions belonging to the nobility on the N. side of the river; one of which, Northumberland house, is yet in existence. The streets S. of the Strand still indicate by their names the site of others that have disappeared. The Royal Exchange was built, and commerce began to flourish. Towards the end of this century water was first conveyed by machinery into private houses, and the New river was projected.

In the 17th century London assumed its present form, with the exception of that part destroyed by the great fire of 1666. Spitalfields was covered with houses; and before 1666 the space N. of the Strand as far as Holborn, and from Temple-bar to St. Martin's-lane, had been extensively built upon. The parts of Westminster, also, from Charing Cross to St. James's palace, had assumed the appearance of a town. The New river had been completed, and each house was supplied with water. Sewers were dug, smooth pavements were laid down for foot passengers, and hackney-coaches became general.

On the 2d of Sept., 1666, the great fire broke out at Pudding-lane, near the spot where the monument was subsequently erected in commemoration of the occurrence. It raged till the 5th, when it ceased, rather by pulling down houses in the line of its course, than by the success of the exertions to extinguish the flames. The ruins, covering 336 acres, comprised 13,200 houses, 90 churches, and many public buildings; the property destroyed being estimated at £10,000,000. But, though productive of great loss, and of much temporary distress and inconvenience, this conflagration was, in its results at least, of signal advantage. It would have been all but impossible, except by some such destructive agency, ever to have got rid of the vast mass of old wooden houses, and narrow and filthy lanes and courts, that had for centuries been the permanent abode of the plague and other pestilential diseases. No doubt it must ever be regretted, that the

designs of Sir Christopher Wren for the renovation of the city were not adopted. But notwithstanding the numerous defects of the new plan, it was a vast improvement on that by which it had been preceded. Though still too narrow, the streets were materially widened; the new houses were constructed of brick instead of wood; party walls were introduced; the old practice of making each story project over that immediately below, was abandoned; obstructions and filth of all sorts were removed; and the sewerage and pavement of the streets were vastly improved. A fire which happened in Southwark ten years afterward, afforded an opportunity for carrying the improvements into that part of the metropolis. The population and trade of the city now increased more rapidly than before. The revocation of the Edict of Nantes occasioned the immigration of a great number of French, who settled in Spitalfields and St. Giles's. The parishes of St. Anne and St. James were formed, the district called the Seven Dials was built, Piccadilly began to extend W., and Soho-square and Golden-square were laid out. In the city the Bank of England was built. St. Paul's Cathedral was almost completed; the parish of Wapping was formed E. of the city; the Penny Post-office was instituted; the number of hackney coaches trebled; and several miscalled asylums (as Alsatia and the Mint), where robbery and crime had been protected, were abolished.

From this period the improvement and increase of London have been constant. In the early part of the 18th century an act was passed for building 50 new churches in and about London, most of which were completed within a few years, and some of them are still among the ornaments of the metropolis. Houses sprang up on every side; and by the middle of this century the W. end of the town, as far as Hyde park, became a compact mass of building, reaching beyond Oxford-street on the N., and extending E. from Portman-square, across Tottenham-court-road, past Montague house and Gray's Inn gardens, through Clerkenwell, Finsbury-square, Spitalfields, and Whitechapel to Wapping. Before this time water-works had been formed at Chelsea, in aid of the supply furnished by the New river. Sewers had become more general, lamps had been fixed in all the principal streets. Westminster bridge was built, St. Paul's completed, and Fleet ditch arched over. In the last half of the century Blackfriars bridge was built, the houses encumbering London bridge were removed, the Mansion house was finished, and Somerset house erected. At the same time many unsightly and inconvenient buildings were removed; overhanging signposts, water-spouts, which occasionally drenched unwary passengers, dirty, stagnant gutters, ash-heaps, and other obstructions to walking the streets, were swept away; the lamps were much increased in number, and lighted during the whole night; raised footways became universal, and the shops, which before were mere stalls, assumed a size and splendour evincing the wealth of their occupiers, and greatly contributing to the ornament of the metropolis.

The citizens of London have, generally speaking, been distinguished by their orderly behaviour and respect for the laws. In 1780, however, the peace, and even, in some degree, the existence of the metropolis, were compromised by the excesses of the mob. Certain concessions made in the course of the previous year to the Roman Catholics, had provoked a good deal of religious excitement in all parts of the kingdom. The contagion spread to London; and the weakness of the government, and the folly, or rather madness, of Lord George Gordon, and other leaders of the ultra Protestant party, led to a dangerous riot. The mob were, in fact, for about two days masters of the city. They took possession of the prisons, and turned the inmates out of doors; destroyed the chapels of the ambassadors of the different Catholic powers; many private houses, including that of Lord Mansfield, were plundered and set on fire; a great distillery, belonging to a Catholic firm, shared the same fate; and an attack was made on the bank, which, however, was happily repelled. At length this formidable riot was effectually put down, though not till a considerable number of the rioters had been killed and wounded. Since this disgraceful epoch, the peace of the city has not been seriously endangered; and the troops in and about town, added to the effective police force that now exists, seem quite adequate, under ordinary circumstances, to ensure the public tranquillity and the safety of the peaceable part of the community.

During the last 30 years London has made greater advances than could reasonably have been expected in an entire century. Within that period four bridges have been built, extensive docks have been excavated, gas has been introduced into every street and alley, steam, both on the river and on railways, has given it an almost unlimited power of intercourse with every part of the kingdom, and of the world; new and handsome markets have been erected, arcades lined with elegant shops have been formed, and wide lines of communication have been opened through close and densely crowded neighbourhoods. A new park, larger and handsomer than any of the other three, has been laid out, and surrounded with houses more resembling palaces than private residences ; an improved police has given additional security both to person and property ; abundant supplies of water have been furnished to every separate dwelling : the paving and sewerage have been greatly improved, especially in districts inhabited by the poor; and the formation of spacious cemeteries in the suburbs is gradually leading to the disuse of interments within the town. At the same time the establishment of colleges and proprietary schools has materially increased the facilities for procuring good education, while the institution of a national gallery and school of design are contributing to improve the national taste, and to add to the innocent pleasures of the people. The spirit of improvement, moreover, is still suggesting extensive and noble works. The act (2 and 3 Vict., c. 80) founded on the Reports of the Metropolis-improvement Committees, furnishes ground for hope that the time is not far distant when several new and large streets shall be constructed "in districts at present secluded from the observation of the wealthy and educated, and exhibiting a state of moral and physical degradation much to be deplored." One of these streets is to form a straight line of connexion between Oxford-street and Holborn ; another is intended to join Long Acre with Broad-street, St. Giles's ; a third is to open a wide line through the alleys about Leicester-square to the W. end of Long-acre ; a fourth is to run from the N. end of Farringdon-street to Clerkenwell-green ; and a fifth will form a direct communication between the London docks and Spitalfields church. "The melioration of the moral condition of the labouring classes" will also be vastly promoted by the project for laying out, on the N.E. and E. sides of London, spacious pleasure grounds and public walks, the enjoyment of which will, there is reason to think, be, at no very distant period, deemed superior by the bulk of the lower classes to the debasing revelry of gin-palaces.

Antiquities.—London possesses few antiquities : its wall is destroyed, its gates have been demolished ; the remains of its monasteries, colleges, and friaries were obliterated at the great fire ; and modern improvements have swept away almost every vestige of olden times. The diligent inquirer, however, may still find a few remnants, though generally either modernised or renovated. The Tower, Westminster Abbey, the Temple church, and St. Saviour's have been already mentioned. The priory of St. Bartholomew, founded in 1102, remains in the parish church, in Smithfield ; the hospital of St. John of Jerusalem has a gateway remaining in Clerkenwell ; the church of the Augustine friars may be seen at Austin friars, near London wall; and portions of the churches E. of the limits of the great fire, may still be seen incorporated in the modern edifices. Several fragments of the wall exist in Cripplegate churchyard, in Allhallows churchyard, Wormwood-street, and near Tower hill. Some ancient houses remain in Bishopsgate-street, the handsomest of which, Crosby hall, has recently been restored to its pristine state, and now exhibits an excellent specimen of a civic palace of the 15th century. London stone, near St. Swithin's church, Cannon-street, is supposed to be the point whence the Romans measured the roads in Britain. Much of it has been chipped off by curious antiquaries, and what remains is enclosed within a niche to prevent farther dilapidation. There are also some ancient crypts, or arched vaults under several private houses in the city ; some of them are handsome, and have been either subterranean chapels, or vaulted chambers belonging to religious houses.

LONDON, p. v., Union t., capital of Madison co., O., 27 m. W. by S. Columbus, 420 W. It contains a brick courthouse, a jail, county offices, several stores, about 80 dwellings, and 560 inhabitants.

LONDON, p. v., capital of Laurel co. ,Ky., 94 m. S.W. Frankfort, 544 W. Situated on a branch of Laurel river, a tributary of Cumberland river. It contains a courthouse, jail, and several stores and dwellings. Nett proceeds of the postoffice, $60.

LONDON, p. t, Monroe co., Mich., 45 m. S.W. Detroit, 502 W. It contains one grist-mill, two saw-mills, one tannery ; four schools, 78 scholars. Pop. 425.

LONDON BRITAIN, t., Chester co., Pa., 35 m. S.W. Philadelphia. Watered by White Clay creek. It has two stores, one grist-mill, three saw-mills, one tannery ; two schools, 100 scholars. Pop. 641.

LONDONDERRY, a marit. co. in the N. of Ireland, prov. Ulster, having N. lough Foyle and the Atlantic ocean ; E. Antrim, from which it is separated by the lower Bann and lough Neagh ; S. Tyrone ; and W. Donegal. Area, 518,270 acres, of which 136,038 are unimproved mountain and bog, and 9,565 water, being mostly included in the portion of lough Neagh, belonging to this county. Surface in some parts mountainous and uneven ; but there is, notwithstanding, a great extent of low, fertile ground. With the exception of lands belonging to the church and to corporations,

the entire property of this county was granted by James I. to 12 of the principal London companies, from whom most part of the land is now held, partly under terminable and partly under interminable leases. Farms vary in size from 2 to 900 acres ; but the average may be from 5 to 20 acres. "Where there has been a perpetuity or a long lease, it is *split*; that is, the children are settled upon divisions of the father's farm ; by which means leases of 40 acres come to be parcelled, in two or three generations, into patches of four or five acres. It seems as if the newly let lands were disposed of under some similar system of parcelling. I could give instances where whole districts are subdivided into patches of six or seven acres, and rarely can boast a farm of 12 or 14." (*Sampson's Survey of Londonderry*, p. 249.) Some landlords have exerted themselves to counteract this wretched system, but hitherto without much effect. It is almost superfluous, seeing the way in which the land is subdivided, to say that agriculture is in a very backward state. Latterly, however, some improvements have been effected. Oats, potatoes, and flax are the principal crops ; but a good deal of wheat is now also raised. Condition of the small farmers and cottiers very unprosperous. Average rent of land, 12s. 2½d. an acre. Various minerals have been discovered, but they are of no great importance. The linen manufacture was, a few years ago, widely diffused, but has latterly been a good deal contracted, the mill-spun yarn being cheaper and better than that spun by hand-wheels. Exclusive of the Bann, the principal rivers are the Foyle, Faughan, and Roe. Principal towns, Londonderry, Coleraine, and Newton-Limavady. The county is divided into six baronies and liberties, and 31 parishes ; and sends four members to the H. of C., being two for the county, one for Londonderry, and one for Coleraine. Registered electors for the county, in 1839–40, 3676. In 1831 Londonderry had 39,077 inhabited houses, 41,239 families, and 222,012 inhabitants, of whom 106,657 were males and 115,355 females.

LONDONDERRY, or DERRY, a city, parl. bor., and river-port of Ireland, cap. co. of same name, and a county by itself, advantageously and beautifully situated on the W. bank of the Foyle, about 5 m. above where it falls into lough Foyle, 121 m. N. by W. Dublin ; lat. 54° 59' N., long. 7° 19' W. Pop. of parl. bor. in 1831, 14,030. The city was originally confined to the hill on which the greater part of it still stands; and which, from its projecting into the river, is called the "Island of Derry." This portion is surrounded by the old city walls, but it is now rapidly extending beyond its former limits, particularly along the river toward the lough. There is also a suburb on the opposite bank of the river, called Waterside. The communication between the latter and the city is kept up by means of a wooden bridge, 1068 feet in length and 40 feet wide, erected in 1789 at an expense of above £16,000, and rebuilt in 1814–15 at a farther cost of £16,801. Derry is well built ; many of the houses in the main streets within the walls are old-fashioned, with high pyramidal gables ; but many modern mansions have, of late years, been erected in this part of the town ; and without the walls rows of mud cabins have been superseded by respectable houses. The principal city streets are broad and clean, well paved, and well lighted ; some of them, however, are inconveniently steep, and there are many narrow lanes and closes. In the centre of the city is an open square space, called the Diamond, from each side of which a handsome street leads to one of the four city gates. The summit of the hill is crowned by the cathedral, courthouse, and bishop's palace.

The cathedral, which is also the parish church, was built in 1633 : it is a large, handsome, Gothic structure, 240 feet long, and has a tower and spire 228 feet high, erected in 1778 ; but this having become dangerous, was taken down in 1802, and was soon after rebuilt, with the addition of Gothic pinnacles. The view from the top is very fine. In the interior is a handsome monument to the late Bishop Knox ; and in it also are displayed the colours taken at the siege of Derry. The bishop's palace is a large, plain building, with extensive pleasure-grounds. There are two other Protestant episcopal places of worship, the chapel of ease and the free church. The latter, which is without the city, was built in 1830, by Bishop Knox, and was intended for the use of the poorer classes, but it is no longer confined to them. There is also a Roman Catholic chapel, and places of worship for Presbyterians, Primitive and Wesleyan Methodists, Seceders, Covenanters, and Independents. Among the public buildings, exclusive of churches and other ecclesiastical edifices, may be specified the Corporation hall, in the centre of Diamond-square ; it was originally constructed in 1692, but received so thorough a repair in 1823, as to be tantamount to a re-erection. The courthouse, adjoining the cathedral, erected in 1813, at an expense of £30,480, is a spacious and a fine building, partly constructed on the model of the temple of Erectheus at Athens. The new jail is a very extensive structure, being 243 feet in front, by 400 feet in depth ; it is

P*

built on the radiating or panoptic principle, and cost above £30,000. Among the chief ornaments of the city is the fluted column, erected, in 1827, in honour of the Rev. George Walker, its heroic defender. It stands on the central W. bastion, and is a well-proportioned pillar, 81 feet in height, bearing a statue of Walker, 9 feet high. It cost £4200, raised by subscriptions.

The principal school, called Foyle college, stands on an eminence near the river; it is a plain, but handsome, building, erected in 1814, having accommodations for 80 resident pupils; it was built by subscriptions from the bishop of the diocese, the Irish society, and other sources, and is maintained by similar means: the head master's salary, from these sources, is about £300 per annum. There is a parish school connected with the church, for the education of 100 boys and 100 girls. The Presbyterian congregation also supports a free school: a school, called St. Columb's national school, was established by the Roman Catholic bishop and clergy, but it is now under the National Board of Education. In 1839 a Mr. John Gwynn left the munificent sum of above £40,000 for the education of as many boys as the funds will afford in the useful parts of a good English education; and, exclusive of the above, there is an infant school, and a number of Sunday and other schools. Among the charitable institutions is the district lunatic asylum for the counties of Londonderry, Donegal, and Tyrone; it was opened in 1829, and cost £25,678; it can accommodate about 190 patients; the number admitted in 1837 was 180, of whom 100 were males and 80 females; the expense of the establishment for that year was £2906, being £15 10s. for each patient. Here is also an infirmary and fever hospital, a dispensary, a charitable loan fund, a mendicity association, and a clergymen's widows' fund, with several minor institutions of a similar description. Among the literary institutions is the Literary Association, with a reading-room and a pretty good library; the Literary Society, in which lectures are given, and discussions take place; a news-room; a mechanics' institute, &c. In 1840 the town had three newspapers. Races take place on a course in its neighbourhood. The citizens of Derry would seem to have but little taste for theatrical entertainments; at all events, the theatre has been converted into a coach-building establishment.

The walls or ramparts by which the city proper is surrounded remain nearly in their original state, except that the ditch has been filled up: they afford a fine broad walk all round the city.

Londonderry was originally granted by Edward II. to Richard de Burgh, earl of Ulster, but the corporation now holds its privileges under a charter granted by James I. in 1613. The government of the city is vested in the mayor, sheriff, recorder, 12 aldermen, and 24 burgesses. The freedom is acquired by birth, by marriage with the daughters of freemen, apprenticeship, and by the gift of the corporation. The city sent two members to the Irish House of Commons; and since the union it has sent one member to the Imperial House of Commons. Previously to the Reform Act the right of voting was in the burgesses and freemen. Registered electors in 1839–40, 1932. The mayor, and aldermen who have filled the office of mayor, are justices of the peace within the liberties. The mayor and recorder hold a court of record every Monday, for pleas, to any amount. A court of general sessions is held quarterly, one of petty sessions weekly; there is also a court of conscience, at which the mayor presides weekly, for debts under £90, and for suits of wages. The assizes for the county and city, and the general sessions for the county, are held here twice a year.

The revenue of the corporation, arising from the tolls of the bridge, and dues on tonnage, quayage, &c., amounted a short while ago to about £7000 a year; but, owing to the expense of improvements, mismanagement, or some other cause, the corporation was involved in the greatest difficulties, and their property has been sequestrated, and mostly made over to other managers.

Manufactures are quite inconsiderable, if we except the conversion of grain into flour. There is a brewery and a distillery in the city, and two distilleries in the vicinity. There was here formerly a sugar-house and a glass-house, but these are now relinquished. A foundry and copper-works were established here in 1821, and have succeeded. Some table-linen is manufactured; and cotton is said to be woven in the parish for the Glasgow manufacturers.

The trade of Londonderry is very extensive, and is rapidly increasing. Its fine river makes it the emporium of a large extent of country; and it is to this that its extensive commerce is principally to be ascribed. Its exports, like those of most other Irish towns, consist principally of agricultural produce, but a good deal of linen is also exported.

Prices were uncommonly low in 1835; and, making allowance for this, and the increase that has since taken place in the trade of the port, we may, perhaps, estimate the present (1840) value of the exports at £1,350,000 or £1,400,000 a year. The vast number and great value of the eggs exported is a peculiar feature in the above account. The imports consist principally of manufactured goods and haberdashery, iron, sugar and tea, timber, wine, coal, glass, earthenware, &c. A great portion of the increase in the trade of the port may be ascribed to the establishment of steam-packets, which now ply regularly between the city and Glasgow and Liverpool. We subjoin an

ACCOUNT of the Quantity and Value of the principal Articles exported from Londonderry in 1835.

Articles Exported.	Quantity.		Estimated Value.		
			£.		
			£.	s.	d.
Corn, meal, and flour	cwts.	416,042	120,676	0	0
Provisions	—	95,890	273,566	0	0
Sugar	—	96	144	0	0
Flax and tow	—	51,190	212,940	0	0
Feathers	—	3	16	0	0
Spirits	galls.	83,680	10,580	0	0
Linen	yards	5,035,862	314,749	0	0
Cotton manufactures	—	968	24	0	0
Oxen and cows	—	855	8,130	0	0
Horses	numb.	73	1,640	0	0
Sheep	—	313	383	0	0
Swine	—	11,108	13,399	0	0
Eggs	—	83,965,000	55,994	0	0
Hides and calf skins untanned	—	24,960	11,226	0	0
Other articles	value	—	21,180	0	0
Total value			1,040,918	0	0

Derry is one of the principal ports for the shipment of emigrants; as many as 5000 to 6000 individuals having of late years frequently sailed for the United States and Canada in the course of a single season. The emigrants are very generally among the best behaved and most industrious, as well as most enterprising, portion of the community.

The gross customs' revenue collected at Londonderry in 1830 amounted to £72,912, and in 1839 to £98,696: the post-office revenue in 1830 was £3038, and in 1836, £3992. The Bank of Ireland, the Belfast, Provincial, Northern, and Agricultural banks, have offices here. A savings' bank, established in 1815, had, in 1835, £16,227 of deposites, contributed by 699 depositors.

Lough Foyle is properly the outer harbour of derry. It is a triangular basin, about 18 m and 10½ m. where widest; but a great part of it is occupied by sand-banks and mud-flats. The navigable channel stretches along the Donegal or Innishowen shore; and by following it, the largest men of war reach the anchorage at Moville, while merchantmen of 500 tons, without difficulty, ascend to the city quays, 5 m. above the lough and 23 m. from the sea. The river is navigable by barges from the city to St. Johnstone, and the Marquis of Abercorn has excavated a canal from the latter to Strabane. A portion of the wooden bridge at the city is constructed so as to open and admit the ascent and descent of vessels. It may be worth mentioning, that both the water and gas for the use of the city are conveyed across this bridge; so that the supply of both is intercepted whenever the bridge is opened.

But notwithstanding its increasing commercial prosperity, there is, we regret to say, much poverty in Derry. The contributors to the savings' bank are mostly menial servants; and the mechanics, tradesmen, and labourers are, in general, very badly off. "Among the labourers great poverty prevails, from the want of steady employment, and their consequent exposure to dissipation, with the total absence of employment for their children. The better class inhabit huts which let for about £3 a year; but the poorer frequently lodge in garrets or out-houses, chiefly in the Bogside, at a rent of about 1s. 3d. a week; and yet, even in these hovels, they contrive to let shares of their rooms at 6d. a week." (Ordnance Memoir of Londonderry, p. 194.)

Derry was colonized and fortified in the reign of James I. by the London companies, who had purchased large tracts of the confiscated estates of the Earl of Tyrone; at which period it took the name of Londonderry. It is famous in Irish history for the memorable siege it sustained in 1689 against the forces of James II. Though ill fortified, and without any disciplined troops, the heroism of the citizens, and the enthusiasm inspired by their brave leader, the Rev. George Walker, enabled them to repel all the attacks of the enemy; and to sustain the more dreadful sufferings occasioned by the pressure of famine. The besiegers lost 8000 men in the course of the siege, which was raised on the 105th day. Derry continued, for a lengthened period after this epoch in its history, to be, as it were, the headquarters of Protestantism, or rather of Orangeism in the N. of Ireland; but even in Derry the Catholics now outnumber the Protestants. (The reader will find, in the Ordnance Memoir referred to above, the most ample details as to all matters connected with the history and present state of Derry; see also, the Boundary and Municipal Reports; the Railway Report, &c.; and Inglis's Ireland, ii. 196.)

LONDONDERRY, p. t., Rockingham co., N. H., 34 m. S.S.E. Concord, 454 W. Bounded N.W. by Merrimac river. First settled in 1719 by a colony of Presbyterians from near Londonderry, in the N. of Ireland, who with their religious teachers, came to New-England in the summer of 1718. It is watered by Beaver river, issuing from Beaver pond, and flowing into Merrimac river, and contains two churches, a Presbyterian and Baptist, Pinkerton academy, named from its founder, who gave to it a fund of $14,000, three stores, six grist-mills, 10 saw-mills; 11 schools, 397 scholars. Pop. 2556.

LONDONDERRY, p. t., Windham co., Vt., 96 m. S. Montpelier, 461 W. Watered by West and Winhall rivers, and Utley brook, which afford water-power. Chartered in 1770, confiscated in 1778, the proprietor having become a tory; re-granted in 1780. It contains three churches, a Congregational, Methodist, and Baptist, three stores, two grist-mills, six saw-mills, two tanneries; 13 schools, 409 scholars. Pop. 1216.

LONDONDERRY, t., Chester co., Pa., 35 m. S.W. Philadelphia. It contains two churches, one store, two grist-mills, two saw-mill; four schools, 96 scholars. Pop. 608.

LONDONDERRY, t., Bedford co., Pa. Drained by Will's creek, which flows into Potomac river. It has one store, one fulling-mill, one saw-mill, one tannery, one distillery, one pottery, one printing-office, one weekly newspaper; one academy, 26 students; one school, 38 scholars. Pop. 869.

LONDONDERRY, t., Dauphin co., Pa., 14 m. S.E. Harrisburg. Drained by Spring and Conewago creeks, the former affording water-power. It contains a Lutheran church, two stores, one fulling-mill, seven grist-mills, one saw-mill; one school, 14 scholars. Pop. 1990.

LONDONDERRY, t., Lebanon co., Pa., 8 m. S.W. Lebanon. Bounded N. by Swatara creek. Drained by Quatapahilla and Conewago creeks. It contains six stores, one furnace, one grist-mill, one saw-mill, three tanneries; five schools, 145 scholars. Pop. 1763.

LONDONDERRY, p. t., Guernsey co., O., 96 m. E. Columbus, 303 W. Named from Londonderry in Ireland, whence some of its inhabitants originally came. It has seven schools, 164 scholars. Pop. 1606.

LONDON GROVE, p. t., Chester co., Pa., 34 m. S.W. Philadelphia, 71 m. E. by S. Harrisburg, 105 W. Drained by White Clay creek and its branches. It contains four stores, one woollen factory, one cotton factory, with 336 spindles, four flouring-mills, four saw-mills, 15 oil-mills, three tanneries, one pottery; three schools, 75 scholars. Pop. 1946.

LONG BRANCH, p. v., Shrewsbury t., Monmouth co., N. J., 39 m. S. New-York, 50 m. E. Trenton, 216 W. A watering place, much frequented for sea air and bathing, and for fishing and gunning. Situated on a peninsular beach on the Atlantic, where the land rises perpendicularly from the beach to the height of 20 feet, and has fine boarding-houses, 20 rods from the water, in front of which are pleasant lawns. It is much frequented in the summer seasons.

LONG ISLAND, N. Y., is the largest island of the United States on the Atlantic coast; and from fort Hamilton at the Narrows, to Montauk point, is about 140 m. long, with a breadth as far east as Peconic bay of from 12 to 20 m., a distance of 90 m. It is bounded N. by Long Island sound, E. by the Atlantic, S. by the Atlantic, and W. by the harbour of New-York, and the strait which connects the harbour with the sound, called East river, from half a mile to two miles wide. The island contains 1500 sq. m., and is divided into three counties, King's co. in the W., Queen's in the middle, and Suffolk on the E. Of these, King's, though the most important and populous, has much the smallest territory, and Suffolk in the E., the largest, comprising two thirds of the whole island. The population of King's co. is 47,613; of Queen's, 30,324; of Suffolk, 32,469; making a total population in 1840, of 170,406. A ridge or chain of hills commences at New Utrecht in the W. part, and extends, with some interruptions, to near Oyster Pond point, in Suffolk county. The highest point of this chain, which is N. of the middle of the island, is Harbour hill, in N. Hempstead, Queen's co., which is 319 feet above the level of the tides. The whole island is underlaid with granitic rock, which rises high in the spine, and breaks out in the form of gneiss along E. river at Hell-gate, and in other places. The surface of the island N. of this ridge is generally rough and broken, while S. of it, it is almost a perfect plain, with scarcely a stone exceeding a few ounces in weight. On the S. side of the island is Great South bay, from one mile to five miles wide, affording an uninterrupted inland navigation for vessels of 60 or 70 tons; and which extends from Hempstead bay, eastwardly for 50 miles. This bay is enclosed by a narrow island or beach, with several inlets, in no part over 3 m. broad. Toward the E. end, the island divides into two parts, the southern consists of a promontory over 20 m. long, and not generally over 1 wide, constituting the township of Easthampton, on the eastern extremity of which is Montauk point, on which is a lighthouse. The N. part, which is less extensive, terminates at Oyster point, and constitutes the township of Southold. Between these two parts is Great Peconic bay and several islands, the largest of which are Gardiner's and Shelter islands. The soil on the N. side of the dividing ridge, generally consists of loam, while through the middle and S. part of the island it consists of sand or gravel. Much of the central part of the island consists of pine forest, in which wild deer are still found. South bay abounds with shell and scale fish, and water fowl of many kinds. A railroad from Brooklyn, 94 m. to Greenport is in progress, which will probably be completed in July 1844.

LONG ISLAND SOUND, extends along the whole length of the island, and separates it from Connecticut on the N. It is a fine body of water, which furnishes great facilities for the coasting trade, being about 110 m. long, and from 2 to 20 m. broad. It has several good harbours, and a number of lighthouses. It communicates with the Atlantic on the E. by a broad and rapid strait called the Race, and with New-York harbour on the E. by the East river, in which is the dangerous pass called Hell-gatt or Hell-gate, which can be safely passed by sail vessels at high or low water, or with a fair wind; and by steamboats at all times.

LONGFORD, an inland co. of Ireland, prov. Leinster, having N. Cavan and Leitrim, E. Westmeath, and W. Roscommon, from which it is separated by the Shannon. Area, 263,645 acres, of which 55,347 are unimproved bog and mountain, and 15,892 lakes. The arable soil is, for the most part, level and fertile. Property mostly in large estates. Tillage farms small, the state of agriculture and the condition of the occupiers being much the same as in the adjoining counties. Grazing, however, is extensively carried on. Average rent of land, 12s. 3d. an acre. It is divided into six baronies and 23 parishes; and sends two members to the House of Commons, both for the county. Registered electors, in 1839-40, 1971. In 1831 it had 19,418 inhabited houses, 20,438 families, and 112,558 inhabitants, of whom 55,310 were males, and 57,248 females.

LONGFORD, an inland town of Ireland, cap. of the above co., prov. Leinster, on the Camlin, an affluent of the Shannon, 65 m. N.N.W. Dublin. Pop., in 1831, 4134. It is "a well-built town, and is increasing rapidly in population and wealth. There is here a very large market for grain; great quantities being exported by the royal canal, a branch of which comes to the town." (Municipal Boundary Report.) It has a parish church, a Roman Catholic chapel, meeting-houses for Presbyterians and Methodists, a market-house, the county courthouse, prison, infirmary, and dispensary, with large cavalry and artillery barracks. The corporation, which, under a charter of Charles II. in 1657, consisted of a sovereign, two bailiffs, 12 burgesses, and a commonalty, sent two members to the Irish House of Commons till the Union, when it was disfranchised. The county assizes and general sessions are held here; and courts for petty causes are held on Mondays and Saturdays. It is a constabulary station. Some linen is manufactured; and there is a tannery, a brewery, and a distillery; but the great business of the town consists in its trade in corn and other raw produce. Markets on Wednesdays and Saturdays: fairs on March 25, June 10, Aug. 19, and Oct. 22. Postoffice revenue, in 1830, £1087; in 1836, £1137. Branches of the Bank of Ireland, and of the Agricultural and National banks, were opened in 1834-35-36.

LONG MEADOW, p. t., Hampden co., Mass., 95 m. W.S.W. Boston, 350 W. Bounded W. by Connecticut river. The principal village is built on a single broad street, one mile E. of Connecticut river, and parallel to it. The township contains three churches, two Congregational, and a Baptist, three stores, one grist-mill, three saw-mills, three tanneries; seven schools, 291 scholars. Pop. 1270.

LONG SWAMP, p. t., Berks co., Pa., 72 m. E. by N. Harrisburg, 165 W. Watered by Little Lehigh river. It contains a church common to Lutherans and Presbyterians, two stores, one furnace, one forge, two grist-mills, two saw-mills, two powder-mills, two tanneries, three distilleries. Pop. 1836.

LONGOBUCCO, a town of the Neapolitan dom., prov. Calabra Citra, 19 m. E.N.E. Cosenza. Pop. about 3000, chiefly employed in working metals and burning charcoal. The horses for hunting in Naples are bred in the neighbourhood.

LONS-LE-SAULNIER, a town of France, dep. Jura, of which it is the cap., in a deep valley, 50 m. S.E. Dijon. Pop., in 1836, 7084. It has no remarkable public buildings; but is generally well built, clean, and furnished with numerous public fountains, one of which, in the Place d'Armes, is ornamented with a statue of Pichegru, in white marble. At the N. extremity of the town is the salt spring from which it derived its ancient name of Ledo Salinaris : this spring continues to yield great quantities of table salt; four pumps are kept constantly at work, and the evaporating houses (*bâtimens de graduation*) are very extensive. Lons has a theatre, a public library with 3000 volumes, a gallery of paintings and antiquities, tribunals of primary jurisdiction and commerce, a communal college, &c. It is the entrepôt of the agricultural produce, iron goods, timber, wines, &c., of the dep., and has a fair on the 15th of every month. (*Huge*, art. *Jura*; *Guide du Voyageur*, &c.)

LOO-CHOO, or LIEOU-KIEOU ISLANDS, a group tributary to the Chinese, in the N. Pacific ocean, nearly midway between Japan and Formosa, and comprised within lat. 26° and 28° N., and long. 127° and 129° E. There are in all about 36 islands; but, excepting the Great Loo-Choo island, towards the centre of the group, 70 m. in length, by from 12 to 15 m. broad, they are mostly of very inferior dimensions. These islands, which are but little known to Europeans, are reported to have a delightful climate, and a soil of great richness, producing the fruits and vegetables of countries the most remote from each other. Rice is cultivated with great care. Cattle, goats, and pigs, are said to be diminutive; but poultry are large and excellent. The islands yield sulphur and salt, and have, it is alleged, rich mines of copper and tin. Conflicting statements have been made by different travellers respecting the civilization, political condition, jurisprudence, &c., of the natives. They appear, however, to be of the same race as the Japanese; and have not merely adopted the costume, but speak the language, of that people. Their religion is a species of Buddhism; and their government, like that of other Asiatic countries, of a despotical character. They are friendly and hospitable; but it is now sufficiently ascertained that Captain Hall was totally mistaken in the estimate he formed of these islanders; who, had his statements been well founded, almost re''.zed the poetical fictions of the golden age. The Loo-Choo islands were for some time subject to Japan, but were conquered by China about 1372. Kintching, the capital, is about 5 m. from its port Napkiang, near the S.W. extremity of Great Loo-Choo, lat. about 26° 14' N., long. 127° 52' E. (*See Hall*, *Macleod*, *and Beechey's Travels*, &c.)

LOOE (EAST and WEST), two contiguous anc. bors. and market towns of England, co. Cornwall, hund. West, on both banks and close to the mouth of the Looe, 12 m. S.E. Bodmin, and 219 m. W. by S. London. United pop., in 1831, 1458. They are mean, wretched places, connected by a narrow, old bridge of 13 arches; and would be unworthy notice, were it not that each of them enjoyed the privilege of sending two members to the House of Commons from the reigns of Edward VI. and Elizabeth down to the passing of the Reform Act, when they were both disfranchised.

LORAIN, county, O. Situated in the N. part of the state, and contains 540 sq. m. Watered by Rocky river and its branches, and by the E. branch of Black river. Organized in 1824. It contained in 1840, 20,931 neat cattle, 19,377 sheep, 18,641 swine; and produced 134,474 bushels of wheat, 3792 of rye, 943,034 of Indian corn, 3869 of buckwheat, 80 291 of oats, 109,577 of potatoes, 326,644 pounds of sugar. It had one commission house in foreign trade, 28 stores, four lumber-yards, one furnace, one fulling-mill, one woollen factory, one flouring-mill, 14 grist-mills, 54 sawmills, eight tanneries, one distillery, three printing-offices, one bindery, three weekly and one semi-weekly newspapers; one college, 498 students; two academies, 135 students; 97 schools, 3123 scholars. Pop. 18,467. Capital, Elyria.

LORAIN, t., Shelby co., O. It has two stores, two sawmills; two schools, 55 scholars. Pop. 904.

LORAINE, p. t., Jefferson co., N. Y., 15 m. S. Watertown, 157 m. N.W. Albany, 408 W. Drained by Sandy creek and its tributaries. It has two stores, two flouring-mills, 10 saw-mills; eight schools, 265 scholars. Pop. 1699.

LORCA (an. *Eliocroca*), a considerable town of Spain, prov. Murcia, cap. of a partido of its own name, on the Guadalentin, a tributary of the Segura, 42 m. W.S.W. Murcia, and 116 m. E.N.E. Granada. Pop., according to Miñano, 40,366. The vale of Lorca is remarkable for picturesque beauty and great fertility; and the town, close under the Sierra del Cano that bounds it on the left, and the fine old castle on a rock hanging over it, adds greatly to the beauty of the picture. This has evidently been a considerable place; but, the lower part of the town being concealed by trees, nothing is seen on approaching it but a number of low houses crowded on the side of the mountain, and from the similarity of colour seeming almost to belong to it. This is the old or Moorish town, and is very irregular and mean in appearance: but the new town, on the plain, is much more regularly laid out, and better built. A collegiate (once episcopal) and seven parish churches, two hospitals, an episcopal palace, and a royal college, are the chief buildings and establishments. Saltpetre is manufactured on a large scale, and soap, thread, and linens, are pro-

duced in small quantities; but the chief resources of the town consist in its great September fair, its markets, and the produce of its neighbourhood, both in flocks and agricultural produce. Mr. Inglis gives a lively picture of the market and its attendants: "All the women here wear a square white woollen shawl, worn like a mantilla: the men are dressed in short white drawers loose at the knee, and instead of stockings use sandals made of rope; and their heads are covered with close-fitting tapering black caps, others from the higher countries being enveloped in blankets of gaudy colours. Among the numerous things exposed for sale, were dried and shell fruits, Catalonian cloths and calicoes, shoes and rope sandals, quantities of Esparto rush and rush-baskets, beads, rosaries, trinkets, &c., in short, everything that one either eats or wears in Murcia. The show of pigs was extremely fine, and nowhere in the world are these animals found in greater perfection than in Spain, fed, as they are, on the ilex nut. The price of a hog weighing 23 stone was 240 reals (£2 8s.), and that of a sucking pig 14 reals. Mutton sells at 12 quartos, a fowl costs 20d., a hare 10d., and bread is 1½d. per lb. The price of labour in the vale of Lorca is 5 reals or 1s. a day." (*Inglis's Spain*, ii., 208.)

Lorca, supposed to be the *Eliocroca* mentioned in Antonine's Itinerary, was exposed to frequent attacks during the contests between the Moors and the crown of Castile, and has at various times sustained sieges. It was nearly destroyed at the commencement of this century. In 1792, a speculator, with the permission of government, collected at a great expense all the waters of the district into a common reservoir (*pantano*) resembling that of Alicant. The basin was said to be "superb," and capable of containing water sufficient to irrigate for years the entire vale of Lorca. Ten years afterward (30th April, 1802) the waters, which had for some time been undermining the reservoir, rushed out with an impetuosity that swept everything before it, men and cattle, public buildings, and even trees and rocks. About 600 houses, a church, two convents, two hospitals, several mills and fountains, were at once swallowed up and disappeared, about 6000 human beings, 24,000 cattle, &c., being at the same time destroyed. The agricultural districts were covered over with sand, rubbish, &c., and the total loss occasioned by the catastrophe is supposed to have exceeded a million sterling. A like disastrous event is said to have destroyed the ancient city of Mareb in Arabia Felix, an account of which will be found in the *Modern Traveller, Arabia*. p. 98. (*Inglis*; *Laborde*, vol. ii.; *Miñano*.)

LORETTO, a town of the Papal States, cap. of a commissariat of the same name, on a bold and commanding eminence, about 3 m. from the Adriatic, and 12 m. S.E. Ancona. Pop. with its suburbs, about 5000. It is surrounded with walls, constructed by Sixtus V., in 1587. This celebrated but poor town is wholly indebted for its fame, and even existence, to its having the good fortune to possess the *Santissima Casa*, or house occupied by the Virgin Mary, in Nazareth, conveyed by angels, first to Tersato in Dalmatia, and thence, by the same agency, in 1294, to its present site! This miraculous edifice is a mean-looking hovel, about 30 feet in length, by 13 or 14 feet in width, and 18 feet in height; apparently built of Apennine limestone, with a modern vault of timber-work. It is incased in a shell of marble, sculptured with bass-reliefs, representing the history of the Virgin; the whole being under the dome of a splendid church, built to protect the sacred edifice. In a niche within the latter, once fenced in with gratings of solid gold, but now with pieces of gilt wood, is the image of the Virgin, affirmed to be the work of St. Luke, to whose talents as an artist it does little credit, being "a little old woman about 4 feet in height, with the features and complexion of a negro." (*Moore*.) Her dress is tawdry, and in the worst possible taste: she literally glitters in jewels and brocade, and reigns "amid the continual glare and smoke of lamps and candles, held by figures of angels." The church, which encloses the *santissima casa* is said to have been designed by Bramante. According to Eustace, it is a "very noble structure;" but it is less favourably spoken of by Woods and others. Its gates, which are of bronze, are embellished with *basso relievos* of the most admirable workmanship; in the area before it, is a handsome marble fountain, and a large statue of Pius VI. The riches formerly accumulated within this sanctuary, were a subject of astonishment to all travellers; and were, most probably, much exaggerated. The popes are believed to have occasionally abstracted some of the gold offerings, and to have substituted false for real gems. But, when the French acquired possession of Loretto, they acted with less reserve; and, undismayed by the sanctity of the place, rifled its repositories, and carried off every article of value, applying them to secular and really useful purposes. It has since, however, received several considerable benefactions

A lucrative trade was formerly carried on at Loretto in

saaries, crucifixes, *agnus Dei*, and such like articles, partly taken off by pilgrims to the shrine, and partly exported. But this trade has now much fallen off. The number of pilgrims, though still very considerable, has also greatly declined; and they are now mostly of the lowest and poorest classes. On their arrival in town, they are received into a hospital, where they are boarded and lodged for three days; and this privilege has probably as much to do as superstition in attracting them to Loretto. (See *Addison's Travels*, p. 94, ed. 1726; *Moore's Italy*, i., 291; *Eustace's Classical Tour*, i., 200; *Forsyth*, 331, &c.)

L'ORIENT, a strongly fortified seaport town of France, dep. Morbihan, cap. arrond., at the confluence of the Scorff with the Blavet, at the head of the bay of Port Louis, about 3 m. from the Atlantic, and 29 m. W. by N. Vannes. Lat. 47° 45′ 11″ N.; long. 3° 21′ 2′ W. Pop., in 1836 (ex. suburbs), 15,138; but, according to recent statements, the population of the town and suburbs may now (1841) amount to 19,000 or 20,000, nearly 5000 of whom are employed in the dock-yard and its appendages. L'Orient is clean, and regularly built: the streets are wide, straight, and well paved, and the houses well constructed and handsome. One of its public squares, the *Place Royale*, is planted with lime-trees, and it has other good promenades. The principal church* is very large, and has a lofty spire, which is a conspicuous landmark. The prefecture, auction-hall, town-hall, and theatre, are handsome edifices. The public slaughter-houses (*abattoir*) are remarkably clean; and the meat, fish, and bread-markets are, next to those of Rennes, the best constructed, and most extensive in Brittany. In the centre of the market-place is a granite column erected to the memory of the commander, Bisson. Some years ago, a bridge, also of granite, was commenced over the Scorff, but the design was abandoned, lest the clearance of the port should be thereby impeded.

The port of L'Orient, about ¾ of a mile in length by nearly ¼ m. in breadth, is secure, commodious, and of easy entrance. It is bordered by fine quays, on which are some extensive buildings and establishments connected with the government dock-yard; an observatory 120 feet in height, which serves also for a telegraph and a lighthouse, and a very handsome public fountain. The naval establishment is on a smaller scale at L'Orient than at Brest; it has no *bagne*, but it has a place of confinement for soldiers guilty of insubordination. More ships of war are now built in the dock-yard of L'Orient than at any other in France, 16, of the estimated cost of 8,652,390 francs, having been constructed here in 1840. L'Orient has slips enough for the construction of 30 vessels of all sizes; frigates are, however, the class of ships chiefly built. Towards the end of 1840, nine slips were occupied each with a frigate in course of active preparation; and, according to the report of an English traveller, 3000 workmen were exclusively engaged on these nine frigates. Many of the subordinate artificers get only 98 sous a day, and few of the better workmen receive more than from 30 to 40 sous a day: but taking into account the cheapness of living in Brittany, they are, perhaps, fully as well off as the workmen in the English dock-yards. The foremen (*chefs d'ateliers*) are not paid by the day, but receive from 1500 to 2000 francs a year, according to circumstances. L'Orient has excellent sheers for masting vessels, &c., and good block sheds, the machinery in which, as well as a portion of that for cable-making, is wrought by steam. At the end of 1840, establishments were in course of being erected for the construction of steam engines for ships of war, and new forges, &c., were about to be commenced. The buildings formerly belonging to the French E. I. Company, are now converted into barracks. The arsenal and naval stores are very extensive, and the artillery barracks are capable of accommodating 1800 men. The lazaret is on a small island to the S., between L'Orient and Port Louis. L'Orient has a school of naval artillery and a spacious artillery ground near the town, a school of hydrography, established 1771, a large and well-arranged commercial college, a preparatory school for training for the government schools, a communal college, gratuitous schools of drawing, geometry, arithmetic, &c., a public and a pretty good naval library, museums of chemistry and mineralogy, an agricultural society, and various educational societies. It is the seat of tribunals of primary jurisdiction and commerce, a chamber of commerce, &c.

The manufactures of L'Orient, chiefly consisting of hats, linens, gold lace, earthenware. &c., are not very important Its trade, though not so flourishing as in 1789, has latterly begun to increase. In 1830, only six merchantmen belonged to the port; it has now (1840) more than four times that number, some of them trading to the French colonies. The chief exports are wax, honey, butter, corn, cattle, and

* Hugo says that this church was originally planned and begun on so immense a scale, that after 30 years labour had been spent in its erection, the builders, despairing of being able to finish the work so commenced, found it necessary to demolish a portion of it to complete the rest.

pilchards, the latter being taken in great quantities on the adjacent coast, are sent to Nantes to be prepared for exportation.

Though at present so little eminent for trade, L'Orient owes its origin and former importance almost wholly to commerce. It was but an insignificant village when, in 1738, the French E. I. Company made it their principal naval *dépôt*; and such was the influence of the change, that in 1738 its population is said to have amounted to 14,000! On the dissolution of the company in 1770, L'Orient was made one of the stations for the French navy, and a free commercial port. (*Hugo*, art. *Morbihan*; *Letter in the Times*, 7th Oct., 1840.)

LORRAINE (Germ. *Lothringen*), one of the largest of the an. provs. of France, in the N.E. part of the kingdom, now distributed among the deps. of Meurthe, Meuse, Mo - selle, Vosges, and Bas-Rhin.

LOSTWITHIEL, or LESTWITHIEL, an an. bor. market-town, and par. of England, co. Cornwall, hund Powder, on the W. bank of the Fowey, 5 m. S.S.E. Bod min, and 211 m. W. by S. London. Area of bor., 120 acres. Population, in 1831, 1548. The houses are chiefly of stone, roofed with slate; but the streets narrow and ill-paved. The church, a curious old building, has a large E. window, and a fine tower and spire at the opposite end: the living is a vicarage, in the gift of Earl Mount-Edgecombe. It has also places of worship for Independents, Wesleyan and Primitive Methodists, with attached Sunday schools. The corporation support a grammar and writing school; another school, for poor children, is slenderly endowed by the trustees of the late Rev. St. John Eliot; and there are a few money bequests. Near the church, an ancient building, supposed to have been either a palace or the house of the Duke of Cornwall, or a court-house for the stannaries, was, till very recently used as a prison during the winter and summer county sessions, now removed to Bodmin. There is a town hall, where petty sessions are held, and under it is a small jail. Tanning and wool-stapling are the principal trades, and the town derives some importance from its situation on the Fowey, by which iron and copper ore, &c. are exported. Lostwithiel was made a free borough by Richard Earl of Cornwall, and incorporated by James I.: it sent two members to the House of Commons, from the 19th Edward II. down to the Reform Act, by which it was disfranchised. It is not included in the Municipal Reform Act; but is now, as formerly, governed by a mayor, six capital, and 17 inferior burgesses. Markets on Friday; cattle fairs, July 10, Sept. 6, and Nov. 13.

About 1 m. N. of Lostwithiel, on the summit of a hill, is Restormel castle, the ancient seat of the baronial family of Cardinan, and subsequently of the earls of Cornwall; it was ruinous even so early as the time of Henry VIII. but was repaired and occupied during the civil war. (*Mun. Corp. and Char. Rep. &c.*)

LOT, a dep. of France, reg. S., chiefly between lat. 44° 13′ N., and long. 1° 20′ E. having N. Corrèze, E. and S.E Cantal and Aveyron, S. Tarn-et-Garonne, and W. Lot-et-Garonne and Dordogne. Area, 525,280 hectares. Population in 1836, 287,003. The dep. is mountainous, with a general slope towards the S.W. Its mountains are ramifications of those of Cantal, and rise in the E. about 2500 feet above the sea. Its chief rivers are the Lot, and the Dordogne; from the first of which it derives its name. The Lot, which rises in Lozère, about lat. 44° 30′ N. long. 3° 45′ (E.)runs with a very tortuous course generally W. through Aveyron, the S. part of Lot, and the centre of Lot-et-Garonne and Gironde; uniting with the Garonne at Aiguillon, about lat. 44° 18′ and long. 0° 19′ E. It is navigable, during four months of the year, for nearly 190 m., Mende, Cahors, and Villeneuve d'Agen are on its banks. There are an immense number of narrow valleys, watered by small rivulets : these have frequently an alluvial soil of great fertility, but the soil in most parts is either calcareous, or stony and gravelly. In 1834, it was estimated that 232,533 hectares were arable. 25,895 in pasture, 54,627 in vineyards, 87,255 in woods, and 71,984 occupied by heaths, wastes. &c. Lot produces more corn than is required for its own consumption; but chestnut-flour forms an important article of food among the rural population. The corn grown is principally wheat, maize, and rye; and the total annual produce of all kinds is estimated at between 1,500.000 and 1,600,000 hectol. Agriculture is extremely backward, and there is a great want of capital, a consequence mainly of the splitting up of the land into an immense number of small properties. In 1835, of 111,948 properties subject to the *contribution foncière*, 50,471 were assessed at less than 5 fr., 18,731 at from five to 10 fr., and 17,652 at from 10 to 90 fr.; while the number of properties assessed at 1000 fr. and upwards amounted to only 18. The plough employed is a fac simile of that described by Virgil, and is drawn by oxen: the spade or hoe is, however, used in the culture of thin soils. The produce of wine amounts to about 600,000 hectol. a year, a third part of which is consumed by the inhabitants, and the rest sold or converted into brandy. The wines known in the market as *vins de Cahors* are strong and very dark-coloured, and are principally employed to give body and colour to other wines, for which purpose they are principally sent to Bordeaux. Tobacco is grown in this dep., and in 1833 about 1850 hectares were appropriated to its culture, and 933,330 kilog. produced. The climate is favourable for the mulberry, but the silkworm does not thrive. A few proprietors have flocks of merino sheep, but the pastures are badly irrigated and attended to, and most black or white live stock are indifferent. The goats' hair of the dep. is, however, highly esteemed. The produce of sheep's wool is estimated at 500,000 kilog. a year. A great many hogs are fattened for sale in the neighbouring deps.; and about 60,000 turkeys and geese are annually exported, preserved in their fat. A considerable proportion of the truffles used in *Pâtés de Périgord* come from this dep. There are some iron and coal mines; but both mining and manufacturing industry are little attended to. A few copper and iron forges, woollen, cotton, and linen cloth factories, paper-mills, and tanneries, are the chief manufacturing establishments: these, however, are so few, that the inhabitants are usually supplied with cloths and leather in exchange for their wool and skins from the adjacent deps. There are nearly 1000 flour-mills in the dep. Lot is divided into three arronds.; chief towns, Cahors, the cap., Gourdon, and Figeac. It sends five members to the Ch. of Dep. Number of electors (1838-39), 1366. Total public revenue, in 1831, 4,765,687 fr.; but the expenditure in the same year amounted to 5,080,509 fr. (*Hugo.* art. *Lot; Official Tables. &c.*)

LOT-ET-GARONNE, a dep. of France, reg. S.W., formerly included in Guienne; chiefly between lat. 44° and 44° 40' N., and long. 0° and 1° E., having N. Dordogne, W. and S.W. Gironde and Landes, S. Gers, and E. Tarn-et-Garonne and Lot. Length and breadth about 50 m. each; area, 530,711 hectares. Population (1836) 346,400. The surface is mostly level, with a slope to the W. The Garonne intersects the dep. from S.E. to N.W., and receives about its centre the Lot from the E. The banks of these rivers may be classed among the most productive portions of France; but 265,496 hectares consist of a chalky soil, and about one eighth part of the surface in the W. of the dep. is composed of *landes*, or sandy plains, sprinkled with marshes, analogous to those in the adjoining deps. of Gironde and Landes. According to the official returns, it comprised, in 1834, 286,100 hectares of arable land, £2,382 do. meadow, 68,349 do. vineyards, and 68,613 do. woods. This dep. is principally agricultural. The corn grown exceeds what is required for home consumption: it is chiefly wheat and maize on the richer lands, and rye on the poorer. The produce of wine is estimated at about 650,000 hectolitres a year, of which nearly a half is exported. The N. part of the dep. produces about 40,000 hectols. a year of chestnuts, from 7000 to 8000 hectols of which are sent to Bordeaux and the neighbouring deps.: 2030 hectares are occupied with tobacco, which produced, in 1833, 746,596 kilogs., valued at about 560,000 fr. Excellent hemp is grown. The prunes of Agen are highly esteemed, and are exported to the value of 600,000 fr. a year: the dried figs of Clairac are also celebrated. On the *landes* are many fir plantations, which furnish about 800,000 kilogs. of resin, and 300,000 kilogs. of turpentine a year, besides pitch, deals, &c. The cork tree grows in a few communes, and its produce is valuable. Artificial pasture lands are rare. According to the official tables, there were in 1830 in this dep. 188,000 head of sheep, and 107,000 head of black cattle. Large flocks of geese are reared, especially near Agen; they are fattened on maize, and preserved in their fat. In 1835, of 122,558 properties subject to the *contribution foncière*, 51,246 were assessed under 5 fr., and 19,780 at from five to 10 fr. Mining industry is insignificant; but some iron ore is smelted by means of charcoal, there being no coal mine in the dep.: there are numerous distilleries. At Tonneins is a royal tobacco manufactory, employing 400 workmen, who produce 400,000 kilogs. of tobacco a year, for the supply of the neighbouring deps. At Nerac, Mézin, Barbaste, &c., are cork factories, which together may employ about 700 hands, and produce 130,000 metrical quintals of corks a year. At Agen is a large sail-cloth factory, with 300 looms, for the service of the French navy; and these are also extensive rope-walks. Lot-et-Garonne has also manufactures of woollen thread, serge, linen and cotton cloths, gloves, paper, starch, glass, and earthenware, besides tanneries, iron works, &c. The dep. is divided into four arrond.; chief towns, Agen, the cap., Marmande, Nerac, and Villeneuve d'Agen. It sends five members to the Chamber of Deputies. Number of electors in 1838-39, 9771. Total public revenue (1831), 7,841,527 fr.; expenditure in the same year, 6,094,709 fr. (*Hugo.* art. *Lot-et-Garonne; Official Tables.*)

LOTHIAN, an extensive, fertile, well cultivated, and rich district of Scotland, lying along the S. shore of the frith of Forth. It is divided into the counties of East Lothian, or Haddington; Mid Lothian, or Edinburgh; and West Lothian, or Linlithgow. We shall make a few observations on each of these, beginning with

1st. *East Lothian.*—This, which, as its name implies, is the most easterly division of the Lothians, has the frith of Forth on the N., the German Ocean on the E., Berwickshire on the S., and Mid Lothian on the W. It is of an elliptical shape, and contains 174,080 acres, of which about four fifths are capable of cultivation. The S. portion of the county is occupied by the Lammermuir hills, which divide the county from Berwick; but with this exception, it is mostly level, or merely undulating; and when viewed from the adjacent heights, appears like an extensive, rich, and beautiful plain, gradually sloping to the sea. The district along its E. coast, comprising about 20,000 acres, has a reddish, loamy, and very fertile soil: the soil gradually becomes more clayey as it recedes from the sea; and, except in the district now referred to, its general character is that of a clay bottom. The climate is comparatively dry and early; but the E. winds, in April and May, are often very severe. This is one of the best cultivated districts of the empire, and is remarkable for the intelligence and skill of its farmers, and their superior husbandry. The best farming is seen in the district along the E. coast, the soil being there adapted alike to the growth of turnips and of wheat. The turnip culture, indeed, is carried on here to a greater extent, on more correct principles, and with better success than in any other part of the empire. In the clayey lands, or those that have a wet, retentive subsoil, summer fallow is extensively practised, and is found to be the best foundation of a profitable system of cultivation. During the late war, when the prices of corn were so enormously high, the raising of corn in this county was carried to an improper extent; and in many parts the land was unduly forced. But this error has since been obviated. The fatting of cattle of all kinds for the butcher is now an important part of the economy of every well conducted farm; and a greater extent of land is kept in grass, and for a longer period. Exclusive of the Lammermuir district, which is principally devoted to the breeding of sheep, the farms in the other parts of the county extend from 60 acres up to 500 acres, or more, the average being about 250 acres. Every farm has a thrashing machine: and of these about 80 are driven by steam, seven by wind, 30 by water, and the rest by horses. Rents are commonly fixed in corn, convertible into money at the *fair* prices of the county. Six bushels of wheat may, perhaps, be taken as the average rent of the wheat lands of the district, which, taking the wheat at 7s. a bushel, will be equivalent to a money rent of 42s. an acre. In 1810, the average rental of the county was 20s. 10d. an acre. Notwithstanding its present highly advanced and improved condition, agriculture was in an extremely backward and depressed state in this county even so late as 1770. The land was then not half tilled; a rotation of crops was comparatively unknown; the stock and implements of husbandry were alike defective; much of the land was injured by the want of drainage; the hinds, or farm labourers, were badly fed and badly clothed; and the ague regularly made its appearance in spring in every hamlet and village, and almost, indeed, in every house. The change in the interval has been most striking and beneficial. Even within the last dozen years many important improvements have been made, principally by the introduction of furrow draining and bone manure, a better rotation of crops, and a more efficient and skilful management. The farm-houses and offices are excellent; but we are sorry to have to add, that while every thing else has been vastly improved, the cottages have not, in the majority of cases, been sensibly ameliorated, and their condition is discreditable alike to the farmers and the landlords. Except, however, as respects their lodging, the labourers are well off; and the hinds, or farm labourers, now receive each 24 bushels of oats a year more than they did previously to the commencement of the improvements. Estates of various sizes; some very valuable. There are about 7,500 acres of wood. The W. division of the county has valuable beds of coal: and limestone is very generally diffused. If we except some considerable distilleries, manufactures are all but unknown. The Tyne, which flows through the centre of the county, is the only considerable stream. The county sends one member to the House of Commons; and the boroughs of Haddington, N. Berwick, and Dunbar, join with Lauder and Jedburgh in returning one member. Registered electors, in 1839-40, 740. In 1831, E. Lothian had 6561 inhabited houses, 8080 families, and 36,145 inhabitants, of whom 17,307 were males, and 18,745 females. Valued rent, £168,874 Scotch. Annual value of real property in 1815, £250,195. (*Robertson's Rural Recollections, passim; New Statistical Account of Scotland.*)

2. *Mid Lothian*, or Edinburghshire, has the firth of Forth on the N.E., Lothian on the E., Roxburgh, Selkirk, Peebles, and Lanark on the S., and W. Lothian on the W. Area, 236,563 acres, of which about two thirds are supposed to be arable. In some parts, especially along its S. border it is rugged and even mountainous; the ridge of the Pentland hills, which approaches within a short distance of Edinburgh, divides its low grounds into two portions, that unite towards the sea. Soil for the most part clayey, and not in general very fertile. Agriculture similar to that of E. Lothian, but inferior; its details being also a good deal modified by the demand of the capital for milk, butter, potatoes, &c. Improvements of all sorts have been prosecuted with great zeal and industry. In 1727, a small field of wheat, within a short distance from Edinburgh, was reckoned so extraordinary a phenomenon that persons came from a great distance to see it! (*Robertson's Recollections*, p. 267.) But, at present (1840), wheat is the principal object of the farmer's attention; and there may be from 18,000 to 20,000 acres under that crop. There are a considerable number of rather large estates; but property is, on the whole, pretty well divided. Average rent of land, in 1810, 24s. 6d. an acre. There are large beds of coal in this county. For details as to its trade, manufactures, literary establishments, &c., the reader is referred to the articles Edinburgh and Leith. This county has, exclusive of Edinburgh, 27 parishes; it returns four members to the House of Commons, viz. one for the county, two for the city of Edinburgh, and one for Leith and Musselburgh. Registered electors for the county, in 1839-40, 2315. In 1831 Mid Lothian had 19,744 inhabited houses, 47,415 families, and 219,345 inhabitants, of whom 99,893 were males, and 119,549 females. Valued rent of county, £191,654, Scotch. Annual value of real property, in 1815, £770,875.

3. *West Lothian*, or Linlithgowshire, the smallest of the divisions of Lothian, has the firth of Forth on the N., Mid Lothian on the E. and S., and Lanark and Stirling on the W. It is of a triangular shape, and contains 76,800 acres, of which about three fourths are arable. Surface varied, with knolls; there are, however, but few hills, and no mountains. In the S. part of the county the ground is moorish, and there are some morasses; but elsewhere it is comparatively fertile. Agriculture similar to that of Mid Lothian, with this difference, that more turnips are raised and fewer potatoes. Estates large; farms of a middle size. Average rent of land, in 1810, 21s. 7d. an acre. Coal is found in most parts of the county. Manufactures of no importance. W. Lothian is divided into 13 parishes: it sends one member to the House of Commons for the county; and the boroughs of Linlithgow and Queensferry join with others in returning representatives. Reg. electors for the county in 1839-40, 716. In 1831 W. Lothian had 3400 inhabited houses, 5014 families, and 23,291 inhabitants, of whom 10,955 were males, and 12,336 females. Valued rent, £73,619 Scotch; annual value of real property, in 1815 £97,397.

LOUDON, county, Va. Situated in the N.E. part of the state, and contains 469 sq. m. Bounded N.E. by Potomac river. Watered by Goose creek and its branches and Broad river. It contained in 1840, 25,690 neat cattle, 31,503 sheep, 38,841 swine; and produced 573,450 bushels of wheat, 81,517 of rye, 891,695 of Indian corn, 6845 of buckwheat, 1472 of barley, 224,706 of oats, 53,947 of potatoes, 1733 pounds of tobacco. It had 79 stores, four fulling-mills, two woollen factories, 55 flouring-mills, 31 grist-mills, 31 saw-mills, 11 tanneries, five distilleries; six academies, 256 students; 33 schools, 1008 scholars. Pop. 762. Of these: whites, 13,840; slaves, 5673, free coloured, 1318; total, 20,431. Capital, Leesburg.

LOUDON, p. t., Merrimac co., N. H., 8 m. N.N.E. Concord, 489 W. Chartered in 1773. Soocook river affords water power. It contains three churches, two Congregational and Methodist, three stores, one woollen factory, three grist-mills, eight saw-mills, one oil-mill, three tanneries, one pottery; 10 schools, 495 scholars. Pop. 1640.

LOUDON, p. t., Seneca co., O., 88 m. N. Columbus, 432 W. Drained by a branch of Sandusky river. It has two stores, one grist-mill; three schools, 94 scholars. Pop. 762.

LOUDUN, a town of France, dep. Vienne, cap. arrond., on a hill, 31 m. N.N.W. Poitiers. Population, in 1836, ex. com., 4428. It was formerly of considerable importance, and has still many large houses and wide streets; but its inhabitants being principally Protestants, it suffered much from the revocation of the edict of Nantes, from the effect of which it has never recovered. It has a hospital, a theatre, the remains of an ancient castle, a tribunal of original jurisdiction, and manufactures of woollen cloth, lace, &c. (*Hugo*, art. *Vienne*.)

This town is famous, or rather infamous, in the history of fanaticism for a judicial murder committed in it, in 1634, when a curate, of the name of Grandier, accused and convicted of sorcery and magic, was burnt alive! The unfortunate curate appears to have had but little respect for that rule of the Roman Catholic religion which enjoins the celibacy of the clergy; and he is said, and we presume truly, to have practised his arts with most success on the nuns belonging to an Ursuline convent in the town. (See *Biographie Universelle*, art. *Grandier*; *Histoire des Diables de Loudon*, passim, &c.)

LOUGHBOROUGH, a market town and par. of England, co. Leicester, hund. W. Goscote, near the left bank of the Soar, 10 m. N. Leicester, and 96 m. N. by W. London. Population of township in 1831, 10,900. It is a clean and respectable looking town, with several streets lined with modern brick houses, meeting the principal avenue on the great London road. The market-place, in which is the town-hall, was formerly narrow and confined, but has been recently laid open by the pulling down of the old market-house. The church, a large and handsome structure in the perpendicular style, has a lofty and well proportioned tower: the living is a rectory (value £1484), in the gift of Emanuel college, Cambridge. There are places of worship, likewise, for Presbyterians, Independents, Baptists, Unitarians, Wesleyan and Primitive Methodists, and the Society of Friends, connected with which are seven Sunday-schools furnishing religious instruction to between 2000 and 3000 children of both sexes. Besides a well-endowed grammar-school, Loughborough has a charity-school for clothing and instructing 80 boys; a subscription-school, attended by 250 boys, and a school of industry with 108 girls. A dispensary and several charitable societies confer essential benefits on the poor, and there is also a large public library and news-room. Fleecy-hosiery and bobbin-net lace are the chief branches of industry, the former occupying nearly 1000 hands in the town and neighbourhood; several persons are employed in making cotton hose and gloves; there are many makers of machinery, and a considerable number of shoemakers, working for the London market. In 1839 there were two worsted-mills, giving employment to 213 persons. The prosperity of the town has been increased by the facility of transit afforded by the Leicester Navigation and Loughborough canal; but much greater benefits are likely to result from the recent opening of the Midland Counties' railway, which brings this town within four hours' distance of the metropolis. Petty sessions every market-day. Loughborough is the election town and principal polling place for the N. division of the county. Markets on Thursday: large fairs for horses, cattle, and sheep, February 14, March 26, April 25, Holy Thursday, August 12, and November 13: cheese fairs, March 24, and September 30.

LOUGHREA, an inland town of Ireland, co. Galway, prov. Connaught, on Loughrea lake, 21 m. E. by S. Galway. Pop., 1831, 4097, mostly Roman Catholic. The town, which was formerly fortified, consists of several irregular streets and lanes. The public buildings are—the parish church, the spire of which was thrown down by lightning in 1832, three Roman Catholic chapels, several large schools, and a barrack. General sessions are held twice a year; petty sessions on Thursdays. It is a constabulary and revenue guard station. Markets on Thursday: fairs on Feb. 11, May 26, Aug. 20, and Dec. 5. Postoffice revenue, in 1830, £568; in 1836, £894. A branch of the National Bank was opened in 1836.

LOUIS (ST.), a town of Western Africa, and the capital of the French possessions in Senegambia, on an island of its own name in the Senegal, about 7 m. from its mouth; lat. 16° 21′ N., long. 16° 13′ 45″ W. Pop., in 1836, exc. garrison, 11,606, of whom 6006 were slaves. It is laid out on a regular plan, nearly a mile in length, by about 206 yards broad. Fort St. Louis, with its esplanade, occupies the centre of the town; and from two of its opposite faces, a street is prolonged, and crossed at right angles by several others. The town has about 250 brick houses, half of which have only a ground floor, and the other half rarely more than an additional story; the other dwellings are mere huts of mud and straw. The chief public buildings are the governor's residence, the barracks, and the new hospital. The last is a superior edifice of its kind for a colony of such inferior rank, and has 122 beds, a number sufficient to accommodate the greatest average number of sick. There is good anchorage in the river on both sides the island, but especially in the E. channel, where ships may lie quite close to the quay. There are neither brooks nor public fountains in St. Louis; and the water for daily use, which has to be brought from the river, is brackish. St. Louis is the seat of a tribunal of primary jurisdiction and of commerce, and a council of appeal. It is also the residence of the apostolic prefect of the colony, and the chief officers of the colonial government. Boat-building and a little weaving are its principal branches of industry. (*Hugo*, art. *Senegal*; *Dict. Géog.*)

LOUISA, county, Va. Situated centrally in the E. part of the state, and contains 570 sq. m. Bounded N. by North

Anna river. Watered by South Anna river. It contained in 1840, 10,796 neat cattle, 12,711 sheep, 20,133 swine; and produced 220,748 bushels of wheat, 899 of rye, 432,661 of Indian corn, 158,731 of oats, 15,395 of potatoes, 2,430,764 pounds of tobacco, 19,120 of cotton. Value of gold produced, $3000. It had 30 stores, three flouring-mills, 25 grist-mills, 16 saw-mills, 11 tanneries; 39 schools, 591 scholars. Pop.: whites, 6047; slaves, 9010; free coloured, 376; total, 15,433. Capital, Louisa C. H.

LOUISA, county, Iowa. Situated towards the S.E. part of the territory, and contains 442 sq. m. Bounded E by Mississippi river. Watered by Iowa river and its tributaries, which afford water-power. It contained in 1840, 2060 neat cattle, 690 sheep, 6813 swine; and produced 10,535 bushels of wheat, 82,695 of Indian corn, 15,725 of oats, 6135 of potatoes. It had three saw-mills. Pop. 1927.

LOUISA, p. v., capital of Lawrence co., Ky., 158 m. E. Frankfort, 436 W. Situated on the W. side of Big Sandy river, and contains a courthouse, jail, and several stores and dwellings. Nett proceeds of the postoffice, $93.

LOUISA, C. H., p. v., capital of Louisa co., Va., 60 m. N.W. Richmond, 103 W. It contains a courthouse, jail, and several stores and dwellings. Nett proceeds of the postoffice, $368.

LOUISBURG. See CAPE BRETON.

LOUISBURG, p. v., capital of Franklin co., N. C., 36 m N.E. Raleigh, 262 W. Situated on the N. side of Tar river. It contains a courthouse, jail, two churches, 10 stores, and about 500 inhabitants. Nett proceeds of the postoffice, $488.

LOUISIANA, the southernmost of the southern United States, is bounded N. by Arkansas and Mississippi; E. by Mississippi from which it is separated by Mississippi r. to the 31° of N. lat., thence E. on that parallel to Pearl r., and down that river to its entrance into the gulf of Mexico; S. E. and S. by the gulf of Mexico; and W. by Texas, from which it is separated by Sabine r. to 32° N. lat., and thence due N. to 33° N. lat., where it meets the S. boundary of Arkansas. It is between 29° and 33° N. lat., and between 89° 40' and 94° 25' W. long. It is 250 m. long, from N. to S. On the gulf of Mexico it is about 300 m. broad, and continues this width for 120 or 130 m. inland, when it suddenly contracts to the width of about 100 m., and on the N. boundary it is 180 m. wide. It contains 48,320 sq. m., or 30,924,800 acres. The population in 1810, was 76,556; in 1820, 153,407; in 1830, 215,575; in 1840, 352,411, of whom 168,452 were slaves. Of the free population 89,747 were white males; 68,710 do. females; 11.526 coloured males; 13,976 do. females. Employed in agriculture, 79,289; in commerce, 8549; in manufactures and trades, 7565; in navigating the ocean, 1322; do., canals, lakes, and rivers, 662; in the learned professions, 1018.

The state is divided into 38 parishes, answering to counties in other states, which with their population in 1840, were as follows:—

Eastern District.		Western District.	
Parishes	Pop.	Parishes.	Pop.
Ascension	6,951	Avoyelles	6,616
Assumption	7,141	Caddo	5,282
Baton Rouge, E.	8,138	Calcasieu	2,057
Baton Rouge, W.	4,638	Caldwell	2,017
Carroll	4,237	Catahoula	4,955
Concordia	9,414	Claiborne,	6,185
Feliciana, East	11,893	Lafayette	7,841
Feliciana, West	10,910	Natchitoches	14,380
Iberville	8,494	Rapides	14,132
Jefferson	10,470	St. Landry	15,233
Lafourche Interior	7,303	St. Martin's	9,674
Livingston	2,315	St. Mary's	5,950
Madison	5,144	Union	1,581
Orleans	102,193	Washita	4,640
Plaquemines	5,060		
Point Coupee	7,998		
St. Bernard	3,237		
St. Charles	4,700		
St. Helena	3,525		
St. James	8,548		
St. John Baptist	5,776		
St. Tammany	4,598		
Terre Bonne	4,410		
Washington	2,649		
Total	249,641	Total	102,770
		Total of State	352,411

New-Orleans, on the N. bank of Mississippi river, which here runs eastwardly, 105 m. from its entrance in the gulf of Mexico, is the seat of government.

The whole southern border of the state, from Pearl river to the Sabine, consists either of sea, marsh or vast prairies, which occupy about one fifth of the surface of the state; and on the borders of the streams are timbered lands. The tract about the mouths of the Mississippi, for 30 m., is one continued swamp, destitute of trees, and covered with a species of coarse reed, 4 or 5 ft. high. The prospect of the country, from the mast of a ship, is an extended and dreary waste. Along the whole border of the gulf of Mexico, a sea marsh extends inland for 20 or 30 m. Back of this the land
240

gradually rises a little, and constitutes the prairies. A large extent of country is annually overflowed by the Mississippi and its various outlets. From lat. 39° to 31° the average width of overflowed land is 20 m.; from lat. 31° to the efflux of La Fourche, the width is about 40 m. All the country below the La Fourche, with little exception, is overflowed. By a survey made by order of the government of the United States, in 1828, it was found that the river overflowed an extent of 5,000,000 of acres, a great proportion of which is at present unfit for cultivation. A part of this is covered by a heavy growth of timber, and an almost impenetrable growth of cane, and other shrubbery. This becomes dry on the retiring of the river to its natural channels, and has a soil of great fertility, and which might, by labour, be rendered fit for cultivation. There are in some parts basins or depressions, in which the water remains until it is evaporated or absorbed by the earth. These by draining might constitute rice fields. The sea marsh is partially overflowed by the tides, and especially when driven in by the equinoxial gales. In the alluvial territory are small bodies of prairie lands, slightly elevated, without timber, and of great fertility. More extended prairies constitute a large portion of the state. The pine woods, which are extensive, have generally a rolling surface, and a poor soil. The greater part of the prairies has a second rate soil, but some parts of those of Opelousas, and particularly of Attakapas have great fertility and feed extensive herds of cattle. More earth is deposited by the Mississippi in its overflow on its immediate margin than further back; and therefore the land is higher adjoining the river than in the rear of its banks. This alluvial margin, of a breadth from 400 yards to 1½ m., is a rich soil; and to prevent the river from inundating the valuable tract in the rear, and which could not be drained, an artificial embankment is raised on the margin of the river, called the Levee. On the E. side of the river, this embankment commences 60 m. above New-Orleans and extends down the river more than 130 m. On the W. shore it commences at Point Coupée, 172 m. above New-Orleans. Along this portion of the river, its sides present many beautiful and finely cultivated plantations, and a continued succession of pleasant residences. The country between the Mississippi, Iberville, and Pearl rivers, in its southern parts, is generally level, and highly productive in cotton, sugar, rice, Indian corn, and indigo. The northern part has an undulating surface, and has a heavy natural growth of white, red, and yellow oak, hickory, black-walnut, sassafras, magnolia, and poplar. In the N.W. part, Red river, after entering the state by a single channel and flowing about 30 m., spreads out into a great number of channels, forming many lakes, islands, and swamps, over a space of 50 m. long and 6 broad. Here the fallen timber, floated down by the stream, has collected and formed the celebrated raft, which formerly extended 160 m., obstructing the navigation of the river. Most of it has been removed by order of the general government, and the remainder will ere long be cleared away, opening this fine river to an extensive steamboat navigation. The bottoms on this river are from one to ten miles wide, and are of great fertility, with a natural growth of willow, cottonwood, honey-locust, papaw and buckeye; on the rich uplands grow elm, ash, hickory, mulberry, black-walnut, with a profusion of grape vines. On the less fertile and sandy uplands of the state are white pitch and yellow pines, and various kinds of oak. The lower courses of Red river have been denominated the paradise of cotton planters.

The staple productions of this state are cotton, sugar and rice. Sugar-cane grows chiefly on the shores of the gulf and the bayous Teche, La Fourche, and Plaquemine, and in some parts of Attakapas S. of 31° N. lat. No cultivation yields a richer harvest, though the labour of the hands is severe. There is a vast amount of sugar lands not brought into cultivation. The quantity of land adapted to sugar has been computed at 250,000 acres; of rice, at 250,000 acres and of cotton, at 2,400,000. Rice is principally confined to the banks of the Mississippi, where irrigation is easy.

There were in this state in 1840, 98,888 horses and mules, 381,248 neat cattle, 98,072 sheep, 323,230 swine; poultry was raised to the value of $283,559. There were produced 60 bushels of wheat, 1812 of rye, 5,952,912 of Indian corn 107,353 of oats, 834,341 of potatoes, 119,824 pounds of tobacco 3,604,534 of rice, 152,555,368 of cotton, 119,947,720 of sugar 24,651 tons of hay, 49,283 pounds of wool, 1012 of wax; the products of the dairy were valued at $133,069, of the orchard, at $11,769, of lumber, at $606,106. There were made 2.884 gallons of wine, and 2233 barrels of tar, pitch, or turpentine.

The climate is mild, though the winters are more severe than in the same latitude on the Atlantic coast. The summers in the wet and marshy parts are unhealthy, and New-Orleans has been frequently visited by the yellow fever. But a considerable portion of the state is healthy.

The Mississippi river divides the state from Mississippi

for a course of 450 m., and enters the state wholly, 230 m. from its mouth, by the course of the channel of the river, and divides into several branches or outlets which, diverging from the main river, wind their way slowly to the gulf of Mexico, carrying off its surplus waters in times of flood, and dividing the southern part of the state into a number of large islands. The Atchafalaya, called here the Chaffalio, leaves the Mississippi on the W. side a little below the mouth of Red river, and is supposed to carry off as much water as Red river brings in; and inclining to the E. of S., it enters Atchafalaya bay in the gulf of Mexico. The outlet Plaquemine leaves the Mississippi 198 miles below the outlet of Atchafalaya, with which the main stream at length unites. Thirty-one miles below the Plaquemine, and 81 above New-Orleans, is the outlet of La Fourche, which communicates with the gulf of Mexico. Below the La Fourche, numerous other smaller streams leave the Mississippi at various points. On the E. side of the Mississippi, the principal outlet from that river is the Iberville, which passes to the gulf of Mexico through lakes Maurepas, Pontchartrain, and Borgne. This outlet on the E. and Atchafalaya on the W., bound what is denominated the Delta of the Mississippi. The Mississippi is navigable for vessels of any size, though the bar at its mouth has on it but 16 or 17 ft. of water. Red river crosses the state in a S.E. direction, and enters the Mississippi 240 m. above New-Orleans. Washita river runs in a S. direction, and enters Red river a little above its entrance into the Mississippi. The other rivers are Black, Tensas, Sabine, Calcasieu, Mermentau, Vermilion, Teche, Pearl, Amite, and Ibberville. The largest lakes are Pontchartrain, Maurepas, Borgne, Chetimaches, Mermentau, Calcasieu, and Sabine.

The vast trade of the valley of the Mississippi centres at New-Orleans, a valley which, for its extent and fertility, has not its like in the world. The exports of this state amounted in 1840, to $34,236,936, but these exports extensively belong to the great and fertile states of the great valley. Its imports were $10,673,190. There were 94 commercial, and 381 commission houses engaged in foreign trade, with a capital of $14,770,000; 2465 retail stores, with a capital of $14,301,694; 597 persons were engaged in the lumber trade, with a capital of $306,045; three persons employed in internal transportation, with 291 butchers, packers. &c., employed a capital of $144,323.

Its manufactures are less considerable. Home made or family manufactures amounted to $65,190; two cotton factories, with 706 spindles, employed 23 persons, producing articles to the amount of $18,900, with a capital of $22,000; six furnaces produced 1400 tons of cast iron, and two forges produced 1366 tons of bar iron, employing 145 persons, and a capital of $357,800; 25 tanneries employed 58 persons, and a capital of $122,025; seven other manufactories of leather, as saddleries, &c., produced articles to the amount of $108,500, with a capital of $49,550; one pottery employed 18 persons, producing articles to the amount of $1000, with a capital of $3000; five sugar refineries produced to the amount of $770,800; 101 persons produced confectionary to the amount of $20,050; machinery was produced to the amount of $5600; and hardware and cutlery to the amount of $30,000; 51 persons produced carriages and wagons to the amount of $40,350, employing a capital of $15,780; mills of various kinds produced articles to the amount of $708,785, employing 972 persons, and a capital of $1,870,795; vessels were built to the amount of $80,500; 190 persons manufactured furniture to the amount of $2300, with a capital of $576,050; five distilleries produced 255,520 gallons, and one brewery 2480 gallons, employing 27 persons, and a capital of $110,000; 75 persons manufactured 2,202,200 pounds of soap, 3,500,030 pounds of tallow candles, and 4000 pounds of wax or spermaceti candles, with a capital of $115,506; 948 brick or stone houses, and 679 wooden houses were built by 1484 persons, and cost $2,735,944; 35 printing-offices, five binderies, 11 daily, 21 weekly, and two semi-weekly newspapers, and three periodicals, employed 399 persons, and a capital of $193,700. The total amount of capital employed in manufactures was $6,438,699.

Louisiana college, at Jackson, was founded in 1825; Jefferson college, at Bringiers, was founded in 1831; St. Charles college, at Grand Coteau, is under the direction of the Roman Catholics; Baton Rouge college, at Baton Rouge was founded in 1838; Franklin college, at Opelousas, was founded in 1839. In these institutions there were, in 1840, 437 students. There were in the state 52 academies, with 1905 students; 179 common and primary schools, with 3573 scholars. There were 4861 white persons, over 20 years of age, who could neither read nor write. In 1835, the legislature granted to three colleges $363,775, to be paid out of the state treasury; viz., $48,775 to Jefferson college, to defray the expense of its buildings, and $15,000 annually for the period of 10 years; to Louisiana college, $15,000 annually for the same period to pay the salaries of their professors, and to lower the rates of tuition and other expenses; and $15,000 also to Franklin college.

The state was originally settled by Roman Catholics, and they are still the most numerous religious denomination. In 1835 they had 27 ministers. Methodists, Baptists, Presbyterians, and Episcopalians exist in considerable numbers, and are increasing.

About the commencement of 1840, the state had 16 banks, with 31 branches, with an aggregate capital of $41,736,769; and a circulation of $4,345,533. At the commencement of 1842, the state debt amounted to $20,830,899. The public debt consists almost entirely of state bonds, issued to the different banks, which bonds have been sold in Europe, and the proceeds constitute the capitals of the banks, which are loaned to the stockholders, on mortgages of their landed property. These mortgages are estimated to be worth $35,400,000.

The constitution of this state was formed in 1812. The governor is elected for four years. The people give their votes for governor at the same time that they vote for senators and representatives, and on the second day of the succeeding session of the general assembly, the two houses, by a joint ballot, elect for governor one of the two candidates who have the greatest number of votes. The governor's term of office commences on the fourth Monday succeeding his election, and continues for four years. He is ineligible for the next four years. The senators are elected for four years, one half being chosen every two years. The present number is 17, chosen by senatorial districts. The representatives are elected for two years, apportioned according to the number of electors, as ascertained by enumeration every four years. The present number is 60. The pay of the members of both houses is four dollars per day. The legislature meets biennially at New-Orleans, on the first Monday of January. The judges of the supreme and inferior courts are appointed by the governor, with the advice and consent of the senate, and hold their offices during good behaviour. The right of suffrage is possessed by every white male citizen of the United States, of 21 years of age or upwards, who has resided in the county where he offers his vote one year next preceding the election, and who in the last preceding six months has paid a state tax.

Several works of internal improvement have been undertaken. Pontchartrain railroad extends from New-Orleans 4½ m. to lake Pontchartrain, and cost originally $900,000, and with its improvements $500,000. West Feliciana railroad extends from St. Francisville on Mississippi river, 90 m., to Woodville, Miss. Orleans-street railroad, through Orleans-street, is 1½ m. long, and connects New-Orleans with the bayou St. Johns, and cost $12,000. New-Orleans and Carrolton railroad extends from New-Orleans, 6½ m., to Carrolton, passing through Lafayette. It has city branches, making its whole length 11½ m. Various other railroads and canals have been projected, and some work has been done upon them, but they are at present suspended.

The river Mississippi was discovered in 1683, by Marquette and Joliette, two French missionaries. In 1682 the country was explored by La Salle, and named Louisiana, in honour of Louis XIV. In 1699 a French settlement was begun at Ibberville, by M. Ibberville, who in the attempt to plant the country lost his life. His efforts were followed up by M. Crozat, a man of wealth, who held the exclusive trade of the country for a number of years. About the year 1717 he transferred his interest to a chartered company, at the head of which was the celebrated John Law, whose national bank and Mississippi speculation involved the ruin of half the French nobility. In 1731 the company resigned the concern to the crown, who, in 1762, ceded the whole of Louisiana to Spain. In 1800, Spain re-conveyed the province to the French, of whom it was purchased by the United States, in 1803, for about $15,000,000. This purchase included all the present territory of the United States, W. of the Mississippi. Soon after the purchase, the present state of Louisiana was separated from the rest of the territory, under the name of the territory of Orleans. In 1812 Louisiana was admitted to the Union as a state, and the part of W. Florida W. of Pearl river was annexed to it In December, 1814, and for several days afterwards, the British made an attack upon New-Orleans, but were repulsed, January 8th, 1815, by the Americans, under Gen. Jackson, with the loss of about 3000 men. killed, wounded and taken prisoners. The Americans lose is stated to have been only seven men killed, and six wounded. Gen. Packenham, the British commander, was killed.

LOUISVILLE, p. t., St. Lawrence co., N. Y., 253 m. N.N.W. Albany, 593 W. Bounded N.W. by St Lawrence river. Watered by Grass river. Several islands in St. Lawrence river belong to it. It has two stores, one gristmill, two saw-mills; 11 schools, 391 scholars. Pop. 1693.

LOUISVILLE, p. v., capital of Jefferson co., Ga., 53 m. E. Milledgeville, 634 W. Situated on Rocky Comfort creek, just above its entrance into Ogeechee river. It contains a courthouse, jail, a church, an academy, nine stores, and about 30 dwellings. The legislature met here from 1795 to 1807. The old state-house has been converted into the courthouse. Here the famous Yazoo acts, conveying several thousand acres of land in western Georgia, now Mississippi, said to have been obtained by fraud, were burned, by order of the legislature, Feb. 13th, 1796, by fire from heaven, kindled by a sun glass. The sum of $500,000 had been paid, which its owners were allowed to withdraw within eight months. Accordingly, $200,000 were withdrawn from the treasury by claimants, and the balance was transferred to the United States government, who engaged to compromise with these claimants. It has been erroneously stated that the money was never refunded.

LOUISVILLE, p. v., capital of Winston co., Miss., 92 m. N.E. Jackson, 918 W. It contains a courthouse, jail, two churches, two academies, and about 180 inhabitants.

LOUISVILLE, city, port of entry, and capital of Jefferson co., Ky., is in 38° 3′ N. lat., and 85° 30′ W. lon. from Greenwich, and 8° 45′ W. lon. from Washington, 52 m. W. by N. Frankfort, 1400 m. by water from New-Orleans, 600 m. by water and 350 m. by land from St. Louis, 650 m. by the river below Pittsburg, and 596 W. Situated on the S. side of Ohio river, at the head of the rapids. The Ohio river opposite to the city is over a mile wide, and for seven miles above is a most beautiful sheet of water, presenting, with its rapids, islands, cliffs, villages, and towns, a delightful landscape. The population in 1778, was estimated at 30 persons ; by the census of 1800, 800 ; in 1810, 1357 ; in 1820, 4012 ; in 1830, 10,336 ; in 1840, 21,210 ; in 1843, by a census taken by order of the mayor and council, 28,643. The city is built on a gentle acclivity, about 75 ft. above low water mark, and is regularly laid out on a slightly undulating plain, with ten broad and straight streets, from 60 to 120 ft. wide, running parallel with the river, nearly E. and W. ; intersected at right angles by 30 cross streets, all 60 ft. wide. The squares are 420 ft. on each street, and most of them are divided by alleys of from 10 to 30 ft. wide. Bear Grass creek, over which bridges have been constructed at convenient distances, divides the upper part of the city ; and the point of land projecting between it and the river renders it an excellent and secure harbour for boats in times of high water, and of descending ice. The public buildings are a city-hall and courthouse, now in a course of completion, on a scale of vastness and magnificence equal to the wants of Philadelphia or New-York ; a city and county jail on the most approved model, combining health, convenience, and security, containing 48 single cells, 6 ft. by 10 ft., and double cells, 10 ft. by 13 ft., all of solid stone, perfectly dry, and thoroughly warmed and ventilated, all opening on interior galleries to the third floor ; a marine hospital, erected in 1830, by a grant of $40,000 from the state ; the lot on which it stands, containing about 7½ acres, was the noble donation of the late Thomas Prather and Cuthbert Bullitt ; the building presents a fine appearance, is well endowed, admirably kept, and is amply sufficient for all the purposes of its establishment ; a medical institute, adapted to its object, well supplied with books and apparatus, selected by an eminent surgeon, sent specially for the purpose to Europe. The institute has been for four years in operation, and for the last two years has been exceeded in the number of its students only by the medical schools at Philadelphia and Lexington ; 25 churches, four Baptist, one Campbellite Baptist, six Methodist, one of which is a Dutch Methodist ; and a Seamen's Bethel, three Presbyterian, two of which are Old School, and one New School ; two Episcopal, one Friends, one Unitarian, one Universalist, two Roman Catholic, one of which is Dutch ; three coloured churches, two of which are Methodist, and one Baptist ; and a free-church, besides a Jews' synagogue. Some of the churches exhibit fine specimens of architecture. There are five banking-houses, four markets, a city workhouse, hospital, and prison, two orphan asylums, a Magdalen Asylum, under the care of the sisters of the Good Shepherd ; a School for the Blind, endowed by the state ; and four large city schoolhouses. Its spacious hotels are among the best in the Union. Many private dwellings, recently erected, exhibit a decided improvement in architectural taste, and add much to the beauty of the city.

The banking facilities are adequate to the wants of the extensive commerce centering at this point. The Bank of Kentucky has a capital of $3,000,000 ; the Bank of Louisville of $1,250,000 ; a branch of the Northern Bank of about $1,000,000 ; and two savings' banks, with about $100,000 each. There are four insurance offices, with capitals of $100,000 each. which, with agencies from offices in other states, are sufficient for all the purposes of marine, fire, and life insurance. The principal business of Louisville is foreign and domestic commerce ; it is the chief wholesale market of the west for dry goods and groceries, and is resorted to by merchants from the river towns above

250

and below ; and from the interior of the adjacent states, as the place at which they can purchase a more complete assortment, and at cheaper rates than at any other place. This commerce is carried on by upward of 300 steamboats, constantly arriving at, and departing from every place accessible to them ; besides numerous flat and keel-boats. Good turnpike roads lead from Louisville to all parts of Kentucky and Indiana, and the hourly arrival and departure of stagecoaches, wagons, and the 100 hacks, and 250 drays and carts employed in the city, present a scene of unusual bustle. It is estimated that, during the year 1843, of the groceries landed at this place, there were 15,000 hogsheads of sugar, and 40,000 bags of coffee, with other articles in proportion ; that the importation of foreign and domestic goods exceeded $10,000,000 ; and that the import and export trade amounted to more than $40,000,000. The exports consist of about 8000 hogsheads of tobacco, 250,000 pieces and coils of bagging and bale-rope, large quantities of pork, lard, bacon, flour, whiskey, feathers, flaxseed, beeswax, ginseng, live-stock, and the usual agricultural products for the southern market. To these should be added steam engines, sugar-mills, and machinery of various kinds.

Manufactures to a considerable extent are already established, and are rapidly increasing. There are six large founderies with machine shops, in which engines and machinery are produced equal to any in the world. During the year 1843, 28 steamboats, many of them of the largest class, were built in Louisville and the adjoining towns, all of which were furnished with engines, &c., from the city founderies. Two extensive steam bagging factories are in full operation, and convert a number of tons of raw hemp into bagging daily ; five ropewalks produce large quantities of cordage and bale-rope. There are three breweries and two distilleries ; an extensive cotton factory ; a woollen factory for the manufacture of jeans and negro-cloth ; four flouring-mills, producing 300 barrels of flour daily ; a mill for the manufacture of hydraulic cement ; starch, white lead, linseed, and lard oil are made in large quantities ; tobacco, snuff, cigars, mustard, cabinet wares, &c., with almost every article produced elsewhere, are now manufactured here. The book trade, printing, bookbinding, and paper-making are carried on with an energy, success, and extent of business unsurpassed by any western city.

The subject of education was one of the first objects to which the attention of the city government was called ; and in 1829 the first free schools west of the mountains were established in the city. There are now 18 schools, of which six are grammar schools, three male and three female, and 12 are primary schools, for both sexes. A college is also connected with the system, in which boys may be fitted, in the most thorough manner, for any occupation or profession. There are also between 40 and 50 private schools, many of which possess high merit. The number of children annually instructed in the public schools is 3080, and in the private schools is 1500. The Mercantile Library Association has a library of 3500 volumes of admirably selected books. The Kentucky Historical Society has a large and valuable collection of books and manuscripts, connected with the settlement and history of the state. There are five daily and an equal number of weekly newspapers, besides several literary, religious, and scientific magazines. An Agricultural and a Horticultural Society have been organized, whose exhibitions would do credit to older places.

The city is abundantly supplied with excellent water from pumps at the intersection of the streets, and by numerous large cisterns, kept constantly full for extinguishing fires. Measures are also in progress for the construction of water-works sufficient to furnish an ample supply of pure water for all domestic and manufacturing purposes. These water-works are expected to be in operation in the course of the year 1844. No city is better lighted. The gas-works, equal to any in the country, furnish an abundant supply of the very best gas, to illuminate brilliantly the streets, shops, and dwellings. The fire-department is well organized, and supplied with all the requisite apparatus ; and, whenever the occasion has called forth its energies, it has been found admirably prompt and efficient.

The falls of the Ohio at this place are caused by a bed of limestone, extending across the river, over which the water pour with an irregular current, and in crooked channels, for two miles, producing a descent in that distance of 24 ft. Previously to the construction of the canal, this was an entire obstruction to the passage of loaded boats for nine months in the year, and for ascending boats for 10 or 11 months. This was a great obstruction to the navigation of the beautiful and important Ohio, above this point. The canal is about 2½ m. long, and has four locks, sufficiently capacious to admit steamboats of the largest class. The canal is 50 ft. wide at the surface, and the entire lockage is 22 ft. With a trifling exception, the entire line is excavated out of compact limestone, to the mean depth of 10 ft. ; everything connected with the canal is of the most substantial kind, and the mechanical execution is excellent. After encountering and overcoming many and great difficulties, this canal was successfully completed in 1833, at an expense of $1,000,000 to the persevering and public-spirited individuals by whom this magnificent work was projected and undertaken; and the company are now reaping the reward their zeal and enterprise so richly deserve. The United States are stockholders to the amount of $290,000; the remainder of the stock is owned by private individuals. The number of steamboat passages through the canal, since it was first opened in 1830, has been 13,550; of keel and flat boats, 5013; and the tolls received have been $1,925,350. The shares were originally $100; they are now selling at $150. When the time shall arrive (and it is probably not remote) for the construction of a permanent dam across the Ohio at the falls, a slack-water navigation can be made for 75 m., to the mouth of Kentucky river, and an incalculable amount of water-power will be obtained; and Louisville may become one of the greatest manufacturing cities in the Union. Recent surveys have made the great water-power which she already possesses apparent, and the contemplated waterworks will show how easily and how cheaply this hitherto neglected element may be made subservient to the wealth and prosperity of the city and surrounding country.

Shippingport and Portland are two villages separated by the canal, about 2 m. below the city. Formerly they were the port of the city below the falls, and goods and passengers from the Mississippi and the lower Ohio were landed at these villages, and brought up to the city by land. But now all steamboats which can run on the Ohio may pass through the canal, and discharge and receive cargoes at the wharves of Louisville.

According to the census of 1840, Louisville had one commercial and 11 commission houses in foreign trade, with a capital of $191,800; 270 retail stores, with a capital of $2,128,400; three lumber-yards, capital $52,000; two flouring-mills, one paper-mill, two tanneries, two breweries, one glass cutting works, one pottery, two rope-walks, seven printing-offices, two binderies, five daily, seven weekly, and three semi-weekly newspapers, and one periodical; total capital employed in manufactures, $713,675; one college, 80 students; 10 academies, 280 students; 11 schools, 386 scholars.

As early as 1778 a fort was built here, and corn was raised on the island opposite to the city, since called Corn island. Gen. George Rogers Clark made it his headquarters, whence he fitted out his celebrated expedition against the British posts of Kaskaskia, Vincennes, &c. For several years after the first settlement, Louisville was harassed by Indian hostilities. In 1780 the legislature of Virginia authorized the laying out of a town at the falls of the Ohio, and named it Louisville, in honour of Louis XVI., the firm ally of the American republic. In 1781, formal possession was taken, a fort was built, and a check was given to the depredations of the Indians. From the peculiar position of the town, in the midst of several swamps and ponds, it was justly considered, until 1823, as excessively unhealthy; but during that and the previous year, all these local causes of disease were removed, and since the fatal autumn of 1822 it is believed that no town or city of the Union has been more uniformly healthy. The average mortality from 1839 to 1842 inclusive, as derived from official sources, has been only two per cent. of the resident population.

LOUISVILLE, or LEWISVILLE, p. v., capital of Clay co., Ill., 112 m. S.E. Springfield, 741 W. Situated on the S.W. side of Little Wabash river. It contains a courthouse, and several stores and dwellings.

LOUTH, a marit. co. of Ireland, on its E. coast, being the most northerly in the province of Leinster, having E. the Irish sea; N. Carlingford bay, which separates it from Down and Armagh; and W. and S. Monaghan and Meath. Area, 208,261 acres, of which 14,916 are unimproved mountain and bog. Surface rugged is the N., but in other parts generally flat or undulating. Soil generally fertile. Estates of a medium size. Farms of all sizes. but the great majority are small. Its crops, agriculture, &c., are similar to those of Meath, which see. Average rent of land, 16s. an acre. Minerals unimportant. The linen manufacture is carried on to a considerable extent. especially at Drogheda, but the business has materially declined. Principal rivers Boyne and Dee. Principal towns Drogheda, Dundalk, and Ardee. Louth is divided into four baronies. and 61 parishes; and sends four members to the House of Commons, viz.. two for the county, and one each for Drogheda and Dundalk. Registered electors for county, in 1830–40, 1192. In 1831, it had 22,040 inhabited houses, 23,585 families, and 124,846 inhabitants, of whom 60,617 were males, and 64,229 females.

LOUTH, a mun. bor. market town, and par. of England, co. Lincoln, in the Wold div. of Louth-Eske hund., paris of Lindsey, 22 m. E.N.E. Lincoln, and 127 m. N. London. Population of borough, in 1831, 6927. The town, agreeably situated in a fertile valley S.E. of the wolds of N. Lincoln,

has of late been much improved, and is well paved and lighted with gas: it has several handsome, and a few elegant buildings, the houses generally being of brick, roofed with slate. The principal public buildings are the mansion-house, town-hall, sessions-house, and a small theatre. The church is a large Gothic structure, with a beautiful E. window, and one of the finest towers in the country, above which rises a light octangular spire, to a height of 290 feet from the ground. The living is a vicarage attached to a prebend. in Lincoln cathedral. A second par. church, once existing, is now destroyed; but its site is marked by the cemetery still used as a place of interment. A new district church has also been erected within the last few years. There is a Roman Catholic chapel; and the Wesleyan and Primitive Methodists, Baptists, and Independents have each places of worship; to which, as well as the church, are attached well attended Sunday schools. The free grammar-school, founded in 1552 by Edward VI., is endowed with landed property producing £700 a year; the half going as salary to the master, the fourth to the usher, and the residue to the support of 12 poor women. A school, established in 1677, provides instruction in English and mathematics to 30 free boys, and 30 pay scholars. There is also a well attended national school; and among the charitable institutions are almshouses, a dispensary, Benevolent society, and Bible society.

"Louth contains little or no manufacture, there being only one establishment of any importance, a carpet and blanket manufactory. The river Ludd flows round a considerable portion of the town: it is not navigable, but feeds a canal beginning at the N.E. extremity of Louth, and communicating with the Humber. It is in a very prosperous state: the principal traffic outwards is that of corn for London, and the W. riding of Yorkshire, the inland freight being chiefly coal. most of which comes down the Humber from York. There is a paper-mill and flour-mill on the river, in addition to wind-mills; and two other mills, one worked by steam, are employed in grinding bones." (Mun. Bound. Rep.) The town has also a soap-boiling establishment, and is famed for its excellent ale. Louth was incorporated in the 5th of Edward VI., whose charter was confirmed by other subsequent monarchs, and lastly, by George IV. Under the Municipal Reform Act, the borough is divided into two wards, and is governed by a mayor, and three other aldermen, with 12 councillors: it has a commission of the peace under a recorder. Corporation revenues, in 1839, £1349 (exclusive of £133 accruing from the sale of property). Louth is also one of the polling-places for the N. or Lindsey division of the county, and the quarter-sessions for the county are held here in January, July, and October, the April sessions being at Spilsby. Markets on Wednesday and Saturday: considerable horse fairs, April 30, 3d Monday after Easter. August 5, and a large cattle fair. November 23. (Mun. Corp. and Bound. Reports.)

LOUVAIN (Dutch Leuven), a town of Belgium, and formerly one of the most populous and industrious in that country, prov. S. Brabant, cap. arrond. and cant., on the Dyle, a tributary of the Scheldt, and on the railway between Brussels and Liege, 14 m. E.N.E. the former; lat. 50° 53′ 26″ N., long. 4° 41′ 46″ E. Population, in 1836, 24,242. It is partly surrounded by walls, and partly by an earth rampart from 80 to 100 feet high, with a deep fosse outside, the total circuit of both being about 7 m.; a great part of the inclosed area consists, however, of fields and gardens. But its fortifications are now cut through by different roads, and are mostly converted into boulevards. The castle, now in ruins, on a hill near the Dyle, is of considerable, but uncertain, antiquity: it was long the residence of the counts of Louvain. The town, which, though regularly laid out, is not generally well built, has several interesting public edifices. The town-hall, begun in 1440, and completed about ten years afterwards, one of the finest specimens of the florid Gothic in Europe, has been recently repaired, or rather restored, with great skill, at the joint expense of the town and government. It is lofty, has six light and elegant minarets, and is most elaborately ornamented. The collegiate church of St. Peter, a curious old edifice probably of the 14th century, with some good paintings, has a finely carved pulpit, and had, formerly, a steeple, blown down in 1604, which is said to have been of the extraordinary height of 533 feet, with two magnificent lateral towers. The university of Louvain was founded by John IV. duke of Brabant in 1425; but it was not till 1431 that it obtained the privilege of teaching theology, for which it was afterwards so celebrated. It had, in the days of its prosperity, more than 40 colleges, some of which were established to help students that had previously belonged to the cloithiers. This famous seminary, after being suppressed by the French in 1797, was re-established in 1817. It has at present 20 colleges, some of which are handsome buildings. Its library, originally the drapers' hall, is richly decorated with antique wooden carvings. Edward III. of

England resided for a year, and the emperor Charles V. was brought up, in the castle of Louvain. The town has five churches, five nunneries, eight hospitals and charitable asylums, a royal college, and a college for ecclesiastics; and is the seat of tribunals of primary jurisdiction and commerce, a chamber of commerce and manufactures, a board of forest inspection, &c.

In the 14th century Louvain was one of the great seats of the woollen and linen manufacture, which supported. It is said, no fewer than 150,000 individuals within the city (*Busching*)! though this, most probably, is a gross exaggeration. But the manufacturers having revolted, in 1382, against the Duke of Brabant, many of them emigrated, on the revolt being suppressed, to foreign countries, and among others to England; where, being hospitably received by Edward III., they assisted in laying the foundations of the woollen manufacture. Louvain seems never to have recovered from this disaster. It has still some inconsiderable woollen fabrics; but it is now principally celebrated for its beer, said to be the best in Belgium. The different breweries produce about 200,000 barrels a year, a large proportion of which is sent to Antwerp and into Flanders. Louvain has also manufactures of lace and cotton yarn, and several dyeing and cotton-printing establishments, with tanneries, distilleries, and glass works, and numerous oil and flour-mills. It is connected with the Demer near Mechlin by the canal of Louvain, navigable for vessels of 150 tons; and has a considerable trade in corn, clover seed, flax, hemp, &c., the produce of the surrounding country. Under the French it was included in the dep. of the Dyle. (See the *Dictionnaire de Martiniere*, for an elaborate article on this city; see also *Busching*; *De Cloet*; *Murray's Handbook*, &c.)

LOUVIERS, a manufacturing town of France, dep. Eure, cap. arrond., on the Eure, and on the road from Rouen to Evereux, 19½ m. N. the latter, and 16 m. S.S.E. the former city. Population in 1836, six. com. 8713. It consists of an old and new town; the former, consisting of three or four principal, united by many smaller streets, is built chiefly of wood; the latter, which is the residence of the principal manufacturers, has a broad and elegant main street, and many well-built brick and stone houses. The Eure, which is navigable from the Seine as far as Louviers, is here crossed by several good bridges. A large church, supposed to have been constructed during the early crusades, a hall built by the Templars towards the end of the 12th century, a theatre, and a public library are the chief public buildings. Louviers was formerly a fortress of some strength; and portions of its wall still exist. It is now, however, distinguished wholly by its industry; and ranks as one of the first seats of the woollen manufacture of France. Fine broadcloths and woollen yarn are its chief products; but of late years, other fine woollen goods have been introduced. Cotton yarn, linen, thread, soap, &c., are made; and there are many dyeing establishments, and bleaching grounds, tanneries, sugar refineries, and factories for looms and other machinery. The woollen manufacture in 1834 employed 6000 hands; and, according to Berghaus, produced, in 1840, goods of the value of between 3,000,000 and 4,000,000 francs. The work-people, according to Villermé, are in pretty much the same condition as at Elbeuf; perhaps their wages are a little lower, but at any rate they are in a tolerably prosperous state, and of pretty correct general habits. In 1836, spinners earned from 1 fr. 60 c. to 3 fr. 56 c. a day; women, as winders, from 1 fr. to 55 c. to ½ fr. 94 c.; and children *rattacheurs* (*drousineurs*), from 60 c. to 80 c. A family of three persons, one of each of the above classes, may gain together 1600 francs a year, or .064, certainly sufficiently to maintain them in a state of comfort.

The peace between Philip Augustus and Richard I., in 1196, was concluded at Louviers. The town was taken and sacked by Edward III. and Henry V. (*Hugo*, art. *Eure*; *Villermé*; *Tableau des Ouvriers*, &c.)

LOVELL, p. t., Oxford co., Me., 63 m. W.S.W. Augusta, 578 W. It contains a large pond which has its outlet into Saco river. A stream issuing from several connected ponds has falls of 40 feet. It has one fulling-mill, two grist-mills, six saw-mills; 10 schools, 410 scholars. Pop. 941.

LOVINGSTON, p. v., capital of Nelson co., Va., 105 m. W. by N. Richmond, 157 W. It contains a courthouse, jail, two churches, several stores, and about 150 dwellings. Nett proceeds of the post-office, $346.

LOWELL, p. t., Orleans co., Vt., 45 m. N. Montpelier, 568 W. Formerly called Kellyvale. Chartered in 1791. Missisque r. rises in a pond in this t., and with its branches, affords water-power. It contains two stores, one grist-mill, two saw-mills; six schools, 149 scholars. Pop. 431.

LOWELL. city, and semi-capital of Middlesex Co., Mass., is in 42° 38' 48" N. lat., and 71° 18' 57" W. long., 25 m. N.N.W. Boston, 37 m. N.N.E. Concord. 37 m. N.E. Worcester, 38 m. S.S.E. Concord. N. H., 444 W. In the rapidity of its growth, and the extent of its manufactures, it stands unri-

valled in the United States, and well deserves the appellation of the "Manchester of America." The town of Lowell was incorporated by the legislature of Massachusetts in March, 1826, with a territory of four miles square, from the E. part of Chelmsford. In 1834, Belvidere village, the W. corner of Tewksbury, was added to it, making a territory of nearly 5 sq. m. The population in the territory in 1820, was less than 200, and the valuation of property not exceeding $100,000; in 1830, the population was 6477; in 1840, 20,796, and the assessors' valuation of property was $12,400,000; the population in 1844, was 25,000. In 1826, Lowell was chartered as a city. It lies on the S. side of Merrimac river, below Pawtucket falls, at the junction of Concord river with the Merrimac; and it possesses a great amount of water-power, easily available. This is produced by a canal, 60 ft. wide, 8 ft. deep, and 1½ m. long, commencing at the head of Pawtucket falls, and extending to Concord river. By locks at its outlet into Concord river, it forms a boatable passage round the falls of the Merrimac. From the main canal, the water is carried by lateral canals to mills and manufactories where it is needed, and is discharged either into Merrimac or Concord river. The entire fall is 31 ft. What 21 years ago was a wild pasture has become a large and flourishing city; a proof of what water-power, seconded by capital and enterprise can do for a place. There are in Lowell 28 cotton factories, with 205,000 spindles; two calico-printing works; two woollen factories; a machine shop, for the manufacture of cotton machinery, loco-motive engines, &c.; two carpet factories, one of which is filled with power-looms; said to be the first and only one of the kind in the world. The capital invested in these works is $10,538,000. In them are employed 2345 males, and 6375 females, making a total of 8720. Less than 250 of these are children under 15 years of age; and these are never employed without certificates of attendance at school for three months of the year, in accordance with the provision of the laws of Massachusetts. The annual amount for the sales of manufactured goods is between 6 and 7,000,000 of dollars. Besides these there are the Lowell bleachery, extensive powder-mills, flannel and blanket-mills, carpet-mills, card and whip factory, iron foundery, grist-mills, saw-mills, &c., employ 600 hands, and a capital of about $900,000. A new cotton factory, much larger than any now in the city, has already been commenced, and extensive additions to others are soon to be made. The number of yards of cloth made annually are 70,275,400. The annual amount of cotton consumed is 22,568,000 pounds, or 56,940 bales of 361 pounds each. The amount of mineral coal consumed per annum, is 12,300 tons; besides 3090 cords of wood; and 69,149 gallons of oil are used. Flour for starch consumed in the mills, print works, &c, is 4000 barrels per annum; and 600,000 bushels of charcoal are used. The average wages, exclusive of board, are, for females, $1 75 per week; and for males, 70 cents per day. The "lock and canals" machine shop can furnish machinery complete for a mill of 5000 spindles in four months. When building mills, they employ directly or indirectly, from 1000 to 1200 hands. Most of the factories, boarding houses, &c., are built according to a contract with this company, and the proprietors of the mill have little to do before it is put in operation, except to see that everything is done according to contract. This arrangement gives ample employment to the resources of the "Proprietors of locks and canals," and secures to the new company the most approved machinery, at the cheapest rates. They also contract with the same company for the necessary water-power and building ground, for which an annual rent is paid.

The public buildings of Lowell are a courthouse, city-hall, market-house, mechanics' hall, hospital, belonging to the factories, and the edifices for the public schools. The mechanics' hall was erected by a Mechanics' Association, incorporated by the legislature as early as 1825. Eight or ten years after, the proprietors of locks and canals gave them a lot of land in the heart of the city, where a spacious and expensive hall has been erected, and where one or more courses of lectures are annually delivered. This building is furnished with a fine library and reading-room, a chemical and philosophical apparatus, and a mineralogical cabinet of several thousands of specimens. The public schools consist of a high school, and six grammar schools, which have spacious and substantial buildings. Besides these there are 30 primary schools. Twenty-four thousand dollars per annum are appropriated to these schools; a very large amount in proportion to its population. There are two newspapers in the place, each published tri-weekly, and two weekly papers from the same office, and three other papers devoted to literature and religion. A periodical is issued called the "Lowell Offering," which has gained deserved celebrity, consisting of original communications by the young women of the factories. This is a new thing in the history of manufacturing cities. There are 23 religious societies of the following denominations: two Episcopal,

LOWER.   LOWESTOFF.

three orthodox Congregational, one Unitarian, two Episcopal Methodist, two Wesleyan Methodist, three Baptist, three Free-will Baptist, two Christian, and two Roman Catholic. There are two banks, with an aggregate capital of $1,050,000, and an institution for savings, having a deposite of $300,000, one third of which belongs to 978 factory young women. Visiters will be agreeably impressed with the neat and respectable appearance of the operatives of this industrious city; and equally so with their moral condition. One third of the entire population of the city is connected with the Sunday schools established by the various religious societies; and there is less intemperance and crime than in any other place of its size.

A railroad extends from Lowell 26 m. to Boston, and was completed in 1835, in a very substantial manner. Andover branch railroad extends from it, 10 m. from the city, and goes to Dover, N. H. The Lowell and Nashua railroad extends 9 m. to New-Hampshire line. The Merrimac canal leaves the Merrimac river, 2 m. above Lowel, and proceeds to Boston harbour.

Lowell is a splendid example of an American manufacturing city, and excites the attention, and in some measure, the jealousy, of Manchester and Glasgow. That a place which, 21 years since, had not a "local habitation and a name," should have become the second place in population in the state, and the fourteenth in the United States, is proof of what manufactures can accomplish. Every true friend of his country must rejoice to see it becoming independent of other countries for the articles of consumption. The manufactures of Lowell are generally of the coarset descriptions; but the same enterprise which has carried them thus far, will extend them farther. Cottons, which a little over 20 years since, would have cost over 30 cents a yard, can now be bought for six cents. And another 20 years may exhibit goods which, for fineness as well as cheapness, will rival those which are now imported. According to statistics of the census of 1840, there were in Lowell 191 retail stores, capital $373,300; five lumber-yards, capital $19,000; one furnace, four fulling-mills, eight woollen-factories, with a capital of $551,300; 36 cotton factories, with 165,000 spindles, and three dyeing and printing works, with a capital of $8,000,000; three powder-mills, capital $150,000; one paper-mill, capital $8000; one flouring mill, three grist-mills, one saw-mill, capital $50,000; two printing-offices, two binderies, three weekly, and two semi-weekly newspapers, and one periodical. capital $10,000: total capital in manufactures, $8,837,460; seven academies, 1311 students; 98 schools, 4036 scholars.

LOWER, t., Cape May co., N. J. More than half of it is covered with sea beach and salt marsh. Cape May island and lighthouse are in the t. It has six stores, three saw-mills; six schools, 940 scholars. Pop. 1133.

LOWER ALLOWAYS CREEK, t., Salem co., N. J., 9 m. S. Salem. It has extensive marshes, secured from overflow by embankment. It has a Methodist and a Friends' church, four stores, one grist-mill. Pop. 1252.

LOWER CHANCEFORD, p. t., York co., Pa., 51 m. S.S.E. Harrisburg. 91 W. Bounded E. by Susquehanna river, across which is McCall's ferry. It contains four stores, one woollen factory, one furnace, one forge, three saw-mills, one paper-mill ; five schools, 125 scholars. Pop. 1991.

LOWER DUBLIN, t., Philadelphia co., Pa., 10 m. N.E. Philadelphia. Pennypack creek affords water-power. It contains five churches, three Baptist, a Methodist, and an Episcopal, 12 stores, one lumber-yard, five grist-mills, one saw-mill ; five academies, 235 students, three schools, 73 scholars. Pop. 3296.

LOWER MACUNGY, t., Lehigh co., Pa. It has five stores, one woollen factory, six grist-mills, three saw-mills, one oil-mill. Pop. 2156.

LOWER MAHANOY, t., Northumberland co., Pa. It has three stores, two saw-mills, two tanneries, four distilleries, two potteries ; four schools, 107 scholars. Pop. 1199.

LOWER MAHANTANGO, p. t., Schuylkill co., Pa., 55 m. N.E. Harrisburg. 165 W. Crossed by Broad and Sharp's mountains. Watered by Deep and Swatara creeks. It contains anthracite coal, and has four stores, five grist-mills, 13 saw-mills, two tanneries ; one school, 30 scholars. Pop. 2465.

LOWER MAKEFIELD, t., Bucks co., Pa., 24 m. N.E. Philadelphia. Delaware canal runs along its E. boundary. It has two stores, two grist-mills, two saw-mills ; fours chools, 232 scholars. Pop. 1550.

LOWER MERION, t., Montgomery co., Pa., 90 m. E. Harrisburg, 150 W. Schuylkill river and Mill and Cobb's creeks afford water-power. It has three churches, nine stores, two woollen factories, three cotton factories, with 1532 spindles, three grist-mills, three saw-mills, seven paper-mills ; six schools, 734 scholars. Pop. 2927.

LOWER MT. BETHEL, t., Northampton co., Pa. Watered by Richmond, Martin's, and Muddy creeks. It contains

a Lutheran church, seven stores, one flouring-mill, eight grist-mills, eight saw-mills, one powder-mill, five tanneries ; five schools, 185 scholars. Pop. 2957.

LOWER NAZARETH, t., Northampton co., Pa. Drained by Manookisy creek. It has two stores ; two schools, 116 scholars. Pop. 1201.

LOWER OXFORD, t., Chester co., Pa., 41 m. S.W. Philadelphia. Watered by Octarara and Elk creeks. It contains a Presbyterian church, four stores, one fulling-mill, two woollen factories, three cotton factories, with 2500 spindles, four grist-mills, six saw-mills, one oil-mill, two paper-mills ; three schools, 75 scholars. Pop. 1922.

LOWER PAXTON, t., Dauphin co., Pa., 6 m. N.E. Harrisburg. Watered by Beaver and Paxton's creeks. It has two stores, one woollen factory, one grist-mill. one tannery, one distillery ; one school, 45 scholars. Pop. 1337.

LOWER PROVIDENCE, t., Montgomery co., Pa. Drained by Perkiomen and Shippack creeks. Lead ore is found on the former. It contains five stores, one flouring-mill, seven grist-mills, one saw-mill ; four schools, 262 scholars. Pop. 1413.

LOWER SALFORD, t., Montgomery co., Pa., 25 m. N.W. Philadelphia. Drained by branches of Perkiomen and Shippack creeks. It contains four stores, four grist-mills, one saw-mill ; five schools, 200 scholars. Pop. 1141.

LOWER SANDUSKY, p. v., capital of Sandusky co., O., 105 m. N. Columbus, 423 W. Situated on the W. bank of Sandusky river, which is navigable to this place for small steamboats. It contains a courthouse, jail, two churches, 10 stores, two warehouses, one grist-mill, one saw-mill. In this t. was fort Stephenson, where Col. Crogan made a gallant defence against the British and Indians during the last war, the remains of which are still visible. Pop. of the t., 1117.

LOWER SAUCON, p. t., Northampton co., Pa., 90 m. E.N.E. Harrisburg, 192 W. Watered by Lehigh river and Saucon creek. Two bridges here cross the Lehigh. It contains the v. of Hellerstown, and has five stores, two fulling-mills, two flouring-mills, seven grist-mills, four saw-mills, one oil-mill, one paper-mill, one tannery, one distillery ; seven schools. 350 scholars. Pop. 9710.

LOWER ST. CLAIR, t., Alleghany co., Pa. 6 m. S.W. Pittsburg. Bounded N. by Monongahela and Ohio rivers. Drained by Chartier's creek. Coal abounds. It contains Birmingham v., a mile below Pittsburg. It has five stores, three furnaces, two flouring-mills, two grist-mills, three saw-mills, one oil-mill, two glass factories ; one academy, 12 students ; 12 schools, 600 scholars. Pop. 4373.

LOWER SWATARA, t., Dauphin co., Pa., 5 m. S.E. Harrisburg. Bounded S.W. by Susquehanna river. Harrisburg lies partly in this t. Watered by Spring and other creeks. It contains three stores, four lumber-yards, capital $31,800, two flouring-mills, one grist-mill, one saw-mill ; two schools, 95 scholars. Pop. 1236.

LOWER SMITHFIELD, t., Monroe co., Pa. It has six stores, one grist-mill, five saw-mills, four tanneries. Pop. 1192.

LOWER WINDSOR, t., York co., Pa., 8 m. E. York b. It contains five stores, one furnace, one forge, one flouring-mill, four grist-mills, three saw-mills, one tannery, two distilleries ; seven schools. 150 scholars. Pop. 1687.

LOWESTOFF, or LOWESTOFT, a market town, seaport, and par. of England. E. coast co. Suffolk, hund. Mutford and Lothingland. 22 m. S.E. Norwich, and 104 m. N.N.E. London ; lat. 52° 29' 10'' N., long. 1° 45' 11' E., being the most easterly land in England. Area of par., 1936 acres. Pop. in 1831, 4238. The town consists of one principal street, which has a gradual descent from N. to S.; and from this main avenue proceed several other streets towards the W.; but, though well paved and lighted, they are narrow and irregular. In the market-place is a building open below, the upper part of which is used for meanbly-rooms and other purposes ; and there is a small theatre. The church is a handsome Gothic building, with a tower and steeple 182 ft. high, the living being a vicarage, in the gift of the bishop of Norwich. There is also a chapel of ease; and the Independents, Baptists, and Wesleyan Methodists have each their places of worship, with attached Sunday schools. A free school furnishes instruction for 40 boys, and there is a good national school. A friendly and benevolent society, a lying-in charity, and dispensary, are the principal charities. Several handsome lodging-houses have been built for visiters coming here for bathing in the summer months; and there are warm baths, reading-rooms, libraries, &c. At the S. end of Lowestoff is a battery, with 13 pieces of cannon, and two others are placed at the N. end, near which latter. on a high point of land. stands a round tower, the upper lighthouse (first built in 1676, and rebuilt in 1778). On the beach, below the cliff, is another lighthouse ; and, by keeping both in a line, vessels are directed safely through the sand banks, which render this coast especially dangerous. The harbour, or rather road, is defended on the E. by the Corton sand; the channel between the latter being marked by a light vessel, and well buoyed. This, which till lately was little better than a little fishing town, engaged in preparing red herrings for the London market, will probably rise to considerable commercial importance. Since 1827, an artificial harbour has been formed on a grand scale at Lowestoff, which communicates with the lake Lothing to the W. of the town; and then, by a short canal, with the Waveney, which is navigable to Beccles. Another canal joins the Waveney with the Yare, which has been rendered navigable for vessels drawing 10 ft. water as far as Norwich. Owing to the flatness of the ground, no locks, except the sea lock at Lowestoff, are required on either line of navigation. This improved communication must be of great service to the country which it intersects, and especially to Beccles and the city of Norwich, on which, indeed, it has conferred most of the advantages of a seaport. (*Priestly on Canals, &c.*, 513.)

Still, however, the chief consequence of Lowestoff, as a port, is owing to its herring fisheries: the quantity of fish annually taken and cured is very large; while, at the same time, their quality is considered superior, and they fetch higher prices in the London market than those sent from Yarmouth. Sail-making, boat-building, and the manufacture of rope and twine are extensively carried on; and several hands are employed in making barrels in which to pack the cured fish previous to their being sent to market or exported. Markets on Wednesday; fairs, May 12, Michaelmas day, and Oct. 10.

The only historical celebrity of Lowestoff is derived from the fact that on the 3d of June, 1665, a sanguinary naval engagement was fought off the coast between the English and Dutch, the fleet of the former being commanded by the Duke of York, afterward James II., and that of the latter by Admiral Opdam, who was killed in the battle.

LOWNDES, co., Ga. Situated in the S. part of the state, and contains 2080 sq. m. Watered by Withlacoochee and Allapaha rivers and their branches. It contained in 1840, 41,003 neat cattle, 9458 sheep, 90,349 swine; and produced 2383 bushels of wheat, 130,198 of Indian corn, 56,285 of buckwheat, 1172 of oats, 25,512 of potatoes, 275,695 pounds of cotton. It had five stores, nine grist-mills, five saw-mills, one distillery; one school, 12 scholars. Pop.: whites, 4394; slaves, 1177; free coloured, 3; total, 5574. Capital, Troupville.

LOWNDES, co., Ala. Situated a little S. of the centre of the state, and contains 1600 sq. m. Bounded N. by Alabama river, by branches of which it is watered. It contained in 1840, 3750 neat cattle, 801 sheep, 97,365 swine; and produced 1421 bushels of wheat, 1736 of rye, 162,540 of Indian corn, 11,113 of oats, 10,353 of potatoes, 4743 pounds of rice, 803,939 of cotton. It had 30 stores, three tanneries; four academies, 947 students; three schools, 100 scholars. Pop.: whites, 6956; slaves, 12,569; free coloured, 14; total, 19,539. Capital, Haynesville.

LOWNDES, co., Miss. Situated in the E. part of the state, and contains 394 sq. m. Watered by Tombigbee river and its branches. It contained in 1840, 11,960 neat cattle, 3069 sheep, 29,463 swine; and produced 13,666 bushels of wheat, 875,140 of Indian corn, 39,915 of oats, 3100 of potatoes, 7,153,056 pounds of cotton. It had 20 stores, nine grist-mills, eight saw-mills, 17 oil-mills, 10 tanneries, two printing-offices, two weekly newspapers; one college, 54 students; two academies, 75 students; eight schools, 221 scholars. Pop.: whites, 5730; slaves, 8771; free coloured, 12; total, 14,513. Capital, Columbus.

LOWVILLE, p. t., Lewis co., N. Y., 137 m. N.W. Albany, 463 W. Bounded E. by Black river, and watered by its tributaries. It contains five churches, a Methodist, two Presbyterian, a Baptist, and Friends'; an academy, a bank, six stores, one fulling-mill, one flouring-mill, three grist-mills, eight saw-mills, three tanneries, one distillery, one printing-office, one weekly newspaper; 16 schools, 601 scholars. Pop. 2047.

LOYALSOCK, t., Lycoming co., Pa. Bounded S. by the W. branch of Susquehanna river, W. by Lycoming creek, E. by Loyalsock creek. It contains Williamsport borough, the capital of the county, exclusive of which it has one store, one furnace, one forge, one grist-mill, four saw-mills, one distillery; five schools, 145 scholars. Pop. 1107.

LOXA, or LOJA, a town of Spain, in Andalusia, prov. Granada, on the Xenil, 26 m. W. Granada, and 92 m. E. by S. Seville. Pop., according to Miñano, 13,866. It stands on the S. side of a rocky gorge, by which the Xenil escapes from the fertile *Vega* of Granada; and "its situation is peculiarly picturesque, the town being built on a steep acclivity, embosomed in groves of fruit trees, and overlooked by a toppling mountain, forming one of the offsets of the Sierra Nevada." It contains three small and shabby parish churches, with two hospitals; and on an eminence at its S. extremity is a ruined Moorish castle, once of great strength and celebrity, but now the residence of a few her-

246

mits. "Loja is proverbially noted for the fertility of its gardens, olive-grounds, and orchards, the abundance and purity of its springs, and the loose and hard features of its rural inhabitants. (*Scott's Ronda and Granada*, ii., 358, 359; *Miñano*.)

LOZERE, a dep. of France, reg. S., between lat. 44° and 45° N., and long. 3° and 4° E., having N., Haute-Loire and Cantal, W. the latter dep. and Aveyron, S. Gard, and E. Gard and Ardeche. Length, N.W. to S.E., 65 m.; greatest breadth nearly 50 m. Area, 514,795 hectares. Pop. (1836) 141,733. This department lies chiefly on the N.W. slope of the Cevennes, with the ramifications of which it is mostly covered. The surface varies from 2500 to 5000 ft. above the level of the sea; but its average elevation may be estimated at 3600 ft. The department derives its name from the mountain Lozere in the S.E., one of the principal summits of the Cevennes, 4886 ft. in height. The rivers Lot, Tarn, Allier, and Gard have their sources within this department, which is not, however, watered by any stream of magnitude. There are several small lakes, one of which appears to occupy the crater of an extinct volcano. The climate is cold; snow remains on the mountains during the greater part of the year, and fogs are frequent. The soil is mostly stony in the N. and S., and calcareous in the centre. In 1834, the surface is said to have been distributed as follows, viz., arable land 908,660 hectares, meadows 35,166 hect., forests 44,580 hect., and heaths, wastes, &c., 179,000 hect. Agriculture is very backward, and little likely to improve, from the sterility of the soil, the remoteness of most parts of the department from great roads, the great subdivision of property, and smallness of the farms. Rye and wheat are raised, but not in sufficient quantities to supply the consumption. In the Cevennes potatoes are pretty extensively cultivated, and form, with chestnuts, the chief food of the inhabitants. About 50,000 hectolitres a year of inferior wine, and some oil and silk are produced; the sharp winds experienced in the department are, however, unfavourable to the silk-worm. Hemp and flax succeed well, but the culture of madder and saffron has been abandoned. The mountain pastures are excellent, and feed many sheep: coarse woollens and serges are made in almost every peasant's family. In 1835, of 43,847 properties subject to the *contribution foncière*, very nearly a half were assessed at less than 5 fr., and 6635 at from 5 to 10 fr.; only seven properties being assessed at more than 1000 fr., and 53 at from 500 to 1000 fr. The department is said to be rich in mineral products, but the mines are but little attended to. Lozere is divided into three arrondissements; chief towns, Mende, the capital, Florac, and Marvejois. It sends three members to the chamber of deputies. Number of electors in 1838-39, 712. Total public revenue in 1831, 2,256,376 fr.; expenditure in the same year, 1,777,870 fr. (*Hugo*, art. *Lozere*; *Official Tables, &c.*)

LUBEC, p. t., and port of entry, Washington co., Me., 180 m. E.N.E. Augusta, 766 W. Situated at the W. entrance of Passamaquoddy bay, and has West Quoddy Head lighthouse on a point of land at the entrance of that bay. It has a good harbour, protected by Grand Menan island, which has a depth of water for vessels of the largest class, and is never obstructed by ice. It has other bays and entrances, and some islands. It contains 12 stores, three grist-mills, nine saw-mills; 12 schools, 847 scholars. Pop. 2307.

LUBECK, a city and republic of N. Germany; the city, which is the capital of the Hanseatic towns, and the seat of their high court of appeal, is situated on the Trave, about 10 m. (direct distance) from Travemunde, at its mouth in the gulf of Lubeck, in the Baltic, 36 m. N.E. Hamburg, and 38 m. S.E. Kiel; lat. 53° 52' 8" N., long. 10° 41' E. Population estimated at rather more than 26,000. The town was built on a gently elevated ridge, on one side of which runs the Trave, and on the other the Wackenitz. The environs are well wooded, and enlivened with cheerful villas, more particularly those along the banks of the Trave. The streets, which are steep, are wider than those of Hamburg. The houses generally appear to be old, and mostly built of stone; like those of Hamburg and Antwerp, their gable ends face the street. They are in general very lofty, six or seven stories not being uncommon. Round the ramparts of the city, in which there are two handsome gateways, is a promenade shaded with fine trees. The principal buildings are the cathedral, four churches, and the town-hall. The cathedral is a curious old building, the spires of which being much out of the perpendicular, momentarily threaten to fall, which also is the case with the towers of St. Mary's and St. Peter's. The church of St. Mary is handsome in the interior, and well worthy a visit, were it only for the sake of the paintings, many of which are curious and of ancient date. Among them is the celebrated picture of the Dance of Death (usually attributed to Holbein, but which belonged to the town for at least 35 years before Holbein's birth). In the same edifice may be seen same fine specimens of sculpture, particularly a representation of the Lord's Supper, carved in white marble, finely designed and beautifully executed." (*Barrow's Excursions in the N. of Europe*, p. 29-24.) Behind the high altar is an old astronomical clock, constructed in 1405, which is the grand curiosity of the place. According to one of the guide-books, it exhibits at a certain hour figures representing the emperor and the seven German electors; but, according to Mr. Barrow, the figures are meant for the Twelve Apostles, who "at mid-day sally forth and march in regular succession, passing a figure of our Saviour, to whom they each face round, and having made a quick and familiar nod of the head, they then march onwards to a door on the opposite side, which closes upon them the moment the twelfth apostle has entered." (P. 94.)

The cathedral, begun in 1170, and finished in 1341, has many monuments of the senatorial families of Lubeck, some of which are well executed, and, among others, a remarkably curious oil picture, by Hans Hemling, dated in 1471, the subject of which is the Passion of Christ, treated in 23 distinct groups. The town-hall, a turretted Gothic building, faces the market-place. It was the place of assembly for the deputies from the cities formerly comprised in the Hanseatic League; but the hall in which they held their meetings was unfortunately destroyed in 1817. Lubeck has a Calvinist and a Roman Catholic church, an exchange, arsenal, and mint, several hospitals and benevolent institutions, a gymnasium, a city-school, which in 1839 had 263 pupils, ecclesiastical and teachers' seminaries, schools of surgery, midwifery, navigation, drawing, swimming, and numerous other schools, a public library of 25,000 volumes, a society of useful sciences and arts, a Bible society, a house of correction and prison, a theatre for operas, &c.

Lubeck, though by no means so prosperous and important as formerly, is still a thriving commercial town. Many of its modern-built houses are on a grand scale. Their basement-stories are used as magazines or warehouses, and they have commonly large court-yards into which the carriages of the proprietors are driven. (*Barrow*, p. 25.) Lubeck and its territory are numerous breweries, distilleries, iron forges, and linen yarn factories; besides manufactures of hats, vinegar, starch, tobacco and snuff, wax lights, paper and cards, musical instruments, with numerous oil and other mills, several printing establishments, and a few woollen, cotton, and golden and silver lace factories. Its trade is principally confined to the N. and W. of Europe. Berghaus states that upwards of 1600 vessels a year enter and leave its port; they are principally Danish, the rest being Russian, Swedish, Lubeck, Dutch, English, and Prussian. Lubeck communicates by means of the Trave and a canal with Hamburg (which see), with which it has an extensive intercourse. The principal article of export is corn: the principal articles of import are wines and silks, from France; cottons, hardware, and other manufactured goods, from England; colonial products, dye stuffs, &c. It has an extensive commission and transit trade, and considerable markets for wool, cattle, horses, &c. Vessels of considerable burden load and unload by means of lighters at Travemunde, at the mouth of the river, which is properly the port of Lubeck. Two steamboats, of small draught of water, ply on the river between the city and its port. Steam-packets sail at fixed periods from the latter for Petersburg, Stockholm, and Copenhagen.

Lubeck has several fire and life insurance companies. Accounts are kept in marks of the value of 1s. 2.67d. each, divided into 16 schellings of 12 pfennigs. The Lubeck rix-dollar, equivalent to three marks, is worth 4s. 6.72d. The lb. = about 18 oz. avoird.; 112 lbs. = 1 centner.

The territory subject to Lubeck consists of a district of about 80 sq. m., immediately adjacent to the city, surrounded by the territories of Mecklenburg, Holstein, and Oldenburg, and the Baltic; of numerous small detached portions of surface enclosed by Holstein; and of the *Vier-länder*, and town of Bergedorf, the sovereignty over which it shares with Hamburg. United area, about 127 square miles. Population, in 1838, 47,900, all Lutherans, except about 300 Calvinists, 400 Roman Catholics, and as many Jews. The land is very productive, yielding good crops of corn, fruits, and kitchen vegetables; but the rearing of live stock is the chief occupation of the rural population. The government is vested in the senate and house of burgesses (bürgerschaft); the former consists of four burgomasters, holding office for life, two syndics, and 16 councillors; and the latter of 12 colleges or companies, only seven of which have, however, the privilege of voting. The house of burgesses has the initiative in all deliberations relative to the public expenditure, foreign treaties, &c.; the senate is entrusted chiefly with the executive duties, but its sanction is necessary to the passing of new laws. Public revenue, in 1838, 748,904 marks; expenditure, 709,882 marks. In 1838, the public debt amounted to

five and a half millions marks; but it has since seen in process of reduction. Lubeck has one vote in the full council of the German confederation, and along with the other Hanse towns, a vote in the committee. It furnishes a contingent of 407 men to the army of the confederation.

It is uncertain when or by whom this city was founded; but no doubt it existed anno 1140. Early in the 13th century, the Emperor Frederick II. made it one of the free towns of the empire; and from 1300 to 1669, Lubeck was the repository of the archives of the powerful association of cities included in the Hanseatic League, and the station of the confederated fleet. The dissolution of the League marked the epoch of the decline of Lubeck. After the battle of Jena, Blucher threw himself into Lubeck, which, after a severe engagement, was taken by the French, and sacked. In 1830, it was made the capital of an arrond. in the department Bouches de l'Elbe; but was restored to its rank, as a free city, by the congress of Vienna in 1815. Sir Godfrey Kneller, the painter, Mosheim, the historian of the Christian religion, Melbomius, and H. Muller, were natives of Lubeck. (*Martinière, Dictionnaire Géographique; Berghaus; Allg. Länder, &c., iv., 465-489; Barrow's Excursions in the N. of Europe.*)

LUBLIN, a city of Russian Poland, cap. of the palatinate of Lublin, in a marshy situation, on the Bistrzyca, a tributary of the Wieprz, 97 m. S.E. Warsaw. Population estimated at 12,500, half of whom are Jews. It is subdivided into the old and new town, the former situated on an eminence, and the latter on the bank of the river. Lublin was fortified by a wall and ditch, till these works were destroyed in the civil wars towards the end of the 18th century. It has still, however, a citadel standing on a high rock, and according to Stein, the ruins of a castle built by Casimir the Great. Its streets are irregular and filthy, and its houses mostly of wood. The principal edifices are a handsome town-hall, the Sobieski palace, the cathedral, the churches of the Dominicans and Carmelites, and that formerly belonging to the Jesuits. There are in all 18 churches and 12 convents, six nunneries, a spacious synagogue, an Episcopal seminary, a Piarist college, several civil and military hospitals, an orphan asylum, and a theatre. Lublin is a bishop's see, and the seat of the second court of appeal in Poland. It has manufactures of coarse woollens; considerable trade in woollen cloth, corn, and Hungarian wines; and three large yearly fairs, each lasting a month, and attended by German, Greek, Armenian, Arabian, Russian, Turkish, and other traders. (*Dict. Géog.; Stein, &c.*)

LUCAS, county, O. Situated in the N.W. part of the state, and contains 600 sq. m. Bounded S.E. by Maumee river, and watered by its tributaries. It contained in 1840, 11,597 neat cattle, 3103 sheep, 18,381 swine; and produced 103,838 bushels of wheat, 5245 of rye, 154,017 of Indian corn, 13,968 of buckwheat, 59,992 of barley, 69,444 of oats, 122,994 of potatoes, 14,061 of sugar. It had six commission houses in foreign trade, 41 retail stores, five grist-mills, 19 saw-mills, one brewery, three printing-offices; 43 schools, 1131 scholars. Pop. 9382. Capital, Maumee.

LUCCA (DUCHY OF), a state of Central Italy, being, excepting San Marino, the smallest of the Italian states; between lat. 43° 46′ and 44° 14′ N., and long. 10° 0′ and 10° 42′ E.; having (except a few small detached portions) N.W. and N. the territories of Modena and the Tuscan Lunigiana, E. and S. Tuscany, and W. the Mediterranean. Length (N. to S.) 26 m.; greatest breadth, 21 m. Area (incl. Montignoso, &c.), 490 sq. m. Pop. (1839), 166,196. The Apennines skirt the N. part of the duchy, two thirds of which they cover with their ramifications; but none of these rise to the height of 4000 feet. The rest of the surface is a low but fertile plain, which becomes marshy towards the coast. The general slope of the country is from N. to S., in which direction it is traversed by the Serchio near its centre. This river is not navigable, but is of great use for irrigation: most of the other streams in the duchy are its tributaries. Near the shore are some small lakes. The mean annual temperature is about 59° Fah.; in the summer it rises to 80°: in winter it rarely freezes in the plain of Lucca. The soil, which is calcareous and stony in the N., is sandy in the S., and rich in the intermediate region. The population is chiefly agricultural, but the corn produced is not sufficient for home consumption; the deficiency being principally supplied by beans, which are largely cultivated, and partly, also, in the mountainous districts, by chestnut flour. The latter is sometimes exported to the neighbouring states, the price varying from 6s. 8d. to 10s. a sack. The culture is extending of all the articles for the production of which the soil and climate afford facilities. The number of mulberry trees has rapidly increased since the peace, and the manufacture of olive oil has been materially improved. The latter is esteemed the best in Italy, and fetches the highest price, especially that grown on high grounds. It is exported to the value of about £38,000 a year: the market price being from 4d. to 5½d. the Lucca pound of 12 oz. Wine is said to give a fair return to the cultivator; hemp and flax are raised, and the produce of silk is very considerable. Lucca, in fact, was early distinguished by her proficiency in the silk manufacture; and in 1319 the culture of the mulberry became an object of public attention. Rice is grown near the coast, in which neighbourhood also most of the cattle in the duchy are reared. There are nearly 25,000 landed proprietors, of whom a large part have necessarily very small properties, and belong to the class of agricultural or manufacturing labourers. The principal causes which have led to this subdivision of the land, as well as to the rapid increase and great density of the population, appear to be the habit of dividing leasehold property equally among the males of a family, the suppression of monasteries, and the abolition of entails.

The *métayer* system of agriculture is not so prevalent here as in Tuscany and elsewhere; but Sismondi represents the peasantry of Lucca as being, notwithstanding, in ' a very depressed condition. "Every day the husbandman is reduced to buy the day's provision. Very rarely has he a reserve of corn; still more rarely, of oil or wine. The former has been sold in the press, and the latter in the tub. He has rarely any provision of salt meat, butter, cheese, or vegetables. All the kitchen utensils are of earthenware; and the whole furniture consists of a table and some wooden chairs, one or two chests, and an indifferent bedstead, on which the father and mother sleep, with their feet in one direction, and the children with their feet against the head-board. When the division under General Vatrain ravaged the districts of the Val di Nievola, in 1799, the peasantry derived this advantage from their indigence, that when they had concealed their clothes, and the gold trinkets of their women, they had scarcely anything left to lose." (*Sismondi, Tableau de l'Agrie. Toscane, &c.*, p. 213, 214.) But, according to Dr. Bowring, this statement must be either too highly coloured, or the condition of the peasantry must have improved in the interval; for he affirms that the labourers, in addition to articles of prime necessity, consume salt provisions, and sometimes fresh meat and colonial products. The ordinary wages of country labourers vary from 5d. to 6½d. a day, with food: farm labourers, who dwell with their masters, get from 45 to 55 francs a year. The mountaineers, who depend almost entirely upon the culture of the chestnut, are said to be in a better condition than the peasantry of the hills and plains. The inhabitants of the districts of Pontito and Sciappa are in particular distinguished by their robust and healthy appearance, and by the beautiful complexion and regular features of the women. This last circumstance is the more remarkable, as, during a great part of the year, these women, have to bear the whole burden of domestic labour; while their husbands, fathers, and brothers emigrate to the Tuscan maremme, and the states of the church, in search of harvest and other work. During winter, about 3000 labourers set out for these territories, Corsica, &c.; and return in summer, bringing with them their small savings, the aggregate of which may amount to £10,000. Most of the Italian image and plaster cast-makers, in other countries of Europe, are emigrants from Lucca. Mining is little or not at all pursued, though copper, iron, and lead ores are met with. Statuary marble, and other fine marbles, are found in great abundance. From 5000 to 6000 hands are employed in the manufacture of silk, wool, and cotton; and there are in the duchy about 30 paper factories, and others of linen cloth, straw and beaver hats, leather, glass, and iron goods. The capital is the chief seat of manufacturing industry.

The value of the exports amounts to about four millions francs a year, more than one fourth part of which is derived from oil and silk. These articles go chiefly to other parts of Italy, and to France, England, and the Levant. Grain, seeds, wine, liqueurs, live stock, lamb-skins, and fresh fish, are sent to Tuscany; and woollen goods to the rest of Italy and the Levant. The imports, which mostly come from Tuscany through Leghorn, consist principally of grain, seeds, rice, fine wines, hemp, flax, cotton, colonial products, salted provisions, pig-iron, &c. British cotton woollen and linen fabrics pay an import duty of 10 per cent. *ad valorem*, cotton twist pays 3 lire, pig and bar iron, 5 lire 15 soldi, and glass wares, 6 lire per 100 lbs. The importation of tobacco and salt is prohibited, except on account of government which has a monopoly of those articles. The lb. (libra) of Lucca is somewhat less than the lb. Troy; the peso grosso = 11 lbs.; the copo of oil = 24 pesi grossi; the stajo of corn = about 5½ gallons. Accounts are kept in lire of 20 soldi and 240 denari. The lira = 7½d.; the scudo = 4s. 5½d.; the gold doubloon or pistole = 14s. 5½d. English. Lucca has only one seaport, Viareggio.

The government is a limited monarchy, under a duke, who exercises the executive power, nominates the minis-

ters, and all other public officers, &c. But an estimate of the amount of the public expenditure is annually laid before the senate, and must be sanctioned by it. This body consists of 35 members, elected from among the four classes of merchants, artisans, scholars, and landed proprietors, called together by the duke for at least a month every year; and without its consent no tax or other public burden can be imposed. The council of state consists of the two state ministers and six additional councillors. Justice is administered by a local commissioner in each commune; and a tribunal of original jurisdiction, a civil and criminal court of appeal, and a court of cassation in the cap. A permanent council of war, and a court of revision, sit to decide in military causes. The military force comprises about 750 men, costing about £10,800 a year. The naval force consists of only a goelette of 12 guns, and three gunboats. The regular and secular clergy amount together to about 1900 persons, under the archbishop of Lucca. There are about 120 communal and other public and private schools, educating 3000 pupils, or one in 53 of the population. The principal establishments of this kind are the college of Charles Louis, with 140 students, the archbishop's seminary, the ducal lyceum with 300 students, and the conservatory of Louis Charlotte with 40 female scholars. There are some extensive charitable institutions, costing the state annually about £12,500. The public revenue and expenditure are about £83,000 a year each. The civil list costs about £25,300. Lucca has no public debt, except that due for pensions, &c.

Lucca, like the rest of Italy, experienced many changes in the middle ages. The capital attained its liberty after the decease of the Countess Matilda, in 1115, when it became an independent republic. In the next century it again fell under feudal authority, and afterwards belonged successively to Louis the Bavarian, and to noble Genoese. In 1370, it again obtained its liberty, by purchase, from the emperor Charles IV., for 100,000 crowns; and from that date to 1805, it was governed by its own *gonfaloniert*. Napoleon united Lucca with Piombino in a principality; the congress of Vienna, in 1814, erected it into a duchy, the greater part of which is hereafter to be united to Tuscany, and the rest to the Modenese territories. (*Bowring's Report on Lucca; Serristori, Statistica d'Italia*, part iii.; *Hoffman; Rampoldi, &c.*)

LUCCA (anc. *Luca*), a city of Italy, cap. of the above duchy, in a plain near the left bank of the Serchio, 11 m. N.E. Pisa, and 38 m. W. Florence; lat. 43° 50′ 49″ N., long. 10° 30′ 49″ E. Pop., in 1839, 24,092. The city is surrounded with walls; which would form, however, but a very feeble defence against an enemy. The towers of the churches, rising above the ramparts, have a fine effect in the rich and beautiful landscape, the view being bounded by vine-clad hills spotted with villas, over which tower the craggy Appennines. On a nearer inspection, the public buildings are less pleasing in their architecture than in their distant effect: yet many of them are very curious structures. According to Mr. Woods, "The churches are all, more or less, imitations of the cathedral at Pisa; smaller, indeed, in size, but some of them are decidedly superior in the proportions and disposition of the parts." (*Letters of an Architect*, ii., 410.) Most of the churches are built of Carrara marble. The cathedral, mostly constructed in the 11th century, has much carved, inlaid, and mosaic work; a rich display of stained glass; a Madonna, by Fra Bartolommeo; and some pictures of the Venetian school. The churches of San Michele, and San Frediano are both ancient. The latter belonged to a monastery restored and enriched towards the close of the seventh century. Frequent notices, both of the monastery and the church, occur in the succeeding centuries, but nothing, it is said, indicates that the latter has been ever rebuilt, or materially altered. Its curious architecture is described by Woods, ii., 411. The ducal palace is a large structure, the exterior of which presents nothing remarkable; but its interior is superbly furnished with articles of Lucca manufacture, the ceilings and walls being also adorned with frescoes by Lucchese artists. The *Palazzo Publico*, the residence of the *gonfaloniere*, in the days of the republic, is described by Forsyth as an immense and august edifice, which makes the city round it look little. There is a small, but handsome theatre. Lucca is generally well built: many of the private houses are very good, though their pointed roofs, gable ends, &c., give it the aspect rather of a Flemish than an Italian city. The streets, though crooked, are broad and well-paved; and the ramparts, planted with trees, form pleasing promenades. It has several colleges, a seminary, founded by Eliza, Princess Bacciochi, sister of Napoleon, for 100 young ladies, a botanic garden, a ducal library with 21,000 volumes, a university library with 16,000 volumes, a *depôt de mendicité*, a *monte di pietà*, and a savings' bank. Forsyth, who visited this city in 1802, speaks of it as silent, dull, and gloomy; it enjoys, however, the title of l'industriosa, and is one of the principal inland commercial towns in Italy. Its manufactures mostly consist of silk and woollen fabrics. The usual wages paid to men vary from one to two francs a day; women and boys earn about half a franc a day. The city has also a considerable trade in olive oil, &c. About 12 or 13 m. up the valley of the Serchio are the baths of Lucca, picturesquely situated, and frequented by numerous visiters. The temperature of the hottest spring is about 120° Fah.

Lucca was colonized by the Romans A.U.C. 575. It was a municipal town, and frequently the head quarters of Cæsar, during his command in Gaul. Traces of a Roman amphitheatre are still discoverable. This city was taken by the French in 1799; and, in 1805, Napoleon made it the capital of a principality he erected for his sister's-husband, Bacciochi. (*Rampoldi; Woods; Forsyth; Cramer's Anc. Italy*, i, 173.)

LUCENA (an. *Eliana*), a town of Spain, in Andalusia, prov. Cordova, 31 m. S.S.E. Cordova, and 68 m. E. Seville. Population, according to Miñano, 19,716. It stands on the slope and at the foot of a hill, comprising some respectable streets, lined with good houses; two squares, and agreeable suburbs. The neighbourhood is distinguished for the abundance of its produce in fruit and grain, which chiefly contributes to the support of the population; but the processes of tillage are of the rudest description, and the resources of the soil are little tried.

LUCERA (an. *Luceria*), a city of S. Italy, Neapol. dom., prov. Capitanata, cap. canton, on a height abrupt towards its N. side, 12 m. W.N.W. Foggia. "The city contains 13,000 inhabitants, apparently in easy circumstances. The houses, which are all tiled, are generally good; but the streets are narrow, ill-paved, and dirty. Some ancient walls, in very bad condition, inclose it; and five gateways open from them to an outward road, which winds entirely round the town. A few gardens and convents are scattered about, and these, with some olive plantations and vineyards, in which the natives have small country houses, contribute greatly to enliven and diversify the prospect. The vines are trained low; and supply the proprietors with a good strong white wine, and a still stronger, but less pleasing red one." (*Craven*, 44, 45.) About ¼ m. from the city, on the edge of the same eminence, is the Castle of Lucera, a ruined Gothic fortress, erected by the emperor, Frederic II. The extent of its walls would almost lead to the belief that they surround a second city; but they at present encircle only an empty area, overgrown with grass. Craven believes that there can be no doubt, from the Roman inscriptions, and pieces of sculpture found within the area of this building, that its situation is identical with that of the citadel of the ancient *Luceria*, taken by the Samnites after the defeat of the Romans at the Caudine Forks, and afterward retaken by L. Papirius. (*Tour in the S. Prove. of Naples*, 48.) This castle is a very conspicuous object; it has a deep moat, a drawbridge, two large round towers, one supporting the telegraph which communicates with Foggia, and the other a piece of masonry, built with consummate skill; the interior of its area are traces of extensive cisterns. The cathedral of Lucera was formerly a Saracenic mosque, and preserves, on the exterior, some marks of its origin. It has a pulpit adorned with that kind of Byzantine mosaic, of which the cathedral of Salerno offers so fine a specimen; but its principal ornaments are 13 beautiful pillars of verd antique, originally found under the cathedral itself, and supposed to have belonged to a temple of Apollo: the capitals are modern. Facing this church is the bishop's palace, considered the finest piece of architecture in Apulia. The tribunal and other public edifices, render the appearance of this part of the city somewhat imposing. The *Tribunals* includes the criminal and civil courts for the province, the register office, the notarial chamber, the residences of the president and judges, and the public prisons. Lucera has a royal college, and an extensive private collection of coins, medals, and antiquities. Great numbers of cattle are kept in its neighbourhood; and its cheese is held in great repute.

Lucera is said to have been founded by Diomed, and was the capital of Dæmia under the Greeks; it afterward became a Roman colony. Having fallen into decay, it was renovated in 1300, by Frederick II., who transported thither a colony of Saracens from Sicily, to whom he gave great privileges. In 1300, however, Charles of Anjou expelled from the Neapolitan dominions such Moors as refused to embrace Christianity, and converted the mosque of Lucera into a church. Numerous antiquities of various ages have been discovered in and about Lucera. (*Craven's Tour, &c.*, p. 43–51; *Swinburne*, 1., 157–160; *Cramer's Ancient Italy*, ii., 286, &c.)

LUCERNE (CANTON OF), a canton of Switzerland, making third in the Confederation, between lat. 46° 47' and 47° 17' N., and long. 7° 50' and 8° 29' E.; having N. Solothurn and Aargau, E. Zug, Schwytz, and Unterwalden,

and S. and W. Bern: length N.E. and S.W. 36 m., breadth varying from 8 to 30 m.; area 587 sq. m. Pop., in 1838, 124,521, all Roman Catholics, except about 50 Calvinists. The surface in the N. is generally plain, undulating in the centre, and rising gradually towards the S., where are several mountain-ranges of considerable height. The principal of these is Mt. Pilate, between Lucerne and Unterwalden, its highest summit, the Tomlishorn being estimated at 7199 feet above the level of the sea. The S. and E. parts of the canton are watered by the Reuss and Little Emmen; the other rivers are the Wigger, Sur, Viñon, &c., all having a N. course, and joining the Aar in Aargau. The lake of Lucerne (which see) forms a part of its E. boundary, and the canton comprises several small lakes, as that of Sempach, 4 m. in length, and memorable for the battle fought on its banks 9th July, 1386 (*see* SEMPACH), those of Baldegg, 3 m. in length, Mauen, &c. The climate is mild, and the soil more favourable to agriculture than that of most of the neighbouring cantons. According to Ebel, more corn is grown than is required for home consumption. Inglis, however, affirms that the surplus is very trifling, and that the greater part of the grain brought to the corn-market in the town of Lucerne comes from the canton of Aargau. "I have never seen," says he, "anywhere more abundant crops than are produced in Lucerne, where time and industry are bestowed upon the lands. In no part of Switzerland might the inhabitants be more at their ease than in this canton; and yet there is not a commune in which paupers are not to be found." Industry is not nearly so active as in the neighbouring cantons of Bern and Zurich. The vine flourishes in some parts, fruit is plentiful, and wine and cider are produced; but the chief occupations of the people are cattle breeding, and dairy-husbandry. The Entlibuch, or valley of the Little Emmen, about 25 m. in length, affords pasturage for about 7000 head of cattle, and 11,000 sheep and goats: the inhabitants make large quantities of cheese, which, though not so good as that of the Emmenthal in Bern, is exported as the produce of the latter district. The inhabitants of this valley are remarkable for their vigour, intelligence, and independent spirit, and are usually richer than those of the rest of the canton; but they are neither so well clothed, nor have such neat cottages as the peasantry of the Emmenthal. (*Coxe's Letters on Switzerland*, let. 33.) Traces of various metals, coal, &c., are met with in this canton, but no mines are wrought. Manufacturing industry is unimportant, and is mostly confined to domestic linen weaving and spinning. The inhabitants are more occupied in the transit trade from N. Switzerland across the St. Gothard, than in any commercial dealings of their own.

The government is vested in the council of one hundred, 50 of whose members are chosen from among the citizens of the capital, and 50 from the inhabitants of other parts of the canton. The 18 arrondissements into which the canton is subdivided, and the three municipalities of Surseo, Sempach, and Villisau, send one member each to the council; and the remaining 29 members from the rural districts, are chosen by the council itself. The council also nominates 40 of the deputies from the town of Lucerne, the remaining 10 being sent by that municipality. The right of election belongs to every native (bourgeois) of the canton 20 years of age, having property to the amount of 400 fr., and who has not been penally condemned, or is bankrupt. Members of the council must be 25 years of age, and pay taxes on property to the amount of 4000 fr., or have rendered important services to the state. A body of 36 members, 30 years of age, chosen from among the council, and holding office for life, form the senate, to which is confided all the executive power. The council meets regularly three times a year, but may be convoked oftener, at the pleasure of the senate. Two avoyers, or presidents, are chosen annually from among the senate, by the council, one to preside at the council and the other in the court of appeal. The latter tribunal is composed of 12 members, chosen from the senate, and has authority in all legal causes, except in cases of capital punishment, when the senate is assembled to pronounce judgment. The council of state for the Swiss Confederation is chosen from among the senate, when Lucerne has the directorial power, which occurs once every three years. In ecclesiastical matters, Lucerne is subordinate to the bishop of Basle; but being at the head of the Roman Catholic cantons of Switzerland, it was the permanent residence of the papal nuncio till 1835, when, in consequence of a dispute with the government, the nuncio removed into the canton of Schwytz. Public instruction is under the direction of a commission of senators: it has been till lately indifferently conducted, but is improving. The public rev. amounted in 1832 to 366,139 Swiss fr., the public expenditure to 359,283 fr. A contingent of 1734 troops is furnished to the army of the confederacy, and 25,000 Swiss fr. money.

LUCERNE, a town of Switzerland, capital of the above canton, and one of the three seats of the Swiss diet, on both sides the Reuss, where it issues from the W. extremity of

the lake of Lucerne, 25 m. S.S.W. Zurich, and 43 m. E.N.E.
Berne. Lat. 47° 3′ 27″, long. 8° 18′ 35″ E. Pop. 7000. Its
situation is highly picturesque, and its environs abound
with pleasant promenades. The town is surrounded by a
circle of watchtowers, and on the land side is inclosed by a
continuous wall. It is pretty well built, and has several
fine public edifices. The cathedral, founded in 695 (*Ebel*),
has a painting of Christ on the Mount of Olives, by Lan-
franc, and an organ with nearly 3000 pipes. The churches
of St. Peter and the Jesuits are handsome buildings; and
there are several convents: that of the Jesuits has, however, been happily converted into a lyceum. The most remarkable objects in Lucerne are the four bridges over the
Reuss, connecting the great and little towns. Some of
these are of considerable length; all of them are covered
and ornamented with pictures illustrative of Swiss and
Scripture history, or copied from the "Dance of Death."
The town-hall, where the diet and cantonal council meet,
erected in 1606, is, though small, a handsome building. In
the arsenal are several suits of ancient armour, including
the coat of mail worn by Leopold of Austria, killed at the
battle of Sempach. Lucerne has two hospitals, an orphan
asylum, a mint, a jail, a theatre, public libraries belonging
to the town, the Jesuits, Cordeliers, Capuchins, &c., and a
lyceum, with 14 professors of theology, law, natural and
moral philosophy, history, mathematics, and the fine arts.
Attached to the lyceum is a large public school. "Into this
school every child until the age of 12 is admitted, upon payment of six francs a year, and is taught reading, writing,
arithmetic, and the first principles of Latin; and this privilege of acquiring, in early years, the rudiments of learning,
is not confined to the city of Lucerne, nor even to the canton; persons may claim admittance from any other of the
Swiss cantons; and even from foreign countries. The college and the school are one establishment; and every one
who has received his education in the school is immediately received as a pupil of the college, and pays nothing for
his instruction there. The original fund for this establishment amounted to 400,000 fr., but has subsequently been
greatly increased by donations." (*Inglis*, 116, 117.) The
institutions for the intellectual and moral improvement of
the inhabitants are on a scale of great liberality, though
education be far from being widely diffused either in the
city or the canton generally. In the town is the celebrated
model in relief of Switzerland, made by General Pfyffer;
and in the Pfyffer Garden, outside the walls, is a monument,
from a design by Thorwaldsen, to commemorate the Swiss
guards who fell at Paris in the memorable attack on the
Tuilleries, on the 10th of August, 1792. "It represents a
lion of colossal size, wounded to death, with a spear stick-
ing in his side, yet endeavouring with his last grasp to protect from injury a shield bearing the fleur-de-lis of the Bourbons, which he holds in his paws. The figure, hewn out
of the sandstone rock, is 28 feet long, and 18 feet high, and
its execution (which is by Ahorn of Constance) merits great
praise." (*Murray's Handbook*.)

The weekly corn market held here is one of the most extensive in Switzerland. Lucerne has a casino and a theatre
open in winter. Dancing is prohibited by the authorities,
except during the last three days of the carnival and on a
few other special occasions. This prohibition is strictly
enforced in Zurich; but it appears less absurd there than in
a Catholic canton.

The city of Lucerne was given by Pepin in 768 to the
abbots of Murbach in Alsace; to whom it belonged till towards the end of the 13th century, when it was sold to the
house of Hapsburg. But, in 1332, the citizens, impatient
of the Austrian yoke, rebelled and joined the three primitive
cantons of the Swiss Confederacy. In less than 30 years
they conquered the territory which now forms the canton.
The town was taken by the French, May 1, 1798, and was
for eight months the capital of the Helvetic government.

LUCERNE (LAKE OF) (Germ. *Walstätter See*, or the lake
of the four forest cantons), a lake of Switzerland, is nearly the centre of that country, between the cantons of Lucerne on the W., Schwytz N., Uri E., and Unterwalden S.
It is the largest and decidedly the finest lake in the interior
of Switzerland, and one of the most picturesque in Europe.
It is of a singular cruciform shape, with an addition to its
E. end, termed the Lake of Uri. Its greatest length is about
25 m.; but the breadth of any of its arms is seldom more
than 2 or 3 m. Area estimated at 43 sq. m.; height of its
surface above the level of the sea, 1380 feet: depth varying
from 300 feet near Lucerne to 900 feet near its E. end. The
Reuss traverses this lake in its entire length, emerging from
it near its W. extremity. Its banks exhibit every gradation
of scenery, from a gently rising and fertile country at its W.
end, to rugged and savage sublimity on the lake of Uri. Its
E. and S. parts are surrounded by mountains rising to many
thousand feet above the sea, the chief of which are mounts
Pilate and Righi. Its shores abound in localities memorable
in early Swiss history. At the N. extremity of what is called
258

the lake of Uri is the little town of Brunnen, where, in
1315, a treaty was entered into by Uri, Schwytz, and Unterwalden, which gave birth to the Helvetic Confederacy.
Like all mountain lakes, it is subject to violent tempests;
and in consequence of the different positions of its different
arms, and the influence of the surrounding mountains, different winds seem to prevail in different parts of its extent at
the same time. A steamboat plies eight times a week in
summer, and five times at other seasons, between Lucerne
at its W. and Fluelen at its E. extremity, calling at the intermediate ports. (*Ebel. Voyageur en Suisse*, 368, 381; *Picot's Statist. de la Suisse*, 212, 233; *Coxe's Switzerland; Inglis's Switzerland, &c.*)

LUCIA (ST.), one of the British W. India islands, belonging to the Windward group, in lat. 14° N., and long. 61°
W., about 90 m. N.N.E. St. Vincent, and 25 m. S. Martinique.
It is of an oblong shape, being nearly 32 m. in length by
about 12 m. in its greatest breadth. Its area has been differently estimated, but may amount to about 300 sq. m. Pop.,
by the last census, 16,017, of whom 13,348 were blacks.
The central and E. parts of the island are occupied by the
table-land, called Capisterre; the W. part, which has a
much less elevation, is called Basseterre. These two districts differ widely in physical aspect; but each, in an eminent degree, is subject to the operation of those agencies
which are supposed to exert a baneful influence on the
health of Europeans in tropical climates. St. Lucia has, in
fact, always been noted for its unhealthiness. Taking an average
of the 20 years, ending with 1836, the annual deaths amounted to upwards of 122 per thousand of the white, and 49 per
thousand of the black troops.

"Basseterre, the best cultivated portion of the island,
abounds in swamps and marshes. Capisterre consists of a
succession of abrupt mountains of the most picturesque and
fantastic shapes, covered to the summit with forest trees
and dense underwood, and intersected by numerous ravines,
which, being too narrow to admit of free ventilation, are at
all times replete with moisture, and choked up with decayed vegetation in every stage of decomposition. The climate
is principally characterized by extreme moisture and variableness. During several months, but particularly in October and November, rain is incessant, and showers are frequent for at least nine months of the year. Cool dry weather
generally sets in about Christmas, and continues three or
four months, at which time the climate is exceedingly
pleasant, though not more healthy, since it is at that period
of the year that the greatest mortality prevails. During the
rest of the year the weather is sometimes dry and sultry, at
others cold and damp, exhibiting a difference of 10 or 12 degrees of temperature in a few hours." (*Tulloch's Report
on Mortality, &c., in the W. Indies*.) The range of the
thermometer is much the same as at Dominica. Nearly
9500 acres are under crops, and 4700 in pasture. The
mountains are feathered to the top with tall forest trees, and
the valleys at their feet abound with excellent timber.

St. Lucia has several good harbours, the chief being the
Carenage on the W. coast, within which 30 ships of the line
may lie in perfect security, without even, as is stated, being
moored. The wish to command this admirable harbour
was, in truth, the motive which made the island be formerly so much coveted by the European powers.

The quantities of the principal articles imported into the
United Kingdom from St. Lucia in 1838 and 1839, were:

Years.	Sugar.	Rum.	Molasses.	Coffee.	Cocoa.
	Cwts.	Galls.	Cwts.	Lbs.	Lbs.
1838	86,691	7,483	4,766	143,288	14,925
1839	80,313	14,051	11,029	34,680	25

The total value of the exports from St. Lucia, in 1836,
amounted to £69,040, and that of the imports in the same
year to £60,344. During that year, 371 ships, of the aggregate burden of 13,044 tons entered, and 379, of the aggregate burden of 13,166 tons, left the ports of the colony.

The island is divided into nine parishes. Castries, the
capital, lies in a low and marshy situation, at the extremity
of a long and winding bay of the same name. The fort,
where most of the troops in the island are stationed, is
built on the summit of a steep hill, called Morne Fortuné,
about 1½ m. from Castries, and 850 ft. above the level of
the sea. Near it is the principal hospital. Another hospital, and some barracks, are erected on Pigeon Island, a
small, conical, and extremely unhealthy islet, near the N.
extremity of the island. St. Lucia is governed by a governor and council, acting under orders from England. The
mutual jealousies of England and France prevented, for a
lengthened period, a permanent settlement being made on
the island, which was then regarded as a sort of neutral
territory. At length it was ceded to the French in 1763;
but, being taken by the English in 1803, it was definitely assigned to us by the treaty of Paris. (*Parliamentary Reports, &c.*)

LUCIA (STA.), a town of Sicily, intend. and distr. of

Messias, cap. canton, in a healthy situation on the declivity of mount Dinnamare, 7 m. S. by E. Milazzo. Pop. in 1831, 6270.

LUCKIPOOR, a town of Hindostan, prov. Bengal, distr. Tiperah, a few miles from the mouth of the Brahmaputra, with which it communicates by a small river, 156 m. E.N.E. Calcutta; lat. 22° 56′ N., long. 90° 43′ E. Coarse cotton cloths of a substantial kind are made here; and the neighbourhood is so fertile and productive that Luckipbor is one of the cheapest towns in British India.

LUCKNOW (Hind. *Lakshmanavate*), a large city of Hindostan, prov. and kingdom of Oude, of which it is the cap., on the Goomty, a tributary of the Ganges, about 150 m. N. W. Benares, and 265 m. S.E. by E. Delhi; lat. 26° 51′ N., long. 80° 50′ E. Pop. formerly estimated at 300,000, but now probably under 200,000. This city is interesting from its being the capital of one of the most powerful native states in Hindostan, with which the British power in its rise and progress in India has been more intimately connected than any other; and it is also one into which European habits have been very extensively introduced by the late reforming sovereign of Oude, Saadet Ali. "When viewed from the summit of a lofty edifice, Lucknow presents a confusion of gilded cupolas and pinnacles, turrets, minarets, and arches, bounded by the winding Goomty, and so thickly interspersed with the richest tropical foliage as apparently to realise the most fantastic visions of oriental splendour. A nearer inspection, however, does not fulfil the anticipations which a bird's-eye survey is calculated to excite. This capital may be divided into three quarters. The first is the city, properly so called, containing the shops and private dwellings of the inhabitants connected with the court and residency. The streets here are sunk 10 or 12 ft. below the surface, and are so narrow that two carts cannot pass, besides being filthy in the extreme. The chowk, and one or two bazaars in its vicinity, are good streets; but, on the whole, this extensive quarter is more meanly built than, perhaps, any city of the same rank in Hindostan. "The second quarter of Lucknow was built mostly by the late nabob, Saadet Ali. It stands near the Goomty, towards the S.E., and consists of one very handsome street, after the European fashion, above a mile in length, with bazaars striking out at right angles, and a well-built new chowk or market-place in the centre, with a lofty gateway at each extremity, which presents a Grecian front on one side, and a Moorish one on the other. The houses that compose the remainder of this street belong to the king, and are occupied by members of his family, or officers of his household. These are, for the most part, in the English style, but with a strange occasional mixture of Eastern architecture. The same remark applies to the palaces, &c., that occupy the space between this street and the river. All these palaces are filled with European furniture and pictures, and may rank with comfortable English houses, but none are on a scale of royal magnificence. The king's peculiar residence only excels the others in being approached through six spacious courts, with reservoirs, fountains, and innumerable pieces of cast statuary, China figures, and other toys that decorate its area. The adjacent buildings of the British residency terminate the great street to the N. At its opposite extremity is the entrance of the Delkusha park, an artificial wilderness of high grass, with which Saadet Ali clothed the arid tract between Lucknow and Constantia, and well stocked with deer, antelopes, and peacocks. "The third quarter of the city adjoins the Goomty to the N.W. being only separated by a wretched bazaar from the second. It consists chiefly of palaces and religious buildings; and, being in a style more purely oriental than the modern portion of the city, is by far the most interesting quarter to a stranger. The magnificent pile of *Imaum-bārah*, with its noble gateway, called the *Roumi-derwazzah*; the new palace built, but never finished, by Saadet Ali, the *Dewlet-khanah*, &c., are the chief ornaments of this division of Lucknow." (*Hamilton's E. I. Gaz.*, ii., 130, 131.) There are many stately *khans*, and some handsome mosques and pagodas scattered in different parts of the wretched alleys, of which the city chiefly consists; but the most striking buildings, as in other Mohammedan capitals, are the royal tombs and mosques. Of these, the *Imaum-bārah*, or tomb and mosque of Asophud Dowlah, is the chief. It is said by Lord Valentia to be the most beautiful building he had seen in India. "The approach to the building is through a very large quadrangle to a garden, elevated a small height; on one side of which is a very beautiful mosque, and on the other the *Bolee* palace. The *Imaum-bārah* itself is built on an elevated terrace, and consists of three long and finely-proportioned apartments, running parallel to each other; in the middle one is the tomb, level with the ground." (*Valentia's Trav.*, i., 157.) The central room is, according to Hamilton, 167 ft. in length, by 32 ft. in breadth, with an octagon room at each end; and

in the rear of the centre, a raised set of rooms, or open arches, with fountains and basins of water under each arch. Lord Valentia says that a range of silver temples or cenotaphs also extends from one end to the other of the room, raised on platforms about 3 ft. from the ground, and each valued at from 50,000 to 100,000 rupees. The *Roumi-Derwazah*, so called from being supposed a copy of one of the gates at Constantinople, is in a light and elegant, though fantastic style, and a mixture of Gothic and Moorish architecture. A good engraving of it and a part of the city is given in Lord Valentia's Travels (i., 173). Near it is the Fivefold palace, a large fortified building, appropriated to the wives, &c., of deceased sovereigns.

Two bridges have been erected over the Goomty at Lucknow; one a heavy bridge of masonry, the other a bridge of boats. The erection of an iron bridge was projected by Saadet Ali, but the materials arrived from England too late for the accomplishment of the work during his lifetime; and his heir, conformably to a prejudice universal among the Mohammedans of Hindostan, declined the unlucky task of completing the unfinished undertaking of a deceased predecessor. About 3 m. from Lucknow is Baroun, a country seat of the last-named chief, built by himself. It is in a Grecian style of architecture, and has, as might be expected, many faults; but it is ornamented by a very fine portico, rising the whole height of the house in front. Near the city is also Constantia, the former residence of a general in the East India Company's service, and erected at an expense of seven lacs of rupees; but this building is in wretched taste, and only imposing at a distance. It has, however, or had, some fine gardens attached to it.

Lucknow is traditionally said to have been founded by *Lakshmas*, the brother of *Rama*; who had his residence here, to extinguish the recollection of which Aurungzebe erected a mosque with two minarets on its site. After the battle of Buxar, Shuja ud Dowlah removed his court from Lucknow to Fyzabad; but on his death, in 1775, his successor made this again the capital of Oude. (*Lord Valentia's Travels*, i., 135–175; *Mod. Trav.*; *Hamilton's Hindostan and E. I. Gaz., &c.*)

LUCKPUT-BUNDER, a town of Hindostan, province of Cutch, of which it is the chief port after Mandavee, on the Khoree, or most easterly branch of the Indus, on the high road from Mandavee to Hyderabad and Tatta, 82 m. S.E. by S. the last-named city, and 67 m. W.N.W. Bhooj. It is defended by a good fort. Early in the present century, it had but 2000 inhabitants, and, owing to the shallowness of the river, could only be approached by very small craft; but, by an earthquake in 1819, the Indus was deepened at Luckput to more than 18 ft. at low water, and there is now 20 ft. water in its channel from the ocean to Busta, 5 m. below this town. (*Geog. Journal*, iii., 119.) This must, no doubt, have contributed to the commercial prosperity of Luckput-Bunder, though we have not learned the particulars.

LUDLOW, a mun. and parl. bor., market town, and par. of England, co. Salop, hund. Munslow, on the Teme, 24 m. S. Shrewsbury, 39 m. W. by S. Birmingham, and 126 m. W.S.W. London. Pop. of parl. bor. in 1831, 5332. The town is neat and well built, and the streets are generally wide, well paved, and lighted. On a bold rock, overhanging the river, at the N.W. angle of the town, stands the castle, supposed to have been built in 1130. The walls and towers which still remain present a mass of extensive and magnificent ruins; and round the castle are public walks, shaded with trees, from which there is a fine prospect of the surrounding country. Near the centre of the town is "the Cross," a handsome stone building, with rooms over it used as a school; and in Castle-street is the market-house, the lower part of which is open, and serves as a corn-market, the upper part comprising several large rooms, used for corporation meetings, assemblies, public balls, &c. The guildhall, where the quarter sessions and court of record are held, is a neat and commodious modern structure; and there is a prison called Gaolford's tower. Performances are given in a small theatre during the races, which are held in the neighbourhood. The church, which stands at the upper end of the town, is a large cruciform building of perpendicular architecture, surmounted by a square embattled tower, rising from the intersection. The interior is very beautiful: lofty pointed arches divide the nave from the aisles; and at the E. end of a very spacious choir is a noble window, entirely filled with painted glass: the whole church is celled with fine oak, and embellished with carving. The S. entrance is peculiar, consisting of a hexagonal porch richly ornamented. The living (valued at £160 a year) is a rectory, in the gift of the lord chancellor. There are three places of worship for dissenters, and to both the churches and chapels Sunday schools are attached, furnishing religious instruction to upwards of 500 children. The grammar school, founded in the reign of Edward VI., is intended to give free instruction, in English and classical learning, to the sons of all residents within the borough. The pupils comprise about 30 free boys, and nearly the same number of pay scholars boarding with the master, who receives a yearly salary of £100, and is assisted by an usher. The master holds, also, the office of preacher, with a salary of £48 a year. A national school, under the superintendence of the rector and a committee, is attended by 150 boys and 100 girls; and is liberally supported, partly by contributions and partly, also, by the funds of a blue-coat charity recently merged into it.

Ludlow, as a place of trade, is of little importance. The glove trade formerly employed several hundred hands; but of late years it has greatly diminished. A small flannel-mill employs about 20 hands, and there is a considerable paper-mill. Malting and tanning are also carried on to some extent; but the chief business is confined to the retailing of goods consumed in the town and neighbourhood. The corporation charter was granted by Edward IV., and has been subsequently confirmed by nine different monarchs. Under the Municipal Reform Act, the government is vested in a recorder, four aldermen, and 12 councillors. Corporation revenue in 1839, £964. The borough has returned two members to the House of Commons since 12 Edward IV.; the right of election, previously to the Reform Act, being nominally vested in the resident burgesses (made so by birth, marriage, or gift), but substantially in the lord of the manor, Earl Powis. The electoral limits were enlarged by the Boundary Act, so as to include, with the old borough, the township of Ludford, and part of the parish of Stanton Lacey. Registered electors in 1839-40, 422. Market (well attended) on Monday; fairs, chiefly for horses, cattle, and pigs, Monday before Feb.13, Tuesday before Easter, Wednesday in Whitsun week, Aug. 21, Sept. 28, and Dec. 6, on the two last of which large quantities of hops are exposed for sale.

The history of Ludlow is closely connected with that of its castle, which, being erected by the barons of Montgomery in the 12th century, continued in a habitable state till the suppression of the council of the marches of Wales by William III. To all lovers of English poetry this castle is interesting, as having been the scene where Milton's "Comus" was performed in 1631, by the family of the earl of Bridgewater.

Ludlow, p. t., Windsor co., Vt., 80 m. S. Montpelier, 672 W. Chartered in 1761, first settled in 1784-5. Watered by Black river. It contains four churches, a Congregational, Baptist, Universalist, and a Union church; a flourishing academy, three stores, three fulling-mills, one woollen factory, one grist-mill, five saw-mills, two tanneries; one academy, 130 students; 16 schools, 488 scholars. Pop. 1363. The township contains beautiful verd. antique marble and excellent iron ore, some of which is magnetic.

Ludlow, p. t., Hampden co., Mass., 82 m. W. by S. Boston, 373 W. Incorporated in 1774. Chickapee river bounds it on the S., by a branch of which it is watered, and which afford water-power. It contains two churches, a Congregational and Methodist; one store, one grist-mill, three saw-mills; 10 schools, 250 scholars. Pop. 1268.

LUDWIGSBURG, a town of Wirtemberg, circ. Neckar, of which it is the cap., on rising ground, about 1 m. W. of the Neckar, and 8 m. N. Stuttgard. Pop. in 1837, 6900. It is one of the best laid out and handsomest towns of the kingdom, but is dull. Charles-street, by which it is traversed from end to end, is 1 m. in length, and like most of the other streets, is lined with rows of trees. From 1727 to 1733, Ludwigsburg was the chief residence of the court; its palace, though now deserted, is one of the largest and finest in Germany; and it has a gallery of old German, Dutch, and Flemish pictures; and a theatre. The palace gardens, formerly celebrated for their beauty, are now falling into disorder from neglect. Ludwigsburg has a Lutheran parish church, three other churches, an arsenal, a military school for 20 officers' sons, a lyceum, an orphan asylum, and workhouse, house of correction for females, school for poor children (kinderrettungsanstalt), a cannon foundry, and manufactures of woollen cloth, earthenware, and buttons. In the neighbourhood are the royal summer palaces of Favourite and Monrepos, and the fine statue of Count Zeppelin, erected by King Frederick of Wirtemburg. About 6 m. distant is Marbach, the birthplace of Schiller, and the mathematician Maver. (Memsinger; Beschreibung von Wurtemburg, 383-84; Berghaus; Stein.)

LUGANO (TOWN AND LAKE OF). The town of Lugano, being with Bettinzoro and Locorno, a cap. of the Swiss canton of Tessin, stands on a bay on the W. bank of the lake of same name, 13 m. N.N.W. Como. Pop. 3800. It is a well built, handsome town, finely situated round the curve of a beautiful bay, surrounded by an amphitheatre of hills, having their slopes studded with villas, vineyards, gardens, and forests; while in the distance are seen the snowy pinnacles and craggy masses of the Alps. Among the principal public buildings are the church or cathedral

of San Lorenzo, on an eminence above the town, commanding a fine view, with a finely-sculptured portal and a façade, said to be by Bramante: the church of the Franciscan friars, remarkable for two paintings of first-rate excellence by Bernardo Luini. It has also some pretty extensive silk manufactures, a large theatre, and a hospital; several establishments for the printing and sale of books newly published or prohibited in Italy; and go fewer than three newspapers, which occasionally advocate doctrines that are but little to the taste of the Austrian and Sardinian governments. Perhaps, however, the town may derive its principal support from its being on the route, and one of the entrepôts, of a considerable portion of the trade carried on between Italy and Switzerland, and Germany by the pass of St. Gothard. Though nominally and politically Swiss, the Luganese are Italians in dress, language, manners, and appearance; in everything, in short, but their greater activity and enterprise; and for this distinction they are mainly indebted to their comparatively free institutions and free press. Monte Caprino, near Lugano, has a great number of natural caverns or grottoes, which, on account of their coolness, are used by the inhabitants in summer as cellars in which to keep their wine, meat, and other provisions.

The lake of Lugano (formerly the lacus Ceresius), is principally within the canton of Tessin, in Switzerland, but partly also in Lombardy, between the lago Maggiore and the lago di Como. It is of an extremely irregular figure: its greatest length, from Porlezza at its N.E. to Porto at its S. extremity, is about 16 m.; but, in addition to its main body, it has two great arms, one stretching S.S.E. to Lugo, and the other N. to Agno. It is nowhere above 2 m. in width, and is mostly surrounded by high mountains, overhanging woods, and bold, abrupt precipices. One of the mountains, San Salvador, on a promontory, washed on two of its sides by the lake, rising to the height of nearly 3000 ft. above its level, is a sublime object from the lake, and commands from its summit a most magnificent and varied prospect. In some parts, however, the banks of the lake slope gently down to the water's edge, and are covered with villages, vineyards, gardens, &c. The bay of Lugano on its W. side, with its surrounding amphitheatre of hills, is particularly fine. Its waters are quite transparent, and so very deep, that in some places no soundings are said to have been attained. It is about 190 ft. above the level of the lakes of Como and Maggiore, into the latter of which the Tresa conveys its surplus waters. (Cendar's Italy. L. 314; Eustace, tr., 60; Coze's Switzerland, iii., 368; Murray's Handbook, &c.)

LUGGERSHALL, or LUDGERSHALL, a decayed bor. market town, and par. of England, co. Wilts, hund. Amesbury, 25 m. N.E. Salisbury, and 68 m. W. by S. London. Area of par. and bor., 1660 acres. Pop. in 1831, 533. The town, now in a wretched and decaying state, contains nothing worth mention, except an old, ruinous church, and a place of worship for Baptists; the inhabitants are chiefly supported by agricultural labour. Formerly, however, it must have been a place of more importance, for a large castle existed here soon after the conquest: it was also one of the most ancient parliamentary boroughs, and, notwithstanding its insignificance in modern times, sent two members to the House of Commons down to the passing of the Reform Act, by which it was disfranchised.

LUGO, a town of Spain, prov. Galicia, and a bishop's see, on the Minho, 47 m. E.S.E. Coruña, and 142 m. N.N.E. Oporto. Pop., according to Miñano, 7909. It occupies an eminence on the E. bank of the river, and is surrounded by an ancient wall of great thickness, with circular projecting towers. The streets are mean and irregularly built: the chief buildings are a Gothic cathedral, four convents, two hospitals, a singular-looking prison, a foundling asylum, and public seminary. The climate is alleged to be colder than that in other parts of Galicia: snow is frequent, and N. winds are common during the winter months. The place appears to be in a languishing condition; the only fabrics are those of thread stockings and Morocco leather. In the neighbourhood are bred great numbers of cattle, horses, mules, sheep, and hogs, which meet with a ready sale at the monthly fairs, and the great fair in October.

Lugo is a place of great antiquity, having been the capital of a conventus, or district, under the Romans, who called it Lucus Augusti. Many monuments of Roman art were existing in the time of Ponz, but they have nearly all been since destroyed. The Roman medicinal baths are still, however, used, and the works formed to protect them from the floods of the Minho may yet be traced. Alonzo the Catholic wrested Lugo from the Moors, and re-established its bishopric. (Miñano, Mod. Trav.)

LUMBERLAND, p. t., Sullivan co., N. Y., 129 m. S.W. Albany, 12 m. S.W. Monticello, 287 W. Bounded S.W. by Delaware river, E. by Mongup river. Drained by tributaries of Delaware river. The Hudson and Delaware canal passes through the t. It contains nine stores, three grist-mills, 30 saw-mills; four schools, 278 scholars. Population, 1905.

LUMBERTON, p. v., capital of Robeson co., N. C., 91 m. S. by W. Raleigh, 379 W. Situated on the E. bank of Lumber river. It contains a courthouse, jail, and several stores and dwellings. Nett proceeds of the postoffice, $85.

LUMPKIN, county, Ga. Situated in the N. part of the state, and contains 730 sq. m. Drained by head waters of Etowah and Chestatee rivers and their branches. It contains the richest gold mines in the country. It had, in 1840, 5490 neat cattle, 3116 sheep, 14,383 swine; and produced 10,061 bushels of wheat, 777 of rye, 214,754 of Indian corn, 9894 of oats, 7813 of potatoes, 4217 pounds of tobacco, 17,812 of cotton. It had 36 stores, 91 smelting-houses, which produced gold to the amount of $74,460, 29 grist-mills, 10 saw-mills, one tannery, 27 distilleries, one pottery; 14 schools, 328 scholars. Pop.: whites, 5143; slaves, 514; free coloured, 12; total, 5671. Capital, Dahlonega.

LUMPKIN, p. v., capital of Stewart co., Ga., 137 m. S.W. Milledgeville, 793 W. Situated on the S. side of Hannahatchee creek, 16 m. E. of its entrance into Chattahoochee river. It contains a courthouse, jail, a Baptist and Methodist church, 24 stores, and about 40 dwellings. Nett proceeds of the postoffice, $640.

LUND, a city of Sweden, near its southern extremity, prov. Malmø, 20 m. N.E. Malmø. Pop. 4120. It is open, and irregularly built, but clean. It is an archbishopric, and has a cathedral, an ancient, irregular building, raised at different periods. But it is chiefly remarkable for its university, founded in 1666. This institution has 22 regular and seven assistant professors, and is attended by about 600 pupils. In 1834 it had 396 pupils, whereof 108 were students of divinity, 130 of law, 50 of medicine, and 160 of philosophy, the sciences, &c. It has a library of 30,000 printed volumes and 1000 MSS., with museums of natural history and mineralogy, antiquities and medals, &c.; an observatory, a chemical laboratory, and a botanical garden. Puffendorf, who, next to Grotius, is the grand authority in matters of public law, was appointed Professor of the Law of Nature and Nations in this university in 1670; and here, in 1672, he published his great work De Jure Naturae et Gentium. "Without," to use the words of a distinguished authority, "the genius of Grotius, and with very inferior learning, he has yet treated this subject with sound sense, with clear method, with extensive and accurate knowledge, and with a copiousness of detail sometimes indeed tedious, but always instructive and satisfactory." (Mackintosh on the Law of Nature and Nations, p. 21.) Linnæus was for some time a pupil in the university of Lund. The town has manufactures of woollen cloths and tobacco, tanneries and sugar refineries, a discount bank, and some foreign trade. The ancient kings of Scania were chosen on the hill of Lyberts, near the town. (Stein; Coxe's Travels, iv., 298; Dict. Geog., &c.)

LUNEBURG, a town of the kingd. of Hanover, cap. of the distr. and principality of Luneburg, on the Ilmenau, 67 m. N.N.E. Hanover, and 27 m. S.E. Hamburg. Population, in 1832, estimated at 11,800. It is surrounded by walls of no great strength, and entered by six gates. It has dark and narrow streets, and old-fashioned houses. The castle, or palace of the prince, the town-hall, council-house, military academy for young nobles, gymnasium, exchange, and cavalry barracks, are the principal public buildings. The military academy has a library of 14,000 volumes, and in the town-hall is another library. Luneburg has four churches, in one of which are the tombs and monuments of many of the ancient dukes of Luneburg, several superior schools, an orphan asylum, and a mont-de-piété.

Luneburg was formerly a Hanse town, was governed by magistrates of its own selection, and had an extensive trade. It took part in the Baltic herring-fishery, and had numerous breweries and manufactures of woollen stuffs, &c., now much fallen off. Lime-burning and the making of salt are at present the chief branches of industry. A large and singular rock of gypsum, rising nearly 170 feet above the Ilmenau, in the immediate vicinity of the town, furnishes abundant materials for the former business. About 20,000 tons of lime a year are sent to Hamburg, Altona, and Holland. About 160,000 centners a year of salt are procured from some adjacent salt-springs: the evaporation is effected by means of turf, and is conducted under a special commission, the government having a monopoly of the article. The price of the undried salt is 40 dollars, and of the dried, 46 dollars the last of 400 lbs. (Berghaus.) Luneburg has some fabrics of woollen and cotton and linen goods, tobacco, paper, cards, and soap; with distilleries, breweries, &c. It also trades in horses; and is the seat of a transit trade between Hamburg and the Elbe, and the interior provinces of Hanover. About 900,000 centners of merchandise are said to have passed in transitu through Luneburg in 1838. (Von Reden's Hanover, ii., 87, &c.; Berghaus; Stein; Hodgskin's N. Germany).

LUNEL, a town of France, dep. Hérault cap. cant. on

the canal of Lunel, 14 m. E.N.E. Montpellier. Pop. in 1836, ex. com., 6091. It has a fine promenade, infantry and cavalry barracks, numerous liqueur and brandy distilleries, and a brisk trade in corn, wines, and raisins. The muscadine wine, produced from vineyards situated on gently rising grounds to the N. of the town, and bearing its name, is reckoned by some connoisseurs as the best of its class, and is rivalled only by the Frontignan. "It is a very delicate wine, of a bright yellow colour, with a less distinct flavour of the grape, and less cloying, than the Frontignan. The vineyard called the Clos-Mazet, which has been long known to afford the first-rate growth, makes about 100 hhds. a year, being one third of the total quantity supplied from the territory of Lunel. Several of the more ordinary muscadine wines, however, come into the market as Frontignan and Lunel; but they may be easily detected by their deeper colour, and the want of the characteristic flavour and perfume." (Henderson on Wines, p. 177.) According to Jullien, the wines of Lunel, "Sont plus précieux et plus fins que ceux de Frontignan; mais ils ont moins de corps, un goût de fruit moins prononcé, et ne conservent pas aussi longtemps." (Topographie de Vignobles, p. 250.)

LUNENBURG, county, Va. Situated in the S. part of the state, and contains 400 sq. m. Bounded N.E. by Nottaway river. Drained by Meherrin river and its branches. It contained, in 1840, 7470 neat cattle, 9004 sheep, 15,805 swine; and produced 28,594 bushels of wheat, 274,547 of Indian corn, 138,945 of oats, 10,136 of potatoes, 2,040,069 pounds of tobacco, 8590 of cotton. It had 30 stores, 28 grist-mills, six saw-mills, three tanneries; 12 schools, 230 scholars. Pop.: whites, 4132; slaves, 6767; free coloured, 216; total, 11,055. Capital, Lewistown.

LUNENBURG, p. t., Essex co., Vt., 58 m. E.N.E. Montpelier, 567 W. Bounded S.E. by Connecticut river. Fifteen Mile falls on that river commence in the S. part of the t. Chartered in 1763. It contains three churches, a Congregational, Methodist, and Baptist, two stores, one fulling-mill, four saw-mills; 11 schools, 377 scholars. Pop. 1130.

LUNENBURG, p. t., Worcester co., Mass., 46 m. N.W. Boston, 428 W. Incorporated in 1728. Watered by branches of Nashua river. It manufactures extensively books and palm-leaf hats. It contains two Congregational churches, one store, one grist-mill, five saw-mills, two tanneries, two printing-offices, one bindery; 10 schools, 494 scholars. Pop. 1272.

LUNENBURG, C. H., called also Lewistown, p. v., capital of Lunenburg co., Va., 78 m. S.W. Richmond, 197 W. It contains a handsome brick courthouse, with a portico of four columns in front, a jail, county clerk's office, two stores, 20 dwellings, and about 100 inhabitants.

LUNEVILLE, a town of France, dep. Meurthe, cap. arrond., on the Vezouze, and on the road from Paris to Strasbourg, 16 m. S.E. Nancy. Pop. in 1836, ex. com., 12,661. It is generally well built, and has a good square, a château erected by Leopold, duke of Lorraine, early in the last century, and long the residence of Stanislaus, king of Poland, a handsome parish church, very extensive cavalry barracks, a parade-ground of 200 hectares, a large covered riding arena, two hospitals, a synagogue, theatre, and manufactures of woollen cloth, woollen and cotton yarn, gloves, &c. Luneville is one of the principal cavalry stations in France. The origin of the town is uncertain, but its name seems to indicate that Diana was anciently worshipped here; and several Roman medals, with the impress of that divinity, have been found near a fountain in the neighbourhood. The peace between France and the German Confederation, in 1801, by which the former acquired the territory on the left bank of the Rhine, was concluded in this town. (Hugo, art. Meurthe; Berghaus; Guide du Voyageur, &c.)

LURAY, p. v., capital of Page co., Va., 130 m. N.W. Richmond, 96 W. Situated on Hawksbill creek. Founded in 1814, when the first house was built. It contains a courthouse, jail, two churches, a Methodist and Baptist, four stores, 50 dwellings, and about 400 inhabitants.

LURGAN, an inland town of Ireland, co. Armagh, prov. Ulster, about 3 m. from the S. border of lough Neagh, and 18 m. W.S.W. Belfast. Pop., in 1831, 2942. It is a neat, clean, and well-built town, consisting principally of one wide street. It has a parish church, a Roman Catholic chapel, meeting-houses for Presbyterians and Quakers, a courthouse, and a bridewell. A manor-court is held every three weeks, and general sessions and petty sessions every Friday. It is a constabulary station; and has two schools on the foundation of Erasmus Smith, and a subscription school. The linen manufacture, particularly that of diapers and damasks, is extensively carried on, as is that of tobacco: there are two breweries and an extensive distillery. Markets on Fridays; fairs, August 5th and November 22d. The nearness of the town to the point where the Lagan and Newry navigation joins lough Neagh, affords great facilities for inland traffic. Postoffice revenue, in 1830, £40¼: in 1836, £893. Branches of the Belfast, Northern, and Provincial banks were opened in 1834 ; of the Agricultural bank in 1836 ; and of the Ulster bank in 1837.

The town is on the estate and in the immediate vicinity of the residence of the Brownlow family, to the head of which it gives the title of baron.

LURGAN, t., Franklin co., Pa., 13 m. N. Chambersburg. Watered by the N. and S. branch of Conedogwinit creek. Bituminous coal is found. It has three stores, five gristmills, six saw-mills, two oil-mills, two tanneries ; seven schools, 250 scholars. Pop. 1143.

LUTON, a market town and par. of England, co. Bedford, hund. Flitt, 16 m. W.N.W. Hertford, and 28 m. N.W. London. Area of par., 15,500 acres. Pop., in 1831, 5693 ; do. of township, 3961. The town, pleasantly situated between two hills in the Chiltern chalk range, is irregularly built, with three long streets, running from a market-place (in which is an old town-hall), in the form of the letter Y. The church is an interesting specimen of Gothic architecture, with a square embattled tower surmounted at the angles by hexagonal pinnacles, and a handsomely decorated W. door : the interior contains, besides some painted windows, a curiously carved font, and some fine old monuments. There are also places of worship for Wesleyan Methodists, Baptists, and the Society of Friends. Three well-attended Sunday schools, a national and Lancastrian school, furnish instruction to the children of the poor ; and there is a well-endowed hospital for lodging and clothing 24 aged widows. The inhabitants are principally engaged in the manufacture of straw hats, and especially of the variety called the Tuscan grass-plait. Lace-making used also to be carried on to a considerable extent ; but this business has been all but extinguished by the rise of the Nottingham frame-lace trade. Two miles E. of the town is Luton Hoo park, a seat belonging to the Bute family, erected by Lord Bute, the favourite of George III. Markets on Monday ; large cattle fairs, April 18th and October 18th.

LUTTERWORTH, a market town and par. of England, co. Leicester, hund. Guthlaxton, 12 m. S. Leicester, and 79 m. N.N.W. London. Area of par., 1890 acres. Pop., in 1831, 2282. The town, situated on the Swift, a tributary of the Avon, comprises one main and well-walled street, with others of inferior size : there are some good houses, but a large proportion of the tenements are mere mud-walled thatched cottages. The church is a large and very handsome structure, in the pointed style, with a high square tower having turrets at the angles : the interior is elegantly fitted up. But it is principally remarkable from having been the scene of the pastoral labours of John Wycliffe, and from its containing his pulpit and portrait. This early and illustrious reformer and eminent divine was appointed rector of Lutterworth in 1374, where he expired 10 years afterward, on the 31st of December, 1384. Luckily, however, his doctrines did not die with him. In 1415 the council of Constance vainly endeavoured to gratify their impotent rage against his memory, by ordering his remains to be disinterred and cast upon a dunghill. This disgraceful sentence was carried into effect : for the bones of Wycliffe, being taken up, were burned, and the ashes thrown into the Swift. "Thus," as Fuller has ingeniously expressed it, " this brook (the Swift) has conveyed his ashes into Avon, Avon into Severn, Severn into the narrow seas, they into the main ocean : and thus *the ashes of Wycliffe are the emblem of his doctrine, which now is dispersed all the world over.*"

Lutterworth has three places of worship for dissenters, four Sunday schools, an endowed free-school, attended by 100 boys, and three smaller subscription schools. Its chief manufacture is that of coarse hosiery, but it is not extensive. It has a considerable trade in farm and dairy produce, chiefly carried on at its seven annual fairs. Markets on Thursday : fairs, Thursday after February 19th, March 10th, April 15th, July 23d, and October 10th ; also on Holy Thursday.

LUTZEN, a town of the Prussian states, prov. Saxony, circ. Merseburg, 12 m. S.W. Leipsic. This town, the population of which is under 1500, would be unworthy notice, were it not that its environs have been the scene of two of the most memorable conflicts of modern times. The first, which occurred on the 16th of November, 1632, took place between the imperialists, under Wallenstein, and the Swedes, under their heroic monarch, Gustavus Adolphus. The latter were victorious ; but the victory was dearly purchased by the death of their king, who fell (it has been alleged by treachery) in the action. Besides their king, the Swedes lost about 3000 men ; but the loss of the imperialists amounted to double that number, and their artillery fell into the hands of the conquerors.

The other great conflict took place nearly on the same ground, on the 2d of May, 1813, between the French, under Napoleon, and the allied army, encouraged by the presence of the emperor Alexander and the king of Prussia. The struggle was most obstinate and bloody ; but in the end victory declared in favour of the French. The allies lost 20,000

men, killed and wounded, and that of the French was also very severe.

LUXEMBURG (GRAND DUCHY and PROVINCE OF), a territory of W. Europe, between lat. 49° 25′ and 50° 28′ N., and long. 5° and 6° 31′ E. ; having N. the Belgian prov. of Leige, W. that of Namur, E. Rhenish Prussia, and S. France. Greatest length and breadth about 65 m. each. Area. 2700 sq. m. Pop., in 1839, 327,665. By the treaty of the 19th of April, 1839, this territory was definitively partitioned between Holland and Belgium ; the E. portion, with an area of about 1000 sq. m., and a population of 160,860, being assigned to the former, and the W. portion, with an area of 1700 sq. m., and a population of about 166,000, to the latter. The title of grand duke of Luxemburg, with the suffrage in the councils of the German confederation, are enjoyed by the king of the Netherlands.

A chain of hills, branching from the Ardennes, traverses the country from S.W. to N.E. It nowhere rises to more than 2000 feet above the sea ; but it forms the dividing line between the basins of the Meuse and the Moselle. The last-named river and the Sur form the E. boundary of the grand duchy ; the other principal streams are the Ourte, Our, Alzette, Semoy, &c., tributaries of either the Meuse or the Moselle. The valleys are fertile, but the rest of the country has mostly a stony and barren soil ; and in some parts, especially about the centre of Belgian Luxemburg, a good deal of the surface is occupied with marshes, heaths, and poor waste land. The entire surface is estimated at 690,000 *hectaires* (a measure nearly answering to hectares), of which about 240,000 are supposed to be in tillage, 211,000 in woods, 127,000 in heaths, wastes, &c., and 112,000 altogether unproductive, or occupied by roads, rivers, &c. It is mostly divided into small properties. Rye, barley, oats, and wheat are the principal corn crops ; and potatoes, with flax, hemp, and beet-root, are raised. The agricultural course almost invariably occupies three years ; the first year wheat, maslin, or rye is sown ; in the second, oats, barley, or potatoes ; and in the third the land is left fallow. The vine is grown on the banks of the Moselle ; and the annual produce of wine was estimated, in 1837, at 75,583 hectol. The chief branch of rural industry is, however, the rearing of cattle for exportation. The sheep yield indifferent wool, but their flesh is excellent. Horses are good. A great many hogs are reared, and, in the first half of 1838, 36,700 were exported to France. The meadow-lands, especially in the valleys of the Alzette, Chiers, and Semois, are carefully irrigated and manured. The woods are an important source of wealth, the annual produce of timber and firewood being estimated, according to Vandermaelen, at nearly 1,100,000 *steres*. Nearly 93,000 hectares of woods belong to communes, there being scarcely a commune without a certain portion of forest-land. There are few countries in which iron is more abundant ; and about 9,200,000 kilog. of metal are produced annually : from the want of coal, it has to be smelted with timber.

Since 1837, however, coal has been admitted into Belgian Luxemburg (where this branch of industry is principally conducted), from Rhenish Prussia, at the reduced duty of one franc per 1000 kilog., and the production of iron is probably on the increase. The slate of Luxemburg is of a superior quality. Viel-Salm, in the N. of Belgian Luxemburg, furnishes about 4,000,000 of slates a year, and in the S. the quarries of Herbenmont and Geripont produce about 10,000,000 a year, mostly exported to the neighbouring countries. Slate-pencils, marble, and a little lead, zinc, copper, and manganese are the other chief mineral products. Next to forges and potteries, woollen cloth, lace, leather, and glass factories, distilleries, and breweries, are the most numerous manufacturing establishments. The commerce of Luxemburg, however, except in iron, slate and cattle, is insignificant. The inhabitants, partly of Saxon extraction, and partly Walloons, are all Roman Catholics. The whole territory is subdivided into three districts ; those of Luxemburg, Diekirch, and Grevenmacher : each has in it a tribunal of original jurisdiction ; and the first, which is identical with the Dutch province, is placed under a Prussian military governor, and a Dutch civil commissary. Belgian Luxemburg is governed in the same way as the other Belgian provinces. Dutch Luxemburg has the 11th place in the German Confederation, with three votes in the full council and one in the committee. It has, since 1839, furnished a contingent of 1850 men to the army of the confederation ; the contingent previously to the division of the duchy having been 2556 men.

LUXEMBURG (Germ. *Lutzelburg*), a town belonging to the kingdom of the Netherlands, the cap. and only place of any importance in the above grand duchy, and one of the strongest fortresses of Europe ; on the Alzette, a tributary of the Sur, 22 m. S.W. Treves, and 77 m. S.S.E. Liege ; lat. 49° 37′ N., long. 6° 7′ 5″ E. Pop., in 1839, 11,242. It is built partly on a steep rocky height, and partly in the valley beneath ; being, consequently, divided into the upper and lower towns, which communicate by flights of steps, and streets running zigzag, so as to be passable for carriages. Both towns are fortified; and the works, which are partly excavated in the solid rock, have been greatly strengthened by the successive possessors of the town—Spaniards, Austrians, French, and Dutch. Great improvements have recently been made in them; and, since 1837, a new fort has been constructed outside the Treves gate. The casemates of that part of the fortifications called *Le Bouc*, resemble those of Gibraltar, and are capable of accommodating 4000 men. Luxemburg is tolerably well built, but has no remarkable public buildings. It has some iron forges, and manufactures of linen fabrics, leather, and tobacco. It is at present garrisoned by 2600 Prussian troops.

The territory of Luxemburg was governed by its own counts from the time of the Carlovingian Frankish kings to 1354, when the emperor, Charles IV., erected it into a duchy. It was taken by the French in 1794, and subdivided among the deps. of Forets, Ardennes, Sambre-et-Meuse, and Ourthe; but, in 1814, it was erected into a grand duchy, and given to the king of Holland, in exchange for the renunciation of his claims upon Nassau. (*Vandermaelen's Luxembourg; Berghaus, &c.*)

LUXEUIL (an. *Luxovium*), a town of France, dep. Haute-Saône, cap. cant., on the Breuchin, 15 m. N.E. Vesoul. Pop. in 1836, ex. com., 3628. It is well built and clean, and has a good town-hall, a large hospital, a communal college, and manufactures of hats, leather, tin and iron goods, &c.; but it is chiefly remarkable for its hot or thermal springs, which are usually frequented by from 500 to 600 visiters. The hot baths of *Luxovium* were known to the Romans, who are said to have decorated them with fine buildings. (*D'Anville, Notice de la Gaule*, p. 430.) The traces of several Roman roads, aqueducts, and edifices, with various statues, medals, &c., have been discovered in and around the town. (*Hugo, art. Haute-Saône; Dict. Géog.*)

LUZERNE, county, Pa. Situated toward the N.E. part of the state, and contains 1350 sq. m. Watered by Susquehanna river. Anthracite coal is extensively found. It contained, in 1840, 30,941 neat cattle, 52,415 sheep, 37,007 swine; and produced 944,389 bushels of wheat, 97,504 of rye, 332,395 of Indian corn, 131,923 of buckwheat, 304,094 of oats, 384,217 of potatoes, 207,878 tons of anthracite coal, of 26 bushels each. It had 132 stores, six furnaces, one forge, 12 fulling-mills, one woollen factory, 30 flouring-mills, 45 grist-mills, 212 saw-mills, one paper-mill, four powdermills, 24 tanneries, five distilleries, two breweries, one pottery, one rope-walk, two printing-offices, two weekly newspapers; five academies, 140 students; 183 schools, 7418 scholars. Pop. 44,006. Capital, Wilkesbarre.

LUZERNE, t., Fayette co., Pa., 12 m. N.W. Uniontown. Bounded N. and W. by Monongahela river. Drained by Dunlap's creek, across which is a bridge connecting it with Brownsville. It has three stores, four flouring-mills, three saw-mills, two tanneries; four schools, 95 scholars. Pop. 1715.

LUZERNE, p. t., Warren co., N. Y., 55 m. N. Albany, 10 m. S.W. Caldwell. Bounded W. by Hudson river. Hadley falls are in the S. part of the t. It contains one church, four stores, one fulling-mill, two grist-mills, 21 saw-mills; 11 schools, 435 scholars. Pop. 1284.

LUZON, the largest and most N. of the Philippine islands, which see.

LYCOMING, county, Pa. Situated a little N. of the centre of the state, and contains 1600 sq. m. Watered by the W. branch of Susquehanna river and its tributaries. The west branch of Pennsylvania canal passes along the N. branch of the rives through the county. It contained, in 1840, 16,713 neat cattle, 34,584 sheep, 26,080 swine; and produced 231,737 bushels of wheat, 116,393 of rye, 177,082 of Indian corn, 62,909 of buckwheat, 208,715 of oats, 194,113 of potatoes, 36,242 pounds of sugar. It had 67 stores, nine lumber-yards, one fulling-mill, nine woollen factories, one flouring-mill, 43 grist-mills, 53 saw-mills, two oil-mills, 20 tanneries, 11 distilleries, one pottery, three printing-offices, two weekly newspapers; three academies, 52 students; 118 schools, 4094 scholars. Pop. 22,649. Capital, Williamsport.

LYCOMING, t., Lycoming co., Pa., 7 m. N.W. Williamsport. Bounded S. by the W. branch of Susquehanna river. Drained by Lycoming creek. It contains four stores, three grist-mills, eight saw-mills, three tanneries; one academy, eight students; six schools, 183 scholars. Pop. 1917.

LYKENS, t., Dauphin co., Pa., 26 m. N. Harrisburg. Watered by Wisconisco, Little, and Mahantango creeks. Superior anthracite coal is abundant, and has an easy access to Susquehanna river. It has three stores, one fulling-mill, five flouring-mills, one grist-mill, seven saw-mills, two tanneries, two distilleries; eight schools, 295 scholars. Pop. 1499.

LYMAN, p. t., York co., Me., 72 m. S.W. Augusta, 523 W. Watered by Saco river, and a head branch of Kenne-

bunk river. It has three stores, one grist-mill, five sawmills; 13 schools, 545 scholars. Pop. 1478.

LYMAN, p. t., Grafton co., N. H., 53 m. N.W. Concord, 498 W. Bounded N.W. by Connecticut river, in which are Fifteen Mile falls, opposite to this t. Watered by Burnham's river, issuing from a pond on Lyman's mountain, 100 rods long and 80 wide. It falls into Ammonoosuc river. It has three stores, one grist-mill, 11 saw-mills; one academy, 100 students; 16 schools, 331 scholars. Pop. 1785.

LYME, p. t., New-London co., Ct., 45 m. S. by E. Hartford, 338 W. Incorporated in 1667. Bounded W. by Connecticut river, over which, at its mouth, is a ferry to Saybrook. It contains six churches, four Congregational and two Baptist, eight stores, one lumber-yard, three fulling-mills, two woollen factories, six grist-mills, six saw-mills, three tanneries; 18 schools, 847 scholars. Pop. 2858.

LYME, t., Jefferson co., N. Y., 12 m. W. Watertown. 172 m. N.W. Albany. Bounded W. by lake Ontario, N.W. by St. Lawrence river. Watered by Chaumont river. It has nine stores, one grist-mill, 13 saw-mills; 14 schools, 466 scholars. Pop. 5472.

LYME, p. t., Huron co., O., 99 m. N. by E. Columbus, 402 W. It has one store, two saw-mills; seven schools, 313 scholars. Pop. 1390.

LYME-REGIS, a parl. and mun. bor., market town, seaport, and par. of England, co. Dorset, in Bridport div. of lib. Loders and Bothenhampton, 20¼ m. S.S.E. Taunton, and 132 m. W.S.W. London. Area of parl. bor., which comprises the two parishes of Lyme and Charmouth, 1980 acres. Pop., in 1831, 3345. "Lyme is a small and irregularly built town, situated among hills, which, by rendering it difficult of access, effectually preclude it from becoming a place of importance. This place, as well as Charmouth, is frequented in the summer as a watering-place, and many respectable families are settled in the neighbourhood; but the streets are very irregular, and not lighted, so that, on the whole, it has the appearance of a poor and inconsiderable place." The pier or cobb (originally erected in the reign of Edward III., and greatly lengthened in 1826, at the expense of government) is 680 feet long and 19 feet broad, furnishing good shelter for shipping between Start point and the isle of Portland; and close to the pier is the custom-house. "The regular trade of the place, however, appears to be altogether inconsiderable; and it is chiefly valuable as a port of refuge for small vessels in bad weather." (*Parl. Boundary Rep.*) In 1836 there belonged to the port 19 vessels, of the burden of 1187 tons: the customs' revenue in 1839 amounted to £2580, indicating a great diminution since the close of the last century, when they amounted to about £16,000 a year. This change is ascribed partly to the decay of its once considerable Newfoundland fishery and Mediterranean trade, and partly also to the separation of Bridport, united with Lyme till 1833. An old church, three places of worship for dissenters, a house used for assemblies, and an old townhall, are the chief public buildings. Two schools for poor children are supported by subscription, and there are almshouses and other charities for the sick and aged.

The borough of Lyme is undoubtedly very ancient, and claims to be one by prescription. Its first charter is dated 12 Edward I.; and its early consequence as a port is shown by the fact, that in the war with France under Edward III. it furnished four ships to serve at the siege of Calais. The municipal borough is now governed by a mayor, three other aldermen, and 12 councillors, but has no commission of the peace. Corporation revenue, in 1839, £374. The borough sent two members to the House of Commons from the reign of Edward I. down to the passing of the Reform Act, which deprived it of one member: previously to that act the right of election was vested in the capital burgesses and freemen. The Boundary Act enlarged its limits, so as to include the entire parishes of Lyme and Charmouth. Registered electors in 1839–40, 277. Markets on Fridays; large cattle fairs, Feb. 13, and Oct. 2.

LYMINGTON, a parl. and mun. bor., seaport and market town of England, co. Hants, in the E. division of the New Forest, close to the mouth of a river of its own name, which falls into the Solent, 23 m. S.S.W. Winchester, and 91 m. W.S.W. London. Pop. of parl. bor. (comprising the par. of Lymington and a part of the par. of Boldre), in 1831, 5361. The town, situated on the W. bank of the river, is well paved and lighted with gas, and consists of one wellbuilt and wide street, crossed by two others of an inferior description. On the E. bank is the village of Undershore, comprising several villas and houses of a superior kind, inhabited by persons of independent fortune; it is connected with the town by a bridge, and clearly forms a suburb of Lymington. Among the public buildings are a town-hall, a neat theatre, and a custom-house; the port, though sufficient for vessels of 300 tons, and provided with wharfs and storehouses, is subordinate to that of Southampton. The church is an irregular building of brick and stone, the living being a curacy dependent on the vicarage of Boldre.

There are likewise three places of worship for Wesleyan Methodists and other dissenters. A free school for both sexes, a girls' national school, and an infant school, provide instruction for the children of the poor, and there are several minor charities.

"The town is considered to be in an improving state, and several large outlays of capital have taken place in the last few years. A company has been formed for the purpose of supplying steam navigation to and from Portsmouth and the isle of Wight; £3000 have been subscribed for the formation of gas-works, and the same sum for the erection of baths; dwelling-houses, also, have been and are now being built on an improved scale, the principal object of these improvements being to induce visiters to resort to the town during the summer. Little or no commerce is carried on here; and the only manufacture of the neighbourhood is that of salt, which some years ago was carried on to a very large extent, but latterly has decreased. A large yearly fair is held for the sale of cheese, exported to various places along the Sussex coast." (*Mun. Corp. Rep.*)

Lymington is a borough by prescription, its corporate officers since the Municipal Reform Act being a mayor, three other aldermen, and 12 councillors; but it has no commission of the peace. Corporation revenues in 1839 (chiefly from quay and river dues), £100.

Lymington has sent two members to the House of Commons since the 27th of Elizabeth, the right of election being vested, till the Reform Act, in the resident burgesses, of whom there were only 38 in 1831. The Boundary Act enlarged the limits of the borough, so as to include the entire parish of Lymington with a part of the parish of Boldre. Registered electors in 1839–2690, 305. Lymington is also a polling place for the S. division of Hampshire. Markets on Saturday; large fairs for cheese, bacon, and cattle, May 12 and Oct. 2.

LYNCHBURG, p. v., Campbell co., Va., 116 m. W. by S. Richmond, 191 W. It is in 37° 36' N. lat., and 79° 22' W. long. Pop. in 1830, 4698; in 1840, 6395. Situated on the S. side of James river, 20 m. below its passage through the Blue ridge. In 1793 it had but five houses. Incorporated in 1805. It contains seven churches, two Presbyterian, two Methodist, a Baptist, an Episcopal, and a Roman Catholic. It has a Friends' church in the vicinity. There are 124 dry goods and grocery stores, four apothecary stores, six public warehouses, in which from 15,000,000 to 20,000,000 pounds of tobacco are annually inspected, 22 tobacco factories, in which from 35,000 to 40,000 boxes of tobacco are manufactured annually, one large cotton factory, one iron foundry, three large flouring-mills, 15 classical schools, besides many common and primary schools, six large hotels, as well as many others, and over 1000 dwellings. The James river and Kanawha canal is in operation from this place to Richmond. Five handsome packet boats, affording fine accommodations for passengers, some of which leave for, and arrive from Richmond every day in the week, except Sundays; and 40 freight boats of from 60 to 100 tons burthen, are employed on the canal. Lynchburg has an extensive trade with the N., the N.W., and the S. parts of the state, together with the adjacent states of Ohio, Kentucky, Tennessee, and North Carolina. It is the fifth place in population in the state.

LYNDEBOROUGH, p. t., Hillsborough co., N. H., 33 m. S.S.W. Concord, 461 W. Incorporated in 1764, first settled in 1750. Watered by branches of Souhegan river. It contains a Baptist church, one store, one fulling-mill, one woollen factory, one grist-mill, six saw-mills, two tanneries; 13 schools, 334 scholars. Pop. 1039.

LYNDON, p. t., Caledonia co., Vt., 44 m. N.E. Montpelier, 554 W. Chartered in 1780, organized in 1791. Watered by Passumsic river, which by two falls affords extensive water-power. It contains two churches, a Methodist and Congregational, and some Baptists, four stores, one fulling-mill, two grist-mills, eight saw-mills, one oil-mill, two tanneries; one academy, 131 students; 16 schools, 607 scholars. Pop. 1753.

LYNN, p. t., Essex co., Mass., 9 m. N.E. Boston, 449 W. Bounded S. by the Atlantic. On the side next the ocean, and on Saugus river, by which it is watered, there is excellent salt marsh. It contains eight churches, three Methodist, two Congregational, a Friends', a Baptist, and a Universalist, two banks with a capital of $150,000 each, besides a savings bank, three insurance companies, 36 stores, three grist-mills, one saw-mill, one rope-walk, two printing-offices, four weekly newspapers; six academies, 133 students; 10 schools, 1035 scholars. Pop. 9367. Lynn is distinguished for the manufacture of ladies' shoes. The various manufactures of leather, consisting chiefly of these articles, amounted, in 1840, to $1,367,500. On the S. of Lynn is the peninsula of Nahant, which has become a fashionable watering place, and is beautifully situated to enjoy the ocean-breezes. The peninsula consists of two parts. The inner portion contains 42 acres. The outer

portion, called Great Nahant, is about 2 m. in length, and in some parts half a mile broad, containing 463 acres, and is finely diversified, rising to elevations of from 40 to 100 feet above the level of the sea. A large and splendid hotel occupies a commanding situation, containing nearly 100 rooms, and surrounded by a double piazza. In the vicinity are several other hotels and boarding houses, and about 30 beautiful cottages erected as country seats by gentlemen of fortune. Great Nahant is connected with Little Nahant by a beach half a mile long. The beach which connects Little Nahant with the mainland is narrow and about a mile and a half long, and barely rises so high as to prevent the water from ordinarily flowing over it. The sand has become so compact from the water dashing on it, that the print of a horse's hoof or a wheel of a carriage is scarcely left in passing over it. Two steamboats are constantly running between this place and Boston in the summer season.

LYNN, t., Lehigh co., Pa. It contains seven stores, one woollen factory, 10 grist-mills, five saw-mills, one powder-mill, four tanneries, 30 distilleries; seven schools, 312 scholars. Pop. 1805.

LYNNFIELD, p. t., Essex co., Mass., 20 m. N. Boston, 460 W. Incorporated as a separate town in 1814. It had previously belonged to Lynn. Bounded N. by Ipswich river. It contains two churches, a Congregational, and a Methodist; three schools, 190 scholars. Pop. 797.

LYNN-REGIS or KING'S LYNN, a parl. and mun. bor., seaport and market town of England, co. Norfolk, locally situated in hund. Freebridge-Lynn, at the mouth and on the E. bank of the Ouse, 36 m. W. by N. Norwich, and 98 m. N. by E. London. Lat. 52° 48' N., long. 26' E. Area of parl. bor., 2690 acres. Pop. in 1831, 11,905. The town, about 1 m. in length, by ¼ m. in breadth, comprising two principal, with other smaller streets, "is generally speaking, well built, and contains many excellent houses, and extensive premises calculated for trade. It is well paved, lighted with gas, supplied with good water, and very clean. The public walks, also, in the E. part of the town deserve notice, for their extent, and the neatness with which they are kept." (*Mun. Bound. Rep.*) Lynn was formerly encompassed on the land-side by a wall and deep wet ditch, defended by nine bastions: these fortifications yet remain, but the wall and bastions are much dilapidated; it is also divided into several parts by four small streams here called *fleets* (from the Dutch *vliet*) over which are 11 bridges. The market place, called by way of distinction the Tuesday's market place, is an area of three acres, situated at the N. end of the town, having a sculptured stone cross in its centre, and surrounded by good houses. A smaller market is held on Saturday in an open space near St. Margaret's church, and outside the town is a cattle-market. The custom-house, built in 1683, and intended for a merchant's exchange, is a handsome building of freestone, with an ornamental front, and a statue of Charles II.: the guild-hall is an old-fashioned building of stone and flint, with suitable apartments for the transacting of municipal business, &c.; and near it is the borough jail, a respectable stone structure, which "seems to be on the whole well regulated, and admits, to a certain extent, of the classification of prisoners." (*Mun. Corp. Rep.*) A new theatre has recently supplied the place of an older one, now converted into warehouses. The parliamentary borough comprises two parishes, that of the St. Margaret's (the living of which is a perpetual curacy in the gift of the dean and chapter of Norwich), and that of All-Saints (a vicarage in the patronage of the bishop of Ely). St. Margaret's church in the N. Lynn, built in the 19th century, is alleged to be one of the largest parish churches in England, and had formerly a lofty steeple blown down by a tempest in 1741: its W. end is still distinguished by two square towers of dissimilar architecture, the upper parts of which are of modern construction. St. Nicholas, a Gothic structure, with a bell-tower and light octangular spire 170 feet high, is a chapel of ease to the above parish church. All-Saints' church, in S. Lynn, is a well-built cruciform edifice, occupying the site of an old convent of White friars. On the opposite side of the Ouse in W. Lynn, but not within the borough, is another parish church, that of St. Peter's. There is also a Roman Catholic chapel; and the Wesleyan Methodists, Independents, Baptists, Unitarians, and the Society of Friends, have their respective places of worship, with large attached Sunday-schools. The grammar school is in the patronage of the corporation, from which its master receives a salary of £63 a year: it has two or three small exhibitions in the University of Cambridge. Various charity-schools have likewise been established, which, with a well-conducted Lancastrian school, furnish instruction to numerous children of both sexes. Gaywood's hospital provides lodging, and a weekly stipend of 5s. to 33 poor widows: there are also three other well-endowed sets of alms-houses, and many minor bequests, &c., for the relief of the aged poor.

"Lynn contains an iron-foundry and four building-yards for ships from 400 tons downward; but there are no other manufactories. A considerable and increasing trade is carried on, coast-wise, in exporting corn, with other natural products of the fens; and importing principally coal: there is also a direct trade with Portugal in wine, fruit, &c.; but this is of much less extent and importance than the coasting-trade." There belonged to the port, in 1836, 120 ships, of the aggregate burden of 15,353 tons; besides which, upward of 2000 coasters, chiefly colliers, come thither, each year, from other ports. Gross customs' revenue, in 1839, £67,253. The harbour is capacious; but the approach to it is rendered both difficult and hazardous by numerous and perpetually shifting sand-banks, occasioned by the action of the tide on the light silt sand forming the bed of the river. The estuary of the Ouse is nearly 1000 feet broad, and there is accommodation in the port for about 300 merchant-ships. Spring-tides rise about 18 feet, and during the prevalence of N. and N.E. winds, are thrown in with such violence and rapidity as sometimes to damage the shipping. "The harbour has also been injured since the completion of the East-brink cut, which has caused a great accumulation of alluvial soil along the King's staith and other quays lining the E. bank of the river; but this evil is now somewhat lessened by the erection of jetties on the opposite shore, which direct the course of the river more to the E. bank, by means whereof these deposits are secured away." (Mun. Bound. Rep.)

King's Lynn, (called Bishop's Lynn before Henry VIII. conferred on it its present name,) received its first charter from King John, in return for valuable services done him by its inhabitants during the baronial wars. Its corporate privileges were confirmed and enlarged by several monarchs, and lastly by Charles II. The borough is now divided into three wards, the municipal officers being a mayor and five other aldermen, with 18 councillors. Quarter and petty sessions are held under a recorder, and a court sits monthly for the recovery of debts under 40s. Corporation revenues, in 1839, £2615. Lynn has sent two members to the House of Commons since the 6th of Edward II., the right of election down to the Reform Act being vested in freemen by birth, servitude, gift, or purchase. Registered electors, in 1839-40, 950. Lynn is also a polling-place for the W. division of Norfolk. Markets, principally on Tuesday, but also on Saturday. Large fairs for London goods, Feb. 14, and five succeeding days, also for cheese a week after old Michaelmas, lasting two days.

LYONS (Fr. *Lyon*; an. *Lugdunum*), a large city of France, being the principal manufacturing town of that kingdom, in the dep. of the Rhône, of which it is the cap., 275 m. E.N.E. Bordeaux, 173 m. N.N.W. Marseilles, 245 m. S.E. Paris, and 70 m. W.S.W. Geneva; lat. 45° 45' 58" N., long. 4° 49' 24" E. Pop. of the city proper, in 1836, 147,923; but, including its suburbs, the pop. is about 200,008. It is situated at the junction of the Rhône and the Saône, chiefly on a tongue of land or peninsula between those two rivers, the length of which is nearly 3 m., and its average breadth about 3 furlongs, though in the N. part of the city increasing to upward of 1 m. Some extensive and important quarters, as St. Just, St. George, St. Irénée, Vaise, &c., are, however, situated on the W. or right bank of the Saône, on and round the hill of Fourvières; and in the E., on the left bank of the Rhône, are the *Faubourg Guillotière* and the *Quartier des Brotteaux*. S. of the city, the handsome and regular suburb of *Perache* is rapidly extending towards the extremity of the peninsula; while on the N., beyond the fortifications, on the declivity of a hill extending from one river to the other, is the municipal commune of La Croix Rousse, comprising the suburbs of Serin and St. Clair. A tower on the hill of Fourvières, 680 feet above the Saône, commands a landscape which combines the rich and the grand in the highest degree. At the spectator's feet is Lyons, with its two noble rivers; its bridges, squares, quays, and public edifices, the vessels that crowd the Saône, and the busy activity that pervades its streets, announcing a highly civilized, prosperous, and opulent community. "Unlike Paris and many other French towns, which stand isolated, as it were, in the country, with ploughed land and meadows coming close up to the barriers, Lyons appears as the nucleus of a vast population, melting gradually by its suburbs into clusters of villages, which break up into smaller villages, hamlets, villas, and manufactories. Even at the distance of 10 m., the country is pretty thickly dotted with buildings, some of which are seen sweetly perched on the S. and W. declivities of the hills which enclose the plain. The high and mountainous land on the W. side of the city is scarcely an exception; for sterile as it seems, it is enlivened by country-houses, villages, and manufactories. Beyond the hills which bound the plain on the N.E., is seen Mount Jura; on the E. we have part of the Alps; above which, at the distance of 100 m. from the town,

Mont Blanc is distinctly seen like a white cloud or a mass of snow." (*Maclaren's Notes*, p. 35.)

The interior of Lyons exhibits little regularity, and chiefly consists of narrow, winding, and dirty streets, rendered dark by the extreme loftiness of the houses. These, though of stone, and solidly built, are old and *triste*, and several of the streets leading up steep declivities are inconvenient for carriages. The *quartier St. George* is disgustingly filthy, and greatly inferior in appearance to the suburbs of Croix Rousse and des Brotteaux, which, like it, are chiefly inhabited by the working classes. But the wretched aspect of some parts of the city is in some degree countervailed by the magnificence of others. Three ranges of quays, two on the Saône, and one on the Rhône, interspersed with 17 bridges, nearly all of modern construction, with the glacis and hill of Fourvières, encompass all that is situated between the two rivers, and form a noble and imposing outline. The Saône, which is far more useful to Lyons in a commercial point of view than the Rhône, is lined with numerous wharfs and landing-places; and along the Rhône from the Faubourg St. Clair to Port Perache, a distance of a league, is a line of elegant public and private edifices, and a public walk, planted with a double row of trees, commanding a fine prospect over the fertile plain to the E. The waters of the Rhône are cold, and clear, and it forms in every respect a remarkable contrast to the Saône, which has a sluggish current, and a muddy. The Rhône is very liable to sudden inundations, to stream the devastating effects of which some extensive embankments have been raised on its left bank. Still the when swollen, frequently does much damage, as was river, evinced in the autumn of 1840, when the inundations carried away some of the bridges, laid a considerable portion of Lyons, and of the surrounding country, under water, and occasioned great damage. There were previously 10 bridges within the city, three of which crossed the Rhône. These were the *Pont de la Guillotière*, originally built in 1190, 530 yards in length by 24 feet wide, with 17 stone arches, but of these only eight are over the water; the *Pont Morand*, constructed of wood in 1774, 283⅓ yards long by 14 wide; and between the two the *Pont Lafayette* (formerly *Charles X.*), a handsome bridge, 235 yards in length, the piers of stone, and the upper part of wood. The bridges over the Saône vary in length from 190 to 140 yards; the principal is the *Pont de Tilsit*, leading from the centre of the city, a stone bridge of five arches, 130 yards long by 15 wide, erected at a cost of 3,000,000 francs, or £120,000 sterling. (*Huge*.) Lyons has 55 *places* or squares, some large and regular, but, as may readily be inferred from their number, the great majority are very much the reverse. The *Place Bellecour* (formerly *Louis-le-Grand*), one of the largest and handsomest in France, and perhaps in Europe, in the very heart of the city, has two of its sides nearly 340 yards in length, the two others measuring 246 and 218 yards. One of the principal streets forms part of its N. face; its two shorter sides consist of symmetrical ranges of handsome buildings; and on its S. side is a fine plantation of linden trees. This square is ornamented with an equestrian bronze statue of Louis XIV., and forms, with the quays, the favourite promenade of all classes. The *Place Louis XVIII.* leads into the *Cours du Midi*, a broad and fine thoroughfare, planted with trees, which separates the city from the new town of Perache. The other principal squares are the *Place des Terreaux*, containing the town-hall, and *Palais des Arts*; *des Cordeliers*, with a fluted column upward of 60 feet in height, supporting a colossal statue of Urania; *de Comédie*, in which is the entrance to the *Grand Théâtre*; *Sathonay*; the *Place Louis XVI.*, in the quartier des Brotteaux, &c. In the N. part of the city a covered arcade has been formed, called the *Galerie de l'Argue*, nearly 500 feet in length, and containing many good shops. Lyons is supplied with water from the Rhône, and has numerous public fountains, but none worth notice.

The town-hall holds the first rank among the public buildings. This edifice, the finest of its kind in France, was erected between 1646 and 1655. It has a front nearly 180 feet in width, flanked with a square tower and dome at either end. Its balustrade is ornamented with two large statues of Hercules and Minerva, and in the centre is a clock tower, surmounted by a cupola, which rises to the height of 157 feet above ground. The depth of the building is 383 yards, at the end of which another handsome front faces the Place de Comédie. Its interior contains a vestibule, in which are two colossal bronze groups emblematical of the Rhone and Saône; a fine staircase, and a saloon, 87 feet long by 40 wide, which formerly contained many fine paintings, destroyed during the Revolution. Of the 18 churches, none is very remarkable either for size or elegance. The cathedral of St. John, on the right bank of the Saône, was begun in the 7th century, but not completed till the reign of Louis XI. It is a Gothic edifice, having at its four corners, four heavy square towers, in one of which

R · 257

is a bell, weighing 35,000 French lbs. The W. entrance is very much ornamented; the interior is characterized chiefly by simplicity. In this church is a remarkable clock, constructed at the end of the 16th century by a native of Basle, which formerly indicated besides the year, month, day, hour, minute, and second, the sun's place, the phase of the moon, and the saints' day, as they occurred. This curious piece of mechanism has been suffered to fall into decay. The church of Ainay, erected on the site of an ancient temple dedicated to the emperor Augustus, has four granite columns and a bass-relief, originally forming parts of that edifice. Several of the churches date from the time of Charlemagne. Here is also a Protestant church and a synagogue.

The hospitals are the largest public buildings in Lyons. The *Hôtel-Dieu*, the most ancient and finest establishment of its kind in France, was founded by Childebert and his queen at the beginning of the 6th century: the present edifice consists of a continuous range of building, extending along the Rhone. It has a noble front, a fine entrance, and two domes, which, as well as the distribution and arrangements of its interior, are generally admired. This establishment receives annually 12,000 in-patients, besides affording medical aid to many persons without its walls. The *Hospice de la Charité*, also on the banks of the Rhone, apparently occupies little less space than the former, and is an asylum for 400 infirm persons of both sexes, besides many orphans, foundlings, and women enceintes. The *Hospice de l'Antiquaille*, for syphilitic and insane patients, stands on the hill of Fourvières, on the site of the Roman palace in which the emperors Claudius and Caracalla were born. The *Hospice de la Providence* has established numerous schools of instruction with the view of checking mendicity, &c.

The prefecture occupies a spacious building, formerly a Dominican convent; its interior is well adapted to its present purpose, and attached to it are some fine gardens. The hall of justice, and the archbishop's palace, present little deserving of notice. The *Palais des Arts*, formerly the Benedictine convent of St. Pierre, consists of four large piles of building, enclosing a square court: different portions of this edifice are devoted to the exchange, and chambers of commerce, the museums of painting, antiquities, and natural history, a cabinet of medals, gallery of casts from the antique, *dépôt* of machinery for the silk manufacture, the academy, schools of drawing and natural history, society of agriculture, &c. The collection of paintings comprises some works of great excellence; and that of antiquities is rich in Roman and middle age specimens of art found in and about Lyons, mosaics, and Egyptian antiquities. The public library, and library of Adamaly (so called from having been presented by a citizen of that name), are deposited in the royal college, and together comprise 100,000 volumes (*Hugo*), among which are some valuable oriental works, and old MSS. The prefecture, Mint, grand theatre (an elegant structure), theatre des Célestins, court of justice, archbishop's palace, new prison, *conditions*, salt-magazine, &c., are among the other chief edifices. The botanic garden is situated within the city, and is a favourite place of public resort. About one and a half miles above Lyons is the beautiful *île Barbe* in the Saône, connected with its left bank by a handsome new suspension bridge.

Manufactures and Commerce.—Lyons is in France what Manchester is in England. And notwithstanding the active competition of Zurich, Basle, Crefeld, and other places on the continent, and of coventry, &c., in England, she still maintains her rank as the first silk manufacturing city of Europe. Her position is peculiarly favourable: she is situated at the point of junction of two large navigable rivers, and has a ready communication with the Mediterranean, on the one hand, at the same time that she is the *entrepôt* of a vast extent of inland country. The districts of France which produce the largest quantities of silk, are immediately adjacent, while Lyons is the natural depôt and place of transit for the silk of Italy, in its way to the great manufacturing countries. Added to which, the manufacture has here had, for centuries, its principal seat: the population have been thoroughly trained and habituated to it; so that, though frequently disturbed by political events, and once or twice nearly annihilated, it has never failed, on tranquillity being restored, to return to its former locality. The silks manufactured here, are distinguished by the equality and perfection of the fabric; the brilliancy, though perhaps not the durability, of their dyes; and by the unrivalled superiority of their patterns, and the taste displayed in the designs. This superiority has been ascribed, with what justice we shall not stop to inquire, to the

School of Arts (*Institution de la Martinière*), and the liberal encouragement of this branch of science by the city authorities, and the government. About 180 students are gratuitously instructed in the various branches of drawing and modelling, and there is a professor, who teaches the "*mise en carte*," that is, the adaptation of designs to the loom. The trade of Lyons, like that of all manufacturing towns, is subject to frequent crises, and periods of distress: a very serious one occurred in 1836-37, which led to formidable riots. But though many workmen implicated in the insurrections have settled in the rival towns of Switzerland, &c., there are, perhaps, were so many looms at work as at present, nor was the manufacture ever more flourishing. The gross produce of the Lyonese looms, in 1838, was estimated at 135 millions of francs, being considerable more than half the estimated value of all the silk goods manufactured in France. (*See* FRANCE, vol. i., p. 931.)

According to M. Villermé, there were, in 1833, in Lyons and its neighbourhood, 40,000 silk looms; 17,000 in the city-proper, 9000 in the suburbs of la Croix Rousse, la Guillotière, and Vaise, 5080 in the neighbouring parts of the department Rhone, and 8920 in the adjacent parts of Loire, Saône-et-Loire, Ain, Isère, and Drôme. Dr. Bowring was furnished with an estimate in 1834, which made the number of looms in the city 16,000, of which 4000 were for figured stuffs; in the suburbs 9000, half for figured silks; and in the country, for 12 or 15 leagues round, 7000, almost wholly for plain silks: making in all 28,000 looms. According to an official estimate in 1835, the master weavers (*chefs d'atelier* or *maître-ouvriers*) in Lyons and its suburbs amounted to about 8000; and the journeymen, or *compagnons*, to 30,000: in all, 38,000 weavers: but the *compagnons* include the wives and children of many of the master weavers. The number of individuals employed in accessory occupations, that is, in the culture of silk, the manufacture of looms, &c., cannot be ascertained; but it has been estimated by M. Villermé and M. Girod de l'Ain at from 27,000 to 30,000. Hugo says that, altogether, 68,000 persons in or about Lyons are supported, directly or indirectly, by the silk manufacture.

Silk weaving at Lyons is not conducted in large buildings or factories belonging to the silk merchants (*fabricans*)[†]; but on the domestic system, in the dwellings of the master weavers, each of whom has usually from two to six or eight looms, which, with the greater portion of their fittings, are his own property. Himself and his family keep as many of these looms at work as they can, and employ *compagnons* for the remainder. The latter are not settled in Lyons; but visit it, and stay a longer or shorter time according to the demand for their labour. Apprentices and *lanceurs* make up the remainder of the working classes. The former are usually apprenticed from the ages of 15 to 18: the latter are children from 9 to 14, who prepare bobbins, and weave fabrics demanding less nicety than others. About three sevenths of the looms are wrought by master weavers, nearly an equal number by *compagnons*, and the remaining seventh by apprentices and children. The *fabricans*, or silk merchants, of whom there are between 300 and 600 in Lyons, supply the patterns and silk to the owners of looms, to whom is entrusted the task of producing the web in a finished state. Half the wages paid by the silk merchants go to the owner of the loom, and half to the labouring weaver. A master weaver may gain by his own labour from two to three and a half francs a day; and he who has three looms is supposed to receive from the two at which he does not himself work, about 900 francs, or £36 a year. His rental may be about 150 francs; the cost of lodging his two journeymen 50 francs; and there remains besides his own labour, a surplus of 670 francs. These weavers are, of course, the most prosperous, who having three or four looms, employ their children to weave on them, and thus receive the whole wages paid by the manufacturer. Three looms will clear to a family from 1500 to 1600 francs (60 to £64) a year.

Wages have risen considerably of late years. In 1838, the price per ell paid for common plain velvets was five and a half francs, for *gros de Naples* 80 to 90 cents, and for common figured silks from one to one and a half francs. A master weaver who made two francs a day in 1834, could make at least two and a half francs in 1838; and the journeymen need never earn less than one franc 75 c., and may frequently get two francs. The hours of work usually vary from 12 to 16 hours; but when the demand is brisk, they reach to 16, 18, and even 20. The weaving population is ill lodged, the master weavers generally having but two rooms at most, and these kept in a disgraceful-

---

* This is an establishment where, by the agency of heat, the unwrought silk is reduced to an equable weight and dryness. In 1831, the weight of silk, submitted to the condition, amounted to 620,000 kilogrammes.

† There is one exception: on the bank of the Saône, opposite the île Barbe, is a factory called la *Sauvagere*, employing from 400 to 500 hands, who may sleep in the building on payment of 30 sous a month, and board there also at a low rate.

ly filthy state. But they live very well; that is, they have abundance of nourishing food, much more than the population of other manufacturing towns in France. Most of the journeymen are boarded by their employers at from 45 to 50 c. a day; and have about one and a half pounds of good bread, quarter litre of wine, a dinner of soup, butchers' meat, &c., with cheese or salad at supper. They rarely save money, and few of the *compagnons* raise themselves to become *chefs d'ateliers*. The weavers speaking generally, are very ignorant; some years since not one fourth part of the children in Lyons could read or write. But after all, according to M. Villermé, there is less prodigacy in Lyons than in many other of the French manufacturing towns. (See *Villermé*, i., 365-69.) The proportion of illegitimate to the total number of births in 1835 was indeed as high as one in three; but, in point of fact, a good many of the connections out of which these births arise are really but little different from matrimonial connections. The weavers, to escape the *octrois*, frequently visit the *cabarets* beyond the barriers, to drink their wine, play billiards, &c., on Sundays and Mondays; but they are not addicted to intoxication or rioting, and it is affirmed, and we believe truly, that they are at present improving in morals, manners, and cleanliness; and certainly they have much room for amendment. Notwithstanding their good wages and liberal supplies of food, the best French authorities admit that the Lyonnese weavers are physically an inferior and degraded race, remarkably subject to scrofulous and scorbutic complaints, spinal diseases, and rheumatism; and according to M. Charles Dupin, half the young men in Lyons liable to military service are exempted on account of weakness, deformity, or deficiency of height, though the standard for recruits or conscripts in the French is considerably below what it is in the English army. (See vol. i., 933.)

Happily, however, the upper and middle classes of Lyons, the latter comprising most part of the shopkeepers, and many of the master weavers, are eminently comfortable, rich and thriving. Mr. Maclaren states that there are three times more villas round Lyons than round Paris; and the number of private and public works erected in and near the city during the last 20 years sufficiently evince the rapid increase of wealth and enterprise. The want of coal is the greatest obstacle to the improvement of the manufactures of the city, and to the extension of its industry. But despite this disadvantage, Mr. Maclaren states that the district of which Lyons is the centre is "advancing with great strides." (P. 26.)

Lyons has numerous dyeing establishments and printing offices, and manufactories of jewellery, liqueurs, &c.; but all these are insignificant compared with its chief branch of industry. It is the seat of a royal court, of tribunals of primary jurisdiction and commerce, a chamber of commerce, one of the five royal libraries of the kingdom, a university, academy, royal college, and academy of sciences, &c.; and has schools of theology, medicine, veterinary medicine, and rural economy; a royal society of agriculture, &c.; societies of medicine, jurisprudence and literature, a Protestant Bible society, deaf and dumb asylum, a *mont-de-piété*, savings' bank, maternity, and many other charitable institutions.

The early history of Lyons is involved in much obscurity. But it appears certain, from the statement of Dion Cassius, that Munatius Plancus, about *anno* 40 B.C., settled in it fugitives from some adjoining towns. (lib. xlvi.) Augustus made Lugdunum the capital of a province, and being embellished and enlarged by succeeding Roman emperors, it became one of the principal cities of the Roman world. The old city was principally built on the hill of Fourvières, which, in fact, is merely a corruption of its ancient name of *Forum Vetus*. (*D'Anville, Notice de la Gaule*, p. 463.) Among the Roman antiquities which still exist at Lyons, are the remains of four aqueducts, several cisterns, a theatre, traces of a palace, and a *naumachia*, recently discovered within the limits of the botanic garden.

From the 5th to the 13th century, Lyons belonged successively to the Burgundians, Saracens, Franks, its feudal archbishops, and its municipal council. In 1312 it was annexed to the crown of France; and in the same century, owing to the immigration of many merchants from Italy, it began to be distinguished by its manufactures. It suffered much during the religious wars of the 16th century; but far more from the revolutionary frenzy of 1793. Its ancient fortifications were then destroyed; but it has been since enclosed on the N. by a line of earth ramparts. Among the distinguished individuals natives of Lyons, were, in antiquity, the emperors Claudius and Caracalla, and Sidonius Apollinarius; and, in modern times, Jussieu, the botanist; J. B. Say, the economist; Jacquard, the inventor of the loom which bears his name; Degerando, the author of the able and elaborate work *Sur la Bienfaisance Publique*, &c., (Hugo, arts. *Rhone; Lyons; French Official Tables;*

Bowring and Symond's Reports; Villermé; Tableau Physique et Moral des Ouvriers; Inglis; Maclaren, &c.)

LYONS, p. v., capital of Wayne co., N. Y., 180 m. W. by N. Albany, 375 W. Organized in 1811. Drained by Clyde river, formed by a junction of Mead creek with Canandaigua outlet. The Erie canal passes through it. It contains a court-house, jail, county clerk's office, six churches, a Presbyterian, two Methodist, an Episcopal, Baptist, and Lutheran, 20 stores, one lumber-yard, one fulling-mill, five flouring-mills, three grist-mills, six saw-mills, three tanneries, two distilleries, one pottery, two printing-offices, two weekly newspapers; eight schools, 972 scholars. Pop. 4302. The village, on the Erie canal, contains about 250 dwellings and 1500 inhabitants.

LYSANDER, p. v., Onondago co., N. Y., 144 m. W. by N. Albany, 364 W. Bounded S. by Seneca river. It has seven stores, two fulling-mills, one woollen factory, one grist-mill, 16 saw-mills, four tanneries; 18 schools, 920 scholars. Pop. 4306.

M.

MAAD (Hung. *Mada*), a town of Hungary, co. Zemplin, in the Heygyalla mountains, about 6. m. N.W. Tokay. Pop. 5646, partly Lutherans and partly Roman Catholics. It is one of the places at which the Tokay wine is grown in the greatest perfection, and near it is the imperial vineyard of Theresienberg.

MAASSLUIS, or MAASLANDLUIS, a town of S. Holland, on a branch of the Maas, 9 m. W. by N. Rotterdam. Pop. 4500. It has manufactures of sailcloth, cordage, leather, &c., and some building docks; and its inhabitants take an active share in the herring and cod fisheries.

MACAO, a seaport town and settlement of the Portuguese in China, province Quang-tong, on a peninsula, projecting from the S.W. corner of the island Macao, on the W. side of the estuary formed at the mouth of the Tigre or Canton river, 84 m. S. by W. Canton; lat. 22° 11' 30" N., long. 113° 32' 30" E. The population is stated in the "Chinese Repository" to amount to upwards of 30,000; but we incline to think that 20,000 is nearer the mark, of whom about 15,000 are Chinese, and the rest chiefly Portuguese and slaves imported from Timor, &c.

The peninsula on which Macao stands is less than 2½ m. in its greatest length from N.E. to S.W., and not 1 m. in its greatest breadth. It is connected with the rest of the island by a long, low, and sandy neck, in one part 400 yards broad, but generally less. Across this isthmus a wall is erected, having in its middle a gate and a guardhouse, called *Casa branca*, for Chinese soldiers; by means of which barrier, all communication between the peninsula and the rest of the island is cut off at the pleasure of the Chinese authorities. The Portuguese inhabitants of Macao are rarely permitted to pass beyond the wall. This town has a very imposing appearance from the sea. It is built chiefly on the declivities of two hills, meeting each other at a right angle, in front of a small semicircular bay forming the harbour. A handsome row of houses faces this bay, with a parade in front embanked with stone to resist the encroachments of the sea, and interrupted once or twice by granite quays with steps leading down to the water. Behind this terrace the houses are arranged in a confused manner, and the gable ends of European residences and the steeples of the churches appear curiously intermixed with Chinese houses and temples. Macao has 12 churches, one of which, that of St. Joseph, is collegiate. There are few other edifices of any note. A spacious senate-house, in the heart of the town, forms a termination to the principal street. The Portuguese governor's residence, near the landing-place, is nowise remarkable, and the contiguous English factory is a plain, commodious building.

The Chinese live chiefly together in the central and back parts of the town, and along the inner harbour; some of them have well furnished shops, and they principally supply Europeans with provisions. Besides the college of St. Joseph, there are in Macao a royal grammar-school and several other Portuguese schools, a female orphan asylum, and other charitable institutions. It is defended by six forts, two of which are placed on a lofty height at either end of the harbour, and it is usually garrisoned by about 400 Portuguese soldiers. At one extremity of the town is a mansion called the *Casa*; in the grounds belonging to which is the celebrated cave of Camoens, sheltered on one side by a lofty rock, and on the other by a grove of bamboos, above which a tower commanding a fine view has been erected. In this sequestered retreat Camoens is said to have composed great part of the *Lusiad*, while holding the office of Portuguese judge at Macao. The land immediately without the town is fertile and is appropriated to vegetable gardens and rice-grounds.

259

The harbour is on the W. side of the town, between it and Priests' island, a small circular island, which formerly belonged to the Jesuits; but it has not depth enough to admit large ships, which accordingly anchor in the roads on the other side of the peninsula, from 5 to 10 m. E. of the town. All foreign vessels coming into the roads send their boats to the custom-house and pay a duty for all goods landed, however trifling. When a ship arrives among the islands, she is generally boarded by a pilot, who reports to the Chinese custom-house officer the nature of her cargo, and obtains a *chop* or permit allowing her to enter the Bogue or Bocca Tigris, with the understanding that she has nothing on board that is contraband. All females must, however, be landed at Macao, as the ship will not be allowed to proceed to Whampoa with them on board. The Chinese regulations do not allow any vessels, except such as belong to Portuguese or Spaniards, to trade at Macao. But the Portuguese inhabitants lend their names for a trifling consideration to such foreigners as wish to be associated with them for the purpose of trading from the port; and vessels of other nations seldom experience any difficulty in obtaining the connivance of the Chinese officers to the landing or receiving of goods in the roads by means of Portuguese boats. Vessels of other nations, if in distress, and not engaged in the contraband trade, are admitted into the harbour for repair, on application to the senate. The latter is composed of the bishop, the chief justice, the military commandant, and several of the chief Portuguese inhabitants; but a Chinese mandarin has substantially the supreme authority in the town. Except during the period of the year when the merchants of Canton are obliged to leave that city and repair to Macao, the latter is said to be dull and uninteresting. At that season, however, the carnival is celebrated with more than its usual sumptuousness in Catholic countries; and balls, masquerades, and concerts follow each other in rapid succession. Macao was given to the Portuguese by the Chinese emperor in 1586, in return for assistance afforded by them against pirates that had infested the coast. (*Hamilton's E. I. Gaz.; Commerce. Dict.; and Private Information.*)

MACASSAR. *See* CELEBES.

MACCLESFIELD, a large manufacturing town, parl. and mun. bor. of England, co. Chester, locally situated in Prestbury div., of the hund. of its own name, on the Bollin, 16 m. S. by E. Manchester, and 143 m. N.N.W. London. Pop. of parl. bor., which includes, with the old bor., parts of the townships of Hurdsfield and Sutton, nearly 30,000 in 1831; and at present the pop. is estimated at 40,000. The town, which is pleasantly situated on a slope near the borders of Macclesfield forest, has greatly increased in size during the last 20 years, and is now about 1¼ m. long, by 1 m. in breadth, consisting of one principal thoroughfare on the London road, crossed by two others leading to numerous subordinate streets. The buildings, in the more conspicuous parts of the town, are of superior construction; the streets also are well lighted, and the inhabitants have plentiful supplies of good water, conveyed from springs in the adjacent hills. An open market-place, with excellent shambles, and a covered corn market, stand near the centre of the town; and the newly erected town hall is a commodious and handsome building, tastefully decorated, and containing, besides courts of justice, offices, &c., a large assembly and concert room. The old church is a large structure, partly Gothic, having a handsome tower, formerly surmounted by a lofty steeple: it was originally erected by Edward I. in 1278, but has, at different times, been almost rebuilt, so that few parts of it can lay claim to any great antiquity. It affords accommodation for about 1700 persons, and has an adjoining chapel containing several interesting monuments: the living is a perpetual curacy, till very lately in the gift of the corporation, but now in private patronage. Christchurch was erected, in 1775, at the private expense of Charles Roe, Esq., who endowed it with £100 a year: it is a regular building, with a neat tower, having, in the interior, an elegant marble monument of the founder, by Bacon. Trinity church, in Hurdsfield, a very recent erection, is beautifully situated on an eminence, and may accommodate about 900 persons. St. George's in Sutton (built in 1823) has accommodation for 1500, and in the S. suburbs of the town, is a fifth church, remarkable for its neat construction, and light spire. There are also several places of worship for Wesleyan Methodists, and other dissenters. A free grammar school, originally founded in 1502, was endowed by Edward VI., with property producing only £95, but now estimated at £1300 a year, and rapidly increasing in value. A head and under-master, who have salaries of £200 and £175 a year (the former having a house and school-field, free of rent and taxes), give instruction in classics, elementary mathematics, history, geography, &c., the average number of scholars being about 60. This school enjoys certain advantages at Brazenose College, Oxford, and it has been proposed to set aside a portion of its property to

260

found four exhibitions of the annual value of £50 each, tenable for four years at any college either in Oxford or Cambridge. The wishes of the trading classes have likewise been consulted by the very recent establishment of a separate school, called the "Modern Free School," endowed with £350 a year, and furnishing good instruction in those branches of knowledge best calculated to enable the scholars to carry on the trades, and support the commerce of Macclesfield. A charity school, national school, and several Sunday schools, with others maintained by voluntary subscription, furnish instruction to the children of the poor; and there are almshouses, various money charities, a dispensary, lying-in charity, and provident society.

" The silk manufacture of Macclesfield affords employment to the largest part of the population; a few, however, are employed in the cotton factories that have been lately established. This place participated deeply in the general distress occasioned by over-trading in 1825, and for several years subsequent to that period the silk trade was in a most depressed state. The effects of that shock, however, seem at length to have subsided, and business has resumed a healthy aspect." (*Mun. Bound. Rep.*) The trade has greatly increased since the date of this report, but it is subject, more than other branches, to sudden shocks, productive of great distress to the working classes. There were at work in the parish of Prestbury (and chiefly in Macclesfield and its immediate vicinity) in 1836, 16 cotton-mills and 68 silk-mills, employing 10,263 hands. The wages of the workpeople employed in these mills vary, at present (1841), from 2s. 6d. to 15s. a week. About 4500 hand-looms are engaged in weaving silk fabrics, chiefly silk handkerchiefs and scarfs of every description, sarcenets, Persians, silk-ferret, and galloon, with a few gros-de-Naples, giving employment altogether to about 9000 persons, whose wages amount to from 6s. to 15s. a week; but there are a few industrious and expert weavers, who can weekly as much as 25s. when in full work. The cotton manufacture, which was introduced only a few years ago, is in a thriving condition, employing a population of about 3000 in factories only; and hatmaking is carried on to some extent. Numerous mechanics, makers of machinery, &c., depend indirectly on the staple trade of the town. The Bollin turns several mills, and the neighbourhood furnishes abundant supplies of excellent coal. Stone and slate also are quarried near the town, and form a considerable object of trade with the surrounding districts. The transit of heavy goods is facilitated by the Macclesfield canal, which connects it N. and S. with the great canal lines of England. There are two private banking establishments, with a savings bank. A newspaper is published weekly, and there is a good newsroom.

Macclesfield, which was incorporated by a charter of Prince Edward, son of Henry III., and subsequently by various sovereigns of England, has been divided by the Mun. Reform Act into six wards, and is now governed by a mayor and 11 other aldermen, with 36 councillors. Corporation revenues in 1839, £2551. It enjoys also a commission of the peace, with petty sessions, under a recorder. This important manufacturing town had no voice in the legislature, till the Reform Act conferred on it the privilege of sending two members to the House of Commons. Registered electors in 1840-41, 894. Macclesfield is also one of the polling places for the N.E. division of Cheshire. Markets on Tuesday and Saturday: cattle, wool, and cloth fairs, May 6, June 22, July 11, October 6, and November 11. (*Parl. Reports, &c.*)

MAC CONNELSVILLE, p. v., Morgan t., capital of Morgan co., O., 73 m. E. by S. Columbus, 330 W. Situated on the E. bank of Muskingum river, on the second bottom, from 10 to 30 feet above high water. It contains a court-house, jail, two churches, a Presbyterian and Baptist, 14 stores, two tanneries, 80 dwellings, and about 500 inhabitants.

MAC CRACKEN, county, Ky. Situated in the W. part of the state, and contains 900 sq. m. Bounded N. and N.W. by Ohio river, W. by Mississippi river, N.E. by Tennessee river, S. by Mayfield's river. Watered by Clark's river and its branches. It contained in 1840, 5091 neat cattle, 2447 sheep, 17,566 swine; and produced 13,455 bushels of wheat, 196,977 of Indian corn, 25,649 of oats, 11,917 of potatoes, 65,643 pounds of tobacco, 9638 of cotton. It had 30 stores, eight grist-mills, one saw-mill, three tanneries, four distilleries, one printing-office, one weekly newspaper; four academies, 115 students; six schools, 112 scholars. Pop. whites, 4064; slaves, 654; free coloured, 27; total, 4745. Capital, Paducah.

MAC DONALD, p. v., capital of Randolph co., Ala. 161 m. E. Tuscaloosa, 757 W. Situated 3 m. E. of Little Tulla poosa river. It contains a courthouse, jail, a male and female academy, and several stores and dwellings.

MAC DONALD, p. v., capital of Barry co., Mo., 200 m. S.W. Jefferson city, 1115 W.

MAC DONOUGH, county. Ill. Situated in the W. part of the state, and contains 576 sq. m. Organized in 1829. Watered by Crooked creek, which flows into Illinois river. It contained in 1840, 5119 neat cattle, 3643 sheep, 15,368 swine; and produced 35,694 bushels of wheat, 699,553 of Indian corn, 41,630 of oats, 14,139 of potatoes. It had nine stores, six grist-mills, 13 saw-mills; 15 schools, 336 scholars. Pop. 5308. Capital, Macomb.

MAC DONOUGH. p. t., Chenango co., N. Y., 119 m. W. Albany. 395 W. Drained by Bowman's and Genoganslette creeks. It contains a Methodist and Baptist church, two stores, one fulling-mill. two grist-mills, 14 saw-mills, three tanneries, one printing-office, one weekly newspaper; 13 schools, 599 scholars. Pop. 1369.

MAC DONOUGH. p. v., capital of Henry co., Ga., 65 m. W.N.W. Milledgeville, 680 W. Situated between Towaliga-ga and S. Ocmulgee rivers. It contains a courthouse which cost $6500, a jail, an academy which cost $9900, two churches, a Baptist and Methodist; eight stores, 50 dwellings, and about 300 inhabitants.

MACEDON, p. t., Wayne co., N. Y., 20 m. E. Lyons, 199 m. W. by N. Albany. 356 W. Watered by Mud creek. It contains nine stores, one fulling-mill, five grist-mills, 10 saw-mills; 13 schools, 396 scholars. Pop. 2306.

MACERATA, a city of Central Italy, Papal States. cap. deleg. of same name, on a hill between Chieti and Potenza, 21 m. S. by W. Ancona, and 170 m. N.E. Rome. Pop., in 1838, 15,690. It is well built, surrounded with walls, and entered by six gates. In the centre of the town is an irregular open space ornamented with several good buildings, including the cathedral, the palace of the delegate, and the theatre. Including the cathedral, there are seven churches, in one of which are some good paintings, 13 convents, several literary associations, and a secondary university for theology, philosophy, and medicine, founded by pope Leo XII. in 1694. This city presents nothing antique, and its most interesting feature is the fine view it commands of the Adriatic, and occasionally of the mountains of Dalmatia.

"Macerata," says Forsyth, "contains a number of palazzi, and therefore a swarm of provincial nobility. The peasants observe an established uniform in dress, of which orange appears the prevailing colour. So constant are the women of this class to local costume, that the female head becomes a kind of geographical index. At Macerata they adhere to the ancient mode of plaiting and coiling the hair, which they transfix with long silver wire tipt at both ends with large knobs, evidently the antique acus crinalis—

"Figat acus tortas emissaeique comas."—Mart.

Macerata is a bishop's see, and the seat of a court of appeal for the delega. Macerata, Camerino, Ancona, Ascoli, Fermo, and Urbino. Its manufactures and commerce are insignificant. Under the French, Macerata was the capital of the dep. Musone. About 2 m. to the N., on the Potenza, are the remains of a theatre of considerable size, with vaults and foundations of other edifices, supposed to indicate the site of Helvia Ricina, colonized by Septimius Severus, and destroyed by the Goths. (Forsyth's Italy, p. 390; Rampoldi.)

MAC HENRY. county. Ill. Situated in the N.E. part of the state, and contains 900 sq. m. Drained by the N. branch of Chicago, by Des Plaines, and Fox rivers. Limestone is abundant. It contained in 1840, 2596 neat cattle, 370 sheep, 4808 swine; and produced 43,957 bushels of wheat, 45,407 of Indian corn, 39,974 of oats, 33,870 of potatoes. It had five stores, seven saw-mills, one distillery; 15 schools, 365 scholars. Pop. 2578. Capital, Mac Henry.

MAC HENRY, p. v., capital of Mac Henry co., Ill., 223 m. N.N.E. Springfield, 758 W. Situated on the W. side of Fox river, 12 m. S. of the line of the state. It contains a courthouse, jail, and several stores and dwellings. Nett proceeds of the postoffice, $118.

MACHIAS, p. t., port of entry, and capital of Washington co., Me., 151 m. E. by N. Augusta, 339 N.E. Boston, by land, and 300 by water, 773 W. First settled in 1762, incorporated in 1784. Situated by the W. branch of Machias river, on the E. side of which the village is situated, at the falls. The river admits vessels of 250 tons within 30 rods of the falls. A great amount of lumber is here produced and exported. It has 10 stores, two grist-mills, 23 saw-mills, two tanneries; 11 schools, 382 scholars. Pop. 1851. Tonnage of the district, 11,847.

MACHIAS, p. t., Cattaraugus co., N. Y., 286 m. W. by S. Albany, 333 W. Drained by Ischua creek, the outlet of Lime lake. It contains a church, three stores, one grist-mill. three saw-mills; nine schools, 310 scholars. Pop. 1085.

MACHIAS PORT, p. v., port of entry, Washington co., Me., 155 m. E. by N. Augusta, 777 W. Situated on Machias river, below the junction of the two branches. It has an excellent harbour, protected by Cross island at the mouth of the bay, and considerable shipping, employed in the lumber trade and the fisheries. Incorporated in 1826. It has

one store, one grist-mill, three saw-mills; 12 schools, 364 scholars. Pop. 834.

MACHYNLLETH, a market town and parl. bor. of N. Wales, co. Montgomery, near the Dyfi, 30 m. W. Montgomery, and 175 m. W.N.W. London. Area of parliamentary borough, about 360 acres. Population of the town in 1831, 858. Machynlleth is an ancient, well-built town, superior to most in N. Wales for cleanness and respectability, the streets being remarkably broad and regular. The townhall, a plain building, was erected by the Wynn family, in whom the manor is vested: the county sessions are held already here and at Montgomery, and the magistrates sit here occasionally in petty sessions for the hundred. The church, a handsome structure, was rebuilt in 1827: the interior is conveniently fitted up, and the W. tower is embattled, and surmounted with crocketted pinnacles. There are places of worship for Independents, Calvinists, and Wesleyan Methodists, with attached Sunday schools, and a well-endowed national school furnishes instruction to poor children of both sexes. There is also a savings bank. "The flannel trade has long existed at Machynlleth, being chiefly carried on at farm-houses: the fabrics are sent for sale to Newtown. Weavers' wages (when on full work) vary from 7s. to 9s. a week, and with respect to their moral condition, it is remarked as being much higher than that of operatives in general, in other districts, but yet neither better nor worse than that of the labouring classes generally within the parish. The truck system is partially practised in this vicinity, being fostered by the improvidence of the weavers, few of whom make any provision for emergencies. The prices of provisions are as follows: flour 1s. for 5 lbs., potatoes 2s. for 40 quarts, mutton 5d. per lb., bacon 9d., butter 11d., and oatmeal 2d. per lb. This town formerly possessed an excellent shipping trade, and was, in fact, the port of Montgomery; but since the canal was brought to Newtown, and facilities were opened direct between Wales and the commercial districts of England, the carrying trade is in barges, and few ships now come to Machynlleth." (Hand-loom Weavers' Report, part 5.) This borough unites with Montgomery and others in sending one member to the House of Commons. Registered electors in 1839–40, 78: ditto in the entire district, 1091. Machynlleth is celebrated in the history of the principality as the place in which Owen Glendwr, in 1402, convoked a parliament, where he was inaugurated Prince of Wales.

MAC INTOSH, co., Ga. Situated in the S.E. part of the state, and contains 600 sq. m. Bounded S.E. by the Atlantic, S.W. by Alatamaha river, N.E. by South Newport river. Drained by Sapelo river. It has Sapelo and Black Bend islands on the coast. It contained in 1840, 7884 neat cattle, 734 sheep, 4692 swine; and produced 30,406 bushels of Indian corn. 40,791 of potatoes, 2,896,903 pounds of rice, 542,877 of cotton, 7300 of sugar. It had one commercial and five commission houses in foreign trade, 17 stores, two grist-mills, four saw-mills, one printing-office: one academy, 65 students; one school, 13 scholars. Pop.: whites, 1343; slaves, 3910; free coloured, 102; total, 5360. Capital, Darien.

MAC KEAN, co., Pa. Situated in the N., towards the W. part of the state, and contains 1470 sq. m. Organized in 1804. Watered by Alleghany river and its tributaries, and by branches of Surnemahoning creek. It abounds with coal, iron, and salt. It contained in 1840, 2977 neat cattle, 3713 sheep, 1781 swine; and produced 8169 bushels of wheat, 1870 of rye, 12,070 of Indian corn, 3136 of buckwheat, 19,378 of oats, 59,911 of potatoes. It had ten stores, one grist-mills, 33 saw-mills, two tanneries, one academy, 30 students; 21 schools, 575 scholars. Pop. 2975. Capital, Smethport.

MAC KEAN, p. v., Erie co., Pa., 9 m. S. Erie, 965 m. N.W. by W. Harrisburg, 336 W. Drained by Walnut creek and its branches, and by Elk creek. It has one fulling-mill, four grist-mills, eight saw-mills; 12 schools, 449 scholars. Pop. 1714.

MACKINAC, or MICHILLIMACINACK, co., Mich. Situated in the upper peninsula of Michigan, and contains 27,684 sq. m. The strait, generally pronounced Mackinaw, which connects lake Michigan and lake Huron, is 40 m. long, and in its narrowest part 4 m. wide, contains several islands, as Mackinac, Bois Blanc, Drummond's, and several smaller ones belonging to this county. The greatest height of Mackinac island, above the level of the lake, is 300 feet. It produces wheat, oats, barley, pease, and beans. It was first settled in 1764, surrendered to the government of the United States in 1796, taken by the British in 1812, but restored by the treaty of Ghent, signed Dec. 24th, 1814. The county contained in 1840, 96 neat cattle, six sheep, 65 swine; and produced 30 bushels of Indian corn, 614 of oats, 2016 of potatoes. It had 11 stores, capital in the lake fisheries $19000, value of skins and furs $21,750. Pop. 923.

MACKINAC, p. v., cap. of Mackinac co., Mich., 300 m. N.N.W. Detroit, 821 W. Situated on the S.E. extremity of an island of the same name, and contains a courthouse, jail, two churches, a Presbyterian and Roman Catholic; a school of the American Board of Commissioners for Foreign Missions, a Roman Catholic missionary school, and a branch of the university of Michigan. Fort Mackinac stands on a rocky eminence, 150 feet above the village, which it commands. The harbour is spacious and safe, capable of accommodating 150 vessels. More than 3000 barrels of trout and white fish are annually exported. It is the seat of an extensive fur trade.

MAC LEAN, co., Ill. Situated a little N.E. of the centre of the state, and contains 1296 sq. m. Drained by Mackinaw river, and Kickapoo and Salt creeks. It contained in 1840, 8049 neat cattle, 6777 sheep, 23.740 swine; and produced 44,471 bushels of wheat, 350,890 of Indian corn, 81,098 of oats, 4596 of potatoes. It had 14 stores, one flouring-mill, 14 grist-mills, 19 saw-mills, three tanneries, six distilleries; three academies, 70 students; 14 schools, 422 scholars. Pop. 6565. Capital, Bloomington.

MAC LEANSBOROUGH, p. v., cap. of Hamilton co., Ill., 165 m. S.S.E. Springfield, 776 W. It contains a court-house, jail, about 25 dwellings, and 150 inhabitants. Nett proceeds of the postoffice, $26.

MAC MINN, co., Tenn. Situated in the S.E. part of the state, and contains 608 sq. m. Watered by Hiwassee river and its tributaries. It contained in 1840, 11,407 neat cattle, 8239 sheep, 37,672 swine; and produced 86,861 bushels of wheat, 3597 of rye, 639,578 of Indian corn, 163,624 of oats, 8833 of potatoes, 7360 pounds of tobacco, 1443 of cotton. It had 90 stores, two cotton factories with 672 spindles, 12 flouring-mills, 23 grist-mills, 29 saw-mills, two oil-mills, one paper-mill, nine tanneries, 51 distilleries, two printing-offices, two weekly newspapers; two academies, 99 students; seven schools, 229 scholars. Pop.: whites, 11,430; slaves, 1941; free coloured, 28; total, 12,719. Capital, Athens.

MAC MINVILLE, p. v., cap. of Warren co., Tenn., 75 m. S.E. Nashville, 694 W. Situated on Collin's river, a branch of Coney fork of Cumberland river. It contains a courthouse, jail, two academies, several stores, five tanneries, a printing-office, and about 800 inhabitants.

MAC NAIRY, co., Tenn. Situated towards the S.W. part of the state, and contains 980 sq. m. Watered by branches of Big Hatchee and Forked Deer rivers. It contained in 1840, 12,252 neat cattle, 6110 sheep, 42,271 swine; and produced 36,958 bushels of wheat, 235,715 of Indian corn, 36,591 of oats, 13,577 of potatoes, 243,120 pounds of tobacco, 42,446 of cotton, 1897 of sugar. It had 10 stores, one woollen-factory, 32 grist-mills, 10 saw-mills, three tanneries, 29 distilleries; one academy, 34 students; 11 schools, 273 scholars. Pop.: whites, 8589; slaves, 763; free coloured, 33; total, 9385. Capital, Purdy.

MACOMB, co., Mich. Situated in the E. part of the state, and contains 485 sq. m. Bounded S.E. by St. Clair lake. Watered by Clinton river and its branches. It contained in 1840, 7193 neat cattle, 8959 sheep, 8069 swine; and produced 81,064 bushels of wheat, 7387 of rye, 71,028 of Indian corn, 8390 of buckwheat, 1344 of barley, 69,792 of oats, 80,881 of potatoes, 32,991 pounds of sugar. It had 19 stores, one fulling-mill, four flouring-mills, one grist-mill, 12 saw-mills, one glass factory, two distilleries, two printing-offices, two weekly newspapers; two academies, 29 students; 76 schools, 1973 scholars. Pop. 9716. Capital, Mount Clemens.

MACOMB, p. v., cap. of Mac Donough co., Ill., 36 m. N.W. Springfield, 859 W. Situated on a fertile prairie, 2 m. S. of Drowning fork of Crooked creek. It contains a court-house, four stores, and about 195 inhabitants.

MACON (an. Matisco), a town of France, dep. Saône-et Loire, of which it is the cap.; on the Saône, here crossed by a bridge of 13 arches, 36 m. N. Lyons; lat. 46° 18' 27" N., long. 4° 50' 8" E. Pop. (1896) 11,944. It is pleasantly situated, but is generally ill built; the streets are narrow, crooked, and paved with rounded pebbles, painful to walk upon; the squares, though clean, are mostly small, and destitute of ornament; and the houses, though mostly of stone, are tristes et mesquines. It was once partially forti-fied, but the works were never completed, and they are now laid out in public walks. A handsome quay borders the Saône, and is continuous with a planted promenade at either extremity. The ancient hôtel de Montrevel, now occupied by the town hall, theatre, and public library, with 9000 volumes; the general hospital, two hospices, some of the churches, the prefecture, and the new prison, are the chief public buildings. Mâcon is the seat of tribunals of primary jurisdiction and commerce, a communal college, schools of mutual instruction and linear design, and of a society of agriculture, arts, and belles lettres; and has manufactures of coverlets, clocks and watches, copper and earthenware, pump machinery, barrels, &c. But Mâcon is principally dependent on its wine trade. The same chain of hills that overhang the rich vineyards of the Côte d'Or extends through the department of the Saône-et-Loire, and

the part of the department of the Rhone called the Beaujolais. But whether it be from some difference of exposure of soil, or other unknown cause, the wines produced in the district now mentioned are, though in many respects excellent, inferior to those of the Côte d'Or. In commerce, the wines both of the Mâconnais or district round Mâcon, and of the Beaujolais, are known by the name of Mâcon wines, from Mâcon being the emporium where they are mostly sold. They are strong and durable, corsés, spiritueux, quelque fois trop fumeux, et toujours agréables (Jullien); and in general may be regarded as ranking next to the Beaune wines. The best growths are those of Torins, Romanèche, Chenas, and Fouilly. (Henderson on Wines, p. 165.) Many Roman antiquities have been found at Mâcon, and the ruins of its cathedral, destroyed during the revolutionary frenzy in 1793, form a very picturesque object. On the opposite bank of the Saône is the flourishing suburb of St. Laurent, the seat of a large corn market. (Hugo, art. Saône-et-Loire; Dict. Géog., &c.)

MACON, co., N. C. Situated in the W. part of the state, and contains 900 sq. m. Drained by head branches of Tennessee river. It contained in 1840, 6925 neat cattle, 3589 sheep, 11,883 swine; and produced 6311 bushels of wheat, 4375 of rye, 125,890 of Indian corn, 39,855 of oats, 8613 of potatoes, 2983 pounds of tobacco. It had five stores, one flouring-mill, 22 grist-mills, five saw-mills, four tanneries. 11 distilleries; three schools, 140 scholars. Pop.: whites, 4446; slaves, 369; free coloured, 55; total, 4820. Capital, Franklin.

MACON, co., Ga. Situated a little S.W. of the centre of the state, and contains 420 sq. m. Watered by Flint river and its tributaries. It contained in 1840, 21,178 neat cattle, 2588 sheep, 19,050 swine; and produced 20,923 bushels of wheat, 1062 of rye, 242,032 of Indian corn, 5985 of oats, 19,883 of potatoes, 1,379,605 pounds of cotton. It had 11 stores, 14 grist-mills, nine saw-mills; three schools, 53 scholars. Pop.: whites, 3553; slaves, 1489; free coloured, 3; total, 5045. Capital, Lanier.

MACON, co., Ala. Situated towards the S.E. part of the state, and contains 970 sq. m. Watered by branches of Tallapoosa river. It contained in 1840, 15,958 neat cattle, 762 sheep, 21,731 swine; and produced 2316 bushels of wheat, 312,313 of Indian corn, 6241 of oats, 29,076 of potatoes, 883,195 pounds of cotton. It had 16 stores, three flouring-mills, 13 grist-mills, 11 saw-mills, two tanneries, one pottery; six academies, 243 students; 14 schools, 368 scholars. Pop.: whites, 5369; slaves, 5851; free coloured, 27; total, 11,247. Capital, Tuskegee.

MACON, co., Ill. Situated near the centre of the state, and contains 1400 sq. m. Watered by Sangamon river. It contained in 1840, 4964 neat cattle, 3546 sheep, 13,985 swine; and produced 21,344 bushels of wheat, 1741 of rye, 173,187 of Indian corn, 44,360 of oats, 5354 of potatoes. It had five stores, one woollen factory, six grist-mills, six saw-mills, one tannery, one distillery; eight schools, 234 scholars. Pop. 3039. Capital, Decatur.

MACON, co., Mo. Situated towards the N. part of the state, and contains 846 sq. m. Watered by Chariton and E. Chariton rivers, and branches of Salt river. In contained in 1840, 6539 neat cattle, 7050 sheep, 30,973 swine; and produced 9105 bushels of wheat, 592,998 of Indian corn, 15,517 of oats, 9581 of potatoes, 57,871 pounds of tobacco. It had 11 stores, six grist-mills, four saw-mills, three tanneries, two distilleries; seven schools, 183 scholars. Pop.: whites, 5806; slaves, 223; free coloured, 1; total, 6034. Capital, Bloomington.

MACON, city, cap. of Bibb co., Ga., 30 m. S.W. Milledgeville, 686 W. Situated at the head of steam navigation on Ocmulgee river, chiefly on the W. side, though it has extended to both sides; connected by a bridge 389 feet long, which was sold to the city for $25,000. The streets run N.W. and S.E., following the course of the river. Wharf street, on the river, is 880 feet wide; the next is 138, and the other streets are from 180 to 180 feet wide. These are crossed at right angles by other wide streets. The public buildings are a courthouse, 93 by 47 feet, and three stories high, with a cupola. In the basement are offices of various descriptions. In the second story are the court rooms; in the third are offices of the clerks; a jail, a market house, five churches, two Presbyterian, a Methodist, Episcopal, and Baptist; and two banks. A great amount of cotton is shipped from this place. Eight or ten steamboats are employed on the Ocmulgee, besides towboats and poleboats. In 1822 a single cabin occupied the site of the city. It is now the third place in population in the state, being exceeded only by Savannah and Augusta. It contained in 1840, nine commercial houses in foreign trade, capital $75,000; 82 retail stores, capital $585,000; nine lumber-yards, capital $7500; one grist-mill, three printing-offices, three weekly newspapers, and one periodical; one college, 150 students; three academies, 200 students; two schools, 75 scholars. Pop. 3927.

MACON, p. v., cap. of Noxubee co., Miss., 125 m. N.E. Jackson, 365 W. Situated on the Noxubee river, and contains a courthouse and several dwellings.

MACON, p. v., Lenawee co., Mich., 65 m. S.W. Detroit, 580 W. Watered by the S. branch of Raisin river. It contains three saw-mills, seven schools, 311 scholars. Pop. 1146.

MACOUPIN, co., Ill. Situated a little S.W. of the centre of the state, and contains 864 sq. m. Organized in 1829. Watered by Macoupin creek and its branches. It contained in 1840, 11,773 neat cattle, 10,531 sheep, 37,917 swine; and produced 42,919 bushels of wheat, 540,980 of Indian corn, 57,685 of oats, 17,179 of potatoes, 11,195 pounds of tobacco. It had 19 stores, two flouring-mills, 16 grist-mills. seven saw-mills, one tannery, three distilleries; one academy, 45 students; 14 schools, 275 scholars. Pop. 7926.

MACQUARRIE RIVER. See AUSTRALIA, vol. i., p. 212, 213.

MACROOM, an inland town of Ireland, co. Cork, prov. Munster, on the Sullane, 20 m. W. Cork. Pop. in 1831, 3028. It is a poor, mean place, consisting of a single street, mostly of cabins. It has a parish church, a Roman Catholic chapel, a large school, a courthouse, market-house, and a constabulary barrack. A manor-court for the recovery of debts to the extent of £3 is held every three weeks. General sessions are held in December, and petty sessions on alternate Tuesdays. Markets on Saturday. Post-office revenue, in 1830, £226; in 1836, £235. Near the town is a large cavern, the interior of which has not been thoroughly explored.

MADAGASCAR, a large island of the Indian ocean, off the E. coast of Africa (from which it is separated by the Mozambique channel), between lat. 12° 2' and 25° 40' N., and long. 44° 90' and 51° 30' E. Length, 930 m.; average breadth, 208 m. Area estimated at about 234,400 sq. m., being somewhat greater than that of France. This country, of which only a few years ago we had scarcely any knowledge, has recently been visited and explored by missionaries and other travellers; so that we now possess very satisfactory information respecting the island and its inhabitants. The coast is generally flat and low; but the interior is considerably diversified, and, though it is not traversed by any continuous chain, many parts, especially the E., N., and S. districts, may be called mountainous, its highest point, Ankaratra, in lat. 19° 40' N., long. 47° 20' E., is about 11,000 ft. above the sea. These mountains consist of granite, syenite, and quartz, covered in the lower parts with clay-slate, primitive limestone, and old red sandstone: volcanic rocks occur in several places, and coal strata, abounding with iron, are widely distributed through the island. Rock-salt and nitre occur near the coast; and iron pyrites, oxide of manganese, and plumbago, have been found in some districts. The rivers of Madagascar are numerous, and many of considerable size, the greater number flowing into the sea on the W. side; but most of them are choked with sand, have frequent falls and rapids, and are almost entirely unnavigable. There are likewise numerous lakes, not only in the central parts of the island, but also in the low alluvial districts near the sea, some of which are remarkable for their size and beauty. The most fertile parts are the valleys, most of which produce rice or other vegetables, or else are clothed with a rich and luxuriant verdure. The climate of Madagascar is extremely diversified; that of the coast being oppressively hot. while in the interior the temperature seldom exceeds 86° Fahr. The heat at Antananarivo, the capital, fluctuates between 46° and 85°; the middle of the day in summer is often extremely sultry, but the mornings and evenings are always pleasant. From May to October (the winter months of this island) the ground is often covered with hoar-frost, and the heat seldom exceeds 44°. At other seasons, however, the fluctuations between heat and cold are extreme and sudden, the temperature in the morning being seldom more than 40°, whereas, in the same day, the afternoon heat often exceeds 80°. The climate of Madagascar is extremely prejudicial to Europeans, in consequence chiefly of the effluvia rising from stagnant lakes and swamps near the coast; but in the central parts, and especially in Ankova, the metropolitan province of the island, the marsh-fever does not exist. The weather on the coast is usually hot and damp or rainy; but in the interior the rains are periodical, in a great measure regulating the divisions or seasons of the year. The trade winds from the E. and S.E. prevail during the greater part of the year; but the rains are often accompanied by violent gales from the N.W., W., and S.W. Earthquakes are occasionally felt, and the capital has more than once suffered considerable damage from such visitations.

Among the animals peculiar to Madagascar, may be mentioned five varieties of the monkey, foxes, wild dogs and cats, hogs, goats, a peculiar kind of cattle and sheep similar to those of the Cape of Good Hope: crocodiles swarm in nearly all the rivers and lakes, and are objects of great

dread to the natives; serpents, also, some of large size, abound in the woods, and lizards, scorpions, and centipedes, are very numerous and troublesome. Birds also, of various kinds, are found in the forests, the principal of which are the parroquet, flamingo, falcon, kite, turtle-dove, pigeon, turkey, and different varieties of land and water fowls. The sea abounds with fish of various kinds, and oysters are numerous on the coast. The soil in many parts is prolific and highly susceptible of improvement, and the island produces numerous and highly valuable plants. The forests yield abundance of trees of varied durability and value; some used as dye-woods, others in building, with ebony, betel, mangrove, dragon-tree, bamboo, sugar-cane, locust-tree, *Urania speciosa*, caoutchouc-tree, plantain, banana, sakana (*Bignonia articulata*), hibiscus, mimosa, castor-oil plant, longoza (*Curcuma cæsaria*), cotton, indigo, and tobacco plants, allspice, pepper, ginger, turmeric, and rice. Various other vegetable productions have been introduced, such as the cocoa-nut, bread-fruit, yam, manioc, lemon, orange, peach, mulberry, quince, fig, and pomegranate. Several varieties of the Cape vine have been found to thrive well, the coffee-plant has been brought from the Mauritius, and the potato is largely cultivated, as well as highly esteemed; but the common European *cerealia* have met with little encouragement. The flora of the country is abundant; but the brilliant aspect usual to the gardens of tropical countries is here missed, in consequence of the rapid alternations of heavy rains and extreme drought.

The husbandry of Madagascar, pursued by a distinct class, consists, in a great measure, of the cultivation of rice, which is conducted with great care and success. Seed time is in September; at which season the grain, after being steeped in water, and subsequently kept in a warm place till it begins to sprout, is very thickly sown in a fine mould, almost covered with water artificially introduced into the fields. The water is afterward drained off, manure is thrown over the seed, and as soon as the sprouts appear above the surface, moisture is again applied. The average produce in inferior grounds is said to be about 50 for 1, but the best cultivated grounds are alleged to produce seventy and even one hundred fold, the harvest being in January and February. Each rice field is separated from those adjoining by banks rising about six inches above the field, and affording great convenience to the labourers. Neither waggon, cart, sledge, nor beast of burden, is used in getting in the harvest, and the threshing is conducted either against a stone, or on the floor, by simply beating the ears with the hand. The secure storing of the grain, however, is an object of special attention: the Ovahs, the prevailing tribe of the island, have underground storehouses, made with extreme ingenuity; but other tribes have granaries above ground, bee-hive shaped, about 16 ft. high, made of thick, clay-built walls, and entered only from the top. Manioc is another great object of farming industry; it is raised from cuttings, and about 18 months elapse between the planting and harvest. The roots, usually about 10 inches in length by three in diameter, are prepared for use by scraping and boiling, and are sometimes made into cakes. Cotton is cultivated to a considerable extent; and the pigeon pea (*cytisus cajan*) is raised for the purpose of rearing silk-worms. The European *cerealia* have been introduced by the missionaries; the plough and harrow have likewise been brought into use, and oxen broken in to cultivate the ground; but the natives prefer their old and imperfect methods of preparing the soil, to the adoption of readier plans. and superior implements. Next to the cultivation of the soil, the working of iron is the most important occupation of the people. In some parts the iron ore is found in large quantities on or near the surface, whence it is gathered in baskets and smelted for use; but when it is dug out of the ground, numerous small pits are made about 6 feet in depth, and no farther attempt is made to explore the riches of the interior. The ore is first crushed, then broken into small pieces, and afterward submitted to the action of a charcoal fire in a rude furnace of stone-work, built up to the height of 2 or 3 feet without mortar, and thickly plastered with clay on the outside, the blast being obtained by means of wooden cylinders, in which a rude sort of piston is fitted to drive the air into a bamboo cane through the fire. The native forges are equally simple; the anvil, about the size of a sledge hammer, is fixed in the ground near the fire, the water-trough is close by, and the smith, when at work, squats on a piece of board while his attendants surround him, armed with large hammers, and ready to strike the metal according to his directions. The articles thus manufactured comprise spears and javelins, knives, hatchets and spades, chisels and hammers, a rude sort of plane irons, files, pots, spoons, lamps, and nails; besides which they have been taught by the English to make hinges, screws, and locks, as well as to draw copper and iron wire. The manufacture of swords and fire-arms was introduced by the French a few years ago, and the native goldsmiths and silversmiths evince considerable ingenuity in making rings, chains, and other gold ornaments, silver dishes, mugs, spoons, &c. The felling of timber (which appears to be a monopoly of the government) employs about 700 men; the pit saw has been brought into general use, and the native carpentry has been so much improved by the application of European tools, that their work may be justly styled "strong, neat, and well finished." (*Ellis*, vol. i., p. 317.) The art of turning wood is practised by the best workmen of the capital; earthenware is made with considerable skill and taste, and many hands are employed in making rope and twine, as well as in tanning leather. The chief occupation of the people, however, next to the cultivation of rice, is the spinning and weaving of silk, cotton, and linen fabrics; but all the processes are so extremely simple, imperfect, and tedious, that it is a matter for surprise that the threads of their cloth are even and well-twisted, the weaving regular, and the patterns regular, and exhibiting fancy and good taste. The art of dyeing is also practised, and several of the native dyes produce bright and durable colours.

The population of Madagascar consists of four chief political divisions, the numbers of which are estimated to be as follows:—

The Ovahs (in the central table-land)	800,000
Sakalavas (W. side of the island)	1,200,000
Betsileos (S. of the Ovahs)	1,500,000
Betanimena and Betsimasarka (on the E. coast)	1,200,000
	4,790,000

It is said, also, that this amount of population is considerably less than it was a few years back, owing chiefly to wars between the different tribes, the prevalence of the slave-trade, &c. It is, also, supposed that the practice of infanticide, which is alleged to have prevailed from time immemorial, has contributed to reduce the population. But most probably the influence of this practice is greatly overrated; and it is evident it cannot at all account for the recent decrease. The inhabitants differ materially in appearance and character, nor is there any doubt, though the people are nominally comprised in one political empire, and speak one language, that they include several distinct and peculiar nations. The distinction of colour separates the population into two great classes, the Ovahs, and a few other tribes, having olive complexions, handsome features, graceful persons, and lank dark hair, whereas the inhabitants of the shore, and indeed the majority of the people greatly resemble the Papuas, being short and stout, almost black, with low foreheads, broad flat faces, large eyes and mouth, and long crisped hair. There are differences also in the languages spoken by various sections of the population, and many of their customs vary so much, as to make it clear that, however amalgamated, they are not one nation, but a combination of several distinct races. With the exception, however, of the Ovahs, they are little better than barbarians, run almost naked, despise a fixed life, are extremely superstitious, and practice most of the vices so generally prevalent among the savages of the neighbouring continent. Circumcision is universal, marriages are formed in very early life, and divorces are very common, and easily effected, the law permits polygamy, restricting the husband to 12 wives; but few have more than two, or at most three. Fidelity to the marriage engagement, however, forms so part of the female character, and modesty is a virtue almost unknown. Their houses are usually of rude construction, except in the capital of the Ovah country, where European improvements have been partially introduced. The diet of the people consists, in great part, of rice and manioc, with smaller portions of beef, poultry, &c., and the cookery is extremely simple.

Pedlary and hawking are favourite, though not profitable, occupations. The markets are favourite places of resort for all classes; and not only is there a daily general market at Tananarivo, but four or five large markets are held in different parts of the province, and well attended by a vast concourse of people from the adjoining districts. Animal and vegetable productions, native and foreign manufactures and cattle are exposed promiscuously; and in no nation are there more clever and persevering bargainers than in Madagascar. The slave-trade, also, which a few years ago was nearly extinct, is now said to be pursued on a large scale. Money has not been coined on the island; dollars are more or less known in all parts, through communication with Europeans; but the trade is more generally carried on by barter. Most goods are sold by measure; rice by the bushel, meat by the eye, snuff by the spoon, fuel by the bundle, &c. Rice, which may here be considered the standard of value, costs about 1s. a bushel; 20 ducks or fowls may be purchased for a dollar, geese cost about 9d. each, and a fine turkey may be got for 1s. A bullock costs from three to eight dollars, sheep average about 1s. 6d. each, and 20 good

pine apples may be had for 3d. Labour is also extremely low, many working for mere food, and others gaining only 2d., or at most 4d. a day. An intercourse has long been carried on with Madagascar by Arabs from Muscat, Indians from the presidency of Bombay, Europeans from the Cape of Good Hope, and Americans from Brazil and the United States. The taste of the people for foreign goods was also on the increase a few years ago; and horses, saddles and bridles, scarlet cloth, gold lace, red satin, purple, green, and yellow silk, silk handkerchiefs, sewing silk, calico and printed goods, hosiery, gloves, finger rings, watches and musical boxes, hardware, salt, and, above all, arrac and rum, fetched very high prices in the markets of Ankova; but since the death, in 1828, of Radama, the most enlightened monarch that ever held sway in Madagascar, the foreign trade has greatly declined, and the policy of the present government seems to threaten the entire cessation of all trade with Europeans.

Madagascar was divided into 28 provinces, all of which have their separate chiefs; but for some years past the Ovahs have been reckoned the prevailing tribe, the chief of which is, in effect, the king of the island, receiving tribute from, and exercising sovereignty over, all the rest. The government is despotic, and the succession to the throne is commonly hereditary, the monarch having the right not only to appoint his immediate successor, but also to settle the line through future generations. He is the father of his kingdom, appoints every subordinate officer, enacts laws and orders their execution, decides cases and raises armies; but he often convokes assemblies of the people, for the purpose of obtaining information or advice on matters requiring mature deliberation, or in cases where the wishes of the aristocracy have to be consulted. The royal family is highly honoured, and no people can be more tenacious of etiquette, and the respect due to rank. The judges, who rank next to the blood royal, hear causes, decide disputes, and are exclusively privileged to communicate between the sovereign and people. Subordinate to these are the *forensies*, the police and tax-gatherers of the country; the *embou-in-jats*, or local magistrates; the *marosoerous*, or military governors of provinces (a very powerful and important body); and the *vadintany*, or royal courtiers, who not only carry government dispatches, but constitute a general patrol for the country. The king receives tithes of all produce, enjoys the monopoly of timber, and is exceedingly rich both in slaves and cattle, receiving also a considerable and valorem duty from the possessors of these valuable articles. The sovereign is also high-priest of the realm, and presides over the great national sacrifices. The religion of the country is a rude species of polytheistic idolatry, and the people almost without exception believe in witchcraft and the efficacy of charms. Christianity was introduced with temporary success by English missionaries, in 1818-1830; but it is believed that at present it is almost powerless, in consequence of the royal edict of 1835, which not only forbad its public profession, but legalised the persecution, and even enslavement, of all natives becoming its adherents.

Madagascar, the earliest accounts of which were given by Marco Polo, from the narrative of others, was discovered, in 1506, by the Portuguese, who established a settlement close to the S. end of the island, and soon after tried, though with little success, to introduce the Roman Catholic religion. It was at first resorted to merely as a place of refuge and provisioning station for ships; but in 1642 an attempt was made by the French to make it one of their colonies, which however proved futile, in consequence of its extreme unhealthiness; and in 1664 most of the colonists removed to the neighbouring island of Bourbon. The Jesuits meanwhile continued to exert themselves in the establishment of Christianity; but owing to the injudicious zeal of Father Stephen, the superior of the mission in Madagascar, the natives were exasperated at the innovations of the foreigners, some of the missionaries were massacred, and the rest were glad to escape from the island. Various attempts have subsequently been made by the French to establish a permanent settlement, and since the general peace of 1815 they have formed four small colonies on the E. coast, as well as on the contiguous island of Madame-St. Mary. The English missionaries were allowed to visit Madagascar in 1818-1825, with full permission to disseminate their moral and religious views, and the sovereign Radama was favourable to the establishment of schools and the introduction of improved methods both of agriculture and manufactures. Since his death, however, there has been a stagnation in the trade with England, the missionaries have been forbidden to approach the island, all possible means have been adopted to destroy the effects which the exertions of Europeans had accomplished in eight years, and Madagascar may now be ranked among the barbarous countries of E. Africa. (*Rochon's Voyage to Madagascar*; *Ellis's Hist. of Madagascar*, vol. i. *passim*; *and valuable private information.*)

MADAWASKA, t., Aroostook co. Me., 220 N.E. by N. Augusta. Bounded N. by St. Johns river. It contains five stores, four grist-mills, one saw-mill; three schools, 56 scholars. Pop. 1584. The settlement lies along the road, following the course of the river. The part N. of the river, by the late treaty, belongs to New-Brunswick, and contains 1876 inhabitants.

MADDALONI, a town of S. Italy, king. of Naples, prov. Terra-di-Lavoro, cap. canton, 14 m. N.N.E. Naples. Pop. 10,500. It has several churches and convents, a house of refuge, a royal college, and a noble aqueduct, which conveys water to the royal palace at Caserta.

MADERA, a great river of S. America, a tributary of the Amazon (which see).

MADEIRA, a famous island in the N. Atlantic ocean, belonging to Portugal, Funchal, its cap., on its S.E. side, being in lat. 32° 38′ N., long. 16° 54′ 26′′ W.: length of Madeira, about 46 m.; breadth about 7 m. Area, estimated at above 300 sq. m. It is a mass of basaltic rock, presenting to those approaching its N. coast, numerous disjointed crags, and tall isolated peaks, interspersed here and there with less elevated spots of verdure, the whole being based on enormous, dark-looking columns, rising perpendicularly several hundred feet from the sea; which is usually so deep, even close in shore, that soundings are not found in less than 50 fathoms and upwards, except in Funchal road where ships anchor in from 30 to 35 fathoms. The cliffs on all sides are very lofty: the Peña d'Ageria (eagle's rock) on its N. coast, a black cone-shaped mass of rock, is upwards of 1800 ft. high; and C. Pargo, at the N.W. extremity of the island, rises 4000 ft. above the sea; but the most curious feature on the coast is the Punta S. Lorenzo, at its E. extremity, a ledge of rock 6 m. in length by 1 m. in breadth, which, though less lofty than other parts, is remarkable for its bold projection into the sea, and its fantastically-broken cliffs and peaks. The rapid declivities of the island are furrowed by deep and narrow valleys, at the bottom of which flow rills of pure spring water; and up their sides vineyards are formed by means of successive terraces, to the height of 2300 ft. above the sea. The mountain scenery of the interior is bold, and highly romantic; one part, a few miles N.W. of Funchal, being called, by way of distinction, "the Switzerland of Madeira. Here is a deep valley, or crater, inclosed on all sides, except seaward, by a range of magnificent precipices, rising upwards of 1000 ft. above the vale, the summits and sides of which are broken into every variety of dark beetling pinnacle, or flattened and tree-clad buttress; while far below smiles a fair region of cultivation and fruitfulness, rich in every species of vegetation, though itself rather more than 3000 ft. above the sea level. The culminating point of the island is Pico Ruivo, rising 5430 ft. above the sea, and covered with vegetation to its summit. Three rivers, or rather torrents, rise on its sides, and cross the island in several directions, contributing greatly to its fertility. The streams are carefully collected, and rendered more available for the purposes of agriculture, by means of artificial channels, or levadas, with sluices, constructed with vast labour. At present (1840) a great work of this kind is in the course of being completed, by which a copious stream, which is precipitated from the top of a cliff 1600 feet in height, will be made subservient to the purposes of irrigation.

The climate of Madeira fluctuates less than that of any country N. of the equator: its mean annual temperature having been found, in a period of 18 years, not to exceed 65° Fahr., that of the hottest months (Aug. and Sept.) being 74°, and that of the coldest, (Dec. and Jan.) 63°, the glass seldom falling below 53° even° in the severest weather. The heat of summer, however, is considerably higher, being increased from 10° to 15° during the prevalence of the hot and parching E. winds (the scirrocco) that blow off the African continent. The temperature of Funchal, however, is considerably higher than that of the island in general: there dews are slight, and the rains few and far between; but in the higher parts of the island, a cool climate is rendered more delicious by frequent dews and rains enriching vegetation, and rendering the air fresh and salubrious. This remarkable equality of climate, not only through the year, but during the days and nights, constitutes the chief recommendation of Madeira to invalids. Persons subject to chronic pulmonary complaints, unattended by any material disorganization, have derived much benefit from a voyage to Madeira; as have others afflicted with diseases of the windpipe; and a still greater number who are the victims of dyspepsia, or other maladies of the stomach, the cure of which is hastened by the regular habits and exercise usually taken by invalid residents in the island. The efficacy of the climate, however, in cases of confirmed tubercular consumption has been absurdly exaggerated. It may then, indeed, lengthen life a little; but it cannot effect a cure. During the last half century, vast numbers of invalids, of whose recovery no rational hope could be enter-

tained, and who should have been left quietly to expire at home, have been hurried off to this island, at an expense which they could often but ill afford, for no purpose unless it were to amuse them with false hopes, or that they might occupy a place in Funchal churchyard. Invalids should not attempt the voyage before the middle of June, nor later than the end of September; spring is a trying season, owing to the prevalence of N.E. winds; and October is the first month of the rainy season of autumn.

Every part of Madeira, not encumbered with rocks, is extremely fertile; the hills are covered with luxuriant vegetation, and the most delicate flowers grow on their summits, which are constantly moistened with dew from the clouds overhanging the island. Trees and shrubs of the finest kinds are everywhere abundant, and tropical plants which have strayed from the gardens soon become naturalized to the soil.

"Here" says Mr. Wilde, "all is sunshine: the green bananas, with their beautiful feathery tops, tell the visiter that he has bid farewell to Europe: the orange trees hold out to him their branches, laden with golden fruit. Plantations of coffee trees fill the spaces between the houses, the splendid coral tree hangs over his head, and the snowy bells of the tulip tree mingle with the scarlet hybiscus. If he wish for exercise, he has the most inviting walks, and the most tempting shades to shelter him: wide-spreading plane trees, and willows of gigantic growth, bend their slender arms over the streams that murmur from the hills. As he begins to ascend from Funchal, the beauty increases, and the sea-view opens to his sight. The roads, though steep, are well paved, and the horses trained to an easy pace. He rides through a perfect vineyard, where in many places the vines are carried on trellises over the road, and large bunches of grapes hang within his reach. Hedges of geraniums, fuschias, and heliotropes, border those narrow paths, and shade him from the sun; the Ficus indicus clothes the cottages, the Salvia fulgens and Guernsey lily are sprinkled over the vineyards, and the Camellia japonica, with its delicate white flower and waxy leaf, adorns every quinta. Higher up grow the yam, prickly pear, dragon-tree, and cedar, the aloe, agave and hydrangea, the sweet potatoe, and the Pharnium tenax; and heaths and pines crown the highest summits of the island." (Narrative, i. 89–91.) Thus it appears, that below the elevation of 1300 ft., many of the most useful tropical plants, as the date, palm, guava, banana, coffee plant, &c., are found, with numerous others peculiar to the warmer part of the temperate zone. Up to 2500 ft., the fruits and grains of Europe, especially maize and corn, are raised; and nearly the whole of this district is covered with vineyards. The chestnut, which is extremely abundant, the beech, and other European trees, with the mahogany, attain an elevation of 3400 ft., above which rise pines, heath, ferns, and grasses. Pasture is scanty: few cows are kept, and the products of the dairy are here expensive luxuries. Horses are little used, their place as beasts of burden being supplied by mules and asses of the Spanish breed. Goats and hogs are very numerous, and are allowed to run wild on the mountains, where also are found large quantities of rabbits. Poultry is abundant, and cheap; and small birds of magnificent plumage occupy the groves. Myriads of finely variegated lizards crowd the gardens and vineyards, occasionally doing much damage to the grapes; but there are no venomous reptiles, and the inhabitants are free from that insect plague that is usually one of the drawbacks of warm countries. The honey bee is abundant, and produces fine honey. Many varieties of fish are caught on the coast, especially tunnies and eels, which are the favourite food of the inhabitants.

Agriculture is chiefly confined to the raising of vines. Land is usually let out in small holdings, varying from 10 to 40 or 50 acres, and the rent is estimated, on the metayer principle, at half the produce, according to a yearly valuation of the crops. Wheat, barley, and rye are produced; but the crops average little more than a third part of the annual consumption. The wheat grown in 1837 is said, though we attach little credit to such statements, to have amounted to 6787 qrs., the barley to 9664 qrs., and the rye to only 570 qrs. Wheat is sown in Oct., and reaped in June, this crop being followed by another of beans or sweet potatoes. Rice is cultivated more as an ornamental grass than for any useful purpose; and Indian corn, which is admirably adapted to the climate, and is much used as an article of food, has till very lately been little grown.

Considerable attention has lately been devoted to the cultivation of the coffee plant, which, should the demand for wine not increase, may, perhaps, become of considerable importance. Fruits and vegetables are raised with little trouble, and the show in the fruit market of Funchal, in a grove of noble palm-trees, would astonish any untravelled European, even from Italy or Spain. Here, besides all the ordinary fruits and garden vegetables of S. Europe, as oranges and lemons, green figs, grapes, pomegranates, water and Valencia melons, pumpkins, &c., one may see bananas and guavas, finer even than those grown in the W. Indies, custard-apples, alligator pears, (the fruit of the *Laurus Persea*), numerous tribes of cucurbits, the exquisitely flavoured fruit of the *Cactus triangularis*, the Cape gooseberry sent as a preserve to Europe, and the *tchoo-tchoo*, said to be "delicious." But its wine is the great glory of Madeira. The grape is not indigenous to the island; and it is said to have received its first plants from Crete, carried thither by order of the famous Prince Henry of Portugal, under whose auspices it was settled by the Portuguese in 1421. Many other varieties of the grape have since been carried to the island, its mild climate and volcanic soil being especially suitable for their growth.

The steepness of the hill-sides, on which the vines chiefly grow, and the necessity of economizing valuable space, have led to the practice of raising the vine-beds on successive terraces, supported by retaining walls. The vines are trellised on bamboo and other supports for the purpose of exposing the grapes to the ripening influence of the sun, and the bunches are frequently of enormous size. The usual method of cultivation is to trench the ground from four to seven feet deep, according to the soil, and to lay a quantity of loose or stony earth at the bottom, to prevent the roots from reaching the clayey soil beneath, which would otherwise hinder their growth. The ground is watered three times, if the summer be very dry, and each time it is thoroughly saturated; but the less it is watered the better is the wine, though the quantity, of course, be diminished.

The N. side of the island, though sufficiently fertile, being the most exposed to cold winds and fogs, is not so favourable to the culture of the vine as the S., where all the finest growths are raised. The best Madeira-malmsey, or *Malvoisia*, is produced on rocky grounds exposed to the full influence of the sun's rays, the grapes being allowed to hang till they are dead ripe. The *Sercial* grape will, also, only succeed on particular spots. The wine made from it is, when new, harsh and austere, and requires to be long kept. The best Madeira wine is produced on the S. side of the island; but it is alleged that not less than two-thirds of the wine grown even in this quarter is of secondary quality; so that in Madeira, as in all wine countries, the first growths (*premiers crus*) are both scarce and dear. The process of making the wine is very simple. The grapes are picked from the stalk, thrown into a vat, pressed, first with the feet, and afterwards with a weighted wooden lever. The proprietor of the land, and the collector of taxes for the crown, both attend at the press; the latter takes out of the tub his *tenth* of the whole *must*, the remainder being equally divided between the landowner and the tenant. Each takes with him a sufficient number of porters to carry away their respective shares, sometimes in barrels, but more frequently in goat skins, *borrachas*, to the cellars in Funchal, where the English merchants have extensive yards and vats for storing the wine, and carrying it through the different processes of fermentation, mixture, &c. They usually advance money beforehand to the growers, to enable them to defray the expenses of cultivation. (*Barrow's Voyage to Cochin China*, p. 22.)

Though naturally strong, a quantity of brandy is added to Madeira wine when racked from the vessels in which it has been fermented, and another portion is added when it is about to be exported. The demand for Madeira wine in the E. and W. Indies, where it is highly esteemed, first led to a knowledge of the improvement it derives from being carried to a warm climate; and it has long been customary for ships outward bound for India and China to touch at Madeira, and take large quantities of wine on board, which they bring home to England. But it must not be supposed that all the Madeira wine that has gone to Calcutta and Canton is necessarily better than any brought direct from the island, as much must obviously depend on the quality of the wine sent to the east. But, if due care be taken in the selection of the wine sent to India and China, it is very much improved and matured by the voyage; and it not only fetches a higher price, but is in all respects superior to the direct importations. Most of the adventitious spirit is dissipated in the course of the Indian voyage, and the full flavour of the wine is evolved.

Madeira wines may be kept for a very long period. "Like the ancient vintages of the Surrentine hills, they are truly *firmissima vina*, retaining their qualities unimpaired in both extremes of climate, suffering no decay, and constantly improving as they advance in age. Indeed, they cannot be pronounced in condition until they have been kept for ten years in the wood, and afterward allowed to mellow nearly twice that time in bottle; and even then they will hardly have reached the utmost perfection of which they are susceptible. When of good quality, and matured as above described, they lose all their original harshness, and acquire that agreeable pungency, that bitter

sweetishness, which was so highly prized in the choicest wines of antiquity; uniting great strength and richness of flavour with an exceedingly fragrant and diffusible aroma. The nutty taste, which is often very marked, is not communicated, as some have imagined, by means of bitter almonds, but is inherent in the wine." (*Henderson*, p. 253.)

The wines of Madeira have fallen of late years into disrepute in England. The growth of the island is very limited, not exceeding 15,000 or 18,000 pipes, of which a considerable quantity goes to the East and West Indies, and America. Hence, when Madeira was a fashionable wine in England, every sort of deception was practiced with respect to it, and large quantities of spurious trash were disposed of for the genuine vintage of the island. This naturally brought the wine into discredit; so that sherry has been for several years the fashionable white wine. It is difficult, however, to imagine that adulteration should ever have been practiced to a greater extent upon Madeira than it is now practiced upon sherry. It is not, therefore, improbable that a reaction may take place in favour of Madeira, which has sunk to a much lower place in the public estimation than it deserves to hold. In 1837, 308,285 gallons Madeira were entered for home consumption, whereas the quantity entered in 1839 amounted to only 118,715 gallons, being less than 5 per cent. of the sherry entered for consumption in the same year! Such is the power of fashion, for there cannot be a question that really good Madeira is one of the very best of wines.

The commerce of Madeira is very considerable; the exports consist principally of wine. Among the minor articles of export are fruits, both fresh and preserved, dragons' blood (the gum of the *Calamus draco*), honey and wax, orchil (a white lichen used in purple dyeing,) tobacco, and provisions for ships. Its imports comprise manufactured goods, sheep, salted provisions, fish (especially herring and cod), oil, corn, and some tropical productions. We subjoin a statement of the vessels, and their tonnage, which arrived at and left Funchal in 1837; specifying the countries whence they came, and for which they cleared out, with the value of their cargoes:—

Countries.	Arrived.			Departed.		
	Ships.	Tons.	Value of Cargoes.	Ships.	Tons.	Value of Cargoes.
			£			£
Gt. Britain	130	29,432	53,296	131	29,570	131,665
Portugal	75	5,539	34,673	75	5,539	6,366
U. States	35	6,324	27,261	36	5,305	71,226
Sardinia	38	4,836	20,451	33	4,896	346
Denmark	10	1,924	6,054	9	1,726	12,669
France	7	915	5,520	7	945	4,365
Holland	3	597	1,549	3	597	11,388
Other countries	14	1,384	9,880	14	1,384	1,809
Total	306	50,271	159,964	307	50,274	240,999

The imports from England in 1838, comprised cottons, woollen and linen fabrics, and haberdashery, to the value of £22,370, with coal, earthenware, butter and cheese, salt meat, rice, sugar, &c., making in all a total value of £35,000. The Americans send timber, whale-oil, salt fish and meat, spermaceti candles, with other articles, in small quantities, to the value, in 1838, of 41,057 dollars.

The government of Madeira has at its head a lieutenant-governor, whose power is so extensive that the comfort and happiness of the inhabitants, especially the British, are greatly dependent on his character and acquaintance with the island. Justice is administered by a tribunal in whose favour little can be said, from which there is an appeal to the courts at Lisbon. The crown revenues are derived partly from a duty of 20 per cent. on all imports, except provisions; but the most productive source is the tithe of wine, with an additional duty per pipe on the quantity exported. A revenue is also derived from the monopoly of snuff, cards, and soap. The revenue is sufficient to defray the expenses of the civil, military, and ecclesiastical establishments; considerable sums are likewise expended in public works, roads, &c., and frequently there remains a surplus which is remitted to Portugal. The number of clergy, including monks and nuns, is stated to be somewhat under 300; they are partly supported by the crown revenues, the tithe on wine being originally intended for their maintenance; but the present government allowance, which they receive in lieu of it is extremely small; so that the monks and clergy traffic in wine, or engage in other secular business, while the nuns gain a considerable income by making artificial flowers of wax and feathers, which are justly admired for their delicacy and beauty, and purchased by visiters and shippers at high prices. It would be well, however, were both nuns and monks suppressed. According to Sir John Barrow, who is by no means disposed to be over censorious, "The clergy are not very rigid in exacting from others the duties of religion, nor in setting an example of pious conduct in their own persons. On the contrary, the lower manners, the intemperate mode of life, and the free conversation of many of the monks, are a disgrace to the sacred office which they hold." (P. 14.)

The population of Madeira, which, including Porto Santo (see *post*), was ascertained by a late census to amount to 112,500, is a mixed race, sprung principally from Portuguese and Moors; but in Funchal many of the labouring classes show, by their English faces and complexions, that there has been a considerable intermixture with British settlers. Negro slaves, also, are still numerous; but they seldom intermarry with those of European origin. On the whole, the natives are a finer and more comely race than the Portuguese; they are of the middle size, well formed, and strongly knit, with masculine features, hair, and complexion. The women are almost universally under the standard height, and, when young, display handsome features, which, however, soon become coarse and unattractive, owing to their laborious field occupations. The men are dressed, somewhat in the costume of English sailors, with large full leather boots, and a little funnel-shaped cap on their heads. This curious head-gear is worn also by the women over the white muslin handkerchief, which covers the head and hangs down over the shoulders; and their gay chintz gowns, and scarlet pelerines give them a light and picturesque appearance. "It is delightful," says Mr. Wilde, " to see groups of these peasantry in companies of eight or ten sitting in some places under the umbrageous palms, eating their morning's meal, or completing their toilet, before entering the town, while others are hastening along, loaded with the various produce of their gardens, consisting of bunches of yellow bananas, strings of crimson pomegranates, &c., or carrying fowl, firewood, and fish to the market of Funchal, each little party preceded by its mandolin-player, who at times accompanies the wire-strung instrument with his voice, and is joined at intervals by the hearty chorus of the whole group." But the condition of a people is not to be learned from such holiday descriptions as this; and the truth is, that the native inhabitants of Funchal are meagre, sallow, and short-lived. "This," as Sir John Barrow has stated, " is not to be attributed to the climate, but to the poverty of their food, which chiefly consists of pumpkins, sour wine, or pernicious spirits; to a life of drudgery and exposure to the great vicissitude of climate, by daily ascending the steep and lofty mountains in search of fuel; and, above all, to a total disregard of cleanliness." In fact, all the natives of the island are infected with a species of itch, which they regard as incurable, and which is accompanied with a great degree of inflammation. (*Barrow*, p. 11.) Among the richer inhabitants are many Portuguese *fidalgos*; but by far the larger part are merchants and private residents belonging to almost every commercial country, especially Great Britain. These hold little intercourse with the other inhabitants, but live either in their town-houses at Funchal, or at their villas or *quintas* higher up the island, where they exercise the most liberal hospitality. A small tax on wine sent to England is levied by our consul, to form a fund for charitable purposes, which is farther increased by the benevolent contributions of the merchants, who also support an English episcopal church, the present minister of which (Mr. Lowe) is at once a zealous clergyman, and a scientific naturalist.

Funchal, the only town of Madeira requiring any special notice, is situated on the S.E. side of the island, and stretches nearly a mile along the margin of the bay. It is irregular, inconvenient, and meanly built, with narrow, crooked, steep, and dirty streets, some of which being paved with sharp-pointed pebbles, are extremely painful to walk upon. Streamlets of water run down some of the streets from the overhanging mountain; but when Sir John Barrow was here, the inhabitants washed their clothes, cleaned their fish, and threw their filth, offals, &c. into these streamlets; so that, in fact, they were rendered rather a nuisance than anything else; but it is stated in late works on the island, that these inconveniences have been in part obviated, and the filth of the town materially abated. The houses are commonly low, not often exceeding one story in height, with white outsides. Those belonging to the fidalgos or rich merchants are comparatively large and handsome, having at the top a *torrinha* or turret, commanding a view of the harbour, used for reconnoitring vessels as they arrive in the offing. The governor's castle is a large clumsy looking Gothic structure, near the beach. The cathedral has a *parvis*, or open space, before its W. door; and beyond it is the *Terreiro da Sé*, a pleasant promenade under several parallel rows of trees, enclosed by a low wall, and overlooked by pretty houses with balconies. In one of the wings of the Franciscan convent is a chamber, the walls and ceiling of which are formed or covered with human skulls and thigh-bones! The English church in the suburbs is an elegant and commodious building, literally embosomed in ever springing roses and white daturas. The convent of *Nossa Senhora do Monte*, amid groves of chesnut-trees half way

up the mountain, commands a very fine view. Funchal roads labour under several disadvantages: the anchorage is in 35 or 40 fathoms; land squalls are often extremely violent; a heavy surf on the beach, especially in spring, makes a landing at all times unpleasant, and sometimes unsafe, except in the shore-boats, in managing which the natives are very skilful. From the autumnal to the vernal equinox, when strong southerly gales throw in a heavy sea, the roads are peculiarly dangerous, and many accidents have then occurred.

A few inconsiderable islands in the vicinity of Madeira are included under its government. Of these Porto Santo, 35 m. N.W., is the only one that is inhabited. It has a parched barren aspect, and has but one fountain of good water. Its products comprise wine of an inferior quality, good barley, water-melons, and other fruits; but it is wholly destitute of wood. The town is insignificant, and is occasionally used as a place of exile from Madeira. The entire population of the island may be about 1400. The little islands called the Desertas, are occasionally visited by a few fishermen and smugglers, and the rest are mere rocks.

Madeira is said to have been discovered in 1344, by Machain, an Englishman, who was wrecked, and cast on its shores. But this story is very doubtful; and it seems most probable, that Juan Gonzales, who had been despatched on a voyage of discovery by Prince Henry of Portugal, and who fell in with this island in 1419, was its real discoverer. When discovered, it was uninhabited, and covered with wood, and was on that account called *Madeira*, that being the Portuguese term for timber. It was settled by the Portuguese in 1421, and has since continued in their possession. (*Robertson's America*, book i.) Its occupation by the English during the late war with France, being merely in order to prevent its falling into the hands of the French, it was restored to Portugal, at the peace of 1814.

MADELEY, a market town and par. of England, co. Salop, franchise Wenlock, on the banks of the Severn, 13 m. E.S.E. Shrewsbury, and 196 m. N.W. London. Area of par. 2750 acres. Pop., in 1841, 7368. The town, which is of considerable antiquity, and celebrated in history as having given refuge to Charles II. after the battle of Worcester, derives its present importance from its proximity to the great coal and iron district of Coalbrookdale. The church is a handsome modern structure, the living being a vicarage in private patronage. The Roman Catholics, Wesleyan, and Primitive Methodists have also their respective places of worship, and there is a meeting-house for the society of Friends. A national school is connected with the church, and there are four Sunday schools. The iron trade, carried on here to a considerable extent, is much facilitated by means of the Shropshire canal, which joins the Birmingham and Liverpool junction canal, and connects Madeley and the Ketley iron-works with the great manufacturing districts of Dudley, Wolverhampton, Birmingham, &c. About 2 m. W. of the town, and near the romantic village of Coalbrookdale, is a cast-iron bridge, erected in 1780, of one arch, 100 ft. in span, 40 ft. above the river, and containing 375 tons of metal, being the first structure of the kind raised in the kingdom. This beautiful rural district, embosomed between high and well-wooded hills, has within the last half century been converted into one of active mining and manufacturing industry, the furnaces now (1841) at work in this vicinity being estimated to produce 20,000 tons of iron a year. At Coalport, about 2 m. from the above mentioned bridge, is a considerable manufactory of china. Markets on Fridays: fairs May 29, and last Tuesday in Oct.

The neighbourhood of Madeley is remarkable for an extraordinary convulsion of the earth, that took place in 1773, when about 30 acres of land were shifted from their site, and broken into irregular chasms, large oak trees were uprooted, and the Severn, blocked up for more than 200 yards by the displaced soil and fallen trees, was compelled to find a new channel, in which it now flows.

MADISON, county, N.Y. Situated a little E. of the centre of the state, and contains 582 sq. m. Watered by head branches of Chenango, Unadilla, Osteic, and Toughnioga rivers. Bounded N. by Oneida lake, into which some of its streams empty. Water-lime and gypsum abound, and marl and bog iron ore are found. The Erie and Chenango canals cross the county. It contained in 1840, 42,191 neat cattle, 204,616 sheep, 20,757 swine; and produced 208,142 bushels of wheat, 2225 of rye, 171,204 of Indian corn, 5996 of buckwheat, 135,635 of barley, 342,307 of oats, 676,549 of potatoes, 107,280 pounds of hops, 215,619 of sugar. It had 160 stores, three furnaces, 26 fulling-mills, 10 woollen factories, two cotton factories with 744 spindles, four flouring-mills, 39 grist-mills, 187 saw-mills, three oil-mills, one paper-mill, one rope-walk, 40 tanneries, seven distilleries, one brewery, nine printing-offices, three binderies, two weekly newspapers, and two periodicals; eight academies, 968 students; 253 schools, 12,297 scholars. Pop. 40,008. Capital, Morrisville.

MADISON, county, Va. Situated a little N.E. of the centre of the state, and contains 330 sq. m. Bounded S. by Rapid Ann r. Drained by Robertson's r. It contained in 1840, 7006 neat cattle, 8964 sheep, 12,851 swine; and produced 100,680 bushels of wheat, 23,627 of rye, 271,980 of Indian corn, 33,005 of oats, 12,796 of potatoes, 148,700 pounds of tobacco. It had 12 stores, 10 flouring-mills, 23 grist-mills, 17 saw-mills, seven tanneries, five distilleries; two academies, 41 students; 17 schools, 356 scholars. Pop.: whites, 3799; slaves, 4308; free coloured 70; total, 8107. Capital, Madison.

MADISON, county, Ga. Situated in the N.E. part of the state, and contains 250 sq. m. Bounded E. by broad river, and drained by its branches. It contained in 1840, 1956 neat cattle, 1571 sheep, 3927 swine; and produced 8239 bushels of wheat, 53,130 of Indian corn, 6657 of oats, 3350 of potatoes, 727,118 pounds of cotton. It had nine stores, three grist-mills, one saw-mill; one academy, 26 students; three schools, 79 scholars. Pop.: whites, 3125; slaves, 1382; free coloured, 3; total, 4510. Capital, Danielsville.

MADISON, county, Flor. Situated in the central part of the territory, extending from Georgia on the N. to the gulf of Mexico on the S. Bounded E. by Suwanne river, W. by Oscilla river. It contained in 1840, 5690 neat cattle, 3069 sheep, 3969 swine; and produced 37,985 bushels of Indian corn, 2305 of oats, 13,913 of potatoes, 1150 pounds of rice, 702,400 of cotton, 1200 of sugar. It had three stores, two grist-mills, one saw-mill, two oil-mills; four schools, 74 scholars. Pop.: whites, 1442; slaves, 1202; total, 2644. Capital, Madison.

MADISON, county, Ala. Situated in the N. part of the state, and contains 760 sq. m. Watered by Flint river and other branches of Tennessee river. It contained in 1840, 23,074 neat cattle, 12,308 sheep, 79,970 swine; and produced 85,009 bushels of wheat, 1,357,808 of Indian corn, 152,816 of oats, 46,371 of potatoes, 15,655 pounds of tobacco, 10,358,897 of cotton. It had 40 stores, 22 grist-mills, 17 saw-mills, 11 tanneries, 11 distilleries, three printing-offices, two weekly newspapers; two academies, 125 students; 33 schools, 1167 scholars. Pop.: whites, 12,297; slaves, 13,265, free coloured, 144; total, 25,706. Capital, Huntsville.

MADISON, county, Miss. Situated a little S.W. of the centre of the state, and contains 548 sq. m. Bounded N.W. by Big Black river, S.E. by Pearl river. It contained in 1840, 21,045 neat cattle, 4906 sheep, 46,029 swine; and produced 1198 bushels of wheat, 877,893 of Indian corn, 79,995 of oats, 152,981 of potatoes, 14,849,153 of cotton. It had 19 stores, one printing-office, one weekly newspaper; one college, 100 students; two academies, 170 students; two schools, 135 scholars. Pop.: whites, 3086; slaves, 11,533; free coloured, 11; total, 15,530. Capital, Canton.

MADISON, parish, La. Situated in the N.E. part of the state, and contains 600 sq. m. Bounded E. by Mississippi r. Watered by Bayou Macon, and Tensas r. It contained in 1840, 6939 neat cattle, 767 sheep, 8641 swine; and produced 190,745 bushels of Indian corn, 13,188 of potatoes, 5,378,610 pounds of cotton. It had eight stores, three saw-mills. Pop.: whites, 1210; slaves, 3693; free coloured, 9; total, 5142. Capital, Richmond.

MADISON, county, Tenn. Situated in the W. part of the state, and contains 670 sq. m. Watered by the S. Fork of Forked Deer r. and its branches, and by the Middle Fork of the same river. It contained in 1840, 12,747 neat cattle, 17,563 sheep, 64,503 swine; and produced 55,178 bushels of wheat, 8530 of rye, 793,215 of Indian corn, 296,604 of oats, 65,236 of potatoes, 136,632 pounds of tobacco, 2,333,039 of cotton. It had 19 stores, one furnace, three woollen factories, two cotton factories, with 1200 spindles, 15 flouring-mills, 21 grist-mills, 13 saw-mills, seven tanneries, 10 distilleries, one printing-office, one weekly newspaper; four academies, 103 students; 13 schools, 512 scholars. Pop.: whites, 10,490; slaves, 6073; free coloured, 37; total, 16,530. Capital, Jackson.

MADISON, county, Ky. Situated centrally towards the E. part of the state, and contains 520 sq. m. Bounded N. by Kentucky r., and drained by its tributaries. It contained in 1840, 8361 neat cattle, 12,906 sheep, 49,104 swine; and produced 41,502 bushels of wheat, 90,902 of rye, 564,684 of Indian corn, 6430 of barley, 88,947 of oats, 16,952 of potatoes, 125,983 pounds of tobacco, 97,171 of sugar. It had 10 stores, two woollen factories, 23 grist-mills, three saw-mills, one tannery, 12 distilleries; eight schools, 234 scholars. Pop.: whites, 10,860; slaves, 5413; free coloured, 82; total, 16,355. Capital, Richmond.

MADISON, county, O. Situated a little S.W. of the centre of the state, and contains 400 sq. m. Watered by Little Darby and Deer creeks. It contained in 1840, 16,177 neat cattle, 15,632 sheep, 19,135 swine; and produced 47,646 bushels of wheat, 1918 of rye, 419,066 of Indian corn, 1287 of buckwheat, 71,173 of oats, 12,566 of potatoes, 1150 pounds of sugar. It had 22 stores, one flouring-mill, nine grist-mills, 15 saw-mills, two potteries; 94 schools, 2917 scholars. Pop. 9025. Capital, London.

MADISON, county, Ia. Situated a little N.E. of the centre of the state, and contains 390 sq. m. White river and its branches afford water-power. It contained in 1840, 5798 neat cattle, 6436 sheep, 21,579 swine; and produced 46,997 bushels of wheat, 1558 of rye, 375,715 of Indian corn, 60,687 of oats, 17,634 of potatoes, 47,887 pounds of sugar. It had two stores, one flouring-mill, seven grist-mills, 13 saw-mills, three tanneries, two distilleries, two potteries; 35 schools, 1120 scholars. Pop. 8874. Capital, Andersontown.

MADISON, county, Ill. Situated toward the S.W. part of the state, and contains 760 sq. m. Bounded W. by Mississippi r. Drained by Cahokia, Silver, and Wood creeks, and their branches. It contained in 1840, 22,030 neat cattle, 13,876 sheep, 53,484 swine; and produced 160,910 bushels of wheat, 1P 735 of rye, 1,370,735 of Indian corn, 12,355 of buckwheat, 72,980 of barley, 209,800 of oats, 121,305 of potatoes, 11,280 pounds of tobacco. It had four commission houses in foreign trade, 67 retail stores, two lumber-yards, seven flouring mills, eight grist-mills, 16 saw-mills, one oil-mill, two tanneries, two distilleries, one brewery, four printing-offices, three weekly newspapers, and one periodical; one college, 101 students; two academies, 144 students; 32 schools, 1445 scholars. Pop. 14,433. Capital, Edwardsville.

MADISON, county, Mo. Situated towards the S.E. part of the state, and contains 780 sq. m. Drained by St. Francis r. and its branches. It contained in 1840, 2453 neat cattle, 2995 sheep, 12,415 swine; and produced 9746 bushels of wheat, 158,510 of Indian corn, 36,331 of oats, 2031 of potatoes, 9850 pounds of tobacco. It had eight smelting houses, producing 1,263,455 pounds of lead, four grist-mills, four saw-mills, two tanneries, eight distilleries; 11 schools, 244 scholars. Pop.: whites, 2763; slaves, 611; free coloured, 23; total, 3395. Capital, Fredericktown.

MADISON, county, Ark. Situated in the N.W. part of the state, and contains 1050 sq. m. Drained by branches of White river. It contained in 1840, 3341 neat cattle, 2875 sheep, 11,634 swine; and produced 4132 bushels of wheat, 146,755 of Indian corn, 4518 of oats, 7949 of potatoes, 14,989 pounds of tobacco, 9690 of cotton, 1097 of sugar. It had five grist-mills, one saw-mill, two tanneries, two distilleries; two schools, 33 scholars. Pop.: whites, 2692; slaves, 63; total, 2775. Capital, Huntsville.

MADISON, p. v., Somerset co., Me., 40 m. N. Augusta, 635 W. Bounded W. by Kennebec r. Incorporated in 1804. It has four stores, two grist-mills, three saw-mills, two tanneries; 16 schools, 697 scholars. Pop. 1701.

MADISON, p. l., New-Haven co., Ct., 56 m. S. Hartford, 298 W. Incorporated in 1826 (formerly East Guilford). Bounded S. by Long Island sound. It contains a Congregational church, Lee's academy, named from its founder; 11 stores, one fulling-mill, one grist-mill; one academy, 60 students; 13 schools, 488 scholars. Pop. 1788.

MADISON, p. v., Madison co., N. Y., 95 m. W. by N. Albany, 365 W. Drained by head-waters of Chenango r. and Oriskany cr. It occupies the summit level of the Chenango canal, which passes through the town. It contains two churches, a Baptist and a Methodist universalist; four stores, one fulling-mill, two grist-mills, nine saw-mills, two tanneries, one distillery; 16 schools, 398 scholars. Pop. 2344.

MADISON, L., Columbia co., Pa., 10 m. N. Danville. Drained by Little Fishing and Mahoning creeks. It has one store, seven saw-mills, three tanneries; six schools, 212 scholars. Pop. 1700.

MADISON, L., Armstrong co., Pa. It has three stores, five grist-mills, four saw-mills; three schools, 64 scholars. Pop. 1365.

MADISON, L., Perry co., Pa. It has three stores, one fulling-mill, one woollen factory, six flouring-mills, 10 saw-mills, two distilleries; seven schools, 315 scholars. Pop. 1369.

MADISON, p. v., capital of Morgan co., Ga., 41 m. N.N.W. Milledgeville, 695 W. It contains a courthouse, jail, a Masonic hall; two academies, a male and female, with a library; a Presbyterian and a Methodist church; several stores, 70 dwellings, and about 450 inhabitants.

MADISON, p. t., Lake co., O., 190 m. N.E. by N. Columbus, 349 W. Watered by Grand r. Bounded N.W. by lake Erie, on which is a harbor. Large quantities of iron are manufactured into hollow ware, mill irons, and other articles, and are exported. It contains nine stores, two grist-mills, nine saw-mills, three tanneries, one distillery; 29 schools, 1250 scholars. Pop. 9800.

MADISON, t., Clark co., O. It has 11 schools, 309 scholars.

MADISON, t., Columbiana co., O. It has three schools, 100 scholars. Pop. 1474.

MADISON, t., Butler co., O. It has four schools, 111 scholars. Pop. 1935.

MADISON, t., Franklin co., O. Watered by Alum and Big Walnut creeks. Pop. 1810.

MADISON, t., Fairfield co., O. It has one store, six grist-mills, eight saw-mills. Pop. 1102.

MADISON, t., Guernsey co., O. It contains Winchester

and Antrim villages, and has one college, 16 students: four schools, 171 scholars. Pop. 1922.

MADISON, t., Licking co., O. Watered by Licking river. The Ohio canal passes through it. It has three schools, 74 scholars. Pop. 1119.

MADISON, t., Highland co., O. It has nine schools, 707 scholars. Pop. 1916.

MADISON, t., Perry co., O. It has one store, one woollen factory, two flouring-mills, four grist-mills, five saw-mills, two distilleries; six schools, 140 scholars. Pop. 1167.

MADISON, p. v. and city, capital of Jefferson co., Ia., 68 m. S.S.E. Indianapolis, 580 W. Situated on a N. bend of the Ohio river, above the reach of the highest floods. It is regularly laid out with broad streets, several of which are paved. A wharf is constructed for the landing of steam-boats. It is handsomely built, mostly with brick, and has a courthouse, jail, market-house, and six churches, a Methodist Episcopal, a Reformed Methodist, a Baptist, an Episcopal, and two Presbyterian; a branch of the state bank, with a neat edifice, a saving's bank, an insurance office, 50 stores, two iron foundaries, a steam engine factory, a cotton factory, a steam flouring-mill, an oil-mill, a grist-mill, a saw-mill, and a boat-yard. Fifteen thousand hogs have been slaughtered here in a year. Many goods are sold at wholesale. It is the second place in population in the state, and in 1840 contained 3798 inhabitants. In the rear of the village are hills 250 feet high, which afford a delightful prospect.

MADISON, t., Putnam co., Ia. It has two grist-mills, three saw-mills. Pop. 1071.

MADISON, p. v., capital of Dane co. and of the territory of Wisconsin. 90 m. N.E. Galena, 90 m. W. of Milwaukie, 847 W. It is pleasantly situated between the third and fourth lakes, of the chain called "the Four Lakes," on a gently rising ground, from which there is a regular descent each way to the water, affording a beautiful water-prospect. It is regularly laid out, with a central square, 914 feet on a side, in the centre of which the statehouse has been erected by the general government. It is a spacious store edifice, two stories high above the basement, with a handsome dome, and can be seen for the distance of ten miles in every direction. The square is surrounded by a fence with 16 gates, for which Congress has appropriated $1000. There is a lot, and several places of worship, and two printing-offices, each of which issues a weekly newspaper, 10 or 12 stores, 60 dwellings, and it had in 1840, 376 inhabitants.

In 1837 the contractor with 40 men and 5 females cut their way, in 11 days, from Milwaukie, through the wilderness, and commenced building the capitol. The distance is now travelled in two days. Since the spring of 1841 the growth of the place has been rapid, and it has a substantial prosperity.

MADISON, C. H., capital of Madison co., Va., 97 m. N.W. Richmond, 98 W. It contains a courthouse, jail, two churches, an Episcopal, and one free to all denominations; six stores, five flouring-mills, two tanneries, 40 dwellings, and about 300 inhabitants.

MADISON. C. H., p. v., capital of Madison co., Flor. It contains a courthouse, and a few dwellings.

MADISON, p. v., capital of Hopkins co., Ky., 127 m. W.S.W. Frankfort, 729 W. It contains a courthouse, jail, and 51 inhabitants.

MADISONVILLE, p. v., capital of Monroe co., Tenn., 172 m. E.S.E. Nashville, 540 W. Situated on Bat cr., which enters into Little Tennessee river. It contains a courthouse, jail, an academy, three churches, six stores, a printing-office, and about 500 inhabitants.

MADISONVILLE, p. v., and seaport, St. Tammany par., La., 35 m. N. New-Orleans, 1137 W. Situated on the W. side of Chifuncier r., 3 m. from its entrance into lake Pontchartrain. It has the best harbour on the lake. Packet schooners ply daily to New-Orleans.

MADRAS (PRESIDENCY OF), an extensive division of British India, being the second in rank and the most southerly of the three presidencies. It comprises, with its tributary states, the whole of Hindostan S. of the river Krishna, the N. Circars, and Canara. It extends from 8° to 20° N. lat., and from 74° to 85° E. long. It is of a triangular shape; the base of the triangle being formed by a line drawn from Ganjam, on the coast of Coromandel, to Sadasharagur, near the 15th degree of lat., on the coast of Malabar, the sides by their coasts, and the apex by Cape Comorin, at the southern extremity of India. It is consequently bounded on two of its sides, the E. and W., by the ocean, while on the third, or N., it has the dominion of the Nizam and the Rajah of Berar, parts of the presidencies of Bengal and Bombay, and the Portuguese territory of Goa.

In greatest length, N. to S., is about 950 m. Its area, population, subdivisions, &c. are specified in the following Table:—

Districts.	Area in sq. m.	Pop. 1836-7.	Pop. to sq. m.	Land Revenue, 1836-7.
				Rupees.
Ganjam	5,700	688,079	118	693,567
Vizagapatam	6,000	1,847,414	187	1,304,008
Rajahmundry	4,680	579,528	184	1,767,157
Masulipatam	4,810	532,039	60	944,877
Guntoor	4,960	19,318	104	1,278,088
Nellore and Ongole	12,000	816,572	113	1,289,614
Arcot, N. division	8,300	506,431	30	1,975,496
— S. division	4,300	650,239	47	1,975,496
Chingleput	2,253	806,319	308	875,352
Madras	30	630,960		65,364
Salem	8,518	905,190	112	1,644,713
Coimbatoor	8,322	807,964	96	2,064,913
Trichinopoly	3,199	554,730	175	1,431,956
Tanjore	3,625	1,129,798	331	3,457,755
Madura	7,456	1,135,411	148	1,750,791
Tinnevelly	5,590	950,891	162	1,660,489
Bellary	12,703	1,112,839	87	2,170,804
Cuddapah	13,783	1,069,164	89	1,912,473
Malabar	6,262	1,140,916	182	1,612,982
Canara	7,477	750,776	94	1,671,216
Total British territ. Tributary States.	138,360	14,804,3o1	113-6	31,527,440
Mysore	29,400	2,671,754	77	
Travancore and Cochin	9,466	1,125,000		
Coorg	2,340	88,000	130	
Grand Total of Madras Presidency	172,028	18,314,805	106-4	

Physical Geography. Mountains.—The surface consists of a central table-land, surrounded on all sides by an undulating or plain country gradually diminishing in elevation as it approaches the sea. The mountain-ranges bounding the table-land on either side are the E. and W. Ghauts, which diverge from each other at the knot of mountains termed the Neilgherries, in about 11° N. lat., and from 78° 30' to about 77° E. long. The W. Ghauts approach much nearer to the sea than the E., so that there is a much greater extent of plain country in the E. than in the W. portion of the presid. The Neilgherry hills, which may be considered the nucleus of the mountain system in S. Hindostan, extend 34 m. E. to W. by 15 m. N. to S., having numerous peaks rising to between 5000 and 6000 feet, and one, Dodabettas, estimated at 8760 feet above the level of the sea. The W. Ghauts are more continuous and generally more elevated than the E.: the latter, even in the district of Salem, where they are highest, seldom attaining to an elevation of 6000 feet, while the former frequently rise 9000 feet higher. The table-land above or between the Ghauts averages in Coorg nearly 5000 feet in elevation, and, in Canara, Balaghaut varies from 3000 to 5000 feet: but it decreases rapidly in height as we proceed E. and N., and even in Mysore, Bangalore is only 2907 feet, and Hurryhur only 1831 feet, above the sea. S. of the Neilgherries, is the Paulghautcherry Pass, in Coimbatoor, 16 m. in width, extending from sea to sea, and forming a complete break in the mountain-system of S. India. S. of this pass, a mountain-chain, little inferior in height to the Neilgherries, stretches nearly due S. to Cape Comorin. This chain separates Cochin and Travancore, on the W., from the district of Madura and Tinnevelly, on the E. The Ghauts elsewhere form the chief line of separation between the British territories and those of the subsidiary states.

The principal rivers are the Godavery and Krishna, with their tributaries; and the Pennar, Palaur, Punnair, Cavery, Coleroon, and Vighey. These have all an E. course, and disembogue on the Coromandel coast. The three principal have been already described. (Vol. i., 588, 1000, ii. 132.) The only other river worthy any particular notice, the Coleroon, is the N. branch of the Cavery, which, having separated from the latter, opposite Trichinopoly, bounds the district of Tanjore on the N., and falls into the sea about lat. 11° 30'. The streams running W. have short courses; the longest is the Ponany, which traverses the Paulghautcherry-pass, but it is of little use for navigation, being very shallow in the dry season. There are no lakes of any importance: that of Colair, in Masulipatam, is the principal. There are numerous salt lagoons, or inlets of the sea, on the Coromandel coast, but they are of little use for navigation; and the whole of the Coromandel coast has a shelving shore, and is beat by so heavy a surf, as to be at all times difficult to reach, and during the monsoon it is quite unapproachable. The inlet of Cochin, on the Malabar coast, is not within the British territory. The Malabar coast within this presidency is also very destitute of good harbours.

The Climate—differs widely in the different portions of this presidency. The W. coast is exposed to all the fury of the S.W. monsoon, during which the rains are excessive, and often accompanied by heavy squalls and thunder storms. On the opposite coast, the rains are, on the contrary, brought in by the N.E. monsoon, a circumstance explained by the fact, that the Ghauts are elevated enough to intercept the passage of the clouds. The N.E. monsoon lasts from October to March: but the monsoon rains are over in December; and much less rain falls on the Coromandel than the Malabar coast, where, as in Canara, the annual fall of rain is sometimes 114 inches. The quantity falling in Coimbatoor, in 1836-7, was only nine inches, and in 1837-8, 22·1 inches. The heat is much more oppressive on the E. than the W. side of S. India, owing to the greater prevalence of dry weather and parching winds. If we may depend on the statement of Berghaus, the average annual temperature of Pondicherry, lat. 11° 55', is no less than 84° 7' Fah. (29·6 centig.), that of Madras being 82°. (*Allg. Länder, &c.*, i. 230.) At the mouth of the Krishna, in the N. Circars, in about 16° lat., the thermometer has been known to stand at 108° Fah. at midnight! (*Hamilton's E. I. Gaz.* i., 418.) The plain country, in the E. part of the presidency, is frequently very unhealthy; but on the Malabar coast this is not the case. The country above the Ghauts, which has a mean temperature many degrees below that of the plains, is decidedly salubrious; it derives rain from both monsoons, having an equable climate, and an atmosphere usually clear, serene, and highly invigorating.

The Geology—of S. India has been noticed with that of the rest of Hindostan. (I. 993.) Syenite, granite, quartz, greenstone, mica, and hornblende are among the chief primitive rocks, in the Ghauts, Neilgherries, &c. The upper soil on the coasts is usually sandy, and not very productive; but in the valleys of the interior, it frequently consists of a rich alluvium or loam. The soil of the Balaghaut districts, N. of Mysore, consists principally of the red and black earth, so prevalent in the Deccan.

Natural products.—Many portions of the soil in the tableland are highly impregnated with carbonate of soda, nitre, and other salts; iron is generally plentiful, and the iron ore of the district of Salem is extremely rich. Copper is found in Nellore, and a few other districts, and diamonds near Cuddapah. The presidency yields no other mineral products of much value. A considerable extent of surface, especially in the upper part of the country, is covered with forests, comprising teak, sandal, ebony, and other valuable timber trees. Teak grows on the E. as well as the W. Ghauts; but that of the Malabar coast is the most available, and best known, in the market; a good deal being floated down to the coast by the small rivers, and sent to Bombay and elsewhere for ship-building. The toddy-palm (*Borassus flabelliformis*), cocoa-nut tree, the products of which form important articles of export from the W. districts; and other palms flourish on the sandy coast lands, which supply few other useful articles. The sugar cane, areca, yam, plantain, tamarind, jack, mango, melons, and various other fruits, ginger, turmeric, cotton, hemp, &c., some of which are indigenous, and pretty generally grown; pepper is an important article of culture on the Malabar coast, and Coimbatoor is celebrated for the excellence of its tobacco. Rice, paddy, wheat, barley, maize, and all the other grains common in India, both wet and dry, are here cultivated: the first is grown chiefly on the plains of the coast; but it forms also the chief export of Coorg, though a high country, and is the great staple of Canara. The Balaghaut districts are almost wholly appropriated to dry grain cultivation.

Animals.—The elephant, tiger, cheetah, bear, bison, elk, spotted deer, antelope, jackal, wild hog, jungle sheep, &c., inhabit this as well as other parts of India; tigers, however, are not so numerous as in the countries watered by the Ganges, and other low and jungly portions of Hindostan. Ivory is a product of some consequence in Coimbatoor; from 700 to 800 elephants being destroyed in that province between 1832 and 1836. Domestic animals are most numerous in the E. and S. districts; Guntoor is celebrated for its cattle; and Coimbatoor for its sheep, which are not hairy and long-legged like those of the Carnatic, but small, yielding good mutton and coarse wool, made into common sorts of clothing, carpets, &c. Live stock, above the Ghauts, is scarce and inferior.

Land-tax, &c.—With the exception of the N. Circars, (see CIRCARS, NORTHERN,) the greater portion of the territories included in the Madras presidency, are assessed for the land-tax on what has been called the *ryotwar* system. It is unnecessary, however, after the copious details we have already given in the article on British India (see *ante*, p. 13), to enter, in this place, into any further investigations with respect to the nature and operation of that settlement. It is sufficient to say that the land-tax in Madras is oppressive in amount; and that the system under which it is assessed, being subject to perpetual changes is, in fact, subversive of the security of property, and consequently of all industry, except what is indispensable to meet immediate wants. We do not mean to impeach the motives, or to depreciate the talents, of Sir Thomas Munro, and the other individuals most instrumental in the establishment of this system; but we do not well see, supposing they had set about devising a scheme for paralysing enterprise, and creating an insuperable obstacle to all improvement, how

they could have hit upon one better fitted to accomplish such objects than the ryotwar assessment. It appears to have every quality that an assessment should not, and not one that it should, have; and it were idle to expect that either the revenue of the presidency, or the industry and condition of the inhabitants, should be materially improved, so long as it is permitted to shed on all sides its withering influence.

Agriculture, &c.—The imposition of an oppressive assessment is the more severely felt, as the land in the Madras presidency is generally much less fertile than in Bengal and many other parts of British India. Tanjore may be said to be the granary of the presidency, and produces the greatest land revenue. The widest breadth of cultivated land is met with in Rajahmundry, Tanjore, and Coimbatoor. The modes of agriculture pursued in the different provinces will be found briefly noticed in the articles which have especial reference to them. Generally, however, it may be said that agriculture is at a very low ebb; that the occupiers, ground down by oppressive taxes, are for the most part miserably poor, and their implements and stock alike bad. Irrigation is extensively practised; and wherever a sufficient supply of water (whether from rivers, tanks, or wells) can be commanded, as in the delta of Tanjore, S. Arcot, &c., the crops of rice are very heavy. The land under dry grains is generally manured; and cow dung used as fuel in this presidency being subject to a tax, it is generally used as manure. Opium is rarely or not at all grown; and indigo only in small quantities, principally in the N. districts. Coimbatoor exports annually upwards of 4000 candies of tobacco to Malabar, Cochin, and Travancore; and large quantities to Trichinopoly and Mysore. The superiority of the tobacco grown in this province is attributed to the soil containing much saltpetre and peroxide of iron, as well as to the attention bestowed on its culture. The exhaustion of the land, from its cultivation, is, however, very great; the ground consequently requires frequent and regular manuring, and is cultivated every other year with dry grains. Tobacco costs on the spot where produced, about 25 rupees per candy. Cotton is a staple product of Tinnevelly; and it and sugar are raised in various other places.

Manufactures, &c.—The chief are those of cotton cloth; and formerly cotton fabrics and other piece goods were largely exported, especially from the N. Circars; latterly, however, the lower price and better quality of British piece goods have enabled them, to a great extent, to supersede those of India in most foreign markets; though the latter are still exported, especially from Tinnevelly to the W. Indies and America. The natives have recently turned their attention to the imitation of English cottons, and in some instances, it is said, with considerable success. The muslins of Chicacole, the woollen carpets of Ellore, and the silks of Berhampore (Ganjam), are of old celebrity; but in general manufacturing industry flourishes most in the S. districts, and the cloths of Madura are highly esteemed for their fine red dye. The state of manufactures appears to depend in a great degree on the state of the roads, and means of communication. In the S. provinces the government has completed several good carriage roads, and in the N. they are also pretty good. Canara, on the other hand, may be said to be wholly without roads, and vehicles are unknown. The Malabar coast has a singular paucity of manufactures: its chief wealth arises from its large export of rice to Arabia and Bombay, and of pepper and other spices, areca, cocoa-nuts, &c. A good deal of iron is made in Tinnevelly; and saltpetre and salt are made in various parts; but the latter are inferior to those of the Bengal presidency. Above the Ghauts the arts are in a very rude state.

Weights and Measures.—At Madras, the maund of 40 seers or 8 ris = 25 lbs. avoird.; the candy of 20 maunds = 500 lbs.; the *garse* for grain = 19·8 mds. At Trichinopoly, the *seer* for metals = 9 oz. 5½ dr. In Malabar, the *tolam* of 40 seers = 23 lbs. 3 oz.; the foot = 10·46 in. At Madras, the *maunzy* = 3400 sq. ft.; the *cawnzy* of 24 *maunzy* = 1·3223 acres. (*Madras Almanack*, 1839.)

The government is vested, as in Bombay, in a governor, subordinate to the governor-general of India. He is assisted by a council of three members, one being the commander in chief, and three secretaries, placed over the revenue and judicial, political, and military departments. In each of the 20 districts there is a European collector, who exerts also the chief magisterial power. Zillah courts are holden in the principal towns of most of the districts; and there are four provincial courts of appeal at Chittoor, Masulipatam, Trichinopoly, and Tellicherry. In Madras is a court of Sudder and Foudjarry Adawlut, an admiralty court, and the high court of judicature for the presidency. The church of England ecclesiastical establishment consists of the bishop and archdeacon of Madras and 19 chaplains, in different parts of the presidency. There are numerous Protestant-dissenting and Roman Catholic chapels, Madras being the see also of a Roman Catholic bishop. According to the government returns of 1836, about 168,600 children were receiving instruction at the schools within the presidency. The Madras military force, according to recent returns, consists of 56,957 men, of whom about 9900 are Europeans.

ACCOUNT of the revenues of the presidency of Madras during the four years ending with 1838-39.

Heads of Revenue.	1835-36.	1836-37.	1837-38.	1838-39.
	Rupees.	Rupees.	Rupees.	Rupees.
Mint duties	16,397		23,807	22,361
Post-office collection	3,01,576	5,80,389	3,12,577	3,50,616
Stamp duties	4,40,543	4,54,688	4,54,480	4,54,013
Madras Town assessment	1,03,069	98,894	1,04,777	2,918
Miscellaneous civil receipts		2,28,9-6	1,54,717	2,96,973
Judicial fees and fines	2,54,797	2,34,506	2,26,795	2,3.',176
Land revenue	3,13,92,179	2,99,26,176	3,14,97,812	3,23,73,016
Abkarry, and small farms and licenses	16,72,945	17,61,896	17,80,291	18,46,620
Moturpha	10,88,466	9,89,047	10,34,885	10,47,397
Miscellaneous receipts in the revenue department	2,19,568	2,40,700	1,86,147	1,60,627
Customs, sea, and inland	44,62,145	46,94,486	46,56,708	60,72,051
Sale of tobacco	7,53,827	7,56,311	8,67,22	8,43,946
Ditto of salt	37,80,972	37,28,946	37,22,073	40,29,584
Marine duties	1,61,200	1,50,682	1,96,617	1,44,173
Profits of the Madras government bank	9,22,415	1,24,288	98,170	1,27,629
Subsidies from Mysore, Travancore, and Cochin	26,93,699	25,95,699	26,95,339	55,36,699
Total gross revenue. Rup.	4,88,98,129	4,61,92,070	4,81,94,324	4,95,30,323
Deduct allowances and assignments, &c., payable out of the revenue	60,94,574	55,77,900	50,56,477	67,07,194
Charges for collecting the revenue	4,39,77,556	4,06,04,118	4,30,97,714	4,24,43,049
	87,45,807	60,96,800	88,97,748	61,29,180
Total nett revenue, after payment of allowances and assignments, and charges of collection	3,52,34,851	3,45,07,912	3,74,70,004	3,76,96,399
Nett sale produce of goods and dead stock	93,729	1,089	14,709	496
Total revenue and receipts. Rup.	3,53,37,779	3,45,08,059	3,74,84,767	3,76,94,394
	3,632,775	2,466,979	3,348,476	3,769,428

EXPENDITURE of the Presidency of Madras during the Four Years ending with 1838-39.

Heads of Expenditure.	1835-36.	1836-37.	1837-38.	1838-39.
	Rupees.	Rupees.	Rupees.	Rupees.
Civil and political charges	34,81,214	34,14,632	32,65,306	34,10,890
Judicial and police do.	29,43,857	28,14,870	26,10,644	26,16,063
Military and miscellaneous do.	2,57,15,428	2,30,71,847	2,48,58,945	2,32,91,398
Interest on debt	6, 0,790	6,51,860	5,76,291	4,65,861
Total charges. Rupees.	3,01,46,349	3,00,53,809	3,27,36,136	2,98,51,619
Do. L.				
Surplus of revenue over expenditure Rupees	51,92,430	44,55,218	52,48,571	48,12,768

History.—In the article INDIA, BRITISH, will be found a table, showing the dates of the successive augmentations to the British possessions in the east. The city of Madras, with a territory 5 m. along shore by 1 m. inland, granted us in 1639, formed the first nucleus of our eastern empire. But we may here notice the chief successive acquisitions under the Madras presidency. The Jaghire, or Chinglepat, was obtained by the E. I. Company from the nabob of Arcot, in 1750 and 1763. In 1799, Malabar, Canara, Coimbatoor, Dindigul, Salem, the Barramahl, &c., were acquired by conquest from the sultans of Mysore; in 1800 the Balaghaut districts were ceded; and in 1801, the remainder of the nabob of Arcot's territories were added to the foregoing. (Madras Almanacks for 1838-39; Parl. Reports and Papers; Hamilton's Hindostan and E. I. Gazetteer.)

MADRAS, a marit. city of Southern India, cap. of the above presidency, is the district of the same name, on the Coromandel coast, 650 m. (direct distance) S.E. Bombay, and 670 m. S.W. Calcutta; lat. 13° 4′ 49″ N., long. 80° 21′ E. The area of the district or collectorate of Madras is only 30 sq. m.; but its pop. in 1836-37 amounted to 638,000; and the pop. of the city and its immediate environs, within a radius of perhaps 2½ m. round fort St. George, is usually estimated at upward of 400,000.

Madras is in all respects badly situated: it is almost wholly unapproachable by sea. "There being no indentation on the coast, nor any island to break off the surge, a heavy swell rolls in throughout the year. Vessels anchor in the open roads; the large ones keeping a mile or two from shore. The swell keeps them pitching and rolling as uncomfortably as when at sea. The danger is so great during the S.W. monsoon that vessels are not allowed to lie here for several months, and the anchorage seems deserted. Cargoes are loaded and unloaded by boats adapted for passing through the surf: these, called catamarans, consist of three flattened timbers, 8 or 10 feet long, tied together horizontally, and sharpened a little at the point. One or two men propel it with a paddle, seated at both ends, and dip first on one side and then on the other. When no boat could live five minutes, these catamarans go about in perfect safety. The men are often washed off, but instantly leap on again without alarm. A waterproof cap, for the carriage of letters to and from newly arrived vessels, is almost their only article of dress. The boats used are large and deep, made without ribs or timbers, of thin wide planks, warped by fire to a proper shape, and fastened together by strong twine. Against the seams straw and mud are fastened strongly by the twine, which ties the planks together. No nails are used, for none would keep a boat together with such thumping. The boatmen display energy and skill scarcely to be surpassed. Keeping time to a rude tune, they now take long, and now short pulls, as the waves run past; they at length push the boat forward on a foaming surf, and she is thrown upon the beach." (Malcolm's S. E. India, i., 53.) These being no pier of any description, passengers and merchandise have all to be landed in the rough way now described.

Madras presents, from the sea, nothing to create great expectations. Only a few public buildings are visible, and not much of the town, as the site is quite level. It is, however, a noble city, and has many fine streets. Fort St. George may be considered the great nucleus and centre of Madras. It is neither so large nor so regular as fort William, at Calcutta; but it is strong, and has the advantages of requiring a smaller garrison, and of being easily relieved by sea. It occupies a semicircular area, rather more than ¼ m. in length, by from 2 to 3 furlongs in width, in a commanding situation, immediately on the beach; and is surrounded by an esplanade traversed by roads, and shaded public walks. Within it were formerly, besides many public offices, some streets of private European dwellings, shops, and stores; but these have been mostly cleared away, and the fort now contains only the barracks, arsenal, a bazaar for the supply of the garrison, the council-house, the old church, the exchange, on which a lighthouse with a lantern 90 feet high is erected. The merchants and tradesmen have mostly removed their establishments to the new streets, opened in the N.E. quarter of the Black-town, and along the skirts of the esplanade. The Black, or native town, which is N. and N.E. the fort and esplanade, is well laid out, and is defended by a substantial brick wall. "The houses are far better, at an average, than those of the natives in Calcutta. Though there are not so many fine residences of rich baboos as in that city, there are some scarcely surpassed in elegance by any in America." (Malcolm, i., 54.) It has probably been improved of late years. Hamilton, in his E. I. Gazetteer says, it is irregular and confused, being a mixture of brick and bamboo houses, and makes a better appearance at a distance, than when closely inspected. A fine range of public edifices, including the custom-house, office for the board of trade, courthouse, granary, and many store-houses, &c., forms its frontage towards the beach, protected from the fury of the surf by a breakwater of massy stones. The front of this terrace, and the drives on the esplanade, form the chief promenades of the inhabitants.

Madras differs from Calcutta, in having properly no European town, except the few houses within the fort. Most of the European settlers reside in suburban houses, and repair in the morning to their offices in the Black-town, returning in the afternoon. Their residences are chiefly on the Choultry plain, a large extent of surface, S.W. of the fort, and separated from it by the river Triplicane, which, in the neighbourhood of the city, is crossed by numerous bridges. The houses all stand in large plots of ground, shaded by trees, and divided by hedges of bamboo or prickly pear. Few are of more than one story, but they are in a pleasing style of architecture, having their porticoes and verandahs supported by stuccoed pillars. According to Heber, the rooms are not quite so large as those of the houses in either Calcutta or Bombay, but they are more elegant and agreeable. On the Choultry plain, near fort St. George, is the governor's residence, a large building, with a spacious banqueting-hall; but opinions vary greatly as to its architectural merit. Heber says, it has some bad paintings of Coote, Cornwallis, Meadows, &c., and one good one of Sir E. Strange, but all are fast going to decay from the moisture of the sea-breeze. Near it are the Chepauk gardens, in which is the residence of the Nabob of the Carnatic; and adjacent to these, is a mosque of grey-stone, with five arches in front, and two handsome minarets, the only Mohammedan structure of any note at Madras. The descendants of the former Portuguese inhabitants chiefly reside at San Thomé, a suburb on the shore, about 3 m. S. from the fort, which has a small cathedral, and two neat chapels under the charge of a Portuguese bishop, and a few priests from Goa. The Protestant places of worship are St. George's cathedral on the Choultry plain, four other episcopal churches and chapels, a Scotch and an Armenian church, and Independent, Wesleyan, and Unitarian chapels. There are also three Roman Catholic churches. The number of native Christians is, however, stated to be very small, though increasing. There are male and female orphan asylums, many schools, and other charitable institutions, conducted in a manner that has been highly eulogised; and numerous missionary establishments, both European and American.

Madras is the seat of all the chief government offices for its presidency, of the supreme court, a board of revenue, marine board, &c. In consequence of its unfortunate maritime position, it has less foreign trade than the capitals of either of the other presidencies. Its commerce is still, however, considerable, as it is the principal emporium of the Coromandel coast, and trades direct with Great Britain, and other European countries, the United States, the South American states, China, the Eastern Islands, the Birman empire, Calcutta, and Ceylon. The principal articles of import are rice, and other grain, chiefly from Bengal; cotton piece-goods, iron, copper, spelter, and other British manufactures; raw silk, from Bengal and China, with betel or areca nut, gold dust, tin, and pepper, from the Malay countries; and rice and pepper from the coast of Malabar, with teak timber from Pegu. The exports consist of plain and printed cottons, cotton-wool, indigo, salt, Ceylon pearls, chank shells, tobacco, soap, natron, some dyeing drugs, and coffee, from the table-land of Mysore, the quantity of which is increasing. The great staples of sugar, rice, opium, saltpetre, and lac dye, of such importance in Bengal, are hardly known as exports here. The importation of sugar from foreign countries is prohibited at Madras. (See *Report on E. India Produce*, 1840.)

In Madras roads, large ships moor in from 7 to 9 fathoms, with the flagstaff of the fort bearing W.N.W. 3 m. from shore. From Oct. to Jan. is generally considered the most unsafe season of the year, in consequence of the prevalence of storms and typhoons. On the 15th of Oct. the flagstaff is struck, and not erected again till the 15th of Dec., during which period a ship coming into the roads, or, indeed, anywhere within soundings on the coast of Coromandel, vitiates her insurance. The light within the fort may be seen from the deck of a large ship at 17 m. distance, or from the mast-head at a distance of 26 m. By the port regulations, no articles are to be shipped or landed without a permit, or after 6 P.M. Any merchandise attempted to be landed without the prescribed forms, or that is not entered in the manifest, is liable to double duty; and where a fraudulent intention shall appear, to confiscation. All goods (except on account of the E. I. Company) are to be shipped or landed at the ghaut opposite to the Custom-house, or pay double duty. Goods exported in British or native vessels are exempted from duty, but they must, nevertheless, pass through the customs' books.

Meat, poultry, fish, and other provisions, are to be procured for shipping at Madras, but they are neither so good nor so cheap as in Bengal. Wood and fuel are rather scarce, and dear in proportion. Water is of very good quality. On account of the dearness of provisions, wages are considerably higher than at Calcutta; and comparatively few servants are kept. The style of living is much the same in Madras as at Calcutta, but visiting is not carried on upon so extensive a scale. In the cool season monthly assemblies are held in the Pantheon, a building erected in the suburb of Vepery, and occasional balls take place throughout the year. During the cool season, also, races are held at St. Thomas's mount, about 7 m. from Madras. The road to the racecourse is certainly the finest in India, and shaded by trees through its whole length. At the foot of mount St. Thomas is the principal cantonment for the artillery of the Madras army, with a noble parade ground, considered one of the best military stations in S. India.

Madras experiences less extreme heat than Calcutta, taking the average of the year, though so much nearer the equator. The minimum temp. in Jan. 1837, was 65° Fah.; the maximum in May of the same year, 99°: the mean annual temp. was 81·7. Several extensive tanks and some swamps surround the city and its territory; but Madras is not said to be particularly unhealthy.

The territory on which Madras is situated formed the first acquisition made on the continent of India by the British, who obtained it by a grant from the rajah of Bijnagur in 1639, with permission to erect a fort thereon. The latter, which was forthwith built, was besieged in 1702 by one of Aurungzebe's generals; and in 1744 by the French

under M. de la Bourdonnais, to whom it surrendered after a bombardment of three days. It was restored to the English at the peace of Aix la Chapelle, and sustained, with credit and success, a memorable siege by the French under Lally in 1758-9; since which it has experienced no hostile attack. (*Hamilton's E. I. Gaz.; Modern Traveller, x.; Malcolm's Travels in S. E. Asia; Madras Almanack; Parl. Reports; Commercial Dict., &c.*)

MADRID, a celebrated city, and the modern cap. of Spain, in the centre of the kingdom, on the Manzanares, a tributary of the Tagus, 39 m. N. by E. Toledo, the former cap. 330 m. E.N.E. Lisbon, and 240 m. S.W. Bayonne; lat. 40° 25' N., long. 3° 38' 15" W. Pop., in 1826, according to Miñano, 181,400, exclusive of about 20,000 occasional residents, designated *forasteros*; but the present pop. is estimated by good authorities to amount to about 210,000, exclusive of 26,000 *forasteros* and foreigners; making a total of 236,000. This city, which, till the time of Philip II., was little more than an obscure country town, stands in a stony barren district, more than 2000 feet above the sea, having no navigable river near it, and scarcely any potable water, and being, at the same time, extremely cold in winter and unbearably hot in summer; the thermometer, at the former season, falling to 18°, and during the latter, rising to 110° or 115° Fahr. This variableness of temperature, combined with the prevalence of piercing E. and N.E. winds during the greater part of the year, renders the climate very unhealthy, and especially prejudicial to persons threatened with pulmonary complaints, some thousands of whom are said to have died during the winter of 1829–30. (*Cook's Sketches*, i., 176.) All authors, indeed, agree that it would have been difficult to fix on a more unfavourable site. "From the Somo-Sierra," says Inglis, "to the gates of Madrid, a distance of nearly 30 m., not a tree, garden, nor country house is to be seen, scarcely an isolated farm-house or cottage, and only three or four very inconsiderable villages. The land is chiefly uncultivated, and even that part under tillage and producing grain is mostly covered with weeds and stones. In the midst of this desert stands Madrid, which is not visible more than two leagues' distance. From this side it appears small and not striking; and although we may count upward of 50 spires and towers, none are so elevated or imposing as to awaken curiosity, like that felt on first discovering the towers of churches in other Spanish cities. Even ¼ m. from the gate, the traveller might still believe himself to be 100 m. from any habitation; the road stretches away, speckled only by a few mules; there are no carriages, no horsemen, scarcely even a pedestrian; there is, in fact, not one sign of vicinity to a great city." (i., 53.)

It occupies a space of nearly 4 sq. m., on a slope inclining S.S.W. towards the Manzanares, usually an insignificant stream crossed by two magnificent bridges, the size and beauty of which contrast so strongly with the river beneath as to have given rise to the saying, that "the kings of Spain should sell the bridge, or should purchase water with the money." The river, however, sometimes swells to a great height, and pours down a magnificent volume of water. (*Swinburne*, ii., 184.) The town is surrounded by a shabby brick wall, in which are 15 stone gates, the handsomest being those of Alcala, San Vincente, and Toledo. The interior comprises an old and a more modern quarter, the former, built before Madrid, was the metropolis of Spain. The E. and more modern part is certainly not devoid of beauty; and its wide and well-paved streets, lined with handsome and lofty houses, chiefly built with brick and gray granite, the extensive and well-planted walks, the squares with their elegant fountains, and the many large and well-built public edifices, remind the traveller that he is in one of the finest, though perhaps the dullest, capitals in Europe. The best entrance to the city is by the Saragossa road, through the gate of Alcala, a noble Ionic structure, with three arches, the central one being 70 feet high. Within the walls, right and left, is the long, wide *Prado*, with its rows of trees stretching in fine perspective for more than ½ m., and in front is the *Calle de Alcala*, reaching into the heart of the city, ¾ m. in length, wider than Regent-street, and flanked by a splendid range of unequal buildings, but all of large size, and good proportions. At its end is the great centre, in which most of the better streets terminate, and, now at least rather inappropriately, designated the *Puerta de Sol*. Here, close to the *Bolsa*, or exchange, is the great morning rendezvous, either for business or pleasure. The best streets uniting in this point are the *Calle Mayor*, the *Calle de la Montera*, and the *Calle de las Carretas*, all busy thoroughfares, with good and showy shops. The *Calle del Arsenal* leads to the palace, and the *Carrera de San Geronimo* is the direct road to the gardens of the Buen Retiro. Among the squares of Madrid, the largest, with the exception of the space fronting the palace, is the *Plaza Mayor*, a rectangular area, 430 feet in length, and 330 feet broad, surrounded by a uniform range of stone buildings, five stories high, the lower part being open in front, and supported by pillars forming a handsome colonnade. The chief streets running into it are those of Atocha and Toledo, the latter passing through the *Plaza de Cebada* (formerly the place of execution for criminals), and through the gate to the bridge of its own name. None of these streets, however, will bear any comparison with the Calle de Alcala: many are good, and very many respectable, tolerably wide, and formed with lofty and well-built houses; but there is no other magnificent street. The bye-streets are narrow and crooked, especially in the S.W. quarter, where decay of material, closeness of building, and extreme filth, are the almost unvarying characteristics.

Among the public buildings, the most conspicuous is the royal palace, occupying, with its gardens, a space of nearly 80 acres, on the E. bank of the river. It stands on the site of the old Alcazar of Philip II., burnt down in 1734, and has four fronts of white stone (each 470 feet in length and 100 feet high), enclosing a spacious quadrangle. The interior is fitted up in a style of costly magnificence, perhaps not surpassed in any palace of Europe. The ceilings are *chefs-d'œuvre* of Menga, Velasquez, Corrado, and Tiepolo; the richest marbles of Spain adorn its walls, and the rooms are hung with paintings by the best masters, and noble mirrors from the manufactory of St. Ildefonso. (*Swinburne*, ii., 168-177.) Many of the best pictures, however, have been removed to the royal picture-gallery in the Prado. Its armoury is especially curious, and presents numerous specimens of arms and accoutrements taken from the Moors by Ferdinand the Catholic and his victorious generals. (*Sir C. Brooke*, ii., 295.) The other chief public buildings are—the custom-house, a handsome range of building, 290 feet in length; the Buena-vista palace, now used as a museum of civil engineering; and the palace of the council of Castile, in the Calle de Alcala; the postoffice, in the Puerta del Sol; the king's printing-office, in the Calle de las Carretas; the duke of Lista's palace, containing a fine collection of pictures, near the gate of St. Bernardino, in the N. quarter of the city; the palace of the duke of Berwick; and the national gallery, in the Prado. Madrid, though a bishop's see, has no cathedral; but there are 67 churches; among which, however, the churches of San Isidro and the Visitation are alone worthy of notice, the rest being externally and internally barbarous. "No mad architect," says Swinburne, "ever dreamt of a distortion of members so capricious, of a twist of pillars, cornices, or pediments so wild and fantastic, but that a real sample of it may be produced in some one or other of the churches of Madrid. They are, with two or three exceptions, small and poor both in marbles and pictures. Their altars are piles of wooden ornaments heaped up to the ceiling and stuck full of wax-lights, which more than once have set fire to the whole church." (*Swinburne*, ii., 164.) Previously to 1834 there were 66 convents; but several have since been pulled down to widen the streets, while others have been converted to different and, no doubt, more useful purposes than the maintenance, in pampered idleness, of hundreds of dissolute monks and nuns. The great walks constitute another grand feature of the city. The *Prado*, or public promenade, is as fashionably attended, especially on Sunday, as Hyde park in London. It is nearly 2 m. long, and comprises a broad walk, called the *salon*, flanked by several of less width, thickly shaded with elm trees: contiguous to it is the garden of the *Buen Retiro*, the palace of that name having been demolished; and still farther S. are the shady gardens called *Las Delicias*, leading to the Canal de Manzanares, which was once intended to connect Madrid with the Tagus at Toledo. These walks, in the afternoons of autumns, are crowded with the most respectable inhabitants, nor can any better idea of the out-of-door appearance of the population be got than by observing them on the Prado. In the spring, however, the scene is varied by visits to Aranjuez, a beautiful park near the Tagus, forming a verdant oasis in the midst of a desert. "The ladies," says Quin (p. 114), "wear, with few exceptions, black silk dresses and shawls, or rather mantillas, of various colours, while their head-dress consists only of a slight veil attached to the hair by a comb, and falling on the shoulder; and the graceful manner in which they wear the mantilla and veil gives to them all a smart and attractive air. The dress of the men is in every respect similar to that of the French or English; but they usually cover their persons with large cloaks, which, from the manner of wearing them, have rather a graceful appearance."

The state of education in Madrid cannot be satisfactorily ascertained. Miñano, indeed, mentions the existence of 166 primary schools and two colleges; but nothing can be inferred from such a statement, as every thing depends on the sort of instruction and the number of the pupils. The schools being generally, however, at least till very recently, under the guidance of the priests, there can be no doubt that the education they afforded was of the very worst description;

and that, instead of expanding and improving the mind, it was only fitted and intended to imbue it with the grossest prejudices. Schools on the Lancastrian system have recently been introduced; but they are opposed by large classes, and at this moment the bulk of the population is involved in the grossest ignorance. The *Colegio Imperial* and the *Seminario de Nobles*, the two schools or colleges frequented by the better classes, are but little superior to the others: no choice is, however, left to parents as to the education of their children, the only alternative being the government school or no school. The instruction given to females is most superficial; reading, writing, and a little geography are taught, in connexion with music and other accomplishments: but few ladies attain to anything like literary distinction, and the majority are "ignorant almost beyond belief." Closely connected with the educational establishments are the various literary and scientific societies, most of which are under the protection of the crown. The Academy of History, which has a handsome mansion in the Plaza Mayor, was instituted in 1735, for the purpose of collecting authentic materials for the history and geography of Spain and her possessions, and has published, among other useful works, an historico-geographical dictionary of Spain and Navarre. The *Academia de la Lengua* has for its object the perfection of the Castilian language, and with this view has published a dictionary, grammar, and other works on Spanish philology. There are also academies of science, the fine arts, medicine, and rural economy, all of which are more or less useful in promoting their respective objects. The public collections comprise, 1. the royal library, with 200,000 printed volumes, besides many valuable Arabic and other MSS., and a fine collection of coins; 2. the library of San Isidro, formerly belonging to the Jesuits, and containing upward of 60,000 vols.; 3. the museum of natural history, in which, besides other good specimens, is the great *megatherium*, described by Cuvier; 4. the botanical garden and library; and, 5. the national picture-gallery, equal in extent, and perhaps little inferior in excellence, to the largest in Europe. "To the lover of the Spanish school," says Inglis, "this gallery possesses attractions which no other can offer. Besides 49 pictures of Murillo, it contains 55 of Velasquez, 29 of Espaноletto, 17 of Juanes, 6 of Alonzo Cano, and many by other native painters; there are also nearly 500 pictures of the Italian schools, and about 300 of the Flemish school; and in the *Sala Reservada* are several *chefs-d'œuvre* of Titian and Rubens. A full description of these pictures is given by Inglis in ch. vi., vii., of his valuable work; also by Cook, vol. i., 166-170.

Several newspapers are now published at Madrid, many of which are violent and abusive in the expression of their political sentiments; but few are sufficiently well conducted to exercise much influence on the public mind. The reprinting of Spanish works has been during some years conducted with great spirit, and translations have been made of popular English and French novels, scientific and elementary works, &c.; and many light writings, with a few more solid productions of unquestionable talent by Castilians of our own day, indicate a gradually increasing taste for literature, which, however, is far from general, even among the better classes.

The theatrical amusements of Madrid are confined to two small establishments, managed by the ayuntamiento or city council. At these theatres, called the *Teatro de la Cruz* and the *Teatro del Príncipe*, Spanish comedy and Italian operas are indiscriminately represented; the musical department is on the whole well conducted; the plays are of the most trifling description, more resembling low farces than regular comedies; but, at any rate, they represent pure and unadulterated pictures of the intrigues and low life of Spain, and exhibit a truth and spirit unknown on any other stage. A large theatre, begun some years ago near the palace, has not been completed. The great and all-absorbing amusement, however, of the people of Madrid (called by their countrymen *Madrileños*), is the bull-fight, held on the Monday afternoons during the season, in a large open amphitheatre, outside the gate of Alcala. Monday in Madrid is always a kind of holyday, and in the afternoon all the avenues leading to the bull-ring are in commotion; the street of Alcala is filled throughout its whole extent with a dense crowd of all ranks, some on foot and others in carriages, all hastening to the same point. The amphitheatre will accommodate 17,000 spectators; the central area has a diameter of 230 ft., and is surrounded by a double fence, behind the exterior of which the benches rise tier above tier to the outer wall, where, at the top of all, and shaded with awnings and blinds, are the boxes occupied by persons of rank and property. The intense interest which the spectators of all classes, women as well as men, feel in this butcher-like sport, is visible throughout, and often loudly expressed; and, says Inglis, "it is certainly a fine spectacle to see thousands of spectators rise simultaneously, as they always do when the interest is intense; the greatest and most crowded theatre in Europe presents nothing half so imposing as this." The expenses of these exhibitions are very heavy; but the receipts are greater, leaving a handsome sum for the general hospital, which, it is said, draws from them a revenue of 300,000 reals, or £3000 sterling.

Inglis says there is less wretchedness in Madrid than in Paris, London, and other great towns of France and England; but the condition of the lower orders has been since much altered for the worse, by the suppression of the convents, on which they were greatly dependent; many of the ecclesiastics have also fallen into the most abject distress, though distress arising from this source, in so far at least as the public are concerned, can only be temporary. There are numerous benevolent institutions for the relief of indigence, and the cure of disease, many of which are supported by handsome endowments. The royal hospital of San Fernando, a very large establishment somewhat like an English workhouse, and the mendicity institution for the reception of beggars, formerly the greatest nuisance in Madrid, are doing much good; and the general hospital not only gives relief to the sick poor, but serves as a practical school for the students of the Academy of Medicine. A *mont de piété*, like that in Paris, lends money on security, with this difference, that at Madrid no *interest* is taken, the expense of the establishment being borne by the government.

Madrid has scarcely any manufacturing industry, nor is it possible, from its situation, at a distance from any navigable river, and in the midst of a stony, unproductive desert, that it can, in this respect, materially improve, even if that love of the *dolce far niente* should be given up which seems to be the *summum bonum* of the Madrilenian. As it is, the workmen of the city are Catalans, Valencians, Aragonese, Asturians, and Galicians: in short, every article in Madrid, whether of manufacturing or farming industry, is exotic. Its fruit comes from a distance of 50 m., butter from Aragon, oranges and lemons from Valencia, and dates from Murcia. A manufactory of porcelain and another of carr us are carried on at the expense of the government, and, most probably with as little profit as the mirror manufactory at St. Ildefonso and the saltpetre works described by Townsend, as entailing a heavy annual loss (vol i., p. 260-278). The consumption of Madrid, in 1825, is stated, by Inglis, to have been as follows: 230,000 sheep, 12,500 oxen, 70,000 hogs, 800,000 bushels of corn, 18,000 bushels of salt, 2,417,357 arrobas of charcoal, 4800 arr. of oil, 13,250 arr. of soap, and 500,000 arr. of wine. (The arroba is equal to 25 lbs. avoird.) The price of provisions, and the general expenses of living, are very high, in consequence of the necessity of bringing almost every article from a distance, and the want of water carriage. The markets are well supplied with meat, poultry, and vegetables; but fish and milk are scarce. Beef and mutton are sold at about 4½d. the lb. of 14 oz., veal fetches 7½d., and pork 5d. per lb.; bread of the best quality (and finer can nowhere be had than in Madrid) is 3½d. per lb., ordinary wine of La Mancha about 6d. the arroba (4½ galls.). Fowls are sold from 2s. to 3s. 6d. the couple, ducks at 2s. each, geese at 3s. 6d., and turkeys from 4s. to 10s., according to the season. Coffee is about one third cheaper than in England; but tea and sugar are scarce, dear, and bad. Fruit is abundant, and very cheap. Fuel is one of the most expensive articles, and lodgings fetch as high rents as those in the best situations in London.

The state of society in Madrid will be best learnt from viewing the habits of the middle classes; for, indeed, it is next to impossible for a stranger, even with good introductions, to know enough of the aristocracy to form a correct judgment of their domestic habits, owing, we believe in a great measure, to the general poverty, which, with the high rate of living in Madrid, is a very effectual bar to hospitality. Almost all families, except those in the very highest ranks, live as in Paris and Edinburgh, in stories or flats, each story being a distinct house. The outer door, which is of enormous strength, has a small window or grating, with a sliding shutter, and the usual salutation from the porter, when one rings for admittance is *Quien es?* to which the proper reply is, *Gente de paz* (people of peace); and the door, in ordinary cases, is opened. This precaution of surveying strangers before admission is, perhaps, attributable to a feeling of personal insecurity, consequent on bad government and religious persecution. A suite of apartments usually consists of a large, well lighted, and respectably furnished saloon, with a recess on one side, in which is a bed, wholly unconcealed, and without curtains; and at another side is a door-way leading into a smaller chamber, similarly furnished to that just described. The lady's boudoir is always handsomely decorated; and the worst rooms in an establishment are invariably the library or study and the dining room, both of which are small, and wretchedly furnished. The apartments are always kept remarkably clean. The manner of living in Madrid is somewhat more generous than in the N. provs. A rather rich soup is usually added to the everlasting *olla*, or *cochido*, which is
274

much better made and more highly seasoned than in the rest of Spain*; and dinner is always followed by cakes, sweetmeats, and fruits, accompanied by a moderate supply of Valdepeñas and other good native wines. The inhabitants, except the tradespeople, rise late, and breakfast on chocolate between 10 and 11. Lounging, reading, or a stroll to the cafés (where, however, they spend nothing) occupies the men, dressing and visiting the ladies, till dinner (about three), after which follows the siesta, a season of almost universal repose in Madrid. The shops then are either shut, or a curtain is drawn before the door: the shutters of every window are closed; scarcely a respectable person is seen in the streets; the stall-keepers spread cloths over their wares, and go to sleep; groups of the poor and idle are seen stretched in the shade; and even the Gallician water-carriers, seized with the general drowsiness, make pillows of their water-casks. The siesta over, the ladies sit in the balconies, and the gentlemen smoke their cigars, till the time for the lounge on the Prado; and then comes the *tertulia*, a very pleasant and social meeting for chit-chat and music, closing the day of Madrid. Dinner parties are seldom or never given, and there are no regular *parties* except balls; and those not frequent, and unaccompanied by any refreshment beyond *agua fresca*. The best national manners are not, as in other countries, to be found in the capital, where thing is sacrificed to the *rage* for imitating the French and English, a feature which distinguishes the *Madrilenes* from all other Spaniards. Morals in all classes, especially the higher, are in the most degraded state. Veils, indeed, are thrown aside, and serenades are rare; but gallantry and intrigue are as active as ever. The men think little of their marriage obligations, and pay no real respect to the other sex; the women make dress and show the business of their lives; court admiration, and are willing victims of unprincipled gallantry. Infidelity in married women is perhaps more frequent than in any of the towns of Italy, scarcely any married lady is without her *cortejo*. The connection, however, if not less sensual, is more lasting than in Italy, and intrigues are usually carried on unknown to the husband, who is generally too proud to connive at his wife's dishonour. Sexual immorality is common also among the lower orders; but there is not that drunkenness, brutality, and insolence which characterise the *canaille* of Paris and London: and the stranger may now walk about the streets in any part of Madrid without fear of being stabbed or plundered, a circumstance attributable more to the improvement of the lower orders than to the excellence of the police, which certainly deserves no eulogium. (*Swinburne*, ii.; *Inglis*, i., 56-240; *Quin's Travels in Spain*; *Cook's Sketches of Spain* in 1829-33, vol. i., c. 8; *Galiano's Lectures on Span. Lit. in the Athenæum of* 1834; *Journ. of Educ.*, vol. ix.; *and Private Information.*)

Madrid occupies the site of the ancient *Mantua Carpetanorum*, a fortified town belonging to the Carpetani. It was afterwards called Majoritum, was taken and sacked in 1083 by the Moors, who gave it its present name. Henry III. repaired and enlarged it at the beginning of the 15th century, and Philip II. made it the capital of Spain. Its subsequent history to the time of the French war is unimportant. On the 23rd of March the city was entered by the French troops under Murat, and the royal family was induced to remove into France. Joseph Bonaparte was then made king; but both he and the French army were, two months afterwards, obliged by the inhabitants, who rose in a body, to evacuate the town. In the December following, Madrid was occupied by Napoleon in person, and his brother Joseph was reinstated. The English troops occupied it for a short time in 1812, and it was again visited in 1823, by the French under the Duc d'Angoulême.

MADRID, t., Franklin co., Me. 58 N.W. Augusta. Incorporated in 1836. It contains four grist-mills, three saw-mills, four tanneries, seven schools, 151 scholars. Pop. 368.

MADRID, p.t., St. Lawrence co., N.Y., 235 N.N.W. Albany. 505 W. Bounded N.W. by St. Lawrence r. Watered by Grass r. and streams flowing into St. Lawrence r. It has several islands in St. Lawrence r. It contains 17 stores, four fulling-mills, eight grist-mills, eight saw-mills, one paper-mill, two tanneries; 27 schools, 1195 scholars. Pop. 4511.

MAD, r., O., a large eastern branch of Great Miami r. which it enters at Dayton. It has a rapid and broken current, and affords extensive water-power, particularly at Dayton.

MAD RIVER, t., Champaign co., O. Watered by Mad r. and Nettle cr. It has two stores, one flouring-mill, four grist-mills, seven saw-mills, one oil-mill, two tanneries, two distilleries. Pop. 1894.

MAD RIVER, t., Clark co., O. The line of Mad r. and Lake Erie railroad is laid through it. Pop. 1340.

MADRIDEJOS, a town of Spain in New Castile, prov. La Mancha, 30 m. N.N.E. Cidnad-Real. and 65 m. S.

* The curious reader is referred to Sir A. C. Brooke's entertaining travels through Spain and Morocco (ii., 345) for a recipe for making this dish in its different varieties.

Madrid. Population according to Miñano, 6900. It is situated in an extensive and exposed plain on the great road from Madrid through Aranjuez to Jaen and Granada, the neighbourhood being rendered not only unhealthy, but also, in some parts, unproductive by the inundations of the Amarguillo, which often greatly injure the town and deprive the people of their means of support. The only public buildings are two parish churches and a hospital; nor are there more than a dozen good houses in the place. A manufactory of serge is the only branch of industry in the town : but the neighbourhood is remarkable for its rich crops of saffron and for extensive sheep-farming. (Miñano.) Inglis describes the inhabitants as "almost a population of beggars," and states that the agriculture of this district is in the lowest state, a great part of the soil being poor and barren, while the indolence and absurd prejudices of the farmers render the rest all but unproductive. (vol. ii. p. 12.)

MADURA and DINDIGUL, a collectorate of British India. presid. Madras, prov. Carnatic, near the S. extremity of Hindostan, between lat. 9° and 10° 45′ N., and long. 77° 10′ and 79° 10′ E., having N. Trichinopoly and Coimbatoor, W. Cochin and Travancore, S. Tinnevelly and the gulf of Manaar, and E. the latter and Tanjore. Area 7856 sq. m. Pop. (1835–37) 1,135,411, chiefly Hindoos of the Sudra caste. The N. and W. parts of this district are mountainous, the S. and E. level. The hilly parts are interspersed with fertile valleys, the principal being that of Dindigul; but the plain country of Madura is by far the most productive portion of the surface. It is intersected by the river Vighey, which rises in this district, and after a course eastward for about 145 m., falls into the gulf of Manaar. A few swamps exist on the shore. The island of Ramiseeram belongs to this district. The climate of the hills is cool and healthy, but the wind often blows with great violence: in the S. it is much warmer, the temperature, in April and May ranging between 70° and 98° Fahr. Different kinds of paddy are grown in the low country, irrigation being there facilitated by plenty of streams and tanks; the husbandry is tolerably good, though not so perfect as in Tanjor. In Dindigul, the dry culture is to the wet as four to one ; and the inhabitants, are in much less comfortable circumstances than those of the S. Property is much subdivided ; some individuals occupy only the 56th part of an acre, and few have more than 135 acres. Madura is celebrated for its piece goods, and its dyers; and its artisans in gold, silver, &c., are in many places much above mediocrity. Its chief exports are piece-goods, cotton, paddy, and chanks; its chief imports, betel nut, chay root, cocoa nuts, and oil seeds. The road, bridges, and other public works in this district, have been of late put into very efficient repair by the government. Total public revenue (1837–38) £219,654, of which the land-tax made £163,363. This district is supposed to be the Regia Pandionis of Ptolemy, having been anciently governed by a Pandian family, and is one of the holy countries of Southern India. It has numerous fine temples, and other monuments of former Hindoo grandeur. It was transferred to the British by the nabob of Arcot in 1801.

MADURA, a town of S. Hindostan, cap. of the preceding district, on the Vighey, 136 m. N.N.E. Cape Comorin, and 270 m. S.W. Madras : lat. 9° 55′ N., long. 78° 14′ E. It is surrounded by a bastioned but dilapidated stone wall ; streets wide and regular. public edifices magnificent, but private dwellings mean and wretched. It has some of the most extraordinary specimens of Hindoo architecture extant. The palace is a vast pile, with a dome 90 ft. in diameter ; but it is much dilapidated ; the great temple, with its spacious areas, choultries, and four colossal porticoes, each a pyramid of 10 stories, covers an extent of ground almost sufficient for the site of a town. In front of the latter is a celebrated choultry, or inn, 312 ft. in length, ornamented with polished green stone columns, and grotesque sculptures. During the Carnatic wars, from 1740 to 1760, Madura underwent many sieges. The British civil station, and seat of the collector, &c., is in a pleasant situation, about 1½ m. S. the town. (Hamilton's E. I. Gaz. ; Madras Almanacks, 1838–39.)

MADURA, an island of the eastern archipelago, immediately adjacent to the N.E. coast of Java. with which island it is politically included, under the Dutch government. (See JAVA.)

MAESE. (See MEUSE.)

MAESTRICHT (an. Trajectus ad Mosam), a fortified town of Holland, prov. Limburg, of which it is the capital, on the Maese, 14 m. N. by E. Liege, and 57 m. E. Brussels : lat. 50° 51′ 7′′ N., long. 5° 41′ E. Population, in 1834, 22,000. It is one of the strongest towns in Holland, being defended by numerous bastions, trenches, &c. : it is well built, with wide, clean, and well-paved streets. The market is held in the great square, the centre of which is occupied by the hôtel-de-ville, built in 1658, and said to be one of the finest structures in the kingdom . .he place-d'armes is also a fine open space planted with rows of trees, and much frequented as a promenade. Among the other public buildings are comprised the exchange, the church of St. Servais, the ci-devant college of Jesuits, the arsenal, and the theatre; and in the town are 10 churches, two hospitals, two orphan-asylums, a lazaretto, athenæum, fine public library, and society of agriculture. Maestricht is the residence of the governor of the province and the seat of a court of assizes and primary jurisdiction, as well as of a chamber of commerce ; and it sends six deputies to the states of the province. The industry of the town comprises the manufacture of woollen cloths and flannels, cotton and woollen yarn, fire-arms, pins, starch, and tobacco ; besides which there are soap factories, tanneries, breweries, and dye-houses. A considerable trade is carried on with various places on the Meuse by means of barges, and packets ply daily between Maestricht, Liege, Namur, &c. These large fairs are held here during the year for horses and cattle. On the other side of the river (crossed here by a stone bridge) is the citadel or fort of Petersberg, in the suburb of Wyk, famous for its extensive subterranean stone quarry, containing numerous intricate galleries and passages, and abounding with curious marine and marine fossils, some specimens of which may be seen in the museum of the Jardin des Plantes at Paris. (Vandermaelen, Dict. de Limburg, &c.)

MAGDALENA, a river of S. America, and next to the Orinoco the principal in the republic of New Granada, through the centre of which it flows, from S. to N., through 9 deg. of lat. It rises in the small lake of Papas, in the Andes, about lat. 2° N., and long. 76° 25′ W., and runs for at least 500 m. between the middle and E. chains of the Cordillera. Its entire course may be estimated at about 800 m. : it enters the Caribbean sea about 65 m. N.E. Carthagena, and 40 m. S.W. Santa Marta. Its principal tributary, the Cauca, flows between the central and W. chains of the Cordillera, and joins it from the W., between 150 and 900 m. from its mouth. Its other affluents are the Sogamoza, Sesar, and Bogota. The towns of Nayva, Honda, and Monpox are on its banks. The descent of the Magdalena is said to be as much as 20 inches a mile (Dict. Geog.) ; and the strength of its waters is such, that they preserve their freshness to a considerable distance from its mouth. The Magdalena is navigable as far as Honda, in lat. 5° 14′ N., near which the navigation is interrupted by cataracts; but its rapidity is such, that a distance of 10 leagues a day is reckoned very good progress in ascending the river, for a champan, or flat-bottomed boat, manned by 24 bogas, or rowers. The oppressive heat of the climate, the abundance of caymans, and the swarms of musquitoes and other insects that infest the river, contribute to render the navigation both dangerous and unpleasant; but the Magdalena is, notwithstanding, the main route for the commercial and other intercourse of the inland province of New Granada with the ocean.

MAGDEBURG, a fortified city of Prussian Saxony, of which province it is the capital, on the Elbe, 74 m. S.W. Berlin, and 50 m. E.S.E. Brunswick, lat. 52° 8′ 4′′ N., long. 11° 38′ 46′′ E. Population in 1837, with its suburbs (exl. garrison), 51,347. Magdeburg is a fortress of the first class, and, from the augmentation and improvement of its defences since the war, it is now considered one of the strongest in Europe. The citadel, on an island in the Elbe, serves also as a state prison, Baron Trenck and Lafayette having, among others, been confined in it.

Magdeburg is divided into the old town, with the suburb Friedrichstadt, together composing the ancient fortress ; and the new town and suburb of Sudenburg. The latter, however, has been for the most part surrounded with walls, and the fortifications are now so extensive that it is said it would require an army of 50,000 men to invest the city. Magdeburg has one good and spacious street, called the Broadway; but all the other streets are narrow and crooked. There are two large public squares, in one of which is the cathedral. This, which is one of the finest Gothic structures of N. Germany, was erected between 1211 and 1363, and has been recently repaired at a cost of £300,000. It has two towers, each 340 feet in height, a lofty vault, a handsome high altar, and numerous tombs and monuments, among which is that of Otho the Great and his empress.

Magdeburg has in all 12 churches, one of which is for Roman Catholics, a synagogue, an ecclesiastical seminary, a female high school, or royal boarding house for the education of girls, a teachers' seminary, with schools for agriculture, commerce, surgery, &c. ; five hospitals, a lunatic asylum, a workhouse, a humane institution a savings' bank, and various charities ; an arsenal, extensive barracks, and other military establishments ; several public libraries, and a theatre. It is a bishop's see, and is the seat of the government, of the board of taxation, the superior courts of justice, the council, and the military commandant of Prussian Saxony. From its position on the Elbe, it is an important entrepôt for the merchandise imported into and exported from the central parts of Germany by that river. In other respects, also, it is very favourably situated for commerce. A canal, commencing about 20 m. below the city, connects the Havel with the Elbe, giving Magdeburg a direct water communication with Berlin and Frankfort on the Oder; and it is also the centre of a number of great roads which lead to all the cities and towns of importance within a radius of 50 m. Its manufactures, which are pretty considerable, consist of silk, linen, cotton, and woollen fabrics; oil-cloth, hats, gloves, tobacco, soap, earthenware, refined sugar, chicory, vinegar, &c., with numerous tanneries, breweries, and distilleries. A large quantity of salt is made in its neighbourhood. It has several native banking establishments, and a branch of the royal bank of Berlin. Several newspapers are published in the town; which has uniformly an air of bustle and activity.

Magdeburg was repaired by Charlemagne, and improved and enlarged by Otho the Great. It has suffered numerous sieges. In 1631 it was taken by assault by the Imperialists under Tilly, by whom it was given up to military execution, and was nearly burned to the ground. It is the birthplace of the celebrated natural philosopher Otto de Guericke, and of the poet Schultz. (*Von Zedlitz, Der Preussische Staat*, iii., 204, 205; *Berghaus, Allg. Länder, &c.,* iv. 655, 656; *Stein Handb.*)

MAGELLAN, or MAGELHAENS (STRAIT OF), a strait at the S. extremity of S. America, separating Patagonia from Tierra del Fuego, Clarence Island, and the Isle of Desolation. It extends from capes de las Virginas and Espiritu Santo, on the Atlantic, to capes Victoria and de los Pilares, on the Pacific ocean, a distance of about 300 m., having a breadth varying from 1½ to 40 m. It has an additional communication with the Pacific by Cockburn channel and Magdalen sound. Its shores are lofty and generally rugged, and its depth is in some parts very great, no bottom having been found with upwards of 1500 feet of line. Some safe and excellent bays communicate with it; but, generally speaking, its passage is extremely dangerous, both from the violence of the currents and the sudden and heavy tempests to which it is subject. It was discovered by Magelhaen, a famous Portuguese navigator in the service of Spain, in 1520. Drake traversed it in his voyage round the world; and it has, since been frequently explored by British navigators.

MAGGIORE (LAGO DI), or Lake of Locarno, (an. *Lacus Verbanus*), a famous lake of N. Italy, lying partly between Piedmont and Lombardy, and partly within the Swiss canton of Tessin. It is long and narrow, stretching above 40 m. from Magadino at its N., to Sesto-Calende at its S. extremity, while in its widest parts, opposite to the mouth of the Toce, it is about 6 m. across, but its ordinary breadth does not exceed from 2 to 3 m. Its general direction is S.S.W. and N.N.E., and it may, in fact, be considered as an expansion of the Tessino, which enters it at its N. and leaves it at its S. extremity. In addition to the Upper or N. Tessino, it receives on its W. the waters of the Toce, and on its E. side those of the Tresa, flowing from the Lago di Lugano. Its only outlet is the Lower or S. Tessino. In some places it is not less than 300 fathoms deep; its waters, which are clear and of a greenish tinge, are well stocked with fish; and, like all Alpine lakes, its navigation is dangerous from sudden squalls.

The scenery of the Lago Maggiore is very varied. That of the upper part is bold and mountainous, its northern branch opening into one of the most beautiful valleys of the Rhætian Alps, which form a magnificent amphitheatre in the back ground. Towards the E. and S., the mountains gradually decline to the plain of Lombardy; and the lower part of the lake is of a more quiet and softened character, yet still very beautiful. Its immediate shores are richly fringed with wood, occasionally broken by picturesque crags, topped with castles and churches, and with numerous villages stretching along the water's edge. Though inferior in wildness and sublimity to the lake of Como, and perhaps, also, to that of Lugano, the softer beauties of this lake are generally allowed to be the more attractive, contrasted as they are, with the distant grandeur of the Alpine chain. (*Conder's Italy,* i., 313.)

The Borromean islands, from which this lake has derived a great portion of its celebrity, are situated in a bay, on its W. side, opposite to the mouths of the Toce. Of these the *Isola Bella* and the *Isola Madre*, are the most famous. They are of small size, and, previously to the middle of the 17th century, were little better than bare rocks; but being the property of Count Vitaliano Borromeo, a descendant of the celebrated St. Carlo Borromeo, he resolved to make them his residence, and to convert them, according to the taste of the time, into a sort of Italian paradise. They were consequently covered with earth brought from the adjoining mainland, formed (especially the Isola Bella) into splendid terraces, lined with trees and statues, and ornamented with superb palaces. Unluckily, however, nothing is natural, all is art.

270

"On ev'ry side you look, behold the wall!
No pleasing intricacies intervene,
No artful wildness to perplex the scene;
Grove nods at grove, each alley has a brother,
And half the platform just reflects the other.
The suff'ring eye inverted nature sees,
Trees cut to statues, statues thick as trees !"
Pope's Moral Essays, iv. l. 114.

For a lengthened period, however, these islands were the theme of universal admiration; but as a simpler and purer taste began to prevail, they came to be regarded with very different feelings, and have latterly, perhaps, been too much depreciated. These are now usually looked upon by Englishmen, at least, as little better than "quarries above ground;" and as evincing only the wealth, extravagance, and bad taste of their founder. (*Eustace's Italy,* vol. iv., 8vo. edit.; *Simons's Italy,* p. 2, &c.)

MAGINDANAO, or MINDANAO, the most S. of the Philippine islands, which see.

MAGNESIA *ad Sipylum* (now MANISA), an ancient town of some celebrity, in Asiatic Turkey, 26 m. N.E. Smyrna. Pop. according to Elliott, about 30,000, of whom 4000 are Greeks, 2000 Armenians, and a few Jews. It is situated near the Kodus, or ancient *Hermus*, embosomed in hills long noted for the production of loadstones, and is one of the cleanest and neatest towns of Asia Minor, being in the width of its streets, and other respects, far superior to Smyrna. The principal buildings are two mosques, with double minarets, indicating a royal foundation, and the interior of each is adorned with paintings, lamps, ivory balls, ostriches' eggs, &c., such as are to be seen in the mosques of Constantinople. There are 23 other mosques, and the Armenians, Greeks, and Jews have their respective places of worship. A Jewish college, lunatic asylum, and the mausoleum of Amurath II. are the only other public edifices, except the khans, which are numerous, and well built. The manufacture of cotton and silk goods, and goats' hair shawls, employs many of the inhabitants, and the town derives some importance from being on the great road between Smyrna and the interior of Asia Minor. (*Elliott,* ii., 56–64; *Chandler,* i., 308.)

Magnesia was in all probability colonized by the Magnesians of Thessaly, not long after the foundation of Cyme and Smyrna, two other Æolian cities. It is celebrated as the scene of a signal victory obtained by the Romans, under the two Scipios, over the forces of Antiochus the Great, who was consequently obliged to retire beyond the chain of Taurus, and leave Asia Minor at the disposal of the conquerors. The inhabitants afterwards displayed great bravery in defending their town against Mithridates. In the reign of Tiberius, A.D. 17, Magnesia, in common with 11 other cities, was all but destroyed by an earthquake, and owed its restoration in a great measure to the emperor's generosity. *Duodecim celebres Asiæ urbes collapsæ nocturno motu terræ; quô improvisior, graviorque pestis fuit: neque solitum in tali casu effugium subveniebat in aperta prorumpendi, quia diductis terris hauriebantur. Asperrima lues in eodem misericordiam traxit: centies sestertium pollicitus Cæsar et quantum ærario pendebant in quinquennium remisit. Magnetes a Sipylo proximi damno ac remedio habiti.* (*Tac. Ann.* ii., 47.)

It was a flourishing city at a late period of the Roman empire, but at the commencement of the 14th century passed into the hands of Sarkhan, sultan of Ionia, and finally was annexed, in 1448, to the dominions of Mahomet II., the conqueror of Constantinople.

The above city must not be confounded with *Magnesia ad Mæandrum*, close to the modern Inek-bazar, and about 50 m. S.S.E. Smyrna, which, though a place of some consequence, was greatly inferior to the Magnesia *ad Sipylum*. It is remarkable, however, for the ruins of a theatre, stadium, and magnificent octastyle Ionic temple, said to have surpassed in the harmony of its proportions even the temple of Diana at Ephesus. (*Leak's Asia Minor,* p. 245.)

MAHABALIPOORAM, or MAVALIPOORAM, a village and a curious assemblage of rock temples in Hindostan, on the Coromandel coast, district Chingleput, about 33 m. S.S.W. Madras; lat. 12° 36' N., long. 80° 16' E. The temples in their general character closely resemble those at Ellora and elsewhere, on the W. side of Hindostan; but, from their being cut in a granite rock, they are in better preservation. They have been chiefly consecrated to Vishnu, whose worship appears to have predominated on this, as that of Siva on the opposite coast of India. At the foot of a hill N. of the village is a pagoda, about 26 ft. high, nearly as long, and about half as broad, hewn from a single rock, and covered with sculptures. Near this temple, the surface of the rock, about 90 ft. in extent, by 30 in height, is covered with bas-reliefs, including a gigantic figure of Krishna, another of his favourite Arjoon, and representations of a number of animals. Opposite to this, and surrounded by a stone wall, are two brick pagodas of great antiquity; adjacent to which are two excavations in the rock, one supported by pillars, in a manner somewhat like the cave at Elephanta, and the other fronting a sculptured group, supposed to represent one of Krishna's adventures. Still proceeding S., the traveller crosses a rocky hill, in which is a spacious excavation,—in the middle compartment of which is a figure of Siva between Brahma and Vishnu; while at one end of the temple is a gigantic figure of Vishnu sleeping upon a cobra-de-capello, and at the other an eight-armed goddess, mounted on a lion, rescuing a human figure from a buffalo-headed demon. Several of the figures are executed in a very superior style. About a mile further S. are other sculptured rocks, said to surpass those already noticed. One pagoda is about 45 ft. in height, by 29 in length and breadth; and another 40 ft. in length and breadth, and 25 ft. in height, but rent, as by some violent convulsion, from top to bottom; besides which there are three smaller structures, and large figures of a lion and an elephant, the last extremely true to nature. E. of the village, and washed by the sea, is an ancient stone pagoda, within which, also, are several sculptured figures. The sea has obviously encroached on this part of the coast, and it has probably submerged many temples that formerly existed here. Mahabalipooram is believed to have been anciently of considerable importance as a metropolis of the kings of the race of Pandion, in Hindoo mythology. (*Goldingham*, in *Asiat. Researches*, v.; *Hobbs*, &c., *passim*.)

MAHADEO TEMPLE, a celebrated place of Hindoo worship in British India, prov. Gundwanah, on the Nerbudda, 60 m. S.E. Hussingabad; lat. 22° 22' N., long. 78° 35' E.

MAHANUDDY (*Maha Nadi*, the great river), a considerable river of Hindostan, having its source in the prov. of Gundwanah; lat. 21° 30' N., long. 81° E., and flowing mostly E. to the bay of Bengal, which it enters by numerous mouths, about lat. 20° N., and between long. 85° 30' and 87° E., after a course of more than 500 m. At Cuttack, about 70 m. from the sea, the river, in the rainy season, has a breadth of about 2 m.; but it is, notwithstanding, fordable at this point from Jan. to June. During the rains it is navigable for a distance of almost 300 m. from the sea. Its deposits consist of a coarse sand, hostile to vegetation, but frequently containing diamonds of the first quality, and which are occasionally of considerable size. (*Hamilton's E. I. Gaz.*)

MAHÉ, a seaport town of Hindostan. It belongs to the French, and was formerly their principal settlement on the coast of Malabar, but is now of little importance. It is admirably situated on rising ground, beside a small river, navigable for boats to a considerable distance inland, 40 m. N.E. Calicut. Pop., in 1825, 2358, nearly all of native races. The town is well built, and has several handsome houses, three churches, &c. Its commerce is, however, small; and nearly confined to cocoa-nuts, pepper, arrack, &c. (*Official Returns.*)

MAHIM, a town of Hindostan, prov. Aurungabad, on the island of Bombay, near its N. extremity, in lat. 19° 2' N., and long. 79° 58' E. It has a Portuguese church and a R. Catholic college, and, in 1816, its population with that of some adjacent villages, amounted to 15,000.

MAHONING, t., Mercer co., Pa., 16 m. S.W. Mercer v. Watered by Mahoning river. It contains 19 stores, three fulling-mills, one woollen factory, two furnaces, 11 grist-mills, 11 saw-mills, five tanneries, one distillery, two potteries; 20 schools, 652 scholars. Pop. 3000.

MAHONING, t., Columbia co., Pa. Bounded S. by Susquehanna river. Drained by Mahoning creek, which flows into Susquehanna river. It contains 10 stores, one fulling-mill, one furnace, one tannery, three distilleries, one brewery, one pottery, one printing office; two academies, 41 students; six schools, 314 scholars. Pop. 1927.

MAHONING, r., O., and Pa., rises in Portage co., O., and passing Warren, in Trumbull co., enters Pa., and falls into Beaver river. The Pennsylvania and Ohio canal passes along this river in its lower part.

MAHONING, p. t., Indiana co., Pa., 173 W. by N. Harrisburg, 296 W. Drained by Great and Little Mahoning creeks, on the latter of which, iron ore is found. It contains seven stores, seven grist-mills, eight saw-mills, two tanneries, three distilleries; one school, 41 scholars. Pop. 2830.

MAIDA, a small town of the Neapolitan dom., prov. Calabria Ultra II., 8 m. S. by E. Nicastro. It is chiefly noted for an engagement fought in its vicinity, on the 4th July, 1806, when an English army under Sir John Stuart entirely defeated a greatly superior French force under Regnier.

MAIDEN CREEK, p. t., Berks co., Pa., 60 m. E. Harrisburg, 152 W. Watered by Schuylkill river, and Maiden creek, its tributary. It contains two churches, a Friends, and one common to Lutherans and Presbyterians, two stores, four grist-mills, four saw-mills. Pop. 1149.

MAIDENHEAD, a mun. bor. and market town of England, co. Berks, hund. Bray, on the S. bank of the Thames, 11½ m. E. by N. Reading, and 27 m. W. London. Pop. of

the bor., in 1831, about 2400. The town consists almost entirely of one street extending from the river about one mile along the high road to Oxford, and lined with numerous respectable and a few handsome houses: it is tolerably well flagged and macadamized, but only partially lighted with gas. The guildhall, in the market place, is a spacious stone building: there is also a handsome chapel of ease, and the Wesleyan Methodists and Baptists have their respective places of worship. A national and infant school, with three Sunday schools, furnish instruction to the children of the poor, and there are almshouses and other charities for the sick and aged. The Bristol, Bath, and Exeter branch of the great western road is here carried over the Thames by a handsome stone bridge of 13 arches, and about 580 yards S. from it is another bridge of three arches, forming part of the Great Western railway, which skirts the town in its whole extent. Maidenhead appears to be in a thriving condition: it has no manufactures, but is in the centre of an opulent neighbourhood, and derives considerable trading importance from its position on one of the most frequented roads of the empire. The borough was first chartered by Edward III., and the corporation now comprises a mayor and three other aldermen, with 12 councillors. Corporation revenue, 2733. Markets on Wednesday; horse and cattle fairs, Whit-Wednesday, Sept. 29, and Nov. 30.

MAIDSTONE, a parl. and mun. bor., market town, and par. of England, co. Kent, hund. of its own name, in the E. div. of the lathe of Aylesford, on the E. bank of the Medway (crossed here by a bridge of five arches), 30½ m. A.S.E. London, and 35 m. W. Canterbury. Area of par. and parl. bor., 4490 acres. Population, in 1831, 15,387. The town which is about one mile in length from N. to S., and three quarters of a mile in breadth, consists principally of a well-built street, leading N.E. from the bridge to a lengthened narrow street, along the road from Rochester to Tenterden; but exclusive of these there are many smaller streets. Among the principal public buildings are the county hall, a modern structure, well adapted for the business of the assizes, the new jail, an immense structure, erected, in 1818, at an expense of £200,000, covering more than 13 acres of land, and ranking as one of the largest and best arranged in England, the barracks near the jail, the county ballrooms, and a small but pretty theatre. The markethouse, the lower part of which is appropriated to the sale of corn, stands in the centre of the town, and behind it is a new market place, conveniently arranged for the sale of provisions. The church, one of the largest in the kingdom, is an extremely handsome embattled edifice, with a lofty tower, formerly surmounted by a spire, destroyed by lightning in 1730: it was made collegiate in the reign of Richard II., and attached to an ecclesiastical college, destroyed with many others at the Reformation: the living is a perpetual curacy, in the gift of the Archbishop of Canterbury. There is also a new district church, erected by the church-building commissioners, at an estimated cost of £13,000, the incumbency of which is in the gift of the curate of Maidstone. Places of worship are also supported by the Wesleyan Methodists, Independents, Baptists, Unitarians, and the Society of Friends. A free grammar school was founded, in the reign of Edward VI., by the corporation, which has two exhibitions at University college, Oxford: freemen have the privilege of sending their sons here gratis, for classical instruction, the master making a charge for other branches of education. This school is not in a very flourishing state, and its inefficiency has led to the establishment of a proprietary school, which is well supported and attended. A blue-coat hospital was founded, in 1711, for the clothing and education of 53 boys and 43 girls, and there are three other endowed charity schools, and a Lancasterian school. Four sets of almshouses furnish lodging, clothing, and money allowances to 20 old women, and various bequests and charities exist for the relief of the sick and aged poor. A philosophical society was instituted here in 1834, a good library and reading-room is established, and a newspaper is published once a week.

"Maidstone is in a very prosperous state, and there is no want of employment. There is a demand for houses of a superior class, and many have been built since the census of 1831; but many of the cottages are unoccupied, owing to the completion of public works, which had been going on for some years. The only manufactory of any importance is that of paper: there are six paper-mills in the parliamentary, employing about 800 hands. The felt, blanket, and hop-bag manufactories are of much less extent. There is a considerable traffic on the river, which has been for many years gradually increasing; and the annual tonnage of vessels passing through Hallington lock, about two miles from the town, is, at present, supposed to average 120,000 tons, on which tolls are paid to the amount of about £3600. The principal articles of merchandlen brought up the river, are coals and timber for the supply of the neighbourhood, and also of Tunbridge, Seven-Oaks, and the whole weald of Kent. A portion of the latter article is imported direct from the Baltic and America. The neighbourhood is celebrated for its abundant produce in hops and fruit, both of which are carried down the river with paper and stone." (*Mun. Corp. Report.*)

Maidstone received its charter of incorporation from Edward VI., in 1549, but forfeited it in the following reign, owing to the connection of its inhabitants with the insurrection of Sir Thomas Wyatt. Qben Elizabeth granted another charter, with increased privileges; but this also became void, by a *quo warranto*, soon after the Revolution of 1688; and a new charter was granted in 1748, by George II. Under the Municipal Reform Act of 1837 the borough is divided into three wards, the corporate officers being a mayor and five other aldermen, with 18 councillors. Corporation revenues in 1839, £4137. exclusive of £39 resulting from the sale of property. The Lent and summer assizes are held here, as also the quarter-sessions for the W. division of Kent. The recorder holds, also, quarter and petty sessions within the borough; and there is a court for the recovery of small debts. This borough has sent two members to the House of Commons from the 6th of Edward VI. Down to the Reform Act, the right of election was vested in the freemen (by birth, apprenticeship, and purchase) not receiving alms. The limits of the borough were not altered by the Boundary Act. In 1839-40, it had 1667 registered electors. Maidstone is also the chief place of election for the members for the W. division of the county. Large markets on Thursday for hops, corn, horses, and cattle: fairs for cattle, &c., first Tuesday in each month, Feb. 13, May 12, June 20, and Oct. 17.

MAIDSTONE, t., Essex co., Vt., 8 m. N. Guildhall, 54 m. N.E. Montpelier. Bounded E. by Connecticut river. Watered by Paul's stream. It has one grist-mill, three schools, 96 scholars. Pop. 271.

MAILCOTTA, a town of Hindostan, prov. Mysore, and a celebrated place of Hindoo worship, on a rocky hill, 17 m. N. Seringapatam; lat. 12° 30′ N., long. 76° 43′ E. The town, which is open and paved, has about 400 good houses, mostly occupied by Brahmins, and several rich pagodas, choultries, &c. The most striking edifice is a temple dedicated to Narasingha (the man-lion), which stands on the highest pinnacle of the mountain, and is approached by a staircase cut in the rock, and ornamented at intervals with smaller temples and arches. It has, besides, a temple to Krishna, a square building of vast dimensions, entirely surrounded by a colonnade, and which is said to be extremely rich in jewels and other articles of value; and held in such esteem that Tippoo did not venture to outrage the prejudices of his Hindoo subjects by plundering it. There is also a large and fine reservoir at Mailcotta, surrounded by numerous buildings for the accommodation of devotees. Near this town the Mahrattas defeated Hyder Ali, in 1772. (*Hamilton's E. I. Gazetteer.*)

MAINE, the northeastern of the United States, is bounded N. by Lower Canada, E. by New-Brunswick, from which it is separated by the St. Croix river and a line due N. from the Monument at the source of the St. Croix river, as designated by the commissioners under the 5th article of the treaty of 1794, between the government of the United States and Great Britain; thence N. following the exploring line marked by the surveyors of the two governments in the years 1817 and 1818, under the 5th article of Treaty of Ghent, to its intersection with the St. John river, and to the middle of the channel thereof; thence up the middle of the channel of the said river St. John, to the mouth of the river St. Francis; thence up the middle of the channel of the said river St. Francis, and through the lakes through which it flows to the outlet of the lake Pohenagamook; thence southwesterly, in a straight line to a point in the N.W. branch of the river St. John. which point shall be 10 miles from the main branch of the St. John, in a straight line, and in the nearest direction; but if the said point shall be found to be less than seven miles from the nearest point or crest of the highlands, that divide the rivers which empty themselves in the river St. Lawrence from those which fall into the river St. John, to a point seven miles in a straight line from the said summit or crest; thence in a straight line in a course about S. 80 W. to the point where the parallel of lat. 46° 25′ N., intersects the S.W. branch of the St. John; thence southerly by the said branch to the source thereof in the highlands at the Mitjarmette portage; thence down along the said highlands which divide the waters which empty themselves into the St. Lawrence from those which fall into the Atlantic ocean, to the head of Hall's stream; thence down the middle of said stream till the line thus run intersects the old line of boundary surveyed and marked by Valentine and Collins previously to the year 1774, as the 45° of N. lat., and which has been known and understood

to be the line of actual division between the states of New-York and Vermont on the one side, and the British province of Lower Canada on the other; and from the said point of intersection W. along the said dividing line, as heretofore known and understood, to the Iroquois, or St. Lawrence river. Such are the terms of the late treaty, now ratified by both governments, and which has happily settled a controversy of the standing of a quarter of a century. The line designated as the old· line, run as the 45° of N. lat., is found to be nearly a mile N. of the true line of 45° N. lat., as is proved by the accurate taking of the latitude, by the more perfect instruments now used. Maine is bounded W. by New-Hampshire, and S. by the Atlantic ocean. It lies between 43° 5′ and 47° 20′ N. lat., and between 66° 50′ and 71° W. long. It is computed to contain 30,000 sq. m., or 19,200,000 acres. The population in 1790, was 96,540; in 1800, 151,719; in 1810, 228,705; in 1820, 298,335; in 1830, 399,955; in 1840, 501,793. Of these 252,959 were white males; 247,449 were females; 790 were coloured males; 635 do. females. Employed in agriculture, 101,630; in commerce, 2921; in manufactures and trades, 21,879; in navigating the ocean, 10,091; in the learned professions, 1899.

The state is divided into 13 counties, which, with their population in 1840, were as follows:

County	Population	County	Population
York	54,034	Somerset	33,912
Cumberland	60,658	Penobscot	45,705
Lincoln	63,517	Waldo	41,509
Hancock	28,605	Piscatiquis	13,138
Washington	28,327	Franklin	20,801
Kennebec	55,923	Aroostook	9,413
Oxford	38,357		

Augusta at the head of sloop navigation on Kennebec river, 50 m. from its mouth, is the seat of government.

Maine is diversified, and has an uneven surface, but is not generally mountainous. On the western side of the state, east of the White mountains in N. H., an irregular chain of high lands commences, and passing N. of the sources of the Kennebec and Penobscot rivers, and S. of the sources of Aroostook rivers extends eastwardly to the eastern boundary of the United States, and terminates at an isolated peak denominated Mars Hill, 1683 feet high. This chain which is not continuous, the British, before the late treaty, claimed as the highlands described in a previous treaty. Katahdin between the E. and W. branches of the Penobscot, 5300 feet above tide-water, is much the highest land in the state, and constitutes a part of the above range, if such it can be called. But the vexed question is now happily settled. The British have ground for their need from Halifax to Quebec, which in fact occasioned all the difficulty, and the United States have received an equivalent. The rest of Maine is hilly, though the hills are not very elevated. The land on the sea coast, for the distance from it of from ten to twenty miles, is not generally very fertile; but further inland, its quality is greatly improved. In the N.W. and S.E. parts, the soil is light and indifferent. Between the Penobscot and Kennebec rivers, is a tract of land not exceeded in fertility by the best portions of the United States. The principal productions are grass, Indian corn, wheat, barley, rye, oats, and flax. The uncleared lands are of great extent, and furnish a vast amount of pine and other lumber, which in the form of masts, plank, boards and timber, is exported to a great extent. Lumber cut and sawed may be regarded as the staple production of the state, and is exported to the amount of about $10,000,000 annually. The state is well adapted to grazing, and the wool produced is estimated at $2,000,000 annually. Lime is manufactured, particularly at Thomaston and the vicinity, to the annual amount of $100,000. A fine building granite is found at Hallowell, and is extensively exported. The Hall of Justice in the city of New-York, is constructed of it. Previously to the year 1807, the wars in Europe gave to the United States much of the carrying trade of the world, and Maine engaged largely in commerce, and neglected her lands for this superior source of wealth. But when the embargo, non-intercourse, and war crippled her commerce, her agricultural resources were developed. Cattle and sheep are raised in great perfection. Maine produced much and good wheat. The crop of Indian corn sometimes suffers from the shortness of the season. Among the fruits, apples, pears, plums and melons succeed well.

According to the census of 1840, there were in the state, 59,208 horses or mules, 327,955 neat cattle, 649,264 sheep, 117,386 swine. There were produced 848,166 bushels of wheat, 137,941 of rye, 950,598 of Indian corn, 355,161 of barley, 1,076,409 of oats, 10,392,280 of potatoes, 681,258 tons of hay, 1,465,551 pounds of wool, 257,464 of sugar. The products of the dairy amounted to $1,496,902; and of lumber to $1,808,683. The amount of wheat and Indian corn had been greater, in some previous years.

Maine has a sea coast of over 230 miles, indented by numerous bays, and protected by numerous islands; and has more good harbors than any other state in the Union. Ships are extensively built, not only for their own use, but for a foreign market. The fisheries employ many of the inhabitants, and are not only a source of wealth, but are a nursery of seamen. Maine in point of shipping is the fourth state in the Union.

The climate of Maine, though subject to great extremes of heat and cold, is generally healthy. The cold of winter, though severe is steady, and is much less trying to health than sudden changes. Near the seashore the heat of summer is greatly tempered by the sea breezes. In July 9th, 1836, the thermometer, in some places, rose to 100 degrees above zero; and in January 26th, 1837, it sunk to 37 degrees below zero. These may be regarded as the extremes of temperature, which are of short continuance. The season of vegetation, at its greatest length, extends from April 21st, to October 16th; though it does not continue in its vigour for more than three months and a half.

Maine has a number of fine rivers. The Penobscot is 250 m. long, and is navigable for large ships to Bangor, 52 m. from the ocean. The tide here rises 17 or 18 feet, and greatly facilitates the entrance and departure of vessels. The Kennebec is 250 m. long, and is navigable for large ships 12 m. to Bath, for sloops of 150 tons, 40 m. to Hallowell, and 2 m. farther to Augusta for vessels of 100 tons, and for boats to Waterville, 18 m. above Augusta. The Androscoggin rises in New-Hampshire, but runs chiefly in Maine. Its numerous falls afford great water-power. It enters the Kennebec, 20 m. from the ocean. The Saco river rises in N. H., but soon enters Maine, and flowing S.E., enters Saco bay, and is navigable 6 m. to the falls. Damariscotta is chiefly an arm of the sea, has a tide of 10 feet, and is navigable 18 m. to Nobleboro'. The Sheepscot is a small river, but has an extensive bay at its mouth, forming the harbour of Wiscasset, one of the best in the state.

It is computed that one tenth of the surface of the state is covered with water. In the interior are many ponds and lakes. The largest, Moosehead, is 50 m. long and 10 or 12 broad. Umbagog, which lies on the border of New-Hampshire, is 18 m. long and 10 broad. The largest island of Maine is Mount Desert, in Frenchman's bay, and is 15 m. long and 12 broad. Long Island, Deer island, and Fox island are on the E. side of Penobscot, bay, There are many others. Penobscot bay is large and open, being 30 m. long and 18 wide, at its mouth. Casco bay extends for 20 m. between cape Elizabeth and cape Small Point.

The most commercial places in the state are Portland, (city,) on Casco bay, the third or fourth in commercial importance in New-England; Bangor (city) on the Penobscot, Bath, Hallowell and Augusta on the Kennebec, Thomaston on St. George river, Belfast on a branch of Penobscot bay, Wiscasset on a bay at the mouth of the Sheepscot, Wells, Gardiner, Brunswick, Frankfort Prospect, Bucksport, Camden, Castine and Eastport.

The exports of Maine, for the year ending September 1841, were $1,078,633, and the imports were $700,961. There were in 1840, 70 commercial and 14 commission houses in foreign trade, employing a capital of $1,646,926; 2230 retail dry goods and other stores, with a capital of $3,573,593; 2068 persons employed in the lumber trade, with a capital of $305,850; 123 persons employed in internal transportation, who with 56 butchers, packers, &c., employed a capital of $95,150; 3610 persons employed in the fisheries, with a capital of $596,967.

The manufactures of Maine are considerable. Homemade or family manufactures amounted to $694,397; 34 woollen manufactories employed 539 persons, producing goods to the amount of $412,366. employing a capital of $316,105; six cotton manufactories, with 29,736 spindles, employed 1414 persons, producing goods to the amount of $970,297, with a capital of $1,308,000; 16 furnaces produced 6122 tons of cast iron, and one forge for bar iron, employed 48 persons, and a capital of $185,950; 15 persons produced 50,000 bushels of salt, with a capital of $95,000; six paper-mills employed 89 persons, producing to the amount of $84,000, with a capital of $90,600; 280 persons produced granite and marble to the amount of $98,720; 37 persons manufactured tobacco to the amount of $18,150, with a capital of $6050; hats and caps were manufactured to the amount of $74,174, and straw bonnets to the value of $2807, together employing 312 persons, and a capital of $26,050; 395 tanneries employed 754 persons, and a capital of $571,793; 530 other manufactories of leather, as saddleries, &c., produced articles to the amount of $443,846, with a capital of $191,717; 21 potteries employed 31 persons, producing to the amount of $90,850, with a capital of $11,353; 864 persons produced bricks and lime to the amount of $261,586, with a capital of $300,622; 330 persons produced machinery to the amount of $99,759; 119 persons produced hardware and cutlery to the amount of $65,555;

four rope-walks employed 34 persons, producing cordage to the amount of $32,680, with a capital of $23,000; 779 persons produced carriages and wagons to the amount of $174,310, with a capital of $75,012; flouring, grist, saw and other mills, employed 3630 persons, producing to the amount of $3,161,502, with a capital of $2,900,565; ships were built to the amount of $1,844,909; furniture was manufactured to the amount of $804,675, employing 1453 persons, and a capital of $663,538; 34 brick, and 1674 wooden houses were erected, employing 3462 persons, and cost $733,067; 34 printing-offices, 14 binderies, three daily, two semi-weekly, and 30 weekly newspapers, and five periodicals, employed 196 persons, and a capital of $468,200. The total amount of capital in manufactures was $7,147,234.

The principal colleges are Bowdoin college at Brunswick, named in honour of its principal benefactor, Hon. James Bowdoin, founded in 1794, and went into operation in 1802; Waterville college, under the direction of the Baptists, founded at Waterville in 1820; Bangor theological seminary, at Bangor, founded in 1816; the Wesleyan seminary, under the direction of the Methodists, founded at Readfield in 1822. These institutions had in 1840, 906 students. There were in the state 86 academies, with 8477 students, 3385 common and primary schools, with 164,477 scholars. There were 3941 persons over 20 years of age, who could neither read nor write.

The principal religious denominations are the Methodists, Baptists and Congregationalists. The Baptists had in 1826, 222 churches, 145 ordained ministers, and 15,000 communicants; the Methodists had 115 travelling preachers, and 15,493 communicants; the Congregationalists had 161 churches, 119 ministers, and 19,370 communicants. There are also some Free-will Baptists, Friends, Episcopalians, Unitarians, Universalists, and Roman Catholics.

On the 1st of January 1842, there were 40 banks, with an aggregate capital of $3,414,000, and a circulation of $1,585,620. At the close of 1840 the state debt amounted to $1,678,367.

Maine has executed several works of internal improvement. The Cumberland and Oxford canal was completed in 1830. It connects Portland with Sebago pond, 20½ m. long; and by a lock in Songo river, the navigation is extended into Brandy and Long ponds, a further distance of 30 m. The whole distance is 50½ m.; 34 feet wide at the surface, and 18 at the bottom, with 26 wooden locks, and cost $250,000. Bangor and Orono railroad was completed in 1836, is 10 m. long, and connects the two places. The Portland, Saco and Portsmouth railroad was incorporated in 1837 and with the Eastern railroad connects Boston with Portland. From Salmon falls, opposite to Portsmouth, its length is 46 miles, and it was completed in 1842, at a cost of $761,509. Several other railroads have been projected.

The government consists of a governor, senate and house of representatives. The governor is elected by the people, and holds his office for one year from the first Wednesday in January. A council of seven persons to advise the governor is elected annually, by the joint ballot of the Legislature. The senate consists of 31 members elected by the people. The house of representatives consists of 151 members, elected annually by the people. The right of suffrage is possessed by every male citizen of the United States, of 21 years of age and upwards, excepting paupers, persons under guardianship, and Indians not taxed, who have resided in the state for three months next preceding an election. The election must be by written ballot. The judiciary is vested in a supreme judicial court, and such other courts as the legislature shall from time to time establish. The judges are appointed by the governor, with the advice and consent of the senate, and hold their offices during good behaviour, or until they are 70 years of age. In a similar manner are appointed the attorney general, the sheriffs, coroners, registers of probate, and notaries public.

The first permanent settlement in Maine was made in Bristol as early as 1625, at Pemaquidpoint. In an old fort once called William Henry, and afterwards Frederick George, built of stone in 1692, and taken by the French in 1696, "are found grave stones of a very early date, and streets regularly laid out and paved, in the vicinity of the fort. On the side of Pemaquid river, opposite to the fort, tan pits have been discovered, the planks of which remain in tolerable preservation, and in other places coffins have been dug up, which bear indubitable evidence of a remote antiquity." In 1635, the district was granted by the British crown to Sir Ferdinando Gorges, and he appointed a governor and council. In 1647 a government was established by the settlers. In 1652 the state of Massachusetts purchased the territory of the heirs or Gorges for $5334. In 1691 it was incorporated with Massachusetts, by a charter of William and Mary, and continued under its jurisdiction until it became an independent state. It had long a sufficient population to become a state, and efforts were made for this purpose in 1785, 1796 and 1802. The inhabitants were averse to a separation. But in 1800 a constitution was formed, and it was admitted to the Union as a sovereign state.

MAINE, a river of W. Germany. (See MAYN.)

MAINE, one of the old provs. of France, now distributed between the deps. Mayenne and Sarthe.

MAINE-ET-LOIRE, a dep. of France, reg. W., formerly comprising the greater part of the prov. of Anjou, chiefly between lat. 47° and 47° 50' N., and long. 0° and 1° W., having N. the deps. Mayenne and Sarthe, E. Indre-et-Loire, S. Vienne, Deux-Sèvres, and Vendée, and W. Loire-Inférieure. Greatest length, E. to W., about 70 m., breadth usually about 40 m. Area, 722,163 hectares. Pop. (1836) 477,970. Surface undulating. The Loire intersects the department from E. to W., dividing it into two nearly equal parts; and is joined within its limits by the Maine, Authion, Thouet, Layon, &c. The Maine is a continuation of the Mayenne, which changes its name after it has been joined by the Sarthe. It passes by Angers, and unites with the Loire about five miles below that city. Its entire length is eight miles, throughout which it is navigable. In 1835, 440,196 hectares of the surface of this department were estimated to be arable, 80.093 in pasture, 38,960 in vineyards, 61,836 in woods, and 48,971 in heaths, wastes, &c. More corn is produced than is required for home consumption. Nearly 2,196,000 hectolitres are said to have been harvested in 1835, of which 1,005,000 were wheat, and 547,680 rye. Agriculture, as in the contiguous departments, is very backward; the lands in lease are all held on the metayer principle, the rent being a certain proportion, usually about half the produce: the occupiers are poor, uninstructed, and, of course, strongly attached to routine practices. Hemp and flax, prunes, melons, walnuts, apples, and various other fruits, are said to be of superior quality. The produce of wine is estimated at about 300,000 hectolitres a year. Some of the white wines are rather well esteemed; but the greater portion of the vintage is either converted into brandy or vinegar. The latter, which enjoys a high reputation, is known in commerce as *vinaigre de Saumur*. Exclusive of wine, this department produces annually from 50,000 to 60,000 hectolitres of cider. The industry of the rural population is, however, chiefly exercised in rearing and fattening cattle for the Paris markets, and in breeding horses. In 1830 there were estimated to be 293,539 head of cattle in Maine-et-Loire—a greater number than in any other department of the W. of France; but, on the other hand, the stock of sheep (180,000) was comparatively small. In 1835, of 140,411 properties subject to the *contribution foncière*, 68,886 were assessed at less than 5 francs, and 21,545 at from 5 to 10 francs. At the same time 332 properties were assessed at more than 1000 francs. This department has the largest and most important slate quarries in France. These are situated near Angers, and are extensive excavations, in one place to the depth of 450 feet below the surface. They employ more than 3000 workmen, and several steam-engines, and are said to yield about 80 millions of slates a year. At Cholet (which see), and other parts, some extensive woollen, cotton, and other manufactures are established, employing a large number of hands, and producing goods of the estimated value of 20,000,000 francs a year. At Angers is a large sail-cloth factory; wooden shoes are made at Mouilcherne; and the department has numerous sugar refineries, breweries, distilleries, paper-mills, dyeing-houses, &c.: and at Angers is one of the two royal schools of arts and trades established in France (the other is at Chalons-sur-Marne), at which about 450 pupils are supported partly or wholly at the expense of government. Maine-et-Loire is divided into five arrondissements: chief towns, Angers, the capital, Baugé, Beaupréau, Saumur, and Segré. It sends seven members to the chamber of deputies. Number of electors (1836-39), 2744. Total public revenue (1831), 11,104,026 francs. (*Hugo*, art. *Maine-et-Loire; Official Tables, &c.*)

MAINLAND. See SHETLAND ISLES.

MAJORCA (Span. *Mallorca*), the largest of the Balearic islands in the Mediterranean sea belonging to Spain, from the E. coast of which it is 110 m. distant, Palmas the chief town being in lat. 39° 35' N., long. 2° 45' E. Greatest length, 48 m.; do. breadth, 42 m.: estimated area, 1340 square miles. Population, according to Miñano, 181,805. Its shape is that of an irregular four-sided figure, the angles of which are formed W. by cape Tramontana, N. by cape Formentor, E. by cape Peri, and S. by cape Salinos. The surface is extremely uneven, and is divided into two pretty equal parts by a range of mountains, the highest of which, the Silla de Torillos, rises 5114 feet above the sea. These mountains are not volcanic, but consist chiefly of granite, syenite, and porphyry, over which lie beds of grauwacke, clay, slate, and coal; lead and iron are found, but not in sufficient abundance for mining purposes. The rivers or rather torrents of Majorca are short, rapid, and very numerous, affording great facilities to irrigation. The climate is exceedingly mild, salubrious, and agreeable; the thermometer during winter scarcely ever falls below 45°, its average height being 65°, and cold and strong N. winds are of rare occurrence. The temperature of summer varies between 84° and 88° Fah.; but the heat is seldom oppressive, owing to the constant sea-breezes. The red, loamy soil of the mountains, though stony, is extremely rich, producing spontaneously great numbers of wild olives, grapes, &c.; in the plains it is much less fertile, owing to the superfluity of moisture, and the absence of any system of drainage. Agriculture is in a very rude and debased state; and the growth of corn, which in wet years totally fails, meets only half the consumption of the island, the annual imports of this article being about 6000 fanegas, chiefly from Catalonia and Valencia. Olives are raised in very large quantities, the crops averaging about 180,000 arrobas yearly; the fruit is smaller than that of Andalusia, but as juicy as the best of the growth of Provence. Wine, both red and white, is abundant, especially near Banalbufar and Falanicbe; considerable quantities are exported, and much is likewise used in the distillation of brandy. Fruit and vegetables, especially oranges, figs, melons, carobs, pumpkins, and cauliflowers, grow plentifully, and attain a large size. Large quantities of saffron also are produced, of preferable quality to that of La Mancha. There is no want of fine pasture in the island; but little attention is paid to cattle-breeding. The sheep are large, and hogs sometimes attain the weight of 600 lbs. or about 30 arrobas. Mules and asses are reared in great numbers, and sent to Valencia and other provinces in the S. of Spain. Hares and rabbits, partridges, quails, snipes, &c., are abundant, and the coast swarms with fish of various kinds and good quality.

The trade of Majorca is, relatively to its size, very considerable, chiefly with Spain, France, and England; its exports comprise oil, wine, brandy, oranges, and other fruits, capers, saffron, wine, mules, and asses, with smaller quantities of home-made goods, as palm brooms and baskets, tannery wares, and water-proof hats for sailors, its imports consisting of wheat, salt beef, iron, sugar, groceries, woollen and cotton goods, hardware, &c., chiefly from France, England, and the N. of Europe; but the precise amount of the trade of Majorca cannot be ascertained.

The inhabitants are described by Fischer as bearing a striking resemblance, "both in their external appearance and general character, to the Catalans, being equally hardy and courageous, equally blunt and jealous of their honour, equally industrious and ingenious, equally good sailors and skilful farmers, with their continental neighbours; and their language is, in fact, nothing but a corrupt dialect of the Catalan."

Majorca comprises only two towns of any importance and 28 villages, the rest being mere hamlets. Numerous detached farms and country houses, however, are scattered over different parts of the island; and in all the fine valleys one may meet with numbers of elegant villas, in which the higher classes, who are usually much attached to a country life, spend the greater part of the year. The roads have also been considerably improved within the last eight years, and there is a tolerably good communication between different parts. The capital of Majorca is Palma (sometimes also called Majorca), situated in a bay of its own name, on the S. side of the island, and having a population (according to Miñano) of 34,343 persons. It is agreeably placed in a delightful country, and is pretty strongly fortified; the houses are large and well built, but the streets, being narrow, dark, and ill paved, give it a wretched and mean appearance. The chief public buildings are the governor's palace, a large structure with extensive gardens, a cathedral, exchange, town-hall, and theatre. The inhabitants are active, enterprising, and laborious; and almost the whole trade of the island is concentrated in its port. The road of Palma affords excellent protection for shipping, except during storms from the S.E.; but the little harbour, called Puerto-Pi, is more secure, and furnishes anchorage for the largest frigates; the port is defended by two well-fortified castles. Among the other towns of Majorca, the largest, with their respective populations, are Llumayor (6530), Campos, remarkable for its mineral waters and salt-pans (4861), Sansenay, celebrated for its stone-quarries (3550), Falaniche, where is made the best brandy of the island (6800), Manacor (8905), Pollenza (7823), and Soller (5614). The small island of Cabrera lies 6 m. S.S.W. of cape Salinas: it is covered with trees, and wholly uninhabited, except by convicts, of whom there is here a small depôt.

The *Balearic Islands*, of which Majorca is the chief, were more anciently known as the Χοιρώδες, so called, probably, from rising out of the sea, like the backs of hogs. The Phœnicians made settlements in them at a very early period; and they were succeeded by the Carthaginians under Hanno, who founded *Mago* (Mahon) and *Jamzen* (Ciudadela), both towns of Minorca. The islanders were celebrated as the most expert slingers in the Carthaginian service during the Punic wars, and were afterward equally noted as successful pirates, till Quintus Metellus subdued them, and hence obtained the surname of *Balearicus*. He was the founder also of two cities in Majorca, *Palma*, the present capital, and *Pollenzia*, now Pollenza. Under the Roman empire, these islands belonged to the judicial district (*conventus juridicus*) of New Carthage in Tarraconensis, and from the reign of Constantine I. to that of Theodosius I., they had their own government. On the breaking up of the W. empire, they became an easy conquest for the Vandals and Huns, from whom they were afterward wrested by the Moors. The people becoming notorious as pirates and robbers on the coast of Christian Europe, Charlemagne headed an expedition against them, and succeeded, not only in taking the islands, but in keeping possession of them for six years, at the end of which they were retaken by the Moors: nor were the latter finally expelled till 1285, when the entire group was formally annexed to the crown of Aragon.

MALABAR. This term is usually applied to designate the whole W. coast of Hindostan from cape Comorin to Bombay, but, strictly speaking, Malabar only extends as far N. as the Malabar language is spoken, or to lat. 12° 30'. The British province of Malabar is a district or collectorate under the Madras presidency, extending between lat. 10° 15' and 12° 15' N., and long. 75° 10' and 76° 50' E., comprising several portions of territory, as Wynaad, &c., not belonging to Hindoo Malabar; and having N. Canara, Coorg, and Mysore, E. Coimbatoor, S. Cochin, and W. the Indian ocean. Length, N.W. to S.E., about 150 m.; average breadth about 42 m. Area; 6802 sq. m. Pop. (1836-37) 1,140,916, of whom 944,186 were Hindoos, 182,037 Mohammedans, and 14,463 Roman Catholics. In the E. the surface is mountainous, comprising a portion of the range called the W. ghauts; the coast is low, and indented by many shallow lakes. Between these two regions the country mostly consists of undulating hills, separated by narrow valleys in general watered by a rivulet. Nearly all the rivers have a W. course. The chief are the Cochin, Beypoor, Baliapatam, Ponany, &c.: the bar of the first is navigable for ships drawing 15 ft. water, and the mouth of the second will admit vessels of 300 tons. Lakes and tanks inconsiderable. The year is divided into three seasons: the hot, from February to May; the wet, from May to October; and the cool, during the remaining months. Dense fogs are rare on the coast, but they usually envelope the ghauts from April to the end of the year. The soil on the coast is sandy, but well adapted for the culture of the cocoa-nut, jack, areca, plantain, cinnamon, and other trees, pepper, coffee, the sweet potato, and other farinaceous roots, garden vegetables, &c. In the interior the soil is of the red kind common in the S. of India, and highly favourable for rice, which frequently yields two and sometimes three crops a year. The rice lands are sown after the first rains in April, and in four months the grain is ripe for the sickle. The second crops are raised by the transplantation of plants a month old, and are reaped in three months. The third crop is assisted by small reservoirs and tanks, and by turning water from streams. About 786 sq. m. are estimated to be under rice, and 190 in gardens and inclosures of productive trees. The sides of the hills are often formed into terraces for cultivation. The rest of the surface, especially in the uplands, is chiefly covered with forests, among which the teak-tree is very prevalent, and an important source of wealth to the district, the teak of Malabar being considered, upon the whole, superior to every other variety. Besides the above articles of culture, the mulberry, mango, tamarind, sugar-cane, ginger, turmeric, mustard, arrow-root, hemp, cotton, &c. are grown, and wheat and barley on the hills. There are few cattle. The elephant and wild hog do great damage on the borders of the forests they inhabit; the tiger, bison, elk, deer, &c. are also met with. Towns are rare in the interior, and villages there are spread over a large space, families usually living separate from each other within gardens inclosed by ditches and high banks. Iron is pretty generally found, and gold, though in small quantities, in the sands of some of the rivers. Coarse cotton cloths are manufactured in a few places from the raw produce of the district; coir is made from the fibrous covering of the cocoa-nut; oil from its kernel, and arrack from the toddy in very large quantities. The chief exports consist of the products of the cocoa palm, amounting to about 805,509 rupees annually. From 10,000 to 15,000 candies of pepper, betel-nut to the value of 350,000 rupees, and cloth from the districts to the E. to the value of from 1,700,000 to 2,300,000 rupees, are annually exported. At Calicut, Tellicherry, Canatore, and Ponany, the chief commercial towns, there are numerous Parsee and other opulent merchants. The roads throughout the

district are in good order, and have convenient bungalows every 10 or 15 m. Public revenue (1836-37), £310,596, of which the land-tax amounted to £161,162. In Malabar, as in S. Canara, inheritance goes by the female line, among the Nairs and other Hindoo castes which inhabit the country. On the coast, a large proportion of the inhabitants are Mohammedans, and many Mopiays, a people originally derived from Arabia. The Christian religion appears to have been planted in this part of India at a very early period, and many churches were found existing by the Portuguese. Malabar was governed by various Nair dynasties, previously to its conquest by Hyder Ali, in 1761, on the fall of Tippoo Saib, it became subsidiary to the British, and was incorporated with the Madras presidency in 1803. (*Madras Almanacks for 1836 and 1839; Hamilton's E. I Gazetteer*, vol. ii.)

MALACCA AND NANING, a British colony, on the W. coast of the Malay peninsula, between lat. 2° and 3° N., and long. 102° and 103° E.; having N.W. the territory of Sangalore, N.E. those of Rumbowe and Johole, S.E. that of Johore, and S.W. the straits of Malacca. Area estimated at 1000 sq. m. Pop. in 1826, 37,708, of whom about 21,800 are Malays, 4680 Chinese, and 2400 Europeans, chiefly English, Dutch, and Portuguese. Surface mostly undulating; the hills are covered with jungle, and the valleys rendered swampy by the rains. The coast also is swampy S. of the town of Malacca, but to the N. it is generally bold and rocky. There are several rivers, but the largest is only navigable by small vessels for 10 or 12 m. from its mouth. Opposite the coast are many small granitic islands, which serve for burial places to the Malay inhabitants of the colony. The country is geologically composed of a granitic formation, overlain by laterite, and this again by a layer of vegetable mould, which becomes thicker the nearer the coast. The soil near the sea shore is very productive, but in the interior it is otherwise; and Naning is much more valuable for its tin mines than for the products of its agriculture. The climate is more salubrious perhaps than that of any other British coast settlement in the East. It has been found that during a period of seven years, the deaths among the troops stationed here amounted to less than two per cent.; and instances of longevity are frequent among both Europeans and natives. The mean annual temperature is about 77° 8' Fah., and there is but little change throughout the year in the barometer, which stands at about 30°. Rain falls continually at intervals of a few days; but as rather more occurs between September and January than at any other time, that period is termed the wet season. Violent squalls and storms of lightning, &c., occur during the S.W. monsoon. The produce of Malacca consists chiefly of rice, jaggery, sago, pepper, rattans, timber, cocoa-nuts, a few nutmegs, cloves, dammer, gambier, gum lac, ivory, gold dust, tin, fruits, poultry, and cattle. A few years ago the rice raised in the colony was scarcely sufficient for four months' consumption, the additional supply being brought from Acheen, Java, and Bengal. A principal cause of this was the former policy of the Dutch, who, while Malacca belonged to them, prohibited the raising of any kind of grain, in the view of rendering the inhabitants wholly dependent for their supplies on Java. The British government, however, has given every encouragement to native agriculture; and, in 1835, the crop of rice amounted to two thirds the annual consumption. Cocoa-nuts form a considerable portion of the food of the lower classes of natives, who also subsist partly by fishing. For the trade of the colony, see *post*. This settlement is included in the presidency of Bengal, and is governed by a resident, with an assistant resident at Malacca, and a superintendent at Naning. The Dutch drew from it a surplus revenue, but since it came into our possession, the expenditure has always exceeded the income by about 100,000 rupees a year. In 1837-38, the revenue only amounted to 53,543 rupees, or £5354.

MALACCA, a town on the W. coast of the Malay peninsula, cap. of the above British colony, at the mouth of the river of the same name, lat. 2° 14' N., long. 102° 12' E., about 106 m. N.W. Singapore, and 230 m. S.S.E. Pinang. Pop. in 1832, 19,190, of whom about 4000 were Chinese, 2600 Malays, 2000 Chuliahs, and 2060 Europeans. "The town of Malacca is divided by the river above mentioned into two parts, connected by a bridge. On the left bank rises the verdant hill of St. Paul, surrounded by vestiges of an old Portuguese fort. Around its base lie the barracks, lines, and most of the houses of the military; the stadthouse, courthouse, jail, church, civil and military hospitals, the site of the old inquisition, convent, the police-office, the school, postoffice, and master attendant's office. On its summit stand the ruins of the ancient church of our *Lady del Monte*, erected by Albuquerque, and the scene of the labours and miracles of that 'Apostle of the East,' St. Francis Xavier; also the light-house and flag-staff. A little to the S. rises the hill of St. John's, and in the rear rises

MALAGA.

that of St. Francis. On these eminences are the remains of batteries erected by the Portuguese and Dutch, commanding the E. and S. entrances to the town. Smaller knolls intervene, covered with the extensive cemeteries of the Chinese. The tombs are white, and constructed with much care, and surrounded by low walls of brick and chunam, in shape resembling a horse-shoe. The bazaars, and by far the greatest part of the town, are situated on the right bank of the river. The anchoring ground in the roads is secure; and though large vessels are obliged to lie at a distance of 2 m. from the shore, accidents have been rarely known to happen. Native craft anchor much nearer, under the lee of one of the islets close in shore." (*Newbold's Malacca*, i., 109–111.)

The principal public institution at Malacca is the Anglo-Chinese college, established in 1818. Its main objects are the cultivation of Chinese literature by Europeans, and of European literature by the Chinese, Malays, and surrounding nations, and the diffusion of Christianity. The college has a library well stocked with European and Chinese books, Siamese MSS., &c.; and attached to it is an English, Chinese, and Malay press. This college was founded by Dr. Morrison, the Chinese scholar, from whom, also, it received a small endowment. But at present it depends almost wholly on the fees paid by the pupils; and its funds are by no means in a prosperous state. Such an institution would, however, appear to be deserving of public support. There are also in the town five Chinese schools, with about 100 scholars, besides several Hindoo and female schools, and schools established by the Malays for their own instruction in English. A full account of the mode of education in the Chinese schools may be seen in Newbold's work on Malacca.

Malacca was formerly a place of considerable trade; but, owing to the superior advantages of Pinang and Singapore, its commerce has rapidly decreased within the last 10 years, and it is now very limited. It exports small quantities of gold dust, balachong, hides, hogs, fowls, jaggery, pepper, dammer, cordage, a little ebony and ivory; iron implements, fire-arms, nails, &c., manufactured by the Chinese smiths at Malacca, with rattans, lac, and aloe-wood. The gold and tin are not the produce of the British territory, but of the adjacent native states, whence they are brought to Malacca by native boats, or overland by coolies. The principal imports are earthenware, iron, rice, sago, opium, nankeens, European and Indian piece-goods, woollens, paper, provisions and liqueurs for the European and Chinese inhabitants; salt, sugar, tea, tobacco, &c., partly for home consumption and partly for re-shipment. The total value of the imports, in 1834–35, amounted to 467,450 dbll.; total do. of exports, 236,192 doll.

Malacca is said to have been founded in 1252, by Iskander Shah, a chief from Singapore, and it soon became a large and flourishing city, its influence extending over all the peninsula and the adjacent islands. It was first visited by the Portuguese in 1508, and captured by them in 1511. In 1641 it was taken by the Dutch, and in 1795 by the English. The latter held it till 1818, when it was restored to the Dutch; but in 1825 the latter finally exchanged it with us for the settlements of Bencoolen, &c., on the coast of Sumatra. But we much doubt whether its possession be of any material advantage, or, at least, whether the advantage be at all adequate to countervail the expense it occasions. (*Newbold's British Settlements in Malacca; Hamilton's E. I. Gaz., &c.*)

MALACCA (STRAITS OF), a channel of the Eastern seas, extending from lat. 1° and 6° N., and long. 98° and 104° E., between the Malay peninsula on the N.E. and the island of Sumatra on the S.W. Its length, N.W. to S.E., may be estimated at about 520 m.; its breadth varies from 25 m. opposite the Naning territory, to nearly 200 m. at its N. extremity. It is the best and most frequented passage from the Indian ocean to the China sea.

MALAGA, an important city and seaport of Spain, kingd. Granada and prov. of its own name, at the bottom of a deep bay, on the Mediterranean, 68 m. N.E. Gibraltar, and 254 m. S. by W. Madrid; lat. 36° 43′ 30″ N., and long. 4° 25′ 7″ W. Pop., according to Miñano, 51,869; but little dependence can be placed on this statement, and the population is believed to amount to near 65,000. It is built along the shore, at the foot of mountains gradually descending towards the sea: westward is the Vega, watered by the great river of Malaga, which delivers a large body of water from the E. end of the Serrania de Ronda; and on the other side rise naked rugged mountains, overhanging the shore, and scarcely leaving room for the town. But the most imposing view of Malaga is from the sea. "It stands in the centre of a wide bay, flanked by lofty mountains, and by the picturesque ruins of its ancient fortifications and castle, which cover the hill rising immediately to the E., and seem, from their great extent, like the remains of a former state." (*Inglis*, ii., 136.) The streets, as in

all Moorish towns, are very narrow, many being only 8 feet wide, with others still narrower, badly paved, and dirty to a proverb: the houses are high and large, built round a court, the interior having a clean and neat appearance, owing to the abundant use of whitewash. There is only one square in the town, and the churches, as well as convents, are so crowded among the houses that their beauty, if they have any, is effectually concealed. "Indeed," says Mr. Inglis, "the only handsome feature of the town is the Alameda, a public walk, the buildings round which are certainly magnificent: the other parts present a labyrinth of narrow, intricate streets, inhabited by the tradespeople." (Vol. ii., p. 139.) The chief public buildings and establishments are a cathedral, with a chapter, four parish churches, a bishop's palace, four hospitals (one of which is for military), a legal seminary, royal college of medicine and surgery, a foundling asylum, a large dépôt for convicts, a custom-house, and two endowed schools. Among these, however, the only edifice worth notice is the cathedral, a large building, having a spire 270 ft. in height: like that of Granada, it is in the transition style, between the Gothic and classic; the roof, instead of being groined, is divided into numerous small circular domes, somewhat like the marigold windows of Gothic architecture; and the modern additions to the building, though not quite in keeping, are, on the whole, designed with good taste. (*Cook's Sketches in Spain*, ii., 100.) The high altar and the pulpit are of flesh-coloured marble: "but the part which most rivets the attention is the choir, called by the biographer Palomino the eighth wonder of the world, and admirable for the perfection of its carved works, representing in very bold relief the twelve apostles, and most distinguished of the saints." (*Townsend*, iii., 12.) On a sharp point of rock commanding the city stands a fine old Moorish castle, in good preservation, called the Gibralfaro (prob. Gebel-al-faro, the great watch-tower), built on the site of a Roman fortress, but still wholly of Arabic architecture: it is altogether, both from its shape and situation, a very curious structure, and, if fortified on the modern system, might be rendered impregnable. Another Moorish building, in tolerable preservation, was formerly the darsena or dock for the ancient galleys, now used as a storehouse. (*Swinburne*, i., 323.) The Alcazaba, an Arabian palace, once occupied a site near the shore; but the greater part of it was pulled down to make room for the custom-house. At a short distance from Malaga is one of the magnificent but unfinished undertakings of Charles III., a bridge and aqueduct over the great river of Malaga, which flows about a league distant from the city; but this work, on which a great outlay was incurred, was rendered useless a few years afterward by a work undertaken by a bishop, who, at his own expense, brought water into the city by a much shorter line. (*Cook*, vol. i., p. 19.)

Malaga is probably entitled to rank as the third or fourth port of Spain; but, owing to the want of official returns, and the prevalence of smuggling, it is impossible to obtain any accurate account of its trade. The principal articles of export are wines and fruit, particularly raisins, almonds, grapes, figs, and lemons: there is likewise a considerable, though smaller, exportation of olive oil, with brandy, anchovies, cummin-seed, anise-seed, barilla, soap, &c. Lead is also brought thither for shipment from the mines of Alora in Granada. The imports comprise salt fish, iron hoops, bar iron, and nails; cotton fabrics, hides, earthenware, &c.; with woollen cloths, all sorts of colonial produce, butter and cheese from Holland and Ireland, linens from Germany, &c. The trade with England has been for some time diminishing, owing to our small demand for Malaga wine; but the trade with America has considerably increased, owing to its pretty large consumption both of the fruit and wine shipped at this port. Mr. Inglis has given the following details with respect to the trade of this port, which may be interesting to some of our readers: "The wines of Malaga are of two sorts, sweet and dry; and of the former of these there are three varieties: 1st, the common 'Malaga,' known and exported under that name, in which there is a certain proportion of burnt wine, which communicates its peculiar taste to the 'Malaga:' the grape from which this wine is made is white, and every butt of Malaga contains no less than 11 gallons of brandy; 2dly, 'Mountain,' made from the same grape as the other, and, like it, containing colouring matter and brandy, the only difference between the two being, that for 'mountain' the grape is allowed to become riper; 3dly, 'Lagrimas,' the richest and finest of the sweet wines of Malaga; it consists of the droppings of the ripe grape hung up, and is obtained without the application of pressure. The dry wine of Malaga is produced from the same grape as the sweet wine, but pressed when greener: in this wine there is one eighth more brandy than in the sweet wine: at least one twelfth part of the dry Malaga being brandy. The whole produce of the Malaga vineyards is estimated at from 35,000 to 40,000 butts; but, owing to the increasing stock of id wine in the cellars, it is impossible to be precise in this calculation. The export of Malaga wines may be stated at about 27,000 butts. The principal markets are in the United States and the states of S. America, to which countries the exports are rather on the increase. The average price of the wines shipped from Malaga does not exceed 35 dollars per butt; but wines are occasionally exported at so high a price as 176 dollars. Many attempts have been made at Malaga to produce sherry, but not with perfect success. The Xeres grape has been reared at Malaga, upon a soil very similar to its native soil; and the sherry made at Malaga might be introduced into the English market as sherry, nor, from its cheapness, could it fail to command a sale. One reason of the very low price of the wines of Malaga, is the cheapness of labour; field labour is paid by 2½ reals a day (4½d.), wages during the fruit and vintage time being about double.

"Next to its wines, the chief exports of Malaga are fruits; as raisins, almonds, grapes, figs, and lemons. During September and October, 1830, the export of raisins amounted to 288,845 boxes, and 31,916 smaller packages. Of this quantity, 195,334 boxes were for the United States; 45,513 for England; the remaining quantity being for France, the West Indies, the Spanish ports, S. America, and Holland. The raisins are of three kinds, muscatel, bloom or sun raisins, and lexias. The muscatel raisin of Malaga is the finest in the world, and is its preparation no art is need, the grape being merely placed in the sun, and frequently turned. The bloom, or sun raisin, is a different grape from the muscatel, but the process of preparing it is the same; like the other, it is merely sun dried. The lexias acquire this name from the liquor in which they are dipped, and which is composed of water, ashes, and oil; these, after being dipped, are also dried in the sun. All muscatel raisins are exported in boxes, and also part of the bloom raisins. In 1830 the number of boxes of muscatel and bloom raisins exported amounted to 330,000, each box containing 25 lbs.; 8,000,000 lbs. in all. This quantity is independent of the export of bloom raisins in casks, and of lexias, the annual export of which does not exceed 35,000 arrobas. The export of raisins to England has fallen off; the export to America has constantly increased. In 1834, 75 ships cleared from Malaga for England, with fruit. In 1836, up to the 1st of November, 34 vessels had cleared out. Of the other fruits exported from Malaga, grapes, almonds, and lemons are the most extensively exported. In the months of September and October, 1830, 11,612 jars of grapes were sent to England; 6499 to America; and 1659 to Russia. During the same period of time 5335 arrobas of almonds (133,375 lbs.) were exported to England; and this constituted nearly the whole export: and during these months, also, there were exported to England 3759 boxes of lemons; to Germany, 4901 boxes; and to Russia, 849 boxes. There is also a large export of oil from Malaga." (Spain, ii., 145-149.)

Malaga has an excellent harbour, formed by a fine mole, 700 yards in length, at the end of which is a lighthouse furnished with a powerful light, revolving once a minute. A shoal that had grown up round the mole-head has been removed by dredging. The harbour, which will accommodate more than 450 merchant ships, may be entered with all winds, and affords perfect shelter. The port dues to a Spanish ship of 300 tons amount to about £11 10s.; those to an English vessel of the same burden being about £31. Goods may be warehoused for any time not exceeding 1/2 months, on paying two per cent. ad valorem in lieu of all charges; but at the end of the year they must be either entered for consumption or reshipped. (For weights and measures, commercial details, &c., see Com. Dict., art. Malaga and Spain.)

Malaga, independently of its export trade, has manufactures of linen and woollen cloths, sail-cloth, ropes, paper, leather, hats, and soap; an iron foundry and a cigar manufactory; but, excepting the latter, they are all on a small scale, and insufficient for the consumption of the inhabitants. Pilchard and anchovy fisheries also give employment to a considerable number of the lower classes. The market is well supplied, the show of fruit in particular being unequalled in Spain. Mr. Inglis quotes the prices of several leading articles, as follows: Beef and mutton, 10 quartos (about 3d.) per lb.; pork, 14 quartos; a fowl, 7 reals (13d.); a duck, 15 reals; a turkey, from 20 to 30 reals; a rabbit, 10 reals; and a partridge, 4 reals. Poultry, however, is here not only exposed whole, but also cut up into joints, like butchers' meat. A barrel of anchovies may be bought for two reals (4½d.), and many other varieties of fish are remarkably cheap and plentiful. Potatoes are sold for seven quartos the 6½ lbs.; and excellent wine may be procured for two reals a bottle. Bread, one of the dearest articles of food, fetches 12 quartos (3½d.) per lb., and eggs are sold for 1d. each. Melons, pomegranates, and prickly pears, which, with fish, constitute the principal food of the lower orders, are so cheap as scarcely to form an article of expenditure.

The general aspect of the population of Malaga is even more Moorish than that of Seville, and affords innumerable pictures of idleness. Hundreds of the lower classes appear in the streets doing nothing, sitting on the ground, lolling against a wall, or lying on the steps of church-doors, wrapped in brown, ragged, and patched cloaks. In fact, Malaga is noted for idleness and demoralization. The necessaries of life being so cheap, there are few motives to industry: begging is very common, and has been encouraged by the ill-judged bounty of the monasteries, which, however, were suppressed in 1835: but, when deprived of their legitimate resources, the mala gente (such being the sobriquet of these people) are at no loss to find, owing to the inefficiency of the police, a few quartos in some other way. The whole of them, indeed, are thieves; and in so degraded a state is public justice in this city, that crimes of the darkest hue pass unpunished. The clasp-knife is in frequent use to gratify revenge, and murders often follow acts of robbery. (Compare Inglis, ii., 138, with Townsend, iii., 18.) This degraded state of morals is attributed, by Sir C. Brooke, partly at least, to the fact that convicts, called presidiarios, are detained here previous to their departure for, and after their release from, the penal settlement of Ceuta, in Africa, which see. (I., 592.) An efficient government would, however, speedily change the whole aspect of society; the impunity that crime has so long enjoyed in this miserable country being the great cause of its prevalence. The more respectable classes of the people are described as agreeable, hospitable, and generally fond of society, the ladies being equally witty and high-spirited with those of Seville, quite as showy in dress, and not a whit more strict in morals. The Italian opera is a favourite resort, and many ladies are good musicians. Numerous foreigners also reside in Malaga, especially English and Americans, who constitute, with a few of the government officers and merchants, the elite of society. Most of these have country seats in the environs, the beauty of which is not surpassed in any part of Andalusia. The weather during summer is intolerably hot, and at this season, especially during the prevalence of the hot southern winds, the inhabitants exclude the sun as much as possible, and remain at home during the day; but when the heat is succeeded by the refreshing coolness of the evening, the whole population is astir, and after nightfall the young people bathe for hours in the sea, a practice quite as conducive to health as pleasure. Nervous and epidemic fevers are still, however, very prevalent, and sometimes carry off great numbers of people.

Malaga, like most other cities of Spain, has had various masters. Built by the Phœnicians, and called by them Malacha, it came successively into the hands of the Carthaginians and Romans, both of whom procured from it considerable supplies of salt fish and provisions. It then passed into the hands of the Goths; and from them, in 714, to the Moors, who were at length driven hence by Ferdinand the Catholic, in 1487. The yellow fever carried off nearly 22,000 of its inhabitants in 1803, and reappeared, though attended with less fatal consequences, in 1813. Malaga was taken by the French in 1810, after an obstinate conflict with a body of Spaniards, officered by monks and commanded by a Capuchin friar; and remained in their possession till 1812. (Inglis's Spain, ii., 136-153; Townsend's Journey, iii., 10-42; Cook's Sketches, i., 18-24; Swinburne, i., 320-327; Sir A. de Capell Brooke's Spain and Morocco, ii., 202-207.)

Malaga, p. t., Monroe co., O., 110 m. E. Columbus, 290 W. It contains one church, two stores; two schools, 55 scholars. Pop. 1442.

MALAY PENINSULA, a long and narrow territory, forming a part of India beyond the Brahmaputra, and the most S. portion of continental Asia, lying chiefly between the 1st and 8th degs. of N. lat., and the 98th and 106th of E. long.: it has N. Lower Siam, with which it is connected by the isthmus of Kraw; and is on all other sides surrounded by the sea, called on the W. and S. the straits of Malacca and Singapore; and on the E. the China sea and gulf of Siam. Length, N.N.W. to S.S.E., 450 m.; breadth varying from 50 to 150 m. Area estimated at 45,000 sq. m.

Exterior Native States.	Pop.	Interior Native States.	Pop.
Quedah and Ligor	50,000	Rumbowe	9,000
Perak	35,000	Sungie-ujong	3,800
Salangore and Calang	12,000	Johole	3,000
Johore (including Sejam at and Muar)	25,000	Jompole	2,000
		Jellabo	2,000
Pahang	40,000	Srimenati	3,000
Komaman	1,000	Aborigines scattered over	
Kalantan	50,000	the Peninsula	8,000
Tringanu	30,000		
Patani	10,000		
	255,000		36,680

British Possessions.				
Malacca and Naning	(1806)			37,706
Province Wellesley	(1535)			48,480
Total Population.	374,266			84,566

MALAY PENINSULA.

As far as lat. 6° S. the country is claimed by the Siamese; but beyond that point the peninsula is subdivided among independent native states and British colonies, which may be enumerated as in the preceding table, with their probable population, as estimated by Lieutenant Newbold.

Physical Geography.—The central and longest of the mountain chains, passing S. from the table land of Yunnan, through the Ultra-Gangetic peninsula, traverses this territory in its entire length. This mountain chain diminishes in height as it approaches the equator; and its highest peaks in Rumbowe and Jahore probably do not exceed 3000 feet in elevation; while many peaks in the N. part of Quedah are supposed to rise to upward of 6000 feet above the sea. M. Ophir, a detached mountain in about lat. 2° 30' N., and long. 102° 30' E., has been roughly estimated at nearly 5700 feet in height, but it is much more lofty than any other summit in the S. part of the peninsula. Between the above mountain chain and the coast the surface is undulating, covered with dense primeval forests, or interspersed with grassy plains, which are by far the most numerous and extensive in the N. An abundance of rivers descend to either coast, in their progress frequently forming marshes and lakes, some of which are of considerable size. Their banks are generally low, swampy, and covered with mangrove and other thickets; and, though several of them are broad, and moderately deep, the sand-banks, coral reefs, &c., at their mouths, usually preclude their navigation by vessels of any magnitude. A number of verdant islets stud the coasts, especially the north-western and the southern.

Geology and Minerals.—The Malay mountain chain, as far as it has been hitherto explored, consists chiefly of gray stanniferous granite and clay-slate. At its S. extremity porphyry occurs; hornblende is met with near Malacca; and quartz is very abundant around M. Ophir, and elsewhere. The geology of the E. coast is almost wholly unknown; but along the W., laterite, similar to that of the Malabar coast, is a very prevalent formation. Clay-slate, sandstone, argillaceous schist, jasper, limestone, and graiwacke are the other most prevalent rocks. Limestone composes a portion of several of the islands off the W. coast, while those off the S. coast are chiefly of granite or syenite. The Elephant rock in the Quedah territory is a mass of calcareous breccia, having many stalagmitic caverns, and interspersed with an abundance of fossil remains. At the S. extremity of the peninsula are evident traces of volcanic action; and submarine thermal springs, scattered over the country, testify the activity of subterranean heat at no great distance below the surface. These are sulphureous and saline. The springs at Ayer-pannas, near Malacca, were found by Newbold to have a temperature of 190° Fahr. at noon, and of 113½° at 6 A.M.

The Malay peninsula produces tin, gold, and iron: tin is, in fact, among its principal articles of export. Mr. Crawfurd observes, that tin, wherever found, has a limited geographical distribution; but that, where tin does exist, it is always in great abundance. The tin of India has, however, a much wider range of distribution than that of any other region, being found in considerable quantity from long. 96° to 107° E., and from lat. 30° N. to 3° S. (*Indian Archip.*, iii., 450.) It has been latterly stated that it is found in abundance at Sakéna, in the interior of Tavoy, lat. 12° 40', and in Siam even as far N. as 14°. At any rate the Malay peninsula appears to be the centre of the region in the eastern seas in which tin is distributed; and, including the island of Junk-ceylon, it has been roughly estimated that its annual produce of this metal amounts to 34,600 piculs of 133⅓ lbs. avoird. The ore of the peninsula is extremely pure, being that which is called stream. The ore of Sunie-ujong, Naning, and Persk is reported to yield 76 per cent. metal, whereas the ores of Cornwall, with all the advantages of European science and ingenuity, do not yield more than 75 per cent. But the process of smelting, as conducted by the Malays, being very defective, and adulteration frequent, the peninsular tin fetches only from 14½ to 15 dollars the picul; while the tin of Banca, wrought by Chinese, sells at from 16 to 16½ dollars. The export of peninsular tin may amount to about 3080 tons a year, including from 400 to 500 tons received from the Malacca straits and Banca. A large portion is brought to England, Malay tin being now very extensively imported for warehousing; large quantities are also carried to Holland, where there are refining houses.

The Malay peninsula does not by any means so well merit the term, *Aurea Chersonesus,* which has been before applied to it, as the neighbouring island of Sumatra. The exports of gold from the S.W. coast of that island amount, according to Marsden and Hamilton, to 26,400 oz. a year, while the annual produce of the peninsula is roughly estimated at less than 20,000 oz. It comes chiefly from the E. coast, and M. Ophir, where it occurs disseminated through quartz, in thin granular veins, and in alluvial deposites. Iron is found in Quedah, but only in small quantities

The Climate is remarkable for its continual moisture, to

which circumstance the perpetual verdure of the peninsula is mainly owing. The year is divided into the wet and dry seasons; but the term "dry season" must not be understood in the same sense as when applied to the climate of Hindostan; for, during its continuance, even three successive days rarely pass without a shower. On the W. coast the dry season comes in with the S.W. monsoon in May; the wet season, with the N.E. monsoon in October. Thunder storms, whirlwinds, waterspouts, and other atmospherical phenomena are frequent, especially during the S.W. monsoon.

Vegetable Products are both numerous and valuable. They include a host of trees, the timber of which is adapted for house and ship building; the finest fruits of tropical climates, bamboos, canes, rattans, &c., of which the jungles are in great part composed; the areca, sago, and gomuti palms, the catechu, dragon's blood, and India rubber plants, the upas of the Javanese, &c. It has been denied that teak is indigenous to the country; but, according to Newbold, the inland Malays affirm that it is occasionally found, and is known under the name of *jati.* The wild nutmeg is a native of the country. The true nutmeg, cinnamon, and clove have been long introduced, and thrive well. Tobacco, coffee, sugar, cotton, and the true indigo (*Indigofera tinctoria*) are cultivated with much success. Mr. Crawfurd (*Embassy to Siam, &c.,* i., 178) estimates that the Malay peninsula produces 28,000 piculs of pepper a year, or about one thirteenth part of the total produce of the E. Rice, and other kinds of grain, are not grown in quantities sufficient for home consumption, and are, therefore, imported chiefly from Bengal and Sumatra.

Elephants roam over the peninsula in great numbers; the rhinoceros, tapir, wild hog, the royal and the spotted black tiger, two kinds of bears, and two species of bison, the axis, plandok, musk-deer, and several other kinds of deer, the vampire, and many varieties of bats, and numerous monkeys, are among the wild animals. The buffalo is a native, and is domesticated; but neither the cow, camel, horse, nor ass, are met with in a state of nature. The great density of the jungles is considered unfavourable to the increase of feathered game; but waterfowl are plentiful, and there are a great many pheasants of the richest plumage. Crocodiles, alligators, and several kinds of formidable serpents, are met with. The dugong, many turtles, and a plentiful supply of fine fish, are caught in the surrounding seas.

People.—The Malays have been ranked by some authors as one of the five great families, or varieties, of the human race. But this opinion is by no means generally entertained. Newbold says, "Both their features, and those of the aborigines in the native states around Malacca are decidedly characterized by the Mongol stamp." (i., 422.) And independent of the Malays having no peculiarity of form or feature to entitle them to be called a distinct variety, there appears to be sufficient evidence to show that they are a mixed race of comparatively recent origin. Antecedent to the 13th century of our era, the coasts of the peninsula, and the adjacent islands, were inhabited, though thinly, by a tribe of *ichthyophagi,* and the interior by a race of negro savages, by whose descendants it is still occupied. In the course of the above century, a body of colonists, the ancestors of the present race of Malays, arrived on the continent, from Menankabowe, in Sumatra: and whether by intermarriage (as traditionally reported) or by conquest, extended their dominion over the whole peninsula. During the succeeding centuries they conquered Sumatra, the Sunda, Philippine, and Moluccа isles, with many smaller groups; and are now found in all those regions, and in Borneo, &c.; but without any centre of unity or power. The chief physical characters of the Malay race consist in a brown colour, varying from a light tawny to a deep brown; black hair, more or less curled, and abundant; the head rather narrow; the bones of the face large and prominent; the nose full, and broad towards the apex; and the mouth large. The average height of the men is about 5 feet 2 inches. A general character can hardly be assigned to a people so widely distributed. The Malay inhabitants of the peninsula are, however, active, restless, and courageous; but their courage is hot of a steady, deliberate character, but is rather a sudden ungovernable impulse, arising from a paroxysm of rage. "To their enemies they are remorseless, to their friends capricious, and to strangers treacherous." (*Hamilton.*) Perhaps their treachery to strangers may, in part at least, be occasioned by the behaviour of the latter, or the antipathy excited against them by the behaviour of former strangers. Malcolm, who remained for some time in this region of Asia, says, that "in their intercourse with each other, domestic and private virtues prevail to a great an extent as among the adjacent nations. A propensity to gambling is a distinguishing trait in the Malay character; and more especially a taste for cock-fighting, to which sport the Malay is so passionately addicted, that his last morsel, the covering of his body, his wife and children, are often staked on the issue of a battle to be fought by his favourite cock." Malcolm admits that a "disregard of human life, revenge, idleness, and piracy, may be considered common to Malays. The universal practice of going armed makes thoughts of murder familiar. The right of private revenge is universally admitted, even by the chiefs; and the taking of life may be atoned for by a small sum of money." (*Travels*, ii., 194.) In the era of peace they are greatly inferior to their neighbours of Java, Japan, Cochin-China, and Siam. The Malay language coincides with monosyllabic tongues in its general construction and analogies, but is properly polysyllabic in its form. It consists chiefly of Polynesian, an intermixture of Sanscrit and Arabic, and a dialect purely Malayan, which last, however, constitutes little more than 1-4th part of the written and spoken language. The literature of the Malays is almost entirely derived from Hindostan, Persia, Arabia, Java, and Siam. Arabic is exclusively their sacred language; and their religion also has been derived from Arabia, all the Malays, with trifling exceptions, being Mohammedans.

The negro tribes which inhabit the interior of the peninsula are called by the Malays *Orang Benua*, men of the soil. They appear to be a distinct variety, differing from and being inferior to both the African and Papuan negro. The average height of the men is only 4 feet 8 inches. The Malay negroes are thinly spread over a considerable extent of territory in and behind Malacca, and thence N. to Mergui; but they probably amount in all to only a few thousands. They are divided into several tribes, some of which are said to dwell altogether in trees or clefts in the mountains. A few have learned a little Malay, and occasionally venture among the adjacent Malay tribes, to purchase tobacco and utensils; but of letters they know nothing. Copious accounts of both the Malays and this people may be found in *Newbold's Malacca, &c.*, vol. ii., ch. 12, 14, 15; and various details respecting the races inhabiting the Malay countries are given in the art. E. ARCHIPELAGO in this Dict. (1. 148.) For the *Commerce* of the British settlements, see MALACCA, SINGAPORE, &c.

The principal articles of export from the native states are tin, gold-dust, spices, elephants' teeth, pepper, sago, sugar, canes, timber for ship and house building, dammer, ebony, bees' wax, betel nut, apees, and eagle-woods, hogs, poultry, buffaloes, tiles, and an immense variety of fruits; in return for which, opium, salt, cotton, cloths, tobacco, rice, and some European manufactures, are the chief imports. The trade is principally with the British and Dutch settlements in the East, Siam, China, and the adjacent parts of the E. archipelago.

In the 15th century a large proportion of this peninsula appears to have been under the sway of the Siamese, but since that time it has been mostly divided into the petty states before enumerated, the historical details of which are destitute of interest. The successive settlements made by the Portuguese, Dutch, and British at Malacca, &c., are elsewhere noticed. The only recent event worthy of mention has been the subjugation of Quedah (or Keddah) by the Siamese, begun in 1821, and completely effected within about 10 years afterward. (*Newbold's Malacca; Crawford, Malacca, Hamilton's E. I. Gaz.; Lawrence's Lectures in Men, &c.*, passim.)

MALDA, a town of Hindostan, prov. Bengal, district Dinagepoor, on the Mahananda, built chiefly of the ruins of Gour, from which it is distant about 10 m. N. Early in the present century it had 3600 houses huddled together along the bank of the river, which, during the rainy season, nearly insulates the town. The E. I. Company established a factory here as early as the 17th century; and there were formerly some prosperous French and Dutch silk and cotton factories in the town; but the trade of Malda has now sunk into irreparable decay, its manufactured goods being unable to withstand the competition of those introduced into India from Europe.

MALDEN, p. t., Middlesex co., Mass., 5 m. N. Boston, 45 W. First settled in 1648. It contains 1000 acres of salt meadow, in the S. part of the t. Malden bridge, over Mystic river, is 9490 feet long, and connects it with Charlestown. It contains four churches, a Congregational, Methodist, Baptist, and Universalist; 13 stores, three grist-mills; eight schools, 650 scholars. Pop. 2514.

MALDIVE ISLANDS, or MALDIVES, a chain of islands in the Indian ocean, extending between the 1st deg. of, and the 7th of N. lat., a distance of about 560 stat. m.; and between 72° 48' and 73° 48' E. long. The Laccadive islands, to the N. of the Maldives, may not improperly be considered a continuation of this island-system. They are of coralline formation, arranged in round or oval groups called *atolis*, separated by several channels, which may be safely navigated by ships of the largest size. The different groups are surrounded by coral reefs, on which the surf beats violently; but between the islands the sea is perfect-

ly smooth, and forms safe harbours for small craft. These islands have been rarely visited by Europeans, though lying in the direct route to India. All that are of any extent are richly clothed with palms and other trees; but no edifice has been seen in sailing past them, whence it may be concluded that none exists higher than a cocoa tree. The Maldives produce millet and other small grains, of which they have two harvests a year; but they are unsuitable for rice and wheat, which are imported. Excellent roots and fruits are found in the greatest profusion; and poultry are extremely abundant, and bred with little or no attention. There are neither horses nor dogs, and but few horned cattle. Fishing is an important occupation, especially that of cowries, a species of shells used as money in small payments in Hindostan and other Asiatic countries, and in extensive districts of Africa. The inhabitants trade with Hindostan, Sumatra, &c., arriving at Belasore and other ports of British India during the S.W. monsoon with cowries, coir, the produce of the cocoa tree, salted fish, tortoise-shell, &c.; and sailing homeward with the N.E. monsoon, taking rice, sugar, manufactured goods, tobacco, &c. The people of the Maldives are Mohammedans, and probably of an Arabic stock. They live under a sultan, who, according to Hamilton, resides in Male, an island about 3 m. in circuit, fortified by walls and batteries, on which above 100 pieces of artillery are mounted. The sultan, however, considers himself dependant on the British government of Ceylon, to which he sends an annual embassy. (*Geog. Journ.*, ii., 79-83; *E. I. Gaz.*)

MALDON, a parl. and mun. bor., river port, and market town of England, co. Essex, hund. Dengey, on the Chelmer, 14½ m. S.W. Colchester, and 37 m. E.N.E. London. Area of parl. bor. (which includes the par. of Heybridge with the old bor.) 4719 acres. Pop., in 1831, 4895. "The town, which is neither paved, lighted, nor watched, and does not appear to be in a flourishing condition, occupies the ridge of a hill on the S. side of the Chelmer, and consists principally of one long street, running parallel to the river, the E. end of this street forming the portion called 'the Hythe;' two other streets, one from the centre of the town, and the other from its W. end, unite at the bottom of the hill, and extend across the Chelmer into an almost insulated flat called 'Potman's Marshes.'" (*Mun. Bound. Rep.*) The town-hall is an old building near the junction of the streets at the W. end of the town, and not far from it is an extensive range of barracks: there is also a small borough jail. Maldon had formerly three parishes; but two of them have been long consolidated. The largest church, that of All Saints, near the town-hall, is an ancient and very large edifice, with a square tower, surmounted by a curious triangular spire. St. Mary's is a spacious building, at the lower end of the town, said to have been founded before the Norman conquest; but the tower and W. end were rebuilt in the reign of Charles I. The united vicarage of All Saints and St. Peter's is in private patronage, the rectory of St. Mary's being in the gift of the dean and chapter of Canterbury. St. Peter's tower is the only part now standing of that disused parish church, and annexed to it is a building formed of the old materials, which has been long used as the depository of a valuable library, containing 5330 volumes, bequeathed to the town, in 1704, by Archdeacon Plume, founder of the Plumian professorship of astronomy in the university of Cambridge: the tower part, which has since been much enlarged, is occupied by the national school, furnishing instruction to about 270 poor children of both sexes. The grammar school, founded in 1621, received an additional endowment from Dr. Plume, who also gave it an exhibition in Christ's College, Cambridge. The estates vested in the hands of trustees yield about £52, which, after some slight deductions for land-tax, and repairs, are paid over to the head-master; six free scholars receive classical instruction gratis, paying a fee for other branches; and there are, besides these, about 12 pay-scholars. Dr. Plume left also a considerable property (the annual produce of which amounts to about £180) for the clothing and instruction of 15 poor boys, and the foundation of a week-day lecture in the church; besides which, he built a workhouse, lately sold under the provisions of the Poor-law Amendment Act. There is also a large Lancastrian school, with two or three minor charities, and money-bequests. (*Char. Comm. 23d Report*, p. 1.) The Roman Catholics, Wesleyan Methodists, and Baptists, have their respective places of worship; attached to which, as well as to the churches, are respectably-attended Sunday-schools.

"Maldon is not a manufacturing town; but it carries on a considerable home trade in coal, iron, chalk, and timber, which it exchanges for corn, and other farming produce. There belonged to the port, in 1836, 136 ships, of the aggregate burden of 9955 tons. Gross customs revenue, in 1840, £6199." The trade of the town, however, is said to be declining; and the principal cause assigned is the new navigation to Chelmsford, which has been carried 1 m. N.

of Maldon along the adjacent village of Heybridge to the Blackwater. (*Mun. Corp. Report.*)

Maldon claims to be a borough by prescription; but its first charter dates as far back as 1155, and was confirmed by Edward I. and subsequent monarchs. The present municipal officers comprise a mayor and three other aldermen, with 12 councillors; a commission of the peace is held under a recorder. Corporation revenues in 1833, £783. Maldon has sent two members to the House of Commons since the reign of Edward I. Down to the Reform Act, the franchise was vested in the resident and non-resident freemen, by birth, marriage, servitude, gift, or purchase. The Boundary Act enlarged the limits of the borough, by including in it the parish of Heybridge. Registered electors, in 1839–40, 876. In cases of succession to burgage tenures, the custom of borough-English prevails here. Markets, well attended, on Saturday; cattle-fairs, Sept. 13 and 14. (*Parl. Rep., &c.*)

MALDONADO, a fortified seaport town of the Banda Oriental in S. America, on the N. bank of the Plata, not far from the mouth of its estuary, and 85 m. E. Monte Video. Its harbour is sheltered from S.E. winds by the small island of Gorriti, but it has little depth. Population uncertain. "Maldonado is a quiet, forlorn, little town, built with the streets running at right angles to each other, and having in the middle a large *plaza* or square, which, from its size, renders the scantiness of the population more evident. It possesses scarcely any trade, the exports being confined to a few hides and live cattle. The inhabitants are chiefly lead-owners, with a few shopkeepers, and the necessary tradesmen, such as blacksmiths and carpenters, who do nearly all the business for a circuit of 50 miles round. The town is separated from the river by a band of sand-hillocks about a mile broad: it is surrounded on all other sides by an open, slightly undulating country, covered by one uniform layer of fine green turf, on which countless herds of cattle, sheep, and horses graze." (*Darwin's Voyage of the Adventure and Beagle*, vol. iii., p. 45.)

MALLOW, an inland town and parl. bor. of Ireland, co. Cork, prov. Munster, on the Blackwater, and on the high road between Cork and Limerick, 18 m. N. by W. the former, and 37 m. S. the latter city. Mallow, properly so called, is built on the N. side of the river, being united by a bridge of 11 arches, with its suburb of Ballydaheen, on the S. side of the river. The latter is included in the parliamentary borough as fixed by the Boundary Act, which comprises an area of 350 acres, and had, in 1831, a population of 7099. It consists principally of one main and well built street, nearly parallel to the river, and has a handsome parish church, a Roman Catholic chapel, two Methodist chapels, an Independent meeting house, a courthouse, a bridewell, barracks, and infirmary, with commodious baths, a public reading-room, library, &c. On its W. side are the ruins of its old castle, the property of the lord of the manor. There are here two schools, one attended by about 300 boys, and the other by about 130 girls, both under the control of the Board of Education. The town is surrounded by thriving plantations, and is situated in a peculiarly rich and well cultivated part of the country. "Though the river be not navigable, and that Mallow has no manufactures, it is yet considered one of the best country towns of its sort in Ireland. It is resorted to in summer on account of the mineral waters that it possesses, the properties of which are much the same as those of Clifton, and in the neighbourhood there is a very unusual number of country gentlemen's houses, occupied by families of the first respectability." (*Parl. Boundary Report.*) Several flour-mills have been established on the Blackwater, and there is a brewery in the town, and a salt work and some quarries in its immediate vicinity; but the great dependance of the inhabitants is on its extensive retail trade, and on the resort of visiters to its Spa. Branches of the provincial and agricultural banks were opened here in 1835. Postoffice revenue in 1830, £700, in 1836, £830. Mallow was incorporated by charter of James I. in 1612, which vested the right of sending two members to the Irish House of Commons in the provost and 12 burgesses. But this charter fell, in no very long time, into disuse; for above a century since the corporate body was extinct, and the right of electing the members for the borough vested in the *freeholders of the manor*, which comprised 1126 acres, and had, in 1831, a population of 7588. Since the union, Mallow has returned one member to the Imperial House of Commons; and the Boundary Act altered the limits of the parliamentary borough, as already stated, by including in it the suburb of Ballydaheen, and excluding the country part of the manor. Registered electors, in 1839–40, 683. The borough has a court leet twice a year, and a court for debts under £2 every third Wednesday. General sessions are held in April, and petty sessions every Tuesday. Markets on Tuesdays and Fridays; fairs on 1st January (for pigs), Shrove Monday, 11th May, 25th July, and 29th October.

MALMEDY a town of Rhenish Prussia, gov. Aix-la-

Chapelle, cap. circ. on the Warge, close to the Belgian frontier, and 20 m. S. Aix-la-Chapelle. Population in 1828, 4212. It has a noble church, formerly belonging to a rich Benedictine abbey, a fine private cabinet of medals, antiques, &c., and is the seat of the council for the circle, a police court, and board of taxation. It has some mineral springs, similar to those at Spa, and manufactures of fine woollen cloth, glue, and soap; but it is chiefly noted for its manufacture of leather for boot soles, with which it supplies a considerable portion of Germany. There are said to be 38 tanneries in active employment: hides are imported principally from S. America, and bark from the forest of Ardennes. (*Berghaus; Schreiber; Guide du Rhin, &c.*)

MALMESBURY, a parl. bor., market town, and par. of England, co. Wilts, hund. of same name, on the Avon, 17½ m. N.N.W. Bath, and 96 m. W. London. Pop. of parl. bor. (which includes, with the old bor., two out-pars., and the several pars. of Brokenborough, Charlton, Garsdon Lea, Great and Little Somerford, Foxley, and Bremhilham), in 1831, 6185. The town, formerly fortified and more extensive, is pleasantly situated on a hill close to the Avon, by which it is nearly encircled, and which is here crossed by six bridges. It consists of three principal streets, two of which running parallel are intersected by the third. In an open space near the centre of the town is the market-cross, an octangular turreted structure, with flying buttresses and a highly carved, supposed to have been erected in the reign of Henry VIII. There appear to have been formerly several churches in Malmesbury; but it now contains only one, the living being a vicarage in the gift of the lord chancellor. The Wesleyan Methodists, Baptists, Unitarians, and Moravians, have likewise their respective places of worship, and there are three Sunday schools. Two free-schools, one of which is conducted on the national system, and well attended, furnish instruction to poor children of both sexes, and there are two sets of almshouses. "Malmesbury is not a place of any trade, nor is it even a considerable thoroughfare. No new buildings are rising in the suburbs, and it contains few houses appearing to be occupied by persons in independent circumstances; indeed, it has altogether the air of a place on the decline, and must now be considered as entirely an agricultural town. (*Parl. Boundary Report.*)

The borough, which is of high antiquity, received its governing charter from William III.; and it was considered too insignificant to be included in the provisions of the Municipal Reform Act. It has sent two members to the House of Commons from the 23 Edward I.: the franchise, previously to the Reform Act, being in the high steward, alderman, and 12 chief burgesses. The Boundary Act enlarged its limits, by including with it the two out-parishes, as above mentioned. Registered electors, in 1839–40, 257. Markets on Saturday, and a cattle market on the last Tuesday of each month, except March, April, and June. Horse and cattle fairs, March 28, April 26, and June 5.

A nunnery was founded here at the close of the 6th century. Other monasteries were formed here in the two following centuries; and it was a place of considerable and rising consequence as the resort of religious votaries, including, among other establishments, an abbey, which afterward attained to high celebrity. The Danes destroyed the town at the close of the ninth century; but monastic wealth and the beneficence of princes soon restored its prosperity, which it enjoyed almost without interruption till the Reformation. The chief monument of Malmesbury's departed greatness is its abbey, the entire buildings of which, with the church, covered about 45 acres. Little, beyond mere foundation walls, is now left except the church, which appears to have been a very magnificent structure, and presents some fine specimens of different eras of architecture, but chiefly of the early English. It was cruciform, with a tower rising at the intersection of the transepts, and another at the W. end, the front of which was exquisitely finished and adorned with sculpture, having also a very fine window filled with painted glass. During the civil wars, however, when Malmesbury was repeatedly besieged both by the royalists and parliamentarians, the church, already partly dismantled, suffered great injury; both its towers were battered down, its cloisters demolished, and now only a fourth part of the building is standing; but the ruins are highly interesting, and the S. porch is one of the finest specimens of its kind in England. In the town are several other remains of ancient monastic and ecclesiastical buildings; and about one mile from it is a field called Came-hills, in which are evident vestiges of a Roman encampment.

Malmesbury claims the honour of having given birth to Aldhelm and Johannes Scotus, William of Malmesbury, second only to the Venerable Bede among our early historians; and Hobbes, so eminent by his metaphysical and political speculations, was a native of Malmesbury, where he first saw the light in 1588. (*Parl. Rep.; Britton's Arch. Antiq. of G. Britain*, vol. i.)

MALMO, a strongly fortified seaport town of Sweden, cap. the len. Malmöhus, on the Sound, nearly opposite Copenhagen, and 110 m. S.W. Christianstadt. Pop. (1836) 8766. It is irregularly built, but has wide streets and a fine market place. It.has a citadel, two churches, two hospitals, manufactures of woollen cloth, stockings, prepared skins, carpets, hats, gloves, tobacco, starch, soap, looking-glasses, &c., and a brisk trade in the products of these establishments, and in corn; its port, however, admits only small vessels.

MALO (ST.), a fortified seaport town of France, dep. Ille-et-Vilaine, cap. arrond., on the British channel, 40 m. N.N.W. Rennes, and 200 m. W. by S. Paris. Lat. 48° 30′ 3″ N., long. 2° 0′ 51″ W. Pop. (1836) 9409. The town is built at the mouth of the Rance, on the peninsula of Aron, connected with the mainland by a causeway. It is defended by strong walls with four bastions constructed by Vauban, and a castle built by Anne, Duchess of Brittany. On its N. side it is inaccessible; but, from the want of outworks, it could not hold out against a regular siege. The town is in many parts well built, and has some excellent houses. Its chief public edifices are a cathedral, bishop's palace, town-hall, theatre, hospital, foundling asylum, communal college, and exchange. The port, on the S. side of the town, is commodious and secure, but is rather difficult of entrance, and dries at low water; though at high water springs it has a depth of above 40 feet. In 1836, however, the French Chamber passed a resolution for the construction here of a floating dock or basin. It has a good roadstead N.W. of the town, and opposite the mouth of the Rance, which is defended by various forts; the principal, La Conchée, being constructed on an all but inaccessible rock, a considerable distance off shore. St. Malo is the seat of tribunals of primary jurisdiction and commerce, a board of artillery, &c., and is the residence of various foreign consuls. It has a hydrographical school of the first class, a chamber of manufactures, a royal tobacco factory, naval rope-walks, and dry docks for the building of vessels of various sizes. It has also manufactures of fishing-nets and hooks, pulleys, and other marine fittings, soap, &c.; a considerable trade in provisions with the French colonies, a brisk coasting trade, and numerous vessels employed in the mackerel, cod, and whale fisheries. St. Malo has given birth to several distinguished persons; among whom may be mentioned the brave admiral Duguay de Trouin, Jacques Cartier, Maupertuis, Le Bourdonnaye, &c. (Hugo, art. Ille-et-Vilaine, &c.)

MALONE, p. t., capital of Franklin co., N.Y., 214 m. N. by W. Albany, 520 W. Watered by Salmon and Trout rivers. It contains six stores, one lumber-yard, one fulling-mill, one forge, two grist-mills, eight saw-mills, two tanneries, one printing-office, one weekly newspaper; 18 schools, 765 scholars. Pop. 2699. The village is situated on both sides of Salmon river, over which is a bridge of a single arch, 80 feet wide, and 65 feet above the surface of the water. It contains a courthouse, jail, county clerk's office, three churches, a Presbyterian, Methodist, and Baptist; Franklin academy, a state arsenal, a large stone cotton-factory with 3000 spindles, and 80 looms, a grist-mill, saw-mill, and various other manufacturing establishments; 80 dwellings, and about 700 inhabitants.

MALPAS, a market town and par. of England, co. Chester, hund. Broxton, 13 m. N.N.W. Chester, and 153 m. N.W. London. Area of parish, 25,140 acres; ditto of township, 2210 acres. Pop. of township in 1831, 1004. The town, which stands on an eminence near the S. extremity of Cheshire, and on the E. side of the valley of the Dee, comprises three tolerably built and well-paved streets. The living is divided into two rectories, in the patronage of the Egerton and Drake families. The church (formerly the chapel to a Cistrian monastery), a structure of unhewn stone, consists of a nave and chancel, without either aisle or steeple; it is highly ornamented, and some its of decorations are supposed to belong to the Saxon æra. There are also two chapels of ease within the parish; and several de nominations of dissenters have their respective places of worship. A grammar school was founded here in the 17th century; but the free instruction is limited to six boys, appointed by Lord Cholmondley. The master's salary from the endowment is £93, with a good house, &c.; and he is also permitted to receive pay scholars, of whom there were 15 in 1836. Alport's school (founded in 1719) has property yielding an income of £119, and furnishes good plain instruction to boys, girls, and recently also, to infants, with clothing for 14 boys. In 1836, there were in attendance 43 infants, 87 boys, and about 50 girls. The other charities comprise an almshouse for six poor women, with an allowance of bread and money; and large sums have been left, at different times, for the relief of the poor. (Cher. Com., 31st Rep.) Malpas is an agricultural town, and derives its chief importance from its large market for cheese, and its position in the centre of a great dairy-farm district. Mar-

kets on Monday: cattle and cheese fairs, April 5, July 23, and Dec. 8.

MALPLAQUET, a small village of France, dep. du Nord, 16 m. N.N.W. Avesnes. This place is celebrated as the scene of one of the bloodiest and most obstinate conflicts of modern times. On the 11th of September, 1709, the allied army, under the Duke of Marlborough and Prince Eugene, attacked the French army under Marshal Villars in their entrenched camp near Malplaquet. The combat was maintained on both sides with undaunted courage and resolution; but in the end the allies succeeded in forcing the entrenchments. The victory, however, was purchased by the sacrifice of above 20,000 men, killed and wounded. Though vanquished, the loss of the French did not exceed half that number, and they effected their retreat in good order. According to Voltaire (Siècle de Louis XIV., cap. 21), who derived his information from Marshal Villars, the army of the allies amounted to 80,000, and that of the French to 70,000, though other accounts represent each army as about 100,000 strong; but, whichever be the more correct statement, there are certainly very few, if any, instances of so great a carnage in an engagement where the defeated army effected an orderly retreat.

MALTA (an. Melita), an island of the Mediterranean sea belonging to Great Britain, 62 m. S.S.W. cape Passaro, in Sicily, and 198 m. N. Tripoli, in Africa. Valetta, its port and capital, being in lat. 35° 54′ 6″ N., long. 14° 31′ 10″ E. Extreme length, 17 m.; do. breadth, 9 m.: area. 98 sq. m. Estimated pop. in 1838 (exclusive of Gozzo), 103,600: pop. of Gozzo, 16,534; the number of British in both islands being estimated at 4550, and that of the other foreigners at 3116. It is of an irregular, oval shape, rising precipitously from the water's edge on the S. and S.W. The surface presents the appearance of an inclined plane, sloping gradually from its highest elevation (about 1200 ft. in the S.W.) to the more level land on the N.E. side, where it dips into the ocean. The substratum consists of soft calcareous sandstone only scantily covered with soil, great part of which has been carried thither from other countries, or artificially created by breaking the surface of the soft rock into small fragments, which crumble by exposure to the air, and in the course of two or three years become good soil. It has neither lake nor river; and from its geological formation, and the absorbent nature of the soil, has no marshy or swampy ground, except, indeed, two spots of very limited extent at the head of Great Harbour and St. Paul's bay, where the sea has receded and left an accumulation of moist soil, from which noxious exhalations have been supposed to emanate. There is no exuberant vegetation, brushwood, or forest; the verdure is scanty, and the greater part of the surface is an arid rock. The climate of Malta, from its being exposed to the winds blowing from the African and Syrian deserts, is unusually hot, especially during summer, when the heat almost equals that experienced in tropical regions. This heat not only lasts during the day, but, owing to the radiation of the caloric absorbed while the sun is up, it continues, with little abatement, throughout the night; so that, by an excess of heat for months together, a feeling is induced among the inhabitants of extreme lassitude and oppression. The medium temperature of the three coolest months (December, January, and February) is 57½° Fahrenheit, the maximum 61½°, and the minimum 33½°; while the medium of the four hot months (June, July, August, and September) is 78°, the maximum 82½°, and the minimum 73½°. Frequent showers occur in September, increasing in frequency during October and November; but from December to February, the rain falls with nearly the same violence as in the tropics, and the atmosphere continues surcharged with moisture till March, when the weather begins to clear; and during the five following months scarcely a drop falls, the sky being generally without a cloud. The most prevalent winds in Malta are from the S.E., S., and N.W.; the first of which, well known as the scirocco, is at once the most prevalent (especially in autumn), and the most disagreeable in its effects on the human frame; neither are there any regular land and sea breezes, as in some southern countries, to modify the temperature. With respect to the salubrity of Malta, the most favourable opinions have been entertained by some writers; but it appears from the following facts, deduced from medical observations and records during sixteen years, that the average deaths between 1819 and 1834, inclusive, amounted to 2552 a year, being about 1 in 39, or nearly 2·5 per cent. for all ages; whereas in England the mortality, at an average of the same years, was only 1 in 47½, or 2·2 per cent.; so that, even as regards the indigenous inhabitants of both countries,[*] Malta would appear to be less healthy than Brit-

---

[*] This is the statement given by Major Tulloch; but to make the comparison between England and Malta quite accurate, allowance should be made for the rapid increase of population in the former, which makes the mortality appear less than it really is.

ain, and seems only to enjoy the average salubrity of the S. of Europe, in which the mortality varies from 1 in 35 to 1 in 40 of the population annually. The mortality, moreover, is sometimes increased by the prevalence of epidemics, and on two late occasions by plague and cholera, the former of which, in 1813, cut off 4500 of the inhabitants, being 80 per cent. of those attacked. For farther particulars as to the climate of Malta, we beg to refer the reader to Major Tulloch's elaborate *Report on the Sickness, Mortality, &c., of Troops in the Mediterranean.*

Cultivation in Malta is pursued with equal diligence and success. In former times the entire surface was but one mass of barren rock; but continued industry has not only rendered a large part of it capable of tillage, but given it fertility. The rock having first been levelled in terraces, the small particles were pulverized and mixed with soil, while the larger masses were employed in erecting walls to sustain these artificial beds. Soil was also, at first, brought from Gozzo, and even Sicily; but after a time this was found unnecessary. Owing to this laborious perseverance, Malta is now, on the whole, a fertile island, the cultivated parts "yielding annual, and often double crops, without a fallow, and frequently 80 or 90 fold." (*Sir R. C. Hoare's Tour*, ii., 286.) Cotton is the principal product both of Malta and the neighbouring island of Gozzo, the annual crops of which average about 4,000,000 lbs. It is sown in May, and gathered before sunrise in October, the chief vent for it being in the ports of Trieste, Leghorn, Genoa, and Marseilles. The corn crops suffice for the supply of the inhabitants with bread during four or five months a year; the remainder is imported from Sicily and the Black sea, the duties on its importation making it rather high priced. The grass of the island, called *sulla*, is similar to saintfoin, and some, though small crops, are raised of cummin and aniseed. The vine has been cultivated with some care; but its produce is very inferior, and wine as well as oil is imported from Sicily. Figs and oranges are very abundant, and of superb flavour. Brydone says, that "the Maltese oranges are deservedly considered the finest in the world. The season continues upward of seven months, from November till the middle of June, during which time the trees are covered with an abundance of delicious fruit. Many of them are of the red kind, and these are certainly the best. They are produced from the common orange bud engrafted on the pomegranate stock, and the juice of the fruit is red as blood." (*Brydone's Tour*, p. 181.) Some good springs of fresh water are made available for the purposes of tillage; and numerous large cisterns and aqueducts are constructed for the purposes of irrigation. Still, however, Malta imports the principal necessaries of life. Sicily and Odessa supply her with corn, oil comes from the ports of Italy, and wine from Naples and Sicily; from which latter, also, snow and ice are brought, no trifling luxuries in an arid climate like that of Malta. Horses and oxen come chiefly from Barbary, but also from Greece and Albania.

Port and Trade.—The central position, excellent port, and great strength of Malta, make it an admirable naval station for the repair and accommodation of the men-of-war and merchant ships frequenting the Mediterranean, and render its possession of material importance to Great Britain. It is also of considerable consequence, particularly during war, as a commercial *dépôt*, where goods may be safely warehoused, and from which they may be sent, when opportunity offers, to any of the ports belonging to the surrounding countries. Malta likewise presents unusual facilities for becoming the *entrepôt* of the corn-trade of the Mediterranean and Black seas. Her *caricatori* for corn are like those of Sicily and Barbary, excavated in the rock, and are, perhaps, the best fitted of any in Europe for the safe keeping of grain. The harbour of Valetta, which lies on the N.E. side of the island, is divided into two sections by a promontory or tongue of land, on which stands the capital, defended by the castle of St. Elmo. The S.E. side, called the Grand Port, is the most frequented, having an entrance about 250 fathoms in width, with an average depth of 10 or 12 fathoms: it runs inward about 1½ m., has deep water and excellent anchorage throughout, the largest men-of-war coming close up to the quays. N.W. fort St. Elmo is port Marsamuscett, which is also a noble harbour, used exclusively by ships performing quarantine: near its centre is an island, on which are built a castle and lazaretto. The Custom-house and storehouses are in the Grand port, and furnish every facility for landing and warehousing goods. An excellent dock-yard, victualling office, naval hospital, &c., have been constructed for the use of the navy. As a trading-port and *entrepôt*, Valetta rose to high distinction during the war with France; but at the general peace, when commerce reverted to its natural channels, the other ports of the Mediterranean took from Malta a large portion of its trade, which was also depressed by the imposition of various oppressive discriminating duties. In 1819 this vexatious system was partially obviated; but it continued to exercise a very pernicious influence till 1837, when, in consequence of a commission of inquiry, the then existing tariffs of customs-duties and port charges were wholly abolished, and a new tariff was substituted, imposing moderate duties, for the sake of revenue only, on a few articles in general demand, without regard to the country whence they came, at the same time that it equalised the tonnage duties, and reduced the warehouse rent on articles in bond to the lowest level. Everything has thus been done that was possible to second the natural advantages enjoyed by Malta for becoming the grand *entrepôt* of the Mediterranean, and there can be little doubt that they will powerfully contribute to bring about the most favourable results.

The following is an official statement of the ships which arrived at and left Malta in 1838, distinguishing those belonging to Great Britain, the British colonies, and the United States.

Places.	Inwards.		Outward.	
	Ships.	Tons.	Ships.	Tons.
Great Britain	113	16,132	66	10,345
British Colonies	11	1,514	13	1,305
United States	14	2,418	2	1,565
Foreign	1,825	179,229	2,607	222,167
Total	1,963	199,500	2,683	234,307

There belonged to Malta, in 1838, 171 vessels, of the aggregate burden of 17,500 tons, and since Malta-built ships were admitted into the ports of the United Kingdom on the same terms as British-built, the trade of ship-building has materially increased. The vessels, which rank among the best in the Mediterranean, are built with oak timber from Dalmatia; the Maltese are diligent, expert shipwrights, and their wages being moderate, Valetta is a favourable place for careening. Owing, however, to the want of a dry dock, all ships, above the size of a sloop-of-war, requiring to have their bottoms examined, are obliged to come to England for that purpose. The articles of export comprise British and foreign manufactures, with colonial produce, chiefly to the ports of the Mediterranean, cotton, both raw and manufactured, of island growth, wool, cigars, grain, and pulse, wine, spirits, of the value of between £296,000 and £400,000 a year: the imports comprise manufactured goods (chiefly from Great Britain), colonial produce, wheat from Sicily and Odessa, wine and spirits, tobacco, and salt-fish, with numerous minor articles of the average value of about £700,000 a year. At an average of the six years, ending with 1838, the value of the exports of the produce and manufactures of the United Kingdom to Malta amounted to £164,632 a year. Latterly, however, the exports have begun to increase, though they are still comparatively trifling. Malta has, within the last few years, become the centre of a very extensive steam-packet system, the steamers from and to England, the Ionian islands, Alexandria, &c., touching here. The French steamers, between Marseilles, Alexandria, and other parts of the Levant, usually perform quarantine at Malta. The industry of the island comprises the manufacture of cotton fabrics, the annual value of which may amount to from £70,000 to £80,000. Cabinet work is made for exportation to Greece and the Ionian islands; soap, leather, macaroni, iron bedsteads, &c., are made on a smaller scale; and the Maltese goldsmiths are remarkable for the elegance of their gold filagree-work, neck chains, &c., the exports of which are valued at about £7000 a year. The currency of Malta consists partly of British silver and copper, introduced in 1825, but partly also of Maltese scudi of the value of 1s. 8d. English, of Spanish dollars valued at 4s. 4d., and of Sicilian dollars at 4s. 2d. each. The weights most in use are the *rottolo* or pound = 12,916 English grains, and the *cantaro*, comprising 100 rottoli, or 147½ lbs. avoirdupois. Corn is measured by the *salma* = 9·221 Winchester bushels, and oil is sold by the *cafiso*, which contains 5½ English gallons. Bills on London are usually drawn at 31 and 60 days' sight; and the deputy commissary-general must, at all times, grant bills on the treasury of Great Britain for British silver tendered to him, at the rate of the £100 bill for every £101 10s. silver, receiving the silver of other countries at a fluctuating rate of exchange. There are two joint-stock banks in Malta, the united capital of which may amount to £20,000: they discount good bills, of short date, at 6 per cent., keep cash without charge, and issue notes payable at sight, which pass current through the island, except in transactions with the government. Any person may establish himself as a merchant, and numerous Englishmen, Frenchmen, and Sicilians carry on an extensive commerce; while among the native traders, perhaps the wealthiest of all are those who speculate in articles of consumption for the island, buying a great variety of goods, in small quantities, for ready money, and realizing large returns, by retail as well as wholesale trade. (For farther particulars as to the trade, port regulations, and tariff, see *Commercial Dictionary*, art. MALTA.)

MALTA.

Government and Garrison.—Malta is a crown colony, the local government of which is conducted by a governor immediately responsible to the secretary of state for the colonies. In legislative matters, however, he is assisted by a council of seven persons, appointed by the crown, and at present consisting of the Roman Catholic bishop of the island, the military officer second in command, the chief justice, and chief secretary, with three unofficial members, appointed by the governor. All orders in council have the force of laws. The principal administrative departments are the chief secretary's office, quarantine department, custom-house, land-revenue department, and audit-office. There are numerous courts of justice, in all of which the procedure is both intricate and expensive; besides which, the laws themselves are frequently contradictory, and generally require revision. The public revenue amounted in 1837 to about £100,000 a year, of which about £23,000 were derived from the rents of government property, £70,000 from customs and quarantine dues, and about £8000 from internal taxes; but it is believed that some reduction in the public burdens has recently taken place, in consequence of the recommendations of the late commission of inquiry. The above revenues not only defrayed the salaries of the various government officers, but the expenses attending the maintenance of the public roads, as well as liberal contributions for the support of schools and public charities. The military force of Malta consists almost entirely of British troops, varying between 2000 and 2500 men. There is also an engineer and artillery corps, the entire maintenance of which, as well as of the army generally, falls on England. There is likewise a native regiment, comprising about 560 men, called the Malta fencibles: but their duties being exclusively local, and rather of a civil than military nature, the maintenance of this body (costing about £11,000 a year) is defrayed out of the revenues of the island.

Religion and Education.—The national religion of the Maltese (secured by the English government) is Roman Catholic, to which the people are strongly attached, scrupulously observing its rites, and celebrating its festivals; but, notwithstanding their sincere adherence to the church of Rome, they entertain little or no jealousy of the Protestants: both parties observe the greatest moderation and deference for the religious opinions of each other. There are in all no fewer than 1000 Roman Catholic clergymen, the church property producing about one fourth part of the rental of the island. The Protestant places of worship comprise the governor's chapel, naval chapel, Church missionary chapel, and Wesleyan mission chapel; besides which, a church has recently been erected at Valetta, for the exclusive use of the English garrison. The total number of Protestants does not, however, exceed 4500. Education did recently has been much neglected; but within the last 15 years several new schools have been established, the principal being the normal free schools at Valetta, Senglea, Notabile, in Malta, and Rabato in Gozzo, giving instruction in 1836 to about 1500 children. Other primary schools are scattered through the villages, and there are about 80 private schools. The university of Valetta, founded in 1771 by the grand master, Pinto, and now occupying the convent of the suppressed Jesuits, is supported by the government, at an expense of between £1000 and £1900 a year, and had (including the lyceum) 375 students in 1836. The bishop has an ecclesiastical seminary at Notabile, giving religious instruction to about 30 boys. Instruction is commonly conveyed in these schools in the Italian language, the mother-tongue of the Maltese (a *patois* of Arabic, mixed with a little Italian), being wholly unwritten, and never applied to the purposes of literature. English is spoken by many of the higher classes, and is making considerable progress even among the lower orders in the cities. In the rural districts, however, Maltese is spoken almost without exception.

Manners of the People.—The Maltese are as dark as the natives of Barbary, but without the Arab features, the men being of middle height but erect stature, robust and active; while the women, though small, and of dark complexions, are graceful, with regular and sometimes handsome features. The working classes are described as laborious and frugal, living on very slender fare, the great bulk of them being employed either in agricultural labour, or quarrying and cutting stone for exportation to Constantinople and Alexandria. The Maltese are celebrated all over the Mediterranean for their good and intrepid seamanship. The dress of the higher orders is similar to that of other Europeans; but among the inferior or working classes the dress of the men is a short loose waistcoat, covering a cotton shirt; short loose trowsers leave the leg bare from the knee, and on the feet are worn *kercks*, a kind of sandals, nearly resembling those of the ancient Romans. The women wear short cotton shifts, blue striped petticoats, corsets with sleeves, and a loose jacket covering the whole. A black veil, called the *faldetta*, is the out-of-doors headdress

of the women; whereas the men wear woollen caps in winter, and straw hats during summer. (*Sir R. C. Hoare's Tour*, ii., 286.) The morals of all classes are much higher than in most parts of S. Europe; and if there be less refinement of manners in the Maltese than among their continental neighbours, there is less vindictiveness and intrigue, while drunkenness and gambling are almost unknown. A few of the aristocratic families, ennobled by the knights of Malta, yet remain, but they form a very small portion of the population, and few of them possess large property.

Cities and Towns.—The principal towns are Valetta, built in 1566, by the famous grand master, John de Valetta, as being more conveniently situated for a capital than the old inland city called Citta Vecchia, the former capital of the island, and identical with the ancient Melita. Valetta, on the N.E. coast, in the centre of a fine double harbour, in lat. 35° 54′ 6″ N., long. 14° 31′ 10″ has a population, including the garrison, and its suburb, Vittoriosa (on the S.E. side of the great harbour), of about 60,000. It is very strongly fortified, and, from its position on a hill, as well as the almost impregnable works and trenches that surround it, has a most imposing appearance; nor is the visiter less struck with its internal beauty. The streets, though generally steep, are wide and well paved with lava, while the public squares and quays along the harbour are of noble proportions, indicative of the former wealth of the knights of Malta. The governor's palace and gardens, lying outside the walls, were formerly occupied by the grand-master: a public library (once belonging to the order) contains upwards of 40,000 volumes; and the general hospital is not only used for the reception of sick troops, but has ample room for stores, and other purposes; the Floriana hospital is also a large building, occupying two sides of a quadrangle; and in the suburb of Vittoriosa is a third military hospital. Other hospitals are open for the relief of the native sick, and among the other public buildings may be mentioned the barracks, prison, theatre, university, collegiate church of St. John, and 19 other churches, including those in the suburbs. Valetta has a bustling animated appearance, from its being the great centre of the industry and commerce of Malta. Citta Vecchia stands on very high ground, overlooking nearly the whole island, about 7 m. W. of Valetta. The rock on which it is built is excavated into large catacombs, "some of which are said to extend 15 m. under ground!" (*Brydone*, p. 186.) This old and decayed city is strongly fortified, and the cathedral is an extremely large and lofty structure, underneath is a grotto in which, as the monks inform us, St. Paul concealed himself for some time after his shipwreck. They have equally authentic legends respecting other localities close to the city. The towns are mere villages, besides which there are about 40 hamlets, chiefly remarkable for their picturesque and well-built churches. The roads, generally speaking, are good, many of them having been recently much improved; but the inland transport is, notwithstanding, chiefly by horses, mules, and asses, nor do the few carts lately introduced meet with much favour from the natives.

Neighbouring Islands.—About 4½ m. W. of Malta is the small island of Gozzo. It produces considerable quantities of cotton, the cultivation of which constitutes the chief occupation of the islanders, who differ in no essential respect from the Maltese. An English garrison is stationed at Chambray, a strong fort elevated about 500 ft. above the sea, and there are other military works well adapted for the defence of the island. Between Gozzo and Malta is another, though very small island, called Cumino, which belongs to a single proprietor, who derives from it the title of a prince palatine.

History.—Malta was probably first discovered by the Phoenicians, who communicated to the Greeks its oldest known appellation of '*Ωγύγια*.* From the Phoenicians it passed to the Carthaginians, from whom it was taken by the Romans in the first Punic war, and made a prefecture subject to the pretor of Sicily. St. Paul, during his voyage from Palestine to Rome, was wrecked here; and being kindly received by the people, performed some miraculous cures, which made him be "honoured with many honours, and, when he departed, laden with such things as were necessary." (*Acts*, xxvii., 39—44; and xxviii., 1-10.) On the decline of the Roman empire, Malta fell under the dominion of the Goths, and afterwards of the Saracens. It was subject to the crown of Sicily from 1190 to 1525, when the emperor Charles V. conferred it on the knights hospitallers of St. John of Jerusalem, who had a short while previously been expelled from Rhodes, giving them

* There are, however, no good grounds for believing that the island was known to the Greeks in the time of Homer, and we regard the notion that either Malta or Gozzo was the island of Calypso, referred to in Homer's *Odyssey*, as unworthy of credit. Several German critics on the *Odyssey* have written learned essays on the subject but a glance at the island, which of course was below their notice, would have sufficed to show the futility of their elaborate trifling.

power to levy taxes, import duties, &c., for the maintenance of the order, on condition that they should wage perpetual war against the Turks and Corsairs. It was besieged by a powerful Turkish armament for four months, in
1565, but without success; the knights, under their heroic
grand master, John de Valette, founder of the city called
by his name, having succeeded in repelling all their attacks,
and compelling them in the end to retreat with vast loss.
During more than 150 years, the island maintained itself
against the Ottoman power; but the order was never sufficiently wealthy to attempt foreign conquests, or equip numerous fleets. At length, however, the inexpediency of the
continuance of the piratical contests, in which the knights
had been so long engaged, became obvious; and, in 1794,
they concluded a truce with the Turks, which secured for
the Maltese in Turkey the same privileges as the French.
The subsequent history of Malta till its surrender to the
French has little worthy of notice. In 1798, a French fleet
of 18 ships of the line, with 16 frigates, and 400 transports,
arrived off Valetta, having Napoleon on board; and the
treachery of the French knights, who desired to be the subjects of France rather than Russia, rendered the capture of
the island, with its capital, no very tedious or difficult task;
and accordingly, after some fighting, the island capitulated
19th July, 1798, one month after the arrival of the fleet,
when the order of Malta was virtually extinguished. In
consequence of the irreligious practices and oppressions of
the French, the Maltese rose *en masse* to expel them; and
compelled them to take refuge in the towns, where they
were closely blockaded for upwards of two years. At
length, the French, being reduced to extremities, surrendered on the 5th September, 1800. The English immediately
took military possession of Valetta, and have since retained it; the treaty of Paris, in 1814, having definitively annexed it to the crown of Great Britain. (*Reports of Commissioners of Inquiry; Brydone, Sir C. Hoare, and other
Travellers; and Private Information.*)

MALTA, p. t., Saratoga co., N. Y., 6 m. S.E. Ballston Spa,
29 m. N. Albany, 400 W. Saratoga lake lies in its N.E.
part, and Round lake in its S.E. part. Watered by Anthony's kill, the outlet of Round lake. Saratoga lake is a beautiful sheet of water, is 9 m. long and 2 m. wide, and lies
partly in Saratoga Springs t. It has three stores, three
fulling-mills, two woollen factories, two grist-mills, three
saw-mills, three tanneries, two distilleries; 10 schools, 357
scholars. Pop. 1457.

MALTA, p. v., Morgan t., Morgan co., O., 72 m. E. by S.
Columbus, 331 W. Situated on the W. side of Muskingum
river, opposite to Mac Connelsville, the capital of the county. It contains three stores, 30 dwelling and about 300 inhabitants.

MALTON (NEW), a parl. bor., market town, and par.
of England, N. riding co. York, wap. Ryedale, on the Derwent, 16 m. N.E. York, and 181 m. N. by W. London.
Area of parliamentary borough, which comprises the parishes of St. Leonard and St. Michael in New Malton, 6640 acres. Pop.,
in 1831, 6602. The town, which occupies an eminence on
the W. bank of the river, is very irregularly laid out; but
the buildings, chiefly of stone, are decidedly improving in
quality: on the opposite side of the Derwent, crossed here
by an ancient bridge, shaped somewhat like an inverted Y,
is the suburb of Norton in the E. riding, a thriving and increasing place; and about 1 m. N.E. of the town is the village of Old Malton, formerly of some consequence, but now
exhibiting all the symptoms of decay. (*Parl. Bound. Rep.*)
The public rooms, theatre, and workhouse, are handsome
modern buildings; and near the bridge stand the remains
of a castle, built by the Vesci family, and destroyed by
Henry II. There are two churches, one of which is surmounted by a tall unfinished spire: the livings are curacies,
dependent on Old Malton, and in the gift of Earl Fitzwilliam. The Wesleyan Methodists, Presbyterians, and the
Society of Friends have their respective places of worship;
and there are three well attended Sunday schools. New
Malton has two subscription schools for children of both
sexes; but the grammar school, founded by Archbishop
Holgate, is at Old Malton. The Derwent being navigable
up to New Malton bridge, is made available for the shipment of large quantities of corn, hams, bacon, and other
farm produce. Malting and tanning are carried on to a considerable extent, and there are two large porter breweries;
but the chief dependence of the town is on its retail trade
with the opulent gentry of the neighbourhood. New Malton is a borough by prescription, governed by a bailiff. It
has returned two members to the House of Commons since
the 23d Edward I. Previously to the Reform Act, the
franchise was vested in the burgage holders and inhabitants, rated to church and poor. The limits of the borough were enlarged by the Boundary Act so as to include
the entire parishes of New Malton, and the parishes of Old
Malton and Norton. Registered electors, 1839–40, 603. New

Malton is also one of the polling-places at elections for the
N. riding; and the petty sessions are held here for the E
div. of wap. Ryedale. Markets on Tuesday and Saturday, but chiefly on the latter for horses and cattle, corn, bacon, and farming implements. Very large cattle fairs, Monday before Easter, day before Whitsunday, and Oct. 11.

MALVERN, GREAT, a town, par., and celebrated watering-place of England, co. Worcester, hund. Pershore, 7½
m. S.S.W. Worcester, and 104 m. W.N.W. London. Area
of parish, 4240 acres. Pop., in 1831, 2010. This town,
which for many years has been a place of fashionable
resort, in consequence of its delightful situation in

"The vale of Severn, Nature's garden wide,
By the blue steeps of distant Malvern wall'd,
Solemnly vast —"   *Dyer's Fleece.*

stands on the E. declivity of the well-known hills bearing
its name, and is neat and well built, comprising, besides
good houses for the tradespeople, several hotels and substantial private residences for visitors. The church, a fine
cruciform structure of Anglo-Norman and pointed architecture (lately renovated in excellent taste), is 171 feet in
length, with an embattled and pinnacled tower rising 124 ft.
above the intersection of the nave and transepts. It formerly belonged to a Benedictine monastery, founded here
in 1083, and long one of the wealthiest and most important
religious establishments in England. At the dissolution of
the monasteries, when the rest of the property was sold, the
church was bought by the inhabitants and made parochial.
Malvern has long been noted for two medicinal springs, the
chief of which (St. Anne's well) is bituminous, and enjoys
a good reputation for the cure of nervous and cutaneous
diseases: the other is a simple chalybeate, and little frequented.

About 3 m. S. is the village of Little Malvern, the road
to which skirts the Malvern Hills, an extensive range composed of greenstone and quartz covered in parts with blue
limestone, and running from N. to S. about 10 m., with an
average breadth of 3 m. The acclivities in many parts
are very gentle: but the summit of the ridge, which attains
a height of 1444 ft., commands magnificent views over
Wales and the counties of Hereford, Worcester, and
Gloucester.

MALWAH, a prov. of Hindostan, chiefly between lat.
22° and 26° N., and long. 74° and 80° E., having N. Rajpootana and Agra, W. Gujrat, E. Allahabad, and S. Gundwanah and Candeish, from which it is separated by the
Nerbuddah. The central part of this province is a table-
land, extending from the Vindhyan mountains, on the S. to
the Chitore and Mokundra ranges on the N., and E. and
W. from Bhopaul to Dohud; but which seldom rises to
more than 2000 ft. above the sea. It declines gently towards the N. in which direction flow most of the principal
rivers, as the Chambul, and its chief affluents, the Kali-
Sind and Betwah, tributaries of the Jumna, and the Mhye,
which falls into the gulf of Cambay. The climate is usually mild and salubrious, except for about two months after
the rains, when fevers are very prevalent. The total fall
of rain from June to September has been estimated at 29
inches. The soil consists either of a loose black loam, or a
more compact ferruginous mould, both noted for their fertility. Wheat, grain, peas, jowares, bajree, mung, and
maize, are among the chief grains cultivated; the first two
furnishing the largest export. Rice is raised only in small
quantities sufficient for home consumption; but opium, sugar, tobacco, cotton, linseed, garlic, tumeric, and ginger, are
grown to a considerable extent. A little indigo, and the rest
of the *Morinda citrifolia*, which supplies a red dye, are also
raised, and fruits, including grapes, flourish in great abundance.

Opium is by far the most valuable product of Malwah,
the soil and climate of which appears singularly well adapted for the cultivation of the poppy. The Malwah opium is
considered by the Chinese, for whose consumption it is
chiefly grown, superior in strength, in the proportion of sev-
en to five, to that of Bahar and Benares, though inferior in
flavour. Since the pacification of central India, the quantity of opium produced in Malwah has increased very rapidly, so much so, that while the total exports of Malwah
opium to China, in 1821, did not amount to 3000 chests, they
amounted to about 21,000 chests in 1839, worth above
£3,000,000. (*Documents relating to the Opium Trade, p.
79.*) In Malwah the culture of opium is freely carried on;
the cultivator paying a proportionally heavy land-tax for
the land occupied in its culture. Previously to 1829, the
Bombay government endeavoured to obtain a monopoly of
the sale of opium exported from the ports under their presidency, but with little success; for 2-3ds of the Malwah
produce were carried to the Portuguese settlement of Damaun and elsewhere, to be exported. But at the above period, the attempted monopoly was abandoned, and a permit,
or transit-duty, similar to that imposed in other states
through which the opium passes, was laid on in its stead.

Since then 9-10ths of the Malwah opium have been shipped at Bombay; and in 1832, the trade yielded to the British government an annual revenue of £900,000 (see *Reports on E. I. Affairs*, 1830–1832), which has since been materially augmented. The tobacco of the province, especially that of the Bilsah district, is also, beyond all comparison, the best in Hindostan.

Malwah is the chief seat of the Bheel race, as it was of the Pindarry and Mahratta powers. It is almost wholly divided among the dominions of native princes, the chief of whom are Scindia, Holkar, and the rajahs of Bhopaul, Kotah, Dewass, &c. Except the Maharajah of the Punjab, Scindia is the only prince in Hindostan who can be called independent of British authority; but his independence has more of semblance than reality, for the power of his dynasty has been completely broken by a succession of reverses: his dominions are surrounded by the territory of the British, or their allies, who are bound to negotiate with foreign states only through the intervention of the British. A stationary British camp is kept up in his neighbourhood; and he is obliged to receive an English resident at his court, and to furnish a contingent of 15,000 men to the Anglo-Indian army. The dominions of Scindia are estimated to comprise 32,940 sq. m., with a population of nearly 4,000,000; and to yield a gross annual revenue of 2,308,000 rupees, out of which the chief derives a nett subsidy of 1,561,000 rupees yearly. The chief cities belonging to Scindia are, Gwalior, his modern, and Oojein, his ancient capital. The states of the other chief native princes of Malwah have been briefly noticed under INDORE, BHOPAUL, KOTAH, &c. (*Pearl. Reports*; *Hamilton's E. I. Gaz.*)

MAMAKATING, t., Sullivan co., N. Y., 12 m. E. Monticello, 101 m. S.S.W. Albany. Watered by Bashe's kill, along which passes the Delaware and Hudson canal. It has twenty-three stores, nine lumber-yards, five grist-mills, 14 saw-mills, six tanneries, three distilleries; one academy, 36 students; 16 schools, 883 scholars. Pop. 3418.

MAMARONECK, p. t., Westchester co., N. Y., 23 m. N.E. New-York, 143 m. S. Albany, 248 W. Bounded S. by Long Island sound. Watered by Mamaroneck creek. It has two churches, an Episcopal and a Methodist, several stores, three schools, 84 scholars. Pop. 1416. Several sloops ply to New-York city.

MAMERS, a town of France, dep. Sarthe, cap. arrond., 24 m. N.N.E. Le Mans. Pop. (1836) 1563. It is indifferently built, but has of late been greatly improved. It is a town of great antiquity, and was surrounded with entrenchments by the Normans, some remains of which are called "the fossés du Robert le Diable." It has a handsome Gothic parish church, a college, a prison, some public baths, a theatre, manufactures of hempen, cotton, and woollen fabrics, pearl buttons, &c., and several tanneries and breweries. (*Guide du Voyageur.*)

MAN, ISLE OF (an. *Mona, Menapia,* or *Monada*), an island belonging to the United Kingdom, in the Irish sea, between lat. 54° 4′ and 54° 27′ N., and long. 4° 17′ and 4° 34′ W.; its N.E. extremity (the point of Ayre), being 17 m. from Burrow-head, in Wigstownshire, its E. coast 34 m. from St. Bee's head in Cumberland, and the town of Peel, on its W. side, 30 m. from Ballyquintin Point, in Ireland: greatest length 35 m.; greatest breadth, about 13 m. Area, 220 sq. m., exclusive of the Calf of Man, a small disjointed fragment of the island, at its S. extremity. Pop., in 1831, 41,000. Its general aspect, as viewed from the sea, is bold and precipitous; a ridge of mountains runs through its whole length, and three of the highest points reach an elevation of more than 1600 ft. above the sea; Snafield. the loftiest, being 2004 ft. high. Several rills and streams flow from the high grounds in different directions; but there are no rivers nor lakes of any considerable size. The prevailing feature in the geology of the island is clay-slate, interspersed with mica-slate; and covered, near the coast, with greywacke and old red sandstone. Limestone also is found on the S. side, near Castleton, intersected in some parts by veins of trap. The clay-slate is quarried at a place called Spanish-head, near Castleton; and stones are raised in blocks averaging about 7 ft. in length, by 1 ft. in breadth, and 6 inches in thickness. Drawing and roofing slates are quarried on the W. side of the island, not far from Peel. Close to Castleton, on the shore, are limestone and marble quarries, which have been worked for many years, and furnished a part of the stone for St. Paul's Cathedral, London. The island also produces lead, zinc, and copper, raised in considerable quantities by the Chester mining company, and by private parties. But mining and quarrying are in a very depressed state; the tools employed are of the rudest description; and not even a common crane is to be seen in the quarries. (*Head's Home Tour*, vol. iii., p. 19.)

The climate of Man is considered milder during winter than that of the adjacent parts of Great Britain and Ireland. Frost and snow are rare; and when they do occur, they are seldom of long continuance. Owing, however, to the frequency of fogs and dews, as well as to the prevalence of E. winds, during many weeks of spring, the summers are deficient in heat, and the harvest is generally rather late. The climate, however, is, on the whole, favourable to health; cases of longevity are frequent, epidemics rare, and agues unknown. The soil is extremely various. Clay and marl, covered with white sand, predominate in the N. and N.W. extremity of the island, which is covered with scanty herbage, affording sheep pasture; but, proceeding S. and E., the quality of the soil improves, and, in the valleys especially, are some tracts, partly sand and loam, and partly stiff clay. No part of Man is, however, very productive; nor are any great pains taken to improve its natural resources. The mountains, commons, and other waste lands, include about 54,000 acres, leaving above 100,000 acres for tillage. Agriculture has considerably improved since the diminution of the herring fishery has made the men turn their attention to farming, which used to be exclusively the occupation of women: wheat, barley, and potatoes are raised in sufficient quantities for exportation, and within the last few years, the turnip husbandry has been introduced with some success by the English and Scotch settlers. The implements, however, are very rude; and the division of land into small farms has combined with the herring fisheries and smuggling to retard improvement. The extent of land under white crops, and the average produce of each in 1835, were estimated as follows:

Wheat 8,000 acres, at 2½ qrs. per acre				21,250 qrs.
Barley 5,080	"	4	"	20,000
Oats 13,000	"	3	"	39,000

This was supposed to leave a surplus of about 5000 qrs. of wheat, and 3000 of barley, over the consumption. Peas are cultivated in the N. parts, clover is a favourite crop, and flax is raised by almost all the farmers for domestic use. The cattle of Man, which at present consist of a mixture of Irish and British breeds, are small and short-horned, running to fat, and not yielding milk till they are six years old. Ayrshire cows have, however, been recently introduced with much advantage. The native sheep, which are small, hardy, and usually of a white or gray colour, are slow feeders, long in coming to maturity, and very coarse-woolled; they are now, however, confined to the hills, the lowlands being mostly stocked with improved breeds. The island yields a race of hardy ponies, capable of much labour, and requiring little food; but for draught and farming purposes other breeds, chiefly Irish, have been imported of larger size and strength. Man had formerly a peculiar breed of hogs, now totally extinct, the animals at the present day being of various kinds, some of which resemble the Chinese variety. Red deer formerly ranged in the mountains, but the game at present consists of hares, rabbits, partridges, snipes, and woodcocks. Foxes and polecats are not found, neither are there any poisonous animals on the island; but weasels and rats are very numerous, and detrimental to the farmers. The Manks tenures are remarkable: the different parishes (of which there are 21), are divided into *treens*, each comprising four quarterlands, varying in size from 60 to 150 acres, and rising in yearly value from £10 to £195; there are 760 quarterlands, and they are esteemed by the islanders as property of the highest nature, in fact strictly entailed estates. Other lands, called *intacks* and cottages, are devisable by will, and, on the whole, considered to be of a far inferior nature. The yeomen are very proud of these little freeholds, which range from 10 to 200 acres, and usually comprise portions of pasture as well as arable land; "but there can be no doubt," says Lord Teignmouth, "that the system is practically vicious, diminishing the wealth both of the farmers themselves, and of the public at large, containing indeed within itself the seeds of its own dissolution." (*Sketches of Scotland and the Isle of Man*, vol. ii., p. 202.) Most of the yeomen have large and expensive families, which the law of Man compels them not only to rear and educate, but to provide for; and hence their estates soon become encumbered, and they are effectually prevented from pursuing any improved system of management, even if they felt inclined to its adoption, which is seldom the case; there being no more obstinate adherents to routine and ancient practices than the Manx husbandmen. Many of them thus become involved in debt, and mortgage their property, the redemption of which being seldom in their power, they are dispossessed of it, and compelled to leave the island, or to resort to trade or predial labour.[*] Hence the class of small proprietors is gradually disappearing: numbers of them having been swallowed up in the extending estates of the Scotch and English residents.

Man used to be one of the principal seats of the herring-fishery: but for several years past it has been comparatively deserted by the herring-shoals, and the fishery has, in

---

[*] The mortgages on land held chiefly by the English, amounted, it is said, in 1835, to 500 000*l.*, which, at 5 per cent., would amount to 10,000*l.* a year.

consequence, become quite inconsiderable, though even now it is the frequent practice of the farmers to purchase a boat, and share in the excitement and profit of the season. This diminution, however, is not to be regretted, as the fishery was carried on from July to October, exactly when the services of the yeomen and others engaged in it were most necessary at home. Being also a kind of lottery, in which by a few weeks' labour, large sums were occasionally realized, it attracted crowds of adventurers, without either capital or skill: while the irregular life led during these pursuits tended to encourage intemperance, and was a main cause of the indolence for which the Manx have been long notorious. There has, in fact, been a material improvement in the habits and industry of the people since the decline of the fishery; and there are, perhaps, few things less to be desired for the island than its revival. The herrings appear off the coast of Man in June, remaining till September, when they seek the E. coast of Ireland to deposit their spawn. The fishing vessels now built are much larger than formerly; they are half-decked, with very short keels, and are good sea-boats, though apt to pitch to a dangerous extent in rough weather: they vary from 18 to 30 tons burden, and are manned by 8 or 10, and sometimes 12 men. Cornish, Welsh, and Irish fishers also visit Man; and, according to Sir J. Rentie's estimate, it appears that out of 250 boats, and 2000 men employed in 1835, only 110 boats with their crews belonged to the island, 100 being Cornish or Welsh, and about 50 Irish. The cod-fishery has been neglected, owing to the want of adequate capital for the supply of proper vessels and lines.

The manufactures are chiefly domestic, and carried on by women, most of whom, when not in the field or farmyard, are employed at their looms or spinning-wheels, producing woollen, linen, and cotton cloth, both for the home and foreign supply, as well as nets for the use of the fisheries. Bleaching is conducted on a large scale in Lazey Glen, stuffs being sent thither from all parts of the island. A paper-mill and brewery are also established in the same neighbourhood. A woollen manufactory is established at Douglas; and hats, made of coarse wool, which cost about 2s., are said to wear extremely well. The exports consist principally of corn, potatoes, eggs, lime, and limestone, lead and copper ore, herrings, lime, sail-cloth, and paper. Owing to the Isle of Man having been formerly independent, a discrepancy has, for a lengthened period, existed between the duties on commodities in it and in Great Britain, the former being considerably lower than the latter. This distinction, which still subsists, has occasioned, at different times, a great deal of smuggling, particularly on those articles on which high duties have been imposed in this country. This, however, is now materially reduced by adopting the plan of allowing only certain quantities of those goods on which the Manx duties are lower than those of England, to be imported into the island (viz., wine, 110 tons; spirits, except British, 80,000 gallons; tea, 75,000 lbs.; coffee, 8000 lbs.; sugar, 10,000 cwt.; and tobacco, 60,000 lbs.): and by maintaining an extra number of customs officers and revenue cruisers for the suppression of smuggling. Nothing, however, can be more impolitic than the continuance of such a system. The public has, at a very heavy expense, purchased all the feudal rights of the Atholl family; and that being the case, it is high time that an end should be put to the anomalous absurdity of having a considerable island lying, as it were, in the very centre of the empire, and in the direct line between some of the principal trading towns, with discriminating duties on many important articles. In making any change, it would of course, be necessary to make the inhabitants some compensation for the temporary inconvenience that it would occasion; but this might be done with advantage to the natives, and without expense to the public, by modifying and improving the internal regulations and legislative policy of the island, which would eventually lose nothing by the change. In 1836 there belonged to the island 242 vessels of the burden of 7229 tons, being a considerable increase on previous years; and as it lies in the line of the steamers plying between Liverpool and Glasgow, most of which touch at Douglas, it has begun to be largely frequented by visitors from these cities and other parts of the empire, whose influx has materially contributed to the improvement of its principal towns. To this cause, indeed, the present improved state of Man may be chiefly ascribed. It is also the residence of numerous half-pay officers, and others, who are induced to live here in consequence of the lower duties on many articles of domestic consumption.

The condition of the labouring population is moderately prosperous. Ordinary labourers receive about 1s. a day; and skilled labourers, if we may so call their clumsy tradesmen, get about 3s. a day, which, considering the low price of provisions, is certainly ample. There is no legal provision for the poor, who have to depend wholly on voluntary

charity. Generally speaking, the cottages are of a very inferior description: they are frequently built of earth or sod, and thatched with straw, having a funnel of sail-cloth, as a substitute for a chimney. There are, however, a few improved cottages, and their number will, no doubt, increase with the spread of improvement.

The feudal sovereignty of Man, which was a kingdom prior to 1504, was held by the Stanleys, afterward Earls of Derby, and their successors, the Dukes of Atholl, from 1406 to 1765, when parliament, conscious of the injury which the revenue and the public generally received from the contiguity of an island only feudally subject to the crown, and hence affording refuge to debtors, outlaws, and smugglers, purchased from the Duke of Atholl, for £70,000, his civil and military rights and patronage, but with certain reservations as to fiscal matters and titular dignity. A further arrangement was made in 1826, and Great Britain now enjoys all the rights and privileges of sovereign of the island. The constitution, however, was left untouched; and for many years, at least, the legislative power has been vested in the House of Keys, a body comprising 24 members, now self-elected, but formerly chosen by the statesmen or owners of entailed estates. Their acts are binding in all cases, and the laws are so few and brief as to admit of being included in a small volume. Attorneys occasionally plead in the courts; but the suitors quite as frequently defend their causes in person: law is cheap, and, as was to be expected, litigation is very common. There are two supreme judges in the island called deemsters, or "wardens of the law," officers of high antiquity, and exercising jurisdiction over all civil and criminal cases; being the presidents (under the crown and governor) of the two courts of chancery and exchequer, each of which is held eight times a year. The former of these has little more to do than to confirm or annul the decisions of the deemsters, who hold a primary court of judicature; and the exclusive business of the latter is to punish offences against the revenue laws. The common-law courts are held at different places for the six different sheadings into which the island is divided, and may be considered as courts of "common pleas," in which all actions personal or real, may be tried, as in the deemsters' court, by a jury of six in real, or of four in personal actions. The appeals from this court are first to the house of keys, afterwards to the governor, and finally to the queen's privy council. A half-yearly jail-delivery is made compulsory, and bailiffs act in the five chief towns to hear and determine cases of debt under 40s. (Feltham's Tour, p. 33–44; Lord Teignmouth, ii., 227–241.)

The established religion is that of the Church of England; all sects, however, enjoy full toleration. The clergy are under the bishop of Sodor and Man, suffragan to the archbishop of York, but holding no English barony, and hence having no voice in the legislature, though privileged to sit in the house of lords. This see has been held by several highly celebrated divines; and among others, by Barrow, Wilson, and Ward. An ecclesiastical court is held twice a year, either by the bishop or his vicars-general, and an archdeacon regulates fabrics and the minor concerns of the 17 parishes. These cures are commonly well attended to by respectable clergymen; but their stipends do not average £90 a year; and the churches, though extremely pretty, are miserably deficient in accommodation. The dissenters have made considerable progress in the present century; but the Methodists comprise even now only one tenth of the population, and the other bodies of dissenters are unimportant. Bishop Wilson and other prelates have done much to promote education, not only by establishing schools, but also by translating the Scriptures and other books into the Manx language. Each parish has its school more or less richly endowed; and while elementary instruction is given in the Manx, every endeavour is made to instruct the natives in the English language. Indeed these can be no doubt, that, at no distant period, the population will be familiarly acquainted with our language; and this will be the surest method to disabuse them of the prejudices which so many entertain against a union with England. A collegiate school was established a few years ago, through the exertions of the late Bishop Ward; and though the funds were far too small for the expected outlay, the establishment, being well conducted, attracted numerous students, and has, on the whole, been successful.

The Manx, like the Welsh, and Scotch Highlanders, belong to the great Celtic family, which probably occupied the whole United Kingdom, previously to the immigration of the Belgæ. Their Celtic origin is clearly evinced by their language, which is a mere dialect of the Irish, Erse, or Gælic. They have a swarthy complexion, stout, with an air of melancholy pervading their countenances. Indolence, and a love of litigation, are distinguishing characteristics of the male part of the population. Even, at present, workmen rest for two hours in the middle of the day, when they may be seen stretched under hedge-rows by the road-sides. The women, however, are extremely industrious; and on them devolve not only the production of domestic manufactures, but also a large share of the labours of agriculture. They are hospitable, superstitiously attached to existing institutions and religious forms, and treat bishops and clergymen as beings of an exalted nature; but they are, notwithstanding, drunken, indelicate, dirty, and addicted to pilfering. Their old habits and prejudices are now, however, gradually giving way; the increasing influx of visiters, during the summer season, having, in this respect, effected an important and beneficial change. The élite of society is composed of the government officers and the large landholders, with a few church dignitaries; the other clergy, the attornies, and medical men being too poor to mingle with the first circle.

The rocky islet, or Calf of Man, already alluded to, at the S. extremity of the island, was formerly the resort of vast numbers of puffins (*Procellaria Puffinus*, Lath.). At present, however, the bird is there entirely unknown. It was supposed to have been driven from this favourite haunt by the too great destruction of its young. These were held in considerable estimation; and Pennant mentions that, in his day, great numbers of them were taken every year by the person who farmed the islet. It appears, however, that rats that had escaped from a vessel wrecked on the coast, were the real exterminators of the birds. (*Quayle's Survey*, p. 8.)

The early history of Man is obscure. It was the Mona of Cæsar, and the Monapia of Pliny; but we know little more of it beyond mere traditions of its being held by the Druids, and subsequently by Norwegian monarchs, till, in 1264, it was purchased by Alexander III. of Scotland, who appointed a viceroy, and made it tributary. The Scotch were soon afterwards expelled by the English, but the power of the latter was not established till the reign of Henry IV., who granted it to the Percys, from whom it fell, by attainder, and thence passed by gift of the same monarch to the Stanley family, by whose heirs it was sold to the British crown.

The chief towns of Man are: 1. Castletown, in which is the college above mentioned, the seat of legislature, and the residence of the governor (pop. in 1831, 2077); 2. Douglas (which see), the chief trading town, with upwards of 7000 inhabitants; 3. Peel, formerly celebrated both as the residence of the earls of Derby and the capital of the kingdom, but now decayed, and having only a population of 1750 persons, which is about the same as that of Ramsey, one of the steam-packet stations between Liverpool and Glasgow, on the N.E. side of the island. (*Quayle's Survey, and Feltham's Tour; Lord Teignmouth's Scotland and I. of Man*, ii., 181–392, and appendix, &c.; *Head's Home Tour*, ii., 1–90.)

MANAAR (GULF OF), an inlet of the Indian ocean, dividing Ceylon from the S. extremity of Hindostan; extending between lat. 7° 30′ and 9° N., and long. 78° and 80° E. It is in general too shallow to be navigated by vessels above the size of sloops; and is separated by the islands Ramisseram and Manaar, and the chain of rocky islands and sandbanks called Adam's bridge, from another inlet of the sea called Palk's strait, also between Ceylon and the continent. The island of Manaar is 18 m. in length, by 2½ m. broad; but has little importance of any kind. For farther particulars, see CEYLON (L., 395).

MANAYUNK, p. v., Roxborough t., Philadelphia co., Pa., 7 m. N.N.W. Philadelphia, 90 m. E. by S. Harrisburg, 144 W. Situated on the E. bank of Schuylkill river, and on Flat rock canal, a part of the Schuylkill chain. It has an extensive water-power, created by the waste water of the canal, with an average fall of 22 feet. It contains five churches, a German Reformed, two Methodist, an Episcopal, and Roman Catholic, 12 or 15 stores, 25 or 30 mills and manufactories, and about 500 dwellings. Two bridges cross the Schuylkill river, and the Schuylkill canal and the Norristown railroad pass through the place.

MANCHA (LA), a prov. of Spain in the S. part of New Castile, bounded E. by Granada, E. by Cuenca and Murcia, and W. by Estremadura. Area about 7500 sq. m. Pop. 250,000 ? This district consists chiefly of lofty and barren plains, upwards of 2000 feet above the sea, and is, without exception, the least picturesque and productive in the whole peninsula. But it produces corn, wine, olives, and saffron; the Val-de-Peñas, a light red wine, is highly esteemed all over Spain. The mules of this province also, are the largest and strongest in the peninsula. La Mancha, however, derives its chief celebrity from the inimitable work of Cervantes; and many of the customs he has depicted are still prevalent in the province. The capital of La Mancha is Ciudad Real, once a flourishing city, but now in decay, and having at present a population of only 8000 persons.

MANCHA (REAL), a town of Spain, in Andalusia, prov. Jaen, 8 m. E. the city of Jaen. Pop., according to Miñano, 4030. It is situated in a spacious plain, and com-

prises some regular-built streets and handsome squares; its chief buildings being a parish church, Carmelite convent, and hospital. Woollen and linen cloths, bedticks, and sacking, are made here, with bricks and tiles in large quantities, for the supply of the province. The neighbourhood is both picturesque and fertile, producing, with little tillage, abundant crops of olives, with smaller quantities of wine and grain.

MANCHE (LA), a marit. dep. and peninsula of France, formerly included in the prov. Normandy, between lat. 48° 40′ and 49° 40′ N., and long. 0° 40′ and 2° W., encircled on the W. and N. sides, and partly on the E., by the English channel (Manche), whence its name; and elsewhere bounded, on the E. by the deps. Calvados and Orne, and S. by Mayenne and Ille-et-Vilaine. Length N. to S. about 85 m.; greatest breadth nearly 40 m. Area, 593,776 hectares. Pop. (1826) 504,392. Surface is generally undulating. A chain of hills, of no great elevation, runs through the dep. N.W. direction, dividing it into two nearly equal parts. Near its N.E. and S.W. extremities are some marshy tracts. The chief rivers are the Vire and the Ouve. The coast is mostly abrupt and rocky, especially in the N., but it has several good roadsteads and commodious harbours, of which Cherbourg is the finest. In 1835, about 360,400 hectares were estimated to be arable, 94,000 in pasture, 24,000 in woods, 30,200 in orchards, and 46,290 in heaths, wastes, &c. Agriculture is better conducted than in many other deputies. The produce of corn, which is chiefly wheat and barley, exceeds the home consumption: potatoes are an important substitute for grain; and, in 1835, the crop amounted to nearly 807,400 hectolitres. Beans, peas, and a good deal of hemp and flax, are raised. The deputy is beyond the limits of the vine-culture; but about 1,000,600 hectolitres of superior cider are annually produced, and some perry. In 1830 there were about 189,000 black cattle in the deputy; and fat cattle and butter are among its principal products. It had also, in the same year, about 291,600 sheep, estimated to yield annually 411,500 kilogr. of wool, though chiefly of inferior quality. There is a considerable traffic in horses and mules. Poultry are reared in great abundance; large quantities of eggs being exported from Cherbourg and Valognes to England and the channel islands. In 1835, of 192,038 properties subject to the *contribution foncière*, 68,792 were assessed at less than 5 fr., and 36,580 at from 5 to 10 fr. The oyster and other fisheries on the coast are important; but, according to Hugo, fish are less plentiful than formerly. Among the mineral products are iron, lead, coal, marble, slate, and granite; which last is found of excellent quality in the Chausey isles, a group of small islands off the coast of this deputy. Salt-works are established at several places on the coast. Manufacturing industry is employed on iron, copper, zinc, woollen, linen, cotton, and various other materials. Cutlery, glass, paper, hair fabrics, lace, &c., are produced; and in some cantons, baskets, panniers, willow sieves, &c., are made, and sent into other parts of Normandy, and into Brittany. But its principal trade is an agricultural produce and fish, fresh or salted. Manche is divided into six arrondissements; chief towns, St. Lô, the capital, Cherbourg, Coutances, Avranches, Valognes, and Mortain. It sends eight members to the chamber of deputies. Number of electors (1836–39), 3368. Total public revenue (1831), 15,145,826 fr. This deputy is rich in Celtic and Roman antiquities. (*Hugo*, art. *Manche; Official Tables, &c.*)

MANCHESTER, a mun. and parl. bor. and parish of England, the great centre of the cotton manufacture of Great Britain, and the principal manufacturing town in the world, co. Lancaster, hund. Salford, on the Irwell, an affluent of the Mersey, 31 m. E. Liverpool, 35 m. S.W. Leeds, 70 m. N. Birmingham, and 163 m. N.N.W. London; lat. 53° 29′ 50″ N., long. 2° 15′ W. The entire parish of Manchester includes an area of 34,260 acres, comprising 30 townships, and had, in 1831, a total population of 271,000 persons.

Manchester and Salford, which, being separated only by the small river Irwell, form a single large town, covering 3000 acres, with a dense mass of buildings, stand on a large plain, encompassed by hills on every side except the W., and dotted with towns and villages, the inhabitants of which are all industriously engaged in the production of woven fabrics. The Irk and the Medlock join the Irwell close to the town, and all three are made extensively useful in moving machinery, and for other purposes. Six bridges connect Salford with Manchester: the handsomest being Victoria bridge, having a single arch of 100 feet span, opened in 1839. The streets are irregularly laid out, and many are narrow and inconvenient, especially in the more central part. Great improvements, however, have been made within the last 30 years: narrow lanes have been pulled down to make way for broad avenues; noble public buildings, which would be ornamental to any capital in the world, have been erected in the chief thoroughfares: factories and warehouses of gigantic proportions have arisen in every direction; confined and mean-looking shops have been replaced by superior establishments, some of which will bear to be compared with the best in London; the paving of the streets, though still in parts very defective, has been much improved; and flagging has been generally introduced, with macadamizing, in the principal streets. The whole town is lighted with gas; but in the poorer districts the lamps are but thinly dispersed, and are extinguished at too early an hour. It is well supplied with water, sufficiently drained (except in some poor districts) by an underground sewerage, and well watched by a day as well as night police. There are three main lines of street, which run in a curve S.E., nearly parallel to each other. The central line, which is the principal thoroughfare of the town, comprises Market-street (formerly a narrow lane, but now vastly improved, having some of the finest shops in town), Piccadilly, and the London road: more to the N., joined to the last mentioned line by Oldham-street, is Great Ancoat-street, with its continuations; and S. is the avenue known in different parts as Quay-street, Peter-street, and the Oxford road, connected with Piccadilly by a handsome line called Mosley-street, and a long narrow street called Deansgate.

The following table exhibits the area, population, and rate of the parliamentary boroughs of Manchester and Salford, the limits of which pretty correctly define the extent of the town and its suburbs:

Townships.	Pop. in 1801.	Pop. in 1831.	Increase per. cent. in 30 yrs.	Houses worth 10l. in 1831.
Bor. of Manchester.				
Manchester, town	70,409	142,026	102·1	9,004
Ardwick	1,762	5,394	215·5	347
Beswick	8	249	4,062·2	7
Bradford	94	101	66·3	10
Cheetham	762	4,025	427·5	688
Chorlton Row	675	20,569	2,947·2	1,905
Harpur-hey	115	468	299·4	9
Hulme	1,677	9,634	473·8	754
Newton	1,296	4,377	235	65
Total (Manchester)	75,776	187,022	143·5	12,729
Bor. of Salford.				
Salford town	13,611	40,786	479·9	968
Broughton	896	1,869	92·4	56
Pendleton	3,511	8,435	135·6	320
Pendlebury— (part of)	497	1,556	202·0	42
Total (Salford)	18,585	52,366	182·7	1,386
Manchester and Salford	95,313	239,388	151·1	14,009

From the above table it appears, that the rate of increase in the two boroughs, during the 30 years ending with 1831, was 151·1 per cent.; and in the borough of Salford, 182·7 per cent., a rate exceeded only by Preston, and one or two other towns. In 1773, the population of the township of Manchester was estimated by Dr. Percival at 22,481, and that of Salford at 4765; making together 27,246; that is, about *one sixth* part of the population in 1831. The present (1841) population of the parliamentary boroughs of Manchester and Salford is certainly not under 300,000.

The public buildings of Manchester are too numerous to admit of individual description; but the following are the largest, best built, and most important. The Exchange, which stands in the market-place, at the lower end of Market-street, is a large but somewhat low semicircular stone building, fronted with Doric pillars. In the lower floor are the commercial room, a magnificent hall, having an area of 6750 sq. feet; and a spacious newsroom (added in 1839 from a space previously occupied by the postoffice, in the rear of the building): there are upper rooms on a corresponding scale, used for public meetings, dinners, &c.: the establishment is supported by subscription, and had, in 1840, about 3000 members. The chief business day is Tuesday, on which, about noon, all the principal manufacturers of Lancashire may be seen in or near this building. The Town-hall, in King-street, which has been recently widened, is of Ionic architecture, and extremely elegant, being formed on the model of the Temple of Erectheus at Athens, with a central octagonal cupola, resembling Andronicus's Tower of the Winds. It cost upwards of £40,000; and comprises, besides rooms for the police business, gas-offices, &c., a spacious and well proportioned public room (ranking among the finest in Europe), 131 feet long, and 38 feet broad. The fresco paintings, however, with which the walls are covered, are said to display little taste, elegance of design, or correctness of execution. Smaller town-halls are situated in Salford and Chorlton, the former of which townships has its separate police establishment, &c. The Corn Exchange, in Hanging-ditch, is a handsome building, erected from a design adapted to it from the temple of Ceres at Athens. Six Ionic columns support the central pediment; and on each side are wings, very slightly projecting, and ornamented with pilasters; between which are the entrances to a square hall, inclosing an area of about

6000 square feet, and affording standing room for 2000 persons. Of the buildings devoted to charitable purposes, to literature, or to public amusement, the following deserve notice, from their architectural beauty. 1. The royal Infirmary and lunatic Asylum, finely situated in Piccadilly, recently faced with stone, and now constituting one of the chief ornaments of Manchester. 2. The Athenæum in Bond-street, a peculiarly elegant structure, designed by Barry, in the Italian style, and completed at an expense of about £13,000. 3. The royal Institution in Mosley-street, built at a cost of £30,000, from Barry's designs, having a noble front, with a portico in the Ionic style, and comprising, besides other apartments, a handsome gallery for the exhibition of pictures, and a theatre for lectures, capable of accommodating 800 persons. 4. The Portico News-room, in the same street as the Institution, having an Ionic portico. 5. The Union Club-house, also in Mosley-street, a fine modern stone building, with internal accommodations equal to those found in the best London establishments of the same description. 6. The Natural History Society's hall in Peter-street, a large square building, having in the principal front a portico supporting a pediment and comprising a fine hall, lighted from a cupola, and different apartments stored with numerous specimens of birds, insects, fossils, shells, &c., and a few quadrupeds. 7. The Concert-hall, near the last mentioned building. 8. The Assembly-room, below which is a billiard and newsroom; and, 9. The asylum for the blind, and the school for the deaf and dumb, at Old Trafford, designed by Mr. Richard Lane, in the Elizabethan style, having a fine frontage of stone, consisting of two wings and a projecting centre formed by the chapel of the two institutions. There are three theatres, one of which has a royal patent; but neither is much patronised.

Among the sacred edifices of Manchester, the collegiate church far surpasses the others, both in size and architectural beauty. It stands close to the Irwell, near Victoria bridge; and was erected in the 15th century, in the ornamental Gothic style, having been frequently since repaired and in part rebuilt. The interior is about 180 feet long and 60 feet broad; but only a portion of the building is fitted up with stalls and pews for choral service, the remainder consisting of private chapels, which cover a large area, and render the form of the edifice exceedingly irregular. The choir is one of the finest in England, and the tabernaclework is unrivalled: the monuments are numerous, and full of interest: the carved figures, with which the church is liberally adorned, are as quaint and grotesque as an antiquary could desire; and there are several beautiful stained-glass windows, with inscriptions and paintings. The college was founded in the reign of Henry VI., dissolved by Edward VI., and again chartered, in 1578, by Queen Elizabeth, who directed that the establishment should comprise a warden, four priests, two chaplains, and eight choristers. This charter was, for the most part, confirmed by Charles I. In the charter dated September 30, 1635, it is ordered that there shall be a warden (appointed by the crown) and four fellows (elected by the warden and fellows), who shall be a body corporate and politic of themselves forever. The charter farther directs that there shall be two chaplains or vicars, and two clerks (one to be in holy orders), four singing men (whether clerks or laymen), and four boys skilled in music. [There are now six singing men and six boys.] It is also ordained that there shall be a subwarden continually in the college, a treasurer, a collector (all of the number of the fellows), a registrar, a master of the choristers, an instructer, an organist, and a bailiff. Whenever either of the sees of Bangor or St. Asaph becomes vacant, they will be united, and Manchester will be raised into a bishopric. Preparatory to this change, the title of warden and fellows has already been changed for that of dean and canons; and this, in due course, will be followed by the appointment of an archdeacon of Manchester to superintend the clergy of the district. A chapel of ease was erected in Salford in 1634: this, St. Anne's, erected in 1712, and St. Mary's, erected in 1755, being the only places of worship in the town till 1760, between which and 1800 eight additional churches were built. Nine other churches, two of which (St. Luke's, Cheetham, and St. George's, Hulme) are very elegant, and cost £14,000 each, have been erected in the present century; so that, on the whole, there are 23 episcopal places of worship in the borough, served by 46 clergymen, and accommodating 29,400 individuals. The Roman Catholics raised their first chapel in 1746, but one belonging to the Presbyterians was opened at an earlier period. The Independents opened a chapel in 1762, and the Wesleyan Methodists in 1780. The dissenting meeting-houses now open comprise, one for seceders from the Scotch kirk, five belonging to Roman Catholics, 11 belonging to Wesleyan Methodists, 16 occupied by various seceders from Methodism, nine belonging to Independents, six to Baptists, four to Unitarians, and six others to Swedenborgians, &c., the whole being calculated to accommodate about 140,000 per-

eums. Three cemeteries have been laid out in Chorlton, Ardwick, and Harpur-hey; and the noxious practice of interring bodies within the town is slowly but gradually going out of use. The two boroughs of Manchester and Salford have above 120 Sunday schools.

The means existing in Manchester, in 1834, for the diffusion of elementary instruction, may be learned from the following summary, drawn up from the report of the Manchester Statistical Society:

Description of Schools	Manchester			Salford		
	Schools.	Scholars.	Per cent. to Pop.	Schools.	Scholars.	Per cent. to Pop.
I. Sunday schools:						
Estab. Church	25	10,304	5·14	9	5,741	4·99
R. Cath.	9	3,060	1·94	2	613	1·11
Dissenters	22	19,602	9·32	20	6,408	11·44
Total of Sun. schools	56	33,196	16·40	31	9,764	17.75
Children returned as day scholars	..	10,611	5·01	..	3,416	6·20
Sund. scholars only	..	22,185	11·40	..	6,344	11·55
II. Free or Subscription day schools	22	4,177	2·05	16	1,776	3·22
III. Evening schools	36	1,483	·79	20	328	·96
IV. Schools supplied by the children's payment						
Primary	460	11,334	5·91	197	3,397	6.11
Higher	114	2,394	1·17	23	663	1·20
Total	721	42,378	21·45	311	14,985	23·43

It would appear from this statement that, in 1834, about 28,000 children were receiving daily instruction in Manchester and Salford, and that 30,000 more were taught on Sundays. But doubts have been entertained as to the accuracy of these returns; and of the day scholars, nearly 19,000 were reported to be educated in private schools, and dames' schools, in two thirds of which the instruction is extremely defective. Within the last half dozen years considerable additions have been made to the better class of schools supported by the church and the manufacturers; still, however, there can be no doubt that the education of the working classes is defective, and that it admits of being very materially improved. From the report of the inspector of "the Chester Diocesan Board of Education," read at a meeting of that body on the 11th of March, 1841, we derive the following particulars relative to the church schools in the town of Manchester:

	On the Books.	Average Attendance.
21 Day schools	3,730	2,300
11 Infant schools	1,900	1,903
10 Evening schools	..	800
30 Sunday schools	15,156	11,519
Total	19,990	15,855

Among the schools deserving particular notice, the first place is due to the grammar school, founded in 1520, by Hugh Oldham, bishop of Exeter. Its revenues amount to upwards of £4500 a year; and in consequence of a decree of Chancery in 1833, its usefulness was increased by the opening of a lower school, and a general augmentation of the establishment. The decree of 1833, however, was ex parte, and as it did not effect all the alterations that were necessary, a suit was instituted to obtain farther reforms, in which a judgment was given by the lord chancellor on 16th Nov., 1840, from which it appears that "the income in 1833 was £4568, and the salary awarded to the head master was £609 per annum; and the salaries of all the masters together was £2050, while the number of scholars, including boarders, was only 196." The lord chancellor concluded his judgment as follows: "I propose, therefore, to declare, that in all future appointments of feoffees and trustees, regard should be had to the qualification required by the statutes; that all children of an age capable of instruction are entitled to be admitted into the school; that no part of the funds of the charity are hereafter to be applied towards paying premiums for exhibitions to boys who are or have been boarders in the house of any of the masters, except in continuing to pay exhibitions already granted; and that such boarders are not in future to derive any benefit from the funds of the charity in any manner by which the expenditure of such funds may be increased; and with these declarations, I shall refer it to the master to approve of such alterations in the scheme contained in the report of 1833 as may be necessary to carry the same into effect, and as the master shall find to be proper for the purpose of more effectually carrying into effect the objects of the charity." We understand, however, that the whole case is in a train for being again brought before the court in another shape. Few establishments in England confer so many university advantages on their alumni. It has 16 exhibitions of £80 a year, tenable for four years; 16 Somerset scholarships at Brasenose college, Oxford, averaging £90 a year; and several others of less value both at Oxford and Cambridge; besides which it derives great, though not exclusive advantages from the valuable Hulme's exhibitions, connected with Brasenose college, and tenable for three years after the degree of B.A. The instruction is efficient; and, should it be placed on the footing directed by the lord chancellor, it will no doubt rank among the best grammar schools in the kingdom. The college, founded by Humphrey Cheetham in 1665, is likewise a wealthy scholastic establishment, comprising, besides lodgings and school-rooms for boys, a valuable library of 25,000 volumes; but it is said that the modern part of this library is deficient, and that it is better suited for the scholar and the antiquary than for men of business: this college has also a museum of curiosities of little real value, but much visited by strangers and holyday people. The number of scholars is restricted to 80, 40 of whom must belong to Manchester and Salford, the rest belonging to Droylsden, Crumpsall, Bolton, and Turton. A plain education is furnished, and the scholars are afterward apprenticed and fitted out in trade. Seven other endowed charities for instructing children are amalgamated with national and other schools, very liberally supported, and conducted in the most efficient manner; and, besides these, the town has an asylum for the blind, erected by public subscription, and supported by an endowment bequeathed by Thomas Henshaw, Esq., of Oldham, and a deaf and dumb school, established in 1823, and remodelled in 1836: there are 80 scholars on the establishment, which is in every respect well conducted. A chapel is connected with the school, to which the public are admitted.

A college (removed by the Unitarians from York) on the plan of King's college, London, under the denomination of the Manchester New college, was opened for the reception of students on the 5th Oct., 1840. Instruction is given by distinguished professors in every department of literature, science, and theology. The erection of another college, to be called the Lancashire Independent college for the education of dissenters, is also in active progress, and will be soon completed.

The charitable institutions of Manchester, for the relief of the sick, disabled, and destitute, comprise an infirmary and dispensary, relieving 18,000 patients yearly, a fever hospital, or "house of recovery," a lying-in hospital, an eye institution, a lock hospital, a night asylum for the destitute poor, a female penitentiary, a provident society, a dispensary for children, a dispensary for diseases of the skin, and four other dispensaries, relieving altogether about 30,000 patients annually, and supported by funds from bequests and subscriptions, amounting to £14,000 a year; besides which, there are various minor charities belonging to Manchester and Salford, the aggregate income of which exceeds £4000 a year: so that upwards of £18,000 are annually expended in the relief of the poor, over and above the expenses of parish support, which in 1838 amounted to £40,000.

The literary and philosophical establishments of Manchester are very numerous, consisting of a philosophical society, instituted in 1781, and numbering among its past and present members Dr. Percival, the three Henrys (father and sons), Dalton, and other eminent men, whose science and discoveries have been of material advantage, not only to the town, but to the world generally: indeed, few provincial societies of the kind have earned so high a reputation; its memoirs (in 16 volumes) have been translated into both the German and French languages: a geological and mining society, founded in 1838, has already upwards of 200 members: a statistical society is actively engaged in collecting local information respecting education, morals, &c.: a botanical and horticultural society, established in 1827, possesses gardens that cover 16 acres: a zoological society has spacious gardens on the Bury road, tastefully laid out, and containing a good and increasing collection of animals: a society of natural history has a good museum, and is supported by about 600 subscribers. The Royal Institution was founded in 1823, for the promotion of literature, science, and the fine arts. We have already noticed its fine building, the principal hall of which has a statue of Dalton, from the chisel of Chantrey. Connected with this institution is a school of design, likely to be of material service to pattern-draughtsmen, machine-makers, and others. The Athenæum, established in the view of affording to the middle classes a suitable resort for reading, study, and conversation, is supported by about 1000 members, and has a library of 4000 volumes: two schools of medicine and surgery are superintended by able teachers, and provided with extensive museums, lecture-rooms, &c. There are two mechanics' institutes, and another establishment for similar purposes called the Christian institute: these institutions are all well provided with libraries, museums, apparatus, &c., and are well attended. There are three lyceums, specially intended for the improvement and recreation of the working classes by furnishing them with books, magazines, newspapers, lectures, and opportunities for friendly intercourse; and there can be no doubt that they have been of very material advantage. The Royal Victoria Gallery has an exhibition of objects in mechanics and science, and courses of lectures. A temperance society, formed in 1835, was the first to inculcate total abstinence from all intoxicating beverages: it has at present 19 branch associations, holding weekly meetings, and about 8000 members.

The banking establishments of Manchester, which are numerous, and conducted on a scale corresponding with the commercial importance of the place, comprise, besides five private banking-houses, mostly of great wealth and respectability, a branch of the Bank of England, and five joint-stocks: viz., the Bank of Manchester, founded 1829; the Manchester and Liverpool District Banking Company (1829); the Union Bank of Manchester (1836); the Manchester and Salford Banking Company (1836); the S. Lancashire Bank (1836); and the Alliance. A savings' bank was opened in 1818; and, from the report of 1839, it appears that it had then deposits to the amount of £237,860, received from 11,862 depositors. Five newspapers are published in Manchester; three of which, the *Guardian*, *Times*, and *Advertiser*, advocate whig and radical politics, the *Courier* and *Chronicle* being conservative. The *Guardian*, which has the largest circulation, and *Chronicle*, are published on Wednesdays and Saturdays; the others appear on Saturdays only.

Manchester possesses several large establishments connected with its internal economy. The workhouse, which occupies an eminence N. of the town, is a very extensive and well conducted establishment, accommodating an average 780 in-door paupers, whose weekly cost is about 2s. 4½d. each for food, and 5½d. for clothing. The house expenditure, in 1839, amounted to about £8750, and £13,600 were distributed among the out-door poor; the total expenditure of the overseers being £41,000. The Salford workhouse, in Greengate, has accommodation for about 350 inmates, whose average cost per head is about 3s. 8d.: the expenditure of the establishment was about £3000, £8000 more being distributed among out-door paupers. The New Bailey prison in Salford, close to the New Bailey bridge, commenced by Howard in 1787, has been since greatly enlarged: it has accommodations for about 800 prisoners, and is well conducted; but, owing to the great increase of population and crime, it is inadequate to the wants of the borough, and is frequently so overcrowded that three persons have to sleep in one cell! A police-office court is held daily within the precincts of the prison, by a stipendiary magistrate, appointed by the chancellor of the duchy of Lancaster, with a salary of £1000 a year. The police of the borough is regulated by a recent temporary act (2 and 3 Vict., c. 87), which provides for the establishment of a police-office, under a chief commissioner, who is to be a justice, and to have an annual salary not exceeding £800; and authorises the appointment of a sufficient number of fit and able men, to act under the commissioner as a day and night police, not only within the borough, but throughout the county of Lancaster. The force consists at present of 304 sergeants and constables, and 24 superior officers, costing in all about £21,700 a year. The police fire-engine establishment is perhaps the most effective in the kingdom, after that of the metropolis: it comprises seven engines, completely furnished with every necessary implement, fire-escapes and water-barrels, and a body of 40 firemen, commanded by a superintendent. The Manchester gas-works are the property of the town, and the profits are applied towards its improvement: the works were established in 1817, but the streets were not generally lighted with gas till 1824. The main pipes extend, in various directions, upwards of 80 m. in length; and the quantity of gas made in 1838 exceeded 160 millions of cubic feet. The Salford gas-works are on a much smaller scale, having only come into operation in 1835. The board of directors of the Manchester establishment paid over to the improvement committee, between 1830 and 1835, £34,000. The Manchester and Salford Water-works Company was originally established in 1808, but assumed its present shape only in 1823, when it received additional powers from parliament. There are two reservoirs, one at Beswick, 110 ft. above the level of the town, and the other at Gorton, 140 ft. above that level: the iron mains extend upward of 70 m., and the daily consumption of water is estimated at 1½ millions of gallons; besides which, 30,000 gallons are daily supplied to the railway companies. The markets of Manchester are not such as a town of great wealth and magnitude might be expected to possess; and this circumstance is most probably owing to the fact, that the tolls are not the property of the town, but belong to the lord of the manor. There are no general markets, like those of Liverpool, Birmingham, and Newcastle; but several are scattered in different parts of the town.

296

In Victoria-street, Swan-street (Smithfield), Camp Field and Deansgate, are markets for butchers' meat and vegetables; and a fish-market was erected near the exchange in 1828. The cattle-market is held every Wednesday, in Cross Lane, Salford; a large area on its sides is fitted up with stalls, filled with various articles both of farming and manufactured produce.

The following table, drawn up from the *Reports of the Manchester Statistical Society*, shows the consumption of butcher's meat in Manchester and its environs (estimated pop. 343,500) in 1836 :—

Description of Meat.	Average weight of Carcasses.	Number of		Quantity to each Person.
		Carcasses.	Pounds.	
Cattle	560 lbs.	40,380	22,568,960	65 lbs. 5 oz.
Sheep	83½ —	105,040	7,212,746	21 —
Lambs	37 —	95,988	3,576,716	10 — 7 —
Calves	90 —	11,791	1,061,190	3 — 1 —
		264,219	34,709,832	101 —
Offal (edible)	. . .	. .	1,387,308	4 — 9 —
Total	. . .	. .	36,097,140	105 — 14 —

The market-days are Tuesday, Thursday, and Saturday, the first-named being the manufacturers' day, and the last the chief market for agricultural produce and provisions. The fairs are held in Easter and Whitsun week, the first week in October, and on November 17. The first of these, called Knot-mill fair, is a mere popular festival, and the rest are cattle fairs.

Manufactures.—Manchester, though situated close to an almost inexhaustible coal-field, and deriving great advantages from the vicinity of three streams, available for machinery, would never, in all probability, have attained to her present magnitude and importance, as the first manufacturing town of the world, but for the invention of the steam engine, and the wonderful improvements made since 1760 in the manufacture of cotton twist and fabrics, through the genius and discoveries of Arkwright, Crompton, Cartwright, Horrocks, and others. How astonishing the revolution effected by the ingenuity and enterprise of a few obscure individuals! Before the spinning frame, which was invented in 1787, came into operation, the imports of cotton wool did not amount to 4,000,000 lbs. a year, and the value of the exports hardly exceeded £200,000. Arkwright's patent was set aside in 1785, and since then the progress of the cotton manufacture of Great Britain, and especially of Manchester, has been rapid beyond all precedent. Previously to 1785, the imports of cotton wool had not reached 12,000,000 lbs. in any single year; but in 1787 they amounted to 23,250,963 lbs.! The progress of the manufacture was not impeded by the late war, to the successful termination of which it contributed more, perhaps, than anything else: and what is not less extraordinary, it has more than quintupled since the peace! The imports of cotton wool, in 1840, amounted to the prodigious sum of about 583,300,000 lbs., of which no fewer than 459,000,000 lbs. were manufactured! In 1803, the value of the exports of cotton goods equalled those of woollen, the long-established and staple manufacture of the country; and they now amount to about £25,000,000 a year, while the exports of woollens do not exceed £6,000,000. Indeed, the cotton manufacture now forms, next to agriculture, the principal business carried on in the country, affording an advantageous field for the accumulation and employment of millions upon millions of capital, and thousands upon thousands of workmen! About 1½ millions of people are supported by spinning and weaving cotton, and the different supplementary employments of the trade; and fabrics of great beauty and excellent quality, which a few years ago were out of the reach of all except the wealthy, have been so much reduced in price as to be within the command of all but absolute beggars. (*For further details, see* ENGLAND AND WALES, vol. i., p. 707.) Of this gigantic manufacture Manchester is the grand centre, absorbing, with its neighbourhood 10 m. round, fully four-fifths of the trade, and comprising, besides spinning-mills, most extensive power-loom factories, and large dyeing and printing establishments. The manufacture of silk goods, also, which was introduced in 1816, has been in a most flourishing state since the removal, in 1826, of the oppressive import duties on raw silk; and this branch of industry in Manchester now exceeds that of Macclesfield. In the infancy of the trade, silk handkerchiefs and mixed goods were principally made; in 1822, gros-de-Naples and figured sarsenets were introduced; and at present (1840) nearly every kind of silk, from the rich brocade to the flimsy Persian, is manufactured, consuming upwards of 1,600,000 lbs. of raw silk, and employing 4000 hand-looms, besides 2000 persons in throwing-mills, and 500 in dyeing and printing houses. Mixed goods of silk and cotton, silk and woollen, and cotton and woollen, occupy several hands; and many hundred persons are engaged in making machinery, and in various branches of handicraft nearly or ere remotely connected with the principal object of industry. The following statement respecting the factories within the borough of Manchester is extracted from the *Parl. Returns* of 1839:—

Description of Mill.	Number.	Engines.	Power.	Hands employed.
Cotton-mills . . .	182	204	6,059	34,090
Worsted . . .	2	2	44	260
Woollen . . .	3	1	16	25
Silk . . .	19	19	300	4,162
Total . .	185	227	7,146	38,440

There are also several flax-mills, and this branch of spinning is extending. In some of these factories the process of spinning only is carried forward; but in many others the whole process is carried on, from the first carding to the ultimate dressing of the woven and bleached fabric. Many of them are buildings of extraordinary size, comprising seven or eight stories, erected at a heavy expense, and filled with machinery costing £30,000 or £40,000. The rooms are kept in the most perfect state of cleanliness, and the strictest order, regularity, and silence prevail throughout the establishments. Several thousands of spindles are at work in each of the principal factories; and in many of them upwards of 800 power-looms are in action, each producing from 15 to 20 pieces of fabric, of 24 yards each, per week. Besides the population connected with the factories, which almost absorb the plain-goods' trade, including jaconets, twilled cloths, and fustians, upwards of 9000 hand-loom weavers are employed in Manchester and the neighbourhood in weaving cotton, silk, and mixed goods. "The cotton fabrics are quiltings, figured waistcoatings, twilled shawls and handkerchiefs, checked and striped ginghams, tape-stripes, dimities, apron-checks, checked handkerchiefs, buff-checks and buff, coarse shirtings and sheetings. The silk fabrics comprise velvets, figured sarcenets, figured and plain levantines, plain satins, plain serges, sarcenets, and gros-de-naples, checked sarcenets, string-persians, ducape handkerchiefs, satin checked cravats, Brussels handkerchiefs, black bandanas, Welsh shawls, romals, turbans, Barcelona handkerchiefs, and grey bandanas. The mixed are chiefly for waistcoatings, handkerchiefs, cravats, shawls, &c. The weaving of each of these fabrics, with a variety of others, may be regarded as a separate branch of the weaving trade; and the earnings of the different weavers employed on each are as various as the fabrics. (*Hand-loom Weavers Rep.*)

We subjoin a statement drawn up by the Manchester statistical society of the amount of steam power employed in the various branches of manufacture within the parliamentary boroughs of Manchester and Salford in 1838:—

Branch of Manufacture.	Horse Power.		
	Manchester.	Salford.	Both Boro.
Cotton-spinning and weaving	5,472	704	6,036
Bleaching, dyeing, printing, &c.	756	521	1,277
Machine-making, foundries, &c.	608	226	734
Silk-throwing and weaving	237½	104	341½
Cotton thread and small wares	270	36	306
Collieries	200	110	308
Saw-mills	141	14	155
Engraving for calico-printing, &c.	75	6	81
Fustian shearing	45	31	76
Breweries	16	62	78
Flax-spinning	—	76	76
Chemical works	56	11	68
Woollen	40	22	56
Various, all of minor importance	408	88	496
Total .	7,924½	1,998	9,921½

The manufacture of machinery is conducted on a most extensive scale, employing many hundred hands. Steam-engines are made of different powers, varying from eight to 100 horse power; and the castings are often of gigantic size, weighing from 30 to 50 tons. The iron-planing and rivetting machines are curious specimens of mechanical ingenuity, and have greatly tended to facilitate the manufacture. Many of the workmen receive from £2 to £3, and few less than 30s. weekly wages. The business of locomotive engine and tool making, also, is most extensively carried on, the largest establishments of this kind being at Fairbairn's; and at Sharp and Robert's; Peel, Williams, and Peel's; Fairbairn's; and Whitworth and Co's, in Manchester; at each of which several hundred men are employed, and the arrangements in every way are most complete and systematic.

The speedy and cheap communication established with the port of Liverpool, and other places, has been at once a cause and a consequence of the increase of manufactures in Manchester. It had become, at the close of the last century, a great centre of internal navigation. Brindley constructed the duke of Bridgewater's canal, uniting with the Mersey at Runcorn, in 1761; the Bury and Bolton canal was projected in 1791; that to Ashton and Oldham in 1792; and that to Rochdale in 1794; and these communicate with other canals, in such a manner as to establish an easy communication with the eastern, central, and southern counties, including the ports of Hull, London, and Bristol, as well as that of Liverpool, which is, *par excellence*, the port of Manchester. Large sums were sunk in excavating these canals; but the returns far exceeded expectation, and the profits to the shareholders were in some cases immense.

The Mersey and Irwell navigation company have recently commenced operations in the river Irwell, which will have a most important influence over the navigation and commerce between the two towns. The river at Manchester is being deepened between the New Bailey and Victoria bridges, in which portion it is not at present navigable: and a report, survey, and plans have been drawn up by Mr. Palmer, (president of the Society of Civil Engineers, London,) in compliance with which it is proposed to open the navigation to vessels of 300 tons burden the whole distance from Manchester to Runcorn. The inhabitants of Manchester have for some time been earnestly soliciting to be allowed the privilege of bonding goods; and bills to that effect have been introduced into the House of Commons. Hitherto, however, the opposition of interested parties has prevented the project being carried into effect; but should the Mersey and Irwell navigation company carry their designs into execution, Manchester will be entitled to claim the privilege of bonding from its being a *port*.

Notwithstanding the rivalry of railways, the tonnage on the canals continues to be very heavy. It was proved before the house of lords, in 1836, that the water-carriage between Manchester, Birmingham, the Potteries, Shrewsbury, &c., amounted annually to 364,100 tons; and if the trade with London, and other southward traffic, be added, it seems probable that the entire canal carriage of Manchester exceeds 700,000 tons a year. But the rapidity and ease of communication have been still more prodigiously increased by the construction of railways, which have brought Manchester within an hour's distance of its great warehouse for the raw material, within four hours of Birmingham, and nine hours of the metropolis! The Liverpool and Manchester railway, opened in 1830, cost £575,000, and the expense of maintenance amounts to £160,000 a year, the annual nett profits being about £103,000 at an average of four years. The passengers on this railway have averaged for some years 480,000 a year; and the receipts from goods, as compared with those from passengers, bear the proportion of three to five nearly. The Grand Junction railway, connecting Manchester and Liverpool with Birmingham, is 82½ m. in length (15 m. of which are on the Liverpool and Manchester line), the original cost amounted to £1,800,000 ; the receipts, in 1839, were £178,000, the expenses averaging 53 per cent. on the returns. The Manchester and Leeds railway was opened the whole distance on the 1st of March, 1841, and in connection with the North Midland and the Midland counties, furnishes another route from Manchester to London. The Sheffield and Manchester railway, and that between Manchester and Birmingham, have been recently commenced, and will probably be completed by the end of 1841.

Manchester has recently received a charter of incorporation; and the municipal borough is divided into 15 wards, the government being vested in a recorder, mayor, 15 aldermen, and 48 councillors. Quarter sessions are held by the recorder; and there is a court for the recovery of debts under £20. Notwithstanding its vast importance, Manchester did not enjoy the privilege of sending representatives to parliament till the Reform Act gave to the manufacturing interests of the country that influence in the legislature to which they had been long entitled. Manchester was then erected into a parliamentary borough, with power to send two members to the House of Commons; its boundaries including, besides Manchester, the eight other townships enumerated at the commencement of this article. Registered electors, in 1839-40, 19,150. The same act conferred on Salford the privilege of sending one member to the House of Commons: its limits comprise two other entire townships, and part of a third. Registered electors, in 1839-40, 2519. Manchester has also been formed into a union under the Poor Law Amendment Act, which came into operation in January, 1841. The union comprises the townships of Blakely, Crumpsall, Cheetham, Moston, Harphur-hey, Bradford, Newton, Failsworth, Great Heaton, Little Heaton, Prestwich, and Manchester.

Condition of the People of Manchester.—The increase of wealth in Manchester, during the last half century, has been quite unprecedented, and there is at present, in proportion to its size, a greater number of opulent capitalists than any other town of the empire. The capital vested in mills, machinery, and stocks of goods, is immense; and, in addition to the vast sums that are thus employed in their peculiar business, the capitalists of Manchester, and the adjoining districts, have been the great promoters of railways in all parts of the empire, and hold a very large proportion of the stock embarked in these undertakings. To achieve such great results, a combination of all those qualities that go to form accomplished men of business has been required; and no where do we find the persevering attention to details, added to the sagacity to distinguish between the doubtful and the certain, and the enterprise to embark in remote and apparently hazardous, though really safe schemes, that characterise the highest class of commercial men, so generally diffused as in Manchester. It is, in fact, the grand arena of industry and enterprise. Every one is striving to raise himself to distinction, and to outstrip his neighbour in the accumulation of wealth. But there are no mean jealousies, or unfair jostlings: there is more than room enough for every one; and every one knows that his success is wholly dependent on his own efforts.

The shopkeepers and middle classes of Manchester are more attached to old habits than those of most other towns. In proof of this we may mention, that by far the greater number of them continue to dine at the primitive and unfashionable hour of one. At no very distant period, indeed, they were accustomed to shut their shops from one till two; and though that be no longer the case, the banks will not, at present, with one or two exceptions, cash cheques sent to them at such a time, or allow their clerks to be interrupted when at dinner!

But it is not so easy to arrive at any very definite conclusions with respect to the condition of the lower classes in this great workshop. On the whole, however, we are inclined to consider it as tolerably satisfactory. No doubt, the condition of the English part of the population has been most injuriously affected by the prodigious influx of Irish immigrants, of whom there are probably not fewer than 65,000 in the town, where they, for the most part, occupy an inferior quarter, called "Little Ireland." The Irish, it is but fair to say, are neither peculiarly disorderly nor peculiarly dishonest; but their competition has depressed wages, or hindered them from rising, and their example has been most pernicious, by accustoming the English to a lower standard of food and comfort. But despite the influence of this fruitful source of degradation, the work-people of Manchester seem, when employed, with the exception of the hand-loom weavers, to be really well off. Unluckily, however, a number of individuals, partly belonging to the town, but mostly new comers from Ireland and other parts of England, are usually without employment, and in a state bordering on destitution. It is unfortunate, too, that so many of the workmen's wives should be employed in factories, as this takes them away from their families, and prevents them from bestowing sufficient pains on the training of their children, and their household affairs. It is singular, indeed, how ignorant workmen's wives, engaged in factories, and brought up as factory girls, are of most matters connected with domestic economy; and how much more comfortable their families might be were they familiar with such details, though their earnings were less. It is not true, however, that the condition of the work-people has been deteriorated, and, in point of fact, it has on the contrary, been very materially improved. Most descriptions of labourers receive good wages; and such skilled labourers as are temperate and industrious are, speaking generally, in decidedly comfortable circumstances.

The lower classes of Manchester live principally in houses above ground, consisting for the most part of cottages, of which many lengthened streets have been built of late years; but, in addition to these, great numbers inhabit cellars or underground floors, sometimes below the cottages, and sometimes below other houses. According to a statement published by the statistical society, the bulk of the population, in 1835, were lodged as stated in the following table:

	Manchester.	Salford.	Total.
Persons occupying houses	94,250	31,693	125,943
" rooms of houses	9,351	3,132	12,483
" boarding with occupants of houses	9,671	2,831	12,502
" occupying cellars	14,214	3,310	17,554
" boarding with occupants of cellars	696	25	711
Total :—Persons resident in dwellings examined	128,292	40,931	169,223
Families resident in dwellings examined	28,186	9,835	37,794
Average persons to a family	4·65	4·29	4·48
Average rent per week of houses, rooms, and cellars, examined	2s. 11½d.	2s. 10d.	2s. 11d.
Gross amount of rent for a year of houses, rooms, and cellars	215,318l.	70,451l.	286,073l.
Number of dwellings comfortable	19,984	7,417	27,391
" uncomfortable	8,302	2,121	10,423

Of the 169,223 persons classified above, 84,333 were adults, 53,699 were children under 12 years of age, and 30,891 were children above 12 years of age, and mostly employed.

It appears, from a statement published by the Manchester statistical society in 1836, which may be regarded as nearly accurate, that 71,799 individuals, comprising nearly all below the rank of shopkeepers, were employed as follows :—

	Individuals.
In cotton factories in Manchester and Salford	16,363
Other factories in ditto	1,433
Hand-loom weavers	3,198
Persons employed in warehouses	7,807
" in manufactures	7,067
" in building trades	4,515
" in clothing trades	6,208
Occupations not classed	23,232
	71,799

The following table, drawn up by the Manchester chamber of commerce, exhibits the average rates of wages paid to the different classes of labourers in and out of factories in Manchester in 1832, since which period no material alteration has taken place :—

	£ s. d.	£ s. d.
Spinners, men	1 0 0	1 9 0
women	0 16 0	0 18 0
Stretchers	1 5 0	1 5 0
Piecers (boys and girls)	0 4 7	0 7 0
Scavengers	0 1 6	0 7 0
In the Card-room :		
Men	0 14 0	0 17 0
Young women	0 9 0	0 9 0
Children	0 6 0	0 7 0
Throstle-spinners	0 6 0	0 9 0
Reelers	0 7 0	0 9 0
Weavers by Power :		
Men	0 13 0	0 16 10
Women	0 9 0	0 12 0
Dressers' men	1 9 0	1 10 0
Winders and warpers	0 9 0	0 11 0
Mechanics	1 4 0	1 6 0
Weaving by Hand :		
Quality. Nankeens, fancy	Woven by men	0 8 0	0 12 0
common	children and women		
best	men	0 10 0	0 12 6
Checks, fancy	men	0 7 0	0 7 6
common	children	0 5 0	0 7 0
Cambrics	all ages	0 6 0	0 13 6
Quiltings	men & women	0 9 0	0 14 0
Fustian cutters	all ages	0 10 0	0 12 0
Machine-makers	men	1 6 0	1 10 0
Iron founders	do.	1 6 0	1 6 0
Dyers and dressers	do.	0 15 0	1 0 0
do. do.	young men	0 12 0	0 14 0
do. do.	boys	0 5 0	0 10 0
Tailors	men	1 0 0	
Porters		0 14 0	0 16 0
Packers		0 16 0	
Shoemakers		0 15 0	0 18 0
Whitesmiths		1 5 0	1 8 0
Sawyers		1 6 0	1 8 0
Carpenters		1 4 0	
Stonemasons		0 18 0	1 5 0
Bricklayers		0 17 0	1 5 0
Bricklayers' labourers		0 12 0	
Painters		0 16 0	
Slaters		3 8 per day	
Plasterers		0 19 0	1 1 0
Spademen		0 10 0	0 15 0

It is greatly to be regretted that effectual provision had not long since been made in Manchester and other large towns for their proper drainage and pavement, and for laying down rules as to the erection of houses. The authorities in Manchester have done all in their power, under the existing laws, to improve the streets; but there is no general building act for the town, and except in certain districts where the commissioners of police are entitled to interfere, each proprietor builds as he pleases. Hence cottages may be seen springing up row behind row, without the streets or alleys between them being of sufficient width, or drained or paved; in some places, indeed, the streets are full of pits filled with stagnant water, the receptacles of all sorts of filth. (Report on Health of Towns, p. 10.) Such a state of things is discreditable alike to the local authorities and the government: and we do not know that any measure is more imperatively necessary, seeing the vast and rapid increase of towns, than the enactment of such regulations as may be required to provide for the proper construction of the streets and houses, and consequently for the health and comfort of the population.

Cellars, however damp and unhealthy, are preferred by a large proportion of the lower classes both here and in Liverpool, not so much from their cheapness, as because they afford facilities for dealing in various sorts of articles, and because their inmates either are or believe themselves to be more independent than if they resided as lodgers in houses rented by other parties.

It is unhappily true, as seen from the previous statements, that many of the dwellings of the lower classes, especially those of the Irish, exhibit a great want of furniture, of cleanliness, and comfort. This, however, is not owing, as many have supposed, to the growth of the factory system, but partly to the poverty, and still more to the perverse habits of the occupiers. In a tract written in Manchester, and published by authority in 1755, long before the factory system had any existence, the houses of the poor are said to be "most wretched," "filthy and nasty" in the extreme, and "noisome and infectious." (See extract from tract in *Manchester as it is*, p. 36.) There is really, therefore, no room or ground for saying, that any portion of the poor are worse lodged now than formerly; while on the other hand, of 37,794 dwellings of the labouring classes, examined by the agents of the Statistical Society, no fewer than 27,281 were decidedly "comfortable;" and as respects the clothing and other accommodations of the poor, they are infinitely superior at present to what they have ever previously been. Their prosperity is evinced by the great average consumption of butcher's meat.

A good deal of fever necessarily prevails at most periods of the year, in the poorer districts of Manchester, especially in those where the streets are in the disgraceful state already noticed. But, on the whole, Manchester is less unhealthy than Glasgow, or than the old town of Edinburgh, which has no manufactures.

The idle and absurd stories that were so industriously propagated with respect to the influence of factory labour on health and morals, are now pretty well exploded. Lately, indeed, there would appear to be a considerable increase of crime; but this increase is apparent only, and is mainly a consequence of the improved state of the police, and of trivial offences that formerly escaped notice being (whether wisely or not, we shall not stop to inquire) now visited with fine or imprisonment. The truth is that, in respect of morality, the labouring population of Manchester has but little to fear from a comparison with that of any large town in the empire. An unexceptionable witness, the Rev. R. Parkinson, canon of the collegiate church, Manchester, in a speech at a public meeting in Feb., 1839, said, "I am aware that an able and well-known poet has said (and the saying has almost passed into a proverb)—

' God made the country, but man made the town ;'

meaning, of course, that the country was the most proper place for man to dwell in, and that the occupations of town life were unnatural. I think, on the contrary, that instead of an agricultural population, the people of this country were meant to be one of a very different character. I have no natural predilections for my present opinions. My birth and early education put me in a very different position from that which I now hold; but being at present an inhabitant of this town, having enjoyed ample opportunities of observing and judging, and being in a position which gives me no motive for a partial judgment, I maintain that, taking an average of all classes of our population and that of other districts, we shall find the morality of this district not below that of the most primitive agricultural population. I have the best authority for saying, that the streets of Manchester, at ten o'clock at night, are as retired as those of most rural districts. When we look at the extent of this parish, containing at least 300,000 souls, being more than the population of half our counties, can we be surprised that there is a great amount of immorality? But a great proportion of that immorality is committed by persons who have been already turned in crime in districts supposed to be more innocent than our own, and who swell our police reports, not so much because we hold out greater facilities for rearing them, as that they are apprehended through the superior vigilance of our police." This is pretty conclusive; and we may add, that the regard paid by females to decency, both of language and deportment, is stated by intelligent witnesses before the factory commissioners of 1833-34 to be greater in Manchester than in most rural districts. It is a fact, too, that the proportion of illegitimate to legitimate children in this country is only 1 in 12, a very low ratio for so dense and varied a population, and not greater than in the purely agricultural counties of Hereford and Salop.

We believe that the doctrines of chartism and ultra-radicalism have made less progress in Manchester than in most other great towns, the metropolis excepted, certainly less than in Glasgow. Stagnations of trade, by occasioning want of employment and reducing wages, necessarily, also, occasion discontent and dissatisfaction; and in such periods demagogues are not wanting to recommend political nostrums of all sorts as infallible remedies for the grievances under which they labour. But the great bulk of the population are, notwithstanding, attached to the principles of the constitution, orderly, and opposed to violence. And in this, no doubt, their opinions are in accordance with their own obvious interests; for, were they to become disorderly, or to cease to respect and uphold the rights of property, the prosperity of Manchester would be instantly terminated: capitalists would withdraw from and shun her as if she were infected with a pestilence, and the mass of the population would sink into a state of squalid and irremediable poverty.

It is needless to observe that the interests of the employers of labour and those of the labourers, though apparently conflicting, are, at bottom, the same; and that neither party can prosper without that prosperity redounding to the advantage of the other. But, notwithstanding this identity of interests, there is, it must be admitted, but little sympathy between the great capitalists and work-people in this or any other large manufacturing town. This is occasioned by the great scale on which labour is now carried on in factories; and by the consequent impossibility of the manufacturer becoming acquainted with the great bulk of the people in their employment. They do not, in fact, so much as know their names; they look only to their conduct when in the mill; and are wholly ignorant of their mode of life when out of it, of the condition of their families, &c. The affections having nothing to do in an intercourse of this kind; every thing is regulated on both sides by the narrowest and most selfish views and considerations; a man and a machine being treated with precisely the same sympathy and regard. It is not to be denied that this is a state of things fraught with considerable danger; and that no society can be in a really sound or healthy state where the bond of connection between the different ranks and orders is such as now prevails at Manchester, and other great towns. In difference, on the one hand, necessarily produces insubordination, and plotting, on the other. However, it is easier to point out a condition of thing? prod than to suggest any means by which it may be obviated. We doubt, indeed whether it admit of any effectual remedy. The whole tendency of society, in modern times, is to make interest, taking the term in its most liberal and sordid sense, the link by which all classes are held together: and should any circumstances occur to make any considerable portion of society conclude that their interest is separate from or opposed to that of the others, there would, we apprehend, be but few other considerations to which to appeal to hinder the dissolution of such society.

In 1838 there were in Manchester 1769 beer-shops, and 625 public-houses, many of the establishments for the sale of spirits vying in splendour with the gin-palaces of the metropolis. Intemperance, however, is not on the increase. Great numbers of coffee-shops have recently been opened; and the influence of the temperance societies has also been most beneficial.

Climate, Temperature, &c.—Manchester, as already seen, is a healthy town; indeed, taking its size and the occupation of its inhabitants into consideration, the mortality is less than in most towns of the north of England; and if means were adopted for the improvement and cleaning the poorer streets and buildings, and for consuming the smoke which at present issues in dense clouds from innumerable factory chimneys, there can be little doubt that its salubrity would be materially increased. The mean annual quantity of rain falling in Manchester (at an average of 35 years) is 36·140 inches, while the mean annual quantity falling in Lancaster (at an average of 20 years) is 30·714 inches; the comparatively slight variations in the temperature likewise contribute greatly to the healthiness of the town. [See table top of next page.]

According to Whitaker, the historian of Manchester, "the Roman invaders of this country fixed a military station in a place since called Castlefield, to which they gave the name Mancunium," whence Manchester has been derived. In the time of the Saxons the old town was deserted, and about 627 another was built on its site. In 920, according to Dr. Aikin, the Saxon king, Edward the Elder, ordered Manchester to be fortified. In Domesday Book the town is called a manor, and is described as having two churches. In the 14th and 15th centuries it received great additions and improvements, so that, in Leland's time, it was reckoned "the fairest, best builded, quickest, and most populous town of Lancashire." Camden also mentions it as being famed in his time for the manufacture of woollen cloths, then called "Manchester cottons," that is, coatings. The first authentic mention of the cotton manufacture in England is made by Lewis Roberts, in his *Treasure of Traffic*, published in 1641, where it is stated, "The town of Manchester in Lancashire must be also herein remembered, and worthily, for their encouragement, commended, who buy the yarn of the Irish in great quantity, and, weaving it, return the same again into Ireland to sell. Neither doth their industry rest here; for they buy cotton wool in London that comes first from Cyprus and Smyrna, and at home work the same, and perfect it into fustians, vermillions, dimities, and other such stuffs; and then return it to London, where the same is vented and sold, and not seldom sent into forrain parts, who have means, at far easier terms, to provide themselves of the said first materials." (*Orig. agr.*, p. 32.) In 1660, the inhabitants of Manchester were reckoned the most industrious in the N. of England. The town was stated to be a mile long, with open and clean streets, and good buildings; and, in 1799, it is described as

The following Synoptical View of the Temperature, &c., of Manchester, during the year 1846, cannot fail to be interesting. It is extracted from the private Diary of Dr. Dalton:—

		Thermometer				Barometer.	Rain.	Aurora.
		Morn.	Noon.	Night.	Mean.		*Inches.*	
Jan.	Mean, &c.	40° 3'	43° 4'	40° 6'	41° 4'	29·41		1
	Greatest	52	54	55		30·32		
	Least	26	31	28		28·68	3·540	
Feb.	Mean, &c.	39° 5'	43° 8'	39° 2'	40° 5'	29·756		8
	Greatest	49	54	50		30·58		
	Least	26	34	20		28·08	1·941	
March,	Mean, &c.	39° 5'	46° 5'	39° 8'	41° 2'	30·42		1
	Greatest	49	52	47		3·66		
	Least	29	39	31		29·72	0·945	
April,	Mean, &c.	50° 5'	58° 2'	50°	52° 9'	30·184		1
	Greatest	63	74	58		30·20		
	Least	39	45	38		29·51	0·790	
May,	Mean, &c.	54° 3'	59° 53'	59° 3'	55° 38'	29·899		8
	Greatest	64	73	61		30·28		
	Least	47	49	46		29·30	3·490	
June,	Mean, &c.	57° 8'	62° 4'	58° 9'	59°	29·942		8
	Greatest	67	76	63		30·17		
	Least	53	54	49		29·62	3·790	
July,	Mean, &c.					29·83		8
	Greatest					30·10		
	Least					29·32	0·049	
Aug.	Mean, &c.	60° 5'	66° 6'	59° 3'	62° 6'	29·917		6
	Greatest	70	77	66		30·26		
	Least	50	54	50		29·00	4·665	
Sept.	Mean, &c.	58° 06'	58° 8'	55° 5'	57°	29·961		1
	Greatest	63	75	63		30·20		
	Least	47	51	45		29·62	3·880	
Oct.	Mean, &c.	48° 01'	52° 5'	46° 7'	49° 4'	29·869		1
	Greatest	54	57	54		30·51		
	Least	34	47	36		29·14	2·405	
Nov.	Mean, &c.	43° 6'	48°	45° 5'	45° 7'	29·53		1
	Greatest	54	58	52		30·35		
	Least	34	40	33		29·30	4·675	
Dec.	Mean, &c.	36° 3'	39° 3'	36° 9'	37° 5'	30·12		2
	Greatest	50	22	44		30·52		
	Least	26	31	28		29·15	0·180	
							34·291	8

" the largest, most rich, populous, and busy *village* in England, having about 34,000 individuals within the parish." Fustians were the earliest article of manufacture, and other fabrics were made soon afterward; but the great increase of population and commercial prosperity did not take place till 1770, when machinery was first introduced into the town. From that year down to the present time Manchester has been a scene of rapidly increasing industry, and has been distinguished by the invention and enterprise of its citizens; its working population supplies every quarter of the world with clothing; and wealth, the reward of successful labour flows in from all sides in a large, rapid, and uninterrupted current. (*Baines's Hist. of Lancaster* (4th ed.), ii., 140-309; *Wheeler's Manchester; Manchester as it is; Parl. Rep.;* but principally *Priv. Inform.*)

MANCHESTER, p. v., Hillsborough co., N. H., 31 m. S.S.E. Concord, 461 W. Incorporated in 1751 by the name of Derryfield, which was, by the legislature, changed to Manchester in 1810. Bounded W. by Merrimac river, E. by Massabesic pond, the outlet of which, Cohass brook, flows into Merrimac river. The canal around Amoskeag falls, with a descent of 45 feet, about a mile long, was completed in 1816, at an expense of over $50,000. It contains five churches, a Congregational, Baptist, Free-will Baptist, Methodist, and Universalist, 31 stores, three lumber-yards, one fulling-mill, one woollen factory, four grist-mills, five saw-mills, one brewery, two printing-offices, three weekly newspapers; eight schools, 950 scholars. Pop. 3235.

MANCHESTER, p. t., semi-capital of Bennington co., Vt., 95 m. S.S.W. Montpelier, 498 W. Watered by Battenkill river and its branches, which afford good water-power. White marble is extensively found and exported. It contains three churches, a Congregational, an Episcopal, and a Baptist, four stores, three fulling-mills, two woollen factories, one grist-mill, 12 saw-mills, one academy, 105 students; 10 schools, 510 scholars. Pop. 1594. The village is pleasantly situated on elevated ground, and handsomely built, chiefly on one street. It contains a courthouse, jail, a Congregational church, several stores, and the Burr seminary, a flourishing academy, founded by a donation of $10,000 by Joseph Burr, Esq., in 1829, who gave property to the amount of $150,000 to different literary and benevolent institutions.

MANCHESTER, p. t., Essex co., Mass., 22 m. N.E. Boston, 463 W. Incorporated in 1645. Bounded S.E. by Massa-chusetts bay. Vessels of 120 tons come to the village, and vessels of any size find good anchorage in the harbour, which is safe. It contains two churches, a Congregational and a Universalist, six commercial houses in foreign trade, two grist-mills, four saw-mills; four schools, 260 scholars. Pop. 1355.

MANCHESTER, p. t., Hartford co., Ct., 10 m. E. Hartford, 346 W. Incorporated from E. Hartford in 1823. Watered by Hockanum river and its branches. It contains two churches, a Congregational and Methodist, seven stores, six fulling-mills, five woollen factories, two cotton factories, with 3600 spindles, two grist-mills, seven paper-mills, two powder-mills; seven schools, 407 scholars. Pop. 1823.

MANCHESTER, p. t., Ontario co., N. Y., 8 m. N. Canandaigua, 209 m. W. Albany, 348 W. Watered by Canandaigua outlet. It contains two churches, a Methodist and Baptist, three stores, three fulling-mills, three woollen factories, two flouring-mills, two grist-mills, five saw-mills, one paper-mill, two tanneries, two distilleries; 10 schools, 975 scholars. Pop. 2912.

MANCHESTER, t., Passaic co., N. J., 18 m. N.W. New York. Organized in 1898. Bounded S. by Passaic river, N.W. by Ramapoo river. Watered by Pompton river. The village lies opposite to Paterson, with which it is connected by two bridges. It has nine stores, seven grist-mills, six saw-mills, two tanneries, one distillery; one academy, 94 students; six schools, 265 scholars. Pop. 3116.

MANCHESTER, p. t., York co., Pa., 18 m. S. Harrisburg, 20 W. It contains four stores, one flouring-mill, eight grist-mills, five saw-mills, two tanneries, 15 distilleries, one pottery. Pop. 2152.

MANCHESTER, t., Morgan co., O. It has one store, one grist-mill, two saw-mills; eight schools, 208 scholars. Pop. 1267.

MANCHESTER, p. v., Chesterfield co., Va., 2 m. S. Richmond, 119 W. Pleasantly situated on the S. side of James river, opposite to Richmond, with which it is connected by Mayo's bridge. It contains two churches, a Methodist and Baptist, six grocery stores, one cotton seed oil mill, one cotton factory, with a capital of $70,000, eight tobacco factories, one flouring-mill, 300 dwellings, and 1500 inhabitants.

MANCHESTER, p. v., capital of Coffee co., Tenn., 66 m. S.E. Nashville, 652 W. Situated on the head waters of Duck river, and contains a courthouse, and several stores and dwellings.

MANCHOORIA (Chin. *Kirin-sola*), an extensive region of N.E. Asia, belonging to China, and the original seat of the present ruling dynasty (Ta-tsing) of the Chinese empire, lying between lat. 41° and 57° N., and between long. 117° and 140° E., bounded N. by the Russian gov. of Yakoutsk, E. by the gulf of Tartary and sea of Japan, S. by China Proper, and W. by the Russian gov. of Irkutsk and Mongolia, from which latter it is separated by a wooden palisade, connected with the great wall of China, and by a line running down the Songari and other rivers to the Daourian range, on the S. of Siberia. Estimated area, 700,000 sq. m. Population unknown. The S. provinces are the only parts of the country that have been visited by Europeans; our knowledge of the remainder being derived only from the doubtful statements of a Chinese geographer. It is, therefore, more than probable, that, should any events lead to the admission of competent travellers into the country, it will be found necessary to make considerable alterations in our maps and descriptions of what is now little better than a *terra incognita*. Manchooria lies chiefly in the great valley formed by the Amur and Songari, with their numerous tributaries, and is bounded by three principal mountain chains, 1, one on the E., running from the peninsula of Corea along the whole line of coast to the N. boundary, and having a probable elevation of 5000 feet; 2, the Daourian mountains (called, by the Chinese, the outer Hing-as-ling), which form the entire N. boundary of Manchooria, but also send out minor offsets into the centre of the country; 3, the inner Hing-as-ling, or Sialkoi chain, which appears to be a continuation of the Shan-see mountains, and to extend, with little interruption, over a great part of Mongolia. Besides the above principal ranges, there are, to the N. of Corea, some chains of inferior importance, bearing several different names; but this part of the country, near the coast, though nominally a part of Manchooria, is inhabited, almost exclusively, by Ainos, a people similar to those inhabiting Jesso and Tarakai, in the empire of Japan. The chief river of Manchooria is the Amur, Segalien or Kwentung (for it is thus variously called), which, measured along its windings, is about 2200 m. in length, and, with its tributaries, drains a territory of about 900,000 sq. m. Several of these streams afford pearls; but the principal pearl-fishery is on the E. coast, in the channel of Tartary. It is a government monopoly, and is carried on by Manchoo soldiers, who are required, annually, to deliver into the imperial coffers, a fixed quantity of pearls. The chief lakes are the Hinkai-nor, a large sheet of water near the source of the Oussouri, in the province of Kirin, and the Hoorun and Pir, which give their names to the most W. district of the province Tsitsihar: there are a few others in different parts of the country, but only of small size.

The nature of the Manchoo soil, and its mineral productions generally, are little known. The people in the N. being chiefly nomads, subsisting by the produce of the chace, pay little attention to tillage; but agriculture is common in the S. districts, and the *cerealia*, as well as hemp and cotton, are extensively cultivated. The staple productions, however, are ginseng and rhubarb, the former being an exclusive government monopoly. The province of Shing-king, on the gulf of Pecheloo, produces corn, millet, and peas, large quantities of which, with ginseng, are sent by sea to the S. provinces of China. The forests, which clothe the sides of most of the mountains, comprise oaks, pines, firs, and birches; lime-trees, maples, oleanders, acacias, &c., being found on the plains towards the S. The domestic animals of central Europe are common in the more cultivated districts; but the cattle are small, and the breed of sheep peculiar to this country, called *argali*, is small, and coarse-wooled. Near the Yablonoi range, reindeer are kept, and camels are to be seen in many parts of the S. provinces. The wild animals comprise the ermine, sable, fox, and bear, hunted for their fur, which are a considerable article of trade with the Russians. Fish, especially salmon, and remarkably fine sturgeons, are abundant in the rivers, and held in high estimation by those living near the banks. The Manchoo territory is divided into three provinces, 1. Shing-king, (comprising the ancient Leaou-tung), near the borders of China; 2. Kirin, occupying the country E. of the Songari: and, 3. Tsitsihar, comprising the whole country W. and N.W. that river. The government of the first of these provinces is conducted by civil officers, on the same plan as in China; but the other provinces are under a government more strictly military than any other portion of the Chinese empire. The governors and magistrates are all military men; and the law makes all males, above 16 years of age, liable to serve under the standards to which they belong by birth, of which there are eight, each being distinguished by its peculiar flag. Kirinoolo is the metropolis of the country, and the residence of the supreme governor. Ningoota, on the Hooka, a tributary of the Songari, is also held in high esteem, in consequence of its having

been the residence in former times of the reigning family of China. With respect to trade, however, both are inferior to Fung-hwang-ching, on the borders of Corea. The seaports frequented by the Chinese junks are Kin-tchou, at the N. end of the gulf of Leantung, and Kaitchou, on the same gulf, E. of that last mentioned. The other cities of Manchooria, except Moukden, the old capital, and still denominated "the affluent metropolis," have no claim to rank higher than villages, though most of them are surrounded by walls, and garrisoned by small bodies of soldiery.

The general history of the Manchoos, or Eastern Tartars, with an account of their physical conformation, has already been given at some length in the article Asia, in this work, (i. 183-185), to which the reader is referred for farther particulars. (See, also, Mongolia. *Ritter's Asien*, i., 85-153, ii., 210-320; *Klaproth's Magasin Asiatique, and Asia Polyglotta, Appendix; Chinese Repository*, vol. i., p. 113-118; and also vols. v. and vi.)

MANDAN, district, a portion of the territory of the United States, bounded N. by the British possessions, E. by Wisconsin, S. by the N. fork of Platte river, and W. by the Rocky mountains. It is about 590 miles from N. to S., and 600 miles from E. to W., containing about 300,000 sq. m. It is an elevated plain or table land, and the soil, though light, bears sufficient grass to support extensive herds of buffaloe, elk, and deer; but some parts of it are barren. It appears that the vast chain of the Rocky mountains is broken through by a pass which admits of a road. The passage is up the N. branch of Platte river, and by the Sweetwater branch, which flows from the Wind river mountains, the highest portion of the Rocky mountains; thence across the height to Snake or Lewis river, a branch of the Columbia. The pass at the dividing ridge is in 42° 24' 32'' N. lat., about 7000 feet above the ocean, but approached by a gradual, and almost imperceptible ascent. A man has actually travelled with a one horse wagon from Hartford, Ct., to the falls of the Columbia. This territory gives rise to the Missouri, Yellowstone, and Platte rivers, and derives its name from the extinct tribe of the Madan Indians, who inhabited a part of it.

MANDAVEE, a town and seaport of Hindostan, being the most populous town, and principal emporium of Cutch, on the S. coast of which it stands, 35 m. S.S.W. Bhooj; lat. 22° 50' N., long. 69° 34' E. Pop. probably 50,000; of whom, upward of 15,000 are Bhattias, 10,000 Banyans, 5000 Brahmins, and the rest Lohannas, Mohammedans, and Hindoos of low caste. "The town is within gun-shot of the beach, and is surrounded with fortifications in the Asiatic style. Its environs are laid out in gardens well stocked with cocoa-nut and other trees. The bed of a river, nearly dry, except in the rains, covers the E. face, and joins the sea, forming the only harbour which Mandavee has. Small boats, laden, can cross the bar at high tides; larger vessels unlade in the roadstead. A brisk trade is kept up with Arabia, Bombay, and the Malabar coast, in which upward of 800 boats, or from 40 to 500 candies tonnage, are employed. The exports are chiefly cotton, *moores* of silk and cotton thread, piece goods of a coarse kind, alum, and ghee. The imports are, bullion from Mocha; ivory, rhinoceros' horns, and hides, from Fowahil; dates, cocoa-nuts, grain, and timber, from Malabar and Demaun. There is a considerable inland trade, by means of *chevres* and other carriers with Marwar and Maiwah." (*Bombay Transac.*, ii., 217; *Geog. Journal*; *Hamilton's E. I. Gaz.*)

MANDURIA, a town of the Neapolitan dom., prov. Otranto, cap. cant., in an arid plain, 22 m. E.S.E. Taranto. Pop. about 5000. It is a straggling but well built town, with wide unpaved streets, many handsome churches, several convents, an orphan asylum, and a large palace, formerly belonging to the Francavilla family. The town during the middle ages, and until 1790, was called Casalnuovo; but at the latter epoch it re-assumed, by royal privilege, the name of the ancient city, on the site of a part of which it is built. When Swinburne visited it, it was noted for nothing except the taste of its inhabitants for dog flesh, the skins of the slaughtered dogs being, at the same time, tanned into an imitation of Turkey leather for the supply of the neighbourhood. It had no other trade or manufacture. The remains of Manduria, destroyed by Fabius Maximus in the second Punic war, consist of its walls, standing several feet above ground, and double, except on the S. side, where the fortifications appear to have been left incomplete. The outer wall, and its ditch, measure 8 yards in breadth; behind this bulwark is a broad space, and than an inner wall, which together measure 14 yards. According to Craven, the walls are no where more than 6 feet in height, having probably been lowered to furnish materials for the construction of the modern town. In the vicinity is a well, mentioned by Pliny, as constantly preserving the same level, whatever quantity of water be added to or taken from it; *lacus ad margines plenus, neque exhaustis aquis minuitur, neque infusis augetur.* (Hist. Nat., lib. ii.) This singular well still exists, and was visited by Swinburne and Craven, (*Craven's Tour, &c.*, p. 165–160; and *Swinburne's Trav.* i., 223. 224.)

MANFREDONIA, a seaport town of the kingdom of Naples, prov. Capitanata, on a bay of the Adriatic, about 19 m. S.W. the promontory (Testa di) of Gargano, and 20 m. N.E. Foggia; lat. 41° 37′ 53″, long. 15° 55′ 40″. Pop. 6000. "In point of symmetry, it may vie with any town in Europe, having been constructed on a regular plan, which never underwent any alteration; and which, notwithstanding the unfinished state of some of the edifices, and the dilapidated aspect of others, gives it an air of grandeur and uniformity very remarkable. It is walled towards both land and sea: from the last a narrow ledge of rocks, almost always under water, divides its bulwarks. One long and wide street runs throughout the city, from one gate to the other; for there are but two gates on the land side, though two others open to the port, which is protected from the effects of the N. wind by a small mole, and commanded by a strong castle, defended by a ditch and drawbridge. The walls are fortified with large round bastions. The harbour is reckoned safe; but its want of depth renders it fit for small vessels only." (*Craven's Tour*, 68, 69.) Four streets run parallel with the principal thoroughfare, and are intersected at right angles by smaller ones. Though narrow, the streets are well kept; and the inhabitants are both cleanly and industrious, in a degree not at all usual in S. Italy.

Vegetables and fish are good, plentiful, and cheap at Manfredonia, but water and wine are indifferent, as are oranges, which form an important article of commerce throughout Apulia. It exports considerable quantities of salt, obtained from the salt lagoons which border the coast of the bay to the S. of the town. It has also a considerable trade in corn, quantities of which are shipped from its port.

About a mile S.W. of the town stood the ancient Sipontum, once a considerable city of *Magna Graecia*, and traditionally said to have been founded or colonised by Diomed. Its site is now principally occupied by a low marsh, abounding with wild fowl, and productive of the malaria which infects Manfredonia. The only remains of the ancient city are its cathedral, and two columns of cipolino marble, both in a dilapidated condition. The former is a small Gothic edifice, with a handsome portico, but little adorned within. It is still the seat of an archiepiscopal see, founded in 1094. Sipontum, which was colonised by the Romans A.U.C. 696, had fallen into such irreparable decay in the 13th century, that Manfred, king of the Two Sicilies, having founded, in 1906, the town which bears his name, but which he called *Novum Sipontum*, removed thither the few inhabitants of Sipontum, bestowing on them many valuable privileges and exemptions. But, though it has always enjoyed some commerce, Manfredonia never attained to the prosperity or celebrity of its ancient predecessor, and has long been stationary. (*Swinburne's Travels in the Two Sicilies*, i., 149–151; *Craven's Tour in the S. Prov. of Naples*, 67, 70; *Cramer's Ancient Italy*.)

MANGALORE, or COREAL BUNDER, a seaport town of Hindostan, prov. Canara, of which it is the cap., on a sandy promontory between a salt lake and the Indian ocean, 440 m. S.S.E. Bombay; lat. 19° 53′ N., long. 74° 57′ E. Early in the present century it had 30,000 inhabitants. The town is well built, and has a fort, now dismantled, which opposed a gallant and successful resistance to Tippoo, in 1783. The port does not admit vessels drawing more than 10 feet water, except at spring tides; but there is good anchorage in the roadstead, in from 5 to 7 fathoms. The exports are chiefly rice, to Muscat, Goa, Bombay, and Malabar; betel nut, black pepper, sandal wood, cassia, and turmeric. Raw silk and sugar are imported from China and Bengal, and oil and ghee from Surat. Mangalore was at an early period much resorted to by Arabian traders; and most of its present inhabitants are of Arabian descent. The vessels employed in its trade belong chiefly to other ports. Salt is made at Mangalore, but it is of bad quality. (*Hamilton's E. I. Gaz.; Parl. Report.*)

MANHATTANVILLE, v., New-York co., N. Y., 8 m. N. of the city hall, N. York. Situated on the E. side of Hudson river, where is a convenient landing and wharf. It contains an Episcopal church, four stores, a white lead factory, 70 dwellings, and about 500 inhabitants. The New-York lunatic asylum is on elevated ground, half a mile S.

MANHEIM, p. t., Herkimer co., N. Y., 64 m. W.N.W. Albany, 397 W. Bounded S. by Mohawk river, East Canada creek, flows near its E. border. It contains seven stores, three grist-mills, nine saw-mills, three tanneries; nine schools, 496 scholars. Pop. 2095.

MANHEIM, p. t., Lancaster co., Pa., 23 m. E.S.E. Harris-

burg, 123 W. Bounded E. by Great Conestoga creek, S.W. by Little Conestoga creek. It contains three stores, three flouring-mills, one grist-mill, three saw-mills; one school, 35 scholars. Pop. 1699.

MANHEIM, t., Schuylkill co., Pa. Watered by Schuylkill river and its tributaries. It contains a Lutheran church, 13 stores, one furnace, one forge, four grist-mills, 16 saw-mills, one powder-mill, two tanneries; five schools, 192 scholars. Pop. 3441.

MANHEIM, t., York co., Pa., 16 m. S.W. York. Drained by Hammer creek, a branch of Codorus creek. It contains five stores, one woollen factory, 11 grist-mills, 12 saw-mills, one oil-mill, one paper-mill, four tanneries, 36 distilleries, one pottery; four schools, 71 scholars. Pop. 1595.

MANILLA (Sp. *Manila*), a fortified seaport city of the Philippine islands, and the cap. of the Spanish settlements in the east, on the E. side of the bay of Manilla, island of Luzon, and on the river Passig, about ½ m. from its mouth; lat. 14° 36′ 8″ N., long. 120° 53′ 30′ E. The pop. of the city and its suburbs was said to amount, in 1818, to from 70,000 to 80,000; and is at present variously estimated at from 100,000 to 150,000, including, besides Tagalos, or natives, from 4000 to 5000 Spaniards and other Europeans, with Chinese, negroes, the descendants of the foregoing races, and foreigners from all parts of the world. The bay and city of Manilla have a very picturesque and imposing aspect from the sea. The former is surrounded by mountains covered with verdure, which, on the E., decline gradually towards the shore. At their feet, on this side, is a small plain, on which the city stands; its buildings consisting almost entirely of the volcanic tufa, of which the plain and its vicinity are geologically constituted. Manilla comprises the city-proper and two suburbs. The former is on the left or S. bank of the Passig, across which it communicates, by a handsome stone bridge of 10 arches, with its important suburb of Bidondo, and those of Tondo, Santa Cruz, &c This bridge, which is about 149 Castilian *veras* (or yards) in length, by 8 in breadth, was founded in 1630; but it has been rebuilt since 1814, when it was for the most part destroyed by an earthquake.[*] The city-proper, little more than 2 m. in circle, is surrounded with strong walls, and a broad ditch, and has not more than 10,000 or 12,000 inhabitants. At the mouth of the river is a small battery, and the town is farther protected by the citadel of Santiago, near its N.W. extremity; but Manilla could not make any effectual resistance to a European force. The city, which is entered by six gates, is regularly laid out; and, according to Meyen, by whom it was visited, in 1831, it is superior in point of appearance to either Lima or Santiago. (*Reise um die Erde*, ii., 207.) The streets have carriage-ways, composed of a mixture of loam and quartz, and are provided with footpaths, and lighted at night. The houses in the city are solidly constructed, though, on account of earthquakes, they are seldom more than one story above the ground floor. The houses in the suburbs, however, are not so substantial. In Bidondo, for example, they are almost wholly composed of bamboo, and are raised from the ground, to the height of 8 or 10 feet, on thick poles, as is customary among ultra-Gangetic nations. Most of the houses are furnished with balconies and verandahs; the place of glass in the windows is supplied by thin semi-transparent plates of shell, which, though more opaque, repel heat better. Bidondo is the most interesting portion of Manilla, and that in which its trade mostly centres. It is principally inhabited by Chinese and Tagalas, and looks very like a Chinese town.

The public edifices are mostly within the walled city. The new *aduana*, or custom-house, is a large fine building, constructed at a great expense; but, like the Dublin custom-house, its size is out of all proportion to the business to be transacted in it. The residence of the Captain-General, and the principal government offices, are also in a huge edifice, occupying one of the sides of the *Plaza Mayor*, or principal square. This square measures about 190 yards either way, and has, in its centre, a bronze statue of Charles IV., on a marble pedestal, presented to the city by Ferdinand VII. in 1824. There are, in Manilla, a vast number of churches and ecclesiastical establishments; and the number of clergymen is said to exceed that of the garrison, which is estimated at about 7000 men! We need not, therefore, be surprised to learn that religious observances are here scrupulously complied with, while real piety and sound morality are at the lowest ebb. The city was erected into an archbishopric in 1598; and the cathedral and archbishop's palace are among its most conspicuous structures. The Augustine, Franciscan, Dominican, and Jesuit convents, the arsenal and cannon foundery in the citadel, the university (founded in 1645), the missionary college, the various schools for natives and Europeans, the hospitals,

* Meyen, *Reise, &c.*, ii., 216. The *Dict. Geog.* says that the bridge was *refaite en 1814, mais en grande partie renversée par le tremblement de terre de 1645.*

310

orphan asylums, and other charities, and the royal cigar manufactory, in which 350 males and 2000 females are said to be employed, include the other principal public buildings and establishments. The promenades round the city are frequented in the evening by the more opulent classes, on horseback, or in their carriages. The neighbourhood is interspersed with orange, areca, tamarind, and mango groves; gardens; coffee, cocoa, and cotton plantations; rice grounds, &c.

The Pasig is navigable for vessels of 600 tons in ballast, or for laden vessels of from 250 to 300 tons, as far as the bridge; and for large shallow boats, drawing from two to three feet water, as far as the lake in which it rises, about nine miles inland. There are 13 feet water, at low ebb, in the channel through the bar at the entrance of the river; for the further deepening of which a steam dredging-boat has been employed since 1837. The rise and fall of the tide in the river is from two to three feet. A lighthouse, at the end of the pier, marks the entrance of the Pasig on the left-hand side.

Ships of all sizes anchor in Manilla roads, at from one to two miles off shore, except during July, August, and September, when the S.W. monsoon throws in a heavy sea, which extends quite to the entrance of the river. At this season, therefore, small vessels load and unload in the river, and large vessels at Cavité, an anchorage sheltered by a neck of land to the S.W., and about six or seven miles by water from the mouth of the river; their cargoes being conveyed, to and from Manilla, in secure decked boats, of from 50 to 70 tons burden.

Manilla is the only port in the Spanish Philippines with which Spanish vessels to or from Europe, or foreign vessels from any quarter, are allowed to trade. Spanish vessels trading to China, Singapore, &c., are, however, allowed to proceed to various outports, and there take on board their outward cargo. The principal articles of export are sugar, which is by far the most important; hemp, and stuffs made of hemp; rice, of which large quantities are sent to China, indigo, sapan and other woods, tobacco, cigars, coffee, cotton, tortoise-shell, hides, ebony, &c. The tobacco of the Philippine islands is excellent, and might be produced in any quantity; but its growth is comparatively limited by its being made a government monopoly. (*See* PHILIPPINE ISLANDS.)

The following is an ACCOUNT of the Quantities and Values of the Principal Articles exported from Manilla in 1837.

Articles.	By Foreign Vessels.		By Spanish Vessels.		Total.		Price.		Gross Amount.		
							Dolls.	Rls.	Dolls.	Rls.	
Sugar	195,852	pic.	96,331	pic.	292,183	pic.	4	2	944,277	6	
Sapan-wood	14,604	—	9,091	—	23,695	—	1	.	23,695		
Hemp	57,363	—	2,104	—	59,467	—	4	2	252,734	6	
Cotton	2,850	—	2,814	—	5,664	—	15	.	84,960		
Coffee	6,206	—	632	—	6,838½	—	13	.	88,894		
Buffalo	7,531	—	2,124½	—	9,655½	—	3	7	37,415	.	10¼
Mother-of-pearl shells	1,004	—	11	—	1,015	—	14	.	14,210		
Hide cuttings	1,417	—	1,531	—	2,958	—	3	.	8,814		
Hemp cordage	875	—	309½	—	1,195½	—	9	.	10,759	4	
Streaked ebony	9,421	—	382	—	9,803	—	1	2	12,253	6	
Roots of Sapan-wood	5,590	—	1,746	—	2,266	—	.	4	1,133		
Pieces of molove (timber)	4,368	—	. . .		4,368	—	5	4	34,924		
Indigo	1,653½ — 99 ctys		142	— 99 ctys.	1,795½	—99 ctys.	66	.	118,503		
Leaf tobacco	35	—	. . .		35	—	12	.	490		
Tortoise-shell	2,706	ctys.	1,910½	ctys.	4,616½	ctys.	7	4	34,623	6	
Rice	45,607	—	76,996	coys.	126,003	coys.	1	2	157,503	6	
Paddy	16,564	pic.	19,048	pic.	35,612	pic.	.	4	17,806		
Coffee in husk	86	—	. . .		86	—	3	6	224	4	
Hats	19,395	in no.	234	in no.	19,629	in no.	.	6	34,350	6	
Cigar cases	5,851	—	70	—	5,921	—	1	4	2,960	4	
Cigars	3,141	boxes.	1,457	boxes	4,598	boxes	25	.	114,950		
Manilla hemp cloth	19,050	pieces	10,000	pieces	29,050	pieces	.	3	10,893	6	
Ditto	4,075¼	—	50	—	4,125¼	—	.	1½	773	3	10
Cocoa-nut oil	876¼	casks	. . .		867¼	casks	2	6	2,410	2	
Ditto	8,768¼	—	. . .		8,768¼	—	2	2	1,728		
Rum	6,951	gallons	132	gallons	7,083	gallons	3	4	2,656	1	
Canes	1,440		. . .		1,440		.	.	720		
Liquid indigo	. . .		230	pic.	230	pic.	4	.	920		
Ditto	. . .		606	casks	606	casks	3	4	2,121		
Cotton canvass	. . .		560	pieces	560	pieces	12	.	6,720		
Mats	. . .		762		762		.	2	190	4	
							Total	.	2,012,638	65	90½

Subjoined is a statement of the import trade of Manilla in 1838, which we have procured direct from that city; but it is right to bear in mind that the official accounts from which it has been drawn up are so defective, that the amounts specified can only be considered as rough approximations; and are, no doubt, under the mark.

ACCOUNT of the Quantities and Values of the principal Articles exported from Manilla in 1838.

Articles.		Quantities.	Value.
			Dollars.
Iron	piculs	21,874	87,495
Cottons, gray	yds.	4,643,375	580,424
Do. white	—	1,176,352	171,050
Ginghams	—	292,190	65,306
Stripes	—	366,179	96,445
Handkerchiefs	doz.	67,364	134,122
Prints	yds.	196,246	24,530
Woollens and worsted	—	157,997	[illegible]
Muslins	—	551,015	187,754
Umbrellas	No.	22,308	16,798
Glass and earthenware	packs	776	27,740
Sundries		. .	214,903
Total value in Sp. dollars			1,666,495

Of the above, goods to the value of 1,145,000 dollars were imported, in Spanish vessels, from China, Singapore, and elsewhere. About 130 ships entered the port of Manilla in 1838; of which, 46 were Spanish, 36 British, 28 American, and 11 Chinese.

The port-charges on foreign vessels consist of a tonnage-duty of two reals, or a quarter-dollar, per register ton; and fees, varying from 15 to 29 dollars, according to the size of the vessel, for port-captain's and health officers' visits, passport, &c. The tariff is bottomed on a custom-house valuation, fixed every five years. Most foreign commodities, imported in foreign vessels, pay an import duty of 14 per cent. *ad valorem*, except wines and spirits, which mostly pay a duty of from 30 to 60 per cent., unless the produce of Spain. Cotton-twist of certain colours, cutlery, ready-made clothes, European fruits, confectionery, and vinegar, pay 40 per cent. if imported in Spanish vessels, and 50 per cent. if in any other. British and other foreign cotton and silk manufactures made in imitation of native cloth, Madras and Senegal cottons, &c., pay 15 per cent. if imported in Spanish, and 25 per cent, if in other ships. Machinery of all sorts for the promotion of industry, cotton-twist of certain colours, gold and silver, plants and seeds, are imported duty free; but tropical products, the same as those of the Philippines, gunpowder, swords, and other warlike stores, &c., are prohibited, unless landed in bond for re-exportation. Exports of nearly all descriptions, by Spanish vessels, pay only from one and a half to two per cent. *ad valorem*, and by foreign vessels double this duty; but manufactured tobacco, rope from Manilla, hemp, and gold and silver, coined or uncoined, if exported to Spain, go duty free. The principal currency of Manilla consists of Spanish dollars, of 8 reals and 96 grains; but S. American dollars are also current. The weights in use are the Spanish lb., which is nearly two per cent. heavier than the English; the arroba = 25½ English lbs. nearly; the quintal = 102 lbs.; and the picul of five arrobas, or one and a quarter cwt. English. The coyan is a measure for rice, &c., varying from 96 to 135 lbs. According to a recent list, there are in Manilla 47 Spanish merchants and 11 foreign firms. The Spanish merchants have a chamber of com-

merce and a joint-stock insurance society. The United
States, France, and Belgium have consuls, and each of the
Canton marine insurance companies has an agent here.
There are, however, neither fire nor life offices nor agen-
cies; nor is any newspaper, price-current, or other periodi-
cal publication issued in Manilla.

Manilla existed as a native town prior to the Spanish invasion; it was taken by the Spaniards, and made the capital of their E. dominions, in 1571. It has frequently suffered very much from earthquakes, especially in 1645 and
1762, and 1824. In 1762, it was taken by the English; but
ransomed by Spain for £1,000,000 sterling. (*Meyen, Reise
um die Erde* ii., 203–213; *Hamilton's E. I. Gazetteer; and
valuable Private Information.*)

MANNHEIM, or MANHEIM, a town of W. Germany,
grand duchy of Baden, lower circ. of the Rhine (*Unter-
rheinkreis*), of which it is the cap., on the Rhine, where it
is joined by the Neckar, 32 m. N. Carlsruhe, and 37 m.
S.S.E. Mayence; lat. 49° 29' 15" N., long. 8° 28' 7" E.
Pop., in 1832, 20,600. It was once strongly fortified, and
has at different times suffered severely from sieges and
bombardments; but towards the end of last century its defences were levelled by the French, and their site is now
laid out in gardens and public walks. Mannheim is a reg-
ularly-constructed, handsome town; though it is, notwithstanding, monotonous and tiresome. It consists of 11
streets, crossed at right angles by 10 others, all perfectly
straight, broad, well paved, and equidistant; and its houses
being uniform, it is difficult for any one, not resident, to distinguish one part of the town from another. It has several handsome public squares, which, though the town be
deficient in good water, have mostly fountains. The spacious *Paradeplatz* and the *Planken*, or principal thoroughfare, both planted with trees, afford pleasant promenades.
The principal public edifice is the palace, a huge structure
of red sandstone, built by the elector palatine when he
made Mannheim his capital, in 1730, but more remarkable
for size than elegance. A part of it is inhabited by the
dowager grand duchess Stephanie, the adopted daughter
of Napoleon and Josephine; and in one wing are museums of antiquities and natural history, the picture-gallery,
with some fine Dutch and Flemish paintings, collections
of plaster casts and engravings, and a library, said to consist of 70,000 volumes. (*Horschelmann's Stein*); but the
other wing, comprising the old theatre, was mostly laid in
ruins during the bombardment of Mannheim in 1795, in
which state it remains. The new theatre, a handsome
fabric, is neatly fitted up, and is rich in scenic decorations.
It is said to have one of the best theatrical companies and
orchestras of Germany; and is celebrated as being the
place at which Schiller's tragedy of the Robbers was originally produced. Opposite the theatre is the house in
which Kotzebue was assassinated. Mannheim has about
an equal number of Lutheran and Roman Catholic churches, of which that formerly belonging to the Jesuits is the
finest. It has also a synagogue, an observatory, with a
tower 115 feet high, and a good collection of instruments,
an arsenal and cannon foundry, an exchange, surrounded
by arcades, several hospitals, a savings' bank, a lyceum,
with schools of drawing, painting, sculpture, surgery, &c.
The Rhine is bordered by a fine terrace in the spacious
grounds belonging to the palace, whence an extensive view
of the surrounding country is obtained; and, like the
Neckar, is crossed by a bridge of boats. Mannheim has
some public baths, and a club called "*The Harmony,*"
with a reading-room, &c. The cheapness of living has attracted a good many English residents.

This town is the seat of the supreme court of justice for
the grand duchy, and of one of the four subordinate courts
of appeal in Baden. It was formerly a manufacturing town
of some importance; and, among other articles, trinkets, of
a compound called Mannheim-gold, were made in large
quantities, but this branch of industry is nearly extinct.
It still, however, produces carpets, linen and silk goods, tobacco, liqueurs, starch, glue, pasteboard, and sealingwax;
and has several coach-building establishments, tanneries,
breweries, and bleaching-grounds. Its neighbourhood produces hops and garden stuff in large quantities; and, besides its traffic in cattle and agricultural products, it has a
considerable transit trade by the Rhine and the Neckar.
Previously to 1606, when it was fortified by the Elector
Frederick IV., Mannheim was a mere village. It soon after received numerous Flemish and other immigrants. In
1777, it was ceded to Bavaria; but, since 1803, has been
again united to Baden. (*Schreiber, Guide du Rhin*. 65, 66;
Berghaus; Allg. Länder, &c., iv., 313; *Stein; Germany
and the Germans, &c.*)

MANITOOWOC, county, Wis. Situated in the N.E.
part of the territory, and contains 468 sq. m. Bounded E.
by lake Michigan. Drained by Manitoowoc river and its
branches. It had, in 1840, 81 neat cattle, ninety swine;
and produced 225 bushels of wheat, 175 of Indian corn,
304

1750 of oats, 1900 of potatoes, 9900 pounds of sugar. It
had one flouring-mill, six saw-mills. Pop. 235. Capital
Manitoowoc.

MANITOOWOC, p. v., capital of Manitoowoc co., Wis., 178
m. N.E. Madison, 900 W. Situated on Manitoowoc river,
at its entrance into lake Michigan. Nett proceeds of the
postoffice, $70.

MANISTEE, an unorganized county of Mich., bounded
W. by lake Michigan. Watered by Manistee river.

MANLIUS, p. t., Onondago co., N. Y., 10 m. E. Syracuse, 191 m. W. by N. Albany, 346 W. Bounded N.E. by
Chittenango creek. Watered by Limestone creek. It
contains six churches, two Presbyterian, two Methodist, an
Episcopal and Baptist, 22 stores, three grist-mills, seven
saw-mills, three tanneries, one printing-office, one weekly
newspaper; three academies, 578 students; 94 schools, 162
scholars. Pop. 5509.

MANNINGTON, t., Salem co., N. J., 6 m. N.E. Salem.
Bounded N. and W. by Salem river. Drained by Mannington creek. It contains two churches, a Methodist and
Baptist, the county poor house, with from 86 to 120 inmates,
a famous nursery, from which, besides other trees, 15,689
peach trees have been sold in a year, one store, one grist
mill; seven schools, 169 scholars. Pop. 2064.

MANOR, p. t., Lancaster co., Pa., 6 m. S.W. Lancaster,
33 m. S.E. Harrisburg, 105 W. Bounded S.E. by Cones-
ga creek, by branches of which it is watered. It has seven
stores, one woollen factory, 10 flouring-mills, 10 grist-mills,
four saw-mills, 18 distilleries; 18 schools, 844 scholars.
Pop. 4152.

MANREZA (*Minorisa*), a town of Spain, prov. Catalonia, 34 m. N.W. Barcelona. Pop. 13,000. It stands on a
rocky height, in the midst of a country irrigated by the Llo-
bregat and its tributaries; is walled and strongly fortified;
has good streets, and comprises among its public buildings
and establishments a collegiate church, with a chapter,
four parish churches, five oratories, a well endowed asylum for female orphans, infantry barracks, free school, and
hospital. The inhabitants rank among the most industrious
in Catalonia, and are pretty equally divided between agriculture and manufacturing pursuits. Cotton and silk fabrics, cotton thread, fine broadcloths, tapes and ribands,
paper, brandy, and gunpowder, are made in considerable
quanties for exportation to Cuba and the West Indies. The
neighbouring district, one of the best cultivated in Spain,
produces corn, hemp, oil, and wine, which, with the goods
above mentioned, find a ready sale at the weekly markets,
and the two fairs held here Sept. 1, and Nov. 30.

MANSFIELD, a market town and par. of England, co.
Nottingham, in the N. div. of wap. Broxtow, 13 m. N.N.W.
Nottingham, and 118 m. N. by W. London. Area of par.
9070 acres. Pop., in 1831, 9498. It is situated in the forest
of Sherwood, near the small river Mann, from which it
takes its name; and, though old-fashioned, and irregularly laid out, it contains many good modern houses, and is
paved and lighted with gas. The chief buildings are the
moot-hall, a structure well adapted for county meetings; a
theatre, and the church, a commodious Gothic edifice, containing some curious monuments, and fine specimens of
painted glass. The Presbyterians, Wesleyan and Calvinist Methodists, and the Society of Friends, have their respective places of worship; to which, as well as the
church, are attached well supported Sunday schools. A
grammar school was established here in 1567, by Queen
Elizabeth, who endowed it with one third part of the church
land of the parish, and founded for it two scholarships, of
£10 each, at Jesus college, Cambridge; but the management appears to have been unsatisfactory, and it had, in
1833, only 27 scholars, including the master's boarders.
(*Char. Comm.,* 25th *Rep.*) There are two other charity-
schools; one of which was founded in 1725, for teaching
and clothing 20 boys and 30 girls, and for paying apprentice fees with the former. Besides the above, there are
several other charities and money-bequests. The inhabitants are chiefly engaged in the hosiery and lace trade, and
in cotton spinning: it had, in 1839, five cotton-mills, which
employed above 400 hands. There are some large iron
foundries, for light castings; and the town has also a considerable trade in corn and malt, as well as in the valuable
building-stone, quarried in its vicinity. A railway connects
it with the Pinxton canal; and, from its proximity to the
N. Midland railway, it seems probable that it will, at no
distant period, be united with that line. Petty sessions for
the hundred are held here; and it is the election-town for
the N. division of the county. Markets on Thursday;
large cattle fairs, 5th April, 10th July, and the 2d Thursday
in October.

About one and a half miles from Mansfield is the village
and township of Mansfield-Woodhouse (population, in 1831,
1850), near which are some curious and pretty perfect remains of two Roman villas. Within a few miles are
Worksop Manor, formerly belonging to the Duke of Nor-

silk, but now the property of the Duke of Newcastle, who has decided on pulling it down; Clumber, the seat of the latter; Thoresby, of Lord Newark; and Welbeck, of the Duke of Portland. Hence, in popular language, this part of the county is called the dukery.

MANSFIELD, t., Lamoille co., Vt., 20 m. E. by N. Burlington, 20 m. N.W. Montpelier. It contains the highest peak of the Green mountains, 4279 feet above tide-water. Drained by Waterbury and Brown's rivers. It has two saw-mills; four schools, 77 scholars. Pop. 283.

MANSFIELD, p. t., Bristol co., Mass., 29 m. S.S.W. Boston, 421 W. Chartered in 1770. Watered by branches of Taunton river, called Rumford, Cocasset and Canoe rivers. It contains two churches, a Congregational and Methodist, five stores, one woollen factory, three grist-mills, three saw-mills; eight schools, 345 scholars. Pop. 1382.

MANSFIELD, p. t., Tolland co., Ct., 24 m. E. Hartford, 360 W. Watered by Willimantic river, Nachaug river and its branches, which afford water-power. It is noted for the production of silk, which has progressed from 1793 to the present time. A large amount of sewing silk is exported. The amount of silk cocoons produced in 1840, was 6151 pounds, reeled silk, 3250 pounds, value $18,050. It contains four churches, two Congregational, a Methodist and Baptist, five stores, one woollen factory, one cotton factory with 1600 spindles, four grist-mills, five saw-mills, three tanneries, one pottery; 17 schools, 613 scholars. Population 2276.

MANSFIELD, p. t., Cattaraugus co., N. Y., 5 m. W. Ellicottville, 300 m. W. by S. Albany, 342 W. Drained by Cattaraugus creek, and tributaries of Alleghany river. It has one store, two saw-mills; six schools, 252 scholars. Pop. 942.

MANSFIELD, p. t., Warren co., N. J., 46 m. N.N.W. Trenton, 207 W. Drained by Musconetcong and Pohatcong creeks. It contains iron ore and a chalybeate spring. The Morris canal winds through the town. Some farmers sell 3000 bushels of wheat, annually. It contains eight stores, one grist mill, three saw-mills; 12 schools, 1037 scholars. Pop. 3057.

MANSFIELD, t., Burlington co., N. J., 7 m. N. Mount Holly. Bounded N.W. by Delaware river. Drained by Black's Craft's and Assiscunk creeks. It contains several villages, and has five stores, one fulling-mill, one grist-mill, one saw-mill, two distilleries, one pottery; 12 schools, 144 scholars. Pop. 2401.

MANSFIELD, p. v., capital of Richland county, O., 69 m. N.N.E. Columbus, 378 W. It is pleasantly situated on elevated ground, near a branch of Mohiccan creek, and contains a courthouse, jail, six churches, 17 stores, two printing-offices, 339 dwellings and 1388 inhabitants.

MANS (LE) (an. Suindinum and Cenomania), a town of France, dep. Sarthe, of which it is the cap., on the Sarthe, here crossed by three bridges, 59 m. N.E. by N. Angers, and 120 m. S.W. Paris. Pop. in 1836, ex. com., 19,103. It stands partly on the declivity of a hill, and partly beside the river. The latter portion is very ill-built, and has narrow crooked streets, impassable for carriages; but the upper town, though irregular, is open, and tolerably well built, its houses being of stone, roofed with slate. A handsome new quarter has been laid out, having a large square in its centre; and there are two good public promenades, one along the bank of the Sarthe. The Romans surrounded the ancient city with walls, a portion of which, on the N.N.E. side, remains nearly perfect; but the modern town is of no strength. Le Mans has several remarkable ecclesiastical structures. Its cathedral, begun in the 9th, but not finished till the 16th century, is a fine Gothic edifice, 416 feet in length, with a large square tower, 212 feet in height, the supports of which in the interior are ornamented with numerous statues. The choir is inferior in elegance only to that of Beauvais; and the stained glass window in the S. arm of the cross is much admired for its richness. The church of St. Julian is an interesting edifice of the 11th century. Another church, built in the 13th century, presents a combination of the Gothic and antique style. The new prefecture, the town-hall, and the theatre, are handsome buildings. Le Mans has two hospitals, a seminary, with a library of 15,000 volumes, a public library, with 45,000 printed volumes and 500 MSS., in excellent preservation; several other libraries, museums of natural history, antiquities, and painting, the latter having several works by Guido, A. Durer, Teniers, Vandyk, &c.; a royal society of arts, a communal college, schools of drawing, midwifery, &c. It has manufactures of linen and coarse woollen stuffs, wax candles, &c.; and a considerable trade in these, and in rags, iron, salt, wine, brandy, and agricultural produce. Le Mans has suffered much from the ravages of war at different periods; and, in 1793, it was the scene of the last struggle between the Republican and Vendean forces. (Hugo, art. Sarthe; Guide du Voyageur, &c.)

MANTINEIA, a celebrated city of ancient Greece, in Arcadia, the ruins of which, close to the wretched hamlet of Palaeopoli, in a marshy plain watered by the Ophis, and enclosed S.E. by the rugged heights of Parthenion and Artemision, are about 7 m. N. Tripolizza, and 17 m. W. by S. Argos. The walls, probably built soon after the battle of Leuctra (B.C. 371), are similar to those of Messene, and enclose an oval space in which the city stood; they have square towers, and the whole exhibits an interesting specimen of Grecian fortification. A ditch, or fosse, round the walls, is supplied by the Ophis; which, at certain seasons, would inundate the plain were it not absorbed by a chasm (κατάβοθρον), through which its waters find a subterraneous vent. Mantineia had eight temples, besides a theatre, stadium, hippodrome, and several other monuments enumerated by Pausanias. (Arcadia, chapter 8-11.) Some imperfect remains of the theatre are still visible, no other ancient building can be identified; and everything, except the enclosing walls, is in a state of total dilapidation. (Dodwell, ii., 422.)

But Mantineia is wholly indebted for its long-continued celebrity to the great battle fought in its vicinity, anno 362 B.C., between the forces of Sparta and Thebes, and their allies; in which Epaminondas, the leader of the Thebans, and the most illustrious, perhaps, of all the warriors of Greece, fell in the moment of victory. Xenophon is very brief in his account of the battle; but it may be collected from his statement that, on the whole, the plan of the Theban general succeeded in all its parts. The charge of the Theban and Thessalian cavalry, which commenced the attack, was completely successful and prepared for the deeper impression made by the column of Theban and Arcadian infantry. But, in the critical moment, when the phalanx of the Lacedaemonians had been broken, and a decisive victory appeared to be secured, Epaminondas received a mortal wound; and, being carried to a rising ground, whence he might view the scene of combat, would not allow the weapon to be extracted till assured that the victory had been won, when he almost immediately expired. But his fall, and the consternation thence arising, paralysed the successful army. They kept the ground they had gained, but did little or nothing more. Hence it was that the result of this great contest disappointed the expectations of those who had supposed that it would be decisive of the fate of Greece. "The Gods," says Xenophon, "decided otherwise. Each party claimed the victory, and neither gained any advantage; territory, town, and dominion was acquired by neither; but indecision, trouble, and confusion, more than ever before prevailed throughout Greece." (Xen. Hell., l. vii., c. 5, ad finem.) This, however, is the statement of a partisan of Sparta, and is not quite fair. The Theban confederacy was, on the whole, decidedly successful. They effectually broke the power and humbled the pride of Sparta; and, by re-establishing the independence of the Messenians, the old and inveterate enemies of the Lacedaemonians, they obtained a new guarantee against any dangerous increase of their power in future. (See Mitford's Greece, sect. viii., cap. 38.)

Mantineia was taken and sacked by Antigonus during the wars of the Achaean league; and its name was changed, in honour of the conqueror, to Antigonia, which it retained till the time of Adrian, who restored its original appellation.

MANTUA (Ital. Mantova), a fortified town of Austrian Italy, prov. Lombardy, cap. deleg. Mantua, on both sides the Mincio, 21 m. S.S.W. Verona, and 37 m. E. by N. Cremona; lat. 45° 9' 10" N.; long. 10° 46' 20" E. Pop., in 1837, 26,865. Its situation is peculiar, being in fact nearly surrounded by lakes, partly natural, and partly formed by damming up the waters of the river. The mounds, or dams constructed for this purpose, are sometimes called bridges, from their being perforated with arches, to allow the superfluous water to escape; and by these the town is connected with the Borgo di Fortezza, or strong citadel of Porto on the N., and with the Borgo di San Giorgio. The latter, as well as the town itself, is surrounded by strong walls; to the S.E. is the outwork of Pradella, and to the S. the fortified island of Cerese, or T., from its alleged resemblance to that letter. The fortifications, though not imposing in their appearance, are very strong, and kept in excellent order; and their strength and the position of the place render it one of the bulwarks of Italy. Mantua has some good streets and squares, but, on the whole, it is ill-built and dirty. Many of the inhabitants live in cellars, its population has declined, and it has a decayed appearance. Its best part is the Piazza Virgiliana, a large square, surrounded with trees, and open to the lake. The climate is subject to great extremes, and in summer the exhalations from the surrounding swamps make it very unhealthy; though, of late years, the Austrian government has exerted itself, by draining part of the marshes, and opening a passage for the stagnant waters, to lessen its insalubrity. Several of the public edifices in Mantua were designed or adorned by Giulio Romano.

But the cathedral, planned by that great artist, is said by Woods to be a bad imitation of the church of Santa María Maggiore, at Rome ; it has double ranges of side isles, and the columns stand very wide apart. The church of St. Andrea, begun in 1470, but not completed till 1782, was designed by Alberti, and is said by Woods to be very superior to the cathedral, and to be, indeed, one of the handsomest churches in Italy : it has fine statues of Faith and Hope, by Canova. The old ducal palace (*Palazzo Vecchio*) is a large imposing building ; and, were it perfect, would be one of the finest palaces in Europe. It is beautifully floored with porcelain, and was formerly splendidly adorned with Flemish and Mantuan tapestry and rich furniture ; and, though repeatedly despoiled, it has still to boast of a room painted in fresco, by G. Romano. But the most celebrated fresco of Romano, "the Fall of the Giants," is in the palace of the T. At the extremity of one of the bridges is a handsome gateway, attributed to Romano, who also erected the open arcade on the bridge over the Mincio, in the heart of the city. Romano inhabited a house opposite the church of St. Barnabas, in which is his tomb. There are numerous convents, a Jews' synagogue, a civil hospital, two orphan asylums, a *monte-di-pietà*, a workhouse, an asylum for 50 poor Jews, an arsenal, cavalry barracks, a large prison, a new and a summer theatre, an imperial academy of arts and sciences, a lyceum, a gymnasium, a public library with 80,000 volumes and many MSS., attached to which are a museum and a fine gallery of sculpture, which has a celebrated bust of Virgil, a botanic garden, and various other scientific and literary institutions. Mantua is a bishop's see, the residence of an Austrian delegate, and the seat of the council, and civil, criminal, and commercial tribunals for the delegation. In the days of her prosperity, and when governed by her own dukes, Mantua is said to have had a population of 50,000, and extensive manufactures ; and, though the latter be greatly fallen off, she still produces limited quantities of silk, woollen, and linen fabrics, with leather, parchment, paper, cordage, &c., and carriages and boats for the navigation of the Po.

Mantua is very ancient, her foundation being probably antecedent to that of Rome. She derives her principal celebrity from her being the native country of Virgil, that great poet having been born in her immediate vicinity, *anno* 70 B.C.

Mantua Musarum domus, atque ad sidera cantu
Evecta Aonio, et Smyrnais æmula pieciris.
Silius Italicus, lib. viii., lin. 595.

Mantua appears, from the contrast, in the first Eclogue, between her and Rome, not to have been a place of much importance in Virgil's time ; and Martial applies to her the epithet of *parva*. (Ep. xiv., 193.) Her unlucky vicinity to Cremona made her territory be divided among the veterans of Augustus. (See art. Cremona, in this work.)

After the conquest of N. Italy by Charlemagne, Mantua became a republic, and continued under that form of government till the 12th century, when the Gonzaga family acquired the supreme direction of its affairs. They were subsequently raised to the title of dukes, and held possession of Mantua till 1707, when it was taken by the Austrians. Under the French, it was the capital of the deputy of the Mincio. (*Forsyth ; Eustace ; Woods ; Oesterr. Nat. Encyc., &c.*)

MANTUA, p. t., Portage co., O., 150 m. N.E. Columbus, 327 W. Watered by Cuyahoga river and its branches. It is distinguished for its fine orchards, and contains a church, an academy, 10 schools, 463 scholars. Pop. 1187.

MANZANARES, a town of Spain, prov. La Mancha, 94 m. E. by N. Ciudad Real, and 100 m. S. Madrid. Pop. 9100. It stands in the loftiest and bleakest part of the province, on the high road between Madrid and Seville ; being, according to Inglis, "a place of considerable size, and proportionate poverty." A parish church of Gothic architecture, a castle, hospital, and cavalry barracks are the only public buildings ; the private houses are better built than in most towns of Spain. The inhabitants are chiefly employed in the production of saffron, for which the neighbourhood is celebrated, and of the Val-de-Peñas wine, highly esteemed all over Castile ; the only other branches of industry being the manufacture of coarse woollens and linens for home supply. Not far from Manzanares are the ruined walls and tower of the ancient *Myrus* ; a city described, in Antonine's *Itinerary*, as being on the road from *Laminium* (Alhambra) to *Toletum* (Toledo).

MANZARES, a small river of Spain, tributary to the Tagus, and flowing by Madrid, which see.

MARACAYBO, MARACAIBO, or NUEVA ZAMORA, a fortified city of Venezuela, cap. dep. Zulia, and prov. Maracaybo ; on the W. shore of the strait connecting the lake of Maracaybo with the sea, 175 m. E.S.E. Santa Marta, and 390 m. W. by N. La Guayra. Lat. 10° 30' N., long. 71° 45' W. In 1801, its population, including a number of Spanish refugees from St. Domingo, was estimated at 94,000 ;

and it may still, perhaps, amount to 20,000. It stands on an arid and sandy soil, partly on the shore of a small inlet of the strait, and partly on a tongue of land which projects into it. Several of its houses are built of a compound of lime and sand, without stone, but they are nearly all thatched with reeds ; and, as the greater number consist wholly of reeds and straw, the town has a mean appearance, and is very subject to fires. A handsome parish church, a chapel, a Franciscan convent, and a hospital, are the only public buildings of which modern travellers make mention. The harbour of Maracaybo, within the bar at the entrance of the straits, has deep water ; and is defended by the three castles of San Carlos, Zapara, and Bajo Seco, situated on the islands of the same names, among the shoals forming the bar. The *Bajo Seco*, or dry shoal, is in advance of the other islands ; and the best channel to the harbour, on the N.W. side, has 13 feet water. The climate of Maracaybo is oppressively hot ; during a part of the year water is scarce ; and in the summer, when violent thunder-storms and earthquakes occur, the city often suffers greatly from very heavy rains. This port has superior facilities for ship-building, and its shipwrights have produced some fine schooners. A brisk traffic is carried on with the interior by the numerous vessels which navigate the lake. The inhabitants are said to be good sailors, and they have generally a taste for a seafaring life. Many, however, devote themselves to the care of cattle, large herds of which are reared in the vicinity. (*Geog. &c. Account of Colombia*, i., 217–225 ; *Mod. Trav.*, xvii. ; *Encyclopædia Americana* ; *Encyc. of Geog.*, American edition.)

MARACAYBO (LAKE OF LAGOON OF), a large lake, or inlet of the sea, in the N. part of S. America, repub. Venezuela, dep. Zulia, prov. Maracaybo. It extends between lat. 8° 5' and 10° 30' N., and long. 71° and 73° 29' W., and is of an oval, or rather "decanter-like" shape ; communicating, at its N. extremity, with the gulf of Maracaybo, by a strait nearly 20 m. in length, and varying in breadth from 5 to 10 m. Length of the lake, N. to S., nearly 100 m. ; greatest breadth, about 70 m. ; circ. probably about 250 m. Inside it has water enough to float the largest vessels ; and, being easily navigated, serves for the conveyance to Maracaybo of the produce of the interior intended for consumption in, or exportation from, that city. But a shifting bar, at the mouth of its strait, where it unites with the sea, in lat. 11° 2', having only 14 feet water, renders it inaccessible to large ships. It receives several considerable rivers, so that its waters are perfectly fresh, sweet, and fit for drinking, except in the spring, when strong N. winds impel inwards a swell from the gulf, which renders them brackish. The lake is not very subject to violent tempests. It abounds with fish and waterfowl ; but tortoises, elsewhere so common in Colombia, are not met with in it. Its banks are in many parts sterile, and only cultivated on the W. side ; and they are, in general, so unhealthy, that the Indians prefer mounting their huts on iron-wood posts in the water, to fixing them on the shore. It was from the Indian villages or towns, built in this way, that the whole country is said to have derived from the Spaniards the name of Venezuela, or Little Venice. Four of these towns are still standing on the E. part of the lake, at considerable distances from each other ; the iron-wood on which they are founded having become a mass of stone, from the petrifying quality of the water. (*Geog. Account of Colombia*, i., 216, 217.)

Towards the N.E. border of the lake is a remarkable mine of asphaltum (*pix montana*) ; " the bituminous vapours of which are so easily inflamed that, during the night, phosphoric fires are continually seen, which, in their effect, resemble lightning. It is remarked that they are more frequent in great heat than in cool weather. They go by the name of the 'Lantern of Maracaybo,' because they serve for lighthouse and compass to the Spaniards and Indians, who, without the assistance of either, navigate the lake." (*Depons, Trav.*, i., 70 ; *Mod. Trav.*, xxvii., 209–211 ; *Geog. Account of Colombia; Blunt's American Coast Pilot*.)

MARAGA (an. *Gamarga*?), a city of Persia, prov. Azerbijan, 50 m. S. by W. Tabreez, and 305 m. W.N.W. Teheran. Pop. about 15,000. It is a well built walled town, in a low valley, at the extremity of a fertile plain, opening to the lake Urumee, which lies 10 m. W. Maraga. The chief buildings are a large and handsome bazaar, spacious public baths, and the tomb of Holaku, one of the most able princes of the dynasty of Jenghis-khan. Maraga is also celebrated for its beautiful and highly productive gardens and plantations, watered by canals drawn from a small river, over which are two bridges, erected in the 11th century. The town has a large manufactory of glass ; but the inhabitants are chiefly employed in the cultivation of the fertile country round the town.

On the top of a mountain rising behind Maraga are the remains of an observatory, built by Holaku, for the use of Naser-a-Deen, one of the most famous Oriental astronomers ; and at the foot of the hill are several cave-temples, similar in form, though not equal either in size or beauty, to those of Hindostan. (*Kinneir's Persia*, p. 156, &c.)

MARAMEE IRON WORKS, p. v., Crawford co., Mo., 63 m. S.W. by W. Jefferson city, 835 W. It is situated at the "Big Spring" of Maramee river, which issues 30,000 cubic feet of water per minute. There is erected on its outlet a saw mill, a grist-mill with two runs of stones, a blast furnace, three forges, capable of manufacturing 1000 tons of bar iron annually. It has been called the head of Maramee river, but two branches, called Water fork and Dry fork, come in on each side of it.

MARANHAM, or SAN LUIS, a city and seaport of N. Brazil, cap. of the prov. Maranham, on the W. coast of the island of the same name, in the bay of Marcos, 390 m. E. by S. Para. Lat. 2° 31' 30" S., long. 44° 16' W. The inhabitants are variously estimated at from 12,000 to 30,000, of which a large proportion are negroes. The city is built on unequal ground, extending inwards about 1½ m. from the water's edge. It is laid out in a straggling manner, with numerous squares and broad streets, the latter being only partially paved. There are many neat and good-looking houses; the better sort consist of a ground floor, and a story above; the lower part being usually employed as a shop, and lodging for servants, and the upper as the apartments of the family. These houses have mostly balconies, and are handsomely fitted up. In the poorer and unpaved streets the houses consist of only a ground floor, and having thatched roofs and unglazed windows, their appearance is extremely mean and shabby. Adjoining the shore is an open space, one side of which is nearly taken up with the governor's palace, town-hall, and prison, which occupy a long, uniform, handsome stone building of one story in height; another of its sides is occupied by the cathedral. This, which was formerly the Jesuit's church, is said to be the finest of any in the maritime cities of Brazil, except that of Para. The Jesuits' college is now the episcopal palace. There are a great number of other churches and convents, a treasury, two hospitals, various public schools, and a custom-house, which, though small, was till recently quite large enough for the business of the place. Latterly, however, its commercial importance has been much increased; and it is the principal port of the empire for the shipment of cotton and rice; the other articles of export consist principally of hides and horns, caoutchouc, isinglass, sarsaparilla, cocoa, &c. We subjoin an

ACCOUNT of the Number and Tonnage of the Ships which cleared out from the port of Maranham in 1837 and 1838, specifying the Countries to which they belonged, and the Value of the Cargoes:

	1837.			1838.		
Countries.	Vessels.	Tonnage.	Value of Cargoes.	Vessels.	Tonnage.	Value of Cargoes.
British	48	6,021	204,452	42	7,090	36,974
Brazilian	54	7,484	44,172	37	4,548	82,312
Portuguese	69	8,696	67,277	17	3,249	52,516
Spanish	36	2,164	84,971	11	466	21,963
French	10	2,550	2,867	4	646	3,360
American	20	1,207	5,714	6	723	7,009
Belgian	1	119	3,280	3	505	3,194
Hamburg	2	280	1,794	4	474	1,156
Danish	2	294	3,444	1	140	
Sardinian				1	219	
Prussian				1	507	
Total	181	28,244	386,752	104	19,165	133,552

We have no authentic information as to the importation of slaves into Maranham; but there can be no doubt it is very considerable, and may, perhaps, be estimated at above 3000 a year.

The harbour of Maranham is rather difficult of access. It is usual for vessels arriving on the coast to make the lighthouse on the island of St. Anna, about 40 m. N.E. Maranham. The harbour of the latter consists of a narrow creek, defended by some indifferent forts. It is so beset with shoals and islets, as to render a pilot always necessary, but with such there is no real danger. It has about 18 feet water at low ebb; but it is said to be filling up, and that the probability is that the port will, at no very distant period, be transferred to Alcantara, on the opposite side of the bay. The latter, indeed, is in all respects a preferable port, being more easily accessible, having deeper water, and greater facilities for getting to sea. The island of Maranham is fertile, and densely peopled; having a number of villages, of which the largest consist of four large timber huts, from 390 to 500 paces in length, and about 20 or 30 feet in depth, each capable of accommodating from 200 to 300 inhabitants.

This city was founded by the French in the early part of the 17th century. (See BRAZIL, in this Dict.; *Mod. Trav.*, XXX., 279, 281; *Encyc. Americana*; *Blunt's American Pilot, &c.*, p. 515.)

MARATHON, p. t., Cortland co., N. Y., 141 m. W. Albany, 318 W. Drained by Toughnioga river and its tribu-

taries. It has two stores, one grist-mill, two saw-mills; seven schools, 350 scholars. Pop. 1063.

MARAZION, or MARKET-JEW, a decayed bor., seaport, market town, and township of England, St. Hilary par., co. Cornwall, E. div. of hund. Penurth, 42 m. S.S.W. Bodmin, and 259 m. W. by S. London. Pop. 1393. It is situated on the shore of St. Mount's bay, on the side of a hill, which shelters it from the cold N. winds. The parish church is 2 m. distant; but it has a chapel of ease, and places of worship for Wesleyan Methodists and other dissenters. An endowed school is held in the guildhall; a national school and three Sunday schools furnish instruction to the children of the poor; and there are a few charities. Its principal trade consists in the importation of timber, coals, and iron, for the supply of the town and neighbouring mines. The market, held on Saturday, is well supplied, especially with ready-made shoes; and two large cattle fairs are held 3d Thursday in Lent and September 29. Though a borough by subscription, this town was chartered by Queen Elizabeth; the corporate officers being a mayor and eight aldermen, with 12 capital burgesses, whose privileges were not interfered with by the late Municipal Reform Act. It is supposed to have sent members to the House of Commons at a former period, but certainly not subsequently to 1658. Its name, Market-Jew, has been supposed to be derived from its having been, in the period of its prosperity, a great trading place for the Jews, but the presumption is unsupported by history; and it appears more rational to conclude that it is a corruption of its ancient name Marghasyon, or Marghasiewe.

MARBELLA (an. *Salduba*), a seaport town of Spain, in Andalusia, prov. Malaga, 30 m. S.W. Malaga, and 36 m. N.E. Gibraltar. Pop., according to Miñano, 6269. "It stands slightly elevated above the sea; and its turreted walls and narrow streets declare it to be thoroughly Moorish. The town is particularly clean, and respectably inhabited; the fishing portion of the population being located more conveniently for their occupation in a large suburb on its E. side." A church, two hospitals, and an old Moorish castle, are its principal public buildings. The trade of Marbella is only trifling: its valuable mines of lead and iron, which formerly secured for it a certain degree of prosperity, have been for many years totally abandoned, its sugar refinery and tan-yards have disappeared, and fishing now forms the chief occupation of the inhabitants. There is no harbour; but vessels find excellent holding-ground, in deep water, near the shore. The landing also is good, on a fine hard sand; and a small pier has lately been constructed. (*Scott's Ronda and Granada*, ii., 378.)

MARBLEHEAD, p. t., port of entry, Essex co., Mass., 18 m. N.E. Boston, 4 m. E. by S. Salem, 458 W. Incorporated in 1649. Situated on a rough and rocky peninsula, extending from 3 to 4 m. into Massachusetts bay. It has a good harbour, defended on the N.E. by Fort Swall. The harbour, in front of the village, is a mile and a half long, and half a mile wide. The place is inhabited chiefly by fishermen, and has about 160 vessels employed in the coasting trade, the fisheries and foreign trade. Its tonnage, in 1840, was 19,478 tons. The t. has five churches, a Congregational, Unitarian, Episcopal, Methodist, and Baptist; two banks, with an aggregate capital of $260,000; two insurance companies, capital $100,000; 29 stores; two academies, 196 students; 20 schools, 896 scholars. Pop. 5575.

MARBLETOWN, p. t., Ulster co., N. Y., 7 m. S.W. Kingston, 64 m. S.S.W. Albany, 318 W. Watered by Esopus and Roudout creeks. The Delaware and Hudson canal passes through it. It contains 15 stores, seven lumber-yards, two fulling-mills, one woollen factory, one flouring-mill, six grist-mills, 10 saw-mills; 12 schools, 687 scholars. Pop. 3815.

MARBURG, a town of Hesse Cassel, cap. circ. Upper Hesse, on the Lahn, a tributary of the Rhine, 50 m. S.W. Cassel, and 36 m. N.E. by E. Coblentz; lat. 50° 48' 41' N., long. 8° 46' 19" E. Pop., including the suburb of Weidenhausen, on the opposite bank of the Lahn, 7700. It is built on the slope of a hill, crowned by a ruined castle; and has narrow and dirty streets, and indifferent houses. Its only building worth notice is the church of St. Elizabeth, an elegant edifice, and one of the earliest existing specimens of the pointed Gothic style, having been commenced in 1235, and finished within the succeeding 48 years. The tomb of St. Elizabeth, in this church, has been long resorted to by pilgrims, and was formerly adorned with numerous gems and articles of value, mostly carried off by the French in 1810. In the transept are several curious monuments of the landgraves of Hesse. The university of Marburg, founded in 1527, has 40 professors, and a good library of 70,000 vols. In 1833, it was attended by 429 students, but, in 1840, the number of pupils had declined to 285. Marburg has also the Wilhelm's Institute, a school of surgery; and a philological seminary, teachers' seminary, botanic garden, school of veterinary medicine, Lutheran and Catholic orphan asylums, a workhouse, a free-school of industry, &c. The inhabitants derive their principal support form the university, and from the manufacture of linen fabrics, stockings, hats, tobacco, and tobacco-pipes, &c. It is the seat of the chief judicial and other state establishments for Upper Hesse. (*Berghaus ; Stein.*)

MARBURG, a town of the Austrian empire, being, next to Grätz, the principal in the prov. of Styria, cap. circ. on the Drave, and on the road from Grätz to Laybach. 26 m. S.S.E. the former city. Pop. in 1837, 4578. Mr. Turnbull says, "it is a good town, and surrounded by a beautiful country, richly planted with vines. The climate here is far more congenial to their growth than on the N. side of the hills, and excellent wine is produced." (*Trav., i., 279.*) Near it, the Archduke John has a vineyard and villa. Marburg has three suburbs, an old castle, a church, in which are several good pictures, a hospital, theatre, gymnasium, military school, swimming school, &c. It is the seat of the council for the circle, furnishes leather and reoogtio, and has some trade in corn, wine, and iron ; but its inhabitants derive their chief subsistence from the active transit trade between Hungary and Croatia and Illyria, (*Turnbull's Austria ; Berghaus ; Oesterr. Nat. Encyc.*)

MARCELLUS. p. t. Onondago co., N. Y., 141 m. W. by S. Albany, 343 W. Drained by Nine-mile creek. It contains three churches, a Presbyterian, Methodist, and an Episcopal, seven stores, four fulling-mills, two woollen factories, two flouring-mills, two grist-mills, six saw-mills, two paper-mills, five tanneries, two distilleries ; 12 schools, 571 scholars. Pop. 2726.

MARCH, a market town, township, and par. of England, belonging to Doddington par., Isle of Ely, hund. Witchford, on the Old Nen, 13 m. N.W. Ely, and 74 m. N. London. Area of township, 20,440 acres. Pop., in 1831, 5117. Excepting the church, which is large and handsome, the town contains nothing worthy of remark ; the streets being generally narrow, and the houses, for the most part, low and meanly built. Its situation on the Nen, which is navigable, makes it the centre of a considerable trade ; corn, hemp, flax, cheese, &c., being shipped here ; and coal, timber, and London goods, imported. Markets on Friday ; fairs, Monday before Whitsuntide, Whit-Monday, and 2d Tuesday in October, chiefly for horses, cattle, and cheese.

MARCY, p. t., Oneida co., N. Y., 6 m. Utica, 96 m. W.N.W. Albany, 343 W. Organized in 1832. Bounded S.W. by Mohawk river and its tributary Nine-mile creek. It contains one fulling-mill, four saw-mills, three tanneries ; nine schools, 674 scholars. Pop. 1799.

MARENGO, a village of N. Italy, Sardinian States, near the Bormida, in an extensive plain, 3½ m. E. by S. Alexandria. This village will be ever memorable for the great battle fought here, on the 14th of June, 1800, between the French under Napoleon, and the Austrians under Melas. Napoleon, believing that the Austrians had withdrawn from the neighbourhood of Marengo, had, on the day previously to the battle, despatched Dessaix with a strong corps to Rivolta. By this means, his army was reduced, when attacked by the Austrians on the following morning, to little more than 20,000 men, whereas the Austrians had nearly 40,000 troops in the field. The contest was most obstinate and bloody ; but, despite a desperate resistance, the Austrians carried the village of Marengo, broke the left wing of the French, and compelled them to retreat. But, at this critical moment, when the fate of the day appeared all but decided, Dessaix, who had returned by a forced march, came upon the field. This gave the French new strength, and inspired them with new courage. The Austrians, exhausted by their previous efforts, were immediately attacked at all points, forced back, and completely defeated, with the loss of all their cannon and baggage, and of a vast number of men left dead on the field and taken prisoners. Dessaix, whose opportune arrival turned the fortune of the day, was killed, charging at the head of his division.

MARENGO, county, Ala. Situated in the W. part of the state, and contains 975 sq. m. Bounded W. by Tombigbee river, and N.W. by its branch, Black Warrior river. It contained in 1840, 15,190 neat cattle, 2175 sheep, 36,819 swine ; and produced 2785 bushels of wheat, 1091 of rye, 649,734 of Indian corn, 22,933 of oats, 30,906 of potatoes, 2815 pounds of rice, 1145 of tobacco, 6,356,699 of cotton. It had 34 stores, one flouring-mill, 18 grist-mills, seven saw-mills, two tanneries, one distillery, one printing-office, one weekly newspaper ; three academies, 131 students ; nine schools, 201 scholars. Pop. : whites, 5350 ; slaves, 11,902 ; free coloured, 12 ; total, 17,264. Capital, Linden.

MARENGO, p. t., Calhoun co., Mich., 100 m. W. by S. Detroit, 572 W. It contains one store, one grist-mill, three saw-mills, one distillery ; 10 schools, 307 scholars. Pop. 872.

MARGARETTA, p. t., Erie co., O., 100 m. N.W. by N. Columbus, 414 W. Bounded N. by Sandusky bay. Wa-

tered by Cold creek. It contains four stores, one flouring-mill, one grist-mill, two saw-mills ; one school, 38 scholars. Pop. 1001.

MARGARITA, an island off the N. coast of S. America, belonging to the republic of Venezuela, and attached to the dep. Cumana. It lies in about lat. 11° N., and long. 64° W., separated from the continent by a channel, 20 m. in width, through which all ships coming from Europe, or windward of Cumana, Barcelona, or La Guayra, must pass in going to to those ports. Length of the island, E. to W., 37½ m.; breadth varying from 5 to 20 m. Population estimated at 15,000. Viewed at a short distance from the N. it appears like two islands, there being a tract of low swampy land in its centre, which is in some parts not more than from 16 to 18 ft. above the level of the sea ; but other parts of the island rise to a considerable elevation ; and Maranco, near its W. extremity, a micaceous schist, is upwards of 3000 ft. in height. The coast-lands are arid and barren ; but the interior is comparatively fertile, producing maize, bananas, and various fruits, with sugar, coffee, cocoa, and other W. Indian products, though not in sufficient quantities for the demands of the inhabitants. A good deal of poultry, and other live stock, is reared, and exported to the continent ; and Margarita has an active fishery, and some salt-works. It was formerly much celebrated for its pearl-fishery ; but this has greatly declined, and the pearls now found are said to be of inferior size and quality. The pearl-fishery was principally conducted at the rocky island of Coche, between Margarita and the main land. The inhabitants have some manufactures of cotton stockings and hammocks, of very good quality. Assumpcion, the capital, and residence of the governor, in the centre of the island, is pretty well built. There are three seaport towns or villages ; one of which, Pampatar, on the S.E. coast, has a pretty good harbour, with anchorage in 7 or 8 fathoms water. (*Blunt's American Coast Pilot, p. 446.*) This island, which is of little value in any other point of view, might, were it occupied by a European power, be of considerable service as a depot for the supply of the adjacent continent. It is better situated for such a purpose than Trinidad. It was discovered by Columbus in 1498, (*Geog. Account of Colombia ; Humboldt's Personal Narrative, &c.*)

MARGATE, a seaport town and much-frequented watering-place of England, co. Kent, in the isle of Thanet, lathe St. Augustine, 16 m. E.N.E. Canterbury, and 65 m. E. London. Area of parish, 3810 acres. Pop., in 1831, 10,339. The town is finely situated, partly along the shore, and partly on the declivities of two hills, one of which presents a bold cliff towards the sea. The older streets are narrow and irregular, lined with inferior-looking houses ; but in the upper parts and outskirts of the town are several handsome streets and squares formed by houses which far surpass, in neatness and regularity of construction would not disgrace the metropolis. The whole is well ·paved, lighted with gas, and plentifully supplied with good water. The town-hall and market-house is a plain but substantial building of recent erection, supported on cast-iron pillars, and fronted by a Tuscan portico. The assembly-rooms in Cecil-square have long ranked among the largest and most elegant in England : a neat theatre stands on the E. side of Hawley-square, where also is a large public library. Numerous bathing-houses line one side of High-street, and near the Parade E. of the town is a very complete establishment formed in the cliff, and furnishing hot and cold baths of a very superior description. There are two churches ; one an old heavy-looking building, with a low square tower ; the other at the opposite side of the town being a very handsome modern Gothic structure, with a light octagonal tower, built at an expense of £36,000. The Roman Catholics, Independents, Baptists, and Society of Friends have also their respective places of worship, to which are attached well-attended Sunday-schools. A national school furnishes instruction to about 250 boys and 180 girls, and there are two other large day-schools. Drapers' alms-houses, founded in 1709, a dispensary, and lying-in charity, are the principal charitable institutions ; and in the immediate vicinity, close to the beach, is a large sea-bathing infirmary, founded in 1792, and since so much enlarged as to furnish accommodation for about 120 patients. The harbour dries at low water. To obviate this defect a stone pier, projecting 900 ft. into the sea. was erected from the designs of the late John Rennie ; still, however, this was insufficient : for the purpose, there not being more than from 4 to 5 ft. water at the pier head at low ebb. Since 1824, a wooden jetty, connected with the pier, has been constructed, which projects into deep water, and may be approached by steamers or other vessels at any time of the tide, except when it blows a gale from the N. or N.N.E. The pier is a favourite promenade for the town's folk and visitors.

Margate enjoys a considerable coasting trade, and has some commerce with Holland and Germany ; but neither these nor its fishery are of any importance compared with the advantages that accrue to it from the thousands of visitors who annually resort thither from the metropolis. The town, indeed, like many others, owes its present importance to the invention of steam; for though prior to 1817 it was a respectable and well-frequented watering-place, the means of access to London were so difficult and tedious, that none but those who could afford a week or two of uninterrupted leisure were ever induced to visit it. But within the last fifteen years, the water-communication with London has been so greatly facilitated, that Margate may now be considered as within five or six hours of the metropolis. Several handsome steamers ply regularly between London bridge and Margate; and for some years past the number of persons landed from these steamers at Margate is supposed to have averaged above 90,000 a year. The fares being extremely reasonable, Margate is frequented chiefly by the families of tradesmen and others belonging to the middle classes, for whose amusement there are numerous bazaars, libraries, &c., with the Tivoli gardens, in the suburbs, very similar to the well-known, but now extinct, Vauxhall of London. Great numbers of persons engaged in business during the week join their families here late on the Saturday, returning to London early on the Monday morning; and it is from the flying visitors that the steam-packet companies derive their chief revenue.

Margate is within the jurisdiction of Dover, by the lordwarden of which the constable of the town is appointed; and as a port, it is subordinate to Ramsgate. It is the chief place of a poor-law union, comprising all the parishes in the Isle of Thanet.

MARIA-THERESIANOPEL, or THERESIENSTADT (Hungar. Szabadka), a royal free town of Hungary, co. Bacs, in the great plain between the Danube and Theiss, 25 m. S.W. Segedin, and 100 m. S.S.E. Pesth. Population said to be about 35,600, chiefly Hungarians and Servians. Its territory, or commune, comprising an area of 300 sq. m., is larger than that of any other town of the Austrian dominions. (Berghaus.) The "National Encyclopædia" says, it is well built, and has numerous handsome public edifices; including several churches, a gymnasium, large barracks, a town-hall, &c. It has manufactures of linen cloth, leather, and tobacco, and a large trade in horses, cattle, sheep, raw hides, and wool.

MARIANNA, an episcopal city of Brazil, prov. Minas Geraes, of which it is the capital, on the Carmo, a tributary of the Doce, 8 m. E.N.E. Villa Rica. Its population, in 1822, was estimated at from 6000 to 7000. (Mawe's Brazil, 258.) It stands principally in a small plain, bounded by rocky hills, the small knolls, and projections of which are crowned by its churches. The city itself is nearly square, and consists principally of two well-paved streets, regularly laid out, and conducting to a kind of square. The houses are whitened, and have a neat appearance. The supply of water is ample, and is of material importance in the cultivation of several extensive gardens; but, being surrounded by lofty eminences, the air is close and hot, and the town unhealthy. There are several churches and a large cathedral. The Carmelite and Franciscan convents, the ecclesiastical college, which has sundry privileges, the bishop's palace, together with fine gardens, and the town-hall, are among the other chief public buildings. It has very little trade, and depends chiefly on the mines and farms in its vicinity. (Mawe's Brazil; Dict. Géog.)

MARIANNA, p. v., capital of Jackson co., Flor., 77 m. N.E.W. Tallahassee, 457 W. Situated on the W. side of Chipola river. It contains a courthouse, and several stores and dwellings. Nett proceeds of the postoffice, $414.

MARIAVILLE, p. t., Hancock co., Me., 163 m. N.E. Augusta, 685 W. Watered by Union river and its branches, which afford water-power. Incorporated in 1836. It has one grist-mill, two saw-mills, one tannery; two schools, 55 scholars. Pop. 273.

MARIAZELL, or MARIANZELL, a village of the Austrian empire, prov. Styria, in a mountainous district, about 55 m. S.W. Vienna. Pop. about 1000. It would be unworthy notice in a work of this kind, but for its celebrated shrine of the Virgin, which renders it the "Loretto" of the Austrian empire, and a principal place of Christian pilgrimage. The town, which stands at an elevation of about 2800 ft. above the sea, is small and mean-looking; and consists principally of inns and alehouses for the accommodation of the visitors, the influx of which only ceases when the roads are impassable by snow. The only building of note is the church, rebuilt, since 1827, on the site of one erected in 1363, by Louis I., king of Hungary, over the chapel, in which the image of the Virgin is placed. The church, as it now stands, is of Roman architecture, except the porch, which is Gothic. It is a spacious edifice, 296 English ft. by 99. inside, and is surrounded by a spire 275 ft. in height. Some of the side altars and chapels are handsomely decorated; but its principal object of curiosity and devotion is the small stone chapel, erected by a mar-

grave of Moravia, in 1202, instead of the wooden hut in which the Gnaden Statue, 'Statue of Grace,' had stood from about 1150, when it was luckily brought thither by a Benedictine monk. This image, like that of Loretto, is ascribed to St. Luke; and, like it, also, is but an indifferent specimen of the apostle's skill in statuary. It is a rudely-carved wooden figure, only 18 in. in height, representing the Virgin, with the Saviour on her knee. Both are as splendid as brocade, gold, gems, and bad taste can make them; their faces are of a negro hue; the effect, perhaps, in part, of the smoke of the solitary lamp kept constantly burning in the dark and gloomy recess in which they are cooped up. The altar and other decorations of the shrine are said to be of solid silver, and the chapel is surrounded by a costly fence of the same metal. A thousand acres of land were assigned for the support of the church; and its treasury was very rich previously to the reign of Joseph II., having received many valuable donations from preceding sovereigns, princes, and private individuals. But Joseph, though he succeeded to the dominions of his mother, inherited none of her superstition: unawed by the sanctity of the place, he did not hesitate to strip the shrine of the greater portion of its wealth; and profusely threw the silver angels that guarded the high-altar, and even the figures of his father and mother, into the melting-pot! The present emperor and empress have, however, made a propitiatory visit to the cell; and have endeavoured, by their pious liberality, to atone, in some measure, for the sacrilegious depredations of their less scrupulous predecessor.

The ecclesiastical establishment of Mariazell consists of about 90 resident priests, deputed from the Abbey of St. Lambricht, who here form a kind of subsidiary Benedictine college, under a pro-rector. During half the year all find abundant employment among the penitents, who arrive here from all parts of the empire. Shortly after the erection of the church, the popes granted the same indulgences to the shrine of Mariazell, as were attached to St. Peter's at Rome; and thenceforward it became crowded with pilgrims. Previously to the reign of Joseph, the pilgrims are stated to have amounted to about 180,600 annually; and it is alleged that, at the celebration of the 18th jubilee of the miraculous image, in 1757, no fewer than 380,000 individuals did homage to the sable Maria ! We confess, however, that we do not attach implicit credit to this statement; but it is, at all events, certain that the number was very great. The Austrian Encyclopædia says that the shrine is, at present, annually visited by 100,600 pilgrims; and, according to Mr. Turnbull, the number is fully 80,000. (Austria, i., 196.) It is customary for the pilgrims from different places to set out together; and, formerly, it was no unusual circumstance for a band of pilgrims from one province or city to have a contest for precedence with those from another; so that disturbances, which frequently ended in bloodshed, were perpetually occurring. The government has, however, put an end to these unseemly brawls by ordering that the pilgrimages from different places should take place at different times. Accordingly, most of the towns of any importance in Upper and Lower Austria, Styria, Bohemia, and Moravia, and some in the W. parts of Hungary, have their stated days on which the devotees assemble, and form their processions of piety and pleasure after the manner described by Chaucer in his Canterbury Tales. In all, about 80 processions take place annually from different parts of the empire. Vienna furnishes four distinct parties, three in June or July, and one in August; the last, which is also the largest, generally consists of about 3000 persons of both sexes, and all ages, travelling chiefly on foot, and performing the journey in four days. In their progress they are jumbled together, without any regularity, until they come within about a mile of the shrine. Here they halt; and some hours are generally occupied in marshalling the confused assemblage into regular devotional order. Banners are unfurled; sacred emblems exposed to view; the maidens and youths are placed in the van of the procession, after whom follow the elder pilgrims, male and female, in distinct parties: and thus they advance to the church, by slow and measured steps, stopping at certain appointed stations on the way, and chanting in their native tongue, whatever it may be, some one of the litanies, in general chorus. Arriving by thousands in a day, they fill to suffocation every inn and house of accommodation within the town; but the larger portion are, notwithstanding, obliged to bivouac in the fields around; where they spend the night in jollity, drinking and singing songs, which are frequently of a kind not especially suited for virgin ears. (Turnbull, i., 197-198; Russel.) It is needless to add, that by far the largest proportion of those who join these processions are but little influenced by religious motives. The enlightened portion of the community despise them as miserable mummeries; and the motley crowd principally consists of the ignorant, the idle, the frolicsome, and the profligate. They are, in fact, an outrage upon religion and morality.

The holy image has been but an indifferent protectress of the village of Marianzell. Six times has it been destroyed by fire, and its population temporarily reduced to ruin. The last conflagration occurred in 1827; when the roof and towers of the church were destroyed, and, out of 111 houses, only 20 escaped. The inhabitants are generally poor. They depend principally on the supply of necessaries, and of rosaries, tapers, relics, and such like articles, to the pilgrims.

The iron foundries, 2 or 3 m. distant from Marianzell, are the most important of the Austrian empire. Every species of casting is executed in them, from the largest cannon and steam-engines, down to trinkets, which are said to rival those of Berlin. Marianzell has also some copper and sulphur works: a great deal of timber is sent from its neighbourhood to Vienna and the Black sea. (*Oesterr. Nat. Encyc.; Turnbull's Austria*, i., 185, 199; *Germany and the Germans*, ii., 291–295; *Horschelmann's Stein, &c.; Russell*, p. 348–354.)

MARIE-GALANTE, one of the French W. India islands. (*See* GUADELOUPE.)

MARIENBURG, a town of the Prussian dom., prov. W. Prussia, cap. circ. Marienburg, on the Nogat, an arm of the Vistula, here crossed by a bridge of boats, 27 m. S.E. Dantsic. Pop. (1837) 5708. This little town is chiefly interesting as having been the seat of the grand masters of the Teutonic order for nearly two centuries. "To the N.E. of the town, and on the summit of a small hill, 50 ft. above the level of the Nogat, and an equal number of feet from the bank of the river, stand the ruins of the Teutonic castle, so often mentioned in the history of chivalrous times. The whole mass is at once imposing and picturesque, bespeaking the grandeur of its former occupants, and the purposes to which it was destined." Most probably this castle had been commenced towards the end of the 12th, or the beginning of the 13th, century. In 1281 it was greatly enlarged, by the addition of that part which was afterwards known as the Old castle; and about the same time, the residence of the grand master was transferred to Marienburg from Venice. Succeeding grand masters built the middle and lower castle (erected, according to Zeditz, chiefly between 1306 and 1309), and the church of Nôtre-Dame, in the immediate vicinity, which is still in existence, and forms a very prominent feature in the landscape of these ruins. In 1644, the Old castle was burnt to the ground; but the rest of the building escaped; and, after undergoing many vicissitudes, was put in complete repair by the present king of Prussia, when crown prince. It comprises a chapel, in which are numerous monuments of the grand masters, cells of the knight-monks, with their halls, dormitories, refectory, subterranean caverns, chapter-house, &c., in tolerable preservation. The chapter-house, by far the most interesting part of the edifice, is a large square apartment, with 20 windows, displaying the arms of the successive grand masters in stained glass. (*Granville's Tour to Petersburg*, i., 341, 342.) An antiquated tower, called the *Butterwelchthurm*, and some singular water-mills in the neighbourhood, are among the other curiosities of the town. It has a Roman Catholic and a Calvinist church, a teachers' seminary, a deaf and dumb school, and numerous other schools: a workhouse, hospital, &c. It is the seat of the council for the circle; and has manufactories of woollen and cotton cloths, stockings, and hats; various breweries, distilleries, and tanneries, and some trade in corn and timber. Marienburg fell into the hands of the Poles by their conquest of the Teutonic knights, in 1457, and was ceded to Prussia, at the treaty of Thorn, in 1466. (*Zedlitz; Der Preuss. Staat.; Berghaus.*)

MARIENWERDER (Slav. *Kwidzin*), a town of the Prussian dom., prov. W. Prussia, cap. of the gov. and circ. of Marienwerder, on the Little Nogat, a tributary of the Vistula; 44 m. S.S.E. Dantzic, and 51½ m. N.N.E. Thorn. Pop., in 1837, 5,590. It stands on elevated ground, is well built, and has four suburbs. Its cathedral, erected in the 13th century, has a steeple 170 ft. in height; and in its interior are the tombs of many church dignitaries and grand masters of the Teutonic order, and some curious mosaics. What remains of the old castle is now appropriated to the judicial courts for the circle and town, and a school of arts. Marienwerder is the seat of the head court of justice for the province of Prussia, and of the provincial council, and agricultural union for W. Prussia. It has a gymnasium, a royal school of agriculture, a school for the improvement of neglected children, a hospital for blind soldiers, to which is attached the *Louisenian*, an institution for the blind widows of soldiers, a large printing establishment, &c. It has, however, few manufactures, and little trade, except in retail; the inhabitants being principally employed in the supply of necessaries to the various public establishments. (*Zedlitz; Der Preussische Staat.*, ii., 483, 484.)

MARIETTA, p. t., capital of Washington co., O., 104 m. E.S.E. Columbus, 174 m. below Pittsburg by the Ohio river,

238 m. above the mouth of the Ohio, 300 W. Pleasantly situated on both sides, but chiefly on the E. side of Muskingum river, at its entrance into Ohio river. First settled in 1788, and is the oldest town in Ohio. The town plat contains 1000 house lots, 90 feet wide, and 180 feet deep, with spacious and airy streets, and extensive commons. It is neatly built, and contains a courthouse, jail, a United States land office, four churches, a bank, a market-house, a public library in a handsome brick edifice, the upper story of which is the hall of the Lyceum, a female academy, and the Marietta Collegiate Institute. The courthouse, bank, and Collegiate Institute, are handsome specimens of architecture. The Marietta Collegiate Institute was founded in 1832, has a president and seven professors or other instructors, 50 students, and 3500 volumes in its libraries. The commencement is on the last Wednesday in July. The edifice is 75 feet long, 50 feet wide, and four stories high, with the basement. Manual labour is connected with the institution. The place has 17 stores, two steam saw-mills, one steam flouring-mill, two carding-machines, an iron foundry, a rope-walk, four tanneries, 250 dwellings, and 1814 inhabitants. According to the census of 1840, the village had one college, 100 students, including a primary department; one academy, 53 students; three schools, 289 scholars. The town had, in addition, six schools, 236 scholars, and 875 inhabitants. It was named in honour of *Marie Antoinette*, queen of France.

MARIETTA, p. v., capital of Cobb co., Ga., 113 m. N.W. Milledgeville, 076 W. Situated 3 m. W. of Chattahoochee river, and contains a courthouse, jail, three churches, two academies, about 30 dwellings, and 299 inhabitants.

MARIGLIANO, a town of the Neapolitan dom., prov. Napoletano, cap. cant., 12 m. N.E Naples. Pop. *estimated* at 5000. It has some ruins, which have been supposed to have formed part of an ancient palace of the Marii.

MARINO (SAN), a town and republic of Italy, under the protection of the Pope; being about the smallest, as well as the most ancient, state in Europe. The territory of the republic, enclosed on all sides by the legation of Urbino, is the Papal states, consists of a craggy mountain, about 2500 ft. high, and some adjacent hillocks, with one town and four villages; comprising an area of about 22 sq. m., and a pop. of 7600. The town stands on the side of the mountain above mentioned, about 15 m. S.W. Rimini, and 26 m. N.N.W. Urbino. Lat. 43° 56′ 21″, long. 19° 27′ 5″. Pop. 5500. It is accessible by only one road, and is irregularly built. It has a principal square, in which is the town-hall; five churches, in one of which are the tomb and statue of St. Marino, the founder of the town; four convents, and three castles. Its inhabitants are chiefly occupied in agriculture and cattle-breeding, or in the manufacture of silk. Most of the wealthy inhabitants of the republic reside in the village of Borgo, at the foot of the hill on which the town is situated.

The legislative powers of the government are vested in a senate, or council, of 60 members, elected for life; 20 from among the nobles, 20 from the citizens, and 20 from the rural population; and in a lesser council, or tribunal of appeal, composed of 12 senators. The executive powers belong to 2 *capitani reggenti*, chosen, every six months, by the inhabitants at large above 25 years of age; the *capitani* preside in the council of 60; and justice is administered by a *commissario*, who must not be a foreigner. Every family is obliged to furnish an individual capable of bearing arms to the military force of the republic, amounting, in all, to between 800 and 900 men; but only about 40 men are ordinarily kept on duty. The state supports a hospital, and four superior and two elementary schools. Public revenue about 6000 *scudi* or crowns a year, which is sufficient to meet the expenditure. The town grew up round a hermitage formed here by an individual of the name of Marinus, or Marino, belonging to Dalmatia, afterwards enrolled in the calendar of saints, in the fifth century; and the insignificance and uninviting character of its territory appear by making it unworthy of attention, to have enabled it to preserve its independence during the disturbed periods of the dark and middle ages. It was occupied by Cæsar Borgia, but for a short period only; and was taken, in 1739, by Cardinal Alberoni; but the pope disavowed the proceeding, and restored San Marino to its privileges. In 1796, Napoleon offered to increase the territory of the republic; but, this being wisely declined, he presented it with four pieces of cannon. (*Serristori, Stat. d'Italia; Addison's Tour in Italy, &c.*)

MARION, district, S.C., situated in the E. part of the state, and contains 1200 sq. m. Drained by Great and Little Pedee rivers and their branches. Bounded S.W. by Lynch's creek. It contained in 1840, 21,909 neat cattle, 8352 sheep, 39,837 swine; and produced 4033 bushels of wheat, 1353 of rye, 377,041 of Indian corn, 16,414 of oats, 61,530 of potatoes, 67,943 pounds of rice, 1852 of tobacco, 603,496 of cotton. It had 12 stores, six flouring-mills, 21

grist-mills, 20 saw-mills, three distilleries; two academies,
71 students; 11 schools, 928 scholars. Pop.: whites, 8892;
slaves, 5951; free coloured, 86; total, 13,939. Capital, Marion, C.H.

MARION, county, Ga., situated towards the S.W. part of
the state, and contains 330 sq. m. Bounded E. by Flint river,
by branches of which it is drained. It contained in 1840,4545
neat cattle, 1846 sheep, 8895 swine; and produced 3634
bushels of wheat, 110,742 of Indian corn, 2694 of oats, 8107
of potatoes, 3,258,232 pounds of cotton. It had four grist-
mills, three saw-mills; one academy, 39 students, four
schools, 105 scholars. Pop.: whites, 3741; slaves, 1070;
free coloured, 1; total, 4812. Capital, Tazewell.

MARION, county, Alabama, situated towards the N.W.
part of the state, and contains 1144 sq. m. Drained by
branches of Tombigbee river and by Bear creek, which
flows into Tennessee river. It contained in 1840, 12,966
neat cattle, 4315 sheep, 13,009 swine; and produced 17,467
bushels of wheat, 2190 of rye, 113,490 of Indian corn, 4395
of oats, 4153 of potatoes, 1100 pounds of tobacco, 118,064 of
cotton. It had eight stores, eight grist-mills, four saw-mills,
two tanneries, three distilleries. Pop.: whites, 3694; slaves,
753; total, 5847. Capital, Pikeville.

MARION, county, Miss., situated in the S. part of the
state, and contains 1476 sq. m. Drained by Pearl and Leaf
rivers, and Black creek. It contained in 1840, 16,785 neat
cattle, 2164 sheep, 17,450 swine; and produced 109,008 of
Indian corn, 4965 of oats, 32,685 of potatoes, 68,390 pounds
of rice, 783,507 of cotton. It had seven stores, 20 grist-mills,
12 saw-mills, two tanneries; one school, 90 scholars. Pop.:
whites, 2121; slaves, 1709; total, 3830. Capital, Columbia.

MARION, county, Tenn. situated towards the S.E. part of
the state, and contains 600 sq. m. Drained by Sequatchie
river and its branches. Bounded S.E. by Tennessee river.
It contained in 1840, 13,051 neat cattle, 8673 sheep, 46,800
swine; and produced 77,763 bushels of wheat, 3580 of rye,
717,617 of Indian corn, 162,014 of oats, 13,073 of potatoes,
14,016 pounds of tobacco, 36,294 of cotton, 1250 of sugar.
It had 31 stores, two flouring-mills, 60 grist-mills, 20 saw-
mills, 19 tanneries, 54 distilleries, one pottery, one printing-office, one bindery; one academy, seven students; 19 schools,
700 scholars. Pop.: whites, 5650; slaves, 390; free coloured, 22; total, 6070. Capital, Jasper.

MARION, county, Ky. situated near the centre of the
state, contains 276 sq. m. Watered by the Rolling Fork
of Salt river and its branches. It contained in 1840, 8963
neat cattle, 13,729 sheep, 31,419 swine; and produced
46,352 bushels of wheat, 16,081 of rye, 324,035 of Indian
corn, 108,167 of oats, 11,561 of potatoes, 185,996 pounds of
tobacco, 62,952 of sugar. It had 16 stores, three woollen-
factories, 17 grist-mills, 11 saw-mills, one oil-mill, six tanneries, 20 distilleries; one college, 247 students; three
academies, 222 students; 12 schools, 359 scholars. Pop.:
whites, 8340; slaves, 2612; free coloured, 80; total, 11,032.
Capital, Lebanon.

MARION, county, O., situated a little N. of the centre of
state, and contains 460 sq. m. Watered by Scioto, Little
Scioto, and Olentangy or Whetstone rivers. Organized in
1824. It contained in 1840, 16,694 neat cattle, 19,541 sheep,
22,637 swine; and produced 147,158 bushels of wheat, 18,476
of rye, 325,410 of Indian corn, 9488 of buckwheat, 158,160
of oats, 77,912 of potatoes, 30,050 pounds of sugar. It had
26 stores, two flouring-mills, 15 grist-mills, 23 saw-mills, one
oil-mill, 10 tanneries, five distilleries, one pottery, one printing-office, one weekly newspaper; 78 schools, 3155 scholars. Pop. 14,765. Capital, Marion.

MARION, county, In., situated in the centre of the state, and
contains 400 sq. m. Watered by by the W. Fork of White
river and its tributaries. It contained in 1840, 15,466 neat
cattle, 15,298 sheep, 38,463 swine; and produced 78,649
bushels of wheat, 2969 of rye, 604,966 of Indian corn, 1068
of buckwheat, 1190 of barley, 148,790 of oats, 39,438 of potatoes, 15,995 pounds of sugar. It had 94 stores, one fulling-
mill, two woollen-factories, three cotton-factories with 1500
spindles, two flouring-mills, 96 grist-mills, 57 saw-mills, four
oil-mills, 13 tanneries, 11 distilleries, three breweries, four
potteries, six printing-offices, four weekly, and two semi-
weekly newspapers; one academy, 75 students; 56 schools,
1595 scholars. Pop. 16,080. Capital, Indianapolis.

MARION, county, Ill., situated S. of the centre of the
state, and contains 576 sq. m. Drained by branches of Kaskaskia river, and by Skillet fork of Little Wabash river. It
contained in 1840, 7589 neat cattle, 4785 sheep, 17,846
swine; and produced 18,697 bushels of wheat, 263,321 of
Indian corn, 42,115 of oats, 5795 of potatoes, 6990 pounds of
tobacco. It had five stores, two woollen-factories, 16 grist-
mills, six saw-mills, two tanneries, one distillery; 17 schools,
417 scholars. Pop. 4742. Capital, Salem.

MARION, county, Mo., situated in the N.E part of the
state, and contains 425 sq. m. Bounded E. by Mississippi
river. Drained by S. Fabius and North rivers. It contained
16,573 neat cattle, 10,104 sheep, 39,857 swine; and produced

38,670 bushels of wheat, 533,785 of Indian corn, 2447 of
buckwheat, 84,835 of oats, 29,654 of potatoes, 38,938 pounds
of tobacco, 10,879 of sugar. It had 39 stores, four flouring-
mills, 11 grist-mills, 19 saw-mills, seven tanneries, four distilleries, three printing offices, three weekly newspapers;
one college, 49 students; 24 schools, 648 scholars. Pop.
whites, 7239; slaves, 2349; free coloured, 42; total, 9623.
Capital, Palmyra.

MARION, county, Ark., situated in the N. part of the
state, and contains 800 sq. m. Watered by White river,
which is navigable for steamboats to the N. part of the
county, and by its branches. It contained in 1840, 3217 neat
cattle, 769 sheep, 5430 swine; and produced 3097 bushels
of wheat, 79,013 of Indian corn, 1951 of oats, 1601 of potatoes, 6127 pounds of tobacco, 3853 of cotton. It had two
stores, four grist-mills, two saw-mills, one tannery, three
distilleries. Pop.: whites, 1291; slaves, 39; free coloured,
65; total, 1395. Capital, Yellville.

MARION, t., Greene co., Pa. It has 12 stores, two tanneries, two printing-offices, one weekly newspaper; two
schools, 61 scholars. Pop. 597.

MARION, p. t., Wayne co., N.Y., 201 m. W. by. N. Albany, 358 W. Drained by a branch of Mud creek. It contains a Presbyterian and a Baptist church, three stores, two
grist-mills, three saw-mills, one tannery; one academy, 32
students; 12 schools, 359 scholars. Pop. 1903.

MARION, v., capital of Marion dist., S.C., 146 m. E. Columbia, 423 W. Situated on the E. side of Catfish creek, a
branch of Great Pedee river. It contains a handsome brick
courthouse, a jail, an academy, 30 dwellings, and about 100
inhabitants.

MARION, p. t., capital of Marion co., O., 44 m. N. Columbus, 416 W. Watered by Little Scioto river and its branches. It contains 13 stores, one flouring-mill, three grist-mills,
five saw mills, two tanneries, one distillery; eight schools,
390 scholars. Pop. 1639. The village contains an elegant
brick courthouse, 68 by 45 feet, a jail, two churches, a Presbyterian and Methodist, 12 stores, a printing-office, issuing a
weekly newspaper, 100 dwelling, and 570 inhabitants.

MARION, t., Athens co., O. Watered by Federal and
Wolf creeks. It contains three stores; six schools, 128
scholars. Pop. 1019.

MARION, t., Hocking co., O. Pop. 940.

MARION, t., Putnam co., Ia. It has two stores, one grist-
mill, two saw-mills; five schools, 451 scholars. Pop.
1030.

MARION, p. v., capital of Smyth co., Va., 275 m. W. by
S. Richmond, 343 W. Situated on the middle fork of Holston river. It contains a courthouse, jail, and several stores
and dwellings. Nett proceeds of the postoffice, $193.

MARION, p. v., capital of Twiggs co., Ga., 41 m. S.W.
Milledgeville, 689 W. Situated 8 m. E. of Ocmulgee river.
It contains a courthouse, jail, six stores, 20 dwellings, and
about 125 inhabitants.

MARION, p. v., capital of Grant co., Ia., 73 m. N.N.E. Indianapolis, 562 W. Situated on Mississinewa river, about
25 m. from its entrance into Wabash river. It contains a
courthouse, jail, a store, and several dwellings. Nett proceeds of the postoffice, $258.

MARION, p. v., capital of Perry co., Ala., 58 m. S. Tuscaloosa, 834 W. It contains three churches, two female academies, one male lyceum with a preparatory school, three
weekly newspapers, and about 1000 inhabitants.

MARION, p. v., capital of Lauderdale co., Miss., 110 m. E.
Jackson, 921 W. It contains a courthouse, jail, and several
stores and dwellings. Nett proceeds of the postoffice, $809.

MARION, p. v., capital of Linn co., Iowa. Situated 4 m.
E. of Cedar river. It contains three stores, a saw-mill, and
several neat dwellings.

MARION CITY, p. v., Marion co., Mo., situated on the W.
bank of Mississippi river, and may be regarded as the port
of Palmyra, the capital of the county. It extends for a
mile and a half along the river, and has two large steam
saw-mills, three other steam-mills, 70 dwellings, and about
400 inhabitants. Most of the merchandise of the county is
landed here.

MARION COLLEGE, p. v., Marion co., Mo., 134 m.
N.N.W. Jefferson city, 938 W. Situated 12 m. W. of Palmyra, the capital of the county. Marion college was founded in
1831, and is a manual labour institution, for which 5000 acres
of the best land have been purchased, and improvements
have been made on them to the amount of $70,000. The
president, professors, and students, are supported from the
products of the soil. Including the preparatory department,
it has a president, six professors or other instructers, 116 students, 43 in the college proper, and 2300 volumes in its libraries. The commencement is on the last Thursday in
September. It has a theological department.

MARKET-BOSWORTH. See BOSWORTH (MARKET).

MARKET-DRAYTON, a market town and par. of England, partly in N. Bradford hund., co. Salop, and partly in
N. Pirehill hund., co. Stafford, on the Tern, a tribute of the

311

Severn, 18 m. N.E. Shrewsbury, and 135 m. N.W. London. Area of par., 13,000 acres. Pop., in 1831, 4619. The town, which stands on the W. side of the river, and in the county Salop, having been recently much improved, is now clean and well built, with tolerably wide streets. The church, originally erected in the reign of Stephen, was all but rebuilt in 1787. There are also places of worship for Roman Catholics, Wesleyan Methodists, and Independents, with attached Sunday-schools. The charitable institutions comprise a free school, founded in the reign of Queen Mary, a national school, and a set of almshouses and dispensary, with a few small money bequests. Drayton was formerly a place of more consequence than at present; its market having been among the largest in England, till the formation of the Liverpool and Birmingham Junction canal gave superior advantages to Stone, in Staffordshire. There are two paper-mills and two horse-hair manufactories close to the town; but most of the inhabitants are engaged either in retail trade or farming pursuits.

Drayton is a borough by prescription, governed by a mayor and two constables, chosen at a court-leet by the lord of the manor; and petty sessions are held here for the Drayton division of Salop. Markets on Wednesdays, chiefly for corn; fairs, for horses and farming-stock, Wednesday before Palm-Sunday, September 19, and October 24.

About one mile from Drayton, on Blore-heath, a battle was fought between the partisans of the houses of York and Lancaster, on the 23d of September 1459. Lord Audley, the Lancastrian general, was slain in the engagement; the spot where he fell being marked by a stone, close to Newcastle road.

MARKET HARBOROUGH. See HARBOROUGH (MARKET).

MARKET-JEW. See MARAZION.

MARKET-RASIN, a small market town and par. of England, Lindsey div., co. Lincoln. wap. Walshcroft, on the river of its own name, a tributary of the Ancholme, 13½ m. N.E. Lincoln, and 130 m. N. London. Area of par., 1290 acres. Pop., in 1831, 1428. This town deserves notice, chiefly on account of its large cattle and sheep fairs, which are attended by persons from almost all parts of the county. The church, an ancient structure, with an embattled tower, has peculiar windows, resembling those of the church at Louth. The living is a vicarage in the gift of the lord chancellor. The Roman Catholics and Wesleyan Methodists have also their respective places of worship; and its only charities are a free school (now incorporated with the national school), and a set of almshouses. Markets on Tuesdays; and fairs on alternate Tuesdays, between Palm Sunday and September 25. About 1½ m. W. Market Rasin is the village of Middle Rasin, remarkable for a small church presenting a most beautiful specimen of early Norman architecture.

MARKSVILLE, p. v., capital of Asoyelles par., La., 225 m. N.W. by W. New-Orleans, 1218 W. It contains a courthouse, jail, and several stores and dwellings. Nett proceeds of the postoffice, $306.

MARLBOROUGH, a parl. and mun. bor. and market town of England, co. Wilts. hund. Selkley, on the Kennet, 27 m. E. Bath, and 70 m. W. London. Area of parl. bor. (which includes, with the old bor., the par. of Preshute), 4380 acres. Pop., in 1831, 4186. The town consists of one broad main street, composed by others of inferior dimensions. The houses are irregularly built, and apparently of great antiquity, having high and curiously carved gables; a portion of the High-street also has a kind of colonnade projecting from the houses. The guildhall is supported on pillars, the lower part being open for the accommodation of the people frequenting the market; above are the council-chamber, sessions-hall, and assembly-rooms. There is also a handsome market-house, the upper part of which is used as a national school. The prison, which serves as a bridewell and house of correction, was built in 1787; but it is too small to admit either of separate confinement or proper classification, and there is no provision for hard labour. There is also a very large hotel, partly built with the materials of the old castle, which once stood at the S. end of High-street. The old church of St. Mary the Virgin, near the guildhall, is of early Norman architecture, with a low square tower: the living is a vicarage, in the gift of the Dean of Salisbury. The other church, which stands at the W. end of High-street, is of more modern construction, and distinguished by its light pinnacled tower: the living is a rectory, in the patronage of the Bishop of Salisbury. The Independents, Wesleyan and Calvinist Methodists have likewise their respective places of worship, with attached Sunday-schools. The national school furnishes gratuitous instruction to 100 boys and the same number of girls; besides which, there are two church Sunday-schools. It has also a free grammar school, founded by Edward VI., and endowed with estates producing about £70 a year: the instruction is almost exclusively classical; and the school

has the privilege of sending an exhibitioner, on the Somerset foundation, to Brasenose college, Oxford.

Marlborough has little trade, and derives its chief importance from being on the great road between London and Bath; but it will soon lose this advantage, as the Great Western railway runs through a line of country considerably N. of the town. Malting and rope-making are extensively pursued. Large quantities of corn and cheese are sent to London and Bristol; their carriage being greatly facilitated by the Kennet and Avon canal, which commences at Newbury, and joins the Avon near Bath, having an entire length of 57 m. "The town has likewise several excellent inns and shops, possesses a large market for the agricultural district, and may be considered in a prosperous state, and highly respectable." (Mun. Bound. Rep., p. 2, Marlbro'.)

The borough, which received its first charter from King John, in 1905, and a subsequent one, in 1577, from Elizabeth, is governed, under the Municipal Reform Act, by a mayor, three other aldermen, and 12 councillors; but it has no separate commission of the peace. Corporation revenue, in 1839, £388 (exc. of £383, accruing from the sale of property). Marlborough has sent two members to the House of Commons since 24 Edw. I.; the right of election, down to the Reform Act, being vested in the mayor and burgesses; but it was, in fact, a mere nomination borough, belonging to the Marquis of Aylesbury. Registered electors, in 1839-40, 269. Markets on Saturday; large fairs, July 10, Aug. 1, and Nov. 23. (Mun. Corp. Rep.; Parl. B. Rep., part vi.; Oxford Calendar; Priv. Inform.)

MARLBOROUGH, district, S.C., situated in the N.E. part of the state, and contains 480 sq. m. Bounded S.W. by Great Pedee river, and drained by its tributaries. It contained in 1840, 9374 neat cattle, 3893 sheep, 15,915 swine; and produced 9076 bushels of wheat, 1417 of rye, 275,399 of Indian corn, 22,161 of oats, 29,547 of potatoes, 2,446,088 pounds of cotton. It had 17 stores, one cotton factory with 2088 spindles, six flouring-mills, 21 grist-mills, 90 saw-mills; three academies, 104 students; 12 schools, 290 scholars. Pop.; whites, 4186; slaves, 4118; free coloured, 103; total, 8408. Capital, Bennetsville.

MARLBOROUGH, p. t., Cheshire co., N.H., 53 m. S.W. Concord, 430 W. Drained by branches of Ashuelot river. Incorporated in 1776. It has one store, one grist-mill, five saw-mills; six schools, 301 scholars. Pop. 831.

MARLBOROUGH, p. t., Windham co., Vt., 130 m. S. Montpelier, 431 W. Chartered in 1761. Drained by the W. branch of West river, by Green river, and by Whetstone brook. It contains three churches, a Congregational, Baptist, and Universalist; one store, two grist-mills, eight saw-mills, one paper-mill; 13 schools, 377 scholars. Pop. 1697.

MARLBOROUGH, p. t., Middlesex co., Mass., 27 m. W. Boston, 414 W. A branch of Concord river affords water-power. It contains four churches, a Congregational, Methodist, Restorationist, and Universalist; six stores, five grist-mills, five saw-mills, two tanneries; one academy, 75 students; 10 schools, 750 scholars. Pop. 2101.

MARLBOROUGH, p. t., Hartford co., Ct., 16 m. S.E. Hartford, 340 W. Chartered in 1803. It contains a Congregational church, two stores, one cotton-factory with 1986 spindles, one grist-mill, four saw-mills; five schools, 185 scholars. Pop. 713.

MARLBOROUGH, p. t., Ulster co., N.Y., 90 m. S. Kingston, 84 m. S.S.W. Albany, 294 W. Bounded E. by Hudson river. Drained by Old Man's kill. It contains three churches, a Presbyterian, Methodist, and Friends'; eight stores, two woollen-factories, one paper-mill; seven schools, 51 scholars. Pop. 2593.

MARLBOROUGH, t., Montgomery co., Pa. Drained by Perkiomer and Swamp creeks. It contains iron ore, and has five stores, one furnace, one forge, one woollen-factory, 34 grist-mills, three saw-mills, 10 powder-mills, one printing-office, one weekly newspaper. Pop. 1140.

MARLBOROUGH, t., Stark co., O., 135 m. N.E. Columbus, 319 W. Pop. 1671.

MARLBOROUGH, t., Delaware co., O. It has five schools, 150 scholars. Pop. 1182.

MARLOW (GREAT), a parl. bor. market-town, and par. of England, co. Bucks, hund. Desborough, on the N. bank of the Thames (here crossed by a handsome suspension-bridge), 29 m. W. London. Area of parl. bor., which includes the several pars. of Great Marlow, Little Marlow, Medmenham, and Bisham (the last being in Berks), 14,910 acres. Pop., in 1831, 6175. The town, formed by several streets, meeting in a large open market-place, is irregularly built; but is well paved and lighted, and contains many substantial houses, and a good town-hall. The parish church, opened in 1835, is a handsome structure, surmounted by a spire. The living is a vicarage, in the gift of the dean and chapter of Gloucester. There are, also, places of worship for Wesleyan Methodists and Baptists, with attached Sunday-schools. A charity school, for 24 boys and

19 girls, a national school, for children of both sexes, and a set of alms-houses, are the principal benevolent foundations. Great Marlow has little trade, except what results from its position, in the midst of a rich and productive country, inhabited by wealthy land-owners. On the Loddon, however, are several paper-mills; and, "on the whole the town is slowly recovering from a state of great depression, consequent to the removal, some years ago, of the military college." (*Parl. Bound. Rep.*) The borough has returned two members to the House of Commons, with some interruptions, since 28 Edw. I.; the right of election being vested, down to the Reform Act, in householders, paying scot and lot. The Boundary Act extended the limits of the parliamentary borough, by including with the old bor. three out-parishes. Registered electors, in 1838-40, 387. Market, well attended, on Saturdays; fairs, for cattle and farming produce. May 1-3 and Oct. 29. (*Parl. Pap., &c.*)

MARLOW, t., Cheshire co., N. H. Drained by Ashuelot river and its branches. It has four stores, one fulling-mill, one grist-mill, four saw-mills: one academy, 12 students; six schools, 189 scholars. Pop. 696.

MARMANDE, a town of France, dep. Lot-et-Garonne, cap. arrond., on the Garonne, here crossed by a bridge of one arch, 30 m. N.W. Agen. Population (1836), ex. com., 4580. It is regularly laid out, well built, and clean; has several good public edifices, and is nearly surrounded by an esplanade, planted with trees. It has a small port, suitable for steamboats, which ascend the Garonne as high as Marmande. It is the seat of courts of original jurisdiction and commerce; and has manufactures of woollen and linen fabrics, cordage and sailcloth, and several brandy distilleries. (*Hugo,* art. *Lot-et-Garonne, &c.*)

MARNE, a dep. of France, reg. N.E., formerly included in the prov. of Champagne, chiefly between lat. 48° 30' and 49° 20' N., and long. 3° 30' and 5° E., having N. Ardennes and Aisne, W. the latter department and Seine-et-Marne, S. Aube, and E. Haute-Marne and Meuse. Length, E. to W., about 70 m.; greatest breadth, nearly as much. Area, 817,037 hectares. Population (1836), 345,945. The hills in this department do not rise to more than 1300 feet above the sea; its general slope is from S.E. to N.W., in which direction nearly all its rivers flow. It derives its name from the Marne, which divides it into two nearly equal parts. This river rises in the department of Haute-Marne, about 3 m. S. Langres; it flows, at first N.W., and afterward generally W., through the departments of Haute-Marne, Marne, Aisne, Seine-et-Marne, Seine-et-Oise, and Seine; and falls into the Seine at Charenton, about 1 m. S.E. Paris, after a course of about 300 m., for 215 of which it is navigable. It has some considerable affluents; and Vitry, Châlons, Château-Thierry, and Meaux are on its banks. About two thirds of this department, including all its central portion, has an arid, barren soil, composed principally of chalk, covered with a thin layer of vegetable mould. But on the borders of this sterile tract are the vineyards which produce the celebrated champaign wine; and surrounding it is a country with a deep and rich alluvial soil. In 1834 the cultivated land was estimated at 614,895 hectares, pasture at 38,454 ditto, vineyards, 13,495 ditto, woods, 78,901 ditto, and heaths, wastes, &c., 16,981 ditto. Considerably more corn is grown than is required for home consumption. In average annual amount has been estimated at 3,000,000 hectol.; but, according to the official tables, 4,570,000 hectol. were harvested in 1835, chiefly wheat, oats, and rye. The culture of the vine is, however, by far the most important branch of industry. The department is supposed to furnish annually from 650,000 to 700,000 hectol. Of this quantity, however, the finest growths, produced in the arronds. of Epernay and Rheims, make but a small portion. The red wines " *se distinguent par beaucoup de finesse, de délicatesse, et d'agrément; ils occupent un rang distingué parmi les meilleurs vins fins du royaume.*" (*Jullien,* 21) But the white wines, which include the finest varieties of champaign, are by far the most celebrated. They are of three sorts, *still, mousseux,* and *grand mousseux.* The *vrais gourmets* prefer the first, or *still* wines, of which Sillery (which *see*) is the best; but the greater number of amateurs prefer the *mousseux,* being that variety of the sparkling wine which merely creams on the surface: the *grand mousseux,* or full frothing wines, are less esteemed. The wine of Ay, the best of the mousseux variety, is an exquisite liquor, worthy, according to the President De Thou, of being called *Vinum Dei!* The best of the red wines are those of Versy, Verzenay, Mailly, Bouzy, St. Basle, Clos-Thierry, &c. The vineyards round Epernay are valued at from 6000 to 10,000, and even 20,900 francs the arpent; and about 5400 *pieces* of wines of the finest growths are produced annually in its arrond. and that of Rheims, about a half of which is exported to foreign countries. Rheims, Epernay, and Avize are the chief seats of the wine trade. Epernay has extensive vaults, excavated in tufa, and admirably fitted for the preservation of wines. (*See* EPERNAY.)

Agriculture, according to Hugo, is in a tolerably advanced state. Near St. Menehould orchards are numerous. More cattle are reared than in any of the adjacent departments, the number, in 1830, being about 120,000. In the same year the department was estimated to have 506,000 sheep, the breeds of which have been much improved by crossing with merinos and English varieties. In 1835, of 179,318 properties subject to the *contribution foncière,* 98,523 were assessed at less than 5 francs, and 24,897 at from 5 to 10 francs; 134 were assessed at 1600 fr. and upward. *Il y a de l'aisance dans le pays, mais il n'y a pas de grandes fortunes. On n'y voit point le contraste affligeant de l'extrême opulence et de la misère; la mendicité n'y règne point.* (*Hugo.*) Marne has but one iron mine; but it furnishes excellent mill-stones, potter's clay, &c. Manufactures of various kinds of woollen fabrics, woollen yarn, &c., are established at Rheims; and hats, silk goods, paper, glass, earthenware, cordage, leather, candles, and soap are made in different places. Marne is divided into five arronds.; chief towns, Chalons-sur-Marne the cap., Epernay, Rheims, St. Menehould, and Vitry-le-Français. It sends six members to the chamber of deputies. Number of electors (1838-39), 2308. Total public revenue (1831), 13,239,636 francs; expenditure, 8,998,985 francs. (*Hugo,* art. *Marne; Official Tables; Jullien, Topographie de Vignobles; Henderson on Wines,* p. 153, &c.,

MARNE (HAUTE), a dep. of France, reg. N.E., between lat. 47° 35' and 48° 40' N., and long. 4° 40' and 6° E., having N. the deps. of Marne and Meuse, E. Vosges and Haute Saône, S. the latter and Cote d'Or, and W. Cote d'Or and Aube. Length, N.N.W. to S.S.E., 80 th.; average breadth, about 30 m. Area, 625,043 hectares. Population (1836), 249,827. The *plateau* of Langres and the Faucilles mountains traverse the S. and E. parts of this department, covering the greater part of its surface with their ramifications. They, however, nowhere rise to any great elevation; Montaigu, the highest point in Haute Marne, being only 1630 feet above the sea. The chief rivers are the Marne, which intersects the department lengthwise; its affluents the Ornain, Blaise, Meuse, and Aube, rise in this department, and have, more or less, a N. course. Surface mostly stony or calcareous: there not being more than 11,000 hectares of rich soil. In 1834, however, the arable land was supposed to comprise 335,611 hectares; pasture land, 35,528 do.; vineyards, 13,136 do.; woods, 174,275 do.: and heaths, wastes, &c., 27,969 do. The farmers devote their attention to the growing of corn, the culture of the vine, and the rearing of live stock. The produce of corn exceeds the demand for home consumption: the annual supply is estimated at nearly 1,800,000 hectolitres, chiefly wheat and oats. The produce of wine amounts to between 400,000 and 500,000 hectols. a year; but the quality is very inferior to that of the wines of Marne. Cherries and walnuts are grown in considerable quantities. The pasture lands are excellent; and, in 1830, there were 84,000 head of cattle, and 221,000 sheep in the department: the annual produce of wool is estimated at 160,000 kilogrs. In some cantons of the arrond. Vassy a good many turkeys are reared. Bees are numerous, and wax and honey are valuable products. This is one of the best wooded departments in France, and St. Dizier has a considerable trade in timber, sent in large quantities to Paris by the Marne. Iron is the only metal found in the department, but the working of the iron mines, and the manufacture of their produce, hold a high rank among the occupations of the people. The department has upward of 50 smelting furnaces (*hauts-fourneaux*), and 100 ordinary forges. Iron plates, rasps, files, and hardware of all kinds are manufactured; and the cutlery of Langres has long enjoyed a high reputation. Chaumont has manufactures of gloves and haberdashery. Linen and cotton thread, wax, candles, leather, brandy, and vinegar are the other chief articles made in the department. In 1833, of 194,714 properties, subject to the *contribution foncière,* 79,094 were assessed at less than 5 francs, and 16,700 at from 5 to 10 francs. Haute Marne is divided into three arronds.; chief towns, Chaumont the cap., Langres, and Vassy. It sends four members to the chamber of deputies. Number of electors (1838-9), 1064. Total public revenue (1831), 6,788,503 francs. (*Hugo,* art. *Haute Marne; Official Tables, &c.*)

MAROS-VASARHELY, or SZEKELY-VARSARHELY (Germ. *Neumarkt,* Wallach. *Oschoref*), a royal free town of Transylvania, the cap. of the Szekler-land, and of the *Stuhle,* or presidency, of Maros; on the Maros, 33 m. N.N.E. Hermanstadt. Pop. 7000. "Although there is nothing very imposing in the wide streets and small houses, of which Maros-Vásárhely is mostly composed, it is rather an important place; and, in winter, many of the gentry in the neighbourhood take up their residence within it. Moreover, both Protestants and Catholics have colleges here: the Protestant has 600, and the Catholic 300, scholars; and these institutions give something of a literary air to its society. Maros-Vásárhely is also the seat of the highest legal tribunal in Transylvania, the Royal Table; and it is, in consequence, the great law-school of the country. Almost all the young nobles who desire to take any part in public business, as well as all the lawyers, after having finished their regular course of study, think it necessary, under the name of *Juraten*, to pass a year or two here in reading law, and attending the court.

"The great pride of the town is the fine library of the Teleki, founded by the Chancellor Teleki, and left to his family, on the condition of its being always open to the public. It contains about 80,000 volumes, which are placed in a very handsome building, and kept in excellent order. It is most rich in choice editions of the Latin and Greek classics." (*Paget's Hungary and Transylv.*, ii., p. 393, 394.) The town has a Roman Catholic gymnasium and seminary, a reformed college, with a library and printing office, two convents, a flourishing *casino*, or literary club, and considerable trade in agricultural produce, particularly tobacco, which is grown in large quantities in this vicinity. (*Paget; Berghaus; Oesterr.; Nat. Encyc.*)

MARQUETTE, county, Wis. Situated in the N. part of the settled portion of the territory. Bounded N.W. by Fox river, which runs through Puckaway lake. It is very little settled, and had, in 1840, 45 neat cattle, 20 swine, and only 18 inhabitants. Capital, Marquette, on the S. side of Nunah river.

MARSALA (an. *Lilybæum*), a city and seaport of Sicily, at its W. extremity, adjacent to Cape Boeo (the *Promontorium Lilybæum*), in the intend. of Trapani, 16 m. S.S.W. Trapani; lat. 37° 48' 10" N.; long. 22° 25' 10" E. Pop., in 1831, 23,388. It is of a square form, and is surrounded by an old wall, flanked at the angles with bastions, but destitute of a glacis. It might be easily rendered a strong military post; but at present, it is without ordnance, quarters, or bomb-proof stores. The town, which is pretty well built, is bisected by a broad and regular street, called the Cassaro, on one side of which is the cathedral, a large edifice, ornamented with 16 fine marble columns of the Corinthian order. It has 16 churches, numerous convents, a *retiro*, or place of retirement under monastic regulations, three abbeys, a gymnasium, a seminary, a hospital, with 70 beds, a *monte-di-pieta*, barracks for cavalry, an old castle, &c. Among its curiosities is a bell towes, which vibrates perceptibly when the bell is rung.

Lilybæum was famous for its port; but, though secure, and well adapted for the use of the gallies of the ancients, it would not have accommodated the larger ships of modern times. Captain Smyth says, that where deepest, the ancient port could not have had more than 14 ft. water. The Romans, in their struggles with the Carthaginians, attempted over and over again to fill up the port, but uniformly without success. This, however, was effected, 1570, by Don John of Austria, who, to prevent the Barbary corsairs from taking refuge here, filled up the port with rubbish. The modern is not, therefore, identical with the ancient harbour, but is about 1 m. S. from the town. It has a mole, constructed by Mr. Woodhouse, for the convenience of the shipment of his wine: large ships anchor S.W. from the city, about 2 m. off shore, in from 8 to 11 fathoms water. The entrance to the port is a good deal encumbered with rocks and reefs; the knowledge of which is as indispensable to the modern as it was to the ancient mariners. (*Smyth's Sicily*, p. 233.)

Marsala is indebted for its importance in modern times to its wine trade; which has grown up, within the last half century, through the skill and intelligence of the Messrs. Woodhouse, Englishmen, who began business here in 1780. The wine, however, did not begin to come into much repute till 1802, when it was supplied, by order of Lord Nelson, to the Mediterranean fleet. It is a dry wine, the best qualities closely resembling the lighter sorts of Madeira; but the extensive demand for it in this country is, no doubt, ascribable more to its cheapness than its quality. It is, however, in all respects, superior to Cape Madeira, with which it principally comes into competition. The success of the Messrs. Woodhouse has led others to embark in the business; and Mr. Ingham and Mr. George Wood have also very large establishments at Marsala. The entire produce of the district is estimated at about 30,000 pipes, of which from 18,000 to 20,000 are exported, partly to the U. States and the W. Indies, as well as to England. In 1823 the entries of Marsala for home consumption in the United Kingdom amounted to 79,686 gallons; whereas, in 1839, they amounted to 369,417 gallons; but, for some years past, the consumption has been pretty stationary.

Besides wine, Marsala exports corn, cattle, oil, salt and soda; but in no great quantities. In 1839, 37 British and nine American ships, of the aggregate burden of 6505 tons, cleared out from Marsala. (*Macgregor's Report on the Commercial Statistics of the Two Sicilies*, passim.; *Commercial Dist.; Official Accounts, &c.*)

Lilybæum, from its proximity to Carthage, and the excellence of its port, was, for a lengthened period, the capital of the Carthaginian possessions in Sicily. It was a place of

great strength; being fortified by strong walls and a deep ditch, into which the sea appears to have flowed (*Polybius*, lib. i., cap. 42); indeed a portion of the ancient ditches still exist in tolerable preservation. (*Heeri's Classical Tour*, ii., 73.) The size of the city may be inferred from the fact of its requiring a garrison of 10,000 men, exclusive of the citizens, for its defence. The successful resistance it opposed to Pyrrhus, by whom it was attacked with great fury, and its defence against the Romans, sufficiently evince its strength and importance. After having ineffectually attempted to carry it by assault, the Romans converted the siege into a blockade; and the city only surrendered at the end of *five* years, when the defeat of Hanno made farther resistance unavailing. (*Ancient Universal History*, xvii. 531, &c., 8vo. ed.) Under the Romans it was the residence of a quæstor; and is called by Cicero, *civitas splendidissima*. (In Verrem, v., cap. 5.) Very few remains now exist of its ancient grandeur; vases, coins, &c., are, however, occasionally dug up; and in the town-hall is a group of two lions destroying a bull, said to be worthy the best period of Grecian art.

MARSEILLES (Fr. *Marseille*, an. *Massilia*), a large commercial city and seaport of France, dep. Bouches-du-Rhone, of which it is the cap., on the E. side of a bay of the gulf of Lyons, 30 m. W.N.W. Toulon, about 176 m. S.S.E. Lyons, and 420 m. S.E. Paris; lat. 43° 17' 49" N., long. 5° 22' 13" E. Pop. of the city, ex. suburbs, in 1836, 120,455; and inc. suburbs, 146,239: but this official statement is said to be under the mark, and it is alleged that, including strangers, the pop. is now (1840) at least 170,000. (*Conducteur dans Marseille.*)

"The situation of Marseilles," says Mr. Maclaren, "is one of the most beautiful I ever saw. It occupies the centre of a basin about 6 m. or 7 m. broad, bounded by lofty precipitous hills. The whole space from the city, back to the hills, is adorned with villas and hamlets; for every merchant or respectable shopkeeper here has his *maison de campagne*. These buildings are showy, sometimes large and splendid. They are called *Bastides*, and I learn that their number is not less than 5,000." (*Notes on France and Italy*, p. 38.) The country around is, however, extremely arid; and the wind called the *mistral* is blighting and noxious in the extreme. The city is somewhat of a horse-shoe shape, and built round its port. It is divided into two parts:—The first, or old town, occupying the site of the ancient Greek city, on rising ground, on the N. side the harbour, is confined, ill-built, with narrow dark streets, or rather lanes, not half ventilated, and inconceivably filthy. The second, or new town, constructed in the modern style, with regular streets, and handsome squares and houses, stands on the S. and E. sides of the port; being separated from the old town by a magnificent street, which extends in a right line from the Porte d'Aix to the Porte de Rome, traversing the city in its entire length N. to S. The middle part of this street, called the *Cours*, is sheltered by trees; the houses on either side are good; it has some handsome fountains, and is one of the chief places of public resort: but the favourite public promenade is the *Rue Cannebiere*, a fine broad street, running at a right angle from the foregoing to the inner extremity of the harbour, and completing the line of demarcation between the old and new town. Marseilles has been fortified at different periods; but its walls were finally destroyed in 1800; and their place is occupied by boulevards planted with trees, beyond which the city is rapidly extending, particularly towards the E. and S. It still, no doubt, is defended by the fort of *Notre Dame de la Garde*, on a steep eminence to the S.; but it is more remarkable for the beauty of its situation than for its strength: the harbour (see *post*) is protected by a fort on either side its entrance, by the Château d'If, on the island of the same name, and by some additional works on the islands of Ratoneau, Pomègue, &c., nearly opposite its mouth.

Marseilles has numerous public edifices, but none merits any detailed notice. The cathedral occupies the site of an ancient temple of Diana; it is extensive but heavy-looking. Its interior is a mixture of various orders; and its ornaments, which are mostly of the 11th and 12th centuries, are in bad taste. In fact none of the churches within the city have any considerable claims to notice. The church of St. Madeline (formerly *des Chartreux*), in the suburbs, an edifice constructed in the 17th century, is far superior to any one else; it has a handsome façade, and 2 steeples (*campaniles*), remarkable for their light appearance. There are in all about 20 Rom. Cath. churches, several chapels, two Greek churches, a Protestant church, and a synagogue. The Prefecture is the finest of the public buildings. The town-hall, on the N. quay, is a heavy edifice, composed of two separate piles of building, connected by a light and elegant arch on the first story. Its ground floor is appropriated to the exchange. There are numerous hospitals, and other charitable institutions. The Hôtel Dieu, one of the first established hospitals in France, was founded in 1188; it has usually from 500 to 600, and is capable of accommodating 750 patients. The

Hôpital de la Charité, founded in 1640, an asylum for aged persons, and for orphans, foundlings, &c., has usually from 800 to 850 inmates. The Lazaretto, one of the largest and most perfect establishments of the kind in Europe, is situated to the N. of the city, and is surrounded by a triple wall. Ships may clear from it while in quarantine. Marseilles has also a lying-in hospital, a *bureau de bienfaisance*, asylums for poor children, a *mont-de-piété*, and a savings' bank. One of the largest public edifices, formerly a Bernardine convent, accommodates the Royal college which has between 300 and 400 students; the Royal society of science, literature and art; the public library of 50,000 printed vols., and 1300 MSS., with cabinets of natural history, medals, and antiquities, &c., and a gallery of paintings, comprising works by Carracci, Salvator Rosa, Rubens, Vandyk, Jordaens, and other artists of the Italian and Flemish schools. The observatory, on the highest point of the old town, has apartments appropriated to schools of navigation, geometry, &c. The Grand theatre, built after the plan of the Odeon in Paris, is spacious and handsome. It has six tiers of boxes; but is in general ill-attended. The *Theatre Français*, a small building, open on Sundays for vaudevilles, and on other days for occasional concerts, is more frequented. The other chief public buildings and establishments are the hall of justice, the new prison, the custom-house, arsenal, barracks, mint, bishop's palace, various public halls, the fish market, &c. Marseilles has a botanic garden, and some excellent public baths. It is well supplied with water from fountains and public wells, but it is not introduced into the houses. At the extremity of the Rue d'Aix is an unfinished triumphal arch, of the Corinthian order, originally erected in honour of the Duc d'Angoulême, after his invasion of Spain in 1823; but it is now intended to commemorate the revolution of 1830, one of the effects of which was to expel the Duc d'Angoulême from the kingdom! Marseilles has but few remains of antiquity. Except a fountain, with an inscription in Greek, an obelisk, and the remains of an aqueduct, none is worthy of mention. It is doubtful, indeed, whether Marseilles possessed any grand or remarkable edifices in antiquity: and if she did, the corroding influence of the sea air, which proves so detrimental to the modern buildings, has been a powerful agent in their destruction. But the Marseillais, for a lengthened period, took little interest in the preservation of the relics of past ages; and, says Hugo, " lorsque l'esprit de conversation entra dans les mœurs, il n'y avait plus rien à conserver."

Marseilles is the see of a bishop, suffragan under Aix; the seat of tribunals of primary jurisdiction and commerce; a chamber of commerce; the residence of a commissary-general and a treasurer of marine; and the head-quarters of the 8th military division of the kingdom. Consuls from all the principal states of Europe and America are resident in it. Besides the public institutions before noticed, it has a diocesan seminary, a royal society of medicine, societies of agriculture and *belles lettres*, a statistical society, an atheneum, and several commercial and other clubs. Three newspapers (two of them daily), and several literary-journals are published in the city; the principal of the former are the *Semaphore de Marseille* and the *Gazette du Midi*.

The port, to which Marseilles is wholly indebted for her early and long continued prosperity, is a fine basin, stretching from W. to E. about 1,000 yards, into the very centre of the city. It has from 15 to 18 ft. water at its entrance, and from 19 to 24 ft. within; so that it is extremely well fitted for moderate-sized merchantmen, of which it will accommodate from 1000 to 1200. The ships come close to the quays, by which it is surrounded on all sides, except at its entrance, which is defended on its N. side by the tower of St. John, a work of the 15th century, and on its S. side by fort St. Nicholas, constructed by Louis XIV. The careening basin, on the right side of the harbour, occupies the site of the ancient necropolis.

Unluckily, this fine basin becomes, from its position, the common sewer, as it were, or receptacle for all the filth of the city; and, as it is not agitated by tides, which are here hardly perceptible, nor by storms, from which it is screened on all sides, nor swept by any current, the water is completely stagnant; and, unless the mud were removed by dredging-machines, it would in no very long time be entirely filled up. But in hot weather the stench arising from this torpid reservoir, and from the detestably filthy streets of the old town, is absolutely intolerable, at least to those not habituated to it; and has doubtless been the cause of Marseilles having suffered so dreadfully on various occasions from the plague, and more recently, from the cholera. Such a state of things is a disgrace to a civilized country. And we agree with Mr. Maclaren in thinking, that, next to the enforcing of proper sanatory regulations in the old town, the best thing that could be done to improve the city would be to cut a canal from the *Anse de Joliette*, on the coast, to the harbour, which would at once create a current, and freshen and agitate the water in the latter. In its present state, Marseilles has been

truly described as a " vast cloaca." We doubt, indeed, whether there be a single water-closet in the city.

There is excellent anchorage ground for men of war and other large ships, about 2 m. W.S.W., between the isles of Ratoneau and Pomegues; which have been connected by a mound. Ships from the Levant perform quarantine at Pomegues; and on Ratoneau island is a hospital for those whose health is dubious. A lighthouse, with a revolving light, 131 ft. in height, is erected on the Isle de Planier, about 10 m from the city, and there is another in Fort St. Jean. Ships having got within a quarter or half a mile of the isle d'If, usually heave to for a pilot. The charge for pilotage is four sous per ton in, and two sous per do. out, for French vessels and vessels belonging to powers having reciprocity treaties with France. With the exception of the above pilotage charges, and the charges on vessels performing quarantine, there are no port charges on ships entering or clearing out from Marseilles.

The trade of Marseilles is very extensive, and is rapidly increasing. She is the grand emporium of the S. of France, and the centre of nine tenths of her commerce with the countries bordering on the Mediterranean. The exports consist principally of silk stuffs, wines, brandies, and liqueurs; woollens and linens; madder, oil, soap, refined sugar, perfumery, stationery, verdigris, gloves, and all sorts of colonial products. Among the principal imports are sugar, coffee, and other colonial products; dye stuffs; corn, from the Black sea and the N. coast of Africa; cotton, from Egypt and America; coal, linen thread, and various descriptions of manufactured goods, from England; hides, wool, tallow, timber, &c. Marseilles engrosses almost the whole trade between France and Algiers. She is now also the principal station for the intercourse carried on by steamers with Malta, Alexandria, and Constantinople; and, besides the steamers employed by the government as packets, she had, in 1839, 12 steam-packets belonging to private companies. Mr. Maclaren says that most of the private steamers have English-made engines, and English engineers; and that they burn English coal, sold here for about 30s. a ton.

The following details set the importance of the trade of Marseilles in a still more striking point of view:—

	Ships.	Tons.
In 1837 there entered the port—		
French ships from foreign ports	1,188	145,190
" " from French colonies	121	30,925
" coasters from the Atlantic to the Mediterranean	404	87,967
" coasters in the Mediterranean trade	3,455	174,708
" cod and other fisheries	64	11,518
Total of French ships	5,232	421,208
Foreign ships from foreign ports	1,802	231,386
Total entries of ships of all sorts	6,934	652,507

We subjoin an

ACCOUNT of the Ships arrived at Marseilles from the Levant during the Four Years ending with 1837; specifying the Ships, and their Tonnage, from each Country.

	1834		1835		1836		1837	
	Ships	Tons	Ships	Tons	Ships	Tons	Ships	Tons
Turkey	119	13,276	113	13,950	111	12,834	134	13,639
Egypt	39	6,589	74	13,278	70	12,972	69	11,400
Greece	18	3,175	17	3,214	9	741	17	2,953
Tunis	37	5,373	47	8,902	26	2,436	30	3,646
Malta	8	709	15	2,405	7	819	13	1,715
Ionian Isles	9	1,197	2	250	3	237	6	—
Total	230	30,160	268	42,069	230	42,079	322	67,234

In this account, the ships from Syria and Candia are included under Egypt. The total value of the imports into Marseilles may be estimated at about 200,000,000 francs, or £8,000,000; and that of the exports at about as much.

The customs' revenue of Marseilles is greater than that of Havre, or any other French port: it amounted in 1836 to 29,806,346, and in 1837 to 30,234,013 francs. A joint stock bank, established here in 1835, is said to have been eminently successful.

Though principally distinguished by its commerce, Marseilles has several important manufacturing establishments. Its soap-works, which are numerous and extensive, employ about 700 workpeople, and consume large quantities of olive oil; but, though soap be exported, by far the greater portion of that produced here is destined for home consumption. The artists of Marseilles prepare and fashion coral into a great variety of articles. Among its other manufactures are woollen stockings and caps, *façon de Tunis*; hats, of which from 30,000 to 50,000 fine, and from 10,000 to 15,000 coarse, are annually exported; morocco and other leather (but the tanneries have fallen off), and sail-cloth. Marseilles has likewise refineries for sugar, sulphur, wax, and borax, with breweries, oil-works, glass-works, brick and tile works, &c.

and furnishes large quantities of vinegar and *liquers*. Another branch of industry is the salting and curing of meat, and the pickling and preparing of capers, olives, and other fruits, and of anchovies and other fish. It has, also, a great variety of trades connected with the building and fitting out of ships, steamers, &c.; and is, in fact, a very prosperous and rapidly increasing town.

There are but few great capitalists in Marseilles. "Here, as in Paris," says Mr. Maclaren, "it is the custom to retire altogether from business as soon as a trader has realized a competency. I was told that there are not a hundred men in Marseilles worth £20,000 each; but there are a great many worth half that sum. The people generally seem stout and well-fed. I went into the Place Royal when it was filled with 400 or 500 of the middle classes, meeting for business. I thought them the tallest, stoutest men I had ever seen. The sailors, porters, and carters, are more tanned than at Paris; but the shopkeepers are not sensibly darker than in the capital. Black eyes, however, are more common; a change invariably observed as we approach the equator. The houses and mode of living resemble those of Paris; but in the new streets, houses with front-doors like our own are common; while in Paris they adhere to the old plan of vast tenements, with a grand gate and open court in the centre."

Marseilles is very ancient, having, according to the best authorities, been founded by a colony from Phocea, a city of Ionia, about 600 years B.C. The Massilians, as the inhabitants were then called, speedily distinguished themselves by their skill as seamen, and the extent of their commerce; and were celebrated for the wisdom of their political institutions, and their civilization. They became, at an early period, allies of Rome; but having espoused the party of Pompey, their city was besieged, and, after an obstinate resistance, taken by Cæsar. But though Marseilles lost her liberty, she preserved her commerce and superior civilization under the Romans; and was highly distinguished as a school of *belles lettres* and philosophy. She is spoken of by Cicero in the highest terms of eulogy. (*Oratio pro L. Flacco*, cap. 26.) At a later period, Agricola was sent thither to be educated; and Tacitus calls her *sedes ac magistra studiorum*. (*Vit. Agricola*, cap. 4.) After the fall of the Roman empire, she underwent many vicissitudes. In the 10th century she was taken and sacked by the Saracens. She was afterwards governed by dukes and counts, and sometimes by her own magistrates, and more recently by the counts of Provence. She was finally united to the crown of France in 1482. During the middle ages she rivalled Venice and Genoa in her trade with the Levant. In 1720 she suffered dreadfully from the plague, which is said to have destroyed from 40,000 to 50,000 of the inhabitants! She also suffered considerably from the revolutionary phrenzy and the anti-commercial policy of Napoleon; but, as already seen, she has risen superior to all these disasters, and is now more populous and flourishing than ever.

Marseilles has given birth to many very distinguished individuals, among whom may be specified Pytheas, one of the most illustrious navigators and astronomers of antiquity who flourished in the 4th century B.C., and Petronius Arbiter, *Auctor purissimæ impuritatis*. Among its modern citizens have been Dumarsais, the grammarian, Mascaron the celebrated preacher, Peyssonnel, the author of a treatise on the commerce of the Black sea and of several other works on the Levant, and Puget, celebrated as a sculptor, painter, and architect.

The famous revolutionary song and air, called the Marseillaise, did not originate in Marseilles, as might be inferred from the name; this was derived from the tune having been played by a body of troops from Marseilles, on their entry into Paris, in 1791. (Besides the authorities already referred to, we have consulted Hugo, art. *Bouches de Rhone*; *Juliany, Essai sur le Commerce de Marseille*; *Dictionnaire du Commerce*; and *Private Information*.)

MARSEILLES, p. v., La Salle co., Ill., 141 m. N.N.E. Springfield, 776 W. Situated at the N. side of Illinois river, at the Grand Rapids, which afford a great water-power. It has several mills, and exports lumber and flour. The Wabash and Erie canal passes through it.

MARSHALL, county, Va. Situated in the N. part of the state, and contains 350 sq. m. Bounded W. by Ohio river. Watered by Grave, Fish, and Wheeling creeks. It contained, in 1840, 4640 neat cattle, 7172 sheep, 9091 swine; and produced 88,890 bushels of wheat, 2086 of rye, 145,892 of Indian corn, 2520 of buckwheat, 163,195 of oats, 29,672 of potatoes, 3896 pounds of sugar. It had six stores, 16 gristmills, six saw-mills, two tanneries; three schools, 70 scholars. Pop.: whites, 6854; slaves, 46; free coloured, 37; total, 6937. Capital, Elizabethtown.

MARSHALL, county, Tenn. Situated S. of the centre of the state, and contains 200 sq. m. Watered by Duck river and its branches. It contained, in 1840, 14,983 neat cattle, 16,594 sheep, 67,729 swine; and produced 69,922 bushels

of wheat, 3405 of rye, 1,019,863 of Indian corn, 121,201 of oats, 16,071 of potatoes, 119,165 pounds of tobacco, 469,631 of cotton. It had 17 stores, nine tanneries, 26 distilleries; 21 schools, 743 scholars. Pop.: whites, 11,466; slaves, 3075; free coloured, 12; total, 14,555. Capital, Lewisburg.

MARSHALL, county, Ala. Situated toward the N.E. part of the state, and contains 600 sq. m. Watered by Tennessee river and its tributaries. It contained, in 1840, 9417 neat cattle, 9408 sheep, 26,399 swine; and produced 8896 bushels of wheat, 362,680 of Indian corn, 23,334 of oats, 14,409 of potatoes, 17,918 pounds of tobacco, 4,385,987 of cotton. It had 13 stores, 73 grist-mills, 16 saw-mills, five tanneries, 15 distilleries; 16 schools, 436 scholars. Pop.: whites, 6688; slaves, 841; free coloured, 24; total, 7553. Capital, Warrenton.

MARSHALL, county, Miss. Situated in the N. part of the state, and contains 800 sq. m. Watered by Tallahatchee and Coldwater rivers and their branches. It contained, in 1840, 14,621 neat cattle, 6044 sheep, 51,276 swine; and produced 19,113 bushels of wheat, 733,514 of Indian corn, 39,540 of oats, 50,550 of potatoes, 9306 pounds of tobacco, 1,064,425 of cotton. It had 26 stores, two flouring-mills, 19 grist-mills, 18 saw-mills, four tanneries, two printing-offices, two weekly newspapers; nine academies, 262 students; 26 schools, 583 scholars. Pop.: whites, 9258; slaves, 6889; free coloured, 8; total, 17,596. Capital, Holly Springs.

MARSHALL, county, Ill. Situated a little N. of the centre of the state, and contains 384 sq. m. Watered by Illinois river. It contained, in 1840, 2422 neat cattle, 1683 sheep, 6495 swine; and produced 31,869 bushels of wheat, 98,689 of Indian corn, 21,405 of oats, 11,990 of potatoes. It had eight stores, one flouring-mill, three grist-mills, eight saw-mills; one academy, 59 students; three schools, 160 scholars. Pop. 1849. Capital, Lacon.

MARSHALL, p.-t., Oneida co., N. Y., 12 m. S.W. Utica, 23 m. N.N.W. Albany, 370 W. Drained by Oriskany cr. The Chenango canal passes through it. It contains three stores, three fulling-mills, one woollen factory, one furnace, one forge, five grist mills, 10 saw-mills, three tanneries, one distillery; 13 schools, 749 scholars. Pop. 2251.

MARSHALL, p.-t., capital of Calhoun co., Mich., 106 m. W. Detroit, 570 W. Watered by Kalamazoo river. It contains 14 stores, two lumber-yards, one furnace, three flouring-mills, one grist-mill, three saw-mills, two printing-offices, two weekly newspapers, and one periodical; one academy, 68 students; eight schools, 238 scholars. Pop. 1703. The village is pleasantly situated on Kalamazoo river, at the confluence of Rice creek. It contains a courthouse, jail, four churches (the Presbyterian a handsome edifice of stone), a bank, 14 stores, several fine hotels, one of which cost from $20,000 to $25,000; was founded in 1831, and is one of the finest villages in the state. It is the seat of Marshall college, recently instituted, which has a president and one professor, 22 students, and 3700 volumes in its libraries. The Detroit and St. Joseph railroad is designed to pass through the v.

MARSHALL, p. v., capital of Clark co., Ill., 123 m. E.S.E. Springfield, 661 W. Pleasantly situated on the National road, and contains a courthouse, jail, and several stores and dwellings. Nett proceeds of the post-office, $652.

MARSHALL, p. v., capital of Saline co., Mo., 87 m. N.W. Jefferson city, 1601 W. It contains a courthouse, a church, four stores, and 150 inhabitants.

MARSHFIELD, p.-t., Washington co., Vt., 15 m. N.E. Montpelier, 500 W. Watered by Onion or Winooski river. It has one store, two grist-mills, one saw-mill; 12 schools, 394 scholars. Pop. 1136.

MARSHFIELD, p.-t., Plymouth co., Mass., 31 m. S.E. Boston, 453 W. Chartered in 1640. Bounded N.E. by Massachusetts bay, and it has a tolerable harbour, and some shipping. Bounded N. by North river, by which, and by South river, it is watered. It contains six churches, three Congregational, a Methodist, Episcopal, and Baptist, six stores, one furnace, six grist-mills, five saw-mills; eight schools, 450 scholars. Pop. 1761.

MARSHPEE, t., Barnstable co., Mass., 19 m. S.E. Barnstable, 65 m. S.E. Boston. This is an ancient Indian settlement, still inhabited chiefly by their descendants, of a mixed blood. It contains about 16 sq. m., or 10,500 acres. It has a church which has had a succession of preachers, some of them Indians, from the earliest times. The land is secured to the Indians in perpetuity, and is held in common, except so much as each one can cultivate. It is bounded S. by the Atlantic, and watered by several inlets and ponds. The inhabitants are agriculturalists or fishermen, and in the

324

latter employment are particularly expert. They are a quiet and orderly people. It has two schools, 60 scholars. Pop. 360.

MARS HILL, an isolated mountain in Me., near the eastern boundary of the United States. It has two peaks, one 1566 feet high, and the other 1363 feet above the waters of St. John's river. The British commissioners, before the late treaty, fixed on it as the commencement of the highlands described in the treaty of 1783. But this claim is now set aside, and Mars Hill will be regarded as of less importance than heretofore. It is an isolated mountain, rather than part of a chain of high lands.

MARTABAN, a town of the Birman empire, cap. of the prov. Martaban, on the Than-lweng (Saluen) river, near its mouth, 30 m. N.W. Maulmain, and 92 m. E.S.E. Rangoon; lat. 16° 28' N., long. 97° 39' E. Population uncertain: in 1826, it was estimated at 9600; but many of the inhabitants were then preparing to emigrate into the British territories, and Mr. Crawfurd (Embassy to Siam, ii., 338) estimates the ordinary population at only 1500. It stands on the E. declivity of a high hill, is more than a mile in length, consisting of two long streets, and is surrounded by a stockade, which separates it from some suburbs. The houses are of wood; it has several conspicuous temples, one of which is upwards of 130 ft. in height. Martaban has an imposing appearance from the water, facing which is a battery on a rocky mound, and a deep wall of masonry with embrasures for cannon, &c., behind the stockade. It was formerly a place of considerable trade; but, early in the course of last century, its navigation was injured by the sinking of vessels in the river by the Birmans in their wars with Pegu; and Maulmain (which see) is at present the emporium of all the adjacent provs. Martaban was taken by the British in 1824. (Crawfurd's Embassy; Hamilton's E. I. Gaz.; Mod. Trav., xi., 170–181.)

MARTHA, or MARTA (SANTA), a seaport town of Colombia, New Granada, dep. Magdalena, cap. prov. Santa Martha, on the Caribbean sea, 195 m. N.E. Cartagena, and 175 m. W.N.W. Maracaybo. Lat. 11° 15' N., long. 74° 18' W. Population estimated at 6000. It has some good houses, a cathedral, which is a conspicuous object in approaching it, both by land and sea, come convents, &c.; but it suffered much from the attacks of the Indians during the revolutionary war, and does not appear to have regained its previous importance. Its harbour, which is one of the best on this coast, having sufficient depth of water and good holding-ground, is defended by several batteries, and by a castle on an insulated rock, commanding both the town and the harbour. Santa Marta was founded in 1525, and made an episcopal city four years afterwards. Before the revolution it had risen to considerable importance as a commercial city, and was the port into which manufactured goods for Bogota were almost exclusively imported. (Cochrane, in Modern Trav., xvii., 297, 298; Blunt's American Coast Pilot.)

MARTHA'S VINEYARD, island, off the S. shore of Massachusetts, S. of Falmouth on the main land, and W.N.W. of Nantucket island. It is 21 m. long, and from 2 to 5 m. broad, and is divided into three towns, Chilmark on the W., Tisbury in the centre, and Edgarton in the E. capital of the island. With several small islands in the vicinity, it constitutes Duke's county, which contains 190 sq. m. The capital is Edgartown. In the W. part of Chilmark is a lighthouse, on a promontory called Gay Head. In the wars with Great Britain the inhabitants have been chiefly without protection. An Indian church was formed here in 1698. The government of the island was at first independent of the other provinces, but in 1664 it was united to New-York, and in 1692 to Massachusetts.

MARTIE, t., Lancaster co., Pa., 10 m. S. Lancaster. Drained by Beaver, Muddy, and Pecquea creeks, which afford water-power. Bounded S.W. by Susquehanna river, across which is a ferry. It has five stores, six flouring-mills, six grist-mills, four saw mills, two tanneries, one distillery; nine schools, 280 scholars. Pop. 3453.

MARTIGUES (LES), a maritime town of France, dep. Bouches-du-Rhone, on an island in the channel between the lagoon of Berre and the Mediterranean, on either bank of which channel are its suburbs of Ferrieres and Jonquieres, 18 m. W.N.W. Marseilles. Population (1836), including its suburbs, 7309. Its situation, amid pools and canals, has made it to be called the Venice of Provence. It is well built, and has several good streets and quays, and handsome buildings; but it is ill supplied with water. Its port is much resorted to by fishing-boats. Merchant vessels are built here; and it has an active trade in olive oil, fish, wine, and salt. (Hugo, art. Bouches-du-Rhone, &c.)

MARTIN, county, N. C. Situated toward the E. part of the state, and contains 431 sq. m. Bounded N.E. by Roanoke river. It contained, in 1840, 4323 neat cattle, 4903 sheep, 17,704 swine; and produced 5156 bushels of wheat, 251,463 of Indian corn, 7475 of oats, 39,639 of potatoes, 37,457 pounds of rice, 291,606 of cotton. It had seven stores, 12 grist-mills,

nine saw-mills; six schools, 130 scholars. Pop.: whites, 4438; slaves, 2816; free coloured, 383; total, 7637. Capital, Williamston.

MARTIN, county, Ia. Situated in the S.W. part of the state, and contains 300 sq. m. Watered by the E. fork of White river and its branches. It contained, in 1840, 4159 neat cattle, 6361 sheep, 11,046 swine; and produced 9684 bushels of wheat, 87,009 of Indian corn, 19,164 of oats, 7433 of potatoes, 15,287 pounds of sugar. It had six stores, nine grist-mills, 10 saw-mills, three tanneries, three distilleries; three schools, 96 scholars. Pop. 3875. Capital, Mount Pleasant.

MARTIN (ST.), one of the Virgin islands, in the W. Indies, belonging partly to the French and partly to the Dutch; about lat. 18° 4' N., and long. 63° 5' W.; between Anguilla and St. Bartholomew, and 12 m. N.W. the latter, and 75 m. N.N.W. Barbuda. Area estimated at 30 sq. m. Though hilly, it has no eminence 9000 ft. in height. It is watered by numerous rivulets; and in the S. are numerous lagoons, from which great quantities of salt are obtained by the Dutch. The coasts, which are deeply indented, afford several good roadsteads, of which Philipsburg and Marigot are the best. The soil is light, strong, and frequently arid; but it is tolerably healthy. The northern, and larger portion of the island belongs to France, forming a commune of the colony of Guadeloupe; and having an area of 5371 hectares, of which 1841 are cultivated, 941 in pasture, 674 in woods, and 2616 unproductive. The annual produce of sugar averages about 900,000 kilogs., syrup about 11,000 kilogs., and rum about 50,000 gallons: a good many cattle are also reared. Population of the French division about 3600, five sixths of whom are slaves. The southern, or Dutch division of the island is less fertile and richly wooded than the French, has been more profitable, on account of the salt it produces, which is sent to the neighbouring islands, and to N. America: it is also estimated to yield annually about 25,000 cwt. of sugar, and 130,000 gallons of rum. (Stein's Handbook, iii., 666.) The Dutch portion is said to be about as populous as the French. Nearly all the white population of St. Martin are of English descent. The Spaniards first colonised this island, but abandoned it in 1650; after which it became a subject of contention between the French and Dutch, who subsequently divided it between them. It has been frequently taken by the English. (Hugo, iii., 209; Notices Statist. sur les Colonies Francaises; Stein, &c.)

MARTINIQUE, one of the Windward islands in the W. Indies, belonging to France; between lat. 14° 23' 43" and 14° 52' 47" N., and long. 60° 46' and 60° 15' W., about 25 m. S.E. Dominica, and 90 m. N. St. Lucia; length, N.W. to S.E., 36 m.; average breadth, about 10 m. Area estimated at 98,782 hectares. Population in 1836, 117,502 of whom 40,000 were white, or free-coloured, and 77,500 slaves. The surface gradually rises on proceeding inland, and mountain-ranges occupy the centre of the country. Their loftiest summits are the Montagne Pelée, towards the N. extremity of the island, and the Piton du Carbet; the former rises to 4429 ft., the latter to 3960 ft. above the sea. These, and other mountains, are evidently extinct volcanoes; having their characteristic conical form, and abounding with lava, and other volcanic products. The flanks of the mountains are mostly covered with a dense and luxuriant forest vegetation, and are in many parts under culture to an elevation of 1200 ft. About a third part of the island consists of pretty level land. It is watered by numerous rivulets; but of these only three or four, which disembogue on the W. coast, are navigable in any part of their extent. At the S. extremity of the island is a small salt-lake. The coasts present many bays and inlets, but the harbours on its E. side are difficult of access, being obstructed by numerous islets, and extensive banks of madrepore. On the S. side is the bay of Marin; and on the W. is that of Fort Royal, forming one of the best harbours in the Antilles; in the N.W. is the roadstead of St. Pierre, where ships ride safely, except during W. winds. The mean annual temp. in the plains is about 81°, the maximum in the shade being 95°, and the minimum 69° Fah.; but the heat is tempered by sea breezes during the day, and land-breezes at night. The moisture of the atmosphere is excessive; and it is estimated that at the level of the sea 85 inches of rain fall annually. Most of this rain descends from July to Oct., a period of the year termed the hivernage; when the hurricanes, from which the island has often suffered severely, are most frequent. The weather for the remaining nine months is generally fine; but Martinique, like the neighbouring island of St. Lucia, is very unhealthy. Mineral springs are abundant, of a chalybeate, saline, or siliceous nature, and useful in cutaneous and liver complaints. The surface consists chiefly of disintegrated pumice-stone, intermixed with vegetable mould, forming a light and very fertile soil. In 1835, it was estimated that 38,390 hectares were under culture; that savannahs and pasture lands occupied 21,773 do.; woods and forests, 23,387 do.; and unproductive lands, 15,203 do. The following official account has been given of the distribution of the cultivated lands, their produce, &c., in 1836 :—

Articles.	Hectares cultivated.	Slaves employed in Culture.	Produce.
Sugar-cane	23,777	34,940 {	Raw sugar . . kil. 53,960,500 Do. . . . — 195,750 Syrup & molas. litr. 6,259,150 Rum . . . — 1,669,920
Coffee	2,317	8,897	Do. . . . kil. 694,807
Cotton	243	—	Do. . . . — 18,703
Cocoa	464	—	Do. . . . — 125,810
Corn, &c.	12,706	8,677	Do. . . . value, fr. 2,986,031
Mulberry	4	—	
Various	.	3,607	
Total .	40,117	55,421	

Of late years, agriculture has made considerable progress in Martinique. The plough has come more into use, and manuring is more extensively practised; and the culture of the sugar-cane, to which the colonists have turned their chief attention since 1820, has been greatly improved. The cane is of two kinds; the Otaheitan variety, and the yellow cane of Batavia. It was first naturalised about 1650. The coffee plant was introduced in 1723; but its culture, like that of most other products, is diminishing in favour of sugar. A few cloves, and some other spices, are grown; and the government has attempted, though hitherto with little success, to introduce the culture of indigo. Martinique formerly produced a pretty large supply of tobacco, but it is now quite insignificant. Manioc, bananas, sweet potatoes, maize, &c., are the principal farinaceous vegetables. The island has about 46,000 head of live stock, of all sorts; about 18,000 being black cattle, and 9000 sheep. In 1836, there were three earthenware and tile factories, and 10 lime-kilns, employing 339 slaves; these are, however, the only manufacturing establishments in Martinique. Carpenters, masons, and such-like workmen are pretty numerous; but there are few of any other description. Most part of the skilled workmen belong to the free coloured class, and only work when they have expended their wages, and are in want of necessaries. A number of hands, both free and slaves, are employed in fishing: and between 400 and 500 are occupied in navigation and the coasting trade. Subjoined is an

ACCOUNT of the Quantities and Values of the principal Articles exported from Martinique in 1836.

Articles.	Quantities.	Value.
		Francs.
Raw Sugar	kilogr.	52,894,754	13,776,852
Molasses	litres	2,433,583	506,962
Raw Cocoa	kilogr.	133,727	120,354
Coffee	—	519,507	931,236
Dye-woods	—	1,299,018	253,879
Rum	litres	144,957	98,995
Cassia	kilogr.	53,006	79,564
Copper	—	40,547	81,594
Cola, gold and silver .			415,180
Other articles . . .			258,299
Total .			16,423,421

The imports consist chiefly of salted meat, butter, and fish, corn, flour, pulse, oils, timber; cotton, linen, and other manufactured goods; wines, soap, candles, hardware, jewellery, apparel, &c.; chiefly from France and the French colonies. In 1836, the value of the imports amounted to 19,480,598 fr. In the same year, 356 French vessels, of the aggregate burden of 48,861 tons, entered, and 353 left, the ports of the island: in addition to which, 495 foreign vessels entered, and 487 cleared out.

The government is vested in a governor, assisted by a privy council, composed of the military commandant, the three principal civil officers of the colony, and three privy councillors nominated by the king; and in a colonial council of 30 members elected for five years. Every individual of French descent, 25 years of age, born or having resided two years in the colony, and paying taxes to the extent of 300 fr. a year, or having property worth 30,000 fr., may be an elector; and inhabitants paying taxes, or possessing property of double the above amount, are eligible to the colonial council. In 1836, there were 819 electors, and 507 individuals eligible to the council. Martinique is divided into the arronds. of Fort Royal and St. Pierre, four cantons' and 96 communes. Justice is administered by a royal court at Fort Royal, courts of assize and primary jurisdiction in each arrond., a justice of the peace in each canton, and a functionary, uniting both the civil and military jurisdiction, in most of the communes. The military force amounts to 2090 men, besides which there is a militia of 4103 men. There are three schools of mutual instruction, two in the capital, and one in St. Pierre; and primary schools in almost every commune. At St. Pierre is a superior female seminary. There are orphan asylums, and various other charities, in the two principal towns; and two newspapers are published, both at St. Pierre. The total expenditure of

318

the colony, in 1837, was estimated at 4,387,266 fr., the total receipts to meet which amounted to only 2,965,711 fr.

Martinique has only three towns worthy of mention. Fort Royal, the capital, and seat of government, on the N. shore of the bay of same name, in the S.W. part of the island; Pop., inc. com., about 11,500. It is well built, its chief public edifices being the parish church, government offices, naval storehouses, arsenal, barracks, hospital, two prisons, and the residence of the préfet apostolique, the superior ecclesiastic of the island. It is defended on the N. by Fort Bourbon, and on the S. by Fort Louis, on a small peninsula, by which it is shut off from its port; but it communicates with the harbour, by a canal, cut within a few years. Near Fort Royal are numerous pleasant country residences. St. Pierre (which see), also on the W. coast, is the largest town in the French W. Indies. La Trinité, on the N. of same name, on the E. side of the island, has a population of about 6500, large warehouses, a prison, some barracks, a hospital, and a handsome church. Its roadstead and harbour are secure; the latter has good holding-ground, but it is difficult of access. Its entrance was formerly protected by a fort, now in ruins.

This island was discovered by the Spaniards in 1493. In 1635 it was settled by the French. In 1762 the English took it from the latter, but restored it in the succeeding year. In 1794 it was again taken by the English, who gave it back in 1802; it came a third time into our possession in 1809, and was finally restored to France in 1815. The Viscount Beauharnais, and his wife Josephine, subsequently espoused by Napoleon, were natives of Martinique. (Notices sur les Colonies Françaises, i., 27-134; Official Tables; Hugo, art. Martinique.)

MARTINSBURG, p. t., capital of Lewis co., N.Y., 134 m. N.W. Albany, 433 W. Drained by Martin's and Whetstone creeks, flowing into Black r., which bounds it on the N.E. It contains seven stores, one fulling-mill, two grist-mills, nine saw-mills, two tanneries, one printing-office, one weekly newspaper; 17 schools, 733 scholars. Pop. 3272. The village situated on elevated ground, contains a court-house and jail in the same building, three churches, a Presbyterian and two Methodist, the Lewis county bank, several stores, mills and manufactories, 75 dwellings and about 350 inhabitants.

MARTINSBURG, p. v., capital of Berkley co., Va., 21 m. N. Harper's Ferry, 160 m. N. by W. Richmond, 77 W. It contains a courthouse, jail, county offices, two academies, 'an alms-house, four churches, an Episcopal, Presbyterian, Lutheran and Roman Catholic, 13 stores, one furnace, one flouring-mill, two grist-mills, three tanneries, one printing-office, one weekly newspaper, 350 dwellings, many of them of brick and neat, and about 1700 inhabitants.

MARTINSBURG, p. v., Clay t., Knox co., O., 57 m. N.N.E. Columbus, 366 W. It contains two churches, a Methodist and Presbyterian, seven stores, two high schools, one for males and one for females, a steam saw-mill, 75 dwellings, and about 450 inhabitants.

MARTINSVILLE, p. v., capital of Henry co., Va., 194 m. S.W. Richmond, 269 W. Situated on Smith's r., a tributary of Dan r. It contains a courthouse, jail, two stores, a tannery, and about 100 inhabitants.

MARTINSVILLE, p. v., capital of Morgan co., Ia., 26 m. S.W. Indianapolis, 599 W. Situated half a mile E. of the E. branch of White r., on a beautiful plain. It contains a courthouse, jail, three stores, and about 125 inhabitants.

MARWAR, a native state of Hindostan. See JOUDPOOR.

MARYBOROUGH, an inland town of Ireland, Queen's co., of which it is the cap.; prov. Leinster, on a branch of the Barrow, 46 m. S.W. Dublin. Pop., in 1821, 2877; in 1831, 3930. "It is a town of very little importance or wealth, possessing a very inconsiderable market, compared to Mount Melick and other towns in the co.; but within the last few years the quantities of grain brought to market have increased." (Mun. Bound. Report.) It is straggling and meanly built. The new co. prison has 75 cells and 36 other prisoners' rooms. The district lunatic asylum has accommodation for 104 patients. It has three schools, one for boys, another for girls, and one for both, partly supported by and connected with the Educational Board, which, in 1839, were attended by 743 children. When the territory of Leix was made shire-ground, at the close of the reign of Philip and Mary, this place, previously a border fortress, was fixed on as the assize town, and named from the reigning queen. It changed masters several times during the war of 1641. It has several good public buildings, among which are the parish church, a Roman Catholic chapel, a convent, Presbyterian and Methodist meeting-house, the infirmary for the co., the district lunatic asylum lately erected for King's and Queen's cos., Westmeath, and Longford, the new co. courthouse and prison. with barracks, schools, &c. It is a constabulary station. The corporation, under a charter of Elizabeth, in 1570, consists of a burgomaster, two bailiffs, and an indefinite number of burgesses and freemen.

MARYLAND.

It returned two members to the Irish House of Commons till the union, when it was disfranchised. The assizes for the co. are held here: and general sessions in April and October, and petty sessions weekly. Markets on Thursdays; fairs, Jan. 1, Feb. 24, March 25, May 12, July 5, Sept. 4, Oct. 26, and Dec. 12. Postoffice revenue, in 1830, £264; in 1836, £437. Rathleague, the seat of Sir Henry Parnell, and Ballyfin, the seat of Sir Charles Coote, are in the immediate vicinity of the town.

MARYLAND, the most southern of the middle United States is bounded N. by Pennsylvania; E. by Delaware and the Atlantic; S. and W. by Virginia. It is separated from Virginia by Potomac river, which pursues a winding course, along its S.E. border. It is between 38° and 39° 44' N. lat. and between 75° 10' and 79° 20' W. long. and between 9° 31' W. and 1° 58' E. long. from W. Its shape is very irregular, and may be considered as 196 m. long and 120 m. broad, which is not its average breadth, containing 13,959 sq. m. or 8,933,760 acres, of which one fifth part is water. The Chesapeake bay runs nearly through the state from S. to N., dividing it into two parts. East of Chesapeake bay, it is called the *Eastern shore*, and W. of it, the *Western shore*.

The population in 1790, was 319,728; in 1800, 345,824; in 1810, 380,546; in 1820, 407,350; in 1830, 446,913; in 1840, 469,232, of whom 89,495 were slaves. Of the free population 152,636 were white males; 150,081 do. females; 29,173 were coloured males; 29,847 do. females. Employed in agriculture, 63,851; in commerce, 3949; in manufactures and trades, 21,325; in navigating the ocean, 731; do. canals, rivers, &c., 1519; in the learned professions, 1647.

The state is divided into 20 counties, which with their population in 1840, were as follows:

Western Shore.

Alleghany	15,690	Frederick	36,405
Anne Arundel	29,532	Harford	17,120
Baltimore	134,379	Montgomery	14,669
Calvert	9,229	Prince George	19,539
Carrol	17,241	St. Mary	13,224
Charles	16,023	Washington	28,850

Eastern Shore.

Caroline	7,806	Queen Anne's	12,633
Cecil	17,232	Somerset	19,508
Dorchester	18,843	Talbot	12,090
Kent	10,842	Worcester	18,377

Annapolis, on Severn r., which enters on the W. side of Chesapeake bay, is the seat of government; though Baltimore is much the largest town.

The land on the eastern shore, with the exception of a small tract at the N., is generally level and low, and in many places, covered with stagnant waters, in the summer and fall, which occasion agues and intermittent fevers, and many of the inhabitants have a sickly appearance. The soil is tolerably fertile, and produces a beautiful white wheat, said to be peculiar to this region, Indian corn, sweet potatoes, and tobacco. Above the falls in the N.W., the country gradually becomes uneven, and in the western part of the state, it is mountainous. Several branches of the Alleghany chain crosses it from Pennsylvania and Virginia. The principal of these are South mountain, North mountain, Sideling hill, Warrior's, Evit's, Willis', and Alleghany mountains. Between these ranges of mountains are many fertile valleys. Wheat and tobacco are the staple productions, cotton of an inferior quality is raised, chiefly for domestic use, and S. of Baltimore a bright tobacco, denominated *kite-foot*. The soil of the state is generally a red clay or loam, and much of it is excellent. Hemp and flax are raised. There are fine orchards, and apples, pears, peaches, plums, and cherries are abundant. The forests abound in nut-bearing trees, which feed great numbers of swine. Beef and mutton are plentiful. The climate in the western part of the state is particularly salubrious.

There were in the state, in 1840, 92,220 horses or mules, 225,714 neat cattle, 257,922 sheep, 416,943 swine; poultry was raised to the value of $218,765. There were produced 3,345,783 bushels of wheat, 732,577 of rye, 8,233,086 of Indian corn. 72,606 of buckwheat, 3504 of barley, 3,534,211 of oats, 1,636,433 of potatoes, 486,301 pounds of wool, 24,816,012 of tobacco, 5673 of cotton, 3674 of wax, 2357 of hops, 2960 of silk cocoons, 36,966 of sugar, 106,687 tons of hay. The products of the dairy amounted to $457,468, of the orchard, to $105,740, of lumber, to $296,977; and 7585 gallons of wine were made.

Iron ore is found in various parts of the state; the bog ore is found in the S. part of the Eastern shore, and is extensively wrought. Bituminous coal abounds between the mountain ranges in the western part of the state, particularly near Cumberland. The Cumberland coal field, extending from Wills' cr. to the head branch of the Potomac, is 60 m. long, and from five to seven miles wide, and from 3 to 15 ft. thick, and covers an area of 460 sq. m., and is of an excellent quality. The Baltimore and Ohio railroad, completed to Cumberland, will make this coal extensively available. Beyond the main ridge of the Alleghany mountains, the Youghiogheny coal field has beds of 20 ft. in thickness. Sulphuret of copper is found in the Monoccy valley, and is easy of reduction. Porcelain clay occurs in the N.E. corner, and excellent clays for stone ware pottery, &c., are extensively found. Red and yellow ochre and chrome ores, alum-earth, and copperas ores are found, in the eastern part of the state.

The state carries on an extensive trade with the other states, the West Indies, and with Europe, principally from Baltimore. The exports consist of flour, tobacco, iron, lumber, Indian corn, pork, flaxseed, beans and fish, particularly shell-fish, with the latter of which Chesapeake bay abounds.

Potomac r., which divides this state from Virginia is 500 m. long, and navigable about 295 miles from the mouth of Chesapeake bay to Washington city. It is 7½ miles wide at its mouth in Chesapeake bay, and 1¼ m. at Alexandria. Susquehanna r. enters the head of Chesapeake bay in this state, is 1½ m. wide at its mouth, and navigable only 5 m.; above which, it is obstructed by falls and rapids. The Patapsco, though a small r., is navigable 14 m. for large ships to Baltimore, and affords above much water-power. The Patuxent is 110 m. long, and navigable for 50 miles for vessels of 250 tons burthen. The other rivers are Elk, Sassafras, Chester, Choptank, Nanticoke, and Pocomoke which flow by broad mouths into the E. side of Chesapeake bay.

Baltimore is much the largest place in the state, and according to the census of 1840, the third in population in the United States. The other principal places are Frederick, Hagerstown, and Annapolis, the capital.

The exports of this state in 1840 amounted to $5,768,768, and the imports to $4,910,746. There were in the state 70 commercial and 117 commission houses in foreign trade, with a capital of $4,414,000; 2569 retail stores, with a capital of $9,246,170; 1330 persons engaged in the lumber trade with a capital of $207,300; 103 persons employed in internal transportation, who with 911 butchers, packers, &c., employed 7900 tons of bar iron, the whole employing 1782 persons, and a capital of $795,650; 17 paper mills produced to the amount of £195,100, and other manufactures of paper to the amount of $3000, the whole employing 171 persons, and a capital of $95,400; 93 persons produced 1,865,949 pounds of soap, 731,446 pounds of tallow candles, and 35,069 of wax or spermaceti candles, and employing a capital of $308,600; 73 distilleries produced 366,213 gallons, and 11 breweries produced 898,140 gallons, the whole employing 199 persons, and a capital of $183,760; hats and caps were manufactured to the amount of $153,456, and straw bonnets to the amount of $13,900, the whole employing 385 persons and a capital of $76,690; 161 tanneries employed 1035 persons, and a capital of $713,855; 406 saddleries, and other manufactories of leather, produced articles to the amount of $1,050,275, with a capital of $434,127; one glass-house employed 37 persons, produced articles to the amount of $25,190; 58 persons produced drugs and paints to the amount of $96,100, with a capital of $85,160; five powder-mills employed 47 persons, and produced 669,125 pounds of gunpowder, with a capital of $46,000; six sugar refineries produced to the amount of $176,000; 102 persons produced confectionary to the amount of $73,450; 378 persons manufactured tobacco to the amount of $639,000, with a capital of $125,100; 947 persons manufactured granite and marble to the amount of $159,750; 1042 persons produced bricks and lime to the amount of $409,456; 723 persons produced machinery to the amount of $348,165; 36 persons manufactured hardware and cutlery to the amount of $15,670; 590 persons produced carriages and wagons to the amount of $357,693, with a capital of $154,955; 180 flouring-mills produced 446,708 barrels of flour, and with other mills employed 898 persons, producing articles to the amount of $3,387,950, and employed a capital of $4,069,671; ships were built to the amount of $979,771; 834 persons manufactured furniture to the amount of $305,360, with a capital of $339,335; 369 brick or stone houses, and 509 wooden houses were built, and employing 2098 persons, at a cost of $1,078,770; 46 printing offices, 15 binderies, seven daily, seven semi-weekly and 28 weekly newspapers, and seven periodicals, employed 736 persons, and a capital of $159,100. The total amount of capital employed in manufactures in the state, amounted to $6,450,284.

St. John's college at Annapolis was founded in 1784. St. Mary's college at Baltimore was founded by the Roman Catholics in 1799. Baltimore Medical school was founded in 1807; and in 1812, there were added to it the faculties of general science, law and divinity, and it received the name of the University of Maryland. Mount St. Mary's college was established near Emmetsburg in 1830, by the Roman Catholics. These institutions had in 1840, 400 students. There were in the state 197 academies or grammar schools, with 4178 students, and 567 common and primary schools, with 16,982 scholars. In the state there were 11,805 white persons over 20 years of age, who could neither read nor write.

The first settlers of this state were Roman Catholics, and they are still numerous. They have an archbishop, who is metropolitan of the United States, and 60 churches. The Episcopalians have 77 ministers. The Presbyterians have 25 ministers. The Baltimore Methodist conference, which extends into neighbouring states, has 172 travelling preachers. The Baptists have 20 ministers. The German Reformed have nine ministers. There are some Lutherans, Friends, Unitarians, &c.

There were in the state in 1840, 13 banks with an aggregate capital of $9,106,031, and a circulation of $2,328,555. At the close of 1841, the state debt amounted to $15,214,761.

The constitution was formed in 1776, but has been frequently amended, since that time. The state is divided into three districts, the Eastern, Southern, and North Western. The governor is elected for three years, by the people from the districts alternately, so that each district is represented in the gubernatorial chair for one term, in each period of nine years. He must have resided in the district for which he is elected for three years next preceding the election. The senate consists of 21 members elected for six years, by the people; one third of the number being elected every two years; and the senators must have resided in the city or county for which they are chosen for three years, next preceding the election. The house of delegates consists of 79 members elected by the people, and must have resided in the county for which they are chosen for one year next preceding the election. All judges are appointed by the governor, with advice and consent of the senate, and hold their offices during good behaviour. Every white male citizen over 21 years of age, who has resided in the state one year next preceding an election, and for six months in the city or county where he offers his vote, enjoys the right of suffrage. The legislature meets annually at Annapolis, on the last Monday of December.

Two of the greatest works of internal improvement in the United States have been projected and commenced in Maryland; the first is the Chesapeake and Ohio canal, which commences at Georgetown, Dist. Columbia, and is designed to extend 341½ m. to Pittsburg. This is the work of a joint stock company, chartered by the states of Maryland, Virginia, and Pennsylvania, and sanctioned by Congress. It was commenced in 1828 and is completed to Hancock, 136 miles. Considerable work has been done between this and Cumberland, where a spacious basin is in process of erection. It is extended 7½ m. to Alexandria, in the S. part. A completion to Cumberland will open a vast and rich coal region. The Baltimore and Ohio railroad is designed to extend from Baltimore to the Ohio r. at Wheeling, 360 miles, and is the second great work. It was incorporated by the legislatures of Maryland, Virginia and Pennsylvania in 1827, and commenced July 4th, 1828. It is completed from Baltimore to Cumberland. There is a side cut, over 2½ m. to Frederick. A railroad extends across the state, passing through Baltimore, and which forms part of the great chain from New-York and Philadelphia to Washington. This road proceeds on the Baltimore and Ohio railroad, 8 m., from the former place. The Baltimore and Susquehanna railroad extends 56 m., from Baltimore, Md., to York, Pa. A railroad 19½ m. long extends from the Baltimore and Washington railroad, to Annapolis. A railroad extends from Frenchtown to New-Castle, Del., connecting the Delaware and Chesapeake bays. The same is effected by a canal, extending from Back cr., a tributary of Elk r., to Delaware city on the Delaware, 42 m., below Philadelphia. It is 66 feet wide at top and 10 feet deep, and affords a passage to vessels requiring that depth of water, completed in 1829, at an expense of $2,750,000.

Maryland was considered as included in the patent of the South Virginia company, until June 20th, 1632, when it was granted to Cecilius Calvert, lord of Baltimore in Ireland, and it received the name of Maryland, in honour of Henrietta Maria, queen of Charles I.. of England. The proprietor offered 50 acres of land in fee to every emigrant, and gave equal privileges to all denominations of Christians. In 1634 the first colony, consisting of 200 Roman Catholics, planted itself on the N. side of Potomac r., at a place called

ed St. Mary's. In 1635 the first legislature was convened at St. Mary's, which divided the territory into baronies and manors, and passed a variety of laws. The next year the legislature passed a law establishing a house of assembly. In 1645 Claiborne, who was the head of a colony, instigated the Indians against the colony of Calvert, and drove him from the province. In 1650 the constitution of Maryland was settled, the legislature was divided into two houses, and the province included three counties. In 1652 parliament violently assumed the government, and intrusted it to commissioners. In 1662 the government reverted to Lord Baltimore, who assumed the administration. Annapolis was made the seat of government in 1699, and has ever since retained it. In 1775 the people were forward to resist the encroachments of parliament, and took an active part in the Revolution. The constitution was formed in August, 1776, but they did not join the confederation till 1781. In convention, April 26th, 1788, they adopted the constitution of the United States, yeas 63, nays 12: majority 51.

MARYLAND, p. t., Otsego co., N.Y. 128 E. by E. Cooperstown, 66 W. by N. Albany, 365 W. Drained by Schoenevus cr. and its branches, which flows into Susquehanna r. It contains a Presbyterian church, six stores, one fulling-mill, three grist-mills, 17 saw-mills, one tannery, one distillery; 16 schools, 763 scholars. Pop. 3085.

MARYPORT, a seaport town of England, co. Cumberland, par. of Cross Canonby, Allerdale Ward, on the Solway frith, close to the mouth of the Ellen, 25 m. S.W. Carlisle. Pop., in 1841, 5211. It is neat and well-built, and, from its salubrity, is much frequented by summer visitors. A modern-built townhall, chapel of ease, and five places of worship for dissenters, are its chief public edifices. A national school furnishes instruction for about 150 children, of both sexes, and there is a school of industry for 90 girls. The present importance of Maryport, which, in 1750, was a mere hamlet, is attributable to the rise of an extensive coal-trade with Ireland and Scotland. It has also three ship-building yards; and sail-cloth, ropes, blocks, &c., are made on a pretty extensive scale. A pier has recently been erected, and there are commodious quays and staiths; but the harbour dries at low water, and has only 19 ft. at high water springs, and 8 ft. at neaps.

MARYSVILLE, p. v., Paris L., capital of Union co. O. Situated on the W. side of Mill cr., a branch of Scioto r. It contains a courthouse of brick, a jail, three stores, 60 dwellings, and 360 inhabitants.

MARYVILLE, p. v., capital of Blount co., Tenn., 18 S.S.W. Knoxville 163 S.S.E. Nashville, 523 W. Founded in 1795. Situated on the great stage road leading from Knoxville to Huntsville, and contains a courthouse, jail, three churches, five stores, three grist-mills, one carding machine, one printing office, various mechanic shops, and about 700 inhabitants. It is the seat of the South Western Theological Seminary, founded by the Presbyterians in 1821, which has two professors, 24 students, 90 have completed their education, and 6000 vols in its libraries.

MASCAL, a town of Sicily, intend. Catania, capital of canton, at the E. base of Mount Ætna, on a small river about 2 m. from the sea, and 10 m. S.W. Taormina. Pop., in 1831, 3083. Its district is exceedingly fertile, and the town was formerly flourishing, but it is now rapidly decaying, while several of its dependent villages are proportionally thriving and increasing, particularly Giarre and Riposto.

MASSA-CARRARA (DUCHY OF). See Modena.

MASON, county, Va. Situated in the W. part of the state, and contains 875 sq. m. Bounded W. and N. by Ohio river. Watered by Great Kanawha river and its tributaries. Salt water is found by digging near Kanawha r. It contained in 1840, 7880 neat cattle. 9209 sheep, 17,169 swine; and produced 66,503 bushels of wheat 1914 of rye; 298,730 of Indian corn, 90,316 of potatoes, 9478 pounds of tobacco, 27,166 of sugar. It had 11 stores, one flouring-mill, four grist mills, five saw-mills, four tanneries; 11 schools, 241 scholars. Pop.: whites,5923; slaves, 808; free coloured 46; total, 6777. Capital, Point Pleasant.

MASON, county, Ky. Situated toward the N.E. part of state, and contains 260 sq. m. Bounded N. by Ohio river. Drained by North Lick cr., a branch of Licking river. It contained in 1840, 11,315 neat cattle, 21,340 sheep; 29,354 swine: and produced 230,763 bushels of wheat, 34,250 of rye, 366,398 of Indian corn, 91,804 of oats, 20,630 of potatoes, 1065 tons of hemp or flax, 1,122.749 pounds of tobacco. It had 11 commission houses in foreign trade, 63 stores, two woollen factories, one cotton factory with 1100 spindles, 13 flouring-mills, 19 grist-mills, four saw-mills, eight tanneries, six distilleries, one brewery, three potteries, four rope-walks, two printing-offices, three weekly and one semi-weekly newspapers; one academy, 70 students; 36 schools, 1695 scholars. Pop.: whites. 11,126; slaves. 4309; free coloured, 278; total, 15,719. Capital, Washington.

MASON, p. t., Hillsborough co., N.H., 44 m. S.S.W. Concord, 450 W. Chartered in 1768, first settled in 1751. Drained by Souhegan river, and branches of Nashua river. It has three churches, a Congregational, Bapdist, and Christian, three stores, one woollen factory, one cotton factory with 9048 spindles, two grist-mills, and two saw-mills; 10 schools, 357 scholars. Pop. 1275.

MASONVILLE, p. t., Delaware co., N.Y., 111 m. S.W. Albany, 313 W. Drained by small branches of Susquehanna river, and of the W. branch of Delaware river. It contains two churches, a Presbyterian and Methodist, two stores, one fulling-mill, one grist-mill, 15 saw-mills; 11 schools, 455 scholars. Pop. 1620.

MASSACHUSETTS, one of the eastern United States, is bounded N. by Vermont and New-Hampshire; E. by the Atlantic; S. by the Atlantic, Rhode Island and Connecticut; and W. by New-York. It lies between 41° 23' and 42° 52' N. lat. and between 69° 50' and 73° 30' W. long. It is about 190 miles long, with an average breadth of 90 miles, and contains about 7500 square miles, or 4,800,000 acres. The population in 1790 was 388,787; in 1800, 422,845; in 1810, 472,040; in 1820, 523,287; in 1830, 610,408; in 1840, 737,699. Of these 369,079 were white males; 368,351 were white females; 6654 were coloured males; 4015 do. females. Employed in agriculture 87,837; in commerce 8063; in Manufactures and trades, 85,176; in navigating the ocean, 97,153; do. rivers and canals, 379; in mining 460; in the learned professions, 3804.

The state is divided into 14 counties, which with their population in 1840, were as follows:

County	Pop.	County	Pop.
Suffolk	95,773	Berkshire	41,745
Essex	94,987	Bristol	60,164
Middlesex	106,611	Plymouth	47,373
Worcester	95,313	Barnstable	32,548
Hampshire	30,897	Dukes	3,958
Hampden	37,366	Nantucket	9,012
Franklin	28,812	Norfolk	53,140

The capital is Boston, the metropolis of New-England, situated on a peninsula in Massachusetts bay.

The face of the country is diversified. The southeastern part is mostly level. There are also level districts of small extent, in the vicinity of Merrimac river in the N.E. Salt marshes are numerous, but not very extensive in the maritime parts. Most of that part bordering on the sea and extending into the interior as far as the county of Worcester may be regarded as level, with slight undulations, but no high hills. Worcester county, which extends across the state, and the three counties west of it present an elevated and various surface diversified with gentle swells with plains and valleys, and with several mountain ranges, and a strong soil, adapted to grazing and most of the purposes of agriculture, and well watered with clear and beautiful streams. Through Berkshire the western county pass two mountain ranges, the Taghkanic, on the western border of the state; and between the Housatonic and Connecticut rivers, the Green mountain range, here called Hoosick mountains. To the east of Connecticut are several mountains with elevated summits. Mount Holyoke near Northampton is more than 1900 feet above the level of the sea, and Wachusett mountain in Princeton in an isolated summit, from 2800 to 3000 feet high. Saddle mountain, in the Taghkanic range, in the N.W. corner of the state, is 4000 feet high, and mount Washington in the same range, in the S.W. corner of the state is about 3000 feet high. The valies of the Connecticut are fertile, as are those of the Housatonic. In no state of the Union has agriculture been more improved than in Massachusetts. The principal productions are grass, Indian corn, rye, wheat, oats and potatoes. Beef, pork, butter and cheese, of an excellent quality, are produced. Apples are found in great abundance, and pears, peaches, plums and cherries are cultivated with success. Marble is produced in various parts of Berkshire county, granite in Chelmsford and Quincy, and iron ore in the B.E. parts of the state. In 1840, there were in the state, 61 484 horses or mules, 289.574 neat cattle, 378,296 sheep, 143 221 swine: poultry was produced to the amount of $178,157; there were produced 157,923 bushels of wheat, 536,014 of rye, 1,809,192 of Indian corn, 87 000 of buckwheat, 165,319 of barley, 1,319,680 of oats, 5,385,652 of potatoes, 941,906 pounds of wool, 254.795 of hops 569,395 tons of hay, 2432 of hemp or flax, 1741 pounds of silk cocoons, 579,227 of sugar. The products of the dairy were valued at $2,373 296; of the orchard at $389,177, and of lumber, at $344,845.

Massachusetts is extensively engaged in the fisheries. There were produced in 1840, 389,715 quintals of dried or smoked fish, 194.755 barrels of pickled fish. 3,630,972 gallons of spermaceti oil, 3,314,725 gallons of whale or other fish oil. In its shipping Massachusetts is the second state in the Union, being inferior only to New-York.

The climate is favourable to health; the extremes of temperature at Cambridge in 1843, were 94° above and 7° below zero of Farenheit; but such extremes are of short continuance.

The principal rivers are the Connecticut, a noble stream, winding for 50 miles across the state, receiving Deerfield and Westfield rivers from the west, and Miller's and Chickapee rivers from the east; Housatonic, which rises in Berkshire county, and flows through the west part of the state; and Merrimac, which rises in New-Hampshire, and has a course of 50 m. in the N.E. part of the state, and enters the ocean below Newbury port. It is navigable for large vessels 15 miles to Haverhill. Besides these, there are Nashua, Concord, Ipswich, Charles, Taunton, and Blackstone rivers.

Massachusetts bay, that gives name to the state, which was formerly called the Bay State, extends from cape Ann on the N., 40 m. to cape Cod on the S., and includes Boston and cape Cod bays. Buzzard's bay on the S. shore, is 30 m. in length. Boston harbour is one of the finest in the world, capacious, safe, easy of entrance and easily defended. New-Bedford on Buzzard's bay, has a fine harbour. The other commercial towns are Salem, Newburyport, Gloucester, and Nantucket. The other principal towns are Lowell, Plymouth, Worcester, Springfield, Pittsfield, and Northampton.

There are several important islands off the S. shore of Massachusetts, belonging to the state. The largest is Nantucket, 15 m. long, and 11 broad, and which constitutes a county of its own name. Martha's vineyard lies W. of Nantucket, is 20 m. long and from 2 to 10 broad, and with other small islands, constitutes Duke's county.

The exports of the state in 1840, amounted to $10,186,961, and the imports to $16.513,868. There were 241 commercial and 123 commission houses engaged in foreign trade, with a capital of $13,881,517; 3625 retail dry goods and other stores, with a capital of $13,705,035; the lumber trade employed 3432 persons, and a capital of $1,029,380; internal transportation employed 799 persons, and with 480 butchers, packers, &c., employed a ca.,ital of $407,850; the fisheries employed 18,000 persons, and a capital of $11,725,850.

The manufactures of Massachusetts are equally distinguished with its commerce. Home-made or family goods were produced to the amount of $231.942; there were 27 fulling-mills and 144 woollen manufactories which employed 3676 persons, producing articles to the amount of $7,982,895, with a capital of $4,179,850; 278 cotton factories, with 665.095 spindles, employing 20,928 persons, producing articles to the amount of $16,553,423, and employing a capital of $17,414,099; 48 furnaces produced 9339 tons of cast iron, and 67 forges, rolling-mills, &c., produced 6004 tons of bar iron, the whole employing 1097 persons, and a capital of $1,239,675; 89 paper-mills employed 967 persons, producing to the amount of $1,659,930, 68 other manufactories of paper produced to the amount of $8,700, and the whole employed a capital of $1,889,800; 463 persons produced salt to the amount of 376,596 bushels, with a capital of $502,980; hats and caps were manufactured to the amount of $918,438, and straw bonnets to the amount of $921,646, the whole employing 6636 persons, and a capital of $609,992; 355 tanneries employed 2446 persons, and a capital of $1,094,699; paints and drugs were produced to the amount of $405,725, and turpentine and varnish to the amount of $25,890; 1532 saddleries, and other manufactories of leather produced various articles to the amount of $3,553,895, and employed a capital of $3,318,544; four glass houses employed 372 persons, producing articles to the amount of $471,000, with a capital of $277,200; 30 potteries employed 71 persons, produced articles to the amount of $44.450, with a capital of $27,975; two sugar refineries produced articles to the amount of $1,025,000; chocolate was manufactured to the amount of $31,508; confectionary was made to the amount of $137,300; 14 powder-mills employed 69 persons, and produced 2,315,215 pounds of gunpowder, with a capital of $825,000; 913 persons produced machinery to the amount of $998,975; 1100 persons produced hardware and cutlery to the amount of $1,881,163; 37 distilleries produced 5,177,910 gallons, and seven breweries produced 429.800 gallons, employing 154 persons, and a capital of $963,100; 397 persons produced 20 cannon and 22,632 small arms; 1402 persons produced carriages and wagons to the amount of $803,999, with a capital of $334,660; 2:4 persons produced granite and marble to the amount of $217,180; 758 persons manufactured bricks and lime to the amount of $316,796; mills of various kinds, flouring, grist and saw, employed 1808 persons, manufactured to the amount of $1,771,185, with a capital of $1,440,139; ships were built to the amount of $1,349,994; 51 ropewalks employed 672 persons, producing cordage to the amount of $632,900, with a capital of $555,100; 2494 persons manufactured furniture to the amount of $1,000,008; 946 persons manufactured musical instruments to the amount of $343,760, with a capital of $555,100; 234 brick or stone houses, and 2349 wooden houses were built, employed 2947 persons, and cost $2,767,134; 104 printing-offices, 72 binderies, 10. daily, 67 weekly, and 14 semi-weekly newspapers, and 14 periodicals, employed 929 persons, and a capital of $416,200. The total amount of capital employed in manufactures was $41,774,446.

Massachusetts has three colleges and two theological seminaries. Harvard university at Cambridge, is the oldest and best endowed seminary in the country, founded in 1638, about 18 years after the first landing on the Rock of Plymouth; Williams college at Williamstown, in the N.W. corner of the state, was founded in 1793 and is flourishing; Amherst college at Amherst, was founded in 1821, and has had an unexampled growth, ranking with the first colleges in New-England. The theological seminary at Andover, under the direction of the Congregationalists, has been munificently endowed by a few individuals, and is one of the oldest and most respectable institutions of the kind in the United States. It was opened for students in the autumn of 1808. The Baptists have a flourishing theological institution at Newtown, founded in 1825. All these institutions had in 1840, 769 students. There were 251 academies and grammar schools in the state, with 16,746 students, and 3362 common and primary schools, with 160,257 scholars. There were 4448 white persons, over 20 years of age who could neither read nor write. These, as in most cases in the states, are principally made up of foreign immigrants.

The principal religious denominations are Congregationalists, of whom some are Unitarians, Baptists, Methodists, Episcopalians and Universalists. In 1836, the Orthodox Congregationalists had about 323 churches, 291 ministers, and 46,950 communicants; the Unitarians had about 120 ministers; the Baptists had 199 churches, 160 ministers, and 20,309 communicants; the Methodists had 87 ministers; the Episcopalians had one Bishop and 37 ministers; the Universalists had 100 congregations and 44 ministers; the Friends had 18 societies; the Roman Catholics had one bishop, and 11 ministers. Besides these there are a few Presbyterians, Christians, Swedenborgians or New Jerusalem, and Shakers.

At the commencement of 1843, the state had 118 banks with an aggregate capital of $31,300,000, and a circulation of about $7,875,392. There is a state penitentiary at Charlestown, substantially on the Auburn plan, and the income more than pays the expenses.

Massachusetts has not been behind her sister states, on works of internal improvement. The Middlesex canal, connecting the Merrimac river, two miles above Lowell, with Boston harbour at Charlestown, is 27 m. long, was completed in 1808, at a cost of $528,000; and it is the first canal of any length, constructed in the U. States. Hampshire and Hampden canal continues the Farmington canal from the N. line of Connecticut, 22 m. to Northampton, making the whole length of the canal from Northampton to New-Haven Ct., 76 m. Blackstone canal, 45 m. long, connects Worcester with Providence, R. I., and was completed in 1828, at a cost of $600,000. Montague canal, around Montague falls in Connecticut river, is three miles long, and overcomes, by eight locks, a fall of 75 feet. South Hadley canal along South Hadley falls in Connecticut river, is two miles long, and has five locks. The railroads are more numerous and important than the canals. Quincy railroad was completed in 1827, from the granite quarry, three miles to Neponset river, and deserves to be mentioned as the first work of the kind built in the United States. The Western railroad commences at the termination of the Boston and Worcester railroad, and extends to the western boundary of the state, 156 m., and is continued to Albany. It cost $7,566,791. A branch of this railroad extends to Hudson, N. Y. Boston and Lowell railroad connects the two places, is 26 m. long, and cost $1,978,256. Boston and Maine railroad, which is continued, after leaving the state to Portland, Me., cost, in the Massachusetts part, $1,960,965. Boston and Providence railroad connects the two places, is 42 m. long, and cost $1,592,831. Boston and Worcester railroad between the two places is 44 m. long, and cost $2,764,396. The Charlestown branch of this road, six miles long, cost $293,145. The Eastern railroad extends to the line of New-Hampshire, 55 m., and cost $2,321,669. The Norwich and Worcester railroad between the two places is 59 m. long, and cost $2,158,562. A steamboat line connects this road with New-York city. Nashua and Lowell railroad is 14 m. long, and cost $380,000. New-Bedford and Taunton railroad is 21 m. long, and cost $495,122. The Taunton branch of this road is 11 m. long. Most of these railroads yield a dividend of from six to eight per cent. At the close of 1840, the debt of the state amounted to $5,149,137.

The government of Massachusetts consists of a governor, lieutenant governor, senate and house of representatives. They are elected annually by the people. The governor must have resided seven years in the state and be worth a freehold of 1000 pounds and declare his belief of the Christian religion. The lieutenant governor must possess the same qualifications. A council of nine persons besides the lieutenant governor are elected annually by the joint ballot of the legislature, and not more than two can be chosen in one congressional district. They rank next to the lieutenant governor. The senate contains 40 members, who must possess a freehold of 300 pounds, and a personal estate of 600 pounds, and must have resided in the state for five years immediately preceding the election. The house of representatives contains 356 members, who must possess a freehold of 100 pounds in the town for which he is chosen, or rateable estate to the value of 200 pounds. The judges and various other officers, as attorney general, &c., are appointed by the governor with the advice and consent of the council. The judges hold their offices during good behaviour. The secretary, treasurer, and receiver-general are appointed annually, by the joint ballot of both houses of the legislature. Every male citizen over 21 years of age (excepting paupers, and persons under guardianship,) who has resided in the state one year, and in the town or district in which he may claim to vote six months next preceding an election, and shall have paid a tax in the Commonwealth within two years, or shall have been exempted from taxation, enjoys the right of suffrage.

Plymouth colony was settled by the Puritans, a part of Mr. Robinson's congregation under Carver and Bradford, Dec. 22nd, 1620. In 1628 the foundation of the Massachusetts colony was laid, and Salem and Charlestown were settled, and Boston in 1630. In 1634 the charter of Plymouth colony was surrendered to the crown, and an attempt was made the same year, and again in 1639 to procure the surrender of the charter of Massachusetts without effect. The patent of the Plymouth colony was, in 1641, transferred to the freemen. In 1643 the four colonies of Massachusetts, Plymouth, Connecticut and New-Haven entered into articles of union, styling themselves the United Colonies of New-England. Rhode Island petitioned to be admitted to the confederacy, but was refused. In 1675 the colony engaged in Philip's war, which lasted after much suffering, in the defeat and extermination of the hostile tribe of the Pequods. In 1645 the troops of Massachusetts, Connecticut and New-Hampshire sailed for cape Breton, and captured Louisburg from the French. In 1765 the colony proposed a general congress, which met at New-York to resist the encroachments of parliament, and sent letters to all the provinces in 1768 to excite them to insist on a redress of grievances. In 1692 the colonies of Plymouth, Massachusetts and Maine were united, by a charter, under the name of Massachusetts. In 1773 the destruction of tea in the harbour of Boston occurred, which was followed the next year, by the shutting up of the port of Boston. The Revolution opened by the battle of Lexington, April 19th, 1775: and of Breed's hill, generally called Bunker hill, June 17th, 1775. Boston was the cradle of American independence, and Massachusetts bore an honourable part in securing it. On the 17th of March, 1776, the British were compelled to evacuate Boston. In 1780, a convention of delegates formed a state constitution. In 1786 Shay's rebellion occurred in the western counties, and was quelled the following year. The state voted in convention, Feb. 6th, 1788 to adopt the constitution of the United States—yeas, 187, nays 168, majority, 19.

MASSA-DUCALE, or DI-CARRARA, a town of N. Italy, belonging to the Modenese dom., cap. duchy of the same name, on the road between Genoa and Leghorn, 3 m. from the Mediterranean, and 28 m. N.W. Lucca. Pop. estimated by Bampoldi at 7000. It is distinguished by the beauty and salubrity of its situation, and is clean and well built, but has few remarkable edifices. Its ancient cathedral was pulled down by Eliza Bacciocchi, sister of Napoleon, when queen of Etruria, on account of its being too near the royal palace. It has an academy of sculpture and architecture, a seminary, college, hospital, public library, and an old castle now used as a prison. It is the see of a bishop; and has manufactures of silk stuffs, and some trade in soap, oil, and other agricultural products, and the fine marble of its vicinity, as to which see CARRARA in this Diet. (Vol. I., p. 539.)

MASSAFRA, a town of the Neapolitan dom., province Otranto, cap. canton, on the road from Bari to Taranto, 19 m. N.W. the last-named city. Pop. about 7000. "Massafra is prettily situated on the slope of a hill interspersed with tufts of trees and shrubs; but when near it, it assumes a most singular appearance. The rock on which it stands is perforated and worked into a thousand fantastic shapes. The houses stand on the brink of a narrow valley, or rather chasm, worked through the rock by the action of running water." (Burgess's Greece, &c., i., 22.) The town is walled, and is conjectured by some authors to occupy the site of the Messapia of antiquity; but others contend that Messagna, between Oria and Brindisi, is the modern representative of that ancient city.

MASSAT, a town of France, dep. Ariège, cap. cant., in a fertile valley, 14 m. W.S.W. Foix. Pop., in 1836, 7193.

There are in its vicinity numerous iron mines, the working of which employs a considerable portion of the people.

MASSENA, p. t., St. Lawrence co., N.Y., 947 m. N.N.W. Albany, 504 W. Bounded N. by St. Lawrence river. Watered by Grass and Racket rivers. It contains several considerable islands in St. Lawrence river, and has a sulphur spring of some celebrity, which emits sulphuretted hydrogen gas, and contains carbonates of soda, lime, magnesia, and sulphur. It contains a Baptist church, nine stores, one fulling-mill, two grist-mills, four saw-mills; 17 schools, 740 scholars. Pop. 2726.

MASSERNE, mountains, sometimes improperly called the Ozark mountains, rise 70 m. S.W. of St. Louis, Mo., and run in a S.W. direction through Missouri and Arkansas into Texas. They are crossed by the Arkansas and Red rivers. It derives its name from mount Cerne, one of its most elevated summits.

MASSILON, p. v., Perry t., Stark co., O., 116 m. N.E. Columbus, 8 m. W. Canton, 100 m. W. Pittsburg, 22 m.-E. Wooster, 450 W. Laid out in 1826 on the Ohio canal, on the E. side of Muskingum river. It contains two churches, an Episcopal and Baptist, a bank, 22 stores, 15 warehouses, one flouring-mill, one woollen-factory, two tanneries, one printing-office, one weekly newspaper, 250 dwellings, and about 2009 inhabitants. Goods are, to some extent, sold here by wholesale. The exports are estimated at over 200,000 annually, and the imports of merchandise amount to over 1,600,000 pounds. Bituminous coal is found in the vicinity, and is exported. In the vicinity are several large flooring-mills.

MASUAH, the principal seaport of Abyssinia, on the Red sea, on an island separated from the continent by the narrow but deep channel of Adowa, 250 m. N.E. Gondar, and 490 m. S. by E. Djidda: lat. 15° 36' 45" N., long. 39° 24' E. Pop. 9900? The island in which Masuah stands is only about half a mile in length, and a quarter of a mile in breadth, one third of its extent being occupied by houses. The town has several stone houses, two stories high; but most of them are in ruins. The other dwellings are mere huts, built, as in Arabia, with poles and bent grass. The only public buildings are four mosques, of small size and rude architecture. Owing to the total absence of springs, water is very scarce, and is collected in large public tanks, that occupy nearly a third part of the island. The harbour, though having a narrow entrance, can accommodate about 50 vessels; and is safe, deep, and easily accessible. The trade carried on between Masuah and the ports of Arabia is of considerable importance. From Djidda are brought many articles of European manufacture, embroidered velvets, arms, glass-ware, silks, and satins; while Mocha furnishes Indian fabrics of every quality, from the finest muslins to the coarse Surat cloth, used as articles of dress in a great part of Africa. The exports comprise a considerable number of slaves, gold-dust, ivory, rhinoceros' horns, and ears, brought from the interior by a large caravan, which arrives in February. The maybe demands 10 per cent., ad valorem, on all exports and imports, and the same amount of duty is levied by the imam. (Valentia's Travels, ii., 44-68.)

MASULIPATAM. See Circars (Northern).

MASULIPATAM, a fortress and town of British India, presid.

Madras, capital of the above district, on the Coromandel coast, 230 m. N.N.E. Madras; lat. 16° 13' N., long. 81° 14' E. The fort is of an oblong figure, 800 yds. by 600, in the midst of a salt morass, and close to a canal communicating with the Krishna. By means of this canal the surrounding country may be entirely inundated, a circumstance constituting the chief strength of the place. The pettah, or native town, is about 1½ m. to the N.W.; it is very extensive, and, for a Hindoo town, tolerably well built. Masulipatam stands on the only part of the Coromandel coast which is not beat with a heavy surf. Its port receives vessels of 300 tons; and it was for a lengthened period a place of considerable trade with Bengal, China, Birmah, Persia, and Arabia. Its commerce is now, however, greatly fallen off, and scarcely extends beyond Calcutta on one side, and Bussorah on the other. Its chief exports are piece-goods and tobacco. The chintzes of Masulipatam, though not equal to those of Europe, have been long, and deservedly, celebrated, and are very generally worn in some parts, especially in Persia. This town is the residence of the district collector and judge. It was conquered by the Bhamanee sovereigns of the Decca, in 1480, ceded to the French in 1751, and taken by the British in 1759.

MATANZAS, a sea-port town of Cuba, ranking next to Havannah, in commercial importance, on the N. coast of the island, at the bottom of a deep bay, 52 m. E. Havannah; lat. 23° 2' 28" N., long. 81° 37' 44" W. Pop., in 1827 (including garrison and strangers, estimated at 3000), 14,341: of the resident population, 6333 were whites, 1141 free people of colour, and 2067 slaves; but, at present (1841), the population is probably not under 20,000. It is pretty well built, has some good streets, and about one third part of its houses are of stone. It has a large hospital, a good parish church, barracks, theatre, two market-places, two printing-offices, a bathing establishment, &c.; and in the neighbourhood is a considerable sugar-refinery, belonging to an English firm. The bay of Matanzas, defended by the castle of San Severino, is extensive, and is exposed only to the N.E. wind. The harbour, in front of the city, is protected by a ledge of rock, 4 ft. below the surface, which serves as a natural breakwater, to defend the vessels at anchor within it from the swell. There are two channels by which to enter, the one by the N., the other by the S. end of the ledge; but the S. channel is fit only for coasting vessels. There are two rivers, one on each side of the town, which deposit so much mud at their mouths as greatly to diminish the extent of the anchorage ground, and render it necessary to load and discharge the shipping by lighters and launches. (Turnbull's Cuba, 217-219; Comm. Dict., &c., passim.)

Matanzas, though situated in one of the most fertile districts of this noble island, was an inconsiderable place till within the last 30 years. Under the old colonial government, it was merely a subsidiary port to the Havannah, and was not allowed to carry on any direct intercourse with foreign countries; but this impolitic restriction being removed in 1809, Matanzas immediately became the centre of a considerable trade; and the town and its commerce have since continued to increase, with the rapidly increasing cultivation of sugar and coffee, and other colonial staples in the adjoining districts. The great importance of the trade of Matanzas will be seen from the subjoined

Account of the Sugar and Coffee exported from the Havannah and Matanzas, in 1838 and 1839.

	Boxes of Sugar.				Arrobas of Coffee.			
	Havannah.		Matanzas.		Havannah.		Matanzas.	
	1838.	1839.	1838.	1839.	1838.	1839.	1838.	1839.
To United States	78,554	65,884	46,521	45,796½	669,460	97,491	145,895½	92,306
Great Britain	13,051	7,191	1,079	2,990	12,727	5,299		
Cowes, &c.	74,992	74,719	33,233	66,340	23,727	15,816½	2,334	8,587
Baltic	39,377	11,304	44,041	13,926	168	496	3,016	
Hamburg and Bremen	53,159	48,463½	42,291	24,039	74,865	169,661½	96,758	23,754
Holland	19,461	20,536	10,493	4,187	1,650	13,148		
Belgium	13,947	5,857	15,513	4,179	9,394	9,769		3,129
France	8,436	10,939	2,575	2,434	88,962	221,580	9,254	8,736
Spain	63,484	69,599	18,998½	19,827	17,400	30,536	2,250½	11,772
Italy	7,399	9,940	2,980	7,095	21,080	45,080	186	10,150
Other ports	2,593	2,706	346	990	1,604½	9,392		492
Total	366,356	326,684	219,069½	191,801½	916,837½	1,204,686	180,504	174,814½

MATARO (an. Illuro), a seaport town of Spain, in Catalonia, 20 m. N.E. Barcelona. Lat. 41° 33' N., long. 2° 30' E. Pop., according to Miñano, 12,949. The more ancient or Moorish portion of the town stands on a slight eminence, at a short distance from the shore, and is surrounded by walls: its streets are narrow and crooked, with the exception of the Riera, which is wide and straight, lined with rows of trees, and forming an agreeable promenade. The new town, which stretches eastward along the seashore, is much larger and more regularly built, with wide streets, and respectable houses. A parish church and a general hospital, with two or three large buildings, formerly used as monasteries, are the only public edifices. The town is celebrated for the excellence of its red wine and brandy, much of which is exported to the United States. Its cloth fabrics, which were favourably noticed by Townsend (t., 109), have much declined; and, since the emancipation of the colonies, its exports of cotton-prints, ribands, and lace, have become quite inconsiderable. The port has a ship-building yard; and there is good anchorage for merchant-ships close in shore. "The neighbourhood is very picturesque, and the country-houses and cottages have an air of greater neatness and comfort; the windows are glazed, and the insides of the dwellings display a good stock of furniture. No beggars and fewer ragged people are seen; industry is evidently active; the ground is better cleared, fences (made of the American aloe) are more general and more neatly constructed; nobody is seen basking in the sun. In short, there is altogether a new order of things, quite different from that seen in any other part of Spain." (*Inglis*, ii., 304.)

MATHURA, or MUTTRA, a celebrated town and place of pilgrimage in Hindostan, prov. Agra, on the Jumna, 30 m. N.W. Agra; lat. 27° 31' N., long. 77° 33' E. It is highly venerated by the Hindoos, from its being the birthplace of their deity Krishna, and consists chiefly of one continued street of temples and ghauts, which, though they do not exhibit the architectural magnificence of similar structures in S. India, have, nevertheless, considerable elegance and richness. Mathura was taken in 1019 by Mahmoud of Ghiznee, who despoiled it of an immense quantity of gold, silver, and gems, threw down many of its temples, and desecrated others by converting them into mosques. Under Acbar and his successors, however, the Hindoos were permitted to rebuild and improve the city; and a temple, erected about that period, is said to have cost 60 lacs of rupees. But this splendid edifice was destroyed by Aurungzebe, who built on the spot a mosque with the materials. Another large mosque, built by a Mohammedan governor, is now in a state of decay. Some extensive cantonments are separated from the town by an interval of broken ground covered with ruins. Mathura has a fort, in which is an observatory, founded by the rajah Jye-Singh of Jyepoor. At the end of the last century it was the head-quarters of the commander of Scindia's infantry: it was, however, taken, without opposition, by the British in 1803. (*Hamilton's E. I. Gaz.*)

MATLOCK, a village and parish of England, celebrated for its mineral waters, hund. Wirksworth, co. Derby, on the Derwent, 14 m. N. by W. Derby, and 135 m. N.N.W. London. Area of parish, 3960 acres. Pop., in 1841, 3782. The town is beautifully situated, partly in a valley and partly on the slope of a hill rising E. from the Derwent, here crossed by a neat stone bridge: the houses are chiefly of stone. The church, picturesquely situated on the brow of a rock, rising perpendicularly above the river, and embosomed in trees, is a small edifice, in the perpendicular English style, with a square tower at its W. end: the living is a rectory (annual value £390), in the gift of the Dean of Lincoln. There are also four places of worship for Wesleyan Methodists, Independents, and other dissenters, with attached Sunday-schools, providing religious instruction for between 400 and 500 children of both sexes. There is also an endowed school, for clothing and educating 30 boys. In 1839, a cotton-mill employed about 300 hands: it has, also, a large paper-mill; and the lead-mines in the neighbouring hills employ about 150 hands. A museum of mineralogy, established here a few years ago, contains a great many valuable specimens of ore, &c., peculiar to this district. Visiters purchase considerable quantities of Derbyshire spar. But the chief dependence of the inhabitants is on the supply of those who come here during summer, to use the mineral waters; which, are considered beneficial in cases of glandular affections, scrofula, bilious disorders, pulmonary complaints, and diabetes. The springs, which first attracted notice, for their medicinal qualities, in 1698, when the first, or old bath, was built, are about 1½ m. S. of Matlock, on the W. bank of the river; and here are the hotels, libraries, and lodging-houses, to which the visiters chiefly resort. Other two springs have been discovered, each of which is now enclosed, with a handsome edifice, conveniently fitted up with baths and pump-rooms. The waters have a temperature of about 66° or 68° Fahrenheit, and bold in solution only a small quantity of carbonate of lime, their specific gravity being less than that of ordinary water: it would hence appear, though having a lower temperature, greatly to resemble the Buxton and Bristol waters.

The scenery of Matlock-dale is peculiarly picturesque and romantic, diversified with rugged, beetling crags, strongly contrasted with the fine verdure of the valley; the most prominent objects being the high Tor and Masson hill. The former rises almost perpendicularly about 300 ft., the upper half of which is a broad mass of naked brown rock, from which fragments often fall into the river which flows immediately below, obstructing the channel, and greatly increasing the impetuosity of the stream after heavy rains. Opposite the high Tor, but of a less bold, though loftier, character, is Masson hill; on the summit of which are the heights of Abraham, rising about 750 ft. above the river, and not only overlooking the whole dale, but commanding an extensive prospect over a considerable part of Derbyshire. Willersley castle, the seat of Mr. Arkwright, son of the great founder of the cotton manufacture, stands on a commanding eminence E. of the Derwent. (*Parl. Rep.; Priv. Inform.*)

MATTAPONY, r., Va., rises by several branches in Spottsylvania county, and in the S.E. part of King William county, unites with Pamunky river, to form York river.

MATTEAWAN, p. v., Fishkill t., Dutchess co., N.Y., situated on Fishkill creek, a mile and a half from the landing on Hudson river. It contains a flourishing boarding school for boys, two churches, a Presbyterian and Episcopal, two cotton-factories with 6040 spindles, an iron foundry and machine shop, which produces every kind of machinery for cotton and woollen manufactories. 200 dwellings, and about 1800 inhabitants. No intoxicating liquors are allowed to be sold to the workmen. The creek has here a fall of 40 feet, affording an extensive water-power.

MATTHEWS, county, Va., situated in the S.E. part of the state, and contains 80 sq. m. It consists of a peninsula between Piankatank and Mobjack bays. Bounded E. by Chesapeake bay. It contained in 1840, 4181 neat cattle, 2336 sheep, 9214 swine; and produced 8875 bushels of wheat, 171,290 of Indian corn, 54,100 of oats, 17,070 of potatoes, 94,500 pounds of cotton. It had 20 stores, 15 grist-mills, two tanneries; two academies, 98 students; 12 schools, 257 scholars. Pop.: whites, 3909; slaves, 3309; free coloured, 164; total, 7442. Capital, Matthews C.H.

MATTHEWS, C.H., p. v., capital of Matthews co., Va., 102 m. E. by S. Richmond, 187 W. It contains a court-house, jail, and several stores and dwellings. Nett proceeds of the postoffice, $250.

MAUCHLINE, a neat village of Scotland, co. Ayr, on an eminence 1 m. N. from the river Ayr, 27 m. S. by W. Glasgow. Pop., in 1831, 1364. The only public buildings are the parish church, and a chapel belonging to the Associate Synod. It has a woollen-mill, which employs about 25 hands: and hand-loom weaving and tambouring for the Glasgow manufacturers employ about 200 hands. It has also a manufacture of beautifully jointed and varnished wooden snuff-boxes, similar to those made at Cumnock and Laurence kirk. There are four schools in the town, of which three are endowed: two subscription libraries, and a savings' bank. The village of Catrine (l., 563), is 3 m. S.E. from the town.

This place, trifling as it is, has been "married to immortal verse." Burns lived for nearly nine years at the farm of Mossgiel, half a mile N. of the village: and Mauchline was the birthplace of "bonnie Jean," and is the scene of two of his inimitable poems, "The Jolly Beggars," and "The Holy Fair."

MAUCH CHUNK, p. t., capital of Carbon co., Pa., 92 m. N.E. by E. Harrisburg, 220 W. The township belongs chiefly to the Lehigh navigation and coal company. Drained by Beaver, Mauch Chunk, Nesquihoning, and Lechs creeks, and Room run. It contains several villages connected with the coal business, which are Mauch Chunk, Coalville, Lousanne, and Nesquihoning. The village of Mauch Chunk is situated on the W. bank of Lehigh river, in a deep and romantic ravine, between rocky mountains, which rise precipitously 800 or 1000 feet above the level of the stream. It contains five churches, a Presbyterian, Methodist, Episcopal, Dutch Reformed, and Roman Catholic, seven stores, 300 dwellings, and about 1800 inhabitants. About 1900 of the inhabitants are employed in mining and shipping coal, and there is little attention paid to agriculture in the vicinity, the provisions being brought from an average distance of 20 miles. An inclined plane 700 feet long, rising 200 feet in that distance, and a railway 9 m. long, extends to the great coal mine. About 30 acres have been worked from this single vein, and have produced more than 1,200,000 tons. Here is located the village of Coalville, containing about 40 dwellings, inhabited chiefly by miners. The township contains seven stores, three lumber-yards, a furnace, an iron foundry, one grist-mill, four saw-mills, a printing-office, one weekly newspaper; six schools, 315 scholars. Pop. 2193.

MAULMAIN, or MOULMEIN, a seaport town of India beyond the Ganges, cap. British prov. Martaban, at the mouth of the great river Than-lŭeng, having N. the Birman town of Martaban, on the opposite side of the river, and W. the island of Balu, which serves as a natural breakwater to defend the port from the heavy seas that would otherwise be thrown in from the W., 100 m. S.S.E. Rangoon, 27 m. N.N.E. Amherst; lat. 16° 30' N., long. 97° 38' E. It was founded so late as 1825, when the site was selected by Sir A. Campbell, as eligible as well for a commercial as a military station. It is about 200 ft. above the level of the river, and extensive and fertile plains stretch eastward from it towards the mountains. Its port is good. and, from its extensive command of internal navigation, it promises to become a considerable emporium. The principal articles of export are teak, timber, and rice; but there is also a considerable export of tobacco, stic-lac, betel-nut, ivory, cutch, cocoa-nut, &c. The imports consist principally of European cotton goods, and marine stores. The principal trade of the place has hitherto been carried on with Calcutta, Madras, Rangoon, and Pinang; but, in 1837, a direct trade was commenced with London. Ship-building is carried on to a considerable extent. We have no recent accounts of the population; but, probably, it is not under 8000 or 10,000. An English newspaper (the *Maulmain Chronicle*), from

	Imports.		
	1836.	1837.	Increase.
	Reals.	*Reals.*	*Reals.*
Calcutta	5,56,819	9,38,891	3,82,072
Madras	77,802	1,39,476	61 674
Straits	51,622	91,137	39,515
Rangoon	1,00,874	1,97,532	96,658
Tenasserim coast	43,730	52,174	8,444
Other ports	34,510	41,455	9,945
Total	8,65,357	14,60,665	5,95,308

" In order to exhibit, at one view, the decided increase of our trade in 1837 over the preceding year, we place the totals to and from each port in juxtaposition.

" From this it will appear, that the imports have increased nearly six lacs of rupees, and the exports two lacs and a half. Among the imports, the increase is found chiefly under the head of marine stores, spirituous liquors, and articles of European and Birmese manufacture. Among the exports, it is found in the staple productions of the country, rice and timber. Under the head of imports, we may notice that five lacs of rupees were received during the year into the government treasury; to which may be added, what does not appear in our statement, about half a lac of rupees, perhaps, from the Mauritius, for the purchase of cargoes of rice and timber. Under the head of exports, also, it would not, we think, be improper to include the estimated value of the vessels built and launched during the year at the several dock-yards. The following is a very rough estimate:

	No.	Tonnage.	Estimated Value.
			Reals.
Vessels launched	9	2,500	250,000
Ditto fitted for sea	.	.	125,000
Total			375,000

" We are not aware that the mode in which our statements are drawn up requires any particular explanation. The word 'sundries' may be, perhaps, thought too comprehensive; but we have divided it among articles of European, Indian, Chinese, and Birmese produce of manufacture.

" The following is a statement of the imports into Maulmain from the neighbouring Shan states during 1837:

	No.	Value.
		Reals.
Elephants	65	13,600
Ponies	145	17,380
Buffaloes	69	980
Cows and bullocks	3,660	45,200
Sundries		4,500
Total		81,610

" Of the exports to those states, we can procure no detailed statement. Little else, however, is taken to them from hence but piece goods, either European or native, the value of which may, perhaps, be estimated at about 60,000 rupees.

" Taking, then, into consideration the various items above alluded to, and which are not brought into our statements, we shall have the following as the amount of imports and exports for the year 1837: imports, 20,22,275 reals; exports, 11,03,410 reals."

MAUMEE, r., O., rises in the N.E. part of Indiana, and is formed by the junction of Little St. Joseph, St. Mary's, and Great and Little Auglaize rivers. It is about 160 m. long, and its average width for 50 m. from its mouth is 70 rods. It enters the W. part of lake Erie, in Maumee bay, and is navigable at all seasons, 18 m. from the lake to Perrysburg, for schooners and steamboats. It is boatable in spring and fall 18 m. above Perrysburg to fort Wayne; but this will be in a great measure superseded by the Wabash and Erie canal, now completed along its borders; and the Maumee canal will extend from it to Cincinnati. It waters a fertile country.

MAUMEE CITY, p. v., Lucas co., O., 124 m. N.N.W. Columbus 455 W. Situated on the W. side of Maumee river, at the foot of the rapids, at the head of navigation for small craft from lake Erie. A side cut here connects the Wabash and Erie canal with the river, and affords extensive water-power. It is extensively laid out, has several stores and warehouses, a considerable number of dwellings, and about 3000 inhabitants. It is in contemplation to remove obstructions in the river, so as to make it accessible to steamboats.

MAURA (SANTA) (an. *Leucas*), one of the Ionian islands, off the W. coast of Greece, and separated from it only by a channel about 100 yards broad, and so shallow, as in some places to be fordable; 48 m. S.E. Corfu and 7

which we borrow the following weak:—

	1836.	
	Reals.	
Calcutta	94,944	2,
Madras	26,900	1,
Straits	49,390	56,
Mauritius		84,
Rangoon	2,04,457	1,48,06
Tenasserim coast	32,614	18,471
Other ports	18,090	28,609
Total	4,94,995	6,05,410

m. N. Cephalonia, its cap. Amaxichi being in lat. 13″ N., long. 20° 43′ E. Length 23 m.; average bre. m. Area, about 180 sq. m. Pop., including troops in 17,385. It is intersected by a chain of mountains, runn. N. and S. through its whole extent, and rising in som. places to the height of 3000 feet, whence secondary ridges branch off in various directions, forming a few small valleys admitting cultivation; but most of the produce is raised on a narrow strip of land, stretching about 90 m. along the N.W. side of the island, and comprising the residences of the greater part of the population. The soil is generally very scanty; and many parts of the surface exhibit nothing but bare rock, interspersed with small patches of verdure: indeed, only about 1-8th part of the surface is capable of cultivation. In the valleys, the soil is either alluvial, or a red loamy earth, tenacious of moisture. There are no rivers; and, though numerous torrents flow from the mountains during the winter months, their channels are quite dry in the summer. There is a winter lake, if we may so call it, about 6 m. S. of Amaxichi, in the bottom of a valley, surrounded by lofty mountains, which dries in summer, and produces rich crops. At the S. end of the island, is a shallow lagoon called the Venetian harbour, now rapidly filling up by the accumulation of sand and mud, the banks of which are said to be exceedingly unhealthy. The temperature of Santa Maura, like that of the other islands, is extremely variable; the thermometer in autumn often rising or falling 20° in 24 hours. In the valleys it seldom falls to the freezing point; but occasionally there is snow on the hills. The quantity of rain, and the seasons in which it falls, are much the same as in the adjacent islands. The low grounds are very unhealthy; and fever usually prevails at Amaxichi during summer, attended with a mortality in some years of 1 in 19 of the population: indeed, most of the natives, except those living on the mountains, present a very sickly appearance. (*Major Tulloch's Reports*.) The quantity of corn raised in the island is barely sufficient for half the consumption of its inhabitants; but wine, olive oil, and several varieties of fruit, are produced in great abundance. The sides of the hills afford excellent pasture, and are grazed by large numbers of sheep and goats. Game is plentiful, and bees form an object of rural economy. The salt-pans near Amaxichi produce annually from 3000 to 6000 tons of basset, which, as well as wine, oil, and cotton, are the chief articles of export. The inhabitants are of Greek origin, and belong to the Greek church. Many of them are employed as fishermen and sailors; while others, especially at harvest-season, cross to the neighbouring continent in quest of agricultural employment.

It has several good ports, and some towns; but none is of any importance, except the capital AMAXICHI (which see).

The ancient *Leucas* once formed a part of the continent; for Homer expressly terms it Ἀκτὴ Ἠπείροιο, in opposition to Ithaca and Cephallenia. So late as the time of Thucydides, the Peloponnesian fleet was more than once conveyed across the isthmus; and Livy informs us, that it had its peninsula shape even in the Macedonian war. *Leucadia, nunc insula et vadoso freto quod perfossum manu est, ab Acarnania divisa, tum peninsula erat occidentis regione arctis faucibus cohærens Acarnaniæ. Quingentos ferme passus longæ fauces erant: lato haud amplius centum et viginti.* In his angustiis *Leucas posita est. coli applicata verso in orientem et Acarnaniam.* (*Hist.* lib. xxxiii., 17.) The cut here mentioned, called Dioryctus, was three stadia in length, and, in Strabo's time, was crossed by a bridge. The famous Leucadian promontory (now *Cape Ducato*) is a low ridge of white marble rocks, projecting S. about 2 m., terminating in a precipice 200 feet high. It was surmounted by a temple of Apollo, and Virgil represents it as an object of dread to mariners:

Mox et Leucatæ nimbosa cacumina montis, Et formidatus nautis apertor Apollo. *Æn.* iii., 274.

but it is wholly indebted for its immortality of renown to its being

The Lover's refuge, and the Lesbian's grave,

—the spot whence Sappho precipitated herself into the ocean, resolved either to recover the affections of Phaon, or to die in the attempt. (See *Ovidii Epist. Sappho Phaoni*, lin. 165, &c.)

Not far from the promontory stood the very ancient town of Nericum, mentioned by Homer as εὐκτιμένον πτολίεθρον, "a well-built city," and of which there are still some small vestiges. The position of the ancient Leucas is fixed by Livy in the above passage close to the narrow strait which divides the island from Acarnania; and Dr. Holland mentions the ruins of an ancient town about 2 m. S. of Amaxichi, exhibiting the remains of massive old Greek walls, ascending a narrow ridge near the sea, and of numerous sepulchres, which appear among the vineyards covering its declivity. (*Travels in Greece*, p. 63.) The modern history of Santa Maura is closely connected with that of the IONIAN ISLANDS generally; and to that article the reader is referred. (*Holland*, ii., 34; see also *Dodwell's Greece*, i., 62, &c.)

MAURICE, r., N. J., rises in Deptford and Franklin townships, and flows into Delaware bay. It is navigable for 20 m. from its mouth, for vessels of 80 or 100 tons burthen. In its upper parts, it affords good water-power. There are fine embanked meadows on its borders, and fine oysters are taken at its mouth.

MAURICE RIVER, t., Cumberland co., N. J., 20 m. S.E. Bridgeton. Bounded W. by Maurice river. Drained by Tuckahoe and Tarkill creeks. It contains six stores, two glass houses, seven grist-mills, four saw-mills; eight schools, 280 scholars. Pop. 2142. It has several villages on the E. bank of Maurice river.

MAURICETOWN, p. v., Downe t., Cumberland co., N. J., 75 m. S. by W. Trenton, 187 W. Situated on the W. side of Maurice river, 10 or 12 m. from its entrance into Delaware bay. It contains a Methodist church, an academy, a store, and about 30 dwellings, some of them of brick, and neat.

MAURITIUS (THE), or ISLE OF FRANCE, an island in the Indian ocean, belonging to Great Britain, situated between 19° 58' and 20° 32' S. lat., and 57° 17' and 57° 46' E. long., from 70 to 80 m. N.E. the Isle de Bourbon, and 500 m. E. Madagascar. It is an irregular oval; length, N.E. to S.W., about 36 m.; breadth varying from 18 to 27 m. Area estimated at nearly 500,000 acres. Pop., in 1836, 92,147, of whom 29,512 were whites and free people of colour, 61,045 apprenticed labourers (blacks); now also free, and 1490 strangers. "From whatever quarter it is approached, the aspect is singularly abrupt and picturesque. The land rises rapidly from the coast to the interior, where it forms three chains of mountains, from 1800 to 2000 feet in height, intersecting the country in different directions. Except towards the summit, these are generally covered with wood, and in many parts cleft into deep ravines, through which numerous rivulets find their way to the low grounds, and terminate in about 20 small rivers, by which the whole line of coast is well watered, from the foot of the mountains to the sea. Though, from its mountainous and rugged character, a great part of the interior is not available for any useful purpose, yet extensive plains, several leagues in circumference, are to be found in the high lands; and in the valleys, as well as along the coast, most of the ground is well adapted either for the ordinary purposes of agriculture, or for raising any description of tropical produce. Extensive forests still cover a considerable portion of the districts of Mahébourg, the Savanna and Flacq, and in the centre of the island are several small lakes. The soil, in many parts, is exceedingly rich, consisting either of a black vegetable mould, or a bed of stiff clay of considerable depth; occasionally the clay is found mixed with iron ore and the *débris* of volcanic rock. In the neighbourhood of port Louis, and generally in the immediate vicinity of the sea, there is but a scanty covering of light, friable soil over a rocky surface of coralline formation. The whole coast is surrounded by reefs of coral, with the exception of a few openings, through which vessels can approach the shore; and at these points the different military posts for the defence of the island have been established. There is a marked difference in the climate of this island in different situations; the windward (or S.E.) side enjoying a lower temperature by several degrees than the leeward (N.W.), owing to the cooling influence of the S.E. breeze, which prevails during most part of the year.

In so far as regards temperature, rain, physical aspect, and diversity of climate, this island exhibits a very striking resemblance to Jamaica; though, being S. of the line, the seasons are reversed; summer extending from Oct. to April, and winter during the rest of the year. The principal rainy season is from the end of December to the beginning of April, but showers are frequent at all times. Hurricanes are of frequent occurrence, and create great devastation, with much loss of life: they principally occur in January, February, and March. "So far as can be ascertained from

the statistical returns of the island, the climate does not exert any prejudicial influence on the health of the white resident population, though it is by no means favourable to the negro race." (*Tulloch's Report on the Sickness, &c., of the Troops in W. Africa, &c.*, p. 3, 4, c.)

Previously to 1825, the sugar and other articles imported from the Mauritius into Great Britain were charged with the same duties that were laid on such articles when imported from India. But, at the epoch now alluded to, the produce of the Mauritius was admitted into our markets at the same duties as W. Indian produce, which were then materially lower than those imposed on the produce of our Eastern possessions. This alteration of the duties gave a great stimulus to cultivation in the Mauritius, particularly to that of sugar, which has since been raised, to the almost total exclusion of coffee, cotton, and indigo, that were previously produced in considerable quantities, the coffee especially being of excellent quality. Wheat and maize are raised in small quantities, with yams, manioc (introduced by the French), potatoes, bananas, and other vegetables. But the island is almost wholly indebted for its supplies of provisions to Hindostan, the cape of Good Hope, Madagascar, the isle de Bourbon, &c. Next to sugar, blackwood, or ebony, of which there is an immense supply, and tortoise-shell, are the principal articles of export.

The exports of sugar from the Mauritius amounted, in 1812, to less than *one* million lbs. In 1814 they amounted to 1,034,294 lbs.; and, in 1818, to 7,908,380 lbs. Since then, but especially since the modification of the duties in 1825, there has been a most extraordinary increase in the export of sugar from the island. It amounted in

		lbs.			lbs.
1820	to	15,594,755	1835	to	64,854,515
1825	.	21,739,766	1836	.	63,357,317
1826	.	42,489,416	1837	.	68,478,874
1830	.	67,926,692			

But the exports of other things are comparatively trifling; having amounted, in 1836, to only 15,819 galls. rum, 23,389 lbs. cloves, and 664,369 lbs. ebony.

In 1837, the total value of the imports amounted to £1,035,783, of which cotton manufactures and other articles from Great Britain made £345,744. The total value of the exports (including £77,799 for imports re-exported) during the same year, amounted to £831,050, of which sugar produced no less than £739,979: of the total export of sugar that year, amounting to 68,478,874 lbs., 57,158,449 lbs. were shipped for Great Britain, 7,585,197 lbs. for New South Wales, 3,896,763 lbs. for the cape of Good Hope, and some small quantities for other places. Among the imports were 54,695,000 lbs. of rice, and nearly 5,000,000 lbs. of wheat, from India; with considerable, though far less extensive, supplies from the cape and Madagascar. During the same year, 433 British ships, of the aggregate burden of 95,831 tons, and 65 foreign ships, of the burden of 16,492 tons, entered the ports of the island.

Nine tenths of the sugar produced in the island, comes to England. We have seen above the distribution of the shipments in 1837; and it appears, from the customs' returns, that we imported, in 1838, 67,874,198 lbs.; and, in 1839, 69,294,960 lbs., from the Mauritius. We believe, however, that the culture of sugar in the Mauritius has attained to a maximum. Indeed, the fair presumption seems to be that, in future, it will decline rather than increase. We found this opinion partly and principally on the greater natural facilities enjoyed by India for the culture of the cane, and partly, also, on the influence of the abolition of slavery, in lessening the supply of labour in the island.

The emancipation of the slaves in the Mauritius does not appear to have been so prejudicial to agriculture as in the W. Indies. By way of supplying the demand for labour, a considerable number of hill coolies have been brought from Hindostan; but it has been contended, and we believe truly, that, despite the regulations under which the coolies were introduced, they would be, in fact, little else than slaves; and their introduction has, in consequence, been stopped. Chinese settlers have also been introduced, but not in any considerable numbers.

The government is vested in a governor, with a salary of £7000 a year, and a colonial legislative council, subordinate to the orders of the sovereign in council. The governor is aided in his duties by an executive council, composed of the military officer second in command, the colonial secretary, and the advocate-general. The legislative council is composed of 15 members, seven of whom hold no official situation. Justice is administered in a supreme civil and criminal court, with three judges, a petty court, from which there is no appeal, and such other minor courts as the governor may see fit. Several provisions of the old French law continue in force. The troops employed in this command have consisted either of two complete, or the service-companies of three, regiments of the line, with one company of sappers and miners, and half a company of artillery. The service-companies of two corps, with the head-quarters of the artillery and sappers and miners, are generally at Port Louis: those of the other corps are distributed between the different stations on the windward side of the island, having their head quarters at Mahébourg. The public revenue, in 1835, amounted to £187,780, and the internal colonial expenditure to £177,740, leaving a surplus of about £10,000, which was to be paid over in aid of the expenditure incurred in Great Britain, in the same year, on account of the colony, amounting to £78,284. (*Parl. Paper*, No. 632, Sess. 1848.) The greater portion of the revenue is derived from the customs' duties received at Port Louis.

Port Louis, or N.W. port, the capital and seat of government, is on the N.W. side of the island, in lat. 20° 9' 6" S., long. 57° 29' 41" E. Pop. 25,000. It is situated at the bottom of a triangular bay, the entrance to which is rather difficult. Every vessel approaching the harbour must hoist her flag and fire two guns; if in the night, a light must be shown, when a pilot comes on board, and steers the ship to the entrance of the port. It is a very convenient port for careening and repairing, but provisions of all sorts are dear. In the hurricane months the anchorage of port Louis is not good, and it can then only accommodate a few vessels. The streets are tolerably regular; but the houses are low, and are principally built of wood. It has extensive, but generally very filthy suburbs. It suffered severely from fire in 1816, and from the cholera in 1819. The town and harbour are pretty strongly fortified. At the W. extremity of the town are some extensive and commodious barracks; and about ½ m. distant is the hospital, on a peninsula of coral rock, jutting into the sea. Mahébourg, is a healthy situation on the S.E. coast, with an excellent harbour, was opened to ships from distant countries in 1836.

The Mauritius has numerous small dependencies between lat. 3° and 20° S., and long. 50° and 70° E. The chief of these are the Seychelles islands, between lat. 4° and 5°, about 830 m. N. from the Mauritius; one of which, Mahé, is 16 m. long, by from 3 to 4 m. broad; fertile, well-watered, very healthy, and having a population of about 7000. Mahé, its chief town, has on its N.E. side about 100 wooden houses, and a garrison of 30 men.

The Mauritius was discovered, in 1505, by the Portuguese. The Dutch took possession of it in 1598, and named it Mauritius in honour of Prince Maurice. They made a settlement in it in 1644, which, however, they abandoned early in the next century. The French having, in 1657, occupied Bourbon, sent occasional settlers to the Mauritius, and, on its evacuation by the Dutch, they established a regular colony in the island in 1715, of which, however, they did not take formal possession till 1721. But the real founder of this important settlement was the justly celebrated M. de la Bourdonnaye, appointed governor in 1734. The isle de France had hitherto been neglected for that of Bourbon, and was, at the arrival of the new governor, in the most impoverished and disordered state imaginable. But M. de la Bourdonnaye immediately perceived the importance of the island, which its two excellent harbours rendered of the greatest consequence to any European power having, or wishing to have, possessions in India; and he set about its improvement with a zeal, sagacity, and success that have rarely been equalled, and never surpassed. Besides extending the culture of the sugar cane, coffee, cotton, and indigo, he introduced the manioc from S. America, and cinnamon, cloves, pepper, &c., from the Dutch islands; though the latter, with the exception of cloves, have not answered his expectations. He fixed the seat of government at Port Louis, which he may be said to have created as well as fortified; and constructed numerous roads, aqueducts, and other useful public works. His administration continued only for 11 years; and in that short space he contrived to change the whole aspect of the country, and rendered it a most prosperous and valuable colony. Even after the possessions of France in India had all fallen into our hands, the Mauritius continued to be of importance to her, and proved how justly M. de la Bourdonnaye had appreciated its capabilities as a naval station. It was estimated that, during the first 10 years of last war, the value of the British ships captured by privateers and other cruisers from the Mauritius amounted to £2,500,000. At length, a formidable armament being sent against it in 1818, it surrendered to our arms, and was definitely ceded to us in 1815.

Every body knows that this island is the scene of St.

* The signal services rendered by M. de la Bourdonnaye here and in India met with a most ungrateful return. On his return to France, in 1748, he was thrown into the Bastile, where he was immured for more than three years, without, as it turned out, there being the smallest foundation for any one of the charges made against him! He died, the victim of this disgraceful treatment, in 1755. (See *Biographie Universelle*, art. *Mahé de la Bourdonnaye*.)

Pierre's admirable tale of Paul and Virginia. The wreck of the St. Geran, so striking and affecting an incident in the story, is a real event, which took place on the 18th of August, 1744. (See *Almanac de l' Isle Maurice pour 1837*; and *Parl. papers*.)

MAURY, county, Tenn., situated a little S.W. of the centre of the state, and contains 750 sq. m. Erected in 1807. Watered by Duck river and its tributaries. It contained in 1840, 22,614 neat cattle, 27,294 sheep, 105,175 swine; and produced 131,027 bushels of wheat, 13,649 of rye, 3,342,682 of Indian corn, 218,856 of oats, 94,616 of potatoes, 51,376 pounds of tobacco, 2,091,876 of cotton. It had 34 stores, two cotton factories with 796 spindles, 23 grist-mills, 20 saw-mills, 10 tanneries, 23 distilleries, two printing-offices, two weekly newspapers; one college, 85 students; 7 academies, 396 students; 98 schools, 703 scholars. Pop.: whites, 17,060; slaves, 11,002; free colored, 94; total, 28,186. Capital, Columbia.

MAXATAWNY, t., Berks co. Pa., ¾ m. N.E. Harrisburg, 165 W. Drained by Maiden creek, and its tributary, Saccony creek, which afford water power. It contains two churches, one Lutheran, and one common to Lutherans and Presbyterians, three stores, one grist-mill, one tannery. Population, 1807.

MAY, cape, N. J. situated at the N.E. side of Delaware bay, in 38° 56' N. lat., and 74° 56' W. lon. It has a light-house on the point. If received its name from Cornelius Jacobus Mey, a distinguished Dutch navigator, who visited the Delaware in 1623.

MAYBOLE, a bor. of barony and market-town of Scotland, co. Ayr, distr. Carrick, of which it is the cap., on the slope of a hill with a S. exposure, 8 m. S. Ayr, and 38 m. S. by W. Glasgow. Pop., in 1837, 4,090. The town consists mainly of an antique well-built street, interspersed with numerous modern buildings. The superiority of the old houses is owing to Maybole having been, in ancient times, the town residence of the aristocracy of Carrick; and the remains of no fewer than 28 baronial mansions are still more or less entire. Of these, the most imposing is "the Castle," once occupied by the Earl of Cassillis, the principal part of which is still standing. The only modern building is the parish church, erected in 1808. There is also a dissenting chapel. Hand-loom weaving, in connection with the Glasgow cotton manufacture, is extensively carried on, employing from 600 to 800 hands. The weavers are mostly Irish; boys and females also engage in the work, and perpetuate the poverty inseparable from the business. A weekly market is held in the town; and it has a bank, and a savings' bank. The parish school and the other schools bear a good character. There are two subscription, and two circulating, libraries. Poor-rates have not been introduced; but, when occasion requires, the inhabitants and land-owners submit to a voluntary contribution to meet the case.

Maybole, being the capital of the bailiery of Carrick, was the seat of the courts of the district previously to the abolition of hereditary jurisdictions in 1747. The remains of Crossraguel abbey are situated close to the town, on the W. A celebrated disputation, which lasted three days, between Quintin Kennedy, one of its abbots, and John Knox, took place, in 1561, in a house in Maybole, now "the Red Lion inn;" it is needless to add, that it ended, as in universally the case with such disputes, in a drawn battle, each party claiming the victory, and conceiving that he had demolished his antagonist. (*M' Crie's John Knox*, p. 241. ed. 1829; *New Statistical Account of Scotland, § Ayrshire*, p. 348.)

MAYENNE, a dep. of France, reg. N.W., formerly comprised in the prov. of Maine; between lat. 47° 49' and 48° 34' N., and long. 0° 5' and 1° 20' W., having N. Manche and Orne, E. Sarthe, S. Maine-et-Loire, and W. Ille-et-Vilaine. Length, N. to S., 55 m.: average breadth about 30 m. Area, 514,803 hectares. Pop. (1836), 352,588. A mountain chain, though of no great height, bounds Mayenne to the N., from which two ranges strike off to the S., one forming the E., and the other a part of the W. boundary of the department. It slopes generally from N. to S., in which direction it is intersected near its centre by its principal river, the Mayenne. The latter rises in the department of Orne, about 12 m. W. Alençon, running at first S.W., and afterwards generally S., through the department of Mayenne and Maine-et-Loire; in the last of which, after receiving the Sarthe and Loire, it assumes the name of the Maine, and falls into the Loire, after an entire course of nearly 130 m. (See also MAINE-ET-LOIRE.) Mayenne, Laval, Château-Gontier, and Angers, are on its banks. There are numerous small lakes in this department. In 1834, it was supposed to comprise 354,298 hectares of arable land; 60,338 do. pasture; 96,379 do. woods; and, 94,629 do. heaths, wastes, &c. More corn is grown than is required for home consumption. In 1835, the total produce was reckoned at upwards of 3,000,000 hectolitres, chiefly wheat, oats, and rye. The annual produce of cider is said by Hugo to be about 600,000 hectolitres. Some inferior wine is produced, but in small quantities only. Flax, hemp, chestnuts, and some other fruits, are the other principal products. Property is very much subdivided; and many of the farms, or rather patches called *closeries*, are so very small that they do not admit of the use of the plough, and are cultivated by the spade only! In 1835, of 36,503 properties subject to the *contribution foncière*, 27,137 were assessed at less than 5 fr., 13,009 at from 5 to 10 fr.; and 13,231 at from 10 to 20 fr., and only 17 were assessed at 1,000 fr. or upwards. We need not, therefore, be surprised to learn that the occupiers are generally destitute of either capital or enterprise, and strongly attached to routine practices. In 1830, there were about 214,000 head of cattle, and 144,000 sheep, in the department; the produce of wool being estimated at 175,000 kilogr. a year. The woods yield excellent timber, a good deal of which is appropriated to ship-building. Some extensive manufactures of linen stuffs and yarn were formerly established at Laval and Château-Gontier. But though these have fallen off, the linen and cotton fabrics, including printed handkerchiefs, &c., of the department, still enjoy a high reputation. The iron trade of Mayenne is of considerable importance; and it also furnishes superior paper. It is divided into three arrondissements; chief towns, Laval the capital, Mayenne, and Château-Gontier. It sends five members to the chamber of deputies: number of electors (1839), 1716. Total public revenue (1831), 6686, 211 fr.; and expenditure, in the same year, 3,791,930 fr. (*Hugo*, art. *Mayenne; Official Tables.*)

MAYENNE, a town of France, in the above dep., cap. arrond., on both sides the Mayenne, 18 m. N.N.E. Laval. Pop. (1836), 8790. The town proper stands on the right; or W., bank of the river; the portion on the opposite bank, though comprising a third of the entire population, being only a suburb. They are connected by a bridge. This is an ill-built town; its streets are steep, irregular, and inconvenient, and its houses old and odd-looking. The castle of Mayenne, founded in the eighth century, made some figure in the wars between England and France; having sustained several sieges, especially one in 1424, when it capitulated to the earl of Salisbury, after resisting four successive assaults. It is now in ruins, and is separated by a planted promenade from the linen-hall of the town. Mayenne has two parish churches, two hospitals, a good town-hall, &c.; and manufactures of linen and cotton fabrics; the former of which has, however, greatly declined of late years, while the latter has increased. (*Hugo*, art. *Mayenne; Guide du Voyageur.*)

MAYFIELD, p. t. Fulton co., N. Y., 8 m. N.E. Johnstown, 58 m. N.W. Albany, 422 W. Watered by Stoney creek, and other branches, of Sacandaga river. It contains a church, four stores, two fulling-mills, two cotton factories, four grist-mills, 18 saw-mills, two paper-mills, two tanneries; 13 schools, 536 scholars. Pop. 9615.

MAYFIELD, p. v., capital of Graves co., Ky., 275 m. W.S.W. Frankfort, 802 W. Situated on a head branch of Mayfield's river. It contains a courthouse and jail, of brick, two churches, a Presbyterian and Methodist, three stores, 25 dwellings, and about 150 inhabitants. Nett proceeds of the postoffice, $118.

MAYFIELD, p. t., Cuyahoga co., O., 161 m. N.N.E. Columbus, 389 W. Watered by Chagrine river, on which is situated the village of Gates' Mills. It contains two stores, one fulling-mill, one woollen factory, two grist-mills, four saw-mills, two tanneries; six schools, 260 scholars. Pop. 851.

MAYN, or MAIN, a river of central Germany, which has its source in Bavaria. It is formed by the union, about 24 m. N.W. Bayreuth, of the White and Red Mayn; the former rising in the Fichtelberg, and the latter in the Frankenjura, about 8 m. S.S.E. Bayreuth. The resulting river flows, with a gentle current, generally W., but with a very tortuous course; first traversing the circles of Upper and Lower Franconia, in Bavaria, then dividing Hesse-Cassel and Nassau, on the N., from Hesse-Darmstadt, on the S., and intersecting the territory of Frankfort; till it ultimately falls into the Rhine, nearly opposite Mentz, after an entire course of about 220 m. Though shallow, it is of equal depth, and is navigable throughout seven eighths of its extent, as far as the confluence of the Regnitz, near Bamberg. The Mayn is of considerable importance as a means of traffic; and Frankfort, especially, owes all its consequence, as a commercial city, to this river. But few rivers, at least in civilised countries, presenting such facilities for improvement, have been more neglected; and, in addition to other inconveniences, the number and amount of the tolls levied on the Mayn oppose serious obstacles to its navigation. A vessel, in passing from the Rhine to Wertheim, in Baden, has to pay tolls amounting to 23 kreutzers per cwt., or 14s. a ton, besides fees, &c. A commission has, however, been recently appointed, for

examining the state of the river, and introducing some improvements. The Mayn will, probably, at no distant period, form a part of the line connecting the Rhine and the Danube; a canal having been already commenced, which is to run from Dietfürth, on the Altmühl, to Bamberg, on the Regnitz. Besides the Regnitz, the Tauber, Mänling, and Gersprenz are its chief affluents from the S., and the Rodach, Saale, Kinzig, and Nidda from the N. Bayreuth, Bamberg, Wurtzburg, Aschaffenburg, Hanau, Offenbach, and Frankfort, are either on, or immediately adjacent to, the banks of the Mayn. (*Dict. Géog.; Berghaus; Private Inform., &c.*)

MAYNOOTH, an inland town of Ireland, prov. Leinster, co. Kildare, on the Lyall Water, an affluent of the Liffey, 14 m. W. Dublin. Pop., in 1831, 2,053. It is without trade, and depends principally for its support on the contiguous college. It has a parish church, a Roman Catholic chapel, and the ruins of a large castle, once occupied by the family of Kildare.

The Royal college of St. Patrick, Maynooth, was founded in 1795, for the education of persons designed for the Roman Catholic ministry in Ireland. It is placed under the direction of a board of trustees, of whom the Roman Catholic archbishops are members *ex officio*, the remainder being selected from the Catholic hierarchy and nobility, in the proportion of seven of the former to six of the latter. An additional board of control was appointed by parliament in 1800, consisting of the lord-chancellor of Ireland, the chief justices of the Queen's Bench, Common Pleas, and Exchequer, the four Roman Catholic archbishops, and the earl of Fingall. This board holds triennial, or, if necessary, more frequent visitations, and has power to examine into all matters connected with the college. The chief functionaries of the establishment are the president, vice-president, and two deans; besides whom, there are three professors of divinity, and five others, giving instruction in various branches of literature and science. The number of students, on the first opening of the classes, in 1795, amounted only to 50; but it has since progressively increased to 450, at which it is limited, by the inadequacy of its funds to admit of further augmentation. The funds consist principally of an annual parliamentary grant of 8,928*l.*, and 2,000*l.* a year are derived from donations and bequests. The income is applied to the maintenance of an order of senior students, nominated from the four ecclesiastical provinces, who not only pursue the higher courses of study, but are also required to take part in the business of instruction. Their number is limited to 20, each of whom has an allowance of 60*l.* a year, half being deducted for board, and other collegiate expenses. Thirty bursaries have likewise been founded, of different annual amounts, from 30*l.* downwards. Two other orders of students have been formed, under the name of pensioners and half-pensioners, the former paying 22*l.* a year, and the latter half that sum, for board, &c. Each free student pays an entrance-fee of eight guineas, but the pensioners pay only four guineas. The free students, of whom there are 250, are appointed, after examination, by the bishops of the several dioceses, and are supplied gratuitously with lodging, commons, and instruction during the courses, which ordinarily occupy seven years. The students belong, with but few exceptions, to the middle and lower classes of Roman Catholic farmers and occupiers. Public examinations are held twice a year. The site of this establishment is a tract of 54 acres, adjoining the town; and the buildings, which form three sides of a quadrangle, comprise the chapel, refectory, library, lecture-rooms, dormitories, and professors' residences. The library contains about 10,000 volumes, chiefly on theological subjects.

"There is much room for doubt, whether the advantages anticipated by government from this institution, have been realized; or, if realized, whether they are not more than countervailed by corresponding disadvantages. There seems to be a very general opinion, among those who have the best means of judging, that the Catholic clergy educated at Maynooth are less intelligent, less liberal, and less gentlemanly, in their manners, than their predecessors; and we should be inclined, on general grounds, to think that this must be the case. The Catholic clergy educated abroad were generally of a superior class to those educated at Maynooth; the scholastic instruction they received at their foreign seminaries was not, perhaps, better than what they now receive at home; but it is pretty certain that their scientific and general education was incomparably superior, and that they were less imbued with sectarian prejudices. The pupils at foreign seminaries mixed much more with the world, especially with the upper classes, than those educated at Maynooth; their opportunities for observation were consequently much greater, and their minds were necessarily liberalised by their familiarity with tastes, habits, and modes of thinking, widely different from their own.

"The students now at Maynooth belong, with but few

exceptions, to the middle and lower classes of Catholic farmers and occupiers. The discipline to which they are subjected appears, judging from its effects, to be well fitted for forming skilful controversial divines, and zealous Catholics: but, in all that tends to expand and liberalise the mind, it is exceedingly defective; and the society of the student during the period of vacation, provided he be allowed to leave the college, is better calculated to increase than to lessen the deficiency. It is difficult to see how this unfavourable state of things is to be amended. Some have proposed opening the general classes of Trinity college to Roman Catholics as well as Protestants; and others have suggested the formation of a new establishment, where clergymen of both sects might be educated, the literary and scientific classes being common to all the students; and there being, at the same time, separate courses of theology by Catholic and Protestant professors, for the students of each persuasion. Either of these plans would, no doubt, materially weaken prejudices, and introduce greater liberality of opinion; but both the one and the other are opposed by great and, we fear, all but insurmountable obstacles." (Statistical Account of British Empire, vol. ii., p. 385-388.)

MAYO, a marit. co. of Ireland, prov. Connaught, of which it occupies the N.W. portion; having N. and W. the Atlantic, E. Sligo and Roscommon, and S. Galway. Area, 1,355,063 acres; of which 625,124 are unimproved mountain and bog, and 37,940 water, consisting principally of loughs Mask, Conn, Carra, &c. The coast-line is extremely irregular, from its being more deeply indented with bays and arms of the sea than any other part of Ireland. On the W. it is fenced with numerous islands, and it has several fine harbours, of which, however, very little use is made. It has every variety of surface, rising, in parts, into high mountains and rugged wastes; but comprising, also, a large extent of comparatively flat and fertile land. The substratum is generally limestone; and, from the thinness of the soil and the humidity and mildness of the climate, it is better suited for grazing than for tillage. Property is, a few hands. There were formerly some very extensive grazing farms in this county, but their number and size have been very greatly diminished within the last half century. Land being here indispensable to existence, the competition for small patches is quite intense; and it is said that any amount of rent that may be asked is sure to be promised! It was formerly usual to let land on the village, or partnership, system; but of late years this practice has, luckily, been getting into disuse. Unfortunately, however, the censers system seems to be rapidly extending; and this is, if possible, worse than the other. (See antè, p. 34.) Hence, notwithstanding the increase of cultivation, the condition of the great bulk of the occupiers of land has deteriorated, and is, at present, as bad as possible. Average rent of land, 8s. 6d. per imperial acre; but the best grazing lands fetch above 40s. per Irish acre. Iron used to be made in this county; the works have, however, been long abandoned, on account of the want of fuel. It has, also, some valuable slate quarries; but its mineral riches have been but very imperfectly explored. The linen manufacture, which had been pretty widely diffused, has materially declined, and its place has not been occupied by any other department of industry. Principal rivers, Moy, Guisbden, Deal, Owenmore, and Robe. Principal towns, Castlebar, Balline, and Westport. Mayo is divided into nine baronies and 66 parishes. It sends two members to the House of Commons, both for the county. Registered electors, in 1839-40, 2,185. In 1831, Mayo had 62,257 inhabited houses, 65,907 families, and 366,328 inhabitants, of whom, 179,595 were males, and 186,733 females.

MAY'S LANDING, p. v., Hamilton t., capital of Atlantic co., N. J., 73 m. S. Trenton, 183 W. Situated at the head of sloop navigation, on both sides of Great Egg Harbor river, 16 m. from the ocean. It contains a courthouse, jail, Methodist church, four stores, and about 30 dwellings. It has considerable trade in cord-wood and lumber, and has some ship-building.

MAYSVILLE, p. v., capital of Buckingham co., Va., 79 m. W. Richmond, 158 W. Situated on the S.E. side of Slate river, which flows into James river. It contains a courthouse, jail, five stores, and about 360 inhabitants.

MAYSVILLE, Morgan co., O. It has two grist-mills, four saw-mills; six schools, 175 scholars. Pop. 1,159.

MAYSVILLE, city, Mason co., Ky., 61 m. N.E. by E. Frankfort, 451 W. Situated on the S. side of Ohio river, just below the entrance of Limestone creek. It was in early times called Limestone. It has a fine harbour for boats, situated on a narrow bottom, between the Ohio and the high hills in its rear. It has three streets running parallel with the river, and four others which cross them at right angles. It contains three churches, and about 500 dwellings; and is the depôt for most of the goods and merchandise for the E. part of Kentucky. It contained in 1840, nine commission houses in foreign trade, capital,

$111,600, 29 retail stores, capital, $133,000, two lumber yards, capital, $10,500, one cotton factory, 1100 spindles, one flouring-mill, one saw-mill, one tannery, one brewery, two printing-offices, three weekly and one semi-weekly newspapers; one academy, 70 students, three schools, 900 scholars. Pop., 2741.

MAYVILLE, p. v., Chautauque t., capital of Chautauque co., N. Y., 21 m. N.W. Jamestown, 334 m. W. by S. Albany, 339 W. Situated on elevated ground at the N. end of Chautauque lake. The buildings are mostly on one broad street and a public square, with a delightful water prospect down the lake. It contains a fine courthouse, which cost $9000, a jail, a fire-proof county clerk's office, all of brick or stone, four churches, a Baptist, Presbyterian, Methodist, and Episcopal, an academy, eight stores, 90 dwellings, and 600 inhabitants. A steamboat plies on the lake, between this place and Jamestown.

MAZAMET, a town of France, dep. Tarn, cap. cant. on the Arnette, a tributary of the Tarn, 32 m. S.S.E. Albi. Pop., 1836, 4438. Its population and prosperity are increasing: it has some manufactures of woollen cloth; several dyeing establishments and paper-mills, and considerable annual fairs for cattle, wool, &c.

MAZANDERAN, a prov. in the N. of Persia (an. a part of Hyrcania), separated from Irak-Adjemi by the lofty ridge of Elburg, and bounded N. by the Caspian Sea, E. by Khorassan, and W. by Ghilan. Length from W. to E., 200 m.; average breadth, 50 m.; area, 10,000 sq. m. Pop., 150,000 (exclusive of the nomad tribes of Kadjars, Kodjavends, and Modanlus). The S. parts of the province are mountainous, abounding with oak-timber, and full of swamps; but the valleys are fertile, producing the finest rice in vast quantities. Besides many smaller streams, the Mazanderan has two principal rivers, both having their sources in the Elburz mountains, and falling into the Caspian sea. With respect to temperature, the province may be divided into a warm and a cold climate, the former being that of the flat country near the sea, and the latter that of the mountain region: in all parts, however, the climate is extremely variable with respect both to temperature and moisture. (Fraser's Caspian Sea, p. 48.) Winter and spring are the healthiest seasons; for during the summer and autumnal heats, such exhalations arise from the fens and marshes overspreading this part of Persia, as to render the air most insalubrious. Agues and dropsies, rheumatism and eye-diseases, are the prevalent disorders, and the natives have generally a sallow and bloated appearance. Heavy rains fall in October, November, and December: snow also falls, but never lies long on the ground; and in spring, the rivers almost invariably overflow. The cultivation of rice is the most important branch of agriculture. Cotton and sugar, also, are raised; but the canes are small, and the produce is dark, moist, and of very inferior quality. Tobacco does not succeed, nor is opium much cultivated, though the poppy grows abundantly. Barley is sown occasionally in spring as a green crop for horses and cattle: it is cut about the middle of May, after which the ground is ploughed, and planted with rice. Wheat is little cultivated, and is of bad quality; but excellent flour is imported from Astrakhan. Unhusked rice is used as dry food for horses and mules. Silk was formerly raised only in small quantities; but it appears that since the government monopoly ceased some few years ago, more attention has been paid to it, and its price has fallen. The trade of the province is chiefly with Russia, in rice, silk, and cotton, which it exchanges for silk, cotton, and woollen fabrics, corn, tobacco, cutlery, &c.

The inhabitants are described as "vain, ignorant, and arrogant, considering themselves as persons of mighty importance, superior to all strangers. Their ignorance of everything beyond their own province, is profound, to a degree hardly credible. Their bigotry in religious matters is excessive, though chiefly confined to forms; for there are few who do not transgress every article of inhibition: all of them drink strong liquors and eat opium." In their appearance and dress, they greatly resemble other Persians, but swarthy and almost black men are of more frequent occurrence than elsewhere. The natives are regarded as the most warlike of the Persians; and, in the time of Timour Bec, they defended their retreats and castles with so much courage and ability, as to secure their independence. This province is also said to have been the grand seat of the war between the Sefeed Deeve (or White Demon) and Rustom, prince of Zablestan; and the relief of his sovereign, who had been besieged in the city of Mazanderan, is one of the most glorious exploits recorded in the life of the Persian hero. The chief cities are Sari, Balfrosh, Ferreabad, and Amul. Most of the towns and villages are open, well-built, and delightfully situated either on verdant hills, or in fertile and well-watered valleys. Among the numerous public works of Shah Abbas the Great, is a magnificent causeway of great length, running nearly parallel to the Caspian. The pavement, even now, is perfect in many places, though it has hardly ever been repaired. In some places it is above 90 yards wide, with ditches on each side; and on it are many bridges, under which the water is conveyed to the rice-fields. (*Fraser's Travels on the Shores of the Caspian; Kinneir, &c.*)

MAZZARA (an. *Emporium*, or *Mazzara*), a seaport town of Sicily, on its W. coast, intend. Trapani, cap. distr., on the Salemi (an. *Mazzara*), at its mouth, 11 m. S.E. Marsala, lat. 37° 30′ 56″ N.; long. 12° 34′ E. Population, in 1831, 8365. It is surrounded by an old wall of Saracenic construction, flanked by small square towers, and has an old ruinous castle at its S.W. angle. The domes of its churches give Mazzara an imposing aspect from the sea; but the contrast, on entering the town, is no less striking. The streets are narrow, unpaved, filthy, and swarm with pigs: the public buildings, for civil purposes, are large, heavy, and mean; and those for ecclesiastical purposes, being very numerous, we need not wonder that it should have become a common saying. "that every house in Mazzara has a priest and a pig;" the latter being by far the, more useful animal of the two. The principal square has a singular appearance. from the antiquated style of its architecture; probably of the 11th century, from its having an equestrian statue of Count Roger destroying a Saracen, over the cathedral gate. Besides the cathedral, an edifice, remarkable for its fine cupola, the principal buildings in this square are the bishop's palace, the senate-house, and the residence of Count Gazziri. "In the cathedral porch are preserved three sarcophagi; the finest of them bears a bas-relief, representing the battle of the Amazons; the second, the rape of Proserpine; the third, and most inferior, the Calydon hunt. At the convent of St. Michael is a Roman tomb, and some marble inscriptions; these, with a small collection of Punic, Saracenic, and Roman coins, are nearly the sum of the antique remains. Nor are there any modern specimens of the fine arts." (*Smythe's Sicily*, 298.) Mazzara has a hospital, a college, and a theatre. Its port, which now, as in antiquity, is formed by the mouth of the river Salemi, is convenient enough for boats and small craft, but larger vessels are obliged to anchor in an exposed roadstead, in from eight to 10 fathoms water. The entrance of the port is ornamented by a statue of St. Vitus, the tutelary saint of the town, in whose honour a festival is held here in August. Notwithstanding the badness of its port, Mazzara enjoys a considerable trade. It has a *carricatore*, for the warehousing of corn, of which it exports considerable quantities; and it also exports pulse, wine, fruit, fish, barilla, madder, oil, and soap. (*Smythe's Sicily*, 294–298, &c.)

Mazzara, or Emporium, was taken by storm by Hannibal, previously to his commencing the siege of Selinus; but it does not appear to have been a place of much importance in antiquity. There can, however, be no doubt, were Sicily subject to a vigorous and enlightened government, capable of developing its gigantic resources, that Mazzara would rise to very considerable distinction as a shipping port. It was here that the Saracens landed when they invaded and conquered Sicily. (*Smythe, ubi supra; Ancient Universal History*, XVII. 360. 8vo. ed.; *Hoare's Classical Tour, &c.*, i. 75–77.)

An extraordinary phenomenon, called the *Marobea*, being a violent agitation of the sea, is witnessed on this part of the Sicilian coast. "Its approach is announced by a stillness in the atmosphere, and a lurid sky; when suddenly the water rises nearly two feet above its usual level, and rushes into the creeks with amazing rapidity; but, in a few minutes, recedes again with equal velocity, disturbing the mud, and occasioning a noisome effluvia: during its continuance, the fish float quite helpless on the turbid surface, and are easily taken. These rapid changes generally continue from half an hour to upwards of two hours, and are succeeded by a breeze from the S., which quickly increases to heavy gusts." Captain Smyth by some speculations as to the cause of this singular phenomenon, for which we beg to refer to his work. (*Smythe's Sicily*, pp. 294–298.)

MEAD, county, Ky. Situated in the N. part of the state, and contains 360 sq. m. Bounded N. and N.W. by Ohio river, and watered by small creeks flowing into it. It contained in 1840, 5714 neat cattle, 7045 sheep, 19,586 swine; and produced 41,597 bushels of wheat, 2308 of rye, 190,223 of Indian corn, 75,529 of oats, 9199 of potatoes, 170,464 pounds of tobacco. It had 17 stores, one fulling-mill, two woollen factories, one cotton factory with 1370 spindles, five flouring-mills. four grist-mills, five saw-mills, two tanneries, three distilleries; 10 schools, 237 scholars. Population—whites, 4369; slaves, 1408, free coloured, 5; total, 5780 Capital, Brandenburg.

MEADVILLE. p. b., capital of Crawford co., Pa., 234 N.W. by W. Harrisburg, 87 m. S. Erie on lake Erie, 94 m. N.W. Franklin on Alleghany river, 307 W. Situated on the E. side of French creek. The townplat gradually rises from the creek to its centre, where is a handsome public

square of five acres, on the E. side of which is an elegant courthouse of brick and cut stone, ornamented with a handsome cupola. It contains seven churches, two Presbyterian, a Methodist, Baptist, Episcopal, Cumberland Presbyterian, and Unitarian, an academy, a state arsenal, 14 stores, one fulling-mill, one grist-mill, one saw-mill, two furnaces, three tanneries, two printing-offices, two weekly newspapers, three schools, 163 scholars. Pop. 1319.

It is the seat of Alleghany college, founded in 1815, which has a president, four professors or other instructers, 160 students, and 8000 volumes in its libraries, chiefly the donation of distinguished gentlemen in Massachusetts. The library contains many rare and valuable works. A portion of the students adopt the manual labour system. The institution is under the direction of the Methodists.

MEADVILLE, p. v., capital of Franklin co., Miss., 80 m. S.W. Jackson, 1090 W. Situated on the W. side of Bayou chitto river. It contains a courthouse, and several stores and dwellings. Nett proceeds of the postoffice, $112.

MEATH, a marit. co. of Ireland, on its E. coast, prov. Leinster, having N. the counties of Louth, Monaghan, and Cavan, W. Westmeath, S. King's county and Kildare, and E. Dublin and the Irish sea. Area, 567,127 acres, of which only about 5000 are said to be unimproved or waste. Surface mostly flat, or only slightly undulating; soil, clay or loam, on limestone or gravel, and generally very fertile. Grazing used formerly to be the principal occupation; but, since the close of the American war, tillage has been gradually extending, and is now spread over more than four-fifths of the county. Notwithstanding the richness of the soil, and the favourable situation of Meath, the state of agriculture, and of the great bulk of the occupiers, are alike bad. A rotation of crops is only beginning to be introduced; corn frequently follows corn for a long series of years; when fallows do occur, they are in general wretchedly executed, so that the land is in general foul and in bad order. Latterly, however, a better system has begun to make its way into the county, and the stock and implements of husbandry have been a good deal ameliorated. Even the better sort of farmers are, for the most part, very badly lodged; and the cabins of the cottiers and labourers are in the last degree filthy and wretched. Potatoes constitute three fourths of the food of the bulk of the people; oatmeal and churned milk are sometimes added, but they rarely taste butchers'-meat, the pig being usually sold to assist in paying the rent. There are some large estates; but property is better divided than in most Irish counties. Tillage farms vary from five to 50, and some few extend to 100 acres. Average rent of land, 18s. an acre, which is higher than that of any other Irish county, except Dublin. Minerals and manufactures of no importance. Irish language pretty generally spoken. Principal river, the Boyne. Principal towns, Navan and Kells. Meath is divided into 12 baronies and 147 parishes. In 1831, it had 28,796 inhabited houses, 31,632 families, and 176,826 inhabitants, of whom 88,993 were males, and 87,833 females.

MEAUX (an. *Jatinum*, afterwards *Meldi*), a town of France, dep. Seine-et-Marne, cap. arrond., on both sides the Marne, which is here crossed by an old stone bridge, 94 m. E.N.E. Paris. Population, in 1836, 7774. It is pleasantly situated, and is tolerably well built. Its ramparts have been laid out in public walks; and it has some good promenades along the river, and a spacious public square. The cathedral, one of the most remarkable Gothic edifices in France, was begun in 1262, but not finished till the 16th century: it is 398 feet in length, 137 feet in breadth; the height of its vault being 136 feet, and that of its tower 213 feet. Its choir and sanctuary are extremely elegant; but it derives its chief interest from its containing the remains and monument of Bossuet; who, having been raised to the bishopric of Meaux in 1681, continued in possession of the see till his death, in 1704. The controversial writings of this great glory of the Gallican church display extraordinary learning and acuteness; but it is to his *Discours sur l'Histoire Universelle*, and his *Oraisons Funèbres*, that he is mainly indebted for his imperishable renown. Exclusive of the cathedral, the public buildings comprise the episcopal palace, in which is the writing-table of Bossuet, a public library, with 14,600 volumes, the college, town-hall, theatre, two asylums, a Protestant and two parish churches, and cavalry barracks. Meaux has manufactures of cotton stuffs, earthenware, and glue. Numerous flour-mills are constructed on the Marne, the produce of which is sent to Paris, and the town has a good deal of trade in this and other articles of farm produce: its traffic is greatly facilitated by the canals of Ourcq and Cornillon. Meaux is very ancient: it was made a bishopric in 375; was twice sacked by the Normans in the ninth century; and was annexed to the French crown by Philippe-le-Bel. The French Protestants first preached publicly in this town, and it was the first to abandon the league, and submit to Henry IV. (*Hugo, art. Seine-et-Marne, &c.*)

MECCA, one of the most famous cities of the eastern world, the birth-place of Mohammed, and the great centre of attraction to all the pilgrims or *Hadjis* of the Mohammedan faith, in Arabia, prov. El-Hedjaz, 51 m. E. from the port of Djidda (which see), on the Red sea, and 270 m. S. by E. Medina; lat. 21° 28' 17" N., long. 40° 15' E. Population 23,000. ? This celebrated city, which, being forbidden ground to Christians, was known to Europeans only through indirect and suspicious information from Mohammedans or African renegades, has a few years ago visited, in disguise, by Burckhardt; who has given a very full though rather tedious description of the localities, climate, inhabitants, government, religious ceremonies, and pilgrim visiters, not only of this city, but also of Medina (which see). Mecca (literally meaning "the place of assembly," but pompously entitled by the Arabs *Om-el-Kora*, "mother of towns," and *Beled-el-Amayn*, "region of the faithful,") stands in a long, narrow, barren, sandy valley, running N. and S., called in the Koran "the valley without seeds." It is a straggling town, nearly 2 m. in length, but nowhere more than about 600 paces in breadth; the streets, which are irregular, unpaved and dusty, are generally wider than those of other eastern cities. The handsomest entrance is from Djidda, the S.W. quarters comprising some of the best houses; but on the other side of the great mosque, which is the nucleus of Mecca, there are three or four older good streets; the best of which, perhaps, is the Mesaa, the great resort of the Turkish pilgrims, and the noisiest, as well as most frequented, part of the town. "Indeed," says Burckhardt, "the Mesaa resembles a Constantinopolitan bazaar. Many shops are kept by Turks from Europe and Asia Minor, for the sale of swords, watches, copies of the Koran, and second-hand Turkish dresses; and there are numerous venders of pies, sweetmeats, &c. Here, too, are numerous coffee-houses, crowded during the Hadj from three in the morning till 11 at night; barbers' shops, auction rooms, &c. W. of the Mesaa branches off a street called Soueyga, or the Little Market, which though narrow, is the neatest street in Mecca, being regularly cleaned and sprinkled with water. Here the rich India merchants offer for sale their piece-goods, Cashmere shawls, muslins, perfumes, Mecca balsam, aloe-wood, civet, &c., strings of coral, necklaces of carnelian, seal-rings, and various kinds of China ware, and Abyssinian slaves. In fact, the Soueyga, being the coolest spot in the town during mid-day, is on that account the most frequented; and here all the gentlemen-*hadjis* take their morning and evening lounge, smoke their pipes, and hear or tell the news." (*Burckhardt's Arabia*, i. 213–215.)

The quarter called Shamye is likewise well built, being chiefly inhabited by merchants or *olemas* (clergy) attached to the mosque, and frequented at the pilgrim time by merchants from Damascus; in whose shops are found silks, cambrics, gold and silver thread, handkerchiefs, carpets, dried fruits, pistachios, &c. Near those last mentioned, also, is another respectably-built quarter, called Gerara, inhabited by some of the wealthiest native merchants. These are certainly the best parts of the town, comprising lofty stone houses, often three stories high, surmounted by handsome terraces, and (what is unusual in eastern towns) having rows of windows fronting the streets. The town, however, is not lighted, is scantily supplied with water, and contains many quarters, which, in filth and closeness, might vie with the dirtiest parts of Constantinople. Though once walled on three sides, Mecca is at present entirely open; but the neighbouring mountains are sufficiently high to form a tolerably strong barrier against an enemy; and on the rising ground S. of the city stands the great castle, a massive square structure, with thick walls and solid towers, comprising a bomb-proof magazine, a reservoir for water, and accommodation for 1000 men. With this exception, Mecca may be said to be almost destitute of public buildings; for the houses belonging to the sherifs, though handsome, are merely private, and not large dwellings, and the *medressa*, or colleges, are now converted either into storehouses or lodgings for pilgrims. This circumstance is, no doubt, attributable to the veneration of the people for their holy house, and this feeling prevents them from erecting any structure which might seem to rival the great object of their affection.

The *Beitullah*, otherwise called *El Haram*, the chief glory of Mecca, and the resort of every pious Mussulman*

who regards the injunction of the Koran, is a building by no means remarkable either for size or beauty; standing on low ground, in an oblong enclosure about 350 feet in length and 200 feet in breadth, formed by colonnades, roofed with numerous small plastered cupolas, supported by 450 pillars, about 20 feet in height, of marble, or Mecca stone. The temple has been so often ruined and repaired, that it has no traces of remote antiquity. The walls, arches, and minarets at the angles of the buildings are gaudily painted in stripes of yellow, red, and blue; but paintings of flowers in the usual Mussulman style are nowhere seen, and the colonnades are very clumsily paved. The Kaaba, or Holy House, which occupies the centre of the enclosure, and is the great attraction for all pilgrims, lays claim to a far more remote origin than that of Mohammedanism; and, though we may safely doubt the alleged fact of its having been built by Abraham and Ishmael, assisted by the angel Gabriel!† there can be no question that its genuine antiquity ascends beyond the Christian era. In all probability, the Kaaba, is alluded to by Diodorus Siculus, when speaking of a temple held in superior sanctity by *all* Arabians. In the second century Maximus Tyrius attributes to the Arabs the worship of a stone; and this, if not identical with, is at any rate analogous to the "black stone" of Mecca, which, as Gibbon justly remarks, is deeply tainted with the reproach of an idolatrous origin. (*Gibbon*, ix. 567.) The Kaaba, which was all but rebuilt in 1627, after having suffered great damage from fire, is an oblong massive structure, 18 paces in length, 14 do. in breadth, and from 35 to 40 feet in height; its door being coated with silver, and embellished with gold ornaments. At the N.E. corner, near the door, is the "Black Stone" previously alluded to, obligingly brought by the angel Gabriel as his contribution to the building: it forms part of the sharp angle of the structure, four or five feet above the ground; being oval-shaped, seven inches in diameter, of a dark brown colour, somewhat resembling lava; and surrounded by a border of cement and silver, to prevent its being worn away by the kisses and touches of the pilgrims. Round the building is a broad marble pavement; and at the S.E. corner is another stone, much revered by all visiters, but of a less noble origin, and less holy than the other. The four sides of the kaaba are covered with a curtain of embroidered black silk stuff, called the *kesona*, annually brought from Cairo at the time of the Hadj, and renewed with some not very decorous ceremonies. The holy fountain of Zem-zem, (said to be that so opportunely found by Hagar, when her son Ishmael was dying of thirst,) which supplies the town with water for *drinking* or *ablution* (its use for other purposes being forbidden), is enclosed in a substantial square building, having a handsome marble-faced entrance, with marble basins for ablution, and a room appropriated to the pilgrims, who come here in crowds to taste the miraculous fountain. From before dawn 'till near midnight, the well-room is constantly filled with visiters; all of whom, if not disposed to buy the services of the attendant, may themselves draw freely from the well. Various stories are afloat respecting the origin and virtues of this sacred spring, which, of course, are all believed by orthodox visiters, few of whom leave Mecca without carrying away some of the water in copper or tin bottles, to give away to their friends, and for their own use during illness, and their ablution after death. These are the chief buildings within the enclosure; but none of them can be considered as consecrated, except during the hour of prayer; for at other times, barbers, and all kinds of retail vendors, porters, idlers, &c., are to be met with at every turn. (*Burckhardt*, vol. i. p. 273.) In several parts of the colonnade, public schools are held for the instruction of young children; while, in other parts, a few professors deliver theological lectures, which meet, however, with little patronage from the pilgrims, who, with all their anxiety to visit the holy house, are, like all Turks, too much attached to the *dolce far niente* to trouble themselves with the doctrines either of theologians or men of science. The exterior of the mosque is adorned with seven quadrangular minarets, from the summits of which a view is obtained of the busy scene around. There are 19 gates, distributed without any order or symmetry; and the outside walls are formed by the surrounding houses, which during the pilgrimage are let, at enormous rents, to the wealthiest hadjis, with whom it is a grand object to be as near as possible in the holy house. The windows of these houses overlook the inclosure; and hence their occupiers are enabled to join in many of the mosque services without stirring from home. The service of the mosque occupies a vast number of people, as the imâms, muftis, officers of the Zem-zem, mueddins, olemas, lamplighters, and menial-servants, all of whom receive regular pay, besides sharing

---

the presents made by the pilgrims. The revenues of the mosque were formerly very extensive; but its wealth has greatly declined, nor does it now possess any treasures except a few golden lamps, the establishment being kept up almost wholly at the expense of the sultan. The first officer of the mosque is the *Nayb el Haram*, or warden, who keeps the key of the Kaaba, receives the pilgrims' contributions, and directs the repairs of the building. Next to him is the Aga of the eunuchs, a body of about 40 negroes who perform the duty of police officers in the temple; preventing disorders, and washing and sweeping the pavement round the Kaaba. In the time of the Ramadan, or great festival, the mosque is particularly brilliant; not only from the number of pilgrims of every age, rank, and nation within the enclosure, but also from the thousands of lamps which illuminate the colonnades. On the termination of the Hadj, however, the temple assumes a very different appearance. Disease and mortality, caused by fatigue, unhealthy lodgings, bad fare, and, in some cases, by absolute destitution, fill the mosque with the sick and dying; all of whom are anxious to have the satisfaction of expiring in view of the Kaaba, of receiving the Imâm's prayers, and of being sprinkled with the sacred water of Zem-zem. Whoever enters Mecca, whether pilgrim or not, is enjoined by the law to visit the temple immediately, and not to attend to any worldly concern before he has discharged that solemn duty. Certain religious rites, such as walking seven times round the Kaaba, and reciting certain prayers, are performed in the interior of the mosque; then comes the ceremony of walking seven times between the hill of Szafa and Merona; and, lastly, the pilgrims must submit to have their heads shaved by the barbers of the mosque. All these ceremonies must be repeated by every Mussulman, who enters Mecca from a journey farther than two days' distance; and they must be again more particularly performed at the time of the pilgrimage to Arafat, a hill about 15 m. E. of Mecca, to which Mohammed used to retire to pray, and which, on this account, is esteemed particularly sacred by all Mohammedans.

The concourse of pilgrims to the holy mount is often immense: Burckhardt says he counted about 3,000 tents dispersed over the surrounding plain; but the greater number of the pilgrims were without tents: between 20,000 and 25,000 camels were to be seen scattered among the pilgrims, whose numbers, he concludes, must have exceeded 70,000. The camp was from three to four miles long, and between one and two miles in breadth. But we suspect that these returns are very decidedly beyond the mark; and the taste for pilgrimages is now rapidly declining throughout the Mohammedan world. A visit to Arafat is indispensable to the pilgrims; none by any chance omit it; nor can the title of Hadji be assumed except by those who have been present at the ceremony. Mecca, like Jerusalem, boasts of many places rendered sacred by tradition. The birth-places of Mohammed and his daughter Fatima, the tomb of his wife Hadija, and the cell where the prophet wrote the Koran, are shown to the pilgrims, who are expected to make contributions for their maintenance. But a visit to these places forms no item of religious duty; and but few depart in any way from the prescribed routine, as such acts would interfere with their profits either as merchants or beggars, and thus frustrate a very important, if not chief, object of the expedition.

The inhabitants of Mecca are, with the exception of a few Hedjaz Bedouins, either foreigners or the offspring of foreigners. The ancient tribe of Koreysh, to which Mohammed belonged, is almost extinct; and there are now in Mecca only three or four Koreysh families, the head of one of which is the Nayb, or keeper of the great mosque. The neighbourhood, however, of the great mart of Djidda, the annual arrival of immense caravans, and the holy house, attract thither vast multitudes of strangers; a portion of whom remain behind, and settle permanently in the city, adopting Arabian habits, and intermarrying with the native population. The most numerous are the descendants of Arabs from Yemen and Hadramaut; next to these in numbers are those of Hindoo, Egyptian, Syrian, African, and Turkish origin; besides whom there are Persians, Afghans, Kurds, and people, in short, of almost every Mohammedan nation, all of whom are careful in preserving a traditional knowledge of their original country. The inhabitants, however, though differing so much, nationally considered, wear the same sort of dress, have the same customs, and care much less for national costume and manners than in any other part of the east. Their colour is a yellowish brown; and in features they closely resemble the Bedouins: the lower classes are generally stout, with muscular limbs, while the higher orders are distinguished by their meagre emaciated forms and black piercing eyes. All classes are fond of dress, and the earnings of the poor are mostly spent on clothes. The women wear Indian silk gowns, with large blue striped trowsers reaching to the ankles, and a white kind of hood to cover the face. There are few families in moderate circumstances that do not keep slaves, most of whom are Nubians and Abyssinians, brought thither from the port of Suakim: many African females are kept as concubines; and, in case of their having issue, the masters usually legitimate the offspring by marrying them. The sale of concubines is confined to the middle and lower classes; the more wealthy regarded the practice as disgraceful. The inhabitants of Mecca, as also of Djidda and Medina, are far more lively and communicative than either the Syrians or Egyptians, and in this respect they resemble the Bedouins. Indeed, vivacity of temper, acute intellect, sagacity, and suavity of manner, are characteristics of almost all the native inhabitants; while, at the same time, their proud, independent spirit, for which they are equally remarkable, is infinitely preferable to the cringing servility of the Turks of Anatolia and Syria. Religion exercises little control over them; and, though they are proud of aping the manners recorded of Mohammed, and thoroughly versed in the Koran, few Mussulmén are so inattentive to the moral duties inculcated by the Prophet. For the most part, indeed, they exhibit great profligacy of character. Drunkenness, gambling, cheating, false-swearing, and the grossest sensuality, are of everyday occurrence; and it is a common saying among the people, " Forbidden things abound in the city forbidden to infidels." Learning and science, which once flourished in Mecca, are now almost wholly neglected. The many *medreses*, or colleges, for which the city was formerly renowned, are turned either into coffee-magazines or lodging-houses for pilgrims; its large libraries have disappeared; the great mosque is at present the only place where teachers of eastern learning are to be found; and the Meccaways themselves, who wish to improve in science, go to Damascus or Cairo.

The employments of the people are in trade and the service of the Beitullah; but there are few employed in the mosque who do not, clandestinely at least, engage in commercial affairs. There are but few artisans, and these much inferior in skill to the same class in Egypt; a few potteries and dye-works are the only manufactories, and the town is wholly dependent on other countries for its necessary supplies. Hence there is a large foreign trade; and the holy city is crowded, during the month of Dhalhajja (the latter end of June and beginning of July), not only with zealous devotees, but opulent merchants, who thus make use of the permission granted them by the prophet: " It shall be no crime in you, if ye seek an increase from your Lord *by trading during the pilgrimage.*" (*Sale's Koran*, ch. ii.) During the whole twelve days that the pilgrims are allowed to remain, a fair or market is held in Mecca and its vicinity; and although the number of pilgrims has greatly declined of late years, it is still a crowded and bustling scene. " Few pilgrims," says Burckhardt, " except the mendicants (a very numerous class), arrive without bringing some productions of their respective countries for sale; and this remark is applicable as well to the merchants, with whom commercial pursuits are the main object, as to those who are actuated by religious zeal; for to the latter the profits derived from selling a few articles at Mecca diminish in some degree the heavy expenses of the journey. The Mograbyns (pilgrims from Morocco and North Africa) bring their red bonnets and woollen cloaks; the European Turks, shoes and slippers, hardware, embroidered stuffs, sweetmeats, amber, trinkets, &c.; the Turks of Anatolia bring carpets, silks, and Angora shawls; the Persians, Cashmere shawls and large silk handkerchiefs; the Afghans, tooth-brushes made of the spongy boughs of a tree growing in Bokhara, beads of yellow soap-stone, and plain coarse shawls manufactured in their own country; the Indians furnish the numerous productions of their rich and extensive region; the people of Yemen, snakes for Persian pipes, sandals, and various other works in leather; and the Africans bring various articles adapted to the slave trade. The pilgrims are, however, frequently disappointed in their expectation of gain, for want of money often obliges them to accept very low prices." The most respectable of the mendicant pilgrims are negroes (called here *Tekrourys*), and these apply themselves to labour immediately on their arrival at Mecca: some serve as porters, for the transport of goods and corn from the ships to the warehouses; some hire themselves to clean the court-yards, fetch wood, carry water, &c.; while others manufacture small baskets and mats of date leaves, or prepare the intoxicating drink called *booza*. The pilgrims are accommodated in lodgings, for which the inhabitants charge a most exorbitant rent; and all, except those of the highest and lowest ranks, live together in a state of freedom and equality, keeping but few servants, and generally dividing among themselves the various duties of housekeeping. The two principal caravans which rendezvous at Mecca are those of Damascus and Cairo; both of which always arrive at fixed periods, generally a day or two before the departure of the Hadj for Arafat. The former of these is very large, and is, at the same time, very well regulated. The caravan of Cairo is much smaller, and its route, along the Red Sea, is more dangerous and fatiguing; but many of the Egyptian and African merchants now come by sea from Suez, Cosseir, and Suakin to Djidda, and thus avoid the weariness of a long land journey. The Persian caravan sets out from Bagdad, and crosses the desert; but it is now of little importance, as all but the poorest Persian pilgrims come round by sea, from Bussorah, between which place and Djidda there is a large and steadily increasing trade.

The climate of Mecca is sultry and unwholesome; especially in August, September, and October, when a hot suffocating wind prevails. The wet season is in December; but the rains are not so continuous as in other tropical countries. Intermittent and inflammatory fevers, dysentery, elephantiasis, and stone, are common diseases; and, with respect to the general health of the town, Burckhardt says, "I seldom enjoyed perfect health while in Mecca: I was twice attacked by fever, attributable chiefly to bad water; and, even on those days when I was free from disease, I felt great lassitude, depression of spirits, and total want of appetite." (Vol. i. p. 450.)

The territories of Mecca, Tayf, Gonfude, and Yembo, were, previously to the Wahabee and Egyptian conquests, under the command of the sherif of Mecca, who held his authority from the grand seignior; but when the Porte was no longer able to send large armies with the Hadj caravans to secure her power in the Hedjaz, the sheriffs became independent. The Wahabees (who are to the Mohammedan religion what the Protestant churches are to Christianity) took possession of Mecca in 1809, and retained it till 1813, when Mehemet Ali restored the holy cities to the nominal protection of the Porte, while at the same time he placed them effectually under his own control. (Burckhardt's Arabia, i. 171., ad finem, and ii. 1-86.)

MECHLIN (Fr. Malines), a city of Belgium. prov. Antwerp. cap. arrond., on the Dyle, a tributary of the Scheldt, and on the road between Antwerp and Brussels, 14 m. S.S.E. the former, and 14 m. N.N.E. the latter. Pop. in 1836, 22,896. The Dyle divides Mechlin into two parts. It is regularly laid out with broad, well paved, and clean streets. Houses grotesque, antiquated, and frequently of a large size; but, being painted in front, they look clean and cheerful. The fortifications were demolished by the French in 1804. The most remarkable public building is the cathedral, a Gothic edifice, commenced in the twelfth century. The body of this building is by no means commensurate with the present altitude of the morisco tower attached to it, and still less with the height to which it was originally intended to be carried. This massive tower, with its truncated steeple, begun in 1452, is 370 feet above ground, being the height of the cross of St. Paul's, London; and, had it been completed according to the original design, it would have been 640 feet high. The Last Supper, the altar-piece, is by Rubens: the heads of the apostles and style of drapery are said to be in his best manner; but the Christ is a failure, and the picture is maldrawn. The church of the Recollects has the famous picture of the Crucifixion, by Vandyke.* "This," says Sir Joshua Reynolds, "is, perhaps, the most capital of all his works, in respect of the variety and extensiveness of the design, and the judicious disposition of the whole. It may be considered as one of the first pictures in the world, and gives the highest idea of Vandyke's powers." (Reynolds's Works, ii. 273. ed. 1819.) There are pictures by Rubens in the cathedral, and some of the other churches, of which the Adoration of the Magi, in the church of St. John, is probably, the best. After the churches, the bishop's palace, town-hall, arsenal and cannon foundry, the Franciscan convent, and the Beguinage, a large asylum for 800 widows, or aged women, are the principal public buildings. Mechlin is the see of an archbishop, who is primate of Belgium, and has a revenue of about 4,000l. a year. It is the seat of a tribunal of primary jurisdiction, and the residence of a military commandant. It has an ecclesiastical seminary, a college, an academy of painting, a society of the fine arts, and a mont-de-piété. It has been long celebrated for the manufacture of lace, of a coarser and stouter kind than that of Brussels; but this has latterly been, to a considerable extent, superseded by the Nottingham lace, and it is said that only eight houses are now engaged in the business. Among its other fabrics are those of cashmere shawls, and gilt leather chairs: the latter were at one time an article of export, and it is said that upwards of 400 workmen are still engaged in their manufacture.

Mechlin furnishes a peculiar species of beer, of a light body, which acquires, by keeping, a vinous flavour and quality. Another delicacy peculiar to this city is the "Déjeûner de Malines," a dish much admired by travellers as

---

* This is Sir Joshua Reynolds's statement! Mr. Murray, on the contrary, says that this is the altar-piece of the cathedral! Non nobis, &c.

---

well as natives, into which pigs' feet and ears enter as important constituents.

Mechlin has an extensive trade in flax, corn, and oil. The tide ascends the Dyle to a league above the city, which is accessible for vessels of considerable burden from the Scheldt. Mechlin is connected with Louvain by a canal navigable by vessels of 160 tons: it would also, but for the absurd opposition of its magistrates, have been the central dépôt of the Belgian railways; but this has been fixed without its boundary, a circumstance which those by whom it was occasioned now deeply regret.

As early as the sixth century, Mechlin appears to have been a place of importance, and the capital of a lordship. It suffered severely from war, plague, and fire in the middle ages; and in modern times has been repeatedly taken by the Spaniards, Dutch, English, and French. (De Cloet; Hasselbing: Belgium. by Emerson Tennent, Esq. M.P.; Murray's Handbook for N. Germany, &c., passim.)

MECHANIC, t., Holmes co., O. Watered by Doughty's fork of Killbuck creek. It has four grist-mills, four saw-mills. Pop. 1,403.

MECHANICSBURG, p. b., Cumberland co., Pa., 9 m. E. Carlisle, 6 m. S.W. Harrisburg, 105 W. Incorporated in 1828. It contains a church, four stores, nine flouring-mills, six saw-mills, one tannery, three distilleries, one printing-office, one weekly newspaper; four schools, 200 scholars. Pop. 670.

MECHIBBES, p. t., Washington co., Me., 157 E. by N. Augusta, 743 m. W. Watered by E. Machias river. It contains Washington academy. which has an edifice 50 by 38 feet, two stories high, a library and philosophical apparatus, and a productive fund of $14,000, 12 stores, one fulling-mill, two grist-mills, 14 saw-mills; one academy above stated, 161 students; eight schools, 460 scholars. Pop. 1,385.

MECKLENBURG, a territory in N. Germany, between lat. 53° and 54° 20' N., and long. 10° 35' and 13° 57' E.; having N. the Baltic, E. and S. the Prussian dom., and W. Hanover, Denmark, and Lubeck. Area, 5,342 sq. m. Pop. about 573,500. It is divided into

1. MECKLENBURG-SCHWERIN (GRAND DUCHY OF), a state of N. Germany, between lat. 53° 7' and 54° 20' N., and long. 10° 37' and 13° 15' E.; having E. Pomerania and Mecklenburg-Strelitz, S. Brandenburg, W. the Hanoverian and Danish provs. of Luneburg and Lauenburg, the principality of Ratzeburg. belonging to M. Strelitz, and the territory of Lubeck, and N. the Baltic. Length, E. to W., about 110 m.; average breadth nearly 45 m. Area, 4,833 sq. m. Pop., in 1840, 482,985. Mecklenburg belongs to the great plain of N. Germany; it is not, however, a dead level, but has an undulating surface, interspersed with some ranges of low hills, one of which, the Ruhneburg, rises to nearly 600 feet above the level of the Baltic. It has several rivers of some size, as the Rechnitz, constituting its N.E. boundary, the Warnow, Stepnitz, &c., flowing to the Baltic, the Elde, a tributary of the Elbe, and others; and a great number of lakes, that of Muritz, which has an area of more than 50 square miles, and is elevated 216 Rhenish feet above the sea. being by far the largest lake in N. Germany; next to it is the lake of Schwerin, having the capital of the duchy on its banks. Notwithstanding its high latitude, this grand duchy has a milder climate than most parts of Germany. The mean temperature of Germany being taken at 51° Fahr. (80·3° R.), that of Mecklenburg will be about 52° 5' F. (90 R.) But the winter is severe, the average temperature of that season being little above the freezing-point; the atmosphere: also, is particularly humid, which, added to the moisture of the soil, renders catarrhs and consumptions frequent. The surface and soil are very various. On either border of the principal range of hills a poor sandy tract extends, covered with heath; and few parts of Germany are worse cultivated or more thinly inhabited than that between Schwerin and Gustrow, a distance of nearly 40 miles. To the south of this district the soil is somewhat better; and beyond Gustrow, towards what was formerly Swedish Pomerania, the sand gradually changes into a fertile loam, well adapted to the growth of rye and wheat. Near the Baltic the soil is, for the most part, a meagre sand, intermixed with stripes of loam. With the exception of the sandy heaths, the face of the country is cheerful and pleasing; the land is mostly enclosed; the woods, which are extensive, are scattered over the country, and on the borders of the lakes good meadow lands occasionally present themselves. Mecklenburg is essentially an agricultural country. It is generally divided into large estates. The demesnes of the sovereign comprise four tenths of the land, and those of the nobility, knights, &c., nearly five tenths; leaving about one tenth in the possession of the municipalities and a few monastic institutions. Farms are generally very extensive; they vary from 300 to 1,000 or 1,200 acres. About one fourth part of the province is cultivated by proprietors resident on their own estates, and who are fre-

quently very intelligent and well-informed ; about a half is occupied by farmers, and a quarter by peasants or boors. The severity of the winter makes it necessary to have farm buildings sufficient to accommodate the live stock, in addition to the corn, hay, &c. ; and hence a gentleman's house has near it, besides labourers' cottages, more than five times the extent of barns, stables, cow and sheep houses, &c., that would be required in England for the same extent of land. Farms, when let, are usually held by tenants on leases, varying from seven to 21 years. The rent varies, of course, according to the quality of the soil, situation, &c. It is uniformly almost paid in money, and the rotation of crops prescribed in the lease must be strictly adhered to. The best farms brought, in 1827, according to Mr. Jacob, about 12s. an acre ; the medium lands from 6s. 4d. up to 8s. 4d., and the sandy districts- in the south from 2s. to 3s. 2d. an acre ; but he thinks that the larger portion of land in the grand duchy did not then produce more than 5s. an acre. Taxes are lower than in most parts of the continent. The various taxes and other out-goings borne by the proprietor may be estimated at about 1d. an acre. The stock on the larger farms usually belongs to the tenants ; but that on the smaller farms, held by peasants, usually belongs to the landlords.

The cultivation of wheat (especially the red variety) has increased greatly of late years, and Mecklenburg is now one of the principal countries of Germany for the export of wheat. Rostock wheat is, however, inferior to either English or Dantzic wheat ; so much so, that while wheat is being shipped from Rostock at 18s. a quarter, it will fetch 27s. or 28s. at Dantzic. Next to corn, peas, beans, potatoes, and turnips are principally grown. Hemp and flax are reared, but in no great quantities, and the culture of tobacco has very much diminished. Of late years horses, instead of oxen, have been employed in field labour. The system of cultivation corresponds with that of Holstein and Sleswick. After a year's fallow, three corn crops, usually rye, barley, and oats, are taken in succession, the land being sown down with grass seeds ; along with the third corn crop, a crop of hay is taken in the fifth year, and the sixth and seventh years the fields are in pasture. (Jacob's Second Report.) The climate is too cold for the vine, though latterly it has been raised to some extent at Crevitz, and some bad wine has been produced. The horses and horned cattle, which are both numerous and excellent, find a ready sale in every part of Germany, and are a source of great profit to the landed proprietors. The breed of horses has been much improved, by means of the grand ducal stud at Redefin and several private studs. Sheep have been a good deal increased of late ; the stock in the grand duchy is now estimated at about 1,300,000, and wool has become a principal article of export. Herds of hogs and flocks of geese are met with in every part of Mecklenburg. The former wander, nearly wild, through the forests, feeding on acorns and roots, and the geese literally cover the banks of the lakes and rivers. The latter supply a considerable part of Europe with quills ; and their breasts, smoked and cured like bacon, are much esteemed as a delicacy.

The population has increased very rapidly within the last 25 years, a consequence partly of the breaking up of the old feudal system, and partly of the introduction of the potato, &c. The number of inhabitants, in 1818, was 377,954, whereas, in 1837, it was 476,499, being an increase of 26 per cent. During the 10 years previous to 1828, the deaths amounted to 10,080, and the births to 16,070 a year. Mecklenburg is still, however, the least populous portion of Germany ; there being only 99 inhabitants to the English sq. m. Till within the last 20 years the peasants were in a state of mitigated slavery. They could acquire, enjoy, and transmit property, but they were adscripti glebæ, and bound to the soil, so as to be sold or let with it. The government took measures, on the restoration of peace, to abolish this relic of the feudal ages ; and, about 1820, all the peasants who still remained in the condition of serfs (for many of the proprietors had previously emancipated those on their estates) were declared free, though their actual manumission did not take place till about 1825. They are now, however, quite free, and may labour where and under whatever conditions they please to stipulate with their employers. Previously to the emancipation of the peasantry, a man was estimated to cost during harvest 1s. 4d., during haymaking, 1s. 1d., and at other periods from 6d. to 1s. a day. Probably but few changes have yet taken place, either in the condition or appearance of the peasantry. The country, at a passing view, seems, from the magnitude of the farm-buildings and the number of enclosures and woods, to be more prosperous than, on a closer examination, is found to be the case. " On a nearer approach: it is scarcely possible to avoid feeling disgust at the miserable filthiness and apparent poverty of the peasants' dwellings and of their clothing ; though a difference may be discerned on the several properties, according to the greater or less degrees of

342

prudence and kindness of the various proprietors." (Jacob's Rep. on Agric.)

The condition of the peasants, of whom a large proportion are noble, appears, in fact, to be much depressed. The author of Germany and the Germans, in 1835-36, says, " The landsman, unlike his neighbour in Holstein, is poor. I sought in vain for the appearance of comfort and plenty which had delighted me in Holstein ; and yet, like the latter, Mecklenburg is one of the most fertile provinces in the N. of Germany, and exports provisions in large quantities to Prussia and Hamburg, while the natives are obliged to content themselves with potatoes, horse-beans, and sauerkraut. During my tour through the more remote villages, I found it impossible to procure a comfortable dinner. Fresh meat was entirely out of the question ; the general bill of fare at the inns consisted of potatoes, bread, butter, and eggs, and those of a superior class added bacon and sauer-kraut. My drink was confined to wretched beer or schnapps ; and, when I demanded wine, they looked at me as if my intellects were deranged. My bed was not unfrequently a straw palliasse, and the only covering a feather bed, enveloped in a gay-coloured cotton cover ; the whole supported on a bedstead, 5 feet long, composed of deal boards nailed together, in form not unlike a packing-box. These miserable arrangements are common in all the remote districts of Mecklenburg, Pomerania, and Prussia.

" As Germany supplies Europe with princes and princesses, it would appear as if Mecklenburg alone were sufficient to furnish it with nobles ; for it is reckoned that the nobility include the half of the population, the possessions of 6-8ths of these dignified persons being limited to their genealogical trees. During my progress through the country, I met with a herr (baron) who exercised the profession of relieving men's chins of what is sometimes considered an incumbrance ; and at one of the inns I found a herr graf (count) for a landlord, a frau gräfinn (countess) for a landlady ; the young herren grasen filled the places of ostler, waiter, and boots, while the fair young fräulein gräfinnen were the cooks and chambermaids. I was informed that, in one village, the whole of the inhabitants were noble except five, and these were married to noble fräuleins." (Germany, &c., i., 43, 44.)

In Mecklenburg the general principle is, that every place shall provide for its own poor, either separately or in common with others. All proprietors are bound to provide for the poor on their estates ; and, in furtherance of that object, are entitled to levy a sum of about 8d. a year from the day-labourers, and 4d. from the maid-servants, &c., on the estate, though but few avail themselves of this privilege. When crown lands are let, there is always a clause in the contract, regulating what the farmer, the dairy farmer, the smith, and the shepherd are to pay for behoof of the poor. The higher classes, public officers, &c., should pay one per cent. of their income to the poor's funds. All poor persons have a legal claim to assistance, and such work must be found them as they can perform.

About 30,000 cwts. of salt are obtained annually ; but, with the exception of lime, minerals are of little importance, and mining is quite neglected. Manufactures are not very considerable. The principal are those of woollen and linen fabrics ; but the former employed, in 1834, only 1128 hands, who produced goods of the value of about 271,000 doll., or about 1-5th part of which was exported. Mecklenburg is famous, even in Germany, for the distillation of corn spirits. Every one may carry on the business of distiller without tax or restriction of any kind ; and this facility has, no doubt, tended to increase that taste for ardent spirits which, unhappily, distinguishes the peasantry. A few cotton, paper, glass, tobacco, soap, and wax-light factories, with breweries and tanneries, complete the list of manufacturing establishments. The government is, however, devoting a good deal of attention to the improvement and diffusion of manufactures, and at least one school of arts and trades is now established in every town. Mecklenburg has an extensive trade in farm produce, which is facilitated by the proximity of the Elbe and the Baltic, especially the former, the principal part of the foreign commerce of the duchy being carried on through Hamburg. The value of the exports is estimated, at an average, at about 4,000,000 dolls. a year, of which, corn, pulse, &c., furnish 2,300,000, wool 800,000, butter and cheese 400,000, and cattle 250,000 dolls. The import trade is also considerable ; but no accurate statement can be made, either of the exports or imports, from the want of custom-house accounts. The commercial policy of Mecklenburg is as liberal as can be desired ; she has no duties on imports, except a trifling excise at her ports ; nor any transit-dues beyond a trifle in the shape of a road-toll, which does not, however, appear to be strictly enforced. The only commercial towns and ports of any consequence are Rostock and Wismar. The other towns have merely a retail trade, a large part of which is in the hands of Jews. In the S. Prussian money and measures are current : but the Hambury measures of length and the Lubeck measures of capacity are in use, in most parts of the grand duchy. The Mecklenburg rod is larger than the Rhenish, in the proportion of 1 to ·809 ; the *morgen* varies from 200 to 400 square rods. The pound is to that of Hamburg as 401 to 408 : the *centner* = 8 *liespfund* = 112-lb. ; but in Rostock the liesp. has 16 lb.

The government is intimately connected with that of Mecklenburg-Strelitz. Each grand duchy has its separate states, which, also, meet separately ; but the states of both grand duchies assemble once a year, alternately at Sternberg and Malchin. The joint assembly has the right, in conjunction with the grand duke of Mecklenburg-Schwerin, to make laws for and impose taxes on the whole of Mecklenburg ; it consists of the landed proprietors among the nobility, and of deputies from towns, &c., in all amounting to between 360 and 600 members. When the states are not assembled, a committee sits at Rostock. The executive power is entrusted to a *directorium*, consisting of eight grandducal councillors, three heads of noble families (*Erb-Landmarschälle*), and a deputy from the town of Rostock, which is itself a sort of little republic, or *imperium in imperio*. The grand duchy is divided into five principal districts—the duchy of Schwerin, or circ. of Mecklenburg, the duchy of Gustrow, or circ. of Wenden ; the district of Rostock, the principality of Schwerin, and the lordship of Wismar ; besides which, there is a small extent of territory, which sends no representatives to the states, and over which three conventual establishments have jurisdiction.

Justice is administered in primary courts in the towns and villages, in patrimonial courts on the estates of the nobility, with courts of secondary jurisdiction at Schwerin, Gustrow, and Rostock, and a high court of appeal at Parchim, which is the supreme legal tribunal for both grand duchies. With the exception of between 3000 and 4000 Jews, the inhab. are nearly all Lutherans. There are upwards of 1000 primary schools, about 40 superior public schools (*Bürgerschulen*), five gymnasia, and the university of Rostock, with several ecclesiastical and other special seminaries. Previously to 1826, there was but one bookselling establishment in the grand duchy ; but, since that period, the diffusion of education and the cultivation of literature has led to the establishment of 11 others, besides 12 printing-offices. (*Berghaus.*)

The public revenues of the grand duchy amount to 4,600,000 fr. per annum, and the public debt to 18,000,000 fr. The dukes of Mecklenburg had formerly five votes in the college of princes, in the diet of the empire. Since 1815, Mecklenburg-Schwerin has held, with Mecklenburg-Strelitz, the 14th rank in the Germanic confederation. They have conjointly two votes in the general assemblies, but only one in the committee of the diet. Mecklenburg-Schwerin furnishes a contingent of 3580 men to the army of the confederation.

Schwerin is the political capital ; but Ludwigslust, a town with a population of about 5000, 14 m. S.W. Parchim, is the usual summer residence of the grand duke. The palace, which is a large fine edifice, has a cabinet of pictures and a collection of Slavonic antiquities ; the surrounding grounds are well laid out ; but the neighbourhood is dull and not very healthy.

2. MECKLENBURG-STRELITZ (GRAND DUCHY OF), a state of N. Germany, consisting of two separate territorial divisions ; the first and largest, or the duchy of Stargard, lying between lat. 53° 9′ and 53° 47′ N., and long. 12° 40′ and 13° 37′ E., having W. Mecklenburg-Schwerin, and surrounded on all other sides by the Prussian territories ; and the second, or principality of Ratzeburg, between lat. 53° 40′ and 53° 61′ N., and long. 10° 29′ and 11° E. United area, 997 sq. m. Pop., in 1846, 99,398. The general features of the country are the same as those described in the above article ; its mean elevation is, however, somewhat less than than of Mecklenburg-Schwerin, though the Helperburg, near Woldeyk, rises to 640 feet above the sea. The chief river is Stargard is the Havel, and in Ratzeburg the Stepnitz. The land is divided among the sovereign, the nobility, and the towns, in the proportions of about 7-10ths to the first, 2-10ths to the nobles, and 1-10th to the municipalities. Nearly 1-4th part of the ducal property consists of forest lands. Agriculture and cattle-breeding are the chief branches of industry here, as in Mecklenburg-Schwerin. The manufactures are even more insignificant than the latter grand duchy, and almost confined to leather, beer, and spirits, with copper wares in Ratzeburg. There is a brisk trade in rural produce. New Brandenburg is, next to Gustrow, the largest wool-market, and Old Strelitz the largest mart for horses, in Mecklenburg. Furstenburgh has some trade in timber and butter.

The government is a limited monarchy, as in Mecklenburg-Schwerin. Justice is administered in eight courts of primary jurisdiction, the superior courts of Ruckscht and Ratzeburg, and the court of chancery in New Strelitz, from

which appeal lies to the supreme tribunal at Parchim. The principal public schools are, the *Gymnasium Carolinum* at New Strelitz, the high schools at New Brandenburg, Friedland, and Ratzeburg, and the school of arts (*Bildungsanstalt*), at Mirow. New Strelitz is the capital and residence of the grand duke. The other chief towns are New Brandenburg, Friedland, and Old Strelitz, Mecklenburg-Strelitz holds, with Mecklenburg-Schwerin, the 14th place in the German confederation, and has also, with it, one vote in the committee, having in the full diet one vote independently. It furnishes 718 men to the army of the confederation.

Mecklenburg has been several times conquered and disposed of by foreign powers ; as by Henry the Lion, in the 12th century by Ferdinand II., who gave it to Wallenstein, and by Napoleon ; but it has always reverted to its original dynasty. The reigning family is the only sovereign house of Slavonian origin, and is one of the most ancient in Europe, with all the principal reigning families of which it has been allied. The separation of Mecklenburg into two states took place in 1701, and both were recognised as grand duchies in 1815. (*Berghaus, Allg. Länder, &c.,* iv., 395–411 ; *Stein, Handbuch der Geog.; Jacob's Second Report on Agriculture ; Parl. Papers,* 1839 ; *Almanack de Gotha,* 1841 ; *Germany and the Germans, &c., passim.*)

MECKLENBURG, county, Va. Situated on the S. part of the state, and contains 640 sq. m. Watered by Roanoke river, formed in the W. part of this county by the union of Staunton and Dan rivers. Bounded N. by Meherrin river. It contained in 1840, 12,938 neat cattle, 14,109 sheep, 31,988 swine ; and produced 77,444 bushels of wheat, 472,345 of Indian corn, 234,107 of oats, 25,107 of potatoes, 4,124,131 pounds of tobacco, 19,054 of cotton. It had 30 stores, one furnace, 17 flouring-mills, 32 grist-mills, 19 saw-mills, four tanneries ; one college, 80 students ; seven academies, 186 students ; 15 schools, 334 scholars. Pop. ; whites, 7754 ; slaves, 11,913 ; free coloured, 1055 ; total, 20,724. Capital, Boydton.

MECKLENBURG, county, N. C. Situated towards the S. part of the state, and contains 900 sq. m. Bounded W. by Catawba river, by branches of which it is watered. It contains several rich gold mines. It had in 1840, 18,541 neat cattle, 19,519 sheep, 33,065 swine ; and produced 78,315 bushels of wheat, 2005 of rye, 586,928 of Indian corn, 61,407 of oats, 14,442 of potatoes, 1,595,327 pounds of cotton. It had 39 stores, two smelting houses for gold, 11 flouring-mills, 23 grist-mills, 32 saw-mills, nine tanneries, 15 distilleries, one printing-office, one bindery, one weekly newspaper ; one college, 81 students ; five academies, 181 students ; 25 schools, 475 scholars. Pop. ; whites, 11,950 ; slaves, 6392 ; free coloured, 101 ; total, 18,273. Capital, Charlotte.

MEDFIELD, p. t., Norfolk co., Mass., 21 m. S.S.W. Boston, 493 W. Bounded W. by Charles river. It contains, one mile E. of the central village, one of the oldest houses in New-England, which was known to be standing in 1676, when the town was burned by the Indians. It has three churches, a Congregational, Unitarian, and Baptist, two stores, six grist-mills, five saw-mills ; three schools, 190 scholars. Pop. 863.

MEDFORD, p. t., Middlesex co., Mass., 5 m. N.W. Boston, 445 W. Watered by Mystic river, which is navigable to this place. Vessels are built here, and floated down the river, without being loaded. The Middlesex canal and the Lowell railroad pass through the town. It contains four churches, a Congregational, Unitarian, Methodist, and Baptist ; nine stores, two lumber-yards, capital $35,000, one grist-mill, two saw-mills, one oil-mill, one tannery, one distillery, one pottery ; one academy, 48 students ; eight schools, 579 scholars. Pop. 9478.

MEDINA, or MEDINÉT-EL-NABI, "the town of the prophet," one of the sacred cities of Arabia, the burial-place of Mohammed, and, next to Mecca, the great centre of attraction to Mohammedan pilgrims, in the prov. el-Hedjaz, 190 m. N.E. its port of Yembo on the Red sea, and 250 m. N. Mecca. Lat. 25° 13′ N., long. 40° 3′ 15″ E. Pop. of town and suburbs, according to Burckhardt, 18,000. This celebrated city stands in a plain, close to a chain of hills which bounds the great desert westward. It is not open, like Mecca, but surrounded by a wall about 40 feet high, and flanked by 30 towers : it was additionally fortified with a ditch by the Wahabees ; but this is in most places nearly filled up. It is entered by three fine gates ; one of which, towards the S., called Bab-el-Masry, is said by Burckhardt to rank second only to the noble gates of Cairo ; a fourth gate, in the S. wall, was closed by the Wahabees, and has not since been re-opened. The houses are well built, of a dark gray stone ; but it has a desolate appearance, owing to the lessened resort of pilgrims. Ruined houses and tottering walls are seen in every part of the town ; and "Medina presents the same disheartening view as most of the eastern towns, which now afford but faint images of their ancient splendour." (*Burckhardt's Arabia,* ii., 150.) The principal street, in

which are most of the shops, leads from the Cairo gate to
the great mosque; another, of respectable size and breadth,
runs from the mosque to the Syrian gate; but many of its
houses are in ruins, and there are few shops. No shops or
bazaars, however, are found in other parts of the town;
and, in this respect, Medina differs from Mecca, which is
one continued market. The suburbs cover more ground
than the city itself, from which they are separated by an
open space, narrow on the S., but widening on the W., before the Cairo gate, where it forms a large public place called *Monakha*, always crowded with camels and Bedouins.
Provisions are sold here in sheds erected for the purpose,
and the numerous coffee huts are beset the whole day with
visitors. The greater part of the suburbs consist of large
court-yards, built round with low houses, tenanted chiefly
by the humbler classes. Each *hoak*, or court-yard, contains
30 or 40 families; the castle belonging to the little community occupy the centre of each; and the only gate of entrance is regularly closed at night. Opposite, however, to
the gate of Cairo are several regular and well built streets,
with houses similar to those within the town; one of these,
called El-Ambarye, comprises some of the handsomest residences in Medina, besides two rather large mosques, all now
remaining, except the great temple, out of 14 mentioned by
the Arabian historians. The town is supplied with good
water, both from wells and open streams.

The glory of Medina, and that which places it, as a sacred
city, almost on a level with Mecca, is the possession of the
tomb containing the remains of the prophet. This tomb,
with the tombs of Abou-Beker and Omar, the friends and
immediate successors of the prophet, are inclosed within the
great mosque, situated at the E. end of the town. Though
smaller than the mosque at Mecca, it is built upon the same
plan, with minarets at the angles, and forms an open square,
surrounded on all sides by covered colonnades. The tombs
are enclosed within a curtain, in a square building of black
stone, detached from the walls of the mosque, and surrounded by a close iron railing. People of rank are admitted
gratis within the sacred precinct, called *El Hedjra*; and
any one, indeed, who has money to spare, finds but little
difficulty in being admitted. The ridiculous stories, long
current in Europe, as to Mohammed's coffin being suspended in the air by a loadstone, are unknown in the East; and
most part of the statements that have been put forth, as to
the richness and magnificence of the tombs and the great
mosque, have been absurdly exaggerated. The tomb of
Fatima, the favourite daughter of the prophet, and the wife
of Ali, is also within the great mosque; but it is doubtful
whether it really encloses her remains. The lofty dome,
which rises above the tombs, is seen at a great distance from
town. The ceremonies observed by persons visiting Medina
are somewhat different from those customary at Mecca; nor
is it absolutely required of the hadjis to visit the prophet's
tomb; hence it is that the enjoined religious duties are considerably less tedious. The building is lighted at night with
lamps and candles, sent either from Cairo or Constantinople.
The mosque has four gates, of which the principal, by
which the pilgrims first enter, called Bab Merouda, is certainly very superior in beauty to any of the gates at Mecca.
The police, cleansing and lighting of the mosque, are entrusted to about 40 eunuchs, somewhat similar to those of the
Beitullah at Mecca, supported, like them, by salaries from
Constantinople, and by fees and presents from the hadjis.
Besides these, and the Imâns, Mueddins, and Olemas, who
are as indispensable here as at Mecca, there are upwards
of 500 inferior servants. The mosque, founded by Mohammed himself, immediately after his flight from Mecca, on
the spot where his camel first rested in the town, was enlarged by Omar after the prophet's entombment, and surrounded with walls by Othman. Subsequent caliphs and
nobles of Arabia greatly embellished it; but the whole edifice was burned down A.D. 1508; and so complete was the
destruction, that only the interior of the tomb was spared.
The present building was erected, in 1514, by Kayd Beg,
then king of Egypt; since whose time only a few imma-
terial improvements have been made by the Othman emperors of Constantinople. (*Burckhardt's Arabia*, ii., 161-
205.) The burial ground of Medina, called *El Bekya*, is another object of extreme veneration, in consequence of its
containing the tombs of Ibrahim, Othman, Abbas, the aunts
of Mohammed, &c. Another place of pilgrim-resort is
Djebel-Ohod, about 2 m. from the town; the scene of a
conflict between the small army of Mohammed, and a very
numerous band of idolatrous Koreysh, under Abu Sofyan.
The prophet's uncle, Hamza, fell in the engagement, with
75 others, all of whom are buried in this mountain, the exact spot being marked by a mosque.

The people of Medina, like those of Mecca, are chiefly
either foreigners or of foreign extraction, drawn thither by
the prophet's tomb, and the gains which it ensures to its
neighbours. The number of sherifs, indeed, descended
from Hassan, the prophet's grandson, is very considerable;

but most of them come from Mecca, or elsewhere, and
nearly all are olemas, or clergymen. The population presents, therefore, as motley a race as that of Mecca; and
Arabians of every district, Egyptians, Africans, Syrians,
and Turks of Anatolia, are found here, more or less naturalised by intermarriage; those long settled being characterised, as at Mecca, by the Arab face, expressive and of
features, and stout thick-set persons.

With respect to commerce, Medina widely differs from
Mecca; for, while the latter is enriched by a transit trade
scarcely inferior to that of any great city in the east, the
trade of the former is merely for the consumption of the
town and its neighbourhood, the articles being chiefly received from Egypt, by way of Yembo. The provision trade
is a lucrative branch of traffic; and the richer merchant
often realise enormous profits when the caravans stay for
any considerable time, and exhaust the stores of the smaller
dealers. The Bedouins supply the town with sheep, butter, honey, and charcoal, taking in return corn and clothing;
but the trade is subject to great fluctuations, in consequence
of continual jealousies between the tribes. The date and
lotus fruit are produced in large quantities in the neighbouring gardens, the former of these being the prime article
of food, and brought thither from all the surrounding country. As respects native industry, Medina is as ill situated
as Mecca; wanting the commonest mechanics, and not
even possessing a pottery. Weaving, dyeing, and tanning
are arts wholly unknown, nor is there a single person in
the whole city capable of making either a nail or a horse-
shoe, unless it be at pilgrim-time, when many of the poorest hadjis endeavour by hard labour to earn the money
necessary for their journey homewards.

The climate of Medina is, during the winter, much colder
than at Mecca. Rain falls irregularly at that season; often
in violent storms, lasting for two or three days, but in some
years so sparingly as to cause a general dearth, from the
want of proper irrigation. The summer heat is alleged to
be greater than in any other part of the Hedjaz; and the
salt-marshes, stagnant pools, and exhalations from the
neighbouring date-groves, are powerful agents in producing
those intermittent fevers, which are so common, and often
fatal, in the city, especially to visiters. The deaths, indeed,
are reckoned by Burckhardt (though, no doubt, very vaguely) at 1200 annually; which, assuming the population at
18,000, is one in 15: and if this be near the truth, it is clear
it must long ago have been depopulated, but for the continued supply of inhabitants from other countries!

Medina, though, probably, not entitled to rank as one of
the cities of what, by the best authorities, is considered the
Hedjaz, has always, since the establishment of Moham-
danism, been considered as a separate principality, and independent even of Mecca. The governor has, till recently,
been appointed by the grand seignior; but, in the absence
of precise information, it is believed that Mehemet Ali now
exercises supreme power over both the holy cities of the
Mohammedan world. (*Burckhardt*, vol. ii., 145-207; *Ab.
Two-, &c.*)

MEDINA, county, O. Situated in the N. part of the state,
and contains 450 sq. m. Drained by head branches of
Black and Rocky rivers. It contained in 1840, 18,325 neat
cattle, 31,450 sheep, 20,969 swine; and produced 215,621
bushels of wheat, 5339 of rye, 188,460 of Indian corn, 5380
of buckwheat, 3175 of barley, 133,563 of oats, 115,464 of
potatoes, 325,501 of sugar. It had 31 stores, one furnace,
eight fulling-mills, one woollen factory, 14 grist-mills, 21
saw-mills, 17 tanneries, two printing-offices, two weekly
newspapers; two academies, 99 students; 97 schools, 2945
scholars. Pop. 18,352. Capital, Medina.

MEDINA, p. t., capital of Medina co., O., 177 m. N.E. by
N. Columbus, 351 W. The village is situated on elevated
ground, and contains a handsome brick courthouse, in which
the county offices are kept, a brick jail, two churches, 3
stores, one fulling mill, three grist-mills, four saw-mills,
one distillery; two schools, 147 scholars. Pop. 675. The
township, exclusive of the village contains one store, two
fulling-mills, one woollen factory; five schools, 199 scholars.
Pop. 896.

MEDINA, p. t., Lenawee co., Mich.. 81 m. S.W. Detroit,
510 W. It has three stores, two grist mills, four saw-mills;
six schools, 165 scholars. Pop. 760.

MEDINO DEL CAMPO, a town of Spain, kingdom of
Leon, prov. Valladolid, on the Zapardiel, a trib. of the
Douro, 27 m. S.S.W. Valladolid, and 83 m. N.W. Madrid.
Pop. 3000. The town on both sides the river (crossed here
by a stone bridge), has a neat square, with a handsome
sculptured fountain in its centre. The houses are most-
very old, and many of them quite in ruins. A cathedral
and six other churches, several monasteries (now suppress-
ted), and two hospitals, one of which has considerable
architectural merit, are its chief public buildings; but most
of them show, by their dilapidated appearance, the degraded
condition of the place. The inhabitants are chiefly em-

ployed in agriculture. It has two weekly markets, and a fair in February, well attended by traders from Toledo, Segovia, Cuenca, &c.

Medina del Campo occupies the site of the ancient *Methymna Campestris*, and was formerly a place of considerable importance; but in the 17th century, after the discovery of America, a large part of its population emigrated, and its decay has since been hastened by the internal troubles of the country. (*Miñano*.)

MEDINA DE RIO SECO, a town of Spain, kingdom of Leon, prov. Valladolid, on the Sequillo, a trib. of the Douro, 32 m. S.S.E. Leon, and 132 m. N.W. Madrid. Pop. 4790. It stands in an open plain, W. of the river, crossed here by three bridges, and has narrow, badly-paved streets, and shabby decaying houses: there are three churches, four monasteries, two hospitals, and a castle; but, with the single exception of the church of St. Maria, all the public buildings are in a ruinous condition. The inhabitants were once so celebrated for their industry and the variety of manufactured goods exhibited at its fairs in April and September, that the district acquired the name of *India chica* (the Little Indies); but every trace of its former prosperity has now disappeared, and the population ranks at present among the most degraded and least industrious in Spain. (*Miñano*.)

MEDINA SIDONIA, a town of Spain, in Andalusia, prov. Cadiz, 22 m. E. by S. Cadiz, and 65 m. S. Seville. Pop., according to Miñano, 9337. It is an old walled town, beautifully situated on the brow of a rocky eminence, looking eastward over a fine champagne country. A castle, two parish churches, six monasteries (now unoccupied), and two hospitals, are the only public buildings. The chief employment of the inhabitants is the manufacture of earthenware, which is conducted on a large scale, furnishing the principal supply for Cadiz, Seville, and, indeed, the whole of Andalusia. The neighbourhood is celebrated for its fine pastures; and the rearing of cattle forms the chief occupation of the rural population.

MEDITERRANEAN SEA (the *Mare Internum* of the ancients, and, more recently, the *Mare Mediterraneum*), a large and very important inland sea, bounded N. by Europe, E. by Asia, and S. by Africa, communicating at its W. extremity, by the straits of Gibraltar, with the N. Atlantic ocean, and at its N.E. extremity, by the Dardanelles and Bosphorus, with the Black or Euxine sea. It extends, in a general sense, from lat. 30° to nearly 46° N., and from long. 5° 54′ W. to 36° 3′ E. Greatest length, 2300 m.; do. breadth, from Venice to the bay of Sidra, 1200 m.; estimated area (according to Stein), nearly 693,000 sq. m. It is of an oblong, but very irregular shape, especially on its N. side, into which project southward the two large peninsulas of Italy and Greece, which thus divide the Mediterranean into three basins, the most westerly of which is included between the straits of Gibraltar and the passage, only 72 m. broad, between C. Bœo in Sicily and C. Bon in Africa; the central part extending eastward from the last-mentioned points to the meridian of C. Matapan in the Morea; while the E. basin, called the Levant, comprises the Grecian Archipelago and the sea that washes the coasts of Karamania, Syria, and Egypt. The principal inlets of the W. basin are the bays of Lyons, Genoa, and Naples: it contains, also, the three large islands of Corsica, Sardinia, and Sicily; the Lipari, and other islands on the W. side of Italy; and the Balearic group, off the coast of Spain. The central basin has a large arm projecting N. under the name of the Adriatic sea; its smaller inlets being the gulfs of Taranto in Italy, Lepanto in Greece, and Cabes and Sidra (an. the two *Syrtes*) in Africa: Malta, the Ionian isles, and the numerous rocky islets skirting the shores of Dalmatia, are its chief islands. The portion of the E. basin or Levant, which stretches N. from the isle of Candia to the coast of Macedonia, is called the Archipelago, and is remarkable, not only for the extreme irregularity of its coast-line, but for the numerous clusters of volcanic islands and rocks that stud its surface; its chief gulfs are those of Egina, Salonika, Contessa, and Smyrna; and its largest islands are Lemnos, Mytilene, Thasos, Scio, and Naxia. The great island of Cyprus lies in the angle between the coasts of Asia Minor and Syria.

The coast of the Mediterranean is as remarkable for difference of altitude as for variety of outline. Its N. shores, as might be inferred from their jagged outline, are generally steep and bold; but in parts, as in Spain and France, near the mouths of the Ebro and the Rhone, and in Lucca, Tuscany, the Papal states, and Naples, as far S. as C. Campanella, the shores are low and gently shelving, varied only by a few bold rocky headlands; the S. side of Sicily and the W. shores of the Adriatic, are, also, with a few exceptions, flat and sandy; but in Istria, Dalmatia, and, in short, all along the E. side of the Adriatic, the coast is bold, broken, and irregular, often presenting cliffs rising between 600 and 700 feet in perpendicular height, with deep sound-

ings close to the shore. The shores of the Archipelago partake, more or less, of the same bold character, except in a few bays, where rivers, by the constant deposition of alluvial soil at their mouths, have formed low beaches, extending considerably beyond the high rocks usual to this coast.

The S coast of Anatolia, which has a less indented line of shore, though by no means low, is much less craggy and precipitous; extremely high promontories here and there stretch out into the deep sea; but beaches, more or less shelving, of shingle, gravel, or sand, are by far the most common on this coast. The cliffs about Iskenderoon are of great height, running round the bay, and furnishing complete security for shipping, except from the E., or land, breezes, which are both violent and dangerous. The shores of Syria are mountainous between Tripoli and Tyre, but present, in many places, a large extent of low and flat coast, especially towards the S. extremity.

Near the mouth of the Nile the country presents a low uninteresting flat, with rocky reefs and shoals, projecting from 5 to 7 m. from the shore; and this continues as far W. as long. 27° E., beyond which a series of not very high cliffs, varied here and there by sandy bays (the largest being those of Sidra and Cabes), marks the whole African coast as far as C. Spartel. Submarine rocks and projecting shoals of mud and sand, not less than the roving piratical habits of the Moors, render the navigation of these shores both difficult and dangerous; and, in this respect, the S. side of this sea presents a striking contrast to the N., where, generally speaking, deep soundings may be had close in shore; while, in parts, particularly between Nice and Genoa, and near Gibraltar, no soundings can be found under 1000 fathoms and upward. The in-shore navigation presents some difficulties, in consequence of a few hidden rocks; but the chief skill of the mariner is required in the Archipelago, where, though there be few hidden dangers, it requires first-rate experience of its shifting winds and currents to guide him safely through its many intricate channels. (*Purdy's Sailing Dir. for the Med.*, part ii., p. 61; *Lyell's Geology*, i., 347.)

It is a curious fact, that, though the Mediterranean generally be so deep that soundings, even where possible, are of no practical utility, except in some of its bays and harbours, the depth of the channel between Sicily and Tunis, according to Sonnini and Smythe, nowhere exceeds 30 fathoms, the average not being greater than the depth of the straits of Dover between England and France. (*Smythe's Hydr. Maps of Sicily and Africa, and Memoir.*) The temperature of its waters is, at an average, from 75° to 76°, or 34° Fah. higher than the W. part of the Atlantic ocean; but it does not appear, from the experiments of Marcet and Woollaston, that its density exceeds that of many ordinary samples of sea-water. (*Lyell's Geology*, ii., 17.) The chief feeders of the Mediterranean are the Ebro, Rhone, Po, and Nile, with the various waters brought from the Black sea by the strong current that sets W. through the Dardanelles. But, notwithstanding this vast supply, the evaporation is so rapid, that water constantly passes in through the straits of Gibraltar, to restore the equilibrium. The Mediterranean has long been considered a tideless sea; but this is not strictly true: for, in the Adriatic, as well as between that sea and the coast of Africa, tides rise from 5 to 7 feet, and their influence is felt, more or less, along the shores of Sicily, and on the W. side of the Morea. The existence of this tide, indeed, may suggest an explanation of the loss of so many vessels in that region of mist and terror, the gulf of Sidra, where there is always a lofty swell and accumulation of waters during the prevalence of N.W. winds. A tide of 8 or 9 feet also ebbs and flows at pretty regular intervals in the smaller gulf of Cabes, on the same coast. In the straits of Messina and Bonifacio, at Naples, in the narrow channel of the Euripus, and on both shores of the straits of Gibraltar, there is an ebb and flow amounting to 3 feet and upward; but whether these movements are to be attributed to lunar influence, or to other causes, has not been determined. (*Purdy*, part ii., p. 9; *Lyell*, i., 375.) The currents peculiar to this great inland sea vary in its different parts: a current sets E. along the African shores, which is turned northward along the coast of Syria, and then westward along that of Cyprus and Karamania; the current in the Archipelago sets almost continually to the S., being increased or retarded, according to the winds; in the Adriatic, the current runs N.W. up the coast of Albania, and S.E. down the Italian shores, bringing with it the waters of the Po. A strong current runs through the Faro of Messina (the Scylla and Charybdis of antiquity), and, by meeting a lateral current, causes numerous eddies and whirlpools. (See SCYLLA and CHARYBDIS.)

But this strait, notwithstanding the statements in the classics, presents no real danger; and, in the late war, it was traversed by the fleet under Lord Nelson. (*Smythe's Sicily*, p. 115–113.) In the straits of Gibraltar, the main

Y　　　　　　　　　　　　　337

current sets eastward, at a rate varying from 3 to 5 m. an hour: it is true that an under-current has long been supposed to run in an opposite direction; but the fallacy of this hypothesis has been fully shown by Mr. Lyell; and it seems that the only outlet for the superfluous water is by the lateral current, which runs westward close to the African shore. (*Geology*, ii., 19.) With respect to the winds of the Mediterranean, it may be observed that the prevalent winds, except during spring, vary between N.W. and N.E., while those in spring are from S.E. to S.W. But the winds are extremely variable, and it is said that three or four vessels may occasionally be seen carrying different, and sometimes opposite, winds at the same time. The *Bora*, a violent N.E. wind in the Adriatic, the *Etesian*, or N.E. winds (called also *Tramontana*), which blow for several months together in the archipelago, and the *scirocco*, or *solano*, are peculiar to this sea. The last of these is described by Capt. Smythe as being extremely troublesome, and producing great dejection and lassitude. "At its commencement," he observes, "the air is dense and hazy, with long white clouds floating just above, and parallel to, the horizon. The thermometer rises to 90° or 95°, sometimes 100°, and the barometer gradually sinks to about 29·60°. It generally continues during three or four days; during which period, such is its influence, that wine cannot be fined, or meat effectually salted; oil-paint laid on during its continuance will seldom harden. But, though blighting in its general effects during summer, it is favourable to the growth of many useful plants in winter, when, indeed, it has few disagreeable qualities." (*Sicily*, p. 6.)

Waterspouts are of very common occurrence; especially on the coast of Asia-Minor, where as many as sixteen have been seen at one time. Many volcanic phenomena have also been observed in this sea; among which may be mentioned the sudden appearance, in 1831, of an island, about 30 m. S.W. of Sciacca in Sicily, and its equally sudden disappearance, three years afterwards. These movements may result from the close proximity of the large igneous region of Italy and Sicily. The presence of electric fluid in the atmosphere is also proved by the play of flame round the mast-heads, called by sailors "the fire of St. Elmo." Several springs of fresh water rise in different parts of the Mediterranean: the largest of these is in the port of Taranto, near the mouth of the Galesus, where the fresh water ascends in such a volume, and with such impetuosity, that it may be taken up at the surface without the least impregnation of salt; but the most celebrated of these fountains is that of Arethusa, in the harbour of Syracuse. (*See* ARETHUSA, I., 146.)

The Mediterranean abounds with fish of many different varieties, as well as with mollusca. The tunny and anchovy fisheries are a source of great profit on the coasts of Italy and Sicily: the sword-fish is very common; and the *murex purpura* supplies the fine Tyrian dye, now, as anciently, celebrated for the brightness of its red colour. Coral is found on many parts of the Barbary coast, and in some of the bays of Corsica and Sardinia. The chief fishery, however, is in the straits of Messina, where there is a coralground upward of 6 m. in length.

In the Scriptures, the Mediterranean is called "the Great sea" (Num. xxxiv., 6). Herodotus calls it (i., 185,) "the Sea;" and Strabo, "the Sea within the Columns" (ʾΗ ἐντός τῶν στηλῶν). It is probable that it witnessed the first rude attempts at navigation. "Having," as Dr. Smith has justly observed, "no (perceptible) tides, nor, consequently, any waves, except such as are caused by the wind only, the Mediterranean was, by the smoothness of its surface, as well as by the multitude of its islands, and the proximity of its neighbouring shores, extremely favourable to the infant navigation of the world; when, from their ignorance of the compass, men were afraid to quit the view of the coast, and, from the imperfection of the art of shipbuilding, to abandon themselves to the boisterous waves of the ocean." (*Wealth of Nations*, book i., cap. 3.) At all events it was navigated, and its islands occupied, in the remotest antiquity: it subsequently was traversed in all directions by the ships of the Phœnicians, and their descendants, the Carthaginians; and, at a later period, by those of the Greeks and Romans. During the middle ages, and down to the discovery of America, it was the grand centre of the commerce and navigation of the old world; and the Venetians and Genoese, by whom its trade was for a while principally engrossed, attained, in consequence, to great wealth and consideration. The discovery of America, and of a route to India by the cape of Good Hope, opened new and far more extensive channels for maritime enterprise. But we incline to think that the depression of the Mediterranean trade, in the 16th, 17th, and 18th centuries, was principally owing to the circumstance of the countries round the Black sea, the Levant, and the whole N. shore of Africa, having been shortly before subjugated by the Turks, the implacable foes of art, civilization, and refine-

ment. Happily, however, their empire has been, to a considerable extent, dismembered; and, within the course of the present century, the trade of the Mediterranean has begun to resume something like its former importance. The opening of the Black sea, and the rise of Odessa and other towns on its shores, the renewed intercourse with India by Alexandria, the occupation of Malta by the English, and of Algiers by the French, the independence of Greece, and the establishment of steamers between the principal ports of the sea, have prodigiously extended its commerce and navigation. And when the old, worn out, imbecile despotism of the Turkish government has been overthrown, and the fine and fertile countries, now under its degrading yoke, have been emancipated, a vast additional stimulus will be given to its commerce; so that the fair presumption seems to be, that the ancient importance of the Mediterranean, as a field for the successful prosecution of commercial navigation, is destined, at no very distant period, to be again equalled, and, most probably, surpassed.

The Mediterranean has on its shores the capital cities of Naples, Palermo, Athens, Tripoli, Tunis, and Algiers. Among its principal emporiums may be specified Marseilles, Genoa, Leghorn, Civita-Vecchia, Venice, Trieste, Syra, Smyrna, Alexandria, Malaga, and Barcelona. Its most important naval stations are those of Malta and Toulon: Gibraltar is, as it were, the key of the sea.

To the scholar and classical traveller the Mediterranean has the most powerful attractions. Her shores were the earliest seats of art, science, and civilization. She has been surrounded and occupied by the most renowned nations of antiquity; and her coasts and islands have still to boast the ruins of some of the noblest and most splendid cities of the ancient world. In short, to use the language of Dr. Johnson, "the grand object of all travelling is to see the shores of the Mediterranean. On those shores were the four great empires of the world; the Assyrian, the Persian, the Grecian, and the Roman. All our religion, almost all our law, almost all our arts, almost all that sets us above savages, has come to us from the shores of the Mediterranean." (*Purdy and Norie's Sailing Directions; Smythe's Sicily, and Hydro. Charts; Beaufort's Karamania, and Charts; Dict. Géog. &c.*)

MEDWAY, an important river of England, which has its embouchure in the estuary of the Thames. It rises in the S.E. corner of Surrey, between the N. and S. chalk ranges; being joined at Penshurst-place by streams from the S. of Sussex. Its course is thence N.E. to Maidstone, and then N. to Rochester and Chatham, about two miles below which it turns nearly E. expanding at the same time into a wide estuary. Interspersed with islands. After prosecuting an easterly course for 8 or 10 m., it turns once more to the N., uniting with the estuary of the Thames at Sheerness. The tide is interrupted by locks, otherwise it would flow up the river to Maidstone. In consequence of works begun in the reign of Charles II. and improved at different periods, it has been rendered navigable as far as Tonbridge; affording a channel of communication of much importance to the surrounding country. From Sheerness to Chatham there is water to float the largest ships; and the ground being soft, and the reaches short, it forms an admirable harbour for men-of-war, many of which are usually laid up here when out of commission. (*See* CHATHAM, I., 605.) The Medway was called by the ancient Britons *Vaga*, to which the Saxons prefixed the syllable *Med*, signifying mid or middle, because it ran through the middle of the kingdom of Kent: hence it came to be called Medweg, and latterly Medway. Considering the shortness of its course, the Medway is one of the deepest of European rivers. (*Hasted's Kent*, I., 278, 8vo ed.; *Statistical Account, &c.*, I., 33.)

MEDWAY, p. t., Norfolk county, Mass., 26 m. S.W. Boston, 418 W. Watered by Charles river. It contains four churches, two Congregational, a Unitarian, and Baptist, eight stores, six cotton factories with 2859 spindles, four paper mills, eight saw-mills; three academies, 150 students, eight schools, 495 scholars. Pop. 2043.

MEERUT, a district of British India, presid. Agra (Bengal), chiefly between lat. 28° 30′ and 29° 30′ N., and long. 77° and 78° E., having N. the collectorate of Mozuffurnuggur, E. that of Moradabad, S. Boolundshahur, and W. Paniput &c. Area, 2930 square miles. Land revenue, in 1833, 14,04,216 rup. The chief towns are Meerut, Sirdhana, Luttoull, and Hustinapoor.

MEERUT, a town of British India, presid. Agra, cap. of the above district, in an extensive grassy plain, 36 m. N.E. Delhi. This, which, like Cawnpoor, is a military station, is a much more agreeable residence than the latter. The town is surrounded by a dilapidated brick wall, and has a ruined fort or citadel. The streets are narrow and mean, and the houses mostly of mud; but it has some good architectural remains of mosques and pagodas; and without the walls are various Mohammedan tombs, built of red stone. A small stream, which swells into a river during the rainy season, is here crossed by a handsome bridge. The cantonments are at some distance N. of the town, from which they are separated by a long and busy bazaar. The barracks are one story in height, and disposed in regular ranges, at intervals, along a space about two miles in length: the bungalows of the officers are surrounded with gardens, enclosed by tall hedge-rows. The church of Meerut is probably the largest in British India, being 150 feet in length, by 84 feet in breadth, and capable of accommodating 3000 people. There is a good free school here, with about 100 native pupils. Meerut was a city of some consequence before the Mohammedan invasion of India. It was taken by Mahmoud, of Ghiznee, in 1018, and by Timour in 1399. It was occupied, with its district, by the British in 1803, and is now the residence of a revenue collector and a judge, and the head-quarters of a force of about 90,000 men, of whom about 3300 are Europeans. (*Hamilton's E. I. Gaz.; Modern Trav.. x., 3, 3.; Parl. Reports, &c.*)

MEHERRIN river, Va., rises in Charlotte county and enters N. C., and unites with Nottaway river, to form Chowan river.

MEIGS, county, O. Situated in the S.E. part of the state, and contains 425 sq. m. Organized in 1819. Watered by Shade and Leading creeks. Bounded E. and S.E. by Ohio river. It contained in 1840, 9747 neat cattle, 16,860 sheep, 12,427 swine; and produced 95,505 bushels of wheat, 193,327 of Indian corn, 75,059 of oats, 34,908 of potatoes. It had 35 stores, one fulling-mill, two flouring-mills, 21 grist-mills, 39 saw-mills, nine tanneries, one distillery; 82 schools, 3601 scholars. Pop. 11,452. Capital, Chester.

MEIGS, county, Tenn. Situated in the S.E. part of the state, and contains 215 sq. m. Bounded N.W. by Tennessee river. It contained in 1840, 5163 neat cattle, 4242 sheep, 17,141 swine; and produced 26,995 bushels of wheat, 2108 of rye, 307,789 of Indian corn, 51,362 of oats, 4038 of potatoes, 19,355 pounds of tobacco, 10,778 of cotton. It had 19 stores, one forge, 15 grist-mills, nine saw-mills, seven tanneries, 28 distilleries; 14 schools, 345 scholars. Pop.: whites, 4498; slaves, 284; free coloured, 12; total, 4794. Capital, Decatur.

MEIGS, t., Adams co., O. Watered by Brush creek. It has five stores, one furnace, one flouring-mill, two grist-mills; one school, 94 scholars. Pop. 1068.

MEIGS, t., Muskingum co., O. It has nine schools, 270 scholars. Pop. 1333.

MEIGSVILLE, t., Morgan co., O. It has six schools, 175 scholars. Pop. 1159.

MEININGEN (SAXE), or SAXE-MEININGEN-HILDBURGHAUSEN (DUCHY OF), an indep. state of Central Germany, consisting of a crescent-shaped territory, between the 50th and 51st degs. of lat., and long. 10° 10' and 11° 27' E.; enclosed on the S. by the territories of Coburg and Bavaria, on the other sides by the dominions of Coburg, Schwartzenberg, Prussia, Hessel-Cassel, and Weimar. Area, 968 sq. m. Pop. (1840), 145,078. This duchy comprises a portion of the Thuringian forest mountains; one of which, the Dollmar, rises to 2370 feet above the sea. The remainder of Saxe-Meiningen is chiefly comprised in the vale of the river Werra, by which it is traversed in a N.W. direction. This last portion of the duchy, though one of the most fertile districts in Germany, does not, however, produce enough of corn for the consumption of its inhabitants. Tobacco, turnips, and fruit are stable products; and the forests and cattle of the duchy are among its most important sources of wealth. Mining is pretty actively pursued; iron, a little copper, coal, alum, vitriol, &c., being produced: it has also marble quarries, and furnishes about 190,000 cwt. of salt a year. The manufacture of hardware and the weaving of linens and woollens are the chief remaining branches of industry; but a number of hands are also employed in making various wooden articles, toys, &c.

The government is a limited monarchy, and, in nearly all respects, similar to that of Saxe-Weimar (which see). The high court of appeal in Jena is the supreme tribunal for this duchy. The inhabitants, who are nearly all Lutherans, are quite as well educated as those of the rest of Saxony. Primary schools are numerous; there are superior schools, or colleges, in all the towns, and the state has a considerable share in the direction and patronage of the university of Jena and of the Prussian gymnasium at Schleusingen, near Erfurt. Public revenue, in 1840, 1,147,860 florins, which was about equal to the expenditure. The public debt, at the same time, amounted to 5,850,000 florins. Saxe-Meiningen has one vote in the full council of the German confederation, and a vote in the committee, conjointly with Saxe-Weimar, Coburg, and Altenburg, together with which it holds the 14th place in the diet. It furnishes 1150 men to the army of the Confederation. Adelaide, queen dowager of Great Britain, relict of

William IV., is a sister of the present sovereign of Saxe Meiningen.

MEININGEN, a town of Central Germany, capital of the above duchy, on the Werra, here crossed by two stone bridges, 31 m. E.N.E. Fulda. Pop. about 6000. It is encircled by wooded hills, is well built and laid out, and surrounded with ramparts and ditches. In the palace, which has been the residence of the dukes since 1681, are collections of paintings, engravings, natural curiosities, the archives of Meiningen and Weimar, and a library of 34,000 volumes. The house of assembly for the states of the duchy, the *herzhordinare*, or gymnasium, the female seminary, riding-school, theatre, and hospital, are the other principal edifices. It has some manufactures of woollen, linen, and mixed fabrics, with tanneries, breweries, &c., and has some of the best public gardens in Germany. (*Berghaus, Allg. Länder, &c., iv., 381-9; Stein's Handb. der Geog.; Alm. de Gotha, 1841.*)

MEISSEN, a town of the kingd. of Saxony, prov. Meissen, on the Elbe, here crossed by a handsome bridge, 14½ m. N.W. Dresden. Pop., in 1837, 7856. "The castle, the Gothic church, and the lofty houses, perched high upon a rocky eminence, have a most imposing effect as you approach Meissen; but the streets are narrow, and the town has internally a gloomy appearance, which is considerably increased by the smoke constantly issuing from the porcelain manufactory." This establishment occupies the castle, built it is said, by the emperor Henry I. "The beautiful pottery ware that goes by the name of Dresden china, is all manufactured here; and though the Meissen potteries are now rivalled by those of Berlin and Vienna, they were at one time the first, and may still be considered the most celebrated, in Europe." (*Strang's Germany in 1831, I., 84, 85.*) The Gothic cathedral is remarkable for the fine open-work of its spire and the elegance of its interior. In it are several antique monuments of the Saxon princes, and some fine old paintings by Albert Durer and Cranach: the latter has introduced into the altar-piece the portraits of Luther, his wife, and his friend, the Elector of Saxony. The neighbouring convent of Afra has been converted into a royal school. The chapter-house, three hospitals, and the orphan asylum, are the other chief public establishments. Besides the porcelain factory, founded by a chemist named Böttcher, in 1710, Meissen has manufactures of hats, stockings, leather, colours, &c., though none is very considerable. (*Berghaus; Stein, &c.*)

MELCOMBE-REGIS. *See* WEYMOUTH.

MELFI, a town of the Neapolitan dom., prov. Basilicata, on a lofty and remarkable volcanic mountain, overlooking the plain of Capitanata, 34 m. S. Foggia. Pop. about 7000. Like most other towns in an elevated situation, Melfi loses somewhat on a nearer approach. It is encircled by old ruined walls. The streets are narrow, ill-paved, and dirty; and most of them, as the town is built on the side of a steep acclivity, are impracticable for carriages. Many of the houses, however, have a respectable appearance; which they owe more to the solidity of their masonry than to their magnitude. The principal object is its castle; which, though partly fallen to decay, and partly restored in very bad taste, retains a venerable and imposing aspect. It stands at the higher extremity of the town, from which it is separated by a kind of platform, commanding a view of the opposite slopes of mount Voltore. Considerable historical interest is attached to it from its being the principal, and probably the first, fortress built by the Normans after their conquest of this portion of the kingdom. The large hall in which the meetings of the Norman confederates were held, and which afterwards accommodated the council of prelates, held here in 1059, and the parliament summoned by Frederick of Swabia, has been converted into a theatre. One only of its towers remains entire; but it affords, in height and solidity, a good specimen of the general structure.

The cathedral has attached to it a high tower, erected in 1151; which, like most of our English belfries, has small Saxon arches for windows. There are eight other churches, with numerous convents, a magnificent bishop's palace, some excellent public cisterns, and a good private collection of the minerals of the neighbourhood. The atmosphere is damp, and the town is said to be at times unhealthy. "The population appears lively, industrious, and active; though there are no particular manufactures. Many of the inhabitants deal in cattle and wine, which last is somewhat less sweet and heavy than the produce of the surrounding towns: it constitutes an abundant article of commerce with the whole of the adjoining province of Capitanata, where, under the name of *Vino di Melfi*, it is sold, and held in general use." (*Craven's Abruzzi, &c., ii., 294; Swinburne, i., 402.*)

MELFORD, LONG, a market town and par. of England, co. Suffolk, hund. Babergh, 17½ m. W. Ipswich, and 51 m. N.E. London. Area of par., 4390 acres. Pop., in 1841,

2597. The town is very pleasantly situated in a picturesque and well wooded country, and consists chiefly of one main street, nearly one mile in length. A handsome Gothic church, two places of worship for dissenters, and an almshouse (founded in 1573) for 12 poor men and two women, are the only public buildings. Spinning, woollen weaving, and retail trade, are the chief occupations of the inhabitants. The weekly market once held here has been discontinued for some years. Cattle and sheep fairs, Tuesday, Wednesday, and Thursday, in Whitsun-week.

MELKSHAM, a market town and par. of England, co. Wilts, hund. its own name, 96 m. N.W. Salisbury, and 86 m. W. London. Area of par., 8090 acres. Pop., in 1841, 6236. The town, formerly much more important than at present, on an acclivity rising from the Avon, consists principally of one long street, with stone houses. The church is a large, old, embattled building, with a central tower and two transepts, both on the S. side: the living is a vicarage, in the gift of the dean and chapter of Salisbury. There are, also, places of worship for Independents, Baptists, and the society of Friends. Two mineral springs, one a strong chalybeate and the other a saline aperient, have been discovered about half a mile from the town; but though they are reported to be as efficacious as those of Cheltenham, they have not led to any great accession of visitors. The staple business of Melksham is the manufacture of woollen cloth; but it has greatly declined, in consequence of the superior facilities enjoyed by the clothing district of Yorkshire. In 1839, two woollen-mills employed 162 hands; besides which there are about 80 hand-loom weavers, whose weekly wages vary from 8s. 2d. to 14s. (*Hand-loom Weaver's Rep.*) Petty sessions for the hundred are held here. Markets for cattle, &c., on alternate Thursdays; fair for horses, cattle, and farming stock, July 27. (*Parl. Papers, &c.*)

MELROSE, a village of Scotland, co. Roxburgh, beautifully situated at the N. foot of the Eildon Hills, on the Tweed, 31 m. S.E. Edinburgh. The village has only about 700 inhabitants, and would be unworthy notice, were it not for its possessing the finest monastic ruin in Scotland. The abbey of Melrose, originally founded by David I., in 1136, for Cistercian monks, was destroyed by the English forces, under Edward II., in 1322. The structure, of which the mutilated remains still attest the grandeur and magnificence, was founded by Robert Bruce, the hero of Bannockburn, in 1326. It was finally completed, in the perpendicular Gothic style, in the reign of James IV.; and must, when entire, have been one of the noblest structures of the kind in the kingdom. This splendid edifice was well nigh demolished by the barbarous zeal of the early reformers. In fact, with the exception of a part of the cloister walls, the abbey has been wholly destroyed; but fortunately a considerable part of the church has been preserved. The great altar or eastern window, 36 feet in height by 16 feet in width, is unrivalled for its fine proportions, the richness of its tracery, and the beauty and delicacy of its workmanship. It has been admirably described by Scott:—

> The moon on the east oriel shone,
> Through slender shafts of shapely stone,
>   By foliag'd tracery combin'd ;
> Thou would'st have thought some fairy hand
> 'Twixt poplars straight the osier wand,
>   In many a freakish knot had twined ;
> Then fram'd a spell when the work was done,
> And chang'd the willow-wreaths to stone.
>
> *Minstrel*, canto ii., st. 11.

The south transept window and door is, perhaps, the most perfect part of the ruin. It is in the decorated style, with crockets and creeping foliage. The compartment of the nave, from the screen work to the cross, was roofed over, and fitted up, in 1618, for the parish church. But this roof does not harmonise with the rest of the fabric; and it is obvious that the stones of which it consists had been quarried from other parts of the building! A great tower rose from the centre of the cross, of which a portion, 84 feet in height, still remains; but the spire by which it was surmounted is entirely gone. The decorated work and masonry of the building have been most admirably executed; the mouldings are still as sharp as if they were fresh from the chisel.

In the centre of the village is a cross, said to be coeval with the abbey: a small property in the village is held on condition of the proprietor keeping up this fabric. The abbey is no longer used as the parish church, a handsome new edifice having been constructed for that purpose some years ago. Abbotsford, the seat and creation of Scott, is situated about three miles W. from Melrose; and Dryburgh abbey, where the great minstrel is buried, is about three miles E. from the village. (See *Monastic Annals of Scotland, &c.*)

MELTON-MOWBRAY, a market town and par. of England, co. Leicester, hund. Framland, on the Wreak (a trib.

of the Soar), crossed here by three stone bridges, 14 m N.E. Leicester, and 92 m. N. by W. London. Area of par., 3570 acres. Pop., in 1841, 3037. The town has of late years been much improved and enlarged; the houses are generally well built, the streets are well paved, watched, and lighted, and there are some excellent hotels. The church is a large and somewhat striking cruciform Gothic building, with a highly ornamented pinnacled tower, rising at the intersection of the nave and transepts: the living is a vicarage, in private patronage. There are places of worship, also, for Wesleyan-Methodists and Independents Three Sunday schools give religious instruction to upwards of 500 children; and two free schools, supported from the town's estate, are attended by about 400 children of both sexes. (*Educ. Rep.*, 1835.) Melton-Mowbray enjoys a small share of the hosiery and bobbin-net trade, which furnishes employment to many thousands in the county; but the chief business and celebrity of the town is attributable to its situation in the centre of a fine hunting country, and to its being, as it were, the hunting metropolis. The hunting season lasts from the beginning of November to the end of March; and during this time the town is frequented by the leading sportsmen of England, who resort thither from all parts of the country, and a few even from the continent. The stabling is excellently arranged, as well are very extensive, there being accommodation for upwards of 800 horses, with their grooms, helpers, &c. The town supports a good subscription library and news-room; and there is a temporary theatre, in which performances are held during the hunting season. Melton has water-communication with Leicester by the Wreak and Soar, and with Oakham by a canal. Petty sessions are held here; and this town is one of the polling places for the N. division of the county. It is likewise the chief place of a poor-law union, comprising 54 parishes, and has a very large and well-arranged workhouse. Markets, well attended, for cattle and provisions, on Tuesday; horse fairs, Monday and Tuesday after Jan. 17; cattle fairs March 13, Holy Thursday, Whit Tuesday, Aug. 21, and Sept. 7. (*Parl. Papers ; Sporting Review*, 1840, &c.)

MELUN (an. *Melodunum*), a town of France, dep. Seine-et-Marne, of which it is the cap.; on an island in and on both sides the Marne, 25 m. S.E. Paris. Pop., in 1836, 6639. It is badly laid out, but is tolerably well built; and, being pleasantly situated, has a prepossessing appearance from without. The different parts of the town are connected by two bridges, one of which admits of the passage of boats. The part of the town built on the island is the most ancient; it has a large central prison for five departments, the most conspicuous edifice in the town; and on the E. side are the remains of a palace, inhabited by several of the French kings. The portion on the right or N. bank of the Marne, called St. Aspais, is the most extensive: it is built on the declivity of a hill, and has a spacious square, an old Gothic parish church, with some fine stained glass, the prefecture, formerly a Benedictine abbey, a theatre, some vapour-baths, and the remains of the abbey of St. Pierre, founded under the Merovingian dynasty. The portion on the left bank of the river is less than either of the others; it comprises the cavalry barracks. Melun has a communal college, a public library, with 10,000 volumes, a society of agriculture, arts, and sciences, a school of drawing, &c., with manufactures of woollen cloths, printed cotton and linen goods, and glass, and some trade in corn and other products destined for the Paris markets. Melodunum is mentioned by Cæsar in his Commentaries: it was taken by his lieutenant Labienus. In the middle ages, it was several times captured by the Normans and English, and was held by the latter from 1419 to 1430. (*Hugo*, art *Seine-et-Marne, &c.*)

MEMEL, a fortified seaport town of the Prussian dom., being the most northerly of any size in the kingdom, and one of the principal shipping ports on the Baltic, prov. Prussia, gov. Königsberg, on the N.E. side of the entrance to the great salt lake, or lagoon called the Kurische Haf, but within the bar. 50 m. N.W. Tilsit, and 74 m. N.N.E. Königsberg ; lat. 55° 42'' N., long. 21° 8' 14'' E. Pop., in 1838, 9034. It consists of the portions called the Old town, New town, and Federick's town, and has several suburbs. It was surrounded by walls in the time of the Teutonic knights, and has a citadel, founded in 1252, now partly used as a prison. It has four churches (two Lutheran, one Calvinist, and one Roman Catholic), a synagogue, arsenal, exchange, theatre, high school, school of industry, female seminary, school for neglected children, a hospital, and various charitable institutions. It is the seat of a council for the circle (Landrath-amt), of judicial tribunals for the circle and town, a board of taxation, police commission, &c.; and b w manufactures of woollen cloth and soap, with docks for ship-building, sawing works, distilleries, &c. The harbour is large and safe, with deep water: but the bar, at the mouth of the Kurische Haff, has seldom more than 17 feet water, and sometimes not more than 12 or 14 feet; so that s...ps drawing more than 15 or 16 feet water are frequently obliged to load and unload a part of their cargoes in the roads, where the anchorage is but indifferent, particularly when the wind is at N. or N.W. A lighthouse, originally 75 feet, but now 100 feet high, has been erected on the N.E. side of the entrance to the harbour. The light, which is fixed and powerful, may be distinguished, in clear weather, at more than 20 m. distance. Timber, particularly oak-plank and fir, of the very finest quality, is the great article of export from Memel; but corn, staves, flax and hemp, linseed for crushing, hides, bones, bristles, wool, &c., are also largely exported. Timber, hemp and flax, and most other articles shipped from this, and, indeed, from most Baltic ports, are brackes; that is, they are inspected, and assorted into three qualities, according to their degrees of goodness, by persons appointed by government for the purpose. (See art. PE-TERSBURG, in this Dict.) Despite the serious difficulties which our corn laws and timber duties throw in the way of the trade with Prussia, we have a pretty considerable intercourse with Memel, especially when there is a demand for foreign corn in England. We send to it small quantities of colonial produce, cotton stuffs and yarn, and cutlery, and considerable quantities of coal, which, however, is reckoned merely as ballast. The trade of ship-building has recently been carried on to a considerable extent at Memel; and, in 1839, 64 vessels, of the aggregate burden of 12,779 lasts of 4800 lbs. each belonged to the port. The present average export of timber is reckoned at about from 75,000 to 88,000 loads fir timber, 5,600 loads oak timber and plank, 700 mill-oak pipe-staves, and about 800,000 fir planks. The exports of other articles are so very fluctuating as hardly to admit of their being reduced to an average. In 1837, of 609 vessels which cleared out from the port, 108 were laden with corn, 50 with linseed, eight with flax and hemp, five with bones, &c., the rest being laden with timber, or in ballast. Of these ships 126 belonged to England. In 1838, 201 English ships cleared out from the port. (Zedlitz, der Preussische Staat.; Bowring's Report on the Prussian Commercial League, &c.)

MEMMINGEN, a town of Bavaria, and formerly a free city of the empire, circ. Swabia, cap. distr. on a tributary of the Iller, 40 m. S.W. Augsburg. Population, 7,000. It is walled, and has a handsome town-hall, an arsenal, barracks, a lyceum, an academy of instrumental and vocal music, &c. It has manufactures of woollen, cotton and linen stuffs, stockings, ribands, oil-cloths, copper and iron wares, &c., with tanneries, linen and cotton printing and dyeing establishments; and an active trade in the products of these, and in salt, wool, corn, hops, &c., which it sends to Switzerland and Italy. (Berghaus, Allg. Länder, iv. 156.)

MEMPHIS, p. v., Shelby co., Tenn., 290 m. W.S.W. Nashville, 915 W. situated on an elevated bluff on the Mississippi, immediately below the mouth of Wolf river. It is built on the site of Old Port Pickering, and is regularly laid out. It contains four churches, a Presbyterian, Methodist, Episcopal, and Roman Catholic, an academy; 53 stores, three printing offices, each issuing a weekly newspaper, 560 dwellings, and 3300 inhabitants. Its commerce is extensive, being equal to that of any town between St. Louis and New-Orleans. Its growth has been exceedingly rapid. A railroad is in operation to Lagrange, intended to be a part of the Charleston and Memphis rail road.

MEMPHREMAGOG, lake, Vt., is from 30 to 40 m. long, and from two to three broad, and lies mostly in Canada. About seven or eight m. of the S. part extend into Vt. It contains about 15 sq. m. in Vt., and receives Clyde, Barton and Black rivers. It discharges its waters through St. Francis river into lake St. Peters, an expansion of St. Lawrence river, 15 m. below the mouth of Richelieu river. On an island, on W. side, at the mouth of Fitch's bay, 9 m. N. of the Canada line, is a quarry of novaculite, known by the name of "Magog oil stone," which have been prepared and vended in various parts of the United States, and is considered not inferior to the Turkey oil stone.

MENALLEN, p. t., Adams co., Pa., 48 m. S.W. Harrisburg, 69 W. It contains six stores, one fulling-mill, eight flouring-mills, eight grist-mills, 23 saw-mills, one oil-mill, one distillery; 11 schools, 440 scholars. Pop. 2260.

MENARD, county, Ill., situated a little N.W. of the centre of the state, and contains 260 sq. m. Watered by Sangamon river and its branches. It contained in 1840, 7,695 neat cattle, 6944, sheep, 23 swine; and produced 21,671 bushels of wheat, 393,900 of Indian corn, 55,150 of oats, 13,352 of potatoes, 2,199 pounds of tobacco. It had six stores, two woollen factories, three flouring-mills, seven grist-mills, 10 saw-mills, three distilleries; 17 schools, 407 scholars. Population, 4,431. Capital Petersburg.

MENAI STRAIT, a strait or channel of N. Wales, separating the island of Anglesea from Caernarvon: it runs N. E. and S.W. about 14 m., varying in width from about 900 yards to about 2 m. Parliament having contributed a sum

of money to assist in improving its navigation, the dangerous rocks, by which it was formerly encumbered, have been removed, so that vessels of moderate burden are able to pass without difficulty through the strait, when it would not be possible for them to double Holyhead. But the Menai Straits is now principally celebrated for the magnificent hanging bridge by which it has been recently crossed. Holyhead being the nearest port to Ireland, and the most convenient place at whish to ship and receive the Dublin mails, it became of great public importance that the access to it should be rendered as safe and expeditious as possible. The usual ferry across the strait was in the vicinity of Bangor; and this being frequently attended with both danger and delay, it was resolved to erect, nearly at the same place, a chain bridge, elevated sufficiently above the water to allow ships to pass freely underneath. This great undertaking was begun in 1819, and completed in 1825. There are seven stone arches, each of 52½ feet span; and the length of the catenary, or chain part, is 579 feet. The bridge cost in all £311,791. Its elegance and solidity reflect great credit on the engineer, Mr. Telford; but it is to the parliamentary commissioners for the Holyhead road, or rather their chairman, Sir Henry Parnell, that the public is mainly indebted for this signal improvement.

MENDE, a town of France, dép. Lozère, of which it is the cap., on the Lot, 48 m. E.N.E. Rodez. Population, in 1836, 5102. It is badly built and laid out, but is well supplied with water by numerous public fountains, and surrounded by a boulevard, forming a public promenade. The cathedral, a Gothic building, has two light spires. The old episcopal palace, now the prefecture has a gallery and hall, enriched with many paintings by Bessard, an artist of the French school. It has also a public library of 6,000 volumes, a communal college, a court of primary jurisdiction, a society of agriculture, science and arts, a chamber of manufactures, &c. Its inhabitants fabricate coarse woollen cloths, called serges de Mende, which are sent into Spain, Italy, and Germany.

Mende is a town of considerable antiquity. For at least 600 years its bishops possessed the privilege of coining money, and other rights of sovereignty, which, however, they began to share with the French kings in 1306. (Hugo, Art. Lozère.)

MENDHAM, p. t., Morris co., N. J., 56 m. N. Trenton, 230 W. Drained by branches of the N. branch of Raritan river and of Whippany river. It contains a Presbyterian church, four stores, one fulling-mill, one woollen factory, one cotton factory with 308 spindles, two grist-mills, one saw-mill, two tanneries, four distilleries; three academies, 95 students; five schools, 183 scholars. Population 1378.

MENDON, p. t., Worcester co., Mass., 33 m. S.W. Boston, 408 W. Watered by Blackstone and Mill rivers. Blackstone canal crosses the S.W. part. It contains nine churches, three Congregational, two Restorationist, one Free-will Baptist, two Friends, one unoccupied; 10 stores, five fulling-mills, four woollen factories, six cotton factories with 19,008 spindles, five grist-mills, six saw-mills, one printing office, one weekly newspaper; two academies, 58 students, 10 schools, 315 scholars. Pop. 2594.

MENDON, p. t., Monroe co., N. Y., 72 m. S. Rochester, 210 W. by N. Albany, 356 W. Drained by Honeoye and Irondequoit creeks. It contains two churches, a Presbyterian and Baptist; seven stores, one fulling-mill, two woollen factories, one furnace, one grist-mill, six saw-mills, two tanneries; two academies, 309 students, 22 schools; 1178 scholars. Pop. 3435.

MENIN (Flemish Meenen), a town of Belgium, prov. W. Flanders, cap. cant. on the Lys, immediately within the Belgian frontier, and 6½ m. S.W. Courtrai. Population, in 1836, 7,394. It is well fortified, and tolerably well built; is the residence of a military commandant; and has manufactures of woollen yarn, and table and other linen cloths, lace, soap, linseed and other oils, &c., with tanneries, breweries, and large bleaching grounds. It has also a considerable trade in horses, cattle, and agricultural produce, and two large annual fairs. (Vandermaelen, Dict. Geog.)

MENTOR, p. t., Lake co., O., 160 m. N.E. Columbus, 355 W. Bounded N. by lake Erie. First settled in 1797. It has two saw-mills; 13 schools, 536 scholars. Population 1,245.

MENOMONEE, r., Mich., has a course of 100 m. and enters Green Bay, bounding the upper peninsula of Michigan on the S.

MENTZ, or MAINZ (Fr. Mayence an. Moguntiacum), a strongly fortified city of Germany, the bulwark of the German Confed. towards the W., in the grand duchy of Hesse-Darmstadt, prov. Rhenish Hesse, of which it is the cap., on the left or W. bank of the Rhine, nearly opposite its junction with the Mayn, 18 m. W.S.W. Frankfort, and 34 m. S.E. Coblentz; lat. 50° 0' 2" N., long., 8° 16' 41" E. Pop., estimated at 40,500, including its garrison of about 6,000 men. It is built partly on level ground, and partly on the declivity of a hill, in the form of a semi-circle, the Rhine forming the basis of the arch. It is surrounded by strongly-built bastioned walls; and is further defended by extensive outworks, including a citadel, lunettes, and six forts. A bridge of boats across the Rhine, 1,666 Rhenish feet in length, protected by a *tête du pont*, connects Mentz with its fortified suburb of Castel, a town of 2,900 inhabitants, near which is an island in the river, that is also strongly fortified. A garrison of 30,000 men would be required for the proper defence of the various works. The city is entered by 10 gates, five on the land side, and five along the river; all which, except on special occasions, are closed at 10 P. M. It has several good streets and squares, which present various indications of improvement; generally, however, it is in most parts irregular, and the streets, which are narrow and dirty, are rendered darker by the loftiness of the houses, many of which have strongly-stanchioned windows: the appearance of the town is, in fact, that of an ancient city, converted into a modern fortress; but it is, notwithstanding, interesting from its antiquity, and its numerous public edifices. The cathedral, built in the massive round-arched Gothic style, was commenced in the 10th, and finished in the 12th, century. Like the cathedral of Worms and Spires, it has a double choir, and a high altar at both the E. and W. extremities. It was nearly destroyed by fire in 1190, and suffered greatly during the siege of the town by the Prussians, in 1793; but, of late years, extensive repairs have been made upon it, by the aid of voluntary contributions; the nave has been newly roofed with slate, and the great E. tower has been surmounted with an iron cupola, 70 feet in height. The interior has numerous monuments of the former archbishops of Mentz, who were sovereign princes, and electors of the empire. It has also monuments of various other historical personages. The side chapels abound in fine old carving; the doors, of solid brass and great height, opening to the market-place, were cast by the founder of the cathedral, and have engraved on them the charter given to the city by Archbishop Adalbert, in 1135. There are six other Roman Catholic parish churches, several conventual churches, and a Calvinist church, most of which deserve notice. On the quay beside the river are two large red buildings; one of which, the ancient electoral palace, has been converted into the custom-house; and the other, the Teutonic-house, once occupied by Napoleon, is now the residence of the military governor. The former palace of the Prince Dalberg, nearly destroyed by fire in 1793, is used for the courts of justice. The arsenal, on the bank of the Rhine, the theatre, a new and handsome edifice, and the episcopal and vice-governor's palaces, are among the remaining principal public buildings. But Mentz derives its principal celebrity from its having been the residence of Gutenberg, and the cradle of the art of printing. The house in which Gutenberg lived has been taken down, and its site is occupied by a casino, belonging to a literary club. In an adjacent court is a statue of Gutenberg, in bronze, from a model by Thorwaldsen; but this work is said (by Chambers) to be clumsy, gigantic, and tasteless; and it is proposed to raise a monument more commensurate with the signal merits of the individual in whose honour it is to be erected. Mentz has a gymnasium, an ecclesiastical seminary, schools of medicine and veterinary surgery, a public library of 90,000 volumes, in which are preserved some of the earliest extant specimens of printing; a museum of natural history, antiquities, coins, &c. Outside the walls are some fine gardens along the bank of the river.

Mentz, formerly the first ecclesiastical city of the empire, is now of importance chiefly as the strongest fortress and principal military post. A mighty revolution has taken place since the visit of Dr. Moore, towards the end of last century, when the abbés, with their handsome equipages, lorded it over the well-behaved troops. "The chapter and the grenadiers have now precisely changed places. You see the meagre occupants of the stalls skulking to mass in threadbare *soutanes*, their looks proclaiming them no longer the monopolisers of the old hock of the neighbourhood; while the Austrian and Prussian soldiers are parading about in the insolence of military superiority. The cafés, the billiard-rooms, the promenades, are thronged with these smoking and swaggering guests, who impart a sort of unhallowed vivacity to the gloomy haunts of superstition and monachism. The university building is converted into barracks, and hospitals and guard-rooms strike one at every corner." (*Autumn on the Rhine*, p. 5.) Mentz is garrisoned by a nearly equal number of Prussian and Austrian troops, and is commanded by a governor, elected alternately every five years from either nation. It is the seat of a military tribunal, and the high court of justice for Rhenish Hesse; its civil authorities being appointed by the government of Hesse-Darmstadt. The town is so environed, on the river side, by its fortifications and other erections, that the Rhine is but little available for commercial purposes, and the ac-

commodation for craft is very inferior. Nevertheless, Mentz is the chief commercial town in the grand duchy (see also HESSE-DARMSTADT, l. 1090.), and, next to Cologne, the chief mart for Rhenish produce in Germany. It has a considerable trade in corn, wine, timber, &c.; and manufactures of leather, soap, hats, glue, vinegar, tobacco, musical instruments, &c.; steam-communications with Mannheim and Holland, and a steam-navigation assurance-company.

Though Mentz abounds in historical associations, its existing vestiges of antiquity are very few. Agrippa, the general of Augustus, established an entrenched camp on the site where Drusus Germanicus, about anno 10 B.C., erected a fort called *Moguntiacum*. Drusus afterwards founded a second fort (*Castellum*) on the opposite bank of the Rhine; and the two were at a subsequent period connected by a Roman bridge, portions of some of the piers of which may still be seen when the water is low. (*Schreiber*.) In the citadel is the Eichelstein, a stone tower, alleged to have been erected by Drusus. At Zahlbach, not far from Mentz, are the remains of an ancient aqueduct; and, between the two, a Roman cemetery has been discovered. The city, which was almost wholly destroyed in the wars at the fall of the Roman empire, was restored by Charlemagne, who erected a church, and rebuilt the bridge with timber. In the 13th and 14th centuries Mentz was a place of some note for literature and the arts. In 1631, it was taken by the Swedes; in 1644, 1688, and 1792, by the French; it was bombarded and taken by the Prussians in 1793; but, being re-taken by the French in 1797, it became during their ascendancy, the cap. of the dep. of Mont-Tonnerre. (*Schreiber, Guide du Rhin*, p. 196–136; *Autumn near the Rhine*, p. 1–13; *Berghaus, Allg. Länder, &c.*, iv. 358, 359, *Chambers's Tour, &c.*)

MENTZ, t., Cayuga co., N. Y., 8 m. N.N.W. Auburn, 156 m. W. Albany. Watered by Owasco inlet, and Seneca river, which bounds it on the W. and N. The Erie canal passes through its centre, crossing Owasco inlet by a stone aqueduct. Montezuma salt works, lie a little N. of the canal, and are connected with it by a side cut. It contains two commission houses in foreign trade, 25 retail stores, four fulling-mills, two woollen factories, three grist-mills, 10 saw mills, two tanneries, one distillery; 29 schools, 1,326 scholars. Pop. 4,215.

MEQUINEZ, a large city of Morocco, and one of the residences of the emperor, 70 m. E. Salee, and 235 m. N.N.E. Morocco; lat. 33° 56' N., long. 5° 39' W. Population, differently estimated, at from 50,000 to above 100,000. It stands in a beautiful, well-watered, and very fruitful valley; and is surrounded by a wall about six feet high, built for a defence against the marauding Berebbers. It owes its present extent and consequence to the late sultan, Muley Ismael; who, after having secured to himself the undisputed sovereignty of the kingdoms now forming the empire of Morocco, made Mequinez one of the capitals, considerably enlarged it, and erected a fine palace, which, owing to its having only one story, is of great apparent extent. In the centre of the enclosure, which contains several well laid-out gardens, is the emperor's harem, formed by a four-sided colonnade, above which are various apartments for the women, eunuchs, and female attendants. The rooms are each about 20 feet long, by 12 feet broad, and 18 feet high; the walls are inlaid with red and blue tiles, and the light is communicated by means of two large folding-doors. Between the chief apartments are paved courts of chequered marble, in the centre of most of which is a fine marble fountain. The houses of Mequinez are neater than those of Morocco; but the streets are not paved; and hence in rains they are infested with mud, and in dry weather with dust. The mellah, or Jews' quarter, is walled round, extensive, and in good repair; but the Negroes' quarter is now a mere ruin. About a century ago a convent was founded here by the king of Spain, for the relief and spiritual comfort of Roman Catholic captives and Christian travellers, but it was deserted by the monks, previously to the accession of the late emperor, Muley Soliman. The inhabitants are described as being more courteous than those more to the S.: they are hospitable to strangers, invite them to their gardens, and entertain them sumptuously. The women are beautiful, and have fair complexions, with black eyes, white teeth, and dark hair; and have a suavity of manners rarely to be met with even in the most polished nations of Europe. (*Jackson's Morocco*, p. 196–199; *Geog. Journ.*, vol. i.)

MERCER, county, N. J., situated in the W. part of the state, and contains 260 sq. m. Bounded S.W. by Delaware river. Watered by Assunpink creek, and a head branch of Millstone river. It contained in 1840, 8098 neat cattle, 8,643 sheep, 19,758 swine; and produced 43,496 bushels of wheat, 48,390 of rye. 179,739 of Indian corn. 39,838 of buckwheat, 1,767 of barley, 216,886 of oats, 57,581 of potatoes. It had 106 stores, eight lumber yards, three furnaces, five fulling-mills, four woollen factories, seven cotton factories with 6500 spindles, four dyeing and printing works, nine flouring-mills, 28 grist-mills, 22 saw-mills, three oil-mills, three paper-mills, one rope-walk, 13 tanneries, 11 distilleries, one brewery, six printing offices, four binderies, three weekly and one semi-weekly newspapers; eight academies, 279 students; 49 schools, 1472 scholars.　Pop. 21,502.　Capital, Trenton.

MERCER, county, Pa., situated in the W. part of the state, and contains 850 sq. m.　Watered by Shenango, Neshannock, and Wolf creeks.　It contained in 1840, 29,896 neat cattle, 63,387 sheep, 47,006 swine; and produced 310,615 bushels of wheat, 89,877 of rye, 254,918 of Indian corn, 119,817 of buckwheat, 1234 of barley, 442,865 of oats, 331,655 of potatoes, 152,314 pounds of sugar.　It had 95 stores, 19 fulling-mills, five woollen factories, one flouring-mill, 73 grist-mills, 126 saw-mills, 25 tanneries, 12 distilleries, three potteries, three printing-offices, three weekly newspapers; two academies, 150 students; 207 schools, 1,746 scholars.　Pop. 32,873.　Capital, Mercer.

MERCER, county, Va., situated toward the S.W. part of the state, and contains 540 sq. m.　Bounded E. by New river and drained by its branches.　It contained in 1840, 2289 neat cattle, 3695 sheep, 5166 swine; and produced 13,239 bushels of wheat, 4711 of rye, 56,254 of Indian corn, 1175 of buckwheat, 98,955 of oats, 7901 of potatoes, 2573 pounds of tobacco, 5842 of sugar.　It had one store, 13 grist-mills; one school, 94 scholars.　Population, whites, 2,127; slaves, 98; free coloured, eight; total, 2,233.　Capital, Princeton.

MERCER, county, Ky., situated centrally toward the E. part of the state, and contains 250 sq. m.　Bounded N.E. by Kentucky river, E. by Dicks river.　Drained by branches of Salt river.　It contained in 1840, 16,778 neat cattle, 29,032 sheep, 49,032 swine; and produced 99,305 bushels of wheat, 36,431 of rye, 3,397,495 of Indian corn, 181,039 of oats, 23,531 of potatoes, 64,747 pounds of tobacco, 19,314 of sugar. It had 25 stores, one fulling-mill, one woollen factory, three cotton factories with 1040 spindles, nine flouring-mills, 49 grist-mills, 21 saw-mills, one powder-mill, nine tanneries, six distilleries, one printing-office, one weekly newspaper; two colleges, 34 students; three academies, 141 students; 30 schools, 726 scholars.　Population, whites, 12,061; slaves, 5996; free coloured, 373; total, 18,790.　Capital, Harrodsburg.

MERCER, county, O., situated in the W. part of the state, and contains 576 sq. m.　Drained by St. Mary's and Wabash rivers, and their tributaries.　It contained in 1840, 4248 neat cattle, 1440 sheep, 6907 swine; and produced 18,996 bushels of wheat, 156,698 of Indian corn, 31,937 of oats, 11,948 of potatoes.　It had 24 stores, one woollen factory, two grist-mills, six saw-mills, two tanneries; 12 schools, 400 scholars.　Pop. 8277.　Capital, Celina.

MERCER, county, Ill., situated toward the N.W. part of the state, and contains 550 sq. m.　Bounded W. by Mississippi river.　Drained by Edwards, and Pope rivers, and the N. fork of Henderson's river.　It contained in 1840, 3072 neat cattle, 1696 sheep, 8339 swine; and produced 27,103 bushels of wheat, 1701 of rye, 111,430 of Indian corn, 20,508 of oats, 11,838 of potatoes, 1735 pounds of sugar.　It had 16 stores, two woollen factories, two grist-mills, 11 saw-mills, one distillery; 15 schools, 390 scholars.　Pop. 2352, Capital, Millersburg.

MERCER, p. t., Somerset co., Me., 34 m. N.N.W. Augusta, 687 W.　Watered by Sandy river.　It has two stores, two grist-mills, three saw-mills; 11 schools, 616 scholars. Pop. 1432.

MERCER, p. b., Springfield t., capital of Mercer co., Pa., 57 m. N. by W. Pittsburg, 234 m. W.N.W. Harrisburg, 277 W.　Incorporated in 1814.　It contains a courthouse, jail, four churches, a Presbyterian, Methodist, Associate Reformed, and a Union, 18 stores, two tanneries, one pottery, two printing-offices, two weekly newspapers; one academy, 100 students, four schools, 126 scholars.　Pop. 781.

MERCER, t., Butler co., Pa.　It has six stores, one woollen factory, four grist-mills, two saw-mills, two tanneries, two distilleries, one pottery; eight schools, 406 scholars.　Pop. 1632.

MERCERSBURG, p. b., Montgomery t., Franklin co., Pa., 15 m. S.W. Chambersburg, 62 m. S.W. Harrisburg, 86 W.　Situated on a branch of Conecocheague creek.　It has four churches, a Presbyterian, Methodist, German Reformed, Associate Reformed, six stores. 200 dwellings and 1143 inhabitants.　It is the seat of Marshall college, founded in 1836, which has four professors or other instructors, and 49 students.　The commencement is on the last Wednesday in August.　The b. has one academy, 54 students, seven schools, 307 scholars.

MEEDIN (an. Mardé), a city of Asiatic Turkey, at the N.W. extremity of the pach. of Bagdad, 50 m. S.E. Diarbekir, lat. 37° 10' N., long. 4° 20' E.　Population about 11,000, of whom 1500 are Armenians, and 200 Jews　It is situated on the slope of the Karadja-dagh, or ancient

Mount Masius, and overlooks a very extensive and fertile tract of country.　It is commanded by a castle, crowning the summit of a rock, and is very difficult of access, the best road to it leading up a steep about 1¼ m. in length. The houses are all built of hewn stone, and appear to be very old; the windows are small, grated with iron, the streets narrow, and the buildings, being on an acclivity, seem to rise one on the top of the other.　The walls are kept in tolerable repair, and a few old pieces of cannon are mounted on the towers of the castle.　Merdin is the frontier town of the pachalic towards Constantinople, and the residence of a mutsellim appointed by the pasha.　The industry of the inhabitants is confined to the manufacture of cotton fabrics and Turkey leather; but it has little external trade, in consequence of not being on any of the great caravan-routes.　The neighbourhood produces an abundance of cotton, grain, and fruits, which find a ready sale in the market of Merdin.　(Olivier, tom. iv., p. 242–245.; Kinneir p. 265.)

MERE, a small market town and par. of England. co. Wilts, hund. its own name, 99 m. W. Salisbury, and 96 m. W. by S. London.　Area of par., 7400 acres.　Population in 1831, 2708, of whom 1482 belong to the town-tything. The town is very indifferently built, having in its centre an ancient cross, the interior of which serves as a market-house.　The church is large, with a square tower at its W. end: the living is a vicarage in the gift of the dean of Salisbury.　The Wesleyan-Methodists, also, support a place of worship and attached Sunday-school.　A silk-mill has recently been erected, and, in 1839, employed 71 hands. Dowlas, also, and bed-ticking, are made here on a small scale; but the town (formerly of considerable importance, having a castle on an adjacent eminence) is now in a miserably decayed condition.

MEREDITH. t., Belknap co., N. H., 29 m. N. Concord. Bounded E. by Winnipesseogee lake.　A part of Great South bay of the lake, extends into its S. part.　It contains two Congregational, two Baptist, and several Free-will Baptist churches. 20 stores, one grist-mill, three saw-mills, three tanneries; 20 schools, 787 scholars.　Pop. 3351.

MEREDITH, p. t., Delaware co., N. Y., 6 m. N. Delhi, 79 m. W.S.W. Albany, 345 W.　Drained by Oulcout creek, flowing into Susquehanna river, and branches of the W. branch of Delaware river.　It contains three churches, two Baptist and a Presbyterian, three stores, one fulling-mill, two grist-mills, three saw-mills; 14 schools, 530 scholars. Pop. 1640.

MERGUI, a town of the Tenasserim coast, in India-beyond-the-Brahmapoutra, capital of the British province of Mergui, on the river Tenasserim, at its mouth, in lat. 12° 12' N., long. 98° 25' E.　Population probably from 6000 to 7000, including natives and British residents, with Chinese, Siamese, Peguans, and descendants of Portuguese.　It is built along the declivity and skirts of a steep hill, and, when taken by the British, was surrounded by a wooden stockade.　But it is not a place of any strength, being accessible to ships, and commanded by a high island in front. The streets are wide, but badly paved; and they would be much dirtier than they are but for the situation of the town on a slope, which facilitates their being cleaned by the rain.　The houses are nearly all of bamboo, reeds, matting, and other fragile materials.　A mean brick gateway stands at the entrance to the town from the river side, which, with some bastions at the angles of the works, a few small pagodas, and some houses erected by Europeans, are the only structures of any solidity.　The harbour is safe for small vessels, having 18 feet water over the bar at low water, with 18 feet rise at springs.　The climate is mild and salubrious: European invalids, sent thither from Rangoon during the war, speedily recovered their health. Mergui was taken by storm by the British in 1824.　(Low. Hist., &c., in Journ. of the Royal Asiat. Soc. ii. 256–259.; Hamilton's E. I. Gaz.)

MERGUI ARCHIPELAGO.　See TENASSERIM PROVINCES.¶

MERIDA (an. Augusta Emerita), a town of Spain, in Estremadura, on the N. bank of the Guadiana, 99 m. E. Badajoz, and 176 m. N.E. Madrid.　Pop., according to Miñano, 4690.　It is situated close to the river, on a slight eminence, in the midst of an open and gently undulating country, naturally very fertile, but almost uncultivated, and unhealthy in summer.　Its chief public buildings are two parish churches, eight ruined monasteries, two hospitals, and a prison: it is, in fact, one of the most decayed towns in the peninsula, and is wholly unimportant, except for its antiquities.　But the remains of the power and magnificence of its Roman masters render it an object of great interest. These are scattered in all directions: in the walls, the houses, the churches, and even in the pavements of the streets are discovered fragments of columns, bases, capitals, friezes, statues, and inscriptions.　Similar vestiges, and in a more perfect condition. are to be seen in the suburbs.　The principal ruins comprise an amphitheatre (used also as a naumachia), circus, theatre, triumphal arch,· baths, &c. The seats of the amphitheatre appear quite perfect; the vaulted dens for the beasts are uninjured; and the conduits by which the arena was filled with water are still distinctly visible. In one of the streets may be seen a large triumphal arch, 150 feet high, but without any inscription or sculpture. The baths are surprisingly perfect, but not large; and round the top of the bathing-rooms runs a cornice of most curious and delicate workmanship, almost as perfect as if it had recently been executed. The bridge over the Guadiana is of stone, and portions of it may be of Roman architecture; but the greater part of the Roman bridge was swept away by a flood in 1610, and the present bridge has been constructed since. Two arches of this structure were blown up, in 1812, by the British troops under the Duke of Wellington. There are likewise two aqueducts, one of Roman, and the other of Moorish architecture, of brick and granite, the former having three, and the latter two tiers of arches.

Augustus Emerita was founded by order of Augustus, anno 25 B.C., who planted in it some of his veterans, called emeriti, whence its ancient and modern names. Though its ancient magnitude appears to have been greatly exaggerated, it was, no doubt, one of the largest Roman cities in the peninsula, and became the metropolis of Lusitania. From the Romans it passed, in 713, to the Moors, who destroyed and altered many of its old buildings. In 1228 it opened its gates to Alphonso IX., after his signal victory over the Moors in the contiguous plain of Matanzas; and from this period downwards it has been attached to the kingdom of Castile and Leon. (Recollections of the Peninsula, p. 178–181; Cook's Spain, i. 142–144; Alfaro; Ancient Universal History, xiii. 492, 8vo. ed.)

MERIDA, a town of S. America, repub. Venezuela, dep. Zulia, cap. the prov. Merida, on the Chama, 330 m. S.W. Caraccas, and 325 m. N.E. Bogota. Previously to 1812, when it was destroyed by an earthquake, this was the largest city in Venezuela, and had a population of 12,000. It continued, for some years, to be little better than a heap of ruins; but it has been again rebuilt, and is now, probably, become more populous than before. It has a cathedral, several chapels, convents, &c., an ecclesiastical seminary, a college for philosophy, civil law, &c. The inhabitants are said to be, for the most part, in good circumstances. The coloured races dye wool and manufacture carpets and other woollen and cotton fabrics. (Geog. Account, 4.o., 270–74; Mod. Trav. xvii.)

MERIDEN, p. t., New Haven co., Ct., 16 m. S. by W. Hartford, 321 W. Incorporated from Wallingford in 1806. Watered by Quinnipiacer r. and its branches. It contains three churches, a Congregational, Episcopal, and Baptist; a bank, eight stores, two fulling-mills, two woollen factories, two grist-mills, nine saw-mills, two paper-mills, one tannery, one distillery; two academies, 50 students; 12 schools, 397 scholars. Pop., 1880.

MERIONETH, or MERIONYDD, a marit. co. of N. Wales, having N. the cos. of Caernarvon and Denbigh, E. and ·S. those of Montgomery and Cardigan, and W. St. George's Channel. It is of a triangular shape, and contains 424,390 acres. This, next to Caernarvon, is the most mountainous county in the principality. Among the principal summits are those of Arran-Fowdy, Cader-Idris, and Snowdon; respectively 2,955, 2,914, and 2,809 feet above the sea. It has, however, some fine vales, especially that of Festiniog, celebrated for its romantic scenery. There are some considerable tracts of low swampy land along the sea-coast; and in parts considerable tracts have been gained by embankments. The soil is very various; but, generally speaking, it is poor, and suited only for pasturage. Oats is the grain principally cultivated; but wheat and barley are also raised, though in no great quantities. Agriculture is in a very backward state; little or no attention is paid to a rotation of crops, and it is a frequent practice here, as well as in Denbigh, to burn the surface for manure. In some parts of the county potatoes are pretty extensively cultivated. The principal dependance of the farmer is, however, on his cattle and sheep, of which great numbers are fed on the mountains, and in the valleys not fitted for husbandry. The small native Welsh ponies, called Merlins, are now rarely met with, except in this county and Montgomery. They are sure-footed and exceedingly hardy. Dairy farming is carried on to a considerable extent. Farms usually small; and being mostly held at will, without any conditions as to management, the low state of agriculture need not be wondered at. Average rent of land, in 1810, 3s. 11d. an acre. Speaking generally, the cottages are wretched in the extreme; though happily they have been a good deal improved in some parts of the county. The minerals seem to be of less consequence than might have been supposed: lead and copper are raised, though in small quantities; large quantities of lime are produced at Corwen, and slates are quarried in different places. The manufactures, which also are unimportant, consist principally of coarse flannels, produced

344

on the domestic system, at Dolgelly, Towyn, and a few other places. The Dee has its source in this county; and it is also watered by the Dyfi, Maw, Dwynwy, &c., flowing W. Bala, the largest lake in the principality (see BALA), is in this county. Harlech is the county town. Merioneth is divided into six hundreds and 37 parishes. It sends one member to the House of Commons for the county. Regis. electors, in 1839–40, 1399. In 1831, the county had 6809 inhabited houses, 7358 families, and 35,315 inhabitants, of whom 17,194 were males, and 18,121 females. Sum expended for the relief of the poor, in 1838–39, 13,107l.

MERMENTAU, r., La., drains the extensive prairies of south western Louisiana. After a southerly course of 220 miles, it enters the gulf of Mexico, 200 m. W. of Mississippi river. It passes through a lake of the same name, and has not more than four feet of water on the bar at its mouth. Live oak is found on its borders.

MERRIMAC, r., N. H., is among the large rivers of New-England. Its longest branch is Pemigewasset river, which rises near the Notch of the White Mountains. After running S. about 70 m., it receives Winnipiseogee river, in Franklin township, the outlet through Great bay of Winnipiseogee lake, and below that it takes the name of Merrimac river. It then flows S. by E. until it enters Massachusetts; it then turns E. and N.E. for 50 m., until it enters the Atlantic, a little below Newburyport. It is navigable for vessels of 200 tons to Haverhill. By means of the Middlesex canal from Chelmsford to Boston harbour, and canals and locks around the falls in the river, a boatable communication is opened from Boston to Concord, N. H. Bow canal, a few miles below Concord, with four locks, overcoming a fall of 25 feet, was completed in 1812, and cost $25,000; 6 m. further down, Hooksett canal, with three locks, overcoming a fall of 17 feet, cost $15,000; 8 m. below Hooksett is Amoskeag canal, with nine locks, overcoming a fall of 45 feet, and cost $50,000. It is the greatest work of the kind on the river. Below Amoskeag canal, the river, for 9 m., is converted into the Union canal, overcoming six distinct falls, with slack-water navigation, and seven locks, and cost $50,000; 5 m. below are Cromwell's falls, made passable by a canal, the expense of which is included in Union canal; 15 m. below, in Massachusetts, Wicasin canal cost $14,000, 3 m. below which is the entrance of the Middlesex canal. A number of bridges cross Merrimac river. Its principal tributaries on the W. side are Contoocook, Piscataquog, Souhegan, and Nashua rivers; and on the E. side are Soucook, Suncook, and Beaver rivers. It has several important places on its banks, the principal of which are Newburyport, Haverhill, and Lowell, in Massachusetts, and Concord in New-Hampshire.

MERRIMAC, co., N. H. Centrally situated toward the S. part of the state, and contains 816 sq. m. Organized in 1823. Watered by Merrimac r. and its branches. It contained, in 1840, 35,911 neat cattle, 90,506 sheep, 12,985 swine; and produced 60,619 bushels of wheat, 46,708 of rye, 189,286 of Indian corn, 8866 of buckwheat, 5951 of barley, 172,232 of oats, 685,370 of potatoes, 64,940 lbs. of sugar. It had 113 stores, 12 fulling-mills, three woollen factories, 3 cotton factories, with 11,768 spindles, one glass factory, 42 grist-mills, 105 saw-mills, one oil-mill, five paper-mills, 35 tanneries, two distilleries, three potteries, 10 printing offices, six binderies, six weekly newspapers, and one periodical; 70 academies, 1984 students; 273 schools, 10,380 scholars. Pop. 36,253. Capital, Concord.

MERRIMAC, p. t., Hillsborough co., N. H., 20 m. S. Concord, 452 W. Bounded E. by Merrimac river. Watered by Souhegan river and its branches, which afford extensive water-power. Incorporated in 1746. It contains a Congregational church, four stores, six grist-mills, six saw-mills; nine schools, 351 scholars. Pop. 1114.

MERRIMAC, t., St. Louis co., Mo. Pop. 1762.

MERRIMAC, t., Crawford co., Mo. Pop. 1111.

MERRIWETHER, co., Ga. Situated in the W. part of the state, and contains 408 sq. m. Bounded E. by Flint river, and drained by its branches and by branches of Chattahoochee river. It contained, in 1840, 18,593 neat cattle, 9121 sheep, 41,544 swine; and produced 86,480 bushels of wheat, 1040 of rye, 356,115 of Indian corn, 58,779 of oats, 20,370 of potatoes, 5,080,229 lbs. of cotton. It had 11 stores, nine grist-mills, five saw-mills, three tanneries; four academies, 155 students; nine schools, 265 scholars. Pop.: whites 8,725; slaves, 5,391; free coloured, 16; total, 14,132. Capital, Greenville.

MERSEBURG, a town of the Prussian dom., prov. Saxony, cap. reg. Merseburg, on the Saale, 56 m. S.S.E. Magdeburg; lat. 51° 22′ 1″ N., long. 12° 0′ 35″ E. Pop., in 1837, 9413. It is walled, and is old and irregularly built. It has several suburbs, a cathedral, a castle, a gymnasium, a hospital, and various other public institutions. Merseburg is the seat of government for the regency, of the council and court of justice for the circle and town, a board of forests, &c.; and has manufactures of woollen and linen cloth, paper, tobacco, and vinegar. The beer of Mersesburg is celebrated as the best in Saxony. (*Von Zedlitz, Der Preussische Staat.*, iii. 223; *Berghaus, &c.*)

MERSEY, a river of England, which has its embouchure on the W. coast of the island, in the Irish sea. Though not large, the Mersey has, from its flowing through the principal manufacturing district of the empire, and giving its name to the gulf or estuary between Lancashire and Cheshire, become. in point of commercial importance, second only to the Thames. It has its sources in the great central ridge, or Pennine chain, on the confines of York-shire, Cheshire, and Derbyshire. After receiving the Goyt from the S., and flowing W. through Stockport, it is joined by its important affluent the Irwell. The latter, which has its source in the Lancashire moors, near Haslingdon, flows S. through Bury to Manchester, where, being joined by two smaller streams, it takes a westerly course, till its confluence with the Mersey. After being still farther increased by the Bodeu from Macclesfield, the Mersey passes Warrington, a little below which it expands into a magnificent estuary, having the great commercial port of Liverpool on its N. side, near its junction with the Irish sea. The Mersey and Irwell have been rendered navigable from Sankey bridge to Manchester; and projects are now on foot for improving and deepening the navigation. (*See MANCHESTER.*)

MERTHYR-TYDVIL, a parl. bor, market-town, and par. of S. Wales. co. Glamorgan, hund. Caerphilly, on the Taff, 19 m. N. by W. Cardiff, and 140 m. W. by N. London. Pop. of parl. bor., which includes nearly all the par. of Merthyr-Tydvil, and the entire par. of Aberdare, with a small portion of the par. of Vainor, 27,460 in 1831, but in 1841, 34,977. "The town lies on the E. side of the valley, down which the Taff descends to Cardiff, scattered in detached masses about the valleys and on the hills, sending forth branches in different directions; and fresh groups are continually rising in the neighborhood of the great iron-works, so that it is somewhat difficult to point out where any collection of houses ends or begins." (*Bound. Rep.*) The houses, generally speaking, are mean-looking, comprising "labourers' cottages, or small ale-houses, beer-shops, or retail shops;" but in the centre of the town there are three tolerably respectable streets, forming a triangle, at one point of which is the parish church, a modern and well-built structure : the living is in the gift of the marquis of Bute. There is, also, a chapel-of-ease, besides several places of worship for dissenters; and the various Sunday schools of the town give religious instruction to nearly 6,000 children of both sexes. National, Lancastrian, and other subscription schools have likewise been formed, and are well attended. A philosophical society and several book-clubs have been established; and there is abundant proof that education is advancing among all classes. A theatre has been built within these few years, and there are two good hotels. In the environs are many handsome seats, belonging to the wealthy iron-masters; and three miles N. of the town, on an insulated hill, stand the ruins of Morlais castle, a very ancient building, demolished during the late civil wars.

The rise of Merthyr has been rapid, almost beyond belief. Towards the middle of last century it was an insignificant village : in proof of this, it is sufficient to state, that, in 1755, the lands and mines for several miles round the village, the seat of the great works now erected, were let for 90 years for 200l. a year. It is wholly indebted for its prosperity to its rich mines of coal, iron-ore, and limestone. The stratum of coal, which is of excellent quality, is accompanied by parallel veins of argillaceous iron, penetrating to a great depth, and yielding, at an average, about 35 per cent. of metal. The iron works are on a vast scale; those of Sir John Guest and Co. at Dowlais, of the Messrs. Crawshay at Cyfarthfa and Hirwain, having actually raised up very populous townships; the Pen-y-darran and Plymouth iron-works are also very extensive. In all, about 150,000 tons of iron a year are produced in the immediate vicinity of the town. Of this, a large proportion goes through the various processes of refinement and rolling into bars, previous to being shipped at Cardiff. The furnaces, refineries, and rolling-mills employ a great many persons; the wages for men ranging from 18s. to 80s.; of women, from 6s. to 10s.; and boys, 7s. to 11s. per week. The trade is of a very fluctuating character, and great numbers of workmen are often thrown out of employment by the stoppage even of two or three furnaces. Distress, however, is less permanent here than in many other districts, as "the work is one requiring less experience than many other manufactures; so that a demand for labour is readily met by a supply; while, on the other hand the labourers feel no great reluctance to transfer themselves to fresh employments." (*Bound. Rep.*) It is said by Mr. Nicholson, in the last edition of the *Cambrian Guide*, that "there is a marked improvement in the houses of the workmen; most of them have good oak chests of drawers, bright as silver; cupboards, with a display of family china cups and glasses; some of the younger women

have a veneered work-box; and all these little things display an attention to the lesser comforts and luxuries of life, of which, a few years ago, they had no idea. On the whole, there is a decided improvement in the general condition and circumstances of our workmen." (P. 421.)

But, notwithstanding their comparatively comfortable condition, and the great increase of temperance societies, it would seem that Chartist doctrines have made a very considerable progress among the labourers in this district, a circumstance for which it is not very easy to account. A large portion of the Chartists engaged in the outbreak at Newport, on the 4th of November, 1839, were understood to be from Merthyr and the adjoining iron-works. It is to be hoped that the severe, but wholesome, castigation they received on that occasion may not have been thrown away; and that it may help to disabuse them of their prejudices, and hinder them from again becoming the dupes of designing demagogues.

The communication with Cardiff is effected by means of the Glamorganshire canal (completed in 1794). which commences at Merthyr, and ends, after a course of 25 m., in the tideway of the Taff, near its entrance into Penarth harbour, the entire descent being 611 ft. (*See* CARDIFF, l. 552.) In 1836, 118,060 tons of iron were sent down this canal. The late improvements of the marquis of Bute in the port of Cardiff have also been of essential advantage to the export trade of Merthyr-Tydvil. The Taff-vale railway, for which the act of parliament was procured in 1836, and which is intended to connect the mining district of Merthyr with Cardiff, was opened as far as Newbridge, 14 m. N. Cardiff, in October, 1840; and it is stated that the traffic, as well as the number of passengers, already exceeds the original estimate. This railway, when completed, will. with its various branches, have a length of 95 m.: the capital is estimated at 620,000l.

The Reform Act created Merthyr a parliamentary borough, with the privilege of sending one member to the House of Commons; the electoral limits comprising the parish of Merthyr-Tydvil, except parts of the hamlets of Forest and Taff and Cynon. the entire parish of Aberdare, and the hamlet of Cefn-Coed-y-Cumner, in the parish of Vainor. Registered electors, in 1839–40, 582. Merthyr is also one of the polling-places at elections for the county. Petty sessions are held here for the upper division of hundred Caerphilly; and a court of requests sits monthly for the recovery of debts not exceeding 5l. An act was passed in 1836 for the better security of life and property in this district; and the three parishes of Merthyr-Tydvil, Aberdare, and Gellygaer are under the superintendence of a stipendiary police magistrate, having a salary of 600l. a year, half of which is levied on the furnaces within the limits of his jurisdiction, and half on the inhabitants of Merthyr alone. Markets on Wednesday and Friday; cattle fairs May 14, the first Monday in July, and the first Monday in August.

MESAGNE, or MESSAGNA, a town of the Neapolitan dom., prov. Otranto, cap. cant., 8 m. S.W. Brindisi. Pop. 5,000 ? It has several convents, an hospital, and a fine palace, belonging to the Francavilla family. It manufactures kitchen utensils: and has some trade in oil and grain, considerable quantities of which are grown in its vicinity. Mesagne is supposed by many Italian authors to be the representative of the ancient town of *Messapia*, but this is doubtful.

MESHED, a city of Persia, the cap. of Persian Khorassan, and esteemed as "holy" from its containing a very superb sepulchre, enclosing the remains of Imâm Reza and the caliph Haroun al Raschid, 455 m. E.-by N. Tcheran; lat 36° 17′ 40′′ N., long. 59° 35′ E. Pop., according to Kinneir, 50,000. It stands in a rich and well-watered plain, is surrounded with a strong wall, and is divided into 12 quarters, of which five are in ruins. The houses are meanly built of sun-dried bricks, and the ark or palace is unworthy of its name. There were formerly 16 medressans or colleges; but most of them are either deserted or in ruins. Indeed, the city has now little worth notice, except its fine and well-supplied bazaar, and the mausoleum of Imâm Reza, the magnificence of which, with its silver gates, jewelled doors, rails once of solid gold, glittering domes and minarets, and handsome arcades, is almost unequalled in Persia. It has, however, been often plundered; and its resources are said by Mr. Fraser to be greatly reduced. Meshed carries on a considerable trade with Bokhara, Balkh, Candahar, Yezd, and Herat; and many of the inhabitants are employed in weaving velvet and making fine pelisses, both of which are much esteemed throughout Persia. (*Kinneir's Persia; Fraser's Khorassan, &c.*)

MESSINA (an. *Zancle* and *Messana*), a celebrated city and seaport of Sicily, ca,. Intendency, near the N.E. extremity of the island, on the strait of its own name, 8 m. N.W. Reggio, 56½ m. N.N.E. Catania, and 120 m. E. by N. Palermo; lat. (of its lighthouse) 38° 11′ 30′′ N., long. 15° 34′ 40′′

E. Pop., in 1831, including that of its canton, 83,772.* The city has a most imposing appearance from the sea, forming a fine circular sweep, about 2 m. in length, on the W. shore of its magnificent harbour, from which it rises in the form of an amphitheatre; and being built of white stone, it strikingly contrasts with the dark forests that cover the mountains in the background. Prior to 1783, the harbour was fronted by a magnificent terrace of lofty houses, called the *Pallazzata*, having in front a broad quay decorated with statues and fountains. But the great earthquake of that year laid the city almost entirely in ruins; and though the the terrace still exists, it is shorn of its former grandeur. The quay in its front, called the *Marina*, has now, as formerly, numerous fountains, and is the favourite promenade. Ruined buildings, and other vestiges of the earthquake, occupy a considerable portion of the city; and few houses have now more than two stories. Swinburne, who visited Messina before 1783, complained that its interior was dirty, with narrow streets, gloomy houses, little bustle of trade, and still less show of luxury. According to Smyth, however, the modern city is regularly built, well paved with square blocks of lava, and several of its streets are wide and handsome, ornamented with numerous churches, convents, statues and fountains. The square in front of the cathedral, and that of *San Giovanni di Malta*, are both well built and handsome. The fountain in the centre of the former ranks with the finest in Sicily. The cathedral, erected soon after the conquest of Sicily by the Normans, has been repeatedly damaged by earthquakes. It is a Gothic building, with a heavy and gloomy exterior: the interior, though devoid of taste, is richly ornamented. The principal entrance is handsome; and the nave is supported by immense granite columns taken from the ruins of a temple of Neptune. The great altar and the roof of the choir are set off with mosaics and precious stones: the carved work of the pulpit is said to be a *chef d'œuvre* of the Sicilian sculptor Gaggini. The church of Monte Virgine has some good paintings in fresco, and that of St. Giorgio is very rich in marbles and inlaid work, and has some tolerable pictures. Adjacent to the viceroy's palace, a noble building at the S. extremity of the city, is a large open space planted and laid out in public walks. The other public buildings include a large hospital, several asylums of various kinds, two theatres, town-hall, exchange, bank, college, &c. Messina is surrounded by an old irregular wall, finished by Charles V. The citadel, a pentagonal fortress, erected on the S. side of the harbour, is constructed according the principles of Vauban; but though well provided with bomb-proof quarters and stores, it is badly situated and commanded in almost every part. Two strong and well-built forts have, however, been constructed on eminences above the town, that would greatly annoy and harass an enemy during any operation against the citadel. The town is further defended by a fort placed so as to command the mouths of the Fiumare, which are the only places where an enemy could land with cannon. The port, to which Messina is wholly indebted for her prosperity, and even existence, is formed by a lengthened curved tongue of land, that might almost be supposed to be an artificial circular mole, projecting first N. E. from the main land, and then bending round to the W. in the form of a sickle.† The entrance on the N., about 700 yards across, is defended on the W., or main-land side, by the bastion of Porto Reale, and at the extremity of the curved promontory by Fort Salvatore. A lighthouse has been constructed on the extreme E. verge of the promontory. The noble basin thus enclosed is about 4 m. in circuit, and, having deep water throughout, is capable of accommodating the largest fleets: it is, in fact, not only the finest harbour in the Mediterranean, but one of the finest of which we have any certain knowledge. Men-of-war moor in the centre of the basin in about 35 fathoms; but merchantmen lie alongside the quay, and have every facility for loading and unloading. The pratique-office, the fish-market, and the custom-house, are all on the Marina. The lazaretto, the best establishment of the kind in Sicily, is in the E. angle of the harbour. The situation of Messina, on the strait between Italy and Sicily, and her admirable port, give her great advantages as a commercial *entrepôt*; and were Sicily and Naples subject to an enlightened government, able and willing to put down abuses, and to call into activity their long dormant energies, Messina would certainly be one of the greatest emporiums of the Mediterranean. Even as it is, her trade is very considerable. Her exports consist principally of oranges and lemons, olive oil, silk, linseed, wines and spirits, shumac, liquorice, rags, corn, salted fish, &c. Almost all the silk exported from Sicily is shipped here. The imports consist of colonial produce, and cotton and woollen fabrics, hides, hardware, &c. (*Macgregor's Re-*

ports on Sicily; Smyth's Sicily, p. 112–123; *Russell's Sicily*, p. 250–261, &c.) We subjoin an

ACCOUNT of the quantities and value of the principal articles exported from Messina in 1832.

Articles.		Quantities.	Total Value.
			£.
Argols and cream of tartar	cwts.	4,941	4,885
Barilla	ditto	5,199	1,545
Brimstone	ditto	8,139	3,950
Cantharides	ditto	26	1,454
Corn, grain, and pulse	qrs.	4,989	9,581
Cotton wool	cwts.	275	689
Essences	lbs.	34,379	7,738
Fish, salted	cwts.	6,685	11,698
Fruit, dry and pickled	ditto	5,198	4,985
Oranges and lemons	boxes	394,894	76,365
Lemon juice	galls.	224,357	5,315
Linseed	qrs.	20,384	34,344
Other seeds	cwts.	11,466	5,729
Liquorice paste	ditto	6,492	14,889
Oil olive	galls.	245,229	38,491
Manna	cwts.	943	2,979
Rags	cwts.	11,215	5,975
Salt	tons	601	944
Shumac	cwts.	23,177	3,145
Silk	lbs.	82,239	62,559
Suines	No.	227,710	7,125
Wine and spirits	galls.	794,349	87,021
Other articles			68,256
Total value			488,345

Messina is the see of an archbishop, and the residence of a Greek *protopapas*, with authority throughout Sicily, but who is nominated by the pope. It is the seat of a royal court of appeal, and of criminal, civil, and commercial tribunals; and has a municipal bank, several *monti di pietà*, or government loan banks, and other benevolent institutions. Next to commerce, its inhabitants are chiefly occupied in the fancy and other fisheries; and in the manufacture of silk stuffs, especially damasks and satins. It has an ecclesiastical seminary, a lyceum, a royal college for law and medicine, and an extensive public library; but Simond says that Messina appears to have made slow progress in refinement, compared with Catania or Palermo. "The education of young people is more neglected; very few in the lower ranks can read; and the nobility do not in general reside in Messina: in short, it is neither fashionable, nor learned, nor rich; nor is it, I think, particularly hospitable." (*Simond's Italy, &c.,* 525.) But, how deficient soever in these respects, Messina has advantages of another sort that entitle her, in the estimation of her citizens, to look down with contempt on most other places. These consist in the possession of an autograph letter of the Virgin Mary, addressed to the Messinians, and assuring them of her especial protection; and what is, if possible, still more precious, a lock of the Virgin's hair, given by her to the persons entrusted with the conveyance of the letter! To question the genuineness of these valuable relics would, in Messina at least, be rather hazardous; and, under such circumstances, we need not wonder that it is firmly believed that, on one occasion, the city was saved from famine by the opportune arrival of a supply of corn, sent by the Virgin. The only wonder is, that she has allowed it to be so often brought to the brink of destruction by earthquakes, and devastated by the plague. A splendid *fête* is annually given in the great square in honour of the exalted protectress and benefactress of the city. Very few vestiges of the ancient city remain; a consequence, no doubt, of the numerous earthquakes by which it has been visited.

The accounts of the origin and early history of Messina differ considerably. It is admitted on all hands to be very ancient; and most probably derived the name it has so long borne from a settlement having been made in it by a body of emigrants from Messene, in Greece. Having been seized by the Mamertini, it became, under them, one of the most populous, wealthy, and powerful cities of Sicily. It was the first town of the island that came into the possession of the Romans. (*Callerii Orbis Antiqui*, i. 973; *Ancient Universal History*, iii. 513., 8vo. ed.)

The principal political events in the history of Messina, in modern times, are its successful resistance to Charles of Anjou, by whom it was besieged, after the Sicilian Vespers; and its revolt against the Spaniards in 1674, followed, in the ensuing year, by the defeat of the latter in its vicinity by a French force. In 1743 the plague broke out in Messina, with the most destructive violence, sweeping off the greater number of the inhabitants.

METHUEN, p. t. Essex co., Mass., 26 m. N. by W. Boston, 454 W. Bounded S.E. by Merrimac river. Watered by Spicket river, which has a fall of 30 feet, two miles above its entrance into Merrimac river, and affords great water power. It contains three churches, a Congregational, Baptist and Universalist, four stores, two cotton factories with 4586 spindles, two grist-mills, two saw-mills, two paper-mills; 10 schools, 650 scholars. Pop. 2851.

---

* According to the official returns, the population, in 1798, was only 46,063, so that it would appear to have nearly doubled in the interval.

† This, in the old Sicilian language, was called Ζαγκλη, whence the original name of the town. (*Thucydides*, lib. vi. cap. 4.)

METZ (an. *Divodurum*, afterwards *Mediomatrici* and *Metis*, whence its present name), a strongly fortified city of France, dep. Moselle, of which it is the capital, at the junction of the Moselle and Seille, 90 m. W.N.W. Strasbourg, and about 180 m. E.N.E. Paris; lat. 49° 7' 5" N., long. 6° 1' 15" E. Population, in 1836, 42,793. "Metz is a fine old city; but, like most fortified places, the streets are narrow, and the houses lofty. Near the river it is more open, the quays are broad, and the bridges magnificent. The river is clear and rapid, and swells to an expanded stream where not confined by the embankments, as it is within the fortifications." (*Jacob's View of Germany, &c.*, p. 436.) Metz was fortified by marshals Vauban and Belle-Isle: it has several strong outworks, and a citadel on the Moselle; but the latter was partly dismantled during the revolution, and its esplanade has been laid out in public walks, which command a noble view of the valley of the Moselle and its bounding hills. The city has nine gates and drawbridges, but only six are in use. The most conspicuous public building is the cathedral, a vast pile, commenced in 1014, but not finished till 1546. It is about 390 feet in length, the height of the nave being about 140 feet, and that of the tower about 400 feet.[*] The latter, which is a fine specimen of Gothic architecture, has in it a bell weighing 25,000 lbs. The whole edifice is remarkable for lightness. Mr. Jacob says that the cathedral of Metz was the most perfect Gothic structure he saw on the continent; and that, though not so old as Westminster Abbey, it may vie in external beauty with that venerable pile. The military hospital, built in the reign of Louis XV., is a noble edifice, consisting of two ranges of building, and capable of easily accommodating 1500 patients. The *Hôtel du Gouvernement*, a large, though rather heavy fabric, fronts the esplanade; it is appropriated to the courts of justice and the city library; the latter has above 30,000 volumes, among which are numerous works printed in the 15th century, and about 800 MSS., some of the 10th century. (*Guide du Voyageur*.) The barracks, military magazines, prefecture, townhall, and mint, several of the churches, the new market, the theatre, with a portico of the Tuscan order, &c., are among the other edifices. The Moselle and the Seille, in and near the city, are crossed by at least 30 bridges. The principal school of artillery and military engineering (*Ecole de Génie d'Application*) in France is established here. Its library has a choice collection of about 10,000 volumes of military and scientific works, with sundry MSS. of Vauban and other distinguished persons. Exclusive of the above, Metz has two other public libraries, with several convents and charitable asylums, a Protestant church, a synagogue, a royal college, a university academy, an ecclesiastical school, and other seminaries; a school for the fine arts, a royal society of arts and *belles lettres*; an agricultural society, a society for the encouragement of primary instruction, and collections in natural history, mineralogy, and chemistry; a botanic garden, a lying-in hospital, a savings' bank, a *mont-de-piété*, &c.

Metz is the see of a bishop, and the seat of a royal court for the deps. of Moselle and Ardennes, and of tribunals of primary jurisdiction and commerce, and a chamber of commerce, and the head-quarters of the third military division of France. "It is also a manufacturing city, in which are made woollen goods of various kinds, hosiery, cotton goods, table-linen, printed paper, musical instruments, starch, and gunpowder; it has, besides, several extensive tanneries. Much trade originates here from the produce of the vines, some portion of which is converted into wine, but more into brandy and vinegar; and Metz is celebrated for the preparation of various kinds of confectionary. It is encircled by hills, covered from the bottom to the top with fruit-gardens and vineyards. The vineyards are mostly in small divisions, and principally cultivated by small proprietors, who are extremely poor, and almost all involved in debt to the capitalists of the city, who take from them their wine, brandy, and vinegar as soon as it is made." (*Jacob's View of the Agric. of Germany, &c.*, 435-436.)

The royal gunpowder factory, on an island in the Moselle, near the city, is superior to most others in the kingdom. Metz has also a royal cannon foundry, a saltpetre refinery, and produces leather, cotton yarn, military and other hats, muslins, beet-root sugar, chicory, nails, and other articles of hardware, cutlery, buttons, glue, &c.

This is a very ancient city. It still possesses several ruins belonging to the Roman period, among which are the remains of an aqueduct, that appears to have conveyed water to a *naumachium* near the S. extremity of the city. The site of the latter is now occupied by outworks belonging to the fortifications. Parts of an amphitheatre and of a Roman palace are still traceable in the city. It suffered considerably, about anno 70, from some excesses of the

troops of Vitellius (*Taciti Hist.*, lib. i., cap. 73), and was nearly destroyed by the savage barbarism of Attila in 452. It had, however, recovered a large portion of its former prosperity in the middle ages, and became the capital of the kingdom of Austrasia. From the 11th century to 1552, when it was taken by Henry II., it was an independent flourishing city. In the same year that it was taken by Henry, it was besieged by the emperor Charles V., with an army of 100,000 men, but the Duke of Guise successfully defended the town, and Charles was obliged to relinquish the siege. It was finally annexed to the French crown by the treaty of Westphalia, in 1648. (*Hugo*, art. *Moselle*; *Guide du Voyageur, &c.*)

MEURTHE, a dep. of France, reg. N.E., formerly part of the prov. of Lorraine, between lat. 48° 20' and 49° N., and long. 5° 40' and 7° 20' E., having N. the dep. Moselle, E. Bas Rhin, S. Vosges, and W. Meuse. Length, E. to W., 74 m.; average breadth, about 35 m. Area, 606,922 hectares. Population, in 1836, 424,304. The Vosges mountains run through the E. part of the dep., the surface of which is mostly covered with their ramifications, though these rise to no great elevation. The dep. belongs almost wholly to the basin of the Moselle, which river intersects its W. part from S. to N., and is joined, within its limits, by the Meurthe. The latter rises in the dep. Vosges, runs generally in a N.W. direction, and, after a course of between 70 and 80 m., unites with the Moselle about 5 m. below Nancy, to which it is navigable. Besides Nancy, St. Dié and Lunéville are on its banks; and it receives the Mortagne, Vezouze, and Mezulle. The Seille and Sarre are the other chief rivers of the dep. There are numerous small lakes, one of which occupies an area of 693 hectares. In 1834, it was estimated that 303,536 hectares of the surface were arable, 71,851 in pasture, 16,371 in vineyards, 116,909 in woods, and 9236 in orchards, &c. The land is very unequal in point of fertility, and is very indifferently farmed; but more corn is grown than is required for home consumption. The total produce of the harvest of 1835, was estimated at 3,434,500 hectolitres, chiefly wheat and oats: in the same year, the crop of potatoes was estimated at 2,149,000 hectolitres. Before the revolution, the culture of the vine was limited to the declivities of hills with a southern aspect; but its culture has since been very much extended, the quality of the produce being less regarded than the quantity. About 550,000 hectolitres of wine are supposed to be produced annually, of which the greater part is consumed in the dep. The wines are generally inferior, though the growths of Pagny, Thiaucourt, Arnaville, Baudonville, and others, may be classed among the secondary qualities of *vins ordinaires*. (*Julion, Topographie*, p. 46.) Dried plums and preserved apricots form important articles of commerce; and the forests, which are more extensive than in most deps., furnish a good deal of timber. The pastures are naturally good, but receive little attention from the farmer. In 1830, there were estimated to be 84,000 head of black cattle, and 167,000 sheep, in the dep., but both are of indifferent quality. The breed of horses has been improved by the fine stud of Rosières. Hogs of an improved breed are numerous, and their flesh and lard are sent to distant parts of France. A great many poultry are reared. Property is much subdivided. In 1835, of 171,092 properties subject to the *contribution foncière*, 160,343 were assessed at less than five francs, and 24,437 at from five to 10 francs. Turf and lime are among the chief mineral products; there are some quarries of marble and alabaster, and a few iron mines; but the latter have been abandoned. The salt mines and springs at Dieuze, Vic, Moyenvic, &c., yield about 45,000,000 kilogrammes of salt, and 1,000,000 do. of soda a year. About 22,000 hands are said to be employed in the manufactures of cotton cloth and yarn, woollen stuffs, glass, and earthenware, and in embroidery, dyeing cotton stuffs, tanning, &c. At Baccarat is a large glass manufactory, employing a great many hands.

Meurthe is divided into five arronds.: chief towns, Nancy the cap., Toul, Château Salins, Sarrebourg, and Lunéville. It sends six members to the chamber of deputies. Number of electors, in 1836-39, 1219. Total public revenue, in 1831, 16,704,392 francs. (*Hugo*, art. *Meurthe*; *Official Tables*.)

MEUSE, a dep. of France, reg. N.E., formerly part of the province of Lorraine, chiefly between lat. 48° 25' N., and 49° 35' N., and long. 5° and 6° E.; having N. Dutch Luxemberg and the deps. Ardennes and Moselle, E. Moselle and Meurthe, S. Vosges and Haute Marne, and W. Marne and Ardennes. Length, N. to S., 80 m.; greatest breadth, about 40 m. Area, 690,555 hectares. Population, in 1836, 314,568. Surface generally hilly, the hills being ramifications of the Vosges and Facilles mountains, with an average height of from 1000 to 1300 feet, though they sometimes reach an elevation of 1600 feet. The Meuse traverses the dep. in its entire length; the other principal rivers are the Ornain, Chiers, and Aire. The plateau, in the E., separating the basins of the Meuse and the Moselle, and other

---

[*] There is some discrepancy in the authorities as to these measures; but the above must be nearly accurate.

portions of the surface, are not very productive; but there are, notwithstanding, about 295,000 hectares of rich soil in the dep., chiefly in the valleys of the Meuse and Ornain. In 1834, according to the official tables, 335,190 hectares were arable, 49,472 do. in meadows, 13,540 do. in vineyards, 7387 do. in orchards. &c., and 137,755 do. in woods. The produce of corn in 1835 was estimated at 2,440,000 hectolitres, of which 1,192,000 were wheat. Potatoes, oleaginous plants, hemp, and flax, are among the other articles of culture. Gooseberries are extensively cultivated in the gardens round Bar and Ligny, and enter largely into the confectionary, for which those towns are celebrated. The produce of wine is estimated at about 400,000 hectolitres a year. The wines of Bar-le-Duc, Bussey-la-Côte, Preue, Ligny, &c., are delicate light wines, ranking in the first class of *vins ordinaires;* but they do not keep above two years, and do not bear carriage. (*Jullien*, 43.) Along the Meuse are rich pasture lands; and at Void, cheese, similar to that of Gruyère, and excellent butter, are made. A good many cattle and sheep are reared in the dep.; but live stock is in general indifferent. The produce of wool is estimated at upwards of 140,000 kilogrammes a year. In 1835, of 157,180 properties subject to the *contribution foncière,* 89,566 were assessed at less than five francs, and 22,169 at from five to 10 francs. Iron, slates, and good building-stone, are the chief mineral products. There are between 90 and

about 1,500,000 kilogrammes of iron a year. About 500,000 kilogrammes a year of cotton yarn are made at Bar-le-Duc, which, also, has fabrics of paper, glue, &c., and is the *entrepôt* of a large trade in timber from the forests of the dep. There are numerous glass-works, with lime-kilns, potteries, beet-root sugar factories, &c. Many working cutlers, shoemakers, and other artisans, emigrate for a part of the year from this into other parts of France, and even to the adjacent foreign countries, with the products of their industry, or in search of employment. Meuse is subdivided into four arronds.; chief-towns, Bar-le-Duc, the cap., Commercy, Montmédy, Verdun. It sends four members to the chamber of deputies. Number of electors, 1838-39, 1188. Total public revenue (1831) 9,071,543 francs. (*Hugo*, art. *Meuse; Official Tables, &c.*)

MEUSE OR MAESE (Dutch *Maas,* an. *Mosa*), a river of W. Europe, flowing through the N.E. part of France, Belgium, and the S. of Holland; its basin being situated between those of the Marne and Scheldt to the W., and of the Moselle to the E. It rises in the dep. of Haute Marne, in France, 10 m. N.E. Langres, in about lat. 48° N., long. 5° 20′ E., and flows at first generally N. through the deps. of Haute Marne, Vosges, Meuse, and Ardennes. Near Charlemont it leaves France, but it continues its previous direction to Nathur, where it receives the Sambre from the W. It here makes a sudden bend to the N.E., in which direction it continues through the provinces of Namur, Liege, and Limburg, to about lat. 51° 30′ N. It afterwards curves to the W., flowing between N. Brabant and Guelderland; and finally at Woudrichem, in lat. 51° 49′, and long. 5°, enters the Rhine or Waal, which loses its own name to assume that of the Maas. (*See* RHINE.) Its entire course may be estimated at 490 m.; nearly the half of which is in France. It is navigable for three fourths of this extent, or as far as Verdun, dep. Meuse. Its chief affluents are the Bar in France, the Lesse, Sambre, and Ourte in Belgium; and the Roer and Niers in Holland. Proceeding from its source to its mouth, the chief cities and towns on its banks are Neufchâteau, Verdun, Sedan, Mézières, Charlemont, and Givet in France; Dinant, Namur, and Liege in Belgium; and Maestricht, Roermond, Venloo, and Grave in Holland, before its junction with the Rhine. The Meuse communicates with the Aisne, and thence with the Seine and Somme by the canal of Ardennes; with the Scheldt by means of the Sambre and the Charleroy canal; and with both the Scheldt and the Rhine by the various branches of the Great North canal; in addition to which, many other canals connected with it are in progress. (*See also* BELGIUM, i. 316-17, 328.)

MEXICO (UNITED STATES OF), a federal republic of N. America, lying between the 15th and 33d parallels of N. lat., and 97° and 113° W. long., being bounded N. and N.E. by the W. districts of the United States of N. America and that wild region called New California, E. by the gulf of Mexico and the Republic of Texas, S. by Guatemala, and W. and S.W. by the Pacific ocean. The line dividing Mexico from Texas commences with the river Nueces, which it follows up to its source, and then runs N. to the head of the Colorado, whence it stretches W., forming the N. boundary of the confederation. The line of separation on the side of Guatemala is very irregular, commencing E. with the river Sarstoon, which it follows to its source, and then takes a N. direction in lat. 17° 30′ N.; it thence runs W. and S.W. to lat. 15° 45′, where it assumes a N.E. course,

including the province of Chiapa. Greatest length from N.W. to S.E., 1700 m.; greatest breadth about 600 m. Area estimated at 1,230,442 sq. m. Nothing can be more unsatisfactory than our acquaintance with this vast country: few even of the principal towns and rivers are correctly laid down, except, indeed, within the small circle personally visited by Humboldt, so that not even the elements of a good map exist; and, with respect to population, and other statistics, the unsettled, disorderly, and almost lawless state of the country makes inquiry all but impossible. The following table has been printed in the American edition of *Murray's Encyclopedia of Geography,* and some other works, and, though little dependence can be placed on it, it is probably as near an approximation to the truth as can, at present, be arrived at:

States.	Extent in sq. m.	Pop. in 1837.	Pop. to sq. m.	Capital Cities.
Chiapas . .	14,760	92,000	4·9	Ciudad de las Casas.
Chihuahua . .	107,500	190,000	1·7	Chihuahua.
Cohahuila	193,600	90,890	0·5	Saltillo.
Durango . .	54,500	150,000	2·6	Victoria.
Guanaxuato	5,000	500,000	62·5	Guanaxuato.
Mexico . .	20,400	1,300,000	48·3	Texcuco.
Michoacan . .	22,466	450,000	20·5	Valladolid.
New Leon . .	21,000	100,040	4·9	Monterey.
Oaxaca . .	32,650	600,000	20·2	Oaxaca.
Puebla . .	18,140	900,000	49·8	La Puebla.
Queretaro . .	7,500	100,000	13·3	Queretaro.
San Luis Potosi .	18,000	300,000	15·9	S. Luis Potosi.
Sonora and Sinaloa .	264,700	300,000	1·1	Villa del Puerto.
Tabasco . .	14,676	73,000	5·1	Villahermosa.
Tamaulihas, or New Santander .	36,100	130,000	4·2	New Santander.
Vera-Cruz . .	27,680	150,600	5·4	Xalapa.
Xalisco . .	70,000	870,000	12·4	Guadalaxara.
Yucatan . .	79,500	570,000	7·2	Merida.
Zacatecas . .	15,250	200,000	10·	Zacatecas.
Federal district . .	·	300,000	·	Mexico.
Total of States .	1,066,442	7,557,000	7·9	
Territories.				
New Mexico . .	300,000	7 60,000	3·2	Santa Fè.
Colima . .	·	7 40,000	·	Colima.
Tlascala . .	·	7 66,000	·	Tlascala.
Total .	1,230,442	7,723,000	6·3	

The most populous cities are Mexico (150,000), Guadalaxara (60,000), San Luis Potosi and la Puebla (55,000 each), Oaxaca and Querétaro (40,000 each), Guanaxuato (34,000), and Merida (20,000).

Of this great tract of country, which is about one third as large as Europe, the portion lying S. of the tropic of Cancer, and comprising a large part of the long and narrow isthmus that connects the American peninsulas, and separates the Atlantic from the Pacific ocean, is by far the most populous and rich, both in mineral and vegetable productions; and nearly all the information gained respecting Mexico has been collected in that part, to which, consequently, it is primarily applicable. The regions N. of the tropic become less populous as we proceed northward; and many large districts claimed by the republic, and divided into states and territories, are almost unknown, being inhabited only by wild Indian tribes, baffling all the attempts of their nominal masters to civilize or subdue them.

The surface of Mexico is extremely varied; and to this circumstance, nearly as much as to the difference of latitude in so extensive a country, may be attributed that singular variety of climate by which it is distinguished from most other regions. The Cordillera, or chain of mountains, generally regarded as a portion of the great chain of the Andes that enters Mexico on the S., where it borders with Guatemala, diverges, as it proceeds N., into two great arms, the the upper part of the letter Y, following the line of the coasts on either side. The most westerly of these chains, or that parallel to the shores of the Pacific ocean, has some very high summits, and preserves its mountainous character till it joins, on the border of the United States, with the Oregon, or Rocky mountains. The other, or eastern arm of the Cordillera, begins to subside after reaching the 21st or 23d degree of latitude, and ultimately subsides, about the 26th or 27th degree of latitude, into the vast plains of Texas. Now, the whole of the vast tract of country between these two great arms, comprising about three fifths of the entire surface of the republic, consists of a central table-land, called the Plateau of Anáhuac, elevated from 6000 to upward of 8000 feet above the level of the sea! Hence, though a large portion of this plateau be within the limits of the torrid zone, it enjoys a temperate climate; inclining, indeed, more to cold than to excess of heat. Some very high mountains are dispersed over the surface of the central table-land; and it is also traversed in parts by pretty well defined ridges, which divide it into extensive sub-plateaus, to which different names have been given. But the surface is interrupted by few transverse valleys; and in some directions it is quite unbroken, either by depressions or by hills. Thus, it is mentioned by Humboldt, that carriages proceed from the capital, in the centre of the plateau, to Santa Fè, in New

Mexico, a distance of 1400 m., without any important deviation from an apparent level. (*Essai sur la Nouvelle Espagne*, i., 254.) The most remarkable tract in this elevated region is the plain of Tenochtitlan (in which is the capital), surrounded by ridges of porphyritic and basaltic rocks, running S.S.E. and N.N.W. It is of an oval form, 55 m. long and 37 m. broad, occupying an area of 1700 sq. m., of which about 160 sq. m. are covered with water. Its S.E. side is that most elevated, and here are seen towering above the plain the volcanoes of Popocatepetl 17,716 feet, Iztaccihuatl 15,760 feet, Citztalepetl or Orizaba 17,380 feet, and Nauhcampaxetl or the Cofre de Perote 13,415 feet above the sea. The waters of the valley are deposited in five principal lakes situated on different levels: that of Tezcuco, which is near the centre of the valley, and covers 70 sq. m., is the least elevated. Farther N. are the lakes of St. Christoval and Tomanitla: while S. is the lake Chalco, occupying an area of 50 sq. m.; and these three are five feet higher than lake Tezcuco. The most elevated, however, of the whole, though the smallest, is the lake Zimpango, the level of which is 30 feet above that of Tezcuco. These lakes are fed by small rivers, and, having no natural outlet, are drained by the Desague of Huchuetoca, an artificial canal cut through the rock, 12 m. in length, 150 feet deep, and 300 feet wide; having its embouchure in the river Panuco, which flows into the gulf of Mexico. This great work, completed in 1789, at an expense of £1,292,000, was undertaken to obviate the frequent inundations, some of which did great damage to the capital. The water of lake Tezcuco is salt, that of the rest is fresh; but from those to the S. sulphuretted hydrogen gas is copiously disengaged, the smell of which is often perceptible at Mexico.

Besides the volcanoes already noticed, those of Tuxtla, Jorullo, and Colima, in the table-land, are at present in a state of activity, and there are several others now extinct. Jorullo, which stands W. of the city of Mexico, first broke out in 1759; when a tract of ground, from 3 to 4 m. square, swelled up like an inflated bladder, emitting flames and fragments of rock through a thousand apertures. These active volcanoes seem to be connected with others parallel to them, and obviously of similar origin. Earthquakes are frequent in Mexico, but they seldom do much mischief.

The geological formation of the Mexican Cordilleras differs considerably from that of the great mountains of Europe and Asia, in which granite is overlaid by gneiss, mica, and clay-slate; for here we seldom meet with granite, as it is covered with porphyry, greenstone, amygdaloid, basalt, obsidian, and other rocks of igneous origin. Granite, however, appears on the surface in the chain bordering the Pacific, and the port of Acapulco is a natural excavation in that species of rock. The great central plateau of Anahuac, between lat. 14° and 20° N., is a mass of porphyry, characterised by the constant presence of hornblende, and the entire absence of quartz; and is it are contained large and valuable deposities of gold and silver. These ores, however, are found in various rocks: in the mines of Comanja rich veins of silver occur in syenite; in those of Guanaxuato, which are the richest in Mexico, the metal lies in a primitive clay-slate passing into talc-slate; and those of Real del Cardonal, Xacala, and Lomo del Toro are situated in a bed of transition limestone. Humboldt says that there were, at the time of his visit, 3000 mines of gold and silver in Mexico; but the ignorance and misrule which prevail in the country have greatly diminished their importance as a source of wealth.

Rivers.—Mexico suffers serious disadvantages from the want of water, and the rivers, as compared with the extent of territory, are few and unimportant. The Rio Grande del Norte, indeed, has a course of more than 1300 m., and the Colorado runs about 700 m. into the gulf of Mexico. The Rio Grande de Santiago, called by the natives Tololotlan, rises in the centre of Mexico, not far from the capital, and, after traversing the lake Chapala, falls into the Pacific at San Blas. The Balsas, or Zacatula, and the Yopez, are the only other rivers on the W. side of the plateau, and on the E. side are the Tula and Tampico and the Tabasco, flowing into the gulf of Mexico; but they have bars at their mouths, which prevent the entrance of large vessels. The other rivers are short, and might more properly be called torrents. The lakes are numerous and extensive; and the principal, besides those in the plateau of Tenochtitlan, already mentioned, are Chapala, in Xalisco, which, according to Humboldt, covers an area of 1300 sq. m.; Pascuara in Michoacan, Mextitlan, Cayman, and Parras, the last two being in the tract called the Bolson de Mapimi.

Climate.—The temperature and climate of Mexico is, of course, extremely various; owing, not only to its great extent from N. to S., but also to the rapidity of the slope both on the E. and W. side. The climates, especially on the E. side, are more distinctly marked by the vegetation. On the ascent from Vera Cruz, says Humboldt, climates succeed each other in in ers; and the traveller passes in review, in

the course of two days, the whole scale of vegetation, from the parasitic plants of the tropics to the pines of the arctic regions. (*Essai Pol. sur la Nouv. Espagne*, i., 270-289.)

Mexico is divided, as respects climate, into the *tierras calientes*, or hot regions, the *tierras templadas*, or temperate regions, and the *tierras frias*, or cold regions. The first, or the *tierras calientes*, include the low grounds, or those under 3000 feet of elevation, on its E. and W. coasts, comprising the greater part of the states of Taumalihas, Vera Cruz, Tabasco, and the peninsula of Yucatan, on the former. The *tierras calientes*, on the W. coast, are less extensive, the eastern arm of the Cordillera approaching nearer to the sea. The mean temperature of this region, or, at least, of that portion of it between the tropics, may be estimated at about 77° Fahr., being from 14° to 16° above the mean temperature of Naples. It is especially suited for the growth and cultivation of sugar, indigo, cotton, and bananas, which flourish in the utmost luxuriance.

This region labours under the serious disadvantage of being nearly inaccessible by sea for half the year, and of being extremely unhealthy during the other half. The winter, on the E. coast, extends from about October to the vernal equinox; and during this season, in the gulf of Mexico, N. or N.W. winds (*los nortes*) are extremely prevalent, blowing with more or less violence. Frequently, especially in the month of March, the N. winds approach to the strength of a hurricane, and continue to blow with the utmost violence, and without intermission, for three, and, sometimes, even for 10 or 12 days together. During the whole of this season the navigation of the gulf is exceedingly dangerous; but on shore the heat is moderate, and the coast free from fever and tolerably healthy. Unluckily, however, it so happens, that during the other half of the year, or from the vernal equinox to October, when the N. winds are comparatively rare, and the ports are easily accessible, the heat is oppressive, a great quantity of rain falls, and the coast becomes the seat of pestilential fevers. A European arriving for the first time at Vera Cruz, or any other part of the coast between the tropics, in August, September, or October, has but little chance of escaping the *vomito prieto*, or yellow fever; and individuals who have merely landed at Vera Cruz, and posted on immediately for Xalapa, have, notwithstanding, caught the infection. The scourge, however, does not extend its ravages beyond the low grounds on the sea-coast; and at the height of 3000 or 3500 feet above the sea it is wholly unknown. The ports of Acapulco and the low grounds along the W. coast are also extremely hot and unhealthy; and, owing to the prevalence of strong gales, approaching to hurricanes, during the months of July, August, September, and down to October, the navigation is then extremely dangerous.

The *tierras templadas*, or temperate regions, which are of comparatively limited extent, occupy the slope of the mountain chains, or barriers, which bound, on either side, the central table-land. It extends from about 2500 to about 5000 feet of elevation. The mean heat of the year is from 68° to 70° Fahr., and the extremes of heat and cold are here equally unknown. The Mexican oak, and most of the fruits and cerealia of Europe, flourish in this genial climate. The cities of Xalapa, on the E., and of Chilpanzingo, on the S.W. slope, are in this region, and are famous for their salubrity and for the abundance of their fruit trees. The frequency of fogs, and the consequent humidity of the atmosphere, is the greatest drawback on the climate of the *tierras templadas*; but this, how injurious soever in some respects, produces great beauty and strength of vegetation.

The *tierras frias*, or cold regions, include all the vast plains elevated 5000 feet and upward above the level of the sea. In the city of Mexico, at an elevation of 7400 feet, the thermometer has sometimes fallen below the freezing point. This, however, is a rare occurrence, and the winters are there usually as mild as in Naples. In the coldest season the mean heat of the day varies from 55° to 70° Fahr.; while in summer the thermometer seldom rises in the shade above 75°. The mean temperature of the city is about 64°, and that of the table-land generally may be taken at about 62°, being nearly equal to that of Rome. But, whenever the table-land rises to more than 8000 feet above the sea, it has, though between the tropics, a rude and disagreeable climate. Under the parallel of Mexico the limit of perpetual snow varies from about 12 to near 15,000 feet. Vegetation in the central plateau is not, owing to the rarity of the air, so vigorous as on the *tierras calientes*, or along the coasts, and the plants of Europe do not succeed so well as in their native soil. In the tropical and central region of Mexico, and as far N. as lat. 28°, there are only two seasons; that of the rains, lasting from July to the middle of September, and the dry season, continuing from October to the end of May. From the 24th to the 30th parallel the rain falls less frequently; but this deficiency is compensated by the abundance of snow during January and February.

The climate of the table-land is, on the whole, favourable to human life. But, though intermittent fevers be of rare occurrence, the natives are occasionally visited by a peculiar epidemic, called by them the *matlasahuatl*; but it owes its origin more to the habits of the people than any other cause. Indeed, famine, and its concomitant privations, have thinned the population more than epidemic complaints. The indolence of the natives prevents all exertions to raise more food than requisite for the wants of a single ordinary season; and no one ever thinks, when there is a surplus, of laying up a stock against future contingencies. Hence, when droughts and severe frosts occur, they are compelled to seek their subsistence in the forests, where roots and wild berries constitute their sole diet; and multitudes are often carried off by hunger and unwholesome food.

Animals.—The zoology of Mexico is but imperfectly known. The domestic animals introduced by the Spaniards have so much increased, that vast herds range wild through these thinly inhabited regions. The wool of the sheep is of inferior quality; but this is attributable more to neglect and mismanagement than to nature: mules are much used in the mining districts. Buffaloes abound in the prairies bordering on the Arkansaw and Red river, and during winter they migrate westward, in quest of pasturage, to the milder climate of the plains along the lower part of the Rio Grande del Norte. Carnivorous animals are not numerous. Bees abound in the low country of Yucatan.

Agriculture.—Mexico, not only from its extent through 21 degrees of latitude, but also from the varying elevation of its surface, and consequent variety of climate, produces most of the plants peculiar to the tropics, as well as those belonging to the temperate regions of S. and middle Europe. "Indeed," says Humboldt, "there is scarcely a plant in the rest of the world which is not susceptible of cultivation in one or other part of Mexico; nor would it be an easy matter for the botanist to obtain even a tolerable acquaintance with the multitudes of plants scattered over the mountains, or crowded together in the vast forests at the foot of the Cordilleras." (*Essai*, tom. ii., p. 370.) The soil also is, in most parts, extraordinarily fertile; and wherever water can be procured for irrigation, the most abundant crops may be raised with very little labour. This, however, is very far, indeed, from being an unmixed advantage; and it is, in fact, more than doubtful whether a very fertile soil and a genial climate, that makes warm clothing and comfortable lodgings of comparatively little importance, be consistent either with active industry and exertion or with a high state of civilization. In most parts of Europe continuous industry is indispensable to existence; but it is otherwise in Mexico and many other countries, and it is found that industry is uniformly proportioned to the strength of the motives by which it is occasioned, and that, wherever the ordinary necessaries and comforts of life may be procured with little labour, the mass of the people are invariably indolent. To suppose, indeed, that they should be otherwise, is to suppose what is contradictory and absurd. This effect of the peculiar nature of the soil and climate was less sensible in Mexico under the Spanish government, because it was then daily receiving adventurers from Europe, imbued with European notions, and anxious to accumulate a fortune. But now that the influx of such parties has nearly ceased, and that there are no such extrinsic and adventitious motives to prompt to activity and enterprise, everything appears to be falling into a state of apathy and languor; and indolence, with its necessary accompaniments of poverty, ignorance, and pride, bid fair to be, for a lengthened period, the distinguishing characteristics of the Mexicans.

We have stated, under the head of climate, how the more useful plants are distributed through the zones into which the country is divided. The banana, which flourishes up to the point where the mean temperature is 75° Fahrenheit, bears the same relations to the Mexicans, in the lower provinces, that the various cerealia bear to the inhabitants of Europe and W. Asia, and the different kinds of rice to the Bengalees and Chinese. About 450,000 sq. m. in the *tierras calientes* are said by Humboldt to be adapted for its cultivation. It is propagated by cuttings; and there is probably no other plant which produces on the same extent of land, and with so little labour, so great a quantity of food. Humboldt affirms that ½ hectare (about an acre) of land, planted with bananas, will furnish food for more than 50 individuals; whereas the same extent of land, if sown with wheat, in Europe, would not support more than two individuals! And all the labour required to raise this enormous produce is to cut of the stems when the fruit is ripe, and to give the earth a slight digging about the roots of the plant once or twice a year. Hence, says Humboldt, nothing strikes an European recently arrived in Mexico with more astonishment than the smallness of the patches of cultivated ground round cabins that swarm with children. It cannot be said of such a country;—

358

But the ease with which subsistence may be procured, and the fewness of their wants, have made the natives in the last degree slothful. Indeed, Humboldt tells us that it has been gravely proposed, in order to stimulate their industry, and rouse their torpid faculties, to grub up and destroy the banana plantations! (*Essai, &c.*, ii., 395.) Such a project is, of course, impracticable and absurd; but the nature of the proposed remedy serves, at all events, to show the violence of the disease.

The same parts of the country which produce the banana produce also the cassava, or manioc, the farina of which yields a very nourishing bread: it requires more care than the banana, somewhat resembles the potato, and arrives at maturity about eight months after the slips have been planted. The culture of maize is scarcely less important in the *tierras calientes* than that of the plants before named; it is not confined, however, to the low lands, but ascends as high even as the plain of Toluca (9190 ft. above the sea), the lowest average temperature favourable to its growth being about 48° Fahr. The plant, under favourable circumstances, rises to the height of 7 or 8 ft., and the returns, in common years, are most abundant; but they are more uncertain than those of any other kind of grain. Maize is the principal food of the people, as well as of most domestic animals; and a deficient harvest, whether from want of rain, or excess of cold, produces a general famine, and compels great numbers of the rural population to seek the deserts in search of wild plants.

There can be no doubt, however, that if agriculture were pursued with any spirit, and the system of irrigation generally introduced on corn lands, or even if there were the slightest degree of providence in the natives, those dearths would not occur that on several occasions have been so fatal, especially in the mining districts. The European *cerealia*, such as wheat, barley, &c., succeed best in the temperate regions, where the mean heat does not exceed 66° Fahrenheit: in fact, in the equinoxial regions of Mexico, these grains are not found under the level of 2380 ft. above the sea. The Mexican wheat is of excellent quality; equal, says Humboldt, to the best of the Andalusian: it is large, white, and nutritive. In well irrigated lands, and on good soils, the produce is said to average 24 for 1; but, since the revolution, this necessary branch of agriculture has been much neglected. Rye and barley resist cold better than wheat, and are cultivated in the highest regions; barley yielding abundant harvests, even where the thermometer indicates a heat during the day of only 5½°. Oats are little cultivated. Among the other alimentary plants, most of which have been introduced by the Europeans, are the potato (confined chiefly to the table-land), the yam, common both to the high and low country, the capsicum, raised in immense quantities for its spice, which is universally used instead of salt for seasoning food, beans, and various other garden vegetables common to Europe and America. Most of the fruits of Europe are common and plentiful; the olive and vine, introduced since the revolution, generally succeed well; and nowhere are there finer pine-apples, pomegranates, guavas, alligator pears, &c. One of the most valuable plants in the country is the maguey (*Agave Americana*), which Humboldt not inaptly terms the vine of Mexico. The Maguey plantations are principally found in the states of La Puebla, Mexico, and Guanajuato; but the plant is very hardy, and occurs in a wild state all over the country. Its growth is slow; but when arrived at maturity, its leaves are from 5 to 8 ft. in length, and the stem often attains the height of 20 or even 30 ft. Its period of flowering is very uncertain, but once in 10 years may be considered a fair average. At the flowering season, when the plant first begins to be useful, the exact time is watched when the stem of the flower is about to shoot up; the top is then cut off, so as to form a hollow, for the reception of the sap, which is regularly drawn off; and a vigorous plant will yield 15 quartillos daily for four or five months successively. The sap, which has a slight sub-acid taste, ferments readily in three or four days, being in its viscous state, called *pulque*, a beverage which somewhat resembles cider, though with a disagreeable smell. Immense quantities of it are drunk by all classes, and many whites as well as Indians use no other liquor. A kind of brandy, called *mexical* (very like whiskey), is made from the distillation of pulque. The maguey is useful, also, in many other ways: its fibres are converted into thread ropes and paper, its prickles serve for pins and needles, and its juice is effective in healing green wounds. Large quantities of sugar are raised in the neighbourhood of the capital, and the crops are very abundant, the lands are cultivated by free labourers, and the farming seems pretty good, though the process of refining be very clumsily conducted. In the commencement of the present century there was a large export of sugar; but this has for some years wholly disappeared, and the present supply is barely sufficient for the home consumption. Vanilla is extensively raised in the *tierras calientes*, E. of the Cordilleras, particularly in the state of Oaxaca. The cultivation of coffee is on the increase, and the quality of that raised on the best soil near the coast is said to be equal to the best produced anywhere else. Tobacco is a government monopoly, and its growth is confined to a small district near Orizava and Cordova. Its quality is inferior to that of Cuba; and, as the consumption exceeds the growth, considerable quantities are imported from the Havannah.

On the whole, it might be fairly concluded, on general grounds, that agriculture in Mexico must have retrograded since the revolution. And such, in point of fact, has been the case, and to an extent that we should hardly have conceived possible. This is evident from the following statement by M. Chevalier[*], who visited Mexico in 1835. "Agriculture," says he, "is neglected. No law, indeed, prevents the planting of the vine and olive tree; not only, however, has no advantage been taken of this change, but the very lands which were cultivated in the time of the Spaniards are now lying fallow. In a circle of a few leagues round Mexico, I have seen large villages almost abandoned. In this delightful climate, the only manure which the land ever requires, is water; this is rather scarce, yet many of the hydraulic constructions raised by the Spaniards at a great cost, are in ruins, and seem likely to remain so. The lands which, by means of this artificial irrigation, were the most fertile in the world, are gradually becoming completely sterile. Their ploughs, and other agricultural instruments, are of the rudest description. No one troubles himself to introduce European improvements, or even to import better tools from the United States. He made the passage from New-Orleans to Vera Cruz with General Arista, who had been exiled in consequence of some insurrection or other in which he was concerned. Wearied with the chances of revolutions, he had determined to devote himself to agriculture. He had scarcely, however, landed at Vera Cruz, when he was thrown into prison, under some vague pretext. He still continues under arrest, and his ploughs, harrows, and winnowing machines, remain under sequestration, suspected, probably, of abetting the general in some subversive designs."

Mining Industry.—The silver and gold mines of Mexico have always been deemed the main sources of its wealth; and, unquestionably, its mineral riches far exceed those of any part of America, except, perhaps, Peru. Before the war of independence, there were, in the 37 mining districts of New-Spain, somewhat more than 3,000 mines, producing annually about 21,000,000 dollars in silver, and about 2,000,000 in gold. Towards the close of the struggle, many of the mines had been deserted, and their produce had declined a half, and does not yet materially exceed that amount. Several of the so-called companies, formed in Great Britain for working the Mexican mines, during the memorable, and, we may add, disgraceful era of 1824–26, were mere swindling engines, and fell to pieces in a very short while. There were others, certainly, that had a more solid foundation; but these were mostly gone into without due consideration, and without any practical knowledge of the country, of the practices that had been followed, and the difficulties to be overcome. We need not, therefore, be surprised at the enormous losses the companies sustained at the outset, and of their want of success in the first instance. But, had the Mexican government been able and willing to repress disorder, and to enforce the observance of contracts, it is probable that the produce of the mines would have been very different at this moment from what it really is. Unluckily, however, no government has yet been established in Mexico with power, even if it had the desire, sufficient to put down disturbances, or to enforce engagements. So long, indeed, as the companies were struggling to put their mines in order, they sustained comparatively little inconvenience from this circumstance; but, as soon as they had succeeded in bringing them once more into a productive state, and were beginning to have some prospect of a return for their enormous outlays, they were annoyed by questions as to title, and by the setting up of claims on the mines, of which they had never heard before. Owing to these causes, as well as the general insecurity of property, the bad condition of the roads, and imperfect mining processes, the results have, on the whole, been very unfavourable, notwithstanding the reduction of the export duty on specie from 10 to 3 per cent.

The following statement of M. Chevalier, as to the insecurity of the miners, in 1835, discovers a state of things

disgraceful to the government; and such indeed as could hardly have been credited upon any inferior authority. "How," asks he, "can the mines be worked with any feeling of security, when it requires a little army to escort the smallest portion of the precious metal to its place of destination? Between the mine of Real del Monte and the village of Temeyuco is a mountain pass, where a grand battle was fought between the miners and the banditti of the country. The former were defeated, overpowered by numbers; but not without having sold their lives as dearly as possible. The mine is now guarded by artillery and grapeshot, and the Englishmen employed there are regularly drilled in the use of the musket." In such a state of things, the wonder is, not that the produce of the mines has declined, but that it continues to be so great as we find it to be. The mineral riches of the country are, however, inexhaustible; and there wants only a government able and willing to afford security, to make the produce of the mines greater than ever. The subjoined statement, which cannot, however, be altogether depended on, exhibits an

Account of the number of dollars coined at the different mints of Mexico during each of the four years ending with 1837; the proportion of gold to silver being as 1 to 26, nearly:—

Mines.	1834	1835	1836	1837
Mexico	964,700	547,145	744,126	596,422
Guanaxuato	2,793,000	2,607,666	2,311,972	3,008,991
Zacatecas	6,527,900	6,184,900	3,449,578	6,325,295
Guadalaxara	718,000	832,870	668,731	879,694
Durango	1,311,000	1,262,985	1,422,718	926,060
San L. Potosi	926,000	1,291,406	1,698,342	1,110,777
Chihuahua	—	—	204,214	225,162
Total	12,540,000	12,668,174	14,469,335	11,496,389

The theory of mining is little understood by the Mexicans, the oldest modes of working being still generally practised, notwithstanding the improvements that have been introduced by the English; and the machinery for draining the mines and raising the ore is of the most primitive description. Indeed, many of the mines have been abandoned, owing to the imperfections of the machinery, which, under more favourable circumstances, might be again worked with profit. The ignorance of the miners is only equalled by their obstinate adherence to old, and elsewhere long exploded, practices. But this should not be matter of surprise, if the testimony of M. Chevalier respecting the education of engineers may be depended on. The school of mines (*Mineria*), the mere building of which cost 190,000*l.*, is at present in the most pitiable condition, although the learned Andres del Rio is still one of the professors. It is unprovided with the means even of the most elementary instruction. It contains a vast chemical laboratory, but without the instruments requisite for the most simple experiments. The collection of minerals is in disorder, badly classed, and very incomplete; the library and the mechanical cabinet are deplorable. The school seems also to have shared the fate of the public treasury—of having been pillaged three or four times over. The very building seems on the point of falling to pieces—an appropriate emblem of the Mexican Republic. But it cannot surely be supposed that the anarchy which has led to such deplorable results is to continue forever. If nothing may be hoped for from within, it is to be wished that foreign interference may rescue this fine country from the barbarism in which it is now involved.

The quantity of silver annually extracted from the mines of Mexico very much exceeds that furnished by all the mines of Europe; but, on the other hand, the gold is not much more abundant than in Hungary and Transylvania; the proportion which the gold of Mexico bears to the silver being as 1 to 26, nearly. Little native silver is found in any of the mines; sulphuretted and black prismatic silver is both very common and exceedingly productive in the veins of Guanaxuato and Zacatecas, two of the richest mining districts: the murias abounds in the mines of Catorce and San Pedro, near San Luis de Potosi; and the martial pyrites of Pachuca yield three marks to the hundred weight. The Mexican ore, however, is poorer than that of Europe, 1600 oz. of ore yielding only about 4 oz. of silver. The gold is produced by washing the earth and sand in some few places; but in the province of Oaxaca occur veins of native gold, usually mingled with the silver veins: the returns, however, seldom exceed 1¼ oz. to the cwt. (*Pehuatt's Notes on Mexico*, p. 296.) Slave labour is not tolerated in the mines; but it would be difficult to find workmen so ignorant, brutalised, and wholly worthless, as the native miners. Indeed, the ill success of the English mining companies is owing, in part at least, to the want of honest and efficient labourers. The business of the mines is followed by native tribes from generation to generation: they lead a migratory life; removing, with their families, to districts where they expect the greatest profit from their labour: they are always paid by a share

naumachia), circus, theatre, triumphal arch, baths, &c. The seats of the amphitheatre appear quite perfect; the vaulted dens for the beasts are uninjured; and the conduits by which the arena was filled with water are still distinctly visible. In one of the streets may be seen a large triumphal arch, 150 feet high, but without any inscription or sculptures. The baths are surprisingly perfect, but not large; and round the top of the bathing-rooms runs a cornice of most curious and delicate workmanship, almost as perfect as if it had recently been executed. The bridge over the Guadiana is of stone, and portions of it may be of Roman architecture; but the greater part of the Roman bridge was swept away by a flood in 1610, and the present bridge has been constructed since. Two arches of this structure were blown up, in 1812, by the British troops under the Duke of Wellington. There are likewise two aqueducts, one of Roman, and the other of Moorish architecture, of brick and granite, the former having three, and the latter two tiers of arches.

Augusta Emerita was founded by order of Augustus, anno 25 B.C., who planted in it some of his veterans, called emeriti, whence its ancient and modern names. Though its ancient magnitude appears to have been greatly exaggerated, it was, no doubt, one of the largest Roman cities in the peninsula, and became the metropolis of Lusitania. From the Romans it passed, in 713, to the Moors, who destroyed and altered many of its old buildings. In 1226 it opened its gates to Alphonso IX., after his signal victory over the Moors in the contiguous plain of Matanzas; and from this period downwards it has been attached to the kingdom of Castile and Leon. (Recollections of the Peninsula, p. 178–181; Cook's Spain, i. 142–144; Miñano; Ancient Universal History, xiii. 493, 8vo. ed.)

MERIDA, a town of S. America, repub. Venezuela, dep. Zulia, cap. the prov. Merida, on the Chama, 330 m. S.W. Caraccas, and 325 m. N.E. Bogota. Previously to 1812, when it was destroyed by an earthquake, this was the largest city in Venezuela, and had a population of 12,000. It continued, for some years, to be little better than a heap of ruins; but it has been again rebuilt, and is now, probably, become more populous than before. It has a cathedral, several chapels, convents, &c., an ecclesiastical seminary, a college for philosophy, civil law, &c. The inhabitants are said to be, for the most part, in good circumstances. The coloured races dye wool and manufacture carpets and other woollen and cotton fabrics. (Geog. Account, &c., 270–74; Mod. Trav. xvii.)

MERIDEN, p. t., New Haven co., Ct., 16 m. S. by W. Hartford, 321 W. Incorporated from Wallingford in 1806. Watered by Quinnipiacer r. and its branches. It contains three churches, a Congregational, Episcopal, and Baptist; a bank, eight stores, two fulling-mills, two woollen factories, two grist-mills, nine saw-mills, two paper-mills, one tannery, one distillery; two academies, 50 students; 13 schools, 397 scholars. Pop. 1880.

MERIONETH, or MERIONYDD, a marit. co. of N. Wales, having N. the cos. of Caernarvon and Denbigh, E. and S. those of Montgomery and Cardigan, and W. St. George's Channel. It is of a triangular shape, and contains 434,360 acres. This, next to Caernarvon, is the most mountainous county in the principality. Among the principal summits are those of Arran-Fowdy, Cader-Idris, and Arrenig; respectively 2,965, 2,914, and 2,809 feet above the sea. It has, however, some fine vales, especially that of Festiniog, celebrated for its romantic scenery. There are some considerable tracts of low swampy land along the sea-coast; and in parts considerable tracts have been gained by embankments. The soil is very various; but, generally speaking, it is poor, and suited only for pasturage. Oats is the grain principally cultivated; but wheat and barley are also raised, though in no great quantities. Agriculture is in a very backward state; little or no attention is paid to a rotation of crops, and it is a frequent practice here, as well as in Denbigh, to burn the surface for manure. In some parts of the county potatoes are pretty extensively cultivated. The principal dependence of the farmer is, however, on his cattle and sheep, of which great numbers are fed on the mountains, and in the valleys not fitted for husbandry. The small native Welsh ponies, called Merlins, are now rarely met with, except in this county and Montgomery. They are sure-footed and exceedingly hardy. Dairy farming is carried on to a considerable extent. Farms usually small; and being mostly held at will, without any conditions as to management, the low state of agriculture itself not be wondered at. Average rent of land, in 1810, 3s. 11d. an acre. Speaking generally, the cottages are wretched in the extreme; though happily they have been a good deal improved in some parts of the county. The minerals seem to be of less consequence than might have been supposed: lead and copper are raised, though in small quantities; large quantities of lime are produced at Corwen, and slates are quarried in different places. The manufactures, which also are unimportant, consist principally of coarse flannels, produced

on the domestic system, at Dolgelly, Towyn, and a few other places. The Dee has its source in this county; and it is also watered by the Dyfi, Maw, Dwyney, &c., flowing W. Bala, the largest lake in the principality (see Bala), is in this county. Harlech is the county town. Merioneth is divided into six hundreds and 37 parishes. It sends one member to the House of Commons for the county. Regd. electors, in 1839–40, 1399. In 1831, the county had 6168 inhabited houses, 7358 families, and 35,315 inhabitants, of whom 17,194 were males, and 18,121 females. Sum expended for the relief of the poor, in 1838–39, 12,107l.

MERMENTAU, r., La., drains the extensive prairies of south western Louisiana. After a southerly course of 220 miles, it enters the gulf of Mexico, 200 m. W. of Mississippi river. It passes through a lake of the same name, and has not more than four feet of water on the bar at its mouth. Live oak is found on its borders.

MERRIMAC, r., N. H., is among the large rivers of New England. Its longest branch is Pemigewasset river, which rises near the Notch of the White Mountains. After running S. about 70 m., it receives Winnipiseogee river, in Franklin township, the outlet through Great bay of Winnipiseogee lake, and below that it takes the name of Merrimac river. It then flows S. by E. until it enters Massachusetts; it then turns E. and N.E. for 50 m., until it enters the Atlantic, a little below Newburyport. It is navigable for vessels of 200 tons to Haverhill. By means of the Middlesex canal from Chelmsford to Boston harbour, and canals and locks around the falls in the river, a boatable communication is opened from Boston to Concord, N. H. Bow canal, a few miles below Concord, with four locks, overcoming a fall of 25 feet, was completed in 1812, and cost $25,000; 6 m. further down, Hooksett canal, with three locks, overcoming a fall of 17 feet, cost $15,000; 8 m. below Hooksett is Amoskeag canal, with nine locks, overcoming a fall of 45 feet, and cost $50,000. It is the greatest work of the kind on the river. Below Amoskeag canal, the river, for 9 m., is converted into the Union canal, overcoming six distinct falls, with slack-water navigation, and seven locks, and cost $50,000; 5 m. below are Cromwell's falls, made passable by a canal, the expense of which is included in Union canal; 15 m. below, in Massachusetts, Wicasse canal cost $14,060, 3 m. below which is the entrance of the Middlesex canal. A number of bridges cross Merrimac river. Its principal tributaries on the W. side are Contoocook, Piscataquog, Souhegan, and Nashua rivers; and on the E. side are Soucook, Suncook, and Beaver rivers. It has several important places on its banks, the principal of which are Newburyport, Haverhill, and Lowell, in Massachusetts, and Concord in New-Hampshire.

MERRIMAC, co., N. H. Centrally situated toward the S. part of the state, and contains 516 sq. m. Organized in 1823. Watered by Merrimac r. and its branches. It contained, in 1840, 33,911 neat cattle, 80,505 sheep, 12,385 swine; and produced 80,612 bushels of wheat, 46,706 of rye, 183,291 of Indian corn, 8988 of buckwheat, 5251 of barley, 173,227 of oats, 685,370 of potatoes, 62,940 lbs. of sugar. It had 113 stores, 12 fulling-mills, three woollen factories, 3 cotton factories, with 11,768 spindles, one glass factory, 42 grist-mills, 105 saw-mills, one oil-mill, five paper-mills, 35 tanneries, two distilleries, three potteries, 26 printing offices, six binderies, six weekly newspapers, and one periodical; 70 academies, 1084 students; 373 schools, 10,360 scholars. Pop. 36,253. Capital, Concord.

MERRIMAC, p. t., Hillsborough co., N. H., 29 m. S. Concord, 452 W. Bounded E. by Merrimac river. Watered by Souhegan river and its branches, which afford extensive water-power. Incorporated in 1746. It contains a Congregational church, four stores, six grist-mills, six saw-mills; nine schools, 351 scholars. Pop. 1114.

MERRIMAC, t., St. Louis co., Mo. Pop. 1782.

MERRIMAC, t., Crawford co., Mo. Pop. 1111.

MERRIWETHER, co., Ga. Situated in the W. part of the state, and contains 606 sq. m. Bounded E. by Flint river, and drained by its branches, and by branches of Chattahoochee river. It contained, in 1840, 18,593 neat cattle, 9121 sheep, 41,844 swine; and produced 86,480 bushels of wheat, 1040 of rye, 256,115 of Indian corn, 58,779 of oats, 20,378 of potatoes, 5,660,229 lbs. of cotton. It had 11 stores, nine grist-mills, five saw-mills, three tanneries; four academies, 155 students; nine schools, 265 scholars. Pop.: whites 8,725; slaves, 5,301; free coloured, 16; total, 14,132. Capital, Greenville.

MERSEBURG, a town of the Prussian dom., prov. Saxony, cap. reg. Merseburg, on the Saale. 56 m. S.S.E. Magdeburg; lat. 51° 22' 1'' N., long. 19° 0' 33'' E. Pop. in 1837, 9413. It is walled, and is old and irregularly built. It has several suburbs, a cathedral, a castle, a gymnasium, a hospital, and various other public institutions. Merseburg is the seat of government for the regency, of the council and court of justice for the circle and town, a board of forests, &c.; and has manufactures of woollen and linen cloth, paper, tobacco, and vinegar. The beer of Mersburg is celebrated as the best in Saxony. (Von Zedlitz, Der Preussische Staat., iii. 223; Borghaus, &c.)

MERSEY, a river of England, which has its embouchure on the W. coast of the island, in the Irish sea. Though not large, the Mersey has, from its flowing through the principal manufacturing district of the empire, and giving its name to the gulf or estuary between Lancashire and Cheshire, become, in point of commercial importance, second only to the Thames. It has its sources in the great central ridge, or Pennine chain, on the confines of Yorkshire, Cheshire, and Derbyshire. After receiving the Goyt from the S., and flowing W. through Stockport, it is joined by its important affluent the Irwell. The latter, which has its source in the Lancashire moors, near Haslingden, flows S. through Bury to Manchester, where, being joined by two smaller streams, it takes a westerly course, till its confluence with the Mersey. After being still farther increased by the Boden from Macclesfield, the Mersey passes Warrington, a little below which it expands into a magnificent estuary, having the great commercial port of Liverpool on its N. side, near its junction with the Irish sea. The Mersey and Irwell have been rendered navigable from Sankey bridge to Manchester; and projects are now on foot for improving and deepening the navigation. (See MANCHESTER.)

MERTHYR-TYDVIL, a parl. bor. market-town, and par. of S. Wales, co. Glamorgan, hund. Caerphilly, on the Taff. 19 m. N. by W. Cardiff, and 140 m. W. by N. London. Pop. of parl. bor., which includes nearly all the par. of Merthyr-Tydvil, and the entire par. of Aberdare, with a small portion of the par. of Vainor, 27,460 in 1831, but in 1841, 34,977. "The town lies on the E. side of the valley, down which the Taff descends to Cardiff, scattered in detached masses about the valleys and on the hills, sending forth branches in different directions; and fresh groups are continually rising in the neighborhood of the great iron-works, so that it is somewhat difficult to point out where any collection of houses ends or begins." (Bound. Rep.) The houses, generally speaking, are mean-looking, comprising "labourers' cottages, or small ale-houses, beer-shops, or retail shops;" but in the centre of the town there are three tolerably respectable streets, forming a triangle, at one point of which is the parish church, a modern and well-built structure: the living is in the gift of the marquis of Bute. There is, also, a chapel-of-ease, besides several places of worship for dissenters; and the various Sunday schools of the town give religious instruction to nearly 6,000 children of both sexes. National, Lancastrian, and other subscription schools have likewise been formed, and are well attended. A philosophical society and several book-clubs have been established; and there is abundant proof that education is advancing among all classes. A theatre has been built within these few years, and there are two good hotels. In the environs are many handsome seats, belonging to the wealthy iron-masters; and three miles N. of the town, on an insulated hill, stand the ruins of Morlais castle, a very ancient building, demolished during the late civil wars.

The rise of Merthyr has been rapid, almost beyond belief. Towards the middle of last century it was an insignificant village: in proof of this, it is sufficient to state, that, in 1755, the lands and mines for several miles round the village, the seat of the great works now erected, were let for 90 years for 200l. a year. It is wholly indebted for its prosperity to its rich mines of coal, iron-ore, and limestone. The stratum of coal, which is of excellent quality, is accompanied by parallel veins of argillaceous iron, penetrating to a great depth, and yielding, at an average, about 35 per cent. of metal. The iron works are on a vast scale; those of Sir John Guest and Co. at Dowlais, of the Messrs. Crawshay at Cyfarthfa and Hirwain, having actually raised up very populous townships; the Pen-y-darran and Plymouth iron-works are also very extensive. In all, about 150,000 tons of iron a year are produced in the immediate vicinity of the town. Of this, a large proportion goes through the various processes of refinement and rolling into bars, previous to being shipped at Cardiff. The furnaces, refineries, and rolling-mills employ a great many persons; the wages for men ranging from 18s. to 60s.; of women, from 6s. to 10s.; and boys, 7s. to 11s. per week. The trade is of a very fluctuating character, and great numbers of workmen are often thrown out of employment by the stoppage even of two or three furnaces. Distress, however, is less permanent here than in many other districts, as "the work is one requiring less experience than many other manufactures; so that a demand for labour is readily met by a supply; while, on the other hand the labourers feel no great reluctance to transfer themselves to fresh employments." (Bound. Rep.) It is said by Mr. Nicholson, in the last edition of the Cambrian Guide, that "there is a marked improvement in the houses of the workmen; most of them have good oak chests of drawers, bright as silver; cupboards, with a display of family china cups and glasses; some of the younger women

have a veneered work-box; and all these little things display an attention to the lesser comforts and luxuries of life, of which, a few years ago, they had no idea. On the whole, there is a decided improvement in the general condition and circumstances of our workmen." (P. 421.)

But, notwithstanding their comparatively comfortable condition, and the great increase of temperance societies, it would seem that Chartist doctrines have made a very considerable progress among the labourers in this district, a circumstance for which it is not very easy to account. A large portion of the Chartists engaged in the outbreak at Newport, on the 4th of November, 1839, were understood to be from Merthyr and the adjoining iron-works. It is to be hoped that the severe, but wholesome, castigation they received on that occasion may not have been thrown away; and that it may help to disabuse them of their prejudices, and hinder them from again becoming the dupes of designing demagogues.

The communication with Cardiff is effected by means of the Glamorganshire canal (completed in 1794). which commences at Merthyr, and ends, after a course of 25 m., in the tideway of the Taff, near its entrance into Penarth harbour, the entire descent being 611 ft. (See CARDIFF, I. 552.) In 1836, 118,060 tons of iron were sent down this canal. The late improvements of the marquis of Bute in the port of Cardiff have also been of essential advantage to the export trade of Merthyr-Tydvil. The Taff-vale railway, for which the act of parliament was procured in 1836, and which is intended to connect the mining district of Merthyr with Cardiff, was opened as far as Newbridge, 14 m. N. Cardiff, in October, 1840; and it is stated that the traffic, as well as the number of passengers, already exceeds the original estimate. This railway, when completed, will, with its various branches, have a length of 25 m.: the capital is estimated at 650,000l.

The Reform Act created Merthyr a parliamentary borough, with the privilege of sending one member to the House of Commons; the electoral limits comprising the parish of Merthyr-Tydvil, except parts of the hamlets of Forest and Taff and Cynon, the entire parish of Aberdare, and the hamlet of Cefn-Coed-y-Cumner, in the parish of Vainor. Registered electors, in 1839-40, 582. Merthyr is also one of the polling-places at elections for the county. Petty sessions are held here for the upper division of hundred Caerphilly; and a court of requests sits monthly for the recovery of debts not exceeding 5l. An act was passed in 1830 for the better security of life and property in this district; and the three parishes of Merthyr-Tydvil, Aberdare, and Gellygare are under the superintendence of a stipendiary police magistrate, having a salary of 600l. a year, half of which is levied on the furnaces within the limits of his jurisdiction, and half on the inhabitants of Merthyr alone. Markets on Wednesday and Friday; cattle fairs May 14, the first Monday in July, and the first Monday in August.

MESAGNE, or MESSAGNA, a town of the Neapolitan dom., prov. Otranto, cap. cant., 8 m. S.W. Brindisi. Pop. 5,000 ? It has several convents, an hospital, and a fine palace, belonging to the Francavilla family. It manufactures kitchen utensils; and has some trade in oil and grain, considerable quantities of which are grown la its vicinity. Mesagne is supposed by many Italian authors to be the representative of the ancient town of Messapia, but this is doubtful.

MESHED, a city of Persia, the cap. of Persian Khorassan, and esteemed as "holy" from its containing a very superb sepulchre, enclosing the remains of Imâm Reza and the caliph Haroun al Raschid, 455 m. E. by N. Teheran; lat. 36° 17′ 40″ N., long. 39° 35′ E. Pop., according to Kinneir, 50,000. It stands in a rich and well-watered plain, is surrounded with a strong wall, and is divided into 12 quarters, of which five are in ruins. The houses are meanly built of sun-dried bricks, and the ark or palace is unworthy of its name. There were formerly 16 medressas or colleges; but most of them are either deserted or in ruins. Indeed, the city has now little worth notice, except its fine and well-supplied bazaar, and the mausoleum of Imâm Reza, the magnificence of which, with its silver gates, jewelled doors, rails once of solid gold, glittering domes and minarets, and handsome arcades, is almost unequalled in Persia. It has, however, been often plundered; and its resources are said by Mr. Fraser to be greatly reduced. Meshed carries on a considerable trade with Bokhara, Balkh, Candahar, Yezd, and Herat; and many of the inhabitants are employed in weaving velvet and making fine pelisses, both of which are much esteemed throughout Persia. (Kinneir's Persia; Fraser's Khorassan, &c.)

MESSINA (an. Zanclè and Messana), a celebrated city and seaport of Sicily, ca., Intendency, near the N.E. extremity of the island, on the strait of its own name, 8 m. N.W. Reggio, 36½ m. N.N.E. Catania, and 190 m. E. by N. Palermo; lat. (of its lighthouse) 38° 11′ 36″ N., long. 15° 34′ 40″

E. Pop., in 1831, including that of its canton, 63,772.[*] The city has a most imposing appearance from the sea, forming a fine circular sweep, about 3 m. in length, on the W. shore of its magnificent harbour, from which it rises in the form of an amphitheatre; and being built of white stone, it strikingly contrasts with the dark forests that cover the mountains in the background. Prior to 1783, the harbour was fronted by a magnificent terrace of lofty houses, called the *Palazzata*, having in front a broad quay decorated with statues and fountains. But the great earthquake of that year laid the city almost entirely in ruins; and though the the terrace still exists, it is shorn of its former grandeur. The quay in its front, called the *Marina*, has now, as formerly, numerous fountains, and is the favourite promenade. Ruined buildings, and other vestiges of the earthquake, occupy a considerable portion of the city; and few houses have now more than two stories. Swinburne, who visited Messina before 1783, complained that its interior was dirty, with narrow streets, gloomy houses, little bustle of trade, and still less show of luxury. According to Smyth, however, the modern city is regularly built, well paved with square blocks of lava, and several of its streets are wide and handsome, ornamented with numerous churches, convents, statues and fountains. The square in front of the cathedral, and that of *San Giovanni di Malta*, are both well built and handsome. The fountain in the centre of the former ranks with the finest in Sicily. The cathedral, erected soon after the conquest of Sicily by the Normans, has been repeatedly damaged by earthquakes. It is a Gothic building, with a heavy and gloomy exterior: the interior, though devoid of taste, is richly ornamented. The principal entrance is handsome; and the nave is supported by immense granite columns taken from the ruins of a temple of Neptune. The great altar and the roof of the choir are set off with mosaics and precious stones: the carved work of the pulpit is said to be a *chef d'œuvre* of the Sicilian sculptor Gaggini. The church of Monte Virgine has some good paintings in fresco, and that of St. Giorgio is very rich in marbles and inlaid work, and has some tolerable pictures. Adjacent to the viceroy's palace, a noble building at the S. extremity of the city, is a large open space planted and laid out in public walks. The other public buildings include a large hospital, several asylums of various kinds, two theatres, town-hall, exchange, bank, college, &c. Messina is surrounded by an old irregular wall, finished by Charles V. The citadel, a pentagonal fortress, erected on the S. side of the harbour, is constructed according the principles of Vauban; but though well provided with bomb-proof quarters and stores, it is badly situated and commanded in almost every part. Two strong and well-built forts have, however, been constructed on eminences above the town, that would greatly annoy and harass an enemy during any operation against the citadel. The town is further defended by a fort placed so as to command the mouths of the Flumare, which are the only places where an enemy could land with cannon. The port, to which Messina is wholly indebted for her prosperity, and even existence, is formed by a lengthened curved tongue of land, that might almost be supposed to be an artificial circular mole, projecting first N. E. from the main land, and then bending round to the W. in the form of a sickle.[†] The entrance on the N., about 700 yards across, is defended on the W., or main-land side, by the bastion of Porto Reale, and at the extremity of the curved promontory by Fort Salvatore. A lighthouse has been constructed on the extreme E. verge of the promontory. The noble basin thus enclosed is about 4 m. in circuit, and, having deep water throughout, is capable of accommodating the largest fleets: it is, in fact, not only the finest harbour in the Mediterranean, but one of the finest of which we have any certain knowledge. Men-of-war moor in the centre of the basin in about 33 fathoms; but merchantmen lie alongside the quay, and have every facility for loading and unloading. The pratique-office, the fish-market, and the custom-house, are all on the Marina. The lazaretto, the best establishment of the kind in Sicily, is in the E. angle of the harbour. The situation of Messina, on the strait between Italy and Sicily, and her admirable port, give her great advantages as a commercial *entrepôt*; and were Sicily and Naples subject to an enlightened government, able and willing to put down abuses, and to call into activity their long dormant energies, Messina would certainly be one of the greatest emporiums of the Mediterranean. Even as it is, her trade is very considerable. Her exports consist principally of oranges and lemons, olive oil, silk, linseed, wines and spirits, shumac, liquorice, rags, corn, salted fish, &c. Almost all the silk exported from Sicily is shipped here. The imports consist of colonial produce, and cotton and woollen fabrics, hides, hardware, &c. (*Macgregor's Re-*

---

ports on Sicily; Smyth's Sicily, p. 113-123; Russell's Sicily, p. 250-261, &c.) We subjoin an

ACCOUNT of the quantities and value of the principal articles exported from Messina in 1836.

Articles.		Quantities.	Total Value.
			£.
Argols and cream of tartar	cwts.	4,341	6,318
Barilla	ditto	5,188	2,845
Brimstone	ditto	2,130	3,068
Cantharides	ditto	98	1,434
Corn, grain, and pulse	qrs.	4,589	9,582
Cotton wool	cwts.	276	698
Essences	lbs.	34,578	7,782
Fish, salted	cwts.	5,688	11,696
Fruits, dry and pickled	ditto	4,150	1,861
Oranges and lemons	boxes	384,284	79,365
Lemon juice	galls.	234,257	6,315
Linseed	qrs.	20,264	34,583
Other seeds	cwts.	11,485	5,728
Liquorice paste	ditto	6,862	14,385
Oil olive	galls.	345,388	58,591
Manna	cwts.	963	1,672
Rags	ditto	11,215	5,972
Salt	tons	288	364
Shumac	cwts.	19,177	3,148
Silks	lbs.	60,038	42,635
Saltas	No.	237,710	7,105
Wine and spirits	galls.	794,249	27,051
Other articles			68,754
Total value			**460,302**

Messina is the see of an archbishop, and the residence of a Greek *protopapas*, with authority throughout Sicily, but who is nominated by the pope. It is the seat of a royal court of appeal, and of criminal, civil, and commercial tribunals; and has a municipal bank, several *monti di pietà*, or government loan banks, and other benevolent institutions. Next to commerce, its inhabitants are chiefly occupied in the tunny and other fisheries; and in the manufacture of silk stuffs, especially damasks and satins. It has an ecclesiastical seminary, a lyceum, a royal college for law and medicine, and an extensive public library; but Simond says that Messina appears to have made slow progress in refinement, compared with Catania or Palermo. "The education of young people is more neglected; very few in the lower ranks can read; and the nobility do not in general reside in Messina: in short, it is neither fashionable, nor learned, nor rich; nor is it, I think, particularly hospitable." (*Simond's Italy, &c.*, 525.) But, how deficient soever in these respects, Messina has advantages of another sort that entitle her, in the estimation of her citizens, to look down with contempt on most other places. These consist in the possession of an autograph letter of the Virgin Mary, addressed to the Messinians, and assuring them of her especial protection; and what is, if possible, still more precious, a lock of the Virgin's hair, given by her to the persons entrusted with the conveyance of the letter! To question the genuineness of these valuable relics would, in Messina at least, be rather hazardous; and, under such circumstances, we need not wonder that is firmly believed that, on one occasion, the city was saved from famine by the opportune arrival of a supply of corn, sent by the Virgin. The only wonder is, that she has allowed it to be so often brought to the brink of destruction by earthquakes, and devastated by the plague. A splendid *fête* is annually given in the great square in honour of the exalted protectress and benefactress of the city. Very few vestiges of the ancient city remain; a consequence, no doubt, of the numerous earthquakes by which it has been visited.

The accounts of the origin and early history of Messina differ considerably. It is admitted on all hands to be very ancient; and most probably derived the name it has so long borne from a settlement having been made in it by a body of emigrants from Messene, in Greece. Having been seized by the Mamertial, it became, under them, one of the most populous, wealthy, and powerful cities of Sicily. It was the first town of the island that came into the possession of the Romans. (*Cellarii Orbis Antiqui*, i. 973; *Ancient Universal History*, iii. 513., 8vo. ed.)

The principal political events in the history of Messina, in modern times, are its successful resistance to Charles of Anjou, by whom it was besieged, after the Sicilian Vespers; and its revolt against the Spaniards in 1674, followed, in the ensuing year, by the defeat of the latter in its vicinity by a French force. In 1743 the plague broke out in Messina, with the most destructive violence, sweeping off the greater number of the inhabitants.

METHUEN, p. t., Essex co., Mass., 26 m. N. by W. Boston, 454 W. Bounded S.E. by Merrimac river. Watered by Spicket river, which has a fall of 30 feet, two miles above its entrance into Merrimac river, and affords great water power. It contains three churches, a Congregational, Baptist and Universalist, four stores, two cotton factories with 4568 spindles, two grist-mills, two saw-mills, two paper-mills; 10 schools, 650 scholars. Pop. 2351.

---

[*] According to the official returns, the population, in 1798, was only 45,052, so that it would appear to have nearly doubled in the interval.

[†] This, in the old Sicilian language, was called *Zanklè*, whence the original name of the town. (*Thucydides*, lib. vi. cap. 4.)

362

METZ (an. *Divodurum*, afterwards *Mediomatrici* and *Metis*, whence its present name), a strongly fortified city of France, dep. Moselle, of which it is the capital, at the junction of the Moselle and Seille, 30 m. W.N.W. Strasbourg, and about 180 m. E.N.E. Paris; lat. 49° 7′ 5″ N., long. 6° 1′ 15″ E. Population, in 1836, 43,793. "Metz is a fine old city; but, like most fortified places, the streets are narrow, and the houses lofty. Near the river it is more open, the quays are broad, and the bridges magnificent. The river is clear and rapid, and swells to an expanded stream where not confined by the embankments, as it is within the fortifications." (*Jacob's View of Germany, &c.*, p. 436.) Metz was fortified by marshals Vauban and Belleisle: it has several strong outworks, and a citadel on the Moselle; but the latter was partly dismantled during the revolution, and its esplanade has been laid out in public walks, which command a noble view of the valley of the Moselle and its bounding hills. The city has nine gates and drawbridges, but only six are in use. The most conspicuous public building is the cathedral, a vast pile, commenced in 1014, but not finished till 1546. It is about 300 feet in length, the height of the nave being about 140 feet, and that of the tower about 400 feet.[*] The latter, which is a fine specimen of Gothic architecture, has in it a bell weighing 25,000 lbs. The whole edifice is remarkable for lightness. Mr. Jacob says that the cathedral of Metz was the most perfect Gothic structure he saw on the continent; and that, though not so old as Westminster Abbey, it may vie in external beauty with that venerable pile. The military hospital, built in the reign of Louis XV., is a noble edifice, consisting of two ranges of building, and capable of easily accommodating 1500 patients. The *Hôtel du Gouvernement*, a large, though rather heavy fabric, fronts the esplanade; it is appropriated to the courts of justice and the city library; the latter has above 30,000 volumes, among which are numerous works printed in the 15th century, and about 800 MSS., some of the 10th century. (*Guide du Voyageur.*) The barracks, military magazines, prefecture, townhall, and mint, several of the churches, the new market, the theatre, with a portico of the Tuscan order, &c., are among the other edifices. The Moselle and the Seille, in and near the city, are crossed by at least 90 bridges. The principal school of artillery and military engineering (*École de Génie d'Application*) in France is established here. Its library has a choice collection of about 10,000 volumes of military and scientific works, with sundry MSS. of Vauban and other distinguished persons. Exclusive of the above, Metz has two other public libraries, with several convents and charitable asylums, a Protestant church, a synagogue, a royal college, a university academy, an ecclesiastical school, and other seminaries; a school for the fine arts, a royal society of arts and *belles lettres*; an agricultural society, a society for the encouragement of primary instruction, and collections in natural history, mineralogy, and chemistry; a botanic garden, a lying-in hospital, a savings' bank, a *mont-de-piété*, &c.

Metz is the see of a bishop, and the seat of a royal court for the deps. of Moselle and Ardennes, and of tribunals of primary jurisdiction and commerce, and a chamber of commerce, and the head-quarters of the third military division of France. "It is also a manufacturing city, in which are made woollen goods of various kinds, hosiery, cotton goods, table-linen, printed paper, musical instruments, starch, and gunpowder; it has, besides, several extensive tanneries. Much trade originates here from the produce of the vines, some portion of which is converted into wine, but more into brandy and vinegar; and Metz is celebrated for the preparation of various kinds of confectionary. It is encircled by hills, covered from the bottom to the top with fruit-gardens and vineyards. The vineyards are mostly in small divisions, and principally cultivated by small proprietors, who are extremely poor, and almost all involved in debt to the capitalists of the city; who take from them their wine, brandy, and vinegar as soon as it is made." (*Jacob's View of the Agric. of Germany, &c.*, 436-438.)

The royal gunpowder factory, on an island in the Moselle, near the city, is superior to most others in the kingdom. Metz has also a royal cannon foundry, a saltpetre refinery, and produces leather, cotton yarn, military and other hardware, smalls, beet-root sugar, chicory, nails, and other articles of hardware, cutlery, buttons, glue, &c.

This is a very ancient city. It still possesses several ruins belonging to the Roman period, among which are the remains of an aqueduct, that appears to have conveyed water to a *naumachia* near the S. extremity of the city. The site of the latter is now occupied by outworks belonging to the fortifications. Parts of an amphitheatre and of a Roman palace are still traceable in the city. It suffered considerably, about anno 70, from some excesses of the

[*] There is some discrepancy in the authorities as to these measures; but the above must be nearly accurate.

troops of Vitellius (*Taciti Hist.*, lib. i., cap. 73), and was nearly destroyed by the savage barbarism of Attila in 452. It had, however, recovered a large portion of its former prosperity in the middle ages, and became the capital of the kingdom of Austrasia. From the 11th century to 1552, when it was taken by Henry II., it was an independent flourishing city. In the same year that it was taken by Henry, it was besieged by the emperor Charles V., with an army of 100,000 men, but the Duke of Guise successfully defended the town, and Charles was obliged to relinquish the siege. It was finally annexed to the French crown by the treaty of Westphalia, in 1648. (*Hugo*, art. *Moselle*; *Guide du Voyageur, &c.*)

MEURTHE, a dep. of France, reg. N.E., formerly part of the prov. of Lorraine, between lat. 48° 26′ and 49° N., and long. 5° 40′ and 7° 20′ E., having N. the dep. Moselle, E. Bas Rhin, S. Vosges, and W. Meuse. Length, E. to W., 74 m.; average breadth, about 35 m. Area, 608,992 hectares. Population, in 1836, 494,366. The Vosges mountains run through the E. part of the dep., the surface of which is mostly covered with their ramifications, though these rise to no great elevation. The dep. belongs almost wholly to the basin of the Moselle, which river intersects its W. part from S. to N., and is joined, within its limits, by the Meurthe. The latter rises in the dep. Vosges, runs generally in a N.W. direction, and, after a course of between 70 and 80 m., unites with the Moselle about 5 m. below Nancy, to which it is navigable. Besides Nancy, St. Dié and Luneville are on its banks; and it receives the Mortagne, Vezouze, and Mezaille. The Seille and Sarre are the other chief rivers of the dep. There are numerous small lakes, one of which occupies an area of 629 hectares. In 1834, it was estimated that 303,636 hectares of the surface were arable, 71,881 in pasture, 16,371 in vineyards, 115,900 in woods, and 6256 in orchards, &c. The land is very unequal in point of fertility, and is very indifferently farmed; but more corn is grown than is required for home consumption. The total produce of the harvest of 1835, was estimated at 3,434,500 hectolitres, chiefly wheat and oats: in the same year, the crop of potatoes was estimated at 2,142,000 hectolitres. Before the revolution, the culture of the vine was limited to the declivities of hills with a southern aspect; but its culture has since been very much extended, the quality of the produce being less regarded than the quantity. About 550,000 hectolitres of wine are supposed to be produced annually, of which the greater part is consumed in the dep. The wines are generally inferior, though the growths of Pagny, Thiaucourt, Arnaville, Baudonville, and others, may be classed among the secondary qualities of *vins ordinaires*. (*Julien, Topographie*, p. 46.) Dried plums and preserved apricots form important articles of commerce; and the forests, which are more extensive than in most deps., furnish a good deal of timber. The pastures are naturally good, but receive little attention from the farmer. In 1830, there were estimated to be 94,000 head of black cattle, and 167,000 sheep, in the dep., but both are of indifferent quality. The breed of horses has been improved by the fine stud of Rosières. Hogs of an improved breed are numerous, and their flesh and lard are sent to distant parts of France. A great many poultry are reared. Property is much subdivided. In 1835, of 171,692 properties subject to the *contribution foncière*, 160,343 were assessed at less than five francs, and 94,437 at from five to 10 francs. Turf and lime are among the chief mineral products; there are some quarries of marble and alabaster, and a few iron mines; but the latter have been abandoned. The salt mines and springs at Dieuze, Vic, Moyenvic, &c., yield about 45,000,000 kilogrammes of salt, and 1,000,000 do. of soda a year. About 22,000 hands are said to be employed in the manufactures of cotton cloth and yarn, woollen stuffs, glass, and earthenware, and in embroidery, dyeing cotton stuffs, tanning, &c. At Baccarat is a large glass manufactory, employing a great many hands.

Meurthe is divided into five arronds.: chief towns, Nancy the cap., Toul, Château Salins, Sarrebourg, and Luneville. It sends six members to the chamber of deputies. Number of electors, in 1836-39, 1919. Total public revenue, in 1831, 16,794,392 francs. (*Hugo*, art. *Meurthe*; *Official Tables*.)

MEUSE, a dep. of France, reg. N.E., formerly part of the province of Lorraine, chiefly between lat. 48° 25′ N., and 49° 35′ N., and long. 5° and 6° E.; having N. Dutch Luxemberg and the dep. Ardennes and Moselle, E. Moselle and Meurthe, S. Vosges and Haute Marne, and W. Marne and Ardennes. Length, N. to S., 80 m.; greatest breadth, about 40 m. Area, 690,555 hectares. Population, in 1836, 314,566. Surface generally hilly, the hills being ramifications of the Vosges and Faucilles mountains, with an average height of from 1000 to 1300 feet, though they sometimes reach an elevation of 1600 feet. The Meuse traverses the dep. in its entire length; the other principal rivers are the Ornaïn, Chiers, and Aire. The plateau, in the E., separating the basins of the Meuse and the Moselle, and other portions of the surface, are not very productive; but there are, notwithstanding, about 225,000 hectares of rich soil in the dep., chiefly in the valleys of the Meuse and Ornain. In 1834, according to the official tables, 335,190 hectares were arable, 49,472 do. in meadows, 13,540 do. in vineyards, 7387 do. in orchards, &c., and 137,755 do. in woods. The produce of corn in 1835 was estimated at 2,440,000 hectolitres, of which 1,192,000 were wheat. Potatoes, oleaginous plants, hemp, and flax, are among the other articles of culture. Gooseberries are extensively cultivated in the gardens round Bar and Ligny, and enter largely into the confectionary, for which those towns are celebrated. The produce of wine is estimated at about 400,000 hectolitres a year. The wines of Bar-le-Duc, Bussey-la-Côte, Preue, Ligny, &c., are delicate light wines, ranking in the first class of *wins ordinaires*; but they do not keep above two years, and do not bear carriage. (*Julien*, 43.) Along the Meuse are rich pasture lands; and at Void, cheese, similar to that of Gruyère, and excellent butter, are made. A good many cattle and sheep are reared in the dep.; but live stock is in general indifferent. The produce of wool is estimated at upwards of 140,000 kilogrammes a year. In 1835, of 157,180 properties subject to the *contribution foncière*, 80,566 were assessed at less than five francs, and 22,169 at from five to 10 francs. Iron, slates, and good building-stone, are the chief mineral products. There are between 90 and 30 iron furnaces (*hauts fourneaux*) in the dep.; and the establishments at Thounelle and Stenay produce each about 1,500,000 kilogrammes of iron a year. About 500,000 kilogrammes a year of cotton yarn are made at Bar-le-Duc, which, also, has fabrics of paper, glue, &c., and is the *entrepôt* of a large trade in timber from the forests of the dep. There are numerous glass-works, with lime-kilns, potteries, beet-root sugar factories, &c. Many working cutlers, shoemakers, and other artisans, emigrate for a part of the year from this into other parts of France, and even to the adjacent foreign countries, with the products of their industry, or in search of employment. Meuse is subdivided into four arronds.; chief towns, Bar-le-Duc, the cap., Commercy, Montmédy, Verdun. It sends four members to the chamber of deputies. Number of electors, 1838-39, 1136. Total public revenue (1831) 9,071,543 francs. (*Hugo*, art. *Meuse; Official Tables, &c.*)

MEUSE OR MAESE (Dutch *Maas*, an. *Mosa*), a river of W. Europe, flowing through the N.E. part of France, Belgium, and the S. of Holland; its basin being situated between those of the Marne and Scheldt to the W., and of the Moselle to the E. It rises in the dep. of Haute Marne, in France, 10 m. N.E. Langres, in about lat. 48° N., long. 5° 20' E., and runs at first generally N. through the deps. of Haute Marne, Vosges, Meuse, and Ardennes. Near Charlemont it leaves France, but it continues its previous direction to Namur, where it receives the Sambre from the W. It here makes a sudden bend to the N.E., in which direction it continues through the provinces of Namur, Liege, and Limburg, to about lat. 51° 30' N. It afterwards carves to the W., flowing between N. Brabant and Guelderland; and finally, at Woudrichem, in lat. 51° 49', and long. 5°, enters the Rhine or Waal, which loses its own name to assume that of the Maas. (*See* RHINE.) Its entire course may be estimated at 480 m.; nearly the half of which is in France. It is navigable for three fourths of this extent, or as far as Verdun, dep. Meuse. Its chief affluents are the Bar in France, the Lesse, Sambre, and Ourte in Belgium; and the Roer and Niers in Holland. Proceeding from its source to its mouth, the chief cities and towns on its banks are Neufchâteau, Verdun, Sedan, Mézières, Charlemont, and Givet in France; Dinant, Namur, and Liege in Belgium; and Maestricht, Roermond, Venloo, and Grave in Holland, before its junction with the Rhine. The Meuse communicates with the Aisne, and thence with the Seine and Somme by the canal of Ardennes; with the Scheldt by means of the Sambre and the Charleroy canal; and with both the Scheldt and the Rhine by the various branches of the Great North canal; in addition to which, many other canals connected with it are in progress. (*See also* BELGIUM, I., 316-17, 328.)

MEXICO (UNITED STATES OF), a federal republic of N. America, lying between the 15th and 33d parallels of N. lat., and 97° and 113° W. long. being bounded N. and N.E. by the W. districts of the United States of N. America and that wild region called New California, E. by the gulf of Mexico and the Republic of Texas, S. by Guatemala, and W. and S.W. by the Pacific ocean. The line dividing Mexico from Texas commences with the river Nueces, which it follows up to its source, and then runs N. to the head of the Colorado, whence it stretches W., forming the N. boundary of the confederation. The line of separation on the side of Guatemala is very irregular, commencing E. with the river Sarstoon, which it follows to its source, and then takes a N. direction to lat. 17° 30' N.; it thence runs W. and S.W. to lat. 15° 45', where it assumes a N.E. course,

including the province of Chiapas. Greatest length from N.W. to S.E., 1700 m.; greatest breadth about 609 m. Area estimated at 1,230,442 sq. m. Nothing can be more unsatisfactory than our acquaintance with this vast country: few even of the principal towns and rivers are correctly laid down, except, indeed, within the small circle personally visited by Humboldt, so that not even the elements of a good map exist; and, with respect to population, and other statistics, the unsettled, disorderly, and almost lawless state of the country makes inquiry all but impossible. The following table has been printed in the American edition of *Murray's Encyclopædia of Geography*, and some other works, and, though little dependence can be placed on it, it is probably as near an approximation to the truth as can, at present, be arrived at:

States.	Extent in sq. m.	Pop. in 1857.	Pop. to sq. m.	Capital Cities.
Chiapas	14,750	94,000	4·5	Ciudad de las Casas
Chihuahua	107,500	190,000	1·2	Chihuahua.
Cohahuila	192,600	90,000	0·5	Saltillo.
Durango	54,500	180,000	2·6	Victoria.
Guanaxuato	9,000	500,000	62·5	Guanaxuato.
Mexico	83,450	1,500,000	42·3	Tezcuco.
Michoacan	22,466	480,000	20·6	Valladolid.
New Leon	21,000	100,000	4·6	Monterey.
Oaxaca	82,650	660,000	20·2	Oaxaca.
Puebla	19,149	900,000	48·6	La Puebla.
Queretaro	7,500	100,560	18·3	Queretaro.
San Luis Potosi	13,000	300,000	23·0	S. Luis Potosi.
Sonora and Sinaloa	254,700	200,000	1·2	Villa del Fuerte.
Tabasco	14,676	75,000	5·1	Villa-hermosa.
Tamaulipas, or New Santander	36,100	156,000	4·6	New Santander.
Vera-Cruz	27,060	156,000	5·4	Xalapa.
Xalisco	70,000	670,000	10·1	Guadalaxara.
Yucatan	79,349	670,000	7·2	Merida.
Zacatecas	18,930	300,000	10·	Zacatecas.
Federal district		200,000		Mexico.
Total of States	1,008,442	7,851,005	7·5	
Territories.				
New Mexico	200,000	? 60,000	0·3	Santa Fe.
Colima		? 40,000		Colima.
Tlascala		? 96,000		Tlascala.
Total	1,230,442	7,728,800	6·3	

The most populous cities are Mexico (130,000), Guadalaxara (60,000), San Luis Potosí and la Puebla (55,000 each), Oaxaca and Querétaro (40,000 each), Guanaxuato (34,000), and Merida (28,000),

Of this great tract of country, which is about one third as large as Europe, the portion lying S. of the tropic of Cancer, and comprising a large part of the long and narrow isthmus that connects the American peninsulas, and separates the Atlantic from the Pacific ocean, is by far the most populous and rich, both in mineral and vegetable productions; and nearly all the information gained respecting Mexico has been collected in that part, to which, consequently, it is primarily applicable. The regions N. of the tropic become less populous as we proceed northward; and many large districts claimed by the republic, and divided into states and territories, are almost unknown, being inhabited only by wild Indian tribes, baffling all the attempts of their nominal masters to civilize or subdue them.

The surface of Mexico is extremely varied; and to this circumstance, nearly as much as to the difference of latitude in so extensive a country, may be attributed that singular variety of climate by which it is distinguished from most other regions. The Cordillera, or chain of mountains, generally regarded as a portion of the great chain of the Andes, that enters Mexico on the S., where it borders with Guatemala, diverges, as it proceeds N., into two great arms, like the upper part of the letter Y, following the line of the coasts on either side. The most westerly of these chains, or that parallel to the shores of the Pacific ocean, has some very high summits; and preserves its mountainous character till it joins, on the border of the United States, with the Oregon, or Rocky mountains. The other, or eastern arm of the Cordillera, begins to subside after reaching the 21st or 22d degree of latitude, and ultimately subsides, about the 26th or 27th degree of latitude, into the vast plains of Texas. Now, the whole of the vast tract of country between these two great arms, comprising about three fifths of the entire surface of the republic, consists of a central table-land, called the Plateau of Anáhuac, elevated from 6000 to upward of 8000 feet above the level of the sea! Hence, though a large portion of this plateau be within the limits of the torrid zone, it enjoys a temperate climate; inclining, indeed, more to cold than to excess of heat. Some very high mountains are dispersed over the surface of the central table-land; and it is also traversed in parts by pretty well defined ridges, which divide it into extensive sub-plateaus, to which different names have been given. But the surface is interrupted by few transverse valleys; and in some directions it is quite unbroken, either by depressions or by hills. Thus, it is mentioned by Humboldt, that carriages proceed from the capital, in the centre of the plateau, to Santa Fé, in New

Mexico, a distance of 1409 m., without any important deviation from an apparent level. (*Essai sur la Nouvelle Espagne*, i., 254.) The most remarkable tract is this elevated region is the plain of Tenochtitlan (in which is the capital), surrounded by ridges of porphyritic and basaltic rocks, running S.S.E. and N.N.W. It is of an oval form, 55 m. long and 37 m. broad, occupying an area of 1700 sq. m., of which about 160 sq. m. are covered with water. Its S.E. side is that most elevated, and here are seen towering above the plain the volcanoes of Popocatepetl 17,716 feet, Iztaccihuatl 15,700 feet, Citlaltepetl or Orizaba 17,390 feet, and Nauhcampatepetl or the Cofre de Perote 13,416 feet above the sea. The waters of the valley are deposited in five principal lakes situated on different levels: that of Tezcuco, which is near the centre of the valley, and covers 70 sq. m., is the least elevated. Farther N. are the lakes of St. Christoval and Tonanitla: while S. is the lake Chalco, occupying an area of 59 sq. m.; and these three are five feet higher than lake Tezcuco. The most elevated, however, of the whole, though the smallest, is the lake Zumpango, the level of which is 30 feet above that of Tezcuco. These lakes are fed by small rivers, and, having no natural outlet, are drained by the Desague of Huehuetoca, an artificial canal cut through the rock, 12 m. in length, 150 feet deep, and 300 feet wide; having its embouchure in the river Panuco, which flows into the gulf of Mexico. This great work, completed in 1789, at an expense of £1,292,000, was undertaken to obviate the frequent inundations, some of which did great damage to the capital. The water of lake Tezcuco is salt, that of the rest is fresh; but from those to the S. sulphuretted hydrogen gas is copiously disengaged, the smell of which is often perceptible at Mexico.

Besides the volcanoes already noticed, those of Tuxtla, Jorullo, and Colima, in the table-land, are at present in a state of activity, and there are several others now extinct. Jorullo, which stands W. of the city of Mexico, first broke out in 1759; when a tract of ground, from 3 to 4 m. square, swelled up like an inflated bladder, emitting flames and fragments of rock through a thousand apertures. These active volcanoes seem to be connected with others parallel to them, and obviously of similar origin. Earthquakes are frequent in Mexico, but they seldom do much mischief.

The geological formation of the Mexican Cordilleras differs considerably from that of the great mountains of Europe and Asia, in which granite is overlaid by gneiss, mica, and clay-slate; for here we seldom meet with granite, as it is covered with porphyry, greenstone, amygdaloid, basalt, obsidian, and other rocks of igneous origin. Granite, however, appears on the surface in the chain bordering the Pacific, and the port of Acapulco is a natural excavation in that species of rock. The great central plateau of Anahuac, between lat. 14° and 20° N., is a mass of porphyry, characterized by the constant presence of hornblende, and the entire absence of quartz; and in it are contained large and valuable deposites of gold and silver. These ores, however, are found in various rocks; in the mines of Comanja rich veins of silver occur in syenite; in those of Guanaxuato, which are the richest in Mexico, the metal lies in a primitive clay-slate passing into talc-slate; and those of Real del Cardonal, Xacala, and Lomo del Toro are situated in a bed of transition limestone. Humboldt says that there were, at the time of his visit, 3000 mines of gold and silver in Mexico; but the ignorance and misrule which prevail in the country have greatly diminished their importance as a source of wealth.

Rivers.—Mexico suffers serious disadvantages from the want of water, and its rivers, as compared with those of the extent of territory, are few and unimportant. The Rio Grande del Norte, indeed, has a course of more than 1200 m., and the Colorado runs about 700 m. into the gulf of Mexico. The Rio Grande de Santiago, called by the natives Tolototlan, rises in the centre of Mexico, not far from the capital, and, after traversing the lake Chapala, falls into the Pacific at San Blas. The Balsas, or Zacatula, and the Yopez, are the only other rivers on the W. side of the plateau, and on the E. side are the Tula and Tampico and the Tabasco, flowing into the gulf of Mexico; but they have bars at their mouths, which prevent the entrance of large vessels. The other rivers are short, and might more properly be called torrents. The lakes are numerous and extensive; and the principal, besides those in the plateau of Tenochtitlan, already mentioned, are Chapala, in Xalisco, which, according to Humboldt, covers an area of 1300 sq. m.; Pascuara in Michoacan, Mextitlan, Cayman, and Parras, the last two being in the tract called the Bolson de Mapimi.

Climate.—The temperature and climate of Mexico is, of course, extremely various; owing, not only to its great extent from N. to S., but also to the rapidity of the slope both on the E. and W. side. The climates, especially on the E. side, are more distinctly marked by the vegetation. On the ascent from Vera Cruz, says Humboldt, climates succeed each other in la ers; and the traveller passes in review, in

the course of two days, the whole scale of vegetation, from the parasitic plants of the tropics to the pines of the arctic regions. (*Essai Pol. sur la Nouv. Espagne*, i., 270–289.) Mexico is divided, as respects climate, into the *tierras calientes*, or hot regions, the *tierras templadas*, or temperate regions, and the *tierras frias*, or cold regions. The first, or the *tierras calientes*, include the low grounds, or those under 3000 feet of elevation, on its E. and W. coasts, comprising the greater part of the states of Taumalihas, Vera Cruz, Tabasco, and the peninsula of Yucatan, on the former. The *tierras calientes*, on the W. coast, are less extensive, the eastern arm of the Cordillera approaching nearer to the sea. The mean temperature of this region, or, at least, of that portion of it between the tropics, may be estimated at about 77° Fahr., being from 14° to 16° above the mean temperature of Naples. It is especially suited for the growth and cultivation of sugar, indigo, cotton, and bananas, which flourish in the utmost luxuriance.

This region labours under the serious disadvantage of being nearly inaccessible by sea for half the year, and of being extremely unhealthy during the other half. The winter, on the E. coast, extends from about October to the vernal equinox; and during this season, in the gulf of Mexico, N. or N.W. winds (*los nortes*) are extremely prevalent, blowing with more or less violence. Frequently, especially in the month of March, the N. winds approach to the strength of a hurricane, and continue to blow with the utmost violence, and without intermission, for three, and, sometimes, even for 10 or 12 days together. During the whole of this season the navigation of the gulf is exceedingly dangerous; but on shore the heat is moderate, and the coast free from fever and tolerably healthy. Unluckily, however, it so happens, that during the other half of the year, or from the vernal equinox to October, when the N. winds are comparatively rare, and the ports are easily accessible, the heat is oppressive, a great quantity of rain falls, and the coast becomes the seat of pestilential fevers. A European arriving for the first time at Vera Cruz, or any other part of the coast between the tropics, in August, September, or October, has but little chance of escaping the *vomito prieto*, or yellow fever; and individuals who have merely landed at Vera Cruz, and posted on immediately for Xalapa, have, notwithstanding, caught the infection. The scourge, however, does not extend its ravages beyond the low grounds on the sea-coast; and at the height of 3000 or 3500 feet above the sea it is wholly unknown. The ports of Acapulco and the low grounds along the W. coast are also extremely hot and unhealthy; and, owing to the prevalence of strong gales, approaching to hurricanes, during the months of July, August, September, and down to October, the navigation is then extremely dangerous.

The *tierras templadas*, or temperate regions which are of comparatively limited extent, occupy the slope of the mountain chains, or barriers, which bound, on either side, the central table-land. It extends from about 2500 to about 5000 feet of elevation. The mean heat of the year is from 68° to 70° Fahr., and the extremes of heat and cold are here equally unknown. The Mexican oak, and most of the fruits and cerealia of Europe, flourish in this genial climate. The cities of Xalapa, on the E., and of Chilpanzingo, on the S.W. slope, are in this region, and are famous for their salubrity and for the abundance of their fruit trees. The frequency of fogs, and the consequent humidity of the atmosphere, is the greatest drawback on the climate of the *tierras templadas*; but this, how injurious soever in some respects, produces great beauty and strength of vegetation.

The *tierras frias*, or cold regions, include all the vast plains elevated 5000 feet and upward above the level of the sea. In the city of Mexico, at an elevation of 7400 feet, the thermometer has sometimes fallen below the freezing point. This, however, is a rare occurrence, and the winters are there usually as mild as in Naples. In the coldest season the mean heat of the day varies from 55° to 70° Fahr.; while in summer the thermometer seldom rises in the shade above 75°. The mean temperature of the city is about 64°, and that of the table-land generally may be taken at about 62°, being nearly equal to that of Rome. But, whenever the table-land rises to more than 8000 feet above the sea, it has, though between the tropics, a rude and disagreeable climate. Under the parallel of Mexico the limit of perpetual snow varies from about 12 to near 15,000 feet. Vegetation in the central plateau is not, owing to the rarity of the air, so vigorous as on the *tierras calientes*, or along the coasts, and the plants of Europe do not succeed so well as in their native soil. In the tropical and central region of Mexico, and as far N. as lat. 28°, there are only two seasons; that of the rains, lasting from July to the middle of September, and the dry season, continuing from October to the end of May. From the 24th to the 30th parallel the rain falls less frequently; but this deficiency is compensated by the abundance of snow during January and February.

The climate of the table-land is, on the whole, favourable to human life. But, though intermittent fevers be of rare occurrence, the natives are occasionally visited by a peculiar epidemic, called by them the *matlazahuatl*; but it owes its origin more to the habits of the people than any other cause. Indeed, famine, and its concomitant privations, have thinned the population more than epidemic complaints. The indolence of the natives prevents all exertions to raise more food than requisite for the wants of a single ordinary season; and no one ever thinks, when there is a surplus, of laying up a stock against future contingencies. Hence, when droughts, and severe frosts occur, they are compelled to seek their subsistence in the forests, where roots and wild berries constitute their sole diet; and multitudes are often carried off by hunger and unwholesome food.

Animals.—The zoology of Mexico is but imperfectly known. The domestic animals introduced by the Spaniards have so much increased, that vast herds range wild through these thinly inhabited regions. The wool of the sheep is of inferior quality; but this is attributable more to neglect and mismanagement than to nature: mules are much used in the mining districts. Buffaloes abound in the prairies bordering on the Arkansaw and Red river, and during winter they migrate westward, in quest of pasturage, to the milder climate of the plains along the lower part of the Rio Grande del Norte. Carnivorous animals are not numerous. Bees abound in the low country of Yucatan.

Agriculture.—Mexico, not only from its extent through 21 degrees of latitude, but also from the varying elevation of its surface, and consequent variety of climate, produces most of the plants peculiar to the tropics, as well as those belonging to the temperate regions of S. and middle Europe. "Indeed," says Humboldt, "there is scarcely a plant in the rest of the world which is not susceptible of cultivation in one or other part of Mexico; nor would it be an easy matter for the botanist to obtain even a tolerable acquaintance with the multitudes of plants scattered over the mountains, or crowded together in the vast forests at the foot of the Cordilleras." (*Essai*, tom. ii., p. 370.) The soil also is, in most parts, extraordinarily fertile; and wherever water can be procured for irrigation, the most abundant crops may be raised with very little labour. This, however, is very far, indeed, from being an unmixed advantage; and it is, in fact, more than doubtful whether a very fertile soil and a genial climate, that makes warm clothing and comfortable lodgings of comparatively little importance, be consistent either with active industry and exertion or with a high state of civilization. In most parts of Europe continuous industry is indispensable to existence; but it is otherwise in Mexico and many other countries, and it is found that industry is uniformly proportioned to the strength of the motives by which it is occasioned, and that, wherever the ordinary necessaries and comforts of life may be procured with little labour, the mass of the people are invariably indolent. To suppose, indeed, that they should be otherwise, is to suppose what is contradictory and absurd. This effect of the peculiar nature of the soil and climate was less sensible in Mexico under the Spanish government, because it was then daily receiving adventurers from Europe, imbued with European notions, and anxious to accumulate a fortune. But now that the influx of such parties has nearly ceased, and that there are no such extrinsic and adventitious motives to prompt to activity and enterprise, everything appears to be falling into a state of apathy and languor; and indolence, with its necessary accompaniments of poverty, ignorance, and pride, bid fair to be, for a lengthened period, the distinguishing characteristics of the Mexicans.

We have stated, under the head of climate, how the more useful plants are distributed through the zones into which the country is divided. The banana, which flourishes up to the point where the mean temperature is 75° Fahrenheit, bears the same relations to the Mexicans, in the lower provinces, that the various cerealia bear to the inhabitants of Europe and W. Asia, and the different kinds of rice to the Bengalese and Chinese. About 450,000 sq. m. in the *tierras calientes* are said by Humboldt to be adapted for its cultivation. It is propagated by cuttings; and there is probably no other plant which produces on the same extent of land, and with so little labour, so great a quantity of food. Humboldt affirms that ½ hectare (about an acre) of land, planted with bananas, will furnish food for more than 50 individuals; whereas the same extent of land, if sown with wheat, in Europe, would not support more than two individuals! And all the labour required to raise this enormous produce is to cut off the stems when the fruit is ripe, and to give the earth a slight digging about the roots of the plant once or twice a year. Hence, says Humboldt, nothing strikes an European recently arrived in Mexico with more astonishment than the smallness of the patches of cultivated ground round cabins that swarm with children. It cannot be said of such a country ;—

Patrr (pod Colendi haud facileam olm viam volait.

But the case with which subsistence may be procured, and the fewness of their wants, have made the natives in the last degree slothful. Indeed, Humboldt tells us that it has been gravely proposed, in order to stimulate their industry, and rouse their torpid faculties, to grub up and destroy the banana plantations! (*Essai, &c.*, ii., 396.) Such a project is, of course, impracticable and absurd; but the nature of the proposed remedy serves, at all events, to show the violence of the disease.

The same parts of the country which produce the banana produce also the cassava, or manioc, the farina of which yields a very nourishing bread: it requires more care than the banana, somewhat resembles the potato, and arrives at maturity about eight months after the slips have been planted. The culture of maize is scarcely less important in the *tierras calientes* than that of the plants before named; it is not confined, however, to the low lands, but ascends as high even as the plain of Toluca (9100 ft. above the sea), the lowest average temperature favourable to its growth being about 48° Fahr. The plant, under favourable circumstances, rises to the height of 7 or 8 ft., and the returns, in common years, are most abundant; but they are more uncertain than those of any other kind of grain. Maize is the principal food of the people, as well as of most domestic animals; and a deficient harvest, whether from want of rain, or excess of cold, produces a general famine, and compels great numbers of the rural population to seek the deserts in search of wild plants.

There can be no doubt, however, that if agriculture were pursued with any spirit, and the system of irrigation generally introduced on corn lands, or even if there were the slightest degree of providence in the natives, those dearths would not occur that on several occasions have been so fatal, especially in the mining districts. The European *cerealia*, such as wheat, barley, &c., succeed best in the temperate regions, where the mean heat does not exceed 66° Fahrenheit: in fact, in the equinoxial regions of Mexico, these grains are not found under the level of 2500 ft. above the sea. The Mexican wheat is of excellent quality; equal, says Humboldt, to the best of the Andalusian: it is large, white, and nutritive. In well irrigated lands, and on good soils, the produce is said to average 24 for 1; but, since the revolution, this necessary branch of agriculture has been much neglected. Rye and barley resist cold better than wheat, and are cultivated in the highest regions; barley yielding abundant harvests, even where the thermometer indicates a heat during the day of only 57°. Oats are little cultivated. Among the other alimentary plants, most of which have been introduced by the Europeans, are the potato (confined chiefly to the table-land), the yam, common both to the high and low country, the capsicum, raised in immense quantities for its spice, which is universally used instead of salt for seasoning food, beans, and various other garden vegetables common to Europe and America. Most of the fruits of Europe are common and plentiful; the olive and vine, introduced since the revolution, generally succeed well; and nowhere are there finer pine-apples, pomegranates, guavas, alligator pears, &c. One of the most valuable plants in the country is the maguey (*Agave Americana*), which Humboldt not inaptly terms the vine of Mexico. The Maguey plantations are principally found in the states of La Puebla, Mexico, and Guanaxuato; but the plant is very hardy, and occurs in a wild state all over the country. Its growth is slow; but when arrived at maturity, its leaves are from 5 to 8 ft. in length, and the stem often attains the height of 20 or even 30 ft. Its period of flowering is very uncertain, but once in 10 years may be considered a fair average. At the flowering season, when the plant first begins to be useful, the exact time is watched when the stem of the flower is about to shoot up; the sap is then cut off, so as to form a hollow, for the reception of the sap, which is regularly drawn off; and a vigorous plant will yield 15 quartillos daily for four or five months successively. The sap, which has a slight sub-acid taste, ferments readily in three or four days, being in its viscous state, called *pulque*, a beverage which somewhat resembles cider, though with a disagreeable smell. Immense quantities of it are drunk by all classes, and many whites as well as Indians use no other liquor. A kind of brandy, called *mexical* (very like whiskey), is made from the distillation of pulque. The maguey is useful, also, in many other ways: its fibres are converted into thread ropes and paper, its prickles serve for pins and needles, and its juice is effective in healing green wounds. Large quantities of sugar are raised in the neighbourhood of the capital, and the crops are very abundant. the lands are cultivated by free labourers, and the farming seems pretty good, though the process of refining be very clumsily conducted. In the commencement of the present century there was a large export of sugar; but this has for some years wholly disappeared, and the present supply is barely sufficient for the home consumption. Vanilla is extensively raised in the *tierras calientes*, E. of the Cordilleras, particularly in the state of Oaxaca. The cultivation of coffee is on the increase, and the quality of that raised on the best soil near the coast is said to be equal to the best produced anywhere else. Tobacco is a government monopoly, and its growth is confined to a small district near Orizava and Cordova. Its quality is inferior to that of Cuba; and, as the consumption exceeds the growth, considerable quantities are imported from the Havannah.

On the whole, it might be fairly concluded, on general grounds, that agriculture in Mexico must have retrograded since the revolution. And such, in point of fact, has been the case, and to an extent that we should hardly have conceived possible. This is evident from the following statement by M. Chevalier*, who visited Mexico in 1835. "Agriculture," says he, "is neglected. No law, indeed, prevents the planting of the vine and olive tree ; not only, however, has no advantage been taken of this change, but the very lands which were cultivated in the time of the Spaniards are now lying fallow. In a circle of a few leagues round Mexico, I have seen large villages almost abandoned. In this delightful climate, the only manure which the land ever requires, is water ; this is rather scarce, yet many of the hydraulic constructions raised by the Spaniards at a great cost, are in ruins, and seem likely to remain so. The lands which, by means of this artificial irrigation, were the most fertile in the world, are gradually becoming completely sterile. Their ploughs, and other agricultural instruments, are of the rudest description. No one troubles himself to introduce European improvements, or even to import better tools from the United States. I made the passage from New-Orleans to Vera Cruz with General Arista, who had been exiled in consequence of some insurrection or other in which he was concerned. Wearied with the chances of revolutions, he had determined to devote himself to agriculture. He had scarcely, however, landed at Vera Cruz, when he was thrown into prison, under some vague pretext. He still continues under arrest, and his ploughs, harrows, and winnowing machines, remain under sequestration, suspected, probably, of abetting the general in some subversive designs."

Mining Industry.—The silver and gold mines of Mexico have always been deemed the main sources of its wealth ; and, unquestionably, its mineral riches far exceed those of any part of America, except, perhaps, Peru. Before the war of independence, there were, in the 37 mining districts of New-Spain, somewhat more than 3,000 mines, producing annually about 21,000,000 dollars in silver, and about 2,000,000 in gold. Towards the close of the struggle, many of the mines had been deserted, and their produce had declined a half, and does not yet materially exceed that amount. Several of the so-called companies, formed in Great Britain for working the Mexican mines, during the memorable, and, we may add, disgraceful æra of 1894-95, were mere swindling engines, and fell to pieces in a very short while. There were others, certainly, that had a more solid foundation ; but these were mostly gone into without due consideration, and without any practical knowledge of the country, of the practices that had been followed, and the difficulties to be overcome. We need not, therefore, be surprised at the enormous losses the companies sustained at the outset, and of their want of success in the first instance. But, had the Mexican government been able and willing to repress disorder, and to enforce the observance of contracts, it is probable that the produce of the mines would have been very different at this moment from what it really is. Unluckily, however, no government has yet been established in Mexico with power, even if it had the desire, sufficient to put down disturbances, or to enforce engagements. So long, indeed, as the companies were struggling to put their mines in order, they sustained comparatively little inconvenience from this circumstance ; but, as soon as they had succeeded in bringing them once more into a productive state, and were beginning to have some prospect of a return for their enormous outlays, they were annoyed by questions as to title, and by the setting up of claims on the mines, of which they had never heard before. Owing to these causes, as well as the general insecurity of property, the bad condition of the roads, and imperfect mining processes, the results have, on the whole, been very unfavourable, notwithstanding the reduction of the export duty on specie from 10 to 3 per cent.

The following statement of M. Chevalier, as to the insecurity of the miners, in 1835, discovers a state of things

* The promised work of this able and intelligent traveller on Mexico has not yet appeared. We have borrowed the following extracts from letters addressed by M. Chevalier to his friends in France, that were published at the time in the Paris journals. The information they contain is by far the most important that has appeared respecting Mexico since the publication of Humboldt's work.

disgraceful to the government ; and such indeed as could hardly have been credited upon any inferior authority. "How," asks he, "can the mines be worked with any feeling of security, when it requires a little army to escort the smallest portion of the precious metal to its place of destination ? Between the mine of Real del Monte and the village of Tezeyuco is a mountain pass, where a grand battle was fought between the miners and the banditti of the country. The former were defeated, overpowered by numbers ; but not without having sold their lives as dearly as possible. The mine is now guarded by artillery and grapeshot, and the Englishmen employed there are regularly drilled in the use of the musket." In such a state of things, the wonder is, not that the produce of the mines has declined, but that it continues to be so great as we find it to be. The mineral riches of the country are, however, inexhaustible ; and there wants only a government able and willing to afford security, to make the produce of the mines greater than ever. The subjoined statement, which cannot, however, be altogether depended on, exhibits an

Account of the number of dollars coined at the different mints of Mexico during each of the four years ending with 1837 ; the proportion of gold to silver being as 1 to 26, nearly :—

Mints	1834.	1835.	1836.	1837.
Mexico	993,300	547,115	764,196	539,422
Guanajuato	2,760,000	2,077,509	1,514,974	3,906,984
Zacatecas	5,557,651	6,187,969	5,456,379	5,354,993
Guadalaxara	715,200	822,250	569,734	479,604
Durango	1,271,500	1,359,965	1,963,728	968,006
San L. Potosi	958,000	1,935,568	1,889,142	1,110,771
Chihuahua	—	—	294,215	225,362
Total	12,040,560	14,005,114	12,466,405	11,878,808

The theory of mining is little understood by the Mexicans, the oldest modes of working being still generally practised, notwithstanding the improvements that have been introduced by the English : and the machinery for draining the mines and raising the ore is of the most primitive description. Indeed, many of the mines have been abandoned, owing to the imperfections of the machinery, which, under more favourable circumstances, might be again worked with profit. The ignorance of the miners is only equalled by their obstinate adherence to old, and elsewhere long exploded, practices. But this should not be matter of surprise, if the testimony of M. Chevalier respecting the education of engineers may be depended on. The school of mines (*Mineria*), the mere building of which cost 190,000l., is at present in the most pitiable condition, although the learned Andres del Rio is still one of the professors. It is unprovided with the means even of the most elementary instruction. It contains a vast chemical laboratory, but without the instruments requisite for the most simple experiments. The collection of minerals is in disorder, badly classed, and very incomplete ; the library and the mechanical cabinet are deplorable. The school seems also to have shared the fate of the public treasury—of having been pillaged three or four times over. The very building seems on the point of falling to pieces—an appropriate emblem of the Mexican Republic. But it cannot surely be supposed that the anarchy which has led to such deplorable results is to continue forever. If nothing may be hoped for from within, it is to be wished that foreign interference may rescue this fine country from the barbarism in which it is now involved.

The quantity of silver annually extracted from the mines of Mexico very much exceeds that furnished by all the mines of Europe ; but, on the other hand, the gold is not much more abundant than in Hungary and Transylvania ; the proportion which the gold of Mexico bears to the silver being as 1 to 26, nearly. Little native silver is found in any of the mines ; sulphuretted and black prismatic silver is both very common and exceedingly productive in the veins of Guanaxuato and Zacatecas, two of the richest mining districts : the muriate abounds in the mines of Catorce and San Pedro, near San Luis de Potosi ; and the martial pyrites of Pachuca yield three marks to the hundred weight. The Mexican ore, however, is poorer than that of Europe, 1600 oz. of ore yielding only about 4 oz. of silver. The gold is produced by washing the earth and sand in some few places ; but in the province of Oaxaca occur veins of native gold, usually mingled with the silver veins : the returns, however, seldom exceed 1½ oz. to the cwt. (*Potsnett's Notes on Mexico*, p. 296.) Slave labour is not tolerated in the mines ; but it would be difficult to find workmen so ignorant, brutalized, and wholly worthless, as the native miners. Indeed, the ill success of the English mining companies is owing, in part at least, to the want of honest and efficient labourers. The business of the mines is followed by the native tribes from generation to generation : they lead a migratory life ; removing, with their families, to districts where they expect the greatest profit from their labour : they are always paid by a share

MEXICO.

in the produce; regular wages, however high, being invariably rejected. The principal mines are in the states of Guanaxuato, Zacatecas, San Luis Potosi, Chihuahua, Durango, Guadalaxara, and Mexico. The richest mineral tract lies between the 21st and 25th parallels of N. lat. Many of the mines have been very imperfectly wrought; and by far the larger part of the richest veins is yet unexplored. It is worthy of remark, also, that the ores appear to increase in richness on proceeding N. The mines in the confines of Durango and Sonora are peculiarly rich, lie near the surface, and hold out, wherever they have been tried, a promise of riches superior to any that Mexico has yet produced.

Iron is found in great abundance in Guadalaxara, Mechoacan, and Zacatecas; but no mines of that metal were worked before 1825. Copper is raised in Mechoacan and Guanaxuato. Large quantities of copper money have been coined in the mint in the city of Mexico, the total value, during the seven years ending with 1837, having amounted to 4,712,000 dollars, or 942,400l. Tin is obtained partly from mines, but principally from washings in the ravines. The lead mines, though rich, are quite neglected. Zinc, antimony, and arsenic, have been found; but neither cobalt nor manganese. A quicksilver mine is wrought in the state of Querétaro. Carbonate of soda, used for smelting the silver ore, is found in great abundance, crystallized on the surface of several lakes.

Manufactures.—The selfish policy of Old Spain, by which she endeavoured to keep her colonies as much as possible dependent on her own markets, or on supplies furnished by her, led to the enactment of laws prohibiting the rearing of silk-worms, and the cultivation of flax, and of the vine and olive. Coarse woollen and cotton fabrics, worth about 1,500,000l. were formerly made; but these have greatly diminished since the revolution. The system on which the cloth and other factories are conducted, is disgraceful to persons having the smallest pretensions to civilisation, and is wholly subversive of all improvement. Each factory is, in fact, a prison, in which the work-people are treated with the greatest rigour, and from which there is no escape; the proprietor, instead of paying his workmen in money, supplies them with spirits, tobacco, and food, at prices fixed by himself. An intelligent German, who resided 40 years in Mexico, states, that the high walls, strong double doors, barred windows, and severe corporal punishments, common to these factories, make them as bad as the worst conducted jail in Europe. Criminals and insolvent debtors are condemned to work in the factories as a punishment. This state of things existed before the revolution, and we regret to say that it has not been at all improved by the free intercourse which the Mexicans have now for several years enjoyed with the manufacturers and capitalists of Europe and the United States. "One might," says M. Chevalier, "have supposed that when the ports were thrown open to the commerce of Europe, manufactories would soon have been established in a country where manual labour is cheap, where the workmen are submissive, and skilful at imitation, where the soil produces the raw cotton, where the Spaniards had multiplied their flocks of sheep to a great extent, and where the rearing of the silkworm might be carried on with astonishing facility. The native Mexicans are, however, destitute of all spirit of enterprise, and strangers cannot attempt any permanent establishment in a country from which, during every session of congress, they are periodically threatened with expulsion. A more than ordinary display of industry would excite the jealousy of the natives; for nothing exasperates a Mexican more than to see Europeans and North-Americans growing rich before his face. A flourishing factory, established by a foreigner, would be very likely to be pillaged during the first popular tumult. Instances of the kind have already occurred. The only European manufactory existing at Mexico, is one founded by M. Duport, a French merchant, for making *mantas*, a coarse cotton stuff much worn in the country. The looms were made at Paterson, near New-York. When the Mexicans had achieved their independence, and were organizing their government, they created a fund for the encouragement of national industry (*banco de avio*), and endowed it with an additional duty of 2½ per cent. on foreign importations. In this way a few hundred thousand piastres were soon procured, which were expended in the vain attempt to establish manufactories. At present, the receipts for this fund are thrown into the abyss of the national deficit, which every year increases in depth, and where they are lost like a drop of water in the sea." Cigars, hats, glass, and earthenware, are produced on a large scale; but the factories are, for the most part, extremely ill-conducted. Mexican leather is very indifferent; paper is of bad quality, and exorbitantly dear: the making of cutlery and hardware is scarcely attempted, and what is done, is badly executed; the use of cast-iron and tin for culinary utensils, is almost

unknown, and a very few years ago there was only one manufacturer of watches and optical instruments in the whole of Mexico. "The Spaniards," says Chevalier, "are bad mechanicians, and no efforts of foreigners have been able to prevail on the Mexicans to deviate from the routine of their forefathers. All their tools are wretched; the common wheelbarrow even is unknown. Some merchants had imported two models, to be used in moving the bales of goods at the custom-house, but the workmen refused to make use of them."

Commerce.—An individual, looking at a map of the world, would be apt to conclude that Mexico is one of the most favourably situated countries for commerce; and, in some respects, this is true. But her trade labours, notwithstanding, under some serious disadvantages. Though washed by the Atlantic and Pacific oceans, neither of her coasts is accessible for several months of the year. On the E. coast, or that bordering the gulf of Mexico, there is not a single good harbour; and during the season when the coasts are accessible they are extremely unhealthy. Owing, also, to the rapid ascent from the shores to the interior, the construction of roads, and the transport of commodities to and from the inner provinces, is alike difficult and expensive. No doubt, however, an efficient government and an industrious people would overcome speedily, in a great measure, overcome these obstacles to an extensive intercourse with foreigners. But Mexico has neither the one nor the other; and, at present, her trade is confined within the narrowest limits. Down to 1778, when the Spanish government relaxed the old prohibitive system, the foreign goods legally imported into Mexico comprised only a few Chinese and European manufactures; the former brought annually in *one galleon* of about 1600 tons, and the latter sent *once in three years* exclusively in ships chartered by government from Seville or Cadiz! On the opening of the trade in 1779, private capitalists engaged in it, and after that period, at an average of 12 years before and after, the returns for exports alone rose from 11,000,000 to 19,000,000 of dollars, the difference being chiefly in the quantity of specie. How much greater would the increase have been, if the trade had not been fettered with vexatious duties, 1. on articles of Spanish produce in the markets of Seville and Cadiz; 2. on shipping for Mexico; 3. at Vera Cruz; and, 4. with an *alcavala*, or transfer duty, at every step, from the merchant to the consumer? On the breaking out of the civil war, the ports of Tampico, Mazatlan and San Blas were opened by the new government; and soon afterward foreign vessels were admitted into all the ports on the same terms as Spaniards. The Spanish capitalists retired to Cuba or Spain; and their places were supplied by British and American merchants, who established themselves in the interior, and supplied the inhabitants in return for dollars with manufactured goods, the superior quality and cheapness of which has, no doubt, had some influence in depressing native manufactures. The jealousy of the natives, however, and the absurd threats of the government against foreign artificers and traders, has tended to prevent their settling in the country, and engaging in any considerable undertaking, other than the mines; and the depressed state of the latter, which have always furnished the principal articles of export, has tended still farther to depress and paralyse commerce. The roads, too, instead of being improved, have been suffered to fall into a state of almost irreparable decay. In this respect, the evidence of M. Chevalier is decisive. "The splendid road which, during the domination of the Spaniards, was constructed across deserts and precipices, by the merchants of Vera Cruz, to the summit of the upper country, is a melancholy instance of the carelessness with which the public interests of the country are directed. During the war of independence, this road was cut up in various points; and, down to this day, the enfranchised Mexicans have not replaced a single stone, nor filled up a single trench, nor even cut down one of the trees, which, in the absence of any considerable traffic, and under the influence of a tropical sun, are rapidly growing up to a magnificent size in the very middle of the road. In the upper country nothing would be more easy than to open noble means of communication. The soil is naturally level; and basaltic rocks, particularly adapted for the construction of roads, are found in great abundance. But even where there are roads, the Mexicans make little use of them. They carry to a yet more extravagant length the inconceivable predilection of the Spanish race in favour of transporting their goods on the backs of animals. You expect to meet with carts and wagons: no such thing; every thing is conveyed on the backs of mules or Indians. Troops of little consumptive donkeys bring into the city in parcels, not much bigger than a man's two fists, the charcoal required for the culinary operations of the inhabitants. The price of every bulky article is thus increased to an enormous degree. The interior districts are so inaccessible as if they were cut off by an enemy's army, and famine frequently ensues."

The following table furnishes an official account of the vessels entering the ports of Vera Cruz, Tampico, San Blas, and Mazatlan during the year 1836, with the invoice value of their cargoes, &c.

Countries.	Inward.			Outward.		
	Ships.	Tons.	Value.	Ships.	Tons.	Value.
			£			£
I. *Vera Cruz:*						
British	6	1,328	167,750	7	1,321	Port blockaded by the French, and value not stated.
Mexican	23	1,500		19	1,306	
American	9	1,450	Not stated	11	1,973	
French	5	1,155		8	2,113	
Others	2	380		2	280	
Total at Vera Cruz	44	5,879	167,750	47	7,217	
II. *Tampico:*						
British	25		235,190	27		1,159,239
Mexican	16	1,300	16,500	16	1,368	800
American	20	2,289	183,400	20	2,316	207,353
Others	9	982	104,900	7	685	7,000
III. *San Blas:*						
British	6	970	110,800	6	970	168,680
American	3	782	70,800	3	783	17,400
Sardinian	2	492	18,000	2	492	2,600
Others	6	613	31,000	4	695	12,380
IV. *Mazatlan:*						
British	6	1,453	57,000	4	1,196	65,200
American	5	1,110	80,800	2	518	42,400
Others	5	2,907	78,600	4	715	
	147	18,018	1,142,870	142	16,876	1,672,661

The above statement, though not complete, shows the comparative trade of different countries with Mexico, and proves that about half her imports come from Great Britain, which also takes off about five sixths of her exports, bullion, the chief article, amounting to about £1,500,000. The nature and amount of the direct trade with this country will be seen by the following statement of the different articles of home and colonial produce sent to Mexico in 1836:—

Articles	Quantities.	Declared Value.
		£
I. *British product and manufactures:*—		
Cotton fabrics	8,438,564 yds.	208,560
hosiery		5,114
yarn and twist	311,800 lbs.	15,207
Linen fabrics	1,395,024 yds.	78,786
Woollen do.	9,697 pieces	37,380
Do.	38,160 yds.	
Silk goods		5,529
Hardware and cutlery	1,120 cwts.	7,505
Iron wares and iron	207 tons.	1,200
Machinery		8,920
Tin and pewter wares		8,064
Earthenware and glass		4,176
Other articles.		7,947
II. *Colonial and foreign produce:*—		
Quicksilver	309,042 lbs.	
Cinnamon	34,941 —	
Raw silk	3,000	280,523
Linen fabrics	2,204 pieces	
Wine	2,150 gallons	
Total of British exports to Mexico		500,620

These returns show a great apparent increase of trade since 1831, when the exports from Great Britain to Mexico amounted only to £109,759; but, instead of going direct, as at present, the imports then were mostly indirect, through Jamaica and other places. The foreign trade is in truth quite insignificant, regard being had to the extent and resources of the country and its population.

The Mexican congress has fixed a tariff regulating the duties on the chief articles of import, and all articles not specified pay an ad valorem duty of 40 per cent.; quicksilver, wooden frames for houses, printed books, maps, and music, philosophical and musical instruments, artificers' tools, agricultural and mining instruments, seeds, and plants, are admitted duty free. All articles which are the growth and produce of Mexico may be exported duty free, except gold, in coin or wrought, which pays 2 per cent. ad valorem, and silver, the duty on which is 3½ per cent. ad valorem. Gold and silver ore, ingots, or dust, are prohibited under penalty of seizure.

Government.—On the resignation of Iturbide, the Mexicans determined on establishing a federal government. The present constitution, dated October 4, 1824, is modelled on that of the United States; the republic was then divided into 19 states, each of which is permitted to manage its own local affairs, while the whole were cemented together in one body politic by fundamental and constituent laws.

The powers of the supreme government are divided into three branches—legislative, executive, and judiciary.

The legislative power was vested in a congress consisting of a house of representatives, a senate, and a president. Representatives, elected by each state at the rate of one member for 80,000 inhabitants, hold their places for two years. The qualifications requisite are 25 years of age, and eight years' residence in the state. The senate consists of two members for each state, of 30 years of age each, who are elected by a plurality of votes in the state congress. The members of both houses receive salaries of 3000 dollars a year. The president and vice-president are elected by the congress of the states, hold office for four years, and cannot be re-elected for four years after. Congress sits annually from January 1 to April 15. A council of government, consisting of the vice-president and half the senate, sits during the recesses of congress. The city of Mexico is the seat of government. The legislatures of the 19 states are similar to that of the republic in general. But the federal has lately been consolidated into a central government with a single legislative body for the entire republic, the states being formed into departments with subordinate councils.

These arrangements appear, however, to be disliked by a large proportion of the people; at all events Mexico has continued, since the establishment of the federal government, to be little better than a theatre for insurrections. The testimony of M. Chevalier is conclusive with respect to the condition of the country in 1835, and there has been no material improvement in the interval. " I have only been two months in Mexico, and already I have witnessed five attempts at revolution. Insurrections have become quite ordinary occurrences here, and their settled forms been gradually established, from which it is not considered fair to deviate. These seem almost as positively fixed as the laws of backgammon or the recipes of domestic cookery. The first act of a revolution is called *pronunciamiento*. An officer of any rank, from a general down to a lieutenant, *pronounces* himself against the established order, or against an institution which displeases him, or against anything else. He gets together a detachment, a company, or a regiment, as the case may be, and these generally, without more ado, place themselves at his disposal. The second act is called the *grito*, or outcry, when two or three articles are drawn up, to state the motives or objects of the insurrection. If the matter is of some importance, the outcry is called a *plan*. At the third act, the insurgents and the partisans of government are opposed to one another, and mutually examine each other's forces. As the fourth act they come to blows; but, according to the improved system introduced, the fighting is carried on in a very distant, moderate, and respectful manner. However, one party is declared victor, and the beaten party *disprononces*. The conquerors march to Mexico, and their triumphal entry into the capital constitutes the fifth act of the play: the vanquished meanwhile embark at Vera Cruz or Tampico with all the honours of war."

The laws are alleged to be mild and just, but they are almost powerless; for nothing can well be conceived more appalling than the state of anarchy described by the very intelligent traveller just quoted.

"With tranquillity, unfortunately, everything else is also lost. There is no longer any security. It is a mere chance if the diligence from Mexico to Vera Cruz proceed the whole way without being stopped and robbed. It requires whole regiments to convey the *conducta* of piastres to Vera Cruz. Travellers who cannot afford to pay for an escort go armed from head to foot, and in little caravans. Here and there, rude crosses erected by the side of the road, and surrounded by heaps of stones, thrown by passers-by, in token of compassion, point out the spot where some wayfarer, and almost always a stranger, has perished by the hand of robbers." "The immediate environs of the most populous cities are infested by malefactors, and even in the interior of cities, not excepting the capital, there is no longer any security. There are numerous instances of people being robbed on a Sunday, and at the hour even when the greatest number of people are abroad, within a league of Mexico. An English chargé d'affaires was lassoed on the Alameda, the public walk, in the middle of the day. In the evening, after sunset, notwithstanding the numerous guardians of the night (*serenos*), notwithstanding the videttes of cavalry at every corner of the streets. notwithstanding the law prohibits the riding on horseback through the streets after eight o'clock, in order to prevent the use of the *lasso*, a man is not safe in Mexico, not even in his own house. If, in the evening at eight or nine o'clock, you visit a friend, before the porter consents to open the enormous gate lined with iron or bronze, there pass as many formalities as if it were a question of letting down the drawbridge of a fortress. Persons on whose words I think I can rely have assured me that as many as 900 dead bodies are yearly deposited in the *morgue* of Mexico."

Revenues.—The amount of the revenue at different dates has been as follows:

1700	.	.	$3,000,000	1827 .	$10,494,299
1763	.	.	5,705,876	1828 .	12,232,285
1802	.	.	20,900,000	1829 .	14,493,189
1825	.	.	10,690,606	1830 .	18,923,299
1826	.	.	13,289,682	1831 .	16,413,060

About half the receipts proceed from the customs' duties: the other sources are, the mint, monopolies of tobacco, salt, pulque, and gunpowder; lotteries, post-office, stamps, tolls, and privileges. The produce of the state lands, none of which have been yet put up to sale, is estimated to be capable of producing from three to four millions of dollars.

Army.—The army consisted, in 1839, of about 30,000 men, exclusive of an active militia of about 30,000. But the troops are without science or a proper feeling of honour, so that they are really worth very little. The military, however, is a favourite service, from the high pay and privileges of the soldier. There are five fortresses—San Juan de Ulloa, Campeche, Perote, Acapulco, and San Blas.

Religion.—The Roman Catholic is the only publicly recognized religion, but others are tolerated. The church establishment consists of the archbishop of Mexico and nine bishops, having an aggregate income of 539,000 dollars, with 3677 parochial clergy. There are also 10 cathedrals, having 168 canons and other dignitaries, and one collegiate church. The regular clergy comprise 1978 monks, chiefly Franciscan; and there are 156 convents. Ecclesiastical property is free from taxation, and they have the sole management of all money bequeathed for pious uses. The annual income of the ecclesiastics is valued at about 12,000,000 dollars. The Spanish monks and priests were expelled during the revolution; and their places are filled by creoles, whose morals are at the lowest ebb. Religion has little influence over the white population, and the hold of the church over the Indians, never complete, is now fast lessening; for they are all, more or less, inclined to idolatry. (*Edwards's History of Texas.*)

Education.—The necessity of education is recognised by the new constitution, which requires that the priests should teach all persons to read and write; but the regulation has little practical effect. Under the old government, botanical pursuits were much encouraged: chemistry and mineralogy were taught in the school of mines; but the progress of science, literature, and the arts have all been checked by the unsettled state of the country since the revolution.

"In fact," says M. Chevalier, "elementary instruction has remained what it was in the time of the Spaniards. The clergy had then the exclusive management of it, and having so still, show but little inclination to enable the poor to read the books published under the régime of a free press. There are even fewer schools than there were, in consequence of the diminution in the number of the clergy. Education of a superior kind is even worse provided for. Under the Spaniards, there existed at Mexico a school for the fine arts, richly endowed: I have been unable to discover its existence now. There is a building called a museum, where

354

I found nothing of interest except a collection of the portraits of the viceroys since the time of Cortez, and a few Aztèque manuscripts. Some years ago, the establishment of a polytechnic school was decreed, but the decree has yet to see the commencement of its execution. There is not even a military school, though the attention of the government is almost exclusively devoted to the army. There is nothing deserving the name of a school of law or medicine; and it may be well imagined that schools of industry or commerce are wholly unknown."

Population.—The amount of the population has been estimated at different periods, both before and after the revolution; but, owing to the jealousy of the old government, and the distracted state of the country, since the declaration of independence, very little credit can be attached to these estimates. The following are those by the best authorities:

1794	.	Humboldt	.	5,200,000
1803	.	Do.	.	6,500,000
1813	.	Poinsett	.	6,368,125
1825	.	Do.	.	6,500,000
1827	.	Ward	.	8,000,000
1835	.	Chevalier	.	7,000,000

The lower estimate of Chevalier may be explained by the emancipation of Texas and California, and by the continuance of the disturbances. The classes of the population are singularly varied and are characterised by distinctions more striking than those in any other country. Four distinct and rival classes may be enumerated: 1. the *Chapetones*, or pure Spaniards, never exceeding 80,000 in the palmy days of New Spain, but now hardly amounting to 34,000, and, politically considered, a degraded class; 2. the *Creoles*, or native whites of European descent, forming the wealthiest and most powerful part of the pop., estimated by Chevalier at 1,300,000; 3. the *Indians*, or native Mexicans, constituting the great mass of the rural labourers, and supposed to amount to 3,800,000; 4. the mixed castes, comprising *Mestizos*, *Mulattoes*, *Zambos*, *Quadroons*, and *Quinteroons*, somewhat exceeding 1,900,000.

The king of Spain formerly exercised a right of conferring the exclusive privileges enjoyed by the white population on individuals of any shade by a decree of the audiencia, *Que se tenga por blanco*—that he be deemed white. These distinctions of colour have been done away with, as far as political privileges are concerned, by the revolution, which admits persons of all colours to the equal enjoyment of civil rights; and hitherto, indeed, this has been by far its best if not its only good effect. The mulattos and zambos principally reside in the low country, the whites on the table land. The Indians are divided into numerous tribes, speaking upwards of twenty languages, totally distinct from each other, and of which fourteen grammars and dictionaries have been published. Their character remains much the same as it is alleged to have been at the time of the conquest. Indolence, blind submission to their superiors, and gross superstition, are as much their characteristics now as formerly. The form of their religion is changed, and that is nearly all: they take the same childish delight in the idle ceremonies and processions of the Catholic church as they once took in the fantastic mummeries of their aboriginal idolatry. They are scattered over the country as labourers, distributed in villages, or else live in the towns as artizans, workmen, or beggars. In a few instances they have accumulated property, and acquired respectability; but, in general, they are indolent, ignorant, and poverty-stricken. We believe them to be wholly incapable of any high degree of civilization; but they might, perhaps, be improved, were measures taken to enforce their education, and to make a fair distribution among them of the many thousands of acres which have been thrown out of cultivation by the consequences of the revolution. They are classed in two great divisions: 1. Mansos, comprising those who have a fixed residence, cultivate the land, adopt the habits of civilized society, and maintain an amicable intercourse with the other races; 2. Bravos, comprising those who live a wandering life, supporting themselves by hunting, and avoiding all intercourse with the other classes, with whom many of their tribes are in a state of perpetual warfare. The latter principally inhabit the N. states along the river Gila, and the extensive and little known mountain ranges on the upper part of the course of the Rio Grande del Norte and the N.W. of Texas, called the Bolxor de Maplmi, from the lake of Man. An independent tribe, called Mayas, inhabits the tract between Yucatan, Tabasco, and central America. It has made some progress in civilization, cultivating maize and cocoa, and wearing garments made of cloth prepared from cotton and the bark of the caoutchouc tree.

"Mexico," says Chevalier, "is a country so rich, that famine scarcely visits even the most indolent. In the *tierras calientes*, and even on the plateau, the natives are content to dwell with their families in a cabin of bamboo trellis work, so slight as scarcely to hide them from the stranger's gaze, and to sleep either on mere mats, or at best on beds made of leaves and brushwood. Their dress consists simply of a pair of drawers, or petticoat, and a *serape* (a dyed woollen garment), which serves for a cloak by day, and a counterpane by night. Each has his horse, a sorry beast, which feeds at large in the open country; and a whole family of Indians is amply supplied with food by bananas, chili, and maize, raised, almost without labour, in a small enclosure round the hut. Labour, indeed, occupies but a trifling portion of the Indian's time, which is chiefly spent in drinking *pulque*, sleep, or singing to his wretched mandolin hymns in honour of Notre Dame de Guadeloupe, and occasionally carrying votive chaplets to deck the altar of his village church. Thus, he passes his life in dreamy indifference, and utterly careless of the ever-reviving *émeutes* by which the peace of Mexico is disturbed. The assassinations and robberies which the almost impotent government allows to be committed with impunity on the public roads, and even in sight of the capital, are to him only matter for conversation, the theme of a tale or ditty. And why should he trouble himself about it? Having nothing in the world but the dress in which he stands, his lance, spurs, and guitar, he has no fear of thieves; nor will the poniard of the assassin touch him, if he himself, drunk with pulque or chingarito, do not use his own."

Antiquities.—Humboldt, Bullock, and other European travellers, have furnished excellent descriptions of numerous ancient monuments, which show that the native Mexicans, before the loss of their independence, had been in some respects a comparatively civilised and ingenious people. Among the most extraordinary are pyramids, somewhat similar in exterior form to those of Egypt, and in some instances even of larger dimensions. The base of the pyramid of Cholula is a square of 1423 ft. on each side, and its height is estimated at 177 ft. A far more elegant building, of similar shape, is situated in the N. part of the state of Vera Cruz; it is formed of large blocks of porphyry, highly polished, and arranged in six stages, diminishing in size according to the elevation, and having all its materials most nicely adjusted. The base is a square of 89 ft. on the sides; it is 65 ft. high: and the ascent to its top is by a flight of 57 stairs: the front is richly adorned with hieroglyphics and curious sculptures. The mountains of Texcuco are nearly covered with the remains of ancient buildings and cities. The ruins of Palenque, near the Rio Chacamas, a branch of the Usumasinta, extend upwards of 20 m. along the ridge of a mountain; and their architecture resembles more that of Europe than Mexico. The remains of an Aztec city, called by the Spaniards La Casa Grande, are to be seen about a league S. of the river Gila, in the state of Occidente. They are spread over a space of more than a square league. In the centre is a teocalli, laid down according to the cardinal points, its sides being 445 ft. by 276 ft. It has three stories and a terrace, but no stairs. Within are five apartments, each 27 ft. long, 11 broad, and 11 high. A wall with towers surrounds the main building. The traces of an artificial canal to the river are visible. The neighbouring plain is strewed with fragments of red, blue, and white earthenware, and pieces of obsidian, which prove that the Aztecs had passed through a country abounding with this volcanic substance before they dwelt on this spot, previously to their final settlement in Mexico. In the W. part of the state of Chihuahua are similar ruins of great extent, which are also considered to have been the site of one of the temporary stations of the Aztecs during their migration southwards. Besides sculptures, vases of elegant form have been found, similar to those of Etruria and Egypt. Roads formed of large hewn blocks of stone may be traced, not only in the neighbourhood of those ruined cities, but at great distances from them.

History.—The first settlers in Mexico are believed to have been the Toltecans, a tribe of Indians from the Rocky mountains, who fixed themselves, after several migrations, near the present city of Mexico, and flourished there for nearly four centuries. Drought, famine, and pestilence at length exterminated them, but not till they had imparted some degree of civilization to the barbarous Chichemecas, who were the next possessors of the soil, and were in their turn displaced by the Aztecans, who, in 1160, migrated southward from a country N. of the gulf of California, and first fixed themselves in the city of Zumpango, in the valley of Mexico, but afterwards in some islands in the lake Tezcuco. Here they maintained themselves by fishing and agriculture, till, in 1325, they founded their chief city on the island of Tenochtitlan, and called it *Mexico*, in honour of their martial deity Mexitli. This nation rapidly increased in power; and, if the remains of monuments and large cities were a just test of civilization, the Aztecans might claim to rank pretty high among the nations of antiquity. But they had invented no alphabet, and nothing better than a rude species of picture writing to record events, and were ignorant even of the useful metals. Their barbarism is sufficiently shown by their custom of sacrificing great numbers of human victims on

coronation fêtes. Montezuma I., the greatest of their sovereigns, extended the Aztec dominions on one side to the gulf of Mexico, and on the other to the Pacific ocean; but it must be stated at the same time, that many tribes within this tract yielded only a reluctant obedience, and some even retained their independence. Such, briefly, was the state of Mexico when Nunez de Balboa first landed on its shores. Its conquest was effected by Fernando Cortes, who sailed thither in 1519 with a small force, comprising, on the whole, only about 700 men. He was met at Vera Cruz by ambassadors from Montezuma the younger, sent to discover his intentions, and to command him to withdraw from the country. But Cortes having refused to return till he had communicated in person with the emperor, at once proceeded to the capital. Here having got possession of the person of Montezuma, Cortes endeavoured by his intervention to effect the subjugation of the empire. But the Mexicans having recovered from the surprise into which they were first thrown by the seizure of the emperor, resolved at all hazards to attempt the expulsion of the Spaniards. Montezuma was soon after killed in a conflict in the city; and Cortes was compelled to retreat to Tlascala. Here having reorganised his small force, secured the co-operation of a large body of Indians, and built brigantines to be employed in the navigation of the lake Tezcuco, he again pushed forward to the city; and having recommenced the siege, took it after an obstinate resistance of 75 days. The fate of the capital decided that of the empire. Province after province submitted; and the power of Spain was extended from Vera Cruz to the Pacific. Cortes, on his return to Spain, was received at first with high honours and liberal rewards; but his court favour soon declined: the emperor refused to appoint him captain-general of Mexico; and after some adventures, suited to his ardent and determined spirit, he died near Seville, in 1554, at the age of 63.

Under the Spanish arrangements Mexico was a subordinate kingdom, governed by a viceroy, with powers nearly equal to those of the sovereign, checked only by the *residencia*, or court of investigation, before which he was liable to be called to account for his administration, on his return home, and by the *audiencia*, or court of final appeal in Mexico. By these arrangements, also, the natives were to be considered as freemen and vassals of the crown; and the Spanish discoverers, settlers, and their posterity, were to have a preference in all civil and ecclesiastical appointments. The natives were thus, in fact, excluded from holding all offices of trust or profit. The great object of the Spanish government was to keep the country in the hands of the European or white population; and the means adopted to effect this object were, 1st, to discourage native manufactures, for the benefit of those belonging to the mother-country; 2dly, to make all the ecclesiastical establishments wholly dependent on the king, without any interference of the pope. The growth of flax, hemp, and saffron was prohibited under severe penalties; that of tobacco was made a government monopoly. The cultivation of the vine and olive was likewise prohibited; that of coffee, cocoa, and indigo tolerated only under certain restrictions, and in such quantities as might suffice for the demands of the mother-country. This system was maintained nearly three centuries; during which Mexico continued to be a blank in the history of nations, and known only by the issue of the precious metals. In 1808, however, the news of the abdication of Charles VI. of Spain gave a shock to the royal authority which it never recovered. The natives and coloured population embraced this opportunity of asserting their claim to the rights of freemen, which was opposed by the audiencia, who also seized on the viceroy, Hurrigarry, and sent him prisoner to Spain, where he was confined till the general amnesty. An open insurrection against the European authorities broke out in 1810, at the head of which were Hidalgo and Morales, two priests of New Spain; and under the auspices of the latter the first national congress assembled at Chilpanzingo in 1813. One of its earliest acts was a declaration of the independence of Mexico.

For several years the history of the revolution is only that of a sanguinary guerilla warfare, leading to no permanent results. At length, in 1821, Iturbide, who had previously been a royalist, declared suddenly in favour of the liberals, and published his celebrated manifesto of Iguala in favour of a constitutional monarchy. His cause was embraced with such enthusiasm by the whole population, that he succeeded not only in putting down the Spanish government, and forming a national congress, but also prevailed on that body to make him emperor of Mexico, under the title of Augustin I. His dissolution of the congress, however, by military force, raised a feeling against him, which, finding it impossible to repress, he abdicated the throne. He was not only allowed to withdraw from the country, but rewarded for his past services by an annual allowance of £5000, accompanied by an edict of outlawry in case of return. In spite, however, of this prohibition, he returned clandestinely, and was soon discovered, apprehended, and executed.

On the expulsion of Iturbide the congress was reassembled, a provisional government formed, and an executive appointed, consisting of Victoria, Bravo, and Negrete, all persons of proved patriotism. The government was modelled on that of the United States; but the hopes then formed of its stability have proved fallacious. Since this epoch repeated attempts at revolution have convulsed the country. During the whole of the struggle for independence, the population had been split into two parties; at first distinguished by the names of imperialists, who adhered to the mother country, and republicans, who asserted its independence: but these parties afterwards merged into those of Centralists and Federalists: the former advocating a single superintending government, and the latter that of the independent government of states, only federally connected. This struggle between the rival parties has now continued for about 17 years, and been a fertile cause of insurrection. Texas and California have already separated from the confederacy, and it is probable that their example will be followed by other states. In fact, there can hardly be said to be anything like regular government. The Centralists are lords of the ascendant to-day; but a successful émeute (as the Parisians term it) may dash all their prospects to-morrow. Meanwhile, all the bonds of society are loosened, property has become almost worthless from its utter insecurity, and life is not safe from assassination and violence. Whether the proposal of the S. American republics to unite their interests with those of Mexico, and form together one grand federation, will be accepted, and whether, if accepted, it would contribute to the improvement of Mexico, is matter for speculation only; but certainly that country, as it exists at present, affords one of the most melancholy examples that modern history has presented of an extensive, fertile, and well situated region being reduced, through an anarchy and misgovernment, to a state bordering on barbarism.

MEXICO, or MEJICO (MEX. Tenochtitlan), the cap. of the united states of Mexico, and anciently the chief city of the empire of Montezuma, 7498 ft. above the sea; lat. 19°25' 40'' N., long. 101° 25' 30'' W. Pop. 150,000? It stands nearly in the centre of an elevated plain, or plateau, surrounded by mountains, and having an area of about 1700 sq. m., one tenth of which is covered by four lakes, the largest of which (Tezcuco), nearest the city, has an area of 77 sq. m. The old city of Mexico, or that taken by Cortes, was built on a group of islands in the lake Tezcuco; but though the modern city occupy its site, it is, owing to the diminution of the waters of the lake, partly originating in natural and partly in artificial causes, situated about ½ m. W. from the lake. The ground on which it stands is, as might be anticipated from the statement now made, low and swampy; the largest buildings are erected on piles, and the roads leading to it are raised 6 or 8 feet above the surrounding flat. Though within the tropics, it is so elevated that its mean temperature is only 65° Fahr., coincident with that of May in England. It is said, by Humboldt, to be "undoubtedly one of the finest cities ever built by Europeans in either hemisphere; being inferior only to Petersburg, Berlin, London, and Philadelphia, as respects the regularity and breadth of its streets, as well as the extent of its public places." The architecture is generally of a very pure style, and many of the buildings are of noble construction, though usually of somewhat plain, exterior. Two sorts of hewn stone, porous amygdaloid and porphyry, are used in the better parts of the city. The balustrades and gates are of Biscay iron, ornamented with bronze; and the houses, which are three or four stories high, have flat-terraced roofs, like those in Italy and other S. countries. (Nouv. Espagne, ii., 51.) The streets are wide, well-paved and flagged, but not lighted or watched at night; so that robberies and assassinations are scarcely less common than in Spain. They run almost uniformly at right angles to each other, many of them being nearly 2 m. in length, perfectly level and straight, and offering, from every point, a view of the mountains that surround the valley. Nearly all the houses are hollow squares, with open courts surrounded by colonnades, and ornamented with plants, &c. The stairs to the interior from the outer gate, and the best apartments, which are showily painted in mosaic and arabesque, generally face the street. (Poinsetts' Notes on Mexico, p. 66.) Numbers of houses are covered with glazed porcelain, in a variety of elegant designs and patterns. The Plaza Mayor, or grand square, is one of the finest to be seen in any metropolis: its E. side is occupied by the cathedral and sagrario, or parish church, and its N. side by the palace, while on the other sides are handsome rows of shops and private dwellings. In its centre is a colossal statue of Charles IV., said to be the finest work of its kind in the new world. The effect of this square, however, is much impaired by the introduction of a paltry building, called the Parian, a large ungainly pile, in one angle, used as a market or bazaar, appropriated to the sale of miscellaneous articles, and the resort of the idlest portion of the

356

inhabitants. The palace, or government-house, a fine building, nearly square, with a front several hundred feet in extent, comprises four large courts, in which are the public offices, barracks, prison, and a large botanic garden; but almost every part of it is falling to decay: the massive tables, staircases, and chandeliers have disappeared, and all is now in the most appalling disorder. (Latrobe's Rambles in Mexico, p. 168.) In this building, also, is the mint of the state of Mexico. The coinage has greatly decreased; for whereas, before the revolution, it amounted to 16 or 18 millions of dollars, it was estimated, in 1822, at 1,920,000 dollars, and, in 1835, at only 547,145 dollars. The cathedral, on the N. side of the square, on the site of the great temple of the god Mexitli, is a heterogeneous edifice; one part of the front is low, and of bad Gothic architecture, while the other and more modern part is in the Italian style, and displays much symmetry and beauty: its two towers are ornamented with pilasters and statues. The interior is imposing, lofty, and magnificent; but the grandeur of the effect is much diminished by the ponderous erections in different parts, and a profusion of massive carved ornaments, pictures, and painted statues. The high-altar and its appendages are inclosed by a massive railing of mixed metal; so valuable, on account of the gold it contains, that a silver-smith of Mexico is alleged to have offered the bishop a new silver rail of equal weight in return for the old metal! (Bullock's Mexico, i., 143.) In the interior, also, are some curious remains, including several idols and a "stone of sacrifice," that is, a stone on which the human victim was placed when the priest tore out his heart! On the outer wall is fixed the Kalenda, a circular stone of basaltic porphyry, covered with hieroglyphic figures, by which the Aztecs, or native Mexicans, used to designate the months of the year, and which is supposed to have formed a kind of perpetual calendar. (Latrobe, p. 177–175; Ward, ii., 68.)

Few monuments of antiquity, however, remain; and we may echo the exclamation of Antonio de Gama, the first among Mexican antiquaries. "Quantos preciosos monumentos de la antiguedad, por falta de intelligencia, habran perecido en esta manera." "How many remains of antiquity have thus perished through ignorance of their value." The church services are celebrated with great magnificence; nor, even in Rome itself, is greater attention paid to the external minutiae of religious observances. Besides the cathedral, there are said to be from 50 to 60 other churches, most of which display, more or less, the barbarous mixture of style that characterised Spanish architecture during the 16th and 17th centuries; there are, also, numerous religious houses, two of which, viz., the Franciscan and Dominican convents, are extensive and wealthy establishments. Opposite to the latter of these is the palace of the Inquisition, now applied to other, and, it is hoped, more useful ends. Bullock describes it as "very elegant, exhibiting little or no appearance of the purposes for which it was intended." This tribunal was abolished by Iturbide, in 1822. The papal religion, however, still maintains its ascendancy; few buildings, whether public or private, are without their patron saint; and the traveller every where meets with statues, pictures, and processions. The Mineria, or college of engineers, was originally a large and handsome building; but, owing either to a want of care in making the foundations, or to the effect of earthquakes, the walls have settled in several parts, and the front is visibly out of the perpendicular. Lectures are given occasionally on the sciences connected with mining; and in one of the rooms is a tolerably good collection of minerals, though generally very inferior to those in European museums, and, as respects a country like Mexico, quite insignificant. In fact, not only the Mineria, but the academy of fine arts, the university, founded in 1595, and public library, are in a state of neglect, disgraceful alike to the government and the people; and we are afraid that the diffusion of elementary instruction since the revolution has not been such as to compensate for the decline of the institutions for the higher branches of instruction. We have seen, in the previous article, the statements of M. Chevalier as to education generally in 1833, than which nothing can be more deplorable. The Acordada, or public prison, is a large substantial structure, fitted to contain about 1200 prisoners; the barracks, also, formerly used as a hospital, are very extensive and well constructed. The theatre is a respectable building, of considerable size; but the establishment has for some years had so little success, that it is very seldom opened. The Plaza de Toros, for the exhibition of bull-fights, consists of a great circular inclosure, fitted up exactly like that of Madrid, and fitted to accommodate from 2000 to 3000 spectators. The great cigar manufactory, which belongs to the government, stands on the S.W. angle of the city, and comprises a very extensive establishment, which supplies the whole legitimate demand of the confederation for cigars. The Alameda, or public walk, at the W. end of the city, somewhat resembles a park, but has the stiff formal appearance of Dutch and French grounds. In the centre is a fountain, supplied with water from the great aqueduct leading from Santa Fé to the city. Another open space, called the Passeo, about 2 m. in length, planted with double rows of trees, is much frequented, on holydays, by persons in carriages and on horseback. In the city, also, are several Portales, or covered colonnades, lined with shops and stalls, and forming a favourite evening promenade long after the Alameda and Passeo have ceased to be frequented. The environs, also, present, on fine dry evenings, a very lively scene of bustle and gaiety; hundreds of canoes of various sizes, mostly with awnings, and crowded with native Indians or Mestizos, are seen passing in every direction along the lake and canals, each boat with its guitar-player at the stern, and some of the party either singing or dancing.

The manufactories are not generally remarkable, either for extent or fineness or of workmanship. Nothing is exposed in the shop-windows, and most of the articles are made in the places where they are offered for sale. Gold and silver lace, trimmings, epaulets, &c., are made in great perfection, and are sold at a much lower rate than in England. Silver-smiths' work is also done on a pretty extensive scale: the ornaments are finished by hand; the chasing is sometimes well executed, but in general the articles are heavy and clumsy. Jewellery employs a few hands; but all precious stones, except rubies, are scarce, and the work is much dearer than in Europe. Cabinet-work is extravagantly dear, and of very inferior quality, made with clumsy tools, and of bad wood; the saw is scarcely known, and the turning-lathe is of the most primitive construction. Coach-making is much better understood; the Mexican vehicles are firmly put together, of handsome shape, and well finished; and, in respect of of painting, gilding, or varnishing, they are but little inferior to those made in Europe, whence the handles and metal furniture are procured. There is a considerable manufacture of hard soap here and at Puebla; but it has greatly fallen off since the time of Humboldt, who states the quantity made in Mexico, in 1802, at 200,000 arrobas. Beaver and felt hats and cotton cloaks are made on a large scale, for the supply of all parts of the union, these being important articles in the internal trade of the country. Woollen clothes are three times as dear as in England, and are uniformly ill-made. Men, not women, are milliners; and it is not uncommon to see some 20 or 30 fellows, who should be porters or coalheavers, employed in decorating ladies' dresses, making flowers, and trimming caps and flounces. (Bullock, L., 262.) The bakehouses are large establishments; and the bread, which is excellent, is made exclusively by slaves, who also perform the work in the cloth factories. (See Manufactures, in art. MEXICO, UNITED STATES OF.) Shops for the sale of pulque (a kind of beer made from the aloe), and native and Spanish brandy, are very common, and have a gay appearance. The markets are well supplied with animal and vegetable productions, brought along the lake and canal of Chalco by crowds of canoes, usually navigated by women. Turkeys, fowls, pigeons, and many varieties of wild water-fowl, are very abundant and cheap; as are hares, rabbits, tortoises, frogs, and salamanders, all of which are esteemed good eating by the inhabitants. The meat market is well supplied with beef, mutton, and pork, but veal is prohibited. The meat, however, is not of the best quality, though, perhaps, this may be owing to its bad preparation by the butcher and cook. There is great variety of vegetables and fruits, and a most enormous consumption in proportion to the population. The vegetable market is larger than Covent garden, but yet unequal to the daily supply; and the ground is entirely covered with bananas, plantains, citrons, shaddocks, melons, pomegranates, dates, mangoes, tomatos, and other vegetable productions of tropical countries.

The greater part of these are cultivated on the chinampas, or floating gardens, of which there are two sorts; one moveable, the other fixed, and attached to the shore. On the marshy banks of the lakes of Xochimilco and Chalco, the water, in the time of the great floods, carries away pieces of earth covered with herbs, and bound together by roots. These, being driven about by the wind, sometimes unite into small islands, which, being taken possession of, are planted with flowers and roots. Artificial chinampas, or islands, are also frequently formed, of reeds, rushes, roots, brushwood, &c., well compacted together, and covered with black mould: these sometimes contain the cottage of the Indian who acts as guard. They are towed or pushed with long poles, and are thus removed from bank to bank. The fixed chinampas are parallelograms from 380 to 400 feet in length, and from 10 to 20 feet in width. They rise about 3 or 4 feet above the water, and afford, from their command of water, beans, small peas, pimento, potatoes, artichokes, cauliflowers, and a great variety of other vegetables.

The population of Mexico is of an extremely mixed character, comprising about 65,000 creoles, or descendants of Spaniards; 26,000 Mestizos, or half-casts between Europeans and Indians, but many of whom are scarcely distin-

guishable by colour from the former; about 35,000 copper-coloured natives; 10,000 mulattoes; and only about 6080 Europeans. There is, or, at all events, used to be, an extreme disparity of wealth in this city. Many of the nobles and successful speculators in mines were excessively rich; but the bulk of the population were at once indolent and indigent. The lower orders are filthy, despite labour of every kind, and are constantly seen lying in the church porches, leaning against the walls, and loitering about the markets. In many respects they bear a striking resemblance to the lazzaroni of Naples; but the latter are not stained with the crimes of robbery and assassination, for which the leperos of Mexico are disgracefully notorious. There is here also a general torpor of the faculties, and the dolce far niente seems to be the summum bonum of all classes. The dress of the higher orders of men closely resembles that of Europeans, the large cloak being as common here as in Spain. The costume of the ladies is universally black, with the veil and mantilla; but, on holydays and public occasions, their dresses are remarkable as well for gayness of colours as for expensiveness of material. Indeed, when in their carriages on the Passeo, they contrast somewhat strangely with the same persons, when seen at home in complete déshabille, without stockings, squatting on the floor, and either pursuing their favourite amusement of cigar-smoking, or eating cakes and capsicums out of the dirty earthenware basins of the country. (Latrobe, p. 150.) The ladies seldom go out during the day: but, after sunset, young and old come forth from their hiding-places, and the Alameda, Passeos, and Portales swarm with the damas and signorinas of the city, chatting and smoking with their gallants. Many gentlemen belonging to the higher classes are intelligent, and a few even fond of literature; but the city is so badly supplied with libraries, and other means of study, as to give little encouragement to such pursuits. There are three or four newspapers; but they are miserable productions, containing little besides the merest chit-chat, copiously interspersed with advertisements. The white creoles are distinguished by their mildness, courtesy, and hospitality: their besetting sin is gambling. Female virtue is on the same low level as in Old Spain; but the Mexican ladies are better educated, and would be agreeable but for the practice of smoking, which is bad enough in men, and intolerable in women. (Poinsett's Notes, p. 160.)

The original city of Mexico, or, as it was called, Tenochtitlan, built, as already stated, on a group of islands in the lake Tezcuco, was founded in 1325: it was connected with the main land by three principal causeways of masonry and earth, about 30 feet in breadth, and extending from 2 to 5 m. over the surrounding marshes. These dikes still exist, and their number has since been increased. They form, at present, paved causeways across the marshy grounds, which were formerly covered with water; and, being of considerable elevation, are useful in securing the city from inundations. The better to preserve the city from the chance of this calamity, the great drain alluded to in the previous article, was commenced in 1607, which has now reduced the lakes of Zimpango and San Christoval within comparatively narrow limits, and prevented their waters in the rainy season from flowing into the lake of Tezcuco, and threatening, as they sometimes did, to submerge the city.

Mexico, when first discovered by the Spaniards, was a rich and populous city; the seat of government, religion, and trade. According to Cortez, it was as large as Seville or Cordova, was well built, and well supplied with various products; but these are the statements of parties naturally disposed to magnify their own services, and should be received with considerable modification. It was taken by the Spaniards in 1521, after a protracted siege, in the course of which it was nearly destroyed. (See previous article.)

MEXICO, p. t., Oswego co., N. Y., 156 m. W.N.W. Albany, 381 W. Bounded N. by lake Ontario. Drained by Salmon creek. It contains three churches, a Presbyterian, Baptist, and Methodist, eight stores, two fulling-mills, one woollen factory, one furnace, five grist-mills, eight saw-mills, one oil mill, four tanneries; one academy, 164 students; 40 schools, 1090 scholars. Pop. 3739.

MEXICO (GULF OF), a large inland sea connected by the Florida channel with the N. Atlantic ocean, and by the channel of Yucatan with the Carribean sea, situated between lat. 18° and 31° N., and between long. 81° and 98° W. Length from E. to W., 1200 m. average breadth, 650 m.; area, about 800,000 sq. m. This sea, which is of an irregular circular shape, is, unlike the Carribean sea, almost clear of shoals and islands, none being found except on the coasts of Yucatan and Florida. Along the coasts of Mexico its soundings are very regular, with 100 fathoms at a distance of 30 m. from the shore. On the N. side, and especially opposite the mouths of the Mississippi, the depth is considerably diminished, and at its E. extremity the navigation is rendered intricate and dangerous by the Tortugas bank, Florida reef, and various other keys, shoals, and lslets, including the great Bahama bank, which surround the N. coast of Cuba. The E. trade winds prevail from April to October, this being usually the wet season : the *Nortes* begin in October, but are not violent till the middle of November, from which time till the end of February they blow with great fury, and are objects of much dread to navigators. These gales last for four or five, and occasionally even ten, days; but their extreme fierceness is usually spent in the first 48 hours. At these times the larger vessels, which cannot enter the shallow harbours of the Mexican coast, are obliged to slip their anchors, and keep as far as possible off shore. Examples are not wanting also of *nortes* happening between May and August, at which time they are particularly furious. Luckily, however, the hurricanes and tornados of the gulf are by no means so fierce and destructive as those in the Carribean sea.

The principal current of the gulf of Mexico, and the only one worth mention, is that which sets W.N.W. between cape St. Antonio and cape Catoche : this runs from 12 to 30 m. a day, and is perceptible even during the *nortes*, except close along the shores of Mexico. At the N.W. extremity of the gulf its course gradually changes, till, at the mouth of the Mississippi, it turns E., and afterward S.E., as it again rushes out into the Atlantic ocean, at the rate of 80 m. in the 24 hours. (This remarkable current, commonly known as the *gulf stream*, is described in the article ATLANTIC OCEAN, i., 290, of this work.) The tides of the gulf of Mexico are of no great importance, they nowhere exceed 3 or 4 feet; but their average rise is not more than 2 feet. The colour of the water is a deep indigo, darker or more intense than that of the ocean : phosphorescent lights shine on it with great brilliancy, and between the coasts of Yucatan and Louisiana great quantities of *fucus natans* occur in parallel lines from S.S.E. to N.N.W., and are carried out in large masses through the straits of Florida. (*Blunt's American Pilot; Purdy's Atlantic Ocean; Darby's Geogr. of United States; Humboldt's Pers. Narr.,* i., 50–69.)

MEZE, a town of France, dep. Hérault, cap. cant., on the lagoon of Thau, 5 m. N.W. Cette. Pop. (1836) 4940. It has a small port, capable of receiving 60 vessels of 40 tons each, and manufactures of brandy and liqueurs. Near it is the abbey of Vallemagne, an edifice of the 13th century, well worth the traveller's notice. (*Guide du Voyageur en France.*)

MEZIERES, a fortified town of France, dep. Ardennes, on the Meuse, which mostly surrounds the town, and is here crossed by two stone bridges, 80 m. N.W. Metz. Lat. 49° 45' 47" N., long. 4° 43' 31" E. Pop., in 1836, 3817. It is walled, and is farther defended by a strong citadel. It is ill built, and has few edifices worth notice, except the townhall, prefecture, the hospital, founded in 1412, and a parish church of considerable antiquity. Mezières, though the nominal capital of the department, has no court of primary jurisdiction, that tribunal being seated at Charleville (which see) : it is, however, the seat of boards of taxation, artillery, and forest inspection, a society of agriculture, &c.; and has tanneries, breweries, and some trade in leather, coarse woollens, and linens. The Chevalier Bayard, with a garrison of only a few thousand men, successfully defended Mezières, in 1520, against a powerful Austrian army; and, in 1815, the town held out for two months against the Prussians. (*Hugo,* art. *Ardennes, &c.*)

MIAKO, a large city, and the ecclesiastical cap. of the Japanese empire, in the island of Niphon, on the Yedogawa, 230 m. W. by S. Yedo; lat. 35° 94' N., and long. 135° 90' E. Pop. (according to the Dutch traders, on whom, however, little reliance can be placed,) 600,000, exclusive of the *Daïri,* or Mikado's court, supposed somewhat to exceed 50,000. It is situated in a spacious plain, enclosed on all sides by high mountains, and almost entirely formed into fine gardens, interspersed with temples, monasteries, and palaces. It is nearly 4 m. in length, and about 3 m. broad, with narrow but regular streets, lined by houses two stories high, built of wood, lime, and clay, most of them being very slightly and poorly constructed. The sacred Mikado, or supreme emperor, emphatically termed, "the Son of Heaven," has his residence on the N. side of the city, in a quarter comprising about a dozen streets, and separated from the rest of the buildings by walls and ditches; but, owing to the great diminution of the revenues furnished by the *sjogûn,* or viceroy (the substantial sovereign), the whole is reported to have a very shabby and dilapidated appearance, little in accordance with the rank of a being more divine than human! On the W. part of the town is another palace, built of stone, and strongly fortified : it belongs to the *sjogûn,* who resides in it when he comes to pay his respects to the emperor. This practice, however, has long been discontinued, and the building is now used for the accommodation of certain functionaries, sent thither from Yedo to watch the proceedings of the *Daïri.* The members of this court, who view themselves as a species of superior beings to the rest of the Japanese, are chiefly engaged in the study of literature and science, the *Daïri* being, in fact, the highest college in Japan for the cultivation of theology, and various other branches of learning. The almanacks, formerly imported from China, are now constructed, including the calculation of eclipses, in the Daïri college; and, at least, ⅔ of all the works, published in Japan, are produced by the *literati* of Miako, some of whom, however, are connected with other colleges and high schools, wholly independent of the Daïri.

This city is likewise the principal manufacturing depot of the empire, every kind of handicraft known in Japan being carried to the greatest perfection. Nearly every house has its attached shop well provided with every description of goods, and the Japanned wares, carved ornaments, &c. of Miako, are unequalled either in Japan or China. Miako is one of the places visited by the Dutch traders, when they, once in four years, pay their respects to the sjogûn at Yedo : they usually spend some days here, which are chiefly occupied in making purchases of Japanese manufactures. Various celebrated samples (of which there are many, though not described,) are freely exhibited to them; and in the gardens attached to one of these buildings, tents are pitched for the purpose, not only of giving a sumptuous entertainment to the *Capitan Holanda* (as they term the Dutch president of the mission), but also of gratifying the curiosity of the natives with a sight of a few strangers from a distant land. (For farther particulars, *see* JAPAN; see also, *Manners and Customs of the Japanese,* p. 140–157; *Siebold,* i. and ii.)

MIAMI, river, O., rises in Hardin co., and flowing southwesterly 100 m., enters Ohio river in the S.W. corner of the state. It is 200 yards wide at its mouth, and is navigated, to a limited extent, for 75 m. It affords extensive water-power, and is connected to Anglaise river, a southern branch of Maume river, by a portage of five miles. But its navigation will be superseded by the Miami canal, constructed along its borders.

MIAMI, county, O. Situated in the W. part of the state, and contains 410 sq. m. Organized in 1807. Watered by Miami river and its branches. It contained in 1840, 14,957, neat cattle, 20,659 sheep, 34,254 swine; and produced 136,150 bushels wheat, 16,998 of rye, 531,132 of Indian corn, 2169 of buckwheat, 3363 of barley, 163,962 of oats, 22,453 of potatoes, 153,238 of sugar. It had thee commission houses in foreign trade, 54 retail stores, three fulling-mills, one woollen factory, 18 flouring-mills, 21 grist-mills, 45 saw-mills, two oil-mills, two tanneries, 11 distilleries, three breweries, one pottery, four printing-offices, one bindery, three weekly newspapers, and one periodical; 67 schools, 3695 scholars. Pop. 19,688. Capital, Troy.

MIAMI, county, Ia. Situated toward the N. part of the state, and contains 380 sq. m. Organized in 1832. Watered by Wabash, Eel and Mississinewa rivers. The Wabash and Erie canal passes through the centre of the county. It contained in 1840, 2910 neat cattle, 2057 sheep, 7564 swine; and produced 6127 bushels of wheat, 132,143 of Indian corn, 12,487 of oats, 19,735 of potatoes, 148 pounds of tobacco, 30,548 of sugar. It had 14 stores, two flouring-mills, three grist-mills, five saw-mills, three tanneries, one printing-office, one weekly, and one semi-weekly newspapers; 12 schools, 356 scholars. Pop. 3048. Capital, Peru.

MIAMI, t., Hamilton co., O. Situated on the north bend of Ohio river, which bounds it on the S. Bounded N.W. by Miami river. Pop. 2189.

MIAMI, t., Clermont co., O. It has two stores, one flouring-mill, one grist-mill, two saw-mills, one tannery, two distilleries; eight school, 587 scholars. Pop. 2982.

MIAMI, t., Greene co., O. Drained by Little Miami river add its branches. It has five stores, two flouring-mills, three saw-mills. Pop. 1938.

MIAMI, t., Logan co., O. It has 13 schools, 1380 scholars. Pop. 3239.

MIAMI, t., Montgomery co., O. It has one cotton factory, eight grist-mills, six saw-mills, six distilleries; 13 schools, 1369 scholars. Pop. 3259.

MIAMISBURG, p. v., Miami t., Montgomery co., O., 12 m. S. by W. Dayton, 78 m. W. by S. Columbus, 472 W. Situated on the E. side of Miami river, on the Miami canal It contains two churches, 17 stores, three warehouses, a market-house, one cotton factory, one iron foundry, one brass foundry, two grist-mills, one steam saw-mill, one brewery, one tannery, and about 150 dwellings. One mile S.E. of the village is one of the largest ancient mounds in the state.

MIAVA, a market town of N.W. Hungary, co. Neutra, on the Miava, a tributary of the Morava, 48 m. N.N.E. Presburg. Pop. 8650, mostly of Sclavonian origin, and Lutheran. It has manufactures of woollen stuffs and begging, several distilleries, and some trade in hemp and flax.

MICHAEL (ST.), an inconsiderable bor. and market town of England, co. Cornwall, in pars. Newlyn and

Border of land. Ryder. This, which is said to have been a town of some importance, previously to the Norman conquest, is now, like others of the Cornish boroughs, an inconsiderable village. It returned two members to the House of Commons from the 6 Edward VI. down to the Reform Act, by which it was disfranchised. The market has long been extinct; but sheep fairs are held here July 20, and Oct. 15.

MICHIGAN, one of the north-western United States, consists of two peninsulas; the principal of which, denominated Michigan proper, is bounded N. by the straits of Michillimackinac, which connect lake Michigan to lake Huron; E. by lake Huron, St. Clair river, lake St. Clair, Detroit river, and lake Erie, which separate it from Upper Canada; S. by Ohio and Indiana; and W. by lake Michigan. This main portion of the state is 289 m. long from N. to S., and has a mean breadth of about 140 m., containing 39,656 sq. m., or 25,387,840 acres. But the upper peninsula of Michigan is a distinct territory, bounded N. by lake Superior; E. by St. Mary's river; S.E. by lake Huron, lake Michigan, and Green bay; and S.W. by Menomonee and Montreal rivers, which separate it from Wisconsin territory. This portion of the state is 394 m. long, with an average breadth of 60 m., containing 20,664 sq. m. The whole territory of both peninsulas is 60,320 sq. m., or 38,736,800 acres of land surface. To this should be added 36,394 sq. m. of water surface. It lies between 41° 39' and 47° 20' N. lat., and between 82° 25' and 90° 39' W. long. and between 5° 23' and 12° 3' W. long. from W. The population in 1810, was 4899; in 1820, 9048; in 1830; 31,639; in 1840, 212,267. Of these, 113,395 were white males; 98,165 were white females; 393 were coloured males; 314 were coloured females. Employed in agriculture, 56,521; in commerce, 728; in manufactures and trades, 6890; in navigating the ocean, 94; &c.-canals, lakes, and rivers, 168; in mining, 40; in the learned professions, 904.

There were in 1840, 32 organized counties, which, with their population, were as follows:

Counties.	Pop. 1840.	Counties.	Pop. 1840.
Allegan	1,783	Lenawee	17,889
Barry	1,078	Livingston	7,430
Berrien	5,011	Macomb	9,716
Branch	5,716	Michilimackinac	923
Calhoun	10,599	Monroe	9,922
Cass	5,710	Oakland	23,646
Chippewa	534	Ocean	208
Clinton	1,614	Ottawa	626
Eaton	2,379	Saginaw	2,168
Genesee	4,268	St. Clair	4,606
Hillsdale	7,240	St. Joseph	7,068
Ingham	2,498	Shiawassee	2,103
Ionia	1,923	Van Buren	1,910
Jackson	13,130	Washtenaw	23,571
Kalamazoo	7,380	Wayne	24,173
Kent	2,587		
Lapeer	4,265	Total	212,267

Detroit, on the W. side of Detroit river, is the seat of government.

The surface of the surveyed part of the southern peninsula is generally level, undulating or rolling, and sometimes broken or hilly. In the eastern part, from the southern boundary to Saginaw bay, the land, to the distance of from 5 to 25 m., is mostly level. Proceeding westward, the land gradually rises to an irregular ridge, in some places 600 or 700 feet high, which divides the waters which flow eastwardly from those which flow westwardly to lake Michigan. This ridge is much nearer the eastern than the western shore. The central counties are somewhat hilly. These hills consist of an irregular assemblage of somewhat conical elevations, sometimes rising to the height of from 150 to 200 feet, though ordinarily not more than from 30 to 40 feet. But the main portion of the central and western part consists of a table land, gradually descending toward the lake, exhibiting a gently undulating, but very rarely a broken surface. The unsettled portions of the state lies N. of a line passing E. and W. through the centre of Saginaw bay. It is but imperfectly known, but it is represented as not very fertile, interspersed with sand ridges and marshes, having a rapid descent from the highlands eastwardly, but a gradual slope to the W. In general, the interior of the state may be regarded as level, but the coasts of lakes Michigan and Huron exhibit high and steep banks; and along the former are bluffs and sand banks, from 100 to 300 feet high. A large part of the soil of the peninsula is fertile, and well adapted to the purposes of agriculture. The forest trees present a great variety; oak, hickory, walnut; ash, linden, sugar maple, soft maple, elm, ash of various kinds, sycamore, hackberry, cotton wood, butternut, box or dogwood, poplar, whitewood, and cherry. On the N.E. border evergreens seem to predominate, as pine, spruce, and hemlock, and in the N. part, large forests of pine and well timbered land extend into the interior. The soil in the settled parts of the peninsula is well adapted to wheat rye, oats, barley, flax,

hemp, garden vegetables, and grasses. No part of the United States is better supplied with fish, aquatic fowls, and wild game. The fish are chiefly white fish and salmon trout, and are extensively taken for exportation. The trout weigh from 10 to 70 pounds, and the white fish are equally large. They constitute a substitute for the codfish in the N.W.

The upper peninsula has been but imperfectly explored; but the geological survey of the state, now in progress, may be expected to make it better known. The lake coast has been estimated at between 700 and 800 m., five tenths of which can be reached by the common lake vessels. It is diversified by mountains, hills, valleys, and plains, rising gradually from lakes Michigan and Superior to the interior. Porcupine mountains, which form the dividing ridge between the waters which flow into lake Superior and those which flow into lake Michigan, have summits towards the western boundary, estimated at from 1800 to 2000 feet high. A greater portion of this peninsula, except the sand plains, consist of millions of acres of white and yellow pine, and a mixture of spruce, hemlock, birch, oak, and aspen, and on the rivers, maple, ash, and elm. Forty large, and 60 smaller rivers flow into the lakes, and will hereafter afford mill-sites, and the means of transportation of a vast amount of lumber. This region does not promise much to agriculture, though there are doubtless fertile tracts; but in minerals it is undoubtedly rich. Iron, copper, and lead, are known to exist, and some surprising masses of native copper have been discovered in Ontonagon river. The climate is cold, but healthy; and though the summers are short, vegetation is exceedingly rapid. The extremes of temperature are 96° above, and 29° below zero of Fahrenheit, which has caused it to be denominated the Siberia of Michigan.

The southern peninsula of Michigan is drained by several large rivers, and many smaller streams. They originate in the dividing ridge, and pass off in an easterly and westerly direction, with some exceptions, to the lakes. Raisin and Huron rivers flow into lake Erie; Rouge into Detroit strait; Clinton and Black rivers into the strait of St. Clair; Saginaw river, formed by the junction of several large branches, enters Saginaw bay; Thunder bay river, Cheborgan, and several smaller streams flow N. into the straits of Mackinaw. But the largest rivers flow W. into lake Michigan. They are St. Joseph, Kalamazoo, Grand, Maskegon, and Manistee rivers. Some of these are navigable, to a considerable extent. The counties of Oakland, Livingston, Washtenaw, Barry, Jackson, and Kalamazoo abound with small clear lakes, well stocked with fish.

There were in 1840, 30,144 horses or mules, 185,190 neat cattle, 99,618 sheep, 295,890 swine; poultry was produced to the value of $42,730. There were produced 2,157,108 bushels of wheat, 127,802 of barley, 2,114,051 of oats, 34,236 of rye, 2,277,039 of Indian corn, 113,592 of buckwheat, 2,109,205 of potatoes, 153,375 pounds of wool, 11,381 of hops, 4533 of wax, 1692 of tobacco, 986 of silk cocoons, 1,329,784 of sugar, 130,805 tons of hay, 755 of hemp or flax. The products of the dairy were valued at $301,829; of the orchard at 16,905; and of lumber at $392,395.

Michigan lake is the largest lake that lies wholly within the United States, being 330 m. long, and on an average 60 m. broad, containing 16,981 sq. m., or 10,868,000 acres. It has Green bay, a large branch on the N.W. The straits of Michillimackinac, 40 m. long, connect lake Michigan with lake Huron. Saginaw bay is a large branch of lake Huron, 60 m. long, and 32 m. wide.

Detroit is much the largest and most commercial place in the state. A large number of steamboats and other vessels ply between this place and Buffalo, and other places on the lakes. The other principal places are Monroe on the river Raisin; Pontiac on Clinton river, 16 m. N.W. from Detroit; Adrian, Ypsilanti, Ann Arbor, Kalamazoo, Marshall, and Jackson in the interior; and St. Joseph and Grand Haven on lake Michigan, at the mouths of St. Joseph and Grand rivers.

The exports of Michigan in 1840, amounted to $162,929, and the imports to $138,610. There were 26 commission houses in foreign trade, with a capital of $177,500; 612 retail dry goods and other stores, with a capital of $2,298,968; 312 persons employed in the lumber trade, with a capital of $45,800; 453 persons employed in the fisheries (lake), with a capital of $28,640.

The amount of home-made or family manufactures was $113,935; 16 fulling-mills and four woollen manufactures employed 37 persons, producing articles to the amount of $9734, with a capital of $34,120; 15 furnaces produced 601 tons of cast iron, employing 99 persons, and a capital of $60,800; one paper-mill employed six persons, produced to the amount of $7000, capital $20,000; 12 persons manufactured tobacco to the amount of $5000, capital $1750; hats and caps were manufactured to the amount of $30,453, and straw bonnets to the amount of $659, employing 42 persons, and a capital of $30,007; 38 tanneries employed 99 persons, and a capital of $70,940; 101 other manufactories of leather, as saddleries, &c., produced articles to the amount of $122,190, with a capital of $79,902; one glass-house employed 34 persons, produced to the amount of $7322, with a capital of $25,000; three potteries employed four persons, and produced to the amount of $1100, with a capital of $1900; three persons produced confectionary to the amount of $3000, with a capital of $1200; 67 persons produced machinery to the amount of $47,000; seven persons produced hardware and cutlery to the amount of $1250; one person manufactured the precious metals to the amount of $5000; six persons manufactured granite and marble to the amount of $7000; 208 persons produced bricks and lime to the amount of $68,913; six persons produced 78,100 pounds of soap, and 57,975 pounds of tallow candles, with a capital of $6000; 34 distilleries produced 337,761 gallons, and 10 breweries produced 308,696 gallons, the whole employing 116 persons, and a capital of $124,900; 59 persons produced carriages and wagons to the amount of $20,075, with a capital of $13,150; 93 flouring-mills produced 202,880 barrels of flour, and, with grist and saw-mills, employed 1144 persons, producing articles to the amount of $1,832,363, with a capital of $2,460,200; vessels were built to the amount of $10,500; 65 persons manufactured furniture to the amount of $22,494, with a capital of $28,050; 39 brick or stone houses, and 1280 wooden houses erected, employed 1978 persons, and cost $571,005; 26 printing-offices, two binderies, six daily and 26 weekly newspapers, and one periodical, employed 119 persons, and a capital of $262,900. The total amount of capital employed in manufactures was $3,112,240.

Michigan university at Ann Arbor has departments of literature, science, and the arts, and of law and medicine. It is designed to have academic branches in different parts of the state, and they have already been established at Detroit, Ann Arbor, Kalamazoo, White Pigeon, and Tecumseh. The institution has been well endowed by 46,000 acres of choice lands, the minimum price of which is established by law at $12 the acre. Marshall college, at Marshall, has been established by the Presbyterians; and St. Philip's college, near Detroit, by the Roman Catholics. The colleges had in 1840, 158 students. There were in the state 12 academies, with 485 students; and 975 common and primary schools, with 29,701 scholars. There were 2173 white persons over 20 years of age, who could neither read nor write.

In 1836, the Presbyterians had 42 churches and 19 ministers; the Baptists had 17 churches and 11 ministers; the Roman Catholics had one bishop and 18 ministers; the Episcopalians had one bishop and four ministers; and the Methodists were considerably numerous.

At the commencement of 1840, the state had nine banks and one branch, with an aggregate capital of $1,929,900, and a circulation of $961,296. At the close of 1840, the state debt amounted to $6,011,000. There is a state penitentiary, on the Auburn plan, at Jackson.

The most important works of internal improvement are the Central railroad, now completed and in operation 80 m. from Detroit to Jackson; the Southern railroad is completed and in operation from Monroe to Adrian, 36 m. The legislature has appropriated the proceeds of both roads for the ironing of the former to Marshall, 30 m. further W., and of the latter to Hillsdale, 28 m. farther W.; and it was expected that both roads would be completed thus far in the course of the year 1843. The legislature has appropriated the proceeds or the value of 150,000 acres of the state lands for the extension of the Central road 34 m. from Marshall, the whole of which is now under contract. This road is designed to extend from Detroit to St. Joseph, on lake Michigan, 194 m. The Erie and Kalamazoo railroad is in operation 30 m. from Toledo to Adrian. The Detroit and Pontiac railroad is in operation 25 m. from Detroit to Pontiac. Other works which have been projected are, for the present, suspended or abandoned.

The governor is elected once in two years by the people. He must be a citizen of the United States, and have resided in the state two years next preceding his election. The lieutenant governor must have similar qualifications, and is elected for the same term and at the same time by the people. He is president of the senate; and in case of the decease, impeachment, resignation, or absence of the governor discharges the duties of the office.

The senate consists of 18 members, elected by the people for two years, one half of the number being elected annually. They must be qualified electors, and reside in the county or district which they represent. The house of representatives consists of 54 members, elected annually by the people, and must have similar qualifications to the senators. The judges of the supreme court are appointed by the governor, with the advice and consent of the senate, and hold their offices for seven years. Judges of county courts, associate judges of circuit courts, and judges of probate are elected by the people for the term of four years. Every white male citizen, above the age of 21 years, who has resided in the state six months next preceding an election, is

permitted to vote only in the township or district in which he resides. The legislature meet annually at Detroit, on the first Monday of January.

Michigan was visited by French traders as early as 1668. Detroit was settled in 1670. At the peace of 1763, this country was ceded by France to Great Britain, and at the close of the Revolutionary war was ceded by Great Britain to the United States. They however held possession of Detroit until 1796, when it was given up to the United States. In 1805 the state was erected into a distinct territory, and received a territorial government. The British had possession of the country in 1812–13, but were soon expelled by the Americans under General Harrison. In 1836 Michigan was admitted into the union as a sovereign state.

MACMEGAN, lake, Mich., one of the five great lakes of North America, and is the second in size; being inferior only to lake Superior and lake Huron. It is between 41° 37′ 58″ and 46 N. lat., and between 84° 40′ and 87° 6′ W. long. In the northern part it communicates with lake Huron by the straits of Michilimackinac, or Mackinac, about 8 m. wide, and 4 in its narrowest part, by which, and its northern part, it separates the two peninsulas of Michigan. (For its dimensions see Michigan state.) It is estimated to be on an average 900 feet deep, and is elevated about 200 feet above tide-water. It has few good harbours. On the west side are those of Chicago, Milwaukie, and Green bay. On the E. side are Michigan city, St. Joseph, at the mouth of St. Joseph river, and Grand Haven, near the mouth of Grand river. It is navigated by many large vessels and several steamboats, which ply from Buffalo to Chicago, stopping at the intermediate places. It affords great facilities for transportation to the N.W. country. Merchants have had their goods carried from New-York over 70 m. up Grand river, through this lake, for a dollar over one hundred pounds. The lake has pure and clear water, and abounds with excellent fish. There are several islands in the N. part. A canal is in progress from Chicago to Illinois river, which is one of the great works of the west, and will probably be soon completed.

MICHIGAN CITY, p. v., La Porte co., Ia., 157 m. N.N.W. Indianapolis, 66 W. Situated on the S. shore of lake Michigan, at the mouth of Trail creek. It was laid out in 1836, and is the only harbour on the lake in the state. The location is in general healthy, and it has a fine elevation for trade. It contains three churches, five large warehouses, 10 stores, one flouring-mill, one iron foundry, a branch of the bank of the state of Indiana, and about 700 inhabitants.

MICHILIMACKINAC, county and strait. (See MACKINAC.)

MIDDLE, t., Cape May co., N.J. Bounded E. by the Atlantic, W. by Delaware bay. Watered by Goshen, Dyer's, Green and Fishing creeks, flowing through marshes into Delaware bay. It contains the village of cape May courthouse, and has 12 stores, two grist-mills, two saw-mills; five schools, 326 scholars. Pop. 1684.

MIDDLEBOROUGH, p. t., Plymouth co., Mass., 49 m. S. by E. Boston, 433 W. Incorporated in 1669. It contains two large ponds, from the bottom of one of which iron ore is taken. It contains eight churches, four Baptists, three Congregational and a Methodist, eight stores, one fulling-mill, two cotton factories with 2500 spindles, one furnace, two forges, four grist-mills, 11 saw-mills; one academy, 125 students; 40 schools, 1582 scholars. Pop. 5085.

MIDDLEBOURNE, p. v., capital of Tyler co., Va., 367 m. N.W. Richmond, 275 W. Situated on the E. side of Middle Island creek, and contains a courthouse, jail, and several stores and dwellings. Nett proceeds of the post-office, $75.

MIDDLEBURG, a town of Holland, prov. Zealand, of which it is the cap., nearly in the centre of the island of Walcheren, 4 m. N. by E. Flushing, and 47 m. S.W. Rotterdam; lat. 51° 30′ 6″ N., long. 3° 37′ 30″ E. Pop. 2586. Though no longer fortified, it preserves its circular mound of earth, divided into bastions and surrounded by a broad and deep ditch. The approaches to Middleburg are somewhat more varied than to most Dutch towns, the roads passing through a number of small plantations and country houses. It is nearly circular; some of its streets are wide and handsome, and the whole are tolerably regular. The market-place forms a spacious square; and part of the town is traversed by canals, crossed by draw-bridges. The whole is extremely clean; the private houses are uniform, and some of the public buildings capacious, particularly the town-house and the Oosthoek (east church); the former is in the Gothic style, and has several statues and paintings. The other objects most worthy of notice are several of the churches, a high spire, commanding a prospect over the whole island, the public walks along the bastions, and the Molenwater, an extensive reservoir or backwater. The chief literary institution is the athenæum, or academy, which affords nearly the same course of instruction as a university, but without the privilege of conferring degrees. It has also a Latin school; a school of design; the Zeeland society of literature, arts, and sciences, which possesses a good library, a collection of medals, &c.; and a society of agriculture.

Middleburg has manufactures of starch, glass, and paper, a cannon foundry, and several saw-mills, and salt refineries. Though four miles from the sea, it has quays of considerable extent, and formerly had a considerable share in the Dutch East India trade. Its other branches of commerce are the importation of wine, chiefly from Bordeaux, and the exportation of corn, brought to its market from the fertile tracts to the eastward of the island. It was the head-quarters of the British army in the unfortunate expedition of 1809. Its atmosphere, like that of the rest of Zeeland, is loaded with moisture, which tends to engender agues and bilious complaints, particularly in autumn.

Middleburg is of considerable antiquity, having been first surrounded with walls in 1122. It was taken by the Dutch from the Spaniards in 1574. In 1795 it was ceded to the French, under whom it was the capital of the department Bouches-de-l'Escaut. It sends eight deputies to the provincial assembly of Zeeland. (*De Cloet; Stein's Handbook; Dict. Géog., &c.*)

MIDDLEBURGH, p. t., Schoharie county, N. Y., 37 m. W. Albany, 378 W. It contains a large pond or marsh called the Vly, the outlet of which flows into Schoharie creek, by which it is bounded on the W. It contains nine churches, five Methodist, a Dutch Reformed, Lutheran, Baptist and Christian, 11 stores, four fulling-mills, six grist-mills, 26 saw-mills, four tanneries; 30 schools, 868 scholars. Population 3643.

MIDDLEBURY, p. t., capital of Addison co., Vt., 33 m. S.S.E. Burlington, 39 m. S.W. Montpelier, 481 W. Watered by Otter creek and Middlebury rivers, which afford extensive water-power. Chartered in 1761. It contains 16 stores, six fulling-mills, two woollen factories, one cotton factory with 3000 spindles, one furnace, one forge, two grist-mills, six saw-mills, two tanneries, two printing-offices, one bindery, two weekly newspapers, and one periodical; two academies, 70 students, 12 schools, 400 scholars. Pop. 3162. The village is situated at the falls on both sides of Otter creek, which are connected by a bridge, and is one of the most flourishing villages in the state. It contains a courthouse, jail, a bank, five churches, a Congregational, Methodist, Episcopal, Baptist and Roman Catholic, and most of the above mills and manufactories. It has a fine marble quarry, which produces to the amount of from $60000 to $85000 annually, and affords some of the finest statuary and other marble. It is the seat of Middlebury college, founded in 1800, which has a president and five professors, or other instructors, has 771 alumni, of whom 945 have been ministers of the gospel, 56 students; and 7054 volumes in its libraries. The commencement is on the third Wednesday in August. The college edifices are two, one of which is of stone, 146 feet long, 40 feet wide and four stories high, containing 48 rooms for students. It has sometimes had over 160 students. The village contains two academies, above mentioned, one male and one female, which are flourishing.

MIDDLEBURY, p. t., New-Haven co., Ct., 32 m. W.S.W. Hartford, 316 W. Watered by branches of Naugatuck river and by Quassepaug pond, the outlet of which affords water-power. It contains two churches, a Congregational and Methodist, two stores, two fulling-mills, four saw-mills, one tannery, three distilleries; one academy, 26 students, six schools, 138 scholars. Pop. 761.

MIDDLEBURY, p. t., Wyoming co., N. Y., 247 m. W, Albany, 305 W. Drained by Allen's creek, a tributary of Tonawanda creek. It contains two churches, a Baptist and Presbyterian, five stores, two fulling-mills, two grist-mills, three saw-mills, two tanneries; one academy, 300 students; 23 schools, 1154 scholars. Pop. 2445.

MIDDLEBURY, p. v., Talmadge t., Summit co., O., 125 m. N.E. Columbus, 380 W. Situated on both sides of Little Cuyahoga river, which affords good water-power. It contains a Presbyterian church, of brick, an academy, seven stores, one furnace, two flouring-mills, one machine shop, two wool-carding and cloth-dressing works, one saw-mill, &c., one large chair and cabinet shop, and various other mechanical establishments, and about 600 inhabitants.

MIDDLEFIELD, p. t., Hampshire co., Mass., 124 m. W. Boston, 368 W. Incorporated in 1783. It lies on the Green mountain range. Watered by a branch of Westfield river. It contains three churches, a Congregational, Baptist and Methodist, four stores, six fulling-mills, two woollen factories, one grist-mill, seven saw-mills; nine schools, 246 scholars. Pop. 1717.

MIDDLEFIELD, p. t., Otsego co., N. Y., 64 m. W. Albany, 387 W. Bounded W. by Otsego lake and outlet, E. by Cherryvalley creek. It contains three churches, a Presbyterian, Methodist, and Baptist, seven stores, three fulling-

mills, five grist-mills, 18 saw-mills, one oil-mill, four tanneries, two distilleries; two academies, 58 students; 17 schools, 820 scholars. Pop. 3319.

MIDDLE PAXTON, t., Dauphin co., Pa., 9 m. N.E. Harrisburg. Drained by Fishing, Stony, and Clark's creeks. Bounded W. by Susquehanna river. It contains anthracite coal, and has a church, four stores, one furnace, three grist-mills, six saw-mills, one tannery, one distillery; one school, 75 scholars. Pop. 1580.

MIDDLESBOROUGH, a river-port, town and par. of England, N. Riding co. York, hund. Langborough, on the Tees, about 3¼ m. from its mouth, 16 m. E. by N. Darlington, and 215 m. N. London. Area of par., 2300 acres. Pop. of township, in 1821, only 40; in 1831, 154: but in 1841 estimated at 4500, an increase attributable to the rapid rise of its coal trade, consequent on the opening of the Stockton and Darlington railway from the collieries of S. Durham. It consists of a main street facing the river, and of another wide avenue running at right angles to it, at the corner of which are the commercial hotel and reading-rooms, surmounted by an observatory. There are also several other respectable streets; and on the whole the town is regularly and substantially built, under the superintendence of a joint-stock building company. The church, erected at an expense of £3500, raised by subscription, and opened in 1846, is a neat Gothic structure, with a rather elegant spire. A national school has been formed, and the church has likewise a well-attended Sunday school. The Wesleyan Methodists, Baptists, and primitive Methodists have also their respective places of worship, with attached Sunday schools, furnishing religious instruction to about 600 children of both sexes.

Middlesborough, the site of which only eleven years ago was occupied by a solitary farm-house, has already become the most considerable part of the Tees, though still regarded as subordinate to Stockton, from which it has taken nearly all its coal-trade, and a large portion of its ship-building. Its rapid rise is owing to its convenient position near the bar of the Tees, and the spirited conduct of Messrs. Pease, Backhouse, and other wealthy coal-owners, who, in connection with other shareholders, have built excellent staiths for loading colliers at the wharfs, and constructed a railway communicating with the important coal-field near Bishop's Auckland. The line was opened to Middlesborough (a distance of 28 m. from Witton Park colliery) in 1839; and the export of coal was stated by J. Pease, Esq., M.P., to amount, in 1839, to 510,000 tons, and the number of passengers in the same year somewhat exceeded 16,000. The entire cost of this undertaking was about £450,000, and the present gross income is upwards of £76,000, the average dividends being about 10½ per cent. The staiths, which are 450 yards in length, and worked by two large steam-engines, are capable of shipping 4000 tons of coal *per diem*. Large docks are also in process of excavation, which, when completed, will comprise a water area of several acres. Steam tug-boats are constantly employed in bringing in and taking out vessels over the bar of the river: steamers run twice or thrice times a week during summer between this port and Sunderland and Newcastle; and there is a weekly steamer to and from London. Two ship-building yards, an extensive pottery, and some sail-cloth and rope manufactories are in active operation. The trade of Middlesborough, independent of coal, is already important; and it bids fair to rival, both in trade and industry, the flourishing towns of Tynemouth and Sunderland. (*Irish Railway Comm. Rep.; Sir G. Head's Home Tour; Granville's Spas of Eng.; Priv. Inf.*)

MIDDLESEX, a co. of England, containing the greater part of the metropolis, having E. the river Lea, which divides it from Essex, N. the co. Hertford, W. Buckingham, and S. the Thames, which separates it from Surry and Kent. It is one of the smallest of the English counties, comprising only 180,480 acres; surface very various. The highest eminences are Hampstead, Highgate, and Harrow-on-the-hill. In some parts along the Thames there are extensive tracts of rich loam; but the higher grounds are mostly gravelly and clayey, and not naturally fertile. There are numerous unenclosed commons in different parts of the county, and Hounslow-heath, on its S.W. angle, is as poor and unimprovable a tract as can well be imagined. Contrary to what might have been expected, agriculture is but little advanced in this county; and, although considerable improvements have been made, the implements and processes of husbandry are still very inferior. But by far the largest portion of the county is in grass, and the business of haymaking is as well understood here as in any part of the kingdom. The rich tract of land along the Thames from Kensington to Isleworth is principally occupied by market gardeners, who send a large supply of fruits and vegetables to the London market. The cows kept for the supply of London with milk are all short-horned. Property is very much divided, and in several districts it is mostly portioned out into villas and pleasure grounds; farms seldom exceed 200 acres, and their average size is supposed to be about 100. Leases pretty common, and mostly for 14 and 21 years. Average rent of land, in 1820, 36s. 8½d. Minerals of no importance; but in the vicinity of London in many places vast quantities of land have been dug up, and converted into bricks. Middlesex is well watered; besides the Thames and the Lea, by which it is bounded; it is intersected and partly bounded on the W. by the Colne; and it is also intersected by the Brent, and by the Grand Junction canal, &c. It is divided, exclusive of the metropolis, into six hundreds and 75 parishes. It returns 14 members to the House of Commons—viz. two for the county, four for the city of London, two for Westminster, two for the Tower hamlets, two for Finsbury, and two for Marylebone. Registered electors for the county in 1839-40, 13,919. In 1841, Middlesex had 907,670 inhabited houses, and 1,576,616 inhabitants; of whom 738,970 were males, and 837,646 females. Sum expended for the relief of the poor in 1838-1839, £386,679.

MIDDLESEX, county, Mass. Situated in the E. part of the state, and contains 806 sq. m. Watered by Merrimac, Charles, Mystic, Sudbury, Concord, and Nashua rivers. The Middlesex canal passes through its N.E. part. It contained in 1840, 33,706 neat cattle, 8351 sheep, 20,391 of swine; and produced 9502 bushels of wheat, 61,105 of rye, 230,473 of Indian corn, 5974 of buckwheat, 28,539 of barley, 86,021 of oats, 741,851 of potatoes. It had 13 commissioned houses in foreign trade, 553 stores, capital $1,388,400, 23 lumber yards, capital $230,000, two furnaces, 36 forges, seven fulling-mills, 18 woollen factories, 35 cotton factories with 189,664 spindles, two flouring-mills, 98 grist-mills, 141 saw-mills, 13 paper-mills, four powder-mills, 34 tanneries, five distilleries one brewery, 13 printing-offices, 19 binderies, eight weekly and two semi-weekly newspapers, and six periodicals. Total capital in manufactures $12,215,055; one college, 341 students; 44 academies, 4014 students; 389 schools, 24,876 scholars. Pop. 106,611. Capitals, Concord, Cambridge, and Lowell.

MIDDLESEX, county, Ct. Centrally situated toward the S. part of the state, and contains 342 sq. m. Connecticut river passes centrally through it, affording great facilities for commerce, and by its small tributaries, affords extensive water-power. It contained in 1840, 15,667 neat cattle, 14,736 sheep, 8494 swine; and produced 5453 bushels of wheat, 54,234 of rye, 103,920 of Indian corn, 9169 of buckwheat, 40,026 of oats, 181,673 of potatoes. It had 105 stores, seven lumber yards, eight fulling-mills, one woollen factory, eight cotton factories with 7646 spindles, one dyeing and printing works, 98 grist-mills, 42 saw-mills, one powdermill, 15 tanneries, four printing-offices, one bindery, two weekly newspapers, and one periodical; one college, 147 students; seven academies, 398 students; 123 schools, 5895 scholars. Pop. 24,879. Capitals, Middletown and Haddam.

MIDDLESEX, county, N. J. Situated near the centre of the state, and contains 339 sq. m. Watered by Raritan river and its branches, and by Millstone and Rahway rivers. Raritan bay forms an excellent harbor, communicating directly with the ocean, and with New-York harbour, through Staten Island sound. It contained in 1840, 12,969 neat cattle, 8310 sheep, 11,658 swine; and produced 35,933 bushels of wheat, 60,736 of rye, 194,474 of Indian corn, 42,334 of buckwheat, 1519 of barley, 185,730 of oats, 86,965 of potatoes. It had 136 stores, five lumber yards, one cotton factory, one flouring-mill, 21 grist-mills, 20 saw-mills, two paper-mills, six tanneries, 15 distilleries, three potteries, four printing-offices, two binderies, one daily and two weekly newspapers; three academies, 113 students; 70 schools, 2340 scholars. Pop. 21,893. Capital, New-Brunswick.

MIDDLESEX, county Va. Situated in the E. part of the state, and contains 170 sq. m. Bounded N.E. by Rappahannock river, S.W. by Piankatank river, and E. by Chesapeake bay. It contained in 1840, 3638 neat cattle, 2603 sheep, 6801 swine; and produced 17,097 bushels of wheat, 1148 of rye, 122,145 of Indian corn, 21,078 of oats, 7597 of potatoes, 1350 pounds of tobacco, 2690 of cotton. It had 14 stores, one flouring-mill, 11 grist-mills, seven saw-mills; 10 schools, 202 scholars. Pop.: whites, 2041; slaves, 2309; free coloured, 242; total, 4392. Capital, Urbanna.

MIDDLESEX, p. t., Washington co., Vt., 6 m. N.E. Montpelier, 519 W. Bounded S. by Onion or Winooski river, by the N. branch of which it is watered. Between this town and Moretown, Onion river passes through a rocky chasm, with perpendicular walls, 20 feet deep, 60 feet wide, and 80 rods long, over which is a bridge. It is a curiosity. It contains a church, one store, one grist-mill, three saw-mills, one oil-mill; 12 schools, 487 scholars. Pop. 1270.

MIDDLESEX, p. t., Yates co., N. Y., 194 m. W. Albany, 393 W. Bounded W. by Canandaigua lake. Drained by West creek. It has one grist-mill, one saw-mill; 12 schools, 610 scholars. Pop. 1439.

MIDDLESEX, t., Butler co., Pa. It has two stores, one fulling-mill, five grist-mills, three saw-mills; nine schools, 345 scholars. Pop. 1692.

MIDDLE SMITHFIELD, t., Monroe co., Pa. It has two grist-mills, two saw-mills. Pop. 1144.

MIDDLETON, a manufacturing market town and par. of England, honour of Clitheroe, hund. Salford, co. Lancaster, 5 m. N.N.E. Manchester, and 165 m. N. by W. London. Area of par., 11,510 acres. Pop. of par. (comprising eight townships), in 1841, 15,488; pop. of township 491. This town, which in 1775 was an inconsiderable village, containing only 300 inhabitants, has, owing to the extension of the cotton-trade, become a large place, with several good streets and well-built houses. The church, rebuilt in 1524, has a low tower, partly of wood, and some fine carvings and painted windows: the living is a rectory, in the gift of Lord Suffield, the lord of the manor. Within the parish are, also, three Episcopal chapels, and seven places of worship for different denominations of dissenters, with 17 attached Sunday schools, furnishing religious instruction to about 2400 children. A free grammar school was founded in 1572; and within the last few years, three subscription schools have been formed for the education of the children of the working classes. The principal employments of Middleton are silk and cotton weaving, cotton spinning, calico and silk printing: there are, within the township, about 2000 silk weavers, and 1000 persons employed in the cotton-mills; besides nearly 500 engaged in subordinate trades. The Rochdale canal, the Manchester and Leeds railway, and the Bolton railway pass through the parish, and afford the greatest facilities for the conveyance both of passengers and goods. The town is governed by the county and manorial constables; and courts leet and baron are held twice a year. Markets on Saturday, 1st Monday after 10th March, ditto after 15th April, and 2d Thursday after 29th Sept. (Butterworth's Stat. of Lanc.; Baines's Lanc., &c.)

MIDDLETON, a market town of Ireland, co. Cork, prov Munster, at the confluence of the Curra and Lewis rivers, at the N.E. extremity of Cork harbour, 14 m. E. Cork. Pop., in 1831, 1946. Middleton, so called from being midway between Cork and Youghal, has a parish church, a Roman Catholic chapel and convent; an endowed grammar school, and two schools partially supported by the commissioners of education, a fever hospital and dispensary, a market-house, courthouse, and bridewell. It is built in a very straggling manner, and is neither lighted nor paved, but it contains several respectable dwelling-houses, and good shops: it is increasing, and is generally admitted to be in a thriving condition. It has two extensive distilleries, and a brewery; and it has the advantage of being in the immediate neighbourhood of the harbour of Ballihacurra, where the merchants ship their commodities, especially flour and agricultural produce. (Mun. Bound. Rep.) The corporation, consisting of a sovereign, two bailiffs, 12 burgesses, and commonalty, returned two members to the Irish House of Commons till the union, when it was disfranchised. Quarter sessions are held in June and Nov.; and it is a constabulary station. Duty was paid, in 1836, on 96,192 bush. malt, and 597,533 galls. whiskey. Markets on Saturday; fairs on the 14th Feb., 14th May, 5th July, 18th Sept., 10th Oct., and 23d Nov. Postoffice revenue, in 1839, £295; in 1836, £486.

MIDDLETON, p. t., Essex co., Mass., 22 m. N. Boston, 469 W. Watered by Ipswich river. It has two stores, one fulling-mill, one woollen factory; four schools, 206 scholars. Pop. 657.

MIDDLETOWN, p. t., Rutland co., Vt., 77 m. S.S.W Montpelier, 441 W. It contains a Congregational and a Baptist church, several stores, nine saw-mills, nine schools, 233 scholars. Pop. 1057.

MIDDLETOWN, t., Newport co., R. I., 28 m. S. by E. Providence. It is the middle of three townships, into which Rhode Island is divided. Incorporated in 1743. It has two grist-mills; five schools, 148 scholars. Pop. 891.

MIDDLETOWN, city, p. t., port of entry, semi-capital of Middlesex co., Ct., 14 m. S. Hartford, 94 m. N.E. New Haven, 35 m. N.W. New-London, 326 W. The township is 9 m. long, with an average breadth of nearly 7 m. A range of hills with a base of granite crosses the Connecticut, a little below the city, where the river is compressed to the width of 35 rods; and a range of greenstone mountains passes through its W. part. Opposite to the city the river is from 50 to 97 rods wide. A horse ferry boat crosses to Chatham. The township, including the city, contains 37 stores, capital $269,500, three lumber yards, capital $40,000, one fulling-mill, one woollen factory, one cotton factory, with 11,000 spindles, one dyeing and printing works, two grist-mills, five saw-mills, one powder-mill, one rope-walk, two tanneries, four printing-offices, one bindery, two weekly newspapers, and one periodical; three academies, 115 students, 33 schools, 1268 scholars. Pop. 7210.

The city is situated on the W. bank of Connecticut river, 31 m. above its mouth in Long Island sound, in 41° 33′ 8″ N. lat., and 72° 39′ W. long. The ground rises gradually from the river, and in the back part attains a commanding elevation, affording a delightful view of the river and of the surrounding country. The principal streets run parallel with the river, and are crossed by other streets running at right angles with them, making over eight in the whole. The principal street is Main-street, elevated from 45 to 50 feet above the river, running N. and S., about a mile in length, is broad and straight, and contains the principal stores and public buildings. The city is at the head of ship navigation, and admits to its wharves any vessels which can pass the bar at the mouth of the river, having ten feet of water. It was first settled in 1651, and was incorporated as a city in 1784, and contains a handsome courthouse, a jail, a fine custom-house of stone, two banks, besides one for savings, seven churches, two Congregational, an Episcopal, Methodist, Baptist, Universalist, and an African, and most of the above mentioned stores and manufactories. The tonnage of the port in 1840, was 14,231. Vessels for Hartford and other towns on the river are registered here. It contains about 480 dwellings, and 3511 inhabitants.

It is the seat of the Wesleyan university, under the direction of the Methodists, founded in 1831, which has a president and nine professors or other instructers, 143 alumni, of whom 49 have been ministers of the gospel, 133 students, and 11,000 vols. in its libraries. The commencement is on the first Wednesday in August. It has a valuable philosophical apparatus, and mineralogical cabinet. The buildings occupy a commanding site in the upper part of the city, over half a mile from the river. Two miles above the city is the village of Upper Middletown or Middletown Upper-houses, which contains two churches, a Congregational and Baptist, and has had considerable ship-building for more than a century. West river, a small stream, enters Connecticut river between this and the city, and affords water-power. The steamboat from Hartford to New-York stops at the city daily in the season of navigation.

Middletown, p. t., Delaware co., N. Y., 79 m. W.S.W. Albany, 335 W. Watered by Papacton, branch of Delaware river. It was organized in 1789, and contains eight stores, two fulling-mills, seven grist-mills, 19 saw-mills, three tanneries; 20 schools, 754 scholars. Pop. 2608.

Middletown, p. t., Monmouth co., N. J., 51 m. E.N.E. Trenton, 220 W. Bounded N. by Sandy-hook and Raritan bays, E. by the Atlantic. The beach forming Sandy-hook runs 6 m. N. from Shrewsbury inlet, on the N. point of which stands Sandy-hook lighthouse. Drained by Swimming and Neversink rivers. It contains three churches, a Dutch Reformed, Episcopal, and Baptist, 33 stores, three lumber-yards, two fulling-mills, 11 grist-mills, six saw-mills, three tanneries, eight distilleries; one academy, 60 students, 19 schools, 1239 scholars. Pop. 6063.

Middletown, t., Bucks co., Pa., 20 m. N.N.E. Philadelphia. Watered by Neshaming creek and its tributary, and Bristol creek. It has four stores; one academy, 27 students; two schools, 164 scholars. Pop. 2124.

Middletown, t., Delaware co., Pa., 20 m. W. Philadelphia. Bounded W. by Chester creek, E. by Ridley creek. It has three stores, two fulling-mills, one woollen factory, four cotton factories, with 4800 spindles, two grist-mills, three saw-mills; five schools, 126 scholars. Pop. 1451.

Middletown, p. v., Frederick co., Va., 158 m. N.N.W. Richmond, 87 W. Situated on Meadow run, on which are several mills. It contains two churches, an academy, five stores, and about 400 inhabitants. Wagons are extensively manufactured.

Middletown, p. v., Lemon t., Butler co., O., 30 m. N. Cincinnati, 90 m. W.S.W. Columbus, 480 W. Situated on the E. side of Miami river. The Miami canal passes through it. It contains three churches, 11 stores, two pork-houses, one woollen factory, one grist-mill; one academy, 150 dwellings, and about 1000 inhabitants. A toll bridge here crosses Miami river.

Middletown, t., Columbiana co., O. It has seven schools, 147 scholars. Pop. 1601.

MIDDLETOWN POINT, p. v., port of delivery, Middletown t., Monmouth co., N. J., 43 m. E. by N. Trenton, 209 W. Situated on Middletown creek, 3 m. from its entrance into Raritan bay. It contains a Presbyterian church, a bank, 16 stores, and about 100 dwellings, many of them neat.

MIDDLEWICH, a market town and par. of England, hund. Northwich, co. Chester, at the confluence of the Dane and Croke, 18 m. E. Chester, and 151 m. N.W. London. Area of par. (which comprises 14 townships), 13,330 acres; pop. of do., in 1841, 4755; pop. of township, 1325. The town, though small, is neat and regularly built, its principal public edifices being a large church, three places of worship for dissenters, and a free school. Middlewich has long been celebrated for its brine-springs, the water of which is alleged to yield ¼ its weight of salt (muriate of soda). The manufacture of salt is hence the chief employment of the inhabitants; but some additional advantages have been derived, within the last 30 years, from the introduction of the cotton trade: there is also a silk-mill, which employed, in 1839, about 40 hands.

Middlewich has an extensive internal navigation by means of the Grand-trunk canal, which passes through the town, and by a branch connecting the town with the Chester canal. It is distant only 2¼ m. from the Winsford station, on the Grand Junction railway, and about 3¼ m. from the yet unfinished Manchester and Birmingham railway. Petty sessions are held here for the hund. of Northwich. Markets on Tuesday; cattle fairs, May 1, Holy Thursday, and Aug. 5.

MIDHURST, a parl. bor., market town, and par. of England, co. Sussex, hund. Easebourne, and rape Chichester, near the Arun, 10 m. N. by E. Chichester, and 46 m. S.W. London. Pop. of parl. bor. (which includes the entire pars. of Midhurst, Easebourne, Heyshot, Chithurst, Graffham, Didling, and Cocking, with portions of parishes. Steep, Bignor, Wool-Lavington, Bepton, Woolbeding, Lynch, Stedham, Sping, Trotton, Sellham, and Lodsworth), in 1831, 5627. The town is small, but particularly clean-looking, and has several good detached houses in its immediate neighbourhood. The church is a small stone building, with a square tower surmounted by a diminutive steeple: the living is a curacy in private patronage. A free grammar school was founded here in 1672, and there is a national and Sunday school for poor children of both sexes. "Midhurst has very little trade, except in corn, large quantities of which are sold at its weekly markets. The surrounding district is entirely agricultural, though formerly iron-works existed within a few miles of it." (*Bound. Rep.*) It is a borough by prescription, and sent two members to the House of Commons from the reign of Edward II. down to 1832, the right of voting being in the holders of burgage tenures. The Reform Act deprived it of one of its members; the electoral limits being, at the same time, so much enlarged as to include, in addition to the parish or old borough of Midhurst, six entire parishes, and portions of 11 others, as above specified. Registered electors, in 1839-40, 251. Petty sessions are held here for the hund. of Easebourne. Markets on Thursday; cattle fairs 5th of April and 29th of Oct.

About ¼ m. E. of Midhurst, and close to the Arun, are the ruins of Coudry house, formerly the residence of the family of Montague, destroyed by fire, with its costly furniture, pictures, books, &c., on the 24th Sept., 1793; the same day that its noble owner was drowned in an attempt to sail down the falls of the Rhine at Schaffhausen.

MIDLAND, county, Mich., recently formed. Bounded E. by Saginaw bay. Watered by Titibawassee river and its branches. Capital, Midland.

MIDNAPORE, a dist. of British India, presid. Bengal, properly belonging to the prov. Orissa, but which has long been attached to that of Bengal; principally between lat. 21° 40′ and 23°, and long. 86° and 88° E.; having N. the Jungle Mehals, E. the Hooghly district and river, S. Cuttack, and W. some zemindaries, tributary to the British. Area, 8260 sq. m. Pop., in 1822, 1,914,060. Notwithstanding this amount of population, a considerable portion of the surface consists of jungles, partially inhabited by a very low caste of Hindoos called *sontals*. The land is generally very fertile, and most part of the articles grown in Bengal are cultivated here; the people, however, are poor and depressed, and it is doubtful whether they ever enjoyed a much higher state of prosperity and civilization than at present. Midnapore has some manufactures of fine calico and gauzes, but of late these have greatly declined. Land revenue, in 1829–30, 1,307,614 rupees. Chief towns, Midnapore, Jellasore, and Pipley. Midnapore, the capital and residence of the judge, collector, &c., of the district, is in lat. 22° 25′ N., long. 87° 25′ E. Its chief buildings are the jail, hospital, and barracks. (*Hamilton's E. I. Gaz.*)

MIFFLIN, county, Pa. Situated near the centre of the state, and contains 900 sq. m. Watered by Juniata river, along which proceeds the Pennsylvania canal. It contained in 1840, 9933 neat cattle, 11,323 sheep, 15,002 swine; and produced 307,696 bushels of wheat, 47,466 of rye, 189,451 of Indian corn, 8619 of buckwheat, 227,321 of oats. 51,499 of potatoes. It had 47 stores, four furnaces, two forges, two fulling-mills, five woollen factories, 24 grist-mills, 61 saw-mills, 14 tanneries, five distilleries, one brewery, two printing-offices, one weekly newspaper; one academy, 27 students; 67 schools, 2125 scholars. Pop. 13,092. Capital, Lewiston.

Mifflin, t., Lycoming co., Pa., 23 m. N.W. Williamsport. It has three grist-mills, five saw-mills, one distillery, one pottery; five schools, 445 scholars. Pop. 1224.

Mifflin, t., Alleghany co., Pa., 8 m. S. Pittsburg. Bounded N.E. and S.E. by Monongahela river, by small tributaries of which it is drained. It contains five stores, four flouring-mills, three saw-mills, four distilleries; five schools, 180 scholars. Pop. 1654.

MIFFLIN, t., Cumberland co., Pa. Bounded S. by Conedogwinit creek. It contains two stores, two fulling-mills, three grist-mills, seven saw mills, two tanneries; eight schools, 310 scholars. Pop. 1412.

MIFFLIN, t., Columbia co., Pa., 17 m. S.E. Danville. It has five stores, one fulling-mill, two flouring-mills, two grist-mills, seven saw-mills, three tanneries; five schools, 148 scholars. Pop. 2150.

MIFFLIN, t., Dauphin co., Pa., 23 m. N.E. Harrisburg. Watered by Wiconisco creek. It contains a Lutheran church, seven stores, one fulling-mill, five flouring-mills, one grist-mill, eight saw-mills, three tanneries, one pottery; seven schools, 220 scholars. Pop. 1781.

MIFFLIN, p. t., Richland co., O., 5 m. E. Mansfield, 72 N. by E. Columbus, 369 W. Pop. 1800.

MIFFLINSBURG, p. b., Buffalo t., Union co., Pa., 76 m. N. Harrisburg, 186 W. Situated on the S. side of Buffalo creek. Incorporated in 1827. It contains two churches, a Lutheran and a Methodist, six stores, two tanneries, two breweries, two potteries; one academy, 25 students; three schools, 180 scholars, 100 dwellings, and 704 inhabitants.

MIFFLINTOWN, p. b., Fermanagh t., capital of Juniata co., Pa., 45 m. N.W. Harrisburg, 136 W. Situated on the N. bank of Juniata river, on the Pennsylvania canal. It contains a courthouse, jail, two churches, a Presbyterian and Lutheran, an academy, several stores, about 100 dwellings, and 450 inhabitants.

MIHIEL (ST.), a town of France, dep. Meuse, cap. cant., on the Meuse, 20 m. N.E. Bar-le-Duc. Pop., in 1836, 5706. It was formerly surrounded with walls, but these were demolished in 1635. It is well laid out, and has several remarkable churches, in one of which is a fine piece of sculpture, representing Christ laid in the sepulchre, the work of L. Richler, a pupil of Michael Angelo. It is the seat of the court of primary jurisdiction for the arrond. of Commercy, and of the court of assize for the dep.; and has a communal college, a public library, and manufactures of cotton cloth and yarn. (Hugo, art. Meuse, &c.)

MILAN (Ital. Milano, Germ. Mailand, Lat. Mediolanum), the principal city of N. Italy, and the cap. of the Austrian dom. S. of the Alps, in a fertile and highly cultivated plain, between the Olona and Lambra, with which rivers it is connected by the Navigilo Grande and other canals. 190 m. W. Venice, and 79 m. E.N.E. Turin; lat. 45° 28' 10" N., long. 9° 11' 30" E. Pop., in 1837, 145,500, including only the inhabitants of the city-proper; but, with the immediate suburbs, 171,968; and including the garrison and strangers, about 185,000. (Berghaus.) It is nearly circular; and is surrounded, except on the N.W., by a bastioned wall of little strength, and broad ramparts, planted with trees, and about 10 m. in circuit. The area thus enclosed comprises, however, not only the city and its suburbs, but a number of gardens and orchards. The city-proper, or closely peopled part in the centre, is surrounded by a canal nearly 5 m. in circumference. Like other old cities, it is irregularly laid out, and most of its streets are narrow and winding; but it has some noble thoroughfares, and is generally extremely well paved. Upon the whole, it is one of the finest and most pleasing cities of Europe. "Milan," says Von Raumer, "stands in a sea of green trees, as Venice in a sea of green waters. In the latter city everything reminds you of the past, as the great and important period; here, on the contrary, the present is full of life, and all that belongs to antiquity is thrown into the back ground. Everything reminds one that Milan is a great central point of wealth and activity. No signs of decay, no unoccupied people, unless in the upper classes, where the possession of fortune invites to the far niente, which, in Venice, goes hand in hand with wretchedness and want. In Venice, and also in Verona, each house is built according to individual fancy or convenience, and the greatest variety of architecture, and the most wanton deviations from all law, order, or harmony, are seen. In Milan, on the contrary, every building is perfectly symmetrical, and scrupulously kept in repair; and not the least symptom is to be seen of a poor or declining population, so evident is everywhere the progress of improvement." (Italy and the Italians, i., 100.)

The principal public edifice is the cathedral; an immense and imposing Gothic structure, inferior in size only to St. Peter's, Rome, and St. Paul's, London. It stands in the centre of a spacious square, nearly in the middle of the city, and is built wholly of white marble. It was begun by John Galeazzo, first duke of Milan, in 1385; but on so large a scale, that it is not yet quite finished; and, from having been continued by many different architects, of adverse tastes, it has a great admixture of styles. Its principal façade has a fine general effect; but it presents the incongruity of Grecian doorways and windows introduced into a Gothic front. The entire building is in the form of a Latin

cross; its length internally is 462 feet; width, 177 feet; total length of the transept, 283 feet 10 inches; height of the nave, 151 feet 11 inches; height to the top of the lantern, 247 feet; do. to the top of the spire and statue, 395 feet. There are 52 piers, 98 pinnacles, and, inside and out, no fewer than 4400 statues. (Wood's Letters of an Architect, p. 207.) In fretwork, carving, and statues, it goes beyond all churches in the world, St. Peter's itself not excepted. "Its double aisles, its clustered pillars, its lofty arches, the lustre of its walls, its numberless niches, all filled with marble figures, give it an appearance novel even in Italy, and singularly majestic." (Classical Tour, &c., iv., 7, 8.) In this cathedral there is no screen, and the chancel is entirely open, and separated from the nave only by its elevation. Neither are there any chapels, properly so called; and the high altar stands, as in the Roman Basilica, and, indeed, in all ancient churches, before the choir, and between the clergy and the people. The pillars, or rather clusters of pillars, which support the vault, though above 90 feet in height, are only 8 feet in diameter, from which comparative thinness they scarcely conceal any part of the interior from the eye. The pavement is of different coloured marbles, disposed in various figures. The dome is surmounted by a tower and obelisk, which last was erected about the middle of the 18th century, adding, however, little to the beauty or magnificence of the edifice. On the top is the figure of the Virgin, to whom the church is dedicated. In a subterraneous chapel immediately beneath the dome is the shrine, enclosing the remains of St. Charles Borromeo, archbishop of Milan in the 16th century, to which numerous pilgrims resort. On the whole, however, the cathedral is, both internally and externally, overladen with ornaments; and there can be no doubt that the removal of 2000 or 3000 of its statues would be a signal improvement; but, with all its faults, it is certainly the finest Gothic edifice in Italy; and, in the opinion of some travellers, the finest church after St. Peter's.

Several other churches in Milan are worthy of notice. The first is that of St. Ambrose, the scene of many ecclesiastical councils and civil conflicts, and in which the German emperors usually received the Lombard crown. It is of high antiquity, and possibly some remains of the original edifice, erected by St. Ambrose towards the end of the 4th century, may form part of the modern building; but the bronze doors, and the court in front, surrounded by arcades, are acknowledged to belong to the 9th century; and the most ancient part of the building, having any character of architecture, appears to be of this same period. This church is divided by arcades into a nave and two aisles, and vaulted in nearly the same manner as the church of the Carthusians at Rome (the great hall of Diocletian's baths). Among its curiosities are the tombs of St. Ambrose and other saints, some Greek mosaics, old paintings in stucco, sarcophagi of considerable antiquity, and a huge brazen serpent, said to be that fabricated by Moses in the wilderness! The churches of St. Victor, St. Mark, San Celso, St. Eustorgio, the Madonne della Grazia, &c., are among the handsomest or most remarkable in Milan, and some of them are adorned with rare works of art. Eustace, however, notwithstanding his disposition to eulogise, says that many of the churches "lose much of their majesty, and even of their beauty, by the profusion of rich and splendid decorations that encumber them. The materials of all are costly, the arrangement of most is tasteless; yet there are few which do not present some object of curiosity worthy of a visit." (Classical Tour, iv., 27.) The steeple of St. Gothard is a curious specimen of the architecture of the 14th century.

In the old Dominican convent is the famous Cænacle, or "Last Supper," by Da Vinci. This magnificent work has suffered severely from damp and age, and, also, as is alleged, through the wantonness of the French soldiers and prisoners when they were quartered in the building. But what better could be expected from common soldiers, when a superior of the convent did not hesitate to cut away the feet of the principal figure, that a doorway might be heightened! It occupies one side of the refectory, and is about 30 feet in length, by 15 in height. It has been so often repaired and retouched, that it is now nearly in the condition of Sir John Cutler's silk stockings: three of the apostles heads are said to be all that remains of the original work, and even they owe their colouring to the pencil of restorers. Morghen's admirable engraving gives now, perhaps, the best idea of the picture and of the genius of the painter. On the wall opposite the "Last Supper" is a fresco, by Mototarro, an artist of the 15th century; more curious on account of its age, than remarkable for beauty.

The Royal palace (Palazzo del Corte), now the residence of the viceroy of Austrian Italy, a noble structure fronting the square of the cathedral, was erected by the French on the site of the old Sforza palace. It has numerous spacious apartments, and some admirable frescoes by Appiani. The floors are beautifully inlaid, and some of the rooms are hung with Gobelin tapestry; but the magnificent paintings, representing the exploits of Napoleon, that formerly decorated the two large saloons, have been removed. The government, judicial, and archiepiscopal palaces, the city-hall and mansion-house, the mint, and the custom-house and treasury, are among the other principal edifices. The large hospital (*Ospedale Grande*), is of much greater extent than Bethlehem hospital in London; being about 880 feet in length, by 369 feet in depth, and inclosing several open courts. It is not remarkable for its architecture, but is under excellent regulations. It was founded by Francis Sforza in the 15th century, and was left by one individual 4,000,000 livres (about £129,000), and by another three fourths of that amount. It is open to all applicants, whatever their country, religion, or disorder: should to it is a dispensary, whence medicines are distributed to the poor *gratis*, on the specification of any physician. The most extensive building in Milan is, however, the Lazaretto, beyond the walls, also founded in the 15th century, for those infected with the plague. It consists of four ranges of building, about 1296 feet each in length, inclosing an area of more than 30 acres. The city abounds in charitable institutions, including several other hospitals, four asylums for poor children, two workhouses, a government loan-bank, in a magnificent edifice built in 1465, &c. (*Oscuov. Nat. Encyc.*)

One of the principal attractions of Milan, especially to strangers, is the famous *Teatro della Scala*. This, which is the largest theatre in the Austrian empire, and, next to San Carlo at Naples, the largest in Italy, has six tiers of boxes, exclusive of the pit, which accommodates 800 visiters. Simond gives the following account of his visit to this theatre. "The house, which is certainly very fine, exceeds perhaps any in Paris or London, and the full band in the orchestra filled it well. Soon, however, the flapping of doors, incessantly opening and shutting, the walking to and fro over that part of the pit which is without seats, and, above all, the universal chattering, overpowered the music. Disappointed in our expectations of hearing this, and finding our attention to what was passing on the stage altogether fruitless, we turned to the spectators, and observed that the boxes, which are little rooms very neatly fitted up, had, by degrees, filled with company; and the lights in some of them (for there were none in the house except the row of lamps on the stage) enabled us to see the people receiving company, taking refreshments, gesticulating in earnest conversation, and laughing. In those boxes where there were no lights, the company remained invisible, and a sort of *chiaro-oscuro* pervaded the fore part of the house. But, when the ballet began, the general hubbub at once ceased, and heads suddenly popped out, cards and conversation being suspended to look at the dancing. This, though much inferior to that of Paris or London, evidently possessed attractions superior to those of music, which was no sooner resumed, after the ballet, than the noise began again as before. A box at the opera, holding eight persons, of whom four only can see, costs 11 francs; and three additional francs are paid by each person for his ticket of admission." (*Travels*, p. 17, 18.) There are eight other theatres, two of which are open for performances in the day-time.

Milan has many spacious and extensive barracks, nearly all of which are in the W. suburbs. The largest, or *Caserna Grande*, occupies an area of about 960 feet in length, by 700 feet in width, having in front, and on either side the *Foro* (*Foro-Bonaparte*), an esplanade, planted with trees, and laid out in elegant public walks. Behind the Caserna is a large open space, called the Place of Arms (*Piazza d'Armi*), from which the Simplon-road opens by the *Arco della Pace*, one of the finest monuments erected in modern times. This arch, commenced in 1807 and finished in 1837, designed and principally completed by the Marquis Cagnola, is altogether of marble, richly adorned with statues and bass-reliefs. It is nearly 73 feet in length, 62 feet in depth, and 74 feet in height; but to the summit of the principal statue is 98 feet. Four fluted Corinthian columns decorate either front; and on the top a bronze herald of victory stands at each angle; and facing the city is a colossal bronze statue of Peace, in a car drawn by six horses. On another side of the *Piazza d'Armi* is the amphitheatre, built by the French in 1806, a poor imitation of the antique structures of the same kind. It is nearly 300 yards in length, by 162 in breadth, and is capable of accommodating 30,000 spectators. It may be made an amphitheatre, a circus, or a *naumachia*, "for charioteers to drive, and athletæ to wrestle, and a navy to give battle on an ocean 4 feet deep; for the area could be held under water at pleasure. The walls of this counterfeit of Roman work are scarcely 25 feet high; and their thin facing of stone, already giving way, shows the rubbish underneath. But the palace annexed to this circus is adorned with columns of red granite, of great size, and each made of a single block. It is, in

every respect, as beautiful as the rest is paltry and contemptible." (*Simond*, p. 19.)

The private palaces of Milan have received little notice from travellers, but some have considerable elegance, as the Palazzo Belgioioso, formerly the villa of Napoleon, and afterward the residence of Prince Eugene Beauharnois, the Serbelloni, Vitti, Marino, and Visconti palaces, &c. Besides the *Arco della Pace*, the city is entered by 10 gates, of which the Porta Orientale is the richest and most remarkable.

Milan, though less striking in its general appearance than Turin or Genoa, is much richer in objects of varied interest, art, and science. The Ambrosian Library, founded in 1609 by Card. F. Borromeo, comprises, according to the Austrian Encycl., 95,000 printed volumes, and 15,000 MSS. Many of the latter are highly valuable, including the note-book of Leonardo da Vinci, some MSS., supposed to date as far back as the 4th century, containing fragments of Cicero's lost orations discovered by Mai. Attached to the library is a hall of painting, with several fine works by Titian, Da Vinci, Luini, Albano, &c., and sketches by Raphael, Pietro de Cortona, Carravaggio, &c. The *Brera*, formerly the principal establishment of the *Umiliati*, is now converted to the use of the Royal academy of arts and science. It has a noble collection of pictures by almost all the first and second-rate masters of Italy, collections of casts and engravings, rooms for the exhibition of the produce of the useful arts, a well-furnished observatory, a good library, and a botanic garden. Many of the private collections in art and literature are excellent; in the Trivulzio palace is a library of 30,000 printed volumes, and many MSS., a considerable collection of coins, and many curious relics of antiquity.

Milan is the seat of government for Austrian Italy, and of the provincial assembly, the court of appeal, and high criminal court of Lombardy. It is the residence of a delegate, and an archbishop's see; and has two lyceums, six gymnasiums, a teachers' seminary, a high female school, many primary schools, a deaf and dumb school, colleges of medicine, midwifery, veterinary surgery, and architecture, a military geographical institute, various societies of literature, agriculture, &c., and a tribunal of commerce.

This city is the centre and most important emporium of the silk trade of Lombardy. Not only do the transactions of the Lombardo-Venetian provinces in silk centre here, but many of the neighbouring states either sell their silk in Milan, or remit it thither in transit to foreign countries; and this is the case, not for raw silk alone, but also for organzine and tram. English houses, in particular, frequently make their advances at Milan to the consignees of silk. The spinning and throwing of silk is also extensively carried on in the city and its immediate neighbourhood, and many of its throwing-mills have steam-engines. (*Bowring's Reports.*) Velvets, silks, ribands, lace, cotton stuffs, carpets, artificial flowers, paper, goldsmiths' wares, glass, felt hats, leather, earthenware, chocolate, &c., are exclusively made in Milan, and it has a royal tobacco manufactory. In addition to silk, Milan has an extensive commerce in rice and Parmesan cheese, and is, next to Venice, the largest book mart in Italy. As a place of residence, it has the advantages of cheap and plentiful provisions, every facility for study and amusement, a well-regulated police, and polite society. Among its drawbacks, are the heats of summer, and the fogs of the autumn; the climate is, however, considered healthy.

Mediolanum, supposed to have been founded by the Insubrian Gauls, was annexed to the Roman dominions by Scipio Napica, *anno* 191 B.C. In the 4th century, it held the rank of the sixth city in the Roman empire; and is one of the few, in Italy, which have survived the devastations of the middle ages, and brought down its celebrity to modern times. It retains, however, but few antiquities; the only good specimen of ancient Roman architecture remaining being a range of 16 beautiful Corinthian columns, with their architrave, before the church of San Lorenzo. In the 12th century, Milan was the capital of a republic, and it subsequently became the capital of a duchy, in the families of Visconti and Sforza. After the taking of Pavia it was held by Spain, until, in 1714, it was ceded to Austria. The French took it in 1796, and again in 1800, after the battle of Marengo. Under their government it was at first the capital of the Cisalpine republic, and, from 1805 to 1814, it was the capital of the kingdom of Italy. Milan has given birth to many distinguished individuals, among whom may be specified the illustrious painter Leonardo da Vinci; the mathematician Cavalieri; Beccaria, the author of the celebrated treatise on Crimes and Punishments; Signora Agnesi, famous for her mathematical and scientific attainments; the poets Parini and Manzoni, &c. There appears to be no foundation for the statement that Valerius Maximus was a native of Milan. (*Eustace; Classical Tour*, iv., 1-36; *Simond's Trav. in Italy*, 9-22; *Wood's Letters of an Architect*, 205-221; *Forsyth; Von Raumer, Italy and the Italians*, i.,

100-122; *Cramer's Anc. Italy*, i., 51, 52; *Berghaus; Oesterr., Nat. Encyc., &c.*)

MILAN, p. t., Duchess co., N. Y., 69 m. S. Albany, 322 W. Drained by a branch of Jansen's creek. It has five stores, one fulling-mill, five grist-mills, five saw-mills; five schools, 366 scholars. Pop. 1725.

MILAN, p. t., Erie co., O., 103 m. N. Columbus, 396 W. Watered by Huron river. It has 11 schools, 608 scholars. Pop. 1531. The village is situated on the S.E. bank of Huron river, 8 m. from lake Erie, on a bluff 60 or 70 feet above the level of the river. A ship canal is designed to connect it with the lake. It contains a brick church, an academy, 10 stores, one grist-mill, two saw-mills, one oil-mill, a carding machine and fulling-mill, and about 800 inhabitants.

MILAZZO, or MELAZZO (an. *Myla*), a fortified seaport town on the N. coast of Sicily, intend. of Messina, cap. canton, on the E. side of an elevated narrow promontory, at the bottom of a bay 25 m. W. by S. the Faro point of Sicily; lat. of lighthouse 38° 15' 58'' N., long. 15° 14' 10'' E. Pop., in 1831, 9206. It is divided into the upper and lower towns, both of which are irregularly built; and though it has a number of large edifices, none of them are remarkable. "The churches, with the exception of that of St. Francis, are generally mean, and the convents poor and dirty ; the prison is a filthy sink ; the public hospital is badly provided ; and the *monte-di-pieta* languishes in bad hands." (*Smyth's Sicily*, p. 103.) The town is principally distinguished by its fortifications ; being so strong, by nature and art, that it may be regarded as the Gibraltar of Sicily. Besides subordinate fortifications, it has a citadel on the highest point of the promontory, 390 feet above the sea, commanding the town and the port. Beneath it is a spacious grotto, called the Cave of Ulysses. The promontory is bounded on all sides by steep rocks, inaccessible from the sea ; and might, according to Captain Smyth, be easily rendered impregnable. In the lower town is the fountain of Mylas, one of those alluded to by Pliny (*Hist. Nat.*, lib. xxxi., cap. 4) as existing in this part of Sicily, the waters of which (in consequence, perhaps, of the melting of snow) are most abundant in summer.

Milazzo is the residence of a military commandant. Its inhabitants are occupied chiefly in the tunny fishery, and in the export of wine, silk, fruit, rags, soap, white and red argols, corn, olive and linseed oils, and *vino colto* ; the last is a cordial made by boiling must with potash. Its trade is principally with Marseilles, Leghorn, and Genoa. Its bay is large and the water deep. Ships may anchor abreast of the town in from 10 to 25 fathom stiff mud, about ¼ m. from the shore.

The gulf of Milazzo (an. *Basilicus Sinus*), between the peninsula on which the town stands and cape Rasacalmo, has been the theatre of some important naval conflicts. The first of these occurred *anno* 261 B.C., when the consul Duillius defeated a Carthaginian fleet, and showed his countrymen how to conquer by sea as well as by land. Another and far more important contest, which influenced, indeed, in no small degree, the fate of the Roman world, took place in this gulf *anno* 31 B.C., when the fleet of the younger Pompey was entirely defeated, and all but destroyed, by Octavius Cæsar, or rather by his general Agrippa. (*Ancient Universal History*, xiii., 450, 8vo ed., and the authorities there referred to.) A third action took place here in 889, between the fleet of the Saracens and that of the Greek emperor Basilius.

MILBORNE-PORT, a decayed borough, market town, and par. of England. co. Somerset, hund. Horethorne, on the Ivel, 28 m. E. by S. Taunton, and 106 m. W.S.W. London. Area of par. 3150 acres. Pop., in 1841, 1748. The town, though considerably enlarged and improved within the last few years, is very irregularly built, consisting chiefly of detached houses, and having the appearance of a mere village. An ancient guildhall stands in the High-street, and near it is the market-house, now converted into warehouses. The church, an ancient cruciform structure, is surrounded by a massive square tower, supported by two pointed and two semicircular arches: the living is a vicarage, in the gift of the Marquis of Anglesey. The Wesleyan Methodists and Baptists have likewise their respective places of worship, with attached Sunday-schools. Milborne-port had formerly considerable manufactures of dowlas, ticking, and sail-cloth, but they have long ceased to exist. The glove trade, however, was introduced here from Yeovil about 25 years ago; and it is stated in *Hull's History of the Glove Trade*, published in 1834 (p. 74), that about 25,000 dozen pairs were then annually produced; and that great efforts had been made by several intelligent and persevering manufacturers to equal the French in this department, and that in some kinds of gloves they had succeeded. The market is disused ; but fairs are held for cattle and pedlary June 5, and October 25.

Milborne-port, which, at the time of the Norman conquest, had a market and 56 burgesses, is a borough by pre-

366

scription, and sent two members to the House of Commons, with some interruption, from the reign of Edward I. down to the Reform Act, by which it was disfranchised. It had been for a lengthened period a mere nomination borough.

MILDENHALL, a market town and par. of England, hund. Lackford, co. Suffolk, on the Lark, 33 m. N.W. Ipswich, and 63 m. N.N.E. London. Area of par. 13,710 acres. Pop., in 1841, 3731. The town is of considerable extent, and well built, consisting of several detached streets, or rows, that form, as it were, a series of little villages. The church is a large and handsome structure, with a rich carved roof and lofty tower : the living is a vicarage, in the gift of Sir H. Bunbury, the chief landowner of the parish. The inhabitants, with the exception of a few retail traders, are chiefly engaged in agricultural pursuits. Petty sessions for the hund. are held here. Markets on Friday ; fair for wool, October 10.

MILES, t., Centre county, Pa. Drained by branches of Penn's creek. It contains two stores, four grist-mills, five saw-mills, one tannery, one distillery ; seven schools, 219 scholars. Pop. 1198.

MILETUS (Gr. Μίλητος), a once famous but now ruined city of Asia Minor, the cap. of Ionia, near the mouth of the Meander (hod. Mendere), 65 m. S. Smyrna. This is a very ancient city, and had borne several names before it received that of Miletus, given to it by Neleus, son of Codrus, king of Athens, who conducted thither a colony of Ionians, *anno* 1100 B.C. Few cities have been more celebrated for their population, wealth, commerce, and civilization. The citizens of Miletus early distinguished themselves by their skill in navigation, and still more by the number of the colonies they had established along the coast of the Hellespont, the Propontis, and the Euxine ; which enabled them to engross the greater part of the trade in slaves, which, in antiquity, were principally furnished by the country round the Euxine, as well as the trade in corn, fish, and furs. She is also famous for her numerous works of art, the magnificence of her festivals, and the luxury, refinement, and opulence of her people. Among her most illustrious citizens were the venerated names of Thales, one of the sages of Greece ; Hecateus, one of the most ancient historians ; the philosophers Anaximander and Anaximenes ; Cadmus, the first who wrote in prose, and Timotheus, a famous musician and poet.[*] She also gave birth to Aspasia, the most accomplished and celebrated of courtesans ; and Venus had nowhere more numerous and beautiful priestesses. Miletus was, in fact, the Athens of Ionia, *urbem quundam Ionis totius belli pacisque artibus principem*. (*Mela*, lib. i., cap. 17.) Near the *Posidæum Promontorium* (hod. cape Arbora), about 12 m. S. by W. Miletus, was an oracle and splendid temple of Apollo, surnamed Didymæus. This temple having been burned down by Xerxes, was rebuilt on a still more magnificent scale, by the Milesians. Part of the ruins yet remain ; and "the columns are so exquisitely fine, the marble mass so vast and noble, that it is impossible, perhaps, to imagine greater beauty and majesty of ruin." (*Chandler*.) Miletus had also within her territory, mount Latmos, famous for the loves of Endymion and Diana ; and the fountain Byblis, so called from the unhappy sister of Apollo, who here expired of love and grief. (*Ovidii Met.*, lib. ix., lin. 454, &c.)

But *quantum mutatus*! Miletus is now a mean deserted place, which still, however, bears the name of Palat, or Palatia, the *Palaces*. The principal existing memorial of her ancient grandeur is a ruined theatre, which must, when entire, have been a magnificent structure. It is 457 feet in front, and is visible at a great distance. The site of the ancient city is encumbered with heaps of rubbish, and overrun with thickets, interspersed with fragments of walls, broken arches, fallen columns, and pedestals. It is evident, from the remains of a number of mosques, that Mohammedanism had once flourished here ; but, with a single exception, the ruins seem to have belonged to mean and paltry structures.

In antiquity, Miletus underwent many vicissitudes. Having joined in the revolt of the Ionian cities, she was besieged and taken by the Persians, in the reign of Darius Hystaspes, *anno* 493 B.C., when the inhabitants were obliged to evacuate their city. But being afterward allowed to return, Miletus again rose to great wealth and distinction. She opposed a vigorous resistance to Alexander the Great ; but, instead of punishing, the conqueror magnanimously restored the city to her ancient freedom. She appears to have been indulgently treated by the Romans ; and continued to be a considerable city, till she fell, in an evil hour, under the ferocious and brutal sway of the Turks, who first sacked, and subsequently destroyed, this ancient glory of Ionia ! To complete her misfortune, her port is now almost filled up

The government of Miletus, and of the other cities of Iona, was usually popular and republican; but, like their mother cities, they were distracted by faction, and frequently subjected to oligarchs or tyrants. Of the Milesian tyrants, the most celebrated was Thrasybulus, whose answer to the inquiry of Periander of Corinth may be seen in *Aristotle's Politics*, lib. iii., cap. 10.

Miletus and the principal states of Ionia, including the islands of Chios and Samos, being connected by the ties of a common origin and interest, were in the habit of sending deputies to a general council or assembly, to debate and determine upon measures for promoting their union and security. This council met at Panionium, so called from the circumstance, on the N. side of mount Mycalè, opposite Samos, about midway between Ephesus and Miletus; the place was regarded as sacred, and was put under the especial protection of Neptune, the chosen guardian and favourite divinity of the Ionians. (*Ibi est Panionium, sacra regio, et ab id eo nomine appellata, quod eam communiter Iones colunt. Mela, ubi supra*; see also *Pliny, Hist. Nat.*, lib. v., cap. 19; and *Herodotus*, lib. i., cap. 148.) Thales, who saw that, without a more intimate union, the Ionians could make no effectual resistance to foreign aggression, advised his countrymen to establish a really federal system of government, and to concert and execute their public measures in common. (*Herod.*, lib. i., cap. 170.) But this judicious advice was not acted upon; and it was only on urgent occasions, such as the invasion of Ionia by the Persians, that a sense of common interest and danger prevailed over their mutual jealousies and antipathies, and made any considerable number of the cities act in unison.

Most commonly the debates and decrees of the assembled deputies seemed to have referred only to matters connected with religion, precedence, or ceremony. This appears evident from the circumstance of the deputies meeting at Panionium, when the Ionian cities were subject to the Persians and others, as well as when they were independent. (for farther information as to Miletus, see *Herodotus*, lib. i., caps. 142, 143, 146, &c. and lib. vi., caps. 18 and 21; *Strabo*, lib. xiv.; *Chandler's Travels in Asia Minor*, cap. 42, 43, and 45; *St. Croix, de l'Etat et du Sort des Anciennes Colonies*, p. 283-289, &c.

MILFORD, p. t., Hillsborough co., N.H., 33 m. S. by W. Concord, 458 W. Watered by Souhegan river, which affords water-power. It contains four stores, one fulling-mill, one grist-mill, one saw-mill, two tanneries; 11 schools, 387 scholars. Pop. 1455.

MILFORD, p. t., Worcester co., Mass., 30 m. S.W. by W. Boston, 400 W. Watered by Charles and Mill rivers. Incorporated in 1780. It has three Congregational churches, eight stores, one woollen factory, five grist-mills, five saw-mills; one academy, 22 students; five schools, 165 scholars. Pop. 1773.

MILFORD, p. t., New-Haven co., Ct., 45 m. S.S.W. Hartford, 291 W. Bounded W. by Housatonic river, S.E. by Long Island sound. Watered by Wepawaug river, which affords water-power, and Indian river. A quarry of beautiful serpentine marble is found here. "A drawbridge 50 rods long crosses Housatonic river, and connects the town to Stratford. The harbour, though not deep, admits vessels of 200 tons burthen. Excellent oysters and clams are taken on the coast, and shad are taken in Housatonic river. An island containing ten acres lies off the shore, and connected with it by a sand-bar, which is bare half the time. A breakwater from Indian point towards the island would form a spacious harbour. It has four churches, two Congregational, an Episcopal, and a Baptist; 11 stores, a lumber-yard, four grist-mills, three saw-mills, one tannery; two academies, 22 students; 14 schools, 764 scholars. Pop. 2455.

MILFORD, p. t., Otsego co., N.Y., 13 m. S. Cooperstown, 77 m. W. Albany, 365 W. Watered by Susquehannah river and its tributaries. It contains a Presbyterian and a Methodist church, four stores, one fulling-mill, one woollen factory, two grist-mills, eight saw-mills, three tanneries; 12 schools, 550 scholars. Pop. 2095.

MILFORD, p. t., capital of Pike co., Pa., 162 m. N.E. by E. Harrisburg, 256 W. The village is situated on a commanding eminence, on the W. side of Delaware river, which is here crossed by a fine bridge. It contains two churches, an academy with 46 students; one school, 36 scholars; 150 dwellings, and 648 inhabitants.

MILFORD, t., Bucks co., Pa., 18 m. N.W. Doylestown, 36 m. N.W. Philadelphia. Drained by Swamp creek, a branch of Perkiomen creek. The village of Charlestown contains a church. It has seven stores, one fulling-mill, 10 grist-mills, 10 saw-mills, one oil-mill, two tanneries, four potteries; two schools, 47 scholars. Pop. 2193.

MILFORD, t., Juniata co., Pa., 10 m. S.E. Lewistown. Bounded S.E. by Tuscarora creek. Watered by Licking creek. It has one academy, 30 students; 10 schools, 423 scholars. Pop. 1884.

MILFORD, t., Somerset co., Pa., 7 m. S.W. Somerset. Watered by Castleman's river and Laurel Hill creek. It contains iron ore and coal, and has three stores, two fulling-mills, five grist-mills, 15 saw-mills, one oil-mill, two tanneries, two distilleries, two potteries; one school, 94 scholars. Pop. 1632.

MILFORD, p. v. and hundred, Kent co., Del., 95 m. S. Philadelphia, 21 S. by E. Dover, 120 W. Situated on the N. side of Mispillion creek, which flows into Delaware bay. The hundred contains 13 stores, six grist-mills, three saw-mills, two tanneries; two academies, 65 students; six schools, 231 scholars. Pop. 2356.

MILFORD HAVEN, is an extensive basin, or inlet of the sea, deeply indenting the S. part of the co. Pembroke, in S. Wales, and forming one of the most capacious and safest asylums for shipping in the British dominions. St. Anne's head, forming the N.W. extremity of the entrance to the haven, lat. 51° 41' N., long. 5° 10' 25" W., is 145 ft. in height, and is surmounted by two lighthouses, with fixed lights, respectively 15 and 45 ft. in height. The entrance is about 1½ m. in width; what may be called the haven is from 10 to 11 m. in length; but it branches out into an immense number of deep bays, creeks, and roads. The water is deep; and being completely land locked, and the anchorage-ground of the very best description, ships ride within the haven as safely as if they were in dock. At springs the tides rise from 28 to 30 ft., affording unusual facilities for the repair of ships, enabling them to get to sea with comparatively little difficulty, and to sail in, even though the wind should be contrary. It may be entered without a pilot as well by night as by day.

MILFORD TOWN, on the N. side of the basin, 6 m. W. by N. St. Anne's head, was founded in 1784. It is finely situated; is especially remarkable for the mildness of its climate, and has some good buildings, including a handsome church. A dock-yard constructed here in 1790 has, however, been removed to Pater-Dock, on the S. side of the haven, and the town has not increased in the degree that was anticipated. It is probable, too, that the advantage of having the rendezvous for the fleets in stations better adapted for watching the coasts of France will always prevent Milford, or any other port on the haven, from attaining that importance as a naval depôt which the excellence of the haven might seem to insure; while the barrenness of the surrounding country, and the want of all internal communication with any considerable manufacturing district, have prevented, and most likely will continue to prevent, any of the places on the haven from becoming of much consequence in a commercial point of view. The mail-packets for Waterford sail from Pater. (*Stat. of the Brit. Empire*, i., 51; *Nicholson's Cambrian Guide*, p. 425.)

MILHAU (an. *Æmilianum*), a town of France, dep. Aveyron, cap. arrond., on the Tarn, 30 m. S.E. Rodez. Pop., in 1836, 9437. It is generally well built, and its streets, though narrow, are regular. It has several squares and public fountains, and a good bridge over the Tarn. Few vestiges exist of its ancient castle and walls; the latter were demolished by Louis XIII. in 1629, after which, says Hugo, *La ville cessa de s'occuper des affaires politiques ou religieuses, et tourna tous ses efforts vers le commerce et l'industrie, qui en ont fait la ville la plus riche et la plus peuplé du département*. It produces woollen cloth, leather and leather gloves, silk twist; and has a considerable trade in cheese, timber, cattle, wool, almonds, wine, and other agricultural products. It is the seat of a court of primary jurisdiction, a tribunal and a chamber of commerce, a communal college, society of agriculture, &c.: it was one of the strongest holds of the Calvinists in the French religious wars.

MILL, t., Tuscarawas co., O. Watered by Big Stillwater creek. It has two commission-houses in foreign trade, three retail stores, one grist-mill, two saw-mills; seven schools, 304 scholars. Pop. 1925.

MILLBURY, p. t., Worcester co., Mass., 43 m. W.S.W. Boston. Bounded N.E. by Blackstone river. The Blackstone canal passes through it. It contains a bank, four churches, three Congregational and a Baptist; five stores, 11 fulling-mills, five woollen factories, three grist-mills, three saw-mills, one paper mill; one academy, 45 students; 10 schools, 462 scholars. Pop. 2171.

MILL CREEK, hund., Newcastle co., Del. It has three schools, 145 scholars. Pop. 3144.

MILL CREEK, t., Hamilton co., O. It lies N. of Cincinnati, and is a suburb of it. It contains one college (Lane Seminary), 32 students; one academy, 125 students; nine schools, 169 scholars. Pop. 6240.

MILLEDGEVILLE, city, capital of Baldwin co., Ga., and of the state, 159 m. N.W. Savannah, 90 m. W.S.W. Augusta, 193 m. N. Darien, 648 W. Situated on the W. bank of Oconee river, at the junction of Fishing creek, at the head of steamboat navigation. The surface is uneven. It contains 10 streets running parallel with the river, crossed by 10 others at right angles with them. The streets are 100 feet wide, excepting Washington, a central street, which is 120 feet wide. There are three public squares, made by the omission of a street in each direction, which are Statehouse-square, Governor's-square, and penitentiary-square, 450 feet each on a side. The statehouse is on an eminence in the centre of Statehouse-square, three fourths of a mile from the river, a tasteful edifice of Gothic architecture, erected in 1838 at a cost of $115,000, with an excellent clock in the cupola which cost $1000. The representatives' hall is 60 by 54 feet, ornamented with full length portraits of General Oglethorpe and La Fayette, and the senate chamber with those of Washington and Jefferson. In the executive office is an old portrait of General Oglethorpe, in an antique dress, and a sitting posture, examining the map of Georgia. The edifice contains offices for the state treasurer, comptroller, surveyor-general, besides clerks' and committee rooms, and several fire-proof rooms for public records. On Statehouse-square are also an academy, a state arsenal, and a powder magazine. The other public buildings are a governor's house, in a range with the N. side of Statehouse-square, three churches, a Presbyterian, Methodist, and Baptist; three banks, and a market-house. A toll-bridge crosses the Oconee, here 550 feet wide, which cost about $20,000. It contains 14 dry goods and 90 grocery stores, wholesale and retail, four printing-offices, four weekly newspapers, 250 dwellings, and 2095 inhabitants. It was laid out in 1803, and the first house was erected of logs in 1804, and the first framed house in 1805. It exports considerable cotton, for which it presents great facilities. It is governed by an intendant and four commissioners. Nett proceeds of the postoffice, $3049. Near to Milledgeville, in Medway village, is Oglethorpe college, founded in 1836, which has a president and five professors or other instructers, 65 students, and 9000 volumes in its libraries.

MILLER, county, Mo., situated in the centre of the state, and contains 555 sq. m. Watered by Osage river. It contained in 1840, 2035 neat cattle, 2320 sheep, 9304 swine; and produced 5005 bushels of wheat, 99,180 of Indian corn, 8163 of oats, 3418 of potatoes, 4601 pounds of tobacco, 1909 of cotton. It had two stores, five grist-mills, one saw-mill, three distilleries; three schools, 65 scholars. Pop.: whites, 2170; slaves, 111; free coloured, 1; total, 2282. Capital, Tuscumbia.

MILLERSBURG, p. v., Hardy t., capital of Holmes co., O., 87 m. N.E. Columbus, 343 W. Situated on the E. side of Killbuck creek. It contains a courthouse, jail, a church, 10 stores, a printing-office, issuing a weekly newspaper, 100 dwellings, and about 600 inhabitants.

MILLVILLE, p. t., Cumberland co., N.J., 67 m. S. by W. Trenton, 178 W. Drained by Maurice river and its tributaries. It contains five stores, five glasshouses, belonging to Quaker, in the group of the central Cyclades, the summit of mount St. Elias; in its S.W. angle, 3036 ft. above the sea, being in lat. 36° 40′ 22″ N., long. 94° 21′ 14″ E. Pop. 2590 ? This island is said by Pliny to be circular (Osmium rotundissima, lib. iv., cap. 12); but it is really of an oblong shape, being about 13 m. in length from E. to W., and, where broadest, about 7 m. across: it is indented on its N. side by a spacious bay, stretching N.W. and S.E. about 6 m., which has deep water throughout, and forms one of the best asylums for shipping in the Levant. This island is obviously of volcanic origin; mount Calamo, indeed, is at this moment a semi-active volcano, emitting smoke and sulphureous vapours; in many places the earth is hot, and there are numerous hot springs, one of which, in a natural grotto, is used by the natives as a sudatory. It also furnishes abundant supplies of iron, alum, sulphur, and salt.

A considerable portion of the surface is rugged and mountainous, and has a naked and sterile appearance; but the valleys and low grounds are extremely fertile, such small portions of them as are cultivated producing corn, wine, oil, cotton, oranges, and other fruits in the greatest profusion. In point of fact, however, Milo is now almost depopulated, and nearly a desert; a result that is partly to be ascribed to the ravages of the plague, the badness of the water, which is generally brackish, and the prevalence of malaria; but far more to the influence of that brutalising despotism under which it has groaned for centuries. Milo, the capital, situated near the bottom of the bay, is rendered unhealthy from the vicinity of salt marshes, and is an inconsiderable, wretched place.

Castro, another town, near the entrance to the harbour, on its E. side, is built on the summit of a conical hill, the houses appearing to rise above the roofs of each other. A little to the S.W. of Castro, near the shore, are the ruins of the ancient city. The remains of a theatre, built of large masses of the finest marble, and fragments of solid walls have been discovered. In the vicinity are numerous catacombs, cut in the porous rock. (Tournefort, Voyage du

Levant, lettre iv.; Olivier, Empire Ottoman, ii., cap. 9; Turner's Tour in the Levant; Purdy's Sailing Directions for the gulf of Venice, &c., p. 97.)

Such is the present state of this once famous island. Melos is said by Thucydides to have been independent 700 years before the Peloponnesian war. The most probable opinion seems to be that the Melians were descended from the Lacedæmonians; but, however that may be, they declined taking any share in that contest, and though pressed by the Athenians to espouse their cause, declared their neutrality. The Athenians, however, having the command of the sea, determined to coerce the Melians into submission to their mandates; and though the first expedition sent against them failed of its object, the second was more successful. Thucydides gives the substance of the speeches made by the Athenian commanders to the Melians previously to their commencing hostilities; and on no occasion has the robber's plea, that whatever the powerful may please to command, the weaker are bound to obey, been more broadly and unequivocally asserted. The sequel of their conduct was worthy of the principle thus laid down; for the Melians having, after a stout resistance, surrendered at discretion, the Athenians put all the full-grown males to the sword, and carried the women and children to Athens, where they were sold as slaves! This detestable atrocity was perpetrated shortly before the Athenians engaged in their expedition against Syracuse; and is related by Thucydides, without note or comment, as if it had been a legitimate and ordinary occurrence! (Thucyd., lib. vi., ad finem.) The fortune of war having, however, soon after turned against the Athenians, the captive Melians were restored to their native country; and the island continued to be comparatively prosperous till, after innumerable vicissitudes, it was seized upon by the Turks, under whom it has been reduced to the abject state in which we now find it. It seems to have retrograded materially between the visit of Tournefort, in 1700, and that of Olivier, at the end of the century. It is now, however, included in the kingdom of Greece, and will, most probably, recover some portion of its former prosperity.

MILO, p. t., Yates co., N. Y., 193 m. W. Albany, 220 W. It contains the v. of Penn Yan, the capital of the county. Bounded W. by Crooked lake, E. by Seneca lake. Watered by the outlet of Crooked lake, which affords water-power. It contains 20 stores, one furnace, four fulling-mills, two grist-mills, nine saw-mills, one oil-mill, one tannery, one distillery; 17 schools, 938 scholars. Pop. 3986.

MILTON, or MILTON-ROYAL, a fishing town and par. of England, lathe Scray, co. Kent, hund. its own name, 11 m. N.E. Maidstone, and 36 m. E. by S. London. Area of par., 2340 acres. Pop., in 1831, 2223. The town, on the declivity of a hill, sloping down to a creek which opens into the channel between the isle of Sheppey and the coast of Kent, is old and irregularly built. It has a market-house and shambles near its centre; and at its N. end is an old courthouse. The church, which stands at a considerable distance from the present town, is a spacious fabric, with a square tower of flint stone laid in even rows. The living is a rectory, in the gift of the dean and chapter of Canterbury. There are places of worship, also, for Wesleyan Methodists and Baptists, with attached Sunday schools; a free school was founded in 1718.

Milton has, for many centuries, almost entirely depended on its oyster fisheries, the produce of which is highly esteemed, and distinguished as the " Milton natives." The right of the fishery, within certain limits, formerly belonging to the abbey of Faversham, and afterwards to the crown, is now held on lease from the lord of the manor by a company of free dredgers, composed of the principal fishermen of the town; and in 1831 there were 119 families immediately dependent on the trade in oysters, which is principally carried on with the metropolis. The town has four wharfs; and, besides oysters, considerable quantities of corn and flour produce are shipped for the London market. Both the town and port are under the jurisdiction of a portreeve, elected annually by the inhabitants paying poor's rates.

Milton disputes with Richborough the honour of having furnished the Roman epicures with the oysters alluded to by Juvenal:—

" Rutupinove edita fundo
Ostrea."　　　　　　　　　Sat. iv. 141.

MILTON, p. t., Strafford co., N. H., 47 m. N.E. Concord, 515 W. Organized from Rochester in 1802. Bounded E. by Salmon Fall river for 13 m., by a branch of which it is watered. It has Teneriffe mount in its E. part. It contains four churches, a Congregational, Methodist, Free-will Baptist and Christian, six stores, three fulling-mills, one woollen factory, three grist-mills, five saw mills, two tanneries, 19 schools, 466 scholars. Pop. 1392.

MILTON, p. t., Chittenden co., Vt., 55 m. N.W. Montpelier, 588 W. Chartered in 1763. Watered by Lamoille river, which affords extensive water-power, having near the v.'s fall of 130 feet in 50 rods. It contains three churches, a a Congregational, Methodist, and Baptist, nine stores, two grist-mills, two saw-mills, one paper-mill, three tanneries; 14 schools, 645 scholars. Pop. 2134.

MILTON, p. t., Norfolk co., Mass., 7 m. S. Boston, 435 W. Chartered in 1662. Watered by Neponset river, which affords water-power. The first paper-mill erected in New-England, was built here in 1728. It contains three churches, a Congregational, Methodist, and Unitarian, two commercial and two commission houses in foreign trade, eight retail stores, two woollen factories, one grist-mill, one saw-mill; two academies, 45 students, five schools, 436 scholars. Pop. 1822.

MILTON, t., Saratoga co., N. Y., 30 m. N. by W. Albany. It contains the v. of Ballston Spa, the capital of the co. Drained by Kayaderosseras creek, and its tributaries. It contains 16 stores, six fulling-mills, three woollen factories, two furnaces, six grist-mills, 10 saw-mills, one oil-mill, one paper-mill, one printing-office, one weekly newspaper; 12 schools, 387 scholars. Pop. 3166.

MILTON, p. b., Turbut t., Northumberland co., Pa., 71 m. N. Harrisburg, 181 W. Situated on the E. bank of the west branch of Susquehanna river, on both sides of Limestone run. It contains three churches, a Presbyterian, Associate Reformed, and German Reformed, an academy, 13 stores, two grist-mills, one saw-mill, one tannery, four distilleries, one brewery, one pottery, two printing-offices, one weekly newspaper, four schools, 250 scholars. Pop. 1508.

MILTON, p. t., Trumbull co., O., 157 m. N.E. Columbus, 316 W. Watered by Mahoning river. It has eight schools, 480 scholars. Pop. 1277.

MILTON, t., Richland co., O. It has 347 scholars in schools. Pop. 1861.

MILTON, t., Wayne co., O. It has two stores, five saw-mills. Pop. 1157.

MILVERTON, a market town and par. of England, co., Somerset, hund. its own name, 6½ m. W. Taunton, and 139 m. W.S.W. London. Area of par. 6400 acres. Pop. in 1831, 2233. The town situated in a richly-wooded and well cultivated country, is small and ancient, consisting chiefly of three irregular streets, with the church, a large building standing on an eminence in the centre. An extensive manufacture of serges and flannels is carried on here; and there is a silk-mill, which in 1839, employed 54 hands. Milverton was formerly a borough, and is still governed by a portreeve, appointed by the lord of the manor. Petty sessions for the hundred are held here. Markets on Friday; cattle fairs, Easter Tuesday, July 25, and October 10.

MILWAUKEE county, Wis. Situated in the W. part of the territory. Watered by Milwaukee, Root, Des Plaines, and Fox rivers. Bounded E. by lake Michigan. It contained in 1840, 5100 neat cattle, 798 sheep, 8814 swine; and produced 34,236 bushels of wheat, 38,820 of Indian corn, 1889 of buckwheat, 26,863 of oats, 64,242 of potatoes, 46,886 pounds of sugar. It had three commission houses in foreign trade, 26 stores, eight grist-mills, 13 saw-mills, two printing-offices, two weekly newspapers; nine schools, 186 scholars. Pop. 5605. Capital, Milwaukee.

MILWAUKEE, p. v., capital of Milwaukee co., Wis., 83 m. E. Madison, 806 W. Situated on both sides of Milwaukee river, near its entrance into lake Michigan. The river affords water-power. Steamboats ply between this place and Buffalo, in the season of navigation. It contains a court-house, jail, a United States land office, two churches, a Presbyterian and Methodist, and in 1842, 2800 inhabitants. It had in 1840, three commission houses in foreign trade, 21 retail stores, one furnace, one saw-mill, two printing-offices, two weekly newspapers; one school, 25 scholars. Pop. 1712.

MINCHIN-HAMPTON, a market town and par. of England, co. Gloucester, hund. Longtree, 12 m. S. by E. Gloucester, and 89 m. W. by N. London. Area of par., 4880 acres. Pop., in 1831, 5114. The town, on the W. escarpment of the Cotswold hills, consists of a long irregular street, extending N. to S. along the road from Gloucester to Chippenham, and crossed by another leading to the parish church, near the market-house. The church is a large cruciform structure, surmounted by an octagonal embattled tower, rising from the intersection of the nave and transepts: the living is a rectory in private patronage. The Wesleyan Methodists and Baptists have also their respective places of worship, with attached Sunday schools; and there is a well-attended national school, for children of both sexes, besides a respectably endowed grammar-school. Minchin-Hampton, which is only 4 m. S.E. of Stroud, the centre of one of the largest clothing districts of the county, has numerous cloth-factories on the banks of the numerous brooks in the vicinity; and in 1839 it had 12 woollen-mills, furnishing employment to 785 hands. Nearly 100 hands were then, also, engaged in hand-loom weaving. Trade, however, has for some years been on the decline, and its fluctuations have caused great distress among the weaving population: the average earnings of each family when in full work amount to 16s. per week, of which 6s. 10d. may be assumed as the produce of hand-loom labour. Markets on Tuesday: fairs for cattle, horses, and cheese, Trinity-Monday and October 29. *See* STROUD. (*Hand-loom Weavers' Rep.*, &c.)

MINCIO (an. *Mincius*), a considerable river of N. Italy, which has its source in the Lago di Garda; and which, flowing S., with many windings, by Mantua, unites with the Po 12 m. S.E. that city. In the upper part of its course, till it approaches Mantua, it is rather rapid; but from near Mantua to the Po it has a sluggish current, and is navigated by the boats that ply on the latter. Virgil, who first saw the light on the banks of this river, has celebrated its praises:

"Tardis ingens ubi flexibus errat
Mincius, et tenera prætexit arundine ripas."

Georg., lib. iii. lin. 14.

MINDEN, a strongly fortified town of the Prussian states, prov. Westphalia, cap. reg. of its own name, on the Weser, here crossed by a bridge 600 ft. in length, near the Hanoverian frontier, 60 m. E.N.E. Munster, lat. 52° 17' 47" N. long. 8° 53' 26" E. Population (1838) 7800. It is irregularly built, and has no remarkable edifice, except a handsome cathedral, and new and good barracks. Minden has a gymnasium or college, a normal school, an orphan asylum, four hospitals, and other public institutions, and is the seat of a court of justice for the town and district, and of a board of taxation; but the court of appeal for the regency is at Paderborn. Manufactures considerable; consisting of woollens, stuffs, linen, hosiery, hats, gloves, tobacco, soap, refined sugar, &c. A number of saw-mills are employed in the preparation of the wood brought down the Weser, and it enjoys a considerable share of the transit trade on this river between Bremen and Prussian Westphalia, Hesse-Cassel, &c. In the neighbourhood are coal mines and salt springs, both very productive. The fortifications of Minden have been much improved since 1815. This town was the residence of several early German emperors, and various diets were held in it.

The French were defeated in the vicinity of Minden in 1759, by the Prussians under prince Ferdinand, brother to Frederick the Great, and the British under Lord George Sackville. The non-compliance of the latter with the orders of the former is said to have saved the French from a complete rout, and gave rise at the time to a great deal of acrimonious discussion. (*Berghaus*, &c.)

MINDEN, p. t., Montgomery co., N. Y., 61 m. W.N.W. Albany, 396 W. Organized in 1798. Bounded N. by Mohawk river. Drained by Otsquake creek. It contains a Presbyterian and Universalist church, 26 stores, two grist-mills, 13 saw-mills, three tanneries, one distillery; 16 schools, 960 scholars. Pop. 3507.

MINERAL POINT, p. v., capital of Iowa co., Wis., 52 m. W.S.W. Madison, 884 W. Situated on the head branch of Pekatonookee river, and contains a courthouse, jail, a United States land office, several stores, and about 700 inhabitants. In the vicinity are three smelting houses for copper, and several for lead.

MINERSVILLE, p. b., Norwegian t., Schuylkill co., Pa., 66 m. N.E. Harrisburg, 176 W. Situated in the valley of the West Branch of Schuylkill river on the West Branch railroad. It contains eight stores, two saw-mills, various mechanic shops, 100 dwellings, and about 600 inhabitants. It is surrounded by anthracite coal mines, of a good quality. Incorporated in 1831.

MINISINK, p. t., Orange co., N. Y., 120 m. S.S.W. Albany, 270 W. On the E. it is covered with the "drowned lands," through which Wallkill river flows, with a sluggish course. It contains six churches, four Presbyterian and two Baptist, 17 stores, two fulling-mills, four woollen-factories, 11 grist-mills, 16 saw-mills, six tanneries, 10 distilleries; four academies, 387 students, 23 schools, 1308 scholars. Pop. 5093.

MINEHEAD, a seaport, decayed bor., and market town of England, hund. Carhampton, co. Somerset, on the Bristol channel, 22 m. N.W. Taunton, and 149 m. W. by S. London. Area of par., 3780 acres. Population, in 1841, 1489. The town comprises three distinct masses of building, forming a triangle, the sides of which are about ½ m. long: the best part, which contains some good houses and inns, being about ¼ m. from the sea. The church, which is large and handsome, stands on the slope of a hill N. of the town: there is also a place of worship for Wesleyan Methodists, and a well attended Sunday-school. A free school for 30 boys is supported by the lord of the manor; besides which, there are several bequests of money-charities for the relief of the poor. Minehead formerly had a considerable share in the herring fisheries: and had a large trade with Ireland, as well as with the Mediterranean and N. America. Its consequence, however, as a port, has greatly declined, notwithstanding its commodious harbour and pier; but it has lately been much frequented as a watering place and, the inhabitants are at present mainly supported by the influx of visitors. Minehead received its charter of incorporation in 1 Eliz., from which time down to the Reform Act, by which it was disfranchised, it returned two members to the House of Commons, the right of election being vested in the resident housekeepers in the parishes of Minehead and Dunster. The corporation was considered too insignificant to be mentioned in the Mun. Reform Act. Markets on Wednesday.

MINORCA (*Balearis Minor*), the second in size of the Balearic islands, belonging to Spain, in the Mediterranean, off the E. coast of Spain, from which it is distant about 140 m., Mahon its cap. being in lat. 39° 51′ 10″ N., long. 4° 18′ 7″ E.; it is of an oblong shape, extending from W.N.W. to E.S.E., but somewhat concave on its S. side. Length 32 m.; average breadth, 9 m.; area about 290 sq. m. Population, according to Miñano, 44,147. The coast is indented on every side, but particularly on the N., with small bays, or deep creeks, and is surrounded with islets, rocks, and shoals. Surface very uneven, with abrupt hills and knolls; but there are no mountains, except El Toro, near its centre, which rises 4793 feet above the sea. Iron, lead, and copper have been found, though in too small quantities to be wrought; but marble is extremely abundant, and of many beautiful varieties, as is seen in the churches and houses of Port Mahon. Water is scarce, and the climate is less mild and agreeable than that of Majorca. The air in winter is damp and raw, owing to the frequency of the N. winds; but snow is seldom seen. The temperature during spring is mild, and the air pure, though somewhat moist: the summer heat is very oppressive, and the autumn is remarkable for its frequent and heavy rains. The soil is in most parts poor, sandy, and unproductive; but on the hill sides are several fertile tracts, on which good crops of corn and wine are raised with little labour. Excepting a few evergreen oaks near the centre of the island, Minorca is almost destitute of trees; a circumstance attributable partly to the devastations of war, and partly to the violent N. winds, which are extremely injurious to plantations. Wheat and barley are the grains chiefly cultivated; both of middling quality, and scarcely sufficient to supply two-thirds of the consumption of the island. Red and white wines are made in large quantities, and about 10,000 arrobas a year are exported, but the olive will not thrive in consequence of the cold N. winds. Capers grow spontaneously, and form an important article of export. Flax, hemp, saffron, and the cotton plant succeed well, but are little attended to. Fruits of various kinds are abundant, though much inferior in flavour to those of Majorca. Vegetables, also, are plentiful, and of good quality. The island is well suited for pasturage, and is well supplied with cattle, sheep, goats, and mules: wool is exported in considerable quantities; and the cheese of Minorca is considered by the Italians as equal, if not superior, to Parmesan. Bees, also, are reared in great abundance, and furnish large supplies of excellent honey and wax. Partridges, quails, and other game, are plentiful. Lizards swarm; and there are several varieties of venomous reptiles, but no beasts of prey. Fish, especially anchovies abound on the coast, and the oysters of Minorca are held in high estimation by the Catalonians.

The trade of Minorca, chiefly carried on at Port Mahon, consists in the export of wine, wool, cheese, capers, honey, and wax, chiefly to Spain, but also to Genoa, Leghorn, and the ports of France. The imports comprise wheat, oil, linen, cotton and woollen fabrics, wood, tobacco, and a variety of manufactured goods and colonial products from Spain, France, and Italy. The possession of Minorca by the British during the greater part of last century did something to awaken a spirit of activity and enterprise among the inhabitants. Since its restoration to Spain, however, its industry and commercial importance have greatly declined. Accounts are kept in Spanish money; but some of the inhabitants still retain the English mode of accounting. "The inhabitants of Minorca," says Fischer, "are ardent, courageous, ingenious, and make excellent sailors. That activity of mind which distinguishes the Mallorcans, they possess, perhaps, in a still higher degree; for they are extremely lively, sociable, and even convivial. As the climate and soil of Minorca are greatly inferior to those of Majorca, the people of the former island are much less opulent than the Mallorcans; but they bear a close affinity to each other in language, manners, and religion." (*Pict. of Valencia*, p. 289.) They are enthusiastically fond of religious processions, and are as bigoted and ignorant as possible. Dancing and playing on the mandolin are their chief amusements. The modern inhabitants are said to be as expert as their ancestors in the use of the sling.

Minorca is divided into the four districts, or *terminos*, of Mahon, Alayor, Mercadel, and Ciudadela, which are the names also of the four largest towns. Mahon, the cap. (an *Portus Magonis*), at the E. end of the island, with a popu-

lation of about 19,000, is on the whole, well built, chiefly in the English style; but the older streets are narrow, crooked, and badly paved. The harbour is one of the best and most capacious in the world: three large squadrons have, more than once, been at anchor in it at the same time, and there is excellent mooring ground in five and six fathoms, sheltered from every wind. It has three rocky islets: on one stands a hospital, on another the lazaretto, and on the third is an arsenal, with naval store-houses, all built by the English. Ciudadela (which see) is the ancient capital, but its population is not above 8000. The other towns are little more than mere villages.

The ancient history of Minorca is nearly identical with that of Majorca. In 1285 the Moors were finally expelled from both islands, which were then formally annexed to the crown of Arragon. In 1708, during the war of the Spanish succession, the English took possession of the island, with the intention of making it a naval station. It was confirmed to the British by the peace of Utrecht, and remained in their possession till 1756, when it was taken by a French fleet and army, after the failure of the attempt to relieve it by Admiral Byng, which led to the memorable trial and death of the latter. At the peace of 1763 Minorca was restored to Great Britain, but in 1782 it was retaken by the Spaniards. It was once more taken by the British in 1798, and was finally ceded to Spain at the peace of Amiens in 1802. (*Fischer's Valencia*, p. 280–289; *Sir G. Temple's Excursions in the Mediterranean*, i. 13–15; *Miñano*.)

MINOT, p. t., Cumberland co., Me., 46 m. S.W. Augusta, 576 W. Bounded E. by Androscoggin river, S. by Little Androscoggin river, which afford extensive water-power. A bridge crosses Androscoggin river to Lewiston. It contains 15 stores, one furnace, two fulling-mills; 26 schools, 1250 scholars. Pop. 3550.

MINSK, a government of Russian Poland, comprising the former palatinate of Minsk, and portions of the palatinates of Polock, Wilna, Novogrodek, &c. It is principally included between the 52d and 56th degs. of N. lat., and the 96th and 30th of E. long., having N. and E. the govs. Witepsk and Moghilev, S. Kief and Volhynia, and W. Grodno and Wilna. Area about 42,000 sq. m. Population, in 1838, 1,034,800, of whom about 100,000 were Jews. Surface mostly level, but in the N. a chain of hills separates the waters flowing towards the Black sea from those that fall into the Baltic. In the S. is a large extent of marsh land, along the banks of the Pripet. Exclusive of this river and the Dnieper, the other principal rivers are their affluents, the Beresina, Styr, Gorin, Pechitsa, &c.: the Dwina forms, for a short distance, the N. and the Niemen the W. boundary of the government. There are a number of small lakes, and in spring a great portion of the country is inundated, so as to form a vast sheet of water. Though one of the poorest and worst cultivated parts of the empire, Minsk produces more corn, principally rye, than is required for home consumption. Hemp and flax are important products, as are potash and tar. The forests are very extensive; and, next to agriculture, sawing and trading in timber is the principal occupation of the population, and numerous large rafts are floated down the rivers to Kherson on the one hand, and to Riga and Koenigsberg on the other. The rearing of live stock is mostly ill-conducted; pasturage is good in some parts, but the sheep yield only inferior wool. A great many bees are reared. Some little iron is obtained. Linen weaving and distilling are pretty general; a little woollen cloth is made; there are some iron forges and glass factories; and at Pinsk, in the S.W., Russia leather is prepared. The trade of the government is chiefly conducted by strangers. In 1832, only 2150 children were receiving public instruction, and there were but three printing establishments in the government. Chief towns, Minsk the capital, Bovhrovich, and Sloutsk.

MINSK, a town of Russian Poland, cap. the above government, and one of the largest in Lithuania, about 400 m. W.S.W., and 150 m. W. by S. Erodno. Lat. 53° 54′ 9″ W.; long. 27° 52′ 15″ E. Pop. 14,600. Its streets are narrow, irregular, and dirty, and its houses nearly all of wood; but the town has, upon the whole, a respectable appearance, and some good buildings, among which are several Greek, Greek-united, and Roman Catholic churches, a synagogue, a gymnasium, founded in 1773, and a handsome theatre. It is the see of a Greek archbishop and a Roman Catholic bishop, and has manufactures of woollen cloths, hats, and leather. Under the Poles, Minsk was the cap. of the palat. of the same name. (*Schnitzler La Russie*, p. 402–403.)

MIRANDOLA, a town of N. Italy, duchy of Modena, cap. cant., on the Burana, 18 m. N.N.E. Modena. Pop. 4600. It is walled, and has a castle, but its fortifications have fallen into decay. Among its principal edifices are a handsome cathedral and numerous churches, a hospital, and a palace belonging to the Cico family. It has manufactures of silk stuffs and twist, and woollen and cotton yarn, and an active trade in these articles and in rice, a good deal of which is grown in the neighbourhood. (*Rampoldi; Dict. Geog., &c.*)

MIRECOURT, a town of France, dep. Vosges, cap. arrond., on the Madon, a tributary of the Moselle, 16 m. N.W. Epinal. Pop., in 1836, 5507. It is ill-built, and has no remarkable public edifice; it is, however, the seat of tribunals of original jurisdiction and commerce, and has a public library of 6500 vols., &c. It is principally noted for its manufactures of violins, guitars, barrel organs, and other musical instruments, which occupy most part of the male population, while the females are employed in making lace. (*Hugo*, art. *Vosges*.)

MIREPOIX, a town of France, dep. Ariège, cap. cant., on the Lers, a tributary of the Ariège, 15 m. N.E. Foix. Pop., in 1836, inc. com., 4060. It is well built and clean, and has a large hospital, a parish church, a town-hall, and a bridge over the Lers, all handsome structures. Its inhabitants manufacture coarse woollen and cotton cloths.

MIRZAPORE, a distr. and town of British India, presid. Bengal. The district is included in the province of Benares, and is in about lat. 25° N., and between long. 82° and 83° E. Area, estimated at 3650 sq. m. Pop. uncertain, but probably about 1,000,000. Land revenue (1829–30), 10,88,291 rupees. The cap., Mirzapore, is on the Ganges, 30 m. S.W. Benares. Lat. 25° 10' N., long. 83° 35' E. It has numerous handsome European and native houses, Hindoo temples and ghauts, and is the chief mart for silk and cotton goods in the British middle provinces. Cotton stuffs and carpets, of a superior kind, are made here; and there are some iron works in the vicinity. (*Hamilton's E. Gaz.*)

MISITRA. *See* SPARTA.

MISKOLCZ, a large market town of Hungary, co. Borsod, of which it is the cap., on the great road from Pesth to Upper Hungary, 92 m. N.E. Erlau. Pop., according to the *Austrian Encyclopedie*, nearly 27,700, principally Protestants. It is well built, and has numerous churches, a Protestant and a Roman Catholic gymnasium, a Greek national school, a synagogue, a Minorite convent, &c. The wine grown in the vicinity is the chief article of trade at Miskolcz.

MISSISQUE, r., Vt., rises in Orleans co., and pursues a northeasterly course into Canada. But it soon returns into Vermont in the N.E. part of Richford. It then runs southwesterly and westerly to Swanton, whence it flows northwesterly until it enters Missisque bay, near the Canada line. It has falls affording good water-power, particularly at Swanton, but the current is generally moderate. It is about 75 miles long by the course of the river, and is navigable for vessels of 50 tons, 6 m., to Swanton falls. It is one of the four large rivers which enter lake Champlain from Vermont.

MISSISSIPPI, one of the southern United States, is bounded N. by Tennessee, E. by Alabama, S. by the gulf of Mexico and Louisiana, and W. by Pearl and Mississippi rivers, which separate it from Louisiana and Arkansas. It is between 30° 10' and 35° N. lat., and between 88° 10' and 91° 35' W. long., and between 11° 5' and 14° 26' W. long. from W. Its extreme length is 338 m., and its breadth 135 m., containing 45,760 sq. m. or 29,286,400 acres. The pop., in 1816, was 45,921; in 1820, 75,448; in 1830, 136,806; in 1840, 375,651, of whom 192,211 were slaves. Of the free population, 97,256 were white males, 81,818 were white females; there were 715 coloured males, 654 were coloured females. Employed in agriculture, 130,794; in commerce, 1303; in manufactures and trades, 4151; in navigating the ocean, 23; do. rivers, canals, &c., 100; in the learned professions, 1506.

Jackson, in Hindes county, a little W. of Pearl river, is the capital of the state. Mississippi has a sea coast of only about 70 miles, with no harbour in this distance which admits large vessels. A chain of low and sandy islands along the coast encloses Pascagoula bay, which is 65 m. long and 7 m. wide, forming an inland navigation between Mobile bay and lake Borgne, which communicates with the gulf of Mexico by a number of entrances, that admit vessels requiring 8 ft. of water. The S. part of the state, for about 100 m. from the gulf of Mexico, is a level country, covered chiefly with pine forests, cypress swamps, open prairies, or inundated marshes. Advancing north, the country becomes more hilly and broken, though there are in the state no elevations that can properly be denominated mountains. From 31° to 35° N. lat., near the Mississippi river, is a range of bluffs, which in a few places reach the margin of the stream, though generally at a little distance from it. They are an extension of the table land which extends over a portion of the state, into the low grounds on the Mississippi river, and have a fertile soil. The alluvial land on the rivers, where not liable to be overflowed, is the most valuable land in the state. The soil in its natural condition is covered with a vast growth of oak, hickory, magnolia, sweet gum, ash, maple, yellow poplar,

pine, and holly, with a great variety of underwood, grape vines, pawpaw, spice wood, &c., and cypress in the swampy alluvions of the Mississippi bottoms. By cultivation, the state produces abundantly cotton, Indian corn, sweet potatoes, tobacco, indigo, peaches, melons, and grapes. The lands watered by the Yazoo, through its whole course in the N.W. part of the state, are very fertile, while much of the land in the state, covered with pine, has a light soil. Cotton is the staple production of the state, and it is found to be more profitable than others to which the soil is also well adapted.

The state is divided into 56 counties, which, with their population in 1840, were as follows:

Counties.	Pop. 1840.	Counties.	Pop. 1840.
Northern District.		Covington,	2,717
Attala,	4,303	Franklin,	4,775
Boliva,	1,356	Greene,	1,636
Carroll,	10,481	Hancock,	3,367
Chickasaw,	2,955	Hinds,	19,098
Choctaw,	6,010	Holmes,	9,452
Coahoma,	1,290	Jackson,	1,965
De Soto,	7,002	Jasper,	3,958
Itawamba,	6,375	Jefferson,	11,650
Lafayette,	6,531	Jones,	1,258
Lowndes,	14,513	Kemper,	7,663
Marshall,	17,526	Lauderdale,	5,358
Monroe,	9,250	Lawrence,	5,920
Noxubee,	9,975	Leake,	2,162
Octibbeha,	4,276	Madison,	15,680
Panola,	4,657	Marion,	3,690
Pontotoc,	4,491	Neshoba,	2,437
Tallahatchie,	2,985	Newton,	2,527
Tippah,	9,444	Perry,	1,887
Tishamingo,	6,681	Pike,	6,151
Tunica,	821	Rankin,	4,631
Winston,	4,650	Scott,	1,653
Yalabusha,	12,218	Simpson,	3,380
		Smith,	1,961
Total,	146,280	Warren,	15,820
		Washington,	7,287
Southern District.		Wayne,	2,120
Adams,	19,434	Wilkinson,	14,193
Amite,	9,511	Yazoo,	10,480
Claiborne,	13,078		
Clarke,	2,986	Total,	229,381
Copiah,	8,954	Total of State,	375,651

There were in this state, in 1840, 109,227 horses or mules, 623,197 neat cattle, 128,367 sheep, 1,001,209 swine; poultry was raised to the value of $369,482. There were produced 196,626 bushels of wheat, 11,444 of rye, 13,161,237 of Indian corn, 1654 of barley, 668,624 of oats, 1,630,100 of potatoes, 175,196 lbs. of wool, 6835 of wax, 53,471 of tobacco, 777,195 of rice, 193,401,577 of cotton. The produce of the dairy was valued at $359,365; of the orchard at $74,458; of lumber at $193,794; tar, pitch, and turpentine, amounted to 9248 bbl.

The climate is mild, but very variable. The extremes of heat and cold at Natchez, for 1840, were from 98° to 94° of Fahrenheit. The sugar cane and orange tree cannot be successfully cultivated N. of lat. 31°.

The Mississippi river washes the entire western border for a distance, by the windings of the stream, for 530 m. A large portion of its bank in this state consists of inundated swamp, covered with cypress, excepting occasional elevated bluffs, which immediately border the river. The Yazoo is the largest river which flows wholly within the state, and enters the Mississippi 12 m. above Vicksburg. It is 100 yds. wide at its mouth, is 200 m. long, and is navigable for large boats for 50 m. It flows through an elevated and healthy country. Big Black is the next river in size, and enters the Mississippi immediately above the Grand Gulf, by a mouth 40 yds. wide. Its length is 200 m., and it is navigable for 50 m. Homochitto is a small river, which enters the Mississippi, 43 m. below Natchez. Tombigbee rises in the N.E. part of the state, and flowing into Alabama, unites at length with Alabama river, to form Mobile river. Pearl river rises near the centre of the state, and drains its southern part, and enters the Rigolets between lake Pontchartrain and lake Borgne. Below the thirty first degree of N. lat., it divides the state from Louisiana. Pascagoula river waters the S.E. part, is formed by the junction of Chickasawha and Leaf rivers, and enters Pascagoula sound. It is navigable for 50 m., for vessels requiring 6 ft. of water, and for boats 100 m. farther; but the bay at its mouth has only 4 ft. of water.

The largest and most commercial place in the state is Natchez, on the Mississippi, situated on a bluff elevated 230 ft. above the surface of the river, 300 m. above New-Orleans, by the course of the river. Vicksburg, 106 m. above Natchez, and 12 m. below the mouth of Yazoo river, has had a rapid growth and is flourishing. The other principal places are Jackson, the capital, Woodville, 18 m. from the Mississippi, in the S.W. part of the state, Port Gibson and Grand Gulf, its port on the Mississippi, Columbus on the Tombigbee, Pontotoc and Hernando in the N., and Mississippi city on the gulf shore.

MISSISSIPPI.

There were in the state, in 1840, seven commercial and 67 commission houses in foreign trade, with a capital of $873,900; 755 retail dry goods and other stores, capital, $5,004,420; 288 persons engaged in the lumber trade, employing a capital of $132,175; 40 persons employed in internal transportation, with 15 butchers, packers, &c., employed a capital of $4250.

The amount of home made or family articles was $082,945; 53 cotton manufactories with 318 spindles, employed 81 persons, produced articles to the amount of $1744, with a capital of $6420; hats and caps were manufactured to the amount of $5140, employing 13 persons, and a capital of $8100; 128 tanneries employed 149 persons, and a capital of $70,870; 48 other manufactories of leather, as saddleries, &c., produced articles to the amount of $118,167, with a capital of $41,945; one pottery, employing two persons, produced to the amount of $1200, with a capital of $200; four persons produced drugs and paints to the amount of $3125, with a capital of $500; two persons produced confectionary to the amount of $10,500; 274 persons produced machinery to the amount of $242,235; 603 persons produced bricks and lime to the amount of 273,870, with a capital of $992,745; there were produced 312,084 lbs. of soap, 31,937 of tallow candles, and 97 of wax or spermaceti candles; 132 persons produced carriages and wagons to the amount of $49,693, with a capital of $34,335; 16 flouring mills produced 1809 bbls. of flour, and with other mills, manufactured articles to the amount of $486,864, with a capital of $1,219,845; vessels were built to the amount of $13,925; 41 persons manufactured furniture to the amount of $34,450, with a capital of $28,610; 14 distilleries produced 3,150 gal., and two breweries 132 gal., employing 12 persons, and a capital of $910; 144 stone or brick houses, and 2247 wooden houses, were built by 2487 persons, and cost $1,175,513; 28 printing-offices, and one bindery, two daily, one semi-weeekly, and 28 weekly newspapers, employed 94 persons, and a capital of $83,510. The total amount of capital employed in manufactures was $1,797,727.

There are three colleges in this state. Jefferson college at Washington, 6 m. E. of Natchez, was founded in 1802, and has been liberally endowed; Oakland college, at Oakland, was founded in 1831, and is prosperous; Centenary college, at Brandon Springs, was founded in 1841, by the Methodists, and is flourishing. The colleges existing in 1840, had about 250 students. There were in the state 71 academies, with 2553 students; and 382 common and primary schools, with 8236 scholars. There were in the state 8360 white persons over 20 years of age who could neither read nor write.

The Methodists and Baptists are the most numerous religious denominations. In 1836, the Methodists had 53 travelling preachers and 9707 communicants; the Baptists had 84 churches, 34 ministers, and 3199 communicants; the Episcopalians had four ministers; the Presbyterians of different descriptions had 32 churches and 26 ministers.

In the beginning of 1840, the state had 23 banks and 15 branches, with an aggregate capital of $30,379,403, and a circulation of $15,171,639. The state debt amounts to $7,000,000, incurred for banking purposes.

A railroad extends from Vicksburg 50 m. to Jackson, and is extended 14 m. farther to Brandon. A railroad extends from Natchez, and is designed to be continued through Jackson to Canton, a small part only of which is completed. Several other railroads have been projected, and some work done on them.

The governor is elected biennially by the people, but is ineligible for more than four years in any term of six years. He must be 25 years of age, and have been a citizen of the United States for 20 years. The senate consists of 30 members, elected for the term of four years, one half of the number being elected every two years, by the people. A senator must be a citizen of the United States, have resided in the state for four years, and in the district for which he is chosen for one year next preceding his election, and be thirty years of age. The house of representatives consists of ninety one members, citizens of the United States, elected biennially by the people. Every representative must have resided in the state for two years, and in the city, town, or district, for which he is chosen, for one year next preceding his election. The judges of the high court of errors and appeals are elected by the people for six years; the judges of the circuit court for the term of four years; the chancellor for the term of six years; the judges of probate for the term of two years. The judges of the high court of appeals and errors must be 30 years of age; and the others be 25 years of age. The secretary of state, the treasurer, and the sheriffs, are elected by the people, for the term of two years. Every white male person of 21 years of age or over, a citizen of the United States, who has resided in the state for one year, and in the county for which he offers his vote four

372

months next preceding an election, enjoys the right of suffrage. The legislature meets biennially at Jackson, on the first Monday in January.

In 1716, the French formed a settlement at the place where the city of Natchez now stands, and laid claim to the country as belonging to Louisiana. This colony was massacred by the Indians in 1729. In 1763, it was ceded to the British, and N. of the thirty first degree of N. latitude was in the chartered limits of Georgia; S. of that, it belonged to W. Florida, which was ceded to the United States in 1798 by Spain. In 1800, this state, with Alabama, was constituted a territory, under the name of the Mississippi territory. In 1817, Mississippi was separated from Alabama, and was admitted to the union as a sovereign state. The constitution was formed in 1817, and revised and amended in 1832.

Mississippi, an immense river of the United States; which, whether we consider its great length, its vast tributaries, the extent of country which it drains, and the distance to which it is navigable, well deserves the title which the Indians give it of the "father of waters." The import in the Algonquin language of *Missi Sepe*, the name by which it is designated, is great river. It drains a country of over 1,000,000 square miles in extent, eminently fertile, and sending through it to its destined market, a vast amount of produce, and receiving in return the productions of other parts of the world. Its extreme source was discovered by School craft in July 13th, 1832, to be Itasca lake, in 47° 10′ N. lat., and 94° 54′ W. long., at an elevation of 1500 feet above the ocean, and 3160 m. from its entrance into the Gulf of Mexico. Itasca lake, or lac la Biche of the French, is a beautiful sheet of water, of an irregular shape, about eight m. long, situated among hills covered with pine forests and fed chiefly by springs. It has its outlet to the N., which is about 10 or 12 ft. wide, and from 10 to 18 in. deep, and flowing northwardly, it passes through lakes Irving and Traverse, and then turns eastwardly, and proceeding through several small lakes, it enters lake Cass. This lake is about 16 m. long, contains several islands, is about 3000 m. from the gulf of Mexico, at an elevation of 1330 ft. above the ocean, and 182 m. below lake Itasca, the source of the Mississippi. It then flows E. to lake Winnipec, and S.E. to Little lake Winnipec, below which it receives Leech lake fork, the outlet of a considerable body of water of a very irregular form, called Leech lake. This was formerly regarded as the source of the Mississippi. The most northerly point obtained by the Mississippi is a little short of 48° N. lat. From the junction of Leech lake fork, the river expands to a hundred feet in width, and increase of depth, and flows with a mean current of one and a half miles per hour, and a descent of three inches in a mile, through a low prairie country covered with wild rice, rushes, sweet grass and other aquatic plants, and is the favourite resort of waterfowls and various amphibious quadrupeds. At the falls of Peckagama, the first rocky stratum, and the first wooded island is seen, at the distance of 685 m. above the falls of St. Anthony. At the fall of Peckagama, the river descends 20 ft. in 300 yards, by a rapid which entirely obstructs navigation. At the head of these falls, the prairies entirely cease, and below, a forest of elm, maple, birch, oak and ash, overshadow the stream. The river now takes a southerly course, curving to the west, and again to the east, to the falls of St. Anthony. The fall of the river above very nearly computed at six inches in a mile, with a current of three miles an hour, exclusive of some rapids, and has some dry prairies on its shores, the resort of the buffalo, elk and deer, and are the only part of the Mississippi in which the buffalo is now found. At the falls of St. Anthony, 843 m. above the mouth of Missouri, the river has a perpendicular descent of a 16 ft. with a formidable rapid above and below. The rapids above the falls, has a descent of 10 feet in the distance of 200 yards, and below the falls, a descent of 15 ft. in the distance of half a mile. An island at the brink of the falls divides the current into two parts, the largest of which is on the west of the island, and immediately below the falls are large fragments of rock, in the interstices of which some alluvial soil has accumulated, supporting a stunted growth of cedars. The whole fall has a descent of about 41 ft. in less than three-fourths of a mile. This all has nothing of the grandeur of Niagara, but the cataract, and the surrounding scenery are widely picturesque and beautiful. In times of high floods it may approach to the sublime. The width of the river above the falls is 500 or 600 yards, and at the falls is 227 yards; but narrows to 200 yards, a short distance below. The portage around the falls, is about 260 rods. In 1805 the United States purchased of the Sioux tribe of Indians a tract of nine miles square, including the falls of St. Anthony for a military post, for the sum of $2000. On this territory, Fort Snelling is situated. A considerable tract of fertile land has been put under cultivation by the garrison. As an indication of the climate, the first green peas were eaten here on the 15th of June, and the first green corn on the 20th of July; and on the last of July, Schoolcraft found much of the corn too hard to be boiled for the table, and some ears might have been selected sufficiently ripe for seed corn. A little below the falls of St. Anthony, the St. Peter's river enters from the W., and is much the largest tributary of the upper Mississippi. For 200 miles above its entrance, Carver found about 100 yards wide, with a great depth, and it receives several important tributaries. It is 150 yards wide at its mouth, and ten or fifteen feet deep. About 44° N. lat. the St. Croix enters on the N.E. side, which is 100 yards wide at its mouth. About 500 yards above its mouth it expands into a lake called St. Croix lake, which is 36 miles long and from one and a half to three miles broad, and the river is said to be navigable for boats for 200 miles. About 15 miles below the mouth of the Croix, the Mississippi expands into a beautiful sheet of water, called lake Pepin, which is 24 miles long, and from two to four miles broad, and is 100 below the falls of St. Anthony. On the E. shore is a range of limestone bluffs, and on the W. an elevated level prairie, covered with a luxuriant growth of grass, and nearly destitute of forest trees, with occasional conical hills, which appear like artificial mounds. At the lower termination of lake Pepin, Chippeway river enters from the N.W., after a course of about 300 miles. In 43° N. lat. Wisconsin river enters on the E. side, forming an easy communication with lake Michigan by Fox river, which enters Green Bay. It is boatable more than 200 miles. A little below this comes in Turkey river from the W., and La Mine river from the E. so called from its traversing the lead mine region of Illinois. A little below 42° N. lat. enters the Webesipinica, a considerable river from the W., and a little down, Rock river, a clear and beautiful stream enters on the E. side. A little above Rock river, are long rapids, which at low water, render it difficult for boats to ascend. A little N. of 40° enters the Iowa, a large river from the W., and below that enter on the same side, Skunk and Des Moines rivers. The latter is perhaps the largest western tributary above the Missouri. It is 150 yards wide at its mouth, and said to be boatable for nearly 300 miles. A few miles above its mouth are rapids, nine miles long, forming a serious impediment to navigation at low water. In about 30° N. lat. comes the Illinois river, a noble, broad and deep stream, and the most considerable tributary of the Mississippi above the Missouri. It is nearly four hundred yards wide at its mouth, is about 400 miles long and navigable for boats more than three hundred miles. A canal is in progress to connect it with lake Michigan. A little below 39° N. lat., comes in from the W. the mighty Missouri, which is longer, and probably discharges more water than the Mississippi; and had it been early explored, it would probably been considered as the parent stream; but it will henceforth be considered only as a tributary. The Mississippi above the junction is a remarkable clear stream, but this is entirely destroyed by the entrance of the turbid Missouri, which communicates its own muddy appearance to the Mississippi through the remainder of its course, thus asserting its lordship over it. (See Missouri river.) Near 38° N. lat. the Kaskaskia river comes in on the E. side. It is 80 yards wide at its mouth, flows through a beautiful country, is over 300 miles long, and is boatable in high water, over 160 miles. Between 36° and 37° N. lat. comes in from the E. the beautiful Ohio, "La Belle Riviere" of the French, and is much the largest eastern tributary; and from the densely populated and highly fertile country on its borders, it is at present, much the most important branch of the Mississippi. The large tributaries, the Cumberland, Tennessee and Wabash, contribute to its magnificence and importance. Between 33° and 34° the St. Francis enters from the W., 200 yds. wide at its mouth, and supposed to be navigable by one of its forks for 300 miles. A little above 34° N. lat., White river enters on the W. side, by a mouth from 300 to 400 yds. wide, and is probably about 1200 m. long. Thirty miles below, between 34° and 23° the Arkansas enters from the W., by a mouth 500 yds. wide, and is supposed to be 2500 m. long. It is next to the Missouri, the largest western tributary, and its waters are of a dark flame color, when the river is full. Between 33° and 32°, a little above the Walnut hills in the state of Mississippi, the Yazoo river enters on the E. side, by a mouth from 200 to 300 yds. wide. A little above 31° Red river enters on the W. side. It is nearly as large as the Arkansas, and discharges about as much water. Here the Mississippi carries its greatest volume of water, as immediately below this, and at intervals after, it sends off several large outlets. Three miles below the mouth of Red passes off on the W. side, the Atchafalaya or Chaffalio Bayou, as it is called, which is supposed to carry off as much water as the Red river brings in. Here the Mississippi has a famous "cut-off" by crossing the isthmus of a large bend, so that the main channel does not pass the mouth of Red river. The Atchafalaya has been

supposed to be the ancient bed of Red river, by which its waters were discharged without mingling with the Mississippi, and much of its water is now discharged by this outlet. A little below Baton Rouge the Ibberville passes off on the E. side, going through lakes Maurepas, Pontchartrain and Borgne into the gulf of Mexico. In times of flood it carries off considerable water. Between this outlet on the E. and the Atahafalaya on the W., is what is called the Delta of the Mississippi. Between Atchafalaya and New-Orleans pass off on the W. side Plaquemine, which joins the Atchafalaya, and La Fourche, which pursues an independent course to the gulf of Mexico. At the distance of 105 miles below New-Orleans by the course of the river and 90 m. in a direct line this majestic river enters the gulf of Mexico by several mouths, the principal of which are called the Balize or north east pass, in 29° 7' 25" N. lat., and 89° 10' W. long. and the south west pass in 29° 8' N. lat. and 89° 25' W. long. The depth of water on the bar at each of these passes is 12 feet, but much greater a little without and within the bar. Most of the vessels enter and leave the river by the N.E. pass. It might naturally be expected from the extent of country drained by the Mississippi, that the spring floods would be great. From the mouth of the Missouri, the flood commences in March, and does not subside before the last of May, at an average height of 15 feet. From the Missouri to the Ohio it rises 25 feet, and for a great distance below the Ohio it rises 50 feet. At every flood it overspreads the country, chiefly on its W. side for 500 miles from its mouth, to the distance of from 10 to 30 miles. From the falls of St. Anthony to the mouth of the Missouri, it has a medial current of two miles an hour; and at every place, except at the rapids of the Des Moines, it has a depth of water of not less than four feet at the lowest stages. Below the Missouri its depth is greatly increased, but its width, except in the forests and swamps, when overflowed, is very little increased. To the mouth of the Ohio it has in the channel six feet of water at its lowest stages, and at its highest, of twenty-five feet. From the mouth of the Ohio to the St. Francis, there are various shoal places, where at low water, pilots are often perplexed to find a sufficient depth of water. Below that point, there is no difficulty for vessels of any draught, except to find the right channel. Above Natchez, the flood begins to decline. At Baton Rouge, it seldom exceeds thirty feet; and at New-Orleans, twelve. This river is exceedingly winding in its course, and sometimes a bend will occur of 30 miles, where the distance across the neck will not exceed a mile. The mighty volume of water often carries away a large mass of earth with its trees, from a projecting point, and frequently endangers vessels. Trees are also often bedded in mud, with their tops projecting, producing snags and sawyers, as they are called, dangerous to navigators. The whirls or eddies which are produced by the tortuous course of the river and its projecting points, render the navigation to a degree difficult and dangerous. Below the Missouri, the medial current is about four miles an hour. It is difficult on viewing it for the first time to have an adequate idea of its grandeur, and the amount of water which it carries. In the spring, when below the mouth of the Ohio its banks are overflowed, although the sheet of water that is making its way to the gulf may be 30 miles wide, yet finding its way through forests and swamps which conceal it from the eye, no expanse of water is seen but that which is included between its wooded banks, which seldom exceeds, but often falls short of a mile in width. But when it is seen from time to time to swallow up many large rivers, it will be easily conceived that it must have a great depth. At the lowest water, at the efflux of the Atchafalaya, at the head of the Delta, it is from 75 to 80 feet deep, at the outlet of La Fourche at Donaldsonville, 130 feet; upwards of 100 feet at New-Orleans, and from 75 to 80 feet, three miles above the main bars. Vessels are often from five to 30 days in ascending from the mouth of the river to New-Orleans; though they will, with a favourable wind, often descend in 12 hours. Before the introduction of steam-boats, it required eight or 10 weeks to go to the Illinois. Boats of 40 tons ascend the river to the falls of St. Anthony, more than 2000 miles from its mouth. The use of steam-boats has nearly superseded all other vessels for ascending the river. Large flat-bottomed boats, denominated arks, which are not designed to return, are extensively used for transportation down the river. The first steamboat on the western waters was built at Pittsburg, in 1811; there are now over 300 on the Mississippi and its tributaries, many of them of great burthen. The passage from Cincinnati to New-Orleans and back, has been made in 19 days. Large ships, moved by sails, rarely ascend above Natchez. There are no tides in the Mississippi. (For the tributaries of the Mississippi, see the several articles, Missouri, Ohio, &c.)

MISSISSIPPI. county, Ark., situated in the N.E. part of the state, and contains 1000 sq. m. Bounded E. by Mississippi river, W. by St. Francis river. Watered by Whitewater river. It contained in 1840, 3042 neat cattle, 76 sheep, 5022 swine; and produced 107,615 bushels of Indian corn, 2908 of potatoes, 22,500 pounds of cotton. It had one store; one school, 25 scholars. Pop. whites, 900, slaves, 510; total, 1410. Capital, Osceola.

MISSISSIPPI CITY, p. v., capital of Harrison co., Miss., 265 S.S.E. Jackson, 1143 W. Situated on the coast of the gulf of Mexico. Its harbor admits small vessels. It contains a courthouse, and several stores and dwellings.

MISSOURI, one of the western United States, is bounded N. by Iowa territory, E. by the states of Illinois, Kentucky, and Tennessee, from which it is separated by Mississippi r., S. by Arkansas, and W. by the Indian territory, from which it is in part separated by Missouri river. It is between 36° and 40° 36' N. lat., and 89° and 95° 45' W. long., and between 11° 59' and 18° 43' W. long. from W. It is 278 m. long, and 235 broad, containing 64,500 square m., or 41,280,000 acres. The population in 1810, was 19,833; in 1820, 66,586; in 1830, 140,074; in 1840, 383,702, of whom 58,240 were slaves. Of the free population, 173,470 were white males; 150,418 were white females; 883 were coloured males; 691 were coloured females. Employed in agriculture, 92,408; in commerce, 2529; in manufactures and trades, 11,100; in mining, 742; in navigating the ocean, 39; do. rivers, canals, and lakes, 1885; in the learned professions, 1496. The state is divided into 62 counties, which, with their population in 1840, were as follows:

Counties.	Pop.	Counties.	Pop.
Audrain	1,949	Miller	2,282
Barry	4,795	Monroe	9,505
Benton	4,205	Morgan	4,407
Boone	13,561	Montgomery	4,371
Buchanan	6,237	New Madrid	4,554
Caldwell	1,458	Newton	3,790
Callaway	11,765	Perry	5,760
Cape Girardeau	9,359	Pettis	2,930
Carroll	2,423	Platte	8,913
Chariton	4,746	Pike	10,646
Clark	6,846	Polk	8,449
Clay	8,282	Pulaski	6,529
Clinton	2,724	Ralls	8,670
Cole	9,286	Randolph	7,198
Cooper	10,484	Ray	6,553
Crawford	3,561	Ripley	2,856
Daviess	2,736	Rives	4,726
Franklin	7,515	St. Clair	7,911
Gasconade	5,380	St. Francois	3,211
Greene	5,372	St. Genevieve	3,148
Howard	13,108	St. Louis	35,979
Jackson	7,612	Saline	5,258
Jefferson	4,296	Scott	5,974
Johnson	4,471	Shelby	3,056
Lafayette	6,815	Stoddard	3,153
Lewis	6,040	Taney	3,264
Lincoln	7,449	Van Buren	4,693
Linn	2,245	Warren	4,253
Livingston	4,325	Washington	7,213
Macon	6,034	Wayne	3,403
Madison	3,395		
Marion	9,623	Total	383,702

Jefferson city, on the S. bank of the Missouri river, 15 m. above the mouth of Osage river, is the seat of government.

The state presents a variety of surface and of soil. South of cape Girardeau, with the exception of some bluffs along the Mississippi, it is alluvial, and a large proportion consists of swamps and inundated lands, most of which are heavily timbered. From thence to the Missouri river, and westward to the dividing ridge between Gasconade and Osage rivers, the country is generally covered with timber, rolling, and in some parts quite hilly; but no parts of the state are properly mountainous. Along the head waters of Gasconade and Big Black rivers, the hills are frequently abrupt and rocky, with fertile alluvion along the water courses. Much of this region abounds with various minerals, as lead, iron ore, gypsum, manganese, zinc, antimony, cobalt, ochres, common salt, nitre, plumbago, porphyry, jasper, chalcedony, buhrstone, marble and free stone. The lead is inexhaustible in quantity and rich in quality. The iron ore of this region is sufficient to supply the whole United States for many thousands of years. Bituminous coal exists in inexhaustible abundance. The difficulty of transporting these products to the market, is the only inconvenience. The western part of this state is divided into prairie and forest land, and much of the soil is fertile. The whole is undulating, and along the Osage it is hilly, abounding with good water, salt springs and limestone. North of the Missouri, the surface is diversified, and divided between timber and prairie land. From the Missouri to Salt river, good springs are scarce, and in several counties artificial wells are dug, to be filled with rain water from the roofs of houses. Between Salt river and Des Moines river is a beautiful country, with a very fertile soil. In the middle counties N. of the Missouri the surface is rolling, and there are some bluffs and hills, with considerable good prairie, and much timber. To the W. of this, and also to the N.,

374

the prairie predominates. Much of the prairie land in this state, is inferior to the same kind of land in Illinois. But independently of some barren and inundated land, the state contains a great proportion of fertile soil. Indian corn, wheat, rye, oats, hemp, tobacco, flax, sweet potatoes, and in the southeastern parts cotton, are produced. The forest trees and underwood, black and white, are walnut, oak of various kinds, locust, ash, cotton wood, papaw, yellow poplar, sycamore, dog wood and grape vines. In the southern part of the state cypress and red cedar are found. On Gasconade river, about 109 m. above its confluence with the Missouri, the timber is principally yellow and white pine, which is very valuable, being rare in the valley of the Mississippi. The state is particularly rich in minerals. The lead region, the centre of which is 70 m. S.W. of Missouri, is 70 m. long, and 45 wide, covering an area of 3150 sq. m. The greatest part of this country is situated in Washington and St. Francis counties, but a part extends into St. Genevieve and Jefferson counties. The ore is of the richest kind. It yields from 80 to 85 per cent of the true metal. Iron ore is equally abundant. In the S.E. part of Washington county is the celebrated "Iron mountain" one mile broad at its base, and 3 m. long, and from 300 to 450 feet high, filled with micaceous oxide of iron, which yields 80 per cent of the pure metal. There is another body of iron ore denominated Pilot Knob, 300 feet high, and a mile and a half wide at the base, which is equally rich. Washington county is a perfect bed of metallic treasures.

There were in 1840, 196,132 horses or mules, 433,875 neat cattle, 348,018 sheep, 1,271,161 swine; and there were produced 1,037,386 bushels of wheat, 65,608 of rye, 17,332,594 of Indian corn, 15,318 of buckwheat, 9801 of barley, 9,234,947 of oats, 783,768 of potatoes, 562,265 pounds of wool, 56,451 of wax, 9,067,913 of tobacco, 121,121 of cotton, 374,853 of sugar, 49,083 tons of hay, 18,010 of hemp or flax; poultry raised at $270,647. The products of the dairy were valued at $100,432, of the orchard at $90,878, of lumber at $76,355.

The climate is healthy, though subject to great extremes of heat and cold. The extreme range of the thermometer is from 100° above zero to 8° below. But these extremes are short, and not of frequent occurrence. The Missouri is frozen so hard as to be safely crossed by loaded wagons, for a number of weeks in winter. The air is generally dry, pure, and salubrious.

The Mississippi winds along the entire boundary of the state, for a distance of 400 miles, and receives the waters of the mighty Missouri, which crosses the state, and deserves to be regarded, on account of its length, and the volume of its waters, as the parent stream. The Missouri is navigable 1800 miles from its mouth in the Mississippi, to the mouth of Yellow Stone river, for four or five months in the year. The Missouri receives La Mine, Osage, and Gasconade rivers on the S. side, and Grand and Chariton rivers on the N. side. Salt river crosses the N.E. part of the state and enters Mississippi river, 85 m. above the mouth of Missouri river. Des Moines river forms a part of the N.E. boundary of the state. Maramee river rises near the head waters of Gasconade river, and after a devious course northeastwardly, enters the Mississippi 18 m. below St. Louis. St. Francis, Current, Big Black, and Whitewater rivers drain its S.E. part.

St. Louis, on the W. bank of Mississippi river, 18 m. below the mouth of Missouri river, is much the largest and most commercial place in the state. St. Genevieve, on the Mississippi, 64 m. below St. Louis, was settled by the French, and has considerable trade, particularly in lead. Potosi, in the mining district, is a flourishing town. Herculaneum is environed by bluffs, which are surmounted by shot towers, and is the principal place of deposite for the lead mines. It is on the Mississippi, 30 miles below St. Louis. New Madrid is the most noted landing place for boats, above Natchez, and Clarksville and Hannibal are noted landing places above St. Louis. St. Charles on the Missouri, 20 m. above St. Louis is a flourishing place, and was the capital of the state until Oct. 1826. Booneville, Lexington, Liberty and Independence are growing places in the W. part of the state.

There were in the state in 1840, three commercial and 29 commission houses engaged in foreign trade, with a capital of $746,500; 1107 retail dry goods and other stores, with a capital of $8,158,802; 345 persons employed in the lumber trade, with a capital of $218,029; 79 persons engaged in internal transportation, who, with 198 butchers, packers, &c., employed a capital of $173,630. Home made or family manufactures amounted to $1,149,544; nine woollen manufactories employed 13 persons, produced articles to the amount of $13,750, with a capital of $5000; two furnaces produced 180 tons of cast iron, and four forges produced 118 tons of bar iron, the whole employing 86 persons, and a capital of $79,900; 21 smelting houses produced 5,285,455 pounds of lead, employed 232 persons, and a capital of

$235,696; 60 persons produced 246,303 bushels of bituminous coal, with a capital of $9466; 36 persons produced 13,150 bushels of salt, with a capital of $5559; 12 potteries produced articles to the amount of $12,175, employing 33 persons, and a capital of $7860; 191 persons produced machinery to the amount of $290,419; 48 persons produced 950 small arms; 19 persons manufactured the precious metals to the amount of $5480; 73 persons produced granite and marble to the amount of $32,030; 671 persons produced bricks and lime to the amount of $185,934; 293 distilleries produced 508,366 gallons, and seven breweries produced 374,700 gallons, the whole employing 365 persons, and a capital of $189,976; 291 persons produced carriages and wagons to the amount of $97,112, with a capital of $45,074; one powder-mill, employing two persons, produced 7500 pounds of gunpowder, with a capital of $1050; eight persons produced drugs and paints to the amount of $13,500, with a capital of $7000; 54 flouring-mills 49,363 barrels of flour, and with other mills, employed 1286 persons, producing articles to the amount of $980,058, with a capital of $1,266,019; 413 brick or stone houses, and 2909 wooden houses were built by 1966 persons, and cost $1,441,573; 40 printing-offices, six daily, five semi-weekly or tri-weekly, and 34 weekly newspapers, employed 743 persons, and a capital of $79,350. The total amount of capital employed in manufactures was $2,704,405.

The university of St. Louis, under the direction of the Roman Catholics, was founded in 1829; Kemper college at St. Louis, under the direction of the Episcopalians, was founded in 1840; St. Mary's college at Barren's, a Roman Catholic institution, was founded in 1830; Marion college, in Marion co., was founded in 1831; St. Charles college, under the direction of the Methodists, was founded in 1830; Missouri university at Columbia, was founded in 1840; Fayette college at Fayette, is a new institution. In the colleges founded before 1839, there were in 1840, 495 students. There were in the state, 47 academies, with 3996 students; and 642 common and primary schools, with 16,788 scholars. There were 19,457 white persons, over 20 years of age, who could neither read nor write.

In 1836, the Methodists had 51 travelling preachers, and 8899 members; the Baptists had 146 churches, 66 ministers, and 4972 communicants; the Presbyterians had 33 churches and 17 ministers; the Roman Catholics had one bishop and 30 ministers; the Episcopalians had three ministers. There were besides, a number of Cumberland and Associate Reformed Presbyterians, and one Unitarian minister.

In 1839 there was one bank and two branches, with an aggregate capital of $1,116,193, and a circulation of $410,740.

At the close of 1842, the state debt amounted to $3,072,361. The governor is elected once in four years by the people, but is ineligible for the next succeeding four years. He must be a natural born citizen of the United States, be at least 35 years of age, and have resided in the state for four years, next preceding his election.

The lieutenant governor is elected at the same time, in the same manner, and must possess similar qualifications. He is president of the senate; and in case of the death, resignation, or removal from office of the governor, discharges the duties of that office, until it is regularly filled.

The senate consists of 18 members, chosen for four years; one half the number being elected biennially. A senator must be at least 30 years of age, a citizen of the United States, have resided in the state for four years, next preceding his election, and for one year in the district for which he is elected, and must have paid a state or county tax.

The house of representatives consists of 49 members, elected biennially by the people. A representative must be at least 24 years of age, have been an inhabitant of the state for two years next preceding his election, and have paid a state or county tax, and be a citizen of the United States.

The governor, with the advice and consent of the senate, appoints the judges of the superior and inferior courts, and the chancellor, who hold their offices during good behaviour. They cannot be appointed before they have attained the age of 30 years, nor hold their offices after the age of 65 years.

Every white male citizen, of 25 years of age or upward, who has resided in the state for one year, next previous to an election, and the last three months in the district in which he offers his vote, is entitled to the right of suffrage.

The general assembly meets biennially at Jefferson city, on the first Monday of December.

The territory of this state was included in Louisiana, purchased by the United States of France, in 1803. The town of St. Louis was settled by the French in 1764, as a trading port with the Indians, and remained such, until it was purchased by the United States. In 1804 Louisiana was divided into the territory of Orleans, extending to the 33° of N. lat., and the residue was styled the district of

Louisiana. In 1805 the district of Louisiana was erected into a territorial government, under the name of the territory of Louisiana, and in 1812 its name was changed to the territory of Missouri. In 1821 a part of this territory was admitted to the union as the state of Missouri, after much debate on the subject of slavery, which was allowed by its constitution, under certain restrictions.

MISSOURI, a large river of the United States, rises in the Rocky mountains, and takes this name after the union of three branches, denominated Jefferson, Gallatin, and Madison, in 45° 19' N. lat., and 110° W. long. The springs which give rise to the Missouri are not more than a mile distant from some of the head waters of the Columbia, which flows W. into the Pacific ocean. At the distance of 411 miles from the extreme point of the navigation of its head branches, are what are denominated the "Gates of the Rocky mountains," which present a view exceedingly grand. For the distance of 5¾ miles the rocks rise perpendicularly from the margin of the river to the height of 1900 feet. The river is compressed to the width of 150 yards, and for the first three miles, there is only one spot, and that only of a few yards, on which a man could stand, between the water and the perpendicular ascent of the mountain. At the distance of 110 miles below this, and 581 m. from its source are the Great falls, 2575 miles above its entrance into the Mississippi. The river descends, by a succession of rapids and falls, 387 feet in about 16¼ miles. The lower and greatest fall has a perpendicular pitch of 87 feet; the second of 19 feet; the third of 47 feet; the fourth of 26 feet. Between and below these falls are continual rapids of from 3 to 18 feet descent. These falls, next to those of Niagara, are the grandest on the continent. The course of the river above, these falls is northerly. The Yellowstone river, 800 yards wide at its mouth, probably the largest tributary of the Missouri, enters it on the S.W. side, 1216 miles from its navigable source, and about 1880 miles from its mouth. This river, at the place of junction, is as large as the Missouri. Steamboats ascend to this place, and could go farther by each branch. Chienne river, 400 yards wide at its mouth, enters the Missouri on the S.W. side, 1310 miles from its mouth, in 44° 20' N. lat. White river, 300 yards wide, enters it on the S.W. side, 1130 miles from its mouth. Big Sioux river, 110 yards wide, enters it 853 miles from its mouth, in 42° 48' N. lat., on the N.E. side. Platte river, 600 yards wide at its mouth, enters it on the S.W. side, 609 miles from its mouth, in 40° 50' N. lat. Kansas river, 233 yards wide at its mouth, enters it on the S.W. side, in 39° 5' N. lat., at the distance of 340 miles from its mouth. Grand river enters it on the N.E. side, 240 miles from its mouth, and is 190 yards wide. La Mine river, 70 yards wide, enters it 200 miles from its mouth. Osage river, 397 yards wide at its mouth, enters it on the S.W. side, in 38° 31' N. lat., 133 miles from its mouth. Gasconade river enters it on the S.W. side, in 38° 45' N. lat., 100 miles from its mouth. The Missouri enters Mississippi river, 3096 miles from its source, which added to 1253 miles, the distance to the gulf of Mexico, makes its whole length 4349 miles; and it is probably the longest river in the world. Through its whole course, there is no substantial obstruction of the navigation, before arriving at the great falls. In principal tributaries are each navigable, from 100 to 800 miles. The alluvial, fertile soil on this stream and its tributaries is not very broad, and back of this are prairies, of vast extent. Through the greater part of its course the Missouri is a rapid and turbid stream, and in the upper part of its course, flows through an arid and sterile country. It is over half a mile wide at its mouth, and through a greater part of its course it is wider. Notwithstanding it drains such an extensive country, and receives so many large tributaries, at certain seasons it is shallow, hardly affording sufficient water for steamboat navigation, owing to its passing through a dry and open country, and being subject to extensive evaporation.

MISTRETTA (an. *Amastra*, or *Mytistratum*) a town of Sicily, intend. Catania; on a high hill, 5 m. S.W. Caronia. Pop. about 8000.

MITCHELSTOWN, an inland town of Ireland, prov. Leinster co. Cork, on an affluent of the Funcheon, 96 m. N.N.E. Cork. Pop., in 1831, 3545. It consists of a well-built square, and two principal streets. It has a parish church, and Roman Catholic chapel, both handsome modern structures; a college, which maintains 12 poor Protestant gentlemen, and 12 gentlewomen, endowed by the Kingston family; a small barrack, and a market-house. A minor court for pleas of the amount of £2 is held every third Monday, and petty sessions every Wednesday. It is a constabulary station. Markets on Thursdays; fairs on the 10th Jan. 25th March, 23d May, 30th July, 12th Nov., and 2d and 6th Dec. Postoffice revenue in 1830, £364; in 1836, £384. Adjoining the town, on the W., is the magnificent seat of the earl of Kingston, erected in 1823.

MITTAU a town of European Russia, cap. Courland, on

the As, 25 m. S.W. Riga, lat. 56° 39′ 10″ N., long. 23° 43′ 30″ E. Pop. 13,000. It is but indifferently built; the houses being mostly of wood, and the streets, with a few exceptions, unpaved. It has a castle, erected in 1739, which served in 1796 as an asylum for Louis XVIII. of France, and is now the residence of the governor and the official authorities. It has a gymnasium and a good library; a theatre, capable of accommodating 3000 spectators, a hospital, a literary society, &c. The nobility and gentry of the province assemble here at stated times for the despatch of business connected with the administration of the province, and many of them reside in town during the winter, when it becomes unusually gay. Its situation is low, sandy, and exposed to inundation. (*Schnitzler, La Russie, &c.*, p. 583; *Granville's Travels*, i., 370.)

MOBILE, bay, Ala., sets up from the gulf of Mexico, and is 30 m. long and 12 m. wide at an average breadth. It has Dauphin island at its mouth, with an entrance on each side. The channel on the W. side has five feet water; that on the E. side has 18 feet of water. The bay has 14 feet of water to the bar in its upper part, on which is 11 feet of water. The channel to the bay is within a few yards of Mobile point, on the E. side. It receives the waters of Alabama river, which is formed by the union of several large rivers, the Tombigbee, Black Warrior, Cahawba, Cocos, and Tallapoosa.

MOBILE river, Ala., is formed by the junction of Alabama and Tombigbee rivers, 40 m. above Mobile city. It enters Mobile bay by two channels, the main or W. channel is called the Mobile, and the E. is called Tensaw river. It is navigable for vessels requiring five or six feet of water by the Tombigbee branch to St. Stephens, 90 m. from the bay, and for steamboats to Tuscaloosa, 285 m. and to Columbus, Miss. The Alabama or W. branch is navigable for vessels of five or six feet draft, 100 m. to Claiborne, and for steamboats to Montgomery, 300 m. by the course of the river. The navigation of these rivers has some obstructions at low water. In time of flood it sometimes rises 50 or 60 feet.

MOBILE, county, Ala. Situated in the S.W. part of the state, and contains 2250 sq. m. Bounded E. by the Tombigbee and Mobile rivers, by branches of which it is watered on the E., and on the W., by branches of Pascagoula river. It contained in 1840, 12,280 neat cattle, 934 sheep, 8069 swine; and produced 31,991 bushels of Indian corn, 32,800 of potatoes. It had 21 commercial, and 93 commission houses in foreign trade, 248 retail stores, 22 grist-mills, 20 saw-mills, five printing-offices, one bindery, three daily, four weekly, and one semi-weekly newspapers; one college, 69 students; 11 academies, 297 students; 11 schools, 170 scholars. Pop.: whites, 11,763; slaves, 6191; free coloured, 787; total, 18,741. Capital, Mobile.

MOBILE, city, port of entry, and capital of Mobile county, Ala., 217 m. S. by W. Tuscaloosa, 1013 W. Situated on the W. side of Mobile river, at its entrance into Mobile bay, 30 m. N. of Mobile point, at the entrance of the bay, 55 m. W. by N. Pensacola, Flor., 10 m. W. by S. Blakeley, 160 E.N.E. New-Orleans. It is in 30° 40′ N. lat. and 88° 21′ W. long. Pop. in 1830, 3194; in 1840, 12,672. It contains a courthouse, jail, custom-house, a U. States naval hospital, a city hospital, three banks, Barton academy, seven churches, a Presbyterian, Episcopal, two Methodist, a Baptist, Roman Catholic and an African. It is pleasantly situated on an extended plain, elevated 15 feet above the highest tides, and has a beautiful prospect of the bay, from which it receives refreshing breezes. Vessels requiring a draft of more than eight feet water, cannot come directly to the city, but pass up Spanish river, six miles round a marshy island, into Mobile river, and then drop down to the city. It has 46 wharves, and next to New-Orleans, is the largest cotton market in the United States; 320,000 large bales have been exported in a year. The exports amount to from 12 to 16 millions of dollars annually. Tonnage of the port in 1840, 17,943. The city is supplied with excellent water, brought in iron pipes for a distance of two miles, and distributed over the city. It is defended by fort Morgan, (formerly fort Bower,) situated on a long low sandy point, at the mouth of the bay, opposite to Dauphin island. It was surrendered to the Americans by Spain in 1813, chartered as a town in 1814, and incorporated as a city in 1819. It has suffered severely by fire. In 1827, 170 buildings were burned, and in 1839, 660 buildings. But it has been rebuilt, with increased convenience and additional beauty. There is a lighthouse on Mobile point, the lantern of which is 55 feet above the level of the sea.

MOCHA, the principal port in the Red sea frequented by Europeans, in that port of Arabia called Yemen, about 48 m. N. from the strait of Bab-el-mandeb; lat. 13° 19′ 30″ N., long. 43° 20′ E. Population variously estimated; but may, perhaps, amount to from 5000 to 7000. It is encircled with walls, and indifferently fortified. Its appearance from the sea is imposing, but internally it is poor and mean.

Mocha is situated on the margin of a dry sandy plain. It is built close to the shore, between two points of land, which project and form a bay. Vessels drawing from 10 to 12 feet water may anchor within this bay at about a mile from the town; but large ships anchor without the bay in the roads, in five or seven fathoms water; the grand mosque bearing E.S.E., and the fort to the S. of the town S. by E., distant about two miles from the shore. The great article of export from Mocha is coffee, which is universally admitted to be of the finest quality. It is not possible to form any very accurate estimate of the quantity exported; but we believe it may be taken at 10,000 tons, or perhaps more. The greater portion is sent to Djidda and Suez; but there is a pretty large export to Bombay and other parts of India, whence some is sent to Europe: occasionally, however, the exports from Mocha and Hodeida, direct for Europe, are very considerable. Besides coffee, the principal articles of export are, dates, adjouse, or paste made of dates, myrrh, gum Arabic, olibanum, senna (*Cassia Senna*), sharks' fins, tragacanth, horns and hides of the rhinoceros, balm of Gilead, ivory, gold dust, civet, aloes, sagapenum, &c. The principal articles of import are, rice, piece goods, iron, and hardware, &c. The ivory, gold dust, and civet, met with at Mocha, are brought from the opposite coast of Abyssinia; whence are also brought slaves, ghee, &c.

MODBURY, an old bor., market town, and par. of England, hund. Ermington, co. Devon, 39 m. S.S.W. Exeter, and 181 m. S.S.W. London. Area of par., 5910 acres. Pop., in 1841, 2048. The town, which is very irregularly laid out, has four principal streets, meeting in a large open market-place. The church is large and well built, having a spire 134 feet high: the living is a vicarage, in the gift of Eton college, to which Henry VI. gave the estates of an alien priory of Benedictines that formerly stood near the church. The Wesleyan Methodists, Independents, Baptists, and the Society of Friends have their respective places of worship; and there were, in 1835, three Sunday schools, furnishing religious instruction to about 300 children. A Lancastrian school is attended by 70 boys; and about 80 children receive instruction in two infant schools.

Modbury had formerly a considerable share in the manufacture of serge, plush, and felt hats; but these branches of industry have long decayed, and the present inhabitants are mostly engaged in agriculture and retail trade. The town, which is a borough, though without an act of incorporation, is governed by a portreeve and other officers; and in the reign of Edward I. it sent two members to the House of Commons; but it afterwards was divested of this privilege, because of its inability to bear the expense! Markets, for corn and other provisions, on Thursday; large cattle fairs, the second Tuesday in each month, and an annual fair, May 14.

MODENA (DUCHY OF), called by the Italians *Stato Estense*, a state of N. Italy, consisting of the united duchies of Modena and Massa-Carrara, principally included between the 70th and 11th degrees of N. lat. and the 44th and 45th of E. long.; having N. Austrian Italy; E. the N. delegs. of the Papal states; W. Parma, and a portion of the Sardinian dominion; and S. Tuscany, Lucca, and the Mediterranean. Area estimated at 2000 sq. m., and the population at 403,000. The N. part of this duchy consists of a portion of the great plain of Lombardy; the S. is traversed from W. to E. by the Appennines, one of the summits of which in this duchy, M. Cimone, rises to upwards of 6500 feet above the sea. The Po constitutes a small portion only of the N. boundary; next to it the principal rivers are its affluents, the Panaro, Secchia, Crostolo, Enza, &c.; which have their sources on the N. slope of the Appennines, and flow N. to the Po; and the Magra and Serchio, which rise on the S. slope of the Appennines, and fall into the Mediterranean. The climate differs on the different sides of the mountains. On the N. it is similar to that of Parma and the Lombardo Venetian kingdom; snow falls and cold weather lasts for several weeks in the winter; while to the S. of the mountains the climate is like that of Genoa, and the olive and orange flourish in the open air. The plain country is very fertile, and abundantly watered by rivulets and canals. In the mountains there are many peasant proprietors, but not in the plains. A great evil here, as in other parts of Lombardy, is the practice of the lords and the possessors of lands in mortmain letting to middlemen, who relet to metayers; under which tenure are all the lands of the duchy. The tenant furnishes half the cattle, and the landlord the other half. Apparently there is not a labourer's house in the country, all being metaying farmers. (*Arthur Young's Trav.* ii., 157.)

The distribution of the land is much the same as in the rest of the plain of Lombardy, where, according to Chateauvieux, very few farms exceed 60 acres. The metayers should receive half the produce for their labour and attention, but the actual quantity falling to their share varies considerably according to circumstances, and in many cases is not more than one third part of the crops produced. The custom of sharing the produce is, however, almost universal; and a lease at a fixed rent being extremely rare, we need not wonder at the low state of industry. (*Chateauvieux, Italy and its Agric., p. 17, &c.*) Rice, wheat, maize, fruits, wine, oil, and hemp are the principal articles of culture; but the supply of corn is notwithstanding inadequate for the consumption. The wine of the duchy is strong, but not of superior quality: the oil S. of the mountains is equal to that of Genoa, but that produced in the N. plains is very inferior. Bees, poultry, and hogs are numerous. Cattle breeding is not very extensively pursued, except in the valley of Garfagnana, where it is almost the sole occupation of the inhabitants. The pasturages on the mountains are excellent; but only the duke and a few of the larger proprietors possess any considerable flocks of sheep. Horses few; oxen of the Lombard and Swiss breeds, and asses, supplying their place for draught, &c., on almost every farm. The declivities of the Appennines are clothed with fine woods of oak, beech, pine, and chestnut; indeed, chestnut flour forms the principal food of the peasantry in the upland region for a considerable portion of the year. The labouring classes, even in the more productive parts of the duchy, live very sparingly; soup of Turkish wheat, or *polenta*, salads, and beans or other pulse, fried in oil, are their ordinary food. Butcher's meat scarcely ever appears except on the tables of the more opulent farmers, and their best beverage is refuse wine, or wine of the second pressure.

Next in importance to rural husbandry, is the culture and manufacture of silk, though the products be of indifferent quality. The other manufactures are on a small scale. They consist principally of canvass, leather, paper, glass, and earthenware. In Garfagnana some iron is forged. The finest statuary marble is found in inexhaustible quantities at Carrara; and amber, petroleum, sulphur, &c., are met with elsewhere.

This duchy is divided into four districts, Modena, Reggio, Garfagnana, and Massa-Carrara. Modena is the capital: the other principal towns are Reggio, Carrara, Massa, and Finale. The government is an absolute monarchy, and perhaps the most despotical in its form of any in Europe. The duke monopolises both the legislative and executive power; but he avails himself of the services of a prime minister, two secretaries of state, and a privy council. The Austrian civil code of laws has been adopted since 1815. There are tribunals of primary jurisdiction in Modena, Reggio, and Massa, and a high court of appeal in Modena. A college is established in each of the principal towns, and the capital has a university and some superior schools; but owing to the jealousy of the government and the restraints laid on the press, public instruction is in a very backward state. The military force consists of 1750 men. It is recruited by volunteers; the recruit receives a premium, and the family of which he is a member is exempted from all personal taxes. The public revenue is estimated at about 1,500,000 florins a year.

The greater part of this territory was erected in 1452 into a duchy, under Borso D'Este, son of Pope Nicholas III. It was taken by the French in 1797, and subdivided into the departments of Panaro and Crostolo. In 1814 it was given to the present ducal family, a collateral branch of the House of Austria.

MODENA (an. *Mutina*), a city of N. Italy, capital of the above duchy, in a fine plain between the Panaro and the Secchia, 24 m. W.N.W. Bologna; lat. 44° 38' 38" N., long. 10° 55' 12" E. Pop. about 27,000. Modena has a citadel, and is surrounded with ramparts which, however, conduce less to its strength than to its beauty. It is regularly laid out, well built, and clean. It has been much improved and embellished within the last fifty or sixty years, and is divided into the new and the old city by the *Strada Maestra*, a part of the Emilian way, which intersects it from end to end. The general architecture of Modena is striking and agreeable; almost all its streets are bordered with arcades over their footways. The ducal palace is the finest public building; it stands isolated in the great square, and, unlike the palace of Parma, it has been completed, and is superbly furnished, and kept up in suitable style. It had formerly a noble collection of paintings; but some of its *chefs d'œuvre* were purchased by the elector of Saxony, and conveyed to Dresden in 1746, and others were taken away by the French: still, however, it is one of the best collections in Italy; it includes works by Raphael, Carlo Dolci, Andrea del Sarto, Guido, Guercino, the Caracci, and Procaccino, the Crucifixion by Pomarancio, a copy of the famous *Notte* by Correggio, &c. The ceiling of the gallery is painted in fresco by Francesconi; and in one of the rooms is a recumbent Cleopatra by Canova. The ducal library, known as the *Biblioteca Estense*, is a valuable collection of 80,000 volumes. Two of the best scholars, and

most laborious, diligent, and able writers of whom Italy has to boast, Muratori and Tiraboschi, were successively librarians during the last century. In the square before the palace is a fine statue of Duke Francis III., the founder of the university. The cathedral is a Gothic edifice of considerable antiquity and imposing appearance, but not in a pure style. It is principally remarkable for a square marble tower, one of the loftiest in Italy, in which is kept the famous bucket, once the cause of a serious feud between Modena and Bologna, and which has been immortalized by Tassoni in the *Secchia rapita*. One of the most celebrated works of Guido, the presentation in the temple, formerly adorned the cathedral, but it was carried off by the French, and has not been restored. (*Condor's Italy,* ii., 45.) The churches are numerous; but few deserve notice, except those of St. Vicenzo, St. Agostino, and the Dominican church, with some colossal statues. The city has several hospitals and asylums, a theatre, some public baths, better inns than most Italian towns, various good scientific collections, and a public library of 80,000 vols., comprising many rare editions of the 15th century, and some valuable MSS. It is well supplied with water by numerous subterranean cisterns; and is united to the Panaro by a canal navigable by boats of 30 tons. Its inhabitants depend principally on the supply of necessaries to the court and government functionaries. Weaving and spinning silk were formerly important branches of industry; but these have greatly declined; and manufactures of hemp, woollen cloths, leather, hats, and glass have, to a great extent, taken their place. It has a large weekly market for agricultural produce, provisions, &c.

Mutina is supposed to have been founded by the Etruscans. It is said, by Livy, to have been colonised by the Romans, A.U.C. 569, (xxxix., 55); and it is styled by Cicero, "*firmissimum et splendidissimum populi Romani colonia.*" (Phil. v., 9.) A few Roman antiquities, mostly tombs, still exist at Modena. It suffered many disasters in the times of Atilla, Odoacer, and the Lombard kings; and was afterwards governed successively by its bishop and magistrates, and belonged to the Popes, Venetians, and the dukes of Milan, Mantua, and Ferrara, before it became the property of the house of Este. Under the French it was the capital of the department Panaro. The learned antiquary Sigonius, the poets Molza and Tassoni, and the celebrated anatomist Fallopius, were natives of Modena. (*Condor's Italy,* ii., 42-46; *Cramer's Anc. Italy,* i., 86, 87; *Thomson; Starke, &c.,* passim.)

MODICA (an. *Motyca*), a town of Sicily, intend. Syracuse, cap. district of its own name, on the Scicli, 31 m. W.S.W. Syracuse. Pop., in 1831, incl. cant., 23,836. It is situated among craggy rocks, and generally ill built. According to Smyth, it has some fine edifices; but it appears to possess few conveniences, for a recent English traveller could find only one inn in the town, which was wretched and dirty in the extreme. Among the public buildings are a castle, numerous churches and convents, a ducal residence, a town-hall, two hospitals, several public schools, and a government loan-bank. The Franciscan convent is said to possess some fine mosaics. In the adjacent valley of Ispica are numerous troglodytic caves, fully described by Smyth (*Sicily,* p. 190) and Russell (*Tour in Sicily,* p. 134-137). In 1833 a good many houses and upwards of 100 persons were buried by the fall of a mountain near Modica.

The district of which this town is the capital has an area of nearly 190.000 acres, with several towns, and a population of about 80,000. It was endowed with peculiar privileges by Roger, king of Sicily, the principal being that its courts of justice should be independent of those of Sicily. These privileges would appear to have been productive of a good effect. "There is a very superior spirit of activity and industry among the natives, attended by greater affluence and comfort than any other agricultural part of Sicily displays, though it is not naturally so fertile as the rest. Modica is, in general, rocky and hilly, with very bad roads; but it boasts several fine plains, and romantic ravines. The soil is mostly loose, calcareous, and dry; but many successful agricultural efforts are made to render it productive, as is testified by the abundant produce of corn, tobacco, oil, wine, cotton, hemp, canary-seed, cheese, butter, and carobs; while, from the attention paid to pasturage, the cattle are in great request. This co. also produces bitumen and salt; and there is so great a quantity of game, as to form an article of export. The trade is principally with Malta, which is supplied from hence with the above necessaries, in exchange for cloth, spirits, hardware, and colonial produce." (*Smyth's Sicily,* p. 191.)

MOFFAT, a village of Scotland, celebrated for its mineral springs, co. Dumfries, on the Annan, at the head of an extensive valley, and bounded, almost immediately on the S., by an amphitheatre of hills, the highest in the S. of Scotland, 45 m. S. Edinburgh. and 20 m. N. by E. Dumfries. Pop. 1600: in summer, during the season of the Spa, the population may be 2000. This, which is an extremely neat, clean, and well-built village, consists principally of a wide street along the line of road from Dumfries to Edinburgh. The public buildings are the parish church, a dissenting chapel, and two good inns. The Earl of Hopetown has a house in the town. The mineral springs, which are sulphureous and chalybeate, rise at no great distance from the town, on the slope of the adjacent hills. One of these springs was discovered in 1633, the other in 1748. They are a good deal resorted to, but less now than formerly.

MOGADORE, or MOGODOR, called by the Moors Sāwera, a seaport town, and the principal emporium of Morocco, on the Atlantic, about 195 m. W. Morocco; lat. 31° 50' N., long. 9° 20' W. Pop. estimated at 10,000. It stands on a patch of granular sandstone rock, which, at high water, is nearly insulated by the sea. The country around is low, flat, and unproductive; so that vegetables have to be brought from gardens from 4 m. to 12 m. inland, and cattle and poultry from a still greater distance. Water is also scarce, and rather dear; being either rain water collected preserved in cisterns, or brought from a river about 2 distant. The white stone buildings give the town an imposing appearance from the sea. It is divided into two contiguous portions, both surrounded by walls: that called the citadel, comprises nearly half the entire town; with the royal palace, the houses of most of the governors and chief officers, the custom-house, the foreign consulates, and a street of well-built shops of red sandstone, formerly occupied by European traders. Beauclerk says, that the houses in this part are well-built and lofty, that the streets are swept, and that it was cleaner than any other town he had seen in the Moorish dominions. It is shut off from the rest of the town, by a high wall, with a strong gate, which is closed at nine o'clock every night. The other portion of Mogadore is not so well laid out, nor so clean, the Jews' quarter, in particular, being excessively filthy; it has, however, a very extensive mosque, with a high square tower, and other public buildings. It is entered by three principal gates; which, with many in the interior, are closed at sunset. To the S. of the citadel, is what is called the port, being an inner roadstead, protected by a rocky island, about 1½ m. in length, 3 m. from the shore. It has not more than 10 feet or 12 feet water at ebb tide, and 23 feet when deepest; it is therefore fit only for small vessels, large ships anchoring outside the harbour, the long battery bearing E., distant 1½ m. The island bounding the harbour is appropriated exclusively to a state prison; and is supposed to be defended by a few crazy pieces of ordnance, ensconced behind mud-wall embrasures. The landing-place is a long stone slip, near the arsenal, protected on the W. by a long battery, mounting several brass cannon, and containing a large tank, and a number of prison cells. The arsenal with which this battery communicates, is a really handsome structure, consisting of a large range of bomb-proof casemates, flanked at either end by an elegant square tower, with turrets at their angles, connected by a battery of two tiers, having in its centre a lofty arched gateway. The long battery defending the whole town on the W., is an extensive lime-wall along the shore, crowded with brass ordnance, and having beneath a range of bomb proof casemates, capable of containing 4000 or 5000 men. On the land side, Mogadore is protected from the attacks of the Arabs by a round tower, furnished with brass cannon. All the fortifications were erected under the superintendence of a European engineer in the last century, and, to an unskilled eye, they appear strong, and well executed; but Beauclerk says they are too flimsy to bear five minutes' breaching. The long battery, offers, however, a fine promenade for enjoying the fresh air of the ocean.

The trade of Mogadore was formerly very extensive; her port was open to the ships of the different European countries, most of whom had consuls here. Most part of the commerce between Europe and Morocco is still carried on through Mogadore; but England and Sardinia are the only states that retain consuls. The principal imports are English woollen and cotton stuffs and hardware, German linens, tin, copper, earthenware, mirrors, glass, sugar, pepper, paper, &c. The exports principally consist of sweet and bitter almonds, gum Arabic, and other gums, bees' wax, cow and calf skins, ivory, ostrich feathers, gold dust, olive oil, dates, &c. Accounts are kept in nutkeels of 10 ounces; the ounce being divided into four blankeels, of 24 fluce each. The blankeel may be valued at 1d., the ounce at 4d., and the nutkeel or ducat, at 3s. 4d. The corn measures are, for the most part, similar to those of Spain. The quintal = 119 lbs. avoird. The market lb., for provisions, = about 1⅓ lb. avoird. The canna, or cubit, = 21 English inches, is the principal long measure. Mogadore has no peculiar manufacture; but a good deal of the excellent woollen cloth of the country is sold in its markets.

The mutton of Mogadore is of a very inferior quality, and the beef but poor. Kids form a great part of the consump-

tion of those who indulge in meat; fowls are very cheap, and a dozen eggs may be bought for 1d. Red-legged partridges are always to be had in abundance. The climate is decidedly healthy, the heats being moderated by the sea breezes. Agues are, however, sometimes prevalent. Mogadore was founded in 1760, by the emperor Sidi Mohammed, who himself worked at its construction, and granted many privileges to mercantile settlers. (Beauclerk's Journey to Morocco, 220–257; Jackson's Morocco.)

MOGHILEF (Pol. Mohylef), a town of Russian Poland. gov. Podólia, cap. circ. on the Dniestr, 38 m. E.S.E. Kamienietz, and 180 m. S.W. Kief. Pop. nearly 7000, of whom many are Jews. From its situation, sheltered on every side by mountains, its climate is milder than that of the rest of Podolia; its fruits are excellent, and the silk-worm thrives well. It has several Greek, Roman Catholic, and Armenian churches, and a Greek convent; and is the residence of an American bishop. It has a brisk trade with Wallachia and the adjacent provinces, in raw produce, and some well-attended fairs. (Schnitzler. La Russie, 502; Malte-Brun, Tableau de la Pologne, 448.)

MOHACZ, a mean but large village of Lower Hungary. on the Danube, co. Barany, 25 m. E. by S. Funfkirchen. Pop. 8300. Near this village, on the 29th of August, 1526, the Turks, under Solyman the Magnificent, obtained a great victory over the Hungarians. Louis, king of Hungary, two archbishops and six bishops, many nobles, and about 22,000 private soldiers, are said to have been killed in the battle and in the pursuit. In 1687, the Turks were themselves defeated in the vicinity of this village by the Imperialists, under the Duke of Lorraine.

MOHAWK, river, N. Y., rises by its extreme source in Lewis county, but chiefly in Oneida county, near the tributaries of Black river. It flows S. about 90 m. to Rome, where it suddenly turns to S.E., and proceeds 37 m. to the Little falls, where the river descends 42 feet. Two miles from its mouth are Cahoos falls, where the river descends perpendicularly about 70 feet, presenting, in high water, a sublime and interesting spectacle. Both the above falls afford extensive water-power. Three quarters of a mile below the Cahoos falls is a bridge across the river, from which is a fine view of the cataract. The river at the falls is 960 feet wide, and the banks below it have a height of 80 or 90 feet. The Erie canal passes along the S. bank of the river as far W. as Rome. The valley of the Mohawk, particularly its river bottoms, is distinguished for its fertility.

MOHAWK, t., Montgomery co., N. Y. It contains the village of Fonda, the capital of the county, 42 m. N.W. Albany, 406 W. Bounded S. by Mohawk river. It contains seven stores, two falling-mills, one flouring-mill, one grist-mill, four saw-mills; nine schools, 936 scholars. Pop. 3132.

MOHILEF, or MOGHILEV (Pol. Mohilow), a gov. of European Russia, formerly included in the gov. of Vitepsk, between the 53d and 55th degrees of N. lat., and the 29th and 33d of E. long.; having N. Vitepsk, E. Smolensk, S. Tchernigov, and W. Minsk. Length, N. to S., 210 m.; average breadth nearly 85 m. Area about 17,470 sq. m. Pop., in 1838, 846,600. The only physical difference between it and the gov. Vitepsk is, that it belongs to the basin of the Dnieper, while the latter government belongs to that of the Dwina. In the N. of the government is a low chain of hills, separating the two river basins; but the rest of the surface is an extended plain, partly covered with forests, and in many parts marshy. The course of the rivers is mostly S.; the principal, next to the Dnieper, are its tributaries, the Soja and Drouets. Small lakes are numerous. The climate is milder and drier than that of Vitepsk. The soil is generally fertile; and though agriculture be extremely backward, nearly four million elements of corn are annually grown, a quantity considerably exceeding the home demand. Rye, barley, oats, hemp, and flax are the principal products; and in the gardens, hops, pulse, &c. The breeds of cattle and horses are very inferior; but latterly the sheep have been improved, by crossing with the breed of Saxony. Goats and hogs are numerous. This is one of the most richly wooded of the Russian governments; and its forests, the produce of which are floated down the rivers to the Black sea, furnish the building-yards of Nicoláeff, Odessa, Sevastopol, &c., with timbers and masts for the largest ships. Only a small proportion of the forest land belongs to the crown. Bog iron is plentiful, but it is dug only by the poorest classes. In respect of manufactures, Mohilef is behind almost all the other governments of the empire. There is no capital, and the inhabitants are without enterprise. The condition of the mass of the population appears to be most wretched. According to Schnitzler, le régime Polonais et l'établissement des Juifs, qui ne sont pas admis à habiter l'intérieur de l'empire Russe, ont produit ici de tristes fruits. Except a few tanneries, all the manufactures are in the hands of the Jews; but, with the exception of some distilleries, and soap and potash works, they are quite unimportant, and did not, in 1830, employ 1000 hands! It is divided into 12 districts; Mohilef, the capital, and Mstislavl are the principal towns. The inhabitants are mostly Russians and Jews, with some Poles, Lithuanians, Moldavians, and Wallacks: their religion is partly that of the Greek and partly of the Roman church, each of which has an archbishop in the government.

MOHILEF, a town of European Russia, capital of the above government, on the Dnieper, 85 m. S.W. Smolensk, and 110 m. E. by S. Minsk: lat. 53° 53´ 49´´ N., long. 30° 24´ 45´´ E. Pop. estimated, in 1834, at 21,080, of whom above 2000 were Jews. It has a better appearance than most Russian towns, many of its houses being of stone or other solid material. It is divided into four quarters, one of which consists of the kreml or castle, built on an eminence, and two of the other quarters are surrounded by ramparts. In the centre of the town is a large octagonal area, with neat stone buildings, including the residence of the Greek archbishop. It has at least 20 churches, three fourths of which are Greek; these are also several convents, a Luther an church, and two synagogues. The government offices and magazines are handsome edifices. Mohilef is the head quarters of the Russian "army of the west," and the seat of Greek and Roman Catholic archbishops, the latter having authority over all the Roman Catholics of Poland and Russia; it has two episcopal seminaries, a gymnasium, a town-school, and various charitable institutions. The business of tanning is extensively carried on; it has an extensive trade with Riga, Königsberg, Dantzic, and Odessa, to which it sends leather, hides, lard, wax, honey, especially the latter, potash, hemp, flax, oil, corn, and other raw products; receiving in return, among other foreign goods, a good deal of thrown silk. The fairs of Mohilef are well frequented. The epoch of its foundation is unknown. After several times changing masters, it was finally annexed to Russia in 1772. (Schnitzler, la Russie, p. 395-400.)

MOISSAC, a town of France, dep. Tarn et Garonne, cap. arrond., on the navigable river Tarn, crossed here by a handsome stone bridge, 14m. W.N.W. Montauban, and 97 m. S.E. Bordeaux. Pop., in 1836 (ex. com.), 6190. It is tolerably well built, and has an elegant fountain in its principal square. The most remarkable feature of the place, however, is a ruined abbey founded in the 11th century, formerly possessing great wealth and importance; the buildings are of great extent, but are for the most part either in ruins, or converted into private dwellings. The church-porch is of high antiquity, and has some curious sculptures; the cloisters are highly interesting; but the church itself is more modern, and of a heavy style. A good deal of corn is ground here for the use of the colonies; and the town has a considerable trade in wheat, oil, saffron, wine, &c. Moissac, founded in the 5th century, appears from its walls to have been formerly much larger than at present. It suffered severely from the religious wars. (Hugo, art. Tarn et Garonne, &c.)

MOLA DI BARI, a seaport town of the Neapolitan dom., prov. Bari, on the Adriatic, 13 m. S.W. Bari. The population, which at the beginning of last century amounted to 13,000, is now reduced to about 6000. It consists of an old and a new division; the former, which has a castle and is surrounded by a wall and ditch, has narrow, crooked, and gloomy streets, and poor houses. The other, or more modern division, is, on the contrary, comparatively well built along the sea-side, and has three creeks, where the small vessels which frequent the port load oil, cotton, and carobs. The traces of an unfinished mole show that this was formerly a place of some commercial importance. The port, between this mole and a rocky reef to the N. is insecure; here there is an open roadstead on either side the town, where vessels may anchor in 10 fathoms water with a sandy bottom. (Croon's Tour, &c., p. 154, 155; Purdy's Sailing Directions.)

MOLD, a market town, dep. and parl. bor., con Flint, hund. of its own name, co. Flint, 10 m. W. by Srth. to ser, and 171 N.W. London. Pop. of par., in 1841, 10,968, ditto of township and parl. bor., 3152. The town, situated in a valley, close to the Alyn, and surrounded by lofty hills, is small and irregularly built; but there is a very handsome town-hall, and, in the environs, are numerous handsome seats and elegant mansions. The church, a large structure of the 16th century, has a highly ornamented embattled tower, and contains some curious monuments. The Wesleyan Methodists, Calvinists, and Baptists, have, likewise, their respective places of worship, with attached Sunday-schools. A cotton-mill, in the town, gave employment, in 1839, to 296 hands; but "there is no other particular branch of trade carried on here, nor is it likely, judging from its present appearance, that it will increase in size or importance; within the parish, however, there are extensive coal-pits, lead and iron mines, which, in 1831, employed 689 labourers. (Bound. Rep. and Parl. Census.) Mold was constituted, by the Reform Act, a parl. bor. contrib. (with six others) to Flint. Registered electors in Mold, in 1839-40,

87. The county assizes are held here. Markets on Wednesday and Saturday. Fairs, February 13, March 21, May 12, August 2, and November 22.

About 1 m. W. from the town is a noted spot called Maes-Garmon, the scene of a victory gained in the 5th century, by the Welsh over the Picts and Saxons; a pillar, with an inscription, commemorates the event. About 1½ m., also, on the Chester road, are some remains of Offa's dike, the ancient boundary between Wales and England. (Hemingway's Pan. of N. Wales; Parl. Papers, &c.)

MOLDAVIA. See WALLACHIA and MOLDAVIA.

MOLDAU, a river of Bohemia, and, next to the Elbe, the principal in that kingdom, through the S. and central parts of which it flows. It rises in the Bohemian forest, about lat. 49° N., and long. 13° 33´ E.; runs at first S.E. to Rosenberg, and thence generally N. to its junction with the Elbe at Melnik, in about lat. 50° 20´, long. 14° 30´, after a course estimated at somewhat more than 200 m. It is properly the head stream of the Elbe, being continuous with the latter in a direct line, and carrying more water to it than the river called the Upper Elbe. It receives the Woltawa, Luschnitz, Sazawa, and Beraun: Rosenberg, Budweis, and Prague are on its banks. The Danube and Elbe have been united by a railway 75 m. in length, completed in 1839, from Linz, in Upper Austria, to Budweis, where the Moldau becomes navigable for boats of from 10 to 15 tons. This railway consists of one line only, and the carriages on it are drawn by horses. The line was rendered unnecessarily expensive through an ill-judged economy, inasmuch as it became necessary to take up the original wooden rails, which were covered with metal plates, and substitute others of cast iron in their stead. The traffic on this road has hitherto chiefly consisted in salt sent from Upper Austria into Bohemia. The nett revenue derived from it amounted, in 1837, to £8120. It belongs to about 13,000 shareholders, at 50 flor. a share; hitherto, however, they have derived no profit from the undertaking. (Oester., Nat. Encyc., &c.)

MOLFETTA (an. Respa), a seaport town of the Neapolitan dom., prov. Bari, cap. canton, on the Adriatic, 16 m. W.N.W. Bari. Lat. 41° 13´ 28´´ N., long. 16° 39´ 36´´ E. Pop. about 12,000. Its appearance from the sea is imposing; and though its streets are narrow and dirty, it has many good houses, among which Mr. Burgess remarked some elevations in a chaste style of architecture, and of a stone almost equal in beauty to white marble. (Greece, &c., i., 13.) It has a cathedral, several other churches, a college, &c., and is the see of a bishop. Its port, formed by a mole, is sheltered from all winds except the N. Opposite to it is a sandbank, which serves as a natural breakwater; the entrances to the harbour being at either extremity of the bank. It has some linen fabrics, a saltpetre manufacture, and some slips for ship-building; and has a considerable trade in the shipping of corn, oil, almonds, &c. (Rampoldi; Schütz, Allg. Erdkunde; Dict. Géog., &c.)

MOLTON (SOUTH), a mun. bor., market town, and par. of England, hund. of its own name, co. Devon, on the Mole, 24 m. N.N.W. Exeter, and 164 m. N. by S. London. Area of par., 6160 acres: pop., in 1831, 3696. The town, occupying an eminence W. of the river, at the union of several high roads, comprises a large market-place, and several well paved and lighted streets, the whole having a peculiarly clean and neat appearance. The guildhall is a handsome and commodious building, near which is a small jail. The church, adjacent to the market-place, is built in the perpendicular style: the living is a perpetual curacy in the gift of the dean and canons of Windsor. The Independents and Wesleyans have also their respective places of worship, and there are three Sunday-schools, attended by about 600 children of both sexes. A grammar-school, founded in 1614, is respectably conducted, and there are two other schools, supported by endowments and subscriptions. S. Molton has a manufacture of serges, shalloons, and felts, employing about 70 families, besides which there is a woollen-mill, which, in 1839, employed 79 hands. The lace manufacture has lately been introduced, but with no great success. The borough is governed (according to the Municipal Reform Act) by a mayor, three other aldermen, and 12 councillors. It is one of the polling places at elections for the N. division of the county. Quarterly and petty sessions are held here, and a court of record sits once in three weeks. Markets on Tuesday, Thursday, and Saturday; that on Saturday being one of the largest in N. Devon. Great markets (not chartered as fairs), Saturday after Feb. 13 and April 27, Wednesday before June 22, and after August 26, Saturday before October 10 and December 12, chiefly for cattle. (Parl. Papers, &c.)

MOLUCCAS, or SPICE ISLAND. See AMBOYNA.

MOMPOX, or MONPOX, a city of S. America, repub. New Granada, and, next to its cap., the most important in the prov. Cartagena; on the Magdalena, about 25 m. above the confluence of the Cauca; lat. 9° 14´ 20´´ N., long. 74° 27´ 30´´ W. Pop. estimated at 10,000, or, with the neighbouring villages, 15,000. "At a distance, on ascending the river, the white houses, with their red roofs, have a neat and clean appearance; but, on a nearer approach, this is exchanged for the general distressed look of Spanish cities. The town is above a mile in length; the streets are of a good breadth, crossing each other at right angles, and some are even furnished with footways. The only decent-looking houses, however, are in the centre of the place, the rest being mere sheds." (*Mollien*.) It has a custom-house and a fine quay, built very high, on account of the floods which take place in December. Several gunboats are stationed here, for the protection of the navigation. Mompox is a place of some commerce. The chief exports are corn, hides, and Brazil wood. Pamplona and Cuença transmit some tobacco, sugar, and chocolate to this *entrepôt*; Antioquia sends gold, and Bogota the produce of the Upper Magdalena. Mompox is surrounded by swamps, and liable to inundations; and alligators come up to the very banks of the river to feed on the offal thrown from the city. "The climate, in the daytime, is burning, the thermometer ranging from 25° to 30° Réaumur; the inhabitants consequently pass the evenings seated in the streets, to breathe the fresh air, and to escape the stings of the mosquitoes. The sky is constantly cloudy, and scarcely a day passes without showers. The nights, on the contrary, are beautifully clear, and truly delicious. It is then a great pleasure to promenade the streets and observe the lively parties which present themselves before the doors of the houses. Loud bursts of laughter are heard on every side, in which the passenger takes part without the least ceremony. Far from this familiarity being offensive, it gives great satisfaction, for the frankest cordiality presides at these meetings. Thus passes the life of the inhabitants of Mompox. The day is spent in their hammocks, the night in the street." This manner of living differs little from that which the inhabitants of the other hot countries in South America have adopted. All classes in Mompox are said to be much addicted to ardent spirits. The surrounding country is wholly in a state of nature. (*Mollien, &c., in Mod. Trav.,* xvii. 301–2; *Geog. Account of Colombia.*)

MONACO, a town and small principality of N. Italy. The principality, which is under the protection of the king of Sardinia, is about 8 m. in length by 6½ in breadth, having W. the div. of Nice, E. that of Genoa, and S. the Mediterranean. Area, 52 sq. m. Pop. about 7000. Being sheltered on the N. by lofty mountains, its climate is very mild, and it produces large quantities of oranges, lemons, and other fruits, from which the revenue of the prince, amounting to about £5000 a year, is mostly derived. The pastures are tolerably good, and cattle numerous. Its inhabitants are occupied almost wholly with agriculture, fishing, and petty coasting trade.

Monaco, the capital (an. *Portus* or *Arx Herculis Monaci*), is built on an elevated promontory stretching into the sea, about 9 m. E.N.E. Nice. Pop. 1900. It is walled and defended by a fort; and has an appearance of strength, but is entirely commanded by an adjacent hill. The largest town in the principality is Mentone, about 6½ m. E.N.E. Monaco, with a tolerable port and 3000 inhabitants.

This principality was founded in the 10th century, and has remained ever since in the Grimaldi family. The reigning prince is a peer of France, with the title of Duc de Valentinois, and usually resides in Paris. (*Almanack de Gotha, &c.*)

MONADNOCK, mountain, commonly called grand Monadnock, Cheshire co., N.H., is 22 m. E. of Connecticut river, extending 5 m. from N. to S., and 3 m. from E. to W. It is an isolated mountain, which rises 3254 feet above the level of the sea, and may be seen far 60 m. in every direction. Its summit, once covered with evergreens, has been by repeated fires converted into a bald rock, and presents a delightful view of the surrounding country, on which the spectator looks down as on a map. Thirty ponds are visible. On ascending it, plats of earth are found sufficient to give growth to the blueberry, cranberry, and ash, and a variety of shrubs. Plumbago or black lead is found in large quantities on its E. side, and there is a mineral spring near its base in Jaffray.

MONAGHAN, an inland co. of Ireland, prov. Ulster, having N. Tyrone, E. Armagh, S. Louth and Meath, and W. Cavan and Fermanagh; area, 397,048 acres, of which 9236 are unimproved mountain and bog, and 7844 water. Surface hilly, but the hills are mostly arable; soil moderately fertile. There are some large, and a great many small, estates. The land is very much subdivided; so much so, that it is said by Mr. Wakefield that the larger class of farms do not average 25, nor the smallest 6 acres! (L. 270), and but little change has taken place, in this respect, in the interval. Con-acre is very general here, and agriculture is in the most depressed state. Principal crops, oats, potatoes, and flax, the latter being very extensively cultivated; but a

good deal of wheat is now grown, and its culture is extending. Considerable improvements have latterly been effected in the breed of cattle; and a good deal of butter is made, though there are no large dairies. Goats are very generally kept by the cottiers for the sake of their milk. A great deal of work is done by the spade. Average rent of land, 13s. 3½d. an acre. The linen manufacture was at one time very widely diffused over the county, most of the small farmers having looms; but this combination of employments, which has been injurious alike to agriculture and manufactures, is now, owing to the greater cheapness of machine-made yarn and fabrics, greatly diminished. The value of the unbleached linen sold in the county in 1824, was estimated at about £126,000. (*Railway Report, Append. B., p.* 31.) The county has vast beds of limestone; and lead ore, and indications of coal have been discovered. There are no rivers of any importance. Monaghan has five baronies, and 19 parishes; and sends two members to the House of Commons, both for the county. Registered electors, in 1839–40, 3421. Principal towns, Monaghan, Clones, Carrickmacross, &c. In 1831 this county had 35,225 inhabited houses, 36,766 families, and 195,536 persons; of whom 95,679 were males, and 99,857 females.

MONAGHAN, an inland town of Ireland, prov. Ulster, co. Monaghan, of which it is the cap. on the main road from Dublin to Londonderry, nearly half way between them. Pop., in 1831, 3948. The town consists of a central square, called the Diamond, with several diverging streets. Its public buildings are the parish church, Roman Catholic chapel, three Presbyterian, and two Methodist meeting-houses; the county jail on the radiating plan, court-house, diocesan school for the sees of Raphoe, Kilmore, and Clogher, a national school, a cavalry barrack, a market-house, and the county infirmary. "It does not appear to possess any important advantages or consequence, except as a market town, chiefly for the sale of agricultural produce, linen, &c." (*Mun. Bound. Report.*) The corporation, consisting of a provost, 12 burgesses, and commonalty, sent two members to the Irish House of Commons down to the union, when it was disfranchised. The assizes for the county are held here, with general sessions four times a year, and petty sessions on Thursdays. It is a constabulary station. It has a considerable linen trade, a large brewery, and is a great mart for agricultural produce. Markets on Mondays, Tuesdays, Wednesdays, and Saturdays; fairs on the 1st Monday in every month. The Ulster canal passes near the town. Post-office revenue in 1830, £758; in 1836, £912. Branches of the provincial and Belfast banks were opened in 1831 and 1833.

MONASTIR, or BITOLIA, a town of European Turkey, prov. Macedonia, cap. sanjak of same name, on the Vetrizza, 82 m. N.N.E. Yamina, and 90 m. W.N.W. Salonika. The population was estimated at 15,000, but we incline to think that this is much beyond the mark; it is the principal entrepôt for merchandise passing from Albania into Roumelia. It suffered great injury from fire in 1806, and was plundered by Ali Pasha. (*Dict. Géog., &c.*)

MONDONEDO (an. *Britonia*), a city of Spain, in Galicia, cap. prov. same name, 30 m. N.N.E. Lugo, and 75 m. W. Oviedo. Pop., according to Miñano, 6074. It is situated on the N. side of the Asturian chain, and is old and irregularly built: its principal public buildings are a cathedral, with 11 dignitaries and 24 canons, a parish church, two convents, now converted into hospitals, and a royal seminary and college. Linen weaving, tanning, and brick-making, are the only branches of manufacturing industry in the town; two large fairs are held in May and October, and the oak-timber of the neighbourhood is better adapted for building than any other in Spain. (*Miñano.*)

MONDOVI, a town of the Sardinian dom. div. Coni, cap. prov. Mondovi, on and round a hill near the Ellero, 12 m. E. by S. Coni. Pop., in 1838, including com., 15,921. It is divided into four parts; the town proper, called the Piazza, on the hill, at an elevation of 1700 ft. above the level of the sea; and the three suburbs of Carassone, Bred, and Piano della Valle, built at its foot. The distance between the upper and lower part of the town is considerable; and the road by which they are connected is inconveniently steep. The town proper has a small citadel, and is surrounded with walls, of no great strength, instead of ramparts. It has a great number of religious houses and churches; the latter including a cathedral, with a handsome altar and sacristy. Its inhabitants are chiefly clergy and country gentry, and it has very little commerce or wealth. The suburbs, on the contrary, are entirely devoted to trade, and have manufactures of woollens and cottons, with tanneries and iron forges; but the chief branch of industry is the spinning of silk. Mondovi is the see of a bishop, and has several seminaries of education. It is comparatively modern, having been founded, according to an inscription on one of the chapels in the cathedral, in the year 1232. It was taken and sacked by the French, under Marshal Soult, in 1799. Beccaria, the natural philosopher, was a native of Mondovi; but he must not be confounded with the Marquis Beccaria, author of the famous treatise on Crimes and Punishments, who was a native of Milan.

MONGHIR, or MUNGGER, a town of British India, prov. Bahar, distr. Bhaugulpore, 89 m. E. Patna; lat. 25° 23' N., long. 86° 20' E. Pop. estimated at 30,000. It is finely situated on a bend of the Ganges, and is of great extent, its ramparts being about 1¼ m. in length by 1 m. in width. The houses, however, are much scattered, and in one quarter only are so close as to resemble a town. Monghir, while a British frontier town, was a station of considerable importance: and when Heber visited it, was in better condition than most native towns. Though the houses are generally small, there are many with an upper story; and the roofs, instead of the flat terrace or thatch, as in Bengal, are generally sloping and covered with reed tiles. The principal edifices are an old Hindoo temple, now occupied by some invalid soldiers; an elegant small mosque; the residence of the commandant and other military officers; barracks for five companies of sepoys; and the remains of a palace built by a brother of Aurungzebe. The shops are numerous; "and I was surprised," says Heber, "at the neatness of the kettles, tea-trays, guns, pistols, toasting-forks, cutlery, and other things of the sort, which may be procured in this tiny Birmingham. I found afterward that the place had been from very early antiquity celebrated for its smiths, who derived their art from the Hindoo Vulcan, who had been solemnly worshipped, and was supposed to have had a workshop here. The only thing which appears to be wanting to make their steel excellent is a better manner of smelting, and a more liberal use of charcoal and the hammer. As it is, their guns are very apt to burst, and their knives to break; precisely the faults which, from want of capital, beset the works of inferior artists in England. The extent, however, to which these people carry on their manufactures, and the closeness with which they imitate English patterns, show plainly how popular those patterns are become among the natives" (i., 296-293). Monghir has also excellent gardeners and tailors. A great deal of clothing for the native army is made here, with shoes in the native and European fashion, furniture, palanquins, carriages, &c. There are several native schools, and the town is a station of the Baptist Missionary Society. (*Hamilton's E. I. Gaz.; Heber's and Valentia's Trav.; Mod. Trav.*, ix.)

MONGOLIA, an extensive tract of country in the N.E. part of Asia, and one of the colonial possessions of China, between the 35th and 53d degrees of N. lat., and the 82d and 123d of E. long.; being bounded N. by the government of Irkutsk, N.E. and E. by Manchooria, S. by China, and W. by Chinese Tartary. Length, from E. to W., about 1700 m.; greatest breadth, 1000 m. area, about 1,400,000 sq. m. The limits, however, are subject, in consequence of wars among the tribes, to constant and great variation. Population conjectured by Timkowsky to be about 2,000,000. Mongolia may be generally described as an elevated plain, almost destitute either of wood or water, enclosed southward by the mountains of Thibet, and northward by various offsets belonging to the great Altaïc range. The central part of Mongolia is occupied by the great sandy desert, or Ta-Gobi, which stretches from S.W. to N.E. about 1900 m., with a breadth, in some parts, of from 500 to 700 m. (See ASIA, i., 163-4.) The most desolate part of the Gobi is called, by the Chinese, the Shamo, or sand sea, from its surface consisting of moveable sand. The desert is, however, intersected by some comparatively fertile tracts, and in other parts a few stunted trees are met with. The chief mountains of this region are, 1. the Altaï and its various subordinate chains, extending eastward, under the names of Tungnon, Khangaï, and Kinte, as far as the banks of the Amour, by which the range is deflected northward and joins the Yablonoï-khrebet; 2. the Tchastajoola and Inchan ranges, which commence in lat. 49° N., long. 107° E., and curve N.N.E. and northward as far as the Amour, in lat. 53° N., where they join the Altaï. The mountains of inner Mongolia are very little known. The rivers are numerous, chiefly in the N., belonging to the basins either of the Irtish or Amour. Connected with the former are the Selenga, Orkhon, and Tula, which unite their streams and flow into lake Baikal: the Keruolun and Onon, which are tributaries of the Amour, rise near each other on opposite sides of the Kente range, and, taking a N.E. course, unite in lat. 53° 30' N., and long. 121° E. In the S. are the Leaou-ho, rising on the eastern slope of the Irchan range, and falling into the gulf of Leaou-tong, and several rivers in the region of Koko-nor, some independent, and connected only with lakes, but others tributary to the Hoang-ho. The chief lakes S. of the great Gobi desert are the Koko-nor, the Oring, and Dzaring, the two latter being near the sources of the Hoang-ho. Inner Mongolia has no lakes of any

importance, and those in the N. region, inhabited by the Kalkas, are of inferior size; but Kobdo, the N.W. district, is a country of lakes as well as mountains, the principal being Upsa-nor, Altaï-nor, and that called the Ike-araï-nor, which receives the waters of the Djabkan, the largest internal river of Mongolia. The air of this country is cold, owing chiefly to its great elevation, but also to the abundance of sulphate of natron, with which the steppes are in many parts covered. Timkowsky reports that the temperature, during October and November, ranged between 2° and 10° Fahrenheit; but he was assured that this was an unusually severe season. Great quantities of snow and rain fall in the Kalkas country; where, also, fogs and heavy dews, with cold mornings, are common in the height of summer. There is no great diversity of soil throughout this vast territory, which is generally sandy, stony, and barren. The banks of the rivers and the mountain valleys abound in good pastures, and in some places there is land fit for tillage. The N. part of the Kalkas region, in particular, is well wooded, and would be very suitable for agriculture; but the people are wholly nomad, and averse to the formation of permanent settlements. The S. sides of the Altaï abound also with gold and silver; but the Mongols are entirely destitute of the knowledge necessary for the working of mines.* S. of Ourga, in lat. 47° N., begin the arid steppes of Gobi: the soil is gravelly, pasturage and water are rare, the grass is short and poor; and yet in these tracts, so little favoured by nature, are to be seen numerous herds of large camels, vigorous horses and oxen, with flocks of sheep and goats, all in good condition. The steppes abound in salt, and the atmosphere is dry and bracing; but there is a total absence of wood, and the ground is quite unfit for agriculture. Caravans are liable to great hardships in passing through the great desert, owing to the want of water and pasturage: the valleys, hills, and mountains offer nothing to the view but a yellow sand. S. of the 39th parallel the arid soil ceases, and is succeeded by lands well watered by rivers, and pretty well adapted to agriculture Wheat is raised by the Mongols of Koko-nor, and also by those living more eastward, in the fertile districts near the great wall of China. The people, however, generally speaking, are too indolent to be good cultivators: they sow millet, barley, and wheat, but in small quantities, and in the most careless manner. Most of them, indeed, pass their whole lives in the open air, on the steppes, and disdain the laborious occupation of cultivating the ground.

"When we asked them," says Du Halde, "why they did not raise even a few vegetables in small enclosures, their prompt reply was, that herbs were the food of animals, whose flesh was the only proper support of man." (*Desc. de la Chine,* iv., 38.) In fact, so great is their love of idleness, that, even in those countries which abound with wood and pasturage, they never make any provision for the winter, except, perhaps, a few stacks of hay: and consequently when there is a heavy fall of snow, and the cold is severe, they sometimes lose nine tenths of their flocks and herds. The quadrupeds of Mongolia are the wild horse, wild boar, stag, goats of various kinds, bears, wolves, hares, foxes, sables, and squirrels: the birds are cranes, wild geese and ducks, moor-fowl, quails, and swans. Of the domestic animals it may be remarked that the horse, though small and shabby looking, is strong and spirited; that the camels have two humps, and that the sheep are white, with long black ears, and furnish very delicate meat. The Mongols have dogs, but very few cats; and mules, as well as asses, are bred in large quantities by the tribe of Karatchin immediately N. of the great wall.

Mongolia is composed of 96 aïmaks, or principalities, all recognising the sovereignty of the emperor of China, and each governed by one of its oldest princes, called *taïdchus.* The division of the Mongol hordes is founded on the necessity of a military administration; but all the officers superintend likewise the direction of civil affairs. According to this military division (introduced by the Manchoos), the whole nation is divided into 135 banners, which are again subdivided into regiments and companies. Each Mongol is bound to serve as a horseman from his 18th to his 60th year. The property of the soil is in the princes, to whom their subjects pay a moderate contribution of cattle, supplying them also with servants and shepherds for guarding their flocks and herds. These princes decide in the last instance all disputes between their subjects, according to the laws established to preserve order in their armies; but the supreme administration is confided to the tribunal of foreign affairs at Pekin, which appoints inspectors-general for the different principalities: these are always chosen from the Manchoo nation. With respect to the attachment of the Mongols to the present Manchoo dynasty of China (Ta-

* Du Halde mentions some tin-mines in the Kalkas country, and Malte-Brun says that there are iron foundries about 50 versts from Kiakhta; but Timkowsky doubts their existence. (Comp. Du Halde, iv., 28; Malte-Brun, ii., 484; and Timkowsky, ii., 364.)

Thsing) it is difficult to speak positively. The Mongols still maintain their ancient hatred of the Chinese; and, though the latter have been enabled to subdue the warlike spirit of these nomads, and to declare them tributary, the court of Pekin sends to Mongolia about 10 times the value of the tribute received from it, under pretext of rewarding the zeal and fidelity of its princes and military officers. Thus, the native Mongol chiefs are bribed into subjection and obedience; but they are, at the same time, vigilantly watched by the Manchoo inspectors, and any misconduct, or show of opposition, is speedily visited by an abridgment or deprivation of their usual presents. The religion of the Mongols is Buddhism, supposed to have been introduced in the 17th century. The temples are not numerous, nor are the lamas much distinguished from the common people by their knowledge and morals. They learn to read Tibetian, because the sacred books and services are copied and printed in those characters; but few of them are even tolerably acquainted with the language, or know the origin and meaning of the religious ceremonies. The lamas observe celibacy, and follow a strictly monastic life: there are also female recluses, who submit to an austere and holy life; but some are married.

The proper or E. Mongols are divided into three great nations; the Kalkas, northward, the Tshakhars, near the wall of China, and the Sunnit, who range over the great desert of Gobi, Their physical conformation, language, general habits, and history, have already been described at some length in the article Asia, in this work (vol. i., p. 184,185), to which the reader is referred for these particulars. The dress of the men is very simple, consisting of a long dark-blue robe, either of cotton or cloth, secured by a leather girdle; their shirts and under garments are of coloured nankeen, their boots of leather, and very thick: in winter they wear pelisses of sheepskin, and fur caps. The costume of the women resembles, in many respects, that of the men. The saddles and bridles of the Mongols are furnished with copper or silver. A bow and arrows, with a short sword, are the arms of a soldier; and muskets or rifles are used only in the chase. Their tents consist, like those of the Khirgiz, of a skeleton of osier, covered with felt, of which there are in winter three layers: the door commonly faces the S.; the hearth is in the centre; and the right side, near the entrance, belongs to the women. The tents of the common people are low, close, and disagreeable; but those of the richer Mongols are spacious and lofty, comprising two or three distinct apartments, the best of which is covered with a Persian or Turkish carpet. Milk, cheese, and butter, with a little mutton and game, form the chief food of this robust and active nation: brick-tea is the principal beverage of the rich as well as the poor. In summer, also, they drink airak, a fermented liquor made from milk, besides koumiss and brandy, purchased from the Chinese. Hunting, horse-racing, wrestling, and archery are their chief amusements: they seem to have no idea of dancing, but their songs are poetical and highly characteristic. The Mongols marry young: a plurality of wives is not forbidden, and divorces are frequent, the least discontent on either side being deemed a sufficient reason for the step. They generally bury their dead, but sometimes burn them, and occasionally even leave them exposed to the birds and wild beasts. Almost every Mongol is a skilful warrior and huntsman; but there are very few workmen or artificers; and, on examining his dress, furniture, and saddle, we find that he is supplied with everything by the Chinese, who give, in exchange for horses, camels, oxen, and sheep, large quantities of brick-tea, tobacco, brandy, silk, cotton, and woollen fabrics, boots, and various utensils in iron, tin, and copper. To carry on this trade, the Chinese go to Mongolia, to the towns of Dolon-nor and Kaigan, or to the great entrepôts of Kiakhta and Ourga, in the country of the Kalkas. The Mongols receive considerable profits from the conveyance of goods through their country: payment is made by the Chinese sometimes in silver, but more frequently in articles of merchandise. (*Timkowsky's Travels through Mongolia,* lii., 207–258; *Du Halde, Desc. de la Chine,* iv., 21–38; *Chinese Rep.,* i., 117–121, &c.)

MONKTON, p. t., Addison co., Vt., 50 m. W. by S. Montpelier, 500 W. Watered by branches of Lewis cr. Excellent iron ore is found, in large quantities, in the S. part. A mile N. of the ore bed is found an extensive bed of kaolin, or porcelain earth. It contains three churches, three stores, one grist-mill, three saw-mills; 12 schools, 555 scholars. Pop. 1310.

MONMOUTH, a marit. co. in the W. of England, adjoining Wales, having N. the cos. of Brecknock and Hereford; E. Gloucester, from which it is separated by the Wye; S. the Bristol Channel; and W. Glamorgan, from which it is separated by the Rumney. Area, 317,440 acres, of which 270,000 are supposed to be arable, meadow and pasture. It is divided into two not very unequal parts by the

Usk, which flows through it from N. to S.; the tract to the W. of that river being comparatively rugged and mountainous, and that to the E. comparatively level and well-wooded. The S. part of both divisions along the Bristol Channel, contains large tracts of marshy land; in some parts of a deep, rich, loamy soil; and in others, of a black peat earth. Large embankments have been raised in different places along the shore, to protect the marsh land from inundation. In other parts of the county, the soil, which is in general good, mostly consists, in the elevated grounds, of a red sandy loam, and in the valleys of a red clay; the substratum is frequently limestone. The arable land is generally clean, and in good order; but the rotation of crops might be a good deal improved. Draining is extensively practised. Cattle principally of the Hereford breed; and inferior only to the same breed in their native county. There are numerous orchards; and, in a few places, hops are cultivated. Stock of sheep estimated at from 170,000 to 180,000. There are some large estates; but property is notwithstanding, a good deal subdivided. The size of farms varies from 60 to 300 acres, 140 acres being supposed to be about the average. They are generally held at will; and the want of leases is much and justly complained of. Average rent of land, in 1810, 12s. 9½d. an acre. Principal minerals, coal, iron, and limestone. The abundance of these has led to the establishment of many very extensive iron works, especially in the N. and W. parts of the county; which were estimated to produce, in 1840, about 300,000 tons of iron. The access to the mines has been facilitated by the formation of canals and railways. A good deal of flannel is made in different parts of the county. Besides the Wye, Usk, and Rumney, it is watered by the Avon, Sirhowey, and Ebwy. Monmouth has six hundreds, and 127 parishes; and sends four members to the House of Commons, viz. two for the county, and two for the borough of Monmouth. Registered electors for the county, in 1839–40, 4,447. In 1831, Monmouth had 18,612 inhabited houses, 19,911 families, and 98,130 persons; of whom 51,286 were males, and 47,035 females. Sum paid for poor-rate, 1839–30, 30,874l. Annual value of real property, in 1815, 286,981l.; profits of trade and professions in ditto, 102,571l.

MONMOUTH, a parl. bor., market-town, and par. of England, cap. of the above co., hund. Skenforth on the Wye, 25 m. N. by W. Bristol, and 112 m. W. by N. London. Pop. of parl. bor. (which includes the par. of Monmouth and a part of that of Dixton), in 1841, 5,446. The town, which is well built, well paved, and lighted with gas, comprises a principal avenue, with other smaller streets, one of which leads to a old stone bridge over the Wye. The guildhall, in the market-place, is a neat and commodious edifice; and at the N. end of the town is a prison, which though externally of imposing appearance, is much too small for the wants of the district. The parish church, partly rebuilt in 1740, has a spire 200 feet in height: the living is a vicarage in the gift of the duke of Beaufort. Another small church stands at the S. W. angle of the town, besides which there are four places of worship for dissenters, with attached Sunday-schools. A free grammar-school was founded here in the reign of James I.; an infant-school has recently been opened, and there is a large almshouse for 20 old men and women. "The town is not flourishing in appearance, and in point of prosperity, is said to be almost stationary. Independently of the conversion of pig iron into bars, and of tin plates, the chief trade of Monmouth consists in the export of bark and timber to Bristol and Ireland, and the general supply of the neighbouring agricultural districts. Coal, for the use of the town, is obtained from the forest of Dean, by means of a railroad: but it is alleged to have been an unprofitable speculation to the projectors." (*Mun. Corp. Rep.*) Monmouth is also a considerable thoroughfare; and from its situation on the romantic banks of the Wye, is likely to maintain its respectability. Its neighbourhood having been selected as the residence of numerous persons of independent fortune. Woollen caps were once largely manufactured in Monmouth. They are referred to by Shakespeare (Henry V., act v., scene 1); and it was ordered by the act 13 Elizabeth, cap. 19, that they should be universally worn on Sundays and holydays.

Monmouth, which was first incorporated in 1539, has been governed, since the passing of the Mun. Reform Act, by a mayor and three other aldermen, with 12 councillors; it has likewise a commission of the peace, under a recorder. The borough, in conjunction with Usk and Newport, has sent one member to the House of Commons since the 27th Henry VIII., the right of election down to the Reform Act being vested in burgesses residing within seven miles of the borough. The electoral limits were left unchanged by the Boundary Act; and, in 1839–40, Monmouth, with its contributary boroughs, had 1304 registered electors. It is also the principal polling-place and election-town for members of the county, as well as the chief town of a poor-law union.

398

Machine on Saturday ; wool fairs, Whit-Tuesday, June 18 and Sept. 4.

MONMOUTH, county. N. J. Situated in the E. part of the state, and contains 1030 sq. m. Bounded E. by the Atlantic, along which are Long Beach and Island Beach, enclosing Little Egg harbour and Barnegat bays. Watered by Neviaink, Shrewsbury, Tom's and Forked rivers, which flow E. into the Atlantic ; and Assunpink and Crosswick's creeks, flowing W. to Delaware river. Marl is found and extensively used as a manure. It contained in 1840, 19,592 neat cattle, 18,604 sheep, 23,941 swine ; and produced 30,368 bushels of wheat, 166,013 of rye, 463,554 of Indian corn, 39,256 of buckwheat, 3411 of barley, 144,066 of oats. 273,980 of potatoes, 1234 pounds of silk cocoons. It had 147 stores, eight lumber yards, seven furnaces, two forges, two fulling mills, 54 grist-mills, 56 saw-mills, one rope-walk, 15 tanneries, 29 distilleries, two printing-offices, two daily newspapers ; 94 schools, 4995 scholars. Pop. 32,909. Capital, Freehold.

MONMOUTH, p. t., Kennebec co., Me., 16 m. S. W. Augusta, 522 W. Watered by Cobbessecontee river. Incorporated in 1792. It contains seven stores, one fulling-mill, three flouring-mills, four grist-mills, one saw-mill; one academy, 164 students. Pop. 1892.

MONMOUTH, p. v., capital of Warren co., Ill., 190 m. N. W. Springfield, 856 W. Situated in a prairie, a little S. of Cedarfork of Henderson's river. It contains a courthouse, jail; six stores and groceries, and about 25 dwellings.

MONONGAHELA t., Pa., rises in Randolph co., Va., at the foot of Laurel mountains, and flowing northwardly for about 300 m., unites with Alleghany river at Pittsburg to form Ohio river. It is nearly 400 yards wide at its mouth, is navigable for light boats, 60 m. to Brownsville, and for small boats nearly 200 m. from its mouth. Its principal tributaries are Youghiogheny and Cheat rivers, which enter it on the E. side.

MONONGAHELA. t., Greene co., Pa., 14 m. S.E. Waynesburg. Bounded E. by Monongahela river. Watered by Whitby creek. It contains six stores, one fulling-mill, one woollen factory, three grist-mills, seven saw-mills, one glass house, two tanneries, four distilleries ; three schools, 75 scholars. Pop. 1178.

MONONGALIA, county, Va. Situated in the N.W. part of the state, and contains 580 sq. m. Watered by Monongahela river, and its branch, Cheat river. It contained in 1840, 16,920 neat cattle, 98,817 sheep, 19,865 swine ; and produced 166,426 bushels of wheat, 6259 of rye, 381,316 of Indian corn. 8936 of buckwheat, 390,928 of oats, 61,723 of potatoes, 14,915 pounds of tobacco, 118,569 of sugar, 167,900 tons of bituminous coal. It had 18 stores, seven furnaces, three fulling-mills, three flouring-mills, 29 grist-mills, 63 saw-mills, one paper-mill, 13 tanneries, 19 distilleries, one pottery, one printing-office, one weekly newspaper ; one academy, 14 students, 29 schools, 639 scholars. Pop.: whites. 16,982 ; slaves, 980 ; free coloured, 146 ; total, 17,368. Capital, Morgantown.

MONOPOLI, a seaport city of the Neapolitan dom., prov. Bari, on the Adriatic, 27 m. S.E. by E. Bari, and 28 m. N.N.E. Taranto ; lat. 40° 57' 19'', long. 17° 18' 58'' E. Pop., in 1823, 15,535. It stands on an eminence surrounded by a wall, and is defended by a castle. Swinburne calls it "a dark, disagreeable town, with narrow crooked streets, and very lofty flat-roofed houses ;" but the account given by Craven is not quite so unfavourable. It is approached from the N. by a newly-built suburb, the small but regular houses of which have each a neat garden. The city has several churches, including a cathedral, which has a fine painting of St. Sebastian by Palma, and a chapel dedicated to the Virgin, and enriched with inlaid marbles of all colours. The town has two ports capable of accommodating vessels of large size ; but the deepest is open to the N., and is consequently exposed to the Bora, or N.E. wind, which often blows in the Adriatic with much violence. Monopoli has manufactures of cotton and linen cloths, and some trade in wine and olives. It is not very ancient, being probably built by the Greeks of the lower empire, partly with the ruins of Egnatia, which stood about three miles S.E., and some traces of which still exist. (Swinburne ; Craven ; Cramer's An. Italy, &c.)

MONREALE, a city of Sicily, intend. Palermo, on a steep hill four m. S.W. Palermo, with which city is communicates by a fine road and causeway, supported by strong buttresses, ornamented with many seats, fountains, urns, &c., laid down at the expense of a late archbishop of Monreale. Pop., in 1831, 19,903. Monreale, though not a fine town, has several remarkable edifices. The cathedral, a large edifice founded in 1174, ranks next after that of the capital: for though heavy, and without symmetry, it has an imposing appearance. Its architecture is a mixture of Lower Greek and Saracenic, and its interior, above the pillars and arches, is wholly incrusted with mosaic work, representing different subjects from the Bible. A destructive fire, in

1811 did great injury to the structure ; but the portions destroyed have been since rebuilt exactly in the former style. An adjoining Benedictine convent has a magnificent cloister, a large library, a collection of coins, and numerous paintings, including one of the finest pictures, of the Sicilian artist, Novelli Monrealese. Near the town is also another rich Benedictine establishment, founded by Pope Gregory the Great. (Smyth's Sicily, p. 90-92.) Monreale is healthy, and commands fine prospects. Its vicinity is very fertile, corn, oil, and fruit being exported from it to Naples, Genoa, and other parts of Italy. (Ortolani, Dizion, della Sicilia ; Smyth's Sicily, &c.)

MONROE, county, N. Y. Situated toward the N.W. part of the state, and contains 607 sq. m. Drained by Genesee river and small streams flowing into lake Ontario, which bounds it on the N., iron ore, gypsum and marl abound, and it has several sulphur springs. The water-power furnished by Genesee river at Rochester, is unsurpassed in the United States. The Erie canal, the Genesee valley canal, and the Auburn, Rochester and Tonawanda railroad pass through it. It contained in 1840, 35,335 neat cattle, 132,970 sheep, 59,309 swine ; and produced 1,074,813 bushels of wheat, 3447 of rye, 406,621 of Indian corn, 37,094 of buckwheat. 61,787 of barley, 523,655 of oats, 721,590 of potatoes, 181,119 pounds of sugar. It had two commercial houses in foreign trade, capital $15,100, 340 retail stores, capital $1,568,840, five lumber yards, capital, $32,300, six furnaces, 20 fulling-mills, 19 woollen-factories, one cotton factory with 3000 spindles, 36 flouring-mills, 15 grist-mills, 69 saw-mills, one oil-mill, two paper-mills, 21 tanneries, seven distilleries, five breweries, two powder-mills, one pottery, nine printing-offices, one bindery, four daily, five weekly, and one semi-weekly newspaper, and one periodical. Total capital in manufactures, $2,633,714 ; 11 academies, 1466 students, 305 schools, 17,656 scholars. Pop. 64,902. Capital, Rochester.

MONROE, county, Pa. Situated in the E. part of the state, and contains 750 sq. m. Organized in 1836. Bounded E. by Delaware river, N.W. by Lehigh river. Watered by Broadhead's, Bushkill, and Tobyhanna creeks. It contained in 1840, 6519 neat cattle, 9422 sheep, 10,642 swine ; and produced 16,961 bushels of wheat, 84,203 of rye, 56,391 of Indian corn, 50,563 of buckwheat, 57,513 of oats, 99,237 of potatoes. It had 19 stores, one flouring-mill, 25 grist-mills, 107 saw-mills, nine tanneries, two printing-offices, two weekly newspapers ; two academies, 31 schools, 794 scholars, Pop. 9879. Capital, Stroudsburg.

MONROE, county, Va. Situated in the W. part of the state, and contains 750 sq. m. Bounded W. by New river. Watered by Greenbrier river. It has an elevated surface, and contained in 1840, 12,456 neat cattle, 20,047 sheep, 13,786 swine ; and produced 67,993 bushels of wheat, 36,873 of rye, 209,498 of Indian corn, 5742 of buckwheat, 194,236 of oats, 22,865 of potatoes, 78,895 pounds of sugar. It had 16 stores, one woollen factory, two flouring mills, 24 grist-mills, 29 saw-mills, three oil-mills, seven tanneries, eight distilleries ; one academy, 80 students, five schools, 119 scholars. Pop.: whites, 7457 ; slaves, 868 ; free coloured, 97 ; total, 8423. Capital, Union.

MONROE, county, Geo. Situated near the centre of the state, and contains 370 sq. m. Bounded N.E. by Ocmulgee river, by branches of which it is drained. It contained in 1840, 10,037 neat cattle, 5341 sheep, 35,903 swine ; and produced 62,651 bushels of wheat, 1308 of rye, 511,941 of Indian corn, 49,931 of oats, 18,001 of potatoes, 4,673,983 pounds of cotton. It had 13 stores, four tanneries ; 17 academies, 752 students, 14 schools, 388 scholars. Pop.: whites, 7804 slaves, 8447 ; free coloured, 24 ; total, 16,275. Capital, Forsyth.

MONROE, county, Flor. Situated in the S. part of the peninsula, having the Florida channel on the S. and the gulf of Mexico W. It contains a part of the Everglades, an inundated region covered with pine and hammock islands, a famous resort of the Indians in the Florida war. In the N.W. part is some fertile land. It contained in 1840, four commission houses in foreign trade, 16 retail stores, one printing-office, one weekly newspaper ; three schools, 28 scholars. Pop. 688. Capital, Key West.

MONROE county, Ala. Situated toward the S. part of the state, and contains 980 sq. m. Watered by Alabama river and its branches. It contained in 1840, 16,189 neat cattle, 3654 sheep, 33,073 swine ; and produced 5335 bushels of wheat, 291,504 of Indian corn, 10,769 of oats, 40,605 of potatoes, 33,351 of rice, 10,000 pounds of tobacco, 986,327 of cotton. It had 20 stores, 32 grist-mills, 15 saw-mills, six tanneries ; 12 schools, 306 scholars. Pop.: whites, 5370 ; slaves, 5292 ; free coloured, 18 ; total, 10,680. Capital, Monroeville.

MONROE, county, Miss. Situated toward the N.E. part of the state, and contains 650 sq. m. Drained by the Tombigbee river and its tributaries. It contained in 1840, 12,097 neat cattle, 9833 sheep, 25,200 swine ; and produced 22,392 bushels of wheat. 331,190 of Indian corn, 30,883 of oats,

MONROE.

19,092 of potatoes, 6797 pounds of tobacco, 712,095 of cotton. It had two commission houses in foreign trade, 24 stores, one flouring-mill, 15 grist-mills, eight saw-mills; one tannery, five distilleries, one pottery, one printing-office, one weekly newspaper; three academies, 93 students; 10 schools, 110 scholars. Pop.: whites, 5146; slaves, 4083; free coloured, 21; total, 9250. Capital, Athens.

MONROE, county, Tenn. Situated in the S.E. part of the state, and contains 750 sq. m. Drained by branches of Tennessee river. It contained in 1840, 6539 neat cattle, 5665 sheep, 17,261 swine; and produced 39,525 bushels of wheat, 2231 of rye, 345,287 of Indian corn, 90,815 of oats, 4750 of potatoes, 1540 pounds of tobacco, 3180 of cotton. It had three stores, one flouring-mills, 10 grist-mills, four saw-mills, two tanneries, nine distilleries, one pottery; two academies, 77 students; 12 schools, 172 scholars. Pop.: whites, 11,125; slaves, 864; free coloured, 67; total, 12,056. Capital, Madisonville.

MONROE, county, Ky. Situated toward the S. part of the state, and contains 375 sq. m. Drained by Big Barren river and its branches. It contained in 1840, 6314 neat cattle, 4952 sheep, 29,396 swine; and produced 33,676 bushels of wheat, 1945 of rye, 311,047 of Indian corn, 77,098 of oats, 11,100 of potatoes, 420,012 pounds of tobacco, 42,959 of cotton, 16,429 of sugar. It had 15 stores, 19 grist-mills, 10 saw-mills, one oil-mill, four tanneries, 30 distilleries; one academy, 25 students; eight schools, 156 scholars. Pop.: whites, 5811; slaves, 703; free coloured, 12; total, 6526. Capital, Tompkinsville.

MONROE, county, O. Situated toward the S.E. part of the state, and contains 520 sq. m. Watered by Little Muskingum river, and Sunfish, Duck and Will's creeks. It contained in 1840, 11,303 neat cattle, 20,518 sheep, 27,354 swine; and produced 168,795 bushels of wheat, 3430 of rye, 428,391 of Indian corn, 7569 of buckwheat, 172,174 of oats, 75,683 of potatoes, 1,098,868 pounds of tobacco, 77,996 of sugar. It had 32 stores, three fulling-mills, one flouring-mill, 35 grist-mills, 37 saw-mills, one oil-mill, 12 tanneries, eight distilleries, one brewery, two potteries; 22 schools, 690 scholars. Pop. 18,521. Capital, Woodsfield.

MONROE, county, Mich. Situated in the S.E. part of the state, and contains 540 sq. m. Bounded S.E. by lake Erie, N.W. by Huron river. Watered by Raisin river, and Swan, Stony, Otter and Bay creeks. It contained in 1840, 8364 neat cattle, 3010 sheep, 9981 swine; and produced 42,856 bushels of wheat, 2103 of rye, 74,407 of Indian corn, 7887 of buckwheat, 2199 of barley, 68,794 of oats, 83,016 of potatoes, 40,697 pounds of sugar. It had 24 stores, one fulling-mill, one woollen factory, two flouring-mills, five grist-mills, 11 saw-mills, one paper-mill, two printing-offices, two weekly newspapers; 30 schools, 777 scholars. Pop. 9922. Capital, Monroe.

MONROE, county, Ia. Situated a little N.E. of the centre of the state, and contains 390 sq. m. Watered by the W. fork of White river and its tributaries, which afford waterpower. It contained in 1840, 4019 neat cattle, 7470 sheep, 19,701 swine; and produced 57,636 bushels of wheat, 2948 of rye, 497,968 of Indian corn, 95,221 of oats, 9780 of potatoes, 20,929 pounds of sugar. It had 20 stores, one flouring-mills, seven woollen factories, six flouring-mills, 12 grist-mills, 10 saw-mills, four oil-mills, 15 tanneries, 13 distilleries, three printing offices, three weekly newspapers; one college, 90 students; one academy, 60 students; 17 schools, 558 scholars. Pop. 10,143. Capital, Andersontown.

MONROE, county, Ill. Situated toward the S.W. part of the state, and contains 360 sq. m. Bounded W. by Mississippi river. Watered by Horse, Prairie de Long and Eagle creeks. It contained in 1840, 8532 neat cattle, 3338 sheep, 16,516 swine; and produced 22,519 bushels of wheat, 903,469 of Indian corn, 31,975 of oats, 10,016 of potatoes. It had 15 stores, two flouring-mills, three grist-mills, four saw-mills, three tanneries, two distilleries; five schools, 168 scholars. Pop. 4481. Capital, Waterloo.

MONROE, county, Mo. Situated towards the N.E. part of the state, and contains 744 sq. m. Watered by Salt river and its branches. It contained in 1840, 5467 neat cattle, 9760 sheep, 29,585 swine; and produced 19,208 bushels of wheat, 5437 of rye, 491,854 of Indian corn, 89,014 of oats, 16,710 of potatoes, 182,414 pounds of tobacco. It had 20 stores, 13 grist-mills, 14 saw-mills, four tanneries, seven distilleries, one pottery; three academies, 125 students; 20 schools, 479 scholars. Pop.: whites, 7813; slaves, 1687; free coloured, 5; total, 9505. Capital, Paris.

MONROE, county, Ark. Situated in the E. part of the state, and contains 1150 sq. m. Watered by Cache and White rivers. It contained in 1840, 2310 neat cattle, 135 sheep, 4074 swine; and produced 54,549 bushels of Indian corn, 3460 of oats, 2706 of potatoes, 26,375 pounds of cotton. It had four stores, four flouring-mills; two schools, 33 scholars. Pop.: whites, 786; slaves, 148; free coloured, two; total, 936. Capital, Lawrenceville.

MONROE, p. t., Waldo co., Me., 54 m. N.E. Augusta, 649

W., Watered by Marsh river, a tributary of Penobscot river. It contains two stores, two grist-mills, seven saw mills; eight schools, 669 scholars. Pop. 1602.

MONROE, p. t., Washington co., Vt., 16 m. N.N.E. Montpelier, 532 W. Chartered in 1781 by the name of Woodbury. Received its present name in 1836. Watered by branches of Onion or Winooski and Lamoille rivers. It has one grist-mill, two saw-mills; 13 schools, 367 scholars. Pop. 1099.

MONROE, p. t., Fairfield co., Ct., 54 m. S.W. Hartford, 292 W. Incorporated in 1823. It contains four churches, a Congregational, Episcopal, Methodist, and Baptist, five stores, one grist-mill, four saw-mills; one academy, 20 students, six schools, eight schools, 293 scholars. Pop. 1351.

MONROE, p. t., Orange co., N. Y., 14 m. S.E. Goshen, 117 m. S. by W. Albany, 279 W. Drained by Ramapo river, which flows into Passaic river. It contains a Presbyterian and a Methodist church, an academy, four stores, one fulling-mill, one grist-mill, one saw-mill; nine schools, 417 scholars. Pop. 2914.

MONROE, t., Middlesex co., N. J. It has eight stores, four grist-mills, five saw-mills, one paper-mill, five distilleries; one academy, 25 students, eight schools, 295 scholars. Pop. 2453.

MONROE, t., Bradford co., Pa. It has five stores, four grist-mills, 18 saw-mills; six schools, 278 scholars. Pop. 1153.

MONROE, t., Cumberland co., Pa., 7 m. S.E. Carlisle. Watered by Yellow Breeches creek, which affords water-power. It has six stores, one fulling-mill, one woollen-factory, four flouring-mills, two saw-mills, four tanneries, one distillery. Pop. 1570.

MONROE, t., Armstrong co., Pa. It has five stores, two fulling-mills, three grist-mills, three saw-mills, two tanneries; five schools, 240 scholars. Pop. 1151.

MONROE, p. v., capital of Walton co., Ga., 61 m. N.N.W. Milledgeville, 635 W. Situated 2½ m. E. of Alcovee river, a branch of Ocmulgee river. It contains a courthouse, jail, two churches, a Baptist and a Methodist, an academy, 11 stores and about 50 dwellings.

MONROE, p. v., capital of Washita par., La., 300 m. N.W by N. New Orleans, 1190 W. Situated at the head of steamboat navigation, on the E. side of Washita river. It contains a courthouse, jail, a U. States land office, an academy, four stores, 100 dwellings and about 600 inhabitants.

MONROE, p. v., Lemoh t., Butler co., O., 95 m. W.S.W Columbus, 479 W. It contains two brick churches, a seminary of the Associate Reformed church, four stores, 30 dwellings, and about 300 inhabitants.

MONROE, t., Clermont co., O. It has seven stores; one academy, 44 students, nine schools, 480 scholars. Pop. 1622.

MONROE, t., Knox co., O., 6 m. N.E. Mount Vernon. Watered by Schenck's creek, which affords water-power. It has six schools, 233 scholars. Pop. 1249.

MONROE, t., Licking co., O. It has seven schools, 277 scholars. Pop. 1156.

MONROE, t., Logan co., O. It has five schools, 155 scholars. Pop. 1903.

MONROE, t., Miami co., O. Bounded E. by Miami river Watered by its W. branch. The Miami canal passes through it. It contains three flouring-mills, one grist-mill, four saw-mills; five schools, 281 scholars. Pop. 1494.

MONROE, t., Richland co., O. Watered by Mohiccan creek Pop. 1624.

MONROE, t., Putnam co., Ia. It has three grist-mills, two saw-mills; two schools, 144 scholars. Pop. 1341.

MONROE, t., Washington co., Ia. It has two stores, one flouring-mill, seven grist-mills, seven saw-mills. Pop. 137.

MONROE, city, capital of Monroe co., Mich., 37 m. S.S.W Detroit, 486 W. Situated on the river Raisin. 2½ m. from its mouth. It contains a courthouse, jail, two banks, a U States land office, seven churches, two Presbyterian, an Episcopal, Methodist, Baptist and two Roman Catholic seven storage and forwarding houses, 22 stores, one woollen factory, one iron foundry and edge tool factory, two flouring-mills, three saw-mills, one paper-mill, two printing offices, two weekly newspapers, a branch of the university of Michigan, two female academies, a reading room, and public library of 12 or 1500 volumes. The courthouse is an elegant edifice of hewn stone, which cost $35,000, and several of the churches are handsome edifices. The waterpower of Raisin river at this place has been estimated by the U. S. engineer, sufficient to propel 350 run of stones. A ship canal, 100 feet wide and 12 feet deep is constructed from the city to the lake. Steamboats from Buffalo to Detroit, stop here in the season of navigation, and many vessels ply on lake Erie. It contains about 380 dwellings and 1703 inhabitants. The town exclusive of the city has 693 inhabitants. It was first settled by the French in 1778, and was increased by the English in 1796. In Jan. 1813, the memorable battle of Frenchtown was fought near this place, when 700 Kentuckians under Gen. Winchester were massacred by the Indians, under the countenance of

MONROEVILLE.

Gen. Procter, who ordered the burning of the place. The order was partially executed, but was stopped by the interference of the celebrated Indian chief Tecumseh.

MONROEVILLE, p. v., capital of Monroe co., Ala., 151 m. S. Tuscaloosa, 943 W. Situated on a branch of Limestone creek, 12 m. E. of Alabama river. It contains a courthouse, jail, and several stores and dwellings. Nett proceeds of the postoffice, $162.

MONS (Flem. Berghen), a town of Belgium, prov. Hainault, of which it is the cap., on the Trouille, by which it is separated into two parts, 32 m. S.W. Brussels, and 30 m. E.N.E. Valenciennes. Lat. 50° 27' N., long. 3° 57' 30' E. Pop., in 1836, 23,081. The town is built partly on level ground, and partly on the declivity of a hill, crowned by a lofty tower, rebuilt in 1662 on the site of an ancient castle, said to have been built by Julius Cæsar. Mons has been, since 1818, when its works were considerably augmented and strengthened, one of the principal Belgian fortresses. Its walls are flanked with 14 bastions, and on its E. sides are two extensive pools, by the aid of which, and the river, its ditches may be easily filled, and the environs laid under water. Without the walls are several suburbs. The town is entered by five gates; several of its streets are steep and winding, but they are in general wide, clean, well paved, and bordered with good houses, many of which are of stone. It has several good squares: of these the Place d'Armes, or great market-place, is the principal, and has in it the government-house, and the hall of the provincial council. The ramparts are planted with trees, forming pleasant promenades; and within the precincts of the citadel is a garden open to the public. The Trouille is here crossed by three bridges, and numerous stone pumps supply the town with water. The town-hall, erected according to Vandermaelen, in 1440, is a large Gothic edifice, surmounted by a fine cupola. The church of St. Waudru, on the site of a chapel founded by that saint in the 7th century, is a fine specimen of Gothic architecture; and the church of St. Elisabeth is also handsome in some of its parts, but it has the incongruity of Gothic pillars supporting Corinthian capitals. The courthouse, the new college, the military hospital designed by Vauban, the arsenal, the new barracks, the theatre, and the academy of arts, are among the most conspicuous public buildings. There are civil, orphan, foundling, and other hospitals, a house of correction, a workhouse, various asylums, a government loan-bank, and other charitable institutions.

Mons is the residence of a civil governor, and of a provincial and a municipal military commandant, and the seat of tribunals of primary jurisdiction and commerce, a chamber of commerce, &c. It had formerly a flourishing manufacture of lace, now much decayed, and several sugar refineries, which have been abandoned. It still, however, produces some woollen and cotton stuffs, gloves, cutlery, hardware, soap, and vinegar; and has copper and lead foundries, flour-mills, &c.; but its chief source of wealth is in the numerous and productive coal mines by which it is surrounded, and which employ a great number of workmen and steam-engines. There are also extensive bleaching grounds in the vicinity. The coal from Mons is sent in part to Paris, by a long line of internal navigation, of which the canal from Mons to Condé forms a part. This canal, commenced by the French in 1807, and finished in 1814, is perfectly straight, 15 m. in length, with seven locks, and at Condé joins the Scheldt. The greater part of it is in the Belgian territory; but a new branch of the Canal d'Antoine has been recently cut from it, avoiding France altogether, and entering the Scheldt not far from Tournay. Mons has sustained many sieges. It was taken in 1691, by Louis XIV., after an obstinate defence; and was occupied by Eugene and Marlborough in 1709. The emperor Joseph II. demolished its former fortifications in 1784. During the French ascendency, it was the capital of the department of Jemmappes. (Vandermaelen, Dict. Geog., De Hainault, &c.)

MONSON, p. t., Hampden co., Mass., 77 m. S.W. by W. Boston, 390 W. Chicopa river and its branches, afford water-power. Incorporated in 1760. It contains four churches, a Congregational, Methodist, Baptist and Universalist, a flourishing academy with 99 students, four stores, three woollen factories, two cotton factories with 3204 spindles, two grist-mills, three saw-mills; 14 schools, 330 scholars. Pop. 2151.

MONTAGUE, p. t., Franklin co., Mass., 85 m. W. by N. Boston, 396 W. Bounded W. and N.W. by Connecticut river, across which is a bridge connecting it with Deerfield. Watered by Saw-mill river. Connecticut river in the W. part of the town has a romantic and majestic fall, called Turner's falls, of 65 feet, which is said strongly to resemble Niagara. The principal pitch is divided by an island. A canal, 3 m. long, 25 feet wide and 12 feet deep, overcomes these falls, by eight locks, with a rise of 75 feet. The water is thrown into the canal by an immense dam, 330 yards long, and 40 feet at its greatest height. Another

MONTBRISON.

immense dam, with a canal, overcomes Miller's falls, four miles above. It contains three churches, a Congregational, Episcopal and Baptist, five stores, one grist-mill, three saw-mills; 11 schools, 385 scholars. Pop. 1255.

MONTAGUE, p. t., Sussex co., N. J., 91 m. N. Trenton, 258 W. Bounded N.W. by Delaware river, and drained by its small tributaries. Across the Delaware is a bridge to Milford, which cost $30,000. It contains one store, four grist-mills, one saw-mill, two tanneries; six schools, 142 scholars. Pop. 1026.

MONTAGNANA, a town of Austrian Italy, deleg. Padua, cap. distr., on the Frassina, 22 m. S.E. Padua. Pop. 6337. It is walled, and has a castle, several churches, a hospital, and a high female school. It has manufactures of woollen, and linen stuffs, hats, and leather, and a brisk trade in agricultural produce. It has several annual fairs, one of which lasts from November 25 to December 24. The hemp grown in the vicinity of this town is esteemed the best in the Lombardo-Venetian kingdom. (Berghaus, &c.)

MONTARGIS, a town of France, dép. Loiret, cap. arrond., at the confluence of the canals of Orleans, Briare and Loing, 39 m. E. by N. Orleans. Pop., in 1836, 7757. Though ill laid out, it is pretty well built; it is in part surrounded by old walls, and has the ruins of a large castle, in which the French kings often held their court. The parish church is remarkable for the elevation and boldness of its pillars and nave. Montargis has two hospitals, a small theatre, and manufactures of coarse woollen cloths. (Hugo, art. Loiret.)

MONTAUBAN, a town of France, dép. Tarn-et-Garonne, of which it is the cap., on an eminence on the banks of the Tarn, crossed here by a brick bridge of seven arches, 122 m. E.S.E. Bordeaux; lat. 44° 1' N., long. 1° 20' 45" E. Pop. in 1836, ex sem., 17,531, one third of whom are Protestants. The town, properly so called, is small, and irregularly laid out, with narrow ill-paved streets, lined by old houses having projecting gables; but the suburbs, which are of considerable extent, present a totally different appearance, having straight, wide, and regular streets, with new, large, and elegant mansions. It has three public squares: that of the Prefecture, the Place-d'Armes, and the Place Royale, the last of which is spacious, and has many handsome houses. The chief public buildings are the cathedral, a cruciform structure with two towers; the town-hall, a large and fine square edifice; the church of St. James, with a lofty brick tower and steeple; the prefecture, bishop's palace, the public library with 11,000 vols., a small theatre, and several hotels. Near the prefecture commences a noble avenue, shaded with six rows of acacias, leading to the terraces of some adjacent promenades, which command extensive prospects of the surrounding country. The beautiful situation of Montauban, the purity of its atmosphere, the good quality as well as abundance of its water, and the cheapness of all the necessaries of life, render it a pleasant and favourite retreat for persons of small fortune. It is the seat of a tribunal of primary jurisdiction, and has a chamber of manufactures, a society of agriculture and science, and a communal college. It has manufactures of serges, flannels, coarse cotton fabrics, and silk stockings, earthenware, soap, brandy, starch, leather, and beer. It likewise carries on a considerable retail trade, and is a large entrepôt for corn.

Montauban was built in the beginning of the 14th century, and owes its foundation to the protection afforded by the Count of Toulouse to the oppressed vassals of certain barons, who claimed, among other privileges, that of prostitution. It afterwards acquired celebrity on account of its early adherence to the cause of the Huguenots, and its great sufferings in their behalf. In 1621, it successfully resisted an army under Louis XIII.; but a few years subsequently, after the siege of Rochelle, it was compelled to open its gates to that monarch. A few years after it was exposed to the dragonnades, that disgraced the reign of Louis XIV. This was the last disastrous event connected with the town, which has since gradually risen to its present importance. (Hugo, art. Tarn-et-Garonne; Guide du Voy. en France, p. 63.)

MONTAUK POINT, East Hampton, Suffolk co., N. Y., 140 E. New-York. It is a high promontory on the E. end of Long Island, against which the waves of the ocean continually dash, and, in a storm, with fearful violence. It contains a lighthouse, the base of which is 46 feet above the ocean, and which is 100 feet high, substantially built of stone. It can be seen for 30 m. at sea, was erected in 1796, and cost $23,996. Near it is a public house, much frequented in the summer season.

MONT BLANC, in Savoy, at once the highest mountain of the Alps and of Europe. (Vol. i., p. 66.)

MONTBRISON (an. Mons Brivo), an ancient town of France, dep. Loire, of which it is the cap., 237 m. S.S.E. Paris; lat. 45° 36' 41" N., long. 4° 4' 23" E. Pop., in 1836, ex. sem., 6030. It was formerly fortified; and is irregularly laid out with narrow streets, and low, shabby houses. A cathedral, founded in 1905, and still in an unfinished state, a prefecture, hospital, college, with a library of 15,000 volumes, theatre, corn-exchange, and infantry barracks are the principal buildings; but the cathedral only has any architectural beauty. Though the capital of a department, and the seat of a tribunal of primary jurisdiction and commerce, and of a society of agriculture, Montbrison is very unimportant, having no manufactures, and only a limited retail trade; indeed, it has been proposed to make the large manufacturing town of St. Etienne, 11 m. S. by E. Montbrison, capital of the department.

MONT-DE-MARSAN, a town of France, cap. dep. Landes, 64 m. S. Bordeaux. Pop., in 1836, ex. com., 3924. It is situated on the side of a declivity close to the navigable river Midouze (crossed here by a stone bridge of two arches), and is clean, well paved, and regularly laid out. the principal buildings being the parish church, town-hall, court of justice, college, public baths, barracks, a small theatre, and a library with 1300 volumes. The suburbs are planted with trees, and laid out in walks. It has manufactures of coarse woollen cloths, blankets, and sail-cloth; and some trade, with Bayonne, in wine and brandy. It is the seat of a tribunal of primary jurisdiction, and of a society of agriculture and commerce.

MONTEFIASCONE, a town of central Italy, Papal States, deleg. Viterbo, on a mountain, 9 m. N.N.W. Viterbo. Pop. about 5500. It has a fine cathedral, and many other religious edifices, but is celebrated principally for its light, white, muscadel wines; but these, as they do not bear carriage, are seldom met with out of the country where they are produced.

MONTELEONE (an. Hipponium and Vibo Valentia), a town of the Neapolitan dom., prov. Calabria Ultra II., cap. of a distr. on a mountain, 27½ m. S.W. Catanzaro. Pop. from 9000 to 10,000. Its commanding situation, with its fine old castle, gives it a fine appearance from without; but its streets are crooked and ill paved, and the houses mostly low and of wood. There are several churches, in which are some good pictures; a royal college, &c. The inhabitants are principally engaged in the tunny fishery, and in trading in silk and oil. According to Strabo (vi., 256), Hipponium was founded by the Locri Epizephyrii. After many vicissitudes, it became a Roman colony; and Cicero calls it illustre et nobile municipium. It had a fine temple of Proserpine, demolished by Count Roger of Sicily, who applied the materials to the construction of the abbey at Mileto, 6 m. distant. (Cramer's Anc. Italy, ii., 419–421; Craven's Tour, p. 419–421; Rampoldi, &c.)

MONTELIMART (an. Acunum?), a town of France, dep. Drôme, on the Jabron, near its confluence with the Rhone, 70 m. S. Lyon, lat. 44° 33' N., long. 4° 45' E. Pop., in 1836, ex. com., 6150. It is surrounded with walls, and is generally well built, the chief street being wide and paved with basalt. It has four handsome gates, and a well-planted public walk along the walls, which adds greatly to its beauty. Near the town is a mineral spring, highly esteemed for its medicinal qualities, and the neighbourhood is remarkable for the abundance and variety of its fruits, &c. The manufacture of figured silks is the only important branch of industry; but it has a considerable retail trade, and is the chief entrepôt of an extensive and highly productive district. It was unsuccessfully besieged by Colognal in 1567. (Hugo, France Pitt., ii., 6; Guide du Voyageur en France.)

MONTEPULCIANO, a town of central Italy, grand duchy of Tuscany, prov. Arezzo, on a lofty hill, 27 m. S.E. Sienna. Pop. from 2000 to 3000. It is surrounded by a wall with battlements, and has numerous ecclesiastical establishments, a college, a hospital, and manufactures of soap, oil, and wine flasks. It is celebrated for its dessert wine, which, with excusable partiality, is preferred by Redi to all other wines:

"Montepulciano d' ogni vino è il re."　　Bacco in Toscana.

MONTEREAU (an. Condate) a town of France, dep. Seine-et-Marne, at the confluence of the Seine and Yonne, each of which is crossed here by a stone bridge, 42 m. S.E. Paris. Pop., in 1836, ex. com., 4379. It has a fine open market-place, and is well built; clean, and respectable; a parish church, town-hall, hospital, and three hotels, are the only public edifices of any importance. It is the seat of a tribunal of commerce, and has an extensive manufacture of earthenware, with some tan-yards, and a considerable trade with Paris, chiefly in corn, flour, and wood for fuel.

MONTEVIDEO, a fortified seaport city of S. America, cap. of the republic of Uruguay, on a peninsula extending into the estuary of the Plata on its N. side, 125 m. E. by S. Buenos Ayres; lat. 34° 54' 11" S., long. 56° 13' 18" W. Population variously estimated, but may probably be about 12,000. It is well fortified, and has a citadel. The houses, which are of stone or brick, are seldom above one story in

height; they are flat-roofed; and timber is so scarce, that their floors consist, for the most part, of brick or bare earth. The streets, being unpaved, are either clouded with dust or loaded with mud, as the weather happens to be dry or wet. The city is ill supplied with water, which has to be brought from a well 2 m. distant, or from pits dug near the sea-side; or is else merely rain-water, collected in cisterns. There are but few public buildings, and those of no great importance; the cathedral is said to be a handsome edifice, but it is badly situated.

The port of Montevideo is the best on the Plata. It is a large circular basin, open to the S.W.; generally the water is shallow, not exceeding from 14 to 19 feet; but, the bottom being soft mud, vessels are seldom damaged by grounding. It should, however, be observed, that the depth of water in the harbour, as well as throughout the whole of the Rio de la Plata, depends very much on the direction and strength of the winds. The harbour is exposed to the impetus of S.W. winds, which sometimes blow with so much force and continuance as to cause the rise of a fathom or more in the depth of water; but they rarely do any damage to vessels properly moored with anchors to the S.W. and S.E. and out to the N. On the opposite side of the bay is a mountain called Montevideo, whence the city has derived its name; on its summit is a lighthouse, having the lantern 476 feet above the sea.

Montevideo has considerable commerce; the imports principally consist of British cottons, woollens, and hardware, flour, wine and spirits, linens, sugar, tobacco, boots and shoes, salt, &c. The great articles of export consist of animal products, which, in 1836, were estimated as follows: Hides, dry, No. 372,019; do. salted, 141,399; horns, No. 593,625; jerked beef, 306,354 cwts.; horse-hair, 18,662 arrobas of 25 lbs.; horse-hides, No. 37,401; wool, 33,908 arrobas; tallow, 43,182 arrobas, &c. In the same year the total value of the exports was estimated at 3,413,867 Spanish dollars; and that of the imports at 3,587,437 dollars. The trade is principally with Brazil, Great Britain, America, France, Sardinia, Spain, and Portugal.

This town was founded by a colony from Buenos Ayres, and its possession was long a matter of dispute between the Spaniards and Portuguese. It was taken by the Brazilians in 1821; and became, in 1828, the capital of the new republic of Uruguay. (Masse, Henderson, &c., in Mod. Trav., xxix.; Parish's Buenos Ayres, Appendix; Supp. to Comm. Dict., &c.)

MONTEZUMA, p. v., capital of Covington co., Mississippi, m. S.E. Tuscaloosa, 914 W. Situated on the E. side of Oconeah river, and contains a courthouse, jail, and several stores and dwellings.

MONTGOMERY, an inland co. of N. Wales, having E. Merioneth and Denbigh, E. Salop, and S. and W. Radnor and Cardigan. It is oval-shaped, and contains 526,360 acres. The Berwyn mountains divide this county from Merioneth; and, with the exception of some considerable valleys, of which that of the Severn is the most extensive, and that of Llangollen, partly in this county, the surface is, for the most part, rugged and mountainous. The soil is very various; but in the vales it is generally clayey, and in parts very fertile; on the whole, however, the land under tillage is not supposed to exceed from 70,000 to 80,000 acres. The Severn has its source at the extreme W. confines of the county, on the skirts of "huge Plynlimmon;" and runs in a N.E. direction, parallel to, and not very distant from, its S. boundary, till it unites with its important affluent, the Vyrnwy, which also belongs to this county, on the borders of Salop. The agriculture of this county, especially in the vales and along the border of Salop, has been a good deal improved; but withal it is extremely similar to, and quite as backward as, that of Denbigh and Merioneth (which see). The climate, though moist, is mild and salubrious. The vales of this county have been long celebrated for a superior breed of horses. Montgomery has also long been, and still continues to be, the best wooded county in Wales. It was formerly regarded as one of the principal sources of the supply of oak timber for the navy; but many of its finest oak woods have been cut down; and, though a good deal of new wood has been planted, it is doubtful whether it be sufficient to supply the place of that which formerly existed. Average rent of land, in 1810, 6s. 51d. an acre. There are a number of fine and commodious farm-houses and offices; but, in general, they are very defective, and the cottages are quite as bad as in Merioneth. Slate is generally diffused over the county, and forms, indeed, the basis of the mountains. Slates are quarried at Llangynog and other places; coal is raised on the borders of Salop; and there are some lead mines, but none that are very productive. Montgomery is the principal seat of the Welsh flannel manufacture, which is extensively carried on at Newtown, Llanidloes, Machynlleth, and Welshpool (which see). The county is divided into nine hund. and 47 parishes. It sends two members to the House of Commons, viz., one for the county and one for the town of Montgomery and its contributary boroughs. Registered electors for the county in 1839–40, 2342. In 1841 Montgomery had 13.650 inhabited houses, and 68.290 inhabitants, of whom 34,252 were males and 34,969 females. Sum expended for the relief of the poor, in 1838–39, £28,241.

MONTGOMERY, a parl. bor., market town, and par. of N. Wales, cap. co. of its own name, 90 m. S.W. Shrewsbury, and 146 m. W.N.W. London. Pop., in 1841, 1208. Though small, it is a clean, well-built town, in a hollow at the foc; of a high hill. The guildhall stands on an eminence near the ruins of an ancient castle, close to which is the county jail, a modern stone building, well adapted for its purpose. The church, a cruciform structure, in the early English style, has a handsome tower, erected in 1816, and an exquisitely carved screen, and some curious monuments: the living is a rectory in the gift of the crown. The Calvinists and Wesleyan Methodists have also their respective places of worship: and there are two Sunday schools, and a small endowed school. No trade or manufacture is carried on in the town; and it deserves notice merely from its being the capital of a county, and a parliamentary borough. It was incorporated by Henry III. under a steward and 19 burgesses; who enjoyed, till the passing of the Reform Act, the privilege of sending one member to the House of Commons. This act, however, made Llanfylline, Llanidloes, Machynlleth, Newtown, and Pool contributary boroughs with Montgomery in the election of the member. Registered electors for the entire district, in 1839–40, 1021, of whom only 116 belonged to Montgomery. The election for the county takes place here; and sessions are held alternately with Newtown. Markets on Tuesday: fairs, March 26th, first Tuesday in May, June 7th, September 4th, and November 14th. Montgomery is very ancient: its castle was built prior to the Norman conquest, and, from its size and strength, was frequently an object of contention during the wars between the English and Welsh. In 1354 it was in the possession of Roger Mortimer, from whom it passed to the crown. In the 15th century the stewardship of the town and castle was granted to the Herberts of Cherbury. The famous Lord Herbert, celebrated alike for his chivalry, wit, and learning, was born here in 1581. It is the birth-place also of the late Dr. Abraham Rees, the learned editor of the voluminous and valuable Cyclopedia which bears his name. (*Nicholson's Comb. Guide ; Parl. Papers.*)

MONTGOMERY, county, N. Y., centrally situated, toward the E. part of the state, and contains 356 sq. m. Watered by Mohawk river and its tributaries. The Erie canal passes through it on the S. side of the Mohawk, and the Schenectady railroad on the N. side. It contained in 1840, 26,306 neat cattle, 36,586 sheep, 29,108 swine; and produced 34,281 bushels of wheat, 40,868 of rye, 90,374 of Indian corn, 38,312 of buckwheat, 193,530 of barley, 422,418 of oats, 559,899 of potatoes, 51,691 pounds of sugar. It had 94 stores, two furnaces, eight fulling-mills, one woollen factory, three flouring-mills, 21 grist-mills, 67 saw-mills, 16 tanneries, six distilleries, two breweries ; three academies, 396 students ; 116 schools, 5655 scholars. Pop. 35,818. Capital, Fonda.

MONTGOMERY, county, Pa., situated in the S.E. part of the state, and contains 425 sq. m. Bounded S.W. by Schuylkill river, which also passes through its S. part. Drained by Perkiomen, Pennypack, and other creeks. On the Schuylkill are seven dams with short canals, which afford extensive water power. It contained in 1840, 31,652 neat cattle, 15,249 sheep, 39,707 swine ; and produced 181.858 bushels of wheat, 275.069 of rye, 503.065 of Indian corn, 452.530 of buckwheat, 642,990 of oats, 239,230 of potatoes. It had 144 stores, 13 lumber yards, two fulling-mills, eight woollen factories, 26 flouring-mills, 100 grist-mills, 77 saw-mills, 12 powder-mills, nine paper-mills, 30 oil-mills, 23 tanneries, one distillery, six potteries, six printing offices, four weekly and two semi-weekly newspapers ; 10 academies, 372 students ; 72 schools, 4465 scholars. Pop. 47,241. Capital, Norristown.

MONTGOMERY, county, Md., situated toward the W. part of the state, and contains 576 sq. m. Bounded S.W. by Potomac river, N.E. by Patuxent river. Drained by tributaries of these rivers. It contained in 1840, 8698 neat cattle, 16,036 sheep, 19,839 swine ; and produced 142,739 bushels of wheat, 27,704 of rye, 308,385 of Indian corn, 2638 of buckwheat, 225,168 of oats, 62,546 of potatoes, 1,098,412 pounds of tobacco. It had 25 stores, one woollen factory, one cotton factory with 790 spindles, one flouring-mill. 18 grist-mills, 16 saw-mills, one oil-mill, four tanneries ; three academies, 172 students ; 17 schools, 514 scholars. Pop.: whites, 8766; slaves, 5377; free coloured, 1313; total, 15,456. Capital, Rockville.

MONTGOMERY, county, Va., situated toward the S.W. part of the state, and contains 600 sq. m. Bounded W. by New river and drained by its branches, and by head waters of Staunton river. It contained in 1840, 9635 neat cattle,

13,439 sheep. 16,634 swine ; and produced 106,258 bushels of wheat, 21,003 of rye, 308,885 of Indian corn, 2615 of buckwheat, 114,365 of oats, 18,071 of potatoes, 941,275 pounds of tobacco, 1000 of sugar. It had 11 stores, four flouring mills, 19 grist-mills, 19 saw-mills, eight tanneries, nine distilleries, two potteries ; one academy, 40 students ; 17 schools, 402 scholars. Pop.: whites, 5625 ; slaves, 1493 ; free coloured, 87 ; total, 7405. Capital, Christiansburg.

MONTGOMERY, county, N. C., situated toward the S.W. part of the state, and contains 500 sq. m. Watered by Yadkin river and its branches. It contained in 1840, 10,603 neat cattle. 8508 sheep, 18,034 swine ; and produced 48,879 bushels of wheat, 1535 of rye, 255,496 of Indian corn, 19,358 of oats, 11,950 of potatoes, 900,597 pounds of tobacco, 2,338,929 of cotton. It had 15 stores, one cotton factory with 528 spindles, 12 flouring-mills, 48 grist mills, 15 saw-mills, eight tanneries, 34 distilleries ; 19 schools, 438 scholars. Pop.: whites, 8321 ; slaves, 2487 ; free coloured, 72 ; total, 10,780. Capital, Lawrenceville.

MONTGOMERY, county, Ga., situated toward the S.E. part of the state, and contains 1100 sq. m. Bounded S.W. by Little Ocmulgee river, N.E. by Great Ohoopee river. Drained by Oconee river and its tributaries, and by Pendleton's creek. It contained in 1840, 6781 neat cattle, 1239 sheep, 6538 swine ; and produced 1119 bushels of wheat, 29,213 of Indian corn, 4000 of potatoes, 18,290 pounds of cotton, 3615 of sugar. It had four grist-mills. Pop.: whites, 1979 ; slaves, 237 ; total, 1616. Capital, Mount Vernon.

MONTGOMERY, county, Ala., situated a little S.E. of the centre of the state, and contains 900 sq. m. Bounded N.W. by Alabama and Coosa rivers. Drained by the Tallapoosa river and its branches. It contained in 1840, 25,325 neat cattle, 9590 sheep, 56,408 swine ; and produced 7907 bushels of wheat, 3059 of rye, 1,353,917 of Indian corn, 72,741 of oats, 87,214 of potatoes, 29,847 pounds of rice, 2296 of tobacco, 14,871,463 of cotton. It had two commission houses in foreign trade, 29 retail stores, one flouring-mill, eight grist-mills, five saw-mills, two tanneries, two printing-offices, two weekly newspapers ; one college, 68 students ; four academies, 212 students ; 21 schools, 678 scholars. Pop.: whites, 8979 ; slaves, 15486 ; free coloured, 116 ; total, 24,574. Capital, Montgomery.

MONTGOMERY, county, Tenn., situated toward the N.W. part of the state, and contains 500 sq. m. Watered by Cumberland river and its branches. It contained in 1840, 17,711 neat cattle, 13,117 sheep, 39,789 swine ; and produced 83,942 bushels of wheat, 2985 of rye, 916,009 of Indian corn, 223,527 of oats, 28,490 of potatoes, 2,549,984 pounds of tobacco, 4502 of cotton. It had 24 stores, six furnaces, three forges, five flouring-mills, 21 grist-mills, 20 saw-mills, seven tanneries, six distilleries ; three academies, 197 students ; 14 schools, 379 scholars. Pop.: whites, 9763 ; slaves, 7059 ; free coloured, 106 ; total, 16,927. Capital, Clarksville.

MONTGOMERY, county, Ky., situated centrally in the E. part of the state, and contains 260 sq. m. Drained by Red river and its tributaries, flowing into Kentucky river, and by branches of Licking river. It contained in 1840, 11,662 neat cattle, 16,189 sheep, 33,183 swine ; and produced 46,191 bushels of wheat, 41,080 of rye, 733,694 of Indian corn, 69,149 of oats, 13,789 of potatoes, 64,212 pounds of sugar. It had 14 stores, 14 grist-mills, 11 saw-mills, one oil-mill, four tanneries, one distillery ; 30 schools, 473 scholars. Pop.: whites, 6409 ; slaves, 2735 ; free coloured, 189 ; total, 9333. Capital, Mount Sterling.

MONTGOMERY, county, O., situated in the S.W. part of the state, and contains 480 sq. m. Watered by Miami river and its branches. It contained in 1840, 16,945 neat cattle, 29,631 sheep, 30,228 swine ; and produced 385,938 bushels of wheat, 54,927 of rye, 814,707 of Indian corn, 3359 of buckwheat, 4729 of barley, 374,481 of oats, 34,068 of potatoes, 192,394 pounds of sugar. It had 130 stores, two lumber yards, three fulling-mills, five woollen factories, six cotton factories with 3530 spindles, 34 flouring-mills, 11 grist-mills, 56 saw-mills, two oil-mills, 12 tanneries, 30 distilleries, two breweries, two potteries, two printing-offices, two bindéries, five weekly newspaper and one periodical ; two academies, 92 students ; 160 schools, 7192 scholars. Pop. 31,938. Capital, Dayton.

MONTGOMERY, county, Ia., situated N.W. of the centre of the state, and contains 504 sq. m. Drained by Sugar, Big and Little Raccoon creeks. It contained in 1840, 12,703 neat cattle, 15,831 sheep, 35,711 swine ; and produced 84,709 bushels of wheat, 3795 of rye, 993,785, of Indian corn, 1771 of buckwheat, 98,681 of oats, 25,650 of potatoes, 17,332 pounds of tobacco, 175,482 of sugar. It had 33 stores, one fulling-mill, 11 woollen factories, 12 flouring-mills, 12 grist-mills, 37 saw-mills, 10 oil mills, 11 tanneries, eight distilleries, one pottery, two printing-offices, two weekly newspapers ; one college, 100 students ; two academies, 243 students ; 29 schools, 685 scholars. Pop. 14,368. Capital, Crawfordsville.

MONTGOMERY, Ill., situated a little S. of the centre of the

state, and contains 684 sq. m. Drained by head branches of Sangamon river and Shoal creek. It contained in 1840, 7464 neat cattle, 7076 sheep, 13,517 swine; and produced 27,500 bushels of wheat, 1222 of rye, 215,592 of Indian corn, 1336 of buckwheat, 57,608 of oats, 9995 of potatoes, 5131 pounds of tobacco. It had nine stores, 25 grist-mills, 17 saw-mills, one oil-mill, two tanneries, three distilleries, one printing office, one weekly newspaper; one academy, 186 students; seven schools, 391 scholars. Pop. 4490. Capital, Hillsboro'.

MONTGOMERY, county, Mo., situated toward the E. part of the state, and contains 576 sq. m. Watered by Cuivre and Loutre creeks. Bounded S. by Missouri river. It contained in 1840, 4858 neat cattle, 5397 sheep, 9054 swine; and produced 12,717 bushels of wheat, 1148 of rye, 122,490 of Indian corn, 26,299 of oats, 76,34 of potatoes, 837,639 pounds of tobacco. It had 12 stores, three grist-mills, two saw-mills, two tanneries. one distillery; one academy, 28 students; 5 schools, 74 scholars. Pop.: whites, 3594; slaves, 887; free coloured, 20; total, 4371. Capital, Danville.

MONTGOMERY, p. t., Orange co., N. Y. S.S.W. Albany, 288 m. W. N. Y. Watered by Wallkill river. It contains three churches, a Presbyterian, Dutch Reformed and Methodist; eight stores, one iron foundry, one grist-mill, one saw-mill; one academy, 61 students; nine schools, 440 scholars. Pop. 4100.

MONTGOMERY, t., Somerset co., N. J., 12 m. S.W. Somerville. Bounded S. by Millstone river, by branches of which it is drained. It has seven stores, one fulling-mill, one woollen-factory, three grist-mills, three saw-mills; seven schools, 212 scholars. Pop. 1482.

MONTGOMERY, t., Franklin co., Pa., 16 m. S.W. Chambersburg. Drained by Conecocheague creek and its tributaries. It has 11 stores, one woollen factory, three flouring-mills, six grist-mills, 14 saw-mills, four tanneries, 10 distilleries, two potteries; one college, 64 students; one academy. 54 students; 21 schools, 795 scholars. Pop. 4360.

MONTGOMERY, t., Montgomery co., Pa., 20 m. N. Philadelphia. Drained by Wissihickon creek, and the W. branch of Neshaminy creek. It contained three stores, one grist-mill, two saw-mills; two schools, 70 scholars. Pop. 1009.

MONTGOMERY, t., Franklin co., O. It contains in its W. part the city of Columbus, the capital of the state. It has a small territory, but is wealthy and populous. Exclusive of Columbus city, the Pop. is 1449.

MONTGOMERY, city, capital of Montgomery co., Ala., 220 m. N.E. Mobile, 112 m. S.E. Tuscaloosa, 389 W. Situated on a high bluff, at the head of steamboat navigation on Alabama river. It contains a courthouse, jail, seven churches, a Presbyterian, Episcopal, Methodist, Baptist, a Methodist Episcopal, Methodist Protestant, a Universalist, and a Roman Catholic; two academies; 30 stores, 260 dwellings and, (in 1842) 2250 inhabitants. There are shipped annually 40,000 bales of cotton, weighing 500 pounds each. The merchandise sold amounts to $500,000, annually. The Montgomery and West Point railroad is in operation 35 m. from Montgomery to Franklin. The city was laid out and began to be settled in 1817. The population according to the census of 1840, was 2179.

MONTICELLO, p. v., Thompson t., capital of Sullivan co., N. Y., 110 m. S.W. Albany, 204 W. Incorporated in 1830. It contains a courthouse and jail of wood, a county clerk's office of brick; two churches, a Presbyterian and Episcopal; seven stores, one grist-mill, two saw-mills, 80 dwellings, mostly of wood, and about 500 inhabitants.

MONTICELLO, p. v., capital of Jasper co., Ga., 35 m. N.N.W. Milledgeville. 542 W. It contains a court-house, jail; two churches, a Methodist and Baptist: a female academy; 19 stores, 60 dwellings and about 375 inhabitants.

MONTICELLO, p. v., capital of Jefferson co., Flor., 20 m. E.N.E. Tallahassee, 925 W. Situated four m. E. of Mickasooky lake. It contains a courthouse, and a few stores and dwellings. Nett proceeds of the postoffice, $553.

MONTICELLO, capital of Lawrence co., Miss., 85 m. S. Jackson, 1004 W. Situated on the W. side of Pearl river, and contains a court-house, jail and several stores and dwellings. Nett proceeds of the post-office, $316.

MONTICELLO, v. v., capital of Wayne co., Ky., 110 m. S. Frankfort, 599 W. Situated on the N. side of Beaver branch of Cumberland river. It contains a courthouse, jail, several stores, and 142 inhabitants.

MONTICELLO, p. v., capital of White co., Ia., 73 m. N.N.W. Indianapolis, 632 W. Situated on the W. side of Tippecanoe river. It contains a courthouse, jail, and several stores and dwellings. Nett proceeds of the post-office, $120.

MONTICELLO, p. v., capital of Piatt co., Ill. Situated E. of Sangamon river, and contains a courthouse, and about 100 inhabitants.

MONTICELLO, p. v., capital of Lewis co., Mo., 145 m. N.E. Jefferson City, 912 W. Situated on the N.E. side of North Fabius river. It contains a courthouse, and several

388

ral stores and dwellings. Nett proceeds of the postoffice, $370.

MONTILLA (an. *Montulia*), a town of Spain, in Andalusia, prov. Cordova, 19 m. S. by E. Cordova. Population, according to Miñano, 12,800. It is well built, and has two parish churches, an orphan asylum, three hospitals, a royal school of Latin and rhetoric, and a bonding warehouse for wine. Its trade is considerable, chiefly with Cordova, both in manufactured goods and farm produce, particularly wine, horses, mules, and horned cattle, which, though small and ungainly in appearance, are very hardy and serviceable. An annual fair is held in September, and well attended.

MONTLUCON, a town of France, dep. Allier, cap. arrond. on the Cher, close to the canal De Berri, in a valley bordered by vine-clad hills, 38 m. W.S.W. Moulins, and 171 m. S. by E. Paris; lat. 46° 20' N., long. 2° 40' E. Pop. in 1836, ex. com., 4280. It was formerly fortified, and is well built and situated. A parish church and hospital are the only public buildings. It produces some coarse woollen and linen fabrics; and has a considerable trade in corn, wine, cheese, and cattle.

MONTMARTRE, a town of France, dep. Seine, only a few furlongs N. of Paris, on a conical hill of the same name, commanding an extensive view of the French metropolis and its suburbs. Pop., in 1836, 6234. It is the favourite resort of the Parisians on Sundays and holydays, and comprises several inns and other houses of entertainment. with some neat-looking villas and private residences. An asylum for 60 old men, a private lunatic establishment, and two schools, have been founded here, and it has oil-cloth manufactories, scagliola-works, and woollen-mills, with mines of gypsum, which supply the whole of Paris with plaster. In 1814 the hill was fortified by the Parisians, who defended it for a day against the allies.

MONTPELIER, p. v., capital of Washington co., and of the state of Vermont. The town contains 22 stores, one furnace, one fulling-mill, three grist-mills, five saw-mills, one paper-mill, six printing-offices, one bindery, two daily and six weekly newspapers, and one periodical; one academy, 101 students; 30 schools, 975 scholars. Pop. 3725. The village is situated in the S.W. part of the town, at the union of the branches of Onion or Winooski river. It is in lat. 44° 16' N., and long. 71° 32' W. Its site is a plain of moderate extent, surrounded by elevated hills. It is 10 m. north-easterly from the geographical centre of the state, and is built on both sides of the river. The great road from Boston to Burlington, which passes through it, makes it a great thoroughfare; and, though this road passes through the most elevated portion of the Green mountains, along the margin of Onion or Winooski river, it passes over no high hills. The village contains a beautiful statehouse, a courthouse, jail, an academy, four churches, two Congregational, a Methodist, and Universalist (there is another Methodist and Friends' in the town), 15 stores, which sell goods annually to the amount of $900,000, several mills and manufactories, and 1790 inhabitants. The statehouse is an honour to the place and to the state, built of granite, 72 feet wide in the centre, with two wings, each 30 feet, making the whole length 150 feet. It has a projecting portico in the centre, of six Grecian Doric columns, 6 feet in diameter and 36 feet high. The centre building is 100 feet deep, and the wings 50 feet deep. The whole is surmounted by a fine dome, 100 feet high from the ground to the top. In the interior are rooms for various offices, and elegant halls for the senate and house of representatives. Its architecture is much admired.

MONTPELLIER (Lat. *Mons. Pessulanus*), a city of France. dep. Hérault, of which it is the cap., on the Lez. about 5 m. from the Mediterranean, and 77 m. W. by N Marseilles; lat. 43° 36' 66" N., long. 3° 52' 45" E. Pop. in 1836, ex. com., 33,864. It is beautifully situated on the declivities of a low hill, commanding views of the Alps, the Pyrenees, the Cevennes, and the sea. It was formerly walled, and a place of considerable strength; but of its ancient fortifications there are now only a few gates, a tower, and some portions of the wall on the N.E. side of the city It still, however, has, at one extremity, a citadel built by Louis XIII.; while at its other extremity is the *Place* or *Promenade de Peyrou*, one of the noblest public walks in Europe. This place is entered by a Doric arch, and ornamented with long lines of balustrades, covered ways, various sculptures, a bronze equestrian statue of Louis XIV. and numerous fountains, including a magnificent hexagonal *château d'eau* of Corinthian architecture. This, like the other public fountains of Montpellier, is supplied by an aqueduct about 8½ m. in length, constructed in the middle of the last century; and which, for a distance of 880 metres, or more than half a mile, is raised on a double row of stone arches, and, in point of elegance, rivals the boasted *Pont du Gard*. Between the town and the ramparts of the citadel is the Esplanade, a fine open space planted with trees and ornamented with reservoirs, &c.; the boulevards sur sounding the town also afford good public walks; and in the outskirts are many newly-built and handsome terraces. The city itself is very ill laid out; its streets are narrow, steep, and winding, and its squares small and irregular; but its houses are generally good, and, according to Hugo and others, it is kept remarkably clean. The public buildings are quite unworthy so considerable a city. Of eight churches, none demands any particular notice; the cathedral is distinguished from the rest only by being larger and uglier; a singular-looking porch, and a tower at three of the angles of the nave, are the principal external ornaments of this edifice. Adjacent to it is the school of medicine, occupying what was formerly the bishop's palace, a large building with several fine apartments. This school, founded by the Arabs driven from Spain in 1180, enjoys a high and well-deserved celebrity, as one of the best conducted establishments of its kind in France; and is all that now remains of the once famous university of Montpellier. It has a new and fine amphitheatre; an examination hall, in which is an antique bronze bust of Hippocrates; a council hall, with portraits of professors from the period of the 13th century, including also a portrait of Rabelais; a library, with 35,000 volumes, including many editions of the 15th century, and 600 valuable MSS. in different European and Asiatic languages; a pretty extensive anatomical museum, several spacious laboratories, &c. The general hospital has accommodation for upward of 600 patients; and there are large and well-conducted lunatic and lying-in hospitals. The botanic garden of Montpellier, which dates from the reign of Henri IV., was the first established in France, and, though small, comprises 8000 species of plants: it is one of the four principal and best arranged botanic gardens in the kingdom, which distinction it owes to its having been the scene of the labours of the late celebrated M. de Candolle. But its greatest interest to Englishmen is derived from its possessing, in one of its most sequestered parts, the tomb of Narcissa, the daughter-in-law of Young, whose funeral the poet has vividly described in "Night the Third."* One of the principal attractions in Montpellier is the museum, founded, in 1825, by the Baron Fabre, a native of the town. It occupies four spacious and well-lighted halls, and comprises collections of paintings, engravings, statues, medals, and other objects of vertu, a library of 15,000 volumes, &c., the whole estimated to be worth 2,000,000 francs. The theatre, built in 1786, is well planned, and capable of accommodating 2600 persons; it is, however, little frequented, and the pit, which is without seats, usually serves as an exchange. The palace of justice, the town-hall, exchange, prefecture, admiralty, barracks, several prisons, including a central prison, with workshops, the Calvinist chapel, synagogue, &c., are the remaining principal buildings, but none deserves especial notice. There are several very good hotels; one of which is said by Inglis to be the best in the S. of France.

Montpellier is a bishop's see, the capital of the ninth military division of the kingdom, and the seat of a royal court for the departments of Aude, Avyyron, Hérault, and Pyrénées-Orientales, a court of original jurisdiction, a tribunal, and a chamber of commerce, boards of taxation, customs, artillery, and engineering, a university, academy, and a royal college. It has schools of veterinary medicine, engineering, drawing, architecture, geometry, and music; societies of agriculture, arts and sciences, medicine, and archæology, a government loan-bank: Protestant Bible societies, a prison society; and numerous other charitable associations of several asylums, &c. It has manufactures of woollen cloths, cotton handkerchiefs, muslins, table and other linens, hats, silk, cotton, and woollen hosiery; with cotton-thread factories, distilleries, sugar refineries, breweries, chemical works, &c. It is connected with its port, Cette, 17 m. S.W., by a railway, and has a brisk trade with it, and with other towns and villages, exporting large quantities of fresh and dried fruits, wool, and other kinds of rural produce, in addition to its manufactured products. It was formerly a place of great resort for English invalids, on account of the alleged salubrity of its climate.

"About twenty-five years ago," says Inglis, "200 English families were sometimes resident here; but since fashion, caprice, or experiment, have sent consumptive patients to die in Madeira or Naples, instead of Montpellier, that number has been reduced to 40 or 50 families, who, indeed, resort thither less for the sake of health than economy. It is undeniable, however, that the air of Montpellier (however little that bustling city may resemble the retired spot pictured by the imagination) is dry and salubrious, possessing the mildness of a southern climate, and yet having its heat tempered by the sea-breeze. It is also a cheap residence, the more so from the diminished influx of strangers. Two well finished rooms may be obtained for about 10s. a week;

* It is stated in Johnson's (Croft's) Life of Young that Narcissa, or Mrs. Temple, died at Lyons, in her way to Nice; but she, in fact, died at Montpellier.

and living is not expensive. Beef and mutton fetch from 5d. to 6d. per lb., fish, of about 20 different sorts, may be had at prices varying from 6 to 20 sous: fruit and vegetables are both cheap and good; wine ranges between one and two francs the bottle."

Montpellier has given birth to many distinguished persons, among whom may be specified Chaptal the chemist, Cambacérès, Daru the historian, &c. It appears to have been founded in the 8th century, and was for awhile dependent on the kings of Majorca. It was acquired by Philip of Valois in 1349, but was not finally annexed to France till the reign of Charles VI. The Calvinists got possession of it under Henry III., and held it till 1622, when it was taken after a long siege by Louis XIII. (*Inglis's Switz. and France*, p. 207; *Frossard, Tableau Pitt. de Nismes et de ses Environs*, ii., 54–63; *Hugo*, art. *Hérault; Guide du Voyageur en France.*)

MONTREAL, a town and river port of British America, and the second city and the chief seat of the commerce of Lower Canada; on the S. side of the island of Montreal, in the St. Lawrence, 142 m. in a direct line S.W. Quebec; lat. 45° 30′ W., long. 73° 25′ N. Population with its suburbs, in 1840, 27,297. Its site is not so commanding as that of Quebec, but it is in every other respect superior to that city. It is not so crowded; and some even of its older streets are of tolerable breadth. Montreal is divided into the Upper and the Lower town; the difference in their elevation is but slight, but the former, being the more modern, is the handsomer division. It has several suburbs, including which it stretches along the river for 2 m. from N. to S., and has, for some distance, a nearly equal breadth inland. The battlemented wall, with which it was formerly surrounded, has fallen into decay, and it is now entirely open, the wooded heights around being covered with villas and pleasure grounds. In the lower town, Paul-street, the chief commercial thoroughfare, extends parallel with the river the whole length of the city; and in the upper town several streets proceed in the same direction, communicating with Paul-street by cross streets. In the upper town and suburbs, which are mostly inhabited by the principal merchants, many of the houses are handsomely and solidly built in the modern style; but in the lower town they are principally of a gloomy looking gray stone, with dark iron window-shutters and tinned roofs. Along the bank of the river is an extensive line of quays and warehouses. Many of the houses in the suburbs are built of wood, but there are no wooden buildings within the space once encompassed by the walls; and this city and Quebec have more of the aspect of old European towns than any other towns in America.

The most remarkable public edifice is the Roman Catholic cathedral, opened in 1829, and superior to any other church in British America. It is of Gothic architecture, 255½ ft. in length by 134½ in breadth. It is faced with stone, and roofed with tin, and has six towers, of which the three belonging to the main front are 220 ft. in height. On the roof is a promenade, 76 ft. in length by 30 in breadth, elevated 190 ft. The principal window is 64 ft. in height, and 32 in breadth. The interior is capable of accommodating from 10,000 to 12,000 persons, who may disperse by numerous outlets in five or six minutes. It comprises seven chapels, and nine spacious aisles. There are several other Roman Catholic churches, mostly belonging to the order of St. Sulpice; to the members of which Montreal chiefly owed its foundation, and who still hold the seigniory of the island upon which it stands. The seminary of St. Sulpice, a large and commodious building adjoining the cathedral, occupies three sides of an oblong area, 132 ft. in length by 90 deep, and is surrounded by spacious gardens. A handsome additional building, 310 ft. by 45 ft., has been lately erected, at an expense of £16,000. In these establishments, students in most of the higher branches of learning are taught at very moderate charges. The principal English church is a handsome building, in the Grecian style, surmounted by a high and beautiful spire. It has also a Scotch kirk, an American Protestant church, and chapels belonging to the Methodists and Scotch dissenters. The Montreal general hospital, erected in 1821-2 by voluntary subscription, a large and well-built edifice, is said to be one of the best regulated institutions of the kind in America. A large conventual structure, the *Hôtel-Dieu*, occupied by a superior-matron and 36 nuns, is appropriated to the reception of the sick and indigent; and the convent of the Gray Sisters partly serves as an asylum for the aged and infirm, the insane, foundlings, &c.

The *Sœurs Noires* have an extensive convent, founded it 1650; its inmates consist of a superior and 60 nuns, whose duties are directed to the education of young girls. The courthouse and prison are substantial buildings, occupying the site of the former college of the Jesuits. The government-house, bank, barracks, ordnance-office, and four market-houses, are among the remaining principal buildings. In one of the squares is a colossal statue of Nelson, placed on a Doric column, the pedestal of which has bas-reliefs representing his principal actions. Besides the educational establishments noticed above, Montreal has a college, with a principal and four professors, a royal grammar school, parochial, union, national, Sunday, and other public schools; and many good private French and English seminaries The university of M'Gill college, endowed by a citizen of Montreal, in 1814, with a valuable estate, and £10,000 in money, and chartered in 1821, is, we believe, not yet opened; but it is to be conducted on a liberal and enlarged scale. Montreal has a penitentiary, a house of industry, a savings' bank, a natural history society, a mechanics institution, a central auxiliary society for promoting education and industry, Bible and tract, agricultural and horticultural societies; several public libraries, an excellent news-room, &c. Several newspapers and other periodical publications issue from the presses of the town. According to Mr. M'Gregor, there is a greater spirit of improvement in this city than in Quebec. There is much activity observable among all classes connected with trade. The position of Montreal at the head of the ship navigation of the St. Lawrence, and near the confluence of that river with the Ottawa, as well as its situation with respect to the U. States, necessarily make it one of the greatest emporiums of America. '(Brit. America, ii., 309.)

The harbour, though not large, is secure, and vessels drawing 15 ft. water may lie close to the shore. Its general depth of water is from 3 to 4½ fathoms. Its chief disadvantage consists in the rapid of St. Mary's, about 1 m. below, which vessels often find it difficult to stem. To obviate the obstructions in the navigation above Montreal, the Lachine canal, 9 m. long, 20 ft. wide, and 5 ft. deep, was undertaken in 1821, and completed at an expense of £130,000. The communication with the opposite sides of the river is carried on by several steam and other vessels; and, during the summer, a regular steamboat communication is kept up with Quebec. At this season, vast rafts of timber come down, and pass the city for Quebec; and scows, bateaux of about six tons, and Durham boats, bring to Montreal the produce of Upper Canada. Neither is the trade of Montreal suspended in winter, like that of Quebec. Thousands of sledges may then be seen coming in from all directions with agricultural produce, frozen carcasses of beef and pork, firewood, and other articles. Montreal is the centre of the commerce between Canada and the U. States, carried on by Lake Champlain and the Hudson; and not only is it the depôt of all the adjacent country, but most of the business done in Quebec, is carried on by branches from the Montreal houses. In 1838, 98 ships, of the aggregate burden of 22,229 tons, entered, and 99 ships, burden 21,901 tons, left the port. Formerly this city was the head quarters of the fur trade, but its interest in it has greatly declined. It has, however, cast-iron foundries; distilleries; breweries; soap, candle, and tobacco manufactories; several ship-building establishments; and machinery for steam-engines. Various articles of hardware, linseed oil, floor-cloth, &c., are made in the town. The markets are abundantly supplied with good butchers' meat, fish, poultry, fruit, vegetables, &c. Mr. M'Gregor says that better accommodations are to be found here than in Quebec; and the society is as good. About three fourths of the pop. are of French descent; the remainder, consisting principally of emigrants from the U. Kingdom, Americans, Iroquois Indians. Montreal, originally called Villemarie, was taken from the French, in 1760. (M'Gregor's British America, ii., 300–317; Encyc. of Geog., Amer. ed. ; Parl. Reports.)

MONTROSE, a royal and parl. bor. and seaport town of Scotland, co. Forfar, at the mouth of the S. Esk, on the N. side of the river, on a projecting tongue of land, between the German ocean, on the E., and the basin of Montrose, on the W., 66 m. N.N.E. Edinburgh; lat. 56° 42' 10" N., long. 2° 27' 15" W. Pop., in 1810, 7974; in 1831, 12,055; but, including the suburbs of Inch (a small island formed by the river) and Ferryden, 12,853, exclusive of sailors, who amount to between 600 and 700. In 1841 the population was 13,532.

The town consists of one wide and regular street extending from N. to S. upwards of ½ m., with numerous closes and subsidiary streets. Many of the houses present their gables to the street, as in the Netherlands. It is a handsome town, well paved, lighted with gas, and supplied with water conveyed, in pipes, from a distance of 3 m. The public buildings are the town-hall, jail, lunatic asylum, academy, trades' school, infirmary, and house of refuge, parish church, with a handsome steeple 200 ft. high. St. John's, a newly erected quoad sacra church, two episcopal chapels, and six dissenting chapels, of which two belong to the Associate Synod and the others to the Methodists, Baptists, Glassites, and Independents. The parish church is collegiate; and one of the two ministers is paid by a tax of 9½ per cent. on the rental of the borough. The narrow downs, provincially links, between the town and the sea, are much resorted to by the inhabitants for golf-playing: a game which is in great favour here and in various other places in Scotland.
300

The most important public structures connected with Montrose is the suspension bridge, completed in 1829, over the principal branch of the South Esk, and uniting the town with the Inch. The distance between the towers of the extremities of the bridge is 432 ft. ; the height of each tower is 71 ft. ; the width of the bridge is 26 ft. within the suspending-rods. The whole cost has been above £25,000; the postage levied amounts to about £1500 a year. The extent of the Inch is less than ¼ m., and the branch of the river on the opposite side is crossed by a drawbridge; so that the communication across the two channels of the South Esk is as perfect as possible.

Montrose has long been celebrated for its schools. It was the first place, in Scotland, in which Greek was taught (M'Cris's Life of Knox, vol. i., App. n. C.); and it has preserved the character which it so nearly (1534) attained. It has at present about 20 schools, and above 1600 pupils, being about a tenth part of the entire population. Of the schools, two are entirely free; one, founded in 1816, by a Mr. White, educates 100 poor children; and another, founded in 1822, by Miss Stratton, educates 42 boys and as many girls. Five schools are partially endowed; the others are voluntary and unendowed seminaries. The Montrose academy, established in 1815, is an excellent seminary. Andrew Melville, who was born in the neighbourhood, was educated at the grammar-school of Montrose. George Wishart, who afterwards suffered martyrdom, was also educated here, and subsequently held the office of teacher in the same school. The celebrated Marquis of Montrose, who made so distinguished a figure in the civil wars in the 17th century, was a native of the town. Archbishop Leighton was descended of a family whose seat was within 2 m. of the borough. There are various subscription libraries, one of which, founded in 1785, has above 8000 vols.; a mechanics' institute, which has occasional lectures on different branches of science; a Natural History and Antiquarian society, to which Lord Panmure has been a liberal benefactor; a Horticultural society; various literary and philosophical societies, and two weekly newspapers.

In addition to the funds bequeathed for the support of schools, there are no less than 12 different bequests, amounting in the aggregate to about £11,000, left for the support of the poor, or for particular classes among them. Poors' rates, however, were introduced in 1836; the average amount of which is about £2800 a year. There is, also, a hospital fund, consisting of certain lands and teinds granted to the town by James VI., in 1587, amounting to about £170 yearly The lunatic asylum, a splendid building, in an airy and healthy situation, was founded in 1779; but, having been greatly enlarged and improved, it was incorporated by royal charter in 1811. It has, at present (1841),77 inmates. The infirmary, a handsome new building, has attached it to a fever hospital, and a dispensary. Average number of patients in the infirmary and free hospital about 21: do. of out-door or dispensary patients, who receive advice and medicines, about 500. A house of refuge was established and endowed in 1838, by a benevolent citizen, William Dorward, Esq., at an expense of £10,000. It is intended to shelter and provide for old and destitute persons of both sexes, and for destitute children. The building is handsome and commodious, and the institution, which is said to be admirably conducted, has at present, about 70 inmates: the children are educated at the different-charity schools. There is, also, a society for the relief of the destitute sick.

The principal business in Montrose is flax-spinning and weaving. The town and its immediate vicinity has at present (1841) six flax-spinning mills; besides one at Logie, and one at Craigo, both connected with the town: it has, also, a power-loom factory, and about 500 hand-looms, three fourths of which are in factories, employed on the fine and heavier linen fabrics, including sheetings, dowlas, sail-cloth, and bagging. There is also a bleach-field at Craigo, and another at Logie.

The hands employed may be estimated as under: viz. employed in spinning, 1650: in heckling, 250; in bleaching, 200; in weaving and manufacturing, 1400: total, 3500. Of the above, 300 spinners and 200 bleachers are employed at the Craigo and Logie works, and the remaining 3000 in Montrose. In 1840 about 53,000 pieces of cloth were woven in Montrose, and 26,000 in the vicinity, making together 79,000: the quantity produced in 1835 did not exceed 47,000 pieces. Montrose has also a small soap factory, one of starch, two rope and sail works, and a machine factory. Ship-building is carried on to a considerable extent, and at present five ships are on the stocks. There is a patent slip for repairing ships. There are five breweries, two tanneries, two candle-works, and a foundry; and a meal and flour-mill driven by steam.

The harbour is one of the best on the E. coast of Scotland. The channel of the river is narrow; but, as it has 15 or 13 ft. water over the bar at low ebb, middling size merchant-men may run in at any time of the tide; and, at high water, it is accessible to the largest ships. A dock is now in the course of being excavated below the present harbour, which will, it is supposed, cost about £50,000; but it may be doubted whether the benefit it will confer on the shipping of the port will be sufficient to countervail the injury arising from the increase of the harbour dues it has occasioned. The basin, immediately W. from the town, has a fine appearance on the map, but is of little use. It is nearly circular, being about 3 m. in diameter: it is shallow, and, excepting the channel of the river, dries at low water. Vessels of 50 or 60 tons, however, reach old Montrose, at the other end of the basin. The entrance to the harbour has, on its N. side, two lighthouses, with fixed lights.

Montrose is a customhouse station; the amount of shipping within the district being 187 ships, of the aggregate burden of 18,900 tons, of which 113 ships, measuring 14,500 tons, belong to the town. Customs revenue, in 1848, £33,494. The trade of the port has more than doubled within the last 15 years; the shore dues, which were under £1000 in 1835, having produced £3076 in 1838: they are now still greater, but the rates have been augmented. The chief imports from foreign countries are flax, hemp, and timber. In 1840, there were imported at Montrose 4680 tons flax and and hemp, 39,604 tons coals, 7400 loads timber, and 11,674 barrels raw herrings; five sixths of the latter being for exportation to London after being cured. During the same year, there were exported, exclusive of manufactures, sent almost wholly to London, 25,232 qrs. of wheat, barley, &c., 1400 boxes salmon, 4647 barrels smoked haddocks, 2275 barrels pork, and 9680 bolls. potatoes (5 cwt. each). Previously to 1839, three or four ships sailed annually to the Greenland whale fishery, but this branch of trade has been abandoned. The majority of the ships belonging to the port are now engaged in the Baltic trade, in which it is believed that about 45 vessels, of the average burden of about 175 tons, are at present employed; whereas, in 1830, only about 10 vessels, of 90 tons each, belonging to this port, were engaged in the trade: The steamers that ply between Leith and the N. of Scotland regularly call at Usan, 2 m. S. of Montrose; and a steamer, belonging to the town, plies twice a week to Leith. There are four banks in the town, a savings' bank, and an insurance society.

Montrose was created a royal br. by David I. in the 12th century. It was here that John Baliol, in 1296, surrendered the Scotch crown to Edward I. Montrose was the first port made by the French fleet, in 1715, with the Pretender on board; and the same personage sailed from it in February, 1716, for France.

The corp. consists of a provost, 3 bailies, and 15 councillors. Municipal revenue, about £3000 a year. Montrose unites with Arbroath, Brechin, Forfar, and Bervie in sending a member to the House of Commons. Registered electors, in 1840-41, 373. (Boundary Returns; Factory Reports, 1839; New Statistical Account of Scotland, § Forfarshire, p. 271-290; and valuable Private Information.)

Montrose, p. b., Bridgewater t., capital of Susquehanna co., Pa., 175 m. N.N.E. Harrisburg, 225 W. Situated on elevated ground and contains a courthouse, jail, four churches, a Presbyterian, Episcopal, Methodist and Baptist, 14 stores, three printing offices, one weekly newspaper; one academy, 110 students; three schools, 135 scholars; 100 dwellings, and 628 inhabitants. It was founded in 1811, and incorporated as a borough in 1824.

MONTSERRAT, one of the British W. Indian islands, a dependency on Antigua, from which it is distant S.W. 27 m.; in lat. 16° 45′ N., long. 61° 6′ W. It is about 12 m. in length, and 5 m. in its greatest breadth. Area, estimated at about 30,000 acres. Pop., in 1838, about 7600, of whom nearly six sevenths were blacks. Montserrat consists of a range of steep abrupt mountains, or rather, perhaps, of one lofty mountain, 2500 ft. high, the summit of which has been broken into a variety of deep precipices and chasms. The upper parts are altogether barren; but the base of the mountain slopes off to the sea by a succession of gentle ridges, admitting of cultivation; and the lower parts are well watered, and very productive. With the exception of the town of Plymouth, unfavourably situated near the S.W. beach, with an amphitheatre of hills in its rear, intercepting the sea breeze, the island enjoys a comparatively high character for salubrity. (Tulloch's Report on the Sickness, &c., in the W. Indies, p. 35.) In 1839, 13,443 cwts. sugar, 29,460 gall. rum, and 18,596 gall. molasses, were imported from Montserrat into the U. Kingdom. The total value of the exports and imports amount to about £4500 each. Amount awarded to the colonial proprietors for the manumission of the slaves, £103,558 16s.; the average value of a slave from 1822 to 1831 having been about £37. This island was discovered in 1493, by Columbus, who gave it its present name. It was colonized by the English in 1632. The French took it in 1664, but restored it at the peace of Utrecht. (Parl. Papers, &c.)

MONTVILLE, p. t., Waldo co., Me., 26 m. E.N.E. Augusta, 689 W. Drained by head branches of Sheepscot river. Incorporated in 1807. It has four stores, two gristmills, five saw-mills; 15 schools, 832 scholars. Pop. 2136.

MONTVILLE, t., New-London co., Ct., 36 m. S.E. Hartford, 360 W. Bounded E. by Thames river, and drained by its branches, which afford water-power. It contains five churches, two Congregational, two Baptist, and one Mohegan Indian, on a reservation of 2700 acres of land, reserved to the remnant of this friendly tribe, who have a school-house as well as a church. The men mostly follow the whaling business, and are only occasionally here. There are in the town four stores, three fulling-mills, five woollen factories, four grist-mills, five saw-mills, one oilmill; 11 schools, 432 scholars. Pop. 1990.

MONZA (an. Modœtia), a town of Austrian Italy, deleg. Milan, on the Lambro, here crossed by three stone bridges, 9 m. N.N.E. Milan. Population, in 1837, 8378. It is regularly laid out, paved with round stones, and tolerably well built. It is interesting from having been the seat of government during the time of the Lombard kingdom; and the iron crown of Lombardy is kept, with other relics, in its cathedral, an edifice supposed to date from the 7th century. The former residence of the Lombard kings is said to have been the building now occupied by the court of justice. Monza has a royal palace, with fine grounds, greatly embellished by Prince Eugene Beauharnois, and which is the usual summer residence of the Austrian viceroy; a gymnasium; two hospitals; a theatre; and manufactures of silk and cotton stuffs, shawls, hats, and leather. (Dict. Geog.; Condor's Italy, i., 347-8.)

MOOERS, p. t., Clinton co., N. Y., 18 m. N. by W. Plattsburg, 198 m. N. Albany, 573 W. Watered by Chazy and English rivers, the latter flowing into Canada. It contains a Congregational and Methodist church, five stores, two grist-mills, 25 saw-mills, two tanneries; 10 schools, 239 scholars. Pop. 1703.

MOOLTAN, or MOULTAN, a city of the Punjab, probably the Malli of Alexander's historians, cap. prov. of same name, on the Chinaub or Acesines, 190 m. S.W. Lahore. Lat. 30° 9′ N.; long. 71° 7′ E. Population about 60,800, one third of whom may be Hindoos; the rest are Mohammedans, the Seiks being confined to the garrison, which does not exceed 500 men. (Burnes's Bokhara, &c., i., 95.) The city is upwards of three miles in circumference, surrounded by a dilapidated wall, and overlooked on the N. by a fortress of some strength. A considerable portion of the two evidently stands on the débris of more ancient buildings. The houses are of burnt brick, with flat roofs; they sometimes rise to the height of six stories, so that the narrow streets are dark and gloomy. The fortress of Mooltan is an irregular hexagon, with a wall of burnt brick, 40 feet high on the outside, and flanked with about 30 towers. In its interior are numerous houses, now uninhabited and falling into ruin, several mosques, and a Hindoo temple of high antiquity. Mooltan has several elegant and highly venerated tombs. Its inhabitants are principally engaged in weaving and dyeing cotton cloths, and silks of a somewhat coarser texture than those of Bahawulpoor, but which are largely exported into the adjacent countries. Many of the fabrics of Mooltan are, as of old, of a purple colour, and interwoven with gold. This city was formerly frequented by a great number of pilgrims, and afforded immense plunder to the Mohammedans in 712. It was captured by Mahmoud, of Ghiznee, in 1010; by Mahomed Ghori, in 1176; by Timour, in 1398; and by Runjeet Singh, in 1818, since which it has belonged to the dominion of Lahore. (Burnes's Bokhara, i., 95-100; Hamilton's E. I. Gaz.)

MOON, p. t., Alleghany co., Pa. Bounded N.E. by Ohio river. It contains Middleton village, nine miles below Pittsburg, and has four stores, two flouring-mills, two gristmills, five saw-mills, two tanneries; seven schools, 245 scholars. Pop. 1981.

MOORE, county, N. C. Situated a little S.W. of the centre of the state, and contains 740 sq. m. Watered by Deep river and its branches, and by sources of Lumber river. It contained in 1840, 3460 neat cattle, 5430 sheep, 5570 swine; and produced 15,643 bushels of wheat, 4195 of rye, 30,385 of Indian corn, 17,341 of oats, 20,194 of potatoes, 750,540 pounds of cotton. It had seven stores, four flouring-mills, 57 grist-mills, 14 saw-mills, three tanneries, three distilleries; 11 schools, 374 scholars. Pop.: whites, 6443; slaves, 1472; free coloured, 73; total, 7988. Capital, Carthage.

MOORE, t., Northampton co., Pa. Drained by head waters of Hockendocque and Manokisey creeks. It contains four stores, two fulling-mills, five grist-mills, five saw-mills, three tanneries, one distillery; four schools, 199 scholars. Pop. 2389.

MOORFIELD, p. v., capital of Hardy co., Va., 178 m. N.W. Richmond, 130 W. Situated on the E. side of the S. branch of Potomac river, at its junction of the S. fork.

It is one of the most fertile valleys in the state, and contains a courthouse, jail, a church, free to all denominations, five stores, two flouring-mills, two carding and fulling mills, 60 dwellings and about 400 inhabitants.

MOORSHEDABAD, a large city of British India, presid. and prov. Bengal, on the Bhagirathi, or most sacred branch of the Ganges, 115 m. N. Calcutta. Lat. 24° 11′ N.; long. 89° 15′ E. Pop. estimated by Hamilton at 165,000. In point of appearance Moorshedabad cannot compare with either of the other great cities of Bengal, but it is not so mean as has been sometimes represented. The houses are principally of mud and straw; the city extends for eight miles along both banks of the river, and a number of brick or chunamed houses are interspersed among the rest, with terraces, small verandahs, flat roofs, &c., "on which you may see the owners, in their Moorish dresses, smoking their hookahs, playing chess, or walking sedately in small parties." A great many small mosques are scattered throughout the city; but a large and fine looking European residence, erected by the British government for the residence of the Bengal nabob, is the only public building worth notice. On the *mootie jheel*, a pool left by a former winding of the river, are the remains of the palace, &c., built by Aliserdi Khan, in the last century, partly with materials from the ruins of Gour. Within the gateway by which the grounds are entered, is a handsome mosque of fine stone, which the zealous frequenters have concealed with thick layers of whitewash. What were formerly gardens are now mere naked fields. Only one fragment of the palace exists, but this is an elegant ruin, consisting of four arches supported by five columns, the whole of beautifully polished black marble.

Moorshedabad is considered unhealthy from the neglected state of the sewers, the closeness and filthiness of the streets, and the rank jungle intermingled with the huts and houses; and pestilential diseases have often raged here with much violence. It is also exposed to the attacks of dacoits and other plunderers, never having been fortified except by an occasional rampart during the Mahratta invasion in the last century. It is the head quarters of one of the six courts of circuit under the Bengal presid.; the seat of a zillah court; the residence of the district collector and other British functionaries, and of the nabob of Bengal; and has a British college, founded in 1826, and endowed with an income of 16,500 rupees a year.

Moorshedabad became the capital of Bengal in 1704, and continued to be the seat of government till the conquest of Bengal by the British in 1756. It was then virtually superseded by Calcutta, to which the revenue-board, collector-general, &c., were transferred in 1751. (*Hamilton's E. I. Gazetteer; Mod. Trav.*, ix., 145–151.)

MOOSEHEAD, lake, Me., is 40 m. long and from 10 to 15 broad, and contains 100,000 acres. Its outlet constitutes the E. branch of Kennebec river.

MOOSEHILLOCK, mountain, N. H., is a lofty eminence in the S.E. part of Benton, on the border of Woodstock, 14 m. E. of Haverhill. It is one of the highest mountains in New-England. The altitude of the N. peak is 4636 feet, and of the S. peak 4526 feet above tide water. It received its name from the moose which formerly inhabited it. Its summit is a bare mass of granite rock, on parts of which snow is found in every month of the year except July. Baker's river rises on its E. side.

MORADABAD, a town and distr. of British India, prov. Delhi. The town, on the Ramgunga, 105 m. E. by N. Delhi, is one of the most populous and flourishing seats of commerce in the upper provinces. It has some good streets, but no public edifice of any importance. The district, or collectorate, is included between the 28th and 30th degrees of N. lat., and 77° 40′ and 79° E. long. Area, 5800 sq. m. Pop., probably 1,500,000. It is well watered, and extensive tracts are very fertile, though a good deal of it be waste. Sugar, cotton, and wheat are the chief products; the latter is almost wholly exported, the food of the population consisting principally of jowares, bajree, &c. At least one fifth part of the land is held rent-free. Total land revenue, (1829–30). 965,110 rupees. (*Parl. Revenue Report*, 1832; *Hamilton's E. I. Gazetteer*.)

MORAT (Germ. *Murten*), a small town of Switzerland, cant. Freiburg, on the S.E. bank of the lake of the same name, and on the high road between Bern and Lausanne, 14 m. W. by S. the former city. Population with its suburbs, 1586. It is partially walled round; and has an ancient castle, now the residence of the *oberamtmann*, a hospital, and an orphan asylum, a Protestant college, a public library, superior, inferior, and commercial schools, and a brisk transit and general trade. The circumstance of several Roman antiquities having been discovered here, has led to the supposition that Morat was anciently one of the suburbs of *Aventicum* (now Avenche). This otherwise insignificant town, owes its celebrity to the great battle fought under its walls on the 22d of June, 1476, in which

the Swiss totally defeated the invading army of Charles the Bold, Duke of Burgundy.

> Moral! the proud, the patriot field! where men
> May gaze on ghastly trophies of the slain,
> Nor blush for those who conquer'd on that plain.
> Here Burgundy bequeath'd his tombless host,
> A bony heap through ages to remain;
> Themselves their monument.
> 　　　　　　　　　　*Childe Harold.*

The loss of the Burgundians was immense; as many as 15,000 soldiers having, it is said, been left on the field, exclusive of those drowned in the lake. The bones of the slain were afterwards collected, in memory of the battle, in a square building, called an ossuary. This singular monument, after standing for more than 300 years, was destroyed in 1798, by the soldiers from Burgundy, in the French army. But though nothing could surpass the gallantry and devotion displayed by the Swiss on this occasion, it is pretty certain that the defeat of Charles at Morat, as well as his previous defeat at Granson, was owing quite as much to his rashness and folly, as to the bravery of his enemies. The principal strength of the duke's army consisted in its cavalry; and yet, on both occasions, he engaged in defiles where they could not act. (*L'Art de vérifier les Dates*, part ii., tom. ii., p. 96.)

MORAVIA (Germ. *Mähren*), an important prov. of the Austrian empire, which, including Austrian Silesia, incorporated with it since 1783, extends between lat. 48° 40′ and 50° 25′ N., and the 15th and 19th degs. of E. long., having N. Prussian Silesia, E. and S.E. Galicia and Hungary, S. the latter country and Austria, and W. and N.W. Bohemia. It is of a rhomboidal shape; greatest length about 185 m.; average breadth, 55 m. Area, about 10,940 sq. m. Pop. in 1836, 2,143,052. In the N. part of the province is a mountainous ridge of no great elevation, stretching W.N.W. and E.S.E., between the Sudetes, bound on the W. and the Jablanka mountains, a branch of the Carpathians on the E., dividing the waters that flow N. into the Oder and the Baltic, on the one hand, from those that flow S. into the Mediterranean on the other. The E. and W. frontiers of the province are also defended by mountain ridges. Excepting in the N., the country is mostly level, or merely undulating, with a gentle slope to the S.; nearly all its great rivers, including the Morawa, by which it is intersected from N. to S., and whence, also, it derives its names, the Iglawa, Thayer, &c., flowing in that direction. The Oder has its sources in the N. ridge. Being sheltered on the N., E., and W., by mountain ranges, and lying in general only from 500 to 1000 feet above the level of the sea, Moravia enjoys a milder climate than most countries in the same lat. The mean temperature of the year at Olmutz is about 48° Fah. The wind is mostly from the S., and the atmosphere clear. A large proportion of the soil is very fertile, and if advantageous markets could be found, large quantities of corn might be raised for exportation; but, in consequence of the want of the latter, the attention of the inhabitants has been of late devoted more to manufactures than to agriculture, and Moravia is no longer a country whence supplies of corn might be drawn, at a short notice, on a very large scale.

An estate of mean size comprises from 850 to 1480 acres of arable land, from 140 to 420 acres of meadow land, and 1000 to 2500 or more of wood, according to the situation. The estates conferring the right of representation in the assembly, and which are only held by knights or nobles, are of all sizes from a few acres to many square miles in extent. These estates can, strictly speaking, be also held by a commoner, but only on his paying a portion of the taxes twice over, and on his renouncing the right to all kinds of patronage and judicial authority. The estates of mean size may be estimated at two thirds of the whole; but about 30 estates exceed 32 English square miles in extent. In purchasing land, a profit of from 4 to 4½ per cent. per annum is generally looked for. The size of the peasant's holdings is very various; in the plains it may be about 28 English acres; but in the hilly parts, where the population is thinner and the soil less productive, it is 30, 40, and in some parts 70 acres. Half holdings, quarter holdings, as well as cottiers with small gardens, are also frequent. It is supposed that of the peasant families, two thirds hold land, and about one third may be considered as mere labourers. The mode of cultivation adopted by the peasants in the low lands consists in a rotation of three crops, viz. wheat, rye, summer corn, fallow; the fallow being only partial. In the hilly parts, the fallows are used for potatoes, turnips, flax, &c.; in the mountains tillage is more irregular. On most of the small estates of the nobles, a better rotation of crops, with clover, green food, and meadows, prevails, according as the soil, or the local advantages of common grazing (which is very extensive) render it necessary. The following rotations, among others, are pursued: 1. potatoes, with manure; 2. barley, or oats, with clover; 3. clover hay; 4. clover as pasture; 5. rye;

6. oats. In heavy soils—1. winter corn with dung; 2. barley, with clover; 3. clover; 4. wheat; 5. green fodder, with manure; 6. wheat; 7. peas and beans; 8. rye. In the low lands millet is a good deal cultivated; in the mountains flax. On the estate of a Moravian nobleman, which is cultivated in a superior manner, but is by no means of a superior quality of soil, as compared with other estates in the same province, the following is the average produce of corn per acre:—

				Maximum.	Minimum.	Mean.
				Bushels.	Bushels.	Bushels.
Wheat	.	.	.	46	14	34½
Rye	.	.	.	26	19½	21
Barley	.	.	.	40	16	24
Oats	.	.	.	46 2-3	17½	28
Potatoes	.	.	.	405	175	240

Distilleries, and even breweries, are commonly established on the low farms; and, within a few years, beet-root sugar manufactories have become frequent.

It is not usual to let lands on lease in this part of the Austrian empire. The few cases in which this mode of tenure occurs, must rather be considered as exceptions than as a rule. From the peasants' holdings the lord usually derives—1st. All that was stipulated on the original cession of the land, whether in the shape of a rent charge in money or otherwise. 2dly. The *Landemium*, or fine on transfer, whether by sale or inheritance (usually five per cent.). 3dly. The *Robot*, or personal service, the maximum of which has been fixed by law. This consists generally in three days' work, with a wagon and horses, weekly, for the peasant's whole holding; the half-holding gives 1½ day's work, and the quarter-holding two or three days' labour, weekly: cottagers give from 10 to 13 days per annum. 4thly. The right of grazing on uncultivated fallows and stubble; which, however, the peasant may exercise upon the land of his lord. 5thly. The great and small tithes, which are often ceded to the church, or have been otherwise transferred. Dominical property (allodial estates) pays in general no tithe. The peasant may cede, or leave by will, his holding to whichever of his sons he pleases; but it is then usually charged with a sum for each of his brothers and sisters. The custom prevails of leaving it to the eldest son; but it is often ceded during the life of the father, who retains a certain proportion of the produce for his own use: this generally happens when the father wishes to free his son from liability to the conscription.

Flax is cultivated in considerable quantities by the descendants of German and Bohemian settlers, in the circles of Brunn and Olmutz: it is celebrated for its fineness and length, and is second only to that of Silesia. In certain favourable situations, the soil and climate of Moravia are well adapted to the grape; and for some time after this was ascertained, the appropriations of land to this kind of culture were so considerable, that government supposed it necessary to interfere, and to issue, in 1803, an order prohibiting the laying out of new vineyards. Wine is mostly grown in the S. circles of Znaym, Brünn, and Hradisch: the average yearly produce is estimated at 436,600 *eimers*; a good deal of which is exported to the adjacent provinces. Large quantities of brandy and beer are also made. Fruits of many kinds are so plentiful, that Moravia is usually styled the orchard of Austria. The forests, formerly much diminished by imprudent waste, are now better attended to. The pasture grounds are extensive in the mountains, and a large proportion of the Austrian heavy cavalry horses is furnished by this province. Cattle are not very extensively bred; considerable numbers are sent thither from Poland to the markets of Olmutz, and from Hungary to those of Austria. In 1827, of 90,007 head of oxen sold in Olmutz, 74,184 were from Galicia. Large flocks of sheep are depastured in the mountainous districts; their numbers having increased with the increase of the woollen manufacture. The breeds have been materially improved by crossing with merinos, &c.; though, from want of proper care, the wool of Moravia is still inferior, and most part of the raw material required is imported from contiguous provinces. Hogs and geese are bred in great numbers for exportation, and game is very abundant. There are valuable mines of iron, copper, lead, and coal; gold and silver mines were wrought previously to the troubles of the 15th and 16th centuries, when they were abandoned, and the works have not since been resumed. Alum, marble, and excellent building stone are found: among other minerals is a species of stone, which, when first dug up, is so soft that it may be moulded with the hand, but which hardens on exposure to the air: a great many pipe-bowls are made from it.

Manufactures and Trade.—Notwithstanding its inland position, this province has made a very considerable progress in manufactures, and has become, since the close of

last century, one of the most thriving portions of the Austrian empire. Woollens, linens, and cottons are all made on a large scale; the first two consuming not only all the wool and flax raised in the country, but requiring a large importation from other parts. Wool is brought from Hungary; flax from Silesia and Austrian Poland. The oldest woollen manufactures are in the neighbourhood of Iglau, in the W.; but those of Brunn are now the most extensive and important. Woollens are also extensively manufactured in other towns; and, exclusive of the goods produced in manufacturing establishments, large quantities are made by the peasants and others for domestic use. Linen and thread are also largely produced; and cotton factories, some of which are on a very extensive scale, have been established, though with but indifferent success, in many parts of the province. Dyeing, especially fine Turkey red colours, is successfully and extensively carried on at Brunn, almost all the cloth made in Moravia being sent thither for that purpose. The other manufactures, such as those of silk, leather, paper, pot-ash, glass, beet-root sugar, &c., are also of considerable importance; and their products are exported to the contiguous countries, and to Hungary, Austria, Italy, and the Levant. The imports consist chiefly of wool, oil, flax, raw cotton, silk, cattle, wine, and hardware. The only navigable river is the Morawa; and hitherto goods have been almost always conveyed in wagons. For these there are two great commercial roads, both leading from Vienna; the one passing by Prague, Znaym, and Iglau, in the west; the other by Brunn and Olmutz in the centre of the province. But the facilities for trade are now in the course of being vastly augmented by the formation of the railway from Vienna to Bochnia, in Galicia, which passes through the valley of the Morawa in this province as far as Magpedi, having branch railways to Brunn and Olmutz. The completion of this great undertaking will be of signal advantage to the province.

The government of Moravia, which is entitled a marquisate, is administered by a governor with direct authority from Vienna. Like the other provinces of the empire, it has its states, or assembly of the clergy, nobility, knights, and citizens; but the power or influence of this assembly is very limited. It meets annually, not to determine the amount of the taxes, but their distribution and mode of collection. The province is subdivided into eight districts or circles, each of which has one or two tribunals of original jurisdiction, and a high court of appeal sits in Brünn.

Education is very generally diffused in this province, and the bulk of the people are comparatively civilised. In 1838, it had 1886 elementary schools, which were attended by no fewer than 272,639 children, being about one eighth part of the population: there is also a great number of superior schools, and the province is well provided with the higher class of seminaries. Among others, it has a university at Olmütz, which, in 1838, was attended by 596 students; it has also faculties of science at Brunn and Nicholsburg: an academy of the provincial states at Olmütz; schools of rural economy at Brünn and Olmütz, attended, in 1838, by 192 pupils; and 11 gymnasia, which, in the same year, had 67 teachers, and 2786 pupils. But, however creditable soever to the government, still it must be borne in mind, that the useful sciences, or those, rather, that are directly subservient to the progress of the arts, and in some degree, also, to the business of administration, are those only that meet with any encouragement, either here or in any other part of the Austrian dominions. All speculative studies, and especially those connected with the principles of political and political economy, are not merely discouraged but proscribed: " Ce que veut avant tout la politique Autrichienne," says M. Saint Marc Girardin, "c'est le calme et le repos; elle veut que le peuple soit tranquille, et pour cela elle veut qu'il soit heureux. Elle veut aussi qu'il ait de l'instruction, mais cette instruction qui apprend à l'homme à mieux se servir de ses forces et de celles de la nature, qui fait les bons ouvriers, les bons laboureurs, et non cette instruction qui agace l'intelligence, qui lui apprend à douter, à raisonner, à examiner. Voulez-vous être mécanicien, manufacturier, agriculteur, architecte? Vous trouverez à cet égard, en Autriche, tout ce qu'il vous faut: écoles, collèges, professeurs, laboratoires, collections. Voulez-vous être avocat, publiciste, homme de lettres, c'est-à-dire raisonner, discuter, douter? Allez ailleurs, allez bien loin; ce n'est point en Autriche que vous trouverez de bonnes écoles pour de pareilles fantaisies. L'utile plutôt que le beau, le pratique plutôt que la théorie, le soin du corps plutôt que le soin de l'intelligence, voilà la maxime fondamentale de l'Autriche. De là suit la mesquinerie des études classiques, et la prospérité des études usuelles, le néant et l'obscurité profonde de l'université de Vienne, et la juste renommée de son institut polytechnique." (*De l'Instruction Publique en Autriche, par un Diplomate Étranger*, 1841, p. 94.)

For a lengthened period after their conversion to Christianity, the Moravians were divided between the Latin and Greek churches; but the doctrines of the Reformation spread widely in this province in the 16th century. The intolerant proceedings of the Austrian government obliged, however, many Protestant families to emigrate into other countries, and many others to embrace the religion of the house of Hapsburg; so that at present the Roman Catholic faith greatly predominates over every other. There is now, however, the most perfect toleration for all sorts of creeds. The archbishopric of Olmütz is, next to the primacy of Hungary, the richest see in the empire; and the chapter of Olmütz enjoy the valuable privilege of choosing this high functionary from among their own members. The Calvinists have their superintendent at Brünn, and the Lutherans theirs at Ingrowitz. The inhabitants are mostly of the Slavonian stock, divided into many different tribes; but among the population there are estimated to be about 450,000 Germans, residing mostly in the towns, 30,000 Jews, and a few Bohemians, Hungarians, &c. This territory was anciently inhabited by the Quadi and Marcomanni. These, or cognate, tribes are said, after the dissolution of the empire of Attila, to have founded a republic here, which maintained a precarious independence for some centuries, and was afterward erected into a kingdom, extending, in the 9th century, over Bohemia, Brandenburg, Silesia, and part of Hungary, &c. Moravia subsequently belonged alternately to the Bohemians and Hungarians: it was finally annexed to Austria, together with Bohemia, in 1527. It was the great theatre of war between the French and Austro-Russian armies, in 1805. (*Oesterr. Nat. Encycl.; Burghaus; Alig. Lander, &c., iv.*)

MORAVIA, p. t., Cayuga .co., N. Y., 158 m. W. Albany, 322 W. Bounded N.W. by Owasco lake. Watered by Owasco Inlet, an E. branch of which has a perpendicular fall of 70 feet. First settled in 1794, organized from Sempronius in 1833. It contains two churches, a Presbyterian and Episcopal, seven stores, two fulling-mills, two grist-mills, six saw-mills, two tanneries; one academy, 177 students, 12 schools, 730 scholars. Pop. 2010.

MORAY, or Elgin, a marit. co. of Scotland, on the S. side of the Moray frith, being the middle district of the old prov. of Moray, having N. the Moray frith, E. the co. Banff, S. Inverness, and W. Inverness and Nairn. It consists of a N. and principal portion, and of a smaller portion on the S., detached from the main body by the intervention of a part of Inverness; and comprises in all 307,200 acres. With the exception of a considerable tract of low, light, arable land along the shore, the rest of the surface is rugged and mountainous. The climate on the coast district is comparatively mild for its latitude; and for a lengthened period wheat has been successfully raised in this district, which occasionally supplies some of the best samples to the London market. This district is also well suited for the turnip husbandry, which has been extensively introduced, and agriculture has been in other respects materially ameliorated; though, on the whole, the progress of improvement has been less rapid in this than in most districts of Scotland. Sheep-farming is not carried on upon a large scale; but the stock of cattle has been improved by crossing with the breeds of Skye and Argyle. Property mostly in large estates. Farms of all sizes; the farm buildings were formerly wretched, but those on the principal farms have been mostly rebuilt, and are now substantial and commodious. Average rent of land, in 1810, 4s. 1¼d. an acre. Manufactures unimportant. Lead, iron, lime, freestone, and slate are met with; but the first two are not wrought, and of the others only the freestone to any extent. It is partly intersected and partly bounded on the E. by the Spey, and on the W., by the Findhorn, and has the Lossie in its centre. The salmon-fisheries, especially those on the Spey, are important and valuable. This county is united with Nairn under one sheriff and in returning one member to the House of Commons. Registered electors for the county, in 1839–40, 639. The boroughs of Elgin and Forres unite with other boroughs in sending two members to the House of Commons. It is divided into 20 parishes; and in 1841 had 8133 inhabited houses, and 34,994 persons, of whom 16,071 were males, and 18,923 females. Valued rent of the county in 1815, £73,988. The county in 1831, £255,600 Scotch. Annual value of real property, in 1815, £73,988.

MORBIHAN, a marit. dep. of France, formerly a part of the province Brittany; between lat. 47° 15' and 48° 15' N., and long. 2° and 3° 45' W.; having N. Côtes du Nord, E. Ille-et-Vilaine, and Loire Inférieure, W. Finisterre, and S. the Atlantic. Length, E. to W., about 70 m.; breadth varying from 30 to 45 m. Area, 699,641 hectares. Pop., in 1836, 433,522. The coast-line is very irregular, presenting many inlets of the sea; from one of which a capacious bay, called by the Bretons *Morbihan*, or the "Little sea," the dep. derives its name. Several islands, including Belle-Isle and Groix, belong to this dep. The N. and centre parts of Morbihan are hilly; but towards the S. are some

tolerably extensive plains. The principal rivers are the Vilaine, with the Oust in the E., and the Blavet, Scorff, &c., in the W. Some of them are navigable for some distance, but none is of any considerable size. The canal of the Blavet, from Hennebon to Pontivy, is wholly included in this dep., and a great part of the canal from Nantes to Brest is within its limits. The climate is mild, but damp, W. winds are most prevalent. The atmosphere is cloudy, and violent storms are frequent in winter. A large proportion of the soil is stony; the vegetable mould is everywhere scanty, but towards the coast it is tolerably fertile. In 1834, it was estimated that 260,971 hectares were arable, 69,062 do. in pasture, 34,462 do. in woods, and 16,380 do. in orchards and gardens; while no fewer than 291,530 do. were occupied by heaths, wastes, &c. Agriculture is extremely backward; but more corn, principally rye, oats, and wheat, is grown than is required for home consumption; and a good deal of rye bread is made for exportation. The cultivation of the potato is on the increase. Turnips, hemp, flax, &c., are grown; and about 700,000 hectol. of cider are produced annually. Near Guer is the model farm of Coetbo, where 300 pupils are instructed in the details of agriculture and the auxiliary sciences, at the expense of government. The rearing of cattle is an important business, and the breeds of both oxen and sheep are in the course of being improved. Butter, both fresh and salted, is an important article of commerce. The annual produce of wool is estimated at 220,000 kilogr. The horses, though small, are strong and good. Bees are very extensively reared; and Hugo states that 450,000 kilogr. of honey, and 30,000 kilogr. of wax, are annually exported, worth together about 375,000 francs. In 1835, of 96,802 properties, subject to the *contribution foncière*, 45,234 were assessed at less than 5 francs, and 13,288 at from 5 to 10 francs: 36 only were assessed at 1600 francs or upward. The conger, oyster, and other fisheries are important; but especially that of pilchards, which employs in the season about 500 boats, manned by 2500 fishermen. About 5-7ths of the fish taken are sold fresh, and the remainder being salted, make up about 15,000 barrels. The nett produce of the pilchard fishery is estimated by Hugo at 1,400,000 francs a year. Manufactures are of considerable importance. The iron works are said to employ, directly and indirectly, from 1500 to 2000 workmen. The woollen cloth factories at Josselin and Malestroit employ together about 900 hands; and the tanneries are supposed to furnish products worth 388,800 francs a year. Paper, glass wares, linen stuffs, cotton yarn, lace, hats, beer, chemical products, &c., are also produced; there are building docks at L'Orient, Vannes, Quiberon, Port Louis, &c.; and salt is made on the coast and islands of the estimated value of 1,000,000 francs a year. Morbihan is divided into four arronds.; chief towns, Vannes, the cap., L'Orient, Ploermel, and Pontivy. It sends six members to the chamber of deputies. No. of electors (1836–39), 1452. Total public revenue (1831) 8,630,117 francs; expenditure in the same year, 11,933,084 francs; the greater portion of which was, however, for the maintenance of the naval and military estab. of the dep. (*Hugo, art. Morbihan; French Official Tables.*)

MOREA (an. *Peloponnesus*), a principal div. of Greece, and the most S. portion of continental Europe, consisting of a peninsula attached to N. Greece by the isthmus of Corinth, between lat. 36° 13' and 38° 90' N., and long. 21° 9' and 23° 30' E.; area estimated by Thiersch at 8890 sq. m. Pop., in 1835, about 380,000. It is said to derive its modern name from the resemblance that it bears to a mulberry leaf; and its coast, which is deeply indented with gulfs and inlets, has numerous headlands, the chief of which are cape Skyllo, eastward; cape Matapan, Gallo, and St. Angelo, southward, and cape Tornese westward. Its surface is extremely diversified, but may be generally described as a lofty table-land, traversed by a main ridge connected northward with the chain of N. Greece, and running southward to cape Matapan, its culminating point (mount Taygetus) rising 5115 feet above the sea. Three branches detach themselves from the main range; one running eastward into the peninsula of Argolis, and another, mount Malevo (an. *Parnon*), running S.S.E., skirting the shore of the Ægean sea; while a third, known in different parts by the different names of Cyllene, Erymanthus, and Olenos, takes a westerly course to cape Tornese. Many of these mountains attain a height of 4000 feet: their geological constitution is of limestone lying on clay-slate, interspersed in a few places by primitive rocks; and their sides are, with a few exceptions, plentifully clothed with pines, firs, oaks, and other deciduous trees. The plains are of no great extent; the largest are those of Tripolizza in Arcadia, of Nisi in Messenia, and of Gastuni in Elis. Numerous rivers and streams run from the mountain-regions in all directions; the Rouphia (*Alpheus*) is by far the largest, having a general N.W. course of more than 70 m., and with its tributaries, the Ladon and Erymanthus, draining nearly 1-3d of the entire peninsula. Next in size is the Gestnal (an. *Peneius*), rising on mount Erymanthus, and flowing, as well as the last-mentioned, into the gulf of Arkhadia: the Iri (an. *Eurotas*), which is the principal river of Laconia, falls into the gulf of Kolokythia: the other streams are mere mountain torrents, rapid in winter, but dried up in summer. Embosomed in the mountains are several lakes; but none deserves any particular mention except Zaraka (the ancient *Stymphalus*), which has two remarkable *katavothra*, or subterraneous caverns, to which its waters are almost confined during summer, and by which it was formerly supposed to connect itself with the little river Erasinus, falling into the gulf of Nauplia near the Lernean lake, now little more than a reedy marsh. (See *Hayed.*, vi., 76.) The atmosphere of the Morea is generally pure, and the climate mild, especially in spring and autumn. The heat of summer is very oppressive in the lower districts; and in winter the country is exposed to hurricanes, and liable to be inundated by heavy rains: fogs, also, are common at that season, and the mountains are covered with snow from December to the end of February. Epidemics, originating in malaria, are common diseases in summer, especially in the neighbourhood of Argos, Corinth, and the whole of the W. coast from Patras to the mouth of the Rouphia, which are the most unhealthy parts of the peninsula. The coldest, and, at the same time, the healthiest region, is the central table-land of Arcadia: the severity of its climate is noticed by many authors; and it probably gave to its inhabitants that robust habit of body which fitted them not only for the pastoral life but for the fatigues of war, and occasioned the old proverb recorded by Athenæus, that "a man should choose his slaves from Phrygia, but his allies from Arcadia." (*Deipn.* l., 27.)

The more elevated regions are devoted to the feeding of sheep and goats, the latter being to the former as about one to four. The wool is coarse; but the ewes afford good milk, butter, and cheese. These flocks suffer much from jackals and wolves, as well as from a disease called the *enlaghia*, or plague. Colonel Leake states the number of sheep and goats in Elis alone to have been, in his time, about 450,000. The uncultivated land serves, also, for the pasture of cattle, which, however, are used only for draught, goat's flesh or mutton being universally preferred for food. The best breeds are said to be found in Corinth; and bulls from this district are often sent to improve the breeds in other parts of the Morea. The valleys and plains are, generally speaking, very fertile, and, with the most imperfect tillage, yield large crops. The produce of Elis comprises wheat, two kinds of *holcus*, called *kalambokhi*, maize, and flax. Wheat, on secondary land, is sown in Oct.; but on the richest, in Nov., Dec., and even so late as Jan.: harvest, on the plains, begins early in June, and is not entirely over till the middle of Aug. The kalambokki is sown in April, and gathered in Sept. Along the N. coast large quantities of the currant grape are raised, and the average yearly production of currants, in the neighbourhood of Patras, is stated by Colonel Leake to amount to 5,000,000 lbs., or about one half the amount of that raised in Zante. Argolis produces extensive crops of rice and cotton, the former being a considerable article of trade between Nauplia and Constantinople. Cotton is likewise raised in Messenia and Laconia, and olive oil, highly esteemed all over Greece. Corn is raised in the irrigated parts of Arcadia; but the greater part of that central district is employed for pasturage. Agriculture, owing to the long-continued insecurity of property, the oppressions under which the peasantry have laboured, and to the obstinate adherence of the farmers to old and imperfect methods of husbandry, is in a most degraded state. Thiersch and Burgess, however, report a marked improvement in the condition of the rural population within the last few years. Land used formerly to be let on the métayer system; the proprietor being at all expenses, and receiving two thirds of the crop, clear of tax; but since Greece has been separated from Turkey, nearly nine tenths of the land has fallen into the hands of government, which offers it for tillage to any one who will agree to pay a quarter part of the produce for rent: the remaining tenth part of the land belongs to individuals, chiefly small proprietors, and is charged with a tax of one tenth of its produce, and the additional burden of obliging the labouring peasant to bring his tithe in kind from a great distance to the town in which it is collected. (*Burgess's Greece and Levant*, i., 131.) The annual produce of corn was, a few years back, estimated by Colonel Leake at 300,000 kilos of 22 okes (each oke = 50½ lbs.). The flora and fauna of the peninsula differ in few respects from those of N. Greece. The manufactures of the Morea are unimportant in amount, but comprise coarse cotton and woollen goods, silk fabrics, leather, and salt. The export trade consists chiefly of wine, oil, currants, rice, fruit, and wool; its chief ports being Nauplia, Patras, Corinth, and Navarin. The Morea, which under the Turks was divided into the two

sandjaks of Tripolizza and Mistra, is now distributed into the five nomes of Argolis and Corinth, Archaia and Elis, Arcadia, Messenia, and Laconia, these being again subdivided into 35 eparchies. Tripolizza was the cap. under the Turks, but recently it has greatly decayed; Nauplia, the modern cap., has about 10,000 inhabitants. The other principal towns are Patras (7000), Modon (6000), Corinth (5000), and Koron (4000).

The most interesting features, however, to the classical traveller are the remains of many ancient cities, existing in the palmy days of Grecian glory. Among the principal are three mentioned by Homer (Il. iv., 51.) Sparta is to be traced only in its ruins; but the beauty of its position, on five hills close to the Eurotas, still recalls the "pleasant Lacedæmon." Various remains of ancient architecture, in the form of dilapidated walls, temples, and forts, exist in different parts of the country, and are described by the general name of *Palaio-Castro*. Of the Cyclopean, or primitive mode of building with uncemented blocks of stone, the chief relics are at Mycenæ and Tiryns. At Mantinea the circuit of the walls is still visible; and the outlines of the celebrated field of battle may be traced. The scene of the Olympic games, though not ascertained with complete certainty, was near the influx of the small river Cladeus into the Alpheus. These interesting ruins are described at some length under their respective heads, to which the curious are referred for further information. The reader will find also some notice of the present inhabitants of the Morea, &c., in the general article GREECE (l. 1014, &c.).

The Peloponnesus, which, before it received that name, was called successively Apia and Argos, received its appellation from the Phrygian Pelops, whose descendants were afterward expelled by the Heraclidæ. Its ancient history forms a part of that of Greece generally. After the destruction of the Achæan league by the Romans, anno B.C., it was formed, with the rest of Greece, into the Roman province of Achaia; and continued, either really or nominally, a portion of that empire during 1350 years. It was taken from the Byzantine emperors by the Franks at the beginning of the 13th century; and in the division of the conquered lands the larger part of the Peloponnesus fell to the Venetians, from whom it received its modern name, either from its supposed resemblance in shape to the leaf of the mulberry (it. *more*), or from the abundance of that fruit in the peninsula. It was repeatedly invaded by the Turks in the 15th, 16th, and 17th centuries, and was finally confirmed to them in 1718, by the treaty of Passarowitz. With the exception of Maina, the Morea, with the rest of Greece, remained under their despotic sway till 1821, when its inhabitants joined in the general struggle for that independence, which, at length, after eight years of anarchy and bloodshed, was established by the treaty of Adrianople in 1829. (*Leake's Morea*, 3 vols., *passim*; *Hughes' Greece and Albania*, i., 167–235; *Burgess's Greece and Levant*, i., 129–265; *Dict. Géog.*, &c.)

MOREAU, p. t., Saratoga co., N. Y., 50 m. N. Albany, 491 W. Bounded N. and E. by Hudson river, which has here a great bend, that includes a part of Baker's falls, Glen's falls, and the great dam at fort Edward, made for the accommodation of the Champlain canal. It has extensive water-power, and contains five stores, two grist-mills, seven saw-mills; nine schools, 319 scholars. Pop. 1576.

MORELAND, t., Montgomery co., Pa., 14 m. N.E. Norristown, 16 m. N. Philadelphia. Watered by Pennypack creek and its tributaries. It contains eight stores, one cotton factory, with 1872 spindles, two flouring-mills, eight grist-mills, four saw-mills, one paper-mill, one printing-office, one weekly newspaper; one academy, 70 students; three schools, 100 scholars. Pop. 2162.

MORETOWN, p. t., Washington co., Vt., 13 m. S.W. Montpelier, 512 W. Chartered in 1763, first settled in 1790. Watered by Mad river, a tributary of Onion or Winooski river, which bounds it on the N. It contains two churches, a Congregational and Methodist, one store, two fulling-mills, three grist-mills, six saw-mills; nine schools, 223 scholars. Pop. 1198.

MORGAN, county, Va. Situated in the N. part of the state, and contains 350 sq. m. Bounded N. by Potomac river. Watered by Cacapon river and Sleepy creek. It contained in 1840, 3691 neat cattle, 3830 sheep, 5669 swine; and produced 37,327 bushels of wheat, 13,812 of rye, 63,101 of Indian corn, 4880 of buckwheat, 41,500 of oats, 17,355 of potatoes, 641 pounds of tobacco, 4298 of sugar. It had 19 stores, one fulling-mill, two woollen factories, five flouring-mills, 12 grist-mills, 25 saw-mills, four tanneries, five distilleries, two potteries; 14 schools, 347 scholars. Pop.: whites, 4113; slaves, 134; free coloured, 6; total, 4253. Capital, Bath.

MORGAN, county, Ga. Situated a little N.E. of the centre of the state, and contains 390 sq. m. Bounded E. by Oconee river, and watered by its branches. It contained in 1840, 11,640 neat cattle, 2360 sheep, 18,565 swine; and produced

25,940 bushels of wheat, 332,116 of Indian corn, 173,700 of oats, 8,947,150 pounds of cotton. Pop.: whites, 3461; slaves, 5646; free coloured, 14; total, 9121. Capital, Madison.

MORGAN, county, Ala. Situated in the N. part of the state, and contains 720 sq. m. Bounded N. by Tennessee river, and drained by its tributaries, Flint river, and Cotaco creek. It contained in 1840, 13,059 neat cattle, 4488 sheep, 35,844 swine; and produced 25,962 bushels of wheat, 788,603 of Indian corn, 1189 of barley, 46,954 of oats, 14,603 of potatoes, 7923 pounds of tobacco, 7,736,974 of cotton. It had one commission house in foreign trade, 19 retail stores, 16 grist-mills, five saw-mills, one tannery, five distilleries, one printing-office, one weekly newspaper; five academies, 131 students; 21 schools, 443 scholars. Pop.: whites, 6580; slaves, 3216; free coloured, 45; total, 9841. Capital, Summerville.

MORGAN, county, Tenn. Situated N.E. of the centre of the state, and contains 640 sq. m. Watered by branches of the S. fork of Cumberland river, and by Emery river and its branches. It contained in 1840, 5321 neat cattle, 1585 sheep, 10,294 swine; and produced 1261 bushels of wheat, 53,954 of Indian corn, 13,561 of oats, 4896 of potatoes. It had three stores, two tanneries, two distilleries. Pop.: whites, 1554; slaves, 84; free coloured, 42; total, 1680. Capital, Montgomery.

MORGAN, county, Ky. Situated in the E. part of the state, and contains 890 sq. m. Drained by Licking river and its branches, by which it is well watered. It contained in 1840, 6594 neat cattle, 9361 sheep, 13,566 swine; and produced 6398 bushels of wheat, 1956 of rye, 193,697 of Indian corn, 34,849 of oats, 9179 of potatoes, 1962 pounds of tobacco, 94,204 of sugar. It had six stores, three grist-mills, three saw-mills, four tanneries. Pop.: whites, 4539; slaves, 61; free coloured, 3; total, 4603. Capital, West Liberty.

MORGAN, county, O. Situated toward the S.E. part of the state, and contains 500 sq. m. Watered by Muskingum river and its branches, and by Duck creek. Salt is produced on the Muskingum, to the amount of 400,000 bushels annually, from wells from 600 to 800 feet deep. It contained in 1840, 13,134 neat cattle, 28,440 sheep, 25,696 swine; and produced 263,860 bushels of wheat, 447,630 of Indian corn, 9496 of buckwheat, 181,447 of oats, 51,991 of potatoes, 49,900 pounds of tobacco, 62,640 of sugar. It had 33 stores, one falling-mill, 20 grist-mills, 44 saw-mills, one oil-mill, 12 tanneries, three distilleries, one brewery, one pottery, two printing-offices, two weekly newspapers; 119 schools, 4449 scholars. Pop. 20,852. Capital, McConnelsville.

MORGAN, county, Ia. Situated a little S.W. of the centre of the state, and contains 453 sq. m. Drained by the W. fork of White river and its branches. It contained in 1840, 10,964 neat cattle, 19,871 sheep, 36,231 swine; and produced 49,217 bushels of wheat, 1078 of rye, 475,753 of Indian corn, 78,757 of oats, 36,522 pounds of sugar. It had 15 stores, one woollen factory, five flouring-mills, eight grist-mills, 16 saw-mills, six tanneries, five distilleries, two potteries; 53 schools, 2030 scholars. Pop. 10,741. Capital, Martinsville.

MORGAN, county, Ill. Situated a little E. of the centre of the state, and contains 510 sq. m. Bounded W. by Illinois river, N. by Sangamon river, and drained by tributaries of these rivers. It contained in 1840, 19,558 neat cattle, 13,037 sheep, 33,568 swine; and produced 60,861 bushels of wheat, 3792 of rye, 799,210 of Indian corn, 8130 of barley, 96,297 of oats, 31,111 of potatoes, 1900 pounds of tobacco. It had 47 stores, 20 grist-mills, 16 saw-mills, one paper-mill, five tanneries, two distilleries, two printing-offices, one bindery, one weekly newspaper, and one periodical; one college, 35 students; six academies, 222 students; 60 schools, 1752 scholars. Pop. 19,547. Capital, Jacksonville.

MORGAN, county, Mo. Situated in the central part of the state, and contains 722 sq. m. Bounded S. by Osage river. Drained by Moreau river, and the S. fork of La Mine river, and their branches. It contained in 1840, 5954 neat cattle, 3913 sheep, 17,487 swine; and produced 9235 bushels of wheat, 163,490 of Indian corn, 27,632 of oats, 7177 of potatoes, 17,831 pounds of tobacco, 2135 of sugar. It had seven stores, eight grist-mills, four saw-mills, one tannery, one distillery; one academy, 30 students; five schools, 121 scholars. Pop.: whites, 3691; slaves, 512; free coloured, 4; total, 4407. Capital, Versailles.

MORGAN, t. Greene co., Pa. It contains two stores, one falling-mill, five grist-mills, five saw-mills, one tannery, two distilleries; five schools, 125 scholars. Pop. 1094.

MORGAN, t, Morgan co., O. It contains McConnelsville village, the capital of the county. Watered by Muskingum river. It contains 10 stores, one saw-mill, two tanneries, one brewery, two printing-offices, two weekly newspapers; five schools, 475 scholars. Pop. 1518.

MORGANTOWN, p. v., capital of Monongalia co., Va., 205 m. N.W. Richmond, 218 W. Situated on the E. side of Monongahela river, at the head of steamboat navigation. It contains a courthouse, jail, two churches, a Presbyterian

and Methodist, a female academy, eight stores, one falling-mill, two flouring-mills, one printing-office, one weekly newspaper, 160 dwellings, and about 700 inhabitants. It is surrounded by an extensive manufacturing country.

MORIAH, p. t., Essex co., N. Y., 115 m. N. Albany, 496 W. The Adirondack mountains which bound it on the W., abound with iron ore. Bounded E. by lake Champlain. Watered by Schroon, branch of Hudson river. It contains a Congregational and Baptist church, 12 stores, four furnaces, five forges; one falling-mill, two grist-mills, 54 saw-mills, three tanneries; two academies, 160 students; 15 schools, 597 scholars. Pop. 2595.

MORLAIX, an ancient town and seaport of France, dep Finisterre, 33 m. E.N.E. Brest, and 263 m. W. by S. Paris; lat. 47° 35' N., long. 30 52' W. Pop. in 1836, &c. sea., 7800. It is situated at the foot of two hills, and at the confluence of two small rivers, forming a considerable estuary and commodious harbour for vessels of 400 tons burden. At the bottom of the harbour stands a well-fortified castle; and hills covered with gardens, formed into terraces, rise immediately above the town, the principal street of which runs parallel with the quays. The principal square (built on arches over the river) comprises many good modern houses, with a very large town-hall, portioned out into government-offices, and a public library. There are two large churches, one an elegant Gothic structure, with a fine tower. A large tobacco manufactory of modern construction, employing between 300 and 400 workmen, a hospital, school of navigation, theatre, and two hotels, are the other chief buildings. Morlaix is the seat of a subprefect, of a tribunal of primary jurisdiction on commerce, and of a society of agriculture: its principal manufactures are those of tobacco and linen cloth; and it enjoys a considerable trade in butter, corn, tallow, honey, and wax.

Morlaix lays claim to considerable antiquity, and was already an important town when taken by the English, near the close of the 14th century. During the two succeeding centuries it suffered greatly from the ravages of civil war.

MOROCCO (EMPIRE OF), (Arab. *Maghreb-ul-Acsa,* "the extreme West"), a tract of country in the N.W. of Africa, between the 28th and 36th degs. of N. lat., and the 2d and 12th degs. of W. long., comprising the *Mauritania Tingitana* of the ancients. It is bounded N. by the Mediterranean sea and the straits of Gibraltar, E. by the Atlas range, which separates it from the Algerine territory and Biled-ul-Jerid, S. by the river Akassa, and Sahara desert, and W. by the Atlantic ocean. Length of coast-line along the Mediterranean, 250 m.; ditto, along the Atlantic ocean, 600 m.; estimated area, 219,300 sq. m., distributed into four kingdoms, the area and population of which are estimated as under:

Kingdoms.	Provinces.	Area in sq. m.	Population.	Chief Towns.
Fez	7	98,657	3,280,000	} Fez, Tangier, Mequinez.
Morocco	7	51,320	3,600,000	Morocco, Mogador.
Suse	3	2-,636	700,000	Agadir.
Tafilet		50,677	1,800,000	
Total of empire		219,390	9,300,000	

Of the above population Gräberg von Hemsö states that 2,550,000 are Moors, 3,750,000 Berebers, and Shelluhs (chiefly devoted to agricultural and pastoral pursuits), 740,000 Bedouin Arabs, 330,560 Jews, 190,000 negroes, and 500 Christians and renegades.

Surface, &c.—Morocco is mostly bounded on the E. by the stupendous chain of the high Atlas, which commences with mount Beni-Ammer, S. of the desert of Angad, on the Algerine frontier, and extends S. as far as capes Geer and Nun. The most elevated parts of the range occur between 30° and and 33° lat.: the highest point, mount Heentet, was estimated by Mr. Jackson at about 26,650 feet, but this is no doubt a gross exaggeration. Captain Washington measured a point called Miltsin, in lat. 31° 12' N., and long. 7° 20 W., which he found to be 11,400 feet above the sea, and this was the highest in the S. portion of the chain. It is probable, however, that the highest summits will be found more to the N. in the province of Tedla: but their height, which has never been ascertained by measurement, cannot well exceed 13,000 feet. (See *Geog. Journ.*, i., p. 140.) A subordinate range, sometimes called the little Atlas, branches N.N.W. and N.W. towards Ceuta, C. Spartel, &c.; and other chains, either continuous or detached, are thinly sprinkled over the country S. of Fez and Mequinez. The geological constitution of these mountains, according to the central ridges, on which are superimposed parts, and even tertiary formations in the less elevated portion of the chain. Silver, iron, and lead mines are little extent. Mineral salt is found in great abundance throughout Morocco, and is a considerable article of export to Soudan. But notwithstanding the gigantic mountains by which it is in part bounded and in part overspread, Morocco has a large extent of comparatively level land. Some of the plains and valleys are of great extent, and extraordinary fertility, especially those of Showiya, Temsena, Ducilla, and Teram between Fez and Morocco. The principal rivers are, 1, the Sebu, rising by several sources on the W. side of the Atlas range, falling into the Atlantic, close to Mehedia, having a probable length of 260 m.; 2, the Wad Oom er-Beg, rising by two principal branches in the high Atlas, and flowing W. and W.N.W. to its mouth at Azamor, after a course of about 300 m.; and 3, the Wad-Tensift, rising about 40 m. E. of Morocco, takes a general course W. by N. to lat. 32° 7′ N., and long. 9° 19′ W., where it falls into the Atlantic ocean. The climate of the country is healthy and genial; the heat is less intense than might be expected from its geographical position, and epidemics are of rare occurrence. The thermometer, even in the hottest season, except during the occasional prevalence of hot winds from the desert, seldom exceeds 92° Reaum. (94° Fahr.); the barometer averages throughout the year 29·30 inches; and the annual fall of rain (chiefly confined to October and November), as calculated on a series of years, amounts to 29 inches. (*G. von Hemsö.* p. 28.) These observations, however, apply chiefly to the N. and W. portions of the empire, or E. of the Atlas range, the heat is intense, and rain seldom falls. The soil is now, as in antiquity, proverbial for its fertility. Mela says of it, *Cæterum solo etiam ditior et adeo fertilis est, ut frugum genera non cum serantur modo benignissimè procreet; sed quædam profundat etiam non sata.* (Lib. iii., cap. 10.) In some favoured spots three crops of corn are reaped in the same year: the soil in many parts is purely alluvial, and in others of clay, sand, and loam, mingled in the most advantageous proportions. (*G. von Hemsö,* 29, 30.)

Agriculture, however, owing, perhaps, to the extreme fertility of the land, which produces luxuriant crops, with little care or attention, is in the most backward state: fallows and rotations of crops are wholly unknown: indeed, the system of culture has remained almost unchanged since the invasion of the Arabs in the 11th century; and it consists of little more, generally speaking, than grubbing up and burning the weeds before the autumnal rains, and afterward ploughing the land about 6 inches deep, with a machine of the most simple description, drawn by a heifer or ass, and in the S. provinces by a camel. Except in the gardens, the Moors never think of using manure or other means of assisting the soil, and consequently, the land near the towns is more impoverished than in less populous districts, where, from the abundance of unemployed land, it is allowed to remain in fallow two or three years, or is the mean time other parts are brought under the plough. The wheat is white, transparent, almost without husk, having a large and exceedingly hard grain, producing a flour superior in fineness and colour to that of the northern countries. A second crop is rarely obtained; but in the S. provinces, when the harvest commences very early, a spontaneous crop springs up. According to Major Beauclerk, "The plains of Dequella alone are capable of producing in one year as much corn as the united kingdoms of Great Britain. Immense crops of corn yearly overstock the markets of Mogador: a bushel of corn may be procured for a partridge, or a coin worth an English shilling: and such is the profusion of grain, that, in many instances, it does not repay the labour of harvesting." (*Journey to Morocco,* p. 286.) Yet, with all this productiveness, so little industry and providence are exercised, that the inhabitants are sometimes, in bad seasons, reduced to the greatest privations, and hundreds of Berbers often die of famine!

Barley is used chiefly for horses and cattle, oats not being raised in any part of the country. Maize and Turkish millet are raised near the towns and along the coast in the S. provinces, and potatoes near Tangiers. On the whole, however, not more than a third part of the arable land is cultivated, and this in so superficial a manner, that the produce might be trebled, or even quadrupled, by a better system of tillage. Holme-oaks, cork and juniper trees are found on the mountains; and immense quantities of date-palms, vines, olive trees, sugar-canes, cotton, tobacco, and the fruits of S. Europe, are found in the level country. Throughout Morocco, however, there is a general scarcity of building timber: the white cedar grows to a tolerable size in the province of Refé; but when large timber is wanted, it is usually imported from Gibraltar. Land is usually rented by the number of oxen required for its cultivation, at the rate of about seven dollars for the yoke of oxen; but in lands belonging to the sultan and allotted out to his soldiery, the same portion of land would be rented at about four dollars a year, and, if sold by auction, would fetch at Tetuan about 200 dollars. (*Sir A. C. Brooke, Spain and Morocco,* i., 400.) The pasture-grounds, also, are extremely rich. the grass often attaining a height unequalled except in the prairies of America. The horses in the country are estimated at 400,000; but the breed once so esteemed under the name of Barbs is greatly deteriorated. A few milk-white, small, and finely proportioned horses, with black manes and tails, are occasionally

to be met with, belonging to the Arab chiefs; but the mass, though active, hardy, and with good action, are poor and meagre-looking; their exportation is entirely forbidden. Mules (of which there are upwards of 1,060,000) are equally well adapted for riding and draught: they are almost universally employed in long journeys, and a good mule, especially if of a bright chestnut colour with a black cross, is valued higher than a horse. Neither the asses nor mules, however, are at all comparable to those of Andalusia. It is said, though we suspect the statement is exaggerated, that about 40,000,000 sheep and 12,000,000 goats are reared; the wool of the former being of the finest quality, and the hides of the latter furnishing the celebrated Morocco leather. Oxen and camels also are bred in great quantities. A duty of about 16 dollars a head is imposed on the exportation of cattle, which being tantamount to a prohibition, the farmer is discouraged from taking any pains farther than to supply his own or his neighbour's wants. The wild animals comprise dogs, hyenas, lions, ounces, panthers, lynxes, gazelles, boars, and different varieties of game; the principal birds being ostriches, storks, quails, snipes, ducks, &c. Fish of many varieties are found in most of the rivers; bees, wasps, and mosquitoes swarm throughout the country, and locusts of large size commit fearful ravages, occasionally devouring every green leaf, and leaving the ground over which they have passed absolutely barren.

Manufactures and Trade are confined within very narrow limits. Except in the principal towns, where the houses are large and square, with a central court and flat roof similar to those of Algiers, the people live almost universally in huts or moveable tents; comparatively destitute of furniture and accommodation. Every woman understands the art of spinning wool or cotton, and the men weave it into cloth. Domestic labour, in short, which is almost wholly performed by women, supplies the principal wants of the inhabitants. Tanning appears to be almost the only exception: leather is made in great quantities all over the empire, but especially in the large towns, that of Fez being red, while that of Tafilelt and Morocco is respectively green and yellow. About 250,000 dozens of goat-skins are annually exported. The red caps, silk fabrics and girdles of Fez are highly esteemed; carpets, chip-baskets, and earthenware are manufactured in different provinces, and in the principal towns may be found skilful saddlers, carpenters, locksmiths, and farriers.

The *Commerce* of Morocco is carried on, 1, with Europe; 2, with the Levant; and 3, with the interior of Africa. The exports of Europe comprise about 2500 cwt. of wax (chiefly to Marseilles, Leghorn, Cadiz, and Lisbon), 1500 cwt. cowhides, 160,000 dozens goat-skins, 2500 cwt. olive-oil, and 4300 cwt. gums, with smaller quantities of wool, dates, honey, indigo, shawls, carpets,&c., to the amount of about 1,000,000 piastres a year; while the imports, chiefly of manufactured and colonial goods, amount only to 750,000 piastres a year. The tariff is regulated by the whim of the sultan, and prohibitions and duties vary at every port. European vessels pay six piastres for harbour dues, and a tax of 10 per cent. is levied on all imported articles. In 1831, the imports were valued at £138,400, the exports at £191,380; and in the same year 64 European ships, of 3670 tons, arrived at, and 98 ships, of 5890 tons, departed from, the different ports of the empire. (*Graberg von Hemsö,* p. 157.)

The trade with the Levant is carried on partly by pedlars, accompanying the pilgrim caravan to Mecca, and partly, also, by feluccas coasting the shores of Africa as far as Alexandria. The communication with the interior of Africa is effected by caravans proceeding from Tafilelt, and crossing the Sahara desert to Timbuctoo, in the manner described in the art. AFRICA (i. 25), where the traders exchange salt, tobacco, cloth caps, girdles, Turkish daggers, &c., for gold-dust, ivory, rhinoceros horns, assafœtida, ostrich feathers, and slaves. Their profits would seem to be immense; since, for 1,000,000 piastres, the value of goods exported, the returns amount at least to 10,000,000 piastres (*G. von H.,* p. 146); but a great part of this excess is swallowed up by the expense of the conveyance of the goods across the desert and back again. Interest on money is forbidden by law; but, notwithstanding, the Jews and others exact sums varying from 7 to 12 per cent. a month. on the security of merchandise. Paper money and bills of exchange are wholly unknown: nor is there any communication by post, for the purpose of facilitating commercial intercourse.

Government, Revenue, &c.—The government of Morocco is a pure despotism. the sultan being the head both of church and state, and the arbiter over the property and lives of his subjects; his chief title is *Emir-el-Mumenin,* "absolute ruler of the true believers." There are not here, as in Turkey, an ulema, the depositary of the national religion, or a mufti, the head of the law, who possesses privileges independent of the sovereign, and may interfere to check his determinations. There is not even a council or divan which he is expected to consult. He has no regular ministers; all is done by his single command, and no subject is supposed to have either life or property but at his disposal. The sultans would appear to consider an adherence to their engagements as an unconstitutional check on their power. "Takest thou me for an infidel," said one of them to a foreigner, "that I must be the slave of my word ? Is it not in my power to say and unsay whenever I shall please ?" (*Chenier's Morocco*, i., 208, Eng. trans.) But, after all, there are here, as in all countries, certain rights which the monarchs dare not touch, and certain duties they must discharge. The sultan cannot safely invade the domestic privacy of his subjects, nor shock any of those customs to which long establishment has given the force of law. He is expected also to give public audience four times a week, when he administers justice to all, even the poorest. Yet prudent persons usually think it more eligible to acquiesce in the sentence of the cadi, than to afford the sultan any insight into their private affairs, of which he might afterward make a not very agreeable use. On these occasions the sovereign appears on horseback, with an umbrella held over his head.

The crown is hereditary, descending to males only, but without the right of primogeniture; and hence it is not unusual for strife and civil war to arise among the children of a deceased sultan. The government has frequently, also, been overthrown by private or public treason. And hence, probably, has arisen the jealous and ferocious character by which the rulers of Morocco have been especially distinguished. Muly Ishmael, who ascended the throne in 1672, a bloodthirsty monster, though not without ability, introduced the system, since kept up, of employing a guard of negro mercenaries, on whose fidelity more reliance may be placed than on that of the Moors.

The most important state officers are the *Mula-et-tai*, or tea-taster, usually the sultan's favourite, and the *Mula-et-tesserid*, or steward of the sultan's household. The sultan sits in public, as already stated, four times a week, to administer justice. The koran is here, as elsewhere in the Mohammedan world, the text-book of justice, and decrees are usually executed immediately after they have been pronounced. For the purposes of civil and military government, Morocco is divided into 28 prefectures, some of which extend over large tracts of country, though others are confined to a single town. The chief provincial officers are the kaids or prefects, who, although removable by the sultan, are despotic governors and commanders of the military forces within their districts. The agricultural tribes have also their respective sheiks; but these are commonly subject to the Moorish governors. The revenue of Morocco in 1822 amounted to 2,600,000 piastres; of which, nearly a half is derived from duties on land, houses, shops, mills, &c., and about a fifth from imported goods. The expenditure of the same year, chiefly for the maintenance of the army and the sultan's household, was only 990,000 piastres; and the yearly surplus goes to enrich the sultan's treasury at Mequinez, which is supposed to contain at least 50,000,000 piastres. (*G. von H.*, p. 226.) The regular army does not exceed 16,000 men; of whom, as already stated, fully a half are negroes. The sultan's body-guard comprises about 3000 infantry, and 2000 cavalry. The Moors are good horsemen, and endure hunger, thirst, fatigue, and every inconvenience. They might therefore make excellent soldiers if they were properly manœuvred and exercised; but they are ignorant of every part of discipline except submission to their superiors. Their standard is the commentary upon the Koran, by Sidi Beccari, the favourite imperial saint, whose book is deposited under a tent in the centre of the army, and is the signal by which they rally. Morocco has 24 fortified and garrisoned towns, the principal of which are Suira, Tangier, Azamor, Salee, and Mazegan. The navy of the empire is quite inconsiderable, comprising only three brigs, mounting 40 cannons, and three shallops.

Population.—The inhabitants may be divided into the classes of Moors, Arabs, Berbers or Berebers, Shellochs, Jews, and Negroes. The Moors are a mixed race, the descendants of the ancient Mauritanians, intermixed with their Arab conquerors, and with the remains of the Vandals, who once ruled over the country; and, with the Moors, expelled from Spain, in the 15th century; but these varieties have been long since obliterated, and the Moors are now moulded into a distinct, peculiar people. They principally inhabit the villages and cities. Their language, called the Occidental Arabic, contains, as might be expected, many words borrowed from the language of the Berebers and Shellochs, and imported from Spain. The Arabs, as distinguished from the Moors, principally inhabit the plains, where, like their ancestors, they mostly lead a wandering life, and follow pastoral pursuits. They occupy *duwars* or moveable villages, composed of tents; and whenever the pastures in the vicinity are exhausted, or the increase of flies and vermin render the tents uninhabitable, they are struck; and placing them, their effects, and children, on panniers on the backs of camels, they set out in search of

some other quarter in which to settle. Their women are not confined; but being subjected to hard labour, tanned by the sun, and sometimes even yoked in the plough with domestic animals, these habits of hardihood, with the loss of all traces of beauty, prove more effectual securities against intrigues than the bolts and bars used in the cities. The mountainous portions of the country is occupied by the Berebers and Shellochs, probably the aboriginal inhabitants of the country. The Berebers, who principally inhabit the country of the lesser Atlas, adjoining the Mediterranean, are nearly white, well formed, of middle size, and athletic; they live mostly in huts of stone and mud, but sometimes, like the Arabs, in tents, and sometimes in caves; they are principally engaged in hunting and pastoral occupations. The Shellochs, who inhabit both sides of the greater Atlas, are less robust than the Berebers, but they are more advanced in civilization, being principally agriculturists and artisans, and enjoying comparatively good houses. A great discrepancy of opinion has been entertained as to whether the languages of those people be radically different, or merely different dialects of the same language; the latter opinion, though accompanied by several difficulties, seems to be, on the whole, the most probable.

The Berebers and Shellochs are sometimes called *amazirghis*, or freemen, a designation to which they have a not ill-founded claim. They have never, in fact, been fully subjected to the Moorish government; they often break out into rebellion; and have carried their arms to the gates of Morocco. Their internal government has even somewhat of a republican form, and they are well trained in the use of arms. The Jews, who are numerous, particularly in the cities, carry on all the mercantile and money transactions; they also act as interpreters, and perform, in the cities, the functions of servants, porters, scavengers, &c. Every species of oppression and contempt, however, is heaped upon them. They are not allowed to mount on horseback, nor to sit before a Moor with their legs crossed. The meanest Moors may insult or maltreat them in the streets, or enter their synagogues for the purpose. They must not read or write Arabic, which being the language of the Koran, is too holy for them ! A worse evil is, that when the emperor, or men in power, happen to be in want of money, they hesitate not to relieve themselves by stripping the Jews of large portions of their wealth, however carefully it may be concealed.

The negroes, who are not very numerous, are imported from Soudan. Sometimes, however, they obtain their liberty; and, as already stated, the emperor has thought fit to select them for his body guard. (The learned and excellent observations of Shaw, as to the different classes of people in Algiers, may be applied, with little modification, to Morocco. See his *Travels*, passim.)

Religion and Education.—The prevailing *religion* is Mohammedanism, and nowhere are its tenets and observances more rigidly enforced. The Jews are universally despised, nor are Christians allowed to reside anywhere except at Tangiers, Mogadore, El-Araich, and Tetuan. There is a Franciscan convent in Tangiers, being the only Christian establishment throughout Morocco.

The *education* of the Moors is, at present, greatly inferior to that of their forefathers in the middle ages, and is almost exclusively confined to learning the Koran by rote, reading, and writing. At the high school of Fez, however, more aspiring students may receive a sort of instruction in grammar, geometry, and the mixed sciences, logic, rhetoric, medicine, and theology. The art of printing is unknown, so that great numbers of persons are employed, in all parts of the empire, copying the koran, &c. Arts and sciences are in the most barbarous states; the literature, and history of foreign countries are wholly unknown; and their only musical instruments are a rude pipe, and, more barbarous, drum.

Manners and Customs.—The Moors are generally a fine looking race of men, of middle stature, and somewhat inclining to corpulence, owing, probably, to their inactive life. The women are pretty when young, blacken their eye-lashes and eye-brows, and stain the tips of their fingers with *henna*. The dress of the country is picturesque and graceful, comprising a shirt with large sleeves, ample drawers of white linen, a *kaftan*, or waistcoat, of yellow or blue cloth, a silk sash, *haick*, or mantle, and slippers, or *bose*, of *yellow* leather. Women, however, wear *red* shoes. The Jews are not allowed to wear colours, and a black cap, with slippers of the same colour, marks their degradation. The usual food throughout the country is a dish, called *kuscus*, composed of mutton or fowls, stewed with vegetables, and served up in large earthenware pans, accompanied with a savoury kind of sauce. Coffee is not used; but tea is a general beverage, always presented to visiters, and highly esteemed by all classes. The Moors do not smoke tobacco, but take large quantities of snuff, and occasionally smoke the hemp-plant, which seems to partake of the interesting qualities of opium: a confection is also made from the hemp-seed, possessed of the same qualities, and to the use of this the natives are much addicted. The distinguishing features of the Moorish character are, a love of idleness, apathy, pride, ignorance, bigotry, and the grossest sensuality. The cities present the same gloomy aspect as in other Moorish states—that of strict seclusion, particularly of the female sex, while habits of gravity and silence prevail among the men, who meet only in the public coffee-houses. Unluckily, their high national pride, and contempt for all other people, is not combined with any sentiments of individual honour. They are not, however, wholly destitute of good qualities, among which may be mentioned their hospitality, and fortitude under misfortune: *Allah-krè*, "God willed it," is their consolation in trouble. They are, also, healthy and long-lived, which could hardly have been expected, considering their habits. The climate is unquestionably good; but leprosy, ophthalmia, hydrocele, and syphilis, originating, most probably, in filthy habits, are not uncommon. Their medicines consist only of a few herbs, and their surgery is such as might be expected among a people without science or arts. The plague visits them about once in 20 years, and carries off thousands of the population. (*Geog. Jour.*, i., 146.)

History.—Morocco, anciently called Mauritania, was inhabited, under the Romans, by a hardy nomadic race, who were never thoroughly subdued by that nation. Early in the seventh century, the country yielded to the Saracens, whose different dynasties disputed for its possession nearly 300 years. At length, in the eleventh century, a chief of Leptana, having acquired so high a reputation for sanctity as to cause all the neighbouring tribes to flock to his standard, overturned the existing government, and extended his dominion all over N. Africa. His son, Joseph Ben-Tessilin, extended the empire by the addition of Fez and the S. provinces of Spain. In 1143, however, another revolution took place, and the Morabites were succeeded by the Almohades, who, in their turn, yielded the empire to more successful adventurers. In this state of anarchy the country remained till the middle of the sixteenth century, when Mohammed-Ben-Achmet, a sheriff and descendant of the Prophet, ascended the throne, which his posterity has ever since continued to occupy. (*Jackson's Travels in Africa; Beauclerk's Journey to Morocco; Sir A. C. Broke's Spain and Morocco,* i.; *Count Gröberg von Hemsö, Storia di Morocco; Geog. Journ.*, i.)

Morocco (Arab. *Marrak'sh*), a large city of N.W. Africa, and the cap. of the above empire, 165 m. E. by N. Mogadar. lat. 31° 37′ 29″ N. long. 7° 36′ W. Pop., according to Captain Washington, about 30,000, of whom 4000 are Shellochs, and 5000 Jews. We doubt, however, whether the population really exceeds 50,000, or at most 60,000[*]. It is beautifully situated about 4 m. S. from the river Tensift, on a plain elevated 1450 ft. above the sea, and is surrounded by a strong wall of lime and mud 30 ft. high and 6 m. in circuit, with square turrets at intervals of 50 paces; but the enclosed area, as in many African and Asiatic cities, comprises, besides houses and streets, many large gardens and open spaces from 20 to 30 acres in extent. The whole town, with slight exceptions, is in bad repair, many parts are in ruins, and it is everywhere filthy in the extreme. It is entered by 11 strong double gates; but the only one worth notice is the *Bab-el-Rom*, a Moorish horse-shoe arch, richly sculptured with arabesque work. Extensive under-ground aqueducts, 10 or 12 ft. deep, surround the walls, and reach across the plain to the foot of Atlas; at present, however, they are mostly in ruins. The houses in the habitable part of the town, a few of which are of stone, but the greater number of mud and lime, are, generally speaking, small, and only one story high, with central courts and flat roofs, the sides fronting the streets being plain and whitewashed, with here and there a narrow opening, unglassed, and scarcely deserving the name of a window. Their interior disposition greatly resembles that of Spanish houses—the doors are of carved cypress-wood, the rooms long and narrow, with scarcely any furniture except a few mats, carpets, and cushions. Most of them, however, being old and in decay, swarm with vermin, especially bugs, scorpions, and snakes. The streets which are unpaved, are frequently so narrow and crooked, that a horse can with difficulty pass through them; and they are in parts so heaved up with accumulations of filth, that the floors of the houses are some feet below the pathway. (*Beauclerk's Morocco,* p. 140.) These inconveniences are further increased by numerous low cross arches and gateways that connect the opposite houses. The city contains several public squares; but, like the streets, they are unpaved, and consequently very dirty when it rains, and covered with dust in dry weather. The sultan's palace

[*] Ali Bey estimated the population at only 30,000, while Jackson, by an absurd exaggeration, carried it up to 270,000!

stands on the S. side of the city, outside the main wall, but is itself enclosed within walls of equal strength. Its precincts consist of a large oblong space about 1300 yds. in length and 800 in width, divided into squares and laid out in gardens, round which are several detached pavilions about 40 ft. square, forming the imperial residence. These have pyramidal roofs covered with glazed tiles, and lighted from four lofty and spacious doors, which are opened according to the position of the sun, the floors being tesselated with variously coloured tiles, and the interior painted in the arabesque style, and ornamented with square compartments, containing passages from the Koran, written in a sort of Arabic short-hand. The luxury and convenience of tables, chairs, and curtains, are unknown, and even the rooms occupied by the sultan are furnished only with a few mats, carpets, cushions, some china and tea equipage, a clock, and arms hung round the walls. The grand pavilion in the middle of the gardens is appropriated to the women: it is a spacious building fitted up in the same simple style as the rest. Near the palace, on the E. side of the enclosure, is the *m'shoar*, or place of audience, an extensive quadrangle, walled in, but open to the sky, in which the sultan gives audience to his subjects, hears their complaints, and administers justice. Attached to the palace, also, are three gardens, each about 15 acres in extent. In two of these, the foreign merchants are allowed to pitch their tents when they visit the sultan, and the third, called *Jenán el Afia,* "the garden of prosperity," is destined for the use of the sultanas. The city has many sanctuaries and mosques; one of these, called El Kontabia, is conspicuous above all by a square tower, 221 ft. high, divided into seven stories, and surmounted by a small lantern. The mosque, Beni-Yusef, next in height and age, has an attached college and a saint's tomb, with a cupola delicately wrought in Saracenic tracery. El Moazin, also, said to be the most ancient mosque in the city, is of great size, comprising several courts opening into each other, and intersected in various directions by highly sculptured horse-shoe arches. Its gates are said to be those of Seville, brought thence by the triumphant Al-Mansor. The mosque of Bel Abbes, the patron saint of Morocco, is built in the shape of a pavilion, surmounted by a cupola covered with green varnished tiles. Attached to it is an immense hospital, said to have accommodation for 1300 patients. Near the S. wall of the city is the *Madrese del Emshis,* a college and mosque, in which are the sepulchres of the sultans of the Moluc Seidia dynasty, once adorned with statues and busts, now defaced.

Morocco, like most other Moorish towns, comprises numerous fountains, several of which have traces of delicate sculpture; and one close to the mosque El-Moazin has a cornice of white marble, still exhibiting the remains of former beauty. Outside the walls are several large cemeteries, one of which, on the E. side is upwards of 100 acres in extent: war, plague, and famine, to which the town owes its present decay, have caused them to be thickly tenanted. In the N. part of the town is the *kaissaria,* or bazaar, a long range of shops, or rather stalls, covered in from the weather, divided into compartments, and serving as a general lounge for all classes of the inhabitants. Here are exposed for sale silk scarfs, shawls, and handkerchiefs, from Fez; carpets and various articles of dress from Duquella; cloth, linen, hardware, tea, and sugar, from England; almonds and raisins, henna and pure spirit, from Suse; corn, beans, &c., from Shragna; dates from Tafilet; and an abundance of boots, slippers, saddles, earthenware, mats and cord, with gold and silver embroidery, in making which the inhabitants particularly excel. A large market is held every Thursday, near the N. gate of the city, and is well supplied with home manufactures: outside the gate, also, is the market for camels, horses, mules, horned-cattle, sheep, &c.; but the display is very indifferent. The tanning of leather is the most important branch of industry in Morocco; and Captain Washington visited one tan-yard, which alone employed 1500 persons. The establishment was extremely defective in order and arrangement; but, in spite of dirt and slovenliness, a bright yellow colour is produced, that has not been successfully imitated in Europe. (*Geog. Jour.*, i., 130.)

The *Millah,* or Jews' quarter, is a walled inclosure about 1½ m. in circuit, at the S.E. angle of the city, very densely peopled, and dirtier even than the parts inhabited by the Moors. The Jews pay a capitation tax to the sultan, and are treated with the utmost contempt; but they are a serviceable body, and are the only goldsmiths, tinmen, and tailors, in Morocco. Shoemaking, carpentry, masonry, smith's work, and the weaving of haiks, are exclusively the occupations of the Moors. Provisions are cheaper even than at Tangiers; but there is very little trade, that which exists being, with the exception of the commerce in leather and salt, confined to the supply of the town. The air about Morocco is generally calm: the neighbouring mountains defend it from the scorching winds that blow from Tafilet and Sahara, while the snow with which the chain is covered nearly all the year, imparts an agreeable coolness to the surrounding atmosphere. In summer, however, the heat during the day is intense, though the nights are cool, and in winter the cold is pretty severe. On the whole, however, the climate is extremely healthy.

Morocco, which is supposed to be situated on or near the spot occupied by the ancient *Bocanum Hemerum*, was founded in 1052 by Abu Tessian, the first Moorish sultan of the Marabou dynasty, and in the following century, during the reign of Ali Ben Yusef, it is said, but no doubt the statement is grossly exaggerated, to have contained 1,000,000 of inhabitants. In later times, its population has greatly fallen off; and, owing to the devastations of successive conquerors, it retains little of its ancient magnificence. At present, it is in many parts little else than a desert; the ruins of houses heaped one upon another serve to harbour thieves and desperadoes of all sorts. Nothing but the wretched government of Morocco could have made so great a city so miserable and so deserted. (*Geog. Journ.*, vol. i.; *Beauclerk's Journey to Morocco*; *Chenier's Hist. of Morocco*, i., 56–63; *Jackson's Morocco*, p. 121–124.)

MORON, a town of Spain, in Andalusia, prov. Seville, on a plain near the foot of a lofty hill, 33 m. S.E. Seville, and 60 m. W.S.W. Cordova. Pop., according to Miñano, 7894. It has some well-built houses, a parish church, and two hospitals. Its inhabitants are almost wholly agricultural, and nearly all the oil used in Seville is raised in the neighbourhood. The existence of Roman inscriptions and other antiquities, has induced some authors to identify Moron with the ancient *Arunci*.

MORPETH, a mun. and parl. bor., market town and par. of England, co. Northumberland, ward of its own name, on the N. bank of the Wansbeck, crossed here by two bridges, 14¼ m. N. by W. Newcastle. Pop. of parl. bor. (which includes the parish of Morpeth, except one detached township, together with the parish of Bedlington), in 1831, 6678.

The town, which is pretty well built, though badly paved, and not lighted, consists principally of two streets, at the junction of which is the market-place, with a high square clock tower, containing a chime of bells. A town-house, a stone structure fronted by a colonnade, and decorated with turrets at the angles, was erected in 1714, at the expense of Lord Carlisle, for the manorial courts, quarter sessions, local business, &c. The county jail, and house of correction, is an extensive and commodious pile erected in 1829, on the S. side of the river. The parish church, a plain brick building, is on Kirkshill, ½ m. S. the town; the living being a rectory in the gift of Earl Carlisle. There is also a chapel of ease. The Roman Catholics, Presbyterians, and Wesleyan Methodists, have their respective places of worship, to which, and the church, are attached Sunday-schools. A slenderly endowed grammar-school was founded here by Edward VI. An English free-school, and infant schools, are supported by the corporation, and there are two subscription schools for boys and girls. The other establishments of the town comprise a provident club, dispensary, mechanics' institute, subscription library, and some minor charities. Morpeth has undergone little change in its condition during the last 50 years: no buildings are in course of erection, and in 1836, there were 115 uninhabited houses. A small woollen manufactory, iron-foundry, and two or three steam corn-mills, are established here, but its chief dependence is on its cattle market, which is one of the largest in the N. of England. Freestone is quarried, and there are four collieries within the parish: but the railway connecting them with the town has been an unprofitable undertaking.

Morpeth is a borough by prescription, and recognized as such by the charter of 15 Charles II.; its municipal officers, since the Municipal Reform Act, being a mayor, three other aldermen, and twelve councillors. Corporation revenue in 1839, 216*l.*, leaving a debt of 903*l.* It has sent two members to the House of Commons since the reign of Mary; the right of election, down to the reform act, having been nominally vested in the bailiff and free burgesses, but substantially in Earl Carlisle, lord of the manor. This act deprived it of one of its members, and the limits of the borough were then also enlarged as stated above. Reg. electors. in 1839–40, 362. Petty sessions are held weekly, and quarter sessions alternately, with three other towns. Markets on Wednesday. Large cattle fairs, the Wednesday, Thursday, and Friday, but one, before Whit-Sunday. Races are held N. of the town on Cottingwood Common, early in September. (*Mun. and Bound. Rep.*, &c.)

It appears from Camden that Morpeth was "burned down by its inhabitants in 1215, out of hatred to King John," that is, with the view of distressing him when on his march to punish the revolt of his barons; and it suffered again from fire in 1689. Its castle, built in 1358, and Cistercian monastery, both mere ruins, and several churches and

baronial residences in the immediate vicinity, are well worth the notice of the antiquary.

MORRIS, county, N. J. Situated toward the N. part of the state, and contains 500 sq. m. Bounded S.E. and E. by Passaic river, N.E. by Pequannock river. Drained by Rockaway and Whippany rivers, tributaries of Passaic river. It abounds with iron ore, which is chiefly magnetic. The Morris canal passes through the county. It contained in 1840, 18,627 neat cattle, 19,147 sheep, 12,719 swine; and produced 27,103 bushels of wheat, 79,551 of rye, 228,249 of Indian corn, 80,175 of buckwheat. 262,385 of oats, 215,895 of potatoes. It had 96 stores, three furnaces, 43 forges, seven fulling-mills, six woollen factories, four cotton factories with 2908 spindles, 37 grist-mills, 57 saw-mills, four paper-mills, 19 tanneries, 50 distilleries, one pottery, three printing-offices, two binderies, two weekly newspapers; 10 academies, 456 students, 103 schools; 4343 scholars. Pop. 25,844. Capital, Morristown.

MORRIS, t., Morris co., N. J., 26 m. W. New-York. Bounded W. and S. by Passaic river, N. by Whippany river. It contains 23 stores, five grist-mills, six saw-mills, two paper-mills, three distilleries; three academies, 116 students, 19 schools, 341 scholars. Pop. 4013.

MORRIS, t., Greene co., Pa. It contains four grist-mills, six saw-mills; three schools, 85 scholars. Pop. 1462.

MORRIS, t., Huntingdon co., Pa. It has one commission house in foreign trade, four stores, one forge, two furnaces, three grist-mills, seven saw-mills. Pop. 1516.

MORRIS, t., Washington co., Pa. It has five grist-mills, one saw-mill; 11 schools, 330 scholars. Pop. 1663.

MORRIS, t., Knox co., O. Watered by three branches of Vernon river, which here unite. It has five schools, 130 scholars. Pop. 1072.

MORRISTOWN, p. t., Lamoille co., Vt., 26 m. N. by W. Montpelier. 544 W. Watered by Lamoille river, which affords water-power. Chartered in 1790, first settled in 1790. It contains three churches, a Congregational, Methodist, and Universalist, a town-house for public meetings, four stores, two fulling-mills, two grist-mills, nine saw-mills, one oil-mill; 12 schools, 423 scholars. Pop. 1583.

MORRISTOWN, p. t., St. Lawrence co., N. Y., 199 m. N.N.W. Albany, 466 W. Bounded N.W. by St. Lawrence river. Black lake, which is 90 m. long, and from 1 to 3½ m. wide, extends through the town. There is a good landing-place on St. Lawrence river. It contains an inexhaustible mine of plumbago or black lead, which yields 60 per cent. of the metal; and has two churches, an Episcopal and a Presbyterian, three stores, one fulling-mill, six saw-mills; 13 schools, 473 scholars. Pop. 2608.

MORRISTOWN, p. v., Morris t., capital of Morris co., N. J., 52 m. N. by E. Trenton, 234 W. Situated on the S. side of Musconetcong river, on an elevated plain, which slopes on two sides to the river, commanding a fine view of the surrounding country. It is regularly laid out with streets crossing each other at right angles, with a public square in the centre, ornamented with trees. It contains a neat courthouse of brick, with a jail in the basement, a bank, an academy, four churches, a Presbyterian, Episcopal, Methodist, and Baptist, 30 stores, one iron-works, one grist-mill, one oil-mill, two paper-mills, five wagon factories, two printing offices, two weekly newspapers, and about 2000 inhabitants. The place is supplied with pure spring water by an aqueduct a mile and a half long. The houses are generally well-built and neatly painted, with fine gardens attached. The Morris and Essex railroad has cars which pass four times daily between this place and Newark, where it connects with the railroad from New-York to Philadelphia. It is noted as a station of the American army, during the revolutionary war, and the ruins of an old fort, overgrown with stately trees, on a hill commanding the town, still mark the spot.

MORRISTOWN, p. v., capital of Henry co., Ill., 141 m. N.N.W. Springfield, 870 W. Situated 4 m. S. of Green river, and contains a courthouse, and several stores and dwellings.

MORRISVILLE, p. v. Eaton t., capital of Madison co., N. Y., 102 m. W. by N. Albany, 350 W. Incorporated in 1819. It contains a courthouse and clerk's office of stone, a jail of wood, three churches, a Baptist, Methodist, and Presbyterian, an academy, a printing-office, issuing a weekly newspaper, five stores, one furnace, one woollen factory, one grist-mill, one saw-mill, one comb factory, 125 dwellings, and about 750 inhabitants. It was founded in 1882.

MORRISVILLE, p. v. Bucks co., Pa., 30 m. N.E. Philadelphia, 125 m. E. Harrisburg, 165 W. Situated on the W. side of Delaware river, opposite to Trenton. Good water-power is obtained from the river. It contains three stores, one grist-mill, one saw-mill, one paper-mill, one button factory; one school, 65 scholars. Pop. 405. Here is a beautiful bridge across the Delaware, suspended from five arches, supported on piers, which is 1100 feet long and 36 ft. wide.

MOSCOW (Russ. *Moskva*), a large city of European

Russia, long the residence of the sovereigns, and still one of the capitals of the empire, on the navigable river Moskva, 400 m. S.E. Petersburg, lat. 55° 45' 13" N., long. 37° 33' E. Pop. in 1838, 304.562.

The total pop. of the city, in 1830, amounting to 305,631, was, according to M. Androsoff, classified as follows:[*]

	Males	Females	Total.
Clergy	1,619	2,127	4,846
Noblemen and superior officers	10,410	11,501	21,944
Raznochintsi	10,167	9,379	19,516
Russian merchants	8,752	7,189	16,210
Do. lesser traders	21,851	25,636	47,267
Artisans and workmen	8,055	4,426	12,484
Foreigners	1,466	1,225	8,651
Peasants, viz —			
1. Of the crown	20,595	8,024	28,619
2. The apanaged	2,094	719	2,847
3. Belonging to individuals	57,791	6,789	41,583
4. Domestic servants	42,336	28,714	70,233
Inferior grades of the army —			
1. On active services	12,300	1,462	13,762
2. On leave of absence	3,345	10,447	13,714
Students, inmates of convents, &c.	4,411	4,240	8,651
Total	144,099	121,533	305,631

This city, which was founded in 1147, is one of the most singular in the world. It is of a circular form, and covers a large extent of ground. The central part, on an eminence on the N. side of the river, is occupied by the kremlin, or citadel, containing the palace of the czars, with cathedrals, monasteries, squares, &c., built at different epochs, and in the most incongruous styles of architecture. The other quarters of the city lie round this central nucleus, increasing in magnitude according as they diverge from it. On the outside of all are the slobodes, or suburbs. The Moskva, which has a very tortuous course through the city, is crossed by various bridges, some of stone, but the greater number of wood.

Previously to the conflagration of 1812, which destroyed two thirds of the city, Moscow presented the most extraordinary contrast—palaces alternating with huts. Asiatic with European buildings, and open fields and gardens with crowded streets. "If I was struck with the irregularity of Smolensk," says Mr. Coxe, "I was all astonishment at the immensity and variety of Moscow; a city so irregular, so uncommon, so extraordinary, and so contrasted, never before claimed my attention. The streets are in general exceedingly long and broad; some are paved; others, particularly those in the suburbs, formed with trunks of trees, or boarded with planks like the floor of a room; wretched hovels are blended with large palaces; cottages of one story stand next to the most stately mansions. Many brick structures are covered with wooden tops; some of the timber houses are painted, others have iron doors and roofs. Numerous churches present themselves in every quarter, built in the oriental style of architecture; some with domes of copper, others of tin, gilt or painted green, and many roofed with wood. In a word, some parts of this vast city have the appearance of a sequestered desert, other quarters of a populous town; some of a contemptible village, others of a great capital." (Travels in the North, i., 383; see also Clarke, i., 60, 8vo. edit.)

There is no longer any question that the conflagration of 1812 was the act of the Russian government, in the view of rendering it impossible for the French to winter in the city. With the exception of the Kremlin, and the quarter (Bielgorod) immediately surrounding it, on the N., the rest of the city was mostly destroyed; and in some quarters the destruction was so complete that the lines of streets could with difficulty be recognised. The Kremlin, too, though it escaped the conflagration, suffered severely from the mines sprung under its walls, by order of Napoleon, on his evacuation by the French. But this wide-spread desolation was repaired in a very few years. Like a phoenix, Moscow has risen from her ashes larger and more beautiful than ever. The streets have been widened, and the buildings are less singular and discordant; still, however, the old and distinctive character of the city is preserved, being at once "beautiful and rich, grotesque and absurd, magnificent and mean." According to M. Androsoff, there were in all 9482 houses in the city in 1834; of which 3127 were of brick and stone, and the residue timber. The erection of a wooden house is an easy matter. A market, held in a large open space in one of the suburbs, exhibits a variety of materials for house building, consisting of trunks of trees cut, shaped, and morticed into each other. The purchaser who wants a dwelling repairs to the spot, explains the number of rooms he requires, examines the different timbers, which are regularly numbered, and bargains for what suits him. The whole is either paid for on the spot, and taken away by the purchaser, or the seller may agree to transport and erect it at the place where it is wanted. A dwelling may thus be bought, transported, reared, and inhabited, within a week!

The Kremlin, which has been completely repaired since 1812, comprises the imperial palace, the archbishop's palace, the cathedral of the Assumption, in which the Russian sovereigns are crowned, with the cathedrals of St. Michael, the Annunciation, &c. It also contains the belfry of Ivan Veliki, a tower 269½ ft. in height, having within it, at different stories, 39 bells, some of them of an immense size: on festivals they are tolled without interruption, the Russians being passionately fond of bell-ringing. The great bell of Moscow, weighing 10,000 poods, or 360,000 lbs. (1600 tons Eng.), is now lying on the ground, at a short distance from the tower of Ivan Veliki. It is said to have been once suspended in a wooden building; but this taking fire, the water thrown upon it, to extinguish the flames, occasioned the large rent now seen in the bell. But the fact of its ever having been suspended is doubtful, and the rent was probably occasioned by some defect in the casting. On festivals, the peasants resort to this bell as they would to a sanctuary. Among the other public buildings, may be mentioned the palace of Arms, in the Kremlin; the foundling hospital; the bazaar, an immense building, containing a great number of shops; the imperial theatre; the hall, for exercising the troops in bad weather, built by the emperor Alexander (of vast dimensions, being 560 ft. in length, 180 in breadth, and 50 in height, the roof not being supported on pillars); the arsenal; the palace of the senate; the university; the postoffice; the Pachkoff palace; the great military hospital. The number of churches, though lessened by the fire, is still immense. In 1831 they amounted, in all to 284, of which seven were for Catholics.

The university founded in 1755, is the most ancient in Russia. It had, in 1835, 120 professors and sub-professors, and 419 pupils. Among the other educational establishments, may be mentioned the gymnasium, dependent on the university; the theological academy, one of the principal in the empire; the medico-chirurgical academy; the military school, or corps de cadets, with 600 pupils; the commercial school, founded in 1804; the establishments of St. Catherine and Alexander, for the education of young ladies; the veterinary school; the institute of Lazarus, so called from its founder, with 80 pupils, and a library particularly rich in Armenian literature, &c. There is, however, a great want of elementary schools for the lower classes; Moscow being, in this respect, far below Petersburg, defective as is the latter.

There are a number of literary societies, libraries, &c. The best library belongs to Count Tolstoi. The university library was partly destroyed in 1812; but it has since been enriched by fresh purchases and donations, and contained, in 1835, about 45,000 volumes. Several nobles have extensive collections of books, pictures, medals, &c.

The foundling hospital is a vast establishment, and is managed in the best possible manner. During the ten years ending with 1831, the admissions were 52,549, and the deaths 34,713. In point of fact, however, a large proportion of the children brought to this, and to all similar establishments, are all but dead before they arrive; and the real objection to such institutions consists not so much in their great mortality, as in the encouragement they hold out to licentiousness, and the desertion of children. The great military hospital has above 1500 beds; and an undoubted judge, Baron Larrey, has declared that its organization is excellent. There is also the hospital of St. Catherine, the hospital of Gallitzen, &c. The population of the hospitals attached to the city in 1830 amounted to 22,227 individuals.

Moscow is the favourite residence of many of the Russian nobles, who pass the winter in the greatest splendour; not being overshadowed, as at Petersburg, by the court. According to M. Lecointe de Laveau (Guide du Voyageur à Moscou), there were reckoned in the city, in 1830, 6396 shops, 476 inns and hotels, 244 restaurateurs, 131 kabaks, or places for the sale of spirits, 58 kabuks for the sale of beer, 135 cellars for the sale of wine, 115 bakehouses, and 25 apothecaries' shops.

Manufactures are prosecuted here on a much larger scale than in Petersburg; but a large proportion of the works on account of the manufacturers and capitalists of Moscow are not in the city, but in the adjoining towns and villages, sometimes at a considerable distance from the capital. The principal establishments are those for the manufacture of cotton, woollen, and silk fabrics, many of which are upon a very large scale, and are fitted up with steam-engines and other improved machinery. Hats, also, are extensively produced; and there are numerous tanneries, breweries, distilleries, &c.

Moscow is the grand entrepôt of the internal commerce of the empire. It has a water communication with Petersburg and Riga, on the Baltic, Astrakhan, on the Caspian, and Odessa, on the Black sea. In spring, or after the breaking up of the ice, the Moskva is navigable for barks, but during the rest of the season it is navigable for rafts only. A great deal of the commercial intercourse between the

---
* Schnitzler La Russie, p. 80.

city and the adjacent and distant provinces is carried on in winter by the sledge-roads.

The same causes which occasion a very great preponderance of males over females at Petersburg (which see), exist in Moscow, though to a less extent, and have a similar result.

MOSELLE, a frontier dep. of France, reg. N.E., formerly a part of Lorraine, chiefly between lat. 49° and 49° 30′ N., and long. 5° 30′ and 7° 40′ E., having N. and E. Luxemburg, Rhenish Prussia, and Rhenish Bavaria; S. the deps. Bas-Rhin and Meurthe, and W. that of the Meuse. Area, 532,796 hectares. Pop., in 1836, 417,003. The E. part of the dep. is covered with ramifications of the Vosges, and the W. with those of the Ardennes mountains; but neither rise to any considerable height. The general slope of the dep. is toward the N., which is the direction followed by its principal rivers, the Moselle in the W., the Sarre in the E., and their affluents, the Ornes and Nied. Some portions of the surface are very marshy. Near the Vosges, where primary formations prevail, the soil is stony; elsewhere it is chalky or clayey, and, in general, of indifferent quality, only about 10,000 hectares being said to be rich land. In 1834, the arable lands were estimated to comprise 303,913 hectares; pastures, 45,807; woods, 92,228; and orchards, gardens, &c., 11,920 hectares. According to Hugo and the Dict. Géog., agriculture has of late made considerable progress in this dep.; and certainly it would seem, from the statements of Mr. Jacob, by whom it was visited in 1819, that, both in this respect and in the condition of the population, there was ample room for improvement. "Through the whole distance," says he, "of 50 m. from the Prussian frontier to Metz, there is not a single good house to be seen, except the convent, and the house of the iron master at Forbach. There is nothing in any of the villages, large and populous as they are, nor within sight of the road, though it is an open country, that looks like a decent farmer's or clergyman's house. All seemed of the same standard; each filthy, dilapidated, and small, with barn and stables adjoining, of corresponding appearance. The horses and wagons look miserable; the cows, few and poor; and I did not see more than fifty sheep, which were of a bad race, and nearly half of them black. The inhabitants were ill clothed, and at every stoppage we were assailed by numerous beggars. The cultivators (near Metz) are all proprietors. They or their parents generally bought the land, at the revolution, for paper money; before which they were bound to their lords in certain feudal services and payments, and were supplied by them with the capital requisite for cultivation. In the worst parts of Germany, where the soil is poorest, and where the feudal power is still in force, the peasantry are better clothed, have more furniture in their houses, and display more new and repaired houses, than are seen in the district from Metz to Verdun. Between those cities, the towns and villages are miserable receptacles of filth and poverty." (Jacob's View of Agriculture in Germany, &c., p., 435, 436.)

Besides wheat, oats, and barley, the other principal articles of culture are turnips, flax, hemp, and oleaginous plants. Moselle furnishes about 180,000 hectolitres a year of second-rate wine, the best of which is the red wine produced in the arrond. of Metz. The white wines are mostly light, and de peu de durée. (Jullien, 42.) Though the pastures are good, all kinds of live stock, except hogs, are said to be inferior. Quills and honey are important articles of rural produce; cantharides are collected in summer near Metz. In 1834, of 153,968 properties subject to the contribution foncière, 64,651 were assessed at less than five francs, and 39,238 at between five and ten francs; only 95 were assessed at 1000 fr. or upward. Iron, which is everywhere abundant, and usually of good quality, is extensively produced and wrought, especially in the arrond. of Thionville. There are many potteries and some glass factories in the dep. Lorraine is famous for its linens; but the value of those produced in this dep. does not exceed 1,800,000 fr. a year. Woollen cloths, lace, paper, glue, and leather are the other principal products. Manufacturing industry may, in fact, be said to have originated and grown up in this dep. since the revolution; but in the interval since that event, has made considerable progress. Moselle is divided into four arronds.: chief towns Metz, the cap., Briey, Thionville, and Sarreguemines. It sends six archers to the chamber of deputies. Number of electors, in 1838-39, 1731. Total public revenue (1831), 13,577,481 fr. (Hugo, art. Moselle; French Official Tables.)

MOSELLE (an. Mosella), a river of W. Europe, flowing through the E. part of France, and the S. part of Rhenish Prussia; its basin being situated between that of the Nahe to the E., and the Meuse to the W. It rises in the Vosges dep. and mountains, about lat. 48° N., long. 7° E., and runs generally in a N.N.E. direction, with a very tortuous course, to Coblentz, in lat. 50° 22′, and long. 7° 33′, where it joins the Rhine. Before entering the Prussian dominions, it

traverses the French deps. Vosges, Meurthe, and Moselle; and separates Dutch Luxemburg from Rhenish Prussia. Its entire course is estimated at nearly 300 m., for about half of which it is navigable. Its average breadth is about 170 yards; its mean depth 6 ft., and its ordinary rate of current about 1½ m. an hour. Its principal affluents are the Madon, Meurthe, Seille, Sarre, and Sure: Epinal, Toul, Metz, Thionville, Treves, Traubach, and Coblentz, are on its banks. The surrounding country is subject to its inundations, which do much damage; but it is of immense utility, as a channel of internal communication, large quantities of timber, slate, coal, charcoal, brandy, salt, potash, oak bark, glass, and earthenware, and wine being sent by it to the Rhine. (Dict. Géog.; Schreiber's Guide du Rhin, &c.)

MOSTAR, a town of European Turkey, pach. Bosnia, and sandjiak Herzegovina, of which it is the cap., on the Narenta, crossed here by a Roman bridge, 48 m. S.W. Bosna Serai; lat. 43° 20′ N., long. 17° 52′ E. Pop. probably under 10,000. It is surrounded by crenellated walls, and its principal streets are on the right bank of the river, about one third of the town being on the other side. It has a celebrated manufacture of swords and fire-arms, besides an extensive traffic in cattle, corn, and wine, brought thither from a great distance. (Stein, Dict. Géog.)

MOSUL, a city of Asiatic Turkey, pach. of Bagdad, chiefly interesting as being near the site of Nineveh, the celebrated capital of the first Assyrian empire. It stands on the W. bank of the Tigris (here very rapid, 300 ft. broad, and crossed by a bridge of boats, as well as an older one of stone), 193 m. N.N.W. Bagdad, lat. 36° 21′ N long. 43° 11′ E. Pop., according to Kinneir, 35,000; of about 9000 are Christians, 1500 Jews, and the rest Arabs, Turks, and Kurds. The city is so near the level of the river, that its streets are often flooded, and, like almost every other town in Turkey and Persia, it is in a declining state, its walls being broken down, and its best buildings crumbling into ruins. It has seven gates, and the castle, now in ruins, occupies a small artificial island in the Tigris. Streets narrow and irregular. Houses built partly of stone, partly of plastered brick, with vaulted roofs and ceilings, surrounded by flat terraces. The mosques, of which there are several that possess considerable beauty, the coffee houses, khans, hummums, and bazaars, are handsomer than in most Turkish towns, and the market is well supplied with provisions from Kurdistan. The Greek Christians have nine churches, and there is a Dominican convent. The principal ornaments of the city are, a college, the tomb of Sheikh Abdul Casim, and the remains of a fine mosque. The minaret of which was built by Noureddin, sultan of Damascus. West of the Tigris, the environs are wholly uncultivated; and this circumstance, combined with the great extent of the cemetery close under the walls, gives it a gloomy and melancholy aspect.

Mosul is under the separate jurisdiction of a pacha of two tails: it formerly had a large caravan trade with all parts of Asia, but has lost much of its commercial importance; it still, however, carries on a trifling trade with Bagdad and Asia Minor; to the former of which it sends, on rafts down the Tigris, gall-nuts and copper, from Kurdistan and Armenia, receiving in return Indian commodities, afterwards forwarded to Diarbekr, Orfah Tokat, Aleppo, &c. Its only manufacture is that of coarse blue cotton cloths, used by the lower orders of the population.

The climate is proverbially healthy, the average temperature of summer not exceeding 66° Fahrenheit: but in spring, during the floods of the Tigris, epidemics are rife. Several sulphur springs are found within a short distance of the town, and are much resorted to for cutaneous diseases. The geological formation of its immediate vicinity consists, according to Ainsworth, of solid beds of massive, compact, and granular calcareous gypsum, arranged in horizontal strata, not fossiliferous, of a bluish white colour, and extensively quarried as marble. Superimposed on the gypsum is a thin formation of a friable limestone, abounding in shells, and forming the common building stone of Mosul, as it probably also formed that of the ancient Nineveh. (Kinneir; see also Nineveh.)

MOULINS, a town of France, cap. dep. Allier, on the river of that name, which is here crossed by a handsome stone bridge of 13 arches, in a fertile plain, 159 m. S.S.E. Paris; lat. 46° 34′ 4″ N., long. 3° 20′ 14″ E. Pop., in 1836, 14,582. Streets narrow and irregular, but clean and well-paved; houses chiefly of brick, but a few also of stone, especially in the principal avenue, the rue de Paris. It has three public squares, that of the Allier being by far the largest and best built. It is well supplied with water from numerous fountains. The chief public edifices are the churches of Notre Dame and the Visitation, the royal college, established in the suppressed convent of the latter, the townhall, hotel of St. Cyr, and a recently erected hall of justice. It has also two large hospitals, barracks, a public library with 90,000 volumes, a museum of natural history, and a small theatre. Several fine walks run in different directions out of the town; and in the neighbourhood are extensive vineyards, and mulberry plantations for breeding silk-worms. Coal and limestone are wrought a short distance from the town, and are articles of considerable trade. Moulins is the seat of a tribunal of primary jurisdiction and commerce, a chamber of manufactures, and a society of agriculture, sciences, and arts: it has a large manufacture of cutlery, especially scissors, which are highly esteemed, and smaller establishments for making coarse cotton and woollen fabrics, haberdashery, and hats, with steam corn-mills, glass-houses, and tan-yards. It has a considerable trade in corn, wine, silk, timber, coal, and cattle, chiefly with Orleans, by the Loire navigation; but also with Bourges, Mâcon, &c.

Moulins, which existed as a town so early as the 11th century, became the residence of the Dukes of Bourbon in 1369. A famous assembly convoked here in 1565, by Catherine de Medici, was followed by the long and sanguinary war of the league. (*Hugo*, art. *Allier*, &c.)

MOULTONBOROUGH, p. t., Carrol co., N.H., 46 m. N. Concord. 527 W. Chartered in 1763, first settled in 1764. Bounded S. by Winnipiseogee lake, into which a considerable neck of land extends. Between this town and Holderness lies Squam lake, 6 m. long, and in its widest part 3 m. wide, containing several islands, the largest of which is a mile long and one third of a mile wide. It is a beautiful sheet of water. It has its outlet into Pemigewasset river. In the N. part of this township are Great Ossipee and Red mountains, over which the wind often blows with great violence. A mineral spring at the base of the Ossipee mountain contains iron and sulphur. There is also a spring 16 ft. in diameter, which discharges water sufficient for a mill stream. Interesting aboriginal remains are found. It contains two churches, a Congregational and Christian, three stores, three grist-mills, three saw-mills; one academy, 11 students; 18 schools, 714 scholars. Pop. 1752.

MOUNT CARMEL, p. v., capital of Wabash co., Ill., 166 m. S.E. Springfield, 714 W. Situated on elevated ground on the W. side of Wabash river. It contains a courthouse and jail, of brick, two churches, a Methodist, and a German Reformed, with which the Evangelical Lutherans are associated; 12 stores, three saw-mills, and one ox tread-mill, one iron foundry and machine shop, and about 1200 inhabitants. Steamboats frequently arrive and depart.

MOUNT DESERT, p. t., Hancock co., Me., 146 m. E. Augusta, 666 W. It is southern of two townships into which mount Desert is divided. This island lies between Frenchman's and Blue Hill bays, and is 45 m. long and 12 broad. The town has excellent harbours, and considerable navigation employed in the coasting trade and the fisheries. It contains six stores, two fulling-mills, two woollen factories, two grist-mills, nine saw-mills; 22 schools, 607 scholars. Pop. 1697.

MOUNT HOLLY, p. t., Rutland co., Vt., 79 m. S. Montpelier, 469 W. Incorporated in 1792. Watered by Mill river, a branch of Otter creek. It has two stores, two grist-mills, four saw-mills; 11 schools, 300 scholars. Pop. 1358.

MOUNT HOLLY, p. v., Northampton t. capital of Burlington co., N.J., 18 m. S. Trenton, 156 W. Situated at the head of tide-water and of navigation on Rancocus creek. It is regularly laid out on seven streets, and contains a brick courthouse, 40 by 60 feet, two stories high, with a cupola, a jail of stone, five churches, an Episcopal, Methodist, Baptist, and two Friends; two female academies, a bank, eight stores, one fulling-mill, one woollen factory, one grist-mill, one saw-mill, one plaster-mill, a large paper-mill, 240 dwellings, many of them of brick and neat, and about 1400 inhabitants. A library company was established here as early as 1765; and, at the period of the revolution, the town had 200 dwellings.

MOUNT HOLYOKE, mt., Hampshire co., Mass., 3 m. S.E. of Northampton. It is 830 feet above the level of Connecticut river, and its top presents the beautiful windings of the river, and altogether one of the finest views in New-England.

MOUNT HOPE, p. t., Orange co., N.Y., 11 m. W. Goshen, 112 m. S.W. Albany, 276 W. Watered by Shawangunk creek. It contains a Presbyterian church, nine stores, one fulling-mill, one woollen factory, five grist-mills, three saw-mills, three tanneries, two distilleries, one printing-office, one periodical; one academy, 100 students; four schools, 213 scholars. Pop. 1565.

MOUNT JOY, t., Adams co., Pa., 6 m. S. Gettysburg. Watered by Rock and Willoways creeks. It has two flouring-mills, two grist-mills, three saw-mills. Pop. 1031.

MOUNT JOY, p. t., Lancaster co., Pa., 25 m. S.E. Harrisburg, 123 W. It has Little Chiques creek on its E. border, and Conewango creek on the N.W. It contains eight

stores, four flouring-mills, three grist-mills; three schools, 93 scholars. Pop. 2275.

MOUNT MARCY, Keene t., Essex co., N.Y., is the highest peak of the Adirondack mountains, 5476 feet above tide water in Hudson river. There are several other peaks in the vicinity, nearly as high, and little inferior in elevation to the celebrated White mountains in New-Hampshire.

MOUNT MORRIS, p. t., Livingston co., N.Y., 10 m. S. Geneseo, 242 m. W. Albany, 353 W. Watered by Genesee river, across which is a dam with a raceway, affording good water-power. It contains 24 stores, one furnace, two fulling-mills, two grist-mills, four saw-mills, one tannery, one pottery, one printing-office, two weekly newspapers; one academy, 60 students; 12 schools, 430 scholars. Pop. 4376.

MOUNT PLEASANT, t., Westchester co., N.Y., 6 m. N. White Plains, 125 m. S. Albany. It has an abundance of marble of a good quality. It contains the village of Singsing, where is a state prison. See Singsing. The township has four academies, 195 students. Pop. 7367.

MOUNT PLEASANT, t., Adams co., Pa., 17 m. N.E. Gettysburg. Watered by branches of Conewago creek. It contains three stores, three flouring-mills, three grist-mills, three saw-mills, one distillery, one pottery; nine schools, 360 scholars. Pop. 1588.

MOUNT PLEASANT, p. t., Westmoreland co., Pa., 170 m. W. Harrisburg, 196 W. Drained by Big Sewickly and Jacob's creeks. It contains two churches, a Methodist and Baptist; 10 stores, one flouring-mill, four tanneries, 13 distilleries; five schools, 120 scholars. Pop. 2123.

MOUNT PLEASANT, t., Wayne co., Pa. Drained by head branches of Lackawaxen, Dyberry, and Equinunk creeks. It has 11 schools, 190 scholars. Pop. 1350.

MOUNT PLEASANT, t., Washington co., Pa. It has four stores, one flouring-mill, one grist-mill, two saw-mills; one academy, 11 students; nine schools, 300 scholars. Pop. 1903.

MOUNT PLEASANT, p. t., Jefferson co., O., 131 m. E. by N. Columbus, 277 W. Bounded N. by Short creek. It contains 10 stores, one woollen factory, three flouring-mills, three saw-mills, two tanneries; one academy, 140 students; three schools, 250 scholars. Pop. 1676. The village contains three churches, a Methodist, Associate Reformed, and Friends, all of brick; a bank, a market-house, several stores, a printing-office, 150 dwellings, and about 700 inhabitants.

MOUNT PLEASANT, p. v., capital of Harlan co., Ky., 165 m. S.E. Frankfort, 473 W. Situated on the N. side of Clover fork of Cumberland river. It contains a courthouse, jail, clerk's office, and several stores and dwellings.

MOUNT PLEASANT, p. v., capital of Martin co., In., 106 m. S.S.W. Indianopolis, 653 W. Situated on the W. side of the E. fork of White river. It contains a courthouse, jail, two churches, four stores, 40 dwellings, and about 250 inhabitants.

MOUNT PLEASANT, p. v., capital of Henry co., Iowa, situated on the N. side of Big creek, and contains a brick courthouse in the centre of a public square, 100 dwellings, and about 500 inhabitants.

MOUNT SORREL (properly, *mount Soar-hill*), a market-town of England, in E. Goscote hund., co. Leicester, parish of Bothley and Barrow-upon-Soar. Pop. of township, in 1841, 1536. It derives its name from being situated close to a steep craggy hill of red granite, which rises immediately from the Soar. On its highest point there formerly stood a fortress, which, being taken by Henry III., was soon after entirely demolished. The town, built along the great road between London and Nottingham, consists chiefly of houses constructed of granite: the principal buildings are, the townhall (in which the petty sessions for the hundred are held), a church, subordinate to that of Barrow, and three places of worship for dissenters.

Mount Sorrel has some share in the hosiery trade of Leicester; and it has derived much benefit from its position on the Great North road; but this advantage it is likely to lose, from the recent opening of the Midland Counties railway. Markets on Monday.

MOUNT STERLING, p. v., capital of Montgomery co., Ky., 59 m. E. Frankfort, 514 W. Situated on a branch of the S. fork of Licking river. It contains a courthouse, jail, 10 stores, and 585 inhabitants.

MOUNT STERLING, p. v., capital of Brown co., Ill., 77 m. W. by N. Springfield, 854 W. It contains a courthouse, and several stores and dwellings.

MOUNT VERNON, p. t., Kennebec co., Me., 15 m. N.W. Augusta, 609 W. Incorporated in 1792. It contains six stores, two fulling-mills, two grist-mills, four saw-mills; 608 scholars in schools. Pop. 1475.

MOUNT VERNON, Fairfax co., Va., 9 m. below Alexandria, on the W. bank of Potomac river. It contains the former mansion-house, and, at present, the tomb of Washington, which is often visited, not only by American citizens, but by foreigners, who take a peculiar interest in the memorials of this great man. "first in war, first in peace, and first in the hearts of his countrymen."

MOUNT VERNON, p. v., capital of Knox co., O., 51 m. N.E. Columbus, 376 W. Situated on Vernon river, and contains a fine courthouse, a jail, four churches, two Presbyterian, a Protestant Methodist, and an Episcopal; 20 stores, three flouring-mills, two saw-mills, one oil-mill, two printing-offices, 250 dwellings, and 2362 inhabitants.

MOUNT VERNON, p v., capital of Posey co., Ia., 188 m. S.W. by S. Indianapolis, 749 W. Situated on a high bank, on a N. bend of Ohio river, of which it commands a view for 16 m. It contains a courthouse. Jail, 15 or 20 stores, a steam flouring-mill and saw-mill, and about 900 inhabitants.

MOUNT VERNON, p. v., capital of Jefferson co., Ill., 129 m. S.S.E. Springfield, 794 W. It contains a courthouse, jail, three churches, an Episcopal, Methodist, and Baptist; nine stores, and about 200 inhabitants.

MOURZOUK. See FEZZAN, 1., 906.

MOYAMENSING, t., Philadelphia co., Pa. It lies S. of the city of Philadelphia, of which it is a suburb, if not a part; though it is under a distinct incorporation, governed by nine commissioners elected for three years, three of whom are elected annually. Incorporated in 1812. Adjoining the city it is compactly built, and contains 127 stores, one cotton factory, one brewery; two academies, 745 students; 14 schools, 2156 scholars. Pop. 14,573.

MOZAMBIQUE, a city and seaport of E. African capital of a colony belonging to the Portuguese, lat. 14° 49' S. long. 40° 45' E. Pop. 3000? It stands on a crescent-shaped island of coral, very low and narrow, and scarcely 1½ m. in length, near the entrance of a deep inlet of the sea, which forms its harbour. The fort, which has six bastions and 80 cannon, is in bad repair; the city comprises a large square, and several narrow, dirty streets, lined with lofty houses fast falling to decay. The governor's palace is an extensive stone building, with a flat lead roof, and a square court in its centre. Three churches, an old town-hall, and hospital, are the only other public edifices. Black town, at its S. extremity, is wholly inhabited by negroes, and consists of bamboo and osier huts.

The administration of the colony is vested in the governor, aided by a council comprising the bishop, the commander of the troops, and the chief civil minister; but it is extremely corrupt, and nearly all the functionaries, both civil and military, are criminals exiled from Portugal. Exclusive of about 500 Portuguese and Creoles, the population consists of free blacks and slaves, with about 900 Arabs and Banyans, chiefly engaged in petty trade and handicraft. The irregular life led by the Europeans, and the general insalubrity of the climate, prevent any increase in the white population; and, at an average, of 100 soldiers, seven only survive a residence of five years. The rural population is in the most degraded state; and, although the soil is naturally rich and productive, the culture of cotton, indigo, sugar, and other articles of commerce is wholly neglected. Rice, millet, and manioc, are raised almost without labour: furnishing, with cocoa nuts, almost the entire food of the slaves. The commerce of Mozambique has greatly decreased, in consequence of our exertions to suppress the traffic in slaves; but, though much diminished, the trade is still carried on to a considerable extent both with Brazil and Arabia. These slaves, who are chiefly of the tribe of Monjores, and brought from the centre of the continent, a distance of 40 or 45 days from the colony, are procured from the native merchants in exchange for salt, shells, tobacco, coarse cloths, &c.; goods costing about two dollars will bring in, as the case may be, either a slave, or an elephant's tusk, weighing from 60 to 80 lbs. of ivory; hippopotamus' tusks, gold dust, columbo root, gums, and amber, are the other chief exports; the imports comprise tea, sugar, coffee, cotton and woollen cloths, with other articles from Diu, Demaun, and Goa. A duty of 16½ crusadores is levied on every slave exported; all other imports and exports are free of duty.

Mozambique was first visited by Vasco da Gama in 1498; and in 1506 Albuquerque made it the centre of the Portuguese possessions in this part of the world, and the seat of the viceroy of the African colonies. When the Portuguese lost their Indian possessions at the commencement of the 17th century, Mozambique began to decline, and has ever since been in a languishing state. The territory, however, still extends from cape Delgado northward to Delagoa bay southward, having a length of coast exceeding 1400 m., and comprising, besides the capital, the several settlements of Ibo, Pomba, Conducia, Mokambo, and Quilluane. The channel between the E. coast of Africa and Madagascar is called the Mozambique channel. (Ritter's Africa, i., 202-204. &c.)

MUDDY CREEK, t., Butler co., Pa. It has five stores, one grit mill, seven saw-mills, two tanneries; 13 schools, 504 scholars. Pop. 1998.

MUHLENBURG, county, Ky., situated centrally towards the W. part of the state, and contains 490 sq. m. Bounded N.E. by Green river, by branches of which it is watered. It contained in 1840, 8866 neat cattle, 7946 sheep, 32,959 swine; and produced 34,900 bushels of wheat, 351,505 of Indian corn, 37,731 of oats, 9396 of potatoes, 286,747 pounds of tobacco. It had 15 stores, one furnace, seven grist-mills, four saw-mills, one oil-mill, five tanneries, ten distilleries; four schools, 98 scholars. Pop.: whites, 5755; slaves, 1196; free coloured, 13; total, 6964. Capital, Greenville.

MULLICA, t., Atlantic co., N.J. It contains 10 stores, one furnace, three grist-mills, seven saw-mills; five schools, 296 scholars. Pop. 1056.

MUHLHAUSEN, a town of the Prussian dom., prov. Saxony, reg. Erfurt, cap. circ., on the Unstrut, 29 m. N.W. Erfurt. Pop., with its suburbs, in 1836, 22,050. (Berghaus.) It is surrounded by a high wall, flanked with towers, and environed by a ditch; has an antiquated appearance, several Lutheran churches, a gymnasium, three hospitals, and an orphan asylum; and is the seat of a judicial court for the town and circuit. It has manufactures of woollen and linen cloths, carpets, &c.; with dyeing-houses, fulling and oil mills, distilleries, breweries, and tanneries, and an active trade in corn and dyeing drugs. (Von Zedlitz; Berghaus.)

MULHAUSEN, or MULHOUSE, a town of France, dep. Haut-Rhin, cap. cant., on the Ill, 22 m. S. Colmar, and 16 m. N.W. Basle. Pop., in 1836, ex. com., 13,789; or with com., 16,932; exclusive of about 7000 individuals, who come daily out of the neighbouring communes to work in the various factories. It is divided into the old and the new town. The former, entirely surrounded by the Ill (here crossed by several bridges), though irregularly laid out, has tolerably broad, well paved, and clean streets, and some good houses. The Protestant and the Roman Catholic parish churches, the synagogue, town-hall, college, arsenal, and hospital, are its principal public edifices. In one of its squares is a column erected to the astronomer Lambert, born here in 1728. The new town, which extends on the S.E. as far as the canal uniting the Rhine and Rhone, is handsomely laid out, and has numerous handsome residences, with the hall of the Society of Industry, the exchange, the chamber of commerce, &c.; it has also a capacious basin on the canal.

Until about the middle of last century, Mulhausen had only a manufacture of woollen cloths; but in 1745 cotton printing was introduced, and it is now one of the principal seats of the cotton manufacture in France. The calico prints and muslins of Mulhausen and its neighbourhood are second only, as respects the perfection and variety of their patterns, to the silk goods of Lyons. (Bowring's Rep. on Switzerland, p. 34.) The manufacturers have, in many instances, branch establishments in other parts of Haut Rhin, and in the neighbouring departments; but Dr. Bowring states that many of their mills and factories are mortgaged to the inhabitants of Basle; and, in fact, Switzerland justly claims considerable capital to the manufactures of Alsace. (See art. RHIN-HAUT in this work.) Hugo says that about one fifth part of its cotton goods are sent out of France. The work-people are badly clothed, dirty, and lodge generally in cellars, or other comfortless dwellings; but of late efforts have been made, by some of the more wealthy manufacturers, to improve the lodgings of the work people in their employ. Wages are good: cotton printers of the first class get from 2½ fr. to 3 fr. 30 c. a day; inferior workmen, from 1 fr. to 1½ fr.; women, an average of 1 fr. 30 c., and children, from 25 to 33 c. a day. Weavers (men and women) and male cotton spinners get from 1½ fr. to 3 fr. a day; females engaged in cotton spinning, from 75 c. to 1 fr 10 c.

Intemperance is not so prevalent here as among the cotton weavers of the department du Nord; and, according to Villermé, the proportion of illegitimate children is but little above the general average of France.

The spinning-mills at Mulhausen are not in a flourishing condition, owing, in part, to their being obliged to use cotton imported by way of Havre or Marseilles. In point of fact, however, they are totally unable to withstand the competition of the English; and since Manchester and Glasgow cotton twist has been admitted at a slight duty, as low as No. 170, they have ceased to spin any high numbers. Woollen cloths, hosiery, straw hats, morocco leather, and beer, are the other principal goods made at Mulhausen, which has also a brisk trade in iron, hardware, and agricultural produce. Before the revolution, this town was the capital of a small republic allied to Switzerland. It was annexed to France in 1798, and has rapidly increased since 1820. (Hugo, art. Haut Rhin; Villermé, Tableau des Ouvriers, i., 14-60; Bowring's and Symon's Reports, &c.)

MULL. See HEBRIDES.

MULLINGAR, an inland town of Ireland, prov. Leinster, co. Westmeath, of which it is the cap., on the Brosna, 44 m. W. by N. Dublin. Pop. in 1821, 3684; in 1831 4245. It consists of one principal street, running E. and W., with several diverging streets and lanes. It has a parish church, a very large Roman Catholic chapel, a convent, a Presbyterian and a Methodist meeting-house, a large barrack, and the prison, courthouse, and infirmary for the county. Though not an incorporated borough, it sent two members to the Irish House of Commons till the union, when it was disfranchised. It has two schools, partly supported by the Educational board, one attended by about 250 boys, and the other by above 400 girls. Courts leet and baron, for small debts, are held every Thursday; and a court of record, with jurisdiction to the amount of £100. The assizes for the county are also held here; with general sessions in January, April, and July, and October; and petty sessions on Saturdays. It is a constabulary and revenue police station. The county prison, built on the radiating plan, contains 100 cells and 13 other rooms for prisoners. It is a large market-town for corn, butter, cattle, and other agricultural produce, having an easy communication-by the royal canal, which nearly encircles it, with Dublin on the one hand, and the Shannon on the other. It has two tanneries and a brewery. Markets on Thursdays; fairs, inferior only to those of Ballinasloe, for the sale of cattle, on April 6, July 4, August 29, and November 11, the last chiefly for horses. Post-office revenue, in 1830, £873; in 1836, £1024.

MUNCY. p. b., 85 m. N. Harrisburg, 195 W. Situated 1 m. E. of Susquehanna river, and contains four churches, a Presbyterian, Episcopal, Methodist, and Baptist; nine stores, one furnace, one tannery, two distilleries, one printing office, one weekly newspaper; two schools, 80 scholars. Pop. 662.

MUNCY CREEK, t., Lycoming co., Pa., 13 m. E. by N. Williamsport. Drained by Big and Little Muncy creek, which flows into the W. branch of Susquehanna river. It contains Muncy borough, exclusive of which it has two stores, two woollen factories, three grist-mills, three saw-mills, one tannery, one distillery; seven schools, 210 scholars. Pop. 1166.

MUNCYTOWN. p. v., capital of Delaware co., In., 58 m N.E. Indianapolis, 529 W. It is on the S. bank of White river, elevated 30 feet above the water of the river. It is laid out in the form of an oblong, having its four principal streets, 60 feet wide, crossed at right angles by other streets, 45 feet wide. It contains a courthouse, jail, three stores, a saw mill, with a fall in the river of 11 feet, and about 400 inhabitants.

MUNICH (Germ. München), a city of S. Germany, the cap. of Bavaria, highly interesting on account of its collections of the fine arts, on the Isar, crossed here by three bridges, about 220 m. W. Vienna, and 118 m. E.S.E. Stuttgard; lat. 4° 8′ 20″ N., long. 11° 34′ 30″ E. Pop., in 1840, 106,537, including military. The city stands in the midst of a plain, which is neither fertile nor picturesque, but is one of the most elevated in Europe, being nearly 1600 feet above the sea. In the last century it was only a second-rate fortified town, with castellated gates, and quaint, ancient-looking houses; but since the beginning of the present century, new quarters and suburbs have so far extended themselves beyond the walls, that the buildings now occupy nearly double the extent of the old town. "Munich," says the author of Germany and the Germans, "has kept pace even with Vienna in the march of modern improvement. This is everywhere visible; for we see new and splendid streets extending in all directions, fine palaces and public edifices, many of them magnificent, surrounded by extensive pleasure-grounds with fine walks and drives: in short, every object shows that it is flourishing beneath the sunshine of peace. Indeed, next to Berlin, Munich is the third city in the Germanic empire; for though Dresden, from its beautiful localities, is more captivating, yet this is the more striking: add to which, the one is dull and stationary, while the other is lively, attractive, and continually advancing in prosperity" (ii., 319). It has, however, an unfinished appearance, in consequence of the open spaces that intervene here and there between the numerous edifices, many of which are still incomplete. The old town comprises numerous streets, diverging from a common central square, called the Haupt-platz, and running towards the walls, which form round it a species of irregular circle. A large and broad street runs from N.W. to S.E., called in different parts, the Neuhausen Strasse, Kaufinger Strasse, and the Thal. The Sendlingen Strasse is another leading avenue, and two narrow lanes, one of which passes through Max.-Joseph's-platz, one of the finest squares in Europe, lead to a fine suburban line of streets. The Ludwig's Strasse, Karl's-platz, and Maximilian's-platz skirt the town on its W. and N. sides. The different public edifices that form the principal glory of Munich are chiefly on its N. side. Here, also, is the Carolinen-platz, in the centre of which is an obelisk, 100 ft. high, formed partly out of cannon taken by the Bavarians in the late war. An equestrian statue of the elector, Max. I., by Thorwaldsen, decorates the Wittelsbacher-platz. The cathedral, a large brick pile, erected at the close of the 15th century, has two tow-

ers, 333 ft. high, and a fine monument of the emperor Louis of Bavaria. The church of the Theatines (so called from Cariffa, bishop of Theate, the founder of the order) is a large structure in the Italian style, with a central dome flanked by two towers. The Jesuit's church, built at the end of the 16th century, is remarkable for its wide roof unsupported by pillars, as well as for two fine porticoes of marble, which form its grand entrance: it is 280 feet in length, and 14 feet wide; has 11 altars and a noble organ, with several monuments, one of which, by Thorwaldsen, to the memory of Prince Eugene Beauharnois, son-in-law of Napoleon, is one of the finest works of its kind. The church of St. Louis, in the Ludwig's-strasse, a brick building faced with marble, in the Byzantine Gothic style, has two towers, 220 ft. above the basement, and the nave is 250 ft. in length. The church of All Saints, recently finished, has some fine carvings and fresco paintings; but is much smaller, and with fewer pretensions to architectural beauty than those before mentioned. There are several other churches, and two have very lately been opened in the suburbs, both on a large scale, and tastefully ornamented with pictures and painted glass windows. A Protestant church, recently built, has a lofty open tower, and is celebrated for a remarkably fine-toned organ. The public cemetery lies outside the S. gate, and is of vast extent; open alike to Protestants and Catholics. Not far from it is the general hospital, a large building having accommodation for nearly 800 patients. The royal palace, or Königs-residenz, consists of an older part, built in the 16th century, and comprising four irregular courtyards, as well as a more modern part, called Neubau, planned on the model of the Pitti palace at Florence, and fitted up in the most sumptuous style, though not yet completed. The apartments already finished are in the style of those seen in Pompeii, and comprise numerous fresco and encaustic paintings, bas-reliefs, richly carved cornices, &c. At the back of the palace is the Hofgarten, a large, square, planted space, surrounded by arcades, with cafés, shops, &c., not unlike those of the Palais royal in Paris, the E. side being occupied by barracks. Connected with the Hofgarten eastward, is a kind of park, called the English garden. The new buildings of the palace face Max.-Joseph's square, on another side of which is the opera-house, opened about 14 years ago, one of the largest and most elegant theatres in Germany; it is fitted to hold about 2500 spectators, and is equalled, as respects its performances, only by those of London, Naples, and Milan. Opposite the palace, in the same square, is the new post-office, copied from that of Rome, and in the centre is a statue of the late king, Max.-Joseph I., by Rauch of Berlin.

The great glory of Munich, however, consists in its fine collections of paintings and sculpture, called respectively the Pinacotheca (from πίναξ, a picture, and θηκή, a repository), and the Glyptotheca (from γλῶπτος, a carving, and θηκή, a repository). The first of these, in the Bäer Strasse, is in the Palladian style, with two wings, and has a front 500 ft. in length: the public entrance is at the E. end, the corridor is adorned with allegorical frescoes in compartments; and the collection, which for specimens of the Flemish and Dutch schools, is one of the finest in the world, occupies seven splendid halls and 23 adjoining cabinets on the first floor, the basement story being devoted to the reception of drawings, enamels, mosaics, &c. The collection of drawings comprises about 9000, including five by Raphael, 30 by Fra Bartolomeo, and several by Rembrandt, Albert Durer, and other Dutch and German artists. The paintings are limited to 1500, and consist of the chefs-d'œuvre from the king's collections, including the galleries of Dusseldorf, Mannheim, Deux-ponts, Schleisheim, and other galleries. Two of the apartments are devoted to the German school, and include the élite of the Boisserée gallery, purchased in 1827 for 375,000 florins, comprising specimens by Albert Durer, J. Von Eyck, Schoreel, Hans Hemling, &c. Three of the rooms contain pictures of the Flemish and Dutch masters, the principal of which are the "Murder of the Innocents," "Fall of the Damned," and other splendid works of Rubens; the "Village Fête" of Teniers; the "Musical Party," by Netcher; the "Girl with the Pitcher" and "The Mountebank," by Gerard Dow; "The Wise and Foolish Virgins," by Schalken, besides numerous highly-coloured works of Wanderwerf and Rembrandt, with various portraits of Vandyck. The specimens of the Italian school, comprised in two apartments, bear no comparison with the invaluable pictures just mentioned; but there are a few fine works by Raphael, Guido, Titian, Domenichino, Annibal Carraci, and Carlo Marratti. Belonging to the Spanish school, also, may be noticed "The Beggars" of Murillo, several works of Espagnoletto, and some portraits by Velasquez. There are also a few paintings of the English school, and among them is the well-known "Reading of the Will," by Sir D. Wilkie. The Glyptothera in the Königs platz is a chaste and elegant structure, in the Ionic style, erected like the last by the

MUNICH.

Baron Von Klenze, and has a noble central portico, the sides being adorned with statues in niches. The collection is distributed in 12 rooms, each of which is devoted to a distinct epoch in the art, and decorated in accordance with its contents. The walls are of scagliola-work, the floors of marble, and the ceilings richly adorned in fresco and stucco work. The marbles from the temple of Jupiter Panhellenius, in Egina, purchased by the present king for 10,000 sequins, occupy an entire room, and are particularly valuable, from their being the only extant specimens of the Eginetan school of statuary. The Ilioneus, however, is said to be "the gem of the collection," and one of the finest existing specimens of ancient art. The Roman hall far surpasses the rest in the splendour of its decorations; but the works that it contains are said to belong rather to the declining stage of the arts. The hall of modern sculpture has, among other works, the Paris and Venus of Canova, copied from that at Florence; the Adonis of Thorwaldsen, and a bust of the king by the same artist.

The Leuchtenberg gallery, formed by the late Prince Eugene Beauharnois, comprises a choice, though not very extensive collection, including, among other chefs d'œuvre, Murillo's famous Virgin and Child, with several cabinet pictures, by Raphael, Vandyck, Rembrandt, and Velasquez, with numerous works of modern French artists, and a few sculptures by Canova. The present king of Bavaria is certainly a liberal, and perhaps, also, a judicious patron of art; and nowhere is the modern German school of painting to be seen to greater advantage than at Munich. Several artists are kept in the king's employ, and an Academy of Arts has a triennial exhibition, supported by government, with salaried professors and pensioned students. This exhibition is encouraged and, in part, supported by a society which devotes annually about 8000 florins to the purchase of modern pictures. Munich was the birthplace of Sennefelder, the inventor of lithography; and it has many eminent professors of that art, who have transferred to stone some of the most celebrated works of the Pinacotheca and Glyptotheca.

The university of Munich, originally founded at Ingoldstadt in 1472, and removed thither in 1826, is the principal school of learning in Bavaria. It comprises 20 professors of four different faculties, with 1300 students, almost exclusively Bavarians, besides a library of about 160,000 volumes. Philological and theological seminaries, as well as two gymnasiums, are attached to the university; and the town has polytechnic, central, and subscription schools. But however high the celebrity of Munich, as respects music and the fine arts, the censorship of the press is fatal to the progress of literature, and to all the higher branches of philosophy. The royal library, lately removed to a noble building of great length, and three stories in height, is equalled only by that of Paris, the best authorities estimating its contents at 540,000 printed books, and 16,000 MSS. The reading-room is open for five hours during three days of the week; but the books may not be withdrawn from the building. The collection of engravings amounts to 300,000; and there are about 10,000 Greek and Roman coins. The museum of natural history is small and poor, containing but few specimens of foreign plants or animals; and the Brazilian collection, made by Spix and Martius during their travels in South America, though originally good and well selected, has been so much neglected as to be unworthy of notice. Munich has no very important manufactures, but comprises establishments for bronze-casting, iron-works, sugar-refineries, silk-throwing mills, and tobacco manufactories. Its telescopes are highly celebrated, and its porcelain is exported, like that of Dresden, to different parts of Europe. The last branch of industry is under royal patronage, and is carried on in a large establishment at Nymphenberg, about 3 m. distant, where also is a handsome palace of the king, with parks, menageries, &c., completed at the end of the 17th century.

Munich owes its present distinguished position, as the Athens of S. Germany, principally to the patronage and encouragement of the reigning monarch. It is very doubtful, however, whether he deserves any considerable portion of the praise that has been lavished upon him on this account. On the contrary, those who are best acquainted with Bavaria, affirm that the embellishment of the capital has been effected at the expense and injury of the rest of the kingdom; and that the vast sums lavished on buildings and pictures would have been far better expended on the improvement of roads, and such like public works.

The immediate environs of Munich abound in taverns and gardens, which are the favourite resort of the middle classes. Beer is the favourite beverage, and waltzes are danced for six or eight hours, without intermission. The beer-houses are exceedingly numerous, and beer is drunk in immense quantities. Some of the breweries are upon a very large scale. (See BAVARIA, i., 307.)

The inhabitants are likewise fond of good cheer in other

MURCIA.

respects, eating and drinking constituting with them the chief business of life. The morals of the inhabitants are alleged to be at a very low ebb; and, according to the government returns of 1634, there were in that year 1991 natural, and only 1330 legitimate children. These returns, however, are probably but little to be depended on, and the probability is that Munich is nearly, in this respect, on a level with the other capitals of Germany, where large garrisons are kept. Seven newspapers and magazines are published in Munich, some daily, one or two weekly, and the rest at longer intervals; besides which, several literary clubs and reading-rooms are established in the city. It has also a yearly festival (Volks-fest) in the early part of October, established by the present king for the purpose of encouraging agriculture, frequented by the farmers and peasants, who bring with them the finest specimens of cattle in competition for prizes offered by government. Poney races and shooting matches take place at this fair; and a high sloping bank, running along the meadow in which it is held, is cut into steps, like a Roman amphitheatre, commanding an excellent view of the whole scene.

Munich was founded by Henry, Duke of Saxony and Bavaria, in 962, on a site belonging to the monks of Schedlar, from whom it takes its name. Otho IV. encircled it with walls in 1157, and in 1632 it surrounded to the Swedes and German Protestants, under Gustavus Adolphus. In the war of 1704, between the Austrian and Bavarians, it fell into the hands of the former, after the battle of Blenheim; and it shared also the vicissitudes of the war of 1740, when the elector made his unsuccessful attempt to attain the imperial crown. In 1796 the French army, under Moreau, approached Munich, and obliged the elector to make a separate treaty. The French again occupied Bavaria in 1800, and from the battle of Hohenlinden till 1813 the country remained in alliance with France. (Murray's Handbook for South Germany, p. 27-49; Strang's Germany, ii., 339-383; Germany and the Germans, ii., 315-328; Berghaus; Schaden, Wegweiser durch München.)

MU'NNEEPOOR, an indep. state of India-beyond-the-Brahmaputra. (See CASSAY.) The small town of the same name, cap. Cassay, is in a fertile valley, about 408 m. N.E. Calcutta: lat. 24° 20' N., long. 94° 30' E.

MUNSON, p. t., Geauga co., O., 166 m. N.E. Columbus, 340 W. It contains five grist-mills, seven saw-mills; seven schools, 301 scholars. Pop. 1983.

MUNSTER, a city of the Prussian states, cap. prov. Westphalia, and of a reg. and circ. of the same name, on the Aa, a tributary of the Ems. Pop., 1838, 19,783. It is pretty well built, is the seat of a R. C. archbishop, of the government, and of the tribunal of appeal for the province. It has a seminary or college for the instruction of Catholics in theology, a gymnasium, a veterinary school, a botanical garden, and a public library. Principal public buildings, the cathedral, the church of St. Lambert, and the episcopal palace, now occupied by the governor. Münster has manufactures of woollen stuffs and starch, with tanneries and breweries; and a considerable commerce in the products of these, and linen fabrics, hams, and other Westphalian produce. It is united by canals with the Ems, and also with the Vechte, flowing into the Zuyder Zee. The treaty of Westphalia was signed in the town-house, in 1648. The famous fanatic Bocold, surnamed John of Leyden, the leader of the Anabaptists, made himself master of this place in 1534; but the town being subsequently taken by the bishop, John of Leyden and two of his accomplices were put to death, after being confined for a while in iron cages, still preserved in the church of St. Lambert. (Berghaus, &c.)

MU'NSTER, one of the four great provinces into which Ireland is divided, comprising the S.W. portion of the island, and counties of Clare, Kerry, Limerick, Cork, Tipperary, and Waterford. (See IRELAND.)

MURCIA, a prov., and formerly a kingd. in the S.E. of Spain, between lat. 37° 20' and 39° 25' N., and long. 0° 46', and 3° 5' W., bounded N. and N.W. by Castile, E. by Valencia, S. by the Mediterranean, and W. by Andalusia. Greatest length, 140 m.; do. breadth, 120 m. Area, about 8000 sq. m. Pop., in 1833, 474,315. It is intersected by numerous ranges of mountains separated by extensive valleys formed by the Segura and its tributaries. The Sierras Segura and Pinoco skirt the country westward, and a chain of mountains runs northward from Carthagena, the highest point in the province being the Sierra España, which rises 5800 ft. above the sea. The mountains are chiefly of limestone, interspersed here and there with other formations. Lead and copper abound; but no mines are wrought. The climate on the sea-coast, and among the mountains, is temperate and delightful, but oppressively hot in the plains. The heat in summer occasionally rises above 108° Fahr. in the shade; and the winters are so mild, that frost is almost unknown. Rain seldom falls, and the sky is usually so clear and blue as to have caused Murcia to be called el cielo serenissimo. The soil, except on the banks of the Segura, is sandy, dry, and unproductive; above two thirds of the surface is incapable of cultivation, and only about one half is fit even for pasture: indeed Murcia is one of the most barren districts in Spain. The huerta, however, which lies close to the Segura, is extremely fertile, producing rich crops of wheat, barley, rye, rice, maize, vegetables, and fruit, particularly oranges, lemons, pomegranates, and melons: mulberry and olive trees are found in great quantities, and evergreen oaks, as well as pines, clothe the sides of the mountains. Silk and oil are extensively produced, with smaller quantities of saffron and wine. The esparto rush grows luxuriantly in the neighbourhood of Carthagena, and with barilla, forms a considerable article of trade. The cattle of Murcia are not numerous, consisting principally of sheep and goats, with only a few horned cattle: the pigs, owing to the abundance of oaks, are almost equal to those of Valencia. Game is abundant, and the coast swarms with a variety of fishes. Wolves, foxes, and wild boar, inhabit the mountains. The manufactures are unimportant, being principally limited to the production of coarse linens, silk-stuffs, and earthenware, and soap. Carthagena, its only port, has a considerable export trade in cutlery, hemp, ribbons, wine, soda, barilla, and saffron; but the roads of the interior are so bad as almost to prevent intercourse. Three large fairs are held in September, at Murcia, Lorca, and Albacete. The inhabitants are proverbial, even in Spain, for pride, apathy, and indolence. Except at Carthagena, the principal inhabitants of which are of French, English, or Italian descent, all classes lead a dull monotonous life, spending their time either in eating and smoking, or else in total inactivity. Agriculture is pursued only from necessity, commerce languishes, and education and science are at the lowest ebb.

Murcia was that part of Spain first conquered by the Carthaginians, who founded Nova Carthago, anno 242 B.C. The country passed, with the rest of the peninsula, into the hands successively of the Romans, Goths, and Moors, the last of whom invaded it in the beginning of the 8th century. It formed a part of the caliphate of Cordova till 1144, when it was annexed to the kingdom of Granada, to which it belonged down to 1226, when it was taken by Alonso X. of Castile, and has since formed one of the provinces of Christian Spain. (Miñano; Mod. Trav.; Inglis, ii., 208–211.)

MURCIA, a city of Spain, cap. of the above prov., on the N. bank of the Segura (crossed here by a "magnificent" stone bridge of two arches), 31 m. N.N.W. Carthagena, and 220 m. S.E. Madrid, lat. 38° N. and long. 1° 14′ W.; pop., according to Miñano, 35,390. It is situated in a vale, which, for beauty and fertility, equals any part of Spain. It was formerly fortified, but is now open on every side, and has narrow, though clean streets, lined with mean houses, sometimes ornamented with grotesque carved-work. Gardens often skirt the streets, as in Seville, and the walls in many parts are overtopped by the heavily laden orange trees and branching palms. There are four considerable squares, the largest of which is used for a bull-ring; the principal public buildings are the cathedral, 11 parish churches, the bishop's palace, five colleges, a town-hall, custom-house, and hospital. The cathedral is of mixed architecture, with a Corinthian portico and Gothic dome. It formerly possessed great riches in plate and jewelry; but these were abstracted during the late war, and it has now only a few pictures. The chief object of attraction at present is its tower, 250 ft. high, which, like that of Seville, may be ascended by a spiral walk or inclined plane, accessible even to horsemen. In the Plaza real is a fine marble column, formerly surrounded by a statue of Ferdinand VI.; and there are four public walks, one of which is formed by a mole or quay skirting the river. The botanic garden is small and ill-arranged. The silk manufacture of Murcia, which once employed some thousand hands, now requires only 400. The silk is prepared by hand labour, and cannot therefore come into competition with that of Valencia, which is for the most part produced by machinery. Considerable quantities of coarse cloth are made for the supply of the poor; and there is a manufactory of saltpetre farmed by government to a company, which makes about 1200 arrobas yearly—only one tenth of the quantity produced at the close of the last century. About 3 m. from the city is a gunpowder-mill, bound to furnish government with 60,000 arrobas a year; but the quantity produced seldom exceeds 32,000 arrobas a year. Most of the inhabitants are supported by agriculture: the land in the vale of Murcia produces two crops a year—wheat and lentils, wheat and maize, or wheat and beans—and may be estimated to return about five per cent. to a purchaser. (Inglis, ii., 214.) Provisions, owing to a heavy market-duty, are somewhat dearer than at Malaga and Seville. The price of labour is from four to five reals a day: female servants receive a dollar per month, and men from one and a half to two dollars.

Murcia is very ill provided with accommodation for trav-

ellers. Miñano mentions sixteen posadas or inns; but they are little better than wretched pot-houses, kept by gitanos or gypsies, a race very thickly scattered over all the S. provinces of Spain, and following the trade of a butcher, tinker, or low innkeeper. The inhabitants of the capital are equally sluggish, gloomy, and reserved, with those in the rest of the province. The African character is more strongly marked in them than in other Spaniards; and the cast of countenance is, in general, very different from that of the Andalusian Moors. (Townsend, iii., 139–159; Cook's Sketches in Spain, i., 35.)

Murcia was little known before the invasion of the Moors, when it was besieged and taken, A.D. 714. It was subject to the caliphate of Cordova from 756 to 1144, when it was annexed to the new Moorish kingdom of Granada. In 1221 it again became subject to Cordova; and, on the dismemberment of that caliphate, was made the capital of a separate kingdom by Hubiel, from whom it was taken in 1266 by Alonso X. of Castile; since which time it has remained in the hands of the Christians.

MURFREESBORO', p. v., capital of Rutherford co., Tenn., situated on the E. side of the W. branch of Stone's r., on elevated ground, surrounded by a rich farming country. It contains a courthouse, jail, three churches, one of them an elegant brick, Presbyterian, an academy, 12 stores, two cotton factories, two cotton gins, one grist-mill, one carding machine, two tanneries, various mechanic shops, 200 dwellings, many of brick and neat, and 1500 inhabitants. It was founded in 1811, incorporated and made the capital of the state in 1817, and continued such until 1826, when the seat of the state government was removed to Nashville.

MURRAY, county, Ga., situated in the N.W. part of the state, and contains 650 sq. m. Drained by branches of Oostanaula r. It contained, in 1840, 5338 neat cattle, 1156 sheep, 12,289 swine; and produced 10,598 bushels of wheat, 174,760 of Indian corn, 10,654 of oats, 5450 of potatoes, 1187 pounds of tobacco, 9875 of cotton. It had nine stores, eight grist-mills, eight saw-mills, three tanneries, five distilleries; 18 schools, 318 scholars. Pop. : whites, 3696; slaves, 798 ; free coloured, 1 ; total, 4695. Capital, Spring Place.

MURRAY, pt., Orleans co., N.Y., 243 W. by N. Albany, 392 W. Bounded N. by Lake Ontario. Drained by Sandy cr. It contains 12 stores, two fulling-mills, one woollen factory, one furnace, three grist-mills, eight saw-mills; two academies, 70 students; 16 schools, 928 scholars. Pop. 2975.

MURVIEDRO (Muri veteres, but more an. Saguntum), a town of Spain, highly interesting on account of its Roman antiquities, prov. Valencia, on the Canales, about 3¼ m. from the Mediterranean, and 15 m. N.N.E. Valencia. Pop., according to Miñano, 6273. It stands at the foot of a mountain of black marble, and at the N.E. extremity of a large and well irrigated plain ; has long, tortuous, narrow streets, and is surrounded by walls flanked with small round towers. The houses in the interior have a mean and gloomy appearance ; but the suburbs are more airy and agreeable, and perfectly level. Two churches, three old convents, and a governor's palace, are its only public buildings. Murviedro formerly exported considerable quantities of brandy ; but its chief dependence, at present, is on the export of the oil, wine, wheat, barley, carobs, and fruit, grown in the adjacent district, sent coastwise to Valencia and other ports of the Mediterranean.

"Murviedro," says Mr. Swinburne, "seems to occupy the same ground as the ancient Roman city ; but, in all probability, the Saguntum of Hannibal was built on the summit of the hill. That the Romans, also, had a fortress on the top is clear, from the large stones and regular masonry on which the Saracens afterwards erected their castle. Half way up the rock are the ruins of the theatre, forming an exact semicircle, about 89 yards in diameter, from outside to outside: the length of the orchestra, or inner diameter, is 24 feet. The seats for the audience, the staircases, and passages of communication, the vomitoria, and the arched porticos, are still easily traced. The back part rests against the hill, and some of the galleries are cut out of the rock. Two walls, going off at an angle, serve to turn off the rain water that washes down from the cliff behind. As the spectators faced the N. and E., and were sheltered from the W. and S., nothing could be more agreeable in this climate than such a place of entertainment, open to every pleasant and salubrious breeze, and defended from all winds that might bring them heat or noxious vapours. It is computed that 9000 persons might be present, without inconvenience, at the exhibitions in this theatre." An attempt was made, at the close of the last century, to repair this noble structure ; and, in 1796, a Spanish comedy was represented within its walls ; but the plan was never carried into execution. The remains of a circus, also, are still discoverable in the orchards outside the town. It extended to a small river, the bed of which only remains, and which was the chord of the segment formed by the circus. When the Saguntines exhibited their mock sea-fights, called *naumachia*, this bed was undoubtedly filled from the neighbouring canals which still exist. A mosaic pavement, 24 ft. in length and 14 ft. wide, in a very perfect state of preservation, was discovered in 1755, at the entrance of the town ; and Ferdinand VI. ordered it to be enclosed, but his orders were not carried into effect, and it was, consequently, soon despoiled. Its fragments may still be seen in several houses of Murviedro. " Indeed," says M. Bourgoing, "the city is full of the remains of its antiquity : the walls of the houses, the city gates, and the doors of the churches and inns, are covered with Roman inscriptions."

The ground occupied by the convent of the Trinitarians was formerly the site of a temple dedicated to Diana. A part of the materials served to build the church, and the rest were sold to build *San Miguel de los Reyes*, near Valencia. The castle on the top of the hill presents some interesting remains of Moorish architecture ; the fortifications divide the hill into several courts, with double and triple walls, erected on huge masses of rock, laid in regular courses by the Romans. (*Swinburne ; Fischer's Valencia*, p. 144 ; *Mod. Trav.* i., 288.)

The prevalent opinion seems to be, that Saguntum was originally founded by colonists from Zacynthus, who were afterwards joined by Rutuli from Ardea. (*Strabo*, lib. iii. ; *Silius Italicus*, ii., 603.) It appears to have early attained to great wealth and distinction ; and being zealously attached to the Romans, it became an object of hostility to the Carthaginians. It was besieged by Hannibal previously to his invasion of Italy ; but the strength of the city, and the determined bravery of the inhabitants, baffled for nearly eight months all the efforts of this great general to effect its subjugation. At length, however, it fell into his hands, *anno* 219 B.C., the inhabitants being in part put to the sword and in part sold as slaves. They had previously thrown a great part of their wealth into the flames ; but the booty was still ample enough to enable Hannibal to reward the valour and devotion of his soldiers, and to facilitate his designs against Italy. (*Polyb.*, lib. iii. ; *Livy*, lib. xxxi., cap. 9.) Having been rebuilt by the Romans, it was afterwards famed for its porcelain, mentioned by Martial (xiv. epig. 108).

" Sume Saguntino pocula ficta luto."

MUSCAT, a city and seaport on the E. side of Arabia, prov. Oman, of which it is the cap., about 96 m. N.W. Cape Ras-el-had, lat. 33° 38' N., long. 58° 37' 30'' E. Pop., estimated by Fraser at from 10,000 to 12,000, of whom 1000 are Hindoos, from Sinde, &c., and the rest a mixed race, the descendants of Arabs, Persians, Kurds, Afghans, Belooches, settled here for the purposes of commerce. The town is situated at the S. extremity of a small cove, shaped like a horse-shoe ; and on either side hills, lined with forts, rise almost perpendicularly 300 ft. from the sea. It is built on a slope, rising gradually from the water, which nearly washes the bases of the houses. On this side it has no defence ; but the other sides are protected by a wall 14 ft. high, with a dry ditch. Its mosques, minarets, and white terraced houses give it an imposing aspect when seen from a distance ; but, on entering, narrow crowded streets, and filthy bazaars, wretched huts, paltry houses, and other tenements more than half fallen to decay, meet the eye in every direction. It has, however, some substantial and even handsome houses ; the palace of the Imaum, those belonging to his mother, the governor's, and several others being of the latter description : their form differs considerably from what is usually seen in the towns of Yemen and the Hedjaz, partaking more of the Persian than Arabian style of architecture. Muscat is supplied with water by means of a deep and strongly guarded well, from which a newly constructed aqueduct conveys it to tanks in the different quarters. During July and August it is excessively hot ; and the fevers then prevalent are especially fatal to Europeans. The country in its immediate vicinity is extremely barren ; but it improves as it recedes inwards. Dates and wheat are the principal articles of produce ; the former being held in high estimation, and largely exported, chiefly to India. A date tree is valued at from seven to ten dollars, and its annual produce from one to one and a half dollars. The value of estates is measured by the number of date trees comprised within the property.

Muscat is a place of considerable importance, being at once the key to, and commanding the trade of, the Persian gulf. The dominions of the imaum are very extensive, and his government is more liberal and intelligent than any other in Arabia or Persia. He has some large ships of war ; and his subjects have some of the finest trading vessels to be met with in the Indian seas. The part of Arabia near Muscat is too poor to have any very considerable direct trade ; but, owing to its favourable position, and the superiority of its ships and seamen, it has become an important entrepôt, and has an extensive transit and carrying trade. Most European ships bound for Bussorah and Bushire touch here ; and more than half the trade of the Persian gulf is carried on in ships belonging to its merchants. But exclusive of the ports on the gulf and the coast of Arabia, ships under the imaum's flag trade to all the ports of British India, to Singapore, Java, the Mauritius, E. Africa, &c.. The pearl trade of the Persian gulf is now, also wholly centered at Muscat. All merchandise passing up the gulf on Arab bottoms pays a duty of ½ per cent. to the imaum. He also rents the islands of Ormuz and Kishmee, the port of Gombroon, and some sulphur mines from the Persian government.

In the magazines of Muscat may be found every species of produce imported into, or exported from, the Persian gulf. Various articles are also imported for the use of the surrounding country, and for the internal consumption of Arabia. Among these, the principal are rice, sugar, coffee from Mocha, cotton and cotton cloth, cocoa nuts, wood for building, slaves from Zanguebar, dates from Bushire and Bussorah, &c. Payment for these is chiefly made in spices and pearls ; but they also export drugs of various descriptions, ivory, gums, hides, ostrich feathers, horses, a sort of earthen jars, called martaban, to Tranquebar, dried fish, an esteemed sweetmeat called *kulwak*, and a few other articles.

The markets of Muscat are abundantly supplied with all sorts of provision. Beef, mutton, and vegetables of good quality may be had at all times, and reasonably cheap. The bay literally swarms with the greatest variety of most excellent fish. Water is excellent, and is conveyed to the beach in such a manner that the casks of a vessel may be filled in her boats while afloat. Firewood is also abundant, and is cheaper than at Bombay.

Mohammedans pay a duty of 2½ per cent. on imports and exports ; and all other nations pay five per cent.

Niebuhr thinks, that Muscat occupies the site of the Mosca of Arrian and other Greek writers (*Voyage en Arabie*, ii. 71, ed. Amst. 1780) ; a conjecture which seems to be confirmed, not merely by the resemblance of the name, but also by the terms applied by Arrian to Mosca being sufficiently descriptive of Muscat ; and as the port is bounded on all sides by rocks, it must now present almost the same appearance as in antiquity. Dr. Vincent, however, though he speaks doubtfully on the subject, is inclined to place Mosca to the west of cape Rasselgate. (*Commerce and Navigation of the Ancients*, ii. 344–347. For further particulars, besides the authorities above referred to, see *Hamilton's New Account of the East Indies*, i. 63 ; *Fraser's Journey to Khorasan*, pp. 5–19 ; *Milburn's Orient. Com.*, &c. The longitude given above is that of *Arrowsmith's Chart of the Persian Gulf*.)

MUSCATINE, county, Iowa. Situated S.E. of the centre of the ter. and contains 440 sq. m. Bounded S.E. by Mississippi river. Watered by Red Cedar river and its tributaries, and Pine creek which affords water-power, and a good harbour for steamboats at its mouth. The Muscatine-slue, a bayou of the Mississippi, about 80 yards wide and four feet deep, with a gentle current, which leaves, and again enters Mississippi river forms a beautiful and fertile island. The county contained in 1840, 1547 neat cattle, 622 sheep, 4920 swine ; and produced 5639 bushels of wheat, 44,345 of Indian corn, 8860 of oats, 8647 of potatoes. It had 15 stores, three grist-mills, four saw-mills ; one pottery ; four schools. 68 scholars. Pop. 1942. Capital Bloomington.

MUSCOGEE, county, Ga. Situated in the W. part of the state, and contains 410 sq. m. Bounded W. by Chattahoochee river, and drained by Upetoy creek and other branches of it. It contained in 1840, 9946 neat cattle, 1807 sheep, 19,828 swine ; and produced 13,386 bushels of wheat, 251,490 of Indian corn, 6665 of oats, 7981 of potatoes, 1,185,139 pounds of cotton. It had six commission houses in foreign trade, 106 retail stores, one woollen factory, one cotton factory with 1600 spindles, 18 grist-mills, 18 saw-mills, three printing-offices, one bindery, three weekly newspapers, and one periodical ; three academies, 60 students, 12 schools, 304 scholars. Pop.: whites, 6930 ; slaves, 4761 ; free coloured. 59 ; total, 11,699. Capital, Columbus.

MUSKINGUM r., O., is one of the largest rivers that run wholly in the state, and is formed by the union at Coshocton, of Tuscarawas river, and Walhonding, or White Woman's river. The former rises in Medina county, near the Cuyahoga river, between which and Tuscarawas river was formerly a portage, which gave name to Portage township and county, and was early used to connect lake Erie with Ohio river, but a better communication is now formed by the Ohio canal. The Muskingum is navigable for large batteaux 100 m. to Coshocton. It is obstructed by falls at Zanesville, overcome by a dam and locks, which afford great water-power. Above Zanesville, a slack water navigation is completed to Dresden, where a side cut, 2½ m. long, connects the river with the Ohio canal. The Muskingum flows with a limpid current, over a pebbly bottom, variegated with thousands of red and white shells. Below Zanesville it has on its borders inexhaustible beds of bituminous coal, and rich deposits of salt. It enters the Ohio at Marietta, by a mouth 225 yards wide.

MUSKINGUM, county, O. Situated toward the S. E. part of the state, and contains 665 sq. m. Drained by Muskingum river and its tributaries. There are about 30 salt works in the county. The brine is obtained by boring in a whitish sandstone, from 350 to 700 feet. Coal is found in every town, but particularly near Muskingum river. Pipe clay and burrh-stone, or cellular quartz, suitable for mill stones, are found. There are numerous ancient mounds. It contained in 1840, 20,994 neat cattle, 37,173 sheep, 51,859 swine; and produced 336,929 bushels of wheat, 3599 of rye, 623,577 of Indian corn, 24,979 of buckwheat, 2098 of barley, 196,408 of oats, 113,149 of potatoes, 38,529 pounds of tobacco, 21,694 of sugar. It had 50 commission houses in foreign trade, 66 retail stores, three fulling-mills, two woollen factories, 22 flouring-mills, 27 grist mills, 72 saw-mills, one paper-mill, eight tanneries, nine distilleries, five breweries, six printing-offices, one bindery, two weekly newspapers, one college, 40 students; four academies, 218 students; 148 schools, 5925 scholars. Pop. 38,749. Capital, Zanesville.

MUSKINGUM, p. t., Muskingum co., O., 61 m. E. Columbus, 346 W. It has three churches, a salt factory; seven schools, 228 scholars. Pop. 1222.

MUSSELBURGH, a parl. bor. and market and seaport town of Scotland, county Mid Lothian, on level ground, at the mouth of the Esk, in the frith of Forth, 5 m. E. Edinburgh. Population in 1831, including Fisher-row, and other environs, 5961. It is well built: the main street runs nearly E. and W., with a slight curvature, parallel to the bay; and it has a great many subsidiary streets. It has numerous villas; which is accounted for by its vicinity to Edinburgh, and its salubrity. Fisher-row, which contains many modern buildings, and new streets, is separated from Musselburgh, properly so called, by the Esk, the communication between them being kept up by means of one wooden and two stone bridges, one of the latter being old and little used. The other stone bridge is a modern and handsome structure, after a design by Rennie. The only public buildings are a jail, an ancient edifice surmounted by a spire, and the parish church of Inveresk (in which parish the town is situated). The latter, on an abrupt eminence ½ m. from the town, is a conspicuous object in every direction. On the shore immediately adjoining the town on the E., are extensive downs or links, used since 1817 for the Edinburgh races, and for the game of golf, which is much practised here.

Musselburgh has two flax-mills, which employ about 790 hands; and about 200 weavers of sail-cloth and other fabrics. The manufacture of hats is, also, carried on to a limited extent; and there are brick-works, a pottery, extensive breweries and distilleries, tanneries, and flour-mills. Fisher-row, along with Newhaven, in the parish of North Leith, virtually monopolizes the supply of Edinburgh with haddocks and other white fish. A branch of the Edinburgh and Dalkeith railroad has its depot at Fisher-row. The country all round the borough abounds with coal, which, by means of the railroad, is conveyed to Edinburgh, and by another branch of the same railway is taken to Leith, for consumption and export. There are two banks in the town. The harbour dries at low water. The exports are coal, spirits, ale, and farm produce.

The schools are numerous and efficient. Gilbert Stuart, author of a history of Scotland and other works, resided at Musselburgh; and New Hailes, the residence of lord Hailes, the annalist and antiquary, is within ½ m. of the town. Poor rates have been introduced: average annual assessment, £1100. In addition to the parish church, it has a quoad sacra church, and chapels belonging to the Episcopalians, Associate Synod, Relief, and Independents.

The chapel of Loretto, to the E. of the town, was, before the reformation, a place of great importance; pilgrimages were often made to it; and, in 1530, James V. performed a pilgrimage thither on foot. About ½ m. S. of Pinkie-house was fought on the 10th Sept. 1547, the battle of Pinkie, in which the English, under the duke of Somerset, totally defeated the Scotch. The battle of Preston Pans, on 21st Sept. 1745, between the forces of Charles Edward and the royal army, took place in this neighbourhood, when the latter were completely defeated.

Musselburgh had no parl. rep. till the passing of the Reform Act. It now unites with Leith and Portobello in sending one member to the House of Commons. Registered voters in 1840–41, 297. Municipal income, £2029. Number of councillors, 12. (Beauties of Scotland. § Edinburghshire; Boundary Reports; Chalmer's Caled.; Private Information, &c.)

MYSORE (Maheshawura), a province of S. Hindostan, forming a state subsidiary to the British, between lat. 11° 20′ and 15° N., and long. 74° 45′ and 78° 40′ E., almost entirely surrounded by the territory of the Madras presidency. Its shape is nearly rhomboidal; greatest length, N.

to S., 240 m. by an extreme breadth of about the same. Area, estimated at 24,750 sq. m.; and population at about 2,500,000. The whole country consists of a table land enclosed on the E., W., and S. by the Ghauts; and varying from 1900 to 4600 feet in elevation above the sea, with a gentle slope towards the N. The Zoongabuddra, Pennar, Colair, Colaroon, &c., all rise within this province, which has, however, no river of much size. The climate is one of the most salubrious within the tropics; the air is temperate and bracing, and the deluging rains, which prevail on either side beneath the Ghauts, are here unknown. The soil, which is mostly of the red and black varieties common in the Deccan, is continually watered by refreshing showers; and produces not only most of the grains and vegetables of other parts of India, but also many of the fruits of Europe. Extensive tracts are overrun with jungle, especially of the date palm, and from the remains of hedges, &c., the province appears at a remote period to have been in a much higher state of cultivation than at present, though it has recovered to a great extent, from its previous devastations, since the period of peace, commencing with the present century. Rice, sesamum, sugar, coffee, betel leaf, castor oil, and cocoa nuts, are the principal articles of produce. Though an inland country, the cocoa palm is almost everywhere abundant, great quantities of salt and soda efflorescing on the soil. About Colar, the poppy is raised, both for making opium and for its seed. Potatoes are grown, and exported to Madras and elsewhere. Tobacco is of inferior quality, and is not much cultivated. From the great imperfection of agricultural implements, and the inferiority of the cattle, the fields are very imperfectly ploughed; but the soil is, in many parts, extremely productive, with the aid of little labour. The cottages of the peasantry are, on the whole, neater and more commodious than in most parts of India. They are almost universally constructed of the red soil of the country, and roofed with tiles; nor are the best habitations of different materials, or otherwise distinguished from the rest than by their size, and from being whitewashed. The inhabitants are nearly all Hindoos.

The Mysore dominions are subdivided into three great districts; chief towns, Mysore, the capital, Bednore, and Chittledroog. The government is nominally in the hands of a native prince, but actually vested in the British resident at Mysore, appointed under the Madras presidency. The subsidiary armed force furnished by Mysore to the British government is undefined, but may be estimated at 4000 men. The present dynasty of the Mysore sovereigns are supposed to have originally emigrated to this province from Gujrat; but no authentic records of them exist previously to the 16th century. From 1760 to 1799, Mysore was governed by Hyder Ali and Tippoo Saib. After a protracted contest with the English, Tippoo lost his crown and life at the taking of Seringapatam. (Parl. Reports; Hamilton, &c.)

MYSORE, a town of S. Hindostan, the capital of the above state, on a lofty hill, 9 m. S. by W. Seringapatam, lat. 12° 19′ N., long. 76° 42′ E. It was suffered to fall into decay by Hyder Ali and Tippoo, but, under the present dynasty, it has been rebuilt, and restored to its ancient importance. It is enclosed by an earthen wall, and consists of the town (pettah) and fort. The latter, which is an extensive work in imitation of a European fortress, is separated from the pettah by an esplanade, and comprises, beside the rajah's palace, the dwellings of the principal merchants and bankers. The architecture of the town is similar to that of Seringapatam, but the houses are larger and better; they are ranged in regular streets, whitened, and intermingled with trees and temples. S. of the fort is a large and good suburb; and on rising ground, near the town, is a British residency. Mysore is well supplied with provisions, and is considered much more healthy than Seringapatam. (Hamilton; E. I. Gazetteer.)

MYTILENE, the ancient Lesbos (insula nobilis et amœna, Tacit. Hist., lib. vi., cap. 3), an island of Turkey in Asia, in the Ægean sea, opposite the coast of Asia Minor, to the north of the entrance to the gulf of Smyrna. It is about 33 m. in length from E. to W., by about 26 in breadth. The strait by which it is separated from the main land varies in breadth from seven to 10 m. Though in part hilly and mountainous, it has notwithstanding a considerable extent of level and very fertile land; and, except in a few places infested with malaria, it is extremely salubrious. The principal products are oil, corn, wine, figs, and other fruits; cotton, timber and pitch, silk, honey, &c. The wines of Lesbos were among the most celebrated of the ancient world. They are said by Athenæus (l. 22) to have deserved the name of ambrosia, rather than of wine, and to have been like nectar when old!

"Hic innocentis pocula Lesbii
Duces sub umbra."—Hor. Od. i. 17.

The wine of the island still continues to preserve some,

though but a slender, portion of its ancient reputation; very little, however, is exported. The figs are excellent, and large quantities of oil of medium quality are annually shipped for Constantinople and other places. The produce of corn is insufficient for the supply of the island. The timber and pitch are derived from the pine forests, with which the mountains are covered. The town of Castro, on the site of the ancient *Mytilene*, stands on the E. coast of the island, on the strait separating it from Anatolia. It contains many fragments of pillars, capitols, friezes, &c., but no considerable ancient ruin: it may have from 5000 to 6000 inhabitants; and has two harbours, but neither is good. The island can, however, boast of two of the finest harbours in the world, Port Jero, or Olivier, and Port Caloni. The former, in its S.E. angle, has a narrow entrance, but the water is deep, and within it expands into a noble basin, capable of containing the largest fleets. Port Caloni, on the S. side of the island, is a basin, similar to the last mentioned, but of more ample dimensions, nearly, in fact, intersecting the island. It has deep water throughout; but the entrance to it being very narrow, it is but little frequented.

Olivier estimated the entire population of the island at about 40,000, half Greeks and half Turks, with a few Jews; but later estimates considerably reduce the number of Turks. There can be no doubt that under an enlightened government Mytilene would speedily recover some portion of her ancient prosperity. Olivier mentions that the singular usage obtains in this island of the eldest daughter succeeding to the paternal property, to the exclusion of her brothers and younger sisters! (*Voyage dans L'Empire Ottoman*, ii., 99.) Most probably the custom has descended from a remote æra; but some modifications have, however, been introduced in modern times in favour of the younger sisters.

Lesbos was one of the most celebrated of the Greek islands. It had several cities, of which Methymna and Mytilene were the most celebrated. The latter was distinguished alike by the magnificence of its buildings, the amenity of its climate, its proficiency in the *belles lettres* and philosophy, the number of its great men, and the luxury and refinement of the inhabitants. Epicurus is said to have read lectures in Mytilene; and Aristotle resided in it for two years to profit by the society and conversation of its learned men. At a later period it became, like Rhodes, a favourite resort of those Romans who preferred quiet enjoyment to the turmoil and bustle of Rome.

Laudabunt alii claram Rhodon aut Mytilenem. *Hor.* Od. i. 17.

Among the illustrious persons who were natives of the city of Mytilene, or of other parts of the island, may be specified. Pittacus, one of the seven sages of Greece; Theophrastus, the scholar and successor of Aristotle; Alcæus, so famous for his odes; Sappho, celebrated alike for her beauty, her poetical talents, her loves and her death; Trepander, who added a seventh string to the lyre; Diophanes, a famous rhetorician, tutor to Tiberius Gracchus, &c. At the same time, however, it must be admitted that the morals of the bulk of the inhabitants were exceedingly corrupt, so much so, that it was common in antiquity to say of a debauchee, that he lived like a Lesbian. (*Cellarii Orbis Antiqui*, ii. 15; Tournefort, i. 36; Ancient Universal History, viii. 290, 8 vo. ed.)

Mytilene was taken and sacked by Julius Cæsar; but Pompey restored it to the full enjoyment of its privileges; and Trajan, who enriched it with several costly buildings, gave it the name of *Trajanopolis*, which, however, it did not retain. Molivo, on the N. coast of the island, is the modern representative of the ancient Methymna.

N.

NAAS, an inland town of Ireland, prov. Leinster, co. Kildare, 18 m. S.W. Dublin. Population, in 1831, 3098. "The town possesses considerable local advantages. Situated in a rich agricultural district, 18 miles from the metropolis, its communication with that city is facilitated by good roads and by means of a branch of the grand canal that enters the town. Its main street, also, presents some appearance of activity, owing to its forming the place of junction of the leading roads from Cork, Limerick, Kilkenny, Waterford, &c., to Dublin. However, far from keeping pace in improvement with the market towns in its vicinity possessed of none of these advantages, its prosperity has been on the decline for the last 15 years. The appearance of the cabins on the outskirts of the town is poor and miserable, many being ruinous." (*Municipal Bound. Report.*) The public edifices are the parish church, a Roman Catholic chapel, a meeting-house for Independents, military and police barracks, a market-house, a fever hospital and dis-

pensary, a courthouse, and a prison. The spring assizes for the county are held here, and the summer assizes at Athy. Its trade is grain, flour, and provisions, is not so considerable as might have been expected. Markets, especially for poultry, on Mondays and Thursdays. Postoffice revenue in 1830, £459; in 1836, £647.

NAKHITCHEVAN, a town of European Russia, on the Don, about 25 miles above where it falls into the sea of Azoff. Population above 10,000, principally Armenians. This and the contiguous town of Rostoff are, as it were, the entrepots of the Don. Except timber, most part of the produce brought down that great river is landed at one or other of these towns, and is thence forwarded by coasters for Taganrog. Nakhitchevan is built in the oriental style, and its inhabitants are distinguished by their commercial enterprise. "The connections they have formed with Astrakhan, Mordok, and Kisliar, also colonies of Armenia, almost annihilates the distance that is between them. They draw from these countries rice, silk, a vast quantity of wine, and about 500 casks of Kisliar brandy, in great esteem in Russia. They receive, moreover, from the Caucasus, all the rough produce of the country. By constantly frequenting the fairs which, in the adjacent towns and villages are very numerous, even to the distance of some hundred versts, the Armenians have formed the means of making themselves masters of the trade of the interior of the S. part of Russia." (*Hagemeister on the trade of the Black sea*, p. 36. Eng. trans.)

NAGPOOR (*nagapura*, "the town of serpents"), a large city of Hindostan, prov. Gundwanah, cap. of the dom. of the Rajah of Berar, between the Wynegunga and Wurdah rivers; lat. 21° 9' N., long. 79° 11' E. Population of the city and suburbs in 1825 estimated at 115,000. Its site is low and swampy, in the rains; and the principal streets with one exception, are narrow, mean, dirty, and intersected by watercourses. The great number of trees interspersed with the huts and houses give it, at a distance, the appearance of a large wood. It presents few good specimens of architecture; the rajah's palace, though an extensive building, has no pretensions to beauty, and has crowded round it a multitude of mean huts of mud and thatch. Some of the principal chiefs and bankers have large houses of brick and mortar, with flat roofs; but these, for the greater part, are old and dilapidated. In 1825, of 27,000 houses in Nagpoor, about 1300 were of mason-work, the rest being principally of mud, thatched or roofed with tiles.

The dom. of which Nagpoor is the cap., extends between the 18th and 23d degs. of N. lat., and long. 78° 30' and 83° E., having N. the Bengal presidency, E. Bengal and the presidency of Madras, and S.W. and W. the dominions of the Nizam. Area estimated, in round numbers, at 64,000 sq. m.; and population at somewhat less than 3,000,000. The general slope of this country is towards the S.E.; the surface is generally mountainous and woody, interspersed with occasional tracts of cultivated land. Principal rivers the Wurdah, Wynegunga, and other tributaries of the Godavery, and the Mahanuddy. The land is assessed on the village system. Wheat, jowaree, and rice; teak and saul timber; cotton, sugar, hemp, tobacco, arrow-root, betel leaf, wild silk, iron, and limestone, are the principal products; some of which are sent in considerable quantities to Bombay, in exchange for European manufactures. Nagpoor had always a large trade with Poonah, though this has very much diminished since the breaking up of the Mahratta empire; and some trade with Benares and Mirzapore in muslins, brocades, &c. The population of the country is nearly all Hindoo, or composed of wild Gonds, and other tribes; and very ignorant, the children of only Brahmins and the mercantile classes being educated. The government is better administered than that of the Nizam dominion. The revenue, which is estimated at between 45 and 47 lacs rups. a year, is collected under the superintendence of British officers. The rajah furnishes a contingent of at least 1000 men to the Anglo-Indian army. (*Hamilton's E. I. Gaz.*; *Evid. of Mr. Jenkins in Parl. Rep.*, 1832; *Report.* 1840. &c.)

NAIRN, a small marit. co. of Scotland, on the shore of the Moray frith, having N. the sea, E. Moray, and S. and W. Inverness: it also includes the detached district of Ferintosh, in the centre of Inverness. Area, 124,000 acres, of which about a third is supposed to be arable. It is in all respects similar to Moray, with which it is united under one sheriff. Along the shore it has a belt of low, flat, sandy soil, mostly suitable for the turnip culture but in parts barren, owing to the sand being dry and movable. The valley watered by the Nairn is generally fertile and well cultivated; but the rest of the county is mostly rugged and mountainous. The statements as to the size of estates and farms, houses, tillage, stock, &c., in the article MORAY, apply equally to this county. The average rent of land, in 1810, 1s. 10½d. an acre. Exclusive of the Nairn, it is watered by the Findhorn (see vol. i. p. 906), and some smaller streams. Except Nairn it has no town of any importance; and it has neither mines nor manufactures. The county which unites with Moray in sending 1 mem. to the House of Commons, had 190 registered voters in 1839–40: the borough of Nairn unites with Inverness, &c., in sending another member to the House of Commons. Nairn is divided into six parishes, and had, in 1831, 9374 inhabited houses, 2946 families, and 9254 inhabitants, of whom 4307 were males and 5047 females. Valued rent, £15,169 Scotch: annual value of real property, in 1815, £14,909.

NAIRN, a royal bor., seaport, and market-town of Scotland, cap. of the above co., near the mouth of the river of the same name, on the public road between Aberdeen and Inverness, 15½ m. N.E. the latter, and 72 m. N.W. by W. the former. Pop., in 1833, inc. the par., 2986. The river is here crossed by a substantial bridge, which, however, was greatly injured by the floods of Aug. 1829. "The principal street is tolerably spacious, but all the others are narrow and confined." (*Boundary Report.*) The only public buildings are the courthouse, jail, established church, two dissenting chapels, and a large inn, built by subscription. The harbour is accessible only for small vessels; and grain, cattle, timber, salmon, herring, and other white fish are exported to London and other markets. Indeed, the fisheries may be said to be the staple branch of industry of the town. The means of education are ample. The town has no fewer than three banks.

Nairn was created a royal borough, by William I., in the 12th century. In the vicinity is Cawdor castle, once a fortress of great strength, but now a ruin. It gave the title of "Thane" to Macbeth, and Shakespeare has made it the scene of the murder of the "gracious Duncan." It now gives the title of earl to a branch of the Campbells of Argyle. Lord Lovat found refuge in a corner of this fortalice, after the battle of Culloden, in 1746. (*Edinburgh Philos. Trans. R.; Beauties of Scotland, § Nairn.*)

NAMUR, (Flem. *Naemen,* Lat. *Namurcum,*) a strongly fortified town of Belgium, cap. prov. of same name, on the Meuse and Sambre at their junction, 33½ m. S.W. Liege, and the same distance S.E. Brussels; lat. 50° 28' 30" N, long. 5° 0' 7" E. Pop., in 1836, 23,176. It is surrounded with good walls, and has been considerably strengthened since the last war: it has strong outworks on both sides the Meuse and Sambre, and is further defended by a citadel, erected in 1817, on the elevated site of a former citadel, demolished by Joseph II. It is well built; many of the streets are broad and clean, and the houses are mostly of bluish stone, roofed with tiles. The cathedral, a modern edifice, with handsome Corinthian portico, and a dome, is principally remarkable for its containing the tomb of Don John of Austria, the conqueror of the Turks at the famous battle of Lepanto. The church of St. Loup, a richly ornamented building, and that of Notre Dame, with some good sculptures; the new townhall, several hospitals, and a theatre are the other principal public edifices. A bridge crosses each river, that over the Meuse having nine arches; a dam has here also been thrown across the Sambre in the view of raising its waters so as to render it navigable; but this design appears to have only partially succeeded. Namur is a bishop's see; the seat of tribunals of primary jurisdiction and commerce, with appeal to the superior courts at Liege, and the residence of the civil governor of the province, a military commandant, a provincial receiver of taxes, &c. It has an episcopal seminary; an athenæum with a library and cabinets of mineralogy, chemistry, &c.; various public and superior private schools, a society for the benefit of the poorer classes, a deaf and dumb and many other asylums, a *mont de piété,* founded in 1619, &c. The situation of Namur is favourable for commerce. Its cutlery is much esteemed on the continent, and it has tanneries, potteries, and brass and iron works; but its manufactures are less flourishing than formerly. The coal and iron mines, and marble quarries of its neighbourhood, are, at present, the principal sources of employment and wealth to its inhabitants. It has four annual fairs, one of which, beginning on July 2d., lasts 15 days.

Namur is supposed to occupy the site of the *Oppidum Aduaticorum,* mentioned by Cæsar. Like other cities in the low countries, it has frequently suffered from the ravages of war. In modern times it has sustained two memorable sieges, one in 1692, when it was taken by Louis XIV.; and the second in 1695, when it was retaken by the Anglo-Dutch under William III. The first is the subject of Boileau's famous ode *Sur la Prise de Namur.* (*Vandermaelen, Dict. de Namur; Dict. Géog.; Murray's Hand-book, &c.*)

NANCY, a town of France, dep. Meurthe, of which it is the cap., in a fine plain, near the Meurthe. 20 m. S. Metz, and 175 m. E. by S. Paris; lat. 48° 41' 55" N., long. 6° 10' 31" E. Pop., in 1836, ex. comm., 29,299. This is one of the handsomest towns of France. It was formerly surrounded with walls, but these were demolished under Louis XIII., and Louis XIV.; and the citadel is its only

existing fortification. It is, however, still entered by several gates, some of which have much beauty. Nancy is divided into the old and new town. The first is, in general, irregularly laid out, though it comprises several good streets and squares, many superior private residences, and most of the principal public edifices. Among the latter, are the remains of the palace of the dukes of Lorraine, now converted into a barrack for the gendarmerie; the church of the Cordeliers, a structure of the 15th century, in which are various interesting monuments; the church of St. Epore, and the ducal chapel. The royal court, the tribunal of commerce, and prefecture are in the *Place Carrière,* a square communicating with the *Place Royale,* in the new town, by a noble triumphal arch. The new town, which, however, is as old as 1603, is remarkable for the elegance and regularity of its streets, which mostly intersect each other at right angles. The *Place Royale* or *Stanislas,* is a square surrounded by edifices, all built on the same plan, comprising the town-hall, the bishop's palace, theatre, &c. Its angles are ornamented with iron gateways and fountains; and in its centre is a bronze statue of Stanislaus, king of Poland and duke of Lorraine, erected by voluntary subscription throughout the duchy, in 1803. Stanislaus, to whom numerous establishments in the town, both scientific and charitable, owe their foundation, is buried, as well as his consort, in the church of *Bon Secours,* which has two marble monuments to their memory. The cathedral is a modern edifice of Corinthian and composite architecture. The remaining principal buildings are the university, with a library of 23,000 vol., the royal college, seminary, civil and military hospitals, a work-house for the déps. Meurthe, and Vosges, and a house of correction. Nancy is a bishop's see, and the seat of a royal court for the dép. Meurthe, tribunals of original jurisdiction and commerce, a board of taxation, a chamber of manufactures, &c. It has an *académie universitaire,* a royal society of science and literature, a royal school of forest economy, a communal college, Protestant, Jewish, and other schools; manufactures of woollen cloth, hosiery, lace, muslins, cotton yarn, liqueurs, chemical products, &c.; with tanneries, dyeing-houses, and refineries of saltpetre for the gunpowder factory at Metz. Nancy is famous for its shot (*boules vulneraires d'acier*). No record exists of this town previously to the 11th century, but in the 13th, it became the capital of Lorraine. It was twice besieged by Charles the Bold, of Burgundy, who was killed under its walls in 1477. (*Huge, art. Meurthe; Guide du Voyageur en France, &c.*)

NANGASAKI, a large town and seaport of Japan, on the S.W. side the island of Kin-sin, and the only place in that empire accessible to Europeans, 600 m. W.S.W. Yedo; lat. 32° 43' 4" N., long. 130° 11' 47" E. Pop. from 60,000 to 70,000. It is situated on the slope of a hill, and, like every other Japanese town, is regularly built, with wide and clean streets. The houses, however, are low, none containing more than one good story, to which is added in some a sort of cockloft; in others, a low cellar; all are constructed of wood and a mixture of clay and chopped straw; but the walls are coated with a cement that gives them the appearance of stone. The height of the street-front, and even the number of the windows are determined by sumptuary laws. Oiled paper supplies the place of glass, and the windows are further protected from the weather by external wooden shutters and Venetian blinds: a verandah, into which the different rooms open, runs round the outside of the houses, to which are invariably attached curiously laid out gardens. Large detached fire-proof storerooms belong to each dwelling, and are so constructed as fully to answer their purpose of preserving the valuables of the inhabitants from the conflagrations so common here and elsewhere in Japan. The chief public buildings are the palaces of the governor and grandees of the empire, some of which cover a considerable extent of ground: there are also in the town and neighbourhood 61 temples, or *pagodas,* usually on commanding eminences, and enclosed in large gardens, the habitual resort of pleasure parties. These buildings are as plain and little ornamented as the private dwellings, and comprise, also, apartments, which are let out to travellers, or used for banqueting rooms, and other purposes. The tea-houses or bagnios are another favorite resort of the natives; and of these, according to Siebold, there are 750 in Nangasaki. The artificial island of Desima, to which the Dutch merchants are rigorously confined, is about 600 ft. in length by 240 ft. in breadth, a few yards from the shore, close to which stands the town, connected with it by a stone bridge, closed by a gate and guard-house, constantly occupied by soldiery. Neither Dutch nor Japanese may pass the gate without being searched; the number of European residents is limited to eleven; and the menial service is performed exclusively by Japanese; all of whom, except courtesans, are compelled to leave the island at sunset. From this imprisonment the Dutch are allowed to escape twice or thrice a year, rather to be exhibited to the great as a curiosity than out of indulgence. A corps of constables and interpreters (the latter of whom form a regular guild, receiving salaries from the *sjogûn.*) are appointed to watch over their minutest actions; and the most degrading servilities are exacted even from the *opperhoofd,* or president of the mission, by the meanest officers of the Japanese government. As respects trade, the Dutch are placed under restrictions elsewhere unparalleled; but these and other particulars have already been detailed in the general article JAPAN, to which the reader is referred. The harbour of Nangasaki extends N.E. and S.W. about 7 m., being in most places less than a mile in width. Ships lie in five or six fathoms water within gunshot of the town, and protected from all winds. (*Siebold,* i. ch. i. 2; *Crawfurd's Hist. of the Indian Archipelago,* iii. 305–308; *Manners and Customs of the Japanese,* 24–57.)

NANKIN, a city of China, in the district of Kiangning-foo, and prov. Keang-soo, near the S. bank of the Yang-tse-kiang, and about 110 m. from its mouth, lat. 32° 4′ N., long. 118° 24′ E. Pop. (acc. to Ellis) about 400,000. The walls, which are of limestone, cemented with sun-baked clay, enclose a very irregular triangular area of about 30 sq. m., and this circuit, as measured by the Jesuits, amounts to 57 *lis,* or nearly 20 m., a fact fully proving the absurdity of the Chinese statement that "if two horsemen should go out in the morning at the same gate and ride round in opposite directions, they would not meet before night!" This enclosure, moreover, comprises groves, fields, and even hills, of considerable extent; less than three fourths of it being covered by the town, which is situated at the S. extremity, and about 6 m. from the river bank. The city has declined much both in size and splendour since the end of the 13th century, when Kublai-Khan removed the imperial residence to Pekin. It now consists of four rather wide and parallel avenues, intersected by six or eight others of less width. The streets are not so broad as those of Pekin, but are, on the whole, handsome, clean, well-paved, and bordered with well-furnished shops. A palace of the emperor which once existed, and many other monuments of grandeur, have now almost disappeared, Nor are the palaces of the mandarins in any respect distinguished from those found in the capitals of other provinces of China; indeed, Nankin possesses no public edifices corresponding to its rank as the second city of the empire, except its famous porcelain tower, belonging to one of the pagodas, several temples, and its gates, some of which are of extreme beauty. The Porcelain tower (called *Paou-gan-sze*) "pagoda of gratitude," which is unquestionably finer than any similar structure throughout China, is an octagonal building, each side 15 ft. wide. It consists of nine equal stories, communicating by a spiral staircase running up the centre of the building, and each comprising one saloon finely painted, gilt, and adorned with idols. The outside wall is white, made of the white bricks commonly used in China: a kind of carved gallery or verandah, ornamented with lightly-tinkling bells, runs round each story, and the whole is surmounted by a gilt conical roof, the height of which from the base somewhat exceeds 200 ft. It was completed in 1432, at a cost of 400,000 taels. An observatory stands about a league N. ward of the pagoda, but though formerly well provided with instruments, &c., it is now almost in ruins. Nankin has extensive manufactures of satin and crape, the quality of the former, both plain and figured, not being equalled by that of any other city in China. The cotton fabric called Nankeen receives its name from this city; but in fact it is made in every part of the province, and scarcely a cottage can be found where the thrifty housewife has not a loom for weaving Nankeen. (*China opened,* i. 79.) The paper of Nankin is highly esteemed; and Indian-ink (as it is called in Europe) is manufactured in large quantities both in the town and neighbourhood, forming an important article of commerce. Nankin is celebrated also for its manufacture of artificial flowers from the pith of a shrub, and so extensive is this branch of industry as to give rise to a large trade. The commerce of the city is very considerable, owing to its position in the centre of the empire, and on the Yang-tse-kiang, which is navigable for small boats to the ports of Soo-cheo-foo and Shang-hae, its great entrepôts for corn, manufactured goods, and other articles. Its communication with Pekin is effected by the imperial canal, which leaves the river about 40 m. below Nankin: the principal traffic with the capital is during April and May, when fast boats, which accomplish the distance in about nine days, are constantly employed in exporting to the imperial court the produce of the Nankin fishery packed in ice. Nankin according to Du Halde, is not less celebrated for literature than commerce: the arts and sciences are studied there with great diligence, and it furnishes more doctors and mandarins than many towns together; its libraries are also extensive and valuable. The booksellers' shops are well provided with the best native publications, and the editions published here are the most esteemed in the empire.

Nankin, which oegan to decrease in the time of Kublai-khan, was further diminished by the removal of the six great tribunals to Pekin, which caused its name Nankin ("Court of the South") to be changed to Kian-nin in all the public acts: in common usage, however, it retains its old appellation. It is still the residence of one of the great viceroys called *Tsong-tuk,* to whose jurisdiction are committed all the judicial affairs, not only of this province but of that also of *Kiang-si* and *Gan-kwey.* The Manchoo-Tartars have here an extensive military dépôt under a general of their own nation, and their quarters are separated from the rest of the town by a lofty wall. (*Du Halde,* i. 149–151.; *Ellis's Journal of a Miss. to China,* 292–306; *Ritter's Erdkunde von Asien,* iii. 681–686; *Gutzlaff's China opened,* i. 76–79; *Private Information.*)

NANTES (an. *Namnetes,* or *Civitas Namnetum*), a large and celebrated commercial city and port of France, dép. Loire-Inférieure, of which it is the capital, on the Loire, where it is joined by the Erdre and Sevre-Nantaise, about 34 m. from its mouth, and 210 m. S.W. Paris; lat. 47° 13′ 6″ N., long. 1° 32′ 44″ W. Pop., in 1836 (ex. comm.), 75,139. "Nantes is a noble city, and its situation can scarcely be excelled. It stands upon the slopes and summit of a gentle hill, half encircled by the Loire, which is broad, clear, and tolerably rapid; and its beauty is greatly increased by several islets which dot the river exactly opposite to the town, and which are covered with pretty country-houses and gardens." (*Inglis,* iv. 338.) The banks of the Erdre too are very agreeable, abounding with chestnut woods, gardens, and country-houses The declivities of the neighbouring hills are in great part covered with vineyards, which add much to the beauty of the scene, though their produce is very inferior. Nantes is built mostly on the N. bank of the Loire, but partly, also, on the islands Feydeau and Gloriette, in which are some of the handsomest quarters. Both the N. bank and the islands are bordered by fine quays, one of which, *Quai de la Fosse,* full half a league in length, is broad, and shaded by fine elms, and bordered with hand-somely nied terraces and warehouses. The Quays *des Brasce* and *Port Maillard* are also planted with trees, being at once well frequented promenades, and the principal seats of commercial activity.

Nantes was formerly fortified, but its ramparts have been mostly demolished, and it is now an open town communicating with four considerable suburbs. Towards the E. end of the city are the *Cours de St. Pierre* and *St. André,* two public walks planted with trees and separated by the square of Louis XVI., in which is a statue of that monarch surmounting a Doric column about 80 ft. in height. Three *Cours* with the *Boulevard,* W. of the Erdré, another fine promenade of the same kind, are on a portion of the site formerly occupied by the fortifications. There are, however, some remains existing of various fortresses erected in the middle ages. In the E. part of the city, skirting the river is the large and imposing castle of the ancient dukes of Brittany, a mass of irregular buildings, surrounded by thick walls flanked with solid round towers. It was founded in the 10th, but it was not until the 15th century that it became a place of any great strength. It is now chiefly dismantled and is the residence of the military governor, and has a powder magazine. Between the Erdre and Loire are some remains of the Chateau de Bouffay, a structure also dating from the 10th century, consisting of some lofty walls surrounding a polygonal tower; and on the bank of the Loire are the ruins of the Tour de Pirmil, erected in 1326. The city is, in general, regularly laid out, and well built and paved. Most of its houses are of stone roofed with slate. There are between 30 and 40 squares, or rather open spaces; the principal of which, the *Place Royale,* is surrounded by handsome shops, and, together with the *quartiers Graslin* and *Feydeau,* may be compared with the best parts of the capital.

The different parts of the city communicate by numerous bridges, several of which are handsome, and one, the *Pont de Pirmil,* 277 yards in length, has 16 arches. The cathedral, though not imposing either without or within, has a fine ornamented with good though mutilated sculptures, and flanked with two towers, 170 feet high: in its interior is a magnificent marble tomb, erected by Anne of Brittany, in memory of her father Francis II., the last duke of that province. No other church demands particular notice. The finest building in Nantes is the Prefecture. It was erected between 1750 and 1777, and was formerly appropriated to the *Cour des Comptes.* It has two noble fronts of the Ionic order, a fine staircase, and several large halls and other good apartments: it is partly used as the depository of the departmental archives. The exchange is a large and convenient building, constructed chiefly within the present century: the theatre, in the *Place Graslin,* built in 1810, is, perhaps, the handsomest provincial theatre in France, after those of Bordeaux and Dijon. The town-hall was commenced in 1607, since which it has received several additions: it has lrse façades, ornamented with Corinthian pilasters, and over its principal front are sculptured figures, emblematical of the Loire and Sevre. The remaining public buildings include the mint, corn exchange, and linen hall; the *Hotel-Dieu*, on the isle Gloriette, erected in 1655, with 670 beds; the *Hospice de Sanitat*, or general infirmary and asylum, with 200 beds; the Hospital of Incurables; the Protestant church, formerly that of the Carmelites; mansion-house, hyper-house, the large prison, public slaughter house, barracks, college; the museum, with an excellent mineralogical collection; the *Salorges*, a general depôt for merchandise, &c. &c.

Nantes is a bishop's see, the capital of the 12th military division of the kingdom, the seat of a Lutheran consistory, tribunals of primary jurisdiction and commerce, a chamber of commerce, &c.; and is the residence of the consuls of many foreign powers. It has a royal college, an academical society, two episcopal seminaries, a public library with 50,000 printed volumes, and many valuable MSS., collections of engravings, paintings, and an observatory, and botanic garden, schools of navigation, medicine, drawing, &c. &c., maternity and Protestant Bible societies, a savings' bank, a *mont de piété*, and a maritime insurance office. The bank of Nantes has a capital of 900,000 francs, in 900 shares.

Ships of 200 tons, in the ordinary state of the river, reach the city quays without difficulty; but vessels of a larger burden have to load and unload in the roads of Paimbœuf, about 24 m. lower down the river. But notwithstanding this disadvantage, the Loire, opposite the city, is crowded with inland craft, and vessels of all nations, but principally from N. Germany, Sweden, Denmark, and Russia. Nantes, Brest, Pontivy, Redon, and other towns in Brittany, will directly communicate with each other on the completion of the canal from Nantes to Brest, now in progress, and which, when finished, will have an entire length of about 230 m.

The manufactures of this city are various, and on the increase. Coarse woollen-cloths and flannels, cambrics, printed cotton goods, handkerchiefs, tickings, and linens, are made on a large scale, besides which there are extensive biscuit-baking houses, chemical works, potteries, rope-walks, copper-foundries, manufactories of iron-cables, cannon, and other stores, with breweries, distilleries, sugar-houses, tanyards, vinegar establishments, and ship-yards for the construction of merchant-ships, corvettes, and smaller vessels. Nantes was formerly famous for her quick sailing vessels; but this is not the case at present, and more ships are now built at Bordeaux. At Indret, near Nantes, on an island towards the mouth of the Loire, the French government has lately founded an establishment for building steam-boats. In 1840, there were here five ships, on some of which, steamboats of from 160 to 220 tons each were in course of being built; the establishment employed altogether 400 workmen, and four marine engineers; and a sum of 200,000 francs was voted by the legislature for its maintenance. (*Ports, &c., of France*, p. 152.) Large naval warehouses are established at Nantes, from which Brest, Lorient, and Rochfort receive supplies both of provisions and ammunition. Previously to the revolution, the foreign trade of Nantes was much larger than at present; and during the time that the slave-trade was carried on, Nantes was more extensively engaged in it than any other French port. Now Marseilles and Havre both rank above her as commercial cities; but she is still the emporium of all the rich and extensive country traversed by the Loire, and has a considerable import and export trade, particularly with the French W. Indies, N. America, and the different ports of Europe. The exports comprise all sorts of French produce, but principally brandy, wine, and vinegar, silk, woollen, and sea goods, refined sugar, wheat, rye, ship-biscuits, &c. The chief imports are sugar, coffee, and other colonial products; cotton, indigo, timber, hemp, &c. Nantes is likewise considerable entrepôt for the commerce of the salt made in the department, chiefly at Noirmoutier and Croisic. (See *OIRE INFÉRIEURE*.) The customs' duties amount to about 5,000,000 francs a year.

Subjoined is a statement of the French and foreign ships that entered and cleared from the port in 1838, specifying the departments in which the French ships were engaged, and the number in each.

Branch of Trade.	Entered		Cleared	
	Ships	Tons	ps	Tons
French ships in foreign trade	301	24,319	137	14,517
" colonial do.	60	11,557	81	19,89
" coasting do.	4,003	133,190	4,303	80,157
" fisheries	18	2,475		
Foreign ships	156	20,183	139	22,703
Steam vessel - chiefly to and from Bordeaux	10	736	13	1,010
	4,463	201,330	4,072	151,343

* Of these, 46 vessels of 4615 tons left in ballast.

Of the vessels engaged in the fisheries, 19 came from the banks of Newfoundland, laden with 15,665 cwt. of codfish and 290 cwts. of oil: three whalers brought in, during the same year, 13,433 cwts. of whale-oil, and 306 cwts. of whale-bone. The pilchard fishery is also carried on with great activity; and employs, in the season, 700 boats, manned by about 3000 seamen. Nantes has two weekly markets, and 12 yearly fairs, one of which, beginning May 25th, lasts 15 days. Living is cheap; and fish of many varieties, as well as the fine fruits of the S. of France, are abundant in the markets. According to Hugo, there are annually killed in Nantes 2700 bullocks, 20,600 calves, 24,300 sheep, and 9000 hogs.

The era of the foundation of Nantes is unknown; but before the conquest of Gaul by the Romans, it was already a considerable city, and the capital of the *Namnetes*, who distinguished themselves by their opposition to Julius Cæsar. In 445, it was unsuccessfully besieged by the Huns, and, in the middle of the 9th century, was sacked by the Normans. In 992, it was added to the possessions of the dukes of Brittany, with whom it remained down to the union of that kingdom with France, by the marriage of Anne of Brittany to Louis XII. But Nantes is chiefly distinguished in history from the famous edict issued here in 1598 by Henry IV., and hence called the Edict of Nantes, which secured to the Protestants the free exercise of their religion, and an equal claim with the Catholics to all offices and dignities. The revocation of this edict by Louis XIV., in 1685, is the grand blot in his reign, and by occasioning the emigration of great numbers of his most industrious subjects, was even more injurious to his kingdom than the victories of Marlborough and Eugene.

During the revolutionary phrenzy, Nantes was the scene of the atrocities of Carrier, the most sanguinary of the republican agents in the reign of terror. Nantes has produced numerous distinguished individuals, including Anne, Duchess of Brittany, the Egyptian traveller Cailliaud; the physiologist Lavenier; Fourche, minister of police, &c. Near it is the Château de Buron celebrated as having been long occupied by Mad. de Sevigne. (*Hugo*, art. *Loire-Inferieure*; *Guide du Voyageur en France*; *Parl. Reports*; *Commercial Dict.*, &c.)

NANSEMOND, county, Va. Situated in the S.E. part of the state, and contains 444 sq. m. Watered by Nansemond river and its branches. The dismal swamp, and lake Drummond, which it contains, lie partly in this co. It contained in 1840, 7239 neat cattle, 4340 sheep, 23,227 swine; and produced 5331 bushels of wheat, 315,672 of Indian corn, 34,514 of oats, 89,330 of potatoes, 153,640 pounds of cotton. It had 42 stores, five lumber yards, one cotton factory with 1040 spindles, 17 grist mills, six saw-mills, one oil-mill, one tannery, one distillery; four academies, 154 students; 13 schools, 270 scholars. Pop.: whites, 4656; slaves, 4530; free coloured, 1407; total, 10,795. Capital, Suffolk.

NANTICOKE, hund., Sussex co., D.L. Watered by Nanticoke river and its branches. It has seven stores, one forge, five grist mills, five saw-mills; one school, 60 scholars. Pop. 1026.

NANTUCKET, island and county, Mass. It is situated in the Atlantic ocean, 10 m. E. of Martha's Vineyard, and 30 m. S. of cape Cod. It is 15 m. long, with an average breadth of 4 m., containing 50 sq. m. It is mostly a sandy plain, almost entirely destitute of trees and shrubbery. The S. part is a plain, elevated not more than 25 feet above the level of the ocean; but in the N part the land rises into hills, which are 40 feet high. The highest land on the island is 60 feet high. The inhabitants hold their land in common, and 500 cows feed in one pasture, and 7000 sheep in one flock. The inhabitants are mostly seamen, engaged especially in the whale-fishery, which they pursue in the Northern sea and the Pacific ocean. S.E. of the island are Nantucket shoals, 50 m. long and 45 wide, where numerous vessels have been wrecked. The island of Nantucket, with four small islands W. of it, two of which are of no value, constitutes the county of the same name. The original right of Nantucket was obtained by Thomas Mayhew of William, Earl of Sterling, in 1641, at New York. It was granted by Governor Mayhew in 1759 to 27 proprietors, many of whom removed to the island in 1660, among whom were three men by the name of Coffin, and Peter Folger, the grandfather of Benjamin Franklin, the father of his mother, who was a man of great influence. Each of the proprietors was at liberty to select a house lot of 60 rods square, of land not taken up. The island is still held as a joint-stock property, and the proprietors have increased to over 3000, mostly Friends. The county contained in 1840, 528 neat cattle, 7500 sheep, 278 swine; and produced 91 bushels of wheat, 521 of Indian corn, 374 of barley, 354 of oats, 4525 of potatoes. It had 33 stores, capital invested in the fisheries, $2,826,000, one fulling-mill, two woollen factories, four rope-walks, three grist mills, two printing-offices, two weekly and one semi-weekly newspapers; five academies, 630 students; 28 schools, 2060 scholars. Pop. 9012.

NANTUCKET, p. v., capital of Nantucket co., Mass., 119 m. S.S.E. Boston, 500 W. Situated at the bottom of a bay on the N. side of the island. It has a good harbour, nearly land locked by two projecting beaches, about three fourths of a mile apart, on one of which, denominated Brant Point, is a lighthouse. Nearly 2 m. N. of the harbour is a bar, with 9 feet of water. About 150 vessels belong to the port, navigated by 3000 men and boys. The village is compactly built with wooden houses and narrow streets, and contains nine churches, a Congregational, Unitarian, two Friends, a Methodist Episcopal, Methodist Reform, Episcopal, two African, one of which is a Baptist, three banks, two insurance companies, an atheneum, with a neat edifice, and a library of over 2000 volumes, a fine museum, and the Coffin school. This last institution was founded by the liberal bequest of Admiral Sir Isaac Coffin of the British navy, who visited the island in 1826, and found that a large part of the inhabitants were remotely related to him. It was incorporated in 1827, and endowed by its founder with a building, and £2500 sterling for its support. The place has daily communication with New-Bedford by a steamboat, stopping at Holmes'-hole, Martha's Vineyard, and Wood's-hole, near Falmouth. The tonnage of the port in 1840, was 31,916 tons.

NANTWICH, a market town and par. of England, co. Chester, and hund. of its own name, on the Weaver, crossed here by a stone bridge, 17½ m. S.W. Chester, and 146 m. W.N.W. London. Area of par. (comprising five townships), 3490 acres. Pop. in 1831, 3357: do. of township, 4886. It is situated in a luxuriant vale near the borders of Staffordshire and Shropshire, is irregularly laid out, and comprises three principal streets, badly paved and lined with mean-looking houses, uniting near the church, a very beautiful cruciform building of red standstone, built in the early English style, and highly ornamental, with an octagonal tower rising from the intersection of the nave and transepts. There are likewise several places of worship for dissenters, with attached Sunday-schools, which, in 1835, furnished instruction to 840 children of both sexes. Two endowed day-schools are attended by about 130 children ; and there are several alms-houses, besides minor charities. A market house and town-hall were built in the last century. Nantwich formerly owed its prosperity to the abundance of its salt-springs; but only one spring is now worked, and nearly the whole trade has been removed to other places. Large quantities of excellent cheese are made in the town and its fertile neighbourhood, besides which, the manufacture of shoes for the London market forms an important branch of industry. Cotton goods, also, are made here in considerable quantities; and in 1829 there was a cotton-mill, employing 111 hands. The glove trade is carried on to some extent. Great facilities of intercourse are furnished by the Birmingham and Liverpool, as well as by the Chester and Ellesmere, canals ; and the Grand Junction canal passes at only a few miles distance. Petty sessions for the hundred are held here ; and the Marquis Cholmondeley, as lord of the manor, holds a baronial court for the recovery of debts under £10. Markets on Saturday ; and fairs, May 15, June 13, September 4, and December 4, 18, 19.

Nantwich, mentioned in Domesday simply as "Wich," was the scene, in 1069, of an unsuccessful attempt by the Cheshiremen to resist the advance of the Normans. In 1438 and 1583, the town suffered considerably from fire ; and during the parliamentary wars, it was besieged by the royalists under Lord Byron, but soon after relieved by Sir Thomas Fairfax. It also deserves notice, from having been the birthplace of General Harrison, one of the regicides, and of Milton's widow, who died here in 1726.

NAPLES (KINGDOM OF); otherwise called the KINGDOM OF THE TWO SICILIES, a European state of the second class, nearly identical with the Magna Graecia of antiquity, comprising the S. portion of Italy, with Sicily and the adjacent islands, included between the 36th and 43d degs. of N. lat., and the 12th and 19th of E. long. It has N. the papal states, E. the Adriatic, and elsewhere the Mediterranean. Its total area may be estimated at about 42,000 sq. m., and its population at nearly eight millions.

The Neapolitan territory is divided into two principal portions, the continental and the insular, the first being called the Dominij al di qua del Faro (the country on this side the Faro, or Straits of Messina), and the latter, the Dominij al di la del Faro (or the country beyond the Faro). The latter portion will be fully treated of under the head SICILY ; we have now only to deal with the continental portion, the area, population, subdivisions, &c., of which are as follow :
(See top of next column.)

Physical Geography.—The continental part of the kingdom of Naples, the limits of which have scarcely varied for the last eight centuries, is about 400 m. in length, N.N.W. to S.S.E., and 130 m. in its greatest breadth, from cape Cam-

Provinces.	Area in sq. m.	Pop. in 1857.	Chief Cities.
Naples (Napoletana)	165	715,654	Naples.
Terra di Lavoro	2,341	684,896	Capua.
Principato Citra	2,618	419,177	Salerno.
Principato Ultra	1,254	379,366	Avellino.
Sannio (Molise)	1,216	342,779	Campobasso.
Abruzzo Citra	1,057	287,692	Chieti.
Ultra L.	1,129	290,257	Teramo.
Ultra IL.	2,195	300,719	Aquila.
Capitanata	3,714	308,662	Foggia.
Terra di Bari	1,711	448,943	Bari.
Otranto	3,069	385,284	Lecce.
Basilicata	3,583	486,270	Potenza.
Calabria Citra	3,532	402,757	Cosenza.
Ultra L.	1,496	272,441	Reggio.
Ultra IL.	1,787	334,915	Catanzaro.
Total	**31,407**	**6,069,266***	

panella to the Garganian promontory. Its shape is very irregular; at its S.E. extremity are the two peninsulas of Calabria and Otranto, forming one the foot and the other the heel of the boot which Italy is supposed to resemble, while to the N.W. of the latter is the conspicuous promontory of Gargano, extending into the Adriatic, representing the spur. On the W. coast also are many promontories and headlands, as those of Sorrento, Gaeta, Baiæ, cape Vaticano, and others, which respectively bound the bay of Naples, the gulfs of Gaeta, Salerno, Policastro, St. Eufemia, &c. The peninsulas of Calabria and Otranto inclose the extensive gulf of Taranto, N. and S. of which are the less spacious gulfs of Manfredonia and Squillace. The coasts are for the most part bold and abrupt, but the W. and S.W. are much more so than the E. and N.E. On either side however the kingdom has several good ports.

The surface of Naples, like that of the greater part of Italy, is mountainous, but it contains extensive and beautiful plains and valleys, which, under the influence of an invariably mild climate present a luxuriance of vegetation, and a beauty of scenery, hardly to be met with in any other part of Europe. The Apennines traverse the kingdom nearly in its centre from end to end. In the Abruzzi, where they reach their greatest altitude, they consist of three lateral ranges, but these unite near Isernia, and the main chain thence proceeds undivided to Monte Caruso, about 14 m. N. Potenza, where it finally bifurcates, the principal range running through Calabria to its furthest extremity, and a less elevated range through the S. part of Apulia. Many important ramifications are given off by the Apennines, both before and after their bifurcation, as that forming the lofty promontory of Sorrento, that of Gargano, the mountain knot of La Sila in Calabria, &c. The Apennines rise to a much greater elevation in S. than in Central Italy. The Gran Sasso d'Italia (M. Corno) in the Abruzzi, reaches the height of 10,185 English feet, and M. Majella, in the same province, that of 9395 feet. There are some separate or detached groups of mountains, of which Vesuvius, though not the largest, is by far the most celebrated. It owes this distinction to its situation close to the city of Naples, and still more to its having long been an active and sometimes a most destructive volcano, Ætnæi ignis imitator.

The largest of the Neapolitan plains is that of Capitanata, having Foggia in its centre. It is mostly appropriated to pasture, and is in part sandy and arid. (See ante, p. 32.) But the most celebrated plain is that of the Terra di Lavoro round Naples, anciently the Campania Felix, an epithet to which it is still well entitled. The choicest gifts of nature have been lavished upon this noble plain. It is above 40 m. in length by from 15 m. to 20 m. in breadth, and, excepting Vesuvius and a ridge between Naples and Misenum, it is everywhere a dead level. The soil, which is deep and loamy, is of the most extraordinary fertility, frequently producing two crops in a season. It is carefully, if not skilfully, cultivated ; the vegetation is most luxuriant ; and being free from malaria, the air mild and genial, and the sky usually clear, it goes far to realise the poetical descriptions of the Elysian fields—Omnium non modo Italia, sed toto orbe terrarum, pulcherrima Campaniæ plaga est. Nihil mollius cælo : denique bis floribus vernat. Nihil uberius solo : ideo Liberi Cereri certamen dicitur. (Florus, lib. i., cap. 16 ; see also anté, p. 51.)

The rivers are numerous, but mostly inconsiderable. The principal are the Garigliano (an. Liris), and the Volturno flowing through Campania to the Mediterranean, the Pescara, Biferno, Candelaro, Cervaro, Ofanto (an. Aufidus, &c., falling into the Adriatic ; and the Bradano, Basiento, Crati, &c., which carry their waters to the gulf of Taranto.

The only lake of any size is that of Celano or Fucino (an. Fucinus), in Abruzzo Ultra, 11 m. in length by 7 m. in its greatest breadth, in a basin surrounded by high mountains. This lake receives the waters of several considerable

---

*In 1838, the population had diminished to 6,987,264, a diminution principally owing to the cholera and febrile diseases, which had been particularly prevalent in the city and province of Naples during the previous year. (Serristori, Statistica d'Italia ; Almanach de Gotha, &c.)

streams, and having no outlet, is apt, occasionally, to over-spread the surrounding country. To obviate this effect, the Emperor Claudius, in imitation of a similar undertaking at the Alban lake (see vol. i., 50, of this work), carried an *emissario*, or aqueduct, 3 m. in length, partly by tunnelling and partly by excavating, from the Liris, through Monte Salviano, to the lake, by which its waters were at once reduced and prevented from again rising, to a higher level. (See *Suetonio in Claud.*, cap. 20; *Tacit. Annal.*, lib. xii., cap. 56, 57.) This great and useful work was, however, in later ages, suffered to fill up and become useless. Latterly, however, it has been renovated under the direction of Afflan de Rivera, in whose work (*Considerazioni sui Mezzi*, &c., i., 305) the reader will find an account of the undertaking.

In respect of *climate*, S. Italy is artificially divided into three regions, according to the elevation of the surface. In the lower parts of the country, the winter is so mild that vegetation is never interrupted; and at the S. extremity of the kingdom, the aloe, palm, and other tropical plants flourish in the open air. In the plains and valleys, near the foot of the mountains, snow seldom falls, and remains only a short time on the ground: but in the higher portions of the Abruzzi, &c., the cold of winter is piercing and long continued. There is a marked difference in the climate on the E. and W. sides of the Appennines. In the plains of Apulia and Bari, for example, rain seldom falls; the ground is, in parts, arid, and almost desert, and, during summer, vegetation is parched up, and the heats are oppressive: but, on the W. coast of the kingdom, and especially in Campania, showers are frequent in summer, and in winter the ground is saturated with moisture. At an average of the kingdom, the annual fall of rain may be estimated at about 29 inches. Except in some marshy tracts on the coast, the climate is in general as healthy as it is genial. The heats of July, August, and Sept. are sometimes, however, rather hazardous, especially when the *scirocco*, from the Syrian and Arabian deserts, exerts its pernicious influence. Its enervating and suffocating blast, notwithstanding the cooling it has sustained in passing over the Mediterranean, becomes positively dangerous, when it lasts more than 20 or thirty hours, and is not followed by a breeze from the N. In Naples and other large cities, the streets are deserted during the prevalence of the *scirocco*, the inhabitants shutting themselves up in their houses, and closing all the doors and windows to exclude the air as much as possible.

Geology, Natural Products, &c. — Granite and gneiss which are absent from the Appennines of central Italy, again appear in the S., and are the predominate rocks in the main chain from the Abruzzi to the end of Calabria. They are intermixed with mica, talc, clay slate, serpentine, &c., and accompanied on either side by the calcareous formations, of which the Italian peninsula is mainly composed. Calcareous or sandy formations constitute the principal portion of the upper soil; but in some parts, especially in the neighbourhood of the capital, the ground consists of volcanic tufa, lava, ashes, &c., which contribute to its prodigious fertility. Vesuvius is at present the only active volcano of continental Naples; but there are numerous extinct volcanoes, the craters of some of which supply sal ammoniac, sulphur, vitriol, alum, and other volcanic products. These with pumice stone from the Lipari islands, salt, lime, and crystals of several kinds, form the chief mineral wealth of the kingdom. Iron is dug up at Stilo in Calabria, and at Vesuvius, but in no great quantity. In Calabria Citra, veins of gold and silver have been detected, but not such as to pay for the expense of working. Salt is procured from salt lakes near Taranto.

The *vegetable products* have been already noticed with those of the rest of Italy (*ante*, p. 55); but in addition to the ordinary forest trees, shrubs, and fruits of Italy, the pistachio, date, sugar cane, &c., flourish, in some places cotton is rather extensively cultivated. Immediately on entering the Terra di Lavoro from the papal states, an increasing luxuriance of vegetation is perceptible, and many new plants common to the tropical climates are added to the *flora* of other parts of Italy.

Bears, wild boars, and an abundance of game, inhabit the forests; and the crested porcupine is an animal said to be peculiar to this part of Italy. The shores abound with fish, and the gulf of Taranto is celebrated now, as of old, for the number and excellence of its shell fish. Among these are the *Pinna marinae*, the silky fibres of which are woven into stockings, gloves, &c.; and the *Murex*, still used, as in antiquity, in dyeing wool. The tarantula and cantharides are met with in the S., which is also occasionally devastated by swarms of locusts.

State of Property, Agriculture. — Notwithstanding the backward state of Naples at the present day, in respect of agriculture and other branches of industry, it has made a very considerable progress during the last century, and especially during the last 30 years. When the Austrian dominion ceased, in 1733, the industry of Naples was at its lowest ebb. The abuses inherent in the feudal system had

then attained to a maximum. The entire property of the country belonged to the crown, the church, the nobility many of whom had vast entailed estates, and the corporations. The bulk of the people were in a state of predial slavery. Every feudal right and service, however onerous and absurd, was rigidly insisted upon. The game laws existed in their utmost extent; justice, if we may so abuse the word, was administered in baronial courts; the services which the peasantry had to render to their lords were not defined, and they could hardly do any thing except with their consent. The roads were neglected, commerce despised; the country, notwithstanding its fertility, frequently suffered from famine; and the people, oppressed, fleeced, and without the means of improving their condition, mostly sank in a state of indolence and apathy, while the more adventurous spirits became bandits and robbers. (See the *Saggio Politico sul Regno delle Due Sicilie*, by Del Rè, p. 16 and passim.)

The Bourbon government laudably exerted itself to suppress some of the worst of these disorders; and it so far succeeded as to introduce a better system of administration, to improve the roads and towns, and to provide, in some degree, for the growth of manufactures and commerce. But the abuses with which the whole frame of society were infected, were too deeply seated to be eradicated by its feeble and timid hands. The wants of the country were set forth by Filangieri, and other able writers, but things continued nearly on their old footing until the French had established their ascendancy in the countries on this side the Faro. We have already seen that Napoleon has an unquestionable title to be regarded as the best friend and benefactor of Italy (see *ante*, p. 57); and this, perhaps, is more especially true of Naples than any other part of the peninsula. In 1806 Joseph Bonaparte, then king of Naples, promulgated, no doubt with the approval of his brother, a law, which overturned the feudal system from its foundations, and produced an extraordinary and most beneficial change in the constitution of property and of society. "The feudal system," says this law, "and all feudal jurisdiction, are abolished; all towns, villages, hamlets, are subjected to the general laws of the country. All feudal dues to the exchequer cease, and feudal estates are subject to the same taxes as others. All feudal burdens, services, and dues of a personal nature, levied from communes or individuals, are abolished without compensation; as are all prohibitive rights or monopolies, wherever they did not originate in purchase. Rivers are public property. The feudality of offices and *fidei commissos* shall cease."

This law was followed by other of an equally decisive and stringent character; and at the same time convents were secularised, the (profligate) drones by which they were occupied were turned out, with only a small stipend, and their estates sold at very low prices. Happily, the changes that these laws had introduced into the distribution of property and the state of society, previously to the revolution that restored the Bourbons to the throne, were such as to render a return to the old state of things impossible. The nobility have, however, recovered some of their old privileges, and have been permitted to establish majorats. The attempts to restore the property and influence of the clergy have been prosecuted with greater zeal, and have been, perhaps, more successful. But, though many convents of monks and nuns have been again established, a comparatively small amount of property is now in their hands; and it is to be hoped that the growing intelligence of the people may at length put a final stop to this most wasteful and vicious mode of employing the public wealth.

The measures now noticed, coupled with the introduction of the law of equal succession in cases of intestacy, have had a much greater influence over the distribution of property than might, *à priori*, have been anticipated. It appears from the returns under the land-tax (*Contribuzione Fondiaria*), that in 1832 there were in continental Naples no fewer than 1,419,191 properties rated to the tax, and that these were held by 1,062,179 individuals, showing that there is, in fact, nearly a proprietor for every family in this part of the kingdom! (*Del Rè, Saggio Politico*, p. 53). As houses as well as land are rated to the land-tax, and as there are no means by which to classify the contributors according to the sums paid by each, we are unable to hazard any estimate of the size and value of the different classes of properties. There can, however, be no doubt that the number of small and middle-sized properties has been prodigiously augmented since 1806. Indeed, the danger now is that the tendency to divide and subdivide estates should be carried too far. When all feudal privileges are abolished, and land is placed on the same footing as other property, the interests of society will be best promoted by adopting a system of primogeniture; and preventing, in as far as practicable, by indirect methods, the division of the land into minute portions.

The method of holding the land by tenants differs in different parts of Naples. In some of the richest and most populous districts round the capital and elsewhere, the land, when not occupied by the proprietor, is mostly let in small lots of four or five acres, on the *metayer* system; but in general this system is but little followed, and in most parts of the country the land is either occupied by proprietors, or by tenants holding under a lease of some years' duration at a fixed rent. The size of farms varies in different districts; and in those where pasture predominates, they are sometimes very large. But notwithstanding the impetus given to industry of late years, agriculture is still extremely backward. The nobility, who are usually involved in debt, care little for agriculture, and their tenants are mostly without either adequate capital or skill. Drainage in some quarters, and irrigation in others, though of essential importance, are generally neglected; and the sowing of artificial grasses, and even the application of manure, are comparatively unknown. "Farming implements, carts, ploughs, and tools of every kind are of the most wretched description." (*Macgregor.*) The more common kinds of grain are wheat, maize, or Indian corn, and, in the older situations, rye. In Calabria, and the more distant provinces, grain is thrashed out by driving cars over it, or by trampling it with the feet of horses and oxen. In many parts few or no ploughs are used in tillage. In the *Campania*, for example, the soil is so very friable, that it is easily turned up even by children; and such is its fertility that no fallows are required in its culture, and but little manure, the gathering of one crop being followed by immediate preparations for another. According to Châteauvieux six crops are obtained round Naples in five years, exclusive of the produce of vines, fruits, and beans, which grow in the same land without prejudice to the corn crops. Most parts, indeed, of the kingdom, except where there is a deficiency of water, are extraordinarily fertile; the crops of wheat and maize are especially most abundant; and there can be no doubt that, with a skilful and careful system of agriculture, the kingdom might afford ample subsistence for four or five times its present population. But here as elsewhere, the fertility of the soil, and the mildness of the climate, contribute, with bad government, to enfeeble the cultivators; and by lessening their wants, and enabling them to supply them with little labour, generate that indolence and apathy that are so universal. In the finest districts numerous families live in the meanest hovels, and in a state that in Great Britain would be reckoned to approach closely to absolute indigence. But under such a climate, and occupying such a soil, they do not often suffer the extreme of want. They rarely, however, partake of any of the enjoyments and luxuries of life, other than those which they share in common with the lower animals. Without ambition, or the desire of rising in the world, and without knowledge of the liberal, and but little of the mechanical arts, they pass their days in a state of brutish apathy and indifference.

The plain of Sorrento, according to Châteauvieux, is almost the only part of the kingdom of Naples where agriculture may be said to be skilfully and actively carried on. The rotation of crops is there, 1st year, maize; 2d, wheat, succeeded by beans; 3d, cotton; 4th, wheat, followed by clover; 5th, melons, followed by peas or beans; making eight crops in five years. Next to the above, rice, barley, rye, onions, and other kitchen vegetables, oil, wine, hemp, flax, tobacco, saffron, and fruits of various kinds, form the principal products; and each might be produced in quantities at least equal to twice the consumption of the inhabitants. The olive is found in all the low and temperate parts of the country, and its culture is widely extended. In Apulia, especially, a great extent of land is covered with olive-trees, which frequently reach the ordinary size of oaks; and throughout the province of Bari, and a part of that of Otranto, a broad belt of olives lines the coast for 100 m. In 1827, a good oil year, the produce of these provinces was estimated at 360.000 *salme* (3,314,000 galls.). In Calabria, also, from Rossano to Gioja, olives grow in great abundance; and the W. shore of the kingdom, from the last-named town to Gaeta, produces large quantities of oil. The average annual produce of Calabria is estimated at 100,000 *salme*. The culture of the olive, and the preparation of the oil, have of late years attracted a great deal more attention than formerly, and have been materially improved; and were it not for the anti-commercial policy of the government, would become of vast importance. At an average of the three years ending with 1817, the exports of olive oil amounted to 2,800,123 staja a year; whereas, at an average of the three years ending with 1832, the exports were 3,008,887 staja a year, or above 6,000,000 imperial gallons. (*Serristori*, p. 61.) Gallipoli is one of the greatest marts in Italy for oil; and that shipped from it is considered equal to that of Genoa or Lucca; a superiority for which it is mainly indebted to the influence of the tufa cisterns in which the oil is purified before being shipped. (See vol. i., p. 959.)

The Neapolitan wines are mostly full bodied and some are of a good flavour; but the principle of the division of labour is very imperfectly understood, and, instead of being

pursued as a separate branch, requiring the undivided attention of the husbandman, the culture of the vine is commonly carried on upon the same farm with that of corn, olives, &c. In the Terra de Lavoro and other parts of the country, elms and poplars are seen, planted in rows for the support of the vines, the intermediate spaces being sown with corn, pulse, and other crops. In parts of the Abruzzi, however, the vines are cut low, and tied to canes, as in France; and in the province of Otranto, they are cut off to about 2 feet above the ground, which is hoed around them with a degree of care by no means general in S. Italy. The wines of Taranto, Gerace, and other parts of Apulia and Calabria, might become very superior by proper treatment. Round Vesuvius some fine growths are obtained, among which is the celebrated *lacryma Christi*: this, which is a red [luscious] wine, is little known except to visiters at the royal table at Naples, the quantity grown being small, and principally purchased for the court. On the whole, though the wines of Naples be probably superior to those of the rest of Italy, they are very inferior to those of France or Spain; are seldom met with out of Italy; and, except the sweet wines, are mostly converted into brandy.

Oranges and lemons are grown in large quantities in Capitanata and about Salerno and Reggio, from which latter town a good many are sent to France and Genoa; tobacco is pretty largely cultivated in S. and Otranto, and suffms in Calabria Ultra, the Abruzzi, and Basilicata. But excepting flax and hemp, which are extensively cultivated, few crops furnish materials for manufactures; these generally speaking, being imported from abroad. Liquorice is grown in Calabria and the Abruzzi, and about 15,000 *cantari* a year of juice are exported. Manna is a product of some importance in Calabria; it exudes from gashes cut in the bark of the manna-ash (*Fraxinusornus*). The product is farmed by the crown; and the labourers who cut the bark and collect the manna are debarred, while so employed, from attending to any other occupation, though they receive only a small pittance, amounting to three *carlini* or about 1s. for every *rotolo* of manna they collect. Manna is also produced near Salerno, where, however, about five *carlini* per *rotolo* are paid for its collection.

The *forests* might become an important source of wealth; but, in consequence of the suppression of the feudal system, the sale of monastic and other domains, and the division of the communal property under the French, numerous woods were felled, and the ground they occupied was subjected to tillage; but, after a few favourable harvests, the soil was exhausted; and it was farther exposed to the redoubled violence of inundations, which by washing away the soil and stones from the mountain declivities, caused great injury to the lands below. For these reasons, in 1819, a special board of superintendence over forests, &c., was appointed; and it was forbidden to appropriate the ground occupied by woods to tillage, except where the site is level and fertile, or under certain other circumstances: neither is any proprietor allowed to fell timber without express permission. These regulations have checked the destruction of the woods; but their preservation has been next to useless; for although many parts of the interior there is abundance of timber admirably adapted for ship-building, the roads and means of conveyance are so bad that its carriage to the coast would more than cover the cost of importing it into Naples from other countries. And, moreover, government discourages the efforts of individuals to improve the roads by prohibiting the export of timber, at the same time that it obliges it to be cut down by laying a heavy duty on the importation of a.

If we except sheep, the rearing of live st ck is not much attended to. There are estimated to be 4,000,000 head of sheep in the kingdom, a great number of which are migratory, being kept on the mountains in summer, and down in winter, into the valleys, and the *tavoliere* of Apulia. The last named region is a tract of about 95 sq. m. belonging to the crown, and which, a few short intervals excepted, has been wholly devoted to pasturage from the time of the Romans, its tillage having, indeed, been prohibited. Under the French, however, a law was passed in 1806, which for a time, completely changed the state of the *tavoliere*; instead of the ground being farmed exclusively as pasture, a fixed rent was substituted; and every one was allowed to apply the land, as he pleased, to pasturage or tillage. But in 1817, this judicious law was repealed, and a new act passed, by the provisions of which (in order to keep up [stock] was alleged, the due proportion between pasture and arable land,) no one was to till more than one fifth part of his land, on penalty of paying tenfold rent, &c. Latterly, however, we believe that this absurd enactment has been in some way modified, it having fortunately been found impossible to enforce its provisions. The horned cattle are estimated at about 300,000 head, exclusive of about 50,000 buffaloes, which wander in large herds over the marshy plains of the N.W. The oxen are of different breeds, and excellent quality, being used both for the plough and for cans. They are mostly stall-fed, on vine leaves, maize, stubble, &c., except in Apulia, or districts where the pasture lands are very extensive. The horses are but indifferent: but a fine small active breed, with dark frizzled hair, is peculiar to Calabria. Mules and asses are the most common beasts of burden. Hogs and goats are very numerous; the former, which are of a large, dark, hairless breed, wander wild in the forest; the milk of the latter is converted into cheese. The herdsmen throughout the kingdom are principally from the Abruzzi.

Except in the S., where the cottages are sometimes built of stone, it is rare that the agricultural labourers inhabit anything but wooden huts, roofed with straw or tiles, in which three or four generations often live together. They live almost wholly on vegetable food; and into this, fruits and pulse enter more largely than corn. Their clothes are coarse, and the dress of the shepherds and poorer peasantry consists almost solely of sheep or goat skins, with the hair outside. The wages of an ordinary field labourer may average about 16 *grani* (or 6d.) a day.

The culture of *silk* is widely diffused, and it now forms a valuable article of export from the kingdom. The greatest quantity of silk is produced in the provinces of Lavoro, Principati, and Calabria, but especially the last. At Reggio, in particular, many families furnish houses expressly for rearing silkworms. The total produce of Naples was estimated, in 1833, at 800,000 lbs., of which 536,000 lbs., worth 2,005,800 ducats, were exported raw and wrought. (*Serristori*, p. 61.) The remainder is consumed in the country. The silk of Calabria is stronger and more compact than that of other parts of the kingdom: a consequence, it is supposed, of the worms being fed with the leaf of the red mulberry peculiar to that province. Much less attention is, however, paid to the culture of the mulberry, both there and throughout the Neapolitan dominion, than in N. Italy. In the vicinity of Naples, where two crops of cocoons are annually collected, the breeders of silkworms are accustomed to buy mulberry leaves at a dear rate, at the same time that they use the poplar as a support for the vines in their farms.

The *fisheries* rank next to the foregoing branches of industry. The tunny (*Scomber thynnus*) frequents the W. coast in large shoals in the early part of the year, and the taking of the fish employs a large number of hands. Large meshed nets, perhaps 1500 feet in length, and divided into several compartments, are laid across the track of the shoals at a considerable distance from the shore, with which they are connected by strong lines and nets of still larger meshes. The fish, having got into the nets, are prevented finding their way out again by the number of compartments; and after the lapse of a day or two, men in flat-bottomed boats surround the nets with harpoons, and kill the larger fish, which are sometimes 8 feet in length. The whole draught is afterward hauled on shore. Anchovies are caught in abundance in the spring; and many mullets are taken, the roes of which are made into *botargo*. A great many of the inhabitants of Taranto employ themselves in taking the shell-fish of the gulf, leaving the adjacent lands to be cultivated by natives of Calabria and Abruzzo. But the quantity of fish taken for food is by no means equal to the home demand, and salted fish is a principal article of import. Coral is raised on different parts of the coast.

Manufactures are, for the most part, domestic; and the majority of the goods made scarcely ever find their way out of the country. Of late years, however, manufactures have been considerably improved, and several considerable manufacturing establishments are now to be met with in different parts of the kingdom. Those of silk are the most extensive; and next to them are those of woollens, principally established in the capital and at Sora and Isola. Linen stuffs are made in several of the provincial towns, and there are numerous paper-mills.

Naples has manufactures of gloves and straw hats, in imitation of those of Tuscany; and the princes of Butera and Gerace, with other partners, have established a considerable glass manufactory at Pausilipo. Reggio has manufactures of gloves, waistcoats, and hosiery; and Avellino, of woollen cloth, sausages, and maccaroni. Campobasso is famed throughout the kingdom for its cutlery. The establishments in which it is made are on a very small scale. "Great was my surprise," says Mr. Craven, "at finding that the various articles of this nature are all fabricated in detached small workshops, containing little more than a common blacksmith's apparatus, and possessing so contracted an assortment of articles that I could with difficulty obtain a selection of a dozen knives and scissors. They all work by commission for dealers in the metropolis and other large towns, with such limited means and capital, that they are entirely incapacitated from executing and maintaining in their laboratories a collection exhibiting anything like variety or choice." (*Excursions in the Abruzzi, &c.* ii., 142.) A miserable cotton factory, a sort of government monopoly established at Salerno, some time ago, and the iron mine and forge at Stilo, comprise nearly all the remaining branches of manufacturing industry. (*Parl. Report*, 1840; *Chateauvieux, Italy, &c.*, p. 187–298; *Von Raumer's Italy*, ii., 248–264: *Rampoldi, Simond, Craven, Swinburne, &c.*, passim.)

Commerce.—The extraordinary fertility of her soil, the variety and superior quality of her raw products, and her admirable situation, give Naples the means of carrying on a very advantageous and extensive commerce. But her natural advantages have been, in this respect, all but nullified by the perverse policy of her government. Her oils, silks, sulphur, corn, wines, &c., would all meet with a ready and advantageous sale abroad, provided the manufactured goods and other products which foreigners have to give in exchange for them, were admitted on anything like reasonable terms into the Neapolitan dominions. Such, however, is not the case. The importation of a great variety of foreign articles is prohibited; and most of those that are admitted are loaded with oppressive duties, varying from 50 to 100 and 150 per cent. *ad valorem.* But this is not all. Not satisfied with attempting to shut out foreign products, the Neapolitan government lays heavy duties on many of the most important articles produced in the kingdom, when exported; and thus by raising their price to the foreigner, and lessening the demand for them, does all in its power to limit and hinder their production! Olive oil, for example, is charged on its exportation with a heavy duty; and as oil is largely produced in Lucca, Tuscany, Genoa, and elsewhere, the effect of this *felo de se* tax is to depress that branch of industry which is most suitable for the country, and to encourage the culture of olives and the oil trade in other parts of the peninsula! After this specimen of the commercial policy of Naples, the reader will not be surprised to hear that the *bonding* of goods is not permitted in any part of the kingdom; in other words, all foreign goods must pay duty on being imported, and no part of this duty is remitted, or drawn back, on their being again re-exported.

Under such multiplied difficulties the wonder is, not that the commerce of the country is confined within comparatively narrow limits, but that it *exists*, and is so extensive as we find it to be. The great superiority enjoyed by Naples in the production of certain articles, and the wants of the people originating in the low state of manufactures, and the demand for colonial products, spices, dye-stuffs, and other indispensable articles, have however proved too strong for the anti-commercial policy of the government, and occasion a considerable intercourse with foreigners. No accurate accounts have, however, been published of the quantities and values of the principal articles imported into, and exported from, the Neapolitan dominions; but, according to the information obtained by Mr Macgregor and others, the value of the exports from the continental portion of the country may, perhaps, be estimated at about £2,000,000, and that of the imports at about as much. The latter consist of cottons, woollens, linens, and other manufactured goods; sugar, coffee, and other colonial products; spices, dye-stuffs, salted fish, iron and hardware, &c. The exports consist principally of olive oil, silk, flax and hemp, wool, wine, corn, linseed, cream of tartar, rags, maccaroni, lamb and kid skins, liquorice, coral, bones, &c. The trade of continental Naples principally centres in the capital; and the reader will find in the following article an account of the principal articles exported from it in 1838.

Subjoined is an ACCOUNT of the Vessels belonging to the different Neapolitan Ports in 1834:—

Provinces.	Emp. in Foreign Trade.	Emp. in Fisheries and Coasting Trade.	Total Vessels.	Burden in Tons.	Men.
Naples	2,669	621	3,280	128,040	24,229
Terra di Lavoro	94	73	167	9,496	1,846
Principato Citra	293	82	275	8,565	2,179
Sannio	15	. .	16	180	98
Capitanata	43	14	57	1,023	258
Bari	509	129	638	14,258	3,123
Otranto	224	7	231	1,167	791
Abruzzo Citra	78	14	92	1,580	653
Ultra I.	24	6	30	414	198
Basilicata	23	. .	23	114	93
Calabria Citra	182	8	190	1,304	1,558
Ultra I.	351	26	377	4,681	1,600
Ultra II.	126	12	138	1,118	1,586
Total	4,565	938	5,493	169,330	36,263

The principal weights in use are, the lb. of 12 *oncie* = 11 oz. avoird.; the *rotolo* = about 31 oz.; and the *cantaro* of 100 rotoli, = 196¼ lbs. avoird. The *tomolo* of corn = 1·45 Winch. bushels; the *barile* of wine, &c. = 9⅔ imp. galls.; the *salma* of oil = 35¼ imp. galls. nearly. The *canna* of 8 palmi = 6 ft. 11 in. Eng.; the *palmo* divided into 12 *oncie* = 10·38 Eng. in. The Neapolitan mile of 7000 palmi = about 1¼ Eng. m.; the *moggia* = ·73 Eng. acre. Accounts are kept in ducats (3s. 5½d. each), divided into 10 carlini and 100 grani. (*Balbi; Serristori, &c.*)

NAPLES (KINGDOM OF).

The Government is a monarchy, hereditary in both the male and female line; and at present in the hands of a branch of the house of Bourbon. The monarchy in respect to the continental portion of the kingdom, was formerly quite unlimited, while Sicily had a parliament of its own. But in 1821 a *consulta* was established for each separate division of the Neapolitan dominion; that for the continental portion consisting of 16 members, and that for Sicily of eight members, appointed by the government from lists of candidates named by the inhabitants of the different provinces. Each *consulta* was presided over by a vice-president nominated by the king; and both assemblies frequently meet in one, termed the *consulta generale*, in which a state minister, also appointed by the king, sat as president. In 1837, these *consulte particolari* were permanently amalgamated into one parliament, which sits at Naples. But the functions of this body are of the most restricted description; and, as is truly observed in the "Quadro del Governo," *non v'ha altra suprema autorità legislativa che quella del monarca; essa è assoluta, ed in qualunque modo faccia egli conoscere la sua volontà debbano i sudditi obbedirvi.* The king is, however, assisted by a cabinet of nine ministers, and a privy council composed principally of noblemen.

Every province is governed by an *intendente* appointed by the king, to whom all the internal administration of the province is confided: at the head of every district (*circondario*) is a sub intendant; and in every commune a syndic or mayor. Each commune has also a body of decurions elected from the heads of families paying taxes of from 12 to 24 ducats a year, according to circumstances, and who consult, under the presidency of the syndic, on the affairs of the commune, fixing the rates, and appointing municipal officers; but their resolutions do not acquire the force of law till they have received the sanction of the *intendente*. Every district has a council of 10 members; and in every province is a council of from 15 to 20 members, nominated by the communes, and chosen by the minister of the interior; which council assembles once a year to examine the proposals of the district councils, to draw up, under the direction of the intendant, the projects proposed for the province, and to examine the provincial accounts, &c. The provincial council must not, however, consult upon any matters but such as are submitted to it; and the entire authority in the province remains with the intendant and his council, which consist of from three to five members, and is formed after the model of the French *conseil de préfecture.*

Justice is administered in a supreme court of cassation in the capital; high civil courts at Naples, Aquila, Trani, and Catanzaro, civil and criminal courts in the capital of each province, by a judge *d'istruzione* in each district, with authority from the provincial criminal court; and by a *conciliatore* in each commune, who decide in cases to the amount of six ducats. In 1834, tribunals of commerce were in operation at Naples, Foggia, and Montéleone. Judges, as well as most other functionaries, usually hold their appointments for three years. Trials are public, and the code of laws, as well as the judicial forms, established by the French, have been generally adopted, except that trials by jury are unknown. Some late statistics and details show that the average of persons accused is as 1 to 1030, and of those convicted as 1 to 1438 of the population. Of 5813 accusations, in a given period, 104 were for offences against religion, 996 for homicide, intentional or otherwise, and 1793 for violations of property. In Capitanata, 1 in 607 of the population was convicted; in Abruzzo Ultra, 1 in 8311. The proportion of the accused, was 1 in 559 of the rural population, 1 in 199 of artisans and servants, 1 in 508 of persons occupied in liberal arts, and 1 in 2819 of landed proprietors. (*Von Raumer's Italy, &c.*, ii., 229–230.) According to these reports, Capitanata is a province distinguished for crime; and both it and Basnio have been, in fact, noted for brigandage on a large scale. Mr. Craven states, that even the favourite amusements of the children, in some districts in these provinces, consist in mock representations of attacks by brigands on travellers, &c., in which the former invariably gain the advantage. The country bordering on the Papal territories is also infamous for robberies. Under the French, the police was well organized, but it is now extremely corrupt and bad. Popular feeling, in the capital at least, is also generally in favour of an offender.

The established religion is the Roman Catholic; but the S. provinces have about 75,600 Greeks, chiefly the descendants of Greek colonists, who settled in S. Italy after the destruction of the Greek empire by the Turks. Jews are few; and there are only about 360 Protestants in the kingdom. In the continental dominions of Naples, there are 30 archbishops and 66 bishops, each of whom receives an income of not less than 3000 ducats. According to Serristori there were altogether, in 1837, 96,304 secular, and 20,906 regular clergy. Under the French, in 1807, about 250 convents were dissolved, their conventual property sold to pay the creditors of the state, and the monks, &c., allowed only a small stipend annually, which sum was afterward considerably diminished. But, as already stated, since 1830, everything has been moving in the contrary direction. Many convents and religious foundations have been restored; many new ecclesiastical fraternities founded; and the Jesuits have been re-established and endowed. According to the *concordant* concluded with the Papal court in 1818, the pope has the sole privilege of confirming the archbishops and bishops of the Neapolitan dominions in their sees, with other important privileges. Still, however, "the Neapolitan government does not allow the publication and application of any Papal rescripts without its own consent, and displays such firmness, nay, sometimes severity, in matters concerning the bishops and clergy, as the court of Rome would scarcely suffer a Protestant sovereign to exercise without reprimand." (*Raumer's Italy, &c.*, ii., 212.)

Public instruction is in the most miserable state. Popular or elementary instruction is confided to the clergy. The Jesuits have, since their re-establishment in 1822, opened day-schools for the gratis instruction of youth, though they have not yet been allowed to re-open their former colleges for lay-boarders. There are grammar schools, as in other parts of Italy; and some attempts have been made to establish Lancastrian schools in Naples, but they seem to have failed. (*Journal of Educ.* v., 85.) There is, or ought to be, a primary school for boys in every commune; but there are few, if any, primary schools for girls, and seminaries for teachers have only just begun to be established. In some of the provinces, scarcely one in 150 or 160 persons learns to read and write.

Naples has a university, with faculties of theology, physics and mathematics, literature, jurisprudence, and medicine, which is attended by about 1500 students; royal lyceums in Naples, Salerno, Bari, Catanzaro, and Aquila; royal colleges in all the other provincial capitals; and 42 secondary schools; in which, however, little is taught beyond reading and writing, and the mere elements of Latin and Greek. But, with the exception of mathematics, antiquities, and perhaps physic, all the higher branches of science and philosophy are in the most degraded state; and even the fine arts have not escaped the general paralysis.

In 1811, a royal society of science was established in Naples, which has been replaced by the royal Borbonic society, with 40 members in three divisions: many other institutions which were founded by the French have disappeared. In 1834, as many as 30 periodical publications appeared in the kingdom; of these, some that were scientific journals had a high character, but the others were mostly indifferent or worthless. The censorship of the press prevents native talent, if it exist, from distinguishing itself; and the oppressive duties on foreign books hinder the people from acquiring that information from abroad which they cannot obtain at home.

Monti frumentarii are established in the different towns for the relief of the poor, in which contributions of corn are received and distributed to the indigent. The hospitals and other charitable foundations were formerly very rich; but they suffered a good deal from the encroachments on their funds by the French government. Their revenue, however, amounts at present (1840) to about 1,500,000 ducats a year: they are generally managed by the clergy. Prisoners in public jails are allowed, but not compelled, to work.

But the provision for the poor is certainly inadequate; and owing partly to this, and partly to the defective state of the police, mendicancy is excessively prevalent. Mr. Maclaren says, that in "all the towns and villages near Naples, strangers are besieged with crowds of mendicants, whose importunities know no bounds. To give anything to the first hive merely excites others to follow you. It is nothing less than a persecution, and is really one of the greatest nuisances a traveller has to endure.

The military force in 1839 amounted to nearly 45,800 men, of whom 29,000 were infantry of the line, 4500 cavalry, 2600 artillery and engineers, and 3600 *gens-d'armes*. In time of war, the effective force amounts to 64,237 men. (*Serristori, Statist. delle Due Sicilie.*)

From the completion of the 18th to that of the 25th year every one (with certain exceptions) is liable to the conscription. A law of 1834 fixes the time of service at five years in the army and five years in the reserve: but gendarmes, artillerymen, and volunteers, serve eight years without reserve. A provincial militia was instituted in 1818; but it was suppressed three years afterward. The soldiers have, at different times, been employed in useful public works, such as paving the streets, &c.; but it must be added, that the troops are ill paid, and deficient in courage, morale, and most of the qualities that constitute good
434

soldiers. In fact, 25,000 English, French, or Austrian troops, would suffice to conquer and retain the kingdom. The principal fortresses are Civitella del Tronto, Pescara, Aquila, Gaeta and Capua, all in the N. provinces: Manfredonia, Brindisi, and Taranto, are only partially fortified.

The naval force consists of two ships of the line, five frigates, two corvettes, and a number of smaller vessels, mounting altogether 496 guns. There are three battalions of marines—two of 1000 men each, and one of 600 men. The principal dockyard is at Castellamare, in the bay of Naples. (*Serristori, Ondinet, &c.*)

The revenues are derived partly from direct and partly from indirect taxes. Of the former the *contribuzione fondiaria*, or land-tax, is by far the most important. Previously to the French *régime*, the estates of the church and the nobility were exempted from direct taxation; but the French made an end of this unjust distinction, and imposed the *contribuzione fondiaria* equally on all descriptions of land, without reference to its proprietors or occupants. The tax was originally fixed at 25 per cent. of the rent, and has not since been changed. The other principal sources of revenue are the customs' duties, the tolls on articles consumed in the different towns, the salt and tobacco monopolies, lottery, &c. The indirect taxes paid by the different districts are insured by companies of the inhabitants, who collect them at a certain per centage. We are not sufficiently informed as to the working of this plan to be able to say whether it be as advantageous as it is ingenious. We subjoin an

Account of the Income and Expenditure of the Kingdom of Naples in 1835.*

Revenue.		Expenditure.	
	Ducats.		Ducats.
Land and direct taxes.	7,442,900	President of the Council of Ministers	54,800
Customs and consumption dues	3,904,000		
Salt	3,871,699	Ministers of Foreign Affairs	250,000
Tobacco	993,800		729,000
Snow, gunpowder, and playing cards	977,800	Justice	60,000
Registrations, &c.	1,257,000	Religion	
Lottery	1,192,00	Finance (including the civil list, interest on debts, &c.	11,833,000
Sinking fund, &c.	704,600		
Deductions from salaries, &c.	1,131,600	War	7,200,000
Quota furnished by Sicily	3,977,000	Marine	1,345,000
Various sources	1,411,000	Police	807,0 0
		the Interior	1,571,000
Total	25,049,000	Total	25,100,000

History.—At a very early period most part of the coasts of Naples and Sicily were occupied by Greek colonists, the founders of some of the greatest and most flourishing cities of the ancient world. They received from this circumstance the name of *Magna Graecia.* Continental Naples was wholly subjugated by the Romans soon after their war with Pyrrhus; and formed a most valuable portion of their empire. In modern times it has undergone many vicissitudes. It was united to Spain in the early part of the 16th century; and continued, as an appanage of that kingdom, to be governed by viceroys sent from Madrid, till the treaty of Utrecht, when it was ceded to Austria. In 1734 it was erected into an independent monarchy, under the Infanta Don Carlos of Spain, who took the name of Charles III. It continued under the Bourbon dynasty till 1798, when it was overrun by the French, who held it from 1803 till 1815.

These circumstances account in part, at least, for the degraded character of the Neapolitans. Down to the invasion of the French they groaned under a succession of tyrannical or imbecile rulers: and with such a government, and with the feudal system in full vigour, a servile and ignorant nobility, a priesthood always ready to protect and absolve every scoundrel who had money and power, it would have been a miracle had the people not become as worthless, as corrupt, and as degraded as their rulers. Had the government of the French been continued for half a century, the regeneration of the country might have been effected. But Naples has again become the prey of dotage and imbecility. And, till a new order of things shall be introduced, a vigorous government established, and the oppressive restrictions on foreign trade and on the circulation of books and papers have been abolished, it would be idle to expect any material improvement in the condition or character of the people.

NAPLES (an. *Parthenope* and *Neapolis*), a famous city and seaport of Italy, cap. of the above kingdom, and the residence of the sovereign, on its W. coast, on the N. side of the bay or gulf of Naples, 118 m. S.E. Rome. In the immediate vicinity of Vesuvius, the lower elevations of which approach to within a little distance of the city, on the S.E.; lat. (Fort St. Elmo) 40° 50' N., long. 14° 15' 50" E. Pop. on the 1st of January, 1837, 351,719; but, owing to the ravages of cholera in the course of that year, the population was reduced, on the 1st of January, 1838, to 336,302. It

---

* Serristori, Statist. delle Due Sicilie
† The public debt in 1838 amounted to 4,857,000 ducats.

may be taken at present (1841) at 350,000. The situation of Naples is one of the finest that can be imagined. Seated partly on the declivity of a hill, and partly on the margin of a spacious bay, it spreads its buildings along the shore, and covers the shelving coasts and adjacent eminences with its villas and gardens. Its suburbs stretch in a magnificent and lengthened sweep, from Portici on the E. to the promontory of Misenum on the W. The bay is extensive, and presents an almost unrivalled assemblage of picturesque and beautiful scenery. On its N.W. side the shores of Pozzuoli rise in a gentle swell from the surface of the water; while on the E. Vesuvius, with its verdant sides and black smoking summit, bounds the prospect: the centre contains the city, with its palaces, churches, and gardens rising one above the other, backed by the heights on which are the royal palace of Capo di Monte, the observatory, and the castle of St. Elmo. The view from the city seaward commands the whole sweep of the bay, bounded on the S. by the promontory of Sorrentum, and having near its mouth the islands of Capri Ischia, &c. The clearness of the atmosphere and the mildness of the climate complete the gratification inspired by the scene, and justify the epithet of *mitis* given to the city by the ancients. The city has an oblong form; but, when viewed from an elevated position, such as the Carthusian monastery, the castle of St. Elmo, or the church of Santa Maria del Parto, it appears irregular, the surrounding country being so studded with houses and villages, that it is impracticable to mark the line of separation between the town and the environs.

But it is principally in respect of its situation that Naples is superior to most other cities. The streets, indeed, are generally straight and well paved, though without footpaths; but they are universally narrow, and, being bordered by lofty houses, have a dark gloomy appearance, that contrasts singularly with the splendour of the surrounding country. The Strada di Toledo, the principal street, having at the one end the Piazza di Mercato, and on the other the royal palace, runs N. and S. for about a mile; but it is only from 40 to 60 feet in width, while the houses on either side are from five to seven stories in height. Few of the other streets are more than 30 feet in width, and many not more than from 15 to 20 feet, and some not so much. The houses are flat-roofed, and covered with a kind of stucco made of Pozzolana sand, which becomes indurated on exposure to the atmosphere. Most of them have balconies in front; and these, and the booths and stalls with which the streets are constantly occupied, make them look narrower than they really are. There are several open spaces or *larghi*; for they cannot be called squares; but they are very irregular both in aspect and plan. The principal are the Largo di Castello, the Largo di Palazzo, and the Piazza di Mercato. Some of the *larghi* are decorated with fountains and obelisks; and the city is, on the whole, pretty well supplied with water.

The houses in Naples bear no analogy to those in London, but correspond pretty closely to those of Paris, except that they are generally on a larger scale. "You see," says Mr. Maclaren, "a vast tenement, with a front as long as that of Edinburgh college, but two stories higher—a grande porte, as large as the college gate, and decorated, too, with columns. This porte opens into a court as long as the building, but perhaps only 30 or 40 feet wide. The tenement, in fact, forms a parallelogram, built all round the court, with wide spacious stairs in each of its interior fronts. The whole of the ground story externally consists of a series of arched cells, probably 10 feet wide, 12 feet high, and 15 or 20 feet long. These are occupied as sale shops, cafés, and workshops. The door is always in three high and narrow divisions; in cold or wet weather the middle only is opened; in mild weather all the three are folded back, and the business is carried on in the open air. In cell No. 1, for instance, you have an oil shop; in No. 2, tripe, sausages, &c.; in No. 3, cloth of some kind; in No. 4, sacks of flour; in No. 5, a copper-smith hammering away; in No. 6 you see half a dozen tailors stitching; in No. 7 you find a confectioner, who is kneading the dough on his counter; in No. 8, a modiste, or dealer in women's dresses; in No. 9, a carpenter; in No. 10, a bookseller; in No. 11, a watchmaker. The cells are all of the same shape and size, and not one front only, but often all the four external fronts of the building are thus arranged and occupied. Such a building is called a *palazzo*, which does not mean a palace, but simply a house, or rather a tenement, in the ground story of which a crowd of shopkeepers and artizans carry on their business, and in the upper part a crowd of other persons live. Naples is almost entirely composed of palazzos, great or small, such as I have described, and they are crowded together amazingly. The ground may be said to bear a crop of houses, as a field bears a crop of corn; for gardens, or open plots of ground for drying clothes, or securing the advantages of light and air, are never dreamed of here, except as appendages to villas in the suburbs. In one thing

419

NAPLES.

Naples is magnificent—its street pavement, which invariably consists of squared blocks of lava, joined as closely and correctly as the flags of our foot pavement. They are said to be laid in mortar, as the old Roman roads were, and hence may be considered as built roads. So firm is the work, that you never see one block an inch higher or lower than another." (*Notes*, p. 51.) There is not, however, a single shop in Naples that would be reckoned handsome in London or Paris. Neither has it any good coffee-house or restaurant.

Notwithstanding his disposition to eulogise, Eustace admits that Naples has but little architectural magnificence. The prevailing taste, if a series of absurd fashions deserve that name, has always been bad. Moresco, Spanish, and Roman, corrupted and intermingled together, destroy all appearance of unity and symmetry, and form a monstrous jumble of discordance. Hence the magnificence of the churches and palaces consists principally in their magnitude, and their paintings, marbles, and other decorations.

The cathedral, built on or near the site of a temple of Apollo, a large Gothic edifice, is overcharged with ornaments in the most discordant style. It is supported by more than 100 granite columns, which originally belonged to the edifice it has replaced. In the subterranean chapel, under the choir, is the body of St. Gennaro, the tutelary saint of Naples, whose blood, carefully preserved in a crystal vase, and miraculously liquefied three times a year, is regarded by the orthodox Neapolitans as the boast of the cathedral, and the great glory and honour of the city. The Santi Apostoli, erected on the ruins of a temple of Mercury, is perhaps the most ancient church in the city, having, it is alleged, been originally erected by Constantine, but subsequently rebuilt with greater magnificence. The churches of St. Paul, St. Filippo Neri, Spirito Santo, and S. Martino, are all well deserving of attention; the latter, indeed, is said to be the most splendid and beautiful church in the city. The church Del Partu, though inferior to most others in size and decorations, deserves notice from the fact of its having been erected and dedicated to the *Virgini parienti*, by Sannazarius, author of the famous Latin poem, *De Partu Virginis*. It contains the remains of its illustrious founder, a native of the city where he expired in 1530, enclosed in a magnificent tomb, with the following distich by Bembo:

" Da sacra cineri flores : hic ille Maroni
Sysperrea * musa proximus, pt tumulo."

In all there are said to be about 300 churches in the city; and the priests compose a large, though certainly not the most valuable, part of the population.

The Neapolitans appear to entertain the most perfect indifference as to the manner in which their mortal remains are disposed of. The great burying-place of the city is alongside the splendid road leading to the *Campo Martio*. It consists of 365 deep cells, dug into the Pozzolana, of which the hill is composed. One of these cells is opened in rotation every morning, and receives all the dead bodies of the day, brought in carts, and tumbled into it, like as much rubbish; this done, it is shut up again for a year, and is then opened to receive a fresh supply of carcasses! But, exclusive of this vast golgotha, a considerable number of funerals take place in churches.

The palaces and mansions of the nobility, like the churches, have little pretensions to purity of architecture; and though in many the apartments are on a grand scale, they are in general too much loaded with ornaments. The kings of Naples have been distinguished by their rage for building. The royal palace (*Palazzo reale*) in the city, near the quay, at the S. extremity of the Strada di Toledo, though a part only of the original design, is a vast building, three stories in height, with four interior courts; the first story is of the Doric, the second the Ionic, and the third the Corinthian order of architecture. Its interior is splendidly fitted up, and it has some good paintings. Another royal residence, the Capo di Monte, finely situated on an eminence outside the town, on the N., commands a magnificent view. It has attached to it some fine gardens; and it possessed, a few years ago, some remains of the famous Parma gallery, including portraits by Raphael, Andrea del Sarto, and Parmegiano, and some fine sculptures; these, however, have been mostly removed to the museum. This palace is now united to the city by a magnificent road, constructed by the French, and called, during their ascendancy, the Strada di Napoleone. The old palace of the Neapolitan monarchs is now occupied by the courts of Justice, and by the archives of the kingdom.

The *Palazzo degli Studii Publici*, erected in the early part of the 17th century, from designs by Fontana, is by far the most interesting building in Naples. It was intended for the university, and was used as such from 1616, when it was completed, down to 1790, when the university was re-

moved to the convent of Gesu-Vecchio, and the Palazzo degli Studii was converted into a great national museum, the *Museo Borbonico*. In addition to a noble library, comprising about 150,000 volumes, and many MSS., this museum contains a matchless collection of bronzes, gems, paintings, household furniture, papyri, and Etruscan vases, from Herculaneum, Pompeii, Stabiæ, Nola, Capua, and other ancient cities: and, in addition to these treasures, which are, in all respects, unique and unrivalled, it contains, exclusive of others, most of the statues and pictures formerly comprised in the Farnese palace at Rome, brought thither when the king of Naples succeeded to the rich inheritance of that family. The collection of statues is, in fact, inferior only to those of the capital and Vatican, and the gallery at Florence, while in paintings it yields only to Rome, Florence, and Bologna. But, despite all the treasures of the south, Naples is not at present either a school or a cradle of art, which is at a lower ebb here than in any other considerable city of Italy.

The university of Naples, founded in 1294, has above 1300 students. It is presided over by a rector, and divided into faculties, under deans, who, with the professors, receive very inadequate salaries from the crown. The professors are frequently chosen with little regard to merit; and the instruction they afford in all the higher branches of science and philosophy is most inefficient and worthless. Public law, moral and political philosophy, and even theology, can hardly be said to be taught at all; or, if taught, the instruction is not intended to expand or enlighten the mind, but to imbue it with the grossest prejudices. There are, besides the university, many superior, as well as inferior, schools; but, speaking generally, they are all miserably bad; and, Turkey excepted, education is nowhere at a lower ebb than in Naples. The censorship is extremely strict; and such foreign works as are admitted, are burdened with an extremely high duty. Under such circumstances, one cannot be so surprised at the gross ignorance prevalent among the mass of the population: the wonder, in fact, is, that there should be so much intelligence among them as is found really to exist.

Naples has a *Societa Reale Borbonica*, or Royal society, divided into the three sections of the fine arts, science, and archeology; and other literary and scientific associations; a military and naval college; a royal medical college, a veterinary college, a royal college of music; a fine botanical garden, constructed by the French; an observatory, in an elevated situation to the N. of the city, &c. One of the most curious institutions in Naples is a school where natives of China are instructed in the principles of Christianity, and qualified to act as missionaries. The number of pupils is small, seldom exceeding six.

Naples has numerous and some very extensive establishments for the support and relief of the poor, including a school for the deaf and dumb, and an asylum for the blind. The two principal hospitals are those, *Degli Incurabili* and *Della Annunziata*; the former, notwithstanding its name, is open to the sick of all descriptions, and has a revenue of about 300,000 ducats a year. The latter is destined to receive foundlings and penitent females. Here, however, as everywhere else, the opening of hospitals for the reception of foundlings is productive of a vast amount of mortality and immorality. Their influence in the latter respect is too obvious to require illustration ; and they are really the most efficient means that could be devised for occasioning the destruction of the children they are intended to preserve. In Naples, for example, in 1837, 2319 children were taken to the different receiving houses attached to the Foundling hospital, of which 1694 died in the course of the year : the greater number being, in fact, in a dead or dying state when they were received ! The truth is, that, instead of discouraging, foundling hospitals really act as a powerful incentive to infanticide. Its *Reclusoria*, or *Albergo de Poveri*, is an immense workhouse, or rather asylum for the destitute poor who are able to work, and for orphans and poor children of both sexes, who are lodged and educated. The hospital of San Gennaro, near the hill of Capo di Monte, is intended for the reception of infirm and aged poor, or poor unable to work. But, despite its hospitals, such is the want of industry, and the defects of the police, that there probably is no other city whose streets are infested by so large a proportion of poor, miserable, wretched mendicants.

Naples has six or seven theatres. That of San Carlo, the largest and finest in Italy, was nearly burnt down in 1815 : but it was soon after repaired, and reopened with more than its original splendour. Among the minor theatres, two or three are wholly devoted to the exhibition of Pulcinella, or Punch, who is here seen in his glory. " What," asks Forsyth, " is a drama in Naples without Punch ? or what is Punch out of Naples ? Here, in his native tongue, and among his own countrymen, Punch is a person of real power : he dresses up and ret ils all the drolleries of the day ; he is the channel, and sometimes the source, of the passing opinions; he can inflict ridicule, he could gain a mob, or keep the whole a kingdom in good humour. Such was De Fiori, the Aristophanes of his nation, immortal in buffoonery." (P. 266.)

The finest promenade is that called the Chiaja, extending along the shore from the Castello dell' Ovo, E. to Virgil's tomb and the hill of Posillippo: it is in part planted and ornamented with statues and fountains, and is, altogether, one of the finest public walks that is anywhere to be met with. The mole also, is a favourite promenade, and the quays that stretch E. from it towards Portici.

Though Naples could offer no effectual resistance to an invading armament, it is not altogether unprovided with the means of defence, having the castle of St. Elmo on a hill on its N.W. side, the Castello Nuovo adjoining the royal palace and the bay, and the Castello dell' Ovo on a rock projecting into the sea. The Castello Nuovo is so situated as to afford a safe retreat to the royal family and court in the event of any disturbance in the capital. The castle of St. Elmo has extensive subterranean bomb-proof works. The arsenal and cannon foundry are situated between the Palazzo Reale and the sea.

A vast number of employments must necessarily be carried on in so great a city; but there are few manufacturing establishments on anything like a large scale. Some woollen, silk, and linen stuffs are, however, produced; as are hats, gloves, earthenware, jewellery, &c.; foundries, wrought on account of government, furnish cannon, fire-arms, iron cables, &c. The preparation of maccaroni may, however, be said to be the distinguishing business of Naples. It forms the principal food of the bulk of the population, and is, therefore, largely produced. The best maccaroni is made of the flour of the hard wheat (grano duro) brought from the Black sea. Being mixed with water, it is kneaded by means of heavy wooden blocks wrought by levers, till it acquires a sufficient degree of tenacity; it is then forced, by simple pressure, through a number of holes, so contrived that it is formed into hollow cylinders. The same gives to the tubes depends on their diameter; those of the largest size being maccaroni, the next to them vermicelli, and the smallest fedelini. When properly prepared and boiled to a nicety, Neapolitan maccaroni assumes a greenish tinge. It is then taken out of the caldron, drained of the water, and being saturated with concentrated meat gravy, and sprinkled with finely grated cheese, forms a dish of which all classes, from the prince to the beggar, are passionately fond. But the maccaroni used by the poor is merely boiled in plain water, and it is rarely eaten with any condiment whatever. When properly prepared, it is nutritious and easy of digestion. The Lazzaroni pique themselves on the dexterity with which they swallow long strings of maccaroni and vermicelli without breaking them.

Commerce.—The harbour of Naples is formed by a mole projecting from the centre of the city, nearly in the form of the letter L, having a lighthouse on its elbow. Immediately within the mole there are from three to four fathoms water, the ground being soft, but only small vessels can approach the town. The water in the bay is deep, and there is no bar, but it is a good deal exposed to the S. westerly winds; and, to guard against their influence, vessels in the bay moor with open hawse in that direction. For remarks on the system under which the trade of Naples and the kingdom generally is conducted, we beg to refer to the previous article. Subjoined is

An ACCOUNT of the Quantities of the principal Articles exported from Naples in 1838.

Articles.	Quantities.			Articles.	Quantities.		
Argols	tons	582	10	1	Lamb and kid skins		145	2	1
Almonds		26	4	2	tons				
Aniseed		1	1	2	Leather gloves	prs.	348,164		
Bones		596	0	0	Do. parings	tons	3	11	3
Bone shaving		162	1	0	Lemon juice	casks	263		
Brandy	bottles	4,324			Linseed paste	tons	92	15	0
Casks used	No.	4,016			Liquorice paste		673	7	0
Cask staves, 6 ft.		4,409			Maccaroni		84	17	1
Do. do. 4 ft.		3,143			Madder root		628	3	1
Do. hoops bundles		10,790			Do. ground		90	13	1
Do. tops & bottoms		3,761			Nut-galls		70	8	3
Cream of tartar lbs.		895,627			Nuts and Walnuts		17	3	0
Coral	tons	11	6	2	Olive oil	galls.	5,074,550		
Do. wrought		0	5	16	Pigeon dung	tons	18	18	2
Essence of orange	lbs.				Rags woollen		545	6	3
		9,797			Raisins, dried		31	1	1
Figs	tons	764	6	0	Silk, raw	lbs.	129,942		
Flax		2	11	0	Do. manufactured		9,374		
Gut strings		0	19	3	Do. dyed		131,176		
Wheat		1,769	3	0	Do. waste		5,643		
Indian Corn		240	13	2	Do. Handkchfs. No.		15,816		
Oats	qrs.	71	2	3	Soft soap	tons	9	15	2
Beans	tons	277	2	1	Hard do.		0	14	0
Haricots	qrs.	1,374			Sulphur		96	5	0
Pease	tons	54	7	2	Tartar, &c.	lbs.	86,174		
Lupins	qrs.	3,592			Tallow	tons	413	12	3
Linseed	tons	1,412	17	2	Do. Candles				
Hay	bundles	3,460			Wool		296	16	0
Leaves of Ind. corn		5,862			Wine in casks		2,636		
Hemp	tons	1,690	3	1	Do. in bottles		5,780		
Rods		7	0	Zaffree	lbs.	5,975		

36

During the same year, there cleared out from the harbour 1227 ships, of which 1051 were Neapolitans, 80 English, 23 French, 43 Sardinian, 90 Tuscan, &c. Naples and Castellamare have about 2200 registered vessels of the burden of nearly 100,000 tons. The duties collected at the Neapolitan custom-house amount, at an average, to about 3,500,000 ducats, or about 660,000l. a year.

There are four or five companies for the insurance of ships, and one for lives. Their terms are generally higher than those of similar establishments in London. Houses are never insured at Naples, their construction rendering fires very rare. The companies are established by royal authority, the shareholders being only liable for the amount of their shares.

The principal merchants of Naples are all, more or less, bankers, inasmuch as they advance money on letters of credit, and deal in foreign exchange, and other financial operations. But the only banking establishment at present in existence, is the Bank of the Two Sicilies, founded by government, and guaranteed by the possession of landed property. It is not a bank for the issue of notes on credit, like the Bank of England, but for their issue on deposits somewhat on the principle of the Bank of Hamburg. Government makes all its payments by means of notes or orders on the bank; and they are issued to individuals for whatever sums they desire, on their paying an equivalent sum of money to the bank. These notes or orders form a considerable part of the circulating medium of Naples; they are paid in cash on demand. The building occupied by the bank is one of the finest in the city.

Government has also established a discount office, where bills endorsed by two persons of good credit, and not at more than three months' date, are discounted at 4 per cent. Goods are universally sold at long credits, mostly from four to eight months, and for manufactured goods sometimes longer. Discount for ready money is at the rate of 6 per cent. per annum. Merchants are arranged by the Chamber of Commerce into five different classes, and a six months' credit is given at the custom-house for duties, to the extent of 60,000, 40,000, 30,000, 20,000, and 15,000 ducats, to individuals, according to the class in which they happen to be enrolled. But this is of little importance, as the transactions of a merchant must be very limited indeed, if the duties he have to pay be not much more than the credit he is allowed. High discriminating duties are charged on all foreign ships entering the port.

Society in *Naples* has undergone many considerable changes during the present century; but its distinguishing features have not materially varied for a lengthened period, and Goldsmith's admirable picture of Italian manners is still more applicable to this than to any other portion of the peninsula :—

> But small the bliss that sense alone bestows,
> And rquual blss is all the nation knows.
> In florid beauty groves and fields appear,
> Man seems the only grow'r that dwindles here.
> Contrasted faults through all his manners reign;
> Though poor, luxurious; though submissive, vain;
> Though grave, yet trifling; zealous, yet untrue;
> And even in penance planning sins anew.

The nobility are exceedingly numerous, and are as fond as ever of splendour and parade. Previously to the occupation of the country by the French, the greater number of them were very poor; and the changes introduced in 1806, and the subdivision of property that has taken place in the interval, have considerably reduced the fortunes of those who had formerly large estates. "Titles are here so common that you find at every corner Precipi or de Precipi without a virtue or a ducat." The rage for carriages and equipages is as great at this moment as it was in the days of Dr. Moore: "Women at all above the lower ranks do not walk; those who cannot afford a carriage, are doomed by pride to perpetual imprisonment in their own houses, or only go to church with one or two poor devils hired for the occasion, who put on antiquated livery, and carry a book or a cushion. I am told that husbands sometimes perform the office, trusting probably that they shall escape recognition under the disguise of a footman, and choosing to gratify vanity at the expense of pride. The roofs of the houses, which are flat, and adorned with flowers and shrubs in boxes, afford air and exercise to the women. Thus living in idle retirement, their mind is exclusively bent on the means of procuring a lover; and the tales of Boccacio and Lafontaine convey a likeness of their moral habits and manners." (Simond's *Italy*, 435.)

The numbers and wealth of the clergy were greatly diminished by the measures adopted by the French; and though both have been increased since the Restoration, their wealth has not been increased in proportion to their numbers. They are, speaking generally, poor, ignorant, and servile; and though the outward forms of religion be respected, it has nowhere so little real influence as at Naples. The lawyers, merchants, physicians, artists, and such-like persons, form the most intelligent and most valuable portion of society.

The *lazzaroni*, so prominent in the descriptions of Naples, formerly included most part of the lower classes, comprising street-porters, hawkers, water-carriers, boatmen, hackney-coachmen, mendicants, &c. Their numbers were loosely estimated at from 30,000 to 40,000, and they were said to constitute a distinct race, immersed in poverty, only half-clothed and not half-fed, without lodgings, and sleeping in the open air in the porches of churches and other public buildings. But it is now admitted that the lazzaroni, properly so called, or the' houseless poor, are merely the dregs of the population, and that they owe their gipsy-like complexion and cast of features to their constant exposure to the sun and air. It is singular that wretches in so destitute a condition, and frequently involved in all but the extremity of want, should, speaking generally, be remarkable for their fine symmetrical and muscular forms, and be distinguished for their vivacity and humour. Great efforts have been made for many years past to lessen the numbers of the lazzaroni; and, under Murat, many of them were drafted into the army. But they are still extremely numerous; Mr. Maclaren says he saw numbers of half-clothed wretches (lazzaroni) asleep, in sunny days, on the pavement of the Chiaja. "They are the refuse of civilization, sunk to the condition of savages. It is said there are individuals among them who do not know their own names; and who go to the priest and confess anonymously, owning sins of whose designation in the decalogue they are ignorant." Unless when pressed by hunger, or under some peculiar and extraordinary excitement, the lazzaroni are neither turbulent nor licentious; but on such occasions they evince all the sanguinary ferocity of savages. They seem, however, to be wholly incapable of any vigorous or long-continued exertion for any public purpose, and may, speaking generally, be regarded as submissive, docile slaves. It used to be a common saying at Naples, that, to keep the populace quiet, three things only were necessary—*festa, farine,* and *forche;* that is, shows, food, and gibbets! And this compendious principle has not yet fallen into disrepute, though certainly they are supplied with but a very scanty portion of *farine.*

Lottery offices are extremely numerous in Naples, and have, as might be expected, a most injurious operation. Tickets are so subdivided, that shares may be purchased for about 2d.: the moral pestilence consequently descends to the very lowest ranks, and even the lazzaroni are speculators. There are, probably, a greater number of pickpockets in Naples than in any other city; and deceit and falsehood are so common as hardly to excite attention. The *donne libere* are also extremely numerous.

Owing principally, no doubt, to its mild climate, a large proportion of the population of Naples may be said to inhabit the streets, and to carry on their business out of doors; and the competition arising among parties so situated, has probably given rise to that universal turmoil and effort to attract notice, that is at once so grotesque and so disgusting to a stranger.

"Naples," says Mr. Forsyth, "in its interior, has no parallel on earth. The crowd of London is uniform and unintelligible; it is a double line in quick motion; it is the crowd of business. The crowd of Naples consists in a general tide, rolling up and down; and in the middle of this tide, a hundred eddies of men. Here you are swept on by the current; there you are wheeled round by the vortex.

Qui vid' io gente, più che altrove troppa
E d'una parte e d'altra con gran' urli,
Portare tutti incontro ——- .   *Dante.*

A diversity of trades dispute with you the streets. You are stopped by a carpenter's bench, you are lost among shoemakers' stools, you dash among the pots of a *maccaroni* stall, and you escape behind a *lazzaroni's* night-basket. In this region of caricature, every bargain sounds like a battle: the popular exhibitions are full of the grotesque : some of their church processions would frighten a war-horse.

"The mole seems, on holidays, an epitome of the town, and exhibits most of its humours. Here stands a methodistical friar, preaching to one row of *lazzaroni;* there, Punch, the representative of the nation, holds forth to a crowd. Yonder, another orator recounts the miracles performed by a sacred wax-work, on which he rubs his *agnuses,* and sells them, thus impregnated with grace, for a grain a piece. Beyond him are quacks in hussar uniforms, exalting their drugs and brandishing their sabres, as if not content with one mode of killing. The next *professore* is a dog of knowledge, great in his own little circle of admirers. Opposite to him stand two jocund old men, in the centres of an oval group, singing alternately to their crazy guitars. Further on is a motley audience, seated on planks, and listening to a tragi-comic *filosofo,* who reads, sings, and gesticulates, old Gothic tales of Orlando and his Paladins.

" If Naples be 'a paradise inhabited by devils' I am sure it is by merry devils. Even the lowest class enjoy every blessing that can make the animal happy—a delicious climate, high spirits, a facility of satisfying every appetite, a conscience which gives no pain, a convenient ignorance of their duty, and a church which ensures heaven to every ruffian that has faith. Here tatters are not misery, for the climate requires little covering ; filth is not misery to them who are born to it ; and a few fingerings of maccaroni can wind up the rattling machine for the day.

"They are, perhaps, the only people on earth that do not pretend to virtue. On their own stage, they suffer the Neapolitan of the drama to be always a rogue. If detected in theft, a lazzaroni will ask you, with impudent surprise, how you could possibly expect a poor man to be an angel. Yet what are these wretches ? Why, men whose persons might stand as models to a sculptor ; whose gestures strike you with the commanding energy of a savage; whose language, gaping and broad as it is, when kindled by passion, bursts into oriental metaphor ; whose ideas are confined, indeed, within a narrow circle—but a circle in which they are invincible. If you attack them there, you are beaten. Their exertion of soul, their humour, their fancy, their quickness of argument, their address at flattery, their rapidity of utterance, their pantomime and grimace, none can resist but a lazzaroni himself.

"These gifts of nature are left to luxuriate unrepressed by education, by any notions of honesty, or habits of labour. Hence their ingenuity is wasted in crooked little views. Intent on the piddling game of cheating only for their own day, they let the great chance lately go by, and left a few immortal patriots to stake their all for posterity, and lose it." (P. 264–267, and 412.)

"The people," says M. Simond, " seem, in general, peaceful and contented, unconscious of want at least ; they consume little, and that little is cheap. For three grains a day (three half-pence sterling) a man has his fill of maccaroni, and for three grains more he may have his fritura (very good fish or vegetables fried in oil) at any of the innumerable stands of itinerant cooks about the streets, which is not the only luxury of the gastronomic kind within his reach. A glass of ice-water costs one sixth of a grain (one twelfth part of a penny sterling), and, if properly seasoned with lime-juice and sugar, two grains. The price of these things is kept down by government, ice or hardened snow being abundantly supplied at the public expense from natural ice-houses, in certain cavernous rocks above Stabiæ and Sorrento, and even on Vesuvius. The ice in baskets is made to slide down the mountain, along light ropes, into boats, which sail across the bay during the night, and land their precarious cargoes before day.

"The lower people have clubs, where they assemble twenty or thirty together, and contribute each one grain for wine of an evening. They elect a president and vice-president. The president calls upon one of the members to drink a glass of wine filled by the vice-president ; but when the member challenged is about to take it, the vice-president has the right to say, I take it for myself, and actually drink it to his health ; a standing joke, which he may repeat as long as he pleases, or as long as he can, but which the disappointed expectant, who has the laugh of the company against him, does not always relish ; and in the end there is sometimes fighting and squabbling." (P. 422.)

The country round Naples is the most beautiful that can well be imagined, and is peculiarly interesting from its classical associations. Virgil was buried in the immediate vicinity of the city ; and the ruins of an ancient mausoleum on the left hand side of the road, leading from the promenade of Chiaja to the grotto of Pausilippo, is said to have contained the remains of the prince of Latin poets. There is, however, no really good foundation for this statement.

The grotto of Pausilippo, now alluded to, is a tunnel cut through the hill of that name, being a part of the road from Naples to Pozzuoli. It is about two thirds of a mile in length, 60 feet in height, and broad enough to serve for a highway. This work is of great, but unknown, antiquity. Seneca, in his 57th epistle, complains bitterly of its length, darkness, and dust. (*Nihil illo carcere longius, nihil illis faucibus obscurius ; etiam si locus haberet lucem, pulvis eam ferret.*) Its dimensions were, however, enlarged in 1537 ; and it is now well paved and lighted with lamps by day as well as by night. (For further information as to the environs of Naples, see the articles BAIÆ, HERCULANEUM, POMPEII, POZZUOLI, VESUVIUS, &c., in this work.)

Historical Notice.—Naples is very ancient. It was founded by the people of Cumæ, a colony from Greece, who gradually spread themselves round the bay of Naples, and was called from this circumstance Neapolis, or the new city. It was also called Parthenope, from its being the burying-place of one of the sirens of that name. (*Falcidius Paterculus,* lib. i., cap. 4 ; *Strabo,* lib. x.) It was, therefore, to all intents and purposes, a Greek city ; its inhabitants spoke the Greek language, and were long distinguished by their attachment to the manners and customs of their ancestors. It was on this account, according to Tacitus, that it was selected by Nero to make his *début* on the stage; such a proceeding being less offensive there, and less repugnant to the prevailing sentiments, than in Rome. (*Tacitus Hist.* lib. xv., cap. 33.) Naples, in truth, was then, as now, a chosen seat of pleasure. Its hot baths were reckoned equal to those of Baiæ; and the number and excellence of its theatres and other places of amusement, its matchless scenery, the mildness of the climate, and the luxury and effeminacy of the inhabitants, made it a favourite retreat of the wealthy and luxurious Romans, and justifies Ovid in calling it *in otia natam Parthenopen*. After the fall of the Roman empire, it underwent many vicissitudes. It, however, early became the capital of the modern kingdom of Naples; and, notwithstanding the calamities it has suffered from war, earthquakes, &c. it has long been the most populous city of Italy, and one of the most interesting that is anywhere to be met with. (Besides the authorities already referred to, see *Serristori's Statistica d'Italia*, *parte settima*, passim; *Official Statement of the Population of Naples as the 1st of January*, 1838; *Stark's and Vallery's Guides*; *Com. Dict.*; *Macgregor's Report*, &c.)

NAPLES, p. t., Ontario co., N. Y., 18 m. S. Canandaigua, 220 m. W. Albany, 395 W. Drained by inlets of Canandaigua and Honeoye lakes. It contains three churches, a Presbyterian, Congregational and Methodist, seven stores, three fulling-mills, three flouring-mills, 16 saw-mills, two tanneries, one distillery, one printing-office, one weekly newspaper; 14 schools, 863 scholars. Pop. 2345.

NAPLES, p. v., Scott county, Ill., 56 m. W. Springfield, 534 W. Situated on the E. bank of Illinois river, 2 m. above the mouth of Mauvaiseterre creek. It contains several stores, some of which are wholesale, three steam saw and grist mills, 190 dwellings and about 600 inhabitants. A ferry here crosses Illinois river. Several hundreds of arrivals and departures of steamboats take place annually.

NAPOLEON, p. t., capital of Henry county, O., 154 m. N.W. Columbus, 485 W. It contains four stores, one grist-mill, one saw-mill; two schools, 59 scholars. Pop. 615. The village is on the N. side of Maumee river, and contains a courthouse, jail, and a few stores and dwellings.

NAPOLEON, p. t., Jackson co., Mich., 66 m. W. Detroit, 542 W. It has three stores, one flouring-mill, two saw-mills; one academy, 86 students; two schools, 131 scholars. Pop. 1698.

NAPOLEON, p. v., Chicot county, Ark., 148 m. S.E. Little Rock, 1097 W. Situated on the W. bank of Mississippi river, at the mouth of Arkansas river. It contains three stores, and about 250 inhabitants.

NAPOLI, p. t., Cattaraugus co., N. Y., 308 m. W. by S. Albany, 341 W. Watered by Coldspring creek. It has one store, two saw-mills; seven schools, 316 scholars. Pop. 1145.

NARBONNE (an. *Narbo Martius*), an ancient city of France, dep. Aude, about 4 m. S. from the Aude, and 7 or 8 m. from the Mediterranean, on a navigable canal that unites it with the sea, on the one hand, and with the river on the other, and which also unites it with the canal du Midi, a little to the N. of the Aude, 32 m. S.W. Montpellier, and 34 m. N. by E. Perpignan; lat. 43° 11' 13'' N., long. 3° 0' 94'' E. Pop., in 1836, ex. com., 10,792. It stands in a fine plain, and is surrounded by a wall flanked with towers, and entered by four gates. Streets narrow and tortuous, and houses mean and ill-built; it is divided by the canal into two nearly equal parts, *la Bourg* and *la Ville*, connected by three bridges; and is plentifully supplied with water by numerous fountains connected with springs outside the walls. The esplanade, or *Place des Barques*, in the centre of the town, is a fine open space; but its beauty is much impaired by the almost total absence of vegetation. Other promenades are formed near the gates; on the banks of the canal is a fine public walk, planted with trees, and the environs generally are extremely beautiful. Narbonne has few buildings worthy of notice, except the cathedral and archbishop's palace. The former, built in the 13th and 14th centuries, is one of the finest specimens of Gothic architecture in Europe: the choir, however, is the only part complete, the nave, commenced in 1708, being unfinished. Two towers rise from its W. end; but they are deficient in that lightness and elegance observable in similar structures of the same æra. The archbishop's palace (celebrated in history, as having been the place where Louis XIII. signed the order for the trial of De Thou and Cinq-Mars), is an ancient castellated building in the Place des Barques, having attached to it a massive square tower, built in the middle ages: in the front court are the remains of a marble altar, erected by the Narbonnese to Augustus Cæsar, and, in the garden, a fine tomb of white marble. The two parish churches are ancient and massive structures, but built in very bad taste,

and remarkable only for some curious sculptures. The other chief buildings are three hospitals, the exchange, arsenal, barracks, prison, a museum, theatre, and public baths. Narbonne is the seat of a tribunal of original jurisdiction and commerce; and has some silk filatures, fabrics of coarse linen cloth, worsted caps, and paper, with numerous distilleries, potteries, chemical works, tanyards, &c. It is the centre of the wine and spirit trade of the department and the principal support of its inhabitants is derived from its trade in wine, corn, brandy, silk, oil, salt (obtained from the neighbouring lagoons), wax, and honey, which it exports, partly to Bordeaux, by the canal du Midi, and partly to Marseilles and other markets on the Mediterranean, by its port of La Nouvelle, at the mouth of the canal on which it is built. The honey of Narbonne is said to be the finest in the world. "Its peculiar excellence is owing to the variety of nourishment for the bees. The hives are moved from one place to another. From the gardens of Narbonne they are carried to the meadows in the neighbourhood; and they are afterwards conveyed 30 or 40 m., as far as the low Pyrenees, so that the treasures of the gardens, meadows, and mountains, are all rifled to produce the honey of Narbonne. In England this system, though doubtless it would be advantageous, could not effect what it accomplishes at Narbonne, because numerous aromatic plants, abounding in the S. of France, are not indigenous to Great Britain. It is of a much higher flavour than any other honey, and so odoriferous that one might fancy himself eating a bouquet." (*Inglis; Switzerland, France, &c.*, p. 311.) Fruit is extremely abundant and cheap. The price of labour do not exceed a franc a day, but the necessaries of life are obtained at the same easy rate. Meat, however, is not reckoned among them by the lower orders; for "at Narbonne we have got so far south, as to discover something of those indolent habits which produce, in still more southern countries, a distaste for all exertion beyond that necessary to preserve existence, and which limit the necessities of life to the natural productions of the soil." (P. 312.) It is worthy of remark, also, that the dress of the peasantry in the neighbourhood bears a striking resemblance to that worn by the Catalonians. The neighbourhood is fertile in corn, but is rendered unhealthy in summer by the salt lagoons fringing the shores of the Mediterranean. There are numerous salt-pans, and marble is quarried near the town.

Narbo, one of the most ancient towns of Gaul, and the chief city of the Volcæ Arecomici, was formed into a Roman colony *anno* 116 B.C.; Julius Cæsar further enlarged it by sending thither the veterans of the tenth legion, and Cicero (Or. pro M. Fonteio, c. i.) terms it *colonia nostrorum civium, specula populi Romani ac propugnaculum.* At the distribution of Gaul into provinces by Augustus it gave its name to the S.W. province, called *Narbonensis:* Mela speaks of it as a place *unde olim terris auxilium nunc et nomen et decus est*, and Strabo designates it as the emporium of all Gaul. Its public buildings, and great commercial wealth, are mentioned by other authors; but the present remains of its ancient grandeur are confined to a few fragments and inscriptions, chiefly incorporated in the walls of the town. It fell into the hands of the Visigoths, A.D. 462, and was shortly after made the capital of their kingdom. In 720 it was taken by the Saracens, and by Pepin-le-bref in 759; after many vicissitudes it was finally annexed to the crown of France in the early part of the 16th century. Its ancient walls were demolished by Simon de Montfort during the wars against the Albigenses: those by which it is now surrounded having been constructed, considerably within the limits of the old walls, by Francis I. It has, at different times, suffered severely from the plague. In the wars of the League, Narbonne embraced the cause of the Huguenots; but in 1591 it submitted to Henry IV. (*Hugo, art. Aude; Guide du Voy. en France; Inglis.*)

NARDO (an. *Neritum*), a town of the Neapolitan dom., prov. Otranto, cap. cant., on the road from Lecce to Gallipoli, 15 m. S.S.W. the former, and 10 m. N.N.E. the latter, city. Pop. about 6000. It is a substantial, flourishing town, neatly built, and well paved. It is a bishop's see; and has numerous churches, a hospital, and manufactures of cotton goods, the raw material of which is grown in its vicinity. It was a place of some note in antiquity, as a city of the Salentines; and was held in esteem as a seminary of learning as late as the middle of the 15th century. (*Orsen's Tour, &c.*, p. 138, 139; *Rampoldi.*)

NARNI (an. *Nequinum* and *Narnia*), a town of central Italy, Papal states, deleg. Spoleto, on a lofty eminence, at the foot of which flows the Nera (an. *Nar*), 44 m. N. by E. Rome. Pop. about 3600. The town has nothing but its antiquity and picturesque appearance to recommend it: it is badly built, with steep and narrow streets, and exhibits every mark of poverty and decay. It has a cathedral, several other churches, numerous convents, a modern aqueduct, which supplies several public fountains, and the ruins of an amphitheatre. But it is principally celebrated for the remains of a noble bridge thrown by Augustus over the Nar, constructed after the Etruscan method, of large blocks of marble without cement: it is supposed to have been originally upwards of 630 ft. in length. Only one of the arches remains perfect, the span of which is above 60 ft.: the piers supporting it are 98 ft. in breadth. Addison styles this bridge "one of the stateliest ruins in Italy;" and few relics of antiquity are better adapted to impress the mind with high ideas of Roman magnificence. Narni was the birth-place of the Emperor Nerva. (*Cramer's Ancient Italy*, i., 277, 278; *Rampoldi*; *Condor's Italy, &c.*)

NARO (supposed to be the an. *Motyum*), a town of Sicily, intend. Girgenti, cap. cant. on the Naro (an. *Hypsa*), 13 m. E. by S. Girgenti, and 21 m. S.W. Caltanisetta. Pop. in 1831, 10,105. It is situated on an eminence, surrounded by picturesque valleys and glens; has a royal college, and a house of refuge, and some trade in oil, wine, and sulphur, which last is very abundant in its vicinity. Many sepulchres, medals, and other vestiges of antiquity, have been found here. (*Smyth's Sicily*, p. 202; *Dict. Geog. Ortolani, Diz. della Sicilia.*)

NARRAGANSET BAY, R. I., is 28 m. long, and from 3 to 12 broad, and sets up from the Atlantic, between point Judith on the W. and Seaconnet rocks on the E. It receives Providence and Pawtucket rivers on the N.W., Taunton river on the N.E. and Pawtuxet river on the W. side. It contains the beautiful islands of Rhode Island, Conanicut, Patience, Prudence and Hope. Mount Hope bay branches from it on the N.E. It has many good harbours, accessible at all seasons. It is a fine naval station.

NARRAINGUNGE, a considerable trading town of British India, prov. Bengal, distr. Dacca-Jelalpore, on a branch of Brahmaputra, 8 m. S.E. Dacca. Lat. 23° 37' N.; long. 90° 35' E. Pop. estimated at 15,000. The inhabitants carry on a large trade in salt, grain, tobacco, and lime; and the town exhibits a scene of bustle and activity seldom witnessed in a community of Bengalese. The banks of the river are studded with indigo factories, and the remains of forts erected to repel former invasions of the Arracanese. (*Hamilton's E. I. Gaz.*)

NARVA, a town of European Russia, gov. of Petersburg, on the Narova, about 8 m. from its mouth, and 81 m. W.S.W. St. Petersburg, lat. 59° 22' 53" N., and long. 28° 24' 40" E. Pop. 3000. It is divided into an old and more recent part; the latter, placed on high ground, is surrounded with fortifications in a good state of repair, and consists of respectable stone houses; the lower and older part comprising only a few wretched wooden tenements, with two churches, one of which belongs to the Greek, the other to the Lutheran, religion. Three other churches, a town-hall, exchange, and the half-ruinous fortress of Ivangorod (built in 1492, by the Czar Ivan III. Vassilievitch), are the only other public edifices.

The place, in fact, would not be worth notice but for the famous battle fought in its vicinity, on the 30th of November, 1700; when Charles XII., king of Sweden, at the head of only 8000 men, attacked and forced the entrenched camp of the Russian army, consisting of about 80,000 men, which had been besieging Narva. The Swedes gained a complete victory. Above 18,000 Russians were killed in their intrenchments, besides a great number drowned in the river: next day above 30,000 Russian troops surrendered to the Swedes, by whom they were disarmed, and dismissed. This extraordinary success did not cost the Swedes above 600 men! On hearing of this disaster, the czar, Peter the Great, said, "*Je sais bien que les Suédois nous battront long temps; mais à la fin il nous apprendront eux-mêmes à les vaincre:*" and the event proved that he was in the right. (See *Voltaire, Histoire de Charles XII.*, liv. ii.)

NASEBY, a decayed market town and par. of England, co. Northampton, hund. Guilsborough, 11½ m. N.N.W. Northampton, and 73 m. N.W. London. Area of par. 3690 acres. Pop. in 1831, 707. This village formerly possessed a market and a worsted manufactory, but they have long ceased to exist; and the market cross, in the centre of the village, is the only extant sign of its past importance.

But, how unimportant soever in other respects, this trifling village will be ever memorable in British history for the battle fought near it on the 14th of June, 1645, between the royalists under Charles I. and the parliamentary army commanded by Cromwell and Fairfax. The action was obstinate and well contested; but in the end the parliamentary leaders gained a complete and decisive victory. The loss in killed on both sides was nearly equal; but the republicans took 500 officers and 4000 soldiers, and all the king's artillery and ammunition. This action may be said to have terminated the civil war.

NASH county, N. C. Situated a little N.E. of the centre of the state, and contains 640 sq. m. Watered by Tar river, and drained by its branches. It contained in 1840, 755 neat cattle, 635 sheep, 2937 swine; and produced 833

---

bushels of wheat, 28,325 of Indian corn, 3582 of oats, 7882 of potatoes, 50,786 pounds of cotton. It had nine schools, 1-8 scholars. Pop.: whites, 4941; slaves, 3697; free coloured 409; total, 9047. Capital, Nashville.

NASHUA, p. t., Hillsborough county, N. H., 12 m. N.W. Lowell, 36 m. S. by E. Concord, 447 W. Bounded E. by Merrimac river, S.E. by Nashua river, which has a fall of 65 ft. in 2 m., producing a great water-power. It contains six churches, two Congregational, a Unitarian, a Free-will Baptist, a Christian and a Universalist, 30 stores, five cotton factories with 34,348 spindles, three saw-mills, one tannery, one pottery, two printing-offices, two weekly newspapers; one academy, 214 students; 36 schools, 1476 scholars. Pop. 6054.

NASHVILLE, city; a capital of Davidson co., and the state of Tennessee is 36° 9' 03" N. lat., and 86° 49' 3" W. long., and in 9° 48' 3" W. long. from W.; 110 m. E. Huntsville, Ala., 183 m. W. Knoxville, Ky., 250 m. S.W. Lexington, Ky., 909 m. S.W. New-York, 648 W. The population in 1830, was 5566; in 1840, 6929; in 1844, 7800 within the city limits, and including its suburbs, 11,000. It is pleasantly situated on the S. side of Cumberland river, 120 m. from its mouth, at the head of steamboat navigation, though keelboats go much farther. The site is undulating and rocky, with unequal elevation, from 50 to 175 feet, interspersed with beautiful groves of cedar. The environs present a rich variety of landscape scenery, and the place is remarkably healthy. The city was originally laid out on 200 acres of ground, in lots of one acre each, with a reservation of four acres for public buildings. The boundaries have since been slightly altered. It has convenient streets, lanes and alleys, crossing each other at right angles, having a good drainage, and many handsome sites for residences. It contains a courthouse, jail, a markethouse, a lunatic hospital, a state penitentiary, three banks, 10 churches, two Presbyterian, an Episcopal, three Methodist, one of which is an African, a Baptist, a Campbellite or Reformed Baptist, a Cumberland Presbyterian, and a Roman Catholic; the halls of the Nashville university, a female academy, besides other schools of a high order. A splendid state-house is about to be erected on the highest hill in the city. The site, consisting of four acres of ground, has been purchased by the city for $30,000, and presented to the state for the purpose. The courthouse stands on the public square, and presents a handsome front of 105 feet, and is 63 feet deep, and three stories high including the basement, surmounted by a dome, 90 feet high, from the bottom to its top. The basement is 11 feet high, and contains a number of public offices. The two principal stories are each 18 feet high. The foundation, and part of the lower story, is of fine hewn stone; and the remainder of brick. The two fronts are ornamented with four white pilasters, each. The dome contains a good city clock, and is supported by eight columns of the Ionic order. The second story besides some offices has two court rooms. Two large rooms in the third story are handsomely fitted up for the accommodation of the state legislature, until the new statehouse shall be prepared for their reception. The market-house, situated on the public square, is one of the finest buildings of the kind in the west. At each end there are spacious rooms, one of which is occupied as a city hall and recorder's office. The Episcopal church is a fine stone building handsomely stuccoed, in the Gothic style of architecture. The basement 9 ft. high, contains a lecture room 40 by 45 ft., besides rooms for the vestry, &c. The body of the church is 45 ft. wide by 69 ft. long and 24 ft. high, is neatly fitted up, and has a fine toned organ. The whole is surmounted with a Gothic cupola, and a bell; and the whole cost about $16,000. The Campbellite Baptist church is a neat brick building 45 by 60 ft., two stories high, with a tall steeple and an excellent bell, and cost $6000. The principal Methodist church is a spacious and elegant building in the form of a parallelogram, 60 ft. in front, and extending 90 ft. back. In the basement are rooms for Sunday schools and class-meetings. In front is a recessed portico of two massive Doric columns. The interior of the building is neatly finished. The whole cost of the building was over $12,000. The Methodists have other convenient churches. The Presbyterian church, erected on the site of one which was burned in 1832, is 91 ft. long, and 60 wide, and is an elegant building. The basement is 11 ft. high, and has a room 64 ft. square for a lecture room and Sunday school. The main room is 65 ft. square, and, with its gallery, will accommodate 1300 persons. The vestibule, approached by a flight of stone steps the whole width of the building, has a portico with six massive brick columns, cased and fluted. The cost of the whole was $16,000. The Cumberland Presbyterian church, is a plain and neat building of brick, 57 by 70 ft., two stories high above the basement story, which is partly below the surface. The front has recessed arches, resting on pillars forming an open vestibule. The main room of the interior is 69 by 55 ft., and with its gallery, will accommodate 1600 persons, and one, exclusive of a steeple, $7000. The Baptist church, recent'y erected, is a neat building in the Gothic style of architecture. The Rom'n Catholic church has a commanding site on the northern declivity of Campbell's hill, and presents a handsome appearance. The female academy, situated in the western suburbs of the city, is a flourishing institution with 200 scholars. There are several other schools for young ladies of the highest order. There are eight excellent preparatory classical schools in the city. The primary schools for both sexes are numerous and good. Probably no city in the union, of its extent, is better supplied with schools and teachers, of every description. The Roman Catholics are particularly active on the subject of schools. The penitentiary, located in the western suburbs, of the city, presents a front of 310 ft., and is 330 ft. deep. The two wings of the front building contain 200 cells. Half of the front building is occupied by the keeper, and the other half is used as a hospital, guard rooms, &c. The yard walls are 8 ft. thick at bottom and 3 ft. at top, with an average height of 20 ft. The cost of the whole was about $50,000. The lunatic hospital is a large and commodious building, 3 stories high including the basement, with an additional tier of rooms in the centre building. The basement and front walls are of stone, and the remainder is of brick. The whole will accommodate over 100 patients. The water works, located on the bank of the river above the city, supply it with water, raised by a steam engine, to a reservoir, 98 ft. above low-water-mark, and 45 ft. above the level of the public square. There may be raised in 24 hours over 200,000 gallons. Vauxhall garden on the southern border of the city, is a fashionable place of resort. It contains a large circular railway, on which the cars are propelled by cranks, turned with the hands of the riders. It contains also a fine assembly room. In the lower suburbs of the city is a strong salt sulphur spring, fitted up with cold and warm baths.

The city is the seat of Nashville university, founded in 1806, has a president and four professors and two tutors, 291 alumni, 100 students, and 10,000 volumes in its libraries. The commencement is on the first Wednesday in October. It has connected with the institution a number of buildings of brick or stone. The main college edifice is 206 ft. long, 50 ft. wide and 3 stories high, containing a chapel, recitation rooms, and 44 rooms for students. To the east of the main building an E. wing has been erected, 76 by 54½ ft. for the library, apparatus and lectures. It is in contemplation to erect a corresponding W. wing. A building of one story, 106 by 40 ft. constitutes the refectory. These buildings are within the college campus, containing eight acres. The president's house is of brick, 55 by 43 ft., with a kitchen in the rear 48 by 22 ft., his 3 stories high, with 31 acres of ground attached to it. The mineralogical cabinet has 20,000 specimens, and the apparatus for illustrating the experimental sciences is very complete. Medical and law faculties are about to be organized, and will commence operations during the present year. 1844.

According to the census of 1840, there were in the city, three commercial and eight commission houses in foreign trade, capital $235,060, 75 retail dry goods and other stores, capital $1,606,406, one forge, one paper-mill, one tannery, four printing-offices, two binderies, one daily, five weekly and three semi-weekly newspapers; total capital in manufactures $151,000; three academies, 390 students; two schools, 122 scholars. Tonnage of the port in 1840, 4733, a considerable portion of which is steamboats. Cumberland river is here navigable at low water for vessels of 30 to 40 tons; and in high water, for vessels of 400 tons.

NASHVILLE, p. v., capital of Nash co., N. C., 44 m. E. by N. Raleigh, 254 W. It contains a courthouse, jail, and a few stores and dwellings. Nett proceeds of the postoffice $110.

NASO (an. Agathyrnum), a town of Sicily, intend. Messina, distr. Palti, cap. cant. on the Naso, near its mouth, in the Mediterranean, 10 m. W. by S. Palti: Pop., in 1831, 6236. It is situated on a hill, in a finely wooded and healthy neighbourhood; is walled, and has some handsome buildings: several warehouses on the sea shore belong to its inhabitants. It acquired some notoriety in 1812, by reason of its population having refused the constitution then promulgated, and armed themselves on behalf of king Ferdinand. (Smyth's Sicily, p. 99, &c.)

NASSAU (DUCHY OF), a state of W. Germany, principally between lat. 50° and 51° N., and long. 7° 22' and 8° 45' E.; having N. and W. Rhenish Prussia, S. Hesse-Darmstadt, and E. the latter, Hesse Cassel, the ter. of Frankfort, and the Prussian circle of Wetzlar. Length, N. to S., 55 m; average breadth, about 33 m. Area, about 1800 sq. m. Pop., in 1838, 386,221. Almost the whole of the surface is hilly with a general slope towards the W. The

Taunus mountains cover the S., and the Westerwald the N. part of the duchy; the Feldberg, the highest point of the former, rises to nearly 2700 ft.; and the Salzburg head (Salzburger Kopf), in the Westerwald, reaches the height of 2600 ft. above the sea. No portion of level surface is sufficiently extensive to be called a plain; and the valleys are generally narrow and confined, though many are highly picturesque.

The Rhine forms a considerable portion of the S. and W boundaries. The Mayn limits the duchy to the S.E., and the Lahn intersects it near its centre, having, for the most part, a S.W. course. The Lahn receives within this duchy the Elz, Ems, Aar, Muhl, &c.; and joins the Rhine at Lahnstein; being navigable as far as Wellburg, 14 leagues from its mouth. There are no lakes: but Nassau has a number of mineral springs, especially in the Taunus, where are Ems, Selters, Schlangenbad, Wiesbaden, &c., among the most frequented spas in Germany. The climate is cold in the mountains, particularly in the Westerwald, but so mild in the sheltered valleys that the vine comes to very considerable perfection. The mean temperature of the year in the Rheingau, S. of Weisbaden, is 10° centigr., or 50° Fahrenheit. Every part of the duchy is tolerably healthy. The soil is nowhere remarkably fertile, but only a small portion of it is barren; it is least productive in the N., where, however, there are good natural pastures. A portion of the soil in Westerwald is volcanic, consisting of basalt and lava; and near Wellburg are traces of an extinct volcano. In 1831, of 1,812,541 morgen of land, 702,004 were arable, 196,190 in meadows, 15,543 in vineyards. 7473 in gardens, no fewer than 736,377 in woods, 6545 occupied with buildings, 106,961 in natural pastures, &c., and 40,947 altogether waste. Agriculture is the principal branch of industry. The land is mostly divided into small parcels, which are not, however, farmed by their actual proprietors. "The whole country, from the Heidelberg to the Rheingau and Hamburg mountains, and from the Rhine to the mountains of the Spessart (which tract includes, besides the S. part of Nassau, part of Hesse-Darmstadt, Frankfort, Hesse Cassel, and Bavaria), presents one uniform face. This plain is divided between large forests of the common pinus silvestris, occasionally interspersed with oak and beech woods, and large flat districts of corn and vegetables, unrelieved by a single tree or hedge. The open fields are divided into small patches, by the difference of culture, which frequently denotes the boundaries of each peasant's little farm. The farms rarely exceed 50 acres: gentlemen farmers, or speculative agriculturists, are unknown; and the opposite extreme, the class of agricultural day-labourers, is very small. The peasant generally holds his little possession at a fixed rent, due to the lord of the soil, which is never increased. He cannot be dispossessed; and his land descends from father to son, subject to this burthen—a tenure much resembling English copyholds. Leibeigenschaft (personal vassalage) is now completely abolished in this, and, indeed, in most parts of Germany. (Autumn near the Rhine, 119, 113.) Wheat is grown in the valleys of the larger rivers; but on the uplands, rye, barley, and oats, are almost the only grains cultivated, with potatoes; and in the Westerwald, buckwheat. The S. declivities of the Taunus are covered with chestnut woods and orchards. In the district called the Rheingau, further S., along the Rhine and Mayn, the culture of the vine occupies a large share of attention. The finest growths of the Rhine, as Hockheim, Steinberger, Rudesheim, and above all Johannisberg (which see) come from this duchy; in which nearly a third part as much land is appropriated to the culture of the vine, as in all Rhenish Prussia. (Berghaus.) The soil of the Rheingau is thin and sandy; but it is well manured and very productive. The Hockheim, properly so called, or hock, is grown at Hockheim, on a little hill behind the ancient deanery, on a space of about eight acres, open to the southern sun, and sheltered from N. winds by the town. Each acre has about 4000 vine plants, valued at a ducat each; and the little hill produces, in good seasons, about 12 large casks (tonneaux) of wine, each of which sells, frequently as soon as made, for 1500 florins (125l.) or upwards. A constant supply of water is afforded to the plants by a small rivulet, and they are protected from too much wet by moveable wooden sheds. (Schreiber, 171.) But there is another vineyard little inferior to the above; and the surrounding lands yield an abundant produce which, as in the case of other wines, often passes for the first growths. The produce of the Steinberger vineyard, which belonged to the suppressed monastery of Eberbach, and is now the property of the grand duke, is the strongest of all the Rhenish wines; and, in favourable years, has much sweetness and delicacy of flavour. The quantity made is about 300 hhds., of which about 60 is first rate, and has occasionally been sold on the spot at 10s. a bottle (Henderson on Wine, 223.) Flax, hemp, fruits, hops, tobacco, turnips, and chicory, are among the other principal kinds of produce. The pastures are well attended to, and a good many crops are grown for fodder, the rearing live stock being an important branch of husbandry, Berghaus estimates that there are 172,600 head of cattle, 196,000 sheep, and 54,000 hogs in the duchy. (Allg. Länder, &c., ill., 513.) The various breeds are said to be improving. Bees are numerous, and game abounds in the woods. Iron, lead, copper, and silver, are the principal mineral products; bovey coal also is found in the Westerwald, and chalk, marble, roofing slate, and potters' clay elsewhere. Mining, forges, &c., are estimated to employ 8000 workmen. Working in metals is, however, the chief branch of manufacturing industry; the other manufactures are mostly domestic. Linen cloths are woven by the peasantry at their own houses; and some cotton-cloths, carpets, woollen yarn and hosiery, morocco leather, sieves, soap, sealing wax, &c., are made: few, however, of the manufactured articles find their way out of the duchy, the exports consisting principally of mineral waters, wine, cattle, wool, mineral products, and hardware and earthenware. The roads are good; and the Rhine, Mayn, and Lahn, present great facilities for commerce; but the trade of the duchy is by no means so flourishing as it might be. Nassau has of late years joined the Prussian commercial league.

Accounts are kept in florins (gulden) of 60 kreutzers, containing four pfennigs each; the florin=1s. 8½d. Engl. The Hessian morgen (steuernormalmorgen)=about eight elevenths of an English acre.

The government is a constitutional monarchy, hereditary in the male line; and is among the most liberal of W. Germany. The landstands, or parliament, of the duchy, consists of two chambers: the first composed of the princes of the ducal house, the heads of eight noble families, and six representatives for the rest of the nobility; the second consisting of 22 members, 15 of whom are representatives of the landed proprietors, and three of the clergy. The states are convoked yearly. The press is free, and personal liberty, the right of petitioning, and eligibility to all public offices are privileges belonging to every subject. Civil justice is administered in a primary court in the capital of each of the 28 districts of the duchy; in secondary courts at Dillenburg and Usingen; and a high court of appeal at Wuisbaden. The principal criminal courts are at Wiesbaden and Dillenburg. There are elementary, royal, and grammar schools at Diez, Usingen, Dillenburg, Hadamar, Wiesbaden, &c., and a gymnasium in Weilburg, besides female schools, many special academies, and deaf and dumb and other charitable schools. The government has taken considerable pains to promote popular education, but it does not seem to be yet much diffused, few children attending the public elementary schools. By an agreement with Hanover, the university of Göttingen has been constituted the high school for the duchy; except in Roman Catholic theology, for which students resort to Marburg, in Hesse Cassel. In respect of religion, about 205,000 of the population are Protestants, 175,000 Roman Catholics, and 6900 Jews: the Lutherans and Calvinists have united in one communion. The military force consists of about 4000 men under arms, with a reserve of 1550 more. Public revenue estimated at 1,800,000 florins a year; public debt about 2,000,000 florins, but this is in process of liquidation. Nassau holds, with Brunswick, the ninth place in the German confederation; it has two votes in the full council; and, with Brunswick, one in the committee. It furnishes to the army of the confederation a contingent of 3098 men.

This country, like Hesse, was anciently inhabited by the Catti. The founder of the reigning house of Nassau was Otho of Laurenburg, brother of the emperor Conrad I. In 1255, two collateral lines were formed; and the descendants of the elder have remained in possession of this territory; while those of the younger (Orange-Nassau) have been seated on the throne of Holland. (Berghaus, Allg. Länder, &c., iv., 431–440; Hörschelmann's Stein; Schreiber, Guida du Rhin, &c.)

NASSAU, county, Florida. Situated in the N.E. part of the territory; and contains 576 sq. m. Bounded N. and W. by St. Mary's river; S.E. by St. John's river. Watered by Nassau river. It contained in 1840, 8836 neat cattle, 436 sheep, 5011 swine; and produced 17,400 bushels of Indian corn, 16,000 of potatoes, 31,500 pounds of rice, 68,425 of cotton. It had two stores, one lumber-yard, one grist-mill, two saw-mills, five schools, 60 scholars. Pop.: whites, 954; slaves, 908; free coloured, 30; total, 1892. Capital, Nassau courthouse.

NASSAU, p. t. Rensselaer co., N. Y., 19 m. S.E. Albany, 385 W. Drained by Kinderhook creek and its tributaries. It contains five churches, two Baptist, two Methodist, and a Presbyterian, 10 stores, three Tulip-mills, six woollen factories, two cotton factories with 3158 spindles, one flouring-mill, six grist-mills, 12 saw-mills, three tanneries; 15 schools, 879 scholars. Pop. 3236.

NATCHEZ, city, and capital of Adams co., Miss., 169 m. S.W. Jackson, 1110 W. Situated on the E. bank of Mississippi river, chiefly on a bluff, elevated over 200 ft. above the level of the river, 322 m. above New-Orleans, by the course of the river. A part of the place called Natchez below the hill lies on the margin of the river, and contains a collection of warehouses, stores for supplying boatmen, and grog shops, is the place where the boatmen principally congregate, and is the seat of much vice. But the greater part of the place is situated on the elevated ground, and is of a different character. It is laid out in the form of a parallelogram, with broad streets, crossing each other at right angles, and ornamented with the China tree. The ground is uneven, and the houses of the more wealthy inhabitants are widely separated, each seeming to occupy a square, surrounded with orange trees, palmetto, and other beautiful shrubbery. The inhabitants are distinguished for their refinement, intelligence, and hospitality. Many of the wealthy planters in the vicinity spend most of their time in the city, which is generally healthy, though sometimes visited with yellow fever, especially when it prevails in New-Orleans. It contains a courthouse, jail, four churches, an Episcopal, a Presbyterian, Methodist, Baptist, and Roman Catholic, three banks, an academy, a female seminary, a hospital, and an orphan asylum, a masonic hall, a theatre, two steam oil mills for manufacturing oil from cotton seed, and 4800 inhabitants. Cotton is extensively raised in the back country, and Natchez is a great cotton mart; and at the proper season, the streets are almost barricaded with cotton bales. Its trade is very extensive. Steamboats are continually arriving and departing, and its elevated situation affords a fine view of the river, with its numerous craft, and of the country to the west, including the village of Concordia on the opposite side of the Mississippi, and a vast region beyond. This place was first settled by the French in 1716, and destroyed by the Indians in 1729. It is now one of the most beautiful places in the valley of the Mississippi. Some of the houses are elegant, but they are mostly of wood, only one story high, with a piazza and balcony. It is much the largest town in the state, and is its commercial metropolis. See the statistics of Adams co.

NATCHITOCHES, parish, Louisiana. Situated in the N.W. part of the state, and contains 4000 sq. m. Bounded S.W. by Sabine river, and drained by its branches, and by Red river and its tributaries. It contained in 1840, 12,917 neat cattle, 1938 sheep, 15,849 swine; and produced 32,325 bushels of Indian corn, 10,080 of oats, 98,563 of potatoes, 114,350 pounds of tobacco, 10,638,709 of cotton. It had 49 stores, 10 grist-mills, 12 saw-mills, four tanneries, two printing-offices, two weekly newspapers; two academies, 129 students; five schools, 58 scholars. Pop.: whites, 768; slaves, 6651; free coloured, 657; total, 14,358. Capital, Natchitoches.

NATCHITOCHES (pronounced Nakitosh), p. v., capital of Natchitoches parish, La., 368 m. N.W. by W. New-Orleans, 1287 W. Situated on the W. side of Red river, 200 m. from its mouth, at the foot of a bluff, and built chiefly on one street. It contains a courthouse, jail, Roman Catholic church, a U. States garrison, two printing-offices, two weekly newspapers, and about 2000 inhabitants. It was settled by the French in 1717, and half of its present inhabitants are of French descent.

NATICK, p. t., Middlesex co., Mass., 17 m. W.S.W. Boston, 423 W. It contains three churches, a Congregational, Unitarian, and Methodist, three stores, two grist-mills, six saw-mills; one academy, 20 students; six schools, 309 scholars. Pop. 1285. The Boston and Worcester railroad passes through it. The first Indian church in New-England was founded here by the Rev. Mr. Elliot, the apostle of the Indians, in 1660.

NATOLIA, ANATOLIA, or ANADOLI (a corruption from anatolé, the East, or Levant), a peninsula of W. Asia, anciently called Asia Minor, and now constituting a pachalik of Asiatic Turkey: it extends between lat. 36° and 42° N., and between long. 26° and 42° E., being bounded N. by the Black sea, E. by Armenia and the Euphrates, S. by Syria and the Mediterranean, and W. by the archipelago. Length, from C. Kara-burun to the Euphrates, 870 m.; breadth from 300 to 440 m.; estimated area, 220,000 sq. m., or about one sixteenth more than that of the Spanish peninsula. Pop. probably about 4,350,000. The coast line is very irregular, especially on its W. and S. sides, where it is deeply indented by the gulphs of Adramyt, Smyrna, Kos, Makry, Adalia, and Scanderoon. Surface very irregular, but may generally be described as a high table-land, dotted with salt lakes, and enclosed by two ranges, detached from the plateau of Armenia, and running nearly parallel to the N. and S. coasts. The latter of these chains, the mons Taurus of the ancients, and Sultan-dagh of the Turks, runs close to the shore in some parts of Karamania, forming a bluff precipitous coast, intersected here and there by narrow gorges, through which numerous tor-sate run into the sea. One of the heights, close to the gulf of Adalia, was ascertained by captain Beaufort to be 7800 ft. high; but there are several summits in the interior, the snow on which descending one fourth the way down their sides, indicates a height of 10,800 ft., or nearly equal to that of Mount Etna. (*Beaufort's Karamania*, p. 37.) The N. range is much less clearly defined, the only snow covered peak being mount Olympus, about 25 m. S. the sea of Marmara. Connected with Olympus westward is the celebrated mount Ida, overlooking the plain of Troy; and the highest summits of which, called *Gargarus* by Homer, and Kaz-dagh by the Turks, rises about 5000 ft. above the sea. About 100 m. S. of Ida runs another range, the *Tmolus* of antiquity, mentioned by Ovid, Virgil, and Seneca, as being celebrated for its excellent vines and rich metallic veins. The central table-land is partly drained by the rivers flowing into the Black sea; but a large portion, lying N. and N.W. the range of Taurus, about 240 m. in length, by 150 m. in breadth, is covered with numerous salt lakes, marshes, and rivers, having no visible outlet. In rainy seasons these lakes overflow, and, but for the ridges that cross the plain and separate it into basins, would submerge nearly 900 sq. m. of the surface. The largest of these is the lake Beishehr, 43 m. W.S.W. Konieh; but by far the most curious in the peninsula is the *Tuzta palus* of antiquity (about 50 m. N. Konieh, and 2500 ft. above the sea), the waters of which, according to Strabo, were so impregnated with brine, that anything immersed in it was soon covered with a saline incrustation: the Turks call it Tuzla, and it still furnishes in abundance the article for which it was anciently famous; but it contains neither fish nor conchiferous animals. (*Geog. Jour.*, x. 299.)

The largest rivers of Natolia flow into the Black sea. The *Halys*, or Kizil-Ermak ("Red River"), rises by two branches on the S. side of mount Erdjik (anciently *Argæus*), and flows by a tortuous course of about 500 m., first N.W., and subsequently N.E., into the Black sea, where it is about as wide as the Seine at Paris. It is the largest river of Asia Minor; and in ancient times was considered the boundary between the Lydian and Median kingdoms, as well as a natural dividing line of the peninsula. (See *Herod.*, i., 72.) E. of the Halys is the *Iris* (now the Sekli Ermak), a much smaller river, rising in the N. range of the table-land, and flowing W. by N. past Tokat into the Black sea, about 20 m. E. Samsoun (an. *Amisus*). In the N.W. part of Natolia is the large and celebrated river *Sangarius* (now Sakaria), the most distant source of which is in the central plateau, about 60 m. S.S.W. Angora; lat. 39° 5′ N., long. 33° 3′ E. After receiving numerous tributaries, it turns northward, near the modern town of Eski-sher (an. *Dorylæum*), and flows into the Black sea, about 50 m. W. by S. Erekli. The three principal rivers flowing into the archipelago are the *Caicus*, the *auro turbidus Hermus* of Virgil (*Georg.*, ii., 137), now the Sarabat, the marshy *Caystros*, at the mouth of which was the "*Asios λειμών*" of Homer (*Il.*, ii., 470), and the *Mæander* (now Menders), by far the largest of the three, and celebrated in antiquity, not only for the sinuosity of its course, but for the fertility of its valleys, and the number of flourishing cities on its banks. It rises by numerous sources in long. 30° 8′ E., and takes a general course, W. by S. about 220 m. to its mouth, near the ruins of *Miletus*. The rivers on the S. side of the peninsula, are, with one or two exceptions, little more than brooks or mountain torrents; and the *Cydnus*, the scene of the splendid pageant of Cleopatra, is at present only 160 ft. wide, and inaccessible to any but the smallest boats. (*Beaufort's Karamania*, 275.)

The geological formation of Natolia partakes in many parts of a volcanic character. The high region of Phrygia, called *catacecaumene*, abounds with lava and other substances, indicating the existence of igneous action at some previous period; earthquakes have frequently visited the W. part of the peninsula, and all but demolished *Laodicea*, *Apamea*, *Cibotus*, *Sardes*, and other cities of antiquity; and it has still numerous thermal and sulphureous springs. The most usual formation, however, is of white limestone, bold cliffs of which rise in Karamania, from 600 ft. to 700 ft. perpendicularly from the sea, exhibiting the most curious contortions of strata. (*Beaufort*, p. 212, 213.) On the N. side of the peninsula the same description of strata prevails, covered with gypsum, and in the highest mountains serpentine is found alternating with the blue mountain limestone. The marble of Asia Minor was extensively used by the wealthy Romans, in building their houses and villas. These mountains abound in mineral riches; copper is wrought to a considerable extent near Trebizond, Samsoun, and Siwas; and the region of the *Chalybes*

———— uvisiuri quæmquæ
Gens Chalybum, duris patiens cui cultus in arvis
Et tonat officina recepet donote ignem massa.
Val. Flac. Arg., iv., 610.

is still an important mining district of the peninsula. Lead

has been found in several places, though not wrought; but rock-alum is procured near Unieh (anciently Œne), and exported in considerable quantities.

The climate of Natolia, owing to the varying elevation and different aspects of its surface, will admit of no general description. On the central plateau, the height of which (exclusive of mountains) varies from 2800 to 3000 ft. above the sea, it is cold, though salubrious, and snow lies, in many parts, for two or three months of winter; but, in July and August, the heat is often intense, and rendered more oppressive by the tendency of the sandy surface to absorb heat. On the W. shores the climate is genial, and the soil very productive; but in some parts, as at Smyrna and elsewhere, epidemics are prevalent, and the plague often makes great ravages among the population. The heat in July is stated by Mr. Addison to range from 84° to 94° Fahrenheit in the shade; rain seldom falls, though the want of it is, in some measure, compensated by heavy dews. (*Damascus and Palmyra*, i., 290, 291.) The climate on the N. side is far more temperate, and rain is frequent. The soil on the coasts is tolerably fertile, producing wines, olives, rice, millet, and other grains; but tillage is much neglected, irrigation and the manuring of land being little practised. The N. shores are covered with forests of oak, ash, larch, beech trees, &c., furnishing abundant supplies of timber for the Turkish navy. The mountains of Karamania are covered principally with pines. Large flocks of sheep and goats graze on the lofty plains of the interior; their wool and hair forming an important article of commerce between Angora and Smyrna.

Natolia is under a pasha or military governor, to whom are subject the respective beglierbegs of Anadoli, Karamania, Marash, Siwas, and Trebizond, the country being further subdivided into 17 sandjaks. The fixed population consists principally of Turks and Greeks, with smaller numbers of Armenians and Jews; besides whom, there are nomadic tribes, both Kurds and Turcomans, employed partly in pastoral, but partly also in marauding occupations. (For farther particulars, *see* TURKEY.)

Natolia, which was first called simply Asia, afterward ἡ μικρὰ 'Ασία, to distinguish it from that more to the E. ἡ ἄνω 'A., was called *Asia propria* or *proconsularis*, by the Romans, and did not receive its appellation of *Asia Minor* earlier than the time of Orosius, in the beginning of the fifth century. With respect to the original inhabitants of this celebrated peninsula, we have little information on which any reliance can be placed; but there is reason to believe that the Phœnicians had settlements, at a very early period, on its S. and W. coasts, and that there were frequent emigrations to it from Thrace, as well as Thessaly, soon after the Trojan war. The great Ionian immigration (composed of colonists from Attica and Achaia) took place anno 1130 B.C.; and, about 80 years afterward, a colony of Dorians, from Megara, Trœzene, and Argos, settled on the S.W. coast, a little S. of those last mentioned. Subsequently to the establishment of these Greek colonies, and during the existence of the Lydian monarchy (which lasted from a period of obscure antiquity down to the overthrow of Crœsus by Cyrus, B.C. 556), Asia Minor was overrun successively by large bodies of Cimmerians and Scythians, who, however, though they penetrated as far as Lydia, and took Sardes, were unable to secure a permanent footing in the country. (*Herod.*, i., 15.) The numerous revolutions, indeed, caused both by conquest and colonization, are sufficiently attested by the statement of Herodotus, that the peninsula, between four and five centuries prior to the Christian era, comprised 30 different nations (ἔθνεα). At the fall of the Lydian kingdom Asia Minor was formed into four satrapies, belonging to the Medo-Persian empire, under which it remained upward of two centuries, though the interior of the country, inhabited by nomadic tribes, was never fully subdued. Notwithstanding the oppressions of the provincial governors, and their occasional struggles with the "Great King," the Greek colonists continued to flourish, and they gradually spread themselves northward, along the Euxine sea, as far as *Trapezus* (now Trebizond), and southward, on the shores of the Mediterranean, to the gulf of Issus, everywhere distinguishing themselves by their industry and commercial activity. In develement, also, and the cultivation of the arts, they were at least equal, if not superior, to their European brethren; at all events, if Asia Minor have not given birth to great warriors and statesmen, she may justly boast of the all but unrivalled excellence of her poets, historians, philosophers, sculptors, architects, and musicians. In poetry the lays claim to Homer, Hesiod, Sappho, Alcæus, and Nicander; in philosophy to Thales, Pythagoras, Anaxagoras, Bias, and Pittacus; and in history to Hecatæus, Hellanicus, Herodotus, Ctesias, and Dionysius of Halicarnassus. The Macedonian succeeded the Persian dominion anno 331 B.C.; from which time, during nearly two centuries, Asia Minor was subject to many vicissitudes consequent on the changing fortunes of Alexander's successors and their descendants, as well as the .formation of several minor kingdoms (as Pontus, Bithynia. &c.) under native princes. During the century immediately preceding the Christian æra the various parts of the peninsula fell, one by one, into the hands of the Romans, under whom it formed a proconsulship; and it attained, during their dominion, not only its most uniform and settled, but also its most prosperous state ; a fact sufficiently proved by the number of large cities built or embellished, and the great works undertaken and completed, during the earlier period of the empire. The decline of the Roman power exposed the peninsula to fresh invasions from the E. ; and at the commencement of the eighth century the Mohammedans began to settle themselves on its E. borders. At the period of the first crusade they had spread over almost the whole peninsula, and reduced it to a state, in many respects, similar to that in which we find it at the present day, except that it was more populous. It was ravaged by the crusaders in the 12th and 13th centuries, and was overrun by the Tartar hordes under Timour after the battle of Angora (1402 A.D.) ; but neither produced any permanent effect on the condition of the population. (*Leake's Asia Minor*, p. 1-144; *Cramer's Asia Minor, passim ; Geog. Journ.*, vi., x.)

NATURAL BRIDGE, Rockbridge co., Va., next to the falls of Niagara, the greatest natural curiosity in the United States, and one of the greatest in the world, 156 m. W. Richmond. 2 m. N. of James river, 12 m. S.S.W. Lexington, the capital of the county, 6 m. from the head of James river and Kanawha canal, at which point starts the Natural Bridge turnpike, passing over the bridge. The bridge consists of a stupendous arch of limestone rock, over an unimportant and small stream; called Cedar creek. The height of the bridge above the stream to the top, is 215 feet, its average width is 80 feet, its extreme length at top is 93 feet, and its thickness from its under to its upper side is 55 feet. The chasm over which it passes is 50 feet wide at bottom, and 90 feet at top. The bridge is covered, to the depth of from 4 to 6 feet, with a clayey earth, with a natural parapet of rocks on its sides, rendered firm by trees and shrubbery. The view from the top, to those who dare take it, is awfully grand ; but the view from below, being divested of terror, is equally sublime, and more interesting. The chasm continues narrow, deep, and straight, for a considerable distance above and below the bridge, and presents a short but pleasing view of North mountain on one side, and Blue ridge on the other, each about 5 m. distant. The bridge is of important use, as it constitutes a safe and convenient passage for a road, over this great chasm, not otherwise possible, for some miles above and below.

NAUMBURG, a town of Prussian Saxony, distr. of its own name, on the Saale, 25 m. S. Halle, and 28 m. S.W. Leipsic ; lat. 51° 8′ N., long. 11° 54′ E. Pop. 12,000. It is situated in a fertile vale, and is tolerably well built, having several good and wide streets, with three suburbs. Its principal public buildings are the citadel, town-hall, and arsenal, a noble Gothic cathedral, five Calvinistic churches, two orphan asylums, six hospitals, a poorhouse, gymnasium, and trade school. It is the seat of a superior and ordinary tribunal for the circle, a council for do., and board of taxation. It has manufactures of woollen cloths, hosiery, and shoes, and large chemical works. The two annual fairs of Naumburg were formerly much celebrated, but have recently declined in importance. Naumburg is celebrated in history as having been besieged in 1432 by the Hussites, under Procopius. This general, irritated by the resistance of the inhabitants, made a vow to put them all to the sword, but was deterred from his purpose by the earnest supplications of the children of the town, who came out in procession and threw themselves at his feet. The anniversary of this event (called the *Kindersfest*, or "Children's Fête"), is still celebrated on the 29th July, and has furnished Kotzebue with the subject of one of his plays. (*Von Zedlitz, Neustrich Press. Staat.*, iii., 285 ; *Berghaus ; Murray's Hand-Book of N. Germ.*, p. 392.)

NAUPLIA, or NAPOLI DI ROMANI, a city and seaport of independent Greece, on the E. side of the Morea, at the extremity of the bay of its own name, 5 m. S.E. Argos, 58 m. W.S.W. Athens ; lat. 37° 33′ 50″ N., long. 22° 47′ 30″ E. Pop., according to Burgess, 16,000. The town, which stands on the N.E. side of a hill, with a tabular summit, and is built in the form of an amphitheatre, has been greatly enlarged and improved since the war of independence.

"Nauplia," says Mr. Burgess, the most recent traveller in Greece, " has no longer any similarity to its former internal appearance (which was that of a filthy and miserable Turkish town). The features of the Palamidi rock, the heights of Itchkali, the low coast sweeping round by Tyrins and the Lerudan marsh, with the citadel of Argos rising out of the plain, the mountainous shores flanking the E. parts, and the headlands jutting beyond the reach of the

naked eye, must endure as long as the landscape remains undissolved ; but everything that man and his institutions can change is now changed at Nauplia" (i., 186). It now comprises several wide streets, regularly laid out, and lined with good houses, in the European style ; some of which, for size and elegance, might pass, in Greece, for minor palaces. The principal public buildings, besides the churches (one of which belongs to the Roman Catholics), are a royal palace, formerly the residence of Count Capo d'Istria, a new court of justice, called the Bouλeυτηριον, and a garrison, occupied by Bavarian troops. The shops are well stored with provisions and other articles ; and there are numerous cafés about the port, and in the chief thoroughfares. A quay faces the harbour, which is commanded by the ancient fort Palamedi, one of the strongest castles in Greece ; at its foot is a stone aqueduct, from which the town is well supplied with water. Nauplia possesses one of the largest government dock-yards in Greece, and extensive storehouses. Its trade is very considerable, the principal exports being oil, wine, gall-nuts, wax, silk, wool, and cotton ; while the imports comprise corn, manufactured and colonial goods, with timber, &c. The commerce is principally carried on in Greek bottoms. In 1836 there arrived 86 vessels of 6085 tons, and 58 vessels of 3726 tons left the port in the same year. The roadstead of Nauplia is W. of the town, in 5 and 9 fathoms ; but within the harbour there are only 3½ fathoms, and in entering it is necessary to keep in mid-channel, to avoid a shoal of 6 ft. water.

Nauplia is, in comparison with the rest of Greece, well provided with literary establishments. They comprise a military academy, school for the middle classes, circulating library, several book societies, two lithographic establishments, and five printing-houses, one of which is the property of the government, and exclusively employed in printing their official paper. The Σωτιρ, or "Saviour," a political and literary newspaper, is published here in Greek and French, and has a wide circulation. The population of Nauplia comprises a considerable number of Germans, French, and Italians : house-rent is high, and the rate of living is not much cheaper than at Paris or Naples. The climate is unhealthy, owing to the miasma from the neighbouring marshes in summer, and the cold, searching N.E. winds that prevail during winter : the town has likewise been frequently ravaged by the plague.

The ancient Nauplia was the port and arsenal of Argos during the flourishing period of Grecian history ; but it was deserted and in ruins when visited by Pausanias, who noticed the vestiges of its walls and docks (λιμενεϛ), the temple of Neptune, and a fountain called Canathus, still existing. The inhabitants had been expelled several centuries before by the Argives, on suspicion of having favoured the Spartans, who in consequence received them into their territory, and established them at Methone in Messenia. The town revived under the Byzantine emperors, and was occupied in the 13th century by the Venetians, who made it their chief settlement in the Morea. It was taken by Sultan Solyman in 1537, but was soon afterward recovered ; nor did the Venetians finally lose possession of it till the treaty of Passarowitz in 1718 secured it to the Porte, which retained it down to the close of the war of independence. Nauplia was the seat of the new government from 1829 to 1834, when it was transferred to Athens. (*Burgess's Greece and the Levant*, i., 185-188 ; *Leake's Morea*, ii., 355-361 ; *Off. Rep. ; Journ. of Educ.*, vol. ix., &c.)

NAUVOO, city, Hancock co., Ill., 124 m. N.W. Springfield, 891 W. Situated on the E. bank of Mississippi river, 18½ m. above the mouth of Illinois river, and inhabited by the Mormons. The river is here about 2 m. wide, and there is a good steamboat landing. The city limits are 4 m. long, and at its greatest width 3 m. wide, bounded W. and S. by the river. At the end of three years from its establishment, it contained 1000 houses, chiefly white-washed log cabins, with a few framed and brick houses. The public buildings are the "Nauvoo house," a spacious hotel, fronting on two streets, 120 feet on each, 40 feet wide, and three stories high above the basement. In this building Joe Smith, the pretended prophet and leader of these "Latter-day Saints," is furnished with a suite of rooms. The *Nauvoo temple*, not yet completed, will be 130 feet long, and 100 feet wide. In the basement is a baptistery, supported on 12 gilded oxen, the model of which is derived from the brazen sea of Solomon. The *Nauvoo legion* consists of from 2000 to 3000 men, with proper officers, armed and disciplined. They have a *university*, which contains a president, a professor of mathematics and English literature, a professor of the learned languages, and a professor of church history. The city is laid out with streets of ample width, crossing each other at right angles. Their property is held as private ; but they have a large farm without the city, which is occupied and cultivated in common. The population within the city limits is about 7000, many of whom are from England, besides about 3000 of the frater-

sky, who reside in the vicinity. The city has a mayor, and is divided into four wards, having two aldermen, four common-council men, and a constable for each of the wards. The city is a curiosity, and the success of its leader has scarcely been paralleled, since that of the prophet of Arabia.

NAVAN, an inland town of Ireland, prov. Leinster, co. Meath, at the confluence of the Blackwater with the Boyne, 16 m. N. by W. Dublin. Pop., in 1831, 4416, and, including its suburbs, about 6000. It has a parish church, a Roman Catholic chapel, a convent, an endowed grammar school, a national school, a preparatory Roman Catholic college, with a chapel, courthouse, bridewell, fever hospital, the infirmary for the county, and cavalry barracks. At one end of the town is a large rath or mote. Owing to the opening of the Boyne navigation to Drogheda, Navan has become a place of considerable trade, especially for all sorts of agricultural produce, sent principally to Drogheda, but partly also to Dublin. It has also five corn or flour mills, two paper-mills, two distilleries, and a tannery. The old corporation sent two members to the Irish House of Commons till the union, when it was disfranchised. Markets on Wednesdays ; fairs on Easter and Trinity Monday, the second Monday in September, and the first Monday in December. Postoffice revenue, in 1830, £608, in 1836, £664. *Railway Rep.; Mun. Boundary Rep.*

NAVARINO, a town and seaport of indep. Greece, on the S.W. crest of the Morea, 136 m. S.W. Athens, and 92 m. S. by W. Patras; lat. 36° 56′ 15′′ N., long. 21° 41′ E. *Pop. 2000 ?* It stands on the S. side of a the semicircular bay of the same name, and is surrounded by walls, and defended by a strong citadel, placed on a lofty rock. Streets narrow, steep, and irregular, lined with small, mean-looking houses, chiefly of stone cemented with mud, and encumbered in many places with the fallen ruins of former habitations. At the opposite extremity of the bay are the remains of Navarino Vecchio, consisting of a fort covering the summit of the hilly peninsula of *Coryphasium*, on the S. slope of which once stood the ancient *Pylos*. The long rocky island of Sphagia (an. *Sphacteria*) stretches about 4 m. from N. to S. across the mouth of the bay, forming a kind of natural breakwater for its protection from the heavy seas that would otherwise be thrown in from the W. The entrance is at the S. side of the island, and the bay is one of the finest asylums for shipping in the Mediterranean. It has water to float the largest ships, and good holding ground. Ships usually moor about one quarter of a mile from the modern town, or behind the island of Marathonisi, near the centre of the harbour. The circular lagoon, on the N. side of the harbour, directly E. Navarino Vecchio, abounds with fish ; but, as it is not mentioned either by Thucydides or Pausanias, it is probably of modern formation.

The ancient Pylos, one of those towns that claimed to be the birthplace of Nestor (called by Homer Νηλήιον ἄστυ), was deserted by its inhabitants after the Messenian war. When the town was restored, we have no information ; but at the time of Pausanias, it was inhabited, and comprised, among other monuments, a temple of Minerva Coryphasia, and a monument of Nestor. (See *Paus. Mess.*, 36, quoted by *Leake*, i., 412.) The island of Sphacteria, which Thucydides (iv., 35–39) has described as " desert, pathless, and covered with wood" (ὑλώδης τε καὶ ἀτριβὴς πᾶσα ὑπ' ἐρημίας), is celebrated in the history of the Peloponnesian war as having been occupied by the Lacedæmonians after the defeat of their fleet by the Athenians, under Demosthenes. They were detained here during 72 days, and were at length compelled to give themselves up as prisoners, and to surrender their fleet in pledge of their fidelity to their engagement. The battle which preceded this blockade took place in the bay of Pylos, or Navarino, which has also obtained celebrity in modern times, during the late war of independence, for the decisive victory gained (October 20, 1827) by the combined fleets of England, France, and Russia, under Sir E. Codrington, over the Turco-Egyptian fleet, commanded by Ibrahim Pacha. Notwithstanding the great preponderance of force and science on the part of the allies, the Turco-Egyptian fleet made an obstinate resistance, but in the end it was almost totally destroyed. A convention was soon after entered into, by which the Turks agreed to vacuate the Morea ; and this battle finally led to the acknowledgment by the Ports of the independence of Greece, in the treaty of Adrianople, in 1829. (*Leake's Morea*, i., 00–415; *Gell's Morea*, p. 19–23, &c.)

NAVARRE (Sp. *Navarra*), a portion of Spain forming a dependent kingdom under that monarchy, on the N.E. side of that peninsula, between lat. 41° 57′ and 43° 12′ N., and long. 0° 41′ and 2° 25′ W.; being bounded N. by France and the Pyrenees, E. by Aragon, S. by Old Castile, and W. by the Basque provinces. Greatest length from S.W. to N.E., 75 m.; breadth, about 50 m. Estimated area, 2440 q. m. Pop. 271,280. The surface consists in a great measure of subordinate mountain ranges, running southward

from the main ridge of the Pyrenees; but the S. part of the province, near the Ebro, has some extensive and fruitful valleys. The principal summits within the limits of the province are Altobiscar, 5380 ft. high ; Adi, 5218 ft. ; and three others rising above 3000 ft. from the sea. The highest points of the Pyrenees, however, are considerably to the E. in the kingdom of Aragon. The principal passes over the Pyrenees from Navarre into France are, proceeding eastward, those of Verra, Maya, and Roncesvalles, the last of which is, according to Bory St. Vincent, 5771 ft. above the sea. The mountains are chiefly of transition and secondary formation, consisting, in a great measure, of the rock called Pyrenean limestone. Jasper and marbles, also, occur in large beds ; and there are several iron-mines, besides one of copper. Rock-salt is quarried at Valtierra, near the Ebro ; and the yearly returns, according to Miñano, amount to 12,000 arrobas ; the province, also, comprises numerous thermal springs. Principal rivers, the Aragon, Zidacos, and Arga, uniting their waters in one channel, which falls into the Ebro opposite Alfaro : the only river flowing into the bay of Biscay is the Bidassoa, which rises in the mountains, forming the Val de Bastau, and has a course E.N.E. of about 45 m., falling into the sea near Fuenterrabia. The climate of the mountainous districts is very severe in winter, and not genial even in summer ; but in the valleys of the Ebro and Aragon the temperature is much higher, and the climate delightful, as well as healthy. The forest trees of the Pyrenees consist chiefly of the pine, large quantities of which are sent down the Ebro to Zaragoza and other places ; but there are also considerable numbers of beeches, deciduous oaks, chestnut-trees, &c. ; and no province furnishes so good supply as Navarre of useful building timber. (*Cook's Sketches*, ii., 286.) The wild animals of the mountains are wolves, wild boars, foxes, and wild cats : game is abundant in every part of the province. Pasturage is extensively followed, especially in the N. districts ; and, according to Miñano, the stock at the last general census included 43,636 oxen, 4616 calves, 95,700 mules, 639,500 sheep, 80,500 goats, and 31,780 hogs ; the produce of wool being estimated at 56,490 arrobas (12,609 cwt.). The higher part of the kingdom, on the frontiers of France, is bleak, cold, and unsuitable for tillage ; but the plains near the Ebro have a rich, productive soil, well watered by numerous streamlets connected with the larger rivers. The principal crops are wheat, maize, barley, and oats. Hemp and flax are also raised, with oil and wine. About the half of the latter, with the greater part of the wool, and about 30,000 bushels of corn are exported, chiefly to France, in return for silk and cotton fabrics, colonial produce, &c. Cider is made in the Val de Bastan, and liquorice is raised in the S. districts for exportation. Agriculture, however, is much neglected, and was recently rendered almost futile, in consequence of this part of the peninsula being the scene of the civil war between the adherents of the present queen and Don Carlos. Manufactures are also inconsiderable, including only those that are most indispensable, and some distilleries. The intercourse with the adjoining provinces is maintained partly by the canal of Aragon, running from Tudela to Zaragoza, and partly, also, by roads intersecting the country in various directions : the great road from Pampluña to Madrid is said by Captain Cook, and other recent travellers, to be little inferior to the modern roads of England.

The kingdom of Navarre is still governed by its separate laws, and has, nominally at least, the same constitution which it enjoyed when it was a separate monarchy ; but its cortes, or estates, have not met since 1713, and cannot be convoked without the authority of the crown. A council, however, representing the cortes, sits permanently at Pampluña, decides on the method of raising the revenue, fixes the tariff, and exercises other commercial privileges. The supreme power is vested in the viceroy, who presides at the royal council (*Consejo Real*), consisting of six judges, an attorney-general, and four *alcaldes* : this is the highest tribunal for civil and criminal causes. The inhabitants of Navarre are tall, and strongly built, resembling the Biscayans in independence of spirit, attachment to their religion, and jealousy of their ancient national privileges. Castilian is the general language of Navarre ; but the Basque is spoken in the N. and W. districts.

Navarre is divided into 17 *partidos*, which are again subdivided into 74 *merindades*, or districts ; and the kingdom comprises nine cities, the principal of which are Pampluña, the capital and seat of government (pop. 15,000), Tudela (8150), the once royal city of Olite (5000), and E tella, the stronghold of the Carlists during the late war (4600). The inhabitants of Navarre, in the time of the Romans, were called *Vascones*, in common with those in the neighbouring parts of the peninsula: and were faithful subjects of the empire till the close of the fifth century, when they were subdued by the Visigoths, under whose away they remained between 200 and 300 years. The Arabs overran the country in the eighth century; but were unable to effect its conquest. Inigo, count of Bigorres, having been elected king in the ninth century, the crown remained upward of five centuries in his family, till in 1590 it became united, through intermarriage, with that of France, the title of whose monarchs, from the time of Henry IV. (with the exception of Napoleon), to that of Charles X., was "King of France and Navarre." In 1512, however, Ferdinand of Aragon united all the country S. of the Pyrenees to the crown of Spain; so that only the small portion N. of that chain remained annexed to the French monarchy: this formed the province of Béarn before the revolution, and is at present included in the department of the Lower Pyrenees. (*Cook's Sketches in Spain*, i., 118–122; ii., 226; *Mignano; Dict. Geog., &c*)

NAXIA (an. *Naxos*), an island of the Grecian archipelago, the largest of the group called the Cyclades, about 5 m. E. Paros, its cap. of the same name being in lat. 37° 7' N., long. 35° 26' E. Shape, oval; circuit, about 48 m.: area, 105 sq. m. Population, according to Burgess, 18,000. The coast is much indented and precipitous, especially on the N.E. side; and the surface is very uneven, comprising several high mountains of primitive formation, on which are superimposed strata of grauwacké and mountain limestone: the culminating point of the island, anciently called the hill of Zeus 3310 feet high), is still called *Zia*: it attained some celebrity from its containing a cavern or grotto, to which, according to tradition, the Bacchantes came to celebrate their mysteries and festivals. Emery is wrought in one part of the island; and, according to Dr. Clarke, almost all the emery of commerce comes from Naxia. Large flocks of sheep feed on the mountain sides; but both their wool and flesh are of inferior quality.

Naxia has few large trees, but is pretty thickly covered with lemon trees, myrtles, oleanders, thorny brooms, the arbutus and labdanum plant, *atractylis gummifera*, the produce of which is chewed by the natives, and various kinds of leguminous plants, besides the olive, fig, and vine, which are extensively cultivated. The wine, however, though characterised by Athenaeus as the "nectar of the gods," is now of very indifferent quality, owing to the want of care in its preparation. The oil, also, is inferior to that produced in most of the other islands. The island was famous in antiquity for its fertility; but agriculture is now so much neglected that the corn raised is sufficient only for six months' consumption of the inhabitants. Vegetables, however, are so abundant, that considerable quantities are sent to Syra. (*Burgess*, ii. 21.) Lime juice also is exported, chiefly to Russia. In the S. of the island is a small saltpan, from which the capital is supplied with salt. The island abounds with game; and field-sports constitute a favourite occupation of the inhabitants.

Naxia, the capital of the island, occupies an eminence close to the sea on the W. coast (which is the only part accessible to shipping), and has 4000 inhabitants. Its narrow streets lined with dilapidated houses, exhibit a profusion of marble; and there is scarcely a dwelling in which there are not ancient inscriptions or other monuments. A castle, built by the Venetians, occupies the summit of a hill above the town. The principal remains of the ancient Naxos are a gate belonging to a temple of Bacchus, on a lofty crag, an aqueduct, and a jetty now under water, but still distinguishable in calm weather. It is the residence of a Greek and Latin archbishop, and there are several churches and convents belonging to both religions. The harbour of Naxia, called the Porto Salice on account of the salt collected there, is exposed to the N. and N.W. winds, and being almost surrounded by hidden rocks, is unfit for the anchorage of large ships. The island comprises 40 villages; and there are numerous country-houses, forming the residences of the nobles and gentry.

Naxos (which, according to Pliny, was called at different times Strongyle, Dia, Dionysias, and Callipolis), was probably first colonised by Carians. The Naxians were among the most steadfast opponents of Persian aggression, and the failure of the expedition undertaken by the Persians against this island at the suggestion of Aristagoras, led to the revolt of the Ionian states. Soon afterwards, Naxos was conquered by the Persian fleet under Datis and Artaphernes, who destroyed the city and enslaved its inhabitants. (*Herod.* v. 28, vi. 96.) The Naxians, however, had sufficiently recovered seven years afterwards to enable them to furnish four well equipped trieremes for the fleet at Salamis. The Athenians, even in the time of Pisistratus, claimed them as colonial dependents; and, after the Persian war, they deprived them of their liberty. Naxos was celebrated in ancient mythology for the worship of Bacchus, who is alleged to have been born in the island. It became tributary to the Romans after the fall of Corinth, 146 B.C., but was ceded by Mark Antony to the Rhodians after the battle of Philippi. The island was afterwards annexed to the possessions of the eastern empire, and subsequently became

the possession of the Venetians, and the capital of a dukedom which embraced most of the other Cyclades. At length, in the reign of Selim II. (A.D. 1570), it was united to the Ottoman empire. The Turks, however, allowed the inhabitants to retain their ancient laws and government, contenting themselves with occasionally sending a *waiwode* to collect the land tax and customs. It now forms a part of the new kingdom of Greece. (*Tournefort*, i. 226–231; *Burgess*, ii. 21; *Cramer's Greece*, iii. 408; *Clarke's Greece, &c.* vi. 90–113.)

NAZARETH, or NASSARA, a small town of European Turkey, in the pachalic of Acre, celebrated as having been the residence, during his youth, of the founder of Christianity, 17 m. E.S.E. Acre, and 70 m. N. by E. Jerusalem. Population 3000, of whom 500 are Turks, and the rest Christians. It stands on the W. slope of a delightful valley, encompassed by rocky mountains of no great height, which rise round it like the edge of a shell, as if to guard it from intrusion. The houses are mostly wretched stone cottages, with mud floors and roofs; nor does it comprise any thing worthy of notice, except a Latin church and convent, with two other churches, belonging respectively to the Maronites and Greek Catholics. The Turks also have a mosque, erected at the beginning of the present century. The Latin convent, belonging to the missionaries of the Terra Santa, at the E. end of the village, is a spacious and commodious building of stone, surrounded by high walls, which enclose a church, cells for the friars, and extensive accommodation for pilgrims and travellers. The church, called that of the Annunciation, is an ill-proportioned and gaudily ornamented building, said to occupy the spot which, according to tradition, the house of Joseph and Mary stood before its miraculous removal to Loretto. The columns and interior walls are hung round with silk damask, and there are two tolerably good organs. Beneath the high altar is the descent to a subterranean cave, in which the Virgin is said to have lived, and which is divided into small grottoes, pointed out as her kitchen, parlour, and bed-room! Here also are two granite columns, each two feet one inch in diameter, and about three feet apart, which are supposed to occupy the very places where the Angel and the Virgin stood at the precise moment of the Annunciation. The innermost pillar is broken through, above the pedestal, and, although it touches the roof, it is represented to be self-supported in the air. "The fact, however," is, says Dr. Clarke, "that the capital, and a piece of the shaft of a grey granite pillar, have been fastened on to the roof of the cave: so clumsily, also, is the rest of the *hocus-pocus* contrived, that what is shown for the lower fragment of the same pillar is not of the same substance, but of *Cipolino marble.*" (*Travels in Greece and the Holy Land*, w. 178.)

It was formerly the custom of the sick, during the prevalence of the plague, to resort thither for the purpose of rubbing themselves against the pillars, believing this to obtain a certain cure; but, within the last few years, a railing has been formed to exclude the patients, who, however, still flock round in hopes of relief from being in its immediate vicinity. Different interesting scenes Christ is pointed out to the pilgrims, such as Joseph's workshop, enclosed in a small chapel, the synagogue in which Christ explained the celebrated passage of Isaiah (*Luke.* iv. 16–22), the table on which Jesus ate his last meal previously to his final departure for Jerusalem; and even the precipice, or "brow of the hill," to which they led him, "that they might cast him down headlong." (*Luke.* iv. 29.) Here however, as at Jerusalem, fancy, and the desire of imposing on the credulity of the devotees, have had a far greater share in fixing these localities, than any regard for authenticity. The chamber containing the fictitious scenes Christ is the favourite resort of all pilgrims, Turks as well as Christians; and to Roman Catholics, who say the Paternoster and Ave Maria in it, the pope grants a plenary indulgence of seven years! The present inhabitants of Nazareth are, with the exception of a few weavers, employed in rural pursuits. Corn is raised abundantly in the neighbourhood, especially by the monks of Terra Santa, who are the chief farmers; and a small portion of it is sent to Acre, which is the chief source of supply for the town. The pasturage of cattle and goats, also, is extensively pursued, from the milk of which is made a large quantity of butter and cheese, both of indifferent quality. (*Turner's Levant*, ii. 130.)

Nazareth is not mentioned in the Old Testament: it was a city of the tribe of Zebulun, and afterwards of the N. portion of Palestine, called Galilee, and was held in so little esteem by the Jews of Jerusalem as to give rise to the exclamation, "Can any good thing come out of Nazareth?" (*John.* i. 46): it owes its entire celebrity to the circumstance of having been the residence of Jesus Christ almost from his birth to the commencement of his ministration. Here the angel Gabriel announced to the Virgin the approaching birth of the Saviour; thither the holy family returned 33 days after his birth at Bethlehem; and during his infancy spent in the house of Joseph the carpenter "the child grew and waxed strong in spirit, filled with wisdom, and increasing in favour with God and man." (*Luke*, ii. 40, 52.) Christ preached here "the acceptable year of the Lord," immediately after the temptation; and found from the unfavourable manner in which he was received, that "no prophet is accepted in his own country." His hearers were filled with wrath, rose up, thrust him out of the city, and led him to a precipice to destroy him; but he passing through the midst of them, went his way (*Luke*, iv. 28–30); and thenceforward Capernaum seems to have been his general residence, though there can be little doubt that he occasionally visited Nazareth to see his mother, and the members of her family. (*Clarke's Travels in Greece and the Holy Land*, iv. 164–183; *Turner's Levant*, ii. 129–132; *Robinson's Palestine*, i. 205–209; *Mod. Trav.*; *Calmet's Dict.* ad vocem.)

NEAGH (LOUGH) a lake of Ireland, the largest in the United Kingdom, in the centre of the province of Ulster, having N. and E. the county of Antrim, S.E. Down, by which it is merely touched, S. Armagh, W. Tyrone, and N.W. Londonderry. It is about 17 m. in length, by about 9 m. in breadth; occupying, inclusive of lough Beg (3551¼ acres), which is joined to it, an area of 61,626 Irish, or 99,822¾ statute acres, at ordinary high-water mark. (*Third Report of Commissioners on Irish Bogs.*) It is fed by several rivers of considerable magnitude, while the Lower Bann is the only channel through which its refluent waters find a passage to the sea. Though by far the largest, it is by no means the most beautiful of the Irish lakes. Its shores consist mostly either of a level sand, or marshy border, liable to frequent floods; and are of course deficient in those varied banks, and bold promontories, without which such extensive sheets of water want picturesque effect, except when their uniformity is broken by islands; and of these there are only two small and uninteresting ones in this lake. Frequent squalls and want of shelter render its navigation rather dangerous for sailing vessels; but these inconveniences will most probably be obviated by the introduction of steam packets. According to Mr. Sampson (*Survey of Londonderry*, p. 116), the mean level of Lough Neah is about 30 feet above that of the sea; and it is said that nearly 10,000 acres of land contiguous to its banks, now annually flooded, might be made available for agricultural purposes, by the outlay of a moderate sum on removal of some obstructions in the channel of the Bann. Its waters are celebrated for their petrifying . (*Statistical Account of British Empire*, ii. 337.)

TH, or NEDD (the an. *Nidum* of Antonine's Itin.), and mun. bor., market town, and par. of S. Wales, the Glamorgan, and hund. its own name, on the Neath (crossed here by a stone bridge), 7 m. E.N.E. Swansea, and 159 m. W. by N. London. Population of parl. bor., in 1831, 4042. The town, situated in a picturesque valley on the E. side of the river, is "indifferent in appearance: the streets are narrow, and badly paved; there are few good houses, and the place is only partially lighted with gas. The cottages of the poor extend irregularly beyond the town, particularly on the Cardiff road." (*Mun. Bound. Report.*) The town hall is a handsome modern building, the lower part of which is used for a corn market: a church, with a lofty square tower, and six places of worship for dissenters, are the only other public edifices. There are two national schools, two Sunday-schools, and two infant schools.

Neath is situated in the immediate vicinity of an extensive copper, iron, and coal district, and it depends in a great measure on the great smelting-houses and foundries that have been established round it, chiefly on the W. bank of the river. "The commerce of Neath is very considerable, and has been steadily increasing for some years; but though vessels of 200 tons can get up to the town, the trade is generally carried on by barge communication with Britten Ferry, which is about 2½ m. lower down the river, and is connected with Neath by a canal running northward, 12 m. higher up the valley. By Britten Ferry, in fact, Neath acts as the vent of all the mineral districts connected with the vale. Another canal joins the last mentioned at Aber-dulais, and terminates in a sea-lock and basin on the E. side of Swansea harbour. The exports are coal, culm, copper, iron, iron castings, fire bricks, oak bark and timber; the imports comprising copper and iron ore, corn and flour, foreign timber, black jack, and general shop goods. (*Parl. Bound. Rep.*)

Neath is a bor. by prescription, and has been governed since the Municipal Reform Act by a mayor and three other aldermen, with 12 councillors: it has also a commission of the peace under a recorder. Before the Reform Act, Neath was a contributary borough to Cardiff; that act annexed it, with Aberavon, Kenfig, and Loughor, to Swansea, which sends one member to the House of Commons. At the same time the electoral limits were so enlarged as to include, with the old borough, that portion of the suburbs on the W. side of the river. Registered electors of Neath in 1839–40, 174; ditto, of entire borough, 1947. Neath is also one of the polling places at elections for the county; and the petty sessions for the hund. are held here, as well as the quarter sessions alternately with three other towns. Markets on Wednesday and Saturday; fairs, July 3, Sept. 12, and the first Thursday after Trinity Sunday.

About one mile from the town, on the low ground bordering the river, are the ruins of Neath Abbey, founded by Richard de Granville in the 12th century: the church is a mere heap of ruins, but the chapter house, a curious specimen of early English architecture, is still in tolerable preservation; and foundations of buildings may be traced to a considerable distance. (*Parl. Rep.*; *Nicholson's Cambrian Guide.*)

NEEDHAM, p. t., Norfolk co., Mass., 12 m. S.W. Boston, 427 W. Bounded S. and N.E. by Charles river which has here two falls, one of them 20 feet, affording good water-power. It has five churches, two Congregational, two Methodist and a Baptist, four stores, one cotton factory with 1200 spindles, two grist-mills, one saw-mill, three paper-mills; three academies, 65 students, six schools, 360 scholars. Pop. 1488.

NEGAPATAM, a decayed town of Hindostan, presid. Madras, distr. Tanjore, and the residence of the British collector for the district, on the shore of the bay of Bengal, 162 m. S. by E. Madras. The European town, which was formerly the capital of the Dutch possessions in the Carnatic, now scarcely exists, Negapatam being seldom resorted to, except by ships, for water and provisions, both of which are plentiful. The native town is tolerably extensive and regular, and on its N. side is a remarkable tower 80 feet high, the origin of which is unknown, but which is very useful as a land-mark. The anchoring ground here is about three miles from shore. Negapatam was taken by the English in 1781. (*Hamilton's E. I. Gaz.*)

NEGOMBO (*Nagambhu*, "the land of serpents.") A seaport town of Ceylon, on the W. coast of that island, 20 m. N. Columbo, and beside the canal, from the latter city to Calpentyn. Lat. 7° 11′ N.; long. 79° 44′. It has a small fort, and several ranges of European buildings; and is principally inhabited by Dutch families in reduced circumstances, attracted thither by the cheapness of provisions, and the salubrity of the climate.

NEGROPONTE, or EGRIPO (an. *Eubœa*), a long, straggling island of the Grecian archipelago, lying close to the E. coast of independent Greece, and forming, with the Sporades, a separate monarchy of its own name. Length, 110 m.; breadth from 5 to 26 m., the widest part being measured from Chalcis to C. Kili: area, 1480 sq. m. Population, in 1836, 60,000. Eubœa is very similar in its mountainous character and geological constitution to the neighbouring continent, from which it seems to have been separated by some sudden convulsion of nature. Grey limestone and clay slate are the chief stratifications, and there are clear indications both of old and more recent volcanic action. The whole country is bold and rugged, with a bluff coast, especially on its E. side, which is dangerous to navigators; the highest points of the mountain range, proceeding from N. to S., through the island, are mount Lithada, 2837 feet high; mount Kandili, 3967 feet; mount Delphi (an. *Dorphessus*), 5725 feet, and St. Elias d'Oro (an. *Ocha*). The soil of the slopes near the shore is very fertile, but only imperfectly cultivated. The orange, citron, almond, and other trees peculiar the climate of Greece, grow abundantly on the lowlands; while the chestnut, oak, and fir, skirt the regions nearer the mountains. The staple produce of the N. part of the island consists of grapes, from which the farmers make large quantities of a thin red wine, very commonly drunk in Greece, and fetching, according to Colonel Leake, about five piastres per barrel. Corn and olives are raised chiefly in the S. districts; but the island has lost the character which it anciently held of being the granary of Greece. (Comp. *Thuc.* i. 2, with *Herod.* v. 77.) Excellent herbage for grazing is found in the more elevated lands; but oxen are bred only for farming purposes. Sheep, however, are numerous, and of an excellent breed, furnishing large quantities both of wool and cheese.

The chief town and port of Eubœa is Chalcis, or Egripos (lat. 38° 30′ N., long. 23° 54′ E.), on the Euripus, or channel of Talanti, where it is only 40 yards wide, and crossed by a bridge, supposed to have been erected by Mahmoud Pacha in 1462. Population 6000. The town (which, according to Strabo, was founded by the Athenians before the Trojan war) is walled and strongly fortified, comprising numerous ancient fragments; but few of them are sufficiently large to be intelligible. It has also two tolerably good harbours, one of which on the N. side, though small, is deep, secure, and capable of containing many merchant ships. The only other town of Eubœa is *Carystus*, or Castel Rosso, a fortified post near its S. extremity, with 2000 inhabitants. There are, also, numerous villages.

The most ancient name of Euboea was Macris; but it was also known, at different times, by the various appellations of Ocha, Ellopia, Asopis and Abanti. Its inhabitants, called Abantes by Homer, were among the earliest navigators of Greece, and, according to Herodotus, joined the Ionian colonists on the coast of Asia Minor, (i. 146). They also founded settlements at a very early period in Illyria, Sicily, and Campania. Soon after the expulsion of the Pisistratidæ, the island became a dependency of Athens, but recovered its liberty, after a hard struggle, in the 21st year of the Peloponnesian war. It afterwards became attached to the Macedonian interests, and was taken by the Romans from Philip, the son of Demetrius. It then gradually declined in population and importance; and Pausanias alludes to its fallen state under the emperors. At the dismemberment of the Eastern empire by the Franca, the Venetians obtained possession of Euboea; but were expelled from it, in 1470, by the Turks, who held it till the formation of the new kingdom of Greece in 1829. (*Leake's N. Greece,* vol. ii.; *Dodwell,* ii. 149–153; *Cramer's Greece,* ii. 123–125, &c.)

NEJIN, a town of Russia in Europe, gov. Tchernigoff, cap. district, on the Oster, 400 m. S. W. Moscow, lat. 51° 2′ 45′ N., long. 31° 49′ 45″ E. Pop. 16,000. It is surrounded by a rampart, most of its houses are of stone, and it is one of the handsomest, best built towns of Little Russia. It has several churches, two convents, a hospital, and a grammar-school founded by Prince Bezborodko. It produces silk, soap, leather, and preserves, and liqueurs that are highly celebrated all over Russia. It is also the entrepôt of a considerable portion of the commerce carried on between the provinces on the Baltic and those on the Black sea. Its merchants are principally Greeks, who enjoy certain peculiar privileges, but they are partly, also, Armenians and Jews. It has several well frequented fairs. (*Schnitzler, la Russie, &c.* p. 464; *Dict. Geog.*)

NEILGHERRY HILLS, or NEILGHERRIES, a collection of mountains of S. Hindostan. (*See* MADRAS PRESID., 245.)

NEISSE, a fortified town of Prussian Silesia, reg. Oppeln, on the river Neisse, which divides the city into two parts, in a marshy district, 48 m. S.S.E. Breslau. Pop. in 1837, 10,787. It is, on the whole, well built, having been greatly enlarged by Frederick II., who also constructed its best fortifications: it is entered by three gates, and comprises among its public buildings a large castle, a commandant's residence, district-hall, seven Catholic and two Calvinist churches, extensive barracks, powder-mills and arsenals, a small theatre, two hospitals, two high schools, a Catholic gymnasium, a poor-school, and an asylum for poor Catholic clergy (*called domus emeritorum*). Neisse is the seat of a council for the circle, a tribune for the principality, a board of taxation, and a consistory court: it has some printing establishments, manufactures of linen and woollen cloths, several distilleries, and a few good hotels. Large yearly fairs are also held here. (*Von Zedlitz; Neukirch; Preuss. Stat,* iii. 139; *Berghaus, &c.*)

NELLORE, a town of British India, presid. Madras. cap. distr. of same name, on the Pennar, 13½ m. from the bay of Bengal, and 100 m. N. by W. Madras. It was, in the last century, a fortress of considerable strength; and is still a populous and busy town, about ⅔ m. in length, full of shops well stocked with commodities, though without a single public or private building of note. The suburbs without the walls are large. The residence of the British collector, &c., is on an elevated ridge S. of the town. A curious discovery was made here in 1787, of a number of Roman gold coins and medals, enclosed in a small pot under the ruins of a Hindoo temple. Many had, unfortunately, been sold and melted: but about 30 were preserved, and found to be of the second century, mostly Trajans, Adrians, and Faustinas. (*Hamilton's E. I. Gazetteer; Madras Almanack.*)

NELSON, county, Va. Situated in the central part of the state, and contains 490 sq. m. Bounded S.E. by James river, N.W. by the Blue Ridge. Watered by Tye and Rockfish rivers. The James river and Kanawha canal passes through it. It contained in 1840, 8139 neat cattle, 7,754 sheep, 20,341 swine; and produced 198,478 bushels of wheat, 35,890 of rye, 397,253 of Indian corn, 90,777 of oats; 18,814 of potatoes; 9,328,817 pounds of tobacco; 994 of cotton. It had 19 stores, one furnace, five flouring mills, 93 grist-mills, 13 saw-mills, six tanneries, eight distilleries; one academy, 70 students; 13 schools, 285 scholars. Pop.: whites, 6,166; slaves, 5,967; free coloured, 152; total 12,287. Capital, Livingston.

NELSON, county, Ky. situated a little N. of the centre of the state, and contains 460 sq. m. Drained by Beechfork, and Rolling fork of salt r. It contained in 1840, 18,508 neat cattle, 20,353 sheep, 46,567 swine; and produced 121,344 bushels of wheat, 56,076 of rye, 473,374 of Indian corn, 3610 of buckwheat, 155,179 of oats, 12,571 of potatoes, 14,711 pounds of sugar. It had 27 stores, six lumber yards, one

furnace, one woollen factory, seven flouring mills, 40 grist-mills, 25 saw-mills, one oil-mill, 65 tanneries, 41 distilleries, three printing-offices, two weekly newspapers; one college, 106 students; three academies, 341 students; one school, 35 scholars. Pop.: whites, 8,878; slaves, 4,543; free coloured, 116; total, 13,637. Capital, Bard-town.

NELSON, p. t., Cheshire co. N.H. 44 m. S.W. Concord, 444 W. Watered by several ponds, which flow into Ashuelot and Contoocook rivers. Chartered in 1774 by the name of Packersfield. It contains one store, one cotton factory with 640 spindles, two grist-mills, four saw-mills; one academy, 37 students; nine schools, 335 scholars. Pop. 833.

NELSON, p. t., Madison co., N.Y., 109 m. W. by N. Albany, 352 W. Drained by Chittenango cr. and branches of Chenango r. It contains three churches, a Methodist, Baptist, and Universalist, three stores, two fulling-mills, two grist-mills, eleven saw-mills, three tanneries; ten schools, 713 scholars. Pop. 2,100.

NELSON, p. t., Portage co., O., 164 m. N.E. Columbus, 339 W. It has nine schools, 391 scholars. Pop. 1,398.

NEMEA, an ancient town of Greece, famous for the games celebrated in its neighbouring grove, but now marked only by the modern village of Agio-Georgio, 19 m. S.W. Corinth, and 10 m. N. by W. Argos. The extant ruins of the town, or village, ((for Pausanias terms it merely a χωρίον), comprise fragments of a temple of Jupiter, a church, and a few blocks and broken Doric pillars, supposed to have formed part of the tomb of Opheltes. Of the temple " three columns only are standing, two of which, belonging to the space between the ante, support their architrave. These columns are 40 5/4 in. in diameter, and nearly 32ft. high, exclusive of their capitals. The temple was hexastyle and peripteral, being supposed by Mr. Wilkins to have had 14 columns on the sides." The lower part of the walls, enclosing the cella, is complete, and the pillars, of which there are numerous fragments, have fallen in such regular order, that the temple appears to have been destroyed by an earthquake, rather than by the lingering and desultory decay of time. Mr. Dodwell says: " I have not seen in Greece any Doric temple, the columns of which are so slender, and the capitals so disproportionately small, as those of Nemea: the whole is of soft calcareous stone, and the columns are coated with a fine stucco." Sir W. Gell mentions, also, " that there are indications of the Nemean theatre at the foot of a neighbouring hill; and probably vestiges of the stadium and hippodrome might be discovered by a search similar to that instituted at Herculaneum and Pompeii." (*Itin. of Morea,* p. 139.)

Nemea was celebrated in mythical history as having been the scene of the first labour of Hercules in destroying the Nemean lion; and the den of this animal was pointed out to travellers even in the time of Pausanias, near the end of the second century of the Christian era. The games are of doubtful origin; but the national mythology ascribes them to the respect entertained for the memory of Ophaltes or Archemorus, son of Lycurgus, a king of Nemea. They were celebrated in the grove of Molorchus, and are thus alluded to by Statius (*Theb.* iv. 159):—

Dal Nemea comlisa et quos in prœlia virus
Sacra Clotoei cognat vincta Molorchi.

With respect to the periods at which these festivals were celebrated, different accounts are given by the old writers; but the most consistent statement is, that they were celebrated triennially, in the Athenian month Boedromion, corresponding with the modern August. The Argives were the judges at these games, which comprised boxing and athletic contests, as well as chariot-races; and the conquerors were crowned with olive till the time of the Persian war, when, in consequence of the losses that the Argolic republic had sustained in that struggle for independence, smallage, a funeral plant, was introduced in its stead. It appears from Polybius and Livy (xxvii. 30), that the games were in a flourishing state in the reign of Philip, son of Demetrius, in the second century preceding the Christian era. It may be inferred, however, from the slight mention that Pausanias makes of the Nemean games, that they had in his time fallen into great neglect. (*Dodwell's Greece,* ii. 208–210; *Cramer's Ancient Greece,* iii. 285; *Burgess's Greece,* i. 171–177.)

NEMI, a village and lake of central Italy, Papal states, in the Comarca di Roma. The village on the N.E. bank of the Lago di Nemi is 2 m. N.E. Albano, and 16 m. S.E. Rome. Nemi, so called from the forest or nemus by which it was anciently surrounded, was famous in antiquity for the worship of the Scythian Diana—nemus glaciale Triveœ—to whom human sacrifices were offered. No remains that can with certainty be ascribed to the temple dedicated to the goddess are now to be met with. The Lago di Nemi is 1022 ft. above the level of the sea; and is now, as of old, beautifully sequestered, and well entitled to its classical epithet of *Speculum Dianæ*. But its principal celebrity in modern times has been derived from the discovery at its bottom, in

1285, of the remains of a very large ship, 680 ft. in length, constructed by one of the early emperors, most probably for some of the *neumachia*, or sham sea-fights, exhibited on the lake. (*Gell's Rome and its Vicinity*, ii. 112 ; *Cramer's Ancient Italy*. ii. 34.)

NEMOURS, a small town of France, dep. Seine-et-Marne, cap. canton, on the Loing, 18 m. S. by E. Melun. Pop. in 1836, 3,635. It is surrounded by the river and the canal du Loing, and inclosed by walls. It is well built, and has a fine old castle, which now serves for several public institutions, including a public library of 10,000 vols. ; several suburbs, a hospital, a small theatre, and a handsome bridge over the Loing. It has some large tanneries and leather factories, and a brisk trade in agricultural produce. The seigniory of Nemours was given to the house of Orleans, by Louis XIV. (*Hugo*, art. *Seine-et-Marne*, &c.)

NENAGH, an inl. town of Ireland, prov. Munster, co. Tipperary, near the Nenagh river, an affluent of the Shannon, within 4½ m. of that river, 82 m. W.S.W. Dublin. Pop. in 1831, 8,448. It is situated in a rich and fertile portion of the co., and was once defended by a strong castle, now in ruins. The principal streets are well and regularly built, and it is decidedly the best town between the cities of Dublin and Limerick. It has a church, a Roman Catholic chapel, Methodist and Independent meeting-houses, an endowed and a national school, a fever hospital, a dispensary, and a large infantry barrack. General sessions are held twice a year, and petty sessions weekly. It is a constabulary station. In contiguity to the Shannon, or rather to its enlargement called Lough Dergh, gives it considerable advantages, and has made it a considerable market for corn and cattle. Markets on Thursdays ; fairs on 24th April, 29th May, 4th July, 4th Sept., 10th Oct., and 1st Nov. Postoffice revenue in 1830, £968, in 1836, £1,026. Branches of the Agricultural and National banks were opened in 1834-5. (*Railway Rep. &c.*)

NEOTS (ST.), a market-town and par. of England, hund. Toseland, co. Huntingdon, on the Ouse, (crossed here by a handsome stone bridge), 8 m. S.S.W. Huntingdon, and 49 m. N. by W. London. Area of par. 4,760 acres. Pop., in 1831, 2,617. The town comprises three or four respectable streets, intersecting each other, with a large market place. The church is a fine building in the perpendicular English style, with large windows of painted glass and an elegant tower 150 ft. high, at its W. end. There are also three places of worship for dissenters, and, in 1834, four Sunday-schools were attended by 280 children. It has also an endowed school for 25 boys, and a large paper-mill; but the principal dependence of the inhab. is on the retail trade with the surrounding district. Markets on Thursday, three large horse and cattle fairs ; and a statute fair on Aug. 1.

NEPAUL (Hind. *Nepala*), an indep. kingdom of N. Hindostan, extending through 8 degs. of long., and comprising a great portion of the S. declivity of the Himelaya chain. It lies between lat. 26° 30' and 30° 30' N., and long. 80° and 88° E., having N. and N.E. the table-land of Thibet, E. the territory of Sikkim, and elsewhere the British territories. Length. E. to W., about 500 m., average breadth, rather more than 100 m. Area may be estimated at 53,000 sq. m. ; and pop. at 2,000,000. This country may be divided into four regions, according to its elevation. The lowest, or *terriani*, is a part of the great plain of Hindostan. In a few places, the British districts reach to the base of the mountains, but in most parts, the Nepaul dominions stretch for about 20 m. into the plain. This region is not wholly level, but undulating, and comprises a good deal of poor land, overgrown with trees and bushes of little value; but there is also a large proportion of rich land, and, upon the whole, the soil is much better than in the adjacent parts of the British territory, the products being, however, nearly the same. The surface here is intersected by numerous small vens, which not only serve for watering the crops, but, in the rainy season, are used for the transit of agricultural produce to the markets of British India, and to float down the floatable timber of the forests. The very name *terriani* (or *regani*), implies, indeed, the country's being navigable. Bounding this region on the N. is another of nearly the same kind, consisting of small hills composed chiefly of clay intermixed, however, with many primary rocks. The lower portion of this region, with a part of the last mentioned, is the grand site of the saul forests, among which many sissoo and toon trees. Higher up the hills are covered with a great variety of trees ; and in the N. are many sals and mimosas, from which catechu is obtained. In a region are many fine valleys, some of which are tolerably cultivated ; while others, though possessing a very good soil, are almost wholly neglected. A few straggling villages are scattered through the woods, the inhabitants of which grow cotton, rice, and other articles with the hoe, clearing the forests away the trees. The third region is that of the mountains, which rise so high as to be covered with snow for a great part of the year, and are divided by valleys,

rising to from 3,000 to 6,000 ft. above the great plain of Hindostan. Of course, these valleys differ very much as to temperature ; some abound with rattans and bamboos, and ripen the sugar-cane and pine-apple ; while others produce only barley, millet, and other grains of cold countries ; and oaks and pines are their only forest trees. The breadth of this belt or region generally may be from 30 to 40 m. N. to S., though further W. it is probably greater. The fourth, or Alpine region, is probably of nearly equal extent, and consists of immense rocks, rising into sharp peaks and tremendous precipices, which, where not perpendicular, are covered with perpetual snow, and almost constantly involved in clouds. The interior, or most lofty chain of the Himalaya, forms the farthest boundary of Nepaul to the N.; through which, however, are several passes into Thibet, while several tributaries of the Ganges, which intersect this country, are supposed to rise on the N. side of the main chain.

The land in the third or mountain region is considered the most valuable in the country, and is that in which all the officers and servants of the crown are paid, and from whence all endowments are made. From the abundance of rain in the warm season (for, the periodical rains extend to Nepaul with nearly the same violence and duration as in Bahar), the land here, considering the inequality of surface, is uncommonly productive of grain. Wherever it can be leveled into terraces, however narrow, it is exceedingly well suited for transplanted rice, which ripens after the rains have ceased, so that the harvest is never injured ; and as most of these terraces can be supplied at pleasure with water from springs, the crops are almost certain. In some parts the same land gives a winter crop of wheat and barley, but in most parts this is judiciously avoided. Where the land is too steep for terraces, it is generally cultivated after fallows with the hoe, and produces rice (sown broadcast), maize, cotton, several kinds of pulse, a kind of mustard, Indian madder, wheat, barley, sugar-cane, and a large species of cardamom ; and in the country between Nepaul Proper (the valley of Catmandoo) and the Kali, ginger is a valuable product ; but transplanted rice may generally be considered as half the entire produce. The sugar-cane is planted in considerable quantities in the valley of Nepaul Proper, and it seems to thrive. Most European kitchen vegetables have been introduced ; but they are only to be found in the gardens of men of distinction, and in very small quantities. From the abundance of rain, the climate is not favourable for many kinds of fruit ; the heats of spring not being sufficient to bring them to maturity before the rainy season sets in, as is the case in Bengal. Peaches grow wild by every rill, but the one side of the fruit is rotted by the rain, while the other is still green. The grapes are also bad from the same cause.

The pasture on the mountains, though not so harsh and watery as that of the low country, is by no means good, and is said to be inferior to that even of the heaths of Scotland. The pastures are in general common. Nothing is paid for pasturage ; but as it is scarce, and as the principal tribes do not employ cattle in agriculture, very few are bred in the country. Buffaloes and goats are imported from the low country ; and horses, yaks, (*Bos grunniens*), shawl-goats, common goats, and sheep, are brought from Thibet, and become tolerably fat on the hills. The buffaloes furnish pretty good beef. The shepherds of some tribes are provided with numerous flocks. In winter they retire to the lower mountains and valleys ; but in summer they ascend to the Alpine regions, and feed their herds in the vicinity of perpetual snow. The sheep which these people possess are very small, and have fine wool, which is woven into a cloth finer than that of Bootan : they give also an abundance of milk, from which is made a kind of cheese.

The lands in Nepaul Proper have long been divided into *khats*, or fields, each of which, in ordinary seasons, produces about 234 bushels of paddy, or rice, in the husk, which, on the supposition that each khat is equivalent to 8½ English acres, would be at the rate of 28 bushels an acre. The arable lands are partly retained as the property of the court for defraying the rajah's household expences ; but the produce of the land so employed is not sold, but serves for the consumption of the court, and for distribution in charity at the temples and among religious mendicants. But by far the greater portion of these lands are let to tenants, or granted in feu for military service ; and the rent of the lands let, as in the former case, forms a principal portion of the rajah's revenue. Landholders who do not cultivate their own estates, in general let them for half the produce. The persons who rent lands from the owners are of two kinds : the *kuriyas*, who occupy free land, and are exempt from any services to government, except the repair of roads, &c. : and the *prajas*, who occupy the crown land, whether that be held by the prince, or granted for military service. Most great proprietors, however, like the rajah, employ stewards, with their servants and slaves, to cultivate land for the supply of their families.

Money rent for land can seldom be procured, and is very low, only from four to twelve annas being paid as a fixed rent in money for land capable of producing a crop, the half of which is worth about 50 annas.

But when the lands are alienated for sale, they fetch from 1600 to 2000 mohurs a khat, which high price is owing to the very small quantity of land that is brought to market. The agricultural implements are very inferior, and almost comprised in an awkward kind of hoe, a weeding-iron and fans for winnowing the corn. In Nepaul, however, they have made a further progress than in India, by the introduction of water-mills for grinding corn.

The mountain region of Nepaul contains a good deal of iron, copper, lead, and some zinc, the first three being found quite on the surface. The copper ore is dug from trenches open above, so that the workmen cannot act in the rainy season, not having sagacity to make a drain. Each mine has attached to it certain families, who seem to be a kind of proprietors, as no one else is allowed to dig. The total quantity of ore dug by each miner may be estimated at 2000 lbs. a year. This is delivered to another set of workmen, by whom it is smelted and wrought, the rajah, to whom the forests mostly belong, furnishing the materials for the charcoal. The ore yields, at an average, 62½ per cent. of metal, 1-3d of which becomes the share of the rajah, 1-3d that of the miner, and 1-5th the share of the smelter; the remainder is divided among the rajah, the miner, and the keeper of the accounts, who usually advances a subsistence to the whole working party, and often furnishes loanavees to the rajah. Iron-ore is found near the surface, and is wrought nearly on the same principle as copper, the miner receiving 1-3d part of the produce. Some of the iron is so excellent that even without being converted into steel, it is made into knives and swords. Only two lead-mines are now wrought; but lead is found in many parts of the country close to the surface, and it contains much silver. There are numerous sulphur mines; but some have been deserted on account of their injurious effects on the workmen. Corundum, here called *Kurren*, is found in great quantities on the hills of Iama and Musikot; but the masses, which always lie close to the surface, are much smaller than those found in the British territory, and seldom exceed four or five pounds in weight.

The most extensive manufacture of Nepaul is that of coarse cotton cloth, woven by the native women of all ranks, and by the men of the Parbatiya caste. These cloths constitute the dress of the middle and lower classes of people, though woollen would be better suited to the temperature of a Nepaul winter. All those, however, who are not very poor, cover themselves with woollen blankets imported from Bootan. The entire dress of the higher ranks is of foreign manufacture, and comprises Chinese silks and shawls, with muslins and callicoes from the low countries. The military alone wear European broadcloth. There are also at Lalita-Patan and Bhatgong extensive manufactures of copper and brass goods, as well as of bells, made from a mixed metal, called *Phul*: these, with iron vessels and lamps, are exported in large quantities to Thibet. A strong paper is made at Bhatgong, from the bark of the *Daphne papyfera*; but the supply is insufficient for the home consumption, and paper is imported from Bootan.

The trade of Nepaul was formerly considerable, though the rajah's territories produce few articles for exportation, except metallic wares and drugs; but at present the badness of the police, and total want of credit, owing partly to the weakness of the law, and partly to the falsehood of the people, operate as a great hinderance to commercial intercourse. The merchants of Cashmere carry their goods, by way of Leh, to different parts of Thibet and W. China, exchanging them for goats' hair, tea, and silks: they also send to China other skins, to the value of about 50,000 rupees a year, procured chiefly from the neighbourhood of Dacca, in Bengal. The merchants of Bootan and Thibet bring to Catmandoo paper, coarse woollen cloths, horses, shawl-goats, sheep, horned cattle, *chaungri*, musk, salt, sal-ammoniac, yellow arsenic, borax, gold dust, silver, and preserved fruit; much of which is again exported to Patna, in exchange for buffaloes and goats, broadcloths, cutlery, glass-ware, and other European articles, Indian cotton cloths, mother of pearl, coral, pepper and other spices, camphor, tobacco, and *phagu*, a red powder thrown about by the Hindoos at their festivals. Most of these articles, with metallic utensils and bells, are sold to the merchants of Thibet. The money of Nepaul consists of *damas*, 4 of which are equal to 1 *paisah*; 4 paisah = 1 ana; and 8 anas = 1 mopur. Gold coins are called *ashruffies*; but the half ashruffy, = 12½ mohurs, is the highest piece now coined: it weighs 84½ grains, and is worth nearly 6s. 3d. at the mint price of Calcutta. The mohur is the common silver coin of the country, and is worth about 4-10ths the Calcutta rupee. The paisah and half-paisah

are the principal copper coins. Grain is sold by measure; 1 muri being equivalent to 2½ Winchester bushels.

The Nepaulese government, which for many years has been monopolised by the tribe called Gorkahs, is essentially despotic, modified, however, by certain observances enjoined by immemorial custom. The *Dharmashastra* forms the basis of jurisprudence both in civil and criminal cases, the principal punishments being by fines, confiscation of property, banishment, degradation of caste, maiming, and death by hanging as well as flaying. Women are never put to death, but are subject to mutilation and torture. The provinces are governed by *subahs*, who are the supreme officers of revenue, justice, and police: each farms the revenue of his own district, and either collects it on his own account, or underlets it to *izaradars*. The amount paid by the subahs, however, forms by no means the whole of the royal revenue; for, besides compulsory presents made by all visiters of the court, a general income-tax is levied on all classes according to the exigencies of the state. Nepaul Proper is governed by a rajah, assisted by the *herzdar*, or council of the 12 great officers of the court; for the support of which Catmandoo pays 18,000 rupees; Lalita Patan, 16,000; Bhatgong, 14,000; and Kirthipoor, 7088. Each farm is assessed at a certain quantity of grain, which may be paid either in kind or in money at the market price. A large proportion of the valley, however, has been alienated either in fee or as charity land. A town called Sanghoo, worth annually 4000 rupees, is the jointure of the queen-regent, and Detwapatan, which is still larger, belongs wholly to certain temples. The religion of the Nepaulese is Buddhism; but in the distinctions of caste and the nature of the priesthood there are essential differences between the religion of the Buddhists of Ava and that professed by those of Nepaul, both of whom are held in equal abhorrence by the Brahmins of Bengal.

The population of Nepaul comprises numerous tribes partly of Mongol and partly of Hindoo descent. The Magars, who occupy the hills in the W. part of the kingdom, form the greater part of the rajah's army, and the Gurungs, who employ themselves either in mining or pasturage. The Newars live in the plain of Catmandoo, and devote themselves to agriculture and the useful arts. They are of middle size, with broad shoulders and chest, flat face, small eyes, and spreading noses, with a sallow complexion. The grand basis of subsistence in Nepaul is rice, with which the poorer classes eat raw garlic, radishes, and lentils: those in more easy circumstances add oil or *ghee*; and the rich eat a great deal of animal food. Even the poorest are occasionally able to sacrifice a pigeon, fowl, or duck, which they afterward eat. The rajpoots of Nepaul, indeed, are so fond of animal food, that, to the astonishment of the Bengalese, they drink the blood of a sacrifice as it flows from the victim. All classes drink spirituous liquors, to which they are excessively addicted. Most of the Nepaulese domestic servants are slaves, the price of which varies between 30 and 40 mohurs. Some are slaves to the rajpoots; but they are not degraded, and are employed in great families either as cooks or in the service of the private chapels. All other ranks are sold as common slaves, and persons of the best families have often been deprived of their caste; but this is not usual, as the Nepaulese are particular in maintaining the distinction of castes. Most of the slaves have been born free: a few perhaps, have been degraded on account of crimes; but by far the greater number have been sold by necessitous parents. The female slaves, even those of the queen, as *donnas libera*, compelled to sell their favours for clothes, no allowance being made to them by their masters except a little rice. Hence they seldom have children, and beauty is the usual lot of the old and infirm. The queen's slaves form her body-guard, and follow her on horseback armed with swords and riding like men. The ordinary language of Nepaul is the Prabratiya, or mountain-Hindoo dialect, which is continually becoming more prevalent, and in some districts has already superseded the language of the native tribes: it is exclusively spoken by the reigning family and the higher castes. The Newars have a language peculiar to themselves, quite different from that of their neighbours, and alleged to possess a copious literature.

Nepaul, which was formerly divided among numerous independent princes, became united by conquest in the middle of the last century under the sovereignty of a chief of the Ghorkas, who in about 40 years subjected all the countries between the Sutledje westward and Bootan on the E. The aggressions of the Ghorkas on the Chinese territory were stopped in 1792 by an army of 70,000 men, who, after many victories, advanced within 25 m. of Catmandoo, and obliged the rajah to make an ignominious peace. The Ghorkas afterward turned their arms against the British, who, after a war of two years, obliged them, in 1816, to cede all the countries between the Sutledje and Kali, as well as to evacuate the territories of the Sikim-rajah. Active symptoms of hostility to the English were displayed in 1839; but these were checked by the events of Afghanistan. (*Hamilton's Nepaul*, passim.)

NEPI (an. *Nepete*), a town of central Italy, Papal states, deleg. Viterbo, 25 m. N.N.W. Rome. Pop. about 1500. It is beautifully situated, and surrounded by a high Gothic wall, partly founded on the original walls erected by the Etruscans. It has numerous churches and convents, and a fine modern aqueduct, but a gloomy and desolate appearance within. Some Roman antiquities exist here. Conjoined with Sutri, Nepi constitutes a bishop's see. (*Gell's Rome, &c.*, ii., 118.)

NERAC, a town of France, dep. Lot-et-Garonne, cap. arrond., on the Baise, a tributary of the Garonne, 16 m. S.W. Agen. Pop., in 1836, ex. com., 3084. It is divided into the old and new town, one on either bank of the river, here crossed by two stone bridges. The old town, on a steep declivity, and partly surrounded with Gothic walls, is ill built and gloomy; but the new town, on a level site, and encircled by promenades, is well laid out, and handsome. Nerac has the remains of an extensive castle, said to have been constructed by the English, a fine parish church, a large hall, and several other good public buildings. One of the promenades has a good statue of Henry IV., who passed most part of his youth in the castle of Nerac. This town has manufactures of coarse woollens, ship biscuit, and corks; and a good deal of trade in linen fabrics, corn, flour, wine, and brandy. Numerous Roman antiquities, including baths and other edifices, medals, inscriptions, &c., have been discovered at Nerac, from which it would appear that it was anciently called *Aqua Nova*, and was either founded or greatly embellished by Tetricus, in the reign of Gallienus. (*Hugo*, art. *Lot-et-Garonne, &c.*)

NERBUDDAH, (*Nermada*, "The bestower of pleasure," called by Ptolemy the *Namadus*,) a river of Hindostan, extending through 9 degs. of long. in the N. part of the Deccan. It rises in the table-land of Gundwanah, lat. 22° 40' N., long. 81° 45' E., near the sources of the Sone and Mahanuddy. It has a general W. direction, with fewer windings than most Indian rivers; and, after a course of about 700 m., falls into the gulf of Cambay, lat. 21° 36', long. 72° 50', 98 m. W. Baroach. It varies considerably in breadth; being 660 yards across, near Jubbulpoor, in long. 80°, and 1200 yards at Mundleystr, 210 m. from its mouth; while above and below Baroach, it sometimes expands to a breadth of 3 m. At its source, the Nerbuddah may be 2460 feet above the level of the sea: its total rate of descent will be therefore nearly 3½ feet in a mile. During its passage it is greatly obstructed by rocks, islands, shallows, and rapids, which render its navigation in most parts difficult or impracticable through the provinces Gundwanah, Malwah, &c.; but, after entering Guzrat, it becomes navigable for small craft for about 160 m. from the sea. The Nerbuddah is joined by no affluent of any consequence. For so considerable a river its basin is remarkably narrow and restricted; it being inclosed on the N., for the most part, by the Vindhyan mountains, and on the S. by the Sautpoora and other parallel ranges, which are seldom more than from 50 to 60 m. from the former. The valley through which it flows, consists of fertile alluvial soil, in which many fossil remains have been found. (See *Journ. of the Asiatic Soc. of Bengal*.) Mundlah, Gurrah Warrah, Hustinagabad, Hindu, Mheysur, and Baroach, are the principal towns on this river. By the war of 1817–18, the British obtained an extent of nearly 30,000 sq. m. of the country watered by this river from the rajah of Berar, which, under the term of "Ceded Districts on the Nerbuddah," has been annexed to the Bengal presidency, and in 1820, produced a total revenue of 1,876,398 rupees. (*Parl. Reports; Asiatic Journals; Hamilton's E. I. Gazetteer.*)

NESCOPECK, p. t., Luzerne co., Pa., 98 m. N.E. Harrisburg, 208 W. Watered by Big and Little Wapwallopen and Nescopeck creeks, flowing into Susquehanna river, which bounds it on the N.W. A covered bridge 1256 feet long, which cost $31,000, here crosses the Susquehanna. It contains two stores, one forge, two flouring-mills, six saw-mills; six schools, 965 scholars. Pop. 1370.

NESHANNOCK, t., Mercer co., Pa., 12 m. S.W. Mercer. Watered by Neshannock creek. It contains two commission houses in foreign trade, three stores, one furnace, one fulling-mill, three woollen factories, seven grist-mills, 12 saw-mills; 13 schools, 555 scholars. Pop. 2068.

NESHOBA, county, Miss. Situated toward the E. part of the state, and contains 600 sq. m. Watered by Pearl river and its head branches. It contained in 1840, 10,003 neat cattle, 677 sheep, 8663 swine; and produced 7163 bushels of wheat, 83,230 of Indian corn, 1810 of oats. 14,506 of potatoes, 1562 pounds of rice, 9918 of tobacco, 1,039,947 of cotton. It had three stores, four grist-mills, three saw-mills, three tanneries; four schools, 65 scholars. Pop.; whites, 1693; slaves, 744; total, 2437. Capital, Philadelphia

NETHER PROVIDENCE, p. t., Delaware co., Pa., 12 m. W. Philadelphia, 90 m. E. by S. Harrisburg, 124 W. Bounded W. by Ridley creek, and E. by Crum creek, both flowing into Delaware river. It contains seven stores, one flouring-mill, three grist-mills, five saw-mills; one school, 144 scholars. Pop. 1085.

NETHERLANDS. *See* HOLLAND.

NETTUNO, a small seaport of southern Italy, in the Campagna and Comarca di Roma, 31½ m. S.S.E. Rome. Pop. about 3000. It seems to have derived its name from an ancient temple dedicated to Neptune, and is built round the bastions of a Papal fortress. It has now but little activity or commerce, owing to the unhealthiness and depopulation of its vicinity; but in antiquity, under the name of Ceano, or Cerio, it was the port of Antium, the capital of the Volsci, some remains of which city exist about 2 m. W. by S. (*Gell's Rome, &c.*, ii., 192.)

NEUBURG, a town of Bavaria, circle Swabia and Neuburg; on the Danube, here crossed by two bridges, 28½ m. N.N.E. Augsburg, and 45 m. W.S.W. Ratisbon. Pop. 6000. It is divided into the upper and lower town; and has some remains of its ancient walls, a royal castle, in which many curiosities are kept, an arsenal, a royal institute, a hospital, a gymnasium, and a teacher's seminary. It is neat and well built; and is the seat of the high court of appeal for the circle. (*Borghaus, &c.*)

NEUCHATEL, or NEUFCHATEL, a canton in the W. of Switzerland, forming a principality belonging to Prussia; between lat. 46° 50' and 47° 10', and long. 8° 25' and 7° 5' E.; having N.E. and E. the canton Bern, S.E. the lake of Neuchatel, S.W. Vaud, and W. and N.W. the dep. of Doubs, in France. Length N.E. to S.W., 33 m.; average breadth about 9 m.; area, 280 sq. m. Pop., in 1837, 56,616, of whom 17,744 were either citizens of other cantons, or foreigners. The Jura chain runs through canton in its entire length, dividing it into two parts, one belonging to the basin of the Rhine, and the other to that of the Rhone. This mountain range often rises to 5000 feet in elevation; and the Chasseral, its highest point within the canton, rises 5285 ft. above the sea. The valleys extend generally in a longitudinal direction, parallel to the mountains. The principal lakes are those of Neuchatel (which see), and a part of that of Bienne; principal rivers, the Doubs, constituting the N.W. boundary; the Reuse, Thielle, Tyon, &c. The climate varies greatly: the vine is cultivated on the banks of the lake of Neuchatel; but in some of the more elevated valleys the winter is very severe, and on many of the mountains snow remains continuously for seven or eight months. The soil is principally calcareous. Of 256,000 *poses* or arpents of land, which the canton is estimated to comprise, 35,600 are arable, 4600 vineyards, 58,000 in artificial, and 60,000 in natural pastures, and 45,000 in forests. There are very few large proprietors: the savings of the labouring population, both agricultural and manufacturing, are generally laid out on the purchase of cottages, with a small portion of adjacent land. Excepting wine and vegetables, this canton does not yield enough of agricultural produce for its own consumption; and the principal part of its supply of corn is imported from the neighbouring cants. of Basle and Bern. Considerable quantities of wine grown around Neuchatel are exported to the neighbouring Swiss cantons. The best wines are those of Cortaillod, Neuchatel, and Boudry. The first in fine years is said to approach pretty closely to Burgundy. Within the last few years the preparation of sparkling wines, sold as champagne, has become a pretty extensive branch of business, from 190,000 to 140,000 bottles being annually exported. A good many cattle are reared, principally cows; and cheese is one of the principal articles of export. Hay is also extensively exported.

Neuchatel is one of the principal manufacturing cantons of Switzerland, especially for watches, printed cottons, and lace. Watch-making, which was introduced early in the 17th century, is carried on to a great extent in the mountainous districts, but particularly in and near Le Locle and Chaux de Fond. It is estimated that from 18,000 to 20,000 hands are employed in this branch of industry, or in manufacturing instruments for the construction of watches. From 100,000 to 120,000 watches are supposed to be annually produced, of which 35,000 are of gold; they are exported to France, Germany, Holland, Italy, Spain, America, Turkey, &c. Mostly all the watches sold in Paris are made in Neuchatel, and the neighbouring Swiss cantons, whence they are either smuggled into France, or regularly imported, the gold watches paying a duty of six and those of silver of 16 per cent. The capital employed in the watch trade has been estimated at 7,000,000 of Swiss francs, or upward of £386,000 sterling. It is difficult to ascertain the ordinary rate of wages, workmen being generally paid by the job, and not by a stipulated salary. But it is believed that a man's wages may be estimated, at an average, at from 1000 to 1500 francs (£55 10s. to £83) a year. Lace making was said by Ebel to occupy from 5000 to 6000 hands; but it has declined, and many persons formerly engaged in it have embraced some branch of watch-making. The printed cotton manufacture was established early in the last century, towards the latter end of which it was in its most flourishing state. At present most of the cotton cloths printed in Neuchatel are furnished by Zürich and other Swiss cantons; and only about 1000 men, women, and children, are employed in this manufacture, who produce annually about 80,000 pieces, each containing about 22 English yards. Of these prints about 30,000 pieces are sent to the Prussian territories, and the remaining 50,000 to Holland, Belgium, and Italy. Hosiery, cutlery, mathematical instruments, and metallic wares of various kinds, are among the other manufactures of the canton. Neuchatel is not a member of the Prussian Commercial League, but its manufactures are admitted into the Prussian dominions at a diminished duty. Watches, &c., pay one half, wine two fifths, and printed cottons pay 20 rix-dollars per cwt. of the ordinary tariff duty. The livre or franc of Neuchatel (of 20 sols of 12 deniers each) = 10 batzen, or about 1s. 8d. English.

Neuchatel recognizes the sovereignty of the king of Prussia, and pays him an annual tribute of 70,000 Neuchatel francs, or nearly £4000. All the administrative functionaries are nominated by the king, without a veto on the part of the legislature. The representative body (audiences générales) consists of the 10 oldest members of the governor's council of state; of 14 members not councillors, chosen for life by the king from lists presented by the inhabitants of the canton; of the head magistrates of the canton, whose number must not exceed 24; and of 30 deputies each at least 25 years of age, elected by all the male inhabitants of more than 22 years, being neither condemned criminals, bankrupts, nor receiving pecuniary relief. This body is convoked and prorogued at the command of the governor, but it must assemble once in two years. No law can be passed, changed, or abrogated, without the consent of the audiences générales; but no resolution of the latter becomes law till it have received the sanction of the king. The budgets are voted by the audiences générales, on whose account the imposts are received, through agents nominated by the king. No custom-houses exist in any part of the canton; and the turnpike dues are much lower than in most of the other Swiss cantons. There is no impediment to the free exercise of any profession, no poll-tax or other direct contribution, duty on raw materials, impost or carriage or communication, or taxes on food or drink, direct or indirect. There are neither stamps nor patents; and the complete absence of all restrictions, and of almost all taxation, is a primary cause of the cheapness of most articles, and the general prosperity of the canton. There are few districts of Europe where so large a proportion of the inhabitants are interested in savings' banks. In 1834, 1 in 18 of the population was a depositor. Paupers are provided for by the communes to which they belong; no general tax can be established for their support. The number of illegitimate children is about 2½ per cent.

The administration of justice is both prompt and economical. There are 21 courts of primary jurisdiction; and two of appeal, at Neuchatel and Vallangin. The laws are, in many respects, similar to those formerly prevalent in Burgundy. The inhabitants speak a French dialect; they are Protestants, except about 3000 individuals under the authority of the bishop of Lausanne. Public instruction is pretty generally diffused, few individuals being ignorant of writing and arithmetic. There are colleges in Neuchatel, the capital, and Chaux de Fond; and schools of watch-making and other arts in those towns and Le Locle. Many societies for instruction, and benevolent purposes, exist. The militia comprises all males between the ages of 18 and 50. Neuchatel furnishes a battalion of light infantry to the Prussian service, and a contingent to the Swiss confederacy. The public revenues, derived from rents, a small tithe or land tax, posts, turnpikes, salt and auction duties, &c., amounted, in 1834, to 362,311 francs, and the expenditure in the same year to 238,153 francs, of which 14,517 francs formed the contribution to the Swiss confederacy.

Neuchatel belonged, in the 11th century, to the German emperors; and was ceded to Burgundy by Rodolph of Hapsburg. In 1406 the town of Neuchatel entered into a treaty with Bern, and soon after allied itself to the Swiss confederacy. In 1707, the last direct inheritor of this territory dying, the states chose the king of Prussia for their sovereign. Napoleon created Neuchatel into a principality, which he conferred on Marshal Berthier; but in 1814 it reverted to Prussia, being constituted, however, at the same time, the 21st among the Swiss cantons. (Ebel; Manuel de la Suisse; Picot, Statistique; Lutz, Geogr., &c.; Bowring's Rep. on Switzerland; Helvetic Almanach, &c.)

NEUCHATEL (Germ. Neuenburg), a town of Switzerland, cap. of the above canton, on the N.W. shore of the lake of

Neuchatel, 17 m. N.W. Freyburg, and 45 m. E.S.E. Besançon. Pop. from 5000 to 6000. It is built upon the steep slope of the Jura mountains, and along a narrow strip of level ground between the hills and the lake. Its objects of curiosity comprise the castle, formerly occupied by the French princes of Neuchatel, but now by the Prussian governor; the church, a Gothic edifice of the 12th century; the town-hall, in which the audiences générales meet; the gymnasium, with a museum of natural history, &c. Its charitable institutions are on a large scale; one hospital and poor-house was founded and endowed with a sum of £166,000 by a townsman; and another, the Hospital Pourtales, is also an extensive establishment, and open to all persons without respect of country. It has, also, an orphan asylum, a house of correction, some public granaries, several good hotels, &c. The extrait d'absinthe is produced here, and it has a considerable traffic in the agricultural and manufactured produce of the canton. (Ebel, &c.)

NEUCHATEL (LAKE OF), otherwise called the lake of Yverdun (Germ. Neuenburger-See), a lake of Switzerland, in the W. part of the confederacy, between the cantons Neuchatel, Vaud, Freyberg, and Bern. It is of an elongated shape; length N.E. to S.W. 24 m.: average breadth nearly 4 m.; area probably 90 sq. m. The elevation of its surface above the sea is estimated at 1390 feet; its greatest depth is 400 feet. Several considerable rivers empty themselves into this lake, which also receives the surplus waters of the lake of Morat. Its own surplus waters are conveyed by the Thiele to the lake of Bienne; and thence to the Aar and the Rhine. Neuchatel, Grauson, Yverdun, Estavayer, and Condrefin, are on its banks. Its scenery is agreeable, but tame in comparison with that of most other Swiss lakes. Its navigation is sometimes dangerous, from its being subject to sudden gusts of wind. A steamer, however, plies on it daily from Neuchatel to Yverdun. (Picot's Suisse; Murray's Handbook for Switzerland.)

NEUILLY, a village of France, dep. Seine, cap. canton, on the Seine, here crossed by a handsome stone bridge, on the road from Paris to St. Germains, 1¼ m. W.N.W. the Barrier de l'Etoile. Pop. in 1836, ex. com., 3732. The bridge of Neuilly, regarded as a chef-d'œuvre of the architect Peronnet, has an entire length of 800 feet (the span across the river being 710 feet), with five arches, each nearly 126 feet in breadth, and 32 feet in height. The château de Neuilly, built in the time of Louis XV., is a favourite summer residence of Louis Philippe, king of the French. The village has manufactures of earthenware and chemical products, and distilleries of ratafia, &c. (Hugo, art. Seine.)

NEUSATZ (Hungar. Uj-Vidék), a royal free town of Hungary, co. Bacs, on the Danube, opposite Peterwardein, with which it is connected by a bridge of boats, 66 m. N.W. Belgrade. Pop. 20,231. (Berghaus.) It consists of long straggling streets, but, being of modern origin, some of them are tolerably well built, and they are generally paved. The Greeks have five churches, the Roman Catholics one church, and the Armenians one; it has also a synagogue, a gymnasium, a Roman Catholic high school, Jewish school, &c. Neusatz is a place of considerable traffic, particularly with Turkey, for which it is chiefly indebted to its position on the Danube, near the influx of its three largest tributaries, the Theiss, Drave, and Save. Its numerous shops are said to be full of "grocery, and clothes, ironmongery, tin-ware, earthenware, wooden bowls, dishes, and trenchers, all of very rude fashion, and jewellery of an ordinary description." (Quin's Voyage down the Danube.) Neusatz is the residence of the Greek bishop of Bacs, and of a protopapas. There are remains of a Roman wall stretching from Neusatz to Csurog on the Theiss, 19 m. N.N.E. (Oesterr., Nat. Encyc.; Berghaus, &c.)

NEUSE, river, N. C., rises in Person and Orange counties, and enters a large estuary connected with Pamlico sound. It is about 400 m. long, is navigable for sea vessels 12 m. above Newbern, and for boats upwards of 200 miles. Much of the land on its borders is fertile, but is liable to inundation.

NEUSOHL, a royal free town of Hungary, beyond the Danube, cap. co. of its own name, on the Gran, at the influx of the Bistricza, 80 m. N. Pesth. Pop. 5814, nearly half being Protestants. "Neüsohl has wide streets, and is a tolerably well built country town, rather imposing in its appearance, because all the houses appear to be in the Italian style, with flat roofs, though probably it is only a high parapet carried up to hide the roof." (Paget, i. 355.) In the parish church (a Gothic structure), is a bell weighing 160 centners. Neüsohl has an old castle, a hospital, several superior schools, &c., and is the residence of a bishop, the seat of a mining council and tribunal. Near it are the mines of Herrengrund, producing 1500 cwts. of copper a year, and some silver; and in the town is the largest smelting-house in Hungary. Neüsohl has also manufactures of sword-blades, and beet-root sugar. (Oesterr., Nat. Encyc.; Paget's Hungary, &c.)

NEUTRA, or NEITRA, an episcopal town of Hungary, cap. co., on the Neutra, in a finely-wooded country, 45 m. E.N.E. Presburg. Pop. 4643. It has a castle, a county hall, a cathedral and bishop's palace, a lyceum, and several high schools; and carries on a considerable traffic in the wine grown in its vicinity.

NEUWIED, a town of Rhenish Prussia, circle Neuwied, of which, and of a mediatized principality, it is the cap., on the Rhine, 7 m. N.N.W. Coblentz. Pop., in 1837, 5766. It was founded early in the last century by a count of Wied, on the broad principle of perfect toleration for all sects; in consequence of which a neat and flourishing town soon sprang up. It is laid out in squares of houses, formed by nine streets intersecting each other at right angles. At its W. extremity, overlooking the Rhine, is a castle, the residence of the princes of Wied: the town has, also, several churches, and other places of worship, a gymnasium, teacher's seminary, hospital, orphan asylum, house of industry, a prosperous Moravian establishment, &c. Its manufactures are of silk, cotton and linen fabrics, and yarn, stockings, iron goods, tobacco pipes, Prussian blue, chicory, potash, and soap: it is the seat of the judicial court for the principality, the circle court, and a mining tribunal. The museums of natural history in the castle and in the Moravian establishment, are worth notice: but the principal object of interest at Neuwied is its collection of antiquities: these were found in the buried Roman city of Victoria, about 2 m. N. the town, supposed to have been destroyed by the Germans towards the end of the 4th century. (A full description of this collection may be found in Schreiber, Guide du Rhin, 290-293; Borghaus; Von Zadlitz, Der Preussiche Staat, &c.)

NEVERS (an. Noviodunum, and Nivernum), a city of France, dep. Nièvre, of which it is the cap.; on the Loire, where it is joined by the Nièvre, and a little above the influx of the Allier: 133 m. S.S.E. Paris; lat. 46° 59′ 17″ N., long. 3° 9′ 31″ E. Pop., in 1836, ex. commune, 13,275. It is agreeably situated on the declivity of a hill facing the S., but is in general ill built and ill laid-out, its streets being narrow, steep, and crooked, and its houses old and gloomy. In its centre, however, is a large and regularly constructed square, on one side of which is the ancient residence of the dukes of Nivernais. Some of the entrances to Nevers are imposing: that from Bourges is ornamented with a triumphal arch, and on the road from Moulins the Loire is crossed by a solid stone bridge of 20 arches. The quays on the river are bordered with good houses, and look clean. The cathedral, on the site of a very ancient church, is an edifice principally constructed between the 19th and 16th centuries. It is large, and has a lofty square tower; in its choir is some fine stained glass. Several other churches, as well as the cathedral, are curious specimens of Gothic architecture. The other public buildings are mostly in a simple but appropriate style: the principal are the barracks, arsenal, prefecture, and public library with 8300 volumes. The park, formerly belonging to the dukes of Nivernais, has now become one of the many public promenades surrounding Nevers. The city preserves but a few remains of its ancient fortifications. It is the see of a bishop, whose diocese extends over the dep. Nièvre; and is the seat of tribunals of primary jurisdiction and commerce, of a chamber of manufactures, a communal college, &c. It has several hospitals, a handsome little theatre, two episcopal seminaries, schools of drawing, geometry, &c., a free school of arts, a commission l'antiquité, and many other scientific establishments. It is also distinguished by its manufacturing industry. It has a royal cannon foundry, in which from 200 to 230 cannons, weighing in all about 550,000 kilogs., are cast annually, besides 50,000 kilogr. weight of other kinds of artillery (Hugo.) It also produces chain cables, iron works for suspension bridges, and other heavy iron goods. Nevers has one for many centuries famous for its china-ware, which, for durability and solidity, is said to be the best made in France; it is sent in large quantities to Paris, and throughout the country watered by the Loire and its tributaries. Its manufacture employs about 700 workmen, whose wages are said to average 1 fr. 75 c. a day. Glass wares, metal buttons, coarse woollen cloths, violin strings, vinegar, fine brandy, and leather are among the other principal manufactures. It has also a considerable trade in timber or ship-building, charcoal, iron and steel, wine, salt, &c., being the great entrepôt for the upper Loire. Its trade is facilitated by a commodious haven at the mouth of the Nièvre. It has nine annual fairs, one of which lasts eight days.

This town existed at the conquest of Gaul by Cæsar; it became a bishopric in 506, and the capital of Nivernais; in 5 it was burned by Hugh Capet; and in the middle ages suffered severely from plague, the inundations of the Loire, the invasions of the English, and religious wars. (Hugo, l. Nièvre; Dict. Géog.)

NEVIS one of the British W. India islands, belonging to the Leeward group; in about lat. 17° 16′, long. 62° 23′ W., separated by a strait 2 m. in breadth, from the S.E. extremity of St. Christopher's. Shape circular; greatest length, N.E. to S.W., 6½ m.; extreme breadth, about the same. Pop. 11,800. It consists of a conical hill, rising from the sea to a height of 2800 feet. Soil mostly a strong tenacious marl, not readily absorbent of moisture: the climate is similar to that of St. Kitt's and Tortola. It is well watered, and, in general, fertile. The inhabitants are nearly all occupied in the raising of the sugar-cane and provisions, and in the preparation of rum and sugar. In 1839, 34,466 cwts. sugar, 39,252 gallons rum, and 3501 cwts. molasses, were imported from Nevis into Great Britain. The total value of the exports amounted, during the same year, to £12,993; that of the imports to £27,183. It is divided into five parishes; Charlestown, the capital, is at its S.W. extremity. This colony is placed under a governor and council, and assembly. It has nine public schools, in which about 480 children are educated. The portion of the compensation for slaves paid to the proprietors of Nevis, amounted to £154,007; the number of slaves by the last registration having been 8722, and the average value of a slave, from 1822 to 1830, £30 4s. Columbus discovered Nevis; which was settled by the English in 1628. (Parl. Papers, &c.)

NEW ALBANY, a city, and capital of Floyd co., Ia., 123 m. S. by E. Indianapolis, 609 W. Situated on the N. bank of the Ohio, 2 m. below the foot of the falls at Louisville, where the river descends 38 ft. in the course of a mile. It is 335 m. above the mouth of the Ohio, and is the largest place in the state. It contains a courthouse, jail, nine churches, two Presbyterian, an Episcopal, two Methodist, two Baptist, a Campbellite Baptist, and a Roman Catholic, a bank, an insurance company, a male and a female seminary, a lyceum, a theological college, four schools, 40 or 50 stores of different kinds, an iron foundry and steam engine factory, a rope-walk, one steam grist-mill, one steam saw-mill, one hemp and bagging factory, and 4296 inhabitants. It is regularly laid out with six streets, running parallel with the river, crossed at right angles by 11 other streets. The streets are from 70 to 80 or 100 feet wide. From 10 to 15 steamboats are built here annually, besides sloops and schooners, to be sold at New-Orleans, and there are several extensive ship-yards. A Macadamized road is finished for 50 m., designed to be continued to St. Louis, Mo. The town, exclusive of the city, has 1396 inhabitants.

NEW ALBION, p. t., Cattaraugus co., N.Y., 307 m. W. by S. Albany, 347 W. Drained by Cattaraugus creek, and branches of Conewango creek. It has one store, one grist-mill, four saw-mills; six schools, 354 scholars. Population, 1016.

NEWARK, a parl. and mun. bor., market town, and par. of England, co. Nottingham, S. div. wap. of its own name, on a lateral stream of the Trent, crossed here by a handsome bridge of seven arches, 16 m. N.E. Nottingham, and 110 m. N. by W. London. Area of parl. bor. 2000 acres. Pop., in 1831, 9557. The approach to Newark from the N. is by a long causeway carried over a flat island formed by the Trent and the Newark branch; and under it are numerous bridges, to give free passage to the waters during the floods. The town, consisting of a principal street on the Nottingham and Lincoln road, crossed by several others, and having a large market-place near its centre, is on the whole well built, paved, lighted with gas, and abundantly supplied with water. Among the public buildings, one of the most interesting, though now in ruins, is the castle near the bridge, called the New Work, from the circumstance of its having been re-edified by Stephen. It comprises a square of large dimensions, with two massive towers, and seems to have had five stories: the interior area is used as a bowling green, but several of the lower rooms are still entire. King John died in this castle, 18th October, 1916. The townhall, in the market-place, a handsome building of stone, erected in 1776, comprises several large apartments for the corporate business, assemblies, balls, &c. It has also a courthouse for the quarter sessions, with a small jail, which, however, is "wholly inadequate and unfit for its purpose." (Pris. Inspect.. 5th Rep.) The church, said by Mr. Rickman to be one of the largest and finest in England, was built in the reign of Henry VI.; it is a cruciform structure, with large aisles, transepts, and chapels, having at its W. end a highly ornamented tower, surmounted by an extremely light steeple, 240 feet in height, round which are niches containing statues of the twelve apostles. Some of the windows have stained glass, representing the history of Jesus Christ; the choir is separated from the rest of the church by a screen of rich oak-carving, and in the interior are several curious monuments. The fabric is kept in repair by the produce of estates belonging to the borough, so that there is no necessity for a church-rate. The living is a vicarage in crown patronage. It has also four places of worship for dissenters, with various Sunday schools, attended by about 100 children. The grammar school was founded in 1539; its endowment, at the time of the Char. Comm. Inquiry, amounted to £2380 a year, and in consequence of a suit in Chancery, the corporation, its trustees, have founded two exhibitions of £80 a year each, tenable for four years at Oxford or Cambridge. The master has a salary of £230 a year with a good house, and £60 a year are paid to an usher; the school is attended by about 40 boys. Two national schools furnish instruction to about 250 children of both sexes, and there are two or three smaller schools wholly or in part supported by subscription. The estates held in trust by the borough for charitable purposes, independently of that above mentioned, are very extensive; and there are several almshouses, a workhouse, and dispensary. A library and small theatre are the only other public establishments.

Newark carries on a considerable trade in malt and corn, and in coal, cattle, and wool. It has also two large brass and iron foundries: bricks and tiles are made here, and large quantities of gypsum and limestone, quarried and prepared in the neighbourhood, are sent by sea to London. Here are two pretty extensive linen manufactories, and two private banks, besides a savings' bank and a branch of the Nottingham banking company. The arm of the Trent on which Newark stands is made navigable by means of a lock close to the town.

Newark was divided by the Municipal Reform Act into three wards; the corporation comprising a mayor and five other aldermen, with 18 councillors. It has, also, a commission of the peace under a recorder, with a court of requests for the recovery of debts under £5. Corporation revenue in 1839, £1250. The quarter sessions for the S.E. division of the county are held here. Newark has sent two members to the House of Commons since the 29th Charles II., the right of election down to the Reform Act being in the mayor, aldermen, and inhabitants paying scot and lot. The electoral limits were not changed by the Boundary Act; registered electors in 1839-40, 1130. Newark is also the election town for the S.E. division of the county. Large markets, especially for corn, on Wednesday; fairs, Friday in Mid-lent, May 14, August 2, November 1, and Monday before December 11.

Newark, which takes its name from the castle, became a place of considerable importance soon after the Norman conquest; but its principal celebrity is owing to the fact of its having been one of the chief garrisons of the royalists during the civil wars of Charles I. It was besieged by the parliamentary forces in 1643; but both the town and castle were held by the royal army till 11th May, 1646, when it was surrendered to the Scotch by command of the king, who was then a prisoner. The castle was at the same time demolished by order of parliament. (*Parl. Rep.; Char. Comm. Rep., &c.*)

Newark, t., Tioga co., N.Y., 8 m. N. Owego, 161 W.S.W. Albany. Drained by East and West Owego creeks. It has five stores, one fulling-mill, five grist-mills, 20 sawmills, three tanneries; 511 scholars in schools. Population, 1616.

Newark, city, port of entry, and capital of Essex co., N.J., 9 m. W. New-York, 49 N.E. Trenton, 215 W. It is in 40° 45' N. lat., and 74° 10' W. long., and 2° 49' W. long. from W. Situated on the W. side of Passaic river, between four and five miles, by the course of the river, from its entrance into Newark bay. It stands on a plain of fertile loam, resting on old red sand-stone, with a rising ground on the W. It is the most populous and flourishing place in the state. The population in 1830 was 10,950; in 1840, 17,290. Of these 237 were employed in agriculture, 906 in commerce, 9494 in manufactures and trades, 12 in navigating the ocean, 47 do. rivers and canals, 101 in the learned professions. It had six academies, 319 students; 30 schools, 1955 scholars.

The Passaic river is navigable to this place for vessels of 100 tons burden. There is a communication twice a day, through a large part of the year, by steamboat, to the city of New-York, and several times a day by railroad. The Morris canal passes through it.

The city is regularly laid out with broad and straight streets, with two large and elegant public grounds, bordered with lofty trees, and bounded by the principal avenues. The ground is elevated 30 or 40 feet above the level of the river, and is open and airy. The city is abundantly supplied with pure water from a spring, over a mile distant, in pipes laid by a joint stock company; and seven miles of iron pipes have been laid for the distribution of the water to the inhabitants. The city is generally well built and neat in its appearance. It contains about 1900 dwellings, many of them of wood, but a considerable number of brick. The wooden houses are generally painted white, and many of the houses are large and elegant. The courthouse is a large edifice, in a commanding position in the W. part of the city, built of brown freestone, in the Egyptian style of architecture. The New-Jersey railroad company have

a large and fine building for a depot, which is an ornament to the place.

There are 17 churches: five Presbyterian, one Episcopal, three Methodist, two Baptist, an Associate Reformed, a Dutch Reformed, an African Methodist, a Roman Catholic, a Bethel, and a Universalist. Of these, four are built of stone, one of brick, and the rest of wood. Several of them exhibit much architectural beauty. There are in the city three banks, with an aggregate capital of $1,400,000, not more than two thirds of which has been paid in. There is an apprentices' library, a circulating library, a mechanics' association for scientific and literary improvement, who have a library and philosophical apparatus, and a mercantile literary association, who support public lectures.

The commerce of Newark is considerable and increasing. The coasting trade employs 65 vessels of 100 tons each. A whaling company was incorporated in 1832, who are successfully prosecuting the business. The tonnage of the port in 1840 was 6687.

According to the census of 1840, there were, in the city, two commercial and two commission houses in foreign trade, with a capital of $15,000; 114 retail drygoods and other stores, capital $321,250; six lumber yards, capital $38,000; 55 men employed in internal transportation, with two butchers and packers, used a capital of $68,000, 50 men produced 12,000 gallons of spermaceti oil, and 80,000 gallons of whale and other fish oil, and whalebone to the amount of $4000, with a capital of $60,000; 16 persons produced machinery to the amount of $7000; 36 persons produced hardware and cutlery to the amount of $30,000; 102 persons manufactured the precious metals to the amount of $154,302; 15 persons manufactured various metals to the amount of $33,000; seven persons manufactured tobacco to the amount of $8000, with a capital of $1250; hats and caps were manufactured to the amount of $476,646, and straw bonnets to the amount of $463, the whole employing 383 persons, and a capital of $121,350; one tannery, employing 425 persons, produced 2000 sides of upper leather, with a capital of $4000; 32 other manufactories of leather, as saddlery, harness, boots and shoes, &c., produced to the amount of $706,946, with a capital of $385,951; 16 persons produced $15,000 pounds of soap, and 115,000 pounds of tallow candles, with a capital of $15,000; two breweries employed 78 men, and produced 54,000 gallons of beer, with a capital of $13,000; four persons produced turpentine and varnish to the amount of $39,000, with a capital of 35,000; one glass-house employed three persons, produced to the amount of $5000, with a capital of $6000; two potteries employed eight persons, produced to the amount of $5200; with a capital of $3200; 483 persons produced carriages and wagons to the amount of $904,900, with a capital of $218,700; vessels were built to the amount of $5000; 247 persons manufactured furniture to the amount of $5200, with a capital of $4000; five printing-offices, two binderies, one daily and three weekly newspapers, and three periodicals, employed 42 persons, and a capital of 32,285; five brick and 15 wooden houses built, employed 71 persons, and cost $44,175. The total amount of capital employed in manufactures was $1,511,339. In addition to the above, there has been recently established an extensive manufactory of fine cutlery, said to equal that of the best imported. For its extent, Newark is a great manufacturing place.

This town was settled in 1666, by a company from Guilford, Branford, Milford, and New-Haven, Conn. It is therefore eminently New-England and puritanic in its origin. They purchased the territory, including several neighbouring towns, of the Indians, for £130 New-England currency. In Indian blankets, and 12 guns. They formed a government, and administered it, often disputing the claims of the proprietors, by holding to an original and superior right, derived from purchase from the natives. Robert Treat, Esq., afterwards governor of Connecticut, was chosen the first recorder, or town clerk.

Newark, p. t., Licking co., O. It contains Newark v., the capital of the co., 39 m. E.N.E. Columbus, 386 W. The The village is situated at the confluence of the three principal branches of Licking river, and on the Ohio canal, which passes along one of its principal streets. It is regularly laid out with streets from six to eight rods wide, crossing each other at right angles, with a large public square in the centre, on which stands an elegant brick courthouse. It has a jail, several churches, a market house, two academies, 15 stores, several large warehouses, two printing offices, 10 schools, 670 scholars, 350 dwellings, and 2785 inhabitants. The town, exclusive of the village, has two schools, 296 scholars. Pop. 1433. It contains some interesting ancient mounds.

New-Athens, p. v., Athens t., Harrison co., O., 115 m. E. by N. Columbus, 291 W. It contains four stores and 300 inhabitants. It is the seat of Franklin college, founded in 1825, which has a president and six professors or other instructors, 84 alumni, 51 students, and 1906 volumes in its libraries. The commencement is on the last Wednesday in September.

NEW-BALTIMORE, p. t., Greene co., N.Y., 15 m. S. Albany, 354 W. Bounded E. by Hudson river. Drained by Dieppe and other creeks flowing into Hudson river. It contains five churches: a Dutch Reformed, Episcopal, Methodist, Baptist, and Christian, seven stores, one fulling-mill, one flouring-mill, three grist-mills, seven saw-mills, one paper-mill, three tanneries; 16 schools, 637 scholars. Pop. 2306.

NEW-BARBADOES, t., Bergen co., N.J. Watered by Hackensack r. It contains Hackensack v., the capital of the county, and has seven stores, six grist-mills, four saw-mills; one academy, 25 students; six schools, 218 scholars. Pop. 2104.

NEW-BEDFORD, p. t., port of entry, and semi-capital of Bristol co., Mass., 26 m. S.E. Taunton, 58 S. Boston, 434 W. It is in 41° 38' 7'' N. lat., and 70° 55' 49'' W. long. Pop, in 1820, 3947; 1830, 7592; 1840, 12,087. Situated on the W. side of an arm of the sea which sets up from Buzzard's bay. The ground rises rapidly from the water, and presents a fine appearance when approached by water, or from the E. It contains many splendid dwellings, in commanding situations. A wooden bridge and causeway, three fourths of a mile long, which cost $30,000, connects it with Fairhaven. It contains a courthouse, jail, four banks, with an aggregate capital of $1,300,000, three insurance offices, total capital $356,000; a savings institution, with an invested amount of $290,000; 14 churches: a Baptist, two Congregational, an Episcopal, two Christian, two Methodist, a Friends, a Unitarian, a Universalist, a Bethel, and an African. The harbour is spacious and safe, though not easy of access. It has from three to four fathoms of water. The business of the place is chiefly devoted to the whale fishery. The tonnage of the port, in 1840, was 89,069, being the second district in the state in this respect. It contained, in 1840, 374 stores, capital $482,350; six lumber-yards, capital, $34,800; capital employed in the fisheries, $4,512,000; salt produced, 13,100 bushels; four grist-mills, two saw-mills, one rope-walk, one paper-mill, three tanneries, three printing-offices, one bindery, two daily and two weekly newspapers; two academies, 118 students; 53 schools, 3455 scholars. A railroad, 24 m. long, connects this place with Taunton.

NEW-BERLIN, p. t., Chenango co., N.Y., 8 m. N.E. Norwich, 86 W. Albany, 347 W. Drained by Unadilla r. It contains seven churches: a Presbyterian, two Methodist, two Baptist, an Episcopal, and a Universalist; 13 stores, one fulling-mill, one cotton factory with 4400 spindles, five grist-mills, 11 saw-mills, one paper-mill, four tanneries, one printing-office, one bindery, one weekly newspaper; 23 schools, 765 scholars. Pop. 3086.

NEWBERN, p. v., port of entry, and capital of Craven co., N.C., 120 m. S.S.E. Raleigh, 100 N.N.E. Wilmington, 348 W. Situated on the S.W. bank of Neuse river, at the junction of Trent river. Pop., in 1830, 3776; in 1840, 3690; It was formerly the capital of the state. The Neuse river is here a mile and a half wide, and the Trent three fourths of a mile wide. It is favourably situated, and more healthy than most places in the state so near the seaboard. It contains a courthouse, jail, a theatre, a masonic hall, two banks, three churches, an Episcopal, Methodist, and Baptist. It is pleasantly situated and handsomely built, mostly of brick. It has considerable commerce, and exports grain, pork, lumber, and naval stores. A steamboat plies to Elizabeth city, and forms a part of the line from Norfolk to Charleston. It had, in 1840, 53 stores, capital $379,410; one flouring-mill, three grist-mills, five saw-mills, two tanneries; exported 83,000 barrels of tar, pitch or turpentine; had one printing-office, one weekly newspaper; four academies, 151 students; three schools, 92 scholars.

NEWBERRY, district, S.C., situated a little N.W. of the centre of the state, and contains 540 sq. m. Bounded N.E. by Broad river, S. by Saluda river. Watered by their branches. It contained, in 1840, 19,600 neat cattle, 7163 sheep, 35,866 swine; and produced 57,350 bushels of wheat, 708 of rye, 635,634 of Indian corn, 1129 of barley, 73,185 of oats, 23,469 of potatoes, 3,105,107 pounds of cotton. It had 34 stores, 15 flouring-mills, 18 grist-mills, 18 saw-mills, four tanneries, three distilleries; three academies, 134 students; 29 schools, 737 scholars. Pop. : whites, 8908; slaves, 9904; free coloured, 238: total, 18,350. Capital, Newberry.

NEWBERRY, C.H., p. v., capital of Newberry district, S.C., 49 m. W.N.W. Columbia, 504 W. Situated on elevated ground, 3 m. E. of Bush river, and has a fine view of the surrounding country. It contains a handsome courthouse, a jail, two academies, a social library, and about 300 dwellings, some of them handsome.

NEWBERRY, t., York co., Pa., 12 m. N. of York. Bounded E. by Susquehanna river, S. by Conewango creek. Watered by Fishing creek. It contains seven stores, six grist-mills, five saw-mills, one oil-mill, two tanneries, seven distilleries, three potteries; 13 schools, 400 scholars. Pop. 1850.

NEW-BLOOMFIELD, p. b., Juniata t., capital of Perry co., Pa., 34 m. W. by N. Harrisburg, 195 W. It contains a brick courthouse and county offices, a stone jail, two churches, a Presbyterian and Methodist, five stores, a printing-office issuing a weekly newspaper, about 60 dwellings, and 400 inhabitants. The borough was incorporated in 1831.

NEW-BOSTON, p. t., Hillsborough co., N.H., 22 m. S. by W. Concord, 467 W. Watered by the S. branch of Piscataquog river. Incorporated in 1763. It contains three churches, a Presbyterian, Baptist and Universalist, three stores, one fulling-mill, four grist-mills, 19 saw-mills, three tanneries; 18 schools, 558 scholars. Pop. 1569.

NEW-BRAINTREE, p. t., Worcester co., Mass., 66 m. W. Boston, 399 W. Bounded N.W. by Ware river. It contains one store, one grist-mill, two saw-mills; 7 schools, 233 scholars. Pop. 752.

NEW-BRIGHTON, v., Castleton t., Richmond co., N.Y. Beautifully situated on the N. shore of Staten Island, fronting N. York city, from which it is six miles distant, connected by a steam ferry. It contains a young ladies' seminary, and a number of beautiful dwellings, on ground which rises abruptly from the water; and near the store, two large and elegant hotels, much resorted to in the summer season. The Sailor's Snug-harbour is one mile W.

NEW-BRIGHTON, b., Beaver co., Pa. Situated on the E. bank of Beaver river, 3 m. from its mouth. It contains six churches: a Methodist, Presbyterian, Associate Reformed, Unionist, and two Friends; a female seminary, the New-Brighton Institute, six stores, one furnace, two fulling-mills, two woollen factories, two flouring-mills, two saw-mills, one printing-office, and one weekly newspaper; two academies, 85 students; three schools, 93 scholars. Pop. 981. A bridge crosses Beaver river at each end of the village. It is regularly laid out, with broad streets crossing each other at right angles, and has a number of beautiful residences.

NEW-BRITAIN, p. v., Berlin t., Hartford co., Conn., 10 m. S.W. Hartford, 296 W. It contains three churches, a Congregational, Methodist, and Baptist, 45 factories, chiefly of brass, employing 700 persons, and a capital of $650,000; and about 1500 inhabitants.

NEW-BRITAIN, p. t., Bucks co., Pa., 25 m. N.W. Philadelphia, 96 E. Harrisburg, 165 W. Drained by branches of Neshaming creek, which afford water-power. It contains 30 stores, five grist-mills, five saw-mills, one oil-mill; two schools, 48 scholars. Pop. 1304.

NEW-BRUNSWICK, a tract of country in British N. America, on the W. side of the gulf of St. Lawrence, between lat. 45° 5' and 48° 5' N., and long. 63° 47 and 67° 53' W., bounded S. by Nova Scotia and the bay of Fundy, N. by Lower Canada, and W. by the state of Maine in the U. States. Extreme length, from N. to S., 180 m.: average breadth, 150 m.: estimated area, 25,931 sq. m. Pop., in 1834, 119,557. This colony, which is divided into 11 counties, has a much less indented coast-line than Canada or Nova Scotia: the surface, however, is broken and undulating, though not mountainous, and considerable rivers intersect it in all directions, the largest being St. John's (which see), Miramichi, and Ristigouche. The principal gulfs are the bay of Chaleur and Miramichi, on its E. coast, and that of Passamaquoddy on the S., into which runs the river St. Croix, which divides the province from Maine, in the U. States. Its geology is very little known; but limestone seems to be the prevailing feature, though clay-slate, granwacke, and even the primitive formations occasionally occur. Coal is abundant, and is wrought, near the Grand lake, by a joint-stock company. Iron and gypsum occur also in considerable quantities. Dense forests cover, by far, the greater part of the surface, and though the soil is generally rich and fertile, except in a few swampy tracts, only one-sixteenth of the whole province has been surveyed and laid open for settlers. The cutting down, and exportation of the fine timber, with which these forests abound, has, however, been extensively pursued for some years, and the quantity of cleared land is progressively increasing. The fauna and flora of the colony nearly resemble those of Nova Scotia, to which, indeed, it formerly belonged. The climate is very similar to that of Canada: winter lasts from November to April, when a sudden change takes place, and vegetation becomes extremely rapid. The temperature in the S. part is milder and more equable; but the prevalence of sea-fogs, on the shores of the bay of Fundy, render the cultivation of wheat near the coast very uncertain, though it does not seem to injure the health of the settlers. Indeed, the climate altogether is uncommonly healthy, and will bear to be compared with that of any part of England. Rheumatism, consumption, and low typhus are the prevalent diseases; but they are in a great measure brought on by exposure to the damp, and the sudden changes of temperature. Agriculture, notwithstanding the rich tracts of alluvial soil skirting the rivers, is considerably less advanced than in

NEW-BRUNSWICK.

Nova Scotia and the Canadas, owing, in part, to its later settlement, but principally to the superior importance attached to its timber trade. Within the last few years, however, great improvements have taken place in these respects; agricultural societies have been formed, new settlers have introduced, in many parts, the more approved systems of husbandry, and emulation has been generally excited by ploughing-matches, cattle-shows, and the distribution of premiums. Wheat, Indian corn, barley, and oats are the principal grain crops; but by far the most important article of produce is the potatoe, the crop of which, in 1835, was estimated at 9,160,000 bushels. Red and white clover are the grasses most cultivated, and beans, peas, turnips, mangel-wurzel, and beetroot thrive well, and are raised in pretty considerable quantities. Pasturage is followed to some extent, and it was estimated that the live-stock of the colony, in 1835, comprised 11,000 horses, 91,000 cattle, 143,000 sheep, and 59,000 hogs. The felling and conveyance of timber constitutes, however, as before observed, the great employment of the labouring classes; but most of the *lumberers* are dissolute and depraved, and the occupation prevents them from paying proper attention to agriculture. Many of the trees, especially the yellow pines, attain a great size, and furnish timber of good quality, though inferior to that of Norway and the Baltic. It is principally conveyed to Great Britain in the log, the remainder being manufactured into deals, boards, staves, &c.

The timber exported from New Brunswick to Great Britain in 1837 was valued at £476,670; that of Lower Canada during the same year having been estimated at £651,783. This trade is wholly forced and artificial, being a consequence of the high discriminating duties imposed in England on Baltic timber. It is contended, however, that the equalization of the duties, how advantageous soever to England, would be injurious to Canada and New Brunswick, by diminishing their trade in timber. But the truth is, that this, so far from being advantageous to either, is distinctly and completely the reverse. The habits which it generates are quite subversive of that sober, steady spirit of industry, so essential to a settler in a rude country: to such a degree, indeed, is this the case, that lumberers have been described as the pests of a colony, "made and kept vicious by the very trade by which they live." Ship-building is pretty extensively carried on, chiefly at St. John's, the capital. In 1839, 164 ships were built, of the aggregate burden of 45,864 tons; but they are of the class called "slop built," and do not enjoy a high character for solidity or durability. Though less deeply indented with fishing bays than Nova Scotia, the coast and rivers of New-Brunswick abound with fish, especially cod, herrings, salmon, and mackerel; the entire value of the exports of fish and fish-oil having amounted, in 1837, to £68,000. The whale fishery, introduced only within the last few years, has already attained considerable importance; the exports of oil and whalebone, in 1837, having been estimated at £37,442.

Except timber, and the produce of its fisheries, the exports of New-Brunswick are quite inconsiderable: the imports consist of corn (chiefly from the United States), of the value, in 1837, of £150,928. British and Irish manufactured goods, valued at £317,391, and various minor articles, making a sum total of £730,563. Subjoined is an account of the ships that entered and left the ports of New-Brunswick in 1839:

Countries.	Inwards.		Outwards.	
	Ships.	Tons.	Ships.	Tons.
Great Britain	576	294,712	596	301,925
British Colonies	1,935	115,176	1,929	118,500
United States	944	64,055	798	33,088
Other foreign ports	37	8,181	4	638
Total	3,492	395,123	3,327	444,051

The colony seems, on the whole, to be improving. The Brunswick land company have done much to promote the immigration of industrious British settlers, and several joint stock companies and banks have been recently established. The premium for bills on England varies from 8 per cent. to 11¼ per cent.; and the difference between the currency and sterling price of money in the province amounts to 11½ per cent. The paper currency, which is very general, consists of the notes of the different banks, and of those issued by the corporation of the city of St. John's. The sum in circulation amounted, in 1839, to about £350,000.

The constitution of New-Brunswick is very similar to that of Nova Scotia, which it resembles in several other particulars, and to which the reader is referred for farther information. The representative body, or parliament, comprises 26 members, and sits at Fredericton, about 90 m. above St. John's. The judiciary courts are the court of chancery, in which the governor presides, the supreme court, directed by four justices, circuit courts, a court of common pleas, and numerous courts for the recovery of small debts. The revenue is extremely variable, and has been much increased of

late years by the sale of unoccupied lands; besides which, a few light taxes are levied for poor rates and other local purposes. After the payment of the local magistracy, &c., the surplus is appropriated to the improvement of the colony, and especially to the formation of roads, that have recently been completed to a very considerable extent. The expense of the regular army is defrayed by the British government; but there is likewise a native militia comprising upward of 20,000 men. The religion of New-Brunswick is similar to that of Nova Scotia; and the diocese of the colonial bishop of that peninsula extends over the province. There are, likewise, Roman Catholics, Presbyterians, Methodists, and Baptists; but the religion of the colonists partakes more of fanaticism than sober, rational worship. As respects education, New-Brunswick enjoys more than ordinary advantages. King's college, at Fredericton, owes its origin to the exertions of Sir Howard Douglas, and has been in active operation for some years. Its maintenance is chiefly provided for by an annual grant of £3000 from the local government, and the mode of instruction nearly resembles that pursued in Oxford; subscription to articles is not, however, required, except from students of divinity: a grammar school also is supported out of the college funds. Nine other grammar schools, which, in 1839, had 321 pupils, are either wholly or partially supported by legislative grants. English schools also are established in all the parishes of the provinces; and, in 1839, there were 477 schools, furnishing elementary instruction to 7019 boys and 5619 girls.

The population of New-Brunswick consists of a mixed race of English, Irish, Welsh, and Scotch; but the last are the less numerous than in the neighbouring colonies. The French also have three small settlements on the E. side of the province. The Indian aborigines have been for many years fast declining in numbers, and all attempts to civilize them or improve their condition have failed: they have a few small villages scattered in different parts, and are all Roman Catholics. In manners and customs the British settlers nearly resemble those of Canada, Nova Scotia, and Cape Breton. The women are handsome; the men generally tall, well made, muscular, and austerely over corpulent. They are remarkably spirited, adventurous, and attached to their country; nor can there be any doubt that they would, if well disciplined, make excellent soldiers.

The country now called New-Brunswick was, in the early part of last century, comprised by the French under the appellation of New France, and viewed as an appendage to Acadia. At the peace of 1763 it was ceded, with the rest of Canada, to the English, and, from that time to 1785, the country, being little more than a mere wilderness, till General Sir Guy Carleton procured for it a royal charter, constituting New-Brunswick a distinct province, with himself as governor. To his exertions it chiefly owes the rapid rise of its prosperity; but it also owes many material improvements in its roads, schools, agriculture, judicial arrangements, &c., to Sir H. Douglas, governor from 1824 to 1831. (*M'Gregor's America*, ii., 5–102; *Murray's B. America*, ii., 234–256, &c.)

NEW-BRUNSWICK, city, capital of Middlesex co., N. J., 29 m. S.W. New-York, 37 m. N.E. Trenton, 102 W. Situated on the S.W. bank of Raritan river, 15 m. from its entrance into Raritan bay at Amboy, by the course of the river. Although it is the capital of Middlesex co., it lies partly in Somerset co., Albany-street being the dividing line between the two counties. The streets on the river are narrow and crooked, and the ground low; those on the upper bank are wide, and many of the dwellings are very neat and elegant, surrounded by fine gardens. The streets are generally paved: this is rendered necessary by the soil which is a red sandy loam, that causes the unpaved streets to become deep with mud in wet weather. From the site of Rutger's college, on the hill, there is a wide prospect, terminated by mountains on the N. and by Raritan bay on the E. A toll-bridge crosses Raritan river, originally built in 1795, and rebuilt in 1811, at a cost of $68,667, which is 1000 feet long, divided into two carriage-ways by a wood partition. It is constructed of wood, and rests on 11 stone piers and abutments. A railroad bridge crosses the river a little above. It was incorporated as a city in 1784, and contains eight churches, a Dutch Reformed, two Presbyterian, an Episcopal, Methodist, Baptist, a Roman Catholic, and a coloured Methodist, two banks, 190 stores, 800 dwellings, and 8693 inhabitants. The first Dutch church was built in 1717; the present was completed in 1836. The first Presbyterian church was erected before 1795, burned by the British during the revolutionary war, and the present one erected in 1784. The Episcopal church was erected in 1742, burned in 1802, and immediately rebuilt. It is the seat of Rutger's college, founded in 1770, which has a president and 10 professors or other instructors, 291 alumni, of whom 77 have been ministers of the gospel, 81 students, and 12,000 volumes in its libraries. The commencement is on the fourth Wednesday in July. Connected with it is the theological department, under the Dutch Reformed church, of 38 students, besides an academic department. The Delaware and Raritan canal commences here, and extends 43 m. to Trenton. It is 75 feet wide and 7 feet deep, admitting the passage of sloops of 75 or 100 tons burden; and, in connexion with Delaware and Raritan rivers, and Staten Island sound, forms a complete inland water communication between the cities of New-York and Philadelphia. The New-Jersey railroad also passes through the city, forming a part of the great railroad communication from New-York to Philadelphia, Baltimore, and Washington. Having these and other facilities of communication, its trade is extensive. The tide sets up in the Raritan, 2 m. above the city, and the river admits vessels of 200 tons burden to its wharves. It has a communication by steamboat with New-York, twice daily.

NEWBURG, p. v., semi-capital of Orange co., N. Y., 60 m. New-York, 84 m. S. by W. Albany, 268 W. It is pleasantly situated on the W. side of Hudson river. The ground rises rapidly from the river, exhibiting the place to great advantage from the water, and in its back parts, 300 feet high, commands a fine view of the river, the highlands, the village of Fishkill opposite, and a beautiful surrounding country. It contains a courthouse, jail, 10 churches, two Presbyterian, two Associate Reformed, a Methodist, Baptist, Episcopal, Dutch Reformed, Roman Catholic, and an African, an incorporated academy, a high school, a theological seminary of the Associate Reformed church, three banks, 100 stores, a large wharf and storehouses, a large iron foundry and machine-shop, one extensive hat factory, a large brewery, two morocco factories, two paper-mills, four plaster-mills, a powder-mill, extensive brick-yards, 350 dwellings, and about 6000 inhabitants. Several sloops and four steamboats are owned here, which trade with New-York city; and a steam ferry-boat crosses the river, here a mile wide, to Fishkill. Some of the shore mills are on Chambers' creek, a little S. of the village. The steamboats from New-York to Albany stop at the place. S. of the village stands an old house, in which Washington had his head quarters in March, 1783, when the famous "Newburg letters," designed to excite the army to mutiny, were anonymously addressed to them by some of the officers, and which Washington, by his great influence, nobly quelled and defeated. Here the American army was disbanded in June 23d, 1783. The town has two academies, 142 students; 23 schools, 1983 scholars. Pop. 8933.

NEWBURG, p. t., Cuyahoga co., O., 150 m. N.E. by N. Columbus, 333 W. Mill creek, on which the village stands, has a great fall, affording extensive water-power. It has one store, one flouring-mill, one saw-mill; eight schools, 457 scholars. Pop. 1342.

NEWBURGH, a seaport and market town of Scotland, co. Fife, on the S. bank of the Tay, 12½ m. S.W. Dundee, and 9 m. S.E. Perth. Pop., in 1831, 2642. It consists chiefly of one street, running E. and W. along the line of the shore, with another at right angles leading down to the harbour. The town is mostly of modern date, particularly toward its outskirts, though many old buildings remain to mark its ancient state. The public buildings are the town-house, with a spire, the parish church, and a dissenting chapel. The splendid mansion-house of Mugdrum is close to the borough on the N.W. The means of education are ample and efficient; from one tenth to one seventh part of the population are at school. The town is lighted with gas.

The harbour is pretty good; but only 10 vessels from 60 to 150 tons, exclusive of fishing-boats, belong to the borough. It has, notwithstanding, a considerable trade, being the port for the greater part of Kinross-shire, Strathearn, and other contiguous districts, both for the export of their agricultural produce, and for importing coals, lime, &c. Most vessels bound for Perth wait here for the flow of the tide; and some of them unload part of their cargo before they can, even at high water, proceed up the river. Newburgh, indeed, is, next to Kirkcaldy, the most important seaport of Fifeshire. The weaving of coarse linens is largely carried on, employing from 550 to 600 looms.

Newburgh existed in the 19th century, and was subject to the neighbouring monastery of Lindores, whose remains are yet pretty entire. In 1631 it was created a royal borough by Charles I.; but, like Falkland, being unable to defray the expenses of its parliamentary representative, it petitioned to be relieved from the burden, which was granted. It has otherwise, however, all the marks of a royal borough, and is governed by two bailies and 15 councillors. Municipal revenue, derived from land, about £170 per annum. There are two curious crosses of remote antiquity in the neighbourhood; one called the Mugdrum cross, the other Macduff's cross. The former is supposed to commemorate a victory over the Danes in the 10th century; the latter was erected as a sanctuary to any of the kindred of Macduff, thane of Fife, who might commit murder. If they fled thither, and paid a certain fixed solatium to their chief, they obtained protection. (*New Stat. Account of Scotland*, § Fifeshire, p. 56-61.)

NEWBURY, a mun. bor., market town, and par. of England, co. Berks, on the Kennett, crossed here by a stone bridge of three arches, 24½ m. S. Oxford, and 53 m. W. by S. London. Population of the municipal borough, which includes, with the parish of Newbury, portions of the parishes of Speen and Greenham, in 1831, 8469. The town, a considerable portion of which on the N. bank of the Kennett is in the hamlet of Speenhamland, consists of two principal, wide, and well-built streets, arranged in the form of the letter T, with smaller and very irregular streets at its S. extremity. The market-place, opposite the church, is a large open square, in which is the guildhall. The church, erected in the reign of Henry VIII. is a large but plain building, with a square tower: the living is a rectory in the patronage of the crown. A district church, in the Gothic style, has recently been erected on the London road, and is remarkable for its extensive catacombs. There are, likewise, five places of worship for dissenters; and the town comprises several Sunday schools, and an endowed free-school, besides numerous and wealthy corporation charities, which, however, had been greatly neglected previously to the passing of the Municipal Reform Act. (*Mun. Corp. Rep.*) The almshouses have accommodation for 98 aged people: there is a small borough jail, and about half a mile S. of the town is a large union workhouse.

"Newbury, which is situated on the main road between London and Bath, has a large posting business, which employs a considerable number of hands, besides occasioning a large importation of horse provender. The Kennett and Avon canal, connecting the former with the Thames, passes through the town, and affords the advantage of water-carriage from London, Bristol, and S. Wales. In the immediate neighbourhood are two silk manufactories and a paper-mill; but they are not considerable. The rapid declivity and copious supply of water in the Kennett, have occasioned the erection of numerous large corn-mills, two of which are within the town. There are also considerable malthouses, and some extensive breweries. The quantity of grain annually exported from Newbury, either as flour, malt, or in its natural state, amounts to upward of 7800 tons, in return for which it imports large quantities of building materials, and various articles of general consumption from the ports of London and Bristol. The town has the appearance of being in a prosperous and improving condition; and there has been a great increase of buildings and population, especially in Speenhamland. High rents are readily obtained in situations favourable to trade. The poor rates are much higher than in the adjoining parishes, and higher also than they should be with reference to the description of the population, a circumstance chiefly attributable to the attractions offered to the lower classes by the numerous charitable foundations attached to the corporation." (*Mun. Corp. Rep.*)

Newbury, which is a borough by prescription, and was afterward chartered in 26 Elizabeth, has been governed, since the Reform Act, by a mayor and three aldermen, with 12 councillors; it has a commission of the peace under a recorder. Corporation revenue in 1839, £1027. The spring quarter-sessions for the county, and petty sessions for the hundred, are held here, and is one of the polling places for the county elections. Large corn markets on Thursday: horse and cattle fairs, Holy Thursday, July 5, September 4, and November 8.

Newbury returned two members to parliament in the reign of Edward I.; and it is not known at what period, or for what cause, it lost the franchise. It was formerly also celebrated for its manufacture of serges, shalloons, &c.; and in the reign of Henry VIII., John Winchcomb, known as Jack of Newbury, kept 100 looms, from the produce of which he became so wealthy as to be able to entertain the king and his retinue during their passage through the town. He was a great benefactor to Newbury; and his house, a large brick structure, is still shown in the High-street, his manufactory being now occupied by a large inn, "the Jack of Newbury." The vicinity is remarkable for two battles fought during the civil wars between the royalist and parliamentary forces, Charles I. commanding his army in person on both occasions. The first was fought on a common called the Wash, on 20th September, 1643; the second on 27th October in the following year; but neither had any decided result. Donnington Castle, a short distance N.W. of Newbury, was the property of Chaucer, the earliest English classic poet, and in it he spent the last two years of his life, which terminated in 1400. (*Mun. Reports, &c.*)

NEWBURY, p. t., Orange co., Vt., 36 m. S.E. Montpelier, 518 W. Situated on the W. side of Connecticut river, opposite to Haverhill, N. H. It has a fine alluvial tract of 450 acres on the river, called the Great Oxbow. A bridge crosses Connecticut river in its N. part, and another in its

S. part, the latter leading to Haverhill Corners. It contains four churches, a Congregational, Methodist, and two others, the Newbury seminary, under the patronage of the Methodists, with a principal and five assistants, seven stores, three fulling-mills, three grist-mills, six saw-mills, one paper-mill, one printing-office, one bindery; one academy, 389 students; 21 schools, 865 scholars. Pop. 2578.

Newbury, t., Essex co., Mass., 31 m. N. by E. Boston, 17 m. N. Salem. Bounded N. by Merrimac river, E. by the Atlantic. Plumb island, 9 m. long, and 1 broad, is partly in this township. Parker river has a fall of 50 feet in 1½ m., affording good water-power. It contains four churches, three Congregational and a Methodist, 15 stores, one fulling-mill, five grist-mills, one saw-mill, three tanneries; four academies, 115 students; 14 schools, 757 scholars. Pop. 3789. Dummer academy, in Byfield parish, was founded in 1756, and is richly endowed.

Newbury, t., York co., Pa., 12 m. N. York. Bounded S. by Conewago creek. Watered by Beaver creek. It contains seven stores, two flouring-mills, one grist-mill, three saw-mills, two tanneries, seven distilleries, three potteries; 13 schools, 400 scholars. Pop. 1815.

NEWBURYPORT, p. t., port of entry, and one of the three capitals of Essex co., Mass., 38 m. N. by E. Boston, 478 W. It is the smallest town in its territorial limits in the state, containing about 647 acres. It is beautifully situated on the S. bank of Merrimac river, 3 m. from its mouth. It is laid out with great regularity, in the form of a parallelogram. The lower street, called Water-street, on which the wharves and docks open, follows the course of the river. Parallel to this, on more elevated ground, High-street extends the whole length of the town, about 3 m., and overlooks a delightful prospect on the opposite bank of the river, the harbour, Plumb island, and the ocean. These streets are crossed nearly at right angles by various other streets. In the centre is a large and convenient market-place, surrounded by brick stores, which is in the vicinity of the principal wharves and docks. The houses are generally neat, and many of them elegant, often surrounded by fine gardens, which give a rural aspect to the place. A bridge crosses Merrimac river, in the N. part of the town, built in 1827, connecting it with Salisbury. The abutment, faced with stone, and filled with sods and gravel, on the Newburyport side is 240 yards, that on the Salisbury side 187 yards. The bridge rests on three abutments and four stone piers, and is supported by chains passing over the tops of pyramids erected on the piers, and under the centre of the arches. It is three sevenths of a mile long, and cost $70,000. A turnpike and bridge connects this town with Plumb island. The road is about 3 m. long, and the bridge about 500 feet in length. This road accommodates parties of pleasure visiting the island, and enables the inhabitants to afford assistance to ship-wrecked mariners. The public buildings and institutions are a brick courthouse, a stone jail, a custom-house of rough granite, with a fine wrought Grecian-Doric portico, with pilasters on the sides, which cost $25,000; eight churches, two Presbyterian, a Congregational, Independent, Episcopal, Methodist, and Baptist; an academy with a fund of $50,000, a brick market-house, an almshouse, a lyceum, three banks with an aggregate capital of $700,000, a savings' bank, and three insurance companies. The harbour is spacious and safe, but difficult of entrance, in consequence of a shifting sand bar at the mouth of the river. The United States erected a breakwater at the mouth of the river in 1830, which cost $30,000. It has numerous vessels, employed chiefly in the coasting trade, and in the cod and mackerel fisheries, in which latter 1000 persons are employed. Its commerce was formerly more extensive than at present. The Middlesex canal has diverted much of the trade from the valley of the Merrimac to Boston. It has 12 commercial and three commission houses in foreign trade, capital $781,000; 116 retail stores, capital $225,200; four lumber-yards, capital $35,000; four cotton factories with 17,736 spindles, two distilleries, three printing-offices, one daily, one semi-weekly, and one weekly newspapers; total capital in manufactures, $647,800; 15 academies, 699 students; 39 schools, 1332 scholars. Population, in 1830, 6388; in 1840, 7161. Tonnage of the port in 1840, 23,965.

A fire on May 31, 1811, destroyed 250 buildings in the fairest part of the town, covering over 16 acres with ruins, including nearly all the dry goods stores, four printing-offices, the postoffice, two insurance offices, four book stores, one church, and the dwellings of more than 90 of the inhabitants. Contributions were made in various places for the relief of the inhabitants, and Boston contributed over $24,000. Mr. Whitefield, the celebrated preacher, died here, September 30, 1770, and his remains repose under the pulpit of the first Presbyterian church, according to his request. An elegant monument of Egyptian and Italian marble, erected to his memory in one corner of the church by an eminent merchant, records, among other things, that "in a ministry of 34 years, he crossed the Atlantic 13 times, and preached more than 18,000 sermons."

NEW-CANAAN, p. t., Fairfield co., Ct., 74 m. S.W. Hartford, 273 W. Incorporated in 1801. It contains three churches, a Congregational, Episcopal, and Methodist, eight stores, two fulling-mills, two grist-mills, six saw-mills, three tanneries; two academies, 63 students; nine schools, 381 scholars. Pop. 2217.

NEW-CASTLE, county, Del. Situated in the N. part of the state, and contains 456 sq. m. Bounded E. by Delaware river, S. by Duck creek. Drained by Brandywine, Christana and other creeks, flowing into Delaware river. The railroad from Philadelphia to Baltimore passes through it. Another railroad extends from New-Castle, Delaware, to Frenchtown, Md. A ship canal connects the Delaware with the Chesapeake bay 13½ m. long, 66 ft. wide at top, and 10 ft. deep, which cost $2,750,000. It contained in 1840, 17,450 neat cattle, 7471 sheep, 14,094 swine; and produced 191,274 bushels of wheat, 3157 of rye, 599,787 of Indian corn, 7528 of buckwheat, 564,015 of oats, 5269 of barley, 84,166 of potatoes. It had 150 stores, 13 lumber yards, two fulling-mills, one woollen factory, 11 cotton factories with 24,492 spindles, 27 flouring-mills, 30 grist-mills, 29 saw-mills, 27 powder-mills, one paper-mill, four tanneries, three breweries, two potteries, three printing-offices, three weekly, and three semi-weekly newspapers; one college, 23 students; 12 academies, 479 students; 60 schools, 2818 scholars. Pop. 33,120. Capital New Castle.

New-Castle, p. t., Lincoln co., Me., 35 m. S.E. Augusta, 603 W. Incorporated in 1735. Bounded E. by Damariscotta river, W. by Sheepscot river. It contains a Congregational and a Roman Catholic church, and the Lincoln academy. The Damariscotta river, which may be considered an arm of the sea, is navigable for large vessels, 15 m. from the sea to Damariscotta bridge, where is a village. It contains two commercial houses in foreign trade, five stores, one fulling-mill, one woollen factory, two flouring-mills, two grist-mills, six saw-mills, two tanneries; one academy, 70 students; 10 schools, 662 scholars. Pop. 1712.

New-Castle, p. v., capital of New-Castle co., Del., 5 m. S.S.W. Wilmington, 42 m. N. Dover, 115 W. Situated on the W. bank of Delaware river, 32 m. S.W. of Philadelphia. It contains a courthouse, jail, a townhouse, market-house, an arsenal, five churches, a Presbyterian, Episcopal, a Methodist, coloured Methodist, and Roman Catholic, a public library of 4000 volumes, 10 stores, 195 dwellings, and 1200 inhabitants. The New-Castle and Frenchtown railroad have here a large establishment for the manufacture of steam engines, locomotives, and other things connected with railroads, including an iron foundry, brass foundry, &c., with a capital of $110,000. There are in the hundred, one academy, 13 students; eight schools, 254 scholars. Pop. 2737. It is surrounded by a beautiful and fertile country. Tonnage of the port in 1840, 3061.

New-Castle, p. v., capital of Henry co., Ia., 47 m. E. by N. Indianapolis, 534 W. Situated on Blue river, and contains a courthouse, jail, an academy, a public library, four stores, 50 dwellings; 10 schools, 359 scholars. Population 528.

NEWCASTLE-UPON-TYNE, a parl. and mun. bor. and river-port of England, locally situated in Castle-ward, co. Northumberland, of which it is the cap., but it is also a county by itself, and is celebrated as the principal British port for the shipment of coal, on the N. bank of the Tyne, about 9½ m. from its mouth, 54 m. E. Carlisle, and 244 m. N. by W. London: lat. 54° 58′ 30″ N., long. 1° 37′ 30″ W. Area of parliamentary borough, which includes, with the town and county, the five townships of Byker, Heaton, Jesmond, Westgate, and Elswick, 9130 acres. Pop. in 1831, 53,613, but at present (1841) probably 65,000. The town (which was formerly fortified, and still has a square Norman castle, with two ancient gates,) and a few remains of the old town wall, occupies the bottom and sides of an acclivity rising somewhat abruptly from the river; and though a few years back it was very irregularly laid out, and consisted, with but few exceptions, of narrow, circuitous, and ill-built lanes, it has been so improved within the last 20 years, (principally through the exertions of Mr. Grainger, a native of the town,) that it is now one of the handsomest towns of England. These improvements have cost nearly a million sterling; and include a great number of new streets and terraces, a handsome square, a market (the largest in England), a central exchange, theatre, dispensary, music-hall, lecture-room, two chapels, two auction marts, &c.

General description.—Grey-street, so called in honour of Earl Grey, the largest of Mr. Grainger's new streets, has now become the principal thoroughfare of the town: it ascends a gentle acclivity, and forms a continuation of Dean-street and the Side, which last reaches nearly to the river. Grey-street is nearly ¼ m. in length, by 80 ft. in breadth, and is lined with substantial stone houses, which, in point of architectural beauty, may vie with Regent-street, in London: the effect is heightened, also, by the curvilinear direction of the street. At the top of Grey-street, where it joins Blackett-street, one of Grainger's earlier and less ambitious works, stands the column dedicated to Earl Grey, 135 ft. high, and surmounted by a colossal statue of that nobleman, by Bailey. Grainger-street, another fine avenue, 380 yards in length, and 66 ft. wide, is on a similar design with Grey-street; and at their junction with Market-street is a large triangular space, on which has been erected the central exchange, a building having three uniform fronts, in the Corinthian style, with circular corners, faced with columns of the same order, and supporting light domes, after those of the temple of Vesta at Tivoli. The outside is formed into handsome shops and ware rooms, enclosing the exchange, which has four large entrances, and is altogether lighted from above: the roof is ingeniously constructed, resting on the external walls, and inwardly on a circular entablature, supported by 14 Ionic columns, enclosing a platform, within which is the news-room, the outer space being open, and used for the promenades and rendezvous of the merchants. Adjoining the exchange is a handsome coffee-room. The establishment is supported by 1600 subscribers, and is managed by a committee; but the property belongs to Mr. Grainger. The proprietor offered it as a free gift to the corporation for a corn-market, but that body thought fit to decline it: the principal commercial business is transacted at the old exchange on the Sandhill. Clayton, Nelson, Nun, and Shakespeare-streets are the other principal thoroughfares in Mr. Grainger's splendid improvements; and besides these, Eldon Square, N. of Blackett-street, Westgate, Percy, and Northumberland-streets, deserve notice. The N. suburbs are open; and being removed from the bustle of town, are occupied by houses suited to the wealthier inhabitants, who have extended their residences into Jesmond township, forming a succession of terraces; and further north is a group of new buildings, called Brandling Place. On Rye-hill, also, W. of Newcastle, terraces and villas are in course of formation; and the same is the case on the road to N. Shields: indeed, it may be said that the town is extending itself in every direction, with marked improvements in architectural taste. In Sandgate, however, and the lower parts of Newcastle, which extend along the banks of the Tyne for nearly 2 m., there are many narrow, inconvenient, and dirty streets, lined with manufactories, ware-houses, &c.; and comprising, also, many lanes and alleys, as filthy, close, and unwholesome, as the very worst of those of Liverpool and Manchester: indeed, "the mind cannot picture a state of greater destitution and misery than what appear in many of these houses; and in Sandgate, E. of the town, the condition of the people seems not much better." (*Rep. to British Association*, 1838.) The communication with the borough of Gateshead (which see), on the S. side of the river, is maintained by means of a handsome stone bridge, of nine elliptical arches; but it has been proposed to supersede this by another so lofty as to admit underneath the free passage of colliers and other masted vessels.

Corporation and Commercial Buildings, &c.—The guildhall, which comprises also the exchange and the court belonging to the incorporated society of boastmen or coalficers (chartered in 1600), is a large building on Sandhill, much enlarged and altered at different periods, but of the most heterogeneous architecture, though at the same time pretty well adapted for business: the rooms contain some valuable portraits of public characters. On the quay, a fine open space faced with stone, and one of the largest in the kingdom, is the custom-house lately faced with stone from a design by Smirke. The Moot-hall, or assize courthouse for the county of Northumberland, is within the precincts of the old Norman castle, and consists of a Grecian building, designed from the Temple of Theseus at Athens, and faced on two sides by Doric porticoes: the interior is well arranged for the business both of the civil and criminal courts. The town-jail, in Carliol-square, is a strong and rather unattractive building, erected in 1827, on the panopticon principle, at a cost of £47,000: but it is reported, that both the site and construction are bad; solitary confinement is impossible, and a better prison might have been built at little more than half the expense. It comprises about 50 cells, and the same number of rooms: the average number of prisoners, not including debtors, amounts to about 80, and the daily cost of each prisoner may average about 1s. 2d. The Trinity-house, in Trinity-chare, Quay-side, is an incorporated institution of great antiquity, intended not only to improve the navigation of the river by the appointment of licensed pilots, but to provide subsistence for poor and decayed brethren, their widows and children: it supports at present about 27 in-pensioners, at 28s. per month; 81 masters or their widows, out-pensioners, at £8 a year, and 37 seamen or their widows, at £6 a year. The Arcade in Pilgrim-street, one of Grainger's

erections, though perhaps the least distinguished in point of taste, consists of an oblong pile of building, with a Corinthian frontage 94 ft. long and 75 ft. high. The N. of England joint-stock bank, and the savings' bank, occupy the front-rooms; and in other parts are the post, stamp, and excise offices, with auction-rooms, shops, and chambers for lawyers, engineers, &c. It was opened in 1832, and cost £40,000. The barracks on the N.W. side of the town, comprise an extensive range of building inclosed within a stone-wall, and accommodate nearly 1600 troops.

Markets.—The Corn-exchange, recently built, seems to be sufficiently capacious and convenient, as is the Fish market. A large butcher-market was built in 1808; but it has been removed; and the only great market now exist ing in Newcastle is that constructed by Grainger, the largest in England, 318 ft. in length, and comprising an area of 9050 sq. yds., exceeding that of St. John's in Liverpool by 9650 sq. yds.: it has 14 entrances, and is lined with 943 shops, besides stalls.

Literary, Scientific Institutions, &c.—The Literary and Philosophical Institution (founded in 1793, chiefly through the exertions of the Rev. W. Turner, a celebrated dissenting minister of Newcastle,) occupies a building of Doric architecture in Westgate-street; and adjoining the library are the meeting-rooms and museums of the Natural History and Antiquarian Societies of Newcastle. A literary, scientific, and mechanical institution has an establishment in Blackett-street. A new Music-hall, beneath which is a large public lecture-room, has been built by Mr. Grainger, in Nelson-street. The old music-hall is now entered from Grey-street, and is occupied as an auction mart. The Newcastle Institution, for the promotion of the fine arts, in Black-street, a handsome building with a Corinthian front, comprises a saloon and octagon gallery well lighted from the top; the establishment is supported by a joint-stock company, and the annual exhibition is in June. Newcastle has six public libraries, and five weekly newspapers. A club, conducted in the modern style, called the Northern Counties' Club, has among its members the most distinguished nobility and gentry of that part of England. The principal places of amusement are the theatre, in Grey-street; the assembly-rooms, in Westgate-street; behind which is the racket-court, the riding-school in the public walk called the Forth, and the baths at the N. end of Northumberland-street. The theatre, built by Grainger in 1835, from designs by Mr. Green, an architect of Newcastle, instead of an older one that he had purchased and pulled down, has a front in Grey-street 190 feet in length, with a portico of six Corinthian columns, supporting a rich pediment; its interior shape is that of a flattened horse-shoe; and in point of size it is surpassed by few or no English theatres, except the Opera-house, and the patent theatres of the metropolis. Races are held in June on the Moor, about 4 m. N. the town; and a good stand has been erected for the accommodation of visitors.

Churches, &c.—Newcastle has four parish churches, the oldest of which is St. Andrew's, a Norman building, at the top of Newgate-street. By far the finest, however, is St. Nicholas (now the parent church) a cruciform structure in the decorated English style, with a choir and nave 240 feet in length, and 74 feet in width, the choir only being enclosed for service; a painted E. window, and a magnificent altar-piece of the Last Supper, decorate the interior; and at the W. end is a tower in the early perpendicular style, surmounted by a crocketted steeple resting on four flying buttresses, the whole being 201 ft. in height. This steeple is said, by Mr. Rickman, "to be as fine a composition as any of its date, and the lightness and boldness of the upper part can hardly be exceeded." St. Giles's in Edinburgh, the college tower in Aberdeen, and St. Dunstan's in the E. of London, are imitations of this steeple, but they all fall far short of the original. A very good library, though chiefly of old or theological books, is attached to this church; the rules of admission are very liberal; and it is, in fact, open to the public free of charge. All Saints' is a modern Grecian building, with an elegant spire 209 feet in height; the interior is of an elliptical shape, and richly fitted up with solid mahogany. St. John's, in Westgate, is a cruciform church, built in the 13th century, having a square embattled tower at its W. end. The chapels of ease are, St. Anne's, on the New road; and another at Barras-bridge, called St. Thomas', in the early English style, and surmounted by a light tower 140 ft. high. An endowed charity school is attached to each of the churches, and a fifth to the chapel of St. Anne. The dissenters here are numerous, and most respectable; out of 30 places of worship, 24 belong to different classes of Methodists, and other dissenters, including Roman Catholics, members of the Church of Scotland, and the Society of Friends. Few of these, however, can have any claim to notice from their architectural beauty. Within the borough are numerous Sunday-schools, furnishing religious instruction to upwards of 5000 children of both sexes. Two public cemeteries have been formed of late years in the suburbs.

Schools and Benevolent Establishments. — The Royal grammar-school of Newcastle was founded by Thomas Horsey, in 1525. Among its pupils have been the late lords Eldon, Stowell, and Collingwood, the poet Akenside, and several other distinguished characters: Dawes, author of the *Miscellanea Critica*, was one of its masters. A Lancastrian school, known as the "Jubilee-school," founded in 1809, has a handsome school-house, with a large library, and is liberally supported by subscription. A second Jubilee-school was founded, to commemorate the 50th year of the preincy of the late Dr. Shute Barrington: it is on the national plan, and supported chiefly by the clergy and lay members of the established church. There are several other endowed and subscription schools, including two infant-schools. According to a report made to the British Association, the number of children receiving instruction of some kind or other, in 1838, amounted to 8239, or to about 51½ per cent. of the population between the ages of 5 and 15. The principal benevolent institutions are, the infirmary, which has accommodation for 800 in-patients; a dispensary; two blind asylums: a small lying-in-hospital; asylum for poor keelmen; Jesus' hospital, for decayed freemen; mendicity society; domestic-guardian institution; and several ranges of almshouses; besides which, there are several minor charities and religious associations. Newcastle has also a large union-workhouse; the expense of maintaining the poor of the borough in 1839 amounted to £13,285.

Coal Trade. — The importance, if not existence, of Newcastle is owing to its convenient situation as a place of shipment for the coal wrought in its neighbourhood. The pits lie on each side the Tyne, from within 2 m. of its mouth to 16 or 18 m. up the river; and it appears from the evidence before the committee of the house of lords on the coal-trade, that, in 1829, there were 93 working collieries on the N. side, and 18 on the S. side of the Tyne. Several more have since been opened, and at present upwards of 50 pits are at work in the neighbourhood of Newcastle, some within ½ m. of the river, but others more than 8m. distant. The coals are conveyed from the pits to the staiths in wooden or cast-iron wagons, brought along railways generally by means of successive inclined planes or locomotive engines. From such staiths (or coal-shipping wharves) as are above Newcastle-bridge, the coal is conveyed in keels (each capable of holding eight chalders or 22 tons) to Wallsend, Jarrow, or Shields, where it is delivered on board the ships; and the strength as well as activity of the Tyne keelmen, is proverbial in the N. of England. Within the last few years, however, the formation of the Brandling Junction, and other railways, to S. Shields, has caused a diminution of the keel navigation. The coal shipped at Newcastle is sent not only to the port of London, but furnishes a large portion of the supply for the E. and S. counties of England, and a considerable quantity is exported, chiefly to France, Holland, and Denmark, as will be seen from the following returns of coal, cinders, and cinders, shipped at the port of Newcastle in 1838 and 1839:—

	London.	Other Ports of the United Kingdom.	Colonies, or Foreign Ports.	Total.
	Tons.	*Tons.*	*Tons.*	*Tons.*
1838	1,167,552	1,378,196	554,175	3,019,903
1839	1,158,356	1,005,465	554,052	2,717,873

Owing to the circumstance of most of the vessels engaged in the coal trade with London and other British ports belonging to Newcastle, her registered tonnage is very nearly equal to that of Liverpool, having amounted, on the fifth of January, 1841, to 1230 ships, of the aggregate burden of 260,000 tons, manned by 13,500 seamen.

The principal exports of Newcastle, besides coal, comprise pig and sheet lead (from 6000 to 7000 tons of the former, and above 2000 tons of the latter) from the mines of Stanhope, glass, and other goods manufactured in the town, hams, grindstones from Gateshead-fell, &c. The gross customs' revenue of the port amounted, in 1840, to £448,951.

The salmon-fishery of the Tyne (once much celebrated) has greatly declined. About 90 steamboats belong to Newcastle, chiefly employed in towing ships up and down the river, or plying for passengers to and from Shields. Other and larger steam-packets ply between Newcastle and Stockton, Hull and Leith; and steamers of still greater size ply to and from London weekly, or even more frequently, during the summer months. The Tyne is navigable, from its mouth up to Newcastle bridge, for vessels of 250 tons, though in some intermediate places the depth, even in the middle of the stream, does not exceed 4 feet at ebb tide. Dredging machines, however, have been in use within the

last few years, and the navigation is said to have been much improved, though the bar at the mouth must always prove a great impediment to the entrance of large ships. It is high water at Newcastle about an hour later than at Tynemouth-bar, the average rise of spring-tides being 11 feet 7 inches, and that of neaps 7 feet 2 inches.

The principal manufactures are those of bottles and window glass, mostly carried on in the township of Byker, of mill-work, steam-engines, &c., and of leather and soap, of which last article 2,532,257 lbs. were made in 1839. Ship and boat building, rope and sail-making employ a considerable number of hands; besides which there are several maulthouses, breweries, iron foundries, lead mills, and chemical works. The gross excise duty collected at Newcastle amounted, in 1839, to £318,331. It has a branch of the bank of England, with the north of England Joint-stock Banking Company, the Newcastle-upon-Tyne Joint-stock Banking Company, Northumberland and the Durham District Banking Company, Newcastle Commercial Banking Company, a private bank, and a saving's bank. The internal communication is maintained not only by the coalpit railways already mentioned, but by the Carlisle and Newcastle railroad (61 m. in length, opened in 1839). The Brandling Junction railway connects the last-mentioned railway with the towns of S. Shields and Sunderland on the S. side of the Tyne; on the N. side of which is the Newcastle and North Shields railway. It is probable, also, that Newcastle will soon be connected with the metropolis by the Great N. of England railway, which is already (1841) open as far as Darlington.

Newcastle was constituted a borough by William the Conqueror, and has received 36 charters from subsequent monarchs. It is divided by the Municipal Reform Act into seven wards, and is governed by a mayor and 13 other aldermen and 48 councillors, and has a commission of the peace under a recorder. Corporation revenue in 1838, £68,475, chiefly derived from ballast-dues, tolls, and rents. The assizes and Epiphany quarter-sessions for the county of Northumberland are held in the Moot-hall, besides which there is a mayor's court and sheriff's court for the recovery of debts to an unlimited amount, a court of conscience for debts under 40s., and a court of conservancy for the river. The town is well-paved, lighted, and cleaned by the corporation, and there is an efficient police. Newcastle has sent two members to the House of Commons since 27 Edward I., the election being vested, down to the Reform Act, in the free burgesses, both resident and non-resident. The electoral limits were enlarged by the Boundary Act, so as to include, with the old borough, the townships of Jesmond, Heaton, Byker, Elswick and Westgate; and in 1839-40 it had 4520 reg. electors. It is one of the polling-places at elections for the county. Markets extremely well supplied, especially with corn, on Tuesday and Saturday: fairs for wollen cloth, hardware, leather, horses, and cattle, Aug. 12, and Oct. 29, each lasting nine days.

The wall of Adrian passes through the town, which is proved, by the numerous antiquities discovered in it, to have been the site of a Roman station; but there is no proof, though a strong presumption, that it was the *Pons Ælii*, mentioned in the "Notitia." Before the conquest it was called Monkchester, from its numerous monastic institutions (of which there are still rather extensive remains), and also from being the resort of pilgrims to the holy well of Jesus' mount (now corrupted into Jesmond). A fortress was built here by Robert, eldest son of William the Conqueror; and it received the name of *Newcastle*, probably, to distinguish it from some more ancient building. In 1080 the walls on the E. side were rebuilt, and in the reign of Edward III. the town was unsuccessfully attacked by David Bruce. Newcastle, at this early period, had become one of the largest commercial ports of the kingdom. It is curious, however, that the first authoritative mention of coal occurs in a charter by Henry III. authorising the burgesses of Newcastle to dig for that mineral. In 1281, the town had a considerable trade in coal, which soon after began to be imported into London; and in 1325 coals were exported to foreign countries. The town furnished, in 1346, 17 ships and 314 mariners for the siege of Calais, a greater force than any port N. of the Thames, except Yarmouth. It continued steadily to increase in commercial importance and mining industry till 1636, when it was visited by the plague, which carried off 5000 of its inhabitants. In the parliamentary wars it warmly espoused the cause of Charles I. With respect to the progress of its coal-trade, it may be stated that in 1703 the masters of the Trinity house of Newcastle reported to the House of Commons that 600 ships, each carrying 80 Newcastle chalders, and navigated by 4500 men and boys, were required for the supply of other ports; and in 1772, 450 keels were employed on the Tyne, the quantity shipped amounting to 351,890 Newcastle chalders. The trade has thence been steadily increasing, and with greater rapidity since the introduction of gas. Lately, however, there has been a slight diminution, owing to the opening of extensive pits in S. Durham, whence large supplies of coal are sent to the ports of Hartlepool and Middlesborough. (*Parl. Papers; Penny Mag., art. Improvements of Newcastle; Scott's R. Com., &c.*)

NEWCASTLE-UNDER-LYNE, or LYME, a parl. and mun. bor., market town, and par. of England, co. Stafford, N. div. hund. Pirehill, 15 m. N.N.W. Stafford, and 135 m. N.W. London. Pop. of parl. bor., in 1831, 8192. The town, which is well paved and lighted, consists of two nearly parallel streets, entered from the London road, and crossed by several others of an inferior description. It is, on the whole, well built, though chiefly consisting of old houses: in the High-street is a large open market place. The guildhall, a respectable-looking building, has good accommodation for the municipal and magisterial business. There are two churches, one of which, with the exception of its square tower, was rebuilt at the beginning of last century: the other, a district church, has recently been erected, the expense being defrayed chiefly by a grant of £4400, from the parliamentary commissioners, but partly, also, by private subscription. A handsome Roman Catholic chapel was built in 1834; and there are places of worship for Wesleyan and other Methodists, Independents, Baptists, and the Society of Friends. Six Sunday schools are attended by upwards of 1500 children; besides which, a national, Lancastrian, infant, and four subscription schools furnish daily instruction to about 800 boys and girls. Newcastle-under-Lyne has, also, a free grammar school, founded in 1602, and is trust of the corporation. The master's salary amounts to £63 a year; but though the sons of freemen may receive gratuitous classical instruction, the school is not much resorted to by the inhabitants of the town, and is seldom attended by more than 19 boys. (*Mun. Corp. Rep.*) Almshouses for 20 aged women were established here in 1637, by the earl of Albemarle; and the town has several benevolent institutions, besides Bible, tract, and missionary associations, &c. A literary and scientific institution was founded in 1836, a public library has 2900 vols., and there is a small theatre, little patronised.

"The manufacture of hats is the chief business carried on at Newcastle, though there are three or four silk-mills and one cotton-mill at work." These mills, however, are not mentioned in the Report of the Factory Inspectors for 1838. "The town, a few years ago, was in some measure regarded as the capital of the pottery district, which includes several towns equal or even superior in population and importance to Newcastle itself. Latterly, however, this connexion has been broken, and the town has suffered much in consequence. Trade has been very languid till within the last year or two, when it somewhat revived; and at present the labouring classes are in full employment." (*Mun. Bound. Rep.*) Newcastle has more recently sustained a serious diminution of its traffic, from the removal of the great line of communication between London, Manchester, &c., to the Grand Junction railway, which passes upwards of 6 m. W. of the town. There is good reason, however, to believe that, on the completion of the Birmingham and Manchester, the Chester and Crew railways, which are intended to run close by it, the losses occasioned by the opening of the former will be more than compensated. Iron-works and collieries are seated in the neighbourhood, and there are considerable tanneries and malthouses, with a paper-mill employed in making tissue paper for the potteries. The town is connected by a branch canal with the Grand Trunk navigation, and has access by a similar line of communication to the coalfield of N. Stafford. A branch of the Manchester and Liverpool District banking company, and a private bank, are established here, and there is a savings' bank. Markets on Monday and Saturday: five yearly cattle markets.

Newcastle-under-Lyne, which received its first charter in the 19 Henry III., was divided by the Municipal Reform Act into two wards, and placed under a mayor and five other aldermen, with 18 councillors; it has also a commission of the peace, under a recorder. The borough has returned two members to the House of Commons from 27 Edward III., the right of voting down to the Reform Act being in the resident freemen (by gift, birth, and servitude). The Boundary Act added to the old borough a small extra-parochial part of the Penkhull-township: registered electors, in 1839-40, 1631. It is also one of the polling-places for the N. division of county of Stafford.

The distinguishing name of Newcastle (*under Lyme*, or *Lyne*) is of doubtful origin; but the best authorities refer it to the fact of its standing near the woodlands, which formed a *lima* (limit) or separating *line* between the county palatine of Chester, and the rest of England. Ashton-under-Lyne, Whitmore-under-Lyne, and Audlem, or Old Lyme, admit of similar explanations. It gives the title of duke to the Pelham-Clinton family. (*Parl. Rep., &c.*)

NEW-DURHAM, p. t., Strafford co., N. H., 33 m. N.E. Concord, 514 W. Chartered in 1762. It contains Merry-meeting pond, which has an outlet in Merrymeeting bay of Winnipiscogee lake. It has a Free-will Baptist church, one store, five grist-mills, four saw-mills; 14 schools, 335 scholars. Pop. 1038.

NEW-ENGLAND, the name commonly given to the N.E. portion of the United States, or to the territory including the states of Vermont, New-Hampshire, Massachusetts, Maine, Rhode Island, and Connecticut.

NEW-FANE, capital of Windham co., Vt., 12 m. N.W. Brattleboro', 100 m. S. Montpelier. Chartered in 1753 by New-Hampshire, charter returned in 1761. Chartered in 1772 by New-York. The land titles are held under this charter. Watered by West river and its tributaries, which afford water-power. The courthouse and jail are in Fayetteville village, in the E. part of the town. The church at the centre of the town is on elevated ground, from which some part of 50 towns may be seen, lying in Vermont, New-Hampshire and Massachusetts. It contains six stores, one fulling-mill, three grist-mills, 12 saw-mills, one oil-mill, two tanneries; 11 schools, 434 scholars. Pop. 1403.

NEW-FANE, p. t., Niagara co., N. Y., 279 m. W. by N. Albany, 416 W. Watered by Eighteen Mile creek. It contains five stores, one furnace, one flouring-mill, one grist-mill, five saw-mills; 13 schools, 549 scholars. Pop. 2372.

NEWFIELD, p. t., York co., Me., 80 m. S.W. by W. Augusta, 531 W. Incorporated in 1794. Watered by Little Ossipee river. It contains one furnace, four grist-mills, two saw-mills; 566 scholars in schools. Pop. 1351.

NEWFIELD, p. t., Tompkins co., N. Y., 283 m. W. Albany, 410 W. Drained by Cayuta creek, and the inlet of Cayuga lake. It contains two churches, a Methodist and Presbyterian, six stores, two fulling-mills, three grist-mills, 26 saw-mills, two tanneries; 27 schools, 1089 scholars. Pop. 3567.

NEWFOUND, lake, Grafton co., N. H., is 6 m. long, and 2¼ broad, lies between Hebron and Bristol, and has its outlet into Pemigewasset river.

NEWFOUNDLAND, a large island of N. America, near the gulf of St. Lawrence, and off the E. coast of Labrador, from which it is separated by the narrow strait of Belleisle, between lat. 46° 30' and 51° 40' N., and long. 52° 15' and 59° 10' W.: Greatest length from N. to S. 359 m.; average breadth, 130 m. Area, 57,000 sq. m. Fixed pop., in 1836, 70,957, exclusive of those who visit the different stations during the fishing season. It may be generally described as of a triangular form, but is broken and indented with broad and deep bays, harbours, coves, rivers, and lagoons, which, besides numerous capes and projecting points of land, form two peninsulas, on one of which, called Avalon, at the S.E. corner of the island, is the town and harbour of Avalon. Its surface is wild and rugged, and its aspect from the sea far from prepossessing. The interior, which, till within the last ten years, was almost unknown, is much broken with water; and lakes, marshes, and scrubby trees, form its general character. The only large and navigable streams are the Humber and that called the river of Exploits. Its prevalent geological constitution is of granite, on which are superimposed in some parts porphyry, quartz, gneiss, mica, and clay-slate, with secondary formations: coal and iron also occur in a few places. The E. half of the interior is generally a low, picturesque country, traversed by hills and lakes, the whole being diversified by trees of humble growth. The country westward is more rugged and mountainous, with little wood, except near the shore; but the mountains are not generally in ridges, each apparently having its own particular base. The highest part of the island is the N. peninsula, lying along the strait of Belleisle; near its centre are flats of considerable extent, swampy, unhealthy, and usually covered with peat or strong wiry grass. Spruce, birch and larch, are the principal forest trees. Pine seldom occurs, and never attains a large size; indeed, there is but little wood of any value, except for fuel and the building of small boats; so that it has scarcely timber enough for its own consumption, much less for exportation. Whortleberry bushes and *wiska capuca* (Indian tea) are the principal plants on high unwooded grounds. The best soil is along the rivers and at the heads of the bays fringing the island; but both the soil and climate generally are unfavourable to the raising of grain, though well adapted for pasturage and the cultivation of potatoes and other green crops. Vast herds of carriboo deer graze in the plains and woods of the interior, and their flesh constitutes nearly the whole food of the Mic-Mac Indians. Beavers are much scarcer than formerly; but foxes are still numerous along the rivers and seacoast. Among the other wild animals are wolves, and bears, hunted by the Indians from Labrador. Insects are numerous in swampy places, especially in hot weather. The best known and most celebrated of the animals belonging to Newfoundland are its dogs, famed for docility, obedience. and attachment to their masters. They are remarkably voracious, and are usually fed on salted fish; but like the aborigines of the country, they endure hunger for a very lengthened period. The true breed has become very scarce, and there are only a few specimens of it in England, the animal so called in this country, though equally sagacious, hardy, and fond of the water, being a breed crossed with the mastiff, or some other English dog. The E. and S. coasts, where the winds blow from the sea, are very humid; and during winter the cold is intense. The harbours on the Atlantic shore are not so long frozen over as those within the gulf of St. Lawrence, where the atmosphere is generally clear, and the climate not unlike that of Lower Canada. During the summer months the days and nights are commonly serene and pleasant; the temperature is very hot during summer, and in winter frequently falls as low as 30° below the freezing point. The island, however, is on the whole extremely healthy; and the inhabitants often attain a great age, attended with more than ordinary bodily as well as mental vigour. Agriculture is progressively increasing; but very few give it their exclusive attention, the population being principally employed in the fisheries. Almost every family, however, has a small quantity of land in cultivation, though tillage be very imperfectly understood.

Newfoundland has long been celebrated for its fisheries, on which, indeed, the inhabitants principally depend. The great bank, on the E. side of the island, is in some places about 200 m. in breadth, and 600 m. in length, the soundings being from 25 to 95 fathoms. There is also an outer bank, lying between lat. 44° 10' and 47° 30' N., and long. 44° 15' and 45° 25' W.; and a continuation of banks extends southward to Nova Scotia. Fogs prevail almost without interruption on these banks, occasioned by the meeting of the waters brought thither by the gulf-stream from the tropics, with the waters carried by the influence of the winds from the polar regions. A counter-current from the N. sweeps, also, along the shore of Labrador, bringing with it large icebergs, and rendering navigation dangerous, especially during foggy weather. The best fishing-grounds on the great bank are between the 43d and 46th parallels; and the principal English settlement, besides St. John's, the capital, are Conception bay, Carbonier, Grace harbour, Trinity harbour, and Placentia, all on the E. side of the island.

The islands of St. Pierre and Mequelon, near the mouth of Fortune bay, on the S. coast, were ceded to France in 1814: the former has a harbour and town of its own name, and is the residence of a governor. The codfishery, which commenced a few years after the discovery of the island, attained so high an importance during the late war with France, that the exports of cod and cod-oil were valued, in 1814, at £2,604,000; but the English fishery has since rapidly declined, so that the average annual value of the fish exported during the years 1837–38–39, amounted only to £507,264. The number of ships employed in the fisheries cannot be ascertained; but, in 1839, there were 6150 boats. The cod-fishery commences early in June; and as the English have for some years abandoned the bank-shoals to the Americans and French, it is principally carried on close to the shore, in small boats, manned by two or four persons. Every fisherman is provided with two lines, each with two hooks, baited with herrings, mackerel, and fish-entrails. In some cases, jiggers, or artificial fish, are used, provided with two strong hooks, which the cod swallows with the bait. Seines are also used, by which multitudes of cod are hauled ashore in coves on the coast of Labrador. So abundant are the fish occasionally, that a couple of cod are hooked on each line before it reaches the bottom; and while one line is running out, the fisherman has only to turn round and pull in the other, with a fish on each hook. As soon as the boat is loaded, which, under favourable circumstances, will be in three or four hours, they proceed to the stage on the shore, where the process of cutting up, salting, and drying takes place; and after having delivered their cargo, return immediately to sea. The cod-fishery, however, is truly precarious. Sometimes the fish is not equally abundant on all parts of the coast, and the fishermen are compelled to go far from the stations, and in some cases, to split and salt the cod in the boat. The incessant labour, also, which attends the curing leaves the shoremen scarcely time during the season to eat their meals, and allows them little more than four hours' sleep. (McGregor's Brit. America, i., 200–207.) The seal-fishery is conducted in vessels varying from 80 to 190 tons, with crews of 20 or 30 men. The season commences early in April: it is principally conducted close to the shore of Labrador, and has become important only within the last 30 years. The cod fishery on the W. coast has been given up to the French; but there is still a small whale-fishery conducted in boats on the S. side of the island. There is likewise a pretty extensive salmon-fishery, the value of which, in 1839, amounted to £11,692.

The trade of Newfoundland consists in the exportation of the products of its fisheries (valued, in 1839, at £944,868), in exchange for manufactured goods, colonial produce, corn, ship-biscuits and a variety of articles for the consumption of the inhabitants. The following table exhibits the number and tonnage of ships that arrived at, and departed from, Newfoundland in 1839:

Countries.	Inwards.		Outwards.	
	Ships.	Tons.	Ships.	Tons.
Great Britain . .	163	16,360	736	16,365
British Colonies .	356	25 004	419	49,217
United States . .	4	5,397	20	1,849
Foreign	94	39,060	129	22,059
Total . .	561	91,691	594	90,386

The government of Newfoundland was long administered by naval commanders appointed to cruise on the fishing station, who returned to Britain in winter. Within the last century, however, it has been deemed more eligible to have a resident governor. In 1832, in consequence of a petition from the inhabitants, a representative government was granted, the election being by almost universal suffrage. This system has hitherto worked very harmoniously, the popular body having been in a state of violent collision, both with the executive and the commercial interests. Great complaints have also been made of the influence of the Roman Catholic clergy in the elections; and the principal merchants have made representations to the government at home, to the effect that trade is injured, and property rendered insecure, by the proceedings of the assembly. The assembly comprises 15 members; and is attached to it is a legislative and executive council. The laws are in English, and administered by circuit courts; but the police is neither numerous nor effective. The militia of the island consisted, in 1838, of 6430 men, including 356 commissioned officers. Five schools are supported by the government; and the Newfoundland School Society has established 15 others: but education is greatly neglected, and in 1839 there were not more than 2000 children receiving any kind of instruction. It is believed, however, that there will be some improvement in this respect, when the Education Act, passed in 1836, has come into full operation. There is no church establishment, all sects having equal privileges; but a titular Roman Catholic bishop resides at St. John's, and a vicar-general at Grace harbour. The Roman Catholics are the prevailing body; but there are also Episcopalians, Presbyterians, and Wesleyan Methodists.

The inhabitants are honest and industrious, but often addicted to drunkenness, and superstitious to a degree almost beyond belief. Capital offences are exceedingly rare, and petty thefts are scarcely known. The people, consisting chiefly of Irish, Scotch, and the inhabitants of Jersey and Guernsey, or their descendants (the Indian aborigines having been long all but extinct), are employed either wholly or occasionally in the fisheries. The pasture of cattle and sheep, and the cultivation of small spots of land, are likewise partial sources of occupation. The women, besides assisting the men in catching and curing the fish, are engaged either in rural occupations, or spinning and knitting worsted stockings, mittens, and socks. In winter much time is occupied in bringing home fuel, building boats, and making or repairing the fishing implements. Marriages and christenings are commonly celebrated at the close of the fishing season, or in winter, and are always times of great festivity and merriment. St. Patrick's and Sheelagh's days are celebrated with riotous mirth by the Irish; and Christmas is a universal holyday, marked by the observance of many customs that are now exploded in England. Celibacy is rare, and families of 10 or 12 children are very common. The fishermen's houses are one story high, built of wood, and covered with boards and shingles, imported from Nova Scotia, New-Brunswick, &c. Their usual diet consists of shipbiscuit, potatoes and fish, salt pork and bohea tea, spruce beer being the common beverage for those living, as most of the people do, on fish and salt meat. Spirits are mixed with the beer, to make the mixture called Callibogus, and rum is so cheap, that the labouring classes are apt to acquire habits of intoxication, which, however, is somewhat obviated by the practice of lugging, somewhat similar to taking the temperance-pledge, either for one or more years, and occasionally for life.

Newfoundland was probably first discovered by the Norwegians, at the beginning of the 11th century, but, if so, it was subsequently forgotten, till John Cabot visited it in the summer of 1497, and gave it its present name. As early as the year 1500 an extensive fishery was carried on, by the Portuguese and French, on the neighbouring banks; but though Sir Walter Raleigh, and others, attempted to form a colony here, no successful settlement was made, till Sir

G. Calvert, afterward Lord Baltimore, in 1623, established himself on the S.E. part of the island, called Avalon, and appointed his son governor. Ten years afterwards a colony was sent over from Ireland, and in 1654 a few English settlers came over, under the authority of a parliamentary grant. The French, who, very early in the 17th century, had formed a station at Placentia, were for many years a constant source of annoyance to the English; and though, by the peace of Utrecht, the possession of the island was confirmed to the English, the subject of fishery rights is still a vexata quæstio between the two nations. With respect to the fishery generally, it was chiefly carried on during the first half of the last century, by the English, Anglo-Americans, and French; but the capture of C. Breton, and other possessions in America, gave a severe blow to the fishery of the latter. The American war divided the British fishery: that portion of it that had previously been carried on from New-England being thereafter merged in that of the United States; but still the English contrived to preserve the largest share. The French were excluded from the fishery during the French war, in consequence of which the English had almost a monopoly of the business; but since the peace it has been carried on chiefly by the French and Americans, that of the English having declined fully three fourths since the peace. (See JOHN'S, ST., and ST. PIERRE.) (M'Gregor's Brit. America, i., 123–216; Murray's Brit. America, ii., 275–298: Comm. Dict.: Parl. Papers &c.)

NEW-HAMPTON, p. t., Belknap co., N. H., 29 m. N. W. Concord, 510 W. Incorporated in 1777. Bounded W. by Pemigewasset r. It contains four stores, two fulling-mills, two woollen factories, five cotton factories with 23,059 spindles, three grist-mills, five saw-mills, four tanneries; one academy, 383 students; 16 schools, 717 scholars. Pop. 1809.

NEW-HANOVER, county, N. C. Situated in the S.E. part of the state, and contains 995 sq. m. Drained by the N.E. branch of Cape Fear r. It contained in 1840, 11,605 neat cattle, 4789 sheep, 34,859 swine; and produced 489 bushels of wheat, 900,423 of Indian corn, 114,014 of potatoes, 1,476,600 pounds of rice, and 28,566 of cotton. It had two commercial and 11 commission houses in foreign trade, 56 stores, 49 grist-mills, 22 saw-mills, one oil-mill, one tannery, two distilleries, two printing-offices, two weekly newspapers; six academies, 216 students; 10 schools, 223 scholars. Pop.: whites, 6371; slaves, 6376; free coloured, 565; total, 13,312. Capital, Wilmington.

NEW-HARTFORD, p. t., Litchfield co., Ct., 20 m. W. by N. Hartford, 344 W. Watered by Farmington r. It contains three churches, two Congregational and a Baptist, six stores, two furnaces, four fulling-mills, one woollen factory, three grist-mills, seven saw-mills, one paper-mill; two academies, 61 students; 11 schools, 494 scholars. Pop. 1703.

NEW-HARTFORD, p. t., Oneida co., N. Y., 4 m. W. Utica, 95 W.N.W. Albany, 364 W. Drained by Sadaquada creek., which affords good water-power. It contains four churches, a Presbyterian, Methodist, Episcopal, and Universalist, six stores, one fulling-mill, one woollen factory, five cotton factories, with 14,164 spindles, three grist-mills, seven saw-mills, one paper-mill; 18 schools, 960 scholars. Pop. 3819.

NEW-HAVEN, county, Ct. Situated in the S. part of the state, and contains 540 sq. m. Bounded S. by Long Island sound, S.W. and W. by Housatonic r. Watered by Naugatuck, Quinnipiac, Pomperaug, West, Mill, and Menunketuck rivers. It contained in 1840, 31,627 neat cattle, 41,392 sheep, 17,234 swine; and produced 7032 bushels of wheat, 122,391 of rye, 234,546 of Indian corn, 30,031 of buckwheat, 2731 of barley, 167,640 of oats, 481,941 of potatoes, 1961 pounds of sugar. It had two commercial and six commission houses in foreign trade, with a capital of $132,000, 342 retail stores, capital $1,222,745, 14 lumber yards. capital $1,149,000, three rope-walks, 26 fulling-mills, 14 woollen factories, four cotton factories with 3356 spindles, two flouring-mills, 46 grist-mills, 93 saw-mills, three paper-mills, 27 tanneries, 22 distilleries, one pottery, nine printing-offices, five binderies, one daily five weekly and one semi-weekly newspapers, and four periodicals; one college 561 students; 32 academies, 1030 students; 242 schools, 9339 scholars. Pop. 48,619. Capital New-Haven. The above statistics as in all cases, include the statistics of all places within the county, both cities and townships.

NEW-HAVEN, city and seaport, capital of New-Haven county, and semi-capital of the state of Connecticut, is beautifully situated round the head of a bay which sets up 4 m. from Long-Island sound, in 41° 18' 30'' N. lat., and 72° 56' 43'' W. long. from Greenwich, and 4° 4' 15'' E. long. from Washington. It is 36 m. S. by W. Hartford, 52 m. W. New-London, 76 m. N.E. New-York, 134 m. W.S.W. Boston, 300 W. The population in 1810, was 5772; in 1820, 7147; in 1830, 10,180; in 1840, 12,960. Of these 180 were employed in agriculture, 474 in commerce, 1653 in manufactures and trades, 255 in navigating the ocean, 51 do. rivers and canals, 945 in the learned professions.

It lies on a plain, with a gentle inclination toward the water, skirted in other directions by an amphitheatre of hills, two of which present, at their termination, bold bluffs of trap rocks, which rise almost perpendicularly to the height of from 330 to 370 feet, which constitute an imposing feature of the scenery. From the tops of these elevations is presented a beautiful view of the city, of Long Island sound, here about 20 m. wide, and of the surrounding country. The harbour of New-Haven is entered by three rivers, Quinnipiac river on the E., Mill river on the N., and West river on the W. The two latter are mere mill streams. Quinnipiac river rises in Farmington, and has a winding course of 35 m.; and towards its mouth furnishes an abundance of fine oysters and clams, the trade in which has principally built up the village of Fairhaven. The city extends about 3 m. from E. to W., and 2 m. from N. to S. It is laid out with great regularity, and consists of two parts; the old town, and the new township. The old town was laid out in the form of a square, half a mile on a side, divided into nine smaller squares, each 52 rods on a side, separated by streets 4 rods wide. These squares have been divided into four parts, by intersecting streets. The central square, however, was reserved for public purposes, and is only divided by Temple street, running northwardly and southwardly through it, and it constitutes one of the most beautiful public grounds in any city in the United States. The eastern part of this square is unoccupied by buildings, but shaded by lofty elms and other trees. Fronting on Temple street, and on its W. side, are three of the finest churches in the country, of various architecture. The Episcopal, or S. church in the row, is a beautiful Gothic structure of stone, and a fine sample of this style of building. North of it is the first Congregational church, built of brick, with a fine Doric portico, with carved oxen's skulls as an ornament of the cornice, and a lofty steeple. Back of the old church which stood on this spot, was a burying ground, containing the monuments of many venerable men, from the time of the first settlement of the place. For a long time the inhabitants refused to have it obliterated, though in the progress of events, it had become a mar to the beauty of the place; but at length the principal monuments were removed to the new burying ground in the outer part of the city, and the ground was levelled and smoothed over. North of the first, is the second Congregational church, a beautiful edifice, built of brick, with a handsome cupola. Near the centre of the western half of the public square is the statehouse, a large and splendid edifice of the Grecian Doric order of architecture, after the model of the Parthenon at Athens. On the N.W. corner of the public square, is the Methodist church, a plain building of brick. On the W. side of the public square, and facing to the E., is the fine row of buildings belonging to Yale college, remarkably symmetrical in their arrangement, with a pleasant yard in front, ornamented by trees. The public square affords a fine prospect, and an open airy situation for the college. The houses of the city are generally built of wood, two stories high, and neatly painted white, surrounded by court yards and gardens, ornamented by shrubbery and fruit trees. As a place for an elegant and quiet residence, it is unsurpassed. Many of the houses recently built are of brick, and generally constructed with elegance and taste. The new township, E. of the old, is regularly laid out, and finely built, and has a handsome public ground, containing 5 acres, called Wooster-square. At the N.W. corner of the old town is the public cemetery, so beautifully laid out and handsomely kept, that a foreign traveller has denominated it the Père-la-Chaise of the United States. It contains 17 acres of ground, intersected by avenues and alleys, at right angles with each other, and divided into family lots, 32 feet long, and 18 ft. broad. The avenues and alleys are bordered by railings painted white, with the names of the owners of the lots inscribed on them. The cemetery contains many elegant monuments, is beautifully ornamented with trees and shrubbery, and deservedly attracts much public attention. The city contains 6 sq. m.; the whole township contains 8; and the small village called Westville, at the foot of West Rock on the W., and about half the village of Fairhaven on the E., are within the town, but not within the limits of the city.

The harbour of New-Haven is safe, but shallow, and gradually filling up with mud. At the time when the first settlers arrived in this town, the northwestern part of the harbour was sufficiently deep for all the purposes of commerce. Ships were built and launched, where are now meadows, gardens, and buildings. In 1765, Long wharf only extended 20 rods from the shore, and yet there was a greater depth of water at its termination than there is now, when it has been extended to 3043 feet in length. There is another wharf which has a basin, in which, by means of flood-gates, the water is always kept at the elevation of high tide. The maritime commerce is extensive; its foreign and coasting trade being both considerable. The sealing business, connected with the China trade, formerly brought considerable wealth into the city. At present its foreign trade is chiefly with the West Indies. The registered and licensed and enrolled tonnage of the port in 1840, was 11,500 tons. A line of three steamboats connects it daily with New-York, and also several lines of packets. The Farmington Canal connects it with Northampton, Mass., and with Connecticut river near it; and a railroad connects it with Hartford. The town contains 20 churches, nine Congregational, three Methodist, three Episcopal, one Baptist, one Roman Catholic, two coloured Methodist, one coloured Congregational. There is also a custom-house, a jail, an almshouse, a museum, three banks and a savings' institution, a young men's institute, and an institution for the support of popular lectures, with a well selected library. The state general hospital was completed in July 1832, and consists of a centre building with two wings; its whole length is 118 feet, and its breadth 48 feet, and two stories high above the basement. It is built of stone, and stuccoed. It stands about half a mile S.W. from the centre of the city. But the most important public institution in the city is Yale college. This institution was founded at Killingworth in 1701, removed to Saybrook in 1707, and permanently established in New-Haven in 1717. There are four college halls, 104 by 40 feet, four stories high, each containing 32 rooms for students. North of these is another hall, devoted to the use of the theological students. Between the colleges are three buildings with the ends in front, finished with cupolas or steeples. These are denominated the chapel, the lyceum, and the atheneum. The chapel is devoted to religious worship and ordinary exhibitions; the lyceum contains the library and the recitation rooms; the atheneum containing rooms for academic purposes. These college buildings, which are all of brick, stand in a uniform row, in the whole 600 feet long, including narrow passage ways, with a spacious yard in front, ornamented with elm and maple trees. There are college buildings of more elegant architecture, but none in the United States which present so imposing a front. In the rear of these, is another range of buildings, consisting of a chemical laboratory, the commons' hall, in the second story of which is a spacious apartment devoted to the most splendid mineralogical cabinet in the United States, consisting of over 16,000 specimens, many of them rare; and a stone building stuccoed, containing a splendid collection of paintings by Col. Trumbull and others. A short distance from these are the buildings devoted to the law and medical departments, the latter of which has a library and anatomical museum. Yale college, with limited funds, has more students than any other college in the United States; in every part of which, its graduates may be found, filling the most important stations. It contained in 1843, 534 students, of whom 376 were undergraduates, 76 in the theological, 52 in the medical, and 30 in the law departments. It has a president and 31 professors, 5189 alumni, of whom 1550 have been ministers of the gospel, and 33,000 vols. in its libraries. The commencement is on the third Thursday in August. The library contains many old and rare, as well as many splendid modern works. A fine new stone building, of handsome architecture, has recently been erected, to contain its libraries.

New-Haven has many subordinate seminaries, both male and female, of high reputation; among which are 11 select schools for males, 10 female seminaries, one Lancastrian school for boys and one for girls, besides several district and common schools.

According to the census of 1840, there were six commercial, and two commission houses in foreign trade, with a capital of $132,000; 204 retail stores, with a capital of $667,600; 10 lumber-yards, with a capital of $198,000; 19 persons were employed in internal transportation, and with 17 butchers, packers, &c., employed a capital of $46,000; 56 persons produced machinery to the amount of $50,000; 66 persons produced hardware and cutlery to the amount of $81,500; eight persons manufactured the precious metals to the amount of $7000; 52 persons manufactured various metals to the amount of $82,000; 10 persons manufactured granite and marble to the amount of $11,500; four persons produced brick and lime to the amount of $3,300; one woollen factory employed 60 persons, produced articles to the amount of $70,000, with a capital of $20,000; 13 persons produced hats and caps to the amount of $9500, with a capital of $3700; five tanneries employed 632 persons, and produced 6000 sides of sole leather, and 9000 sides of upper leather, with a capital of $50,000; 37 other manufactories of leather, as saddleries, &c., produced to the amount of $383,300, with a capital of $101,700; 11 persons produced 110,000 pounds of soap, and 144,000 pounds of tallow candles, with a capital of $18,000; four persons produced turpentine and varnish to the amount of $19,000, with a capital of $21,000; one pottery employed three persons, producing $1500 with a capital of $3000; seven persons pro-
448

duced confectionary to the amount of $12,520, with a capital of $7300; one rope-walk employed eight persons, producing cordage to the amount of $4000, with a capital of $9000; 211 persons produced carriages and wagons to the amount of $234,031, with a capital of $101,200; one gristmill, employing two persons, produced to the amount of $2500, with a capital of $9000; vessels were built to the amount of $30,000; 93 persons manufactured furniture to the amount of $83,900, with a capital of $39,900, nine printing-offices, five binderies, one daily, two weekly, and one semi-weekly, newspapers, and four periodicals, employed 39 persons, and cost $119,000. The total capital employed in manufactures was $921,206. There were one college with 561 students; 13 academies with 385 students; 27 schools, 1119 scholars. The above statistics are those of the city, exclusive of the township.

New-Haven was settled in 1638 by a colony under Theophilus Eaton, the first governor, and John Davenport, the first minister, whom Cotton Mather denominated the "Moses and Aaron of the settlement." Its original Indian name was Quinnipiack; and in 1640, it was decreed that the plantation at Quinnipiack should be called New-Haven. In 1639, it was ordered that a meeting-house be erected, of 50 ft. sq. In 1643, every planter was required to give in his name, the number of his family, and the amount of his estate; when there were found to be 419 persons in the colony, with a total estate of £32,196. In 1647, a ship of 150 tons was despatched for England, which was lost. The judges of Charles I., Goffe and Whalley, arrived at New-Haven, March 7th, 1661, and they were secreted, part of the time in a cave on West Rock, where the "judges' cave" is still shown. In 1679, they were joined by John Dixwell, another of the judges. In the old burying ground at New-Haven, there were stones supposed to mark their graves, inscribed with the initials of their names, and the date of their deaths, the last in 1688-9. In 1665, the colony of New-Haven was united by a royal charter to the colony of Connecticut. In 1667, Rev. Mr. Davenport removed to Boston, on a call from a church there, where he died, March 5th, 1670, in the 73rd year of his age. In 1684, Rev. James Pierpont was settled as minister of New-Haven, and continued there until his death in 1714. In 1716, Yale college was removed from Saybrook to New-Haven, which produced great excitement. It was named in honor of Elihu Yale, Esq., of London, who bequeathed to it £300 in money, and an equal amount in goods, the latter of which were never received. About this time the first college building was erected, 170 feet long, 22 feet wide, and three stories high, containing near 50 studies, besides the hall, library, and kitchen, and cost about £1000. In 1753, the building of the brick church, where the first Congregational church now stands, was commenced, which was completed before 1757. Difficulties having previously occurred in the church, in 1759 the White-Haven society was separated from the old church. This now constitutes the second Congregational church. About 1750, the Episcopal church was founded. On July 5th, 1779, New-Haven was entered by the British under Gov. Tryon, who in the short time in which they held the town, committed many outrages. They were at length driven off by the prompt gathering of the militia. The loss of property was estimated by commissioners, appointed by the general assembly for the purpose, at $54,566, for which the sufferers were compensated by lands in the western part of the Connecticut Western Reserve, in the N. part of the state of Ohio.

NEW-HAVEN, p. t., Addison co., Vt., 60 m. W. Montpelier. 480 W. Chartered in 1761, first settled in 1769. Bounded W. by Otter creek. Watered by Little Otter creek and New-Haven rivers, which afford water-power. It contains a Congregational church, three stores, two fulling-mills, one woollen factory, two grist-mills, two saw-mills, two nurseries, one pottery; 12 schools, 630 scholars. Pop. 1562.

NEW-HAVEN, p. t., Oswego co., N. Y., 10 m. E. Oswego. 161 m. W.N.W. Albany, 383 W. Bounded N. by Lake Ontario. Drained by Catfish river flowing into it. It contains one store, two grist-mills, three saw-mills; 10 schools, 665 scholars. Pop. 1738.

NEW-HAVEN, p. t., Huron co., O., 92 m. N. Columbus, 394 W. It has six stores; six schools, 262 scholars. Pop. 1278.

NEW-HAMPSHIRE, one of the Northern United States is bounded N. by Lower Canada; E. by Maine; S.E. by the Atlantic, on which it has a sea coast of 18 miles; S. by Massachusetts; and W. by Vermont, from which it is separated by the western bank of Connecticut river. It is between 42° 41' and 45° 11' N. lat., and between 70° 40' and 72° 28' W. long. It is 168 miles long, and from 19 to 90 broad, containing 9491 square miles, or 6,074,240 acres, of which 110,000 acres are computed to be water. The population in 1790, was 141,885; in 1800, 183,858; in 1810, 214,460; in 1820, 244,161; in 1830, 269,328; in 1840, 284,574. Of these, 139,004 were white males; 145,032 were white females; 265 were coloured males; 290 were coloured females. Employed in agriculture, 77,949; in commerce, 1379; in manufactures and trades, 17,826; in navigating the ocean 455; do. lakes, rivers and canals, 198; in the learned professions, 1640.

The state is divided into 10 counties, which, with their population in 1840, were as follows:

Counties.	Pop. 1840.	Counties.	Pop. 1840.
Rockingham,	45,771	Strafford,	23,118
Merrimack,	36,273	Belknap,	17,988
Hillsborough,	42,104	Carroll,	19,873
Cheshire,	28,429	Grafton,	6,531
Sullivan,	20,340	Coos,	9,449
		Total.	284,574

Concord, on Merrimac river, is the seat of government.

The shore of the sea coast in the S.E. part, is, in most places, a sandy beach, bordered with salt marshes, penetrated by creeks and coves, furnishing harbours for small craft, and affording but one harbour for ships, which is that of Portsmouth, at the mouth of Piscataqua r. This harbour is one of the best in the United States. For the distance of 20 or 30 miles back from the sea, the country is level, occasionally diversified with hills and valleys. Beyond this the hills increase in number, and in many parts of the state rise into lofty mountains, particularly in the N., where are situated the White mountains, which are among the highest in the United States, E. of the Rocky mountains; and others along the heights between Merrimac and Connecticut rivers. In different parts are some picturesque lakes, and fine waterfalls; and the beauty and grandeur of the scenery presented by its mountains and lakes has caused it to be denominated the Switzerland of America." The principal mountain peaks are Grand Monadnock, toward the S.W. part of the state, 3254 ft. above the level of the sea; Sunapee mountain, near Sunapee lake, Kearsarge mountain between Sutton and Salisbury, 2461 feet; Carr's mountain in Ellsworth; Moosehillock in Benton, 4636 feet high; and mount Washington, the highest peak of the White mountains, 6428 feet high. These mountains, though not a continuous range, are sometimes regarded as a continuation of the Alleghany range. The Notch in the White mountains is justly regarded as a curiosity, being in some places not more than 22 feet wide, with lofty precipices on both sides, affording some of the wildest and grandest scenery in nature. A road passes through this Notch, being the only place in which the mountain can be passed. By this road, the produce of the N. part of New-Hampshire, and the N.E. part of Vermont, finds a market at Portland, Me.; and so important is this communication regarded by Maine, that its legislature has at sometimes made grants for its improvement.

The soil of New-Hampshire is generally fertile, particularly on the margins of the rivers, and especially on Connecticut r. And the hills have a moist and warm soil, particularly adapted to grazing. Indian corn, wheat, rye, oats, barley and flax are produced, and pork, beef, mutton, poultry, butter and cheese are annually exported. Apples, pears, plums, and cherries are produced in abundance. Among the natural growth are oak, elm, birch, maple, pine, and hemlock. Sugar is extensively made from the hard maple tree.

According to the census of 1840, there were 43,892 horses or mules, 275,563 neat cattle, 617,390 sheep, 121,871 swine. Poultry was raised to the value of $107,092. There were produced 422,124 bushels of wheat, 308,148 of rye, 105,103 of buckwheat, 1,162,572, of Indian corn, 721,899 of barley, 1,296,114 of oats, 6,906,608 of potatoes, 1,260, 517 pounds of wool, 243,425 of hops, 1,162,368 of sugar, 496,107 tons of hay. The produce of the dairy was $1,633,543, of the orchard, $239,973, of lumber, $433,917.

The climate is subject to great extremes of heat and cold, the thermometer ranging from 95° above to 15° below zero of Farenheit. Greater extremes occur very rarely. The air is salubrious, as is proved by the longevity of the inhabitants. In 90 years since 1732, 75 persons have died over 100 years of age, one of whom was 116, and another 120. In the month of November, the streams are generally frozen over; and snow commonly lies until April, and in the mountainous parts, until May.

Connecticut, the largest river, is made boatable to the fifteen mile falls in Bath, 250 m. above Hartford, Ct.; Merrimac river is boatable by means of locks and short canals to Concord. The Saco, Androscoggin and Piscataqua rise and run partly in this state. The other rivers are Upper and Lower Ammonoosuc, Sugar and Ashuelot, which are tributaries of the Connecticut, and Contoocook and Nashua which flow into the Merrimac. Among the beautiful lakes are Winnipiseogee, in the centre of the state, 23 m. long, and from 2 to 10 broad, Umbagog, which lies partly in Maine, Ossipee, Sunapee, Squam, and Newfound lakes.

Portsmouth is the largest and most commercial town in the state. Its harbour is unsurpassed in the world, being safe, easily defended, and having 40 feet of water at low tide. The other principal places are Dover, Concord, Na-

shua, Keene, Exeter, Manchester, Peterborough, Walpole, Claremont, Gilmanton, Meredith, Hanover, and Haverhill.

The recent geological survey of this state by Dr. Charles F. Jackson, has resulted in the discovery of extensive copper and iron mines. A copper mine in Coos co. yields an ore of 33 per cent. of pure copper, which can be transported to the Boston market for one cent per pound.

The exports for the year ending Sept. 30th, 1841, were $16,394; and the imports were $73,701. The tonnage entered was 11,129, and cleared was 3405. There were in 1840, 18 commercial and six commission houses in foreign trade, with a capital of $1,330,600; 1075 retail dry goods and other stores, with a capital of $2,602,422; 117 persons employed in internal transportation, who, with 38 butchers, packers &c., employed a capital of $54,120; 626 persons employed in the lumber trade, with a capital of $29,000; 399 persons employed in the fisheries, with a capital of $50,680.

Home-made or family goods were manufactured to the amount of $538,303. There were 66 woollen manufactories, and 152 fulling-mills, employing 893 persons, producing goods to the amount of $795,784, with a capital of $740,345; 34 cotton factories, with 195,173 spindles, employed 6991 persons, producing goods to the amount of $4,142,304, with a capital of $5,523,300; 15 furnaces produced 1320 tons of cast iron, and two forges produced 185 tons of bar iron, together employing 121 persons, and a capital of $96,900; one smelting house, employing two persons, produced 1000 pounds of lead; 13 paper mills produced articles to the amount of $150,600, and other manufactories of paper to the amount of $1500, the whole employing 11 persons, and a capital of $104,300; hats and caps were manufactured, to the amount of $190,526, and straw bonnets to the amount of $9679, together employing 2048 persons, and a capital of $48,458; 17 persons manufactured tobacco to the amount of $10,500, with a capital of $2100; 251 tanneries employed 776 persons, and a capital of $386,402; 2131 other manufactories of leather, as saddles, boots and shoes, &c., produced articles to the amount of $712,151, with a capital of $234,649; five distilleries produced 51,244 gallons of distilled spirits, and one brewery 3000 gallons of beer, together employing seven persons, and a capital of $15,998; three glass houses employed 85 persons, producing to the amount of $47,000 with a capital of $44,000; 14 potteries employed 29 persons, producing $19,100, with a capital of $6840; 20 persons manufactured 10,900 pounds of soap, 28,845 pounds of tallow candles, and 50,000 pounds of spermaceti or wax candles, with a capital of $13,550; 191 persons produced machinery to the value of $106,814; 47 persons produced musical instruments to the amount of $28,750, with a capital of $14,050; 197 persons manufactured hardware and cutlery to the amount of $124,460; 55 persons produced granite and marble to the amount of $21,918; 236 persons produced bricks and lime to the amount of $63,166; 450 persons produced carriages and wagons to the amount of $239,240, employing a capital of $114,762; seven powder-mills, employing 11 persons, produced 185,000 pounds of gunpowder, with a capital of $58,000; mills of various kinds, employed 1296 persons, and produced articles to the value of $758,260, with a capital of $1,149,193; ships were built to the amount of $78,000; 233 persons manufactured furniture to the amount of $105,827, with a capital of $50,964; 90 brick and 434 wooden houses were built by 935 persons, and cost $470,715; 36 printing offices, 23 binderies, 27 weekly newspapers, and six periodicals, employed 256 persons, and a capital of $110,850. The total capital employed in manufactures was $9,252,448.

The principal literary institution of the state is Dartmouth college in Hanover, founded in 1770, to which is attached a flourishing medical department. The Gilmanton Theological seminary, at Gilmanton was founded in 1835, under the direction of the Congregationalists. In these institutions there were in 1840, 433 students. There were in the state, 68 academies, with 5799 students; 2127 common and primary schools, with 86,632 scholars. There were 942 white persons over 20 years of age, who could neither read nor write.

The principal religious denominations are the Congregationalists, Baptists, and Methodists. In 1836 the Congregationalists had 159 churches, 142 ministers, and 18,982 communicants; the Baptists had 90 churches, 64 ordained ministers, and 6505 communicants; the Free-will Baptists had 100 congregations, and 81 ministers; the Methodists had 75 ministers. Besides these there are Presbyterians, Unitarians, Episcopalians, Universalists, and some Roman Catholics, and two societies of Shakers.

In June 1839 there were in the state 26 banks, with an aggregate capital of $2,939,508, having a circulation of $1,430,519. The state has no public debt. There is an asylum for the Insane at Concord, which went into operation in October 1842. There is also a state prison at Concord.

The constitution was formed in 1784, and in 1792, was altered to its present form. The governor is elected annually by the people on the second Tuesday in March. He must have resided in the state for seven years next preceding his election, be 35 years of age, and possess property to the amount of 500 pounds, one half of which must be a freehold within the state. The council consists of five members, chosen by the people, who must have resided in the state for seven years, and possess property to the amount of £500, and be 30 years of age. The legislature consisting of the senate and house of representatives, is denominated the general court of New-Hampshire. The senate consists of 12 members, elected annually by the people, who must be at least 30 years of age, have resided in the state for seven years preceding the election, and possess property within the state, to the amount of £900. The house of representatives consists of 250 members, elected annually by the people. A member must have resided two years in the state, next preceding his election, be 30 years of age, and possess property to the amount of £100, within the district which he represents, one half of which must be a freehold. All judicial officers are nominated and appointed by the governor and council, and hold their offices during good behaviour; but are removable by the governor, with the consent of the council, at the representation of both houses of the legislature. No judicial officer can hold office after he is 70 years of age. The secretary and treasurer are elected by the joint ballot of both houses of the legislature. Every male inhabitant of 21 years of age or over, excepting paupers, and persons excused from paying taxes at their own request, have the right of suffrage.

The navigation of Merrimac has been improved by dams, locks, and short canals. They are Bow falls, 3 m. below Concord three fourths of a mile long; Hooksel falls, one eighth of a mile; Amoskeag falls, one mile; Union falls, nine miles; and Sewall's falls, one quarter of a mile. By means of the Middlesex canal, there is a boatable communication from Concord to Boston. The Eastern railroad extends from the Massachusetts line 15¼ miles to Portsmouth, where it passes into Maine, and is continued to Portland. A railroad extends from Concord to Nashua, and connects with the Boston and Lowell railroad.

New-Hampshire was discovered by Capt. John Smith, an English navigator, in 1614. It was first granted to Ferdinando Gorges and John Mason in 1622. The first settlements were made at Dover and Portsmouth, in 1623. In 1629, Rev. John Wheelwright purchased of several sachems the country between Merrimac and Piscataqua rivers; and in 1638, with a small colony commenced the settlement of Exeter. In the same year Hampton was settled, and these were the first settled towns in the state. In 1641 all the settlements, by a voluntary act submitted to Massachusetts, but were made a separate province by an act of Charles II., in 1679. In 1689 it again united with Massachusetts, but revived and established a separate government, in 1692. It was several times afterward connected with Massachusetts, until 1741, since which it has remained a separate state. A temporary government was established in 1776 to continue during the war with Great Britain. The state constitution was established in 1784. In June 21st, 1788, the state in convention adopted the constitution of the United States, yeas, 57, nays, 46; majority 11.

NEW-HOPE, p. b., Solebury t., Bucks co., Pa., 110 m. E. by N. Harrisburg, 179 W. Situated on the W. side of Delaware r. It has six stores, two cotton factories with 7000 spindles, two flouring-mills, one saw-mill; one academy, 50 students; two schools, 100 scholars. Pop. 890. A covered bridge here crosses Delaware river, supported on nine piers, which cost $100,000, built by a company who possess, by charter, banking privileges.

NEW-HUDSON, t., Alleghany co., N. Y., 14 m. W. Angelica, 270 W.S.W. Albany. Drained by Black cr. It has three stores, eight saw-mills; two schools, 44 scholars. Pop. 1502.

NEW-IPSWICH, p. t., Hillsborough co., N. H., 47 m. S. S.W. Concord, 447 W. Chartered in 1762. Watered by Souhegan river and its branches. It contains three churches, a Congregational, Methodist and Baptist, four stores, three cotton factories with 3600 spindles, two grist-mills, six saw-mills; one academy, 186 students; 11 schools, 493 scholars. Pop. 1578.

NEW-JERSEY, one of the middle United States, is bounded N. by New-York; E. by Hudson river, separating it from New York, and by the Atlantic; S. by the Atlantic; and W. by Delaware bay and river, separating it from Delaware and Pennsylvania. It is between 38° 58' and 41° 21' N. lat., and between 73° 58' and 75° 29' W. lon., and between 1° 22' and 3° 7' E. lon. from W., containing 7,276 sq. m., or 4.656,330 acres. It is 164 m. long, and from 40 to 75 m. wide. The population in 1790 was 184,189; in 1800, 211,149; in 1810, 245,592; in 1820, 277,575; in 1830, 390,779; in 1840, 373,306. Of these 177,055 were white males; 174,533 were white females; 10,780 were free colored males; 10,964 were free colored females. Employed in agriculture, 56,701; in commerce, 2,253; in manufactures and trades, 27,604; in mining, 365; in navigating the ocean, 450

1,143; do. rivers and canals, 1,695; in the learned professions, 1,027.

The state is divided into 18 counties, which with their population in 1840, were as follows:

Counties.	Pop. 1840.	Counties.	Pop. 1840.
Atlantic,	8,726	Mercer,	21,502
Bergen,	13,123	Middlesex,	21,893
Burlington,	32,831	Monmouth,	32,919
Cape May,	5,324	Morris,	25,844
Cumberland,	14,374	Passaic,	16,734
Essex,	44,621	Salem,	16,024
Gloucester,	25,438	Somerset,	17,456
Hudson,	9,485	Sussex,	21,770
Hunterdon,	24,789	Warren,	20,366
			3,4,20c

Trenton on the E. side of Delaware river, 30 m. above Philadelphia, is the seat of government.

The southern part of the state is level and sandy, and naturally barren, excepting where it is fertilized by the use of marl, which is extensively found. The natural growth is shrub oaks and yellow pine; and the extensive use of the latter for steam boats and glass houses, has contributed to raise the value of this land, formerly considered of little value. In the swamps the white cedar is found, which is valuable for fencing. The central portion of the state has an undulating surface and a fertile soil, which produces wheat, rye, Indian corn, barley, oats and potatoes, and excellent fruit, as apples, pears, peaches, plums and cherries. The northern parts of the state are hilly, and even mountainous; the Blue ridge, and other ranges crossing it. It is well adapted to grazing, and has a fertile soil. The mountainous region abounds with iron ore. The cities of New-York and Philadelphia on its eastern and western borders, furnish an extensive market for its agricultural productions, and its fruits. The state exports wheat, flour, horses, cattle, hams, cider, lumber, flax-seed, leather and iron. There were in the state in 1840, 70,503 horses or mules, 220,202 neat cattle, 219,285 sheep, 261,443 swine, and poultry was produced to the value of $336,953. There were produced 774,903 bushels of wheat, 1,665,820 of rye, 4,361,975 of Indian corn, 3,083,594 of oats, 12,501 of barley, 858,117 of buckwheat, 2,072,069 of potatoes, 397,907 pounds of wool, 4,321 of hops, 10,051 of wax, 1,966 of silk cocoons, 334,861 tons of hay, 2,165 of flax or hemp. The products of the dairy amounted to $1,328,032; of the orchard to $454,605; of lumber to $271,501; 9,416 gallons of wine were made, and 2,880 barrels of tar, pitch, turpentine and rosin, were produced.

A large part of the state is open to the influence of the sea, and has a mild climate; but in the mountainous region in the N. part, the cold in the winter is severe. The extreme range of the thermometer at Trenton in 1842 was from 13° to 87° above zero of Fahrenheit.

Hudson river flows on the E. side of the state, and Delaware river on the W. side. Besides these, the Raritan enters Raritan bay at Perth Amboy, and is navigable 15 miles to New-Brunswick; the Passaic enters Newark bay, and is navigable 15 miles to the port of Aequackanock. It has several important falls, affording great water-power particularly at Paterson. The Hackensack river is from 35 to 40 miles long, affords good water power, and is navigable 15 miles from its mouth in Newark bay to Hackensack, where it meets the tide water. Great Egg Harbor river enters the Atlantic, and is navigable 20 miles for small vessels. The principal bays in this state are Newark bay, N. of Staten Island, 7 miles long and 2 wide; and Raritan bay, between Staten Island and Sandy Hook, 14 miles long and 2 miles wide at Amboy Point, but increases in width below. The principal entrance has from 24 to 28 feet of water. Perth Amboy, at the head of this bay, is the principal seaport in the state. Delaware bay lies partly in this state. The principal capes are cape May, at the N. entrance of Delaware bay, and Sandy Hook, which is a low sandy island, about 3 miles long, S. of the outer harbor of New-York. The principal towns are Newark, New-Brunswick, Paterson, Trenton, Burlington, Bordentown, Elizabethtown, and Perth Amboy.

The commerce of the state is principally carried on through the two great cities on its borders. Its exports in 1841 were $19,166, and its imports were $22,315. The tonnage entered was 132; and cleared 2732. But this gives an imperfect view of the commerce of the state. It had in 1840 two commercial and eight commission houses in foreign trade, with a capital of $99,000; 1,504 retail dry goods and other stores, with a capital of $4,113,947; 1280 persons engaged in the lumber trade, with a capital of $416,570; 493 persons employed in internal transportation, who with 30 butchers, packers, &c., employed a capital of $304,380; 179 persons were engaged in the fisheries, with a capital of $93,275.

The home-made, or family manufactures amounted to $901,695; 31 woollen manufactories, and 49 fulling-mills, employed 427 persons, produced goods to the amount of $440,716, with a capital of $314,650; 43 cotton factories, with 63,744 spindles, employed 2,408 persons, manufactured articles to the amount of $2,086,104, with a capital of $1,722.870; 96 furnaces produced 11,114 tons of cast iron, and 80 forges produced 7,171 tons of bar iron, the whole employing 2,056 persons, and a capital of $1.721,820; 41 paper-mills produced articles to the amount of $568,260, and other manufactories of paper produced $7,800, the whole employing 400 persons, and a capital of $460,100. Hats and caps were manufactured to the amount of $1,181,562, and straw bonnets to the amount of $22,230, the whole employing 957 persons, and a capital of $882,629; 130 tanneries employed 1090 persons, and a capital of $415,739; 478 other manufactories of leather, as saddles, boots and shoes, &c., produced articles to the amount of $1,582,746; 93 glass-houses and four glass cutting works, employed 1,075 persons, produced articles to the amount of $664,700, with a capital of $569,800; 22 potteries employed 129 persons, produced articles to the amount of $256,207, with a capital of $125,850; 232 persons produced machinery to the amount of $255,650; 219 distilleries produced 334,017 gallons of distilled spirits, and six breweries produced 906,375 gallons of beer, the whole employing 304 persons, and a capital of $230,870; 123 persons produced hardware and cutlery to the amount of $83,575; 71 persons manufactured 2,010 small arms; 70 persons manufactured drugs, medicines and paints to the amount of $127,480, and turpentine and varnish to the amount of $43,000, with a capital of $140,890; 1,234 persons produced carriages and wagons to the amount of $1,307,140, with a capital of $644,928; 64 flouring-mills manufactured 168,797 barrels of flour, and with other mills, employed 1,986 persons, and a capital of $2,641,300; eight rope-walks, employed 80 persons, and produced cordage to the amount of $63,075, with a capital of $37,305; ships were built to the amount of $334,946; 517 persons manufactured furniture to the amount of $176,586, with a capital of $130,525; 573 persons produced bricks and lime to the amount of $576,985; 205 brick or stone, and 581 wooden houses were built, employing 2,966 persons, and cost $1,092.052; 40 printing offices, 30 binderies, four daily, one semi-weekly and 21 weekly newspapers, and four periodicals, employed 1,203 persons, and a capital of $479,080. The total amount of capital employed in manufactures, was $11,517,582.

The college of New-Jersey, or Nassau Hall, at Princeton, is one of the oldest and most distinguished in the country, and has educated many eminent men. At the same place is the Theological seminary of the Presbyterian church, more recently founded, but equally distinguished. Rutgers college, (formerly Queens college,) in New-Brunswick, was founded in 1770, and latterly has been flourishing. Connected with it is the Theological seminary of the Reformed Dutch church, founded in 1784, which is respectable. In these institutions there were in 1840, 423 students; there were in the state 66 academies, with 3,027 students, and 1,207 common and primary schools, with 32,583 scholars. There were 6,385 white persons over 20 years of age, who could neither read nor write.

In 1835 the Presbyterians had 100 churches, and 105 ministers; the Dutch Reformed had 48 churches, and 49 ministers; the Episcopalians had 35 churches, one bishop and 29 ministers; the Methodists had 64 ministers, and about twice as many congregations; the Baptists in 1839, had 61 churches and 54 ordained ministers; the Friends had 67 meetings; the Roman Catholics had four ministers. There are also some Congregationalists, Universalists, and others.

There is a state prison at Lamberton, whose income exceeds the expenses.

There are several important works of internal improvement. The Morris canal proceeds from Jersey city opposite to New-York westwardly 101½ miles to Easton, Pa., and cost $3,100,800. The Delaware and Hudson canal commences at Bordentown, proceeds N.W. to Trenton, and thence N.E. and E. to New-Brunswick, on the Raritan river. It cost $2,500,000. The Camden and Amboy rail-road commences at Camden, opposite to Philadelphia, and proceeds N.E. to Amboy, 61 miles. It has a branch from Bordentown eight miles to Trenton, and another from Crafts Creek, 13 miles to Jobstown. The Paterson and Hudson railroad proceeds from Jersey city 16½ miles to Paterson. The New-Jersey railroad proceeds from Jersey city through Newark, 34 miles to New-Brunswick, and cost $2,000,000. Here it joins the New-Brunswick and Trenton rail-road, which connects the two places, 27 miles distant. The Morris and Essex railroad extends from Newark 22 miles to Morristown. A railroad extends from Elizabeth Port 25 miles to Somerville. The Camden and Amboy rail-road had in 1841 a nett income of $372,682, and the Raritan canal a nett income of $26,034, equal to about 7½ per cent. interest on the capital stock.

The government consists of a governor, council, and house of assembly. The governor is elected annually, by the joint ballot of the council and the assembly. The council consists of 18 members, elected annually by the people. A councillor must have resided one year in the state, and possess personal and real estate of £1000 in the county for which he is chosen. The council elects from its body a president, who performs the duty of a lieutenant governor in the other states. The assembly is composed of 58 members, elected annually by the people. A member must have resided one year in the state, and possess property to the amount of £500, within the county for which he is elected. The judges of the supreme court are elected for seven years, and of the inferior courts for five years, by the joint ballot of the council and assembly. All persons of the age of 21 years, who are worth £50 proclamation money, and have resided in the state one year, immediately preceding an election, enjoy the right of suffrage.

The settlement of New-Jersey was commenced by the Dutch from New-York between 1617 and 1620, in Bergen county. In 1628, a colony of Swedes and Finns settled on the Delaware, and purchased of the Indians the land on both sides of the river to the falls. In 1664 Charles II. granted to his brother the Duke of York, all the territory between Connecticut and Delaware rivers, which included New-Jersey; and the same year sent a squadron, which conquered it from the Dutch, and the duke conveyed the territory of New-Jersey to Lord Berkely and Sir George Carteret. The two proprietors formed a constitution of government, and appointed Philip Carteret, Esq., governor, who came over in 1665, and fixed the seat of his government at Elizabethtown. In 1672 the people assumed the government, and appointed James Carteret, son of Philip, their governor. In 1675 Lord Berkely disposed of his property, and shortly after the territory was divided by the proprietors into East and West Jersey, and the eastern portion was assigned to Carteret. In 1681 the governor of West Jersey summoned a general assembly, and several laws were established, securing the rights of the people, and defining the power of rulers. In 1682, East Jersey passed from Carteret to William Penn and 23 associates, mostly Friends; and Robert Barclay, author of the "Apology for the Quakers" was appointed governor. Active measures were taken to promote emigration, and many people came over, particularly from Scotland. Many disputes arose between the settlers and the proprietaries, and in 1702, the government was surrendered to the crown, and East and West Jersey were united; and the governor of New-York was appointed governor of New-York and New-Jersey. In 1706 the inhabitants, by petition, requested of the English government that they might have a separate government, which was granted, and a governor was soon after appointed. The constitution of this state was formed in 1776. The state for several years, during the Revolutionary war, was occupied by the American and British armies, and several important battles were fought in its territory, particularly those of Trenton, of Princeton, and of Monmouth, and the inhabitants bore their full share of toil and suffering during that memorable period. On the 19th of December, 1787 this state in convention adopted the constitution of the United States by a unanimous vote.

NEW-KENT, county, Va. Situated toward the E. part of the state, and contains 295 sq. m. Bounded N.E. by Pamunky r., S.W. by Chickahominy r. It contained in 1840, 4139 neat cattle, 2518 sheep, 9379 swine; and produced 22,131 bushels of wheat, 139,784 of Indian corn, 51,307 of oats, 8496 of potatoes, 4138 pounds of cotton. It had 10 stores, 12 grist-mills, one tannery, 16 distilleries; 12 schools, 287 scholars. Pop.: whites, 2473; slaves, 3385; free coloured, 373; total. 6930. Capital, New-Kent, C. H.

NEW-KENT, C. H., capital of New-Kent co., Va., 30 m. E. Richmond, 147 W. It contains a courthouse, jail, six stores, and about 15 dwellings.

NEW-LEBANON, p. t., Columbia co., N. Y., 23 m. S.E. Albany, 368 W. It contains the village of New-Lebanon Springs and a large Shaker settlement, and has 11 stores, three fulling-mills, five woollen factories, two grist-mills, six saw-mills, three tanneries; one academy 75 students, 16 schools, 819 scholars. Pop. 2538. The springs are warm, having a temperature of 730 Fareuheit, and are useful for rheumatism, salt-rheum, and cutaneous affections. The spring emits nitrogen gas. The country around is beautiful and picturesque, and the springs are much frequented. The Shaker settlement contains 600 inhabitants, in a state of celibacy, inhabiting 10 large dwellings, occupied by the society in common. They have also 3000 acres of land, held in common, and highly cultivated. They have a large church 30 by 61 feet, with a portico, and a domical roof covered with tin, and shops for many manufactures of things useful, in which they are distinguished for peculiar neatness.

NEW-LISBON, p. t., Otsego co., N. Y., 90 m. W. Albany, 345 W. Drained by Butternut and Otsego creeks. It contains four stores, two fulling-mills, one woollen factory, one cotton factory with 110 spindles, three grist-mills, six saw-mills; 14 schools, 515 scholars. Pop. 1909.

NEW-LISBON, p v., Centre t., capital of Columbiana co., O., 150 m. E.N.E. Columbus, 35 m. S.W. Warren, 280 W. Situated on the Middle fork of Little Beaver river, and on the Sandy and Beaver canal. It contains a courthouse, jail, a bank, six churches, 10 stores, three printing offices, 300 dwellings, mostly of brick and about 2000 inhabitants. A number of the streets are paved or Macadamized, and the side walks are laid with brick. In the immediate vicinity are one furnace, four flouring-mills, four saw-mills, one paper-mill, one fulling-mill, and one carding machine, and an abundance of bituminous coal. It is a flourishing and growing village.

NEW-LONDON, county, Ct. Situated in the S.E. part of the state, and contains 560 sq. m. Bounded S. by Long Island sound. Watered by Thames river, formed by the junction of Shetucket and Yantic rivers, which afford extensive water-power in Norwich. The Shetucket receives the Quinnebag, as large as itself, in the N. part of the co. Hawcatuck r. bounds it on the E., and separates it from the state of R. Island. It contained in 1840, 27,441 neat cattle, 62,395 sheep, 14,256 swine; and produced 10,138 bushels of wheat, 27,633 of rye, 186,587 of Indian corn, 16,966 of buck wheat, 21,875 of barley, 122,355 of oats, 394,301 of potatoes. It had one commission house in foreign trade, with a capital of $30,000, 247 retail stores, capital, $770,450, 11 lumber-yards, capital, $76,500, capital invested in the fisheries $1,190,000, 24 fulling-mills, 23 woollen factories, 16 cotton factories with 29,013 spindles, 57 grist-mills, 64 saw-mills, three oil-mills, five paper-mills, seven rope-walks, 20 tanneries, two potteries, four printing-offices, three binderies, three weekly newspapers; 17 academies, 747 students; 231 schools, 10,790 scholars. Pop. 44,463. Capitals, New-London and Norwich.

NEW-LONDON, city, port of entry, and semi-capital of New-London co., Ct., is situated in 41° 22' N. lat., and 72° 9' W. long. from Greenwich, and 4° 59' E. long. from W.; 44 m. S.E. Hartford, 52 m. E. New-Haven, 55 m. S.W. Providence R. I., 353 W.' The population in 1810, was 3238; in 1820, 3330; in 1830, 4356; in 1840, 5519.

It stands on the W. bank of Thames river, 3 m. from its mouth, and is built on a declivity, which descends to the S. and E. Back of the city the ground rises to a considerable height, presenting, from its more elevated parts, a delightful view of the harbour, of Long Island sound, of the Atlantic, and of the surrounding country. The site of the city is uneven and rough, and overspread with granite rocks. The difficulty of laying it out with much regularity prevented the attempt, but it affords many commanding sites for dwellings, and is generally well built. Some of the houses recently erected are neat and elegant.

The county courts are held alternately in this city and in Norwich. It contains a courthouse, jail, a custom-house, three banks, an almshouse, two markets, nine churches, two Congregational, two Methodist, two Baptist, one Episcopal, one Universalist, and one Roman Catholic, 83 stores, one grist-mill, three manufactories of hardware, three sperm candle and oil manufactories, two tanneries, three printing-offices, two binderies, 750 dwellings, and about 7000 inhabitants, in the year 1844.

The harbour of New-London is the best in the state, and one of the best in the United States, easy of access, spacious and safe, having a depth of water of 30 feet, entirely sufficient for the largest ships of war. It is rarely obstructed by ice, and serves as the port of Connecticut river, which is not navigable for vessels of a large class, and, for some months of the year, not at all. This in some measure compensates for the limited back country which naturally flows to New-London for a market.

New-London became a place of considerable importance even before the Revolutionary war, when a large amount of shipping was employed in the European and West India trade, consisting of 130 vessels. The commerce, however, dwindled away, and was entirely extinguished by the last war with Great Britain. A few years after the peace, the merchants of this place turned their attention to the whale fisheries, which are now very extensively carried on. There are belonging to this place some 50 ships, besides some smaller vessels, employed in this business, employing over 1600 seamen, and a capital of $1,500,000. Besides these, there is a considerable number of vessels employed in the shore fisheries, which supply the market of New-York, and most of the neighbouring cities, with fish. There are also a few vessels employed in the coasting trade, some of which visit the southern states. The tonnage of the port in 1840, registered, licensed and enrolled, was 44,829, which is more than that of any other port in the state. The harbour is defended by two forts. Fort Trumbull is situated on a point projecting from the W. side of the harbour, about a mile below the city, and is generally garrisoned by soldiers of the United States. The other fortification is fort Griswold, on

an eminence on the opposite side of the river, in Groton. The lighthouse is on a projecting point of land, which divides the harbour from Long Island sound, 3 m. below the city.

There were in the city according to the census of 1840, 49 retail stores, with a capital of $220,000; three lumber-yards, with a capital of $30,000; capital employed in the fisheries $630,000; machinery was produced to the amount of $40,000; hardware and cutlery, to the amount of $61,000; one tannery employed a capital of $3000; three rope-walks employed a capital of $10,000; one printing office, one bindery, one weekly newspaper. Total capital employed in manufactures $91,300. It had four academies, 131 students; 14 schools, 787 scholars.

The first settlement of New-London was made in 1646, and the township was laid out into lots in 1648. The Indian name of the place was Nameaug. It was the seat of Seasacus, grand sachem of Long Island, a part of Connecticut and Narragansett, and belonging to the Pequot tribe of Indians. It is the smallest town in the state, being only 4 m. long, and at an average three fourths of a mile wide. On the 16th of September, 1781 a large portion of the place was burned by the British, under the traitor Benedict Arnold, and fort Griswold on the Groton side was captured, and the greater part of the garrison, consisting of militia hastily collected, was barbarously put to the sword, mostly after they had surrendered. A granite obelisk, 125 ft. high erected near the spot, commemorates the event; and on a tablet are inscribed the names of those who fell on that occasion. The loss of property by the burning of New-London was ascertained by commissioners appointed by the general assembly of the state in 1783, to be $485,980; as a compensation for which, the sufferers received a grant of land in the Connecticut Western Reserve, in the state of Ohio. During the last war, a British squadron, under the command of Commodore Hardy, chased two American frigates into the harbour of New-London, which they blockaded for a considerable time. But the fortifications were strengthened and well garrisoned, and the British never ventured to attack the place, but made an assault upon Stonington, a much weaker place in the vicinity, and were repulsed.

NEW-LONDON, t., Chester co., Pa., 36 m. W. Philadelphia. Bounded W. by Elk cr. Watered by Clay cr. It contains two churches, seven stores, two woollen factories, one forge, six grist-mills, seven saw-mills; one academy, 80 students, five schools, 131 scholars. Pop. 1533.

NEW-MADRID, county, Mo. Situated in the S.E. part of the state, and contains 1625 sq. m. The surface is level, and in some parts liable to be overflowed. Bounded E. by Mississippi r. It contained in 1840, 6635 neat cattle, 993 sheep, 21,406 swine; and produced 9503 bushels of wheat, 461,110 of Indian corn, 11,055 of oats, 9527 of potatoes, 1194 pounds of cotton. It had 15 stores, four grist-mills, four saw-mills; two academies, 25 students, four schools, 99 scholars. Pop.: Whites, 3748; slaves, 801; free coloured, 5; total, 4554. Capital, New Madrid.

NEW-MADRID, p. v., capital of New-Madrid co., Mo., 271 m. S.E. Jefferson city, 911 W. Situated on the W. bank of Mississippi river, which here has a bend, placing it on the N.W. side. The bank of the river, somewhat elevated, has been undermined by the current of the river. The place was visited by earthquakes in 1811–12. It contains a court-house, jail, Roman Catholic church, a nunnery, a female seminary, 10 stores, and about 500 inhabitants. It has extensive exports of corn and lumber. About 75,000 bushels of corn are shipped annually. It is well situated for shipping produce, at all seasons of the year.

NEWMARKET, a market town of England, celebrated for its races, partly in Suffolk, partly in hund. Cheveley, co. Cambridge, and partly in hund. Lackford, co. Suffolk, 13 m. E.N.E. Cambridge, and 55 m. N. by E. London. Area of its two parishes, 570 acres. Pop., in 1831, 2648. It comprises one large and wide street, lined with respectable shops, handsome private residences, numerous hotels and inns for the accommodation of the nobility and others who flock thither during the races. It has some handsome public buildings, among which may be specified the new rooms belonging to the Jocky Club. The stables are most extensive, and are fitted up with every convenience. Of the two parish churches, that of St. Mary's is by far the most handsome, and has a tower and steeple that form a prominent feature when seen from a distance. The Wesleyan Methodists and Baptists have also their places of worship, with attached Sunday schools. A subscription charity school is attended by 73 boys and 53 girls, the number on Sunday amounting to 176 children of both sexes. There are numerous other small charities.

Horse-racing, though now so favourite a diversion, is of rather late origin in England, and does not appear to have been much practised till the latter part of the reign of Queen Elizabeth. In the following reign, however, James I. was a distinguished patron of the turf, and imported Arabian horses for the improvement of the native breeds. In the early part of the reign of Charles I., Newmarket became celebrated for its races; and Charles II., who was still more zealously attached to this new resort of the sporting world, regularly attended its races, and repaired and enlarged the house in the town that had been occasionally occupied by his father and grandfather. From this epoch Newmarket has been the racing metropolis of the empire, and has always had to boast of the most distinguished patronage. "Newmarket fame and judgment in a bet" being an object of the highest ambition with many nobles and wealthy commoners of our own day, as well as with those of the days of Pope. The racecourse, on the heath to the W. of the town, is probably the finest in England. It it apportioned into different distances, corresponding with the ages and supposed powers of the horses, the longest course being 4 m. 1 furlong and 138 yards, and the shortest 2 furlongs and 47 yds. The grand stand has every accommodation for spectators, with a betting-room, coffee-room, &c. There are seven race meetings during the year, instituted at different periods, each lasting three days: the earliest is the Craven meeting, on Easter-Monday; then follow the two spring meetings; a fourth takes place in July; and there are three others in October, the last being called "the Houghton meeting." The sovereign gives three plates annually; one is provided from a fund left for the purpose, and others are given by the nobility or subscribed for by the members of the turf. The training-ground, on a slope S. of the town, is considered superior even to the course for trying the mettle, wind, and speed of the horses. About two thirds of the adult male population are trainers, stable-keepers, grooms, &c.; and, in fact, the town is wholly dependent for support on the races, and the training of horses. Markets on Tuesday: fairs, Whit-Tuesday and Nov. 8., chiefly for horses and sheep. Newmarket was nearly burnt down in 1683, and again at the commencement of last century. (*Blaine's Encyc. of Rural Sports*, p. 238, 373.)

NEW-MARKET, p. t., Rockingham co., N. H., 38 m. S.E. Concord, 488 W. Bounded N.E. by Lamprey river, S.W. by Exeter r. Incorporated in 1727. It contains four churches, two Methodist, a Free-will Baptist, and a Universalist, 30 stores, three cotton factories, with 14,000 spindles, one furnace, two grist-mills, two saw-mills, two tanneries; one academy, 46 students; 13 schools, 764 scholars. Pop. 2730. Vessels of 80 or 100 tons come up to the village on Lamprey r.

NEW-MARKET, p. v., Shenandoah co., Va., 139 m. N.W. Richmond, 112 W. Situated a mile E. of the N. fork of Shenandoah r. It contains three churches, a Lutheran, Methodist and Baptist, an academy, five stores, a variety of manufacturing establishments, 150 dwellings, and about 800 inhabitants.

NEW-MARLBOROUGH, p. t., Berkshire county., Mass., 131 m. S.W. by W. Boston, 20 m. S. by E. Lenox, 364 W. Granted in 1736, incorporated in 1759. Watered by Konkapot river and its branches. It contains a Congregational church, and some Methodists, four saw-mills, four fulling-mill, one forge, four grist-mills, 15 saw-mills, one powder-mill, three paper-mills; 11 schools, 430 scholars. Pop. 1682.

NEW-MILFORD, p. t., Litchfield co., Ct., 51 m. W.S.W. Hartford, 308 W. Watered by Housatonic river and its tributaries. It contains nine churches, two Congregational, two Episcopal, two Methodist, two Baptist and one Friends, seven stores, one cotton factory with 1500 spindles, two grist-mills, two saw-mills, five tanneries; 23 schools, 879 scholars. Pop. 3974. Marble is found here. Two bridges cross Housatonic r.

NEW-MILFORD, p. t., Susquehanna co., Pa., 185 m. N. Harrisburg, 299 W. Drained by Mitchell's and Salt Lick creeks. It contains an Episcopal church, three stores, one fulling-mill, one grist-mill, four saw-mills, one oil-mill, two tanneries; eight schools, 240 scholars. Pop. 1148.

NEW-ORLEANS, city, port of entry, and capital of Louisiana, in the parish of Orleans, is the fourth in population, and the third in commerce in the United States; and is in 29° 57' 30" N. lat., and 90° 0' W. long. from Greenwich, and about 13° 7' W. long. from W. It is 109 miles from the mouth of the river, following its course, by not more than 90 in a direct line, 953 below the mouth of the Ohio, 1149 below the mouth of the Missouri, 1397 S.W. New-York, 1612 S.W. Boston. 1173 by post-road from Washington. The population in 1810 was. 17,242; in 1820, 27,176; in 1830, 46,310; in 1840, 102,193; of whom 23,448 were slaves. Employed in agriculture, 1,430; in commerce, 7,392; in manufactures and trades, 4,593; in navigating the ocean, 1,315; do. rivers, lakes, &c., 285; in the learned professions, 438.

It is situated on the left bank of the Mississippi river, which by a singular bend causes the city to be on its N.W. side, facing the S.E. The city is built on an inclined plane, descending gently from the river towards the swamp in the rear; so that when the Mississippi is full, the streets are three or four feet below the surface of the river. To pre-

vent inundation an artificial embankment called the *Levee*, has been raised at a great expense, extending from fort Plaquemine, 43 miles below the city, to 120 miles above it, which is 15 feet wide and 4 feet high. Directly in front of the city, it affords a very pleasant walk. The position of New-Orleans as a vast commercial emporium is unrivalled; for the Mississippi, with its numerous tributaries, brings to it for a market, the products of 20,000 miles of navigation, and the immense resources of the great valley are yet but partially developed. The city proper is in the form of a parallelogram, running along the river 1320 yards, and extending back 700 yards. This portion of the city is traversed by 32 streets, forming 84 principal and 14 minor squares. The whole extent of the city, including its incorporated faubourgs, is not less than five miles parallel with the river, and it extends perpendicularly to it, from a quarter to three quarters of a mile; and to the Bayou St. John, two miles. The houses are principally of brick, except some of the ancient and dilapidated dwellings in the heart of the city, and some new ones in the outskirts. The modern buildings, particularly in the upper parts of the city, or Second Municipality, are generally three and four stories high, with elegant and substantial granite fronts. Many of the houses in the outer parts, are surrounded with gardens, and ornamented with orange trees. The view of the city from the river, in ascending or descending, is beautiful and on entering it, the stranger finds it difficult to believe that he has arrived at an American city. This remark applies especially to the central and lower parts, where the older buildings are ancient and of foreign construction, where the manners, customs and language are so various; the population being very nearly equally made up of Americans, French, Creoles and Spaniards, with a mixture of almost every nation on the globe. During the business season, extending from the first of November until July, the *Levee*, in its whole extent, is crowded with vessels of all sizes, from all quarters of the world; with hundreds of large and splendid steamboats, and numerous barges and flat boats, &c. Nothing can present a more busy, bustling scene than the levee at this time, the loading and unloading of vessels and steamboats, with 1500 drays transporting tobacco, cotton, sugar, and the various and immense produce of the far west. In 1836 the legislature passed an act dividing the city into three municipalities, ranking them according to their population. The first includes the city proper, extending, with that width, from the river back to lake Pontchartrain, and occupying the centre: the second adjoining it above, and the third below, both extending from the river to the lake. Each municipality has a distinct council for the management of its internal affairs, which do not encroach on the general government.

Among the public buildings, the Cathedral or Church of St. Louis on the Place d'Armes or Parade-square, strikes the stranger forcibly by its venerable and antique air. It was founded in 1792, and to a certain extent completed in 1794. The lower story is of the rustic order, flanked at each of the front angles by hexagonal towers, projecting one half of their diameter, crowned by low spires. The second story is of the Roman Doric order. Above, on the apex of the pediment of this story, rises the principal turret, square below, about 20 feet high, and hexagonal above, with a belfry with apertures on each side to let out the sound, with an elevated pinnacle above. Every Saturday evening, by the conditions of its erection, masses are offered for the soul of its founder, Don André, and at sunset on that day, the tolling of its bell recalls his memory to the citizens. On the right of it, looking from the square, is a large building of the Tuscan and Ionic orders, two stories high, occupied by various offices in the lower stories, and in the second story by the court-rooms of the parish, district and criminal courts, with the offices for their clerks. On the left of the cathedral is a building, corresponding to the one last described, the lower story containing the city guard room and the police prison, and above the offices of the mayor, the city treasurer, the comptroller, and the common council room. The second Presbyterian church is finely located, fronting on La Fayette-square, the handsomest public place in the city. It is of the Grecian Doric order, with a fine portico. The basement is of granite, the remainder of brick, plastered to imitate stone. It was completed in July, 1835, and cost $55,000. In the court in front, is a neat obelisk, erected to the memory of Rev. Sylvester Larned, the first Presbyterian pastor in the city, who died in 1820, at the early age of 24, deservedly lamented. The new Methodist Episcopal church, corner of Poydras and Carondelet streets, is of the Grecian Doric order, copied from the temple of Theseus at Athens. It has a fine portico, and a steeple rising from an octangular obelisk, resting on a lofty pedestal of Egyptian architecture combining novelty, grandeur and beauty. The steeple is 170 feet high from the side-walk, and the building was completed in 1837, at an expense of $50,000. The first Congregational church is a plain brick edifice in the Gothic style of architecture, finished in 1819, at an expense, including the cost of the ground, of $70,000. Rev. Sylvester Larned, its first pastor, died of the epidemic in 1820. St. Antoine's or the Mortuary chapel, at the corner of Conti and Rampart streets, was erected as a place for the exhibition of the bodies of the dead, and the performance of the funeral ceremonies according to the Roman Catholic ritual. It is a neat edifice of the Gothic composite order, and cost $16,000. All the funeral ceremonies of the Roman Catholics are performed here. The Ursuline chapel, is a building in the quaint old style of architecture, erected, according to a Spanish inscription on a marble tablet, in 1787, and is an interesting monument of former times. The State-house, formerly the Charity hospital, was purchased by the state in 1834. It consists of a centre building and two detached wings. It occupies a whole square between Canal and other streets. The entrance from Canal-street is through ground laid out and ornamented as a pleasure-ground, and neatly kept. The principal building is occupied by the chambers for the senate and house of representatives, and offices for the clerks and others. The wings are occupied with offices for the governor, the secretary of state, the treasurer, and other public officers. The new Charity hospital is a large building, completed in 1834, at an expense, including ground, of $149,571. The old Charity hospital (now the State-house) was purchased by the state for $125,000, for bonds payable in 50 years at five per cent interest. The new Charity hospital is 290 feet long and three stories high, and is entered from Common-street under a Doric portico. The cupola presents a magnificent view of the city and its environs. The lower story is occupied by the library, the physicians' and surgeons' rooms, a lecture room for the Medical college, &c., and the second and third stories are divided into wards for the patients, and rooms for the accommodation of the Sisters of Charity, who devote themselves to an attendance on the indigent sick. It is calculated to contain 540 patients. The grounds around it are handsomely improved and neatly kept. It is a noble charity, rendered peculiarly necessary by the sickness, which often prevails at certain seasons in New-Orleans, particularly among strangers, who are often cured in this place, while others are carried from it to their long home. The Franklin Infirmary is a private hospital, fronting on the Pontchartrain railroad. It is a beautiful building, 65 by 55 feet, and two stories high, with an imposing portico in front, surrounded by handsome shrubbery, having attached to it a variety of buildings, and can accommodate 100 patients. Several of the markets are large and commodious structures. Poydras-street market is 402 feet long and 42 feet wide. The vegetable market is 172 feet long, and cost $25,800. The meat market is built of brick, on the levee, and extends from Ann to Main streets, is a striking object as the city is approached by water, and cost about $50,000. St. Mary's market is in the rusticated Doric style, in the second municipality, built of brick and plastered to imitate granite, 480 feet long and 42 wide, and cost $47,000. It was rented in 1838 for $34,650. Washington market, in the third municipality, is designed to be a fine structure, and is but partially completed. The theatres are among the prominent buildings of New Orleans. The most magnificent of these is St. Charles theatre, completed in Nov. 1835, 132 feet long by 175 deep. In the centre of the dome is suspended a magnificent chandelier, 12 feet high and 36 in circumference, weighing 4800 pounds, lighted by 176 gas lights. It contains 23,300 cut glass drops, which weigh about 900 pounds. The interior and exterior of this building have a corresponding magnificence. It cost $350,000, and stands on the E. side of Charles-street. It is said to be now converted to other uses. The Orleans theatre is a spacious edifice, of the Roman Doric, and a mixture of the Corinthian and Composite orders, and cost $180,000. The performances are in the French language. The Camp-street, or American theatre is 60 feet wide and 160 deep, and cost, with its ground and furniture $130,000. It is in the second municipality, and can accommodate 1100 persons. Several cotton presses are among the imposing structures of the city. The Orleans Cotton Press is on ground 639 feet long and 308 wide, which is nearly covered with buildings. It contains a centre building three stories high, surmounted by a cupola, which affords a fine view of the city. The wings are two stories high, and very extensive. It presses on an average 150,000 bales of cotton annually, but its capacity is much greater. There are other cotton presses. Several of the banks have fine buildings, and some of the hotels are magnificent. Two of these hotels, one of which contains the exchange, cost $500,000 each. The United States Branch Mint has an edifice 2x2 feet long and 108 feet deep, with two wings 29 by 81 feet, the whole three stories high, which cost $182,000. The city is supplied with water raised by a steam engine from the Mississippi river, into a reservoir, constructed on an artificial mound, 21 feet high at its base. The reservoir is 250 feet

square, built of bricks, and plastered with hydraulic cement. It is divided into four compartments, to allow the water to settle before it is distributed over the city in cast iron pipes, which are laid to the aggregate length of 18 miles. The Water-works belong to the Commercial bank, and cost $722,004. The City Water-works have a pipe a mile long, to furnish running water in hot weather, through the gutters of the city, and cost $110,000. A draining company, with a capital of $640,000 has two steam engines for draining the marshes of 35 square miles in extent between the city and lake Pontchartrain. There are in the city a custom house, a U. S. branch mint, a U. S. land-office, 16 banks with an aggregate capital of over $40,000,000, some of which are not in good credit, and 12 insurance companies, with an aggregate capital of $3,600,000. There are fewer churches than in most large American cities. There are three Roman Catholic, two Episcopal, and a Congregational, Presbyterian, Methodist and Baptist, one each, and a Mariner's church. There are packet lines to several large cities; to New-York of nine ships, one sailing weekly; to Boston of five ships, sailing twice a month; to Philadelphia, seven vessels, sailing once a week; to Baltimore, four vessels, sailing twice a month; to Charleston, four vessels, twice a month; and one steam-packet to Texas, twice a month.

New-Orleans is often familiarly called the Crescent city, from its form; for though the streets are straight, those which follow the river have two turns at large angles, giving it something of this form. The river opposite to the city is half a mile wide, and from 100 to 160 feet deep, and it preserves the same width to near its entrance into the gulf of Mexico. On the bar at its mouth it has a depth of from 13½ to 16 feet of water, with a soft muddy bottom. Large and powerful steam tow-boats, some of which will tow six large vessels, are constantly employed to facilitate the passage of vessels to and from the gulf. A canal 4½ miles long leads from a basin within the city to lake Pontchartrain, through the bayou St. John. Through this canal the trade of the country bordering on lake Pontchartrain and Borgne, and all the coast of the N. part of the gulf of Mexico as far as Florida, comes to the city, and a considerable fleet of sloops is often seen in the basin. A railroad also, 4½ miles long, connects this city with lake Pontchartrain, which will probably supersede the use of the canal. A harbour is formed in the lake at the termination of the railroad, and a considerable village is there springing up. The facilities for trade are great, and well improved. Its exports, including the foreign and coasting trade, are not less than $40,000,000, which are greater than those of any other American city, except New-York; but its imports are vastly less. Much of the western country, which exports its produce by the way of New-Orleans, imports its goods from New-York. In 1842, 740,967 bales of cotton were exported to foreign ports and countries. New-Orleans is growing rapidly, but will never probably equal New-York; though it is very likely to become the second city in the Union. The licensed and enrolled tonnage in 1846, was 196,613. Its unhealthiness is against it, though this has often been exaggerated; and the same is true of its morals. It is said to be an orderly and peaceable city, and its inhabitants are distinguished for their politeness, hospitality, and kindness to the distressed.

According to the census of 1840, there were eight commercial and 375 commission houses in foreign trade, with a capital of $16,490,000; 1881 retail stores, with a capital of $11,018,225; 22 lumber yards, with a capital of $67,500; six furnaces, with a capital of $355,000; hardware was manufactured to the amount of $30,000; one cotton factory with 700 spindles employed a capital of 20,000; tobacco manufactures employed a capital of $60,000; one tannery had a capital of $50,000; two distilleries employed a capital of $56,000; three sugar refineries produced to the amount of $700,000; three steam saw-mills had a capital of $173,800; 18 printing-offices, 5 binderies, 9 daily, 6 weekly, and 2 semi-weekly newspapers employed a capital of $162,300; 201 brick or stone houses, and 210 wooden houses were built, at a cost of $2,934,300. The total capital employed in manufactures was $1,774,300. There were two colleges with 105 students; 10 academies, with 440 students; 25 schools with 975 scholars.

In 1718, Bienville, then governor of Louisiana, selected a spot for the chief settlement of the province, which had hitherto been at Biloxi, and fixed on the present site of New-Orleans, and left 50 men to clear the ground, and erect the necessary buildings. In 1719 the Mississippi rose to an extraordinary height; and as the company were not able to erect dykes the spot was overflowed, and it was for a time abandoned. In 1721, De Pauger completed a survey of the passes of the Mississippi. He found a bar at its mouth consisting of a deposite of mud, 300 feet wide, and twice that in length, having about 11 feet of water. In Nov. 1722, Delorme removed the principal deposite to New Orleans, pursuant to orders. The next year, Charlevoix reached New-Orleans from Canada by the way of the river, and found at New-Orleans 100 cabins without much order, a large wooden warehouse, two or three dwelling houses, a miserable store-house, which had been used as a chapel, a shed being converted into a house of prayer, and a population not to exceed 200. A negro was at this period sold for $126; rice at $3 the barrel ; and brandy at $30 the quarter cask. A company of Germans, disappointed by the failure of the financier Law, descended the river to New-Orleans, with a view to return to France, but were induced to remain on small allotments of land made to them at what is now called the German coast, and supplied the city with vegetables. Their descendants still cultivate the land on a larger scale. In September of this year a terrible hurricane levelled the church, hospital, and 30 houses, drove three vessels which were in the harbour, ashore, destroyed the crops and gardens, and produced a scarcity of provisions, and several of the inhabitants thought of abandoning the colony. In 1727 the Jesuits and Ursuline nuns arrived, and were accommodated on a tract of land in the lowest part of the faubourg St. Mary. The nuns removed to a house erected for them, in 1730. This property became in time very valuable, and was sold ; and the nuns removed to a new convent in 1824, two miles below the city. In 1763, Clement XIII. expelled the Jesuits from the dominions of the kings of France, Spain and Naples, and they were obliged to leave Louisiana, and their property was seized and sold for about $180,000. The same property, with its improvements, is now worth 15,000,000. In 1764 British vessels began to visit the Mississippi. They would sail past the city, make fast to a tree opposite the present city of Lafayette, and trade with the citizens. The exports during the last year of its subjection to France was $250,000, and the population of the city was 3190. The commerce suffered by the restrictions of the Spanish. In 1785 the population of the city, exclusive of the settlements in the vicinity, was 4980. A more liberal course of the Spanish government revived the trade of New-Orleans, and French, British and American vessels began to visit New-Orleans. In 1788 a fire consumed 900 houses. In 1791 the first company of French comedians arrived from Cape François, having fled from the massacre at St. Domingo ; other emigrants opened academies, the education of youth having been previously in the hands of priests and nuns. In 1792 Baron Carondelet arrived. He divided the city into four wards, and recommended lighting it, and employing watchmen. The revenue of the city did not amount to $7000, and the lighting it required a tax of 1¼ cents on every chimney. He erected new fortifications, and had the militia trained. In 1794, the first newspaper was published in Louisiana. In 1795, permission was granted by the king to the citizens of the United States to deposite their merchandise at New Orleans, during a period of ten years. In 1796 the canal Carondelet was completed. In March 21st, 1801, Louisiana was ceded by Spain to the French republic, and in April 30th, 1803, Bonaparte, as first consul, sold it to the United States, for about $15,000,000, and it was taken possession of on the 30th of November. The population of the city did not then exceed 8,056, and of the province but 49,473 ; 42,000 of whom were within the present bounds of Louisiana. The duties of the custom-house the year preceding the cession, amounted to $117,515, which would have been greater, but for the corruption of the officers. The Roman Catholic religion was the only one publicly allowed. The revenues of the city in 1802, were $19,278. There entered the Mississippi this year 256 vessels, of which 18 were public armed vessels ; of American, 46 ships, 63 brigs, 50 schooners, and nine sloops ; of Spanish, 14 ships, 17 brigs, 4 polacres, 64 schooners, and 1 sloop ; of French 1 brig. In 1804 New-Orleans was made a port of entry and delivery, and the bayou St. John, a port of delivery. A city charter was granted New-Orleans in 1805. Jan. 10th, 1812, the first steamboat arrived at the city from Pittsburg, having descended in 259 hours. In August, a hurricane did great damage to the houses and shipping, which has not been an unfrequent occurrence.

Early in December 1814, the British approached New-Orleans with about 8000 men by the way of lakes Borgne and Pontchartrain. Their passage into the lake was opposed by a squadron of gun-boats, under Lieut. Jones. After a spirited conflict, in which the killed and wounded of the enemy exceeded the whole American force, he was compelled to surrender to superior numbers. Dec. 21st, 4000 militia arrived from Tennessee. On the 22nd, the enemy, having previously landed, took a position near the main channel of the river, eight miles below the city. On the evening of the 23rd, Gen. Jackson made a furious attack upon their camp; they were thrown into disorder, but rallied ; and Gen. Jackson withdrew his troops, and fortified a strong position four miles below the city, supported by batteries on the W. side of the river. These fortifications

were unsuccessfully assailed on the 28th of Dec. and the 1st of Jan. In the meantime both armies received reinforcements ; and on the 8th of Jan. the British prepared to storm the works. In the night, a regiment was transported across the river to storm the works on the western bank. Early in the morning, the main body of the British, consisting of 7 or 8000 men, marched from their camp to assault the American works. Many were killed by grapeshot as they approached. When they came within musket shot, a stream of fire burst forth from the American lines. Gen. Jackson had placed his troops in two lines, where those in the rear loaded for those in front, which caused the fire to be incessant, which from Kentucky and Tennessee marksmen, must have been deadly. While leading to the walls the regiment which bore the ladders, Gen. Packenham, the chief in command, was killed; Gen. Gibbs, the second in command, was mortally wounded ; and Gen. Keene, severely. Without officers to direct them, the troops halted, fell back, and soon fled in disorder to their camp. In a little more than an hour, 2000 of the British lay prostrate on the field, while only seven Americans were killed and six wounded ; a disproportion unparalleled in the history of warfare. The men on the west side of the river fled before an inferior force, but the events on the east side caused the British to cross the river and retire to their entrenchments. Gen. Lambert, upon whom the command devolved, despairing of success, retired with his troops on board the fleet; and Gen. Jackson, being resolved to hazard nothing, suffered him to retreat unmolested. Immediately after the event, news arrived of peace having been concluded between the United States and Great Britain, which had in fact taken place a short time before the battle, though the news of it did not arrive till after.

In May 1816, the levee, nine miles above New-Orleans, broke through, and inundated the back part of the city, from three to five feet deep, and destroyed several plantations. The crevasse was finally closed, principally by the exertions of Gov. Clairborne, by sinking a vessel in the breach.

NEW-PALTZ, p. t., Ulster co., N. Y., 74 m. S. by W. Albany, 306 W. Watered by Wallkill r. Bounded E. by Hudson r. It contains two Dutch Reformed churches, seven stores, three fulling-mills, one woollen factory, three grist-mills, four saw-mills, two tanneries ; one academy, 35 students ; 26 schools, 1100 scholars. Pop. 5408. The inhabitants are mostly Dutch descent.

NEW-PHILADELPHIA, p. v., Goshen t., capital of Tuscarawas co., O., 115 m. E.N.E. Columbus, 313 W. Situated on the E. bank of Tuscarawas river opposite the mouth of Sugar creek. It is on a beautiful plain, and contains a court-house, jail, several stores, a printing office, and various mechanic shops, 80 dwellings, and about 500 inhabitants.

NEWPORT, a parl. bor., market-town, and par. of England, in the centre of the Isle of Wight, of which it is the cap., on the Medina (crossed here by an old stone bridge), 14 m. S.S.E. Southampton, and 75 m. S.W. London. Pop. of parl. bor., which comprises, with the old bor., a portion of the par. of Carisbrooke, in 1831, 6700. The town has one principal street, with two or three others meeting it at right angles, and forming spacious market places. The best streets are well built, paved, and lighted with gas; but there are several inferior houses on the N. side of the town and along the river. The market-house is an old building, open in the lower part, the upper story being formed into apartments for the corporation business, &c. The church is a large edifice, having three aisles, divided from each other by pointed arches, and an embattled tower at its W. end : the living is a vicarage, subordinate to Carisbrooke. The Roman Catholics, Wesleyan Methodists, Baptists, Unitarians, and the Society of Friends have their respective places of worship; and there are three Sunday-schools. A grammar-school was founded in 1619, by James I. In its school-room, a venerable-looking structure of gray stone, Charles I. and the parliamentary commissioners carried on the negotiation which ended so fatally for the former. There is also a girls' charity school. The Literary institution, assembly-rooms, and the-atre are the other principal public establishments. Within the borough is a jail, built at the joint expense of the borough and the island generally : it has rooms for the separate confinement of male and female tried and untried prisoners; but there is little further classification. The old castle of Carisbrooke (which see), occupies an eminence, about 1½ m. S.W. the town. About 1 m. N. is a workhouse for the poor of the entire island, and near it is Parkhurst military depôt and hospital, erected in 1780, and furnishing accommodation for upwards of 3000 troops. "The town formerly derived much benefit from the presence of a large military force; but the barracks have been all but deserted since 1825, and the withdrawal of the stimulus has been seriously felt. There is a luce manufactory close to the town, employing from 600 to 700 hands, and another of less consequence at the distance of about 2 m. The present importance of Newport depends principally on its being a market-town in the centre of the island, which is an active agricultural and grazing district. The markets are said to be somewhat injured by the existence of the tolls; but the general prosperity of the town seems to be neither increasing nor diminishing to any material extent." (*Mun. Corp. Rep.*)

The bor. of Newport is supposed to have been incorporated in the reign of Henry II., but its principal charter was granted by James I. Under the Mun. Reform Act it is divided into two wards, and is governed by a mayor, five aldermen, and 18 councillors: it enjoys, also, a commission of the peace, under a recorder. Corporation revenue, in 1839, £559. An ancient court, the *Curia militum*, consisting of freeholders, is held once in three weeks at the town-hall, and exercises jurisdiction over all the island, except the borough. Newport has sent two members to the House of Commons, since the 23 Edward I. A portion of the parish of Carisbrooke was added to the old borough, by the Boundary Act: registered electors in 1839–40, 669. Newport is also the election-town for the Isle of White. which, under the Reform Act, sends one member to the House of Commons. Markets on Wednesday and Saturday; fairs on Whit-Monday and two following days. (*Parl. Papers; Corp. Rep., &c.*)

NEWPORT (Welsh, *Castell-newydd*) a parl. bor., market town, and river port of England, hund. Wentloog, co. Monmouth, on the W. bank of the Usk, crossed here by a stone bridge of five arches, and about 4 m. from its mouth, 20 m. S.S.W. Monmouth, and 19½ m. W. by N. London. Pop. of parl. bor., in 1831, 7097. The town comprises a narrow and crooked main street, bifurcating at its S. extremity, and crossed by others still meaner and more irregular. On an eminence S. from the town is the old parish church of St. Woollos, with a square tower, apparently of Norman architecture, though much altered at different periods; the living is a vicarage, in the gift of the bishop of Gloucester. There are places of worship for Wesleyan Methodists, Calvinists, Baptists, and Roman Catholics. National and Lancastrian schools are established here, and the Sunday-schools are attended by about 900 children. Near the bridge are some interesting remains of a baronial castle, said to have been erected by Robert Fitzroy, son of Henry I.; and not far off are the ruins of an old monastery. "Newport is extensively engaged in the iron and tin trade, and in the export of coals. It is connected with Pontypool and Crumlin by the Monmouthshire canal. Iron and coal are brought from the former, coals only from the latter. Tram-roads also connect Newport with the Rumney, Tredegar, Sirhowey, Ebbervale, and Beaufort iron-works. It may, indeed, be considered a very thriving place: new docks and wharfs are building, or in contemplation, and the town is rapidly increasing." (*Mun. Bound. Rep.*) In 1839, 483,855 tons of coal were shipped from this port, of which, 13,035 tons were sent to foreign countries. The gross customs' revenue in 1839 amounted to £18,250, but in 1840 it fell off to £10,530. The river is navigable for sea-going ships close up to the town, and ship-building is carried on to a considerable extent. Between 60 and 70 ships belong to the port. The iron-foundries are on a large scale, and there are nail factories, roperies, breweries, and a pretty extensive pottery.

Newport, which received its earliest charter in the reign of Edward III., was divided by the Municipal Reform Act into two wards; its municipal officers being a mayor, and five other aldermen, with 18 councillors. It has also a commission of the peace, under a recorder. Corporation revenue in 1839, £617. In conjunction with Monmouth and Usk, Newport has sent one member to the House of Commons since the 27th Henry VIII., the right of election down to the Reform Act being vested in the resident burgesses. The electoral limits were enlarged by the Boundary Act, so as to include, with the old borough, additional portions of the parishes of St. Woollos and Christ-church. Registered electors for the united boroughs, in 1839–40, 1304. It is also one of the polling-places at elections for the county, and the principal town of a poor-law union, comprising 30 parishes and townships. Markets on Saturday; cattle markets the 3d Monday in each month; fairs, Holy Thursday, Whit-Thursday, 13th Aug., and 6th Nov. (*Parl. Mun. Bound. Rep., &c.*)

NEWPORT, a market town and par. of England, S. Bradford, hund. co. Salop, near its E. limit, 16½ m. W.N.W. Shrewsbury, and 126 m. N.W. London. Area of parishes 800 acres. Pop., in 1831, 2745. It consists principally of a main street, on the road between Shrewsbury and Stafford, in the centre of which stands the parish church: the living is a perpetual curacy, in the gift of the lord-chancellor. A grammar-school, founded in 1565, is endowed with lands producing about £1000 a year, and funded property to the amount of £12,450: it has eight exhibitions at Oxford and Cambridge, and is conducted by two masters. An English school is supported out of the funds of the same charity, and there are two sets of almshouses. The town com-

prises also an old, but well-built market-hall. The chief business of Newport is its retail trade for the supply of the neighbourhood. Malting is carried on pretty extensively; and it derives some advantages from its situation on a branch canal connecting the Shrewsbury canal with the Liverpool and Birmingham Junction canal. A private bank and savings' bank are established here. Markets on Saturday; cattle and sheep fairs, first Tuesday in February, Saturday before Palm Sunday, May 28, July 27, Sept. 25, and Dec. 10.

NEWPORT, a decayed bor. and market town of England, co. Cornwall, N. div. hund. East, separated from Launceston, of which it is a suburb, by a small rivulet. Though it had for many years been quite insignificant, this borough sent, from the reign of Edward VI., two members to the House of Commons (nominees of the Duke of Northumberland) down to the passing of the Reform Act, by which it was disfranchised.

NEWPORT, county, R. I. Situated in the S.E. part of the state, and contains, exclusive of water, 136 sq. m. It consists of several islands in Narraganset bay, and a portion of the main land. The principal island is Rhode Island, from which the state takes its name. which is 15 m. long, and 3½ broad. It also contains Conanicut, and Prudence islands in Narraganset bay, Block island in the Atlantic, and the townships of Tiverton and Little Compton on the main land. It possesses great facilities for navigation, and especially for the fisheries. It contained in 1840, 6633 neat cattle, 29,230 sheep, 5679 swine; and produced 408 bushels of wheat, 3500 of rye, 98,058 of Indian corn, 33,005 of barley, 62,607 of oats, 142,218 of potatoes. It had seven houses in foreign trade, 133 retail stores, five lumber yards, capital invested in the fisheries $301,557, two fulling-mills, five woollen factories, 19 cotton factories with 21,583 spindles, one dyeing and printing works, 94 grist mills, four saw-mills, three rope-walks, four tanneries, one distillery, one brewery, three printing-offices, two binderies, three weekly newspapers; three academies, 100 students; 62 schools, 2471 scholars. Pop. 16,874. Capital, Newport.

NEWPORT, p. t., Penobscot co., Me., 54 m. N.E. Augusta, 649 W. It has four stores, one grist-mill, one saw-mill; eight schools, 471 scholars. Pop. 1136.

NEWPORT, p. t., capital of Sullivan co., N. H., 39 m. W. by N. Concord. 474 W. Chartered in 1761. Watered by Sugar river and its branches, which afford water-power. It contains four churches, a Congregational, Baptist, Free-will Baptist, and Universalist, seven stores, two fulling-mills, one woollen factory, two grist-mills, 10 saw mills, two tanneries, one printing-office, one weekly newspaper; 19 schools, 650 scholars. Pop. 1958. It has a courthouse, jail, and an academy.

NEWPORT, p. t., capital of Newport co., and semi-capital of the state of Rhode Island. Situated on the S.W. side of Rhode Island, 30 m. S. by E. Providence, 75 m. S.W. Boston, 45 m. E. by N. New-London, 408 W. It is in 41° 29' N. lat., and 71° 19' 19" W. long. The harbour is enclosed by Brenton's point on the S.W., and Goat island in front, and is spacious, safe, and well defended, having a depth of water sufficient for the largest ships. The harbour is defended by three forts. Fort Green is on the N. side of the town; fort Adams is on Brenton's point, 2 m. N.W of the town; and fort Woolcott is on Goat island, in front of the town. On this island there is a military hospital belonging to the United States. Newport is finely situated for commerce. It has some trade with the East Indies, and the north of Europe, and considerable with the West Indies, particularly the island of Cuba. The coasting trade, particularly with the middle and southern states, is considerably extensive; and the fisheries have been successfully prosecuted. In the whaling business seven ships, and three brigs are employed. Of merchant vessels, there are two ships, two barques, 12 brigs, 16 schooners, and 19 sloops. Packet lines ply to New-York, and to other places on the coast. The site of the town is beautiful. The ground rises by a gentle acclivity from the shore, causing the place to appear finely from the water. The principal street is over a mile in length, and there is a public square called Washington square, on which the state house stands. The houses are neatly built, and those which are old are kept in good repair. The beauty of its situation, and the salubrity of its climate, have made it a favourite place of summer resort, particularly from the southern states. Its fish market is unrivalled, having nearly 68 different kinds of scale and shell fish, and in great abundance. It contains a state house, a jail, seven banks, with an aggregate capital of $715,000, 12 churches, three Baptist, a Congregational, Methodist, Friends, Moravian, two Episcopal, two Unitarian, and a coloured, besides a Jews' synagogue, 1280 dwellings, and 8333 inhabitants. The Redwood library was founded in 1747, has a neat edifice, and contains 4000 volumes, many of them rare old folios. There is also a Mechanics' library, and a Marine society. The place has several cotton manufactories. The Coddington mills have 10,200 spindles, 344 looms, and employ 170 hands; the Newport manufacturing company have 5090 spindles, 116 looms, and employ 140 hands; the Perry manufacturing company have 8000 spindles, 220 looms, and employ 217 hands. According to the census of 1840, Newport had five commercial and two commission houses in foreign trade, with a capital of $126,700; 104 retail stores, with a capital of $346,515; three lumber yards, capital $26,800; one fulling-mill, two woollen factories, four cotton factories, with 20,290 spindles, seven grist-mills, three tanneries, one distillery, one brewery, three printing-offices, two binderies, three weekly newspapers. Total capital in manufactures $798,983; two academies, 100 students; eight schools, 265 scholars. The registered, and licensed and enrolled tonnage of the district in 1840, was 10,924.

NEWPORT, p. t., Herkimer co., N. Y., 86 m. N.W. Albany, 405 W. Watered by West Canada creek. It contains one Episcopal and Baptist church, seven stores, two fulling-mills, two cotton factories with 1636 spindles, two grist mills, six saw-mills, two tanneries, two distilleries; 10 schools, 422 scholars. Pop. 2020.

NEWPORT, t., Luzerne co., Pa., 8 m. S.W. Wilkesbarre. Watered by Nanticoke river. Anthracite coal and bog iron ore abound. It has two stores, three flouring-mills, four saw-mills; five schools. 225 scholars. Pop. 1009.

NEWPORT, p. v., capital of Campbell co., Ky., 86 m. N.N.E. Frankfort, 493 W. Situated on the S. side of Ohio river, immediately above the mouth of Licking river, directly opposite to Cincinnati. It contains a courthouse, jail, a market-house, an academy endowed by the state with 6000 acres of land, and a United States arsenal. It occupies an elevated plain, which commands a fine view of Cincinnati.

NEWPORT, p. v., capital of Cocke co., Tenn., 232 m. E. by S. Nashville, 465 W. Situated on the S.W. side of French Broad river. It contains two churches, a Methodist and Presbyterian, an academy, and 150 inhabitants.

NEWPORT, p. t., Washington co., O., 118 m. E.S.E. Columbus, 263 W. Bounded S.E. by Ohio river. It has a church, two stores, 10 schools, 484 scholars. Population 1228.

NEWPORT, p. v., capital of Vermillion co., Ia., 78 m. W. Indianapolis, 650 W. Situated on the S. side of Vermillion river, 2 m. above its junction with Wabash river. It contains a courthouse of brick, a jail, four stores, two tanneries; one school, 30 scholars. Pop. 192.

NEW-PORTLAND, p. t., Somerset co., Me., 56 m. N.N.W. Augusta, 650 W. It has seven stores, one fulling-mill, three grist-mills, three saw-mills; 17 schools, 597 scholars. Pop. 1620.

NEWPORT-PAGNELL, a market-town and par. of England, at the N. extremity on. Buckingham, hund. of its own name, near the junction of the Ouse and Ousel (crossed here by two stone bridges, and one of iron), 13 m. E.N.E. Buckingham, and 46¼ m. N.W. London. Area of par., 3220 acres. Pop., in 1841, 3509. The town is straggling, ill-built, and only occasionally lighted with gas. The church, which has lately been thoroughly repaired, is a large building of considerable antiquity, occupying an eminence which commands an extensive view of the surrounding rich country : the living is a vicarage, valued at £230 a year, and in crown patronage. The Roman Catholics, Wesleyan Methodists, and Independents have also their respective places of worship, with attached Sunday schools. National, Lancastrian, and infant schools are supported by subscription ; and there are two endowed charity schools for girls. A mechanics' institute was established here a few years since; and there is a theological academy for training Independent ministers. Revis's almshouses provide lodging, clothes, fuel, and a stipend of £10 a year to seven aged persons, and Queen Anne's hospital (founded by Anne, consort of James I) is appropriated to the maintenance of six poor men and women, an allowance, also, of £10 a year being made to the vicar as its master. There are several other minor charities and bequests belonging to the parish, and in trust of the vicar and churchwardens. Newport-Pagnell had formerly a very extensive manufacture of bone-lace, which, though greatly injured by the competition of the machine-lace of Nottingham, still forms the staple trade of the town. The petty sessions for the hund. are held here. Markets well supplied with corn on Saturday ; cattle and lace fairs, April 22, June 22, and October 22.

NEW RICHMOND, p. v., Ohio t., Clermont co., O. 116 m. S.W. Columbus, 487 W. Situated on the N bank of Ohio river. It contains two churches, a Methodist and Presbyterian, 10 stores, a large steam grist and steam sawmill. 130 dwellings. and about 900 inhabitants.

NEW-ROCHELLE, p. t., Westchester co., N. Y., 18 m. N. E. New-York. 145 m. S. Albany, 244 W. Bounded S. by Long island sound, on which is a pleasant village. It was early settled by the Huguenots, from Rochelle in

France. It has two academies, 38 students; one school, 50 scholars. Pop. 1816.

NEW ROSS, a parl. bor. and river port of Ireland, co. Wexford, prov. Leinster, on the declivity of a steep hill, on the E. side the Barrow, 13 m. N.E. Waterford. Pop., in 1831, 6284. It is in general pretty well built, but its appearance does not indicate prosperity : this, however, is said to be owing to the difficulty that has existed of late years in obtaining land from the proprietor, Mr. Tottenham, on leases of sufficient length to encourage building. (Boundary Report.) Its public buildings are the parish church, a chapel of ease, a Roman Catholic chapel, a friary, and a nunnery, with a chapel to each ; meeting-houses for Quakers, Methodists, and the Christian brethren ; several endowed schools, an infirmary, a fever hospital, and dispensary, a lying in hospital, the Trinity hospital, and other almshouses, with several minor charitable institutions ; a market-house, and corn-market ; a barrack, the borough courthouse, sessions-house, and bridewell. A wooden bridge, 500 feet in length, with a drawbridge for the passage of vessels, leads across the river to the suburb of Rossbercon, in the county Kilkenny. The corporation, which received its first charter in the reign of Edward I., consisted of a sovereign, burgesses, and commonalty. It returned two members to the Irish House of Commons till the union, since which it has sent one member to the Imperial House of Commons. The electoral limits, as fixed by the Boundary Act, comprise about 326 acres. Registered electors, in 1839-40, 229. General sessions are held at Easter and Michaelmas ; petty sessions every fortnight. The town is a constabulary station, and it has three breweries and distilleries. Markets on Wednesdays and Saturdays ; fairs, 10th Jan., 10th Feb., 17th March, Easter Monday, 3d May, Whit-Monday, 10th June, 10th July, 10th Aug., 10th Sept., 18th Oct., 10th Nov., and 8th Dec.

New Ross is extremely well situated for trade : vessels of 900 tons reach it at all times of the tide, and those of 600 tons at high springs : the river is also navigable for barges to Athy, where it unites with the Grand canal, communicating with Dublin on the one hand, and the Shannon on the other. In 1835, its exports, consisting principally of grain, provisions, and live stock, were valued at £59,074. It imports fish from Newfoundland, and timber from N. America and the Baltic. Postoffice revenue in 1830, £885 ; 1836, £1035. Branches of the Agricultural and National banks were opened in 1835. (Boundary Rep. ; Railway Rep.)

Though called New, Ross is really an old town. A sanguinary conflict took place here on the 4th of June, 1798, between the insurgent Irishmen and the military. The former repeatedly forced their way into the town ; but were in the end repulsed with great loss.

NEWRY, a parl. bor. river-port, and town of Ireland, prov. Ulster, on the Newry Water, about 6 m. above where it falls into Carlingford bay, and on the line of the Newry canal, which gives it a navigable communication with the bay on the one hand, and with lough Neagh on the other, 34 m. S.W. Belfast, and 56 m. N. Dublin. Area of parl. bor., 2500 acres. Pop., in 1831, 13,369. It is divided by the river into two unequal portions, the largest of which, on its W. side, is in the county Down, and the other in Armagh ; the communication between them being kept up by four bridges, two of which are handsome structures.

It is a well-built thriving town. (Mun. Bound. Rep.) Its more ancient part, on the declivity of a hill, has narrow and ill-arranged streets ; but the modern portion, on the low ground along the river and canal, has wide airy streets, with good houses, mostly of granite. The principal public buildings are two Protestant Episcopal churches; St. Patrick's, originally built in 1578, and rebuilt after the revolution, and St. Mary's, a handsome structure, erected in 1812, with a spire 190 feet in height. It has also two Roman Catholic chapels, one of which, of large dimensions, in the Gothic style of architecture, is regarded as the cathedral of the see of Dromore. A convent of the order of St. Clare has also a chapel attached to it. The Presbyterians have three places of worship, the Methodists two, and the Independents and Kellyites one each. The institutions for education comprise a preparatory seminary for Maynooth college, a school attached to the convent of St. Clare, and three schools connected with the board of national education, in which and in other minor schools, about 1700 pupils are instructed. Some of the apartments of a suite of assembly-rooms, erected in 1794, are now used as public offices and for a savings' bank. It has also a mendicity association, and some almshouses: a hospital, with accommodation for 40 patients ; a good custom-house ; and barracks for 700 men.

The environs, which are very beautiful, are studded with numerous seats, surrounded by well-wooded demesnes. The town is paved, cleaned, lighted with gas, and watched, under the management of a board of commissioners: the

assessment for these purposes amounts to about £1100 a
year. The supply of water is wholly derived from numerous private springs.

The *Lordship of Newry*, of which the town forms part,
extends over about 21,000 acres. It formerly was attached
to a monastery, and enjoyed very extensive privileges,
which, after the dissolution of religious houses in the reign
of Henry VIII., were vested in the Bagnal family, of which
the Earl of Kilmorey is the present representative; his
lordship being lay rector and impropriator of the tithes.
The borough was incorporated by James I. in 1613; but
the corporation under this charter having, from some cause
or other, been extinguished, the seneschal, appointed by
Lord Kilmorey, became the ruling officer in the town, holding a manor-court every third Wednesday for sums not exceeding £10, and a weekly court of record on Mondays for
pleas to the amount of £3 6s. 8d. Irish. The general sessions for the county of Down are held here twice a year,
as are those for the W. division of Armagh, in which the
town is partly situated. Here is a bridewell, in which prisoners are confined until transmitted to the county prisons
at Downpatrick or Armagh. The borough returned two
members to the Irish House of Commons; and since the
union it has returned one member to the Imperial House
of Commons. The charter restricted the right of voting to
the provost and 12 burgesses; but on its extinction the
franchise came to be enjoyed by the inhabitants at large.
It was, however, confined by the act 35 G. 3, cap. 29, to the
occupiers of houses rated at the annual value of £5. The
Boundary Act did not change the limits of the borough.
Registered electors, in 1839-40, 1669.

Though not distinguished by its manufactures, it has two
foundries, a flint-glass factory, a distillery, three breweries,
and some large corn-mills; and in its vicinity are two large
flax-mills and a cotton-mill. The opening of the Newry
canal connecting Carlingford bay with lough Neagh, has
been of great advantage to the town, having made it the
entrepôt of a very considerable district, and the seat of an
extensive commerce: it is the principal port in the kingdom
for the shipment of butter. Vessels of small burden come
up to its quays, but those of larger burden load and unload
at Warren's point, about 4 m. lower down. Subjoined is

An ACCOUNT of the Quantity and Value of the Principal
Articles shipped from Newry in 1835.

Articles.		Quantity.	Estimated value.
			L.
Corn, meal, and flour	cwts.	497,347	204,537
Provisions	"	21,233	54,620
Potatoes	"	5,340	802
Flax and tow	"	26,940	77,820
Feathers	"	34	340
Tobacco	lbs.	8,735	1,526
Spirits	gallons	5,800	595
Linen	yards	3,686,950	184,311
Eggs	number	2,293,000	4,682
Cows and oxen	head	8,551	17,755
Horses	—	898	8,936
Sheep	—	30	30
Swine	—	15,565	15,595
Other articles	value	—	15,500
Total			616,436

Exclusive of its cross-channel trade with Great Britain,
especially that with Liverpool and Glasgow, which is by
far the most extensive, it has some trade with North
America, the ports on the Baltic, and other foreign countries. The gross customs' duty received at the port amounted, in 1840, to £44,040; and it has about 160 ships of the
aggregate burden of about 9000 tons, exclusive of two or
three steamers. A general market, and a market for linens,
which are extensively produced in the neighbourhood, are
held every Thursday; and a market for grain on Tuesdays,
and for meat on Saturdays. Postoffice revenue, in 1830,
£2841; do., in 1836, £3739.

Newry was early of considerable importance, and had a
castle. It suffered in the war of 1641; and was nearly destroyed by the Duke of Berwick in 1689. It is now one of
the most thriving towns in the N. of Ireland.

NEW-SALEM, p. t., Franklin co., Mass., 73 m. W.N.W.
Boston, 403 W. Bounded N. by Miller's river, by a branch
of which it is watered. Incorporated in 1753. It contains
three churches, two Congregational, and a Baptist, four
stores, one fulling-mill, two grist-mills, eight saw-mills,
two tanneries; one academy, 103 students; 12 schools, 360
scholars. Pop. 1305.

NEW-SCOTLAND, p. t., Albany co., N. Y., 9 m. W.
Albany, 379 W. Drained by Coyeman's creek, and a branch
of Normanskill river. It contains a Presbyterian church,
three stores, two grist-mills, four saw-mills. Population
2912.

NEW-SEWICKLY, t., Beaver co., Pa., 5 m. E. Beaver.
Watered by Beaver river, which affords extensive water-
power. It has one store, two grist-mills, two saw-mills,
456

one tannery, one distillery, one pottery; four schools, 132
scholars. Pop. 1740.

NEW-SHARON, p. t., Franklin co., Me., 28 m. N.W.
Augusta, 621 W. Watered by Sandy river, which flows
into Kennebec river. Incorporated in 1794. It contains
four stores, one fulling-mill, two grist-mills, two saw-mills;
17 schools, 729 scholars. Pop. 1829.

NEW-SHOREHAM, p. t., Newport co., R. I., 15 m. S.S.W.
Point Judith, 13 m. N.E. Montauk-point, Long Island. It
comprises Block island in the Atlantic, 8 m. long, and from
2 to 4 broad. About one seventh part of the surface is
covered with ponds. The only fuel is peat, as there are
no forests. The inhabitants are chiefly employed in the
cod and mackerel fisheries. It has no ship harbour, and
the boats are hauled ashore in bad weather. It contains a
Baptist church, four stores, one grist-mill; four schools,
374 scholars. Pop. 1069.

NEWSTEAD, a village of England, co. Nottingham, being
a liberty of the par. of Paplewick, in the N. div. wap. Bros-
tow, 8 m. N. by W. Nottingham. Pop., in 1831, 159. The
village would be wholly unworthy of notice were it not
for its proximity to Newstead Abbey; a structure, the fame
of which will be as immortal as the English language.
The abbey was formerly a priory of Black Canons, founded
by Henry II., and granted at the dissolution to Sir John
Byron, the ancestor of the illustrious poet, to whom it is
wholly indebted for its celebrity. The part now inhabited
consists principally of the rooms and offices of the priory,
the church, except the S. aisle, having fallen entirely into
decay. The front has a noble and majestic appearance,
being built in the form of the W. end of a cathedral, adorned with rich carving and lofty pinnacles. The clusters
exactly resemble those of Westminster Abbey, only on a
smaller scale, but possessing, if possible, a more venerable
appearance. The cloister-court has a basin in the centre;
and many of the ancient occupants of this noble pile lie
under its flagged pavement. The chapel is still entire.
The abbey stands

　　" —— embosom'd in a happy valley,
　　Crown'd by high woodlands ——."

And the ivy-covered ruins of the Gothic church, with its
" mighty window" and tower, strikingly contrast with the
castellated mansion and its offices.

　　" Before the mansion lies a lucid lake,
　　Broad as transparent, deep, and freshly fed
　　By a river," ——

with woods sloping down to its banks. The apartments
are spacious and superbly furnished in the old style; and
the venerable fabric, with its remnants of monastic and
baronial magnificence, its sombre appearance and sequestered situation, seems to have harmonized well with the
moody mind of the " noble Childe." An antique cross of
red sandstone stands in the courtyard, and a Gothic greenhouse leads into a beautiful garden, formerly the cemetery
of the priory, where is a pedestal of white marble erected
by the poet over a Newfoundland dog that had saved his
life. The remains of Lord Byron are interred in Hucknall
churchyard, a few miles from the abbey, which has passed
from the family.

NEWSTEAD, p. t., Erie co., N. Y., 20 m. N.E. Buffalo,
266 m. W. Albany, 386 W. Bounded N. by Tonawant
creek. It has four stores, two fulling-mills, one grist-mill,
six saw-mills, two tanneries; 15 schools, 702 scholars. Pop.
3653.

NEWTON, county, Ga. Situated N.W. of the central
part of the state, and contains 480 sq. m. Drained by
branches of Ocmulgee river. It contained in 1840, 6611
neat cattle, 4212 sheep, 17,983 swine; and produced 46,562
bushels of wheat, 901 of rye, 465,329 of Indian corn, 34,789
of oats, 13,990 of potatoes, 747 pounds of tobacco, 4,115,691
of cotton. It had 26 stores, one cotton factory, with 792
spindles, four flouring-mills, 12 grist-mills, eight saw-mills,
two tanneries, three distilleries; one academy, 25 students;
one school, 30 scholars. Pop.: whites, 7880; slaves, 3720;
free coloured, 18; total, 11,698. Capital, Covington.

NEWTON, county, Miss. Situated a little S.E. of the centre of the state, and contains 540 sq. m. Drained by head
branches of Chickasawha river. It contained in 1840, 2236
neat cattle, 1387 sheep, 14,190 swine; and produced 1984
bushels of wheat, 95,751 of Indian corn, 10,319 of potatoes,
1670 pounds of rice, 273,899 of cotton. It had six grist-mills,
six saw-mills; one college; two schools, 46 scholars. Pop.:
whites, 1980; slaves, 546; free coloured, 1: total, 2527
Capital, Decatur.

NEWTON, county, Mo. Situated in the S.W. corner of
the state, and contains 1150 sq. m. Drained by branches of
Neosho river. It contained in 1840, 6694 neat cattle, 3721
sheep, 17,390 swine; and produced 615] bushels of wheat,
264,116 of Indian corn, 5918 of oats, 9661 of potatoes, 9170
pounds of tobacco, 1588 of cotton, 1752 of sugar. It had 12
stores, six grist-mills, four saw-mills, four tanneries, four
distilleries; six schools, 114 scholars. Population: whites,

3616; slaves, 169; free coloured, 5; total, 3790. Capital, Neosho.

NEWTON, p. t., Middlesex co., Mass., 7 m. W. Boston. 433 W. It is bounded on three sides by Charles river, which, by two considerable falls, affords extensive water-power. The Boston and Worcester railroad passes through it. It contains the Newton Theological seminary, under the direction of the Baptists, founded in 1825, has three professors, 33 students, 137 who have completed the course, and 4000 vols. in its library. It has a brick edifice 85 feet long, 49 wide, and three stories high above the basement, which cost $10,000. There are also houses for the professors, and a mansion house for boarding the students. The township contains five churches, two Congregational, two Baptist, and an Episcopal, 15 stores, one cotton factory, with 5712 spindles, two grist-mills, three paper-mills; four academies, 114 students; 11 schools, 509 scholars. Pop. 3351.

NEWTON, p. t., capital of Sussex co., N. J., 70 m. N. Trenton, 238 W. The borough is on Paulinskill river, has several streets, and a large public square in the centre, on which stands the courthouse, jail, and county offices. It contains three churches, a Presbyterian, Episcopal, and Methodist, an academy, a high school, a bank, a lyceum, a public library, eight stores, two printing-offices, each issuing a weekly newspaper, 150 dwellings, and 900 inhabitants. The township contains 15 stores, one furnace, one fulling-mill, two woollen factories, five grist-mills, seven saw-mills; three academies, 91 students; 15 schools, 537 scholars. Pop. 3657.

NEWTON, t., Gloucester co., N. J., 6 m. N.E. Woodbury. Watered by Cooper's and Newton creeks. It contains nine stores, one fulling-mill, one woollen factory, three grist-mills, one tannery, one pottery; three academies, 155 students, five schools. Pop. 1863.

NEWTON, t., Cumberland co., Pa., 16 m. W. Carlisle. Watered by Yellow Breeches creek. It has two commission houses in foreign trade, two stores, one furnace, two fulling-mills, two woollen factories, three flouring-mills, six saw-mills, two distilleries; seven schools, 317 scholars. Pop. 1499.

NEWTON, p. t., Muskingum co., O., 60 m. E. Columbus, 345 W. It contains three churches, and has several salt works. It has 15 schools, 522 scholars. Pop. 2568.

NEWTON FALLS, p. v., Newton t., Trumbull co., O., 162 m. N.E. Columbus, 311 W. Situated at the junction of two branches of Mahoning river, and on the Pennsylvania and Ohio canal. It possesses a good water-power, and contains two churches, four stores, three warehouses, two woollen factories, two grist-mills, three saw-mills, one paper-mill, and about 450 inhabitants. Iron ore is found in the vicinity.

NEWTON-IN-THE-WILLOWS, otherwise called Newton-in-Macresfield, a bor., market-town, and township of England. W. Derby, hund. co. Lancaster, 15 m. W. by S. Manchester, and 166 m. N.W. London. Area of township, 3101 acres. Pop., in 1831, 2139; but probably much increased, owing to the recent erection of iron-foundries, engine-manufactories, and glass-works. It comprises one main and rather long street, conveniently situated near the point where the Manchester and Liverpool railway unites with the Grand Junction and Union railways: it has also a large depôt and station. Two episcopal chapels have been built here within the last seven years, and there are two or three Sunday schools. A free school, founded in 1699, is endowed with £55 a year. Horse-races take place annually on the common N. of the town. Its market, long disused, was re-established in 1836, and is held on Saturday. Fairs, May 17 and 18, August 11 and 12. Newton returned two members to the House of Commons, nominees of the lord of the manor, from 1st Eliz. down to the Reform Act, by which it was disfranchised. (Butterworth's Stat. of Lancashire.)

NEWTON-LIMAVADY, an inland town of Ireland, co. Londonderry, prov. Ulster, near the Roe, about 1½ m. E. from lough Foyle, and 15 m. N.E. Londonderry. Pop., in 1831, 2428. "It is agreeably situated on the E. bank of the river Roe, in a fertile and well-cultivated district. The town is rapidly increasing and improving, arising from the impulse which of late years has been given to husbandry in its vicinity. Wheat is now raised in considerable quantities, though not many years ago it was imported for home consumption. But the great increase in the culture of flax is the principal cause of its prosperity; and such has been the extent of this produce, that it was found expedient to open a market in the town for its disposal within the last twelve months, where the average weekly sales now amount to from 15 to 17 tons, and to the value of £1000 sterling; and it is now considered that this amount will rather increase than diminish. Since the municipal commissioners visited the town, two banks, or branches of banks, have been established, with a fair proportion of business." (Municipal Bound. Rep.) The public buildings

comprise the parish church, three Presbyterian meeting-houses, and one Methodist do.; a dispensary sessions-house, a market-house, and a bridewell.

The corporation, under a charter of James I., in 1613, consisted of a provost, 12 burgesses, and a commonalty, and returned two members to the Irish House of Commons till the union, when it was disfranchised. General sessions are held in June and December; petty sessions on alternate Tuesdays. The town is a constabulary station. Markets for corn are held on Tuesdays and Fridays, and for general sales on Mondays. Fairs on the second Monday in February, 28th March, 13th June, 12th July, and 29th October. Postoffice revenue, in 1830, £478; in 1836, £469. The banks referred to above were the Belfast and Northern banks opened in 1835.

NEWTON-STEWART, a market-town of Scotland, co. Wigtown, beautifully situated in the vale of the Cree, mostly on level ground, on the banks of that river, and on the high road from Dumfries to Portpatrick, 36 m. W. the former, and 25 m. E. by N. the latter, direct distance. Pop. of town, 2241; or, including the parish, 3461. It chiefly consists of one main street along the road. A suburb, called Cree Bridge, on the opposite side of the river, and in the stewartry of Kirkcudbright, is connected with the town by a handsome granite bridge.

A large cotton-mill was erected here about 45 years ago, but the speculation not succeeding, the premises were sold in 1826, for a fifth part of the original cost, and have since been pulled down. A few hand-loom weavers are employed by the Glasgow manufacturers; and it has a brewery and tan-work. But the inhabitants are chiefly dependent for support on its retail trade with the surrounding country and its markets. Pork, of the value of about £6000 a year, is cured here, chiefly for the English market. Vessels of 70 or 80 tons come up the Cree (which falls into Wigtown bay) to Carty, within 1 m. of the town.

The parish church, designed by Mr. Burn, architect, Edinburgh, and now (1841) about being finished, is the handsomest Gothic edifice for religious purposes in the S. of Scotland. It occupies an elevated situation on the outskirts of the town, has a fine light spire, and is altogether extremely elegant, and in the best taste. It cost nearly £7000. Here are also chapels belonging respectively to the Associate Synod, the Relief, and the Roman Catholics. There are nine schools in the parish, of which the most important is the Douglas school, founded and endowed by a gentleman of that name, a native of the parish, who died in Jamaica in 1799. The teacher has a salary of £80 a year, and is allowed to charge school fees. Dr. Alexander Murray, the celebrated orientalist, was born (1775) in the neighbourhood, where a granite monument, 89 ft. in height, has recently been erected to his memory. Adjoining the town is Kirrouchtree, the seat of the Heron family. (New Stat. Account of Scotland, & Wigtownshire, 167-195.)

NEWTONARDES, a town and seaport of Ireland, co. Down, at the N. extremity of lough Strangford, 10 m. E. Belfast. Pop., in 1831, 4442. It has a large square and several good streets, in which are the parish church, a small Roman Catholic chapel, three Presbyterian meeting-houses, two for Methodists, and one each for Seceders and Covenanters, a large school on the foundation of Erasmus Smith, a spacious town-hall, a courthouse, and a house of industry. "It is increasing very much in extent; many houses have been built within the five years ending with 1835; and others are building, but they are generally of a small description." (Mun. Bound. Report.) The corporation, which, under a charter of James I., in 1613, consisted of a provost, 12 burgesses, and a commonalty, returned two members to the Irish House of Commons till the union, when it was disfranchised. A manor court sits every third Saturday for the recovery of debts to the amount of £10. General sessions are held in June and December, and petty sessions on the first and third Saturday of every month. A constabulary force is stationed here. The weaving and embroidery of damask muslins are carried on to a considerable extent for the Glasgow manufacturers; and there is a large brewery. Markets on Saturdays; fairs on the second Saturday of every month, and on the 23d January, 14th May, and 23d September. Postoffice revenue, in 1830, £220; in 1836, £552. A branch of the Belfast bank was opened here in 1836.

NEWTOWN, a parl. bor., manufacturing and market-town, and par. of N. Wales, hund. of Newton, co. Montgomery, on the Severn, crossed here by a handsome stone bridge, close to the upper end of the Montgomery canal, 26½ m. S.W Shrewsbury, and 152 m. N.W. London. Pop. of par., in 1831, 4550; in 1841, 3990. The town consists of a number of small streets, lined with mean-looking houses of lath and plaster. The town-hall is of brick, and there is a handsome modern cloth-hall. The church, an ancient structure in the English style, has a low square tower, surmounted by a wooden belfry; and another church is at present in the course of being erected. There are also several places of worship for dissenters, and numerous Sunday-schools. "There is not a single day school, however, in which the rising poor can receive gratuitous instruction." (*Hand-loom Weavers' Report*, v. 550.) "Newtown is one of the most considerable towns in Montgomeryshire; and appears, indeed, to be more flourishing, and rising into greater importance, than any other town in N. Wales, owing to the number of flannel manufactures carried on in the town and its neighbourhood. The greater quantity of the Welsh flannel is made here; and the peculiar quality of the water is one of the causes assigned for the excellence of its woollen articles. Land in the neighbourhood lets at a high rent." (*Bound. Rep.*) "The flannel markets (removed thither from Welshpool in 1832) are held on alternate Thursdays, and the quantity in the mart averages 400 pieces, valued at £10 each, every market-day. The supply comes from every part of the country, except Llanidloes; and from many districts the flannel is sent in the rough, and finished or dressed at Newtown, where there are greater facilities of machinery and water. There are about 700 hand-looms in the town. The labour is performed principally by male weavers, but also by women and children, the average nett wages amounting to 9s. 6d. per week. The best weavers are never out of employ; but a great number of the middling hands are thrown out of work by the slightest depression of the trade. Considerable distress prevailed a few years ago among the weavers, owing to the failure of numerous small manufacturers, but the trade has now returned to a wholesome channel." (*Hand-loom Weavers' Report*.)

In 1839, Newtown had four woollen-mills, employing 91 hands. Machinery is made on a considerable scale, and there are foundries, potteries, tanyards, and malthouses, besides two joint-stock banks. From the extent of its trade, it is designated "the Leeds of Wales." Its communications are facilitated by the Montgomery canal, which comes close up to the town, and connects it with the internal navigation of the central and northern districts. The Reform Act made Newtown a parliamentary borough, contributory with Llanidloes, Welshpool, Machynlleth, and Llanfyllin, to Montgomery. The Boundary Act included with the parish the townships of Hendidley and Gwestydd. Registered electors of Newtown, in 1839-40, 384; and of the united boroughs, 1021. Provision markets on Tuesday and Saturday; fairs, first Tuesday in February, last Tuesdays in March and August, June 24, October 24, and December 16. (*Hand-loom Weavers' and Bound. Reports*; *Nicholson's Camb. Guide, &c.*)

NEWTOWN, a decayed bor. and town of England, in the isle of Wight, on the river of the same name, 5 m. W. by N. Newport, and 100 m. W.S.W. London. Pop. in 1831, 66. It was anciently called Frankville, and is supposed to have been of some importance previously to its being burnt down by the French in the reign of Richard II. Notwithstanding its decayed condition, it sent two members to the House of Commons from the 27th Elizabeth down to the Reform Act, by which it was disfranchised.

NEWTOWN, p. t., Fairfield co., Ct., 62 m. S.W. Hartford, 263 W. Incorporated in 1708. Bounded N.E. by Housatonic river. Watered by Powtatuck river, a tributary of the Housatonic. It contains five churches, two Episcopal, a Congregational, Methodist, and Baptist; 16 stores, three woollen factories, one cotton factory with 300 spindles, six grist-mills, three saw-mills, three tanneries; two academies, 45 students; 18 schools. 583 scholars. Pop. 3189. The Housatonic railroad passes through it.

NEWTOWN, p. t., Queens co., N.Y., 8 m. E. New-York, 152 m. S. Albany, 228 W. Bounded W. by East river, N.E. by Flushing bay, S.W. by Newtown creek, which receives several branches from the township. Peat is considerably used for fuel, taken from an extensive bog near the village. The village contains four churches; a Presbyterian, Episcopal, Methodist, and Dutch Reformed; a townhouse, 80 dwellings, and about 500 inhabitants. The township contains 16 stores, one woollen factory, two grist-mills; one academy, 55 students; 10 schools, 466 scholars. Pop. 5034.

NEWTOWN, p. t., Bucks co., Pa., 90 m. N.E. Philadelphia. 118 m. E. Harrisburg, 160 W. Newtown creek, a branch of Neshaminy creek, affords water-power. It contains a Presbyterian and a Friends' church; one store, two grist-mills, one saw-mill; one academy, 50 students; six schools, 436 scholars. Pop. 1414.

NEW-UTRECHT, p. t., Kings co., N.Y., 7 m. S. New-York, 151 S. Albany, 231 W. It occupies the extreme W. end of Long Island. The "Narrows" bound it on the W. It contains the "Bath house," a celebrated watering-place. It has one Dutch Reformed church, four stores; one academy. 35 students; two schools, 107 scholars. Pop. 1283.

NEW-WINDSOR, p. t., Orange co., N.Y., 86 m. S. by W. Albany, 699 W. Bounded E. by Hudson river. It contains nine schools, 641 scholars. Pop. 2489.

NEW-YORK, the most northern of the middle United States, and the most populous in the union, is bounded N. by lake Ontario, the river St. Lawrence and Lower Canada; E. by Vermont, Massachusetts and Connecticut; S. by the Atlantic, New-Jersey and Pennsylvania; W. and N.W. by Pennsylvania, lake Erie and Niagara river. It is between 40° 30' and 45° N. lat., and between 71° 56' and 79° 58' W. long.; and between 5° 05' E., and 2° 55' W. long. from W. Exclusive of Long Island, it is 340 miles long and 310 broad, containing, in the whole state, 46,000 square miles, or 11,040,000 acres. The population in 1790, was 340,120, in 1800, 586,050; in 1810, 959,049; in 1820, 1,372,810; in 1830, 1,913,508; in 1840, 2,428,921. Of these 853,929 were white males; 816,976 were white females; 6435 were coloured males; 6428 were coloured females. Employed in agriculture, 455,954; in commerce, 28,408; in manufactures and trades, 173,193; in mining, 1898; in navigating the ocean, 3511; do. rivers lakes and canals, 10,167; in the learned professions, 14,111.

The state is divided into 59 counties, which, with their population in 1840, were as follows:

Counties.	Pop. 1840.	Counties.	Pop. 1840.
Northern District.		Rensselaer,	60,295
Albany,	68,593	Saratoga,	40,553
Allegany,	40,976	Schenectady,	17,387
Broome,	22,338	Schoharie,	32,358
Cattaraugus,	28,872	Seneca,	24,854
Cayuga,	50,338	St. Lawrence,	56,706
Chautauqua,	47,975	Steuben,	46,137
Chemung,	20,732	Tioga,	20,527
Chenango,	40,785	Tompkins,	37,948
Clinton,	28,157	Warren,	13,422
Cortland,	24,607	Washington,	41,080
Delaware,	35,396	Wayne,	42,057
Erie,	62,465	Yates,	20,444
Essex,	23,634		
Franklin,	16,518	Total,	1,385,465
Fulton,	15,049		
Genesee,	59,587	*Southern District.*	
Hamilton,	1,907	Columbia,	43,252
Herkimer,	37,477	Dutchess,	52,398
Jefferson,	60,984	Greene,	30,446
Lewis,	17,830	Kings,	47,613
Livingston,	35,140	New-York,	312,712
Madison,	40,008	Orange,	49,135
Monroe,	64,902	Putnam,	12,825
Montgomery,	36,518	Queen's,	30,326
Niagara,	31,132	Richmond,	10,965
Oneida,	85,310	Rockland,	11,975
Onondaga,	67,911	Suffolk,	32,469
Ontario,	43,501	Sullivan,	15,629
Orleans,	25,127	Ulster,	45,882
Oswego,	43,619	Westchester,	48,686
Otsego,	49,628	Total,	743,893

Wyoming county has been formed since the taking of the census and is not included in the above. The counties are divided into 635 townships, nine cities, and include 145 incorporated villages.

Albany, on the W. bank of Hudson river, 145 m. N. of New-York, is the capital. The surface is various. Two principal chains of high lands, rising to mountains, cross the eastern part of the state. One of these comes from New Jersey and crosses the Hudson at West Point, where on each side of the river the mountains, in places coming to the water's edge, and rising to the height of from 1800 to 1680 feet, constitute some of the grandest and most interesting scenery in the country, known as the "Highlands." These mountains are from 15 to 20 miles wide, and after crossing Hudson river proceed in a northerly direction, but not in a continuous range, form in the N. part, the Taghkannic mountains, which divide the waters which fall into Hudson river from those which flow into Housatonic river, and thence into Long Island sound. Another range comes from the N.W. part of New Jersey and constitutes the Shawangunk mountains. A third range comes from Pennsylvania, and proceeding N. through Sullivan, Ulster and Greene counties, constitutes the Catskill or Kaatsberg mountains, having in their highest parts an elevation of 3800 feet, and proceeding N. at a less and varied elevation, crosses the Mohawk river at Little Falls. The Adirondack mountains in the N.E. part, and S.W. of lake Champlain, are the loftiest mountains in the state, mount Marcy, the highest peak, being 5460 feet high, and little inferior to the White mountains in N. H. These mountains are exceedingly rich in iron ore of the best quality. The country in the eastern part of the state is generally hilly, where it is not mountainous. In the western part of the state it is generally level, excepting on the S. towards Pennsylvania, where it is uneven and rough. The soil is generally good, and in some parts exceedingly fertile. The eastern part is best adapted to grazing, and the western part to grain. It produces wheat, Indian corn, grass, rye, oats, buckwheat, barley and potatoes. Beef and pork, butter and cheese, horses and cattle, pot and pearl ashes, flour, flaxseed, potash, beans and lumber, are extensively exported. Apples, pears, plums and peaches, are the most common fruits. The produce of the state will be best understood by the following statistics. There were in the state, according to the census of 1840, 473,543 horses or mules, 1,911,244 neat cattle, 5,118,777 sheep, 1,900,065 swine; poultry to the value of $1,153,413. There were produced 12,286,418 bushels of wheat, 2,979,323 of rye, 10,972,284 of Indian corn, 2,520,060 of barley, 2,287,885 of buckwheat, 30,123,484 of potatoes; 9,845,295 pounds of wool, 417,250 of hops, 1735 of silk cocoons, 10,048,109 of sugar, 3,127,947 tons of hay, 1130 of hemp or flax. The products of the dairy amounted to $10,496,021, of the orchard to $1,701,935, of lumber to $3,891,302. There were produced 6799 gallons of wine, 7613 tons of pot and pearl ashes, 402 barrels of tar, pitch or turpentine.

The climate of New-York is various. In the S. part the winters are mild, but changeable. In the N.E. part they are severe, but more uniform. In the level country W. of the mountains, the climate is more mild than in the same latitude in the E. part. The extremes of temperature at Albany, are 16° below and 93° above zero of Farenheit; at Flatbush, Long Island, 4° below and 87° above zero; at Canandaigua 8° below and 87° above zero. These may be regarded as representing the northern, southern and western divisions of the state, with some exceptions.

The principal rivers are the Hudson, 324 miles long, which enters New-York bay, and is navigable for sloops 151 miles to Troy; the Mohawk, 135 miles long, which enters the Hudson a little above Troy; the Genesee, 125 miles long, which enters lake Ontario, with falls at Rochester of 226 feet in three miles, having three perpendicular falls of 96, 78, and 20 feet, affording a vast water-power; Black river, 120 miles long, which flows into lake Ontario; Saranac, 65 miles long, which enters lake Champlain at Plattsburg; the Ausable, which after a course of 75 miles, enters lake Champlain; the Oswegatchie which flows 100 miles, and enters the St. Lawrence; the Oswego which proceeds from Oneida lake, 40 miles to lake Ontario The St. Lawrence forms a part of its N.W. boundary.

Lakes Erie, Ontario and Champlain lie partly within the state. Of the lakes which lie wholly within it, are lake George, 33 miles long, with an average breadth of two miles, having its outlet into lake Champlain, and surrounded by much picturesque scenery; Oneida lake, 20 miles long and 3½ wide; Skaneateles lake, 15 miles long, and from 1 to 1½ m. broad; Owasco lake, 11 m. long, and from one to two in width: Cayuga lake, 38 m. long, and from one to four broad; Seneca lake, 35 m. in length, and from two to four broad; Crooked lake, 18 m. long, and from 1 to 1½ m. broad; Canandaigua lake, 14 miles long, and one broad. Excepting lake George, all these small lakes discharge their waters into lake Ontario. In the extreme W. part of the state is Chautauque lake, 18 miles long, and from one to three broad, situated within a few miles of lake Erie, but discharging its waters S. into Alleghany river.

There are several important islands in New-York. Long Island is 120 miles long from E. to W., with an average breadth of 10 miles, and contains the counties of Kings, Queens and Suffolk; Staten Island, S.W. of New-York harbour, is 18 miles long and eight wide, and contains the county of Richmond; Manhattan or New-York island is 15 miles long, with an average breadth of 1½ miles, and contains the city and county of New-York. At the E. end of Long Island are Fisher's island, Shelter island, and Robbins's island, all, excepting the first, small. Grand Island in Niagara river is 12 miles long, and from two to seven wide, and extends within 1½ miles of Niagara falls.

The harbours are New-York, one of the finest in the United States, which extends eight miles above the Narrows, and is 25 miles in circumference, which is safe, spacious, well defended, and accessible at all seasons of the year, and has from 21 to 27 feet of water on the bar at its entrance. The Hudson river which enters this harbour, is navigable for large ships, about 130 miles to Hudson. Brooklyn on the W. end of Long Island has a good harbour, as has Sag Harbour, at its E. end. Sacketts harbour on lake Ontario is a good natural harbour, and Oswego harbour has been made good by artificial means. Buffalo and Dunkirk have good harbours on lake Erie.

New-York is the chief commercial city in the state, and in the United States, its facilities for commerce are unrivalled, and next to London, it is the greatest commercial city in the world. Brooklyn, opposite to New-York on Long Island, is an appendage of the larger city, and the second place in population in the state. Albany, Rochester, Troy. Buffalo and Utica are large and flourishing cities, Poughkeepsie, Newburgh. Hudson, Catskill and Lansingburgh on the Hudson; Schenectady on the Mohawk; Geneva, Syracuse, Auburn, Lockport, and Ithaca in the W., and Plattsburgh in the N. part of the state, are flourishing places.

The commerce of the state of New-York greatly surpasses that of any other state in the union. The exports of 1841 were $33,130,833, and the imports were $75,713,496; the tonnage entered was 1,111,880; the tonnage cleared was 965,548.

In 1840, there were 469 commercial, and 1044 commission houses in foreign trade, with a capital of $49,583,001; 12,207 retail dry goods and other stores, employing a capital of $42,130,795; 9502 persons engaged in the lumber trade, with a capital of $3,694,170; 7593 persons engaged in internal transportation, and 804 butchers, packers, &c., the whole employing a capital of $2,833,916; 1998 persons employed in the fisheries, with a capital of $849,250.

The manufactures of the state of New-York are also extensive. Home-made or family manufactures amounted to $4,636,547; 393 woollen manufactories, with 800 fulling-mills, employed 4636 persons, produced articles to. the amount of $3,537,337, with a capital of $3,469,349; 117 cotton-factories, with 211,659 spindles, employed 7487 persons, and a capital of $4,900,774; 339 persons produced 2,867,884 bushels of salt, with a capital of $5,601,000; 186 furnaces produced 29,088 tons of cast iron, and 190 forges, produced 53,693 tons of bar iron, consumed 193,677 tons of fuel, employed 3456 persons, and a capital of $2,193,418; nine smelting houses produced 670,000 pounds of lead, employing 333 persons, and a capital of $221,000; 77 paper-mills produced articles to the amount of $673,121, and other manufactories of paper produced $89,637, the whole employing 749 persons, and a capital of $703,550; hats and caps were manufactured to the amount of $2,914,117, and straw bonnets to the amount of $160,248, the whole employing 3680 persons, and a capital of $1,676,550; 1216 tanneries employed 5579 persons, and a capital of $3,907,348; other manufactories of leather, as saddles, boots and shoes, &c., produced articles to the amount of $6,239,924, with a capital of $3,743,765; 13 glass houses, and 11 glass cutting works employed 498 persons, produced articles to the amount of $411,371, and employed a capital of $204,700; 47 potteries employed 197 persons, producing articles to the amount of $150,292, with a capital of $98,450; machinery was produced to the amount of $2,893,517, employing 3631 persons; 692 persons produced hardware and cutlery, to the value of $1,566,974; 112 cannon and 8308 small arms were made by 903 persons; 1713 persons manufactured the precious metals to the amount of $1,106,243; 1447 persons manufactured granite and marble to the amount of $943,980; 489 persons made 11,939,834 pounds of soap, 4,029,783 pounds of tallow candles, and 533,000 pounds of spermaceti or wax candles, with a capital of $618,875; 669 persons manufactured tobacco to the amount of $531,570, with a capital of $795,580; 219 distilleries produced 11,973,815 gallons of spirits, and 83 breweries produced 6,059,192 gallons of beer, the whole employing 1486 persons, and a capital of $3,107,066; 4710 persons manufactured carriages and wagons to the amount of $2,364,461, with a capital of $1,485,023; 338 flouring-mills produced 1,861,385 barrels of flour, and with other mills, produced articles to the amount of $16,953,280, employing 10,807 persons, and a capital of $14,648,814; vessels were built to the amount of $797,317; 3660 persons manufactured furniture to the amount of $1,971,776, with a capital of $1,610,810; 3160 persons produced bricks and lime to the amount of $1,198,327; 1883 brick or stone houses, and 5198 wooden houses were built by 16,768 persons, and cost $7,265,844; 321 printing-offices, 107 binderies, 34 daily, 13 semi-weekly or tri weekly, and 198 weekly newspapers, and 57 periodicals, employed 3931 persons, and a capital of $1,876,540 The total amount of capital employed in manufactures was $55,252,779.

This state has several important literary institutions. Columbia college, formerly King's college, was founded in New-York city in 1754, and is under the direction, though not exclusively, of the Episcopalians; Union college at Schenectady was founded in 1795; Hamilton college in Clinton, was founded in 1812; Geneva college, conducted by the Episcopalians, was founded in Geneva in 1823; the University of the city of New York was founded in 1831. The Hamilton literary and theological seminary was founded by the Baptists in 1819; the Theological Institute of the Episcopal church was founded in 1819, in the city of New-York, by the Episcopalians. The New-York theological seminary, connected with the University, was founded in 1826; the theological seminary at Auburn was founded by the Presbyterians in 1821; the Hartwick seminary was founded at Hartwick, in Otsego county, by the Lutherans, in 1816; the theological seminary of the Associate Reformed Church was founded at Newburgh, in 1836; the college of physicians and surgeons in the city of New-York was founded in 1807; the Albany medical college was founded in 1839. In the above named institutions, there were in 1840, 1985 students. There were in the state 505 academies, with 34,715 students, and 10,539 common and primary schools, with 502,367 scholars. There were 44,452 white persons over 20 years of age, who could neither read nor write.

In 1838, the Presbyterians, with a few Congregationalists,

had 564 ministers, and 86,000 communicants; the Dutch Reformed had 142 ministers, and 15,800 communicants; the Methodists had 591 ministers, and 30,700 communicants; the Baptists had 483 ministers and 67,183 communicants; the Episcopalians had 207 ministers, and about 10,000 communicants; the Associate Reformed had 30 ministers; the Lutherans had 27 ministers; the Roman Catholics had 32 ministers; the Universalists had 25 ministers; the Unitarians had eight ministers; and there were a few others.

There were in the state, Jan. 1st, 1843, 85 chartered and 46 free banks, with an aggregate capital of $43,950,137, and a circulation of $14,520,843. The state debt, over available funds in hand for its redemption, was, in Sept. 30th, 1843, $23,330,083. The valuation of real estate in 1842, was $504,254,029, and of personal estate was $116,595,233, making a total of both of $620,849,262.

There are two penitentiaries in the state, one at Auburn, 77 m. W. of Utica, which is considered as a model for all similar institutions; the other at Sing Sing on Hudson river, 35 m. above the city of New-York, on a similar plan.

New-York has taken the lead in the United States in works of internal improvement. The Erie canal extends from Albany 363 miles to Buffalo. It was commenced in July, 1817, and completed in 1825, and cost $7,143,790. The widening of the Erie canal has progressed considerably, but is at present suspended. If completed, it would more than double the expense. The Champlain canal, extending from its junction with the Erie canal at Waterford, 64 m. to Whitehall, on lake Champlain, cost $1,257,604. The Oswego canal extends from its junction with the Erie canal at Syracuse, 38 m. to Oswego on Lake Ontario, and cost $565,437. The Cayuga and Seneca canal extends from Geneva, at the outlet of Seneca lake, 21 m. to Montezuma, part the way by Seneca river, and cost $236,805. The Chemung canal connects the head of Seneca lake with Elmira on Chemung river, is 39 m. long, and cost $331,694. Crooked lake canal connects Crooked lake, near Penn Yan, with Seneca lake at Dresden, is 8 m. long, and cost $156,777. Chenango canal extends from the Erie canal at Utica to Binghampton on Susquehanna river, is 97 m. long, and cost $2,270,605. These canals are all completed. The Genesee Valley canal is to extend from Rochester 108½ m. to Olean on Alleghany river, below which it is navigable to Ohio river. It is finished 52 m. to Dansville, and has cost $1,309,292. Most of it is under contract, and the whole, it is estimated, will cost not far from $5,000,000. Black river canal is to extend from the Erie canal at Rome, to the High falls on Black river, 35 m., whence the navigation of the river is to be improved for 42¼ m. A navigable feeder of 10 m. extends from Black river to the canal at Boonville. The canal, with the feeder, is completed to the summit level, 14 m., and cost $446,841. The whole expense of the canal, aside from the river navigation, is estimated at $1,760,046, a small part of which is under contract. The Delaware and Hudson canal is the work of a private company, but the state loaned $800,000 to the company, for which it took stock. It extends from Rondout on Rondout creek in Ulster county, to Delaware river, 84 m., and thence extends 25 m. to Honesdale, and 16¼ m. by railroad, to the coal mines at Carbondale. The whole length is 125¼ m., part of which lies in Pennsylvania. The cost of the whole work was $1,875,000.

A line of railroads, under different names and companies, extends from Albany to Buffalo. A railroad extends from Buffalo to Niagara falls, and is extended to Lewiston, where it connects with a steamboat line to Oswego. A branch of this railroad extends to Lockport. A railroad extends from Schenectady to Ballston spa. Another extends from Troy to Saratoga, and another from Troy to Schenectady. A railroad of 16 m. in New-York and of 24 m. in Pennsylvania, extends from Corning in Steuben county, N. Y., to the Blossburg coal mines in Pa., with a total length of 40 m.; and more than 40,000 tons of bituminous coal are transported on this railroad annually, which, according to the demand, may be indefinitely increased. The New-York and Erie railroad is designed to extend 446 m., from Piermont on Hudson river, to Dunkirk on lake Erie. It is finished 46 m. to Goshen; and though something has been done on other parts, it is, at present, suspended for want of funds. The Harlem railroad extends from the city hall, New York, to Bronx river, 14½ m.: and it is designed to extend it to the Housatonic railroad in Connecticut, unless it can be continued directly to Albany. Unless a railroad can be extended from the city of New-York to Albany, much of the trade of the west will be diverted to Boston, which is connected by a railroad with Albany.

The governor is elected biennially by the people. He must be 30 years of age, be a native-born citizen of the United States, and have resided five years in the state. The lieutenant is elected in like manner, and must possess similar qualifications. He is president of the senate: and in case of the impeachment, removal, death, or absence, of the governor, discharges the duties of the office. The senate consists of 32 members, who are chosen for five years, one fourth of whom are elected every year. The assembly consists of 128 members, elected annually by the people. The governor nominates all judicial officers, except justices of the peace, and has the power of appointment, with the consent of the senate; and the judges hold their offices during good behaviour, or until they are 60 years of age. Every male white citizen of full 21 years of age, who has resided for one year in the state, and for six months preceding the election, in the county where he offers his vote, enjoys the right of suffrage. Persons of colour are allowed to vote, who have resided five years in the state, and who possess a freehold of £250, and have held it for one year previous to the election, and paid a tax upon it.

Henry Hudson, an Englishman in the service of the Dutch East India company, discovered Hudson river, and Manhattan island, in 1609. He entered the river, and sailed with his ship above the highlands, and sent a party of men in a boat to explore the river farther, who ascended above Albany. The Dutch merchants sent a company to trade with the natives, who in 1614 built a fort on Manhattan island, and another called fort Orange, on an island, just below Albany. These establishments were made for the purpose of trade with the Indians. In 1621 the Dutch West India company was incorporated, and directed their attention to forming settlements for the purpose of trade, and they laid claim to the country from Delaware to Connecticut river, on both of which they built forts, which involved them in controversy with the Swedes and the English. They also had severe contests with the Indians. Peter Stuyvesant, the most intelligent of the Dutch governors, established, in 1650, with commissioners of the New-England colonies, the boundaries between the Dutch and English colonies. The former relinquished all claim to Connecticut, excepting the lands which they actually occupied, and the latter retained the eastern part of Long Island. In 1664, Charles II. granted the country to his brother, the Duke of York, and the English, under Col. Robert Nichols, took possession of the country. Manhattan assumed the name of New-York, and fort Orange that of Albany, in honor of James, Duke of York and Albany. In 1673, an expedition from Holland captured the country from the English, but it was restored in 1674, by the treaty of Westphalia. From this time to the American Revolution, it was under governors appointed by the crown of Great Britain, with whom the people were often in controversy, and whose measures, when arbitrary, they were prepared to resist. The inhabitants bore an important part in the French and revolutionary wars. In September, 1776, after the disastrous battle of Flatbush on Long Island, the city fell into the hands of the British, who held it until the peace of 1783, when Washington marched the American army into it, in triumph. On the 26th of July, 1788, this state in convention adopted the constitution of the United States,—yeas 30, nays 25, majority 5.

NEW-YORK, city, the principal city of the state of New-York, and in population, wealth, and commerce, the largest city in the United States, deserves to be denominated the London of America. The City-hall is in 40° 42' 40" N. lat. and 74° 1' 8" W. long. from Greenwich, and 3° 0' 16" E. long. from W. It is 86 m. N.E. Philadelphia, 210 m. S.W. Boston, 225 m. N.E. Washington, 670 m. N.E. Charleston, 145 m. S. Albany, 372 m. S. Montreal, 1370 m. N.E. New-Orleans. The population in 1790, was 33,131; in 1800, 60,489; in 1810, 96,373; in 1820, 123,706; in 1830, 202,589; in 1840, 312,710. Employed in commerce, 11,365; in manufactures and trades, 43,390; in navigating the ocean, 2786; do. rivers and canals, 716; in the learned professions, [illegible].

The compact part of the city is situated on the S. end of New-York or Manhattan island, at the confluence of Hudson or North river, with a strait called East river, which connects Long Island sound with the harbour of New-York. The chartered limits of the city embrace the whole island, which is of the same extent with the county. The island extends from the Battery, on the S. point of the island, 13 m. to Kingsbridge, in its N. part; and has an average breadth of 1 m. and three-fifths. The greatest breadth is on the line of 88th street, where its breadth is about 2¼ m., and it contains about 14,000 acres. It is bounded on the N. by Harlem river or strait, the western part of which, from Kingsbridge to the Hudson, was named by the Dutch Spuyten Duyvel creek; on the E. by East river or strait, which separates it from Long Island; on the S. by the harbour; and on the W. by Hudson river, which separates it from New-Jersey. It is connected with the main land on the N. by three bridges, Harlem bridge, Macomb's bridge, and King's bridge. It is connected with Long Island by six ferries, four of which proceed to Brooklyn, and three to Williamsburg. Of the ferries to Brooklyn, the South ferry from Whitehall, New-York, to Atlantic-street, Brooklyn, is 1300 yards wide; the Fulton ferry, from Fulton-street, New-York, to Fulton-street, Brooklyn, is 731 yards wide ; the Catherine ferry, from Catherine-street, New-York, to Main-street, Brooklyn, is 736 yards wide ; the Navy Yard ferry connects Walnut-street, New-York, with Jackson-street, Brooklyn, and is 707 yards wide. Of the three ferries to Williamsburg, one is about 950 yards wide, and another, crossing East river obliquely, is about a mile and a half in length. Three ferries connect the city with New-Jersey, one to Jersey city, which is about a mile wide, and two to Hoboken, which are wider. On two ferries, boats are continually plying to Staten Island. Thus the insular situation of New-York proves no serious disadvantage. The ferries to Brooklyn are by far the most important, as many persons who do business in New-York reside in that city. These ferries are crossed in from four to six minutes, and at the charge of two cents, which most persons would prefer paying, to walking over a bridge.

The harbour is spacious and safe, the inner harbour extending 8 m. from the Narrows to the city, and several miles farther up both the North and East rivers, but particularly the latter. It is about 25 m. in circumference, and the largest vessels come to its wharves. Besides this, it has an outer harbour, extending from the Narrows to Sandy Hook, consisting of Raritan bay. Sandy Hook, on which is a light house, is 18 m. from the city ; and at this point, there are 27 feet of water on the bar at high tide, and 21 feet at low tide. Within Sandy Hook, there is a good anchorage. The harbour is entered not only from the ocean at Sandy Hook and through the Narrows, but on the N.E. from Long Island sound, and on the S.W. through the Kills and Staten Island sound. By an accurate enumeration made March 16th, 1844, there were found to be 1611 vessels in the harbour of New-York, viz., 121 ships, 43 barques, 101 brigs, two galliots, 208 schooners, and 536 ordinary sloops and schooners, all of which are licensed at the custom-house, lying at a total extent of about 7 m. of wharves. To these should be added, when the Hudson river opens, about 80 steamboats, 75 tow boats of from 100 to 400 tons burden, and 100 canal boats ; making a grand total of 1316 vessels of different descriptions, which will be increased as the season opens. Several islands within the inner harbour are attached to the city, which are Governor's, Bedlows, and Ellis's islands, on all of which are strong fortifications ; and Blackwell's, Great Barn, and Randall's islands, in the East river. Governor's island is 3080 feet distant from the city at the Battery, and includes 70 acres of ground. It contains fort Columbus, in the form of a star on the S. side of the island, and castle Williams on the N.W. point, which is a round tower, 600 feet in circumference and 60 feet high, with three tiers of guns. There is also a battery on the S.W. side, commanding the entrance through Buttermilk channel. At the Narrows, on the eastern shore of Long Island, are fort Hamilton and fort Lafayette, the latter of which (formerly called fort Diamond) is built on a reef of rocks, 200 yards from the shore. The Narrows, here about one third of a mile wide, has on Staten Island, on its W. side, fort Tompkins and fort Richmond, which are strong fortifications. The entrance from the sound on East river is defended by fort Schuyler, on Throg's neck. On Governor's island are houses for the officers, and barracks, which are occupied by a considerable garrison. On the whole, the harbour of New-York must be considered as well defended, and in a very different state from what it was during the last war with Great Britain. If even then they did not venture to approach New-York, much less would they do it now, in case of war, which it is hoped may not occur. After passing the bar at Sandy Hook, the channel to the city has a depth of water of from 35 to 50 feet. The average tide at the wharves is from 6 to 7 feet. Steamboats are constantly employed in towing vessels to and from the ocean. The safest and best portion of the harbour, and where the vessels chiefly lie, is along East river, where there is rarely any obstruction from ice. The tide sweeps through this part with a strong and rapid current ; and in the winter of 1844, when the harbour of Boston, and the W. part of Long Island sound were much obstructed by ice, no inconvenience was felt at New-York. The excellence of its harbour, and other great natural advantages, have contributed to make it the second commercial city on the globe.

The surface of the island was originally uneven and rough, as is now the case in the northern parts, with occasional low valleys and marshy swamps ; but the hills in the southern part of the island have been levelled, and the swamps and marshes filled up. Many creeks and inlets on the margins of the rivers have also disappeared, and the large ledge of rocks which occupied the site of the Battery, has long since been buried beneath the water ledge which constitutes that beautiful promenade. The water line has been materially altered from what it originally was. A large part of Water, Front, and South streets on the East river, and of Greenwich, Washington, and West streets on the Hudson river in the S. part of the city, occupy ground which has been made by filling in these rivers. The most elevated ground on the island is 238 feet above tide water. The city is built, extends over 3 m. on each river, and the compact part has a circumference of over 9 m. But its limits are continually extending, and will soon greatly surpass these boundaries. The streets were originally laid out according to the make of the ground, and some of them were crooked ; and in imitation of European cities, many of them were narrow. But in later times, they have been widened and improved, at a great expense. It seems a little barbarous at the time, to cut off 20 feet from valuable houses by the row ; but when these improvements have been effected, the proprietors themselves find occasion to be well satisfied with the change. By these improvements, many fine streets have been made of those which before were unsightly and inconvenient. As instances of this among others, John-street and West Broadway may be particularly mentioned, as well as many others. Had the forecast of William Penn been concerned in originally laying out the city of New-York, millions of dollars might have been saved. But it should be recollected that he had a more smooth and even surface to work upon. He expected that he was laying out a great city for posterity ; but the first inhabitants of New-York never expected that it would be what it now is, and is likely to become. In latter times care has been taken to lay out the streets straight, and of an ample width. This is particularly true of all the N. part of the city, which was laid out under the direction of Governor Morris, De Witt Clinton, and others, commissioners appointed by the legislature for this purpose, and surveyed by Mr. John Randall, Jr., completed in 1821, after having occupied ten years. No city can exhibit a more beautiful plan than this portion of the city of New-York, which extends to 154th street, about 10 m. N. of the battery.

Broadway extends from the Battery nearly 3 m. to Union-square, where it joins the fourth-avenue. It is 80 feet wide, perfectly straight, occupies the height of land between the two rivers, and has generally, particularly in its S. part, an excellent drainage. It is well built, with many fine houses and large retail stores. It is the great promenade of the city, and much resorted to in pleasant weather by the gay and fashionable. Pearl-street, between Broadway and the East river, is in a crescent form, over a mile long, and is the principal seat of the wholesale dry goods and hardware business, which has also extended into Cedar, Pine, and other adjacent streets. Water and Front streets, between Pearl-street and the East river, is occupied chiefly by wholesale grocers, commission merchants, and mechanics connected with the shipping business. South-street, extending along the margin of East river contains the warehouses and offices of the principal shipping merchants. In front of it is, at all times, a dense forest of masts. Wall-street extends from Broadway to the East river, and is occupied by banks, insurance offices, newspaper and broker's offices ; has Trinity church at its head, the Custom-house and Merchant's exchange, and many fine granite buildings, which has caused it to be sometimes denominated the granite street ; and is the centre of the heaviest money transactions in the United States. The Bowery is a wide and extensive street E. of Broadway, running N. and S., connected with the third avenue, which is Macadamized to Harlem, and forms the principal entrance to the city from the N.E. East Broadway, and parallel to it Madison, Henry, and Monroe streets, running a little N. of E. and S. of W., are broad and straight streets, and handsomely built. Bleecker, Bond, and other streets in the N. part of the city are beautifully built, and have become a fashionable place for residences. Canal-street, half a mile N. of the City-hall, and now much below the centre of the population, is a wide street with a huge covered canal under it, is occupied extensively by stores, and is the seat of an extensive retail business. It crosses Broadway, nearly at right angles, and extends to Hudson river. In the year 1800, the site of this street was extensively occupied by a large pond, called Fresh Water, or the Collect, which received the drainage of 400 acres of ground. This was the northern limit of Broadway in 1801, and much beyond the thickly settled part of the city. There are other streets which deserve a particular notice, especially Greenwich street, a long, wide, and nearly straight street, extending N. from the battery nearly two miles and a half, parallel to Hudson river, which has many stores and fine buildings ; and Hudson-street, W. of it, and parallel to it, which is wide and straight, extending from Chambers-street to the ninth avenue, over a mile and three quarters long, and well built in many of its parts. Chatham-street, named in honour of the Earl of Chatham, extending from Broadway to Chatham-square, at the commencement of the Bowery, is a great thoroughfare, and particularly distinguished for its numerous clothing stores. Grand, Broome, and Houston streets are extensive and important streets, N. of the centre of the city. But it is impossible to notice all the important streets.

In the harbour of New-York may be found vessels not

only from all parts of the United States, but from the principal commercial nations on the globe. Although it has a great extent of harbour, the vessels usually lie at the wharves along South-street, several tiers deep. The tonnage for the port entered in 1833, was 494,493; cleared 398,426. The imports in 1841 were $75,968,015; the exports were $30,731,519; and the duties paid at the port were $10,802,119. The imports and exports in 1843 were less, and the duties were more. In 1843 the imports were $50,308,590; the exports were $23,440,336; and the duties collected were $11,300,407. There is a line of steam-packets to Liverpool, consisting of the Great Western, to which will be added the Great Britain, a large iron steamer, in the spring of 1844. There are besides these, in several lines, 24 ships of the largest class, with a great capacity for freight, and elegant accommodations for passengers, one vessel of which sails for Liverpool every five days. Two lines of 12 large ships sail for London, one vessel every ten days. Two lines, one of 19 and the other of four large ships, making 16 in the whole, sail for Havre, France, every eight days. Lines are also established to all the principal ports of the United States, the West Indies, and South America. There are also steamboat lines to Hartford, New-Haven, Bridgeport, Norwich, and New-London, Stonington, and Providence, besides the numerous steamboats to Albany, Troy, and the various places on Hudson river. The arrivals from foreign ports in 1843 were 1832. Of these 409 were American ships, 183 barques, 515 brigs, and 298 schooners; eight were British ships, 18 barques, 184 brigs, and 56 schooners; 16 were Bremen ships, 25 barques, nine brigs, and three schooners. The remainder were from Sweden, Hamburg, France, and various other countries, making in the whole six steamers, 439 ships, 239 barques, 789 brigs, eight galliots, 355 schooners, and three sloops. The number of passengers in 1843 was 46,302, of whom 341 were Americans returning to their own country. In addition, there were in 1843, 4737 arrivals of coasters, which, added to 1832 arrivals from foreign countries, makes the whole number of arrivals in the year to be 6566; which is 801 more than in the year 1842.

The city must be considered somewhat deficient in public grounds or places, but it has several important ones. In addition to several triangular areas, as Hanover-square, Franklin-square, and Chatham-square, as they are denominated, with some others of a like description, there are several more important public places. *The Battery*, at the south-eastern end of the island, is situated at the junction of Hudson and East rivers. It is in the form of a crescent, and contains about 11 acres of ground, beautifully laid out with grass-plats and gravelled walks, and shaded with trees. On the side next to the city, is an iron-railing, and from this ground is a fine view of the bay, with its islands, and the adjacent shores of New-Jersey, and Long and Staten islands; and this view is generally enlivened by shipping under sail. *Castle Garden* is built on a mole, and connected with the battery by a bridge. It was originally erected as a fortification, and having become unnecessary for this purpose, was ceded by the United States to the corporation of the city, in 1823. Within its walls, 10,000 people can be accommodated in a great amphitheatre; and it is used for public meetings and exhibitions.

The *Bowling Green*, at the southern termination of Broadway, is an elliptical area, 290 feet long and 140 broad, enclosed by an iron fence. It was established before the American Revolution, and formerly contained a leaden statue of George III., which was converted into bullets at that period. It contains in its centre a public fountain, supplied by the Croton Waterworks. In the centre of a large basin, a rude pile of huge flat stones, in a somewhat circular form, about 15 feet in diameter, and as many feet high, has a jet from its centre, and small jets around it; and when it is in action, presents, by the water pouring down its sides, a wild and picturesque appearance.

The *Park*, called in early times the *Commons*, is a triangular area of about 10½ acres, lying between Broadway, Chatham, and Chamber streets, is laid out with trees, and planted with trees, and surrounded by an iron fence, which cost over $15,653. It contains the City-hall, the New City-hall or old Almshouse, and the Postoffice. It has also toward its S. part, a public fountain, within a basin about 100 feet in diameter, which has a variety of jets, which are occasionally changed. When the water is thrown in a single stream, it ascends to the height of about 70 feet, presenting a majestic and interesting appearance. It is designed to be made yet more ornamental.

Hudson-square, or *St. John's Park*, is private property, belonging to Trinity church, which, however, has been reserved as a permanent public square. It is between Beach, Laight, Varick, and Hudson streets, is encompassed by elegant buildings, beautifully laid out with walks, and shaded with trees, and surrounded by an iron fence, which cost $36,000. It contains about four acres of ground, and has on its E. side St. John's Episcopal church, of beautiful propor-

tions, with a lofty spire. It has a public fountain, and is, perhaps, the most ornamental spot in the city.

Washington-square, a mile and a half N. of the City-hall, between Wooster and M'Dougal streets, contains about 9½ acres of ground. Two thirds of this area was the Potter's field until 1827, and the remainder was purchased by the city for about $78,000, and the whole was enclosed by a wooden fence, which cost nearly $3000, laid out in walks, and planted with trees. On the E. side it has the fine building of the New-York university, and an elegant Dutch church, both handsome specimens of Gothic architecture. On the N. and S. sides it has blocks of fine buildings, and is destined, when the trees shall be fully grown, to be a highly ornamental ground.

Union-place, at the northern termination of Broadway, is in an elliptical form, enclosed with a fine iron fence, having a public fountain in the centre, with ornamental jets; and when the vicinity shall be more densely settled, will be a delightful breathing place to the inhabitants. All these public grounds are much frequented in the summer season. Farther up the city are other public squares, as Madison-square, Hamilton-square, and others, not yet regulated. On the E. are Tompkins-square and Bellevue, the latter the seat of the new almshouse.

The city of New-York has some superb public buildings. The most splendid of these is the *Merchant's Exchange*, which covers the whole space between Wall, William, Exchange, and South William streets. It has a somewhat confined situation, and shows to less advantage than if it were surrounded by open grounds. It is built in the most substantial form of blue Quincy granite, and is 200 feet long by 171 to 144 feet wide, 77 feet high to the top of the cornice, and 124 to the top of the dome. The front on Wall-street has a recessed portico of 18 massive Grecian-Ionic columns, 38 feet high and 4 feet 4 inches in diameter, each formed from a solid block of stone, and weighing 43 tons. It required the best application of the mechanical power, aided by horses, to raise these enormous masses. Besides numerous other rooms for various purposes, the Exchange in the centre is in a circular form 80 feet in diameter, with four recesses, making the length and breadth each 100 feet, the whole 80 feet high, surmounted with a dome, resting in part on eight Corinthian columns of Italian marble, 41 feet high, and lighted by a skylight, 25 feet in diameter. On the S. side of the roof is a telegraph, which communicates with another on Staten Island; and an hourly report is sent down from the telegraph to the newsroom, for public inspection. When it is recollected that this fine building has been erected in the place of an elegant exchange building burned in the great fire in 1835, it is a matter of congratulation that this building is absolutely incombustible, having no wood but the doors and window frames used in its construction. The cost of this building, including the ground, is estimated at $1,800,000.

The *Custom-house* is a splendid building, constructed in the Doric order of Grecian architecture. It is built in the most substantial manner of white marble, something after the model of the Parthenon at Athens, at the head of Broad-street, on the corner of Nassau and Wall streets. It occupies the site of the old Federal hall, in the open gallery of which General Washington was inaugurated; and nearly over the front door is the place where he stood, when the oath of office, as first president of the United States, was administered to him by Chancellor Livingston, April 30, 1789. Would not this be a fine spot in which to place a group of statuary, representing this interesting transaction? The building is 200 feet long, 90 feet wide, and 80 feet high. At the S. end on Wall-street, is a portico of eight pure Grecian-Doric columns, 5 feet 8 inches in diameter and 32 feet high; and on the N. end on Pine-street, is a corresponding portico of similar columns. The front portico is ascended by 18 marble steps, and the rear portico on Pine-street by only three or four marble steps. It is two lofty stories high above the basement story. The great business hall is a splendid circular room, 60 feet in diameter, with recesses and galleries, making it 80 feet in diameter, surmounted by a dome, supported by 16 beautiful Corinthian columns, 30 feet high, ornamented in the dome with stucco, and at top with a skylight. On each side, on the outside, are 13 pilasters, in perfect keeping with the pillars on the two fronts. The cost of the building and its furniture was $950,000, and including the ground, $1,175,000. It has other large apartments than those specified, for various purposes connected with the business of the Custom-house. It is absolutely incombustible, and may be expected to stand to a late posterity, an honour to the commercial capital of the nation.

The *City-hall*, heretofore regarded as much the finest building in the city, and one of the finest in the United States, has a commanding situation in the middle of the Park, though somewhat in the rear, and shows to greater advantage than either of the fine buildings already described. It has more ornament than either the Exchange or the Custom-house, but less simple grandeur; though with its furniture, it is, perhaps, the most interesting building in the city. It is 216 feet long and 105 wide. The front and ends are constructed of white marble, and the rear of brown freestone. It is two stories high above the basement, with a third or attic story in the centre building; and there rises from the centre a lofty cupola, containing a city clock, of fine workmanship, and on the top, a colossal statue of Justice. In the upper part of this cupola is a room occupied by a man whose business it is to give alarm in cases of fires; and from this elevated position, he is able to overlook the whole city. Behind this is another less elevated cupola, with eight beautiful Ionic columns, which contains the City-hall bell, weighing 6910 pounds, whose deep and solemn tones often sound the knell of property, and by the different number of strokes, indicate the district of the city in which a fire occurs. The front of the City-hall is ornamented with columns and pilasters of the Ionic, Corinthian, and Composite orders, rising above each other in regular gradation. The building is entered in front by a flight of 12 marble steps. In the centre is a double staircase, ascended by marble steps, at the top of which is a circular gallery, floored with marble, from which 10 marble columns of the Corinthian order ascend to the ceiling. where is a handsome panelled dome ornamented with stucco, and a skylight, which gives light to the interior of the building. There are halls which lead from the centre to each end of the building in each story. In the basement and the stories there are 29 offices and other public rooms, the most conspicuous of which are the Governor's room, and the chambers of the common council and assistant aldermen. The *Governor's room* is appropriated to the use of the governor of the state when he visits the city, and has been used as a reception room for other distinguished men who have occasionally been here. It is 52 by 30 feet. The walls of the room are hung with a fine collection of portraits, including the governors of the state, the mayors of the city since the Revolution, some of the Dutch governors, and the principal military and naval heroes of the late war, all which are regarded as excellent likenesses, and many of them are full-length portraits. The *Common Council room* is 42 by 30 feet, and the president occupies the identical chair occupied by General Washington when he presided over the first American Congress, which assembled in New-York. It is surmounted, as is meet, by a canopy. The seats of the aldermen are ranged in a semicircular form, in the centre of which is a table for the clerk The room contains several fine full-length portraits painted by Trumbull, of which that of Washington is thought to be the best in existence, when he was in the prime of life. The room of the assistant aldermen is handsomely fitted up. The *Superior Court room* is very neat and convenient, 42 by 30 feet, and neatly fitted up for its purpose. But he who looks over the several apartments of this building will obtain a higher idea of it, than he can from any concise description. The building was commenced in 1809 and completed in 1812, and cost $538,734. To the E. of the City-hall, in the park, is the Hall of Records, two stories high, with a lofty portico of four Ionic pillars on each front; and in the rear of it is the New City-hall, which was the old almshouse, and contains 15 offices, besides the Marine Court-room, and the rooms of the American Institute, the latter containing a valuable library and various interesting models of machines.

The *Hall of Justice* occupies the whole space between Centre, Elm, Leonard, and Franklin streets, and is a unique and beautiful building of the Egyptian architecture. It is 253 feet long and 200 feet wide, the front of which is occupied by the main building, and other outer portions of which consist of lofty walls, with apartments in some of their parts. Within this enclosure, and toward its back part, is the house of detention or prison, 142 feet long and 45 feet wide, which is entirely separate from the outer wall and building, and consists of 148 cells for different classes of prisoners. The court of sessions occupies a part projecting back from the front building, the roof of which is supported by lofty Egyptian columns, and it has a gallery. The front of the building is entered by eight steps, leading to a portico of four massive Egyptian columns. From this there is an ascent by twelve steps, between two massive columns, to an open area of 50 feet square, which has eight large columns supporting the ceiling above. From this area there is an entrance to the various offices and apartments of the building. The windows, which extend to the height of two stories, have massive frames, and cornices ornamented with the winged globe and serpents. The two fronts on Franklin and Leonard streets have each two entrances, with two massive columns each; and the back entrance forms a carriage way, for taking prisoners to and from the house of detention. This building, though handsome of its kind, has a heavy and gloomy aspect, which has acquired for it the name of the *Egyptian Tombs*. It is constructed of a light coloured granite, from Hallowell, Me.

Among the churches, some deserve to be particularly noticed, on account of their architecture. The new Trinity Episcopal church will, when completed, be one of the finest buildings of the city, and the most complete Gothic structure in the United States. It stands at the head of Wall-street, which it fronts. It is in the finest style of English church architecture, built of a light-brown freestone, with much beautifully ornamented sculpture in its various parts; is 192 feet long upon the outside, 84 feet wide, and the steeple, when completed, will be 264 feet high, built of stone to the top. Above the first story is a roof which considerably narrows the building in the second story, as is common in the old churches of England. The inside will be even more imposing than the outside. The situation is fine, and it is only to be regretted that it could not have stood farther back from the street. St. Paul's Episcopal chapel is situated on Broadway, between Fulton and Vesey streets, and the burying ground extends W. to Church-street, and thus includes the whole block, 400 by 180 feet, surrounded by a handsome iron railing. The body of the church is 90 by 78 feet, with a beautiful spire, 200 feet high, painted and overcast with sand to resemble brown freestone. The church was erected in 1765, and the steeple in 1794. It has a beautiful portico of four Roman Ionic fluted columns of brown stone, supporting a pediment with a niche in the centre, containing a statue of St. Paul, painted white. St. John's Episcopal chapel is situated on Varick street, directly opposite the centre of St. John's-square, is a building of fine proportions and beautiful appearance, was finished in 1810, and cost over $200,000. It is 111 feet long and 73 wide, built of stone, and has an admirably proportioned steeple, 220 feet high, at present the finest in the city. It has in front a splendid portico of four Corinthian columns. St. Thomas's Episcopal church in Broadway, corner of Houston-street, is a large and imposing building, 62 feet wide by 113 feet deep, built of stone. It is in the Gothic style, and has two octagon towers, one on each of the front corners, 80 feet high, with pointed turrets, and an immense Gothic window in front between them. One or two other Episcopal churches, recently erected in the upper part of the city, deserve particular notice. Of the Presbyterian churches, though many of them are peculiarly neat and convenient, few of them have anything imposing in their external appearance. The Scotch Presbyterian church, corner of Grand and Crosby streets, is a stone building, 95 feet long and 67 broad, with a fine Ionic portico of six stone columns, and cost $114,000. The venerable brick church, corner of Nassau and Beekman streets, built in 1767, has been recently and thoroughly repaired, at a great expense, and has a lofty and well-proportioned steeple. The Rutgers-street church is a fine stone edifice, with a lofty square tower. The Duane-street church has an imposing Ionic portico. The first Baptist-church in Broome-street, corner of Elizabeth-street, is a fine stone edifice of Gothic architecture, is from 88 to 110 feet long, and from 75 to 87 feet wide, with two octagon towers on the front corners, and a pointed window between them, 62 ft. wide and 41 ft. high. The interior is more imposing than the exterior. St. Peter's Roman Catholic church in Barclay-street, corner of Church, is a large and substantial granite structure, with a very imposing Ionic portico of six granite columns, and a statue of St. Peter in a niche in the pediment. The French Protestant Episcopal church, corner of Franklin and Church streets, is built of white marble, and has a portico with a double row of four marble columns of the Ionic order. The Reformed Dutch church on Washington-square is a large and imposing structure of Gothic architecture, and appears well, even by the side of the splendid New-York university. St. Patrick's Roman Catholic cathedral is of stone, 120 feet long and 80 feet wide, but is more distinguished for its magnitude than for its elegance, though it is an imposing structure. Some churches recently built in the upper part of the city have a fine appearance, and are expensive and commodious buildings. Perhaps there is no circumstance that more strikingly shows the progress of the population toward the N. part of the city than the removal of churches. Murray-street church was built in 1812, was 92 by 77 feet, with a lofty and beautiful cupola, and was an ornament to the lower part of the city. It was in good repair, and would have stood, without alteration, well for 100 years. Yet this building was taken down in 1842, and removed a mile and a half N. of its former site; and with such care, that it presents a front and cupola which is at once recognised, and with pleasure, by those who have formerly admired its beautiful proportions, while the inside has been much improved. A new and splendid Episcopal church has been commenced to supply the place of Grace church in Broadway, near Trinity, two miles and a half N. of its present site; and the present church will be taken down, though a good building, when the new one is completed. And it is said that Wall-street church, and perhaps others, in the lower part of the city, will, before long, share the same fate. But in all cases, the new structures will greatly exceed the old, in convenience and magnificence.

New-York contains several important literary institutions. The oldest is Columbia college, chartered by George II. in 1754, by the name of King's college, and confirmed, with the necessary alterations, by the legislature of New-York, in 1787. It has a president and 10 professors, 1170 alumni, 100 students, and 14,000 volumes in its libraries. The commencement is on the day after the first Monday in October. The building is situated on a beautiful square between Chapel, Church, and Murray streets, is 200 feet long and 50 feet wide, with two projecting wings, one at each end, in which are accommodations for the families of the professors. It contains a chapel, lecture rooms, hall, museum, and an extensive philosophical and chemical apparatus. The students do not reside in the building, as is common in many other colleges. They are considered safer under the immediate eye of their parents or guardians. The funds amount to about $200,000; and the annual income to $7000 or $8000. There is a flourishing grammar school attached to the institution, over which one of the professors presides, as rector. The university of the city of New-York stands on Washington-square, between Washington-place and Waverley-place, has a fine edifice of white marble in the Gothic style of English collegiate architecture, and is an ornament to the upper part of the city. It is 180 feet long and 100 wide, with a centre building and wings, and an octangular turret on each of the four corners. The chapel, a highly finished room, receives its light from a window of stained glass in the W. front, 24 feet wide and 50 feet high. The principal entrance is through a richly moulded and deeply recessed portal, with doors of oak richly panelled and ornamented. The wings are four stories high, and the corner towers five stories high. The whole building is a fine specimen of architecture, and presents a beautiful appearance, especially from the large square which spreads out before it, and shows it to great advantage. This institution was founded in 1831, has a president and 11 professors, has in the collegiate department 145 students, and a valuable library and philosophical apparatus. Connected with it, is an extensive grammar school, superintended by a professor, and a flourishing medical department, the whole of which contain 680 students, viz., undergraduates, 145; medical, 323; grammar school, 212. The general theological seminary of the Episcopal church, corner of ninth avenue and 21st street, contains two handsome buildings of stone, was founded in 1819, has five professors, 74 students, and 7200 volumes in its library. The New-York Theological seminary was organized in 1836, has six professors, 104 students, and 12,000 volumes in its library. The Public School society had on May 1st, 1843, 16 schools, with male, female, and primary departments, besides two schools for coloured children; and 48 primary schools, besides five for coloured children. The number of scholars in all these schools was 20,186. Besides these, under the late law of the state, there are established public district schools, in different parts of the city, in which are numerous children, and these schools are spoken of by the superintendent of schools, as well taught and flourishing. What the final result of these two separate organizations will be, remains yet to be seen. The Rutger's Female Institute in Madison-street, has a fine granite building, a valuable library and philosophical apparatus, and 450 students, ranged in a number of distinct apartments, under separate teachers. The Mechanics' school in Crosby-street has a number of teachers, and 550 pupils. The Protestant Episcopal school has several teachers, and a large number of scholars.

The College of physicians and surgeons of the city of New-York has a handsome edifice in Crosby-street, near Spring-street, and was founded in 1807, has eight professors and about 100 students. The lectures commence on the first Monday in November, and continue four months. Degrees are conferred by the regents of the university, at the recommendation of the board of trustees. The whole expense of the course is about $100. The New-York Hospital has a collection of extensive buildings in a handsome situation, with a fine yard, on Broadway, at the junction of Pearl-street, has a large number of physicians and surgeons attached to it, and over 200 patients. The Eye Infirmary has four surgeons attached to it, relieves a large number of applicants, and is a useful institution. The New-York Lunatic Asylum, connected with the New-York Hospital, is located at Bloomingdale, near Hudson river, has a large and fine building, attached to which are 40 acres of ground, tastefully laid out in gardens, pleasure grounds, and gravelled walks, and is one of the most elevated sites on the island, from which is a fine view of Hudson river and the surrounding country. The principal edifice is of stone, 210 feet long and 60 feet wide, cost, with its grounds, over $200,000, and contains about 150 patients. The Deaf and Dumb Asylum is on 50th street, near the fourth avenue or Harlem railroad, and has a building 110 feet long and 60

feet wide, which will accommodate a large number of pupils, with the teachers and family of the principal. It had, January 1st, 1844, 173 pupils. It is under a principal and eight professors. The Institution for the Blind is on the eighth avenue, and has about 70 pupils.

The New-York Society Library is an old institution, founded in 1754, has a handsome and substantial edifice of brown freestone on Broadway, corner of Leonard-street, with six Ionic columns resting on the basement story. It is among the interesting buildings of the city, and besides spacious accommodations for the library, has a handsome and commodious lecture room, and the rooms of the Academy of Design. The library is open on every week day with a few exceptions, and contains nearly 40,000 volumes. The Historical Society, located at the university, has a valuable library of 12,000 volumes, besides a collection of coins and medals. It has published several volumes of historical collections. The National Academy of Design, instituted in 1826, has purchased the statuary of the Academy of Fine Arts, and exhibits annually a large collection of paintings by living artists, which are visited by great numbers of persons. The same painting is not allowed to be exhibited twice, so that the collection is always new. Clinton-hall Association was founded in 1830, for the promotion of literature, science, and the arts, and is the proprietor of Clinton-hall, in which the Mercantile Library is located. The Mercantile Library Association was formed for the special benefit of merchants' clerks, and is one of the most useful institutions of its kind in the city. It has a library, after deducting for volumes lost, of 20,567 volumes. About 4500 volumes have been lost since its establishment in 1820. It has a beautiful reading room, open on week days, well warmed and lighted, and much frequented. It sustains in the winter season an interesting course of literary and scientific lectures, for which some of the best talents in the country are put in requisition. The Apprentices Library in Crosby-street has a library of 19,000 volumes, read by 1800 apprentices, and exceedingly useful to young mechanics. The American Institute, incorporated in 1829, for the encouragement of agriculture, commerce, manufactures, and the arts, occupies rooms at the W. end of the second story of the New City-hall, where it has a valuable library and reading room, and interesting models of machinery. It holds an annual fair at Niblo's garden, where is exhibited a splendid array of the fruits of American ingenuity and industry. It is one of the most interesting exhibitions in the city, and is visited by not fewer than 20,000 persons yearly. By the distribution of premiums, it gives great encouragement to agriculture and the mechanic arts. The New-York Lyceum, founded in 1838, sustains, in the winter season, an able course of lectures.

There are many religious charitable institutions which have their centre in New-York. The American Bible Society, founded in 1816, received for the year ending May 24, 1843, $196,448, and has a large edifice in Nassau-street; the American Tract Society, founded in 1816, received for the year ending April 15th, 1843, $96,940. This society has a large brick building in Nassau-street. The American Board of Commissioners for Foreign Missions, received $244,681; the American and Foreign Bible Society (Baptist), received $23,638; the Baptist Home Missionary Society, received $40,583. Most of these societies hold an anniversary in the city in May.

There are 185 churches in New-York, viz., 31 Presbyterian, 29 Episcopal, 21 Baptist, 26 Methodist Episcopal, 9 Dutch Reformed, three Lutheran, five Congregational, two Protestant Methodist, four Friends, three Welsh, two Unitarian, four Universalist, 13 Roman Catholic, nine African, six Jews' synagogues, and 13 various.

There are 26 banks in the city of New-York, with an aggregate capital of $27,903,200; several marine insurance companies, with a total capital of about $3,008,800; 22 fire insurance companies, with an aggregate capital of about $6,000,000, besides several mutual insurance companies. There are four savings banks. There are 15 markets, of which Fulton, Washington, Catharine, Essex, Clinton, Tompkins, and Franklin, are the principal. There are six theatres, of which the Park, Bowery, and Chatham are the principal, and two museums, the American and Peale's. There is also a circus, and there are occasionally performances at Niblo's garden, somewhat resembling a theatre.

There are in New-York several splendid hotels, of which the Astor House is the most remarkable, a mammoth granite building, said to be superior to anything of the kind in London or Paris; Howard's Hotel, little less extensive; the City Hotel, a very extensive brick building; the Franklin House; the United States Hotel, of white marble, six stories high; the American Hotel; and many others equally commodious and extensive, some of them elegant hotels and boarding houses united.

The number of buildings erected in the city in 1843, was 1773.

In 1842, the aggregate of the city expenses was $1,043,719;

of which there were for the public schools, $96,671; for the poor, $228,000; for the watch, $295,567; for the fire department, $51,887. The city debt was $14,790,424. The amount collected by taxation for city purposes, was $1,476,876.

A large number of the streets, stores, and other buildings are lighted with gas. The works of the New-York Gas Light Company are situated at the corner of Canal and Centre streets, and nearly 30 m. of iron pipes are laid down. The expense of gas for a store with four lights is about $60 a year. But the most splendid and expensive public work undertaken by the city is the Croton Waterworks. It was at first estimated that it would cost five or six millions of dollars; and at the city charter election of 1835, the citizens were required to vote for or against supplying the city in this way. The whole number of votes given was 17,330, of whom 5983 were against it, and 11,367 in favour of it. It is probably happy that they did not then know how much it would ultimately cost; but after having experienced its great advantages, no one now regrets its construction. The aqueduct commences at the Croton river, 5 m. from Hudson river, in Westchester county. The dam is 250 feet long, 70 feet wide at bottom, and 7 feet at top, and 40 feet high, built of stone and cement. It creates a pond 5 m. long, covering a surface of 400 acres, and contains 500,000 gallons of water. From the dam the aqueduct proceeds, sometimes tunnelling through solid rocks, crossing valleys by embankments, and brooks by culverts, until it reaches Harlem river, a distance of 33 m. It is built of stone, brick, and cement, arched over and under, 6 feet 3 inches wide at bottom, 7 feet 8 inches at the top of the side walls, and 8 feet 5 inches high, has a descent of 13¼ inches per mile, and will discharge 60 millions of gallons in 24 hours. It will cross Harlem river on a magnificent bridge of stone, 1450 feet long, with 14 piers, eight of them 80 feet span, and seven of 50 feet span, 114 feet above tide-water to the top, and will cost $900,000. This bridge is in progress; and for the present the water is brought across the river in an iron pipe, laid as an inverted syphon. The receiving reservoir is at 86th street, 38 m. from the Croton dam, and covers 35 acres, and contains 150 millions of gallons. The water is conveyed to the distributing reservoir on Murray's hill, in 40th street, in iron pipes. The distributing reservoir covers 4 acres, and is constructed with stone and cement, 43 feet high above the street, and holds 20 millions of gallons. Thence the water is distributed over the city in iron pipes, laid so deep underground as to be secure from frost. The whole cost of the work will be about $12,000,000. The water is of the purest kind of river water. There are laid below the distributing reservoir in 40th street, 150 m. and 2685 feet of pipe, from 6 to 36 inches in diameter, the majority of which is from 6 to 12 inches in diameter; and free hydrants are opened in most of the streets, besides the fire hydrants. There are 1490 fire, and 690 free hydrants. But little inconvenience has been felt during the cold weather of the past winter to the hydrants, and that inconvenience can be easily remedied. No city in the world is now better supplied with pure and wholesome water than the city of New-York, and the supply would be abundant, if the population were five times its present number. The water can be freely expended for washing the streets, which will contribute to the health of the population, who are not now satisfied with the water from the pumps. You might as easily persuade the drunkard from his cups, as to choke the citizens off from the Croton water.

The Harlem railroad extends from the City-hall through Centre-street to Broome-street, where it turns at right angles to the Bowery, and there it again turns nearly at right angles, and follows the Bowery to the fourth avenue, on which it proceeds to Harlem, 8 m., and crossing Harlem r., it is extended into Westchester county as far as where the line crosses Bronx river. A part of this course is a deep cut through solid rock, with a tunnel 595 feet long, 24 feet wide, and 21 high to the crown of the arch, and a high embankment. It has a double track the whole length; and so Harlem cost $137,500 per mile, being by far the most expensive railroad for its length now in the United States. Within the thickly settled parts of the city the cars are moved by horses, and beyond that, by locomotives. The cars start from the City-hall once in 15 minutes, and it affords a pleasant ride, as through a considerable part of its course it presents a fine view of the East river, and of the surrounding country.

There are not more than five or six cities in Europe more populous than New-York, viz., London, Paris, Constantinople, St. Petersburgh, Naples, and perhaps Vienna.

According to the census of 1840, there were in the city 417 commercial and 918 commission houses engaged in foreign trade, with a capital of $45,941,300; 3620 retail dry goods and other stores, with a capital of $14,648,595; four furnaces employed 56 persons, consumed 4500 tons of fuel, produced 8900 tons of cast iron, with a capital of $23,000; the pro-

ducts of the dairy were $22,400, of the garden for market, $66,640, of the nursery and florists, $24,650; 61 lumberyards employed 2605 persons, and a capital of $731,500; 398 persons employed in internal transportation, with 43 butchers, packers, &c., employed a capital of $648,780; 1419 persons produced machinery to the amount of $1,150,000; 145 persons produced hardware and cutlery to the amount of $135,300; 30 persons produced 1462 small arms; 542 persons manufactured the precious metals to the amount of $202,700; 798 persons manufactured various metals to the amount of $1,087,800; 472 persons manufactured granite and marble to the amount of $403,850; 18 persons produced bricks and lime to the amount of $97,000; 248 persons engaged in the manufacture of wool employed a capital of $30,000; 18 cotton manufactories, and two dyeing and printing works, employed 290 persons, produced to the amount of $150,708, with a capital of $61,300; mixed manufactures employed 1653 persons, produced to the amount of $1,201,700, with a capital of $507,050; 309 persons manufactured tobacco, producing to the amount of $187,700, with a capital of $53,055; hats and caps were manufactured to the amount of $1,031,346, and straw bonnets to the amount of $198,200, the whole employing 1361 persons, and a capital of $444,300; one tannery produced 2000 sides of sole leather, and 3500 sides of upper leather, employing 507 persons, and a capital of $30,000; 217 other manufactories of leather, as saddleries, &c., produced to the amount of $1,509,156, with a capital of $545,730; 229 persons produced 7,813,700 pounds of soap, 2,003,400 pounds of tallow candles, and 250,000 pounds of wax or spermaceti candles, with a capital of $277,600; 11 distilleries produced 2,973,278 gallons of distilled spirits, and 15 breweries, 1,205,490 gallons of beer, the whole employing 274 persons, and a capital of $575,076; 293 persons produced paints and drugs to the amount of $225,050, and turpentine and varnish to the amount of $161,360, with a capital of $648,650; three glass houses, and six glass-cutting establishments, employed 104 persons, produced to the amount of $127,131, with a capital of $53,000; one pottery employed 12 persons, produced to the amount of $4000, with a capital of $3000; seven sugar refineries produced to the amount of $385,000, and confectionaries produced to the amount of $249,242, the whole employing 397 persons, and a capital of $425,706; one paper mill produced to the amount of $25,000, and other manufactories of paper, as playing cards, boxes, &c., produced to the amount of $20,137, the whole employing 51 persons, and a capital of $27,900; six rope-walks, employing 61 persons, produced cordage to the amount of $24,000, with a capital of $9996; 281 persons produced musical instruments to the amount of $353,531, with a capital of $207,874; 297 persons produced carriages and wagons to the amount of $207,874, with a capital of $90,950; two grist-mills and eight saw-mills, employed 63 persons, produced to the amount of $331,800, with a capital of $146,800; ships were built to the amount of $354,000; 1319 persons manufactured furniture to the amount of $916,675, with a capital of $596,150: 542 brick or stone houses, and 50 wooden houses were built, employing 4033 persons, and cost $1,889,100; 113 printing-offices, 43 binderies, 18 daily, 45 weekly, and five semi-weekly or tri-weekly newspapers, and 28 periodicals, employing 2025 persons, and a capital of $1,965,280. The total amount of capital employed in manufactures was $11,228,894.

There were four colleges, with 430 students, 146 academies or grammar schools, with 7207 students, and 209 common or primary schools, with 32,867 scholars.

Hudson river, and Manhattan or New-York island, were discovered by Henry Hudson, an intrepid English navigator, September 3d, 1609. He had previously explored the North sea, in the fruitless attempt to discover a north-west passage, and Hudson's bay received its name from him. Although Sebastian Cabot had previously discovered the coast, he knew nothing of Hudson river. Henry Hudson set sail from the Texel in a vessel called the Half Moon, navigated by a crew of 20 men, English and Dutch; and, after doubling the cape of Norway, proceeded toward Nova Zembla, until, being impeded by ice, he determined to proceed S. toward Virginia, in doing which he discovered and explored the harbour of New-York, and the river which bears his name, which he penetrated with his ship as far as he thought prudent, and thence in a boat above Albany. He returned to Dartmouth, in England, November 7th, 1609, whence he sent an account of his discoveries to the Dutch West India company, in whose employ he sailed. The point of the island where New-York now stands, he found possessed by the Manhattans, a brave and savage tribe: the Indians on the E., or Jersey shore, were more friendly; but were the deadly enemies of the Manhattans. The Dutch West India company sent a second vessel to Hudson river for trade in 1610: and, finding the Indians more friendly in that quarter, they obtained permission of the natives to build a small fort on an island lying a little below Albany, on the W. side of the river. In 1612 the Dutch had a fort on York island, which consisted of a redoubt near the corner of Garden-street and Broadway, overlooking Hudson river. In 1614 an expedition from Virginia, under Captain Argal, took possession of New-Amsterdam, as New-York was then called: there were then but four houses outside of the fort. An arrangement was soon after made with the English government, by which the Dutch remained in the peaceable possession of the place for 50 years. The establishment was made for the purpose of trade, which they successfully prosecuted with the Indians, receiving furs in exchange for beads, trinkets, blankets, and hatchets. The Dutch had frequent quarrels with the New-England colonies on Connecticut river, and the Swedes on Delaware river; the former claiming all the country between these two rivers. The Dutch were not able to obtain permission of the Manhattans to build a fort on the island for some time; but in 1623 they obtained leave to build a better one than that which previously existed, and made a purchase of the present site of the S. portion of New-York, and erected a fort. Most of the settlers resided in the fort; but, the colony increasing, some houses were built on the outside near it, which formed the commencement of Pearl-street. The fort was in a square form, with four bastions, at the junction of Hudson and East rivers, near the present site of the Bowling-green and the N. part of the Battery. It was from time to time strengthened, by building additional walls on the outside of the first wall. It contained the houses of the Dutch director-general, the commandant, and other officers. The Dutch resolving to establish a permanent colony at New-Amsterdam, in 1629 appointed Wouter Van Twiller governor, who held the office for nine years. In 1635 he erected a more substantial fort, with four bastions, which mounted 42 cannons, mostly brass, 12 and 18 pounders. In 1643 a church was built in the S.E. corner of the fort. This church was 72 feet long, 52 wide, and 16 feet high. The governor's house, also within the fort, was 100 feet long, 50 feet wide, and 24 feet high. These buildings were burned in 1741, and not afterward rebuilt. It appears that, in 1638, tobacco was produced to a considerable extent on the island, and negro slavery had been introduced. In 1644 the City-hall, or Stadt-house, or tavern, was built, on the corner of Pearl-street and Coenties-slip, and was a very important house in those days, being the place where the courts and the public meetings of the citizens were held. May 11th, 1647, Governor Stuyvesant, the last of the Dutch governors, arrived, and held the office for 17 years, until the colony was captured in 1664. He was a military character, and had lost a leg in the capture of Tobago. In 1652 the first public school was established in the city. In 1653 a wall of earth and stones was built from Hudson river to the East river, running between Wall and Pine streets, with a gate near the present corner of Wall and Pearl streets, called the water-gate, and another in Broadway, called the land-gate. The walls and palisades were designed as a defence against the Indians. In 1665 Governor Stuyvesant captured fort Casimer, now Newcastle, from the Swedes on Delaware river, then called South river, whence probably Hudson river received the name of North river. In 1656 a market house was built at the present corner of Pearl and Broad streets, then called by other names. The city had 120 houses and 1000 inhabitants, including the garrison. In 1658 the first public wharf was built by the burgomasters of the city, where Whitehall-street now is. The governor's house stood opposite, at the beginning of Water-street. In 1660 the first map of the city was sent to Holland by Governor Stuyvesant. In 1662 a wind-mill was erected near the site of the present City Hotel. In 1664 a patent conveying the colony to the duke of York, was issued, and Colonel Nichols, with four frigates and 300 soldiers, arrived from England, where he had been appointed governor of New-York and New-Jersey, and the city surrendered to him without resistance, though it was done with great reluctance on the part of Governor Stuyvesant. The property of the Dutch West India trading company was all confiscated. The style of the government of New-York was altered from scout, burgomaster, and schepen, to mayor, aldermen, and sheriff. Twelve hundred guilders were raised for the support of the ministry in New-York. In 1669 the governor permitted the Lutherans to settle Jacobus Fabricius as their minister. In 1668 a carriage-road was ordered to be made to Harlem, there being none before. A race-course was established at Hempstead, for the purpose of improving the breed of horses; and subscriptions were taken for those who were willing to run for a crown in silver, or a bushel of good wheat. In 1672 the first Friend or Quaker preached in New-York. In 1673 the post-rider began his trips between New-York and Boston, once in three weeks. In July, 1673, the Dutch retook the city, and the fort was surrendered by its commander, Captain Manning, without firing a gun: but in the next year it was restored to the English, and Manning was tried by a court martial for treachery and cowardice, and sentenced to have

his sword broken over his head. All the inhabitants were required to take the oath of allegiance to the English government. In 1675 it was ordered that the land in the city convenient to be built on, if the owners did not choose to do it, might be valued, and sold to those who were willing to build. The streets were ordered to be cleaned every Saturday, or oftener if necessary, and the cartmen obliged, under the penalty of forfeiting their license, to carry away the dirt. In 1676 a law was passed to pave streets; also to fill up, level, and pave Heeren Gracht or Broad-street on each side of the canal which occupied its centre. At this time the water came up to Garden-street, where the ferry-boats landed. It was not lawful to sell liquor to the Indians; and if any were found in the street drunk, without knowing at whose tavern they obtained the liquor, the whole street was liable to a fine. No grain was allowed to be distilled, except that which was unfit for other purposes. In 1677 there were found by the assessments to be 12 streets and 384 houses in the city. In 1683 there belonged to the city three barques, three brigantines, 26 sloops, and 46 open boats. Twenty cartmen were allowed. At this time New-York had, by law, the exclusive right of bolting and packing flour and meal, which formed the main business of two thirds of its inhabitants. This was regarded as a grievance by the country people. In 1684 the first watch was appointed, consisting of 8 persons, at 12 pence each per night. In 1685 the assessed valuation of property in the several wards amounted to £75,694, and a tax of three farthings on the pound was laid. In 1686 James II. of England abolished the representative system; likewise the use of printing presses. The city paid the governor £300 for the charter, and £24 to the secretary, which had to be borrowed at 10 per cent.: this charter, with some alterations, has been continued to the present time. Chimney-sweepers were appointed, and ordered to "cry and make a noise." Houses with two chimneys were to have one fire-bucket; with more, three buckets. In 1688 the inlet in Broad-street was limited by a frame-work to the width of 16 feet, with a cartway on each side of 28 feet, making in the whole 72 feet, the present average width of the street.

The assessors' valuation of property in the several wards denominated West, North, South, East, and Dock wards, with Harlem and the Bowery, was £78,231; of this £22,254 were in South ward.

In 1690 a meeting of commissioners (called a congress) of the several colonies took place at New-York. In 1694 there belonged to the city 60 ships, 25 sloops, and 40 boats. Out of 983 houses, the inhabitants of 600 of them depended on bolting flour or meal for subsistence. In 1696 Trinity church was built, afterward enlarged, and burned in 1776. Ordered that a city-hall be built, valued at £3000. In 1697 a city watch of four sober men was ordered. It was also ordered, November 23d, that lights be put in the windows of the houses fronting on the streets, under a penalty of nine pence for each night of default; and, on December next following, that every seventh house hang a lighted lantern on a pole, and that the seven houses bear an equal portion of the expense. In 1698 the old City-hall was sold to John Rodman, merchant, by "public outcry," for £920, situated in what is now Pearl-street, at the head of Coenties-slip. In 1701 the docks and slips of the city were rented for £25 per annum. In 1703 Wall-street was paved from Smith (now William-street) to the English church. In 1707 Governor Cornbury prohibited Presbyterians from preaching, and two ministers were arrested and tried, but acquitted, on paying $220 costs. In 1710 several hundred Palatines arrived from Germany by the way of England, fleeing from persecution; they built a Lutheran church, on the present site of Grace church. In 1711 a slave-market was established in Wall-street, near the East river. In 1712 an insurrection of the negroes took place, who fired the city in several places, and killed some of its inhabitants. Nineteen of them were executed. In 1718 a rope-walk was established in Broadway, opposite to the Park (then called the Commons, and covered with brush and underwood). In 1719 a Presbyterian church was built in Wall-street. In 1720 a duty of two per cent. was laid on European goods imported, the first regular tariff mentioned in the early history of the city. In 1725 the New-York Gazette, a weekly newspaper, was established. In 1729 the society in London for propagating the gospel in foreign parts gave notice of a present to the city of 1642 volumes, belonging to the library of the late Dr. Millington; the books arrived, and were arranged in a room in the City-hall, appropriated to their keeping; and the thanks of the corporation were returned for the munificent gift. Three pence per foot given for land on the west side of Broadway, near the Battery. The Middle Dutch church built. In 1731 the boundaries of the colony were finally adjusted with Connecticut. In 1732 the first stage began to run between New-York and Boston once a month, being 14 days on the journey. In 1733 a law was passed to preserve the fish in Fresh Water pond, now

Canal-street, and contiguous streets. House of correction or bridewell instituted. In 1736 Water-street first mentioned as extending from Maiden-lane to Countess' Key and Rodman's wharf. In 1737 the town of Brooklyn disputed the right of the corporation of New-York to the ferry, and the city retained two counsel in the case, at a doubloon each. A market-house was erected in Broadway, opposite Crown-street, now Liberty-street. The city contained 1416 houses, only 16 having been built in seven years. In 1740 the New-York Society Library was founded. In 1741 a severe fire broke out in the fort, which destroyed the secretary's office and the old Dutch church. In this and the following year the yellow fever prevailed to an alarming extent. In 1741 occurred the celebrated "negro plot," when the city contained 12,000 inhabitants, one sixth of whom were slaves. A plot no doubt existed, but the account of it was greatly exaggerated, and the fears of the inhabitants excited by repeated fires and robberies. Some Irish Catholics were implicated with the negroes: 154 negroes and 20 white persons were committed to prison, of whom 55 were convicted and 78 confessed; 13 negroes were burned at the stake, at the present intersection of Pearl and Chatham streets, then out of town; 20 were hung, one in chains, on an island in Fresh Water pond, where the arsenal now is in Elm-street; 78 were transported to foreign parts, and 50 discharged. Thirty-six watchmen were appointed, and divided into three divisions, to watch alternately. In 1742 wheat was quoted at 3s. 6d. a bushel. The yellow fever prevailed near the tan-vats and docks. Coal was imported from England, as cheaper fuel than wood. In 1745 Lady Murray owned the only coach in New-York. In 1746 the city contained 1834 houses, having increased 418 in 11 years. In 1750 a theatre was established. Dey-street opened, regulated, and paved from Broadway to high-water mark in Hudson river, having a descent of 96 feet 2 inches. Beekman-street laid out and paved. In 1751 the Moravian church in Fulton-street built. In 1752 St. George's church in Beekman-street built. Also an exchange at the lower end of Broad-street, on the W. side, by private subscription, to which the corporation gave £100. In the winters of 1753-4-5 sloops passed from Albany to New-York in the months of January and February. In 1755 the exchange was let for one year, for £30. One thousand stand of arms were imported from England by the corporation for £3008, and deposited in the City-hall; the corporation petitioned for a lottery, to discharge this "excessive and alarming" debt. In 1759 Chatham-street began to be laid out, and a few houses erected. Thirty pounds per acre paid for land in the outer ward. In 1761 Vesey-street regulated and paved. Lamps and lamp-posts were purchased. The city contained 60 firemen. In 1763 the first Methodist chapel was erected, by the successful preaching of Lieutenant Webb of the army, assisted by some friends of the cause. In 1765 St. Paul's church was built. A Congress met at New-York, composed of delegates from the colonies. Great excitement existed on account of the Stamp Act. The governor and the devil, holding the Stamp Act, were burned in effigy, after having been paraded through the streets. In 1767 the Brick church in Beekman-street was built, on a triangular piece of ground granted by the corporation for a rent of £50 per annum. In 1768 the first Methodist church in America was built in John-street. In 1769 the North Dutch church in William-street built. Five church buildings erected before the revolution are now standing—the South and North Dutch churches in William-street, the Lutheran, now Coloured, church corner of William and Frankfort streets, the Brick church in Beekman-street, and St. Paul's church in Broadway. Most of these are now in fine repair. The New-York hospital was founded by subscription. In 1770 the expense of lighting the city was £760 per annum. In 1771 an iron railing was made round the Bowling-green, which cost £800. Warren-street laid out and regulated. In 1776, August 26th, by the disastrous battle of Long Island, the city fell into the hands of the British. On the 21st of September a great fire consumed 493 houses, nearly one eighth part of the city. Before the fire it contained 4200 houses and 30,000 inhabitants. In 1780 the winter was intensely cold, and two cakes of ice completely blocked up the ferry from Powles Hook to Courtland-street; hundreds of citizens and loaded teams and artillery passed on the bridge of ice, which continued for a considerable time. Hudson river, measured on the ice at this place, was found to be 2000 yards wide. On the 25th of November, 1783, the British evacuated New-York, after having held it since 1776, and General Washington, at the head of the American army, entered it. The British left their flag flying at the Battery, and had greased the flag-staff, so that it was with difficulty hauled down, and the American flag was raised in its place. A large number of loyalists and tories left with the retreating army. The British had erected works across the island near Duane-street. All the churches, excepting the Episcopal, had been destroyed, or used for hospitals, barracks, or riding-schools. The schools and college had been shut up. The city did not then extend N. of Murray-street. The books and accounts of the corporation during the revolutionary war were taken away by Mr. Cruger, treasurer, who joined the British army and left the country. In 1784 the Exchange in Broad-street was converted into a market place. Much difficulty was found in tracing out and securing the public property, of every description. At different dates La Fayette, John Jay, lately arrived from Europe, Baron Steuben, and especially General Washington, received the freedom of the city, and the latter an address of congratulation and thanks. The streets were cleaned for £150 per annum, and wells and pumps repaired for £140 per annum. Lot 116 Chatham-street was leased for 21 years for £6 per annum, and lot No. 18 of the same street, for the same term, for £4 per annum. The corporation offered any accommodation in their power to the Federal Congress. In 1785 the first Congress of the United States after the war, was organized in the City-hall, corner of Wall and Nassau streets. The Bank of New-York went into operation. In 1786 St. Peter's, the first Roman Catholic church, was built in Barclay-street. The state, until the present year, presented no instance of divorce in any case whatever. In 1788 the New-York city library was kept in a room in the City-hall. The adoption of the new constitution of the United States was celebrated by a grand federal procession. In 1789, April 30th, Gen. Washington was inaugurated in the open gallery of the old City-hall, facing Broad-street; and at the conclusion of the ceremony the collected thousands shouted with one heart, "long live George Washington." Broadway opened through the fort to the Battery. The City-hall was repaired and enlarged for the accommodation of Congress, at a great expense for that day, the whole done under the direction of Major L'Enfant, who received the thanks of the corporation, the freedom of the city, and an offer of 10 acres of land of the public property, which last he politely declined. The salary of the mayor commuted for £600 per annum. In 1790 the salary of the mayor was £700 per annum. The population of the city, December 11th, was 29,906. In 1791 the city was divided into seven wards. One hundred lots of ground in Broadway and adjacent streets in the vicinity of the New-York hospital, 25 by 100 feet, were offered for sale at £95 per lot. In 1792 the General Society of Mechanics and Tradesmen was incorporated, and Mechanics-hall built. Mayor's salary £800 per annum. A museum was allowed in the City-hall. In 1795 the new almshouse on Chambers-street was built, and contained 622 paupers, who were supported at an expense of £8319 15s. 7d. per annum. South-street laid out 70 feet wide, and ordered that no water lots be farther laid out, and no more buildings be erected in that direction. The Park theatre erected. Powles Hook ferry leased for £250 per annum. Water-street was laid out, which limited the city on the East river. In 1795 a lot on the S.W. corner of Broad and Wall streets was purchased by the corporation for £800. All the printing of the corporation done for £35 per annum. In 1797 the Brooklyn ferry leased for £2000 per annum. Free schools were established. In 1798 the Park theatre was completed, and the proprietors petitioned for the erection of a portico over the side-walk, which was not granted. A street commissioner was appointed. The chamber of commerce and citizens petitioned the corporation to fortify the city, and £50,000 were appropriated and expended for the purpose. The yellow fever prevailed from July to November, and 2086 persons died.

The Manhattan company in 1799 received an unlimited charter for supplying the city with pure and wholesome water, with a capital for the purpose, with the privilege of using their surplus funds in banking operations, and an exclusive use of the springs on the island for a supply. What this company have never been able to do, has been effectually done by the Croton water works of the city. The old Exchange in Broad-street was ordered to be taken down. December 20th the news of the death of General Washington was received, the bells of the churches were ordered to be muffled and tolled from 12 to 1 o'clock, until the 24th; the citizens were recommended to wear crape for six weeks, and a funeral oration was delivered by Governeur Morris in St. Paul's church. In 1800, eight lots of ground, a part of the present Washington-square, purchased by the Corporation for $1000. In 1801 the United States Navy Yard at the Wallabogt, Brooklyn, was established. The Brooklyn ferry at Fulton-street was leased for $2600 per annum. Broadway ordered to be continued and opened through Thomas Randall's land, called the Sailors Snug harbour, to meet the Bowery, and the hills levelled and carted into Fresh Water pond, (now Canal-street,) which to this time was the northern limit of this street, and far beyond the settled parts of the city. The total valuation of real estate in the city was $21,964,037. A City-hall was voted to be erected, and after much doubt and hesitation, the sum of $250,000 was devoted to the object, and contracts were entered into, and the foundation stone was laid September 20th, 1803, with due ceremony by Edward Livingston, Mayor, and by the corporation, though the prevalence of an epidemic, in some measure damped the ardor of the citizens. In 1804, July 11, the duel between Colonel Burr and General Hamilton occurred, in which the latter was mortally wounded, and died the next day, to the great grief of the citizens. Colonel Burr after this event fled as a fugitive to France, and after many years returned to the United States, to be neglected. December 18th a great fire destroyed 40 stores and dwellings, 15 on Wall-street, 17 on Front-street, and eight on Water-street, with a loss of between one and two millions of dollars. It was supposed to be the work of incendiaries. In 1805 the New-York Free school was incorporated. The upper part of Broadway was regulated and paved. The yellow fever prevailed, in the summer, and 280 persons died. The inhabitants of the city numbered 75,770, one third of whom left their dwellings. In 1806 the first successful attempt at navigation by steamboats by Fulton and Livingston took place on Hudson river. In 1809 the Historical society was established. In 1811 a great fire in Chatham-street consumed from 80 to 100 houses. The brick church and the jail, narrowly escaped. July 4th the Corporation met in the new City-hall, in the Mayor's room. In 1812 the old City-hall in Wall-street was ordered to be sold, and the new City-hall was finished. June 20th war was declared with Great Britain. November 12, the Brooklyn Fulton ferry was leased to Robert Fulton for $4000 per annum for seven years, on condition of establishing new steamboats upon it. In August an experiment was made with gas lights in the Park. In 1814 there were 3212 free holders; owners of personal estate over $150, 5512; tenants, 13,804; jurors, 4138; aliens, 3495; slaves, 976. The population was 92,448, which was less by 2312 than in 1810. The Literary and Philosophical society was instituted. October 29, the steam frigate launched. The interments this year amounted to 1794. In 1815 the news of peace with Great Britain was celebrated with great rejoicings. In 1816 the duties on merchandise imported amounted to $16,000,000. It July 1817 the Erie canal was begun near Utica. In 1818 the public wharves, piers, docks and slips sold for one year for $42,750. In 1821 Mr. John Randall, Jr., finished his maps and surveys of the N. part of the city and island, having been engaged in it, under the direction of the commissioners, for ten years, at a cost of $32,485. In January the harbour was closed by ice for the first time since 1780. The citizens crossed on the ice to Powles Hook, and some to Staten Island. The distance from Courtland-street to the Jersey shore, was found to be a few feet over a mile. In 1822, July, the yellow fever appeared, and most of the city S. of the City-hall was vacated, and the infected district fenced in; 388 died of the fever. November 25, burials in Trinity churchyard discontinued. In 1823 interments were forbidden S. of Canal-street. Washington-square formed and regulated. The New-York Gas-light company incorporated. In 1824 1600 houses were erected. In 1825 the Merchants' Exchange commenced in Wall-street. The city was divided into 12 wards. May 11th gas pipes were laid in Broadway, from Canal-street to the Battery, on both sides. October 26th the completion of the Erie canal was announced by the firing of cannon through the whole line, from Buffalo and back in 12 hours. November 4th the first canal boat arrived, and was greeted with great rejoicing. In 1827 the Merchant's Exchange was completed. In 1829 the American Institute in the city of New-York was instituted. In 1832 the cholera swept off a great number of the inhabitants. The whole number of deaths in July was 2467, in August, 2206; during the year, 10,359. In 1833 the number of pupils taught in the public schools was 6140 boys, 4390 girls, total, 10,460. In 1834 the number of inmates at the Almshouse at Bellevue in January was 2011, of whom 1051 were natives, and 960 foreigners. On the night of the 16th of December, 1835, occurred the great fire, which swept over between 30 and 40 acres, of the most valuable part of the city, covered with stores and filled with rich merchandise. The number of buildings burned was 648, and the amount of property destroyed was estimated by a committee appointed for the purpose at nearly $18,000,000. The Merchant's Exchange, and the South Dutch church were burned. It is proof of the great wealth of New-York that they were able to bear such a loss, without feeling it more. Few failures resulted from it. The burnt district was immediately rebuilt, with additional convenience and beauty. With this great event the Editor concludes this hasty glance at the annals of the commercial metropolis of the nation.

NIAGARA RIVER and FALLS. The river commences at the N.E. part of lake Erie; and discharges the waters of the great upper lakes Superior, 320 miles long and 130 broad, Huron, 240 miles long and 150 broad, Michigan, 330

470

miles long and 60 broad, and Erie, 240 miles long and 60 broad, besides smaller lakes and numerous rivers which enter them. These great inland seas as they may be called, contain nearly half the fresh water on the surface of the globe. Niagara river, as it flows from lake Erie, is about three fourths of a mile wide, and from 20 to 40 feet deep; and has, for three miles a rapid current, and then becomes smooth and placid, till within one mile of the falls. Five miles from lake Erie, the river begins to expand, till it becomes more than eight miles in width, measured across Grand island, and embraces, before it reaches the falls, about 40 islands. Grand island is 9 miles long, and in its widest part, 6 miles wide, contains 17,381 acres, and belongs to the state of New York, the channel being on its W. side. Navy island contains 304 acres, belongs to Canada, and was famous during the Canadian rebellion, in 1837-8, having been for a time occupied by the insurgents, headed by William L. Mackenzie. Grand island commences five miles from lake Erie, and extends to within three miles of the falls. Among the other islands are Bird island, between Buffalo and fort Erie; Squaw island opposite to Black Rock, containing 131 acres; Strawberry island, containing 100 acres; Beaver island of 30 acres; Rattlesnake island, containing 62 acres; Tonawanda island, of 60 acres; Cayuga island, 4 miles above the falls, containing 100 acres; Buckhorn island, containing 146 acres; besides a number of smaller islands; and Goat island which belongs to a description of the falls. Below the termination of Grand and Navy islands, the river is compressed to the breadth of two and a half miles, and proceeds with an accelerated motion. Three fourths of a mile above the falls commence the rapids, which have a descent of from 52 to 57 feet, the greatest descent being on the British side, more than a mile in width, with white crested breakers, and a dashing and foaming torrent, tossing from 10 to 30 feet above the main current, until they cover to the tremendous cataract of Niagara falls. Woe will to the man or animal who falls into these rapids, by which numbers have been hurried to destruction; while a few almost by a miracle, have escaped. The falls of Niagara, are about 21 miles below lake Erie, and 14 miles above lake Ontario, and are generally regarded as the greatest natural curiosity in the world. The river, which constitutes the great outlet of the upper lakes, is here precipitated over a precipice 160 feet high, with a solemn and tremendous roar, which is ordinarily heard from 5 to 20 miles; but has in some instances been heard at Toronto, 45 m. distant. And yet, at the village on the American shore, near the cataract, there is little to give notice of its awful proximity. In consequence of a bend in the river, the principal weight of the water, supposed to be seven eighths of the whole, is thrown on the Canadian side, down what is called the Horse-shoe fall, which name has become inappropriate, as the edge of the precipice has ceased to be a curve, and forms a moderately acute angle. Near the middle of the fall, Goat island, containing 75 acres, extends to the brow and the bottom of the precipice, dividing the river into two parts; and Luna island, at a little distance from Goat island, again divides the cataract on the American side. Goat island at its lower extremity, presents a mass of perpendicular rock, reaching from the bottom to the top of the precipice, and creating two distinct falls. A bridge has been constructed to Bath island. This island is 24 rods in length, and contains about two acres. The bridge to Bath island is 28 rods long from the shore, and another bridge of 16 rods long extends to Goat island. The view of the rapids, foaming and tossing among the rocks, is fearfully beautiful, as seen from these bridges. On Bath island is a toll-house, where travellers enter their names, paying a small fee, can see curiosities, and frequently obtain walking canes, cut on the island, which they often retain as memorials of their visit to the falls. On this island is also a large paper mill. From the W. part of Goat island, a bridge has been constructed, 300 feet long, projecting ten feet over the great Horse-shoe or principal fall; and near the termination of this bridge, in the water and on the rocks, at the very verge of the precipice, a stone tower has been erected, 45 feet high, ascended by winding steps on the inside, and as has balustrade around its open gallery at the top. It requires some strength of nerve, and a little reflection, to feel safe in looking down into this awful abyss, and to survey the tumbling confusion and tremendous roar of the mass of waters, from a situation so near the brow of the precipice. But the unnumbered initials of names, inscribed on every part of the building will convince the spectator, that thousands have been there before him. The distance across the fall, from the American shore to Goat island is 65 rods; across the front of Goat island, is 78 rods; around the Horse-shoe fall, on the Canada side, 144 rods; directly across the Horse-shoe fall, 74 rods. The height of the fall near the American shore is 163 feet; near Goat island on the same side, 158 feet; near Goat island on the Canada side, 154 feet. Table rock, a shelving projection on the Canadian side, at the edge of the precipice, is 150 feet high. This last place is thought to present the finest view of the fall; and taken as a whole, this is doubtless true. But if the spectator wishes to view the crescent or Horse-shoe fall, in all its grandeur, and will visit the stone tower on Goat island, a little after sunrise, when the whole cavity is illuminated by the full strength of the solar rays poured into it, discovering it nearly to the bottom, and spanning it with a perfect rainbow, it will leave him nothing further to desire, in regard to this part of the fall. At the lower end of Goat island, about one third across it, is the Biddle-stair-case, named from the gentleman who erected it in 1829, to afford visiters an opportunity of descending to the bottom of the fall, and passing for a considerable distance behind the two main sheets of water. The descent from the top of the island to the margin of the river is 185 feet. A flight of common steps leads down 40 ft., to the perpendicular spiral steps, 90 in number, enclosed in a hexagon. resting on a firm foundation at the bottom. There is another staircase, about six rods below the falls, leading down the bank on the American side, to the ferry, where is a fine view of the American fall, and a safe passage by the ferry boat to the Canada side, where many fine views of the falls are obtained. The American fall, though sublime, inclines to the beautiful; while the Canadian fall, though beautiful, is characterised by an overpowering sublimity. There are good hotels on both sides of the river, and small villages are laid out. On the American side, the water power is immense, and easily available; and but for its expeadneas in case of a war, would probably be soon, and extensively occupied. It is computed that 100 millions of tons water are discharged over the precipice every hour. The river at the falls is a little over three fourths of a mile wide, but below, it is immediately compressed to less than one fourth of a mile wide, and, as ascertained by sounding is about 250 feet deep. One of the best general views of the falls is from a projecting rock, 900 feet high above the river, about a mile below the village. About two miles below the falls the river is comparatively smooth, and thence to Lewiston it flows with amazing velocity. While the river makes a constant descent, the banks have a gradual ascent for six miles: and some have supposed that the falls have receded from Queenston to their present situation; but they are known now to occupy the same situation that they did 200 years since. About two miles below the falls on the American side, is a mineral spring, containing sulphuric and muriatic acids, lime and magnesia, useful in scrofulous, rheumatic, and cutaneous complaints. One mile farther down is a terrific whirlpool, almost as tremendous as the Maëlstrom of Norway, where logs and trees are whirled round for days in its outer circles, and finally drawn down perpendicularly with great force, and shot out again at the distance of many rods. A mile below the whirlpool is an excavation from the side of the bank, containing about two acres, and 150 feet deep, called the De vil's hole. The high banks of the river extend nearly to Lewiston, 7 miles below the falls, when they become level as above the falls. From Lewiston, a steam boat proceeds to lake Ontario, and through it to Oswego. The Welland canal affords a passage for sloops and schooners of 125 tons burden around the falls of Niagara, and connects lake Erie with lake Ontario. It is 42 miles long, 56 feet wide and from 8½ to 16 feet deep. The whole descent from one lake to the other is 334 feet, which is accomplished by 37 locks. It has a deep cut through the mountain ridge, 45 feet deep, where an immense amount of earth and rocks were removed. This canal was completed in 1829, at cost of $1,000,000. A canal has also been projected on the American side, but it is doubtful if it will ever be undertaken.

The number of visiters at the falls is from 12 to 15 thousand annually, and the number is increasing. While curiosity constitutes an attribute of the human character, these falls will be frequented by admiring and delighted visiters, as one of the grandest exhibitions in nature. The fashionable, the opulent, and the learned here congregate, in the summer season, from the principal cities of the country; from the southern and western states, South America, the West Indies, the Canadas, and various parts of Europe, and indeed from all parts of the civilized world. An American poetess has well said of Niagara.

"Flow on forever, in thy glorious robe
Of terror and of beauty! God hath set
His rainbow on thy forehead, and the cloud
Mantles around thy feet. And he doth give
The voice of thunder, power to speak of Him
Eternally—bidding the lip of man
Keep silence, and upon the rocky altar pour
Incense of awe-stricken praise."

NICARAGUA (LAKE OF), the most considerable lake of central America, comprised within the state of Nicaragua, and extending principally between the 11th and 19th degs. of N. lat., and the 84th and 86th of W. long., about 12 m. in a direct line from the Pacific, and 90 m. from the Caribbean sea. It is of an oval shape: length, N.W. to S.E., about 130 m.; average breadth, perhaps, about 40 m. It

has numerous creeks and harbours, and several islands. It receives a good many rivers, especially along its N., N.E., and W. sides; its surplus waters are carried to the Caribbean sea by the Rio San Juan, which issues from its E. extremity, and is said to be navigable, during the rains, throughout its whole extent.

The project for uniting the Atlantic and Pacific oceans, by means of the river San Juan and of a canal from the lake of Nicaragua to the Pacific, has been often mooted; and the country certainly presents greater facilities for effecting this great work than any other part of Central America, or than the isthmus of Panama. The river San Juan, notwithstanding it is in one part impeded by cataracts, is said, by Thompson, to be available for craft drawing 3 or 4 ft. water, at all times of the year, and for vessels drawing from 10 to 12 ft. water to from 20 to 35 leagues from the sea. The lake itself is adapted for ships of the largest burden, being 15 fathoms deep. The distance between its S.W. shore and the gulf of Papayago, in the Pacific, is only 29,280 yards, or 15½ m.; and though the intervening country be laid down in many maps as mountainous, the greatest actual height of any part of it above the level of the lake is only 19 ft.; as was proved by a series of 347 levels, about 100 yards apart, taken in 1781. (See *Thompson's Guatemala*, Append., p. 519–520.) The surface of this lake is about 134 ft., and its bottom 42½ ft. (Engl.) above the level of the Pacific; but the ascent might probably be overcome by a succession of locks. The difference in the level of the two oceans, formerly supposed to be so serious an obstacle to the undertaking, was ascertained by Humboldt not to exceed 20, or, at most, 22 ft. (*Pol. Essay.* i., 31.) At its W. extremity, the lake of Nicaragua is connected by a small river with the lake of Leon. The latter, 50 m. in length by nearly 30 in breadth, is said to be also of sufficient depth for the largest ships. It is but 13 m. from the Pacific, and 5 from the river Tosta, which enters that ocean; and the height of the intervening ground between it and the Pacific is not more than 51 ft. above its own surface, which last is only 3 ft. higher than that of the Tosta. In this direction, also, a communication has been contemplated. At one period, a British company proposed connecting the two oceans, and it was afterward said that the Dutch government had undertaken it; we believe, however, that no active steps have been yet taken to carry either of the above plans into execution. (*Thompson's Guatemala; Geog. Journ.*, vi., &c.)

NICASTRO, a town of the Neapolitan dom., prov. Calabria Ultra II., on the W. side of the Apennines, 19 m. S. by W. Cosenza. "It is a straggling town of 5000 inhabitants, and the seat of a bishop and a sub-intendant. Its houses are mean, and all roofed with red tiles. A ruined castle, on a conical hill, rising from amid all these modern buildings, is that in which Henry, the eldest son of the emperor Frederic II., was confined for some time." (*Craven. Tour,* p. 33.) It has some trade in oil, and there are many mineral baths in its vicinity.

NICE (Ital. *Nizza*, an. *Nicæa*), a city and seaport of the Sardinian dom. in Italy, cap. div. and prov. of its own name; on the Mediterranean, about 5 m. E. from the bar, the boundary of the French territory, 95 m. S.W. Genoa, and 98 m. S. by W. Turin; lat. 43° 41' 16'' N., long. 7° 16' 27'' E. Pop. in 1838, of the town and canton (ex. garrison), 33,811. It is beautifully situated in a small plain at the foot of the maritime Alps, by which it is protected from the N. and E. winds; while the cool sea-breeze, which prevails every day with a regularity almost equal to that of a tropical climate, moderates the summer heat. The principal disadvantage of its situation is that, being open on the W., it is exposed, with but little protection, to the influence of the *mistral*, or *vent de Bize*, which is often keen and piercing. It is encircled by bastioned walls; and has on the E. the steep rocky hill of Monte Albano, surmounted by the ruins of an old castle. The view from this hill is very fine, and at sunrise and sunset the island of Corsica is sometimes clearly distinguished, though it is some 70 or 80 m. distant. The port, which is small and protected by a pier, admits vessels of 300 tons burden, and is visited by the steamers from Marseilles to Genoa. Nice is divided into two parts by the river Paglione, here crossed by a good stone bridge The old town has narrow and crooked streets, which, however, are kept very clean. The new town to the W. of the river is well laid out and handsome: it has a square surrounded by open arcades, and some of the houses near the sea, and in the vicinity, are very superior. The cathedral, several convents, three hospitals, the governor's residence, college, public library, theatre, and a fine arch erected in honour of Victor Amadeus III., are the principal public buildings: it has several bath-establishments, and some good hotels; and Dr. Farr states that the rents of houses and apartments are lower here than in any other place of general resort on the continent. It has manufactures of silk twist, snuff, soap, essences, perfumery, and paper, a fishery of anchovies, and a considerable trade in the export of oil, wine, oranges, hemp, &c., and in the importation of corn from the Black sea, salt fish, manufactured goods, and colonial produce. In 1839, the value of the imports amounted to 12,343,450 fr., and that of the exports to nearly as much. It is a bishop's see, and the seat of a royal council, and of the head court of justice for its division.

Nice, in common with Montpelier, enjoys the reputation of having a peculiarly genial climate, and is accordingly resorted to by numerous invalids, especially from England, during the months of November, December, and January. Mr. Forsyth says, that when he arrived at Nice on Christmas day; "a soft and balmy air, oranges growing in every garden, lodgings without a chimney, and beds with mosquito-curtains, presented the first signs of Italy." But at other seasons it is less suitable for invalids. In February, the *vent de Bise* begins to blow; and it is very trying to persons with delicate constitutions. This explains the singular discrepancies in the accounts of different travellers as to the climate of Nice. Dr. Farr and Sir James Clark, especially the former, give full and satisfactory information as to the climate of Nice, and its surrounding localities, and the classes of invalids most likely to be benefitted by a residence in it. A noble road, constructed at a vast expense, leads over the Maritime Alps from Nice to Turin. Another road, begun by Napoleon, but not completed till 1827, leads along the sea coast from Nice to Genoa; and a third road is now about being opened from Lyons to Nice, which will be a shorter and better way of entering Italy than by Mount Cenis.

Nice is said to have been founded by colonists from Marseilles. Under the Romans, it was originally the seat of a naval arsenal; but, under Augustus, the latter was transferred to Frejus. Under the French, it was the capital of the department *Alpes-Maritimes*. Among the celebrated individuals to whom it has given birth, are the painter, Vanloo, the astronomer, Cassini, and marshal Messena, one of Napoleon's ablest generals. (*See Dr. Farr's admirable Guide to Nice*, passim; *Clark on Climate*, third ed. p. 202.)

NICHOLAS, county, Va. Centrally situated toward the W. part of the state, and contains 1430 sq. m. Bounded S.W. by New river. Watered by Gauley river and its branches. It contained in 1840, 4095 neat cattle, 4863 sheep, 5832 swine; and produced 4454 bushels of wheat, 3222 of rye, 56,297 of Indian corn, 37,733 of oats, 11,354 of potatoes, 3013 pounds of tobacco. It had three stores, 20 grist-mills, five saw-mills, two tanneries; three schools, 77 scholars. Pop.: whites, 2440; slaves, 72; free coloured, 3; total, 2515. Capital, Summerville.

NICHOLAS, county, Ky. Situated in the N.E. part of the state, and contains 350 sq. m. Watered by Licking river and its branches. It contained in 1840, 8121 neat cattle, 19,067 sheep, 29,427 swine; and produced 60,765 bushels of wheat, 25,315 of rye, 613,804 of Indian corn, 77,086 of oats, 8949 of potatoes, 47,638 pounds of tobacco, 57,020 of sugar. It had 16 stores, one woollen factory, 19 grist-mills, 10 saw-mills, five tanneries, three distilleries; 22 schools, 501 scholars. Pop.: whites, 7310; slaves, 1253; free coloured, 182; total, 8745. Capital, Carlisle.

NICHOLAS, C. H., capital of Nicholas county, Va., 210 m. W. by N. Richmond, 322 W. It contains a courthouse, jail, and several stores and dwellings. Nett proceeds of the postoffice, $78.

NICHOLS, p. t., Tioga co., N. Y., 8 m. S.W. Oswego, 170 m. W.S.W. Albany, 268 W. Bounded N. by Susquehanna river, and drained by its small tributaries. It contains a Methodist and a Baptist church, five stores, three fulling-mills, eight grist-mills, 26 saw-mills; 849 scholars in schools. Pop. 1986.

NICOBAR ISLANDS, a group in the Indian Ocean, between the 3rd and 10th degrees of N. lat., and the 93rd and 94th of E. long., about midway between the N.W. point of Sumatra and the Andaman islands, and from 100 to 130 m. from each. Sambelong and Carnicobar, the former at the S. and the latter at the N. extremity of the group, are the principal; there are, however, about half a dozen other islands of some consequence, and a number of small islets. Most of these islands are hilly, and all are covered more or less with dense woods of cocoa-nut, areca-palm, and various timber trees. The climate is extremely unhealthy to Europeans, and is supposed to owe this quality, in great part, to the extensive spontaneous decomposition of vegetable matter. These islands are inhabited by a race of natives of the Indo-Chinese stock, whose inoffensive character contrasts strongly with the wild ferocity of their neighbours of the Andaman Islands. Their chief occupations are fishing, rearing hogs and poultry, a little agriculture, and trafficking among themselves, and with foreigners who touch at the Nicobars. Cocoa and betel-nuts are met with in immense quantities, and most of the Indian ships bound eastward, call here to take in a cargo of the former, which they obtain at the rate of 4 nuts for a leaf of tobacco, and 100 for a yard

472

of blue cloth. The natives also exchange fowls, hogs, birds nests, ambergris, tortoise shell, wild cinnamon, sessafras &c., for iron, tobacco, cloth, silver coin, and other European goods. They live under a number of petty chiefs; but little is known of their internal economy, customs, &c., the great insalubrity of the climate having successively broken up all the establishments formed on the Nicobars by the Danes, the British missionaries, &c., in the latter half of the last century. (*Hamilton's E. I. Gaz., &c.*)

NICOLAEFF, a town and river port of European Russia, gov. Kherson, at the confluence of the Ingul with the Bug, about 20 m. above where the latter falls into the estuary or liman of the Dniepr, lat. 46° 58' 21" N., long. 30° 0' 21" E. Pop. 8500. Nicolaeff was founded in 1790; and was intended to be a great naval depôt, and the station of the Russian fleet in the Black sea. It stands in an elevated, healthy situation, covers a large extent of ground, and is extremely well built. The streets are wide, and regularly laid out, and the private houses, which are mostly of brick, have a handsome appearance. Among the numerous public buildings may be specified the new church or cathedral, the admiralty, the town-house, the marine barracks, the naval hospital, &c. In the vicinity is an observatory. The admiral commanding the fleet in the Black sea resides here; and here, also, are the various offices connected with this department of the service, with schools for the instruction of pilots, ship-builders, naval artillery, &c.

Nicolaeff owes its existence to its river, which has its entrance without the bar of the Dniepr, and water sufficient to float large ships up to the town. There are extensive docks and yards for the building of ships; but the latter are, notwithstanding, mostly constructed at Kherson, being sent thither to be laid up, or, when necessary, repaired. Still, however, Nicolaeff has not, as its founders anticipated, become a large, thriving, town, and latterly, indeed, it has been either stationary or has retrograded. This is ascribable partly to the want of good water[a], and the scarcity and high price of fuel caused by there being no timber in its vicinity; partly to its harbour being, though very superior to that of Kherson, decidedly inferior to that of Sevastopol in the Crimea, at which a part of the fleet is now always stationed; and partly, and principally perhaps, to the great advantages enjoyed by Odessa as a commercial emporium. Nicolaeff is, in fact, nearly deserted by all the mercantile class, and depends entirely on the employment afforded by government. (*Clarke's Travels*, ii. 350, 8vo. ed.; *Lyall*, i. 201; *Schnitzler, La Russie, &c.*, p. 723; *Pinkerton's Russia* p. 160.)

NICOLAS (ST.), a town of Belgium, prov. E. Flanders, cap. canton, on the high road between Ghent and Antwerp, 19 m. E.N.E. the former and 12 m. W.S.W. the latter. Pop in 1836, 16,153. It is well built, and handsome, and has its habitants generally opulent. It has a fine town-hall, a parish church, in which are some good Flemish paintings, a hospital, two orphan asylums, a convent, a prison, and a large market-place, partially planted with trees. It is one of the most flourishing towns of Belgium, and has manufactures of woollen, cotton, and silk fabrics, hats, paper, soap, tobacco, chocolate, &c.; with salt-refineries, tanneries, breweries, dye-houses, and potteries. It has, perhaps, the largest market for flax in Europe, and large annual fairs for cattle and horses. It is the seat of a tribunal of commerce; has academies of music, drawing, &c., and sends one deputy to the states of the province. (*Vandermaelen, Fland. Orient.*)

NICOPOLIS, a town of Turkey in Europe, prov. Bulgaria, cap. sanjack, on the Danube, 100 m. E. by S. Widin. Pop. estimated at 10,000. It has an imposing appearance, being situated on a range of hills above a bay of the river, and surrounded by strong ramparts mounted with cannon. It is further defended by an ancient castle, and has several suburbs, in which the Greek and Bulgarian inhabitants principally reside. Generally it is ill-built, but has some large houses, and several handsome mosques and public baths. It is the seat of a Greek archbishop and a Roman Catholic bishop: its position on the Danube gives it some commercial importance; it is, however, said to be in a state of decay. (*Elliott's Trav.* i. 175.) Nicopolis was founded by Trajan, and some portions of its ancient walls are said still to exist. But it is chiefly memorable, at least in modern times, for the great battle fought in its vicinity, on the 28th of September, 1396, between the Ottoman army under Bajazet, and that of the Hungarians and their allies under their king Sigismund. The latter sustained a complete defeat, ascribable as much to the rashness and presumption of the count de Nevers and other French leaders, as to the bravery and superior discipline of the Turks. (*Gibbon*, cap. 64.)

---

[a] Dr. Lyall says that this deficiency has been supplied by the construction of a reservoir, the existence of which has, however, been doubted, from its not having been mentioned by the accurate Dr. Pinkerton. But it may however been overlooked by the latter, or may not have sufficiently answered the purpose for which it was intended.

NICOSIA (an. *Tremitus?*), the principal city of the land of Cyprus, near its centre, on the small river Pedia; lat. 35° 13' 11" N., long. 33° 26' 45" E. Pop., according to Turner, about 17,500, of whom about two thirds are Turks. It stands in a low fertile plain, near the S. foot of a range of high mountains, and is surrounded by walls in the shape of a hexagon, flanked by 13 bastions. The ground of the enclosure is very unequal, being in some parts elevated to the height of the walls, and in others forming a deep valley. The streets are in general not more than 10 and 15 ft. in breadth; and, being unpaved, are always filthy, and, in inter almost impassable. Having been the residence of the principal Venetian families during the period that the island was subject to Venice, it has many fine houses, which are now, however, mostly in ruins; and at present consists principally of brick and mud huts. The bazaar, though tolerably well supplied, is not even arched, but roofed with reeds and mats, which admit the rain in all directions. Most houses have gardens, which abound with olive, lemon, and pomegranate-trees; and hence the first view of the city is very pleasing, from the contrast between the foliage and the dark mountains to the N. There are eight mosques, all of which were once churches, the principal having been the cathedral church of St. Sophia, built by the Venetians; is in the Gothic style, of an oblong shape, with a pentagonal projection at the end opposite the entrance, for the reception of the altar. The interior is laid out in three aisles, divided by clumsy white-washed Corinthian columns. On the two belfries the Turks have erected two high and handsome minarets. There are still six Greek churches, one Roman Catholic, and several Greek convents. The city has also four public baths, and a large, but ruined caravaneral. It has some manufactures of carpets, printed cottons, and red morocco leather, and exports wine and cotton.

Nicosia is supposed to occupy the site of the ancient *Tvimus*, or *Trimithus*, mentioned as a place of some note by the Byzantine historians. When Richard I. of England took Cyprus in 1191, and conferred it on Guy de Lusignan, it was made the capital of the new kingdom, and greatly enlarged. It fell, in 1480, to the Venetians, who built the present walls, and several churches and handsome palaces; and who held it, with the island, till 1571, when it was taken from them by the Turks, under whose brutal and destructive sway it has since continued. (*Turner's Levant*, . 544–547; *Kinneir's Asia Minor*, 178, 189; *Drummond's Travels, &c.*)

NICOSIA, a city of Sicily, intend. Catania, district of its own name, on two hills, 14 m. N.E. Castrogiovanni. Pop., n 1831, 13,151. Like other towns in the interior of the island, it is remarkable for nothing but the number of its churches and convents. It has few manufactures, and hardly any export trade, but a considerable traffic in the corn and cattle of the surrounding country, which is very fertile. Its situation is such as to afford a strong military position; and it is supposed to be the ancient *Herbita*, founded in the earliest period of Sicilian history.

NIÈVRE, a dep. of France, reg. centre, nearly co-extensive with the old prov. of Nivernais, between lat. 46° 40' and 47° 35' N., and the 3rd and 4th degrees of E. long.; having N. Yonne, E. Côte d'Or and Saône-et-Loire, S. the latter and Allier, and W. Cher. Area, 681,093 hectares. Pop. (1836), 282,521. A mountain chain runs from S.E. to N.W. through its centre, dividing the basin of the Loire from that of the Seine; the culminating point of the chain in this department being 3000 ft. above the sea. The Loire and Allier bound Nièvre on the W.: the other principal river is the Yonne. The Loire and Yonne are united by the canal du Nivernais, which, commencing at Decize on the former river is continued through the departments Nièvre and Yonne, for a distance of above 100 m.; but the work is not yet completed. The Nièvre, whence the department has its name, flows through its W. part, and, after a course of about 25 m., generally southward, joins the Loire at Nevers. It turns many mills, but it is navigable only for rafts or small boats. The soil is not, in general, very fertile. In 1834, 295,991 hectares were estimated to be in cultivation, 67,366 in meadows, 9900 in vineyards, 3607 in orchards and gardens, and 239,561 in woods. In 1835, of 3,851 properties subject to the *contribution foncière*, 43,650 were assessed at less than 5 fr., and 13,996 at from 5 to 6r. The fertile portions of the surface are comparatively well cultivated, and sufficient corn is produced for home consumption. The annual produce, in wine, is estimated at about 260,000 hectolitres; of which the white wines of Pouilly are the best. *Ils ont du corps du spiritueux, un léger parfum de pierre à fusil, et un gout fort agréable; ils ne sont pas sujets à jaunir, et conservent assez long-temps leur douceur.* (Jullien, p. 156.) In 1830 there were supposed to be about 132,000 oxen and cows, and 315,000 sheep in the department; but the breeds are not particularly good. The chief resources of Nièvre are in its forests and mines.

Most of the small rivers, which are not navigable, have been adapted to floating down rafts of timber and fire-wood, a good deal of the latter being sent down the Yonne and Seine to Paris. The coal wrought near Decize is principally destined for the supply of Paris and Orleans. Lead, copper, and some other metals are found, but iron is by far the most important metallic product, and its yearly value in the shape of pig iron, iron plates, anchors, files, &c., is estimated at 8,728,000 fr. Hardware and cutlery, at Cosne and La Charité, glass, and earthenware, especially at Nevers, linen and woollen cloths, and musical strings, are among the principal goods manufactured. Nièvre is divided into four arronds.: chief towns, Nevers the cap., Château Chinon, Clamecy, and Cosne. It sends four members to the chamber of deputies. Number of electors (1838–39), 1379. Total public revenue (1831), 6,256,756 fr. (*Hugo*, art. *Nièvre; French Official Tables.*)

NIGER, JOLIBA, or QUORRA, a celebrated river of central Africa, having its remote sources near the extreme W. coast of the continent, in the country of the Mandingoes, in about 8° N. lat., and 6° W. long. It thence pursues a course N.W. and N. to the 10th deg. of lat., and then follows a general N.E. course to Timbuctoo, below which it turns S.E., and afterwards S. and S.W., to its mouth, in the gulf of Benin. Supposed length about 2300 m. The upper part of the Niger, called by the natives the Joliba, was first discovered in modern times by Mungo Park, who was sent out in 1795 by the African association: he describes it at Sego, the cap. of Bambarra, as "glittering in the morning sun, broad as the Thames at Westminster, and flowing slowly to the eastward." (*Travels*, p. 220.) He succeeded in ascending it as far as Bammakoo, 250 m. above Sego, the cap. of Bambarra. From Cabra he sailed down the stream to Boussa, where, unfortunately, he was killed by the natives. Major Laing concluded, from information obtained in the neighbourhood, that the sources of the river were on the N. side of the mountains of Kong, at a height of 1600 ft. above the sea, in lat. 8° 20' N., and long. 9° 10' W.; but Mr. Macqueen conjectures that the Allmar, its principal source, rises farther to the E. than Laing supposed. Lander, the servant of captain Clapperton (who was murdered near Saccatoo), sailed from Boussa, with the stream, to the mouth of the river, previously called the Nun, in the bight of Benin; and thus finally identified the Niger and the Quorra, and put an end to all the doubts and theories that previously existed as to the course and termination of the former. It hence appears that the length of the Niger, measured along its banks, exceeds 2300 m.; and it is probable that its basin is nearly, if not quite, as extensive as that of the Nile. According to Caillié, it is navigable for large canoes within 100 m. of its source: for 200 m, below that point it has not been navigated by Europeans; but from Bammakoo to Timbuctoo it has been pretty accurately laid down, both by Mungo Park and Caillié. The river valley is here of considerable width, fertile, and comprising numerous towns and villages on either bank. The current of the river is not strong; and both travellers saw flotillas of canoes of 60 tons and upwards, frequently passing up and down the river, which in the rainy season is flooded on both banks to a considerable distance. (*Caillié*, ii. 34.) In about lat. 16° N., the stream expands, forming a lake, called Debo, which measures about 10 m. from N. to S., is from 12 to 15 ft. deep, calm, transparent, and surrounded by extensive marshes. (*Caillié*, ii., 20.) Hence to Timbuctoo the valley becomes still wider; the pasturage of cattle, the tillage of rice, millet, maize, &c., is extensively pursued, and along the banks are numerous villages, which export rural produce. In lat. 17° 30' N. and long. 3° 10' W., the river bifurcates, and on the N. and narrower branch is Cabra, the port of Timbuctoo: these branches, however, seem to unite a few miles lower down. It has already been observed that the course of the river below Timbuctoo was traversed by Mungo Park as far down as Boussa; but, as that traveller was killed there, and his papers were lost, we know nothing of this portion of the river, except that it is navigable for vessels of considerable size.

The highest point of what may be called the lower Niger hitherto visited by Europeans is the neighbourhood of Yauri (lat. 11° 20' N. and 5° E.), which point Lander reached in 1830. Here the river leaves the great plain of Soudan, and enters the defiles of a mountain range crossing this part of Africa from E. to W., and probably connected, on one side, with the Djebel-el-Kumri, and on the other with the mountains of Kong. The direction of the stream from Yauri, for about 150 m., is nearly due S.; but it is full of rocks, sand-banks, &c., and wholly unnavigable, except at the time of the rains, and immediately after. Below Boussa, the banks on both sides are generally high and rocky; cultivated plains intervene in many places between the river and the mountains, but in others the offsets come close down to the water's edge. From Boussa downwards, the Niger is navigable for moderate-sized vessels; and in lat. 6° N., a little below Atta, it leaves the hilly country, and enters an alluvial plain, the lower part of which is an unhealthy swamp, covered with jungle: many branches here diverge from the main stream, and at the mouth is an extensive delta, which, however, is, as yet, very imperfectly known. At Atta, the river is about 2 m. wide, and near Bubba, in lat. 8° 45', it attains a width of 5 m.; but its breadth, close to the mouth, is somewhat less than a mile. The tide is said to extend within about 30 m. of Atta, or about 190 m. from the sea. The only branch of the Niger hitherto explored is the Chadda, which joins it on the left bank in lat. 7° 52' N., 32 m. above Atta. Captain Allen and Mr. Laird sailed about 100 m. up this tributary, and inform us that it is quite equal in width, though not in depth, to the parent river. It has many shoals and sand-banks. It has been conjectured that this river has its source in the great lake of Tchad, discovered by Messrs. Clapperton and Denham; but the more probable opinion seems to be, that it has its sources on the N. declivity of the Gebel el Kumri, not far, perhaps, from the sources of the Bahr-el-Abiad, or W. arm of the Nile. The only other known tributaries of the lower Niger are the Saccatoo, Mayarrow, and Coodonia, all joining it on the left or E. bank: the former of these was discovered by Clapperton, and from the course which it pursues, it may possibly be the same river that joins the Niger near Yauri. (See *Arrowsmith's new map of Negroland.*) Both rivers flow from a range of mountains, running N.W. through Houssa, and forming the watershed between the tributaries of lake Tchad and the Niger.

In the article AFRICA, (vol. i. 41), we have given a succinct account of the successive modern expeditions that have been fitted out for the purpose of exploring the course of this river, so long involved in doubt and obscurity; and though much still remains to be accomplished, its general course and leading features have been pretty well ascertained. This, however, has not been done without a great sacrifice of human life. The inhabitants of the countries in the lower part of its course are among the most degraded in the scale of human beings: the slave trade is extensively carried on; and wars being continually waged between the different tribes, travellers are exposed to the greatest dangers. (*Ritter's Africa,* ii. 110, and *Buxton on the Slave Trade,* pp. 41, 42.) The climate, also, is extremely unhealthy, so much so, that out of 40 persons who sailed in 1832, on a commercial expedition up the Niger, only 11 survived. Three steamers, well fitted up for the purpose, were despatched (June 1841) to this river, by the British government, with full powers to form commercial treaties with the natives, and to concert measures for the extinction of the slave trade. But did our limits permit, it might be easily shown that there are no rational grounds on which to anticipate any considerable success from this expedition, and that the barbarism of Africa seems to depend on natural and irremediable causes. (*Caillié's Travels in Central Africa,* ii., with *M. Jomard's Remarks; Ritter's Africa,* ii. 47–173; *Lander's Exped.* 3 vols.; *Geog. Journal,* vols. ii. and viii.; *Laird and Oldfield's Exped. into Africa,* &c.)

The *history of the Niger* is involved in extreme obscurity. Herodotus was informed by the Greeks of Cyrene, that in the interior of the African continent, a city had been reached by some Nasamon travellers, which was inhabited by negroes, and stood on the banks of a river containing crocodiles, and flowing from the W. eastward (ἀπὸ ἑσπέρης πρὸς ἥλιον ἀνατέλλοντα, ii. 32), which he conjectured to be the Nile. Now, as the Bahr-el-Abiad, or W. arm of the Nile flows from W. to E., and is certainly more likely to have been reached by the Nasamons than the Niger, the conjecture of the venerable father of history, that the river which they encountered, was in fact the Nile, seems to be more consistent with probability than that of D'Anville, Rennell, and other learned moderns, who suppose that the city visited by the Nasamons was Timbuctoo, and the river the Joliba of Mungo Park. The latter theory has, however, so far prevailed, that the name Niger is that which is now usually given to the river discovered and explored by Park and Lander. The word Niger, or Nigris, is first used by Pliny (*Nat. Hist.* v. i. 1–9.), from whose somewhat confused account it would appear that there were supposed to be two rivers of that name, one in Mauritania, 8. of the great chain of the Atlas, and the other in Æthiopia, thus briefly described:—"*Nigri fluvio eadem natura quæ Nilo: calamum et papyrum et easdem gignit animantes, iisdemque temporibus augescit.*" He seems, also, to have conceived that the Niger and Nile were united, and that there was a large water-system, having many branches in the interior of Africa. The poet Claudian also entertained the idea of a similar connexion:—

        Gir notissimus amnis
    Æthiopum, similli mentitus gurgite Nilum.

Ptolemy furnishes a somewhat more detailed account of the

river, and assumes that there are two separate streams in the interior of Africa, both having many branches (ἐκτροπαί), and connected with lakes; the river most eastward he terms the Gir (Γείρ), that to the W. being the Nigir (Νίγειρ), communicating with the lake Libye, which may, perhaps, be identical with the lake Tchad, discovered by Denham and Clapperton. Ptolemy says nothing, however, respecting the course of the river, though he seems to have been of opinion that its waters were absorbed in lakes, or lost by evaporation. Edrisi, Abulfeda, and other Arabian geographers, conceived that the Niger (by them called *Nil-el-Abid,* "Nigris Nileo") flowed westward, discharging its waters either into the Atlantic or some lake of the interior; and they represented it as rising from the same source as the Nile, and identified with it in the upper part of its course: this, indeed, is the opinion still maintained by the natives; and it is far from improbable that some of the affluents of the W. Nile may be connected, during the period of the inundation, with some of the affluents of the Niger. Such, in a few words, seem to be the leading statements of the more celebrated of the old geographers respecting the Niger. It is doubtful, perhaps, whether the Greek and Roman writers really possessed any authentic information as to the rivers and lakes S. of the Great Desert; and, at all events, the statements now referred to, if they really apply to that part of the continent, are at once extremely limited and extremely vague. That the caravans, which appear from a very remote period to have maintained an intercourse between the countries to the N. and those to the S. of the Great Desert, should have fallen in with and had some knowledge of the Joliba, is far from improbable; and perhaps, had any remains of the literature of Carthage come down to our times, they might have thrown considerable light on the question as to its identity with the Niger: but, with our existing means of information, it would appear, notwithstanding the learning and ingenuity that have been brought to its investigation, to be all but insoluble. The notices of the ancients are too obscure to admit of any certain inferences being deduced from them; and supposing (against the opinion of the learned Baron Walcknaer, *Recherches sur l'Afrique,* 409.) that the Niger is to be looked for to the S. of the Great Desert, the Bahr-el-Abiad, or western arm of the Nile, the Yeo, and other considerable rivers falling into the lake Tchad, correspond quite as well with their statements as the Joliba.

NIJAR, a town of Spain, in Andalusia, prov. Granada, 15 m. E.N.E. Almeria, and 78 m. E.S.E. Granada. Pop. according to Miñano, 5,792. It has two parish churches: its chief branch of industry is the manufacture of house-cloths.

NIJIR-EGYHAZA, a large market town of Hungary, co. Szaboles, 28 m. N. Debreczin. Pop. 15,640, principally Protestants. It has a saltpetre refinery; but by far the greater part of its inhabitants are agriculturists.

NIJNII-NOVGOROD, vulgarly *Nijegorod,* that is Lower Novgorod, a government in the central part of European Russia, on both sides the Wolga, between lat. 54° 28' and 57° 6' N., long. 41° 40' and 46° 38' E., having N. the government of Kostroma, E. Kasan and Simbirsk, S. Penza and Tambof, and W. Vladimir; area, 18,740 sq. m. Pop. (1838) 1,071,000. Surface flat or gently undulating; the soil, which consists principally of sand and black friable mould, is exceedingly fertile; and being (for Russia) well cultivated, this is one of the most productive provinces of the empire. Exclusive of the Wolga, several of its affluents, including the Oka, Betlouga, Piana, &c., traverse different parts of the government, which is well watered, at the same time that it is not marshy. There are some very large forests, those of the crown amounting to about 1,390,000 desiatines. The produce of the corn crops considerably exceeds the consumption. Hemp and flax are very extensively cultivated. Great numbers of cattle and horses are bred; and government is taking the most effectual measures to improve the latter. This is a considerable manufacturing, as well as a rich agricultural district. Coarse linen, canvas, and cordage, are the principal manufactured products; there are, also, some iron-works, with numerous distilleries and tanneries, soap-works, glass-works, &c. Commerce extensive and growing. The exports consist of corn and flour, cattle, horses, leather and tallow; the manufactured articles specified above, with iron, timber, potash, mats, glass, &c.

NIJNII NOVGOROD, NIJEGOROD, or NIJNII, the cap. of the above government, in the angle formed by the confluence of the Oka with the Wolga; lat. 56° 19' 49" N., long. 44° 37' 30" E. Stationary pop., 25,000. It stands partly on a steep hill, about 400 ft. in height, the summit of which is occupied by the Kremlin or citadel, and partly on the low ground along the sides of the rivers. The citadel, from the ramparts of which there is a noble view of the Wolga, Oka, and surrounding country, contains the government offices, two cathedrals, built after the model of Moscow: an obelisk 75 ft. in height, erected in honour of the deliverers of their country, the patriotic citizens, Minin, and Prince

ojarski; and other public buildings. The upper part of the
own has several good streets; and being ornamented by
numerous churches, placed in conspicuous situations, has
a imposing appearance. The lower town consists princially of a very long street, bordering the Wolga. With the
xception of the principal public buildings, and a few pri-
fic houses, the rest of the city is constructed of wood.
mong the establishments are three convents, a bazaar, a
mnasium, and four primary schools, an ecclesiastical
eminary, and a large military school. The town is an-
ent, having been founded in 1222. The Kremlin was sur-
ounded by strong walls and towers in 1508.

A bridge of pontoons leads across the Oka to the splen-
d new bazaars erected on the left bank of that river for
te exhibition and sale of merchandise brought to the fair.
These, which are divided into parallel rows or streets, are
onstructed of stone, roofed with iron, having covered galle-
es in front, supported by 8,000 iron pillars. They are built
a piles, and to guard against the danger of inundation, the
round on which they stand was raised about 20 ft. Being
nclosed on three sides by canals, and on the fourth by a
avigable inlet of the Oka, there is every facility for the
elivery and shipment of merchandise. The establishment
of very great extent, comprising above 2,500 booths; and
admitted on all hands to be-at once the largest and
ost perfect of its kind that is any where to be met with.
cluding the church, dedicated to St. Macarius, the patron
f the fair, it is said to have cost in all about 11,000,000 rou-
les.

Nijni Novgorod has various manufactures, but it owes its
reat importance almost entirely to its commerce. It is the
rand entrepôt for the trade of the interior of the empire,
nd has, in fact, a greater command of internavigation than
ny other city of the old world. Besides the corn, cattle,
nd other products of the surrounding country, the Kama,
he principal affluent of the Wolga, conveys to Nijni the
alt of Perm; the gold, silver, copper, and other metallic
reasures of the Oural mountains; the furs &c., of Siberia;
nd even the teas of China. The silks, shawls, and other
merchandise of central Asia, and the fish, caviar, &c., of
outhern Russia come up the river from Astrakhan; while
he manufactured goods of England and western Europe,
he wines of France, the cotton of America and the sugar
f Brazil, are conveyed to her from Petersburg and Archan-
el, with both of which, as well as Moscow, she is connect-
d by navigable rivers and canals. These advantages,
oined to her situation in a fertile country in the centre of
he monarchy, were so highly appreciated by Peter the
reat, that it is said he at one time intended to have made
Nijni the capital of his empire; and it is, perhaps, to be re-
retted that he did not carry this project into effect.

Latterly the commercial importance of Nijni has been
astly increased. Previously to 1817, the great fair, now
eld here, was held in a less convenient situation at Makna-
ief, lower down the Wolga. But the buildings for the ac-
ommodation of the merchants at Makarieff having been
ccidentally burnt down in 1816, government took advan-
age of the circumstance to remove the fair to Nijni. It
egins on the 1st of July, and continues for a month or six
eeks, and is well known, not only over all Russia, but over
ost other countries of Europe and Asia. It is carried on
rithin the bazaars already noticed, which were constructed
y government for the accommodation of the traders, to
hom they are let at moderate rents. The produce dispos-
d of is classified as follows, viz. 1st., Russian produce. raw
nd manufactured; 2nd, merchandise from the rest of Eu-
ope, consisting principally of manufactured and colonial
roducts; and 3d, products of China, Bokhara, the Kirghi-
es. and other Asiatic nations. The estimated value of the
roduce belonging to each of these classes, exposed to sale
t 1827, 1838, and 1839, has been as under:—

	First Class.	Second Class.	Third Class.	Totals.
	Roubles.	Roubles.	Roubles.	Roubles.
1827	67,800,000	14,000,000	22,000,000	105,000,000
1838	88,400,000	17,000,000	14,700,000	125,200,000
1839	122,267,000	16,035,000	23,005,000	140,597,000

In 1830, Russia sent to the fair silk goods valued at
,500,000 roubles; hides. tanned and raw, 3,000,000 roub.;
ry and salted fish, 1,800,000 roub.; cotton stuffs and yarn,
9,000,000 roub.; woollen stuffs. 500,000 roub.; furs and
eltries, 8,000,000 roub.; and 2,000,000 pouds iron. During
e same year there were sent to the fair by foreigners,
roollens of the value of 2,900,000 roub.; 33,368 boxes of
ea; 306,570 lbs. of silk, &c. Every sort of article is to be
ound in one or other of the different bazaars. In 1839, the
ottons exposed to sale were valued at 28,544,000 roub.; and
e metals and metallic goods at 22,390,000 roub. The con-
ourse of strangers during the fair is quite immense; so
uch so, that the population is then increased. according
o the lowest estimates by from 150,000 to 300,000 individu-
ls. Here are seen dealers from India, China, Tartary, Bo-

khara, Persia, Circassia, Armenia, and Turkey; and from
Italy, Poland, Germany, France, England, and even America. Amusement, as well as business is attended to: theatrical representations, shows of wild beasts, and other Bar-
tholomew-fair diversions, being got up for the entertainment
of the multitude. (Schnitzler, La Russie, &c., 114–190;
Lyall, ii. 329–355; Possart, &c., Das Europäische Russland,
Sch. &c.

NIKOLSBURG, a town of Moravia, circ. Brünn, from
which city it is 28 m. S. Pop. about 8,500, a third part of
whom are Jews. It has a fine castle and grounds belonging to Prince Dietrichstein, a philosophical academy, a gymnasium, and several other superior schools; and in the castle is an extensive library, comprising many valuable MSS.
The town is dirty and wretched; it has however, manufactures of woollen cloth and other stuffs, and some trade in
wine and marble, both produced in its vicinity. (Oesterr.
Nat. Encycl.; Berghaus.)

NILE (Lat. Nilus, Gr. Νεῖλος, from νέα ἰλύς, "new
mud," because it brings down vast quantities of slime or
mud*). a large and famous river of N.E. Africa, flowing N.
through Abyssynia, Nubia, and Egypt, to the Mediterranean
sea, celebrated alike for its magnitude, the inexhaustible
fertility which it confers on the "land of Egypt," the uncertainty of its origin, its connection with some of the most
interesting events in remotest periods of authentic history,
the great cities that were early built on its banks, and the
stupendous monuments that still attest the wealth and
power of their founders. The discovery of its real source
was an object of intense curiosity to the ancients, as it still
remains to the travellers and geographers of modern days;
the words of Tibullus,

Nile pater, quanam te possum dicere causa
Aut quibus in terris occuluisse caput,   —

being nearly as applicable now as in his time.

The Nile is formed by the junction, at 15° 34' N. lat. and
39° 30' 58" E. long., of two great arms, the *Bahr-el-Azrek*,
(the *Astapus* of the ancients), or Blue river, from the S. E.,
and the *Bahr-el-Abiad*, or White river, from the S.W. The
sources of the former, which derives its name from the dark
colour of its water, were discovered and described by Paez,
in 1618, and were subsequently visited by Bruce, who ridiculously pretended to have, for the first time, ascertained
the true sources of the Nile, and thus solved a problem that
had for ages occupied the attention of the learned world!
This E. branch rises from two fountains near Gersh in Gojam, in Abyssinia, at an elevation of about 10,000 ft. above
the level of the sea, in lat. 10° 59' 25" N., long. 36° 55' 30"
E. It thence flows N. to the lake of Dembea, or Tzana, a
large sheet of water which receives many other streams;
but the Nile is said to preserve its waters with little intermixture with those of the lake, across which its current is
always visible. Escaping from this lake, it sweeps in a
southerly direction round the E. frontier of the province of
Gojam and Damot, till, between the 9th and 10th deg. N.
lat., it takes a N.W. direction. which it preserves till, at
Khartoom, it unites with the other great arm, the Bahr-el-
Abiad, flowing from the S.W. The Bahr-el-Azrek receives
in its course several important tributaries, as is in several
parts interrupted by cataracts, one series of which has a fall
of 280 ft. At the point of junction with the other great
arm, it is about ½ m. in breadth, and has a rapid current;
but, during half the year, its waters are low.

The W. arm, Bahr-el-Abiad, or White river, derives its
name from the fine whitish clay usually suspended in, and
colouring its waters. It is broader and deeper than the E.
arm, brings down a larger volume of water, and appears to
have been regarded in antiquity as the true Nile. (Herod.
ii. caps. 30, 31. ; *Memoire de D'Anville* ; *Mémoires de l'Académie des Inscriptions*, &c., xxvi. 46.) If, however, the derivation of the name previously given be correct, the Bahr-
el-Azrek would seem to have the best right to be considered
the genuine Nile, inasmuch as it carries down the greater
portion of that mud whence its name has been derived, and
the deposits of which have. in the lapse of ages, formed
the land of Egypt. But without insisting further on this
point, though the sources of the Bahr-el-Abiad have not
hitherto been explored,

Nec licuit populos parvum te, Nile, videre,   '

its course was traced in 1827, by Linant, for about 160 m.
from its confluence with the Bahr-el-Azrek. (*Geog. Journal*, ii. 171–187.) And a party sent by the pacha of Egypt
on a slaving expedition have since traced it to a much greater distance, or to a point supposed by Col. Leake to be in
about the 10th deg. of N. lat. and 29th deg. of E. long. ; and
at this point no mountains were in sight, the river being also

---

* This is the derivation given by Servius in his notes to Georg., lib. iv., v.
291; but many other derivations have been proposed, and perhaps the Nile
may, like the Hebrew Nahhal, merely mean the river, or river par excellence. (See *Dictionnaire de Trevoux*, art. Nil.)

of great breadth, full of islands, and shallow. Perhaps, however, we have not be far wrong in fixing its sources on the N. slope of the Gebel-el-Kumri, or Mountains of the Moon, in about the 6th deg. of N. lat., and between the 22d and 30th deg. of E. long. But whether its confluents form themselves into lakes, as was conjectured by Ptolemy, or fall successively into the main stream, are questions as to which no information can be given. The course of the Bahr-el-Abiad, so far as it has been explored, to its junction with the Bahr-el-Azrek, is pretty uniformly N.N.E.; it receives many tributaries, and forms numerous islands. "At the point of confluence, the Bahr-el-Abiad is only about 1,800 ft. across; but a little above it enlarges much, its banks being frequently 3 or 4 m. apart, and in some places during the inundations the waters extend 21 m. from side to side. In its ordinary state, and in mid channel, it has here from 3 to 4 fathoms water." (*Geog. Journal*, ii. 187.)

After the junction of its two great arms, the united river, or Nile, takes a generally N. direction, but with almost innumerable windings. Not far below the point of confluence is a low range of mountains, through which the river rushes in a narrow gorge, forming what is called the sixth cataract; and thence deflecting eastward, through extensive and verdant plains, it passes the capital of Shendy and the ruins of the ancient Meroë. It receives, close to the town of Addamer (lat. 17° 45' N.), the waters of its important tributary the Tacazzé (the *Astaboras* of the ancients), which has its sources in the high lands of Lasta, in Abyssinia, in lat. 11° 40' N. long. 39° 40' E., about 2½ deg. E. of lake Dembea, pursuing thence a pretty uniform course N.N.W. to its junction with the Nile. From this point to its embouchure, a distance of about 1350 m., the Nile receives no affluent whatever, either on its E. or the W. bank, a solitary instance, as Humboldt has remarked, in the hydrographic history of the globe. (*Per. Narr.*, v. 744.)

At Abu Hamed, in about 19¾ deg. N. lat., and 33 deg. E. long., the river, which had previously been following a northerly course, turns suddenly to the W., and thence pursues a south-westerly course to Edab, in the province of Dongola, in the 18th deg. of lat., where it again curves round to the N. This deflexion is called the great bend of the Nile. In this course through Dongola, the valley on each side is very circumscribed. The river enters Lower Nubia in about 19° 40' N., where it is precipitated over a ledge of granite rocks, forming what is commonly called the 3d cataract. Under the 22d parallel the 2d cataract, of Wady-Halfa. The first, or lowest cataract, is that of Assouan (an. *Syene*), near the island of Elephantine, where the river has cut its way through a ridge of granite rocks. (*See* vol. i., 810.) It must be observed, however, that the term "cataract," as applied to the broken course of the Nile, bears no analogy to the great cataracts of Niagara, the Pisse-Vache, &c.; for most of them scarcely exceed a few feet in height, and are, in fact, rather rapids than cataracts. In a portion of Lower Nubia the river-valley is very much contracted: the rocks on both sides approach the shore so closely as to allow little space for the deposit of alluvium; and in other places on the Libyan side, the sand covers the whole level space between the hill and the bank. At Kalabsheh, the an. *Talmis* (which has a temple bearing a close resemblance to the temple of Tentyra, Edfou, and Phile), the river rises from 30 ft. to 40 ft. during the floods; and after their subsidence in Feb., the stream flows at a rate of two or three nautical miles an hour. (*Geog. Journ.*, vol. ix.. part 3.) The Nile after entering the boundaries of Egypt at Philæ, six miles from Assouan, runs in a quiet and very tortuous stream, though generally northward, through the whole length of the country, enriching it by its waters, and its deposits, which, indeed, not only give to Egypt its fertility, but make it habitable. But, with the exception of the district of Fayoum (vol. i., 901), the valley of the Nile in upper and central Egypt is of very contracted dimensions, the mountains and the burning sands of the desert encroaching so closely upon it, that it seldom exceeds 10 m. in width, and is frequently not half so much. But how limited soever, this narrow stripe is of extraordinary beauty and fertility, and contains the magnificent remains of some of the noblest and most populous cities of the ancient world. But we beg to refer the reader to the article EGYPT for farther particulars as to the past and present state of the bed of the river, its inundations, and its delta. In antiquity, the Nile seems to have poured its waters into the sea by seven mouths; but it has now only two mouths, those of Rosetta and Damietta. The former, or most westerly, has a breadth of 1800 ft., with a depth of about 5 ft. in the dry season. The Damietta mouth is only 900 ft. wide; but its depth averages between 7 ft. and 8 ft. when the river is lowest. The greatest breadth of the Delta is about 85 m. from E. to W.; the distance of its apex from the sea being rather more than 90 m. Great changes have, however, taken

place in it during the lapse of ages; the soil has not only been elevated many feet by alluvial deposits, but its shape and the position of its apex have greatly altered even within the period of modern history. The river begins to swell in its higher parts in April, and even earlier in the Bahr-el-Abiad; but at Cairo no increase occurs till the beginning of June, its greatest height at that city being in September, when the Delta is almost entirely under water. The waters subside in November, leaving a rich alluvium, which is the great source of the fertility of Lower Egypt. *Quotannis certis diebus, præcipue circa solstitium æstivum, aucto magno per totam spatiatus Ægyptum, terram pluviis omnibus destitutam aquis suis irrigat, limo tegit, et fæcundissimam aficit. Unde unica spes Ægyptii in Nilo posita est, quia fertilis aut sterilis annus est, prout ille magnus aut parcior fuit.* (Cicero, *De Nat. Deor.*, L. cap. 38.)

We need not, under such circumstances, feel surprised that the ancient Egyptians regarded the Nile as a god to whom they paid divine honours. The greatest breadth of the river may be estimated at 2000 ft., or about twice the width of the Thames at London bridge. Its average current does not exceed 3 m. an hour: the water is always muddy; and even in April and May, when it is clearest, it has a cloudy hue. When it overflows, the colour is of a dirty red, consisting chiefly, we believe, of the red-clay deposit of the Bahr-el-Azrek; for, as already stated, the Bahr-el-Abiad brings down only a fine whitish clay. The Nile abounds with a great variety of fish, such as the *Labrus Niloticus*, or white trout, the *Murœna anguilla*, and a large species of salmon. The *Oxyrynchus* of this river, so famed in the antiquities of Egypt is, according to D'Anville, the fish now called *Kesheo*. None of the fish, however, except eels, have any very close resemblance to those of Europe. Among the waterfowl of the Nile, the most characteristic is the Turkey-goose, or *Anas Niloticus*, the flesh of which is both palatable and salubrious. From Assouan down to Cairo, about 360 m., the banks, except in the rocky parts, present no native plant, but abound with all sorts of succulent vegetables, raised by the industry of the inhabitants in this peculiarly fertile soil. Cultivation, however, is more common on the E. than on the W. bank of the river. Hippopotami are found in Nubia, but not in Egypt; the crocodiles, also, are greatly reduced in number, and are now confined to the district above Assiut.

NILES, p. t., Cayuga co., N. Y., 10 m. S.E. Auburn, 169 m. W. Albany, 329 W. Bounded E. by Skeneateles lake, W. by Owasco lake. It has a Presbyterian church, four stores, one fulling-mill, two grist-mills, 16 saw-mills, two tanneries, one distillery; 11 schools, 629 scholars. Population 2234.

NILES, p. t., Berrien co., Mich., 182 m. W.S.W. Detroit, 697 W. Watered by St. Joseph's river. It contains two churches, a Presbyterian and Episcopal, three commission houses in foreign trade, 11 stores, two flouring-mills, two grist-mills, four saw-mills, two tanneries, two distilleries, two printing-offices, two weekly newspapers; one college, 38 students; four schools, 94 scholars. Pop. 1622.

NIMEGUEN, or NYMEGEN (probably the an. *Noviomagus*), a town of Holland, prov. Gueldertand, cap. arrond., on the Waal, here crossed by a flying bridge, 34 m. S. by W. Arnheim, and 53 m. S.E. Amsterdam. Population about 14,000. It stands on several small but steep hills, and is pretty strongly fortified. Though not ill built, it has an irregular appearance, the streets being narrow; and, on account of the abrupt elevation from the river, the windows of one range of houses overlook the chimneys of another. Among the public buildings worth notice, are an old edifice, said to have been raised by the Romans, and now forming part of the fortifications; the old castle of Valkenof, believed to have been built by Charlemagne; and the town-house, an edifice of considerable beauty. Several of the churches are likewise entitled to attention; and a high tower, called the Belvidere, is much resorted to by visitors, on account of the extensive view which it commands of the course of the river and the surrounding country. Nimeguen is the seat of tribunals of primary jurisdiction and commerce, and the residence of a military commandant and a receiver of taxes. It has a branch of the Society of Public Good, a commission of agriculture, and a Latin school. It produces Prussian blue, glass, &c., and has some tanneries; but the only article for which it is celebrated is its pale beer, sent to almost every part of the Netherlands. Nimeguen is known in history from the treaty concluded here, in 1678, by Spain, France, and Holland. It was taken by the French on the 8th September, 1794, after a severe action, in which the allies were defeated. Various Roman antiquities have been discovered in and about the town. (*De Cloet; Dict. Geog.; Barrow's Tour in Holland.*)

NIMES, or NISMES (an. *Nemausus*), a city of the S. of France, dep. Gard, of which it is the cap., in an extensive and fertile plain, near the Vistre, 23 m. W.S.W. Avignon, d 30 m. N.E. Montpellier: lat. 43° 50′ 8″ N., long. 4° 21′ ″ E. Pop., in 1836, ex. com., 41,194. The distant view Nimes is not imposing. Notwithstanding its numerous e edifices, it has only the *Tourmagne* to render it con- icuous at a distance. The city-proper, which is sur- unded by boulevards, on the site of the ancient fortifica- as, is confused and irregular with narrow streets and ill- uilt houses. But the boulevards and suburbs, which mprise three fourths of the houses, are regularly laid t, clean, and have numerous handsome modern build- gs and fine public promenades.

Nimes is principally interesting on account of its remains antiquity, of which it probably possesses more than any her city of Europe, Rome excepted. The most classical, ough not the most extensive, of these is the oblong tem- e, absurdly called the *Maison-carrée*, nearly in the cen- e of the city. This edifice was supposed, from an in- ription discovered on its frieze, to have been built in onour of Caius and Lucius Cæsar, grandsons of Augus- s; but, from subsequent discoveries, it would appear to ave been erected to the adopted sons of Antoninus Pius. t any rate, it dates from the finest period of Roman art, nd is one of its most perfect remains. It is raised on a latform ascended by 15 steps, and has 30 Corinthian col- mns, six in the front and at the back, and nine on each de, exclusive of those at the angles. The portico, which of ample dimensions, is supported by six detached col- mns in front, and two on either side: the other columns n the sides and back of the building are sunk half way to the walls. The capitals of the columns, and the frieze, ornice, and other parts of the building, are profusely dorned, in the most exquisite taste. The measurements f this edifice are as follow: length, 82½ ft.; breadth and eight, 40½ ft. each; height of the platform on which it tands, 18½ ft.; height of the stylobate, 9½ ft.; height of the oorway, 23¼ ft.; breadth of do., 10½ ft. The columns, which are about 30 ft. in height, have a height equal to 0½ diameters. (*Frossard, Tableau Pittor. de Nismes*, ii., 71.)

The *maison-carrée* was considerably injured in the mid- le ages; but it is protected from future spoliation by being nclosed within an iron palisade, and since 1823 it has been mployed as a museum of paintings and antiques. But it would have been more consistent with good taste to have reserved it untouched and unoccupied, in its ancient sim- licity.

The amphitheatre of Nimes is admitted to be the most erfect structure of its kind extant, after that of Verona. t stands on one of the boulevards, surrounded by a large pen space, on which no buildings are allowed to be erect- d. It is said to have been founded by Antoninus Pius. Its ongest external diameter is 437 ft.; its shortest 332½ ft.: it as 32, or, according to some authorities, 35 ranges of eats, and is variously estimated as having sufficient ac- ommodation for from 17,000 to 23,000 spectators; the eight of the building outside is from 68 to 104 ft., and its otal external circumference is 1174½ ft. (*Frossard*, i., 135.) Though it was occupied by the Visigoths, and afterwards he Saracens, as a fortress for their defence against the Franks, the outer wall is still nearly entire. It consists of wo stories, each having 60 arches, and an attic story, and s entered by four gates, one at each of the cardinal points, he principal being on the N. side. The arcades of the round story are separated by pilasters, those of the upper y columns, in an irregular Tuscan or Doric style. The nterior is in many parts dilapidated and overgrown with egetation; but it still serves for bull-baits, jousts, and dra- atic entertainments, to which the modern inhabitants of Nimes are as much addicted as their ancestors were to the ore barbarous exhibitions of gladiators.

A few portions of the ancient walls still remain, princi- ally in the *Portes d'Auguste* and *De France*: the first, which, in the time of the Romans, was the principal gate f the city, consists of two large and two smaller arches: he former, which are in the middle, have between them small Ionic column, respecting which there has been uch controversy, all the other decorations of this gate eing of the Corinthian order. The *Porte d'Auguste* is laborately ornamented with sculptures, which constitute ne of the principal points in which it differs from the *Porte de France*. In the N.W. part of Nimes is a ruined *nymphæum*, or Roman bath, of considerable size, improper- y termed the Temple of Diana. Near this, on a height verlooking the city, is the *Tour magne* (turris magna), a ower supposed to have been built by the Greek colonists f the city before the Roman invasion; but the original urpose of which has not been correctly ascertained. It is n the Doric style; its lower part being heptagonal, its up- er, octagonal. It is in great part ruined; but being still 00 ft. in height, and in a conspicuous position, it is used o support a telegraph. The above are the principal ob- ects of architectural interest in the city. The Vandals, and other barbarians, are said to have destroyed the basili- ca of Plotinus, the temples of Apollo, Ceres, Augustus, &c.; but the still existing memorials of antiquity are more than sufficient to evince the almost unequalled magnifi- cence of the ancient city.

Nimes does not, however, owe its sole interest to its an- tiquities. It has several large, and some good modern, ed- ifices. The cathedral, begun in the 11th, but principally constructed in the 16th and 17th centuries, has little to re- commend it, except its occupying the site of the temple of Augustus, but the *Palais de Justice* on the esplanade, the *Hotel-Dieu*, principally rebuilt in 1830, the general hospital, the new theatre, several of the churches, and the public library, are handsome, well-contrived buildings. A large fortress to the N. of the city was constructed by Vauban, on the site previously occupied by the basins that received the water brought thither by the aqueduct, of which the *Pont du Gard* forms a part. It is now the central prison for the S. departments of France, and has usually about 1200 inmates. The bishop's palace, Episcopal seminary, college, and large barracks, are the other principal pub- lic buildings. The esplanade contiguous to the amphi- theatre, and the *Cours Neuf*, are among the finest prom- enades. The last named extends quite through the W. part of Nimes from N. to S.; and leads to the fine and extensive *Jardin de la Fontaine*. This garden derives its name from a large and handsome fountain, and has in it many statues and other Roman antiquities, besides the *nymphæum* mentioned above.

The *Pont du Gard*, above alluded to, formed part of a superb Roman aqueduct, 25½ m. in length, which convey- ed a supply of water from the neighbourhood of Uzes to Nimes. We have no certain details as to the founders of this great work, the æra of its construction, or the pur- pose for which the water brought by it was employed. Some antiquaries have ascribed its erection to Agrippa, son-in-law of Augustus, about *anno* 19 B.C., while others have ascribed it to Adrian, or his successor Antoninus, who derived his origin, by the father's side, from Nemau- sus. But, by whomsoever constructed, it was worthy the most brilliant æra of Roman power. The Pont du Gard consists of that part of the aqueduct which was thrown across the river Gardon, in a wild defile, 11 m. N.E. Nimes. It consists of three rows of arches, or, as it were, three different bridges, raised the one above the other, the whole being constructed of large stones, without cement. The first, or lower tier or bridge, has a length of 599 English ft., and a height of 65½ ft.; and consists of six arches of unequal size, the breadth of the largest, through which the Gardon usually flows, being 82½ ft. The second, or middle tier, is 846 ft. in length, and 62½ ft. in height: it consists of 11 arches, generally smaller than those of the first tier, but like them of unequal size. The third or upper tier, 870 ft. in length, and 23½ ft. in height, has 35 arches, which of course are much smaller than those of the other tiers, being respectively only 13½ ft. in width. The entire height of the structure is 188 ft.; its width or thickness, which is 19½ ft. at its base, diminishes as it as- cends; on its summit is the watercourse, 4½ ft. in depth and 4 ft. in breadth, and through it a person may now pass with ease from one end of the structure to the other. About the middle of last century, a carriage road was built up against the bridge as high as the base of the sec- ond tier of arches. The *Pont du Gard* is in the Tuscan style; it is very little ornamented, but is a highly pictu- resque object. With singular good fortune, it escaped di- lapidation during the dark ages; and the greatest injury it experienced was in 1600, from the Duke de Rohan, who broke away a portion of the second tier of arches to facil itate the passage of his artillery; but the breach was af terwards repaired at the expense of the states of Lan- guedoc.

Nimes is a bishop's see, the seat of a royal court for the departments Gard, Lozere, and Vaucluse, courts of prima- ry jurisdiction and commerce, a chamber of commerce, *conseil de prud'hommes*, a university academy, the royal academy of Gard, a royal college, &c. It has schools of drawing and chemistry, as applied to the arts, societies of agriculture, medicine, &c., a Bible society, a commission of antiquities, an athenæum, an extensive public library, and a cabinet of natural history.

Nimes is further distinguished by its manufacturing in- dustry. It is one of the principal seats of the silk manu- facture of France; ranking, in this respect, immediately after Lyons and (perhaps) St. Etienne. Its manufactures are principally silk hosiery and shawls; and silk stuffs mixed with cotton, linen, and woollen. There are, alto- gether, between 7000 and 8000 looms at work in Nimes, many of which are Jacquard looms. All the weavers work with their families at their own homes, there being no large factories except for dyeing, or for printing silk stuffs; which latter branch of industry has greatly augmented since 1836, when it employed from 600 to 700 hands, exclusive of children. But, though the silk manufactures of Nîmes be extensive, the goods produced are not much esteemed by the upper and middle classes, being mostly mere imitations of those of Lyons, and of inferior quality. From this and other causes the export trade of Nîmes is small; its industry is not progressive, and its population often experience distressing *crises*. The weavers employ about 11 hours a day at the loom; the wages of a man being estimated by Villermé at an average of 30 sous, those of a woman at 12 sous, and of children from 5 to 12 sous. These low wages being barely sufficient to provide current necessaries, the weavers are almost all wretchedly clothed, dirty, and ill provided with fuel in winter. According to Villermé, they are intelligent and laborious, and not addicted to drunkenness or other kinds of profligacy; but they have neither economy nor foresight, with the exception of the silk stocking weavers, who being employed on articles less subject to the caprice of fashion, are less affected by *crises* than the rest. These form, in fact, a separate class, distinguished for economy and prosperity, notwithstanding that their wages are smaller than that of most other artisans. The besetting fault of the working population of Nîmes is a want of perseverance. Few are able to write and read; and many spend a considerable part of the week in idling. Besides silks, Nîmes has manufactures of cotton goods, gloves, leather, brandy and vinegar, and a good deal of trade in wine, essences, drugs, colonial produce, &c. It is also the principal entrepôt for the raw silk produced in the S. of France, of which material almost all its own silk manufactures are made. Of the population of Nîmes and its suburbs, about 32,000 are Roman Catholics, and 12,000 Protestants; and in few towns is there so much acrimonious party-spirit and violence displayed on account of religion. This violence broke out, soon after the downfall of Napoleon, into the most atrocious acts on the part of the Catholics, which might easily have been suppressed by a vigorous government, but which were, in fact, rather encouraged by the imbecile bigots then at the head of affairs in France.

Nemausus is supposed to have been founded by a colony of Phocians; it was subjugated by the Romans, anno 121 B.C. In the middle ages it belonged successively to its own viscounts, the counts of Thoulouse, and the kings of Aragon, by one of whom it was ceded to Louis IX., in 1258. Nîmes has given birth to many distinguished persons, among whom may be specified the Court de Gebelin, author of the *Monde Primitif*, and M. Guizot, now (1841) minister for foreign affairs, and author of the able and original works on the progress of civilisation in France and Europe, &c. (*Histoire de la Civilisation en France*, and *Histoire Générale de la Civilisation en Europe, &c.*) This illustrious individual, to whose enlightened and rational patriotism, integrity, and good sense, France and Europe are under the greatest obligations, first saw the light on the 4th of October, 1787: he is a Protestant, and the simplicity of his character, and perfect freedom from all sorts of pretension, give additional lustre to his talents and eloquence. (*Hugo*, art. *Gard; Tableau de Nîmes; Villermé, Tableau des Ouvriers, &c.*, i., 406–417; *Bowring's Report, Inglis, &c.*)

NEMISHILLEN, t., Stark co., O. Watered by Nemishillen creek. It has one school, 30 scholars. Pop. 1927.

NINEVEH, a great and famous city of the ancient world, the cap. of the Assyrian empire, is supposed to have stood on the E. bank of the Tigris, opposite to the modern city of Mosul (which see). There is every reason to think that its site was identical with that of the village of Nunia, and the "tomb of Jonah," about ½ m. from the river, which stand upon and are surrounded by vast heaps of ruins; lat. 36° 20' 17" N., long. 43° 10' 17" E. Herodotus (i., 185,) and other profane writers ascribe its foundation to Ninus, son of Belus, and first monarch of the Assyrian empire; but, according to the Bible (Gen. x., 11), "Ashur (the grandson of Cush) went forth out of the land of Shinar, and builded Nineveh." Its history is lost in the obscurity of succeeding ages; but it was, no doubt, a very large city nine centuries before the Christian era, for at that period Jonah described it as "an exceeding great city of three days' journey." (iii., 3.) Strabo says (l. xvi,) that it was larger even than Babylon; the circuit of which he estimated at 385 stadia; and, according to Diodorus Siculus (l. ii.), it was of an oblong shape, 150 stadia in length, and 90 in breadth; that is, above 54 m. in circuit. Very little dependence can, however, be placed on these statements; and it is, at the same time, admitted that the walls included a large extent of well cultivated gardens, and pasture grounds. The description of its walls, given by Diodorus, is too obviously exaggerated to require any notice. The prophet Jonah says that Nineveh "had more than six score thousand persons that could not distinguish between their right hand and their left." (Jonah, iv., 11.) This expression, the import of which is by no means clear, has been

generally understood to refer to the children; and, taking it in this sense, and including under the term children the younger persons under nine years of age, they might be taken at about one fourth part of the population, which, consequently, would be 480,000. But if we suppose, as some critics have done, that the children referred to by the prophet could not well exceed five years of age, they might be taken at between one sixth and one seventh part of the population, which would, consequently, amount to from 720,000 to 840,000. It is plain, however, that these statements are far too vague to be entitled to any considerable weight.

Nineveh was the residence of the Assyrian kings, and a city of such commercial importance, that Nahum apostrophises her: "Thou hast multiplied thy merchants above the stars of heaven." (iii., 16.) She was besieged and taken by Arbaces the Mede in the 6th century B.C., but appears to have been regarded as the capital of the Assyrian empire down to anno 612 B.C., nearly three centuries after Jonah's prophecy of her destruction, when she fell, after a protracted siege, into the hands of Ahasuerus, or Cyaxares, king of Media, who took "spoil of silver and gold, and none end of the store and glory out of all the pleasant furniture," making her "empty, and void, and waste." (Nahum, ii., 9, 10.) The spoil was taken to Ecbatana, the citizens were dispersed in villages, and the Assyrian empire, which for four centuries had been the glory of the eastern world, gave way to that of the Medes and Persians. It seems certain, however, either that the city had not been wholly destroyed, or, which is most probable, that a new and inferior city had, at a subsequent period, grown out of the ruins of the more ancient city; and the latter, no doubt, is that referred to by Tacitus. *Annal.* xii., 13, and Ammianus Marcellinus, xxiii., 22. Benelr supposes that the present remains, comprising an oblong rampart and fosse, about 4 m. in circuit, with a mound-covered wall about 20 feet in height, are those of the more recent city. The ruins at first sight present a range of hills, from which large stones are constantly dug out; and the bridge over the Tigris seems to have been entirely built of stones from the ruins, which have, no doubt, furnished as large a supply for modern buildings as the quarries near Mosul. The tomb of Jonah occupies one of the hills above mentioned, and is covered by a mosque, held in high veneration all over the East. Bricks, entire as well as in fragments, and pieces of gypsum, with inscriptions in the wedge-formed character, are found here, closely resembling those of Babylon (*Kinneir's Persia*, p. 256–259; *Olivier, Voyen Turquie*, iv., 265–278; *Ainsworth's Assyria, &c.*, p. 138–258; *Calmet, Dict. de la Bible*, ad voc. *Nineveh*.)

NING-PO, a city of China of the first rank, prov. Che-Keang, at the confluence of the rivers Kin and Yoeu, near their mouth in the harbour of Chusan, 46 m. E. by S. Hang-tcheou, and about 180 m. S.E. Nankin; lat. 29° 57' N. long. 121° 17' E. Pop. estimated at from 280,000 to 400,000. It is surrounded by walls and bastions, now in ruins, and is entered by five gates; the streets are broad and long, and the shops surpass those of Canton in elegance and splendour. It is intersected by numerous canals; a fine brig bridge crosses the inlet; and there are several pagoda, government warehouses, and other public buildings. The suburbs are flat, presenting rich fields and rice-gardens; but at the back, skirting the sea-shore, are dark-looking barren hills. Ning-po may be considered the third or fourth emporium of the Chinese empire; and its trade to the I. and S. districts of China, as well as to Siam, is of the highest importance. In the neighbourhood are very extensive salt works, and salt is exported in considerable quantities. The town is accessible by vessels of 300 tons; but large ships unload at Chinhae, a fortified town at the entrance of the inlet.

The English formerly traded to Ning-po, and the ruins of the British factory are still perceptible near the harbour of Chusan. They were compelled, however, by the Chinese, in the 17th century, to confine themselves to Macao, at the same time that similar restrictions were imposed on the Portuguese. (*China Opened*, i., 115.)

NIORT, a town of France, dep. Deux-Sèvres, of which it is the cap., on the Sèvre-Niortaise, 34 m. E.N.E. La Rochelle, and 43 m. W.S.W. Poitiers; lat. 46° 19' 2" N. long. 0° 19' 12" W. Pop., in 1836, ex. com., 18,915. It is pleasantly situated on the declivities of two hills, and is surrounded by planted promenades. It was formerly ill-built, but has been greatly improved since the Revolution, many new and good streets having been constructed on the site of the ancient fortifications. The castle of Niort, which has been long converted into a prison, was the birth-

---

* The well-informed authors of the learned and valuable work, *L'Art de Vérifier les Dates*, have made a singular blunder in noticing this subject. They say that the children of six years of age and under do not exceed the 1-20th part of the pop. of a city; and that, consequently, the population of Nineveh must have amounted to about 2,400,000. (*Tom. ii., 363, 3me. ed.*)

place of Mad. de Maintenon. The town has two good parish churches, one of which was built by the English, two hospitals, some good barracks, public baths and public halls, a handsome arcade (*galerie vitrée*), a theatre, a public library with 20,000 vols., including some rare MSS.; and a botanic garden, having attached to it a large horticultural school. It is the seat of tribunals of-primary jurisdiction and commerce, a royal athenæum, a council *des prud'hommes*, a society of agriculture, and a communal college. It has manufacture of leather, gloves, shoes, woollen stuffs, wooden and horn articles, &c.; and is an entrepôt for the wines of the Gironde, and for timber, wool, hides, and cattle. It is also celebrated for its confectionery (*confitures d'angelique*). (*Hugo*, art. *Deux-Sèvres*.)

NIPHON. *See* JAPAN.

NISHAPOOR, a town of Persia, in Khorassan, cap. district of its own name, 46 m. W. by S. Meshed, lat. 38° 55' N., long. 38° 8' E. Pop., according to Captain Conolly, 8060. The town has a poor appearance, being confined within a mud wall and ditch, without either minarets or domes; the only building that appears above the wall being a shapeless mosque. The circuit of the present wall does not exceed 4000 paces, and the greater part of the enclosed area is covered with ruins. The houses now inhabited, of which there are about 12,000, are meanly built, chiefly of mud. A tolerably large bazaar is well filled with goods for the consumption of the town, and provisions are alleged to be cheap and of good quality.

Nishapoor has few manufactures, and cannot boast of a single branch of foreign trade, except that of turquoises, from which, owing to the exactions of the government, and clumsy mode of working, it derives little benefit. The turquoise mines (from which exclusively are derived our supplies of this valuable gem), are about eight or nine in number, principally situated in a hill about 40 m. W.S.W. Nishapoor; of these, however, some have been abandoned, and others are so imperfectly wrought, as scarcely to pay the miners' expenses. The gems are usually found in a reddish brown rock, but occasionally also in a firm quartzose rock of a whitish grey colour, abounding with veins of specular iron. The produce of the mines would be very great under proper management; but nothing can be more inartificial than the process now adopted by the peasant-farmers, no skill or ingenuity being exerted, and no sort of contrivance used to lessen labour, or economise time and material. This defective management is mainly attributable to the wretched government, and the consequent insecurity of property from the oppressions of the local authorities. The mines are rented from the crown for about 2000 tomans annually, and wrought almost exclusively by the inhabitants of the surrounding villages. The produce is either sold to merchants resorting thither, or sent for sale to Meshed; but the miners practise every possible deception on purchasers; and the gems cannot, according to Mr. Fraser, be procured at a rate which would yield any considerable profit on a sale in Europe. Iron and rock-salt are also wrought within the district. Agriculture is little understood; the soil is tilled only once in three or four years, the ground being left fallow during the intervening time; one fifth of the produce is claimed as the property of the shah.

Nishapoor lays claim to high antiquity. It is said to have been destroyed by Alexander the Great, and rebuilt by Shapoor: afterward, during the Seljuk dynasty, it was one of the four royal cities of Khorassan; but in 1209 it was destroyed by the Tartars, who massacred most of its inhabitants. It was again pillaged by Jhengiz-khan; and more recently, in 1740, by Nadir Shah, from whose ravages it has never recovered. (*J. B. Fraser's Khorassan*, p. 395–423; *Conolly's Travels to the N. India*, i., 211–214; *Kinneir's Persia*, p. 185.)

NIVELLES (Flem. *Nyvel*), a town of Belgium, prov. S. Brabant, cap. arrond., on the Thienne, 17 m. S. Brussels. Pop., in 1836, 7814. It is said to have had, in the 16th century, a pop. of 30,000; and it is still half a league in circuit, exclusive of its suburbs. It is not well built; but it has a remarkable church, in which are two finely carved pulpits, and on the tower is a colossal statue, called Jean de Nivelles, which strikes the hours. It is the seat of a court of primary jurisdiction, and the residence of a receiver of taxes; with manufactures of woollen stuffs, coarse lace, cotton, and linen cloths, hats, paper, and oil; and sends two deputies to the states of the province. It originated from a remarkable Benedictine convent, founded by St. Gertrude in 645, the abbesses of which enjoyed the title of princesses of Nivelles. (*Dict. Géog.*)

NOANAGUR (*Nowanagara*, "The New City"), a town of Hindostan, prov. Gujrat, dom. of the Guicowar, and cap. of the most considerable subordinate chieftain in the Guj-rat peninsula; on the Nagal, near its mouth, in the gulf of Cutch. 63 m. N.W. Joonagur. It is 4 m. in circuit, and is defended by a stone wall, of no great strength, and a ditch.

Many of its inhabitants are weavers, and manufacture large quantities of coarse and fine cloths, which are sent to Surat, and various other commercial towns. The water of the river is supposed to be peculiarly adapted to dyeing cloth, for which branch of industry this town is also celebrated.

NOBLE, county, Ia. Situated in the N.E. part of the state, and contains 432 sq. m. Watered by Elkhart and Tippecanoe rivers. It contained in 1840, 3802 neat cattle, 1585 sheep, 6954 swine; and produced 18,319 bushels of wheat, 66,716 of Indian corn, 90,551 of oats, 23,327 of potatoes. It had seven stores, four grist-mills, eight saw-mills, one pottery; eight schools, 111 scholars. Pop. 2702. Capital, Augusta.

NOBLEBOROUGH, p. t., Lincoln co., Me., 32 m. S.E. Augusta, 600 W. Bounded W. by Damariscotta pond and river. At the bridge, at the head of navigation on the river, is a considerable village. Ship building is extensively carried on. It contains 23 stores, nine lumber yards, 11 saw-mills, two tanneries; nine schools, 873 scholars. Pop. 2210.

NOBLESVILLE, p. v., capital of Hamilton co., Ia., 20 m. N.E. Indianapolis, 574 W. Situated on the E. bank of the W. fork of White river, and contains a courthouse, jail, a Methodist church, three stores, and about 200 inhabitants. Nett proceeds of the postoffice, $204.

NOCERA DEI PAGANI (an. *Nuceria Alfaterna*), a town of the Neapolitan dom., prov. Principato Citra, on the Sarno, 8 m. N.W. Salerno. Pop. 7000. The walls and citadel of the ancient city are on a hill above the present town, which consists of detached groups of houses, interspersed with trees and gardens. Nocera is the see of a bishop; it has some fine cavalry barracks, several public schools, and manufactures of linen and other fabrics. Nuceria was of great antiquity, and is said to have been founded by the Pelasgian inhabitants of Italy. (*Cramer's Anc. Italy*, ii., 212.) It was sacked and burned by Hannibal in the 2d Punic war. It is supposed to have derived its surname of Pagani, from a colony of Saracens, settled in it by the Emperor Frederick II. (*Swinburne's Tour*, ii., 109; *Craven*, p. 46; *Rampoldi*, ii., 930, &c.)

NOCKAMIXON, p. t., Bucks co., Pa., 14 m. N. Doyles-town, 40 m. N. Philadelphia. Bounded E. by Delaware river. It has four stores, two grist-mills, one saw-mill, seven potteries; two schools, 71 scholars. Pop. 2055.

NOGENT-LE-ROTROU, a town of France, dep. Eure-et-Loir, cap. arrond., on the Huisne, 32 m. W.S.W. Chartres. Pop., in 1836, 5813. It stands at the foot of a mount, on which is the château, formerly the residence of the virtuous minister of Henry IV., the famous Maximilian de Bethune, duc de Sully; to whose memory a monument has been erected in the town.

Nogent is well built, and has manufactures of woollen fabrics and cotton thread, with dyeing houses and tanneries.

NOIRMOUTIERS, an island off the W. coast of France, dep. Vendée, of which it forms a canton; in about lat. 47° N., long. 2° 13' 45'' W.; separated from the main land by a channel about 1 m. in breadth, but which at ebb tide may be passed by horses and vehicles. Area of the island about 70 sq. m. Pop., in 1826, 7027. It is in no part much above, and in many parts below high water mark; being protected against inundations on the W. by a range of natural sand-hills or *dunes*, and on the S. by artificial embankments. A portion of the surface is very fertile; and corn, beans, &c., are grown for exportation: a little wine is also grown, but the chief product of the island is salt, from extensive marshes and salt-pans. In 1837, 16,250,000 kilogr. of salt, 12,313 hect. of wheat, and 4621 hect. of beans, were sent from the island, mostly to other parts of France. The town of Noirmoutiers, with about 2200 inhabitants, is on the E. side of the island. It is tolerably well built and paved; defended by an old castle founded in 830, and several adjacent batteries; and has a harbour capable of receiving vessels of from 50 to 60 tons. (*Hugo*, &c.)

NOLA, a town of the Neapolitan dom., prov. Neapolitano, in a wide and fertile plain, the *Campania Felix* of the ancients, 14 m. E.N.E. Naples. Pop. 9000. Though ill built and dirty, it has numerous churches and convents, a hospital, a college, and public seminary, large cavalry barracks, an old palace of the counts of Nola, and a good market-place.

In antiquity Nola was one of the most considerable cities of Magna Græcia. It is said by Pliny (lib. iii., cap. 5), and by Silius Italicus, to have been founded by a colony of Chalcidians:

> Hinc ad Chalcidicam transfert citus agmina Nolam.
> Campo Nola sedet, crebris circumdata in urbem
> Territore, et celso facinus bumlar adori
> Planitiem vallo —　　　　　　*Punics*, lib. xii., v. 161.

But Velleius Paterculus (lib. i., cap. 7) states that Nola was founded, along with Capua, by the Tuscans; and the many fine Etruscan vases that have been found here seem

479

to corroborate this statement. It was besieged by Hannibal soon after the battle of Cannæ; but Marcellus, who had thrown himself into the town, having made an unexpected assault upon the Carthaginian army, Hannibal withdrew from the siege. It is, however, principally celebrated in ancient history from its having been the place where Marcus Agrippa, the faithful friend and successful general of Augustus, breathed his last, anno 12 B.C.; and where Augustus himself expired, A.D. 14, in the 75th year of his age. But, with the exception of its vases, it has now but few remains of antiquity. In the days of its prosperity it had two marble amphitheatres; of which, however, nothing now remains but the brick walls, the marble having been taken away to be employed in the construction of modern edifices. (Swinburne, i., 97; Ancient Universal History, xiv., 39, 8vo. ed., &c.)

The famous Giordano Bruno was a native of Nola, where he was born about the middle of the 16th century. He appears, at a very early period, to have become dissatisfied alike with the prevailing systems of philosophy and religion, and attempted to innovate in both. In 1583 he came to London, where he published, in 1584, his most celebrated work, Spaccio della Bestia Trionfante, dedicated to Sir Philip Sydney, of which there is a very flimsy notice in the 389th number of the Spectator. Having returned to the continent, he resided some time in Germany; but, being anxious to revisit his native country, he arrived at Venice in 1598/. Here he was arrested and thrown into prison, on the convenient charge of heresy and atheism. From Venice he was transferred to Rome; where, sentence having been pronounced against him, he was committed to the flames on the 17th of Feb., 1600! An elaborate estimate of the philosophy of this victim of the implacable hatred of the Inquisition may be found in the Historia Critica Philosophiæ of Brucker (vol. v., cap. 2), and in Enfield's compendium of the same work. (See, also, Biographie Universelle, art. Bruno; and Tiraboschi, tomo vii., p. 476–483.)

NOLACHUCKY, r., N.C. and Tenn., rises in Buncombe co., N.C., on the W. side of the Blue ridge, and enters French Broad river in Tenn., 40 m. above its junction with Holston river.

NORCIA, a town of Central Italy, Papal states, deleg. Spoleto, in a high valley near the source of the Nar, and 17½ m. E.N.E. Spoleto. Pop. 4000. It has a brisk trade in wine, oil, truffles, turnips, and other rural produce. It is identical with the ancient Nursia, noted for the coldness of its climate:

"Qui Tiberim, Fabarisque bibuat, quos frigida misit, Nursia."   Æneid, vii., 715.

NORD (DEP. DU), or Department of the North, so called from its being the most N. dep. of France, lying principally between the 50th and 51st degs. of N. lat., and the 2d and 4th of E. long., having N. and E. the North sea and Belgium, and S. and W. the deps. Aisne and Pas-de-Calais. Shape very irregular; length N.W. to S.E. 115 m., by a breadth varying from 4 to nearly 40 m. Area 567,963 hect. Pop., in 1836, 989,938, it being the most populous of all the French deps. Surface almost an uninterrupted plain; the highest hill being no more than 360 feet above the sea. The shore is bordered with sandy downs (dunes), as in Belgium and Holland. The Aa and Yser water the N., the Lys and Scheldt the central, and the Sanubre the S. parts of the dep. The arrond. of Dunkirk (Dunkergue), has a good deal of marsh land, called the Wateringues, and the Motres; but it has been mostly drained, and rendered cultivable. The soil, except along the coast, is generally very fertile. In 1834, the arable lands were estimated to comprise 359,570 hectares, meadow lands 95,832 hectares, orchards 16,334 hectares, and woods, &c., 35,627 hectares. This dep. is among the best cultivated in France. The properties, as elsewhere throughout that country, are, in general, small, since of 221,559, subject to the contrib. foncière, in 1835, 102,776 were assessed at less than 5 francs; but it has, notwithstanding, more large properties than most other deps. The largest farms are round Douai; the smallest generally about Lille. In the wooded tracts they run mostly from 13 to 22 hectares; but in the marshy region called the Wateringues, they vary up to between 60 and 70 hectares. Leases are seldom for more than nine years, except in the arrond. of Avesnes, where they are frequently from 18 to 27 years, or even longer. (Hugo.) On the large farms horses are used for the plough; but spade husbandry is common on all the smaller holdings, and nearly universal on the lands appropriated to flax, hops, tobacco, or potatoes. Fallows are rare, and the cultivators are not here, as in most parts of France, so addicted to routine practices as to reject all new and improved methods of culture. All kinds of corn are cultivated, but principally wheat and oats. In 1835, nearly 6,000,000 hectolitres of grain were harvested, besides about 1,960,000

hectolitres of potatoes; but from the density of the population but little more corn is usually grown than is required for the home demand. Kitchen vegetables are good and plentiful; and beet root, oleaginous grains, hops, chicory, flax, hemp, wood, and fruits, are also extensively raised. Hugo states that there are 500 oil mills in the dep., which annually produce 470,000 hectolitres of oil; and that the produce of flax is 3,285,000 kilog. a year. Tobacco is variously estimated at from 3 to 4 millions kilogr. a year. The pastures are very good, especially on the Sambre and in the N. According to the official tables, there were, in 1836, about 214,000 black cattle, and 193,000 sheep in the dep. The cows are of the fine Flemish breed, and it is estimated that they supply 7,000,000 kilogr. butter, and 1,500,000 kilog. cheese a year. The annual produce of wool is about 745,000 kilog.; a good deal is of very fair quality, the sheep being partly merinos, and partly of the long and fine woolled Flanders breed. The inhabitants of the coast are actively employed in the herring fishery, and at Dunkirk and Gravelines many vessels are fitted out for the cod and whale fisheries.

Dunkirk is the centre of the maritime trade; but every week three or four vessels leave Gravelines with a cargo for the English market of from 500,000 to 600,000 eggs, produced in this and the neighbouring departments. Iron, marble, building stone, &c., are found here; but the principal mineral product is coal, of which about 6,000,000 quintals a year are raised. Manufactures highly important. Nearly half the beet-root sugar produced in France is raised in this department: the quantity made in it being estimated, in 1836, at 21,172,000 kilogr. Lille is one of the chief seats of the French cotton trade; and it also occupies the population of Roubaix and Turcoing.

Lace and linen fabrics at Valenciennes; carpets, stuffs of hemp, cordage, arms, at Maubeuge, Cambrai, &c.; hardware, cutlery, glass, and earthenware, hats, paper, soap, chemical products, barrels, tiles, and bricks, are among the other chief manufactures. A great many distilleries, breweries, sugar and salt refineries, dyeing and bleaching establishments, and tanneries, are spread over the department. No portion of France has its commerce so much facilitated by navigable rivers, canals, and good roads. The department is divided into seven arronds.; chief towns, Lille (Lisle), the capital, Avesnes, Cambrai, Douai, Dunkirk, Hazebroucke, and Valenciennes. It sends 12 members to the chamber of deputies. Number of electors (1836–9), 6667. Total public revenue (1831), 38,810,234 fr. This department was annexed to the French crown by Louis XIV. (Hugo, art. Du Nord; French Official Tables; Dict. Geog. Univ.)

NORDHAUSEN, a town of Prussian Saxony, gov. Erfurt, cap. circ., on the Zorge, 49 m. W. Halle. Pop., in 1838, 12,163. It is surrounded with old walls flanked with towers, and is generally built in an antiquated style. It has several churches, in one of which are two paintings by L. Cranach; three hospitals, a gymnasium, an orphan asylum, a theatre, &c., and is the seat of a circ. council, a board of taxation, and judicial courts for the town and circ. It is, for its extent, one of the most flourishing commercial towns in the Prussian dominions, having numerous distilleries, the refuse of which support great numbers of hogs and cattle. Woollen cloth, sealing-wax, vitriol, soap, mineral waters, and cream of tartar, are made at Nordhausen, which is farther noted for its peculiar manufacture of fuming sulphuric acid. It has also numerous oil-mills, some marble-works, and a considerable trade in corn, produced in its vicinity. It was the native place of the celebrated philologist Wolf. (Von Zedlitz, Der Preussische Staat, ii. 943; Berghaus, &c.)

NORDKOPING (Swed. Norrköping), a town and port of Sweden, lan Linköping, on the Motala, near its mouth in the Baltic, 85 m. S.W. Stockholm. Pop., in 1836, 11,449. After Stockholm, it covers more ground than any other Swedish town, but it has no public building worthy of notice. It has straight and broad streets, and is well situated for trade, having a commodious quay, close to which vessels can lie. It has several churches, a synagogue, public school, house of correction, and savings' bank, and numerous factures of brass and hardware goods, linen, cotton, and coarse woollen fabrics, gloves, starch, paper, leather, &c., and several sugar refineries. A profitable salmon fishery is also carried on here in the river. (Forsell; Stein's Handbook.)

NORDLINGEN, a town of Bavaria, circ. Middle Franconia; on the Eger, 48 m. S.W. Nuremberg. Pop. 6388. It is surrounded with old bastioned ramparts. The cathedral, a handsome Gothic edifice, has some curious monuments and paintings, and a tower 345 ft. in height. The town-hall is ornamented with fresco paintings of the battle of Nordlingen, in 1634; in which, after an obstinate and doubtful conflict, the Austrians and Bavarians, under the Archduke Ferdinand, defeated the Swedes and their allies, under the famous Bernard, Duke of Weimar. The town has flourishing carpet factories, and a considerable trade in feathers, geese, and hogs. (*Berghaus*.)

NORFOLK, a marit. co. of England on its E. coast, having N. and E. the German ocean, S. Suffolk, and W. Cambridge, a point of Lincoln, and the inlet of the sea called the Wash. It is of a circular shape, and contains 1,295,360 acres, of which about 1,200,000 are supposed to be arable, meadow, and pasture. Surface generally flat, and, where most diversified, merely undulating. Soil very various : in the W. parts of the county, contiguous to Cambridge, and the bottom of the Wash, there is a considerable tract of marsh-land included within the great level of the Fens ; and there is also some marsh-land in the S.E. corner of the county contiguous to Yarmouth. But with these exceptions, the rest of the county consists principally of a light sandy loam, especially suitable for the turnip and barley husbandry. Climate dry and early ; but in spring, the E. winds are often very severe. Few counties in the empire have been so much improved as this. Little more than a century ago, the greater portion of it consisted of wastes, commons, sheep-walks, and warrens of little or no value. But, through the judicious application of marl, which is found in the greatest abundance in all parts of the county, and the extension of the turnip husbandry, introduced by Lord Viscount Townshend in the reign of George II., followed up by the introduction of the drill husbandry, and an improved rotation of crops, it is now, perhaps, the best farmed county in England, and is a striking example of what may be accomplished by intelligence, perseverance, and industry. The usual rotation in the turnip land is, 1st, turnips ; 2d, barley ; 3d, clover, or clover and rye grass ; and 4th, wheat. Turnips form the basis of the system, and are said, with marl, to "have made the county." On some estates no oats are allowed to be raised, and barley is, in all respects, the leading corn crop. Tenants are strictly prohibited from taking two white crops in succession, and the land is kept remarkably clean, and is not injured by over-cropping. Ploughing is wholly executed, as in Scotland, by ploughs drawn by two horses or two oxen. The grazing husbandry of Norfolk is very inferior to the arable, though it has been latterly a good deal improved. Great numbers of galloways, and other Scotch cattle, are purchased at the great fairs in the county to be turnip fed, and otherwise fattened, for the market of the metropolis. The stock of sheep is very large, amounting to between 700,000 and 800,000 head. Vast quantities of turkeys are raised in this county and Suffolk, which furnish the greater part of those supplied to London, especially at Christmas. Estates of all sizes, from £40,000 a year downwards. Farms mostly large ; and, in fact, the great improvement of which Norfolk has been the theatre never could have been affected by small farmers. Leases vary from 7 to 14, and in a few instances to 21 years. Farm-buildings generally good ; barns very large. Average rent of land, in 1810, 14s. 4½d. an acre. Minerals, with the exception of marl, of no importance. The woollen manufacture, especially the worsted branch, has been long extensively carried on in this county, especially at Norwich, where various descriptions of shawls, crapes, silks, &c., are also manufactured. (*See* NORWICH.) But owing to the superior facilities for the successful prosecution of manufacturing industry enjoyed by Bradford, Paisley, and other towns in the N. engaged in the same departments, the manufactures of Norfolk are rather on the decline. Principal rivers, Great and Little Ouse, Nen, Waveney, Yare, Wensume, &c. A navigable communication admitting vessels drawing 10 ft. water has recently been effected between Norwich and Lowestoft. (*See* LOWESTOFF.) Norfolk has no fewer than 33 hundreds and 713 parishes. Principal towns, Norwich, Yarmouth, and King's Lynn. It sends 12 members to the House of Commons ; viz. four for the county, two for the city of Norwich, and two each for the boroughs of King's Lynn, Thetford, and Yarmouth. Registered electors for the county, in 1839-40, 16,033, being 8474 for the E., and 7559 for the W. division. In 1841, Norfolk had 85,922 inhabited houses, and 412,821 inhabitants, of whom 199,055 were males, and 213,566 females. Sum expended for the relief of the poor, in 1838-39, £177,567. Annual value of real property in 1815, £1,516,651 ; profits of trade and professions in do., £593,011.

NORFOLK, county, Mass., situated in the E. part of the state, and contains 400 sq. m. Watered by Neponset and Charles rivers. A fine blue granite is found in Quincy, and transported to Neponset river, by the first railroad built in the United States, 3 m. long. The Boston and Providence railroad passes through the county. It contained in 1840, 15,110 neat cattle, 2297 sheep, 12,411 swine ; and produced 3341 bushels of wheat, 19,866 of rye, 99,193 of Indian corn, 2678 of buckwheat, 18,903 of barley, 17,063 oats, 495,981 of potatoes. It had 15 commercial and 22 commission houses in foreign trade, capital $696,000 ; 331 retail stores, capital $1,737,650 ; 17 lumber-yards, capital

$135,400 ; value of hardware and cutlery manufactured, $191,000 ; 9400 bushels of salt produced ; four furnaces, five forges. four fulling mills, 15 woollen manufactories, 38 cotton factories with 30,237 spindles, one flouring mill, 44 grist-mills, 59 saw-mills, 13 paper-mills, 21 tanneries, one pottery, six rope-walks, seven printing-offices, three binderies, four weekly newspapers. Total capital in manufactures, $2,834,180 ; 46 academies, 1473 students ; 197 schools, 11,776 scholars. Pop. 53,140. Capital, Dedham.

NORFOLK, county, Virginia, situated in the S.E. part of the state, and contains 544 sq. m. Drained by Elizabeth and Pasquotank rivers. It contains a part of the Dismal Swamp. Hampton Roads lies on its N. border. It contained in 1840, 8248 neat cattle, 2943 sheep, 18,715 swine ; and produced 2797 bushels of wheat, 260,215 of Indian corn, 34,745 of oats, 35,490 of potatoes, 1000 pounds of cotton. It had nine commercial and eight commission houses engaged in foreign trade, capital $204,500 ; 116 retail stores, capital $1,781,350 ; 10 grist-mills, two saw-mills, four printing-offices, one bindery, two daily, four weekly, and three semi-weekly newspapers ; 19 academies, 545 students : 21 schools, 535 scholars. Pop.: whites, 11,280 ; slaves, 7845 ; free coloured, 1967 ; total, 21,092. Capital, Norfolk.

NORFOLK, p. t., Litchfield co., Ct., 36 m. W.N.W. Hartford, 344 W. Drained by Blackberry river, which has a fall of 30 ft., affording water-power. It contains a Congregational church, three stores, one fulling-mill, one woollen factory, one grist-mill, eight saw-mills, four tanneries ; one academy, 139 students ; 11 schools, 371 scholars. Pop. 1393.

NORFOLK, p. t., St. Lawrence co., N.Y., 234 m. N.N.W. Albany, 513 W. Watered by Racket river and its tributaries. It contains two churches, a Presbyterian and Methodist ; four stores, two furnaces, two fulling-mills, one woollen factory, two grist-mills, six saw-mills ; 13 schools, 644 scholars. Pop. 1798.

NORFOLK, p. b., port of entry, capital of Norfolk co., Va., 110 m. by water below City point, 106 m. E.S.E. Richmond, 230 W. Situated on the N.E. bank of Elizabeth river, just below the confluence of its two branches, 8 m. above its entrance into Hampton Roads, and 32 m. from the ocean. It is in 36° 50' 50" N. lat., and in 76° 18' 47" W. long. Pop in 1810, 9193 ; in 1820, 8478 ; in 1830, 9816 ; in 1840, 10,920.

The site of the place is low, and there are marshes in the vicinity ; the principal streets are well paved, lighted, and clean ; but others are irregular, crooked, and less commodious ; and most of the houses are not distinguished for elegance. It contains a courthouse, jail, a market-house, a theatre, two banks, two insurance offices, an academy, an orphan asylum, and an atheneum with a respectable library. There are eight churches, two Episcopal, two Methodist, a Presbyterian, Baptist, Roman Catholic, and an African. The harbour is deep, spacious, safe, and easy of access. The entrance, between Old Point Comfort and a sandbar called the Rip Raps, is over a mile wide, defended by fort Monroe and fort Calhoun. Fort Monroe, on Old Point Comfort, exclusive of the ditch or moat, covers 56 acres ; and, including the ditch, 70 acres. The whole peninsula ceded by the state to the United States, contains 250 acres. The work, when completed, will mount 335 guns, generally 24's, 32's, and 42's, and about 130 of them are under bomb-proof covers. Fort Calhoun, 1 m. distant from fort Monroe, on the opposite side of the river, is on a flat near the channel, and covers about seven acres. The foundation has been made by throwing in stones, and suffering them to settle for several years, to render it sufficient to sustain the superstructure. It will mount 265 guns. 24's and 32's, and nearly all under cover. These fortifications will completely command the entrance from Hampton Roads. Opposite to Norfolk is Portsmouth, immediately above which is Gosport, the seat of one of the most important navy-yards in the United States, which has a splendid dry-dock, constructed of hewn granite, which cost $974,356. On Washington point, between the E. and W. branches of Elizabeth river, about a mile from Norfolk, is the United States marine hospital, a handsome brick edifice. The tonnage of the port, in 1840, was 19,079. The Dismal Swamp canal connects the waters of Chesapeake bay with Albemarle sound, through Pasquotank river, and opens an extensive water communication from Norfolk to the south. It has more foreign commerce than any other place in Virginia. There were in 1840, eight commercial and eight commission houses in foreign trade, with a capital of $202,000 ; 35 retail stores, capital $1,500,500 ; two printing-offices, one bindery, two daily, and one semi-weekly newspapers. Total capital in manufactures, $178,300 ; 16 academies, 515 students : 17 schools, 604 scholars.

NORMANDY, one of the provs. of France under the old regime, now distributed among the departments of Seine Inférieure. Eure, Orne, Calvados, and La Manche.

NORRIDGEWOCK, p. t., capital of Somerset co., Me., 32 m. N. Augusta, 697 W. It lies on both sides of Kennebec river. The village is chiefly on the N. side, but connected by a bridge with a village on the S. side. It contains a courthouse, jail, and Congregational church on the N. side, and a female academy on the S. side. The streets are pleasantly shaded by trees. The town contains seven stores, one fulling-mill, one grist-mill, one saw-mill, one oil-mill, two tanneries; one academy, 25 students; 20 schools, 835 scholars. Pop. 1865.

NORRISTOWN, p. b.. Norriston t., capital of Montgomery co., Pa., 91 m. E. by S. Harrisburg, 154 W. Situated on the N. side of Schuylkill river. It contains a courthouse, jail, county offices, an academy of brick, two churches, a Presbyterian and Episcopal, both of stone. a public library of 1200 volumes, a literary society with a cabinet of natural history, and about 400 dwellings. A bridge crosses Schuylkill river, 800 feet long, and with the abutments, 1050 feet, which cost $41,200. The streets are handsomely graded, and the sidewalks extensively laid with bricks and flagging stones. A dam across the Schuylkill creates great waterpower. It has 14 stores, two lumber-yards, one forge, three cotton factories with 19,064 spindles, two flouring-mills, one saw-mill, two printing-offices, two semi-weekly newspapers; one academy, 33 students; six schools, 600 scholars. Pop. 2937.

NORTHALLERTON, a parl. bor., market-town, and par. of England, in the liberty of Allertonshire, N. riding, co. York, on a small tributary of the Whisk, 13¼ m. S.S.E. Darlington, and 31 m. N.W. York. Area of parl. bor., which comprises the townships of Northallerton, Romanby, and Brompton, 2340 acres. Pop., in 1841, 5273. The town, which is on level ground, consists almost entirely of a main street along the great N. road from London to Edinburgh. It is wide, well paved, and lighted with gas : a market-house stands near the centre of the town, and at its N. extremity is a fine open space, in which are the church and churchyard. The former is a large cruciform structure, of considerable beauty, with a square tower at its W. end : the living is a vicarage, in the gift of the dean and chapter of Durham. A grammar-school has been founded here under the same patronage, and there is a large national school for children of both sexes. There is also a place of worship for Wesleyan Methodists, with an attached Sunday-school. The register office for the N. riding of the county was built here in 1736; and there is a courthouse, in which the general county sessions of the peace are held. A jail has also been built, within the present century, on the plan of Howard, the discipline and arrangements of which are said to be, on the whole, very efficient : the number of prisoners averages about 60, and the cost of each is 1s. 5d. per diem. (Jail Returns, and Prison Inspectors 4th and 5th Reports.) "Northallerton is not a corporate town, and is under the jurisdiction of the county magistrates. No manufactures are carried on, nor are there any local advantages likely to attract them. Linen weaving, however, employs a portion of the population, both in Northallerton and the surrounding villages, its chief seat being at Brompton." (Parl. Bound. Rep.) A branch of the Darlington Joint-Stock Banking company, a private bank, and a savings' bank are established here. The town has been a great thoroughfare ; but it has lost this advantage from the recent opening of the Great N. of England railway, which passes at some distance westward. It has, however, very large weekly cattle and corn markets on Wednesdays, and large fairs for horses, cattle, sheep, and cheese, February 14, May 5, September 5, October 3, and second Wednesday in October. Northallerton sent two members to the House of Commons from the 15th Charles I. down to the passing of the Reform Act, which deprived it of one of its members. "The elective franchise was formerly attached to about 210 burgage-houses, mixed up and conjoined with the other buildings from one end of the town to the other." (Bound. Rep.) The electoral limits were enlarged, as above mentioned, by the Boundary Act, and in 1839–40, there were 281 registered electors.

At a short distance from Northallerton is Standard hill, celebrated as having been the scene, in 1138, of a sanguinary conflict between the Scotch, under David I., and the English, under the Earls of Albemarle and Ferrers. It was called the battle of the Standard, from the circumstance of the victory of the English being attributed to their possessing a standard whence were displayed the banners of St. Peter of York, St. John of Beverley, and St. Wilfred of Rippon, the whole being surmounted by a consecrated host ; but the true cause of the defeat of the Scotch was their consternation at the supposed death of their king. (Parl. Papers ; Priv. Inform., &c.)

NORTHAMPTON, a central county of England, having at its N. extremity the county of Lincoln ; on its E. and S.E. side, Cambridge, Huntingdon, Bedford, and Buckingham ; S. Oxford ; and W. and N.W. Warwick, Leicester, and Rutland. It stretches N.E. and S.W. from Banbury to near Crowland, a distance of 66 m. Area, 630,340 acres, of

which about 580,000 are supposed to be arable, meadow, and pasture. Surface beautifully diversified with gently rising hills, fine valleys, and extensive woods : it is traversed nearly in its whole extent by the Nen, which rises near Daventry. Though of various qualities, the soil is in general very fertile, and is, in many parts, strong and well adapted for the culture of wheat and beans, which are the principal crops. The climate is mild and salubrious, and there are more gentlemen's seats in this than in most other counties. Agriculture, though still capable of material improvement, is, on the whole, in a comparatively advanced state. About half the county is in grass ; and great numbers of heavy horses, and of cattle, mostly short-horns, and sheep, are bred. Estates are generally large ; but there are few large farms ; and the circumstances of their being let only from year to year tends to perpetuate the routine practice that keep their ground in this and other counties. Farmhouses and offices are mostly inferior, and inconveniently placed ; and this is also true of cottages. This is one of the counties in which there is a great waste of horse labour, five horses being usually employed to do the same work that might be as well done by two, or at most three. Average rent of land, in 1810, 21s. 5d. an acre. The woodlands are very extensive ; and a good deal of wood is used as fuel. Except limestone, which is very abundant, and slates, dug up at Collyweston, minerals are of little importance. Boots and shoes are extensively produced in the town of Northampton, and in Wellingborough, and other places ; but the want of coal is an all but insurmountable difficulty to the progress of manufacturing industry. Exclusive of the Nen, the Ouse and Welland have their sources in Northamptonshire. Principal towns, Northampton, Peterborough, and Wellingborough. This county is divided into 20 hundreds and 306 parishes, and sends eight members to the House of Commons ; viz. four for the county, two for Northampton, and two for Peterborough. Registered electors for the county in 1839–40, 8724, being 1287 for the N., and 4607 for the S. division. In 1841, Northampton had 40,903 inhabited houses, and 199,061 inhabitants, of whom 98,886 were males, and 100,175 females. Sum assessed for the relief of the poor in 1838–39, £83,163. Annual value of real property in 1815, £947,578 ; profits of trade and professions in do., £185,904.

NORTHAMPTON, a parl. and mun. bor., market and manufacturing town of England, cap. of the above co., hea. dist. hoc, on the great N. road. and on the N. bank of the Nen, crossed here by a stone bridge of three arches, and two others of inferior size, 90 m. S.S.E. Leicester, and 38 n. N. W. London. Area of parl. and mun. bor. (which comprises four par.), 1580 acres. Pop. in 1881, 10,844 ; ditto in 1841, 21,242. The town, which comprises four principal street, meeting in a very large open market-place, is well built, paved, and lighted with gas : the houses in the principal street along the line of the great north-road are of stone, large, and substantial ; but in the smaller streets are many inferior houses, almost entirely occupied by journeyman shoemakers, and other workmen employed in shoe-making. The pars. of All Saints' and St. Giles's comprise the principal portion of the respectable classes of society. St. Peter's is a small par., inhabited principally by the inferior tradespeople and working classes. St. Sepulchre's is extensive, but chiefly occupied by artisans and labourers. (Mun. Bound. Rep.) There were formerly seven parish churches, of which four still remain. That of All Saints, in the centre of the town, (rebuilt, in 1680, on the site of one destroyed by fire,) is a large and handsome, though somewhat incongruous, building, with a central cupola supported by four Ionic columns, and a tower at its W. end, rising above an Ionic portico : a fine organ, and a full length statue of the late Spencer Percival, are the principal ornaments in the interior. St. Giles's at the E. end of the town, is a large cruciform structure, partly of Norman, and partly of later English architecture, with a square tower rising from the intersection of the nave and transepts. St. Peter's, at the W. end of the town near the castle, erected shortly after the Norman conquest, consists of a nave, with side aisles separated from it by piers and arches, with a square west ern tower, and is altogether "a remarkably fine and curious specimen of enriched Norman." (Rickman, p. 214.) St. Sepulchre's, an almost equally ancient edifice, built by the Knights-Templar, at the N. end of the town, comprises a circular part, forming the body of the church, a square chancel with side-aisles, and a square tower surmounted by a spire at its W. end. The remains of the old church of St. Gregory is a school-house but the two others have entirely disappeared, and of the numerous religious houses existing in Northampton before the Reformation, two only, St. Thomas's and St. John's, both in the later English style, now remain, having been converted into almshouses for the aged poor. The Wesleyan Methodists, Baptists, Presbyterians (now Unitarians), Rom. Catholics, and the Society of Friends, have their respective places of worship, the castle

NORTHAMPTON.

hill meeting-house was, for 22 years, the scene of Dr. Dodd-ridge's ministrations, during which period he was also master of the Presbyterian academy in this town. Attached to the various churches and chapels, are numerous Sunday schools, furnishing religious instruction to between 2000 and and 3000 children of both sexes. A central national school, serving as a model-school for the co., is attended by about 400 boys and girls ; a Lancasterian school, by upwards of 500 children ; two infant schools (one of which is supported by the Wesleyan Methodists), have 250 children ; besides which, the corporation charity-school, Dryden's charity-school, and the girls' school in King's-well-street, provide clothing and education for 120 children of both sexes. The free grammar-school, in Marefare, was founded in 1542. Among the other buildings of the town, by far the most handsome is the Shire-hall, on the S. side of the market-square : it is of Grecian architecture, and comprises two large courts, and other apartments well adapted for the business of the assizes. The town-hall is an ancient structure of brick, adjoining All Saints' church ; and near it is the bor. jail, now disused. On the E. side of the town is a large co. jail, built in 1794 on the plan of Howard ; but notwithstanding its size, the cells are not sufficiently numerous to allow of the entire seclusion of prisoners. The silent system, however, accompanied by hard labour, is strictly enjoined ; and the management has been greatly improved within the last few years. The yearly expense of each prisoner to the co. amounts to £7. This jail is used also, by agreement, between the county and borough magistrates, as a place of confinement for prisoners belonging to the bor. (Prison Inspector's 3d Report, lt. 80.) The theatre in Marefare, built at the beginning of the present century, is a neat building, and, though small, is sufficiently large for a town in which dramatic entertainments are little relished. The barracks form a large enclosure on the W. side of the road leading to Leicester ; and in the E. suburbs is the infirmary, a large and respectably-built edifice, completed in 1793, and furnishing excellent accommodation for patients. A race-course was formed N. of the town in 1778, and the races, which take place in autumn, are invariably well attended. About ½ m. S. on the London road is an ancient cross, one of those erected at the halting-places of the funeral of Queen Eleanor, on its passage from Hardeby, in Lincolnshire, to Westminster Abbey.

"The bor. is evidently in a flourishing condition. Its fairs and markets are resorted to by the inhabs. of the agricultural districts ; and the shops are numerous, respectable, and thriving. The principal manufacture is that of boots and shoes ; and a large proportion of the lower orders, men, women, and children, are employed in this craft, which has thriven and increased during the last 30 years, without being affected by the various changes which have occurred within that period. In 1831, upwards of 1300 men (exclusive of women and children) were employed in this trade ; and we are assured that the wages of the journeymen at present amount to the weekly sum of £2000. The fixed prosperity of this trade has been assigned as the cause why the pop. of the bor. was nearly doubled during the 30 years preceding 1831, and is stated to be still rapidly increasing." (Mun. Bound. Rep.) These shoes are sent in large quantities to London, and furnish the chief supply of the shops that deal in cheap, ready-made shoes ; they are, also, extensively exported. Leather currying and saddlery are extensively carried on ; but the stocking and lace trades, once very considerable, have greatly declined since the introduction of machinery at Leicester and Nottingham. There are also three founderies, and the manufacture of light brass and iron work is prosecuted on rather an extensive scale. The Northampton Union Bank, Northampton banking Company, a private bank, and savings' bank, are established here ; and the town has two weekly newspapers, one of which is among the oldest provincial papers in England. The prosperity of Northampton was materially promoted by its situation on the great mail-coach road between London, Leicester, Nottingham, &c., and its coach and posting establishments are very considerable. These, however, are now all but extinguished, in consequence of the opening of the Birmingham and other railways, leading to the great towns in the midland and N. counties. About 5 m. W. the town is the Blisworth station of the Birmingham railway, 62½ m. from London. by which the metropolis may be reached from Northampton in about three hours ; and it is supposed by many that the facility of communication thus afforded will o far to indemnify the town for the loss of its business as a osting station. The Nen, also, and the numerous canals uniting with that river, give to Northampton the advantage f a water communication with the German ocean, London. Liverpool. Manchester, and Bristol.

Northampton is a bor. by prescription, and has received several royal charters, the last being granted in the 36th en. III. By the Mun. Reform Act it has been divided into three wards, and is governed by a mayor and five other aldermen, with 18 councillors ; having also a commission of the peace under a recorder, as well as a court of record for civil suits. Corporation revenue in 1839, £4805, exclusive of £83 accruing from the sale of property. The assizes for the co. are held here in spring and summer ; and the quarter sessions take place in Jan., April, July, and Oct. The bor. has sent two mems. to the H. of C. since the reign of Edward I. ; the right of voting, previously to the Reform Act, being in inhabitant-householders occupying a distinct dwelling for six months previously to the election, and not having received alms for 12 months. (Bound. Rep.) The electoral limits were left untouched by the Boundary Act, and in 1839–40, it had 2057 reg. electors. Northampton is likewise the principal polling-place and election town for the S. div. of the county. A large cattle market is held every Saturday, and there are smaller markets on two other days. Extensive horse and cattle fairs attended by jobbers from all parts of England, Feb. 20th, April 6th, May 4th, and Aug. 5th : other fairs, August 26th, Septem. 19th, Novem. 28th, and Dec. 19th.

After the Norman conquest North-Hamtune, which according to the Domesday Survey, had then only 40 burgesses, was given by William I. to Simon St. Liz, who built a castle here (now marked only by an earth-mound, on the W. side of the present town). Numerous synods and parliaments met here during the succeeding reigns ; and at the beginning of the 13th century, Northampton was considered of sufficient importance to have a mint. In the reign of Henry III. an attempt was made to establish a university here, consisting of emigrant students from Oxford and Cambridge ; but though the scheme was at first sanctioned by the king, a mandate was afterwards issued to compel the students to return to their old seminaries, and to forbid the continuance of the establishment. In the wars of the Roses, its neighbourhood was the scene of a great battle, (fought 10th July, 1460), between Henry VI. and the Earl of March (afterwards Edward IV.), in which the former was defeated, and taken prisoner. In 1642, the town was seized by Lord Brooke, who fortified it for the parliament. In 1663, Northampton suffered greatly from the flood, and in 1675, was nearly destroyed by fire, the loss of property being estimated at £150,000. To this calamity, however, may be attributed the increased width and regular arrangement of the streets, for which it is remarkable above most other provincial towns. (Parl. Reports ; Private Information.)

NORTHAMPTON, county, Pa. Situated in the E. part of the state, and contains 600 sq. m. Bounded E. by Delaware river. Watered by Lehigh river. It contained in 1840, 19,471 neat cattle, 19,207 sheep, 36,163 swine ; and produced 289,595 bushels of wheat, 595,157 of rye, 378,772 of Indian corn, 82,191 of buckwheat, 12,650 of barley, 244,760 of oats, 203,600 of potatoes. It had 163 stores, 20 lumber yards, six furnaces, four forges, eight fulling-mills, one woollen factory, 65 grist-mills, 69 saw-mills, five oil-mills, one paper-mill, three powder-mills, 24 tanneries, 16 distilleries, four breweries, four printing-offices, two binderies, five weekly newspapers ; two colleges, 68 students ; nine academies, 601 students ; 76 schools, 3445 scholars. Pop. 40,996. Capital, Easton.

NORTHAMPTON, county, Va. Situated on the S. part of the Eastern shore, and contains 390 sq. m. Bounded E. by the Atlantic, W. by Chesapeake bay. It contained in 1840, 4574 neat cattle, 5453 sheep, 12,209 swine ; and produced 279 bushels of wheat, 296,718 of Indian corn, 197,058 of oats, 51,456 of potatoes, 6003 pounds of cotton, 665 of sugar. It had 16 stores, 96 grist-mills, one saw-mill, eight oil-mills ; seven schools, 166 scholars. Pop. whites, 3341 ; slaves, 3899 ; free coloured, 754 ; total, 7715. Capital, Eastville.

NORTHAMPTON, county, N. C. Situated towards the N.E. part of the state, and contains 546 sq. m. Bounded N.W. by Roanoke river. Watered by Meherrin river. It contained in 1840, 12,627 neat cattle, 5965 sheep, 40,545 swine ; and produced 20,306 bushels of wheat, 3125 of rye, 716,050 of Indian corn, 78,650 of oats, 38,600 of potatoes, 66,944 pounds of tobacco, 5,210,724 of cotton. It had 16 stores, 15 grist-mills, three saw-mills, one oil-mill ; one academy, 22 students ; 10 schools, 147 scholars. Pop.: whites, 5818 ; slaves, 6759 ; free coloured, 792 ; total, 13,369. Capital, Jackson.

NORTHAMPTON, p. t., capital of Hampshire co., Mass., 17 m. N. by W. Springfield, 93 m. W. Boston, 67 m. E. Albany, 39 m. N. Hartford, 380 W. It is in 42° 19′ 8″ N. lat. and 72° 36′ 21″ W. long. Pop. in 1810, 2631 ; in 1820, 2854 ; in 1830, 3613 ; in 1840, 3750. The Indian name was Nonotuck ; incorporated in 1654. It was the third town settled on the river, in this state. A body of fine meadow land, of between 3000 and 4000 acres lies on Connecticut river, near the village. A bridge, 1080 feet long, 26 feet wide, resting on six stone piers, and two abutments, which in the deepest water are 40 feet high from the bottom, was rebuilt in 1826. The floor of the bridge is 21 feet above low water mark, and it connects the town with Hadley. The village

NORTH BEAVER.

is regarded as one of the most pleasant for an elegant residence in New-England. It is irregularly but handsomely laid out, about a mile W. of Connecticut river, a little elevated above the adjoining meadows, presents many pleasant sites for dwellings, and is well built. In the W. part of the village, Round Hill, a considerable elevation, in a very regular form, is the seat of the celebrated Round Hill seminary, on the plan of a German gymnasium, and a number of elegant residences. The village contains an elegant brick courthouse, a brick county house and jail, a handsome townhall, two banks, six churches, two Congregational, a Methodist, Baptist, Unitarian, and an Episcopal, a female seminary, two high schools, one for boys and one for girls, over 10 years of age, numerous well conducted district schools, for children under 10 years of age, 600 dwellings and about 4000 inhabitants. The first Congregational church is a handsome edifice, 100 feet long and 75 feet wide, so constructed that the voice of the speaker easily fills its large dimensions. The female seminary is a handsome Gothic edifice, and is patronized from all parts of the United States; and the Round Hill seminary is of the highest order. The scenery around this place is highly picturesque, including the beautiful valley of Connecticut river and mount Holyoke, 830 feet high, on the opposite side of the river, whose top commands one of the finest views in this part of the United States. A stream passes through the centre of the town which affords good water-power. A canal, which here joins Connecticut river, connects Northampton with New-Haven, Connecticut. According to the census of 1840, there were in the town 34 stores, with a capital of $125,700; two persons produced machinery to the amount of $1900, one person manufactured the precious metals to the amount of $2900; nine persons manufactured various metals to the amount of $7500; two persons produced granite and marble to the amount of $1900; eight persons produced bricks and lime to the amount of $1700; two fulling-mills, and two woollen factories, employed 156 persons, produced to the amount of $900,000, with a capital of $110,000; 12 males and 38 females produced 3007 pounds of reeled silk, valued at $24,056, with a capital of $50,000; five persons manufactured hats and caps to the amount of 4000, with a capital of $2000; one tannery, employing four persons, produced 1000 sides of sole leather and 300 sides of upper leather, with a capital of $5000; eight other various manufactories of leather, as saddleries, boots, shoes, &c., produced $13,500, with a capital of $3800; one person produced 6000 pounds of tallow candles, with a capital of $900; one paper-mill employed 10 persons, produced to the amount of $20,000, with a capital of $40,000; nine persons produced carriages and wagons to the amount of $1500, with a capital of $2500; one flouring, three grist, and 11 saw-mills, employed 15 persons, produced to the amount of $12,500, with a capital of $18,600; nine persons produced furniture to the amount of $4500, with a capital of $3000; three printing-offices, two binderies, and three weekly newspapers, employed 25 persons, and a capital of $9700; six wooden houses built, employed 18 men, and cost $12,000. The total capital employed in manufactures was $254,800. There was one academy, with 56 students; 21 schools, with 937 scholars.

NORTHAMPTON, p. t., Fulton co., N. Y., 18 m. N.E. Johnstown, 47 m. N.W. Albany, 415 W. Watered by Sacandaga river and its branches. It contains 10 stores, one fulling-mill, one woollen factory, two grist-mills, 12 saw-mills; eight schools, 278 scholars. Pop. 1596.

NORTHAMPTON, t., Burlington co., N.J. Drained by Rancocus creek, and tributaries of Little Egg harbour river. It has 29 stores, two fulling-mills, one woollen factory, one cotton factory with 9500 spindles, seven grist-mills, 11 saw-mills, two paper-mills, two tanneries, five distilleries, one pottery, two printing-offices, two binderies, two weekly newspapers; four academies, 91 students; six schools, 386 scholars. Pop. 6813.

NORTHAMPTON, t., Bucks co., Pa., 10 m. S.E. Doylestown, 20 m. N. Philadelphia. Drained by branches of Neshaminy creek. It has three stores, four grist-mills; three schools, 114 scholars. Pop. 1094.

NORTH BEAVER, t., Beaver co., Pa. It has two stores, one flouring-mill, six grist-mills, seven saw-mills, four tanneries, five distilleries; 14 schools, 431 scholars. Pop. 2293.

NORTH BERWICK, p. t., York co., Me., 89 m. S.W. Augusta, 506 W. It has four stores, four fulling-mills, one woollen factory, two grist-mills, five saw-mills, three tanneries; 17 schools, 566 scholars. Pop. 1461.

NORTHBOROUGH, p. t., Worcester co., Mass., 33 m. W. Boston, 406 W. Watered by Assabet river. Incorporated in 1766. It contains three churches, two Congregational, and a Baptist; three stores, two cotton factories with 2059 spindles, six grist-mills, five saw-mills; two academies, 45 students, six schools, 366 scholars. Pop. 1348.

NORTHBRIDGE, p. t., Worcester co., Mass., 39 m S.W. by W. Boston. 406 W. Watered by Blackstone and Mumford

NORTH CAROLINA.

rivers. The Blackstone canal passes through it. It contains four churches, two Congregational, a Methodist and Friends, three stores, six cotton factories with 10,108 spindles, one grist-mill, three saw-mills; eight schools, 250 scholars. Pop. 1449.

NORTH BRIDGEWATER, p. t., Plymouth co., Mass., 22 m. S. by E. Boston, 439 W. It contains four churches, a Congregational, Unitarian, a Methodist, and a New Jerusalem, 16 stores, three grist-mills, two saw-mills, two academies. 66 students, nine schools, 488 scholars. Pop. 2616.

NORTH BROOKFIELD, p. t., Worcester co., Mass., 60 m. W. Boston, 392 W. It contains a Congregational and a Methodist church, four stores, one fulling-mill, one woollen factory, two grist-mills, four saw-mills; 11 schools, 383 scholars. Pop. 1485.

NORTH BRUNSWICK, t., Middlesex co., N. J. Bounded N. by Raritan river, N.E. by South river. Drained by Lawrence's brook. It contains most of the city of New-Brunswick, and has 79 stores, one flouring-mill, two grist-mills, one saw-mill, three tanneries, one distillery, two potteries, three printing-offices, two binderies, two weekly newspapers; one academy, 45 students, 18 schools, 666 scholars. Pop. 5866.

NORTH CAROLINA, one of the southern United States, is bounded N. by Virginia, E. and S.E. by the Atlantic, S. by South Carolina and Georgia, and W. by Tennessee. It lies between 33° 50′ and 36° 30′ N. lat., and between 75° 45′ and 84° W. long., and between 6° 59′ W. and 10 17′ E. long. from W. It is 430 m. long, with a breadth varying from 20 to 180 miles, containing about 50,000 square miles, or 32,000,000 acres. The population, in 1790, was 393,754 : in 1800, 478,103; in 1810, 555,500; in 1890, 638,829; in 1830, 738,470; in 1840, 753,419, of whom 245,817 were slaves. Of the free population, 240,047 were white males; 244,823 were white females; 11,296 were coloured males; 18,505 were coloured females. Employed in agriculture, 217,095; in commerce, 1734; in manufactures and trades, 14,322; in navigating the ocean, 327; do. canals, rivers, &c., 379; in the learned professions, 1086.

The state is divided into 68 counties, which, with their population in 1840, were as follows:

Counties.	Pop. 1840	Counties.	Pop. 1840
Anson,	15,077	Jones,	4,945
Ashe,	7,467	Lenoir,	7,946
Beaufort,	12,225	Lincoln,	25,160
Bertie,	12,175	Macon,	4,420
Bladen,	8,022	Martin,	7,47
Brunswick	5,265	Mecklenburg,	18,273
Buncombe,	10,084	Montgomery,	10,780
Burke,	15,799	Moore,	7,286
Cabarrus,	9,259	Nash,	8,042
Camden,	5,663	New Hanover,	13,312
Carteret,	6,591	Northampton,	12,381
Caswell,	14,693	Onslow,	7,527
Chatham,	16,943	Orange,	24,356
Cherokee,	3,427	Pasquotank,	8,514
Chowan,	6,690	Perquimans,	7,346
Columbus,	3,941	Person,	9,790
Craven,	13,438	Pitt,	11,461
Cumberland,	15,284	Randolph,	12,875
Currituck,	6,703	Richmond,	5,889
Davidson,	14,606	Robeson,	8,378
Davie,	7,574	Rockingham,	13,442
Duplin,	11,182	Rowan,	12,409
Edgecombe,	15,708	Rutherford,	16,487
Franklin,	10,890	Sampson,	10,437
Gates,	8,161	Stokes,	16,305
Granville,	18,517	Surry,	15,653
Greene,	6,106	Tyrrel,	4,689
Guilford,	19,170	Wake,	21,118
Halifax,	16,965	Warren,	10,865
Haywood,	4,975	Washington,	4,631
Henderson,	5,129	Wayne,	10,413
Hertford,	7,484	Wilkes,	12,577
Hyde,	6,458	Yancey,	8,205
Iredell,	15,685		
Johnston,	10,599	Total,	753,419

Raleigh, near the centre of the state, 6 m. W. of Neuse river, is the seat of government.

The whole eastern coast of North Carolina consists of a ridge of sand and low islands, separated from the main land in some parts by narrow, and in other parts by broad sounds and bays, entered by various inlets, generally shallow and of dangerous navigation. Ocracoke inlet is the only one N. of Cape Fear, through which vessels pass. For the distance of from 60 to 80 m. from the shore, the country is a dead level, the streams are sluggish and muddy, and there are many swamps and marshes. The soil is generally sandy and poor, excepting on the margins of the streams, where it is fertile. The natural growth of this region is almost universally the pitch-pine, which grows much larger than the same tree in the northern states, and yields extraordinary tar, pitch, turpentine, and lumber, which constitutes an important portion of the exports of the state. Turpentine is merely the resinous sap of the pine tree, obtained by incisions, and flows from the middle of March to the end of Oc-

484

tober. It is received in boxes, which are emptied five or six times during the season, and 40 trees will yield about a barrel of turpentine. Oil or spirits of turpentine are produced by distillation, and the residuum is rosin. Large quantities are manufactured in Brooklyn and New-York, from pitch imported from North Carolina. Tar is produced from billets of pitch-pine wood, burned in pits covered with sods or earth, and the tar is caused, by a slow combustion, to flow out in a trench dug in the earth. Its value depends much upon the cleanness with which it is manufactured, and that from the north of Europe is said, in this respect, to be superior to that made in the United States. Pitch is made by boiling tar down to dryness. In the swamps of this region, fine rice is produced. Back of the lower country, and extending to the lower falls of the rivers, is a tract of country, 40 miles wide, which has a moderately uneven surface, of a sandy soil, in which the pitch-pine is the prevailing natural growth. Above the falls the country is uneven, the streams have a more rapid current, and the soil is more fertile, producing wheat, rye, oats, barley, flax, tobacco, and Indian corn. The western part of the state is an elevated table land, and in some places rises into elevated mountains. In Yancey county is the highest land in the United States E. of the Rocky mountains. Black mountain is 6476 feet above the level of the sea, which is 242 feet higher than the highest peak of the celebrated White mountains in N. H. Roan mountain is 6038 feet high, and Grandfather mountain is 5556 feet high. It is but recently that the elevation of these mountains has been correctly ascertained. The Blue Ridge constitutes the main range through the western part of the state, and on the extreme western part is a chain called by different names—as, Smoky mountain, Unka mountain, Bald mountain, Iron mountain, and Stone mountain. In this range Roan mountain is situated. Between these mountain ranges the soil is fertile. Throughout the state Indian corn is raised, and in some parts considerable cotton. In the low country, grapes, plums, blackberries, and strawberries, grow spontaneously; and, on the intervals, canes grow luxuriantly—and, continuing green through the winter, furnish food for cattle. In the low country the climate is somewhat unhealthy, but in the elevated parts it is salubrious. In the elevated parts, the natural growth is oak, walnut, lime, and cherry, which are often large. In the northern part, extending into Virginia, is the Great Dismal swamp, 30 m. long and 10 broad, containing 130,000 acres; and on the Virginia line is lake Drummond, 15 m. in circumference. This swamp is thickly wooded with pine, juniper, cypress, and, in the drier parts, with white and red oak. Between Albemarle and Pamlico sounds is the Alligator or Little Dismal swamp, which also has a lake in the centre. It is computed that there are 2,500,000 acres of swamp within the state, which is capable of being drained, at a moderate expense, and made to produce cotton, tobacco, rice, and Indian corn.

This state contained, according to the census of 1840, 166,608 horses and mules, 617,371 neat cattle, 538,279 sheep, 1,649,716 swine, and produces poultry to the value of $544,125. There were produced 1,960,885 bushels of wheat, 15,391 of buckwheat, 913,971 of rye, 23,893,763 of Indian corn, 3574 of barley, 3,193,941 of oats, 2,609,239 of potatoes, 625,044 pounds of wool, 16,772,359 of tobacco, 2,880,388 of rice, 51,926,190 of cotton, 7163 of sugar, 3014 of silk cocoons. 101,369 tons of hay, 9879 of hemp or flax. The products of the dairy were valued at $674,349; of the orchard, at $398,696; of lumber, at $506,766. There were made 98,752 gallons of wine.

Gold and iron are found in this state. The gold region lies on both sides of the Blue Ridge, and extends to the east of the Yadkin river. It exists in grains, in small masses and lumps, and in veins. Many persons have commenced digging for gold, and a considerable amount is sent annually to the mint of the United States.

Cape Hatteras, cape Fear, and cape Lookout, on the coast of this state, are a terror to navigators, and have caused many shipwrecks. The principal rivers are the Chowan and Roanoke, both of which rise in Virginia and flow into Albemarle sound; Neuse, which also enters Albemarle sound; Cape Fear river, the longest which runs wholly in the state, 280 miles long, which is navigable with 11 feet of water to Wilmington, and has from 10 to 14 feet of water on the bar at its mouth; the Yadkin, which forms the Great Pedee in South Carolina. The sluggishness of the rivers as they approach the sea, and the sandy character of the coast, cause them to be obstructed by sand-bars at their mouths, and the state has few good harbours. Much of its commerce is carried on through the neighbouring states. Wilmington, on Cape Fear river, 40 m. from the sea, is the most commercial place in the state. Newbern, on Neuse river, 30 m. from its mouth in Pamlico sound, has some commerce, and Fayetteville, at the head of boat navigation on Cape Fear river, has considerable trade.

The exports of the state, in 1840, amounted to $387,484;

and the imports to $252,532. There were four commercial and 46 commission houses engaged in foreign trade, with a capital of $151,300; 1068 retail stores, with a capital of $5,082,835; 432 persons employed in the lumber trade, with a capital of $46,000; 213 persons employed in internal transportation, who, with 24 butchers, packers, &c., employed a capital of $9000; 1784 persons were employed in the fisheries, with a capital of $213,502.

The amount of home-made or family manufactures was $1,413,342; three woollen manufactories, with one fulling-mill, produced articles to the amount of $3900, with a capital of $9800; 25 cotton manufactories, with 47,934 spindles, employed 1219 persons, and produced articles to the amount of $438,900, with a capital of $1095,300; eight furnaces produced 068 tons of cast iron, and 43 forges produced 963 tons of bar iron, the whole employing 468 persons, and a capital of $94,981; two smelting houses, employing 30 persons, produced 10,000 pounds of lead; 10 smelting houses, employing 389 persons, produced gold to the amount of $255,618, with a capital of $8632; two paper-mills produced articles to the amount of $9785, with a capital of $5000; hats and caps were manufactured to the amount of $38,167, and straw bonnets to the amount of $1700, the whole employing 142 persons, and a capital of $13,141; 353 tanneries employed 645 persons, with a capital of $271,979; 238 other manufactories of leather, as saddleries, &c., produced articles to the amount of $185,387, with a capital of $76,163; 16 potteries employed 21 persons, and produced articles to the amount of $9200, with a capital of $1531; 89 persons produced machinery to the amount of $43,395; 43 persons manufactured hardware and cutlery to the amount of $1200; 608 persons manufactured carriages and wagons to the amount of $301,601, with a capital of $173,318; 323 flouring-mills produced 87,641 barrels of flour, and, with other mills, employed 1830 persons, producing articles to the amount of $1,552,096, with a capital of $1,670,228; vessels were built to the amount of $62,800; 223 persons manufactured furniture to the amount of $55,002, with a capital of $57,980; 40 persons manufactured 1988 small arms; 15 persons manufactured granite and marble to the amount of $1063; 276 persons produced bricks and lime to the amount of $58,336; 367 persons manufactured 1,612,825 pounds of soap, 148,546 pounds of tallow candles, and 335 pounds of wax or spermaceti candles, with a capital of $4754; 2802 distilleries produced 1,051,979 gallons of distilled spirits, and breweries 17,431 gallons of beer, employing 1422 persons, and a capital of $180,200; 38 brick or stone houses, and 1822 wooden houses were built, employing 1707 persons, and cost $410,984; 26 printing-offices, four binderies, 26 weekly and one semi-weekly newspapers, and two periodicals, employed 103 persons, and a capital of $55,400. The total amount of capital employed in manufactures was $3,836,900.

The University of North Carolina, at Chapel Hill, 37 m. W.N.W. Raleigh, was founded in 1791; Davidson college, in Mecklenburgh, was founded in 1837. In these institutions there were, in 1840, 158 students. There were in the state 141 academies, with 4398 students; and 632 common and primary schools, with 14,937 scholars. There were in the state 56,609 white persons, over 20 years of age, who could neither read nor write.

There were in the state, near January 1st, 1840, one bank and five branches, with a capital of $1,500,000, and a circulation of $1,165,857. The state has no public debt.

The Methodists and Baptists are the most numerous religious denominations in the low country. In the elevated country, in the W. part of the state, are many Presbyterians. The Methodists and Baptists have each about 20,000 communicants; the Presbyterians about 11,000 communicants. The Episcopalians have a Bishop, and about 20 ministers; the Lutherans have 18 ministers, 38 congregations, and 1886 communicants. Besides these, there are some Moravians, Friends, and Roman Catholics. The above numbers are increased at the present time.

A railroad extends from Wilmington, 161 m., to Weldon, on Roanoke river. A railroad extends from Raleigh, 85 m., to Gaston, on Roanoke river These railroads unite with railroads from Virginia. The Dismal Swamp canal of Virginia extends into N. Carolina. (See Virginia.) The Weldon canal extends 5 m. round the falls in the Roanoke.

The governor is elected by the qualified voters for the house of commons, once in two years, but cannot hold the office more than four years in six. He must be 35 years of age, possess a freehold estate to the value of £1000, and have resided in the state for five years. The council consists of seven persons, elected for two years by the general assembly. The senate is composed of 50 members, elected once in two years by the people. A senator must have a residence, and possession for one year previous to the election, of 300 acres of land, in the county for which he is chosen. The house of commons consists of 120 members, chosen once in two years by the people. A member must have a residence, and possession for one year previors to the election, of land to the amount of 100 acres, in the county for which he is chosen. The general assembly, by joint ballot, appoint the judges of the supreme courts of law and equity, judges of admiralty, and the attorney general. The judges hold their offices during good behaviour, and the attorney general for four years. Every person of 21 years of age or upwards, who has resided in one county one year previously to an election, and paid taxes, is entitled to vote for members of the house of commons. In addition to this, to be entitled to vote for senator, he must possess 50 acres of land. Free negroes, and persons of a mixed blood from negro ancestors, to the fourth generation, are excluded from the right of suffrage. The legislature meets once in two years at Raleigh, on the second Monday of November.

The first permanent settlement in North Carolina was made on the eastern bank of Chowan river, N. of Albemarle sound, and called Albemarle, by a company of emigrants who fled from religious persecution in Virginia. After several other attempts at settlement, the province was granted, in 1763, to Lord Clarendon and others, who caused a constitution of government to be prepared for it by the celebrated John Locke. The chief magistrate was called " the Palatine," and there was a hereditary nobility. The legislature was called a parliament. This constitution was found to be so defective in practice, that it was abolished in 1693. In 1729, the crown purchased the whole of the Carolinas, for £17,500; and the king divided it into the two provinces of North and South Carolina, which have ever since continued separate.

In 1769, this province successfully resisted the oppression of the British ministry. Two years after, 1500 of the inhabitants, assuming the name of *Regulators*, rose in rebellion, but were defeated by Gov. Tryon. Three hundred were killed in battle; and of those who were taken, 12 were condemned for high treason, and six were executed. During the war of the revolution, the inhabitants were the devoted friends of American freedom. A kind of congress, composed of militia officers, assembled at Charlotte, county of Mecklenburgh, of which county they were inhabitants, May 19th, 1775, and put forth a public declaration breathing the spirit, and to some extent using the language, of the subsequent Declaration of Independence by the American Congress. It is equally bold and uncompromising in its character. (See American Almanac for 1835, p. 226.) Several battles of the revolution were fought in this state, particularly the severe one of Guilford courthouse, on the 25th of March, 1781. In 1776, early in the revolutionary war, this state formed a constitution, which, with some recent modifications, continues to the present time. In convention, November 27th, 1789, this state adopted the constitution of the United States: yeas 193, nays 75; majority 118.

NORTH CASTLE, p. t., Westchester co., N. Y., 6 m. N. White Plains, 132 m. S. Albany, 263 W. Drained by Byram river. It has four schools, 110 scholars. Pop. 2058.

NORTH EAST, p. t., Dutchess co., N. Y., 33 m. S.S.E. Albany, 332 W. The Taghkanic range of mountains passes through it. It contains some lead ore, and has five stores, one fulling-mill, one woollen factory, three grist-mills, two saw-mills; six schools, 144 scholars. Pop. 1385.

NORTH EAST, p. t., Erie co., Pa., 10 m. N.E. Erie, 261 m. N.W. by W. Harrisburg, 358 W. Bounded N. by lake Erie, E. by the state of New-York. It contains two fulling-mills, four grist-mills, nine saw-mills, one paper-mill; two schools, 66 scholars. Pop. 339. The town, exclusive of the borough, has three stores, 17 schools, 468 scholars. Pop. 1793.

NORTHFIELD, p. t., Merrimac co., N. H., 15 m. N. by W. Concord, 406 W. Incorporated in 1780. Bounded W. by Merrimac river. It contains a Methodist church, two stores, one fulling-mill, one cotton factory with 1000 spindles, one grist-mill; three saw-mills, three tanneries, 11 schools, 407 scholars. Pop. 1413.

NORTHFIELD, p. t., Franklin co., Mass., 22 m. W.N.W. Boston, 412 W. Situated on both sides of Connecticut river. It contains two Congregational churches, three stores, one fulling-mill, four grist-mills, six saw-mills; one academy, 85 students; 13 schools, 538 scholars. Pop. 1673.

NORTHFIELD, p. t., Washington co., Vt., 10 m. S.S.W. Montpelier, 506 W. Watered by Dog river, which affords water-power. It contains five churches, three Methodist, a Congregational, and Free-will Baptist; three stores, 16 fulling-mills, two woollen factories, six grist-mills, 12 saw-mills; one academy, 30 students; 17 schools, 599 scholars. Pop. 2013.

NORTHFIELD, t., Richmond co., N. Y., 9° N. Richmond, 156' S. Albany. Bounded N. and W. by Staten Island sound, and the Kills, along which are broad marshes. It has 17 stores, eight flouring-mills, four grist-mills, one saw-mill; six schools, 222 scholars. Pop. 2745.

NORTHFIELD, p. t., Summit co., O., 143 m. N.E. Columbus, 342 W. The Cuyahoga river and the Ohio canal cross

486

the S.W. part. A small tributary of Cuyahoga river affords water-power, on which is a woollen factory and saw-mill. It contains six schools, 217 scholars. Pop. 1031.

NORTHFLEET. *See* GRAVESEND.

NORTH HAVEN, p. t., New-Haven co., Ct., 30 m. S.S. W. Hartford, 306 W. Watered by Quinnipiac river, on which are extensive salt marshes. It contains a Congregational and an Episcopal church, three stores, two grist-mills, three saw-mills, eight schools, 302 scholars. Pop. 1349.

NORTH HEMPSTEAD, t., capital of Queens co., N. Y., 168 m. S. Albany, 20 m. E. New-York, 248 W. Bounded N. by Long Island sound. It contains Harbour hill, 319 feet above the ocean, the highest land on Long Island. The village contains the courthouse and jail in the same building, and a few stores and dwellings. The town contains 10 stores, one fulling-mill, two woollen factories, one paper-mill, one glass house, two flouring-mills, four grist-mills, three saw-mills; one academy, 35 students; nine schools, 353 scholars. Pop. 3691.

NORTH HERO, p. t., capital of Grand Isle co., Vt., 71 m. N.W. Montpelier, 26 m. N. Burlington, 544 W. It comprises the smaller of the two largest islands on lake Champlain, and contains 6273 acres. A narrow strait separates it from the S. island. It contains a stone courthouse and jail, a store, and a few dwellings.

NORTH KINGSTON, t., Washington co., R. I., 29 m. S. by W. Providence. Bounded E. by Narragansett bay. It contains the village of Wickford on a branch of Narragansett bay, which has a good harbour, and considerable navigation, employed in the coasting trade, and the fisheries. It contains a Baptist and an Episcopal church. The town has 24 stores, four woollen factories, five cotton factories with 5756 spindles, four grist-mills; one academy, 79 students; 14 schools, 624 scholars. Pop. 2909.

NORTH MIDDLETON, t., Cumberland co., Pa. It has one store, one fulling-mill, one flouring-mill, five saw-mills, three distilleries. Pop. 1909.

NORTHPORT, p. t., Waldo co., Me., 50 m. E. Augusta, 643 W. Bounded N.E. and E. by Penobscot bay, and its branch, Belfast bay. It is well situated for navigation, and has four stores, one saw-mill. Pop. 1207.

NORTH PROVIDENCE, t., Providence co., R. I. 4 m. N. Providence. The village of Pawtucket is a large manufacturing place, partly in Massachusetts, being on the line between the two states, on Seekonk river. (See Pawtucket.) The town was incorporated in 1767, and contains 38 stores, two furnaces, two fulling-mills, 20 cotton factories with 30,000 spindles, five grist-mills, one saw-mill, two tanneries, two printing-offices, one bindery, one weekly newspaper; two academies, 80 students; eight schools, 265 scholars. Pop. 4207.

NORTH SALEM, p. t., Westchester co. N. Y., 54 m. S. White Plains, 113 m. S. Albany, 261 W. Bounded W. by Croton river. Drained by Titus's creek, flowing into it. It contains three churches, a Presbyterian, Episcopal, and Universalist, three stores, one grist-mill, one paper-mill; one academy, 48 students, five schools, 87 scholars. Pop. 1161.

NORTH SEWICKLY, p. t., Beaver co. Pa., 232 m. W. by N. Harrisburg, 963 W. Bounded W. by Beaver river. Drained by Conequenessing creek. It has four stores, eight grist-mills, seven saw-mills, two tanneries; also schools, 354 scholars. Pop. 2292.

NORTH STONINGTON, p. t., New-London co., Ct., 53 m. S.E. Hartford, 366 W. It contains four churches, three Baptist and a Congregational; five stores, two grist-mills, seven saw-mills; one academy, 15 students, 13 schools, 544 scholars. Pop. 2299.

NORTH STRABANE, t., Washington co., Pa. Drained by Chartier's creek. It has two grist-mills, three saw-mills; six schools, 300 scholars. Pop. 1207.

NORTHUMBERLAND, a marit. co. of England, being the most northerly of the kingdom; having N. a small detached portion of Durham, by which it is separated from Scotland, E. the German ocean, S. Durham, and W. Cumberland, and the co. of Roxburgh and Berwick, in Scotland. Area, 1,197,440 acres, of which about 500,000 are supposed to be arable, meadow, and pasture. It exhibits every variety of surface and soil. It is divided from Scotland and Cumberland by the Cheviot hills, and a portion of the Pennine, or great central range of mountains, which stretch out into extensive moors, and cover a large portion of the W. parts of the co. with their ramifications. There are, however, very extensive tracts of low land along the coast, and in the vales of the Coquet, Tyne, and other rivers, the soil of which consists, for the most part, of a strong clay loam, and is very fertile. The Cheviot hills are mostly covered with fine verdure, affording excellent pasture for the peculiar and valuable breed of sheep, called by their name, and now so widely diffused : but the mountains and other offsets belonging to the Pennine range are mostly covered with peat earth; and are bleak, dreary, covered with heath, and interspersed with swampy morasses. The climate varies with the elevation and nature of the soil; but along the coast and in the vales it is dry and early. Northumberland is distinguished by its improvements, and is now one of the best cultivated cos. of the empire. Wheat and oats are the principal corn crops; but barley, beans, and peas are, also, extensively raised. Turnips are an important crop in the coast district; they are universally drilled, and their culture is no where better understood. Cattle are of various breeds; but the improved short-horns are now, perhaps, the greatest favourites. Estates of all sizes, but mostly large. Farms, also, large, and their occupiers distinguished by their superior intelligence and enterprise. Farms mostly held on leases, varying from 7 to 14 and 21 years. Farm-houses and cottages good. Average rent of land in 1810, 15s. 1½d. an acre. With the exception of those carried on at Newcastle (which see), manufactures are of no importance. Pit-coal forms the staple produce of Northumberland, and is raised and shipped in vast quantities from the Tyne, for the supply of London and other ports on the E. coast, and for exportation. In proof of this, we may mention that of 7,475,577 tons of coal shipped coastwise from the different ports of the United Kingdom in the year 1840, 2,981,343 were shipped from Newcastle. (Parl. Paper, No. 269. Sess. 1841.) The pitmen, who are a numerous and important class, receive wages varying from 15s. to 25s. a week, and are honourably distinguished among the working classes by their superior comforts and enjoyments. Their houses are generally clean, roomy, and well furnished; they live well, are but little influenced by political agitation, and are more orderly and decidedly less addicted to ardent spirits, cock-fighting, and such like demoralizing sports, than they were 30 years ago. (Minutes of Committee of Council on Education, p. 61.) Exclusive of its coal, Northumberland has mines of lead and iron, and abundant supplies of limestone and sandstone, the quarries of the latter at Gateshead Fell supplying the "Newcastle grindstones," so famous in most parts of the world. Principal rivers, Tyne, Coquet, Alne, Blythe, Wansbeck, and Till. Principal towns, Newcastle, Tynemouth, N. Shields, Morpeth, &c. It returns (inc. Berwick) 10 mem. to the H. of Com., viz. four for the county, two each for the bors. of Berwick and Newcastle, and one each for Morpeth and Tynemouth. Registered electors for the co. in 1838–39, 8012; being 2742 for the N., and 5270 for the S. division. The co. is divided into six wards and 88 pars.; and had, in 1841, 48,704 inhab. houses, and 250,268 inhab.; of whom 121,271 were males, and 128,997 females. Sum expended for the relief of the poor in 1838–39, £61,918. Annual value of real property in 1815, £1,991,413. Profits of trade and professions in do. £436,404.

NORTHUMBERLAND, co., Pa. Situated near the centre of the state, and contains 440 sq. ms. Watered by the Susquehanna r. It contained in 1840, 11,693 neat cattle, 17,409 sheep, 18,855 swine; and produced 927,927 bushels of wheat, 141,016 of rye, 165,790 of Indian corn, 54,549 of buckwheat, 160,190 of oats, 115,985 of potatoes, 27,305 pounds of tobacco. It had 55 stores, 99 grist-mills, 26 saw-mills, one oil-mill, 17 tanneries, 14 distilleries, one brewery, six potteries, three printing-offices, one weekly newspaper; 61 schools, 2854 scholars. Pop. 20,027. Capital, Sunbury.

NORTHUMBERLAND, co., Va. Situated in the E. part of the state, and contains 240 sq. ms. Bounded N.E. by the mouth of Potomac river, S.W. in part by Rappahannock river. It contained in 1840, 6300 neat cattle, 4057 sheep, 12,035 swine; and produced 98,038 bushels of wheat, 178,972 of Indian corn, 54,594 of oats, 26,396 of potatoes, 11,808 pounds of cotton, 45,190 of sugar. It had 17 stores, 23 grist-mills, six saw-mills, three tanneries, one distillery; four academies, 106 students; eight schools, 180 scholars. Pop.: whites, 4094; slaves, 3343; free coloured, 547; total, 7984. Capital, Heathsville.

NORTHUMBERLAND, p. t., Saratoga co., N.Y., 15 m. N.E. Ballston Spa, 28 m. N. Albany, 408 m. W. Bounded E. by Hudson river. It has four stores, two fulling-mills, one woollen-factory, one grist-mill, five saw-mills; six schools, 198 scholars. Pop. 1672.

NORTHUMBERLAND, p. b., Northumberland co., Pa., 59 m. N. Harrisburg, 169 m. W. Situated at the confluence of the N. and W. branches of Susquehanna river. A bridge across the N. branch, connects it with Sunbury. Another over the W. branch leads to Union co. The Shamokin dam across the river here is 9½ ft. above the bottom of the river, and 3783 ft. long. The chute through it for the passage of boats and rafts, is 64 ft. wide, and 650 ft. long. The bor. contains three churches, a Presbyterian, Methodist, and German Reformed, an academy, a market-house, a town-house of brick, a bank, six stores; three schools, 190 scholars, and 150 dwellings. Pop. 926. Incorporated in 1828. The Susquehanna and N. and W. branch canals meet here.

NORTH WHITEHALL, p. t., Lehigh co., Pa., 95 m. E.

N.E. Harrisburg, 188 m. W. Drained by tributaries of Lehigh river. It contains two churches, eight stores, seven grist-mills, four saw mills, five tanneries, 13 distilleries; seven schools, 980 scholars. Pop. 3333.

NORTHWICH, a market-town and township of England, par. of Great Budworth, hund. of its own name, co. Chester, on the Weaver, 16¼ m. E.N.E. Chester, and 155 m. N.W. London. Area of township, 200 acres. Pop. in 1841, 1368. It has an antiquated appearance, with badly paved streets. The church, which is subordinate to that of Great Budworth, is a large building, with a semicircular choir, remarkable for the curious decorations on the roof of the nave. There are places of worship, also, for Wesleyan Methodists and Independents, with attached Sunday-schools. A grammar-school was founded in 1558; it is handsomely endowed, and the government is vested in twelve trustees, who appoint both the masters and the free scholars. There is, also, a charity-school for twelve poor children. Northwich is one of the wiches or salt-towns of Cheshire, and vast quantities of salt are annually produced in the town and its vicinity. The salt mines are very extensive; they have been wrought since 1670; and the quantity of salt obtained from them is greater, probably, than is obtained from any other salt-mines in the world. In its solid state, when dug from the mines, the salt is not sufficiently pure for use, and is sent to Frodsham and other places on the S. side of the Mersey, where it is refined, by being dissolved in sea-water, and afterwards separated by evaporation and crystallization. By far the largest quantity, however, of the salt now produced in Cheshire is obtained from the brine springs. The brine is first pumped up, principally by means of steam-engines, from very deep wells, and is collected in reservoirs, where it is sometimes saturated or strengthened by an admixture of crushed rock-salt. The business has greatly increased within the last few years, and it is estimated that above 230,000 tons are annually produced in Northwich and its vicinity. A considerable number of the inhab. are also employed in the cotton manufacture. It has every facility for water-carriage by its position on the Grand Trunk navigation, and it is close to the Grand-Junction railway. It is one of the polling places at elections for the N. div. of Cheshire. Markets on Friday; a large cattle fair, Apr. 10; other fairs Aug. 2 and Dec. 6. (Parl. Papers, &c.; Stat. Account of the British Empire.)

NORTHWOOD, t., Rockingham co., N. H., 20 m. E. Concord. It has six ponds, which give rise to Isinglass, Suncook, and Lamprey rivers. First settled in 1763. It has three churches a Congregational, Baptist, and Free-will Baptist; six stores, two-saw-mills; eight schools, 435 scholars. Pop. 1172.

NORTH YARMOUTH, p. t., Cumberland co., Me., 10 m. N. Portland, 40 m. S.W. Augusta, 555 m. W. Bounded S. E. by Casco Bay, affording facilities for navigation, of which it has about 4000 tons employed in the lumber trade and the fisheries. It has four churches, three Congregational and a Baptist; 15 stores, two fulling-mills, three grist-mills, three saw-mills, five tanneries, three potteries; an academy, 265 students; 18 schools, 1100 scholars. Pop. 3894.

NORTON, CHIPPING. See CHIPPING-NORTON.

NORTON, p. t., Bristol co., Mass., 33 m. S. Boston, 428 m. W. Incorporated in 1711. Watered by Rumford, Cocasset, and Canoe rivers, branches of Taunton river, which afford good water-power. It has nine stores, one furnace, four cotton-factories with 1964 spindles, three grist-mills, nine saw-mills; one academy, 190 students; eleven schools, 377 scholars. Pop. 1545.

NORTON, t., Summit co., O. The Ohio canal crosses its S.E. corner, along the Tuscarawas river. It has seven schools, 263 scholars. Pop. 1479.

NORWALK, p. t., Fairfield co., Ct., 68 m. S.W. Hartford, 269 m. W. Bounded S. by Long Island sound. The v. is situated on both sides of Norwalk river, over which is a bridge, to which vessels come which require six feet of water. It contains two churches, a Congregational and Episcopal; a bank, 24 stores, two printing-offices, two weekly newspapers, and over 100 dwellings. One and a half miles S. of the bor. is South Norwalk where the steamboats land, which ply to New York. The town contains three grist-mills, four saw-mills, two tanneries, two potteries; two academies, 67 students, nine schools, 927 scholars. Pop. 3863.

NORWALK, p. t., cap. of Huron co., O., 99 m. N. by E. Columbus, 392 m. W. The village has an elevated situation, and contains four churches, an Episcopal, Methodist, Baptist, and Presbyterian, a Methodist seminary, a High-school, an insurance company, a public library and reading-room, a steam paper-mill, two printing-offices, two weekly newspapers. 900 dwellings, and about 1800 inhabitants. The town has three saw-mills, one tannery, one distillery; 12 schools, 460 scholars. Pop. 2613.

NORWAY, (Norw. Norge, Germ. Norwegen), a country

NORWAY.

of N. Europe, forming the W. portion of the great Scandinavian peninsula, and at present united to the crown of Sweden. It extends, including Norwegian Lapland, between the 58th and 71st degs. of N. lat., and the 5th and 31st of E. long.; having E. Russian Lapland and Sweden, S. the Skagerrack, separating it from Denmark, and W. and N. the North sea, and the Atlantic and Arctic oceans. Its entire length from the Naze, its most S. promontory, to the North cape, is upwards of 1100 m. Its breadth varies greatly; in Nurrland, near its N. extremity, it may average about 50 m.; but towards the S. it is as much as 250 m. The area, pop., &c., of Norway, have been estimated as follows:—

Dioceses.	Provinces.	Area in sq. m.	Pop. 1833.	Capitals.
Aggerhuus, or Christiania	Aggerhuus	1,983	90,316	Christiania.
	Smaalehnen	1,566	64,941	Moss.
	Hedemarken	9,516	77,929	
	Christian	9,418	90,903	
	Buskerud	4,787	76,669	Drammen.
	Jarsberg & Laurvig	966	54,516	Laurvig.
Christiansand	Bradsberg	5,560	63,136	
	Nedenees	4,256	45,842	Arendal.
	Mandahl-Stavanger	2,012	54,252	Christiansand.
	S. Bergenhuus	3,005	64,858	Stavanger.
Bergen		6,495	104,471	
	N. Bergenhuus	7,515	66,778	Bergen.
Trondhjem	Romsdal	5,933	70,174	Romsdal.
	S. Trondhjem	7,094	77,724	Trondhjem.
	N. Trondhjem	8,069	57,422	
Norrland	Norrland	15,082	57,791	Bodoe.
	Findmark (Norw. Lapland)	27,470	33,394	Tromsoe.
Totals		121,725	1,150,000	

In 1835, the pop. amounted to 1,194,827, viz. 585,381 males, and 609,446 females, the ratio of the two sexes being as 100 to 104. The increase between 1825 and 1835 was 13·6 per cent., or about 1½ per cent. annually. The rural pop. in 1835 amounted to 1,065,825; the remaining 129,002 individuals lived in towns of 3000 inhab. and upwards, of which there are only 11. The increase of the rural pop. during the previous ten years had been 14 per cent.; that of the town pop. 11 per cent. (*Statistical Journal*, July, 1839.)

Physical Geography.—The chief physical characteristics of Norway are its *fjelds and fjords*; the first being lofty mountain plateaux in the interior, and the second deep indentations or arms of the sea all round the coast. Nearly the whole of the country is covered with mountains. The main chain, called the Kiölen (or keel), forms the line of separation between Norway and Sweden, as far S. as lat. 63°; but thenceforward it tends to the S.W., under the names of Dovrefjeld, Langefjeld, &c., forming the watershed between the rivers flowing into the Skagerrack on the S.E., and the North sea and Atlantic on the W. Many of the Norwegian mountains rise to from 6000 to 8000 ft. above the level of the sea. The Sneehætta, (lat. 69° 35′, long. 9° 40′) 8190 ft. in height, has been long considered the most elevated point of land; but it is now supposed that the Harunger Fjeld, in the prov. Bergen, overtops the former by at least 700 ft. (*Laing*, p. 53.) The *Fjords* have been sometimes compared to the Scottish *friths*; but they are generally smaller than the latter, and rather resemble the Scottish salt-water *lochs*. They are most numerous on the W. coast, where the Sogne and Hardanger fiords, with their continuation, stretch inland for at least 100 m. in a direct line; and are of the greatest use as means of communication. Norway has numerous rivers, some of which, as the Glommen, Lougen, Drammen Nid, &c., all taking a S.S.E. direction, are of large size; but their courses are so beset with cataracts, that they are of little service for navigation. Lakes are numerous in the E. half of the country; but none of them can be compared in respect of extent to the lakes of Sweden. The W. coast is lined in its entire extent by a vast number of islands. The principal of these are the Loffoden group (which see *ante*, p. 208). The shores of Norway (like the W. coast of almost all countries in high latitudes) are ironbound, and difficult of access; and at the S. extremity of the Loffoden Isles is the celebrated Maelstrom, which inspires the Norwegian fishermen with as much terror as Charybdis did the ancient navigators of the Mediterranean; and perhaps, with more reason.

Geology.—The formations of Norway are for the most part primary. The mountains were long supposed to consist almost exclusively of granite, but in reality this rock is far from common. The most abundant rock is gneiss; next to which, though by no means so widely diffused, is mica slate, resting upon and alternating with the gneiss; and in beds subordinate to both, are limestone, quartz, and hornblende. Upon the high table lands, the ground is often covered with blocks of a conglomerate rock, in which pebbles of quartz, felspar, and other crystallised substances are embedded, and which, being smooth and rounded, have ev-

idently been, during a remote but lengthened period, subject to violent friction. Mr. Lyell (*Princ. of Geology*, i. 336) denies the occurrence of volcanic action and earthquakes in the Scandinavian peninsula; but his opinion, though entitled to great weight, differs from the statements of some recent travellers, from which it would seem that earthquakes are not unfrequent, and that the physical appearance of the country, especially of its fiords, almost demonstrate that it has at a distant period been upheaved by volcanic action. (See *Laing's Norway*, p. 75, 76–114, &c.)

The climate must, of course, vary greatly, according to the elevation of the surface, as well as the difference of latitude; but generally the summers are short, and the changes sudden and extreme. From lat. 58° to 59° the average temp. is about 45° Fah.; and there is no constant snow-region.

The same vegetables and fruits grow as in England, except apricots and peaches. Beech woods cease at 59°. From 59° to 60° the average temperature is about 44° Fahr.; all kinds of grain grow here on the best soils, and the same fruit trees as before, but at 60° the plum ceases to ripen. From 60° to 61° the average temperature on the coast is 43°; in the interior, 41°. In this division the pine and Norway fir become the predominant forest trees, with birch, hazel, and aspen. The elm ceases; and beyond 61° the oak is not seen in perfection. The principal crops are rye, oats, flax, and hemp; but wheat ripens in favourable situations. Between 61° and 62°, the average temperature is about 40°; all the common fruits still ripen; as will wheat, in certain places; but this grain is very precarious and little cultivated. N. of 62° the ash is scarcely seen. The region between 62° and 63° comprises the highest land in the country, and the upper 5000 ft. of the Doone Fjeld is covered with perpetual snow. The average temperature of the valleys in this zone is about 39° Fahr. Beyond 63°, peas begin to be precarious, cabbage ceases to come to perfection, flax scarcely ripens, and wheat is not seen, except near the sea coast in small quantities; but the pine and fir tribes, birch, mountain-ash, and aspen flourish. From 63° to 64°, the hardier frosts ripen in sheltered situations only, and oats begin to be a precarious crop. From 64° to 65°, rye, oats, and barley ripen; but beyond 65°, neither oats nor any fruit, except currants, succeed; and the pine begins to degenerate. Respecting the climate of the country further N., *see* LAPLAND, *ante*, p. 147.

Stoves begin to be lighted in Christiania in the middle of September, and cannot be dispensed with till the middle of May; the summer then advances rapidly, and the thermometer, in July, often rises at noon to above 80° Fahr.; but the heats are of short duration; frosts frequently occurring in the latter end of August. The W. coast, though generally rainy and damp, is not unhealthy: in the interior, the atmosphere is usually dry and bracing. In some places vegetation is so quick that the corn is sown and cut within six weeks.

Land, Agriculture, &c.—Norway is essentially an agricultural and pastoral country. In 1835, of a total male population of 585,381, of whom 434,987 were above 16 years of age, 309,000 were connected with agriculture, either as proprietors, farmers, or farm-servants, journeymen, country people, &c. 28,903 were estimated to be engaged in navigation and the fisheries; 23,145 in commerce and manufactures; 1192 were government officers. 2104 pensioners, and 4739 pauper-inhabitants of towns. Only about 100th part of the entire surface is supposed to be under culture, or otherwise productive. As Norway is a country where the feudal system was never established, the land is mostly the property of those who cultivate it. Such land is termed *udal*, a word in its origin probably the same with the German word *edel*, or noble, since it carries an equivalent meaning in all its applications. Udal land is noble land, held from or under no superior, not even the king; but by the same right by which the crown itself is held. It is possessed, consequently, without charter, and is not subject to fines, escheats, forfeitures, nor personal suit or service; nor to any of the burdens affecting land held by feudal tenure direct from the sovereign, or from his superior vassal. The succession to land is not vested in the oldest male heir. On the contrary, all the kindred of the udalman in possession are what is called *odelsbaarn* to his land, and have, in order of consanguinity, a certain interest in it, called *odelsbaarn ret*. Hence, if the udalman in possession should sell or alienate his land, the next of kin is entitled to redeem it on repaying the purchase money; and if he should decline to do so, it is in the power of the one next him to claim his odelsbaarn ret. Formerly the power to redeem estates was unlimited in point of time; but as a power of this sort, by rendering the title of the occupier insecure, prevented him from making any improvements, the right of redemption has latterly been limited to within five years of the sale; and it has also been ordered that the purchaser shall be indemnified for his improvements. But though this be an improvement on the former law, no doubt the better plan would be to abolish

486

altogether the privilege of redemption, by making the sale absolute and final.

In 1835 the number of taxed agriculturists was 103,192, of whom 72,694 were proprietors, enjoying the *odels ret*, leaving 30,568 individuals, with only the limited possession or use of a farm.

Farms generally consist of three divisions; the in-field, or acres inclosed for the crops and best hay; the mark, or out-field, also enclosed for pasturing the cattle; and the *seater*, a tract of unmeasured grass land, which is sometimes 30 or 40 m. distant; and on which chalets are erected, and the cattle are pastured for three or four months in summer. A farm of average size is stated by Laing to have comprised 280 acres, exclusive of the *seater*. Of this extent, 148 acres, comprising the in-field, were cleared; only about one third, however, yielded corn and potatoes, the remainder being always in grass for hay. The out-field is usually half cleared, being fenced off and ploughed in patches; and it is in this division that the housemen or cottiers, paying from three or four dollars each of rent, and working at about eight shillings (3d.) a day, with their food, have their houses and their fenced pieces of land. The farm referred to above, supported 20 cows, seven horses, and a score or two of sheep and goats. The accommodations for the cattle were very good, the cow house being floored with timber, and lighted by glass windows: the cows were tended by a woman. The rent was 2500 dollars a year; the taxes, including tithe, poor rate, and all other direct assessments, amounted in all to about 36 dollars; the indirect taxes, including excise and other duties, were inconsiderable. A property like this is reckoned worth about 4000 dollars: and the prices of ordinary estates vary from 2500 to 4500 dollars. Almost all the houses are of wood: they are generally comfortable; and owing to the facility with which they may be constructed, there is but little difference between the residence of a public functionary, a clergyman, or a gentleman of large property, and that of a *bonde* or peasant proprietor. The division of property among children prevents the erection of any splendid mansions, or any thing more expensive than is proportioned to the property upon which it stands.

Except in a few favoured spots the arable land is, generally speaking, sandy and poor. Hence, if a few days of warm sunshine succeed each other without rain, as is frequently the case in the early part of summer, the roots of the corn and grass are apt to be burned up. In autumn, on the contrary, the decreased warmth prevents the corn from ripening, and not unusually, even in favourable seasons, it is injured by violent autumnal rains. Mr. Malthus says there are three nights about the end of August, distinguished by the name of *iron nights*, on account of their sometimes blasting the promise of the fairest harvests. (i. p. 375.) The crops are, in consequence, extremely precarious. Even in the best years a considerable supply of corn has to be imported; and in bad years the inhabitants, especially in the interior, have to sustain the greatest privations.

In addition to the depressing influence of the soil and climate the peasantry are said to be deficient in industry; and wedded to routine practices; and a considerable influence is also ascribed, in the production of dearths, to the great consumption of corn in distilleries. We believe, however, that the latter complaint is wholly without foundation. The demand for corn for distillation makes, no doubt, a greater quantity be sown in ordinary years than if it were prohibited; so that in bad years, when distillation almost wholly ceases, there is a greater supply to meet the necessities of the people. Rye is the crop most extensively cultivated, and next to it oats, flax, and potatoes. The agricultural implements, which are usually made by the peasants themselves, are better than could, under such circumstances, be expected: even threshing machines are pretty common.

All over Norway there are corn magazines, to which the farmers may send their surplus produce, and whence, also, they may be supplied with loans of corn; the depositors receiving at the rate of 19½ per cent. of increase on the corn deposited for a twelvemonth; and the borrowers replacing the quantities advanced at the expiration of the same period, with 25 per cent. increase. These depositories are found to be useful in consequence of the extreme precariousness of the crops. The difference between the increase allowed, on the corn received, and that charged on the corn given out, pays the expenses. In the north, and even in other parts, in years of scarcity, the inner rind of the fir tree, kiln-dried, and ground, is used, together with corn meal, for bread. Some travellers have, strangely enough, and without sufficient information, denied this fact; but, according to Laing, the use of this material is more extensive than is generally supposed. The inner rind next the wood is taken off in flakes, like foolscap paper, steeped in warm water, and hung to dry in the sun. When dry it is pounded in small pieces, mixed with corn, and ground on the hand-mill. The extended cultivation of the potato since the peace, has probably placed the inhabitants of the lower country beyond the necessity of generally using it; but those in the higher parts use it, more or less, every year. It is not unpalatable, but is costly. The value of the tree, left to perish, would buy a sack of flour, if the English market were open. "The Norwegians starve, and we shiver in our dwellings, though each country has the means of relieving the other with advantage to itself; and all for the sake of supporting colonies and other interests, which add little to the well-being of the people of Great Britain." (*Laing*, 340, 341.)

The most profitable branch of rural industry is cattle breeding. The cattle are small in the bone, thin skinned, usually red or white, and obviously of the same stock with the common unimproved breed in England, France, and Germany. The cows give excellent milk, and dairy produce enters largely into the food of every family. Goats are a favourite stock, and on every farm they appear more numerous than sheep. Hogs are not numerous. The horses are, in general, inferior to those of Sweden. The real Norwegian pony, however, met with in the N. of the country, is an admirable little animal, fast going, hardy, and fit for a great deal of work. A few are occasionally imported into Scotland. The live stock suffers frequently from wolves and bears, the hunting of both of which is actively pursued; but that of the latter not so much as formerly, the price of bear skins having greatly fallen. The elk, and many kinds of game, are found; and in the N. large herds of rein-deer constitute the chief wealth of the Laplanders. Aquatic birds are so abundant, that the search after their eggs occupies a large share of the attention of the inhabitants of the coast.

From the want of markets, and of other facilities for commerce, the Norwegian farmer is seldom able to convert his surplus produce or savings into money. His object, indeed, is not to raise produce for sale, but to supply himself with the various materials required for the food, drink, and clothing of his family. "The food of the labourers who work for gentlemen or large farmers, consists of black rye bread and salted butter or cheese, for breakfast; and boiled barley and a herring or some other fish, with beer, for dinner. Once a week, and sometimes twice, they have fresh meat. The common people live nearly in the same way, only not quite so well; and some who have large families are often in great distress." (*Clarke's Travels*, x. 448.) Mr. Laing says that the labourers get frequently at their meals an allowance of home-made potatoe or corn spirit. The latter article is especially abundant, being distilled, without let or hindrance, on every farm. Great quantities are drunk, its price being only about 14d. a gallon. The farm labourers, called housemen, live, as has been stated, in cottages on the mark or out-field, at a fixed rent for two lives, under the obligation of furnishing a certain number of days' work on the main farm, at a certain rate of wages. A system, in some respects similar, prevails in some of the best cultivated districts of Scotland; but, according to Mr. Laing, the Norwegian houseman is better off than the Scottish married farm servant. Land, he says, being of less value in Norway, the houseman has more of it: in fact, it constitutes a complete little farm, keeping generally two cows and some sheep, and producing a full subsistence for a family.[*] The law of the country has especially favoured the class of housemen. In default of a written agreement registered in the parish court, the houseman is presumed to hold his possession for his own life and that of his wife, at the rent last paid by him. He may give up his land and remove, on giving six months' notice, before the ordinary term, and is entitled to the value of the buildings put up at his own expense, which he may have left; but the landlord cannot remove him or his widow so long as the stipulated rent and work are paid.

Fisheries.— Above the parallel of lat. 65°, agriculture and cattle rearing cease to be the primary occupations of the population. The inhabitants of Norrland and Finmark, amounting to 91,000 persons, subsist chiefly by fishing, when they are not supported on the produce of herds of rein-deer. The Loffoden islands (see *ante*, p. 208) are the principal seats of the cod-fishery; and the average value of the fish caught there, during the winter, has been estimated at £86,500. The winter fishery lasts from February to April; after which the fishermen are either employed by the Russian merchants, or retire to their homes to begin the herring fishery. Besides these general fisheries, in every fiord, even at 100 m. from the sea, an abundance of cod, whiting, haddocks, flounders, herrings, &c., is caught daily for use and for sale by the seafaring peasantry.

The Forests and Mines of Norway might be rendered two of its principal sources of national wealth. Fir timber, deals, &c., are among the chief exports. But the want of navigable rivers, canals and roads occasions great difficulties in the conveyance of timber to the coast; for it is only dur-

---

[*] Dr. Clark, however, says, that in the neighbourhood of Christiania, the housemen have seldom land to keep a cow (x. 450); and we understand that this is frequently the case in other parts.

ing the spring thaws that the rivers or torrents are deep enough to float the timber down to the fjords. No doubt, however, were the timber trade of this country placed on a proper footing, by doing away the impolitic preference given to American timber, a great stimulus would be given to its importation from Norway; and the advantages thence arising would, it may be fairly presumed, lead to the formation of improved means for supplying the shipping ports with timber and deals. The manufacture of the latter is the principal branch of industry carried on in the country. They are mostly shipped from Christiania, Drammen, &c. "Their forests are of the most essential service to the Norwegians, who apply their products to an infinite variety of purposes. Their *summum bonum* seems to consist in the produce of the fir. This affords materials for building their houses, churches, and bridges—for every article of their household furniture—for constructing sledges, carts, and boats—besides fuel for their hearths. With its leaves they strew their floors, and afterwards burn them, and collect the ashes for manure. The birch affords, in its leaves and tender twigs, a grateful fodder for their cattle, and bark for covering their houses. The bark of the elm, in powder, is boiled up with other food, to fatten hogs; sometimes also, though rarely, it is used in the composition of their bread." (*Clarke*, x. 344, 8vo. ed.)

No coal has yet been discovered in Norway; but Berendiah, between the N. cape and Spitzbergen, appears to consist principally of that mineral. Some is occasionally brought thence by Tromsoe and Hammerfest whalers; and, were the forests raised to their due importance by better means of conveyance, it is probable that coal might be supplied to the country in quantities which would render the cutting down of the trees for fuel in a great measure superfluous. The iron of Norway, though inferior to that of Sweden, is of excellent quality, and very generally found. Copper is found at Rœraas; and near Kongsberg a silver mine, which has been wrought for upwards of 200 years, was, about the middle of last century, accounted the richest in Europe. In 1768, it produced ore to the value of £79,000; it has since, however, materially declined. Lead, cobalt, arsenic, and a little gold are met with in various places. At Waldö is a salt mine, producing about 20,000 tons a year. Alum, asbestos, marble, slate, building stone, &c., are among the other mineral products.

Manufactures are almost wholly domestic, the division of labour being carried to a less extent in Norway than in, perhaps, any other European country. The *bonder*, or agricultural peasantry, build their own houses, make their own chairs, tables, ploughs, carts, harness, iron-work, basket-work, and wood-work; in short, except the window glass, cast iron ware, and pottery, every thing about their houses is of their own make. The Norwegian peasant, indeed, unites most trades in his own person, his principal tool for executing all kinds of work in wood being the knife he carries in his girdle. The shoemaker and tailor go round and cobble and sew for a few weeks at each village, getting their maintenance, and being commonly paid over or above, in potatoes, meal, butter, or other produce. Spinning-wheels and looms are at work in every cottage and house in the country; the farmers and country people spinning their own flax and wool, and weaving their own linen and woollen clothes. An official report, in 1829, made the total number of manufacturing establishments 327, but of these 128 were distilleries and 80 tobacco-factories principally in Christiania, Drammen, and Bergen. There are, in fact, very few fabrics of clothing materials; and few Manchester or Glasgow fabrics are seen. No doubt, however, were greater facilities afforded to the Norway timber trade by Great Britain, our manufactures would, at no very distant period, supersede the ruder fabrics of the Norwegian peasantry.

Commerce.—Foreign trade is in a very depressed state from various causes, but principally from impolitic restraints. Bergen and Trondhjem were formerly members of the Hanseatic Association, on the decline of which these towns retained, and still hold separately, the same privileges they enjoyed in conjunction with the other members of that body, though Christiansand and some other minor towns have succeeded, after a long struggle, in obtaining a share of the commercial monopoly. The merchants and shopkeepers in Norway are all licensed burgesses of Bergen, Trondhjem, or other privileged towns, to which they pay a certain tax; and each has a certain tract or circle belonging to his factory, within which no other person is entitled to buy or sell. The imports consist principally of coffee, sugar, tobacco, corn, spices, brandy, wines, tea, &c., and the exports of fish, timber, and other native produce. The trade of Norrland and Finmark is, however, different from that of the rest of the kingdom. The privileged traders do not supply the inhabitants of these provinces with necessaries, except during the winter fishing season: and as no other Norwegian dare interfere, the trade of these provinces is now almost wholly in the hands of the Russians, whose ships have been, since

1838, allowed admission, free of duty, into every port N. of Tromsoe.

Owing to the thinness of the population, and the bad state of the roads and other means of communication, there is but little internal trade. "Even in the largest towns, such as Christiania and Trondhjem, there is nothing that can be called a market. It is extremely difficult to get a joint of fresh meat; and a pound of fresh butter is an article not to be purchased even in the midst of summer. Fairs are held at certain seasons of the year, and stores of all kinds of provisions that will keep are laid in at these times; and, if this care be neglected, great inconveniences are suffered, as scarcely any thing is to be bought retail. Persons who make a temporary residence in the country, as small merchants, not possessed of farms, complain heavily of this inconvenience." (*Malthus*, i. 372.) Latterly, however, some improvements have been made in the facilities of interchange: and the inconveniences depicted by Mr. Malthus have been in so far diminished.

The *Bank of Norway*, established in 1816, has its head office in Trondhjem, and branches in Bergen, Christiania and Christiansand. Its principal business consists in advancing its own notes, upon first securities over land, any sum not exceeding two thirds the value of the property, at 4 per cent. interest. The notes of this bank are at only a trifling discount when compared with silver; and its paper is in general use in Norway instead of silver for sums above a mark. The mark or ort of 24s. = 9⅗ English; 5 marks = 1 specie-dollar. Money being scarce, internal traffic is almost wholly conducted by barter. Provisions are generally cheap; and a dollar has been, in this respect, estimated as equivalent to a pound sterling in England; but it should be added, that the Norwegians, especially the innkeepers, never omit an opportunity of overcharging travellers. The most usual mode of travelling in this country is with the *carriola*, a little carriage formed somewhat like a shell, and slung between the shafts and two cross bars, horsed as in Sweden, by the farmers along the road, at the order of station-masters. The price of posting in this manner is about 4d. a mile. Steamers ply along the coast between Christiania and Bergen, but there is no similar communication farther N.

Government.—Though Norway be under the same crown with Sweden, she is no more connected with that country than Hanover was formerly with Great Britain. The constitution differs from that of Sweden in many important respects. The Swedish government is in part aristocratical; that of Norway is a hereditary monarchy, with a democratic assembly only. This, which is called the *Storthing*, consists of a certain number of members, between 75 and 100; about one third of whom are returned by the towns, and the rest by the rural districts. Every native Norwegian of 25 years of age who is a burgess of any town, or possesses property, or the life-rent of land to the value of £30, is entitled to elect and be elected: but for the latter privilege he must not be less than 30 years of age: nor an officer of the crown (which has no representative or organ in the Norwegian storthing); and he must have resided in Norway for 10 years. The country is divided into election districts and sub-districts, according to their population. The mode of election is double, being performed through the intervention of election-men. In the towns one election-man is chosen by every 50 voters; in the rural sub-districts by every 100 voters: the choosing of these takes place in the parish church at the end of every third year. The election-men afterwards meet at the place appointed for the district or provincial election, and there elect among themselves, or from among the other qualified voters of the district, the representatives to the storthing, in the proportion of one fourth of the number of election-men for the towns, and one tenth of those for the rural sub-districts. Substitutes (being those who have the next number of votes) take the places of both election-men and members of the storthing, in the event of their unavoidable absence from duty. The storthing meets for three months once in three years, *suo jure*, and not by any writ from the king or the executive. It may be convened at other times, but in that case it can pass only temporary acts, which must be ratified during the next ordinary session, otherwise they do not become law. Each storthing settles the taxes for the ensuing three years; enacts, repeals, or alters laws; opens loans on the credit of the state; fixes the administration of the revenue; impeaches and tries before a section of its own body all state ministers, judges, and its own members, &c. This body, when elected, divides itself into two houses. One, called the *lagthing*, has functions corresponding generally to those of our house of lords, and is composed of one fourth of the total number of members of the storthing: the other three fourths constitute the *odelsthing*, or lower house; and all proposed enactments must originate in this division. A bill which has passed both houses usually becomes law, by receiving the sanction of the king. But the Norwegian storthing enjoys a right which no other legislative assembly in Europe possesses.

If a bill pass through both divisions in three successive storthings, on the third occasion it becomes a law of the land without the royal assent; and this right was exerted when the Norwegians abolished their hereditary nobility in 1821. Each member of the storthing has an allowance of 1½ dollar a day during its session.

The mode of assembling the people in the country for public business is simple, but curious. A budstick, or message-stick, about the size and shape of a constable's baton, with a spike at one end, is made hollow to hold a piece of paper, on which are written the official notice to meet, with the time, place, and object. This is delivered from the court-house of the district to the nearest house-holder who is bound by law to carry it, within a certain time, to his nearest neighbour; he must transmit it to the next; and so on. If the owner be not at home the bearer is to stick it " in the house-father's great chair by the fire-side;" and if the door be locked he must fasten it to the outside. He who, by neglect in passing the budstick, has prevented others from attending, pays a fine for every person so absent.

Justice, &c.—The Norwegian peasantry were never adscripti globæ, subject to local judicatories, as in feudal countries, but subordinate only to the general jurisdiction of the country. The small kings, expelled in the ninth century by Harold Haarfager, seem never to have attained the powers of the great feudal lords in other countries, but were always in some degree dependent upon the general things, or courts, of the people. Trial by jury is a very ancient institution in Norway; but many of the details in the administration of justice originated with the Danes. The latter instituted the courts of mutual agreement, one of which exists in each parish, the arbitrators being chosen by the householders every third year. Norway is divided, for legal purposes into four stifts and 64 sorenskriveries. In each of the latter divisions is a legal court, which sits once a quarter, and in which the sorenskriver, who presides, has only a vote as a member of the jury, a majority of whom decides the case. The stifts amts, consisting each of three judges, with assessors, and established in the chief town of each stift, are the courts of appeal from the foregoing. The hoiests ret, in Christiania, composed of a president and eight assessors, is the highest court, and one of final resort. The special courts are the regris-ret, or lagthing, the ecclesiastical, and the military tribunals. Judges are responsible in damages for their decisions. Capital punishment has been abolished; slavery in chains, for a longer or shorter period, being the ordinary sentence for all kinds of crimes.

The religion is the Lutheran; but much ceremony still remains in the forms of worship. Norway is divided into five bishoprics and 336 parishes: the latter divisions are very extensive, but several are frequently under the cure of one priest. The incomes of the parish priests amount to from 800 to 1600 dollars, those of the bishops to 4000 dollars a year. (Laing, p. 180.) The former are paid by means of rents from glebe lands, a small tithe of corn from each farm, or of fish in some parts, and fees, and other unfixed sources of revenue. There are no dissenters; all sects of Christians are, however, tolerated, but Jews are excluded from settling in Norway, nor are even suffered to remain in the country for more than a few hours at a time.

In 1837, 176,733 persons, or about one seventh part of the population were receiving public instruction. Schoolmasters are settled in each parish, who live either in fixed residences, or move at stated intervals from one place to another, and who frequently attend different schools, devoting one day only in the week to each. They are paid by a small tax levied on householders, besides a personal payment from each scholar, amounting, in the case of agricultural servants, to about eight skills, or half a day's wages in the year. Instruction in the primary schools is limited to reading, writing, arithmetic, and singing, with sometimes the rudiments of grammar and geography. Almost every town supports a superior school; and in 13 of the principal towns is a lærde skole, or college, the instruction in which includes theology, Latin, Greek, Norwegian, German, French, English, mathematics, history, geography, &c. Christiania has a university, founded by the Danes, in 1811, which is modelled on the system of the German universities, but differs from them in the professors not receiving fees; and in which the number of students varies from 600 to 800. (Brønner.) There are, also, schools of drawing and architecture, commerce and navigation, and other special schools. Sunday-schools have been widely established; and the Society of Public Good maintains a public library in most parishes of the kingdom.

The press in Norway is altogether free. Every man is at liberty to print and publish what he pleases, being responsible, however, for what he does print. No tax exists on newspapers; and somewhat more than 20 are published in the kingdom, besides several scientific journals. But notwithstanding these aids to science and advancement, Norwegian literature is not in a very flourishing state, and can

by no means bear to be compared with that of Denmark or Sweden.

The army of Norway consists of about 10,000 infantry, 1600 cavalry, 1000 artillerymen, and 150 engineers; in all, 12,150 men. A militia is raised throughout the interior, into which all males, between 27 and 30 years of age, must enter; and on the sea coast there is a kind of marine militia, in which all seafaring men, and inhabitants of sea ports of a certain age, must be enrolled. The naval force consists of five brigs and 117 gun boats. (Alm. de Goths, 1841.)

The public revenue, for the three years 1839–41, was estimated in the budget of the former year at 2,130,090 dollars; the expenditure for the same period is fixed at 2,415,936 dollars; the deficiency will be made up from the reserve funds in the hands of the government.

People, &c.—The Norwegians are members of the widely spread Teutonic race. The men are, in general, rather small in stature, but well made, and appear to have great muscular power. The Gudbrandsdal peasants are said to be the most athletic, but they are decidedly, as a body, shorter and slighter of limb than the mountaineers of Delecarlia, in Sweden. Their complexions, hair, &c., are fair, and resemble more of the Danes, and other N. German tribes, than the Swedes. The dress of the men varies greatly in the different districts, being, for the most part, more gay and fanciful than that of the women: in the towns, however, the upper classes have fully adopted the costume common in the rest of Europe. "The peasants possess much spirit and fire in their manner, are frank and undaunted, yet not insolent; never fawning on their superiors, yet paying proper respect to those above them. The principal mode of salute is by offering the hand; and when we gave them a title, instead of returning thanks by a word or a bow, they shook our hands with great frankness and cordiality." (Case's Travels in the North of Europe, v. 7.) They are generally addicted to drinking, and the standard of morals is said to be, in other respects, higher in Sweden than in Norway. Women are very generally employed in field-labour; and beggars are numerous, especially in the towns. The average number of illegitimate births is about 1 in 5; and in one district it was, from 1826 to 1830, as much as 1 in 3½. (Laing, 151.) But illegitimate children are most commonly legitimized by a legal act, and are seldom or never abandoned by their parents. The Norwegians are extremely fond of dancing, music, and dramatic entertainments, which are the principal amusements introduced at their festivities.

History.—Norway is interesting as the original seat of the Northmen, who made such frequent descents on the coasts of England and France in the dark ages, and who were the ancestors of that remarkable people the Normans, who conquered and carried their institutions to England and other of the fairest portions of Europe.

Little is known of the history of Norway before the end of the 9th century, when Harold Haarfager united the whole country under his dominion. Christianity was introduced by Olaf I. in the succeeding century. In 1387 Norway was annexed to Denmark, to which it remained attached till 1814, when the allied powers gave it to the Swedes in indemnity for Finland. The Norwegians, indignant at the transfer, took arms, and elected Prince Christian Frederick of Denmark for their king; but the latter resigned the crown in the same year, and the country has since been united to Sweden; and this union will probably be maintained without difficulty so long as the Swedish cabinet attempts no rash or violent changes in the internal administration of the country. (Official Rep.; Laing's Norway is an able work, but its statements are obviously much too favourable, and must be received with great modification; Brønner's Excursions, vol. i.; Coxe, Inglis, Clarke, Barrow, &c., passim; Dict. Geog.)

NORWAY, p. t., Oxford co., Me., 44 m. W. by S. Augusta, 588 W. It has a large pond with its outlet into Little Androscoggin river, Incorporated in 1797. It contains four stores, two fulling-mills, one flouring-mill, three grist-mills, four saw-mills, two tanneries; one academy, and 662 scholars in schools. Pop. 1786.

NORWAY, p. t., Herkimer co., N. Y., 86 m. W.N.W. Albany, 411 W. Drained by tributaries of West Canada creek. It contains three churches, a Presbyterian, Episcopal, and Baptist, two stores, one fulling-mill, one grist-mill, five sawmills; eight schools, 285 scholars. Pop. 1046.

NORWEGIAN, t., Schuylkill co., Pa. It is an important centre of the coal mines and trade. Watered by Schuylkill river and its branches. It contains 14 stores, one flouring-mill, one powder-mill, 14 saw-mills; eight schools, 309 scholars. Pop. 2812.

NORWICH, a city of England, being a county of itself, and an important manufacturing town, locally situated in hund. Humbleyard, co. Norfolk, of which it is the cap., on the navigable river Wensum (crossed here by 10 bridges), 56 m. N.E. Cambridge, and 96 m. N.N.E. London; lat. 52° 7′ N., long. 1° 16′ E. Area of the city and co. 5930 acres: population, in 1811, 37,256: ditto, in 1821, 50,288; ditto,

NORWICH.

in 1831, 61,116; and in 1841, 62,344. The county of the city is an irregular circular form, with an average diameter of about 5½ m., the city itself standing a little E. of its centre on the slope and summit of a hill, gently rising from the river. The buildings are, in a great measure, circumscribed by the remains of the ancient fortifications which still exist, particularly on the W. and N. sides of the city. The streets, with the exception of Giles-street, and one or two more, are narrow, and so irregularly laid out, as to preclude the possibility of any general description. It has, however, many handsome houses, open spaces, &c., and is well paved, watched, and lighted with gas; and its appearance from a distance is remarkably striking. The castle and cathedral are the principal public buildings; but it has no fewer than 36 parish churches, besides chapels and other edifices. The castle (supposed to have been built at intervals between the 10th and 12th centuries, by Canute, Roger Bigod, and others,) occupies a commanding eminence near the cattle-market in the centre of the town, and is a very imposing object at a distance: the part now extant forms a large square, on the E. side of which is an entrance tower, recently restored on the original plan. The entire building formerly occupied an area of 23 acres, and had three nearly circular and concentric lines of defence formed by a wall and ditch: the inner ditch, now laid out in gardens, still remains, inclosing the inner ballium, and is crossed by a semicircular bridge of one arch, 40 ft. in diameter, forming one of the largest and most perfect Saxon arches in the kingdom. In 1793 a county jail was commenced on the Castle-hill; and at the same time the ditch was enclosed with iron palisades and gates. Within the precincts, also, a new county hall has recently been erected in the Tudor style. These modern additions, however, are quite incongruous with the ancient and venerable appearance of the original Norman fabric. The cathedral, originally built in 1096, but subsequently so repaired and enlarged that it did not assume its present form till the 16th century, is one of the largest and finest ecclesiastical edifices in the kingdom. The plan is almost wholly Norman. It consists of a nave, with side aisles, two transepts, and a choir with a semicircular E. end: the whole length from W. to E. is 411 ft., that of the transepts from N. to S. being 191 ft., and the breadth of the nave and choir, 72 ft. The cloisters form a square of 174 ft. within the walls adjoining the S. side of the nave. From the intersection of the cross formed by the nave, choir, and transepts, springs a lofty Anglo-Norman tower of four stories, highly ornamented and surmounted by an elegant spire, rising 317 ft. from the basement of the church. The W. entrance is extremely beautiful, and is the best point of view from which the cathedral can be seen; but the friable nature of the stone used in its construction has caused a decay of the more salient ornaments, and thus greatly diminished the external effect. The appearance of the interior is, on the whole, grand and imposing; the architecture, however, is of various eras, from the Anglo-Norman to the English-perpendicular style; and modern alterations and additions have not always been in the best taste. The ecclesiastical establishment consists of a dean and six prebendaries, now five, and having a nett revenue of £5240, besides eight minor canons with separate allowances; but it is to be subjected to various retrenchments. The bishop's diocess comprises the whole of Norfolk, with part of Suffolk, and the revenue amounted, in 1838, to £4465. The epis palace stands N. of the cathedral, on the site of that built by the founder: it was erected in 1318, and, after undergoing repairs, and receiving considerable enlargements from successive prelates since the Restoration, has become a tolerably commodious residence, attached to which is a large and well-laid out garden, comprising some ruins of the hall belonging to the ancient palace. Near the W. front of the church is an ancient chapel, dedicated to St. John the Evangelist, which had underneath a charnel-house; it is now used as a free grammar-school. Near it are the two ancient gates of St. Ethelbert and Erpingham; the former is in the decorated English, and the latter a fine specimen of the late perpendicular style. Among the churches, which are here more numerous than in any city except the metropolis, a few deserve notice as good specimens of ancient architecture. St. Peter's, Mancroft, at the corner of the market-place, is a large edifice in the perpendicular style, surmounted at its W. end by a lofty tower; the inside is remarkably light and elegant, and it has a fine altar-piece and E. painted window. The churches of St. Andrew. St. George Colegate, St. Lawrence, St. Saviour, present similar architectural features, having high towers either of stone or flint. Those of St. Etheldred, St. Benedict and St. Julian, have round towers, and belong apparently to the early Norman era; but they have been much altered and mutilated. Norwich abounds, also, with the remains of other ecclesiastical edifices. The common-hall, in St. Andrew's parish, consists of the nave of a church attached to a monastery of Black Friars; the workhouse till very lately occupied an old Flemish convent, near which is the Dutch church, now used

as a chapel to the workhouse, and St. Giles's hospital comprises portions of the former church of St. Helen's. There are two Roman Catholic chapels, seven places of worship for Baptists, three for Independents, two for Calvinist-Methodists, five for Wesleyan or Primitive Methodists, one for Swedenborgians, one for Unitarians; besides which the Society of Friends have two meeting-houses, and the Jews a synagogue. Attached to the various places of worship are numerous Sunday Schools, of which about a half are supported by the established church, and the rest by dissenters; the whole furnishing religious instruction to upwards of 7000 children; besides which, there are several endowed charity-schools, with national, Lancastrian, and infant schools, either wholly or in part supported by subscription, and attended by about 3500 children of both sexes.

The free grammar-school, founded in 1318, and restored by Edward VI., is maintained out of the funds of a corporation charity, called the Great hospital: it has an upper and under master, and possesses fellowships and exhibitions at Caius college, Cambridge. The boys' and girls' hospitals, founded in the 17th century, are supported by the produce of estates in trust of the corporation, and furnish clothing, and instruction to upwards of 100 children. Doughty's hospital, established in 1517, and under the same patronage, provides for numerous infirm and aged persons; but the principal corporation charity is St. Giles's hospital, near the cathedral, otherwise known as the "Great" or "Old Man's hospital," maintained by rents and other property, averaging £7000 a year, and providing clothing, food, and a small stipend for 165 inmates, besides servants. It appears however, that till very recently these trusts were most extensively abused for political purposes. (Comp. Char. Comm. 27th and 29th Report, with Mun. Corp. Report. App. iv.)

Among the charitable institutions of Norwich supported by subscriptions, the first place is due to the Norfolk and Norwich hospital, occupying a large brick building, erected in 1771, and enlarged in 1802: it has accommodation for about 190 in-patients, and has about the same number of out-patients. Bethlehem hospital is a well-endowed lunatic establishment, established in 1713; and at Thorpe, about 2 m. distant, is the county lunatic asylum. The other principal institutions of this kind are the dispensary, eye-infirmary, magdalen asylum, lying-in charity, and blind asylum, with numerous minor benevolent associations, Bible and tract societies, provident clubs, &c.

The buildings devoted to the purposes of municipal or civil jurisdiction comprise the guildhall, a large old building of the 45th century, but subsequently much altered and enlarged, though even now it is little worthy of so large a town; 2. St. Andrew's, or the new hall, a noble fabric, previously mentioned as having formed part of an old Dominican church; 3. the county hall, in the castle precincts, a fine commodious building of perpendicular architecture and recent construction; 4. the new city jail and bridewell, a modern and well constructed edifice outside the walls, near St. Giles's gate; and, 5. the county jail and house of correction, on the castle-platform, a large but plain building, well adapted for its purpose, the establishment being conducted on the system of silence, separate confinement, and hard labour; the criminal prisoners averages about 80, the weekly cost of each being 2s. 11d. (Pris. Inspec. 4th Rep.). A commodious corn exchange was erected in 1826; and the cavalry barracks in Pockthorpe are substantially built of red brick, enclosing an area of 17 acres. Norwich has, also, a large workhouse, belonging to the united parishes: the sum expended for the relief of the poor amounted in 1839 to £14,976.

Among the literary establishments is the public library, originally formed in 1784, and now occupying a handsome structure in the Grecian style, lately erected on the site of the old city jail; the Norfolk and Norwich literary institution, occupies a building of recent erection, and has a good library and a numerous body of subscribers. In the same building is a museum, but not connected with the above society. A society called the Norfolk and Norwich art-union, has occasional exhibitions. Concerts are held in the common hall in St. Andrew's parish, and the musical festivals are held in St. Andrew's hall. Norwich has also a neat modern theatre, and assembly-rooms, two news-rooms, and a mechanics' institute. Two newspapers, (the "Norfolk Chronicle" and "Norwich Mercury,") are published every Saturday.

Norwich has been celebrated for its manufactures since the era of Henry I., when the Flemings first settled here, and introduced the spinning and weaving of long woollen stuffs, called worsteds, from the name of the village in which the business was first established: the worsted and bombazine trade was also greatly increased during the 16th century, by the immigration of Flemish weavers from the low countries. Norwich, however, appears to have attained its greatest prosperity at the close of the last century, when the value of its goods exported to the E. Indies

508

Russia, and other places abroad (consisting chiefly of camlets and pamletees, callamancoes, worsted satins, figured stuffs, lastings, damasks, and shawls), have been estimated to amount to £1,000,000 a year, or to one fourteenth part of the British manufactured goods exported at that period. (*Handloom Weavers' Report*, part ii., p. 303.) We believe, however, that this estimate is beyond the mark; and since then the manufactures of Norwich have materially declined, or rather, perhaps, have not kept pace with their progress in Lancashire and the West Riding of Yorkshire, &c.; the greater facilities enjoyed by the latter, in the command of coal, the absence of corporation privileges, and the greater scope given to competition and improvement, have enabled them to produce, at cheaper rates, several articles that were at one time peculiar to Norwich. In fact, the greater part of the yarn now used in making Norwich fabrics is spun at Bradford, in Yorkshire; and the worsted manufacture of the West Riding is now decidedly more extensive and valuable than that of Norfolk. The principal fabrics that are at present manufactured in and about Norwich comprise bandanas, bombazines and paramattas, fillovers, or ornamental shawls and shawl borders, gauzes and crapes, princettes (a fabric of mixed warp, with a worsted shoot), silk, silk shawls, woollen shawls, jacquard, coachlace, lustres, shallis and mousselines-de-laine, fringes, &c., with sacking and horse-hair. In 1830, there were in the city and its vicinity 5075 looms, of which 1021 were unemployed; and of the 4054 looms then at work, there were 3398 in the weavers' houses, and 656 in shops and factories: indeed, by far the greater part of the looms belong to families having only one or two. The labourers at these looms comprise 2311 men, and 1648 women, with 195 children. In the same year, two silk-mills employed 731 hands, three worsted-mills 365 hands, two woollen-mills 30 hands, and one cotton-mill 120 hands, making a total of eight mills wrought by a power (chiefly steam) of 151½ horses, and employing 1285 persons. With respect to the weavers' wages, no general reduction has been made since 1829; and the rates of payment for the various fabrics are, on the whole, pretty fairly adjusted according to the labour: if there were full employment at the present scale, the weavers would be able to secure for themselves a comfortable maintenance; but they are usually out of employ four months a year, so that wages are really a third less than they appear to be. The gross wages of the weavers when fully employed, range between 8s. and 25s., those engaged on fillovers, shallis, and fine fabrics, earning from 15s. to 25s. a week; but when "play-time" and expenses have been deducted, the average nett wages of the hand-loom weavers are said to amount to only 7s. 9d. a week; but this, we believe, is speaking generally, below the mark. A power-loom factory of mohair (or Angora wool) has been established at Lakenham (one of the out-townships), and employs about 400 hands, chiefly children. As respects the health of the weavers, Dr. Mitchell reported to the hand-loom commissioners, that Norwich is most favourably situated, the ground being on a bed of gravel over a substratum of chalk; that the working people, weavers included, have a fresh and healthy complexion; and that the physical condition of the weavers is much better than that of the same class in Spitalfields. Epidemics, however, occasionally prevail, as in Bethnal Green; but not to the same extent, or so continually, owing to the better air and house-ventilation usual in Norwich. The prevalent diseases are dyspepsia, consumption, female diseases, and those belonging to children, which last are attended with a large mortality. (*Handloom Weavers' Rep.*, ii., 399.) On Sunday, the work-people, especially the women, are well dressed; in general they attend divine service, and drunkenness prevails less than in most large towns. Frugality, however, is said to be little practised; few save any thing when they have the means; and when work is scarce, they are in the greatest distress. The industry and morals of the people have suffered materially from their frequent strikes and riots; from the disunion among the master-manufacturers; the party spirit which pervades all classes, and the electioneering abuses, for which Norwich is pre-eminent. (*Handloom W. Rep.*, ii., 339-338.)

Besides its worsted and silk manufactures, Norwich has iron and brass foundries, snuff-mills, vinegar-works, malt-houses, breweries, oil, mustard, and corn-mills; but they are not on an extensive scale. The excise duty collected in this district amounted, in 1839, to £183,193. The trade of the town consists in the exportation of its manufactures, chiefly to London and other English ports, but partly also abroad, in exchange for corn, coal, and various other articles of consumption. The town has, since 1833, had the important advantage of being accessible to vessels drawing 10 ft. water, by means of the canals connected with the Lowestoft navigation (*see* Lowestoft); and its trade will also, no doubt, be materially promoted by the Eastern Counties' railway. A branch bank of the Bank of England

is established here, and there is a joint-stock bank, called the East of England banking company, besides two private banks and a savings's bank. The Norwich Union insurance company is an establishment of great importance; but which has been shown by a late investigation to have been grossly mismanaged.

Norwich, which claims to be a borough by prescription, and received its principal charter constituting it a separate county in 1403, was governed before the Municipal Reform Act by a mayor, 24 aldermen, and 60 common councilmen; but by the provisions of that act it is divided into eight wards, and has a mayor, with 15 other aldermen, and 48 councillors. The borough has also a commission of the peace under a recorder, and a sheriff's court for the recovery of debts to any amount: the assizes and quarter sessions for the county are also held here. Norwich has returned two members to the House of Commons since the 25th Edward I., the right of election, down to the Reform Act, being vested in the freemen and feeholders not receiving alms. The electoral limits were left untouched by the Boundary Act, except that the castle precinct was included. Registered electors, in 1839-40, 4234. Norwich is likewise a polling-place and principal election-town for the E. division of Norfolk. Markets on Wednesday and Saturday, but chiefly on the latter, for corn and cattle; large horse and sheep fairs, day before Good Friday, Easter-Monday, and Whit-Monday.

Norwich is supposed to have risen from the decay of an old Roman town, now known as Castor St. Edmond's, probably the *Venta Icenorum* of antiquity. A royal fortress was erected here by the East-Angles in the 6th century, and a town was gradually formed round it, which, even before the Norman conquest, was so important as to have a mint and 25 parish churches, with 1320 burgesses. William the conqueror bestowed the castle on Roger Bigod, one of his Norman followers, who probably erected the present keep. It continued in the possession of his descendants till the reign of King John, when it was seized by the king, and finally surrendered to the crown in 1294. In the reign of Henry I., a colony of Flemings came over, who were joined by a still greater number of immigrants in 1336, from which time Norwich became an important seat of manufactures. In 1403, Henry IV. separated the city from the county, and made it a county of itself with peculiar privileges. Its prosperity, however, owing to plague, scarcity, and frequent fires, had begun to decline, when, in 1566, a fresh immigration took place of 4000 Flemings, who had fled from the persecutions of the Duke of Alva. In the civil wars of Charles I., Norwich declared for the parliament, and was occupied by its forces till Cromwell became protector. It is remarkable in ecclesiastical history for its numerous convents and other religious establishments, the funds of which have in most cases been converted to charitable uses, and placed in the trust of the corporation. Among other distinguished persons Norwich has given birth to Matthew Parker, archbishop of Canterbury in the reign of Elizabeth; John Cosin, bishop of Durham; Dr. Kaye, one of the founders of Gonville-and-Caius college, Cambridge; Dr. Samuel Clarke, the author of the famous work on the Attributes; and Beloe, the translator of Herodotus. (*Parl. Papers; Comm. Rep. &c.*)

NORWICH, p. t., Windsor co., Vt., 46 m. S.S.E. Montpelier, 490 W. Chartered in 1761, first settled in 1763. Organized in 1768. Bounded E. by Connecticut river, which is here from 30 to 40 rods wide. Watered by Ompompanoosuc river, and small tributaries of Connecticut river which afford water-power. A bridge crosses the Connecticut to Hanover. It contains five churches, two Congregational, two Methodist and a Baptist, five stores, three grist-mills, two saw-mills; one college, 70 students; one academy, 70 students; 20 schools, 747 scholars. Pop. 2318. It is the seat of the Norwich university, originally established in 1820 by Capt. Alden Partridge as the American Literary, Scientific and Military academy, and conducted by him and his associate instructers for a number of years. In 1834 it was chartered as the Norwich university, and Capt. Partridge returned, after the discontinuance of his academy at Middletown, Ct., (now the Wesleyan university,) and became its president. It has besides its president, six professors or other instructers, and 40 students. The commencement is on the third Thursday in August. It has no regular term for its collegiate course; but the candidates for degrees are examined as to their qualifications. Everything sectarian is carefully excluded.

Norwich, city and semi-capital of New-London co., Ct., is in 41° 33' N. lat., and 72° 7' W. long.; 13 m. N. New-London, 39 m. S.E. Hartford, 50 m. E.N.E. New-Haven, 38 m. S.W. Providence, R. I., 357 W. Bounded E. by Shetucket and Thames river. Watered by Yantic river which affords extensive water-power. The place consists of three parts and two villages. The city, (formerly called Chelsea or the landing,) the town, 2 m. N. of it; and Westville, (formerly Bean hill,) 1 m. N.W. of the town; and the villages of Yanticville, and Greenville. The city is situated on a steep acclivity, at the head of navigation on Thames river, which is formed by the junction of Shetucket river, with a cove at the mouth of Yantic river. The streets rise as in terraces; and the houses in the rear overlook those in front. The situation is peculiarly romantic. It contains four banks, with an aggregate capital of over $1,000,000, and two insurance companies with a total capital of over $100,000, and a bank for savings, a high school for boys, and a female academy. The town is 2 m. N.W. of the city, in a pleasant valley, surrounded by hills, which on the E. rise rapidly from the street, affording many handsome sites for residences. Near the church is a pleasant triangular area or public ground, surrounded by dwellings and stores. Between the city and town is a beautiful plain, with many fine buildings. N.W. of the town, on the road to Hartford, is Westville, which contains a number of manufacturing establishments and pleasant residences. At the head of the cove, which sets up over a mile from Thames river, enters Yantic river, by a romantic cataract affording great water-power. A bridge, which serves partially as a wharf, crosses the mouth of the cove, and around the head of it is Yanticville, a flourishing manufacturing village. The falls at this place are a curiosity, being singularly wild and picturesque; and in the rocks at its bed, which are exposed at low water are circular holes, often six feet deep, worn with all the regularity of a work of art, by stones whirled round within them by the force of the water. From a high projecting rock which overhangs the foot of these falls, the Mohegan Indians formerly plunged to destruction, rather than fall into the hands of the Narragansetts who were pursuing them. A mile E. of the city on the Shetucket river, is Greenville, a flourishing manufacturing village, possessing great water-power, by means of a dam across the river. The whole town contains a courthouse, jail, eight churches, three Congregational, two Methodist, an Episcopal, Baptist, and a Universalist. The city has 97 stores, with a capital of $337,000; five lumber-yards, capital $39,000; hardware was manufactured to the amount of $58,000; one fulling-mill, one woollen factory, capital 35,000; one cotton factory with 4000 spindles, and a capital of $100,000; two grist-mills, one oil-mill, two paper-mills, two rope-walks, one pottery, three printing-offices, two binderies, two weekly newspapers; total capital in manufactures $408,700; three academies, 71 students; 13 schools, 906 scholars. In the t., exclusive of the city, are 14 stores, capital $36,000; six fulling-mills, five woollen factories, one cotton factory with 4096 spindles, 11 grist-mills, two paper-mills, one tannery, one pottery; total capital in manufactures $453,500; two academies, 90 students; 11 schools, 871 scholars. Pop. in 1830, in the city 3144; whole t., 5179; 1840, city, 4200, excepting the city 3039; total in the whole t. 7239.

NORWICH, p. t., capital of Chenango co., N. Y., 132 m. W. Albany, 336 W. Watered by Chenango river. It contains a courthouse, jail, county clerk's office, bank, four churches, an Episcopal, Congregational, Methodist, and Baptist, 20 stores, five fulling-mills, one woollen factory, six grist-mills, 90 saw-mills, one oil-mill, four tanneries, two printing-offices, two weekly newspapers, and two periodicals; three academies, 82 students; 34 schools, 1049 scholars. Pop. 4145.

NORWOOD, a populous village of England, hunds. Brixton and Wallington, co. Surrey, on the top and sides of a steep range of hills, 5½ m. S. London. Pop. estimated at about 6080. It is very irregularly laid out, chiefly on a wide and elevated common, commanding an extensive view of the metropolis-northward, and of the plains of Surrey southward. The neighbourhood is studded with villas, belonging partly to merchants and others engaged in business in the city, and partly to persons retired from active pursuits. Of late years, Norwood has been much frequented in consequence of the discovery of a mineral spring at a place called Beau-lieu, or Beulah; where large gardens, laid out, with terraces, plantations, &c., have been opened to the public for fêtes, picnic parties, &c. On the N. acclivity of the hill is a handsome church, opened in 1825 (subordinate to Lambeth), with a Corinthian portico and steeple. There are, also, places of worship for Wesleyan Methodists and Independents, with attached Sunday-schools. A school of industry, established here in 1812, furnishes instruction in reading, writing, and needle-work, to nearly 300 girls. Here also is a large national school for children of both sexes, and a Lancastrian school, attended by about 200 boys. But the principal ornament of Norwood is the South-Metropolitan Cemetery, lately opened. It comprises about 40 acres, has two fine chapels, is beautifully laid out, and cost in all from £70,000 to £75,000.

NOTO, a city of Sicily, intend. Syracuse, cap. distr., on a hill near the Noto, and 16 m. S.W. Syracuse. Pop., in 1831, 11,156. It stands near the site of the ancient Netum, now called Vecchio Noto, the surviving inhabitants of which removed thither after the destruction of their city by an earthquake in 1693. Noto has large squares and regular streets, and is one of the best built, most agreeable cities of the island. Besides many handsome private residences, it has various ecclesiastical buildings, a council-house, lyceum, hospital, &c. Some, however, of its public buildings, being on two magnificent and expansive a scale for a provincial town, are unfinished. There is in this city an excellent private museum, especially of medals and coins; and also of antiquities, minerals, &c. The ruins of an amphitheatre and of a gymnasium are the principal remains of the ancient city, which stood about 4 m. N.W. the modern town. It is rather unhealthy, but is surrounded by a very fertile tract of country, in the produce of which it has an active trade. (Smyth's Sicily, p. 177; Russell's Tour, 1381-40; Ortolani, Dizionario delle Sicilia, &c.)

NOTTINGHAM, a central co. of England, in the bosom of the Trent, having N. the cos. of York and Lincoln, E. the latter and Leicester, and W. Derby. It is oval-shaped. Length, N. to S., 50 m. Area, 536,690 acres; of which about 470,000 are arable, meadow, and pasture. The Trent partly traverses and partly bounds the county on the E., and it is also traversed by its important tributary, the Idle. Excepting the vales of the Trent and Belvoir, the surface is mostly hilly and uneven; but the hills do not rise to any considerable height. The soil in the vales is either a sandy or a clayey loam, and is very fertile; elsewhere it is principally sandy and gravelly. The climate is reckoned peculiarly dry and good. The ancient forest of Sherwood, the scene of the exploits of Robin Hood and his companions, anciently covered the greater part of the hilly portion of this county along its W. side; but it has long since been disforested, and now contains some magnificent seats and parks. Agriculture, though still susceptible of material improvement, is, on the whole, good. The vale of the Trent is famous for its crops of oats; but wheat, barley, beans, peas, and cabbages are also extensively grown. There is a considerable extent of grass and meadow land; and irrigation has been extensively practised, particularly on the estate belonging to the Duke of Portland. The breeding of heavy black horses is pursued to some extent. Cattle principally of the short-horned variety. Estates of all sizes; many small. Farms generally small, and mostly held at will. Average rent of land, 19s. 11½d. an acre. Coal is abundant in the W. parts of the county. Nottinghamshire is the grand seat of the manufacture of bobbinet, or Nottingham lace, and also of the manufacture of cotton and silk stockings. It is divided into six wards, one liberty, and 283 parishes. It returns 10 members to the House of Commons, viz. four for the county, and two each for the boroughs of Nottingham, Newark, and East Retford, which are its principal towns. Registered electors for the county in 1839-40, 7360; being 3746 for the N., and 3614 for the S. division. In 1841, the county had 50,541 inhabited houses, and 249,773 inhabitants; of whom 121,699 were males, and 128,113 females. Sum expended for the relief of the poor in 1838-39, £53,973. Annual value of real property in 1815, £751,696. Profits of trade and professions in do., £314,561.

NOTTINGHAM, a parl. and mun. bor., and extensive manufacturing town of England, and co. of itself, locally situated in the above county, of which it is the cap., hund. Broxtow, on the Leen, about ½ m. from its junction with the Trent, crossed here by an old bridge of 19 arches, 14 m. E. by S. Derby, and 123 m. N.N.W. London. Area of parl. bor., which is co-extensive with the co. of the town, and comprises three pars., 9010 acres. Pop. in 1821, 40,415; in 1831, 50,680; and in 1841, 53,091. The town stands partly at the bottom and partly on the sides of a steep red sandstone rock, the summit of which is crowned by a modern building, called the castle, occupying the site of a castle built by the son of William the Conqueror, and demolished by order of Charles II. The streets, many of which rise above each other in successive terraces, are narrow, and irregularly laid out; two long thoroughfares run very nearly parallel N. and S., crossed at right angles by other streets; and considerable improvements have taken place in the N. part of the town. All the streets are well paved, and lighted with gas; there is a good supply of water from two companies, and the police is tolerably effective. The market-place, which is spacious, and surrounded by handsome buildings, has, at its E. end, the new exc'range, a quadrangular building of four stories, erected at the beginning of last century, and much improved within the last few years: the lower part comprises shops, butchers' stalls, &c.. the upper stories being used for assemblies and public business. The county hall, on the high pavement, near St. Mary's church, is another very conspicuous edifice, comprising two law-courts, a grand-jury room, and other apartments, for the business of the assizes. Behind it, and connected by a long covered passage, is the county prison, built on the edge of a rock, below which, at a depth of 70 ft., is the densely crowded and low quarter, called the Marsh. "The building has been altered and enlarged at various times, but is even now very ill suited for carrying on any efficient system of prison discipline: the supervision and control of the prisoners is hence no easy task; and the entire management is susceptible of great improvement." The borough house of correction, or "St. John's" prison, so called from occupying the site of an old monastery, is conveniently situated for its purpose; and shortly after the passing of the Jail Acts was re-constructed, on the principles of classification. It comprises two sides of a square, each three stories high, possesses considerable capabilities, and is in an efficient state,—both as respects discipline and general management. At Lenton, about 1½ m. from the town, is a small jail, belonging to the Honour of Peveril; but "it is small, insecure, and totally unfit for its purpose: indeed, its abolition would be a public benefit." (*Prison Inspectors' 3d Rep.*, ii. 36–39.) The town-hall is a large building, three stories high, of which the lower part is used as a bridewell, while the upper apartments are used for corporate business, and other purposes. The other public buildings consist of a small theatre, little frequented, the cavalry-barracks in the castle-park, the foot-barracks, the yeomanry riding-house, now used as a circus, and the grand stand, on the race-course N. of the town: with eight churches and chapels, besides numerous places of worship for dissenters.

St. Mary's church, standing on a bold eminence, 170 ft. above the level of the adjacent meadows, is a cruciform structure, in the perpendicular style, with an elegant square tower, rising from the intersection of the nave and transepts. A few years ago it underwent a thorough repair, on a plan consistent with its original architecture, and is now the handsomest church of the town: the living is a vicarage, of the annual value of 700l., with a glebe-house, in the gift of Earl Manvers. St. Peter's, near the market-place, is a building of mixed architecture, partly Saxon and partly Gothic, with tasteless modern additions, being remarkable chiefly for its lofty spire. St. Nicholas, erected in 1678, on the site of a more ancient edifice pulled down during the parliamentary wars on account of its proximity to the castle, is of brick, with stone corners: it comprises a nave with two side aisles, and has a light appearance outside, as well as good interior accommodation. St. James' church, on Standard hill, in the district called the Park, is a modern edifice in the perpendicular style, with a low embattled tower. St. Paul's church, of Grecian architecture, with a Doric portico, is capable of accommodating upwards of 2308 persons. It was formerly subordinate to St. Mary's, but has recently been made an independent district parish church. Trinity church, a handsome Gothic structure, with a tower and spire, fitted to accommodate 1400 persons, is now nearly completed. There are numerous places of worship for dissenters, including a Roman Catholic chapel, meeting-house for the society of Friends, and Jews' synagogue, connected with which, as well as the churches, are above 30 Sunday-schools, attended by above 5000 children of both sexes. The blue-coat school furnishes clothing and instruction to 60 boys and 20 girls; a national school had, in 1834, 570, and three Lancastrian schools, 484 children: there are, also, four other subscription-schools, and five infant schools. The free grammar-school was founded in 1512, and before the close of the last century, had fallen into disuse; but in 1807 the establishment was revived, and it now furnishes the means of a respectable classical education to between 50 and 60 boys.

The other charities of Nottingham comprise, 1. Plumtree hospital (founded in the reign of Richard II., and subsequently enlarged) for 13 aged widows, besides out pensioners; 2. Collins' hospital, which provides ample accommodation for 24 poor men and women, with a stipend and allowance of coals; and, 3. Lambly hospital for decayed burgesses or their widows; besides which, several other charities confer essential benefits on the infirm and aged of both sexes. On Standard hill is the general hospital or infirmary, standing in a spacious enclosure, and comprising a centre and two wings, with large, airy wards for patients, about 1300 of whom are relieved, on an average, every year. The lunatic asylum, opened in 1812, is in New Sneinton, and has good accommodation not only for pauper but other patients. Nottingham has likewise two dispensaries, and several other benevolent institutions, with bible, tract societies. &c.; and there are few towns in which so much is expended in the relief of the sick and necessitous of the working classes. The three parishes of Nottingham are formed into a poor-law union: the maintenance of the poor, in 1839, cost £18,556.

The chief literary establishment of the town is the public library and news-room in the Market-place, which has a collection of more than 8000 volumes, a museum of mineralogy, lecture-rooms, &c., with an attached literary society.

A mechanics' institute, established in 1824, has a considerable library, with apparatus, &c. Nottingham, also, issues three weekly newspapers, the Journal, Review, and Mercury.

Nottingham is celebrated as being the great centre of the bobbin-net and lace manufacture, besides which it enjoys, in common with Derby and Leicester, a large share of the hosiery business. The first attempt at the manufacture of lace by machinery dates as early as 1768; but though this was followed by many subsequent attempts to shorten the tedious process of making lace on the pillow, it was not till 1809 that Mr. Heathcoat, of Tiverton, discovered the correct principle of the bobbin-net frame, and obtained a patent for his invention. Steam-power, first introduced in 1816, and becoming general in 1822-23, gave a great stimulus to the trade, which was further increased on the expiration of Heathcoat's patent. Prices fell in proportion to the increased production; and the Nottingham lace-frame soon became the organ of general supply, rivalling and supplanting, in plain nets, the most finished productions of France and the Netherlands: so much so that large quantities were smuggled into those very countries from which lace was formerly smuggled into England. But the great object of the manufacturer is not so much to produce very fine, and, consequently, high priced lace, for which the demand must at all times be very limited, as to improve the fabric and lower the cost of the inferior qualities for which the demand is comparatively extensive. At present (1841) there are supposed to be about 1680 bobbin-net and warp-lace frames employed in the town and its immediate vicinity: the wages paid to the individuals engaged in the trade vary from 1s. 6d. to 5s. per week for children, 7s. to 15s. per do. for young people and inferior men, 15s. to 30s. per do. for power and handframe men on plain work; and 20s. to 40s. per do. for men working hand-frames, weaving patterns.

The health of the power-machine workmen is said to be, on the whole, good: the factories are neither hot nor confined; and the workmen have only to superintend, not work the machines. Hand machine labour is much heavier; but as it is the custom to work by shift, each man is seldom more than six hours a day at the frame. Hand machines used to be let out at a weekly rent to the workmen by capitalists; but they are now much reduced in numbers; and nearly all those that now exist are the property of those who work them, either by their own hands, or by the aid of journeymen.

Subordinate to its other departments, the town had, in 1839, three cotton-mills, two worsted-mills, and three silk-mills; employing, in all, nearly 900 hands. Nottingham has likewise very extensive establishments for making bobbin-net and stocking frame machinery, large bleaching-works, malthouses, and breweries: the Nottingham ale has obtained considerable celebrity. There are two banking companies, and three private-banks. The Nottingham canal connects the town northward with the Codnor iron and coal district, and southward with the Trent, and the great canal system of the N. midland counties. It has also very extensive railway communication, by means of the Midland Counties railway (opened in June, 1840), which is united southward with the Birmingham line at Rugby, and with the North Midland railway at Derby. It is 57½ m. long, and cost about £1,300,000.

Nottingham claims to be a borough by prescription, but received charters from Henry II., and many subsequent monarchs, Henry VI. having granted to it the additional privilege of being a county of itself. It is divided, under the Municipal Reform Act, into seven wards; and is governed by a mayor, 13 other aldermen, and 42 councillors: it has likewise a commission of the peace, under a recorder. Corporative revenue, in 1839, £13,718, exclusive of £3440 accruing from the sale of property. A court for the recovery of debts under 40s. is held monthly; the assizes, both for the borough and county are held in spring and summer; and quarter sessions for the S. division of the county, in January, April, June, and October. Nottingham has sent two members to the House of Commons since the 12th Edward I., the right of election down to the Reform Act being in the freemen. (by birth, apprenticeship, and purchase), and in freeholders to the amount of 40s. The electoral limits were not altered by the Boundary Act. Registered electors, in 1820-40, 5436. Nottingham is also a polling-place for the N.W. division of the county. Markets on Wednesday and Saturday, but principally on the latter. Large fairs for cattle, cheese, &c., March 7th, 8th, and 9th; Oct. 2d, 3d, and 4th; three other smaller fairs.

The origin of Nottingham is involved in obscurity; but so early as in the time of Alfred, it was of sufficient importance to give its name to the county. The castle was built by William Peverill, the natural son of William the Conqueror. Edward III. held several parliaments here, in one of which were enacted the laws relating to the settlement of the Flemish manufactures. Nottingham was the chief place of rendezvous for the troops of Edward IV. and Richard III. during the wars of the Roses; and it was here, in 1642, that Charles I. formally raised his standard against the parliament. The inhabitants, however, being attached to the republican cause, the king was soon compelled to abandon the town and castle to the parliamentary forces. Nottingham has been the scene in more recent times of disturbances among the working classes. In 1811, when considerable distress prevailed among the weavers in consequence of our exclusion from the continental markets, combinations were formed among the workmen for the purposes of breaking the frames, which they erroneously supposed had thrown them out of employment; and to such an extent did they proceed, as to call for the most vigorous interference of the legislature. Disturbances of a minor nature have occurred several times since that period; but the only serious riot of late years took place on the 8th Oct. 1831, during the agitation preceding the passing of the Reform Act, when the rioters burnt down the castle. (*Parl. Papers; Private Information.*)

NOTTINGHAM, p. t., Rockingham co.. N. H. 25 m. E.S.E. Concord, 489 W. Incorporated in 1722, first settled in 1727. Drained by North river, a branch of Lamprey river. It contains a Free-will Baptist church, five stores, three grist-mills, six saw-mills; 11 schools, 467 scholars. Pop. 1193.

NOTTINGHAM L, Mercer co., N. J., 17 m. N.E. Mount Holly. Watered by Assunpink and Crosswick's creeks. It contains a Presbyterian and a Baptist church, 28 stores, three lumber yards, three furnaces, three fulling-mills, three woollen factories, seven cotton factories with 6500 spindles, 10 grist-mills, seven saw-mills, one oil-mill, five tanneries, four distilleries, one printing-office, one bindery; one academy, 36 students; 13 schools, 348 scholars. Pop. 5109.

NOTTOWAY, county, Va. Situated S.E. of the centre of the state, and contains 290 sq. m. Bounded S.W. by Nottoway river. Drained by tributaries of Appomattox river. It contained in 1840, 5540 neat cattle, 6627 sheep, 10,294 swine; and produced 42,145 bushels of wheat, 248,863 of Indian corn, 70,130 of oats, 8367 of potatoes, 2,212,950 pounds of tobacco, 20,682 of cotton. It had seven stores, eight flouring-mills, 17 grist-mills, five tanneries; 10 schools, 195 scholars. Pop.: whites, 2490; slaves, 7071; free coloured, 158; total. 9719. Capital, Nottaway, C. H.

NOTTOWAY, p. t., St. Joseph co., Mich., 137 m. W S.W. Detroit, 592 W. Watered by St. Joseph's river. It contains six stores, one flouring-mill, two saw-mills; 10 schools, 201 scholars. Pop. 1226.

NOTTAWAY, C. H., p. v., capital of Nottaway co., Va, 67 m. S.W. Richmond, 186 W. Situated on the N.W. side of Nottoway river. It contains a courthouse, jail, clerk's office, one store, one flouring-mill, 15 dwellings, and about 90 inhabitants.

NOVARIA (an. *Novaria*), a city of the Sardinian continental dom., cap. division, prov. and mand. of its own name, on an eminence between the Gogna and Terdoppia, 32 m. N.E. Turin, and 27 m. W. by S. Milan. Population, in 1838 (ex. garr.), 18,584. It is surrounded by ramparts and ditches, and defended by a castle. Though the streets are mostly narrow, it is, on the whole, tolerably well built, and has many handsome residences. The cathedral, the Dominican church, and that of St. Gandenzio, and the large barracks, are the principal public edifices. Novaria has numerous convents, several hospitals and colleges, a theatre, and a government loan bank. It is a bishop's see; and has manufactures of silk and linen fabrics, leather, &c., and two large annual fairs. Under the French it was the cap. dep. Agrogna. (*Dict. Géog.*)

NOVA SCOTIA, one of the British colonies of N. America, consisting of an oblong shaped peninsula, between lat. 43° and 46° N., and long. 61° and 67° W.; connected with New Brunswick by a low sandy isthmus, only 14 m. across, and separated from Cape Breton by the narrow strait called the Gut of Canscaw. It is about 300 m. in length, and of very various breadth. Area according to Haliburton, 15,620 sq. m., about 1·5th part of which consists of lakes, rivers, and salt-water inlets. Population, in 1838, 155,000. The coast-line is extremely irregular, forming numerous capes and bays. Capes George and Canscaw are the chief promontories on the N.E. side, and at the S. extremity is cape Sable. The basin of Minas is a deep inlet on the N.W. side of the peninsula, forming a part of the bay of Fundy, which separates Nova Scotia from New Brunswick. St. Mary's and Argyle bays are on its S.W. side; Pictou, Antigonishe, and Chedebucto bays form the chief irregularities on the N. coast; and the E. coast, from cape Canscaw to cape Sable, is indented with almost innumerable small bays, harbours, and rivers. Rocks and islands fringe its shores, and the aspect of the entire Atlantic coast is exceedingly picturesque. Deep water is found, almost without exception, close to the rocks and islands; and the peninsula

presents towards the bay of Fundy bold and almost precipitous cliffs. The interior is intersected in almost every direction by streams, rivers, and lakes; but, with the exception of Annapolis river and lake Rossignol, connected with the sea by the river Mersey, most of them are of very inferior size. The peninsula has no elevations deserving the name of mountains; its highest point, mount Ardoise, between Windsor and Halifax, not rising more than 700 feet above the sea. A pretty high ridge of hills skirts the shore of the bay of Fundy.

As respects geological constitution, "the greater part of Nova Scotia may be described as a low range running from S.W. to N.E., resting on solid rocks of granite, trap, and slate alternately. Towards the E. end of the peninsula are beds of sandstone, graywacké, gypsum, limestone, porphyry, and many other kinds of rock; and on these strata there is usually a rich deep soil. The barren tracts are chiefly of sand or clay; and in these parts, especially about Pictou, are the great coal-fields of the peninsula. Iron is abundantly interspersed among the coal strata; and different varieties of lead and copper ore are met with, though in smaller quantities. Near Pictou are several brine springs, one of which is saturated with salt in the proportion of 12 to 88 of water." (*Report on Brit. N. America*, App. B., p. 140.)

The climate of Nova Scotia, with respect to temperature, bears a general resemblance to that of Lower Canada, and is subject to the same great and sudden variations. The greatest degree of heat observed at Halifax by Captain Moorsom was 95° Fah. and the extreme of cold 10°. The difference of temperature within 24 hours often exceeds 50°, and a difference of 69° has been known to occur within the same period. These changes, however, are seldom so frequent or extreme in the interior, or in those parts of the province less immediately on the Atlantic. Notwithstanding the occasional lowness of temperature, the maritime situation of Nova Scotia tends to abridge the duration of the frost. The severe weather usually sets in about the middle of December, and it is not uncommon for the frost to break up at the end of January. The quantity of snow not only varies greatly from year to year, but is also very unequally distributed throughout the province. The snow-storms are very heavy, some having been known to continue for 60 or 70 hours without intermission. (*Moorsom, p. 162.*) The severity of winter ends late in March, when cold, damp, east and north-east winds succeed, caused by the breaking up and passage along the coast of vast fields of ice from the gulf of St. Lawrence. Hence the most disagreeable season in this country is from the vernal equinox to the end of April. Spring approaches tardily and irregularly, the close of May often arriving before the fields are fully clothed with verdure. A very warm summer occupies three months, dating from the early part of June. May and June are marked by the prevalence of fogs, especially on the eastern coast, while July and August are usually remarkable for a continuance of calm serene weather. Autumn, the most beautiful season of the year, may vie with that of any country. September and October are very similar to the same months in England; but in November, and even December, there are days which, for beauty, warmth, and mildness, are equal to the loveliest months of an English May. (*Moorsom, p. 167.*) Westerly and S. W. winds are most prevalent; the fine bear to the wet days a proportion of eight to five. The extreme variations of temperature common in this country have not that injurious influence on health which one might naturally expect. Rheumatic and inflammatory complaints are far more prevalent than any other; and a considerable annual mortality occurs from pulmonary consumption. Intermittent fevers, however, so common in Canada and the United States, are here wholly unknown; typhus occurs only in a mitigated form, and the ravages of the yellow fever have never been felt. Nova Scotia, therefore, may, on the whole, be termed a healthy country. Its inhabitants often live to extreme age; many attain ninety and even a hundred years. (*Murray's Brit. America*, N. 119.)

As respects agriculture, Nova Scotia is estimated to comprise somewhat more than 5,000,000 acres of land available for tillage; the proportion of land under cultivation at present being to the wilderness as 1 to 26. The first large public grants of land appear to have been made in 1783, and in less than 13 years from that time, nearly 6,000,000 acres (including the whole of Prince Edward's Island, then a part of Nova Scotia) were granted in lots, ranging from 20,000 to 150,000 acres, to individuals, or companies in England. These grants contained conditions of improvement but the grantees, after having incurred some expense in trying to settle their extensive properties, abandoned the land to its few inhabitants, or suffered it to remain absolutely waste. Efforts made to escheat these lands to the crown were repeatedly baffled by the influence of the absentee proprietors; and thus the province was effectually closed against immigration, either from England or the neighbouring colonies. Large grants of escheated land were, however, made on the breaking out of the American war to refugee royalists; but these were seldom occupied, and are now, for the most part, uncultivated, opposing serious obstacles to the cultivation of the lands around them. Licensed occupiers, however, and squatters, have improved some portions of these tracts; and to them the progress made by the colony in population and agriculture is almost entirely ascribable. The system of selling in lots not exceeding 1000 acres was introduced in 1827; and the average price of unimproved land, in 1839, amounted to 2s. 3d. an acre. The largest portion of it, however, has been acquired, not by actual or intending settlers, but by speculators, who, tempted by the low price, have purchased, on account of the timber, or with a view to profit from a future sale. Indeed, out of 3,750,000 acres that have been granted in Nova Scotia, only about 400,000 are under cultivation. (*Report on British North America.* App. B., pp. 12, 13, and 129.)

The total quantity of land still ungranted in Nova Scotia was estimated by Mr. Morris, in 1838, at about 2,500,000 acres; but of these not above one-eighth part is fit for tillage. The country, as respects the quality of land and the state of agriculture, may be divided into three distinct sections. The E. division, formed by a line from the mouth of the river Philip to that of the St. Mary, presents a strong upland soil, well adapted for grain, and varied with strips of rich intervale land along the sides of its rivers. The upland consists principally of a strong, loamy clay, intermixed more or less with sand and gravel, the soil of the intervale being a rich, sandy, alluvial loam. The lands about Pictou are very rich and productive, seven successive crops of wheat being frequently raised without the use of manure. Agriculture, however, is only imperfectly understood, and no proper use is made of the resources of the soil. In the S. district the land is almost wholly upland, with very little intervale or marsh: the soil is extremely rocky, varying from a strong loam to a light sand. Good varieties of wheat and the coarser grains are obtained in some places; but the state of the farms generally exhibits the very reverse of intelligence. The unskilful use of manure, the indiscriminate employment of sea-weed, and, in many instances, the total neglect of any manure whatever, have retained those lands in a poor and backward state which better management would have rendered comparatively productive. (*Moorsom,* p. 183.) The N.W. division comprises upland, intervale, and marsh land; the first two being poor, and scarcely susceptible of any improvement. The marsh land is of two kinds,—one, called salt-marsh, being little more than a flat surface of spongy soil, overflowed at spring-tides, and covered with a long rank grass, sometimes converted into hay; the other, called the dyke-marsh, owes its formation to the impetuosity of the tide in the bay of Fundy, which brings along with it fine loamy particles, which it leaves behind as it recedes, and thus, in course of time, a succession of layers raises the surface to the level of spring-tides, when an embankment or dyke, called an *aboiteau,* is formed to prevent any farther overflow. A newly-enclosed marsh is usually left untouched for the first three or four years: in the third year it is fit to receive the plough, and is then sown with wheat, the first crop averaging about 60 bushels an acre; and on long cultivated marshes the returns average about 40 bushels of wheat, and 2½ tons an acre of hay. The crops usually cultivated are wheat, oats, and barley, with smaller quantities of peas, buckwheat, and rye. Potatoes are universally cultivated, and form the staple article of food throughout the province. Crops of beans or cabbage are rarely seen, and horticulture meets with very little attention.

The average produce per acre of medium farm land is estimated at 25 bushels of wheat, 40 of oats, 200 of potatoes, and two tons of hay; but all calculations of this nature are very vague, and little to be depended on. Good dairy-farms are found in the N.W. division. Hire-labour is difficult to procure, and too expensive to allow of its adoption, except by the more wealthy. Labourers (who do not exist here as a separate class, but comprise the more indigent of the new settlers) are usually hired during the six months of summer, for which they receive from £15 to £18, with board and lodging; but a part of the payment is made in produce. (For further information, see Moorsom's work, p. 176-222, from which the above remarks are mostly derived.)

The forests of Nova Scotia abound with good timber: pine and birch, oak, beech, ash, and maple, are the most common trees; and many of the inhabitants have for years been supported by the timber trade. The exports of timber in 1837 were valued at £143,726. The principal wild animals of the province are the moose-deer, carriboo, bear, loup-cervier, fox, martin, otter, mink, and squirrel. Hunting and trapping were once extensively pursued; but in proportion as the country has become settled, the number of

animals has gradually but rapidly decreased, so that the exports of furs in 1837 were not estimated at more than £4330. The rivers abound with many varieties of freshwater fish; besides which, cod, herrings, mackerel, halibut, and other kinds of seafish, are found in the deep bays of the coast. Chedabucto bay and Annapolis basin are the principal stations for the herring and mackerel fishery: but the inhabitants share, also, in the whale, seal, and cod fisheries; and this branch of industry has for some years been on the increase. The fish of all sorts (chiefly cod) exported in 1837 was valued at £181,960; besides which, the exports of train-oil were estimated at £30,980. The fisheries are said to employ about a third part of the inhabitants; but this is, no doubt, an exaggerated statement. Another important branch of employment in Nova Scotia is mining. Coal and iron are abundant, and are pretty extensively wrought by the general mining association, to which all the mines have been let. The total value of the coal produced in Nova Scotia and C. Breton amounted, in 1839, to £25,000.

Gypsum, which abounds in the W. districts, is highly prized in the United States as manure, and the quantity exported thither from Nova Scotia in 1837 amounted to 22,396 tons, valued at £6738. A stone is found in many parts of the province extremely well adapted for grindstones, which are celebrated all over America under the appellation of "Nova Scotia blue grits:" the exports to the United States were valued, in 1837, at £12,085. The manufactures are quite unimportant: the weaving of coarse woollen cloths, called *homespuns,* is pretty general throughout the colony; and carding-mills are established in some parts. Carpets, also, are woven in small quantities, and ropes are made of hemp imported from N. Europe. Grist and saw-mills are very numerous; besides which, there are several breweries and tan-yards. The position of Nova Scotia, on the extreme W. side of N. America, gives it great commercial advantages; and its trade, especially with the U. States, has been for some years steadily on the increase. The exports, chiefly to Canada, the U. States, and Great Britain, consist of fish and fish-oil, timber, coals, &c.; the whole being valued, in 1837, at £478,461. The imports during the same year comprised corn and flour, British manufactures, colonial produce, &c., and were valued at £790,765. The trade principally centres in Halifax, which *see.* Subjoined is an

ACCOUNT of the Ships and their Tonnage that entered and left the Ports of Nova Scotia in 1839:

Counties.	Inwards.		Outwards.	
	Ships.	Tons.	Ships.	Tons.
Great Britain,	97	27,448	102	28,739
British Colonies,	2,517	169,631	2,815	179,712
United States,	1,211	198,540	1,266	138,477
Foreign,	181	19,039	49	6,209
Total,	4,006	314,136	4,232	354,177

The means of internal communication have been improved within the last few years, and some of the roads are stated by Moorsom to be equal to the secondary roads in England. They are partly supported by annual grants from the house of assembly, and the inhabitants of each district are compelled to furnish, either personally or by substitute, a certain quantity of labour for the same purpose: this system, however, has not been found successful; and large tracts are still left uncultivated, owing either to the absence or wretched condition of the roads. (*Report, App. B.* p. 135-140.) A water communication has been effected between Halifax and Windsor; but the want of any such communication is severely felt by those whose settlements are at a distance from Halifax, the chief market in the colony for agricultural produce.

The constitution of Nova Scotia is a representative provincial government. The lieutenant-governor, who is subordinate to the governor-general of British N. America, is commander within the province, and the supreme civil, as well as military authority. Under him is a council of 12 members, of whom the bishop and chief justice are members *ex-officio,* and the rest appointed by the crown. The legislative assembly is a body of 41 members, elected by 40s. freeholders. It is elected, like the British House of Commons, for seven years, but may be prorogued or dissolved by the lieutenant-governor. It meets every year, and all money bills must originate in this assembly: other bills require the consent of the governor and council before they become law. For the purposes of election, Nova Scotia is divided into 10 counties (including Cape Breton). The counties have two members each, and the other representatives are returned by the towns. Justice is administered by a court of queen's bench, sitting at Halifax, and by district courts in the different counties. The common and statute laws of England are in force, together with statutes passed by the local legislature, and approved by the queen in council. The laws, according to M'Gregor, are, on the whole, judicious; and, as far as they go, calculated to pro-

mote the prosperity of the colony. But there is too frequent a recurrence to courts, and the harmony of society is
often broken by a love of litigation.

The provincial revenue, amounting to somewhat more
than 60,000l., is raised by duties of 2½ per cent. *ad valorem*
on property generally, with an additional rate on wine and
spirits, and by duties on imported goods, lighthouse dues,
rents, &c. Taxation, however, is extremely light; the cost
of defence being defrayed by Great Britain, and the inhabitants being burdened only with the civil government and
local improvements. The military force consists of three
regiments of the line, the expense of whose maintenance to
England is estimated at about £120,000 a year.

The Church of England is the established religion, and
the colony is divided into 32 parishes, each of which has a
rector salaried by the crown or the Society for the Propagation of the Gospel. Nova Scotia was made a bishopric
in 1787, the diocese extending over New-Brunswick and
Prince Edward Island, Newfoundland, and the Bermudas.
The bishop draws no revenues from the colony, and holds
spiritual jurisdiction only over the members of his own
church. The Presbyterians, however, are the most numerous body, and a synod of 17 members meets annually at
Halifax. There are numerous Roman Catholics, consisting
principally of the Acadians and Irish settlers. The Baptists
and Wesleyan Methodists are also an important body; and
a complete toleration is granted to all religious denominations.

Among the establishments devoted to education, the principal is Windsor college, partly supported by the provincial
government and partly by subscription; but it has not met
with much success. The Presbyterians have an academy at
Picton, and the Baptists have another at Horton, attended
by about 50 students. There are eight schools at Halifax,
furnishing instruction to 1000 children. An Episcopal school
at St. George's is attended by 190 boys and 100 girls; and
there is a large grammar school at Sydney. The province
comprises, also, 600 common schools, and 30 combined English and Latin schools, attended altogether by about 20,000
children. These schools are supported by grants, subscriptions, &c., and 40 other schools are maintained by the Society for the Propagation of the Gospel.

Of the population of the province, the Indians do not now
exceed 600; there are about 6600 Acadians (or descendants
of French settlers, before the country was ceded to the
British), and about 2000 free negroes: the remainder of the
population consists of Germans, or their descendants, British
emigrants, chiefly from the N. of England and Scotland, a
few Irish, and the descendants of refugee loyalists from the
U. States. The Acadians congregate in settlements of their
own, mixing little with the other colonists.

Nova Scotia was discovered, by John Cabot, in 1497. It
was first settled by the French, who called it Acadia. It
subsequently fell under the English, having been, in 1627,
granted by James I. to Sir W. Alexander, and named Nova
Scotia. In 1632, it was restored to France by the treaty of
St. Jermain's; but it several times subsequently changed
masters, and was not finally established in the quiet possession of the British till 1758. At the peace of 1763, the
boundaries of this colony were so defined, as to include
New-Brunswick and Cape Breton; but, in 1784, the former
was made a separate government. HALIFAX (which see)
is its capital and the seat of government. (*M'Gregor's
Brit. America*, i. 250–427; *Moorsom's Letters from Nova
Scotia*, passim; *Murray's Brit. America*, vol. ii.)

NOVELLARA, a town of N. Italy, duchy of Modena, in
the plain of the Po, 16 m. N.W. Modena. Pop. 4070. It is
the capital of a principality annexed to Modena in 1737, and
has some silk and leather manufactures.

NOVGOROD, a government of Russia in Europe, between the 57th and 61st degrees of N. lat., and the 30th and
49d of E. long.: having E. the government of Vologda, S.
those of Jaroslavl, Tver, and Pskof; W. the latter and Petersburg, and N. the last named and Olonets. Length. N.E.
to S.W., about 400 m.; breadth, varying from 40 to 100 m.
Area, estimated at 43,880 sq. m. Pop., in 1838, 895,400. The
surface, which in the N. is low and level, rises gradually
towards the S.W., where the Valdaï plateau reaches an
elevation of 1000 feet above the sea. The government is
well watered: principal rivers, Volkhof, Msta, Chezna, Mo-
logda, Lovat, &c., some of which run towards the Wolga,
and others towards the lake of Ladoga. Among the lakes
are those of Bielo-Osero, Voje, and Ilmen. The climate,
especially in the N., is more severe than in the government
of Petersburg, not being tempered by the sea breezes. Except in a few districts, the soil is not eminent for fertility,
and eight-frosts often spoil the crops. Scarcely any orchard
trees are met with, but hemp and flax are grown for exportation; and in 1832, 864,000 *chetwerts* of corn, principally
rye, oats, and barley, were raised. Timber is an important
product; a large part of the government is covered with
forests, those belonging to the crown covering 2,727,200 de-

ciatines. Few cattle are reared. Next to agriculture, fishing is a principal occupation. The salt-springs of Staraïa-
Roussa furnish an adequate supply of salt for this government and that of Tver. Manufacturing industry is very
backward; there are a few copper, glass, tile, leather, woul-
len cloth, and other factories; but in 1832 there were not,
in all, 50 manufacturing establishments in the government;
the population have, however, a turn for commerce, and the
different fairs and markets are well attended. Novgorod
is divided into 10 districts: Novgorod, Tikhvine, and Valdaï
are among its chief towns. Except some Lutherans among
the Finnish inhabitants, the population is principally of the
Greek church. Education is very little diffused. The capital has a gymnasium; and there are schools there and in
other parts of the government; but the young persons of all
descriptions at school in 1835 amounted to only 1086! The
government is not supposed to possess a printing-press.
Civil public revenue estimated at 2,733,000 roubles. The
territory was made a separate government in 1776.

Novgorod, (called *Veliki*, or "the Great,") a city of Russia, and formerly the most important in that empire, capital
of the above government, on the Volkhof, near its exit from
the Lake Ilmen, 100 m. S S E. Petersburg, and 305 m. N.W.
Moscow. Lat. 58° 31' 33" N.; long. 31° 16' 24" E. Its population, which, in 1830, amounted to only 8534, was estimated to have amounted, in the height of its prosperity, in
the 15th century, to 400,000, though this, probably, is much
beyond the mark. At this period, Novgorod, with London,
Bergen, and Bruges, constituted the four principal foreign
depôts of the Hanseatic League; but the fall of the League,
and still more the massacres perpetrated by the bloodthirsty
barbarian Ivan Vassilievitch II., in 1570, proved fatal to this
great emporium; and it soon after fell into all but irremediable decay. La Motraye, who visited it early in the last
century, gives the following description, which will apply
nearly as well in the present day. "Nothing is more deceitful than the view of Novgorod from a distance: its extent,
and the number and height of its towers and spires, seem
to announce one of the finest cities in Europe; but, on near-
ing it, the traveller perceives that its walls and houses are
only of wood; and on entering, he finds it ill built and
wretchedly paved. Only the churches and a very few private residences are of stone or brick. There may be from
80 to 85 churches, including those of the monasteries: besides which, the castle, a large fortress bristling with artillery, is the remaining principal edifice." (*La Motraye, in
Schnitzler, la Russie*, p. 70.) The town, in fact, though
comprising a large space, consists principally of scattered
groups of miserable habitations, separated by ruins or by
fields, which, it is evident, had once been covered with
houses. It is divided into two parts by the Volkhof, here
crossed by a handsome bridge of 11 arches, which is about
the only modern structure in the city. The piles, &c., of
this bridge are of granite; the rest is chiefly of timber: its
entire length is 270 yards, and the breadth of its central
arch 85 feet. In the *Torgovatz*, or market-town, are the
governor's residence, an ancient palace of the czars, and
most of the shops and warehouses. The *Sefnakaia*, on the
opposite bank of the Volkhof, is about 1½ m. in circuit, and
surrounded by an earth rampart and a ditch. In it are the
kreml, or citadel, the cathedral of St. Sophia, the archbishop's palace, and the various tribunals. The citadel is
in many respects similar to the kremlin of Moscow, having
a stone wall, flanked with many round and square towers.
The cathedral, built between 1044 and 1051, and repaired
in 1832, is of stone, somewhat on the model of St. Sophia at
Constantinople. It has some remarkable bronze gates, with
sculptures in *alto-relievo*, representing passages from scripture history; and many of the paintings on its walls are
curious, being said to date from a period previously to the
revival of the arts in Italy. Novgorod is the seat of a military governor, whose authority extends over the adjacent
province of Tver. It has a few manufactures of sail-cloth,
leather, and vinegar, and some trade in corn. Though not
the original capital of Rurick, it became the seat of the Russian government in 864. In the beginning of the 11th century, the inhabitants obtained considerable privileges that
laid the foundation of their liberty and prosperity; and as
the city and its contiguous territory increased in population
and wealth, they gradually usurped an almost absolute independency; so that, in effect, Novgorod, in the middle ages,
should rather be considered a republic, under the jurisdiction of an elective magistrate, than a state subject to a regular line of hereditary monarchs. During the 12th, 13th,
and 14th centuries, Novgorod formed the grand entrepôt between the countries E. of Poland and the Hanseatic cities;
and its wealth and power seemed so great and well established, and the city itself so impregnable, as to give rise to
the proverb—

" *Quis contra Deos et magnum Novogardiam ?"*
" Who can resist the Gods and Great Novgorod ?"

But in 1477 it was obliged to submit to Ivan I., great duke

of Russia. In 1554 it was visited by the famous Richard Chancellour, who describes it as the "great mart town of all Muscovie, and in greatnesse beyond Moscow." But not long after, it was subjected, as already stated, to the scourge of the destroyer, and fell, never to rise again. The foundation of Petersburg took from it all hope of ever recovering any portion of its ancient prosperity. (*Schnitzler, La Russie*, 152–174 ; *Coxe's Travels in the N. of Europe*, ii., 77–90, 8vo. ed.)

NOVI, a town of N. Italy, Sardinina dom., div. Genoa, cap. prov. and mand.; in the fertile plain of Marengo, at the foot of the Appenines, 14 m. S.E. Alessandria. Pop., in 1838, 10,278. Few remains exist of its old castle; its streets are narrow and ill-paved, and its public edifices undeserving of notice. It has, however, a handsome square. It is the seat of civil and commercial tribunals, and has a college and hospital, and manufactures of the best silk twist in the divisions. It is also an *entrepôt* for goods passing between Italy and Germany. On the 16th of August, 1799, an obstinate conflict took place near this town, when a French army, under Joubert, who fell in the action, was defeated by the Austro-Russian army, under Suwarrow.

NOVI, p. t., Oakland co., Mich., 25 m. N.W. Detroit, 547 W. Watered by the W. branch of Rouge river. It has two stores : 10 schools, 255 scholars. Pop. 1351.

NOVI-BAZAR, a town of Turkey in Europe, province Bosnia, capital Sanjiak, on the Rachka, 130 m. S.E. Bosna-Serai. Pop. estimated at from 8000 to 10,000. It is a town of considerable traffic, the residence of a pacha and a Roman Catholic bishop, and has some warm baths. Our acquaintance with it is, however, very limited, as it is seldom or never visited by travellers from W. Europe.

NOXUBEE, county, Miss., situated in the E. part of the state, and contains 680 sq. m. Drained by the W. fork of the Tombigbee river. It contained, in 1840, 10,075 neat cattle, 1461 sheep, 21,140 swine ; and produced 4072 bushels of wheat, 400,750 of Indian corn, 15,036 of oats, 11,638 of potatoes, 1,606,345 pounds of cotton. It had seven stores, two flouring-mills, four grist-mills, four saw-mills, one tannery, one printing-office, one weekly newspaper ; seven academies, 170 students ; four schools, 70 scholars. Population : whites, 1817 ; slaves, 6157 ; free coloured, 1 : total, 6975. Capital, Macon.

NOYON, (an. *Noviomagus Veromanduorum*,) a town of France, dep. Oise, cap. Canton, on the Verse, a tributary of the Oise, 42 m E.N.E. Beauvais. Pop., in 1836, 3473. It is well built, and surrounded with numerous gardens. The cathedral, erected under Pepin and Charlemagne, is 340 feet in length, its main entrance being flanked by two towers upwards of 200 feet in height. Noyon has manufactures of fine linens. talile, hosiery, leather, copperas, &c., and a brisk general trade. It was erected into a bishopric in 531. Charlemagne held his court in this town for a considerable period, and in it Hugh Capet was proclaimed king. But it is chiefly remarkable for its having been the birth-place of the famous reformer John Calvin, born here on the 18th of July, 1509. (*Hugo ; Guida du Voyageur*.)

NUBIA, (an. *Æthiopia*,) an extensive tract of E. Africa, between the S. boundary of Egypt and the N. limit of Abyssinia ; bounded E. by the Red sea, W. by the desert of Libya, between lat. 13° and 24° N., and long. 33° and 36° E. Estimated area, 360,000 sq. m. Pop. unknown. The country is divided into Lower Nubia, or Nubia Proper, extending from Egypt to the N. frontier of Dongola, and thence to the junction of the river Atbara or Tacazzé with the Nile ; and Upper Nubia, which includes Shendy, Halfay (an. *Meroë*), and Sennaar. (*Encyc. Britannica*, art. *Nubia ; Ritter's Africa, &c*.)

Nubia is situated almost entirely in the basin of the Nile. Rocks and mountains are the characteristics of *Lower Nubia* ; and the mountains here press so closely on the river, that there would be but little ground left for tillage, if they were not interrupted by lateral plains, the productiveness of which, however, is diminished by the continual encroachments of the deserts. Numerous rocky islands dot the stream, and in some places congregate so as to form rapids, hardly deserving the name of "cataracts," by which they are usually designated. Some of these islets are rendered productive (like the high banks of the Nile), by means of the artificial irrigation effected by *sakeas*, or Persian water-wheels. Between the river and the Red sea extends the stony and sandy Nubian desert, interspersed here and there with small fertile spots, or oases. On the coast of the latter are a few inconsiderable towns. In *Upper Nubia* the country wears a somewhat different aspect. Instead of one river, several streams flow through it to pour their waters into the majestic Nile. The land is also much more elevated, being situated on the lowest of the three plateaux on which, according to Ritter, this part of Africa is placed. The S. extremity of Nubia has an elevation of 4000 feet above the level of the sea : but northwards the elevation gradually lessens, and the Nubian desert forms

the gradual transition from the lower course of the Nile to the higher and more southern lands of Africa.

Mountains.—Ranges of mountains, forming a continuation of the range traversing Egypt, skirt the entire Nubian coast of the Red sea ; but they are not nearly so high nor so important as some travellers have stated. Those through which the Nile forces its course are figured in most maps as running parallel to its bed, as in Egypt ; but the numerous so-called cataracts, and the many valleys which intersect the hills, prove that these ranges traverse the Nile parallel to each other from E. to W., crossing the current of the river, instead of accompanying its course from S. to N. *Gebel Smigré* (*Chiggré*, in Bruce) and *Gebel Safeha* take the former direction. From Faka, on the river Atbara, to Suakim, on the Red sea, stretches another lateral chain, called the *Orbay Langoy*. Several inconsiderable chains and detached rocky hills, off-shoots from this chain, are distributed over the E. desert skirting the Red sea.

The *Climate*, of lower Nubia, though intensely hot, is healthy, on account of its dryness ; the plague has never been known to reach beyond the second cataract ; but the higher districts are subject to those violent tropical rains which contribute in some degree to the regular swelling of the Nile ; the N. limit of these rains is between lat. 17° and 17° 30' N. In Berber and Shendy they continue throughout March and the two succeeding months. The deserts E. and W. of the Nile are subject to violent storms of wind. The *geological structure* of the rocks in lower Nubia consists of granite and syenite, interspersed with black marble, of which last the second cataract is formed. Slate (in the E. desert), porphyry, sandstone, and limestone, have also been enumerated. In the upper countries coarse grey granite, primitive quartz, and mica-slate are likewise often mentioned by travellers. Along the coast of the Red sea gold and silver mines are said to exist ; but the pachs of Egypt has made more than one unsuccessful attempt to open them. Neither have the "Emerald" mountains, which pass the Egyptian frontier, yielded any treasure to modern adventurers.

Animals.—The S. parts of Meroë seem to be the N. boundary of the natural habitation of the African elephant. Tygers and lions have been seen in the valleys of Shendy, where crocodiles also abound. Wild dogs and foxes are exceedingly numerous, Ruppel having discovered four new species. The hippopotamus seldom ascends the Nile higher than Dongola. Antelopes, of three species, occupy the banks of the White Nile and the desert W. of Dongola. The giraffe (*zerafa*, " the elegant,") inhabits the mountains of Dender, near the Atbara. The principal birds of Nubia are the ocellated vulture, the red-throated shrike, and several curious specimens of the family of great-legged thrushes. Bustards are also abundant, with partridges, quails, and several other species of game.

Mirbel having, very properly, classed Nubia with Barbary and Egypt, and placed them in the S. transition zone, the *botany* of Nubia assimulates very nearly with that of those countries. (*See* EGYPT, *sec. vegetable productions*, t. 744.) The great enemies to vegetation here, as in other hot countries, are locusts, clouds of which sometimes darken the air, and settling on the land, strip it of every remnant of verdure ; on these occasions, the inhabitants catch and eat them, " out of self-defence."

Description of Lower Nubia.—The Nubian valley of the Nile, which ascends as high as the 7th cataract, and ranges between the 9th and 24th degrees N. lat., comprises 13 states, each governed by its melek, or chief subordinate to the pacha of Egypt. Ten of these states are in Lower, and three in Upper Nubia. Between the 1st and 2nd cataracts, in the state of Kenous and Wady Nubia, the Nile flows through a rocky bed, and precipices enclose the river within very narrow limits, scarcely allowing of cultivation on either side ; but at short intervals occur those excavated monuments which will hereafter be more minutely described. El-Kalatsbeh, the largest village on the W. bank, occupies the site of the ancient *Talmis* ; and opposite to it is that of *Contra-Talmis*, the ruins of which have occupied the attention of modern travellers.. At Sebou, lat. 22° 50' N., the river inclines to the N.W., flowing past Derr. which, though a mere village of 200 houses, is the capital of the five states N of Dongola. Ipsambool, with its well-known excavated temples, is near the centre of the state. called Wady Nubia, on the W. bank of the Nile, in lat. 22° 20' N. The second cataract, which occurs about 35 m. below Ipsambool, is formed by numerous rocky islets intercepting the stream, on each side of which in this vicinity stretches an extensive and not unfertile plain.

Through the district of Batn-el-Hadjar, the Nile passes between a chain of syenite hills, those on the W. side having at their foot many deserted villages and monasteries : only the E. side of the valley is now inhabited. The district of *Sakket* has many poor villages on both banks of the river, possesses numerous antiquities, and is joined southward by *Mahass*, where the most cultivated spots, hitherto situated on the E. bank of the river, are transferred to the opposite shore. Remains of castles, churches, and houses afford evidence that this district was formerly well peopled. The course of the Nile here is tortuous; but S. of the 3d cataract, forming the N. boundary of Dongola, it runs in a pretty direct channel as far as Old Dongola, in lat. 18° 10′ N. The stream then takes a sweep to the N.E., preserving that direction for about 100 m., through the highly fertile district of Sheygya, ascending beyond the 4th cataract to the island of Mokrat, which divides the stream. The state of Berber commences southward of the 5th cataract, and in this district the villages stand at a considerable distance from the river. Berber, or El-Mekhair, the capital, is near the E. bank, about 17 m. below the junction of the Tacazzé with the Nile.

Upper Nubia, is a triangular tract lying chiefly between the White Nile, or Bahr-el-Abiad, the Blue Nile, or Bahr-el-Azrek, and the Tacazzé or Atbara. It is now divided into the three states of Shendy, Halfay, and Sennaar. From the Berber frontier, for some considerable distance southward, the soil of Shendy consists of immense fertile plains, stretching out from both sides of the Nile on elevated ground, at some distance from the river. Shendy el-Garb "on the W. bank" is a large and not ill-built village, with about 6000 inhabitants: Shendy "on the E. bank" is the capital of the province; and being a place of rest for the caravans from Sennaar, possesses regular and well-stocked markets. N. of Shendy are some ruins, supposed to be those of the ancient *Meroë*. Halfay lies between Shendy and Sennaar; and, before the Egyptian conquest by Ismael, the son of Mehemet Ali, belonged to the melek of Sennaar: its chief town, having the same name, lies N. of the confluence of the White with the Blue Nile, which takes place at Khartum. W. of the Bahr-el-Azrek is the district of Sennaar, or Fungi; it is a flat and fertile tract, with some large villages, mostly composed of conical houses, similar to those of the S. African tribes. Six days' march S. of Khartum is Sennaar, the *entrepôt* of the caravan merchandise for Kordofan, Darfur, and Abyssinia. Its environs are wide plains, with a long ragged mountain, about 15 m. W. of the town (Sennaar). The most considerable port upon the Red sea is Suakim, whence merchants embark their goods for Arabia, &c.

Population, Languages, &c.—The inhabitants of the different parts of Nubia differ considerably in personal appearance; and those southward are much darker than those in the states bordering on Egypt. The marked features of the whole race, however, are long oval countenances, curved noses, somewhat rounded towards the top, rather thick lips, but not so far protruding as those of the negroes. retreating chins, scanty beards, lively dark eyes, strongly frizzled hair, and well-knit, muscular bodies. The Noubas, properly so called, are about the best looking of the race; both men and women have good features and well-proportioned persons, their disposition and character also being, according to Burckhardt, more susceptible of improvement than those of the Dongolese, who are described as dirty, idle, and ferocious. (*See* DONGOLA, i. 764.) The people inhabiting the valley of Sheygya, E. of Dongola, are the most powerful of the Nubian tribes N. of Sennaar. They are good horse-soldiers, and were employed as such by Ismael Pacha, on his expedition against the negroes of the S. The common people are almost naked, wearing nothing but a hip-cloth. They usually speak the Arabic language; and the learned caste among them cultivate most branches of Mohammedan literature. The Berbers present, perhaps, the worst specimens of Nubian character: treachery, dishonesty, and drunkenness, are prevailing features among the men; and the women, who in the better parts of the country are modest and observant of conjugal fidelity, here indulge in the greatest profligacy, and pay no attention to the marriage vow. The inhabitants of upper Nubia are of Arabic descent, speak the language of the Arabs, and resemble them in their love of a restless, roving life. A pastoral population inhabits the banks of the Tacazzé, which, also, are visited by mountaineers, when in search of pasturage, during the dry season. The E. desert is infested by wild nomadic tribes, constantly at war with each other, and remarkable for addiction to thieving and treachery towards strangers.

Productive Industry and Commerce.—The cultivated portions of the Nubian valley being, on account of the height of its banks, beyond the inundation of the Nile, the land can only be watered by artificial means. Even in the lateral valleys, the few canals cut through them are rarely full: and the water, both from them and the Nile, is raised by Persian wheels. Dhourra is reaped in December and January; next follows a crop of barley, and then dhourra again. Tobacco is universally raised. Although the S. districts present some excellent land, agriculture offers few charms for the inhabitants; and Sennaar and Shendy are celebrated only for being the *entrepôts* of the chief commerce

of E. Africa. The town of Shendy, having Soudan and Abyssinia to the S., Egypt and the Arabian gulf to the N. and E., and Darfur to the W., is the centre of much of the trade with those countries. Markets are regularly held there twice a week; and at one of them Burckhardt saw from 4000 to 5000 cows, as many camels, nearly 100 asses, and several horses for sale. In Shendy are several forges for iron and silver. The merchants from the W. pay regular visits to Sennaar; they exchange Indian goods for gold, which they transport to Djidda and the E. The price of gold at Sennaar is estimated at $12 an ounce, and at Shendy $16. Every two months merchant caravans arrive at both these places, frequently consisting of 500 or 600 camels laden with dhourra; others, comprising about 100 camels, trade in various products, as well as slaves. The traffic in slaves is extensively carried on, upwards of 5000 being annually imported from the interior of Africa; of these 2500 are disposed of in Arabia, 1500 in Egypt, and 1000 in Dongola and other parts of Lower Nubia. The Arabs of the Desert supply the caravans with senna of the best quality, ostrich feathers, and charcoal.

History, Government, &c.—It has been supposed that the country of the Ethiopians was among the earliest in which advances were made towards civilization, and that the arts descended from Meroë to Egypt. But we have little or no authentic information respecting the state of this country in antiquity; and it was not till the 6th century that the wandering ancestors of the Nubians appear to have settled under a regular government. At that period mention is made of Silco, king of the Nubetes and the Euhtopians (*Letronne, Journal des Savans*, 1825); under whom they were converted to Christianity, the country divided into ecclesiastical districts, and the whole subjected to the patriarch of Alexandria. After the loss of Abyssinia, the kings of the Noubas resided at Dongola; but in the 14th century their power ceased, and Nubia was divided into several petty states. In the succeeding century the Mohammedan conquerors reached and subdued the country, Christianity was suppressed, and Mohammedanism took its place.

Down to the year 1821 the people of Nubia were independent, living under their own meleks, or chiefs; but at that period Ibrahim Pacha reduced them to a dependency on Egypt. This change is so far fortunate for travellers, that with the permission of Mehemet Ali, the whole country is open to their researches, and no danger is to be apprehended, except from the climate and the dishonesty of the natives. The same system of military despotism and oppressive taxation that exists in Egypt has been extended to Nubia: but it is a question whether the people be now more heavily taxed than formerly by their petty chiefs, while, in other respects, their condition is improved.

Monumental Remains of Nubia.—Ipsamboul.—Of all the relics of ancient art with which the valley of the Nile abounds over the whole distance from Meroë to Memphis, none have excited more admiration than the excavated temples at Ipsamboul, lat. 22° 19′ 47″ N.; long. 31° 38′ 54″ E. According to Champollion, the great temple, "*est une merveille, qui serait une fort belle chose même à Thèbes*!" It is wholly cut out of the solid rock, and presents a façade, supported by four seated colossi, of exquisite workmanship, and not less than 61 ft. in height. They represent Rameses the Great, and are all portraits, for the faces bear a perfect resemblance to the figures of that king at Memphis and elsewhere. The interior is not less grand than the entrance: 16 apartments have been enumerated; the first of these is sustained by eight pillars, against which rest the backs of as many figures of Rameses, each 30 ft. in height. The walls of this immense hall are covered with innumerable bas-reliefs on historical subjects, the most striking portraying the conquests of the same prince in Africa. The other apartments afford some curious particulars that supply many conjectures relative to Nubian and Egyptian religious history, which it remains for future students in hieroglyphics to verify. The whole is terminated by a sanctuary, at the back of which are seated five statues, representing Amon, Ra, Phré, Phtah, with the never-absent Rameses the Great. The smaller of these excavations is a temple dedicated to Hathor by Nofre-Ari, wife of Rameses the Great, whose façade has six colossi, each 35 ft. high, carved out of the rock. They represent Rameses and his wife, having at their feet statues of their sons and daughters, all of whom have their names and titles. The front of this temple is free from sand, and access is much easier to its interior than to that of the greater. A passage leads to the pronaos, which is 35 by 3½ ft., supported by six square pillars, three on each side; to this chamber succeeds a vestibule, which leads to the adytum, or sanctuary, containing the remains of a sitting statue cut in the rock, which, however, is not in such good preservation as the rest of the structure. The bas-reliefs adorning the sanctuary are painted, the figures yellow, and are enclosed by a border of these colours: the colour of the ceiling is blue.

We are indebted to Berckhardt and Belzoni for bringing these splendid temples to light. The entrance of the great temple is so blocked up with sand that it is only passable by a person divesting himself of nearly all his clothing, and creeping on his hands and knees; and then the heat within is more intense than that of a Turkish bath, the want of air being almost insufferable.

Besides the excavated temples of Nubia, of which Ipsambohl does not present the only specimen, there are others, partly hewn out of the solid rock, and partly built. Those of Girshe (lat. 23° 31' 45" N.; long. 39° 56' 55" E.), Sebona, Dendera, and Gebel-el-Birkel (lat. 18° 31' 41" N. Ruppell) are of this class. The interior of these temples is cut out of the solid rock, while the exterior chambers and appendages are formed of stone-work. From the primitive character of the masonry, the rudeness and decay of the sculptures, and the decomposition of the walls, it has been concluded that the temple of Gebel-el-Birkel is older than many of the temples of Egypt, or even of Nubia. This she is also remarkable for 13 pyramids, lying in the desert to the E. of the town, differing from those previously known, their sides presenting small temples with gateways and enclosures. Opposite to Birkel, on the other side of the Nile, at Nouri, is another assemblage of pyramids. The age of all these vast masses of stone, many of them exhibiting little else to the modern traveller than mounds of debris, no doubt belongs to the remotest antiquity. At Senumeh and Dendera, we find specimens of a more perfect class of temples than those before mentioned, and which belong to the last epochs of Nubian art. That at the latter place has the proportions of Grecian structures, and in the pillars have been recognised a mixture of the Greek and Egyptian styles.

Ruins of Meroë.—The tract of country enclosed by the Nile and the Tacazzé, or Atbara, and terminating at the confluence of these rivers, was the island of Meroë of ancient geographers; and near Assur on the Nile, in the prov. of Shendy, the ruins of the ancient capital of Ethiopia have been recognised. Nothing remains but the Necropolis; which consists of a vast assemblages of pyramids, similar in every respect to those of Birkel. (For a particular account of which we refer the reader to *Hoskins' Travels in Ethiopia*, p. 66, et seq.) We must not, however, omit to notice the inferences drawn from these and other Nubian monuments by those whose acquaintance with antiquities entitles their opinions to respect. They conclude that arts and civilization, instead of ascending the Nile from Egypt, descended to it from Ethiopia. The decay in which the mounds of Meroë are now found, produced entirely by the slow hand of time, the sculptures of their interior, exhibiting religious rites of a purer and simpler stamp than those of Egypt, and other circumstances which nice observers have supposed they have detected, prove, according to the authorities referred to, that they have been the models of the more stupendous Egyptian structures. The excavated temples, too, furnish, it is alleged, proofs of the remotest attempts at architecture. But, how plausible soever, these conclusions amount to no more than probabilities; and it would not be difficult, were this the proper place, to show that they must be received with great limitations and modifications.

NUDDEA, a district of British India, presid. Bengal, chiefly between lat. 23° and 24° N., and long. 88° and 89° E.; having N. the districts Moorshedabad and Rajeshaye, E. Jessore, W. Beerbhoom, Burdwan, and Hooghly, and S. Calcutta and the 24 Pergunnahs. Length, N. to S., about 80 m.; average breadth, nearly 40 m. Area, 3,105 sq. m. Pop., in 1822, estimated at 1,187,160. Its natural features are the same with those of the rest of the delta of the Ganges, by many arms of which it is intersected. The culture of the soil has greatly increased since the establishment of the perpetual settlement: total land revenue, in 1829-30, 11,66,951 rupees. Gang-robbery formerly prevailed to a great extent in Nuddea; but under the British rule, it has decreased so as to be now of rare occurrence.

NUDDEA, a town of British India, presid. and prov. Bengal, cap. of the above distr., at the commencement of the Hooghly river, 80 m. N. by W. Calcutta. It is the residence of the collector and judge for the district, and was formerly the capital of a rajahship, and a celebrated seat of Hindoo learning, but it has now fallen into decay. (*Parl. Reports, &c.*)

NUNDA, t., Alleghany co., N. Y., 20 m. N. Angelica, 253 m. W. by S. Albany. 352 W. Drained by Canaseraga and Cashaqua creeks. It contains three churches, a Presbyterian, Baptist, and Universalist. 13 stores, one fulling-mill, two flouring-mills, six saw-mills; one academy, 47 students: 13 schools, 1064 scholars. Pop. 2637.

NUNDYDROOG, a celebrated hill-fortress of Hindostan, prov. Mysore, on a hill 1700 feet in height. 100 m. N.E. Mysore; lat. 13° 22' S., long. 77° 44' E. The hill on which it stands is inaccessible, except on one side: the fort has within it several barracks, magazines, &c., besides a Hindoo temple, in which worship is paid to the bull Nundy, whence the name of the fortress. Nundydroog was taken by the British in 1791, after an obstinate defence of three weeks.

NUNEATON, a market town and par. of England, Atherstone div., hund. Hemlingford, co. Warwick, on the Anker, 8½ m. N. by E. Coventry, and 90 m. N.W. London. Area of par., 7090 acres. Pop., in 1841, 7105. The town is large and well built; consisting principally of a long main street, whence another diverges, in which is the market-place. The church is a Gothic structure, with a square tower; the living is a vicarage, in crown patronage. There is also a modern-built chapel-of-ease; and the Wesleyan Methodists, Independents, and Baptists have their respective places of worship. A free school was founded, by the inhabitants, in the reign of Edward VI.; and there is another endowed school, called "Smith's charity school," besides which there are two or three other day schools and Sunday schools. The inhabitants are principally engaged in riband weaving, and, in 1838, there were 2800 looms, nearly all employed. The engine trade is confined to four firms; and there are scarcely any engine looms in the place that are not their property, though some are worked on the premises of the weavers. Floret gauze ribands are the staple article of manufacture; but they are occasionally laid aside for figured sattins, sarsnets, and lustrings. (*Hand-loom Weavers' Report*.) Malting is carried on to a considerable extent; and there is a silk factory, which, however, was unemployed in 1839. Coal is procured in the neighbourhood, where are also some extensive stone quarries. The government of the town is vested in a permanent constable and three others, annually elected at a court-leet. It is one of the polling-places at elections for the N. division of the county. Markets on Saturday: fairs, May 14th, February 16th, and October 31st, for horses and cattle. (*Parl. and Comm. Reports, &c.*)

NUREMBERG (Germ. *Nürnberg*), a city of Bavaria, circ. Middle Franconia, on the Pegnitz, a tributary of the Regnitz, 93 m. N.N.W. Munich. Lat. 49° 27' 31" N., long. 11° 6' 40" E. Pop., in 1838, with its suburbs, estimated at 40,400, of whom about one tenth part are Roman Catholics. It stands in a sandy but fertile plain, at an elevation of about 1000 feet above the sea; and is divided by the Pegnitz into two nearly equal parts, the Sebald side and the Lawrence side, each deriving its name from its principal church. Nuremberg covers more ground than any other city of Bavaria, and is, next to the capital, the most populous. "It is surrounded by feudal walls and turrets, and these are inclosed by a ditch 100 feet wide and 50 feet deep, lined throughout with masonry. Its arched gates are flanked by four massive cylindrical watch-towers, no longer of use as fortifications, but picturesque in a high degree, and serving to complete the coronet of antique towers which encircle the city, as seen from a distance. The stranger arrived within its walls might fancy himself carried back to a distant century, as he treads its irregular streets, and examines its quaint gable-faced houses. Its churches and other public edifices are singularly perfect, having escaped unharmed the storm of war, sieges, and even of the Reformation, which its inhabitants adopted at an early period, without any outbreak of fanaticism or iconoclasm. Its private buildings, including the palace-like mansions of its patrician citizens and merchant nobles, having been built of stone, are equally well preserved, and many are still inhabited by the families whose forefathers originally constructed them. Though built with narrow but highly ornamented fronts and acutely pointed gables, they are often of large size, inclosing two or three courts, and extending back from one street to another." (*Murray's Handb. for S. Germany*.) The most elevated position within the town, near its N. extremity, is occupied by the Reichsveste, or imperial castle, a building of great antiquity, and a frequent residence of the German emperors in the middle ages. A portion of this castle is fitted up for the accommodation of the king of Bavaria, when he visits the town; and comprises a picture gallery, the paintings in which, however, except one by A. Durer, are generally of the most ordinary merit. The two principal churches are highly deserving of notice. That of St. Sebald, a fine Gothic edifice, with an elegant choir, built in 1337, has numerous sculptures and carvings by Adam Kraft and V. Stoss, many old paintings and stained glass windows and the remarkable shrine of St. Sebald. This, which still stands in the centre of the church, though the latter is devoted to the Lutheran service, is the masterpiece of the celebrated artist Peter Vischer, who, with his five sons, was employed on it for 13 years. "It is a miniature Gothic chapel, entirely of bronze, consisting of a rich fretwork canopy, supported on pillars, beneath which the relics of the saint repose in an oaken chest, incased with silver plates. The workmanship is most elaborate. The figures of the 12 apostles occupy the niches around the shrine, and are truly first-rate works of art. Above them are 12 smaller figures of fathers of the church; while about 70 fanciful representations of cupids, mermen, animals, &c., distributed among flowers and foliage, are scattered over the other parts. The miracles of the saint are subjects of the bas-reliefs under the coffin. In a niche below, at one end, is an admirable statue of the artist himself, in a mason's dress, and at the opposite end is a figure, equally excellent, of St. Sebald." (*Handb. for S. Germ.*) The church of St. Laurence, founded in 1274, is the largest in the town; and has some very handsome entrances, fine stained glass, curious carvings, &c.; and above all, a repository for the sacramental wafer, a tapering spire of Gothic open-work, 64 feet in height, executed by A. Kraft, with a minuteness more commonly bestowed on ivory than on stone. The church of St. Giles, erected in 1718, in the Italian style, has a fine altar-piece by Vandyk, various bas-reliefs, escutcheons, &c.; the Roman Catholic church, finished in 1361, and distinguished for its rich decorations; and the church of the Teutonic knights, begun in 1784, are the other principal ecclesiastical edifices: the Gothic chapel of St. Maurice, constructed in 1313, has been converted into a picture gallery, and filled with rejected paintings from the gallery of Munich, &c. The *Rathhaus*, or town-hall, in the Italian style, is one of the most remarkable edifices in the city; it was chiefly built in 1619, but includes the ancient town hall, dating from 1340. In the latter are the great hall and the council-chamber; the walls of the former apartment being ornamented with several oil paintings by Albert Durer, and those of the latter having many concealed doors "leading to subterranean passages, which extend from the Rathhaus under the streets to the town ditch, beyond the walls." (*Handbook*, p. 56.) Nuremburg has a gymnasium, founded by the famous reformer, Melancthon, whose statue is placed in its front; an arsenal, barracks, a theatre, many hospitals and asylums, a savings' bank, a charity for distributing food to the poor, a house of correction, &c. It has also several fountains, some of which are worthy notice, especially the "Beautiful Fountain" (*Schöner Brunnen*), in the great market-place; a Gothic obelisk, or spire of open-work, with statues of various historical characters. Among the other remarkable objects in and near the city are the house of Albert Durer, now occupied by a society of artists; St. John's churchyard, in which is Durer's tomb, together with those of many distinguished natives; a succession of stone pillars between the cemetery and the city, ornamented with curious bas-reliefs, &c. Nuremberg is the seat of a high police court, a civil court of justice, a commercial court of appeal, and a forest board. It has a royal and other high schools, several Latin and numerous inferior schools, a teachers' seminary, an academy of arts, a polytechnic, and a high commercial academy (*Handlungs Institut*); a number of public libraries, including the city library of 40,000 printed volumes, and 800 MSS. (*Adrien in Statist. Journ.*, 1841); societies of national industry, and medical and natural science, an agricultural union, and collections of every description in the arts and sciences. There are but few pictures by the celebrated native artist A. Durer; but those by other artists are very numerous.

Nuremberg has given birth to many distinguished men, including, among others, the famous painter Albert Durer, born here in 1471. Several important inventions in the arts are said to have been made in this city. The famous machine for drawing wire is supposed to have been constructed by Rudolph, a native of this city. (*Beckmann's Hist. of Inventions*, ii., 236.) Gun-locks are supposed to have been first fabricated here in 1517; and Beckmann says that the circumstance is probable, though he doubts whether the locks were of the present construction. (iv., 608.) Owing partly to these inventions, but more to the freedom and industry of its inhabitants, Nuremberg early rose to great eminence as a manufacturing and commercial town. It was, in fact, the continental Birmingham of the middle ages, during a portion of which period it is said to have had 70,000 inhabitants. Cannon are said to have been cast here as early as 1356; and in the same century it furnished paper and playing-cards. It had also a very extensive commerce, being a principal entrepot for the produce of both the N. and S. of Europe. It is still, and has long been, celebrated for an extensive manufacture of wooden clocks and toys, which it exports to all parts of the world. It also produces various species of metallic goods and jewellery, with telescopes, mirrors, mathematical and musical instruments, sealing-wax, and lacquered wares; lead pencils, alabaster, horn, and ivory articles; brushes, woollen yarn, lawn, paper, parchment, brandy and liqueurs, chicory, &c. Printing is also carried on to some extent.

Though considerably declined, it still ranks as one of the principal commercial cities of Bavaria; and its commerce will probably receive some augmentation from the opening of the canal between the Danube and Rhine, now in progress. The first railroad for steam-carriages in Germany was completed in 1835-36, between Nuremberg and Fürth, a distance of 4½ m., now traversed in 15 minutes.

Nuremberg, supposed to have been founded in the sixth century, became, in 938, the seat of the first Germanic diet. Until 1417, it had a *burggraf*, or resident governor, appointed by the emperor, and the ancestors of the present royal family of Prussia make their first appearance in history in that capacity. It was subsequently governed much in the same way as Venice, by a merchant aristocracy, consisting of about 30 families, who appointed the executive officers among themselves. It was at the summit of its prosperity in the 15th and 16th centuries. The famous Æneas Sylvius, afterward Pope Pius II., who had travelled over the greater part of Europe, celebrates the wealth of this city, and says in his work *De Morib. Germ.*, published in the 15th century, that the kings of Scotland would wish to be as well lodged as the meanest burgesses of Nuremberg. *Cuperent tam egregie Scotorum reges, quam mediocres Nurembergæ cives habitare* (p. 1055).

Nuremberg early embraced the doctrines of the Reformers, and is celebrated in the history of the Reformation. A diet assembled here in 1524, was of great service to the cause of the Reformers; and here, on the 23d of July, 1532, a treaty was signed, by which full toleration was granted to those professing the new doctrines. The city preserved its privileges as a free town of the empire to the peace of Presburg in 1805, when it was annexed to Bavaria by Napoleon. (*Berghaus, Allg. Länder- und Völk.*, iv., 145-147; *Murray's Handbook for S. Germany*, 54-68; *Dict. Geog.; Stein, Cannabich, &c.*)

O.

OAKHAM, a market town and par. of England, hund. of same name, co. Rutland, of which it is the cap., 17 m. E. by N. Leicester, and 83 in. N. by W. London. Area of par., 3130 acres. Population in 1841, 2726. It is tolerably well built. The chief public buildings are the county hall; serving the only remaining part of a castle built in the reign of William the Conqueror), a fine church with a lofty spire, and a large edifice belonging to the Rutland Agricultural association. The free school, founded in 1584, and closely connected with that established at Uppingham, is under the control of 14 official governors; it is well endowed, and has 34 exhibitions at Oxford and Cambridge; it has two masters, and the school is open gratuitously to the children of the poor inhabitants. A hospital for old men was incorporated with it by Queen Elizabeth, and endowed with alienated church property, now producing above £368 a year. Another hospital once existed here, but it has fallen into decay. A boys' national school is established, and there is a well-attended Sunday school. Oakham is of very little importance with respect to trade, its chief dependence being on its markets and the retailing of goods for domestic consumption. It is connected by a canal with Melton-Mowbray, and has a considerable traffic in coal. The assizes, quarter and petty sessions are held here; and Oakham is the election town for the county. Markets on Saturday; fairs for cattle and sheep, March 15th, second Saturday in April, May 9th, Saturday in Whit-week, Saturday after October 10th, and December 15th.

OAKHAM, p. t., Worcester co., Mass., 64 m. W. Boston, 416 W. Watered by branches of Chickapee river. Chartered from Rutland in 1762. It contains three churches, two Congregational and a Baptist, one store, three grist-mills, two saw-mills; nine schools, 404 scholars. Pop. 1408.

OAKHAMPTON, a decayed bor., market town, and par. of England, hund. Lifton, co. Devon, on the Oke, a trib. of the Torridge, and near the N. border of Dartmoor, 20 m. W. Exeter. Area of parish (including the villages of Okesacot, Meldon, and Kegbear), 19,570 acres. Population in 1841, 2194. It is old and irregularly built. The church stands on rising ground about 1 m. westward; and there is an ancient chantry chapel in the market-place, with places of worship for Wesleyan Methodists and Independents. It has a small endowed free school and two subscription schools, with minor charities for the relief of the aged poor. "The town is not flourishing:" the inhabitants depend chiefly on retail trade, a few, however, being supported by serge weaving. It also derives some advantage from its situation on the great road between Exeter and Falmouth. (*Mun. Corp. Report.*) The borough was not incorporated till the 21 James I., and, having fallen to decay, it was considered too insignificant to be included in the provisions of the Municipal Reform Act. It, however, sent two members to the House of Commons from the reign of Charles I. down to the Reform Act, when it was disfranchised. Markets on Saturday; fairs, second Thursday after March 11th, May 17th, first Wednesday after July 5th; August 5th, and Saturday after Christmas day.

OAKLAND, county, Mich. Situated in the E. part of the peninsula, and contains 900 sq. m. Watered by Clinton, Huron, Flint, Rouge. and Shiawassee rivers. It contained in 1840, 20,361 neat cattle, 19,656 sheep, 39,313 swine; and produced 264,965 bushels of wheat, 8157 of rye, 254,909 of Indian corn. 238,005 of oats, 399,807 of potatoes, 90,614 pounds of sugar. It had 62 stores, one furnace, six fulling-mills, one woollen factory, 24 flouring-mills, 14 grist-mills, 41 saw-mills, three tanneries, six distilleries, two printing-offices, two weekly newspapers; 117 schools, 4800 scholars. Pop. 23,646. Capital, Pontiac, ;

OAKLAND COLLEGE. p. v., Claiborne co., Miss., 25 m. N.N.E. Natchez, 87 m. S.W. Jackson, 1097 W. It is under the patronage of the Presbyterians. was founded in 1831, has a president and five professors, or other instructers, and 160 students.

OAXACA, or GUAXACA, a city of Mexico, the cap. of the state of the same name, on the Rio Verde. 205 m. S.S.E. Mexico, and 160 m. S.S.W. Vera Cruz ; lat. 17° 5' N., long. 27° 8' W. Estimated population, 40,000. It is built in the form of a parallelogram, about 2 m. in length and 1½ m. in breadth, including its suburbs, which are laid out in gardens and planted with nopal trees. The streets, which are broad, straight, and well paved, are lined with good houses of a greenish kind of stone ; and, on the whole, it is the neatest, cleanest, and most regularly built city of Mexico. The public buildings are in general handsome, solidly constructed, and richly decorated : the town-hall, cathedral, and bishop's palace form three sides of the principal square. There are several churches and convents ; and numerous fountains are supplied with water conveyed by aqueducts across the valley from the neighbouring hills of St. Felipe. The climate is peculiarly good, the thermometer seldom falling below 63° or rising higher than 78° ; but it is exposed to earthquakes, and suffered considerably during the last that happened in Mexico. Oaxaca was founded by Nuno del Mercado, one of the companions of Cortez, and received its name from the trees called guazas that abound in its neighbourhood.

The state of which Oaxaca is the capital is remarkable for its extreme fertility, and for the richness and variety of its products. The cereals and the sugar-cane are raised with great facility, and cochineal is extensively cultivated. Considerable attention is likewise paid to the culture of silk. The mineral riches of the state have been very little explored. (*Ward's Mexico*, ii., 389 ; *Mod. Trav., &c.*)

OBAN, a parl. bor. and seaport of Scotland, co. Argyle, on a bay of the same name, in a secluded but beautiful situation, 61½ m. N.W. Glasgow. Pop. 1480. The only public buildings are a new quoad sacra church connected with the establishment, and a dissenting chapel. It has no manufactures, and no trade, except in such articles as the limited consumption of the place and neighbouring district require. It is visited by the steamboats between Glasgow and Inverness, and those that ply between either of these places and Staffa, Iona, &c. The harbour is excellent ; and the inhabitants engage extensively in fishing. The magnificent ruins of the royal palace of Dunstaffnage stand on a promontory 3 m. N. the town. The town had no parliamentary representative till the passing of the Reform Act. In 1832 ; which united it with Campbelton, Inverary, Irvine, and Ayr, in sending one member to the House of Commons. Registered voters, in 1840–41, 51.

OBERLIN, p. v., Russia t., Lorain co., O., 110 m. N.N.E. Columbus, 379 W. It is the seat of the Oberlin Collegiate Institute, which contains a male and female department, on the manual labour system. It was founded in 1832, has a president and 10 professors, or other instructers, and 70 students, in the collegiate department. It has a theological department, with 58 students. The Oberlin Evangelist, a religious periodical, is published here.

OBI, a large river of Asiatic Russia, in the governments of Tomsk and Tobolsk, rising by two principal sources on the N.W. side of the Altai chain near the frontiers of the Chinese empire ; lat. 51° N., long. 89° E. ; flowing first N.W., and then N., into the gulf of Obi, after a course of about 2700 m. ; but if the Irtish, which joins it in lat. 60° 50' N., and is the longest and widest stream and most direct from the source, be considered the main river, its length will exceed 3000 m. ; the area of its entire basin has been estimated at 1,357,000 sq. m. The Obi, which is the eastern branch, has numerous affluents, the principal of which are the Tom, Tchelim. and Ket, joining it on the E. or right bank. After its junction with the Irtish, it attains a breadth in some places of nearly 20 m., with a depth varying from 2 to 7 fathoms, and has a very rapid current, forming in the lower part of its course numerous islands, and flowing over rocky ledges that greatly impede navigation during the few months that the river is free from ice. The Irtish rises within the Chinese empire, in lat. 47° N., long. 89° 10' E., on the W. side of the Great Altai chain, and pursues a course nearly W.N.W. of 240 m. to lake Tsigan, through which it flows, and then turns northward for about 100 m., after which it has a general N.W. direction, passing Semi-

polatinsk and Omsk, as far as Tobolsk Below this point it makes a curve northeastward of about 300 m., and joins the Obi at Samarova. Both the Obi and Irtish abound with fish, which might be made a lucrative article of trade, as there is a free navigation during the greater part of the year along the Northern ocean to Archangel. (*Stein's Geographie*, lii., 87 ; *Dict. Geog.*)

OBION, county, Tenn. Situated in the N.W. part of the state, and contains 700 sq. m. Bounded W. by Mississippi river. Watered by Obion and Reelfoot rivers. Along the Mississippi it is liable to be overflowed. It contained in 1840, 7580 neat cattle, 1490 sheep, 23,409 swine ; and produced 10,246 bushels of wheat, $35,715 of Indian corn. 14,137 of oats, 13,577 of potatoes, 943,190 pounds of tobacco, 42.446 of cotton, 1897 of sugar. It had eight stores, six grist-mills, one saw-mill, two tanneries, three distilleries ; six academies. 160 students : 10 schools, 223 scholars. Pop. : whites, 4219 ; slaves, 587 ; free coloured, 8 ; total, 4814. Capital, Troy.

OCANA, a town of Spain in New Castile, prov. Toledo, 26 m. E. Toledo, and 34 m. S. by E. Madrid, on the great road leading from Madrid to Granada. Population according to Miñano, 5013. It is an ancient town of considerable size, surrounded by ruined walls, situated on the summit and sides of a steep hill. Streets generally narrow and ill-built ; but there are two or three squares which give it a tolerably respectable appearance. It has four parish churches, three decayed monasteries, a hospital, cavalry barracks, and a school of primary instruction ; but the only object worth notice is the Fuenta vieja, a fountain and aqueduct of stone, on 19 arches, supposed to have been constructed by the Romans, which supplies the town with excellent water. Ocana, in the days of its prosperity under the Grand Masters of the Order of St. Ingo, established here in the 12th and 13th centuries, carried on a considerable trade in gloves ; but its industry at present is confined to the manufacture of hard soap, the tanning of leather, and the weaving of coarse woollen and linen cloths. A festival and fair is held on the 8th of September and eight following days, which is much frequented, especially by Jewish traders. During the peninsular war Ocaña was the scene of a disastrous and obstinately contested battle (Nov. 19th, 1809), between the Spaniards under Areizaga, and the French under Mortier and Victor, which terminated in the total defeat of the former. (*Miñano : Sir A. Brooke's Spain and Morocco,* ii., 291 ; *Mod. Trav.*)

OCMULGEE. r Ga., rises in Gwinnet and De Kalb counties, runs S.S.E. for 200 m., and unites with Oconee r. on the S. border of Montgomery co., to form Alatamaha r. It is navigable for steamboats to Macon. The Oconee is navigible for steamboats to Milledgeville.

OCRACOKE INLET, N. C. is the pass from the Atlantic ocean into Pamlico sound. It has 14 feet of water on the bar at low tide : but has dangerous shoals on each side. It is 22 m. S.W. cape Hatteras, in 34° 55' N. lat., and 75° 39' W. long.

ODENSEE, a town of the Danish dom., isl. Funen, of which it is the cap., on a small river, about 2 m. from the bottom of Stegestrand bay. a deep gulf to which it has been united by a navigable canal. 88 m. W. by S. Copenhagen; lat. 550 24', long. 10° 24' E. Pop. 8700. It is well built ; and has one of the finest cathedrals in Denmark, in which many of the Danish kings are buried, an old episcopal palace, with a library of 6000 vols., a gymnasium, a church seminary, and a convent with an extensive library of Danish books, the collection of which commenced with the introduction of printing into the kingdom. It is the residence of the governor and of the bishop, and has a patriotic society. Most of the gentry of the island reside here for a part of the year, and the inhabitants are said to be the best educated and informed of his Danish majesty's subjects. (*Inglis*.) It is celebrated for its manufacture of gloves and leather accoutrements : it has also manufactures of cloth, with extensive breweries and distilleries. soap works, &c. It is the most ancient town of Denmark ; and was a place of great note long before Copenhagen was in existence. (*Stein ; Inglis, Coxe, &c.*)

ODER, a large and important river of Germany, traversing the centre of the Prussian dom. It rises in Moravia, about 15 m. E. Olmutz, lat. 49° 35' N., long. 17° 25' E., at an elevation of 1800 ft. above the sea. It runs, at first generally N.E. to Oderberg, near which it leaves the Austrian dom. ; it thence flows in general N.W. to near Oderberg in the Middlemark of Brandenburg, from which point its course is mostly N.N.E. to the great Haaff, an inlet of the Baltic, which it enters by numerous mouths near Stettin. In the lower part of its course it forms numerous islands. Its principal tributaries are the two Neisses, the Oppa, Katzbach, and Bober, on its W., and the Malapane, Bartsche, and Netz with the Wartha, on its E. side ; the Wartha being by far its most considerable affluent. It is subject to sudden floods, and frequently inundates the plain country through

ODESSA.

which it flows. It is navigable for small boats as far as Ratibor in Prussian Silesia, and for barges of from 40 to 50 tons as high as Breslau. Next to this city, Frankfort, Stettin, Oppeln, Glogau, Crossen, Kustrin, and Schweidt, are the principal towns on its banks. It is connected with the Havel and Elbe by the Fluow canal, with the Spree by Frederick-William's canal, and with the Vistula by means of the canals from Nakel on the Netz to Bromberg. It is of the highest commercial advantage to the country through which it flows; as to which, see article PROSSIA, in this vol. (*Dict. Geog.; Berghaus, &c.*)

ODESSA, a celebrated city, sea-port, and emporium of S. Russia, gov. of Kherson, on the N.W. coast of the Black sea, about half way between the mouths of the Dniestr and Bugg; lat. 46° 28' 54'' N., long. 30° 43' 22'' E. Pop., in 1838, 69,023. The rise of this emporium has been quite extraordinary; its foundations having been laid, by order of the Empress Catherine, so late as 1792, after the peace of Jassy. It was intended to serve as an entrepôt for the commerce of the Russian dominions on the Black sea, and has, in a great measure, answered the intention of its founders. It has been said, indeed, that a better locality might have been chosen; and in proof of this, it is stated that there are no springs nor fresh water within 3 m. of the town; that the vicinity is comparatively barren and without wood; and that not being on or near the mouth of any great navigable river, its communications with the interior are difficult and expensive. That these considerations have great weight is clear; but, on the other hand, the situation has the advantage of being central and salubrious: the bay, or roadstead, which is generally open and easy of access, is extensive, the water deep, and the anchorage good; the port, which is artificial, being formed by two moles, is fitted to accommodate above 200 ships, and has a lazaretto on the model of that of Marseilles; the inconvenience arising from the want of water has been obviated by the cutting of a canal, by which it is conveyed to the town; and, on the whole, we doubt whether any position could have been chosen so well suited to serve as an entrepôt. The vicinity is by no means so barren as has been represented. Latterly, indeed, it has been signally improved by the formation of many gardens, and by the planting of extensive vineyards.

The town is well built of soft calcareous stone; but the houses being, for the most part, detached from each other, there are few handsome streets. But a more serious defect is, that the streets are generally unpaved; and after rain the ground is so deep that, according to Mr. Elliott, "it is not uncommon for gentlemen to be obliged to leave their carriages in quagmires in the middle of the streets, and to send oxen to drag them out!" (*Trav. in Austria, Russia, &c.*, i. 256.) But some of the principal streets are now either paved or macadamized; and in this respect the city has been materially improved. The warehouses for corn are very extensive. The city is defended towards the sea by some batteries, and on its E. side is a citadel, which commands the town and port. The space, comprising the city and a small surrounding district, to which the franchise of the port extends, is bounded by a rampart. Though it cannot be called a manufacturing town, Odessa has some fabrics of coarse woollen and silk goods; and has extensive tallow refineries, breweries, distilleries, rope-walks, &c.

Among the public buildings may be specified the church or cathedral of St. Nicholas, with a cupola, the exchange, palace of the governor, theatre, barracks, Roman Catholic church, a hospital two stories in height, a large and fine building, with public baths, large hotels, &c. On the quay facing the port, in the centre of the esplanade, is a statue in bronze in honour of the Duc de Richlieu, to whose enlightened administration much of the prosperity of the city is ascribable. Of the various institutions which the city owes to the duke, the Lyceum, which bears his name, founded in 1817, is one of the most important. Its organization has been modified of late years; and at present it is divided into the faculties of philosophy and jurisprudence, and has attached to it a gymnasium with four, and a primary school with three classes. In 1835 it had 259 pupils, and a library with about 7000 vols. There are also schools for the education of young ladies, founded in 1829 and 1835; a Jews' school, attended by about 400 pupils; an institution for the the study of the eastern languages; schools of navigation and commerce; an orphan school, &c. The inhabitants as in other commercial towns that have had a rapid rise, are a very motley race, consisting of Russians, Greeks, Jews, Poles, Italians, Germans, French, &c.

In 1817, a ukase conferred on Odessa, for a period of 30 years, the important privilege of being a free port; and her commerce has since rapidly increased. Not being at the mouth of any great river, nor having any considerable manufactures; she is not a port for the exportation of what may be called articles of native growth; but in consequence of her convenient situation, excellent port, and the privileges

which she enjoys, she is, as already remarked, the emporium where most of the produce of southern Russia destined for foreign countries is collected for exportation, and where most of the foreign articles required for home consumption are primarily imported. The shallowness of the water at Taganrog, and the short period during which the sea of Asoff is navigable, tends to hinder foreign vessels of considerable burden from entering the straits of Yenikale, and occasions the shipment of a considerable portion of the produce brought down the Don in lighters to Caffa and Odessa, especially the latter. All the products brought down the Dniestr, the Bug, and the Dniepr, are exported from Odessa; but owing to the difficult navigation of the first and last mentioned rivers, by far the greater part of the corn brought to Odessa from Podolia, and the Ukraine, &c., is conveyed to the town in carts drawn by oxen. The roads traversed by these carts are only practicable at certain seasons of the year; and nothing would contribute so much to increase the commerce of the port, and the prosperity of S. Russia, as the opening of improved communications with the interior; whether by removing obstructions in the channels of the rivers, constructing canals, or railways, or good common roads.

Among the articles of export from Odessa, corn, especially wheat, occupies, as every one knows, a high rank; but tallow is also a most important article; and next to it are linseed, wool, iron, hides, copper, wax, caviar, potash, beef, furs, cordage, sail-cloth, tar, butter, isinglass, &c. We subjoin an official

Account of the Value of the Exports from, and Imports into, the Port of Odessa in the different Years from 1802 to 1839 inclusive.

Years.	Exports. Rbls.	Imports. Rbls.	Years.	Exports. Rbls.	Imports. Rbls.
1802	1,554,000	719,600	1825	30,025,000	
1803	9,339,000	1,125,500	1826	14,351,000	
1805	3,980,000	2,184,000	1827	18,419,000	
1812	1,854,000	2,465,400	1828	1,978,000	
1814	7,125,000	4,596,000	1829	9,754,000	
1815	11,664,000	4,316,000	1830	27,891,000	
1816	37,717,000	4,291,000	1831	90,659,000	
1817	41,506,000	19,210,000	1832	35,480,000	
1818	20,534,000	11,191,000	1833	24,500,000	
1819	15,223,000	7,748,000	1834	15,273,000	
1820	16,541,000	7,729,000	1835	23,341,000	
1821	16,085,000	6,085,000	1836	34,263,000	
1822	13,034,000	7,710,000	1837	33,426,000	
1823	15,910,000	8,551,000	1838	38,256,000	
1824	13,039,000	6,946,000	1839	45,000,000	

The great amount of the exports in 1816 and 1817 is ascribable to the failure of the corn crops in Italy and western Europe during those two years, and to the consequent high price, and ready demand for the wheat of Odessa, Dantzic, &c. The small amount of the trade in 1828, and 1829 is accounted for by the war with Turkey having interrupted all communication with the port by the channel of Constantinople. Subjoined is an

Account of the arrivals of wheat at Odessa since the beginning of the present century.

1801	arrived	53,142 chetwerts.*
1802	—	285,106 —
1803	—	494,838 —
1804–13	—	1,898,367 average 184,836 chets. a year.
1814–23	—	6,800,000 average 680,000 chets. da.
1824–33	—	7,279,000 average 729,900 chets. do.
1834	—	691,000 —
1835	—	378,100 —
1836	—	878,700 —
1837	—	950,498 —
1838	—	1,941,000 —
1839	—	1,159,000 —
1840	—	680,000 —

The small amount of wheat brought to Odessa in 1835 was owing to the almost total failure of the crops in S. Russia in the course of that year. The price of the best wheat in Odessa is rarely under 25s. or 28s. a quarter; and, during the three years ending with 1840, it was 34s. 6d. free on board; the freight and other charges on importing a quarter of wheat from Odessa to England may be estimated at about 16s. a quarter. Constantinople, Genoa, Leghorn, Marseilles, and Malta are the principal markets for Odessa wheat; but, when our crops are deficient, considerable quantities are shipped for England. In 1839, the exports from Odessa comprised, among other things, 1,910,232 chetwerts of wheat, nearly 900,000 do. rye, oats, &c., 155,194 do. linseed, 115,669 pouds of wool, and 933,192 pouds tallow. The subjoined account of the merchants belonging to Odessa inscribed in the different guilds affords an additional illustration of the progress of its commerce since 1808.

* A chetwert is equivalent to about 6 imperial bushels, or 3-4ths of a quarter.

504

	In 1st Guild.	2d Guild.	3d Guild.	Total.
In 1808	39	30	135	204
1813	53	13	190	256
1823	87	11	213	311
1833	64	34	304	342
1838	67	64	644	765

The great articles of import into Odessa consists of sugar, coffee, and other colonial products; cottons, silks, woollens, and other manufactured goods; oil, wines, and spirits; spices, and dye-stuffs; cotton-twist and raw cotton; lemon-juice; tin and tin-plates; cutlery, timber for building and firewood, &c. About 800 ships from foreign parts enter and leave Odessa in ordinary years. Including Maltese and Ionians, nearly 900 ships under English colours has arrived in the port in a single season; but their number is very fluctuating, depending essentially on the state of the corn trade.

A tribunal of commerce was established at Odessa in 1804, whose jurisdiction extends over all disputes connected with trade. There is no appeal from its decisions except to the senate. There are 12 sworn brokers, approved and licensed by the tribunal of commerce, who have deputies appointed by themselves. They register all transactions, and receive one half per cent. from each party as commission. There is a discount or loan bank, established in 1826, and marine and fire insurance societies. Most articles of provision are cheap; and fish, which cost next to nothing, is excellent. Fuel, however, is scarce and dear. (*Official Reports; Private Information; Schnitzler, La Russie,* p. 794, &c.)

ODEYPOOR, or OUDEPORE, a city and rajahship of Hindostan, prov. Rajpootana, the city standing in a basin surrounded with rugged hills, 135 m. S.S.W. Ajmere, and 165 m. N. W. Oojein; lat. 24° 35′ N., long. 73° 44′ E. It has, at a distance, an imposing appearance. On the W. it skirts a large lake, the palaces and garden residences on the brink of which are all of marble, with sculpture that are both highly finished and display considerable taste. (*Hamilton.*) It is protected from inundation by an extensive embankment stretching along the lake. Images, toys, and other articles in marble, crystal, &c., are sent from Odeypoor into the neighbouring provinces. The rajahship, or principality, of which this city is the capital called also Mewar, or Chittore, holds a high rank among the Rajpoot states. It has N. Joudpoor; E. the territories of Kotah, Sindia, &c.; S. many small principalities of Malwah, Guzrah, &c.; and W. Carowy. Area estimated at 11,784 sq. m.; and population at 300,000 ?

The surface is hilly and well watered, producing sugar, indigo, tobacco, rice, wheat, barley, &c. Fuel is abundant; and there are mines of iron, copper, lead and sulphur; the last mentioned product being, however, of inferior quality. The pop. consists principally of Rajpoots, Jauts, Brahmins, Bheels, and Meenas. The rana, or chief, claims to be of the purest dynasty in India, and is held in great reverence by the Mohammedans, because of his supposed descent from the Persian sovereign, Noshirvan. In prosperity and power, however, this state is much inferior to those of Jeypoor and Joudpoor; and for a lengthened period previously to its becoming subsidiary to the British, it had been wretchedly mismanaged. The treaty of 1818 secured to the British, as the price of their protection, three eighths of the public revenue.

Chittore, the ancient cap., is the only other town in this principality worth notice. It is on the summit of a scarped rock, 68 m. E.N.E. Odeypoor. Heber says, "It is still what would be called in England a tolerably large market town, with a good many pagodas, and a meanly built, but apparently busy bazaar." It was, however, formerly famous for its splendour and riches, and has many interesting Hindoo temples, palaces, and other buildings. It was several times taken by the Mohammedans, and by Akbar, after a siege, an account of which is given in the second volume of the *Miscell. Trans. from Oriental Languages.* See also *Dow's Hist. of Hindostan,* ii., 356,357; *Heber's Journal,* 274–284.

ODIHAM, a market town and par. of England, co. Southampton, hund. of its own name, 21 m. N.W. Winchester, and 40 m. W.S.W. London. Area of par., 7550 acres. Pop., m 1831, 2647. It is pleasantly situated on the N. side of a chalk down, and comprises a principal and well built street, met by two others of inferior size. The church, a large brick structure, has a square tower at its W. end; the living is a vicarage, in the gift of the chancellor of Salisbury Cathedral. It has, also, two places of worship for dissenters, with attached Sunday-schools, a free school for 20 boys, and a large national school, and alms-house for 12 poor persons. Odiham has a considerable retail trade, and some of the inhab. are supported by spinning worsted and winding silk. It also derives some advantages from its situation on the Basingstoke canal. Petty sessions are held here, and

43

it is one of the polling-places at elections for the N. division of Hampshire. Markets on Friday; cattle fairs, March 23 and July 31.

About one mile from the town are the remains of an old castle, in which David I., king of Scotland, was imprisoned; and close to the town is a ruined gate, the only existing portion of a royal palace. Lilly, the celebrated grammarian and first master of St. Paul's school, London, was a native of Odiham.

ODENBURG (Hung. *Soprony,* an. *Sopronium*), a royal free town of Hungary, cap. co. of its own name; in a wide and fertile plain near the S.W. border of the Neusiedl lake, 49 m. W. Raab, and 37 m. S.S.E. Vienna. Pop. in 1837, 12,500, principally of German extraction. It is generally well built: the town-proper, which is not extensive, is regularly laid out, and tolerably well paved; and the suburbs are in every respect much superior. The only relic of its ancient fortifications is a huge watch tower, which, according to the Austrian *Nat. Encycl.,* is the loftiest in Hungary. It has several Roman Catholic churches, some of which are interesting specimens of Gothic architecture; a Calvinist church; Dominican and Ursuline convents; Roman Catholic and Lutheran high schools, two hospitals, two large barracks with a good riding school, a military academy, theatre, &c. It is the residence of the superintendant of the Calvinist church for Hungary beyond the Danube. The inhabitants refine sugar, weave cotton and woollen goods, manufacture potash, and saltpetre: and trade in wine grown in the vicinity), corn, tobacco, wax, honey, and cattle, for which it is an extensive market. Numerous Roman antiquities have been discovered in its vicinity. (*Oesterr. Nat. Encyc.; Berghaus, &c.*)

OELAND, an island of the Baltic, belonging to Sweden, near its S.E. extremity, being separated from the prov. of Calmar, in which it is included, by the straits of Calmar, a channel varying from 2 to about 20 m. in breadth. It is long and narrow, extending between lat. 56° 13′ and 57° 22′ N., and long. 16° 20′ and 17° 10′. Area estimated at 300 sq. m.; and the population at 31,000. (*Horschelmann's Stein.*) The W. shore of the island is low, the E. hilly; in the centre is a plateau, elevated about 150 ft. above the sea, principally of a calcareous or sandy formation. It is principally appropriated to pasturage, only a small portion of the land round the coast being under culture. Fishing and navigation form the principal occupations of the inhabitants, who send their fish, butter, cattle, &c., to the mainland, receiving corn, manufactured goods, &c., in return. The forests are rather extensive; and the deer, roebuck, and wild boar, are pretty abundant. About 300 hands are said to be employed in alum mine, the produce of which is supposed to be worth 50,000 dollars a year. Borgholm, on the W. side of the island, is its chief town and seat of commerce. A royal edict of 1820 conferred the freedom to pursue any trade or calling, without authority from any guild or company, on all handicraftsmen settling in this town. (*Horschelmann's Stein,* i. 589; *Dict. Géog.*)

OELS, a town of Prussian Silesia, gov. Breslau, cap. circ., and principality of Oels; on the river of the same name, a tributary of the Oder, 17 m. N.E. by E. Breslau. Pop., in 1838, 5680. It was formerly fortified, but is now merely enclosed by a lofty wall. It has a large ducal castle, in which are some extensive collections in art and science, several churches and hospitals, a theatre, and numerous public schools and charitable institutions. It has manufactures of woollen and linen fabrics. (*Von Zedlitz, Der Preuss. Staat, &c.*)

OESEL, an island of the Baltic, belonging to Russia, and included in the gov. Livonia or Riga, extending across the mouth of the gulf of Riga, principally between lat. 56° and 58° 40′ N., and long. 21° 40′ and 23° E. Area, estimated at 1150 sq. m. Pop., including the inhabitants of the adjacent islands of Moen and Runoe, about 35,000, all Esthonians except some German landed proprietors, and a few Swedes. The coasts are bold; the island is well watered, and its climate is milder than that of the neighbouring continent. The soil is mostly stony, calcareous, or loamy; but with manuring it becomes tolerably fertile, producing wheat, rye, barley, oats, peas, &c.: a considerable extent of the surface is covered with forests. Rearing cattle and fishing are the principal occupation of the inhabitants; and the seal fisheries are of some importance. Manufactures quite insignificant. People all Lutherans. Arensburg, on the S.E. coast, with about 1800 inhabitants, is a bishop's see, and the centre of the commerce of Oesel. This island belonged to the Teutonic knights, when their order possessed Livonia; it afterward belonged to Denmark and Sweden, but was ceded to Russia with the rest of Livonia in 1721. (*Schnitzler, La Russie; Dict. Géog. Universelle.*)

OFEN. See BUDA.

OFFENBACH, a town of Central Germany, being the principal manufacturing town of the Grand Duchy of Hesse-Darmstadt, prov. Starkenberg, on the Mayn, 5 m. E. by S.

2 1*

505

Frankfort, and 17 m. N. by E. Darmstadt. Pop., in 1838, 7600. It is well built, and has a castle, four churches, and a synagogue. Its manufactures consist of silk and cotton stockings; cotton fabrics; carriages, and other vehicles; tobacco and snuff; lacquered iron ware, sealing-wax, jewellery, toys, umbrellas and parasols, a few carpets, and other woollen fabrics, &c. Next to Mayence (*Mentz*) it has the largest general trade of any town in the Grand Duchy. Some good wine is grown in its environs. (*Berghaus, &c.*)

OGDENSBURG, p. v., port of entry, Oswegatchie t., St. Lawrence co., N. Y., 210 m. N.N.W. Albany, 477 W. Situated on the S.E. side of St. Lawrence river, at the mouth of Oswegatchie river. It lies on a beautiful plain, and is regularly laid out and well built. It contains five churches, an Episcopal, Presbyterian, Methodist, Baptist, and Roman Catholic; an academy, a bank, with a capital of $100,000; three forwarding houses, 60 stores and groceries, one flouring-mill, one saw-mill, one furnace, a machine shop, two tanneries, one distillery, one brewery, one carding machine and fulling-mill, 425 dwellings, and 2526 inhabitants. A steamboat plies to Prescott, in Canada, on the opposite side of the river. Several steamboats owned here ply to lake Ontario, and one goes down the river 40 m., to the long rapids. A dam across Oswegatchie river, has a fall of 14 feet. A canal, 1900 feet long, with a lock of 12 feet lift, connects the river below with the pool above the dam. A canal has been projected from this place to lake Champlain, 130 m. distant, but nothing has been done towards it, though it is thought to be feasible.

OGLE, county, Ill. Situated in the N. part of the state, and contains 625 sq. m. Watered by Rock river, and its branches. It contained in 1840, 3333 neat cattle, 597 sheep, 10,329 swine; and produced 69,250 bushels of wheat, 106,868 of Indian corn, 41,850 of oats, 46,130 of potatoes, 300 pounds of tobacco, 370 of sugar. It had 10 stores, one flouring-mill, three grist-mills, 12 saw-mills, two distilleries; one academy; seven schools, 215 scholars. Pop. 3497. Capital, Oregon city.

OGLETHORPE, county, Ga. Situated towards the N.E. part of the state, and contains 490 sq. m. Bounded N.E. by Broad river, by branches of which it is drained. It contained in 1840, 11,812 neat cattle, 7479 sheep, 27,461 swine; and produced 33,919 bushels of wheat, 1180 of rye, 490,510 of Indian corn, 66,229 of oats, 19,496 of potatoes, 1002 pounds of tobacco, 2,639,655 of cotton. It had 19 stores, five flouring-mills, 23 grist-mills, 13 saw-mills, one tannery, one distillery; three academies, 107 students; 10 schools, 278 scholars. Pop. whites, 4506; slaves, 6331; free coloured 31; total, 10,868. Capital, Lexington.

OHIO, the northeasternmost of the north-western states, is between 38° 30' and 42° N. lat., between 80° 35' and 84° 42' W. long., and between 3° 34' and 7° 41' W. long. from W. It is about 200 m. from N. to S., and 220 from E. to W., containing about 44,400 sq. m., or 28,416,000 acres. It is bounded N. by Michigan and lake Erie; E. by Pennsylvania and Virginia, from which latter it is separated by Ohio river; S. by Ohio river, which separates it from Kentucky; and W. by Indiana. The Ohio river, by its various windings, bounds this state for a distance of 436 m. The population in 1790 was 3000; in 1800, 45,365; in 1810, 230,760; in 1820, 581,434; in 1830, 937,637; in 1840, 1,519,467; being the third in population in the United States. Of these 775,360 were white males, 736,762 were white females; 8740 were coloured males, 8602 were coloured females. Employed in agriculture, 272,579; in commerce, 9201; in manufactures and trades, 66,265; in navigating the ocean, 919; do. rivers, canals, and lakes, 3393; in mining, 704; in the learned professions, 5663.

The interior and northern parts of the country bordering upon lake Erie are generally level, and in some places marshy. About a quarter or a third of the eastern and southeastern part of the state bordering on Ohio river, is very hilly and broken. There is nothing in the state which deserves the name of a mountain; but the state is an elevated table land, rising from 600 to 1000 feet above the level of the sea. A ridge of slightly elevated high lands divides the waters which enter lake Erie from those which flow into the Ohio, which is situated much nearer to the lake than to Ohio river; and the waters which flow into lake Erie are more rapid in their course, and more frequently broken by falls than those which flow in the opposite direction. The interior of the state which borders on the Scioto river, which divides the state into two nearly equal parts, and on the Great and Little Miami rivers, contains the most extensive bodies of level and fertile land in the state. On the head waters of the Muskingum and Scioto rivers, and between the sources of the two Miamis, are extensive prairies. On these prairies no timber grows, excepting occasionally a few scattering trees. Some of these prairies are low and marshy, while others are elevated and dry. The latter are frequently called barrens, but not always from

their sterility; for they are often tolerably fertile. On the dry prairies the grass grows, but not luxuriantly; but the wet prairies yield spontaneously a coarse grass, from two to five feet in height, which is of a tolerably good quality. The forest trees are black walnut, oak of various species, hickory, maple of different kinds, beech, birch, poplar, sycamore, ash of several species, papaw, buckeye, cherry, and whitewood, which last is extensively used as a substitute for pine, that is very scarce. Wheat may be regarded as the staple production of the state, and immense quantities are annually exported. Indian corn is also extensively produced, and from 70 to 100 bushels to the acre have been frequently raised. Rye, oats, buckwheat, and potatoes are also abundantly produced. There is very little waste land in Ohio. Nine tenths of the surface is susceptible of cultivation, and nearly three fourths of it are eminently fertile. The traveller is surprised on passing through the state to see so large a portion of the original forest undisturbed. There is no doubt that the state could support three times its present population without any difficulty, and much more than that, if it were necessary.

The summers are warm and pretty uniform, but subject at times to severe drought. The winters are generally mild in the southern part, though attended with cold and piercing winds on the borders of lake Erie. For the distance of 30 m. from lake Erie, there are generally several weeks of good sleighing in the winter; but in the south part the quantity and duration of snow are not sufficient to produce much good sleighing. The climate is generally healthy, excepting in the neighbourhood of low and marshy ground, where fevers and agues, and bilious fevers, sometimes prevail.

The state is divided into 79 counties, which, with their population in 1840, were as follows:

Counties.	Pop. 1840.	Counties.	Pop. 1840.
Adams	13,163	Licking	35,096
Allen	9,079	Logan	14,015
Ashtabula	23,721	Lorain	18,467
Athens	19,108	Lucas	9,382
Belmont	30,901	Madison	9,025
Brown	22,715	Marion	18,352
Butler	28,173	Medina	18,352
Carroll	18,108	Meigs	11,452
Champaign	16,721	Mercer	5,277
Clark	16,402	Miami	19,480
Clermont	23,106	Monroe	18,551
Clinton	15,719	Montgomery	31,412
Columbiana	40,378	Morgan	20,412
Coshocton	21,590	Muskingum	38,749
Crawford	13,152	Ottawa	2,248
Cuyahoga	26,506	Paulding	1,034
Darke	13,282	Perry	19,344
Delaware	22,060	Pickaway	19,725
Erie	12,669	Pike	7,536
Fairfield	31,924	Preble	19,482
Fayette	10,984	Portage	22,965
Franklin	23,919	Putnam	5,189
Gallia	13,444	Richland	44,532
Geauga	16,297	Ross	27,460
Greene	17,528	Sandusky	10,182
Guernsey	27,743	Scioto	11,192
Hamilton	80,145	Seneca	18,139
Hancock	9,986	Shelby	12,134
Hardin	4,598	Stark	34,603
Harrison	20,099	Summit	22,560
Henry	2,503	Trumbull	38,107
Highland	22,269	Tuscarawas	25,631
Hocking	9,741	Union	8,488
Holmes	18,088	Van Wert	1,577
Huron	23,933	Warren	23,141
Jackson	9,744	Washington	20,823
Jefferson	25,030	Wayne	23,333
Knox	29,579	Williams	4,465
Lake	13,719	Wood	5,357
Lawrence	13,719	Total	1,519,467

Columbus, on Scioto river, just below the confluence of Olentangay or Whetstone river, near the centre of the state, is the capital.

There were in this state in 1840, 430,597 horses and mules, 1,217,874 neat cattle, 2,028,401 sheep, 2,099,746 swine; poultry was produced to the value of $551,193. There were produced 16,571,661 bushels of wheat, 814,205 of rye, 33,668,144 of Indian corn, 212,440 of barley, 633,139 of buckwheat, 14,363,103 of oats, 5,805,021 of potatoes, 3,895,313 pounds of wool, 5,943,275 of tobacco, 6,363,386 of sugar, 69,195 of hops, 38,050 of wax, 4317 of silk cocoons, 1,022,037 tons of hay, 9,080 of hemp or flax; the products of the dairy were valued at $1,848,869; of the orchard at $475,371; of lumber at $292,891. There were made 11,524 gallons of wine; 6809 tons of pot and pearl ashes.

Salt springs have been found on Yellow creek in Jefferson county; on the waters of Killbuck creek, in Wayne county; on Muskingum river, near Zanesville, and at various other places. Bituminous coal is found in great quantities in the eastern part of the state, particularly near Massillon in Stark county, and in Tallmadge in Summit county. This coal is delivered to consumers in Cleveland at 15 cents a bushel. Iron ore is found in various places,

OHIO.

particularly at the falls of the Licking, 4 m. W. of Zanesville, and on Brush creek, in Adams county.

Ohio river, which gives name to the state, washes much of the eastern, and the entire southern border. This river is 959 m. long from Pittsburgh to its mouth in the Mississippi, though in a straight line it is only 614. It has a gentle current, and no fall excepting a rocky rapid of 22½ feet descent in 2 m., at Louisville. Around this is a canal, 2½ m. long, with four locks, sufficiently capacious to admit steamboats of the largest class. For about half the year, the Ohio has a depth of water admitting of navigation by large steamboats through its whole course. The Muskingum is the first river which flows wholly in this state. It is formed by the junction of the Tuscarawas and Walholding rivers, and flows into the Ohio at Marietta. It is navigable for boats 100 m., to Coshocton, and is 225 yards wide at its mouth. Scioto, the next river in magnitude, is 200 m. long, and enters Ohio river at Portsmouth. It is navigable for boats 130 m., and is 150 yards wide at its mouth. Its largest branch is Olentangy or Whetstone river, which enters it immediately above Columbus. The Great Miami river is a very rapid stream, 100 miles long, and enters the Ohio in the S.W. corner of the state. It is 200 yards wide at its mouth. The Little Miami is 70 m. long, and enters the Ohio 7 m. above Cincinnati. The Maumee rises in Indiana, is 100 m. long, and enters lake Erie in Maumee bay. It is navigable for steamboats 18 m. to Perrysburgh, and above the rapids is boatable for a considerable distance. The Sandusky river rises in the northern part of the state, and after a course of 80 m., enters Sandusky bay of lake Erie. The Cuyahoga river rises in the N. part of the state, and after a broad curve to the S., flows N. into lake Erie at Cleveland. It has a number of fine falls, and affords valuable water-power. It is about 60 m. long, and forms the fine harbour of Cleveland at its mouth. Huron, Vermillion, Black, Grand, and Ashtabula rivers flow into lake Erie.

Cincinnati, on Ohio river, is much the largest and most commercial place in the state. Next in importance is Cleveland, at the mouth of the Cuyahoga river, at the northern termination of the Ohio canal. Dayton, Columbus, Steubenville, Zanesville, Chilicothe, Lancaster, Newark, Circleville, and Massillon, are considerable and flourishing places.

The exports of the state in 1841, were $793,114; and the imports were $11,318. This includes but a small portion of its trade, having relation only to its foreign commerce. There were in 1840, 53 commercial and 241 commission houses in foreign trade, with a capital of $8,998,900; 4805 retail dry goods and other stores, with a capital of $21,999,295; 2891 persons engaged in the lumber trade, with a capital of $373,968; 854 persons engaged in internal transportation, who, with 1061 butchers, packers, &c., employed a capital of $4,617,570.

House-made or family goods manufactured to the amount of $1,853,937; 130 woollen manufactories, and 206 fulling-mills, employed 935 persons, produced articles to the value of $685,757, with a capital of $537,965; eight cotton manufactories with 13,754 spindles, employed 946 persons, produced articles to the amount of $139,378, with a capital of $113,500; 72 furnaces manufactured 35,936 tons of cast iron, and 19 forges produced 7466 tons of bar iron, consuming 104,312 tons of fuel, employing 2968 persons, and a capital of $1,161,900; 434 persons produced 3,513,409 bushels of bituminous coal, with a capital of $45,595; 14 paper-mills employed 305 persons, and produced articles to the amount of $370,902, with a capital of $808,900; 21 persons manufactured hemp or flax, producing to the amount of $11,737, with a capital of $942; hats and caps were manufactured to the amount of $796,513, and straw bonnets to the amount of $3028, the whole employing 963 persons, and a capital of $369,637; 812 tanneries employed 1790 persons, and a capital of $957,393; 1180 other manufactories of leather, as saddleries, &c., produced articles to the amount of $1,986,146, with a capital of $917,245; 187 persons manufactured tobacco to the amount of $212,818, with a capital of $68,810; 99 potteries employed 199 persons, manufactured to the amount of $89,734, with a capital of $43,450, 858 persons produced machinery to the value of $675,731; 289 persons produced hardware and cutlery to the amount of $303,500; 70 persons produced three cannon and 2450 small arms; 37 persons manufactured the precious metals to the amount of $53,125; 589 persons manufactured other metals to the amount of $782,901; 70 persons produced drugs and paints to the amount of $101,880, with a capital of $126,335; 401 persons manufactured granite and marble to the amount of $256,131; 1469 persons produced bricks and lime to the amount of $712,927, two powder-mills employed 13 persons, and produced 222,500 pounds of gunpowder, with a capital of $18,000; 195 persons manufactured 3,603,036 pounds of soap, 2,318,456 pounds of tallow candles, and 151 pounds of wax or spermaceti candles, with a capital of $385,780; 390 distilleries produced 6,329,467 gallons of distilled spirits, and 59 breweries produced 1,422,584 gallons of beer, the whole employing 789 persons, and a capital of $693,119; 21 rope-walks employed 66 persons, and produced cordage to the amount of $29,750, with a capital of $37,675; 11 persons produced musical instruments to the amount of $8454, with a capital of $5000; 1490 persons manufactured carriages and wagons to the amount of $701,528, with a capital of $290,540; 536 flouring-mills produced 1,311,954 barrels of flour, and, with other mills, employed 4661 persons, producing articles to the amount of $6,868,213, with a capital of $4,931,034; vessels were built to the amount of $522,855; 1928 persons manufactured furniture amounting to $761,146, with a capital of $553,317; 970 brick or stone houses, and 2764 wooden houses were built, employing 6060 persons, and cost $3,776,893; 159 printing-offices, 41 binderies, nine daily, seven semi-weekly, and 107 weekly newspapers, and 20 periodicals, employed 1175 persons, and a capital of $446,790. The total amount of capital employed in manufactures was $18,905,957.

The university of Ohio at Athens was founded in 1821; the Miami university at Oxford was founded in 1809. These institutions have been publicly endowed with large grants of lands. Franklin college, at New-Athens, was founded in 1825; the Western Reserve college, at Hudson, was founded in 1826; Kenyon college, at Gambier (Episcopal), was founded in 1826; Granville college, at Granville (Baptist), was founded in 1832; Marietta college, at Marietta, was founded in 1834; the Oberlin collegiate institute, at Oberlin, was founded in 1834; Cincinnati college, at Cincinnati, was founded in 1819; as was also Woodward college, at the same place. Willoughby university, at Willoughby, is a medical institution, with a college charter. Lane theological seminary, at Cincinnati, was founded in 1829. There are theological departments in Kenyon, Western Reserve, and Granville colleges, and in the Oberlin institute; a Lutheran theological school at Columbus; and two medical and one law school at Cincinnati. At all these institutions, there were in 1840, 1717 students. There were in the state, 73 academies, with 4310 students, and 5186 common and primary schools, with 218,609 scholars. There were in the state 35,394 white persons over 20 years of age, who could neither read nor write.

In 1836, the Presbyterians had 247 ministers; the Methodists had 900 ministers; the Baptists had 170 ministers; the Lutherans had 47 ministers; the Episcopalians had one bishop and 25 ministers; the German Reformed had 26 ministers; and there were besides, a considerable number of Friends, Roman Catholics, and a few others.

In January 1st, 1840, there were in the state 36 banks and one branch, with an aggregate capital of $10,507,591; and a circulation of $4,607,137. The state debt of every description, in Dec. 1842, amounted to $19,947,325. The whole exports of the state exceed its imports by about $12,000,000. There is a state penitentiary at Columbus.

This state has some important works of internal improvement. The Ohio canal extends from Cleveland on lake Erie, 307 m., to Portsmouth on the Ohio. It has the following navigable branches: 14 m. to Zanesville, 10 m. to Columbus, 9 m. to Lancaster, 50 m. to Athens, the Walholding branch of 23 m., Eastport branch of 4 m., and one of 9 m. to Dresden. This important work was begun in 1825, and finished in 1839, at an expense of $5,000,000. The Miami canal extends from Cincinnati, 178 m., to Defiance, where it meets the Wabash and Erie canal, thus completing a second line of canal from lake Erie to Ohio river. This canal is estimated to cost $3,750,000. The whole distance from lake Erie is 265 m. The Mahoning canal extends from the Ohio canal 88 m., eight of which are in Pa., to Beaver river, and cost $764,372. The Sandy and Beaver canal extends from the Ohio canal at Bolivar, 76 m., to Ohio river, at the mouth of Little Beaver creek, and is but partially completed. The Milan canal extends from Huron, 3 m., to Milan, to which steamboats now ascend. The Mad river and Sandusky city railroad extends from Tiffin, 36 m., to Sandusky city, and is designed to be continued to Cincinnati; but is finished only 28 m. from Cincinnati, and the remainder of this great work is not likely to be soon completed. Other railroads have been projected.

The governor is elected biennially by the people, but is not eligible for more than six years in eight. He must have been a citizen of the United States for 12 years, and an inhabitant of the state for four years next preceding his election, and must be at least 30 years of age. The senate consists of 36 members, chosen biennially by the people, one half being elected annually, must be citizens of the United States, and have resided in the county or district for which they are elected for two years immediately preceding the election, and have paid a state or county tax. The house of representatives consists of 72 members, elected annually by the people, and must have resided in the county for which elected, for one year next preceding the election, and have paid a state or county tax. The judges of the supreme court, and of the courts of common pleas, are appointed by the joint ballot of the two houses of the general assembly, and hold their offices for seven years. The secretary of state, treasurer, and auditor, are elected in like manner, for three years. The general assembly meets annually at Columbus on the first Monday of December. All white male inhabitants of full 21 years of age, who have resided in the state for one year next preceding an election, and have paid, or are charged with a state or county tax, enjoy the right of suffrage.

The first permanent settlement in Ohio was made, April 7th, 1788, at Marietta, and the first judicial court was held in September of the same year, under an act of Congress passed in 1786. The next settlement was at Columbus, 6 m. above Cincinnati, in 1789. The next was made by French emigrants, at Gallipolis, in 1791. The next was made at Cleveland and Conneaut on lake Erie, in 1796, by emigrants from New-England. In 1799, the first territorial legislature met at Cincinnati, and organized the government. Early in 1800, Connecticut relinquished her jurisdiction over the Western Reserve, and received a title to the land, which she sold, to constitute her large school fund. In 1802, Ohio formed her state constitution, and was admitted to the Union. Thus, in a little over 50 years, this state has risen up, which in 1840, contained 1,519,467 inhabitants, with taxable property to the amount of $112,037,801. For the early events, see the accounts of Cincinnati and Cleveland.

OHIO, a large river of the United States, which separates the states of Virginia and Kentucky from Ohio, Indiana and Illinois is second in importance only to the Mississippi, is formed by the confluence of Alleghany river from the N., and Monongahela from the S., at Pittsburg, in the western part of Pennsylvania. The Alleghany river rises in Potter co., Pa., on the W. side of the Alleghany mountains, flows into the state of New-York, and returns into Pennsylvania, and is the most important tributary of the Ohio. It is navigable for boats of a hundred tons, and of a light draft to Olean, Cattaraugus county, New-York, 270 m. from its mouth in the Ohio, 600 feet above the level of the river at Pittsburg, 1920 feet above the level of the ocean, and 2500 m. from the gulf of Mexico. The Monongahela rises in Virginia, and where it unites with the Alleghany, is more than 400 yards wide. It is navigable at a good stage of the water for large boats, 100 m. from its mouth. The The Alleghany, though not larger than the Monongahela at the junction, is the more important stream. Immediately below the junction, the Ohio is over 600 yards wide, and is a placid and beautiful stream. At Pittsburg it is 680 feet above tide water; at the mouth of the Muskingum, 541 feet; at the mouth of the Scioto, 464 feet; at Cincinnati, 414 feet; at its mouth in the Mississippi, 300 feet. Its length from Pittsburg to Cairo, according to the Western Pilot, is 959 m.; but the distance in a direct course is about 614 m. Its average descent is not quite five inches in a mile. The French called it la belle rivière, or the beautiful river; but its name, according to Heckewelder, is derived from the Indian word Ohiopekhanne, meaning a very white stream, alluding to the white caps with which its gentle surface is covered in a high wind, omitting all but its first part for ease of pronunciation. The Ohio for some distance below Pittsburg is rapid, and the navigation interrupted at low water, by chains of rocks extending across the bed of the river. The scenery is exceedingly beautiful though deficient in grandeur, exhibiting great sameness. The hills, two or three hundred feet high, approach the river, and confine it on either side. Their tops have usually a rounded and graceful form, and are covered with the verdure of an almost unbroken forest. Approaching Cincinnati, the scenery becomes still more monotonous. The hills recede from the river, and are less elevated. Heavy forests cover the banks, and limit the prospect, but exhibiting a beautiful verdure, and often exuberant with blossoms. The river exhibits the same scenery, as we continue to descend it, except that the hills become less bold and rocky. Many villages and farm houses are passed, through the whole course of the river; but as the bottom lands on its immediate margin are liable to be overflowed, the inhabitants prefer to settle a little back from the river, so that the dwellings in view, do not correctly exhibit the population in the vicinity. Between Pittsburg and the mouth of the Ohio there are as many as 100 considerable islands, besides a great number of sand bars and tow-heads. These last are low sandy islands incapable of cultivation, and covered with willows. Some of the islands are of exquisite beauty, and furnish desirable situations for a retired residence. The principal tributaries of the Ohio are the Muskingum, Great Kanawha, Big Sandy, Scioto, Great Miami, Kentucky, Green, Wabash, Cumberland and Tennessee. The last three are the most important, of which the last is the largest. One remarkable circumstance respecting the Ohio as well as other western rivers, is its great elevations and depressions. In the summer and autumnal months, it often

dwindles to a small stream, affording limited facilities for navigation. Among the hills of Pennsylvania and Virginia, it is seen rippling over chains of rocks, through which a passage is barely afforded to boats of the lightest burthen. Farther down, sand bars either extend across the stream or project into the bed of the river. Steamboats are at sometimes grounded on the bars, where they are obliged to wait in peril, for the periodical rise of the river. The lowest water is generally in the months of July, August, and September. The melting of the snows in the spring, and heavy rains in autumn or winter fill the river to overflowing, and many of its islands and the bottoms on its margin are covered with water. These rises are generally gradual, and attended with no danger. As the waters rise, trade and navigation are quickened into activity; the largest steamboats, often of 600 tons burthen, now float in security. The average rise of the water from low water mark, is 39 feet, but in the year 1832 an extraordinary flood was experienced. The river began to rise early in February, and on the 18th of that month it was 63 feet above low water mark, and the lower parts of Cincinnati and Covington were flooded. The river here is 1006 feet wide, and the velocity of its stream at its height 6½ miles per hour. The water discharged by the rise of the river above low water alone, would fill a lake of one square mile in surface. 107 feet deep, in one hour. The surface drained by the Ohio and its numerous tributaries is about 77,000 sq. m.; and water 6 inches in depth on this surface would be sufficient to maintain the river at the above height and velocity for fourteen days. Such a flood as this has scarcely been known, since the first settlement of the country. There are no considerable falls in the river, excepting at Louisville, Ky., where it descends 22½ feet in the course of 2 miles. Even over these, boats pass in high water. But they have been obviated by a canal around them, which admits of the passage of the largest steamboats. The current of the Ohio is very gentle; at the mean height of the river the current is about 3 miles an hour, at high water it is more, but at low water not more than 3 miles. During five or six weeks in winter, the navigation is obstructed by floating ice. The Ohio and its tributaries have not less than 5000 miles of navigable waters. The following distances have been derived from the Western Pilot, and are doubtless correct. From Pittsburg to Steubenville, O., is 70 miles; to Wheeling, Va., 92 miles; to Marietta, O., 174 miles; to Gallipolis, O., 264½ miles; to Portsmouth, O., 349 miles; to Maysville, Ky., 397 miles; to Cincinnati, O., 455½ miles; to Lawrenceburg, Ia., 479½ miles; to Louisville, Ky., 587 miles; to New-Albany, Ia., 591 miles; to the mouth of Cumberland river, Ky., 900 miles; mouth of Tennessee river, Ky., 911½ miles; mouth of Ohio, 959 miles.

OHIO, county, Va. Situated in the N.W. part of the state, and contains 195 sq. m. Bounded W. by Ohio river, which extends along its border for 36 m. It contained in 1840, 4252 neat cattle, 96,669 sheep, 9651 swine; and produced 194,096 bushels of wheat, 2276 of rye, 393,337 of Indian corn, 1434 of buckwheat, 145,865 of oats, 43,058 of potatoes, 3909 pounds of sugar, 1,178,800 bushels of bituminous coal. It had 91 stores, four furnaces, one forge, two woollen factories, one cotton factory with 408 spindles, 12 flouring-mills, seven grist-mills, 11 saw-mills, four paper mills, seven tanneries, one distillery, three glass-houses, two potteries, three printing-offices, three weekly and two semi-weekly newspapers, and one periodical; two academies, 90 students; 29 schools, 999 scholars. Pop.: whites 12,842; slaves, 212; free coloured, 303; total, 13,357. Capital, Wheeling.

OHIO, county, Ky. Centrally situated toward the N.W. part of the state, and contains 576 sq. m. Bounded S. by Green river, and drained by its branches. It contained in 1840, 7028 neat cattle, 6511 sheep, 25,022 swine; and produced 30,646 bushels of wheat, 1905 of rye, 346,015 of Indian corn, 42,091 of oats, 6790 of potatoes, 954,780 pounds of tobacco. It had 17 stores, one cotton factory with 39 spindles, two flouring-mills, seven grist-mills, four saw-mills, four tanneries, three distilleries; one school, 25 scholars. Pop.: whites, 5747; slaves, 822; free coloured, 22; total, 6592. Capital Hartford.

OHIO t., Alleghany co., Pa., 11 m. N.W. Pittsburg. Drained by Great and Little Sewickly creeks. It has three stores, four grist-mills, seven saw-mills; two academies, 71 students; nine schools, 302 scholars. Pop. 1631.

OHIO, L. Beaver co., Pa., 10 m. S.W. Beaver. Watered by Little Beaver creek, flowing into Ohio river, which has ter bounds it on the S. It has four stores, one grist-mill, three saw mills; seven schools, 229 scholars. Pop. 1223.

OHIO CITY, p. v., Brooklyn t., Cuyahoga co., O., 145 m. N.N.E. Columbus, 360 W. Situated on lake Erie, at the mouth of Cuyahoga river, opposite to Cleveland. The ground on which it is built is uneven, but presents fine situations which overlook the lake, the city of Cleveland and the surrounding country. It contains an elegant Episcopal church of stone of the Gothic architecture, a Presbyterian and other churches, seven commission houses in foreign trade, seven retail stores, two furnaces; machinery manufactured to the amount of $40,000, and 1577 inhabitants. A bridge across the Cuyahoga connects it with Cleveland, with which place it enjoys a harbour in common. From its situation, it is likely to constitute virtually one great city with Cleveland.

OHLAU, a town of Prussian Silesia, gov. Breslau, cap. circ. on the Oder, 17 m. S.E. Breslau. Pop., in 1838, 4000. It was formerly one of the strongest fortresses of Silesia, but its works were, in great part, demolished after its cession to Frederick the Great, in 1741. It has a royal palace, with a gallery of paintings, several Lutheran and Roman Catholic churches, a hospital, orphan asylum, workhouse, manufactures of woollen cloth, &c. A good deal of tobacco is grown in its neighbourhood. (Von Zedlitz; Borghaus.)

OISE, a dep. of France, reg. N., formerly comprised in the Île-de-France; between lat. 49° 5' and 49° 45' N., and .ong. 1° 40' and 3° 10' E.; having N. the dep. Somme, E. Aisne, S. Seine-et-Marne and Seine-et-Oise, and W. Eure and Seine Inférieure. Length, E. to W., 63 m.; average breadth. about 35 m. Area, 582,560 hectares. Pop., in 1836, 397,725. Surface gently undulating. A range of hills traverses the dep., dividing the basin of the Somme from that of the Seine, but no summit rises to any considerable height. Principal rivers, the Oise, Terrain, and Epte, all of which have a S. direction. The Oise, whence the name of the dep., rises in the Belgian province of Hainault, near the frontier of Ardennes; and after a generally S.W. course of about 160 m., through the deps. Du Nord, Aisne, Oise, and Seine-et-Oise, joins the Seine at Conflans St. Honorien, about 12 m. N.W. Paris. It is navigable from Chauny in Aisne to its mouth, a distance of 75 m. Some pools and marshes exist in the E. and S.E. of the dep. In 1836, the arable lands of this dep. were estimated to comprise 359,486 hectares; meadows, 92,997 hect.; vineyards, 3501 hect.; orchards, gardens, &c., 15,388 hect.; forests, 80,578 hect. and heaths, waste, &c., 15,709 hect. In 1835, of 214,149 properties subject to the contrib. foncière, 109,027 were assessed at less than 5 francs; and 34,606 at from 5 to 10 francs. The number of large properties is, however, greater than in most deps. Soil principally calcareous, everywhere requiring manure; agriculture is considerably advanced, and is improving. Fallows are decreasing, and agricultural implements are made more effective. More corn is grown than is required for home consumption; it is principally oats and wheat. Peas, beans, &c., are raised in large quantities for the Paris markets. Pear and apple orchards are numerous, and a great deal of cider of good quality is made. Some wine is grown, but of indifferent quality. The rearing of cattle is an important branch of rural economy; and the fat calves, known in Paris as veaus de Pontoise, are from this dep. In 1830, the stock of horned cattle, calves, &c., was estimated at about 96,500 head, and that of sheep at 536,000. The latter have been improved by crossing with the merino, Southdown, and Leicester breeds, and yield annually about 800,000 kilogr. wool. (Hugo.) Butter and cheese, including the cheese of Songeons, are valuable products. Hogs and poultry are numerous. Mineral products, excepting limestone, are few, and of little importance. Oise is distinguished for its manufacturing industry. Woollen fabrics, especially at Beauvais and Crèvecœur; table-linen, cotton, and hempen cloths, woollen and cotton yarn, cotton stockings, lace, metallic and glass wares, and fans, horn, wooden and ivory articles at Méru, &c., are among the principal goods manufactured. This dep. is divided into four arronds.; chief towns, Beauvais, the cap., Clermont, Compiègne, and Senlis. It sends five members to the chamber of deputies. Registered electors in 1838–39, 3105. Total public revenue in 1831, 14,099,295 francs. (Hugo, art. Oise; Dict. Geog.; French Official Tables.)

OKTIBBEHA, county, Miss. Situated N.E. of the centre of the state, and contains 625 sq. m. Watered by Oktibbeha creek and other branches of Tombigbee river. It contained in 1840, 8447 neat cattle, 764 sheep, 10,042 swine; and produced 2475 bushels of wheat, 158,990 of Indian corn, 16,961 of oats, 10,808 of potatoes, 1900 pounds of tobacco, 1,634,444 of cotton. It had four stores, two grist-mills, three saw-mills; 13 scholars. Pop.: whites, 1984; slaves, 2197; free coloured, 15; total, 4276. Capital, Starksville.

OLBERA (an. Ilipa), a town of Spain, in Andalusia, prov. Seville, in a mountainous district, 48 m. S.E. Seville. 'op., acc. to Miñano, 6900. It is, acc. to Captain Scott, "a wretched place, containing some of the rudest-looking and most scrupulous inhabitants of the Serrania de Ronda." A ariah church, hospital, three decayed convents, and a Moorish castle, are its principal buildings; and the view of the last is very commanding over a great extent of mountains, intersected by well wooded valleys. A few

oil-mills are established here; but the population is almost wholly occupied in rearing hogs for the Seville market. (Miñano; Scott's Ronda and Grenada, ii., 151.)

OLD CODORUS, t., York co., Pa., 10 m. S.W. York. Drained by Codorus creek and its branches. It has four grist-mills, three saw mills, one tannery, 28 distilleries; three schools, 97 scholars. Pop. 1131.

OLDENBURG (GRAND DUCHY OF), a state of N.W. Germany, consisting exclusive of some portions of country inclosed by the Duchy of Holstein, of an oblong-shaped territory, between lat. 52° 30' and 53° 43' N., and long. 7° 35' and 8° 50' E.; having N. the North sea, and surrounded on all other sides by the Hanoverian dom., except on a small portion of its E. frontier, where it adjoins the territory of Bremen. Length, N. to S., 80 m.; breadth varying from 10 to nearly 50 m. It is divided into three provinces, as follows:

	Area in sq. m.	Pop., 1837.	Chief towns.
Oldenburg	2,104	217,087	Oldenburg.
Lubeck	166	20,082	Eutin.
Birkenfeld	143	27,651	Birkenfeld.
Total	2,413	265,370	

The natural features, climate, &c., of this duchy are similar to those of the adjacent kingdom of Hanover. (See p. 1080.) It is almost a perfect level, except towards the S., where are some hills, though none rises above 300 or 400 feet. The principal rivers are the Weser, on the N.E. boundary, its tributaries, the Hunto, Haase, Leda, Jahde, &c. There are many small lakes, the principal being the Drummersee, in the S. The coast is so low that dykes are necessary, as in Holland and Friesland, to prevent inundations of the sea. Here, and on the banks of the rivers, the soil is alluvial and rich; but in most parts of the grand duchy it is either sandy or sandy; and the country does not produce sufficient corn for home consumption, the deficiency being mostly made up by potatoes and pulse. The industry of the inhabitants is, however, principally rural; flax, hemp, hops, and rape-seed; together with cattle, horses, salt beef, butter, and bacon, are the chief exports. The horses and cattle are of superior breeds; large flocks of sheep are pastured on the heath lands, but their wool is of inferior quality. In this district, also, a good many bees are kept. Next to tillage and grazing, taking fish, with which the rivers abound, is a chief employment of the population. Timber, fit for ship-building and carpenter's work, grows in the hilly district in the S. of the duchy, where the forests are estimated to cover an extent of nearly 170 sq. m.; but in other parts the fuel used consists almost wholly of turf, which is very abundant in the marshes. Iron is the only other mineral product of much utility. The spinning of linen yarn, and the domestic weaving of linen and woollen stuffs, are the chief branches of manufacturing industry; but these are pursued only as auxiliary occupations by the agricultural population. Though the country produces oily seeds, animal fat, &c., in considerable quantities, neither candles, soap, nor oil are made to any extent, all being imported from foreign countries, to which the raw materials are sent. Neither is the trade of the grand duchy at all extensive; it has but a small seafaring population, and its commerce is principally confined to a coasting traffic with the neighbouring countries of Denmark, Hanover, Holland, Lubeck, &c.

The government is an unlimited monarchy, except in respect to the distribution of the taxes, which is under the control of the states, composed of deputies of the nobles, citizens, and peasantry. For administrative purposes, the Grand Duchy (exclusive of Lubeck and Birkenfeld) is divided into six circles and 28 districts, each of which has its own judicial courts. The court of chancery, and other high courts of appeal, are in Oldenburg or Jever, except for the principality of Lubeck, which has its own superior courts in Eutin. Total public revenue estimated at 850,000 rix dollars a year. The population is principally Lutheran, but there are about 68,600 Roman Catholics, and a few Calvinists and Jews.

Oldenburg holds the 10th place in the full diet of the Germ. Confed., in which it has one vote; and the 15th in the committee, in which it has a vote in conjunction with Anhalt and Schwartzburg. Its contingent to the army of the Confed. consists of 2229 men. The house of Oldenburg is connected with the reigning families of Denmark and Russia. The duchy was erected into a sovereign state in 1773, but Birkenfeld was not united to it till 1813.

OLDENBURG, a town of N.W. Germany, cap. of the above Grand Duchy, and residence of its sovereign, circle of same name, on the Hunte, a tributary of the Weser, 24 m. W.N.W. Bremen. Lat. 53° 8' 24" N., long. 8° 13' E. Pop., in 1837, 5564. It is fortified, and divided into the Old and the New Town, the latter being pretty well built. The ducal castle is an imposing building, with a fine park. The chancery-chamber, and other buildings for the use of the government, St. Lambert's church, in which the sovereigns of Oldenburg are interred, some other places of worship, the observatory, and the barracks, are the principal public edifices. It has a gymnasium, a military school, and a ducal library of 24,000 volumes. Its manufacturing industry is quite insignificant; but it has some trade in wool, timber, &c. (*Berghaus, Allg. Länder und Völkerk,* iv. 419-430; *Dict. Geog.*)

OLDHAM, a parl. bor., market-town, and township of England, par. of Prestwich, hund. Salford, co. Lancaster, near the source of the Irk, and not far from its junction with the Medlock, 6½ m. N.E. Manchester, and 162 m. N.N.W. London. Area of parliamentary borough (which includes, with Oldham, the townships of Chadderton, Crompton, and Royton), 11,180 acres. Pop., in 1841, 42,595. The town has entirely risen since 1760, when it comprised only about 60 thatched tenements; it now consists of many well-built streets, extending on the side of a hill on the road from Manchester to Leeds; and is well paved, lighted with gas, and well supplied with water. The chief thoroughfare runs from E. to W., and is crossed by two or three others in an opposite direction. The principal edifices are the town-hall, built in 1810, a small theatre, the "Terrace Buildings," comprising a public room, a market-house, &c., and a large gas establishment. The church, which is subordinate to that of Prestwich, is a modern Gothic structure, completed in 1830, at a cost of £22,000; there are also two chapels of ease within the town, the livings of which are in the gift of the rector of Prestwich. There is a Rom. Catholic chapel; and 11 other places of worship belong to different denominations of dissenters, among whom Methodists are the prevailing body: and attached to the churches and chapels are numerous Sunday schools, which, in 1839, furnished religious instruction to 12,827 children of both sexes. The town has a small endowed grammar school, and a large blue-coat school, founded, in 1807, by the late Mr. Henshaw, hat-manufacturer. This school, however, owing to a long chancery suit respecting the property, which amounted to £40,000, was not opened till 1833. The schoolhouse is at Oldham-edge, and the establishment supports, clothes, and educates 110 boys. There are likewise two large national schools, and several Lancastrian as well as infant schools, wholly, or in part, supported by subscription. The other public institutions consist of three benevolent societies, a bible and tract association, subscription library, lyceum (with schools, news-rooms, &c.), and two mechanics' institutes.

Oldham owes its present consequence entirely to the cotton manufacture, which it was early a considerable seat. In 1785 there were within the chapelry six cotton-mills; but such and so rapid has been the increase of the manufacture in the interval, that in 1839 it had 200 manufactories, set in motion by a steam-power equal to 2942 horses, and employing 15,291 hands. It had then, also, 200 hand-loom weavers, chiefly of silks similar to those manufactured in Manchester. (*Factory, Comm. Rep.*) Hat-making (once the staple manufacture of Oldham) is also very extensively carried on, employing 2800 hands; and about 1400 men are engaged in collieries, that have been opened during the last few years within the chapelry. The beds vary from 3 to 5 ft. in thickness; the coal is of excellent quality, and furnishes the chief supplies for Manchester, Ashton, Rochdale, and other manufacturing towns, with which it is connected by the Oldham and Rochdale canals, which latter also communicates with the Ayr and Calder navigation. The Manchester and Leeds railway passes within 2 m. of Oldham, and a branch railway up to the town is in course of construction. Besides the Oldham banking company, it has branches of one or two other banks, and a savings' bank. The affairs of the township are regulated by commissioners, according to an act passed in 1826. Petty sessions are held twice a week; and there is a court of requests for the recovery of small debts, once a month. The Reform Act conferred on Oldham, for the first time, the important privilege of returning two members to the House of Commons. The electoral limits comprise with the townships three out-townships, as above mentioned. Registered electors, in 1839-40, 1402. Markets on Saturday: fairs May 2d, July 8, and first Wednesday after October 12. (*Butterworth's Stat. of Lanc.; Baine's Hist. of Lancaster; Parl. Rep.*)

OLDHAM, county, Ky., situated in the N. part of the state, and contains 920 sq. m. Bounded N.W. by Ohio river. It contained in 1840, 5803 neat cattle, 9672 sheep, 19,144 swine; and produced 87,846 bushels of wheat, 2967 of rye, 419,997 of Indian corn, 72,761 of oats, 8163 of potatoes, 162,078 pounds of tobacco, 2191 of sugar. It had 14 stores, three flouring-mills, 11 grist-mills, 13 saw-mills, two tanneries, 16 distilleries, one pottery; one academy, 20 students; 14 schools, 330 scholars. Pop.: whites 4858; slaves, 2377; free colored, 145; total, 7380. Capital, La Grange.

OLD TOWN, p. t., Penobscot co., Me., 80 m. N.E. Au-

gusta, 675 W. Bounded E. by Penobscot river, and contains a part of Marsh island in Penobscot river. On an island in Penobscot river, about a mile above Old town, is the settlement of the remnant of the Penobscot tribe of Indians. The settlement is called Indian Old town, and contains 95 families and 388 individuals, and they own all the islands in the river as far up as the forks. They receive a considerable annuity from the state, for lands which they have sold. They have a small Roman Catholic church, with a bell, containing the Latin inscription, "Deus pro nobis, quis sit contra." They have a number of small dwellings for themselves. The island is fertile, and many shad are taken here in the spring of the year. The township contains 11 stores, one grist-mill, 35 saw-mills; eight schools, 699 scholars. Pop. 2342.

OLEAN, p. t., Cattaraugus co., N.Y., 293 W. by S. Albany, 90 m. S.E. Ellicottville, 307 W. Watered by Allegany river and Oil creek. It contains a Presbyterian and Episcopal church, five stores, one grist-mill, three saw-mills; two schools, 100 scholars. Pop. 638. The village on the Alleghany river is the point where the Genesee valley canal is to terminate, whence the river is navigable to Pittsburg.

OLERON (ISLE OF), an island off the W. coast of France, dep. Charente-Inférieure, opposite the mouth of the Charente, lat. 46° N., long. 0° 20' W., 7 m. S. the Isle of Ré, and 2 m. from the nearest point of the continent. Area estimated at about 100 sq. m. Pop., in 1836, 14,481. It is tolerably fertile, producing various kinds of corn, timber, red and white vines (a portion of which is converted into brandy), and considerable quantities of salt, from salt-pans along the coast. Château d'Oleron, the cap., on its S.E. side, is a fortified town, with a population of about 3600.

OLEY, p. t., Berks co., Pa., 62 m. E. Harrisburg, 135 W. Watered by head branches of Manatawny and Maxatawny creeks. It contains two churches, a German Presbyterian, and a Lutheran; five stores, two furnaces, three forges, six grist-mills, eight saw-mills, two oil-mills, two paper-mills, two tanneries. Pop. 1877.

OLIVA, a town of Spain, prov. Valencia, 38 m. S.S.E. Valencia, and 218 m. S.E. Madrid, built amphitheatre-wise on the side of a hill, 1½ m. from the Mediterranean, in a well watered and productive district. Pop., according to Miñano, 5587. Its chief public buildings are two churches, one of which has a collegiate establishment, a hospital, ancient palace, and two prisons. Its manufactures are confined to hemp and linen fabrics. In the neighbouring river Molinet are found fine eels and leeches, the latter of which are exported in considerable quantities to France. (*Miñano.*)

OLIVENZA, a town of Spain, prov. Estremadura, close to the frontiers of Portugal, and about 6 m. from the left bank of the Guadiana, 14 m. S.S.W. Badajoz, and 221 m. W.S.W. Madrid. Pop., according to Miñano, 76,447. It is surrounded with walls, and strongly fortified; it has three parish churches, seven convents (now applied to secular uses), three hospitals, and a poorhouse. The surrounding country, though very imperfectly cultivated, produces abundant crops of wheat, barley, and other grain, with pulse, wine, &c.; and the town is much resorted to by traders from Alentejo, who come to exchange manufactured goods for farm produce. Olivenza was attached to Portugal till 1801, when it was ceded to Spain, to which it is still attached, notwithstanding the order for its restitution by the Congress of 1815. (*Miñano.*)

OLIVER, t., Mifflin co., Pa. It contains three stores, one woollen factory, one furnace, one forge, three grist-mills, 10 saw-mills; 10 schools, 323 scholars. Pop. 1997.

OLMUTZ, a town of Moravia, of which it was formerly the cap., being one of the strongest fortresses of the Austrian dom., cap. circ. Olmütz, on the March, 40 m. N E. Brünn. Pop., in 1837 (ex. garrison), 12,782. It is well built, but the loftiness of its buildings darkens the streets. The cathedral is a fine Gothic edifice, in which its founder, Wenceslaus III. of Bohemia, is buried: some of the other churches also deserve notice. The archbishop's palace, deanery, town-hall, theatre, arsenal, barracks, a military hospital, a hospital for lying-in women and orphans, and a large conventual establishment comprise the other chief public buildings. Olmütz is the seat of a university, founded in 1581, and restored in 1827. In 1833 it had, in all, 682 students. It still possesses a library of about 50,000 printed volumes, and many hundred MSS.; it had formerly a rich library of Slavonic literature, but this was carried away by the Swedes, and lost towards the end of last century. It has a gymnasium, an Episcopal seminary, an academy of nobles, a military school, and numerous inferior schools, and is the residence of the high military authorities, and the seat of the superior judicial courts for the circle. It has some manufactures of woollen, linen, and cotton fabrics, earthenware, leather, and vinegar; and an active transit trade with the neighbouring Austrian provinces, Prussia, Poland, &c., especially in cattle. Olmütz was taken by the Swedes in the 30 years' war; and was besieged unsuccessfully by Frederick the Great in 1758. Lafayette was confined here in 1794. (*Oesterr. N.t. Encyc.; Berghaus.*)

OLNEY, a market-town and par. of England, co. Buckingham, hund. Newport, on the W. side of the Ouse, crossed here by a bridge of four arches, 16 m. E.N.E. Buckingham, and 50 m. N.W. London. Area of par. 3140 acres. Pop. in 1831, 2344. The town consists of one long street, lined with stone houses. The church is a large Gothic structure, with a spire 185 ft. high, seen from a great distance: the living is a vicarage in the patronage of the Earl of Dartmouth. The Baptists and Independents have their respective places of worship, which, as well as the church, have well attended Sunday schools. Almshouses for the necessitous females are supported at the sole expense of a benevolent Quaker lady. Lace-making was long the chief employment of the inhabitants; but it has been for many years declining. More recently silk-weaving and the manufacture of hosiery were attempted on a small scale, but they also have been abandoned. Olney derives its principal celebrity from its having been for a lengthened period the residence of the poet Cowper: the "substantial brick-built house" in which he resided still stands near the centre of the town, and the arbour in which he studied is in excellent preservation, and is an object of great attraction. The latter years of the poet's melancholy existence were not, however, passed here, but at East Dereham, in Norfolk. It is rather singular that though the vicarage of Olney be not worth £100 a year, it has been held by some rather distinguished persons, including Moses Browne, Scott the biblical commentator, and Newton the friend of Cowper.

OLONETZ, a gov. of European Russia, between the 60th and 65th degs. of N. lat., and the 30th and 43d of E. long.; having N. and N.E. the gov. Archangel, S.E. and S. Vologda, Novgorod, and Petersburg, and W. lake Ladoga and Finland. Area (including lake Onega) estimated at nearly 62,400 sq. m. Pop., in 1836, 239,000. The W. part of this gov. resembles Finland, in being alternately mountainous and marshy, or covered with lakes. Of the latter, Onega (which see) is by far the largest. Principal rivers, the Onega, by which the lake Latcha discharges itself into the White sea, Vodla, Svir, Suna, &c. For 23 weeks in the year the mean temperature is below 39° Fah., and mercury sometimes freezes. Bleak winds are almost constant; but the country is tolerably healthy. Soil thin, stony, and not very fertile. Except in the district of Kargopol, into which some improvements have been introduced, agriculture is very backward. The produce of corn, in 1832, was estimated at 269,000 *chetworts*, a quantity insufficient for the wants of the population. The peasantry are supported chiefly on turnips, carrots, and other vegetables, of which their bread partly consists, and on the produce of the chase, fisheries, &c. Hemp and flax are grown for exportation; but the principal source of wealth consists in the forests, which are of great extent, those belonging to the crown covering 6,956,795 deciatines. Pasturage is not abundant, and few cattle are reared. Marble, granite, serpentine, alabaster, &c., are found; and there are mines of iron, copper, and even silver, though they are but little wrought. The poverty of the country obliges many of the inhabitants to emigrate annually into the adjacent governments to take charge of cattle, hew millstones, &c.; and in summer the number of absentees is estimated at about a third part of the entire population. These circumstances are hostile to manufacturing industry; and, exclusive of the royal cannon foundry at Petrozavodsk, it has only a few tanneries and iron forges. It exports raw produce to Petersburg and Archangel; from which cities, corn, salt, spirits, and colonial and manufactured goods are imported. Oionetz is under the same military jurisdiction as Archangel. Its inhabitants are principally of the Greek church, and subordinate to the archbishop of Novgorod. Total public revenue estimated at only 298,110 rubles. (*Schnitzler, La Russie; Possart; Russland, &c.*)

OLORON, or OLERON (an. *Iluro*), a town of France, dep. Basses-Pyrénées, cap. arrond., on the summit and declivity of a hill behind the Oleron, across which it communicates with the town of Ste. Marie by a lofty bridge, 13 m. S.W. Pau. Pop., in 1836, 6037. It has a court of primary jurisdiction, a board of customs, and a chamber of manufactures; with manufactures of woollen cloths, yarn, hosiery, paper, and leather; and an active trade in French and Spanish wool, sheep-skins, *jambons de Bayonne*, and other salted meats, cattle, and timber. It is the general depôt for the timber of the Pyrenees destined for the dockyard of Bayonne.

OLOT, a town of Spain, in Catalonia, near the small river Fluvia, 53 m. N.N.E. Barcelona; lat. 42° 12' N., long. 2° 45' E. Pop., according to Miñano, 13,845. It is tolerably well built on level ground, at the foot of a range of volcanic

hills, and has several squares and streets adorned with fountains; its public buildings consist of two parish churches, cavalry barracks, and a hospital. It has considerable manufactures of cotton cloths and woollen caps, with extensive tanneries and soap factories, and some paper-mills. Well-attended markets are held twice a week; and Olot is one of the most thriving towns of Catalonia. Its neighbourhood is peculiarly interesting to geologists, on account of the extinct volcanoes with which it abounds, scattered over a tract measuring about 15 m. from N. to S., and about 6 m. from E. to W. Mr. Lyell, who visited it in 1830, says, "There are about 14 distinct cones with craters in the vicinity of Olot; and the largest, called Santa Margarita, is 455 ft. deep, and about a mile in circumference. These volcanic rocks, also, have often a cavernous structure; and at the base of the same hill, adjoining the town, are the mouths of about 12 subterraneous caverns, here called *bufadors*. In 1421 the whole of Olot, except a single house, was thrown down by an earthquake; but this calamity may, perhaps, be ascribed more to the cavernous nature of the subjacent rocks than to the extraordinary violence of the movements on that spot; for Catalonia is beyond the line of those European earthquakes which have within the period of history destroyed towns throughout extensive areas." (*Miñano; Lyell's Geology*, iv., 36–49.)

OMER (ST.), a strongly-fortified town of France, dep. Pas de Calais, cap. arrond., on the Aa, and at the union of several great roads, 40 m. N.W. Arras, and 29 m. E. by N. Boulogne. Lat. 50° 44' 46" N.; long. 2° 25' 18" E. Pop., in 1836, 18,789. It is partly built on a hill, but principally in the low and marshy plain at its foot. The circ. of its ramparts is about 2¼ m.; beyond its moats and glacis are several strong and extensive outworks; and from the town being half surrounded by marshes, the greater part of its vicinity may be readily laid under water. Its streets are broad and regular, but being lined generally with measlooking houses of yellow brick, it has a dull appearance. It is, however, well furnished with public fountains. The cathedral of Notre-Dame, an edifice completed towards the middle of the 15th century, is of Gothic architecture, and 373 ft. in length internally. In it are several colossal statues, a fine painting by Rubens, the tomb of St. Omer, a superior organ, &c. The abbey of St. Bertin, in which the last of the Merovingian kings died, was destroyed during the phrensy of the revolution, and only some ruins of its church exist. The college, formerly the Jesuit's church; the military hospital, occupying the building formerly a seminary for the English and Irish Roman Catholic clergy; the town-hall, arsenal, and powder magazines, several hospitals and prisons, the theatre, and some convents, are the other principal public buildings. The ramparts are planted with elms, and form fine promenades; as do the banks of the canal of Neuf Fossé, which connects St. Omer with Aire, and the Aa with the Lys. It is the seat of a sub-prefecture, and has courts of primary jurisdiction and commerce, a chamber of manufactures, a communal college, and a public library of 16,000 volumes. Its manufactures consist of common woollen cloths, woollen yarn, lace, basket-work, fishing-nets, soap, starch, glue, tobacco-pipes, &c.; it has also many distilleries, breweries, paper-mills, and tanneries, and an active trade in corn, wine, oils, flax, coal, &c. Beyond the walls are two suburbs, the inhabitants of which are principally gardeners. Near the town is a lake, on which are some curious floating islands, held together principally by the trees which grow on them, and affording pasturage for sheep and cattle.

This town was originally called Sithiu; it took its present name from St. Omer, who founded its cathedral about 645. It was walled at the end of the 9th century; and was long an object of contention between the Burgundians and French, to whom it finally fell in 1677. (*Hugo*, art. *Pas de Calais; Dict. Géog.*)

ONATE, a town of Spain, in Biscay, prov. Guipuzcoa, 28 m. E.S.E. Bilbao, and 194 m. N.N.E. Madrid. Pop., according to Miñano, 12,000. It stands on the side of a hill in the valley of its own name, and is well built with regular streets, most of them terminating in a large square, which has a remarkably fine town-hall, a parish church, with a tower 190 ft. high, and a large building with a Doric portico, formerly used as a convent of Jesuits: in the centre of the square is an elaborately ornamented fountain. There are two other parish churches, and several deserted convents, a well supported hospital, and a college of handsome architecture, attended by between 150 and 200 students. Iron is extensively wrought in the neighbouring mountains, and within the town are iron-foundries, nail factories, &c. The late civil war, however, which chiefly raged in the Basque provinces, gave a great shock to its industry, from which it is only slowly recovering. The surrounding district is extremely productive, and has numerous mineral springs and quarries of jasper and limestone. (*See* BISCAY.)

ONEGA (LAKE) a considerable lake of Russia, being

ONEIDA.   ONTENIENTE.

next to that of Ladoga, the largest in Europe, in the centre of the gov. of Olonetz, between lat. 60° 50′ and 62° 50′ N., and long. 34° 20′ and 36° 20′ E. Length, N.W. to S.E. 130 m.; breadth varying from 30 to 45 m. Area variously estimated at from 3300 to 4300 sq. m. It receives numerous rivers, and at its S.W. extremity discharges itself into the lake Ladoga by the Svir. Its shores, which are generally rocky, present several deep bays and gulfs; and there are numerous islands near its N. extremity. Its navigation is impeded by sandbanks, but it is less subject to storms than lake Ladoga. Principal towns on its bank, Petrozavodsk and Povienetz. (*Schnitzler, La Russie; Possart; &c.*)

ONEIDA, county, N.Y., situated N.E. of the centre of the state, and contains 1101 sq. m. Drained by Black and Mohawk rivers. The Erie and the Chenango canals, and the western line of railroads pass through it. It contained in 1840, 92,669 neat cattle, 177,070 sheep, 66,543 swine; and produced 238,159 bushels of wheat, 6064 of rye, 364,075 of Indian corn, 30,240 of buckwheat, 98,331 of barley, 657,952 of oats, 1,574,109 of potatoes, 38,794 pounds of hops, 286,502 of sugar. It had two commercial and three commission houses in foreign trade, 342 retail stores, 14 furnaces, one forge, 40 fulling-mills, 23 woollen factories, 13 cotton factories with 37,316 spindles, two dyeing and printing works, five flouring mills, 57 grist-mills, 248 saw-mills, two oil-mills, six paper-mills, two rope-walks, three glass-houses, 61 tanneries, 11 distilleries, four breweries, two potteries, nine printing-offices, nine weekly newspapers; two colleges, 190 students; 24 academies, 2035 students; 441 schools, 20,166 scholars. Pop. 85,310. Capital, Utica, Rome, and Whitesboro'.

ONEIDA, lake, N.Y., borders on Oneida, Madison, Onondaga, and Oswego counties, and is 22 m. long, and from to 6 m. broad. It receives Wood, Oneida, and Chittenango creeks, and has its outlets into Oswego river.

ONEONTA, p. t., Otsego co., N.Y., 90 m. S. Cooperstown, 9 m. W. Albany, 349 W. Watered by Susquehanna river and its tributaries. It contains two churches, a Presbyterian and Baptist, eight stores, two furnaces, one fulling-mill, two grist-mills, seven saw-mills, one printing-office, one weekly newspaper; one academy, 35 students; 13 schools, 555 scholars. Pop. 1936.

ONION, or Winooski river, Vt., rises in Cabot, and flows S. and S.W. to Montpelier, where it receives a large branch, then flows northwesterly, until it enters lake Champlain m. N. of Burlington village on the dividing line between Burlington and Colchester townships. Along the margin of this river is the most level road through the Green mountain range. On each side of it in some parts the mountains tower to a great height, leaving on the bank of the river, barely sufficient room for a road. It is one of the four large rivers of Vermont on the W. side of the mountains, which enter lake Champlain, and is about 70 m. long. Winooski is the Indian name; and though this river has been long known by the name of Onion river; Thompson, in his Gazetteer of Vermont, has adopted the name Winooski, and it may perhaps prevail. Winooski signifies the land of onions. The turnpike road from Burlington to Montpelier, along this river, was long since chartered by the name of the Winooski turnpike. The river has falls, which afford good water-power.

ONLOW, county, N.C., situated in the S.E. part of the state, and contains 720 sq. m. Watered by New river and its branches. It contained in 1840, 9003 neat cattle, 5161 sheep, 23,128 swine; and produced 2117 bushels of wheat, 1556 of rye, 228,759 of Indian corn, 6126 of oats, 74,924 of potatoes, 4411 pounds of rice, 218,104 of cotton. It had four stores, three flouring-mills, 18 grist-mills, 10 saw-mills; three schools, 71 scholars. Pop.: whites, 4675; slaves, 2739; free coloured, 113; total, 7527. Capital, Onslow C.H.

ONONDAGA, county, N.Y., centrally situated towards the N. part of the state, and contains 711 sq. m. It contains Skeneateles and Otisco lakes, and has Oneida lake on its N.E. border. Watered by Oneida and Seneca rivers, which unite to form Oswego river. The salt springs of this county are owned by the state, and yield to it a large revenue. In the villages of Salina, Syracuse, Liverpool, and Geddes, which are near to each other, and the centres of the salt manufacture and trade, there were produced in 1840, 2,692,335 bushels of salt, yielding a revenue to the state of $162,404. The Erie canal passes through it; and the Oswego canal extends from it at Syracuse to lake Ontario. It contained in 1840, 46,020 neat cattle, 159,650 sheep, 61,733 swine; and produced 655,799 bushels of wheat, 3593 of rye, 401,303 of Indian corn, 14,420 of buckwheat, 348,615 of barley, 534,762 of oats, 800,317 of potatoes, 178,590 pounds of sugar. It had 264 stores; 2,864,634 bushels of domestic salt were produced; one furnace, 23 fulling-mills, 10 woollen factories, three cotton factories with 3793 spindles, 23 flouring-mills, 29 grist-mills, 114 saw-mills, one oil-mill, two paper-mills, 43 tanneries, seven distilleries, one brewery,

518

six printing-offices, one daily and five weekly newspapers, and one periodical; 12 academies, 1539 students; 333 schools, 17,670 scholars. Pop. 67,911. Capital, Salina.

ONONDAGA, p. t., Onondaga co., N.Y., 5 m. S. Syracuse, 133 W. by N. Albany, 346 W. Drained by Onondaga creek. It contains five churches, three Presbyterian, an Episcopal, and a Methodist; one academy, 294 students; 25 schools, 1748 scholars. Pop. 5658.

ONONDAGA, lake, N.Y., is situated in the N. part of Onondaga co., is 8 m. long, and from 2 to 4 m. broad, and is celebrated for the salt springs on its borders, the largest and best in the United States, though its own waters are fresh. It has its outlet into Seneca river, and receives from the S. Onondaga creek.

ONSLOW, C.H., p. v., capital of Onslow co., N.C., 145 nf. S.E. Raleigh, 372 W. Situated W. of New river. It contains a courthouse, jail, and several stores and dwellings. Nett proceeds of the postoffice in 1842, $57.

ONTARIO, lake, N.Y., the easternmost and smallest in extent of the five great lakes of North America. It is between 43° 10′ and 44° N. lat., and between 76° and 80° W. lon. It receives Niagara river, the great outlet of the upper lakes, in its S.W. part, and has its outlet by St. Lawrence river in its N.E. part; in which, immediately below the lake, is the cluster denominated "the Thousand Islands." Its shape approaches to a long and narrow ellipse, being 190 miles long, and 55 miles wide in its widest part, and about 480 miles in circumference. It is very deep, being in some places over 600 feet, so that its bottom is considerably below the level of the Atlantic. It is 334 feet below the level of lake Erie, and 231 feet above tide-water. In every part, it has sufficient depth of water for the largest vessels. It has many good harbours, and is rarely frozen, excepting in shallow places near the shore. The principal rivers which enter it on the S. side are the Genesee, Oswego, and Black rivers, and a large number of creeks. The bay of Quinte is a long and irregular body of water in its E. part, which receives a considerable river, the outlet of several small lakes; and Burlington bay is in its W. part. Both of these bays are in Canada. It has several important places on its shores, the principal of which are Kingston, Toronto, and Coburg, in Canada; and Oswego, Secket's Harbour, port Genesee or Charlotte, in the United States. It is subject to violent storms and heavy swells; but the numerous steamboats which navigate it pass quietly through it, having a great draught of water. It is connected with the Erie canal and Hudson river by the Oswego and Erie canals, and much of the trade of New-York for the west passes through it, and through the Welland canal, which is 36 m. long, with 34 locks; and admits the passage of the largest vessels which navigate the lakes. The canal-boats generally have decks, which fit them for navigating the lakes. This canal commences at Sherbrooke, near the mouth of Grand river on lake Erie, and terminates at port Dalhousie on lake Ontario, 9 m. W. of Niagara village. Its entrance being a considerable distance W. of the outlet of lake Erie, it is open earlier than the Erie canal at Buffalo, where the ice often accumulates in the spring.

ONTARIO, county, N.Y., situated centrally in the W. part of the state, and contains 617 sq. m. It has Seneca lake on its E. border, Canandaigua lake in its central and S. part, and Honeyoye and Hemlock lakes in its W. part. The outlet of Canandaigua lake flows into Seneca river. Iron ore, gypsum and marl, are abundant. It has a spring which emits carburetted hydrogen or inflammable gas, and several sulphur springs. The Erie canal touches its N. border, and the Auburn and Rochester railroad passes through it. It contained in 1840, 34,300 neat cattle, 172,199 sheep, 46,837 swine; and produced 770,235 bushels of wheat, 6162 of rye, 246,018 of Indian corn, 16,961 of buckwheat, 117,000 of barley, 462,266 of oats, 395,844 of potatoes, 183,273 pounds of sugar. It had 136 stores, 2; fulling-mills, 10 woollen manufactories, 23 flouring-mills, 35 grist-mills, 22 saw-mills, one oil-mill, one paper-mill, 20 tanneries, 14 distilleries, three breweries, three potteries, six printing-offices, three binderies, four weekly newspapers; one college, 146 students; 18 academies, 1375 students; 383 schools, 12,427 scholars. Pop. 43,500. Capital, Canandaigua.

ONTARIO, p. t., Wayne co., N.Y., 22 m. W. Lyons, 235 m. W. by N. Albany, 369 W. Bounded N. by lake Ontario. It contains one store, one furnace, one forge, one grist-mill, 11 saw-mills; 10 schools, 423 scholars. Pop. 1879.

ONTENIENTE (an. *Fontenientis*), a town of Spain, prov. Valencia, on the Clariano (a trib. of the Jucar), crossed here by a handsome stone bridge, 35 m. N. by W Alicante, and 47 m. S.S.W. Valencia. Pop. acc. to Miñano, 12,000. It is tolerably well built, with a fine central square, and several wide though steep streets; its principal public buildings being three parish churches, five decayed convents, a hospital and college now partly in ruins. It is a place of considerable industry, chiefly exerted in weaing linens and woollen fabrics; besides which there are several failing, corn, oil, and paper-mills. In the neighbourhood is an extensive *huerta*, or irrigated tract, which is very productive. A great fair is held here in November.

OOCH, a town of N.W. Hindostan, prov. Mooltan, rajahship of Bhawalpoor, in a fertile plain 4 m. E. the Chenab (an. *Acesines*), where it is joined by the Garra, or united Suttlej and Beas; lat. 29° 11' N., long. 70° 50' E. Pop. 20,000. † "It is formed of three distinct towns, a few hundred yards apart from each other, and each has been encompassed by a wall of brick, now in ruins. It is a mean place: the streets are narrow, and covered with mats as a protection from the sun. It is highly celebrated in the surrounding countries for the tombs of two saints of Bokhara and Bagdad. These are handsome edifices, about 500 years old; but an inundation of the Acesines, some years back, swept away one half of the principal tomb, with a part of the town." (*Burns's Trav. to Bokhara, &c.*, i., 70-82, &c.) Ooch is built on an elevated mound of clay, apparently composed of the debris of former houses, it being a place of high antiquity.

OOJEIN (Hindoo *Ujjayini* or *Avanti*, the *Ozene* of Ptolemy and the Periplus), a city of Central India, prov. Malwah, and the former cap. of Scindia's dom., on the Siprah, a tributary of the Chumbul, 34 m. N. by W. Indore, and 1695 ft. above the sea; lat. 23° 11' N., long. 75° 51' E. It is of an oblong form, about 6 m. in circuit, and fortified with a stone wall and towers. Within this space is some waste ground, but the greater part of the surface is thickly covered with buildings and very populous. The streets are broad, airy, paved, and clean; the houses of brick or wood, and tiled or terraced. Four mosques, some mausoleums, Scindia's palace, an extensive and commodious edifice, but without any claim to magnificence, an ancient Hindoo gate, and some Hindoo pagodas, are the principal public edifices. In a temple to Mahadeo is an extraordinary sculptured image of the bull Nandi. The population of Oojein includes a great many Mohammedans, who are actively engaged in trade. The imports are principally fine white cloths, turbans, and dyed goods; European and Chinese produce from Surat; assafœtida, &c., from Sinde; cotton, coarse cloths, opium and other drugs, are exported, and diamonds in transit from Bundlecund to Surat.

Ancient Oojein, which stood about 1 m. northward, was destroyed at an uncertain period by some physical catastrophe. On digging to a depth of 15 or 16 ft., brick walls, stone pillars, and other antiquities have been discovered, frequently in good preservation. Adjoining these subterraneous ruins, is what has been called the cave of Bhirtry, a palace now in great part buried by an accumulation of the surrounding earth, but of which many portions remain entire, including a long gallery, supported by pillars curiously embellished with figures in relief. Elsewhere around Oojein, there are various temples, &c., worthy of notice; and about 4 m. N. is an elegant summer palace, cooled by artificial cascades, built in the 16th century, and but little injured by time. (*Forbes, Hunter, &c., in Mod. Trav.*, x., 220-223; *E. I. Gaz.*)

OOSTERHOUT, a town of Holland, prov. N. Brabant, arrond. Breda, cap. canton, 5 m. N.E. Breda. Pop. about 6300. It has numerous potteries and brick-kilns, and is the seat of three large annual fairs, at each of which the sale of woollen and linen fabrics, shoes, &c., is estimated to amount to 40,000 florins.

OPELOUSAS, p. v., capital of St. Landry par. La., 217 m. W.N.W. New-Orleans, 1242 W. Situated on a branch of Teshe river, and surrounded by a level and pleasant country. It contains a courthouse, jail, U. States' land-office, three stores, and about 500 inhabitants It is the seat of Franklin college, founded in 1839, which has a president and three professors or other instructers, and 70 students. The commencement is on the 1st of November.

OPORTO, or PORTO, an important commercial city and seaport of Portugal, on the N. bank of the Douro, about 2 m. from its mouth, 174 m. N. by E. Lisbon, lat. 41° 10' 30" N., long. 8° 37' 18" W. Pop., in 1827, including Villanova and Gaya, on the opposite side the river, 80,000. The town has four suburbs, which, with the city itself cover an area of about 2 sq. m. It is built amphitheatre-wise, partly on the sides and tops of two hills, but partly also on a plain near the river, from which it has a strikingly beautiful appearance. A wall, flanked at intervals, with towers, encircles the town, which is farther protected by a fort; but these fortifications have not been kept in good repair, owing to the city being naturally secure against an attack by sea, and one on the land side not being apprehended. An elevated quay extends the whole length of the town: it has on one side a row of houses, and on the other a strong stone wall, with rings, booms, &c., for securing vessels during the heavy swells of the river. It is generally well built, the houses are all white-washed, and though it has many narrow and dirty streets, it is said to be (which,

however, is no great recommendation) the cleanest and most agreeable town in Portugal. From the quay rises a broad well-paved street, flagged on both sides, and leading to two cross streets of equally fine proportions; but the streets on the slope of the hill are mostly irregular, contracted, steep, and dirty. At the E. end of the city the houses, which overhang the side of the river, are built on so steep an acclivity as to be accessible only by steps cut out of the rock. On the summit of the hills are several fine broad streets, lined with good houses, with gardens attached, occupied by some of the principal merchants. Oporto has several squares, the largest of which are the *Praça da Constituição* and the *Campo da Cordaria*, lined with three rows of trees, and much frequented as a public promenade. There are nine parish churches, and a great many other churches and chapels. The cathedral, built in 1105, is a large and fine, though rather heavy edifice: the church, *das Clerigos*, has the highest steeple in Portugal, except that of Mafra: the rest exhibit no features worth notice. There are also 17 convents, now luckily unoccupied, or applied to secular uses. The other public edifices comprise a modern-built episcopal palace, a town-hall (*senado da Camera*), courthouse, with attached prisons, royal hospital, *casapia*, or pawnbroking establishment, and a very pretty theatre, with extensive warehouses belonging both to the Oporto company and the British merchants. The English factory is a handsome building, in one of the principal streets, comprising reading-rooms, ball-rooms, &c., and a residence for the British consul. Oporto has several establishments for public instruction, the principal of which are the academy of navigation and commerce, the school of medicine and surgery attached to the hospital *de la Misericordia*, the episcopal seminary, school for foundlings, and four colleges, with numerous schools for primary and higher instruction. It has a large tobacco factory, a soapboiling establishment, with roperies, tanneries, and fabrics of cotton, silk, linen, and wool: besides which there are ship-building yards for the construction and repair of merchant-vessels; but in none of these establishments is there any great display of activity. The harbour within the bar, across the mouth of the Douro, can only be entered, at least by vessels of considerable burden, at high water; and it is rarely practicable at any period of the tide for any vessel drawing more than 16 ft. On the N. side the entrance is the castle of St. Joao de Foz, near which, on high ground, is a lighthouse, with a fixed light. The ordinary rise of spring tides is from 10 to 12 ft., and of neaps from 8 to 8 ft. The bar being liable, from the action of the tides, and from sudden swellings and *freshes* in the river, to perpetual alterations, should never be attempted by any vessel without a pilot.

The swellings or *freshes*, now alluded to, most commonly occur in spring, and are caused by heavy rains, and the melting of the snow on the mountains. The rise at such times often amounts to 40 ft.; and the rapidity and strength of the current are so great that no dependence can be placed on anchors in the stream. Fortunately, a fresh never occurs without timely warning, and it is then the practice to moor with a cable made fast to stone-pillars erected on the quay for that purpose. On the opposite side of the river, but connected by a bridge of boats, are the towns of Gaya and Villanova, which may be considered as suburbs of Oporto. The former of these is said to occupy the site of the ancient Cala: more eastward is Villanova, inhabited chiefly by coopers, porters, and other labourers, employed by the merchants; and between these towns are immense vaults or warehouses for storing wine previously to its shipment.

Commerce.—Owing to her situation on the Douro, which is navigable partly by barges and partly by boats about 100 m. inland, Oporto is the emporium of a large portion of Portugal, and enjoys a pretty extensive commerce. The famous and well-known red wine called port is produced on the banks of the Douro, about 50 m. above Oporto, and has derived its name from its being exclusively shipped at this city. The exports of port, which is the great article of trade, have varied during the last 10 years from about 16,500 to above 40,000 pipes. England is by far the largest consumer of port. The high discriminating duties on French wine, imposed in the reign of William III., originally introduced port into the British markets, and gave it a preference, to which, though an excellent wine, it had no natural claim: this preference first generated, and its long continuance has since so confirmed, the taste for port among the great bulk of the population, that it bids fair to maintain its ascendancy as an after-dinner wine, notwithstanding the late equalization of the duties. In 1840, for example, 2,088,534 gallons port were entered for consumption in the United Kingdom; whereas the entire entries of French wines during the same year, including champagne, claret, sauterne, &c., amounted to only 341,841 gallons! At an average of the three years ending with 1840, the shipments of port wine from Oporto for England were 25,964 pipes a year. Next to England, Brazil, the U. States of America, and Hamburg, are the principal importers of port. The other articles of export from Oporto are oil, oranges, and other fruits; wool, refined sugar, cream of tartar, shumac, leather, cork, &c. The imports are sugar, coffee, and other colonial products, principally from Brazil; corn, rice, beef, salt fish, and other articles of provision; cotton and woollen goods, hardware, tin plates, &c., from England; hemp, flax, and deals from the Baltic, &c. Subjoined is a

STATEMENT of the Shipments of Port from Oporto during each of the Three Years ending with 1840, specifying the Quantity shipped for each Country.

	1838.	1839.	1840.	Annual Average of the 3 Years.
	Pipes.	Pipes.	Pipes.	Pipes.
Great Britain	26,057	26,159	25,878	25,364
Brazil	7,181	1,913	2,758	3,903
United States	2,629	3,471	1,400	2,498
Hamburg	803	223	645	580
Other countries	1,305	1,949	2,574	1,783
Total	37,975	39,205	33,196	34,790

The climate of Oporto is damp and foggy in winter, less from the vicinity of the Atlantic than from its position in the midst of woods and mountains. The cold is severe for the latitude, though it seldom freezes; and in summer, on the other hand, the heat would be intense, if not moderated by winds blowing regularly from the E. in the morning, S. at noon, and W. at night. The soil in the vicinity is not fertile, nor is Oporto supplied with provisions from its own immediate neighbourhood; but there are many beautiful and pleasant gardens, producing, according to their exposure or elevation, the fruits of N. or S. Europe. The neighbouring mountains exhibit many traces of metallic ores; and along the S. bank of the river are veins of copper and beds of coal. (Dalrymple's Travels in Portugal, 196-198.)

Oporto was occcasionally the residence of the ancient earls of Portugal, till Alphonso I., in 1774, wrested Lisbon from the Almoraves, and made it the permanent capital of his kingdom. The city received many important privileges from John II. at the close of the 15th century; but most of them were withdrawn, in consequence of an insurrection of its inhabitants in 1757. In 1805 it was taken and sacked by the French, who retained it till 1809, when the British crossed the Douro, and compelled them to retreat. It afterwards became, in 1831-39, the scene of an obstinate and long-protracted conflict between the late Don Pedro and his brother Miguel. The siege of Oporto lasted upwards of a year, during which a considerable portion of the town was battered down by Don Pedro's artillery, a great deal of property was wantonly destroyed by Miguel's troops, many of the wine-stores were blown up, and several of the wealthiest merchants were ruined by the annihilation of all trade. (Dalrymple's Portugal; Balbi, Essai Stat. de Portugal; Dict. Géog.; Mod. Trav.)

OPPELN (Slav. Oppolse), a town of Prussian Silesia, cap. reg. and circ. Oppeln, on the Oder, 51 m. S.E. Breslau. Pop., in 1838, 6921. It is walled, and has, in general, lofty and massive houses, with an old Gothic cathedral, several other Roman Catholic and Protestant churches, various schools, a royal salt magazine, and several good hotels. It is the seat of government for the regency, of a board of taxation, a municipal tribunal, &c.; and has a gymnasium and a society for the promotion of the public good. (Von Zedlitz; Berghaus.)

OPPENHEIM, p. t., Fulton co., N. Y., 64 m. N.W. Albany, 402 W. Bounded W. by East Canada creek. Drained by branches of Mohawk river. It has a Dutch Reformed church, one store, 17 saw-mills, one oil-mill, three tanneries; 13 schools, 601 scholars. Pop. 2169.

OPPIDO, a town of the Neapolitan dom., prov. Calabria Ultra, cap. cant., on a hill close to mount Aspromonte, and 14 m. N.E. Reggio. Pop. about 6000. The old town of Oppido, supposed by Cluverius to occupy the site of the ancient Mamertium, was utterly ruined by the great earthquake of 1783. The modern town, which is also a bishop's see, was built in the vicinity of the former.

Oppido is the name of another Neapolitan town, prov. Basilicata, 13 m. N.E. Potenza. Pop. 5000. (Croom's Tour, &c., p. 275-76.)

ORANGE (an. Arausio), a town of France, dep. Vaucluse, cap. arrond. on the Meyne, a tributary of the Rhone, in a fine plain about 5 m. E. the latter river, and 12 m. N. Avignon. Pop. in 1836, 5697. It has several parish churches, a Protestant church, a communal college, hospital, &c., and many good private houses and public fountains; but the widest thoroughfare being only 12 or 13 ft. across, scarcely any vehicles enter its streets; and the road from

514

Lyons to Avignon, instead of passing through, goes round the town. It has manufactures of handkerchiefs, coloured linens called toiles d'Orange, serge and silk twist.

Orange is indebted for its celebrity to its Roman antiquities. The principal of these is a splendid triumphal arch, situated a short way out of the town. It has been conjectured that this structure was erected by Marius; but from its profusion of ornament it would seem to date from a much later period; and, according to Woods, is probably not prior to the age of Hadrian. (Letters of an Architect, i., 146.) In many of its details it presents more of the Grecian than the Roman style of architecture; but from the absence of any inscription, its date is wholly conjectural. It is about 64 ft. in length and breadth, and rather more in height. It has three arched passages, the central and largest of which is 26½ ft. in height. The archways are flanked by fluted Corinthian columns, and the whole structure is profusely covered with groups of figures and other sculptured ornaments. This monument was a good deal injured in the middle ages, from having been converted into a fortress; but it is, notwithstanding, in a state of tolerable preservation; and of late years it has been repaired, and surrounded by a palisade. Extensive remains of a theatre, and the traces of several other Roman buildings, also exist here.

Orange was long the capital of a small principality of the same name that gave the title of Orange to the family which now occupies the thanes of Holland and Nassau. The king of Holland, however, retains merely the title of Prince of Orange; the town and principality having been ceded to Louis XIV. at the peace of Utrecht. The principality, 12 m. in length by 9 m. in breadth, is very productive of wine, oil, saffron, madder, fruits, and numerous plants; its inhabitants are distinguished by their industrious habits. (Hugo; Guide du Voyageur; Prosvost, Itinéraire de Nismes, &c.; Artaud, Art d'Orange; Woods, &c.)

ORANGE, county, Vt. Situated in the E. part of the state, and contains 650 sq. m. Watered by Ompompanoosuc river, Waits' river, and branches of White river. Connecticut river bounds it on the E. Large quantities of sulphuret of iron are found in Thetford, and some lead ore in Strafford. It contained in 1840, 36,833 neat cattle, 136,862 sheep, 22,516 swine; and produced 69,565 bushels of wheat, 11,933 of rye, 190,543 of Indian corn, 30,144 of buckwheat, 5260 of barley, 945,678 of oats, 1,055,379 of potatoes, 628,639 pounds of sugar. It had 60 stores, 22 fulling-mills, five woollen factories, 33 grist-mills, 109 saw-mills, two oil-mills, two paper-mills, 36 tanneries, one printing-office; six academies, 994 students; 287 schools, 9682 scholars. Pop. 27,873. Capital, Chelsea.

ORANGE, county, N. Y. Situated in the S.E. part of the state, and contains 760 sq. m. Shawangunk mountains is in its W. part, and the highlands in the S.E. part. Drained by Delaware and Wallkill rivers. Bounded E. by Hudson river. The Delaware and Hudson canal crosses its W. part, and the Erie railroad is finished and in operation, 61 m. from Piermont, on Hudson river. It contained in 1840, 54,799 neat cattle, 50,318 sheep, 47,664 swine; and produced 94,774 bushels of wheat, 396,566 of rye, 416,194 of Indian corn, 119,883 of buckwheat, 1870 of barley, 417,761 of oats, 350,563 of potatoes, 1845 pounds of hops. It is particularly celebrated for its butter and cheese. The products of the dairy were valued at $1,191,395; 21 lumber yards, capital $23,650; three furnaces, 14 fulling-mills, 13 woollen factories, ten cotton factories with 3390 spindles, four flouring-mills, 83 grist-mills, 100 saw-mills, four paper-mills, 28 tanneries, 6 distilleries, one brewery, three potteries, six printing-offices, one bindery, five weekly newspapers and two periodicals; total capital in manufactures $1,515,415; 21 academies, 1409 students; 170 schools, 8787 scholars. Pop. 50,739. Capital, Goshen and Newburgh.

ORANGE, county, Va. Situated in the central part of the state, and contains 380 sq. m. Drained by branches of North Anna and Rivanna rivers. It contained in 1840, 7399 neat cattle, 10,708 sheep, 15,095 swine; and produced 97,747 bushels of wheat, 8450 of rye, 294,784 of Indian corn, 91,671 of oats, 20,897 of potatoes, 416,395 pounds of tobacco, 2051 of cotton. It had 15 stores, eight smoking-houses, producing gold to the amount of $64,400, six flouring-mills, 24 grist-mills, 20 saw-mills, seven tanneries; two academies, 41 students; 26 schools, 347 scholars. Pop.: whites, 3575; slaves, 5584; free coloured, 186; total, 9235. Capital, Orange C. H.

ORANGE, county, N. C. Centrally situated towards the N. part of the state, and contains 1300 sq. m. Watered by Neuse and Haw rivers. It contained in 1840, 9297 neat cattle, 46,340 sheep, 36,131 swine; and produced 97,530 bushels of wheat, 9843 of rye, 388,426 of Indian corn, 81,633 of oats, 14,847 of potatoes, 969,869 pounds of tobacco, 253,437 of cotton. It had 19 stores, three cotton factories with 2360 spindles, 18 flouring-mills, 27 grist-mills, 25 saw-

mills, three oil-mills, 11 tanneries, 11 distilleries, one pottery; five academies, 909 students; seven schools, 157 scholars. Pop.: whites, 16,771; slaves, 6954; free coloured, 631;
total, 94,356. Capital, Hillsboro'.

ORANGE, county Ia. Centrally situated towards S. part
of the state, and contains 378 sq. m. Watered by Lost
river and Lick and Patoka creeks, which afford water-power. It contained in 1840, 7508 neat cattle, 15,678 sheep,
22,220 swine; and produced 140,864 bushels of wheat,
1991 of rye. 529,997 of Indian corn, 140,796 of oats, 9068 of
potatoes, 908,697 pounds of tobacco, 22,968 of sugar. It
had 27 stores, 14 grist-mills, five saw-mills, two oil-mills,
five tanneries, four distilleries, three potteries, one printing-
office, one weekly newspaper, and one periodical; 40
schools, 974 scholars. Pop. 9008. Capital, Paoli.

ORANGE, p. t., Franklin co., Mass., 74 m. W. by N. Boston, 415 W. Watered by Miller's river, which affords
water-power. Incorporated in 1783. It contains a Baptist
church, six stores, two grist-mills, seven saw-mills; 12
schools, 494 scholars. Pop. 1501.

ORANGE, t. New-Haven co., Ct., 4 m. S.W. New-Haven.
Organized in 1822 from New-Haven and Milford. It contains three churches, two Congregational and an Episcopal,
three stores, four grist-mills, three saw-mills; eight schools,
493 scholars. Pop. 1309.

ORANGE, p. t., Steuben co., N. Y., 201 m. W. by S. Albany. 313 W. Drained by Mead's creek. It contains two
stores, six saw-mills, one tannery; 10 schools, 325 scholars.
Pop. 1894.

ORANGE, p. t., Essex co., N. J., 44 m. N.W. Newark, 53
m. N.E. Trenton. 219 W. Drained by Second river, and
branches of Rahway river. It contains five churches,
three Presbyterian, an Episcopal and Methodist, 10 stores,
one tannery; eight schools, 330 scholars. Pop. 3952.

ORANGE, t. Cuyahoga co., O., 13 m. E.S.E. Cleveland,
13 m. S. Willoughby. The E. branch of Chagrin river
here has a fall of 28 ft. perpendicularly, affording good
water-power. It contains four stores, one fulling-mill, one
woollen factory, one flouring-mill, one grist-mill, five saw-
mills; eight schools, 330 scholars. Pop. 1113.

ORANGEBURG, dist., S. C. Situated a little S. of the
centre of the state, and contains 1824 sq. m. Bounded N.
by Congaree river, E. by Santee river, S.W. by South Edisto river. Watered by North Edisto river. It contained in
1840, 22,507 neat cattle, 3593 sheep, 33,383 swine; and produced 12,490 bushels of wheat, 477,011 of Indian corn,
12,357 of oats, 74,940 of potatoes, 510,670 of rice, 878,370 of
cotton. It had 14 stores, 67 grist-mills, 78 saw-mills; one
academy, 40 students; 11 schools, 256 scholars. Pop.:
whites, 6321; slaves, 11,934; free coloured, 264; total,
18,519. Capital, Orangeburg.

ORANGEBURGH, C. H., p. v., capital of Orange dist.
S. C., 44 m. S. by E. Columbia, 550 W. Situated on the
E. side of North Edisto river, and contains a courthouse,
jail, an academy, several stores, and about 500 inhabitants.

ORANGE, C. H., p. v., capital of Orange co., Va., 84 m.
N.W. Richmond, 94 W. It is situated at the foot of S.W.
mountain, and contains a courthouse, jail, two churches, a
Methodist and Presbyterian, a female academy, nine stores,
a printing-office, issuing a weekly newspaper, about 60
dwellings, mostly of brick, and 500 inhabitants.

ORANGETOWN, t., Rockland co., N. Y., 123 m. S. Albany. Bounded E. by Hudson river, on which is Piermont
village, where the New York and Erie railroad terminates.
It contains 30 stores, four lumber-yards, one cotton factory
with 576 spindles, seven grist-mills, four saw-mills, two
tanneries, two distilleries; eight schools, 312 scholars. Pop.
2771.

ORANGEVILLE, p. t., Genesee co., N. Y., 257 m. W.
Albany, 369 W. Drained by Tonawanda creek and its tributaries. It has two stores, one grist-mill, five saw-mills;
16 schools, 659 scholars. Pop. 1949.

OREGON TERRITORY, consists of a large extent of
country lying between the Rocky mountains and the Pacific ocean, and drained by the Columbia river and its tributaries, and supposed to contain about 350,000 sq. m. The
boundaries of the territory are in dispute between the United States and Great Britain, which it is hoped may be
amicably adjusted by a pending negociation.

This territory is divided into three belts or sections, separated by ranges of mountains running very nearly par-
llel with the shore of the Pacific. The first or western
section lies between the Pacific ocean and the Cascade
mountains or President's range. The second or middle
section is between the Cascade mountains and the Blue
mountains. The third or eastern belt is between the range
f the Blue mountains and the great range of the Rocky
mountains. These sections have a distinction of soil, climate and productions. The first range of mountains is
mountainous, from 100 to 150 miles from the coast, and has
any high conical peaks of from 12,600 to 14,000 ft. above
e level of the sea, that are above the region of perpetual

snow, which is here 6500 ft. high. The Blue mountains are
irregular in their course and occasionally interrupted, but
generally proceed from the east of north to the south of
west. On the S. part of the territory is the Klamet range,
running on the parallel of 42° N. lat., and dividing the ter
ritory from California. The climate of the region between
the ocean and the first range, though not unhealthy, is not
in general very favorable to agriculture. The climate is
mild throughout the year, neither experiencing the severe
cold of winter nor heat of summer. The mean temperature is 54° of Farenheit. The winter is supposed to last
from the last of December until February; the rains begin
to fall in November and last till March, but they are not
heavy, though frequent. Snow sometimes falls, but it seldom lies longer than three days. The frosts are early,
occurring in the latter part of August. This however is
owing to the proximity of the mountains, as the winds
from them always cause a fall in the temperature. The
nights are cold, and affect vegetation so far that Indian corn
will not ripen. Fruit trees blossom early in April, at Van
couver and Nisqually; at which latter place pease were
a foot high on the 12th of May, strawberries were in full
blossom, and salad had gone to seed. The country is in
general well timbered with pine, firs, spruce, oaks, (red
and white,) ash, arbutus, arbor vitæ, cedar, poplar, maple,
willow, cherry and tew; with an undergrowth of hazel,
rubus, roses, &c. Near the coast, the trees grow to an astonishing height. A fir tree growing near Astoria, 8 m.
from the sea, was 46 ft. in circumference, 10 ft. from the
ground, and 153 ft. in length before giving off a single
branch, and not less than 300 ft. in its whole height. Another tree of the same species, on the banks of the Umqua,
was 57 ft. in circumference, and 216 ft. in length below its
branches; and sound pines, from 200 to 280 ft. in height,
and from 20 to 40 ft. in circumference, are not uncommon.
These enormous trees are an impediment to cultivation
Near the foot of the Cascade range, the climate and soil
are adapted to all kinds of grain, wheat, rye, oats, barley,
pease, &c.; and apples and pears succeed well. The low
grounds of the eastern section are well adapted to grazing,
and cattle subsist on the green or dried grass through the
year, which favours the rearing of great numbers of horses
and horned cattle. The valley of the Willamette or Mult
nomah river is thought to contain the finest land in Oregon
This river has a course of about 100 m., nearly N. and en
ters the Columbia on the S. side. The wheat of this valley
is of a superior quality, and yields from 20 to 30 bushels to
the acre. The soil of the second or middle section is general
ly a light sandy loam; in the valleys a rich alluvion, and
barren on the hills. The third or eastern section of Oregon between the Blue and Rocky mountains, is a rocky
broken, and barren country. Stupendous mountain spurs
traverse it in all directions, affording little level ground;
and in its elevated parts, snow lies nearly through the year.
It rarely rains, and no dew falls. The difference of temperature, at sunrise and at noon is often 40 degrees.

The Columbia is the great river of this territory. Its
northern branch rises in the Rocky mountains in 50° N
lat., and 116° W. long., and thence pursues a northern
route to near McGillivray's pass in the Rocky mountains
Here the river is 3800 ft. above the level of the sea, and
receives Canoe river, it then turns S., and receiving many
tributaries, among which are Kootanie, or Flat Bow, and
the Flat Head or Clark's river from the E., proceeds to fort
Colville. The Columbia is thus far surrounded by high
mountains, and often expands into a line of lakes. At Col
ville it is 2049 ft. above the level of the sea, having fallen
550 ft. in 290 m. To the S. of this it tends to the W., receiving Spokan river from the east. Thence it pursues a
westerly course for 60 m., and bending to the S. receives
Okanagan river, which has its source in a line of lakes,
extensively susceptible of canoe navigation. The Colum
bia thence passes to the southward until it reaches Wallawalla, in 45° N. lat., and receives Saptin or Lewis river.
This river has many rapids, which greatly obstruct canoe
navigation, and receives the Kooskooske, Salmon and several other rivers from the E. and W. The length of Lew
is river to its junction with the Columbia is 590 m. from its
rise in the Rocky mountains. The Columbia at Wallawalla is 1286 ft. above the level of the sea, and is 3500 ft
wide. It now takes its last turn to the westward, receiving
Umatilla, Quisnel's, John Day's and Chute rivers from the
south, and Cathlatate's from the N., pursuing a rapid course
for 80 m., previously to passing through the range of the
Cascade mountains, where is a series of falls and rapids
that form an insurmountable obstacle to passage of boats,
even in time of floods, which are overcome by portages
From thence there is a still-water navigation for 40 m.,
when it is again obstructed by rapids. Thence to the
ocean, 120 m., it is navigable for vessels requiring 12 ft. of
water, at the lowest state, though obstructed by many
sand bars. In this part, it receives the Willamette from

the S., and the Cowelitz from the N. The Willamette is navigable for small vessels to within 3 m. of its falls. The Columbia is greatly increased in its width within 20 m. of the ocean, which it enters between cape Disappointment and point Adams, 7 m. apart, from each of which a sand bar extends, which renders its entrance dangerous. Fraser's river rises in the Rocky mountains near the sources of the Canoe river, and flowing northwesterly for 80 m., it turns to the southward and receives Stewart's river, which rises in a chain of lakes near the N. boundary of the territory. It then pursues a southerly course, receiving several rivers on the E. and W. side, and under the parallel of 49° M. lat. it passes the Cascade range in a succession of falls and rapids; and after a westerly course of 70 m. it empties into the gulf of Georgia in 49° 7' N. lat. Through this last portion, it is navigable for vessels requiring 12 ft. of water; its whole length being 350 m. The lower part of this river, and its precise outlet were unknown in 1819. There are many small lakes in various parts of the territory. The harbors on the coast at the mouth of the rivers are generally obstructed by sand bars; and even the entrance of the Columbia is impracticable for two thirds of the year, during which time it is equally dangerous to leave it. Within the straits of Juan de Fuca, there is a great number of the finest harbors for the largest class of vessels; and the rise and fall of the tides is 18 ft. Here are found the greatest portion of the valuable harbors of the territory.

Among the most striking features of the territory are the passes through that immense barrier, the Rocky mountains. These mountains are in general a continuous chain; and though they are often 16,000 ft. or more in height, yet in several places they are so interrupted as to allow of a road for crossing them. At fort Boisais on Lewis's river, 8 m. N. of the mouth of Reed's river, Farnham in his tour through Oregon says: " Among the curiosities of this establishment were the fore wheels, axletree and thills of a one-horse wagon, said to have been run by the American missionaries from the state of Connecticut through the mountains, thus far toward the mouth of the Columbia. It was left under the belief that it could not be taken through the Blue mountains. But fortunately for the next that shall attempt to cross the continent, a safe and easy passage has lately been discovered by which vehicles of this kind may be drawn through to Wallawalla." There are three places where the great chain of the Rocky mountains has been passed. The north one is that passed by Lewis and Clark, in about 46° 30' N. lat. The second is in 44° 30', where a road is practicable. But the best, probably, is the south pass, in about 42° 20' following up the north fork of Platte river, and its branch the Sweetwater river at the foot of the Wind River mountains and W. of them. over to the head waters of Lewis's river. This is believed to have been the route of the wagon above spoken of.

The various tribes of Indians inhabiting this territory amount probably to about 20,000. Of the whites, Canadians and half breeds, there are between 700 and 800, about 150 of whom are Americans; the rest are settlers, and officers and servants of the Hudson's bay company. The American board of commissioners for foreign missions have several missionary stations, and the Methodists have a flourishing station on the Willamette river. Fort Vancouver, on the N. bank of the Columbia, 90 m. from the ocean, is the principal seat of the British fur trade, though they have a number of other posts, which are at present reaping nearly all the advantages of the fur trade of the territory.

On the 7th of May, 1792, Captain Robert Gray, in the ship Columbia, of Boston, discovered and entered the Columbia river, to which he gave the name of his vessel. He was the first person who established the fact of the existence of this great river, and this gives to the United States the title from discovery. In 1804–5, Captains Lewis and Clarke, under the direction of the government of the United States, explored the country from the mouth of the Missouri to the mouth of the Columbia, and spent the winter of 1805–6 at the mouth of the Columbia. This exploration of the river, the first ever made by white men, constitutes another ground of the claim of the United States to the territory. In 1808, the Missouri fur company, established a trading house on Lewis's river, the first ever formed on any of the waters of the Columbia. In 1810, the Pacific fur company, under John Jacob Astor of New-York, was formed; and in 1811, Astoria was founded by them, at the mouth of the Columbia. In consequence of the exposure of Astoria by the last war with Great Britain, the post was sold out to the Hudson's bay company; but it was restored to its original proprietors, by order of the British government, at the close of the war, agreeably to the first article of the treaty of Ghent. See a Memoir of the North West coast of North America, by Robert Greenhow, translator and librarian to the department of state;

516

presented, by a select committee, to the senate of the United States, Feb. 10th, 1840; a work which clearly exhibits the mass of historical facts that bear upon the great question of title to this territory, which it is hoped will be soon settled by negociation.

OREBRO, or ÖREBRO, a town of Sweden, in the centre of the country, lap. län. Orebro, at the W. extremity of the lake Hjelmar, 100 m. W. Stockholm, lat. 59° 17' 12", long. 15° 13' 20" E. Pop., in 1836, 4135. " The houses are built chiefly of wood, not merely of logs clumsily put together, as is the case in Russia, but of logs covered with boards neatly finished, the workmanship of which would not disgrace the tools of any of our English carpenters. Their exterior planking is invariably painted a deep red colour, with white doors and window-frames. The roofs are generally covered with turf; but there are several well-built brick houses stuccoed white." (Barrow's Excur. in the N. Europe, 162.) Streets wide and clean, and some of them are paved with granite. In the principal church is a monument in honour of Englehardt. From its central position, Orebro becomes sometimes the seat of the Swedish diet. The inhabitants manufacture woollen cloths, stockings, &c., and carry on an active trade with Stockholm, by the lake Hjelmar, the canal of Arboga, and the lake Melar. The town also is an entrepôt for the iron of the interior. (Stein's Hand-Book; Barrow, &c.)

OREL, a central prov. of European Russia, to the S. of Kalouga and Tula. Area, 16,780 sq. m. Pop., 1,366,688. Surface undulating; soil, extremely rich and fertile. Principal rivers, Desna, an affluent of the Dniepr; Sosna, an affluent of the Don; and Oka, an affluent of the Wolga. Forests very extensive; occupying nearly a third part of the surface. Agriculture is the principal dependence of the inhabitants; and owing to the excellence of the soil, the return, notwithstanding the bad husbandry, is frequently 7, and sometimes 10 times the seed. Manufactures have made little progress, but distilleries are numerous. Instruction very defective; there being in the entire government, in 1832, only 30 schools, and 4986 scholars. The public revenue amounts to about 10,500,000 roubles, of which the spirit and beer duties yield 6,336,963 roubles! (Schnitzler, La Russie, &c., 346, &c.)

OREL, a city of European Russia, cap. of the above gov., on the Oka, where it receives the Orlyk; lat. 52° 57' 37" N., long. 35° 57' 15" E. Pop., according to the official accounts, in 1830, 31,406; having been only 20,000 in 1825. This rapid increase is owing to its favourable situation for commerce, in the centre of a rich country, on a navigable river communicating, partly by the aid of canals, with the Baltic on the one hand, and the Black sea and the Caspian on the other. Orel may be reckoned the entrepôt of the commerce between Russia, Little Russia, and the Crimea; and at the same time the depôt for corn, both of its own and the adjacent fertile government. The principal articles of commerce are corn, hemp, tallow, butter, bristles, leather, wax, honey, cloth, horned cattle, &c., which its merchants chiefly buy in the southern provinces. Immense quantities of most of these articles are shipped upon the Oka, partly to be landed at Serpuchof for the consumption of Moscow, and partly to be forwarded to Petersburg. It has an ecclesiastical seminary, attended by a great number of pupils; a gymnasium, &c. It has also establishments for the spinning of cotton, manufactures of cloth and coarse linen, with tanneries, distilleries, tallow-melting houses, rope-works, &c.; and is the seat of some important fairs. It is built of wood, and palisadoed; and the inhabitants are distinguished for their industry and wealth. (Schnitzler, La Russie, &c., p. 351; Lyall's Travels, i., 29, &c.)

ORENBOURG, a very extensive government of the Russian empire, mostly in Europe, but partly in Asia, lying between the governments of Astrakhan on the W., and that of Tobolsk on the E., and having Persia on the S., and the country of the independent Kirghesse on the S. Area estimated at about 138,000 sq. m. Pop. 1,771,008. (Koppen.) It is divided into two unequal portions by the Oural mountains; and the river Oural has its source and termination in this government. It is also traversed by the Bielia and other affluents of the Wolga, and is bounded on the E. by the Emes. Soil very various; in part mountainous, in part arid, saline steppes, but the larger portion is decidedly fertile. Forests very extensive. Climate in extremes; being exceedingly hot in summer, and proportionally cold in winter, particularly to the E. of the Oural mountains. Notwithstanding the backwardness of agriculture, the produce of corn is estimated at about 4,000,000 chetwerts. The principal wealth of the inhabitants consists in their horses, cattle, and sheep; of all which, but especially the first two, they have vast numbers. They have also camels, hogs, &c. The river Oural teems with fish, which are taken in great numbers, and are said to furnish the best caviar. Mines important and valuable, yielding some gold, with large quantities of iron and copper, in the working of which many individuals are employed. A manufactory of arms was established in the district of Troitsk, in this government, by workmen from Rhenish Prussia. The salt-mines of Iletsk furnish annually about 500,000 poods, and a large supply is obtained from the salt-lakes. A considerable commerce is carried on with the Kirghises and other people to the S. of Orenbourg, which principally centres in the town of that name. In 1833 there arrived at the latter 14 caravans, with 2547 camels; and during the same year there were despatched 13 caravans, with 4769 camels and 64 draught-horses. The value of the imports was 3,551,198 roubles, and of the exports 3,577,921 do. (*Schnitzler, La Russie, &c.*, p. 704.)

ORENBOURG, the principal city of the above government, and the residence of the military governor, on the N. bank of the river Oural, lat. 51° 46′ N., long. 55° 4′ 45″ E. Pop. 8,000 or 22,000.* It is well built, and regularly fortified. Principal edifices, cathedral and custom-house. In 1825 a school was established here for the special use of the Mohammedans, and the study of their language. The seat of the commerce alluded to in the preceding article is on the southern side of the river, in a vast bazaar erected exclusively for that purpose, and protected by a camp of Cossacks.

ORFORD (formerly *Ore-ford*), a decayed bor., market-town, and par. of England, hund. Plomesgate, co. Suffolk, at the confluence of the Alde with the Ore; 16 m. E. by N. Ipswich, and 90 m. N.E. London. Area of bor., 2740 acres. Pop., in 1841, 1038. It was formerly of much greater importance than at present, having, in 1359, sent three ships and 62 men to Edward III. at the siege of Calais. Its decay is attributed to the loss of its port, the sea having receded from this part of the coast. It is now, having lost its privilege of sending members to the House of Commons, sunk into insignificance; though its large ancient-church, decayed town-hall, assembly-houses, and fine old castle, attest its former consequence. Its present inhabitants are chiefly supported by the oyster fisheries in the neighbouring rivers. Orford claims to be a borough by prescription, but has received several royal charters. Its corporate officers are a mayor, eight portmen, and 12 capital burgesses; but it was considered too unimportant to be included in the provisions of the Municipal Reform Act. It sent two members to the House of Commons from the reign of Richard III. down to the Reform Act, by which it was disfranchised. Sudborne hall, a seat of the Marquis of Hertford, is about 1 m. N.E. from the town. Markets on Monday; fair, June 24.

ORFORD, a p. t., Grafton co., N.H., 30 m. N.N.W. Concord, 17 m. N. Hanover, 505 W. Bounded W. by Connecticut river, over which is a bridge to Fairlee, and drained by its small tributaries, and by headwaters of Baker's river, which flows to the Pemigewasset and Merrimac. Limestone and a fine quality of soapstone are extensively found. It contains three churches, two Congregational and a Universalist, four stores, two fulling-mills, one woollen factory, four saw-mills, one tannery, one pottery; 17 schools, 560 scholars. Pop. 1797.

ORIA (an. *Hyria*), a town of the Neapolitan dom., prov. Otranto, cap. canton, 21 m. E. by N. Taranto. Pop. 5000, principally of Greek origin. It is a "city romantically situated upon three hills, in the centre of the plains. The castle and cathedral stand boldly on the highest points." (*Swinburne.*) Oria is a bishopric, and is very ancient. (*Cramer's Anc. Italy*, ii., 310, &c.)

ORIHUELA, an Episcopal city of Spain, prov. Valencia, on both sides the Segura, crossed here by two bridges, 14 m. E.N.E. Murcia, and 26 m. S.S.W. Valencia. Pop., according to Miñano, 25,551. It is situated at the foot of a ridge of bare rocks, near the head of a very fruitful *vega* or *ale*, forming a continuation of the *huerta* of Murcia. The streets are broad, but not paved; and there is not a single fountain in the town. Its principal buildings are, a cathedral (with an attached chapter of five dignitaries and 17 canons), five parish churches, two of which are in the suburbs, and five dependent churches (*anejos*). two hospitals, a mendicant asylum, public granary, and cavalry barracks. The inhabitants are remarkable for their superstition, devotion to civilization, idleness, and poverty. The greater part are engaged in agriculture; and the town is a considerable mart for corn and oil. It produces linen and hats, and has numerous tanneries, corn and oil mills, soap-houses, and other manufactories. A large fair is held in October. The neighbouring huerta, about 17 m. in length by 5 m. in breadth, is scarcely to be exceeded in beauty and fertility. "Even the vale of Murcia yields in this respect to that of Orihuela, because the latter is so abundantly supplied with water as to be independent of rain. The cypress, silver elm, and

pomegranate are here seen mingled with the mulberry, orange, and fig; and here, also, the palm, rising in rich clusters, lends novelty as well as beauty to the enchanting scene." (*Inglis*, ii., 220.) The huerta yields also very plentiful crops of wheat, barley, and canary seed, hemp, flax, garden vegetables, &c., and is deservedly called the "garden of Spain."

Orihuela, the an. *Orcelis*, is supposed to have existed prior even to the Carthaginians, from whom it passed successively to the Romans, Goths, and Moors. In 1057 it was made the capital of a small kingdom subordinate to the caliphate of Cordova. In 1228 it became annexed to the Moorish kingdom of Murcia, and finally, in 1265, fell by conquest into the hands of James I., the Christian king of Aragon. (*Inglis*, ii., 216-220; *Med. Trav.; Miñano.*)

ORINOCO, a large river of S. America, in Columbia, for a knowledge of which we are chiefly indebted to Humboldt, who explored the greater part of its course in 1800-1802. Its sources have not been traced, but are supposed to lie in about lat. 3° 30′ N., long. 64° W.: it has a circular course, running first W., then N., and latterly E., to its embouchure in the Atlantic, opposite to and S. of Trinidad: its length, including windings, is estimated at 1380 m., being nearly equal to that of the Danube. In lat. 3° 10′ the river runs over a ledge of rocks, forming a cataract called the Raudal de Guaharibos, about 50 m. above the station of Esmeralda, the highest point attained by Humboldt. About 15 m. below this station it is joined on the S., or left bank, by the Cassiquiari river ("two or three times broader than the Seine nearer the Jardin des Plantes"), which unites with the Rio Negro, an affluent of the Amazon; and thus connects the Orinoco and the latter by a navigable water communication. Its course from this point is W.N.W. to the junction of the Guaviare, after which it becomes a broad and deep river, flowing N. by E. over a rocky bed, in which are the two large cataracts called the *Raudalito*, or rapids, of Maypures and Atures, joining together an archipelago of islands, which fill the bed of the river for several miles, and in some places do not leave a free passage of 20 or 30 ft. for its navigation, though its breadth at this point exceeds 8000 ft. (*Humboldt's Pers. Narr.*, v., 138.) About 50 m. below these falls the Orinoco receives from the W. bank the waters of one of its principal affluents, the Meta; and about 70 and 180 m. lower still, the large rivers Arauca and Apure. At the junction of the Apure, where the rocky country terminates, the main stream deflects eastward, and runs E. by N. past Angostura, to the delta at its mouth, the only considerable affluents in this part of its course being the Caura and Carony, joining it from the S. The delta has its apex about 130 m. from the sea: the S., or principal channel, called *Boces de Navios*, runs eastward into the Atlantic, and is divided for a distance of about 45 m. into two channels, by the island of Imataca, the E. end of which is about 35 m. from point Barima, at the mouth of the river, in lat. 8° 45′ N., long. 60° W. The N., or smaller channel, divides itself into a number of branches, called *Bocces chicas*, or small mouths; most of them are sufficiently deep for vessels of considerable burden; but they cannot be navigated without the aid of the neighbouring Indians, who alone are acquainted with the deep and safe channels formed in the alluvial soil near the mouth of the river. The greatest distance between the mouths of the Orinoco is estimated by Humboldt at 47 nautical leagues, or 140 m.: two of the northern mouths fall into the gulf of Paria.

The affluents of the Orinoco are very numerous, many of them contributing an immense volume of water to the principal river. Those on the W. and N. banks, however, are the only rivers available for navigation, except in the lower part of its course, where it receives Caura and Carony. The Guaviare, which is supposed to connect itself with the Rio Negro by a natural channel, is the same way as the latter river is connected with the Orinoco, appears to have a navigable course of more than 200 m. The Meta rises in the Andes, E. of Bogota, and is navigable for about 500 m. The Apure rises in the Andes, by several sources, between the 6th and 8th parallels of N. lat.; and after a course of nearly 500 m., enters a low and swampy district, through which it forms many different channels, in the neighbourhood of which are extensive *llanos*, furnishing very rich pasturage.

The tides of the Orinoco, at the lowest season, in March, are perceptible as far as Angostura, about 280 m. from Ft. Barima; but the rise is not material above the juncture of the Carony, about 180 m. from the mouth. The annual swell of the Orinoco commences in April and ends in September, during which it remains with the vast body of water which it has acquired the five preceding months, and presents an astonishingly grand spectacle. At the distance of 1300 m. from the ocean, the rise is equal to 13 fathoms. In the beginning of October the water begins to fall, imperceptibly leaving the plains, and exposing in its bed a num-

* The official returns quoted by Schnitzler make the population only 800; but this is no doubt an error, and most probably refers either to houses or families. According to the same returns, the births in 1833 were 805, and the deaths 658, showing that the population cannot differ materially from that is stated above.

44   517

ber of rocks and islands. At the beginning of February it is at its lowest ebb, and continues in this state till the beginning of April. It abounds in fish of various descriptions. Amphibious animals are also found in great numbers on its shores; caymans, or round-mouthed crocodiles, are met with in great abundance throughout the river, and are, not without justice, an object of dread to the natives. Scorpions and mosquitoes are stated, likewise, to be so abundant as to offer the greatest hindrances to European travellers. The Indian tribes above Angostura are described as a miserable, savage race, little improved by the efforts of the Jesuit missionaries; and the ferocity of the natives along the upper part of the river prevented Humboldt and Schomburgk from ascertaining its farthest sources. (*Humboldt's Pers. Narr.*, v.; *Geog. Journal*, x., 249–247.)

ORISSA, a prov. of Hindostan, now wholly included in the British presid. of Bengal and Madras, between the 18th and 23d degs. of N. lat., and the 83d and 87th of E. long.; having N. the provinces Bengal and Bahar, W. Guhdwanah, S. the Godavery, and E. the bay of Bengal. Length, N.E. to S.W., about 400 m.; average, breadth, 70 m. Area, 28,000 sq. m. Population uncertain. Orissa proper is almost wholly included in the British district of Cuttack, which see (i., 722; *see* also *Asiatic Researches*, xv., for a detailed description of the prov.) The shore of Orissa is in general low and sandy; the interior remains in a very wild state, being composed of rugged hills and uninhabited jungles, pervaded by a pestilential atmosphere. The population consists, for the most part, of castes considered impure by other Hindoos; including Ooreas, and other hill tribes, quite distinct in language, features, and manners, from the Hindoos of the plains. Principal towns, Cuttack, Juggernauth, and Balasore. Orissa has been continually subject to foreigners since 1558, when it was conquered by the Afghans. In 1578 it was annexed to the Mogul empire; in 1756 it was transferred to the Mahrattas, and in 1803–4 it was occupied by the British. (*Hamilton's E. I. Gazetteer.*)

ORISTANO, a town of the island of Sardinia, div. Cagliari, cap. prefecture, near the mouth of the Tirsi (an. *Thirsus*), in the bay of Oristano, on the W. shore of the island, 30 m. N.N.W. Cagliary. Pop., in 1838, 5791. It is in a fertile, but unhealthy plain, and is not fortified. Its steeples and turrets give it a tolerable appearance from the sea; but it is straggling, unpaved, and ill supplied with water. It has a cathedral, a spacious edifice, with a detached octangular belfry, one of the most striking objects in the town. There are several other churches and convents, a hospital, a Tridentane seminary, and a Piarist college. The Tirsi is crossed here by a bridge of three arches. The bay of Oristano, the mouth of which is 5 m. across, affords excellent anchorage during the prevalence of winds blowing off shore, but those from the W. throw in a heavy sea. The E. shore of the bay is shoal, but near its N. side vessels anchor in six or seven fathoms water. Many of the inhabitants are occupied in the manufacture of salt, and the tunny fishery; and some tolerable wine is grown near the town. Oristano was founded about 1070. (*Smyth's Sardinia*, 394, 395, &c.)

ORIZABA, a town of Mexico, in the state of Vera Cruz, in a valley remarkable for its fertility, 70 m. W.S.W. Vera Cruz, and 140 m. E.S.E. Mexico. Pop. between 8000 and 10,000, including whites and Indians. It is laid out in wide, neat, and well-paved streets; though so great is the power of vegetation, that grass grows in almost every part of the town. Coarse cloths are made here in small quantities, and there are several tanneries. The valley in which Orizaba is situated is well clothed with forest trees, above which rises the now extinct and snow-covered volcano of Orizaba, to the height of 17,380 ft. above the Atlantic. The neighbourhood produces all the tobacco consumed in Mexico; and within the town is a large government manufactory of that article.

ORKNEY AND SHETLAND ISLANDS. These islands, which are most probably the *Thule* of the ancients, lie in two groups to the N. of Scotland, and form between them a county. The Orkneys (*Orcades*), the most southerly group, are separated from the county of Caithness by the Pentland frith, about 6 m. in breadth. They are comprised between the parallels of 58° 44' and 59° 24' N. lat., and 2° 25' and 3° 26' W. long. There are about a dozen principal islands; Pomona, or Mainland, being decidedly the largest. But, including the smaller islands (provincially *holms*) and islets, the total number is estimated at about 67, of which about 40 are uninhabited. They are supposed to comprise an area of about 281,000 acres, and had, in 1831, a population of 28,847.

The Shetland, or Zetland, isles, the most northerly group (perhaps the *Ultima Thule* of the ancients), are separated from the Orkneys by a channel 48 m. across, and lie principally between the parallels of 59° 52' and 60° 50' N. lat., and 30' and 1° 40' W. long. Including islets, they are supposed to exceed 100 in number; but the mainland or princi-

pal island is a good deal more extensive than all the others put together. Between 30 and 40 are inhabited. They are about twice as extensive as the Orkneys; their total area being estimated at 563,200 acres. In 1831 they had a population of 29,392.

The aspect of these islands is pretty similar; but the Shetland group is the more wet and barren of the two. They are generally fenced, particularly on the W. side, with high, black, precipitous cliffs, against which the sea, when vexed by storms, dashes with astonishing fury. They are destitute of high mountains; the altitude of mount Rona, in Shetland, the highest, not exceeding 919 ft. Their general appearance is that of dreary, heathy wastes, interspersed with rocks, varied sometimes with swamps and lakes, and, in a few places, with beds of moveable sand. In some parts, however, particularly in Orkney, the land is abundantly fertile, producing good crops of corn and luxuriant herbage. Some of the islets, or holms, appear like gigantic pillars, rising perpendicularly from the sea: these are the resort of vast numbers of sea-fowl; and in the breeding season hunting for eggs and young birds forms one of the principal and most dangerous employments of the natives. Climate similar to that of the outer Hebrides, except that the days are a little longer in summer and shorter in winter. During the latter, the aurora borealis is uncommonly brilliant. The cultivated lands bear but a very small proportion to the others, being supposed not to exceed 25,000 acres in Orkney, and 22,000 in Shetland. Farms generally very small; few having more than 10 acres of arable land, and many not nearly so much. Agriculture is considered, particularly in Shetland, of subordinate importance, and, though a good deal improved, is still very backward. In Shetland, most part of the ground is turned over with the spade; but in Orkney ploughs are in pretty general use. Oats, and bere or bigg (*Hordeum hexastichon*), are the only white crops cultivated; and, except on a few improved farms, they follow each other alternately as long as the land will bear any thing, which it does for a very long time when well manured with sea weed. The barley of Orkney is a great deal more abundant, and of a much better quality, than could have been anticipated; and, besides supplying the home demand, considerable quantities are exported. Potatoes are cultivated in all the islands, and form an important part of the food of the people. Turnips have also been planted, and have succeeded very well. At present, no trees can be made to grow, and hardly a shrub is to be met with; which is the more singular, as the trunks of large trees are not unfrequently found imbedded in moss and sand, both in Orkney and Shetland. The hardy, spirited little horses, known by the name of *Shelties*, are bred in Shetland, and are exported in considerable numbers. The stock kept in the islands is estimated at from 10,000 to 12,000; they are never housed, nor receive any food, except what they gather for themselves. Some of them are exceedingly well proportioned, active, and strong for their size. The horses of Orkney are inferior in estimation. Cattle very small, sometimes not weighing more than from 35 to 40 lbs. a quarter; they are shaggy, and not well shaped; but they are hardy, feed easily, and, when fattened, their beef is fine and tender. The stocks in both groups of islands is supposed to amount to about 45,000 head. The native sheep are of the small dun-faced kind: they yield short wool, which, though generally soft and fine, is sometimes as hairy as that of a goat. Recently the black-faced and Cheviot breeds, and even pure merinos, have been introduced into Orkney with considerable success. The stock in both groups of islands is believed to exceed 135,000. A small breed of swine is very abundant: they roam at large, and are not a little destructive. Rabbits are abundant in both sets of islands, but particularly in the Orkneys; as many as 36,000 skins have been exported in a single season from the port of Stromness. Fowls are plentiful, and large quantities of eggs are exported from Orkney. The fisheries, however, in Orkney as well as Shetland, are the grand object of pursuit. The islands are periodically visited by vast shoals of herrings; while the surrounding bays and seas are uniformly well supplied with cod and other species of white fish. *Divitiae eis sunt a mari, ab omni parte summa piscandi commoditate objecta.* (*Buch.*, lib. i., § 50.) Bressay sound, in Shetland, has always been one of the principal stations of the Dutch herring fishers; but the fishing there is now principally carried on by the islanders. During the year ended the 5th of April, 1840, 19,396 barrels herrings, and 9988 cwt. cod were cured in Orkney; and 25,625 barrels herring, and 42,38 cwt. cod, in Shetland. About 100,000 lobsters are supposed to be annually shipped from the Orkneys for London. Vessels from British ports bound for the N. whale fishery mostly touch at Orkney or Shetland; and, besides taking on board supplies of provisions, usually complete their crews with seamen belonging to the islands, whom they put on shore on their way back. Rye straw grown in Orkney has been found peculiarly well fitted to serve as a substitute for the straw used in Italian plait; and the manufacture of this straw into plait has been carried on for several years to a considerable extent, and with pretty good success. Kelp, though comparatively unprofitable, continues to be extensively produced. Woollen stockings and gloves, sometimes of extraordinary fineness, are exported from Shetland. Messrs. Anderson, in their valuable work on the Highlands, give the following account of the Sums received in Orkney, in 1833, for Farm Produce, Manufactures, Fish, &c., exported.

	L.	s.	d.
Bere, or bigg, 5178 bolls, at 15s. per boll of 6 bushels, or 20s. per quarter	3883	10	0
White oats, 1515 bolls, at 12s. do. of do., or 16s. per quarter	909	0	0
Malt, 10,698 bushels, sent to Leith, say 3s. per bushel	1604	8	0
Peas, 234 do., at 3s.	35	2	0
Oatmeal, 40 bolls, at 14s. per boll of 140 lbs. Imperial	28	0	0
Horses, cows, and oxen, 954, 1 quarter horses, at £3; 3 quarter cows, at £3.	4280	0	0
Do. not entered at custom-house, about 946, at do	1104	0	0
Eggs sent to Leith, 50 tons, 100 doz. per cwt., 100,000 doz. at 6d.	2500	0	0
Sheep and swine, 40 of each at £1.	80	0	0
Butter, about £2000; hides, about £700.	2700	0	0
Rabbit-skins, more than 2000 doz., at 5s. 6d. per doz.	600	0	0
Feathers, about	250	0	0
Kelp, supposed scarcely to exceed 500 tons, at £4 10s.	2250	0	0
Straw manufacture, including rent, cutting, planting, &c.	4800	0	0
Herrings, 34,000 barrels, at 10s. per barrel	17,000	0	0
Cod, fished by about 40 sloops of 30 tons, 14 tons each, at £13 per ton	7280	0	0
Lobsters, caught by 432 men, in 216 boats	1800	0	0
Whale fishing, about 25 ships, taking 20 men each, 500 men, at £15	7500	0	0
Hudson's Bay Company pay annually for the wages of men employed in Hudson's Bay, about	1500	0	0
Total	**66,114**	**0**	**0**

There is no similar account for Shetland; but the sums received by its inhabitants do not differ materially from those received by their neighbours in Orkney. The shipping of the islands is considerable. In 1836 there belonged to the Orkneys 77 ships, carrying 4218 tons and 383 men; and there belonged at the same time to Shetland 101 ships, carrying 3967 tons and 744 men, exclusive of a great number of boats engaged in the fisheries.

The people of these remote islands being of Scandinavian and not Celtic origin, neither the Gaelic dress nor language has ever prevailed among them. All of them now speak English; but, of old, Norse was the prevalent language. The cottages of the poorer ranks are in general miserable hovels, affording accommodation in winter to cows and fowls, as well as to the family. Owing to the scarcity or exhaustion of moss, the want of fuel is in some islands very severely felt. On the whole, however, the inhabitants are decidedly better off than those of the Outer Hebrides, being comparatively industrious, civilized, and well fed. Kirkwall in Orkney, and Lerwick in Shetland—the only towns of consequence in the islands—had, in 1831, the former a population of 3065: and the latter 2750. The society in both is good, and the inhabitants hospitable.

Shetland and some parts of Orkney suffer much from the exaction of tithes. They are not only charged upon the produce of the land, but on that of the fisheries; and being generally farmed, they are rigidly collected, are productive of much irritation, and are a formidable obstacle to improvement. Seeing the good effects that have resulted everywhere else in Scotland from the abolition of tithes, it is singular that they have not been commuted here. The feu duties, payable to the crown, or rather to its donatory, Lord Zetland, have also contributed materially to check improvement. Marl, though neglected, is common in Orkney. Lead ore also has been met with, and limestone is of frequent occurrence. These islands are divided into 40 parishes. The gross rental of Orkney amounted, in 1810, to £9495; and that of Shetland to £6741.

The Orkneys are divided into 18, and the Shetlands into 22 parishes. They send one member to the House of Commons. Registered electors in 1839-40, Orkney, 374; Shetland, 152: making together a constituency of 526. Inhabited houses in both islands, in 1831, 10,298; families, 11,815. Population, 58,239. Annual value of real property, in 1815, £90,930. (For farther information, see *Wallace's Descrip-*

tion of the Orkney Islands, 8vo., 1700; *Barry's Hist. of Orkney*; and *Sheriff's Agricultural Survey of Orkney*; *Edmonston's View of Zetland*, 2 vols., 8vo.; *Hibbert's Description of the Shetland Isles*; and *Sheriff's Agricultural Survey of Shetland*, &c.)

ORLAND. p. t., Hancock co., Me., 61 m. E. by N. Augusta, 657 W. Bounded W. by Penobscot river. Opposite to it is Orphan island. Incorporated in 1800. It possesses great facilities for navigation, and has 15 schools, 693 scholars. Pop. 1381.

ORLEANS (an. *Genabum*, and afterward *Aurelianum*), a city of France, in the centre of the kingdom, cap. dep. Loiret, on the Loire, 34 m. N.E. Blois, and 68 m. S.S.W. Paris. Lat. 47° 54' 19'' N.; long. 2° 45' 40'' E. Pop., in 1836, 40,272. "Orleans is a large, but not a beautiful city; and its environs, though rich and highly cultivated, are less agreeable than the country round Tours or Blois. The city itself has few good streets; but there is one spacious and elegant avenue, terminating in a noble bridge. The great square is also magnificent." (*Inglis's Tour*, p. 365.) The bridge across the Loire, the foundations of which were laid in 1751, is 354 yards in length, and has nine arches, the central one being 104 feet in width. On either side the river are spacious quays; and from the bridge, the *Rue Royale*, one of the handsomest streets in France, leads in a direct line to the *Place du Martroy*. In this square is the monument erected to Joan of Arc; consisting of a bronze statue of that heroine, 8 feet in height, on a marble pedestal, upon the sides of which are four base-reliefs in bronze, representing the principal actions of her life. A few remains of the ancient fortifications of Orleans exist, but their place is now principally occupied with plantations and public walks, one of which is a fine promenade called the *Mail*. In the old parts of the city the houses are chiefly of timber, and the public thoroughfares narrow, dirty, and wretchedly paved; but several new and tolerable streets have been opened of late years, and various improvements are in progress. (*Hugo*.) The cathedral, one of the finest Gothic edifices in France, is in a great measure hidden by the surrounding houses. It was begun in the 13th century; partly destroyed by the Huguenots, but rebuilt by Henry IV. It has a fine western portal, flanked with two towers, built by Louis XV. in the most gorgeous style. At the intersection of the nave and transepts is an elegant spire. The side entrances, the lofty vaults, the high altar, and the carving of some parts of the interior also possess great beauty. Some of the other churches and chapels are handsome; but, though still numerous, several of them have been converted into warehouses. The old town-hall, an edifice of the 15th century, is now appropriated to the museum; in its court-yard is an old tower, serving to support a telegraph. The Palace of Justice, a handsome edifice erected in 1821, the theatre, *abattoir*, prison, large infantry barracks, corn-hall, intendency, and general hospital, are the other principal public edifices. There are several private buildings, curious for their Gothic architecture and decorations; the most remarkable of these are the houses of Agnes Sorel and Francis I. The city is surrounded by extensive suburbs, and its vicinity is sprinkled with numerous villas.

Orleans is the seat of a bishopric, of a royal court for the deps. Loir-et-Cher, Indre-et-Loire, and Loiret; of tribunals of primary jurisdiction and commerce, a court of assize, the forest-direction for the basin of the Loire, a *conseil de prud'hommes*, and a chamber of commerce. It has a royal college; an *académie universitaire* (for the three deps. specified above); a society for the promotion of science, *belles-lettres*, and art; a public library of 25,000 vols.; a museum, with an extensive collection of paintings of the French school; cabinets of natural history, &c.; a botanic garden; courses of medicine, &c.; at the general hospital; of drawing, architecture, botany, &c.; maternity and Bible societies; schools of mutual instruction, &c.; and a departmental assurance company. Its former university, founded in 1512, had, among its illustrious students, De Thou, Erasmus, Calvin, and Theodore Beza.

Orleans is well situated for commerce, but its trade is less flourishing than before the revolution. It has declined, while Havre and Paris have risen as commercial towns. Its manufactures comprise fine woollen cloths, flannels, woollen yarn, hosiery, cotton yarn, refined sugar, vinegar, and wax candles; and besides its trade in these, Orleans deals extensively in corn, wines, timber, wool, cheese, and colonial produce. It has a large general fair in June, which lasts 15 days, and one in November, lasting eight days.

D'Anville has shown conclusively that Orleans occupies the site of the an. *Genabum*, the emporium of the Carnutes, taken and burned by Cæsar. (*Notice de l'Ancienne Gaule*, p. 345.) It subsequently rose to great eminence, and was unsuccessfully besieged by Attila and Odoacer. It became the cap. of the first kingdom of Burgundy, under the first race of French kings. Since the time of Philip of Valois, in the 14th century, it has usually given the title of duke to a member of the royal family. It was besieged by the English in 1428-29, who were ultimately obliged, through the efforts of Joan of Arc, to raise the siege and retire. In 1563 it was besieged by the Catholics; and during the progress of this siege the Duke of Guise was assassinated. (*Hugo*, art. *Loiret*; *Dict. Géog.*, &c.)

ORLEANS, county, Vt. Situated in the N. part of the state, and contains 700 sq. m. The S. part of lake Memphremagog lies in its N. part. Drained by Barton, Black, and Clyde rivers, flowing into lake Memphremagog, and by head branches of Mississippi, and Onion or Winooski rivers. It contained in 1840, 18,259 neat cattle, 46,669 sheep, 9750 swine; and produced 33,315 bushels of wheat, 2400 of rye, 90,896 of Indian corn, 90,067 of buckwheat, 10,997 of barley, 133,301 of oats, 596,895 of potatoes, 507,446 pounds of sugar. It had 38 stores, one furnace, one forge, 13 fulling-mills, four woollen factories, 19 grist-mills, 53 saw-mills, one oil-mill, 11 tanneries, one distillery; two academies, 111 students; 122 schools, 3916 scholars. Pop. 13,634. Capital, Irasburg.

ORLEANS, county, N. Y. Situated in the N.W. part of the state, and contains 372 sq. m. Bounded N. by lake Ontario, and drained by Oak, Orchard, Johnson's, and other creeks flowing into it. The Erie canal passes through it. It contained in 1840, 18,123 neat cattle, 60,563 of sheep, 27,933 of swine; and produced 791,212 bushels of wheat, 198,966 of Indian corn, 10,047 of buckwheat, 30,798 of barley, 180,581 of oats, 303,314 of potatoes, 1523 pounds of hops, 150,786 of sugar. It had 77 stores, five furnaces, 12 fulling-mills, four woollen factories, five flouring-mills, 15 grist-mills, 53 saw-mills, 10 tanneries, one brewery, four printing-offices, four weekly newspapers; seven academies, 750 students; 163 schools, 8247 scholars. Pop. 25,127. Capital, Albion.

ORLEANS, parish, La. Situated in the S.E. part of the state, and contains 160 sq. m. Bounded N. by lake Pontchartrain and the Rigolets, E. by lake Borgne, S. by Mississippi river. It has much low and marshy land, and produces cotton, rice, Indian corn, garden vegetables, and oranges, figs, and peaches. It contained in 1840, 9639 neat cattle, 1807 sheep, 1824 swine; and produced 3100 bushels of Indian corn, 19,000 pounds of cotton, 10,000 of sugar. It had eight commercial and 375 commission houses in foreign trade, capital $16,490,000; 1881 retail stores, capital $11,018,295; 39 lumber yards, capital $67,800, three saw-mills, one tannery, two distilleries, 18 printing-offices, nine daily, six weekly, and two semi-weekly newspapers; two colleges, 165 students; 10 academies, 440 students; 25 schools, 975 scholars. Pop.: whites, 59,519; slaves, 23,448; free coloured, 19,226; total, 102,193. Capital, New-Orleans; the statistics of which are, of course, included in the above, as is always the case in the United States census.

ORLEANS, p. t., Barnstable co., Mass., 95 m. S.E. Boston, 496 W. It extends across the narrow part of cape Cod. The inhabitants are extensively employed in the fisheries, and the manufacture of salt. It has four churches, a Congregational, Methodist, Baptist, and Universalist, 10 stores, four grist-mills; 10 schools, 765 scholars. Pop. 1974.

ORLEANS, t., Jefferson co., N. Y., 16 m. N. Watertown, 172 m. N.N.W. Albany. Drained by Chaumont and Perch rivers. It has a church, five stores, two grist-mills, three saw mills; 11 schools, 271 scholars. Pop. 3001.

ORMSKIRK, a market-town and par. of England, hund. W. Derby, co. Lancaster, 11½ m. N.N.E. Liverpool, and 180 m. N.W. London. Area of par., which comprises six townships, 30,083 acres. Pop. of Ormskirk township, in 1841, 14,608. The town is well-built, paved, and lighted with gas, consisting of four principal streets, meeting each other at right angles in a large market-place, in which is the town-hall, built in 1779. The church is a large edifice (rebuilt in 1790), with a tower and steeple, detached from each other, and standing side by side: the living is a vicarage, in the gift of Earl Derby, lord of the manor. The out-townships have three district churches. The Wesleyan Methodists. Independents, and Unitarians have their respective places of worship; and at Scarisbrick is a Roman Catholic chapel. Attached to the churches and chapels are seven Sunday-schools, three of which are also national day-schools. A grammar school, endowed in 1614, is supported by an income of about £140; and there are three infant-schools. The other charitable institutions consist of Lathom's almshouses, and apprentice-fund, three benevolent societies, a savings' bank, and a dispensary, opened in 1797. The principal business of the inhabitants is in weaving light cottons and silks, silk-winding, hat and rope-making: in 1836, one cotton-mill employed 58 hands. Within the parish, also, are considerable coal-mines, the produce of which is sent to Liverpool and other places by the Druylin Navigation, and by the Leeds and Liverpool canal, which passes within 3 m. of the town. The local government of Ormskirk is in the county and manorial

police; and courts-leet are held by the lord of the manor (Earl Derby) once a year. Petty sessions, also, are held here; and it is one of the polling-places at elections for the S. division of Lancashire. Markets on Thursday; large cattle-fairs, Whit-Monday and Tuesday, and 10th Sept.

About 3 m. E. Ormskirk is Lathom House, once the seat of the Stanleys, Earls of Derby, and celebrated for the siege which it sustained under the Countess of Derby in the civil wars of the 17th century: it is now the property of Lord Skelmersdale. A battle was fought near the town in 1644, between the royalists and parliamentarians, when the former were defeated with great loss. (*Parl. Papers*, *Butterworth's Stat. of Lancashire*.)

ORMUZ (an. *Oxyris*), an island situated at the mouth of the gulf of Persia, in lat. 27° 12' N., long. 56° 23' E., about 15 m. in circ. It resembles, when viewed from the sea, a mass of rocks and shells violently thrown up from the bottom to the surface of the ocean. The fort, which is in tolerable repair, is built on a narrow projecting neck of land; and this, with a wretched suburb, has a population of not more than 500 persons. The remains of aqueducts, walls, &c., on a plain near the suburb, mark the seat of the former capital. The harbour is sheltered on three sides by land, and has good anchorage ground. A range of hills intersects the island from E. to W.; and the rocks consist almost entirely of fine crystallised salt, which might be exported in unlimited quantities. The geological formation of the island indicates the former existence of volcanic action, and sulphur, iron, and copper, are found in large quantities, though few attempts have yet been made to apply these mineral riches to any useful purpose. (*Kinneir's Persia*, p. 13, 13.)

This island, at present so inconsiderable, would not have been worth notice but for its former celebrity and importance. It had, however, owing to its advantageous situation, become, previously to the appearance of the Portuguese in the east, a great emporium, being, in fact, the centre of the trade of the Persian gulf, and of the contiguous countries, and possessed great wealth, population, and prosperity. It was taken by Albuquerque, the Portuguese viceroy, in 1515; and was held by the Portuguese till 1622, when it was wrested from them by Shah Abbas, assisted by an English fleet. The booty acquired by the captors on this occasion is said to have amounted to two millions sterling! Subsequently the trade of the island was diverted to Gombroon and other places; and this once rich and flourishing emporium gradually fell into that state of irreparable decay in which we now find it. (*Modern Universal History*, ix., 301, &c., 8vo., ed.)

ORNE, a dep. of France, reg. N.W., formerly included in the prov. of Normandy and Perche; between lat. 48° 12' and 48° 48' N., and long. 1° E. and 1° W., having 3. the deps. Calvados and Eure, E. the latter and Eure-et-Loir, S. Sarthe and Mayenne, and W. Manche. Length, E. to W., 80 m.; breadth very variable. Area, 610,381 hectares. Pop., in 1836, 441,881. A chain of hills runs E. to W. through this dep., separating the basins of the Orne and Seine from that of the Loire; but its summits do not reach a height of more than from 1900 to 2000 feet: the hills are mostly covered with thick woods. The dep. is abundantly watered. Principal rivers, the Orne, Dive, Ta, &c., running N., and the Sarthe, Mayenne, and Huisne, S. The Orne, whence the name of the dep., has its source near Siez, and flows generally N. through the deps. Orne and Calvados to the English channel; which it enters, after an entire course of about 90 m., 15 m. below Caen, from which city it is navigable. Small lakes are supposed to occupy 1300 hectares; and there are numerous marshes. The soil is very various; and in several places there are distinct traces of volcanic action. In 1834, 333,469 hectares were estimated to be arable, 131,045 in meadows, 11,121 in orchards, 72,000 in woods, and 18,953 in heaths, waste, &c. In 1835, of 147,135 properties, subject to the contrib. foncière, 63,854 were assessed at five francs and under, and 26,128 at from 5 to 10 francs. Agriculture is extremely backward. On the small farms, which are extremely numerous, spade husbandry is very general. Except oats, not enough of corn is produced for home consumption, and the deficiency is, in part, made up by potatoes and buckwheat. Hemp and flax are among the principal products: in some cantons beet-root for sugar is grown. Large quantities of cider and perry are made, from a portion of which brandy is distilled. The best horses of Normandy are reared in this dep. Cattle, hogs, and poultry are fattened for the Paris markets, and honey is an important product. The sheep, which are of an inferior breed, are supposed to yield 450,000 kilog. wool a year. Iron mines are wrought in some parts; manganese, building and other stone, and porcelain clay, being the other principal mineral products. Metallic and linen goods are those chiefly manufactured. L'Aigle is celebrated throughout France for its needles and pins, copper and brass wire. The coarse linen cloths made at Montagne amount annually to about 19,400 pieces, of from 80 to 100 ells each; and Alençon is particularly famous for a fine and highly prized species of lace, termed points d'Alençon. Muslins, calicoes, hair cloths, paper, glass, and beet-root sugar are among the other manufactures. Orne is divided into four arronds.; chief towns, Alençon, the cap., Argentan, Domfront, and Mortagne. It sends seven members to the chamber of deputies. Registered electors in 1836-39, 2312. Total public revenue in 1831, 10,414,389 francs. (*Hugo*, art. *Orne*; *Dict. Géog.*; *Official Tables*.)

ORONO, p. t., Penobscot co., Me., 75 m. N.E. Augusta, 670 W. Bounded E. by Penobscot river, in which is Marsh Island, belonging to this t., containing 5000 acres. The falls in the Penobscot afford great water-power. Incorporated in 1806. A railroad, 12 m. long, extends from Bangor to Stillwater village in this township, and to Oldtown, which cost $35,000. It contains six stores, two lumber-yards, one grist-mill, 36 saw-mills; four schools, 350 scholars. Pop. 1521.

ORONTES, (Arab. *El-Aassy*, "the rebellious,") a river of Asiatic Turkey, in Syria, which rises in a natural rocky basin on the E. side of the mountain chain of Antilibanus, near the village of El-ras, within the pach. of Damascus, about 30 m. N. of that city. It runs N.N.E. as far as the lake Kadez, through which it flows, and then takes a N.N.W. direction through the beautiful vale of El-Ghab, as far as lat. 36° 15', where it receives the waters of lake Antakiah, near the city of that name (the ancient *Antioch*), and then suddenly deflects westward, falling into the Mediterranean, near Soveidin, or Seleucia, after a course of about 240 m. At its mouth is a bar, over which there is from 3½ to 9 feet water during winter. (*Geog. Journ.* viii., 230.) "The Orontes in the winter season inundates a part of the low grounds, through which it flows in the upper part of its course, thus insulating the villages and cutting off all communication between them, except by boats. In summer the inundation subsides; but the lakes remain half dried up, and give birth to swarms of gnats and flies, which, coupled with the exhalations from the marshes, oblige the inhabitants to retire into the mountains with their cattle, goods, and chattels." (*Robinson's Pal. and Syria,* ii., 347.) This river is not navigable; "and the rapidity of the stream in many parts of its course, its sudden and numerous windings, its frequent shallows, its various bridges, and the many changes to which it is subjected in the vicissitudes of the seasons, appear to be insuperable obstacles to any plan for making it navigable, or for using it to any considerable extent for trading purposes. In fact, the Orontes is scarcely available at all, even for small craft; and to reach Antioch in a steamer would be a work of consummate difficulty, and, when accomplished, by no means worth the trouble and expense incurred." (*Bowring's Stat. of Syria,* p. 49.) Its use, therefore, is chiefly confined to the irrigation of the surrounding country, which is effected by means of water-wheels similar to those described in the article HAMAH, *ante,* p. 10.

The river abounds with fish, and produces a species of eels much in request with the Greeks; they are salted and sent in every direction to serve during the fasts before Easter. They are said to produce 60,000 piastres a year to the proprietors of the nets at Antioch, in passing through which they are taken. (*Barker's Rep. in App. to Bowring's Stat.*) The valley of the Orontes has on several occasions been visited by earthquakes, the last of which, in January, 1837, nearly destroyed several cities, and occasioned the loss of many thousand lives. (*Robinson's Syr. and Pal.,* ii.; *Geog. Journal,* vii. and viii.; *Mod. Trav.*)

ORRINGTON, p. t., Penobscot co., Me., 70 m. N.E. Augusta, 686 W. Bounded W. by Penobscot river. It has two stores, one fulling-mill, one flouring-mill, six saw-mills; 10 schools, 645 scholars. Pop. 1590.

ORTHEZ, or ORTHES, a town of France, dep. Basses-Pyrénées, cap. arrond., on the Gave de Pau, across which it communicates with a suburb by an old bridge of two arches, 24 m. N.W. Pau. Pop., 1836, 5967. It is well laid out and built, but ill supplied with water. It has manufactures of woollen stuffs, brass and iron wire, and copper wares; with dyeing houses, tanneries, &c.; and an extensive trade in hams of a superior kind, improperly termed *jambons de Bayonne,* goose-feathers, and cattle. It suffered much during the religious wars. One of its governors, a Viscount d'Orthez, is justly famous for being one of the few who refused to carry into effect the orders of the court for the detestable massacre of St. Bartholomew.

Near this town, on the 27th of February, 1814, the Anglo-Spanish army, under the Duke of Wellington, defeated a French force under Marshal Soult. The action was well contested and bloody: the French lost nearly 4000 men, killed, wounded, and prisoners, and the allies, 2300. (*Napier's Peninsular War,* vi., 564.)

ORTONA a seaport town of central Italy, kingdom of

Naples, prov. Abruzzo Citra, cap. canton, on the Adriatic, 11 m. E.N.E. Chieti. Pop., in 1832, 6000. Its chief edifices are its cathedral and other churches, convents, &c.; and a palace, once the winter residence of Margaret, daughter of the Emperor Charles V. It was anciently the principal port and naval arsenal of the Frentani (*Strabo,* v., 241; *Pliny,* iii., 11, &c.), and it has still a few insignificant remains of antiquity; but its harbour has now ceased to exist. Vessels anchor in soft ground about half a league from shore, in from 10 to 15 fathoms water, or nearer if requisite, in less water; but the station is exposed to N. and E. winds; and there are various rocks and shallows. (*Nurie's Sailing Directions.*)

ORVIETO (the an. *Herbanum,* afterward *Urbs Vetus?*), a town of central Italy, Papal States, cap. deleg. of same name, close to the junction of the Paglia and Chiane, about 5 m. from their union with the Tiber, 13 m. N.E. the lake of Bolsena, and 59 m. N.N.W. Rome. Pop. about 7000. It stands on an isolated and scarped tufa rock; and is clean, well built, and embellished with fine palaces. Among the latter are the bishop's palace; the Gualterio, with frescoes by Domenichino, Albano, L. Signorelli (?), &c.; and the *palazzo* Petrangeli, with paintings by Pietro Perugino. The cathedral, founded in 1900, is a remarkable Gothic edifice, very rich in bas-reliefs, mosaics, paintings, and statuary, with a large and handsome circular window. It has several other churches, a Jesuits' college, and a large well, dug by order of Clement VII., which is shown as a curiosity. Various Etruscan antiquities have been discovered here; and a light white wine is grown near the town, which has acquired greater celebrity than it deserves. The inhabitants carry on some trade in cattle, wine, and silk. (*Dict. Géog.*; *Wood's Letters of an Architect,* i., 319, 322.)

ORWELL, p. t., Rutland co., Vt., 70 m. S.W. Montpelier, 463 W. Bounded W. by lake Champlain, lies opposite to Ticonderoga, N.Y., and contains mount Independence, both of which were celebrated in the Revolutionary war. Chartered in 1763, organized in 1787. Drained by East creek and Lemonfair river, which afford water-power. It is one of the best farming towns in the state, and contains two churches, a Congregationalist and Baptist, three stores, one fulling-mill, one grist-mill, one saw-mill; 10 schools, 473 scholars. Pop. 1504.

ORWELL, p. t., Bradford co., Pa., 146 m. N. Harrisburg, 226 W. Watered by Wysox creek and its tributaries. It contains one store, one woollen factory, one grist-mill, four saw-mills, two tanneries; eight schools, 225 scholars. Pop. 1037.

ORWIGSBURG, p. b., capital of Schuylkill co., Pa., 62 m. N.E. Harrisburg, 179 W. Situated on a small creek flowing into Schuylkill river, and surrounded by beautiful hills, cultivated to their tops. It contains 150 dwellings, many of them handsome brick houses, three stories high. It has a courthouse surmounted by a cupola, and the county offices, all of brick, a jail, an academy endowed by the state with $60000, a spacious Lutheran church; seven stores, one saw-mill, one printing-office, one weekly newspaper, two academies, 87 students; two schools, 80 scholars. Pop. 779.

OSAGE, r., Mo., rises in the Indian territory in about 38° 30' N. lat., and 97° W. long., and flows into the state of Missouri, and falls into Missouri river 133 m. above the Mississippi. It has a very winding course, is 397 yards wide at its mouth, and is navigable for steamboats for 900 miles. There is much fertile land on its borders. Its length is supposed to be about 350 miles.

OSIMO (an. *Auximum*), a town of central Italy, Papal States, deleg. of Ancona, 8½ m. S.S.W. the city of that name. Pop. about 7000. It is well built, having a handsome bishop's palace, a cathedral, several churches and convents, a college, &c. In antiquity this was one of the most important towns of Picenum. It was included among the cities of the Pentapolis, and was taken by Belisarius from the Goths, after an obstinate defence. (*Cramer's Anc. Italy; Dict. Géog.*)

OSNABURG (Germ. *Osnabrück*), a town of the kingdom of Hanover, cap. prov. and principality of its own name, on the Hase, a tributary of the Ems, 63 m. W. by S. Hanover. Pop., in 1833, 11,500. It is walled and divided into the Old and New town. "The palace, the townhouse (in which the treaty of Westphalia was concluded in 1648), the court of justice, and the cathedral (in which some relics supposed to have belonged to Charlemagne are kept), are all good buildings; and there are a great many good-looking private houses, belonging to merchants. Though not the largest, it is undoubtedly the best situated and the handsomest town of the Hanoverian dominion." (*Hodgskin's Trav. in the N. of Germ.,* 1819, i., 310.) But other authorities represent the town as irregularly and ill built. There are two Lutheran and two Roman Catholic churches, a Lutheran orphan-house, four hospitals, a workhouse, a Catholic and a Lutheran gymnasium, and a house of correction. Osna-

OSSIPEE.   OSTIA.

burg is a place of considerable trade, from being in the centre of a country where great quantities of the linen cloths termed Osnaburgs are made, and which are brought thither for inspection, stamping, and sale. (*See* HANOVER, i., 1062.) But it is in a great measure indebted for its trade in these fabrics, and in cattle, to its position on the high road between Bremen and the Lower Rhine. It has, also, manufactories of woollen cloths, tobacco, chicory, soap, paper, leather, &c. No court has been kept up in Osnaburg since the time of Ernest Augustus, father of George I.; but the nobility of the province generally reside here; and without either having a university or being a royal residence, it is in some degree celebrated for the literature and polish of its inhabitants. It is the seat of a Roman Catholic bishop; and its civil governor, nominated by the King of Hanover, is called, though without having any ecclesiastical duties to discharge, the Prince-bishop of Osnaburg: this dignity was held by the second son of George III. (*Berghaus, Allg. Länder, &c.,* iv., 234; *Hodgskin's Trav. in N. Germ.,* i., 306, 318, &c.)

OSNABURG, p. t., Stark co., O., 129 m. N.E. Columbus, 306 W. Watered by a branch of Sandy river. The village contains a church, three stores, two tanneries, three distilleries, 70 dwellings, and about 400 inhabitants. Pop. of the township, 2333.

OSSIPEE, p. t., capital of Carrol co., N.H., 52 m. N.N.E. Concord, 533 W. Bounded N.E. by Ossipee lake, covering about 7050 acres. Incorporated in 1780. Watered by Bear Camp river. It contains a courthouse, jail, four churches, a Congregational and three Freewill Baptist; seven stores, five grist-mills, 10 saw-mills, three tanneries; 22 schools, 878 scholars. Pop. 2170.

OSSUNA, a town of Spain, in Andalusia, prov. Seville, 42 m. E. Seville. Pop., according to Miñano, 16,000. It is built amphitheatre-wise on the declivity of a lofty hill, on the top of which stands the parish and collegiate church. It has, also, four hospitals and two sets of barracks. The town formerly possessed a university, which attained considerable celebrity in the time of Cervantes; but at the close of last century it was in a state of decay, and was suppressed in 1824. Osuña is neat and pretty, surrounded by orchards, among which are some fine public walks; but it has an insufficient supply of water. The climate is good, except in summer, when, during the prevalence of the *Solano*, or E. wind, the thermometer often rises to 111° Fahr. The inhabitants are principally employed in agriculture, and the neighbourhood has the reputation of being one of the most productive grain districts of Andalusia. It is, also, celebrated for its capers, large quantities of which are pickled and sent to Seville and Cadiz. (*Miñano.*)

OSTEND, a fortified seaport town of Belgium, province of W. Flanders, capital of the canton, on the shore of the North sea, 14 m. W. by N. Bruges, 27 m. E.N.E. Dunkirk, and about 60 m. E. the North Foreland. Lat. 51° 13′ 57″ N., long. 2° 55′ 8″ E. Pop., in 1836, 12,161. Few English travellers speak in favourable terms of Ostend. It is, however, neatly and regularly built, and has a lively appearance, the houses being painted of different colours. (*Barrow's Tour,* 285.) It is also a favourite watering place of the Belgians, and is occasionally resorted to by the royal family. It has no public edifice worth notice, except a good bathing-house, with reading-rooms, &c., on the *levée*, a sloping glacis of stone-work, originally erected to serve as a dyke, having on its summit a favourite promenade. Ostend is strongly fortified by ramparts, a broad ditch, and a citadel; but it is ill supplied with water, which is assigned as a reason for its being in parts rather dirty. The interior harbour, which is large and commodious, is bordered by a broad quay; but ships of considerable burden can only enter the port at high water, and in strong off-shore winds is difficult of access. There are two lighthouses which, when brought in line, mark the channel that leads into the port. The exports consist of corn, clover seed, cattle, and other farm produce; and the imports of sugar, coffee, and other colonial products, wines, spices, English manufactured goods, &c.

The cod and herring fisheries, especially the former, are carried on to a considerable extent from Ostend. In 1836 the produce of this fishery amounted to 7841 tons salt fish, and in 1827 to 8799 tons. The aggregate value of the imports by sea, in 1840, amounted to 9,383,557 fr. Exclusive of its trade, Ostend has some sugar and salt refineries, and sailcloth, soap, tobacco, and other factories, with rope-walks, building-docks, distilleries, &c. It is connected by a canal and railroad with Bruges, and by the great Belgian railroad with Antwerp and other cities of the interior. It is the seat of a tribunal and chamber of commerce, and the residence of a military commandant and of an English consul.

During the ever-memorable struggle made by the Dutch to emancipate themselves from the blind and brutal despotism of Old Spain, Ostend sustained one of the most celebrated sieges of which history has preserved any account. It continued from the 4th of July, 1601 to the 28th of September, 1604, when the garrison capitulated, on honourable terms, to the ablest of the Spanish leaders, the famous Marquis of Spinola. This siege is supposed to have cost the contending parties the lives of nearly 100,000 men! (*See Watson's Philip III.,* i., 99–188, 8vo ed.; *Official Returns, &c.*) We subjoin an

ACCOUNT of the Ships that entered and left the Port of Ostend, in 1836 and 1837; specifying the Countries to which they respectively belonged.

	1836.		1837.	
	Entered.	Left.	Entered.	Left.
Belgian	947	919	227	221
English	130	134	134	115
French	92	96	34	25
Norwegian	51	51	96	96
Hanoverian	71	84	30	66
Oldenburgh	4	6	6	4
Mecklenburgh	7	7	6	6
Prussian	3	3	14	15
Hamburgh	1	1		1
Danish	10	12	6	10
Bremen	1		2	1
Knaphausen	1		2	2
Russia	1		2	
Spanish	1	1		
Total	**548**	**548**	**557**	**597**

And this number is exclusive of the steamers conveying the mails to and from England.

OSTERODE, a town of central Germany, kingdom of Hanover, princip. Grubenhagen, on the Söse, a tributary of the Leine, at the foot of the Harz, 49 m. S.S.E. Hanover. Pop., in 1837, 4600. It is walled, and has several churches, a hospital, a gymnasium, &c., but its principal public edifices are the royal granaries, which supply about 58,000 scheffel of corn annually, at 16 groschen (2s.) the scheffel, to the miners and other labourers of the Harz. Osterode has manufactures of woollen and cotton goods, table linens and long cloths, hats, tobacco, soap, white lead, copper and wooden articles, with breweries, distilleries, tanneries, &c. (*Hoeselmann's Stein; Berghaus.*)

OSTIA, a decayed town and seaport of Italy, Papal State, on the left or S. arm of the Tiber, a little below where it divides into two branches to enclose the Isola Sacra, about 3 m. from its mouth, and 15 m. W.S.W. Rome, lat. 41° 45′ 35″ N., long. 12° 16′ 35″ E. This miserable little town, which has scarcely 100 inhabitants, and which, in fact, is all but uninhabitable from malaria, was in antiquity a flourishing emporium. It was for a lengthened period the seaport of Rome; and was founded by Ancus Martius in the view, who, it is, also, said to have constructed the salt-works in its vicinity. (*In ora Tiberis Ostia urbs condidit, salina circa facta : Livius in Anco,* lib. i., cap. 33.) In the course of time Ostia rose, with the rise of Rome, to be a place of great wealth, population, and importance. It was taken by Marius, who appears to have treated it with great severity. (*Livii Epit.,* lib. lxxix.) But it soon recovered from this disaster, and continued for a lengthened period to engross the whole trade of Rome carried on by sea. But its port had never been good; and, owing to the gradual accumulation of the mud and other deposits brought down by the river, it ultimately became inaccessible to ships of considerable burden, who were obliged to anchor on the coast at an exposed and hazardous situation. Many efforts were made at different periods to obviate these inconveniences, but apparently without much success; and at length the Emperor Claudius determined to construct a new port *Portus* at the mouth of the N. or right arm of the Tiber. This harbour was wholly artificial, and was formed at a vast expense by moles projecting into the sea. (*Suetou. in Claud.,* cap. 20; *Dio Cassius,* lib. ix.) Mannert, in his article on Ostia, says that the port constructed by Claudius was repaired by Trajan, and continued to be the port of Rome as long as the Roman empire existed! This statement is, however, in all respects, wide of the mark. The truth is, that the same circumstances (the accumulation of sand and mud at the mouth of the river) that had destroyed the harbour of Ostia, very soon began to choke up the new port; and instead of attempting to improve the latter, Trajan judged it more expedient to construct a totally new harbour at Centumcellæ, now Civita Vecchia, though the latter was more than double the distance of the former from Rome! (*Plinii Epist.* lib. vi., ep. 31, and art. CIVITA VECCHIA in this work.) The harbour formed by Trajan is at this moment the best by far on the W. coast of central Italy; the great works, the construction of which is described by Pliny, still remain entire, and evince the superior discernment and power of its illustrious founder. The port of Claudius, as well as Ostia itself, is now at a considerable distance from the sea; and its harbour (which, according to Bergier (*Hist. des Grands Chemins,* ii., 356), could not have been executed by any European monarch, is a shallow, noisome pool :—

"Tantum evi longinqua' valet mutare vetustas."

OSTUNI, a town of the Neapolitan dom., prov. Otranto, ap. canton, on the brow of a steep hill, 21 m. W.N.W. Brindisi. Pop. about 6060. It is a bishop's see, but remarkable for little more than the number of its churches and convents. Its climate is said to be highly salubrious. *Craven's Tour, &c.*, 123, 194.)

OSWEGATCHIE, t. St. Lawrence co., N.Y., 16 m. W. Canton, 200 m. N.N.W. Albany. Bounded N.W. by St. Lawrence r. Watered by Oswegatchie r. It contains a part of Black lake in its S.W. part. The village of Ogdensburg lies in its N. part. It contains five churches, an Episcopal, Presbyterian, Methodist, Baptist, and Roman Catholic; two commission houses in foreign trade, 57 stores, three furnaces, three fulling-mills, two grist-mills, seven saw-mills, four tanneries, two printing-offices, one bindery, two weekly newspapers; 19 schools, 683 scholars. Pop. 3193.

OSWEGO, r., N.Y., is formed by the union of Seneca and Oneida rivers. Seneca river receives the waters of Seneca, Cayuga, Crooked, Canandaigua, Owasco, Skeneateles, and Cross lakes. Oneida river forms the outlet of Oneida lake. After the junction, Oswego river pursues a N.W. course for 24 m., and enters lake Ontario at Oswego village. It forms for some distance the Oswego canal, its falls being overcome by a number of locks. The harbour of Oswego at its mouth has been much improved by artificial structures. The village affords good water-power. It is not navigable for lake vessels, having falls near its mouth.

OSWEGO, county, N.Y., situated in the N. part of the state, and contains 923 sq. m. Bounded N.W. by lake Ontario, S.E. by Oneida lake. Watered by Oswego river. The Oswego canal passes through it. It contained in 1840, 35,369 neat cattle, 63,843 sheep, 39,233 swine; and produced 138,602 bushels of wheat, 1676 of rye, 189,327 of Indian corn. 41,618 of buckwheat, 11,961 of barley, 215,177 of oats, 599,137 of potatoes, 264,980 pounds of sugar. It had seven commission houses in foreign trade, 107 retail stores, three lumber-yards, six furnaces, 16 fulling-mills, two woollen factories, 10 flouring-mills, 99 grist-mills, 135 saw-mills, two oil mills, one paper-mill, 25 tanneries, five printing-offices, five weekly newspapers; five academies, 468 students; 325 schools, 19,193 scholars. Pop. 43,619. Capitals, Oswego and Pulaski.

Oswego, p. v., port of entry, and semi-capital of Oswego co., N.Y., 160 m. W.N.W. Albany, 372 W. The village lies in the townships of Oswego and Scriba. Incorporated in 1828. It is regularly laid out with broad streets, crossing each other at right angles. The two parts of the village are connected by a bridge across Oswego river, 700 feet long, and which cost $80000. It contains a courthouse, jail, seven churches, two Presbyterian, an Episcopal, a Tabernacle (Union), Methodist, Baptist, and Roman Catholic; two banks, with an aggregate capital of $400,000, an academy, and a flourishing female seminary. The harbour, next to that of Sackett's harbour, is the best on the S. side of lake Ontario. It is formed by a pier or mole filled with stone 1259 feet long on the W. side of the harbour, and 900 feet on the E. side, with an entrance between them. The water within the pier has a depth of from 12 to 20 feet. The cost of this work was $93,000. On the end of the W. pier there is a light, and a lighthouse on the hill on the E. side of the harbour near the port. More than 70 vessels, including steamboats, are owned at this port, besides a large number of canal-boats, built in the most substantial manner, generally having decks, and capable of being towed through the lakes. Not far from one fourth of the trade from New-York city with the west, goes through Oswego and the Welland canal, which passes round Niagara falls into lake Erie. The salt from Salina, destined for the west, generally passes this way. The tonnage of the port in 1840, was 8346. Oswego has great manufacturing advantages. A feeder dam, 7½ feet high, three fourths of a mile above the village, furnishes an abundance of water taken from the canal on the E. side of the river, under a fall of 19 feet. A canal has also been constructed on the W. side, by a wall on the margin of the river 18 feet high, forming a canal along the bank 62 feet wide, 7 feet deep, with a fall of 19 feet, which cost $75,000.

Oswego village contains three commercial and four commission houses in foreign trade, with a capital of $346,000; 46 retail stores, with a capital of about $100,000; two cotton factories, one iron foundry, eight flouring-mills, one saw-mill, six machine shops, one plaster-mill, one large tannery, one morocco factory, one snuff factory. There are a marine railway and three ship-yards. The market building, a large edifice of brick, contains also a town-hall, the corporation chamber and offices, the postoffice, county clerk's office, a room for the Mechanics association, and in the attic a spacious room 133 feet long, for various public occasions.

The cost of the building was nearly $25,000. There is nearly a mile of wharves and dockage in the harbour, and many extensive warehouses and forwarding establishments. The population is about 6000. Daily lines of steamboats for the conveyance of passengers, run from Oswego to Lewiston, Sackett's Harbour, and Ogdensburg, and to Kingston in Canada. On the E. side of the harbour, on rising ground ceded to the United States, is fort Oswego, for the defence of the harbour, which occupies nearly the site of an old fort, famous in the French and Revolutionary wars. The town fell into the hands of the British during the late war, but they were driven from it in 12 hours, with loss.

OSWESTRY (corrupted from *Oswaldestrae*), a market-town and parish of England, hund. of its own name, co. Salop, on the borders of Wales, 16 m. N.W. Shrewsbury. Area of par., 13,680 acres: pop. of township, in 1831, 4478. The town, which was formerly surrounded with walls, is well paved and lighted; its chief public buildings are a town-hall, prison, theatre, and a fine old church, remarkable for its lofty, ivy-mantled tower. The living is a vicarage, in the gift of Lord Clive. The Independents, Baptists, Wesleyan Methodists, Welsh Calvinists, and Primitive Methodists, have their respective places of worship, to which are attached Sunday schools, furnishing religious instruction to upward of 700 children. A grammar school has been established here, and "is free for all boys born in the parish to be instructed in grammatical learning in the English, Latin, and Greek languages. The present annual value of the endowment is £230; but the number of free scholars seldom exceeds 20, besides whom there are about 94 pay-boys residing with the master. There is also an extremely well-regulated national school, attended by about 230 boys and 150 girls." (*Mun. Corp. Report.*) Oswestry, which from 1461 to 1691 was the great mart for woollens called Welsh webs, has still a few manufactures of flannel and coarse linen cloth; but its principal dependance is on its retail trade with an extensive agricultural district. The borough, which received its charter in 22 Richard I., is divided, under the Municipal Reform Act, into two wards, and the corporation consists of a mayor, five other aldermen, and 18 councillors. Corporation revenue, in 1839, £411. Quarter and petty sessions are held under a recorder, and there is a court for the recovery of small debts. Oswestry is one of the polling-places at elections for the N. division of Salop. Races are held near the town in September. Markets on Wednesday: large cattle fairs, 13th March, 12th May, Wednesday before 24th June, 15th August, and 10th December.

Oswaldestrae (more anciently called *Maserfield*) is supposed to have derived its name from Oswald, king of Northumbria, killed here in battle in 642, and subsequently canonized. It was surrounded by walls by Edward II., in 1277, and became highly important as one of the keys to the principality of Wales. At the W. end of the town, on a lofty hill, are some remains of its castle, supposed to have been built about the time of the Norman conquest. (*Pennant's Tour in Wales*, i., 333, &c.; *Parl. Reports.*)

OTAHEITE. (See POLYNESIA.)

OTEGO, p. t., Otsego co., N.Y., 22 m. S.W. Cooperstown, 86 m. W. Albany, 340 W. Bounded S.E. by Susquehanna river, and drained by its tributaries. It contains three churches, a Presbyterian, Episcopal, and Baptist; five stores, two fulling-mills, two grist-mills, eight saw-mills; 13 schools, 571 scholars. Pop. 1919.

OTIS, p. t., Berkshire co., Mass., 15 m. S.E. Lenox, 119 W. Boston, 368 W. Drained by branches of Farmington and Westfield rivers. It contains four churches, a Congregational, Episcopal, Methodist, and Baptist; three stores, one forge, two grist-mills, six saw-mills, two tanneries; one academy, 90 students; 10 schools, 305 scholars. Pop. 1177.

OTISCO, p. t., Onondaga co., N.Y., 15 m. S. Syracuse, 138 m. W. by N. Albany, 325 W. Bounded N.W. by Otisco lake. It contains two churches, a Methodist and Presbyterian; two stores, one grist-mill, four saw-mills, two tanneries; 16 schools, 782 scholars. Pop. 1906.

OTISFIELD, p. t., Cumberland co., Me., 71 m. S.S.W. Augusta, 578 W. Bounded W. by Crooked river, which flows into Sebago lake or pond. It contains four stores, one fulling-mill, two grist-mills, two saw-mills. Pop. 1307.

OTLEY, a manufacturing and market-town, parish and township of England, W. Riding, co. York, upper div. wap. Skyrack, on the Wharfe, 9½ m. N.W. Leeds. Area of par., comprising 12 townships, 23,060 acres. Pop. of Otley township in 1831, 3161. The town, though small, is well built, and delightfully situated in a picturesque river valley. The church is large, but has few remains of its original architecture: the living is a vicarage in the gift of the lord chancellor. There are places of worship for Independents, Wesleyan and Primitive Methodists, with attached Sunday schools. A grammar school was founded here in 1611, and there is a national school for children of both sexes. Otley formerly enjoyed a considerable share in the woollen trade;

but it has long since been removed to situations nearer to the coal districts, and better placed for inland navigation. Within the parish, however, there were, in 1839, two cotton-mills, four woollen-mills, and 10 worsted-mills, moved chiefly by water-power, and employing 1182 hands, chiefly in the townships of Grisley, Yeadon, and Rawden. Tanning and malting are carried on in the town to a pretty considerable extent, and it has large markets and fairs for corn and cattle, besides an agricultural show held in April. The Archbishop of York is lord of the manor, and holds courts-baron for the recovery of small debts. Petty and quarter sessions are held here by the magistrates under the archbishop's commission. Markets on Friday: cattle fairs, Wednesday in Easter week, and every fortnight after till Whit-Sunday, and then every three weeks till 1st August.

OTRANTO (an. *Hydruntum*), a seaport town of Italy, near its S.E. extremity, kingdom of Naples, prov. of its own name, cap. canton, on the strait of Otranto, close to the point of Italy nearest to the Greek peninsula. 24 m. S.E. Lecce, and 44 m. W.S.W. cape Linguetta, in Albania. This petty town, of 1600 inhabitants, has little to interest the English reader, except the celebrated "Castle of Otranto," with the name of which every lover of romance is familiar. "The castle, however, is far from realizing the expectations created by the perusal of the work bearing its appellation. It is now, what it ever was, the citadel of the town, a fort of no considerable extent or power, but not entirely deficient in picturesque beauty, especially on the land side. Two large circular towers rise from the rich foliage of the trees which fill the town ditch, and among which a very high palm is conspicuous." (*Craven's Tour.*) The castle, which comprises prisons, stables, a mill, a chapel, &c., was built by Alphonso of Aragon, who otherwise fortified the town, as a bulwark against the Turks. Otranto has a very ancient cathedral, in which are some columns taken from the temple of Minerva in the vicinity; an archbishop's palace and a few Roman antiquities. In 1480 it was taken and sacked by the Turks. Under Napoleon, it gave the title of duke to Fouché. (*Craven's Tour in the S. Provinces of Naples*, 143, 144; *Burgess's Greece*, i., 26, 30; *Dict. Géog.*)

OTSEGO, lake, N.Y., is 9 m. long, and from 1 to 2 broad. It is situated in the N. part of Otsego county and is a beautiful sheet of water, furnishing a variety of excellent fish. Its outlet forms the E. branch of Susquehanna river, and on it, as it leaves the lake, is the beautiful village of Cooperstown, which commands a fine view of the lake to the N.

OTSEGO, county, N.Y. Situated centrally towards the E. part of the state, and contains 892 sq. m. It contains Otsego and Canaderaga lakes, the outlets of both of which flow into Susquehanna river. Bounded W. by Unadilla river. Watered by Susquehanna river, and Butternuts, Otsego, and Cherryvalley creeks. In Burlington, the stones abound with marine petrifactions. It contained in 1840, 66,035 neat cattle, 235,979 sheep, 47,637 swine; and produced 148,880 bushels of wheat, 68,936 of rye, 122,388 of Indian corn, 45,059 of buckwheat, 116,715 of barley, 693,989 of oats, 1,293,109 of potatoes, 168,705 pounds of hops, 351,748 of sugar. It had 139 stores, seven furnaces, 43 fulling-mills, four woollen-factories, eight cotton-factories with 17,502 spindles, one dying and printing works, 85 grist-mills, 292 saw-mills, three oil-mills, one paper-mill, 47 tanneries, nine distilleries, one brewery, five printing-offices, four weekly newspapers and one periodical; nine academies, 385 students; 300 schools, 13,392 scholars. Pop. 49,628. Capital, Cooperstown.

OTSEGO, t., Otsego county, N.Y., 66 m. W. of Albany. Bounded E. by Otsego lake, N.W. by Canaderaga lake, and Oak creek; the outlet of the latter, affords water-power. It contains 26 stores, two cotton-factories with 4000 spindles, one furnace, six grist-mills, 12 saw-mills, one paper-mill, 4 tanneries, one brewery, three printing-offices, two weekly newspapers and one periodical; three academies, 197 students; 34 schools, 1093 scholars. Pop. 4190.

OTSELIC, p. t., Chenango county, N.Y., 15 m. N.W. Norwich, 86 m. W. by S, Albany, 344 m. W. Drained by Otselic creek. It has one store, one grist-mill, seven saw-mills, two tanneries; 14 schools, 581 scholars. Pop. 1691.

OTTAJANO, a town of the Neapolitan dom., prov. Napoletano, at the E. foot of mount Vesuvius, 19 m. E. Naples. Pop. estimated at 6000. It has three churches, a castle, and some other public buildings. Its inhabitants are principally engaged in agriculture, having but little taste for commerce. Several antiquities found here are supposed to have formed part of a palace anciently belonging to the Octavian family, from whom the town is conjectured to have derived its name.

OTTER CREEK, Vt., the longest river which runs wholly in the state; rises in mount Tabor, Peru, and Dorset, and flowing by a general course of N.N.W. for 90 miles, it enters its most fertile alluvial borders; and at Middlebury, Weybridge, and especially at Vergennes, it affords extensive water-power. It is naviga-

ble for the largest lake vessels and steamboats, 8 m. to Vergennes. The large vessels of McDonough's squadron, and several of the largest steamboats on the lake, have been built at Vergennes.

OTTER, Peaks of, Va., constitute the highest summits of the Blue ridge, and the highest land in the state. The E. peak is 3104 feet high, and the W. 2946 feet high. Other statements make the greatest elevation over 4000 feet. They are situated between Bedford and Botetourt counties, 20 m. W. by N. of Lynchburg.

OTTERY ST. MARY, a market-town near of England, so. Devon, hund. its own name, on the Otter (whence it derives its name), 11½ m. E. by N. Exeter. Area of par. 9470 acres. Pop. in 1841, 4194. It is large but irregularly built; containing many vestiges of antiquity, among which is a house formerly inhabited by Sir Walter Raleigh. The church is a large and curious structure, built like Exeter cathedral, with two towers opening into the body of the church, and serving as transepts; a ladye-chapel occupies the E. end, and in the interior is a fine arched monument. The living is a vicarage, in the gift of the lord chancellor. The Independents and Wesleyan Methodists have likewise their respective places of worship, with attached Sunday-schools. A grammar-school was founded here by Henry VIII.; but it has for many years been an almost useless appendage to the town. Two day-schools are supported by subscription, and an infant-school is attended by about 60 children. Almshouses are established here, and there are numerous minor charities for the relief of the poor within the parish. Ottery St. Mary, which had formerly a considerable share in the manufacture of serges and coarse woollen-cloths, is now chiefly supported by agriculture and retail trade. A silk-mill, however, has been established within the last few years, which in 1839, employed 385 hands. Petty sessions for the hundred are held here. Markets on Tuesday; fairs, Tuesday before Palm-Sunday; Whit-Tuesday, and Aug. 5, for cattle and sheep.

OTTO, p. t., Cattaraugus co., N.Y., 10 N.W. Ellicottville, 306 m. W. by S. Albany, 348 m. W. Bounded N. by Cattaraugus creek, by branches of which it is drained. It contains three stores, one fulling-mill, two grist-mills, five saw-mills; 12 schools, 633 scholars. Pop. 2132.

OTTAWA, county, O. Situated towards the N.W. part of the state, and contains 350 sq. m. Drained by Carrying or Portage river, and Toussaint and Crane creeks. It contained in 1840, 327 neat cattle, 1489 sheep, 5074 swine; and produced 14,506 bushels of wheat, 1693 of rye, 36,737 of Indian corn, 1522 of buckwheat, 3465 of oats, 15,734 of potatoes, 9449 pounds of sugar. It had five stores, two grist-mills, two saw-mills; four schools, 115 scholars. Pop. 2248. Capital, Port Clinton. It includes Cunningham's, and the Bass's islands in lake Erie.

OTTAWA, county, Mich. Settled in the N.W. part of the settled portion of the peninsula. Watered by Grand river and its tributaries, and Muskegon river. Bounded W. by lake Michigan. It contained in 1840, 289 neat cattle, 60 swine; and produced 1226 bushels of wheat, 3936 of Indian corn, 2525 of oats, 7741 of potatoes, 13,229 pounds of sugar. It had two commission-houses in foreign trade, two retail stores, 12 saw-mills, one tannery. Pop. 208. Capital, Grand Haven.

OTTAWA, p. v. cap. of La Salle co., Ill., 133 m. N.N.E. Springfield, 773 m. W. Situated on both sides of Illinois river, at the junction of Fox river. For the distance of eight or nine miles below, there are rapids, which are an impediment to navigation, except in high-water. The place has a convenient steamboat landing with deep water. It has ten stores; about 80 dwellings, and 500 inhabitants.

OUDE, (*Ayodhya*), a prov. and kingdom of Hindustan, under the protection of the British, between the 26th and 28th degs. of N. lat., and the 79th and 83rd of E. long. It has N. Nepaul, but is every where else surrounded by the territories of the Bengal and Agra presidencies, having W. the provinces of Delhi and Agra, S. Allahabad, and E. Behar. Area estimated at about 25,300 sq. m. Pop. probably 5,000,000. This country is an extended plain, bounded northward by the lower Himalaya ranges, and W. and S. by the Ganges, being well watered by several tributaries of the latter. When properly cultivated, the soil is extremely productive, yielding crops of wheat, barley, rice, and other grains, sugar, indigo, opium, and all the richest crops raised in India; and Heber states, that from Lucknow to Sundee the country is as populous and well cultivated as most of the Company's provinces. But Oude was for a lengthened period so wretchedly mismanaged by its native authorities, that, from being one of the richest states of Hindustan, it had become, a few years ago, one of the poorest and most miserable, being peculiarly distinguished for anarchy and disorder. The total revenue is estimated at between 18 and 19 million rupees a year, and its collection is farmed out to the highest bidders. Large tracts of the country are also in jaghire, or feudally conferred for military service; the aumils, or revenue collectors, and the zemindars, are so independent of the royal authority, that they yield only a very imperfect obedience, and ae late as 1830 Oude was distracted by civil war between the sovereign and his military chieftains. A full account of the condition of the country at that period may be found in the *Report on E. I. Affairs*, 1832. (*Political Appendix*, p. 396–420.) Under the new king, however, great ameliorations have been effected, and the standing army has been reduced from 50,000 men to less than half that number. The subsidiary troops furnished by Oude to the Anglo-Indian army amount to 10,000 men. The foreign relations and treaties of the state are wholly conducted through the British resident at Lucknow. It has been secured ever since 1765, by the British government from foreign aggression; and it is to be regretted that we did not also so far interfere in its internal concerns as to introduce regularity and efficiency into the administration, and to repress disorders. *See Crawfurd's Letter in Rep. Polit. Append.*, p. 93, 94.)

OUDE, a town of Hindostan, in the above prov. and kingdom, of which it was the former capital; on the Gogra, across which an iron bridge, the materials having been brought from England, is said to have been very recently thrown, 74 m. E. Lucknow; lat. 26° 48' N., long. 82° 4' E. It extends for a considerable distance along the banks of the river, stretching as far as Fyzabad. It is said by Hamilton to be tolerably populous; but except along the river's brink, it consists wholly of ruins and jungle, among which are the remains of various celebrated Hindoo temples. Hindoo pilgrims still visit Oude; and did so in great numbers, until Aurungzebe demolished most of their places of resort. A mosque erected by that monarch, and two tombs, greatly venerated by Mohammedans, are now the principal and almost sole remaining public edifices. (*Med. Tres.* ix. 312–315, &c.)

OUDENARD, or AUDENARD, a town of Belgium, prov. E. Flanders, cap. arrond. on the Scheldt, 14 m. S.S.W. Ghent. Pop. in 1836, 5539. It is generally well built, and has one of the handsomest town-halls in the Netherlands, several churches, a hospital, two orphan asylums, a convent, a college, and other schools, including one for spinning yarn. It has some manufactures of cotton and woollen fabrics, with breweries and tanneries. On the 11th of July, 1708, a powerful French army, commanded by the Dukes of Burgundy and Vendôme, was defeated in the vicinity of this town, and obliged to make a disorderly retreat, by the allied army under the Duke of Marlborough and Prince Eugene.

OUNDLE, a market-town and parish of England, county Northampton, and hundred Polebrook, on the Nen (crossed here by two bridges, one of which has five arches), 25 m. N.E. Northampton, and 67 N. by W. London. Area of par., 5300 acres; pop. in 1831, 2308. The town, though small, is neat and well built, having a good market-house. The church is large and handsome, with a square-tower, having octagonal turrets at the angles, and surmounted by a lofty spire; the living is a vicarage, in the gift of the lord chancellor. A grammar-school, established in 1541, is attended by about 60 boys. Two charity-schools furnish clothing and instruction to 42 boys; and a national-school is attended by about 130 children of both sexes. There are also two almshouses. Petty sessions for the hundred are held here once a fortnight. Markets on Thursday; fairs, Feb. 25, Whit-Monday, August 21, and October 12, for horses, sheep, &c.

OURAL, or YAIK, the *Rhymnus* of the ancients, a large river of the Russian empire. It rises in the Oural mountains, whence its name, in the district of Troitsk; and after a lengthened south-westerly coarse past Orsk, Orenbourg, and Ouralsk, pours its waters by various mouths into the northern part of the Caspian sea. It is shallow, and of little use for navigation; but teems with fish, which, at the proper season, afford a rich harvest to the Cossacks of the Oural. It is reckoned one of the bulwarks of the Russian empire against the incursions of the nomades of the Tartar steppes.

OURAL, or OURAL MOUNTAINS, an extensive mountain-chain, extending, including its subsidiary portions, nearly under the same parallel from the N. border of the sea of Aral to the shores of the Arctic ocean, or from about the 48th to about the 69th degree of N. lat. It forms, during the greater part of its course, the boundary between Europe and Asia. Where highest, it attains to an elevation about 6400 ft. above the level of the sea; but the ascent to the summit, particularly on the European side, is so very gradual, that its height does not appear nearly so great as it really is. Its breadth varies from one to five geographical miles. It is very productive both of the precious and useful metals; being estimated to afford, at an average, about 300 pouds of gold, 290,000 do. of copper, 5,500,000 do. of forged, and 8,500,000 do. of cast iron. (*See* art. RUSSIA; *Schnitzler, La Russie, &c.*, p. 652; *Humboldt's Fragmens de Géologie, &c.*, ii. 315, &c.)

OURFA, (an. *Edessa* or *Callirrhoë*) a fortified city of Asia Minor, pach. Diarbeker, 81 m. S.W. the city of that name, and 112 m. N.E. Aleppo. Pop. 30,000, 3-4ths of whom are Turks and Arabs, and the rest Armenians and Jews. It occupies the slopes of two hills, in the valley between which is a fountain and large basin containing a number of fish accounted sacred by the inhabitants. The houses are substantially built of hewn stone, and surmounted by terraces; gutters two or three feet in width run through the middle of the streets, and on each side are tolerably clean pavements. The mosques, which are numerous, have all lofty and many of them handsome minarets; there are churches also for the adherents of the Greek and Armenian creeds. The bazaars are of tolerable size, and covered in from the weather, some being occupied by cloth merchants, others by goldsmiths and other artizans.

Ourfa is a place of considerable industry; large quantities of cotton-fabrics are made in it; its goldsmiths' work and morocco leather are highly esteemed, and the produce of the neighbourhood, especially wheat and barley, is sent to Aleppo and the N. of Syria, and, by way of Bir, across the Euphrates. "The general trade with Aleppo is carried on almost entirely by about twenty Turkish and Christian merchants. They employ a capital not exceeding 50,000 dollars; but they trade on credit, both at Ourfa and Aleppo, for a much greater amount. Three fifths of this are for British manufactures, principally cotton twist, callicoes, a few prints, tin-plates, and nankeens; the remaining two fifths being for colonial produce, and different articles in the country. The inland duty paid at Ourfa by the rayas is invariably 5 per cent. on the invoice cost; and the price of carriage from Aleppo ranges between 80 and 100 piasters per Aleppo cantar on every kind of goods." (*Bowring's Report on Syria*, p. 44.)

The ancient *Edessa* was for a lengthened period the cap. of the prov. Osroene, in Mesopotamia. It is said to have been one of the numerous cities built by Seleucus Nicator, and continued under his successors as long as they remained possessors of Syria. It was taken by the Arabs during the caliphate of Abubekr, in the seventh century, and, after many vicissitudes, Selim I. united it, in 1517, to the Ottoman empire. (*Olivier, Voyage dans l'Empire Oth.*, iv. 210–223.)

OUSE, a river of England, co. York, one of the principal affluents of the estuary of the Humber, and which, indeed, may be considered as representing the various rivers that join it before it falls into that great estuary. The Swale, the remotest branch of the Ouse, has its source in the mountain of Shunnor Fell, on the borders of Yorkshire and Westmoreland, one of the most elevated in the great central range. Pursuing a S.E. course, the Swale is joined a little below Boroughbridge by the Ure, from Askrig, Middleham, and Ripon. A little lower down, the united river takes the name of the Ouse, and flowing past York, receives at Cawood its important tributary, the Wharfe, which, flowing through Tadcaster, has its source near Arncliffe. From Cawood the Ouse flows S.E., with many windings, to Goole, where it unites with the Humber, receiving the Derwent from the N., and the Aire and Don from the S. The Ouse is itself navigable for considerable vessels as far as York, and for barges to Linton; and the Aire, Don, and Derwent, have been partly by improvement in their channels, and partly by canals, rendered navigable to a great distance. The Ouse, is in fact, connected not only with the ports on the Humber but by an internal navigation, with Liverpool, London, and Bristol.

Exclusive of the above, another river, called the Great Ouse, rises in Northamptonshire, near Brackley; its course at first is E., a little inclining to the N., through Bucks; it then passes Olney and Harold, and, after many windings, reaches Bedford, where it becomes navigable. It then traverses the counties of Huntingdon and Cambridge, and the N.W. corner of Norfolk, falling into the estuary of the Wash at King's Lynn. From Huntingdon Sluice to Denver Sluice, a distance of about 20 miles in a direct line, the Ouse is now called the New Bedford river, from the greater part of its water flowing in the great channel or drain of that name, dug in the reign of Charles II. The principal affluents of the Great Ouse are the Nen, Cam, Little Ouse, Lake, Wissey or Stoke, and Nar.

OVERTON, a pari. bor. of N. Wales, co. Flint, hund. Maylor, on the Dee, (crossed here by a handsome stone bridge), 14 m. S. Chester, and 156 m. N.W. London. Pop. of parl. bor. and par., 1746. The town is situated on rising ground above the river, and is on the whole well built. The church is a venerable structure, picturesquely situated; and in the church-yard are some yews which, for size and beauty, are ranked among the wonders of the principality. The town has little trade; and would be unworthy of notice, were it not that it enjoys the privilege, in connexion with Flint and six other towns, of returning one member to the House of Commons. Registered electors for the united boroughs in 1839–46, 1053, ditto for Overton, 44.

OVIEDO, (*Ovetum*), an ancient city of Spain, cap. of the prov. of Asturias, in a plain at the confluence of the two small rivers Ovia and Nora, 60 m. N. by W. Leon, lat. 43° 19' N., long. 5° 57' W. Pop., according to Miñano, 10,476. It is an old-fashioned city, with many narrow and irregular streets; but it has several good squares, that forming the market-place being large and handsome. The town is supplied with water by a magnificent aqueduct of 41 arches, communicating with the fountains in the public squares. The principal public buildings are the cathedral, the collegiate church, and three parish churches, besides a district church, three hospitals, and four colleges. The cathedral (supposed to have been founded in the 8th century) is a large structure of Gothic architecture, and one of the most elegant in Spain, very similar to that of Toledo, though much smaller; it is surmounted by a beautiful, though unfinished, tower; and at the W. end is a noble open porch. (*Cook's Sketches in Spain*, ii. 99.) It contained many valuable vases, &c., taken away during the Peninsular war: it has also a large mausoleum, in which are deposited the remains of 14 kings and queens of Asturias. The university, founded in 1580, is well endowed, and has a large library: the university buildings are among the finest in the town. Oviedo has a weekly market, and three annual fairs; but its trade is chiefly confined to the neighbourhood. A few tanyards, manufactories of hats, horn combs, and metal buttons are established here, and domestic weaving is carried on to a considerable extent. A manufacture of arms was, till lately, supported by the government; but within the last few years it has been abandoned.

Oviedo is supposed to have been founded about A.D. 759. It afterwards became a place of refuge, during the persecutions of the Moors, for great numbers of Christian clergy, and hence acquired the name, *Civitas Episcoporum*. The pope in 901 made Oviedo an archbishop's see; but afterwards this honour was transferred to St. Jago, since which time the bishops of this city have been mere suffragans. The foundation of the university improved the condition of the inhabitants; and for upwards of 150 years Oviedo was a popular resort for literary monks and others desiring to avail themselves of the advantages offered by the ecclesiastical seminaries. When Mr. Townsend visited the town, it swarmed with mendicants, encouraged by a wealthy clergy; but since the French war, and the suppression of the conventual establishments the importance both of the town and university has very much diminished. (*Townsend*, ii. 1–14; *Mod. Trav.*; *Miñano*.)

OVERTON, county, Tenn. Situated in the N. part of the state, and contains 625 sq. m. Drained by Obies river, and its branches. It contained in 1840, 12,213 neat cattle, 11,011 sheep, 42,034 swine; and produced 36,452 bushels of wheat, 3713 of rye, 541,647 of Indian corn, 77,681 of oats, 19,464 of potatoes, 261,160 pounds of tobacco, 18,949 of cotton, 31,330 of sugar. It had 15 stores, two forges, 28 grist-mills, 14 saw-mills, one oil-mill, two powder-mills, five tanneries, 53 distilleries; 20 schools, 648 scholars. Pop.: whites, 8334; slaves, 889; free coloured, 56; total, 9279. Capital, Monroe.

OVID, p. t., semi-capital of Seneca co., N.Y., 189 m. W. Albany, 323 m. W. Situated about mid-way between Seneca and Cayuga lakes. It contains a courthouse, jail, county clerk's office, six churches, a Presbyterian, Episcopal, Methodist, two Baptist, and a Dutch Reformed, 12 stores, a furnace, a steam grist-mill, one printing-office, one weekly newspaper; one academy, 48 students; eight schools, 308 scholars. Pop. 2721.

OWASCO, p.t., Cayuga co., N.Y., 3 m. S.E. Auburn, 162 m. W. Albany, 334 m. W. Owasco lake lies on its S.W. border. It contains a Presbyterian church, one store, one fulling-mill, one grist-mill, four saw-mills; two schools, 831 scholars. Pop. 1319.

OWEGO, p. t., cap. of Tioga co., N.Y., 161 m. W.S.W. Albany, 275 m. W. Watered by Susquehanna river and Owego creek. It contains a court-house, jail, county clerk's office, a bank, four churches, a Presbyterian, Episcopal, Methodist, and Baptist, 36 stores, eight grist-mills, 52 saw-mills, two oil-mills, two printing-offices, two weekly newspapers, one academy, 268 students; 1309 scholars in schools. Pop. 5340.

OWEN, county, Ky. Situated in the N. part of the state, and contains 390 sq. m. Bounded S.W. by Kentucky river. It contained in 1840, 5,796 neat cattle, 11,810 sheep, 19,900 swine; and produced 61,030 bushels of wheat, 3913 of rye, 368,575 of Indian corn, 59,965 of oats, 8357 of potatoes, 776,969 pounds of tobacco, 18,457 of sugar. It had 13 stores one flouring-mill, eleven grist-mills, 12 saw-mills, two tanneries; 17 schools, 511 scholars. Pop.: whites, 6915; slaves, 1981; free coloured, 36; total, 8932. Capital, New Liberty.

OWEN, county, Ia. Situated a little S.W. of the centre of the state, and contains 360 sq. m. Watered by the W. fork of White river, and its tributaries, and by the S. fork of Eel river, which affords good water-power. It contained in 1840, 8450 neat cattle, 9578 sheep, 24,971 swine; and pro-

duced 47,082 bushels of wheat,1990 of rye, 357,030 Indian corn, 45,965 of oats, 11,969 of potatoes, 41,744 pounds of tobacco, 69131 of sugar. It had 19 stores, one woollen factory, one flouring-mill, 25 grist-mills, 16 saw-mills, one oil-mill, four tanneries, eight distilleries; 19 schools, 558 scholars. Pop. 8350. Capital, Spencer.

OWENSBORO', p. v., cap. of Daviess co., Ky., 168 m. W.S.W. Frankfort, 693 m. W. Situated on the S. side of Ohio river, and contains a courthouse, jail, a church, five stores, and about 200 inhabitants.

OWINGSVILLE, p. v., capital of Bath co., Ky., 73 m. E. Frankfort, 499 m. W. Situated a little W. of Slate creek, a branch of Licking river. It contains a courthouse, jail, two stores, and 251 inhabitants.

OWHYHEE. See POLYNESIA.

OXFORD, an inland and central co. of England, of a very irregular shape, bounded S. and S.W. by the Thames, by which it is separated from Berks, and having W. Gloucester, N. Warwick, N.E. Northampton, and E. Buckingham. Area 483,846 acres; of which above 405,000 are said to be arable, meadow, and pasture. Surface, a good deal diversified. The S. division of the county is traversed by the range of the Chiltern hills; the rest elsewhere it is mostly flat, or merely undulating. Soil very various; in the N. it consists of a deep, red, fertile loam; in the middle district it is comparatively sandy, gravelly, and poor; and in the S., thin and chalky soil predominates. The county is extremely well watered; for, besides being bounded, as already stated, for a lengthened distance by the Thames, it is traversed by the Windrush, Evenlode, Cherwell, and Thame. Oxford is principally in tillage; but though numerous improvements have been effected of late years, its agriculture is far from being in a very advanced state. This is accounted for principally, perhaps, from the tenants not being bound to follow any particular mode of husbandry, and not being restricted in the sowing of wheat, so that the land is frequently foul and over-wrought. There is, also, in many parts, a great want of drainage. The soil is particularly suitable for barley, which is the principal crop; but large quantities of wheat are also raised. Turnips are extensively cultivated. Dairy husbandry is, in some districts, carried on upon a large scale, and the sheep stock is supposed to exceed 300,000 head. There are but few large estates, and farms are generally small; it is common to grant leases for 7 and 14 years. Average rent of land in 1810, 20s. 6½d. an acre. Manufactures and minerals of no importance. Principal town, Oxford. The county is divided into 14 hundreds, and 217 parishes. It sends nine members to the House of Commons, viz. three for the county, two for the city, and two for the university of Oxford; and one each for the boroughs of Banbury and Woodstock. Registered electors for the county in 1839–40, 5721. In 1841, Oxford had 39,141 inhabited houses, and 161,573 persons, of whom 80,357 were males, and 81,190 females. Sum expended for the relief of the poor, in 1838–9, £79,847. Annual value of real property in 1815, £790,886; profits of trade and professions in ditto, 312,809.

OXFORD, a parl. bor. and city of England, cap. of the above county, and the seat of one of the two great English universities, on the left bank of the Isis, near its confluence with the Cherwell, which are both crossed by numerous bridges, (one of which on the London road, is a handsome stone structure, of five arches,) 52 m. W.N.W. London, lat. (Observatory) 51° 45' 38" N., long. 1° 15' 29" W. Pop. of parl. bor. (which includes with the old borough, the parishes of St. Clement's and a portion of that of Cowley) in 1841, 21,345.

The city stands on a plain, in the midst of meadows thickly planted with trees, and is surrounded on three sides by the above mentioned rivers: it has an imposing external appearance from whatever side it may be viewed; but more especially from the adjacent high ground on the London and Abingdon roads. The High-street is one of the finest in England; not only for its width and regular arrangement, but for the beauty and magnificence of the churches and collegiate edifices lining it on both sides: the towers of Magdalen and All Souls' colleges, the noble fronts of University and Queen's colleges, and the university church, are its chief and most admired features. Three other streets meet it at its W. end; one of which, called the Corn-market, leads northward to the airy suburbs of St. Giles's; and the second passes southward by the town-hall, and the noble building of Christ-church, towards Abingdon; while the third, called Queen-street, runs westward in continuation of High-street, into the low and densely peopled parishes of St. Ebbe and St. Thomas. Parallel to and N. of High-street, is another fine, though not long line, called Broad-street, in which are Baliol, Trinity, and Exeter colleges, the Ashmolean museum, Clarendon-rooms, Sheldonian theatre, &c.; and between High-street and Broad-street is an oblong space, occupied by a quadrangular building, forming a hollow square, round which are the academical "schools," the upper stories being occupied by the Bodleian Library and Picture Gallery. Between the schools and St. Mary's church is the Radcliffe Library, a circular Grecian edifice, surmounted by a dome, and contrasting rather strangely with the Gothic structures by which it is surrounded. St. Giles's and Beaumont-streets are lined with substantial private dwellings; besides which, St. John's and Worcester colleges, and the Taylor-building, a large structure of Corinthian architecture, now in course of construction, greatly contribute to the embellishment of this part of Oxford. The other streets are mostly narrow, irregular, and crooked. Some new streets, however, with good substantial houses, an entire new suburb, and several hundred smaller tenements, have been erected within the last 20 years, and the city has thus been greatly improved. (*Municipal Bound. Report.*) The principal thoroughfares are well paved, cleaned, lighted with gas, and plentifully supplied with water. The police, a very efficient body, is regulated and maintained by the university. The town-hall, a long stone building, with little pretension to architectural elegance, is open below for the accommodation of farmers and corn-factors, the upper stories being divided off into court-rooms, and apartments for municipal and judicial business. A large and commodious new sessions-house, however, is in course of erection near the old castle, which has been converted into a modern jail. The arrangements of the county prison are very imperfect; its size does not admit of classification or solitary confinement to any great extent, and hard labour is only partially enforced. There is likewise a city bridewell, comprising about 50 cells; but its management is said to be unsatisfactory. (*Prison Inspector's 4th Rep.*, part iii. p. 193–201.) The market-house is a modern range of buildings, entered from the High-street, and, with its aisles, arcades, and shops, forms one of the greatest improvements made in the city. The Radcliffe infirmary erected towards the middle of the last century by the trustees of the fund left by Dr. Radcliffe, is in the N. suburb, not far from the observatory, which owes its origin to the same founders. It is a plain stone building, within a spacious enclosure, capable of accommodating between 150 and 200 patients; besides medical officers, &c. The only other edifices, exclusive of those devoted to public worship, are the house of industry, a large structure near the infirmary, built for the reception of the poor belonging to 11 united parishes (the expense of whose maintenance amounted, in 1830, to £3195), a small music-hall, and two sets of alms-houses.

The city is divided into 14 parishes, and is the seat of a bishopric. The cathedral church connected with Christ-church college, presents the styles of different ages, from the 12th to the 16th century; it is built in the form of a cross, and measures, from E. to W., 154 ft.; the length of the transepts being 102 ft., and the height from the floor to the roof, 42 ft. At the intersection of the nave and transepts rises a tower, surmounted by a spire 146 ft. in height. The carvings of the choir are very elaborate, though somewhat heavy; and in some of the windows are fine specimens of painted glass. Unfortunately it is so hemmed in by college buildings and gardens, that no view of the whole can well be obtained. The cathedral establishment is identical with that of Christ-church college; and the sum annually divided by the dean and eight canons amounts to £13,550, besides which each has a handsome residence. The income of the bishopric amounted at an average of the three years ending with 1831, to £2548 a year; but it is to be increased to about £4000 a year. St. Mary's church (used by the university for the academical sermons, Bampton lectures, &c.) is a fine structure, in the perpendicular style, surmounted by an elaborately ornamented tower and spire, 180 ft. high. The side towards the High-street, however, is disfigured by an incongruous porch, with twisted columns. The interior was renovated in 1826, and is handsomely fitted up. It is likewise a parish church, the living being in the gift of Oriel college. Carfax, or St. Martin's, the corporation-church, facing the W. end of High-street, is an oblong stone building, with a low tower. All-Saints, in the High-street, is in the Grecian style: the roof is entirely supported on the side walls, and the whole is surmounted by a tower and spire: the living is in the gift of Lincoln college. St. Peter's-in-the-East, near Queen's college, is the oldest church in Oxford. Mr. Rickman says that its original portions are Norman; but it has had many introductions and alterations, mostly in the perpendicular style, which have greatly altered its exterior appearance. The interior has recently been restored and beautified, so as to correspond with that of the original building; the living is in the gift of Merton college. St. Mary Magdalen at the juncture of the Corn-market with Broad-street, is in the decorated style; it has recently undergone a thorough repair, and is about to be enlarged by the addition of a large aisle and lofty tower, to be erected by subscription, in commemoration of the martyrs, Cranmer, Hooper, and Ridley. The other churches of Oxford deserve no particular description. The Roman Catholics have a small

chapel: and there are places of worship for Wesleyan and Calvinist Methodists, Independents, and Baptists. Most of the churches have their attached charity-schools, besides which there are various Sunday schools, attended by great numbers of children of both sexes. The diocesan national schools give instruction to about 600 boys and 300 girls; and a school of industry is attended by 200 girls. A few schools, also, are supported by dissenters. Most of the parishes have considerable funds for the relief of the aged and sick poor; And there are dispensaries, lying-in charities, clothing-societies, provident clubs, &c., to aid the numerous indigent persons in the town and neighbourhood.

The great glory of Oxford, however, consists in its buildings devoted to collegiate education; which far surpass those of Cambridge in number, and in extent and beauty. Most of them are built in the style peculiar to the 15th and 16th centuries; but a few, as Queen's and Worcester colleges, with parts of Christ-church and Magdalen colleges, partake more of the Grecian style, introduced late in the 17th century. They are chiefly built in hollow squares, round which are the member's rooms; and the quadrangles of Christ-church, All-Souls, Magdalen, New, and Brasenose colleges are very large and imposing. The chapels, halls, libraries, and gardens of these establishments are likewise extremely beautiful; nor must we omit to mention the shady promenades, called the Christ-church Meadows and Magdalen Walks, which are of great extent and beauty.

Oxford is a very great degree depends, and has during many centuries depended, for its prosperity on the university. Till the opening of the Birmingham and Great Western railways, it enjoyed considerable advantages from being on the great roads leading northward to Birmingham, Shrewsbury, &c., and westward to Cheltenham, Gloucester, and South Wales. Between 20 and 30 coaches used daily to pass through the town, and its inns were among the largest in England: but this source of wealth is now almost extinct, and owing to the opposition of the university, no railway has hitherto been brought near the city. It is believed however, that this will speedily be the case; and there can be no doubt that Oxford will gain incomparably more by such a speedy means of communicating with the metropolis and other great towns than she has lost by the annihilation of her posting business. There are no manufactories; and the trade of the place is chiefly confined to the supply of the academic population. It has the advantage of a canal navigation by the Isis to London, and by the Oxford canal northward, which channels supply it with coal, and all the more bulky articles of domestic consumption. (*Parl. and Mun. Bound. Rep.*) It is also the mart for an extensive agricultural district, and its weekly corn-market is one of the largest in the midland counties. Oxford has four private banks, with a savings' bank; and two weekly newspapers, the "Oxford Journal," and the "Oxford University Herald." Races are held during August in the Port-meadow, W. of the town.

The corporation of Oxford claims to exist by prescription, but it has also received many charters, the last of which was granted in 3 James I. It is divided, according to the Mun. Reform Act, into five wards, and is governed by a mayor, nine other aldermen, and 30 councillors. Corp. revenue, in 1839, £2914, exclusive of £61 accruing from the sale of property. Quarter and petty sessions are held by the recorder; besides which, there is a mayor's court for the recovery of small debts, and a court of hustings. The assizes for the county are held here: the quarter sessions take place on the Mondays after January 4, April 5, June 28, and October 18. The vice-chancellor of Oxford is a magistrate *ex officio* within the borough, and exercises jurisdiction over the town as well as the university. The city has sent two members to the House of Commons since the reign of Edward I.; the right of election down to the Reform Act having been in the free burgesses (becoming so by birth, apprenticeship, purchase, or gift). The limits of the borough were enlarged by the Boundary Act so as to include with the old borough, the parish of St. Clement's and a part of Cowley parish E. of the Cherwell. Registered electors, in 1839–40, 2773. Oxford is likewise the election town for the county. Markets on Wednesday and Saturday, but chiefly on the latter. Fairs, May 3, Monday after September 1, and Thursday before Michaelmas-day.

Oxford (originally called Oxnaford, or Oxneford) lays claim to very high antiquity. It suffered much during the ravages of the Danes, and was the residence of Canute, and of his son Harold Harefoot. William the Conqueror stormed the town in 1067. Soon after, the castle (remains of which are still existing contiguous to the county jail) was built, by Robert de Oilgi, one of the Norman barons. Henry I. built a palace here, which continued to be a favourite regal residence during several successive reigns; but it was pulled down at the dissolution of the religious houses. Oxford had a share in the civil wars of Stephen and Henry II., which were terminated by a council held in it in 1154. The history of the city is henceforward closely connected with that of the university, which now began to attain a high celebrity. Hot disputes on points of scholastic doctrine prevailed between the reigns of Henry III. and Edward III.; and in the middle of the 14th century a large body of the students removed to Stamford, in Lincolnshire. (See Gutch's *Annals*, anno 1348.) Pestilence at the same time made great ravages; the city was almost deserted, and the university all but ruined. The introduction of the doctrines of Wycliffe, at the close of the 14th century, occasioned a great commotion in the academic body; the city suffered much during the wars of the Roses, and Oxford was again visited by plague in the reign of Henry VII. The troubles of the Reformation, and the spoliation of the academic houses by Henry VIII. drove many of the students from their habitations; but that monarch may be said to have resuscitated the university by the establishment of the cathedral of Christ-church, as well as by the foundation of professorships for the learned faculties. The forms of popery were restored under Queen Mary; and during this period Oxford acquired an unenviable notoriety by the martyrdom of the three great reformers, Cranmer, Ridley, and Latimer, in 1555–56. In the following reigns the city and university gradually recovered from their previous depression, and the latter received from James I. and Charles I. many important privileges. Oxford warmly espoused the cause of the royalists during the reign of Charles I., who made it his head-quarters after the battle of Edgehill. For a lengthened period after the Revolution, Oxford was attached to the party of the Jacobites; and since the accession of George III. down to the present time, the university has, speaking generally, supported what are called high church and high tory principles. (*Ackerman's History of the City and University of Oxford; Ant. Wood's Antiq. of Oxford; and Athenæ Oxon.; Brewer's Account of Oxfordshire*, art. *Oxford; Parl. Papers; Priv. Inform.*)

OXFORD (UNIVERSITY OF). This celebrated university lays claim to very high antiquity; but its exact origin is unknown. Tradition has assigned its foundation to King Alfred, about the year 800; and University college is supposed to have been the nucleus round which, in the course of nine centuries, have been formed the present assemblage of colleges and halls. Mr. Hallam and Mr. Dyer countenance this opinion, which, however, seems to rest on no very solid grounds. But there can be no question that Oxford was known as a school of ancient learning as early as the reign of Edward the Confessor, for Ingulphus, abbot of Croyland, says of himself, that "he was sent to study at Oxford, where he made greater progress in the Aristotelian philosophy than most of his contemporaries, and became well acquainted with the rhetoric of Cicero." (See *Conringius de Antiq. Academ. Diss.*, iii. ch. 7.) During the reigns of Henry I. and II., Oxford appears to have comprised a theological school of some note, and civil law was studied in it as early as the middle of the 12th century, about which time doctors both of divinity and law were first created; but we do not find it designated as a university till the 3 John, 1201, which is of earlier date than the application of the same title either to Paris or Cambridge. The earliest charter was granted by John, and its privileges were confirmed and extended by subsequent monarchs, the act by which it was created a corporate body, by the style or title of "The Chancellor, Masters, and Scholars of the university of Oxford," having been passed in the 13th Elizabeth, 1570. The statutes, however, by which the university is now either nominally or really governed were drawn up during the chancellorship of Archbishop Laud, and received the sanction of Charles I. in 1636. (*Ant. à Wood's Antiq. of Oxf.*, transl. by Gutch, ii. 403.) The university was sanctioned, also, by Papal authority; and Oxford is mentioned in the Constitutions of Clement V. (A.D. 1311), with Paris, Bologna, and Salamanca. Collegiate foundations date from a very early period; and University, Baliol, Merton and Oriel colleges, were founded prior to the reign of Edward III. The number of *colleges* or endowed establishments, however, was for some centuries small, in comparison with that of the *halls* or *inns*, in which the students lived, chiefly at their own expense, under the supervision of a tutor, or principal. For the establishment of these halls, of which there were about 300 in the early part of the 14th century, nothing more was necessary than the agreement of a number of students to form themselves into a society, under a doctor or master of their own choice; for the chancellor could not refuse his sanction to the establishment. Pestilence, civil war, the decline of the scholastic philosophy, and other causes, led to a diminution in the number of students, in consequence of which, also, the halls decreased in number. The Reformation still further thinned the ranks of the students, and at the beginning of the 16th century the university was almost entirely deserted, and the halls fell to decay; a circumstance which gave to the secular colleges a preponderating weight, and allowed them so to extend their circuit, and increase their numbers, that they were subsequently

able to comprise within their walls nearly the whole academical population, though, previously to the 15th century, these endowed establishments appear to have rarely, if ever, admitted independent members. (See *Edin. Review*, lii. 411.) In 1546, the number of halls had fallen to only *eight*, and Antony Wood informs us that in 1551 the ancient halls were "either laid waste, or had become the receptacles of poor religious people, turned out of their cloisters." Many of these buildings were purchased by the colleges, which were thus considerably extended, and began to provide for the accommodation of members *not on the foundation*. Six colleges were founded in the 16th century, chiefly on the sites of old halls or deserted houses. After this period, one fresh college (Wadham) was founded; and three out of the eight surviving halls (namely, Broadgates, Gloucester, and Hert halls) were changed, by endowment, into Pembroke, Worcester, and Hertford colleges, of which, however, the last is now extinct. The Earl of Leicester, chancellor of Oxford during the reign of Elizabeth, obtained from the university the privilege of nominating the principals of the halls; and this right, which was, in effect, a veto on the institution of new halls, was vested by statute in his successors. Of the five still existing halls, Magdalen and St. Edmund's are the best attended by students; and they are the only examples in the country of unendowed academical houses; for the establishments called *halls* at Cambridge differ in no respect from the colleges. In the 15th century an enactment was made compelling all students to become members of some college or hall; and by the regulations of Leicester (which were confirmed by Laud), it was made necessary for them to enter under a particular resident tutor. The business of instruction was originally carried on by the doctors and masters of arts (all of whom had the liberty of teaching), or else by the salaried professors of the university. The tutor, therefore, was at first rather a moral guardian than a professed teacher; and his duties did not consist in teaching the sciences constituting the trivium and quadrivium, but in imbuing his pupils with good principles, instructing them in the rudiments of religion, especially in the doctrines of the 39 articles, and making them conform generally to the statutory regulations of the university. These duties the tutor still performs; but he has, also, become an acknowledged teacher, giving daily instruction in languages, science, &c., to those under his charge; and bye-statutes enacted at different times have rendered an attendance on the professors' lectures merely optional, and wholly unnecessary either for the acquisition of the university degrees, or for a participation in academical honours. Indeed, out of 28 professors, only 12 now give lectures; and, excepting those of one or two professors, they receive little attention from the great body of students. The following are the existing collegiate institutions of Oxford, with the dates of their foundation, and the number of foundation-members, independent members, &c., in 1841. (See *Mr. Malden's* valuable work *On the Orig. of Universities*, p. 80–92.) [See Table, next page.]

The buildings belonging to all or most of these societies, are of great extent and beauty. Christ-church, New college, All-Souls, Magdalen, and Queen's colleges, are very large, comprising several quadrangles, and lay claim to considerable architectural elegance. The libraries and dining-halls of these establishments are on a large scale, and the rooms within the colleges are capable of accommodating several hundred students. Oxford, indeed, from the great number of its fine academic buildings, has a very imposing appearance when seen either near or at a distance; and it has been not inappropriately called a city of palaces, vying in external and internal beauty with the finest cities.

Each college, as in Cambridge, is governed by its own statutes; and its principal or head must, in most cases, in orders, as a living commonly forms the chief part of his *solatium*. The direction of the college is vested in the principal and senior fellows (technically called the *seniority*); but in matters affecting discipline the principal is the supreme arbiter, and he appoints the deans and tutors, who are immediately responsible to him for the conduct of the members *in statu pupillari*. The fellowships and scholarships are, in most instances, reserved for the natives of certain towns and counties, or for those who have been educated at certain schools; so that birth or interest, more than any positive amount of scholarship, usually procures the candidate's election. At Baliol, Oriel, Lincoln, and Wadham colleges, however, the fellowships and scholarships have been either wholly or in part thrown open to general competition, and the candidates for these usually comprise the most promising students of the university. Most of the colleges have *exhibitioners*, or students, receiving annual allowances from charities held in trust by the fellows, by city companies, trustees of schools, &c.; and at Christ-church there are *servitors*, an inferior class of students, somewhat resembling the *sizars* of Cambridge. All Souls' college has no under-graduate members, except its bible-clerks, and cannot be considered as an establishment

Colleges and Halls.	Date of Foundation	Founders.	Visitors.	Details of Foundation.	Members in 1841. M A.	Total.
University . .	872 ? 1940	King Alfred and William of Durham.	The crown.	A master, 13 fellows, 2 travelling fellows, 17 scholars. (8 open), and 4 exhibitioners : 10 benefices	119	238
Baliol . . .	1263 ?	John Baliol, father of John Baliol, king of Scotland.	The Archbishop of Canterbury (elected by the College).	A master, 13 fellows, and 14 scholars, (all open) : several exhibitions for natives of Scotland : 20 benefices	151	233
Merton . . .	1264	Walter De Merton, Bishop of Rochester, who removed it from Maldon in Surrey, in 1204.	The Archbishop of Canterbury.	A warden, 24 fellows, 14 post-masters (portionistae), 3 chaplains, and 8 bible clerks : 18 benefices	70	149
Exeter (originally called Stapledon Hall) . .	1314	Walter De Stapledon, Bishop of Rochester.	The Bishop of Exeter.	A rector, 25 fellows, 19 scholars (3 open) : 13 benefices	132	245
Oriel . . .	1326	Edward II., at the suggestion of Adam De Brome, his almoner.	The Crown.	A provost, 18 fellows, and 17 exhibitioners (all open) : 13 benefices	149	252
Queen's . .	1340	R. Eggesfield, confessor to Philippa, consort of Edward III.	The Archbishop of York.	A provost, 12 fellows, and 6 scholars on the old foundation ; 3 fellows, 4 scholars, and 4 exhibitioners (all open), on the Michael foundation : 20 benefices	130	306
New . . .	1386	William of Wykeham, Bishop of Rochester, founder of Winchester School.	The Bishop of Winchester.	A warden, 70 fellows and scholars, 10 chaplains, 3 bible clerks, and 16 choristers : 35 benefices	76	150
Lincoln . .	1427	Richard Fleming, Bp. of Lincoln.	The Bishop of Lincoln.	A rector, 12 fellows, 8 scholars, 12 exhibitioners, on Lord Crewe's foundation : 10 benefices	71	161
All Souls . .	1437	Henry Chichele, Archbishop of Canterbury.	The Archbishop of Canterbury.	A warden, 40 fellows, 2 chaplains, and 4 bible clerks	88	107
Magdalen . .	1456	William of Wykeham ? Bishop of Winchester and Lord Chancellor of England.	The Bishop of Winchester.	A president, 40 fellows, 30 scholars, called demies, 8 schoolmasters, 4 chaplains, 8 bible clerks, and 16 choristers : 36 benefices	136	174
Brazen nose Col. and King's Hall	1509	William Smith, Bishop of Lincoln, and Sir R. Sutton of Prestbury, in Cheshire.	The Bishop of Lincoln.	A principal, 20 fellows, 17 scholars, 7 bible clerks, and 15 exhibitioners, on a separate foundation by W. Hulme ; 42 benefices	228	360
Corpus-Christi .	1516	Richard Fox, Bishop of Winchester, and Lord Privy Seal.	The Bishop of Winchester.	A president, 20 fellows, 20 scholars, 2 chaplains, and 4 exhibitioners : 23 benefices	92	127
Christ-Church .	1525 and 1532	Cardinal Wolsey : King Henry VIII. refounded it and gave it the name of Henry VIII's College. It was made a cathedral church in 1546.	The Crown and Chancellor of the University.	A dean, 8 canons, 101 students, 8 chaplains, 8 bible clerks, and 8 choristers : 91 benefices	509	914
Trinity . .	1554	Sir Thomas Pope of Titenhanger, in Hertfordshire.	The Bishop of Winchester.	A president, 12 fellows, and 12 scholars (all open) : 9 benefices	122	204
St. John's . .	1555	Sir Thomas White, alderman of London, and founder of Merchant-Tailors' School.	Ditto.	A president, 50 fellows and scholars, 3 chaplains, 3 bible clerks, 6 singing men, and 6 choristers : 29 benefices	136	260
Jesus . . .	1571	Queen Elizabeth, on the endowment of Dr. Price.	The Earl of Pembroke (hereditary).	A principal, 19 fellows, and 18 scholars, all natives of Wales : 24 benefices	85	156
Wadham . .	1613	Nicholas Wadham of Thinfield in Somersetshire, and Dorothy his wife.	The Bishop of Bath and Wells.	A warden, 15 fellows, and 15 scholars (all open), 10 exhibitioners on Dr. Hody's foundation, 2 chaplains, and 2 bible clerks : 9 benefices	98	207
Pembroke . .	1624	Thomas Tesdale, and Richard Wightwick, B.D.	The Chancellor of the University.	A master, 14 fellows, and 25 scholars or exhibitioners, and 4 bible clerks : 13 benefices	109	120
Worcester . .	1714	Sir Thomas Cookes of Bentley, in Worcestershire.	The Bishops of Oxford and Worcester, and the Vice Chancellor of Oxford.	A provost, 21 fellows, 16 scholars, 3 exhibitioners, and 2 bible clerks : 5 benefices	113	247
St. Mary Hall .	1333	Oriel College.	The Chancellor of Oxford, who appoints the Principals.	None	23	74
Magdalen Hall	1487	Bishop Waynflete, the founder of Magd. College.		None	49	178
New Inn Hall .	1392	William of Wykeham, who gave it in New College.		None	4	64
St. Alban Hall .	1547	Merton College.		None	8	28
St. Edmund Hall .	1289 and 1659	The canons of Osney ; afterward William Denyer, provost of Queen's College, which society has the appointment of the Principal.	The Provost of Queen's College.	None	52	90
19 Colleges, 5 Halls }	. .	. .	. .	. .	2,780	5,215

(Oxford Calendar for 1841 ; and Chalmer's History of the Colleges and Halls of Oxford.)

for education, though it has 40 fellows, bene nati, bene vestiti, et in arte cantandi mediocriter docti.. New college is connected with Winchester school, which owes its origin to the same founder, and, like King's college, Cambridge, has little connexion with the rest of the university ; but it has not, like the last-named college, the privilege of examining its own members for academical degrees. Residence in college, which, at Cambridge, is to a certain degree optional, is compulsory at Oxford on all under-graduates who have not passed their examination for the B. A. deg. ; and hence the out-college men at Oxford comprise the senior, not the junior portion of the academic population. Attendance at chapel and the tutors' lectures is enjoined on each student, and omissions in either case are visited by impositions, rustication (temporary banishment), and other penalties. At the end of each term, also, examinations are held within the colleges on the subjects of the tutors' lectures : these trials (called collections) have no necessary connexion with the public examinations, though they in some measure familiarize the student with the method pursued in the academic schools. With respect to college lectures, however, it seems to be a generally received, and, we are inclined to think, a well-founded opinion, that, as at present conducted, they confer but little benefit on the student, who is indebted for his degree and distinction, should he acquire any, almost entirely to his own exertions, or to the assistance of a tutor, procured at

an expense of from £50 to £70 a year. The degree of attainment requisite for passing the university examinations has, since 1801, been determined by statute ; and every candidate must belong to some college or hall, and must have been under collegiate instruction. It is usual, also, for the students to consult their tutors before they put down their names in the list of those to be examined. On admission to the university, each member is required to subscribe his assent to the 39 articles of the Church of England, the act of supremacy, and certain sections of the university statutes ; and hence Oxford is exclusively resorted to by parties professing the established religion ; whereas, at Cambridge, the education furnished by the university, though not the degrees, may be participated in by all parties, dissenters as well as churchmen.

There are at Oxford four terms in each year, two of which (Michaelmas and Hilary terms) last nine, and two others (Easter and Trinity terms) last from three to four weeks each. By statute, however, the two first mentioned terms may be kept by six weeks' residence, and the two latter by three weeks each ; though by those who have taken the first degree in arts, or have passed the examination for that degree (having previously resided 12 terms), any term may be kept by a residence of three weeks. Sixteen terms (occupying the same time as the twelve terms at Cambridge) are required for the degree of B. A. from all, except the sons of British and Irish peers, baronets, knights, and their sons, *if so entered in the university books;* but of these the term of matriculation is reckoned as one, the day of admission to the degree as another, and two others are dispensed with by a regular order of the governing body; so that, in fact, a residence of 12 terms in Oxford, as of nine terms in Cambridge, is all that is necessary for the B. A. degree. The examinations for this, the lowest degree, are conducted in a large square building, known as "the schools," and consist of two trials, the first, or preliminary examination, called the *little-go,* or *responsions,* and the second the examination for, the B. A. degree, both under masters of arts appointed by the vice-chancellor and proctors. There are seven public examiners of candidates for the B. A. degree, and three examiners, or masters as they are called, of the *little-go* schools. The *little-go* examination, which commonly takes place at the completion of the eighth term from matriculation, comprises a ner; grammatical and elementary examination, offering few impediments except to the dullest or idlest students. To have failed in this examination is in Oxford phraseology, said to have been *plucked;* and three successive failures are considered as tantamount to a disqualification from further university pursuits. The next examination, or that for the degree of B. A., is the last to which the student is subject during his probationary residence: the lowest acquirements for the degree comprise a knowledge of the rudiments of religion, sacred history, the doctrines of the thirty-nine articles, the *literæ humaniores,* including, at least, two works of Latin and two of Greek authors, (usually Herodotus or Thucydides, with a few Greek plays, and portions of Livy or Tacitus, with Virgil or Horace,) with a fair knowledge either of Aldrich's Logic or Euclid's Elements. For honours, however, (of which, since 1830, there have been *four* instead of *three* classes as previously,) a very extended course of reading is necessary; the number of classical works on the candidate's list (each being compelled to send in an account of them) often amounts to twelve or sixteen, of which Aristotle's Ethics and Rhetoric commonly form two; and the amount of historical and metaphysical knowledge requisite for the honourable distinction of a *first-class* man is so great as to require intense labour during the three years' probation. It has been alleged, however, that the education of first-class men at Oxford is more extensive than solid, owing to its not being bottomed on any sound philological basis: but it may be doubted whether there be any good foundation for this statement. Classics are the favourite studies of the under-graduates, and till recently the pursuit of physical science was, if not discouraged, at least not encouraged. Hence mathematics form an unimportant part of the general examination for the B. A. degree; but for the honours *in disciplinis mathematicis et physicis* an amount of knowledge is required, varying from that of pure mathematics, (including differential equations,) to a somewhat extended acquaintance with physics, astronomy, &c. The highest mathematical degrees, however, may be generally obtained by persons of less attainments than the *wranglers* of Cambridge. (See *Prof. B. Powell on the State of Mathem. Studies in the Univ. of Oxford; Journ. of Education,* vii., 46; and *Stat. Account of the Brit. Empire,* vol. ii.) Of those members who, during the last six years, have taken the degree of B. A., the *passmen* (those not ambitious of honours) have averaged at each examination about 100, those receiving honours in classics, 78, and in mathematics only 18. A few of the students aim at distinction both in classics and mathematics; and there are occasional instances of men having attained a *double first-class,* the highest honour that the university can bestow. The examinations for the degrees are held twice a year, during the Easter and Michaelmas terms, and last about three days. A large part of the examination is now carried on by written questions and answers, the oral examination being principally confined to theology. About 300 students pass at each of these examinations.

The annual prizes of the university, which are subjects of considerable competition among those *in statu pupillari,* comprise three of £20 each, given by the chancellor for the best compositions in Latin verse, Latin prose, and English prose; the first being confined to under-graduates, and the others to those who have exceeded four, but not completed seven years. Sir Roger Newdigate, in 1806, left property for an annual prize for English verses on ancient sculpture, painting, or architecture, confined to under-graduates; a prize of 90 guineas was founded, in 1825, by Dr. Ellerton, for the best English essay, by bachelors of arts, on the doctrine or duties of religion; and two other theological prizes of £30 each were founded by Mrs. Denyer, in 1836, for the best discourses written on selected subjects by clerical members of the university under the standing of 10 years. The university has likewise several public scholarships. The principal of these are the Vinerian scholarship and fellowship, five Craven scholarships for undergraduates, four scholarships of £30 a year each, established by

Dean Ireland, and tenable for four years, the Boden scholarships for the encouragement of Sanscrit literature, three mathematical scholarships founded in 1831, the Kennicott scholarships for proficients in the Hebrew language, and the Eldon scholarships of £200 a year, tenable for three years by bachelors of arts studying the profession of the law. Several scholarships have, also, within the last few years, been established with the view of encouraging mathematical studies; and they are usually held by those who have attained honours in the mathematical schools of the university. These prizes and scholarships are contested for with great spirit, and the holders of them are usually among the ablest of those *in statu pupillari.* The prize compositions are in most cases recited in the Academical or Sheldon theatre, at the Commemoration or Act held in Trinity term for conferring honorary degrees. After the degree of B. A. has been taken, there are no further examinations except for degrees in medicine; but certain exercises, now merely nominal, are performed in the schools, and the candidate must have had his name on the books of some college or hall for a certain number of terms, during some of which also he must reside in Oxford. Subjoined is a summary of the different periods at which the several degrees, &c., may be acquired in the course of a student at Oxford:

	Terms	In residence
B. A.	16	12 in residence.
M. A.	28	13 or one after B. A.
B. C. L.	28	17 —
B. M.	28	17 —
B. D.	56	14 —
D. M.	40	or one after B.A.
D. C. L.	36	but usually honorary.
D. D.	68	—
D. Mus.		chiefly honorary, but dependent on the performance of a musical exercise.
B. Mus.		

The orders in the different colleges rank as follows: 1. The heads, or principals, some of whom are D.D. 2. The fellows, D.D., M.A., or B.A. 3. Noblemen who have graduated. 4. Graduate members not on the foundation. 5. Under-graduate noblemen. 6. Gentlemen-commoners. 7. Scholars, 8. Commoners. 9. Bible-clerks and servitors.

The expenses of academical residence vary greatly according to the taste and habits of the student; but about £200 a year may be assumed as the average outgoings of the most economical commoner, and an additional expense of £30 a year is incurred by those who engage the services of a private tutor. The payments to the college for living, rooms, &c., are usually called *battels,* and in the case of commoners range from £70 to £90 a year; but those of noblemen and gentlemen-commoners are on a considerably higher scale, and their annual expenses are consequently much greater.

The University of Oxford is governed, as before stated, by the *Corpus statutorum,* drawn up by Archbishop Laud. The highest officer is the chancellor, anciently elected for three years, but since 1434 for life. This office, however, as well as that of the high steward or seneschal, is little more than an honorary dignity conferred on some distinguished nobleman; and the chancellor's duties are ordinarily performed by the vice-chancellor, who is, in fact, the supreme executive and judicial authority resident in the university. He is annually selected by the chancellor from the heads of houses, and approved by convocation; but in practice the office is held for four years, and four deputies are appointed, called pro-vice-chancellors, to take the duties of the office in case of the vice-chancellor's absence or illness. The proctors (two masters of arts, of at least four and not more than 10 years standing) are the conservators of the peace and discipline of the university; they rank next to the vice-chancellor, and have an extensive police jurisdiction over the town. They are assisted in their duties by four pro-proctors, and have at their command a large constabulary force. They are usually nominated by the colleges, each college taking in turn, according to a cycle fixed by the statutes. The business of the university, in its corporate capacity, is transacted by the doctors and masters at large, in two distinct assemblies, called *congregation* and *convocation.* The former consists of regents either *necessary* or *ad placitum,* including resident doctors, heads of houses, professors and tutors of colleges, its business being chiefly confined to the granting of degrees and dispensations: the vice-chancellor has a negative on its proceedings, and the proctors *conjointly* have the same privilege. Convocation is the legislative assembly of the university, comprising all doctors and masters resident or not, whose names are on the books of a college or hall; and its business is unlimited, extending to all subjects affecting the credit, interest, and welfare of the corporate body. The statutes, however, cannot be altered, nor any new laws be enacted, before the matter has been referred to the *hebdomadal meeting* of the vice-chancellor, proctors, and heads of houses, who, if they approve of the measure, draw up the terms in which it must be promulgated in convocation.

The hebdomadal meeting was first instituted in the reign of Charles I.

The public professorships of Oxford are of two classes, those established by royal foundation, and those supported by private endowment. The regius-professorships of divinity, civil law, medicine, Greek, and Hebrew, were founded by Henry VIII., and canonries in Christchurch cathedral are attached to the chairs of divinity and Hebrew. George I. also founded a regius-professorship of modern history in 1794, which was confirmed by George II. in 1738. There is also, as at Cambridge, a margaret-professor of divinity. The other professorships are, 1. of natural philosophy, founded by Sir W. Sedley, in 1618; 2, geometry and astronomy, established by Sir H. Savile; 3, moral philosophy, by Dr. White, in 1621; 4, ancient history, by W. Camden, in 1622; 5, anatomy, in 1626; 6, Arabic, by Archbishop Laud, in 1636; 7, botany, in 1632; 8, poetry, in 1708; 9, Anglo-Saxon, in 1750; 10, common law, by Charles Viner, in 1755; 11, clinical medicine, in 1772; 12, 13, 14, anatomy, practice of medicine, and chemistry, by Dr. Aldrich, in 1803; 15, political economy, in 1835; and, 16, Sanscrit, by Colonel Boden, in 1830. There are also lectureships or readerships of Arabic, anatomy, experimental philosophy, mineralogy, and geology; there being, in all, 26 professors or readers. The lectures are delivered either in the public schools, or in a building near them, formerly the university printing-office. An edifice, however, is in course of erection, the funds for which are provided by the munificent bequest of the late Michael Angelo Taylor: it will be, when completed, one of the finest buildings belonging to the university, and will comprise several large lecturing theatres for the professors, besides a noble picture gallery for the reception of the numerous portraits and other pictures belonging to the university.

The public orator, who delivers the Crewelan oration alternately with the professor of poetry, is chosen by convocation; and his office is to write public letters and make addresses on grand occasions in the name of the university. The archives are kept by a registrar, elected also by convocation; this office was first established in 1634. The Clarendon press is superintended by delegates, of whom the vice-chancellor and proctors form three ex officio; the rest are heads of houses. The present building, opened in 1829, is of great extent, the bible department is on a magnificent scale, and the editions of classical and other works printed at this establishment are celebrated both for beauty and accuracy. The Bodleian library, founded by Sir Thomas Bodley, is the property of the university, and its affairs are regulated by the vice-chancellor, proctors, and the five regius-professors, its officers being a librarian, two under-librarians, and two assistants. It has received many valuable additions from the libraries of Selden, Archbishop Laud, Bishop Tanner, Browne Willis, Hearne, Gough, Malone, &c.; and it now comprises, exclusive of about 300,000 printed books, a great number of valuable MSS.: it is entitled, also, to a copy of all new works published in the United Kingdom. It is said, however, to be of less utility than might have been supposed. Owing to a justifiable apprehension of fire, the library is very inadequately warmed, and is very uncomfortable in winter; and the books are not allowed to be removed from the library. But though this be a very proper regulation as respects the rare and more valuable works, all the more common works might be lent out here as in Edinburgh, on lodging a deposit equal to their value, without any loss to the library, and with very great advantage to the students. The Radcliffe library, founded by Dr. Radcliffe, in 1718, and erected at an expense of £40,000, is under private trustees, and has little or no connexion with the university. The books in this collection are principally on medicine and natural history. An observatory was erected in 1772, out of the funds left by the same munificent individual, and the observer (commonly the Savilian professor of astronomy) is appointed by the Radcliffe trustees. The Ashmolean museum was built in 1683, for the accommodation of a rich collection of natural objects and articles of virtù, brought together by Elias Ashmole: large additions are annually made to it; and in the department of natural history, this museum is inferior only to the British Museum and that of the Zoological Society. It is under the care of official visiters, appointed by Ashmole's will.

The few remarks we ventured to make, under the article ABERDEEN, on the system of education followed in that university, apply with little variation to Oxford. This, like a sister institution, is essentially a scholastic establishment, and is well fitted to make good Greek and Latin scholars, and perhaps good divines. But it is obviously quite unsuited, as a place of instruction, for the élite of the youth of such a country as this. It must be entirely changed before either send forth legislators capable of appreciating the various interests of this vast empire, or individuals capable of promoting and extending those manufacturing and commercial pursuits, to the success of which we are mainly in-

debted for our unparalleled increase in wealth and population. The university of Oxford received, in 1603, the privilege of sending two representatives to the House of Commons; the right of election is vested in the vice-chancellor, doctors, and other members of convocation, of whom there were, in 1840, 2799. (Gutch's ed. of Wood's Antiquities of the University of Oxford; Ackermann's Illustrat. Hist. of Oxford, 3 vols.; Malden on the Origin of Universities; Ingram's Memorials of Oxford, 3 vols.; Oxford Calendar; Edinburgh Review, vols. 31, 35, and 53; and Private Information.)

OXFORD, county, Me. Situated in the W. part of the state, and contains 1600 sq. m. It contains a number of lakes in its N, part, connected together, which discharge their waters into Umbagog lake, and thence into Androscoggin river, by which, and by Saco river, it is watered. It contained in 1840, 24,955 neat cattle, 63,507 sheep, 8299 swine; and produced 66,696 bushels of wheat, 29,342 of rye, 83,339 of Indian corn, 4852 of buckwheat, 7075 of barley, 110,172 of oats, 609,684 of potatoes. 19,158 pounds of sugar. It had 59 stores, one furnace, 10 fulling-mills, one woollen factory, six flouring-mills, 37 grist-mills, 63 saw mills, three oil-mills, one paper-mill, 26 tanneries, one printing-office, one weekly newspaper; eight academies, 339 students; 309 schools, 14,330 scholars. Pop. 38,351. Capital, Paris.

OXFORD, p. t., Oxford co., Me., 50 m. S.W. Augusta, 563 W. Watered by Little Androscoggin river, which receives the outlet of a considerable pond in its S. part. It contains five stores, one flouring-mill, two grist-mills, three saw-mills; 12 schools, 516 scholars. Pop. 1354.

OXFORD, p. t., Worcester co., Mass., 52 m. W. Boston, 394 W. Watered by French river, a branch of Quinnebaug river, which affords water-power. It contains four churches, two Congregational, a Methodist, and Baptist, nine stores, three woollen factories, four grist-mills, six saw-mills; one academy, 30 students; 11 schools, 500 scholars. Pop. 1742.

OXFORD, p. t., New-Haven co., Ct., 35 m. S.W. Hartford, 309 W. Bounded S.W. by Housatonic river. Drained by Nangatuck river, which affords water-power. It contains two churches, a Congregational and an Episcopal, five stores, four fulling-mills, three woollen factories, two grist-mills, seven saw-mills, two tanneries; one academy, 25 students; 13 schools, 482 scholars. Pop. 1695.

OXFORD, p. t., Chenango co., N. Y., 10 m. S. Norwich, 109 W. Albany, 328 W. Watered by Chenango river, along which passes the Chenango canal. It contains six churches, a Presbyterian, Episcopal, a Methodist, two Baptist, and a Universalist; 17 stores, five fulling-mills, one woollen factory; two academies, 389 students; 25 schools, 909 scholars. Pop. 3179.

OXFORD, t., Warren co., N. Y. Drained by Pequest creek. It contains 12 stores, two grist-mills, one saw-mill, two tanneries, four distilleries; 12 schools, 822 scholars. Pop. 2853.

OXFORD, t., Philadelphia co., Pa., 4 m. N.E. Philadelphia. Bounded S.E. by Delaware river. It contains Frankford v. Watered by Tacony creek. It contains an insane asylum, a U. States arsenal, 40 stores, three lumber-yards, three cotton factories with 9472 spindles, one tannery, one brewery, one pottery; six schools, 313 scholars. Pop. 1582.

OXFORD, p. t., Butler co., O., 105 m. W.S.W. Columbus, 502 W. The land belongs in fee simple mostly to the Miami university, located in the t., granted by the Congress of the United States, and yields an income of $4500 annually, which is increasing. The university was founded in 1809, has a president and five professors or other instructers, has 309 alumni, of whom seven have been ministers of the gospel, 105 students, and 4352 volumes in its libraries. The commencement is on the second Thursday in August. The t. has one academy, 162 students; nine schools, 264 scholars. Pop. 3388.

OXFORD, p. v., capital of Granville co., N. C., 45 m. N. Raleigh, 256 W. Situated 8 m. E. of Tar river. It contains a courthouse, jail, market-house, a hospital for males, and also for females, three churches, a Presbyterian, Episcopal, and Methodist, and about 450 inhabitants.

OXUS (called by the natives Amoo or Jihoun), a river of central Asia, flowing westward through the territories of Budukshan, Kunduz, Bokhara, Khiva, &c., into the Aral sea, and extending between long. 58° and 74° E.; estimated length, 1300 m. This great river was, in 1838, traced up to its source by Lieut. Wood, who ascertained that it rises in the mountain-lake of Sir-i-kol, within the district of Pamir, lat. 37° 27' N., long. 73° 40' E., at an elevation of 15,600 ft. above the sea. (Wood's Journey to the Oxus, p. 354.) Its course hence is S.W. for about 70 m. to Langer Kish, where it turns westward. In long. 71° 40', it passes the ruby mines of Budukshan, near the town of Ishhaam, and is deflected northward by a large offset of the Western Himalaya chain. After another turn southward, its course is pretty regularly W.N.W. through extensive plains, and at the point where Sir A. Burnes crossed it on his way to Bokhere, he found it to be upwards of 800 yards in width, about 20 feet in depth, with muddy waters, and a current of about 3½ m. an hour, and from Kharjoo downwards, for 300 m., it is made available for commercial communication. (*Burnes' Travels*, ii., 214, and *Geog. Journal*, iv., 309.) The river passes about 20 m. N.E. Khiva, which is situated in a verdant plain, irrigated by numerous canals supplied from its waters. It forms at its mouth a pretty extensive delta, the apex of which is about 50 m. from its principal and only navigable embouchure in the Aral sea, the breadth of coast from the W. to the E. mouth being about 45 m. The Oxus has numerous tributaries; few of which, however, have been satisfactorily explored. A large river, called the Kokcha, rises in the Hindoo-Koosh, near the celebrated lapislazuli mines of Budukshan, and, flowing N.W., joins it at Kijapack on the S. bank. About 75 m. lower its waters are further augmented by the Ghori, an important stream rising in the Hindoo-Koosh, near the celebrated pass of Bamian, and having a general direction northward, passing in its course the large cities of Ghori and Kunduz. The only other affluent explored by Europeans is the Kulm, partly traced by Moorcroft, and joining the Oxus on its S. side, about 30 m. below the tributary last mentioned. Several tributaries flow in from the N. bank, bringing considerable volumes of water; but their extent is almost wholly unknown. The Oxus, according to Moorcroft, begins to rise in April, and remains full till July, when it again falls. When at its height it inundates the plain on either side, but especially on the right bank, the extent of the floods being marked by a belt of sedge, weeds, &c., and then by a thick jungle of dwarf trees and brushwood. (*Moorcroft and Trebeck's Travels*, ii., 496.)

The Oxus, regarded by some critics as the Araxes mentioned by Herodotus as flowing through the territories of the Massagetæ (i., 201–205; iv., 11), was supposed by Strabo and Ptolemy to fall into the Caspian; and the traces of a valley, nearly resembling the dry bed of a river, have induced some modern geographers to adopt the opinion, that in the course of ages the Oxus formed for itself a new channel, running into the Aral sea. But, however confused our information respecting this river, it undoubtedly formed the boundary line between the more civilized and settled nations of W. Asia and the wandering hordes of Tartary. The Oxus was the northern limit of the territories subdued by Cyrus and Alexander, and it seems to have been used, at a very early period, as a channel for commercial intercourse between India and the countries bordering on the Caspian and Euxine. The Ochus is mentioned by Strabo as one of its principal affluents; but his account is inconsistent, and unworthy of credit. (*Strabo*, xi.)

OYSTER BAY, p. t., Queen's co., N. Y., 189 m. S. by E. Albany, 282 W. It extends across Long Island. It has Oyster bay of Long Island sound on the N., and Great South bay of the Atlantic on its S. part. It contains three churches, an Episcopal, Baptist, and Friends, 14 stores, four fulling-mills, three woollen factories, eight grist-mills, two sawmills; one academy, 30 students; 20 schools, 809 scholars. Pop. 5865.

P.

PACIFIC OCEAN (THE), a vast expanse of water, extending between Asia and America (sometimes, though improperly, called the SOUTH SEA), and covering a large portion of the surface of the globe. Its extreme S. limit is the Antarctic circle, from which it stretches northward through 138 degrees of lat. to Behring's straits, which separate it from the Arctic ocean. Its greatest breadth from E. to W., measured along the equator, is about 10,100 m. Its shape is very irregular; but it becomes gradually narrower, as it extends northward, till at length the sea of Kamtschatka has a breadth of only 170 m. The American coast is pretty uniform, though high and bold, presenting the long range of the Andes close down to the shore. Its chief indentations are the gulf of California and bay of Panama; besides which, at the N. and S. extremities, it is broken and rugged, forming numerous islands and fiords, similar to those of other high latitudes. The coast-line of Asia, on the contrary, is extremely irregular, formed into deep bays, and subdivided by groups of islands into separate gulfs or seas, as the sea of Okhotsk, separating Kamtschatka from Siberia, the channel of Tartary dividing Saghalien from the main land, and the Yellow sea separating the peninsula of Corea from China; besides which numerous straits are formed between the islands of the Asiatic archipelago, as the straits of Sunda, between Sumatra and Java; the straits of Macassar, between Borneo and Celebes; Torres straits, between New Guinea and Australia; Bass's strait, between Australia and Van Diemen's Land, &c. The equator divides this vast expanse of water into the two grand portions of the N.

and S. Pacific oceans, both being remarkable for the numerous groups of small coralline and volcanic islands with which they are studded, and which constitute a separate portion of the world, entitled POLYNESIA, to which the reader is referred for further particulars. These numerous islands form several archipelagos, in which are reefs and sandbanks, that render the navigation extremely difficult and dangerous. The reefs are sometimes of great extent, stretching from island to island, upwards of 600 m. Earthquakes are felt in most of the islands; and all the archipelagos seem to be the seat of extensive volcanic action. (*Lyell's Geology*, iii., 236–239.)

The general motion of the Pacific ocean is from W. to E., or from the coast of America to that of Asia; and this movement is very powerful in the vast and uninterrupted extent of its waters, though it gradually decreases as it approaches the shores of Asia, while its temperature increases: its average velocity is stated by Captain Beechy to be about 28 m. a day. (*Geog. Journal*, i., 210.) Near cape Corrientes, in Colombia, the sea, owing to this cause, appears to flow constantly from the land; and from Acapulco, in Mexico, ships are carried with great celerity to the Phillippine islands. In returning, however, it is found advisable to take a course N. of the tropics, in order to have the advantage of the variable winds and polar currents, as well as of a counter-current, which sets eastward in about lat. 18° N. In the S. Pacific, the Polar currents being less interrupted by land, proceed with less deviation from their general course than those in the N. hemisphere; and carry icebergs nearer to the tropical regions than is usual N. of the equator. The equatorial current, as it approaches the shores of Asia, is interrupted and broken by the vast chain of islands, shoals, and submarine banks, which stretch from China to New Zealand. The general direction is changed and modified by the form of these lands, and the vast mass of New Holland is one cause of those dangerous currents around its shores, noticed by Cook, La Perouse, and Flinders. A current, also, sets eastward in the lat. of the Japanese islands, but turns northward about 150 m. from the shore, and probably joins the stream that runs N.N.E. through Behring's strait; besides which there is a variable current on the E. side of Australia, setting southward from August to April, and northward during the rest of the year. The N.E. trade wind prevails uninterruptedly between lat. 5° and 23° N.; and, with the currents, enables vessels to sail from America to Asia with great rapidity, and almost without changing the sails. The S.E. trade wind, which is not met with near the American coast, varies in its extent at different seasons; but it commonly prevails between the equator and 26° S., so that the region of calms in the Pacific extends over only five degrees of lat., or somewhat less than in the Atlantic. In this region, however, there are excessively severe storms, attended with lightning and heavy rain. (See *Bennett's Whaling Voyage*, i., 180.) These winds are still stronger in the numerous straits of the Asiatic archipelago, and in the neighbourhood of the Philippine islands, and immediately N. of Australia their violence becomes extreme, and even dangerous. The attraction of great masses of heated land also causes local variations in the wind, as is the case in New Holland, on the W. side of which there prevails a constant W. wind. Every island has, likewise, its land and sea breezes. In lat. 40°, on both sides the equator, tempests and variable winds prevail; and it may be remarked generally, that N. of lat. 40° N., winds from W. and N.W. are more prevalent than any others, whereas in the region S. of the trade winds, the prevailing winds are from S.W., and often extremely violent. Winds from the S., however, are found along the coast of Peru, and may be attributed in some measure, at least, to the strength of the polar current in the S. hemisphere. They are generally light, though steady; and N. of Guayaquil they always blow from S.S.E., extending westward as far as the Galapagos. (*Tuall's Phys. Geography*, 95, 198; *Malte-Brun*, i., 337, 395; and *Hall's America*, ii., App.)

Vessels in sailing northward from the coast of Chili are favoured both by wind and tide, so that they may safely run near the shore; but those going in the contrary direction sail south-westward, by means of the trade winds, till they arrive in the region of the variable winds, and are obliged to run as far as about lat. 28° S., before they can reach a port. Navigators traversing the ocean between Asia and America, sail westward from Mexico, touching at the islands of Luzon and Formosa; but from the ports of S. America the ordinary track is westward, between the Marquesas and Society islands, beyond which it assumes a W.N.W. direction, and joins the former in lat. 20° N., and long. 125° E. The voyage from Asia to America is effected by seeking the region of the variable winds N. of lat. 30°, and making the coast of California; but from Sydney the course is pretty direct E. as far as the coast of America, where the winds and currents are favourable for reaching

PADANG. PADUA.

its principal ports. One track for vessels sailing from Sydney to India is through the islands N. of New Guinea ; the other, however, by Bass's straits, is more common, and is the only one used in making the passage from India to New South Wales. (See *Berghaus's Physical Map of the Pacific Ocean*, and vol. i., 574, of his *Geography*.)

This ocean, which received its name *Pacific* from Magellan, in consequence of the prosperous weather with which he met while navigating its surface, was not known to the ancients, nor was the existence of so vast an ocean at all suspected by Europeans, till, in 1513, Vasco de Balboa beheld it from the summit of the mountains near the isthmus of Panama. Magellan traversed it from America to Asia in 1521, and at the close of the same century Sir Francis Drake explored a great portion of the W. coast of America. In the view of ascertaining whether this ocean had any other communications with the Atlantic than by the straits of Magellan and round cape Horn. The Pacific was pretty extensively explored during the 18th century ; and to the observations of Behring, Anson, Byron, Bougainville, Cooke, Vancouver, Broughton, and La Pérouse, we are principally indebted for the grand outlines of our best maps of this ocean. These navigators have been succeeded in the present century by Entrecasteaux, Krusenstern, Beechey, Fitzroy, Bennett, &c., and means are now provided for the formation of a pretty accurate chart of this sea so thickly studded with rocks and islands. Meanwhile the intercourse of the islanders with Europeans, and the efforts of European missionaries, have introduced among some of them the arts of civilized life ; trade has gradually extended itself along the American shore as well as in the different islands ; and in Australia, and very recently in New Zealand, the British have established numerous and very flourishing colonies. For particulars respecting the trade of its various ports, the reader is referred to the articles MANILLA, CANTON, and NANGASAKI, on the coast of Asia ; to ACAPULCO, PANAMA, GUAYAQUIL, CALLAO, VALPARAISO, on the W. side of America ; and to SYDNEY, and ZEALAND (New), in Australia. (*Malte-Brun's Geog.*, i. and iii. ; *Dict. Geog.* ; *Hall's S. America*, vol. ii., Appendix ; *Geog. Journal*, i., 190–222 ; *Bennett's Whaling Voyage*, &c.)

PADANG. *See* SUMATRA.

PADERBORN, a town of Prussian Westphalia, reg. Minden, cap. circ. at the source of the Pader, a tributary of the Lippe, 52 m. E.S.E. Munster. Pop., in 1838, 7895, principally R. Catholics. It is walled, is tolerably well built, and has a good cathedral and several other Roman Catholic churches, a Lutheran church, synagogue, gymnasium, Episcopal seminary, female teachers' seminary, and numerous almshouses, and other charities. It has a few manufactures of starch, leather, &c. ; but its trade is insignificant. It was erected into a bishopric by Charlemagne, who is said to have made it his head-quarters during his wars with the Saxons. It was the temporary residence of several succeeding emperors, and the palace they occupied still exists. Paderborn was subsequently one of the Hanse towns. In 1622 it was taken and pillaged by the Duke of Brunswick ; and in 1802 it was annexed to Prussia. (*Berg.* ; *Dict. Geog.*, &c.)

PADIHAM, a town and chapelry of England, par. of Whalley, co. Lancaster, and upper div. of hund. Blackburn, on the Calder, a tributary of the Ribble, 15½ m. E. Preston. Area of township, 1700 acres : pop. in 1831, 3529. The town, though small, is respectably built, and has an appearance of considerable activity. The church, subordinate to that of Whalley, was rebuilt in 1776 ; but its old tower, built at the close of the 15th century, is still remaining. The Wesleyan Methodists and Unitarians have their respective places of worship ; Sunday-schools are established. The inhabitants are principally employed in the manufacture of cotton goods. A market once held here has been for some years discontinued. Fairs, on the 8th May, and 26th September.

PADSTOW (corrupted from *Patrickstow*), a seaport, market-town, and par. of England, co. Cornwall, and hund. Pyder, on the W. side of the estuary of the Camel, 11 m. N.W. Bodmin, and 230 m. W. by S. London. Area of par. 3270 acres ; pop. in 1841, 2145. The town, which is situated in a richly cultivated vale, sheltered by bold rocks and hills, has been considerably improved by the erection of new houses ; but the streets are inconveniently narrow, and many of the buildings are antiquated. The church is in the perpendicular style : the living is a vicarage in the patronage of the descendants of Dr. Prideaux, the learned author of the famous historical work on the "Connection" of the Old and New Testaments, a native of the town, where he was born in 1648. The Wesleyan Methodists have also a place of worship, and there are two Sunday-schools, besides a well endowed national school. Facing the river are good quays and a custom-house, the gross amount of customs' duty in 1840 being £2018. The entrance to the harbour is between Stepper Point, on the W., and Pentire Point on the E., close to the former. The passage

is narrow, and rather difficult, especially with N.W. winds ; it has from 13 to 18 ft. water at spring ebbs. This is the only harbour between the Land's End and Hartland Point. (*Purdy's Sailing Directions for the English Channel*.) In the reign of Eward I., Padstow furnished two ships for the siege of Calais ; and in the time of Leland it carried on a considerable trade with Ireland and Wales ; at present about 80 ships, of the aggregate burden of about 4500 tons, belong to the port. The town was incorporated by Queen Elizabeth ; but the charter has lapsed by desuetude.

PADUA. (Ital. *Padova*, an *Patavium*), a city of Austrian Italy, gov. Venice, cap. deleg. of its own name, in a low and rather marshy situation, between the Brenta and Bacchiglione, at the termination of the canal of Monselice, 94 m. W. Venice ; lat. 45° 24' 7" N., long. 11° 52' 15" E. Pop. circa, 43,000. (*Austrian Encyc.*) It is of a triangular shape, is surrounded with walls and a broad ditch, and intersected by canals. Mr. Rose speaks very slightingly of Padua, and represents it as a city which, beyond all others, disappoints the expectations of the traveller. (*Letters*, i. 51.) It is, certainly, dull, damp, and gloomy, having numerous narrow, dirty, monotonous streets, bordered by arcades, without any leading thoroughfare ; there are three or four squares or open spaces, which, however, are all of very limited dimensions, excepting the Prato della Valle, the principal public promenade. This, which occupies what was once a marsh, bears some resemblance to a London square, but the interior is differently laid out ; being surrounded by a circular stream of running water, the banks of which are fringed with a double row of statues representing distinguished natives of Padua. The houses, though old, are generally well built and lofty. The principal public buildings are the churches, of which there are said to be nearly 100. The cathedral, a large brick edifice of Grecian architecture, was intended to have a stone front, which, has not yet been built. There is in it little remarkable ; except a monument to Petrarch, his portrait, and some Madonnas, one of which is by Titian. The church of St. Anthony, begun in 1250 and finished in 1234, 396 ft. in length by 160 ft. in width, is a vast ugly pile, exhibiting seven domes, a small octagonal tower above the gable of the front, two high octagonal towers, near the choir, and a lofty cone in the centre surmounted by an angel. (*Wood's Letters of an Architect*, i., 246.) The splendid shrine of the saint, with *mezzo-reliefs* in white marble, and two fine bronze panels, by Riccio, are the principal objects of interest within. The church of S. Giustina, begun and finished during the 16th century, is partly modelled on the foregoing, but is far handsomer. It is of brick, 367 ft. in length, by 232 ft. in the transept, and 88 ft. in height inside. It was built from a design by Palladio ; its interior is generally admired. Forsyth says, " it is rich in the bones of 3000 saints, and disputed bodies of two apostles ;" and it possesses a less questionable, if not so precious a relict, in a fine painting of Paul Veronese. The Benedictine abbey, to which this church was attached, is now converted into a barrack. The churches of the Eremitani ; the Annunziata, with some fine frescos by Giotto ; la Madre dolente ; S. Gaetano, &c., have all valuable works of art, or are remarkable for their architecture. The palace of Justice, or town-hall, is one of the most striking edifices in Padua : it has a saloon, 276 ft. in length, 86 in breadth, and 75 in height, being one of the largest in Europe, unsupported by columns. The roof is of dark carved wood, shaped like a reversed keel, and sustained by a number of iron ties. The walls are ornamented with frescoes, originally the work of Giotto. In the hall is a monument in honour of Livy, a native of Padua ; and at the entrance are two basalt statues, brought from Egypt by Belzoni, who also belonged to the city. The tower of Ezzelin, still used as an observatory ; the theatre, the museum of antiquities, &c. ; the mayor's and several other official and private palaces ; the *caffé Pedrocchi*, one of the oldest and best establishments of its kind in Europe ; several good hotels ; and the university buildings, are among the most conspicuous of the remaining public edifices ; but, according to Mr. Woods, the last mentioned structure hardly surpasses mediocrity.

The university of Padua, founded in the 12th century, was in the height of its popularity during the 15th and 16th centuries, when it was not only frequented by vast numbers of students from all parts of Europe, but even by some from Mohammedan countries. Its medical school was particularly celebrated. Fallopius, Fabricius ab Aquapendente, Morgagni, &c., have been among its medical teachers ;[a] and Galileo, Guglielmini, &c., among its professors of philosophy. Dante, Petrarch, and Tasso were of the number of pupils. Harvey took his doctor's degree here in 1602, Evelyn also studied here in 1645, and it was resorted to by many

<hr>

[a] It is said in *Corder's Italy* (ii. 132), that Vesalius was professor of anatomy at Padua from 1537 to 1542, but this is an error. He was offered the chair of anatomy, vacant by the death of Fallopius, in 1564, the same year in which he died. (*Biographie Universelle*, art. *Vesale*.)

other distinguished foreigners. Defects of discipline and the quarrels of the students seem to have been the first causes of the decline of the university, which has for more than a century been in a languishing state.' But it has still to boast of several distinguished professors and ranks as the second seminary of its kind in Italy, that of Pavia being the first. It has faculties of theology, law, medicine, and philosophy, and 35 professors, with between 400 and 500 students. It is governed by a senate, composed of a rector and 12 other individuals chosen from its general assembly; which includes, with the directors, deans, and professors, all the doctors who have graduated at Padua, and reside in the city. The university library comprises about 70,000 vols,. and it has a fine botanic garden, one of the oldest in Europe. Padua has a celebrated society of arts and sciences an Episcopal seminary, with an extensive library, formerly belonging to the Benedictine abbey, a city-school, two gymnasia, a high female school, agricultural, veterinary, and various other schools, a famous chemical laboratory and cabinet of mineralogy, and several libraries and museums of the arts, &c., this city being the seat of one of the five sections of the literary union of Austrian Italy. Among the charitable institutions are a civil and military hospital, a workhouse, foundling and orphan asylums, a monte di pietà, &c. Padua is a bishop's see, and the seat of the council and superior judicial courts for the deleg. It has been celebrated, both in ancient and modern times, for its woollen manufactures; but these have greatly declined since the time of the Venetian republic, to which they once supplied a considerable revenue.

It has still, however, manufactures of woollen cloth, broad silks, silk ribands and leather, and an extensive trade in wine, oil, cattle, garden vegetables. The fair of St. Anthony, which last 15 days, from June 13th, renders the city for a time a scene of bustle and gaiety; and the inhabitants derive some benefit from Padua being, for a part of the year, the residence of the Venetian nobility. It is very ancient, being said to have been founded by Antenor, after the siege of Troy—

" Hic tamen ille urbem Patavia, sedesque locavit
Teucrorum, et genti nomen dedit ;"	Æneid., l. 248.

and Mela enumerates the Patavium Antenoris among the principal cities of that part of Italy. (Lib. ii. cap. 4.) The historian Livy was a native of Padua; and the alleged patavinity of his style has long been a topic for critical discussion. Padua was taken by Alaric, Attila, and the Lombards; but being restored by Charlemagne to something like its former grandeur, it became, under his successors, flourishing and independent. In 1316, it came into the possession of the Carrara family; and in 1405 was united to the Venetian territory. Under the French, it was the cap. dep. Brenta. (Œst. Nat. Encycl.; Barghaus; Von Raumer's Ita. l, 197, 198; Eustace, Class. Tour, l. 144-158; Wood's Letters of an Archit. i. 245-250; Ross's Lett. from the N. of Italy.)

PADUCAH, p. v., capital of Mc Cracken co., Ky., 284 m. W.S.W. Frankfort, 816 W. Situated on the S. bank of Ohio r., immediately below the entrance of Tennessee r. It contains a courthouse, jail, 15 or 20 stores, and about 1000 inhabitants.

PAGE, county, Va. Situated toward the N.E. part of the state, and contains 160 sq. m. Bounded E. by the Blue Ridge. Drained by S. fork of Shenandoah r. It contained in 1840, 5300 neat cattle, 5993 sheep, 12,517 swine; and produced 105,199 bushels of wheat, 29,898 of rye, 155,784 of Indian corn, 29,116 of oats, 14,742 of potatoes; 6589 pounds of tobacco. It had 11 stores, two furnaces, five forges, 21 flouring-mills, 14 grist-mills, 40 saw-mills, one oil-mill, five tanneries, two distilleries; 11 schools, 527 scholars. · Pop. : whites, 5197; slaves, 781; free coloured, 216; total, 6194. Capital, Luray.

PAINSVILLE, p. t., capital of Lake co., O., 30 m. E. Cleve land, 179 m. N.E. Columbus, 349 W. Watered by Grand r., at the mouth of which are the villages, called cities, of Fairport and Richmond, which have good harbours. The village is situated on a high bank of Grand river, near the centre, and contains a courthouse, jail, five churches, a Presbyterian, Congregational, Episcopal, Methodist and Universalist. The town contains 53 stores, about 90.of which are in the village, one furnace, two grist-mills, five saw-mills, two printing offices, three weekly newspapers; one academy, 190 students; nine schools, 975 scholars. Pop. 2880.

· PAINT, t. Holmes co., O. Watered by a branch of Sugar er. It has one grist-mill, two saw-mills. Pop. 1361.

PAINT, t. Highland co., O. It has six schools, 410 scholars. Pop. 2580.

PAINT, t., Fayette co., O. It has three stores, one saw-mill; nine schools, 360 scholars. Pop. 1910.

PAINT, t., Ross co., O Watered by Paint cr. Pop. 1260.
PAINT, t., Wayne co., O., 95 m. N.E. Columbus. It has six stores, one flouring-mill, five saw-mills, three tanneries. Pop. 1610.
534

PAIMBŒUF, a seaport town of France, dep. Loire Inférieure. cap. arrond., on the Loire, 22 m. direct distance W. Nantes, of which it is, in fact, the deep-water harbour Pop., in 1836, 3850. It consists principally of one good street, fronting the quays which border the river. It has a fine moist 300 ft. in length, a school of navigation a communal college, and court of primary jurisdiction. Vessels of more than 200 tons trading with the port of Nantes are here to load or unload their cargoes. (Hugo, art. Loire Inférieure.)

PAINSWICK, a market town and par. of England, co Gloucester. hund. Brisley, on the S. declivity of Sponehed hill, 6 m. S. Gloucester, and 90 m. W, by N. London. Area of par., 6510 acres. Pop., in 1841, 3730. The town is small and irregularly built, the streets being neither paved nor lighed. The church, which is large, has at its W. end a fine tower and spire 174 ft. in height; but the building is rendered unsightly by the strange admixture of Doric and Ionic pillars, with the more ancient architecture in the Gothic style. There are, also, two places of worship for dissenters, and six Sunday schools, furnishing religious instruction to about 500 children of both sexes. The town has an endowed free school for 26 boys besides which three subscription schools, and an infant school, are attended by 300 children. The inhabitants are principally employed in the manufacture of woollen cloth, especially broadcloth and kerseymeres. In 1839 there were nine mills at work, employing 319 hands. . About 70 families are employed in hand-loom weaving, the weekly wages of a family averaging 10s. 10d., in 1838. Markets on Tuesday; fairs, Whit-Tuesday and September 19 for sheep and oxen.

At the top of Sponebed hill is an ancient fortification, called Kinsbury castle; its form is that of a parallelogram, enclosing about three acres within a double trench, and commanding the adjacent country. The discovery of numerous coins and other antiquities have led to the supposition that this fortress is of Roman origin.

PAISLEY, a parl. bor., market and manufacturing town of Scotland, co. Renfrew, partly on an eminence and partly on a plain, on both sides the White Cart, 3 m. S. Renfrew Ferry, on the frith of Clyde, and 6 m. W. by S. Glasgow. The surrounding country is nearly a dead level, except on the S., in which direction the "Braes of Gleniffer" rise to a height of 760 ft. within little more than 1 m. of the town. Population of the town and parish in 1755, 6799; in 1821, 31,179; in 1831, 57,466. Population of the town only, in 1831, 45,199; in 1841, 48,628.

Paisley, inc. its suburbs, is spread over a tract of ground comprising an area of about 2¼ sq.m. ; but the boundary of the parl. bor. embraces an area of about 6 sq. m. Its main street runs from E. to W. for nearly 2 m., and forms part of the road from Glasgow to Beith, and the towns on the coast of Ayrshire. Another long line of road passes through it from N. to S. That part which lies E. of the river is called the new town ; the first houses in this important addition to the borough having been erected in 1779. But though well built, Paisley is not so handsome as some of the larger Scottish towns. Of late years, however, its appearance has been greatly improved by the substitution of numerous substantial houses for low thatched cottages. Improvements of this description are in rapid progress ; but still, in passing along the streets, one observes a singular alternation of handsome with mean edifices. The streets are generally paved, and are lighted with gas; and the town is well supplied with water from the Gleniffer Braes, by means of reservoirs constructed under an act passed in 1838. The police is efficient. In the neighbourhood are many elegant villas and baronial seats. The most important of the public edifices is that for the civil business of the town and county, including the jail and bridewell, erected in 1820 at an expense of £98,000. It is a quadrangular building, in the castellated style. The original parish of Paisley has been divided into four distinct parishes, with seven parishes quoad sacra, having no fewer than seven places of worship connected with the estab. church. Of these Abbey church, which is a collegiate charge, is the most imposing and magnificent. It consists of the nave of an ancient monastery, being the only portion of that once splendid building which now remains. The High church, on an eminence in the old town or "the borough," as it is commonly called, is an elegant building, with a lofty spire. The other churches require no particular notice. If we except St. George's, a spacious Grecian structure; the Episcopal chapel, a handsome building of chaste Gothic ; and one of the secession churches, an elegant Grecian edifice. The new town is connected with the borough by three bridges: and the river is also crossed by the line of the Glasgow, Paisley, and Greenock railway, which passes through the town. About two years ago, barracks were erected in one of the suburbs for the accommodation of a battalion of infantry.

In addition to the eleven churches (one of which is a

Gaelic chapel), there are no fewer than seven Presbyterian dissenting churches, besides two Baptist places of worship, two Methodist chapels, and one each, belonging to the Independents, Roman Catholics, Episcopalians, Unitarians, and New Jerusalemites; and there are two or three additional dissenting chapels in Johnston; a large village within the parish. Within the parl. bor., 19,812 persons belonged, in 1836, to the established church; 22,490 to other religious denominations; the remainder not being known to belong to any Christian denomination. Paisley is the seat of two seminaries for theological instruction; one connected with the Reformed Presbyterian Synod, the other with the Relief Synod: the professor of divinity in each being a minister in the town. The number of students at both seminaries may average 40. Each hall has attached to it an extensive collection of theological books. (*New Stat. Acc. of Scotland—Renfrewshire*, p. 947, 948.)

The grammar-school is a royal foundation (though the endowments have nearly disappeared), established by James VI., in the 16th century, and confirmed by subsequent royal deeds: it is an efficient seminary. There are, exclusive of Sunday schools, 64 other schools in the town and parish, none of which are parochial or endowed, with some trifling exceptions. A philosophical institution, similar to a mechanics' institution, was founded here in 1808, for the delivery of courses of lectures on different branches of science and literature. A small library and a museum are attached to it; but the institution is not in a very flourishing condition. There are various printing presses; and the town has two weekly newspapers. Among the eminent characters that Paisley has produced may be named Alexander Wilson, the celebrated American ornithologist, and Robert Tannahill, the Scottish poet. Dr. Witherspoon, author of various theological works, and afterwards president of the college of New-Jersey, was, for ten years, one of the ministers of the town. The Public Subscription library contains about 5000 volumes, and is supported by 200 subscribers. The Trades library, supported chiefly by workmen, is a valuable collection. There are various others, generally of religious books.

Poor rates were introduced into the borough so early as 1740; and into the Abbey parish in 1785. The annual assessment in each is nearly £3000. The number of paupers in both, including occasional and permanent poor, is about 1800, including children. The whole sums left for charitable purposes do not exceed £3000. There is a hospital or poor-house in the borough erected in 1752. The number of inmates, old and young, is about 200. The children are boarded and educated in the country. A public infirmary, which can accommodate about 45 patients, was erected about 30 years ago. The town was visited by cholera asphyxia in 1832: number of cases, 796; of deaths, 446. Cholera reappeared in 1834: number of deaths, 140.

Manufactures.—Paisley was early distinguished by its manufactures. The first impulse given to this department was by pedlars or travelling merchants, who, soon after the union, bought the goods made here, and sold them in England; and a good many of whom, having made some money, settled in the town. The articles then manufactured were striped linen cloths, handkerchiefs, and Bengals: these were succeeded by plain lawns, some of them chequered with cotton, and others ornamented with a great variety of figures; by linen gauze; and by white sewing thread, known by the name of *Ossnas* or *Nun's thread*: in fact, this species of thread, so long as it was made in Scotland, was principally produced in Paisley.

In 1760 the making of silk gauze was first attempted, in imitation of that of Spitalfields; and it soon attained to great importance, both in the town and villages round to the distance of 20 m. This trade afterward declined; but not till the Spitalfields manufacturers had been driven out of the market, and some of them had transferred their establishments to Paisley. It has again revived; and "Paisley now furnishes nearly all the silk gauzes used in the kingdom, with the exception of those imported from France." (*New Stat. Account, ut supra,* p. 267.)

In 1785, when the silk gauze trade experienced a temporary interruption, many of the principal houses in the town, entered into the muslin manufacture, which rose to a great height of prosperity. This branch has considerably declined; but the fabrics, which are chiefly designed for the London market, are of first-rate excellence in point of taste and elegance of execution. The ornamenting of muslins by fine needle-work has lately become a considerable employment.

The shawl manufacture is one of the staple branches carried on in Paisley. Imitation shawls of all kinds have, at different times, been made here,—such as Thibet shawls, Cashmere ditto, and zebras; the last being so called from their resemblance to the skin of the zebra. The genuine Cashmere wool is imported for making the Cashmere shawls; and the first Cashmere shawl made in Britain was manufactured in Paisley. It is a curious and not easily explained fact, that the yarn is generally spun in France, and that the attempts to produce it here have not been very successful. Edinburgh had long the lead in this manufacture, but it has been nearly beat out of the field; and Paisley is at present without a rival in this department. Damask and embroidered shawls are also extensively manufactured; and a beautiful and ingenious kind of shawl, *Chenille* (caterpillar), from its variegated colour and the softness of its feel. This shawl is often labelled in shops with the words *velour du soie* (silk velvet), a name descriptive of its appearance. But the shawls now chiefly made are, 1st, those wholly of silk; 2nd, half silk and half cotton; and 3d, wholly cotton. The manufacture of these varieties has been increasing with astonishing rapidity for some years past; several makers effecting sales to the amount of from £40,000 to £60,000 a year. The total sales in 1834 were estimated at £1,000,000 sterling; and the trade has greatly increased since. Machinery has been advantageously employed in finishing the shawls, in the operation of *clipping*, which was formerly performed in a comparatively clumsy way, by the hand. The Jacquard loom has also been recently introduced. All the trades depending on and subordinate to the shawl branch have necessarily increased; in particular that of dyeing. "Fifteen years ago," to quote from the *New Statistical Account,* written in 1837, "perhaps 40 or 50 hands were employed as dyers; now, ten times that number at least are engaged." (P. 267.) Crape dresses are manufactured to a considerable extent. Paisley has, in fact, been long the centre of the manufacture of ornamental or fancy goods. (*Wilson's Survey of Renfrewshire,* p. 255.)

Instead of the linen-thread formerly made in this town, a pretty extensive cotton-thread trade has sprung up, which employs several factories; and the annual value of the thread may be about £100,000 sterling. But while so large a quantity of cotton-yarn is exported from Paisley, it imports from Lancashire a considerable portion of the yarn used in making the finer muslins; and yarn is not unfrequently sent thither by the Manchester manufacturers to be woven into those fancy muslins for which Paisley is so famous. The cotton manufacture employs altogether about 13 mills, with nearly 3000 hands. There is also a silk-mill, with about 380 hands.

"The number of looms in Paisley at present (1837) is ascertained to be about 6000: of these, 5700 are employed by Paisley manufacturers; the remaining 300 work to Glasgow houses. About 3000 looms are employed in the country by Paisley capital, chiefly in the neighbouring villages. The number of apprentices to the looms in Paisley is at present 728; the number of harness weavers is 5350; plain-weavers, 650; female weavers, 40; in all, exclusive of apprentices, 6040." (*New Stat. Acc.,* p. 268.) Besides the above, there is in the new town a power-loom factory for cotton cloth used in printing. The printing of silks and other fabrics, has lately been attempted, but on a small scale.

The town has four iron-foundries, three brass foundries, one large tan-work, three breweries, two distilleries, one large soap-work (which has been in operation for nearly 70 years), seven extensive bleachfields in the neighbourhood, and various other minor branches of business. The foregoing accounts, generally speaking, are confined to Paisley and its suburbs, and do not include Johnston, Elderslie, and other villages at some distance, though within the limits of the original parish of Paisley, (*Vide* Johnston, in this work.) Still, however, we regret to have to state that Paisley, during the last 10 years, has done little more than keep its ground, and that neither its population nor its manufacturing industry has materially increased. During that period, its manufacturing pop. has frequently, indeed, been involved in extreme distress; occasioned sometimes by fluctuations of demand, but more frequently, perhaps, originating in the improvident loans made by the banks to individuals without capital; which has tempted the latter to engage in the most hazardous speculations, generally to their own ruin, and in most instances, also, to the great injury of the town.

Renfrew, or Renfrew Ferry, 3 m. from the town, is, properly speaking, the port of Paisley; but the White Cart, which falls into the Clyde, 3 m. from the borough, and only a few hundred yards W. of Renfrew, is navigable to Paisley for vessels of 80 tons. Much has been done to improve the navigation of this river. A railway has recently been opened between the town and Renfrew Ferry, on which locomotive wagons regularly ply. This railway has a rise of 16 ft. in the whole distance (3½ m.). A railway from Glasgow, recently opened, passes through the town, where it divides itself into two branches, one going to Ayr, with a subsidiary branch to Kilmarnock; the other leading to Greenock. The Glasgow, Paisley, and Johnston canal, opened in 1811, commences at Port Eglinton, near Glasgow, passes Paisley, and terminates at Johnston, a distance of 11 m. There were conveyed along this canal, in 1840, 396,248 passengers, and 76,393 tons of goods. But it

535

will, no doubt, be injured by the opening of the railway. The first light iron passage-boats were established on this canal in 1831. In addition to the Paisley Commercial Banking Company, there are three branch banks; and a savings' bank, or provident bank, instituted in 1815. The neighbourhood of the town produces coal, ironstone, fire-clay, and potters' clay; and there are manufactures of sulphate of iron, or copperas, alum, muriate of potash, and sulphate of ammonia. (New Stat. Acc., p. 157-159.)

Previously to the passing of the Reform Act in 1832, Paisley, notwithstanding its great wealth and importance, had no parliamentary representative; but the act in question conferred on it the important privilege of sending one member to the House of Commons. Registered voters in 1840-41, 1257. Municipal revenue, £3997 13s. 4d. Number of councillors, 16. The sheriff courts of the county were transferred from Renfrew to the capital of the shire, to Paisley, so early as 1705.

Paisley is very ancient, and is supposed to occupy the site of the Roman station, Vanduaria. In 1164, Walter, son of Allan, lord high steward of Scotland, founded a monastery here, of which nothing remains but the nave and its collateral appendages, now used as the parish church. This abbey, the precincts of which were enclosed with a wall about one mile in circuit, was the burial-place of this noble family till they became kings of Scotland. At the Reformation, this property passed into the hands of a branch of the noble house of Hamilton, now represented by the Marquis of Abercorn, in whose possession (with a slight interruption) it has ever since remained. Paisley, in 1483, was regularly constituted under the jurisdiction of the abbot. The "Black Book of Paisley" has lately been ascertained to be simply a MS. copy of Fordun's "Scotichronicon." The "Chartulary of Paisley" was printed in 1832 by the Maitland Club of Glasgow. Sir William Wallace is said to have been born at Elderslie, 2 m. S.W. of the town. (In addition to the works already quoted, see Chalmer's Caledonia, vol. iii.; Crawfurd's Hist. of Renfrewshire, 3d ed., 1818; Bound. Reports, and Priv. Inform.)

PALATINE, p. t., Montgomery co., N. Y., 56 m. W.N.W. Albany, 390 W. Bounded S. by Mohawk river. Watered by Garoga creek. It contains a Presbyterian and a Lutheran church, two stores, one grist-mill, two saw-mills; three schools, 115 scholars. Pop. 2823.

PALAWAN, an island of the E. archipelago, 4th div., about midway between Borneo and the N. Philippines. It is long and narrow, extending between lat 8° and 11° N., and long. 117° and 120° F.; having N.W. the China sea, and S.E. that of Sooloo. Length, N. to S., 275 m.; average breadth about 32 m. It is little known to Europeans; but is W. appears to be loftier than its E. side, and inhabited by a savage people, who seldom approach the coast. Palawan produces cowries, wax, tortoise-shell, ebony, and lake wood; the sea-slug abounds around its shores.

PALEMBANG. See Sumatra.

PALENCIA, a city of Spain, k. of Leon, cap. prov. of its own name, on the Carrion (crossed here by two bridges), 57 m. S.E. Leon, and 116 m. N.N.W. Madrid. Pop., according to Miñano, 10,813. It is agreeably situated in an extensive valley near the canal of Castile, and comprises several straight and pretty wide streets, with a spacious square, having colonnades on two of its sides. In the environs, also, are several well-planted walks, or alamedas; its principal public buildings are the cathedral (one of the largest Gothic structures in Spain), five parish churches, a well-endowed hospital, a poor-house, (formerly a palace, built by the Cid,) foundling asylum, and bishop's palace. Palencia had a university prior to the establishment of that at Salamanca; and it still possesses a superior seminary, with about 60 students, of grammar and philosophy. It has manufactures of woollen goods, blankets, coverlets, and serge, which meet with a ready sale throughout Spain; and of hats and earthenware, with tanneries, &c. A fair is held annually in September.

PALERMO (an. Panormus, from παν, all, and ὅρμος, a station for ships, from the number of vessels that frequented its port), the cap. city, and principal seaport of Sicily, on its N. coast towards its W. extremity; lat. (observatory) 38° 6' 44" N., long. 13° 20' 15" E. Pop., in 1831, 173,478; but in consequence of the prevalence of cholera, it is now (1846) said to be under 140,000. It is built along the S.W. side of an extensive bay, in a plain which, from its luxuriance, and from being surrounded by mountains on three sides, has been termed the "golden shell" (conca d'oro). In front of the city, the numerous steeples, cupolas, and towers of which give it a noble appearance from the sea, is the Marina, a raised platform or terrace, extending above 1 m. along the bay, and about 80 paces in breadth. At the E. extremity of this walk is the Flora, a public garden, laid out in walks, interspersed with statues, fountains, and summer-houses. People of all ranks are admitted, and in fine evenings it appears the rendezvous of the whole city.

Adjoining the Flora is the botanical garden, at the entrance of which is a building similar to an ancient temple, in which botanical lectures are delivered. The garden is well laid out, and contains an extensive collection of valuable plants. On the W., Palermo extends to the foot of the rocky and abrupt mountain Pellegrino, but on the E. a reach of well cultivated grounds ascends gradually to cape Zaffarano, which bounds the bay on that side. The city is surrounded by an old wall, of little or no strength, some of the bastions being occupied by gardens, while others have been cut away to increase the breadth of the Marina. It is, however, defended by a citadel and several other forts, which are tolerably strong towards the sea; though from being much scattered they would require a large garrison, and could not hold out against a force investing the city by land.

Palermo is regularly built, and, if better finished, might be esteemed an elegant city. Two large streets, the Cassaro and Strada Nuovo, each upward of a mile in length, intersect each other at right angles, dividing the city into four equal parts, and each leading to one of the four principal gates. These streets are well paved with large flat blocks of lava, and are faced throughout their whole length with handsome buildings. The central space where they meet is an octagon (Piazza Ottangolare); each of its sides consists of an edifice three stories in height, combining the Doric, Ionic, and Corinthian orders; and it is besides enriched with statues and fountains. A coup-d'œil of similar magnificence is that enjoyed from this piazza is, perhaps, not to be met with in any other city of Europe. (Russell's Sicily, 43.)

There are several other public places or squares adorned with obelisks, jets-d'eau, and sculpture, of which the principal are the column of St. Dominic, and the superb fountain opposite the praetorian palace. But all the streets, except those above named, are irregularly laid out, narrow, and ill built. The houses are almost all high, and a number of them have balconies with iron railings. These projections lessen the symmetry of the architecture, but this is more than countervailed by the convenience they afford of enjoying the cool evening breeze in so warm a climate. Almost every house has a common stair; and each story of apartments forms, as in the old houses of Paris and Edinburgh, the separate residence of a family. Several of the mansions of the nobility are admired for their architecture, but their interior is usually deformed by a multiplicity of false ornaments. Many have marble columns, either in front, or in the large court, which they generally inclose; but their effect is frequently destroyed by the meanness of the adjoining buildings. Altogether Palermo presents an incongruous mixture of pomp and poverty, exemplified in noble ranges of palaces disgraced at their bases by shops and stalls, and in showy equipages parading the same streets with sturdy mendicants vociferously demanding food, or sluggishly taking their siesta on the pavement. Swarms of priests, nobles, officers, and other loungers, yawn on chairs before the coffee-house; and artisans of every kind at their respective employments outside their shop doors, usurp the sides of the street, obliging foot passengers to walk in the centre among the numerous carriages. The constant calling out this occasions on the part of the coachmen, added to the hurry of business, and the groups round the ice-water stalls, form an animated and singular if not a pleasing scene.

The supply of water is peculiarly abundant, and most of the houses have fountains, even in their second and third stories; hence the city is in general clean, except after heavy rains, when, from the lowness of its site, it becomes extremely muddy, and recourse is sometimes had to moveable iron bridges for crossing the streets. There is an excellent supply of provisions of every description; and during the absence of the moon, the principal streets are tolerably well lighted. The city, excepting on the site of the ancient port, where malaria is generated in autumn, is healthy. The temperature of winter seldom falls below 39° Fah. In summer, however, the thermometer keeps for months between 80° and 90°; and then the inhabitants generally shut up their houses and shops a little before noon, keeping them shut for three or four hours, an interval during which all is silence and stagnation. The Scirocco is very oppressive, but fortunately not of frequent occurrence.

Palermo has a great number of public edifices and institutions. Convents and churches are particularly numerous; of the former there are even said to be nearly 70! Most of the churches are sumptuous; but they discover no taste, and offend the eye by a profusion of ornament. A striking monotony reigns in their construction, being generally built with an elevated façade, a large nave, and two side aisles, bounded by lateral chapels, dedicated to various saints, and decorated with pillars, paintings, statues, flowers, and candelabra. Some, however, as that formerly belonging to the Jesuits, must be exempted from this censure. The cathedral, erected about 1180, by Archbishop Walter, an

Englishman, is externally of Gothic architecture; and, though not in the best taste, is a tolerable specimen of the style of the 15th century. It has, however, been spoiled by the modern addition of a cupola; and its interior has been somewhat recently altered to the Greek style. Within are many fine red porphyry sarcophagi, of considerable antiquity, in which have been deposited the remains of different sovereigns of the island, including Roger the founder of the Norman kingdom of Sicily, the emperor Frederick II., &c. The church of St. Giuseppe, also on the Cassaro, is profusely and richly ornamented, and has some fine columns of grey Sicilian marble, nearly 60 feet in height. The royal palace, the residence of the viceroy, is a spacious building of mixed Arabic and Norman architecture. It has many spacious apartments, a gallery with some good paintings, and a neat armoury: on its summit is the observatory erected in 1748, whence Piazza discovered the planet Ceres. Attached to the palace is the beautiful little church of St. Peter, which, with its crypt and superb mosaics, forms one of the most complete specimens of Saracenic magnificence extant. The square in front has a statue of Philip IV. of Sicily, surrounded by four other statues. The tribunal of justice and the custom-house occupy a large edifice on the Marina, formerly the palace of the Inquisition, abolished in 1782. The public prison, in one of the main streets, built round a large courtyard, though well supplied with water, is dirty, and in many respects badly provided. The Jesuit's college, a magnificent edifice in the Cassaro, with various schools, and a fine library, in which the Sicilian parliament formerly held their sittings; the university, the archbishop's palace, and the principal government pawn-bank, a spacious building, with a neat portico, are among the remaining most remarkable edifices. There are several theatres, but they are generally ill constructed, and not to be compared to those of Naples, Paris, or London.

At the N.W. extremity of the city is the arsenal, from which a fine mole, fully 1-4th m. in length, having a light-house and battery at its extremity, projects S. into 9 or 10 fathoms water, forming a convenient port, capable of accommodating a great number of vessels. This important work cost about £1,000,000 sterling; but the lighthouse, though a splendid structure, is said to be very ill lighted. Ships that do not mean to go within the mole may anchor about half a mile from it, in from 16 to 23 fathoms. There is an inner port, reserved for the use of the arsenal, with large naval magazines, prisons for galley-slaves, &c. There is also a small cove in front of the town, called the Cala Felice, the representative of the two ancient harbours, and capable of accommodating vessels of from 150 to 200 tons. On its E. side is the pratique office; the lazaretto, a dirty and inefficient establishment, is in a rocky bay at the back of the mole.

Few indications exist of the ancient splendour of the city, except the remains of a naumachia, and some vestiges of an amphitheatre. In the senatorial hall are preserved fragments of various marbles, &c.; and in the royal palace are two ancient bronze rams, brought thither from Syracuse, and said to have been made by Archimedes! (*Smyth's Sicily*, p. 73.)

In the neighbourhood are many fine specimens of Moorish architecture; the principal being the Saracenic fortress of Kaba, now used as cavalry barracks; and the Zisa, a palace erected in the 9th or 10th century, still in good repair, and occasionally used as a royal residence. Near the latter is a Capuchin convent, with a cadavery, or receptacle for the reception of dead bodies. A royal residence, in the Chinese style, stands outside the walls, near M. Pellegrino; and about 16 m. E. Palermo, near the bay, is La Bagaria, the favourite residence of many Sicilian nobles. Several of the villas of the nobility are richly adorned, both by nature and art: that of Prince Palagonia, however, is chiefly noted for its statues of all sorts of monsters.

Palermo is the see of an archbishop, who is primate of Sicily; the seat of an intendant and council of intendency; a departmental council; a supreme court of justice, with 14 judges; a civil and criminal court for the intendency, and a tribunal of commerce. It has a university, the second in the Neapolitan dominion, attended by about 600 students, comprising several eminent names among its professors. It has a library of upwards of 30,000 vols., a printing press, several museums; but only a few lectures are given, and the education is worthless in the extreme. Palermo has also a high female seminary, second to that of Naples; a college of nobles, an episcopal seminary, and many inferior schools; numerous charities, including two large hospitals, a lunatic and a foundling asylum, houses of industry for mendicants, &c.; public baths, and several public libraries and scientific associations. Silk manufactures were established here in the 11th century, and they still form the chief branch of manufacturing industry, though much less flourishing than formerly. Cotton fabrics are also produced,

with oil-cloth, leather, &c.; and there is here a glass-work, the only one in Sicily. The tunny fishery employs from 900 to 1000 boats, and 3500 fishermen. But the principal resources of the inhabitants depend on Palermo being the residence of the viceroy and the seat of government, and on her trade. The latter, indeed, is but trifling, compared to what it would be were Sicily under an enlightened government, capable of calling forth its vast resources. But even at present it is far from inconsiderable. The great articles of export are shumac, fruits of various sorts, including oranges and lemons, wine, manna, brimstone, &c. Subjoined is a

STATEMENT of the Quantities and Values of the principal Articles exported from Palermo in 1839.

Articles.		Quantities	Value in Pounds Sterling.
			L.
Argols and cream of tartar	cwts.	4,799	4,607
Barilla	—	6,385	2,989
Brimstone	—	39,305	16,811
Cantharides	—	62	1,394
Corn, grain, pulse, and rice	qrs.	942	942
Essences	lbs.	6,735	1,063
Fish, salted	cwts.	1,475	1,884
Fruits:—			
Dry and pickled	boxes	86,848	142,119
Orange and lemons	do.	164,186	32,986
Linseed	qrs.	5,797	9,561
Other seeds	cwts.	3,478	1,597
Liquorice paste	—	4,385	5,318
Manna	—	3,797	41,107
Oils:—			
Linseed	galls.	10,346	1,056
Olive	—	77,984	7,304
Rags	cwts.	19,342	9,967
Shumac	—	217,489	256,794
Silks	lbs.	2,510	2,810
Skins	No.	86,000	1,874
Wine and spirits	galls.	804,440	66,115
Other articles			14,928
Total value			**624,964**

Of the above, goods to the amount of £290,122 went to the U. States of America; £198,562 to Great Britain and her colonies; £59,446 to France; £25,674 to the Baltic; £19,737 to Belgium and Germany; £37,418 to the Italian states; and £1,005 to other countries.

The imports consist principally of sugar and other colonial products; cotton, linen, silk and woollen fabrics; earthenware, hardware, and other manufactured goods; dye stuffs, spices, &c.

The city funds, derived principally from landed property, the land-tax, and taxes on consumption, are said to amount to about £100,000 a year; but for many years past the expenditure has exceeded this sum, and the city is now deeply in debt.

Sicilian writers have made many absurd and ridiculous statements concerning the foundation of Palermo; but the most rational and generally received opinion, confirmed by the authority of Thucydides and Polybius, is, that it was founded by a colony of Phœnicians; the beauty of the situation, and the convenience of the port, whence, as already seen, it derived its name, being powerful inducements to a trading people, to make it a settlement. (*Tucyd.* lib. vi.; *Polybius,* lib. i. cap. 38.) It subsequently fell into the hands of the Carthaginians, who made it the capital of their Sicilian dominions. Soon after the beginning of the first Punic war, it passed into the hands of the Romans, who established a colony in it (*Strabo.* lib. vi.), conferred on it various privileges, and allowed it to be governed by its own laws. In a subsequent age, the Saracens made it the capital of their Sicilian territories; and since their time, with the exception of some short intervals, it has been the capital of Sicily. It was the residence of the court of Naples during their exclusion from that city from 1806 to 1815. (*Smyth's Sicily,* 70–89, and Append.; *Swinburne; Brydone; Russell; Simond; Von Raumer's Italy,* ii.; *Parl. Reports, &c.*)

PALERMO, p. t., Waldo co., Me., 19 E.N.E. Augusta, 614 W. Incorporated in 1804. Watered by head branches of Sheepscot river, one of which issues from a pond. It has two stores, one grist-mill, two saw-mills; 12 schools, 587 scholars. Pop. 1694.

PALERMO, p. t., Oswego co., N. Y., 15 S.E. Oswego, 157 W.N.W. Albany, 380 W. Drained by Catfish creek. It has three stores, 11 saw-mills; 15 schools, 696 scholars. Pop. 1928.

PALESTINE, *See* SYRIA and PALESTINE.

PALESTINE, p. v., capital of Crawford co., Ill., 155 S.E. Springfield, 680 W. Situated 3 ms. from Wabash river, and contains a courthouse, jail, a U. States land-office, two churches, eight stores, various mechanic shops, and about 300 inhabitants.

PALHANPOOR, a fortified town of Hindostan, prov. Gujrat, cap. of a Mohammedan principality, 88 m. N.N.W. Ahmedabad. Lat. 24° 11'; long. 72° 20' E. Pop. 30,000 (?)

It is about 1½ m. in circ., and is surrounded by a brick wall, flanked with towers. the gates being defended by outworks, mounted with small cannon. It is of considerable political importance, being a frontier town on the desert separating Gujrat from Sinde and Cutch, and on the main route from Rajpootana southward. The state, of which it is the cap., extends from 30 to 40 m. round, including two towns, and about 130 villages, and producing a yearly revenue of about 2½ lacs rup. (*Hamilton, E. I. Gaz.*)

PALMA. See MAJORCA.

PALMAS, the principal town of the Canary islands, which see.

PALME, or PALMI, a town of the Neapolitan dom., prov. Calabria Ultra I., cap. distr., on the gulf of Ginja, 21½ m. N.E. Reggio. Pop. about 7000. It was partially destroyed by the earthquake of 1783, but has since been restored. It is well built; its streets being regular, and its houses mostly of stone, and in good taste. In its centre is an elegantly sculptured and well supplied fountain. It has some manufactures of silken and woollen fabrics, and trades in oil, liqueurs, &c. (*Craven's Tour, 292.*)

PALMER, p. t. Hampden co., Mass., 81 W. Boston, 391 W. Bounded W. by Swift river, S. by Chicapee river. Watered by Ware river. It contains a Congregational and a Baptist church, 8 stores, 3 grist-mills, 3 saw-mills; 14 schools, 350 scholars. Pop. 2139.

PALMYRA, (the Tadmor of the Scriptures, by which name it has always been designated by the Arabs,) a celebrated city of antiquity, and the cap. of the region of Palmyrene in Syria, in an oasis in the midst of deserts in the modern pach. of Damascus, 147 m. S.E. Aleppo, and 187 m. S.S.W. Damascus, lat. 34° 29' N., long. 38° 48' E.

This once famous city is now all but deserted, not having more than 100 inhabitants, and it derives its whole importance from its classical associations and the number and magnificence of its ruins. These, which stand near the E declivity of a mountain range running from N. to S., may occupy a space of about 3 sq. ma., though it is probable that the ancient city extended over a larger area, exclusive of the tombs on the tops and sides of the adjacent hills. The oasis, in which the city is situated, is traversed by two streams, which, though hot and sulphureous, are said to be wholesome, and not disagreeable. But the water used in the ancient city was of the best quality, being brought from a considerable distance by a large subterranean aqueduct, of which there are still some remains. The first view of the city is described by all travellers as extremely magnificent. "On opening upon the ruins," says Captain Mangles, "as seen from the Valley of the Tombs, we were much struck with the picturesque effect of the whole, presenting altogether a most imposing sight. It was rendered doubly interesting by our having travelled through a wilderness destitute of a single building, from which we suddenly opened on the innumerable columns and other ruins, the snow-white appearance of which, contrasted with the yellow sand, produced a very striking effect." (*Irby and Mangle's Travels, p. 262.*) The ruins are not, however, to be compared, as respects the size of the gates, columns, and temples, with those of Baalbec and Thebes; but they are more remarkable than either for their vast extent, and they are less encumbered with modern fabrics than most other ancient remains.

The ruins now extant comprise the fragments of two or three temples, several gateways, (one of which is more perfect than the rest), colonnades, sepulchres, &c. With respect to the antiquity of these ruins, it is difficult to form a conjecture: the tombs are evidently the oldest, but even these do not date as far back as the Christian era. The other buildings are considerably more recent, and most of the fine and expensive edifices appear to have been constructed during the three centuries ending with the reign of Diocletian.

On approaching the city, a ruined mosque, built by the Saracens, introduces the stranger to a fine gateway, having a lofty central arch, flanked by two others of smaller size, which lead directly to a grand-avenue, which, from the remains, must have been nearly one mile in length, and bordered on either side by rows of Corinthian columns, of which, however, only 114 now remain. This avenue leads to a gateway. beyond which are ranges of pillars supporting a frieze and entablature, supposed by Mr. Addison to be the ruins of two noble gateways, that may have led from the central avenue to other colonnades now entirely destroyed. A circular colonnade, of which eighteen columns only are now standing, has in its centre a small but richly ornamented building, with niches for statues; and immediately beyond it are the prostrate remains of a magnificent building, constructed of a species of marble superior to that found in other parts of the ruins. It appears to have comprised two very large rooms; but whether it were a temple or palace, is difficult to determine. By far the most extensive ruin, however, is the Temple of the Sun, the grand entrance to which was supported by four fluted Ionic pillars, and adorned with rich carvings of vine-leaves and clusters of grapes in bold and spirited relief, beautifully chiselled. The outer precinct, which encloses a quadrangular space of 220 sq. yards, is formed by a lofty wall. adorned with pilasters both within and without. Inside this court are the remains of two rows of noble marble pillars, each 37 ft. in height, and another row of columns 50 ft. in begin. appears to have encircled the temple, which, however, was only 100 ft. in length by 45 ft. in breadth: it has since been converted into a mosque, and its interior is disfigured by passages from the Koran written round the walls.

The sepulchres, which are, perhaps, the most interesting of all the ruins, occupy the tops and sides of the surrounding eminence, some presenting mere heaps of rubbish; others half fallen, exposing their shattered chambers; while one or two still exist almost entire. They are built in the shape of square towers, from three to four stories in height, each forming a sepulchral chamber, with recesses divided into compartments for the reception of the bodies. Some of the chambers are ornamented with Corinthian pilasters and sculptures, in almost perfect preservation, executed in high relief; the walls are of white stucco, and the ceilings are divided into diamond-shaped compartments, delicately ornamented with white stars on a blue ground: over the doorways are tablets with inscriptions both in Greek and Palmyrene. A few of the streets may be traced with some difficulty, and the foundations of houses are distinguishable in some places; but not a vestige remains of the old walls destroyed by Aurelian, though a wall still exists that has been made of materials from the sepulchres, and was probably erected soon after the demolition of the older fortifications. The inscriptions are both in Greek and in the unknown Palmyrene language; all of those on the columns are honorary, generally to the effect, that the senate and people inscribed them in honour of an individual whose pedigree is given through several generations. The inscriptions on the tombs are in Greek, and tolerably perfect. Fac-simile copies of them are given in the great work of Messrs. Wood and Dawkins, which also contains drawings of all the principal buildings of Palmyra.

History.—The earliest accounts of the existence of Palmyra are derived from the sacred writings, which state that "Solomon built Tadmor in the wilderness, and all the stone cities which he built in Hamath" (2 Chron. viii. 3, 4, and his motive for thus founding it was, according to Josephus, "because in that place were fountains and wells of water. He gave it the name of Tadmor, which is still prevalent among the Syrians; but the Greeks name it Palmyra." (*Ant. Jud. i. viii. ch. 6.*)

Pliny has noticed the city, and the peculiarities in its situation to which it owed its rise and importance: *Palmyra urbs nobilis situ, divitiis soli et aquis amœnis; vasto undique arenis circulo includit agros; ac velut terra exempta a rerum natura, privata sorte inter duo imperia summa, Romanorum Parthorumque; et prima in discordia semper utrimque cura. (Hist. Nat., lib. v. cap. 25.)* The fertility of the oasis round Palmyra made it a suitable situation for a small town; but its position in other respects was still more advantageous, from its being the resting-place of the caravans between the Persian gulf and the great cities on the Euphrates and Tigris, and Aleppo, Damascus, and the ports on the Mediterranean. Palmyra thus became a principal emporium of the commerce between the eastern and western worlds; and to this, no doubt, is to be ascribed the wealth and importance to which she early attained. Being situated between the empires of Rome and Parthia, it was an object of great importance with the Palmyrenians to preserve a strict neutrality, and to keep on good terms with them both. But after the victories of Trajan had established the unquestionable preponderance of the Roman arms, Palmyra became a dependency of Rome, and assumed to the rank of a colony. "It was during that peaceful period, if we may judge from a few remaining inscriptions, that the Palmyrenians constructed those temples, palaces, and porticoes* of Grecian architecture, whose ruins, scattered over an extent of several miles, have deserved the curiosity of our travellers." (*Gibbon, chap. 9.*)

The most splendid period of the history of Palmyra was that which immediately proceeded her fall. Valerian, emperor of Rome, having been made prisoner by Sapor, king of Persia, Odenathus, a citizen of Palmyra, who had obtained to the principal direction of her affairs, joined the Roman forces, and had a large share in avenging the insult offered to the majesty of Rome. He attacked the Persians, drove them beyond the Euphrates, penetrated as far as their capital city Ctesiphon, and captured the treasures and women of the great king. For these services, the sen-

---

* According to Stephanus Byzantius, these were mostly erected by the emperor Adrian; but there is no evidence that such was really the fact, though he may have done so to some extent.

ste; with the approbation and applause of the Roman world, conferred on Odenathus the title of Augustus, and associated him in the empire with Gallienus. These honours, however, he enjoyed only for a brief period, being soon after (A.D. 263) assassinated by his nephew. The vacant throne was seized by his young, warlike, and beautiful widow, the famous Zenobia, who broke the alliance with the imbecile Gallienus, and assumed the title of Augusta, queen of the east. The accounts that have come down to us of this extraordinary woman are so very flattering that we may not unreasonably suspect them of being exaggerated, in the view, perhaps, of enhancing the merit of her conqueror Aurelian. But that she was highly accomplished there can be no doubt. "Her manly understanding was strengthened and adorned by study. She was not ignorant of the Latin tongue, but possessed, in equal perfection, the Greek, the Syriac, and the Egyptian languages. She had drawn up, for her own use, an epitome of oriental history, and familiarly compared the beauties of Homer and Plato, under the tuition of the sublime Longinus. The success of Odenathus was, in a great measure, ascribed to her incomparable prudence and fortitude." (Gibbon.)

Zenobia, who boasted of being the descendant of Cleopatra and the Ptolemies, sent, after the death of her husband, on pretence of this relationship, an army into Egypt, which she annexed to her dominions. But her troops were unequal to a contest with the disciplined legions of Aurelian. After being defeated in two great battles, Zenobia shut herself up in Palmyra. But, seeing that it must fall into the hands of Aurelian, she attempted to make her escape; and being intercepted in her flight, the city soon after surrendered. The victor sullied the glory of his conquest by ordering the execution of Longinus, author of the famous treatise on the sublime, and other advisers of the unfortunate queen; but, in other respects, the city was treated with great lenity. Unhappily, however, as soon as it was understood in Palmyra that the emperor, with his captive princess, had crossed the Hellespont, the citizens rose in rebellion, and, having massacred the Roman governor and garrison, proclaimed their independence. The instant Aurelian heard of this revolt, he at once, without a moment's hesitation, began to retrace his steps; and hastened to the ill-fated city with an irresistible force, and an insatiable thirst for vengeance. The sequel may be learned from his own words:—Mulieribus non pepercimus, infantes occidimus, senes jugulavimus, rusticos interemimus: cui terras, cui urbem deinceps relinquemus? Parcendum est iis qui remanserunt. (Flavius Vopiscus in Hist. Augusti., p. 218.) At the same time the walls of the city were rased to the ground, and, in the words of Gibbon, "the seat of commerce, of arts, and of Zenobia, gradually sunk into an obscure town, a trifling fortress, and at length, a miserable village. 'Zenobia herself was taken to Rome to grace the triumph of Aurelian, who, however, behaved towards her with a generous clemency seldom exercised by the ancient conquerors, and presented her with an elegant villa at Tibur, where the Syrian queen insensibly sunk into a Roman matron, her daughters married into noble families, and her race was not yet extinct in the fifth century." (Decline and Fall, ii., 44–48.) Palmyra afterwards fell with the surrounding country under the power of the Mohammedans; but history is entirely silent respecting the causes and period of its total desolation. (Wood and Dawkins on the Ruins of Palmyra; Addison's Damascus and Palmyra, ii., 294–323; Irby and Mangle's Travels, p. 269–287; Mod. Trav.)

PALMYRA, p. t., Somerset co., Me., 49 m. N.N.E. Augusta, 644 W. Drained by Sebasticook river which affords water-power. It has two stores, one saw-mill; 11 schools, 689 scholars. Pop. 1500.

PALMYRA, p. t., Wayne co., N. Y., 195 m. W. by N. Albany. 382 W. Watered by Mud creek. The Erie canal passes through it. It contains four churches, a Presbyterian, Episcopal, Methodist and Baptist, 20 stores, one fulling-mill, three grist-mills, five saw-mills, two tanneries, one printing-office, one weekly newspaper; two academies, 77 students; 15 schools, 668 scholars. Pop. 3549.

PALMYRA, p. v., capital of Marion co., Mo., 111 m. N.N.W. Jefferson city. 995 W. Situated 8 m. from Missouri river. It contains a courthouse, three handsome brick churches, a Presbyterian, Methodist and Baptist, a U. States' land office. two printing-offices, two weekly newspapers, and several stores. Marion college is 12 m. W., and Lower college 6 m. S. Both are manual labour institutions.

PAMELIA, t., Jefferson co., N. Y., 4 m. N. Watertown, 168 m. N.W. Albany. Bounded S. by Black river. Watered by Perch river. It contains six stores, two fulling-mills, one cotton factory with 2434 spindles, two grist-mills, eight saw-mills, two tanneries; 16 schools, 780 scholars. Pop. 2104.

PAMIERS, a town of France, dep. Ariège, cap. arrond.,

on the Ariège, 11 m. Foix. Pop., in 1836, 5972. It is well situated, and is generally well built and laid out. The cathedral, several other churches, the bishop's palace, a Carmelite convent, the courthouse, and a large civil hospital, are its principal buildings. No remains exist of its castle, built during the crusades, and called Apamea, from the Syrian town of that name, whence, by corruption, the present name of this town. (Hugo, &c.)

PAMLICO, river and sound, N. C. The river is the broad estuary of Tar river, below Washington, Beaufort co.; it is 40 m. long, from 1 m. to 8 m. broad, and has a depth of water sufficient for any vessels which navigate the sound. Pamlico sound is a shallow body of water, 80 m. long, and from 8 m. to 30 m. broad, separated from the Atlantic by low sandy islands, scarcely a mile wide, covered with bushes. The outer island of one of these islands constitutes cape Hatteras. The principal entrance to the sound is Ocracoke inlet, which admits vessels requiring 14 feet of water. In its N. part it connects with Albemarle sound. The land around it is low, and in some places marshy. Pamlico sound receives Neuse and Tar rivers.

PAMPELUNA, or PAMPLONA, a fortified city of Spain, cap. kingd. of Navarre, on a hill near the left bank of the Arga. 48 m. S. Bayonne, and 195 m. N.E. Madrid; lat. 42° 49' 57" N., long. 1° 39' 45" W. Pop., according to Miñano, 15,000. It is surrounded by a strong wall, with bastions, but derives its principal defence from two castles, one within and the other outside the walls, the latter, the citadel, being situated on a rock (of which the only accessible part is covered by a morass), and encircled by a deep ditch. The interior comprises several wide and straight streets, lined on both sides with trottoirs; three public squares, in the largest of which bull-fights are held; six public fountains, supplied with water from a fine aqueduct 3 m. in length; and the Taceners, a public walk. Outside the walls are three other planted walks, and six bridges across the river, connecting the town with the suburbs. The houses are irregularly built; and the public edifices, which comprise a cathedral, four parish churches, two palaces, a prison, poor asylum, and small theatre, are more remarkable for antiquity than beauty. Within the citadel are extensive barracks and magazines, and a curious corn-mill, turning five sets of stones, and capable of grinding 360 quintals of wheat a day. Pampeluna is a place of little industry, confined chiefly to the manufacture of coarse linen cloths, parchment, and white wax. Its trade also has long been in a languishing condition; and the town exhibits few signs of activity, except at its fair in July, which is much frequented both by the French and Spaniards. The surrounding country abounds with many varieties of grain and fruit; but agriculture is much neglected, and is only slowly recovering from the ravages inflicted during the late civil war.

Pampeluna, supposed to have been built by Pompey, after the defeat of Sertorius, and called by him Pompeiopolis, was taken in the fifth century by the Goths, from whom it passed to the Moors. After the foundation of the kingdom of Navarre, it was made its capital city, and sustained several sieges. The most memorable event connected with the town, however, is the contest that took place for its possession between the English and French at the close of the peninsular war. In June, 1813, on the sudden retreat of the French army from Vittoria, the road to Pampeluna was alone open, and this fortress was hastily garrisoned and provisioned. It was forthwith invested by the British; but the approach of Marshal Soult, with an army, toward the close of July, promised it an early deliverance. It was in the vicinity of Pampeluna that the obstinate conflicts of the 27th and 29th of July took place; and, the French being compelled to repass the Pyrenees with great loss, Pampeluna was cut off from all supplies, and surrendered on the 31st of October.

PANAMA, or PANAMA (ISTHMUS OF), the narrow neck of land which connects the continents of N. and S. America, forming a province of the Colombian republic of New Granada, between the 8th and 10th degrees of N. lat., and the 77th and 81st of W. long., having E. the Colombian province Choco, W. that of Veragua, N. the Atlantic, and S. the Pacific ocean. Its shape is that of an arc, the convex side facing the N.; length, W. to E., about 300 m.; general breadth about 40 m., but, where narrowest, not more than 28 m. from sea to sea. Population of province in 1835, 79,665. The Cordillera, or chain of the Andes, is here interrupted by several remarkable breaks of low and level land, through which it has been proposed to carry a canal or railway: but near the city of Panama its summits rise to 1000 or 1100 feet of elevation, and farther E. they are considerably more lofty, and are generally covered with dense forests. The isthmus is extremely well watered; and, though without any river of considerable length, several of its streams are partially navigable. The dry season lasts from December to April, and the wet during the rest of the year. The quantity of rain is prodigious; but a very remarkable phenomenon occurs throughout the isthmus, in the height of the rainy season, of which no satisfactory explanation has yet been offered. On the 20th of June the rain ceases for five or six days, and the sun shines out during the whole day with the utmost splendour; nor is any instance known of irregularity in the recurrence of this singular break in the ordinary course of the seasons. (*Geog. Journ.*, i., 78.) The temperature and salubrity vary greatly. Porto Bello is one of the hottest and most unhealthy places in the world. On the opposite coast, at Panama, the thermometer in the rainy season does not rise higher in the day time than 87°, and though at other times it is very sultry, it can hardly be called unhealthy. Rice, maize, coffee, cocoa, and some sugar, are cultivated; but most part of the sugar used is imported in skins from central America, or Cauca. Storax, caoutchouc, various dyeing drugs, and the finest timber trees, abound in the forests.

Near Panama is a considerable extent of cultivated land; but round Porto Bello, and on the E. coast, most part of the surface is uncultivated. Elsewhere the landlords keep their estates chiefly in grass, to save trouble; few of the inhabitants are industrious; and many, indeed, depend almost wholly on the chase. Droves of wild hogs, deer, and a variety of other wild animals, are met with; monkeys are frequently used as food, as are sharks, guanas, &c. Horses are small, but hardy; mules are, however, the favourite beasts of burden, and fetch sometimes 120 dollars each.

The isthmus was formerly famous for its gold mines; but these are now all but exhausted and abandoned. The pearl fishery in the bay of Panama is still carried on, and with some success. The trade of the isthmus, notwithstanding its favourable position, is at a very low ebb. On the Pacific some little traffic is carried on with various ports both N. and S.; but, on the Atlantic, almost the only communication is with Jamaica and Cartagena. The inhabitants are said to be less advanced in civilisation even than their neighbours; and their education and morals seem to be alike bad. The isthmus is divided into seven cantons; chief towns, Panama, the capital, Chagres, Porto Bello, Natà, and Carreto. (*Geog. Journal*, i., 66–101.)

The isthmus of Panama, from its narrowness, appears, on the map at least, to be the most advantageous point for establishing a direct communication between the Atlantic and Pacific oceans. At first it was supposed that this might be effected by a canal, but the difficulties in the way of such an undertaking would not be easily surmounted; and the general opinion seems to be, that the preferable plan would be to construct a railway from Chagre, on the Caribean sea, across to Panama. It is believed, however, that the project for connecting the two oceans by means of the river San Juan and the lake Nicaragua, presents, on the whole, greater facilities. *See* NICARAGUA (LAKE OF).

Colony of Darien.—The place marked New-Edinburgh in Arrowsmith's map, on the W. coast of the gulf of Darien, derived its name from its being the site where, in 1698, the Scotch attempted to form a settlement. This colony was projected by a Scotch gentleman of the name of Paterson, the founder of the Bank of England, and was zealously patronised by all classes of his countrymen, who formed a joint stock company, and subscribed large sums to carry the project into effect. It was, however, extremely ill suited for a country in the then situation of Scotland; and provoked the well-founded hostility of the Spaniards, and the bitter, though unreasonable and unfounded, jealousy of the English West India merchants and ship-owners, who either were, or pretended to be, much alarmed lest this new settlement, in an unoccupied and unhealthy country, should seriously injure their commerce and navigation. The selfish opposition of these interested parties to the project having been abetted by the English parliament, the king disavowed the company, and even issued orders to the governors of the West India and American colonies, charging them not to permit any intercourse with the Scotch at Darien! In consequence of these vindictive measures, and of the threatened hostilities of the French and Spaniards, the settlement was abandoned. This event was most acutely felt by the Scotch, whose pride was mortified by the failure of a scheme, of the success of which they had formed the most exaggerated expectations; and many of whom were ruined by the loss of the sums they had embarked in the project. It farther inflamed the existing prejudices against the English, and against the projected union of the two kingdoms, which happily however, was not long after effected. (*Laing's History of Scotland*, iv., 261–277; *Burnett's History of his Own Times*, iii., 299, &c., ed. 1753.)

PANAMA, a city and seaport of Colombia, repub. New-Granada, cap. of the above prov., on the Pacific, 43 m. S.E. Chagres, and 480 m. N.W. Bogota: Lat. 8° 57′ N., long. 79° 30′ W. Pop. about 11,000. It stands on a rocky peninsula, projecting into the bay of Panama, and has a very imposing aspect from the sea. Its plan is not strictly regu-

lar, but the streets are tolerably well ventilated, and it is said to be cleaner than most Spanish American cities. It is encircled by irregular and not very strong fortifications, constructed at different periods. The houses are partly of wood, straw, and other fragile materials; but many are substantially built of stone, the larger having court-yards, or patios, in the old Spanish style. The public edifices are a cathedral, four convents, a nunnery, and a college; but most of them are falling into ruin, and a large and fine Jesuits' college is in a state of total dilapidation. Its roadstead is exposed to northerly gales; but Mr. Lloyd says there are a number of islands a short distance from the main land, which afford secure anchorage, and from which supplies of provisions, including excellent water, may easily be obtained.

Panama is still the centre of some trade, not only with the ports on the Pacific, but also with the W. India islands, &c. Previously to 1740, when the trade with the Pacific first began to be carried on round cape Horn, it was the principal entrepôt of trade between Europe and W. America. From that period, however, it has fallen off; and its decay has been peculiarly rapid since the independence of S. America, and the opening of the other ports of the Pacific. Its situation is, however, highly favourable; and should a canal or railway be carried across the isthmus, it will probably attain to greater commercial distinction than ever.

Old Panama, founded by the Spaniards in 1518, stood about 3 m. to the E. of the present town. It was destroyed by the buccaneer Morgan, in 1670; shortly after which the existing city was commenced. (*Lloyd in Geog. Journal*, i., 85, 86; *Hall's S. America, &c.*)

PANIANY, a commercial town and seaport of British India, presid. Madras, prov. Malabar, on the sea of British, 38 m. S. Calicut. It has numerous mosques, being principally inhabited by Moplays, or fishermen of Arabian descent. Before Tippoo Saib captured this town its trade was very considerable, its merchants trading direct with Surat, Mocha, Madras, and Bengal. It still exports teak, cocoa-nuts, iron, and rice; and imports wheat, pulse, sugar, salt, catichu, and spices; but the mouth of its river is closed by a bar which only admits boats of small burden.

PANOLA, county, Miss. Situated in the N.W. part of the state, and contains 760 sq. m. Watered by Tallahatchie river. It contained in 1840, 7353 neat cattle, 1119 sheep, 15,099 swine; and produced 9041 bushels of wheat, 26,931 of Indian corn, 6125 of oats, 3000 of potatoes, 699 pounds of tobacco, 463,979 of cotton. It had 13 stores, five grist-mills, two saw-mills, one tannery; four schools, 93 scholars. Pop.: whites, 2237; slaves, 2415; free coloured, 5; total, 4667. Capital, Panola.

PANOLA, p. v., capital of Panola co., Miss., 161 m. N. Jackson, 939 W. Situated on the S. side of Tallahatchie river, and contains a courthouse and about 100 inhabitants.

PAOLI, p. v., capital of Orange co., la., 34 m. S. by W. Indianapolis, 695 W. Situated on Lick creek, a beautiful mill-stream. It contains a courthouse, jail, a brick academy, six stores, one fulling-mill, two flouring-mills, a cotton factory, two oil-mills, two carding machines, various mechanic shops, and about 450 inhabitants. It is surrounded by a fertile and highly cultivated country.

PAPA, a considerable market town of Hungary, beyond the Danube, co. Wesprim, 88 m. S.E. Vienna. Population in 1837, 13,939. It was formerly fortified, and has a large castle belonging to the Esterhazy family. It has also numerous churches, one of which is a very handsome edifice, built with immense blocks of stone; Roman Catholic, Lutheran, and Calvinist colleges; manufactures of earthenware, glass, and paper, and an active trade in agricultural produce. (*Berghaus; Conver. Nat. Encyc.*)

PAPAL STATES (THE), STATES OF THE CHURCH, or POPEDOM, an independent country of Europe, occupying the greater part of central, with a portion of N. Italy, being principally comprised between lat. 41° and 45° N., and long. 11° and 14° E.; having N. Austrian Italy, from which it is separated by the Po; W. Modena, Tuscany, and the Mediterranean; S. and S.E. the Neapolitan dominions; and N.E. the Adriatic. It is very irregularly shaped; the length of a line drawn from its N. to its S. extremity may be about 570 m. Its breadth is very various. Per its area, population (exclusive of about 10,000 Jews), subdivisions, &c., see the annexed table.

The Apennines, which intersect the papal territories nearly in their centre, have here an average height of about 4000 feet; but Monte della Sibilla rises to 7210 feet (five guisire); and several other peaks are not greatly inferior in elevation. The provinces of Perugia, Spoleto, Camerino, and the others constituting what was formerly called the March of Ancona, are those principally covered with the ramifications of the Apennines, which, in this part of Italy, approach more nearly to the Adriatic than the Mediterranean, leaving, however, an extensive plain on either side

Legations, Delegations, &c.	Area in sq. m.	Pop. in 1833	Chief Cities.
Comarca di Roma	546	263,436	Rome.
Legations of Bologna	1,105	321,225	Bologna.
Ferrara	1,005	210,383	Ferrara.
Forli	1,149	164,390	Forli.
Ravenna	901	156,353	Ravenna.
Urbino	1,895	225,956	Urbino.
Velletri	686	86,504	Velletri.
Deleg. of Ancona	641	153,550	Ancona.
Macerata	1,003	245,120	Macerata.
Camerino	342	39,502	Camerino.
Ascoli	477	70,046	Ascoli.
Fermo	462	99,491	Fermo.
Perugia	1,721	204,680	Perugia.
Spoleto	1,362	146,539	Spoleto.
Rieti	650	49,254	Rieti.
Orvieto	1,698	21,877	Orvieto.
Viterbo		115,041	Viterbo.
Civita Vecchia	174	18,611	Civita Vecchia.
Frosinone and Ponte-corvo	905	139,479	Frosinone.
Benevento	89	23,049	Benevento.
Total	17,218	2,720,036	

pagna to its present all but desert state. (See *ante*, p. 51.) In antiquity it was bordered along the shore by dense forests; and it is believed by many that the destruction of the woods has been a principal cause of the increase of malaria. "The ancients," says M. Simond, "planted, or preserved, these woods, under an idea, probably erroneous, that they screened them from certain winds carrying noxious vapours; but, though mistaken as to their real mode of agency, they were quite right in supposing them useful. To the destruction of the woods the increase of solstitial fevers has been clearly traced; the one having uniformly followed the other. During the decline, also, and after the fall of the Roman empire, those stupendous aqueducts, which in earlier times brought whole rivers to Rome, having been broken and overturned, in some places poured their waters over the land, which became a marsh; and the population, diminished by wars, was farther and still more reduced by pestilence. The country became more unhealthy as it was less inhabited; in the course of a few centuries the millions of ancient Rome dwindled down to 30,000; and it was not before the 16th century, under Leo X., that the scanty population grew more numerous. Another cause of the increase of malaria is, that sandy ridge gradually thrown up on both sides the mouth of the Tiber for many leagues; various outlets, natural and artificial, are thus choked up; and hence the Pontine Marshes, formerly confined to a narrow space near the promontory of M. Circello, now extend under other names all along the coast." (*Tour in Italy*, p. 350-350.)

Agriculture.—It has been estimated that of 590,000 *rubbi* of productive land (about 2,655,000 acres, or less than ¼ part of the entire surface), 242,000 are arable, 169,000 in pasture, 14,800 in vineyards, 1400 in gardens, &c., and 170,000 in woods. In the March of Ancona, and other mountainous districts, and round the towns and villages, both the properties and farms are small; but it is otherwise in the Campagna and the plain of Bologna. The whole of the Campagna is divided into about 600 estates, varying from 500 to 1000 hectares and upwards each. The largest of these vast estates, which are mostly held in mortmain, belongs to the chapter of St. Peter. The value of land is very various; the rent in the Campagna varies from ½ to 4 scudi per hectare. Beyond the maremme, as the population increases in density, the rental rises to perhaps 90 scudi per hectare, for lands on which there are olive, vine, or mulberry plantations; or where there are adjacent markets for corn. In the neighbourhood of Rome, where the land is rented at a fixed price, it readily sells for 40 years' purchase; while land farmed on the metayer principle do not generally sell, owing to the greater difficulty of collecting the rent, for more than 33 or 35 years' purchase. Lands in the maremme are frequently rented by middlemen, who underlet them in smaller portions to the actual cultivators. But speaking generally, land is everywhere held under the *metayer* system, the occupier paying a certain proportion, generally a half of the produce, to the owner. The soil is mostly fertile; but owing to the badness of the government, which oppresses the occupiers with injudicious taxes; the want of capital, skill, industry, and markets; the ignorance of the cultivators, the number of holidays, and the prevalence of the metayer system, agriculture is in the most wretched state. The implements of husbandry made in the country are as rude as those described by Virgil; and heavy customhouse duties prevent the introduction of improved implements from abroad. The cultivated part of the maremme produces wheat, maize, beans, and vines; but the lands often lie fallow from three to seven years; and Mr. Maclaren states that, from what fell under his observation, not one acre in eight is under the plough or hoe. (*Notes on Italy*, 66.) In the more populous and best cultivated parts there is usually an annual change from spring grasses to corn produce; but by far the greater part of this region is in a state of nature. Formerly all the farms were let with a considerable stock of horses, cattle, &c.; but the proprietors, when in want of money, parted with them. In whatever direction the traveller may enter the Campagna from Rome, he would pass over at least from 20 to 30, and frequently from 50 to 60 m., without meeting with a single field cultivated by resident inhabitants. In fact, though it embrace an area of about 4000 sq. m., or 2,560,000 acres, it is not supposed to have a resident population of more than 16,000 or 18,000 inhabitants, mostly wandering shepherds. There is on each estate a *casale*, or large building, where the implements of husbandry are kept; but neither bakehouses nor kitchen gardens exist throughout the whole tract, the labourers being wholly supplied at a few scattered depôts with provisions, sent thither from Rome. The shepherds are in about as depressed a condition as possible; they have a sickly, cadaverous appearance; their clothing consists principally of sheepskins, worn with the wool outside; and they sleep either on the ground in the open air, or in some of the ruins with which the country is strewed. They are

that on the N., between the Po and the Adriatic, comprises the legation of Ferrara, and the greater part of the legations of Bologna, Ravenna, &c. It includes the *Valli di Comacchio*, a very extensive marsh, but, with this exception, is highly fertile and productive. The plain to the S. of the Apennines is of still more ample dimensions, embracing all the vast undulating tract known by the names of the Campagna and Maremmo, extending between the declivity of the mountains and the sea from the frontiers of Tuscany on the one hand, to those of Naples on the other. The S. portion of this great plain, or that next Naples, consists of the district called the Pontine Marshes (an. *Pomptinae Paludes*), which, notwithstanding the vast sums expended upon it, is still very imperfectly drained. We have elsewhere fully described the present state of this vast and naturally fruitful plain, famous in antiquity for its fertility, but now, unhappily, the seat of pestilence and death. (See *ante*, p. 51.)

The Po, which forms their N. boundary, is the largest river of the Papal states; but by far the most celebrated is the Tiber. The latter rises at St. Albongio in Tuscany, and runs generally S. or S.E., but with a very tortuous course, to within about 25 m. N.N.E. Rome, whence it flows mostly S.S.W. to its mouth in the Mediterranean, 15 m. below Rome, after an entire course of about 200 m. Before entering the sea the Tiber divides into two arms, enclosing the small island of *Isola sacra*. At Rome the greatest breadth of the Tiber is only about 400 feet, or scarcely one third part of the breadth of the Thames at Blackfriars bridge, and nearly approaching that of the Seine at Paris. It is justly entitled to its ancient epithet *flavus*, being almost constantly loaded with yellow mud, from the crumbling and disintegration of its banks. Its principal affluents are the Topino, Nar, and Teverone on the left or E., and the Chiana on the right bank. It is navigable for boats to near Perugia. Except the Tiber, no river of any consequence falls into the Mediterranean in this part of Italy. The country to the E. of the Apennines has, however, a great number of rivers, though none of them be of any very considerable magnitude, falling partly into the Po and partly into the Adriatic. Several of the most celebrated Italian lakes are in the Papal states, as those of Perugia (an. *Lacus Thrasimenus*), Bolsena, and Bracciano (which see). The lakes of Vico, Albano, Nemi, Gabii, &c., though insignificant in point of size, are interesting from the classical associations with which they are connected. They are situated in a mountain region, and evidently occupy the craters of extinct volcanoes.

Geology.—The primary rocks in the Apennine region consist mostly of serpentine, mica, clay-slate, and quartz. Gneiss is met with in various places along the coast. Mountain limestone is frequent, and indeed a large portion of the country consists of calcareous formations; but the region round the capital is of volcanic origin, and abounds with volcanic products, as sulphur, alum, &c. Rome is principally built of volcanic tufa, which composes the general soil of the Campagna. Some sulphur springs also exist *r* Poretta, N. of the Apennines, and various kinds of mineral springs are common elsewhere.

Climate.—In the legations N. of the Apennines the thermometer often sinks in winter to 10° Reaumur, and oranges, lemons, &c., do not flourish in the open air. But the greater portion of the Papal states is situated within the second Italian region. Vegetation is here scarcely interrupted at any period of the year. The air in the mountain districts pure and salubrious; but the plains of Ferrara and Bologna, the Campagna di Roma, and the Pontine Marshes, are at that season very unhealthy: the latter especially are subject to malaria. The origin of malaria has been a subject of much dispute, and we have already noticed some of the circumstances that have conspired to reduce the Cam-

paid, not in money, but in cattle pastured with those of the farmer. The harvests in the Campagna are reaped by peasants from distant mountainous districts, who come to it in companies of from 20 to 100 individuals. Even in favourable seasons, ¼ or ⅓ part of their number are attacked by fever; and in unhealthy seasons, the proportion is much larger. Many die in the hospitals of Rome, or in the Campagna; others perish on the road home; and others again return condemned to pass the remainder of their days a prey to intermittent fever, or other diseases brought on by the climate: and yet such is the poverty of the population in the mountainous districts, that the chance of realising a few scudi continually tempts new adventurers to undergo the same risks.

In 1800, on the estate of Prince Rospigliosi at Zagarolo, land was let out at a low fixed rate in lots of a rubbio each (about 4½ acres) to the peasantry to cultivate as they pleased; and this plan, it is alleged, had the best results. Cultivation extended for a considerable distance round Zagarolo into the plain beneath; and the climate of the neighbourhood was improved. Were such a plan followed round the other centres of civilization, a considerable portion of the Campagna would, probably, soon lose a portion of its desolate and pestilential character, but neither the proprietors nor farmers show, in general, much disposition for change or improvement; and till the government and public economy of the state be completely changed, it were idle to expect that they should evince any such disposition. In order to arrest the depopulation of the Campagna, Pope Pius VII., in 1802, laid an impost of five *pauli* per rubbio on the uncultivated land immediately round the towns, and deducted five pauli per rubbio from the tax on cultivated lands. But this miserable attempt to extend industry by fiscal regulations (though it appears to be approved by M. Sismondi) necessarily failed. The regulation, which never was acted upon, speedily became obsolete; and the peasantry of the Campagna generally remain in the same condition as before the French invasion. (*Sismondi, Etudes sur l'Economie Politique*, ii. 19–198.)

In the mountainous parts of the Papal states, where the country is divided into small farms, and rather thickly inhabited, pease, beans, and kitchen vegetables, which form a large proportion of the food of the peasantry, occupy most part of the land; the remainder being appropriated to wheat, maize, &c. Little skill is evinced in agriculture; the crops being generally raised only for the supply of the cultivators, no one thinks of raising those products for which his land may be the best fitted, till after he has provided an adequate supply of grain or other produce for the use of his family. In the mountains near Rome, white crops are taken from the grounds for two or three years successively, without any manure being applied to the land; three crops of wheat may be succeeded by maize or kidney beans for two years, and once in five or six years, a crop of hemp or flax is perhaps raised. The grain is trodden out by horses, and winnowed by hand, immediately after harvest. The wages of a man in harvest time, amount at Poli, to about two paula a day, with bread and *piquette*, or weak wine; but they are generally higher the nearer the district to the capital. The herdsmen in the Apennines take charge of the cattle belonging to many different persons, and tend them on the mountains, night and day, receiving at the end of the season payment from each proprietor, at the rate of two scudi per month for every score of cattle. Besides bread and *piquette*, the food of the peasantry in the mountains principally consists of cheese from goats' or ewes' milk, onions, garlick, and other vegetables, and *polenta*, a kind of hasty pudding, made with maize, pulse, &c. Goats' flesh and pork are sometimes eaten by the labourers, but very little other animal food. (*Graham's Three Months in the Mountains*, 7–56; &c.)

The provinces of Ferrara, Bologna, Romagna, and those forming the March of Ancona, produce rye, wheat, barley, and maize in abundance. Rice is grown in the legations of Ferrara and Bologna, but there only. Hemp and flax are cultivated along the Adriatic; and saffron, coriander, aniseed, woad, and great quantities of kitchen vegetables in the same districts and the N. provinces. The export of hemp is supposed to average 30,000,000 lbs. a year. Tobacco is grown in several places, especially at Chiaravalle, near Ancona; but being a government monopoly, its culture is confined within narrow limits; Serristori states that for 300,000 lbs. of tobacco exported, 1,000,000 lbs. are imported. The sugar-cane, indigo, and cotton, are cultivated near Terracina, though neither are grown to any great extent. Olive plantations were long among the most productive investments; but they are now less so than those of the white mulberry. The olive is abundant in the S. provinces; and though the Roman oil be badly made, and mostly consumed at home, a million lbs. have occasionally been exported in favourable years. Vineyards are said not to yield returns proportioned to the outlay. The vine is

tolerably well cultivated in the vicinity of Velletri; the plants in regular lines being tied to trellisages of large reeds; but the most esteemed growths are the light, white, muscadel wines of Orvieto and Montefiascone, near Viterbo: they do not, however, bear transport well, and are seldom met with out of the country. The timber of the dense forests in the deleg. of Viterbo is not turned to much account from the distance of markets, and is cut principally for smelting iron ore, making charcoal, &c. Cork trees abound in the country about Cisterna, Velletri, &c.

The rearing of live stock is, as has been said, the principal branch of rural industry. The number of sheep in the Papal states is estimated at 2,000,000. There are two varieties; the *negretti*, a small, short-legged variety, in every respect resembling the breed of Dauphiny, except that their wool, though good, is chocolate-coloured; and the *pecus*, a variety with a wool of a whiteness almost equalling that of the breed of Aragon. Still, however, it would seem from the statements of Serristori that the total quantity of all sorts of wool exported from the Papal states does not exceed 800,000 lbs. a year, sent to Tuscany, Piedmont, France, and England. The ewes are mostly kept for their milk, and the greater part of the lambs are killed, the mutton not being good. About 1,000,000 lbs. cheese and 400,000 lamb-skins are annually exported, principally to the other states of Italy. Cattle breeding is extensively carried on in the marshes of the Po, in the provinces Bologna, Perugia, &c.; and about 40,000 oxen are exported. Many buffaloes and hogs are kept in the marshes; and about 100,000 ox and buffalo skins are annually exported. The horses are mostly of good breeds, and are exported to Tuscany and Lombardy. Goats are extensively reared, their flesh and lamb being the principal animal food. In Perugia and other delega. great numbers of poultry, and in Forli, Macerata, &c., many bees are kept. (*Châteauvieux, Italy and its Agric.; Bowring's Report on the Roman States; Graham; Sismondi; Sismondi; Etudes sur l'Economie Politique; Serristori, Statist. d'Italia*, part. vi.)

The fisheries on the coast are almost wholly conducted by Neapolitan fishermen. Mining industry is also at a very low ebb. The government works the mines of alum at Tolfa, but the rest are left to private speculators. Iron ore is pretty abundant in some places, but only a few traces of other metals have been discovered. About 4,000,000 lbs. of sulphur are raised in Romagna at Pesaro, &c.; and 100,000 lbs. of vitriol at Viterbo, half of which is exported. Lime, building stone, potters' clay, variegated and statuary marbles, fuller's earth, bitumen, naphtha, and coal are met with; but the last, though under the French it was raised in considerable quantities, is no longer made use of. From 70 to 80 million pounds of salt are annually made at Cervia, Commachio, Corneto, and Ostia, rather more than the half of which is sent to the adjacent states.

Manufactures, though in the most depressed and backward state, serve almost entirely for home consumption. Woollen fabrics are the principal, and include cloths, cassimeres, serges, woollen caps, blankets, and carpets. Rome, Spoleto, Matelica, Perugia, Norcia, &c., are the chief places in which these are made; but since their manufacture has ceased to be bolstered up by government premiums, their production has greatly diminished, and their total yearly value does not exceed 300,000 scudi. Hats, of the value of 200,000 scudi, are made principally at Rome; good felt cloth at Fabriano; silk goods at Rome, Bologna, Camerino, Perugia, Pesaro, &c.; leather and gloves at Rome; and paper, about 3,600,000 lbs. a year, are the other most prominent manufactures. Bologna was formerly famous for its crapes, but the value of the exports of these does not now exceed 30,000 sc. a year. The iron furnaces are estimated to yield 18,000,000 lbs. pig. and about 2,000,000 lbs. a year bar iron; rasps, files, nails, needles, pins, screws, &c., are made in various towns; glassware, to the value of 90,000 scudi, copper goods to 80,000 do.; earthenware to about 150,000 do., &c. Roman musical strings enjoy a high and deserved celebrity, and are exported to most countries of Europe. The most flourishing branch of manufacture is the refining of sulphur, a product which, under a free system, might be supplied in unlimited quantities. (*Serristori, Statistica; Bowring, &c.*) Such is the meagre catalogue of Papal manufactures. "Many a town of Great Britain, of only 30,000 inhabitants, produces a greater quantity of manufactured goods than the 3,000,000 inhabitants of the Pontifical states! Notwithstanding the enormous sacrifices made by the Papal government, the protections, the prohibitions, the premiums given for the encouragement of what is called native industry, scarcely a valuable discovery has been introduced: the woollen spinning, in some cases by hand, in others by machinery, is far behind the state in England, Belgium, Prussia, or France. The looms, such as were generally employed during the 14th century, are little better than those used by the Indians of the Deccan: the rowing and carding are wholly done by solitary workmen and with the ancient teasels and hand cards ; the shearing with the antique hand shears, such as have been employed from immemorial time ; and in some places the fulling is performed by men employed to trample on the cloth, a process probably not now to be found in any other part of the civilised world." (*Bowring's Rep.*, p. 94.)

Commerce.—From the circumstance of the Apennines dividing the country into two portions, between which there is little communication, some provinces are exporting while others are importing the same kinds of produce. The N. provinces have generally a superabundance of corn, while in the S. provinces the crops are insufficient for home consumption. On the other hand, oil is exported from the S., while in the N. legations, the Marche, &c., 3,000,000 lbs. are annually imported from S. Italy and Tuscany. Besides the articles of export previously specified, about 3,000,000 lbs. rags are sent every year from the N. legations, partly to Genoa as measure for orange trees ; planks are sent to Spain, France, and America ; organzined silk, about 200,000 lbs., chiefly to France and England ; about 450,000 lbs. linseed ; 2,800,000 lbs. charcoal ; 1,000,000 lbs. potash, with vinegar, cork bark, tartar, woad, tallow, bones, honey, works of art and antiquities, are the other principal exports. The imports, in addition to those already noticed, are raw sugar, about 10,800,000 lbs. a year, and other colonial products from England and France, coarse wool for mattresses, cheese and butter from Lombardy, salt fish, pilchards, &c., from England, to the amount of 8,700,000 lbs. a year ; about 2,080,000 lbs. of iron ore from Elba, and the same quantity of lead annually from England, and metals and manufactured goods of most kinds from N. and W. Europe. (*Serristori, Statistica d'Italia.*) Serristori estimates the total annual value of the imports at nearly 7,000,000 and that of the exports at above 5,000,000 scudi. The importation of salt, tobacco, alum, and some other kinds of native produce, including wheat when the price is under 14 scudi the rubbio on the Mediterranean, or 12 scudi on the Adriatic, is prohibited ; as is the export of hempseed and wheat, when the price is above 16 scudi in the Mediterranean, and 14 scudi in the Adriatic ports, and other grains in proportion. The importation of such books as would communicate any useful information, as to politics, political economy, or philosophy, is of course absolutely forbidden. Notwithstanding the low state of manufacturing industry, high duties are levied on manufactured goods when imported. Woollen cloth, woollen and cotton fabrics, and cambrics, pay 100 scudi ; dye or stamped cottons, 50 scudi ; and porcelain, 20 scudi per 100 lbs. The principal seats of the foreign trade are Ancona and Civita Vecchia. In 1838, 1922 ships, of the aggregate burden of 66,838 tons, with cargoes of the value of 1,109,300 scudi, cleared from the former port ; and in 1837, 1280 ships, burden 133,402 tons, cleared out of the latter. (*Parl. Reports*, 1838 ; *Bowring's Reports, &c.*)

Accounts are kept in *scudi* (crowns), = 4s. 3½d. each, and divided into 10 paoli and 100 bajocchi. The Roman *libbra* or pound of 12 oncie and 288 danari = nearly 12 oz. avoirdupois. The *barile* of wine, of 32 boccali, and 128 foglietts = about 13 gallons ; the barile of oil contains only 26 boccali. The Roman foot or 11·7 English inches ; the palmo of architects = about 8¾ inches ; the Roman mile = 1628 English yards. Generally the measures of Rome have less capacity than those of the N. legations.

The Government is wholly ecclesiastical, no one being eligible to fill any civil office who has not attained the rank of abbot. The pope enacts all laws, and nominates to all clerical appointments. He is assisted, however, by the high college of cardinals, comprising about 70 members; and the different branches of the government are conducted each by *Congregations*, with a cardinal at its head. Each legat. and *deleg.* is governed by a cardinal, assisted by two assessors, and a council of four individuals nominated by the pope, half of whom are changed every five years. The jurisdiction of the temporal nobles is provisionally retained in some provinces, but all the judicial officers of the nobility must be confirmed by the pope, and are subject to the general laws. In each cap. of a deleg. there is a tribunal of primary jurisdiction, which also decides in appeal on certain matters that first come before the district officers. The proceedings in these courts are public, but there is no jury. There are four courts of appeal, one at Bologna, one at Macerata, and two in Rome ; and a supreme tribunal of final resort, that of *l' Uditore Santissimo*, with a single judge ; nine tribunals of commerce also exist. Every town has its own jurisdiction and magistracy, and a municipal court of from 1 to 48 members, according to the population. The laws in force are nominally those of the Justinian code ; but the pope has power to alter or annul any previous law, and, which is incomparably worse, the provincial judges have extensive discretionary powers. Criminal proceedings in the Papal states are very dilatory ; and in all cases the accused is thrown into prison, whence there is no liberation by bail. In 1832, 2708 criminals were confined in the vari-

ous prisons, of whom 580 had been convicted of homicide, 384 of other offences against the person, 295 of burglary, and 1072 of other offences against property, and 76 of state offences. Brigandage is less frequent than formerly ; and the government has stationed five military posts along the road from Velletro to Terracina, for the protection of travellers. Still, however, the police and the law are equally defective ; and assassinations and other crimes of violence are daily taking place without the perpetrators being ever brought to justice. The whole frame of government is, in fact, a tissue of abuses.

On the fall of Napoleon, the alienation of church domains was confirmed ; but the compensation since made to their former owners, and the restoration of suppressed churches and convents, have cost the government prodigious sums, and are the principal causes of the wretched state of the finances. (*Von Raumer's Italy.*) Within the limits of the Papal states there are no fewer than eight archbishops' and 59 bishops' sees ; and it is estimated that in Rome there is a clergyman for every ten families. It is needless to add that this superabundance of priests, instead of promoting religion and morality, is, in fact, a principal cause of their low state in the city. The outward deportment of the papal court is, however, at present highly decorous. "Those times so disastrous and disgraceful, when the popes had so many nephews, and those nephews built so many splendid palaces and villas, called by the Romans, in derision, 'miracles of St. Peter,' are now almost as much forgotten in Rome, as the times when horses were made consuls, and eunuchs emperors." (*Lyman's Polit. State of Italy.*)

Public Instruction.—There are two chief universities—in Rome and in Bologna—each having at least 38 professorships ; and six universities of secondary rank—at Ferrara, Perugia, Camerino, Macerata, Fermo, and Urbino—each at least with 17 professors. The university of Rome was, in 1839, attended by 843 students (*Serristori*) ; that of Bologna, which ranks, in Italy, second only to Pavia, is usually attended by from 500 to 800 students ; that of Perugia by about 300 ; and those of Camerino, Macerata, and Urbina, by about 200 each. Altogether, upwards of 2000 students annually attend the universities. (*Journal of Education*, viii., 209.) There are various other high colleges in Rome, &c., the principal of which is the Gregorian (*see* Rome). Secondary schools exist in most towns ; but there is no general system of elementary instruction, and it has been estimated that only 1 in 60 of the population attend public schools. The truth is, that education in the Papal states is in the most degraded state imaginable. It is wholly in the hands of the clergy ; and is conducted on the principle, if we may so call it, of imbuing the pupils with the grossest prejudices, and of proscribing every study or pursuit that might tend to expand or enlighten their minds, or make them acquainted with their rights and duties. The university education, excepting, perhaps, in so far as respects medicine, is altogether contemptible. Even theology is not efficiently taught ; and philosophy, politics, and political economy are as little relished in Rome as in Morocco.

The censorship of the press is severe in the extreme ; and the gazettes published in the different towns insert nothing not approved by the censors. The journals, of which there are several, devoted to *belles lettres*, antiquities, the fine arts, &c., being under a less severe surveillance, occasionally display originality and learning ; but the literature of the Roman states is like their government, emasculated and imbecile. "The Eternal city prohibits all the best works on mental philosophy. She has not one eminent man of science ; and if she had a Cuvier or a Buckland, she would not permit him to lecture or to publish his discoveries to the world till they had been subjected to the pruning knife of some ignorant censor. The apathy and timidity, the dread of independent thinking and free inquiry manifested by the Papal government, seems, however, to admit of easy explanation. Its dogmas, its rites, its principles of action were framed in accordance with the opinions of the 12th century. It does make some changes silently, by dropping a few untenable pretensions ; but it can make no great and marked change without abandoning its professed character of being the depository of immutable truth. The rulers of Rome, therefore, finding themselves unable to raise up their old institutions to the level of modern knowledge, endeavour to keep down this knowledge to the level of their institutions. They see the props and stays of their system dropping off, and one source of influence failing after another, and their prudence counsels them to shut out, as far as they can, the light which is sapping their authority, and to look upon innovations, even of the most harmless kind, with suspicion. They are religious *Conservatives* in the strictest sense of the term." (*Maclaren's Notes on Italy*, p. 97.)

Charitable Institutions.—The Papal states are literally overrun with all kinds of *charitable institutions*. In Rome, especially, the sums expended on charitable foundations

PAPAL STATES.

are, in proportion to its extent, twice as large as in Paris; so that, as Serristori exclaims, "*Dovrebbe credersi che negli Stati Pontefici, e molto più in Roma non esistesse mendicità.*" (*Statist.* p. 39.) But nowhere are the pernicious consequences of indiscriminate charity better exemplified than in the Roman states, where mendicity, wretchedness, and want prevail to a frightful extent. The universality of beggary removes all sense of shame; and a large proportion of the population are degraded enough to prefer subsisting on alms to making any attempts to provide for themselves. There is a *monti-di-pietà*, or government pawn-broking establishment, with a capital of 230,000 scudi, in Rome; and others in most of the principal cities; where also savings banks have been established with considerable success. (*Bowring's Rep.*, p. 98-90.)

The army, if so it may be called, is under the direction of a cardinal-president, and a board of three general officers; and consisted, in 1840, of a permanent force of 14,680 men. (9300 infantry, 800 cavalry, &c.); and a body of reserve and national guard, together amounting to 9000 men.

The principal fortresses garrisoned by the pope are those of Rome, Civita Vecchia, Urbino, castle Franco, Terracina, and Ancona: by the treaty of Vienna, the emperor of Austria has the right to garrison Ferrara and Commacchio. The naval force consists of a solitary ship of war, manned by 33 men. (*Almanach de Gotha*, 1840; *Oudinot, Italie et ses Forces Milit.; &c.*)

The taxes are very heavy, and are imposed in the worst possible manner. The principal consists of a land tax; heavy duties are also laid on most articles consumed in towns and villages; and all sorts of grain, except rye, maize, barley, and oats, pay a heavy tax when ground at the mill. Salt, tobacco, alum, and vitriol, are monopolies in the hands of government. The customs' duties are probably, however, on the whole, the most oppressive and injurious. The lottery, also, notwithstanding its demoralising influence, is a fertile source of revenue; and contributes, in fact, about one tenth part of the entire public revenue !

The public revenue and expenditure was as follows in 1835 :

Revenues.	Scudi.	Expenditure.	Scudi.
Direct taxes	2,643,356	Treasury charges	226,377
Customs, &c.	4,851,038	Customs do	590,329
Stamps and registry	877,910	Stamps, &c. do	79,071
Postoffice	396,065	Postoffice	162,469
Lotteries	896,396	Lotteries	596,011
Various sources	48,334	Papal household	185,026
	8,312,961	Interest on debt	2,547,565
		Government and justice	1,344,564
		Public instruction	109,061
		Public works and charities	906,491
		Army and police	1,323,146
		Public health	284,070
		Various	233,344
			9,039,759
			8,312,961
		Deficit in the Revenue	616,898

But according to the *Alm. de Gotha*, 1841, the accounts for the preceding year were more satisfactory.

History.—The rise of the Popedom as a temporal power, dates from 755, when Pepin, king of the Franks, invested the pope with the exarchate of Ravenna; to which Charlemagne added the provinces of Perugia and Spoleto. Benevento was given to the pope by the emperor Henry III. in 1053; and in 1102 the marchioness Matilda of Tuscany bequeathed to the holy see the provinces forming the "Patrimony of St. Peter." In 1297, Forli and the rest of Romagna, and, in 1364, Bologna, became portions of the Papal dominion; and at the end of the 14th century the pope acquired full jurisdiction over Rome and Sabina. Ferrara was acquired in 1598, Urbino in 1626, and Orvieto, in 1649.

The French invaded the states of the church in 1797, after which the N. legations were annexed to the Cisalpine republic. In 1798, Rome was taken by the French, and in 1810 the whole of the Papal states were included in the kingdom of Italy. Since 1815, most part of the pope's former dominions have been restored; but his authority, especially in the N. legations, is far from being well established, and his power as a temporal prince, depends wholly on the support given him by Austria. (*Percival, Hist. of Italy; Sismondi; Maclaren, Notes on Italy; Von Raumer's Italy and the Italians, &c.*)

PAPUA, or NEW GUINEA, a very large island, or perhaps a dense cluster of islands, in the E. archipelago, third division; between the equator and the 9th deg. of S. lat., and the 130th and 150th degs. of E. long.; having N. and E. the Pacific ocean, W. and S.W. the sea in which Gilolo, Ceram, the Aroo isles &c., are situated, and S. Torres

---

straits, separating it from Australia. It is indented by several deep bays; but even its coast line is in many parts unknown, and its interior has been but little explored, and is, in fact, a *terra incognita*. The coast, viewed from the sea, rises gradually into hills of considerable elevation; but no mountains of any remarkable height have yet been discovered. The whole island being covered with palm trees and timber of a large size, little can be said respecting its soil, which, however, is presumed to be fertile. The cocoa-nut, the two species of the bread-fruit tree, pine-apples, and plantains are found here: nutmeg-trees also grow wild; but it is not known whether they produce good spice. It is said that there are no quadrupeds in Papua, except dogs, wild cats, and hogs; and that to the E. of Gilolo no horned animals of any description are found. The woods abound with wild hogs, which the natives kill with spears and bows and arrows, in the use of which they are very expert. There is reason to believe that gold is found in the interior of the island. The natives of Papua appear to consist of two distinct races; those in the W. being identical with the negroes of the E. archipelago, while the inhabitants of the E. part of the island belong rather to the sallow complexioned long-haired natives of the South sea islands (*see* POLYNESIA). The Papuan negroes, of whom a brief notice is given in the art. ARCHIPELAGO, EASTERN (1. 1-62), continue, for the most part, in their original state of nakedness and barbarism, devoid of homes or clothing, and subsisting principally on the precarious produce of the chase, or on the spontaneous products of the forests. On the N.W. coast, which has been the most frequently visited by Europeans, the dwellings of the natives are raised on posts, as in other parts of the archipelago and among the ultra-Gangetic nations of the Asiatic continent. These tenements accommodate many families, who live in cabins on either side of a wide common hall that occupies the centre of the building. The cabins are miserably furnished; a mat or two, a fire-place, an earthen pot, with perhaps a china plate or basin, and some sago flour. As they cook in each cabin, and have no chimney, the smoke issues at every part of the roof; and at a distance the whole building seems to be on fire. Their clothes are very scanty, but they contrive to bedizen themselves so as to attract the attention of European observers. Their hair is not so short, close, and woolly as that of the African negro, and they wear it bushed out round their heads to the circumference of 2½ and 3 ft.; and, to make it more extensive, comb it out horizontally, occasionally adorning it with feathers.

The men in general wear a portion of the inner bark of the cocoa-nut tree, resembling a coarse kind of cloth, fastened round the middle; and the women use blue Surat India in a similar manner. Boys and girls go entirely naked till puberty. All are fond of glass and coloured china beads, and wear them about their wrist, &c. The women, as generally happens among savages, lead a laborious life; and Forrest says that he has often seen them labouring hard in fixing posts in the ground for stages, in making mats, or in forming pieces of clay into earthen pots, while the men were sauntering about.

In the interior the inhabitants are supposed to practise gardening and some sort of agriculture, as they supply the inhabitants on the coast with food, in exchange for axes, knives, and other coarse cutlery. The natives on the coast purchase these from the Malays and the Chinese, particularly the latter, from whom they also buy blue and red cloths. In exchange, the Chinese carry back mussoy bark, slaves, ambergris, sea slug (*bêche de mer*), tortoise-shell, small pearls, birds of paradise, and many other species of dead birds, which the Papuas have a particular method of dressing.

The Dutch may have some trade with Papua; but Mr Earl says that no intercourse whatever takes place between it and the British settlements in Australia, Singapore, or elsewhere in the east.

The inhabitants of the more westerly islands of the E. archipelago buy the Papuans for slaves; and the natives of the W. coast of Papua make slaves of those of the E. and sell them to strangers. With a similar view, probably, they were formerly, and perhaps still are, accustomed to assemble in great numbers, and make war on the inhabitants of Gilolo, Ceram, Amboio, and other islands still farther W.

The Arabians, in their early voyages, appear to have come into contact with the Papuans, whom they constantly describe as cannibals. Papua was discovered by Europeans in 1511, and frequently resorted to by the Portuguese during the 16th century. Toward the end of the 18th century Forrest, McCluer, and other British navigators visited it; since which it has been but little noticed.

PARA, formerly called *Belem*, a city and seaport of Brazil, cap. prov. of same name at the confluence of a considerable river, with the great estuary of the Tocoantins, or Rio Para, on its S. side, opposite the island of Joannes or

544

Marajo, estimated about 60 m. from the Atlantic, and 300 m. W.N.W. Maranham ; lat. 1° 30' S., long. 48° 22' 32' W. Population estimated at 20,000, including comparatively few negroes. It stands in a fertile plain, and is one of the finest Brazilian cities. Its streets being straight, and the houses almost all of stone, and both solid and elegant. The cathedral and governor's palace are said to be magnificent edifices. There are several other churches, but only two convents appear to exist at present, that of the Mercenarios having been converted into barracks, and the Jesuit's college into the episcopal palace and seminary. Para has a judicial tribanal, royal college, botanic garden, hospital, theatre. and arsenal. The harbour is confined, and is said to be diminishing in depth ; the approach from the ocean is also rather difficult, and it is always expedient to take on board a pilot at the mouth of the estuary. The principal articles of export are cocoa. of which it exports above 60,000 bags ; caoutchouc, of which it is the principal mart ; with laugiass, rice, drugs, and cotton, amounting in all from £250,000 to £350,000 a year. The sugar grown in the neighbourhood is bad, the soil not being favourable for the cane. The communication with Great Britain is principally with Liverpool. Ships of war have been built here ; and timber used to be exported to Lisbon for the use of the arsenals. The climate of Para is very hot ; and thunder-storms occur almost daily. It was formerly deemed very unhealthy, but in this respect it has latterly been materially improved. (*Mawe's Tres. in Brazil*, 403, 404 ; *Mod. Trav.*, xxx. ; *Dict. Geog.*)

PARADISE. t., York co., Pa. It has five stores, two fulling-mills, one woollen factory, six grist-mills, two saw-mills, nine distilleries ; three schools, 90 scholars. Pop. 2117.

PARAGUAY, an indep. state of S. America, principally between the 21st and 27th degs. of S. lat., and the 54th and 58th of W. long. ; having N. and B. Brazil ; S.E. and S. the territory of La Plata; and W. the latter repub. and Bolivia. Shape nearly oblong ; length N. to S. about 460 m. Estimated area, 74,000 sq. m. Estimated pop , 300,000 (*American Almanac*, 1841) ; principally whites of Spanish descent, native Indians, negroes, and mixed races originating from the foregoing, those of the Indo-Spanish descent greatly preponderating. Paraguay is an inland peninsula, inclosed E. and S. by the Parana and its tributary the Yajuari, and W. and N.W. by the rivers Paraguay and Blanco. A mountain chain, the Sierra Amambahy, enters Paraguay on the N., runs through it near its centre to near lat. 26°, and then divides into two branches, inclosing the basin of the Tibiquari. From the undivided chain many rivers flow on either side to join the Parana or Paraguay ; but none of them require any special notice, though in the rainy season they are all swollen so as to inundate a considerable extent of country. Respecting the Parana and Paraguay, see PLATA (RIO DE LA). There is but one lake, that of Ypoa, worth mention ; extensive marshes, however, abound in the W. The climate is temperate, but damp. Paraguay, in point of fertility, forms a striking and favourable contrast to the adjacent parts of the Argentine republic. It is well wooded, and diversified with undulating hills and verdant vales. Mr. Robertson, who entered it at Neembucú, states, "I was glad to meet with much more frequent traces of cultivation and industry than were to be found in the solitary tracts over which I had heretofore sped my monotonous way. Whitewashed cottages often peeped from among the trees, and around them were considerable fields of the cotton, yucca, and tobacco plants. The Indian corn and sugar-cane were also frequently to be seen in the vicinity of the farm-houses of a better character than the cottages; and there was abundance of wood and of the prickly pear. With the latter, the cultivated country, as well as the potreros or paddocks, were invariably well fenced." (*Letters from Paraguay*, i. 259, 260.)

Almost half the entire territory is national property. It consists of pasturage lands and forests, which have never been granted to individuals, the estates of the Jesuit missions, and other religious corporations ; and a great number of country houses and farming establishments confiscated by the dictator. The latter has paid great attention since the commencement of his reign to the improvement of agriculture, and to rendering the government property productive; and has, by so doing, created a branch of revenue which, aided by time and a wise government, may be found sufficient of itself for all the wants of the state. He has let a part of these lands at a very moderate rent, and for an unmited period, under the single but indefinite condition, that they shall be cultivated, or turned into pasturage. On other parts of these lands he has established large farms, here thousands of cattle and horses are bred. These supply his cavalry with horses, and his troops with provisions; besides which, they also furnish great numbers of oxen as; for the consumption of the capital. For these the dictator charges a high price, and will allow no one to undersell m. The farming establishments are objects of peculiar

solicitude to the dictator; and every month the master herdsmen are obliged to make a detailed report concerning them. (*Rengger and Longchamp's Reign of Francia in Parag.*, 174–6.)

The arbitrary measures of Francia have produced certainly a salutary result on farming economy at large throughout Paraguay. Before the establishment of his sway, the farmers never thought of cultivating any article beyond tobacco, the sugar cane, and yucca-root; while the gathering of the mate or Paraguay tea engrossed almost all hands. In 1820, an extent of country, 80 leagues in circumference, was devasted by locusts, and a famine impended over the inhabitants. To avert this, the dictator compelled the proprietors to sow a second time, a large portion of the land which had been laid waste, and the harvest that followed was most abundant. On the complete success of this experiment, Francia determined to extend his measures to the whole country, so that, ultimately, every farmer was under the necessity of employing himself in that particular branch of agriculture which the dictator pointed out! By these violent regulations, which, perhaps, were, at the outset, the best suited for the country, a total change in its rural economy has been produced ; it is no longer customary to import common necessaries from Buenos Ayres and the adjacent provinces. The migration of the rural population has been forbidden. Rice, maize, yucca, kitchen vegetables, are now cultivated on a more extended scale ; and the growth of cotton, which had formerly been wholly received from Corrientes, suffices for the consumption. The breeding of horses and horned cattle has been equally encouraged ; and instead of receiving cattle from Entre-Rios, as previously, the farmers have new a surplus stock. The prohibition to interfere with the forests, and the total suspension of the intercourse between Paraguay and other countries, contributed very much to produce these results, as they turned to the cultivation of the soil all the industry which used to be applied to navigation, the cutting down of timber, and the collection of the *yerba mate*.

The latter, or Paraguay tea, is the leaf of the *Ilex Paraguaensis*, an evergreen about the size of an orange-tree, growing wild and in great abundance in the dense forests in the N. and E. provinces, to which the natives resort in great numbers for its collection. It is difficult to penetrate the country where it is found ; but the profits derived from the article are ample, Paraguay tea being in as general demand through all La Plata, Chili, and many parts of Peru, as the teas of China are in Europe. Its collection is undertaken by merchants in Assumption, who each employ a master-workman or *abilitador*, and from about 20 to 50 peons, the master providing axes, knives, tobacco, mules, bulls for slaughter, and other provisions, with money advanced to him by the merchant. The boughs of the yerba, with the leaves attached, are first hewn down and scorched; the leaves being then roughly removed, 'and dried by being placed over a wide arch of wood work, underneath which a large fire is kindled ; and, together with the small twigs, they are afterward ground to powder by a rude wooden mill. The tea is next weighed and stored by the overseer, who pays the peons for it, at the rate of 2 rials or 1s. each arroba of 25 lbs. It is next rammed tightly into bags of bull's hide, which are left to dry in the sun, and contain from 900 lbs. to 220 lbs. each; and in this state it goes to market. Mr. Robertson estimates that for six months' work, the peon may obtain about £57 in wages; but he has run in debt to his master perhaps £12 before entering the woods, and as much more while employed there, for neither of which sums he has got half the value. Of the remaining balance of £33, he spends perhaps £12 in ornaments for his horse. £5 more in personal decorations, and the rest in gambling, to which all are very much addicted. "In a month the peon re-sells his horse-furniture and personal apparel; and in a fortnight after that he is left without a farthing; and in week more he is to be found again naked in the *yerbales*." (*Robertson's Letters*, i. 134–150.)

Manufacture received a considerable impulse from the dictator's prohibition of foreign commerce. The people had previously imported cotton, woollen, and almost all other manufactured goods ; and there used to be no such thing as a good workman in Paraguay. But the exercise of ingenuity was excited, not only by necessity, but by terror. The dictator caused a gibbet to be erected, and threatened a poor shoemaker to hang him up, because he had not made some belts of the size he required ; and once he sentenced an unfortunate smith to hard labour, because he had improperly placed the *sight* of a cannon! (*Rengger, &c.*, p. 50.)

Francia, next to personal aggrandizement, appears to be actuated by the vulgar, short-sighted, barbarous policy of wishing to render Paraguay dependent solely on her own internal resources, and wholly unconnected in any way with any other South American state. Except in special cases, he permits no ingress or egress of individuals or merchandise to and from Paraguay ;* and this system of exclusion will probably be kept up as long as he lives. While Paraguay remained a Spanish province, the yearly value of its exported produce fell little short of 1,500,000 dollars. Of Paraguay tea, 8,000,000 lbs. were annually sent to Santa Fé and Buenos Ayres, besides 1,000,000 lbs. tobacco, large quantities of timber, cotton, sugar, molasses, spirits, &c. But the only trade, if so it may be called, which has been carried on of late years, has been upon account of the dictator. "When he wants an assortment of foreign goods, a permit is sent over to the adjoining province of Corrientes for a vessel to proceed to the opposite port of Neembucú. On her arrival there, the invoice of the cargo is immediately forwarded to him at Assumption, from which place, after selecting such articles as he requires, he orders a quantity of yerba-maté to be sent on board in payment. There is no appeal from his own valuation: no one is allowed to go on shore, and the ship is sent back as soon as the yerba is delivered. This article is in such demand, from his having stopped the trade in it, that the people of Corrientes are glad to get it upon his own terms. In the same manner, for a short period, he allowed a peddling traffic to be carried on between the Brazilian missions beyond the river Uruguay and the port of Ytapua, but that trade he altogether stopped about 10 years ago." (*Parish's Buenos Ayres, &c.*, 234, 235.)

The government of Paraguay is an anomaly to the present times. It approaches as near to an absolute despotism as can well be conceived, the dictatorship of Sylla in ancient Rome being the only model with which it may be compared. The state is nominally republican, having a so styled congress of several hundred members; but the entire power centres in the dictator, who is not only commander-in-chief, but head of the church, the law, and every other branch of the administration. The country is divided into 20 sections, or *commandancias*, exclusive of a territory in the S.E., called the Missions, occupying 600 sq. leagues, and governed by a special officer. Besides Assumption, the cap., there are but four towns in Paraguay; the other collections of houses being mere villages. There is no law save what is dictated by the caprice of the dictator; and his punishments are as barbarous as his policy is tyrannical and oppressive. The military force comprises about 3000 men, principally cavalry; besides which, there is a militia, comprising every free male citizen, 17 years of age, and capable of bearing arms. The naval force consists of only a few brigantines and gunboats. The amount of the public revenue is uncertain: it is derived from state property, the greater part of which has been confiscated by the dictator; tithes in kind upon all articles of produce, the right to levy which is sold each year to the best bidder; taxes upon shops and storehouses in the capital; the *droit d'aubaine*, or right to the property of all foreigners dying in Paraguay; fines, postage, sale, stamp and commercial dues, &c. The principal state expenditure is for war-stores, and the support of the army. There is no public debt. Public education is not much encouraged by the dictator; but there are many primary schools for male children, and, according to Reugger and Longchamps, "it is a rare occurrence in this country, where no printing-press exists, to find a free man who cannot read and write." Morals are at a very low ebb.

Paraguay was discovered in 1526 by Sebastian Cabot. The Jesuits afterward established many missions in the S. part of the country; and were supposed to have effected astonishing improvements in the condition and habits of the natives; but no sooner had they been expelled in 1768, than the fabric they had been so long in raising fell straightway to pieces, and the Indians relapsed into their former barbarism. In 1776 Paraguay became a province of the viceroyalty of Buenos Ayres. In 1810, the Buenos Ayres revolutionary troops were defeated by the Paraguayans; but the latter soon afterward deposed their governor, and, in 1812, proclaimed Paraguay a republic under two consuls. In 1814 the second consul, Dr. Francia, found means to get himself made sole dictator for three years, and at the expiration of that term, for life. In 1826 Francia declared Paraguay independent, and its independence was formally recognised by the emperor of Brazil in 1827. (*Robertson's Letters on Paraguay; Reugger and Longchamps; the Reign of Dr. Francia; Parish's Buenos Ayres, &c.*)

PARAMARIBO. *See* GUIANA (DUTCH).

PARGA, a fortified town and seaport of European Turkey, in Albania, sanjack Delvino, on the Ionian sea, near the mouth of the Fanar (an. *Acheron*). 48 m. S.W. Yannina, and 13 m. E. Paxo; lat. 39° 15' 45" N., long. 20° 24' S.E. Pop. 4000. It is built amphitheatrewise on the side of a steep rock, surrounded on three sides by the sea, the summit of which is crowned by an almost impregnable fortress, commanding a magnificent view of the surrounding coast

and country. It is surrounded by strong walls, and has a double harbour. The streets are narrow, steep, and dirty; it has no public buildings of importance, and many of the houses are in ruins. The inhabitants export oil, tobacco, different kinds of fruit, and some tolerably good wine, all being the produce of the fertile and well-watered tracts surrounding the town. Sir J. C. Hobhouse states that the Pargiotes were among the worst of the Albanians, and that their connexion with the Christian states had taught them only the vices of civilization without diminishing their ferocity. (*Albania and Turkey*, p. 169.) The opinions of General Campbell and Colonel Leake are much more favourable; and Colonel de Bosset, who had excellent opportunities of estimating their character, pronounced them to be spirited and independent, though at the same time temperate, docile, and, if well treated, easy of command. The Albanians, however, mostly withdrew from the town on its being ceded to Ali Pacha, and the present inhabitants are principally Turks.

Parga is hardly mentioned in history until 1401, when it entered into an alliance with Venice, which continued nearly four centuries, until the subversion of the latter in 1797. Parga, being independent of Ali Pacha, tyrant of Albania, afforded an asylum to refugees from his violence, and was the seat of frequent cabals against his government; so that it became an object of importance for him to annex it to his dominions. In 1814, it was besieged by Ali, but being assisted by the British in Corfu, Ali was obliged to give up the siege, and the Pargiots had reason to believe that they would be incorporated with the republic of the Ionian islands. The British government did not, however, agree to this arrangement: they felt the importance of Parga to Corfu; but the dread of continued dissensions with the Albanians led to a negotiation for its surrender, on Ali paying a pecuniary indemnity to such of the inhabitants as should refuse to remain after a change of government. In consequence of this agreement, which was severely, and perhaps justly, censured, most of the Pargiots withdrew to the Ionian islands; and Ali had to pay, in all, about £336,000 by way of compensation. The cession took place in the year 1819.

PARIS (an. *Lutetia* or *Lucutecia*), a celebrated city of W. Europe, the metropolis of France, being the next European city to London in magnitude and importance, in the dep. of the Seine, of which, with its suburbs, it comprises the largest portion, on the Seine, about 110 m. direct dist. from its mouth, 210 m. S.S.E. London, and 170 m. S.S.W. Brussels; lat. (observatory) 48° 50' 14" N., long. 2° 20' 15" E. Pop., in 1896 (exclusive of troops and temporary residents), 890,431, and, in 1836, 909,126. The city stands in a plain surrounded on several sides, but especially N. and N.E., by considerable eminences; and the geological constitution of the district is so peculiar that the French geologists have called it the Paris basin, in the same way that the English have called the tertiary formations near the English metropolis the London basin. Here are found alternate strata, abounding with marine and fresh-water shells, and containing also many fossil remains of extinct animals. Gypsum (known in England as *plaster of Paris*) is found in large quantities; and S. of the Seine is quarried good building-stone, of which, indeed, some of the principal edifices of Paris are formed: the older quarries, all of which were subterranean, have been converted into catacombs, or repositories for the bones of the dead, removed from the public graves that once abounded, greatly to the injury of the health of the city.

Paris, like London, is situated on both sides a considerable river, which runs through it from S.E. to N.W., and divides it into two parts, of which the largest is on the N. side; the most ancient part of the city being, however, confined to the small islands within the channel of the river. In the course of centuries it has so extended itself that it now occupies an area of about 14 sq. m., including the Champs Elysées, and other open spaces at its W. extremity. Many of the best streets are parallel to the river, and the open spaces, or quays along its banks, present an agreeable feature of which London is almost wholly destitute. A few of the streets more recently built are wide, and lined on each side with *trottoirs*; but, generally speaking, the streets are narrower, and less regular, than those of the British metropolis. The style of building, however, is the best; streets, is probably superior to that of London. The houses are very high, and many of them comprise seven stories, including the ground-floor; for there are no such stories. All the tenements have rich heavy cornices over each story below the roof, and the fronts are invariably coated with plaster, and repainted from time to time. The town has, therefore, in its better parts, a gayer and handsomer appearance than London; but externally, the houses (which are of great extent, inhabited by many families, and, in some cases, formed round internal courtyards, accessible by *porte-cochères*), want the many comforts and conveniences which are found in English houses." (*Maclaren's Notes,* p. 12.) As in London, the fashionable part of Paris is at its W. end, while the districts of an opposite character are mostly in the E. and S. The *boulevards,* a succession of open, circular roads, similar to the "Circular Road" which surrounds Dublin, encircle the more densely-peopled portion of the city. They occupy the site of the old fortifications built in the reign of Louis XIII., are from 60 to 70 yards in width, and, being planted with trees, form agreeable places of resort for all classes of the inhabitants.

The city was originally divided into four quarters (*quartiers*), but as it increased, new allotments became necessary, though the old name was retained; and hence we find that there are at present 48 *quartiers.* For electoral and municipal purposes, however, Paris is divided into 12 arrondissements, each comprising four *quartiers,* and each sending one member to the chamber of deputies.

The following table exhibits the population belonging, in 1826 and 1836, to the different arrondissements, with the names of the *quartiers* comprised in each, the order of succession being from W. to E. on each side the Seine.

Arrondissements.	Quartiers.	1826.	1836.
1. N. of the Seine.			
I.	Quartiers du Roule, des Champs Élysées, de la Place Vendôme, des Tuileries	72,101	82,796
II.	Quartiers du Chaussée d'Antin, du Feydeau, du Palais Royal, du Faubourg Montmartre	73,699	80,226
III.	Quartiers Poissonnière, de Montmartre, de St. Eustache, du Mail	54,197	57,060
IV.	Quartiers de St. Honoré, des Marchés, de Louvre, de la Banque de France	51,769	59,129
V.	Quartiers des Bonnes-nouvelles, du Faubourg St. Denis, du Faub. St. Martin, de Montorgueil	75,569	82,994
VI.	Quartiers du Temple, des Lombards, de la Porte St. Denis, de St. Martin des Champs	90,651	94,106
VII.	Quartiers des Arcis, du Ste. Avoye, du Mont-de-Piété, de Marché St. Jean	73,903	68,407
VIII.	Quartiers des Quinze-vingts, de Popincourt, du Faub. St. Antoine, du Marais	72,375	82,084
IX.	Quartiers de la Cité, de l'Arsenal, de l'isle St. Louis, de l'Hôtel-de-Ville	57,795	71,780
2. S. of the Seine.			
X.	Quartiers des Invalides, de la Monnaie, de St. Thomas d'Aquin, du Faub. St. Germain	90,699	88,172
XI.	Quartiers du Luxembourg, du Palais de Justice, de l'École de Justice, de la Sorbonne	65,749	58,797
XII.	Quartiers de l'Observatoire de St. Jacques, du Jardin des Plantes de St. Marcel	97,922	82,261
		890,431	909,126

"Paris, however," observes Mr. H. Lytton Bulwer, "is divided into quarters as well by its manners as its laws, and these different districts differ as widely one from the other in the ideas, habits, and appearance of their inhabitants, as in the height and size of their buildings, or the width and cleanliness of their streets. The Chaussée d'Antin breathes the atmosphere of the Bourse, and the Palais Royal is the district of bankers, stock-brokers, generals of the empire, and rich trades-people; and it is the quarter fullest of life, most animated, most rife with the spirit of progress, change, luxury, and elegance. Here are all the new buildings, arcades, and shops, and here are given the richest and most splendid balls. How different is the *quartier* St. Germain, the district of the long and silent street, of the meagre repast, and the large, well-trimmed garden, of the great courtyard, of the broad and dark staircase, inhabited by the administrations and the old nobility, manifesting no signs of change, no widening of streets, no piercing of arcades or passages: it hardly possesses a restaurant of note, and has but one unfrequented theatre. Farther E., on the same side of the Seine, is the *quartier* of the students, at once poor and popular, inhabited by those eloquent and illustrious professors who give to France its literary glory. Then there is the Marais, the retreat of old-fashioned judges and merchants, where the manners have been changed almost as little as the houses by the philosophy of the 18th century: here are no carriages, no equipages: all is still and silent; you are carried back to the customs of the grand hotels in the time of Louis XIII. Then there is the Faubourg St. Antoine, the residence of those immense masses that

reigned under Robespierre, and which Bonaparte, after Waterloo, refused to summon to his assistance. And behold the ancient city of Paris surrounded by the Seine, and filled by a vast and wretched population; there, proud and with sordid roofs around them, rise the splendid towers of Notre Dame, that temple of the 12th century which, in spite of the Madeleine, has not been surpassed in the 19th; there is the Hôtel-Dieu, the antique hospital as old as the time of Philip Augustus, and there is the Palais de Justice, where sat the parliament of Broussel, remarkable in the chronicle of De Retz!" (*France, Social, Lit., and Pol.,* i., 44–47.)

Barrières, Boulevards, &c.—Paris, as defined by the walls erected in the reign of Louis XVI., is of an irregular oval shape, its greatest length from N.W. to S.E. being 4½ m., and its greatest breadth from the Barrière de la Villette northward to the Barrière d'Enfer southward about 3½ m. In these walls are 58 gates, at each of which is a toll-house for the collection of the *octrois,* or local dues on goods entering the city; and on the outer side of the walls are well-planted walks, called "the outer boulevards," abounding with *guinguettes,* wine-shops, &c., the favourite resort of the lower orders, the wine drunk here not being subject to the town dues. Between the outer and the inner or great boulevards, already noticed, are the suburbs or *faubourgs,* forming some of the best built quarters of Paris.

General Condition of Streets, Houses, &c.—The streets in the interior of Paris, except those of more modern construction in the fashionable quarters N.W. the Tuileries, have been formed more or less on the model of the narrow lanes and alleys constructed before the general introduction of carriages, at a time when the absence of police and the frequency of popular tumults rendered it necessary to defend the streets at night with cross barriers or chains. Much attention, however, and vast sums of money have lately been devoted to the improvements of the great thoroughfares; *trottoirs* of basaltic stone from Auvergne have been laid down in many of the streets; and more recently the asphalte-pavement has been successfully introduced in the Rue Rivoli and on the Boulevards. Gas has been provided by two or three chartered companies; and the quarters of the Tuileries, Chaussée d'Antin, and Palais Royal, as well as the arcades and principal shops N. of the Seine, are now brilliantly lighted with gas. But in the streets of other quarters, and even in the best streets S. of the river, may still be seen the melancholy oil-lamps, or *reverbères,* suspended by a cord across the street, emitting only sufficient light to make darkness visible, till midnight, when all the lights are extinguished, and the town is plunged in obscurity, rendered more annoying and dangerous from the inefficiency of the night police. (*Maclaren's Notes,* p. 22.) With respect to cleanliness, Paris, though greatly improved, is still very far from what it should be. The sewers, which were begun at the commencement of the 15th century, were at first wholly open, running through the middle of the streets, either directly into the Seine, or into its tributaries, the brook Ménil-montant northward, and the Bièvre southward. In some of the closer and less wealthy districts this nuisance still remains; but by far the larger part of Paris is now supplied with subterraneous drains, arched with stone, some of them being of great size. Drainage, however, for separate houses is still far from being common; there is a general want of water-closets; and domestic filth is in many cases allowed to accumulate, greatly to the injury of the public health, and in spite of police regulations not strictly enforced. The houses, also, which are always five, and often seven, stories in height, though *without cellars,* consist, as in Edinburgh, of separate stories; and sometimes a single story, or floor, is divided into different sets of apartments, occupied by different individuals or families, the access to the different stories being from a common staircase, usually very dirty, and often, indeed, filthy to a degree revolting to an Englishman: and as water is seldom introduced into the houses by pipes, but is brought by porters from the public fountains in the streets, we need not be surprised at the deficiency of cleanliness. The more modern houses, however, are provided with better means for insuring this advantage; and in the new quarters we do not observe that *melange* of inmates, from the prince *au premier* to the poor seamstress *au sixième,* that distinguishes the houses of Paris from those of most other capitals of Europe. The broader streets have externally a pleasing appearance, owing to the regularity of the houses, the great number of windows (for which there is scarcely any tax), and the general use of balconies and "external shutters called *persiennes,* formed by thin bars of wood, turning on hinges, and folding back on the walls when not in use." (*Maclaren,* p. 23.) The streets S. of the Seine are gloomy and monotonous: in the chief hotels of the nobility few windows face the street, and large *porte-cochères* lead to an inner court yard, round which the building is arranged.

Principal Streets, Places and Parks.—The leading street

of Paris, corresponding with the Strand or Oxford-street of London, is the *Rue St. Honoré*, running westward from the *Marché des innocens*, and forming, with its continuation, the *Rue Faubourg St. Honoré*, a line of streets very nearly 2 m. in length. The houses in this faubourg are large and handsome, belonging chiefly to the higher classes; but those in the part nearer the centre of the city consist principally of shops and residences of persons in trade. The Rue St. Honoré is connected eastward with the *Rue St. Antoine*, terminating in the *Place de la Bastille*, in which is the model of the projected *Fontaine de l'Eléphante*. These streets entirely intersect the capital from W.N.W. to E.S.E., and the continuous line measured from the *Barriere du Route* westward to the *Barrière du Trône* eastward, is exactly 5 m. in length. From N. to S. runs another and almost equally important avenue, formed N. of the Seine by the Fauxbourg and Rue St. Martin, crossing the river and isle of Paris by two bridges (the *Pont Notre Dame* and *Petit Pont*), the part S. of the river consisting of the *Rue* and *Faubourg St. Jacques*, terminating in the *Barrière d'Arceuil*, from which to the N. end of the line at the Barriére de la Vilette is a distance of 3¾ m. The Rue St. Denis runs parallel to the Rue St. Martin, connecting the Pont-au-Change with the N. Boulevards at the Porte St. Denis; and more westward, running in the same direction, are the *Rue de Richelieu*, *Rue de Castiglione*, and *Rue de la Paix* connected with the *Place Vendome*, the *Rue de Luxembourg*, and the *Rue Royale* and *Tronchet*, which run into the square containing the church of *La Madeleine*. These again are crossed by other streets running from W. to E., the principal being the Rue St. Augustin, which connects the Bourse with the Boulevard des Capuchins; and opposite to the gardens of the Tuileries is the *Rue de Rivoli*, a noble well-paved street, lined on its N. side by government buildings and fine hotels; it is about ¾ m. in length; and lined throughout its whole extent with colonnades. The chief streets S. of the Seine and parallel to its banks are in the Faubourg St. Germaine, comprising the *Rue de Grenelle*, a handsome avenue lined with several large and handsome government-buildings, the *Rue St. Dominique*, *Rue de l'Université*, and *Rue de Bourbon*, the last running close to and in a line with the *Quai d'Orsay*. Most of the streets at the E. end of Paris are narrow and irregularly built: the *Quartier Latin*, S. of the Seine, comprises several collegiate edifices; but the streets are confined and crooked, in fact, little better than mere lanes and alleys. Indeed, the only handsome streets of Paris, except the Boulevards, are to the W. of the Palais Royal on both sides the river. The quarters of the Tuileries, the Place Vendôme, and the Chaussée d'Antin, are the most fashionable districts N. of the Seine; but the houses, or *hôtels*, of the noblesse are chiefly in the Faubourg St. Germain. Besides the streets just mentioned, Paris has about seventy squares, or *places*, the principal of which are—1. the *Place de la Concorde*, an open space W. of the garden of the Tuileries, in the centre of which is the obelisk lately brought from Luxor, in Egypt; 2. the *Place Vendôme*, an octagon surrounded on three sides by handsome buildings, with Corinthian fronts, and having in its centre a noble column, formed on the model of that of Trajan at Rome, covered with bronze castings, representing the achievements of the grand army in 1805, and surmounted by a statue of Napoleon; 3. the *Place des Victoires*, originally formed in 1685, and having in its centre an equestrian statue of Louis XIV. on a marble pedestal, with bas-reliefs; 4. the *Place de Grève*, where public executions were formerly carried into effect, and having on one side the *Hôtel de Ville*; 5. the *Parvis Notre Dame*, in front of the cathedral of that name, and having on its S. side the *Hôtel-Dieu*; and 6. the *Place de la Bastille*, in the Faubourg St. Antoine, occupying the site of the Bastille destroyed 14th July, 1789: in its centre is the *Colonne de Juillet*, a large Doric column, erected in commemoration of the revolution of 1830, 130 ft. in height, surmounted by a colossal figure of the Genius of France. Paris has also 65 fountains, communicating by pipes with the Seine and the canal de l'Ourcq: some of them, as the fountains of St. Sulpice, St. Eustache, Grenelle, the *Châteaux d'Eau* in the Boulevard de Bondi, and that in the Marché des Innocens, are worthy of notice from their architectural beauty; though the supply of water from them be not only insufficient, but of bad quality. In consequence of the great depth to which the *calcaire grossier* of the Paris basin penetrates, there are nó springs in the city itself, though aqueducts bring pure water from a distance: but pipes, leading as in London to each house, are of very modern date, and only partially introduced.

Exclusive of several handsome gates and barriers, Paris has four splendid triumphal arches; those of St. Denis and St. Martin in the N. Boulevards, erected in honour of Louis XIV.; the *Arc de Carrousel*, forming the principal entrance to the palace of the Tuileries, built in 1806 on the plan of that of Septimius Severus at Rome, being 60 ft. in width, by 45 ft. in height; and, the *Arc de l'Etoile*, at the W. end of the *Avenue de Neuilly*, commenced by Napoleon in 1806, and only recently completed. The latter is a most magnificent monument, and is, indeed, by far the most stupendous structure of the kind ever erected, either in ancient or modern times. It consists of a single arch 96½ ft. in height, 48 ft. in width, and 73 ft. in depth, and of two smaller transverse arches; the whole structure being 147 ft. in length by 73 ft. in depth, and 162½ ft. in height! It stands quite separate from any other building, so that it is seen to the best advantage. It has numerous colossal groups of sculpture, depicting most of the great battles gained by the French during the revolutionary war. The effect of this prodigious structure is grand in the extreme, and is worthy the genius and magnificence of its founder.

Contiguous to the *Arc de l'Etoile* is the garden of the Tuileries, an enclosed space of sixty-seven acres, laid out by the celebrated Le Nôtre in broad walks and angular beds, and profusely ornamented with vases, statues, &c. It is a favourite resort of the Parisians, and is separated by the Place de la Concorde from the Champs Elysées, an open space about 1000 yds. in length by 400 yds. in breadth, planted by Colbert in 1670, with pavilions along the sides provided with seats and entertainments. These parks, for so they may be called, constitute with the *Avenue de Neuilly* the Hyde-park of Paris, and, like it, are thronged on Sundays and festival days. S. of the Seine is the *Champ de Mars*, an oblong space bordered by a double avenue of trees, and used for reviewing troops, horse-racing, &c. It was the scene of the celebrated *Fête de la Féderation*, 14th July, 1790, as well as of the Champ de Mai during the 100 days. Races are held in it in May and September; but English sportsmen describe them as very inferior. Several minor gardens are dispersed in the different faubourgs; besides which, near the E. and W. suburbs respectively are the *Parc des Vincennes*, about 2¼ m. in length by 2 m. in breadth, and the *Bois de Boulogne*, a favourite resort of carriage company, as well as of duelists and suicides. Duels, however, have become much less frequent since the enactment of the law, allowing damages to the family of the deceased party.

Palaces and Government Buildings.—Paris contains four royal palaces; but only one of these, that of the Tuileries, is inhabited by the royal family; the Louvre has not been the residence of a French monarch since the minority of Louis XV., and is now formed into a national museum and picture gallery; the Palais Royal, built by Cardinal Richelieu, and the favourite abode of Louis XIII. and Anne of Austria, consists principally of shops, cafés, restaurateurs, and *estaminets*, crowded at all times, day and night, by almost every class of the Parisians; and the Palais du Luxembourg built for Marie de Medicis, widow of Henry IV., is now devoted to the use of the chamber of peers. Palace of the Tuileries, erected in the 16th century, on the site of a manufactory of tiles (*tuiles*, whence its name), was greatly enlarged by Henry IV., Louis XIII. and XIV.; and additions have since been made to it by Napoleon and Louis Philippe. Its architecture is of a somewhat mixed character; but the earlier parts may be taken as a good specimen of the revived Italian style. Wings extended from the main building on the side towards the Place du Carrousel, one of which, on the S. side, connects it with the museum of the Louvre; and on the garden side are arcades extending through the central portion of the building, at the sides of which are handsome pavilions formed into state apartments, remarkable for their lofty windows, flanked by Corinthian pilasters. The general effect is extremely grand, especially on the garden side; but its grandeur results more from its great length and the variety of outline it presents, than from any excellence or congruity in the details. The state-rooms are on the first floor, running the whole length of the garden front, the principal being the *Salle du Conseil*; the state dining-room, known as the *Galerie de Diane*, from which other rooms lead to a vast saloon and state ball-room in the centre of the building, called the *Salle des Maréchaux*, adorned with portraits of the great marshals of France, and unquestionably one of the finest rooms of the kind in Europe. The court of the Tuileries, on the E. side of the palace, was formed chiefly by Napoleon, and forms a wide space, separated by an iron railing from the *Place du Carrousel*, and now used for the inspection and review of the troops on duty in different parts of Paris. S. of the Place du Carrousel is the long gallery of the *Louvre*, connecting it with the Tuileries. The pictures are deposited in a splendid range of rooms on the first floor, facing the river, above ½ m. in length; but nearly the whole interior of the palace, which forms a hollow square, is appropriated to the reception of museums, &c., which will be subsequently noticed. A portion of the basement story, however, in the S. wing, is divided into apartments for the residence of officers attendants, &c. As respects its external architecture, the

Louvre, is undoubtedly one of the finest regal structures in Europe. Its E. front, facing the *Place du Louvre*, consists of a magnificent colonnade formed by 28 coupled Corinthian columns, rising above the basement story, and surmounted by a beautiful cornice and line of balustrades. The S., or r___ though much less ornamented, is still extrem___ ___ome, being faced with 40 Corinthian pilasters, ___ich rise a balustrade and central pediment; the ___ fronts are quite plain, and form a striking ___ver ___those just described. The internal courtyard of the ___ely ___is a perfect square, each side being 400 ft. in ___above___buildings surrounding it are of the Corin- N., and ___opposite orders, highly adorned with sculpture. ___is *Royal*, which stands to the N. of the mass of buildings just noticed, has towards the Rue St. Honoré a front with two wings, united by a screen which encloses a court-yard somewhat resembling the horse guards or admiralty of London, not inelegant, though of a very impure style. Round the oblong space, at the back of the palace, the father of the present king of France erected large houses and handsome colonnades, occupied by jewellers, tailors, *marchands-de-mode*, shoemakers, printsellers, *restaurants* (the principal of which are Véry, Vefour, and the frères Provençaux), keepers of *cafés*, *estaminets*, or smoking-rooms, &c. The gardens are tastefully laid out, the whole being "brilliantly illuminated with gas;" and hence the Palais Royal is the perpetual rendezvous of the idle and curious, as well as of the little *rentiers* of the capital. The *Palais du Luxembourg*, where, during the republic, the directory held its sittings, and which, since the re-establishment of the monarchy, has been used by the chamber of peers, is a stately edifice, facing the *Rue de Vaugirard*, with two wings, connected by a screen and gateway; being remarkable for strength and solidity, as well as for the beauty of its proportions. The interior comprises several handsome apartments, the most interesting being the *Salle des Séances*, a semi-circular chamber of no great size, round which are arranged chairs for the peers, while the flat side is occupied by the president's seat, and tables for clerks, &c.; this room, however, is much too small for its present purpose, and a larger and more commodious building is now being erected. The gallery of the Luxembourg, which once boasted of a fine collection of old pictures, since removed to the Louvre, is now appropriated to the reception of works by living French artists; the gem of the modern gallery being "the bathing nymph," by Julien. The gardens behind the palace, laid out in the old French style, have a sheet of water in the centre. Among the buildings devoted to the use of the government and legislature, the first place is due to the Palais Bourbon, erected chiefly by the Prince de Condé, between 1722 and 1789: it was occupied during the revolution by the *Council of Five Hundred*; but at the restoration of the Bourbons was restored to the Condé family, with a proviso for the accommodation of the deputies in that portion of the building previously occupied by the council. The chamber of deputies was purchased from the family in 1829; and a treaty has lately been completed for ceding the entire property of the Palais Bourbon for the use of the legislature. The Corinthian portico fronting the *Pont Louis XVI*., the pediment of which has lately been completed and exposed to public view, is of fine proportions; but, when compared with the building to which it is the entrance, it is much too large, and leads to no apartments of any great size, except the chamber itself, a semicircular room ornamented with 20 Ionic columns of white marble, having gilt capitals. The president's chair and the tribune form the centre of the axis of the semicircle, round which rise successively the seats and desks of the 459 deputies, to the height of the basement supporting the columns. The walls are adorned with pictures and statues; and a spacious double gallery, capable of accommodating 700 persons, runs round the semicircular part of the chamber, fitted up with tribunes for the royal family, corps diplomatique, &c., and seats for the public. The place of each deputy is marked at the beginning, and retained to the end of each session; but when a member addresses the assembly, he does not, as in the British House of Commons, speak from his place, but ascends the tribune near the president's chair. The sittings are held chiefly by daylight. The library of the deputies, a long and handsome gallery, contains about 44,000 vols., chiefly reports and law-books, both French and English, including also a few rare MSS.

Some of the government offices are extremely handsome edifices, especially the *Hôtel des Finances*, an insulated structure of vast size in the Rue Rivoli; and the *Hôtel du Quai d'Orsay*, a noble stone building, on the plan of the Farnese palace at Rome, and unquestionably one of the most magnificent in Paris, comprising about 800 rooms, allotted into offices for the council of state, the Cour des Comptes, and for the departments of the Interior and Public Instruction. The *Hôtel des Affaires Etrangères*, in the Rue

des Capucines, is a building of considerable extent, but with few pretensions to architectural elegance. The *Hôtel des Monnaies*, or Mint, S. of the Seine, a little W. the Pont Neuf, built in 1771, has two fronts, the principal of which facing the river, has a length of 300, with a height of 78 ft. All the operations of coining are carried on within this building; and it is the place of assay for all gold and silver articles made in Paris. In one of the apartments is a superb collection of medals and casts belonging to all eras of French history. Among other buildings nearly or more remotely connected with the government of Paris are the following: 1. The *Hôtel de Ville*, in the *Place* of the same name, commenced in 1549, but not completed till 1605, having a singularly uncouth front, with two side pavilions, higher than the rest of the edifice, and two gates leading to a quadrangle, in which is a bronze statue of Louis XIV. It comprises some fine apartments, particularly the *Grand Salle*, at one of the windows of which Lafayette, in 1830, introduced Louis Philippe to the populace as the best of Republicans! Its exterior is now undergoing extensive repairs. 2. The *Palais de Justice*, in the Isle du Palais, an edifice in a mixed style, erected between the 14th and 18th centuries, on the site of a still-more ancient structure; in the interior a central staircase leads to a grand saloon, called the *Halle des Pas Perdus*, which comprises apartments for the court of cassation, the Cour Royale, and Cour d'Assize. 3. The *Hôtel de la Legion d'Honneur*, on the Quai d'Orsay. Paris has eight prisons, among which those of St. Pélagie and St. Lazare are the most extensive and best managed. The prison for juvenile offenders in the Rue de la Roquette, is built on the *panopticon* principle. The *Conciergerie* and *Abbaye* are small, and very inefficiently regulated the latter being now exclusively employed for the detention of military offenders.

Religious Edifices.—The sacred buildings of Paris, like those devoted to secular purposes, exhibit a great variety of styles; but from the close of the 16th century downwards, which the Grecian has prevailed. The first place, however, both as respects antiquity and grandeur, must be ceded to the *Cathédrale de Notre Dame*, erected between 1010 and 1407: it is a cruciform-structure, with an octagonal E. end, and double aisles surrounding the choir and nave; a third aisle also being occupied by a series of seven external chapels. At the W. end are two lofty towers, evidently intended to have been the bases for steeples; but the tower usually seen at the intersection of the nave and transepts was destroyed during the revolution. The length of the church externally is 442 ft., the breadth 163 ft., and the length of transepts 392 ft.: the towers are 235 ft. high. The exterior, though not without beauty, is heavy, owing to the absence of steeples, pinnacles, &c., which give a light appearance to the majority of Gothic buildings. The W. front, however, with its three large gates and circular window, and the noble gateway on the W. side of the church, are highly worthy of admiration. The inside of the church has a very splendid and imposing appearance, owing to its numerous aisles and chapels; but the uniformity of effect has been entirely destroyed by the embellishments of the choir, which, though in themselves beautiful, are wholly unsuited to the rest of the building. The church of *St. Germain des Prés*, built about 50 years after Notre Dame, is cruciform, with a circular E. end. A considerable portion of the old building has gone to decay; and, out of three towers, only one remains; but the interior contains some good modern decorations, valuable pictures, old monuments, &c. There are six other churches, either wholly or in part of Gothic architecture; the most interesting is the church of *St. Germain l'Auxerrois*, not only from its structure, but its associations with the less peaceful periods of the history of Paris, especially those of the massacre of St. Bartholemai. Among the more modern churches in the Grecian style, the largest and most splendid is that of St. Geneviève, now called the *Panthéon*, in the quarter of the university. It was commenced in 1764 by Louis XV. The portico is composed of 22 fluted Corinthian columns, 60 ft. in height, supporting a triangular pediment 120 ft. broad by 24 ft. in height, in which is a sculptured composition, by David, representing the genius of France (a colossal figure 14 ft. high), surrounded by the great men of the nation: on the frieze beneath is inscribed in gold letters—

"AUX GRANDS HOMMES, LA PATRIE RECONNOISSANTE."

The plan of the church is a Greek or equilateral cross, the exterior having no windows, and being ornamented only by a frieze and cornice. In the interior a gallery and colonnade line the nave and transepts on both sides, forming so many smaller naves and aisles. Semicircular windows rise above the colonnades, throwing a strong light into all parts of the building; and from the centre of the cross rises a dome 282 ft. in height, the lower part of which is encircled by a Corinthian peristyle of 32 columns, each 36 ft. high. The inside is now perfectly empty, without any embellishments, except its architectural decorations; but it is intended that it shall be enriched with statues of Voltaire, Rousseau, Lagrange, and other illustrious individuals, whose remains have been deposited in the spacious vaults beneath the pavement. The total length of the Pantheon, including the portico, is 352 ft.; interior length from E. to W. 295 ft.; length of transept 265 ft.; uniform breadth 104 ft. On the whole, this church is a work of great merit: the general proportions are good, and there is much grace and elegance in the outline, as well as grandeur and simplicity in the design; but it is by no means entitled, either from size or composition, to be compared with St. Paul's. Among the other churches, the most deservedly celebrated is the Madeleine, at the N. end of the Rue Royale, on the model of the Parthenon of Athens, but larger, being 328 ft. in length, and 138 in breadth, while its archetype is only 228 by 100 ft. It is altogether a very noble structure, and is remarkable for purity and elegance of design. Paris has, in all, twelve parish churches, corresponding with the arrondissements, and 29 district churches, besides six others unattached, and several belonging to hospitals, convents, &c.; but none of these, except that attached to the *Hôpital des Invalides* (for which see *Hospitals*), requires any particular description. There are also several places of worship for dissenters from the Roman Catholic religion; two of which belong to the French Lutherans, two to Calvinists, and five to independents; besides which, there are three English churches and five chapels, two American chapels, two Jews' synagogues, and a Greek church; but none of them have any claims to architectural beauty except the church of the Visitation, in the Rue St. Antoine, and the Gothic church belonging to the British embassy, in the Rue d'Aguesseur. The Roman catholic clergy of Paris comprise an archbishop, 12 vicars-general, 3 metropolitan and 4 diocesan officials, 16 canons of Nôtre Dame, 34 honorary canons, with curés and vicars to the different churches.* The Protestant clergy comprise 3 Lutheran and 5 Calvinist pastors, 4 French independent ministers, an English bishop, 5 English Episcopal clergymen, and several ministers of other denominations. Paris still comprises several convents for females; but those of the present day bear but a very slight resemblance to those the old nunneries: and are now little more than religious boarding schools for young ladies, or lodging houses for the numerous *sœurs de charité*, who devote themselves to the nursing of the sick in the hospitals, &c. There are, also, 14 societies, some of which are liberally supported, for the promotion of religion at home and abroad, as well as of religious education.

Judging from the statements of the most intelligent travellers, it would seem, whatever may be the other wants of the French capital, that an increase of church accommodation is not one of them. "Had I not looked into the almanack, I should never have found out which day was Sunday. The churches are open every day, and of course afford no criterion. The shops are open too; and cars and carriages are plying on the streets, and placards invite you to vaudevilles at the theatres and ballets at the opera. Your first impression is that Sunday has been blotted out of the French calendar. On closer inspection, you discover that there is a difference between this and the other days in the week, though I am sorry to say it is a small one. In making a circuit about 12 o'clock through the Palais Royal, the Rue Vivienne, Boulevard des Italiens, Rue de la Paix, and Rue Rivoli, I found about one shop in 20 shut or half shut. At 4 o'clock, on a shorter tour, I found about half of them shut, and at 6 o'clock, three fourths. The thoroughfare of carriages is perhaps also rather less, and that of loaded carts decidedly so. Some of the working classes, I understand, rest on Sunday, going to church perhaps in the morning, and in the evening to a theatre with their wives, or to a cheap café, and playing at dominos.

"On Sunday-week I went to the once celebrated Café de Mille Colonnes, (now sunk to the character of an *estaminet*, or smoking-house,) to get a cup of coffee. It was about seven; I found two or three parties playing at billiards, and a score of little groups, of two, or three, or six individuals, busy with dominos. Two of the parties near me consisted each of a man, with his wife and daughter. The greater proportion, however, of the working classes, ply their labours on Sunday till dinner-time, then rest in the afternoon; and that they may not want their holyday, go beyond the barriers, where wine is cheap, and spend the Monday in drinking and dancing. The over-rigid observance of the Sunday in Scotland, which sometimes disgusts young minds with religion altogether, is a light evil, when compared with this." (*Maclaren's Notes*, p. 17.)

Cemeteries.—The Parisians formerly interred their dead like the ancient Romans, along the sides of the roads leading out of the city; but, as the population increased, and its

* The cures in France are the incumbents of livings, the *vicaires* being the merely salaried servants of the curés; in the same way as the stipendiary curates of England are the dependants of the rectors and vicars, though provided by the bishop.

boundaries were extended, these grave-yards became included within its precincts, and were at length almost in the centre of the town. They were, however, both few and small; so that the inhabitants were compelled to have recourse to other modes of interment; and accordingly, huge trenches (similar to those opened during the prevalence of the plague in London) were dug for the reception of corpses thrown in till the holes were filled, when they were covered over, and others opened close to them. This disgusting method of burying necessarily rendered the neighbourhood of these cemeteries extremely unhealthy; and at length the government issued a prohibition against all funerals within the town, and ordered the formation of spacious cemeteries at a mile distance from the city-walls, at the same time ordering that the bones in the old grave-yards should be deposited in the subterraneous quarries or catacombs under the Quartier St. Germain. Paris has now five large and well-laid out cemeteries, similar in many respects to those which have since been formed on the same model near London, Liverpool, Leeds, and other large towns of England. The Père-la-Chaise, outside the E. barrier, is the finest of the Parisian cemeteries; and its advantageous situation on the slope of a hill, the number, as well as beauty of its monuments, and the celebrity of many of those whose remains have been brought thither, make it one of the most interesting sights in the French metropolis. The catacombs are very extensive, running under about one third part of Paris S. of the Seine: they are arranged into galleries lined with piles of bones, and the entrance is near the Barrière de l'Enfer; but, being deemed unsafe, they are no longer shown to visiters. (*Dulaure*, ix. 211—240.)

Hospitals and benevolent Institutions.—Hospitals for the relief of the sick, and *hospices* for the aged, infirm, or foundlings, existed in Paris from a very early period; but being exclusively under the direction of ecclesiastics, the objects of the founders were grossly perverted, and the revenues of these establishments applied to improper uses. No improvement took place till the revolution; when, by a decree of convention in 1793, the two old and only remaining hospitals were ordered to surrender a portion of the inmates of their crowded and unhealthy wards to the convents and other houses that had become national property. An administration, consisting of a general council and administrative committee, was formed in 1801, for the purpose of improving the condition of the public charities of Paris: taxes on places of amusement and graves in the cemeteries were applied to defray the necessary expenses; and from that time to the present these institutions have progressively increased both in number and utility; so that at present there are 13 hospitals and 21 *hospices*, having altogether about 20,000 beds, and supported in 1837 by a revenue of 18,696,390 fr., or £747,852 sterling, chiefly arising from portions of the *octroi* dues, contributions from the theatres, dues from the *Mont-de-Piété*, interest of funded property, &c. The *Hôtel-Dieu* is entitled to the first notice, on account of its antiquity; for it is known to have existed in the middle of the 18th century, and even at that early period to have had some valuable endowments. It was enlarged between the reigns of Hénry IV. and Louis XVI.; and since the revolution the buildings have been so much improved, that at present the Hôtel Dieu, with its subordinate establishment in the *Rue de Faubourg St. Antoine*, furnishes accommodation for upwards of 1000 in-patients. In 1837, 26,412 patients were admitted into this establishment, the average mortality being one in 8·6. It is in every respect extremely well appointed, and has among its medical officers the most celebrated physicians and surgeons of Paris: indeed, the Hôtel-Dieu may be considered as the great normal hospital of France. The hospitals next in importance are those of *La Pitié, La Charité, St. Antoine, Beaujon, des Enfans Malades* and *des Vénériens;* the whole number of these establishments under the civil administration amounting to 16, exclusive of three military hospitals, regulated by the minister of war. These hospitals, however, are, with two or three exceptions, situated in close neighbourhoods, and, from the antiquity of the buildings, ill-planned; but the interior management is extremely good, and may bear a comparison with that of the first London hospitals. Among the *hospices*, the principal are the *Bicêtre*, for infirm old men, the *Salpetrière*, for aged women, two *hospices des incurables*, and one for orphans and foundlings. There are also numerous *maisons de santé* in and about Paris, which receive patients at certain fixed scales of payment, and may therefore be called hospitals for the middle classes. A great number of minor institutions for the relief of the sick and poor are supported by private subscription.

Of all the establishments, however, in the French capital devoted to the support of the aged and infirm, by far the most important, both on account of the grandeur of its buildings and the benefits which it confers on its inmates, is the *Hôtel des Invalides*, intended for the support of disabled officers and soldiers, or those who have been in active service upwards of 30 years. The edifice, situated in the S. end of an avenue leading over the Pont des Invalides from the Champs Elysees, and begun in 1675, is a conspicuous object from a distance on account of its gilded dome, lantern, and spire rising to a height of 383 ft. above the floor. It is composed of 5 courts of equal form and size, surrounded by buildings 5 stories in height, and covers a space of nearly 7 acres; and is, on the whole, a very heavy building, without any exterior beauty. The church of the establishment is indebted for its noble appearance principally to its magnificent dome supported by 20 pairs of Composite pillars. Besides about 170 pensioned officers, there are about 3080 sub-officers and privates, who are boarded, lodged, and clothed, and receive a monthly stipend varying according to rank. The dormitories contain each from 30 to 60 beds; besides which there are large infirmaries for the sick. All, except field-officers, mess at public tables, and all wear the same uniform. Their only duty is to mount guard within the precincts of the hotel; and, when the king comes within its walls, they have the exclusive privilege of guarding his person. On the whole, the Hôtel des Invalides, though by no means so beautiful a building as Greenwich Hospital, near London, is entitled to rank with it as one of the grandest national institutions of Europe. And it is now more than ever deserving of notice from its containing the remains of the emperor Napoleon, perhaps the greatest general, and perhaps, also, the greatest man of modern times.

Commercial Establishments:—Paris, till 1826, though abounding with fine public edifices, had no structure specially devoted to the transacting of commercial business. The merchants previously met in the Hôtel Mazarin, and afterwards in the Palais Royal; but the inconvenience to which they were subject, led, in 1808, to the formation of a plan for constructing an exchange sufficiently large for the multifarious business of so great a capital. The form of the *Bourse*, which stands in a spacious square at the E. end of the Rue St. Augustin, is a parallelogram 212 ft. in length by 196 ft. in width, surrounded by a peristyle of 66 Corinthian columns. The *Salle de la Bourse*, or great hall, on the ground-floor of the building, 116 ft. in length by 76 ft. in breadth, is surrounded by arcades of Doric architecture. A grand staircase leads to a spacious gallery supported by Doric columns, and to the hall of the Tribunal of Commerce. Corridors run round both the upper and lower hall, communicating with various rooms devoted to commercial purposes; and on the whole the arrangements are of the most complete description. The hours for transacting business are from one to five; but the galleries and corridors are open from nine till five. The *Banque de France*, erected by Mansard in 1720, possesses little architectural beauty. The present establishment was founded in 1803, and received the exclusive privilege for 40 years of issuing notes payable to bearer. Its capital consists of 90,000,000 francs, in shares of 1000 francs each. The notes issued are for 1000 and 500 francs. The customary rate of discount varies according to circumstances, but averages 4 per cent.: the bank, however, discounts no bills that have more than three months to run. It opens, also, *comptes courants* with all requiring them, and charges no commission, its only remuneration for such transactions arising out of the use of money placed in its hands. The government of the bank is vested in a council of 30 elected by the 200 largest proprietors; the governor and deputy-governor are appointed by the king. The institution is flourishing, and enjoys unlimited credit. The public establishment connected with wholesale trade are called *halles*, the principal of which is the *Halle au Blé*, or corn-market, a circular building, completed in 1767. The *Halle aux Vins*, on the Quai St. Bernard, S. of the Seine, near the Jardin du Roi, is an immense enclosure, having an area of 31,100 sq. yards, walled on three sides and fenced towards the quay by an iron railing about 850 yards in length. This great market is divided into streets called after the principal French wines; there are seven in piles of buildings, four in front and three behind, one of which is fire-proof, and used solely as a store-house for spirits. The warehouses and vaults will contain 400,000 casks. Wines entering this *dépôt* are not charged with the octroi till taken out for consumption; but they pay one franc per cask for warehouse room, &c. The hall is open from six to six in summer, and from seven to five in winter; and the counting-houses of some of the principal wine merchants are within the premises. The quantity of wine taken out of the hall in 1836 amounted to 932,402 hectolitres, and of spirits to 36,910 hects. The other wholesale markets are the *Halle aux Draps*, and the *Halle au Cuir*, the *Halle aux Veaux*, near the Quai de la Tournelle, being now exclusively used by the chiffonniers for the sale and exchange of rags, &c. The *Mont-de-Piété* of Paris is a government establishment, enjoying the exclusive privilege of lending money on moveable effects at the rate of 9 per cent. a year, or ¾ per cent. a month. The money which it lends is borrowed from government at the rate of 3 per cent.: and the whole is under the management of a board of commissioners.

Markets, &c.—Paris has 36 markets, the principal of which is that of St. Germain, opening on the Rue de Seine, and constructed from the designs of Blondel, which has served as a model for all the others since built. It is a parallelogram, 500 feet in length, by 480 feet in breadth. One of the most celebrated markets is the *Marché des Innocens*, in the centre of which, as already observed, is one of the noblest fountains in the capital: this, in fact, is the Covent Garden market of Paris, and has at least an equally fine and abundant show of fruit and vegetables; but connected with it, and in its immediate neighbourhood, are several other markets for fish, cheese, eggs, &c.; and indeed it may be termed a "quarter of markets." The *Marché du Vieux Linge* (old clothes market), built in 1809, partly on the site of the old *Temple* (the prison of Louis XVI. immediately prior to his execution), comprises four galleries containing 1888 stalls or shops, in which are exhibited for sale all kinds of old clothes, shoes, iron, tools, &c.; and is, on the whole, not unlike the Rag Fair or Monmouth-street of London. The other markets are, generally speaking, commodious, but they require no particular description. Paris has, also, five *abattoirs* (built in consequence of a decree of Napoleon, 9th Feb., 1810), where the animals necessary for the consumption of its inhabitants are killed. The abattoir of Montmartre is about 1074 feet long, and 364 feet broad, and that of Menil-montant is nearly as large; the other three (two of which are S. of the Seine) are inferior both in size and arrangement.

Internal Consumption.—All estimates respecting the consumption of provisions in a populous city must, of course, be extremely vague; but, perhaps, with regard to Paris, there is less uncertainty, owing to the octrois or duties levied on most articles coming across the barrier. It is very difficult to form any estimate of the consumption of bread, on account of the permitted exportation of wheat into the neighbouring districts, whenever the price outside the barriers exceeds that of the *Halle au Blé*, but the daily consumption of flour is supposed to amount to about 1700 sacks. Cattle, sheep, &c., chiefly come from Normandy, the isle of France, and Limousin. It appears, however, that, though, in 1826, the average consumption of each inhabitant was estimated at 63·4 kilogr., the allowance to each in 1836 amounted to only 59·4 kilogr., a result ascribable wholly, or almost wholly, to the increased price of butcher's meat, occasioned by the oppressive duties laid on foreign cattle when imported, and the inadequacy of the home supply. (*See* art. FRANCE.)

The following table, though not complete, gives a tolerable idea of the consumption of Paris in 1836.

Wines	922,363 hect.	Sausages,	
Spirits	36,441 —	hams, &c.	3,301,518 kilog.
Cider and		Cheese, dry	1,944,527 —
perry	18,123 —	Grapes	618,995 —
Vinegar	17,541 —	Sea-fish	4,771,363 fr. val.
Beer	111,811 —	Oysters	1,219,650 —
Oxen	73,230 head.	Fresh-water	
Cows	17,442 —	fish	541,745 —
Calves	77,563 —	Poultry and	
Sheep	378,476 —	game	8,387,276 —
Pigs	91,999 —	Butter	11,539,080 —
Pies and prepa-		Eggs	4,935,864 —
red meats	210,773 kilog.	Hay	7,842,393 bottes.
Meat, coarse	819,921 —	Straw	11,959,813 —
Offal	1,233,779 kilog.	Oats	1,003,945 hect.

Among other articles of consumption, the tobacco sold in Paris during the same year amounted to 708,793 kilog., which, it should be seen, is here one of the most costly articles of domestic expenditure, consists principally of wood; but large quantities of coal and charcoal are also made use of.

Industry, Commerce, and Trade.—Paris, besides being the political cap. of France, is one of the chief seats of the national industry and commerce. Many branches of industry are conducted on an extensive scale; the advantages resulting from the greater subdivision of employment, the greater command of scientific assistance, and of skilled workmen, being more than sufficient to countervail the higher wages and heavier expenses in other respects attending their prosecution in so great a city. Still, however, it is ludicrous to suppose that a city like Paris, without coal, and without the command of water-power, should ever be able to come into successful competition with such places as Manchester, Glasgow, or Birmingham. The articles produced in Paris are, in fact, chiefly those of vertû, jewellery, the fine arts, or those immediately ministering to the luxurious wants of a great capital. In these, however, a great increase has taken place during the last 10 years; and in 1836 the prefect of the department estimated the value of its manufactures exported to foreign countries at 80,000,000 francs, and of those sent into other parts of France at an equal sum. Two large manufacturing establishments belong to the government, and, like all similar establishments, are carried on at a heavy loss. One of these, the *Manufacture Royale des Gobelins*, so called from the place where it is carried on having originally belonged to a family of the name of Gobelin, who amassed great wealth as dyers; but the property having changed hands, it was converted into an establishment for weaving tapestry; and, becoming celebrated for the beauty of its products, was purchased by Colbert for Louis XIV., in 1662, since which it has been a government monopoly. The pieces of tapestry are most exquisitely executed, and the effect of some of them is scarcely inferior to that of the best oil paintings. The manufacture of a single piece frequently occupies three or four years, costing from 15,000 to 18,000 francs. Some splendid carpets are likewise manufactured in this establishment. Its products are chiefly destined for the royal palaces and public buildings, or for presents by the king to other monarchs: a few of the inferior pieces of tapestry are allowed to be sold, but the sale of carpets is forbidden. A drawing school is attached to the manufactory, and lectures are annually delivered by the most celebrated chemists on the chemical principles of dyeing. The royal manufactory of tobacco is conducted on a very extensive scale, in a handsome modern house on the Quai des Invalides. The quantity of tobacco purchased by the government, in 1838, amounted to 5,885,000 kilogs., besides 135,000,000 cigars; and about one fifth part of the entire produce of tobacco in France is manufactured in this establishment. The royal manufactory of Sèvres porcelain, though about 6 m. from Paris, and not in the department of the Seine, may be noticed in this place: it has been the property of the crown since 1759. Some of the articles furnished by this manufactory bring very high prices, and are esteemed alike for elegance of form, and the beauty and brilliancy of the paintings. An exhibition takes place in September, when there is an extensive show of foreign china and earthenware, as well as of specimens in different stages of progress. The *salo-dépôt*, in Paris, is in the Rue Rivoli. Among the other manufactures of the capital, jewellery, works in gold, silver, bronze, and steel, watch-making, the manufacture of chemical products, hats, carpets, artificial flowers, and the compounding of all kinds of bonbons and sweetmeats, furnish employment to a vast number of persons; besides which, Paris has 98 woollen factories, employing upward of 1900 hands, and producing broad-cloth, cashmere shawls, *schalis*, flannels, &c., one establishment for weaving silk shawls, and 24 factories for cotton or mixed goods of cotton and worsted. The manufacture of both silk and cotton stockings is likewise conducted on a most extensive scale. With respect to the wages of workmen in Paris, there has been a considerable fall since the revolution of 1830: stone-cutters, carpenters, masons, &c., receive from 3 to 4 francs a day, and a few of those engaged in the more delicate branches of handicraft may earn from 5 to 6 francs per the day of 12 hours. Those engaged in the cotton and woollen factories receive only about 1½ francs per diem, and work for 14 or 15 hours: and the wages of females, in whatever branch they are employed, scarcely exceeds that rate. Young women in shops receive their food, washing, and lodging, with wages varying from 160 to 400 francs a year. The expenses of living to these classes range between 17 and 28 sous a day, and lodging may cost about 100 francs a year. The Parisian workmen scarcely ever work on Sunday, and they are quite as fond as those of London of keeping holyday on Monday, devoting both these days to amusements in the guinguettes, theatres, &c. A great number of workmen belong to benefit societies, of which there are about 170; and the moral condition of the labouring classes has been greatly improved by the establishment of fire and life insurance companies, savings' banks, infant and primary schools, as well as by the abolition of lotteries and gambling-houses. In 1838, 26,697,968 francs were paid into the Paris Savings' bank, and 21,379,500 francs were taken out.

The export trade of Paris consists chiefly in the transmission to foreign countries of its different manufactures. The following is an official statement of the value of the exports from Paris from 1828 to 1837 inclusive:—

1828	66,972,467 fr.	1833 . . . 95,274,381 fr.
1829	64,737,731 -	1834 . . . 98,315,090 -
1830	64,231,106 -	1835 . . . 119,441,922 -
1831	66,758,574 -	1836 . . . 134,495,449 -
1832	66,911,055 -	1837 . . . 137,000,000 -

The retail trade of Paris is on a very extensive scale, and it is estimated that there are 6500 retailers of food and drink, that 5000 are employed in making and selling articles of dress or ornament, 3000 in building and furnishing houses, 850 in printing, publishing, and selling books, &c., 1570 in trades connected with the fine arts, 250 in trades connected with mechanics, optics, &c., and about 650 in keeping hotels, restaurants, cafés, &c.; but these statements are all very vague, and not to be relied on. The tradesmen's licences, issued in Paris, during 1838, amounted to 73,26, including hawkers; and the returns to government from this source were estimated at 8,321,254 francs.

Paris has also 42 banking firms, 198 stock and insurance brokers, 1000 physicians and surgeons, 250 apothecaries, and about 400 persons keeping seminaries and *pensions* for children of both sexes.

Seine, Bridges, Quays, and Navigation.—Paris has not like London, a deep broad river, navigable to the city by sea-borne vessels of large burden; but the Seine is, notwithstanding, a striking feature in Paris on account of its bridges and quays, as well as advantageous from its extensive boat-navigation. It enters Paris from the E.S.E. about 3¼ m. below its junction with the Marne, at Charenton, and in its course forms a slight curve northward, its whole length from the *Barrières de la Rapée* at the E. end, to the *Barrière de la Grenelle* at its W. extremity, being 8000 metres, or nearly 5 m., in which space it forms three islets, the smallest but highest up the stream being the *Isle Louvier*, used as a dépôt for wood-fuel, the isle of St. Louis, about 700 yards in length, and the isle du Palais, the site of the ancient Lutetia, about 5 furlongs in length, by 2 do. in breadth. The river is crossed by 24 bridges, of which five are on the suspension plan, three of iron and stone, one of wood, and the rest of stone. These structures, though usually on a level with the quays, and on the whole convenient, will not bear to be compared with the bridges of Sta. Trinità at Florence, or St. Angelo at Rome, much less with the noble bridges crossing the Thames. "The Parisians," says Mr. Woods, "boast of their bridges, but without great reason: the *Pont d'Austerlitz*, sometimes called *Pont du Jardin du Roi*, is fine for an iron bridge; the *Pont Neuf*, which crosses two branches of the Seine, and has 12 arches, has little pretension to beauty: the *Pont des Arts* is a light, not to say slight construction of iron, for foot passengers; the *Pont Royal* is a well constructed stone bridge of five arches, but hardly a handsome one; the *Pont de la Concorde* is a stone structure of five very ugly looking flat arches, and the *Pont de Jena* is a caricature of flattened elliptical arches, and apparent lightness, its entire merit being confined to some ingenuity in the construction in order to obtain this effect, which, nevertheless, is certainly a blemish." (*Letters of an Architect*, i., 22.) The islets in the river are connected with the N. and S. banks by 10 bridges, some of stone and others of wood, of inferior size and little beauty. The banks of the Seine are not blocked up, like those of the Thames, with coal-wharves, warehouses, and irregularly built houses, running close down to the water's edge, but have fine open quays, affording uninterrupted walks, extending on both sides the river from one end of the city to the other, Paris being in this respect greatly superior to London. Wharves and landing-places are formed in different parts, particularly towards the E. end of the city. *Dépôts*, for fire-wood are to be found along the river, and on all the outskirts of the town, and the boats along the wharves on both sides the isle du Palais furnish supplies of wood and charcoal. The navigation of the river is effected by large boats called *coches d'eau*, by barks, and within the last few years by steamers, the number of which is progressively increasing. From the higher parts of the river about 11,000 boats arrive every year with fruit, corn, and flour, hay, wine, paving-stones, bricks, &c., besides about 4000 barks laden with timber, charcoal, and fire-wood. Barges of from 40 to 60 tons burden come from Rouen with colonial produce, cotton goods, cider, salt, foreign corn, &c., and two steamboats ply daily between Rouen and Paris, and two others during summer between Paris and Montereau. The river does not fail, and do not average more than 700,000 francs a year, exclusive of the octroi.

Canals and Railways.—The canals N. of the Seine consist of two or three branches, connected with an undertaking for uniting the waters of the Seine and Ourcq, with the view not only of making an inland navigation, but also of supplying the capital with water of better quality than that of the Seine. The canal de l'Ourcq receives the waters of the Ourcq about 98 m. N.E. Paris, and after collecting several minor streams, falls into a basin in the N.E. suburb of the city, from which branch several canals almost encircling the N. side of Paris. These canals, however, are less used for navigation than for supplying the city with water. Several railroads are in course of construction from Paris; those to St. Germain and Versailles are already open, are much frequented, and will, apparently, be profitable undertakings. The principal projected railways are intended to connect Paris with Orleans, and with Rouen and Havre. Other railways will, in all probability, be formed within a few years; and there is reason to hope that the country may, at no distant period, possess some considerable portion of the facility and speed of communication which distinguish England and Belgium from the other countries of Europe.

Establishments for Education.—The university of Paris, which appears to have been established early in the 13th century, though some writers have traced its antiquity up to the time of Charlemagne, comprised, before the revolution of 1789, 10 great and 36 minor colleges, which had each faculties of divinity, law, physic, and arts. The professors appear to have been attached to colleges rather than to the university as an abstract body, and the number of students living in the colleges was very considerable. The income of the university arose out of a 28th part of the rent paid by the farmer-general of the royal posts and messageries, which, with the students' fees, made a large income; and the extensive buildings, still remaining, testify the importance once attached to these institutions, at the same time that the literary annals of France prove that, speaking generally, the various functionaries performed their duties pretty efficiently. At the revolution, however, the colleges were abolished, their estates confiscated, and the whole establishment was remodelled. Further alterations were made by Napoleon; but the present system dates from the restoration of the Bourbons. The university of Paris, as now constituted, is in fact the central establishment for education in France, and has under its direction all the faculties, colleges, and schools of the country: indeed, it may be considered as a mere government establishment, since the minister of public instruction, is *ex-officio* rector of the university, which has jurisdiction over the five royal and two private colleges of Paris, the royal college of Versailles, and all the *institutions* and *pensions* within certain limits. There are four faculties for higher degrees granted by the academic council; but no students can be admitted to them without having previously taken the degree of bachelor of letters. Professors are attached to each faculty, and deliver gratuitous courses of lectures, a certain attendance on which, as well as other exercises, is necessary for obtaining degrees. Candidates for the degree of bachelor of letters must produce certificates of having studied at least one year in one of the colleges of the university, in some *authorised institution*, or under the charge of a regular tutor. Examinations are publicly held four times a year under four professors appointed by the academic council: they include numerous oral questions on philosophy, literature, history, and the elements of mathematics, with papers for written composition and the translation of Greek and Latin authors. The examination of each candidate must last 1½ hours, but may be protracted at the pleasure of the professors. All the higher degrees are granted only after severe trials, and numerous candidates are annually rejected. The number of regularly-entered students in law is about 3200, and in medicine 2500. There are seven colleges in Paris, all of which have a certain number of attached bursaries, and receive boarders as well as day-pupils. The following table exhibits the numbers and classes of the students in the different colleges in 1840-41:

Colleges.	Bursars.	Free-board-ers.	Out-door Pupils.	Total.
Collège Louis le Grand	162	431	663	1176
de Henri IV.	165	438	325	920
de St. Louis	91	214	175	434
de Charlemagne			794	794
de Bourbon			980	980
Stanislas	36	144	44	255
Rollin	47	242		309
Total	245	1726	3181	5279

At the end of each academical year, in August, a grand distribution of prizes takes place in the public hall of the Sorbonne to the most deserving pupils of these colleges, and the degree of emulation thus excited among more than 5000 students is immense. The minister of public instruction presides at this ceremony, the professors deliver Latin orations, and the company comprises the most distinguished functionaries and literati of France.

Paris, besides its colleges, has 38 *institutions* and 93 *pensions* for boys, with 22 institutions and 48 pensions for girls. These are similar in all respects, except in size; the institutions being on a larger scale, and the course of study more general and more advanced in them than in the pensions. The conductors of these establishments must be at least graduates of letters, and are bound by law not only to follow a course of study prescribed by the university, but to send their pupils daily to one of the royal colleges in order to attend the professors' lectures. It is their duty, also, to assist them in their studies, and prepare them for the lessons to be gone through in the public class-rooms of the college. All the permanent students of the colleges wear a military-looking uniform, and are summoned to their duties by beat of drum. There are no colleges for girls, and their education varies according to the system followed in each particular establishment; the mistresses, however, of the different schools are obliged to pass an examination before persons authorised by the university. (See a valuable paper on the *Present State of superior Education in Paris*, in the *Statistical Journal of London*, iv., 50-66.)

Among the other establishments for education, the highest is the college of France, founded by Francis I. in 1530, and augmented at different periods. It consists of professors only, called *lecteurs du Roi*, among which are Biot, Thènard, Elie de Beaumont, Michelet, Lacroix, Jaubert, St. Hilaire, Chevalier, and other distinguished scholars and men of science. All their lectures are gratuitous, and open to every body. The museum of natural history, in the *Jardin des Plantes*, has likewise an attached corps of 13 professors, who deliver courses of lectures on different departments of natural history, chemistry, agriculture, &c. These lectures are, also, perfectly open and gratuitous, as are those delivered on Oriental literature at the *Bibliothèque du Roi*; on painting, sculpture, &c., at the royal school of fine arts; and on various branches of practical science at the *Conservatoire des Arts des Métiers*. One of the best of the educational institutions to which the revolution gave rise is the *Ecole Polytechnique*, established in 1794-95, for the promotion of mathematical and physical science and the graphic arts, and the preparation of pupils for the artillery, engineering, and mining departments. The school is under the control of the minister of war; but the details of management and instruction are left to a general council. The most distinguished masters in every branch of science are hired by government; and no students are admitted without having previously undergone an examination, to prove their competency in the classics and elementary mathematics. The pupils, of whom there are about 300, study two, sometimes three years; and no one can enter the higher departments of the military service without a certificate of attendance at this establishment. The present king has founded 24 scholarships, twelve of which are in the gift of the minister of war, eight of the minister of the interior, and four of the minister of *marine*. The establishment is supported by government; but the pupils pay an annual sum for board and lodging. The mathematical education at this institution is excellent, much superior indeed to what it is in most other institutions of the same kind. Another seminary of great importance, and closely connected with Paris, though not within the capital, is the *Maison Royale* of St. Denis, established by Napoleon, and furnishing an excellent education to between 600 and 700 young ladies, the daughters, sisters, and nieces of members of the legion of honour; of whom 400 receive their instruction gratuitously. This institution is in every respect admirably conducted, and might serve as a model for a large college of females; it has, also, two succursal houses, in which 400 pupils are gratuitously instructed. There are also several normal schools, with lectures, &c., for the purpose of forming teachers in the primary schools. The adult primary schools in the department of the Seine are attended by about 8500 pupils; and the primary schools for children were attended, in 1838, by 19,650 boys, and 11,350 girls; the expense of these establishments being estimated at more than half a million francs a year. There are at present, also, between twenty and thirty infant schools in Paris, supported by subscription, and attended by about 5000 children.

Literary Institutions.—Among the many chartered and private literary societies of Paris, the highest place is due to the Institute, unquestionably the first establishment of the kind in Europe. A decree of the convention, in 1793, annihilated the old *académies*, including among others, the *Académie des Inscriptions et des Belles Lettres*, the most celebrated academical institution that ever existed. In 1795, however, the *Institute* was formed, in the view of combining the literary and scientific academies into one body. In 1803, Napoleon divided the Institute into four classes; and in 1832, a fifth was formed of moral and political science. The titular members, of whom there are 217, receive pensions from government, ranging between 1200 and 1500 francs a year, besides whom there are 45 free academicians, 51 associates, and 219 corresponding members. The magnificent building in which this learned body meets is on the Quai Conti, near the Pont Neuf: it was built at the close of the 17th century, after the designs of Levau, at an expense of 2,000,000 francs, left by Cardinal Mazarin, who intended it should be a college for natives of four provinces then recently annexed to the crown of France. The principal room of the palace (formerly used as a church) is now appropriated to the sittings of the different classes, and fitted up with benches forming a semicircle facing the president's chair. The Mazarine library comprises 160,000 printed vols., with 4500 MSS.; and the more recently formed library of the Institute has upwards of 100,000 vols., chiefly scientific works, At one end of this library is Pigalle's celebrated marble statue of Voltaire. The interior is adorned with busts, bas-reliefs, &c.; and this palace is, on the whole, one of the finest public edifices in Paris. The *Bureau des Longitudes* is another important public body, formed in 1795, for the discovery of the best methods of ascertaining the longitude, and for the general improvement of navigation: its meetings are held at the Observatory (near the Barrière d'Arceuil,) a building well suited for astronomical observations, and furnished with every description of philosophical instruments, and a good library of scientific works. The bureau produces annually the celestial almanack, called the *Connoissance des Temps*; for the use of navigators; and another work of a more general character, the *Annuaire du Bureau des Longitudes*. The Royal Academy of Medicine, formed in 1776, and restored in 1820, is charged with making reports to government on all matters of public health: this establishment, as definitively organized in 1835, consists of 175 resident, and 25 country members, with 22 foreign associates. The following are among the principal learned societies of Paris supported by private subscription:—

* Soc. Royale des Antiquaires,
— Royale d'Agriculture.
* — Géologique de France.
* — de Géographie.
— Grammaticale.
* — Philomathique.
— Philotechnique.
* — de Statisque Univ.
— d'Horticulture.
* — Asiatique.

Société Entomologique de France.
—Soc. d'Historie Naturelle.
—Athénée Royale de Paris.
* Institut Historique.
—Athénée des Arts.
—Académie de l'Industrie Française.
* Soc. pour l'Encouragement de l'Industrie Nationale.

Libraries, Museums, and Picture Galleries.—By far the most celebrated library of Paris, and probably the largest and most valuable that anywhere exists, is the *Bibliothèque du Roi*, or royal library. This vast collection of books is deposited in the old Hôtel de Nevers, a long, inelegant-looking edifice in the Rue Richelieu. It was begun in the middle of the 16th century; and at the death of Louis XIV. it had 70,000 vols.: it was afterwards greatly increased by the addition of MSS. and printed books from the suppressed convents; and it is said to comprise at present about 700,000 books and pamphlets, 80,000 MSS., 100,000 medals, 1,500,000 engravings, and 300,000 maps or plans; though there is reason to think that these numbers are in some measure, considerably overrated. It is open from ten to three daily (except Sundays, and during a recess of six weeks in September and October), and every facility is given for literary research, except that the books may not be removed from the building. Among the libraries attached to public establishments, the following are the largest:—That of St. Geneviève, comprising 200,000 printed vols. and 30,000 MSS.; the two libraries of the Institute, consisting together of 360,000 vols. and 4500 MSS., that of the Arsenal 190,000 vols. and 6300 MSS.; that of the chamber of deputies (50,000 vols.); and that of the Bibliothèque de la Ville (50,000), all open to the public. Paris has also several valuable museums, belonging either to the government or the university. The museum of natural history, in the *Jardin des Plantes*, may be said to stand at the head of every institution of the kind, not only in France, but elsewhere; it is conducted at an annual expense of 300,000 francs, and gives employment to 160 persons. The botanic garden, exclusive of a large collection of exotic plants from all climates, comprises buildings fitted up as dens for carnivorous animals, with menageries for foreign birds and beasts, which are all provided with habitations suitable to their modes of life. This collection of living animals, perhaps the largest in Europe, has enriched the museum with many new species, and enabled zoologists to improve the study of comparative physiology. Large additions to the *ménagerie* have recently been made, and the collection is constantly on the increase. The amphitheatre of anatomy and museum of natural history occupy a large space at the end of the garden: on the first floor of the latter is a superb collection of reptiles and birds; the second floor is devoted to the reception of mammalia, birds, insects, mollusca, &c., arranged according to the system of Cuvier (who here pursued those studies that have gained him an imperishable fame); and on the third story is arranged a general herbal, comprising upwards of 50,000 species, besides special collections amassed by Tournefort, Jussieu, Humboldt, Decandolle, &c., the entire number of specimens amounting to nearly 400,000. Along the E. side of the garden runs a long gallery, in which is deposited a noble museum of geology and mineralogy, only recently formed, and even now second to none in Europe. The museum of comparative anatomy is kept in a detached building W. of the garden; and the number of preparations, which fill 15 rooms, considerably exceeds 15,000. The menagerie is open every day: and the museums may be seen by tickets twice, or three times a week. Paris has many

* Those marked with an asterisk publish memoirs and bulletins of their proceedings.

564

minor collections of anatomy, mineralogy, &c., attached to the school of medicine, and to its numerous literary societies; but none of them are sufficiently extensive to require description. The *Conservatoire des Arts et des Métiers*, in the Rue St. Martin, deserves notice, both on account of its great extent and the astonishing variety of instruments and machines, specimens of manufacture, models of patents, &c. deposited therein; in fact, this gallery of practical science is one of the most interesting exhibitions in Paris; and having been re-arranged and newly catalogued, may now be seen to great advantage. It is open during two days in the week; but the library is accessible seven or eight hours a day.

The great glory of Paris, however, as respects the fine arts, is the gallery of the Louvre, comprising a most extensive and valuable collection of pictures and statues. During the latter years of the reign of Napoleon, this gallery was the richest and most magnificent by far of any that has ever existed, having then to boast of the *chefs-d'œuvre* of Rome, Florence, and, in fact, of the greater part of continental Europe. But victory having deserted the eagles of Napoleon, these treasures were again restored to their former possessors, and the Louvre has no longer to glory in the Apollo Belvidere, the Venus de Medici, and other matchless productions. Still, however, the collection is a very noble one. Eighteen large halls, on the ground-floor, are filled with pieces of sculpture, including the choicest treasures of the Villa-Borghese, and many works that once embellished ancient Rome. Many of them are of great value, especially the *Diane à la Biche*, standing on a pedestal adorned with most exquisite bas-reliefs; a statue of Mars, supposed to have been modelled from a picture by Zeuxis, once in the temple of Concord, at Rome; the celebrated Fighting Gladiator, by Agasias of Ephesus; the Hermaphrodite of the Villa-Borghese, a statue of Jason, erroneously called Cincinnatus; and the magnificent group of Silenus and the infant Bacchus discovered in the garden of Sallust at Pompeii. Five other rooms in the basement story are devoted to the reception of works by modern sculptors, a large apartment was filled in 1830 with a collection of Egyptian antiquities, and a large gallery has since been opened, called the *Musée de la Marine*, comprising models and sections of vessels, plans of ports, and other naval curiosities. The great picture gallery, which is on the first floor, is approached by a grand staircase painted by native artists, and comprises a suit of nine apartments, the walls of which are lined by upwards of 1900 pictures by artists belonging to the French, Flemish and Dutch, Italian and Spanish schools. Among the pictures of the French school are 15 admirable landscapes by Claude, the best of which is the well known "Disembarkation of Cleopatra;" 16 compositions, chiefly scriptural, by N. Poussin, among which may be distinguished an "Assumption" and "Holy Family," and 17 beautifully coloured marine paintings by Vernet. Among the Flemish and Dutch pictures, which, like those of the French school, occupy three apartments, the most distinguished specimens are "Gerard Dow's Dropsical Woman;" several pictures by the Vandycks, particularly a sketch of the "Dead Saviour in the Virgin's arms;" 14 fine studies by Rembrandt, including his "Venus and Cupid;" and 38 pictures by Rubens, the principal of which are his "Flight into Egypt," and a composition known to connoisseurs as the *Vierge aux Anges*. The schools of Italy and Spain occupy the three remaining apartments, which, indeed, contain the gems of the gallery. The following have been specified as those more particularly entitled to notice:—Raphael's "*Belle Jardinière*," and "Holy Family;" his portrait for Francis I., who paid for it upwards of 45,000 francs; Leonardo da Vinci's portrait of "Monalisa;" Carrage's group of "Jupiter and Antiope;" Domenichino's "St. Cecilia;" Guido's "Infant Saviour on the Virgin's knee;" Guercino's "Repentance of St. Peter;" a magnificent battle-piece, and the "Witch of Endor," by Salvator Rosa; Murillo's "Infant Jesus;" and Espagnoletto's "Adoration of the Shepherds."

The Louvre is open every day except Mondays and on Sundays the concourse of visiters is particularly great. Owing to the want of room, the pictures of the ancient masters are removed from the grand gallery, or are covered over, from the 1st of February to June, to make way for the annual exhibition of the works of the modern artists. This arrangement has been much objected to, and is, perhaps, the most defective of any connected with the fine arts in Paris. It is to be hoped that it may speedily be obviated, by providing some other place for the modern exhibitions. The French metropolis comprises several valuable private collections, especially that of Marshal Soult, which has some fine specimens of the Spanish school: the gallery of the Palais Royal consists chiefly of works by modern artists.

Literature and Periodical Press.—Paris is the great centre from which emanate all the most important publications of France, and in which is congregated all the most distinguished French writers. The increase of publications appears to have been regularly progressive from 1817. And within the last 90 years a great number of highly important and valuable new works, especially in history, philosophy, and science, have issued from the French press, as well as several republications of old standard works. These large undertakings are mostly conducted at the expense and risk, not of one, but of several publishers, on a joint stock principle. The periodical press of Paris is well conducted, and has great influence. In 1836, 947 periodical works were published, exclusive of 66 political papers. The principal magazines are the Revue de Paris (which appears weekly), and the Revue des Deux Mondes, Revue du Nord, and Revue Britannique, published monthly.

The following table exhibits the circulation, in 1836, of the nine principal newspapers of Paris:—

Le Siècle	11,667
La Presse	9,700
La Constitutionel	5,833
Le Moniteur Parisien	5,300
La Gazette de France	5,006
La Quotidienne	3,333
Le National	3,333
Le Journal des Débats	3,166
Le Journal du Commerce	3,100

The sale, however, of the "Presse" and "Constitutionel," has considerably diminished since this return. The Moniteur Universel" is the official morning journal of the government. "Galignani's Messenger," a well conducted daily paper, in English, is extensively circulated in the principal cities and towns of continental Europe. The royal printing-office of Paris employs about 400 workmen. The censorship of the press, which was early introduced, and was exercised, though with considerable indulgence, down to the revolution of 1789, was finally abolished after the revolution of 1830.

Theatres and other Amusements.—Paris may be regarded as the dramatic capital of Europe. Every Parisian, even of the lowest class, esteems himself, more or less, a critic of the drama; and the fondness for this species of entertainments makes the 14 theatres be almost nightly crowded to excess: indeed, the receipts of the theatres have for some years been steadily on the increase, and amount at present nearly 9,000,000 francs a year, of which a tenth goes to the support of public hospitals and charities. The most fashionable spectacles of Paris are the Italian and French operas. The former of these (which has been held in the Salle Favart, since the destruction by fire of the old Salle Favart, but will soon be removed to the Théâtre de la Renaissance) strictly confined to the representation of Italian operas, and its administration for many years has secured an almost precedented amount of vocal and instrumental talent. The French opera house, belonging to the *Académie Royale de Musique*, is partly supported by the government; the operas are represented by the pupils of the academy, the dancers in the ballet are the first in Europe, and the stage mechanism is brought to matchless perfection. These theatres are supported chiefly by the higher classes, and as fashionable resorts may be considered analogous to the Italian opera in London. The other theatres, five or six of which are nightly crowded almost to suffocation, are supported by the middle classes, the small *rentiers*, and althy tradespeople. Vaudevilles and musical farces are the most popular entertainments; but among the lower classes frequenting the Porte St. Martin and Ambigu Comique, there is a marked predilection for the horrible, as depicted by Victor Hugo, Dumas, &c. The following is a list of the theatres now open in Paris, with the amount of accommodation in each:—

...démie Royale de usique	1,940	Théâtre du Palais Royal	930
...on (Ital. Opera)	1,800	Théâtre de la Porte St. Martin	1,800
...âtre Français	1,520	Théâtre Gaieté	1,800
— de l'Opéra ...mique	1,900	— des Folies Dramatiques	1,400
...âtre des Variétés	1,240	Théâtre de la Porte St. Antoine	1,930
— de la Renaissance	1,800	Gymnase Dramatique	1,300
...âtre de l'Ambigu ...mique	1,900	Cirque Olympique, (for horsemanship)	1,800

...e theatres, however, are by no means the only amusement of the Parisians; for they seem to be equally attached to their concerts, balls, and guinguettes, which abound in every part of the French metropolis. The concerts Musard in high estimation, and those of Strauss in the *Jardin* ..., near the Porte St. Martin, are almost equally celebrated. The *Bal masqué* of the opera deserves notice also, being the nearest approach made by the French to the almost insane revelries of an Italian carnival.

Hotels, Clubs, &c.—Paris abounds with excellent hotels, cafés, and restaurants; about a dozen of which may vie in respectability and amount of accommodation with similar establishments in the W. end of London. They are uniformly clean, and in many cases elegantly furnished. "Indeed," says Mr. Maclaren, "in the interior of these establishments, two peculiarities immediately strike the eye of a stranger—the profuse use of mirrors and marble. Many apartments are furnished with three or four mirrors of very large size; and you will see a restaurant panelled all round with mirrors, each 3 or 4 ft. broad. The tables in these places are almost invariably slabs of marble, and the tops of cabinets and even basin-stands are of the same substance. In the cafés here, also, there are no fires visible, as in the London coffee-houses; but the stoves are so managed as always to maintain an agreeable warmth throughout the apartments. (*Notes*, p. 14 and 90.) The *restaurants*, or dining-houses, are frequented by all classes of the inhabitants, female as well as male. In fact, however, it may be accounted for, whether it have originated in the greater advantageousness in an economical point of view, in the taste of the people for society, or whatever else, home, in the English sense of the word, has but few charms for the bulk of the Parisians, who may be said to live in public, dining in restaurants, spending their evenings on the Boulevards, or in the theatres or cafés, and, appropriately it must be allowed, ending their days in the public hospitals! At the principal restaurants (such as Véry's, Véfours, the Frères Provençaux, and the Rocher de Cancale) the bills of fare include hundreds of dishes, and the charges are necessarily high; but at many houses of great respectability dinners may be had for two or three francs, including half a bottle of *vin ordinaire*. Beaune and Pomard, however, are the wines commonly drunk by persons in good circumstances. The houses of the *traiteurs* are frequented almost exclusively by the tradespeople and lower classes; but they also supply dinners to people at their own houses at a fixed price for each dish. Clubs, similar to those of London, have been established within the last few years; the largest and most respectable is the Club Anglais, the habitual resort of the leading men in the fashionable and diplomatic circles. The Jockey Club is frequented by sporting men, and a still more heterogeneous assemblage may be found at the Cercle. It may be remarked, also, that these clubs have been much more numerously attended since the suppression of the salon and other *licensed* gambling-houses, which till very recently infested the metropolis of France.

Local Government.—Paris, with its environs, forms the small department of the Seine, of which the form is nearly circular, and the diameter about 15 m. At the head of it is a prefect, under whom are twelve mayors, one for each of the twelve divisions of the town, and two sub-prefects for the country quarter. As to the administration of justice, the courts of Paris are less comprehensive than those of London, their jurisdiction comprising only the capital and seven adjoining depts.: but in all other respects Paris is as much the common centre of public business for France, as London is for England. The court of cassation is the supreme court of appeal from all the tribunals of France, and the *Cour des Comptes* has authority to examine all the public accounts of the kingdom. The *Cour Royale* confines itself to the trial of criminals in the dep. of the Seine; besides which there is tribunal *de première instance*. Paris is likewise the permanent residence of the sovereign and royal family; the place of meeting for the legislature; the seat of all the ministerial bureaux, and of the public offices generally. It is the seat of an archbishop, and the head-quarters both of the royal guards and of the first of the twenty-two great military divisions of France. It has also a numerous corps of national guards, or volunteers, composed of twelve legions, comprising about 30,000 men. For mercantile purposes it has a chamber and several courts of commerce. Finally, it is the centre of almost all associations for public purposes, such as those for the promotion of national industry, for the management of prisons, for the diffusion of vaccine inoculation, &c. As already stated, Paris sends 12 members to the chamber of deputies, one for each arrondissement. Total number of electors, in 1838–39, 16,871.

Population, Health, &c.—In 1836, when the population amounted, excl. troops and foreigners, to 909,126, there were 28,949 births and 24,057 deaths, showing an increase during the year of 1-186th part of the gross population. Of the births, 14,645 were males and 12,043 females: the illegitimate being, in the same year, to the legitimate births as one to three, while the average throughout France is as one to 13. This gives but an indifferent view of the morality of the French capital; but it is favourable compared with the fact, that of the whole number of children born, 4792 were abandoned by their parents, and sent to the Foundling Hospital! We are glad, however, to have to state, that the admissions to this hospital have been of late years progressively diminishing; and that, in this respect, there has been a material improvement. Of the children produced in 1836, no fewer than 4772 were born in hospitals, and these, of course, were mostly abandoned. Of the deaths in 1836, 9060, or between ⅓ and ⅕ of the whole, took place in the public hospitals, 64 in prisons, and 289 were found dead, and deposited in the Morgue, or public dead-house, in the Isle de la Cité. The suicides in Paris, during the last 15 years, have averaged about 240 a year, or about one in 93 of the deaths during that period, a proportion about double that of London. The most densely peopled arrondissements are the 2d, 8th, and 12th; those most thinly peopled being the 4th and 9th. Of the entire population of the city it is supposed that nearly a half are working people, the rest being composed of tradesmen, professional men, and persons of independent property. There are about 80,000 servants, and nearly the same number of paupers; but the latter have been gradually on the decrease for some years.

Nearly 15,000 patients are constantly in the hospitals, and about 60,000 pass through them annually, of whom about 4-5ths are discharged cured. The foundlings may average about 20,000; and about the same number of aged and infirm persons are supported by public charity. The population of the prisons is very fluctuating, but may be taken at about 4000 at an average. The climate of Paris is not so variable as that of London, but the winters are sometimes very severe; snow does not lie long, fogs are not frequent, and on the whole the climate is favourable. There has been a great improvement in the health of the inhabitants since the revolution, though it be far short of the improvement that has taken place in London during the same period. As was to be expected, the least densely peopled arrondissements, and those occupied by the wealthier classes, are by far the healthiest.

Fortifications.—During the course of the present year (1841), it has been, after much opposition, determined to fortify Paris, by surrounding it on all sides with a continuous chain of fortifications. The sum of 140 millions francs (£5,600,000) has been voted for the execution of the works, which have been already commenced. We believe, however, that most military men are of opinion that the fortifications in question will not oppose any effectual resistance to an invading army; and that it is, in fact, impossible successfully to defend so great a city. It is true, the fortifications may be serviceable as a means of overawing the Parisians; and it has been suspected that this has been with many the real motive that made them approve of their construction.

History.—When Gaul was invaded by Cæsar, Paris, then called *Lutetia*, was the chief town of the Parisii, a Belgic tribe, and was afterwards included by Augustus in the province of *Lugdunensis quarta* or *Senonia*. It attained no importance, however, till the middle of the 4th century, when it took the name of *Parisii*, and became the see of an archbishop. It was the favourite residence of the Emperor Julian, who, in his *Misopogon*, terms it his φίλη Λευχετια; being taken by the Franks, under Clovis, in 494, it became the capital of the new kingdom. As late as the close of the 9th century, the walled part of Paris was still limited to the island of la Cité, though considerable suburbs were extending themselves along both banks of the Seine. It was greatly enlarged by Louis VI. and VII.; still more so by Philip Augustus; and after the battle of Poitiers, in 1356, new walls were raised on the N. side of the Seine. The treaty of Troyes, in 1420, gave Paris into the hands of the English, who held it till 1436, when it was recaptured by the French, and the English garrison put to the sword. The population of Paris, in the 15th century, is estimated by Dulaure at 150,000, and great architectural improvements had been gradually taking place; but the police was so bad, that both life and property were insecure, and morals were almost at their lowest possible ebb. The city was farther improved by Francis I., and the circuit of the walls was greatly enlarged by Charles IX. and Henry IV., under whose reigns the entire suburb of St. Germains was rebuilt. In the religious wars of the 16th century, Paris was the scene of a revolt against the troops of Henry III., known as "the day of the barricades." It was held by the Leaguers, from 1585 to 1594, when it surrendered to Henry IV. The palace of the Luxembourg and the Palais Royal were built in the reign of Louis XIII., and the walls were so extended as to include nearly the whole space within the present boulevards; but the police was still wretchedly inefficient, and disorders were of almost daily occurrence, particularly in the turbulent quarter of the university. In the 17th century, Paris was the principal scene of the tumults of the *Frondeurs*, supported by the inhabitants, against the French and Swiss guards; but notwithstanding these disturbances, the city still increased, churches, convents, and hospitals were built, the palace of the Tuileries was finished, the quays and boulevards were laid out, sewers formed, and other improvements effected at a great expense. The most

memorable scenes connected with the history of the French revolution, from the destruction of the Bastile, in 1789, to the assumption of imperial power by Napoleon, were enacted in the metropolis, which has long had a preponderating, though not always a beneficial influence, over the affairs of the kingdom. Under the government of Napoleon, Paris was greatly improved, and many of those scientific and other establishments were either formed or remodelled, which have contributed to increase its literary and scientific renown. The work of improvement proceeded slowly during the reign of Louis XVIII. and Charles X.; but since the revolution of 1830, which placed the Duke of Orleans on the throne, great activity has been evinced by the legislature in raising solid as well as splendid monuments, some of which, as the Madeleine, the Arc de l'Etoile, and the ministerial hotel on the Quai d'Orsay, may vie in magnificence with the finest European structures. At the same time, new pavements, bridges, sewers, markets, public gardens, and prisons, attest that no department of the metropolis is neglected by the government; and whenever gas is generally introduced, and water more generally diffused in private houses, Paris will be one of the most comfortable as well as handsomest and most luxurious capitals of Europe. (*Dulaure, Histoire de Paris; Paris and its Historical Scenes, i.; Metropolitan Magazine.* 1837; *Galignani's New Paris Guide: Planta's Picture of Paris; Dictionnaire de la Conversation et de la Lecture,* art. *Paris,* and *Private Information.*)

PARIS, p. t., capital of Oxford co., Me., 39 m. W. by S. Augusta, 593 W. Watered by Little Androscoggin river, which affords water-power. It has a courthouse, jail, 10 stores, one woollen factory, three grist-mills, three saw-mills, one paper-mill, four tanneries, one printing-office, one weekly newspaper; 170 schools, 1050 scholars. Pop. 2454.

PARIS, p. t., Oneida co., N.Y., 8 m. S. Utica, 96 m. W.N.W. Albany, 378 W. Drained by Sadaquada cr. It contains five churches, two Presbyterian, an Episcopal, Methodist, and Baptist, seven stores, two fulling-mills, three cotton factories with 4152 spindles, five grist-mills, 12 saw-mills, two paper-mills, two tanneries, two distilleries; one academy, 47 students; 15 schools, 864 scholars. Pop. 2944.

PARIS, p. v., capital of Henry co., Tenn., 96 m. W. Nashville, 784 W. Situated on Bally fork of Little Sandy river, and contains a courthouse, jail, two churches, an academy, 10 stores, various mechanic shops, and about 900 inhabitants.

PARIS, p. v., capital of Bourbon co., Ky., 36 m. E. Frankfort, 507 W. Situated on the S. fork of Licking river, and contains a courthouse, jail, two churches, 16 stores; three academies, 110 students, two schools; 98 scholars. Pop. 1197.

PARIS, t., Stark t. and co., O. Drained by Sandy creek. Pop. 2474.

PARIS, p. v., capital of Edgar co., Ill., 114 m. E. Springfield, 665 W. It contains a brick courthouse, nine stores, 60 dwellings, and about 350 inhabitants.

PARIS, p. t., Oswego co., N.Y., 22 m. E. Oswego, 16 m. W.N.W. Albany, 376 W. Drained by Salmon creek. It contains a church, two stores, one fulling-mill, one grist-mill, four saw-mills; 16 schools, 404 scholars. Pop. 1561.

PARISHVILLE, p. t., St. Lawrence co., N.Y., 16 m. E. Canton, 221 m. N.W. by N., Albany, 505 W. Watered by St. Regis, Racket, and Oswegatchie rivers. It contains two churches, a Congregational and Baptist, five stores, one furnace, two forge, one grist-mill, six saw-mills, one tannery, one distillery; 15 schools, 800 scholars. Pop. 2228.

PARKERSBURG, p. v., capital of Wood co., Va., 285 m. N.W. by W. Richmond, 383 W. Situated on the N. side of Little Kanawha river, at it junction with Ohio river. It contains a courthouse and 10 stores.

PARKE, county, Ia. Situated in the W. part of the state, and contains 450 sq. m. Drained by Big and Little Raccoon, and Sugar creeks, tributaries of Wabash river. It contained in 1840, 14,255 neat cattle, 18,858 sheep, 9,135 swine; and produced 107,188 bushels of wheat, 456 of rye, 949,850 of Indian corn, 137,140 of oats, 23,067 of potatoes, 15,150 pounds of tobacco, 195,576 of sugar. It had 29 stores, one furnace, two fulling-mills, nine flouring-mills, 26 grist-mills, 44 saw-mills, four oil-mills, 11 tanneries, 22 distilleries, one pottery, one printing-office, one weekly newspaper; two academies, 150 students; 54 schools, 1841 scholars. Pop. 13,449. Capital, Rockville.

PARKMAN, p. t., Piscataquis co., Me., 71 m. N. by E. Augusta, 666 W. Incorporated in 1822. It has three stores, one grist-mill, three saw-mills; 12 schools, 546 scholars. Pop. 1905.

PARKMAN, p. t., Geauga co., O., 162 N.E. Columbus, 327 W. Drained by head waters of Grand river. It has four saw-mills; one academy, 18 students, three schools. 258 scholars. Pop. 1181.

PARMA (DUCHY OF), and indep. state of N. Italy, between lat. 44° 20' and 45° 8' N., and long. 9° 20' and 10° 38

E (exc. the detached distr. Guastalla), having N. Austrian Italy, from which it is separated by the Po, W. the Sardinian dom. S. the latter and Tuscan Lunigiana, and E. Modena. Area estimated at 2268 sq. m. Pop., in 1833, 15,673. This duchy, lying between the Po on the N., and the Apennines on the S., is partly and principally included the great plain of Lombardy; but a large portion of its surface is covered with the ramifications and slope of the Apennines, the crest of the ridge, here about 4000 feet in height, forming its S. boundary. It slopes to the N., and all rivers fall into the Po. The soil, which is very fertile the plain, where it is watered by numerous canals, is dry and arid in the mountainous districts. It is principally cultivated by metayers: but while in the surrounding states such agreements are entered into between the metayer and his landlord as they may approve, in this duchy the law interferes to protect the tenant, who might be inclined to submit to pecuniarly onerous conditions. Thus, a landlord stipulates for more than half, but not for more than ⅔ of the produce. If the landlord furnish the cattle, the tenant has a right to, at least, 1-3d of the profit thence arising; and if the tenant furnish the cattle, no lease is binding that does leave him 9-3ds the produce. (Von Raumer, i., 307, 308.) Along the Po, the farther we advance E., the soil becomes deeper, richer, better watered, more fertile. The dairy is here the great object of attention, and the famous cheese which derives its name from this duchy, is still made to a great extent in the neighbourhood of Pavia, Lodi, &c. Parma are very small, and the dairy inferior to that of Modena. (See ITALY (Austrian &.)

corn of the numerous oaks, along the banks of the an immense number of hogs, which are generally sent to those of any other part of Italy; in consequence sent to Tuscany and the Papal states, whither, also to Genoa, great numbers of fat oxen are sent. p of the duchy are indifferent: the superior flocks in the Farnese Apennines in summer, belong to Tuscany, and other adjacent states. Both poultry are abundant. Maize, wheat, beans, tobacco, wine, are the principal products of the plain country: only raw materials manufactured are silk and according to Serristori, the produce of the former is to 100,000 lbs. a year. Rice is grown in the Guastalla, and near Parma. In the mountains but is grown; and the climate is too cold for the trout and skimmed milk, with cheese made of goats and ewes, form the principal food station. Potatoes were not introduced into the districts when this duchy was visited by Châteauvieux; inhabitants of these districts make a good deal but their principal revenue is derived from cows: for all the active inhabitants quit their favourable season, to work in Lombardy and to e money they gain and save from this source the capital circulating among them. (Châteauvieux and its Agricul. Trans., p. 55.)

vitriol, and petroleum, are found in the t the principal mineral product is salt, of 300 quintals are made annually. Manufactures such as are domestic, are of trifling importances are the principal, and are made in all the There are several iron forges; and hats, are, paper, and gunpowder are made in, and San Donnino. The value of the imexist mostly of woollen and linen cloths. d articles of luxury, is estimated at 748,000 value of the exports, including 8600 head 50 hogs, is estimated at only 166,500 lire a Statist.) But it is hardly necessary to in fact, be no such discrepancy between the one statement or the other, or percompletely erroneous. Except the Po, able river, nor is there any canal, except in this duchy.

is an unlimited monarchy, hereditary in t present vested in a female, this duchy d, by the treaty of Vienna, to the Archus, ex-empress of France. At her death, it, under the existing arrangements, to or his heirs. The administration is conof six ministers. The duchy is divided ief towns, Parma, the capital, Piacenza, Borgo Taro, and Guastalla. There are arbitration, consisting of a single judge, -y jurisdiction and appeal in Parma and trials take place in public, but without a jury; the judges composing the court of votes. The code of Parma, promulbottomed upon the Code Napoleon, but uliar to itself. Secret societies are prolems for definite objects, if consisting of

more than 20 members, require to be authorized by government. All games of chance are prohibited, under penalty of fine and imprisonment. Public provision is made for the poor; and beggars are either punished with imprisonment, or sent to a workhouse. If death ensue from a duel, the challenger is liable to imprisonment for from 10 to 20 years, and the challenged from 8 to 10. (Von Raumer, i., 314–316.) The Roman Catholic is the state religion; but others are tolerated. Public instruction is furnished by about 380 elementary schools, attended by nearly 10,000 pupils; there are also secondary schools in most of the towns; and superior academies in Parma and Piacenza. The military force consists of about 1300 men, chiefly infantry. The public revenue, derived from state property, territorial and personal taxes, patents, communal and river dues, custom duties, octrois, in Parma and Piacenza, amounted, in 1834, to 6,895,840 lire. The public debt, in 1833, amounted to 10,700,000 do. (Serristori, Statist. d'Italia.)

This territory anciently formed part of Cisalpine Gaul. Charlemagne gave it to the Holy See; but during the quarrels between the popes and the emperors, Parma and Piacenza became independent republics. They afterwards fell successively under the dukes of Milan, the popes, the Farnese family, and the Spaniards. By far the most celebrated of its native sovereigns was the famous Alexander Farnese. Though he served one of the most odious tyrants, Philip II., Farnese was alike generous and brave, and was certainly the most skilful and accomplished general of his age. In 1801, Parma was ceded to the French; and in 1805, it was principally included in the dep. Taro, belonging to the kingdom of Italy. (Serristori, Statistica d'Italia; Rampoldi; Châteauvieux; Raumer, &c.)

PARMA, a city of N. Italy, the cap. of the above duchy, on the little river of its own name, a tributary of the Po, here crossed by three bridges, 35 m. S.E. Piacenza, and 33 m. S.W. Mantua: lat. 44° 48' 1" N., long. 10° 20' 43" E. Pop. circa 36,000. The walls are between 3 m. and 4 m. in circumference; it is surrounded by a glacis which forms a favourite public promenade. It is well built and laid out, especially its principal thoroughfare, which forms a part of the Flaminian way (via Flaminia). It has many good public and private edifices; they are mostly, however, of brick, and none is remarkable for beauty: many of them are, also, in a decaying condition, and its streets are dull and dreary. The Farnese palace, though an immense pile, is little more than half the original design; it is raised on open arcades; and, though partly in a ruinous state, it serves for the residence of the archduchess, and accommodates the academy of arts. In the new picture gallery belonging to the latter are several masterpieces by Correggio, Parmegiano, Raphael, the Carracci, &c. The academy has also a museum, in which are many interesting antiquities from the buried city of Velleia (18 m. S. Piacenza), and an extensive and well arranged library. Attached to the palace is the large Farnese theatre, designed by Vignola on the model of the ancient theatres, 300 feet in length, and capable of accommodating some thousand persons. It is built entirely of wood, and is well constructed for hearing; but it has been long disused, and is said to be in a very dilapidated and ruinous state. Another, but smaller theatre, exists on the same floor; and a third, built by the present archduchess, was finished in 1830.

The cathedral, an edifice of the 11th century, though in a mixed and semi-barbarous style, is, on the whole, a magnificent building: its fine dome is ornamented with one of the last and most celebrated productions of Correggio. The city has a great number of other churches, several of which possess some fine works of art. It is a bishop's see, and a seat of the high court of revision for the duchy, besides several inferior courts; and was, till 1809, the seat of a university; it has now a superior school of divinity, medicine, and philosophy, attended by about 420 students; a college of nobles, founded in 1600; an episcopal seminary, some inferior schools, and several hospitals and other charities. The principal and most extensive establishment in the city is the famous printing-office of the Bodoni, established in 1765: it has produced some of the best specimens of typography, especially some of the most splendid editions of the classics of which modern Europe has to boast. The city is supplied with water by a conduit, said to be 50 m. (?) in length. The Palazzo Giardino, and a large public cemetery, are situated without the walls. Parma has some silk and other fabrics; but its manufactures and commerce are comparatively insignificant.

Parma became a Roman colony A.U.C. 569. It suffered greatly in the civil war between Antony and Augustus; and was colonized anew by the latter. from whom it received the name of Julia Augusta Colonia. It was anciently much celebrated for its wool.

Velleiabus prima Apulia, Parma secundis
habalos.   Martial, xiv. Ep. 85.

(Crador's Italy, ii., 26–39; Rampoldi; Dict. Géog., &c.)

PARMA, p. t., Monroe co., N. Y., 10 m. N.W. Rochester, 231 m. W. by N. Albany, 380 W. Bounded N. by lake Ontario. Drained by Salmon and Little Salmon creeks. It contains two churches, three stores, two fulling-mills, two grist-mills, five saw-mills, two tanneries; 17 schools, 914 scholars. Pop. 9652.

PARNASSUS, a mountain-chain of Independent Greece, prov. E. Hellas, and nom. Phocis, famous in Grecian poetry and mythology, the favourite resort of Apollo and the muses, and especially sacred to Bacchus. It runs from W.N.W. to E.S.E., forming the connecting link between mount Pindus and mount Helicon; but the only part of it that requires any particular mention is its culminating point N. of Delphi, now called Liakura, lat. 38° 31' 57" N.; long. 22° 38' 36" E. According to M. Peytier, it rises 8068 ft. above the sea, and being covered with snow during the greater part of the year, would have been rather an uncomfortable residence for the muses, who inhabited its lower regions, especially the laurel groves in the vicinity of the Castalian fountain. Dr. Clarke, who ascended this celebrated mountain, describes its summit as somewhat resembling that of Cader Idres in N. Wales; and adds that "after having been for years engaged in visiting the tops of mountains, he must still confess that he never saw any thing to compare with the view from the summit of Parnassus. The gulf of Corinth had long looked like an ordinary lake, and it was now reduced to a pond. Northwards, beyond all the plains of Thessaly, appeared Olympus, with its many tops, clad in shining snow, and expanding its vast breadth distinctly to view. The other mountains of Greece, like the surface of the ocean in a rolling calm, rose in vast heaps according to their different altitudes; but the eye ranged over every one of them. Helicon was one of these; and it is certainly inferior in height to Parnassus. One of the principal mountains in the Morea, now called Tricala, not far from Patras, made a great figure in that mountainous territory; it was covered with snow, even the lower ridges not being destitute of it. We looked down on Achaia, Argolis, Elis, and Arcadia, as upon a model. The higher region of the mountain is of limestone, bleak and destitute of all herbage, except a few alpine plants." (Travels, vii., 261.) From the sacred town and temple of Delphi the mountain appears to have two summits, one of which was sacred to Phœbus and one to Bacchus—

" —— Parnassus gemino petit æthera colle,
Mons Phœbo Bromioque sacer." Lucan, v. 72.

Running down the cleft between these two summits is the famous Fons Castalius, the genuine source of poetical inspiration. It is thus alluded to by Virgil, in connection with the neighbouring mountain—

" Sed me Parnassi deserta per ardua dulcis
Raptat amor. Juvat ire jugs, qua nulla priorum
Castaliam molli devertitur orbita clivo." Georg. iii., 293.

Even at present it is by no means unworthy its ancient renown. Mr. Dodwell says, "it is clear, and forms an excellent beverage. The fountain is ornamented with pendent ivy, and overshadowed by a large fig-tree." Higher up the mountain is the Corycian cave, which, during the Persian war, afforded a safe retreat to the less adventurous Greeks after the battle of Thermopylæ. (Herod. viii., 36.) It is described by Mr. Raikes, the first modern traveller who has visited it, as a chamber 330 ft, in length, and nearly 200 ft. in breadth, with a roof studded with stalactites. Above this cave, and at a distance of about 80 stadia from Delphi, stood the town of Tithorea, taken and burnt by the army of Xerxes at the close of the Persian war. The ruins were found by Dr. Clarke, near the modern village of Velitza. For further particulars see DELPHI.

PAROS, a famous island of the Grecian archipelago, group of the central Cyclades, to the W. of Naxos, from which it is separated by a strait, 5 m. across; mount St. Elias, the most elevated point in the island, being in lat. 47° 2' 46" N.; long. 25° 11' 25" E. Pop. 8000 ? It is oval-shaped, being about 12 m. in length by 8 in breadth. Though rugged and uneven, it is, speaking generally, extremely fertile; and, if well cultivated, would support four or five times its present population. It produces considerable quantities of cotton, with corn, wine, oil, &c. Tournefort says that the butchers' meat is good, that there are a great number of hogs, and that pigeons and partridges are extremely abundant. (i., 303, 4to. ed.) Port Naussa, on the N.E. coast of the island, is one of the best harbours in the archipelago, and was used, in 1770, as the rendezvous of the Russian fleet. Parecchia, on the site of the ancient city of Paros, on the W. coast of the island, is the capital. Its harbour is open to the W., and there are some sunken rocks in its vicinity, on one of which the Superbe, a French line-of-battle ship, was lost, in 1833. The present town consists of mean houses, which, however, are interesting, from their chiefly consisting of fragments of the old city, including portions of the shafts and
538

capitals of columns, &c.: the cathedral church is said to be the best in the archipelago.

Paros was famous in antiquity for its beautiful snow-white marble, whence Virgil has called the island niveus Paros. (Æneid. iii., v. 126.) The finest of the ancient statues, including the Venus de Medici, the Apollo Belvidere, and the Antinous, were formed out of this material. Indeed, the best sculptors used no other, comes antem tantum candido marmore usi sunt e Paro insula. (Plin. Hist. Nat., lib. xxxvi., cap. 5.) The quarries were situated about 4 m. from the city of Paros, and remain exactly in the state in which they were left by the ancients. Dr. Clarke says they had been wrought with infinite skill; and that the blocks had been cut out with such precision that there was not the smallest waste. (vi., 138, 8vo. ed.)

According to Thucydides (lib. i.), Paros was originally settled by Phœnicians. It early attained to great wealth and consideration, and established colonies in Thasos and other islands. During the first Persian war, it sided with the Persians; and, after the defeat of the latter at Marathon, the city of Paros was unsuccessfully besieged by Miltiades. Themistocles, however, rendered it tributary to Athens. It produced several distinguished individuals, among whom may be specified Archilochus, the inventor of Iambics.*

In modern times, the only event of importance connected with the history of Paros is the discovery of the " Parian, or Arundelian Chronicle." This is a chronological account, cut in marble, of the principal events in the history of Greece during the period of 1318 years, beginning with Cecrops, and ending with the archonship of Diognetus, anno 264 B.C. The chronicle for the last 90 years is, however, obliterated; and the inscription is in many parts a good deal defaced.

The marble slab on which this chronicle is cut was purchased on the spot, in 1624, for the earl of Arundel, whence it is now frequently called the Arundelian chronicle; and being brought to England in 1627 the inscription was soon after copied, translated, and published by Selden and other eminent scholars. Unfortunately the marble afterwards met with the most barbarous treatment, having been broken, and a part of it employed, as is alleged, to repair a chimney in Arundel house. The portion that escaped this worse than Gothic usage was presented in 1667 to the university of Oxford, of which it is one of the most precious relics. (Robertson on the Parian Chronicle, p. 43–48.)

For a lengthened period the Parian chronicle was regarded as of unquestionable authority, and was referred to as such by all inquirers into ancient history. In 1788, however, its authenticity was assailed, in a singularly clear, able, and ingenious dissertation, by the Rev. John Robertson, who contended that it was altogether spurious, and had been fabricated in modern times. As was to be expected, this dissertation elicited various answers, by Mr. Hewlett; Porson; the celebrated Greek scholar; Gough, the antiquary, &c.; and at present it seems to be generally concluded by the ablest critics and scholars, that the objections of Robertson have been satisfactorily disposed of, and that there is no good or sufficient reason for doubting that the Parian Chronicle was really compiled about 264 years B.C.

PARSONSFIELD, p. t., York co., Me., 86 m. W.S.W. Augusta, 536 W. Bounded N. by Great Ossipee river. It has seven stores, one fulling-mill, four grist-mills, four saw-mills, one oil mill, four tanneries; one academy, 28 students; 20 schools, 473 scholars. Pop. 2442.

PARSONSTOWN, or BIRR, an inland town of Ireland, King's co., on the confines of Tipperary, on the Birr, a branch of the Lesser Brosna, 7½ m. above its confluence with the Shannon, and 69 m. W.S.W. Dublin. Pop. in 1821, 5408; in 1831, 6925. It has a large square, in which is a pillar surmounted by a statue of the duke of Cumberland, erected in commemoration of the victory of Culloden, in 1746, and some good streets. Its public buildings are the parish church, a fine Roman Catholic chapel, the cathedral of the see of Killaloe, three meeting-houses for Independents, one for Quakers, and two for Methodists, a fever hospital, a dispensary, a courthouse, and a bridewell. Near it are large barracks. It has various schools, and is the seat of a manor court, general sessions and petty sessions: it is also a constabulary station, and has two distilleries, a brewery, and an extensive retail trade. The river is navigable for 2 m. from the Shannon, for barges. Markets on Saturdays; fairs, Feb. 11, May 5, Aug. 23, and Dec. 10. Postoffice revenue in 1830, £967; in 1836, £1267. Branches of the provincial and agricultural banks were opened in 1833 and 1836.

---

* The Dictionnaire Geographique says that the famous sculptors Phidias and Praxiteles were natives of Paros. In point of fact, however, the former was born at Athens, and the birthplace of the latter is unknown.

Close to the town is Birr castle, the seat of the earl of Rosse, the head of the noble family of Parsons, whence the town has its name; and to whom it is greatly indebted. The castle, which is of considerable antiquity, has been completely modernised, and greatly improved by its present proprietor.

PASCAGOULA, river, Miss. is formed by the union of Leaf and Chickasawhay rivers, and enters Pascagoula sound of the gulf of Mexico. It is navigable 50 m. for vessels drawing 6 ft. of water, and 150 m. for boats. It is 200 m. long.

PASCAGOULA, sound, Miss. and Ala. Situated off the mouth of Pascagoula river, separated from the gulf of Mexico by low islands, and communicates on the E. with Mobile bay. It is 55 m. long, with an average breadth of 8 m.

PASCO, or CERRO DI PASCO, the principal mining town of Peru, dep. Junin, prov. Huanaco, in an irregular hollow on the table land of Bombon, nearly 14,000 ft. above the level of the sea, and 60 m. S. by W. Huanaco. Population varying at different seasons from 4000 to perhaps 12,000. It is a mean, wretched place, which, previously to the establishment of the Peruvian Mining Company, in 1825, had not a house with a chimney, fire-place, or glazed window; and even now its dwellings are principally covered with thatch, a frequent cause of destructive fires. The town—of which the very adobes, or unburned bricks, partly used in some of the houses, contain silver—is so burrowed under, that a person is in no small danger of inadvertently falling into old mines, or rather pits, sometimes superficial, sometimes deep and fathomless, and half filled with water. There are several hundred well-known mines, from which silver has been and still could be extracted in large quantities, provided a perfect drainage were effected. But during the revolution a great many of the mines were allowed to fill up with water, and only about 36 are now wrought for eight months a year. From 1825 to 1836 inclusive, 2,180,555 marcs of silver were reduced to bars in the foundry at Cerro Pasco; the produce in the latter year having been 237,849 marcs. These mines have the advantage of being near a coal mine, which has of late years been opened; but turf, dung, timber, &c., are the kinds of fuel most commonly used. The miners choose whether they will be paid in money or a proportion of the ore. In the former case they get four reals, or 2s. a day; but they prefer of course payment in ore, if the mine be productive; and sometimes realise, in this way, very high wages. But the gambling nature of the pursuit has the worst effect on all parties engaged in it. The miners are almost universally prodigate, and involved in debt; and but few of the undertakers have made fortunes. (Smith's Peru as it is, i. 1–20.)

PAS-DE-CALAIS, a dep. of France, reg. N., formerly comprised in the provs. of Artois and Picardy, between the 50th and 51st degs. N. lat., and 1° 35′ and 3° 10′ E. long.; having N.E. and E. the dep. Du Nord, S. Somme, W. the English channel, and N. the strait of Dover, or Pas-de-Calais, whence its name. Area, 655,645 hectares. Pop. is 538, 655,515. There are several chains of hills, but none of any considerable height. The Scarp, Lys, and Aa, rise in this department; besides which the principal rivers are the Lisane, Canche, and Authie, having mostly a N.W. course. Except about Boulogne, the coasts are generally low, and in some places bordered with sandy downs, which are, however, prevented from increasing to an inconvenient extent by being carefully planted. The soil is, for the most part, good; and agriculture is on the whole well conducted. Of the surface, in 1834, 499,374 hectares are supposed to be arable; 46,219 in pasture; 21,852 in orchards, gardens, &c.; 43,167 in woods, and 18,845 in paths and wastes. Near Boulogne, farms vary in size from 85 to 250 acres; but, in general, they do not exceed 50 acres. Few are cultivated by the proprietors, being usually let to farmers who pay a money rent, and are also charged with the payment of the land tax. All kinds of corn, but principally wheat and maslin, and large quantities of beans, peas, and oleaginous seeds, are raised. About 1,522,000 hectolitres of potatoes were grown in 1835; and a good deal of land is devoted to the growth of beet-root. The annual produce of beer is estimated at 1,000 hectolitres; of cider, at 36,000 hectolitres; and of it spirits, at 11,000 hectolitres. In 1830, there were estimated to be nearly 200,000 oxen and cows, and 300,000 sheep in the department; the produce of wool averages out 662,000 kilogr. a year. The farmers, though not sperous, are contented with their condition; and there few paupers requiring permanent relief. In 1835, of 0.002 estates subject to the contributions foncières, 101,918 are assessed at less than 5 fr., and 39,402 at from 5 to 10. Some coal is met with; but the greater part of that in use of in the department is brought from Belgium; wood and turf are the principal species of fuel. About

8,964,000 kilogr. of beet-root sugar were made in this department in 1836, a greater quantity than in any other French department, that of the north excepted. Arras is famous for lace and gingerbread. A portion of the population of Boulogne and Calais is occupied in the manufacture of tulles; in the arrondissement of Bethune many hundreds are employed in making linen stuffs and yarn; and manufactures of cotton stuffs and yarn are pretty general. Spirits, leather, gunpowder, soap, glass and earthenware, tobacco pipes, &c., are also produced. Artesian wells (so named from the province Artois) originated in this department. The Pas-de-Calais is divided into six arrondissements; chief towns, Arras, the capital; Bethune, Boulogne, Montreus, St. Omer, and St. Pol. Calais and Boulogne are the principal sea-ports, and have a considerable coasting trade, and share in the cod, herring, and mackerel fisheries. The department sends eight members to the chamber of deputies; number of electors in 1838–39, 4512. Total public revenue (1831), 18,813,372 fr. (Hugo, art. Pas-de-Calais; Dict. Géogr.; French Official Tables; and Parl. Reports on Agriculture, 1834.)

PASQUOTANK, river, N. C., rises in the S. part of the Dismal Swamp, and enters Albemarle sound by a broad estuary, including which it is 40 m. long. It admits ordinary coasting vessels to Elizabeth city. It derives its chief importance from the Dismal Swamp canal, which extends from Deep creek, a tributary of Chesapeake bay, 23 m. to Joice's creek, a branch of Pasquotank river. This canal is 45 ft. wide, and at intervals of a quarter of a mile, 96 ft. wide for turnout stations, 6½ ft. deep, with six locks 100 by 20 ft. It has a navigable feeder of 5 m. from lake Drummond, in the Dismal Swamp; and it forms an important channel of communication from Norfolk to the south.

PASQUOTANK, county, N. C. Situated in the N.E. part of the state, and contains 300 sq. m. Bounded N.E. by Pasquotank river, S.W. by Little river. It contained 1840, 5719 neat cattle, 3415 sheep, 15,605 swine; and produced 19,130 bushels of wheat, 473,970 of Indian corn, 45,101 of oats, 21,065 of potatoes. It had 96 stores, 10 grist-mills, two saw-mills, one tannery, one printing-office, one weekly newspaper; two academies, 78 students; five schools, 109 scholars. Pop.: whites, 4650; slaves, 2788; free coloured, 1076; total, 8514. Capital, Elizabeth city.

PASSAIC river, N. J., rises in Mendham, Morris co., and flows through a mountainous country, with a quiet and sluggish course, excepting at two falls. At the Little falls it descends by two leaps and a rocky rapid 31 ft. in the distance of a half a mile. Five and a half miles below are the Great falls at Paterson. Here the river pours itself in one unbroken cataract of 60 ft. wide, 50 ft. descent, and a total fall of 70 ft., affording an immense water-power. It finally enters Newark bay, and is navigable 2 m. to Acquackanonck. It flows through marshes near its mouth, and is about 70 m. long, and is the longest river which runs wholly in the state.

PASSAIC, county, N. J. Situated in the N.E. part of the state, and contains 180 sq. m. Watered by Passaic river and its tributaries. It contained in 1840, 5532 neat cattle, 5075 sheep, 4555 swine; and produced 5190 bushels of wheat, 36,721 of rye, 70,874 of Indian corn, 24,012 of buckwheat, 45,945 of oats, 78,886 of potatoes. It had 126 stores, four lumber-yards, one furnace, 18 forges, four fulling-mills, one woollen factory, 20 cotton factories with 40,056 spindles, two dyeing and printing-works, 13 grist-mills, 21 saw-mills, four paper-mills, seven tanneries, two distilleries, one brewery, two printing-offices, two weekly newspapers; three academies, 198 students; 41 schools, 1987 scholars. Pop. 16,734. Capital, Paterson.

PASSAMAQUODDY, bay, lies partly in the state of Maine, and partly in the British province of New-Brunswick, and is 6 m. wide, and 19 m. long. It has a sufficient depth of water for the largest vessels, and is never closed by ice. It abounds with cod, mackerel, herring, and other fish. The boundary of the United States passes through it on its W. side, into St. Croix river, which enters its N.W. part.

PASSAU (an. Castra Batava), a fortified frontier city, principally belonging to Bavaria, circ. Lower Bavaria, of which it is the cap., on the Danube, where it is joined by the Inn, and also by the small river Ilz, 68 m. E.S.E. Ratisbon: lat. 48° 34′ 28″ N., long. 13° 28′ 5″ E. Pop., circa 9000. It consists of the city proper, built in the angle between the Danube and Inn, and of three smaller portions beyond the Danube, the Inn, and the Ilz, the latter being within the Austrian dominions. These different parts are connected by bridges, and surrounded with fortifications; and are further defended by two citadels, and some inferior forts; this being, in fact, one of the most important fortresses in the line of the Danube. The defile, in which the town is situated, is highly picturesque; and it has a striking appearance from the river, though not generally well built. The cathedral, however, is a magnificent modern edifice, in the Italian style, and several of the other churches are handsome : the old Jesuit's college, now a lyceum, the bishop's palace, several hospitals. an orphan asylum, and the postoffice are the other principal public buildings. On a hill adjacent to the Instadt, is the shrine of *Maria-hilf*, a celebrated place of Roman Catholic pilgrimage. Passau is the seat of circle, police and taxation boards, and has an Episcopal seminary, a school of industry, manufactures of leather, tobacco. and pottery-ware, docks for ship building, and an active trade both up and .down the Danube. It was long the capital of an ecclesiastical principality, secularised in 1805. Here, in 1552, a treaty was concluded between Maurice, elector of Saxony, on the one hand, and Ferdinand, king of the Romans, on the part of the emperor Charles V., on the other, by which the latter agreed to set the landgrave of Hesse at liberty, and to allow the Protestants full freedom of conscience. (*Berghaus; Sketches of Ger.*, ii., 101, 102, &c.)

PASSYUNK, t., Philadelphia co., Pa. It includes the W. part of the tongue of land included between the Delaware and Schuylkill rivers, below the city of Philadelphia. It contains League Island in the Delaware, and is covered with meadows and gardens. Pop. 1504.

PATAGONIA, an extensive country of S. America, comprising nearly the whole of but one continent S. of lat. 38° S., and having N. the territories of La Plata and Chili, S. the strait of Magellan, separating it from Terra del Fuego, E. the Atlantic, and W. the Pacific. Little is known respecting this region beyond its coast outline. The Andes in Patagonia appear to consist of but one cordillera, the mean height of which may be estimated at 3000 ft. ; but opposite Chiloe there are some mountains probably from 5000 to 6000 ft. in height. (*Geog. Journ.*, i., 157.) The W. coast is abrupt, very much broken, and skirted with a great number of irregularly shaped rocky islands. The E. coast has been most explored. The surface of the country appears to rise from the Atlantic to the Andes, in a succession of terraces, all of which are alike arid and sterile, the upper soil consisting chiefly of marine gravelly deposites, covered with coarse wiry grass. No wood is seen larger than a small thorny shrub, fit only for the purpose of fuel, except on the banks of a few of the rivers subject to inundation. where herbage and some trees are occasionally found. This sterility prevails throughout the whole plain country of Patagonia, the complete similarity of which, in almost every part, is one of its most striking characteristics. It is stated, however, by the Indians on the Rio Negro, which forms the N. boundary of Patagonia, that near the Andes, wheat, maize, beans, lentils, pease, &c., are raised. This latter region is not, however, placed under the same circumstances as the country more to the eastward, nor is it subject to the causes which mainly occasion its sterility.

A great deal of rain falls in the Andes, and the country immediately E. of the mountains is thickly wooded, and is injured by too much rain. This results from the moisture which the W. winds, that prevail throughout most of the year, bring with them from the Pacific, being condensed and precipitated in the mountains and immediately adjacent territory; so that after passing these regions the winds are quite dry ; and E. winds, which are very rare in Patagonia, are those only which convey any moisture to the desert E. of the Andes. Porphyry, basalt, sandstone, containing numerous organic remains, and a friable rock, greatly resembling, but not tinted with, chalk; are among the mineral formations hitherto remarked as the most prevalent in E. Patagonia. The zoology of the country is as limited as its *flora*. Guanacoes are met with sometimes in herds of several hundreds; and their enemy the puma, and a small kind of fox, are almost the only other wild quadrupeds at all abundant, except mice. The latter are of many species, and so numerous that, according to Mr. Darwin, Patagonia, poor as she is in some respects, can, perhaps, boast of a greater stock of small *rodentia* than any other country in the world. (*Voyages of the Adventure and Beagle*, iii., 215.) The condor and the cassowary are included among the few species of birds. The reptile and insect tribes present nothing remarkable.

The Patagonian Indians are tall and bulky, and, though not absolutely gigantic, they may be said, after rejecting the exaggerations of the early and the contradictory statements of later travellers, to be the tallest people of whom we have any accounts, the average height of the men being probably not under six feet. Their heads and features are large, but their hands and feet small ; and their limbs are neither so muscular nor so large-boned as their height and apparent stoutness would induce one to suppose. Colour a dark copper brown ; hair black, lank, and coarse, and tied above the temples by a fillet of plaited or twisted sinews. A large mantle of guanaco skins loosely gathered about them and hanging from the shoulders to the ankles, is, with a kind of drawers and loose buskins, almost their only article of dress, and

adds much to the bulkiness of their appearance. They neither pierce the nose nor lips, but disfigure themselves greatly with paint. They lead a nomadic life, living in tents formed of poles and skins, and subsisting on the flesh of the wild animals they catch. Both men and women ride on horseback, and are often furnished with saddles, bridles, stirrups, spurs, and Spanish goods of various kinds, which they obtain from Valdivia and other places in S. Chili. Their arms consist generally of a long tapering lance, a knife or scimitar if one can be procured, and the *bolas*, a missile weapon of a singular kind, carried in the girdle, and consisting of two round stones, covered with leather, each weighing about a pound. These, which are fastened to the two ends of a string, about 8 ft. in length, used as a sling, one stone being kept in the hand, and the other whirled round the head till it is supposed to have acquired sufficient force, when they are together discharged at the object. The Patagonians are so expert at the management of this double-headed shot, that they will hit a mark not bigger than a shilling with both the stones at a distance of 15 yards. It is not customary with them, however, to strike either the guanaco or the ostrich with them; but to discharge them so that the cord comes against the legs of the ostrich, or the fore legs of the guanaco, and is twisted round them by the force and swing of the balls; so that the animal being unable to run, becomes an easy prey to the hunters. These people live under various petty chiefs, who, however, seem to possess but little authority.

Patagonia was discovered by Magellan in 1519. The baldness of its harbours, the mostly difficult and dangerous of access, and afford little or no security for vessels above the size of a brig, has hindered the formation of any European settlement, except at Port St. Julian, about lat. 49° 10′ S., and long. 67° 40′ W., where the Spaniards settled about 1779, but speedily abandoned the establishment. A few expeditions have been undertaken to the interior in the last century, and latterly by the officers of the Adventure, principally up the larger rivers ; but the coasts are rarely frequented by any other than whaling vessels, and the nature of the country is not such as to hold out any hope of its ever emerging from its present state of savage barbarism. (*Parish's Buenos Ayres*, &c., 56–95; *Voyage of the Adventure and Beagle*; *Geog. Journ.*, i. vi.)

PATERSON, city, capital of Passaic co., N. J., is situated on Passaic river, near the great falls, 4 m. from this river, 13 m. N. Newark, 17 m. N.N.W. New-York city. 3 m. N.E. by N. Trenton, 214 W. A society for the establishment of manufactures, projected by Alexander Hamilton, was formed by the subscription of $200,000, about half only of which was fully paid in, in 1791. The contributors were incorporated by the Legislature in November 22d, 1791, and authorised to increase their capital to $1,000,000. After several other places had been proposed, the company located near Passaic Falls at Paterson. At this period there were not more than 10 houses in the place. This was one of the first attempts to establish cotton manufactures in the United States, and had to contend with the obstacles to which such establishments were then incident ; the object of the company in a great measure failed, and were abandoned. But more propitious times have enabled their successors to raise up here a large manufacturing city. The city of Paterson was incorporated November 22d, 1791, and lies partly in Bergen county and partly in Passaic county, is on both sides of the river, and includes 36 sq. m. By a dam in the river, 4½ feet high, and a canal around the falls, a vast water-power is created, sufficient for the supply of numerous manufactories. It contains a courthouse, jail, a bank, and many spacious edifices for manufactories, built chiefly of stone. There are nine churches, a Presbyterian, two Dutch Reformed, a Methodist, Reformed Presbyterian, Episcopal, Baptist, True Reformed Dutch, and a Roman Catholic. There is a philosophical society of young men, who have a respectable library, and a Mechanics' society for the advancement of science and the mechanic arts, with a library and philosophical apparatus. The Morris canal passes near the city, a little to the S. of it, and the railroad to Jersey City gives an easy access to the city of New-York. Passaic Falls are celebrated for their picturesque beauty, and are often visited as a curiosity.

There were in the city in 1840, 104 stores, with a capital of $199,950, machinery was manufactured to the amount of $607,000, four fulling-mills, one woollen factory, with a capital of $20,000, 19 cotton factories, with 45,936 spindles, two dyeing and printing works, with a capital of $256,400, one grist-mill, one saw-mill, two paper-mills, with a capital of $92,000, one tannery, two printing-offices, two binderies, two weekly newspapers. Total capital in manufactures, $1,799,500. It has one academy, 80 students; 14 schools, 1006 scholars. Pop. in 1825, 5084 ; in 1830, 7731 ; in 1840, 7596.

The city is governed by a mayor, recorder, common council, &c. There are many manufactories not enumerated above, and it is now one of the most respectable manufacturing places of its population in the country.

PATMOS (hod. *Patmo*), a small island of the Grecian archipelago, belonging to the Sporades, celebrated in ecclesiastical history as the place of St. John's exile, during Domitian's persecution, 11 m. N. W. Lero, and 90 m. S. the W extremity of Samos; its chief town being in lat. 37° 17' l' N., long. 26° 35' 14" E. Pop., according to Burgess, about 1,000. It is of very irregular shape, about 10 m. in length, 5 m. in breadth, and 28 m. in circ. Tournefort, whose authority is entitled to the highest respect, describes Patmos as the most barren rock (*meckant éceuil*) of the archipelago (t. 438); others, however, extol its fertility; and Dr. Clarke says, that from all he could collect on the subject, it is about as fertile as any of the neighbouring islands, were it not for the danger to which property is exposed from the continual incursions of pirates. (vi. 66., 8vo. ed.) It has numerous harbours, of which that of La Scala, on the E. side, deeply indenting the island, is the principal.

Above the landing-place is a small village, comprising about 50 houses and shops. On the ridge of a mountain, overlooking the port, is the cap. of the island, comprising about 400 substantial stone houses; its streets however, are steep, ill paved, and extremely narrow, few being more than 8 ft. wide. The monastery of St. John's, on a mountain close to the town of Patmos, built in the commencement of the 12th century, is, in fact, a pretty strong fortress, and commands a noble and extensive view of the sea and surrounding islands. It is peopled by about 30 monks, and has an attached church, and a library containing some early printed books and numerous MSS., some of which were purchased by Dr. Clarke. The famous grotto, or cavern, covered by a chapel, where St. John is said to have written the Apocalypse to the dictates of the Holy Spirit, is situated on the face of the hill, about half way between the town and the port. Clarke says that it is not spacious enough to have afforded a habitation even for a hermit; but the monks, to quash all doubts as to its being really the retreat of St. John, show the crevices in the rock through which, as they allege, the divine commands were communicated to the apostle! In point of fact, however, there is not a sentence in the apocalypse to warrant the inference that it was written in a cave; and if there had, any other cave in the island would have answered the purpose quite as well as this.

The island produces only a few grapes, and is furnished with corn and other articles of subsistence from the Black sea, Samos, and Smyrna. The male inhabitants are chiefly seamen, and, from their pretty extensive intercourse with different European nations, have become more enlightened than the generality of Greeks. The women (who, according to Tournefort, are generally pretty, though much disfigured by their strange costume) are chiefly employed in knitting cotton stockings. Tournefort mentions that for every man on the island there are at least twenty women; but this disproportion was, no doubt, occasioned by the men being mostly at sea, and is not mentioned by later writers. *Tournefort, Voyage du Levant.* i., 439; *Clarke*, vi., 27, &c.; *Burgess's Greece and Levant*, ii., 24–28.)

PATNA, a city of British India, presid. Bengal, prov. Bahar, of which it is the cap., on the S. bank of the Ganges, 44 m. E.N.E. Benares, and about 300 m. N.W. Calcutta; at. 25° 37' N., long. 85° 15' E. Population estimated at upward of 300,000. (*Hamilton*.) Within the walls it is ot much more than 1½ m. in length, by three fourths of a mile in average breadth; but, including its straggling suburbs, it extends for 9 m. along the banks of the river, and 2 m. inland. The town itself is very closely built, and surrounded with fortifications in the Hindoo style, which are ow, however, completely decayed. At a short distance it as a very striking appearance; being full of large buildings, with remains of old walls and towers, and bastions projecting into the river, and backed by irregularly elevated land. has, however, but one wide street, all the other thoroughfares being narrow, crooked, and mean. The houses are artly built of brick, and many have terraced roofs and balmies; but the greater number are of mud, with tiled roofs. of the city is a large suburb, with many well-built storehouses, and in the same direction are the gardens belonging the palace of Jaffier Khan, two or three miles in circuit. he suburbs of Bankipoor, where the East India company is most of its offices, and where are most of the residences its servants, is W. of the city; here is also a remarkable difice erected during the governorship of Warren Hastings, d originally intended for a grain magazine, but now used a depôt for military stores. Patna has a small citadel, t there do not appear to be any other public buildings orth notice; though the Hindoos, Mohammedans, Seiks, ., have many religious temples. The Portuguese inhabitants have a Roman Catholic church; and there is a itish seminary with about 100 pupils. But though Patna the seat of one of the six courts of circuit in the Bengal

presidency, the residence of a zillah-judge, a collector, a commercial resident, and an opium agent, it has but few English inhabitants. This city is not celebrated for any particular manufacture; but most articles of foreign and domestic produce may be procured in its bazaars. The neighbouring country produces the finest opium and saltpetre, and great quantities of wheat and other grains, sugar, indigo, &c. The opium and saltpetre are monopolised by government, and produce a large revenue; but a considerable trade is carried on in the other articles. It was at this city that the English first established a factory in the eastern provinces of Hindustan; and it is indebted to the European trade for most part of its growth and prosperity. (*Hamilton's E. I. Gaz.; Heber, &c., in Mod. Trav.*, ix.)

PATRAS (an. *Patra*), a town and seaport of independent Greece, on the N.N.W. coast of the Morea, and on the E. side of the gulf of its own name, about 5 m. S.W. the mouth of the gulf of Lepanto, 107 m. W. by N. Athens; lat. 38° 14' N., long. 21° 46' 40" E. Population according to Burgess, about 4000, exhibiting a decrease of nearly two thirds since the commencement of the revolution. It is built amphitheatre-wise on the side of a hill rising from the shore, which has at its summit the acropolis, commanding a fine view of the surrounding coast and country. The fortifications are in good repair, and have been recently much enlarged. The interior comprises one pretty broad and well-built street, with numerous narrow lanes and alleys lined with mean wooden tenements, the overhanging eaves of which nearly meet over the street. The Greeks have a few good houses; but those of the European consuls are the best. Every considerable house is surrounded with a garden well stocked with orange, fig, pomegranate, and other fruit trees, which give the town an agreeable aspect, and conceal the greater part of the poorer habitations. The only public buildings are two hospitals and several churches; the remains of ancient buildings are but inconsiderable.

The bay in front of the town being unsafe, and exposed to heavy seas, particularly in winter, vessels go a little further up the gulf, where the port is situated, and where there is a mole for their security. Patras has a pretty extensive trade. The principal exports are currants (by far the most important article), oil, valonia, wine, raw silk and cotton, wool, skins, wax, &c. The imports here, as at the other Greek ports, consist principally of colonial produce, manufactured goods, salt fish, cordage, hemp, deals, &c., chiefly from the Ionian islands and Great Britain, Venice, Trieste, Leghorn, and Marseilles. The imports from Great Britain amounted, in 1838, to £13,490, the shipments to Great Britain in the same year being valued at £171,340; but this discrepancy is apparent only, the excess of exports being balanced by the imports of British produce from the Ionian islands and other places. According to the tariff now in force, the rate of duty (without distinction of foreign or native flags) is 10 per cent. *ad valorem* on imports, and 6 per cent. *ad valorem* on exports.

The ancient Patra is supposed to have been founded by the Ionians. Herodotus enumerates it among the 12 towns of Achaia. (I. 146.) Its inhabitants took an active part, and the town suffered greatly, in the Achæan war. After the battle of Actium, however, it was raised to its former flourishing condition by Augustus, who made it a colony by establishing in it some of his veterans. In Strabo's time it was a large and populous town; and in the beginning of the second century it was still prosperous, though remarkable for the dissoluteness of its inhabitants. (*Pausanias, Achaia*, c. 21.) It was the seat of a dukedom under the Greek emperors, and in 1408 was bought by the Venetians, from whom it was taken by the Turks in 1446. It was pillaged by the Albanians in 1770, and was the stronghold of the Ottomans from 1821 down to the period of the emancipation of Greece. (*Burgess's Greece and Levant*, i., 194–198; *Mod. Trav.; Cramer's Greece.* iii., 66–69.)

PATRICK, county, Va. Situated in the S. part of the state, and contains 541 sq. m. Drained by head branches of Smith's, S. Mayo, and Little Dan rivers. Bounded N.W. by the Blue Ridge. It contained in 1840, 6988 neat cattle, 6047 sheep, 24,292 swine; and produced 28,469 bushels of wheat, 3190 of rye, 222,964 of Indian corn, 63,940 of oats, 13,114 of potatoes, 618,384 pounds of tobacco. It had seven stores, one flouring-mill, 31 grist-mills, 12 saw-mills, five tanneries; five schools, 190 scholars. Pop.: whites, 6087; slaves, 1842; free coloured, 103; total, 8032. Capital, Taylorsville.

PATTERSON, p. t., Putnam co., N. Y., 92 m. S. Albany, 295 W. Drained by Croton river. It contains two churches, a Presbyterian and an Episcopal, five stores, two fulling-mills, three grist-mills, six saw-mills, two tanneries; 10 schools, 394 scholars. Pop. 1349.

PATUXENT, river, Md., rises on the border of Montgomery and Anne Arundel counties, and, flowing S.E. 40 m., it turns to the S., flowing 50 m. nearly parallel with Chesapeake bay, and enters the W. side of the bay by a large æstuary. It is navigable 50 m., to Nottingham, for vessels of 250 tons.

PAU, a town of France, dep. Basses-Pyrénées, of which it is the cap., on the Pau, here crossed by a fine stone bridge of seven arches, in a fertile though marshy plain, 56 m. E. by S. Bayonne. Lat. 43° 17′ 29″ N., long. 0° 22′ 30″ W. Population in 1836 (ex. com.), 11,959. It is regularly laid out and well built, consisting principally of one long and broad main street. It has, however, several squares, or open spaces, and is environed by public walks. Its principal, and by far the most interesting, edifice is the castle, in which Henry IV. first saw the light on the 13th of December, 1553 ; it was founded by the princes of Bearn in the 10th century, is situated on a commanding height to the W. of the town, and forms an irregular collection of massive towers, having a fine terrace on the side fronting the river. It was much injured during the revolution. Having been converted into military quarters ; but it has since been completely repaired and renovated.

The chamber, memorable as the birth-place of Henry IV., retains its ancient portraits and furniture, the tortoise-shell cradle of the king, &c. There is a marble statue of Henry in the vestibule of the castle, and another statue in bronze in the *Place Royale*. The other principal buildings comprise the prefecture, hall of justice, college, and one or two hotels.

Pau has a royal court of tribunals of primary jurisdiction and commerce, boards of taxation and forest economy, a royal college, an *académie universitaire*, a society of agriculture, a school of design and gallery of paintings, a public library of 18,000 volumes, &c. Its manufactures include cotton stuffs, linen cloths or *toiles de Bearn*, &c., and it has considerable dyeing establishments and tanneries : it has also a pretty extensive trade in manufactured products, and in wines, Bayonne hams, salted geese, &c. It has two large weekly markets, and three important annual fairs. Inglis says, " Pau has always enjoyed the reputation of being one of the most interesting cities of the S. of France. It is clean, airy, and abounds in every convenience and in most luxuries. It is a great resort for strangers, particularly English ; and, excepting Bayonne, is probably the most desirable of any of the towns selected by foreigners as a residence. Excellent houses are to be obtained at a very moderate expense, and the markets are both abundant and cheap. There are generally 50 or 60 English families in Pau and its neighbourhood ; and the number, I understand, is upon the increase." (*Switzerland, the Pyrenees, &c.*, 309, 310.)

Besides the "great" Henry, Pau has produced several distinguished persons, among whom may be specified Marshal Bernadotte, now king of Sweden ; and Viscount Orthez, governor of Bayonne, who nobly refused to execute the orders issued by Charles IX. for the massacre of St. Bartholomew. (*Hugo*, art. *Basses-Pyrénées* ; *Guide du Voyageur, &c.*)

PAULDING, county, Ga. Situated toward the N.W. part of the state, and contains 600 sq. m. Drained by head branches of Tallapoosa river, and tributaries of Etowah river. It contained in 1840, 3804 neat cattle, 995 sheep, 10,937 swine ; and produced 11,263 bushels of wheat, 199,170 of Indian corn, 6012 of oats, 1556 of potatoes, 1630 pounds of tobacco, 118,793 of cotton. It had three stores, two grist-mills, two saw-mills, two distilleries, three breweries ; three schools, 68 scholars. Pop. : whites, 2102 ; slaves, 454 ; total, 2556. Capital, Van Wert.

PAULDING, county, O. Situated in the N.W. part of the state, and contains 433 sq. m. Watered by Maumee and Auglaize rivers. It contained in 1840, 201 neat cattle, 85 sheep, 736 swine ; and produced 3859 bushels of wheat, 8420 of Indian corn, 2749 of oats, 2518 of potatoes, 3316 pounds of sugar. It had two stores ; one school, 28 scholars. Pop. 1034. Capital, Paulding.

PAVIA (an. *Ticinum*), a frontier city of Austrian Italy, Lombardy, cap. deleg. of its own name, on the Ticino, 19 m. S.S.W. Milan ; lat. 45° 10′ 47″ N., long. 9° 9′ 48″ E. Pop. in 1837 (ex. com.), 23,531. It is surrounded with old walls, and communicates with a suburb across the Ticino by a bridge of seven arches, constructed in 1351. This structure, partly of stone, but principally of brick, is one of the most singular monuments of the 14th century : it is 300 feet in length by 12 in breadth, and is covered with a curious roof, supported on 100 pillars of rough granite. At one end is the Austrian and at the other the Sardinian custom-house, the Ticino separating their respective territories. This is a magnificent venerable city ; but its buildings and its fame belong to another age, and it has long been in a state of paralysis and decay. From the bridge, the *Strada Nuova* or *Corso* extends through the centre of the city to a superb gate, begun under the viceroyalty of Prince Eugene. In this street the principal palaces of the Pavian nobility, mouldering and dismantled, are mingled with shops, churches, colleges, cafés, theatres, and hospitals. From the main street others of greater antiquity branch off at right angles,

some terminating in *piazzas* opening before vast and cumbrous palaces, now half ruinous and dismantled. From its numerous public edifices, Pavia was formerly called the "City of a hundred Towers ;" but these are now greatly diminished. It has, however, a ruined castle, once the residence of the Lombard kings, and several other buildings traced up to the time of the Lombards, particularly the church of St. Michael, in a barbarous and grotesque style. The church of *San Pietro in Cielo d'Oro*, which is said, though on doubtful authority, to contain the remains of St. Augustine, and which certainly contains those of Boëthius, is in the same early and rude style ; but, *tempore mutantur*, the venerable edifice is now converted into a granary or barn ! The cathedral has little that is remarkable ; it was begun in 1485, and is of large dimensions ; but it yields in interest to the churches of the Carmine, San Francesco, S. Salvador, &c. The palace of Theodoric was destroyed in a popular tumult in the 11th century ; and the tower in which Boëthius was confined and wrote his famous treatise, *De Consolatione Philosophiæ*, no longer exists. On the site of the latter, however, is the Malespina palace, at the entrance of which is a marble monument and bust of the philosopher. Previously to the reign of the emperor Joseph II., Pavia had 46 wealthy convents ; but few, if any, exist at the present day. The theatre and the university buildings are almost the only other structures worth notice : the interior of the former is rendered dark and gloomy by the black marble of which it is constructed, and the latter, according to Mr. Woods, are magnificent rather by their extent than by any merit in their architecture.

The university of Pavia, the first and most frequented in Italy, was founded by Charlemagne, and restored by Galeazzo Visconti in the 14th century ; but it owes its present form and institutions to the Empress Maria Theresa, and her minister Count Firmian. It has faculties of law, medicine, and philosophy ; being particularly celebrated as a school of medicine. It has no faculty of theology, but in every other respect its constitution is similar to that of Padua (which see). It has at present 38 professors, three adjuncts, and 11 assessors ; and in 1837 had 1307 students, of which 287 belonged to the philosophical, 436 to the legal, and 582 to the medical faculty. Its revenues are derived principally from imperial treasury grants, legacies, municipal and communal funds, and fees paid by students on obtaining degrees ; which last average about 150,000 *lire* a year. The whole expenditure for the university amounted, in 1832, to 259,669 *lire*. The professors have annual salaries of from 3000 to 6080 *lire* (£120 to £240 sterling) : and enjoy, by special privilege, the distinction of personal nobility. Several of the most distinguished names in the history of Italian literature and science, have been professors in this university. Among others may be specified Vesalius, Cardan (a native of the city), Spallanzani, Volta, Scarpa, Tamburini, &c. ; and its has still to boast of many eminent teachers. The university has an extensive library, which it chiefly owes to Count Firmian ; a fine botanic garden, instituted by the French ; a valuable collection of natural history, physical and anatomical museums, &c. Students are lodged and boarded gratuitously, in three colleges attached to the university. The *Collegio Caccia*, founded by a noble family of Novarra, receives 30 boarders from that city and district ; the *Collegio Borromeo*, a stately and venerable edifice, founded by the famous Cardinal Borromeo, supports 36 students ; and the *Collegio Ghislieri*, founded by Pope Pius V., receives 60 students ; but the greater number of the students are extra collegians. Pavia has also a gymnasium, a high school and female school, two hospitals, numerous asylums, and charitable institutions, a *monte di pietà*, &c. It is a bishop's see, the seat of the superior court of the delegation, and a chamber of commerce. It has some silk manufactures, and a considerable trade in agricultural produce with Milan and the cities on the Po. It is connected with Milan by a navigable canal, traversed by boats like the Dutch treckschuyts. A good deal of the cheese, called Parmesan, is made in this neighbourhood. About 4 m. N. Pavia is the celebrated Certosa, the most magnificent of Italian monasteries, founded by John Galeazzo Visconti in 1396 ; and dissolved, and in part stripped by the French, in 1794.

Ticinum, which was an important city under Augustus, began to be called *Papia* (whence its present name), during the Gothic dominion in Italy. In modern times it has sustained numerous sieges ; but it is principally distinguished, in a historical point of view, by the great battle that took place in its vicinity on the 24th of Feb. 1525, between the French army, that had undertaken the siege of Pavia, under Francis I., and the Imperialists, under the viceroy Lannoy. The French were totally defeated. Francis, who had displayed the greatest heroism, and the king of Navarre, were taken prisoners ; and exclusive of many generals and persons of distinction, between 9000 and 10,000 private soldiers were left on the field of battle. The French army was, in fact, entirely destroyed ; and there was quite as much of truth as of point in the laconic epistle addressed by Francis to his mother after the battle—" Madam, we have lost all except our honour." (*Austr. Nat. Encyc.*; *Von Raumer's Italy; Journal of Education; Wood's Letters; Conder's Italy; Robertson's Charles V.*, book iv., &c.)

PAWCATUCK, r., R. I., is formed by the union of Wood and Charles rivers, which afford extensive water-power. In its lower part it forms the dividing line between the states of Rhode Island and Connecticut, and enters the Atlantic ocean.

PAWLET, p. t., Rutland co., Vt., 87 S.S.W. Montpelier, 431 W. Chartered in 1761. Watered by Pawlet river. It contains a Congregational and a Baptist church, seven stores, five fulling-mills, four woollen factories, one cotton factory with 1076 spindles, one furnace, one grist-mill, three saw-mills, one oil-mill; 10 schools, 495 scholars. Pop. 1748.

PAWLINGS, p. t., Dutchess co., N. Y., 90 E. Poughkeepsie, 89 S. Albany, 298 W. Croton river has its rise in this township. It contains two churches, an academy, four stores, one grist-mill, one saw-mill; three schools, 96 scholars. Pop. 1571.

PAWTUCKET, r., R. I., rises in Worcester co., Mass., with the name of Blackstone river. It enters Rhode Island in the N.E. part of the state, and is called Pawtucket river, until it enters Seecoak river, at Pawtucket, and finally becomes Providence river, at and below Providence. It affords extensive water-power, particularly at Pawtucket. The Blackstone canal from Providence, R. I., to Worcester, Mass., passes along this river.

PAWTUCKET, p. v., North Providence t., R. I., 4 N. Providence, 404 W. Situated on both sides of Pawtucket river, and lies partly in North Providence t., R. I., and partly in Pawtucket t., Mass. In both parts of the village there are seven churches, two Baptist, a Congregationalist, Episcopal, Methodist, Free-will Baptist, and Universalist, three banks with an aggregate capital of $236,550, besides a savings bank, 38 stores of different kinds. There were at the close of 1842, 17 cotton factories with 94,368 spindles, 531 looms, employing 523 hands, and producing 86,870 yards weekly; of cotton cloths and printed goods, besides yarn. At Central falls, half a mile above, there are 11 cotton factories, with 16,344 spindles, 380 looms, employing 373 hands, and producing 57,700 yards, weekly, of coarse cottons and printed goods, besides considerable thread. The village contains about 8000 inhabitants, and has the honor of being the place where the cotton manufacture first successfully commenced in the United States. The fall in the river within a short distance, is 50 feet. A bridge unites the two parts of the village.

PAWTUCKET, t., Bristol co., Mass., 36 S. Boston, 4 N. Providence. Situated on the E. side of Pawtucket river. Organized in 1828. It contains a part of Pawtucket village, and had in 1840, 11 stores, nine cotton factories with 16,362 spindles, two grist-mills, one saw-mill; one academy, 60 students, five schools, 260 scholars. Pop. 2184.

PAWTUXET, r., R. I., rises by several branches, and flowing E. enters Providence r., 5 miles below Providence. It has on it many mills-seats and numerous cotton manufactories, for a particular account of which see Kent co. R. I.

PAWTUXET, p. v., port of entry, Warwick t., Kent co., R. I. Situated on both sides of Pawtuxet river, at its mouth, and lies partly in Cranston t., Providence co. Vessels of from 90 to 50 tons come up to the village, the two parts of which are connected by a bridge. It contains two Baptist churches, six stores, one cotton factory with 1500 spindles, 150 dwellings, and about 1200 inhabitants. The river has here four feet fall.

PAXO (anc. *Paxos*), the smallest of the seven principal Ionian islands, in the Ionian sea, 10 m. S. by E. Corfu, and about the same distance W. the main land of Greece; its N. point being in lat. 39° 14' N., long. 29° 9' E. Area, 26 sq. m. Pop., in 1836, 5287. It is oval-shaped, and extremely mountainous: its soil being stony, and so destitute of moisture, that in summer the inhabitants are obliged to procure fresh water from the neighbouring continent. The climate is extremely mild and agreeable; but the island produces little else than olives, almonds, and vines, the quantity of corn raised being altogether insignificant. Mules and goats are abundant on the coast. Port Gai, the principal port on its E. side, affords good anchorage for a few vessels; but a more secure harbour is formed by the channel between this and the neighbouring islet of Antipaxo. The town has a population of 4000 persons, but in appearance is little better than a mean village.

PAZ (LA), a city of Bolivia, cap. dep, of its own name, 195 m. N.N.W. Chuquisaca, lat. 17° 30' N., long. 68° 25' W. Pop. 20,000 ? It is situated on the E. declivity of the Andes, at an elevation of 12,170 ft. above the sea, and at no great distance from the sources of the Beni, a principal affluent of the Amazon. It has a cathedral, four other churches, several conventual establishments, and is a

bishop's see, with very considerable revenues. It is the centre of a considerable trade in Paraguay tea.

LA PAZ was founded in 1548, and received its name in commemoration of the peace that ensued after the defeat of Gonzalo Pizarro and his associates. It suffered considerably a few years ago during a revolt of the Indians, but still ranks as a city of some wealth and importance. (*Dict. Géog., &c.*)

PEACHAM, p. t., Caledonia co., Vt., 36 E. by N. Montpelier, 538 W. Chartered in 1763, first settled in 1773, organized in 1783. Onion river pond of 300 acres in the W. part of the town, gives rise to one branch of Union or Winooski river. A bog meadow in the E. part contains an inexhaustible quantity of shell marl from which lime is made. Limestone is found. It contains a Congregational church, an old and flourishing academy, three stores, one fulling mill, two woollen factories, one grist-mill, six saw-mills, two tanneries; one academy, 195 students; 12 schools, 548 scholars. Pop. 1443.

PEACH BOTTOM, p. t., York co., Pa., 95 S.E. York, 62 S. E. Harrisburg, 81 W. Drained by Fishing creek entering Muddy creek, which bounds it on the N. It contains seven stores, two fulling mills, one woollen factory, one flouring mill, five grist-mills, one saw-mill; three schools, 27 scholars. Pop. 1074.

PEARL, r., Miss. and La., rises by several branches near the centre of Mississippi, and enters by several mouths the Rigolets, which forms a communication between lakes Pontchartrain and Borgne. From lat. 31 N. it forms the boundary between Alabama and Louisiana. The navigation is obstructed by sand bars, shallows and rafts of timber. It is navigable for small craft to Jackson, the capital of Mississippi.

PEDÉE, GREAT, r., N. C. and S. C., rises chiefly in Wilkes co., N. C., and bears, for a considerable distance, the name of Yadkin river. After it enters South Carolina, it is called Great Pedee river, and flowing southeasterly, it receives Lynches creek from the W., and Little Pedee river, 32 m. above its mouth, on the E. side, and Waccamaw river on the same side, at its entrance into Winyaw bay, and enters the Atlantic 12 m. below Georgetown. It is navigable for boats of 60 or 70 tons for about 200 miles.

PEEBLES, an inland co., of Scotland, having N. Mid-Lothian, E. Selkirk, S. Dumfries, and W. Lanark. . Area, 304,160 acres, of which a comparatively small portion only is arable. This is almost wholly a pastoral district; the surface consisting of mountain, moor, and bog, with the exception of a limited extent of low, level land along the banks of the Tweed, which rises in and runs through the county. The highest mountains are in the S. part of the county adjoining Dumfriesshire, where the Tweed has its source.. The summit of Broadlaw rises 2741 ft. above the level of the sea; and this, which is about 100 ft. above the altitude of the contiguous summit of Hartfell, is the highest elevation in Scotland S. of the frith of Forth. The hills are generally smooth, and afford good sound sheep pasture. In the low parts of the county, agriculture has been very much improved; but it is now pretty generally believed that tillage had been too much extended during the late war. The buildings on farms of any importance have been entirely renovated, principally within the last 30 years. The black faced breed of sheep were, at no very distant epoch, diffused over the whole county, to the exclusion of every other; but about 1795 Cheviots began to be introduced; and their numbers have since so rapidly increased, that, even in the parish of Tweedsmuir, which is the wildest and most exposed, there are now three Cheviots to one black-faced sheep. The total sheep-stock at present in the county may be estimated at above 100,000. Property in a few hands, farms being very large; average rent of land in 1810, 5s. 7d. an acre. Neither minerals nor manufactures of any importance. Peebles is divided into 16 parishes, and returns one member to the House of Commons. Registered electors in 1839–40, 863. Peebles is the only town of any importance. In 1831, this county had 1789 inhabited houses, 2079 families, and 10,578 inhabitants, of whom 5342 were males, and 5,236 females; and in 1841 10.520 inhabitants. Valued rent, £51,938 Scotch. Annual value of real property in 1815, £64,182.

PEEBLES, a royal bor. and market town of Scotland, co. Peebles, of which it is the capital, and the only town, romantically situated in a mountainous pastoral district, on the Tweed, which is here crossed by a bridge of five arches, at the point where it is joined by the Eddleston, 21½ m. S. Edinburgh. Pop. in 1831, 2100. It is divided by the Eddleston into the Old and New towns. The main street runs E. and W., in a line nearly parallel with the Tweed. The houses are unusually substantial and good for so small a town. It has a parish church, with a handsome spire, two Presbyterian dissenting chapels, an Episcopal chapel, a town-house, and jail. The grammar-school enjoys a high reputation, and is well attended. A scientific association established in the town, at which lectures are given, has an average attendance of no fewer than 190 members! Though the town has no manufactures, it supports two banks. It is regarded as peculiarly salubrious; and is much resorted to in summer as a favourite country residence.

Peebles was long a hunting residence of the Scottish kings, particularly of Alex. III., who founded in it a monastery for Red Friars, in 1260, of which the ruins are still pretty entire. The poem "Peblis to the Play" was written by James I. of Scotland. Neidpath Castle, inhabited by the Earls of March till 1778, stands on a rocky promontory overhanging the Tweed, ½ m. W. of the town. Mungo Park, the traveller, practised as a surgeon in Peebles for some time previously to his second mission (1805) to Africa. Before the passing of the Reform Act, Peebles was united with Falkirk, Linlithlow, and Lanark, in sending a member to the house of congress; but it was then merged in the county constituency. (*Pennecuik's Description of Tweeddale*, 1815 ; *New Stat. Acc. of Scotland ; Boundary Returns*.)

PEEKSKILL, p. v., Cortlandt t., Winchester co., N. Y., 46 N. New-York, 106 S. Albany, 269 W. Situated on the E. side of Hudson river, at the entrance of Peekskill creek. The village proper is half a mile from the landing, on a hill 200 feet high. It contains seven churches, two Dutch Reformed, a Presbyterian, Episcopal, Methodist, and two Friends, a bank, an academy, with an edifice which cost $7000, 25 stores, two large iron founderies, a printing-office, issuing a weekly newspaper, and, including the landing, 250 dwellings, and about 1800 inhabitants. One steamboat and six sloops ply between this place and New-York for transportation, and one steamboat for passengers. Steamboats from New-York to Albany land and receive passengers at this place.

PEGU, a former kingdom of India-beyond-the-Brahmaputra, forming at present the S. portion of the Birmese empire. (*See* BIRMAH.)

PEGU, a decayed city of the Birmese empire, and the ancient capital of the above kingdom, on the Pegu river, a tributary of the Irrawadi, 50 m. Rangoon. Lat. 17° 40′ N., long. 96° 12′ E. While it remained the capital of an independent state, it is said to have had a population of 150,000; but being taken in 1757 by the Birmese monarch, Alompra, he razed most of its buildings to the ground, and reduced its inhabitants to a state of slavery. At present it consists of only two streets, one parallel to the river, and the other leading to the celebrated Shoe-madoo, or great pagoda. This, the most famous edifice in the Birman empire, boasts of high antiquity, and is raised on successive terraces in a manner similar to the religious structures of the Mexicans, as described by Humboldt. It stands on an apparently artificial hill, the sides of which are sloped into two terraces, and ascended by steps of hewn stone. The lower and greater terrace forms an exact parallelogram, and is about 10 ft. in height; the upper and smaller terrace is of similar shape, and rises about 20 ft. above the lower terrace, or 30 ft. above the level of the ground. According to Col. Symes a side of the lower terrace is 1391 ft. in length, and of the upper, 684 ft. The earth forming the terraces appears to have been taken from the ditch which formerly surrounded the city, and which may still be traced, enclosing a quadrangular space 1½ m. on each side. The brick walls sustaining the sides of the terraces were formerly covered with plaster, wrought into various figures, but they are now in a ruinous state. The area of the lower terrace is strewed with fragments of decayed buildings, but the upper is kept free from filth, and is in tolerable order. On the second terrace is the pagoda, a pyramidal building of brick and and mortar, without excavation or aperture of any sort, octagonal at the base, each side measuring 162 ft., and diminishing in breadth abruptly till it becomes of a spiral form. Its entire height from the ground is about 360 ft.: it is surrounded by two rows of small spires, a great variety of mouldings, ornaments in stucco, &c.; the whole being crowned with the *tee*, a sort of umbrella of open iron work, gilt, 56 ft. in circumference, surrounded by a number of small bells, and from the centre of which, when Symes saw it, rose a rod with a gilded pennant. At the time of his visit, also, there were temples in miniature of the larger, at all the angles of the upper terrace, various saloons embellished with carving and gilding, numerous sculptures in masonry, idols, flying pennants, three large bells used by devotees on the N. side of the temple, many dwellings for priests on the lower terrace, &c.: probably most of these still exist, for a recent traveller states that the pagoda, with its appendages, is in tolerable preservation. (*Malcolm*, i. 89.) Pegu has several other temples, but they are mostly in ruins ; and the site of the ancient city is almost wholly under water, probably from neglect of the drains and sluices. Mindjeree Praw, king of Birmah, in 1790, endeavoured to restore to Pegu a portion of its former importance, by transferring thither the provincial government from Rangoon, but he did not succeed. Zangnomang,

however, on the opposite side of the river, is a prosperous town, and adjacent to it, for many miles, is a succession of thriving villages. (*Malcolm's Travels in S. E. Asia*, i., 88, 89 ; *Symes in Mod. Trav.; Asiatic Researches, &c.*)

PEKING, or PEKIN (Chin. *Pih-king*, meaning "the northern capital"), the modern metropolis of the Chinese empire, prov. Pecheilee, in a vast, sandy plain, between the Pei-Ho (which has its embouchure in the gulf of Pecheilee), and its important affluent the Hoen-Ho, within about 5 or 6 m. of each, and being united to the Pei-Ho by a canal, 562 m. N. by W. Nankin, and 100 m. W.N.W. the gulf of Pecheilee, in the Yellow sea ; lat. (observatory) 39° 54′ 13″ N., long. 116° 27′ 45″ E. We know nothing certain of the population, except that it is very great : some writers estimate it at two millions, and others at three millions; we believe, however, that even the smallest of these estimates is beyond the mark, and that probably it does not much exceed, if it be not under, 1,500,000. Klaproth estimates it at 1,300,000. A large portion of the space within the walls is occupied by gardens and enclosures; and there is no keeping up of one family above another as in European towns. The city is divided into two principal portions, exclusive of the suburbs. The most northerly portion, which is nearly a perfect square, is called *nei-tching*, or the inner city ; it contains the palace of the emperor and the principal government officers, and is mostly occupied by Manchous, whence it is sometimes called the "Imperial," and sometimes the "Tartar city." The other, or more southerly portion, denominated the *wai-tching*, or outer city, is a quadrilateral rectangle, entirely occupied by Chinese, and is at once the seat of business and the residence of the great bulk of the population. Both divisions are surrounded by walls, the extent of which may be about 18 m. The walls of the Chinese city are 30 ft. in height, and 25 ft. broad at the base, and 12 ft. on the top, the slope being mostly on the inner side ; but those of the Imperial city are 40 ft. in height. Square towers project from the outer side at intervals of about 70 yards from each other, and each of the 16 city gates is surmounted by a tower nine stories in height, with port-holes for cannon.

The principal streets are of great width, and perfectly straight (*tirées au cordeau*, Duhalde), running between opposite gates in the divisions of the city to which they respectively belong ; those in the northern being, for the most part, better built, and preferable to those in the southern division. The other streets, however, are very narrow, and are, in fact, mere lanes branching off at right angles from the principal thoroughfares. It is singular, that though the great roads leading to the capital be paved with large blocks of granite, the streets are not paved, which makes them dusty and disagreeable in hot, and dirty, and, in parts, all but impassable in wet weather. "En général," says M. Hyacinthe, "l'inégalité, le mauvais entretien des rues, ou plutôt des sentiers, qu'on est obligé du suivre dans les rues, est un juste sujet de blâme contre la police Chinoise;" p. 12. The houses, which rarely exceed a story in height, are built of brick, and covered with tiles ; but, according to Sir John Barrow, "they are void of taste, grandeur, beauty, solidity, and convenience :" none but the great shops have either windows or openings in the front wall ; but most of them have a sort of terrace, with a raised balcony or parapet-wall, on which are placed pots of flowers, shrubs, or stunted trees. The houses in the smaller streets or lanes, many of which are occupied by public functionaries, are very similar to those in the larger streets ; and the regularity with which the streets intersect each other, the uniformity in the size and appearance of the houses, and the absence of towers, spires, domes, and even of chimneys (of which not one is to be seen), give the city an extremely monotonous appearance, resembling, in fact, a vast encampment. (*Barrow*, 93.)

"The shops in the principal streets make an ostentatious display of painting and gilding. Sky-blue and green, mixed with gold, are the prevailing colours upon the walls. The goods are not only displayed within, but exposed in heaps in front of the houses. Before these are generally erected wooden pillars, whose tops are much higher than the roofs of the houses, bearing inscriptions in gilt characters describing the goods to be sold, and assuring the buyer he will not be cheated : To attract more notice, they are generally hung with various coloured flags, streamers, and ribands, exhibiting the appearance of a line of shipping, dressed in their different colours. Lanterns of horn, muslin, silk, or paper, are arranged before the doors, and exhibit such variety of form, that the Chinese appear to have exhausted on them all the powers of imagination. The streets are peculiarly crowded, in consequence of the number of trades that are carried on in the open air. The numerous moveable workshops of tinkers and barbers, cobblers and blacksmiths ; the tents and booths where tea, fruit, rice, and other eatables, are exposed to sale ; the wares and merchandise arrayed before the doors ; the troops of dromedaries laden with coals from Tartary; the wheelbarrows and handcarts stuffed with vegetables, leave in the broadest streets only a very narrow space unoccupied." Room, indeed, is scarcely allowed for the frequent processions of men in office, with their numerous retinues and strange insignia, or for the pompous trains which attend at funerals and marriages. With the confused voices of the multitude buying and selling their various commodities, are mixed the cries of jugglers, conjurors, fortune-tellers, mountebanks, quack doctors, comedians, and musicians. It is, however, a curious fact, that the crowd and bustle are wholly confined to the great thoroughfares: the cross streets and lanes being perfectly still and quiet. "Women are frequently seen among the crowd, either walking, or riding on horses, which they bestride like men; but they are all Tartar females, whose manners alone admit of such exposure: the Chinese ladies being more rigidly confined to the house in the capital than in the rest of the empire." (*Barrow's Travels*, p. 94-96.)

At the four points, where the great streets intersect each other, are singular looking erections, somewhat resembling triumphal arches, but, in fact, monuments to the memory of those who had deserved well of their country, or who had attained an unusual longevity! They are built sometimes of stone, but more generally of wood, and consist invariably of a large central gateway, with a smaller one on each side, covered with a narrow roof, and painted, gilt, and varnished.

The northern city, which comprises the residence of the emperor and the principal government offices, consists of three enclosures—an outer, a middle, and an inner. The latter contains the imperial palace and the houses of the different members of the imperial household; the second is chiefly inhabited by Chinese merchants; and the third, or outermost enclosure, constitutes the open city. The inner portion, or that comprising the imperial palace and its dependencies, including gardens and pleasure-grounds, occupies an oblong space, about 2½ miles in circuit. This sacred enclosure, or "forbidden city," is surrounded by a high wall, similar to that surrounding the city, being, like it, flanked with towers, and faced with yellow tiles. Each side of the wall has a large gate surmounted by a tower; the walls, leading to the principal halls, being paved with large slabs of white and gray stone. The "Meridian gate," through which alone the emperor can pass, is by far the most splendid of all the approaches to the palace: here he distributes presents to foreign ambassadors, views the captives that may have been taken by his "invincibles," and shows himself whenever he dispenses mercy. In the *Tse-ho-men*, or "gate of extensive peace," which is a superb building of white marble, 110 ft. in height, the emperor receives congratulatory visits of ceremony from the various officers of his court: but by far the most sacred, as well as richest and most magnificent portion of the palace, is the *Kaen-tsing-kiong*, or "tranquil palace of heaven," the emperor's private retreat, which none may approach without special permission. It is used also as a cabinet, where the great officers of state assemble for consultation, and where candidates for office receive their appointments. The palace of the empress is also very extensive; and beyond it is a gate leading to the grand flower-garden, laid out in walks, filled with pavilions, temples, and groves, and interspersed with canals, fountains, lakes, and beds of flowers. Near the empress's palace is a library, alleged by the Chinese to comprise a collection of most books published in the empire. Within the precincts also is a temple, to which the emperor comes on certain stated occasions to obtain blessings from the manes of his ancestors, and to show his filial piety. Six palaces are occupied by the princesses of the imperial family; and other ranges of buildings constitute the residences of the emperor's stewards, &c.; besides which there are halls for councils, courts, &c., and a large printing establishment. (See *Gutzlaff's China Opened*, i., 62, 63.)

The reader, however, would form a very inaccurate notion of these buildings, if he supposed they bore any considerable resemblance to European palaces, or that the magnificence of the buildings at all corresponded with their imposing names. The truth is, that there is but little of pomp or splendour in the imperial residence. "The buildings that compose the palace, and the furniture within them, if we except the paint, the gilding, and the varnish that appear on the houses even of plebeians, are equally void of unnecessary and expensive ornaments. Those who should rely on the florid relations, in which the missionaries and some travellers have indulged, in their descriptions of the palaces of Pekin, and those of Yuen-min-yuen, would experience, on visiting them, a woful disappointment. These buildings, the common habitations of the country, are all modelled like after the form of a tent, and are magnificent only by a comparison with the others, and by their number, which is sufficient, indeed, to form a town of themselves. Their walls are higher than those of ordinary houses, their wood-

en columns of greater diameter, their roofs are immense, and a greater variety of painting and gilding may be bestowed on the different parts: but none of them exceeds one story in height, and they are jumbled and surrounded with mean and insignificant hovels. Some writer has observed, that the king of England is worse lodged at St. James's palace than any other sovereign in Europe. Were I to compare some of the imperial palaces in China to any royal residence in Europe, it would certainly be St. James's; but the apartments, the furniture, and conveniences of the latter, bad as they are, infinitely transcend any of those in China. The stone or clay floors are, indeed, sometimes covered with a carpet of English broadcloth, and the walls papered; but they have no glass in the windows, no stoves, fireplaces, or fire-grates in the rooms; no sofas, bureaux, chandeliers, nor looking glasses; no bookcases, prints, nor paintings. They have neither curtains nor sheets to their beds; a bench of wood, or a platform of brickwork is raised in an alcove, on which are mats or stuffed mattresses, hard pillows or cushions, according to the season of the year; instead of doors, they have usually screens, made of the fibres of the bamboo. In short, the wretched lodgings of the state officers at the court of Versailles, in the time of the French monarchy, were princely palaces in comparison of those allotted to the first ministers of the Emperor of China, at the capital as well as at Yuen-min-yuen." (*Barrow*, 194.)

The second enclosure, in the northern city, is called *Hoang-tching*, or the "august city," about 6 m. in circ., surrounded by walls 20 feet in height, and entered by four large, and three smaller gates. This section of the city comprises several idol temples, a depository of military stores, extensive public granaries, and a military seminary. It has also an artificial mountain in the centre of an extensive park. The third enclosure, or that called the "imperial city," contains the offices of the six superior tribunals of the empire. The Russian mission, the temple of *Yung-ho-Kung*, or "of eternal peace," the largest and most sacred edifice in the city, having connected with it an institution for the instruction of lamas for the service of Thibet. Here also is the National college, in which is concentrated all the learning and literature in China. All the literati of the empire, all the colleges and principal schools, are subordinate to this establishment, which nominates the examiners of the compositions required of candidates for civil offices. Manchoo, Chinese, and Russian literature meets here with equal attention, and all religions are sanctioned within its precincts. (*Gutzlaff's China Opened*, i., 65.) Indeed, it is somewhat strange that Pekin, the capital of the most exclusive empire in the world, should comprise, besides its numerous temples and pagodas, a magnificent mosque, a Greek church and convent, and a Roman Catholic chapel!

The S. division of the city is the grand emporium of all the merchandise brought for sale from other provinces; and as this portion is not subject, like the other, to the rigour of military discipline, it is frequented by those who are in search of business, amusement, or repose. Its buildings do not, however, require any special notice. But it should be stated that it contains an enclosure, where sacrifices are offered up to the god of agriculture, and where the imposing ceremony of the emperor holding a plough annually takes place.

There are suburbs round most of the gates of the city, some of which extend more than a mile from the wall, and comprise several large temples, with a few other public buildings.

The streets are not lighted at night. Sir John Barrow says that the cross-lanes were generally watered, but that that did not appear to be the case in the main streets. A large sheet of water, comprising several acres, within the N. division, furnishes an abundant supply to that part of the city, and to the palace; and a small stream, which runs along the W. wall supplies that neighbourhood. There are, besides, numerous wells; but the water of some of these is dreadfully nauseous; and, when mixed with tea, the well water is, to Europeans at least, particularly disgusting. But good potable water is brought from beyond the barriers. (*Hyacinthe, Ville de Pékin*, p. 13.)

"Although Pekin cannot boast, like ancient Rome or modern London, of the conveniences of common sewers to carry off the filth and dregs that must necessarily accumulate in so large a city, it enjoys one important advantage rarely found in capitals out of England: no kind of filth or nastiness, creating offensive smells, is thrown out into the streets, a piece of cleanliness that perhaps may be attributed rather to the scarcity and value of manure than to the exertions of the police officers. Each family has a large earthen jar, in which is carefully collected everything that may be used as manure; when the jar is full, there is no difficulty of coverting its contents into money, or of exchanging them for vegetables. The same small boxed carts, with one wheel, which supply the city with vegetables, invariably return to the gardens with a load of this liquid manure. Between the palace of Yuen-min-yuen and Pekin, I have met many hundreds of these carts. They are generally dragged by one person and pushed on by another; and they leave upon the road an odour that continues, without intermission, for many miles. Thus, though the city be cleared of its filth, it seldom loses its fragrance. In fact, a constant disgusting odour remains in and about all the houses the whole day long, from the fermentation of the heterogeneous mixtures kept above ground, which in our great cities are carried off in drains. To counteract these offensive smells, they make use of a variety of perfumes, and strongly scented-woods and compositions." (*Barrow*, 98.) This statement is completely borne out by that of Hyacinthe, who speaks of the *puanteur insupportable d'urine* felt in walking along the streets.

About 10 m. from Peking is a large park, belonging to the emperor, having an extent of at least 12 sq. m.; it exhibits all the great features of nature, lakes and rivers, mountains, rocks, and forests, thrown together in the boldest and most irregular manner. It comprises 30 distinct palaces, and a village of no inconsiderable size; but these palaces are ill-arranged, falling to decay, and wholly unworthy of the name.

The highest class of inhabitants is composed of the Manchoo troops and officers, most of whom are in poor circumstances, though a few possess considerable property. Next to these rank the Chinese merchants, many of whom are extremely wealthy; and below these are the artisans and other labourers, most of whom come from the provinces to procure employment. The poor are employed in cleaning and watering the streets, gardens, &c., and in cultivating the ground; but, notwithstanding the general discouragement of pauperism, and the severity of the police, it is alleged that there are in Peking many thousand persons, who, being without employment, have recourse to robbery and cheating. The cross streets are shut, and the others are patrolled at night; and in consequence, says Duhalde, *la paix, le silence, et la sûreté régnent dans toute la ville.*

Hired carriage and sedan-chairs are common in all the public thoroughfares; but the males of the higher classes almost universally ride on horseback, though many of them keep their private carriages.

Peking is indebted for its importance to its being the residence of the emperor and the seat of government; and a very large proportion of its inhabitants depend for subsistence on employment in one or other of the departments connected with the army, the administration, or the court. It is to China in respect of literature what Paris and London are to France and England. The printing and bookselling business is very extensive. A great many works, especially upon history, issue from the imperial press, and are sold at a low price to the booksellers. It is not distinguished by any peculiar manufacture, unless it be that of coloured glass; nor has it any foreign commerce or trade, other than that directed to the supply of its own wants. This, however, is necessarily very considerable. The country round the city being sandy and poor, a large portion of its supplies are brought from a distance, partly from sea by the Pei-ho, but principally by the Grand canal and the Eu-ho, which connect it with Nankin, and most of the E. provinces. Mutton and beef, however, which constitute the principal food of the Manchoos, are brought principally from Mongolia. The Chinese prefer pork; and hogs, consequently, form a principal article of import. Geese, ducks, and chickens, are the common domestic fowls; and in winter the shops are well supplied with partridges, pheasants, and other game.

A considerable portion of the taxes imposed on the different provinces is paid in kind; and a part of the rice and other grain so collected, being sent by canal to the capital, is stored in public granaries, whence it is issued to the troops, and others engaged in the public service. But notwithstanding this resource, it not unfrequently happens that the supply of corn proves deficient, and that numbers of the inhabitants are reduced to the greatest privations. Tea, of excellent quality, is the common beverage; but they also use a strong spirit made of rice.

The Pei-ho is navigable for vessels of considerable burden to Tiensing, nearly 40 m. from its mouth; and it may be ascended in flat-bottomed boats to within 12 m. or 20 m. of the capital. Peking might, consequently, be easily attacked from the sea. There are no forts or other obstacles to the navigation of the river, which might be effected, with the greatest facility, by the aid of steam tugs.

The early history of Peking is involved in obscurity; but it is generally regarded by native authorities as one of their most ancient cities. It is clear, however, from the statements of Marco Polo, who describes Peking under the name of Kambalu, that the N., Imperial, or Tartar city, was either built or restored by Kublai Khan. Marco Polo describes it, as it now exists, as having perfectly straight

streets, lined on each side with booths and shops. The Mongul dynasty, founded by Kublai, continued to occupy this city till it was expelled from China, in 1367. In 1421, the third emperor of the Chinese dynasty of Ming transferred his residence thither from Nankin, since which it has been the capital of the empire. (*Marco Polo, by Marsden,* lib. ii., cap. 7.)

PELEW ISLANDS. *See* POLYNESIA.

PELHAM, t., Hillsborough co., N. H., 37 m. S. by E. Concord, 32 m. S.W. Portsmouth. Chartered in 1748. Watered by Beaver river, a tributary of Merrimac river. It contains a Congregational church, two stores, one fulling-mill, two grist-mills, four saw-mills; six schools, 268 scholars. Pop. 1003.

PEMBERTON, p. v., Northampton t., Burlington co., N. J., 22 m. S. Trenton, 162 W. Situated on the N. branch of Rancocus creek, and contains two churches, a Methodist and Baptist, five stores, one cotton factory, one cupola furnace, one grist-mill, one saw-mill, 190 dwellings, and about 750 inhabitants.

PEMBROKE, a maritime co. of S. Wales, the most westerly in the principality, having N. St. George's channel and the co. of Cardigan, E. the latter and Caermarthen, and S. and W. the Bristol channel and St. George's channel. The coast line is very irregular, being deeply indented with arms of the sea, including Milford Haven and St. Bride's bay. Area estimated at 390,400 acres. In the N. part of the county the highest point of the Precelly mountains rises to the height of 1754 feet above the level of the sea; but with this exception the surface elsewhere is merely undulating. It is watered by the Cleddeu, Cleddy, and other streams, and owing to the number of its deep bays, it has, in most parts, every convenience for water-carriage. The soil is very various: in the S. it rests on a limestone and sandstone bottom, and is, speaking generally, very fertile: elsewhere the soil rests mostly on a slaty rock, and though not so fertile as the other, it is still, with few exceptions, far from unproductive. Principal crops, wheat, barley, and oats. Lime, shelly-sand, or marl, may almost everywhere be had; and, in fact, were this county well farmed, it would be one of the most productive districts of the empire; but we regret to have to add that its agriculture is very far behind. There is a great want of drainage, and of a proper rotation of crops: the land is often foul and exhausted and the implements of husbandry, and the mode of using them, are alike bad. Latterly, however, some improvements have been introduced. Leases for 14 years have been substituted for leases for three lives; and clauses have been inserted in the leases for the preservation of over-cropping. But a vast deal still remains to be done before agriculture in this and the adjoining Welsh counties attains to even a medium state of advancement. Owing to the great mildness and humidity of the climate, and the nature of the soil, this county is extremely well suited for grazing and dairying; and a good deal is done in both departments; the cows used in the dairies are now generally a cross with the Ayrshire breed. There are some large estates; but property is, notwithstanding, a good deal subdivided: farms are of all sizes, but mostly rather small. The modern farm-houses and offices are generally good and commodious, but many are still very inferior, and very inconveniently situated. Not a few of the older farm-houses and many of the cottages have mud walls, about 5 feet in height, with a "wattle-and-daub" chimney, and are both mean and miserable dwellings. Luckily, however, they are gradually becoming less numerous, and will, it is to be hoped, be at no distant period wholly extirpated. Average rent of land in 1810, 8s. 2½d. an acre. Anthracite coal, slate, and limestone are found in large quantities. Manufactures unimportant. Principal towns, Pembroke, Tenby, Haverfordwest, and St. David's. It is divided into seven hundreds, and 148 parishes, and returns three members to the House of Commons, one for the county, and one each for the Haverfordwest and the Pembroke districts of boroughs. Registered electors for the county in 1839-40, 3297. In 1841, Pembroke had 18,532 inhabited houses, and 88,522 inhabitants, of whom 40,343 were males, and 47,919 females. Sum contributed to the relief of the poor in 1838-39, £33,119. Annual value of real property, in 1815, £580,291; profit of trades and professions in ditto, £43,102.

PEMBROKE, a parl. bor., market-town, and seaport of S. Wales, co. its own name, of which it is the cap., on the margin of Downpool, a creek on the S. side of Milford Haven, 29 m. W.S.W. Caermarthen, and 265 m. W. London. Pop. of parl. bor., which includes the connected town of Paterchurch, 6511. The town is built on a tongue of land, dividing the creek into two branches, one of which runs on the N. side, while the smaller branch bends southward under the suburb of Monkton. It consists of one long street, running along the ridge of a hill, on which are sloping gardens; and, though it be the county town, its small size and general quietness give it more the appearance of a village. The public buildings comprise a town-hall, custom-house, and three churches, one of which is in the suburbs. St. Michael's, at the E. end of the town, is a cruciform structure of Norman architecture; St. Mary's is in the pointed style, and somewhat more ornamental: St. Nicholas', the parent church, is in the W. suburb of Monkton. The livings are united in a single vicarage, in the gift of Sir John Owen. There are several places of worship for dissenters, most of which, as well as the churches, have Sunday-schools for children of both sexes. Pembroke has also a small endowed grammar school. On a high and rocky eminence W. of the town, is the castle, an octagonal structure, nearly surrounded by water, which, on account both of its extent and beauty, ranks among the most splendid monuments of military architecture in the principality: it was built in the 11th century, and dismantled in 1649, after a brave defence by its royalist garrison. The keep is 75 feet high to the dome, and 183 feet in circumference at its base, the mean thickness of its walls being 14 feet. It consists of four stories, and is still covered in with a vaulted stone roof. About a mile N.W. the town, and within St. Mary's parish, is Paterchurch, now more commonly called Pembroke or Paterdock, from the government dock-yard transferred thither from Milford in 1814. It is pretty regularly built, and contains the houses of the persons connected with the dock-yard. It has a handsome market-place, and many good shops, most of which, however, belong to the tradesmen of Pembroke.

The dock-yard, which is defended westward by a strong battery, occupies about 60 acres of land; and when the improvements now in progress are completed, it will be one of the finest building-yards in the kingdom, capable of having on the stocks, at once, five or six first-rate ships, and several others of smaller size. There is also a small private dockyard. A very fine jetty has recently been completed, and as ships of any burden may come up to the dock, there can be little doubt that the importance of the place will rapidly increase, especially as the packet establishment has lately been removed here from Milford. "The inhabitants of Pembroke consist, at present, of shop-keepers, people of small independent fortunes, and a few persons whose business is at Pembroke dock. Pembroke serves, in a great measure, as a dépôt for the neighbouring district. Stone coal is brought from a distance of about 6 m. eastward, and bituminous coal from Swansea, Llanelly, Newport, and the S. coast in general. The articles of export are confined to cattle, corn, and butter; the imports consist chiefly of articles of ordinary consumption." (*Mun. Corp. and Bound. Rep.*) The borough of Pembroke was incorporated in the 10 Henry II.; and is divided into two wards, under a mayor, five other aldermen, and 18 councillors. It has likewise a commission of the peace, under a recorder. Corporation revenue, in 1839, £125, exclusive of £38 accruing from the sale of property. Pembroke returns one member to the House of Commons, in connexion with Tenby, Wiston and Milford. Registered electors of the united boroughs, in 1839-40, 1179. The electoral limits of the borough were left untouched by the Boundary Act, and, in 1839-40, it had 782 registered electors. It is also one of the polling places at elections for the county. Markets on Saturday: fairs, April 12, Trinity Monday, July 16, Oct. 10, and Nov. 30. (*Nicholson's Camb. Guide; Corp. Bound. Reports, &c.*)

PEMBROKE, p. t., Washington co., Me., 180 m. E.N.E. Augusta, 766 W. Bounded S.E. by Cobscook bay. It has two stores, one furnace, one forge, one grist-mill, five saw-mills; eight schools, 376 scholars. Pop. 1050.

PEMBROKE, p. t., Merrimac co., N. H., 5 m. S.E. Concord, 490 W. Bounded W. by Merrimac river, S.E. by Suncook river. It contains three churches, a Congregational, Methodist, and Freewill Baptist, five stores, one glass factory, three grist-mills, four saw-mills; one academy, 295 students; nine schools, 333 scholars. Pop. 1336.

PEMBROKE, p. t., Plymouth co., Mass, 26 m. S.E. Boston, 445 W. Watered by North river, which flows into Scituate harbour. It contains a Congregational, Methodist, and Friends church, six stores, one woollen factory, four grist-mills, 10 saw-mills; one academy 25 students, eight schools, 288 scholars. Pop. 1258.

PEMBROKE, p. t., Genesee co., N. Y., 262 m. W. Albany, 387 W. Watered by Tonawanda creek. It contains a Presbyterian church, four stores, one furnace, eight saw-mills; 13 schools, 381 scholars. Pop. 1970.

PEMIGEWASSET, r., N. H., the principal constituent stream of Merrimac river, rises in Franconia, and flowing S. It retains this name until it receives Winnipiseogee river, the outlet of Winnipiseogee lake, when it assumes the name of Merrimac river to its entrance into the Atlantic ocean.

PENDLETON, county, Va. Situated toward the N. part of the state, and contains 999 sq. m. Its surface is an elevated table land between two ridges of the Alleghany

mountains. The south branch of Potomac river flows through it, from S.W. to N.E. It contained in 1840, 14,161 neat cattle, 20,973 sheep, 12,777 swine; and produced 65,795 bushels of wheat, 33,547 of rye, 130,010 of Indian corn, 8189 of buckwheat, 51,168 of oats, 35,645 of potatoes, 112,151 pounds of sugar. It had 11 stores, one fulling-mill, one cotton factory with 500 spindles, three flouring-mills, 31 grist-mills, 46 saw-mills, one oil-mill, four powder-mills, eight tanneries, 44 distilleries, one pottery; 12 schools, 225 scholars. Pop.: whites, 6445; slaves, 468; free coloured; 23; total, 6490. Capital, Franklin.

PENDLETON, county, Ky. Situated in the N. part of the state, and contains 450 sq. m. Watered by Licking river and its branches. It contained in 1840, 4,136 neat cattle, 6475 sheep, 13,108 swine; and produced 31,175 bushels of wheat, 5637 of rye, 225,021 of Indian corn, 31,232 of oats, 3562 of potatoes, 107,573 pounds of tobacco. It had six stores, one flouring-mill, eight grist-mills, four saw-mills, two tanneries, two distilleries; 10 schools, 265 scholars. Pop.: whites, 4013; slaves, 437; free coloured, 5; total, 4455. Capital, Falmouth.

PENDLETON, p. t., Niagara co., N. Y., 7 m. S.W. Lockport, 284 m. W. by N. Albany, 406 W. Bounded S. by Tonawanda creek. The Erie canal passes through it. It contains three stores, one saw-mill; seven schools, 230 scholars. Pop. 1098.

PENDLETON, p. v., Anderson dist., S. C. Situated on Eighteen mile creek, a branch of Seneca river. It was the former capital of Pendleton dist. It contains three churches, an Episcopal, Methodist, and Presbyterian, two academies, seven stores, one printing-office, one weekly newspaper, and about 633 inhabitants.

PENFIELD, p. t., Monroe co., N. Y., 8 m. E. Rochester, 222 m. W. by N. Albany, 371 W. Drained by Irondequoit creek. It contains six churches, two Presbyterian, a Methodist an Episcopal, and two Baptist, an academy called the Pennfield Lyceum, five stores, one fulling-mill, one flouring-mill, two grist-mills, two saw-mills; one academy, 170 students; 16 schools, 943 scholars. Pop. 3942.

PENKRIDGE, a market town and par. of England, co. Stafford, E. div., hund. Cuttlestone, on the Penk (a tributary of the Trent), crossed here by a stone bridge, 5 m. S. Stafford. Area of township, 14,500 acres. Pop. in 1841, 3129. The town, which is very ancient, is supposed by some to be the Pennocrucium of the Romans. The church, formerly collegiate, is a large building, in the early English style, with a square tower: the living is a curacy in the gift of Lord Littleton. The Wesleyan Methodists and Baptists have likewise their respective places of worship, with attached Sunday-schools. There is also an excellent charity school, in which 12 boys and eight girls are boarded, clothed, and instructed; and a national school, established in 1816, furnishes instruction to about 200 children of both sexes. Petty sessions for the hund. are held here; and Penkridge is one of the polling-places at elections for the S. division of the county. It is also the chief place of a poor-law union, comprising 21 parishes. Market disused; fairs, April 30, and first Monday in September, a very large horse fair.

PENNINGTON, p. v., Hopewell t., Mercer co., N. J., 8 m. N. Trenton, 177 W. It contains two churches, a Presbyterian and Methodist, an academy, a public library, three stores, about 50 dwellings, and 300 inhabitants.

PENN'S NECK, LOWER, t., Salem co., N. J., 5 m. N.W. Salem. Bounded W. and S.W. by Delaware river. It contains three churches, a Methodist, Episcopal, and Presbyterian, four stores, one grist-mill; five schools, 185 scholars. Pop. 1919.

PENN'S NECK, UPPER, t., Salem co., N. J. It has two churches, a Methodist and Friends, four stores; two schools, 95 scholars. Pop. 1854.

PENNSYLVANIA, one of the middle U. States, is bounded N. by New-York and lake Erie; E. by New-York and New-Jersey, from which it is separated by Delaware river; S. by Delaware, Maryland, and Virginia; W. by Virginia and Ohio. It is between 39° 43' and 42° 15' N. lat., and between 74° 44' and 80° 34' W. long.; and between 2° 17' E., and 3° 31' W. long. from W. It is 310 m. long, and 160 broad, containing nearly 47,000 sq. m., or 30,080,000 acres. The population in 1790 was 434,373; in 1800, 602,545; in 1810, 810,091; in 1820, 1,049,313; in 1830, 1,346,672; in 1840, 1,724,033. Of these, 884,770 were white males; 831,345 were white females; 22,75₤ were coloured males; 25,109 were coloured females. Employed in agriculture, 207,533; in commerce, 15,338; in manufactures and trades, 105,883; in mining, 4603; in navigating the ocean, 1815; do. canals, rivers, and lakes, 3951; in the learned professions, 6706. The state is divided into 58 counties, three of which have been formed since the census of 1840, viz., *Clarion* formed in 1839, *Carbon* in 1843, *Elk* in 1843. The following were the counties, and their population, according to the census of 1840.

Counties.	Pop. 1840.	Counties.	Pop. 1840.
Eastern District.		Cambria .	11,256
Adams .	23,044	Centre .	20,492
Berks .	64,569	Clearfield .	7,834
Bucks .	48,107	Clinton .	8,323
Chester .	57,515	Columbia .	24,267
Cumberland .	30,953	Crawford .	31,724
Dauphin .	30,118	Erie .	31,344
Delaware .	19,791	Fayette .	33,574
Franklin .	37,793	Greene .	19,147
Lancaster .	84,203	Huntingdon	35,484
Lebanon .	21,872	Indiana .	20,782
Lehigh .	25,786	Jefferson .	7,253
Monroe .	9,879	Juniata .	11,080
Montgomery .	47,241	Luzerne .	44,006
Northampton .	40,996	Lycoming .	22,649
Perry .	17,096	M'Kean .	2,975
Philadelphia .	258,037	Mercer .	32,873
Pike .	3,832	Mifflin .	13,092
Schuylkill .	29,053	Northumberland	20,027
Wayne .	11,848	Potter .	3,371
York .	47,010	Somerset .	19,650
Total .	905,744	Susquehanna .	21,195
		Tioga .	15,498
Western District.		Union .	22,787
Alleghany .	81,235	Venango .	17,900
Armstrong .	28,365	Warren .	9,278
Beaver .	29,368	Washington .	41,279
Bedford .	29,335	Westmoreland .	42,699
Bradford .	32,769	Total .	815,289
Butler .	22,378	Total of the State .	1,724,033

Harrisburg, on the E. bank of Susquehanna river, 98 m. W.N.W. from Philadelphia, is the capital, though Philadelphia is much the largest city in the state.

The surface of Pennsylvania is greatly diversified. There are few large tracts of level land in the state. The southeastern counties, though they can scarcely be denominated hilly, have an undulating and variable surface. South mountain extends from the Delaware below Easton in a S.W. direction through the state, to the borders of Maryland in Adams county. Next to this, Kittatiny, or Blue mountain, extends from the Delaware Water Gap, and proceeds S.W. with a regular elevation of from 700 to 1900 ft. above the level of the streams at its base, and terminates at Parnell's knob, an elevated and picturesque summit in Franklin county, near the S. border of the state. N. of the Blue mountain, and between the Lehigh and Susquehanna, is a wild, mountainous region, where the anthracite coal is found. This region is interspersed with high and barren ridges in close succession, interlocking with each other, and enclosing long and pointed valleys between them. The Second and Sharp mountains are between Kittatiny and the first coal basin. Next comes broad mountain, an irregular elevation, with a broad and barren table-land at its top. E. of the Susquehanna are several ridges with various names. The valley of Wyoming is enclosed by a chain of lofty mountains, known by many local names. Between Kittatiny, or Blue mountain, and the Alleghany, is what has been called the Appalachian chain, which consists of high and nearly parallel ridges, sometimes approaching near to each other, and at other times with valleys between them of 20 miles in breadth, frequently divided by smaller ridges. The elevated range, called the great Alleghany, extends nearly across the whole state, presenting on its southeastern side a steep ascent, but extending north and west with a gentle descent, and consists of an elevated and undulating table-land. Beyond the Alleghany are Laurel hill and Chesnut ridge, which are high ridges running parallel with the Alleghany ridge.

Most of the country W. of the Alleghany mountains is a hilly country, with many irregular and abrupt elevations, not disposed in chains. The soil of the state is generally good, and much of it is of a superior quality; the richest tract is on the S.E., on both sides of Susquehanna river. This part of the state has been long settled, and is under excellent cultivation. In the country W. of the Alleghany mountains there is much fertile land. For some distance from the mountain, the country is hilly and rough. The more level tracts, particularly along the streams, are highly fertile. Between the Alleghany river and lake Erie, and the western border of the state, the soil is excellent. By far the most important production of the state is wheat, and next in value to that is Indian corn. Rye, oats, barley, buckwheat, hemp, and flax, are also extensively cultivated. Apples, cherries, and peaches are abundant, and much cider is made. Although the state, as a whole, is better adapted to grain than to grazing, yet in many parts there are large dairies; and fine horses and cattle are raised.

There were in the state in 1840, 361,558 horses or mules, 1,161,576 neat cattle, 1,755,597 sheep, 1,485,360 swine. Poultry was produced to the value of $081,979. There were produced 12,993,218 bushels of wheat, 6,544,654 of rye, 2,096,916 of buckwheat, 14,077,363 of Indian corn, 206,658 of barley, 20,485,747 of oats, 9,477,343 of potatoes; 3,028,657 pounds of wool, 48,694 of hops, 32,708 of wax,

325,018 of tobacco, 2,965,755 of sugar, 7262 of silk cocoons, 1,302,685 tons of hay, 2644 of hemp or flax. The products of the dairy were valued at $3,152,987; of the orchard at $610,512; of lumber at $1,146,355. There were made 14,298 gallons of wine.

The mineral wealth of Pennsylvania is very great, consisting of coal, iron, and salt, which are abundant. The immense coal regions form the most interesting feature of the mineral resources of the state. Bituminous coal of an excellent quality, and inexhaustible in quantity, is almost everywhere found west of the Alleghany mountains, and in the S. part of the state, to the E. of them. In Pittsburg and the vicinity, it is extensively used for manufacturing purposes. In this region salt springs occur, which afford a very strong brine. The *anthracite* coal region, with some few exceptions, is bounded on the N.W. by the N. branch of Susquehanna river, extending in a N.E. direction for over 60 m., and divided into the southern, middle, and northern coal-fields. The southern coal-field is divided into four mining districts; the Lehigh, Schuylkill, Swatara, and Susquehanna. The middle coal field includes the Beaver Meadow, Hazelton, Mahanoy, and Shamokin districts. Towards the eastern part, both the southern and middle coal-fields afford a hard, shining, and very compact coal, difficult of ignition, but affording an intense heat; while, farther west, the coal is less dense and more easy of ignition, and yielding both white and red ashes. The northern coal-field is, like the others, about 65 m. long, and 5 or 6 broad, and from 1 to 30 feet in thickness, and includes the Wyoming and Lackawanna coal. The latter is sent by the Delaware and Hudson canal to Hudson river, and thence to New-York, Albany, &c. This coal is lighter than most of the other kinds of anthracite coal, and more easy of ignition, and yields a good heat, though less intense than the Lehigh, and some others. The three great deposites of anthracite coal have been calculated to contain 975 sq. m., or 624,000 acres, in some places 50 or 60 feet deep; and as each cubic yard in the ground is calculated to yield a ton of coal, it is easy to conceive that the quantity must be immensely great. The following is the amount produced in the several mining districts in 1842. Lehigh, Beaver Meadow, &c., 272,126 tons; Schuylkill, 540,892; Swatara, 39,381; Lyken's valley, 4864; Shamokin, 10,000; Wyoming, 47,346; Lackawanna, 205,253; making a total in 1842 of 1,112,862 tons, though the business of mining is yet but in its infancy. In 1820, only 365 tons of anthracite coal were sent to market. The *bituminous* coal region of Pennsylvania has been estimated at 21,000 sq. m., or 13,440,000 acres, over which it is scattered. It has been computed that nearly 8,000,000 bushels, or 980,000 tons of this coal are annually consumed at Pittsburg alone. This coal constitutes a great resource for the steamboats on the western waters.

The climate of Pennsylvania, though healthy and temperate, is variable and inconstant. The extremes of temperature are from 20° below zero of Fahrenheit to 98° above; but such extremes are of short continuance. The mean temperature is from 44 to 52 degrees. The greatest heat usually occurs in July, and the greatest cold in January. In the southern and southwestern parts, winter sets in in severity in the latter part of December, and snow disappears in the latter part of February or the first of March. In the northern and more elevated parts, winter commences early in December, snows are deep, and generally cover the ground until the latter part of March. The heat of summer during the day is as intense here as in the less elevated parts of the state, but the mornings and evenings are much cooler. Frosts often appear in September. The autumn is usually a delightful season in all parts of the state.

The Delaware river washes the eastern border of the state, and is navigable for ships of the line to Philadelphia. The Lehigh, after a course of 75 m., enters it at Easton. The Schuylkill is about 130 m. long, and unites with the Delaware, 6 m. below Philadelphia. The Susquehanna river rises in the state of New-York, and flows S. through this state, and enters Chesapeake bay in Maryland. It is much obstructed by rapids and falls, but furnishes a descending navigation for boats and rafts, in time of high water. The Juniata rises among the Alleghany mountains, and, after a course of 180 m., enters the Susquehanna, 11 m. above Harrisburg. The Alleghany river, 400 m. long, from the N.W.; and the Monongahela, 300 m. long, from the S., unite at Pittsburg to form Ohio river, which latter flows a short distance in this state. The Youghiogheny is a small river which enters the Monongahela on the E. side.

Philadelphia, between the Delaware and Schuylkill rivers, and Pittsburg, at the junction of the Monongahela and Alleghany rivers, are the two most commercial places in the state. The other principal towns are Lancaster, Reading, Harrisburg, Easton, York, Carlisle, Alleghany, and Erie.

The exports for the year ending September 30th, 1841

$5,152,501, and of the imports $10,346,698. The tonnentered was 99,385; and cleared, 83,383. There were in 1840, 153 commercial and 178 commission s engaged in foreign trade, with a capital of $3,682,811; retail dry goods and other stores, with a capital of 9,178; 5964 persons engaged in the lumber trade, employed in d transportation, who, with 446 butchers, packers, aployed a capital of $737,830; 58 persons employed potteries, with a capital of $16,460.

manufactures of Pennsylvania are extensive. According to the census of 1840, home-made or family manufactures amounted to $1,392,439; 235 woollen manufactories—237 fulling mills employed 2909 persons, producing to the amount of $2,338,061, employing a capital of 46; 108 cotton manufactories with 146,494 spindles, and 5389 persons, producing articles to the amount of 17, with a capital of $3,325,400; 2997 persons producing 59,686 tons of anthracite coal, with a capital of 2, 1798 persons produced 11,620,654 bushels of us coal, with a capital of $300,416; 213 furnaces 98,395 tons of cast iron, and 169 forges, &c., producing 144 tons of bar iron, employing 11,522 persons, and of $7,781,471; 87 paper-mills produced to the $792,225, and other manufactories of paper to t of $65,500, the whole employing 704 persons, al of $581,800; hats and caps were manufactured unt of $919,431, and straw bonnets to the amount employing 1487 persons, and a capital of $449,107; ries employed 3393 persons, and a capital of 2132 other manufactories of leather, as saddlerduced articles to the amount of $3,453,243, employed $1,249,923; 255 persons produced 549,478 dt. with a capital of $191,435; 30 powder-mills d 1,184,225 pounds of gunpowder, employing nd a capital of $66,800; drugs, paints, &c., employed persons, produced to the amount of $2,100,074, ne and varnish to the amount of $7865, the ring a capital of $2,179,625; 28 glass-houses, cutting establishments employed 835 persons, cles to the amount of $772,400, with a capital 182 potteries employed 322 persons, producing amount of $157,902, with a capital of $75,562; produced machinery to the value of $1,993,752; produced hardware and cutlery to the amount 28 persons produced five cannon, and 21,571 5 persons manufactured the precious metals amount of $2,670,075; 536 persons manufactured and marble to the amount of $443,610; produced bricks and lime to the amount of persons manufactured carriages and wagt of $1,903,732, with a capital of $559,831; produced 6,238,768 gallons of distilled spirles produced 12,765,974 gallons of beer, the 1601 persons, and a capital of $1,585,771; produced 1,181,530 barrels of flour, and, mployed 7916 persons, producing articles 9,232,513, with a capital of $7,773,784; ectured 5,097,090 pounds of soap, 3,316,843 indles, and 5002 pounds of spermaceti or rhole employing a capital of $394,442; o the amount of $908,015; 2357 persons ure to the amount of $1,151,187, with a 1991 brick houses, and 9406 wooden employing 9881 persons, and cost ring-offices, 46 binderies, 12 daily, 10 2 weekly newspapers, and 49 periodical ersons, and a capital of $680,340. The nal employed in state manufactures.

imerous. The university of Pennsylvania n Philadelphia, in 1755; Dickinson 1783; Jefferson college in Cannons gton college in Washington, in 1806; leadville, in 1815; Pennsylvania college 1832; La Fayette college in Easton, ge at Mercersburg, in 1836. Besides spartment of the university of Penniladelphia, in 1765; Jefferson medibhia, in 1824; the medical department oliege in Philadelphia, in 1839; the ' the Lutheran church at Gettysainary of the German Reformed 5; the Western theological semi-828; the theological seminary at theological seminary at Pittsburg. here were, in 1840, 2034 students. 90 academies, with 13,970 students, ury schools, with 179,989 scholars. 13,940 white persons over 20 years ' read nor write.

inations, the Presbyterians, inclu-

ding the Associate Reformed, had in 1836 460 ministers; the Methodists, 250; the Baptists, 140; the German Reformed, 73; the Episcopalians, 70; the Friends, 150 congregations. Besides these are several other denominations, less numerous.

In January, 1840, there were 49 banks, with an aggregate capital of $24,286,405, and a circulation of $9,338,636. The state debt amounted, in January, 1843, to $37,937,788.

By a splendid course of internal improvements, Pennsylvania has greatly extended and facilitated her trade, but has contracted the largest debt of any state in the Union, which she is abundantly able to discharge, but can only do it by a moderate taxation. The central division of the Pennsylvania canal commences at the termination of the Columbia and Philadelphia railroad at Columbia, and extends along the Susquehanna and Juniata rivers, 172 m. to Hollidaysburg, where it passes over the Alleghany mountain by a railroad. The western division of the Pennsylvania canal extends from Johnstown to Pittsburg, 104 m. This completes the line of railroads and canals from Philadelphia to Pittsburg, 395 m. A canal extends from the Pennsylvania canal at the mouth of the Juniata river, and proceeds 39 m. to Northumberland, where it connects with the North and West branch canals. The West branch canal extends from Northumberland along the W. branch of Susquehanna r., 73 m. to Farrandsville, in Clinton county, reaching the bituminous coal region in that vicinity. The North branch division extends from Northumberland, 73 m. to a little above Wilkesbarre. The Delaware division of the Pennsylvania canal extends from tide water at Bristol, 60 m. above Philadelphia, to Easton, at the mouth of the Lehigh, where it joins the navigation of the Lehigh company, extending to the coal region, 25 m. The Schuylkill navigation commences at the Fairmount dam near Philadelphia, and extends to port Carbon in Schuylkill county, the heart of the coal region. The Union canal extends from the Schuylkill near Reading to Middletown on the Susquehanna, 82 m. It has a navigable feeder of 23 m. on Swatara creek, which communicates with the coal region. The Susquehanna or Tidewater canal extends from Wrightsville opposite to Columbia, and extends 45 m. to Havre de Grace in Maryland, and connects the Pennsylvania canal with the tidewater of Chesapeake bay. The Philadelphia and Reading railroad extends from the W. side of the Schuylkill near Philadelphia, to Pottsville in Schuylkill county, 90 m., giving a ready access to the coal region. There are many minor railroads which have relation to the transportation of coal. The Alleghany Portage railroad extends from Hollidaysburg to Johnstown, 36½ m., and connects the eastern and western divisions of the Pennsylvania canal. The Philadelphia and Columbia railroad, one of the most important in the state, extends from Philadelphia, 82 m., to Columbia on the Susquehanna. A railroad extends through the S.E. part of the state, leading from New-York to Baltimore and Washington.

The governor is chosen by the people for three years, but cannot hold the office more than six years in nine. He must be 30 years of age, and have resided in the state for seven years. The senate consists of 33 members, elected by the people for three years, one third being chosen annually. A member must be 25 years of age, and have resided four years in the state, and the last year in the district for which he is chosen. The house of representatives consists of 100 members elected annually by the people. A member must be 21 years of age, have resided in the state three years next preceding his election, and the last year in the district from which he is chosen. All judicial offices are appointed by the governor, with the advice and consent of the senate. The judges of the supreme court hold their offices for 15 years. The president judges of the court of common pleas hold their offices for 10 years. The associate judges of the court of common pleas hold their offices for five years. The secretary of state is appointed by the governor, and holds office during his pleasure. The treasurer is elected annually, by the joint ballot of both houses of the legislature. Every white person of the age of 21 years, who has resided in the state for one year next preceding an election, and ten days in the district where he offers his vote, and has paid a state or county tax, enjoys the right of suffrage. The legislature meets annually at Harrisburg on the first Tuesday of January.

In 1638 Pennsylvania was first settled by the Swedes, who purchased from the natives the land upon the western shore of Delaware bay and river, from cape Henlopen to the falls opposite the present city of Trenton. In 1642, the Swedish governor erected a handsome house for himself on an island just below the mouth of the Schuylkill, and caused a church to be built, which was consecrated in 1646. The Dutch government at New Netherlands conquered the Swedes in 1654. When the English conquered New Netherlands in 1664, the Dutch possessions on Delaware river fell into their hands, and for several years remained subject to the governors of New-York. In 1681 Pennsylvania was granted by Charles II. to William Penn, a member of the society of Friends, in consideration of the services of his father as a British admiral. Four years from the grant of the charter, the province contained 22 settlements, and Philadelphia had 2000 inhabitants. In 1684 Penn returned to England, and appointed five commissioners, with a president, to administer the government during his absence. He returned to the state in 1699, having been confirmed in his proprietary rights. In 1768, Mason and Dixon's line was drawn, to mark the boundary between this state and Maryland. Penn died in 1718, leaving his interest in Pennsylvania as an inheritance to his children, who continued to possess it until the revolutionary war, when their claim was purchased by the commonwealth for £130,000 sterling, or $580,000. In 1784, the last remaining portion of the state, not previously purchased, was bought of the Indians, lying in its N.W. part. In 1776 a state constitution was formed, which continued in operation until 1790, when another constitution was adopted, which remained until 1838, when the present constitution was adopted. The United States constitution was adopted in convention December 13th, 1787 ; yeas 46, nays 23 ; majority, 23. Philadelphia remained the seat of the United States government until 1800, when it was removed to Washington.

PENN TOWNSHIP, t., Philadelphia co., Pa. Situated N.W. of the city of Philadelphia, to which it joins, and of which it is a suburb. Bounded W. by Schuylkill river. Drained by Falls run, which flows into Schuylkill river, and affords some water-power. It is the seat of Girard orphan college, founded by the bequest of $2,000,000, for a description of which *see* PHILADELPHIA.

PENN-YAN, p. v., Milo t., capital of Yates co., N. Y., 192 m. W. Albany, 329 W. Pleasantly situated on the outlet of Crooked lake, which affords water-power. It is half a mile below the foot of the E. arm of the lake, and contains a courthouse of brick with four Doric columns in front, a jail of stone, a fire-proof county clerk's office, five churches, a Presbyterian, Congregational, Episcopal, Methodist and Baptist, a bank with a capital of $100,000, an academy; 20 stores, two flouring-mills, two saw-mills, one furnace, various mechanic establishments, two printing-offices, one bindery, one weekly 'newspaper, about 300 dwellings, and 1800 inhabitants. The Crooked lake canal, 8 m. long, passes through the place to Seneca lake, and thence to the Erie canal.

PENOBSCOT, river, the largest river of Maine, rises by two main branches. The western branch which is the largest rises in the high lands near the northern boundary of the United States, and not far from some of the head waters of the Chaudière river which flows into St. Lawrence river. After flowing a considerable distance E., it enters and passes through Chesuncook lake, whence flowing S.E. it in the like manner passes through Pemadumcook lake, and continuing S.E. it unites with the eastern branch, 54 m. in a right line N. by E. of Bangor. The eastern branch, called Sebools river, rises near the sources of the Aroostook river, and flows nearly S. to the junction. It continues S.E. and 60 m. above Bangor it receives the Mattawankeag, its largest eastern branch. It then proceeds S.S.W. until the entrance of the Piscatiquis, its largest eastern branch, and thence S. by W. to its entrance into Penobscot bay, at Owl's Head. Before reaching tide-water it has many falls and rapids, affording extensive water-power. Its whole length· from its source to the ocean is about 275 m. It is navigable 50 m. from the ocean to Bangor, for vessels of the largest class, and for boats some distance above that. The tide generally rises at Bangor 17 or 18 ft. This river is the medium of an immense lumber trade, which extensively centres at Bangor. There are several important towns on the river. On the E. side from its mouth including the bay, are Castine, Bucksport, and Orrington ; and on the W. side are Thomaston, Camden, Belfast, Prospect, Frankfort, Hamden, Bangor and Orono. *Penobscot bay* is a fine body of water, extending from the ocean at Owl's Head, 20 m. to Belfast bay ; and across the mouth of the bay from Owl's Head to Burnt Coat island, it is about 30 m. wide. It contains a number of islands, the principal of which are Deer island, Fox islands, Isle au Haut, Long island and others. From an elevated summit back of Camden, and from other points, this bay with its beautiful islands, and numerous vessels under sail, presents a delightful prospect. The bay and river contain many fine harbors, the principal of which are those of Castine, Belfast, Frankfort, Bucksport, Bangor, and other. The total tonnage of the district above Belfast in 1840, was 37,130 ; of Belfast, 38,218.

PENOBSCOT, county, Me. Situated in the N. part of the state, and is of great extent, contain 3282 sq. m. The S. part only is thickly settled, and the N. part remains a wilderness, but not for want of a fertile soil. Watered by Penobscot river and its branches. It contained in 1840, 19,416

neat cattle, 35,319 sheep, 9449 swine ; and produced 131,041 bushels of wheat, 9767 of rye, 35,694 of Indian corn. 350s of buckwheat, 7919 of barley, 103,596 of oats, 922,699 of potatoes, 12,303 pounds of sugar. It had 11 commercial and commission houses in foreign trade, 226 retail stores, two lumber-yards, 10 fulling-mills, two woollen factories, four flouring-mills, 28 grist-mills, 942 saw-mills, one paper mill, 21 tanneries, one pottery, three printing offices, two binderies, one daily and two semi-weekly newspapers. four academies, 275 students, 270 schools, 14,797 scholars. Pop. 45,705. Capital, Bangor.

PENOBSCOT, p. t., Hancock co., Me., 8 m. N. by E. Castine, 78 m. E. by N. Augusta, 674 W. Bounded W. by Penobscot river. An arm of Penobscot bay extends into the t., affording facilities for navigation. It exports lumber and has two stores, three grist-mills, three saw-mills ; 13 schools, 526 scholars. Pop. 1474.

PENRITH, a market town and par. of England, county Cumberland, ward Leath, in a valley watered by the Eamont and Lowther, which unite their streams about 1 m. below the town, 17 m. S.S.E. Carlisle, and 43 m. N. Lancaster. Area of parish 6640 acres. Population, in 1841 was 6429. The town, which mostly consists of a principal street along the line of road from Kendall to Carlisle, is clean and neat, built chiefly of red free-stone, much improved of late years. The church is a large and handsome structure, nearly rebuilt in the beginning of last century on its walls are many curious old inscriptions, and in the church-yard is a rude monument called the Giant's Tomb, consisting of two stone pillars 10 ft. high, and 13 ft. apart. The living is a vicarage in the gift of the bishop of Carlisle. The Independents, Wesleyan and Primitive Methodists, and the Secession church of Scotland have also their respective places of worship, with attached Sunday schools. A free-school was founded here by queen Elizabeth, and there are several charity-schools. On an eminence W. of the town are the remains of Penrith castle, a square structure, surrounded by a deep foss and rampart. It is supposed to have been built during the wars of the roses and was destroyed in the time of the Commonwealth. Northward is an excellent enclosed race-course, on which races take place at the beginning of October. The inhabitants are principally employed in agriculture and lace-weaving ; but the town has also a considerable retail trade, and large markets. Its situation, too, on the great road to the W. of Scotland, and in the neighbourhood of the lakes, occasions a large influx of visiters. Quarter and petty sessions are held here ; besides which there is a court for the recovery of debts under 40s. Penrith is also one of the polling-places at elections for the E. division of the county. Markets on Tuesday and Saturday : fairs April 25th and 26th, Sept. 27th, and Nov. 11th, for horses, cattle, &c.

Penrith is a town of considerable antiquity, and often suffered in the border wars. The chief objects of interest in its vicinity are Brougham hall, Eden hall, Greystoke and Dacre castle, the Giant's cave, and king Arthur's round table, with other British antiquities scattered over the district. (*Tottersall's Guide to the Lakes*, p. 95-96.)

PENRYN, a parl. and mun. bor., market town, and township of England, par. of St. Glavias, co. Cornwall, and E. div. hund. Kerrier, on the slope of a hill at the mouth of a small river running into Falmouth harbour, 4 m. N.W. Falmouth (of which, indeed, it may almost be considered a suburb), and 78 m. S.W. Exeter. Pop. of new mun. bor. in 1831, 3042 ; and of parl. bor. (which includes the neighbouring bor. of Falmouth), 11,531. The town consists chiefly of one wide street, crossed by three or four others of inferior size, its principal buildings being the town-hall (with a small attached jail) and a custom-house. The church is a large plain building the living is a curacy subordinate to the vicarage of St. Glavias, the church of which is on the opposite side of the river. The Wesleyan Methodists and Baptists have, also, their respective places of worship ; and there are three Sunday schools. " Penryn possesses no manufacture of consequence, nor is the general trade such as to warrant the expectation that the town will increase materially in wealth or importance. The commodities imported are confined to those required for the consumption of the town, and for the use of the mines in the immediate neighbourhood. The principal and almost only export is the granite which is quarried on the moors a few miles from the town ; and this trade has of late been on the decline. There appears, however, to be a considerable general trade ; and the shops are very numerous." (*Parl. Boun. and Corp. Reps.*)

The borough of Penryn was incorporated in 18 James I. Its municipal boundaries were considerably enlarged by the Municipal Reform Act ; under which its corporate officers consist of a mayor, three other aldermen, and 12 councillors. Corporation revenue in 1839, £2264. Penryn sent two members to the House of Commons since the n of James I., the right of election down to the Reform having been in freeholders, resident leaseholders for ears, and householders, after a residence of six months. Boundary Act enlarged the electoral limits of the parentry borough so as to include/with the old borough ntire parish of Falmouth, with portions of the parish Glivias and Budock. Registered electors in 1839—40, Markets on Wednesday and Saturday. Fairs, May ily 7, and Dec. 21.

\8ACOLA, bay, Florida. Sets up from the gulf of o. and 11 m, from its mouth divides into three parts, Escambia bay, Yellow Water bay, and East bay; last is 7 m. long, and admits frigates of the largest and is entirely sheltered from all winds. Its enbetween fort Barancas, and the W. end of St. Rosa is about three fourths of a mile wide, and is well nd. The bar has 22 ft. of water at low tide, the ' is completely land locked, and is the best on the Mexico.

iCOLA, city, port of entry, and capital of Escam-Flor., 962 m. W. Tallahassee, 1080 W. Situated and sandy plain, 40 or 50 ft. above the level of the , Pensacola bay, 10 m. from its entrance into the lexico. It is regularly laid out with streets crossother at right angles. It has two public squares by 360 ft., and contains a courthouse, jail, two a Roman Catholic and an Episcopal, a marketustom-house, a public storehouse, and about 9000 . A wharf extends 500 or 600 feet into the bay. d States navy yard is 8 m. from the city, and 5 ntrance of the harbour, and covers 80 acres of losed by a high brick wall. It contains houses cers, and a naval store and other buildings, the wants of the establishment. The grounds nely laid out. The tonnage of the port in 1840,

i government of Russia in Europe, between 52° V. lat., and 43° 26' and 46° 41' E. long., having nment of Nijnii Novgorod, E. Simbirsk, S. Sarv, Tamboff; area, 14,350 sq. m. Pop. (1838), rhce, flat or feebly undulating; soil, extremely ite, mild. Rivers numerous, but except the Mokcha, affluents of the Wolga, the others mportance. Produce of the corn-crops esti-9,000,000 to 10,000,000 chetwerts, of which is are exported. Considerable attention is ding of cattle, sheep, and horses, particularly reats very extensive and valuable. There on mines near Troitsk; and in some parts ries of mill-stones. Large quantities of l woollen stuffs are prepared in the cottages r; and there are besides several consideruifactories; with tan-works, soap-works, a foundries, &c.; in 1838, there were six beet-root sugar. (Schnitzler, La Russie,

. of the above government, on its S.E. ' Sura; lat. 53° 11' N., long. 45° 38' E. Houses and government offices of wood; a large building, and some of the churches. Tanning and the manufacture of soap rried on; and it has a considerable com-

mun. bor., seaport, market town, and nd, par. Madron, near the W. extremity l on the N.W. side of Mount's bay, only nd's End, and 96 m. W.S.W. Exeter. The town consists chiefly of four ight angles, in the market place; they i, and, for the most part, lined with s. A handsome town-hall, chapel of the parish of Madron, and places of aspectively to Methodists, Presbyteriof Friends, are its chief public buildhool, three Sunday schools, a dispenr society, and other societies, as well um, belonging to the geological socie-i has its head quarters in this town. vantages, arising from soil, climate, e is the market town of an extensive orn which the produce of the neigh-iberties is exported in considerable role, the town is fast increasing both r, and many good houses have been t three or four years." (Mws. and d copper are extensively wrought in shery of pilchards, whitings, &c., is activity. The gross customs' duties n 1840, amounted to £29,562. The except for the smaller class of vesvater springs being only 13 ft., and at

neaps only 9 ft. (Purday's English Channel, p. 86.) The pier is upward of 600 ft. in length, having a lighthouse at its extremity. About 100 ships, of the aggregate burden of about 5000 tons, belong to the port. The mildness of the climate, and the fertility as well as beauty of the surrounding district, render it a desirable residence for invalids, many of whom are settled here, and for whose accommodation baths, libraries, boarding-houses, &c., have been established within the last few years. The scenery of Mount's bay is also extremely fine, and on its N.E. side is St. Michael's mount, a rock of conical form, having a base of nearly a mile in extent, and gradually diminishing to the summit, which is crowned with a chapel, its tower being 230 ft. above low-water mark.

The borough of Penzance was incorporated in the 12 James I., when it was also made one of the coinage-towns of the duchy of Cornwall. The Municipal Reform Act considerably enlarged its limits, and, at the same time, it was divided into two wards, its municipal officers being a mayor, five other aldermen, and 18 councillors. It has a commission of the peace, under a recorder, and a civil court for the recovery of debts under £50. Corporation revenue in 1839, £3164, exclusive of £4500 money borrowed.

Petty sessions for the W. division of the hundred are held here, and Penzance is one of the polling places at elections for the W. division of Cornwall. Among other distinguished citizens, Penzance has to boast of Sir Humphry Davy, born here on the 17th of December, 1778. He also received his early education, and served an apprenticeship as surgeon, in the town. Markets on Thursday and Saturday; large fairs, Thursday before Advent, and Thursday after Trinity Sunday, for cattle and farming produce. (Municipal Corp. and Boundary Reports, &c.)

PEORIA, county, Ill. Situated toward the W. part of the state, and contains 648 sq. m. Organized in 1825. Bounded E. by Illinois river, which here has the expansion called Peoria lake, wider than the river, and 20 m. long. Watered by Spoon river, and Kickapoo creek. It contained in 1840, 4848 neat cattle, 3554 sheep, 24,077 swine; and produced 43,240 bushels of wheat, 215,540 of Indian corn, 5010 of barley, 68,416 of oats, 30,039 of potatoes. It had two commission houses in foreign trade, 19 retail stores, one tannery, one brewery, two printing-offices, two weekly newspapers; 37 schools, 1161 scholars. Pop. 6153. Capital, Peoria.

PEORIA, p. v., capital of Peoria co., Ill., 70 m. N. Springfield, 784 W. Situated on the W. bank of Illinois river, at the bottom of Peoria lake, and formerly called fort Clark. The situation is beautiful. The first bank is from 6 to 12 ft. above high water mark and extends west a quarter of a mile, gradually rising from the river, when it rises 5 or 6 ft. to the second bank. This extends nearly on a level to the bluffs, which are from 80 to 100 ft. high. The ascent is steep, and on the bluffs the surface again becomes level. From these heights, the prospect is beautiful, embracing a fine prairie in front and the river and lake. It contains a courthouse, jail, an academy, six churches, two Presbyterian, a Methodist, Baptist, Episcopal, and Unitarian, 21 stores, two steam saw-mills, one tannery, one brewery, two printing-offices, two weekly newspapers; four schools, 236 scholars. Pop. 1467.

PEPPERELL, p. t., Middlesex co., Mass., 38 m. N.W. Boston, 434 W. Bounded E. by Nashua river by a branch of which it is drained; and both afford water-power. It contains two churches, a Congregational and Unitarian, a private insane asylum, three stores, three grist-mills, five saw-mills, three paper-mills; one academy, 60 students; eight schools, 468 scholars. Pop. 1571.

PEQUANNOCK, t., Morris co., N. J., 10 m. N. Morristown. Bounded N.E. by Pequannock creek; E. by Pompton river. Drained by a branch of Rockaway river. The Morris canal passes through its S. part. It contains a Dutch Reformed church, 13 stores, two woollen factories, six grist-mills, nine saw-mills, six tanneries, two distilleries; 22 schools, 873 scholars. Pop. 5190.

PERIGUEUX (an. Vesuna), a town of France, dep Dordogne, of which it is the cap., on the Isle, here crossed by a handsome bridge, 68 m. E.N.E. Bordeaux, lat. 45° 11' 8" N., long. 0° 43' 24'' W. Pop., in 1836, 9399. It consists of the city proper, and Puy-St.-Front, which, previously to which it is the cap., on the Isle, here crossed of the town, and an elegant public garden bequeathed to the city by a wealthy citizen. The cathedral of St. Front
571

is probably one of the most ancient churches in France, if not in Christendom. It appears to have been founded towards the end of the 4th century, and restored about the beginning of the 6th; and portions of the edifice are still supposed to date from these remote epochs. Its architecture is partly Roman and partly Gothic, and, though it have little elegance to boast of, it is altogether a bold and majestic structure. (*Guide du Voyageur, &c.*) A church, formerly belonging to the Jesuits, and having a remarkable piece of carving, the prefecture, town-hall, hospital, barracks, a handsome theatre, are the other principal buildings. Perigueux is a bishop's see, the seat of tribunals of primary jurisdiction and commerce, and has a communal college, a museum of antiquities, a botanic garden, and a public library of 16,000 vols., with manufactures of coarse woollens, hosiery, and liqueurs, and a considerable trade in cattle, poultry, game, *patés-à-la-Périgord, &c.* Its hog-market is considered the largest in France. Here are several Roman antiquities, including the remains of a more extensive amphitheatre than·that of Nîmes. The town continued long in the possession of the English; and was a stronghold of the Calvinists during the religious wars. (*Hugo, art. Dordogne; Dict. Geog., &c.*)

PERM, an extensive government of the Russian empire, extending from the 56th to the 62d deg. of lat., and from 59¼ to 64 deg. E. long., having W. the government of Viatka, and E. that of Toboisk. It is divided by the Oural chain into two unequal portions, the larger, or that on the W. side being in Europe, and the smaller, or that on the E. side, in Asia. Neither the area nor the population has been exactly ascertained, but the former may be taken at about 58,000 sq. m., and the latter at 1,450,000 or 1,500,000 individuals. More than three fourths of the surface is covered with dense forests. The W. side of the government is traversed by the Kama, one of the principal affluents of the Wolga: the rivers on the E. side of the Oural mountains fall into the Obi. The climate is very severe, and the soil beyond the 60th deg. of lat. is hardly susceptible of cultivation. The mines·in the Oural mountains furnish employment to about 100,000 work people, and yield large quantities of gold, silver, copper, iron, salt, &c. (*See Russia.*) The corn produced in the government is insufficient for the consumption of the inhabitants. ·

Perm, the chief town of the government, is situated on the Kama, lat. 58° 1′ N., long. 56° 26′ 15″ E. Pop. 10,000. It is built entirely of wood, and is the seat of an archbishopric, and has several public establishments. The inhabitants are principally employed in smelting the iron, copper, and other products, of the adjacent mines.

PERNAMBUCO, a city and seaport of Brazil, inferior only to Rio Janeiro and Bahia in commercial importance, cap. prov. of its own name, on the Atlantic, at the mouth of the Capabaribe, 210 m. N.E. Bahia; lat. 8° 4′ 7″ S., long. 34° 51′ 44″ W. Pop. estimated at 62,000. It consists of the separate towns of Olinda, Recife, Boa-Vista, and St. Antonio; the first of which is on the mainland, and the others lie S. from it on a succession of low sandy banks, separated by saltwater creeks and different arms of the river, but connected with each other by two bridges. Recife, or Pernambuca proper, the most southerly, about 4 m. S.W. Olinda, is defended by the principal forts, and comprises the dock-yard and the larger merchants' warehouses. Most of its streets are narrow; its houses are chiefly of brick, and sometimes from three to five stories in height, but usually less lofty. Several churches, the port-admiral's residence, and the custom-house, are among its most conspicuous public buildings.

St. Antonio, the residence of the greater number of of the provincial authorities, has broad streets and large houses, the ground floors of which are appropriated to shops, warehouses, stables, &c. When Koster visited Pernambuco, the shops were without windows, the only light being admitted by the door. There existed, also, very little distinction of trades, all kinds of goods being sold by the same person; the bridges of the city were crowded with shops; neither the streets of this town or those of Boa-Vista were paved, but it is probable that since that period many important improvements may have taken place. In St. Antonio are the governor's palace, formerly the Jesuits' convent; the treasury, town-hall, prison, barracks, several convents and churches handsomely decorated within, and several good squares; and it has a gay and lively appearance. A long embankment connects this town with the main land. Boa-Vista is extensive, but irregularly laid out; it has one handsome street and comprises the residences of many of the richer inhabitants of the city, surrounded with gardens, various churches and convents, &c. Olinda is beautifully situated on small hills, the sides of which are alternately either broken down abruptly, or covered with a most luxuriant tropical vegetation; amid which, the white cathedral, the convents and churches, the bishop's palace, and numerous villas, have a most picturesque effect. But Olinda is

in a state of decay, having been deserted by many of its population for Recife and the other parts of the city.

The harbour of Pernambuco is defended from the swell of the ocean by an extensive reef (*recife*); which seeming to Koster, continues along the whole coast from Maranham, at a variable distance from the shore, and has numberless breaks, through which ships approach the land. This reef, which is said to be of coral, "is scarcely 16 ft. broad at top; it slopes off more rapidly than the Plymouth breakwater, to a great depth on the outside, and is perpendicular within to many fathoms." (*Graham in Modern Traveller, xxx. 298.*) This natural breakwater forms the harbour; for though at high-water the waves beat over it, they strike the quays and buildings of the town with diminished force. Along the sandy neck of land between Olinda and Boa-Vista, however, which is uncovered by the reef, the surf is very violent; and the harbour itself is not considered very safe. It consists of two parts: the Poco, capable of receiving vessels of 400 tons and upwards, entered across a bar at which there are from 17ft. to 30 ft. water; and the Mosqueras, much better protected than the former, but on the bar of which there are but 7 ft. water at ebb tide. Vessels trading with Pernambuco ought not to draw more than from 10 ft. to 12 ft. water. (*Blunt's Am. Coast Pilot, 318.*) The harbour is defended by several strong military works, the principal being the stone forts of Do Barreo, and Do Brus, and has at its entrance a lighthouse with a revolving light.

The exports consist principally of cotton to the extent of from 60,000 to 70,000 bags a year; sugar to the extent of from 20,000 to 25,000 cases; and hides and dye-wood. At an average, the value of the exports may amount to from £600,000 to £700,000 a year. In 1835, 347 vessels from foreign ports arrived at Pernambuco, of which 39, of the burden of 11,954 tons, were from England. The value of the imports from England, in the course of that year, was estimated at about £500,000, consisting principally of cotton and linen manufactures, hardware, earthenware, &c.

PERNAU, a fortified seaport town of Russia, gov. Livonia, at the confluence of the Pernan with the N. angle of the gulf of Riga, lat. 58° 21′ 20″ N., long. 36° 36′ 5″ E. Pop. 9000. It comprises an old and a new town, and two suburbs; and has three ch urches and a Latin school. There is a bar at the mouth of the river, which can only be crossed by the smaller class of vessels; those of larger burden having to load and unload in the roads, where the anchorage is not very good. It has a considerable export trade, especially in corn, hemp, and flax, linseed, &c.; the aggregate value of the exports amounting to about 3,000,000 roubles or 3,000,000 roubles a year.

PERONNE, a fortified town of France, dép. Somme, cap. arrond. on the Somme, 29 m. E. by N. Amiens. Pop. in 1836, 4029. It has a handsome town-hall, a Gothic parish church, in which one of the Merovingian kings is interred; but the edifice possessing the greatest interest is the old castle, in a tower of which Charles the Simple was confined by a count of Vermandois, and subsequently Louis XI. by Charles the Bold, duke of Burgundy. The latter circumstance forms the basis of one of the finest episodes in Scott's novel of *Quentin Durward.*

Peronne has been frequently besieged by the Spaniards, but unsuccessfully; and it had not been taken till it was stormed by the British about a week after the battle of Waterloo. At the castle of Applincourt, near Peronne, the famous "league" was concluded, in 1576, between Henry III. and the Duke de Guise. (*Hugo art. Somme; Guide de Voy.*)

PERPIGNAN, a strongly fortified town of France, dép. Pyrénées-Orientales, of which it is the cap. on the Tet, where it is joined by the Basse, 80 m. S.W. Montpellier, lat. 42° 42′ 3″ N., long. 2° 49′ 9″ E. Pop. in 1836, 14,721. It is built partly on a declivity, and partly in the plain beneath; and is separated by the Basse from *les Marguerites*, or the new town, and by the Tet from a suburb. Each river is crossed by a bridge, that across the Tet consisting of seven arches. The fortifications of Perpignan have been improved considerably since 1815; and it is now one of the best fortified towns of France. It is surrounded with walls of brick and stone, flanked by several bastions, and enclosed by ditches, beyond which are numerous outworks. It is further defended by a citadel, with a double line of ramparts, beside outworks; within which are the barracks and the castle, formerly the residence successively of the counts of Roussillon, and of the kings of Arragon and Majorca. The town is not well built, though improving. There are a few good streets and squares, and some pleasant public promenades; but the public thoroughfares are designed for pedestrians, from being paved with small round stones. The cathedral is a handsome Gothic edifice, of the 14th and 15th centuries, 255 ft. in length, by 64 ft. in breadth, remarkable for the beauty and boldness of its nave and vault, which last is sustained without the aid of any column rising from the area of the building. The old church of St. John, an edifice of the 11th century, and several other

e; the Castillar, a defensive work of uncertain date,
lar to structures erected by the Moors in Spain; the
ll, mint, arsenal, (formerly a large convent), theatre,
ustice, and prison, are the other chief public build-
srppean is a bishop's see, the seat of tribunals of
jurisdiction and commerce, and of boards of artil-
users, and customs. It has a communal college,
a, a botanic garden, and a library of 13,000 vols.,
ul. cturers of woollen stuffs, lace, corks, soap, and
urda, and is an extensive entrepôt for the wines,
liqueurs, wool, silk, oil, and other products of the
succ. (*Hays*, art. *Pyrén.-Orient.* ; *Guide du*
)

!MANS, county, N.C. Situated in the N.E. part
a. and contains 175 sq. m. Bounded N.E. by
r, 8 by Albemarle sound. Watered by Perqui-
r It contained in 1840, 6335 neat cattle, 4756
23 swine; and produced 33,540 bushels of wheat,
Indian corn, 22,068 of oats, 40,223 of potatoes,
ds of cotton. It had nine store, ten grist-mills,
mills: five schools, 82 scholars. Pop.: whites,
, 2943; free-coloured, 307 ; total, 7346. Capital,

'ON, t., Monroe co., N.Y. It has five stores,
mills, one woollen factory, one flouring-mill, one
ur saw-mills; 16 schools, 740 scholars. Pop.

uaty, Pa. Situated a little S. of the centre of
d contains 540 sq. m. Watered by Juniata
rman's creek. On the latter, 11 m. N. Car-
s. W. Harrisburg, is a warm medicinal spring.
ineous complaints, which discharges 90 gal-
ute. It contained in 1840, 15,043 neat cattle,
1,484 swine ; and produced 200,638 bushels
19 of rye, 150,095 of Indian corn, 37,052 of
2.258 of oats, 80,369 of potatoes. It had 57
illing-mills, five woollen factories, eight fur-
es, 24 flouring-mills, 98 grist-mills, 120 saw-
ies, 13 distilleries, four potteries, two print-
weekly newspapers ; one academy, 90 stu-
ls, 3829 scholars. Pop. 17,096. Capital,

y, Ala. Situated a little S W. of the centre
contains 936 sq. m. Watered by Cahawba r.
. It contained in 1840, 23,813 neat cattle,
2 swine ; and produced 32,694 bushels of
e, 792,384 of Indian corn, 99,297 of oats,
, 12,680,177 pounds of cotton. It had nine
w-mills, two tanneries, one distillery ; two
dents ; six schools, 136 scholars. Pop.:
s, 10,343 ; free coloured, 22 ; total, 19,088.

Miss. Situated in the S.E. part of the
1044 sq. m. Watered by Leaf river and
r Black creek, a tributary of Pascagoula.
in 1840, 13,645 neat cattle, 1415 sheep,
produced 42,520 bushels of Indian corn,
15,520 pounds of rice, 68,446 of cotton.
six grist-mills, three saw-mills; six
Pop.: whites, 1425 ; slaves 454 ; free-
879. Capital, Augusta.
in. Centrally situated toward the W.
d contains 780 sq. m. Watered by
Buffalo-river. It contained in 1840.
sheep, 13,072 swine ; and produced
, 58,291 of Indian corn, 18,679 of oats,
5 pounds of tobacco. 4787 of cotton.
urnaces, ten grist-mills, 20 saw-mills,
istilleries ; four academies, 113 stu-
300 scholars. Pop.: whites, 6713 ;
d 8 ; total, 7419. Capital, Perrysburg.
Situated in the S.E. part of the state,
Watered by the north and middle
. It contained in 1840, 6001 neat
i swine; and produced 2981 bushels
an corn, 18,441 of oats, 4421 of pota-
acco, 3761 of sugar. It had four
saw-mill, three tanneries, 48 dis-
2923 ; slaves, 143 ; free-coloured,
Hazard.
uated toward the S.E. part of the
g. m. Watered by tributaries of
gum rivers. It contained in 1840,
heep, 23,968 swine ; and produced
19 ; 3 of rye, 326,312 of Indian corn,
159 of oats, 50,194 of potatoes,
91,732 of sugar. It had 36 stores,
sn-factories, seven flouring-mills,
, one oil mill, 11 tanneries, four
wo printing-offices, two weekly
, 80 students; 92 schools, 2930
spital, Somerset.

PERRY, county, La. Situated in the S. part of the state,
and contains 400 sq. m. Bounded S. and S.E. by Ohio river. Drained by Anderson's and Oil creeks. It contained in
1840, 4611 neat cattle, 5073 sheep, 14,780 swine ; and produced 13,452 bushels of wheat, 170,295 of Indian corn, 29,736
of oats, 8622 of potatoes, 21,419 pounds of tobacco, 6076 of
sugar. It had 12 stores, two grist-mills, four saw-mills, six
tanneries, five distilleries, one pottery ; eight schools, 194
scholars. Pop. 4655. Capital, Rome.

PERRY, county, Ill. Situated toward the S.W. part of the
state, and contains 432 sq. m. Drained by Beaucoup and
Little Muddy creeks. It contained in 1840, 16,666 neat cattle, 4113 sheep, 10,223 swine ; and produced 14,667 bushels
of wheat, 160,445 of Indian corn, 31,754 of oats, 4556 of potatoes, 3140 pounds of tobacco. It had four stores, three
grist-mills, one saw-mill ; one academy, nine students ; 13
schools, 335 scholars. Pop. 3222. Capital, Pinkneyville.

PERRY, county, Mo. Situated toward the S.E. part of the
state, and contains 400 sq. m. Bounded N.E. by Mississippi
river, and drained by several fine mill streams. It contained in 1840, 3690 neat cattle, 3116 sheep, 8530 swine ; and
produced 11,684 bushels of wheat, 184,320 of Indian corn,
9415 of oats, 5361 of potatoes, 11,400 pounds of tobacco,
2500 of cotton. It had one commercial house in foreign
trade. 14 stores, seven grist-mills, four saw-mills, four tanner
ies, four distilleries ; one college, 75 students ; eight schools,
169 scholars. Pop.: whites, 4968 ; slaves, 778 ; free-coloured, 14 ; total, 5760. Capital, Perryville.

PERRY, p. t., Washington co., Me., 5 m. N.W. Eastport,
191 m. E.N.E. Augusta, 777 m. W. Bounded E. by Passamaquoddy bay, S. by Cobscook bay. It is connected to
Eastport by a bridge. It has one grist-mill, five saw-mills,
11 schools, 356 scholars. Pop. 1008.

PERRY, p. t., Wyoming co., N.Y., 243 m. W. Albany, 380
m. W. Sliver lake lies partly in its S. part, by the outlet of
which it is watered. It contains four churches, a Presbyterian, Methodist, Baptist, and a Universalist, ten stores, two
fulling-mills, one flouring-mill, three grist-mills, 11 saw-
mills, one oil-mill, two tanneries, one printing-office, two
weekly newspapers ; eight schools, 918 scholars. Pop. 3082.

PERRY, t., Union co., Pa. It has two stores, two fulling-
mills, five grist-mills, nine saw-mills. Pop. 1254.

PERRY, t., Armstrong co., Pa., 19 m. N. Kittaning.
Watered by Alleghany river, and its tributary, Clarion river.
It contains iron ore, and has two stores, two fulling-mills,
five grist-mills, nine saw-mills. Pop. 1122.

PERRY, t., Jefferson co., Pa., 16 m. S. Bookville. Watered by Little Sandy Lick, and Mahoning creeks. It has two
stores, one fulling-mill, two grist-mills, four saw-mills.
Pop. 1078.

PERRY, t., Fayette co., Pa. It has seven stores, two
flouring-mills, one grist-mill, six saw-mills, two tanneries ;
four schools, 156 scholars. Pop. 1350.

PERRY, p. t., Lake co., O., 182 m. N.E. Columbus, 356 m.
W. Bounded N. by lake Erie. It has two churches ; 10
schools, 627 scholars. Pop. 1339.

PERRY, t., Brown co., O. It has four schools, 92 scholars.
Pop. 1869.

PERRY, t., Carroll co., O. It has two schools, 55 scholars.
Pop. 1344.

PERRY, t., Coshocton co., O. It has three stores, one grist-
mill, two saw-mills, two tanneries ; four schools, 111 scholars. Pop., 1339.

PERRY, t., Columbiana co., O. It contains Salem village.
Pop. 1530.

PERRY, t., Franklin co., O. Bounded W. by Scioto river.
It has two schools, 65 scholars. Pop. 1037.

PERRY, t., Fairfield co., O. Pop. 1172.

PERRY, t., Logan co., O. Watered by Rush creek. It
contains East Liberty village, and has six schools, 210 scholars. Pop. 1014.

PERRY, t., Montgomery co., O. It has six schools, 305
scholars. Pop. 1881.

PERRY, t., Muskingum co ., O. Pop. 1061.

PERRY, t., Pickaway co., O. It has three stores, one fulling-mill ; one school, 40 scholars. Pop. 1277.

PERRY, t., Richland co., O. Pop. 1853.

PERRY, t., Stark co., O. It has two schools, 40 scholars.
Pop. 2209.

PERRY, t., Tuscarawas co., O. It has four stores ; six
schools, 556 scholars. Pop. 1381.

PERRY, t., Marion co., Ia. It has two grist-mills, three
saw-mills, one distillery ; 9 schools, 281 scholars. Pop. 1510.

PERRY, p. v., capital of Houston county, Ga. Situated on
the N. bank of Big Indian creek, between Flint and Ocmul
gie rivers. It contains a court-house, jail, an academy, a
Methodist and a Baptist church, eight stores, and about 40
dwellings.

PERRYSBURG, p. t., Cattaraugus co., N.Y., 306 m. W. by
S. Albany, 358 m. W. Bounded N. by Cattaraugus creek
It has three stores, one grist-mill, five saw-mills ; nine
schools 400 scholars. Pop. 1680.

PERRYSBURG, p. v., capital of Wood co., O., 123 m. N.N. W. Columbus, 454 m. W. Pleasantly situated at the head of steamboat navigation on Maumee river, 18 m. from the lighthouse on the lake shore, and 12 m. from the entrance of the river into Maumee bay. It is on the S.E. bank of the river, elevated 60 ft. above its surface, commanding a fine view of the river and the surrounding country. It contains a handsome courthouse, a jail, three churches, a Presbyterian, Methodist, and a Universalist, 2) stores, three large warehouses, two steam saw-mills, a ship-yard, a printing-office issuing a weekly newspaper, 350 dwellings, and about 2000 inhabitants. It has 1000 ft. of permanent wharf, and was laid out in 1817. The Wabash and Erie, and the Maumee canals will add much to its importance as a place of business. It was named in honour of Commodore Oliver H. Perry, the hero of lake Erie; and appropriately, as it is not far distant from the scene of his conflict with, and triumph over the British fleet.

PERSEPOLIS, (the Istakhar of the an. Persians), a celebrated city of antiquity, and during a considerable period the capital of Persia, and residence of its monarchs, province Farsistan, in a fine plain at the foot of a mountain, against which it abuts, near the Araxes, or Bundemir, 30 m. N.E. Shiras; lat. 29° 59' 30'' N., long. 53° 20' E. The city appears to have covered a large surface; bricks, fragments of walls, and rubbish being found widely scattered; but the only extant ruins of any interest or importance are those of a vast building, or rather series of buildings, supposed to have been the palace of Darius, burned by Alexander the Great.

It would be impossible, without the aid of plates, to give any intelligible description of these vast ruins. We may, however, state, that they occupy the summit of a platform about 1430 ft. in length, 803 ft. in width on the S., and 926 ft. on the N. side, and about 50 ft. in height, formed by levelling the summit of a marble rock. This platform is ascended by easy flights of steps, cut into the rock on its W. side, and, when entire, consisted of three fronts or terraces, the mountain forming its E. side.

The ruins consist of the remains of vast portals or gateways (one of which is formed of two enormous sphinxes), pillars, walls, on which, but especially on the sides of the staircases, figures are cut in basso relievo, which are highly interesting as illustrating the costume and armour of the ancient Persians. Some of the remaining columns are 60 ft. in height, and though their capitals and form be very different from what we have been accustomed to consider classical, they are extremely beautiful, and many of the sculptures are executed with infinite spirit. Numerous tombs have been cut into the mountain, on which, as already stated, the palace abuts. (Porter's Travels, i. 582–583; Rich's Babylon and Persepolis; Niebuhr, Voyage en Arabie, ii. 98–131; Mod. Trav., &c.)

Perhaps, however, the most curious portion of the ruins of this vast palace consists of the inscription in arrow-headed or cuneiform characters, similar to those on the bricks of Babylon, found in great profusion on most parts of the walls. Very discordant opinions have been entertained with respect to the nature and value of these characters; but Dr. Grotefend, who has bestowed the greatest pains on the subject, has shown that the cuneiform marks are real alphabetical letters; that every inscription is treble, (the first being in Zend, and the others in different Persian dialects); that the inscriptions are to be read from left to right; and that all of them belong to the period between Cyrus and Alexander. Heeren considers that these characters are the 'Ασσυρια γραμματα, mentioned by Herodotus (iv. 87); but, in point of fact, notwithstanding the investigations of Grotefend, we are still far from being well informed as to the true meaning of these inscriptions.

The history of Persepolis is, for the most part, hidden in obscurity; but it seems to be an established fact that this city is identical with the Istakhar of Persian historians, the foundation of which is ascribed to Cyrus the Great, the Jemsheed of Oriental writers (whence the modern name of the ruins, Takhti-Jemsheed). Herodotus, Ctesias, and the older Greek writers do not mention it, and it may not then have been a permanent royal residence. The inscriptions, however, (if they have been correctly interpreted), show that it must, occasionally at least, have been visited by Darius, and the several monarchs called Xerxes. It is, at all events, certain that this city was the residence of the unfortunate Darius Codomannus, who, with his court, fled from it after his defeat at Arbela or Guagamela (anno 331 B.C.) by Alexander the Great. The conqueror soon after took the city, and gave it up to military execution. Alexander himself set the palace on fire. Ancient circumstances, which, if we may believe Diodorus, have been accurately as well as admirably depicted in Dryden's noble Ode. But Arrian, a far less questionable authority, has given a very different account of the matter. He states that Alexander destroyed this palace contrary to the advice of Parmenio, not in a

574

drunken frolic, but in cold blood, and on principle, in retaliation of the destruction of the Greek temples by the Persians. (Arrian, lib. iii. cap. 18.) From the few notices that now exist, it appears that Istakhar was an important city under the Sassanian dynasty. In the 7th century it was taken by the Mahommedans, who, having founded Shiras, made it the capital of Persia; and Persepolis, long on the decline, rapidly sunk to a state of total decay. It may be right to mention that Persepolis has been regarded by some writers as identical with Pasargadæ; and it may be fairly inferred, from the statement of Arrian (lib. iii. cap. 18), that the palace destroyed by Alexander belonged to the latter; and though the question as to their identity be not free from difficulty, we are, on the whole, inclined to prefer the conclusion to the contrary theories of later writers.

PERSERIN, a considerable town of Turkey in Europe, pach. Albania, sanjak Scutari, near the Drin, at the foot of a mountain, 72 m. E.N.E. Scutari. Population, according the Dict. Géog., 16,000; but in Horschelman's Stat. it is stated at only 5000! Its inhabitants are principally Arnaouts, but partly, also, Mohammedans, and partly Christians. It is the residence of a Greek bishop, and of a military governor, who occupies a fort of no great importance. Albania is said to be principally supplied with firearms from a factories.

PERSHORE, a market town of England, county Worcester, hund. its own name, on the W. bank of the navigable river Avon, crossed here by a stone bridge, 7 m. S.E. Worcester. Population of its two parishes (exclusive of the out-townships), in 1831, was 2536. The town, which consists of one principal street, nearly 1 m. in length, is well built and paved, having many respectable and a few handsome houses. The church of Holy-cross, formerly attached to a Benedictine monastery, of which there are still some remains, is a large cruciform structure, with a lofty square tower. That of St. Andrew is small and mean looking; both livings are in the patronage of the dean and chapter of Westminster. The Wesleyan Methodists and Baptists have their respective places of worship; and there are three Sunday schools and a national school. Some of the inhabitants are employed in the manufacture of stockings; but the town depends principally on its retail trade for the supply of the neighbouring gentry. The petty sessions for the hundred are held here; and Pershore is the chief place of a poor-law union, comprising 40 parishes; the expense of maintaining its own poor having been £970 in 1839. It is also one of the polling places at elections for the E. division of the county. Markets on Tuesday; fairs Easter Tuesday, June 26, and last Tuesday in October.

PERSIA, a celebrated and very extensive country of central Asia, between the 39th and 25th degs. N. lat. and the 44th and 62d deg. of E. long. The political boundaries of the country have varied at different times with the character or exploits of its monarchs: sometimes embracing Armenia and Georgia on the W., Khàrezm and Bokhara on the N., and Affghanistan on the E.; and sometimes being reduced to less than its natural limits. The latter on the S. are the Indian ocean and the Persian gulf; on the S.W. and W. the Tigris; on the N. the Aras, the Caspian sea, and an indefinite line in the desert that separates Persian Khorasan from the territories of Kharezm or Khiva, stretching from the Attruck, which falls into the Caspian, to about the 36th deg. of N. lat., and the 61st deg. of E. long.; whence a waving and undefined line, drawn southwards, separates the Persian territories on the E. from those of Cabul and Affghanistan. At present, however, the actual limits of Persia are much more circumscribed. The extensive province of Beloochistan, along the Indian ocean, is quite independent. (See BELOOCHISTAN.) The Turkish mountains embrace a large portion of country to the E. of the Tigris; and the country of Talash, to the S. of the Aras, belongs to the Russians, but still, even with these deductions, its area probably exceeds 450,000 sq. m.; though, from the vast extent of its deserts, the badness of the government, and the want of industry, the population does not probably exceed 9 or 10 millions.

Name.—The most ancient name of this extensive region is that of Elam (Genesis x. 22). The name of Persia, by which it was afterwards known in Europe, appears to have been derived from that of the province of Fars, or Phars, which being changed by the Greeks to παρσις, was applied by them to the whole country. This designation has not, however, been adopted in the east; the Persians, both in ancient and modern times, having styled their country Iran. The countries occasionally subject to the Persian monarchs beyond the Gihon, or Oxus, have usually been called Amiran or Tooran, that is beyond Iran. (Ancient Universal History, v. 49, 8vo. ed.)

Face of the Country.—Persia may be considered as an elevated plateau, diversified by many clusters of hills, chains of rocky mountains, extensive plains, and barren deserts, with two extensive declivities, or lower tracts—one along;

shores of the Persian gulf and the banks of the Shat-el-Arab, and the other along the shores of the Caspian. The southerly portion of the former consists of a succession sandy or gravelly plains, where water is so scanty that station is only seen in patches where a well or a rivulet gives the inhabitants to irrigate some portions of the soil. region is called by the natives *Dushtistan* and *Gurm-* that is, the hot country; and according to Morier, *wureness, solitude, and heat,*" are its principal character-: but in the province of Kuzistan, to the E. of the -el-Arab, this low tract is comparatively well watered umerous streams, and its upper portion is naturally very active. The low country along the banks of the Caspian is extremely well watered, and is covered with forests verdure; it unites on the E. with the desert, which tches from the E. shore of that sea to the Tartarian as.

plateau, or elevated space which lies between these wer slopes, and which rests, as it were, on two great of mountains, may rise to an elevation of from 2500 ft. above the sea, and on this again chains of moun-ar themselves to various altitudes, seldom, however, ing 7000 or 8000 ft. above the sea, and including aces between their ranges valleys of various dimen-od sometimes rather appearing as islands in the ex-plain.

most striking feature of Persia are its chains of rocky ns; its long arid valleys without rivers; and, above as salt or sandy deserts.

tains.—There are two great chains of mountains, while they support the plateau of Persia on the N. seem to be the stocks from whence all the minor cceed. The most northerly of these, striking off Caucasus, crosses the Kur, to the W. of the plains ., and from Ardebeel runs parallel with the south-of the Caspian sea to Asterabad. It thence passes direction to Meshed, and, stretching S. of Balkh to o Koosh, is lost in the range of the Himalaya, and ndous central knot of mountains where the lar-of Asia take their rise.

mense chain, in its extent of more than 20 degs. is forth everywhere a number of branches, which aces sink into the plains or deserts on the E. of l sometimes connect themselves with other ele-Of these, the Sahund mountains, striking off from umeah in a N.E. direction, spread themselves jan, and connect more or less with the spurs s of that extensive aggregation of mountains in Euphrates, Tigris, Zab, and other large rivers, urces, and whence they derive their supplies. f the Taurus may be considered as a great this central knot, which spreading out in all vers the pachalics of Diarbekir, Erzeroom, and Koordistan, with piles of immense moun-rise to a great height between the lakes Van particularly to the W. of the latter, where ewar are supposed to attain an altitude of 3 ft. above the sea. From this mass a chain varying in height and breadth, runs S.E. istan, bounding at a distance the valley of ning the high lands of Louristan and the Buchtiarees, and giving birth to the rivers Abzool, &c. After passing to the S. of Shi-ll more to the E., and following at uncer-the line of the Persian gulf and Indian sionally almost disappearing, joins the ran-tan and Mekran, and finally sinks into the or is lost in the high grounds which di-countains of Afghanistan.

wo ranges may probably be traced every mountains that cover Persia as with a net-es ; though many even of those which at-ee of altitude appear almost insulated. t peaks may be mentioned that of Dema-Tehran, from 12,000 to 13,000 ft. above near Hamadan, nearly as high ; Sahund, e Koh-i-Zerd, near Ispahan, &c., which sured. These mountains include among system of valleys and plains, differing in eness, according to their nature and cli-water abounds they are fertile, but mois-which nature is the least liberal in Per-ne happy regions, even streamlets are rivers scarcely any are navigable beyond m their mouths.

the Tigris, being assigned as one of the of Persia, can scarcely be set down as a are many large streams which descend ountains to feed it. Of these the prin-n. supposed by D'Anville to be the Cho-Herodotus, the Ulai of sacred writ; but hat the Kerah, or Kerau, represents the

Choaspes ; and that the greater and lesser Zab, which have their sources in the chain of Zagros and the mountainous districts of Koordistan, are the ancient *Capros* and *Zabolus*. On the N. the Arras, or Araxes of classical writers, though a boundary line of the present Persia, derives much of its waters from Persian Koordistan ; and the salt lake of Uru-meah receives from the same hills, and from the Sahund mountains, a number of streams. The provinces bordering the Caspian are as remarkable for their moisture as the rest of the country is for its aridity ; but their rivers are chiefly torrents, sometimes full and foaming, at other times nearly dry. The Kizzilozein, which rises in Ardelan, and passing by Miana falls into the Caspian in the province of Ghilan ; the Herauz, which flows through Amol ; and the Tejen, which passes by Saree, both in Mazunderan, are the largest. The Attruck and Goorgan, both considerable streams, falling into the S.E. corner of the Caspian, are fed from the N. face of the Elbruz of Astrabad. It is a singu-lar fact, as Sir J. Malcolm remarks, that from the mouths of the Indus to those of the Karoon and Euphrates, there is not found one river navigable more than a few miles from the ocean ; and, in fact, the rivers that fall into the Persian gulf and the Indian ocean, on the shores of Laris-tan and Kerman, are mere torrents, almost dry during the long period of the summer and autumn heats.

Lakes.—In a country so arid there can be but few sheets of standing water, and those which do exist are chiefly salt. Of these the lake Urumeah or Shahee, in Azerbijan, near the frontiers of Turkey, 36 m. W. Tabreez, is the most remarkable, both for size and intensity of saltness. Ac-cording to Kinneir, it may be about 300 m. in circumference, and it has several islands. So saturated is the water with the salt it holds in solution, that immense quantities are deposited, assuming the appearance of a pavement under the shallow water near the brink, and its shores are covered with saline efflorescence. Its waters, like those of the sea, appear to be dark blue, streaked with green, and are pellu-cid in the highest degree. The lake of Baktegan, near Shiraz, is another of these sheets of salt water, but on a smaller scale ; as is the lake Zurrah, in Seistan. Except-ing small pools among the mountains, which are the well-heads of streams, there are no lakes of any considerable size ; but on the banks of the Caspian sea, the beating of the surf, by damming up the mouths of streams, has given birth to some extensive lagoons.

Deserts.—Those which are the most striking of its phy-sical features, Persia shares with a large portion of central Asia and Africa : they consist of salt deserts, called by the natives *Kuveer*, and sandy wastes called *Sahra*. The great *Deria Kuveer*, or salt sea, as it is called, is of prodigious dimensions, and may be said to be to Persia what the great desert of Sahara is to Africa. It commences on the N. at the foot of the Elburz mountains, in about the 36th deg. N. lat. ; and, uniting with the desert of Kerman, extends S. to about the 30th deg.: on the other hand, it extends from about the 51st to the 60th deg. of long., occupying all the central and eastern portion of the country. It has a few oases, or cultivated spots ; but they do not amount to 5 per cent. of its extent. The cultivated portions of the country lie round the margin, as it were, of this vast desert plateau, principally to the N.W., W., and S.W., but partly also, as already seen, to the N., along the Caspian. S. of the desert is Beloochistan, and E. Affghanistan.

The nature of this desert varies in different places. In some the surface is dry, and even produces a few salugi-nous plants ; in others, it is covered with a crackling crust of earth, white with saline efflorescence. A considerable portion is marshy ; and during winter the melting of the snow and the increase of the torrents, cause an accumula-tion of water in its lower parts, which, being evaporated in the hot months, leaves behind a saline incrustation in cakes upon a bed of mud. In extensive tracts sand pre-dominates, either in the shape of level plains or wave-like hillocks, easily drifted by the wind, and sometimes so light and impalpable as to be carried to a vast distance by tem-pests. In some places the plain surface is broken by ridges of bare black rocks. Nothing can be more dreary than these dismal wastes. When the traveller has advanced some distance into them, the boundless expanse around, blasted with utter barrenness, and hoary with bitter salt, glistening and baking in the rays of a fervid sun, only broken here and there by masses of dark rock, distorted by the powerful refraction into a thousand wild and varying forms, impress him with a sense of desolation that cannot be described.

Forests and Appearance of the Country.—Although the greater part of Persia is bare of vegetation, there are a few tracts exempted from this sterility. Among these are the provinces of Ghilan, Mazunderan, and Asterabad, bordering the Caspian sea. The strip of low land constituting these provinces, with the N. face of the lofty mountains by which it is overhung, is covered with dense forests of oak, elm,

PERSIA.

beech, sycamore, and all sorts of fruit trees, among which the vine grows with the greatest luxuriance. The swamps and back waters are bordered with alders of magnificent size, and among the underwood is found abundance of box, of a magnitude unknown in Europe. In the S. the chain of mount Zagros, including Persian Koordistan, Louristan, and the Buchtiaree mountains, is partially covered, and in many places densely, with forests of oak, which, however, does not attain any great size; and parts of Kuzistan are overrun with low jungle, the haunt of wild beasts. The district of Bebahan is rich in wood; and Kinneir praises highly the beauty of the finely-wooded vale of Ram-Hormuz in Upper Kuzistan. But except in those districts, which bear but a small proportion even to the inhabited portions of the country, its appearance is dreary in the extreme, and lacks almost everything that gives interest and beauty to European landscapes. It has no green plains or grassy slopes, no parks nor inclosures, no hedges nor woods, no magnificent seats nor comfortable-looking cottages, and, excepting in spring, even the portions cultivated round the villages can hardly be distinguished from the brown, arid expanse, that everywhere meets and fatigues the eye of the traveller. And if the reader will farther figure to himself towns and villages, consisting mostly of mud houses, partly in a state of decay, and many of them wholly deserted; roads, if so we may call wretched paths, wholly impracticable for carriages, and unsafe even for horsemen; property insecure, and tyranny and rapine everywhere lords of the ascendant, he will be able to form a pretty accurate notion of the state of this celebrated country. (*Fraser's Persia; Edinburgh Cabinet Library*, p. 29; *Kinneir's Persia, passim., &c.*)

Soil.—Lime in various shapes abounds everywhere, and being mingled in the glens and valleys with the remains of decayed vegetables and other detritus, forms a loamy soil of inexhaustible fertility. Indurated clay is often found to mingle with the calcareous matter. Artificial irrigation is here, almost everywhere, essential to the raising of crops. It is, in fact, the great business of the Persian agriculturist: and is well understood, having been practised from the remotest antiquity. Wherever, indeed, it is neglected, the land is, for the most part, barren and unproductive.

Climate.—This is found to vary to almost the greatest possible extent in different provinces; and the statement of the Younger Cyrus, that one extremity of his father's dominions stretched into those climates that were uninhabitable through heat, and the other into those uninhabitable through cold, is nearer the truth than might be supposed. The summer heats in the S. provinces are almost insupportable; while the cold of winter in those of the N. rivals that of Canada or Russia. In summer, however, even in the N., the heat is so great that all who can leave the towns and villages of the plains in the months of June, July, August, and September, resort to temporary lodgings or tents in the mountains. In the low provinces on the Caspian, the heat, though great in summer, is not so excessive as in the S., partly perhaps from the evaporation that takes place, as well as from the breezes from the sea: but the climate is here extremely unhealthy, and in the end of autumn putrid and intermittent fevers prevail to a great degree.

Minerals.—The mineral riches of Persia are almost wholly unexplored. Iron, copper, and lead are, however, known to abound in all the great mountain ranges. The first is not largely produced, and much of the required supply is imported from Russia. Copper has been worked in several places, particularly in Khorasan and Azerbijan; but the distracted state of the country has hitherto prevented much progress being made in such undertakings. Of late, however, an enterprising native, whose mind has been enlarged and his knowledge improved by a residence in England, has commenced working mines in Karadang under favourable appearances of success; and, from the connexions he has formed, he may perhaps escape the extortions to which others of his countrymen would probably be exposed. The mines of Fars and Kerman supply the greater part of the demand for lead, though some is also brought from India. Antimony is found, but is little used. Gold and silver are said to exist, but there are no mines of either worth notice. The turquoise is almost the only gem found in Persia, to which it is peculiar, the mines near Nishapour yielding this precious stone in an abundance and of a beauty unknown in any other part of the world. It is found disseminated in veins, nodules, and irregular masses, in beds of porphyritic conglomerates or limestone deeply tinged with iron, and often veined with micaceous iron ore. Garnets are also found in various parts, especially near Hamadan, of great size and beauty. Rock salt is very abundant all over the country; and the mines of Khameer furnish abundant supplies of sulphur, which is also found in other places. Coal has been discovered in Azerbijan, and naptha is abundant, cheap, and useful.

Vegetable and Animal Productions.—As Persia embraces a variety of climates, its vegetable productions necessarily vary in different parts. The climate of many of the northern provinces resembles that of Europe, so that most European fruits and vegetables are found there in great perfection and abundance, with several belonging to more southern latitudes. The forests of European trees, that cover the Caspian provinces, and the woods that more thinly cover the southern mountains of Louristan and the Buchtaree, have already been noticed; and to these may be added the stately chinar, or *Platanus orientalis*, the Lombardy poplar, willow, jujube tree, and, in the warmer parts, the cypress and pinaster. The plains are covered with a stunted and prickly herbage, among which the camelthorn, wild liquorice, wild rue, and many aromatic plants, are conspicuous. Among the rest, the tall stalk of the gum-ammoniac rears itself on most of the gravelly plains of Irak and Khorasan, dropping its bitter tears upon the waste. The assafœtida plant abounds in parts of Khorasan. The orchards of Persia are rich in all the fruits of Europe: cotton, tobacco, the opium poppy, figs, vines, and the mulberry, abound everywhere. The Palma Christ (castor-oil plant) is reared for lamp oil in the warm districts, and the manna-bearing tamarisk is found in many low moist spots.

Among the animals are found most species common to Europe, with the addition of the camel, wild ass, wild sheep (or Argali), lions, tigers (rarely), leopards, hunting leopards, tiger-cats, lynxes, and hyænas. There are many celebrated breeds of horses, of which those of the Toorkman plains and the Chaab district are held in highest repute. They have been a good deal improved by crossing with Arab horses, and though not handsome, at least in the estimation of Englishmen, have great strength, speed, and the most extraordinary powers of enduring fatigue. The Persians are extremely fond of, and take great care of their horses. They are clothed with the greatest attention, according to the season of the year; and in warm weather are put into the stable during the day, and taken out at night. Next to camels and dromedaries, mules are in greatest repute as beasts of burden, and form the bulk of the caravans employed in transporting goods from one part of the country to another. There is a great variety of birds, and the Caspian is well supplied with fish.

Persia is not, in general, much infested by reptiles or insects; but the black scorpion and large centipedes are met with in various parts, and the plains in some places swarm in summer with immense Phalangii and enormous spiders, the bite of which is venomous. The stories of the poisonous bug of Miana are believed to be grossly exaggerated; but no words can exaggerate the swarms of gadflies and other stinging insects which set upon the traveller who enters the jungles of Mazunderan in summer, near the clouds of mosquitoes which are bred in its swamps and other low marshy parts of the country. Hosts of locusts, too, occasionally visit the land, destroying every green thing, and themselves supplying food to myriads of wild fowl.

Tenures of Land, and Agriculture.—Property in land is of a fourfold description: 1st, *Khalisas*, or crown lands which, since the confiscations of Nadir Shah, have become very extensive; 3d, those which belong to private individuals; 3d, those granted to charitable or religious institutions; 4th, those granted by the king for military service, or in payment of salaries or annuities. Persons may become proprietors of land by inheritance, by purchase, by gift from the crown, or by reclaiming it from waste by producing the means of irrigation. In any of these cases, except the third, the proprietor's right (he not being the occupant) amounts to the privilege of exacting from the cultivator a tenth part of the produce. In the third case, that of being an assignee of crown lands—he may exact 3-10ths, which include all government dues, and what he can get from the farmers. If the assignment be on the estate of another, he can only demand 2-10ths. If the proprietor be the occupier of his own land, he makes what bargain he pleases with his cultivators; but the regulations for the protection of the husbandman have little or no practical influence. Almost the whole expense of government has to be defrayed by taxes on the land, the amount of which is perpetually varying, not only with the measures of the government, but with the character and disposition of the governors of the different provinces. The state of the country is such that "the cultivator rarely expects to reap the fruit of his labours. His lands and houses are liable to be plundered by the retainers of every petty chief; and he and his family may, in an instant, be deprived of their little capital, and reduced to beggary and want." (*Kinneir*, 37.) Under such circumstances, agriculture must necessarily be in the most depressed possible state. No improvement is ever dreamed of; only the most easily worked portions of the soil are cultivated, and the instruments of husbandry are of the rudest and most primitive onstruction. Such, however, is the fertility of the land, that, despite the want of skill and attention on the part of the husbandmen, wherever the means of irrigation can be commanded, from 10 to 30 returns of the seed are said though we attach little weight to such statements) to be usually reaped, and in many places the produce reaches y and sixty fold. The grains chiefly cultivated are wheat, ley, maize, and rice; the latter being most abundant in low and well watered provinces of Mazunderan and ilan. Oats are very little, if at all, raised. In the greater ber of provinces there are two sorts of cultivation, wet dry, or by means of irrigation and without this assist-s. In the former, which is also by far the most exten-, the ground is roughly turned up by a wooden plough, etimes slightly shod with iron, and drawn generally by oxen; and the seed being harrowed in, the fields of individual are laid under water as frequently as may quired, or as he has a right to; for the water of each u is portioned out by time into shares, which are t and sold as property. The water is derived either natural rivulets or from under-ground canals, con-d with great skill, and carried to a great distance. are the property of those who construct them; the I streams belong to the sovereign, or to those who urchased them, or on whom they may have been ed. All disputes about irrigation, like those on other i, are settled by the kethhods (magistrate) or elders illage. The dry cultivation is conducted, as to til-nearly the same way as the other, but the grain is nourished by rain only; so that, in this arid coun-ust be confined to particular spots which experience vation have pointed out as fit for it. Manure is dy applied to corn-land. Near cities, the melon, r, and vegetable grounds are enriched with manure soil of the caravanseras, &c.; and in the neigh-of Ispahan pigeon dung is so highly valued in e of the fine melons, for which that district is , that pigeon-houses are built for the purpose of this manure, which sells at an enormous price; re almost the only instances in which the land ly assisted. The use of lime as a manure is un-this, as in other Asiatic countries; and fallows be the only means resorted to for the improve-land.

articularly the large-tailed variety, are every abundant. Their flesh is almost the only meat . exclusive of poultry and game; beef not being ed, nor of good quality. Sheep's' wool is unif for clothing, and sheep skins with the wool s and cloaks. The property of the wandering la consists of sheep, with cattle, horses, camels, large flocks and herds. Wool might become le article of export. In Kerman they have a which yield a down, not much inferior to the ol, which might be greatly increased all over ne parts of the country. Manufactures are l so articles of home consumption. Till late-ne clothing of the poorer classes, both cotton was home-made; but for some years past rse chintzes and printed cottons in use have rom England and Russia by way of India, - Caspian. Still a great deal of cotton cloth in and printed, some of which goes to Rus-. In silk, also, a good deal is done, the fab-our groe-de-naples, taffetas, satins, velvets, erchiefs, &c., produced at Cashan, Yezd, d, Ispahan, and many other places, are not be in request for export to Russia, and us for home consumption. Besides these, fnctured at Kerman, of the wool of that ion of those of Cashmere, and which, n softness and fineness, are still a hand-great request both for home use and ex-s and felts for sitting on are also made in best of the former being produced at He-tral districts of Irak, and generally by the The best are made in Khorasan; but use in the country, few being exported. ufactures of arms, swords, daggers, guns, scissors and knives. The former are nz and Mushed; the guns of Kerman-y, are highly prized, and the cutlery of has obtained some celebrity in Persia; cissors of Birmingham are so far supe-d quality, that there is no great demand . China ware, and all superior pottery, ttle coarse glass is manufactured in the also a manufactory of leather and sad-

country where there are no made roads, canals, and scarcely any seaports or there is but little security of property,

it might be inferred that there can be no commerce. Yet, with all these disadvantages, Persia has a good deal of trade, and there is no inconsiderable degree of commercial spirit among its inhabitants. The means of land-carriage, as over most of the E., is by caravans of camels, mules, and small horses, called yaboos, for there is not a wheel carriage in the country. The two latter are particularly suited to the stony roads and high mountain passes which occur in al-most every day's journey : and in this way is brought the whole merchandise from Bokhara, Caubul, much of that from India, and from all the nearer countries, to the vari-ous marts; the returns being transported in like manner. The only seaports are Bushire, Bunderabbas, or Gombroon, Congoon, and some still smaller places on the Persian gulf; and Enzellee, Balfroosh, and Asterabad, upon the Caspian. Of these Bushire, on the gulf, and Balfroosh and Enzellee on the Caspian, are the most considerable, the former being the mart of all the trade with India, and the two latter of that with Russia ; and from these the communication with the interior is kept up by caravans, as already mentioned.

The principal trade of Persia is with India, Turkey, Rus-sia, Bokhara, and Afghanistan, and of late direct with England. From the first the imports are chiefly indigo, chintzes, muslins, and calicoes, gold and silver brocades, precious stones, china, and earthenware. Sugar and sugar-candy, cashmere shawls, iron, lead, copper. From Turkey, European goods from the Levant, and specie. From Rus-sia, iron, broadcloth, coarse and fine printed calicoes, gold lace and metal buttons. Trunks of all sizes ; tea, coarse cutlery, leather, glass-ware, tea-urns, and copper in sheets, quicksilver, furs, paper, in great quantity ; cochineal, &c. From Bokhara black lamb-skins, raw hides, dried prunes, ru-bies, and other gems ; shawls, china ware, camblet ; Indian and Russian produce by that route. From England, broad-cloths, and narrow woollens of all sorts ; cotton manufac-tures, imitation shawls, jewellery, arms, cutlery, watches, spectacles, earthen and glass-ware, iron, tin, copper, &c. From Arabia coffee, pearls, horses, &c.

The exports are chiefly to England, silk, gall-nuts, a little wool, madder, yellow berries, occasionally a few pearls and precious stones, and specie to be converted into bills at Con-stantinople. To India, specie, dried fruits, tobacco, wine, drugs, dates, sulphur, torquoises, Kerman shawls, rose-water, swords, horses, greyhounds, raw silk ; copper (from Turkey), saffron, &c. To Turkey, grain, raw silk, tobacco and pipe-sticks, cotton, lamb and fox skins, carpets, silk manufactures, cotton do., salt, sheep ; besides foreign ar ticles in transit from India and Bokhara. To Bagdad much the same as to Turkey ; to the Uzbecks and Turkmans, Kerman shawls, and woollens ; silk stuffs, gold embroidery ; copper ware from Cashan ; cotton goods, arms, Hamadan leather shoes, and cloths, torquoises ; sugar, raw and refined, opium, and some Indian goods; to Arabia, wheat, dates, dried fruits, rose-water, cloaks.

To Russia, raw silk and cotton, rice, grain, timber, tobac-co, raw hides, lamb-skins, fish, gall-nuts, naphtha, drugs, torquoises, Kerman shawls, silk and cotton goods, brocades, besides foreign produce, as Cashmere shawls, pearls, &c.

Where no official records are preserved, where every art is put in practice to disguise the true amount of imports and exports, and where, also, the existing state of things is al-ways changing from external or internal causes, it is impos-sible to arrive at any just estimate of the value of com-merce. In 1890 the export trade of Persia was estimated by Mr. Fraser at about £1,925,000 a year. Since then the imports have undoubtedly increased, in spite of great dis-tress in the country from three years of plague and famine ; for in 1735, from the best materials that can be collected on the subject, they appear to have nearly amounted to 3¼ mil-lions sterling, and the imports of 1835 to Trebizond of Euro-pean produce, the greater part of which finds its way to Persia, exceeded a million sterling; and there is every reason to expect a gradual increase from all quarters, though doubtless liable to occasional fluctuation.

Shipping.—Scarcely any vessels belong to Persian own-ers. The trade between India and the gulf is carried on in bottoms belonging to Indian, Arab, or Armenian merchants resident at the seaports; and that of the Caspian, in vessels from Astrakhan. The Russian government discouraging any vessels in that trade which is owned by one of their own subjects, the Persian or Armenian merchants who have embarked in it generally become subjects of Rus-sia, which leads to their ultimate residence in Astrakhan. There are, however, some vessels, of from 50 to 150 tons, built at Enzellee.

Money.—The coins chiefly current in Persia are bajo-glees, or Persian ducats ; sahebkerans, commonly called koroonees, a silver coin, nine of which go to the bajoglee ; and copper coins, called pool-e-siah, or black money. There are also several pieces of one or more abbasees or shahees, the abbasees or shahee being the fourth or fifth part of a

49   2 O   577

hstoonce; there are about four pool-e-siahs in an abbassee. The old tomân of gold, and real or rupee of silver, are now seldom seen, although used at times in calculation; the tomân being equal to 10 koroonees, or 8 reals. Except the Russian or Austrian ducat, which is in common use, there is little foreign coir now current.

Races—Population.—The ancient Persian stock has been much intermixed in the course of ages by the settlement of other races in the country, especially by the influx of Greeks during and after the conquest of Alexander the Great; and more recently by that of Arabs and Turks. Still, however, the distinguishing characteristics of the family appear to be pretty well preserved. The complexion is fair, but not transparent, and there is little or no colour in the cheek. Hair long, straight, and almost always jet-black; beard abundant, bushy, generally black, but now and then with a reddish tinge. Features regular and handsome, though generally minute, and, excepting the beard, rather effeminate. Stature little short of the European standard; body gracefully, but not very strongly formed, being, altogether, less robust than that of the European. Though early civilized, they have made no considerable progress in arts, science, or arms; and though remarkably clever, and not deficient in bravery, they have never been able to establish anything like a free system of government, or to set any limits to the caprice and tyranny of their rulers. They have occasionally gained some advantages over other Asiatic nations, but they have never been able to oppose any effectual resistance to Europeans. In antiquity, a small army of Greeks overthrew the Persian empire when in the zenith of his power; and in more modern times, it has been overrun by the Arabs, and even the Affghans. At present it owes its existence to no intrinsic vigour of its own, but merely to the forbearance and jealousies of its enemies. The Persians have, in fact, contributed nothing to the improvement or civilisation of mankind; and excepting Zoroaster, have not produced a single benefactor of his species known to history.

At present the population of Persia may be divided into two distinct classes, the fixed and the nomadic. The first comprise all who live in towns and villages, and have fixed habitations; the second consists of the various tribes, indigenous and of foreign extraction, who lead a pastoral and erratic life, having no regular habitations. But the whole body of the people may more conveniently be divided into four classes: first, those who are attached to the metropolitan and provincial courts, including the functionaries of government and military; second, the inhabitants of towns, comprising merchants, shopkeepers, artisans, with men of the religious orders, of business or of learning; third those employed in agriculture; and fourth, the tribes, including the *Eeliaats*, or *Illyats* (dwellers in tents), or nomades.

The officers of court are more remarkable for skill in business, versatility, politeness, and courtesy, than for probity, honesty, or good principles. Forced, in self-defence, to dissemble and control their feelings, they do so successfully, and, looking to wealth as the best means of purchasing favour in the day of adversity, as well as of enjoyment in prosperity, they stick at no means by which it may be acquired. Accordingly, they become, in general, great intriguers; and are at once deceitful, sensual, venal, treacherous, and, when they dare, arrogant and overbearing. Ministers of state are generally selected from among the men of business or meezas, who, though less arrogant than the nobles, are equally corrupt and immoral: they do not assume so much state as military chiefs, and are distinguished by a roll of paper stuck in their girdle, instead of a sword or dagger. One remarkable class of court dependants are the royal gholaums or body guards, the confidential and devoted guardians of the monarch's person; whence the name *gholaum*, or slave. They are usually either Georgian captives or sons of respectable families; and resemble somewhat the *mousquetaires* of the old French government. They are employed in lucrative and confidential services, and the situation is much sought after; but their tyranny and dissoluteness know no limits, and the arrival of a gholaum-e-shah in a district, creates a sensation not unlike the attack of a pestilence.

The towns-people, *shehrees*, as they are called, are a mingled race of all those which have ever conquered or had intercourse with Persia, grafted on the original stock—Turks, Tartars, Arabs, Armenians, Georgians. They are a more industrious and less depraved class than the first; but being nurtured in falsehood and deceit, they are adepts in these vices, being at the same time, however, cheerful, polite, sociable, kind masters, and good servants. The merchants are numerous and often wealthy, and, having more intercourse with foreign nations, are usually of more cultivated and enlarged minds than others of their countrymen. The shopkeepers are, of course, a grade lower.

The ecclesiastical body, which is also numerous, is, with some rare exceptions, more remarkable for hypocrisy and profligacy than for piety and morality; originating, most probably, in the want of a suitable provision to live on, and the consequent necessity of practising fraud and imposition. The cultivators of the soil are those on whom the tyranny of their rulers falls most heavily; yet it cannot be said that they exhibit much misery. They are themselves, as well as their wives and children, for the most part sufficiently, though poorly, clad, and have abundance of wholesome, though coarse, food, as wheat or barley bread, cheese, sour milk, rice, &c. Extortion and tyranny are met, as usual, by cunning and deceit; and, as the peasantry are active and intelligent, they contrive to avoid being completely fleeced.

The fourth class is an interesting and extensive one. It consists not only of the native nomades of Persia, who occupied the south-western and southern ranges of mountains long before the Mohammedan conquest, but of all those of nomadic origin who came with the various conquerors that have overrun the country since that era, as the Arabs, Ghiznavedes, Seljook Toorkmans, Mughuls, Toorks, Usbecks, &c. But the greater number consists of those of Arab and Turkish origin, particularly the latter. It may be remarked, as a singular anomaly, that these nomadic tribes supply not only the principal military force of the country, but, as a consequence, probably, its only hereditary aristocracy, and, generally, its sovereign himself. Of these tribes, a portion is always approximating more nearly to the habits of fixed life; but the greater part by far are strictly nomadic, living in tents, which they shift from place to place, according as lack of pasture for their flocks and herds, or change of season, suggests. In these their wealth consists; and though many of them cultivate a little grain, they live by the sale of the surplus of their stock, and by the produce in milk, wool, and flesh. Their character and habits are every where much the same. Being poor, they are frugal and abstemious; and unaccustomed to more civilized manners, they are rude and blunt, fond of independence, and passionately fond of martial exercises, of the chase and war. Predatory both from inclination and education; but hospitable, comparatively honest when their faith is pledged, and brave. Their chiefs, seen among their own people and in their own country, appear to great advantage, as frank, liberal, and generous, though hasty and passionate; at court they are constrained to assume somewhat of the manners of the place, and do not shine so much as at home.

The Koords come under the denomination of the "tribes," though less erratic in their habits. They claim a high descent: some pretending to be descendants of the great of the air by terrestrial women, and others the progeny of certain persons saved from the tyranny of Zohaak. But their antiquity is unquestionable, and, probably, they may be descendants of the Carduchii described by Xenophon.

The Toorkman tribes, inhabiting the desert on the N. of Khorasan, are likewise to be reckoned among the nomades of Persia. They are wholly addicted to robbery and pillage, their chief occupation being that of making plundering parties, which destroy whole villages, carry off the inhabitants into slavery, and their cattle and property. But to enumerate, far more to describe, the various nomades of Persia would greatly surpass our limits; and we must refer our readers, on this interesting subject, to works whereon it is treated of at greater length.

National Character.—In general, it may be said of the Persians that they "are handsome, active, and robust; of lively imagination, quick apprehension, and agreeable and prepossessing manners. As a nation they may be termed brave; though the valour they have displayed, like that of every other people in a similar state of society, has, in a great degree, depended on the character of their leaders and the nature of the objects for which they have fought." (*Malcolm's History of Persia*, ii., 638.) Unhappily, however, their vices are far more prominent than their virtues. Though the despotism to which they are subject is similar to that which weighs down all the eastern nations, they have a peculiar and distinctive character. As compared with the Turks, they are not unlike what the Irish are as compared with the English or the Scotch, being gayer, livelier, more active, more versatile, and less to be depended on. Though easily inflamed into passion, and when under no influence abusive in the highest degree, they are, speaking generally, courteous, affable, and polite. They flatter with equal skill and profusion of compliments. Their language is extravagantly hyperbolical; and a stranger, ignorant of their character, would suppose them ready to devote their fortune and life to his service. A foreigner, therefore, on his first arrival, can hardly avoid receiving the most favourable impression of their friendly disposition. A longer acquaintance, however, proves that their flattery is nothing to their insincerity. However it may be accounted for, whether it be ascribed to the despotic nature of their government and the frequency of revolutions, the influence of their religion, or whatever cause, all travellers are agreed that the Persians have reduced dissimulation and falsehood to a system, and have practised them so long and so universally, that it would be difficult for them, even if they intended it, to speak the truth. Their whole conduct is a tissue of fraud and artifice; and they rarely think of fair dealing till they find they have to do with one who sees through their impostures. "There is no deceit, degradation, or crime in which they will not stoop for gain; and no habit of falsehood so inveterate, that untruths flow, if it were spontaneously from their tongue, even without any apparent motive." (Fraser's Khorasan, p. 174.) Mr. Kinir's estimate of their character is, if possible, still more favourable. "They are," says he, "haughty to their inferiors, obsequious to their superiors, cruel, vindictive, treacherous and avaricious, without faith, friendship, gratitude, or honour." (Memoir, p. 22.) Presents, a necessary instrument of business over all the east, are expected in Persia with peculiar avidity. Without presents no inferior can approach a superior, or any individual ask a favour from another; and the donation, being supposed to confer honour, is made in the most public place and manner possible. ... are said to be, with few exceptions, incorrigible spendthrifts: their dress, horses, harems, &c., are generally kept up on a scale exceeding their means, and intended for ostentation; and the difficulties in which they are thus involved make them resort to any expedient, however mean and discreditable, for raising money.

These statements must, however be understood as applying more particularly to the sedentary population, and especially to the inhabitants of cities and towns. "The Eelleats ... the virtues and vices of their condition: are sincere, noble, and brave; but rude, violent, and rapacious. They are not in need of falsehood and deceit, and therefore, which in the habit of practising them; but, if they have vices than the citizens of Persia, it is evidently the effect of temptation, and the ignorance of luxury and refinement, which give them all the superiority they boast; as remarked that they never settle in towns, or enter into visiters, without exceeding the inhabitants in every kind of prodigacy." (Malcolm.)

The natives of Persia do not recline on cushions, in the manner of Turks; but sit in an erect posture on what is called a nummud. They have seldom, if ever, fires in apartments, even in the coldest season, and, in order to obtain, fold themselves in a fur pelisse on a barraxane, or a handsome robe of crimson cloth, lined with velvet. Like other oriental nations, they rise early; and having dressed and said their prayers, take coffee, or, perhaps, some fruit. They then enter on business of the day, if they have any; and if not, converse until about 11 o'clock, at which time they have their breakfast, and then retire into the harem they remain until about 3 o'clock, when they again, see company, and finish their business; so that people the most important affairs are disturbed transacted in public. Between 9 and 10, the principal meal, is served up. This chiefly consists, and of mutton and fowl, dressed in various ways; however, they eat but moderately. Wine is before company; although, in private, they are notorious drunkards, and invariably drink between. They are passionately fond of tobacco, which most incessantly from the moment they rise for them to retire to rest: it constitutes individual source of amusement to a man of fortune if not for his caleean. I am at a loss to imagine how spend his time. In this respect, indeed, there is something peculiarly inconsistent in the character. When without an inducement to exert himself entirely to luxury and ease; and the one, with his caleean in his mouth, would appear day in a state of stupor, when roused into action on his horse, will ride for days and without termination. Hunting and hawking as well as rustic exercises, are favourite amusements. By these means their bodies become hardened and as they are taught to ride from their youth to their houses with great boldness and frequently use the 'warm bath,' but seldom ." (Kinneir's Persian Empire, 245.) These, at least those of the sedentary part are for the most part closely concealed. They great pass their time in visiting their and amusing themselves with diversions of one kind and with intrigues. The bath is, however, one of their enjoyment and relaxation, where, options, they give full scope to merri-... They differ equally from us in their mode of taste. Large, soft, and languishing ... in their opinion, the perfection of beau-... figure their natural charms by painting sometimes also by tatooing their skins of

various colours, while constant smoking spoils their teeth and mouths. Many of the women of Shiraz and other cities are as fair as those of Europe; but they want, owing to their confinement, the bloom so essential, in our estimation, to female loveliness. The Persian ladies would seem to be totally devoid of delicacy and refinement. "Their language," says Mr. Scott Waring, "is often gross and disgusting, nor do they feel more hesitation in expressing themselves before men than they would before their female associates. Their terms of abuse and reproach are indelicate to the utmost degree. I will not disgust the reader by noticing any of them; but I may safely avow, that it is not possible for language to express, or the imagination to conceive, more indecent or grosser images. When they leave the house, they put on a cloak which descends from their head to their feet, and their faces are carefully veiled, holes only being left for their eyes. It is curious to see a number of tall and elegantly formed figures walking in the streets, and presenting nothing to your view but a pair of sparkling black eyes, which seem to enjoy the curiosity they excite. The veil appears to be essential to their virtue; for so long as they conceal their face, they care not how much they expose the rest of their person." Like the Mohammedans, the Persians are restricted to four legitimate wives, but they may have as many concubines as they please; the latter being acquired by purchase or hire. Few, however, unless they belong to the richer classes, indulge in the luxury of a plurality of wives, or keep concubines. Marriages are usually celebrated with great splendour, and often entail a ruinous expense on the parties.

Amount of Population. — There are no existing data on which to found any thing like an accurate estimate of the amount of the population. Pinkerton supposed it might amount to about 10 millions, which Sir John Malcolm thinks may be a pretty close approximation to the truth. Another writer (Fraser) has set down the fixed population at about seven millions, and the migratory population at from two to three millions, thus approaching to Pinkerton's estimate. But besides the loss of territory which Persia has since suffered from her wars with Russia, a great depopulation has taken place within the last ten years, from plague, famine, and various prevailing maladies, and there is reason, besides, to believe that this estimate of the migratory population was much beyond the mark, so that it is probable the population of the countries subject to the shah does not exceed 8, and is certainly under 10 millions. We subjoin from the *Weimar Almanack* an estimate of the area and population of the different provinces; but it is needless to say that it is but little to be depended on.

Provinces.	Area in sq. m.	Population.
Irak-Ajemi	83,376	2,150 000
Thaburistan	6,984	130 000
Mazanderan	7,547	850 000
Ghilan	6,215	250,000
Azerbijan	30,337	2,000 000
Kourdistan	14,282	450,000
Kermian	99,356	900,000
Fars	126,160	1,700 000
Kerman	65,496	600,000
Kohistan	23,574	150,000
Khorasan	81,132	1,700 000
Total	484,180	11,230,000

The government of Persia, like that of most eastern countries, is, in principle at least, an absolute despotism. The shah is regarded as the vicegerent of the prophet, and, as such, is entitled to implicit obedience. His word is law; he is absolute master of the lives and properties of his subjects; and the first man in the empire may, at his command, be instantly stripped of all his dignities, bastinadoed, or strangled! the only control on his actions being the risk of provoking rebellion or assassination. The two principal ministers are the grand visier, or *Visier Azem*, and the lord high-treasurer, or *Amen a Doulah*. The former superintends every thing connected with foreign relations, and, in the absence of the sovereign, commands the armies; while the latter, who is subordinate to the other, superintends the internal arrangements, the collection of the revenue, &c. The whole executive government is in the hands of these two functionaries, whose authority, so long as they continue in power, is as absolute as that of their master; but their greatness, being built on the favour of a tyrant, is of the most unstable kind, and they are very often precipitated from their slippery elevation.

The duties of a monarch, who either regards his own safety or the well-being of his people, are numerous and weighty. One of the most important is the distribution of justice. The Mohammedan law, both civil and criminal, is founded on the precepts of the Koran and the oral commentaries and sayings of the prophet's immediate successors. This is called the *Sherrah*, or written law, and is the rule in all regular courts, where persons of the ecclesiastical order, such as moonshtehehds, preside. But there is also the *Urf*, or customary law, administered by secular magistrates, having the king as their head. It is more arbitrary, and the judgments of the king and his lieutenants are more summary, than those of the other court, and enforced with greater vigour. But there is an appeal to the superior functionaries, and the power of life and death rests with his majesty, who seldom delegates it, except to princes of the blood royal, or to governors of remote provinces. Theft is always punished with extreme severity.

The system of civil government is simple. Each province, or important district of a province, including some large city, has a *Beglerbeg*, or governor, usually a prince of the blood or nobleman of rank, who appoints his lieutenants, or *Hakims*, over the districts and subdivisions: and each village has its *Ketkhoda*, or magistrate, generally one of the elders or more respectable inhabitants, who is the organ by which communication is kept up with government. There are also governors of cities and towns, *Daroghas*, or lieutenants of police, and *Ketorenters*, or chief magistrates of cities, in which each *muhuleh*, or parish, has its ketkhodah, or head, who are in general practically chosen by the people, and who look to the kelounlee as their head.

The beglerbegs, like the Turkish pachas, are, at the expiration of a certain period clied to court, where, admitting their conduct to have been ever so irreproachable, persons are not wanting to accuse them of injustice and mal-administration; and unless the demands and avarice of the court be completely gratified, their eyes are put out, and their property confiscated! Conscious, therefore, of the necessity of amassing a sufficient sum of money to answer the rapacity of the king and his ministers, and aware, at the same time, that, provided the money be forthcoming, no inquiry will be made respecting the manner in which it has been acquired, the same mode is applied by the beglerbegs to the hakims and other subordinate authorities; who, in turn, oppress the head of villages and the cultivators, so that the land becomes the prey of a subordination of vultures, and venality and extortion pervade every class from the throne to the cottage, (*Kinneir's Memoir*, p. 31.)

But, after all, the principal evil under which the country labours consists in the perpetual insurrections and the sudden changes of sovereigns and dynasties. The insecurity, devastation, and proscriptions, to which this state of things has led, have necessarily gone far to extinguish all industry, and many provinces that were formerly well peopled and well cultivated are now all but deserts. It has, also, prevented any idea of stability being associated with the existing state of things; and has made change, and the insecurity and falsehood inseparable from it, almost a necessary state of existence. Neither under existing circumstances does it seem reasonable to expect that any reform can be effected from within; and its subjugation by a European power would certainly be the most desirable event that could happen for the country.

The revenue of the shah has been variously estimated, but does not probably amount to more than £1,500,000 or £2,000,000. As already stated, it is principally derived from taxes on lands and farms, capitation taxes, duties on imports and exports, tributes paid by the nomadic tribes, &c.

Religion.—The Persians are Mohammedans of the sect called Schiites, or Sheahs, or of those who look upon Ali, the son-in-law of the prophet, as his legitimate successor. They repudiate the first three caliphs, Abubekr, Omar, and Othman, and their successors, as usurpers of the right of their patron Ali, holding that of his sons Hassan and Hossein to the caliphat as indefeasible, and acknowledging their 12 immediate descendants as the 12 high priests, or imáms, of their religion; the last of whom, Imaum Mehdee, they consider as still alive, though (*ghaib*) concealed for a time, so that no other can exercise the office. This doctrine is quite opposed to that of the Turks, who belong to the sect of Sunnites, and between whom and the Persians the most rancorous and irreconcilable animosity exists as to religious doctrines.

The priesthood consists of many orders, of which the mooshtehed is now the chief. There are seldom above four or five of this dignity, and these are elected as much by the public voice as by that of their brother mooshtehds, by whom they must be declared, for the shah has no voice in their appointment. The sheikh-ul-Islam, or ruler of the faith, is next in rank, but he receives his appointment and a salary from the shah, and there is one in all large towns. In every mosque of consequence and at every considerable shrine or place of pilgrimage, there are at least three regular ecclesiastical officers: the mooturelie, who manages its temporal affairs; the muezzin, or cregree to prayers; and the mollah, who conducts the ceremonial. If the establishment be rich, there are several mollahs, from among whom are selected a *peesh namâz*, who recites the prayers and goes through the motions and genuflections to guide the congregation. They also preach occasionally sermons from texts of the Korân. Of all these, except the sheikh-ul-Islam the income and means of life depends chiefly on the cele-

brity of the individual for wisdom, virtue, and religious sanctity; so that there are no means of estimating the income of individuals: but most mosques and shrines have large property in land and villages, the gift of the crown or of pious individuals, and out of this the priesthood attached respectively to each is maintained. Besides those above enumerated, there are in every city, and in every seminary of learning, a crowd of mollahs who live by their wits waiting for the chance of employment, but having little of the priest but the name. They practise astrology, write letters and contracts for those who cannot do it for themselves, and descend to all manner of meanness and vice for a livelihood.

The Persians, though there are many enthusiasts and bigots among them, are not generally intolerant: they listen without anger to the professions or arguments of those who hold a different belief, and do not allow this circumstance to cause any interruption of social intercourse. The only exception is in the case of the Guebres, or fire worshippers, who are probably rendered odious to the modern rulers of Persia by connecting with their faith an attachment to its ancient laws and political system. This unfortunate race is now almost entirely extirpated, only a small remnant being found in Yezd, and other cities of Kerman. Indifference, scepticism, and free-thinking, are, however, making a rapid progress. This last, which may be identified with what is called Soofieism, extends every day. There is, if we may so speak, a religious and a sceptical Soofieism; the former is a sort of a mystical or fanatical aspiration after the mysteries of divine love, but without laying any or much stress on the rules and regulations of the Koran: the latter is of a bolder character, and approaches nearer to the European notions of free-thinking; its votaries affect no particular respect for religion, but are a species of metaphysical Deists, regarding the Koran merely as an elegant work, embodying sound moral doctrines, but not otherwise entitled to attention. Hence all who profess or are suspected of Soofieism are hated or persecuted by the mollahs. But Mohammedanism in Persia, as in other countries in which it is professed, appears decidedly on the decline, and Soofieism is likely to be one great instrument of its overthrow.

The ancient religion of the Persians, which is not yet entirely extirpated, was materially reformed and renovated by the famous legislator Zoroaster, or Zerdusht. The life and even the epoch of the birth, of this great reformer, are involved in the utmost obscurity; but the preferable opinion seems to be that he flourished about the 6th century B.C. He inculcated the doctrine of an eternal, self-existing, supreme Being, from whom every thing else has its origin, and from whom are sprung two antagonist powers, Ormusd, the source of all good, and Ahriman, the source of all evil of which, however, the former, though this point be extremely obscure, is destined, in the end, to obtain the ascendancy. The doctrines of original sin, the immortality of the soul, the happiness of the good and the misery of the bad in another life, are all laid down by Zoroaster. But the distinctive feature in the religion of the Persians was the extreme veneration paid to fire, light, and heat, which is regarded as symbolical of the divinity. *C'est par les qui tout respire: la terre lui doit sa fécondité; l'animal, son existence; l'arbre, sa végétation. Nous seulement il anime les êtres, il forme encore leurs rapports, et son action, par conséquent, n'est pas moins ancienne que le monde.* (Pastoret, Zoroaster, Confucius, &c., p. 30.) Herodotus says, that the ancient Persians neither erected temples nor statues to their gods, but sacrificed to them on the tops of mountains, or other high places. (*Herod.*, lib. i. cap. 131.) They had, also, the singular, and, as it appears to us, barbarous custom of exposing the bodies of the dead to be devoured by birds; (*Herod.*, lib. i., cap. 140); and Niebuhr distinctly states that this custom was observed in his time by the Parsees at Bombay. (*Voyage en Arabie*, ii. 30.) The magi, or priests, established by the Persians, had great influence.[*]

Education.—In former reigns, particularly in the time of the Suffaveans, when literature was more encouraged than now, considerable attention was paid to education. Medressas, or colleges were built and endowed, in which mollahs and teachers of suitable abilities were placed to instruct the students. These buildings consisted of a quadrangle, the interior sides of which were pierced with small cell-like apartments, like those of a caravanserai, in each of which a student lived. But these institutions were rather for students of more advanced age; for younger pupils of the lower classes there are schools kept by private persons, where reading and writing are taught, and some knowledge of the practice of religion is imparted, with perhaps, to some who are destined to become "men of the pen," a little in-

---

[*] The *Zend Avesta*, the most important work on the religion of the Persians, was translated and published in 3 vols. 4to. by Anquetil du Perron in 1771; but it is believed by some of the ablest critics that the most ancient portion of this work is long posterior to the age of Zoroaster.

periodical instruction in logic and grammar. The children of the higher orders are taught at home by *moolisms* and *lal-lals*, or learn, from his infancy, to the extremes of the duties of their religion, and teach them to read the Koran, with such works as are calculated to impress them with a strong regard for Sunnite doctrines. Next come the works of Saadi and Hafiz, with a superficial course of grammar, logic, and philosophy. All this time athletic exercises, riding, hunting, and the use of arms, are not neglected; and from the earliest age every boy is carefully schooled in all that regards the ceremonial of social intercourse. He is taught how to sit down and rise up, and to stand in the presence of his elders or superiors; and so much stress is laid on these matters, that it is not uncommon to observe the least deviation from due etiquette even in children of five or six years of age. But the whole system is artificial, more showy than solid, and tends in no small degree to nurse up the rising generations that disposition to deception and hypocrisy which marks strongly the national character.

Military Force and Resources.—"Frugal in his diet, robust in his constitution, capable of enduring astonishing fatigue, and inured, from his infancy, to the extremes of heat and cold, to hunger and thirst, nature seems to have formed the Persian for a soldier. But as, according to the ancient customs of this people it is deemed degrading to a man, who has money sufficient to purchase a horse, to travel on foot, the infantry of Persia has been, from the earliest times, contemptible: whilst her numerous bodies of irregular cavalry have more than once carried terror and defeat amongst the disciplined legions of Rome." (*Kinneir's Mem.*) These forces, however, both cavalry and foot, have varied in amount and efficiency with the varying abilities and martial skill and daring of the different monarchs. Until lately no attempts have been made to form regular corps, disciplined after the European fashion, the army has mostly consisted of levies of irregular cavalry, furnished by the chiefs of the different wandering tribes, according to their assessed numbers and strength, and also by the different settled towns, on a plan corresponding in many respects to the feudal levies of the middle ages in European countries. The troops thus collected, though brave, are totally deficient in organisation and discipline, and could make no impression on a body of European troops. Inasmuch as the arms and horses on which the horsemen mounted do not belong to the state, but to the individuals who frequently constitute their whole property, they are apt to prefer their safety to other considerations; a circumstance which, on more than one occasion, has proved the ruin reputation of the Persian army. The whole force might thus be collected on an emergency might, amount to 100,000 or 150,000 men. In the late first attempt was made to introduce European discipline and tactics among the Persian soldiers. The prince royal, Abbas Mirza, obtained leave from his father, and offered to the E. I. company to raise and discipline a body for Azerbijan, with a view of opposing the Russians, strengthening his internal government; and he did form a standing, with cavalry and artillery, to about 12,000 infantry; but after the peace of Goolistan with Russia, proposing this corps were unwisely permitted to return to their homes, mustering only occasionally; the remaining useless appendages of state at the court of Abbas. On the commencement of the war with Turkey, as British officers could not serve against a power in amity terms with Great Britain, they were dismissed. The army of 35,000 men, regular and irregular, which the prince marched against the Turks, was victorious, dispersed by the cholera; and from that time there were only one or two British officers retained to command the serbás or regular troops, formed and maintained in Azerbijan. When the prince royal, proposed to subdue the rebellious chiefs, and to reduce some of the other more remote quarters, he felt the want of more efficient aid to military force, and applied to the British minister for a supply of commissioned and non-commissioned officers, who only reached Persia after his death. His son, Mahomed Mirza, the present shah, inherits his father's plans, but with less ability and though desirous to increase and improve has never employed these officers in the way services most useful.

In the disastrous campaigns in Khorasan, and particularly last, the army suffered much from sickness, and desertion; so that no calculation can be made as to state. In 1837, however, when the shah made his noble effort to bring a large force against Herat, the army did not certainly exceed 35,000 men, and on. The falling off in the amount of the Persian war of 1827-28 is very striking; for

the prince royal had then a well-appointed army of 40,000 men, with all its complement of baggage, cattle, and attendants on the frontiers; while the shah was at Choee with another army of 50,000. It may be doubted, however, whether the attempts to introduce European tactics and discipline into such a country, and under such a government as that of Persia, can ever succeed; and whether it would not have been better policy to trust now, as of old, mainly to levies of cavalry, and endeavouring to improve and amend the defects in their constitution and discipline. What may be called the household troops of Persia, consist of a kind of militia of about 10,000, quartered in the capital and its vicinity, and liable to be called out at a moment's warning. The *gholauns*, royal slaves, or body guard, have been already noticed.

Arts, Language, Sciences, and Literature.—Of Persian proficiency in these, previously to the Mohammedan conquest, little or nothing is known, all that may have existed having been destroyed by the Moslems. But we may infer, from the relics of sculpture of the Sassanian era that remain, and from accounts of contemporary authors of other countries, that some of the arts, at least, were then successfully cultivated. In the days of the Suffavean painting appears to have received some attention, and architecture still more; but though attempts at depicting the human form, as well as animals and landscapes, are numerous among the Persians of this day, they are but rude and unsuccessful, the total absence of all drawing and perspective rendering their performances ludicrous, if not disgusting. In fact, being quite without models for either painting or sculpture to copy from, excellence is scarcely to be looked for, especially in a country where the tyrannical spirit of the government and nobility would render such attainments dangerous rather than profitable to the owner. Their most successful performances are the inkstands and small boxes, made chiefly at Shiraz and Ispahan, which are ornamented with figures of boys and girls, birds and flowers, finished with surprising minuteness and accuracy. The stone and seal cutters of the same cities are also famous for their workmanship.

When the Arabs overran Persia, about the middle of the seventh century, three languages were spoken in the country, the Parsee, Pehlvi, and Deri, exclusive of the Zend, or language dedicated to religion. The first of these languages has superseded the rest, which are now only known by name, and become the universal language of the country. It is of simple structure; and, like the English or French, has few or no inflections, prepositions governing its cases, and auxiliary verbs its tenses and modes. Many of its roots can be readily traced to the Sanscrit; and, in the course of time, it has received a large accession of Arabic words. All the existing literature of the Persians dates from the Arabic conquest, and mostly, indeed, belongs to the 15th and 16th centuries.

In science the Persians are scarcely more advanced than in art. Astronomy, judicial astrology, metaphysics, logic, mathematics, and physic are among those professedly cultivated. But their efforts in the first are contemptible; their theories, founded on the Ptolemaic system mixed up with fantastic notions of their own, are utterly useless, unless to aid their dreams of astrology. Their firm belief in this science is universal, and no Persian will undertake the most trivial affair without consulting some professor of its mysteries for a lucky hour. Their metaphysics and logic are scarcely less puerile. The first consists of little more than a collection of disputations, sophisms turning on wild and unprofitable paradoxes; the second is an ingenious method of playing upon words, the object being not so much to arrive at truth, as to display quickness of mind and readiness of reply, in the discussion of plausible hypotheses. Geography is no better understood. Their knowledge of countries, and their relative positions, is extremely confused; nor can they describe, with any exactness, even those places or regions with which they are most familiar.

Mathematics, though not much more beneficially applied, are taught upon better principles, for the Persians are acquainted with the works of Euclid. Chemistry is unknown; but alchymy is a favourite study, and the search after the philosopher's stone continues to be eagerly prosecuted. In medicine, though they profess themselves pupils of Galen and Hippocrates (*Jalenoos* and *Boerat*), they practise only the most wretched empyricism, united with the exhibition of a few simples, the qualities of which experience has taught them. Diseases are classed into hot and cold, moist and dry, upon no apparent principle, and each disease is combated by a remedy supposed, as vaguely, to be of an opposite quality. They are quite ignorant of anatomy, and even of the circulation of the blood, so that their knowledge of surgery is no greater than that of medicine. Yet, though they admire the skill of Europeans, and eagerly possess themselves of their remedies, they adhere obstinately to their own practice; and all the persuasion of the British mission, and its medical men, were for ten years exerted in vain to introduce vaccination, although the ravages of the smallpox were frequently dreadful. The profits of science are confined to those who are regarded as proficients in divinity, astrology, and physic. The two former, when combined, thrive best.

The Persians make high, and, in a few respects, not ill-founded, pretensions to literature. Their treatises on the sciences now mentioned are in a great measure borrowed from the Arabians; and little improvement has been made of late in these branches. Their historical works are of a higher cast, and include some of considerable merit; but these belong chiefly to the earlier and brighter times of the empire. Among the more modern may be noticed a History of Nadir Shah, a flowery but authentic record of the life of that extraordinary monarch. But it is in poetry the Persians claim peculiar excellence; and they, no doubt, can produce the names of more eminent authors in this department than any nation of the east. From the highest to the lowest, they possess an exquisite relish for poetical compositions: from the men of letters to the lowest groom they recite passages from their heroic poets, or chant odes of Hafiz; and if you find fault with a tent pitcher, he probably replies with a stanza from Rudiki, or a moral apothegm from Saadi. It is singular, however, that the moral lessons inculcated by their poets and learned, and, as it should seem, admired, by the people, should be wholly inoperative in practice: the Persians being certainly as corrupt, sensual, and immoral as any people of Asia.

Their poetry may be divided into epic and narrative, moral and lyric. Of the first class Ferdousi is the father, though Dukeekee did compose about 1000 verses of the Shahnameh, in which the former is also said to have been assisted by Asidi. Next to Ferdousi ranks Nizâmi, who composed a poetic life of Alexander the Great; but this, like the Yusseeff and Zuleika of Jâmi, another on the same subject by Ferdousi; the Leila and Mignoon of Hatife; Khoosroo, Shireeu, and others, recited with rapture all over Persia, are, in fact, poetic romances, called Musnavees.

Of the didactic poets the chief, without question, is Saadi, whose Goolistân and Bostâm abound in beautiful maxims and fine moral precepts. Sheikh Saadi was born at Sheraz (A. D. 1194), and in his youth was a great traveller. While in Syria he was taken by the Crusaders, and actually compelled to labour as a slave at the fortifications of Tripoli. From this condition he was relieved by a merchant of Aleppo, who not only paid ten crowns for his ransom, but gave him his daughter with 100 for her dowry. The lady, however, proved a shrew, and Saadi, in several parts of his works, gives vent to the chagrin caused by this marriage. Among other taunts, she is said to have reproached him with having been bought from the Christians by her father for ten crowns: "Yes," replied the unhappy moralist with a sigh, "and he sold me to you for a hundred." He died in his native city at the extreme age of 120 lunar, or 116 solar years; and his tomb is still to be seen near the place of his birth—a small mosque-like edifice, within an enclosure, in which are some fine old fir trees and some cypresses.

In the mystic and lyrical strain there is none who can come into competition with Hafiz, to whom also Shiraz had the honour of giving birth. He flourished in the time of Tamerlane or Timour Bec, who, when he came after the defeat of Shah Mansora to the place where the poet dwelt, desired to see and converse with him. With feigned or real displeasure, the conqueror demanded to know how he dared to dispose of his two noble cities of Samarcand and Bokhara, which, in a beautiful stanza,* he declared he would give for a mole on the cheek of his mistress: "Can the gifts of Hafiz ever impoverish Timour?" was the reply, which changed the monarch's wrath into admiration, and elicited reward instead of punishment. The poetry of Hafiz is considered by Persian scholars as of a singularly original character—simple and unaffected, yet possessing a wild and peculiar sublimity. Like most lyrical effusions, his odes will not brook translation, so that his beauties can never be comprehended by the mere English reader. In his own country, however, he is fully appreciated; and perhaps no poet of any country ever attained greater popularity among those for whom he wrote than the khanjeh of Shiraz. His mortal remains rest near the city whose praises he has celebrated, not far from the tomb of Saadi, and near his favourite stream of Rôknabad. The tomb is in a small enclosure, whither the people of the place resort to sit under the shade of the old cypresses, recite the odes of their favourite bard, and draw omens from the pages of his works.

Next to Hafiz, in celebrity, has been placed Abdul Rahman Janie, a famous doctor of laws, and no less famous soofise, whose Dîwân, or collection of odes, is in small estimation with the enthusiasts of his sect. His wit is said to have been equal to his poetic genius.

To these already mentioned might be added many names

* Beautifully, though freely, translated by Sir William Jones.
598

scarcely less celebrated, whose works it would require too much space to particularize or describe. But it is not to be imagined that their perusal would give any pleasure to European readers. They contain, it is true, many beautiful thoughts, and their diction is frequently euphonious and expressive; but they have the vice of most eastern works, that is, of being disfigured by the wildest extravagance and bombast, and by an endless repetition of metaphors and similes.

History.—Modern Persia comprises the countries known in antiquity by the names of Media, Susiana, Caramana, Hyrcania, and Persia proper. Its ancient history is intimately connected with that of Greece and Rome. In some modern times it has been the theatre of endless civil wars, revolutions, and changes devoid of all interest to foreigners. Toward the end of the 16th century, however, order was restored, and Persia rose to distinction under the government of Shah Abbas, surnamed the Great, who defeated the Turks in several battles, taking from them the city of Taurus and the province of Georgia, and Ormuz from the Portuguese. Abbas was succeeded by a series of imbecile tyrants; and, in 1727, the country was overrun by the Afghans. At length the famous Thamas Kouli Khan, a brigand chief, was raised to the throne by the title of Nadir Shah, and distinguished himself alike by his victories and his ferocity. Nadir being assassinated, in 1743, his death was followed by a long-continued civil war. After a vast deal of blood had been spilt, the eunuch Mehemet Khan succeeded, by his superior ability and good fortune, in establishing his authority over most of the provinces now comprised in Persia; and transmitted his authority to his nephew, Futieh Ali Shah.

This prince waged an unsuccessful war with Russia, who stript him of a large territory in Armenia, and obliged him to pay £2,500,000 as an indemnity for the expenses she had been put to in the contest: Futieh Ali kept an enormous harem; and it was his practice to disperse his sons over the empire, as governors of provinces, towns, &c., of which, speaking generally, they were the scourges. On the death of Futieh, in 1835, his grandson, the present sovereign, son of the prince royal, Abbas Mirza, succeeded to the throne in terms of his grandfather's will. A few of his uncles, who were reckoned most dangerous, were deprived of sight; but on the whole the succession was unusually tranquil and bloodless. This sovereign is understood to be favourable to the interests of Russia, or rather, perhaps, he thinks it most prudent to keep on good terms with those who have the greatest means of injuring him. His unsuccessful expedition against Herat is said to have been undertaken at the instigation of Russia.

PERSIAN GULF, an extensive arm of the Indian ocean, separating Persia from Arabia, between the 24th and 30th degs. N. lat., and the 47th and 57th E. long., uniting with the Indian ocean by the strait, about 32 m. across, between cape Musseldom (lat. 26° 19' N., long. 56° 30' E.) and the opposite coast. This gulf has somewhat of an oval shape, extending about 550 m. N.W. and S.E., with an average breadth of about 160 m.; but toward its S.E. end it is upward of 330 m. in width, though it soon afterward, on taking its northern bend, previously to its junction with the coast, becomes much narrower. It receives at its N.W. end the united waters of the Euphrates and Tigris, above 70 m. below Bussorah; but it has few or no other affluents of any importance. These streams, however, assisted in some measure also by the shape of the gulf itself, tend to diminish the height of the tides, which is considerably less than in the Red sea. (*Traill's Phys. Geog.*, p. 116.) The climate round the shores of this gulf is extremely hot; and, notwithstanding the prevalence of N.W. winds, the thermometer in some parts stands at a higher elevation than in almost any other locality with which we are acquainted. Owing to the number of small islands, and the number and extent of its reefs, the navigation of this sea, especially along the coast of Arabia, is hazardous, difficult, and dangerous: it is less encumbered along the Persian coast. The trade carried on in the ports on or connected with the gulf, is very considerable. Bussorah is the principal inlet through which Indian and other eastern products find their way into the Turkish empire, and Bushire, in the Persian territory, is the chief entrepôt of the trade between that country and Bombay, whence it receives the products of Europe, China, and the eastern archipelago. The chief interest, however, that attaches to the Persian gulf, is its pearl fishery, on which, indeed, the inhabitants of the S. coast mainly depend, as the land produces only a few dates, and is insufficient to support the population. (For details as to these fisheries, see the article BAHREIN ISLANDS; and see also BUSSORAH and BUSHIRE, in this Dict.) This sea was surveyed between 1821 and 1829; but, although much information has come to us through charts and memoirs in the *Geog. Journal* (vols. v. and viii.), we are far from possessing any satisfactory information respecting its islands, which

, in all probability, more numerous and important than
a hitherto been supposed. The ancient importance of
Persia gulf is principally owing to its connexion with
conquests of Alexander, and its commercial intercourse
h India. Were the scheme for the steam navigation of
Euphrates to succeed, this sea might again become, as
as during a certain period of antiquity, a thoroughfare
he commerce between the E. and W. worlds; but the
usages in this respect enjoyed by the route by the Red
and through Egypt, are so very superior, that we have
oubt it will continue to engross by far the larger portion
e trade not carried on by the cape of Good Hope The
is and shores of the Red sea have been at all times a
urite resort of pirates. At present, however, they are,
consequence of the exertions of the British government,
r extirpated.

2SON, county, N. C. Situated in the N. part of this
and contains 640 sq. m. Drained by branches of Dan
and by head branches of Neuse river. It contained
1 9271 neat cattle, 3027 sheep, 7858 swine; and pro-
18,718 bushels of wheat, 12,638 of Indian corn, 36,610
4994 of potatoes, 1,036,363 pounds of tobacco, 107,900
n. It had five stores, five flouring-mills, 15 grist-
l saw-mills; one academy, 90 students; 7 schools,
lars. Pop.: slaves, 3299; slaves, 4351; free colour-
toral, 2780. Capital, Roxboro'.

H. one of the largest and most important counties
nd, nearly in the centre of that part of the U. King-
communicating by the frith of Tay with the Ger-
un, having N. the counties of Iverness and Aber-
Forfar, S Fife (from which it is mostly separated
th of Tay), Kinross, Clackmannan, the frith of
l Stirling, and W. Dumbarton and Argyle. Ex-
a small detached portion on the frith of Forth, it
mpact circular form. Area 1,688,390 acres, of
109 are water. This great county comprises with-
most all that is peculiar to, or characteristic of,
having every variety of surface and soil, from
rile mountains, to low, level, fertile vales. Its
ivers are also on a grand and varied scale; and
s as different as its surface, being severe in the
ted, and mild and early in the lower districts.
st is the inhabitants is equally great: the Celt
on the mountains and the Saxon on the plains,
fering widely from the other in language, dress,
s. Perth is naturally divided into highlands
s: all the country, including the Ochill and Sid-
m in S. frontier to the foot of the Grampians,
d in the lowlands, and the remainder in the
he part of the Grampian chain in this county
ne of the highest mountains in Scotland,
 may be specified Ben Lawers, 3945; Ben
Ben Gloe, 3690; Schichallion, 3550; Ben
B; and Ben Ledi, 2863 feet above the level
esides the mountains and hilly districts, there
alve, though progressively diminishing, tracts
and bog. There is also a large extent of
nd plantations. The latter were much ex-
operations of the late duke of Atholl, who
15,000 acres! But, notwithstanding these
cultivated land is estimated at from 530,000
l, or at about a third part of the entire sur-
t valuable tract of low land is denominated
wrie, being the district bounded by the Tay
'., the Sidlaw hills on the N., and Forfar-
Its soil is mostly a deep rich clay loam;
ertility, it is not, perhaps, surpassed by any
own. The lower part of Strathearn, from
confluence of the Earn and Tay, consist
and is hardly less fertile. Exclusive of
r lands along the Tay, above Perth, there
of the Teith, Forth, and other rivers, ex-
:arse land, and of sandy, gravelly loam.
l is, indeed, predominant in Perthshire.
ery large estates; but there is also a fair
naller class of proprietors. Arable farms
50 to 500 acres. The same plan that
in Argyle (which see) of holding lands
led throughout the highlands of Perth-
s of it are, at present, river in the latter
Farms in the lower districts are uni-
se, generally for 19 years; large stock
a lease; but some of the small highland
year to year. Buildings and other ac-
e farmers, in the lower districts, are for
ntial and excellent; but in some of the
ey are still, in many instances, bad and
ind beans, of excellent quality, are the
itivation in the carse of Gowrie, Strath-
nore, the valley of the Forth and Teith,
d districts barley, and, in the higher,
pal crops. Potatoes everywhere culti-

vated, largely consumed, and recently exported in large
quantities to the London market. Turnip culture extensively prosecuted. Considerable quantities of fruit, as apples, pears, &c., are produced in the vales, particularly in
Gowrie. Breeds of cattle various, but none peculiar to the
country; the stock differs with the varying quality of the
land on which it is pastured. Number of sheep vastly increased within the last 40 years, and the Cheviot breed now
generally diffused. "It begins to be generally understood
that the land cannot stand a constant system of cropping
without intervals of rest. The sheep husbandry is, therefore, daily gaining ground; and the breed of cattle has been
improved both in size and earliness of maturity." (New
Statistical Acc. of Scotland, Perthshire, p. 196.) Roads signally improved; as much so, certainly, as in any other Scotch
county. Coal is found in the S. part of the county, contiguous to the frith of Forth; and limestone and freestone are
pretty generally diffused. Average rent of land, in 1810,
5s. 6d. an acre. The linen and cotton manufacture has
been introduced, particularly into the city of Perth; but
neither has had much success; so that, on the whole, Perthshire may be regarded as an essentially agricultural district;
and we are glad to say that the progress of agriculture du
ring the present century has been most satisfactory. The
following statements by the author of the account of the
parish of Wester Foulis in the New Statistical Account of
Scotland, are generally applicable to the whole county
"Since the date of the last statistical account great improvements have been made in rural economy. Waste
ground has been planted and brought into cultivation. The
roads are in a superior condition, and new ones have been
constructed. Farm steadings were then covered with
thatch, and indifferent in building and accommodation;
but they are now all slated, well built, and adapted for
every necessary purpose. Wheat, potatoes, turnips, and
artificial grasses are cultivated in a much greater breadth.
More manure is laid on the soil, and it is meliorated by
fences, cleaning, and draining. Horses and harness, the
different breeds of cattle and sheep, and all the implements
of husbandry, are much improved." Principal rivers, Tay,
Forth, Earn, Teith, Lyon, Garry, Tummel, &c. Fisheries
on the Tay about the most valuable in the kingdom. Perth
is divided into 80 parishes, and returns two members to the
House of Commons, one for the county and one for the city
of Perth. Registered electors for the county, in 1839–40,
4294. Some parishes in the S.W. part of the county are
joined, for election purposes, with the counties of Kinross
and Clackmannan; and the borough of Culross unites with
that of Inverkeithing, Dunfermline, &c., in returning one
member to the House of Commons. Principal towns, Perth,
Crieff, and Dunblane. In 1831 Perth had 23,809 inhabited
houses, 31,749 families, and 142,894 inhabitants, of whom
68,565 were males and 74,329 females. Valued rent,
£339,809 Scotch. Annual value of real property, in 1815,
£555,539.

PERTH, a royal and parl. bor. and manufacturing town
of Scotland, co. Perth, of which it is the cap., on a plain on
the right bank of the Tay, 33 m. N. by W. Edinburgh, lat.
56° 23' 40" N., long. 3° 26' 20" W. The town is surrounded,
except on the line of the Tay, with gently rising, verdant,
or richly wooded hills. Pop., in 1801, 14,878; in 1831,
20,016. It is connected by a handsome bridge of nine
arches, 880 feet in length (built by Smeaton in 1771, at an
expense of £26,632), with the village of Bridgend, on the
left bank of the Tay, included in the parl. bor. The main
street runs N. and S., nearly parallel to the river; and it
and the other streets are for the most part straight. Many
of the more modern streets and crescents are of freestone,
and altogether the town is remarkably neat, clean, and
well-built, and has a substantial, wealthy appearance. It
is lighted with gas, and the side pavements are good. The
inhabitants are well supplied with water, raised by a steam-
engine from the river into a reservoir, whence it is conducted
in pipes through the streets. The expense of these, which
were constructed in 1830, was £13,609. The assessment
is limited to 5 per cent. on rental. North and south of the
town are the two large public greens, called the North and
South Inches. The former, which is flanked on the W.
by Athole Crescent, has the race-course; the latter is surrounded by stately trees and elegant villas.

In addition to the gas and water-works, the public edifices are the county buildings and jail, of Grecian architecture, fronting the river, erected in 1819, at a cost of £32,000;
the church of St. John, a building of ancient but unascer-
tained date, surmounted by a pyramidal spire of wood covered with lead, and divided into three places of worship,
appropriated to three distinct parishes; St. Paul's church,
built in 1807, at an expense of £7000; the academy, also
erected in 1807, at a cost of £6000; lunatic asylum; the
theatre; barracks; Marshall monument (built in commemoration of a late lord provost), containing the public library
and the museum of the Perth Antiquarian society, and the

town-hall. A little way S. from the town, an extensive military prison, capable of accommodating 7000 captives, was constructed, in 1812, at an expense of £130,000; but it is now (1841) about being converted into the central and model prison for Scotland.

The town has four parishes; and three *quoad sacra* churches (in one of which the service is performed in Gaelic) connected with the establishment, have recently been erected. There are no fewer than 14 dissenting chapels; three of which belong to the United Secession; two to the Old Light Burghers; two to the Relief; and one respectively to the Independents, Baptists, General Baptists, Methodists, Glassites, Episcopalians, and Roman Catholics.

The grammar school of Perth was at an early period the most celebrated in Scotland, being attended by pupils from every quarter of the kingdom. It was the first seminary in Scotland in which Hebrew was taught. (*M'Crie's Life of Knox*, ii., 14–16.) Its eminence may be said still to continue. The academy, which embraces the most ample course of instruction, scientific, literary, and commercial, was founded in 1760; its first rector, Dr. Robert Hamilton, afterward of Aberdeen, is well known by his able work on "The National Debt." These two seminaries are endowed; and there are, besides, about 28 other schools, of which six are endowed; in addition to which a large seminary has recently been erected, partly by public subscription, and partly by a grant from government, for the education of 400 poor children. Perth has six public libraries, one of which contains 6000 volumes; a literary and antiquarian society; a reading-room; and four weekly newspapers. Printing and the publishing of literary works have been carried on here to a greater extent than in any town of a similar size in Scotland, perhaps in the empire. Among other products of the Perth press may be specified the *Encyclopædia Perthensis*, and Heron's *History of Scotland*, 6 vols., 8vo.; this branch has now, however, materially declined.

Poor laws have been introduced into the town, the assessment being about £2200 a year, in addition to the church collections, which average about £700. The number of paupers ranges between 700 and 800. The incorporated trades, in addition to this, give an annual allowance of £2270 to their poor members, whose number is generally about 200. A hospital for behoof of the poor was founded by James VI. This institution has a nett revenue of £397 8s. 6d.; distributed among 61 paupers, who are out-pensioners, the building being let. There are two dispensaries and an infirmary, in addition to the lunatic asylum above noticed.

Perth is not eminent for trade or manufactures. About 80 years ago, however, it had an extensive trade in gloves; Perth-made gloves having a preference throughout the kingdom. Latterly, however, Dundee has quite superseded it in this department. In consequence of this, the business of tanning, which principally depended on the glove-trade, has been greatly narrowed. Its manufactures consist at present principally of coloured cottons, especially for umbrellas. A great quantity of handkerchiefs, checked and striped ginghams, imitation India shawls, scarfs, trimmings, &c., are also woven. The aggregate number of weavers is about 1600, some of whom are employed by Glasgow and Paisley houses. A flax-spinning mill has recently been erected, which employs about 70 hands. A large bleachfield at Tulloch, 2 m. W. of the city, employs 250 individuals. There are in the town several distilleries, breweries, and corn-mills, and three considerable iron-founderies. The salmon-fisheries on the Tay, belonging to the parish, bring a rent of £1350 per annum, of which £830 goes to the city. The quantity of salmon, including grilses, shipped from Perth for London, amounts, at an average, to about 4500 boxes, or 225 tons a year.

The Tay is navigable to North for vessels of considerable burden; but the navigation is much obstructed, and a great deal is required to be done for the improvement of the navigation. In this view an act was obtained in 1834 for deepening the bed of the river, and forming a new harbour and wet dock, and the works are now proceeding. There belong to the port about 100 vessels of the aggregate burden of about 8000 tons; and about 800 coasting-vessels annually enter the port. The exports consist of manufactured goods, corn, and potatoes. The quantity of potatoes shipped for the London market is often as high as 100,000 bolls: corn of all kinds, between 40,000 and 50,000 quarters. The town supports two native banks and four branch banks; it has also a savings bank, established in 1815; two weekly markets and five annual fairs. Gross customs revenue, from 5 July, 1840, to 5 Jan., 1841, £10,766.

Perth is very ancient; and some authorities refer its origin to the Romans. It was a borough at least, as early as 1106. Its church being consecrated to John the Baptist, it was long called St. Johnstoun. Prior to the reign of James II., Perth was the capital of Scotland; and from its central situation it is, perhaps, to be regretted that it did

not continue to enjoy that distinction. The kings were crowned at Scone, about 2 m. N. of the city, and had a residence in the town. The "Parliament house," of Perth remained standing till 1818, when it was pulled down. Scone palace, long a royal residence, has been rebuilt by its noble owner, Earl Mansfield, and is now a splendid modern residence. The famous stone, reckoned the palladium of Scotland, and on which the Scottish kings were crowned, was transferred from Dunstaffnage, in the 9th century, to Scone, whence it was removed by Edward I., in 1296, to Westminster Abbey; since which time it has been enclosed within the frame-work of the regal chair on which the British sovereigns have been crowned since Edward II. There were no fewer than four monasteries in Perth, exclusive of one in Scone, two nunneries, and eight chapels. But the violence of the Presbyterians at the Reformation may be said not to have left a relic of these buildings. James I. was murdered in the Blackfriars' Monastery here in 1437, and was buried in the Carthusian Monastery, which he had founded. In August, 1600, Perth was the scene of that most mysterious incident in Scotish history, entitled the "Raid of Ruthven," or the *Gowrie conspiracy*. Gowrie palace, the residence of the noble family of that name, was pulled down within the last 25 years, and its site is now occupied by the county buildings. Perth is associated with many important events in history; but we limit ourselves to mentioning that, in 1644, it was taken by Montrose, after his victory at Tippermuir, in the neighbourhood; in 1651 it capitulated to Cromwell; and was occupied by the insurgents in the rebellions of 1715 and 1745.

Before the passing of the Reform Act, Perth was united with four other boroughs in sending one member to the House of Commons; but that act conferred on it the important privilege of returning a representative for itself. Registered voters in 1840-41, 949. Perth is an opulent borough; the corporation revenue for 1839-40 amounted to £6226; number of councillors, 16.

Perth, though it never was the see of a bishop, is called a city; and, in the rolls of the Scottish parliament, it held rank next to Edinburgh. Its chief magistrate has the title of lord-provost. (*New Stat. Account of Scotland*, § *Perthshire*, i., 162; *Boundary Report*, &c.)

The situation of Perth is one of the finest in Scotland. Close to the city, on the E., is the hill of Kinnoul; the summit of which, of easy access, commands one of the noblest prospects that is anywhere to be met with. Towards the S. and E. is the valley of the Tay, and the confluence of the Tay and Earn; to the W. is a finely variegated country, and to the N. the prospect is bounded by the many girdle of the Grampians. The country round Perth is among the most fertile in Scotland.

PERTH AMBOY, city, and port of entry, Middlesex co., N. J., 46 m. N.E. Trenton, 212 W. Situated at the confluence of Raritan river with Arthur Kill, or Staten Island sound, at the head of Raritan bay. It was laid out in 1683 by the proprietaries of East Jersey, and designed for the capital of the province. The present city charter was given in 1784. The harbour is spacious, safe, and easy of access, having in the estuary of the river 12 feet of water, and from 34 to 36 in the main-channel. It contains three churches, a Presbyterian, Episcopal, and Baptist, nine schools, capital $38,500, one pottery; two schools, 96 scholars. Pop. 1303. The collection district, containing much of the S.E. part of New-Jersey, had a tonnage in 1840, of 17,343. It is a pleasant place of summer resort.

PERU, a celebrated country of South America, formerly one of the most valuable possessions of the Spanish crown. It then included the modern republic of Bolivia (South or Upper Peru); but at present the term is restricted to the republic of Lower or Lower Peru, lying chiefly between lat. 3° and 21° S., and long. 65° and 81° W.; having N. the Columbian republic of Ecuador, E. Brazil, S.E. and S. Bolivia, and W. the Pacific. Extreme length, S.S.E. to N.N.W., about 1500 m.; breadth varying from 40 to 660 miles. Estimated area and population, subdivisions, &c., as follows:

Departments.	Area in sq. m.	Population.	Chief Town.
Lima Truxillo Junin Huancabelica Ayacucho Cusco Puno Arequipa	500,000 ?	1,300,000 ?	Lima. Truxillo. Huamanga. &c. Huancabelica. Cusco. Puno. Arequipa.

Physical Geography, &c.—The country is naturally divided into three regions; that between the coast and the Andes; that occupied by the latter; and the region E. of the Andes, forming a part of the basin of the Amazon. All these divisions differ widely in their physical character,

The coast region from Tumbes, on the N. frontier to the river Leche, is mostly a desert; and wherever, in fact, the coast region is not traversed by streams, or is unsusceptible f irrigation, it consists principally of arid sandy wastes, and is in the last degree barren. The Andes and their ramifications have been roughly estimated to cover, in Peru, an extent of 300,000 sq. m. They consist here, as in Bolivia, of two main chains, or Cordilleras, connected in various parts by cross ranges, and inclosing several extensive and lofty valleys. Round Cusco is a vast knot of mountains, occupying about three times the extent of Switzerland; round Pasco, in lat. 10° S., is another knot surrounding plain of Bombon, 13,500 feet above the sea level, and which are the rich silver mines of Cerro Pasco. The Peruvian Andes are not in general so elevated as the Bolivian; though many of their peaks rise far above the limits of perpetual snow. The loftiest summits are towards the N., where the Nevado de Chuquibamba (about lat. 15°) rises to 21,000 feet in height; and several other mountains surrounding the valley of Desaguadero, which belongs to Peru (see BOLIVIA, i., 396,) may at least approach elevation.

Bolivia the E., but in Peru the W., Cordillera is the st. At the mountain-knot of Pasco, the Andes separate into three collateral chains, which, proceeding N., form the basins of the Maranon, Huallaga, and Ucayale. The first range of the Andes to the E., in Peru, extends to the 6th and 13th parallels of lat., at a distance between 200 and 400 m. from the Pacific, and separate basin of the Ucayale from those of the Yavari, and other affluents of the Amazon. Probably no this range rises above 10,000 feet in height. (See vol. i.)

space enclosed between the gigantic ridges of the E. Cordillera, called the Sierra, is partly occupied by barren and naked rocks, partly by table-lands yielding grass, and extensive hilly pasture-ground, very general outline, to the Highlands of Scotland, though of heath, and partly by extensive and fertile valleys once supported a much larger amount of population; third region, or country E. of the Cordilleras, is known: it is mostly covered by all but interminable forest, and a large portion of it can scarcely be said to Peru, since only a few Roman Catholic missions and there scattered over its surface, the rest of being in the exclusive possession of the native

s birth to some of the largest rivers in the world. agua, generally regarded as the proper source Amazon, and its vast confluents the d Ucayale (the latter formed by the junction mac and Paro), have their sources on the E. W. Cordillera, between 10½° and 16° S. lat.; though with many windings, a northerly course the country. These great rivers are mostly not, with the assistance of steam-navigation, at some future period, carry the riches of region across the continent to the ports on the great lake of Titicaca (which see) is mostly excepting it, there is no other large lake: ever, some smaller lakes, one of which, the ocha, to the N. of the Cerro de Pasco, gives guaragua. (Geog. Journ., v., vi., viii.; Poeppig, &c., ii.; Meyen, Reise um die Erde, &c.,

throughout rugged and lofty. In the N. miles of a loose sandy desert intervene in 'een the high lands and the ocean; but in its approach close to the shore, which has an extent of 1000 m. a dozen secure harbours. e of Callao, Payta, Sechura, Salina, Pisco, and a few others. Truxillo and Lambaeque open roadsteads. The water being almost camels are obliged to approach within ½ m. re they can anchor; and the prodigious in unbroken from the Pacific occasions a ous surf. The operation of landing is, excess, at once difficult and hazardous; it is of balsas, or platforms raised on inflated in shape in different parts of the coast. Captain Hall off Mollendo, " was made skins inflated, placed side by side, and pieces of wood, and strong lashings of platform of cane mats forms a sort of in width, and 6 or 8 feet in length., At who manages the balsa kneels down, double-bladed paddle, which he holds by 'ikes alternately on each side, moves it passengers, or goods, being placed on the m. All the goods which go to the interior of the coast, are landed in this manner. river, and the bags of dollars, also, which are shipped in return for the merchandise landed, pass through the surf, on these tender, though secure conveyances." (Hall's S. America, i., 205, 206.)

Climate.—The year may be divided into two seasons; the wet and the dry. From June to Oct., the coast lands in all the S. and central provinces are covered during the night and morning with a dense fog, the only moisture supplied by nature to this part of the country. These fogs diminish as we proceed N., and in the N. province of Piora, which is celebrated for its dry atmosphere, rains occasionally occur; and when such is the case, the arenales, or arid sands, are speedily clothed with the most exuberant vegetation. While, however, the dry season prevails on the coast, and especially from January to March, heavy rains fall in the mountains, frequently accompanied with thunder, which never occurs along the coast. The extensive valleys between the Cordilleras, such as that of Cusco, 10,000 feet above the level of the sea, enjoy an admirable climate; and though between the tropics have in consequence of their elevation, all the advantages of the best climates of the temperate zone, with but few of their disadvantages. Beyond this, and at the level of about 14,000 feet, commences the limit of perpetual snow. Even in the coast region the temperature is not so high as might be supposed from the latitude; cooling S. winds being uniformly prevalent, and the sea-breezes by day alternating with others blowing from the land at night. The mean temperature of the year in Lima is about 70° Fahr. Wheat, and other European cerulis, though but little cultivated, succeed admirably in the elevated valleys of the Sierra; potatoes thrive best at an elevation of from 11,000 to 13,000 feet. The country is, on the whole, salubrious. Colics, bilious and inflammatory diseases, small-pox, and hydrophobia are common; but in Lima, and many other parts of the country, individuals frequently live to an advanced age.

Peru is more subject, perhaps, than any other country, to the tremendous visitation of earthquakes. Shocks are felt every year; and they occasionally become so violent as to be productive of the most disastrous consequences. The earthquake which occurred in 1746 swallowed up the entire seaport town of Callao, and destroyed the greater part of the city of Lima. The years 1687, 1806, and 1828, were, also, distinguished by the occurrence of severe and most destructive shocks.

Minerals and mines.—Peru, like Mexico, is famous for her mineral products; and we have been long accustomed to associate her name with the almost unlimited abundance of gold and silver. But, though the most exaggerated notions of the value and importance of the Peruvian mines were long, and, perhaps, are still prevalent in Europe, they have, no doubt, furnished vast supplies of the precious metals. The famous mine of Potosi in Bolivia, or Upper Peru, was discovered by accident in 1545: it produced, for a lengthened series of years, vast quantities of silver; but it is now comparatively neglected, and is supposed to be nearly exhausted. The greater number, as well as the most productive of the mines that are at present wrought, are situated in the Cerro di Pasco, in the dep. of Junin. They were, like Potosi, accidentally discovered in 1630.

The produce of the Peruvian, like that of the Mexican mines, has materially declined since the commencement of the revolutionary struggle. Humboldt, who had the best means of obtaining accurate information, estimated the annual value of the gold and silver produced in Peru at the commencement of the present century, at 6,240,000 dollars (£1,248,000). But at present (1841), owing to the anarchy and insecurity that has prevailed, their value is probably not much above half this amount. The silver smelted in Peru, from 1830 to 1834, is said to have amounted to 3,906,573 marcs, worth £3,804,175.

The following is a statement of the quantities of silver reduced to bars at the different smelting houses of Peru in 1834 (Board of Trade Table for 1839):

Departments.	Marcs.	Departments.	Marcs.
Lima	15,321	Ayacucho	2,417
Truxillo	15,367	Puno	31,379
Junin (Pasco)	272,568	Arequipa	4,362
		Total	341,404 marcs, worth 851,088l.

It is estimated that over and above the quantities given above, 1-3d part more is actually raised in the different departments, except in Junin, where 1-5th part of excess is produced, and smuggled out of the country in the state of plata piña, or native silver. In 1838, 900 marcs of gold, and 287,258 oz. of silver were coined at the several mints of Peru. Perhaps, on the whole, allowing for smuggling, &c., the total produce of the gold and silver mines of Peru, including Bolivia, may now (1841), amount to from £70,000 to £750,000 a year.

Huancavelica has one of the richest quicksilver mines in the world, one portion of which (St. Barbara,) furnished

5800 quintals a year of quicksilver, for two centuries. It is said that the metal might be procured here at an expense of 65 dollars the quintal; though in 1837–38, quicksilver was so scarce in Peru as to cost from 200 to 230 dollars per quintal. These mines were then, of course, unwrought; but a private company has since been formed to work them, which has proceeded to a considerable extent with its operations. (*Smith's Peru*, ii., 24.)

Exclusive of the above, Peru produces iron, copper, tin, coal, saltpetre, &c. The latter, indeed, under the name of nitrate of soda has, within the last few years, become an important article of export.

Vegetable products are numerous, and very dissimilar. Sugar, rice, tobacco, yams, sweet potatoes, and cocoa, are raised in the warmest situations; the vine, wheat, and quinoa (*chenopodium quinoa*) are planted in colder places, and potatoes on the highest cultivated grounds. The grapes are well-flavoured, but the wine made from them is inferior. The sugar cane is mostly the creole species. Three sorts of maize are cultivated, and this grain, which forms the principal farinaceous food of the modern inhabitants, appears to have been also the principal formerly in use among the Indians, large quantities having been discovered in subterranean granaries, where it had probably remained from a period previously to the Spanish conquest. Cotton is grown in almost every part of Peru, and the Peruvian ranks immediately after the Sea Island and Egyptian cotton in the English markets. Except in the prov. Piura, it is all short stapled. The culture is rapidly increasing, and it may now be considered one of the staple products of the country. Lucern is a good deal grown for provender: it reaches the height of three feet, and is cut five times a year. Culinary vegetables abound; beans, with potatoes, forming the principal food of the lower classes in the uplands. Olives succeed well in certain districts, and oil is extensively produced, but it is inferior to that of France or Italy. Plantain, bananas, guavas, and other tropical fruits, with oranges, lemons, nectarines, plums, and others common to Europe, are found in great profusion. Bark, and numerous medicinal plants, with cedar, ebony, mahogany, walnut, and other valuable timber trees abound in the forests. The valuable and well known drug called Peruvian bark consists of the rind of the *cinchona lancifolia*, *oblongifolia*, *cordifolia*, &c., and is, consequently, of several varieties. The genus *cinchona* is supposed to be confined to the Andean ranges between lat. 19° N. and 22° S., where it grows luxuriantly to the height of 10,000 feet above the sea. The prov. of Loxa, Ecuador, and Huanuco, are those in which the bark is principally obtained; and during a lengthened period after its first introduction into Europe in 1640, it was called Loxa bark, from the former of these provinces. Its collection begins in May and continues till November. The trees are cut down close to the roots; the stems are then divided into pieces of uniform length; and about three or four days afterwards the bark is taken off in broad strips. The price which the article fetches depends on the rapidity with which it is dried, which is effected by exposing the bark with the least possible delay to a hot sun, which makes the pieces roll up and sometimes form a solid cylinder, without any cavity in the centre. It is afterwards carefully packed in bales of 4 or 5 arrobas each, and exported in chests closely covered with skins. (*Poeppig, Com. to Bot. Mag.*, i. 249.) Coca, the dried leaf of the *Erythroxylon coca*, is largely used by the Peruvians for chewing, much in the same way as betel in the east. Poeppig says that indulgence in its use brings on a gloomy kind of mania; but other authorities deny that it has any such effect.

Agriculture, which was never in a prosperous state, has, like every other branch of industry, been greatly neglected since the revolution. Dr. Smith gives an account of the agriculture of a district between the Cordilleras, which he says may serve for that of the *Sierra* generally. "The agriculture of Huanuco, though alluring to the eye of the ordinary traveller, who only glances at its rich and waving fields of maize, inclosed within tapias or fences of mud, and hedges of the Indian fig and aloe or maguey plants, is in every way defective. The fields owe their luxuriance to nature rather than man, except in the single advantage of water, which he often directs and supplies to them. Manure is a thing never thought of; and the implements of husbandry are of the rudest kind. The plough, slight and single handed, is constructed merely of wood, and without a mould-board. The ploughshare is a thick iron blade (or, where iron is not at hand, a piece of hard iron-wood), only tied, when required for use, by a piece of thong, or lasso, on the point of the plough, which divides the earth very superficially. Harrows they have, properly speaking, none; but sometimes use, instead, large clumsy rakes, or a cross bough dragged over the sown ground, with a weight upon it to make it scratch the soil. Instead of the roller, they break down the earth intended for cane-plants, after it has got eight or ten ploughings and cross-ploughings, with the

heel of a short-handed hoe. For smoothing down the clods of earth, some Indians use a soft, flat, round stone, about the size of a small cheese, which has had a hole bored through its centre by dint of blows with a harder and [illegible] stone. To the stone thus perforated they fix a long handle; and as they swing it about they do great execution in the work of *cuspiando* or field levelling. Lucern, or *alfalfa*, is cut down, and used green, cattle and working oxen for the plough and sugar-mills being fed on it; yet the scythe is not in use among the great planters, who find it necessary to keep two or three individuals at the sickle to cut down food for herds, which, in the day time, are fed on irrigated pastures, but at night in corrales or pens. The inhabitants are accustomed to break up potatoe grounds on the face of steeps with deep narrow spades having long handles. In the same manner the soil is turned up by those who have neither plough nor oxen, but who yet sow maize on the temperate flats on the hill sides. People thus circumstanced make holes in the ground with a sharp pointed stick, when they bury the seed. The Indian sows the white grained maize in preference to the yellow, as he considers that when toasted it makes the best "*cancha*," or substitute for bread; and that when boiled it makes the best "*mote*," or simply boiled maize: it has moreover the credit of making the most savoury *chica* or beer, which they house have whenever they have a little surplus grain. They also make a kind of beer from the fermented juice of the maize stalks compressed between small rollers of wood turned by the hand. Dry maize leaves and stubble are most used in the foddering of cattle. The sugar mills in the valley of Huanuco are mostly made of wood, and wrought by oxen. On the larger estates small brass rollers are used; but water-power is not thought of, the proprietors adhering to the old practice of working with oxen day and night throughout the year, barring accidents, and frosts, and holidays." (*Peru as it is*, ii., 60–66.)

The wild animals include the puma, or American lion: the uturuncu (*felis onca*, L.), a kind of tiger cat; the acumari (*ursus Americ.*, L.), a black bear inhabiting the mountains; the *anas*, or skunk; great numbers of deer, wild bears, armadillos, &c., which are objects of the chase, and several varieties of animals, as the llama, alpaca, guanaco, and vicuna, used, especially the llama, by the native Peruvians, previously to the Spanish invasion, as beasts of burden. (Copious descriptions of these animals may be found in Stevenson's *Trav.*, ii., 85–92, and *Smith's Peru*, &c., ii., Appendix.) Four varieties of the condor are included among the native birds. Alligators are met with in the rivers; but neither the reptile nor the insect tribes appear to infest Peru so much as the country around Guayaquil and some other regions within the tropics.

Of the *foreign quadrupeds* acclimated in Peru, sheep appear to have succeeded best. They have increased in an amazing degree on the great commons or pastures of the Andes, at an elevation of 12,000 or 14,000 feet above the sea. Few sheep are bred on the coast; but during certain months, large flocks are driven from the interior, and fattened for the Lima market. Many of the ewes are in lamb, and the common bargain between the drover and the farmer is to give the lambs for the pasturage, the farmer calculating on receiving 150 lambs for every 100 ewes. Besides this increase, which is greater than in England, the ewes bear twice a year, generally in June and December. (*Stevenson*, ii., 2.) Little attention has been hitherto paid to breeding sheep, so as to improve the wool; but as the latter is now becoming an increasing article of export, more care will doubtless be bestowed on this object. The largest quantity of wool exported are from Islay, and are chiefly produced in the neighborhood of Lampa, Puno, and Cuzco. It is soft and similar in appearance to English wool; but, being badly cleaned, it does not fetch more than 9d. or 10d. per lb.; that from the mountains between Lima and Pasco being better cleaned, usually brings 1s. per lb. additional. The wool produced on the coast is of very inferior quality. Vicuna wool is exported, but only in small quantities. In the high region, cattle, horses, asses, &c., are of a standard size; but in the valleys and on the coast they are large, spirited, and showy.

"The cattle of Peru are not so large as those of Lincolnshire; but at an average, they are as large as the English, French, and Spanish cattle: when fed on lucern, the meat is well flavoured, fat, and juicy, and the bones very small." (*Ibid.*, i.) The black cattle of the Sierra do not agree with the climate of the coast, and when brought there speedily die. Ordinary horses and mules fetch from 45 to 50 dollars each. Piura is noted for its excellent breed of the latter, and many mules are taken thence to Truxillo, Lima, &c., where they sometimes fetch 250 dollars each. The same province is also famous for its goats. A good many pigs are reared in Peru; they are considered fit for slaughter at from 10 to 16 months old, when they sell at from six to nine dollars each, if of a good breed.

PERU.

The *Population* consists principally of native Indians, Spaniards, negroes, and the races of mixed origin derived from the foregoing; but of the number of each we have no authentic estimate. The accounts of the Indians given by recent travellers are in many respects conflicting and various; we believe, however, that the statements of Ulloa may, on the whole, be safely depended on. That excellent observer represents them as in the lowest stage of civilization, without any desire for the comforts and conveniences of civilized life, immersed in sloth and apathy, from which they can rarely be roused, except when they have an opportunity of indulging to excess in ardent spirits, of which they are excessively fond. (*Voyage* II., liv. vi., cap. 6.) With the exception of Mr. Stevenson, most recent travellers agree that they are dirty in the extreme, seldom taking off their clothes even to sleep, and still more rarely using water. (*Modern Traveller, Peru,* p. 205.) Their habitations are miserable hovels, destitute of every convenience and accommodation, and disgustingly filthy. Their dress is coarse, and mean, and their food coarse and scanty. Their religion is still mixed with the superstition of their forefathers; but they are great observers of the external rites and ceremonies of the church, and spend large sums of money in masses and processions; a species of profusion to which they are excited and encouraged by their priests, who profit by it. We have previously made some statements illustrative of their attainments in the arts at the time of the Spanish invasion. (See vol. i., 81.) The opinions to which they have since been subject have probably sunk them to a lower point in the scale of civilization than they then occupied; and, no doubt, it would be possible were proper care taken, materially to improve their mind and condition. A good deal, too, of their apathy and little progress in arts and industry, must be ascribed to the physical circumstances under which they have been placed —the mildness of the climate and the fertility of the soil, which, on the one hand, by diminishing their wants, and, on the other, by enabling them to supply those which they do feel with comparatively little exertion, take away and greatly weaken some of the most powerful motives that prompt to labour and invention. Still, however, we are well convinced, notwithstanding the statements and reasonings to the contrary of M. d'Orbigny (see the learned work entitled *L'Homme Americain,* passim), and others, that the Indians are naturally an inferior race, and, indeed, wholly incapable of any degree of civilization.

The principal burden to which the Indians were subject under the Spanish government, was that of the *mita,* or compulsory labour in the mines. All male Indians from 16 to 50 years of age were compelled, during a certain specified period, to undergo this servitude. Its severity had, however, been materially abated previously to the revolution, and it was then entirely suppressed.

Manufactures are in a very backward state, though many of the natives evince considerable ingenuity. In Tarma they make *ponchos,* or loose cloaks, of great beauty and fineness, and on the colder table lands warmer and coarser blankets, ponchos, &c. In the valleys, goat skins are made into cordovans, cow hides into saddle-bags, and travelling cases for bed and bedding, and mats for carpeting from rushes. Cordage for packing is manufactured from the maguey in Piura; and at Guamanga is made the fine filigree silver work, for which inland Peru is celebrated. But in general, the manufactures of Europe have, in the larger towns, superseded those of the natives, and are supplied to Peru in exchange for raw produce.

ESTIMATE RETURN of the Quantity and Value of the different Articles of Peruvian and Bolivian Produce exported, in 1837, to Europe and the United States.

Articles.		Quantities exported.	Average Price free on Board.		Total Value in Dollars.	Total Value in Pounds sterling.
			In Dollars.	In Pounds sterling.		
			Dolls.	*L. s. d.*	*Dolls.*	*L. s. d.*
...	Quintals	9,191	98 per qtl.	5 14 0	251,548·4	52,389 14 0
... crown, &c.		1,800	20 —	6 0 0	30,000	6,000 0 0
... and specie :—						
5. Peru to Great Britain	Dollars	1,197,238			1,197,238	227,448 12 0
... Lima		2,315,200			2,315,200	463,040 0 0
... Lambayeque and Paita	—	56,000			56,000	14,800 0 0
... in bills for supplies to foreign						
... of war and public agents	—	100,000			100,000	20,000 0 0
7. Peru to the U. States	—	1,200,000			1,200,000	240,000 0 0
Do. France	—	400,000			400,000	80,000 0 0
Do. Germany		500,000			500,000	100,000 9 0
Do. Italy	—	}				
Do. Spain	—					
... skins	Dozen	7,376	4·4 per doz.	0 18 0	35,172	7,084 2 0
... rilla	Quintals	6,846	9 per qtl.	1 16 0	75,126	15,025 3 0
... bark	—	131	14 —	2 16 9	1,834	368 16 0
... Islay		619				
...		940	} 14 —	2 16 0	208,705	44,125 4 0
... Nepaum	—	2,300				
...		12,000				
... or in the seed from other						
the coast	—	2,000	8 —	1 9 0	15,000	3,300 0 0
... cow :—						
... and Islay	No.	607				
...		1,800	} 3 each.	0 8 8	21,364	4,250 16 0
...		1,200				
... to the N.		1,000	2·4 —	0 10 0	2,000	600 0 0
...	Quintals	148,380	2·2 per qtl.	0 9 0	273,039·2	74,416 1 0
...	—	12,000	4·2 —	0 17 0	52,632·3	10,560 12 0
...	—	5,670	14·4 —	2 18 0	81,730	16,326 0 0
... Peru	—	17,800	} 16 —	3 0 0	245,000	58,000 0 0
...		6,000				
... Peru	—	3,000	20 —	4 0 0	60,000	12,000 0 0
Total value					7,397,548·0	1,485,509 18 0

... through the Bolivian port of Cobija, or La Mar, chiefly consisting in bullion and copper ore, are not included in this return.

he above articles of produce there were British account as returns for goods sold value of Equator. New Granada and central account of £51,964. The exports of bark, Bolivia, Chinchilla skins, salt-petre, &c., considerable increase in the above over the export of seal skins had diminished. ous great and indiscriminate slaughter cannot become an article of very considerable through the greater part of the tanned &c., is imported from Equator. yet from ing timber on the coast and table land, ily used as a substitute for wood. The s of native produce, except the precious uty.

be not distinguished from Bolivian produce, the exports are almost wholly from Peru, the foreign trade of Bolivia being very insignificant, owing chiefly to her great deficiency of coast and sea ports. The imports into Peru and Bolivia are supposed to average about £1,500,000 a year, those from Great Britain, making about one third part of the whole. Of the total imports, more than two thirds in value go to Peru, and the rest to Bolivia. They consist principally of plain and coloured cottons and woollen goods from England and the U. States; linens from Germany, France, Ireland, &c.; French and China silks, hats, glass, earthenware, hardwares, &c., chiefly from European countries.

The remains of the Incas' road, extending through the centre of Peru from Quito to Cuzco, a distance of 1500 m., may, according to Humboldt, be compared with the finest Roman roads; though, when it is recollected that the Peruvians were ignorant of the arch, and that their bridges were made of osier ropes, this statement will probably appear not a little extravagant. Various passes were also cut in the steep acclivities of the Andes by the Peruvians before the Spanish conquest. The roads laid down by the European masters of the country bear no comparison with the foregoing. They consist, indeed, with a few exceptions in the vicinity of the large cities, only of foot tracks for horses or mules : and, in point of fact, goods are exclusively conveyed on the backs of the latter. In the department Junin, however, Dr. Smith says that laudable efforts have been lately made for improving the roads; but no regular post-houses are yet any where established, and at this moment the want of roads and of improved means of communication, is the greatest obstacle to the improvement of the country. So much is this the case, that Lima and some other towns along the coast find it more convenient to receive supplies of corn from Chili by sea than from the interior of their own country! We have elsewhere described the sort of bridges in use in the Andes (see art. ANDES, vol. i., 107, 108).

The *Government* is popular and representative, the sovereignty, in theory at least, emanating from the people. Peru has a senate or chamber of deputies, consisting of an uncertain number of members, which delegates the executive power to a president and the other high authorities of the state. The chamber of deputies consists of representatives elected by the electoral colleges of provinces and parishes. The parochial electoral colleges consist of all the citizens resident in a parish, for every 200 of whom an elector is nominated; and in every village with an amount of population entitling it to name an elector, a municipal body is established, subject to the approbation of the departmental *juntas*. The electoral colleges of provinces are composed of parochial electors constituted according to law, who elect deputies to congress in the proportion of one for every 20,000 inhabitants. The province, however, in which the whole population does not come up to 10,000, may nevertheless send a deputy. The government of every department is vested in a prefect, that of a province in a sub-prefect, that of a district in a governor, and that of every town or Indian village in an *alcalde*, who is entrusted with the command of the local police. To fill the foregoing appointments, it is required that the candidate should be an active and approved citizen above 30 years of age. The prefects are charged with the economical administration of their respective departments, but are strictly prohibited from interfering with the course of popular elections, or the functions of departmental juntas. The latter are bodies sitting in the capital of each department, composed of two members from each province, elected in the same manner as the members of the senate, and whose functions include the assessing of taxes, examining the accounts, and determining the military force, of the department.

Justice, in all the departments, is administered in the name of the republic ; and in every town are justices of the peace, whose business is to endeavour to bring about an amicable termination without a formal law suit, few suits being, in fact, admitted without some preliminary attempt at settlement. In some provinces the functions of the judge are exercised by the sub-prefect. Justice is not said to be positively corrupt, but, the law being ill understood by many judicial functionaries, civil suits especially have been frequently decided on erroneous principles. Few of the municipalities have revenues adequate to the maintenance of a sufficient police ; the latter is said to be better in Junin than in the other departments. The province prisons are bad and insecure. Slavery is extinguished, except as respects those who were slaves previously to the declaration of Peruvian independence in 1820. Except those individuals, a traffic in whom still goes on in the interior, every one enjoys the right of citizenship, excepting only vagrants, gamblers, drunkards, and those who, without cause, abandon their wives, or are divorced on account of misconduct.

The state religion is the Roman Catholic ; and, Peru having been the country in which the direct influence of Spain was perhaps more felt than in any other of her transatlantic possessions, a great deal of intolerance was formerly shown toward individuals of a different creed ; though we believe a considerable portion of this has disappeared since the establishment of the republic. Lima is the seat of an archbishop, who holds the chief ecclesiastical authority. The Jesuits in the 17th century, and afterwards the Franciscan monks, established various Indian missions in the E. parts of the country. But these have almost all gone to decay ; and the former missionary college of Ocopa, about 12 m. S.E. Tarma, suppressed at the revolution, but afterwards restored, is by no means flourishing, and many Indians of the interior are relapsing into paganism. The clergy are said to be careless of their duties, and lax in their morals. " The Indians and curates are often seen chattering and driving hard bargains in rela-

588

tion to first fruits (for tithes are collected by the state), marriages, burials, and religious festivals, which latter are closely interwoven with the entire social system of the country. The Sierra curates are men commonly much worn out in constitution at the age of 40. These gentlemen, when their home becomes irksome, start off, swayed by some sudden impulse, to the nearest town of white inhabitants, where they enjoy a finer climate and more gratifying company. The curate not unfrequently resorts to a mining village, under pretext, perhaps, of selling his 'prancia,' or first fruits, in grain, gambling with the miners day and night, till the premicia be all swallowed up ; and the poor residentiary returns home involved in a debt which he cannot pay for the next six months, even should his curacy be worth 4000 or 5000 dollars a year, though it be oftener much less." (*Peru as it is*, i., 211, 212.)

Schools for reading, on the Lancastrian plan, are common in the capital, and exist in the larger provincial towns ; and all the white children are taught the elements of instruction. Lima has a university and several other colleges ; but the former has seldom more than 50 students, and the latter establishments have mostly dwindled into insignificance. Superior education is confined to a very few among the whites, and ornamental almost universally takes precedence of useful instruction : the negroes and Indians have rarely any education except what is necessarily acquired in the ordinary intercourse between man and man. There are some good libraries in the capital, and a medical college ; but medical science generally is at a very low ebb. In the rural districts especially, what is called medicine is the grossest quackery ; and other branches of general science are not in a much better condition. There are few hospitals or other charitable institutions, such foundations having been mostly suffered to fall into decay.

The constitution provides that a national militia shall be raised throughout the country ; but in most of the provinces it can hardly be said to exist, except in name. The standing army is estimated at 3000 men, and the naval force is quite insignificant. (*Smith, Peru as it is*, i., 65–142.) We cannot form any clear estimate of the public revenue and debt of this state : the *Weimar Almanach* roughly states the probable amount of the latter at 20,000,000 piastres.

'*History*.—When the Spaniards under Pizarro and Almagro arrived in Peru, in 1532, they found that country under the dominion of the incas, who, according to the traditions of the Indians, had held the sovereignty about four centuries. The first inca, Manco Capac, had either immigrated from some distant country or been a person of very superior acquirements. He pretended that his sister, Oello, whom he married, and himself were children of the sun, and that they were sent to instruct rude and barbarous natives in the duties of religion, the arts and civilisation. He made Cuzco the capital of his dominions ; and, having erected a temple to the sun in that city, appointed 12 virgins of the blood royal to act as priestesses to the divinity, and became both the high-priest and law-giver of his people. The government and manners of the Peruvians were, as compared with those of the Mexicans, mild in the extreme. Still, however, a considerable number of the attendants of the incas were sacrificed on their death, and interred with them, that they might appear in the next world with their former dignity, and be served with the same respect. The remains of the roads, aqueducts, palaces, temples, and other structures, scattered over the country, attest the advanced state of civilisation at which the Peruvians, as compared with most other Americans, had arrived. The empire of the incas fell an easy conquest to Pizarro and his bloodthirsty comrades. It continued in possession of the Spaniards till 1821, when Lima, having submitted to a Chilian army under San Martin, its independence was declared on the 28th of July. Since that time Peru has been, like the other ci-devant Spanish colonies, involved in all but perpetually occurring vicissitudes. (*Ulloa; Robertson's America; Stevenson's Residence in Peru; Smith, Peru as it is; Humboldt; Hall; Miller; Geog. Journal, &c.*)

PERU, p. t., Oxford co., Me., 40 m. W. by N. Augusta. 627 W. Bounded N. by Androscoggin river. Incorporated in 1821. It has one store, two grist-mills, four saw-mills ; eight schools, 209 scholars. Pop. 1002.

PERU, p. t., Clinton co., N. Y., 10 m. S. Plattsburg, 132 m. N. Albany, 303 W. Bounded E. by lake Champlain. Watered by Ausable river. It contains 12 churches, five Methodist, two Presbyterian, a Congregational, two Friends, a Baptist and Roman Catholic, nine stores, three forges, one woollen factory, one grist-mill, 19 saw-mills, two tanneries ; 13 schools, 305 scholars. Pop. 3134.

PERU, p. t., Huron co., O., 93 m. N. Columbus. 386 W. Watered by a branch of Huron river. It has 14 schools, 554 scholars. Pop. 9000.

PERU, p. t., capital of Miami co., Ia., 90 m. N. Indianapalis, 599 W. Situated on the N. bank of Wabash river. The Wabash and Erie canal passes through it. It contains .. courthouse, jail, nine stores, two grist-mills, two saw- ..ills, one printing-office, one weekly and one semi-weekly newspaper, four schools, 112 scholars. Pop. 961.

PERUGIA (an. *Perusia*), a town of the Papal states, .. being of the same name, nearly in the centre of the Italian peninsula, and on the main route between Rome and Florence, about equidistant from the Mediterranean and Adriatic, and 85 m. N. Rome; lat. 43° 6' 46'' N., long. ..27' 13'' E. Pop. 30,000. It stands on the summit and ..livity of a hill, 700 or 800 ft. in height. It is fortified, ..ugh not strongly; being defended, exclusive of its walls, .. castle, erected by Pope Paul III., in 1543. It is irregu-..ly laid out, but well-built, and has several public build- .. and remains of antiquity that are worth notice. The ..edral is a large Gothic edifice, which would be hand- .., were it not so party-coloured. Like many of the .. churches, it is rich in works of art, having paintings ..rocci, Guido, and Perugino, exclusive of four famous ..res by Raphael—the Annunciation, Circumcision, As- ..tion, and Adoration of the Magi. The churches of ..inico and St. Francisco are interesting; the last be- very handsome specimen of early Italian architec- ..The *palazzo publico* is a remarkable specimen of Gothic. Among the antiquities are an arch, report- ..ave been built by Augustus, though, according to ..ood (*Letters of an Architect*, ii., 104), it is probably A circular building, still tolerably perfect, which is ..ly of Roman origin. There are numerous public ..s; one of which, in the principal square, has been ..ted with bas-reliefs and statues by Arnolfo de ..Perugia has a university, with about 200 students, academies, numerous convents, two theatres, a and an admirable ground for playing *pallone*. ..is the residence of a Papal delegate, a bishop's seat of a tribunal of primary jurisdiction, and a ..police, and the residence of the directing engineer ..and roads for the delegates of Perugia, Spolito, ..It has manufactures of carpets, silk goods, pre- ..s, hats, cream of tartar, soap, and wax candles, ..trade in wines, oil, and other agricultural pro-

is scarcely inferior in antiquity to Cortona, and ..al in rank among the cities of Etruria. Antony ..t himself up in the city, it was taken, after a ..stance, by Octavius Cæsar, who dismissed An- ..he city was hardly dealt with, more, as Velleius ..h the irritation of the soldiers than the inclina- ..reneral. (lib. ii., cap. 74.) It was annexed to ..ominion by Julius II., in 1512. The famous ..acci. surnamed Perugino, was a native of this ..'s *Letters of an Architect* ; *Simond, &c.*)

..an. *Pisaurum*), a coast town of central Italy, ..leg. Urbino. on the Toglia. near its mouth, ..20 m. N. by E. Urbino ; lat. 43° 55' 10'' N., ..'' E. Pop. about 15,000. It is surrounded ..ons, and is well built. The streets are clean ..it has in general a neater appearance than ..Italy. Its market-place is ornamented with ..arble statue of pope Urban VIII. Being the ..p, it has, of course, its cathedral, and the ..ent of churches and convents. It has no ..erely an open roadstead. Some of the ..markable for their paintings, as are several ..f the higher ranks. The palace of the .. Urbino is now occupied by the papal le- ..nany handsome private residences. two bo- ..g asylum, a good theatre, &c. The aque- ..eys water to different parts of the town, is ..work of the Romans; and there are the ..ncient bridge and theatre. Silk and cot- ..ottles, cream of tartar, &c.. are produced ..but the inhabitants are principally em- .. in agricultural produce, the surrounding ..rich and well cultivated, producing the ..with wine, olives, silk, &c. There are ..villas in the vicinity; one of which was ..19, by Queen Caroline of England, Pi- ..an colony. It is noticed by Catullus, ..nda sede Pisauri.'' Carm. 82.

..ts climate, which made it be so charac- ..o a considerable extent obviated by the ..ljacent marshes. The famous musical ..s a native of, and resident in this town. ..*Starke, &c., passim.*

..ne "Advanced Post''), a considerable ..cap. of a principality of the same name. .. belonged to Runjeet Singh, but which ..cluded in the dominion of the Caubul ..s in a nearly circular plain, about 35 m.

5J

in diameter, and watered by many branches of the Caubul river, 140 m. E. by S. Caubul, and 236 m. N.W. Lahore; lat. 34° 6' N., long. 71° 13' E. Pop., formerly estimated by Mr. Elphinstone at 100,000, but it is now certainly much less. It is built on uneven ground, and is upwards of 5 m. in circ. The houses are mostly built of unburnt brick, enclosed in wooden frames, and are commonly three stories high; the lower story being usually occupied by shops. The streets, though narrow, are paved, and have a kennel in the middle. Two or three brooks run through the town, skirted with willows and mulberry-trees, and crossed by bridges. There are many mosques, but no public building is said to deserve public notice, except the *Bala Hissar*, or citadel, and a fine caravanseral. The former, a castle of no strength, on a hill N. of the town, commands an extensive and beautiful view, and when visited by Mr. Elphinstone in 1809, had some fine halls, and was surrounded with spacious gardens, being the occasional residence of the king of Caubul; but when Moorcroft saw it about a dozen years afterward, it was a heap of rubbish, and the only use made of it by its Seikh rulers was as a quarry whence to procure materials for dwellings of their own erection. Many of the houses in the city were also, at the latter period, untenanted and in ruins; in the plain numerous villages were deserted; and in the immediate vicinity of the town the Seikhs had inflicted more mischief than many years' labour could remedy, by destroying gardens and orchards, and demolishing the wells, and channels of irrigation. (*Moorcroft's Trav.*, ii., 337.)

It is probable, however, that Peshawer has since recovered a portion of its previous prosperity. It is well situated for trade; and should we take possession of the Punjab, or the Indus be extensively navigated by British vessels, Peshawer would most likely become a considerable entrepôt for the trade between India, and Afghanistan, Khorassan, and the countries N. of the Hindoo Koosh. The inhabitants are very mixed, but principally of Indian origin, and occupied in commerce. The shops are well supplied with fruits and other provisions, saddlery, boots and shoes, woollens, hardware, books, and other manufactured goods.

It is said by some authorities to have been founded by Acber; but the district of Peshawer is mentioned in the histories of the 10th century; and it is more probable that it should have taken its name from a city already existing, than the reverse. Peshawer was, however, greatly improved and enlarged by Acber, in the 16th century. (*Elphinstone's Caubul, &c.,* 79-81; *Moorcroft's Travels*, ii., 337, 338; *Burnes's Bokhara, &c.*)

PESTH, a city of Hungary, on the E. side of the Danube, immediately opposite to Buda, with which it is connected by a bridge of boats, being the seat of the chief judicial tribunals of the country, as Buda is the residence of the viceroy and other chief state authorities, 135 m. E.S.E. Vienna. Pop. (excluding garrison and strangers) about 63,000, principally Roman Catholics; or, together with Buda (which see) and its suburbs, probably 110,000. (*Berghaus.*) Pesth, with its suburbs, occupies a space nearly 4 m. in circ. It stands on level ground; and being almost wholly of modern date, is much more regularly laid out and handsomely built than Buda. The streets, which are mostly wide and straight, are paved, and partially furnished with trottoirs, some of them being, in the splendour of their shops and their elaborately painted signs, little inferior to those of Vienna. "After a fashion, once common with us, and of which one or two specimens still exist in London, every shop has a name and sign; so that you may buy your cigars at the *Young Prince*, your cravats at the *Three Graces*, and bonbons at the *English Lord*; and for the instruction of those who do not read, or to attract the attention of those who do, these subjects are all illustrated by large paintings, in a style by no means contemptible." (*Paget's Hungary*, i., 243-4.) The squares are generally very well built; but, from the want of some object in the centre, they look bare and deserted: besides affording room for the accumulation of those heaps of sand with which the city is infested, but which might probably be prevented by planting round the outskirts. The growth of Pesth within the last few years has been most rapid. Along the river-side, which, a few years ago, was nothing but a marsh, is now a wide quay, partially paved and walled in, and lined for upwards of 1 m. by a succession of handsome buildings. Near the centre of these are the new theatre, and *Redonten-saal*, or public ball-room; and at one end, ornamented with a portico like the last named edifice, is the National Casino, an institution similar to the clubs of London, recently established by Count Szechenyi. It is open to strangers, who may use it during their stay in Pesth, and its reading-room is furnished with the leading newspapers and magazines of Europe. A part of the establishment is appropriated as a casino for tradesmen; and in the centre of the building is a very fine ball-room. Among the most conspicuous of the public edifices is the *Neugebauds*, in the suburb of Leopolstadt, begun by Joseph II. in 1786; a structure of immense size, four stories in height, ranged round four spacious areas, and now used as a barrack and artillery *depôt*. Pesth has only a small number of churches in proportion to its size and population, and none is particularly distinguished for architectural beauty. Service is performed in them according to the United Greek, Roman Catholic, Dissenting Greek, Lutheran, and Calvinistic rituals; and in the German, Hungarian, Slavonic, and Greek languages.

There are also several synagogues. Besides the large theatre on the quay, an elegant new national theatre, destined solely for Hungarian performances, has recently been completed by the aid of a grant from the diet. The grenadiers' barracks, county hall, Jesuit college, and two or three of the hospitals, are worth notice. But, in respect of public buildings, Pesth must yield to Buda; it has, however, many handsome noble palaces and other private residences, and excellent hotels and coffee-houses.

This city is also distinguished by its establishments devoted to the higher branches of scientific instruction. Its university, established at Tyrnau in 1633, and transferred thither from Buda, by Joseph II., in 1784, is the only one in Hungary, and one of the most richly endowed in Europe. The instruction is entirely gratuitous; and it has about 50 professors, an observatory on the Blocksberg in Buda, a large botanic garden, a veterinary hospital, and a library of 60,000 volumes. In 1834–35, it had 1610 students; 72 in divinity, 184 in law, 665 in medicine and surgery, and 403 in philosophy, &c. The National museum, founded in 1802, has a fine library, rich in Hungarian MSS.; a complete collection of Hungarian coins from the 10th century; collections of minerals, fossils, antiquities (principally Roman, and others, found in Hungary and Transylvania); historical relics, specimens of manufactures, &c. The Hungarian academy of sciences, originally founded for the cultivation of the Magyar language, has received many munificent donations, and has an income of about £3000 a year. It publishes transactions, and gives annual prizes for the best works in Hungarian. Pesth has also a gymnasium, Roman Catholic and Lutheran seminaries, an English conventual school for noble ladies (*Englisch Fraüleinstift*), a teacher's seminary, many primary, and Greek, Protestant, and Jewish schools, a blind school, &c. The chief judicial tribunals are the *Curia Regis*, or Royal Table, and the *Septemviral Tafel*, so termed because it originally consisted of seven members, but is now extended so as to include the palatine, four prelates, nine magnates, and seven other nobles. There are various charitable institutions. Several newspapers are published in the Hungarian language; the principal of which has a circulation of about 4000. Though near the extreme verge of European civilisation towards the E., Pesth has all the appearance and conveniences of a city of W. Europe. It is well furnished with provisions of every kind, and, in some respects, its inhabitants are better off than those even of the capitals of England and France! "The *fiacre* is none of the heavy, shabby, slow coaches found on the stands of London; but a very clean, smart, open calèche, with two high-bred little horses, which whisk along at a famous rate; and a driver as far superior in sharpness and wit to his wooden-shod confrère of Paris, as the equipage is to that of London. In winter, instead of the open calèche, a neat, close chariot takes its place; for he is a very poor *fiacre* in Pesth, who has not a winter and a summer carriage." (*Paget*, i., 230.)

The greatest variety of costume may be seen in Pesth, especially at the four great annual fairs, which are attended by at least 90,000 strangers, many of them from very distant parts. The business transacted at these fairs is very extensive. (See *Bright's Travels in Lower Hungary*, p. 917–923.)

Pesth manufactures silk and woollen fabrics, leather, straw hats, oil, tobacco, &c.; but its principal manufacture is that of *meerschaum* pipe-bowls. These, which consist of the species of earth called *kaf-kil*, dug in the Crimea, are first rudely fashioned in Constantinople, but are finished for the German markets in Pesth. They are thence conveyed to Vienna, and ultimately to the fairs of Leipsic, Frankfort, Mannheim, &c.; where the best fetch from £3 to £5, and even £7 sterling. A considerable intercourse is kept up between Pesth and Buda; and though a large proportion of the population be exempted from payment, by their privilege of nobility or otherwise, the tolls on the bridge of boats across the Danube are estimated to amount to 16,000 florins a year. But the no lies have agreed to relinquish this offensive privilege in favour of the project for building a suspension bridge across the river. (See BUDA.)

The amusements of both cities differ little from those of the German capitals. The theatres, coffee-houses, and public gardens in the neighbourhood are the favourite places of resort. Immediately beyond the barriers of Pesth is the *Rákos Mezö* a wide plain on which the diets of Hungary were held for many centuries; and on a part of which
540

horse races, somewhat after the English fashion, are now held yearly in May or June.

Pesth is supposed to occupy the site of the *Transaquinum* of the Romans, on the ruins of which a town was afterward built by Arpad. Bela IV., in the 13th century, surrounded the town with walls; and it subsequently rose to considerable commercial importance. It was held by the Turks 160 years. The present town, one of the best built and handsomest in the Austrian dominions, may be said to have wholly grown up since the reign of Maria Theresa. It suffered severely in 1838 from an inundation of the Danube, which is said to have destroyed 1200 houses in the older part of the city. (*Œst. Nat. Encyol.*; *Berghaus, Allg. Länder, &c.*; *Paget's Hungary*, i., 229–245; *Bright's Travels in Lower Hungary, &c.*)

PETERBOROUGH, an episcopal city and parl. bor. of England, in liberty of its own name, co. Northampton, on the navigable river Nen, crossed here by a wooden bridge, 37 m. N.E. Northampton, and 75 m. N. by E. London. Area of parl. bor. (which includes the entire parish of St. John the Baptist, with the Minster precincts), 1439 acres. Pop., in 1841, 13,066. This small city consists of several streets close to the N. bank of the river, regularly laid out, well paved, and lighted with gas, the houses generally being well built, and some of recent erection. The principal public building is the cathedral, formerly attached to a Benedictine monastery, founded here in 870, and regarded at the dissolution as one of the most magnificent abbeys in the kingdom. It is a regular cruciform structure of Norman and early English architecture, erected during the 12th century. The dimensions of the interior are, length, 476 ft.; breadth of nave and aisles, 78 ft.; breadth, including the great transept, 203 ft.; breadth of transepts, 69 ft.; height of ceiling, 78 ft.; ditto of lantern, 135 ft.; length of the W. front, 156 ft.; height of the central tower, 150 ft. A tower and spire once stood over the N.W. transept; but the latter has been taken down. The approach to the cathedral has a very monastic appearance. "Passing under a Norman gate, with later additions, a court is entered, the right side of which is formed by the domestic buildings of the abbey; and at the end is the noble front of the church, consisting of three fine early English arches; but their beauty is much impaired by the small chapel or porch, which, in another place, would have been very beautiful. The E. end is circular, and the aisles are made out square by perpendicular additions. The choir has a wooden groined roof of very inferior workmanship; a handsome stone screen has recently been substituted for one of wood, and the fittings of the choir have been entirely renewed, under the direction of Mr. Blore. The nave is a very good description of Norman work which has its [...] of shafts; the proportions are good, and the [...] of that [...] ance is fine, without that overwhelming [...] those edifices where the great circular [...] There are few monuments, shrines, or chantry chapels in the [...] parliamentary troops having plundered the church of most of its ornaments of this description. Catherine of Arragon and Mary, queen of Scotland, were interred here; but their graves are not marked by any sepulchral monument. " The remains of the monastic buildings in the court fronting the cathedral are of somewhat varied style, but present, on the whole, a valuable specimen of bold and varied outline both of plan and elevation. In the same court, also, leading to the cemetery on the N. side of the church, is a late perpendicular gate, remarkably rich in ornament, and near the S. transept are some remains that may have been the refectory, or infirmary chapel of the ancient monastery." (*Rickman's Gothic Architecture*, p. 209.) The cathedral corporation consists of the dean and six prebendaries, who divide among them a nett revenue of £5118: there are also four minor canons and a precentor. Peterborough was erected into a bishop's see by Henry VIII., and the diocese now comprises the counties of Northampton, Rutland, and Leicester. The nett revenue of the see amounted, at an average of the three years ending with 1831, to £3103 a year; but it has been since augmented by an order of council to £4500, the deficiency being supplied from the surplus fund contributed by the larger sees.

The parish church is a spacious building, lately remodelled and put in very complete repair: the living is a vicarage in the gift of the bishop. The Independents, Baptists, and Wesleyan Methodists, have also their respective places of worship, and there are Sunday schools attended by upwards of 900 children of both sexes. The cathedral grammar school, founded by Henry VIII., is attended by about 30 boys, and endowed with three scholarships and a fellowship at St. John's college, Cambridge. There are two charity schools for boys, and a national school is attended by about 300 boys and 180 girls. There are numerous other charities, including a pretty large infirmary and a dispensary. The town-hall is a small but neat structure, the area beneath being used as a market-place. The jail and house of correction for the city and liberties are both "small and miserably deficient in accommodation." (*Prisons' Report.*) There is also a small theatre, open during the summer months.

The trade of Peterborough arises chiefly from the transit of corn and malt, large quantities of which are brought down the Nen from the interior; it also imports coal, timber, bricks, stone, and other goods. The city is not incorporated, but is under the jurisdiction of the dean and chapter, whose steward holds a court for the trial of civil actions within the city. Quarter sessions are likewise held here for the liberty of Peterborough, and it is one of the polling places at elections for the N. division of the county. Peterborough has returned two members to the House of Commons since the 1 Edward VI., the right of election down to the Reform Act being in the inhabitants paying scot and lot. The electoral limits were enlarged by the Boundary Act, so as to include with the old borough the remainder of the parish, and the extra-parochial precincts of the cathedral. Regist. electors, in 1839–40, 569. Markets on Saturday: fairs, July 1 and Oct. 1, chiefly for cattle.

PETERBOROUGH, p. t., Hillsborough co., N.H., 41 m. .W. Concord, 447 W. Chartered in 1738. Drained by ontoocook river and its branches, which afford waterower. It contains five churches, a Presbyterian, Methost, Baptist, and two Universalist; six stores, two fullingills, two woollen factories, five cotton factories with 6064 indles, one furnace, six grist-mills, seven saw-mills, one per-mill, two tanneries; one academy, 90 students; 13 hools, 671 scholars. Pop. 2163.

PETERHEAD, a parl. and mun. bor. and seaport of otland, co. Aberdeen, on the point of a flat, rocky promtory, projecting into the German ocean, 27 m. N.N.E. erdeen; lat. 57° 32' N., long. 1° 47' W., being the most terly point of land in Scotland. Pop., in 1831, 5112. erhead was erected into a borough of barony by the ily of Keith, earls marischal, on whose estate it was lt, in 1593. On the attainder of that family, it was pursed by the York Buildings company, who sold it to the ernors of the Merchant Maidens hospital of Edinburgh, are now the superiors of the town, and have always its liberal and munificent patrons. It did not, how-, attain to any distinction till about 1770, soon after ch the famous engineer, Smeaton, was employed to truct a harbour on the S. side of the promontory on h the town is built. This harbour, though on a small , was sufficient to demonstrate the importance of the , and the advantage that would result, not merely to own itself, but to the shipping employed on the E. of Scotland, from the improvement of the harbour. nsequence, measures have been undertaken and carnto effect during the present century for excavating cks that obstructed the S. harbour, and for constructnew and extensive harbour and graving dock on the e of the peninsula. These great works have since ompleted, at an expense of above £30,000, and Peterhas, in consequence, been rendered one of the best irs on the E. coast of Scotland. The area of the S. r is 6·6, and that of the N. nearly 11 acres. They med by strong moles projecting into the sea. The r-dues, which, in 1808, amounted to only £367, had ed, in 1837, to £2879, and are now about £3000.

general appearance of the town is airy and health- streets are wide and well paved; and the houses t of the fine red granite which abounds in the neighd, and is an object of traffic." (*Parl. Bound. Re* The streets are well lighted with gas, and the town ied with excellent water, brought from a distance 2¼ m. Among the public buildings may be specitownhouse, with a handsome spire 125 ft. in height, parish church, a respectable building, 118 ft. in A handsome cross, consisting of a Tuscan pillar e surmounted by the arms of the earl-marischal, ted in 1832. The town has also a *quoad sacra* parhurch; an Episcopal chapel, with chapels for the d Dissenters, Independents, and Methodists. The al establishments are not such as one might exd in so large and thriving a town; but measures, e, are now in progress for having a grammar academy organized on an adequate scale. The a scientific association, a news' room, a valuable elonging to Mr. Arbuthnot, two public libraries, is friendly societies. There are mineral springs borough that used to be a good deal resorted to: re now comparatively neglected. Except roped ship-building, Peterhead has no manufactures. re belonged to Peterhead 82 vessels of the agrden of 11,022 tons. The inhabitants early ene N. sea whale fishery, and carried it on for a period with great vigour and success. In 1823, longing to this port brought home 2217 tons of however, was the maximum of prosperity, and the business has since progressively declined, and is now nearly extinct. But the loss of the whale-fishery has been compensated by the extension of the herring-fishery and the success which has attended it. During the year ending the 5th of April, 1840, 53,077 barrels of herrings were cured here: the cod or white fishery is also prosecuted to a considerable extent.

The products of the fishery form, of course, a considerable portion of the exports from the port; but, exclusive of these, very considerable quantities of corn, butter, and other agricultural products, are exported. In 1836, for example, there were shipped 27,164 quarters of corn, 14,424 bolls meal, 3343 cwts. butter, 1,730,000 eggs, 3268 cwt. pork, &c. Great quantities of granite are also occasionally exported.

The Reform Act conferred on Peterhead the privilege of sending a member to the House of Commons in conjunction with the boroughs of Elgin, Banff, Cullen, Inverury, and Kintore. Registered electors in the united boroughs, in 1839-40, 648; in Peterhead, only 241. The borough has 12 councillors. (*New Statistical Account of Scotland, par. Peterhead; Boundary Report, &c.*)

PETER-LE-PORT (ST.). *See* GUERNSEY.

PETERS, t., Franklin co., Pa., 12 m. S.W. Chambersburg. Watered by the W. branch of Conecocheague creek. It contains three stores, one fulling mill, one woollen factory, one furnace, three forges, three flouring-mills, one grist-mill, six saw-mills, one oil-mill, four tanneries; 13 schools, 385 scholars. Pop. 1939.

PETERSBURG, a government of Russia in Europe, being that in which the capital is situated, between 58° and 60° 30' N. lat., and 28° and 34° E. long.; having N. the gulf of Finland, the gov. of Wyborg, lake Ladoga, and the gov. of Olonetz; E. and S.E. Novgorod; S. Pskof; and W. the lake Peipus, and the gov. Revel. Area, estimated at about 15,000 sq. m. Pop., in 1838, 585,000. The country is generally flat; but in the N. and S.E. are a few undulating hills. The general slope is towards the N.W.: all the rivers, the principal of which is the Neva, flow to the gulf of Finland, or lake Ladoga. The soil is mostly sandy and thin; and the climate damp, severe, and unhealthy. At an average, frost prevails, more or less, for 160 days in the year. Rye, barley, oats, and some wheat are grown, but the climate is unfavourable to the culture of corn; and nearly 2-3ds of the province is covered with wood, marshes, and lakes. Timber, indeed, forms the chief source of what may be called wealth, deals and masts being the great articles of export, and the villagers subsisting chiefly by making woodwork of different kinds. The trade is limited, in a great measure, to the capital. The best agriculturists are German colonists, who raise flax, hemp, and, above all, kitchen vegetables, for the supply of the capital. Russians form the majority of the inhabitants; the remainder is composed of Finns, Carelians, Ijors, Germans, &c., most of whom are Lutherans. This government is divided into nine districts; Petersburg, the capital, is the only place of importance.

PETERSBURG, the modern metropolis of the Russian empire, and one of the largest cities of Europe, at the E. extremity of the gulf of Finland, where it receives the river Neva, by which the city is intersected, lat. 59° 56' 31" N., long. 30° 18¾' E. Pop., in 1838, 469,720. The city, which is of a circular form, stands partly on the main land, on the S. side of the Neva, and partly on islands formed by its branches. It owes its existence to the genius and power of Peter the Great, by whom its foundations were laid in 1703. At first all the public buildings and houses were of wood, and were huddled together without regard either to regularity or convenience. But brick and stone buildings were soon after introduced; and the streets were laid out on a regular plan, crossing each other at right angles. This was greatly facilitated by the ravages of destructive fires in 1736 and 1737, which having destroyed some thousands of the old houses, enabled government to lay down judicious regulations for their reconstruction. The empress Elizabeth did much to improve the city; but it is chiefly indebted for its regularity, beauty, and magnificence to the empress Catharine II. Under this princess the principal channel of the Neva was faced by noble granite quays; several new streets and canals were opened; and seven of the finest public buildings and monuments were either rebuilt on an improved plan, or constructed of wood. The late and present emperors have also distinguished themselves by their efforts to improve and embellish the city. It is now certainly one of the finest in Europe, and is unmatched for the width and regularity of its streets, the length and magnificence of its quays, and the elegance of its squares and public buildings.

Among the latter, which are principally situated on the quays bordering the main channel of the Neva, and in the street entitled the Nefski Perspective, may be specified the Winter palace, or ordinary residence of the emperor, a vast but heavy building. It communicates by a gallery with the Hermitage, another palace, long the residence of Catherine II.; the latter has attached to it the court theatre, and contains a noble picture gallery, a valuable library, and an extraordinary rich collection of engraved stones, jewels, &c. There are also the Marble palace, the palace of Anitchkoff, formerly occupied by the reigning monarch, the Taurida palace, built by Catherine II. for her favourite Potemkin, &c. The admirality, an immense brick building, occupies the centre of the city; it contains store-houses, docks for the construction of men-of-war, and a very extensive collection of objects connected with navigation and natural history. The high gilt tower of the admirality, erected by the empress Anne in 1734, is one of the most striking objects in approaching Petersburg. Among the other public buildings are the hotel of the academy of the fine arts, accounted one of the finest in the city; the exchange; the palace of the senate; the hotel *de l'état Major,* a magnificent building; the barracks for the guards; the new theatre; the exchange bank; imperial library; foundling hospital; hotel of the land cadets, and a host of others.

The citadel, founded by Peter the Great, but since reconstructed, stands on an island in the centre of the city. It is a regular hexagon, and has a tower 360 feet in height. Among the churches may be specified the cathedral of our lady of Kasan, consecrated in 1811. It is built on the model of St. Peter's at Rome, and is one of the finest ornaments of the capital. The cathedral of St. Peter and St. Paul was founded by Peter the Great in 1712: it is of large dimensions; but it derives its principal interest from its containing the tombs of all the Russian sovereigns, from its illustrious founder down to the late emperor, with the exception of that of Peter II., interred at Moscow. The church of the convent of St. Alexander Nefski, at the end of the magnificent street to which it has given its name, occupies the third rank among the churches of Russia. The mausoleum of the saint is of solid silver; several distinguished persons are interred in the church; and there is attached to it a grammar school on a very large scale; having generally about 800 pupils. The cathedral of St. Isaac, now in course of being built, will be one of the finest churches in the city. The total number of churches in 1836 amounted to 58, of which 43 belonged to the established Greek faith, 11 to Catholics and Protestants, and four to dissenters. There were also, at the same epoch, 91 Greek chapels in private houses, and one monastery.

Petersburg contains some noble monuments: at the head of these may be placed the magnificent equestrian statue of Peter the Great, by Falconet. "The monarch is represented in the attitude of mounting a precipice, the summit of which he has nearly attained. His head is uncovered, and crowned with laurel; he wears a loose vest, in the Asiatic style, with half boots, and sits on a housing of bear skin; the right hand is stretched out, as in the act of giving benediction to his people, and the left holds the reins. The design is masterly, and the attitude bold and spirited. The horse is springing upon the hind legs, and the tail, which is full and flowing, appears slightly to touch a serpent, artfully contrived to assist in supporting the vast weight. The contrast between the composure of Peter and the fire of the horse, eager to press forward, is very striking. The simplicity of the inscription corresponds to the sublimity of the design—PETRO PRIMO, CATHARINA SECUNDA, 1782." (*Coxe's Trav.,* ii., p. 112.) The pedestal on which this noble statue is erected is a gigantic rough block of granite. It was found at a distance of several miles from the capital, and its conveyance thither was a work of extraordinary difficulty.

The column erected in honour of the late emperor Alexander is one of the finest of its kind. It is 150 feet in height; the pedestal is of granite and bronze; the shaft of the column consists of a single piece of red granite, 84 feet in length, and 14 feet in diameter. This, which is the largest monolithe in the world, was cut from the quarries of Pyterlar, in Finland, several miles from Petersburg. The column is surmounted by a capital and a small dome in bronze, on which is placed a statue emblematical of religion.

The cottage occupied by Peter the Great during the foundation of Petersburg is still preserved, and will be regarded by most persons as the most interesting monument in the city. It is built of wood, painted in the Dutch style, and is not 20 feet high.

Petersburg has a great variety of institutions for the promotion of education and literature. The university, founded in 1819, has already attained to considerable distinction, and had, in 1839, 413 pupils and 48 professors. The *medico-chirurgical academy,* founded by Peter the Great, and re-organized by the emperor Alexander, for the instruction of medical men, enjoys a high reputation; two hospitals, on a large scale, are attached to it; the instruction is gratuitous, and the number of pupils may amount to 590. A sum of 396,290 roubles a year is appropriated to the support of this establishment. The education is good, and the examinations strict. Among the educational institutions may be

specified the military schools (*see* art. RUSSIA); the theological academy; the school for training and instructing schoolmasters; the Oriental institution, founded in 1823. the school of commerce and navigation; the technological institution; the Protestant school of St. Peter, at which above 500 children of both sexes are educated; two gymnasiums or colleges; the schools of Sendiuol and St. Catherine, for the instruction of nearly 700 young ladies; the military orphan school; the grammar school of St. Alexander Nefski, already alluded to, and a number of others.

According to the official returns, the pupils at the various schools and educational institutions in the city, in 1836, amounted, in all, to 11,293. Now, if we take the stationary population of the city at 370,000 (*see post*), this would give the proportion of scholars to the population at about one in 33. But when everybody is educated, the proportion at school cannot well be less than one in 16. The number of children in Petersburg is, no doubt, less in proportion to the population than in most other great towns; but it is, notwithstanding, sufficiently obvious that the means of elementary instruction are very insufficient.

Petersburg has some noble libraries, and scientific and literary collections. The Imperial library, one of the largest and finest in the world, contains about 425,000 printed vols. The academy of sciences, founded by Peter the Great in 1724, has long occupied a distinguished place among such bodies. It contains a valuable library, an Asiatic museum, with cabinets of medals, natural history, &c. It is also furnished with an observatory, whence the Russian geographers reckon their first meridian. The imperial Russian academy, the academy of fine arts, &c., are celebrated all over Europe; and there are besides a great number of similar institutions. The botanic garden is extensive, and has a valuable collection of exotics. There were in the city, in 1836, 32 printing and lithographical establishments belonging to the crown, and 31 to individuals.

The hospitals are numerous, and well endowed. The most celebrated is the foundling hospital, founded by Catherine II., and much improved and enlarged by the late benevolent empress, Maria Feodorowna. But, however well intended, it is pretty certain that this foundation has always been productive of infinitely more evil than good. The mortality used to be enormous: and, notwithstanding the improvements that have been introduced, it continues, necessarily perhaps, to be very great. From 1822 to 1831, the admissions were 39,114, and the deaths 31,779! The establishment costs about 1,000,000 roubles a year. The city hospital, and the Imperial hospital, for sick poor, are both on a large scale: there is also an institution for deaf and dumb persons, a blind asylum, &c.

The Neva is deep, rapid, and its waters as clear as crystal. The main stream, which is broader than the Thames at London, is crossed by two bridges, and its branches by three. These are all of boats, and are removed in the beginning of winter, before the river is frozen over. The granite quay, along the S. side of the Neva, is a stupendous work, above 3 m. in length. The S. side of the town is intersected by canals, of which the Fontanka is the finest.

The streets are, for the most part, paved with stone; a few, however, and still floored with logs of timber: and recently some have been macadamised. The foot pavements are generally good; but the carriage ways, in wet weather become very dirty.

Many of the brick houses are stuccoed and painted, so as to have the appearance of stone; but the number of genuine stone houses is very limited, indeed, and wooden houses still predominate. According to the official account, there were in the city, in 1836, 8641 houses, of which 3024 were brick and stone, and 5617 wood. The better class of houses are covered with iron or copper, and the inferior with tiles. The principal houses have arched door-ways, under which carriages enter, and spacious courtyards, in which wood for fuel, &c., is stored up.

Owing to the barren nature of the soil round the city, most part of the provisions required for its consumption have to be brought from a great distance by canal or by sledges. An extraordinary market for butchers' meat, game, fish, &c., is held at the conclusion of the fast, ending the 28th Dec. (O.S.), at which a large proportion of the inhabitants supply themselves with provisions for the remainder of the winter. The carcasses of the vast quantities of oxen, sheep, hogs, fowls, &c., brought to this market, are all frozen: the smaller animals are piled up in pyramids, but the larger ones, which are skinned, and set on their legs in rows, along side each other, have a ghastly and frightful appearance to persons not familiar with such exhibitions. They are cut in pieces with hatchets; and when carried home are preserved in snow cellars, of which each house has one. Previously to its being dressed, the frozen meat is thawed in cold water, but it loses much of its flavour and all families of condition use fresh killed meat. The length and severity of the winter necessarily occasions a great demand for fuel. This consists partly of coal from England, but chiefly of wood; and as the neighbourhood does not afford a sufficient supply, vast quantities are brought from the interior. The barges, too, which arrive from the sttier are almost all broken up, and used either as timter to the construction of houses, or as fire-wood. Flour and bullet magazines have been erected by government, for the accommodation of the poor in scarce and severe seasons.

There were in 1836, within the city, six manufacturing tablishments and workshops carried on upon account of e crown, and 218 by private individuals. Among those longing to the crown are manufactures of gunpowder, weary, and porcelain, and a cannon foundry. Among private manufactures are those of silks, cottons, sailth, woollens, paper, leather, stuffs, tobacco, wax-cloth, p. types, &c.; there are extensive glass-works at Oserski, r the city; and there are great numbers of watchmakers jewellers, coachmakers, mathematical and musical instrument makers, carpenters, &c.

tersborg has the most extensive foreign trade of any in the N. of Europe. This arises not so much from its population as from its being the only great maritime on the gulf of Finland, and from its vast and various iunications with the interior. By means partly of s, but principally of rivers, Petersburg is connected the Caspian sea, goods being conveyed from the latter capital, a distance of 1434 m., without once landing The iron and furs of Siberia and the teas of China reived at Petersburg in the same way; but, owing to at distance of these countries, and the short period which the rivers and canals are navigable, they ree years in their transit by water. Immense quanthe less bulky and more valuable species of goods brought to the city during the winter upon the ice

rincipal article of export is tallow; and, next to it, p and flax, iron, copper, grain, particularly wheat; potashes, canvass and coarse linen; linseed and d; linseed and hempseed oils; furs, hides, leather. fox, hare, and squirrel skins; cordage, caviare, gl es, tar, &c. The principal imports are sugar r colonial products; cotton yarn, raw cotton, and stuff; dye stuffs, wines, silks, woollens, hardware, from Holland, Silesia, &c.; salt, lead, tin,.coal.

he year 1836, the total value of the exports 1,693 roubles; the principal articles being, tal ,350; hemp, 19,221,395; flax, 6,291.808; copper, iron, 6,860,329; canvass and coarse linen, ristics, 5,316,052; hides, 2,518,190; Russia leath 7; and potashes, 2,134,660. During the same sports amounted to 180,913,930 roubles; the icles being, cotton yarn (almost entirely from 4.418,476; raw cotton, 5,902,980; raw sugar from Havanah), 37,343,544; coffee, 4,316,995; 2; woollens, 6,174,868; cotton stuffs, 3,344,434; 85; and gold and silver, 2,948,450 roubles. duties during the same year amounted to bles.

trade is principally conducted by foreigners, English. Cronstadt, 20 m. lower down the y the port of Petersburg. All ships drawing 9 feet water stop at the former, their cargoes l to and from the city by means of lighters. dt is included in the Petersburg customneele clear indifferently for the one or the umber of vessels entering the port varies 20 a year, of which the English are by far ons. The arrivals in 1836 were 1105 ships, of 108.612½ lasts. The Russians have few A commercial bank was established here urther details as to commerce, &c., see Rus-

towns females predominate over males; erwise in this capital: of 451,974 individuhe population in 1836, 330,564 were males, females! This extraordinary discrepancy partly by the great number of soldiers in of whom are married, partly by the great ried male domestics in great houses, and that the wives and families of many of established in the districts of the country long. Deducting military and strangers. ulation of the city does not exceed 370,000

vere 2185 marriages, 10,167 births, and 1e city is not unhealthy. In 1430, of 6348 1nls belonging to the Greek church, 3314 by colic, 2037 by fever, 1152 by phthisis, 376 by dropsy, and 209 died suddenly.

Suicides are rare, but accidental drowning is not uncommon.

Though well situated for commerce, the position of the city is, in most other respects, far from good. The ground on which it stands is low and swampy, and the surrounding country partakes of the same character, and is covered with forests. No one less bold and enterprising than Peter the Great would have dreamed of building a large city in such a situation, and no one with less gigantic means at his disposal would have been able to carry such a project into effect. But no art can ever overcome the defects inherent in its situation. The climate is severe, foggy, wet, and variable in an unusual degree. The sub-soil is so very porous and saturated with water, that it is hardly possible to excavate a cellar any where in the city, and there is the greatest difficulty in constructing sewers to carry off filth and other impurities. But the great drawback on Petersburg consists in its little elevation above the level of the sea and the river, and its consequent exposure to the most dreadful inundations. These are generally occasioned by a W. or S.W. wind, accumulating the water of the gulf at the mouth of the river, and preventing the free exit of the latter. The years 1726, 1752, 1777, and 1824 have been particularly distinguished by these inundations. The last of these visitations was the most appalling and destructive. The whole city was laid under water; above 8000 individuals perished, and property to a vast amount was destroyed.

The amusements of Petersburg are modified by the religion and the climate. There are four theatres, and plays are acted in Russian, German, French, and sometimes English. The actors are all paid by government, and do not, in any degree, depend on their audiences. The personnel of the theatres, in 1836, amounted to 1196 individuals. During winter the national amusements are sledge-driving and sliding down artificial elevations, similar to those called in Paris the Montagnes Russes. All classes use the vapour bath: in 1836 the public and private baths in the city were estimated at 350. During the same year there were open 18 public markets, 2517 shops attached to houses, 118 magazins de modes, 5 great magazines for merchandise, 40 live-boats for the sale of live fish. 53 eating-houses, 29 furnished hotels, 37 coffee-houses, 36 confectioners, 46 restaurateurs, 59 do. for the lower classes, 308 wine-merchants, 98 taverns, 5 spirit magazines, 70 dépôts for the sale of spirits, 197 bakehouses, 247 forges, &c. (For further particulars as to the climate, manners, and classes of the people, police, &c., see RUSSIA.)

The country in the vicinity of Petersburg being flat and marshy, presents few rural beauties. The imperial family have country residences at Oranienbaum, on the gulf of Finland, Kammanoi, Osteof, Peterhof, Czarkoselo, &c. Czarkoselo, situated on the Neva, about 16 m. from Petersburg, is an exception to the general rule, inasmuch as it stands on an eminence rising 280 ft. above the river; Paulook, in the vicinity of Czarkoselo, is also considerably elevated.

The palace of Czarkoselo is reckoned the finest summer residence belonging to the Russian monarchs. The town, though small, is handsome and improving. It has a lyceum, with 14 professors, a forest school, and is the capital of a circle. Paulook has also a pretty little town dependent on it. These have been, for some time past, favourite resorts of the citizens of Petersburg; and the presumption is that they will be more resorted to now than ever, a railroad, the first work of the kind constructed in the empire, having been opened between them and the capital.

We have consulted in drawing up this article the Travels of Coxe, the Voyage de Deux François dans le Nord de l'Europe (an accurate and valuable work), Storch's Picture of Petersburg, &c.; but we are principally indebted to the excellent account of the city in Schnitzler's work La Russie, La Pologne, &c., p. 187–301, and the official accounts published by the Russian government.

PETERSBURG, p. t., Rensellaer co., N. Y., 26 E. Albany, 309 W. Watered by Little Hoosic creek. It contains five stores, one grist-mill, five saw-mills, one oil-mill, two tanneries; 12 schools, 627 scholars. Pop. 1801.

PETERSBURG, p. t., port of entry, Dinwiddie co., Va., 22 S. by E. Richmond, 140 W. Situated in 37° 13' 54" N. lat., and 77° 20' W. long. Population in 1830, 8322; in 1840, 11,136. It is on the S. bank of Appomattox river, 12 m. above its entrance into James river, at City Point. The falls of the river here afford extensive water-power, and there is a canal around them for the purpose of navigation. Vessels of 100 tons come to this place. The borough contains besides Petersburg, the villages of Blandford, in Prince George county, and Pocahontas in Chesterfield county. It has a courthouse, jail, a masonic hall, two banks, an insurance office, seven churches, a Presbyterian, Methodist, Baptist, Episcopal, and two African, six commercial and eight commission houses in foreign trade. with a capital of $275,000; 191 retail stores, capital $1,096,250; two lumber yards, capi-

PETERSFIELD.

tal £6,000; one furnace, six forges, one woollen factory, six cotton factories with 10,000 spindles, two flouring mills, one grist mill, two saw-mills, two rope walks, one pottery, two printing-offices, one semi-weekly newspaper; total capital in manufactures, £726,555; eight academies, 386 students, seven schools, 125 scholars. In July 1815, a disastrous fire destroyed 400 buildings, and property to the amount of about £2,000,000. The place has been rebuilt with brick, in an improved form. It exports extensively flour, tobacco, and cotton. Vessels requiring more than seven feet of water, discharge and receive their cargoes at City Point, at the mouth of Appomattox river, at its entrance into James river.

PETERSBURG, p. v., capital of Pike co., Ia., 138 S.W. Indianapolis, 677 W. Situated one mile S. of White river, and three miles below the junction of its E. and W. branches. It is on elevated ground, and contains a courthouse, jail, six stores, and about 400 inhabitants.

PETERSBURG, p. v., capital of Menard co., Ill., 21 N.W. Springfield, 801 W. Situated on the W. side of Sangamon river, and contains a courthouse, jail, six stores, one steam saw-mill and grist-mill, and about 30 dwellings.

PETERSFIELD, a parl. bor., market town and parish of England, co. Southampton, and hund. Finch-dean, near the Loddon, 15 m. N.N.E. Portsmouth. Pop. of parl. borough (which includes with the old bor. the tithing of Sheet, the several pars. of Buriton, Lyn. and Froxfield, three tithings belonging to the parish of East Meon and the parish of Steep, with the exception of two tithings), in 1831, 4321; ditto, of town, 1495. It is a small, clean, country-town, its principal street crossing the Portsmouth and London road, nearly at right angles. The church is a large brick structure, with a low, square tower: the living is a curacy subordinate to the rectory of Buriton. Near the church is an equestrian statue in bronze of William III., erected at the expense of the late W. Jolliffe, Esq., the father of the present lord of the manor; but it has a shabby appearance, and is any thing but an ornament. The Wesleyan Methodists and Baptists have their respective places of worship, with attached Sunday schools. The endowed charity school, called Churcher's college, was founded in 1722, and furnishes clothing and food, with general and mathematical instruction, to 10 or 12 boys. "Petersfield has but little trade, and any consequence which it possesses arises entirely from its position on the high road between London and Portsmouth," an advantage, however, which it will not enjoy after the completion of the Portsmouth branch of the south-western railway. (Parl. Bound. Rep.) The borough was incorporated in the reign of Elizabeth, and is governed by a titular mayor and common council, chosen annually at the court-leet of the lord of the manor. It returned two members to the house of congress, from the reign of Edward I. down to the Reform Act, the right of election being in the freeholders in general within the borough, though, in point of fact, it had been for many years a close borough, belonging to the lord of the manor. The Reform act deprived it of one of its members, and the Boundary act enlarged the electoral limits, so as to include with the old bor. the additional pars. and tithings mentioned above. Registered electors in 1839-40, 343. It is also one of the polling-places at elections for the N. division of Hampshire. Markets on Saturday; and fortnightly cattle fairs on alternate Wednesdays; other fairs March 5, July 10, and Dec. 11, for sheep and horses.

PETERSHAM, p. t., Worcester co., Mass., 64 W. by N. Boston, 469 W. Swift river affords water-power. Incorporated in 1754. It contains three churches, two Congregational and a Baptist, three stores, two grist-mills, seven saw-mills, two tanneries; 13 schools, 388 scholars. Pop. 1775.

PETTIS, co., Mo. Situated near the W. part of the state, and contains 680 sq. m. Watered by La Mine river and its branches. It contained in 1840, 4821 neat cattle, 2331 sheep, 12,068 swine; and produced 6579 bushels of wheat, 162,145 of Indian corn, 38,640 of oats, 6380 of potatoes, 11,475 pounds of tobacco. It had five stores, five grist-mills, three saw-mills, five distilleries. Pop.: whites, 2377; slaves, 552; free coloured, 1; total, 2930. Capital, Georgetown.

PETERWARDEIN, or PETER-VARA, the cap. town of the Sclavonian military frontier, and one of the strongest fortresses in the Austrian empire, on a scarped rock, on the S. bank of the Danube, opposite Neusatz, 46 m. N.W. Belgrade. Pop., with suburbs, about 4800, exclusive of its garrison, which is usually about 3000 strong, though there are barracks adapted for 10,000. Peterwardein has several churches, schools, &c., and communicates with Neusatz by a bridge of boats defended by a strong tête du pont. It is the residence of the general commandant of the Sclavonian military frontier and several subordinate military authorities. It is supposed to occupy the site of the ancient Acumincum, and derives its present name from Peter the Hermit.

PHILADELPHIA.

mit, who marshalled here the soldiers of the first crusade. (Anst. Nat. Encyc.; Berghaus; Slade; Payot, &c.)

PETHERTON, NORTH, a market-town and parish of England, co. Somerset, hund. its own name, 7½ m. N.E. Taunton. Area of parish, 11,080 acres. Pop. in 1841, 3733. The town consists of one long street, in which are many well-built houses and a spacious market-place. The principal buildings are, the parish church, a handsome edifice, a chapel-of-ease, and an endowed school for 20 boys. The inhabitants are chiefly employed in retail trade and agriculture. A large corn-market was formerly held here; but it has long since declined in favour of that held at Bridgewater. Fairs, May 1, and the Monday before November 13.

PETWORTH, a market-town and parish of England, rape of Arundel, co. Sussex, hund. Rotherbridge, on the Arun, 13 m. N.N.E. Chichester, and 42 m. S.W. London. Area of parish, 6148 acres: pop., in 1841, 3364. The streets are very irregular, but there are many well-built and even handsome houses, besides a fine market and sessions-house near the centre of the town. The church, a neat stone edifice recently enlarged, has a square tower: the living is a rectory in the gift of the Earl of Egremont, the lord of the manor. The Wesleyan Methodists and Independents have also their respective places of worship, and there are various Sunday schools and a national school. Here are almshouses, a hospital, and a house of correction on the plan of Howard. A considerable retail trade is carried on for the supply of the neighbouring gentry; but most of the inhabitants are engaged in agriculture. The Epiphany and Easter quarter-sessions for the W. division of Sussex are held here, and petty sessions on alternate Saturdays. Petworth is one of the polling-places at elections for the W. division of the county. Markets on Saturday; fairs, Holy Thursday, July 20, and Oct. 2.

Close to the town is Petworth house, the seat of the Earl of Egremont, which, both for the elegance of its exterior and the sumptuousness of its internal fittings, may vie with the finest mansions of the English nobility: the park, which is enclosed by a wall, 12 m. in circumference, is beautifully laid out, and commands many picturesque as well as extensive views.

PEZENAS, (an. Piscena), a town of France, dép. Hérault, cap. cant., near the junction of the Hérault and Peine, 24 m. W.S.W. Montpelier. Pop., in 1836, 7,496. It is finely situated, and its old castle commands a magnificent prospect. Several of its streets are wide and lined with good houses. It has a handsome parish church, and had formerly many conventual churches; but one of these has been converted into a distillery, another into an hotel, and a third into a theatre! Pezenas is the seat of a tribunal and chamber of commerce, and a comm. coll. It has manufactures of woollens, cotton stuffs and yarn, linens, &c. But it is principally celebrated for the great fair, held here in September, which is attended by dealers from all parts of the S. of France. A great deal of business is then transacted in wool, woollens, cottons, and other fabrics. (Hugo, art. Hérault, &c.)

PHELPS, p. t., Ontario co., N.Y. 15 E. Canandaigua, 195 W. Albany, 350 W. Watered by Canandaigua outlet and Flint creek. Gypsum abounds in the vicinity. It contains 18 stores, 3 flouring mills, 8 flouring mills, 4 grist-mills, 11 saw-mills, 5 distilleries, 1 printing-office, 1 weekly newspaper; 2 academies, 90 students, 30 schools, 1416 scholars. Pop. 5563.

PHILADELPHIA, co., Pa. Situated in the S.E. part of the state, and contains 120 sq. m. Bounded E. by Delaware river. Watered by Schuylkill river, and Pennypack, Pequessing, Frankford, and Darby creeks. These streams, especially the Schuylkill, afford good water-power. It has five colleges, 737 students; 120 academies, 8897 students; 948 schools, 16,212 scholars. Pop. 258,037. For the other statistics, see Philadelphia city, its capital.

PHILADELPHIA, the largest city of Pennsylvania, and, next to New-York, the largest city in the United States, is situated in lat. 39° 57' 9" N., long. 75° 10' 37" W., and 1° 59' 47" W. long. from Washington, as taken at the High School observatory. It is 300 m. S.W. Boston, 86 m. S.W. New-York, 97 m. N.E. Baltimore, 98 m. E. by S. Harrisburg, 138 N.E. Washington. The population in 1800 was 70,287; in 1810, 96,287; in 1820, 119,325; in 1836, 162,325; in 1840, 258,691. Of these, in 1840, 693 were engaged in agriculture, 7912 in commerce, 34,900 in manufactures and trades, 1403 in navigating the ocean; ditto rivers and canals, 595; in the learned professions, 1553.

The city is situated between the Delaware and Schuylkill rivers, extending from the one to the other, 3 m. above their junction. The distance between the two rivers, on Market-street, is about 2 m.; but, as the rivers curve toward each other, the distance between them is something more, above and below this point. It is 120 m. distant from the ocean by the course of the river, and about 55 m. in a

PHILADELPHIA.'

ect line S.E. Its name is derived from that of a city in
in Minor; and the Greek words of which it is compound-
signify "brotherly love," a name which accorded well
h the pacific principles of William Penn, its founder.
ording to the original design of the proprietary, it would
e covered an area of more than 12 sq. m.; but, by its char-
of 1701, it was contracted to the limits of Delaware river
he E., Schuylkill river on the W., Vine-street on the N.,
Cedar-street on the S., the present limits of the city
er. But there are five adjoining districts which proper-
long to Philadelphia, though they have incorporations
municipal authorities entirely distinct from the city
er, and from each other. They are the Northern Lib-
s, Kensington, and Spring Garden on the N., and South-
and Moyamensing on the S. All these are included
e above account of the population. In 1840 the city
r had 93,665 inhabitants. The district of the *Northern*
ties was incorporated in 1803, and the act was amend-
1819. It is governed by 21 commissioners, elected for
ears, one third of them being chosen annually. The
ipal officers are chosen by the commissioners, but no
ssioner can be appointed to any office to which a
isation is attached. *Spring Garden* lies W. of the
rn Liberties, was incorporated in 1813, and is govern-
15 commissioners, who serve for three years, five
s being elected annually. The district of *Southwark*
orporated in 1794, and is governed by 15 commission-
serve for three years, five being elected annually.
sing was incorporated in 1812, and is governed by
missioners, elected for three years, three of them
cted annually.

ty is built on a plain, slightly ascending from each
highest point of which is elevated 64 feet above
er mark in the river. The streets of the city vary
130 feet in width: originally the city proper con-
ine streets running E. and W., from the Delaware
lkill, and 25 running N. and S., from Vine to
ets. Numerous small streets, alleys, and courts,
which run partially across a block, divide the dif-
res. The whole number of streets in the city
ng districts is about 1280. High or Market street,
from river to river, a little N. of the middle of
oper, is 100 feet wide; and Broad-street, which
t right angles, a little W. of the middle, is 130
Front-street, on both rivers, is 60 feet wide;
is 66 feet wide; and most of the other streets
wide. The city was originally laid out with
egularity, all the streets crossing each other at
which is still the case, with the exception of
which is crooked, occupying the site of a for-
le creek, that is now obliterated, having been
The adjoining districts, though tolerably regu-
the same regularity in their arrangement as
r; nor does the whole connect in one consis-
orm plan. While the city proper had very
of surface, above this on the Delaware, and
low on the Schuylkill, there are many situa-
manding prospects, and much varied scenery.
nays in the streets are paved with rounded
d in gravel, forming a dry and durable, but
h, surface. Paving with wooden blocks has
n experiment; but, though pleasing to the
ble to the traveller when first laid, they
been found here, as in New-York, to decay
o lead to the abandonment of the plan.
from Fourth to Sixth streets, has been re-
ith square blocks of stone, which, though
it, possess smoothness and durability, and
iperior to any other mode yet tried. The
most universally paved with bricks, but is
with smooth, well-dressed flagging stones.
ed by common sewers, which are arched
work, constructed under most of the main
sely-built parts of the city and districts
f about 8½ m., and its length on the Dela-

re chiefly of brick, and the style of archi-
d neat, rather than showy and ornamen-
e, which abounds in this vicinity, is gen-
or-steps, window-sills, &c., and the base-
is faced with this material. A number
es present an entire exterior of marble,
of a fine quality, from the quarries in
es of Montgomery and Chester, which
uch to the embellishment of the city.
provided with many public squares,
recnst of its founder. In the original
in the centre was designed to contain
each quarter of the city to contain
space between Front street and Dela-
igned to be left open. Although these
ly carried out, yet there are many pub-

lic squares, which are finely ornamented, though none of
them are very large. Penn-square, at the intersection of
Market and Broad streets, near the centre of the city, was
designed by the proprietor to be surrounded by the public
buildings; but in early times it was too far removed from
the centre of population, and hence Independence-square
was procured. Penn-square, however, before the removal
of the water-works to Fairmount, was the most ornamental
spot in the city. It was laid out in the form of a circle, had
a neat marble temple in the centre, containing the steam
engine for raising the water of the Schuylkill, had an elegant fountain with a beautiful piece of statuary, and was
handsomely laid out with walks and shaded by trees. But
its ornaments have all been removed, and it is now divided
into four parts by running Market and Broad streets through
it, which has added to the convenience of the city, has
destroyed the beauty of this public ground. Independence-
square is between Chestnut and Walnut streets, and Fifth
and Sixth streets. It is surrounded by a solid brick wall,
3 or 4 feet high above the adjacent streets, surmounted by
an iron railing. It is laid out with walks and grass-plots,
and shaded by majestic trees. It is in the rear of the old
state-house or Independence hall, a venerable building,
which deserves to be denominated the cradle of American
Independence, as Faneuil hall in Boston is called the cradle
of American Liberty. In this building the "declaration of
independence" was adopted, and from its steps it was read
to the multitude assembled in this square to hear it; and it
is still a place where public meetings are held when the
audience is too large to be accommodated in the hall. It
is an inviting spot on its own account, as well as from its
associations. The northern part of this ground was purchased, and the state-house erected, in 1735; the southern
part, fronting on Walnut-street, was purchased in 1760.
Washington-square, between Walnut, Locust, Sixth, and
Washington streets, is a little S.W. of Independence-square,
and is one of the most handsome and frequented public
grounds in the city. It is finely ornamented with walks,
trees, and shrubbery, and surrounded by a handsome iron
railing. For many years it constituted the Potter's fields,
a place of burial for the poor and for strangers. When the
yellow fever prevailed in the city, many interments took
place here. About the year 1795 it was closed by the city
authorities against future interments, and about 1815 it was
regulated as a public square, laid out with walks, planted
with trees, and made to put on its present attractive appearance. It is in contemplation to erect a monument to
Washington in the centre of this square, of which the corner stone was laid in 1833. Franklin-square, between Race,
Vine, Sixth, and Franklin streets, in 632 feet long by 550 feet
broad, handsomely embellished, and has a splendid fountain in its centre, surrounded by a marble basin. A part
of it was formerly used as a cemetery by the German society. It is now an attractive public promenade. Logan-
square, between Race and Vine streets, and Rittenhouse-
square, between Walnut and Locust streets, both on
Schuylkill Fifth-street, are enclosed and planted with trees,
and promise to become attractive public grounds, and a
resort of the inhabitants of the western part of the city.
Among the public buildings, the old state-house, now
called Independence hall, deserves, from its historical associations and venerable appearance, the first place in the
enumeration. It fronts on Chestnut-street, having, as already mentioned, Independence-square in its rear. It was
begun in 1729, and finished in 1735. The wings, extending
from the main building to Fifth and Sixth streets, are of
modern construction. The wood work of the steeple of the
main building was so much decayed in 1774, that it was
taken down, leaving only a small belfry to cover the bell
for the use of the town clock. The bell for the first steeple
was imported from England in 1752, but was broken by accident when first hung up. A new one was cast in Philadelphia, under the direction of Isaac Norris, then speaker
of the colonial assembly, who caused to be inscribed on it
a passage from Leviticus xxv. 10, as if prophetic of its future
use, "*Proclaim* LIBERTY *throughout this land, unto all the
inhabitants thereof.*" Accordingly, its joyous tones rang
loud and clear to proclaim to anxious thousands the declaration of American Independence. The chamber in which
the declaration was signed is on the first floor, at the eastern end of the centre or original building. Some years
since the wood work of the interior was removed, to give
place to more modern decorations. But, with great good
taste, the people demanded its restoration; and it now presents the same appearance that it did when it contained
that venerable body whom William Pitt, the friend of America pronounced, in the British parliament, to be the most
distinguished for wisdom of any body of men of whom he
had read, in ancient or modern times. In 1828 the present
steeple was erected, to correspond, as nearly as possible, to
the original structure. A statue of Washington in Independence hall, carved in wood, by Rush, is said to be an excel-

lent likeness, and several fine paintings adorn this venerable room. The west room is occupied by the mayor's court, and the second story by the district and circuit courts of the United States. Until time shall destroy them, it is hoped that it will be regarded as sacrilege to alter materially this building or Faneuil hall at Boston. They are noble monuments of revolutionary times and characters. In the steeple is an excellent clock, with a ground glass dial-plate, capable of being illuminated by night, and a heavy bell to give alarm in cases of fire. A watchman, stationed in the steeple, indicates, by the number of strokes, the district in which the fire takes place. The old bell, an interesting relic, is carefully preserved in the cupola. On each side of the state-house are two wings, occupied as offices for the law courts, and depositories for the records and documents in relation to the county of Philadelphia. The United States bank is a chaste specimen of the Grecian Doric architecture, in imitation of the Parthenon at Athens, with the omission of the colonnades on the sides. It is 87 feet in front by 161 feet deep, with marble steps ascending to the portico, which consists of eight columns, 4 feet 6 inches in diameter, supporting a plain entablature and pediment. The large banking room in the centre of the building is 81 feet long and 48 feet wide, richly ornamented with fluted Ionic columns and sculptured embellishments. The whole interior arrangements are admirably adapted to its purpose: it was commenced in 1819, and completed in 1824, at a cost of about $500,000. The bank of Pennsylvania is a beautiful edifice, of white marble, and is a fine specimen of Grecian architecture. It has a portico on each front, of six Ionic columns, supporting entablatures and pediments. It is 121 feet long and 21 feet wide; and has an enclosure neatly ornamented with plants and shrubbery, surrounded by an iron railing. The Girard bank was purchased of the old bank of the United States, by Stephen Girard, Esq., and used by him as a banking house. It has a marble front, with the sides and rear of the building of brick, and is enriched by a portico of six Corinthian columns. Several other banking houses are spacious and handsome buildings, but of a plainer style of architecture. The Merchants' exchange is a splendid building, erected on the triangular space between Third, Walnut, and Dock streets, in the centre of commercial and financial transactions. The eastern front, on Dock-street, presents a semicircular piazza, supported by six Corinthian columns, on a basement about 12 feet high, presenting a beautiful appearance, and producing an imposing effect. In the basement are insurance and brokers' offices. The first floor is divided into several apartments; that on the eastern side is devoted to the use of the subscribers, who assemble during the business hours of the day. The rotunda has four appropriate columns, and adjoining it is a spacious reading room, well furnished with the current literature of the day. Some of the churches deserve attention on account of their architecture. St. Stephen's Episcopal church is a fine specimen of Gothic architecture, on Tenth-street, between Market and Chestnut streets. It is 102 feet long and 50 feet wide, and on its front corners has two octagonal towers, 86 feet high. The upper parts of the windows are embellished with cherubims, in white glass on a blue ground; and the sashes are filled with diamond-shaped glass of various colours, ornamented in the same manner. St. John's Roman Catholic church is an elegant Gothic structure, situated on Thirteenth-street, between Chestnut and Market streets, with square towers on each of its front corners. The windows are composed of stained glass, and the interior is decorated with several appropriate pictures. The first Presbyterian church, fronting on the S. side of Washington-square, is the most elegant Presbyterian church in the city. It is after the model of a temple on the Ilissus, near Athens, and is of the Ionic order of Grecian architecture, having a portico of six Ionic columns in front. The walls are of brick, covered with mortar, and painted to resemble marble. It is surmounted by a neat cupola. The fifth Presbyterian church, on Arch-street, above Tenth-street, deserves to be mentioned, on account of the beauty of its architecture. Among the buildings of Philadelphia distinguished for their architecture, the Girard college deserves a particular description. It was founded by the bequest of the late Stephen Girard, of $2,000,000, and a farther sum, if necessary, for the erection of the buildings and the education of *orphan children*, for whose exclusive use it was designed. This immense sum would have been sufficient to found and enclose a most splendid college. Over $1,000,000 have already been expended upon the buildings alone, and a farther sum will be required to complete them; and it is questionable whether these splendid preparations are necessary to promote the object in view. Less expensive buildings would be better calculated to teach the orphans economy, to say nothing more. The Girard college, being on high ground, occupies a commanding position, about a mile N.W. of the limits of the city proper. The lot on which it stands contains about 45 acres of ground,

bequeathed for the purpose by the founder of the institution. The college buildings consist of a centre building, to be exclusively devoted to the purposes of education, and four other buildings, two on each side, for the residence and accommodation of the professors, teachers, and students. The centre building is one of the most splendid edifices in the United States. It is 218 feet long from N. to S., and 160 feet wide from E. to W. It is surrounded by 34 columns of the Corinthian order, with beautiful capitals, supporting an entablature. Each column, including its capital and base, is 55 feet high and 6 feet in diameter, having a base 3 feet high and 9 feet in diameter, and leaving a space of 15 feet between the columns and the body of the building. At each end is a door or entrance, 16 feet wide and 32 feet high, decorated with massive architraves, surmounted by a sculptured cornice. Each of these doors opens into a vestibule 35 feet wide and 48 feet long, the ceiling of which is supported by eight marble columns. In the second story, over these vestibules, are lobbies of similar dimensions, having their ceilings supported by Corinthian columns. In each corner of the building are marble stair-ways, lighted from the roof. On each floor are four rooms, each 50 feet square, and the third is lighted by a sky-light, which does not rise above the roof. No wood is used in the construction of the building, excepting for the doors, so that it is entirely incombustible. The building is warmed by furnaces built in the cellar, with flues in the interior walls. The remaining four buildings, two on each side of the main building, are each 52 feet wide by 125 feet long, and two stories high above the basement. The most eastern of these is divided so as to constitute four distinct houses for the professors. The other buildings are devoted to the residence and accommodation of the pupils. The benevolent and charitable institutions are very numerous. One of the oldest and most respectable of these is the Pennsylvania hospital. It was founded chiefly by the exertions of Dr. Franklin and Dr. Bond, in 1738. It has received, besides, individual grants, and donations from the legislature, until its endowment is ample. The building, grounds, and garden occupy the entire square between Eighth, Ninth, Spruce, and Pine streets. The principal front is on Pine-street, and consists of a centre building and two wings. The E. wing was erected in 1755, the W. wing in 1796, and the centre building in 1804. Between the east and the building is a handsome area planted with lofty trees, and laid out in grass-plots, in the centre of which is a colossal statue of William Penn, bronzed. In the rear of the lot, fronting on Spruce-street, is a small neat edifice, erected for the purpose of containing the picture, by West, of Christ healing the sick, presented to the institution by its author, the exhibition of which adds something to the funds of the institution. This spacious building affords accommodation to indigent patients as well as to those who pay. It contains a fine anatomical museum, and a valuable library of books, and is eminently useful to the numerous medical students in the city. From its roof a fine view of the city and its environs is obtained. The insane asylum is a branch of the Pennsylvania hospital, about 2 m. W. of the Schuylkill, and is surrounded by 111 acres of ground, 41 of which are enclosed by a substantial wall, and laid out in pleasure-grounds for the patients. The building has a front of 236 feet, is two stories high above the basement, and contains 904 rooms for the patients and their attendants. The United States Marine hospital or Naval asylum has an extensive and elegant building on the E. bank of the Schuylkill, below Cedar-street. It is intended for invalid seamen, and officers disabled in the service. It has a front of 386 feet, is built of white marble, having a centre building 146 feet in front and 175 feet deep, and two extensive wings. The centre building has a fine Ionic portico of eight columns, projecting in front and rear beyond the wing, which have balconies extending their whole length, and resting on iron pillars. The basement contains a large refectory with cooking apparatus. The whole building can lodge about 400 persons, and is partly occupied by a naval school. The almshouse is on the W. side of Schuylkill river, opposite to Cedar-street, and extends its benefits to the poor of the city and the adjoining districts of Southwark, Northern Liberties, Kensington, Spring Garden, and Penn township; the other districts of the county provide for their own poor. The edifice is an immense structure, consisting of four main buildings in the form of a parallelogram, covering and enclosing an area of 10 acres of ground. The front, on Schuylkill river, presents a handsome appearance: the main building has a portico of eight massive pillars 30 feet in height, and, the site being elevated above the sloping bank of the river, presents a fine view of the city and the adjacent country. It contains workshops for such of the inmates as are capable of employment, furnished with a steam engine for propelling machinery. The average number of paupers maintained in this establishment for the year ending May 15th, 1842, was 1546, of whom

730 were men, 630 women, and 157 children. The number is augmented on the approach of winter, and diminished on the return of spring. The tax assessed for its support in 1842, amounted to $184,694. The Pennsylvania institution for the deaf and dumb is at the corner of Broad and Pine streets, having extensive buildings adapted to the purposes of the establishment. The system of instruction is similar to that of Abbe de L'Epee and Abbe Sicard of Paris. In addition to literary and moral instruction, the pupils are taught some mechanical trade, by which they may be able to provide for themselves in after-life. Most of them are supported by funds derived from the state; some of them by the states of Maryland and New-Jersey; and some by the friends of the institution. The Pennsylvania institution for the instruction of the blind is situated in Race-street, near Schuylkill Third-street. The main edifice is built of brick, stuccoed to imitate marble, and occupies a lot 247 feet on Race-street and 220 feet on Third-street. In front the ground is decorated with flower gardens, and in the rear are play grounds. A commodious building is erected for workshops, and the pursuit of such trades as are carried on by the pupils. They are taught not only reading, writing, arithmetic, geography, and music, but also to manufacture a great variety of useful and ornamental articles. Among the other charitable institutions may be mentioned the orphans' asylum, the asylum for indigent widows and aged women, Wills' hospital for the lame and blind, Preston-street, the Magdalen asylum, Foster home, a shelter for coloured orphans, institute for coloured youth, St. John's and St. Joseph's orphan asylum, Christ church hospital, Friends' almshouse, Friends' lunatic asylum near Frankford, Philadelphia northern and southern dispensaries, for supply of medicine to the poor; house of industry, union relief association, fuel savings' society, seaman's friend society, humane society, society for alleviating the miseries of public prisons, and many others. There are societies for the relief of distressed emigrants from foreign countries, and a number of moral relief societies, and other associations of a benevolent purpose. Indeed Philadelphia is distinguished among its sister cities for a provision for the relief and varied forms of human suffering. Temperance societies are numerous, exercising a happy influence on the city, and contributing to dry up this great source of suffering and crime.

Of the literary institutions, the university of Pennsylvania is one of the oldest and most considerable. It was first a charity school in 1751, chartered and endowed, erected into a college in 1755, and into a university. It comprises three departments, the academical, collegiate, and the medical. The usual course in the department occupies a term of four years, in instruction is given in the usual branches taught in academies and universities in the United States, and the degrees are conferred. The medical school connected with the university is the oldest and most flourishing in the country, usually from 480 to 560 students, and from its foundation it had 3390 graduates. It has an extensive anatomical museum and cabinet of natural history, and an chemical and philosophical apparatus. Three professors, two male and one female, are supported by the university. The university buildings are situated, between Market and Chestnut streets. They have a handsome edifice 85 feet front by 112 feet deep, surmounted by an open area, separated from the street by an iron railing. Jefferson medical college was founded in a flourishing institution, having usually about It was formerly connected with the college buildings, but is now an independent institution. Its edifice, between Walnut and Chestnut streets, has accommodations for lectures and anatomical, with a museum of preparations in the science connected with medical studies. Pennsylvania college is connected with Pennsylvania university, was founded in 1839, and empowered to confer degrees. It is located in Filbert-street above and is a young and flourishing institution. An interesting portion of the literary institutions is its public schools. By a law passed in the county of Philadelphia were erected into the first School district of the state, denominated the First School district of The schools have been divided into eleven districts, which are in the city proper, and the outer districts. At the head of them stands the High model school, and the arrangement of the model school, and to constitute one of the most perfect systems of education to be found in any country. More than two-thirds or three fifths of the whole population of 5 and 15, attend the public schools, and for admittances are more than can be remembered of schools in the district is 214, or the High school, 40 are grammar schools, 18 primary schools, and 80 in the outer sections

not classified. The number of teachers, including the professors of the High school, is 499, of whom 87 are males, and 412 females; and the number of pupils in all the schools was, in 1843, 33,130, which is 5392 more than in 1842. The whole expenses of the board of the district, for all purposes excepting the erection and fitting up of school-houses, including tuition, fuel, books, stationery, &c., was $192,511 per annum, making an average to each scholar, including the pupils of the High school, the sum of $5 61 cents. As some of these schools are without the city and incorporated districts, the following, from the report of 1843, will show the proportion. In the city and incorporated districts there were 62 primary schools, with 9342 pupils; 11 secondary schools, with 2597 pupils; 19 grammar schools, with 8445 pupils; and one high school, with 307 pupils. In the outer sections of the district there were 92 schools, with 6347 pupils. The whole were accommodated in 96 substantial schoolhouses, generally three stories high, and capable of accommodating from 600 to 1000 scholars each, worth, with the lots on which they stand, $540,000, and the value of school furniture, libraries, &c., was $40,800. The High school, from the number and ability of its professors, and the extent and duration of its course of studies, deserves to be denominated the people's college. It has a principal and 10 professors, a professor of belles-lettres, of ancient languages, of modern languages, of theoretical mathematics, of practical mathematics, of natural history, of natural philosophy, of chemistry, of mental, moral, and political science, and of graphics. Attached to the High school, is an astronomical observatory, furnished with fine instruments; and for practical astronomy, it ranks high among the literary institutions of the country. The first or English course in this institution occupies two years; the second or principal course four years; and the classical course four years.

Among the literary and scientific institutions is the American philosophical society. It was founded in 1743, principally through the exertions of Dr. Franklin, by the union of two previous societies. It has had among its members many distinguished men in the United States and in Europe. Its hall is in South Fifth-street, below Chestnut-street. It has a library of 15,800 volumes of rare books, and a collection of minerals, fossils, and ancient relics. It has published several volumes of its transactions. The Franklin institute was incorporated in 1824 for the promotion of manufactures and the mechanic and useful arts, by popular lectures, and the formation of a cabinet of models, minerals, and a library. It has over 2000 members, a library of about 3000 volumes, and an extensive reading room. The Academy of Natural Sciences occupies a spacious building in Broad-street, between Chestnut and Walnut streets, has a library of 9000 volumes of valuable books, and a cabinet of natural science. The Pennsylvania Academy of Fine Arts was founded in 1805, and in Chestnut-street above Tenth-street, has an extensive collection of paintings, statues, and other works of art, always open to students, and exhibited to visitors. The Artists' Fund society has its hall in Chestnut-street, in front of the Academy of Fine Arts, and exhibits its productions for the benefit of decayed artists. The Philadelphia library was founded in 1731 by the influence of Dr. Franklin, containing upwards of 30,000 volumes of books on almost every branch of knowledge. To this has been added the Loganian library of 11,000 volumes, of rare and valuable books, mostly classical. The Atheneum was incorporated in 1815, and contains the periodical journals of the day, and a library of several thousand volumes. The Mercantile library was established in 1822, and has a collection of about 6000 volumes, chiefly relating to commerce, and its kindred subjects. It sustains popular lectures on commerce, commercial law, the arts and sciences, and general literature. The Apprentices' library was instituted in 1819 by voluntary contribution from the citizens, and contains about 14,000 volumes, read by about 1380 young persons of both sexes. The literary, scientific, and benevolent institutions of Philadelphia are worthy of the city of William Penn and Benjamin Franklin.

The United States mint was founded in 1790, and the business of coining commenced in 1793, in the building occupied at present by the Apprentices' library. It was removed in 1830 to the fine building it now occupies in Chestnut-street above Olive-street. The edifice is of white marble, and the north front opposite to Penn-square is 123 feet long, with a portico 60 feet long, of six Ionic columns, and the south front on Chestnut-street has a similar portico. The whole amount of coinage since its establishment in 1793, to December, 1842, has been 235,067,171 pieces, to the value of $65,573,059¼. In 1842 there were coined 7,463,169 pieces, to the amount of $9,496,351½. The amount of gold deposited for coinage at the mint at Philadelphia, since its institution, with the exception of $27,117, was from the mines in the states of Virginia, North Carolina, South Carolina, Georgia, Tennessee, and Alabama, and amounted to

$6,135,531. The first deposit from these states was made in 1824.

The eastern penitentiary in the N.W. part of the city is at the corner of Coates-street and College-avenue, S. of Girard college, and its towers, battlements, and massy walls, resembles a baronial castle of the middle ages. It occupies a square area of 10 acres, which is surrounded by a wall, 30 feet high, and surrounded at the angles, and on each side of the front entrance, by watch towers, which command a view of the inside and outside of the external walls. It is constructed on the principle of strictly solitary confinement in separate cells, and is admirably calculated for the security and the health, and, so far as consistent, with the comfort of its occupants. In the centre of the whole enclosure is an observatory or watch house, from which long passages radiate in different directions, on the sides of which the cells are ranged so that the watchman at the observatory commands a view of all the passages and the entrance to all the cells. These cells are 12 feet long, 8 feet wide, and 10 feet high, made perfectly secure, and lighted by a convex glass at top. Food is conveyed to the prisoner by means of a cast iron drawer, which forms a table on the inside when the prisoner takes his meals. The cells are well cleansed and ventilated, and to each of them is attached a yard, 18 feet by 8, surrounded by walls 12 feet high, which are overlooked from the observatory at the centre of the prison. As a reward for good behaviour, the Bible, and some other religious and moral books are given to the prisoners, and some suitable employment is granted; but they are withheld as a punishment. A suitable person administers moral and religious instruction to the convicts, standing near the centre of the passage on which the cells are ranged, where he can be heard, but not seen from the cells. This institution is well calculated to make a trial of the principle of solitary confinement.

The county prison on Passyunk road, below Federal-street, is appropriated to the confinement of persons awaiting trial, or those which are sentenced for short periods. The building is spacious, and presents an imposing front of Gothic architecture, and is admirably adapted to solitary confinement, for which it is intended. The debtors prison, adjoining the county prison on the N., is constructed of red sand stone, in a style of massive Egyptian architecture.

The house of refuge is at the corner of Coates-street and the Ridge road, not far from the penitentiary, and is appropriated to the confinement of vicious and abandoned juvenile offenders, of both sexes; where, in addition to their moral culture, they are taught the various elementary branches of an English education, and various mechanical employments. At the expiration of their terms of confinement they are apprenticed, the boys to respectable mechanics or farmers, and the girls to families to perform the work of domestics.

There are in the city 144 churches of different denominations, of which the Presbyterians have 23, the Protestant Episcopal 22, Methodist Episcopal 19, Methodist Protestant 5, Reformed Presbyterian 3, Associate Reformed 2, Baptist 14, Lutheran 6, German Reformed 3, Dutch Reformed 2, Friends, Orthodox, 4, Hicksite 3, Universalist 3, Independent, Moravian, Unitarian, Swedenborgelan, Menonist, Disciples of Christ, Bible Christians, Protestant (Mariners' chapel), and Mormons, 1 each, Roman Catholic 11, Jews 3; African, Presbyterian 3, Methodist Episcopal 4, Protestant Episcopal 1, Baptist 4. Attached to these churches are 163 Sunday schools, with 28,390 scholars.

The city is supplied with water derived from the Schuylkill river. Before the present water-works were constructed at Fairmount, the water was raised by two steam engines, one of which was situated on Penn-square, where the water was raised into a reservoir 36 feet high, and distributed in wooden pipes over the city. The supply was expensive and inadequate, and after having expended $657,396, these works were abandoned in 1815, and the Fairmount water-works were put in operation. But in 1818, after having expended upon them $390,699, the plan of operation was changed. A dam is constructed across the Schuylkill, and by means of wheels moved by the water, which work forcing pumps, by which the water of the river is raised into a reservoir, 102 feet above its level, and 56 feet above the highest ground in the city. There are four reservoirs to let the water settle, the whole containing 22,000,000 gallons. The dam is 1600 feet long, and the race upwards of 400 feet long and 90 feet wide, cut in solid rock. The mill house is of stone, 238 feet long and 56 feet wide, and capable of containing eight wheels, and each pump will raise about a million and a quarter of gallons in 24 hours. The reservoirs contain an area of over 6 acres, and they are 12 feet deep, lined with stone and paved with bricks, laid on a bed of clay, in strong lime cement, and made water-tight. The water is conveyed from them into the city by iron mains, one 20 and the other 22 inches in diameter, and distributed through the streets in pipes from 6 to 10 inches in diameter,

and from these smaller ones lead into the yards and dwellings. The daily consumption of the water in the city and districts in the year 1843 was 4,422,400 gallons, supplied to 26,549 tenants. There are laid in the city and districts 171 m. of iron pipes; and there are 1070 fire-plugs. The whole cost of these works, including those which have been abandoned, is $1,584,594; and they are inferior only to the Croton water-works of New-York, while their expense has been greatly less. Three thousand families in Philadelphia are supplied by the public pumps, which receive their water from cisterns, filled by the water-works of the Schuylkill. These constitute a part of the above named tenants.

The principal streets, most of the stores, hotels, churches, public institutions, and many of the private dwellings, are lighted with gas, supplied by the city gas-works, situated on Schuylkill river, between Market and Filbert streets, which have a retort house capable of containing 120 retorts with their connected apparatus. On the eastern part of the lot are eight gazometers of 50 feet diameter each, the whole capable of containing 280,000 cubic feet of gas. The total length of the street mains, Jan. 18th, 1844, was 45½ m., and the length of the service pipe connected, was 10 m. The number of customers was 4429, the number of private lights 30,658. There were 837 public lamps in the streets, and 62 in the public squares, making the whole number of lights supplied, 31,557. There were consumed 154,009 bushels of coal, 2500 barrels of rosin for distillation, and 54,602 bushels of lime for purifying the gas. The coal used produces, as a residuum of coke, more than sufficient for heating the retorts. The average consumption of gas in the city is 135,000 cubic feet per day. There is a gas company in the Northern Liberties for supplying that district and Kensington, with a capital of $150,000, with a line of mains 33,725 feet in length, supplying 633 consumers, and 146 public lamps. There were consumed in 1843, 42,776 cubic feet of gas, and the works could supply much more.

There are in the city 14 banks, with an aggregate capital of nearly $12,000,000, besides a number of saving banks and institutions, and over 20 insurance companies with a total capital of about $6,000,000.

Philadelphia is well protected against fire, by about 60 engine and hose companies, greatly aided by the Schuylkill water, and superior engines. The expense of the fire apparatus is borne by the members of the companies, aided by appropriations from the treasuries of the city and districts.

There are several theatres in the city, the principal of which are the Chestnut-street theatre, the Arch-street theatre, and the Walnut-street theatre. The Philadelphia Museum, founded by Charles Wilson Peale in 1784, and long one of the most distinguished institutions of the kind in the United States, is alike a place of entertainment and instruction. It was removed in 1839 to convenient rooms at the corner of Ninth and George streets. It occupies the upper story of an edifice 238 feet long, 70 feet wide, and two stories high. The lower story, formerly occupied by the Chinese Museum, is now used for temporary exhibitions and public meetings. Philadelphia is celebrated for its excellent markets, supplied not only from the parts of states adjacent, but extensively from the state of New-Jersey opposite. The principal market is in High or Market-street, extending from Delaware river to Eighth-street, and the end of which near the river is a fish market. There are five other markets in different parts of the place, as in Market-street from Schuylkill Seventh to Schuylkill Eighth streets; Callowhill-street, from Fourth to Seventh-street; in Spring Garden, from Marshall to Ninth-street; in North Second-street, from Coates-street to Poplar-lane; in South Second-street, from Pine to South-street; in Moyamensing, Passyunk, &c. The regular market days in High-street market are Wednesday and Saturday, and those in Second-street market on Tuesday and Friday; but the markets are well supplied with provisions on other days of the week, and they are strictly regulated. The United States navy yard is situated at the southern end of Swanson-street, and contains an area of about 12 acres, which was purchased in 1801, for $37,500, but is now considered worth over $70,000. The area is enclosed on three sides by a high and substantial brick wall, and on the fourth side by Delaware river. It contains every preparation for building vessels of war, and has means of barracks sufficient for accommodating 150 men, with handsome quarters for the officers. The shears for fixing the masts are 120 feet high, said to be the most complete in the United States. The Pennsylvania, one of the largest and finest vessels in the world, pierced for 140 guns, was built at this yard. Her masts are about 150 feet high, her main yard is 120 feet long, and her largest anchor weighs 14,609 lbs. Other large vessels have been built at this yard. Philadelphia has made provision for an ornamental cemetery, located on Laurel Hill, on the E. bank of the Schuylkill, about three and a half miles from the city, on the Ridge road, containing over 20 acres, with every variety of surface and prospect. The entrance contains a fine example of the

PHILADELPHIA.

Doric architecture, 216 feet long, on the Ridge road. It is handsomely laid out, and nature and art have combined to render it a highly inviting spot. Such ornamental cemeteries have become common in our principal cities, and they alike bear witness to the respect for the dead, and the kind feelings of the living. It thus becomes peculiarly pleasant to the living to visit the tombs of those whom they have loved in life.

Philadelphia has an extensive foreign, but a still greater domestic trade. The registered tonnage entered from foreign ports in 1843, was 63,369; and the registered tonnage cleared for foreign ports for the same period, was 58,894. The value of the imports was $4,916,535; and of the exports was $3,643,390. The United States custom-house is situated in Second, below Dock-street. The front of the basement is of marble, and the remainder of brick. It stands back from the street, and has an iron gate in the brick wall before it. A niche in front contains a statue, emblematical of commerce.

Delaware bay, though a fine body of water, having a path in its principal channel sufficient for the largest vessels, is still much obstructed with sand bars and shoals, and unless a skilful pilot to navigate it with safety. Hence, in case of war, it furnishes much to deter an invading force, therefore Philadelphia, in the last war, felt much more apprehensive of an attack by the way of the Chesapeake than the Delaware bay. The principal harbour of Philadelphia is on Delaware river, where large vessels come up, foreign commerce centres; but the Schuylkill affords a convenient harbour for small vessels, and has become especially important since the extensive opening of the coal trade. There are many wharves on the Schuylkill below the bridge at Market-street, at which a large number of vessels are generally found. A permanent bridge was completed across the Schuylkill, at the end of Market-street, in which cost $300,000. It was originally a toll, but is now a free bridge. A light and beautiful wire bridge has been erected at Fairmount, which is also a free bridge. There is a railroad bridge at Gray's ferry, below the city, which also accommodates vehicles and foot passengers; another one of the Columbia railroad, with similar connection, above the city. By means of railroads and Philadelphia possesses facilities for communication great extent of country, and shares with New-York a lion in the trade of the great West. Many canals, turnpikes, and others, greatly facilitate and extend trade. The most important road is the Philadelphia and Columbia railroad, 82 m. long, to Columbia, on the Susquehanna, which connects with the eastern division of the Pennsylvania canal; and forms the principal route to the West, which cost $3,983,302.

Philadelphia is distinguished for its foreign and domestic trade, but more for its manufactures. According to the census of the United States of 1840, there were 186 commercial and 63 commission houses engaged in foreign trade, with a capital of $2,119,501; 2078 retail dry goods and other stores, with a capital of $17,385,993; 60 lumber-yards, with capital of $1,252,000; three furnaces and four forges with 25 persons, and a capital of $314,050; 10 persons required granite and marble to the amount of $4600, with capital of $1300; 1084 persons manufactured machinery to the amount of $1,088,864; 345 persons produced hardware and cutlery to the amount of $217,445; 13 persons manufactured 500 small-arms; 215 persons manufactured precious metals to the amount of $2,651,510; 156 persons manufactured various metals to the amount of 15 persons produced granite and marble to the amount of $266,733; 588 persons produced bricks and lime to the amount of $278,804; 16 fulling-mills and 39 woollen manufactories employed 1156 persons, producing to the amount with a capital of $576,600; 45 cotton factories, spindles, and 392 dyeing and printing works, employed persons, producing to the amount of $3,157,119, capital of $1,923,600; 32 men and 15 women or thereabouts produced 1912 lbs. of reeled silk, valued at $12,560, capital of $76,600; 100 persons manufactured flax to the amount of $59,100, with a capital of $51,000; mixed textiles employed 2880 persons, producing to the amount of 57,820, with a capital of $1,359,488; 390 persons manufactured tobacco to the amount of $312,600, with 584,900; 584 persons produced hats and caps of $313,620, and straw bonnets to the amount with a capital of $153,700; 10 tanneries, employing persons, produced 48,300 sides of sole leather, and other leather, with a capital of $198,000; 271 other persons of leather, as saddleries, &c., manufactured to $1,504,034, with a capital of $475,216; 152 persons 1,181,845 pounds of soap, 1,367,200 pounds of candles and 5000 pounds of wax or spermaceti candles, with a capital of $254,300; 11 distilleries produced distilled spirits, and 19 breweries 11,345,920, the whole employing 181 persons, and a

capital of $418,600; 368 persons manufactured paints and drugs to the amount of $1,823,600, and turpentine and varnish to the amount of $3900, with a capital of $1,865,630; one glass house, and one glass-cutting works, employed 106 persons, and produced to the amount of $70,000, with a capital of $23,500; seven potteries employed 39 persons, produced to the amount of $45,300, with a capital of $24,600; 12 sugar refineries produced $890,000, chocolate factories $6000, and confectionaries $156,300, the whole employing 134 persons, and a capital of $228,700; 10 paper-mills produced $97,250, and other manufactories of paper as a material, $59,500, employing 189 persons, and a capital of $50,750; 20 rope-walks employed 151 persons, producing to the amount of $120,800, with a capital of $87,050; 97 persons manufactured musical instruments to the amount of $27,000, with a capital of $32,500; 491 persons manufactured carriages and wagons to the amount of $291,290, with a capital of $127,670; 17 flouring-mills produced 113,810 barrels of flour, and 13 grist-mills, 13 saw-mills, and one oil-mill, employed 103 persons, and produced to the amount of $454,068, with a capital of $249,575;. vessels were built to the amount of $467,480; 736 persons manufactured furniture to the amount of $547,960, with a capital of $396,700; 47 printing-offices, 13 binderies, eight daily, 17 weekly, and a fair view of what substantially belong to Philadelphia and its immediate vicinity. A small portion belongs to the towns in the vicinity, Blockley, Bristol, Germantown, Kingsessing, Lower Dublin, Moreland, Oxford, Northern Liberties (unincorporated), Roxborough, and Penn. township. They contain a population of 37,814 out of 258,037, and a much smaller proportion of stores and manufactories.

The government of the city proper is in the hands of a mayor, a select council of 12, and a common council of 90 members. One third of the select, and the whole of the common council, are chosen annually by the people, and the councils elect a mayor. The aldermen, who are 15 in number, are appointed by the governor to act with the mayor and recorder, as judges, during good behaviour; and the aldermen act as justices of the peace. The whole legislative power is in the councils, of which the select council is a kind of senate. The incorporated districts are governed by commissioners, as already noticed.

Philadelphia was first laid out at the close of 1682, by Thomas Holme, the first surveyor general of the province. The ground selected was claimed by three Swedes by the name of Swenson, to the amount of about 800 acres, granted by the Dutch governor of New-York, in 1664. This claim was extinguished by Penn on his arrival, by giving them in exchange a tract on the Schuylkill above Fairmount. Shortly after his arrival, Penn held a treaty with the Indians under a large elm tree in Shackamaxon (now Kensington), on the bank of the Delaware, at which certain grants of land were confirmed, and the most amicable relations were settled between the colonists and the Indians. This venerable tree was regarded and preserved with great reverence, as a memorial of this interesting transaction, until it was blown down in 1810. A marble monument, with an appropriate inscription has been erected on the spot. By the close of 1682, 23 ships had arrived with passengers, and numbers followed in the succeeding year, so that when Penn departed for England in 1684, the city contained 360 dwellings, and 2500 inhabitants. It seems that from the first the idea was entertained of building up a large place. Dean Prideaux, in his celebrated work entitled "The Connexion of the Old and New Testament," after presenting the plan of the ancient city Babylon, observes, "Much according to this model hath William Penn, the Quaker, laid out the ground for his city of Philadelphia, in Pennsylvania: and were it all built according to that design, it would be the fairest city in America, and not much behind any other in the whole world." In 1689, the first public school was founded by the Friends, and placed under the care of George Keith, a native of Aberdeen in Scotland. The Revolution of 1688 having driven James II. from the throne, and placed William the Third and Mary, the daughter of James, on the throne, Penn, from a supposed adherence to the former, was, in 1693, deprived of his authority over Pennsylvania, and Colonel Fletcher, governor of New-York, succeeded to the government. But in 1694, the warrant by which he was deprived of his proprietary right was revoked, and he was reinstated in his rights and authority over the province.

PHILADELPHIA.

In Oct. 25th, 1701, the city of Philadelphia received a charter from the proprietary, and Penn sailed for England. The same year the customs on goods chiefly imported through Philadelphia amounted to £8000 sterling. In 1702 eight vessels were loaded with tobacco for England, with 80 or 90 hogsheads each. Tobacco seems to have been early much cultivated here. In 1704, the first Presbyterian church was erected. In 1706, 800 hogsheads of tobacco and 25 or 30 tons of furs were exported. In 1707, British goods were consumed to the amount of £14,000 or £15,000 sterling. In 1718, William Penn died in England, and left his interest in Pennsylvania as an inheritance to his children, which they and their heirs possessed until the Revolution. In 1719 the first weekly Gazette, entitled "The American Weekly Mercury," was published by Andrew Bradford. The mayor and alderman Hill, in conjunction with the regulators, were requested to employ Jacob Taylor to run out the seven streets of the city, and cause the same to be staked out, to prevent any encroachment in building, for the want thereof. Some foot pavements of brick, and crossing places of stone, were laid in the streets. In 1727, Benjamin Franklin published "The Pennsylvania Gazette." In 1730, the first Roman Catholic church was founded. In 1730, a fire consumed a number of stores and dwellings, to the loss of £5000 pounds sterling. Soon after, the corporation paid nine shillings a piece for 100 fire buckets; and in addition, 250 fire buckets and two fire engines were imported from England: the fire buckets were hung up in the courthouse, and houses were provided for the engines. In 1738, Benjamin Franklin instituted the first fire-company in Philadelphia. In 1739, Rev. George Whitefield preached to 15,000 people, assembled on Society Hill, near the flag-staff. Aug. 17th, 1741, to prevent riotous assemblages of negroes and others, the constables were charged to disperse all such meetings at half an hour after sunset, and all persons who refused to depart were to be brought before the magistrates. In 1742, the first Moravian church was built. May 4th, 1743, stalls erected for the sale of goods in the market were ordered to be removed, that the market might be kept open and clear. During this year the first Lutheran church was built; and the first Dutch Reformed in 1747. In 1749, the grand jury presented the city watch as defective; and only five or six persons were employed, who went the rounds in company. Second-street from High-street to Chestnut-street first paved; a horse having been mired there, and the rider, a respectable man by the name of Wharton, having been thrown and broken his leg. The city this year contained between two and three thousand houses, and 15,000 inhabitants. In 1751, an act passed for a nightly watch, and enlightening the city; the last of which was completed in October of this year. Previously to this the constables acted as a city watch. In 1754, a theatre was erected in Southwark, much against the remonstrance of the citizens. In 1755, Fifth-street might be regarded as the western extremity of the city. In 1756, the first stage to New York, through Trenton and Perth Amboy, was established, going through in three days. St. Paul's, the first Episcopal church, was founded in 1760. This year the Pennsylvania hospital was founded, and Dr. Franklin wrote an inscription, which was engraved on its corner stone. The first public library was founded in Philadelphia by the influence of Dr. Franklin; 38 persons agreed to pay 40 shillings each, and ten shillings annually afterwards. In 1761–2, the streets were first paved in the middle, and it was at first accomplished by a lottery. In 1765, a scavenger was appointed to see the streets cleaned once a week. The use of curb-stones was not then introduced, and the gutters were often in the middle of the streets. A second line of stages was established to New-York, twice a week, going through in three days, at two pence a mile. It was a covered wagon without springs. In the same year a line of vessels and wagons, once a week, was established to Baltimore. In 1766, a third line of stages to New-York was established, called the "Flying Machine," going through in two days. The first Methodist church was founded in 1773, the first stage coaches were established from Philadelphia to New-York, going through in two days, with a fare of four dollars for each passenger. In 1777, Gen. Howe, the British commander, was in possession of the city; many of the inhabitants were gone, and there were 3508 houses, of which 287 were untenanted; there were 287 stores, and 21,767 inhabitants, exclusive of the army and strangers. In 1780, the bank of Pennsylvania was established for the purpose of supplying the army of the United States for two months, by a subscription of £300,000, by 90 persons, the principal of whom were Robert Morris and Blair McClennachan, who subscribed £10,000 each. The bank of North America was founded by the United States' Congress in 1781. April 19th, 1790, Dr. Benjamin Franklin died, leaving, among other benefactions, £1000 sterling, or $4444·44 to be loaned to young unmarried artificers under 25 years of age, who had served an apprenticeship in the city, and fulfilled their indentures,

PHILIPPINE ISLANDS.

in sums of not more than £60 nor less than £15 at 5 per cent. interest, with sufficient sureties. This constitutes what is called the Franklin fund; and on the first of January, 1830, it amounted, with the interest due, to $32,192. Southwark district was incorporated in 1794. In 1800, the seat of the United States' government was removed from Philadelphia to Washington, District of Columbia. March 28th, 1803, the district of the Northern Liberties was incorporated, and the act was amended March 16th, 1829. In March 22d, 1813, the district of Spring Garden was incorporated. In March 24th, 1812, the district of Moyamensing was incorporated; and on March 6th, 1820, the district of Kensington was incorporated. These districts, with the city proper, make up what is virtually the city of Philadelphia. The police of Philadelphia would probably be more efficient if it were under a single incorporation.

PHILADELPHIA, p. t., Jefferson co., N. Y., 16 N.E. from Watertown, 179 N.W. Albany, 432 W. Drained by Indian river and its tributaries. Taken from Le Roy in 1821, first settled from Pennsylvania in 1813. It contains one Friends' church, four stores, one grist-mill, six saw-mills; 18 schools, 709 scholars. Pop. 1,868.

PHILADELPHIA. See ALA-SHEHR.

PHILIPPINE ISLANDS, a group of islands of the eastern archipelago, principally included in its 5th division, forming, Cuba excepted, the most valuable colonial possession still belonging to Spain, between the 5th and 20th deg. N. lat., and the 117th and 126th deg. E. long., having N. the Balintang channel, which divides it from the Batanes and Bashee islands; E. the Pacific ocean; S. the strait of Basilan, separating it from the Soolos archipelago to the E. of Borneo; and W. the Chinese sea. Aggregate area (according to Berghaus), including Palawan, 134,115 sq. m. Of these islands, ten are large, and the rest of very inferior size, their extent and populations, in 1857 (from an official return), being respectively as follows:

Islands and Provinces	Area in sq. m.	Pop. in 1857.
Luzon (13 provs.)	56,664	2,484,687
Mindoro	4,155	28,672
Panay (3 provs.)	4,516	656,684
Isla de Negros	3,774	26,662
Zebu	2,162	209,287
Leyte	4,196	92,165
Samar	4,198	99,295
Masbate	1,915	2,99
Palawan	7,998	11,809
Magdanao (2 provs.)	35,887	74,969
Smaller islands, including the Islas de Calamianes	8,926	15,009
Total of 30 provs.	134,115	3,509,008

The islands of Palawan and Magdanao, however, are but very imperfectly known, portions of them being only subject to the colonial government of Manilla, the rest being governed by chiefs of native and often hostile tribes. LUZON, the largest of these islands, and that also which is best known to Europeans, is of extremely irregular shape, but may be described as a long and narrow island, running N. and S., with a peninsula (called Camarines) stretching out at its S.E. side, its length from point Calcunga northward, to point Calaan southward, being about 450 m., and its breadth ranging from 10 to nearly 140 m. The coast generally is rocky, and indented with numerous bays and gulfs; on the E. side is the Seno de Luzon, a deep narrow inlet, nearly separating the peninsula of Camarines from the rest of the island, and on the W. side are the gulf of Lingayen, the bay of Manilla, the Ensenada de Balayan, and the Seno de Ragay. A large portion of the surface is covered with mountains; and N. of Manilla the chains are divided by the Cagayan, the largest river of the island, into two nearly parallel ranges, called Sierra Madre and Sierra de los Caravallos, that terminates respectively in the points Cabicunga and St. Vincent, the extreme N. points of the island. The latter range is the most elevated; but even its highest peaks do not rise more than 4000 ft. above the sea. S. of Manilla the chain may be traced into the peninsula of Camarines, a large portion of which it covers, and finally terminates in the Punta Calaan. The plain of Pampanga N. of the city of Manilla, extends northward nearly as far as the gulf Lingayen, from which it is separated by a ridge of rather lofty hills: it is about 50 m. in length by about 30 m. in breadth, and being watered by a river falling into the bay of Manilla, is extremely fertile, and is covered with plantations, and densely peopled. (Berghaus, Memoir zur Erklärung der Karte von den Philippinen Inseln, p. 16.) S. of Manilla is the Laguna de Bahia, about 20 m. in length and 10 m. in breadth, the waters of which are conveyed into the bay of Manilla by a wide and pretty deep stream, the Pasig, flowing through the cap. A few miles farther S. is the Laguna de Taal which communicates with the Senada de Balayan, by the

* This estimate includes only the Spanish portion of these islands.

short but deep river Bombon; it is nearly circular, being about 11 m. in diameter, and contains the island of Taal, in which is a volcano, with two active craters. About 90 m. E. is another volcano; and more to the S., in the peninsula of Camarines, are ten volcanos, one of which has frequent eruptions: in fact, throughout Luzon and most of the Philippines, the igneous formations have been found in constant connexion with the primitive rocks, and there can be no doubt that they form a part of the great volcanic band extending from Kamtschatka, through Japan and Formosa into Borneo, Java, and Sumatra. Gold, iron, and copper, have been found in the mountains of Luzon, and rock-salt is so abundant in some parts as to be an article of export. Luzon is separated from Mindoro by the strait of that name, about 5 m. broad, and from Samar by the Embocadero de San Bernardino, the common passage for vessels navigating the Pacific on their way to China.

The Bisayan group, which lies S.W. and S. of the great island last mentioned, comprises about eight or nine considerable islands, the most westerly of which is Mindoro, and that most eastward Samar. With the exception of Panay, which is triangular, these islands are generally long and narrow; Mindoro, Negros, Samar and Masbate, are very mountainous, and only moderately productive; but Panay and Zebre contain much good level land, and are, on the whole, the most important islands of the Philippine group. The Palawan, the extreme W. island of the Philippines, extends from N.E. to S.W. about 250 m., with a breadth of not more than one tenth its length: it is reported to be extremely mountainous; but the greater part i inhabited by savages, and it has been little visited by Europeans. (Berghaus, p. 65.) A small portion only at a N. extremity belongs to Spain.

Magindanao, or Mindanao, the largest of the Philippine lands after Luzon, is of very irregular shape, having a peninsula stretching 150 m. from the main part of the land: length from N. to S. about 330 m.: average breadth, exclusive of the peninsula, 95 m. Dampier and Forest are the principal authorities from whom we derive any knowledge of this still imperfectly known island. In the interior near the bay of Iliano, is a considerable lake, which, according to Forest, is between 15 and 20 m. in width. The coast is mountainous, and numberless hills occur in the inland S. districts, which are generally well-covered with timber: large tracts, however, are found in some parts destitute of trees, and covered, like the savannahs of America, with long, rank grass. So large an island is not necessarily have large rivers; but only two of them are known, one on the N. side, flowing into the Bay of Bus, and another, called the Pelangy, flowing westward by the bay of Iliano, opposite the island of Bunwrit. The Spanish settlements, which are chiefly on the N. side, form provinces in the capitanato of the Philippines. The interior is inhabited by Horaforas, who are treated as slaves by the Malays occupying all the S. coast; they acknowledge the supremacy of a native sultan.

The climate of the Philippines, owing to their extent, is more variable than in the other groups of islands lying so near the equator. In and about Manilla, the district usually visited by Europeans, the mean temperature of the hot season (from August to October) may average about 83° and that of the cold season, usually preceding the hot, about 70° Fah. The year, as in other tropical countries, is divided into a wet and dry season, here depending on the monsoons: the former lasts from May till the commencement of November, and the latter during the rest of the year. In the S.W. monsoon immense quantities of rain falls, and the rain frequently lasts for 12 or 14 days without intermission. A large part of Luzon is within the region of the typhoons, which are as formidable as in the tropics of the W. Indies: they last from May to December, but seldom continue more than six or eight hours at a time. In Luzon (with which we are better acquainted than the rest of the group), the inhabitants suffer from ague and dysentery, in consequence of the great extent of low shy and low grounds inundated during the rainy season, and exhaling pestilential vapours during the rest of the year. (Meyen, Reise um die Erde, ii., 261.)

The agricultural products of the Philippine islands are rice, millet, and maize; sugar, indigo, hemp, tobacco, and cotton, with a great variety of other articles of importance; but with the modes of culture we are little acquainted, though, according to Meyen (Malayan Archip., the Chinese implements are pretty generally used in agriculture. Rice is the chief support of the population, and is raised in large quantities throughout the the marshy nature of the country in many parts being favourable to its cultivation. In other districts, however, upland rice is cultivated. The sugar-cane is raised on the plain of Pampanga and in the island of Panay: the mode of extracting the sugar is defective, but the sugar is excellent; its culture is also rapidly extending, and it

now forms by far the most important article of export from the Philippines. Tobacco grows well, and might be produced on a very extensive scale; but its growth is limited from the manufacture of cigars, in which form alone tobacco is exported, being a government monopoly. Indigo, also, is pretty extensively cultivated, but is inferior in quality to that of Bengal. Sapan wood constitutes the chief timber of the hilly districts, and is exported in large quantities to China. The coffee-plant was introduced by the Spaniards at the close of the last century, and grows wild on the W. side of Luzon, though not in the other islands. The coffee exported from Manilla is almost wholly procured from these wild plants, and is alleged to be almost equal to that of Bourbon or the Mauritius. The banana is found on nearly all the island, and hemp is produced very abundantly in the neighbourhood of Manilla, as well as in Panay and Zebre, furnishing materials both for cordage and a strong coarse cloth woven by the Malays. (Meyen, ii., 278.) Excellent sago grows in most parts of Luzon, and the cocoa-nut, which was introduced from Guatimala, is very abundant, and superior in quality even to that of Peru. The shrub-cotton (gossypium herbaceum) thrives well; but, owing to some imperfection in the mode of breaking the pod, the cotton is of inferior quality, and little is exported. Cinnamon, the betel-pepper, and the clove-tree are found wild in many of the islands, and the bread-fruit, mango, and shaddock are raised very abundantly and with little labour. The other fruit-trees are few in number and of very indifferent quality. The mountains produce excellent timber for building both houses and ships; and the bamboos, used in the construction of the houses of the Malays, are very long, some being as thick as a man's thigh. The fauna of the Philippines comprises buffaloes of uncommon size and strength, a small, but hardy, breed of horses, introduced by the Spaniards, goats, pigs, and a few sheep, with immense numbers of ducks and fowls both wild and domestic. Land-tortoises are plentiful in most of the islands, and their shells constitute an important article among the exports. There are no beasts of prey; but caymans are found in most of the rivers and lakes, and are particularly numerous in the Laguna de Bahia. Among the birds may be noticed the swallows which supply the edible nests. Fish abounds on the coasts, and the native fishermen are equally expert with the other islanders of the E. archipelago. Pearl-oysters also are found in large quantities, and the shells are exported to China. The sea-slug, or holothuria, is also an important article of commerce.

Manufactures are of very little importance. The plaiting of straw and chips of wood into hats, cigar cases and matting is carried on pretty extensively, and the hats are highly prized by the Spaniards. Domestic weaving occupies most of the females; and cotton cloth was till recently an article of export to Mexico. Considerable quantities of earthenware are made in Luzon, but of very inferior quality to that of China. The manufacture of cigars is a government monopoly; and, according to Meyen, ii., 211, the royal manufactory at Manilla employs about 2000 persons, two thirds of whom are females.

Commerce.—Considering the great fertility and varied productions of the Philippines, and their peculiarly favourable situation for carrying on commerce, the limited extent of their trade, even with its late increase, may excite surprise. This, however, is entirely a consequence of the wretched policy of the Spanish government, which persevered until very recently in excluding all foreign ships from the ports of the Philippines, confining the trade between them and Mexico and South America to a single ship! Even ships and settlers from China were excluded. "Provisions," says La Pérouse, "of all kinds are in the greatest abundance here, and extremely cheap; but clothing, European hardware, and furniture, bear an excessively high price. The want of competition, together with prohibitions and restraints of every kind laid on commerce, render the productions and merchandise of India and China at least as dear as Europe!" Happily, however, this miserable policy, the effects of which have been admirably depicted by M. de la Pérouse, has been materially modified within the last few years. The events of the late war destroyed for ever the old colonial system of Spain; and the ships of all nations are now freely admitted into Manilla and the other ports in the Philippines. An unprecedented stimulus has, in consequence, been given to all sorts of industry; and its progress will, no doubt, become more rapid, according as a wider experience and acquaintance with foreigners makes the natives better aware of the advantages of commerce and industry, and disabuses them of the prejudices of which they have been so long the slaves.

In 1837, among other articles exported from Manilla, were 222,183 piculs (about 13,250 tons) sugar; 50,457 piculs hemp; 126,603 coyangs rice; 4598 boxes cigars, with considerable quantities of indigo, hempen cloth, sapan wood, and a great number of other articles; the total value of the exports amounting in the course of that year to above *two* million dollars. Cotton manufactured goods constitute the great article of import. The foreign trade of the islands centres almost entirely in Manilla, which see. It is, in fact, the only port in the Philippines to which either Spanish ships to or from Europe and foreign vessels generally are allowed to trade; but Spanish vessels trading with China and Singapore, may proceed to the outports of Pangasinau, Ylvers, and other places, to take on board their outward-bound cargoes. (*See* MANILLA.)

The *population* of the Philippines is extremely various; and, independently of foreign settlers, the natives consist of a great number of distinct tribes, partly of Malay, partly of Papuan origin, and speaking several distinct languages or dialects. Some of the natives still adhere to the polytheism they professed before the arrival of the Spaniards; but a large proportion have been converted to the Catholic faith, which, indeed, is the common bond between them and their new masters, and the principal means by which the latter have so long been able to maintain their ascendancy. The natives are said to be the most active, bold, and energetic of any belonging to the E. archipelago. "These people," says M. de la Pérouse, "appear in no respect inferior to those of Europe. They cultivate the earth like men of understanding; are carpenters, joiners, smiths, gold-smiths, weavers, masons, &c. I have walked through their villages, and found them kind, hospitable, and communicative; and though the Spaniards speak of them and treat them with contempt, I perceived that the vices with which they are charged, ought rather to be imputed to the colonial government." (*Voyage*, chap. 15.) The people here described, however, are the Malays, who have pretty generally acknowledged the supremacy of the Spaniards, by whom they are treated as free subjects, and allowed to be proprietors of land; the Papuans, who chiefly occupy the higher parts of the country and less frequented islands, are miserable savages, incapable of civilization, and avoiding all communication with foreigners. The rest of the population comprises European and Creole Spaniards, Spanish and Indian mestizos. Mohammedans from the E. Indies, and Chinese. The Spaniards, however, do not, perhaps, exceed 4000 or 5000.

The *seat of government* is at Manilla, the residence of the captain-general of the Philippines; but there are lieutenant-governors in the most important of the other islands, and alcaides in each of the provinces, which also are subdivided into *pueblos*, having their separate intendants. The revenues of the Philippines are principally derived from *ad valorem* duties on imports and exports, and from a capitation tax, and the tobacco monopoly. Foreign commodities, imported in foreign vessels, pay 14 per cent., and in national vessels from 7 to 9 per cent.; but there are some exceptions, and wines of all sorts, except Champagne, pay from 40 to 50 per cent., according as they are brought in native or foreign bottoms. Spanish products, imported by Spanish vessels, pay 3, and, by foreign vessels 8 per cent. Exports (in which are comprised commodities produced in the island, and imports from foreign countries for consumption) pay from 1 to 1½ per cent. *ad valorem*, by Spanish, but from 2 to 3 by foreign ships. Tobacco, however, and hempen rope, made at Manilla, may be exported free of duty. Spaniards may export rice free of duty; but foreigners are charged 4½ per cent. The capitation tax, or annual tribute, is charged at certain rates on all the inhabitants except European mestizos. The Chinese, who constitute the chief portion of the shopkeepers, traders, mechanics, coolies, and household servants are divided into four classes, paying rates, varying from 12 doll. 6 rs. to 120 doll. 6 rs., to which they are subject from the age of 16 to 60; but no Chinese is allowed to settle on the islands after the age of 40 years. Chinese and Indian mestizos pay 1 doll. 3 rs. annually, from the time of puberty till death, and Indians of both sexes pay 5 rs. 6 gr. each, during the same period. The number of tribute payers amounted, in 1837, to 1,305,142, of whom 901,924 belonged to the 17 provinces comprised in the island of Luzon. The army consists of about 7000 men, of whom only 700 are Spaniards, and the rest Malays. The cavalry are chiefly European; but the Malays are said to be good soldiers, and occupy most of the fortresses in Luzon.

The Philippines were discovered by Magellan in 1521, but were not claimed by the Spanish till 1565, soon after which Manilla was constituted the capital of their possessions in this part of the world. The islands received their present appellation in honour of Philip II. when king of Spain. The Dutch and Chinese tried to make settlements here in the 17th century, though without success, owing to the determined and jealous opposition of the Spaniards. In 1762, Manilla was taken by the English, who gave it back to Spain in 1764; since which they have held it without interruption. The government, however, is inefficient

and unable to protect the islanders from the ravages of the Sooloo and other pirates, who capture vessels, plunder villages, massacre or enslave the inhabitants, and commit other enormities with impunity. (*Berghaus's Asia; Meyen*, ii., 210-264; *Moor's Malayan Archipelago*, p. 74-65; *Mavor's Hist. View of the Philipp. Islands; Hamilton's Gaz.*)

PHILLIPS, co., Ark. Situated in the E. part of the state, and contains 730 sq. m. Watered by St. Francis river Bounded E. by Mississippi river. It contained in 1840, 7634 neat cattle, 557 sheep, 11,415 swine; and produced 151 bushels of wheat, 160,675 of Indian corn, 3830 of oats, 15,256 of potatoes, 1455 lbs. of tobacco, 146,409 of cotton. It had 10 stores, 22 grist-mills, five saw-mills, two tanneries, two distilleries, two printing-offices, two weekly newspapers; five schools, 109 scholars. Pop.: whites, 925; slaves, 905; free coloured, 17; total, 3547. Capital, Helena.

PHILLIPSTOWN, t., Putnam co., N. Y.. 14 W. Carmel 96 S. Albany. Bounded W. by Hudson's river. It contains some of the most elevated summits of the highlands. Iron ore abounds in the vicinity. It has four churches, a Presbyterian, Methodist, Episcopal and Baptist, 18 stores, one fulling-mill, one furnace, one flouring-mill, four grist mills, four saw-mills, one paper mill; nine schools, 547 scholars. Pop. 3814.

PHIPPSBURG, p. t., Lincoln co., Me.. 44 S. Augusta, 294 W. Situated on a peninsula on the W. side of Kennebec river, at its mouth. It has cape Small Point at its S. extremity, on the Atlantic. It contains seven stores, one grist-mill, 33 saw-mills; nine schools, 654 scholars. Pop. 1655.

PIACENZA (anc. *Placentia*), a city of N. Italy, duchy Parma, cap. prov. and duchy of its own name, on the Italian way, near the Po, where it is joined by the Trebbia. 37 m. W.N.W. Parma, and 37 m. S.E. Milan: lat. 45° 7' 44', long. 9° 42' 32". Pop. 30,000. It is of an oblong form, surrounded with ramparts, now partly converted into public promenades, and defended by a citadel garrisoned by Austrian troops. Its streets are wide and regular, especially the principal, called the *Stradone*, which is one of the handsomest in Italy; but they are dull and deserted. The houses are built of brick. The principal square is mostly surrounded with old and mean buildings, but includes the town house, with the prison, an antique structure, with fine Gothic tracery-work; the governor's palace, an old building, with new front, raised by the French; and a large church: in the centre of the square are the bronze equestrian statues of Alexander Farnese and his son. The Farnese palace, an unfinished edifice of singular architecture, designed by Vignola, is now converted into a barrack, its walls were formerly adorned with the works of Raphael, Correggio, Parmegiano, &c.; but these were mostly removed when the last duke of Parma changed his brick palace of Piacenza for the throne of Naples. Placenza abounds with churches. The cathedral is a heavy-looking building of the 12th century; but its cupola is ornamented with fine frescoes by Guercino and Procaccchini; and it has as altarpiece of high merit, by Procaccchini, with other paintings by that artist, the Caracci, Parmegiano, &c. Many of the other churches, also, boast of fine paintings; but that formerly belonging to the Augustine convent, an elegant structure by Vignola, has been converted into a granary. The city has a pretty theatre, good hotels, a public library, and to comprise 30,000 vols.; two orphan asylums, &c., and about ½ m. distant, a bridge crosses the Po. Placenza is a bishop's see, the seat of the high court of appeal for the duchy, two inferior tribunals, an Episcopal seminary, with about 200 students, schools of the fine arts, architecture, a high school for young ladies, &c. It has a few manufactures of silk twist, woollen stuffs, stockings, hats, earthenware, &c.; but its chief trade is in agricultural produce. It has a large fair in April.

When colonized by the Romans, *anno* B.C. 219, Placentia was the most important and strongest city in Cisalpine Gaul; and it afforded a secure retreat to the Romans, after the unfortunate battles of the Ticinus and Trebbia. In fine amphitheatre, beyond the walls, was destroyed in the war between Otho and Vitellius. Piacenza, with its territory, was taken possession of by Pope Julius II., and given by Paul III. to L. Farnese. It has since mostly followed the fortunes of Parma. Pope Gregory X., Cardinal Alberoni, Pallavicini, Landi, &c., were among the remarkable natives of this city. (*Rampoldi; Cramer's Anc. Italy*, i. 72-82; *Conder's Italy*, ii. 18-23, &c.)

PIATT, co., Ill. Situated a little N.E. of the centre of the state, and contains 440 sq. m. Watered by Sangamon river Capital, Monticello. Organized since the census of 1840.

PIAZZA, a town of Sicily, near its centre, intend. of Caltanisetta, 18 m. E.S.E. city of that name. Pop. in 1831, 13,229. It is built upon an isolated eminence. There is nothing remarkable in its buildings; but it is admired for the richness of its territory, and the great beauty of the contiguous country. The *chiesa madri*, a good church eral convents, and a college, are the principal edifices. s a bishop's see; but, from the little attention paid to it classical writers, and the absence of any vestige of antiquity, it was evidently a place of no great consequence in ly times. (*Heere, Sir R. C., Classical Tour*, ii. 253,

/CARDY, one of the former provs. of France, now subded among the deps. Aisne, Somme, Pas-de-Calais, Arnes, and Oise.

ICKAWAY, co., O. Situated a little S. of the centre of state, and contains 470 sq. m. Watered by Scioto river its tributaries. It contained in 1840, 19,373 neat cattle, 58 sheep, 37,895 swine; and produced 946,661 bushels vheat, 10,414 of rye, 1,323,684 of Indian corn, 1342 of kwheat, 169,838 of oats, 18,067 of potatoes, 1198 lbs. of cco, 11,859 of sugar. It had three commission houses oreign trade. 39 retail stores, one furnace, four fulling s, two woollen factories, eight flouring mills, 15 grists. 36 saw-mills, one oil-mill, 12 tanneries, eight distilleone brewery, two potteries, three printing-offices, two kly and one semi-weekly newspapers and four periods-; one academy, 15 students; 54 schools, 1523 scholars. 19,785. Capital Circleville.

CKAWAY, t., Pickaway co., O. It has one flouring-mill, saw-mills; six schools, 176 scholars. Pop. 1579.

CKENS, dist., S. C. Situated in the N. W. part of the , and contains 1050 sq. m. Drained by Seneca river its branches. It contained in 1840, 14,873 neat cattle, sheep, 33,566 swine; and produced 43,815 bushels of t. 3440 of rye, 513,215 of Indian corn, 91,539 of oats, 8 of potatoes, 13,613 lbs. of tobacco, 365,523 of cotton. d 16 stores, one fulling-mill, one woollen factory, nine ng-mills, 72 grist-mills, 25 saw-mills, seven tanneries, stilleries; one academy. 51 students, 25 schools, 579 ars. Pop.: whites, 11,548; slaves, 2715; free coloured, otal, 14,356. Capital Pickens, C. H.

KENS, co., Ala. Situated in the W. part of the state, ontains 720 sq. m. Drained by Tombigbee river and anches. It has three academies, 190 students, 93 is. 609 scholars. Pop.: whites, 9347; slaves, 7764; -loured, 7; total, 17,118. Capital, Carrollton.

KERING, a market-town and parish of England, N. co. York. W. divis. of wap. Pickering-Lythe, on the e of the Egton Moors, 18 m. S.W. Whitby, and 23 m. York. Area of parish, including five townships, acres: pop. of township in 1831, 2555. The town, d on a gentle eminence near a small tributary of the nt, is old and straggling. Near its W. extremity are ns of a castle, in which Richard II. was confined af-depostion, and prior to his final removal to Ponte-The church is a fine old building, with a lofty spire: ing is a vicarage in the gift of the dean of York. Wesleyan Methodists, Independents, and other dis-have their respective places of worship; and there ious Sunday schools and a well-endowed charity for 150 children of both sexes. The town sent two rs to the House of Commons in 23 Edw. I.; but the te was discontinued in the same reign, and has not stored. It belongs to the duchy of Lancaster, and es in its jurisdiction several neighbouring villages, nle forming what is called "the Honour of Picker-A railway, 24 m. in length, connects this town with . A manor-court is held here at Easter and Mias, for the recovery of debts under 40s., and petty are held on alternate Mondays. Markets on Monrs the Mondays before Feb. 14, July 5, and Oct. 11, or cattle.

MONT, a country of N. Italy, composing the principion of the continental dom. of the kingdom of Sardice SARDINIA, KINGDOM OF.)

MONT, p. v., Orange t., Rockland co., N. Y. 24 N. rk, 135 S. by W. Albany, 958 W. A thriving vilthe W. side of Hudson river. Here, by a pier in the arly a mile long, the Erie railroad commences; and highlands terminate by an abrupt hill.

₹E (ST.) *See* MARTINIQUE.

co., Pa. Situated toward the N.E. part of the d contains 729 sq. m. Bounded N.E. and S.E. by e river. Watered by Lackawaxen river, along the which the canal of the Hudson and Delaware passes. It contained in 1840, 2935 neat cattle, ep. 3491 swine; and produced 4469 bushels of 2,994 of rye, 21,759 of Indian corn, 19,317 of buck-5,019 of oats, 51,614 of potatoes. It had 11 stores, llen factory, seven grist'mills, 46 saw-mills, four ; one academy, 46 students, 18 schools, 496 schol-, 3839. Capital, Milford.

o., Ga. Situated a little W. of the centre of the contains 470 sq. m. Bounded W. by Flint river, ted by its branches. It contained in 1840, 11,746 e. 4475 sheep. 24,397 swine; and produced 47,171 of wheat, 296,289 of Indian corn, 57,516 of oats,

14,715 of potatoes, 3947 lbs. of tobacco, 6,074,950 of cotton. It had 11 stores, three grist-mills, five saw-mills, two oil-mills; one academy, 25 students, eight schools, 256 scholars. Pop.: whites, 6505; slaves, 2549; free coloured, 22; total, 9176. Capital, Zebulon.

PIKE, co., Ala. Situated toward the S.E. part of the state, and contains 1100 sq. m. Bounded E. by Pea river. Drained by Conecuh river and its branches. It contained in 1840, 98.375 neat cattle, 3796 sheep, 36,340 swine; and produced 3090 bushels of wheat, 395,993 of Indian corn. 8173 of oats, 39,380 of potatoes, 733,366 lbs. of cotton. It had eight stores, 12 grist-mills, seven saw-mills; 19 schools, 412 scholars. Pop.: whites, 7987; slaves, 2111; free coloured, 10; total, 10,108. Capital, Troy.

PIKE, co., Miss. Situated toward the S.W. part of the state, and contains 864 sq. m. Drained by Bouge Chitto river and its branches. It contained in 1840, 9491 neat cattle, 9482 sheep, 19591 swine; and produced 367 bushels of wheat, 180,345 of Indian corn,. 9198 of oats, 26,818 of potatoes, 169,600 lbs. of rice, 1,416,751 of cotton. It had three stores; 12 schools, 342 scholars. Pop.: whites, 3.756; slaves, 2374; free coloured, 21; total, 6151. Capital, Holmesville.

PIKE, co., Ky. Situated in the S.E. part of the state, and contains 400 sq. m. Bounded S.E. by Cumberland mountains. Watered by the W. fork of Big Sandy river. It contained in 1840, 2917 neat cattle, 2964 sheep, 11,886 swine; and produced 4986 bushels of wheat, 174,396 of Indian corn, 21,292 of oats, 8767 of potatoes. 11,369 lbs. of tobacco, 18,860 of sugar. It had 12 stores, 48 grist-mills. one sawmill; one school, 25 scholars. Pop.: whites, 3469; slaves, 85; free coloured, 13; total, 3567. Capital, Piketon.

PIKE, co., O. Situated in the S. part of the state, and contains 421 sq. m. Watered by Scioto river. It contained in 1840, 9559 neat cattle, 8706 sheep, 19,362 swine; and produced 82,470 bushels of wheat, 356,434 of Indian corn, 108,334 of oats, 4596 of potatoes. It had 23 stores, three flouring mills, five grist-mills. 19 saw-mills; 37 schools, 893 scholars. Pop. 7536. Capital, Chillicothe.

PIKE, co., Ia. Situated in the S.W. part of the state, and contains 325 sq. m. Organized in 1816. Bounded N. by White river. Drained by Patoka river. It contained in 1840, 4733 neat cattle, 5152 sheep, 18,578 swine; and produced 26,947 bushels of wheat, 239,037 of Indian corn, 31,934 of oats, 9366 of potatoes, 11,440 lbs. of tobacco, 6732 of sugar. It had 15 stores, eight grist-mills, three saw-mills, three tanneries, three distilleries; one academy, 15 students, 19 schools, 349 scholars. Pop. 4769. Capital, Petersburg.

PIKE, co., Ill. Situated in the W. part of the state, and contains 800 sq. m. Bounded E. by Illinois river, S.W. by Mississippi river, and drained by their small tributaries, which afford water-power. A salt spring, 20 ft. in diameter, is found on McKee's creek. It contained in 1840, 12,597 neat cattle, 6987 sheep, 23,413 swine; and produced 81,879 bushels of wheat, 1221 of rye, 359,945 of Indian corn, 1740 of buckwheat, 30,809 of oats, 31,615 of potatoes, 12,065 lbs. of tobacco, 2940 of sugar. It had 20 stores, one flouring mill, 11 grist-mills, 18 saw-mills, 14 tanneries, one distillery; two academies, 105 students; 41 schools, 656 scholars. Pop. 11,728. Capital, Pittsfield.

PIKE, co., Mo. Situated toward the N.E. part of the state, and contains 720 sq. m. Bounded N.E. by Mississippi river, Watered by Salt and Cuivre rivers. It contained in 1840, 13,515 neat cattle, 14312 sheep, 23,940 swine; and produced 42,971 bushels of wheat, 2306 of rye, 513,160 of Indian corn, 73,980 of oats, 21,641 of potatoes, 829,922 lbs. of tobacco, 20,045 of sugar. It had one commercial house in foreign trade, 36 retail stores, three woollen factories, two flouring-mills, 15 grist mills, eight saw-mills, 11 tanneries, nine distilleries, two potteries, one printing-office, one weekly newspaper; 27 schools, 1048 scholars. Pop.: whites, 8157; slaves, 2479; free coloured, 17; total, 10,646. Capital, Bowling Green.

PIKE, co., Ark. Situated toward the S.W. part of the state, and contains 500 sq. m. It contained in 1840, 1772 neat cattle, 448 sheep, 5744 swine; and produced 653 bushels of wheat, 79,115 of Indian corn, 2670 of potatoes, 1018 lbs. of tobacco, 11,998 of cotton. It had one grist-mill, one saw-mill. Pop.: whites, 860; slaves 109; total, 969. Capital, Murfreesboro'.

PIKE, p. t, Alleghany co., N. Y., 20 N. Angelica, 237 W. by S. Albany, 350 W. Drained by Westkoy and Eastkoy creeks, tributaries of Alleghany river. It contains three churches, a Methodist, Baptist and Presbyterian, nine stores, one furnace, three fulling-mills, two grist-mills, six saw-mills, three tanneries, one distillery; nine schools, 289 scholars. Pop. 2176.

PIKE, p. t., Bradford co., Pa., 158 N. Harrisburg, 268 W. Drained by Wyalusing creek and its tributaries. It has three stores, two fulling-mills, one woollen factory, one grist-mill, nine saw-mills, two tanneries; 11 schools, 185 scholars. Pop. 1518.

PIKE, t., Perry co.. O. It has six stores, two flouring-mills, one grist-mill, four saw-mills, two tanneries, one pottery ; seven schools, 285 scholars. Pop. 1608.

PIKE RUN, p, t., Washington co., Pa., 18 S.E. Washington b., 194 W. Harrisburg, 210 W. Bounded E. by Monongahela river. It contains six stores, one saw-mill, one distillery ; two schools, 58 scholars. Pop. 2187.

PIKETON, p. v. capital of Pike co., O., 19 S. Chillicothe, 64 S. Columbus, 398 W. Situated on the E. side of Scioto river. It contains a courthouse, jail, six stores, 80 dwellings, and about 500 inhabitants.

PILLAU, a seaport town of the Prussian states, prov. E. Prussia, at the point of a tongue of land, on the N. side of the opening from the Baltic into the large maritime inlet, called the Frische Huff, lat. 54° 33′ 39″ N., long. 19° 52′ 39″ E. Pop. 3600. Pillau has a pretty good port, but the water is rather shallow, not exceeding 11 or 12 ft. in depth, so that vessels of large burden must anchor outside the bar. A light-house, having the lantern elevated 90 ft. above the level of the sea, has been erected on the S. side of the town, contiguous to the port. Pillau is properly the seaport of Conigsberg and Elbing, and is, in consequence, largely frequented by shipping.

PILES GROVE, t., Salem co., N.J., 10 N.E Salem. Salem creek affords water-power. It contains three churches, a Methodist, Friends, and an African Methodist. 10 stores, four grist-mills, two saw-mills, two tanneries ; eight schools, 319 scholars. Pop. 2477.

PILSEN, a town of Bohemia, cap. circ. Pilsen, on the Beraun, a tributary of the Elbe, 53 m. S.W. by W. Prague. Pop. in 1834, 8906. It is one of the best built towns in the kingdom, and has a fine Gothic parish church, and town-hall, a gymnasium, military and other schools, a theatre, &c. Its manufactures of woollen goods are flourishing, and it has others of Morocco leather, iron and horn wares, and alum. Being on the high route from Prague to S. and central Germany, it has a considerable transit trade ; and a large annual fair is also held at Pilsen, attended by traders from every part of Bohemia. (*Austr. Nat. Encyc.*; *Berghaus.*)

PINE, t., Alleghany co., Pa., 11 N. Pittsburg. Drained by tributaries of Alleghany river. It has one store, two fulling-mills, five grist-mills, five saw-mills, three tanneries ; nine schools, 349 scholars. Pop. 1770.

PINE GROVE, p. t., Schuylkill co., Pa., 43 N.E. Harrisburg, 153 W. Watered by Swatara creek, along which runs a navigable feeder of the Union canal. It contains six stores, two furnaces, one forge, three grist-mills, 12 saw-mills ; one school, 25 scholars. Pop. 1695.

PINE PLAINS, p. t., Dutchess co., N. Y. 25 N.E. Poughkeepsie, 68 S. Albany, 397 W. Drained by head branches of Wappinger's creek, issuing from several ponds. It contains a Presbyterian church, six stores, two fulling-mills, one flooring-mill, four grist-mills, three saw-mills ; one school, 18 scholars. Pop. 1534.

PINEROLO (Fr. *Pignerol*), a town of the Sardinian dom., in Piedmont, divis. Turin, cap. prov., on the Clusone, near the foot of the Alps, 21 m. S.W. Turin. Pop. in 1838, 13,501. It was formerly a place of strength ; but on its cession to Savoy in 1713, its fortifications were blown up by the French ; and at present it is surrounded only by a slight wall. It is neither regularly nor well built, but contains a spacious piece of arms, with a handsome hospital and cavalry barracks. It has a fine cathedral, and numerous convents. The manufactures comprise coarse woollens, silk, twist, paper, and leather ; and the inhabitants have a considerable trade in these articles, and in corn, wine, spirits, and fire-wood.

PIQUA. p. v. Miami co., O., 25 N. Dayton, 73 W. Columbus, 467 W. Situated on the S.W. side of Miami river, on the line of the Miami canal. It is regularly laid out with spacious streets, crossing each other at right angles. It contains five churches, a market-house, 16 stores, two grist-mills, four saw-mills, a printing-office, issuing a weekly newspaper, 250 dwellings, mostly of brick, and many of them handsome, five schools, 169 scholars. Pop. 1481.

PISA (an. *Pisæ*), a famous city of central Italy, the cap. of one of its most celebrated republics, and now the cap. of the prov. of its own name in the grand duchy of Tuscany, in a fertile, though rather marshy plain on the Arno, about eight m. from its mouth, 13 m. N.N.E. Leghorn, 50 m. W. Florence, and 12 m. S.W. Lucca ; lat. 43° 43′ 11″ N., long. 10° 24′ E. The pop., which, in the 13th century was estimated, though probably far beyond the mark, at 150,000, was, in 1836, only 20,943! The walls of the city are nearly five m. in circuit. In the days of its prosperity it was celebrated for the strength of its fortifications, its patrician towers, its profusion of marble, and its grave magnificence ; but it is now only " the shell of a great city." (*Addison.*) Its ancient gravity has degenerated into dulness; its powers, however, though no longer a mark of nobility, may be traced in its modernised houses; and it can still boast of

many marble edifices, and of one of the finest marble bridges in Europe. Its streets, though crooked, are wide, and paved, as those in Florence, with large flag stones ; the river is embanked with stone quays ; and a street, the *Lung-Arno*, which extends along both its banks, has a most majestic appearance. Some of the houses have curious old fronts, and one street is wholly bordered with arcades. In a large grass-grown square, at the N. angle of the city, are four remarkable buildings—the cathedral, baptistery, leaning tower, and *Campo Santo* ; " all," says Forsyth, " built of the same marble, all varieties of the same architecture, all venerable with years, and fortunate both in their society and their solitude." The cathedral is an edifice of the 11th century, and principally interesting as a specimen of the style then prevalent in Italy. Its length is 397 ft., breadth 106 ft.; and it has a front of 177 ft. in height. Internally it is divided into five aisles by 68 insulated columns of Corinthian or Composite architecture ; and four piers support an elliptical cupola. There are some beautiful altars, three magnificent bronze doors with sculptures in relief by John of Bologna, and many fine paintings and bas-reliefs by some of the first Italian masters. But incongruities are numerous: "The marble pulpit is supported by a naked figure of the most gross design ; Bacchanals and Meleager's hunt are encrusted on the walls ; an ancient statue of Mars is worshipped under the name of St. Peter ; and the heads of satyrs are carved on a cardinal's tomb !" (*Forsyth.*) Among the 80 other churches are many which can boast of rare works of art. The baptistery, opposite the cathedral, was built between 1152 and 1154, when Pisa was so populous and rich, that a voluntary contribution of one florin from each family is said to have sufficed to pay for its erection. This building is an immense polygon, above 100 ft. in diameter and 176 in height, surmounted by a cupola and a cone terminated by a figure of St. John the Baptist. Nearly the whole of the exterior is of marble, and the interior is handsome ; but, according to Mr. Woods, the building wants finish, and is not altogether well proportioned. The famous *Campanile*, or Leaning Tower, is an edifice of little actual beauty, but rendered extraordinary by its inclination from the perpendicular. It was begun about 1174, but not finished till the middle of the 14th century : it consists of two concentric walls, each two ft. thick, the diameter of the circular well in the centre being 22 ft.: it is eight stories, or 190 ft. in height, with outside galleries projecting seven ft. The topmost story overhangs the base on one side about 15 ft.; and to a spectator looking down from the top, the effect is terrific ; though as the centre of gravity is still 10 ft. within the base, it is perfectly safe. The view from the summit is alike extensive and beautiful. It has been supposed by some that the inclination is not accidental, but intentional, and that it was so constructed originally ; but the more probable opinion seems to be, that it is a consequence of the sinking of the foundation. The observatory and baptistery have also a slight inclination, which is, no doubt, owing to the same cause.

The *campo santo*, or cemetery, is the most beautiful edifice at Pisa, and unique in its kind. It is an oblong or rhomboidal court, 383 ft. in length, by 127 ft. in breadth, surrounded by arcades of white marble, adorned with ancient Etruscan, Greek, and Roman bas-reliefs, busts, and other sculptures; and the walls covered with fresco-paintings, &c., by the earliest Italian masters. In its centre is an enormous mound of earth, said to have been brought thither from Palestine during the crusades, and formerly used as a burial ground. Pisa has a grand ducal residence, and several other palaces, which, with the nobility's clubhouse and bank, are fine buildings : the three bridges are handsome ; and the hospitals, theatre, modern aqueduct four m. in length, &c., are well adapted to their purpose. Various remains of antiquity exist, as those of the aqueduct of Caldanelli, of the *hypocaustum*, &c., supposed to have been constructed by Nero, but especially a *sudatorium*, or vapour-bath, near the Lucca gate. About 3½ m. distant on the Lucca road, are the *Bagni de Pisa*, supposed to be the baths mentioned by Strabo and Pliny, and still frequented by numerous visitors.

Pisa has a university, formerly among the most celebrated in Italy, and remarkable for its tolerance ; its degrees, except in divinity and canon law, being attainable by persons of all creeds. It is still the head university of Tuscany; and, in 1836, had 545 students, 255 in jurisprudence. 173 in medicine and surgery, 30 in physical and mathematical sciences, and 96 in theology. Its library comprises 25,000 vols.; and attached to it are the Ferdinand college, a fine botanical garden, cabinet of natural history, chemical laboratory, and observatory. This university was one of the first to revive the study of the civil law ; though there be no good foundation for the common story that this revival was in consequence of the Pisans having found a copy of the Pandects among the spoils of Amalphi, sacked by them in 1140. The university has had many illustrious names in he soil of her professors, including, among others, Galileo, Farricelli, Redi, Malpighi, Thomas Dempster, Borelli, Castelli, Gronovius, &c. Besides the university, Pisa has an episcopal seminary, with about 80 students, several congregational female schools, normal and Lancasterian schools, &c.; and is an archbishop's see. Its manufactures, which re unimportant, consist chiefly of soap, white lead, vitriol, glass, and a few other articles; and its trade has sunk in proportion to the rise of that of Leghorn. The city is connected by navigable canals both with Leghorn and the Arno; but it is a curious fact that no vestige can now be found of its ancient port or roadstead at the mouth of the river, though, in the 13th century, it accommodated large fleets. This singular result is supposed to have been brought about by a change in the course and embouchure of the Arno.

Pisa derives some advantage from its being the winter resort of the grand duke and of the Tuscan court, as well as numerous invalids, attracted thither by the mildness of its climate. On the shore near the city, amidst an extensive forest of isles, is a farm belonging to the grand duke, where a number of camels have been reared, it is said, in the time of the crusades; and whence most of the zoological collections in Europe are supplied with these animals.

Morals seem to be at a lower ebb in this, than even in most other Italian cities.

> "All evils have contaminate the mind,
> That opulence departed leaves behind."

According to Simond, idleness, ignorance, and profligacy, n the general character of the inhabitants. "Every day our disgusting stories of meanness and dirty artifice in ry transaction of life." Foreigners cannot hire a house, make a bargain of any sort without being cheated. most of the ladies have their *cavalieri servienti*.

Most writers attribute the foundation of this city to colonists from Pisa in the Peloponnesus soon after the Trojan

> "Alpheus ab origine Pisa
> Urbs Etrusca solo."   *Æneid, lib. x. v. 179.*

became a Roman colony, and its port was, in the time of Strabo, an important naval station. It did not however, attain to any great distinction till the 10th century, when it took the lead among the commercial republics of Italy. In 11th century, its fleet of galleys maintained a superiority the Mediterranean, commanding the coasts of Sicily, Sardinia, Corsica, and Barbary, and assisting the French in the crusades. But in the 13th century, a struggle commenced with Genoa, which, after many vicissitudes, ended the total ruin of the Pisans. The latter were defeated at the island of Meloria, in 1284, in a great naval engagement by the Genoese, with the loss of the greater part of fleet, and above 16,000 men killed and taken prisoners. Pisa subsequently became the prey of various petty states, and was finally united to Florence in 1406. (*Eustace*, ii. 440-461; *Wood's Letters of an Architect*, ii. 393-; *Matthews*; *Williams*; *Forsyth*; *Condor's Italy*, iii. &c.)

CATAQUA. r., N. H., rises in Wakefield t., and runs S.E. about 40 m., falls into the Atlantic below Portsmouth, where it forms one of the best harbours in the United In the upper part of its course it is called Salmon Fiver, and including this, it constitutes, through most of its course, the boundary between New Hampshire and

Through Great bay which enters it on the S. side, receives Lamprey, Exeter, and several other streams on the S. and W. It is navigable, with its tributaries, for to South Berwick, Dover, New-Market, Durham, Exeter. Great Bay unites with it at Hilton's Point, les above Portsmouth, and is, in some places, four wide. The tide flows up the river at Portsmouth, powerful current, which prevents this fine harbour ng much obstructed by ice.

TAQUA. t., Middlesex county, N. J., 5 N. Newick. Bounded S. and S.W. by Raritan river, N.W. n Brook. It contains an Episcopal church, seven ve grist-mills, three saw-mills, two tanneries, eight 229 scholars. Pop. 2328.

ATAQUIS, r., Me., is a large western branch of of river, is about 100 m. long, and enters the Penobn. below the junction of Mattawamkeag river, the stern branch.

TIQUIS, co., Me. Situated in the N. part of the d contains a great amount of territory, extending to nada line. The N. part is a wilderness, only a small f the S. part being settled. Moose-head lake forms its W. boundary, and it has many other lakes. by the W. branch of Penobscot river, and by Pischer river and its branches. It contains Katadin mountainsolated summit, 5300 ft. above the level of the sea. ned in 1840, 8642 neat cattle, 17,830 sheep, 4730

swine; and produced 59,298 bushels of wheat, 2769 of rye, 12,983 of Indian corn, 4229 of barley, 57,675 of oats, 445,376 of potatoes, 8454 lbs. of sugar. It had 10 stores, three fulling-mills, one woollen factory, three flouring-mills, 19 grist-mills, 31 saw-mills, one oil mill, nine tanneries, two printing-offices, two weekly newspapers; one academy, 100 students; 143 schools, 5578 scholars. Pop. 13,138. Capital, Dover.

PISTOJA, (anc. *Pistoria*), a city of central Italy, grand duchy of Tuscany, prov. Florence, on the Ombrone, a tributary of the Arno, at the foot of the Apennines, 20 m. N.W. Florence. Pop. in 1836, 11,266. It is between 2 and 3 m. in circuit; is surrounded by old walls said to have been originally constructed by Desiderius, the last of the Lombard kings, and is further defended by a citadel built in 1352. It is clean, handsome, and well built, with unusually broad streets, and many curious and splendid edifices; but, like many other Italian towns, is dull, monotonous, and silent. Its cathedral is in the same style of architecture as that of Pisa and that of Lucca, but inferior to either; it has, however, some interesting monuments. The baptistery, constructed by Andrea Pisano, in 1337, is a small and handsome octagonal edifice. Several of the inferior churches are remarkable for their style of architecture or works of art: and the courthouse is a fine old building. Instead of the suppressed Jesuits' college, there is a seminary for priests; and a large hospital is kept in good order.

Pistoja has a well supplied market, a museum, two small public libraries, a large theatre, assembly-rooms, and a race ground. Provisions are cheap and good; the climate is cool and healthy; and the city is the residence of many noble and respectable families. The manufactures, which are on a small scale, include silk twist, straw hats, paper, musket barrels, cutlery, nails, iron ware, &c.; and near the city are some tolerably extensive iron-works.

Pistoja is with Prato a bishop's see, and has an Episcopal college, a superior private lyceum, with normal, girls', Lancasterian, and various inferior schools. In the middle ages it was the cap. of a republic which became subject to Florence about the same time with Pisa. (*Rampoldi; Bowring's Report; Condor's Italy*.)

PITCAIRN'S ISLAND. See POLYNESIA.

PITCHER, p. t., Chenango co., N. Y., 196 W. Albany, 332 W. Watered by Ostelic river. It contains two churches, a Presbyterian and Baptist, three stores, two fulling-mills, one woollen factory, four grist-mills, three saw-mills, two tanneries; one academy, 26 students, 16 schools, 871 scholars. Pop. 1562.

PITT, co., N. C. Situated toward the E. part of the state, and contains 800 sq. m. Bounded S. by Neuse river. Watered by Tar river. It contained in 1840, 5984 neat cattle, 4369 sheep, 31,037 swine; and produced 9751 bushels of wheat, 2570 of rye, 376.455 of Indian corn, 6424 of oats, 73,316 of potatoes, 510,329 lbs. of cotton. It had 15 commission houses in foreign trade, nine retail stores, one flouring-mill, six grist-mills, five saw-mills, 14 distilleries; one academy, 30 students; 4 schools, 63 scholars. Pop.: whites, 6128; slaves, 5648; free coloured, 30; total, 11,806.

PITT. t., Alleghany co., Pa. Bounded N. by Alleghany river, S. and W. by Monongahela river. The city of Pittsburg and the villages of East Liberty and Lawrenceville are within its limits. It contains an abundance of coal. Exclusive of Pittsburg, it contains 17 stores, one furnace, one forge, four glass-houses, two saw-mills, five tanneries; one academy, 20 students; 10 schools, 961 scholars. Pop. 6002.

PITTENWEEM, a royal and parl. bor. and seaport of Scotland, co. Fife, on the N. shore of the frith of Forth, 26½ m. N.E. by N. Edinburgh. Pop. 1317. It was created a bor. in 1537, and bears the marks of antiquity and decay. In addition to the parish church, the Episcopalians and the Relief synod have each a chapel. Pittenweem has a small harbour; but the only business connected with it is that of fishing to a limited extent. Here are the remains of a monastery of Augustine friars. Dr. Douglas, bishop of Salisbury,

> "The scourge of impostors, the terror of quacks."

was born here in 1721.

Pittenweem unites, with St. Andrew's, the two Anstruthers, Crail, Cupar, and Kilrenny, in sending one member to the House of Commons. Registered electors in this bor. in 1839-40, 49. Corporation revenue, £448. Number of councillors, 24.

PITTSBORO', p. v., capital of Chatham co., N. C., 34 m. W. by S. Raleigh, 332 W. Situated on Robinson's creek, a tributary of Haw river, and contains a courthouse, jail, two churches, an Episcopal and Methodist, an academy, and about 300 inhabitants.

PITTSBURG, city, port of entry, and capital of Alleghany co., Pa., is situated in 40° 30' N. lat., and 80° 2' W. long., 230 m. W.N.W. Baltimore. 297 m. W. by N. Philadelphia, 200 m. W.N.W. Harrisburg, 296 W. The population in

1816, was 4768; in 1820, 7248; in 1830, 12,542; in 1840, 21,115, being the second city in population in the state, and the thirteenth in the United States. With the adjoining cities and boroughs of Alleghany, Manchester, Birmingham, and Lawrenceville, and other suburbs of Pittsburg, there is a population of not less than 40,000.

The city is situated at the confluence of the rivers Alleghany, 400 yards wide, and Monongahela, 450 yards wide, which by their union here form the Ohio river, which is at this place a quarter of a mile wide, and navigable by steamboats to the gulf of Mexico, through the Mississippi river. It is built on a beautiful plain between the two rivers of a triangular form. About a mile back of the point it is surrounded by Grant's, Ayers's, and Quarry hills. It is compactly built with many handsome edifices, chiefly of brick, to which however a dingy appearance is given by the dust of the bituminous coal, so extensively used in its manufactories and otherwise. It was laid out in 1765 on the N.E. bank of the Monongahela river, with streets running parallel with the river, and crossed by others at right angles with them. The streets on the Alleghany river also run parallel with the river, and are crossed at right angles with other streets. The cross streets meet each other obliquely a few streets back from Alleghany river. Three covered bridges cross the Alleghany river, one of which has a walk on the top of the roof. A bridge also crosses the Monongahela river, which cost $102,000. The Pennsylvania canal crosses the Alleghany river in an aqueduct, and several ferries cross the Monongahela. The harbour of Pittsburg is chiefly on the Monongahela, where the water is deeper than on the Alleghany. Eighty-nine steamboats employed on the Ohio and connected rivers, with an aggregate of 12,436 tons are owned either wholly or in part in Pittsburg. The hills with which the city is surrounded are filled with bituminous coal, which can easily be brought to the city, and affords unequalled facilities for manufacturing.

Among the public buildings of Pittsburg, the new courthouse is a splendid edifice, at an elevation which commands an extensive view of the three rivers, with the city and the surrounding country, adorned with hills and valleys, villages and steamboats. It is 165 feet long and 100 feet wide, of the Grecian Doric architecture, having in the rear a spacious and well conducted prison. It is surmounted by a dome, 37 feet in diameter at the base, and 148 feet high above the level of the street, and cost about $200,000. A splendid Roman Catholic cathedral is located on Grant's hill. The Western university of Pennsylvania is also situated near Grant's hill. The third Presbyterian church, several of the banks, and two splendid hotels, the Exchange hotel and the Monongahela house, which are the largest in the western country, have much that is imposing in their appearance. There is a museum which contains many aboriginal curiosities, three market-houses, and several literary societies, with small libraries, which would probably be more efficient if united in one large institution. The city is supplied with water raised from the Alleghany river, a stream which yields fine water, by a steam engine, which raises 3,000,000 gallons daily, and which is sent over the city in pipes, which have a total length of over 9¼ miles. The water works cost over $250,000. The city is lighted with gas, extended through its principal streets, by gas works which were completed in 1836. The bituminous coal, which abounds in the vicinity, furnishes great facilities for the production of gas, at a little expense, and which affords a brilliant light.

There are 35 churches in Pittsburg, five Presbyterian, one Reformed Presbyterian, four Methodist, one Protestant Methodist, one Cumberland Presbyterian, three Baptist, two Episcopal, five Scots Presbyterian, two Lutheran, two Congregational, three Welsh Methodists, one Unitarian, one Disciples, three Roman Catholic and one African. There are four banks, with an aggregate capital of $3,000,000; and three insurance companies, with a capital of $540,000; and a bank for savings. The tonnage of the port in 1840 was 12,000. It had according to the census of 1840, seven commercial, and 32 commission houses in foreign trade, with a capital of $1,241,110; 408 retail stores, with a capital of $4,165,190; 17 lumber-yards with a capital of $167,000; 23 furnaces and five forges, with a capital of $1,319,000; machinery was manufactured to the amount of $443,500; hardware and cutlery to the amount of $276,500; five cannon and 1350 small arms were manufactured; the precious metals were manufactured, producing $14,860; various metals, $196,700; one fulling-mill, and one woollen manufactory employed a capital of $10,000; two cotton factories with 3000 spindles, four tanneries, five breweries; paints and drugs were manufactured with a capital of $203,300; four glass houses, two glass-cutting works, two flouringmills, five saw-mills, one oil-mill; 18 printing-offices, seven binderies, four daily and 11 weekly newspapers; 53 brick or stone houses, and 15 wooden houses were built, which cost $161,900. The total capital employed in manufactures

was $2,037,952. There were one college with 50 students; nine academies with 755 students, and 18 schools with 259 scholars.

Several places in the vicinity of Pittsburg deserve to be described with it, although they are under separate incorporations. The most important of these is Alleghany city, on the opposite side of Alleghany river. It contains 15 churches, two Presbyterian, two Reformed Presbyterian, two Cumberland Presbyterian, two Methodist, two Protestant Methodist, two Scots Presbyterian, a Baptist, an Episcopal, a German Reformed, and a Disciples. It contains the Western Theological seminary, founded in 1822, which has two professors, 20 students, 182 have been educated, and has 6000 volumes in its libraries. It is under the direction of the Presbyterians. The edifice occupies a commanding eminence, is 140 feet long and 50 feet wide; the central part being four stories, and the wings, three stories high. It is also the seat of the theological seminary of the Associate Reformed church. Alleghany city presents many fine residences, in commanding situations, occupied by many persons doing business in Pittsburg, who are pleased to retire from the bustle, smoke and coal dust of the principal city. Its commercial and manufacturing business, properly its own, is considerable. It contained in 1840, 52 stores, with a capital of $63,400; nine lumber-yards with a capital of $50,000; one furnace, and one forge, with a capital of $216,000; value of hardware and cutlery manufactured, $50,000, three cotton factories, with 14,270 spindles, two tanneries, one brewery, one rope-walk; total capital in manufactures $726,640; one theological college, 31 students; three academies, with 169 students; 12 schools, 709 scholars.

Birmingham, borough, the postoffice of which is called Buchanan, is another considerable suburb of Pittsburg. It occupies the S. side of Monongahela river, at the distance of about a mile from the centre of the city. It is connected with Pittsburg by a bridge 1500 feet long, and by a ferry, and has important manufactures of glass and iron. It contains two churches, a Presbyterian and Methodist, and a market-house. It had in 1840, 10 stores, capital $14,300; two lumber-yards, capital $9300; one forge, six glass-houses, four glass-cutting works, capital $140,000, one pottery; total capital in manufactures $153,750; four schools, 115 scholars. Pop. 1554. There are a number of other flourishing villages in the vicinity of Pittsburg.

Pittsburg occupies the site of the former French fort Du Quesne, which remained in their possession from 1754 to 1758, and whence they instigated the Indians to hostilities against the frontier settlements of Pennsylvania. In 1755 General Braddock lost his life, and the army suffered a defeat in attempting to drive the French from the post. The youthful Washington displayed his military skill and gave promise of his future greatness, by conducting in a masterly manner the retreat of the shattered forces. A more formidable force was collected at Carlisle, under General Forbes, for the capture of fort Du Quesne. Major Grant, who was sent forward with a force of 800 men, was surrounded by the French and Indians, on what is now called Grant hill in Pittsburg, and captured, after losing about 300 men in killed and prisoners, himself being among the captured. General Forbes, however, pressed forward, and before he reached the place, the French burned their fort and retired. Near this a temporary stockaded fort was built on the bank of the Monongahela, which was called fort Pitt, in honour of the distinguished British statesman. In 1759 General Stanwix commenced the construction of a more formidable work, nearer the site of fort Du Quesne, which cost the British government £60,000 sterling. This fort was unsuccessfully attacked by the Indians in 1763. In 1764 peace having been restored between the British and French, a brick redoubt was built, remains of which are still visible. Until the close of the revolutionary war, Pittsburg advanced slowly. In 1773 the number of dwelling houses, within the present limits of the city, amounted to from 25 to 30. The ground on which it stands belonged to Penn's manor, and was the property of the family. In 1784 it was laid out into town lots, which sold rapidly. In 1786 the first number of the Pennsylvania Gazette was published, and in it, it was stated that the number of houses in the village was about 100. In 1788 Alleghany county was constituted, and in 1791 Pittsburg was made its capital, and a lot was purchased for the site of a courthouse and jail. Among those who came in 1794 to quell the whiskey insurrection, there were many enterprising young men who were pleased with the situation of Pittsburg, and resolved to make it their future abode. The large sums expended by the army in this expedition and in the Indian wars, had a favourable influence on the business of the place.

In the Pittsburg Gazette it is stated Jan. 9th, 1796, that by a census just taken by the assessors, the population amounted to 1395. This is the earliest authentic account of the population of the place. During this year Lewis ippe, the present king of the French, visited Pittsburg, upon a considerable time in the place. In 1811 the steamboat constructed on the western waters, named Orleans, was built at Pittsburg. This invention has ly changed the character of the trade of the west. its introduction it was the work of a whole season. te a trip to New-Orleans and to return; now many de in a season. In 1814, the manufactures of Pittsmployed 1637 hands, and amounted to $2,265,366. ansylvania canal constructed between 1825 and 1836, of great importance to Pittsburg and the west. It ured the transportation from Philadelphia to Pittsm 3 or 4 dollars to less than one dollar for a hundred besides greatly facilitating the travel on this route, ing the time and reducing the expense to less than former amount. The Pennsylvania canal crosses ghany river by a splendid aqueduct 1200 feet long. the state at an expense of $104,000. After enterburg it passes through a tunnel under Grant's hill, rs the Monongahela river. Another branch of the see through Alleghany city and enters Alleghany w hundred yards above the head of Ohio river. se manufactures, mechanical products, and sales ds, foreign and domestic, in Pittsburg, were estifrom $20,000,000 to $25,000,000 of dollars. Pittsburg red deservedly the name of the Birmingham of U It is a great work shop, and industrious hive; ay who once toiled in its manufactories, now live , and ride in coaches, still practising the economy heir wealth has been acquired.

[ELD, p. t., Merrimac co., N.H., 15 m. N.E. Con-. Drained by Suncook r., which affords waterntains three churches, a Congregational, Methose-will Baptist; five stores, one cotton factory andles, one grist-mill, two saw-mills, three taaacademies, 168 students; 11 schools, 461 schol-'16.

o, p. t., Berkshire co., Mass., 131 m. W. Boston, x, 33 m. E.S.E. Albany, 376 W. Drained by Housatonic river, which afford water-power. from Boston to Albany passes through it. as been greatly improved here. The village e, is the largest in the county, and the most is part of the state. The houses are mostly in fine taste, and neatly painted white, with , ornamented with shrubbery. It contains a Congregational, Episcopal, Methodist, and ik, and the Berkshire medical institution. is connected with Williams college, was 3, has five professors, 103 students, and 473 lectures commence on the first Thursday The village contains a male and a female ting-office, 350 dwellings, and 2500 inhabicentral square, of 4 acres, ornamented by a emains of the original forest. The townnumber of stores, one furnace, one cotton 0 spindles, two grist-mills, eight saw-mills, one brewery, one printing-office, one weekne academy, 68 students; 15 schools, 696 747.

t., Otsego co., N.Y., 18 m. W. by S. m. W. Albany, 352 W. Bounded W. by id drained by its tributaries. It contains 'ulling-mill, one cotton factory with 3500 t-mill, nine saw-mills; five schools, 160 95.

., capital of Pike co., Ill., 70 m. W.S.W. ., Situated on elevated ground on a rollv between Illinois and Mississippi rivers, thouse, jail, five stores, various mechanic 50 inhabitants.

t., Rutland co., Vt., 70 m. S.W. Moataartered in 1761; first settled in 1769. creek, and its tributary Furnace brook, r-power. It abounds with iron ore, and oarse grained, and slightly flexible. The s also found. It contains three churches, tional, and Methodist; four stores, two ies; 18 schools, 650 scholars. Pop. 1927.

Monroe co., N.Y., 6 m. E. Rochester, bany, 362 W. Drained by Irondequoit ios. The Erie canal, and the Auburn ad, pass through it. It contains three erina, Baptist, and Methodist; three mill, one tannery, one brewery; 13 Pop. 1983.

t., Salem co., N.J., 16 m. E. Salem. of Maurice river, and head waters of creeks. It contains four churches, sbyterian, and Baptist; six stores, one grist-mills, three saw-mills; seven Pop. 2390.

PITTSTON, p. t., Kennebec co., Me., 7 m. S. by E. Augusta, 595 W. Bounded W. by Kennebec river, small tributaries of which afford water-power. Incorporated in 1779. Lumber is exported. It contains nine stores, three gristmills, 11 saw-mills; 16 schools, 696 scholars. Pop. 2460.

PITTSTOWN, p. t., Rensselaer co., N.Y., 12 m. N.E. Troy, 25 m. N.E. Albany, 393 W. Bounded N. by Hoosick river. Drained by Tomhannock creek. It contains three churches, a Presbyterian, Methodist, and Baptist; 10 stores, three fulling-mills, one woollen factory, three cotton factories with 2890 spindles, one flouring-mill, five grist-mills, 13 saw-mills, one oil-mill, two tanneries; 20 schools, 857 scholars. Pop. 3784.

PITTSYLVANIA, county, Va. Situated in the S. part of the state, and contains 891 sq. m. Bounded N. by Staunton river. Watered by Dan and Bannister rivers. It contained in 1840, 19,451 neat cattle, 19,277 sheep, 42,513 swine; and produced 143.178 bushels of wheat, 5747 of rye, 679,319 of Indian corn, 233,763 of oats, 24,409 of potatoes, 6,438,777 pounds of tobacco, 18.499 of cotton. It had 25 stores, four flouring-mills, 50 grist-mills, 37 saw-mills, 13 tanneries, four distilleries, one printing-office, one weekly newspaper; four academies; 195 students; 38 schools, 817 scholars. Pop., whites, 14,263; slaves, 11,55d; free coloured, 557; total, 26,398. Capital, Competition.

PITTSYLVANIA, C. H. (called also Competition), p. v., capital of Pittsylvania co., Va. Situated on Cherrystone branch of Bannister river, and contains a courthouse, jail, a Methodist church, about 125 dwellings, many of them of brick, and neat, and about 300 inhabitants.

PLAIN, p. t., Wayne co., O., 92 m. N.E. Columbus, 349 W. Pop. 2134.

PLAINFIELD, p. t., Sullivan co., N.H., 62 m. N.W. Concord, 478 W. Bounded W. by Connecticut r. Queechy falls, in this river, are opposite to the town, and a bridge crosses the river. Hart's island in Connecticut river, containing 19 acres, belongs to the town. Union Academy, founded in 1813, to aid young men for the gospel ministry, received a bequest from Hon. Daniel Kimball, of $40,000, and is flourishing. It contains three churches, two Congregational, and a Baptist; three stores, two grist-mills, five saw-mills, two tanneries; one academy, 175 students; 11 schools, 613 scholars. Pop. 1552.

PLAINFIELD, p. t., Windham co., Ct., 45 m. E. Hartford, 372 W. Bounded W. by Quinnebaug river, and watered by its tributary, Moosup river, which affords water-power. The Norwich and Worcester railroad passes through it, a little W. of the village. The village is built chiefly on one extended street, on elevated ground, and contains a Congregational church, an academy, founded in 1783, and a number of neat dwellings. The town contains four churches, a Congregational, two Baptist, and a Friends; six stores, two. woollen factories, seven cotton factories with 15,900 spindles, four grist-mills, five saw-mills, nine oil-mills; 12 schools, 692 scholars. Pop. 2383.

PLAINFIELD, p. t., Otsego co., N.Y., 15 m. N.W. Cooperstown, 84 m. W. Albany, 372 W. Watered by tributaries of Unadilla river, which bounds it on the W. It contains a Presbyterian church, two stores, two fulling-mills, two gristmills, four saw-mills; 16 schools, 491 scholars. Pop. 1458.

PLAINFIELD, p. v., Westfield t., Essex co., N.J., 30 m. N.E. Trenton, 206 W. Situated on Green brook, and contains five churches, a Presbyterian, Methodist, Baptist, and two Friends; a bank, four stores, several mills and manufactories, an insurance company, a ladies' library, an apprentices library, 150 dwellings, and about 1000 inhabitants.

PLAINFIELD, t., Northampton co., Pa. Drained by the E. branch of Bushkill creek. It has four stores, three gristmills, four saw-mills, one oil-mill, two tanneries. Pop. 1501.

PLAQUEMINE, parish, La. Situated in the S.E. part of the state, and includes the mouths of the Mississippi proper. It contains about 2500 sq. m. It is level and low, the most elevated parts being not more than 10 feet above the level of the gulf of Mexico. Much of it is liable to submersion. The arable soil is extremely fertile, and is confined to the margin of the Mississippi. It contained in 1840, 4194 neat cattle, 1832 sheep, 224 swine; and produced 100,185 bushels of Indian corn, 985,500 pounds of rice, 10,320,000 of sugar. It had 15 stores, one distillery, one brewery; one college, 38 students. Pop.: whites, 1551; slaves, 2385; free coloured, 324; total, 5060. Capital, Fort Jackson.

PLAQUEMINE, p. v., capital of Ibberville par., La., 112 m. W.N.W. New-Orleans, 1192 W. Situated on the S. side of Plaquemine bay, at its efflux from Mississippi river. It contains a courthouse, 15 stores, and about 250 inhabitants.

PLASENCIA, a fortified city of Spain, in Estremadura, on a peninsula almost surrounded by the Jertes, a tributary of the Tagus (crossed here by three bridges), 102 m. N. by E. Badajos, and 190 m. W. by S. Madrid. Pop., according to Miñano, 6787. It stands in a plain, surrounded N. and

E. by high mountains, is surrounded by strong walls, entered by six gates, and has several pretty wide, level, and well-paved streets, with seven parish churches, a cathedral, an episcopal palace, five hospitals, a fine old aqueduct of 80 arches, which conveys water to the town, and a private collection of antiquities. The cathedral, a modern Gothic structure, is not completed: the chapter comprises a bishop, eight dignitaries, and 24 canons. The manufactures comprise leather, hats, woollen, linen, and hempen cloths, and the surrounding plain, or *puerta*, is extremely fertile, producing large quantities of grain, fruits, oil, &c.

Plasencia, though not the *Ambracia* of the Romans, is proved, by the numerous antiquities found in it, to be of remote origin. The present city, however, was built near the end of the 12th century, by Alphonso IX. of Castile. It was formerly possessed by its own lords, and gave title to a duchy; but, in 1442, it was united to the crown of Castile. (*Dillon's Travels*, p. 282; *Mod. Trav.; Miñano.*)

PLASSEY, a village of Bengal, on the Hooghly river, 80 m. N. Calcutta, and 30 m. S. Moorshedabad. This village will be ever memorable in Indian history, for its having been the scene of the great victory gained by Lord (then Colonel) Clive, on the 23d of June, 1757, over Suraja Dowla, soubahdar of Bengal. Clive's army consisted of only 900 Europeans, 2100 Sepoys, and 100 Topasses; yet, with this small force, he did not hesitate to attack the soubahdar's army, of 50,000 foot, and 18,000 horse, supported by a formidable train of artillery! This, however, was not so Quixotic a proceeding as it may at first sight appear; for Clive knew that the native troops had no confidence in their general, and, in fact, they instantly gave way, so that the victory was at once complete, and easily won. The result of this contest threw Bengal into the hands of the English, and laid the foundation of our empire in India. (*Mill's India*, iii., 152.)

PLATA, LA (REPUBLIC OF), or ARGENTINE REPUBLIC, an independent state, or rather confederation of states, in S. America, extending between the 22d and 41st degrees of S. lat., and the 54th and 72d of W. long., having N. Bolivia; E. Paraguay, Brasil, and the Banda Oriental, from all which it is separated by the Paraguay, Parana, and Uruguay rivers; S. the Atlantic ocean and Patagonia; and W. Chili and Bolivia. Sir Woodbine Parish gives the following estimate of the area, population, and subdivisions of this territory:

Provinces.	Area in sq. m.	Estim. Pop 1836-37.	Chief Towns.
Buenos Ayres		160,000 to 200,000	Buenos Ayres
Santa Fe		15,000 — 20,000	Santa Fe
Entre Rios		25,000 — 30,000	Parana
Corrientes		35,000 — 40,000	Corrientes
Cordova		80,000 — 85,000	Cordova
Santiago		45,000 — 50,000	Santiago
Tucuman	726,000	40,000 — 45,000	Tucuman
Salta		50,000 — 60,000	Salta
Catamarca		30,000 — 35,000	Catamarca
La Rioja		15,000 — 20,000	Rioja
San Luis		30,000 — 35,000	San Luis
Mendoza		35,000 — 40,000	Mendoza
San Juan		22,000 — 25,000	San Juan
Total	726,000	600,000 to 675,000	

Inasmuch, however, as the population of the city of Buenos Ayres may itself be estimated at about 100,000, we are inclined to think that this estimate is under the mark, and that the population may be set down at about a million. It has sometimes, indeed, been estimated at two millions and upwards; but there can be no doubt that this is very far beyond the mark. The Indians, who are not included in any of the above estimates, are, probably, under 50,000.

On the W. this territory is bounded by the great Cordilleras of the Andes; and the N.W. prov. of Salta is almost wholly mountainous, as are extensive portions of the adjacent provinces of Catamarca and Tucuman. Some points of the Despoblado chain in Salta rise probably to the height of 13,000 feet; and in Cordova are isolated chains, which any where but in the neighbourhood of the Andes would be called mountains. Still, however, after allowing for these and other deductions, five sixths of the country consist of plains, several of which are of vast extent. But notwithstanding its freedom from mountains, and the number and magnitude of its rivers, it is far from being the fertile region that is so generally supposed; and a large proportion of its surface is, in fact, condemned to perpetual sterility. In the N. is the S. portion of the immense tract known by the name of the *Gran Chaco*, a vast plain, occupying the whole triangular space between Bolivia on the N., and the great rivers Paraguay on the E., and Salado on the W. This immense plain, which probably covers from 110,000 to 120,000 sq. m., is but little known; in the N. it is covered with extensive forests, but its more S. portion, between the Vermejo and Salado, is, in fact, a sandy, arid, and uninhabitable desert. This, also, is the character of the extensive tract between the Salado and the Rio Dulce, and W. from the latter, as far as the Sierra Velasco, in about the 68 degree

of W. long., extends the great Salt Desert of Salinas, in great part covered with saline efflorescence, and extremely hot. The great southern plain, or that which extends over the whole country S. of the 33d degree of lat., is, fortunately, of a very different character from those previously noticed. This vast tract, which includes an area of above 300,000 sq. m., is called the *Pampas*. It may, taking its vast size into account, be regarded almost as a dead level, its slope towards the E. being gradual and imperceptible. It is interspersed with innumerable lakes; but these, as well as most of the rivers by which they are fed are brackish, the soil through which they flow being strongly impregnated with salt. Perfectly fresh and portable water is, however, found at the depth of from 30 to 50 feet. Substantially, however, the Pampas are divided into several regions, differing in climate and produce, though under the same latitude. "On leaving Buenos Ayres, the first of these regions is covered for 180 m. with clover and thistles alternately; the 2d region, which extends for 450 m., produces long grass; and the 3d, which reaches the base of the Cordillera, is a grove of low trees and shrubs. The 2d and 3d of these regions have nearly the same appearance throughout the year, for the trees and shrubs are evergreens, and the immense plain of grass only changes its colour from green to brown; but the 1st region varies with the four seasons of the year in a most extraordinary manner. In winter the leaves of the thistles are large and luxuriant, and the whole surface of the country has the rough appearance of a turnip field. The clover in this season is extremely rich and strong; and the sight of the wild cattle grazing in full liberty on such pasture is very beautiful. In spring the clover has vanished, the leaves of the thistles have extended along the ground, and the country still looks like a rough crop of turnips. In less than a month the whole region becomes a luxuriant wood of enormous thistles, which have suddenly shot up to a height of 10 or 11 feet, and are all in full bloom. The path is hemmed in on both sides; the view is completely obstructed; not an animal is to be seen; and the mass of the thistles are so close to each other, and so strong, that, independent of the prickles with which they are armed, they form an impenetrable barrier. The sudden growth of these plants is quite astonishing; and though it would be an unusual misfortune in military history, yet it is really possible that an invading army, unacquainted with the country, might be imprisoned by these thistles before it has had time to escape from them. The summer is set over before the scene undergoes another rapid change. the thistles suddenly lose their sap and verdure, their heads droop, the leaves shrink and fade, the stems become black and dead, and they remain rustling with the breeze one next another until the violence of the pampero, or hurricane, levels them with the ground, where they rapidly decompose and disappear; the clover rushes up, and the scene is again verdant. The vast region of grass in the Pampas for 450 m. is without a weed, and the region of wood is equally extraordinary. The trees are not crowded, but in their growth such beautiful order is observed, that one may gallop between them in every direction. The whole country is in such beautiful order, that if cities and millions of inhabitants, could suddenly be planted at proper intervals and situations, the people would have nothing to do but to drive out their cattle to grass, and without any previous preparation, to plough whatever quantity of ground their wants may require.

"The climate of the Pampas is subject to a great difference of temperature in winter and summer, though the changes are very regular. The winter is about as cold as our month of November, and the ground at sunrise is always covered with white frost, but the ice is seldom more than 1-10th of an inch thick. In summer the sun is oppressively hot. The difference, however, between the atmosphere of Mendoza, San Luis, and Buenos Ayres, which are all nearly under the same latitude, is very great; in the two former, or in the regions of wood and grass, the air is extremely dry; there is no dew at night; in the hottest weather there is apparently very little perspiration, and the dead animals lie on the plain dried up in their skins. But in the province of Buenos Ayres, or in the region of thistles and clover, vegetation clearly announces the humidity of the climate, and the dead animals on the plain are in a rapid state of putrefaction. On arriving at Buenos Ayres, the walls of the houses are so damp that it is cheerless to enter them; and sugar, as also all deliquescent salts, are there found nearly dissolved. This dampness, however, does not appear to be unhealthy. The S. part of the Pampas is inhabited by Indians, who have no fixed abode, but wander from place to place as the herbage are and these becomes consumed by their cattle. The N. part and the rest of the provinces of La Plata are inhabited by a few straggling individuals, and a few small groups of people who live together only because they were born together. The travelling across the Pampas is really a very astonishing effort. The country has no road but a tract which is constantly changed. The huts, termed posts, are at different distances, but, upon an average, about 20 m. from each other; and in travelling with carriages, it is necessary to send a man before to request the peaches to collect their horses. The country is intersected with streams, rivulets, and rivers, with pantanos (marshes), &c., through which it is absolutely necessary to drive. In one instance, the carriage, strange as it may seem, goes through a lake, which of course is not deep. The banks of the rivulets are often very precipitous, and I constantly remarked that we drove over and through places which, in Europe, any military officer would, I believe, without hesitation, report as impassable. The most independent way of travelling is, however, on horseback, without baggage, and without an attendant. In this case the traveller has to saddle his own horse, and to sleep at night upon the ground on his saddle; and as he is unable to carry any provisions, he must throw himself completely on the feeble resources of the country, and live on little else than beef and water." (*Head's Journeys cross the Pampas*, p. 9–10, 46–50.)

Many of the minor plains are of a very different character from either this or the Gran Chaco; and some, as those of Tucuman, yield corn and maize, rice, tobacco, the sugarcane, &c., in the greatest abundance. The provinces of Cordova, Salta, &c., are also in parts very fertile. In general, the N.W. provinces are the most productive of grain, like the E. provinces, or those between the Parana and the Uruguay, and the S.E. provinces, abound most in cattle, and furnish the greater portion of the exports from Buenos Ayres.

The Argentine Republic, excepting a small portion towards its S. extremity, watered by the Rio Colorado, and a few smaller rivers, is wholly comprised in the basin of the Plata. (*See next art.*) Its vast branches supply the most extensive means of internal communication; though the utility of some of them is at present a good deal impaired by the prohibitory measures of Dr. Francia, dictator of Paraguay. Many large rivers water the great plains; but several of these lose themselves in the considerable lakes variously noticed, without finding their way to the sea. The principal of these is the Rio Dolce, which intersects the provinces of Tucuman and Santiago, and falls into the lake *ios Porongos*, or great salt lake. 35 m. in length, by 20 m. breadth. Several lakes and swamps, of a size little inferior to this, are laid down in maps as existing in the Pampas; but the largest lake, or lagoon, is that of Ybera, in Corrientes, which extends over 1000 sq. m., and supplies four considerable rivers. It is probable that the Parana formerly its course through this lake: at present no stream runs in it, and it is supposed to derive its waters through some underground drainage.

Geology and Minerals.—The N. E. and S.W. shores of Rio de la Plata present the greatest contrast in their geological features. The N. shore is elevated, and, like the bed in the river's bed above Buenos Ayres, composed of gneiss, and clay-slate; while, on the S. side, every sort of rock is entirely lost, and for hundreds of miles inland even the smallest pebble is to be met with. As we are yet acquainted with it, the whole of the vast forming the Pampas appears to be one immense bed of fluvial sand, quietly deposited, during the lapse of ages, what was anciently a gulf of the Atlantic, of which the story of the Plata is now the only remaining portion. The process appears to be, at present, going on here also: the estuary, which, in the 16th century, is reported to have been deep enough for ships of any burden, is gradually filling up, and we may expect that, at some future period, instead of discharging itself by a wide mouth as at present, the Plata, probably, in the most ancient times, the Nile, will enter the ocean by a delta, like the Amazon, the Ganges, or the Nile of our day. In the alluvium of the Plata, vast quantities of marine shells, and the remains of *Megatherium clyppidon*, &c., have been found, and, according to Mr. Darwin, its whole area is one wide sepulchre of extinct quadrupeds. (*Voyage of the Adventure and Beagle*, iii., 155, &c.; *Parish's Buenos Ayres*, 164, 165.)

The precious metals, with copper, lead, iron, &c., are found in different parts of the country; but, speaking generally, its mineral riches have been very imperfectly explored. Gold and silver have, however, been obtained in considerable quantities, in various districts in the N.W. and provinces connected with the Andes, especially at Salta, in Rioja, where the ores of silver are said to be rich. Humboldt estimated the total value of the gold and silver obtained from mines and washings in the captainship of Buenos Ayres, at the commencement of the century, at nearly 5 millions of dollars a year. But the latest portion of what is now called Bolivia was included within the limits of the government; and as seen no late accounts of the produce of the mines, it would be safe to place the smallest reliance:

perhaps it does not, at present (1841), amount to £400,000 a year. Iron is also believed to exist in the Chaco, in extensive veins, intermixed with small proportions of nickel and cobalt; and Sir W. Parish considers it extremely probable, that the immense mass of metal presented by him to the British Museum, and considered meteoric, is rather a production of the soil. (*Buenos Ayres*, 256–263.)

Salt is the most abundant mineral, and exists in a state of efflorescence over the surface of immense tracts, in a multitude of brackish springs and pools, and in mines of rock salt. Epsom and Glauber salts, limestone, gypsum, alum, mineral pitch, and an abundance of sulphur, are to be met with along the Cordillera, besides bituminous shale, with appearances of coal in many places, and it is said, though the statement requires confirmation, that there are extensive beds of coal in the extreme S.W. angle of the country.

The *vegetable products* in the N. provinces include most of those which flourish between the tropics; while in the S. they are in general similar to those of S. Europe. But even as far S. as Corrientes, cotton, tobacco, rice, sugar cane, indigo, and many other articles of primary importance in the markets of Europe, may be produced to almost any extent; and a large extent of country is extremely well adapted to the culture of wheat, maize, and other grains. Wheat, which, till of late, was little cultivated, has now become an article of export. But the implements and processes of agriculture are still in the rudest possible state. "In many towns of the interior, a common wheelbarrow is as yet unheard of; while in the capital itself, the first pump ever seen in a private house was put up a very few years ago by an Englishman!" (*Parish's Buenos Ayres*, p. 357.)

The immense tract annually inundated by the Plata, now wholly in a state of nature, might, it is said, be made available for the culture of rice on a most extensive scale. The inhabitants of Arauco, a department of La Rioja, are principally employed in the culture of the vine, and make annually from 5000 to 10,000 small barrels of a strong sweet wine, which is sent to Cordova and the neighbouring provinces. A good deal of strong and full-bodied wine and brandy is also sent from Mendoza to Buenos Ayres. The demand for sugar in the inland provs. is not sufficient to induce the country people to attend much to the cane; but tobacco is pretty largely cultivated, and finds a ready sale in Tucuman and the adjacent provs. Catamarca supplies all the surrounding states with cotton of a superior quality, for their domestic manufactures; and exports large quantities of red pepper to Buenos Ayres. The cactus, which feeds the cochineal insect, grows in Santiago, Salta, and elsewhere in the W., and to an unusual size: and from the first named province from 8000 to 10,000 lbs. a year of cochineal were formerly sent to Chili and Peru. Aloes are equally abundant, and from the macerated fibres the Indians of the Chaco make yarn, ropes, fishing-nets, and a variety of bags and pouches, for which there is always a demand among their more civilised neighbours; these articles are dyed in indelible colours, prepared by the Indians from native plants. The cocos plant (*Erythroxylon peruvianum*), which, when mixed with lime, forms a stimulant chewed by the Peruvian, as the betel in S. Asia, grows plentifully in Salta. Bonpland found three new species of indigo in Corrientes; madder is indigenous in several places; and on one species of shrub a small insect called the *clavile* is found, which affords a most beautiful green dye. The E. flanks of the Andes, and the banks of the W. affluents of the Paraguay, are covered with dense forests; which are, however, for the most part, useless, being at so great a distance from the Plata and seaports of the republic. The trees are mostly of the mimosa family; and from the fruit of the algaroba, mixed with maize, the Indians make cakes; and by fermentation, produce their chica, a strong intoxicating spirit, in very general use. The *guinquina*, or Peruvian bark tree, various palms, and the *yerba maté*, or Paraguay tea, are indigenous in Salta, &c. Figs, oranges, peaches, walnuts, apples, and other fruits, have been introduced by and Europeans, flourish with great luxuriance in many of the central and S. provinces; the want of population being, for the most part, the only great drawback to the development of the vast natural resources of the country.

But the chief source of wealth is the immense herds of horned cattle which wander over the widely-extended plains of the Pampas. Formerly, the greater number of these were unappropriated, like the land they roamed over, and were lassoed and destroyed at pleasure for their hides or fat, though sometimes, also, for their flesh, and sometimes for their tongues only, the rest of the body being left to beasts and birds of prey. But the Pampas are no longer unappropriated; a large part has been carefully measured by the government officers, and allotted to individuals, the extent and boundaries of whose estancias, or estates, are duly registered. Every animal a year old is branded with the mark of the owner, and that mark, being registered by the authorities, entitles him to claim his property wherever found. According to Sir W. Parish, it has been estimated that in the single province of Buenos Ayres there are from three to four million head of cattle, and above another million in the other provinces. The number must certainly be immense; for hundreds of thousands have, on some occasions, perished through inundation and drought without sensibly affecting the supply for the market.* But all estimates of the number of cattle in these provinces are purely conjectural, and entitled to little weight; though we incline to think that Sir W. Parish has much underrated the numbers of cattle in the provinces other than Buenos Ayres. In Entre Rios, before the revolution, "an estate of 3 leagues in length, by 2½ in breadth, that is, of 12½ sq. leagues, might have had upon it generally about 8,000 head of horned cattle, and 15,000 horses. The price of it, with stock, &c., might be—the horned cattle at 2s. each = £800; the horses at 6d. each = £375, and fixtures £100; cost, therefore, of stock and fixtures, £1,275, leaving the estate of 37½ sq. m. as a bonus to the purchaser!" (Robertson's Letters on Paraguay, ii. 215, 216.)

Cattle-rearing estates are frequently of vast size; and Candioti, probably the largest landholder of La Plata, is said, by Mr. Robertson, to have been the proprietor of 300 sq. leagues of territory, with 250,000 head of horned cattle, and 30,000 horses and mules. The annual increase upon a well-regulated estancia varies from 30 to 40 per cent., which yields an enormous profit to the proprietor, whose expenses are comparatively trifling. And since the revolution, which has thrown open the trade of S. America, the cattle, which were previously not worth more than 3s. or 4s. a head, are now worth 20s. and for these 20s. the farmer can buy double the quantity of the necessaries and luxuries (his own commodity of cattle always excepted), which he could procure for them before. The estancieros of Santa Fé were formerly among the richest in the viceroyalty, and furnished by far the greater part of the 50,000 mules yearly sent to Salta for the service of Peru; but the stoppage of the trade with the latter country and Paraguay, and the attacks of the Indians, have impoverished and depopulated that province almost to the last degree. The mules were commonly sent when two or three years old, to a periodical fair, near Salta, to which the purchasers from Peru repaired, and bought them in droves, at the rate of from 14 to 16 dollars each. The struggle for independence stopped this traffic; for Peru being to the last in the possession of the royalists, all intercourse with Salta was long cut off, and not having been renewed to any great extent, the breeding of mules has declined considerably in La Plata.

The horses of the Pampas are similar to the common Spanish horse, and of all colours; like the cattle, the original stock was introduced from Europe. They wander wild in immense herds, being caught indiscriminately by the gauchos, whenever they are required. Though as serviceable to the gaucho as to the Arab, the former, owing, no doubt, to the animal being raised without any attention on his part, cares very little for his horse; he goads it in the most unmerciful manner, and when it drops with fatigue, he forthwith lassoes and mounts another, abandoning the exhausted horse to the wild animals. Mares are better treated by the Spaniards; but among the Indians mares' flesh is the ordinary animal food. It may be supposed that mutton is neither very excellent nor dear in this country, since, by an old royal edict, the inhabitants were forbidden to drive sheep alive into the brick-kilns for fuel! The wool, also, a few years ago, was hardly worth the expense of cleaning, but it has latterly much improved in quality; and the export of this article, which in 1830, reached only 19,444 lbs., amounted in 1837 to 2,907,951 lbs. Still, however, the breed of sheep, like that of goats and hogs, is very inferior. The singular animal, the coypou, which furnishes the skins known in commerce by the name of nutria, is abundant in Buenos Ayres, its skins forming a principal article of export; the chinchilla also abounds in various districts. Along the Andes are found guanacoes, llamas, vicunnas, &c. Wild boars, deer, with jaguars, pumas, armadillos, &c., are also met with. The biscachis, a rodent quadruped, which makes travelling over the Pampas dangerous from its numerous burrowings, is very abundant; and condors, vultures, and numerous birds of rich plumage, inhabit the country. Its greatest pests are giant ants, locusts, immense bugs, musquitos, and other insect tribes.

The gauchos, or native peons, are the descendants of European colonists, and many of them have sprung from the best families of Spain. They are at once the most active and the most indolent of human beings; living, when not on horseback (which they generally are), in the rudest manner in mud huts. They are without agriculture, subsisting almost wholly on the flesh of oxen and game of various kinds,

which they catch by means of two singular weapons, in the use of which they are extremely dexterous, the lasso and the bolas. The former, used by most natives of La Plata and Chili, is a strong plaited thong of green hide, about 40 ft. in length, with an iron ring at one end forming a running noose, the other end being fixed by the peon on horseback to his saddle-girth. The gaucho, when about to close an animal, whirls the noose with a portion of the thong horizontally round his head, holding the rest of the lasso coiled up in his left hand; and, when near enough to the object, at a precise point of its rotation, flings off the noose, which seldom or never fails to secure the animal. If a horse, it invariably falls over the neck; if an ox, over the horns. As soon as the rider has succeeded in his aim, he suddenly turns his horse, which sets his legs in a position to resist successfully the pull of the entrapped animal. The dexterity evinced in this operation, and the certainty with which an animal running at full speed, is caught, is very striking. The bolas, used also by the Indians, is briefly described in the article Patagonia, in this volume.

The Indians are of Araucanian origin (see Chili, vol. i. p. 690.), living, like the gauchos, chiefly on horseback, but partly in moveable tents made of hides. To their main food, the flesh of mares and colts, if they add any thing it is maize, obtained from the Spaniards in exchange for salt, cattle, and blankets, made by their women. They live together in different tribes, each governed by a cacique. Some are friendly to the whites, but the greater part are bitterly hostile; and the two races maintain against each other an exterminating warfare. A few Indians in the provinces of Salta, &c., are employed by the whites in agriculture; receiving in payment for their services some coarse woollen cloths, beads, baubles, and a few other articles of dress; but in general, the independent tribes wearing only a poncho, or short cloak, boots of colt-skin, and other articles of domestic manufacture, place little value on European goods. Numerous settlements were made in the 16th century by the Jesuits in the Missiones, S.E. of Corrientes, and civilization is said to have made some progress among the Indians of this quarter; but after the expulsion of the Jesuits, in 1767, they speedily lapsed into their previous barbarism.

La Plata has scarcely any manufactures. Ponchos, saddle-cloths, blankets, &c., are made by the Indian women, and sold in great numbers to the people of Tucuman and Salta. Cordova is the principal manufacturing town; but the above kind of goods and morocco leather, with wooden bowls and dishes, comprise almost all the articles made there. The foreign imports consist principally of European goods for the white colonists; mostly from Great Britain. "The gaucho is everywhere clothed in British manufactures. Take his whole equipment—examine every thing about him—and what is there (not of raw hide) that is not British? If his wife has a gown, ten to one, it is made at Manchester; the camp-kettle in which he cooks his food, the earthenware he eats from, the knife, spurs, bit, are all imported from England." (Parish, 338.) The foreign trade of the republic is increasing; but, as it wholly centres in Buenos Ayres, the reader is referred for an account of it to the article on that city in this Dictionary, i. 448.

Education, as may be supposed, is not very flourishing in the provinces; but in the chief towns it is not, upon the whole, so backward as in some other parts of S. America. Cordova has a university, and Buenos Ayres a good public library. Newspapers are published in several towns, and the press is quite free.

The Government is nominally a representative republic or confederation, each of the provinces being, to a certain degree, independent of the rest, as is the United States, Mexico, &c. But, in 1835, General Rosas, who was unanimously called to the presidency, refused to act, unless invested for a period with extraordinary powers. These were accordingly granted him, so that at present the government is a nearly absolute dictatorship, presenting, however, a favourable contrast to the dictatorship in the neighbouring state of Paraguay. There is a junta or parliament of 44 deputies, half annually renewed by popular election; and a senate of two deputies from each province. The provincial government consists of the popular assembly, which nominate the governor. But though democratic in theory, they are quite otherwise in practice; the lower classes knowing with obsequious deference to the higher classes the upper and "if any appeal to the people be ever made by the latter, it is generally from the necessity of supporting by a show of stration of brute force the pretensions of some particular candidate."

The powers of the president are constitutionally very extensive: he appoints to all civil, military, and judicial offices; but he, as well as his ministers, is responsible for every new measure, and liable to impeachment before the senate and house of representatives. The military force is of no certain amount; but during the late war with Brazil, about 10,000 troops were collected, besides a numerous seamen.

---

* In the last drought of 1831-32, it was estimated that from 1 and 1½ to 2 millions of animals died:—the borders of the lakes and streamlets in the province of Buenos Ayres were long afterwards white with their bones. (Parish's Buenos Ayres, p.371.)

The province of Buenos Ayres alone supports the government expenditure; the other provinces contributing nothing to the general expenses of the confederation. The public revenue was estimated, in 1837, at 12,000,000 dollars, a sum insufficient to meet the ordinary expenditure. The public debt, at the same period, amounted to 35,917,166 dollars, bearing 6 per cent. interest.

This country was first discovered in 1517, and settled by the Spaniards in 1553. It was long dependent on Peru; but, in 1778, was erected into a viceroyalty, comprising together with La Plata, Bolivia, Paraguay, and Uruguay. The English made an unsuccessful attempt on this country in 1807. In 1810, the struggle began between the inhabitants of Buenos Ayres and Spain, which terminated in 1816, in the independence of the former. The first congress met at Tucuman, but the federal capital was soon transferred to Buenos Ayres. In 1827 a war broke out between the republic and Brazil, respecting the possession of Uruguay (Banda Oriental), established as an independent state in 1828; and more recently La Plata has been involved in disputes both with Bolivia and France. These wars have contributed to retard the march of her prosperity; but with all her accumulated difficulties, La Plata, appears only to require a few years of repose to develop her abundant natural resources, and to become a comparatively flourishing country. (*Rush's Buenos Ayres ; Head's Journ. across the Pampas ; Andrew's Journey in S. Am. ; Robertson's Letters on Paraguay ; Scarlett ; Miers ; Darwin ; Am. Alman. &c. passim.*)

PLATA (RIO DE LA) (*River of Silver*), a large river S. America, draining with its numerous affluents the outer part of the states of La Plata, Banda Oriental, and Paraguay, with smaller portions of Bolivia and Brazil. It is formed by the union of two important branches, the Paraná and Uruguay; and, gradually increasing in width, becomes a very large estuary, entering the S. Atlantic ocean between Punta Negra (lat. 34° 55' S., long. 55° 5' W.) on N. E., and cape St. Antonio (lat. 36° 21' S., long. 56° 42' on the S.W.; having on its N. bank the city and port Monte Video and the colony of San Sacramento, while, the opposite side, 124 m. from its mouth, is Buenos Ayres. The basin of this great river is estimated to occupy at 1,250,000 sq. m., being inferior in extent only to those of Amazon and Mississippi: its length, from the source of Paraguay to its mouth, is about 2450 miles.

The longest and most direct river, and that of the largest size, belonging to this great water system, is the Paraná, which, on receiving the waters of the Paraná at Corrientes (27° 30' S. lat.), assumes the name of that branch. As its sources between lat. 13° and 14° S., in the lofty ranges connecting the great mountains of Peru and Brazil, that constitute the water-shed between the affluents of the Amazon and those of the Rio de la Plata. Many navigable streams join it from the E., as it passes through Brazil; but on the W. side, though not so numerous, are much more extensive. Between the 17th and 19th degs. of S. lat. that wide region of swamps called the lake of Xarayes which, during the periodical inundations of the river, extended so extensively as to form a great inland sea, running from E. to W. between 200 and 300 m., and from N. upwards of 100 m., with a depth of 10 or 12 feet. At the close of the rainy season, these waters are carried off by Paraguay, which is navigable from this point to its mouth for vessels of 40 or 50 tons, a direct distance of 1800 m. The other western affluents are the Pilcomayo and Vermejo, which fall into it between Assumption and Corrientes, both having their sources in Bolivia, and flowing through the great chaco, or desert. The Pilcomayo, in a course of 1000 m., enters the main stream by two months, about 60 m. apart; it is shallow, and not navigable by canoes. The Vermejo, which falls into the main stream about 135 m. below that last mentioned, rises on the E. side of the Andes, and is navigable for large boats through the whole of the level country for nearly 700 m.

The Paraná (which, as we have before observed, joins Paraguay at Corrientes, and gives its own name to its part) rises in the table land of Brazil, in lat. 21° S., 820 m. from the shores of the Atlantic. It flows S. and curves westward, separating Brazil from Paraguay, and lower down, divides the latter country from the states of La Plata. It has numerous affluents, but though the stream be upwards of 1000 m. in length, it is not navigable for more than 100 m., owing to the saltos, or falls, the chief of which, close to the island of Apipe, is in lat. 27° 26' and 47° 47' W. From this point the river at once becomes navigable for vessels of 300 tons. The most important, however, is considerably higher up the stream, in lat. 27° S., being upwards of 50 ft. in height. From Corrientes the united river, now from 1½ m. to 2 m. in width, through a vast channel, much broken by islands, with trees, and subject to inundation. The only considerable tributary of the Paraná below Corrientes is the Salado, which rises in the E. Cordillera of the Andes, and

after a devious course through the mountains, runs south-eastward through the Pampas, to its junction with the main river, near Santa Fe, in lat. 31° 40' S. Here the Paraná divides into numerous branches, formed by pretty large islands, becoming more frequent lower down the stream, which at length opens into the estuary of La Plata, by a long, but narrow delta, having two principal branches. The depth at the mouth is seldom less than two fathoms, and there is an uninterrupted navigation throughout the year for vessels of 300 tons from Assumption, upwards of 800 m. from the mouth. It has been estimated, says Mr. Darwin, that " the river, at its source, has only a fall of 1 ft. per mile, and much less lower down in its course ; indeed, a rise of 7 ft. at Buenos Ayres may be perceived 180 m. from the mouth of the Paraná. But notwithstanding these advantages, we met during our descent very few vessels. One of the best gifts of nature seems here wilfully thrown away ; so grand a channel of communication being left nearly unoccupied : a river so which ships might navigate from a temperate country, as surprisingly abundant in some productions as destitute of others, to another, possessing a tropical climate, and a soil, perhaps unequalled in fertility in any part of the world. How different would have been the aspect of this country if English, instead of Spanish colonists had, by good fortune, first sailed up this splendid river !" (*Voyage of the Adv. and Beagle*, iii. 164.) The inundations of the Paraguay and Paraná bear a close analogy to those of the Nile. "Both rivers," says Sir Woodbine Parish, " rise in the torrid zone, nearly at the same distance from the equator ; and both, though holding their courses towards opposite poles, disembogue by deltas in about the same lat.: both are navigable for very long distances, and both have their periodical risings, bursting over their natural bounds, and inundating immense tracts of country." The Paraná begins to rise about the end of Dec., soon after the commencement of the rainy season in the S. tropic, and increases gradually till April, when it begins to fall somewhat more rapidly till the beginning of July ; a second rising, called repunta, is caused by the winter rains S. of the tropic of Capricorn ; but it seldom overflows the banks.

The ordinary average of the increase below Corrientes is 12 ft.; but at Assumption, where the river is more confined, the rise is said to be sometimes as much as 5 or 6 fathoms. Occasionally, however, these floods are much higher; penetrating into the jungles of the interior, and drowning numbers of wild animals, the carcasses of which poison the air for months afterwards. The river at these times is exceedingly turbid, from the great quantity of vegetable substances and mud brought down by it ; the velocity of the stream in the higher and narrower parts at first prevents their deposition; but as it approaches the lower lands, or pampas, they are spread over the face of the country, forming a grey slimy soil, which increases vegetation in a surprising degree. The extent of ground thus covered during the inundations is estimated at 30,080 sq. m.

The Uruguay, the other great branch of the estuary of La Plata, takes its name from the numerous falls and rapids which mark its course. It is upwards of 800 m. in length, rising in lat. 27° 30', on the Sierra de S. Catherina, in the province of that name, only about 75 m. W. the Atlantic ocean. Its course is at first nearly due W., but it afterwards turns southward by a mountain range, separating it from the Paraná. It receives several important affluents, of which the Negro, the principal river of Banda Oriental, is the chief. It joins the estuary of La Plata about 50 m. below the junction of the latter ; and its clear blue waters may be distinguished from the muddy streams of the Paraná for miles after their junction. The country through which the Uruguay flows is of a very uneven and rocky character; its navigation of which the navigation is broken by many reefs and falls, only passable during the periodical floods. Of these, the lowest are the Saltos Grande and Chico in lat. 31° 30', about 190 m. above its mouth.

The estuary of the Rio de la Plata, the recipient of these great rivers, is about 185 m. in length ; its breadth at the mouth being about 130 m., though it gradually becomes narrower, till, opposite Buenos Ayres it has a width of only 30 miles. The coast on the N. side is in general high and rocky ; whereas on the opposite side the shores are low, extending inwards in immense pampas. The depth of the river increases towards the mouth, where it averages 10 fathoms ; but at Monte Video it scarcely exceeds 3 fathoms, and gradually lessens, so that vessels drawing more than 16 ft. water cannot ascend above Buenos Ayres. E. of Monte Video is an immense bank of sand and shells, called the English Bank ; besides which there are many other sand-banks, covered when the river is low with only about 8 ft. water, one of which, called the Ortiz, is in some parts between 11 and 12 m. in width. The currents are extremely irregular, both in rate and direction, a consequence of the immense volume of water brought down at certain seasons by the Paraná, as well as of the influence of the winds at the mouth of the river: indeed, this variability of the winds and currents constitutes one of the chief difficulties in navigating the Plata, which on this account, has been termed "*Il Inferno de los Marineros.*"

In calm weather the currents are generally very slack, and almost as regular as tides, setting up and down the river alternately. The effect produced by the *pamperos*, or S.W. gales, so called from their blowing over the pampas S. of Buenos Ayres, is remarkable from the singular fluctuations in the depth of the water before and after their occurrence; the river being always higher than usual when they begin, whereas, after they have continued for a few hours, the water is forced out to sea, so that the sand-banks begin to appear, and, on some occasions, even the anchoring grounds have been laid bare! The tides are so much disturbed, and, as it were, hidden by the currents, that it has been affirmed they have no existence; but, according to the *Amer. Coast Pilot*, they are clearly discernible in calm weather, though their rise seldom exceeds 6 ft. (*Am. Coast Pilot.*)

The Rio de la Plata, which, with its affluents, furnishes an internal navigation of many thousand miles, must, of course, even its present neglected state, have a considerable commerce, of which BUENOS AYRES and MONTE VIDEO are the great *entrepôts*, and to which the reader is referred for further details. The river trade, however, is far less extensive at present than is generally supposed; the intercourse with Chili and Peru having greatly decreased since the establishment of Independence at Buenos Ayres, and that with Paraguay having been all but annihilated by the exclusive and despotic policy adopted by its present dictator, Dr. Francia. The river is pre-eminently well adapted for steam navigation, and, under more happy circumstances, might become the medium of a trade inferior only to that of the Ganges and the Mississippi. (*Sir W. Parish's Buenos Ayres and La Plata;* p. 179–905; *Voyages of the Adventure and Beagle,* ii. and iii.; *Amer. C. Pilot, &c.*)

PLATÆA, a considerable city of ancient Greece, now wholly in ruins, in Bœotia, at the northern foot of mount Cithæron, about 7 m. S.W. Thebes. This town has acquired an immortality of renown from its having given its name to the great battle fought in its vicinity, on the 2nd of September, anno 479 B.C., between the combined Greek forces under Pausanias and the Persian army under Mardonius, generalissimo of the forces left by Xerxes in Greece. The Grecians gained a most complete victory. Mardonius was killed in the action; and the camp to which the fugitives retreated, having been forced, a prodigious slaughter took place. In fact, with the exception of about 40,000 horse, who escaped under Artabazus, the entire Persian army, said to have been nearly 300,000 strong, was all but entirely annihilated. (*Herod.* lib. ix. cap. 60.) The victorious Greeks, besides securing the independence of their country, found immense booty in the camp of the Persians. A tenth part of the spoil was given to Pausanias, the general, whose great talents materially contributed to the success of the day; and another tenth was set apart as offerings to the gods. From the produce of the latter was presented to the shrine of Delphi a golden tripod, supported on a brazen pillar formed of three serpents twisted together. And it is a singular and curious fact, that this identical pillar, having been carried to Constantinople, still exists in the Hippodrome of that city! (*Herod.* lib. ix. cap. 80; and the arts. CONSTANTINOPLE, and DELPHI, in this work.)

Notwithstanding the services the Platæans had rendered to the common cause in this great struggle, their city was, at a subsequent period, (anno 374 B.C.), taken and razed by the Spartans. But she was afterwards restored, and her walls rebuilt, by Alexander the Great. The existing remains of the city date from the era of that conqueror. (*Dodwell's Greece,* i. 278.)

PLATTE, r., Indian Ter., one of the largest tributaries of the Missouri, rises by two branches called the North and South forks, in the Rocky mountains, which unite about 400 miles from their source. Flowing eastwardly about 400 miles farther, it unites with Missouri river, in 41° 3′ 13″ N. lat., about 700 miles from the Mississippi. It is about 600 yards wide at its mouth, is broad and shallow, being from one to three miles broad, and at low water may be forded at almost any place. Its principal tributaries are the Elkhorn entering it near its mouth, and Loup fork, 90 m. above. By the N. fork or this river, following the Sweetwater branch, the best passage is obtained across the Rocky mountains, and a wagon has been actually driven across, to the Columbia river.

PLATTSBURG, p. t., capital of Clinton co., N.Y., 163 m. N. Albany. 338 W. Bounded E. by Lake Champlain. The village is situated on both sides of Saranac r., at its entrance into Cumberland bay of lake Champlain. Incorporated in 1815. It contains a courthouse, jail, county clerk's office, a bank, an academy, a lyceum, four churches, a Presbyterian, Episcopal, Methodist, and Roman Catholic. The town contains 43 stores, four fulling-mills, three woollen factories, two cotton factories with 12,000 spindles, one furnace, four

forges, one flouring-mill, three grist-mills, 25 saw-mills, five tanneries, two printing offices, two weekly newspapers; two academies, 111 students, 19 schools, 646 scholars. Pop. 6416.

This place is celebrated for a famous battle in the last war, on Sunday, September 11th, 1814, between the British under Sir George Prevost, who commanded 14,000 of their best land forces, and Commodore Downie who commanded the naval force; and the American General Macomb, who commanded the land force, and Commodore Macdonough who commanded the American naval force. The land force of the Americans consisted of 1500 regulars and 2500 militia. The British naval force rather exceeded the American, in guns and men. After a hard fought battle of over two hours, the British naval force surrendered to Commodore Macdonough. Commodore Downie was killed early in the action. The British had possession of the village on the N. side of the Saranac, but were unable to cross the river; and the object of the expedition having failed by the loss of the fleet, they retired to Canada.

PLAUEN, a town of the k. of Saxony, circ. Zurichau, cap. districts Plauen and Pausa: on the White Elster, 38 m. S. by W. Leipsic. Pop., in 1837, 9485. It is built on uneven ground, walled, and has several churches and hospitals, a gymnasium, and a royal castle. It is a thriving town, with manufactures of linens, and cotton goods and yarn: it has also considerable markets for wool. Pearls are occasionally found in this part of the river; and there is a r yal pearl fishery at Oelsnitz, in the neighbourhood. (*See Berghaus; Murray's Handb. for S. Germany*)

PLEASANT VALLEY, p. t., Dutchess co., N.Y., 7 m. N.E. Poughkeepsie, 79 m. S. Albany, 306 W. Drained by Wappinger's cr. It contains three churches, a Presbyterian, Methodist and Friends, nine stores, one fulling-mill, one woollen factory, one cotton factory with 2256 spindles, four flouring-mills, four grist-mills, three saw-mills; five schools, 202 scholars. Pop. 2219.

PLUMB, t., Alleghany co., Pa., 14 m. E. Pittsburg. Bounded N.W. By Alleghany r. It has one academy, 40 students; four schools, 156 scholars. Pop. 1953.

PLUMB CREEK, t., Armstrong co., Pa. It has five stores one fulling-mill, seven grist-mills, 11 saw-mills, one tannery, two distilleries; 17 schools, 1618 scholars. Pop. 2216.

PLUMB ISLAND, Southold, t., Suffolk co., N.Y. Separated from Oyster Pond point at the E. end of Long Island, by Plumb island gut, one mile wide. It is 3 miles long and 1 mile broad, and has a lighthouse on its W. end.

PLUMSTEAD, p. t., Bucks co., Pa., 30 m. N. Philadelphia, 106 m. E. Harrisburg, 168 W. Drained by Tohickon and Neshaming creeks. It has eight stores, two grist-mills, one saw-mill, two oil-mills, one pottery. Pop. 1872.

PLYMOUTH and DEVONPORT (for, though separated in their municipal and political privileges, they constitute, in fact, parts of the same town), two parl. bors. and seaport towns of England, co. Devon, and hund. Roborough, standing together the principal naval port of Great Britain: on Portsmouth, at the bottom or N. end of Plymouth sound on a kind of rocky promontory between the Tamar and Plym. 36 m. S.W. Exeter, and 193 m. W. by S. London; lat. old church, 50° 22′ 14″ N., long. 4° 7′ 32″ W. The pop. of the different towns and districts, popularly included under the term Plymouth, was, in 1801, &c., as follows:—

	1801.	1811.	1821.	1831.	1841.
Plymouth	16,040	20,405	21,594	31,080	36,486
Devonport	23,747	30,923	33,578	34,883	33,820
East Stonehouse	3,407	5,174	6,045	9,571	9,711
	43,194	56,602	61,212	76,534	80,048

These towns had together, in 1841, 8799 inhabited houses. There is a large excess of females, the latter having amounted in the same year to 45,115, and the males to only 35,945. The towns are built on rather unequal ground: Plymouth being on the E., and Devonport on the W. side of the space (3 m. broad) between the two rivers: the medial suburb of Stonehouse connects (with the help of a bridge over Stonehouse-pool) the two towns.

Plymouth is old and irregularly laid out; several of its streets are narrow and ill-built, a few also being steep. Many improvements, however, have been made within the last few years, and it has now several handsome streets and good squares, lined with substantial stone houses. Devonport, formerly called Dock, may be said to be the new town of Plymouth, having been almost entirely built since 1700; most of its streets are straight and wide, and the older houses are being gradually replaced by handsomer and more substantial buildings. Its wide, handsome streets, which cross each other at right angles, are paved with limestone quarried in the neighbourhood. Both towns are well lighted with gas; and water is abundantly supplied to Plymouth by the corporation, and to Devonport by a joint stock company.

The principal public buildings of Plymouth are, a modern, though by no means handsome, guildhall, with a small attached borough jail; an exchange, and custom-house; a very elegant edifice, called the Athenæum, belonging to the Plymouth Literary Institution; a public library; a splendid hotel and theatre, built by the corporation, at an expense of nearly £40,000; the royal Union Baths, and a plain stone structure called the Free-masons' Hall. A new and extensive market-place has been formed, at an expense of more than £10,000; and it has several barracks, hospitals, and others belonging to government. The naval hospital, an alehouse, is of great extent, and admirably arranged; and close to the water-entry of this hospital is the royal military hospital, with an arcade of 41 arches, supporting a terrace, and covered promenade below: there is, also, a large military prison at Mill-bay, capable of accommodating nearly 2000 persons. One of the most striking features of Plymouth is its citadel, erected on a commanding eminence in ?: it has five bastions, and is surrounded on three sides by a deep ditch and counterscarp: the interior comprises a governor's house, residences for numerous military officers, and extensive barracks. The victualling office, erected some 10 years since at Duval's or Devil's point, S. of Stonehouse, is of large size, and replete with every convenience. ' the citadel is the elevated walk called the Hoe, which commands a fine view of the sound and the surrounding country, including mount Edgcumbe, with the rich hills in the W., the high land of Dartmoor in the distance, and Saltram, its neighbourhood, and distant town in the ?. Plymouth is divided into two parishes, St. Andrew the Martyr, of the annual value of £920 and respectively: both have been till very recently in the gift of the corporation. St. Andrew's church, erected at different periods between the 12th and 15th centuries, is a large and rather handsome structure, with a square embattled tower; its interior comprises accommodation for 2500 and is, on the whole, elegantly arranged. Charles's church, built toward the close of the 17th century, and named after Charles I., is a neat building, with a square tower, surmounted by a light steeple. There are two chapels ?ne in each parish, besides the chapel in the citadel the mariners' church, for which a new structure, Trinity church, is in course of erection. There are four chapels, and two others belong to the sect denominated "Plymouth Brethren." The Wesleyan, Associated Bryanite Methodists, Independents, Presbyterians, Unitarians, have each one or more chapels; and a meeting-house for the Society of Friends, and a Five Sunday-schools are attached to the established church; and religious instruction is furnished by the ? to several hundred children of both sexes. A school, in the patronage of the corporation, is attended by 90 to 30 boys; and another school, formerly ? furnishes a good general education to about 50 or more of each sex. Dame Rogers's charity gives ? instruction to 52 girls; and a Lancastrian school, added by 175 boys, and 120 girls. An orphan asylum, established some few years ago; and there are almshouses, besides the "South Devon, and Hospital," and "Public Dispensary," for proor with gratuitous medical aid; and other beneficial religious institutions supported by voluntary The literary establishments comprise the institution, or Athenæum, for the promotion of literature; the National History society of Cornwall; a public library with 6000 vols.; a ?, established in 1794; a law library, established a theological library; and a mechanics' institute are five weekly newspapers, of which two ?outh, and three to Devonport. ?like Plymouth, is a regularly fortified town, lines commenced in the reign of George II., improved. It is also defended by the for-Vise, between the town and the harbour on ? fort on mount Pleasant to the N., and a ?sk Point on the peninsula of mount Edg-?pposite sides of the harbour. Inasmuch, ?ral of the adjacent heights command the ? oppose any effectual resistance to an army ?nd; but it is quite secure from any attack ? the fort of mount Wise is the governor's ? stone building, fronted by a parade, form-?romenade: on its N. side is the port ad-recent and capacious stone building. A ?ected a few years ago, and other edifices ?es have been built since the grant of the Opposite the town-hall (which is a neat ?tands a fluted column, intended to com-?ming of Devonport, in 1824; its summit ? view of the harbour and surrounding ?r. church is at Stoke-Damerell, nearly 1 m.

N.E. the town; but there are three Episcopal chapels: ' ease, one of which is attached to the dock-yard. The Baptists, Independents, Wesleyan Methodists, and Unitarians, have each places of worship, with attached Sunday schools: there are also several endowed and subcription schools, furnishing instruction to nearly 5000 children of both sexes. A public dispensary and savings bank have also been established. Devonport, however, is mainly dependent on its dock-yard, which, indeed, constitutes, by far the most interesting feature of the united town.

Plymouth has been, for a lengthened period, the second naval harbour of Great Britain. The dock-yard at Devonport, commenced in the reign of William III., is one of the finest establishments of its kind in Europe. It extends along the shore of Hamoaze, 3500 ft., with an extreme breadth of 1300 ft., including an area of 73 acres; the entrance on the land side is from Fore street, one of the principal thoroughfares of Devonport. The basin of the dock is only 250 ft. in length, and 180 ft. in width; but the excellence of Hamoaze as a natural harbour renders a larger basin of less importance. The wharf wall extends along the shore: and the depth of water is such as to allow the largest ships of the line to come close up to the principal jetties to take in stores.

The dock-yard is divided into two pretty equal parts by a canal (similar to that in Portsmouth), which furnishes easy access for boats to the storehouses, roperies, smithies, &c. In the N. half, and facing the harbour, are two dry docks for ships of 120 guns, with jetties for their accommodation after having been undocked. A fine double dock, for ships of 74 guns, communicates directly with the harbour, and a smaller dock opens into the older basin, in addition to which a capacious new basin is now being formed in the S.W. part of the yard. The roofs of the docks are extraordinary specimens of architectural skill, each being formed of a single arch, unsupported either by buttress or pillars. Workshops and sheds are erected in various parts, and a quadrangular range of fire-proof stone buildings comprises magazines of stores, sails, rigging, &c. At the N. end, also, is a range of handsome houses with attached gardens, forming the residence of the principal officers. On the S. side of the yard are building-slips for large vessels, and others for those of inferior size: near these is a building in which planks are steamed when required to receive any particular curvature; and in this part, also, are extensive timber-berths, saw pits, and mast-houses, besides a large pond in which masts, yards, &c., are kept soaked to prevent their cracking by exposure to the sun. The blacksmith's shop, about 300 ft. square, comprises many forges. Anchors are made in it of the largest size. The ropery, which is the finest in the kingdom, comprises two ranges of building, each 1200 ft. in length, and three stories in height, built entirely of stone and iron, as a security against fire; contiguous to them is the hemp-magazine. In this part of the yard, also, is the model-loft, in which are the patterns of the various parts of ships ordered to be constructed by the admiralty.

The gun-wharf, or arsenal, separated from the dock-yard by North Corner-street, includes within the walls about 4½ acres: it has two principal warehouses for muskets, pistols, and other small arms, with sheds for gun carriages, a powder magazine, and a cooperage.

A great number of men are employed in the different departments of the dock-yard, especially during war, or when a fleet has to be fitted out; and the scale on which the various operations are conducted, the diversity of the employments, the perfection of the machinery, and the activity prevalent throughout the establishment, are all objects of admiration. But no individual who is not in uniform, or is not well known by the authorities, is allowed to enter the dock-yard, unless by special permission.

The harbour of Plymouth is double, being formed by the estuaries of the Plym and Tamar, opening into the N.E. and N.W. angles of Plymouth sound: the estuary of the Plym forms the Catwater, a convenient and capacious harbour for merchant vessels; and that of the Tamar expands into the noble road or harbour of Hamoaze, about 4 m. in length, by about ⅓ m. in width; it is almost completely land-locked, and has moorings for 100 sail of the line, with secure anchoring-ground for a still larger number: its average depth is 9 fathoms at ebb tide, and the largest ships float close to the quays. Subordinate to the harbour of Catwater is Sutton-pool, a small tide-basin, to the E. of the citadel, surrounded by quays for the convenience of colliers, coasting vessels, and fishing smacks, by which it is almost exclusively frequented. - In Mill-bay, also, to the W. of Plymouth, a pier is in course of being constructed for the accommodation of the largest steamers at all times of the tide. Spring tides rise from 15 ft. to 18 ft., and neaps from 6 ft. to 8 ft.

The bay or arm of the sea, called Plymouth sound, into which these harbours open, is used for the accommodation of the ships that have been refitted in the dock-yards, and as a safe asylum for all sorts of ships in stormy weather. Owing, however, to the heavy swell thrown in from the S., it was formerly a very unsafe place for anchoring; and, to obviate this inconvenience, a stupendous breakwater, or mole (similar to that of Cherbourg, or, rather, to that of Civita Vecchia, constructed by the emperor Trojan, see CIVITA VECCHIA), has been formed in the middle of the sound, stretching in a line, straight in the middle, but inclined inwards at either extremity, about 1 m. in length, between Cawsand bay on the one side, and Bovisand bay on the other. From the commencement of the work to the 31st of July, 1841, 3,777,063 tons of stone, procured from quarries on the banks of the Plym, consisting of rough cubical blocks, each weighing from 1¼ to 2 tons and upwards, had been employed in this great work. The top presents a flat surface, about ten yards in width, whence it slopes on both sides to the bottom, the principal slope being on the side next the sea. A lighthouse is now being erected on its W. extremity.

This great national undertaking has cost a very large sum, but the important object in view in its construction has been completely attained. That part of the sound within the breakwater has been rendered one of the very best roadsteads in the world: it is accessible on either side, and is sufficiently capacious to admit the largest fleets, which ride under cover of this immense bulwark during the stormiest weather in perfect safety. The Eddystone lighthouse (which see) stands about 14 m. S. by W. the breakwater, and is an important appendage to the harbour, the entrance to which would, but for this beacon, be comparatively dangerous, in consequence of the hidden rocks on which it is placed. Within the breakwater, and opposite to and commanding the entrance to Hamoaze, is St. Nicholas island, which has been strongly fortified, and constitutes, with the redoubt at Staddon height, above Bovisand quay (near which is formed the reservoir for supplying H. M.'s ships with water), the principal defence of the town and harbour on the side of the sea.

The Hamoaze is bounded on the S., directly opposite Devonport, by the beautiful peninsula of mount Edgcumbe, the seat of the noble family of Edgcumbe. On the upper part of the Hamoaze, and on its W. side, is the town of Saltash (which see).

In Catwater harbour there are wet and dry docks, formerly suited to the construction of seventy-four-gun ships; but for many years they have been used exclusively for merchant-vessels. On the E. side of the Catwater are the villages of Oreston and Turnchapel, mostly occupied by persons engaged in the dock-yards, or otherwise connected with the trade of the port. In this direction, also, are the finely-situated villages of Upper and Lower Hooe. Mount Batten, at the S.W. extremity of the Catwater, opposite Sutton-pool, is a very picturesque object. It is surmounted by an ancient castle.

The trade of Plymouth is of considerable importance: the gross custom duties amounted in 1840 to £135,931. There belong to the port about 370 vessels, of the aggregate burden of 30,096 tons, new measurement.

A large part of the trade of the town depends on the dock yard, and timber is, of course, a principal article of import. Steamers touch here almost daily, on their passage between London, Dublin, Belfast, &c., and boats run twice or thrice a week to and from Southampton. The manufacturing establishments of Plymouth comprise a very extensive sailcloth factory, a sugar refinery, glasshouse, starch factory, and a soap-boiling establishment, which produced, in 1839, 2,904,145 lbs. hard soap. The communication with the country E. of the Plym is effected by an iron bridge of five arches, raised on granite piers, built at the sole expense of the late Earl of Morley; and mailcoaches, carriages, horses, passengers, &c., are ferried over to Cornwall by a kind of floating steam-bridge, running every quarter of an hour between Torpoint and Devonport. A railway, 24 m. in length, connects Plymouth with Prince-town, or Dartmoor; and as a line of railway is in progress, uniting Exeter with the metropolis, there can be little doubt of its being continued thither. The bank of England has a branch here; and the Western District banking company, the Devon and Cornwall banking company, and the National and Provincial bank of England, have establishments both at Plymouth and Devonport; besides which there are the "Naval bank," and two private banks.

Plymouth was incorporated in 13 Henry VI. Its present municipal officers are a mayor, 11 other aldermen, and 36 councillors, the borough being divided into six wards. Corporation revenue, in 1839, £11,431, exclusive of £1297, accruing from the sale of property. Quarter and petty sessions are held under a recorder; and there is also a borough court for the trial of civil actions, and a local court of requests, established by a recent act for the recovery of small debts, embracing a jurisdiction extending, in some directions, nearly 12 m. Devonport has also very recently,

on petition, been made a corporate town, and divided into wards; its municipal officers being a mayor, aldermen, and councillors. It has a commission of the peace under a recorder. Corp. rev., in 1839, £1818. Plymouth has regularly sent two members to the House of Commons since the reign of Henry IV.; but it occasionally exercised the franchise at an earlier period. The electoral boundaries were enlarged by the Reform Act, so as to include within the old borough a small portion of land N.E. the town, on the Exeter road. Registered electors, in 1839-40, 1907. The Reform Act constituted Devonport a parliamentary borough, conferring on it the privilege of returning two members to the House of Commons. The electoral limits comprise the parish of Stoke-Damerell and township of Stonehouse. Registered electors, in 1839-40, 2121. Neither Plymouth nor Devonport is attached to poor-law unions; but the expenditure of the united boroughs in 1839 amounted to £78,582, of which £9390 were expended on the poor of Plymouth. Markets in Plymouth, on Monday, Thursday, and Saturday; in Devonport, on Tuesday, Thursday, and Saturday; both abundantly supplied with every kind of provisions.

Plymouth, originally called Tamearwerth, and afterward Sutton (or South-town), received its present name at the period of its incorporation, in the reign of Henry VI. The town received a considerable accession of wealth on the dissolution of the monasteries, and in the reign of Elizabeth was greatly benefited by a supply of good water, conveyed by a channel, 24 m. in length, from Dartmoor, planned by the famous Sir Francis Drake, a native of the town. During the same reign, Plymouth sound was the rendezvous of the fleet opposed to the Armada, and also of the fleet sent against Cadiz. The town suffered greatly, on three occasions, from the plague, which, in 1626, carried off 2000 persons. During the parliamentary wars it embraced the cause of the parliament, and was besieged by Prince Maurice and the royalists, though without success.

Devonport, as is already stated, is quite a modern town, which owes its rise to the rapidly increasing importance of the dock-yard during the American and French wars. Stonehouse, which connects Plymouth with Devonport, is still more modern, and has been wholly built within the present century. All these towns suffered most severely in 1832-33 from the cholera. (Parl. Bound. and Mun. Reports; Private Inf.)

PLYMOUTH, county, Mass., situated in the S.E. part of the state, and contains 608 sq. m. Watered by North river and branches of Taunton river. Iron ore is abundant. It contained in 1840, 14,663 neat cattle, 12,293 sheep, 979 swine; and produced 10,765 bushels of wheat, 25,505 of rye, 125,999 of Indian corn, 5831 of barley, 39,100 of oats, 392,802 of potatoes. It had nine commercial houses in foreign trade, 230 retail stores, seven lumber-yards; capital invested in the fisheries, $775,950; salt produced, 14,948 bushels; 11 furnaces, 16 forges; value of hardware and cutlery manufactured, $1,079,603; one fulling-mill, four woollen factories, 14 cotton factories with 47,994 spindles, 60 grist-mills, 103 saw-mills, one oil-mill, two paper mills, even rope-walks, 14 tanneries, one pottery, four printing-offices, four weekly newspapers; 12 academies, 699 students; 198 schools, 11,541 scholars. Pop. 47,373. Capital, Plymouth.

PLYMOUTH, p. t., semi-capital of Grafton co., N.H., 40 m. N. Concord, 521 W. Bounded E. by Pemigewasset river. Watered by Baker's river. It contains a courthouse, jail, three churches, a Congregational and two Methodist, six stores, one grist-mill, two saw-mills, three tanneries, three potteries; one academy, 66 students; 12 schools, 489 scholars. Pop. 1281.

PLYMOUTH, p. t., Windsor co., Vt., 73 m. S. Montpelier, 475 W. Watered by branches of Queechee river. Black river affords water-power. Limestone and steatite, or soapstone, are extensively found. It contains three churches, a Congregational, Methodist, and Christian; two stores, three grist-mills, 13 saw-mills; 16 schools, 477 scholars. Pop. 1417.

PLYMOUTH, p. t., port of entry and capital of Plymouth co., Mass., 38 m. S.E. Boston, 447 W. The township is of great extent, and contains 50 ponds, covering 3000 acres, and many small streams. The village is in the N. part of the town, on Plymouth bay, a branch of Massachusetts bay. It contains a courthouse, jail, six churches, two Congregational, a Unitarian, Methodist, Baptist, and Universalist; two banks with an aggregate capital of $200,000, a marine insurance company, capital $100,000, and Pilgrim-hall. The harbour is spacious, but has not sufficient depth of water for the largest vessels. There are 45 vessels employed in the cod and mackerel fisheries, several vessels employed in the whale fishery, and others in the coasting and West India trade. Pilgrim-hall is 70 feet long by 40 feet wide, and 33 feet high, forming two stories. It is built of unhewn granite. In front is a Grecian Doric portico of six columns. 20 feet high. The Pilgrim society commemorates the landing of the pilgrims, which took place on December 22d, 1620, O.S., the first English settlement in New-England. In Pilgrim-hall is a large painting representing the landing of the pilgrims from the Mayflower, the chair of Governor Carver, the sword-blade of Captain Miles Standish, and various aboriginal curiosities. The Pilgrim society, at its annual celebration, meets at Pilgrim-hall. The rock on which the pilgrims first landed, was conveyed in 1774 to the centre of the village. In the whole township there are, besides the above churches, two Congregational and a Baptist; five commercial houses in foreign trade, 46 retail stores, four cotton factories with 40,000 spindles, four grist-mills, one saw-mill, two printing-offices, two weekly newspapers; two academies, 123 students; 51 schools, 1,378 scholars. Pop. 5281.

PLYMOUTH, p. t., Litchfield co., Ct., 23 m. W. Hartford, 330 W. Drained by Naugatuc river, which affords good water-power. This town is celebrated for its manufacture of clocks, which have been widely diffused. It contains two churches, a Congregational and an Episcopal; seven stores, one fulling-mill, one woollen factory, one cotton factory with 9650 spindles, two furnaces, two grist-mills, eight saw-mills; 13 schools, 505 scholars. Pop. 2205.

PLYMOUTH, p. t., Chenango co., N.Y., 7 m. N.W. Norwich, 120 m. W. Albany, 344 W. Drained by Canasawacta creek, a tributary of Chenango river. It has one store, one fulling-mill, two grist-mills, eight saw-mills; 19 schools, 511 scholars. Pop. 1623.

PLYMOUTH, p. t., Luzerne co., Pa., 124 m. N.E. Harrisburg. 298 W. Drained by Harvey's and Toby's creeks. It abounds with coal, and has five stores, five grist-mills, eight saw-mills; six schools, 282 scholars. Pop. 1765.

PLYMOUTH. t. Montgomery co., Pa., 14 m. N.N.W. Philadelphia. Inhabited chiefly by Friends, who have a church and a female boarding school. It contains three stores, one grist-mill, one saw-mill; five schools, 315 scholars. Pop. 1417.

PLYMOUTH, p. v., capital of Washington co., 162 m. E. Raleigh, 286 W. Situated on the S. side of Roanoke river, 8 m. from its mouth in Albemarle sound. It contains courthouse, jail, a church, and about 778 inhabitants.

PLYMOUTH, p. t., Wayne co., Mich., 25 m. S.W. Detroit, 29 W. Drained by the W. branch of Rouge river. It contains a Presbyterian church, nine stores, six flouring-mills, five saw-mills, one distillery, one pottery; 12 schools, 16 scholars. Pop. 2163.

PLYMPTON EARLE, or PLYMPTON MAURICE, a decayed bor., market town, and par. of England, co. Devon, and. its own name, 4½ m. E. Plymouth. Area of borough and parish, which are co-extensive, 170 acres. Pop. in 1841, 933. The town comprises two streets disposed in the rm of the letter T.; and it has an old guildhall, under which is held the corn-market. The church is small, and a living is a curacy subordinate to the adjoining parish of Plympton St. Mary. The Wesleyan Methodists and Independents have places of worship. North of the town are ruins of a once magnificent castle, occupying nearly 10 acres of ground: it was built by Richard de Rivers, earl of Devonshire, to whom the town was granted, with the honour of Plympton," by Henry I. Plympton Earle came to be a borough by prescription, but received a royal charter in the 12th Edward III. It was one of the stannary towns; but for many years it has been in a decaying condition, and the borough was not considered of sufficient importance to be included in the provisions of the Municipal Corm Act. But, decayed as it is, it sent two members to House of Commons, with some interruptions, from the reign of Edward I. down to the Reform Act, by which it is most properly disfranchised. Markets on Saturday: fairs for cattle and woollen cloth, February 25, April 5, August 12, and October 28.

D. See ITALY.

POCAHONTAS, co., Va., situated a little N.W. of the centre of the state, and contains 710 sq. m. Drained by Greenbrier river. It contained in 1840, 6734 neat cattle, 6465 sheep, 4944 swine; and produced 17,846 bushels of wheat, 21,083 of rye, 41,380 of Indian corn, 4638 of buckwheat, 50,111 of oats, 21,040 of potatoes. It had three mills, one fulling-mill, 22 grist-mills, 22 saw-mills, one oil-five tanneries, eight distilleries; six schools, 122 scholars. Pop.: whites, 2684; slaves, 219; free coloured, 19; 2922. Capital, Huntersville.

POCKLINGTON, a market town, par., and township of England, E. riding co. York, Wilton-Beacon, div. of wapentake hill, on a small tributary of the Derwent, 12½ m. E. by N. York. Area of parish, including four townships, 4380 acres. Pop. of township, in 1841, 2559. It is a good country town; and the market-place, though small, is conveniently arranged. The church, in the centre of the town, a large cruciform structure, with a handsome tower at its W. end, and a chancel containing some carved stalls: the living is a vicarage in the gift of the dean of York. A grammar school was founded here in 1526, and endow-

ed with lands that now yield upwards of £1000 a year: the master is appointed by the master and fellows of St. John's college, Cambridge. A national school also was established in 1819. It has no manufactures, but a considerable trade has grown up, chiefly in corn, flour, timber, coal, and general goods, since the completion of the canal from E. Cottingwith on the Derwent, to Street Bridge, about a mile from the town. Markets on Saturday: fairs for horses, cattle, &c., March 7, May 6, Aug. 5, and Nov. 8. Great show of horses, Feb. 24 and Dec. 17.

PODOLIA, a government of Russia in Europe, chiefly between the 48th and 50th degs. of N. lat., and the 96th and 31st degs. of E. long.; having N Volhynia, N.E. Kiev, S.E. Kherson, S.W. Bessarabia, and N.W. Galicia. Length, N.W. to S.E., 240 m. Area estimated at 12,300 sq. m. Pop., in 1838, 1,548,000, principally Poles, but including some Russians, and about 150,000 Jews. The greater part of the country is flat; but a low branch of the Carpathians extends through it in an easterly direction. The general slope is towards the S.E. Principal rivers, Bug and Dneister, which last forms the S.W. boundary. The climate is healthy, and mild enough for the vine and mulberry to flourish in the open air. Soil stony, but in general very fertile, and Podolia formerly ranked among the most valuable provinces of Poland, as it now does of the Russian empire. Corn is produced in abundance; the produce of wheat, in 1820, was estimated at 6,000,000 chetwerts, a quantity exceeding the home consumption by one third. Hemp, flax, tobacco, hops, beans, and various fruits are grown. The culture of the vine, though on the increase, is not yet of any importance; and orchard and garden husbandry is conducted in a negligent manner. Pastures luxuriant. Cattle rearing is an important business; and many head of cattle are sent into Germany, where they are much prized for their beauty and excellence. The sheep yield but indifferent wool. A good many hogs are kept, as well as poultry and bees. The forests are estimated to cover 991,443 decatines, or nearly three millions acres; only a small proportion of which belongs to the crown. Game is scarce, but the fisheries are highly productive. Saltpetre, lime, and alabaster, are the principal mineral products. Manufactures are quite insignificant; except distilleries, there are only a few woollen cloth, leather, potash, and saltpetre factories. The trade, which consists mostly of the export of grain to Odessa, and cattle to Gallicia and Germany, is wholly in the hands of the Jews. Podolia is divided into 12 districts; its capital is Kaminietz. It is one of the 10 governments privileged with respect to its judicial administration and the distillation of spirits: it is subordinate to the government of Kiev, both as to military affairs and public instruction. Most of its inhabitants belong to the Greek church.

Podolia was long governed by its own princes: but, in 1569, it was united to Poland, who erected it into the two voivodes of Podolia and Braczlaw. It has belonged to Russia since 1793. (Schnitzler, La Russie; Possart, &c.)

POINT COUPEE, parish, La., situated in the S.E. part of the state, and contains 600 sq. m. Bounded N.E. by Mississippi river, W. by Atchafalaya river. Much of it is liable to subversion. It contained in 1840, 6983 neat cattle, 5211 sheep, 3200 swine; and produced 93,500 bushels of Indian corn, 3985 of potatoes, 6,294,796 pounds of cotton, 411,000 of sugar. It had 95 stores, two saw-mills; one college, 51 students; five schools, 83 scholars. Pop.: whites, 2087; slaves, 5430; free coloured, 381; total, 7898. Capital, Point Coupee.

POINT COUPEE, p. v., capital of Point Coupee par., La., 140 m. W.N.W. New-Orleans, 1174 W. Situated on the S. side of Mississippi river, opposite to St. Franceville, and contains a courthouse and a jail; the land in the vicinity is very fertile, and laid out in cotton plantations. Here commences the levees, or embankment along the river, which extends to New-Orleans. The inhabitants are chiefly of French descent.

POINT PLEASANT, p. v., capital of Mason co., Va., 379 m. W.N.W. Richmond, 361 W. Situated on the E. side of Ohio river, at the mouth of Great Kanawha river. It contains a courthouse, jail, 50 dwellings, and about 200 inhabitants.

POITIERS, or POICTIERS (an. Limonum, and afterwards Pictavi), a city of France, dep. Vienne, of which it is the cap.; on the Clain, a tributary of the Vienne, 58 m. S.S.E. Tours, and 78 m. N.E. by E. La Rochelle; lat. 46° 35' N., long. 0° 20' 20" E. Pop., in 1836, 22,000. It is surrounded by old walls, flanked with towers. Few French cities occupy a greater extent of ground: but a large space within the walls consists of fields and gardens. The streets are inconveniently steep, ill-paved, and gloomy, and the city generally is ill-built, the houses being without either taste or dignity. It has but one good square, the Place Royale, in which was formerly a statue of Louis XIV. Previously to the revolution, few towns in France had so many churches; and though much diminished, many of them still exist; but there are few other public edifices worth notice. The cathedral is a large, though rather low Gothic edifice, said to have been founded by Henry II. of England. The church of St. Radegonde is much more ancient, being said to owe its origin to the wife of Clotaire, in 587. The crypt containing her tomb, and some other portions of the original edifice, are still extant, but the rest of the building mostly dates from the 11th century. The church of Nôtre Dame is very handsome, and several of the other churches have some curious tombs and monuments. The hall of justice, public library, with 12,000 volumes, bishop's palace, theatre, cavalry barracks, and baths, are the other most conspicuous buildings. Poitiers is the seat of a royal court, of tribunals of original jurisdiction and commerce, a university academy, faculty of law, and royal college; and has societies of agriculture, arts, and sciences, a departmental nursery-ground, botanic garden, &c. It is the see of a bishop, whose diocese comprises the departments Vienne and Deux Sèvres. Its manufactures, though not extensive, comprise very various articles, as coarse woollen cloths, blanketing, hosiery, cotton netting, lace, hats, prepared sheep-skins, and goose-down. It has a considerable trade in agricultural produce, and six annual fairs.

Poitiers is one of the most ancient towns in Gaul. The vestige of a Roman palace, an aqueduct, and an amphitheatre, are still visible. The Saracens were totally defeated in 732 by Charles Martel, in a great battle between this city and Tours. But Poitiers is chiefly memorable for the signal victory obtained in its vicinity on the 19th of September, 1356, by an English army commanded by Edward the Black Prince, over a vastly superior French force commanded by King John. The French army was wholly dispersed; and, besides many thousand common soldiers, a vast number of persons of distinction were killed or taken prisoners, the king and one of his sons being among the latter. (*Hugo*, art. *Vienne; Dist. Géog.*)

POITOU, the name of an extensive prov. of France, previously to the revolution: it is now distributed among the departments of Vienne, Deux Sèvres, and Vendée.

POLA, in antiquity, a splendid city, but now a poor, decayed seaport town of the Austrian empire, gov. Trieste, on the W. side, and near the S. extremity of the peninsula of Istria; at the bottom of a bay of the same name; lat. 44° 52′ 18″ N., long. 13° 50′ E. Pop., about 900. It is surrounded by walls flanked with towers constructed by the Venetians in the 15th century, is the seat of a bishopric, has a castle, a cathedral, a Greek church, and three convents. The harbour is one of the best on the Adriatic. The entrance to it is narrow, but the water is deep, and within it expands into a large basin, land-locked and safe. It might easily be rendered an excellent station for a fleet intended to command the Adriatic. The chief occupation of the inhabitants is fishing. The sand used in the Venetian glass-works is brought from its environs.

Pola owes all its celebrity to its ancient greatness, and to the magnificent remains of antiquity of which it has still to boast. The principal of these is a noble amphitheatre, standing outside the town, and near the bay. This splendid monument is in a very perfect state of preservation, and is scarcely exceeded in magnificence by that of the Colosseum at Rome, while, in point of dimensions, it is in a very small degree only inferior to the amphitheatre of Verona. It is in the form of an ellipsis, its longest diameter being 436 ft. 6 in., its shortest 346 ft. 2 in., and its height, in the most perfect parts, 97 ft.* It is estimated to have been capable of accommodating above 20,000 spectators. The height is divided into three stories, and the whole circ. into 72 arches. It is constructed of Istrian stone of a very superior quality, and which, in appearance and durability, is equal to the purest marble.

Within the town are two temples which, when perfect, must have been exactly similar, and worthy the best period of the Augustan age. The best preserved is dedicated to Rome and Augustus Cæsar. There is, also, an arch raised, as the inscription announces, by a Roman lady, in testimony of her affection for her husband. It is a beautiful and elegant structure, admired for its simplicity and admirable proportions. Part of a Roman gateway, containing three arches, was discovered by Messrs. Stanhope and Allason, in 1816. The cathedral has apparently been built on the site of an ancient temple.

These ruins sufficiently attest the former magnitude and wealth of this now miserable place. Strabo, Pliny, and Mela, say that it was built by a colony from Colchis; and of its great antiquity there can be no doubt. It became a Roman colony, and was for a lengthened period the principal town of Istria. *Pola quondam a Colchis, ut ferunt,*

habitata (in quantum res transeunt!) nunc Romana colonia (*Mela.* lib. ii. cap. 3.)

Malte-Brun says that it was destroyed by Cæsar for its devotion to Pompey, and rebuilt by Augustus, at the intercession of his daughter Julia. But there is no evidence whatever of its having been so destroyed, or of the restoration in the way now mentioned, other than what may be derived from the fact of its having been sometimes called *Pietas Julia!*

At present it is very unhealthy in summer, owing to marshes near the town; but these, it is said, might be easily drained. (See *Allason's Picturesque Views of the Antiquities of Pola; Murray's Handbook*, p. 295; *Spon at Wheeler, Voyage d'Italie, &c.*, L., 48, 12mo edit.)

POLAND (Lat. *Sarmatia; Pol. Polsk*, signifying a plain country), formerly the same of an independent and extensive country of E. Europe, comprising the territories between the 48th and 58th degs. N. lat., and the 15th and 33d degs. E. long.; including, with Poland Proper, Lithuania, Samogitia, Courland, the Ukraine, Podolia, and other provinces now belonging to Russia, with Galicia, belonging to Austria, the province of Posen, and some other districts in Prussia. But the existing kingdom of Poland, constituted by the congress of Vienna in 1815, and now united to the Russian empire, is of comparatively limited dimensions, extending only between the 50th and 55th degs. of N. lat., and the 18th and 24th degs. of E. long.; having N. Prussia Proper and the Russian government of Wilna; E. the governments of Wilna, Grodno, Bialystok, and Volhynia; S. Austrian Poland and the territory of Cracow; and W. Prussia, Poland (the grand duchy of Posen) and Silesia. Area, population, subdivisions, &c., as follows:

Governments.	Area in sq. m.	Pop. in 1 34	Chief Towns.
Cracow	4,451	438,7-7	Xa'wt.
Sandomir	5,984	415,^96	Radom.
Kalisz	6,412	549,384	Ka'm.
Lublin	6,728	514,980	Lublin.
Płock	6,148	496, 07	Płock
Masovia	3,947	700,3 14	Warsaw.
Podlachia	4,533	361,^08	Siedlce.
Augustow	6,359	58^,405	Suwałki.
City of Warsaw		136,104	
Total	**50,737**	**4,364,968**	

Of the population about three fourths consist of Poles, one tenth of Jews, and the remainder principally of Russians, Germans, Tartars, and gipsies.

The whole country, except in the S., where are some scattered offsets from the Carpathian mountains, is an extended plain, with a general slope toward the Baltic, in which its principal rivers have their embouchure. These are the Vistula, with its tributaries, the Wieprz, Bug, Narew, Pilitza, &c., the Niemen, and the Warta. The Vistula, after bounding the kingdom for a lengthened distance on the S., traverses its centre, leaving it near Thorn. The Niemen, Bobr, and Bug bound nearly all the E.; and the Prussa, a tributary of the Warta, a considerable part of the western frontier. These rivers are all more or less navigable. There are innumerable smaller streams, Poland being an extremely well-watered country; and in the N., E. and W. are a great number of lakes and many very extensive marshes. The surface, though flat, is abundantly diversified, presenting alternately fertile corn lands, savage steppes, rich pastures, sandy wastes, dense forests, and dreary swamps. The climate is rigorous; the cold of winter is often as great as in Sweden, in a lat. 10 degs. higher; and in 1799 the thermometer descended to 27° below zero (Réaum.). In summer, however, the heat sometimes rises to 120° (Fah.). The mean temperature of the year at Warsaw is about 46° Fah. The atmosphere is humid, rainy and cloudy days occupying half the year. Between the Vistula and the Prussian frontier the soil is generally fertile, the most productive districts being in the governments of Cracow and Sandomir, and the neighbourhood of Warsaw. In the N.E. are also some very fertile tracts; but there, and in the governments of Płock, Lublin, &c., the surface is in great part waste.

"The traveller in Poland sometimes finds himself in an expanse of surface, almost without a house, a tree, or any single object large enough to attract his notice. Soon, however, are descried the skirts of some vast forest fringing the distant horizon; and, on entering it, we proceed for 9 or 10 m., more or less, winding with the road, through lofty pines, &c., precluded from the sight of all objects but trees and shrubs. Sometimes, in the midst of a forest, we meet with a small spot of ground (for example, of 10 or 20 acres) cleared and cultivated; its sides prettily fenced by the green surrounding woods. Sometimes a small lake is found thus situated, its borders ornamented in a similar manner: and these, generally speaking, are the prettiest scenes which Poland furnishes. These forests, in some places, are 15 and even 20, m. in all directions. Indeed, if we exclude

orasses and the level pasture lands, perhaps not more
ian half of the country, speaking generally, is cleared. At
rtant intervals are found plains of some extent, affording
ch pasturage. The best are those contiguous to the Vis-
la, some of which are periodically overflowed by that
ver. Such are those in the neighbourhood of Warsaw,
hich supply that town with good butchers' meat." (*Bur-
tt's View of Poland*, p. 29–52.) This description was writ-
n early in the present century; and, though a considerable
oportion of forest land has been cleared in the interval, it
still substantially accurate. Of 741,000 *wloka* of land
mprised in the kingdom, 255,000 *wloka* are supposed to be
ble; 205,000 in forest; 171,000 in natural pastures, rivers,
d marshes; 46,000 in meadows; 39,000 occupied with
ds and buildings, and 25,000 in gardens. Poland has, for
engthened period, been the granary of a great part of Eu-
e. But Volhynia, Podolia, and Galicia, formerly included
the Polish dominion, were the principal corn-growing
vinces; and in the existing kingdom of Poland, with the
eption of Sandomir and Cracow, the land, according to
Jacob, is so poor that it can scarcely be made to yield
iedium crop of wheat more than once in nine years.
soil is mostly thin, sandy, or sandy loam, resting chiefly
bed of granite, through which the heavy rains gradu-
percolate. S. of the Pilitza, however, the appearance
he land and the f:ce of the country improve; and as
proceed southward to the Vistula, the surface becomes
undulating, and the soil stronger and more tenacious.
his quarter there are extensive tracts of clayey loam,
iring three or four horses to plough it, and yielding,
o tolerably well managed, excellent crops of wheat and
Where, in this district, anything like a system of ro-
n is adopted, the crops are very heavy.

me of the estates belonging to the nobility of the high-
ink are of enormous extent; and, not long since, those
ince Czartoryski and Count Zamoyski, taken together,
ied a space nearly equal to half the extent of England!
times of the republic the former contributed 20,000
he latter 10,000 men to the army. Owing, however,
practice of dividing the land equally among the chil-
unless a majorat be established in favour of the eldest
rhich is sometimes the case, much of it is possessed
aller allotments. These, however, we should still
arge, for they mostly vary from 5000 or 6000 up to
or 40,000 acres each. The rent and price of land is
ally low; depending much more on the number of
its than the extent of the farm. The crown lands,
sing one third part of the whole surface, or about
000 acres, include perhaps 2,800,000 acres of wood,
mainder being chiefly arable land, leased to tenants.
consequence, acquire right to the services that may
lly demanded from the peasantry. The tenants of
wn are exempted as well as their peasants, from
xes to which the other occupiers of land are subject,
consequence, the crown estates are better stocked
amts. With this freedom from taxation and supply of labourers, the rent of 8,000,000 acres of land
by Mr. Jacob, from whom we have borrowed these
to have amounted, in 1827, to no more than 4,000,000
about £93,000 sterling, or somewhat less than 3d.
lish acre. But a large extent of land is included in
rage that is literally of no value; so that, according
acob, the rent of the cultivable land may be fairly
n at from 6d. to 14d. per acre. It might, however,
curred to Mr. Jacob that, in point of fact, the money
land, in a country like Poland, without town and
a market for its produce, affords no test whatever
al value. Lands belonging to private individuals
ly, indeed, ever let, except for services to be per-
on the other parts of the same estates; and the
the land is to be determined not by the amount of
oney rent it will bring, but by the amount of sub-
it affords, or the number of individuals it will main-
a verage state of comfort, according to the customs
ts of the society. Nothing, indeed, can be more
inconsequential and absurd than to set about
g the value of land in such a country as Poland by
standard by which it is measured in England.
the whole lands of the republic were the proper-
nobility or gentry, and could not be held by any

The possession of land was, in fact, of itself, a
obility; and the owner of an estate of three acres
voted in the elections of nuncios, and, in respect
il rights and privileges, was on a level with the
bleman in the country. But this state of things
iolly changed. Landed property is no longer the
of a particular class; but may be indifferently
bles, burghers, and peasants. Jews only are pro-
m becoming proprietors of the soil, though they
erous mortgages thereon. When they foreclose,
must consequently be sold; and as the Jews,
ss the greater part of the money capital of the

country, cannot become purchasers, the prices they yield
are very trifling. Latterly, however, some modifications
have been made in the regulations respecting the Jews,
and various privileges have been conceded to them.

The most numerous class of cultivators are peasants,
who are a species of *quasi* proprietors of the lands they occupy, holding them under condition of working a stipulated
number of days in each week on their lord's demesne, and
paying him, in addition, specified quantities of poultry, eggs,
yarn, &c. The extent of their holdings varies according to
the quality of the land, the quantity of work to be performed, and of payments in kind to be made. On a large property examined by Mr. Jacob, the peasants had each about
48 acres of land, for which they were bound to work two
days a week with a pair of oxen. If their farther labour
was required, they were paid at the rate of 3d. a day for
two days more, and if beyond that number, they received
6d. a day. On another property the peasants had about 36
acres, for which they worked two days a week with two
oxen; when called upon for extra labour, they were paid
6d. a day for themselves and their oxen for the next two
days, or, without the oxen, 3d.

Under the republic the Polish peasants were slaves, and
did not, in fact, enjoy any greater consideration than the
blacks of Carolina and Georgia in the present day. They
were the absolute property of their masters. Down to 1768,
a lord who had killed his slave was merely amerced in a
small fine: and, though in that year the offence was made
capital, such an accumulation of evidence was required to
prove the fact, that the enactment was rendered quite nugatory. (*Coxe*, i., 113.) It was customary to make the slaves
work five days a week on the estates of their lords; the latter also might seize on whatever wealth the slaves had accumulated, might inflict on them corporal punishment, and
might sell them as if they had been so many head of cattle.
The boasted freedom of Poland was, in truth and reality,
merely the license of the gentry to trample under foot the
mass of the people, to browbeat their sovereign, and sell
their votes. It is due, however, to the nobility to state that
some among them, as the Zamoyskis, the Czartoryskis, and
others, perceived the miserable consequences of such a
state of society, and were most anxious for the improvement
of the peasantry on their estates, of whom they emancipated considerable numbers. Generally, however, the Polish gentry were not inclined to establish or give efficacy to
any regulations in favour of the peasantry, whom they
scarcely considered as belonging to the same race of beings
as themselves, or as entitled to the common rights of humanity. Under these circumstances, no one will be surprised to learn that the Polish peasantry, at the dismemberment of the republic, were in the lowest state of degradation,
being at once ignorant, indolent, addicted to drunkenness,
poor and improvident in the extreme. (*Coxe's Travels*, i.,
14; *Voyage de deux Français dans le Nord de l'Europe*, v.,
105; *Bursching's Geog.*; *Introduc. to Account of Poland*,
§4; *Connor's State of Poland*, passim; Coyer, *Vie de So-
biaski*, i., 199; Malte Brun, *Tableau*, &c.)

The servitude of the peasants was modified by the constitution of 1791, and it was wholly abolished in the grand
duchy of Warsaw, nearly identical with the existing kingdom, in 1807; the labour and services due by the peasants
to their lords having been since regulated and defined by
law. Owing to the ignorance of the peasantry, the influence of this great and salutary change was for a lengthened
period less considerable than might have been supposed.
Though the peasants may now leave one part of the country to settle in another, they must first pay off any debt that
may be owing their lords; and from inability to do this, and
various other circumstances, they do not often quit the estates on which they were born. When a young peasant
marries, his lord assigns him a certain quantity of land, sufficient for his maintenance and that of his family in the way
in which they have been accustomed to live. Should the
family grow numerous, some little addition is made to the
farm. At the same time, the young couple obtain also a
few cattle, as a cow or two, with steers to plough their
land. These are fed in the stubble, or in the open places
in the woods, as the season admits. The master also provides them with a cottage, with implements of husbandry;
in short, with all their little moveable property. Owing to
the powerful influence of old habits, but few peasants improve the little stock committed to their management; their
conduct, according to Mr. Jacob, being most frequently marked by carelessness and a want of forecast. This, however,
is by no means uniformly the case: there have been many
instances of accumulation; indeed, several of the peasants
have become proprietors, while others have hired a larger
extent of land. But it will require the lapse of a lengthened series of years before any very general change be made
in the habits and condition of the bulk of the people.

Speaking generally, the houses of the Polish peasantry
are miserable hovels. They are all built of wood; even

" those of the better class have merely the ground floor. On the exterior they are, in every point of view, humble, very often mean in appearance : the interior is occasionally somewhat better. though an Englishman looks in vain for anything like comfort. There are usually two or three ordinary rooms, whitewashed, though only one serves, for the most part, as a sitting-room. The floors are sometimes of earth only, but more frequently planked. A bed stands almost always in every room." (*Burnett's Travels*, p. 196.) The villages, which are of the most wretched description, are thinly scattered, rather along the skirts than in the midst of the forests, and sometimes in vast bare heaths, where no other object is to be seen. They consist of from 10 to 50 miserable huts, rudely constructed of timber, and covered over with straw, turf, or shingles; and afford so imperfect a shelter, that the inhabitants are glad to stop up the chimneys in winter, and to be half smothered with smoke, rather than die of cold. Each of these huts consists generally of only one apartment, with a stove, round which the inhabitants and their cattle crowd together. Bad as these villages are, you may travel 10 m., even in the clear part of the country, without seeing one, or indeed beholding any human habitation. The common diet of the peasantry is cabbage; potatoes sometimes, but not generally; pease, black bread, and soup, or rather gruel, without the addition of butter or meat. Their chief beverage is the cheap whiskey of the country, which they drink in quantities that would astonish the best customers of the gin-palaces of England. Their houses scarcely have little that merits the name of furniture; and their clothing is at once coarse and disgustingly filthy. These, however, are only their general characteristics. The condition of the peasantry depends much on the character of their lords, and upon the more or less embarrassed state of the property on which they may be settled. On the estates of opulent and enlightened landlords it is wholly different from what it is on the estates of those of an opposite description, and may, indeed, be said to be decidedly comfortable.

It is, perhaps, hardly necessary to state that, from the labour applied to the lords' estates being rendered as compulsory service, it is performed in the most negligent and slovenly manner possible. Mr. Jacob says that all the operations of husbandry are very ill executed; the ploughing is shallow and irregular; the harrows, with wooden tines, do not penetrate sufficiently to root up weeds in fallowing; so that the land is always foul, and in bad order. The same want of attention prevails in thrashing. In short, the natural effects of the system of duty-labour are strikingly visible in the whole administration of most of the large estates where it is followed; and is hardly even prevented from exhibiting itself on the estates belonging to the few proprietors who have intelligent and active managers, and are free from pecuniary embarrassments. The common course of crops is the old system of a whole year's fallow, followed by winter corn, and then by summer corn, and then a fallow again, so that one third part of the land bears nothing. The winter crop, in the N. of Poland, consists of wheat and rye, the latter being to the former nearly as nine to one, the little manure that is preserved being laid out on the wheat land. In the S. part of the kingdom the wheat bears a larger proportion to the rye, amounting, on the more tenacious soils, to one fifth, and in some cases to one fourth part, or upward. On a well-managed farm in the province of Lublin, the quantities of seed and produce are said by Mr. Jacob to have been as follows: Potatoes, about 90 bushels to the acre planted, and about 900 bushels raised; wheat, two bushels sown, and from 16 to 90 reaped; rye, two bushels sown, and from 12 to 15 reaped; buckwheat, three bushels sown, and from 10 to 15 reaped. The barley and oats scarcely yield four times the seed. Manure is applied after potatoes for the wheat, the farmer having the benefit of fallowing. This farm was one of the few in which all the labour, except that of the oxen and their drivers, was paid for in money, and not in produce. The common plan of thrashing is to give the thrasher a certain proportion of the corn, varying, according to circumstances, from the 14th to the 18th bushel. In the generality of farms the increase is considerably less than the above; the average produce of wheat being estimated at not more than 14 or 15 bushels; rye, 10 or 12; barley, 14 or 16; and oats and buckwheat, from 8 to 12 the acre; or at not more than half the average produce of similar crops in England. In the S. part of Sandomir and Cracow the crops are more than usually heavy; but they are celebrated more for the excellent quality of their corn than for its greater produce. Sandomir, a narrow district about 60 m. in length, extending along the Vistula, produces the heavy and fine grain known in London as Dantzic white wheat, but the average growth is rarely beyond 90 bushels an acre.

The stock of cattle is small in proportion to the extent of land and the number of the inhabitants. The Polish horses, formerly held in high estimation, have much degenerated,

and a good breed is to be met with only in a few studs. A miserable race of colts is employed to transport merchandise, and field labour is almost wholly performed by oxen or cows. The latter are small, and generally kept in bad condition, both as to food and cleanliness. They are mostly stall-fed, but, from negligence, yield very little butter, and no good cheese. The common breed of the country may be worth from 27s. to 30s. a head; but considerable numbers of a superior breed are annually imported from the Ukraine, which may be worth £3 or upward a head. Previously to the late revolution, the total number of sheep in Poland was roughly estimated at about 3,000,000; but, though the country be extremely well adapted for sheep breeding, the Polish breeds were greatly inferior to those of Saxony, and there were very few flocks of fine-wooled sheep. Latterly, however, the Polish wool has improved very much in point of quality; and is now sent in large quantities to the markets of Leipsic, Berlin, and Bremen, where it sometimes brings a very high price. Hogs, though not very numerous, are of a good breed, originally from Hungary.

The burdens laid directly on the land are not very heavy. Tithes are moderate, and principally compounded for at fixed rates. A small sum is levied in each district for the repair of roads, bridges, and other local purposes; but the and the land-tax do not exceed 25 per cent. on the presumed annual value of the land, which is usually far below its real value. The other taxes fall equally on the different classes of the community. That on beer is let to farm by the government to the brewers. Heavy duties are laid on foreign commodities, such as sugar, coffee, wine, &c. The great mass of the population cannot, however, afford to purchase such luxuries, but content themselves with honey, dried chicory, and whiskey.

The forests are highly important, and in the governments of Augustow and Plock they cover more than a third part of the surface; though in some of the other governments they have been much neglected, and wantonly cut down, especially in the government of Cracow, where, however, especially the place of wood-fuel is supplied by coal. Scotch pine, black fir, alder, aspen, oak, beech, ash, maple, linden, and elm, are the principal forest trees, and the Polish oak and fir timber is decidedly preferable to that of America. Most of the larger forests belong to the crown, and are felled in portions annually, so as to cut them every 30 years. Mr. Jacob states, that the wood cut in one year, on the forest land belonging to the crown, produced £42,000 sterling, being at the rate of 5½d. the acre on the whole of the woods, or 24s. on the part actually cut. (*Jacob's Report on the Agric. of Poland; Burnett's View of Poland; Oot's Trav., &c.*)

Among the wild animals may be specified the bison (*Pol. Zubr*), found in the vast forests of the province of Plock, traversed by the Narew. The Emperor Alexander prohibited the chase of the bison, of which, perhaps, the only remnant in Europe is now to be found in Plock and the adjoining Russian province of Bialystock. (*Malte Brun, Tableau de la Pologne*, p. 55.) The other wild animals include the elk, roebuck, wild boar, badgers, foxes, hares, &c., the skins of which last form articles of export.

Minerals are more numerous and valuable than might have been expected in so flat a country. Bog iron is found almost everywhere; but the principal mining districts are in the S., in the governments of Cracow and Sandomir. Coal is raised in considerable quantities at Bendzine, Sieden, Niemcy, &c. Zinc, which is exported in considerable quantities, is found in the vicinity of Cracow; lead at Olkusz; and copper at Kielce. Iron of excellent quality is also mined in Sandomir.

The domestic *manufacture* of woollen and other stuffs is universal throughout Poland, almost every agricultural family having a loom for the manufacture of the coarse cloths required for their consumption. The yarn used to be partly imported from foreign countries, but lately a large spinning factory has been established at Girardow, which occupies 500 hands. and produces, besides yarn, a quantity of linen cloth. In 1829, the woollen cloth made in the country was estimated at 7,000,000 Polish ells, worth upwards of 70 millions florins, about a tenth part of which was sent into Russia. During the disturbed period which followed, the production of Polish woollens sunk to one third of what it had previously been; but it has lately revived in consequence of the importation of Polish cloths into Russia. duty free, where they are in extensive demand for the clothing of the troops, and other purposes. They are, also, sent in considerable quantities to Kiachta, on the borders of Chinese Tartary. Leather is the manufacture next in importance; and then follow linen and cotton fabrics, sailcloth, paper, bleached wax and wax candles, glass and other chemical products, glass, printing types, jewellery, carriages, &c. Generally, however, these articles are produced on a very small scale; and notwithstanding the eapness of labour, they are mostly, from the want of ill on the part of the workmen, at once high-priced and ferior. Poland, in fact, is an agricultural country; and. cept a few of the more bulky and coarser articles, it ould, were the citizens permitted to resort to the cheap t markets, derive almost all its manufactures and arti- es of luxury from other countries, in exchange for corn, ool, timber, tallow, flax, spirits, and such like articles. irits are distilled in every village from rye and potatoes. t their sale is still, as formerly, a manorial right, each d of a manor having the exclusive sale of spirits within domain. There are breweries in Warsaw, and in some er large towns; and mead, and drinks made from rasp- ries, cherries, &c., principally in the S. provinces, are ourite beverages of the people. Of late years several t-root sugar factories have been established.

The *trade* of Poland is almost wholly in the hands of the ws. The internal commerce is carried on chiefly by ans of fairs, at which, also, a considerable portion of foreign trade is conducted. The latter principally with sia, Prussia, and the republic of Cracow: in 1830 the owing was the value of the several exports to, and im- ts from each of these states, according to the *Petersburg ercial Gazette* for 1833.

	Exports.	Imports.
	Polish Florins.	Polish Florins.
ossia	14,514,496	19,234,599
russia	27,596,921	12,590,106
acria	259,879	3,996,395
racow	2,703,841	1925,184
Totals	45,073,137	47,713,684

uring the revolution the exports decreased greatly, le the imports were considerably augmented. Since period, however, the balance has been in a great mea- restored. England, Holland, France, and sometimes America, take off, through Dantzic, most of the corn h Poland has to export. But in years when prices are in England, and when, consequently, there is a great nd for corn in Dantzic, a good deal of the supplies ght to that port come from Galicia. The customs and duties produced in 1830, 8,794,231 florins, and in 9,238,003 do. Goods are conveyed in summer by y wagons, and in winter by sledges; but the roads are ally bad, and during the late insurrection were much ip; latterly, however, government has been exerting for their improvement. Steam navigation is but in fancy; and merchandise is at present forwarded the rivers by flat-bottomed boats to the Prussian

But Russia seems to be endeavouring to put a stop intercourse between Poland and the Prussian ports Baltic, by constructing a great commercial road from W. angle of Poland to the Baltic; and a railway has planned, for which the capital is already subscribed, e works begun, to convey to the harbours of Windaos ibau the goods which formerly went to Tilsit or Me- r by the Pregel to Königsberg. (*Russia under Nico- p.* 194, 125.) A similar purpose is served by the ca- Augustowo, connecting the Narew and Vistula with emen, and which is to be continued to the Baltic by Indau canal, in the government of Wilna. The ca- Augustowo is 96 m. in length, from 5 to 6 ft. in depth, sufficient breadth for two large boats to pass each vith ease. It has 17 locks, and several convenient in different parts of its course. It was wholly com- between 1821 and 1829, and is now the means of an traffic. These, however, are all violent and unnat- easures, and can have no useful or permanent re- No one can take up a map without being satisfied at e that the Vistula is the proper highway of the , and Dantzic her proper shipping port.

unts in Poland are kept in *zlots, groschen,* and flo- e florin = about 9d., is divided into 30 gr. The Po- is about equal to 14 oz. avoird.: the ship pound =

-w nent.—Previously to 1831, Poland had its two leg- chambers, those of the deputies and the senate; e the unhappy attempt at a revolution that then t, Russia has suppressed these chambers, and Po- overned nearly in the same way as the other por- the empire. The council of administration for the consists of three directors-general (of the interi- ce, and finance), a comptroller-general, and other appointed by the sovereign. The reports of this are submitted to the emperor by a secretary of state nd residing in Petersburg. There is also in that a department for Polish affairs, established since which the government of Poland is confided. The e power is vested in the sovereign, and the pro- ws for this kingdom are submitted for his sanction ussian council of state. The local administration d is exercised by civil governors, with the same

powers as those established in the different governments of Russia.

The civil and commercial codes at present in force are, for the most part, the same as in France: the criminal code is modelled on that of Prussia and Austria. Personal and religious liberty are nominally guaranteed; and those who do not interfere with politics are as secure in Poland as any where else. But those who wish to enjoy this se- curity must have a care not to find any fault with any ac- tion of the government. The press is under the control of censors, who are stricter here than even in Russia. Jus- tices of the peace decide in civil causes up to the amount of 500 florins; above which the latter come before the tri- bunals of original jurisdiction in the capitals of the several governments. At Warsaw, besides a court of appeal, there is a supreme court of cassation, and commercial tribunals are established in all the principal towns. Criminal caus- es are tried in separate tribunals, of which there are four in the kingdom. Political offences come under the cogni- zance of a council of war, or a commission specially ap- pointed.

Religion.—Until lately, upwards of three fourths of the Poles belonged to the Roman Catholic, or the United Greek church, the Greco-Russian communicants being but few in number. But of late the Russian government has, by eve- ry means, been endeavouring to shake the spiritual de- pendance on the court of Rome, not only of the Poles, but of the United Greeks throughout the empire; and its measures, in this respect, appear to have been attended with so much success, that, in 1839, from three to four mil- lions of the United Greeks, including most of those of Po- land, had joined the orthodox Greek church. Until 1832, the Greco-Russians had no prelate in Poland; but at that period an archimandrite was appointed, who resides at Warsaw. The bishop of the United Greek church resides at Heline, in Lublin. The Roman Catholics have an arch- bishop and eight bishops, nominated by the pope on the re- commendation of the emperor of Russia; and the bishop of the government of Cracow exercises authority over the free city of that name, and its territory. There are a num- ber of convents possessing territorial revenues; but the secular clergy receive a regular stipend from the govern- ment, the landed possessions formerly belonging to them being now public property. The parish priests, however, receive tithes, the amount of which is sometimes very considerable. The Lutherans and Calvinists, amounting together to about 290,000 persons, are principally Germans. There are a few Mennonites and Moravians, and some Mo- hammedans.

Public Instruction.—Previously to 1380, education was scarcely diffused at all, except among the nobility and up- per classes residing in the towns, and the total number of persons receiving instruction at that period is said not to have exceeded 16,000, or about 1 in 980 of the population. After the suppression of the insurrection, the schools were shut for several months, and when re-opened, were or- ganized upon the same plan as those of Russia. Private schools are subject to the same inspection on the part of the government as public schools. In 1834 there were, in all, 43,794 pupils in public and private establishments; and in 1839 the number, at 1159 schools of all kinds, was estimated at 70,000 or 1 to every 69 individuals. (*Russia under Nicholas I.,* p. 188.) In 1836 an order was issued by the Russian government, directing that there shall be a teacher of the Russian language in every primary school; and that all children attending such schools shall be obliged to learn the Russian language: it was, also, at the same time, ordered that no individual should be employed as a tutor unless he possessed a testimonial granted by the proper authorities, certifying his ability to give instruction in the Russian language; and that no person unacquaint- ed with Russian should be promoted to any civil or military employment.

This regulation, as was to be expected, gave much of- fence to the Poles, and has been the theme of much idle declamation in this and other countries. Russia, no doubt, wishes to secure her hold over Poland; and every thing that tends to Russianise the latter, and to give her people the same tastes, habits, and modes of thinking as the Rus- sians, must necessarily contribute to this end; and we be- lieve it will be generally admitted, that of all the means to bring about this consummation, the gradual substitution of the Russian for the Polish language will be one of the most effectual. If there were any rational prospect of Poland being able to emancipate herself from the yoke of Russia, one might regret the measure. But as there is no such prospect, the interests of both countries will be best promoted by their being, as far as possible, consolidated into one people.

The Polish army, which before 1831 amounted, in time of peace, to 35,000 men, is now amalgamated with that of Russia.

POLAND.

The Poles are a remarkably fine race of people, being well formed, strong, and active. In their general appearance, they are said to resemble the western Asiatics rather than the Europeans, and are, most probably, of Tartar origin. The gentry are haughty and brave, but, at the same time, frank and generous. The peasantry, however, bowed down by continual oppression, are cringing and servile; their whole behaviour evincing the state of abject servility from which they are now being emancipated. The nobility are very numerous in Poland, amounting at present to not less than 283,490 individuals. According to the old laws of the republic, the nobles were *terrigena*; every person who possessed a freehold estate, how small soever, or who could prove his descent from ancestors formerly possessed of such an estate, and who had not debased himself (!) by engaging in any sort of manufacture or commerce, was a nobleman or gentleman, the terms being in Poland synonymous. The gentry were all held to be equal to each other, the titles of prince, count, &c., which some of them enjoyed, not being supposed to add any thing to their real dignity. Under the republic, the nobility were every thing, and the rest of the people nothing. The former were the absolute lords of their estates, and of the boors by whom they were occupied. They enjoyed the royal privilege of maintaining troops, and constructing fortresses; and they only could elect the sovereigns. No noble could be arrested without previous conviction, except in cases of high-treason, murder, or robbery on the highway; and then only provided he were taken in the fact! His house was a secure asylum to all to whom he chose to extend his protection, whatever might be their crimes. Even his vassals could not be arrested, nor their effects seized; they were exempted from all payment of tolls and direct duties; and though the king might bestow titles, he had no power to create a nobleman or gentleman, that being the exclusive privilege of the diet. Happily, however, this state of things has been wholly changed. Under the vigorous governments of Russia, Austria, and Prussia, the oppressive privileges of the nobles have been suppressed; they can no longer trample with impunity on their inferiors, nor commit offences without subjecting themselves to the full penalty of the law; and a poor gentleman no longer considers it a degradation to engage in some department of industry. (Busching, *Introduction to Poland*, § 4; Malte Brun, *Tableau de la Pologne*, 266; Coxe, i., 102, &c.)

Though modernised in a considerable degree, the richer Polish nobles continue to live in large castles, in a state of rude hospitality, entertaining great numbers of their dependants and such strangers as may happen to visit them. At these feasts the practice of sitting below the salt is still kept up, the best dishes and the best wines being appropriated by the *élite* of the guests.

Jews are more numerous in Poland than in any other European country, having amounted, in 1838, to 411,307 individuals, of whom 338,677 lived in towns, and 72,630 in villages, or in the country. They are, as already stated, in the exclusive possession of the commerce of the country; they, also, are the great manufacturers and sellers of spirituous and fermented liquors; advance money on lands and goods; are the only jewellers and silversmiths; and carry on all pecuniary dealings. Those in the towns are mostly all burgesses; and they may be said to engross all the most lucrative business. But notwithstanding all this, the majority of the Israelites are extremely poor. They seem, also, to be in a lower state of civilization than any other class. Even the richer individuals, though they occupy the best houses in the towns, appear to care little for cleanliness or comfort; and the lower orders live in a state of filth and discomfort that would be intolerable any where else.

There are in Poland many instances of longevity, and, on the whole, the country may be said to be healthy; but the people are, notwithstanding, especially liable to endemical diseases, such as small pox and fevers, which frequently make great havoc. Among the diseases peculiar to, or nearly so, to Poland and Lithuania, the *plica polonica* is the most remarkable. This is a disease of the head, which terminates by affecting the hair, which it dilates, softens, and clots into one undistinguished mass. This disgusting malady spares neither age nor sex, gentry nor peasants, though it be more frequent among the latter than the former. Various theories have been formed to account for its origin: most probably it is occasioned by the bad water, unwholesome food, and filth of the people.

Poland suffered much from the outbreak in 1831, in consequence partly of the destruction of property, and partly of the proscriptions and oppressive measures which it occasioned. Within the last few years, however, the country has again begun to revive.

The municipal revenues of Warsaw amounted, in 1834, to 4,094,000 florins, an increase over their amount in 1829 of nearly one fourth: and the value of insured buildings, throughout Poland, in 1835, was estimated at 585,187,000 florins, being an increase of 112,204,000 florins over that of the buildings insured in 1829. Population has increased still more rapidly; having amounted, in 1828, to 4,088,98?, and to 4,296,963 in 1838. The old roads, also have been materially improved, at the same time that several new ones have been undertaken; so that, on the whole, however depressed in some respects, the country is certainly advancing in improvement.

History.—The history of Poland commences from the 10th century. In 1139, Boleslaus, under whom Christianity had been introduced into the country, divided the kingdom among his four sons, which was the source of a lengthened series of civil wars, and of all sorts of disasters. At length these different portions were united under one sovereign, in 1296. The reign of Cassimir the Great, which began in 1333, and terminated in 1370, is the most brilliant in the Polish annals; still, however, the foundations were laid in it of that anarchy that destroyed the kingdom. Casimir, having no children of his own, and being anxious that the crown should devolve, at his death, on his nephew Louis, king of Hungary, in preference to the legitimate heirs, obtained, for that purpose, the sanction of a general assembly of the nobles, and Louis agreed to the conditions under which they offered him the crown (Koch, *Tableau des Revolutions*, i., 284); establishing, in this way, a precedent for the like interference on future occasions. On the death of Louis the grand-duchy of Lithuania was united to the crown by the *marriage* of Jagellon, its grand-duke, to the daughter of Louis, who had succeeded to the throne. The house of Jagellon continued to occupy the Polish throne for about two centuries; but at each change of a sovereign an assembly of the nobles or *diet* was held, at which the new sovereign was formally elected to the throne. On the death of the last of the Jagellons, in 1572, the throne of Poland became, substantially as well as formally, elective, and it was called not a kingdom but a republic. Henceforth, on the death of a sovereign, the nobility or gentry repaired in vast numbers, sometimes to the amount of 100,000, on horseback, and armed, with crowds of attendants, to a sort of camp in the neighbourhood of Warsaw, to elect his successor, who had to subscribe, and make oath to observe, the *pacta conventa*, or conditions under which he had been elected. These were such as to reduce the royal authority within the narrowest limits, to secure and extend the privileges of the nobility and clergy, and to perpetuate the degradation of the people, who, being slaves in the fullest extent of the term, were not supposed, in fact, to have any legal existence! The famous John Sobieski, the deliverer of Vienna, elected to the throne in 1674, was the last of the great monarchs of Poland. But in the latter part even of his reign the vices of the Polish constitution, and their fatal operation, became obvious; and they thence continued to increase in number and virulence till its total subversion. (See the *Histoire de Jean Sobieski*, by the Abbé Coyer, *passim*.)

Exclusive of the diets for the election of the sovereign, ordinary diets were held, at least, once every two years, at which all matters connected with the government of the country were discussed and decided upon. It is easy to see, from what has been already stated, that this form of government could not fail to produce great party contests and disorders, and that it must have afforded every facility to the surrounding powers for acquiring a preponderating influence in the diet. Probably, however, the abuses already noticed might have been repaired, but for the principle, if we may so call it, first introduced in 1652, that no decision could be come to upon any matter submitted for consideration, unless the diet were unanimous. Hence the singular and extraordinary privilege of the *liberum veto*, by which any single member of the diet was permitted to interpose his absolute veto, and, by doing so, could nullify its whole proceedings! And, which is even more extraordinary, this absurd privilege, which allowed the whim, caprice, or bad faith of an individual to prevent the adoption of any measure, however necessary and however generally approved, were, for a lengthened period, regarded by the Poles as the palladium of their liberties! (*Abregé de l'Histoire du Nord*, ii., 683.)

It is plain, from these statements, that latterly the whole powers of the state were engrossed by the nobles, or gentry, many of whom, though enjoying the same political rights and franchises as the others, were miserably poor. In consequence, corruption, intimidation, and such like arts had full scope in the Polish diets, particularly in those held for the election of sovereigns; and latterly the crown was

---

* The diets consisted, 1, of the senate, composed of the bishops, palatines, or perpetual governors of provinces, castellans, or governors of towns, and the grand officers of the crown; and, 2, of the nuncios, or representatives of the nobles or gentry. These bodies did not, however, deliberate separately, but together; and, as will be immediately seen, they could come to no resolution without being unanimous.

fact, either sold to the highest bidder, or the election
as decided under the influence of foreign force. And if,
hile the government was in this state of abasement, we
ar in mind that the whole people, with the exception of
ie nobles or gentry, were slaves, on whom every indigni-
ought be practised by their masters, it will be seen that
ere is but little to regret in the subversion of such a state
things.

Indeed, the only wonder is, that Poland was not sooner
ased from the list of nations. Its partition had, in fact,
en proposed by the Swedes in the reign of Casimir V., a
ort while previously to the election of John Sobieski, as
e only method by which the disorders that agitated the
untry could be put an end to, and the inconvenience
ence arising to the surrounding states be obviated.
uhibère, *Anarchie de Pologne*, i., 68.) But it was not
more than a century after that the first partition was
reed upon, in 1772, by the emperor of Austria, the em-
ess of Russia, and the king of Prussia, by which about a
rd part of the kingdom was dismembered, and added to
: dominions of the partitioning powers! But it was not
be supposed that having once begun to share in so rich
poil, these powers would rest satisfied with this acquisi-
i. The pretexts for farther interference still continued
liminished. Poland, as before, remained a prey to all
s of disorders, and the Russian ambassador, and not the
g. was the real sovereign.

1791 the majority of the nobility and gentry then as-
bled in a diet, which had been made permanent, being
rous to raise their country from the miserable state
which it had fallen, and stimulated by the events
ected with the French revolution, drew up the *projet*
new constitution on a more liberal and broader basis,
shing the *liberum veto*, and making the crown heredi-
on the demise of the king, in the Saxon family. This
titution was accepted by the king; but the great bulk
e nation did not, and could not, take any interest in
change; and the government were wholly without the
ns of supporting the new order of things. Russia had
difficulty in fomenting fresh disorders; and the un-
nate Poles, with an imbecile sovereign, without forces,
abandoned and betrayed by their pretended allies.
again compelled to submit to a fresh dismemberment
eir country.

voked by these repeated indignities the Poles, under
usko, rose in rebellion in 1794. But their means were
v inadequate to the struggle in which they had en-
l; after displaying prodigies of valour, Kosciusko was
ted and taken prisoner (10th October, 1794), and Pra-
e suburb of Warsaw, being taken by storm, that city
vith surrendered; and there being no longer any ob-
in the way, a dismemberment of the remaining terri-
of the republic took place in 1795, and Poland was
obliterated from the map of Europe.

powers who dismembered Poland had, in reality,
g better to allege, in justification of their measures,
he robber's plea, that the power to commit an act
it at once right and expedient! But, how objection-
never the motives by which they were influenced,
w dangerous soever the precedent which they estab-
there can be no reasonable doubt that their mo-
iave been decidedly advantageous to the great bulk
Polish people. The vices inherent in Polish society
uch that it is idle to suppose they could have been
ted by any remodelling of the constitution. There
middle class (or none worth notice) in the country;
: between nobles, jealous of their rank and privi-
m the one hand, and newly emancipated slaves,
ed and degraded by a long course of oppression, on
er. To restrain the first within the limits prescribed
and to raise the second class, was a work that
nly be undertaken by a powerful government, such
were no means of forming out of native materials.
e regretted that Russia obtained the lion's share of
l; but even in Russian Poland the condition of the
as been very decidedly changed for the better; and
ian and Prussian Poland, the improvement in their
n has been signal and extraordinary.

xisting kingdom of Poland originated in the grand
f Warsaw, established by Napoleon in 1807. It
gned to Russia by the congress of Vienna, and ob-
rom the emperor Alexander a representative con
. Unfortunately, however, the disgust occasioned
rutality of the grand duke Constantine, command-
ef of the Russian forces in the kingdom, conspir-
the excitement produced by the French revolution
and the abuse of Russia in intemperate and ill-
peeches in the H. of C. and chamber of deputies,
r de it be believed that England and France were
assail that power, precipitated the Poles into an
ion. The result was such as all men of sense an-
from the outset. The Poles made a gallant stand

in defence of their liberties; but in the end every vestige
of their independence was totally destroyed. The name
of the kingdom remains; but its peculiar privileges have
been subverted, and it is now substantially and in fact a
part of the Russian empire.

POLAND, p. t., Cumberland co., Me., 51 m. S.W. Augusta,
574 W. Incorporated in 1795. Bounded N. by Little An
droscoggin river. It contains a small village of Shakers,
who possess 600 acres of finely cultivated land. It has
five stores, one grist-mill, one saw-mill, one oil-mill; 34
schools, 1052 scholars. Pop. 2360.

POLAND, t., Chautauque co., N. Y., 28 m. S.E. Mayville,
317 m. W. by S. Albany. Drained by Conewango creek
and Chautauque outlet. It contains two stores, one grist-
mill, 15 saw-mills, seven tanneries; 10 schools, 400 scholars. Pop. 1087.

POLAND, p. t., Trumbull co., O., 73 m. N.E. Columbus,
284 W. Watered by Mahoning river. Pop. 1583.

POL-DE-LEON (St.), a seaport town of France, dep.
Finisterre, cap. cant., on the channel, 10 m. N.W. Morlaix.
Pop., inc. com., in 1836, 6451. Though ill-built, it is clean,
well paved, and remarkable for its antique edifices. Its
cathedral, a structure of the 15th century, has some excellent carving; and a tomb, said to be that of the first Breton
king. The church of Kreizker, built in the 14th century,
has a spire 394 ft. in height, and one of the handsomest in
France. (*Guide du Voyageur*.) St. Pol has some trade in
cattle, horses, linen, hemp, &c., but to no great extent.

POLIGNY (an. *Castrum Olinum*), a town of France,
dep. Jura, cap. arrond., on elevated ground, 13 m. N.E.
Lons-le-Saulnier. Pop., in 1836, ex. comm., 6366. It consists principally of four long parallel streets, and is well
built, clean, and ornamented with several handsome fountains. Among its public edifices is a well-constructed
slaughter-house. Poligny was formerly a place of importance, and a favourite residence of the sovereigns of Burgundy. It has a few manufactures of common earthenware, saltpetre, glue, &c.; and is a sub-prefecture, though
the superior courts for the arrondissement sit at Arbois.
(*Hugo*, art. *Jura*, &c.)

POLK, county, Tenn. Situated in the S.E. part of the
state, and contains 330 sq. m. Bounded N. by Hiwassee
river, by branches of which it is drained. It contained in
1840, 3541 neat cattle, 1799 sheep, 10,899 swine; and produced 8710 bushels of wheat, 220,294 of Indian corn, 94,489
of oats, 5362 of potatoes, 3680 pounds of tobacco, 14,883 of
cotton, 1060 of sugar. It had seven stores, six grist-mills,
one saw-mill, one tannery, four distilleries; two schools, 38
scholars. Pop.: whites, 3949; slaves, 304; free coloured,
17; total, 3570. Capital, Bentonville.

POLK, county, Mo. Situated S.W. of the centre of the
state, and contains 768 sq. m. Drained by Pomme de Terre, Sac, and Niangua rivers. It contained in 1840, 11,402
neat cattle, 6411 sheep, 30,515 swine; and produced 12,143
bushels of wheat, 392,830 of Indian corn, 19,042 of oats,
17,903 of potatoes, 25,650 pounds of tobacco, 29,823 of cotton. It had 14 stores, 10 grist-mills, three saw-mills, two
tanneries, four distilleries; five schools, 93 scholars. Pop.:
whites, 7978; slaves, 463; free coloured, 9; total, 8449.
Capital, Bolivar.

POLTAVA, a gov. of European Russia, lying along the
E. side of the Dniepr, by which it is separated from the
governments of Kherson and Kieff, having the government
of Tchernigoff on the N., and those of Kharkoff, and
Ekaterinoslaff on the E. and S. Area estimated at 22,500
sq. m. Pop., in 1838, 1,622,000. (*Koppen*.) Surface quite
flat; soil excellent: in some parts there is a scarcity of
wood. Besides the Dniepr, the principal rivers are its
affluents, the Vorskla, Piriol, and Sula. This and the surrounding governments constitute what may be called the
granary of Russia. It is one of the best cultivated districts
of the empire: the return of the corn crops is said to be as
six to one, the total produce being about 6,506,000 chetwerts, of which about 1,500,000 are exported. The grazing
grounds are excellent, affording pasturage for large herds of
the fine Ukraine breed of oxen, and for immense flocks of
sheep, the breed of which has latterly been much improved Some peasants have above 100 bee-hives. Manufacturing industry has not made much progress; but there
are fabrics of cloth and linen, with numerous distilleries,
and establishments for the preparation of tallow, candles,
&c. Large quantities of corn, tallow, and other products,
are annually sent from this government to Odessa, and
oxen to Moscow, Petersburg, &c. (*Schnit., La Russie, &c.*,
p. 465.)

POLTAVA, the cap. of the above gov., on the Vorskla;
lat. 49° 35' 4" N.; long. 34° 41' 15" E. Pop. 9500. It
stands on an eminence, and is built principally of wood,
with broad and straight streets. There is a good square,
with brick houses, embellished with a granite monument
in honour of its deliverer, and the regenerator of Russia,
Peter the Great. It is surrounded by a rampart, and has

a cathedral, gymnasium, convent, &c., with three great fairs annually, and a considerable commerce.

Charles XII. of Sweden having besieged this town in 1709, Peter the Great marched to its relief; and in its vicinity, on the 27th of June of the same year, was fought the famous battle of Poltava. The Russians gained a complete victory. The Swedish army was entirely destroyed: it lost above 9000 men left dead on the field of battle, and from 2000 to 3000 made prisoners in the pursuit; while the residue, consisting of about 14,000 men, under General Lewenhaupt, after escaping from the battle, were compelled to lay down their arms and surrender on the 12th of July. Charles, with only a small escort, effected his retreat across the Bug, and took refuge in Turkey. This great victory established the power of Peter on a solid foundation, and secured not merely his empire, but the success of his vast projects and plans for the civilization and improvement of his people. To use the words of Voltaire—" Ce qui est le plus important dans cet bataille, c'est que de toutes celles qui ont jamais ensanglantée la terre, c'est la seule, qui au lieu de ne produire que la destruction, ait servi au bonheur du genre humain, puisqu'elle a donnée au Czar la liberté de policer une grande partie du monde." (Histoire de la Russie sous Pierre le Grand, part i., cap. 18.)

POLYNESIA, " the region of many islands," a collective term used by geographers to designate the numerous groups of volcanic and coralline islands scattered over a great part of the Pacific, but especially between the tropics, extending eastward from the Philippine islands and New Guinea to the coast of S. America. Supposed aggregate population, 1,500,000; but all calculations of this kind are merely conjectural, as there are few or no data. The equator forms a convenient dividing line between these groups, which may accordingly be comprised under the heads of N. and S. Polynesia, as follows, the order being from W. to E.

Names of Insular Groups.		Situation.	
		Lat.	Long.
I. N. Polynesia.			
Pelew Islands	about	8° N.	135° E.
Ladrone —	.	17°	147°
Caroline —	.	7°—17°	135°—70°
Radick —	.	0°—10°	164°—172°
Sandwich —	.	20°	156°— W.
Gallapagos —	.	2°	9.0°
II S. Polynesia.			
Admiralty Islands	.	2° S.	148° E.
New Ireland, New Britain, and New Hanover	.	6°—9°	152°
Louisiade (little known)	.	10°	144° ?
Solomon's Islands	.	9°	157°—164°
New Hebrides and New Caledonia	.	19°—22°	164°—170°
Q. Charlotte's Islands	.	12°	166°—170°
Friendly —	.	15°—21°	173° E.—185° W.
Navigators' —	.	13°	173°
Society —	.	13°—21°	150°—143°
Dangerous Archipelago	.	14°—19°	141°—132°
Marquesas —	.	9°—11°	140°
Pitcairn Island	.	20°	133°
Eastern —	.	27°	109°

The whole of Polynesia may be considered as a series of submarine mountain ranges; for no portion of the earth's surface has more numerous inequalities, and nowhere, except in America, have the chains so marked a course from N. to S. Indeed, all the archipelagos have, more or less, this direction, and it not unfrequently happens that the small chains are individually terminated by an island of larger size than the others with which it is connected. Many of the larger islands, and particularly those which shoot up to a considerable elevation from the sea, consist of basalt, as well as other igneous formations; and in many of them are distinct traces of volcanic action, with a few active volcanoes. To this class belong the Friendly islands (the largest of which, called Otaheite, has a mountain rising to the height of 10,000 ft. above the sea), the Marquesas, and Sandwich islands, in the last of which are several, both extinct and active, volcanoes, rising from 12,000 to 16,000 ft. above the sea. The desert group of the Galapagos is likewise in a state of igneous action, and the whole is a mere mass of lava and similar productions. The numerous small islands that stud the Pacific S. of the equator, and W. of the Friendly islands, and particularly those that rise but a short distance from the level of the sea, are based on reefs of coral rock. Of those examined by Captain Beechey, none were more than 30 m. in diameter; they were of various shapes, chiefly formed of living coral, or at any rate encompassed by a reef of that substance. Most of them have lagoons in their centres, the bottoms and sides of which are likewise formed of coral; and the generally circular form of these islands, the existence of these lagoons, and shelving conical form of the submarine mountains, has led to the supposition that they are nothing more than the crests of submarine volcanoes, having the rims and bottoms of their craters overgrown

with coral. It is also well known, that the Pacific is a great theatre of volcanic action, and every island, yet examined in Polynesia, consists either of volcanic rocks or coral limestone, and in many instances of basalt and lava, having a girdle of coral. (Lyell's Geology, iii., 236-233.) The formation of coral, which, according to Captain Beechey, is very gradual, ceases as soon as it reaches the surface of the water; but it serves as a basis for a vegetable soil, which in these regions is soon covered with plants, cocoa-nut and other trees. The larger islands of Polynesia are indented with deep bays, furnishing tolerably good harbours for shipping; but the circumference of the largest is less than 200 m., so that their extent does not admit of the formation of any important lakes or rivers. By far the largest portion of Polynesia is between the tropics; but the small extent of the islands procures for them the temperature of the ocean, and a succession of light sea and land breezes. Hence the heat never becomes oppressive, even to Europeans. Hurricanes and earthquakes are of rare occurrence.

The numerous islands of the Pacific afford an extremely diversified vegetation; and among the many plants covering their surface are some of high utility for human support, especially the bread fruit tree (a favourite article of food among the islanders), cocoa-nut, yam, the root of the Arum esculentum, the banana, plantain, and sugar-cane, which last grows naturally in the Sandwich islands, and succeeds better than in any other part of the world. A native chestnut, called Rata (Tuscarpus edulis), furnishes the natives with a sweet nut, that forms an agreeable substitute for bread-fruit; besides which, the Ahia (Eugenia malaccensis) bears a pulpy fruit, shaped like an apple. The Ti-root (Dracæna terminalis) furnishes an inferior spirit, called Ava, the preparation of which is pretty well known by most of the islanders, and has produced very demoralising effects. The use of foreign spirits, however, has, in the groups best known to Europeans, almost superseded the necessity of drinking this nauseous stuff. The tropical productions of the American continent have been successfully introduced by European navigators and missionaries; so that many of the islands, besides their indigenous productions, bear an abundance of oranges, citrons, shaddocks, pine-apples, guavas, figs, and cape mulberries. The vine also was introduced, but was destroyed by the natives in their wars: fresh plants have more recently been imported, and thrive well in some of the islands. The growth of corn has been more than once attempted, without success; owing more, however, to the imperfect means of tillage than the unfitness either of the soil or climate. Pumpkins, melons, cucumbers, cabbages, and kidney-beans flourish better, and with less trouble, than any other foreign vegetables. The hills of the more elevated islands are clothed with forests of stately trees; the most valuable of which are the apapa and faifea, which yield excellent timber for canoes, and the candle-nut (Aleurites triloba), the oil of which is used for domestic illumination. The principal trees growing on the plains are the tamanu (Callophyllum inophyllum), and the Hibiscus tiliaceus (highly valued as materials for furniture, canoes, &c.), the Chinese paper mulberry (Morus papyrifera), and the sandal-tree, the timber from which being so exported from the Sandwich islands to Europe and China.

The fauna of Polynesia is characterized, like that of Australia and its surrounding islands, by the absence of beasts of prey, and indeed, all the larger animals. Hogs, dogs, rats, lizards, were the only quadrupeds originally found on the islands. The native hog is different from the European breed since introduced, and has now almost disappeared: the present breed is reared in great numbers, and pork constitutes the favourite food of the natives. The flesh of the dog is also esteemed a luxury, especially by the Sandwich islanders. Rats were occasionally eaten unconsidered by the Friendly islanders; but that practice has been discontinued. Cats have been domesticated in most of the houses. Horses, asses, horned cattle, goats, and sheep have since been added, and thrive exceedingly well. The oxen are a fine breed, chiefly imported from New South Wales. The horses come from S. America, and, being used only for the saddle, are never shod. Among the indigenous water birds may be noticed the albatross, tropic bird, several kinds of petrels, herons, and wild ducks. Woodpeckers, turtle-doves, and pigeons are common; besides which, the parroquet (of the species trichoglossus) is abundantly distributed over all the islands; and in most of them the domestic, and several species of wild fowl appear to be indigenous. Fish are numerous on the coasts and in the lagoons of all parts of Polynesia: the albicore, bonito, ray, and shark are eaten by the natives; in addition to which the shores abound with a great variety of delicate rock-fish. The fresh water streams also swarm with salmon, eels, &c., and many varieties of molluscous crabs, &c., with turtles, are caught on the coral rocks. Among the whales that frequent the coasts of Polynesia, the largest and most valuable is the *cachelot*, or sperm-whale (the male of which, according to Mr. Bennett, sometimes yields from 70 to 90 barrels of oil, and about 15 barrels of spermaceti); the cape-whale, hump-back, and black-fish; besides which, porpoises and other small cetaceous animals are extremely abundant. (*Ellis's Polynesian Researches*, i. 30–77; *Moerenhout*, i. 367–382; *Bennett's Whaling Voyages*, i. 36, and append.)

Polynesia has, of late years, been much visited by Europeans, partly curious to inquire into the habits of people differing essentially from those of the rest of the world, and partly anxious to communicate to them the arts of civilization and a knowledge of the Christian religion. Its isolated inhabitants were found to possess many interesting features of character, but at the same time to be plunged into the grossest barbarism, sensuality, and idolatry. Cook traced among some of the islanders, a confused notion of a supreme intelligent Deity; but they almost exclusively worship a number of inferior gods, particularly marine and aërial deities, demons, birds and fishes. Their rude idols were supposed to exercise a powerful influence; their temples were polluted with human sacrifices; and divination, witchcraft, &c., were practised by the chiefs as political engines for overawing their subjects. (*Ellis's Pol. Researches*, i. ch. 13, 14.) At the same time, morality, as understood by Europeans, had no existence among this wild people, and the grossest animal appetites had full sway, their gratification being encouraged, also, by their religious institutions. The law of the strongest prevailed: fierce and bloody wars frequently took place between the different insular tribes on the most trivial pretext, and conquest was generally followed by acts of the most horrible cruelty, including the extermination of the vanquished tribe, and occasionally even an indulgence in cannibalism. The female sex, too, was found in as licentious, degraded, and oppressed a state as in the wildest districts of Africa. A kind of civil marriage appears to have been generally observed; but as the sex was too much despised to allow the existence of affection, the wives were repudiated on the slightest pretext, or else neglected for more fascinating concubines. Polygamy was common in nearly all the islands, and in some groups it is still prevalent. Moerenhout, indeed, tells us, though the statement savours strongly of exaggeration, that some chiefs in the Fidjee islands have as many as 300 wives. (*Voyage au G. Ocean*, ii. 66–69.) Female virtue was formerly wholly unknown; and notwithstanding the labours of numerous missionaries during upwards of twenty years, chastity is still, we apprehend, extremely rare. Sexual indulgences, and even infanticide, were encouraged by a singular institution called the Areoi society, the baneful influence of which appears to have been pretty generally diffused over the islands of the Pacific. The missionaries state that about two thirds of the children born were destroyed by their parents; and notwithstanding the introduction of Christianity, the practice still prevails: but we are not disposed to attach implicit credit to these statements, which, no doubt, go far beyond the mark.

Cannibalism is still practised in the Marquesas, and some of the other groups. The islanders, however, do not live in the rude independence of savage life, but acknowledge the arbitrary sway of hereditary chiefs, whose power is controlled only by those subordinate to them in particular districts. Some attempts have been made to introduce governments of a more liberal character; but they have signally failed.

The habits of the natives are still in many respects those of barbarians. Their houses, almost without exception, are confined to a single story; usually, though not always, of oblong shape, and very simply constructed with stakes of the bread-fruit tree, driven into the earth for the purpose of supporting the roof, which is commonly thatched with the leaves of the *Pandanus odoratissimus*, or cocoa-tree. The framework of the walls is composed of bamboo or hibiscus rods, and a large portion of one side is open, being covered only at night with a kind of cloth curtain. The interior comprises only a single apartment, sometimes with a board-ed floor; and the furniture consists simply of a few mats and cotton-stuffed pillows spread on the floor, a few low wooden stools, a trough and stone for preparing their favourite paste called *poe* (made from the *arum esculentum*), some cocoa-nut shells, used as cups, with a fishing-spear, and perhaps a musket. Bunches of fruit hang from the walls; and occasionally may be seen a work with a litter of pigs occupying a space in one corner, railed off for her accommodation. A separate shed is employed for cooking; and in the more advanced islands a plot of enclosed ground, planted with useful vegetables or favourite flowers, surrounds many of the houses. (*Bennett's Whaling Voyage*, i. 100; and *Moerenhout*, ii. 86–89.) Some of these huts are exceedingly large. Mr. Ellis mentions one belonging to a native chief that was nearly 400 ft. in length; and houses 100 ft. in length are by no means infrequent. (*Polyn. Researches*,

i. 175.) The domestic habits of most of the natives are not unsocial, but irregular, as respects refreshment and sleep, labour, and amusement. Their meals are arranged according to their avocations or the supply of their provisions. They usually eat in the forenoon; but their principal meal is in the evening, when, if well supplied, they eat to excess. They rise early, and go late to rest; but the men are often buried in sleep during many hours of the day, while every species of household drudgery is performed by the females, who are oppressed and degraded in the last degree. The habit of frequent bathing prevails throughout Polynesia, and the natives are remarkable for cleanliness, and most of them pay great attention to personal ornament: indeed, says Mr. Ellis, "their appearance on public occasions is in a high degree imposing." The hair, in particular, is regarded as an object of great attention by both sexes, and the females commonly appear in loose ringlets entwined with flowers. (*Researches*, i. 133–135.)

The dresses of the islanders originally consisted of cloths woven by the women from the bark of trees, and wrapped loosely round the body, leaving a large part of it uncovered; but since their connection with Europeans, they have introduced a very droll *mélange* of native and European costume. The practice of tattooing the body prevails more or less through all the islands, though attempts were made a few years ago to abolish the barbarous custom in Otaheite. Tattooing is performed during childhood; and in the Society islands at the early age of 8 or 10. The patterns vary in the different groups; but nowhere is the body so extensively disfigured as in the Marquesas, the inhabitants of which have a most hideous appearance. In some of the islands the face is left in its natural state, the legs, arms, and breast being the only part tattooed. (*Moerenhout*, ii. 121–4.) The natives of all the best known groups, except the Sandwich islands, are, like the half-civilized inhabitants of most tropical countries, extremely indolent, having, in fact, little occasion for industry, owing to the abundance with which the fruits of the earth are spontaneously produced. Their principal employments are agriculture, fishing, canoe-building, and the manufacture of cloth. Agriculture, as previously observed, is in the rudest state that can well be conceived, the only tillage that the earth receives being by a rude iron-shod stick, about as broad as a European chisel. Fishing is a far more favourite employment, and the methods used are numerous and sometimes highly ingenious. The fish are sometimes caught in circular fences, built up in the shallow parts of their lakes, and simply taken out with a hand net; these enclosures are also excellent preserves for fish not wanted for immediate use. Large nets, made of the twisted bark of the hibiscus, are used for fishing salmon, herrings, &c.; and on most of the islands the natives exhibit a surprising dexterity in the use of the fishing spear. In fact, nowhere are there more skilful fishermen; and considering that before their intercourse with Europeans they were entirely destitute of iron, their various modes of fishing apparatus was astonishing. The situation of these islanders necessarily imparts a maritime character to their habits, and much of their attention is, accordingly, devoted to the building and management of their vessels. Their canoes are of various size, as well as shape, and are either double or single. The largest of those seen in the Society islands are nearly 70 ft. in length, with very high sterns and sterns, but only 2 ft. wide. Those used in war are between 40 and 60 ft. in length, firmly built, of rather elegant shape, highly ornamented with carving, and when in use decorated with gay-coloured flags and streamers. In the double canoes (which are merely single canoes lashed together), planks are thrown across to form a kind of deck for the accommodation of passengers, and over it is sometimes spread an awning of platted cocoa-nut leaves. The paddles, made of the hibiscus, are not heavy; but as they are used alternately on each side of the boats, the labour of rowing is by no means inconsiderable. The canoes used in fishing on the reefs are single, and are commonly the excavated trunks of trees; they seldom carry more than two persons. Many of the canoes have moveable masts, which are only raised when the sails are used. The latter, of which there are sometimes two, but more frequently only one to each canoe, are made with matting of the pandanus leaf, in the shape of a half oval, and the rigging is of the simplest description. On the whole, the canoes of the Society islands are decidedly superior to those of the other groups: those of the Sandwich islanders are large and strong, but less elegant: those used in most of the other islands are of smaller size, and less skilfully constructed. The dexterity of the natives in managing these frail barks when out at sea is also, perhaps, unparalleled among the inhabitants of savage countries. (*Ellis*, i. 128–170.)

The manufacture of cloth, which is more or less carried on in all the islands, is almost exclusively conducted by females, the materials commonly employed being the inner fibres of the bark from the branches of the bread fruit tree.

POLYNESIA.

These fibres, after having been macerated, are beaten on a long board with a grooved mallet, the blows from which cause the moist fibres to interlace with each other, and to assume the appearance of woven cloth. By this process bales are sometimes made containing upwards of 200 yards of cloth, 4 yds. wide. The colour of the unbleached cloth is a darkish brown; but it is always either bleached or coloured with vegetable dyes. Skill in the manufacture of this fabric was formerly highly prized by females of all ranks; but since the introduction of European cloth, it has been made in much smaller quantities, and its use (in the Sandwich, Society, and Friendly islands) is now confined chiefly to women, children, and the lower classes. The missionaries tried to introduce the spinning and weaving of cotton, but with little success; and this pursuit has been since abandoned. A kind of delicate matting is made in some of the islands from the bark of the hibæcus; the fabrics thus formed being either bound over the loins, or worn as mantles on the back. In the manufacture of this last article the islanders of the Palliser group far excel all others. A coarser kind of matting, also, is made of palm-leaves, for bedding, and the sails of canoes.

As respects foreign trade, it may be said to have had no existence in these islands till a late period; the intercourse is chiefly kept up by means of the whaling ships, and the number of vessels touching at the Sandwich islands may average about 80 annually. Trading intercourse has made the islanders aware of the value both of goods and money; and beads, looking-glasses, and buttons have wholly lost their former commercial value.

Most of the islanders of Polynesia are of a lively excitable disposition: hence, when not employed in the graver pursuits of fishing, canoe building, or war, they give themselves up with great ardour to a variety of amusements, among which dancing is, perhaps, the most prominent, being common on all occasions, not merely of pleasure, but also of religion and state ceremony. Some of these dances are stated by the missionaries to have been very objectionable; while others were of a graceful and more dignified character. The exchange of a Christian profession for a debasing polytheistic idolatry has diminished the frequency of these exhibitions, though they still occasionally take place. The musical instruments of the islanders consist of a long narrow drum, a trumpet formed of a species of murex, into which is inserted a bamboo cane for a mouth-piece, and a flute of bamboo, about 16 inches long and about 1 inch in diameter. Boxing and wrestling, also, used to be favourite amusements; but these exercises, as well as many other national entertainments, have been all but abandoned since the introduction of Christianity, though there is, after all, but too much reason to suppose, that the efforts of the missionaries have produced little radical change for the better in the morals of the mass of the population.

The islanders of the Pacific, as respects physical character, may be divided into two distinct classes. The most ancient tribe is composed of Papuan negroes, who are distinguished by darkness of skin, smallness of stature, and black woolly or crisped hair: they chiefly inhabit the Admiralty islands, New Britain, New Ireland, New Hebrides, New Caledonia, and the Solomon islands. The other tribe, which is far more widely dispersed over the numerous groups of this great ocean, exhibits many of the features belonging to the Malays and aboriginal Americans, but is, in some respects, so different, as to form a separate and intermediate race. The people of each cluster, also, are distinguished by minor peculiarities. The inhabitants of the Society islands are of good stature, and well made; of olive complexion, with open, prepossessing features, with a facial angle as perpendicular as in the European head; a bright, full, and jet-black eye, placed under well-arched eyebrows, a straight or aquiline nose, well-formed mouth, and moderately high forehead, with straight though not wiry hair, either black or brown. The Sandwich islanders have more firm and muscular limbs, but in other respects bear a close resemblance to those just described: and they are generally active in their movements, graceful and stately in their gait, and perfectly unembarrassed in their address. Both sexes incline to corpulency in advanced life. It is remarkable, also, that the chiefs, and persons of hereditary rank, throughout the islands, are almost without exception, superior to the common people, in stateliness, dignified deportment, and physical strength: indeed, so great is the difference, that Bougainville, and others, have supposed them to be a distinct race, whose ancestors at some remote period had brought the aborigines into permanent subjection. (*Ellis's Pol. Research.*, i. 78-84; and *Moerenhout*, ii. 247-253.)

With respect to the languages of Polynesia, Marsden first ascertained that there is one general language pervading the whole of the South Sea islands, and extending, with its different dialects, from the E. Indian archipelago to the E. extremity of the Polynesian groups: "Indeed," says M. Moerenhout, "it is impossible to avoid observing the close

analogy between the dialects spoken in the many different islands. So striking a similarity is there between the languages of the Society islands and New Zealand, that the natives mutually understand each other: the inhabitants of the Sandwich, Marquesas, and Society islands converse after only a few days' practice, and the occupants even of the far distant Easter island are intelligible to the whole of the other islanders both N. and S. of the equator. It has been believed by some linguists that all these dialects are branches, more or less, of the Malay language; and many words certainly bear some analogy to those in the Malay vocabulary; but, in fact, "there is no living language either of Asia or America, which can be denominated the parent-stock of the great Polynesian language." (*Crawford's Indian Archipelago*, ii. 80-86; *Moerenhout*, i. 395-398.)

1. *The Caroline Islands* (sometimes also called the New Philippines) extend over about 20 degrees of longitude, and are divided by captain Lütké into 46 groups, comprising several hundred islands, a few of which are high and rising in peaks, but by far the greater number are low, and merely of coralline formation. They were discovered in 1686 by a Spaniard, who named them after Charles II. king of Spain. The productions of these islands are very similar to those of the Sandwich and Society islands; but the bread-fruit is found only in a few of the groups, and the hog is wholly absent; hence these islanders live chiefly on fish. The inhabitants are reputed to be the most expert sailors and fishermen of Polynesia; and, notwithstanding the tempestuous sea by which they are surrounded, they have a considerable trading intercourse in canoes with the Ladrone islands, and the E. archipelago.

2. *The Sandwich Islands* (10 in number, of which 8 are inhabited) form a group many hundred miles distant from a. the rest. The area of the largest island, Owhyhee (or, more properly, Hawaii), is estimated at 4300 sq. m., being more than half that of the entire group, and, is indeed, by far the largest island of Polynesia. It rises in high and towering cones to an elevation of nearly 16,000 ft. from the sea; and not only is the gigantic volcano of Kiraoca, with its immense crater, 2 m. in length by nearly 1 m. in width, and several hundred feet deep, in a constant state of terrific ebullition, but the whole island is one complete mass of lava; and, being perforated with innumerable apertures in the shape of craters, may be considered as forming a hollow cone over a vast furnace in the heart of a stupendous submarine mountain. (*Ellis*, iv. 236, 950.) The population of the Sandwich islands is estimated by the American missionaries at 150,000, more than half of whom reside in Owhyhee. The natives are of a darker complexion than those in the Society islands; the females have coarse and disagreeable features, and both sexes are gloomy and reserved. The natives generally are remarkable for their attention to the arts of industry, and have distinguished themselves above all others by their efforts to introduce European civilization.

Christianity was introduced by the American missionaries in 1820, and is now the religion of the state: schools have been established, churches have been built, and the forms of religion are, at least, pretty generally observed. European usages have also become fashionable, and the costume of the better classes, women as well as men, closely resembles that of the Anglo-Americans. Honoraru, in the island of Oahu, is the capital of the group, and has a population of about 6500. Some of the houses are built of stone, but the natives still prefer living in their huts, so that the town is grotesquely irregular. The principal public building is the "English school," in which instruction in English reading and writing is given to children of both sexes: there are, also, two churches, numerous boarding houses, and many well-stocked shops. The harbour (formed by a barrier-reef of coral, having a single opening) has accommodation for between 60 and 70 vessels of 500 tons: in 1831, there belonged to the islands 14 ships of 9630 tons, of which 4 brigs and 7 sloops were the property of the natives. In 1836, from the 1st of July to the 13th of December, there were 154 arrivals at the port, of which 80 were brigs and schooners belonging to the country, 56 from the United States, and 17 from England. An English newspaper has been established in the town; and in some numbers, which we have seen, we have noticed advertisements of ladies' shoes from Paris, eau-de cologne, ices and so forth. The government has associated commercial treaties on a liberal footing with England, the United States, &c. (*Sandwich Islands' Gazette; Bennett's Whaling Voyage*, i. 190-236.

3. *The Society Islands*, with which Europeans became earlier acquainted than the other groups, consist of six larger and several smaller islands. The principal of them, called Otaheite (or, more properly, Tahiti), is 180 m. in circ. and has a population of about 7000 persons. It is extremely mountainous, some parts attaining an elevation of nearly 7000 feet; but "extensive as well as fertile vales open on every side towards the sea, and the entire land is clothed from the water's edge to its topmost heights with a

694

perennial verdure, which for luxuriance and picturesque effect is certainly unparalleled." (Bennett, i., 62.) Next in importance to Otaheite, and about 130 m. N.W. that island, is Ulietea, or Raïatèa, nearly 60 m. in circ., encircled by a reef of coral, bordered by numerous islands: it has a bold, mountainous appearance, and is scarcely less picturesque than that last mentioned. Eïmèo is another mountainous island, with an abrupt rocky coast, and is chiefly distinguished as the central station of the missionaries on this group: a school and printing-office are established here, but both on a confined scale. The Society islanders are light-hearted, merry, and fond of social enjoyment; but, at the same time, indolent, deceitful, thievish, and addicted to the excessive use of ardent spirits. The forms of Christian worship are enforced here as rigidly as in the Sandwich islands; but civilization is considerably less advanced, and European costume considerably less prevalent. (Bennett, i., 79.) The seat of government and principal port of Otaheite is Pápéta, which exhibits the same combination of European houses and native huts as the capital of the Sandwich islands. The harbour is a capacious sheet of smooth water, of a circular shape, and so completely landlocked as rather to resemble a large dock-basin than a natural harbour. The commerce, consisting in the exportation of pearl-shells, sugar, cocoa-nut oil, and arrow-root, in exchange for European manufactures, chiefly cloth and hardware, is carried on exclusively by foreigners, as the natives have no vessels larger than their double canoes. This port is also frequently visited by whalers coming here to refit or to obtain supplies. Otaheite is not and never can be so important a commercial station as Oahu, in the Sandwich islands.

4. The Marquesas, which were discovered by the Spanish in 1595, consist of 13 islands, extending about 200 m. from N.W. to S.E. The largest, Nunhiva, is about 70 m. in circumference, and is the only one generally frequented by shipping. The coast-scenery is neither picturesque nor inviting, its principal features being black naked cliffs and barren hills; but in the interior are many fertile vales, and very picturesque scenery. The inhabitants, with regard to personal beauty, are superior to most others of the Polynesian tribes; and the women, though short in stature, are well-proportioned, and sometimes even handsome. In civilization, however, they are far behind the Sandwich islanders; and are generally characterized by covetousness, irascibility, love of revenge, and gross sensuality. Cannibalism was practised by them within a very recent period; and they have steadily resisted all attempts to convert them to Christianity. Polygamy, however, though not forbidden by law, has fallen into disuse.

5. The Friendly islands are low, encircled by dangerous coral reefs. The soil is almost throughout exceedingly rich, producing, with very little care, the banana, breadfruit, and yam. The population may amount to about 50,000; but the natives, though favourably mentioned by Captain Cook, appear to be as treacherous, savage, and superstitious as any in the worst parts of Polynesia. The Wesleyan missionaries established themselves in these islands in 1821, and are reported to have met with considerable success.

6. Pitcairn island, which stands alone, near the E. extremity of Polynesia, is chiefly interesting on account of its having been the refuge of the mutinous crew of Captain Bligh's ship, the Bounty. The mutineers, after having turned their captain and a few of the crew out into an open boat, tried to make a settlement in the Society islands, but afterward fixed themselves in this isolated spot, where a few of them, with their descendants, were found by Captain Beechy, in 1825. It is not more than 7 m. in circumference, with an abrupt rocky coast, and rises about 1050 feet above the sea. The present population comprises about 80 persons, who (being the descendants of Europeans and native women) form an interesting link in person, intellect, and habits between the European and Polynesian races. They are tall and robust, though not handsome, with black glossy hair, and frank, honest, goodhumoured dispositions. In morals, also, they present a striking contrast to the inhabitants of the other islands; for they are a simple, inoffensive race; on the whole industrious, and strictly observant both of morality and religion.

7. The New Hebrides (discovered in 1506, and so named by Captain Cook, who surveyed the entire group) are considerably hilly, though well clothed with fine timber; and the valleys are extremely abundant, producing figs, nutmegs, and oranges, besides the fruits common to the rest of Polynesia. The inhabitants present about the most ugly specimen extant of the Papuan race: the men live almost in a state of nudity; and the women, who are used as mere beasts of burden, wear only a petticoat made from the plantain-leaf. Their canoes are more rudely fashioned than in most of the other islands; and, on the whole, these

people seem to be among the most degraded of the islanders of the Pacific.

8. New Caledonia, which is the largest island of the W. groups, is far less fertile, and produces a smaller variety of fruits and vegetables than any of the islands yet mentioned. The natives closely resemble in habits, and the total absence of civilization, those of the New Hebrides; though, owing to the ravages caused by famine, they are infinitely more wretched. The Pelew islands, in about lat. 7½ N., long. 135° E., are chiefly known from the accounts of Captain Wilson, who was wrecked on them in 1783. He describes the inhabitants as hospitable, friendly, and humane. Breeds of cattle, goats, poultry, &c., were subsequently sent to the islands, and have succeeded extremely well. They have now a considerable trade with China.

Our knowledge of the many extensive groups of islands comprised in Polynesia is still very far from being satisfactory. Magellan began the work of discovery early in the 16th century, and he was followed at the close of the same century by Mendana and other Spanish navigators. The Dutch made farther discoveries in the 17th century; but to England mainly belongs the honour of having explored and laid down the exact position of the principal groups of the South sea islands, and the names of Byron, Wallis, Cook, Vancouver, and Beechey, must ever rank high in the estimation of geographers: great credit is also due to La Pérouse, D'Entrecasteaux, Freycinet, and other learned navigators, sent out on exploring voyages by the French government. In consequence of the labours of these and other navigators, aided also by the information gained from missionaries resident in the islands, we have gained a pretty intimate acquaintance with the condition of the natives in the principal groups of Polynesia; but there remains a great number of islands, especially on the W. side of the Pacific, that have very seldom been visited by Europeans, and are occupied by people as savage and uncivilized as the Sandwich islanders of the last century.

POMERANIA, a large prov. of the Prussian states, lying along the S. coast of the Baltic, from long. 12° 20′ to 18° 2′ E., having E. and S. West Prussia and Brandenburg, and W. Mecklenburg. Its form is oblong, its length (from E. to W.) being above 200 m.; while its breadth varies from 30 to 60 and 80 m. Area, 12,179 sq. m. Pop. in 1837, 970,117, of whom 956,334 were Protestants. It is divided into three regencies, and these again into 25 circles. Surface flat. Principal towns, Stettin, Stralsund, Greifswald, Stargard, Stolpe, &c. Principal rivers, Oder, Lebs, Stolpe, Rega, Persante, Ucker, Peene, and Ihna. The Haff is a large bay, or rather lagoon, of an irregular form, which communicates with the Baltic by the mouths of the Oder. Along the sea the land is in many parts so low that it would be overflowed, were it not protected by ranges of sand hills, and where these are wanting, by dykes, as in Holland. Soil mostly sandy, and unproductive, except along the rivers and lakes, where it is marshy and comparatively fertile. A large part of the country is covered with forests and heaths, and there are also many shallow lakes. Agriculture is in a very backward state; but since the abolition of vassalage in 1811, it has made considerable progress. Wheat and barley are grown; but rye, buckwheat, and oats, are the principal corn crops; potatoes are largely cultivated, and form, indeed, a principal part of the food of the people; flax and hemp are also grown, with tobacco, &c. The breeding of cattle, sheep, and hogs, is a favourite occupation. Geese are reared in immense numbers; and besides those consumed in the province, large quantities are exported smoked and dried. Very few farms are occupied by tenants holding under a lease, but are chiefly farmed by or on account of the proprietors. The peasantry live in mean dirty cottages of wood or clay; their clothes are all of home manufacture. Common agricultural labourers earn from 6 to 8 silver groschen a day, exclusive of subsistence. Minerals unimportant. The woollen manufacture in a domestic state is pretty generally diffused; and linen, leather, &c., are also produced. There are numerous breweries and distilleries, and dram-drinking is as prevalent here as in other parts of the monarchy. The fishery, particularly of sturgeon and salmon, is carried on along the coast, and in the creeks and rivers.

POMFRET, p. t., Windsor co., Vt., 49 m. S. Montpelier, 490 W. White river passes through its N.E. corner, and Queechee river through its S.E. corner, by small branches of which it is drained. It contains a Congregational church, and has some Baptists, Methodists, and Christians, two stores, three fulling-mills, two woollen factories, one sawmill, three tanneries; 31 schools, 610 scholars. Population 1774.

POMFRET, p. t., Windham co., Ct., 41 m. E. Hartford, 377 W. Bounded E. by Quinnebaug river, by branches of which it is drained. It contains five churches, two Congregational, a Baptist, Episcopal, and Friends', seven stores, three cotton factories, with 3550 spindles, five grist mill, nine saw-mills, two tanneries; one academy, 35 students; 11 schools, 486 scholars. Pop. 1868.

POMFRET, t., Chautauque co., N. J., 12 m. N.E. Mayville, 315 m. W. Albany. Watered by Canadawa creek. It contains the villages of Fredonia, Vanburen, and Dunkirk. At the latter, the New-York and Erie railroad is to terminate. Bounded N. by lake Erie. It contains five commission houses in foreign trade, 28 retail stores, four fulling mills, one woollen factory, one flouring-mill, four grist-mills, 15 saw-mills, one oil-mill, one paper-mill, five tanneries, one distillery, two printing-offices, two weekly newspapers, and one periodical; one academy, 250 students; 27 schools, 1411 scholars. Pop. 4566.

POMONA, one of the Orkney islands, which see.

POMPEII (called by the Greeks *Pompeia*), a long-buried city of ancient Italy, in Campania, not much celebrated in ancient history, but now an object of the greatest interest in consequence of its rather recent discovery and exhumation. It was originally close to the sea; but it is now nearly 1½ m. inland, and is about 5 m. S. by W. the crater of Vesuvius, and about 15 m. S.E. Naples. The æra of its foundation, as well as the greater part of its early history, is involved in obscurity; but the presumption is, that it was settled by Osci and Pelasgi prior to the establishment upon this coast of the Greek colonies from Euboea. About anno 440 B.C., it fell into the hands of the Samnites, from whom it was taken, with their other possessions, by the Romans about 80 years afterward. Pompeii revolted, with the other Campanian towns, during the social war; and little more is known of it till it was visited by an earthquake (A.D. 63), which occasioned great devastation. *Motu terræ, celebre Campaniæ oppidum, Pompeii, magna ex parte proruit.* (*Tacit. Annal. lib. xv., cap. 22.*) The repairs consequent to this disaster were incomplete, as is seen by the state of the excavated ruins, when the city, with Herculaneum, Stabiæ, and other towns in its vicinity, was wholly overwhelmed by an eruption of Vesuvius, A.D. 79. This tremendous calamity has been admirably described by the younger Pliny, whose uncle was one of the sufferers, in a letter to the historian Tacitus; "*Processerat per multos dies tremor terræ, minus formidolosus, quia Campaniæ solitus; illa verò nocte ita invaluit ut non moveri omnia sed everti crederentur. Mare in se resorberi et tremore terræ quasi repelli videbamus. Certè processerat littus, multaque animalia maris in siccis arenis detinebat. Ab altero latere nubes atra et horrenda ignei spiritus tortis vibratisque discursibus rupta in longas flammarum figuras dehiscebat; fulgoribus illæ et similes et majores erant. . . . Mox audiere ululatus fœminarum, infantium quæritatus, clamores virorum: alii parentes, alii liberos, alii conjuges vocibus requirebant, vocibus noscitabant: hi suum casum, illi suorum miserabantur: erant qui metu mortis mortem precarentur. Multi ad deos manus tollere. Nec defuerunt qui fictis mentitisque terroribus vera pericula augerent. . . . Mox dies verus, sol etiam effulsit, luridus tamen: occurrebant trepidantibus adhuc oculis mutata omnia, altoque cinere tanquam nive abducta.*" (*Plin. Epist. vi., 20.*) It seems probable, however, from the small number of skeletons discovered, here and at Herculaneum, that the inhabitants of both cities not only found time to escape during the confusion, but also to carry with them their most valuable effects.

From this time forward, for about 1669 years, Pompeii continued buried under the ashes, pumice-stone, and other volcanic matter by which she had been overwhelmed, and even her situation was matter of doubt and conjecture. It is surprising however, that her ruins did not sooner attract attention; for, in 1592, the celebrated architect and engineer, Dominico Fontana, having been employed to construct an aqueduct to convey water to Torre, fell in with the remains of the buried city. But this discovery appears to have attracted little or no attention; and it was not till 1748, that peasants employed in cutting a ditch fell in with the ruins of the city, that they became an object of interest and attention. The excavations were commenced in 1755; and have since been pretty constantly, though not very vigorously, prosecuted. Not having been overwhelmed by lava, but with tufa, ashes, and scoriæ, the excavations are much more easily effected here than at Herculaneum. (*Description de Pompeii, par Bonucci*, p. 31, &c.)

Pompeii, to borrow the words of an intelligent observer, is "the most wonderful of the antiquities of Italy; and it is perhaps the only one which never disappoints a traveller who is even moderately acquainted with the history of ancient Rome. The impression which it gives of the actual presence of a Roman town, in all the circumstantial reality of its existence 2000 years ago, is so vivid and intense, that it requires but a small effort of imagination to place yourself among the multitude which once thronged its streets and theatres, and occupied its now voiceless chambers. The expression so often used, that you expect to see the inhabitants walk out of their homes to salute you, is scarcely

a figure of speech. Many things, in fact, concur to foster the illusion. You see a street before you carefully paved and well worn, and bordered with *trottoirs*, in good preservation, as if it had been in use on the preceding day. The houses generally extend in unbroken lines, and even the dilapidation is in some measure concealed by the small modern roofs placed over the walls to protect them from farther waste by the weather. The doors and windows, indeed, are all open; but so they generally are in the modern houses of Italy; and the sombre brown tints of the walls is not very different from what is seen in the decayed towns of the same country at the present day. You turn to the right and the left, and wander from street to street, and still you have the perfect image of a town before you, except that no inhabitants appear, and these you may suppose have only left it a few days before. We have detached public buildings of many kinds elsewhere; but here we have a Roman forum, with all its accompaniments of temples, porticoes, curiæ. &c., not indeed perfect, but only so injured that what is missing can be replaced, and what is mutilated restored. We have also many shops, with their utensils of trade in them, and about a hundred private houses of all descriptions, from the poor cottage to the patrician mansion, enabling us, for the first time, to obtain a distinct idea of the forth and arrangement of a Roman house, and giving us, as it were, a glimpse of the domestic life and manners of the people. The public baths here, which were almost entire, have thrown new light on the structure of those buildings. Lastly, the *tout-ensemble* of the walls, gates, streets, forum, houses, temples, fountains, theatres, associated as they are with each other, give us a conception of a Roman town incomparably more clear and satisfactory than any number of such objects scattered over distant localities could have furnished." (*Maclaren's Notes*, p. 100, 103.)

It seems evident on an examination of the superincumbent strata, which consists of various layers, that intervals had taken place in the original eruption, which lasted for three days; and it is farther probable that some of the uppermost layers may have been the result of subsequent eruptions.

This resuscitated city, of which about 1-4th part is now laid open to public view, is of a somewhat oval form, ¼ m. in breadth, and ¾ m. in length, covering an area of 160 acres, or about 3-4ths of the new town of Edinburgh, and considering the narrowness of the streets, the nature of the houses, and the mode in which the slave population generally were lodged, the entire population could scarcely have exceeded 25,000 inhabitants. The walls, which have been traced on every side, except towards the sea, are about 20 feet thick, and nearly equally high, being faced with blocks of lava inside and outside. There are six gates, and many towers, rising high above the ramparts, and pierced with arches. The best approach to Pompeii is by the Appian way to "the Gate of Herculaneum," which is nearest the sea, and on the N.W. angle of the city. Along each side of the road, approaching this gate, extends a line of tombs, many of which remain perfectly entire, their angles being so sharp, their inscriptions so legible, and their whole appearance almost as fresh as if they had been erected only a few years ago. The monuments vary greatly in size, pattern, and material: many are mere cenotaphs, while others have niches for urns; and a pretty common form is that of a small oblong temple, adorned with columns or pilasters. On the whole, these tombs are not unlike the more ambitious monuments in our own churchyards; but there is nothing resembling our single upright slabs, or flat gravestones. The gate of Herculaneum consisted, like the others, of a large central and two smaller side-gates, not unlike those of Temple Bar in London, the central archway being about 20 feet in height and 15 feet in width. The streets appear to have been arranged pretty regularly in parallelograms; but they are very narrow, the most usual breadth being 18 or 20 feet, of which 1-3d is occupied by the trottoirs, which invariably line both sides of the highway. An exception, however, must be made in favour of the "Street of the Silversmiths," which is 48 feet in width, and decidedly the finest in Pompeii. The middle of the road is paved, like the Via Appia, with masses of lava of irregular shapes, and from one to two feet in diameter, the most level surfaces being placed uppermost; and so many parts the ruts produced by the wheels are still obvious. The trottoirs, which are raised about 16 inches above the rest of the street, average about 4 feet in width: they are generally made of a sort of compound of lime, sand, and gravel, not unlike the asphalte used for modern pavements in London and Paris.

The largest excavated space in Pompeii, and that which exhibits most architectural magnificence, is the Forum, an extensive oblong area, once paved with large slabs of marble. The feelings of a classic traveller, on beholding, for the first time, such a monument of antiquity, are well described by Mr. Maclaren. "I felt that it was not a trifling incident in my life, to stand in a veritable Roman forum. There it lay distinctly before me, rifled of the greater part of its marble pavement, of its statues, and some of its columns, yet retaining enough of its ancient lineaments to give a perfect idea of its form, extent, and distribution of its parts. It had been terribly injured by the earthquake A.D. 63, and was rebuilding when the great catastrophe occurred. The forum was the great place of public resort: the idle came here to inquire after news, the busy to talk of business, friends to keep appointments, patrons to meet their clients, suitors to attend the courts, candidates for office to solicit votes; here the orators harangued, and the people shouted, the magistrates met in council, and the tax-gatherers collected the revenues: here the decrees of the senate were promulgated *viva voce*. and plays, festivals, and gladiatorial shows, were advertised by short notices badly spelt, painted on the walls in rudely-formed letters" (p. 131). The entrance from the N.W. corner (that nearest the gate of Herculaneum), is by a flight of steps leading downwards through a brick arch into an oblong area about 490 ft. in length by 114 ft. in width, surrounded by columns, and the ruins of temples, triumphal arches, and other public edifices, the uses of which can in general only be conjectured. The red masses of brick divested of their marble casings, the brown and yellow tints of the tufa, the fragments of white stucco attached to the shattered walls of the different edifices, and the pedestals which once supported statues in honour of illustrious patriots, are all that now remain to attest its former beauty and magnificence." (*Pompeii. Library of Entertaining Knowledge*. i., 100.) A Doric colonnade ran round three of its sides, and the fourth was occupied by a temple of Jupiter. The columns formed a species of covered gallery, raised above the central area, and a second row of smaller columns, placed on the top of the first, formed a second gallery, which would afford a view of everything passing in the area. The temple of Jupiter has a prostyle portico, supported by 12 very beautiful Corinthian columns, and its total length, from the front steps to the back wall of the *cella*, is 120 ft., the uniform breadth being 45 ft. On the W. side of the enclosure stood the prisons and public granaries, a peripteral temple of Venus, having a façade looking southward, of nine Corinthian columns.

a Basilica, or court of justice, which is the largest building in Pompeii. It is oblong shaped, 220 ft. in length and 80 ft. in breadth, and is entered through a vestibule having five doorways of masonry. The roof of the interior was supported by a peristyle of 28 Ionic fluted columns, and at the farther end are some remains of what was once the praetor's tribunal. At the S. end of the Forum, which was also ornamented by a triumphal arch, are the remains of three buildings of nearly equal size, and similar shape, that may have been curiæ, or places of assembly for the magistrates: these, however, are of very confined dimensions, and possess little interest. On the E. side, opposite to the Basilica, and flanked on one side by the Street of the Silversmiths, is a large enclosed building, in the shape of a parallelogram. within which was an oblong peristyle of white marble Corinthian columns: it is commonly called the Chalcidicum, and was built by Eumachia (whose statue is still standing), but its former use seems to be quite conjectural, though Gell and Donaldson seem to think, from some of the pictures and other remains found there, that it may have been a kind of cloth-hall. Adjoining it, and fronting the Forum, is a small temple of Mercury, in the court of which is an altar of white marble, beautifully sculptured in bass-relief, representing a sacrifice. What the next building may have been is very doubtful; but as it has an altar, without a cella, we incline to think that it may have been a *sacarium* or hall of meeting for the town-council. Lastly, the space close to the N.E. angle of the Forum is occupied by a very large enclosed peripteral structure. supposed to have been a pantheon dedicated to the twelve Dii Consentes of Roman mythology, and comprising, besides an ædicula or raised chapel, numerous cells for the accommodation of the priests. Under the colonnades of the Forum close in front of the Pantheon, the entrance to which is by a rather narrow vestibule, are the remains of seven recesses or shops, in some of which the pedestals of the table are still visible; these may possibly have been the *taberna argentaria*, common in most Greek and Roman fora. (For farther particulars respecting these several buildings, and the monuments contained in them, the curious reader is referred to the accurate and very instructive little work on *Pompeii*, in the *Library of Entertaining Knowledge*.)

Next in importance to the excavation of the Forum, is that of the quarter occupied by the theatres. Its best approach from the Forum is by the "Street of the Silversmiths;" the space cleared comprises two theatres of unequal size, a square usually called the soldiers' quarter, and temples, with other buildings of minor importance.

The theatres will not bear to be compared, in point either of size or splendour, with the magnificent structures at Rome; but still they have the remains of considerable beauty, and the largest, at least, would be considered of large size in any modern city. It has six entrances, leading to different parts of the building, and six inner doors, or *vomitoria*, opened on an equal number of staircases running down from the external circular corridor to all parts of the house. The benches were about 1 ft. 3 in. in height, and 2 ft. 4 in. in width, and it would appear that they may have been capable of accommodating about 5000 males, chiefly of the middle classes; those of high rank sat on chairs in the orchestra, and the women occupied a gallery running round the top of the building over the corridor. All the benches as well as the orchestra seem to have been entirely covered with marble, of which, however, there are now but few vestiges. Like the Coliseum and other ancient theatres, it was open at top; but on the outside wall may still be seen the iron rings inserted to receive the masts supporting the awning. Of the scene itself, enough remains to show that the three chief doors were situated in deep recesses, and that behind them was the postscenium. The smaller theatre. which communicates with that last mentioned, is built on the same plan, and similarly arranged, having had accommodation for about 1500 spectators; but, unlike the other, it seems to have been permanently roofed. Its shape, also, is rather elliptical than semicircular. Close to the theatre is a large open space, supposed to have been soldiers' barracks; and near it is a Doric temple of Hercules, the oldest in the city, and said to have been erected at least 800 years B.C. The great amphitheatre occupies a large space at the E. angle of the town, quite separate, and at some distance from the rest of the excavations. Like other amphitheatres, it is oval-shaped, the extreme outside length being 430 ft., and do. breadth 335 ft. The seats rise above each other in 24 successive rows, and must have accommodated upwards of 10,000 persons.

The baths, which occupy a space not far from the Forum, of about 100 ft. square, are interesting, not so much from their size as from the simplicity of their construction, which makes their arrangements more intelligible than in the complicated buildings of this kind in Rome and elsewhere. As, however, it would be impossible without a diagram to give a satisfactory account even of those at Pompeii, we pass at once to the domestic architecture of the city, which, indeed, is one of its most interesting features. Mr. Maclaren closely examined about a dozen of the private houses, which, he says, are so different from those either of Britain or modern Italy, that it is not easy to comprehend the use of their several parts. Indeed, most of them are so dilapidated that they could not be understood at all without the aid of Pliny and Vitruvius. "The present condition of the houses and shops resembles what we see in our tenements after the occurrence of a fire. The roof, upper floors, doors, and all the woodwork, have disappeared, the furniture has been carried off, and nothing remains but the half-dilapidated walls, the pavement of mosaic on the ground flats, columns entire or in fragments, stone counters, and a few bulky or heavy articles of too small value to be worth removing. The apartments, however, have been carefully cleaned out; and not only the houses but the streets were completely free of foreign matter, except a thin covering of ashes and scoriæ, deposited by the recent eruption of Vesuvius, and which was easily removed by a besom or whisp of straw." The paintings, also, are still on the walls, and remarkably fresh. The house of Pansa, though not the largest, is better calculated than any other to convey to the reader an idea of a private town residence; and. taking this as a guide, we may remark that the houses consisted generally of a square or oblong enclosure surrounded by blind walls, the central court being open, and chambers formed round it, over which sheds (*compluvia*) projected inwards, which discharged the rain water into a stone or marble basin (the *impluvium*), in the centre of the court. The larger houses, however, have a second court, with its corresponding impluvium, surrounded by columns; and hence, while the outer court is called simply *atrium*, the inner one (divided from it by a square apartment, called the *tablinum*), from being usually adorned with pictures and statues) is denominated the peristyle and was devoted to the use of the family. Bed-rooms and parlours run round both courts, a garden extends some way at the back of most of the houses, and the front entrance, in Pompeii at least, is by a passage 8 or 10 ft. wide. Paper, horn, mica, and even pretty thick glass, were pretty generally used; and in cold or hot weather awnings were commonly thrown over the impluvium; but no fireplaces were seen in the houses, and, like those in modern Italy, they seem to have been wholly built for summer use. The exterior of the houses is generally of brick, coated with plaster, and formed into panels: in the interior, also, the walls are covered with fine plaster, which serves as a ground for frescoes, which are found pretty abundantly in all parts of the larger houses throughout the town: and these paintings, if not equal to other extant specimens of Roman art, are highly valuable for the light which they throw on the costumes, habits, and amusements of the ancient inhabitants. The shops, like those of Naples, seem to have been extremely small, scarcely exceeding 14 ft. square, and wholly open in front, with the exception of a low counter, being closed by shutters at night, somewhat in the same manner as the butchers' stalls and shops in England. Some of the implements of trade still remain, such as earthen jars, ovens, mills, cooking-pans, &c.; and we have reason to believe that the ancient inhabitants pretty well understood the division of trades. Most of the shops and other places of public entertainment, not excepting those belonging to the *Domus Liberæ*, had images or figures over the doors, serving, like the signs in modern towns, to indicate the profession or business of the occupants. The household furniture and domestic utensils found in the excavations both here and at Herculaneum have, with a few slight exceptions, been removed to the Museo Borbonico at Naples; and the number of articles of every kind and material is truly immense. "Among these," says Mr. Maclaren, "are several iron chairs, like our garden-chairs; braziers for burning charcoal or wood, keys and locks, metallic mirrors, pots and pans, glass bottles and drinking vessels, lamps of copper and earthenware, vases and urns, marble statues and bas-reliefs, ancient armour, seals, styles and inkstands, bells, moulds for bread and pastry, glazed plates for the table, scales and steelyards, spoons, ear-rings, and similar articles." The discovery of Pompeii has, in fact, thrown a strong and steady light on many points connected with the private life and economy of the ancients that were previously involved in the greatest obscurity. An acquaintance with its remains is indispensable to the classical student; and he cannot study them to more advantage than in the volumes already referred to in this article, or in the more elaborate and better illustrated works of Gell, Donaldson, David and Maréchal. (*Pompeii*, 2 vols. passim; *Gell's Pompei. and Illust.; Map of Pompeii, Soc. of Use. Knowl.; Mecl. Notes, p. 99–132; Lyell's Geol. ii. 95–103.*)

POMPEY, p. t. Onondaga., N. Y., 14 m. S.E. Syracuse, 125 W. by N. Albany, 337 W. Watered by Limestone cr., which has two falls of 70 feet perpendicular each, within 90 rods of each other, which afford extensive water-power. It contains two churches, a Presbyterian and Baptist, ten stores, two fulling-mills, one flouring-mill, three grist-mills, five saw-mills; one academy, 127 students; 98 schools, 1350 scholars. Pop. 4371.

POMPTON, t., Passaic co., N. J., 43 m. N.W. Hackensack. Bounded S. by Pompton r. Watered by Ringwood r. It contains a Dutch Reformed church, an academy, two stores, one furnace, eight forges, three grist-mills, six saw-mills; five schools, 186 scholars. Pop. 1437.

PONDICHERRY (Fr. *Pondichéry*), a town of Hindostan, and the principal French settlement on the Asiatic continent; on the Coromandel coast, 83 m. S.S.W. Madras; lat. 11° 57' N., long. 95° 54' E. Pop., in 1835, 32,197, of whom 696 were Europeans. Standing on a flat sandy plain, near the shore it it has a very imposing appearance from the sea; and it is in reality a handsome regularly laid out town. The streets in the European quarter are of uniform breadth, built with remarkable regularity, and intersecting each other at right angles. The houses, which are of a good height, have flat terraced roofs, the walls being stuccoed white and yellow, and not intermixed with native huts. Nearly in the centre is a spacious square, laid out in walks, shaded by rows of trees, with the government house on the N. side, and open on the E. to the sea. The black, or native town, to the W. of the former, and separated from it by a canal, crossed by several bridges, is laid out with nearly the same regularity as the European town, though the houses are very inferior. Pondicherry was formerly strongly fortified; but the only portion of its works that now remains is an old brick tower, on which the flag is hoisted. The chief buildings are the government house, an edifice of a single story, adorned with columns, and surmounted by a balustrade; the church, built by the Jesuits, and a good market-place. It has a college for Europeans, and a school for the Indians, a botanic garden, and a government pawn bank.

The French possessions in India, comprising Pondicherry, Chandernagore, Karical in the Carnatic, Mahé in Malabar, and Yanaon in Orissa, with the territory attached to each, have a total population of about 166,000, of whom 1000 are whites. The territory attached to Pondicherry is considerably larger than the rest, and had, in 1835, 10,613 hectares under culture, producing 6,486,640 kilogr. rice, 6,734,000 kilogr. other grains, 6,900,000 cocoa nuts; with some betel, and a little indigo, tobacco, and cotton. The total value of the imports into these settlements, in the same year, amounted to 1,744,960 fr., and that of the exports to 5,309,619 fr. The trade, by far the greater part of which centres in Pondicherry, is chiefly with the rest of the Coromandel

coast, Sumatra, the Isle de Bourbon, the Mauritius, and Senegal. The governor of the French settlements in India is usually a peer of France, and resides at Pondicherry, where also is the royal court for these colonies, a tribunal of primary jurisdiction, police court, &c.

This town was purchased by the French from the Beapoor sovereign, in 1672. It was successively taken by the British, in 1761, 1778, 1793, and 1803; but was finally restored at the treaty of Paris, in 1815. (*Hugo; Official Reports; Hamilton's E. I. Gaz., &c.*)

PONT-A-MOUSSON, a town of France, dep. Meurthe, cap. canton, on the Moselle, by which it is intersected; 16 m. N. Nancy. Pop., in 1836, ex. com., 7008. It derives its name from a bridge of eight arches which here crosses the river, and led formerly to the old castle of Mousson, now in ruins, on an eminence E. of the town. It is surrounded by planted boulevards, and has several good edifices, including a Gothic church, built towards the end of the 13th century, a town-hall, St. Mary's Abbey, now converted into a seminary, large cavalry barracks, a good hospital, and a building termed the *Maison des Sept Peches Capetaux*, from its front being ornamented with old sculptures intended to represent the cardinal sins. This town has manufactures of coarse woollen stuffs, printing-types, earthenware, tobacco-pipes, and beet-root sugar. (*Hugo, Guide du Voyag. &c.*)

PONTECORVO, a town of S. Italy, the cap. of a detached territory, surrounded by the Neapolitan dom., but belonging to the Papal states, deleg. Frosinone; on the Garigliano, 20 m. S.E. Frosinone, and 37 m. N.W. Capua. Pop., about 5500, chiefly occupied in agriculture. It has several churches and convents and an old castle. Napoleon conferred on Marshal Bernadotte the title of Prince of Pontecorvo, which he enjoyed till he became king of Sweden. Near it are some considerable ruins, supposed to indicate the site of the ancient *Interamna ad Lirim*.

PONTCHARTRAIN, lake, La. is 40 m. long and 25 m. broad, communicates with lake Maurepas on the W., and lake Borgne, through the Rigolets on the E., and with New Orleans through St. John bayou, and a canal. The water is about 16 or 18 feet deep in its deepest parts, but near the shore it is not more than 9 feet deep. It receives several rivers from the N., and is the outlet of considerable commerce from New Orleans.

PONTE-DELGADA, a seaport, and the largest and most populous city (though not the cap.) of the Azores, on the S. side of the island of St. Michael; lat. 37° 45' 10" N. long. 25° 41' 15" W. Pop. estimated at 22,000. Its appearance from the sea is far from picturesque, exhibiting a compact uniform mass of bright-looking buildings, backed by a few conical hills, some of which, however, are covered with a luxuriant vegetation. It had, when visited by Captain Boid, in 1831, six churches, eight monasteries, and four convents; but the latter, which were celebrated alike for their artificial flowers made of birds' feathers, their sweetmeats, and the easy virtue of the vestals by which they were occupied, have since luckily been suppressed. The houses are substantial; but the streets are very ill-paved, and disgustingly filthy. Its markets are abundantly supplied with fish, poultry, eggs, and vegetables of all sorts, including Spanish beans, yams, sweet and common potatoes, oranges, lemons, &c.; and every thing is remarkably cheap. When contrasted with the other towns in the Azores, it displays considerable wealth, activity, and industry. The exports consist of oranges, wines, brandy, rocella, &c. A mole has been constructed for the accommodation of the smaller class of vessels; but those of considerable burden have to anchor in an open roadstead. The town and harbour are defended on the W. by the castle and fort of St. Braz, mounting 28 guns of cannon, and on the E. by the forts Sam Pedro and Santo de Cao. The governor of the island St. Michael and St. Mary resides at Ponte-Delgada. (*Boid's Azores, 119–118; Dict. Géog.*)

PONTEFRACT (vulgo *Pomfret*), a parl. and mun. bor., market town and parish of England, W. riding county York, upper div. wap. Osgoldcross, on a commanding eminence about 2 m. S.W. the Aire, 10 m. E. Wakefield and 21 m. S.W. York. Pop. of parl. bor. (which includes, with the old bor., the extra par. dist. of Pontefract park, the castle precincts, and also the several townships of Tanshelf, Monkhill, Knottingley, Ferrybridge, and Carleton), in 1831, 9675. The town, which is well paved, and has been lighted with gas since 1828, is well-built, with open spaces, and clean streets, lined by handsome houses, chiefly of brick, the principal thoroughfare running N.E. past the ruins of the old castle, at the N. end of the town. The principal public buildings are a modern-built town-hall, with an attached jail, "now used only for debtors and prisoners under remand," and a courthouse for the quarter sessions of the W. riding, a well-constructed market-house, &c. The parish church, originally built in the reign of Henry I., has subsequently been so altered, that little remains of the olden edifice: a more ancient, and once parochial church is now in ruins;

but the tower at the intersection of the nave and transepts is nearly entire, and is worthy, according to Rickman, of attentive examination. The living is a vicarage in the gift of the chancellor of the duchy of Lancaster. There are also places of worship for Roman Catholics, Wesleyans and Primitive Methodists, Independents, and the society of Friends.

A grammar school, founded in the reign of Edward VI. was revived and rechartered in 1792: it is one of 12 schools privileged to send candidates for Lady E. Hastings' exhibitions at Queen's college, Oxford. Another charity-school for children of both sexes, endowed with £95 a year, has been incorporated with a national school; and a neat building, formerly the theatre, has been converted into schoolrooms capable of accommodating 400 children. The town has four Sunday schools, and six hospitals, or almshouses, of various foundations, furnishing lodging, clothing, and a small stipend to 53 aged people of both sexes. A subscription library, mechanics' library, and news-room have recently been formed; but the races, formerly held here in September, are now extinct.

"Pontefract is not a manufacturing, but a very substantial and respectable country town. A considerable trade is carried on in malt, which is said to be increasing. The corn markets generally are also improving; but there does not appear any prospect of a material change taking place in the condition of the borough. (Mun. Corp. Rep.) The neighbouring village of Ferrybridge within the borough is on the Aire (crossed there by a stone bridge of three arches), and till recently enjoyed considerable advantages from its position on the Great North Road: it is now in a state of decay. The vicinity of Pontefract is famed for its gardens and nurseries, which furnish vegetables in great abundance for the markets of York, Leeds, Wakefield, Doncaster, &c. The deep loamy soil around it is also well adapted for the cultivation of liquorice (glycyrhiza glabra), which is grown here in large quantities, and supplied extensively to London and other large towns. Filtering stones are quarried on the castle-hill, and are in great request in all parts of the kingdom. Pontefract received its charter of incorporation in 2 Richard III. Under the Municipal Reform Act, it is governed by a mayor and three other aldermen, with 12 councillors, having also a commission of the peace under a recorder. Corporation revenue in 1839, £588. The borough has sent two members to the House of Commons, with some interruptions, since 23 Edw. I.; the right of voting down to the Reform Act having been in the inhabitant householders within the borough. Registered electors in 1839-40, 815. The spring quarter sessions for the W. riding are held here in Easter week. Markets on Saturday, and large fortnightly fairs for cattle, besides eight other annual fairs.

The principal celebrity of Pontefract is owing to its castle, once of great extent, but now a mere ruin, its site having in a great measure been converted into garden-ground. It appears to have consisted of several towers, with intervening walls and other buildings, the round-tower, or dongon keep, having occupied an eminence at its W. extremity. It was finished in 1080. In the beginning of the 14th century it became, by marriage, the property of the Earls of Lancaster, and in the reign of Henry IV. was attached with the rest of the duchy to the crown. For the space of many centuries it stood the ornament and terror of the surrounding country, till the civil wars of Charles I., when, after sustaining three successive and desperate sieges, it was finally taken by the parliamentary army in 1649, when it was unroofed and demolished by order of parliament. This castle has been the scene of various tragical events in English history. In the reign of Edward II., Thomas, Earl of Lancaster, was brought a prisoner here, and detained till the day of his execution. Richard II. was confined, and eventually murdered here; and is it, also, Anthony Woodville, Earl of Rivers, Richard Lord Grey, Sir Thomas Vaughan, and Sir Richard Hawse, were put to death by order of Richard III., without even the form of a trial. (Mun. and Bound. Rep.; Inspector of Prisons, 2d and 4th Reports.)

PONTINE MARSHES. See ITALY.

PONTIAC, p. t., capital of Oakland co., Mich., 25 m. N.W. Detroit, 549 W. Watered by Clinton river. It contains 18 stores, two fulling-mills, one woollen factory, one furnace; nine schools, 456 scholars. Pop. 1904. The village, on Clinton river, contains a courthouse, jail, a branch of the Michigan university, a bank, a Congregational church, 13 stores and groceries, two saw-mills, one fulling-mill, one iron foundry, two printing-offices, two weekly newspapers. A railroad connects it with Detroit.

PONTOISE (an. Briva Isura), a town of France, dep. Seine-et-Oise, cap. arrond., on the Oise, from a bridge laid by the Viosne, 20 m. N. Versailles. Pop., in 1836, ex. comm., 4990. It stands upon an abrupt rocky hill, and was formerly surrounded with walls, portions of which still exist. The lower part of the town is tolerably well-built and paved, but has no building of importance except a new and good

hospital. The rivers here turn numerous flour-mills; and in addition to these Pontoise has iron and copper works, tanneries, and manufactures of watches, jewellery, cotton yarn, &c. It was frequently taken and retaken in the wars between England and France; and the parliament of Paris sat here several times during the 17th and 18th centuries. (Dict. Geog., &c.)

PONTOTOC, county, Miss. Situated toward the N.E. part of the state, and contains 900 sq. m. Drained by Tallahatchee river and its branches, and by branches of the W. fork of Toombigbee river. It contained in 1840, 6203 neat cattle, 962 sheep, 12,365 swine; and produced 8783 bushels of wheat, 167,845 of Indian corn, 8140 of oats, 4130 of potatoes, 145,953 pounds of cotton. It had seven stores, 10 gristmills, seven saw-mills, two tanneries, one printing-office, one weekly newspaper, one academy, 30 students; three schools, 86 scholars. Pop., whites, 2695; slaves, 1593; free coloured, 3; total, 4491. Capital, Pontotoc.

PONTOTOC, p. v., capital of Pontotoc co., Miss., 175 m. N.N.E. Jackson, 888 W. It contains a courthouse, jail, a United States land office, and about 500 inhabitants.

PONT-ST.-ESPRIT, a town of France, dep. Gard, cap. cant., on the Rhine, 31 m. N.N.E. Nismes. Pop., in 1836, 4160. It derives its name from its famous bridge over the river, constructed between 1265 and 1309, 875 yards in length; it has 23 arches, but is only about 12 feet in breadth. It is kept in perfect repair, and has of late been made practicable for carriages. The town, which is ill-built, with narrow, winding streets, has a small port on the Rhone, and a considerable traffic in corn, wine, oil, and silk. (Hugo, &c.)

PONTYPOOL (corrupted from Pont ap Howell), a market town and township of E 'and, par. Trevethin. hund. Abergavenny, co. Monmouth, m. S.W. Monmouth, and 125 m. W. by N. London. Area of par. and township, 18,460 acres. Pop., in 1831, 10,980. The town, occupying the side of a steep hill, near a rivulet tributary to the Usk, and nearly surrounded on all sides by mountains, is large and straggling, with two principal streets, lined with neat houses and numerous shops. The church, on an eminence near the town, has an embattled square tower; and there are two chapels of ease, besides places of worship for Wesleyan Methodists and Independents, with attached Sunday-schools.

Pontypool was formerly celebrated for its manufacture of japanned goods, introduced by Thomas Allgood, their inventor, in the reign of Charles II.; but this branch of industry has greatly declined, owing to the successful competition of Birmingham, which now nearly monopolises the trade. Its present importance is derived from the iron and tin mines in the surrounding district, chiefly belonging to Capel Hanbury Leigh, Esq., and from the iron-works, situated a little S.W. from the town, and conducted upon a very large scale. It is connected partly by railway and partly by canal with Newport, its shipping place for the produce of the mines. Petty sessions for the hund. are held here. Markets on Saturday, and cattle fairs on the last Monday of each month, besides others, April 22, July 5, and Oct. 16.

Pontypool park, the seat of the Hanbury family, is at a short distance N.E. the town; it is finely wooded with oak and other forest trees; and the house, on a perpendicular cliff above the Avon-Llwyd, commands a fine view of the surrounding country. (Nicholson's Cambrian Guide, &c.)

POOLE, a parl. and mun. bor., seaport, market town, and par. of England, and a co. of itself, with separate jurisdiction, on a peninsula on the N. side of the extensive and almost landlocked harbour, whence it derives its name, 20 m. E. Dorchester, and 97 m. W.S.W. London. Pop. of parl. bor., which includes, with the old bor., small portions of pars. of Canford and Hamworthy, in 1831, 7059. The town consists of several streets intersecting each other at different angles, the principal running in a N.E. direction through the market-place, in which is the town-hall, a convenient building, with attached shambles. The modern houses are generally substantial, and regularly built; but the older parts of the town have a mean, shabby appearance. The church is of modern erection, and there is a chapel of ease, both being built in the gift of the parishioners. The Wesleyan Methodists and Independents have their respective places of worship, and the town has four Sunday-schools, a free grammar school, several charity schools, and two or three well-endowed almshouses. There is a small jail, and the parish workhouse has lately been formed into a union with seven other parishes, the expense of supporting the poor of this parish having been £2031 in 1839. "The whole town, with its inhabitants and trade, may be described as in a prosperous and increasing state. Several vessels from this port are engaged in the Newfoundland fisheries; besides which it has a large and increasing coasting-trade, the exports being principally Purbeck clay (for the Staffordshire potteries) in exchange for coal." (Parl. Bound. Rep.) The port has also a considerable foreign trade: there belonged to it in 1840 about 160 ships, of the aggregate burden of nearly 16,000 tons. Gross customs duty, in 1840, £12,666. The entrance to Poole harbour, about ¼ m. in width, has a shifting bar, over which there are seldom more than 15 feet, even at high water. Vessels drawing 14 feet water may, however, come up to the quays. "It is a considerable and singular advantage to Poole harbour, that the tide ebbs and flows twice in 12 hours. It first flows regularly 6 hours, and ebbs for 1½ hours: it then flows for 1½ hours, and ebbs during the remaining 3 hours. The second flood seems to arise from the peculiar situation of the entrance; for, being in a bay facing the E., the tide of ebb from between the isle of Wight and the main, falls into that bay, forcing its way so as to raise the water for 1½ hours, at which period the water without the bar, by its falling to a lower level than that within, produces a second ebb till low water." (*Purdy's Sailing Directions for the English Channel*, p. 29.)

Near the mouth of the harbour is a bank, from which large quantities of oysters are taken, to be fattened in the creeks of Essex and Kent.

Poole, which claims to be a borough by prescription, has received several new charters, the principal being in the 10th Eliz. Under the Mun. Reform Act, it is divided into two wards, and is governed by a mayor, five other aldermen, and 18 councillors. Corporation revenue in 1839, 3451. The borough has returned two members to the House of Commons since the reign of Edward III., the right of election down to the Reform Act being in the members of the corporation, resident and non-resident. The electoral limits were enlarged as above mentioned by the Boundary Act; and in 1839–40, it had 543 registered electors. Markets on Monday and Thursday; fairs, May 1 and Nov. 2.

POONAH, a distr. of British India, presid. Bombay, principally between the 17th and 19th degs. of N. lat., and the 74th and 76th of E. long.; having N. Ahmednuggur, W. the Concan, S. the Sattarah dom., and E. those of the Nizam. Area, 8281 sq. m. Pop. about 558,000. The face of the country is mountainous and irregular but interspersed with many fertile and well-watered valleys. The climate is good and invigorating, and more suitable to Europeans than most parts of India. A good deal of the surface consists of the black and red cotton soils common in the S. of India: rice grounds comprise about 1-16th part of the land in cultivation, and gardens about 1-10th. The land is assessed on the village system: land revenue, in 1827-8, 15,16,323 rupees. Poonah is the only city; but there are several other considerable towns, at which coarse woollen, cotton, and silk fabrics, and metallic ornaments, are made. The celebrated cave-temples of Carlee, and several places of Hindoo pilgrimage, are in this district. (*Parl. Reports ; Hamilton's E. I. Geog.*)

POONAH, a city of British India, presid. Bombay, formerly the cap. of the Mahratta dom., but now the cap. of the above distr.; on the Moolla, a tributary of the Beemah. 89 m. E.S.E. Bombay. Lat. 18° 30' N.; long. 74° 2' E. Pop. estimated at about 110,000. It stands in an extensive and bare plain, about 2000 feet above the sea, at the foot of a small insulated hill, crowned with a pagoda. It is without walls, and can neither lay claim to antiquity nor beauty; it is very irregularly built and paved, with mean bazaars, deep ruinous streets, and no large or striking edifices. Heber says, "that it has as few evidences as can well be conceived of its having been, till lately, the residence of a powerful sovereign." But according to Hamilton, "its principal street is wide and handsome; and the mixture of rude paintings, in illustration of mythological legends, with the carved frame-work of dark-coloured wood, gives the fronts of the houses a fantastic and cheerful appearance." The principal palace is surrounded by high and thick walls, with four round towers, and is entered through a pointed archway. There are several other palaces, but they are small and insignificant. A little west of the city is the British cantonment, on an elevated site, with wide streets, a spacious church, a good station-library for the soldiers, and another library for the officers, and regimental schools, supported by subscription. The city has a Hindoo college, established in 1821, for 100 students, with classes for Hindoo divinity, medicine, metaphysics, mathematics, and astronomy, law, logic, rhetoric, grammar, &c., which costs 15,250 rupees a year. Poonah is the residence of the British collector and judge for the district, and has a well conducted district jail, several British schools, a Roman Catholic church, &c. East of the city is an excavated temple, apparently dedicated to Siva. Poonah is first noticed in history early in the 17th century; but it did not become the permanent residence of the Mahratta sovereign till the middle of the 18th century. It came into our possession in 1818. (*Hamilton; Heber, in Mod. Trav., &c.*)

POPAYAN, a city of Colombia, cap. of the gov. of Cauca, New Granada, on an extensive table-land, nearly 6000 feet above the sea, having the Cauca river, about a league distant, on the N., and a mountain named M, from its resem-

blance to that letter, on the E.; 230 m. S.W. Bogota, and 235 m. N.E. Quito; lat. 2° 28' 30" N., long. 76° 31' 30" W. Pop. estimated at 25,000. Like most other Spanish-American cities, it is laid out on a perfectly regular plan, its broad streets being bordered with stone footways. The houses have for the most part only one story, and are usually built of unburned brick; but, according to Mollien, some of them would not discredit a European capital; and that traveller would have preferred this to most South American cities, had it not been that the myriads of disgusting insects rendered a residence in it all but intolerable. It has several squares, one of which is spacious and handsome; a cathedral and other churches; numerous conventual buildings, some of which are now converted into barracks, or appropriated to other uses, and was formerly the seat of a royal mint and of a tribunal of finance. Two bridges are thrown across the Molina, a tributary of the Cauca, which runs rapidly through the city, and drains it of its filth. Popayan is principally inhabited by Negroes and Mulattoes, the number of whom a few years since, was double that of the whites. It was formerly the *entrepôt* of the trade between Bogota and Quito, and had a large traffic in the precious metals; but the revolution, by turning the trade into other channels, gave a blow to its prosperity, from which it has not hitherto recovered. It has still, however, some trade in woollen stuffs, salt, flour, sugar, cocoa, &c.; and its markets are always well supplied with provisions. Popayan was the first city built by Europeans in this part of the New World, having been founded by Benalcazar in 1537. A considerable portion of the city was destroyed by an earthquake in 1827. (*Geog. Account of Colombia; Mollien, in Mod. Trav., xxvii.; Dict. Geog.*)

POPE, county, Ill. Situated in the S.E. part of the state, and contains 576 sq. m. Bounded S.E. and S. by Ohio river. Drained by Big Bay, Great Pierre, and Rock creeks. It contained in 1840, 5936 neat cattle, 5350 sheep, 6041 swine; and produced 1246 bushels of wheat, 36,908 of Indian corn, 6080 of oats, 1444 of potatoes, 15,700 pounds of tobacco. It had six stores, 10 grist-mills, three saw-mills, two tanneries, 21 schools, 468 scholars. Pop. 4094. Capital, Golconda.

POPE, county, Ark. Situated in the N.W. part of the state, and contains 720 sq. m. Watered by Arkansas river and its small tributaries. It contained in 1840, 6143 neat cattle, 1178 sheep, 10,356 swine; and produced 3418 bushels of wheat, 121,874 of Indian corn, 2736 of oats, 7168 of potatoes, 13,576 pounds of tobacco, 52,344 of cotton. It had six stores, two grist-mills, four saw-mills, three tanneries, one distillery; four schools, 78 scholars. Pop., whites, 366. slaves, 215; free coloured, 9; total, 2650. Capital, Dover.

POPERINGEN, or POPERINGHE, a town of Belgium, prov. W. Flanders, cap. cant., on the Schipvaert canal, near the French frontier, 7 m. W. by N. Ypres. Pop. about 1000. It has several churches, a handsome town-hall and college, some rather extensive woollen manufactures, with oil-mills, &c. It has also a considerable trade in hops. It sends along with Thielt, three members to the states of the prov. (*De Cloet ; Dict. Geog.*)

PORTAGE, county, O. Situated in the N.E. part of the state, and contains 500 sq. m. Watered by Cuyahoga river and head branches of Mahoning river. The Pennsylvania and Ohio canal passes through it. It contained in 1840, 21,922 neat cattle, 26,894 sheep, 8550 swine; and produced 112,144 bushels of wheat, 9497 of rye, 113,370 of Indian corn, 6301 of buckwheat, 119,694 of oats, 119,742 of potatoes, 258,038 pounds of sugar. It had 38 stores, one fulling-mill, one woollen factory, nine grist-mills, 41 saw-mills, 11 tanneries, two distilleries, one pottery, two printing-offices, one weekly newspaper; two academies, 101 students; 143 schools, 5077 scholars. Pop. 22,965. Capital, Ravenna.

PORTAGE, county, Wis. Situated in the N. part of the ter., and contains 850 sq. m. Bounded W. by Wisconsin river, and head branches of rock river. It contained in 1840, 372 neat cattle, 127 swine; and produced 1813 bushels of wheat, 293 of Indian corn, 1018 of oats, 5539 of potatoes. It had nine stores, 14 saw-mills, one school, 19 scholars. Pop. 1623. Capital, Fort Winnebago.

PORTAGE, t., Allegheny co., N. Y., 18 m. N. Angelica, 247 W. by S. Albany. Watered by Genesee river, which has falls of 300 feet in the distance of 2 m., producing a great water-power. The Genesee valley canal passes through it. It has an Episcopal church, 18 stores, two fulling-mills, one flouring-mill, three grist-mills, 18 saw-mills, three tanneries, one distillery; 17 schools, 1071 scholars. Pop. 6781.

PORTAGE, t., Summit co., O. Akron village, the capital of the co., lies in its S part. Here was a former portage between Cuyahoga and Tuscarawas river, whence it derives its name. See AKRON.

PORTER, county, Ia. Situated in the N.W. part of the state, and contains 415 sq. m. Bounded N. by lake Michigan. Watered by Grand Calumet and Kankakee rivers. It contained in 1840, 3092 neat cattle, 1198 sheep, 7220 swine; and produced 30,712 bushels of wheat, 60,444 of Indian corn, 1740 of buckwheat, 53,312 of oats, 16,673 of potatoes, 16,406 pounds of sugar. It had four stores, two grist-mills, seven saw-mills, two tanneries ; eight schools, 189 scholars. Pop. 2162. Capital, Valparaiso.

PORTER, t., Niagara co., N. Y., 16 m. N.W. Lockport, 7 m. N. Lewiston, 300 m. W. Albany. Bounded N. by lake Ontario. Drained by Tesearors creek. It contains old fort Niagara, opposite to fort George in Canada, at the mouth of Niagara river, which bounds it on the W. It contains 11 stores, one saw-mill ; one academy, 22 students ; eight schools, 481 scholars. Pop. 2177.

PORT-GIBSON, p. v., capital of Claiborne co., Miss., 45 m. N. Natchez, 73 m. S.W. Jackson. 1082 W. Situated on Pierre Bayou, 30 m. above its entrance into Mississippi river. It contains a courthouse, jail, and several stores and dwellings.

PORT-GLASGOW, a parl. bor. and seaport town of Scotland, co. Renfrew, on the S. side of the frith of Clyde, 16 m. W.N.W. Glasgow, and 2 m. E. Greenock. On the W. the town is flanked by a steep range of hills, about 400 feet in height ; and is, in fact, so much overshaded by these hills, that the rays of the sun do not reach it for about six weeks in winter. Pop. in 1790, 4036 ; in 1835, 6018. The town is neat and regular ; the streets, which are straight, for the most part cross each other at right angles ; while the houses, which are generally lofty and substantial, have a pretty uniform appearance. The only modern buildings worth notice are the town-house and parish church. The latter is ornamented in front with a portico, resting on four massy fluted pillars, and is surmounted with a handsome spire, rising from the centre. In addition to the parish church, there is a quoad sacra chapel belonging to the establishment, and a chapel in connexion with the Associate Synod. There are eight schools, one of which is parochial : and another an endowed seminary, called Be iton's school, from the name of its founder. There are three public libraries, and a reading-room. A legal assessment for the support of the poor has not been introduced ; but voluntary contributions to the extent of above £600 are raised for the poor, including church collections and the usual parish dues. A savings' bank, under the designation of a Provident Bank, has existed here since 1818 ; and there are various friendly societies.

Two sugar-refining houses employ about 50 men ; and a rope-manufactory about 45. But the most important branches of business are ship-building, and the manufacture of canvass for sail-cloth and coarse lioen fabrics. The former gives employment to about 200 men, exclusive of apprentices ; wages £1 or £1 1s. a week. The building of steam-boats, some of them of the largest class, is extensively carried on. A flax-mill employs about 440 hands.

Port-Glasgow, as every one knows, was the seaport or deep-water harbour of the city of Glasgow ; and was long regarded as a mere dependency of the latter. It has two capacious harbours, furnished with ample quay and shed room, together with a graving dock, the oldest in Scotland. A large and commodious wet-dock has recently been erected. Formerly the trade of this place was almost entirely carried on in ships belonging to merchants resident in Glasgow. Of late years, however, the people of Port-Glasgow have themselves become ship-owners, and at present about 1-4th part of the shipping belonging to the port, or about 7000 tons, is owned by residents in the town. It is the principal port on the Clyde for the importation of N. American timber, the quantity imported having varied during the last 15 years, from 16,000 to 30,000 tons a year. Owing to the great improvements that have recently been effected in the navigation of the Clyde, the greater part of the trade belonging to Glasgow that formerly centered in this port, has been transferred to the former. The town, consequently, is not improving. The tonnage belonging to the port engaged in foreign trade, has ranged, for several years back, from 21,000 to 23,090 tons ; but the customs' revenue has very materially decreased ; a consequence of a large proportion of the goods formerly warehoused here being now carried direct to Glasgow. Thus, the customs' revenue which, in 1830, amounted to £343,349, had sunk, in 1840, to £84,936. More than half the trade of the port is with the British N. American possessions ; about a fourth with the W. Indies ; and the remainder with the E. Indies, the Mediterranean and the U. States. There is now little coasting trade between the town and Glasgow ; but the numerous steamboats that navigate the Clyde, except those that ply to comparatively distant ports, touch here in passing and re-passing.

The Glasgow, Paisley, and Greenock railway, recently opened, passes close on the W. of Port-Glasgow. We have lately seen (Vol. I., 994) that the magistrates of Glasgow, in order to provide a deep water port for their city, purchased, in 1662, the ground on which Port-Glasgow stands, and laid the foundations of the town and harbour. The town supports four branch banks. The intercourse between Glasgow and its newly acquired port was at first carried on

principally by land ; but the improvements effected in the Clyde navigation have been such as to make Glasgow nearly independent of this or any other port, except its own. The ruins of the castle of Newark, which originally formed the seat of the proprietor of the estate on which Port-Glasgow is built, stands on the shore, immediately contiguous to the town on the E. In 1775 it was created a borough of barony, and a municipal constitution was conferred on it. The Reform Act united it with Renfrew, Rutherglen, Dumbarton, and Kilmarnock, in sending one member to the House of Commons. Registered voters in 1840-1, 218. Municipal revenue, £2616. (Vide the article GLASGOW, in this work ; New Stat. Account of Scotland. § Renfrewshire, p. 69-74 ; Bound. Reports ; Factory Returns, &c.)

PORT JACKSON. See SYDNEY.

PORT KENT. p. v., Chesterfield t., Essex co., N. Y. 12 S. Plattsburg. 151 N. by E. Albany, 398 W. Situated on the W. side of lake Champlain, and contains about 250 inhabitants. A steam ferry boat plies to Burlington, 11 miles distant, of which place it has a delightful view.

PORT LOUIS. See MAURITIUS.

PORT MAHON, a seaport of the island of Minorca, which see.

PORTPATRICK, a seaport town of Scotland, co. Wigtown, on a gentle declivity on the E. coast of the Irish channel ; bounded on the land side by hills which suddenly rise, in a romantic semicircular form, to the height of 200 or 300 ft. ; 109 m. S.W. Edinburgh, and 21 m. N E. Donaghadee, Ireland. Pop. in 1791, 512 ; in 1832, 1010, of whom about a third are Irish, or of Irish descent.

The principal street is in the form of a crescent, parallel to the bay ; and there are three smaller streets connected with it, stretching at right angles towards the hills. The houses are in general, well built, comfortable, and covered with slate. With the exception of the parish church, erected in 1629, there are no public buildings ; nor is there any other place of public worship, though the dissenters of all kinds in the town and parish exceed 400. Education is at a low ebb ; lower, perhaps, than in any other place of similar size in Scotland. There is a small parish library ; and a still smaller one connected with the Sunday school. There are no manufactures ; but the cod fishery is carried on to some extent.

Two government steam-packets, in the service of the postoffice, ply between Portpatrick and Donaghadee ; this having been long one of the principal lines of communication between Great Britain and Ireland. The fare of cabin passengers is 5s., and of deck passengers 2s. The number of the former is Ireland, annually, is about 500 ; from Ireland, about 450 : the number of the latter to Ireland, about 5000 ; from Ireland, about 4500. This includes the last 14 years ; but before that time, that is, before the general introduction of steam navigation, by which the most ready conveyance is obtained from all the great ports of Ireland to similar places in England and Scotland, Portpatrick, in consequence of the shortness of the passage from Donaghadee, formed the principal port of entry from and to Ireland, for cattle and horses. For example, 17,275 horses and cattle were imported here in 1701 ; 20,000 in 1819 ; but only 1060 in 1837 ; and the number has considerably diminished since that year.

Formerly the harbour of Portpatrick was a mere inlet between two ridges of rocks, and was one of the worst and most dangerous on the W. of Scotland. Whenever a vessel approached the harbour, the inhabitants assembled to draw her to the beach, there being no quay or creek to afford shelter from the waves. But a quay and reflecting lighthouse were built about 60 years ago ; and it having been determined to make the place a steam-packet station, a new harbour has been constructed, proceeded by two piers, curved to resemble a horseshoe, and furnished with jetties near their extremities, by which the entrance is contracted to 180 ft. ; the area of the basin thus formed being about seven acres. In accomplishing this great work, rock to an immense extent required to be excavated, which was effected by means of puddle-dikes and the diving-bell ; the works are not yet quite finished. The original estimate was £120,000 ; it is, however, supposed that the total expense will not fall much short of £200,000. But the entrance to the harbour is, after all, very difficult in rough weather, and it is doubtful whether Portpatrick should have been selected for a packet-station ; indeed, notwithstanding the outlay on the harbour, it has been already proposed to abandon it, and to select Finnart, 8 m. N. from Stranraer, on the confines of Ayrshire and Wigtownshire, as a more suitable place, on the alleged ground that the harbour of Portpatrick can never be made sufficiently accessible or secure, and that it is too remote from the central parts of Scotland.

Portpatrick was long resorted to as the Gretna Green for Ireland, and was celebrated for its runaway, or irregular marriages. The lowest sum charged was £10, payable to the parochial clergyman, who performed the marriage ceremony, and £1 to the session clerk. The practice was abandoned in 1826, owing to the interference of the church courts; but in the records of these marriages during the preceding period of 50 years, there occur the names of 198 gentlemen, 15 officers of the army or navy, and 13 noblemen. (*New Stat. Acc. of Scotland, § Wigtownshire*, p. 129-161, and *Priv. Inform.*)

PORT-AU-PRINCE, now called PORT REPUBLICAIN, a city and seaport, and the modern cap. of the republic of Hayti, on the W. coast of the island, at the bottom of the bay of Gonaives, 90 m. S. cape Haytien, and 165 m. W. St. Domingo; lat. 18° 33' 42" N., long. 72° 27' 11" W. Pop. variously estimated, probably from 18,000 to 20,000. It is partially fortified on the land side, and the harbour is protected by a battery on a small island near the shore. The streets are laid out with great precision, crossing each other at nearly right angles; but the town is irregularly built; the houses are principally of wood, and as they seldom exceed two stories in height, have a paltry appearance. Except the palace, which is a large building, with a handsome flight of steps leading to good reception rooms, there are no public buildings of any importance. The arsenal, church, mint, lyceum, military hospital, and courts of law, are all insignificant. The adjacent country is low and marshy; and the heat in the summer months being excessive, the climate is then exceedingly unhealthy. The entrance to the harbour is between White island and the S. shore. The depth of water varies from about 18 ft. at ebb, to 20 ft. at full tide. It is customary, but not compulsory, to employ a pilot in entering the harbour. Ships moor head and stern at from 100 to 500 yards from the shore; loading and unloading by means of boats, as there are neither docks nor quays to facilitate these operations. The harbour is perfectly safe, except during hurricanes, which may be expected from August to November. The markets are tolerably well supplied with beef, mutton, fowl, fruit, and vegetables; but the supply of fish is uncertain; and such is the indolence of the inhabitants, and their neglect of the most obvious resources, that though turtle abound in the bay, they are rarely found in the markets.

Port-au-Prince is the seat of government, the supreme court of justice, court of cassation, a tribunal of original jurisdiction, &c. It is also the residence of the principal foreign consuls in Hayti, and the grand entrepôt of the commerce of the island, which, however, is trifling, compared to what it was previously to the commencement of the disturbances and atrocities that devastated this fine colony. (*See* HAYTI.) It is of late origin, having been founded in 1749. It is very subject to earthquakes, by one of which it was nearly destroyed in 1770. (*Mackenzie's Notes on Haiti*, i. 5-25; *Encyc. Amer.; Comm. Dict.*)

PORT ROYAL, a town and seaport, and formerly the commercial cap. of Jamaica; at the extremity of a narrow point of land, bounding Kingston harbour on the S. and E., about 5 m. S.S.W. Kingston. It formerly had 2000 houses, and was handsomely built; but having been in great part destroyed by an earthquake in 1692, and having subsequently suffered severely by fires and hurricanes, its public offices were transferred to Kingston, and it is now insignificant as a town. It is still, however, strongly fortified, and is the seat of the royal navy yard, the naval hospital and of some regimental barracks.

PORTARLINGTON, a parl. bor. and inland town of Ireland, prov. Leinster, on the Barrow, which divides it into two portions, the larger of which is in Queen's, and the smaller in King's co., 40 m. S.W. by W. Dublin. Area of parl. bor., 933 acres, pop. of ditto, in 1831, 3000. It consists principally of a single street, nearly 2 m. in length, extending on both sides the river, which is here crossed by a stone bridge. This is, perhaps, the best built, and cleanest country town in Ireland. "Few towns of its size have so respectable an appearance, which arises, not from its trade or manufactures, for of these it has none beyond the retail trade consequent on its population, but from the unusual number of its resident gentry." (*Mun. Bound. Report.*) A considerable part, perhaps, of the distinguishing features of this town may be ascribable to the fact of a colony of French Protestant refugees having been settled in it by William III. It has two churches, in one of which, frequented by the refugees and their descendants, divine service was performed in the French language within the course of the present century. It has, also, two Roman Catholic chapels, a Methodist meeting-house, a market-house, and a dispensary. The schools belonging to this town have long enjoyed a high reputation, especially those for French; and in them two of the most illustrious individuals of whom Ireland has had to boast, the Duke of Wellington, and his brother the Marquis Wellesley, received the rudiments of their education. Under the charter granted by Charles II., in 1667, the corporation consisted of a sovereign, two portreeves, 12 burgesses, and a commonalty, who returned two members to the Irish House of Commons down to the

union. Since then it has returned one member to the Imperial House of Commons. The Boundary Act fixed the limits of the parl. bor. as stated above. Registered electors in 1839-40, 171. An obelisk on a hill adjoining the town commands an extensive prospect. Manor courts and petty sessions are held here; and it has two constabulary stations, one in the portion of the town in each county. Markets on Wednesdays and Saturdays; fairs on the 5th of Jan. 1st March, Easter Monday, 22d May, 4th July, 1st Sept. 12th Oct., and 23d Nov. The town, with an extensive surrounding district, was granted, in the reign of Charles II. to Lord Arlington (Sir H. Bennett; the Eliab of Dryden's Absalom and Architophel), one of the famous CABAL. The town, which was previously called Cultodry, took the name of its new owner, the prefix Port being given to it in consequence of its having a small landing-place on the Barrow. Lord Arlington, however, soon afterwards sold the property; and, after passing through various hands, it was acquired by Mr. Dawson, an ancestor of the present Earl of Portarlington. Emo-house, the residence of this noble family, is about 4 m. S. from the town. (*Mun. and Parl. Boundary Reports; Fraser's Guide to Ireland, &c.*)

PORTLAND, city, port of entry, and capital of Cumberland co., Me., in point of wealth population and commerce, the largest town in Maine, and until 1829 the capital of the state, is in 43° 39' 28" N. lat. and 70° 20' 30" W. long. from Greenwich, and 6° 40' 54" E. long. from Washington. It is 54 N.N.E. Portsmouth, 102 N.N.E. Boston, 59 S.S.E. Augusta, 545 N.E. Washington. The population in 1790 was 2240; in 1800, 3677; in 1810, 7169; in 1820, 8521; in 1830, 12,601; in 1840, 15,218. Of these, 397 were engaged in commerce; 1032 in manufactures and trades; 736 in navigating the ocean; 101 in the learned professions.

It is pleasantly situated on a peninsula in Casco bay, having Back cove on the N., and a branch of the ocean called Fore or Casco river on the S. The ground on which it is built rises at its E. and W. ends, and the city presents a beautiful appearance as approached from the sea, rising like an amphitheatre between two hills. It is regularly laid out, and extends for 3 m. from E. to W., with an average width of three fourths of a mile. The houses are mostly of brick, many of them are neat, and some are elegant. The public buildings are a courthouse of brick, a spacious city hall, a jail, a custom house, six banks, an Athenæum, containing a library of nearly 5000 volumes, a market house, and 12 churches, three Congregational, an Episcopal, two Unitarian, two Baptist, a Free-will Baptist, three Methodist, a Swedenborgian, a Universalist, a Bethel, and an African.

The harbour is one of the best in the United States; it is near the ocean, easy of access, of a sufficient depth of water for the largest vessels, spacious and safe. At its entrance, on a point called Portland head, is a lighthouse of stone, 72 ft. high, exclusive of the lantern, erected in 1790. By several islands it is completely land locked. The harbour is defended by two strong forts on two islands; fort Preble on Bang's island, and fort Scammel on House island. On an eminence in the N.E. part of the town, is an observatory, about 70 ft. high, which affords a fine view of the harbour, with its beautiful islands and variegated scenery, and the city and surrounding country, extending to the White mountains in New-Hampshire. The harbour is not much obstructed by ice, except in unusually cold weather. By the Cumberland and Oxford canal, completed in 1830, a communication is opened between Portland and Sebago pond, and is extended to Long pond, the whole being 38 m. in length. This city receives much of the trade of the N. part of New-Hampshire, and of the N.E. part of Vermont, through a road which leads through a notch in the White mountains; and so important has this road been regarded to the prosperity of Portland, that the legislature of Maine has at times made considerable appropriations for its improvement. The Eastern railroad extending from Boston, has been completed, and was opened for travel in 1842, being 105 m. long between the two places. The total number of vessels cleared from the port in 1841, was 969, with crews amounting to 1910 men, and a tonnage of 42,153. The number of vessels which entered the port in 1841, was 774, with crews of 1250 men, and a tonnage of 27,303. The total amount of the registered licensed and enrolled tonnage of the district at the same period, was 56,135 tons. The total valuation of estate in the city, according to the assessment of 1842, was $3,106,514. The expenses of the city government in 1842 were $74,377; of which $9614 were for the public schools; $4000 were for the poor; $1498 for the city watch; $1389 were for the fire departments. The amount raised by tax for city purposes, was $50,143. The city debt amounted to $193,306. The public schools of Portland receive much attention, and there are a number of respectable schools of the higher order. The principal articles of export from the city are lumber, fish, beef, butter, &c. It has an extensive coasting trade, particularly with Boston. The tonnage employed in the coasting trade in 1841, was 13,385;

and in the cod and mackerel fisheries, was 3651 tons. Its trade is also extensive to the West Indies, and considerable to several countries of Europe. In the season of navigation, steamboats ply regularly to Boston, and packets to various ports.

According to the census of 1840, there were 40 commercial and eight commission houses in foreign trade, with a capital of $658,500; 256 retail stores, with a capital of $574,450; two lumber yards, capital $4000; the fisheries employed a capital of $11,300; machinery was produced to the amount of $3000; one furnace employed a capital of $5000; two tanneries a capital of $6000; two potteries a capital of $4000; two rope-walks a capital of $18,000; nine printing-offices, five binderies, two daily, seven weekly, and three semi-weekly newspapers, and three periodicals, employed 94 persons, and a capital of $34,500. The total amount of capital employed in manufactures was $615,350. There were 11 academies or grammar schools, with 1118 students, 32 common and primary schools, with 1976 scholars.

The settlement of Portland commenced in 1632, and the place was purchased by Gorges, the proprietor, in 1637. It was destroyed in the Indian war in 1675, and again in 1690 by the same enemy. It was rebuilt in 1715, when the foundations of the present city were laid. In 20 years after this, its trade in lumber was extensive, and it particularly supplied the British navy with masts and spars, which were chiefly transported by foreign ships. At the commencement of the Revolutionary war, the population was 1900, and the tonnage belonging to the port was 2555. It had 230 houses, and a Congregational and an Episcopal church. In 1775, it was bombarded by the British fleet, 136 of the principal houses were destroyed, among which were the townhouse, courthouse, and the Episcopal church. The total loss was about $340,000. The place was then called Falmouth. From the close of the Revolutionary war to the year 1807, this place, in common with many others on the shores of Maine, especially during the wars in Europe, when the carrying trade of the world was extensively in the hands of the Americans, experienced great prosperity. Subsequently the embargo, non-intercourse and war, put a severe check upon this prosperity. In 1789 the tonnage was 5000, and the duties received $6000; and in 1830, it was $39,000, and the duties were 346,000. In 1786 the place was incorporated by the name of Portland; and in 1832, it received a charter as a city, and is governed by a mayor, seven aldermen, and 21 common councilmen.

PORTLAND, p. t., Middlesex co., Ct., 15 S. Hartford, 297 W. Bounded W. by Connecticut river, across which is a ferry to Middletown. It was recently taken from the N. part of Chatham.

PORTLAND, p. t., Chautauque co., N. Y., 7 S. Mayville, 332 W. by S. Albany, 352 W. Bounded N. by lake Erie, on which it has a good harbour. It contains two stores, one fulling mill, two grist-mills, 12 saw-mills, two tanneries, one distillery; 14 schools, 716 scholars. Pop. 2136.

PORTO BELLO, a famous seaport and town of Columbia, on the Caribbean sea, on the N. side of the isthmus uniting the two great continents of N. and S. America, at the point where it is less than 40 m. across; lat. 9° 24' 29" N., long. 79° 43' 35" W. The town, now greatly fallen off, is built along the shore, at the foot of a mountain range which surrounds and shelters the harbour. But this barrier, at the same time that it protects the port, prevents the circulation of the air, and, combined with the heavy periodical rains, the influence of the surrounding forests, and the excessive heat, renders this locality a favourite seat of yellow fever, and one of the most unhealthy places in the world. Owing, however, to the excellence of its port, which is one of the finest that can be imagined, and its contiguity to Panama, on the opposite side of the isthmus, it was, for a lengthened period, an important commercial entrepôt. Down to 1740, the galleons from Old Spain, with merchandise for the Spanish Main, Peru, and the W. coast of America, used to rendezvous at Porto Bello, about the same time that the Peru fleet arrived at Panama, the gold and silver, and other produce brought by the latter, being conveyed across the isthmus by means of oxen, and conversely. As soon as the galleons had unloaded, and the merchandise from Panama had arrived, a fair was held, which was attended by a great concourse of strangers, and when a great deal of business was transacted. But in 1740 the galleons ceased to resort to Porto Bello, the commerce with Peru and W. America having been since carried on direct by vessels that sailed round cape Horn. In consequence, the importance of Porto Bello rapidly declined; and the advantages of its port not being sufficient to countervail the unhealthiness of its climate, it is now comparatively deserted. But should the project for effecting a communication across the isthmus, by canal or railway, take effect, it is probable that Porto Bello may recover some portion of its former importance. The climate is said to have

been improved by an opening made in the mountains that encircle the town, and by the cutting down of a portion of the adjacent forests.

Notwithstanding Porto Bello was formerly very strongly fortified, it was taken, with little loss, by Admiral Vernon in 1739. The importance of this exploit, and the abilities of the admiral, were, at the time, much overrated; and it was supposed that if he were furnished with an adequate force he would have little difficulty in reducing all the Spanish settlements in this quarter. But the events that took place during the next two years, and especially the failure of the attack on Carthagena, undeceived the public.

Porto Bello was discovered in 1509, by Columbus, its name being derived from the excellence of its harbour. (*Alcedo's Dictionary; Geog. Account of Columbia,* i. 303; *Smollet's Hist. of England,* cap. 18, &c.)

PORTO BELLO, a parl. bor. and seaport of Scotland, co. Mid Lothian, in a plain on the S. bank of the frith of Forth, 2 m. E. Edinburgh. Pop. in 1831, 2781; but in summer its population, owing to the influx of visiters from Edinburgh, is much larger. It has no public buildings, except a *quoad sacra* church connected with the establishment, and chapels belonging to the Episcopalians, Independents, and Relief. The Episcopal chapel is an elegant building. The main street lies along the line of the public road running E. and W.; a number of cross streets diverge from it, leading down to the sea beach, or stretching in the opposite direction. Some of these consist of lines of detached villas. Separate villas, also, abound throughout the town, some of them fronting the sea. No fixed plan, in truth, has been observed in laying out the town, which has a straggling appearance, and some of the older parts are very mean.

Adjacent to Porto Bello is the village of Joppa, now almost a part of it. There is no harbour at either place; but it has long been proposed to construct one at Porto Bello. There are manufactories of bricks, tiles, earthenware, glass, and crystal. Fire-clay abounds on the E. point of Joppa, where fire bricks are manufactured to a considerable extent. A bank has recently been established in Porto Bello. It is estimated that the average number of visiters for sea-bathing in the town, from May to October, amounts to 500. And, owing to the beauty and salubrity of the situation, many families resort to it as an eligible permanent residence. A branch, leading to Leith, of the Edinburgh and Dalkeith railway passes the town on the S.; and stage coaches ply between Porto Bello and Edinburgh at least every hour.

Porto Bello is quite a modern town. It derives its name from the first house having been built, about a century ago, by an individual who had been with Admiral Vernon, in 1739, at "Porto Bello." The Reform Act united it with Leith and Musselburgh in sending one member to the House of Commons. Registered electors in 1840-41, 243. Municipal councillors, nine.

PORTO FERRAJO, a seaport and principal town of the island of Elba, which see.

PORTO RICO (Span. *Puerto Rico*), one of the West Indian islands belonging to Spain, being the smallest and most E. of the greater Antilles; chiefly between lat. 17° 55' and 18° 30' N., and long. 650° 40' and 670° 20' W.; having N. the Atlantic, and S. the Caribbean sea, separated on the E. from the Virgin islands by the Virgin passage, and from Hayti on the W. by Mona passage, 80 m. across. Its shape is that of a parallelogram, the length E. to W. being about 100 m., and the breadth about 36 m. Area, 3700 sq. m. Under the old colonial system of Spain, in 1788, the population did not exceed 80,650; whereas it amounted, in 1836, according to the official returns, to 357,086, of whom only 41,818 were slaves; and it is now (1841) probably little, if at all, short of 380,000 or 390,000. A mountain chain runs E. and W. through the centre of the island; the highest summit of which, at the N.E. extremity, is about 3000 ft. in height. Numerous rivers have their sources in this chain, flowing on either side to the sea, some of which are navigable for two or three leagues from their mouth, for schooners and coasting vessels. The coast line is indented with numerous bays and creeks, some of which form excellent harbours for ships of large burden. The surface, which is finely diversified, is well watered, and the soil is generally rich and fertile. The climate is supposed to be less unhealthy, and better adapted to Europeans, than in most of the Antilles: it differs widely, however, in different parts; the N. coast being especially subject to heavy rains, and the S. to droughts. Violent hurricanes often do immense damage. Porto Rico is singularly destitute of wild animals. "There are almost no indigenous quadrupeds; and scarcely any of the feathered tribe are to be found in the forests. The birds are few both in number and species; you may travel whole leagues without seeing a bird, or even hearing their chirp. On the rivers there are a few water-fowl, and in the forests the green parrot. Almost every other island in the West Indies is infested by snakes, and other noxious reptiles. Here are none. But rats of an enormous size, and in great numbers, infest the country, and sometimes commit dreadful ravages on the sugar canes; and although continually persecuted, their numbers do not decrease." (*Flinter's Puerto Rico*, 53.)

The resources of Porto Rico are wholly agricultural; no manufactures exist, nor have any mines of gold or silver, or other mineral products, been hitherto explored. In 1830, the lands under cultivation in the island were not supposed to exceed 110,000 acres, not being more than 15th or 16th part of the land suitable for that purpose. According to an official return the lands under the different species of culture in 1830, and their produce, were as follows:—

Articles of Culture.	Acres in Culture.	Produce.	
Sugar-cane	14,803	Sugar (musc.)	414,806 cwts.
		Molasses	1,807,748 galls.
		Rum	12,165 puchas
Plantains	36,760	Plantains	617,925 loads.
Maize	16,194	Maize	63,750 fanegas.
Rice	14,960	Rice	—
Tobacco	2,589	Tobacco (cured)	34,640 cwts.
Manioc	1,150	Cassava bread	30,419 loads.
Sweet potatoes	1,294	Sweet potatoes	28,570 cwts.
Yams	6,696	Yams	7,450 —
Pulse	1,100	Pulse	4,870 —
Coffee	16,904	Coffee	250,000 —
Cotton	3,079	Cotton	?
Fruit-trees and gardens	140		
Total acres in culture	108,547		

But during the last 10 years there has been a great increase of pop., and consequently, also, of production. Many thousand acres of land have been cleared and brought under tillage; and a very great increase has taken place of the staple products of the island. Indeed, it will be seen from the account of the exports in 1830, that the sugar and molasses exported in the course of that year very materially exceeded the entire quantity grown in 1830, as given in the above return. Indeed, we have little doubt, supposing the previous return to be nearly accurate, that the lands under cultivation in Porto Rico may at present (1841) be estimated at above 200,000 acres.

In 1828, 1,437,285 acres of land were held by 19,140 proprietors, giving to each, at an average, 75½ acres. At the same time 423 individuals were proprietors of estates regularly established with slaves; 275 being sugar estates, and 148 coffee plantations. The remaining 17,440 proprietors bred cattle, and raised provisions and small quantities of coffee. In consequence of the small capital required, the inhabitants formerly applied themselves almost exclusively to the latter branches of industry: so much so was this the case, that at the commencement of the present century there were but 29 sugar estates on the island; and in 1802 the total value of the exports did not exceed 57,5000 dollars! At present, however, the case is widely different; the S. coast of the island is generally occupied with sugar plantations, and in 1839 the exports of sugar amounted to no less than 692,458 cwts., or 34,629 tons, worth above 3,400,000 dollars. But, with the exception of the S. coast, most parts of the island exhibit a promiscuous cultivation, plantations of sugar being intermixed with those of coffee, and with field rice, maize, plantains, tobacco, and pasture. Artificial irrigation is nowhere practised; but, notwithstanding the drought which prevails in the S., plenty of water for the cane is found at about 2 feet below the surface. The average produce of sugar per acre, for the whole of Porto Rico, is estimated by Flinter at 30 cwt., being more than double the quantity raised on the best lands in the most favoured of the British Antilles! (*Puerto Rico*, 180.) The coffee is of a peculiarly good quality, much care being taken in plucking and preparing it. The trees, which mostly belong to small proprietors, grow to a great height, and sometimes yield from 20 to 40 lbs. each. Every poor family has 20 or 30 trees; and even in the woods trees are to be found in a wild state, laden with coffee. The free labourers at the harvest come into the market, some with 50 lbs., others with a cwt., and so on, being the surplus of their little crops, after leaving enough for the use of their families for the season. This they sell to the merchants, to purchase articles of clothing. The plant on the large estates, cultivated by slaves, is pruned and cut low, and yields, at an average, one pound per tree. Flinter estimates that, in 1830, the 148 coffee estates may have produced 300 cwt. each, or altogether 44,400 cwt.; and that the remaining 205,600 cwt. of the estimated produce (250,000 cwt.) of that year was the growth of free labour. The tobacco of Porto Rico, which is but little inferior to that of Cuba, is wholly the produce of free labour. Poor families, white and black, plant a half or quarter of an acre, and cure and dispose of the produce to the shopkeepers in the villages, who are generally agents for the more extensive merchants of the capital. These shopkeepers furnish clothes and money in advance, at an enormous interest, to the cultivators, from whom they receive their crop at a certain price, generally less than half

its value. The soil in many places is particularly suitable for the growth of cotton, the culture of which has been very greatly extended within the last half dozen years. Indigo grows spontaneously, but is not cultivated; and few European vegetables are raised, though the greater number might be produced without much difficulty.

The pasture lands in the N. and E. are superior to any in the W. Indies for breeding and fattening cattle; the total number of which in Porto Rico, in 1830, was estimated at about 200,000. Cattle breeding is, perhaps, more profitable here than any other branch of agriculture; but, owing to the subdivision of property, few individuals possess so many as 1000 head of cattle. In the breeding districts, where there are no arable lands, the cattle are permitted to roam at large, as on the American continent, but on a smaller scale elsewhere they are penned up in enclosed meadows. They are mostly reared in the valleys distant from the coast, whence the carriage of sugar and coffee would be so expensive for the cultivator. The larger proprietors, who have from 100 to 150 head, if they have not sufficient pasture land of their own, divide their stock among the poorer land-owners, giving them cows, and calves already weaned, to be taken care of; and dividing the produce of the animals, when sold, with them. in a certain proportion. The cattle are turned into the fattening pastures at a year old, and in three or four months are fit for sale. A considerable traffic has long been carried on in cattle from this island with the French, English, and Danish W. India islands; for, in consequence of their being tamer, they are easier managed on board ship, and are not so liable to die or fall off at sea as the wilder cattle of the American continent they fetch about 33½ dollars per head. The home supplies of cheese and butter are insufficient for the consumption. The island is not adapted for wool growing, the fleece of the sheep degenerating into a species of hair; but the mutton is excellent. The numbers of sheep, goats, hogs, mules, and asses are, however, very limited; but there are, probably, above 80,000 horses of a tolerably good breed. The value of the live stock in 1830 was estimated at about 7,856,000 dollars; of which the cattle made 4,680,000 and the horses 3,900,000. The forests in the interior supply timber of the best quality for ship and house building; and, to prevent their decrease, the government has ordered that three trees should be planted for every one cut down.

Previously to 1815, Porto Rico, being excluded from all direct intercourse with other countries excepting Old Spain, was either stationary or but slowly progressive, the entire value of the exports in that year having amounted to only 63,274 dollars! But at that epoch a royal decree appeared, which exempted the trade between Spain and the Spanish colonies and Porto Rico from all duties for 15 years; and she was then also permitted to carry on a free trade, under reasonable duties, with other countries. In consequence principally of these wise and liberal measures, but partly also of a considerable immigration of rich Spanish colonists from S. America, Porto Rico has latterly made a most extraordinary progress. Great improvements have been effected in the police and internal administration, and roads have been constructed in all parts of the island.

Probably, however, the most advantageous circumstance in the condition of this island is the decided preponderance of the white and free coloured population. According to the census of 1836, the inhabitants consisted of

Whites	. . . . . .	188,869
Free mulattoes	. . . .	241,575
Free blacks	. . . .	26,194
Slaves	. . . . .	41,818
Total		357,085

This is a better distribution of classes than prevails in any other part of the W. Indies, Cuba not excepted, and the deep-rooted antipathy that exists between the white and black races in the other islands is nearly unknown in Porto Rico. The slaves have always been particularly well treated in the Spanish colonies; and here they are decidedly comfortable, being well clothed, well lodged, and not over-worked.

"The necessaries, and many of the comforts, of life are enjoyed by the great majority of the inhabitants of Porto Rico. The Xivaros, a name applied to all the whites below the better classes, swing themselves to and fro in their hammocks all day long, smoking their cigars, and scraping a guitar. A few coffee plants and plantain trees, a cow and a horse, an acre of land, in corn or sweet potatoes, constitute the property of what would be denominated a comfortable Xivaro; who, mounted on his meagre and hardworked horse, with his long sword protruding from his basket, dressed in a broad-brimmed straw hat, cotton jacket, clean shirt, and check pantaloons, sallies forth from his cabin to mass, to a cock fight, or to a dance, thinking himself the most independent and happy being in existence.

"The houses of all classes, in the country, are usually

built of wood. The windows have no glass: they are shut with sliding boards; so that when it rains, or when the wind blows with violence, the family remains in darkness. The roofs of the better class of houses are covered with wooden shingles. There are no inns for travellers either in the towns or country." (*Flinter*, p. 242–251.)

The following statements as to the trade of Porto Rico in 1839, taken from the *Balanza Mercantil*, published in the capital, afford the most conclusive proof of the advancing prosperity of the island:

ACCOUNT of the Value of the Imports into, and of the Exports from, Porto Rico, in 1839, specifying the Values of those imported and exported under different flags.

Flags.	Imports.		Exports.	
	Dolls.	Cwts.	Dols.	Cwts.
Spanish (from Spain)	725,740	40	600,401	16
Spanish ca es from				
St Thomas . . .	1,041,817	81	414,996	81
American . . .	1,102,679	87	2,5 8,1·2	84
English . . .	145,½86	90	347,994	81
French . . .	86,362	40	296,0 ·4	90
Hamburg, Altona, and				
Bremen . . .	223,671	58	478,453	7
Dutch . . .	8,615	47	10,988	
Portuguese . .	895	75	984	95
Swedish . . .	—		119	
	4,380,355	5	4,734,19·1	2
In transit . .	1,111,346	93	984,413	57
Total . . .	5,601,800	98	5,516,811	60

ACCOUNT specifying the Quantities and Values of the principal Articles exported from Porto Rico in 1839.

Articles.	Quantities	Value
		Dollars.
Cotton . . .	11,920 cwts.	126,431·76
Sugar . . .	892,451 —	2,472 60/·10
Coffee . . .	86 3 44 —	863,036·½0
Molasses . . .	3,311,722 gallons.	496 757 90
Tobacco . . .	43,;03 cwts.	172,813 58
Rum . . .	650 punchs.	16,241 86
Hides . . .	6,736 cwts.	60,844 92
Horses and mules .	293	11,363 00
Cattle . . .	3,598	120,303·06
Timber . . .		21,256 10
Rice, salt, &c. .		28,904·31
Specie . . .		130 . 89 00
		4,508,531·06
Cocoa, cottons, and other imported goods and exported		968,079·72
		5,516,611·00

ARRIVALS of Shipping in Porto Rico in 1839.

Flags.	Vessels.	Tons.
Spanish	675	22,976
Americans	439	62,939
English	114	5,336
French	86	6,904
Danish	47	4,577
Swedish	3	81
Dutch	9	449
Hans Town . . .	16	5,705
Portuguese . . .	2	187
	1,394	116,356
Custom-house duties on imports . .	734,781·81 dollars.
Do. exports . .	261,062 08 do.
Do. dues on shipping . .	28,756 85 do.
	Total	1,024,562 94

The government, laws, and institutions are nearly similar to those established by Spain in the rest of her Transatlantic colonies. Porto Rico is governed by a captain-general, whose authority is supreme in military affairs, and who is president of the royal *audiencia* for civil matters. The latter court is composed of the captain-general, a regent, three judges, a fiscal, two reporters, and a marshal; and is superior to all other constituted authorities, including the ecclesiastical tribunal. The captain-general has a *junta*, or council of the principal military officers. In the seven towns and villages, which are the capitals of departments, justice is administered by the mayors: in the smaller towns and villages by inferior magistrates, called lieutenants, who determine debts under 100 dollars, act as justices of the peace, collect the duty of subsidy, receiving six per cent. on the collections, &c. They are appointed by the captain-general, who also appoints the clergy to their different livings, on the recommendation of the bishop. Public instruction is very backward; but schools, though few, are increasing. The island is divided into seven military departments, each under the command of a Spanish colonel. The regular military force comprises about 10,000, and the militia about 46,000 men. The naval force consists of only a man-of-war, a schooner, and about a dozen gun-boats. The fueros and alcavala duties were abolished in 1815, and the subsidy, or direct contribution on landed property, established instead. In 1832 the total revenue of the island amounted to 798,400 dollars; the direct taxes producing 207,140 dollars, the custom-house duties 493,930 dollars, and the in-

direct taxes (on cock-pits, billiard-rooms, lotteries, stamps, saleable offices, legacies, bulls, &c.) 97,000 dollars. The duties on imports, which, in 1839, produced 734,781 dollars, consist of certain rates charged on different articles, according to the presumed value. On some articles they are very heavy, and there is, in consequence, a considerable contraband trade. Articles imported in Spanish vessels pay one third less duty, and Spanish produce imported in Spanish bottoms pay a half less. The chief towns of the island are San Juan de Porto Rico, the capital, Mayaguez, Ponce, Aguadilla, and Faxardo.

Porto Rico, when discovered by Columbus, in 1493, is supposed, though probably on no very good grounds, to have had, 600,000 inhabitants, who were, in no very long time, almost exterminated. In the latter part of the 17th century it was taken by the English: but, from the prevalence of dysentery, they were soon after obliged to abandon it; since which it has been mostly in the quiet possession of Spain. A revolutionary movement, which broke out in 1820, was happily put an end to in 1823. (*Flinter's Present State of Porto Rico; and Official Returns.*)

PORTO RICO (SAN JUAN DE), the principal city and seaport of the above island, of which it is the capital, on rising ground, at the extremity of a peninsula, joined to the land by a narrow isthmus; lat. 18° 29′ 10″ N., long. 66° 13′ 15″ W. Pop. 30,000 ? The town, which lies along the E. side of the harbour, is strongly fortified. The streets cross each other at right angles: being on a declivity, it is well drained, and may be considered as one of the best and healthiest towns in the W. Indies. In the earlier part of the present century most of the houses were of wood; but at present, except in the suburbs, not a wooden house is to be seen, and they are principally two stories high. There are some good public buildings; including the bishop's palace, and seminary; the royal military hospital, with 350 beds; public jail, house of correction, a handsome theatre, town-house, with a magnificent public hall, several convents, &c. The government house, though old and sombre looking, has some fine apartments. The cathedral is a large, unfinished, heavy fabric; there are several other churches, with a custom-house, arsenal, &c. The harbour has a striking resemblance to that of the Havannah, to which it is but little inferior. Its entrance, about 300 fathoms in width, has the Morro castle, at the N.W. corner of the city, on its E. side, and is defended on the W., or opposite side, by forts erected on two small islands. Within it expands into a capacious basin, the depth of water varying from 5 to 6 and 7 fathoms. On its W. side, opposite to the city, there are extensive sand-banks; but the entrance to the harbour, and the harbour itself, is unobstructed by any bar or shallow. Porto Rico is the residence of the governor, and the seat of the superior courts, &c., for the island. It has a society for the promotion of the fine arts, with numerous public schools, hospitals, &c. It engrosses the larger portion, by far, of the commerce of the island, and has, in consequence, attained to considerable distinction among the emporiums of the W. Indies. (*Flinter's Porto Rico; Comm. Dict., &c.*)

PORTSMOUTH, including its suburb of Portsea, a parl. bor., and a celebrated seaport town of England, being the principal naval arsenal of Great Britain, and the grand station of the fleet, hund. Portsdown, co. Hants, on the W. side of Portsea island, at the entrance to Portsmouth harbour, opposite Gosport, and on the N. side of the channel separating the isle of Wight from the main land: 16 m. S.E. Southampton, and 65 m. S.W. London; lat. 50° 48′ 3″ N., long. 6° 5′ 59″ W. The area of the parl. bor. and the pop. at the under-mentioned decennial periods have been as follows:—

Parishes.	Area in stat. acres.	Pop. in 1801.	Pop. in 1811.	Pop. in 1821.	Pop. in 1831.	Pop. in 1841.
Portsmouth	170	7,839	7,103	7,269	8,083	9,354
Portsea . *Portsea Guildable	4,920	24,327	31,595	34,795	38,199	43,673
		1,069	2,969	3,894	4,107	
Total	5,090	33,266	40,567	45,645	50,349	53,027

Portsea island, which has Portsmouth at its S.W. extremity, is about 4 m. in length (N. and S.), and from 2 to 3 m. in breadth, between Portsmouth harbour on the one side, and Langstone harbour on the other: it is connected with the main land, at its N. extremity, by a stone bridge, and is generally fertile and well cultivated, producing excellent crops of corn, and large quantities of particularly fine garden vegetables. Besides the above towns, Portsea island has several villages; and its coasts are well defended, at numerous points, by strong military works, including, together with the fortifications of Portsmouth itself, fort Cumberland, Southsea castle, a long line of intrenched works at

* Portsea Guildable is that part of the parish of Portsea not within the jurisdiction of the borough of Portsmouth previously to the passing of the Municipal Reform Act.

Hilsea, &c. The external appearance of Portsmouth and Portsea is greatly embellished by the fine trees which ornament their ramparts; and few towns exhibit so imposing an approach as Portsea at its principal entrance from London. The entrances to Portsmouth, the older and more southerly part of the parliamentary borough, are much less striking; but its interior is far superior to that of its neighbour. Portsmouth may be generally described as consisting of three or four parallel streets, crossed at right angles by two or three other lines of thoroughfares. High-street, the principal, with its angular continuation, Broad, or Point-street, runs entirely through the centre of the town; it is wide and handsome, having on either side many large and excellent public buildings, and some very superior hotels. It has also been much improved by the removal, in 1836, of the old town-hall, an unsightly brick building, which previously stood about its middle, blocking up the greater part of the coach-road. Many good private houses are to be met with in some of the other streets, and on the Grand-parade, a spacious open area, used for garrison inspections, and for the daily muster of the several guards; but in general the private buildings are of an inferior character, and the back streets, particularly those at the Point and toward the N. part of the town, are of the lowest character, and, which is worse, disgustingly filthy. The *Point* is a small peninsula stretching W. to form the mouth of the harbour, and mostly beyond the walls of the town. It is, with the opposite part of Portsea, the principal seat of naval traffic, most of the ship agents, brokers, &c., having their offices here, and, in time of war especially, it presents a scene of the greatest activity.

Portsea, which has entirely grown up since the beginning of last century, on a tract formerly called Portsmouth Common, N. of Portsmouth, now greatly surpasses the latter in extent and population. It is divided into two nearly equal parts by its main thoroughfare. Queen-street, which runs for about ½ m. in a direct line from Lion-gate, and is lined on each side with shops, many of which emulate those of the Strand or Fleet-street. Some few of the other streets, as St. James's-street, King-street, the Common Hard, &c., are tolerably broad and well built; but except these, none of the others approach even to mediocrity. The houses in Portsea are rarely more than two stories in height, and their fronts are but seldom stuccoed. It has but one handsome open space (St. George's-square), and few public buildings; and, indeed, till within the last few years, it had scarcely an hotel deserving the name. Both towns are well paved, well supplied with water, and well lighted with gas, but their police is said to be very defective.

The suburbs beyond the walls comprise at least half the parliamentary borough, their more densely peopled portion extending over a space fully as large, if not larger, than that occupied by the two towns. Of these suburbs, Southsea alone has any pretensions to beauty. It consists of a succession of well built terraces facing the sea, and the E. ramparts of Portsmouth, being inhabited principally by naval, military, and government civil officers, and visitors resorting thither during the summer season. Behind these terraces are a number of tolerable streets, and some new squares, &c.

Elm Grove and Somerstown are sections of this suburb; the former, a little farther eastward, is a series of elegant detached villas, surrounded by fine plantations and gardens, commanding prospects of Spithead, the isle of Wight, &c., and inhabited by opulent individuals. Landport, Flathouse, &c., immediately to the E. and N. of Portsea, have nothing, in point of appearance, to recommend them; their inhabitants are, in a great measure, retail tradesmen and workmen employed by government. Newtown (Mile End), Kingston, Buckland, &c., E. of the foregoing, are cheerful and agreeable suburbs, principally extending along the London road, and inhabited by much the same classes as those residing in Southsea.

Immediately without the walls of Portsmouth, stretching along the shore, is Southsea Common, a fine large open space, used for reviews and military inspections, and a favourite place of public recreation.

The importance of Portsmouth, like that of Plymouth, depends wholly on the excellence of her harbour, and on her convenient situation as a place for the outfit and rendezvous of the fleets in the channel, or of those cruising off the coasts of France and Spain. It is this that has made her be selected as the principal station of the navy, and has consequently advanced her to the highest destination as a naval depôt.

The harbour, which is unequalled in Great Britain, and surpassed but by few in the world, has a narrow entrance, not exceeding 220 yards in width, between Portsmouth and Gosport; but within its width increases, and it expands into a noble basin capable of containing the greater part of the navy of Great Britain. There is a bar outside the entrance to the harbour; but as it has about 13 feet water

over it, even lowest ebbs, it can hardly be said to be any obstruc... to navigation; and within the harbour there is water sufficient to float the largest men-of-war at any time of the tide. The anchoring ground is excellent; and, being free from sunken rocks, or other obstructions, ships lie as securely in it as if they were in dock.

The dock-yard, which comprises about 120 acres, lies along the E. side of the harbour. It comprises all the establishments necessary for the construction and repair of ships of war, and for their outfit with the greatest despatch, including numerous building and graving docks, partly opening into the harbour, and partly into a large basin, which communicates with the latter. Along the quay, fronting the harbour, extends a noble line of warehouses, having in its centre a handsome octagonal observatory, on the top of which is a telegraph connected, by a line of signal stations, with the admiralty in London. In the ropehouse, nearly 1200 feet in length, cables are twisted that are sometimes 30 inches in circumference; and the anchor forge produces anchors of the largest size. The iron and copper-mills, the copper foundry, where the copper is rolled into sheets for sheathing by steam-power, the rigging and mast-houses, timber berths, saw-pits, seasoning sheds, mast-ponds, &c., are all on the most extensive scale. Probably, however, the most interesting machinery is that invented, or, at all events, vastly improved by Sir Isambard Brunel, for cutting blocks. It is exceedingly ingenious, and has been productive of a vast saving of labour. During the late war upward of 4000 working men were employed in Portsmouth dock-yard, of whom 1500 were shipwrights and caulkers; but in time of peace the numbers are greatly reduced.

In the dock-yard are the navy pay-office, the residence of the port-admiral, the admiral-superintendent, and the heads of the principal departments of the establishment. The port-admiral's residence, formerly that of the commissioner (whose duties are now performed by the superintendent), is an elegant edifice of white brick, surrounded by gardens. Near it is the Royal Naval college, a spacious dark brick edifice, erected in 1729, its centre surmounted by a cupola and observatory well furnished with instruments. Here youths intended for the navy were formerly instructed in navigation, &c.; but, in 1839, the college was remodelled, and is now appropriated to the instruction of junior naval and marine officers in the higher branches of science connected with their profession, and especially the principles and practice of naval gunnery. The officers belonging to this establishment are boarded and lodged in the college, but are borne on the books as part of the complement of a ship of the line in the harbour. Immediately facing the residence of the port-admiral, is a handsome white brick building, intended originally for a school or college of naval architecture, for the education of a "superior class of shipwrights;" a plan which, though on no very satisfactory grounds, has since been abandoned. On the green, in front of the last mentioned building, is a bronze leaden statue of William III. Adjacent to the college is a neat chapel-of-ease for persons attached to the dock-yard. The latter was, during last century, the scene of several conflagrations. One of these, in 1776, was clearly the work of an incendiary, who was convicted and executed for the offence. The dock-yard is, however, daily open to the inspection of visitors who apply for admission at the gate.

To the S. of the dock-yard, and nearer the mouth of the harbour, is the "Gunwharf," or arsenal for ordnance stores. This is an extensive and very complete establishment. As a depôt for cannon, it is inferior to the arsenal at Woolwich, but, in most other respects, it is very superior to the latter. It comprises many extensive and handsome storehouses, filled with all kinds of ammunition; a neat armoury roofed with copper, and containing small arms for 25,000 men; a laboratory, and various other offices, spacious quays along the harbour, and a terrace of excellent residences for its officers, fronted by a finely planted inclosure. This establishment is separated into the two divisions of the old and new Gunwharf by the Mill-pond; a dammed up creek between Portsmouth and Portsea, which supplies the moats of both, and also turns a considerable flour-mill formerly attached to the victualling office, but now in private hands. The victualling department, which formerly occupied some large piles of building within the town of Portsmouth, was, in 1828, removed to Weovil (see GOSPORT); and its former storehouses have partly been purchased by merchants of the town, and partly given place to a handsome row of modern dwellings, the new alms-houses, and the building of the Philosophical Society. The Custom-house is an insignificant building, but is a convenient situation. The functions of governor are exercised by a lieut. gov., who occupies a noble mansion in High-street, formerly the residence of the port admiral.

Portsmouth appears to have been originally fortified by

Edward V.' Its works were greatly augmented and improved during the reigns immediately succeeding, and in those of Elizabeth, Charles II., and James II. Under William III. they were completed, nearly as at present, the town being almost wholly inclosed within a bastioned enceinte, the ramparts faced with masonry, and encircled with broad moats, with a glacis beyond. But, owing to the rise of Portsea, the N. side of these works soon became useless; and, in 1770, the government began also to surround Portsea with works on a still more extensive scale. At present a continuous line of ramparts extends round both towns, and the belt of fortification is completed by the works surrounding Gosport, on the opposite of the harbour. The ramparts, being planted with elms and poplars, form the favourite promenades of the inhabitants: and facing the sea is the Platform, a fine stone battery, mounting 93 pieces of cannon, and commanding an extensive and beautiful view. Portsmouth is entered by four, and Portsea by two carriage gateways, some having considerable architectural beauty. Besides the town batteries, Spithead, and the approaches to the harbour, are defended by Southsea castle, and forts Cumberland, Blockhouse, and Monkton. Southsea castle, founded by Henry VIII., about 1 m. S.E. Portsmouth, is built almost wholly of stone; as are forts Monkton and Blockhouse on the main land. Fort Cumberland, on the E. extremity of Portsea island, a structure of the last century, covers a very large space, and has earthen ramparts faced with brick, and barrack-room for 3000 men. The town, its suburbs, and auxiliary fortresses are garrisoned by the Portsmouth divisions of royal marines, and marine artillery, and a certain number of infantry of the line. Within the town are several capacious and excellent barracks; and there are others in the gunwharf, at Tipner, Hilsea, &c. Portsmouth has a military hospital, and a marine infirmary situated between the custom-house and the gunwharf. But Haslar hospital for the reception of sick and wounded seamen, the principal establishment of its kind in the kingdom, is on the opposite side of the harbour at Gosport (which see). The chief engineering department for the S. and W. of England, and the residences of the commandants of the marines and engineers, are among the other principal government buildings.

The parish church of Portsmouth, founded about 1220, but principally rebuilt in 1693, is a spacious stone edifice with a square tower, 190 feet in height, surmounted by a cupola and vane, which forms an important landmark. Among other monuments, it has one to Villiers, duke of Buckingham, assassinated here in 1628. The vicarage, value £555 a year, is in the patronage of Winchester college. Adjoining the grand parade is the garrison chapel and burying-ground. In Portsea are two chapels of ease, St. John's and St. George's. St. Paul's, Southsea, and All Saints', Newtown, are elegant Gothic edifices of similar architecture, the former built, in 1822, at an expense of £15,329, and the latter, in 1827, at a cost of £19,464. Two new churches are in progress of erection, one at Landport, and the other in the town of Portsea. Portsea parish church is an antiquated edifice at Kingston, about 1½ m. from the town, and surrounded by a very extensive cemetery. The living is a vicarage in the patronage of St. Mary's college. Winchester, value £606 a year. There are numerous places of worship for Independents, Baptists, Wesleyans, Bryanites, &c.; the Scotch, Presbyterians, Unitarians, Roman Catholics, and General Baptists have each a chapel, and the Jews a synagogue. The town-hall and jail together occupy a large edifice; the latter, which is clean, convenient, and well-conducted, is under the jurisdiction of the borough magistrates. A new market-house and exchange, the general dispensary, savings' bank, workhouses, female penitentiary, beneficial society's hall, literary and philosophical institution, with a handsome edifice, and a good museum and library; Hampshire library with 5000 volumes, the King's rooms at Southsea, with an excellent bathing establishment. Green row and York rooms used for balls, &c., and a theatre, are the other principal buildings of public interest. There are Lancastrian and national schools, an endowed free grammar school for 50 boys, and St. Paul's school, a joint-stock subscription academy for superior classical and mathematical instruction; besides several good private academies, a mechanic's institute, forensic, philharmonic, &c., societies, and various charities. On the London road, about 1½ m. from the town, is a new and spacious public cemetery, with an elegant entrance, a chapel, officiating minister, &c.

In addition to its other conveniences, Portsmouth harbour enjoys the important advantage of opening into the celebrated road of Spithead, between the Hampshire coast and the Isle of Wight. It derives its name from a sandbank called the Spit, extending about 3 m. in a S.E. direction, from the narrow neck or tongue of land on which Gosport is built. A ship of war was formerly kept moored, as a guard or receiving ship, at the head of this bank; but

since the peace this practice has been discontinued, and the roadstead is merely marked by buoys placed at regular intervals. It is here that ships fitted out in the docks and harbour rendezvous before going to sea; and it is also a secure and convenient asylum for the channel fleet and other vessels, during the occurrence of storms. From its safety and capaciousness this roadstead is called by sailors, "the king's bed-chamber." "Thus," to use the words of Dr. Campbell, "it appears that Portsmouth derives from nature all the prerogatives the most fertile wits and most intelligent judges could devise or desire; and that these have been well seconded by art, without consideration of expense, which, in national improvements, is little to be regarded. Add to all these the striking excellence of its situation, which is such as if Providence had expressly determined it for that use to which we see it applied, the bridling the power of France, and, if I may so speak, the peculiar residence of Neptune." (Survey of Great Britain, i., 370.)

Inasmuch as Portsmouth depends for support on its being a great naval port and arsenal, its prosperity is necessarily greatest during war. At present, however, if we compare it with previous periods of peace, it may be said to be flourishing. It necessarily has a considerable trade in the importation of the various articles required in its numerous establishments, and for the supply of the inhabitants, and the victualling of the fleet. Above 900 merchant-vessels, of the aggregate burden of above 12,000 tons, belong to the port. The gross customs revenue received here in 1840, amounted to £70,188. Excepting some extensive breweries, no manufacture, unless it be that of ships of war, is carried on within the town. About 3 m. N.E. were some extensive salt-works, so very ancient as to be referred to in Doomsday-book; but they have been abandoned within these few years.

Portsmouth is connected with Arundel and London by a navigable canal, and communicates with Gosport by a floating bridge for passengers, carriages, &c., the property of a company incorporated in 1838; by which also the town is placed in communication with a branch of the South Western railroad. (See GOSPORT.) Many coaches connect Portsmouth with London, Brighton, Southampton, Bristol, Salisbury, Oxford, &c.; and steam packets ply during the year to Ryde, Cowes, Lymington, Havre, Plymouth, Dublin, London, New-York, &c. A branch of the bank of England, and five private banks, are established here; and it has a weekly newspaper.

Portsmouth received its first charter from Richard I., which was confirmed by various subsequent monarchs. Under the Municipal Reform Act it is divided into seven wards; its municipal officers being a mayor, 13 aldermen, and 42 councillors. Corporation revenue in 1839, £4211. It has a commission of the peace under a recorder; and the boundaries of the municipal and parliamentary borough are co-extensive. A court of record is held here for the recovery of debts under £5; and petty sessions are held three times a week. Portsmouth has sent two members to the House of Commons since the 23d Edward I.: the right of election down to the Reform Act being vested in the mayor and corporation, the number of which seldom exceeded 60. The electoral limits were enlarged by the Boundary Act, so as to include all the parish of Portsea with the old borough; and, in 1839-40, there were 1837 registered electors. Portsmouth is one of the polling places at elections for the S. division of Hampshire. Markets on Tuesdays, Thursdays, and Saturdays. The charter of Richard I. established a fair in the town, called the "Free Mart;" which lasts for 15 days from July 10, and is succeeded by a three days' fair on Portsdown hill, attended by a great concourse of people.

The earliest mention of Portsmouth occurs in the Saxon chronicle, which states that it existed in 501. It probably owes its origin to the sea having retired from Portchester, probably the ancient Portus Adurni at the upper end of the harbour; on which account the inhabitants removed thither, and built a town at the mouth of the port. It was taken and burnt by the French, in 1377, but was soon recovered; and in the reign of Henry VIII. had become the principal naval arsenal of England. (Boundary, Municipal, &c., Reports; Charpentier's New Portsmouth Guide; Private Information.)

PORTSMOUTH, p. t., port of entry, Rockingham co., N. H., is in 43° 4' 35" N. lat., and 70° 45' 50" W. long., and 60° 15' 34" E. long. from W. It is 14 m. E.N.E. Exeter, 24 m. N. Newburyport, 45 m. E.S.E. Concord, 54 m. S.S.W. Portland, 54 m. N. Boston, and 493 W. The population in 1810, was 6934; in 1820, 7327; in 1830, 8082; in 1840, 7887. It is the largest town, and the only seaport in the state, built on a peninsula on the S. side of Piscataqua river, 3 m. from the ocean. The situation is healthy and pleasant, the land gradually rising from the harbour. It is not laid out with great regularity, but is handsomely built; many of the houses are neat, and some of them are elegant. There are in the town 63 streets, 41 lanes, 13 roads, and three squares. It is connected with Kittery in Maine by two bridges, one 480 feet in length, on 20 piers; the other 1750 feet long, on 70 piers. For 900 feet under this last bridge, the water is from 45 to 53 feet deep at low tide, and the piers are from 61 to 72 feet high. The tide is here very rapid, and fears were entertained that it would not stand, which have happily been disappointed. It was completed in less than six months from its commencement, at a cost of only $22,000. A bridge also connects Great island, comprising the township of New-Castle, with Portsmouth, erected in 1821. It contains seven churches, three Congregational, an Episcopal, Methodist, Baptist, and a Universalist, seven banking houses, two market houses, an academy, an athenæum, an almshouse, and a state lunatic hospital. The Episcopal church is a spacious and elegant brick edifice, from the cupola of which is a fine view of the town, the river, the islands, and the surrounding country, affording a prospect exceedingly various and beautiful. A brick market-house, erected in 1800 near the centre of the town, is 80 feet long and 30 feet wide, and has in the upper story a large room called Jefferson hall, well adapted to public uses. The athenæum was incorporated in 1817, and has a library of more than 5000 volumes, besides cabinets of minerals and of natural history. On Navy island, which is on the E. side of the river, and within the bounds of Kittery in Maine, though it is owned by the general government, is the navy yard belonging to the United States, containing every convenience for the construction of vessels of the largest class, having three ship houses. One of these ship houses is 240 feet long, and 131 feet wide, the roof of which is covered with slate, which weighs in the whole 130 tons. Several large vessels have been built here, of which two were 74 gun ships. Ship building also for the merchant service has been, to a considerable extent, carried on, for which the fine white oak timber of New-Hampshire furnishes great facilities. In 1782, the America of 74 guns, the first of this size ever constructed in the United States, was built here and was presented by congress to the king of France. Several other large armed vessels were built here in still earlier times. Good water is brought from a spring in Newington, 3 miles distant, and distributed in pipes, through almost every street in the place, by a company formed in 1799; and a reservoir is filled for the supply of the shipping, at a cheap rate. Portsmouth has one of the best harbours in the United States. In the main channel it has 40 feet depth of water at low tide, and 30 feet at the wharves. The tide here rises 10 feet, and sweeps through the harbour with a powerful current, which prevents it from being obstructed by ice. The harbour is completely land-locked, and protected from winds by a number of islands, the most considerable of which is Great island, comprising the township of Castleton. This island contains 450 acres, and on the N.E. point of it is a lighthouse, and on the N.W. point is fort Constitution. The main entrance to the harbour passes on the E. side of this island, and opposite in Kittery, is fort McClary. Fort Sullivan and fort Washington, on two other islands, are not garrisoned in time of peace. There is another entrance on the S. side of Great island, called Little harbour; but the water is shallow. The harbour is already well defended, and is capable of being made impregnable. Portsmouth contains much wealth, and has considerable trade, though its back country is somewhat limited. The trade of the N. part of the state flows naturally to Portland, and much of the central and eastern part goes through Merrimac river and the Middlesex canal to Boston. This will furnish a reason why the population has diminished for the last 10 years. Its unobstructed and excellent harbour, however, gives it great advantages. The tonnage of the port in 1840, was 27,375. There are 35 ships belonging to the port, employed in the freighting business; 100 vessels in the cod and mackerel fisheries; 52 in the coasting trade; three in the West India trade; and one in the whaling business. The railroad from Boston to Portland, passes through the place.

According to the United States census of 1840, there were in the place 18 commercial and six commission houses in foreign trade, with a capital of $1,251,500; 137 retail stores, with a capital of $278,500; six lumber-yards, with a capital of $11,000; the fisheries employed a capital of $10,600; four furnaces had a capital of $46,000; one woollen factory employed a capital of $30,000; five tanneries had a capital of $15,700; two distilleries a capital of $5000; two flouring-mills, two grist-mills, and one saw-mill, employed a capital of $15,000; one rope-walk, a capital of $5000; three printing offices, three binderies, and two weekly newspapers, employed 18 persons, and a capital of $9100. The total amount of capital employed in manufactures was $187,101. There were three academies, 188 students; 16 schools, 2222 scholars.

Portsmouth was first settled in 1623, under the auspices

of Gorges and Mason, and was incorporated in 1633. Secured on three sides by water, in early times a stockade fence was constructed across the isthmus, as a defence against the savages; the population was easily considerable, and it was therefore exposed to less annoyance than many other places. The first church (an Episcopal), was erected about 1638. The first Congregational church was erected in 1671. The second Congregational church was erected in 1715. The first Baptist church was founded in 1826. The first Methodist church was founded in 1809. The first Universalist society was constituted in 1816. The year 1802, 102 buildings were burned; in December 1813, 14 buildings, including St. John's Episcopal church, were destroyed. In December 1813 a more desolating fire over 15 acres, destroying 397 buildings, among which were over 100 dwelling houses. The parts burned have been rebuilt of brick, with additional convenience, and increased beauty.

PORTSMOUTH, p. t., Newport co., R. I., 7 m. N.W. Newport, 24 m. S.E. Providence, 417 W. It occupies the N. part of the island of Rhode island, is surrounded on three sides by Narragansett bay, and enjoys great facilities for navigation. Many of its inhabitants are employed in the fisheries. It contains mineral coal, and has four stores, one fulling-mill, one woollen factory, five grist-mills. Pop. 1706. It is connected with Tiverton by a bridge. Several islands are connected with it, the largest of which is Prudence, 6 m. long, and three fourths of a mile wide.

PORTSMOUTH, p. v., port of entry with Norfolk, and capital of Norfolk co., Va., 105 m. E.S.E. Richmond, 223 W. Situated at the mouth of Elizabeth river, on the W. side, opposite to Norfolk, 1 m. distant, and has one of the best harbours in the United States. The largest vessels come to its wharves. It contains a courthouse, jail, five churches, a Presbyterian, Episcopal, Methodist, Baptist, and Roman Catholic, a branch of the bank of Virginia, a masonic hall, a theatre, a military academy, 50 stores, and 6500 inhabitants. A United States navy yard, dry dock, and naval hospital, are located at Gosport, a suburb of Portsmouth. By means of the Portsmouth and Roanoke railroad, and the Wilmington railroad, a daily line of communication is formed with steamboats plying to Charleston.

PORTSMOUTH, p. v., capital of Scioto co., O., 90 m. S. Columbus, 424 W. Situated on the E. bank of Scioto river, at its entrance into the Ohio river, and at the termination of the Ohio canal. It contains a courthouse, jail, a market house, four churches, 30 stores, seven warehouses, 300 dwellings, and about 1500 inhabitants. It contains a number of mills and manufactories, and occupies an important position. Iron ore, coal and building stone abound in the vicinity.

PORT TOBACCO, p. v., capital of Charles co., Md., 30 m. S.W. Annapolis, 32 W. Situated at the head of a bay which sets up from Potomac river, and contains a courthouse, jail, a church, several stores, 120 dwellings, and about 600 inhabitants.

PORTUGAL (KINGDOM OF), see Lusitania. Great W. state of Europe, occupying the greater part of the W. portion of the Spanish peninsula, between the 36th and 42d degs. N. lat., and the 6th and 10th W. long., having E. and N. Spain, and S. and W. the Atlantic. Length, 350 m.; average breadth, rather more than 100 m. Area, pop., subdivisions, &c., as follow:—

Provinces.	Area in sq. m.	Pop. 1826.	Pop. to sq. m.	Chief Cities.
Minho	2,671	874,400	335·5	Oporto
Tras-os-Montes .	4,065	331,900	81·5	Villa Real.
Beira, Upper . .	8,626	998,350	120·7	Coimbra.
" Lower . .		100,420		
Estremadura . .	10,945	700,700	89·5	Lisbon.
Alentejo	10,945	314,210	28·6	Evora.
Algarve	2,089	135,380	64·4	Faro.
Total . . .	38,510	3,445,630	89·5	

Geographically considered, Portugal can be regarded in no other light than as a dependency on, or portion of Spain; and, in fact, all the mountain chains and great rivers by which she is traversed originate in the eastern and more extensive portion of the peninsula. The principal mountain chain, the Sierra de Estrella, runs S.W. and N.E. from the Spanish frontier, near Almaida, to cape Roca, near Lisbon, the most westerly land in Europe, lat. 38° 47′ N., long. 9° 30′ 24″ W. The culminating point of this chain, near Covilha, is 7524 feet above the sea. Another chain, the Sierra Monchique, runs across the province of Algarve, the most southerly in the kingdom, terminating in cape St. Vincent. There are a great number of inferior chains, and the provinces to the N. of the Douro are especially encumbered with mountains.

The great rivers, the Tagus, Douro, Minho, and Guadiana, have their sources in Spain, though they are joined by some considerable affluents in their passage through

Portugal. (*See* the names.) There are but few lakes, and these of no importance; but mineral and hot springs are not uncommon. Water, in many districts, particularly in the S. is both scarce and bad; and, in consequence, extensive tracts in the great plain of Alentejo and other provinces are nearly uninhabitable. The climate is, in general, milder and more agreeable than in Spain, owing to the height of the mountains, and the great extent of coast. In the rugged tracts of the N.E. (Tras-os-Montes) the air is in many parts keen. In the valleys, and in the S. part of the kingdom, the case is generally very different; but all along the coast, the heat is tempered by the sea breeze. Snow seldom lies on the low ground; but the rains of winter are often heavy and long continued; and at this season the vicinity of Lisbon and other parts of the country are very subject to earthquakes. Violent hurricanes are also of frequent occurrence.

The general aspect of Portugal is similar to that of Spain, and even more luxuriant.

> It is a goodly sight to see
> What heaven hath done for this delicious land!
> What fruits of fragrance blush on every tree!
> What goodly prospects o'er the hills expand!
> *Childe Harold*, canto i., st. xv.

The *Vegetable Products* are very various, as well from difference of latitude, as from the great variety of elevation. Wheat, barley, oats, flax, hemp, and other products of a northern latitude, are raised in the high grounds; vines and maize in those of warmer temperature; and rice in the low grounds. The chief fruits are olives, oranges, and lemons; but the last two grow only in the warm and sheltered valleys of the S. and central parts of the kingdom. The woods are extensive: in the N. they consist principally of oak; in the central provinces of chestnut, and in the S. of the pine, kermes, and cork trees. Algarve produces the American aloe, date, and other intertropical products: and Portugal is supposed to have a greater number of indigenous plants than any other part of Europe. (*Balbi, Essai Statistique*, i., 143.) Silk is produced of very good quality; and, in general, any deficiency, whether in the vegetable or animal products, is to be imputed not to the soil or climate, but to the indolence and unskilfulness of the people.

The *Agriculture* of Portugal, though recently it has begun to improve, is still, speaking generally, in the most backward and degraded state imaginable. A variety of circumstances have conspired to bring about this result; among which, the heat of the climate and the want of water, especially in the southern provinces, have, no doubt, a very considerable influence. Probably, however, the mildness of the climate has been still more injurious than its aridity, for this has at once encouraged the indolence, and lessened the wants of the people. And if to these powerful physical causes we add the pernicious practice of exempting the clergy and nobility from those direct taxes which were made to fall with their full weight on the cultivators; the vast amount of property in mortmain, and prevented from coming into the hands of those who would turn it to the best account; the want of a proper method of keeping, and the consequent insecurity of the occupiers; the want of a manufacturing population, and of great towns, that is, of markets for agricultural produce; the extreme badness of the roads, and the difficulties in the way of internal communication; the number of Saints' days, fasts, and other superstitious observances; and the ignorance of the people; we shall certainly have little cause to wonder at the low state of agriculture in most parts of the country.

In the greater portion of the kingdom the farmers are quite unacquainted with the rotation of crops, and, one would be almost disposed to conclude, of the differences of soil, inasmuch as they continue to raise the same crops indiscriminately from all sorts of land. Their implements are of the clumsiest and rudest description; the harrow and the hoe were, till lately, nearly unknown, and threshing was usually performed by trampling the grain under the feet of horses and cattle. Though, in so dry a country, the command of water and the irrigation of the lands be indispensable, this, in many extensive districts, is quite neglected. In consequence, the country is in parts but little occupied, and the traveller sometimes proceeds a distance of 15 or 20 m., without discovering as many houses. To show the deficiency of the means of communication, it may be sufficient to state, that on travelling from Abrantes to the Spanish frontier, along the N. side of the Tagus, a distance of about 100 m., there are six rivers to cross without a single bridge, though they are fordable only in dry weather.

It must not, however, be supposed that these statements apply equally to the whole country. The inhabitants of the greater part of the provinces of Entre-Douro-e-Minho and Tras-os-Montes, to the N. of the Douro, and of the adjoining portions of Beira, participate, to a considerable ex-

tent, in the industrious qualities of their neighbours the Galicians. (*See* GALICIA.) An abundant supply of water is here provided, partly from natural streams, but principally from wells dug in the sides of the mountains; and, in consequence, good corn crops are raised in the lower grounds, while the hills are covered with vineyards, and olives and other fruits are also extensively raised.

But, with these exceptions, agriculture in Portugal is, at this moment, probably in a more backward state even than in Spain or any other European country. We incline, however, to think that this will not long be the case. The more intelligent classes have at length become aware of the vicious nature of the institutions which have so long prevented the development of industry; and, of late years, most important changes have been effected in the tenures under which landed property is held, and in its distribution. The feudal rights of the nobility and other landed proprietors have been suppressed; an equal system of direct taxation has been introduced; and a large extent of crown property and of estates belonging to monasteries, sold at low prices, has mostly found its way into the hands of industrious proprietors. Hence, though the want of capital, the ignorance and indolence of the peasantry, be most formidable obstacles to the rapid spread of improvement, it has notwithstanding already made a considerable progress. In proof of this we may mention that, despite the facilities afforded for the importation of corn and other bulky products from the interior into Lisbon, by means of the Tagus, which runs through the centre of the kingdom, that city was long indebted to foreign countries for a considerable portion of her supplies of corn; but this, we are glad to say, is no longer the case; and, in 1839, for the first time for centuries, considerable quantities of Portuguese corn were shipped from the Tagus! Flax, hemp, and potatoes are grown only to a small extent; and, owing to the want of due care and attention, olive oil is of an inferior quality.

Wine, however, is the staple produce of Portugal, and that by which she is best known in this country. The red wine, called port, from its being all shipped from Oporto, is produced in the Upper Douro, about 50 m. above Oporto, on a succession of low hills on both sides the river, having the finest soil and exposure. The produce of this district is generally divided into two sorts of wine, the *vinho de feitoria* or *Factory-wine*, for exportation; and the *vinho do ramo*, an inferior wine for home consumption and distillation. Great complaints having been made, about the middle of last century, by the merchants in England and their agents in Oporto, of the bad quality and adulteration of the wine, the matter came under the notice of the Portuguese government; and the method which it took to redress the evil is singularly illustrative of its sagacity and principles of action. Instead of leaving the matter to be adjusted between the growers of the wine and the merchants, or, at farthest, contenting itself with confiscating such wines as were found to be adulterated, it made over the whole district to a joint stock company, and invested them with almost despotic privileges! Thus, the agents of the company were authorised to class the wines belonging to individuals, and to fix their maximum price; so that the company became, in fact, the sole buyer, at its own price, of the wines produced within the limits of its charter.

But for the rooted taste for the wines of the Upper Douro established in this country through the influence of the long-continued high discriminating duty on French wines, it is probable that the institution of the company would have destroyed the Portuguese wine trade. It has, however, owing to the continued demand for the British markets, continued to keep its ground, or rather to increase, though not nearly to the extent that, under other circumstances, might have been anticipated. We have elsewhere seen (art. OPORTO) that, at an average of the three years ending with 1840, the shipments of port wine from Portugal amounted to 34,790 pipes a year. It is a curious fact, that the Oporto wine company, after being suppressed by Don Pedro, as a nuisance, has been re-established, though with less oppressive privileges. (*Henderson on Wines*, 196-214.) In addition to port, considerable quantities of Lisbon, Calcavella, and other white wines, are exported. Some red wine is also exported from Figueiras.

The pastures in Minho, and in the Sierra Estrella and some other parts, are excellent. But throughout most part of the kingdom they are very much neglected. In consequence of the great number of fast days enjoined by the church, few cattle were reared; and a large proportion of those required for the consumption of the principal towns were imported from Galicia and other adjacent Spanish provinces. Horses are scarce; oxen being commonly used for draught, except in towns. Mules, however, are numerous, and of an excellent breed; and, with asses, are generally preferred, on account of the rugged nature of the country, to horses for travelling. Sheep breeding is principally conducted in Beira whence large flocks are sent to winter in Alemtejo. The wool of the Portuguese sheep might, no doubt, with a little attention, be rendered equal to that of the Spanish sheep; but no pains have been taken to improve its quality. Goats and hogs are numerous; and the latter are of a superior kind, and yield excellent hams. The fisheries, which were formerly important, are now insignificant: except in Algarve, where tunny, pilchards, &c., are taken, Galicia and England supply the greater part of the fish made use of.

The *mineral* products are considerable, though few mines except those of iron have been wrought, in consequence partly of the scarcity of fuel, and partly of the supply of minerals (chiefly copper and lead) from Brazil. The mountains abound in fine marble, and contain traces of gold and silver. Large quantities of salt of a very superior quality are produced in bays along the coast by natural evaporation, especially at Setubal or St. Ubes, whence it is largely exported.

Gold dust is obtained by washings; and in antiquity the Tagus was famous for its golden sands: *Tagus auriferis arenis celebratur.* (*Plin. Hist. Nat.*, lib. iv., cap. 22.) Coal is found near Oporto and elsewhere; and lead, antimony, &c., are raised, but not to any considerable amount.

Manufactures and Trades.—An Englishman can with difficulty form an idea of the backward state of manufactures in Portugal: they are in general carried on in separate cottages, like the coarse woollens of North Wales, or the linen of Normandy, and are found on the primitive plan of every family manufacturing for its own consumption. Manufacturing establishments are but few in number, and are principally for the production of woollens, silk, and earthenware. Cotton has, also, been attempted of late years; and paper, glass, and gunpowder are made in a few places.

One might be disposed to think from the pompous enumeration of factories made by Balbi, and his vindication of the Portuguese from what he is pleased to call the *reproche banal* of their being without manufactures, that on the contrary, manufacturing industry was in a very flourishing state among them. He even goes so far as to say that the cotton yarn produced at Thomar is at least equal to that of England! It was hardly necessary, perhaps, to notice such ridiculous statements. The cottons, woollens, linens, hardware, and earthenware of England, are all largely imported into Portugal, and are used by all but the very lowest classes. To suppose, indeed, that there should be any thing like really successful manufacturing establishments in a country like Portugal, is to suppose what is all out contradictory and absurd.

"A very superficial knowledge even of some of the commonest arts exists. A carpenter is here the most awkward and clumsy artisan; and the way in which the doors and wood-work belonging to the good houses are finished would have suited the rudest ages. Their carriages of all kinds, more particularly their wagons and carts, their agricultural implements and management, their cutlery, locks, and keys, are ludicrously bad. The chief *forte* of the Portuguese appears to lie in ship-building and stone-masonry; they also excel in embroidery, and make good artificial flowers, and lace." (*Baillie's Lisbon.* i., 17, &c.)

The navigation and commercial intercourse of Portugal are of more importance; and though even in the times of Emanuel and Albuquerque, they were by no means so extensive as is generally supposed, they were, notwithstanding, very considerable, and appeared immense from the small amount of the shipping and trade of other nations. For a long time past, the import and export trade of the country has been conducted chiefly by foreign merchants, particularly British, settled in Lisbon and Oporto. The exports consist almost entirely of raw produce, or of wine, oil, salt, wool, fruits, cork, &c. The imports have hitherto included corn and flour, but, as already seen, these will now probably be dispensed with; the other articles are cotton goods (by far the most important article), hardware, woollens, fish, linens, earthenware, drugs, tea, coal, &c. We subjoin an

Account of the Quantities of the Principal Articles of Foreign and Colonial Merchandise imported into the United Kingdom from Portugal, in 1838.

Articles.	Quantities.	Articles.	Quantities.
Figs . . cwts.	1,702	Brandy . . galls	2,537
Lemons and oranges . chests	98,292	Sheep's wool . lbs.	627,522
Olive oil . galls.	407,048	Wine (Portuguese)	
Shumac . cwts	1,760	. galls	3,055,543
Goat-skins (undressed No.	23,580	Cork (unmanufactured,) . cwts.	47,304

Nearly all the foreign trade centres in Lisbon and Oporto, which see. It is principally carried on with England and Brazil. At an average of the six years, ending with 1838, the real value of the British produce and manufactures exported to Portugal amounted to £1,949,114 a year; but of

this a considerable portion was subsequently re-exported for, or rather smuggled into Spain. Perhaps, at an average, the value of the imports and exports of Portugal may each amount to about two millions a year. Formerly, Lisbon had about 400 large ships employed in the trade with S. America, but now she has not more than 50 or 60 ships in all departments of her foreign trade, and those, too, of comparatively small burden.

The internal traffic of such a country is, as may be supposed, inconsiderable. There is no navigable canal and till of late years, not a single road in Portugal was practicable for carriages for more than 20 or 30 m. from Lisbon. In fact the only mode of travelling by land from Lisbon to Oporto is in a litter, or on the back of a mule or horse; and in the wine country of the Douro, or in the province of Minho, two oxen sometimes take a whole day to convey a pipe of wine 5 or 6 m.; and to prevent the cart from being overturned, it is attended by two men. Accounts are kept in reis and milreis; the milrea contains 1000 reis, and is worth about 4s. 6d. The *dobros*, or *doubloons* = £3 6s. 6d.; the *crusado* = about 2s. 6d. The lb. = 459 kilogr., or rather more than the lb. avoird.; the arroba = 32 lbs.; the quintal = 4 arrobas. The *mayo* for grain &c., = 24 bushels; the almudo = 4½ gallons. The Portuguese foot is a little longer than the English.

Constitution.—Like the peninsula kingdoms of Castile and Arragon, Portugal had anciently her Cortes or assemblies of the states. One of these assemblies, held at Lamego in 1141, conferred the title of king on Alfonso Henriquez, who had two years previously defeated the Moors in the great battle of Ourique. The Cortes at the same time enacted a law for regulating the succession to the throne; in which, among other things, it is laid down that females shall not be eligible to the crown, though in the direct order of succession, if they have married a foreigner, and that their marrying a foreigner when on the throne shall be considered equivalent to an act of abdication. The powers of the Cortes corresponded, in other respects, with those of similar assemblies in other countries; but their privileges and those of the sovereign were very ill defined: and the latter contrived, in the course of time, to engross all the powers of the state; the government of Portugal became, in all respects, as despotical as that of Spain: and the last convocation of the Cortes took place in 1697. (See *L'Art de verifier les Dates*, part ii., tom. vii., 1–40. 8vo. ed.

From this period down to the administration of the Marquis de Pombal (1750–1776), every abuse continued to multiply, and Portugal was distinguished only by the imbecility of her government, the power and profligacy of her nobility and clergy, and the poverty and indolence of her people. The Marquis de Pombal suppressed the order of the Jesuits and confiscated their estates; he also suppressed some of the more oppressive privileges of the nobles and clergy; and effected various important reforms in several departments of the administration. In other respects, however, his policy evinced the narrowest and most illiberal views; and on his dismissal from power, most part of the old abuses in the government revived, and the country continued in its former state of apathy and abasement.

The events connected with the late war in the peninsula, the emigration of the court to Brazil, the long continuance of the English armies in the country, the organisation of the Portuguese army on an improved footing, and the influence of the change in Spain, laid the foundations of a new order of things. The nation was dissatisfied with the continued residence of the court in Brazil, which, in fact, made Portugal a dependency of the latter, and the wish for some fundamental changes in the frame of the government became general. At length, in August 1820, a revolution broke out, and a free constitution was soon after established, having for its basis the abolition of privileges, the legal equality of all classes, the freedom of the press, and the formation of a representative body in one chamber. Our limits preclude our attempting to trace the obscure history of the Portuguese government from this period. Suffice it to say, that the constitution of 1820, which was violently opposed by the clergy and privileged classes, having been suppressed, in 1823, was replaced in 1826, by a constitutional charter, granted by Dom Pedro; and that, in 1836, the latter was suppressed, and the existing constitution established in its stead.

Under this constitution, the government is an hereditary monarchy, with an upper and a lower representative chamber, both of which are elective, the franchise being vested in the holders of a certain small amount of fixed property. The Cortes meet and dissolve at specified periods, subject the intervention of the sovereign, and the latter has no veto on a law passed twice by both houses. At present, therefore, the constitution of Portugal is decidedly liberal, probably too much so for a country in its peculiar situation. Each province has a governor, to whom the details of its government are entrusted. Justice is administered in the first instance by the Juizes de Fora: and appeals are made to the corregidors of the provs., and from these to the Casa de supplicaçaos at Lisbon, and the Relaçaos do porto at Oporto. All these tribunals are, however, subordinate to the royal court in Lisbon. Great abuses are said to exist in almost every department, both in the judicial and administrative branches, the inadequacy of the salaries leading to the acceptance of bribes. Assassination is more frequent here than it has ever been in Italy; the law and the police being impotent alike either to secure property or life. It is stated in an official return published by the Cortes, that in 1837 there were no fewer than 176 assassinations, and 296 robberies in the district of Oporto; and 166 assassinations, and 234 robberies, in that of Guarda! The country, in fact, teems with ruffians and crime.

The religion of Portugal is the Roman Catholic, unalloyed with any taint of Protestantism, and contributing but little, if anything, to the morality of the people. The inquisition no longer interferes with the freedom of conscience, having been abolished in 1821. The Portuguese church is under the jurisdiction of a patriarch, with extensive powers; three archbishops, and 14 bishops.

Notwithstanding the hostility of the Marquis de Pombal to the monks, who used to say of them that they were la vermine la plus dangereuse qui puisse ronger un état, their numbers previous to the late revolution were estimated at about 8000, and the revenue of the conventual establishments was supposed to exceed £1,200,000 a year! Luckily, however, an end has been put to this preposterous state of things. the monks have been turned out of their establishments, to support themselves as they best may, on a small stipend that has been allowed them, so that the public wealth is no longer wasted in maintaining hordes of idle, profligate drones. The nunneries will, it is most probable, share the same fate.

It is said that infidelity is now very prevalent among the more intelligent classes in Portugal; and it is natural that such should be the case. They have never seen religion except in association with all that was most revolting. In Portugal its ministrations have consisted only in idle and unmeaning ceremonies; and its priests have uniformly been the enemies of every thing like popular rights, and the supporters of the most oppressive and offensive privileges. But now that they are reduced to their proper sphere, and are no longer used as instruments for the debasement of the people, it is probable that juster religious views will begin to be entertained; and that the profligacy and corruption of the clergy will cease to react upon religion itself.

The language of Portugal is merely a dialect of the Spanish, differing but little more from the latter than Scotch from English. Education is, at once, little diffused and of bad quality. There is a university at Coimbra; besides which, 17 high seminaries, and numerous schools exist, affording instruction to about 33,000 pupils. With the exception of Camoens, few Portuguese authors are known beyond the limits of their country. The army consists of about 28,000 men; 21,500 infantry, 3700 cavalry, and 2400 artillerymen. The forces of Portugal, whether naval or military, have in general been very inefficient. A partial stimulus was given, in 1760, to the Portuguese army, by a German commander, the Count de la Lippe; but after his death his plans were not followed up; and it was not till 1809, that Portuguese troops, recruited by British funds, and disciplined by British officers, became worthy the ancient renown of their country. It remains, however, to be seen whether if left to their own resources, and without the example and assistance of foreigners, they would preserve the laurels they gained during the late war. The navy comprises only two ships of the line, four frigates, six brigs, and some smaller vessels.

The public revenue and expenditure for the year ending with June, 1837, were as follows:

Revenues.		Expenditure.	
	Milreas.		Milreas.
Rent of gov. property .	676 096	Royal family, Civil-	
Direct taxes .	1,670,439	list, &c .	4,786,727
Indirect do. .	4,244,140	Ministry of interior .	1,094,727
Various sources .	1,865,630	" justice, &c.	972,362
	143,986	" finance	663,966
Insular, &c. rev. .	1,071,313	" war.	3,191,751
		" marine	1,062,463
		" foreign	
		affairs .	945,694
		Insular, &c. expend. .	1,073,235
Total. .	9,491,986		13,077,651
Deficit			3,585,792

The public debt, in 1834, amounted to 5,689,229 milreas. The Portuguese are but little indebted to the accounts given of them by travellers. But their character, as drawn by Dca Chatelet, though not very flattering, is, we believe, nearly correct. "Il est, je pense, peu de peuple plus laid que celui de Portugal. Il est petit, basané, mal conformé. L'intérieur répond, en général, assez à cette repoussante enve-

loppe, sur-tout, à Lisbonne; où les hommes paroissent réunir tous les vices de l'âme et du corps. Il y a, au reste, entre la capitale et le nord de ce royaume, une différence marquée sous ces deux rapports. Dans les provinces septentrionales, les hommes sont moins noirs et moins laids, plus francs, plus llans dans la société, biens plus braves, et plus laborieux; mais encore plus asservis, s'il est possible, aux préjugés. Cette différence existe également pour les femmes; elles sont beaucoup plus blanches que celles du sud.

"Les Portugais, considérés en général, sont vindicatifs, bas, vain, railleurs, presomptueux à l'excès, jaloux, et ignorans. Après avoir retracé les défauts que j'ai cru appercevoir en eux, je serois injuste si je me taisois sur leurs bonnes qualités. Ils sont attachés à leur patrie, amis genereux, fidèles, sobres, charitables. Ils seroient bons chrétiens, si le fanatisme ne les aveugloit pas. Ils sont si accoutumés aux pratiques de la religion, qu'ils sont plus superstitieux que dévots. Les hidalgos, ou les grands de Portugal, sont très bornés dans leur éducation; orgueilleux et insolens; vivant dans la plus grande ignorance, ils ne sortent presque jamais de leur pays pour aller voir les autres peuples." (Voyage en Portugal, i. 69–71.)

The Spaniards and Portuguese regard each other with a deep-rooted national antipathy.

"Well doth the Spanish hind the difference know
Tw'xt him and Lusian slave, the lowest of the low."

"Strip a Spaniard of all his virtues, and you make a good Portuguese of him," says the Spanish proverb. "I have heard it more truly said," says Dr. Southey, "add hypocrisy to a Spaniard's vices, and you have the Portuguese character." The two nations differ, perhaps purposely, in many of their habits. Almost, every man in Spain smokes, the Portuguese never smoke, but most of them take snuff. None of the Spaniards will use a wheelbarrow: none of the Portuguese will carry a burden: the one says, 'it is fit only for beasts to draw carriages;' the other, that 'it is fit only for beasts to carry burdens.'" (Southey's Letters, ii. 64.) In one respect, however, their tastes are identical, bull-fights being quite as popular among the Portuguese as among the Spaniards. Semple's statements, as to the Portuguese character, coincide with those of Du Chatelet. "The Portuguese are generally dark-complexioned and thin, with black hair, irascible and revengeful in their tempers, and eager in their gestures on trivial occasions. They are also said to be indolent, deceitful and cowardly; but they are temperate in diet, and that may be classed at the head of their virtues, if, indeed, they have many more. They have no public spirit, and, consequently, no national character, An Englishman, or a Frenchman, may be distinguished in foreign countries by an air and manners peculiar to his nation; but any meagre swarthy man may pass for a Portuguese." All classes seem to despise cleanliness; and Lisbon, and the Portuguese towns generally, are, certainly, entitled to the not very enviable distinction, of being about the filthiest in Europe. The morals of both sexes are lax in the extreme; and, as already stated, assassination is a common offence. On the whole, we incline to think, that, owing to vicious institutions, the Portuguese rank about as low in the social scale as any people of Christendom. But the fair presumption is, that, under the beneficial influence of the new constitutional arrangements, the abuses that have depressed and degraded the nation will be extirpated; and that the Portuguese will once more recover their ancient place among European nations.

History.—This country, anciently called *Lusitania*, was taken possession of by the Romans, about anno 200 B.C.; previously to which some Phœnician and Carthagenian colonies are supposed to have been planted on its shores. It remained a Roman province till the 5th century, when it was invaded by the Suevi, Visigoths, &c.

The Moors landed in the S. of Spain in the early part of the 8th century, and meeting with little resistance from its thinly spread population, easily overran the greater part of Portugal; but the nature of the country favouring the operations of the inhabitants, they were not long in recovering possession of its more northern and mountainous portion. The name of Lusitania seems to have been exchanged about this period for that of *Portucale*, subsequently changed into Portugal, from the circumstance of Oporto, the principal of the strong-hold of the Christians, being then called *Calls* or *Porto Calls*. (D'Anville, *Etats formés en Europe*, p. 192.) In the 11th century Portugal became an earldom, under the kingdom of Leon and Castile; and during the 12th it was erected into an independent kingdom. Its power now rapidly increased; and by the acquisition of Algarve, in 1249, it arrived at its present limits. In the latter half of the 14th century the voyages and discoveries commenced which have shed immortal lustre on the Portuguese name. During the 15th century, Madeira, the Canaries, and Azores were discovered, and colonized; and in 1498, Vasco de Gama doubled the Cape of Good Hope, and opened a new route to India.

In the following century the Portuguese explored the coasts of Newfoundland and America; took possession of Brazil; made important acquisitions in India and the Persian gulf, and discovered the Moluccas; by which successful enterprises they monopolized the commerce of the east, and a great share of that of the west. But the prosperity of Portugal was short-lived. After the disastrous defeat and death of king Sebastian, in Africa, in 1578, Philip II. of Spain seized on the kingdom, which remained a Spanish prov. from 1580 to 1640; and when she regained her independence; the greater part of her commerce, and her foreign possessions, were in the hands of the Dutch.

But, notwithstanding the emancipation of Brazil, Portugal still preserves the Azores, Madeira, Cape de Verd, and Guinea islands; the settlements of Angola and Mozambique, in Africa, and those of Goa, Diu (Timor), Macao, &c., in Asia. In 1807, Portugal was invaded by the French, when the royal family removed to Brazil. John VI. dying in 1826, Dom Miguel usurped the throne in 1827, which he held till 1833; when, after a lengthened contest, Donna Maria II. was established in its possession. (*Balbi, Essai Statistique sur le Royaume de Portugal, &c.; Du Chatelet, Voyage en Portugal; Baillie's Lisbon; Semple, Southey, Murphy, &c.*, in *Mod. Trav.* xix.; *Journal Statistique*, 1838; *Ursulla's Geog.; Convers. Lexicon. &c.*)

POSEN, a prov. of the Prussian monarchy, comprising the portion of Poland assigned to Prussia by the treaty of Vienna in 1815, having N. the prov. of Prussia and Brandenburg, E. Poland, and S. and W. Silesia and Brandenburg. It is of a triangular shape. Area, 11,374 sq. m. Pop., in 1827, 1,158,608, of whom 739,527 were Catholics, 344,853 Protestants, and 74,194 Jews. Principal towns, Posen, Bromberg, Nakel, &c. It is divided into two regencies, and these again into six circles. Surface generally flat, and in part occupied by extensive marshes and forests. Principal rivers, Warta, Netze, and Obra. Soil various, but generally clay and black loam intermixed with sand, and naturally very fertile. Principal products, corn, timber, wool, honey, &c. Minerals and manufactures unimportant. A vast number of leeches are taken in this country, especially in the circle of Bomster; above 1,300,000, worth, on the spot, about 65,000 rix-dollars, were exported in 1840; and, in 1839, the exports were still more extensive. This is the most backward of the Prussian provinces. When it first came into the possession of Prussia, in 1792, the great bulk of its inhabitants were in a state of predial slavery, and were as ignorant and brutalized as can well be imagined. The vigorous and enlightened government of Prussia at once put down the excesses of the nobles, and has exerted itself by introducing an improved judicial system, establishing schools and otherwise, to improve the habits and condition of the people. These efforts, combined with the total abolition of servitude (*see* Prussia). have had the best effects; though a lengthened period will still have to elapse before the vices and habits engendered by centuries of slavery and degradation be completely eradicated, and the population become as intelligent and industrious as in the more advanced provinces.

POSEN, a city of Prussia, cap. prov. and reg. of the same name, at the confluence of the Prosna with the Warta, 147 E. by S. Berlin; lat. 52° 29' N., long. 16° 53' E. Pop., in 1826, 32,456. Since the peace of 1815 its fortifications have been vastly improved, and it is now one of the bulwarks of the kingdom on the side of Russia. Though an old town it is pretty well built. Principal edifices, a cathedral and townhouse. It is the residence of the provincial authorities. and of an archbishop; and has a court of appeal, a gymnasium, or college, a theological seminary for the education of Catholics, a school of arts, &c. The business of watch-making, is carried on to some extent; and there are manufactures of leather, linen, fire-arms, &c. It has three great annual fairs. Here, as in the rest of Poland, the buying and selling of goods is chiefly managed by Jews, who occupy a particular quarter of the town.

POSEY, county, In. Situated in the S.W. corner of the state, and contains 450 sq. m. Bounded S. by the Ohio river, W. by Wabash river. Drained by Big and Flat creeks. It contained in 1840, 9973 neat cattle. 9521 sheep, 35,818 swine; and produced 55,103 bushels of wheat, 1821 of rye, 607,799 of Indian corn, 1509 of barley, 66,157 of oats, 7841 of potatoes. It had 28 stores, seven flouring-mills, 10 grist-mills, nine saw-mills, four tanneries, three distilleries, two printing-offices, two weekly newspapers; 10 schools, 300 scholars. Pop. 9683. Capital, Mount Vernon.

POTENZA, (an. *Potentia*), a city of the Neapolitan dom., prov. Basilicata, of which it is the capital, on a hill in a wild and rocky tract, near the source of the Basiento, 54 m. E. Salerno. Pop., circa 9,000. It is fortified, and has a cathedral. several other churches, and convents, a seminary, a royal college, lyceum, hospital, government pawn-bank, c.; but, speaking generally, it is poor and meanly built. It is a bishop's see, and the seat of the superior judicial courts for the province. It suffered greatly from earthquakes,

especially in 1694, and 1812. (*Rampol., Corografia; Dict. Geog.*)

POTOMAC, r., Md. and Va. rises by two main branches, the N. and S. branches near the Alleghany mountains, and forms, by the N. branch, and through the greater part of its whole course, the boundary between the states of Virginia and Maryland. It is 550 miles long, and at its mouth 10 miles wide. At Alexandria, it is a mile and a quarter wide, 290 miles from the ocean. It is navigable for ships of the line, 300 miles, to the navy-yard at Washington. The greatest tributary is the Shenandoah, 200 miles long, that carries nearly as much water as the main stream which it enters from Virginia, on the S. side, shortly before its passage through the Blue ridge. This passage forms a grand and picturesque scenery, to view which Mr. Jefferson pronounces worth a voyage across the Atlantic. Chesapeake and Ohio canal is constructed along the river as far as Harper's Ferry. The principal The the E. side of the river are Georgetown, Washington and Tobacco, and Leonard Town; and on the W. side Harper's Ferry, and Alexandria. The Potomac is 42 ft. deep at its mouth, and 18 ft. deep at Alexandria.

POTOSI (SAN LUIS DE), a city of Mexico, cap. of the state of same name, near the source of the river Tampico, 165 m. W. Tampico, and 75 m. N.N.E. Guanaxuato; lat. 22° N., long. 103° 1' W. Population of the city itself about 15,000 (*Poinsett*), but including the *barrios*, or suburbs, which cover a large extent of ground, it may amount to from 50,000 to 60,000 (*Ward*). "It presents a fine appearance: the churches are lofty, and some of them very handsome. The streets are well built. very clean, and intersect each other at right angles. The houses in the square, and in the principal avenues leading to it, are of stone, and two stories high; those in the suburbs are low, and of adobe (sun dried bricks). The government house, in the square, is not yet completed; but the front, which is of cut stone, and ornamented with Ionic pilasters, would do credit to any city in Europe. The market-place is well supplied with meat, fruits and vegetables. Pedlars hawk up and down the coarse manufactures of the country. Stalls are erected and set out with *mantas*, blankets, leather breeches and leggings, saddles, bridles, huge wooden stirrups, iron spurs, weighing at least 2 lbs., and a great variety of manufactures from the filaments of the Agave; ropes, cord, twine, and thread, matting, bags, saddle-cloths, &c., &c. Here, as in every part of Mexico, the vendors were satisfied with one half of their asking price, and frequently with one third part of what they, the instant before, had sworn on their consciences the article was worth." (*Poinsett's Notes on Mexico*, pp. 243-246.)

From its situation, this city is the natural *depôt* of the trade of Tampico with the N. and W. Mexican states. In foreign trade it is at present almost wholly in the hands of natives of Old Spain or of the United States. The European imports consist principally of French brandies, wines, silks, and cloths, English hardware and printed cotton goods, with some mantas or ordinary cotton manufactures from the United States. In addition to its foreign trade, San Luis supplies the neighbouring states of Leon and Cohahuila with home made goods of various kinds. The town abounds in tailors, hatters, leather dressers, and smiths; and the whole population seems industrious. (*Ward's Mexico.* ii. 227.) The people are better dressed, and there are fewer beggars here, than in almost any other part of Mexico. The mines in the neighbourhood have long ceased to be wrought, from exhaustion of the ores: they were however, formerly very productive. A college, founded by voluntary subscription, and in a flourishing state. affords gratuitous instruction to poor students in Latin, jurisprudence, theology, and constitutional rights. The city was founded in 1586. (*Poinsett's Notes on Mexico*, p. 242-246; *Ward's Mexico*, ii., 226-230.)

Potosi, a city of Upper Peru or Bolivia, famous for its rich silver mines, on the N. declivity of the Cerro di Potosi, a mountain belonging to the Andes, lat. 19° 35' S., long. 67° 21' 45" W.; 54 m. S.W. Chuquisaca. Early in the 17th century this city is said to have had 150,000 inhabitants; but it is now almost deserted. It is built on uneven ground, and has a spacious square in the centre. The government house, the town house, and the jail, under the same roof, occupy one side; the treasury and government offices another; a convent, and an unfinished church the third; and private houses the fourth. Extensive suburbs, once tenanted by Indians and miners, are now without an inhabitant. and the vestiges of the streets are all that remain. Among the most remarkable public edifices is the mint, substantially built of stone, in 1731, at a cost of 1,148,000 dollars. In the principal square an obelisk 60 ft. high was erected in honour of Bolivar, in 1825. The houses of Potosi, generally, are of stone or brick, and of only one story, with wooden balconies, but without chimneys. The country round is perfectly barren, and the climate disagreeable; the rays of the sun are scorching at noon, while at night the air is piercingly cold. The market is well supplied; though, from many articles having to be brought from a considerable distance, the necessaries, as well as the luxuries of life, are very dear.

The Cerro di Potosi, which is 18 miles in circuit, and rises to the height of 16,037 ft., is supposed to be a solid mass either of the ores or the matrix of the precious metals, of which it has produced a vast quantity. Viewed from the city, it appears dyed all over with numerous tints, green, orange, yellow, grey, and rose colour. The discovery of its wealth was made by an Indian, who, in hunting some goats, slipped, and, to save himself, took hold of a shrub, which, in coming away from the ground, laid bare the silver at its root. The mines were first wrought systematically in 1545, from which time till 1803 they are said to have produced 1,095,500,000 piastres, or £237,358,334, worth of silver on which duty was paid; and during the same period they also produced a large quantity of gold; at the same time that great quantities of both metals were smuggled, or put into circulation without payment of the duty. About 5000 openings are said to have been made in the mountain; but the number of mines wrought during the present century has rarely exceeded 100. At one time, the mines yielded about 30,000 ducats a day; and for a lengthened period they produced about 9,000,000 dollars a year. But they had begun to decline long previously to the revolution; and since then they have been, whether from their exhaustion, defects in the mode of working, or the want of capital, nearly unproductive. The ore is pulverised in water-mills, worked with overshot wheels, at from 1 to 10 miles from the city; but according to Helms, both the mining and reduction of the ore were conducted in the most bungling manner. (*Andrews*; *Müller*; *Mod. Trav.*, xxviii.; *Geog. Journal*, v.; *Encycl. Americana*, &c.)

POTSDAM, a town of the Prussian states, prov. Brandenburg, cap. reg., at the confluence of the Rathe with the Havel, on an island formed by the two rivers, a canal and lakes, 17 m. S.W. Berlin. Pop., in 1837, 25,560. Potsdam has been appropriately termed the Versailles of Prussia. It is a favourite royal residence; streets straight, broad, and well paved; and the houses, though frequently small, and not very commodious within, have, for the most part, splendid fronts. It is encompassed by walls and palisades; has numerous gates and bridges, some of which are highly ornamental, and is divided into three parts, the old and new towns, and Frederickstadt. The most remarkable edifices are, the palace a magnificent structure on the Havel, having attached to it a theatre, menagerie, and spacious stables; the church of the garrison, in which are deposited the remains of Frederick the Great; the church of St. Nicholas; the great military orphan hospital, &c. In the old market place is an obelisk of red Silesian marble, 75 ft. high, on a pedestal of white Italian marble; on the base are inscribed the names of the great elector and his successors. A large garrison is always kept at Potsdam, so that the barracks are very extensive. There is a lyceum, a military school, with various public schools of inferior note, and sundry literary establishments. There are also, exclusive of the military orphan hospital, mentioned above, an infirmary, workhouse, &c. Potsdam was formerly said to be more of a barrack than of a town; but for a good many years past it has been distinguished in various branches of manufacture, such as that of silk, hardware articles, arms, &c. Being situated on a navigable river, communicating by canals, &c. with many large towns, and with the Elbe and the Oder, it has a good deal of commerce.

Potsdam is a very old town, having existed in the 8th century: it did not, however, become a place of any importance till the elector Frederick William selected it for a residence, and began the palace. It was materially improved by the king, Frederick William 1.; but, like Berlin, it owes its principal embellishments to the taste and liberality of Frederick the Great. In its environs is Sans Souci, the favourite residence of that illustrious prince, and the place where he expired, on the 17th of August, 1785. The new palace and the marble palace are also in its vicinity.

POTSDAM, p. t., St. Lawrence co., N. Y., 212 m. N.N.W. Albany, 496 W. Watered by Racket and Grass rivers. It contains six churches, a Presbyterian, Congregational, Episcopal, Methodist and Universalist, a large academy, occupying two buildings, each four stories high, 12 stores, one fulling-mill, one woollen factory, one furnace, four grist-mills, five saw-mills, three tanneries; one academy, 240 students; 36 schools, 1363 scholars. Pop. 4473. The village is situated on the falls of Racket river, and to it most of the above statistics belong.

POTTER, county, Pa. Situated in the N. of the state, and contains 1110 sq. m. Drained by head branches of Alleghany and Genesee rivers, and by Sinnemahoning, Pine and Kettle creeks. Coal and iron ore are found. It contained in 1840, 7315 neat cattle, 4960 sheep, 2230 swine; and produced 11,410 bushels of wheat, 2363 of rye, 6091 of

Indian corn, 3818 of buckwheat, 24,739 of oats, 60,571 of potatoes, 103,199 pounds of sugar. It had six stores, eight grist-mills, 30 saw-mills, one printing-office, one weekly newspaper; 23 schools, 476 scholars. Pop. 3371. Capital, Condersport.

POTTER, p. t., Yates co., N. Y., 201 m. W. Albany, 296 W. Drained by Flint creek. It contains seven stores, two fulling-mills, one furnace, two grist-mills, six saw-mills; 14 schools, 830 scholars. Pop. 2845.

POTTER, t., Centre co., Pa., 10 m. S.E. Bellefonte, 71 m. N.W. Harrisburg. It has two stores, one fulling-mill, one woollen factory, five grist-mills, eight saw-mills, one tannery, one distillery; eight schools, 240 scholars. Pop. 1787.

POTTSTOWN, p. v., Pottsgrove t., Montgomery co., Pa., 37 m. N. Philadelphia, 71 m. E. Harrisburg, 164 W. Situated on the E. side of Manatawny creek, near its entrance into Schuylkill river, and is chiefly built on one broad street containing many brick or stone houses. It contains three stores, one flouring mill, one grist-mill, one saw-mill, one oil-mill, one printing-office, one weekly newspaper; two academies, 45 students, 100 dwellings, and 721 inhabitants.

POTTSVILLE, p. b., Norwegian t., Schuylkill co., Pa., 99 m. N.W. Philadelphia, 68 m. E. Harrisburg, 173 W. Situated at the termination of the Schuylkill canal. It contains two churches, a bank, 33 stores, one furnace, one flouring-mill, two tanneries, three breweries, three printing-offices, three weekly newspapers; one academy, 40 students; eight schools, 170 scholars. Pop. 4345. In 1824 there were but five houses on the spot. The immense coal trade has produced its rapid growth.

POUGHKEEPSIE, p. t., capital of Dutchess co., N. Y., 71 m. S. Albany, 299 W. Bounded W. by Hudson river, E. by Wappinger's creek. Fall creek enters Hudson river at the village by a succession of cascades, with a fall in the whole of over 160 ft. The town contains 54 stores, eight lumber-yards, three cotton factories with 4088 spindles, four flouring-mills, two grist-mills, one saw-mill, two tanneries, one brewery, two potteries, three printing-offices, two weekly newspapers, and two periodicals; two academies, 290 students; 14 schools, 1077 scholars. Pop. 10,006. The village is one of the handsomest and most flourishing in the state, situated about midway between N. York city and Albany, on an elevated plain, 200 ft. above the level of the river, from which it is concealed from the view by the abrupt bank. It is handsomely laid out, and contains 79 streets, the principal of which are paved, with convenient side walks, and compactly built with houses and stores which would do honour to a large city. It contains a court-house, jail, a collegiate school with a fine edifice, 115 ft. long and 35 ft. wide, after the model of the Parthenon at Athens, Dutchess academy, four seminaries for young ladies, three banks with an aggregate capital of $100,000, besides a bank for savings, two whaling companies, with a total capital of $400,000, which employ four ships, two market houses, a lyceum, 14 churches, a Presbyterian, Congregational, two Episcopal, two Methodists, a Primitive Methodist, a Dutch Reformed, Baptist, two Friends, a Universalist, Roman Catholic, and an African, 92 stores and groceries, and a large variety of mechanical and manufacturing establishments. The village is supplied with water from the vicinity brought into a reservoir, and distributed through the streets, at an expense of about $30,000. It contained in 1844, 1251 dwellings and 8158 inhabitants. It has a rich back country, extending to the state of Connecticut. A silk company has been formed, with a capital of $200,000, who have erected a building 100 ft. long, 36 ft. wide, and four stories high. Several steamboats belong to the place, which tow barges to New-York, and the steamboats from the latter place to Albany stop daily at the landing. A number of sloops are owned here. The convention of the state of New-York which adopted the constitution of the United States, assembled at this place in 1788, and previously, the state legislature frequently assembled here.

POULTNEY, t., Rutland co., Vt., 60 m. S.W. Montpelier, 46 m. N. Bennington. Watered by Poultney river. It contains four churches, a Congregational, Methodist, Episcopal and Baptist, a bank, five stores, one fulling-mill, one woollen factory, three furnaces, five saw-mills, three tanneries; one academy, 112 students; 15 schools, 648 scholars. Pop. 1878.

POWHATTAN, county, Va. Situated near the centre of the state, and contains 300 sq. m. Bounded N. by James river, S. by Appomattox river. It contained in 1840, 4759 neat cattle, 6594 sheep, 9160 swine; and produced 53,935 bushels of wheat, 186,810 of Indian corn, 126 273 of oats, 6737 of potatoes, 1,849,750 pounds of tobacco. It had 13 stores, seven flouring-mills, 16 grist-mills, four saw-mills, two tanneries, 10 distilleries; five academies, 125 students; four schools, 94 scholars. Pop. whites, 2432; slaves, 5129; free coloured, 363; total, 7924. Capital, Scottsville.

POWNAL, p. t., Cumberland co., Me., 18 m. N.E. Port-

land, 42 m. S.W. Augusta, 563 W. Incorporated in 1808.
It contains five stores, two saw-mills. Pop. 1210.

POWNAL, p. t., Bennington county. Vt., 126 m. S. by W.
Montpelier, 396 W. Watered by Hoosic river, which affords water-power. Chartered in 1760. It contains a Baptist church, and some Methodists, two stores, one woollen
factory, one cotton factory, two grist-mills, five saw-mills ;
10 schools, 349 scholars. Pop. 1613.

PRAGA, a suburb of WARSAW, which see.

PRAGUE, a city of Bohemia, of which it is the cap.,
near the centre of the kingdom, on the Moldau, by which
it is intersected, 73 m. S.S.E. Dresden, and 152 m. N.E. Vienna ; lat. 50° 5′ 18″ N., long. 14° 25′ 15″ E. Pop., in 1837,
107,338, ex. garrison, inhabs. of the citadel, and strangers ;
whose united numbers raise it to upwards of 190,000. It
stands in a basin, surrounded on all sides by rocks and eminences, upon the slopes of which the buildings rise tier
after tier, as they recede from the water's edge ; and few,
if any, cities of Germany, or, indeed, of any country, have
so grand and imposing an external appearance. It is divided into four quarters, of which two, the Altstadt and
Neustadt, are on the right, and the others, the Kleinseite
and Hradschin, on the left bank of the Moldau. The Altstadt, or most ancient part of the city, stretches along the
margin of the river, and for a considerable distance up the
ascending ground ; it comprises the university and the
archbishop's palace, the municipality, the principal churches and public edifices, the theatre, and all the superior
shops. It is the district of commerce and general activity ;
and is crowded with a dense and active population. Its
streets are generally narrow, dark, and winding ; the principal edifices massive and gloomy ; and the private buildings, usually of stuccoed brick, are black with age and dirt,
and so lofty as to exclude the light from the avenues between them ; but, on the other hand, there is an air of antiquity, and a singularity of architecture about many of the
edifices, public and private, that renders them at once venerable and interesting. The open places are often surrounded by low heavy arcades, beside which are the churches
or public buildings, exhibiting a fantastic mixture of Gothic
and Italian decorations ; while at every turn the eye is met
by some memorial of historical events. Beyond the Altstadt, surrounding it on three sides, and separated from it
only by a large wide street termed the *Graben*, from its
having been formerly the city ditch, is the Neustadt (new
city), founded by the emperor Charles IV., the most splendid of Bavarian monarchs ; the streets of which are much
more open and spacious, and are generally rectangular. Here
are the vast convents, hospitals, and other public buildings,
which owe their magnificence to the Jesuits ; but the
houses are poor, and the inhabitants chiefly mechanics, artisans, and traders of the lower class. At one extremity of
the Neustadt, up the river, is the fortress and arsenal of
the Wissehrad, erected on a bluff rock, and connected with
the line of works which extends in a curve behind the old
and new city, embracing them both, and descending to the
river at each extremity. On the opposite bank of the Moldau, the surface of ground is for a small space comparatively even, behind which arises a range of high, bold, craggy hills. On the even space, and partly up the ascent, is
built the Kleinseite (small side) : this is the quarter of the
aristocracy ; in it are the palaces of the ancient Bohemian
nobles, with attached gardens and shrubberies, which often extend high up the irregular ascent behind.

The lofty ridge above the Hradschin forms a magnificent
termination of the prospect, as viewed from the bridge below or from the opposite side of the river. Here, on the
summit of a long bold eminence, is the vast palace of the
old Bohemian monarchs; and close behind it rise the choir
and tower of the cathedral. Further on, along the hill, are
groups of stately edifices ; and beyond these again may be
seen, on a loftier point, the fine Premonstratensian monastery of Strahow, with its lofty towers and dark thick groves
overhanging the river. The quarters of Prague on the left,
like those on the right bank of the Moldau, are inclosed by
fortifications ; but these are of little strength, and were
raised by Charles IV., merely to give employment to the
working population, as the chance of invasion was then
but inconsiderable.

The bridge which connects Altstadt with the Kleinseite,
the only one hitherto constructed within the limits of the
city, is the longest in Germany. It is a ponderous structure of stone, 1780 ft. in length, and 35 in breadth ; with a
lofty tower at each extremity, and colossal stone statues,
single and in groups, among which is pre-eminent that of
St. John Nepomuck, the tutelar saint of the city. Not far
from the bridge, and attached to the Altstadt, is the Judenstadt, a district allotted to the Jews, whose number is about
8000, living, as usual, in crowded filthy abodes, forming a
labyrinth of narrow winding streets. (*Turnbull's Austria,
&c.*, i., 93–96, and *Reeve's Sketches of Bohemia*, No. ii.)

The Hradschin, or palace on the hill, is a vast pile, more

remarkable, however, for extent than beauty. It is said to
be larger than the palace at Vienna, and to comprise 440
apartments, including the hall of Landtaus, Imperial audience-room, hall of assembly for the states, &c. On a
narrow terrace immediately below the palace, two obelisks
mark the spot where the imperial commissioners, and their
secretary, sent thither with the most intolerant edicts
against the Bohemian Protestants, were indignantly thrown
out of the windows of the green chamber, by the deputies
of the kingdom, in 1618. Notwithstanding the great height
of the windows whence they were ejected, the commissioners escaped unhurt, by falling, as is said, on a dunghill.
This event may be regarded as the commencement of the
30 years' contest, ended by the peace of Westphalia in
1648, which, while it secured the liberties of the rest of
Germany, unfortunately consummated the slavery of Bohemia, which had long been foremost in freedom and toleration. The cathedral, begun in 1344 and finished in 1474,
is within the precincts of the Hradschin. It is a fine specimen of Gothic architecture, and is surmounted by a lantern-crown similar to that on the tower of St. Giles, Edinburgh.
The choir, built by Charles IV., and the unfinished chapels
that surround it are much admired. In the cathedral are
the tombs of many Bohemian sovereigns and other distinguished individuals ; a fine altar piece and other paintings
mosaics, and the costly shrine of St. John Nepomuck, &c.
Others of the numerous churches, as that of the *Theinkirche*, in which is the tomb of the famous astronomer,
Tycho Brahé, who died here on the 13th October, 1601, are
interesting for their monuments. Prague had formerly a
great number of convents ; but Joseph II. secularised most
of these establishments. Among the chapels is one which
is an exact representation of that of Loretto.

The town-hall, arsenal, military hospital, military orphan asylum, lying-in hospital, principal workhouse, and
theatre are among the finest of the numerous public edifices. Of the private palaces, that built by the famous
Wallenstein, duke of Friedland, is the most remarkable.
Though unfinished, it is of immense extent, 100 horses
having been pulled down to make room for its site. It
still belongs to a collateral branch of Wallenstein's family : the apartments and furniture, which are said to remain in their original state, are shown to strangers, and
the park attached to the palace is thrown open to the public ; but the residences of the nobility in the Kleinseite are
mostly deserted. "They are generally large ugly buildings, some, however, with a good deal of architectural decoration ; and the dirty rubbishy appearance of their brick
walls, half covered with worn-out stucco, conveys the
idea of prisons or poorhouses rather than of mansions of
distinguished nobles. Their proprietors have transferred
themselves and their wealth to the Austrian capital ; leaving to the Bohemians these sad memorials of times when
the court of Prague might have looked with scorn on the
inferior splendour of Vienna. Yet, in some of these desolate abodes, covered with dust and rubbish, we found immense collections of books. The Lobkowitz library is
said to comprise more than 70,000 vols., the Kinsky 60,000,
the Klebelsberg 18,000, the Klam Martinitz 21,000, and
others equal or superior numbers. In some of these palaces
a few rooms are fitted up and occupied during winter by a
minor branch of the family, and in many of them one office for the stewards and managers of the Bohemian estates ; but when, on particular occasions, as, for instance,
at the coronation of a sovereign, it is requisite for the proprietors to visit Prague, they usually occupy apartments in
some hotel, their own palace being quite unfit for their reception : there are, however, a few exceptions to this general emigration. Here and there may be found the mansion of some great noble, who still upholds the local dignity of his ancestors ; and below these highest magnates are
a considerable body of resident nobles, inferior in wealth,
though perhaps not in blood, who take a part in the provincial administration, and who form among themselves in
the winter season an agreeable and elegant society."
Turnbull's Austria, &c., i., 101.) In the palace of Count
Nostitz is a gallery rich in cabinet pictures of the Dutch
and Flemish masters ; and in that of Count Sternberg is
the national museum, comprising extensive collections of
paintings, books, fossils, and natural objects. The library,
in the Strahow monastery, one of the finest apartments of
its kind in Germany, comprises a collection of about 30,000
volumes.

The university of Prague, founded by Charles IV. in 1348,
occupies a large edifice termed the *Carolinum*, and is remarkable as the first great public school established in Germany. The students were formerly divided into four nations, and are said, though there can hardly be a doubt
that the statement is grossly exaggerated, to have amounted, early in the 15th century, to 40,000 ! In consequence,
however, of a measure proposed, in 1409, by Huss, who
was then rector, to abridge the privileges of the foreign

students, more than half the pupils attending the university withdrew to Leipsic, Heidelberg, Cracow, and other seminaries. The Carolinum is now exclusively devoted to instruction in medicine, law, and the sciences: while education in theology is conducted in the *Clementinum*, an immense building, founded by Ferdinand III., in 1653, as a convent and seminary for Jesuits. The university library, in the latter, comprises about 150,000 volumes: it has also an observatory, botanic garden, and various museums; and was attended, in 1840, by about 1700 students; but it must be borne in mind that the censorship of the press, the prohibition of foreign works, and the jealousy of the government, oppose the most formidable obstacles to the diffusion of knowledge, and hinder any real progress being made in political or even philosophical science. The same causes render the newspapers and other journals published here of little interest or importance. There are three gymnasia, preparatory to the university, with several other high schools, ecclesiastical, teachers', and Jewish schools; a polytechnic institute, conservatory of music, academy of the fine arts; and many orphan and deaf and dumb asylums, and other charities. The Jews settled here at a very early period, and have an infirmary and orphan asylum of their own, and as many as nine synagogues, one of which is very ancient.

Prague has manufactures of printed cotton, linen, silk and woollen stuffs; leather, hats, liquours, earthenware, refined sugar, &c.; and is the grand centre, not merely of the commerce of Bohemia, but of an extensive and rapidly increasing transit trade. It owes this distinction to its situation on the Moldau, or principal arm of the Elbe, which is navigable by large boats to Budweis (80 m. direct distance from the city), where it is joined by a railway from Linitz, on the Danube. Prague is thus rendered the centre, as it were, of the communication between Hamburg on the one hand, and Vienna on the other; and is, besides, intimately connected with Dresden, Leipsic, and other German cities. Several annual fairs are held here, including a large wool fair in June.

"Owing to the number of its palaces, churches, public buildings, and other splendid remains of its ancient grandeur, Prague is more imposing than Vienna, and far preferable as a residence; the situation being much more salubrious, and the climate more mild and equable, the cold in winter rarely exceeding 24° Reaum., and generally averaging between 7° and 10°; while, during the greatest heat of summer, the thermometer seldom rises above 23°. Dr. Stulzt, a celebrated German physician, who has written upon the relative salubrity of German towns, considers Prague as one of the most healthy in the empire, and affirms that it is no uncommon occurrence for the inhabitants to attain the age of 100, and even, sometimes, 115. Provisions are good and cheap; and an excellent red wine resembling Burgundy is produced in the neighbourhood. The theatre equals that of Vienna, and the musical department and orchestra cannot be too highly praised. Public and private concerts are also very frequent; and, except Vienna, there is no town in Germany where music is cultivated with so much success; indeed, this taste may in the Bohemians be termed truly national, for they excel both in vocal and instrumental music; and not a few of the natives travel to Italy, acquire the language, Italianise their names, and make large fortunes in Vienna. The harp appears to a stranger their native instrument; for we meet with itinerant harpists in every part of the country, whose strains generally accompany the midday repast at every inn, however small, whether in the capital or the provinces. Their language, which is rich and expressive, is also musical, and sounds as pleasing as the Italian when wedded to melody." (*Spencer's Germany and the Germans*, i., 207–8.)

Jerome, the friend of the great Bohemian reformer John Huss, was a native of this city, and was thence surnamed "of Prague." He suffered the same fate as his illustrious friend, having been burnt alive, in pursuance of a sentence of the council of Constance (see CONSTANCE), on the 30th of May, 1416. A very interesting account of the unfair and barbarous treatment he experienced on his trial, and of the singular ability, courage, and eloquence with which he defended himself, and exposed the malignity and bad faith of his adversaries, is given in a letter of Poggio-Bracciolini, who was present on the occasion. (This interesting document may be seen in *Shepherd's Life of Poggio*, or in the art. *Jerome*, in Aikin's *Biographical Dictionary*.)

Prague is very ancient; but the date of its foundation is uncertain.

Bohemia is one of the few countries in which persecution has been successful. At one time the Protestant faith was that of a great and decided majority, not of the citizens of Prague only, but of the whole kingdom. But the sovereigns of Austria having succeeded, in 1596, to the Bohemian crown, succeeded, after a lengthened struggle, partly by force of arms and partly by the most atrocious persecution of which history has preserved any account, in exterminating every germ of the reformed faith, and in totally subverting the free institutions of the country. For more than 150 years the language of the people was proscribed, their spirit was broken, and they were subjected to every species of insult and indignity. But since the middle of last century, and especially since the reign of Joseph II., the government has been conducted in a more tolerant and liberal spirit: and Prague, with the rest of Bohemia, is now advancing as rapidly in prosperity as any part of the Austrian monarchy. Prague was taken by the Prussians under Frederick the Great in 1741, but they were soon after obliged to evacuate the city, and it has ever since been held by the Austrians. (*Oesterr. Nat. Encyc.; Berghaus; Reeve's Sketches of Bohemia, in Metrop. Mag.,* 1837; *Turnbull, Strang, Gleig, Murray, &c.,* passim.)

PRAIRIE DU CHIEN, p. v., capital of Crawford co., Wis., 125 m. W. Madison, 957 W. Situated on the E. bank of the Mississippi river, 300 m. below the falls of St. Anthony, 600 m. above St. Louis. It contains a courthouse, jail, and three churches, a Methodist, Presbyterian, and Roman Catholic; an academy, nine stores, 150 dwellings, and 1500 inhabitants. Fort-Crawford, a United States military port, is situated immediately S. of the village. There are many ancient mounds in the vicinity.

PRATO, a town of the grand duchy of Tuscany, cap. distr. on the Bisenzio, a tributary of the Arno, 10 m. N.E. Florence. Pop. in 1836, 10,849. It is surrounded with a wall and ditch; the streets are regular, and the houses generally good. It has several squares, of which the best is the Piazza Mercatale; but the chief ornament of the town is the cathedral, a fine edifice of white marble, with ornamental parts of dark serpentine. Several other churches are handsome, and worth notice. Prato has two workhouses, several hospitals, the Cicognini college for secular instruction, normal Lancastrian and infant schools, &c. The manufacture of straw hats and bonnets employs nearly 1000 females; and it has also manufactories of woollen stuffs and caps, the latter for exportation to the Levant; with iron and copper works, paper-mills, a rope-walk, a glass factory, &c. The average wages of the working classes at Prato may be reckoned at 2¼ paoli, or 10½d. a day. (*Bowring's Rep. on Tuscany,* p. 34, 35.) In the middle ages, Prato was the capital of a republic, conquered by the Florentines in 1353. The poet Casti was a native of the town. (*Woods; Williams; Condor's Italy; Bowring's Rep.; Geog. Dict.*)

PRATTSBURG, p. t., Steuben co., N. Y., 209 m. W. by S. Albany, 315 W. It contains two churches, a Presbyterian and Methodist; four stores, two fulling-mills, three grist-mills, 22 saw-mills; one academy, 304 students; 18 schools, 993 scholars. Pop. 3455.

PRATTSVILLE, p. t. Greene co., N. Y., 38 m. W. Catskill, 54 m. S.W. Albany, 360 W. Watered by Schoharie creek. It contains two churches, a Dutch Reformed and Methodist; seven stores, one flouring-mill, one woollen factory, two grist-mills, five saw-mills, four tanneries; one academy, 25 students; six schools, 150 scholars. Pop. 1613.

PRAYA (PORTO) a seaport town of the Cape de Verd islands, which see.

PREBLE, county, O. Situated toward the S.W. part of the state, and contains 432 sq. m. Drained by St. Clair's and Twin creeks, tributaries of Great Miami river. It contained in 1840, 15,865 neat cattle, 29,255 sheep, 40,785 swine; and produced 198,107 bushels of wheat, 8955 of rye, 1,110,611 of Indian corn, 1340 of buckwheat, 3571 of barley, 270,804 of oats, 94,775 of potatoes, 49,996 pounds of sugar. It had 37 stores, one fulling-mill, three woollen factories, 16 flouring-mills, 18 grist-mills, 34 saw-mills, one oil-mill, nine tanneries, five distilleries; two printing-offices, two weekly newspapers. Pop. 19,489.

PREBLE, p. t., Cortland co., N. Y., 131 m. W. Albany, 294 W. Drained by Tioughnioga river. It contains two churches, a Presbyterian and Methodist; one store, one fulling-mill, one grist-mill, one saw-mill; 10 schools, 464 scholars. Pop. 1247.

PRENZLOW, a town of the Prussian states, the chief place in that part of Brandenburg, called the Ucker Mark, at the point where the river Ucker escapes from the lake of that name, 32 m. W.S.W. Stettin. Pop., in 1837, 10,508. It is thriving and well built; has various churches, schools, and hospitals, a valuable public library, and manufactures of linen, woollens, and tobacco, with breweries, tanneries, &c. In 1806, a conflict took place in the suburbs of this town, which ended in the surrender to the French of 90,000 Prussian troops, escaped from the battle of Jena.

PRESBURG, or PRESSBURG (Hungar. *Posony*, an. *Posonium*), a royal free town, and formerly the cap. of Hungary, and still the seat of the diet, immediately within its W. frontier, cap. co. of its own name, on the N. bank of the Danube, 34 m. E. by S. Vienna; lat. 48° 0' 59" N., long. 17° 6′ 15″ E. Pop. in 1837 (excluding garrison and strangers), 37,380. Travellers differ greatly in their statements as to the appearance of Presburg; but the latest and best authority, Mr. Paget, says it is prettily situated along the banks of the Danube, and, for a town of its size, offers a greater number of handsome buildings than are often seen together. It has, however, more of the characteristics of a German than a Hungarian city, and has few public buildings worth notice. The most conspicuous of the latter is the castle, a huge square brick structure, built upon a height above the town. It is now a ruin, having been burned in 1811, by an Italian regiment in the French service; it is, however, memorable as the scene of the appeal made in 1741 by Maria Theresa to the Hungarian states, which was so generously responded to by the latter. The hall of the diet, or *Landhaus*, is a plain, unpretending edifice, both externally and internally. The two chambers, which constitute the Hungarian diet, meet in apartments furnished simply with long tables, round which are benches for the members who speak, as in the House of Commons, from their places, and not, as in France, from a tribune. The lower chamber, visited by Paget, " is a long, plain hall, traversed in nearly its whole length by two tables covered with green baize, at which the deputies sit, with pens, ink, and paper before them. At the upper end is a raised part occupied by the president, vice-president, and secretary ; and behind these sit the judges of the royal table." In the intervals between the sessions of the diet, this apartment is used as a concert-room. The members attend the diet armed, in full national costume : and, since 1835, the debates have not been carried on in Latin, but in Hungarian. They are sometimes very stormy. The cathedral, a Gothic edifice, supposed to date from the 11th century, and in which the kings of Hungary are crowned ; the county-hall, German theatre, barracks, and archbishop's palace, are the other principal public buildings. There are several handsome noble residences, but they are seldom occupied, for Presburg is not a favourite place of abode with the Hungarian nobility ; and most of the members of the diet, when assembled, live in lodgings, dining at one of the numerous *restaurateurs* in the town. Presburg was formerly surrounded with walls, but it has long outgrown these, and they are now mostly demolished. It is a bishop's see, and the residence of the archbishop of Gran, primate of Hungary. It has a Roman Catholic academy and a Calvinist lyceam, both possessing good libraries ; a Catholic high gymnasium, Catholic seminary, college for poor students, and various other public schools ; five hospitals, including one supported by the Jews, who are very numerous here, and have a quarter to themselves ; and many charitable institutions. A fine library, belonging to Count Appony, is open to the public ; and, according to Mr. Paget (i., 93-4), the booksellers' establishments are well supplied with good works. The manufactures, which are various, include silk and woollen goods, saltpetre, rosoglio, tobacco, &c. ; and the town has a large transit trade in corn, linen, and Hungarian wines. Immediately outside the town is the *königsberg* (king's mountain), a small circular mound to which the king of Hungary proceeds to perform an important ceremony, immediately after his coronation. A very beautiful and fertile country extends along the other bank of the Danube, opposite Presburg ; and on that side are the favourite resorts of the inhabitants ; the promenade in the Au : public gardens ; arena, or theatre in the open air for national performances, &c. ; the communication across the river being kept up by a bridge of boats. Presburg is very ancient. Joseph II. transferred its previous title of capital of Hungary to Buda. The treaty which gave Venice to the French and the Tyrol to Bavaria was concluded here in 1805. (*Austr. Nat. Encycl.; Berghaus; Paget's Hungary, &c.*, i., 1-98; *Gleig, Elliot, &c.*, passim.)

PRESCOT, a market town, par., and township of England, hund. W. Derby, co. Lancaster, 8 m. E. by N. Liverpool, and 93 m. W. by S. Manchester. Area of par., 34,940 acres ; do. of township, 940 acres. Pop. of township in 1831, 5655. It is situated on high ground, over a large and rich coal field, and consists of one long principal street, well paved and lighted with gas, on the turnpike road between Liverpool and Manchester, about 1 m. N, the railway between those towns. The principal public buildings are a town hall, sessions-house, prison, and mechanics' institute. The parish church, an ancient structure, has a modern tower and steeple 156 feet in height, forming a conspicuous object to the surrounding country : the living is a vicarage, of the annual value of £593 in the gift of King's college, Cambridge, to which the manor belongs. In the out-township are eight district churches, the patronage of six of which is with the vicar. There are places of worship in the town for Wesleyan and Primitive Methodists, Independents, and Unitarians, besides which there are within the parish three Roman Catholic chapels, and nine meeting-houses for different denominations of dissenters. A grammar school, with an endowment producing £160 a year,

has about 30 foundation boys (sons of inhabitants, with other pay-scholars, those born and educated in the parish having a preference to seven fellowships and several good exhibitions at Brasennose college, Oxford. Several alms-houses, erected in 1708, furnish lodging for 19 old women, and there are numerous money-charities. The other benevolent institutions are a ladies' charity, bible society, and savings' bank.

Prescot has long been celebrated for its manufacture of watch tools and movements, in both of which branches it greatly excels : files, also, of first-rate quality, and engravers' tools, are made here. In 1836, 100 men were employed in file-making, and 500 in the construction of watch tools, motion work, &c. The manufacture of coarse earthenware has for many years been carried on, the clay of the neighbourhood being well adapted for such a purpose. Cotton spinning is carried on in two mills, employing, in 1838, 99 hands; and there is a small flax-mill. Coal mines are wrought in every direction round the town ; it is estimated that upwards of 2000 men are employed in the collieries within the parish ; and Liverpool receives from Prescot its chief supply of coal. Many of the out-townships are very populous, St. Helen's and Eccleston having attained to some importance as manufacturing towns. Prescot has petty sessions, and a baronial court for the recovery of small debts. Markets on Saturday, and fairs on alternate Tuesdays. (*See* ST. HELEN's.) (*Baines's Hist. of Lancaster. Butterworth's Stat. of Lancaster, &c.*)

PRESTEIGN, a pari. bor. and market town of S. Wales. hund. New-Radnor, co. Radnor, near the S. bank of the Lug, in a fertile and well-cultivated valley, adjoining the confines of Herefordshire, 19 m. E. by N. Leominster. The parish of Presteign, which extends partly into Wigmore hund., co. Hereford, had, in 1831, 2282 inhabitants, of which the parl. bor. had 1699. This is a well-built town, and, notwithstanding its limited size, is the capital of the county, the assizes and quarter sessions being held in it. " A woollen manufacture was once carried on here ; but that is now given up, and there is no manufacture at present in the town, unless it be that of malt, which is made in large quantities. It is a sort of metropolis of the county for 5 m. round, supplying it with groceries, drapery, mercery,' and shop goods in general. It has also a considerable trade in timber. The inhabitants are principally professional men, tradesmen, mechanics, handicraftsmen, and a few farmers. There are also some persons of small independent incomes, who reside here for the sake of the cheapness of the place." (*Parl. Bound. Report.*) The church, which is very ancient, has some curious monuments and fine old tapestry. The living, a rectory of the annual value of about £890, is in the gift of the Earl of Oxford. The Wesleyan and Primitive Methodists, and Baptists, have places of worship. The county hall is said to be a handsome structure, and there is also a county jail and a free school. The latter, founded and liberally endowed in the reign of Elizabeth by a clothier of the town, furnishes a plain English education to between 50 and 60 boys; and there are other minor schools and Sunday-schools attached to the Church and the Wesleyan connexion. At the N. end of the town is a fine circular mound, laid out in public walks, presented to the inhabitants by the Earl of Oxford. Presteign unites with New-Radnor and other small boroughs in sending one member to the House of Commons. Registered electors in the whole borough in 1839-40, 599. It is governed by a bailiff and constables.

The Rev. Richard Lucas, author of the " Enquiry after Happiness," was a native of Presteign, having been born here in 1648. The work now referred to, which has passed through a great number of editions, and is still held in considerable estimation, was written after the author had become blind. (*Aikin's Biographical Dict.*)

PRESTON, a pari. and mun. bor., market town, and par. of England, in its own div. of hund. Amounderness, co. Lancaster, on the N. bank of the Ribble, crossed here by three handsome bridges, 19 m. S. Lancaster, and 29 m. N.N.E. Liverpool. Area of parl. bor. (which comprises Preston and Fishwick townships), 2560 acres. Pop. in 1831, 33,871; in 1841, 50,073. The town, which consists of a broad principal street, running N.E. from the river, crossed by several others in different directions, is " well built, well paved, with handsome dwelling-houses and thriving factories ; having a perfect drainage, and good roads leading from it." (*Bound. Rep.*) The streets are well lighted with gas, and there is an abundant supply of water. The market-place, at the junction of Fishergate and Friargate, contains about 3000 sq. yards. The public buildings comprise an elegant courthouse, erected in 1825, an exchange or market house, town-hall, assembly-rooms, theatre, borough prison, house of recovery, and a large county penitentiary. The church, originally erected in the 16th century, was rebuilt in 1770 : the living is a vicarage, of the annual value of £685, in the gift of the trustees of

Holme's charity. There are also seven district churches, chiefly of modern erection, and four others have been built in the out townships. The places of worship for dissenters comprise four for Roman Catholics (a numerous and increasing body), two each for Wesleyan Methodists and Independents, with others for Unitarians, Huntingdonians, Primitive Methodists, Baptists, Sandemanians, and the Society of Friends; there being, in all, within the town, eight churches and 17 dissenting chapels. National schools are attached to the several churches; and all or most of the dissenting places of worship have large Sunday-schools, furnishing religious instruction to between 7000 and 8000 children. A grammar school, founded prior to 1686, has an endowment of £50 a year; there is, also, a partially endowed blue-coat school, and several day and infant schools, supported by subscription. The other charities comprise nine almshouses, and several money bequests; a dispensary, house of recovery, built in 1809; provident society, workhouses, and savings' bank. Among the literary establishments, the first place is due to the Preston Institution for Diffusing Useful Knowledge, which has a library of about 2300 volumes, and an excellent museum. The Palatine and Dr. Shepherd's libraries are open to all classes, and the town has a public law-library. An agricultural society was founded in 1811. Avenham Walk, on the summit of the hill which rises from the banks of the Ribble, is a favourite promenade, and is kept in order at the cost of the corporation. Two newspapers are published, each once a week.

Preston, from its central position, its vicinity to an important coal district, and its extensive means of communication with the interior by canals and railways, united to the skill and enterprise of its citizens, has of late years rapidly increased in wealth and population, and is now, in fact, one of the great seats of the cotton manufacture. In 1838, there belonged to the town 35 cotton-mills, impelled by a steam power of 1339 horses, and employing above 7000 hands. It has, also, numerous hand-loom weavers, though owing to the competition of the power-loom, their wages had fallen in 1838 to about 7s. a week at an average. The manufacture of linen cloth, formerly the principal branch of industry in Preston, is still pretty extensively carried on; and in 1838, six flax-mills employed about 1400 hands. It has also numerous iron foundries, and other establishments for making machinery and other articles. Leather is tanned in considerable quantities, and there is a small fishery on the Ribble, which abounds with salmon, smelts, and eels. Two banking companies have been recently established, and there are three private banks. The Ribble is navigable at spring tides, as far as Preston-marsh, for vessels of 150 tons; but, being ill adapted for sea-borne vessels, it is frequented almost exclusively by coasters. The navigation, however, which is impeded by sandbanks, is in course of being improved by a company incorporated in 1837-8; and it is expected that a depth of 18 feet water, at ordinary spring tides, will be attained. The river does produce about £1000 a year. The Lancaster canal, formed in 1796, passes the town; and it is connected with other parts of the county, and of England generally, by the North Union railway (32¾ m. in length), which crosses the Ribble on a viaduct of five arches, 68 feet above the river, the Lancaster and Preston railway (20¼ m. in length), the Preston and Longridge railway (7 m. long), the Bolton and Preston railway, not yet completed, and the Preston and Wyre railway, which last connects it with the new seaport of Fleetwood, at the mouth of Wyre harbour, now rapidly rising in importance. Large markets on Saturday, with others on Wednesday and Friday for fish, butter, and vegetables. Great fairs in Jan., March, Aug., and Nov.; the first of which, called the "Great Saturday," is celebrated for its show of horses.

Preston is a borough by prescription, and received its first charter from Henry II. By a subsequent charter, granted by Henry III., the officers of the borough were authorized to hold a guild merchant for the renewing of the freedom of the burgesses, and other purposes. This privilege is made the occasion of great festivity. For a long time after their first institution, the guilds were held at irregular periods; but they have now for more than a century been uniformly celebrated every twentieth year, commencing on the Monday next after the decollation of St. John, which generally happens in the last week of August. The last was held in 1832. Processions of the corporation, and the different trades in characteristic dresses, as well as of ladies, and females employed in the factories, take place on two of the days; and the amusements, which are varied and interesting, continue for a fortnight. But for civic purposes, the guild books are open for an entire month. Under the Mun. Reform Act, Preston is divided into six wards, its municipal officers being a mayor and 12 aldermen, with 36 councillors. Corporation revenue in 1839 (exclusive of £670 accruing from the sale of property), £4867. Quarter

and petty sessions are held under a recorder; and there is a court for the recovery of debts to any amount contracted within the borough. Quarter sessions for the hunds. of Amounderness, Blackburn, and Leyland, are held here, and there is a monthly county court. Preston has sent two members to the House of Commons since the reign of Edward VI., the right of voting, down to the Reform Act, being in the inhabitants at large. The Boundary Act enlarged the electoral limits, so as include the township of Fishwick with the old borough. Registered electors in 1839-40, 6633.

Preston is supposed to have risen on the decay of the ancient *Rerigonium*, or Ribchester, a city now reduced to the condition of a mere village, about 11 m. higher up the river, and it derived its name of *Priest's-town* from the number of religious houses established here, and of which there are still some remains. It was partly destroyed by Robert Bruce, in 1322. In the parliamentary wars of Charles I., its inhabitants declared for the king, and it was besieged and taken by Sir Thomas Fairfax. In 1715, the Jacobite insurgents took possession of the town, and erected barricades for its defence; but, after a brave resistance, they were compelled to surrender to the royalist force under General Willes. In 1745, Preston was visited by the Pretender, on his retreat; but he was compelled to withdraw on the approach of the Duke of Cumberland. (*Baines's Hist. of Lancashire; Butterworth's Stat. Sketch of Lancashire*, p. 109–113; *Parl. and Bound. Rep.*)

PRESTON, county, Va. Situated in the N. part of the state, and contains 501 sq. m. Watered by Cheat river, the principal constituent of Monongahela river. It contained in 1840, 7459 neat cattle, 12,249 sheep, 9069 swine; and produced 2801 bushels of wheat, 17,877 of rye, 42,587 of Indian corn, 16,057 of buckwheat, 128,966 of oats, 35,209 of potatoes, 3992 pounds of tobacco, 27,132 of sugar. It had 11 stores, two fulling mills, two woollen factories, two cotton factories with 90 spindles, 29 grist-mills, 28 saw-mills, one oil-mill, 11 tanneries, 22 distilleries, one printing-office, one weekly newspaper. Pop., whites, 6743; slaves, 91; free coloured, 32; total, 6866. Capital, Kingwood.

PRESTON, p. t., New-London co., Ct., 45 m. S.E. Hartford, 383 W. Bounded W. and N.W. by Quinnebaug, Shetucket, and Thames river. It contains three churches, a Congregational, Episcopal, and Baptist; nine stores, one cotton factory with 200 spindles, two grist-mills, four saw-mills; 13 schools, 469 scholars. Pop. 1727.

PRESTONPANS, a bor. of barony, and seaport town of Scotland, co. Haddington, on the shore of the frith of Forth, 7½ m. E. Edinburgh. In 1835, the parish, which is one of the smallest in Scotland, had a population of 2467, of whom the borough may have about 1900. It is straggling and ill-built, consisting principally of a single street parallel to the frith of Forth. It derives its name from its having, for a lengthened period, had a number of salt-works or pans for the production of salt by the evaporation of the sea water, and for the refining of rock salt. The latter branch of the business is now the only one that is carried on. It has, also, a brewery, at which some of the best Scotch ale brought to the Edinburgh market is produced, a soap work, and two works for the manufacture of coarse pottery. There are extensive oyster beds in the vicinity of the town, whence the Edinburgh markets derive a large proportion of their supply. Morrison's Haven, the port of Prestonpans, about ¾ m. W. from the town, is a small creek, with not more than 10 feet water at springs. (*New Statistical Account of Scotland*, par. of *Prestonpans*.)

Near this village, on the 21st of September, 1745, the royal army, under Sir John Cope, consisting of about 2100 regular troops, was totally defeated and dispersed with great loss, by the Highlanders, who were but little superior in point of numbers, under the Pretender. The king's troops being panic struck, threw away their arms, and fled at the first fire, and were cut down almost without resistance. (*Johnston's Memoirs of the Rebellion*, p. 34–36.)

PRESTWICH-CUM-OLDHAM. (See OLDHAM.)

PREVESA, a town of Turkey in Europe, prov. Albania, at the entrance of the Ambracian gulf, 56 m. S.S.W. Yanina. Pop. 4000? It is ill built, badly paved, and dirty; but it is in a good situation for commerce, and was formerly the entrepôt of the trade of Epirus. On the isthmus, connecting the peninsula upon which it stands with the main land, are the remains of Nicopolis, consisting of the ancient walls, a theatre, some baths, and various other vestiges of antiquity. (*Burgess's Greece, &c.*, i., 93–96; *Hobhouse, Hughes, &c.*)

PRINCE EDWARD'S ISLAND, (formerly ST. JOHN's), an island of N. America, belonging to Great Britain; in the gulf of St. Lawrence, principally between the 46th and 47th degs. of N. lat., and the 62d and 64th of W. long., from 90 to 25 m. E. New-Brunswick. Length, of a curved line, passing through its centre E. to W. about 140 m.; greatest breadth 34 m. Area estimated at 1,369,709 acres, or about

2150 sq. m. Pop. about 33,000, principally Highland, Scotch, Irish, and Acadian French, with some English, Dutch, Americans, and Swedes. A chain of hills of moderate height partly intersects it; but the surface in general is level, or at most only undulating. It is well watered, and its shores are deeply indented with numerous bays. The climate is milder than in any of the surrounding British colonies, and appears to be favourable to longevity. The atmosphere is nearly free from the fogs prevalent in cape Breton and the adjacent countries. Below the thin vegetable mould the soil is generally clay or loam, resting on a base of sandstone: there are some swamps, and pine barrens; but these bear only a small proportion to the whole surface. The island in general is well wooded, the principal trees being spruce, fir, beech, birch, and maple. Oak, ash, larch, &c., are scarce, and the quality of the first is very inferior. All kinds of grain and vegetables raised in England come to perfection. Wheat is grown in abundance, and the surplus is exported to Nova Scotia. Barley, oats, and beans, are said to be equal in weight and quality to any met with in the English markets. Flax of excellent quality is raised, and manufactured into linen for domestic use. Hemp will grow, but not to the same perfection as in the adjacent colonies. It is said, though we apprehend the statement must be taken with large qualifications, that had the natural advantages of this island been turned to proper account, it might at this time have been the granary of the British colonies, instead of barely supporting a poor and limited population. Of nearly 1,400,000 acres contained in the island, only 10,000 are said to be unfit for the plough; but only 100,000 are now under cultivation. The origin of this state of things is ascribed, in Lord Durham's Report, to the injudicious grants made to absentee proprietors, under conditions that have been totally disregarded. "The absent proprietors neither improve the land, nor will let others improve it. They retain it, and keep it in a state of wilderness." (Report, p. 70–86.) What land is under the plough, is cultivated in a very slovenly manner; though the establishment of an agricultural society of late years has done something to improve husbandry.

Pastures are good, and suitable for cattle and sheep; owing to the want of proper attention, hogs are said not to thrive so well as the former. Live stock used to suffer greatly from the ravages of bears, loup-cerviers, and other wild animals; but these are much less numerous now than formerly. The island has no mines. Its fisheries are, however, of some importance; the value of the fish cured in 1839 having amounted to £5340 sterling. But owing to the want of capital and industry, the Americans have an ascendancy in the fisheries. A good many ships are built in the colony; 69 vessels of the aggregate burden of 9934 tons having been built and registered in 1839. Total value of exports from the colony, in 1837, £7871, timber, deals, &c., making £7074 of this amount. Total value of imports, £1946.

The constitution is nearly similar to that of Nova Scotia, and in all civil matters independent of any jurisdiction in America. The government and legislature is vested in a lieut. governor, a council of nine members, and a House of Assembly of 18 representatives, elected by the people. The governor is chancellor of the court of chancery; the chief justice and attorney-general are appointed by the sovereign; and the high sheriff is appointed annually by the local government. In the supreme court of judicature all criminal and civil matters of consequence are tried by jury. Cases of petty debt and breaches of the peace are decided by special magistrates and justices of the peace. There are two superior schools in Charlotte town, and about 40 district schools. Charlotte town, the capital and seat of government, on Hillsborough river, near the S. coast, has one of the best harbours in the gulf of St. Lawrence. The town, on gently rising ground, is regularly built, and clean, with about 3500 inhabitants. The courthouse, Episcopal and Scotch churches, several chapels, the barracks, and the fort, are its only conspicuous public buildings.

This island was taken from the French in 1758. It was annexed with cape Breton to the government of Nova Scotia in 1763, but since 1768 has formed a separate colony. (Macgregor's America; Durham's Report on N. America, &c.)

PRINCE EDWARD, co., Va. Situated centrally toward the S. part of the state, and contains 375 sq. m. Bounded N. by Appomattox river, by branches of which it is drained. It contains Hampden Sidney college, 1 m. from the courthouse, which was founded in 1783, has a president and four professors, or other instructers, 63 students, and 8000 volumes in its libraries. The commencement is on the fourth Wednesday in September. A preparatory academy is connected with the institution. There were in the county in 1840, 7634 neat cattle, 11,698 sheep, 15,428 swine; and produced 57,993 bushels of wheat, 303,997 of Indian corn, 139,210 of oats, 13,147 of potatoes, 3,106,950 pounds of to-

648

bacco, 11,121 of cotton. It had 33 stores, 22 flouring-mills, 28 grist-mills, eight saw-mills, three tanneries; two colleges, 66 students; three academies, 52 students; 26 schools, 465 scholars. Pop.: whites, 4923; slaves, t576; free coloured, 570; total, 14,069. Capital, Prince Edward, C. H.

PRINCE FREDERICKTOWN, p. v., capital of Calvert co., Md., 46 m. S. by W. Annapolis, 76 W. Situated on Parker's creek, which flows into Chesapeake bay. It contains a courthouse, jail, a church, several stores and about 500 inhabitants.

PRINCE GEORGE, county, Va. Situated toward the E. part of the state, and contains 305 sq. m. Bounded N.E. by James river, N.W. by Appomattox river. Drained by Blackwater river. It contained in 1840, 3695 neat cattle, 2727 sheep, 5911 swine; and produced 31,439 bushels of wheat, 176,640 of Indian corn, 35,231 of oats, 6453 of potatoes, 115,000 pounds of tobacco, 23,349 of cotton. It had eight stores, nine grist-mills, two saw-mills, seven distilleries; six schools, 117 scholars. Pop.: whites, 2692; slaves, 4014; free coloured, 469; total, 7175. Capital, Prince George, C. H.

PRINCE GEORGE'S, county, Md. Situated in the W. part of the state, and contains 575 sq. m. Bounded E. by Patuxent river, W. by Potomac river. A part of the district of Columbia was taken from this county. It contained in 1840, 10,482 neat cattle, 13,633 sheep, 34,901 swine; and produced 80,147 bushels of wheat, 38,209 of rye, 507,366 of Indian corn, 107,070 of oats, 21,570 of potatoes, 9,259,493 pounds of tobacco. It had 31 stores, one flouring-mill, 19 grist-mills, six saw-mills, one printing-office, one weekly newspaper; four academies, 43 students; 12 schools, 253 scholars. Pop.: whites, 7893; slaves, 10,636; free coloured, 1080; total, 19,539. Capital, Upper Marlboro'.

PRINCE OF WALES'S ISLAND (native Pulo Pinang, 'the Areca island'). An island and British settlement in the eastern seas, about 2 m. from the W. coast of the Malay peninsula, lat. 5° 15' N., long. 100° 25' E. Length, N. to S., nearly 16 m.; breadth varying from 6 to 12 m. Area about 160 sq. m. Pop., in 1836, 38,450, principally Malays, Chinese, and Chuliahs, the Europeans being under 800. The N. part of the island is mountainous, and a range of hills runs through its centre, declining in height as it approaches the S.W. extremity. But two thirds of the whole surface are level, or of gentle inclination, and like the hills, covered with woods. The thermometer, in the plains, ranges between 76° and 90° Fahr., and on the higher hills at from 64° to 76°. Except in a few places, Pinang is considered very healthy. Refreshing showers fall at short intervals throughout the year. The climate of the high lands is said to resemble that of Funchal, in Madeira. The geological formations are primitive. Nearly all the hills are of granite; and the subsoil, where not alluvial, is principally the detritus of that rock. Tin ore is found at the base of the mountains. The island produces a good deal of timber, well adapted for ship-building and masts; and fine fruits; and it is well adapted for the growth of spices. As a commercial mart, this settlement is much inferior to Singapore; but, according to Mr. Newbold, "it bids fair by its plantations of nutmegs and cloves, to render us independent of the spice islands which we have given up to Holland." (Malacca, &c.)

The attention of the agriculturists is now almost exclusively directed to the improvement and extension of the spice-plantations, and dry waste land for that purpose may be obtained from the government on leases of 40 years, at a small quit-rent. The annual produce of the staple articles is estimated as follows:

Articles.		Quantities.	Values in Spanish Dollars.
Nutmegs	.	.	piculs.	400	12,400
Mace	.	.	—	170	34,400
Cloves	.	.	—	320	4,300
Pepper	.	.	—	16,080	64,300
Total value					114,400

The value of the nutmegs, mace, and cloves, exported in the years 1836-37, amounted to 156,800 rupees. A good many cocoa-nuts are grown; and gambir, indigo, cotton, areca, and tobacco in small quantities. Coffee, sugar-cane, betel nut and leaf, rice, cotton, and ginger, are also raised. The trade of Pinang is chiefly that of transit, between Great Britain and British India, on the one hand, and the Malay peninsula and Tenasserim, provinces Siam, Anam, Borneo, Java, Sumatra, China, &c., on the other. A considerable trade in cotton cloths is kept up by the Chuliahs with the Coromondel coast.

The revenue of Pinang and province Wellesley is derived from lands, customs, licences, and the sale of the government monopolies of opium, spirits, pork, betel-leaf, and the market. In 1835-36 it amounted to 178,909 rupees; but the disbursements, not including the expenses of the military and the convict establishment, amounted to 253,388 rupees, leaving a deficit of 74,398 rupees.

Province Wellesley, a dependency of this settlement, is a strip of coast land on the Malay peninsula, opposite Pinang, 35 m. in length, by about 4 m. in breadth. Area, 140 sq. m. Pop., in 1836, 47.555, of whom 43,000 were Malay. It has an undulating surface, chiefly of alluvial soil, and a healthy climate. Its principal products are rice, pepper, sugar, and cocoa-nuts; and the gross value of its surplus agricultural produce is estimated at 80,000 Spanish dollars. It supplies Pinang with cattle and poultry. These settlements are under the Bengal presidency, and governed by a resident at George Town, and an assistant resident in province Wellesley.

George Town, the capital, at the N.E. extremity of Pinang, has a population of about 13,000. It is built on level ground, and consists of a long and broad street, intersected by others of inferior dimensions. It has a fort, a handsome church, an Armenian chapel, two Roman Catholic chapels, a courthouse, jail, public school, poor-house, the governor's offices, and the civil and military hospitals. An English weekly newspaper is published in the town, and a few shops are kept by Europeans, but the major portion by Chinese. There are cantonments for the native troops near the town. Penang was purchased by the East India company in 1786, and province Wellesley in 1800. *Newbold's Malacca, &c.*)

PRINCESS ANN, county, Va. Situated in the S.E. part of the state, and contains 374 sq. m. Bounded E. by the Atlantic, N. by Chesapeake bay. Back bay, a branch of Currituck sound, sets up in its S. part. It contained in 1840, 10.982 neat cattle, 6623 sheep, 20,761 swine; and produced 6655 bushels of wheat, 298,960 of Indian corn, 84,771 of oats, 37,029 of potatoes, 951 pounds of cotton. It had five stores. 16 grist-mills, three tanneries, 10 distilleries; one academy, 49 students; eight schools, 179 scholars. Pop., whites, 3996; slaves, 3087; free coloured, 202; total, 72·5. Capital, Princess Ann C. H.

PRINCESS ANN, p. v., capital of Somerset co., Md., 110 m. S.S.E. Annapolis, 150 W. Situated on the E. side of Manokin river, 18 m. from its entrance into Chesapeake bay, and is near the head of tide water. It contains a courthouse, jail, a bank, three churches, a Presbyterian, Episcopal, and Methodist, and about 120 dwellings. It is laid out with streets crossing each other at right angles, and neatly built, with brick or with wood houses, painted white. Washington academy is in the vicinity.

PRINCESS ANN, C. H., p. v., capital of Princess Ann co., Va., 132 m. S.E. Richmond, 256 W. It contains a courthouse, two Methodist churches, one store, 20 dwellings, and about 130 inhabitants.

PRINCETON. p. t., Worcester co., Mass., 16 m. N. by W. Worcester, 47 m. W. by N. Boston, 417 W. Drained by branches of Nashua river. It contains Wachusett mountain, an isolated summit, 3000 feet above tide water, and 1900 feet above the surrounding country. The prospect from its top is extensive, various and grand. The township contains four churches, a Congregational, Presbyterian, Baptist and Universalist, three stores, three grist-mills, une saw-mills, two tanneries; 10 schools, 404 scholars. Pop. 1347.

PRINCETON, p. b., Mercer co., N. J., 11 m. N.E. Trenton, 77 W. It is a neat and pleasant village, chiefly built on one extended street, and contains four churches, two Presbyterian, an Episcopal, and an African, a number of stores, 00 dwellings, and about 1400 inhabitants, exclusive of those connected with the literary institutions. The Delaware and Raritan canal passes within a mile of the bor., and have the office of the company in the village. It is elevated ground, in a healthy location, and distinguished a pleasant place of residence. The society is intelligent id refined. The college of New-Jersey was founded at Elizabethtown in 1746, was removed to Newark in 1748, nd it was permanently located at Princeton in 1756. It s had a list of venerable presidents, and has educated any of the distinguished men of the country. It has a esident and 12 professors or other instructors, 2554 alumni, whom 483 have been ministers of the gospel, 190 students, and 12.500 volumes in its libraries. The commencement is on the last Wednesday in September. Its buildings are convenient, spacious and neat. The Theological minary of the Presbyterian church is also located here, nded in 1812, has four professors, 109 students, 753 have npleted their course, and there are 7000 volumes in its aries. It has spacious and convenient buildings, and is of the most respectable institutions of the kind in the ntry. Princeton township according to the United tes census of 1840, had nine stores, capital $47,600; one iber-yard, capital $2500; two grist-mills, one saw-mill, printing offices, one bindery, one weekly newspaper; colleges, 328 students; two academies, 90 scholars; schools, 110 scholars. Pop. 3055. In the immediate hbourhood of this place, was fought the memorable le of Jan. 3rd, 1777, in which the British army was

roused by the Americans, under General Washington, immediately subsequent to the battle of Trenton, and when General Mercer was mortally wounded. In the college chapel is a large painting, commemorative of this event.

PRINCE WILLIAM, county, Va. Situated in the N.E. part of the state, and contains 370 sq. m. Bounded N.E. by Occoquan river, E. by Potomac river. It contained in 1840, 6614 neat cattle, 6502 sheep, 8900 swine; and produced 47,471 bushels of wheat, 3704 of rye, 180,483 of Indian corn, 3181 of buckwheat, 105,376 of oats, 6476 of potatoes, 4974 pounds of tobacco. It had 18 stores, one cotton factory with 1088 spindles, 10 flouring-mills, 13 grist-mills, 10 saw-mills, five distilleries; five schools, 118 scholars. Pop.: whites, 4867; slaves, 2767; free coloured, 510; total, 8144. Capital, Brentsville.

PROSPECT, p. t., Waldo co., Me., 53 m. E. by N. Augusta, 649 W. Situated on the W. side of Penobscot river, at its entrance into Penobscot bay. Incorporated in 1794. It has 15 stores, one fulling-mill, two grist-mills. 12 saw-mills, two tanneries; 90 schools, 1416 scholars. Pop. 3492. Considerable ship-building is carried on.

PROVENCE, one of the former provs. of France, in the S.E. part of the kingdom, now subdivided into the deps. Basses-Alpes, Bouches-du-Rhone, Var, and a portion of Vaucluse.

PROVIDENCE, county, R. I. Situated in the N. part of the state, and contains 380 sq. m. Watered by Pawtucket, or Blackstone, Wanasquatucket, Masbassck, and Pawtuxet rivers, which afford extensive water-power. It is particularly distinguished for its manufactures. It contained in 1840, 13,157 neat cattle, 11,661 sheep, 10,669 swine; and produced 641 bushels of wheat, 16,870 of rye, 157,577 of Indian corn, 1573 of buckwheat, 13,374 of barley, 16,949 of oats, 347,269 of potatoes. It had 78 houses in foreign trade, capital $1,588,850; 550 retail stores, capital $1,987,200; 19 lumber-yards, capital $165,500; capital invested in the fisheries, $130,000; five furnaces, 18 fulling-mills, six woollen factories, 130 cotton factories, with 367,251 spindles, 15 dyeing and printing works, 59 grist-mills, 73 saw-mills, two paper-mills, two rope-walks, eight tanneries, two distilleries, two breweries, 10 printing-offices, six binderies, two daily, five weekly, and four semi-weekly newspapers, and two periodicals. Total capital in manufactures $7,165,887; two colleges, 324 students; 37 academies, 2935 students; 209 schools, 8079 scholars. Pop. 50,073. Capital, Providence. The above statistics, of course, include those of Providence city.

PROVIDENCE, city, port of entry and semi-capital of Rhode island, is situated in 41° 49' 22'' N. lat., and 71° 24' 45'' W. long., and in 1° 23' 24'' E. long. from W., 30 m. N. Newport, 43 m. S.S.W. Boston, 55 m. N.E. New-London, 70 m. E. Hartford, 173 m. E. New-York, 396 W. The population in 1820 was 11,767; in 1830, 16,832; in 1840, 23,171. Of these 999 were employed in commerce; 3948 in manufactures and trades; 422 in navigating the ocean; 90 do. rivers and canals; 165 in the learned professions.

Its present limits contain an area of about 9 square miles. Its compact part is nearly equally divided by Providence river and cove, which is here in fact an arm of Narragansett bay, across which are convenient bridges, connecting the two parts of the town. On the east side of the place flows Seekonk river, which is the estuary of Pawtucket or Blackstone river. Over this estuary are two bridges, the one called Indian bridge, and the other Central bridge, connecting the place with Seekonk, in Massachusetts. It was, at the census of 1840, the second city in population in New-England. The city is handsomely laid out, though not with entire regularity. On the E. side are three principal streets running parallel with the river and cove, which are Water-street, Main-street, and Benefit-street. On Main-street stand a number of public buildings, and a considerable proportion of the stores and offices. This is an ancient and populous street, and contains many elegant, stately and substantial brick edifices. Market-square extends from this street to a bridge which connects the two parts of the town. On this side, the ground rises abruptly, and the cross streets have a steep ascent. Benefit-street has an elevated and sightly situation, and east of it the city is laid out with considerable regularity, the streets generally running in an east and west direction, crossed by others, nearly at right angles. On this side of the water are the governor's official residence, the buildings of Brown university, the alms-house, most of the banks and insurance offices, and many elegant residences of the wealthy citizens, enjoying fine prospects. On the west side of the river, the ground is more level, the streets are broad, and the plan of them more regular. On this side the two principal streets are Westminster and Weybosset-streets, the latter somewhat serpentine in its direction. Numerous other streets are well-built and populous.

Among the public buildings of Providence are a state-house built of brick, of not very imposing appearance, a city hall, jail, and state prison, hospital, three market-houses, a custom-house, athenæum, the buildings of Brown university, a high school, theatre, six public school buildings, 30 churches, seven Baptist, three Congregational, two Unitarian, four Methodist, two Episcopal, a New Jerusalem, Free-will Baptist, a Universalist, Friends, Christian, two Roman Catholic, coloured, a Methodist, Baptist, and Episcopal, a coloured Bethel, and a seamen's Bethel. Several of the churches have fine edifices, with lofty spires. The first Baptist, two Unitarian, one Congregational with its chaste portico and dome, and one Episcopal with its square Gothic tower and pinnacles, add much to the beauty of the city. The buildings of Brown university occupy a commanding situation on Prospect-street, at the head of College-street, on the E. side of the river. They are four in number, University hall, Hope college, Manning hall, and Rhode Island hall. Manning hall has a Doric portico in front, and is occupied by the chapel and library. This institution which derived its name in 1804, from Nicholas Brown, Esq., its principal benefactor, was originally founded in Warren in 1766, and removed to Providence in 1770. The president and a majority of the trustees are required to be Baptists. It has a president and eight professors, 1496 alumni, of whom 474 have been ministers of the gospel, 167 students, and 17,200 volumes in its libraries. The commencement is on the first Wednesday in September. It has an extensive philosophical and chemical apparatus; and the cabinet of mineralogy and natural history are very complete, and occupy Rhode Island hall, which has also a number of spacious lecture rooms. A university grammar school is connected with the institution. The athenæum is a handsome Grecian Doric edifice of 40 feet front and 78 feet deep, and is occupied in the basement by the Franklin and Historical societies, the first of which has an extensive collection of models, philosophical instruments and curiosities; and the latter, a valuable collection of books, papers, and records. The upper story is occupied by the athenæum society, with its valuable library, reading room, librarian's office and committee room. The cost of the building and its furniture was about $90,000, and of the library about $30,000. The arcade, though not strictly a public building, deserves to be mentioned as one of the finest of its kind in New-England. It extends from Weybosset-street on the one front to Westminster-street on the other, having an imposing Doric portico of six granite columns of one entire stone, 3 feet in diameter and 25 feet high, on each front. It is 225 feet long, 80 feet broad, and 72 feet high, divided into three stories, containing 82 shops or warerooms in the interior, and the whole is lighted by a glass roof from above. It is built of granite, and cost $130,000. The Friends boarding school, belonging to the yearly meeting of this denomination in New-England, is a flourishing high school, situated three quarters of a mile N.E. of the university, has a spacious edifice of stone and brick in a commanding situation, has 10 instructers, and about 200 pupils.

Providence has 26 religious and benevolent societies of various kinds, some of which are very efficient. There are in the place 21 banks, with an aggregate capital of $7,089,750, and three insurance companies, with a total capital of $390,000, besides three mutual insurance companies without a definite capital.

Providence possesses great commercial facilities, which have been well improved. The harbour, at the head of Narragansett bay, 23 m. from the ocean, is spacious, and has a sufficient depth of water for the largest ships. There belong to the port 106 vessels, viz., 17 ships, 14 barques, 34 brigs, 27 schooners, and 24 sloops. The registered licensed and enrolled tonnage of the port in 1840, was 16,610 tons. Formerly the East India trade was extensive, but is now less considerable than formerly. The trade is principally with parts of Europe, with the West Indies, particularly the island of Cuba, and with the southern states. The extensive cotton manufactories in the vicinity require a large supply of cotton, and furnish a great amount of cotton goods to be sent abroad. There are two lines of packets which ply regularly to New-York; two lines to Albany; one line to Philadelphia, and one to Baltimore. These several lines contain 23 vessels, of different descriptions. A railroad 41 m. long extends to Boston, and with this is connected a daily line of steamboats to the city of New-York, during the season of navigation, which is most of the year. A railroad extends from Providence to Stonington, which has also a connected line, by steamboats to New-York. When the Long Island railroad shall be completed, as it is expected to be by the fourth of July 1843, that, with the railroad from Stonington through Providence to Boston, will constitute the most speedy route between New-York and Boston. The Blackstone canal extends from Worcester to Providence, 45 m., passing through a region of manufactories, scattered along the whole course of the Blackstone and Pawtucket rivers.

According to the census of 1840, there were in Providence

22 commercial and 55 commission houses in foreign trade with a capital of $1,592,850 ; 392 retail stores, wit i a capital of $1,758,040 ; 18 lumber-yards, with a capital of $170,5?? the fisheries employed a capital of $130,000 ; machinery was manufactured to the amount of $270,000 : the precious metals to the amount of $147,550 ; one fulling-mill and one woollen factory employed a capital of $10,000 ; 21 cotton factories with 76,554 spindles, and eight dyeing and printing works, employed a capital of $1,448,000 ; three tanneries employed a capital of $22,000 ; two distilleries and two breweries, a capital of $51,000 ; paints and drugs employed a capital of $30,000 ; one paper-mill produced : the amount of $15,000 ; five grist-mills employed a capital of $6500 ; eight printing-offices, five binderies, two daily, three weekly, and four semi-weekly newspapers, and ?? periodicals, employed a capital of $23,100. The total amount of capital employed in manufactures was $3,812,5??. There were two colleges, 394 students ; 21 academies, 259 students ; 42 schools, 6639 scholars.

Providence was first settled in 1636 by Roger Williams, who was banished from Massachusetts on account of his religious opinions, and who adopted in his new establishment the principles of universal toleration. It was incorporated as a town in 1649. It suffered considerably in the Indian war of 1675. In September 23rd, 1815, a southeast early storm drove an unusual tide into the harbour, raising the water 12 feet higher than the usual spring tides, spreading devastation and ruin along the wharves, destroying the bridge, demolishing one church, overturning houses and stores, and doing immense damage to the shipping. The total loss was estimated at over a million of dollars. In 1831 Providence was chartered as a city, and is governed by a mayor, six aldermen, and 24 common council-men.

PROVIDENCE, p. t., Saratoga co., N. Y., 15 m. N.W. Ballston Spa, 41 m. N.N.W. Albany, 409 W. Watered by Sacandaga river. It has two stores, one woollen factory, three grist-mills, 30 saw-mills, two tanneries ; nine schools, 313 scholars. Pop. 1507.

PROVIDENCE, p. t., Luzerne co., Pa., 146 m. N.E. Harrisburg, 250 W. Watered by Lackawannock creek. It abounds with anthracite coal, and contains six stores, three flouring-mills, eight saw-mills. Pop. 1143.

PROVIDENCE, one of the Bahama islands, which see.

PROVINCETOWN, p. t., Barnstable co., Mass., 123 m. S.E. Boston, 340 W. Occupying the extreme N.W. part of cape Cod. It has a good harbour within the point of the cape, with a sufficient depth of water for the largest ships. The inhabitants are chiefly employed in the cod and mackerel fisheries. Salt is extensively manufactured. It contains 15 stores, two lumber-yards ; two academies, 109 students ; eight schools, 667 scholars. Pop. 2122.

PRUSSIA, an important European kingdom, between the 49th and 56th deg. N. lat., and the 6th and 23d deg. E. long. The principal part of the Prussian dominions lies continuously along the S. shore of the Baltic, between Russia and Mecklenburg, comprising the N. part of what was formerly Poland, and most part of the N. of Germany. The inland frontier of this part of the monarchy on the E. and S. is sufficiently connected ; but on the W. side it is very rugged, some small independent states being almost entirely surrounded by the Prussian dominions. But exclusive of this principal portion, there is an extensive Prussian territory on both sides the Rhine, divided into the provinces of Westphalia and Rhine. This portion is separated from the rest of the monarchy, or from what may be called the Eastern states, by Hesse Cassel, part of Hanover, Brunswick, &c. The Canton of Neufchatel, in Switzerland, and some detached territories in Saxony, also belong to Prussia.

Eastern Prussia has on the N. the Baltic ; on the E. Russia, Russian Poland, and Cracow ; and on the S. and W. the Austrian states of Galicia, Moravia, and Bohemia, with Saxony and other German states. West Prussia, or the provinces on the Rhine, have on the N. and E. Hanover and other German states; on the S. France ; and on the W. Belgium and the Netherlands. From the extreme eastern frontier of Prussia to Aix-la-Chapelle, in an E.N.E and W.S.W. direction, the distance is about 775 m. and from the promontory on the Baltic, above Stralsund, to the extreme southern frontier of Silesia, in a N.E. and S.W. direction, the distance is 404 m. Owing, however, to the irregularity of the frontier, and the intervention of other countries, these measurements give no information as to the extent of the monarchy.

The disjointed state of the dominions of Prussia detracts materially from her power. The possession of Warsaw gives Russia a position whence she may attack the very centre of the monarchy. An alliance with Saxony would bring an Austrian army within a few days' march of Berlin ; and the Rhenish provinces are exposed to be overrun by France. The government, aware of the weakness occasioned by the circumstances now alluded to, have systematically laboured to give a more compact form to the dominions. But they have only partially succeeded; and it is, in consequence, necessary for the public security that the military establishment should be placed on a very imposing footing.

Aggrandisement of Prussia.—The rise of the Prussian power has been rapid and extraordinary. The kings of Prussia are descended from petty German princes, who, in the 14th century, were burgraves of Nuremberg. In 1415, Sigismund, emperor of Germany, sold the marquisate (afterwards electorate) of Brandenburg to Frederick, one of these burgraves, for 400,000 ducats, who, by this purchase, laid the foundation of the future grandeur of his family. (Pfeffel, *Histoire d'Allemagne*, anno 1417.) In 1515, Albert, margrave of Brandenburg, was elected grand master of the Teutonic knights, who then possessed ducal Prussia; and by a treaty concluded in 1525, this territory was secularised and erected into a dachy in favour of Albert and his successors. (Pfeffel, 1525.) In 1657, Prussia was acknowledged by Poland to be a free and independent state; and, after other aggrandisements, it was advanced to the dignity of a kingdom in 1700. Part of Pomerania was not long after added to Prussia.

But notwithstanding these acquisitions, when Frederick the Great ascended the throne, in 1740, his disjointed dominions did not contain 2,500,000 inhabitants, who had made but little progress in the arts, or in the accumulation of wealth. But this extraordinary man, with no extrinsic assistance, and by mere dint of superior talent, wrested, in the early part of his reign, the valuable and extensive province of Silesia from the house of Austria. He afterwards defended himself, during the seven years' war, against the combined efforts of Austria, Russia, and France, and forced these powers to conclude a treaty, by which Silesia was solemnly guaranteed to Prussia. In the latter part of his reign, in conjunction with Russia and Austria, Frederick planned, and partly carried into effect, the partition of Poland, acquiring as his share the western parts of Prussia, and secured in addition to the increase of territory, an unrestricted communication between the different great divisions of his kingdom.

By these different acquisitions, Prussia, at the death of Frederick, in 1786, had been increased in size nearly a half; while, owing to the superior fertility of the conjoined provinces, and the improvement effected in every part of his dominions, after the peace of 1763, the population had increased, according to the estimate of the Prussian writers, to about 6,000,000.

Prussia acquired by the subsequent partition of Poland in 1792, and its final dismemberment in 1795, a great extension of territory, including the important city of Dantzic, and upward of 2,000,000 inhabitants. In addition to this, she acquired the bishopric of Paderborn and the principalities of Bayreuth and Anspach, with several lesser districts in Germany; so that in 1805, according to the statement of Krug, she contained nearly 5000 geographical square miles of territory, and a population of 9,640,000.

Her disastrous contest with France in 1806, and her subsequent humiliation, are well known. But the spirit of the people was not subdued; and after Napoleon's campaign in Russia, the population rose *en masse*, and the zeal and bravery of the Prussians were mainly instrumental in effecting the final overthrow of Napoleon. At the general peace of 1815, Prussia became more powerful than ever. She recovered all her former possessions, except a portion of her Polish dominions assigned to the kingdom of Poland; but this was more than compensated by valuable acquisitions in Saxony, Pomerania, and the Rhenish provinces.

Account of the Provinces and Regencies of the Prussian Monarchy in 1837, specifying their Extent in Geographical and British Square Miles; their Population in 1816, 1825, and 1837, and the Density of the Population per Geographical Square Mile in 1837.

Provinces and Regencies.	Area in Geog. sq. m.	Area in British sq. miles, 69·15 to a degree.	Pop. 1816.	Pop. 1825.	Pop. 1837.	Pop. per Geog. sq. mile, 1837.
Prussia :						
Königsberg	409·13		583,104	673,268	735,868	1,810
Gumbinnen	298·21		351,058	478,640	556,066	1,865
Dantzic	152·28		333,058	310,244	341,975	2,246
Marienwerder	319·41		395,184	497,117	491,696	1,539
	1,178·03	24,974	1,432,404	1,829,269	2,125,535	1,804
Posen :						
Posen	321·68		570,758	706,396	779,895	2,494
Bromberg	214·83		243,190	325,529	379,913	1,764
	536·51	11,374	813,948	1,031,925	1,159,608	2,160
Brandenburg :						
Potsdam with Berlin	382·51		688,300	802,038	968,272	2,531
Frankfort	348·43		565,876	634,882	725,770	2,082
	730·94	15,496	1,254,176	1,436,920	1,694,042	2,318
Pomerania :						
Stettin	236·88		310,952	389,412	452,387	1,910
Cöslin	258·56		234,421	298,218	360,634	1,395
Stralsund	79·02		125,988	142,312	157,096	1,988
	574·46	12,179	671,361	829,942	970,117	1,689
Silesia :						
Breslau	248·14		764,822	903,404	1,010,639	4,072
Oppeln	243·06		516,619	647,399	798,209	3,284
Liegnitz	250·54		632,652	729,818	836,318	3,338
	741·74	15,725	1,914,093	2,280,621	2,645,166	3,561
Saxony :						
Magdeburg	210·13		460,405	520,272	589,686	2,806
Merseburg	188·76		485,531	558,584	643,779	3,411
Erfurt	61·74		234,477	263,231	305,888	4,954
	460·63	9,765	1,180,413	1,342,087	1,539,353	3,342
Westphalia :						
Munster	132·17		347,537	380,054	402,144	3,043
Minden	95·68		335,609	369,204	412,587	4,312
Arnsberg	140·11		374,713	427,652	502,810	3,589
	367·96	7,801	1,057,859	1,176,910	1,317,541	3,581
Rhine Provinces :						
Cologne	72·40		324,632	363,896	418,481	5,780
Dusseldorf	96·32		587,922	652,875	750,156	7,791
Coblentz	109·64		337,478	392,573	449,195	4,096
Treves	131·13		288,289	342,684	437,383	3,335
Aix-la-Chapelle	75·65		307,324	336,025	369,103	4,879
	487·14	10,397	1,845,645	2,087,983	2,433,250	5,995
Total of civil inhab.			10,169,899	12,075,657	13,883,612	2,734
Military .			179,132	181,063	214,513	
Total of civil and military	5,077·41	107,641	10,349,031	12,256,725	14,098,125	2,776
Neufchatel	13·95	296	53,000	52,223	59,448	4,261
Gross Totals	5,091·36	107,937	10,402,031	12,308,948	14,157,573	2,781

PRUSSIA.

Considering, however, the importance of Prussia, as a counterpoise to Russia on the one hand, and to France on the other, it were to be wished that her territories were both more extensive and more compact. One of the greatest faults committed by the congress of Vienna was the diminution of the acquisitions of Prussia in Poland. This was to be regretted, as much on account of the general interests of Europe, which requires that Prussia should be a power of the first order, as of the inhabitants of the provinces separated from her dominions.

Divisions and Extent of the Country. Population.—The Prussian monarchy is divided into eight provinces, and these again into 25 regencies, which are farther subdivided into 335 circles. Each regency takes its name from its principal city, as is seen in the preceding Table, which also shows their extent and population at different epochs.

Face of the Country. Mountains.—The surface of the Prussian states is generally flat. With the exception, indeed, of part of the Hartz mountains, in the province of Saxony, the Teutoburgher Wald, and some other mountains in Westphalia and Saxony, the volcanic districts in it and the lower Rhine, and the Riesengebirge, or Giant's mountains, on the S.W. confines of Silesia, there is no other tract that is more than hilly. Prussia is, in fact, a country of vast plains; and is in most parts so very level, that many marshes and small lakes have been formed by the inundations of the rivers. The eastern, or principal part of the monarchy slopes imperceptibly from the S. frontier toward the Baltic, the shore of which is low and sandy. From this circumstance, and the nature of the soil, which in many places consists of little else than mere loose sand, some geologists have supposed that the sea had at one time overspread the greater part of its surface; and there would seem to be considerable plausibility in the supposition. At a comparatively recent period the country was covered in most parts with immense forests, of which there are still very extensive remains. These, when they belong to the crown, are under the control of the administration of forests.

Soil.—The quality of the soil is very various. In Brandenburg and Pomerania it is generally poor : in many parts, indeed, it consists of tracts of loose barren sand, diversified with extensive heaths and moors ; but, in other parts, particularly along the rivers and lakes, there is a good deal of meadow, marsh, and other comparatively rich land. In Ducal Prussia and Prussian Poland, including the province of Posen, the soil consists generally of black earth and sand, and is in many parts very superior. But Silesia, and the Saxon and Rhenish provinces, are, naturally, perhaps, the most productive. The plain of Magdeburg, on the left bank of the Elbe, is, perhaps, the most fertile and best cultivated district of the monarchy.

Rivers and Lakes.—Prussia is extremely well watered. The Rhenish provinces are traversed by the Rhine, while their E. frontier is partly formed by the Weser. The Elbe traverses the Saxon provinces ; the Oder, which is almost entirely a Prussian river, runs through the whole extent of the monarchy, from the S. frontier of Silesia to the isle of Usedom, where it falls into the Baltic. Polish Prussia (or Posen) is watered by the Wartha ; West Prussia by the Vistula ; and Ducal Prussia by the Pregel and Niemen. And, besides the above, there are many other large rivers, as the Ems, Moselle, Spree, Havel, Netz, &c.

Owing to the flatness of the country through which they flow, none of the great rivers are interrupted by cataracts, and they are all navigable—the Rhine, Elbe, and Vistula throughout their whole course in the Prussian dominions : the Oder is navigable, for barges, as far as Ratibor in S. Silesia ; and the Pregel and Niemen to a considerable distance inland. The establishment of steam-packets on these rivers, and the freeing of the navigation of the Rhine and the Elbe from the oppressive tolls and regulations by which it was formerly obstructed, have already been, and will no doubt continue to be, of vast service to the country. Canals have also been constructed, connecting the Elbe, the Oder, and the Vistula ; so that goods shipped at Hamburg may be conveyed by water to Dantzic, and conversely. (*See* the accounts of the above rivers, under their different names.)

Lakes are exceedingly numerous, particularly in Ducal Prussia and Pomerania. There are also along the coast several large bays, or rather lagoons, communicating with the sea by narrow mouths, and possessing more of the character of fresh-water lakes than of arms of the sea. They are denominated *Hafs*, the principal being the Curische Haff and the Frische Haff, on the coast of Ducal Prussia, and the Haff at the mouth of the Oder.

Seaports.—The principal seaports are Memel, Königsberg, or rather Pillau, Dantzic, Stettin, and Stralsund. With the exception of Stettin, or rather of its outport, Swinemunde, the water at these ports is rather shallow, seldom exceeding from 10 feet to 12 feet. But at Swinemunde there are from 19 feet to 21 feet.

Climate.—The climate of Prussia is not less various than

the soil. Along the Baltic it is moist, and, in Ducal Prussia especially, the winter is long and severe. It is also severe in the S. parts of Silesia, contiguous to the Carpathian mountains. In N. Silesia, Brandenburg, and the Saxon and Danish provinces, it is comparatively mild.

Minerals.—The Prussian monarchy is richer in minerals than might have been anticipated from its flatness. Iron is the most generally diffused. It is very extensively wrought in Silesia, principally on account of the crown, but also by private individuals. The iron-works in the Rhine provinces, near Dortmund, Solingen, Iserlohn, &c., and those near Schmiedeberg, Tarnowitz, Sprottau, &c., in Silesia, are very extensive. Coals are very abundant in the Rhenish provinces, Saxony, and parts of Silesia, and large quantities are annually produced. Salt, which is a government monopoly, is produced principally in the Saxon provinces, which also yield considerable quantities of copper, and some silver. Silesia furnishes annually large quantities of zinc, lead, and tin ; but the last-mentioned metal is partly, also, supplied by Brandenburg. Amber has long been known as a product of Ducal Prussia. It is principally found along the low narrow tongue of land between the Curische Haff and the sea.

If we distinguish the mineral products into, I. metals ; II. combustible minerals ; III. stones ; IV. clay, sand, earth; V. salt, alum, &c.—we have in Prussia,

I. Metals.—*Silver*, in Saxony (Maarfeld). Westphalia (Siegen), Silesia. *Copper*, Saxony (Maarfeld). Westphalia (Siegen), Silesia. *Lead*, Silesia, Rhine, Westphalia, Saxony. *Iron and steel*, in every province, but principally in the mountains of Silesia, Westphalia, and Rhine. *Cobalt*, Westphalia (Siegen) and Saxony. *Arsenic*, Silesia. *Calamine and zinc*, Silesia, Rhine, and Westphalia.

II. Combustible Minerals.—*Sulphur*, Silesia, Saxony. Prussia. *Pit-coal*, Silesia, Westphalia, Rhine, Saxony. *Brown-coal*, Saxony and Rhine. *Turf*, in every province, principally in Brandenburg.

III. Stones.—*Amethyst, agate*, in Silesia. *Alabaster*, Saxony. *Marble*, Westphalia, Saxony, Rhine, Silesia. *Volcan tophus*, Rhine province, and very important. *Serpentine stone*, Silesia. Of *sandstone, mill-stones*, in Silesia, Saxony, Westphalia, Rhine. *Grinding*, or *whet-stones*, in Westphalia, Silesia, Saxony. *Limestone*, in Silesia, Westphalia, Rhine, Saxony, Brandenburg. *Gypsum* in the same provinces as limestone. *Slate*, Westphalia and the Rhine.

IV. Clay, Sand, Earths.—*Porcelain earth*, in Saxony, near Halle. *Pipe-clay* and *fuller's earth*, Silesia. *Sand*, suitable for the fabrication of glass, in all provinces. *Brick-clay* and *marl*, in all provinces.

V. Salt.—*Kitchen salt*, in Saxony, Westphalia, Pomerania, Rhine. *Alum*, Silesia, Saxony, Westphalia, Rhine, Brandenburg. *Saltpetre*, in some provinces.

In 1835 there were produced in the monarchy ... cwt. zinc, 1,653,297 cwt. iron, 10,896,433 tons coal, ... salt, &c. The total value of the minerals produced in that year has been estimated at 9,196,365 rix-dollars.

(*See*, for farther details, the valuable and important work of the able statistical writer, Von Dieterici, *Statistische Uebersicht der wichtigsten Gegenstände des Verkehrs und Verbrauchs im Preussischen Staate*. Berlin, 1838.)

Vegetable and Animal Productions.—These do not differ materially in Prussia and Great Britain. Rye and wheat, with buckwheat, oats, barley, potatoes (now very extensively cultivated), and flax and hemp, are the principal products of Prussian agriculture. About 700,000 cases of very fair wine are made in the Rhenish provinces. It is mostly consumed in the country, the exports being so very trifling as not to exceed from 3600 to 4000 casks a year. The average price of Prussian wine may be estimated at from 15 to 50 rix-dollars per eimer. The animals of Prussia are the same with those of this country, except that wolves and wild boars, which were long since exterminated in Great Britain, continue to exist in considerable numbers in the Prussian territories.

Agriculture.—Down to a comparatively recent period, the state of landed property in Prussia, and the condition of the occupiers of the soil, was similar to its state and their condition in most parts of continental Europe. The country was mostly divided into pretty considerable estates ; and down to 1807, none but nobles or privileged persons could acquire landed property. Such parts of an estate as were not in the immediate possession of the lord, were held by occupiers, in a sort of predial slavery, on condition of their paying a certain rent, consisting sometimes of service to be performed on the lord's land, sometimes of the delivery of a certain proportion (generally a *half*) of the produce, and more frequently, perhaps, of both the one and the other. In some places the tenants had acquired a sort of hereditary right to their possessions, on their making the accustomed payments ; but in other parts the title to the lands they occupied was only for life or for a certain number of years ; though, by a most absurd regulation, the proprietor could

>t then resume the lands into his own hands, but was
obliged to relet them to an occupier of the same grade as
the one who had left them ! In 1807, however, the regulation which prevented peasants, tradesmen, &c., from acquiring land was abolished ; and in 1811 appeared the famous edict which enacted that all the peasants who held
perpetual leases, on condition of paying certain quantities
of produce, or of performing certain services on account of
the proprietors, should, upon giving up *one third* part of the
land held by them, become the unconditional proprietors
of the other *two thirds* ! And with respect to the other
classes of peasants, or those who occupied lands upon life-
leases, or leases for a term of years, it was enacted that they
should, upon giving up *half* their farms, become the unconditional proprietors of the *other half* ! This edict certainly
effected the greatest and most sweeping change that was
ever peaceably effected in the distribution of property in
any great country. It was regarded at the time, and in
some respects justly, as a dangerous interference with the
rights of individuals. But the abuses which it went to
eradicate were so injurious to the public welfare, and were,
at the same time, so deeply seated, that they could not
have been extirpated by any less powerful means. It has
given a wonderful stimulus to improvement. The peasantry, relieved from the burdens and services to which they
were previously subjected, and placed, in respect of political
privileges, on a level with their lords, have begun to display
a spirit of enterprise and industry that was formerly unknown. Formerly, also, there were in Prussia, as there
are been in England and most other countries, a great extent of land belonging to towns and villages, and occupied
in common by the inhabitants. While under this tenure,
these lands rarely produce a third or fourth part of what
they would produce, were they divided into separate properties, and assigned to individuals, each reaping all the advantages resulting from superior industry and exertions.
The Prussian government, being aware of this, has succeeded in effecting the division of a vast number of common properties, and has thus totally changed the appearance
of a large extent of country, and created several thousand
new proprietors. The want of capital and the force of old
habits rendered the influence of these changes at the outset
as striking than many anticipated : but these retarding
circumstances have daily diminished in power ; and it may
safely affirmed, that the country has made a greater
progress since 1815 than it did during the preceding hun-
-d years.

The frequency of mortgages, and the embarrassed state
great numbers of the landed proprietors, are still, how-
-r, loudly complained of. Perhaps the extent of the
l is exaggerated ; and we incline to think that it is in no
considerable degree ascribable to the institution of land-
-ks, or rather of associations for the lending of money
the security of land. These associations were establish-
-with the most laudable intentions : but the facilities
y have afforded of contracting debt, coupled with the
-e risk there is of the principal ever being demanded,
-ided interest be paid, have tempted many individuals
-dulge in inconsiderate expenditure, and have made
-n injurious rather than otherwise.

-ye used to be in Prussia an article of universal con-
-ption, occupying the same place there that wheat oc-
-es in England, and potatoes in Ireland. Latterly, how-
-, potatoes have been gaining upon it, and now form a
important part of the food of the people. The usual
-se is to fallow every third year, taking either first a
of rye, and then wheat, or conversely. The greater
of the wheat shipped at Dantzic is brought from the
-h provinces under Russia and Austria. Flax and
-p are cultivated for domestic use, and also for sale, in
-arts of the monarchy, but especially in Silesia. The
-flax is raised from foreign seed, the seed produced at
-e being used to make oil-cake. Tobacco, hops, mad-
-nd other plants, used in dyeing, are also raised. Chic-
-ry largely cultivated. Beet-root plantations are very
-sive, and have recently made great progress, especial-
-Saxony and Silesia. Their produce may be estimated
-m 100,000 to 120,000 quintals, being about a fourth
f the sugar consumed in the monarchy.

-ept on the crown estates, there are few farms. Most
-erable landed proprietors are accustomed to manage
-estates by stewards ; and, as already seen, the smaller
-ers are mostly all proprietors. It is impossible to
-ny accurate estimate of the productiveness of the
-ffering as it does so very widely in quality, culture,
-n the most fruitful and best cultivated districts, as
-f Magdeburg, the produce of wheat is reckoned at from
16 *scheffel* the *morgen* (1 scheffel = 1·6 bushel ; 1
-= 1·59 acre) ; whereas, in Brandenburg and Pom-
-the produce of wheat is not reckoned at more than
to 10 scheffel the morgen. The produce of rye va-
-the best districts from 4 to 8 sch. per morgen. But

in the sandy and sterile portions of Pomerania, Brandenburg, and W. Prussia, the produce is often not more than 2
or three sch. per morgen. The rent of cultivated land is
as various as the crops, being dependent partly on soil and
partly on situation. In the immediate neighbourhood of
Berlin, the best lands yield a rent of 15 riz dollars per morgen ; in the country around Magdeburg the rent is in general about 10 r.-d. per do. ; in Brandenburg, if it be not in
the neighbourhood of Berlin, the rent is seldom more than
2½ r.-d. per morgen ; and in the sandy and sterile tracts, the
rent is sometimes not more than 1 or 2 silver groschen per
morgen.

Horses, cattle, and sheep are raised everywhere throughout Prussia. The growth of the latter has been of late
years an object of much attention, particularly in Brandenburg, Saxony, and Silesia. In consequence of the improvements effected by the introduction of merino sheep into
Germany, the wool of Saxony, Silesia, and some other
provinces, has become superior even to that of Spain. The
fall in the prices of corn, subsequent to 1815, gave a great
stimulus to this branch of industry. Wool now constitutes, in fact, the principal article of export from Germany,
and has been productive of much wealth to many Saxon
and Silesian proprietors, as well as to many in other provinces. The number of sheep in the Prussian dominions is
estimated at about 15,000,000, and the total produce of wool
at from 80,000 to 100,000 quintals, of which about a third
part is exported. Hogs are very extensively reared in
Westphalia, and immense numbers of geese are bred in
Pomerania and the N. part of Brandenburg. We subjoin
An Account of the number of Horses, Cattle, Sheep, and
Hogs, in the different Provinces of Prussia, as determined
by an enumeration made in 1837 :

Provinces.	Horses and Foals.	Black Cattle.	Sheep.	Goats.	Hogs.
Prussia	442,591	879,969	2,149,577	8,464	551,583
Posen	144,124	416,364	3,165,070	5,524	232,386
Brandenburg	179,968	591,914	4,982,943	43,770	475,194
Pomerania	135,915	401,456	2,114,609	12,749	139,847
Silesia	311,650	891,457	2,769,199	82,999	124,252
Saxony	141,749	457,736	2,170,752	71,484	253,323
Westphalia	129,409	578,896	953,304	79,956	287,966
Rhine prov.	122,134	775,913	605,753	52,134	275,884
Total	1,474,964	4,838,622	15,911,862	307,328	1,796,365

Manufactures.—Though more of an agricultural than a
manufacturing country, Prussia has greatly distinguished
herself, particularly of late years, in various branches of
manufacture. The Rhenish provinces, Saxony and Silesia, are the districts most prominent in this department.
Linens and coarse woollens for domestic consumption are
made in every village, and, indeed, in most cottages throughout the kingdom. The linens that are exported are chiefly
produced in Silesia, Westphalia, and the Ermeland, or portion of Ducal Prussia containing the circles of Braunsberg,
Heilsberg, Rossell, and Allerstein. The manufacture in Silesia was formerly very extensive, but latterly it has been,
to a considerable extent, interfered with by that of cotton.
The total value of the linen stuffs annually manufactured
is estimated at from 25 to 30 millions rix-dollars, of which
a fourth part is exported. Hirchberg, and the adjacent
towns and villages, are the principal seats of Silesian manufactures. Large quantities of silk and cotton goods, linens, &c., are produced in Elberfeld, and other towns in the
Rhenish provinces. Very superior broad cloth is largely
manufactured at Eupen, Malmedy, Berlin, Aix-la-Chapelle,
&c. Prussia occupies a respectable rank in the production
of the useful metals. The total quantity of bar and pig
iron annually produced may (1840) be estimated at about
130,000 tons. The articles of hardware, cutlery, &c., made
at Iserlohn, Hagen, Solingen, Olpe, Essen, &c., enjoy a
high reputation ; but in this respect they are very inferior
to the cast-iron articles, whether of fancy, ornament, or
utility, produced at Berlin : these, as regards beauty and
delicacy of execution, are unequalled by any made either in
England or any other country. Porcelain, jewellery, watches, coaches, &c., are largely produced at Berlin and other
towns. Vast numbers of books annually issue from the
presses of Berlin and Halle. Beer and spirits are very extensively produced, and consumed in all parts of the monarchy. The consumption of spirits amounts, in fact, to
from 160,000,000 to 180,000,000 quarts, or from about 40 to
45 millions imperial gallons ! (3·9 Prussian quarts = 1
imp. gall.) Now, it may be worth while observing, as illustrative of the habits of the people, that the entire quantity of British and foreign spirits entered for home consumption in the United Kingdom in 1840, amounted to only
25,517,396 gallons, notwithstanding our population is about
double that of Prussia. Indeed, if we take Prussia for a
standard, the people of the United Kingdom may be said
to be temperate in the extreme ; for, while the consumption
of spirits in Prussia amounts, at an average, to about three

gallons to each individual, the consumption in Great Britain and Ireland is only about three fourths of a gallon! and we believe that the consumption of beer in Prussia exceeds its consumption in the United Kingdom in a corresponding proportion.

The principal manufacturing district of Prussia, and probably of the continent, is in the Rhenish provinces on the Wupper, having Elberfeld and Sollingen for its principal towns. It is well supplied with coal and water-power; and the inhabitants are alike industrious and inventive. The population of Elberfeld has increased during the present century from 11,790 to 38,162; and the progress of many of the other towns and villages in the vicinity has been hardly less remarkable.

Some of the manufacturing establishments in the Rhine district are on a large scale, employing from 400 to 500 work-people. The first steam-engine used in Prussia was set up in 1780. In 1840, there were about 450 steam-engines in the monarchy, of which above 30 were a loan. The wages of the work-people employed in manufactures vary according to the work to be performed, the expense of living, &c. At Elberfeld, weavers earned in 1836 from 1 to 3 rix-dollars, or from 6s. to 9s. a week, working from 9 to 12 hours a day.. Children employed in factories earn from 3d. to 4d. a day.

We subjoin an official return of the looms at work in the different provinces of the monarchy in 1837, specifying the departments in which they were employed, and whether they were employed as a principal or secondary business.

Provinces	Looms :— Principal Business.						Looms :— Secondary Business.			
	Cloths and Stuffs of all kinds.				Stockings.	Ribbands.	Linen.	Coarse Woolen Stuffs.	Other Stuff.
	Silk or mixed Silk.	Cotton or mixed Cotton.	Wool or mixed Wool.	Linen.					
Prussia . .		16	557	710	36	13	9a,548	2,196	3
Posen . .		25	997	1,236	5	5	24,945	179	
Brandenburg .	2,315	4,998	4,621	5,844	306	119	94,377	146	39
Pomerania .		30	523	2,298	14	4	35,268	466	17
Silesia . .	78	17,730	2,617	12,347	940	223	11,2.8	435	1,1a
Saxony . .	129	2,775	2,606	4,237	351	731	13,303	54	38
Westphalia .	116	2,947	564	3,431	363	102	20,990	57	76
Rhine provinces	11,475	9,564	4,457	3,906	892	2,422	72,974	576	1,32?
Totals .	14,111	36,234	16,937	35,a77	2,118	4,340	346,294	4,806	1,9a4

OFFICIAL ACCOUNT of the Number of certain Fabrics and Manufactures in the different Provinces of Prussia in 1837.

Provinces.	Type Foundries.	Book and Music Printing Offices.	Copper and Steel Engraving Offices.	Lithographic Institutions.	Iron Furnaces.	Other Furnaces driven by Water.	Copper Works.	Chemical Fabrics.	Sugar Refineries.	Beet-root Sugar Factories.	Starch and Fine Meal Factories.	Pot and Wood Ash Factories.	Glass Works.	Porcelain and Earthenware Factories.	Limekilns.	Sawing and Tile Mills.	Tar Works.
Prussia . .	1	83	1	12	63	5	7	3	11	7	3	19	11		134	38	
Posen . .		8	1	14	2		1	3		5	12	19	3		47	46	4
Brandenburg .	8	61	11	80	95	10	9	18	92	9	33	9	19	19	103	32	31
Pomerania .		14	1	12	6		4	1	2	9	11		11		5	20	19
Silesia . .	1	80	2	37	263	117	11	90	5	14	56	60	25	9	171	73	1
Saxony . .	3	79	8	41	44	28	6	90	13	27	36	25	5	3	475	39	4
Westphalia .		44	2	33	401	345	11	30	4		10	212	14	3	345	571	1
Rhine provinces	4	112	11	85	288	276	24	49	31	6	10	415	19	87	463	135	1
Totals . .	17	401	37	264	1,062	779	74	144	89	76	170	750	100	99	1,741	1,90?	78

Commerce.—The exports from Prussia consist principally of corn, wool, timber, Westphalian hams, zinc, flax, bristles, salted provisions, and other articles of raw produce; with linen and woollen cloths, silk wares, iron and hardware, jewellery, watches, and wooden clocks, Prussian blue, spirits, beer, &c. The imports consist chiefly of sugar, coffee, and other colonial products, raw cotton, and cotton twist and stuffs, indigo and other dye stuffs, spices, French and other wines, cash for the use of the ports on the Baltic, salt, &c. The amount of the trade of Prussia cannot, owing to the free system of internal commerce now established in Germany, be ascertained with any precision; but it is very considerable, and is rapidly increasing. Except in dear years, when we are large importers of corn, our trade with Prussia is carried on at second-hand through Hamburg, Bremen, and the Netherland ports. But we have elsewhere shown that the real value of our exports to Prussia may be fairly estimated at nearly three millions a year. (*Supp. to Commercial Dictionary.*) Besides the facilities afforded to internal commerce by the rivers and canals already alluded to, others of a very important nature have recently been furnished. Previously to 1815, the roads in Prussia were, with few exceptions, about the very worst in Europe. They were, in fact, mere pathways, without any artificial construction; and owing to the loose sandy nature of the soil, the wheels not unfrequently sunk in them to the axle, and the carriage was drawn rather through than over the ground. But since the period alluded to, a very great change has been effected. New roads, constructed on the most approved principles, macadamised, and equal to any in England, are now carried from Berlin to all the most distant parts of the monarchy. The cross roads have also been materially improved; and every day is adding to the means of communication thus afforded. In all, about 1900 German miles of road have been constructed since 1815, three fourths of which have been made at the expense of government, and the rest by the country districts. Mail coaches, which travel at the rate of about six miles an hour, are established along the principal lines of road. They are under the orders of the government, and are well conducted. All travelling charges are regulated by a tariff fixed by the police. Railways have been opened, or are about to be, from Berlin to Potsdam, and from Cologne to Aix-la-Chapelle; and others

either have been opened, or are in the course of being constructed.

Within the last few years Prussia has prevailed on far the greater number of the secondary and smaller German states to enter into a commercial league, by adopting a uniform tariff of duties on imports, and establishing a free system of internal commerce. Previously to the adoption of this plan, each petty state had its own customhouses, and its own system of duties and revenue laws; these frequently differed widely from those of its neighbours, so that the internal trade of the country was subjected to all the vexatious restrictions that are usually laid on the intercourse between distant and independent states, and was, in consequence, comparatively trifling. But these restraints are now entirely got rid of. Internal custom-houses and separate custom duties no longer exist. Each state participates, in proportion to its population, in the amount of the duties collected at the frontiers of the league; and a commodity admitted at any one of the external custom-houses may be subsequently conveyed, without let or hindrance from Aix-la-Chapelle to Tilsit, and from Stettin and Dantzic to the frontiers of Switzerland and Bohemia. It has been supposed by many that this system threatened to be very injurious to the trade of Britain and Germany; but there seems to be no ground whatever for any such apprehension. The freedom of internal commerce will do more to promote the prosperity of the countries included within the league, than any other measure their rulers could have adopted; and, as population increases, and the inhabitants become more wealthy, there will, no doubt, be an increased demand for foreign products. Generally speaking, the duties are moderate. It is, indeed, obvious that were we to attempt made to raise them to an exorbitant height, the facilities for smuggling along the frontiers of the league are so very great, that its territories would very speedily be deluged with overtaxed products. And in addition to this signal reform, the tolls and other impediments that formerly obstructed the free navigation of the Rhine to the Elbe are now mostly removed; and there are no longer any exclusive companies, or incorporated guilds or bodies, to obstruct the general freedom of industry and competition.

Shipping.—Considering the extent of sea-coast possessed by Prussia, and the facilities she enjoys for ship-building, the shipping is not very considerable. In 1825 there belonged to the different Prussian ports 576 ships, of the burden of 38,077 lasts of 4000 lbs. each. In 1838, Prussia had 901 ship, of 73,696 lasts. Stettin possesses the largest amount of shipping, and next to it Dantzic. Prussia has entered into reciprocity treaties with most foreign powers.

Money, &c.—Accounts are kept in rix-dollars, or thalers, of 30 silver groschen. Each rix-dollar should contain 557·69 gr. fine silver, and is worth 2s. 11½d., but is generally taken at 3s. The centner, or quintal of 110 lbs. is equal to 113·381 lbs. avoirdupois. The last, by which ships' tonnage and freights are estimated, contains 4000 Prussian lbs. One Prussian mile is equivalent to 4·68 English miles. The morgen is equal to 1·52 Imp. acres.

Races; Population.—The people of Prussia belong mostly to the great German and Slavonian families; the Poles in Posen, W. Prussia, &c., belonging to the latter, and the great bulk of the inhabitants in the other districts to the former. German is the language of the court, and of all the better informed classes in all parts of the kingdom. We have seen that the population of the monarchy, exclusive of Neufchâtel, has increased from 10,349,031, in 1816, to 14,098,125 in 1837. As only 35,029 of this augmentation is due to an enlargement of territory (the Duchy of Lichtenberg, acquired in 1834), it is a more rapid rate of natural increase than has taken place during the same period in any other country; and the condition of the people being, at the same time, materially improved, it affords the most unequivocal evidence of the increase of prosperity.

No doubt a considerable portion of this increase may be ascribed to the change already noticed in the distribution of property. It is more than doubtful, however, whether the influence of this change will continue to be as advantageous as it has been hitherto. With few exceptions, there are neither entails nor majorats in Prussia; but when an individual dies, his property, whether it consists of land or moveables, is equally divided among his children, without respect to sex or seniority. The influence of this law in causing the splitting of property was limited, previously to 1807, by the regulation that none but nobles could hold land. But, as it may now be held by any one, the system of equal partition will exert its full force; and there is but too much reason to fear that, in the end, it may occasion the too great subdivision of landed property, and an excess of agricultural population.

We regard this as being, in fact, by far the most formidable of the unfavourable contingencies to which Prussia would seem to be exposed; and, unless it be met by the organization of such a system of poor-laws as will make it w the interest of the proprietors to oppose themselves to a very great subdivision of the land, and a too rapid increase of population, the consequences may be most disastrous. Already, indeed, there are in several quarters symptoms of a excess of population. This is a matter to which the attention of government cannot be too early and steadily directed.

The population, in 1837, was divided as follows:—

Age.	Males.	Females.
To 14 (completed) years of age	2,4·7,154	2,16,745
From 14 to 60 (completed)	4,153,610	4,191,5·8
Above 60 years	409,459	450,571
Total	**7,009,225**	**7,658,902**

At an average of the three years ending with 1837, the number of births, marriages, and deaths, in the different provinces of the monarchy, was as follows:—

Province.	Births.	Marriages.	Deaths.
...ssia	94,492	21,411	74,498
...sen	49,603	11,244	37,036
Brandenburg	64,044	15,379	44,842
Pomerania	34,543	8,679	26,082
...ssia	111,812	25,·04	94,314
...ony	57,±31	13,510	44,030
Westphalia	49,115	10,083	41,150
...Rhine province	92,414	90,349	76,024
Total	**557,495**	**125,022**	**438,905**

There are in Prussia, exclusive of Neufchâtel, 535 towns a population of 2000 and upward; the inhabitants of which amount in all to 3,060,315, or to about three fourths of the population.

Government.—The government of Prussia is monarchical, the exercise of the royal prerogative being modified by the ages of the different ranks and orders of the people, and more by their intelligence and the power resulting from military organization. The succession is hereditary direct male line. The king is assisted by a council e ; and there are nine ministers for particular departments, each having a salary of 12.000 rix-dollars (£1800). i have been established in the different provinces by of an ordinance issued in 1823. Projects of all laws during their peculiar provinces, or which involve changes ts, or of the rights of property or individuals, are subject to their deliberation, and are rarely passed without

their sanction; but they have no power directly to originate any measures, though they may effect this indirectly by making representations to the king. The electors in the provinces of Silesia, Saxony, Westphalia, and the Rhine, are divided into four classes or states: consisting, 1st, of mediatised princes and other superior nobles; 2d, of inferior nobles and great landholders; 3d, of burgesses; and, 4th, of peasants, little landholders, and fee-farmers: the electors in the provinces of Prussia, Posen, Brandenburg, and Pomerania are divided into three classes or states, consisting, 1st, of nobles; 2d, burgesses; 3d, peasants, little landholders, and free-farmers. The deputies are chosen by, and represent, these states. The following table shows the number of representatives for each state in each province:

Provinces.	Classes of States.			Total Rep.
	I.	II.	III.	
Prussia	45	23	23	91
Posen	26	16	8	49
Brandenburg	35	23	12	70
Pomerania	25	16	8	49
Total	**130**	**83**	**50**	**263**

Provinces.	Classes of State.				Total Rep.
	I.	II.	III.	IV	
Silesia	10	30	30	16	96
Saxony	6	29	24	13	72
Westphalia	11	20	20	50	71
Rhine province	5	25	25	25	80
Total	**32**	**110**	**99**	**74**	**315**
And the 4 other provinces	142				
	130		83	50	263
Gross total	**272**	**1·2**	**1·2**	**124**	**578**

It is next to certain, however, that very material modifications will speedily have to be made in this system of representation. Considerable disappointment was experienced at the outset of the reign of his present majesty from his not at once conceding an extension of popular rights and privileges; and the general opinion seems to be that such concessions cannot long be withheld. The city of Breslau, and other important bodies, have already petitioned the king on the subject; and should the wish for an extension of popular privileges become general, and be vigorously pressed upon the government, it will, no doubt, have to be granted.

It is common in this country to call the government of Prussia a despotism; but if we mean by a despotism a government in which the monarch may do as he pleases, with little, if any, regard for the feelings and interests of his subjects, there is, perhaps, no government less a despotism than that of Prussia. It, in fact, may be said to be self governed; for, though the king be all-powerful so long as his conduct is approved by the bulk of the people, he would, on the contrary, be quite powerless were he to lose their confidence and support. The king of Prussia has no extrinsic resources, or peculiar interests, on which to fall back in the event of his becoming unpopular. There is neither a powerful church nor a powerful aristocracy in Prussia: the army, too, is merely a portion of the citizens; and where every man is educated and every man is a soldier, the acts of the government cannot, speaking generally, be other than the acts of the public. But it is most probable, as stated above, that the Prussians will install upon a direct control over the measures of government: but whether that should be so or not, no one acquainted with the state of society in Prussia can doubt that public opinion is there all but omnipotent.

The deputies are elected for six years, and must be 30 years of age, and the electors 25. The king nominates a president and vice president of each state, chosen from the first or second order, and fixes the duration of the session. A majority of two thirds is necessary on all questions proposed by the king, but on other matters a mere majority suffices. Their sittings are not public, and the results only of their proceedings are published. All Prussians are entitled to address themselves by petition directly to the sovereign; and there is no instance of any petition, however humble the party presenting it, remaining without a distinct answer. Political writings are subject to a censorship.

A president, with powers similar to those of a French *préfet*, and a salary of 6000 rix-dollars (£900), is placed at the head of each of the eight provinces into which the kingdom is divided. Each province has also a military commandant, a superior court of justice, a provincial director of taxes, a provincial consistory, all appointed by the king. The last is divided into three sections—one having the superintendence of schools, another of ecclesiastical affairs, and another of the public health. The provinces are subdivided into regencies or counties, and these again into *kreise*, or circles (arrondissements), and the latter into *gemeinden*, or parishes (communes). Each regency has a president and an administrative board or council; and the farther subdivisions have also their local authorities. The municipal organization of the towns is more complicated than that of the communes. The principal functionaries are all elective; but the elections must be confirmed by the king or the authorities.

The system of law principally in force in the E. states of the Prussian monarchy is embodied in the well-digested code entitled *Landrecht für die Preussischen Staaten*, which received the royal sanction in 1791, and became a law in 1794; but it is occasionally modified by custom; and Polish, Swedish, and German laws are still in force in certain parts of the monarchy. The Rhenish provinces follow, with some exceptions, the rules laid down in the *Code Napoleon*. The primary proceedings in judicial matters take place before local courts established in the circles and towns; thence they may be carried before the provincial courts (*oberlandes gerichte*); and in the last resort before the supreme tribunals at Berlin. The judges are independent; and justice is purely and cheaply administered. Juries are employed in the Rhenish provinces, but not in the other parts of the monarchy. Tribunals of arbitration have also been established in the provinces of Brandenburg, Pomerania, Prussia, Silesia, and Saxony, similar to those established in Denmark (which see), and with the same beneficial results.

In 1836 of 13,831 cases brought before these tribunals in the above provinces (excepting Prussia, from which there are no returns), 11,835 were settled, and only 1996 sent to the higher courts. In no other country, perhaps, is patronage of so little consequence, and merit so sure to lead to distinction and advancement. Candidates for public employment go through a course of education appropriate to the functions they are desirous to fill; and, before being appointed, have to submit to a severe examination as to their knowledge, conduct, and fitness for the office. "*La Prusse*," says M. Cousin, "*tous les fonctionnaires sont nommés; et comme ils n'arrivent à aucune fonction qu'après des examens sévères, tous sont éclairés; et comme de plus ils sont pris dans toutes les classes, ils portent dans l'exercice de leurs emplois l'esprit général du pays, en même tems qu'ils y contractent l'habitude du gouvernement.*"

The king of Prussia is, next to the emperor of Austria, the most important member of the Germanic Confederation. His contingent to the army of the Confederation is 79,234 men.

Revenue and Expenditure. — The following is an official statement of the Revenue and expenditure of the Prussian monarchy for 1835:—

REVENUE.		Rix-d.	Rix-d.	EXPENDITURE.		Rix-d.	Rix-d.
Produce of the administration of domains and forests, after deducting the revenue reserved for the feoffee trust of the crown	. . .		4,212,000	*National Debt, viz.:* Interest of the general and provincial debt, and expenses of management		6,397,000 2,430,000	
Of the sales of domains, &c., for the more speedy paying off of the national debt	. . .		1,000,000	Sinking fund			
						8,827,000	
Of mines, forges, and salt-mines, and of the royal china manufactory at Berlin	. . .		717,000	Interest and sinking fund of debts of different provinces		41,000	
Of the post-office	. . .		1,200,000				8,968,000
Of the lottery	. . .		699,000	*Pensions and Life-Rents, viz.:*			
Of the Administration of Taxes, viz.:				Pensions to servants of the state, and their widows and children, and other charities		966,000	
(a) Taxes on landed and house property	9,735,000			Pensions for life to members of suppressed spiritual corporations, &c.		1,584,000	
(b) *Classensteuer* (poll-taxes)	6,401,000						2,550,000
(c) Taxes on trade (patents)	1,976,000			*Perpetual Rents, viz.:*			
		18,112,000		Indemnification for abolished claims and usufructs		260,000	
(d) Importation, exportation, and transit duty, on the consumption of home products: tolls;				For capitals and securities called in		622,000	
duty on navigation, &c., the use of harbours, canals, and bridges, and other communications; and stamp duty	20,052,000			For the cabinet, the office of the minister of state, accountant-general's office, and the administration of the treasury, the mint, archive of the state, and provincial archives, the secretary of state's office, the account-office, the chancellory royal, and the statistical office			982,000
(e) Revenue from salt	5,906,000			Ministry of spiritual, educational, and medical affairs			396,000
			45,530,000	Ministry of the interior and police			2,000,000
Surplus revenue from the principality of Lichtenberg	. . .		80,000	Ministry of the interior for trade and general commissions			2,216,000
Various items not comprised under the above	. . .		382,000	Administration of commerce and manufactories, including land and aquatic buildings, but exclusive of roads			172,000
				Maintenance of old and making new roads, including interest of capital raised for making roads, &c.			9,500
				Ministry of foreign affairs			2,000,000
				Ministry of war, including the supplies for the great military orphan asylum at Potsdam, and its affiliated institutions			494,000
				Central Administration of Finances, viz.:			22,000,000
				(a) The finance office and for the state treasury offices		150,000	
				(b) The general administrations of the domains and forests		94,000	
							224,000
				Ministry of Justice			2,000,000
				Chief president's and provincial courts (Regierungen)			1,700,000
				Breeding of horses (Haupt- und Landgestüte)			97,000
				Revenues not recovered, extra expenses, and improvements			1,200,000
Total	. . .		51,740,000	**Total**			51,740,000

In 1834 the revenue amounted to 52,651,000 rix-dollars, which, as above, exactly balanced the expenses.

It should, however, be observed, that the above statement represents only the nett amount of revenue after the expenses of collection are deducted. These are estimated at from 3 to 4 per cent. on the produce of the direct taxes, and at from 10 to 12 per cent. on the produce of those that are indirect. It will also be observed, that in the account of expenditure nothing is set down for the civil list, or for the support of the monarch and his court; a sum of 2,500,000 dolls. is, however, specially appropriated for this purpose.

The tax on classes is a kind of poll-tax imposed on the inhabitants of the country and towns, according to the amount of their property, and the population of the towns. The tax on industry or trades consists of the sums charged for patents or licences for leave to carry on trades, &c.

National Debt. — Frederick the Great effected all his extraordinary achievements without contracting any debt, and left at his death a large sum in the coffers of the state. Though his successors were not so economical, still, at the breaking out of the war with France, Prussia was free from debt. In the course of that contest she was compelled to borrow pretty large sums, which, however, have been considerably reduced since the peace. The public debt of Prussia amounted, in 1840, to about 130,000,000 rix-dollars, or £19,500,000.

Religion. — The royal family belongs to the reformed v Protestant religion; but all denominations of Christians enjoy the same privileges, and are equally eligible to places of trust or emolument. In 1837 the population consisted of 8,604,748 Protestants, 5,294,603 Catholics, 183,579 Jews, and 14,495 Mennonites. The Protestants predominate very decidedly in Brandenburg, Pomerania, Saxony, and Ducal Prussia; while the Catholics predominate in the Rhenish provinces and Westphalia, in the regency of Oppeln, in Silesia, and in Posen. When Silesia was acquired by Prussia, the mass of the population were Catholics; but at present the Protestants predominate in the regencies of Breslau and Liegnitz, particularly the last.

The Protestant church is governed by consistories, or boards appointed by government, one for each province. There are also synods in most circles and provinces, but no general synod has yet been held. The constitution of the Catholic church differs in different provinces. In the Rhenish provinces it is fixed by the concordat entered into between the French government and Pope Pius VII. But in every part of the monarchy the crown has wisely reserved to itself a control over the election of bishops and priests In the entire kingdom there were, in 1837, 5748 ordained Protestant clergymen, and 140 assistant do.; there were, at the same time, 3510 Catholic priests, and 2033 vicars, chaplains, &c.; so that the proportion of Catholic clergymen rather exceeds that of Protestants.

The incomes of the clergy are very different. The higher Catholic clergy are paid by the state, the archbishop of Breslau receiving £1700 a year, and the other bishops about £1135. The incomes of the parochial clergy, of both sects, mostly arise from peculiar endowments. Generally government does not guaranty the stipend either of Protestant or Catholic clergymen; but in many parishes the clergy enjoy a public provision from the state. This is peculiarly the case in the Rhenish provinces, in virtue of the concordat already alluded to. Proselytism, or the attempting to induce a person to change his religion, whether by force or by persuasion, is prohibited by law; and all controversial sermons, or peculiar displays of religious zeal, would certainly attract the notice, and incur the displeasure, of the authorities. If we except the Rhine province and Westphalia, the population of which are bigoted Catholics, perhaps there is in no country less of religious acrimony and contention than in Prussia. For farther details as to the religious state of Prussia, *see* the parliamentary paper drawn up by Mr. Lewis from the official returns furnished by the Prussian government.

Education.—Prussia can boast of possessing a more perfectly organized and complete system of national education, than has ever existed in any country. Frederick the Great has the merit of having introduced the system into Silesia, after he had wrested it from Austria; and such of our readers as are curious about these matters, will find in "*Adam's Letters on Silesia*" a very full and interesting account of the plans of Frederick in relation to this important subject, of the obstacles he had to encounter in carrying them into effect, and of their result. From Silesia the system has been gradually extended to the other provinces, and is now in full vigour in every part of the monarchy. Attendance at school is *enforced by law.* Every child, whether male or female, rich or poor, must attend a public school from the age of five years complete, till such time as the clergyman of the parish affirms that the child has acquired all the education prescribed by law for an individual in his station: generally speaking, the school time extends from 6 to 14 years complete. Should a child not attend, its parents or guardians must satisfy the public authorities that it is receiving an appropriate education at home or in a private seminary. The school fees are exceedingly moderate; and the children of such poor persons as are unable to pay them, are instructed gratuitously at the public expense. Accord-

ing to the census of 1837, the population of the monarchy exclusive of Neufchâtel, was 14,098,125. It is calculated that of this number 2,830,398 were between the ages of 6 and 14, complete; and it appears, from the official returns, that *of this number* 2,171,745 were actually at school, or were otherwise receiving sufficient instruction. It must not, however, be supposed that the 658,583 (2,830,328—2,171,745) apparent excess of children are not instructed. The 2,830,328 includes, as already stated, all the children between 6 and 14 years complete; but the school education of a good many children is completed by the time they complete their 12th or 13th year; and, owing to the delicate state of their health, and other circumstances, a good many do not go to school till they have attained their seventh or eighth year. And, allowing for these circumstances, we believe it may be safely affirmed of Prussia, with the exception of a few districts in Posen, that every child born within her limits is educated! In so far, too, as we are able to judge, Prussian education is of the most excellent quality; and is, in this respect, as much superior to that of the lower and middle classes in England and Scotland as can well be imagined. The late king of Prussia deserved the esteem and gratitude, not of his own subjects only, but of every friend to humanity in every part of the world, for the zeal, perseverance, and discretion he displayed in maturing this system. Nothing has been omitted that could render it perfect. In the schools for the instruction of the masters, the examinations to which they are subjected, and the supervision exercised over every part, the utmost anxiety is evinced to render it as perfect as possible.

No particular religious creed is allowed to be taught in any school; but on particular days, set apart for the purpose, the children are instructed by the clergymen of the different sects to which they belong. Their religious instruction is not, therefore, neglected; while the intermixture of the different sects from their earliest years, on a perfect footing of equality, removes all asperities and religious animosities. All matters relative to the public schools are managed in each province by a public board appointed for that purpose; and the expense is defrayed by government. (For farther information as to the Prussian schools, see the Report of M. Cousin, *Sur l'Instruction Publique en Prusse, &c.*)

Exclusive of the gymnasiums and superior schools, Prussia has *six* universities, and the two semi-universities of Munster and Braunsberg. These are placed directly under the control of the minister of public instruction; and much pains has been taken to render them efficient and useful. The number of professors and subordinate teachers in the Prussian universities is very great; and we should incline to think that they had been needlessly multiplied. The following statement exhibits the attendance at the Prussian universities in 1834 and 1837.

Account of the Number of Pupils attending the different Universities and Semi-Universities of Prussia during the Summer Sessions of 1834 and 1837 respectively.

Universities, &c.	Year.	Prot. Theology.	R. Cath Theology.	Law.	Medicine.	Philosophy.	Total.	Total of whom.	
								Natives.	Foreigners
Berlin	1834	578	. .	504	402	289	1,863	1,343	520
	1837	430	. .	475	356	324	1,585	1,183	402
Breslau	1834	217	215	257	118	113	920	905	15
	1837	158	191	118	128	126	721	703	18
Bonn	1834	110	196	278	154	116	854	737	117
	1837	71	108	217	150	102	657	571	86
Halle	1834	505	. .	127	109	60	801	655	146
	1837	354	. .	87	128	69	638	521	117
Königsberg	1834	151	. .	83	83	105	422	384	38
	1837	140	. .	64	65	110	379	363	16
Griefswald	1834	93	. .	43	72	19	230	207	13
	1837	43	. .	14	58	102	218	190	28
Munster	1834	. .	181	. .	. .	61	242	200	42
	1837	. .	138	. .	. .	68	206	166	40
Lyceum Hosianum, Braunsberg	1834	. .	17	. .	. .	9	26	26	. .
	1837	. .	23	. .	. .	5	27	27	. .
Total	1834	1,654	609	1,362	938	765	5,348	4,457	891
	1837	1,196	459	975	895	906	4,431	3,724	707

Military Force.—The situation of Prussia, surrounded by powerful monarchies, and with a disjointed territory, requires for her security a large military force. But as the revenue of the country is comparatively limited, it became indispensable to endeavour to organise the army so that it might embrace the maximum of force with the minimum of expense. The Prussian government seems to have pretty satisfactorily solved this important and difficult problem. The obligation of military service is universal; every man being obliged to enter the army of the line, or the *landwehr*

(provincial army), between the ages of 20 and 39, as a private, and to serve in the one or the other for *three years*. At the end of this period he may enter the army of the line for a limited period as a volunteer, of which description of recruits it chiefly consists, or return home to prosecute some civil employment, his service in the army having secured for him various privileges. But the individuals whose period of service in the army of the line, or the *landwehr*, has been completed continue bound, on returning home, to serve in the second band of the landwehr till they be 39 years of age. They are seldom, however, called on to be exercised, and then only for limited periods. The *landsturm*, or levy *en masse*, consists of all the men not in the army or the landwehr up to the age of 50, and of young men between 17 and 20. This force is only called out in case of invasion. The army of the line may at present be estimated at 192,000 men, distributed as follows, viz.: infantry 83,000, cavalry 20,000, artillery 12,000, engineers 2000, and permanent landwehr 5000. The reserve and the first ban of the landwehr amount to 228,000 men, and the 2d ban of the landwehr comprises 180,000, making the total military force of the country about 530,000 men. Of these the regular army and the first ban of the landwehr, amounting to 350,000, are ready to act immediately against an enemy. There are numerous institutions, at Berlin and elsewhere, for military education. The greatest pains are taken to have the officers well instructed, and equal to their duties. Their appointments are good. The condition of the common soldier has also been vastly improved. He is seldom subjected to corporal punishment; and his pay and rations have been nearly doubled since 1806. The Marquis de Carsman estimates the total annual expense of a Prussian foot soldier, including rations, at 63½ rix-dollars (£9 9s. 6d.) a year; and that of a cavalry soldier, including the keep of his horse, at 213 rix-dollars (£31 19s.) a year. These allowances, considering the cheapness of provisions in Prussia, place the Prussian soldiers on a level, in point of comfort, with those of England or of any other country. Nothing, in fact, is omitted that seems calculated to benefit the army, and to keep up the martial spirit of the people.

This system imposes, no doubt, severe privations on the public. But there is a general conviction of its being necessary; and its universality, and the impartiality of the laws, prevent it from being felt as a peculiar burden by any particular class. At all events, it is clear, whatever difference of opinion may exist as to its merits in other respects, that it has *completely nationalised the army*; which must be always actuated by the sentiments and feelings that prevail among the mass of the people from which it is taken, and to which it is constantly being returned. When, therefore, it is said that Prussia is a "military monarchy," it is necessary to bear in mind that the army is not composed of mercenary troops, but of citizens serving for a limited period, and that it has very little analogy indeed to most other European armies. (For farther and ample information as to the Prussian army, see the excellent Essay of the Marquis Carsman, *Sur l'Organisation Militaire de la Pruss.* passim.) Prussia has no ships of war.

Provision for the Poor.—The question as to the provision for the poor has only become of importance since the abolition of vassalage in 1810. Previously to that epoch they were provided for by their lords. At present poor persons, or those unable to provide for themselves, have a legal claim to support. But it is rarely necessary to resort to compulsory proceedings to enforce this claim, the really necessitous being generally provided for by voluntary subscriptions. There are workhouses in most large towns. That of Berlin had, in 1838, an average population of from 850 to 900 persons; which cost at an average 30 rix-dollars a year each.

ACCOUNT of the Crimes judicially prosecuted in Prussia in 1836; the Rhine Province, Neufchâtel, and Out Pomerania excepted.

Provinces.	Number of inhab. belonging to the circle	High Treason.	Tumults, or Resistance to the Authorities.	Official Offences by Civil Officers.	Duels.	Murder and Manslaughter.	Infanticide.	Concealed Pregnancy.	Rape, &c.	Theft and Robbery (Petard Thefts excepted.)	Highway Robbery and Murder.	Linger., Forgery, Usury.	Bankrupt.	Quackery.	Incendiarism.	Real Injuries.	Wood and Forest Theft., Poaching.	Other Crimes.	Total of Truth in Number of Crimes.
Prussia	3,095,374	16	969	106	1	41	5	27	56	4,537	26	197	4	27	131	411	27,636	29	34,059
Posen	1,125,591	26	646	71	.	21	6	21	49	3,125	37	184	2	60	80	197	13,205	45	18,060
Brandenburg	1,640,835	46	1,091	67	3	26	10	13	75	4,926	36	469	16	28	51	458	42,454	111	19,328
Pomerania	768,742	3	307	34	1	13	2	11	13	1,103	4	86	6	15	89	189	9,397	77	13,172
Silesia	2,515,551	41	980	90	.	59	14	80	117	4,845	161	349	16	44	163	551	51,629	45	34,328
Saxony	1,471,400	27	1,001	80	2	28	6	14	52	5,408	31	347	10	34	84	416	19,864	206	23,467
Westphalia	1,482,507	35	546	75	2	27	4	15	36	2,102	94	187	9	35	86	445	22,913	194	27,769
Total	11,606,390	19	5,439	519	9	225	57	181	439	25,948	380	1,747	61	233	535	2,557	185,406	944	205,408
Rhine Provinces in 1837	2,298,526	30	1,171	62	.	20	12	2	42	2,637	23	285	40	74	44	2,117	58,464	. .	58,178

ACCOUNT of the Number of Deaths in the Prussian Provinces in 1837, specifying the Numbers that died of each Disease.

Provinces.	Of Old Age.	Violent Deaths.		In Childbed.	Small-pox.	Hydrophobia.	Inflammatory Fevers.	Linger-ing Fevers, &c.	Hæmorrhage and Apoplexy.	External Diseases not determined Wounds.	Diseases not determined.	Born dead.	Total.
		Suicides.	Accidents.										
Prussia	7,355	192	1,290	948	376	9	41,917	24,344	3,170	1,391	.	5,503	3,794
Posen	4,149	75	440	548	381	4	11,841	5,596	2,730	873	.	4,300	1,778
Brandenburg	5,979	374	697	490	366	1	11,368	17,942	2,668	772	.	3,704	2,821
Pomerania	2,742	113	471	383	90	1	6,038	8,433	1,882	390	.	1,904	1,421
Silesia	10,431	303	966	969	307	18	20,637	39,345	9,737	1,948	.	4,507	4,770
Saxony	6,990	279	597	544	90	2	9,882	18,994	3,977	708	.	2,624	2,356
Westphalia	4,191	67	419	406	244	.	10,782	19,025	1,621	544	.	2,341	1,465
Rhine province	10,549	94	609	671	491	1	16,911	30,017	3,368	955	.	1,233	2,444
Totals	53,256	1,507	5,881	4,956	2,195	36	103,306	67,163	31,312	7,571	.	30,857	21,120

Food, Dress, Diversions, and Habits of the Bulk of the People.—The food, dress, diversions, and habits of the people are very different in different provinces. In general, rye is the ordinary bread-corn. About half a century since the consumption was estimated at from 5 to 6 scheffel per head, but now it is not more than 3 scheffel; the deficit being made up by the increased consumption of potatoes. In many districts, indeed, potatoes are now almost the only vegetable food. Besides rye-bread and potatoes, the people use barley, buckwheat, and milk. In the Rhenish provinces wheat is more extensively used than in the other provinces. The wheat cultivated in Prussia Proper and Posen is rather for sale than for inland consumption. Beans and pease are extensively used in most parts of the monarchy. Coffee, mixed with chicory, is extensively used; and on Sundays it is used by all classes: ten is but little used. The consumption of sugar has rapidly increased; and now amounts, at an average, to about 4 lbs. per individual, notwithstanding the substitution of honey for sugar in very extensive districts. The consumption of butchers' meat is considerable, principally in towns. It is estimated over the whole monarchy at from 36 to 37 lbs. a year to each individual. (Dieterici, *Statistische Übersicht*, s. 234.) But in the towns it may be from 40 to 50 lbs., and in the country, perhaps, not more than from 20 to 30 lbs. In Berlin the consumption of butchers' meat exceeds 100 lbs. per individual: and this, in all cases, is exclusive of poultry, fish, and game. Game,

indeed, is only used by the richer families; but fish is an ordinary food of the peasantry in many districts. Poultry, especially geese, are largely used by the common people in Pomerania. There is also no want of butter for the common people, but the consumption of cheese is greater.

We have already seen that the consumption of spirits is quite immense; it is less, however, in the Rhenish provinces, where wine is extensively used, than in other parts of the monarchy. Beer also is an ordinary beverage, and the consumption is very great. The consumption of tobacco is estimated at 3½ lbs. per individual. The common people (males) use, in winter, a woollen great-coat, and in summer a linen coat. Women now begin to be pretty extensive consumers of cotton stuffs; it is still, however, customary for the young female peasants to prepare stocks of linen in anticipation of their marriage. Wooden shoes are worn, but are not so frequent as in France or Ireland; and in bad weather almost every body is well shod with leather boots and shoes: but in fine weather the common people often go barefoot. Silks are only used by the wealthier classes.

It is usual for the peasants to go to church regularly on the Sunday forenoon; and it is customary for them to spend the evening of the day in diversions of some sort or other, frequently in the ale-house, and in fiddling in dancing-parties. Most little towns have shooting-places; and the meetings of the landwehr for exercise are looked forward to with much satisfaction. (*From Official Doc. and Priv. Inform.*)

PRUSSIA PROPER.   PULASKI.

PRUSSIA (PROPER), an extensive and important prov. of the Prussian dominions, formerly divided into the provinces of East or Ducal Prussia, and W. Prussia, having N. the Baltic, E. and S. Russia, Poland, and the prov. of Posen, and W. Brandenburg and Pomerania. Area, 24,974 sq. m. Pop., in 1837, 2,125,335, of whom 1,532,396 were Protestants, 555,930 Catholics, and the rest Jews and Mennonites. It is divided into four regencies, and 57 circles. Principal towns, Konigsburg, Dantzic, Elbing, Tilsit, Marienwerder, &c. It consists principally of an immense plain, traversed by the Vistula, Pregel, Passarge, and other rivers, and interspersed with numerous lakes and immense forests. Soil various, but generally fertile, particularly the delta of the Vistula and the country watered by the Niemen. Principal products, wheat and other sorts of corn, timber, hemp, and flax, provisions, wool, ashes, &c. Rye is more cultivated than any other sort of grain; wheat is also largely cultivated, but more for exportation than for internal consumption, rye being the ordinary bread corn. Oats, barley, and peas, are also raised; and latterly potatoes have been extensively cultivated. Farming implements defective and ill constructed; the harrows are made of wood, without any iron even for teeth. With the exception of the crown estates, which are let on lease, but little land is rented. In the circles of Dantzic, Elbing, and Marienwerder, good land fetches, when let, from about 4s. 6d. to 5s. an acre, the tenant bearing all taxes. But in other districts the rent of the cultivated land does not exceed from 1s. 3d. to 3s. 6d. an acre. The wages of farm labour vary in summer from 4d. to 8d., and in winter from 3d. to 5d., without food; but with a cottage free, or at a low rent, a garden, and pasture for a cow. In harvest the wages are a good deal higher. The peasantry live in wooden or clay cottages, with but few luxuries, principally on potatoes, rye bread, and milk, rarely tasting animal food, but drinking great quantities of spirits and beer. Linen, from flax of their own growth, and woollen spun in their cottages, furnish their clothing. Domestic servants get butchers' meat, generally pork, once or twice a week. Wood and turf are almost every where abundant, and are the principal articles of fuel. Amber is found along the sea-shore, but otherwise minerals are quite unimportant. There are many distilleries and breweries, but, with the exception of Posen, this is the least manufacturing province in Prussia. (Jacob's First Rep., p. 13–22; Parl. Pap., No. 84, Sess. 1836, &c.—See PRUSSIA and KONIGSBERG.)

PSKOF, a government of European Russia, chiefly between the 56th and 58th degs. N. lat., and the 28th and 32d E. long., having N. Petersburg and Novgorod, of each of which govs. it formerly made a part; E. Tver and Smolensk; S. Witepsk, and W. Livonia. Area estimated at 22,154 sq. m. Pop., in 1838, 705,300. The surface is nearly flat, with a slope to the N., the direction taken by most of the rivers. None of these are of considerable size; but the gov. is, notwithstanding, well watered. At the N. W. extremity is the lake of Pskof, connected by a strait with that of Peipus. Marshes are numerous. The atmosphere is usually damp, though, on the whole, the climate is far from unhealthy. Soil thin, and not very fertile; but owing to the fewness of the inhabitants, more corn is grown than is required for home consumption. The produce averages 2,500,000 chetwerts a year, of which upward of one million may be exported. It consists chiefly of rye, barley, and oats, the proportion of wheat being small. A good deal of hemp and flax is raised. The forests are extensive, and abound with game. Cattle are not of great importance, and bees are less reared than in most provinces. Manufactures have increased during the present century, but they are still of no great consequence. The leather of this government is much esteemed, but its principal wealth consists in its corn and natural produce. Pskof is divided into eight districts; chief towns, Pskof, the capital, Toropetz, and Velikie-Louki. Its population consists mainly of Russians, with some Lithuanians and Finns. Public education is little extended, in 1831, only 2110 individuals were receiving instruction, or 1 in 302 of the population; and only one printing-press existed in the government.

PSKOF, or PLESKOW, a town of European Russia, cap. of the above gov., on the Velikaia, 165 m. S.W. Petersburg. Pop., in 1831, 8731. It covers a large space of ground, and is divided into three parts, the Kremlin or citadel, the Middle Town, and the Greater Town, all surrounded with an earthen mound. It is mostly built of wood, but has two good edifices belonging to the archbishop and the consistory; a number of churches, two convents, and a high school. Its only manufactures are of leather; but it has a considerable trade in the export of the products of the country. A large annual fair is held here in February, at which large quantities of woollen, silk, and cotton fabrics, leather, books, jewellery, &c., are sold. This town is rather ancient, being mentioned in history as early as 903. (Schnitz. La Russie.)

PUEBLA (LA), a city of Mexico, cap. of the state of its own name, on the declivity of a hill, 76 m. E.S.E. Mexico,

and 125 m. W. by S. Vera Cruz, lat. 19° 0' 15" N., long. 98° 2' 30" W. Pop. 50,000. (Ward.) It is compactly and uniformly built. The streets, which, though not very wide, are straight, and intersect each other at right angles, are paved with large diamond shaped stones, with broad and well kept footpaths on either side. The houses, of stone, are generally two stories high, with flat roofs, having mostly a court in the centre, surrounded with open galleries, and a fountain of fine water, conveyed thither by earthen pipes, Many have iron balconies toward the street, and their fronts are inlaid with highly glazed tiles, or else gaudily and fantastically painted. The apartments are spacious: according to Mr. Ward, they are mostly paved with porcelain, carpets not being in use; and their walls are adorned with paintings in fresco. The family of the proprietor usually resides in the upper story, the ground-floor being occupied with shops, warehouses, or offices, and the second story by servants.

This would seem to be a perfect hotbed of priests; when Bullock visited Mexico, Puebla had no fewer than 69 churches, nine monasteries, 13 nunneries, and 23 colleges. He says of the churches that they were the most sumptuous he had ever seen. "Those of Milan, Genoa, and Rome are built in better taste; but in the expensive interior decorations, the quantity and value of the ornaments of the altar, and the richness of the vestments, they are far surpassed by the churches of Puebla and Mexico." The cathedral, which forms one side of the principal square, has nothing remarkable in its exterior, but its interior is very rich. The high altar, which, however, is too large for the building, is particularly splendid. Several of the other churches are handsome, and, with the cathedral, abound in gold and silver ornaments, paintings, statues, &c. The bishop's palace has a library 200 ft. in length, which has a tolerable collection of Spanish and French books.

La Puebla is governed by 4 alcaldes and 16 subordinate magistrates. Its market is well supplied with all sorts of provisions, except fish. Many of the inhabitants are wealthy, and have handsome carriages drawn by mules; but, like Mexico, the city swarms with beggars, a consequence of the want of industry, occasioned partly and principally by the mildness of the climate, but partly, also, by the distribution of provisions at the convent doors. It was formerly famous for its manufactures of coarse woollens, cottons, glass, earthenware, soap, &c.; but most of these have declined with the decrease of the trade formerly carried on with Acapulco, Callao, and the other ports on the Pacific. The manufactures of glass and earthenware however, keep up their reputation, and the soap made here is sent to most parts of Mexico.

La Puebla was founded by the Spaniards. The state of which it is the capital comprises the towns of Cholula, Tlascala, Huetxocingo, and other ancient Mexican cities; it also includes Popocatepetl. 17,716 ft. above the sea, being the highest mountain in North America. (Ward's Mexico, ii., 72–74; Poinsett's Notes, &c., p. 51–58; Humboldt, Bullock, &c. in Med. Trav., xxv.)

PUERTA DE STA. MARIA. See ST. MARY's.

PUERTO-REAL, a town of Spain in Andalusia, prov. Cadiz, and on the bay of that name, 5 m. E. Cadiz, and 60 m. S.S.W. Seville. Pop., in 1826, according to Miñano, 5000; but according to Inglis, in 1830, 12,000. It is tolerably well built, with straight, regularly formed streets, crossing each other mostly at right angles, and lined with good houses. The market-place, an oblong space of considerable extent, has covered passages for provision-stalls, &c. There are two other public squares, both of which, as well as that last mentioned, are ornamented with handsome fountains. The only public buildings are a parish church and two hospitals. Fronting the sea is a wharf rising about 1½ yard above the level of the highest tides; and a mole, 50 yards broad, runs out 300 yards to low-water mark, having steps on each side, for embarkation at all times of the tide; the whole is of stone, and has a handsome appearance. Near the town are extensive salt works, the annual produce of which was estimated by Laborde at 21,300,000 quintals; and hence salt is an important article of commerce. The process of manufacture is by evaporation in wide and deep basins, communicating with the sea by locks. There are three series of these basins, and the salt undergoes a three-fold evaporation before it is considered sufficiently crystallized.

Puerto-Real is of no great antiquity; it was taken by the French in 1809, but soon abandoned; and during the war of independence it was the great depôt for the French troops during their investment of Cadiz. (Miñano; Laborde, &c.)

PULASKI, county, Ga. Situated a little S. of the centre of the state, and contains 680 sq. m. Watered by Ockmulgee r. and its branches. It contained in 1840, 16,904 neat cattle, 9299 sheep, 15,115 swine; and produced 7439 bushels of wheat, 1433 of rye, 153,764 of Indian corn, 5369 of oats, 21,778 of potatoes, 1,735,783 pounds of cotton. It had 14

660

stores, 10 grist-mills, six saw-mills.   Pop.: whites, 2972; slaves, 2385; free coloured, 32; total, 5389.   Capital, Hawkineville.

PULASKI, county, Va. Situated toward the S.W. part of the state, and contains 350 sq. m.   Watered by New r. It contained in 1840, 6920 neat cattle, 9653 sheep, 11,752 swine; and produced 46,008 bushels of wheat, 16,940 of rye, 144,087 of Indian corn, 2446 of buckwheat, 80,170 of oats, 15,064 of potatoes. It had 10 stores, seven grist-mills, five saw-mills, one oil mill, one paper-mill, six tanneries, 20 distilleries; seven schools, 136 scholars.   Pop.: whites, 2768; slaves, 954; free coloured, 17; total, 3739.   Capital, Newbern.

PULASKI, county, Ky. Situated in the S.E. part of the state, and contains 800 sq. m.   Bounded S. by Cumberland r., by branches of which it is drained. It contained in 1840, 10,862 neat cattle, 13,366 sheep, 19,490 swine; and produced 43,985 bushels of wheat, 120,301 of Indian corn, 94,655 of oats, 4504 of potatoes. It had 10 stores, four tanneries; 29 schools, 754 scholars.   Pop.: whites, 8583; slaves, 1019; free coloured, 18; total, 9620.   Capital, Somerset.

PULASKI, county, Ia. Situated toward the N. part of the state, and contains 342 sq. m.   Watered by Tippecanoe r. It contained in 1840, 591 neat cattle, 270 sheep, 1589 swine; and produced 1399 bushels of wheat, 13,075 of Indian corn, 1825 of oats, 3381 of potatoes, 2131 pounds of sugar. It had one store, one saw-mill, three potteries. Pop. 561. Capital, Winamac.

PULASKI, county, Mo. Situated toward the S. part of the state, and contains 1332 sq. m. Drained by head branches of Gasconade r., and by branches of Osage r. It contained in 1840, 10,513 neat cattle, 6600 sheep, 25,131 swine; and produced 13,680 bushels of wheat, 385,860 of Indian corn, 23,143 of oats, 11,662 of potatoes, 19,091 pounds of tobacco, 7727 of cotton, 2602 of sugar. It had 10 stores, 22 grist-mills, 15 saw-mills, one tannery, 11 distilleries; six schools, 116 scholars. Pop.: whites, 6338; slaves, 190; free coloured, 1; total, 6529. Capital, Waynesville.

PULASKI, county, Ark. Situated near the centre of the state, and contains 2050 sq. m. Watered by Arkansas r. and its small tributaries. It contained in 1840, 7935 neat cattle, 949 sheep, 12,031 swine; and produced 559 bushels of wheat, 164,324 of Indian corn, 6920 of oats, 10,312 of potatoes, 7809 pounds of cotton. It had 20 stores, 12 grist-mills, 10 saw-mills, three printing-offices, three semi-weekly newspapers; one school, 48 scholars. Pop.: whites, 3961; slaves, 1244; free coloured, 105; total, 5350. Capital Little Rock.

PULTNEY, p. t., Steuben co., N. Y., 15 m. N.E. Bath, 213 m. W. by S. Albany, 316 W. Bounded E. by the W. branch of Crooked lake, into which its streams flow. It has two stores, one grist-mill, eight saw-mills; 12 schools, 1174 scholars. Pop. 1784.

PUNJAB (THE), (country of the *Five Rivers*), or LAHORE, a nominally indep. territory of N.W. Hindostan, consisting of most part of the region watered by the five great arms of the Indus: the Sinde, or Indus proper ; the Jhylum, or Behui (an. *Hydaspes*) ; the Chenab (an. *Acasines*) ; the Ravee (an. *Hydraotes*) and the Sutledje (an. *Hysudrus*), with its tributary, the Beas (an. *Hyphasis*). It extends principally between the 29th and 34th degs. of N. lat., and the 70th and 77th E. long. ; and formed, with some adjacent states, the extensive dom. of the late Maharajah Runjeet Singh. It is of a triangular shape, the apex directed S.W. ; it has S.E. Rajpootana and the Bahawulpoor territories, W. Beloochistan and Caubul, and N.E. the ridge of the Himalaya, separating it from Cashmere, Ladakh, and Little Thibet. Area estimated at 80,000 sq. m. Pop. probably 4,000,000, chiefly Seiks, Jauts, Rajpoots, Hindoos of low caste, and Mussulmen.

Nearly the whole country is flat: it is in many parts fertile, especially along the banks of the larger rivers ; but it also comprises some wide, sandy, and barren tracts, especially between the Indus and Hydaspes. Cultivation generally increases and improves as we proceed eastward. Of the four divisions of the Punjab E. of the Hydaspes, the two nearest to that river are principally depastured by herds of oxen and cattle ; and that most to the E. is the best cultivated. Sir A. Burnes says, " there is, perhaps, no inland country which possesses greater facilities for commerce than the Punjab, and there are few better supplied with the products of the mineral, vegetable, and animal kingdoms. These relieve it from any great dependence on external resources. The wheat and barley of the plains are expended within the country ; and such is the number of horses, that gram, bajree, and other grains reared in a dry soil, are imported. Rice is exuberantly produced under the mountains ; but it is not a diet which suits the palate of the people. The cane thrives luxuriantly, and sugar is manufactured for exportation. The smallness of its stalk is remarkable ; but it is said to produce the most saccharine fluid, and is preferred to the thicker canes of India. Indigo is reared E. of Lahore, and exported to the Mohammedan countries westward. A valuable oil is extracted from the *siroya*, or sesamium plant,

and used both for the lamp and culinary purposes. Esculent vegetables, such as turnips, carrots, &c., are produced everywhere ; and most of the vines and fruit trees common to Europe may be seen in the mountains. The climate is not very favourable to the cotton shrub ; it is produced in the ' donb' between the Sutledje and Beas rivers, but it is also imported from the dry country S. of the former river. The mineral resources of the Punjab have been very imperfectly explored. A range of hills, extending from the Indus to the Hydaspes, formed entirely of rock-salt, furnishes an inexhaustible supply, and, being a close monopoly, contributes to enrich the ruler. It is in general use throughout the country, and most extensively exported, till it meets the salt of the Sambre lake in Rajpootana and the Company's territories. There is another deposit of salt on the verge of the mountains towards Mundi, but of an inferior description. In the same vicinity, it is said, some mines of coal have been discovered ; and there are also extensive mines of iron. The salt range, and the other high lands, yield alum and sulphur ; and nitre is gathered in large quantities from the plains." (*Bokhara, &c.*, iii., 316–392.) But these statements must be received with considerable limitation. Vegetable products are abundant only in the central parts of the country, through which Burnes travelled. The climate in the N., though hot in summer, is as cold in winter as that of France and central Europe, and never sufficiently warm to mature the most valuable products of Hindostan. Rice is grown in the valleys, but in limited quantities, the usual food of the population being wheat or peas, made into a thick soup, and, according to Mr. Trevelyan (*Parl. Report on India*, 1840), the Punjab does not produce sufficient sugar for its own consumption, but imports it from British India. Mr. Elphinstone, who travelled both in the N. and S. parts of this country, states that not one third part of the surface seen by him was under tillage ; and there can be no doubt of the correctness of his statement, that, except near the rivers, no part of the Punjab will bear a comparison for productiveness with British Hindostan. (*Elphinstone's Caubul*, i. 109.)

The plains, which are diagonally intersected by so many rivers, might be successfully irrigated by canals ; as is proved by the existence of some, and the remains of many others, the work of the Mogul emperors. The country abounds in cattle and horses, though the former is small and ill-conditioned, and no attention is paid to the breed of either. The salt mines, which were opened at a very early date, are one of the most productive sources of revenue. Burnes states that they yield about 800,000 maunds a year. The salt is sold at 2 rupees the maund, or at a third part of the price of that of Bengal ; the working the mines cost ½ lac rupees a year ; but the profit to the government amounts, notwithstanding, to 1,100 per cent. ! (*Burnes*.) The remainder of the public revenue, which amounts to about 3½ crore rup., is principally derived from exorbitant taxes on land and agricultural stock. Moorcroft mentions a peculiar method of assessing the land tax, adopted, in his time between the Beas and Sutledje, by a collector who had been chief financial minister to Runjeet Singh. " A given quantity of earth was put into a fine muslin sieve, and washed with water until all the mould was carried through, and nothing but the sand left, and, according to its proportion to the whole, a deduction was made from the assessment. Four rupees for 2 begas was the fixed rate for the rich soil : 3 if it contained one fourth of sand ; 2 if it had a half ; and 1 where the sand was three fourths the quantity. The general character of the soil of the Punjab, composed chiefly of mould and sand, renders this mode of appreciating its assessment more correct than might be supposed, and it was at any rate, preferable to the old plan of assessing the land according to the estimated out-turn of the standing crop." (*Moorcroft's Trav. in the Himalayas*, vol. i., p. 121.)

As respects the commerce of the Punjab, the staple commodities are the shawls of Cashmere, which reach India and Europe wholly through this channel. The annual revenue from the shawl manufacture, exclusive of every expense, is rated at 18 lacs of rupees, though, from frauds of all kinds, this sum greatly exceeds the amount that actually reaches the treasury. It is a curious fact, that the silk-worm is unknown in the Punjab, though the silks of the immediately adjacent state of Mooltan have a high reputation in India. The natives in the E. of the Punjab excel in the manufacture of cotton, and their looms furnish white cloth at from ⅓ to 1 rupee a yard, which, though inferior in appearance to that of British manufacture, is stronger and more durable. There is a considerable demand for foreign copper, brass, tin, and lead ; all kinds of British hardware and woollens are much prized. There is a considerable importation of European articles ; and British chintzes have wholly superseded those of Mooltan.

The Punjab is interesting to the classical scholar, from its being the theatre of Alexander the Great's Indian victories. Mr. Elphinstone supposes the scene of the defeat of Porus to

676

have been at Jelalpoor, on the Hydaspes, while, according to Burnes, it was most probably at Jelum, about 25 m. higher up the river. Burnes imagines he has discovered on the opposite sides of the Jhylum, about lat. 32° 40' N.. long. 73° 40' E., the sites of Nicæa (victory town), built by Alexander at the point where he crossed the Hydaspes, and of Bucephalia, built in commemoration of his favourite horse, Bucephalus, which expired in this region. (*Mitford,* viii. 200.) Burnes joins Major Wilford in identifying the neighbourhood of the celebrated tope of Manikyala, between the Indus and the Hydaspes, with the site of the anc. Taxila. There are, however, very few Greek remains in the Punjab; and the statements of the historians of Alexander, as to the places where the great events in his Indian expedition occurred, are far too indistinct to allow of any certain conclusions being deduced from them. The seiks, now the ruling race in this quarter, originated as a Hindoo sect, about the middle of the 15th century, and remained in a turbulent feudal condition till early in the present century. About that period, Runjeet Singh, having subdued the other Seik chieftains, established a despotism; which, though far behind the governments of Europe, was yet far in advance of most native governments in the east. He maintained an army of about 25,000 regular infantry, drilled as Europeans, 5000 regular cavalry and artillery, and perhaps, 50,.00 irregular horsemen, who were paid by assignments of land for military service; and by their means he made himself feared and respected by his neighbours. His government was vigorous. without being either cruel or unnecessarily severe. But, since his death, which occurred in 1839, no successor adequate to the task of government seems to have appeared; and it is probable that this territory will, at no distant period, be definitively incorporated with the British dominions. (*Burnes's Travels,* 3 vols.; *Elphinstone's Caubul,&c.; Moorcroft; Perl. Reports.*)

PUTIWL, or POUTIVL, a town of European Russia, gov. Koursk, cap. distr., on the Seim, a tributary of the Delepr, 110 m. W.S.W. Koursk. Pop. 9000. It has a high school, and a great number of churches, most of which are, however, only wooden structures. Its general trade is considerable, and it has a large annual fair. (*Schnitzler La Russie.*)

PUTNAM, county, N. Y. Situated in the S.E. part of the state, and contains 216 sq. m. Bounded W. by Hudson r. It is very mountainous, containing the Highlands. Drained by head branches of Croton river. It contains iron ore. It had in 1840, 14,971 neat cattle, 14,945 sheep, 12,888 swine; and produced 19,250 bushels of wheat, 35,367 of rye, 86,679 of Indian corn, 37,090 of buckwheat, 86,431 of oats, 149,564 of potatoes. It had 47 stores, one furnace, two forges, 11 fulling-mills, one woollen factory, two flouring-mills, 27 grist-mills, 36 saw-mills, two paper-mills; 63 schools, 2935 scholars. Pop. 12,825. Capital, Carmel.

PUTNAM, county, Ga. Situated a little N. of the centre of the state, and contains 340 sq. m. Bounded E. by Oconee river, by branches of which it is drained. It contained in 1840, 9497 neat cattle, 7071 sheep, 29,991 swine; and produced 35,666 bushels of wheat, 1174 of rye, 360,320 of Indian corn, 21,332 of oats, 13,693 of potatoes, 3,279,145 pounds of cotton. It had 15 stores, one cotton factory with 2000 spindles, eight flouring-mills, 16 grist-mills, 14 saw-mills, two tanneries; five academies, 210 students; five schools, 174 scholars. Pop.: whites, 3741; slaves, 6482; free coloured. 37; total, 10,260. Capital, Eatonton.

PUTNAM, county, O. Situated in the N.W. part of the state, and contains 376 sq. m. Drained by Auglaize river, and its tributaries. It contained in 1840, 5570 neat cattle, 2365 sheep. 11,799 swine; and produced 43,999 bushels of wheat, 1903 of rye, 136,485 of Indian corn, 20,587 of oats, 30,115 of potatoes, 45,979 pounds of sugar. It had 18 stores, eight grist-mills, 11 saw-mills, five tanneries, two printing-offices. Pop. 5189. Capital, Kalida.

PUTNAM, county, Ia. Situated in the W. part of the state, and contains 486 sq. m. Watered by Eel r. It had in 1840, 16,022 neat cattle, 21,077 sheep, 42,534 swine; and produced 72,274 bushels of wheat, 2212 of rye, 894,792 of Indian corn, 98,206 of oats, 25,183 of potatoes, 37,582 pounds of tobacco, 292,075 of sugar. It had 30 stores, one fulling-mill, three flouring-mills, 33 grist-mills, 29 saw-mills, two oil-mills, 13 tanneries, 10 distilleries, three potteries, one printing-office, one weekly newspaper, and one periodical; one academy, 27 students; 44 schools, 2414 scholars. Pop. 16,842. Capital, Greencastle.

PUTNAM, county, Ill. Situated toward the N. part of the state, and contains 325 sq. m. Watered by Illinois and Spoon rivers. It contained in 1840, 2975 neat cattle, 1732 sheep, 6410 swine; and produced 40,885 bushels of wheat, 71,223. of Indian corn, 46,572 of oats, 18,818 of potatoes, 11,030 pounds of sugar. It had nine stores, two saw-mills, two printing-offices, two weekly newspapers; 12 schools, 675 scholars. Pop. 2131. Capital, Hennepin.

PUTNAM, p. t., Muskingum co., O., 55 m. E. Columbus

340 W. Situated on Muskingum river, opposite to Zanesville. It has one academy, 154 students, one school, 54 scholars. Pop. 1071.

PUTNAM VALLEY, p. t., Putnam co., N. Y., 9 m. W. Carmel, 100 S. Albany. It is mostly covered by the Highlands. Iron ore is extensively found. It has three stores, two forges, one fulling-mill, five grist-mills, eight saw-mills; 10 schools, 461 scholars. Pop. 1659.

PUTNEY, p. t., Windham co. Vt., 111 m. S. Montpelier, 432 W. Bounded E. by Connecticut r. Sacket's brook has a fall of 75 feet in 80 rods, and affords extensive water-power. It contains a Congregational and a Baptist church, three stores, two fulling-mills, two woollen factories, two paper-mills, three grist-mills, four saw-mills, two tanneries; 12 schools, 462 scholars. Pop. 1383.

PUY (LE) (an. *Revessio* and *Vellavi*), a town of France, dep. Haute-Loire, of which it is the cap., on the Borne, here crossed by a bridge of eight arches, 36 m. S.W. St. Etienne. Pop., in 1836, 14,738. It stands on a steep acclivity, and has, when seen from a distance, a picturesque appearance; but in reality, it is ill-built, and the streets, which are narrow, dirty, and ill-paved with smooth pieces of lava, are frequently impracticable for vehicles of any kind, and even dangerous for foot-passengers. Lately, however, the commerce of Le Puy has revived, and the town has been a good deal improved. The cathedral, in a very conspicuous situation, a Gothic structure of the 10th century, has a richly ornamented altar, on which is a statue of the Virgin, brought by Louis IX. (St. Louis) from Egypt, and presented by him to the city in 1254, on his return from his unfortunate expedition to that country. Adjacent to the town is a very singular needle-shaped rock, about 300 ft. in height, on the summit of which another church is erected. There are several other churches; and it is said that the remains of the constable Duguesclin, originally deposited in the abbey of St. Denis (*Biographie Universelle*), have been transferred within these few years to the church of St. Laurence, in this city. (*Hugo; Guide du Voyageur.*) Among the other public buildings are the prefecture, a handsome new edifice; a public library, with 5000 vols.; town-hall, bishop's palace, seminary, college, hall of the tribunal of commerce, two hospitals, barracks, and theatre. Le Puy has a tribunal of original jurisdiction, a chamber of manufactures, a commercial college, normal school, a society of agriculture, science, and arts, gratuitous courses of geometry and mechanics, applied to the arts, and a small museum of antiquities and paintings. It is celebrated for its manufacture of white and black lace. Le Puy is very ancient, but its origin is uncertain. It is called *Pollavenses* by Gregory of Tours; and in the middle ages several ecclesiastical councils were held in it. It suffered considerably in the wars of the League: and till lately its trade and prosperity were much depressed. (*Hugo,* art. *Haute-Loire,&c.; Guide du Voy.; D'Anville, L'Ancienne Gaule,* 545.)

PUY-DE-DOME, a central dep. of France, formerly a part of Auvergne, between lat. 45° 17 and 46° 15' N., and long. 2° 20' and 4° E.; having N. Allier, E. Loire, S. Haute-Loire and Cantal, and W. Corrèze and Creuse. Area, 797,238 hectares. Pop., in 1836, 573,106. This dep. is mostly covered with mountains of volcanic formation, the highest of which, the Pic-de-Sancy, is 6293 ft. in height: the Puy-de-Dome, whence the dep. takes its name, has an elevation of 4842 ft.[*] Principal rivers, Allier, Dore, Sioule, and Dordogne, most of which have a N. direction; the Allier, intersects the dep. about its centre, and its valley, the Limange of Auvergne, is exceedingly fertile. In general, however, the soil is sandy; stony, and otherwise indifferent; and agriculture, owing to the poverty and ignorance of the natives, is in a more backward state than in most other parts of France. Great numbers of the inhabitants annually emigrate in search of field labour into the adjacent provinces. In 1834, 366,339 hectares of the surface were said to be arable, 90,131 hect. in pastures, 99,152 in vineyards, 82,389 in woods, and 199,112 in heaths and wastes. Rye, oats, and wheat, are the principal grains cultivated, and in 1835 the total produce of corn was estimated at 2,965,000 hectolitres, in addition to which 1,900,900 hect. of potatoes were grown. The annual produce of wine is estimated at from 400,000 to 500,000 hect.; it is mostly of inferior quality, cannot be conveyed from place to place without the risk of being spoiled, and, generally speaking, is good for little, unless it be mixed with other wines. (*Jullien,* 173.) The annual produce of wool is estimated at 900,000 kilogr. Chestnuts, timber, honey, walnut oil, and

---

[*] We may remark. by the way, that it was here, on the 19th of Sept. 1648. that the decisive experiments recommended by Pascal, for determining the weight of the air, and its influence on the height of the mercury in the barometer. were carried into effect, by conveying the instrument to the summit of the mountain. The result was precisely such as Pascal had anticipated, the mercury falling as the barometer reached a higher elevation. (See *Montucla, Histoire des Mathematiques,* ii., 303; and *Biographie Universelle,* art. *Pascal.*)

cheese, are among the principal sources of its wealth. In 1835, of 225,172 properties subject to the *Contribution foncière*, 113,095 were assessed at less than 5 fr., and 37,485 at from 5 to 10 fr. Puy de Dome yields lead, antimony, coal, granite, &c. Its manufactures, which are mostly confined to the arrond. of Thiers and Ambert, consists in the former principally of cutlery and hardware, paper, leather, wax lights, &c.; and in the latter, of woollen, linen, and cotton goods, lace, and paper. The dep. is divided into 5 arronds.; chief towns, Clermont-Ferrand, Ambert, Issoire, Riom, and Thiers. It sends seven members to the chamber of deputies. Registered electors in 1839–39, 2106. Total public revenue in 1831, 10,057,196 fr. (*Hugo*, art. *Puy de Dome*; *French Official Tables; Dict. Geog.*)

PYRENEES, a lofty chain of mountains in S.W. Europe, separating Spain from France, and which, taken in its largest extent, may be regarded as stretching from cape Creux, in Spain, on the Mediterranean, near the frontier of France, W. to the coast of Galicia, a distance of about 650 m. It is more usual, however, to confine the term to that portion of the chain which separates Spain from France:—

> " Pyrene celsa nimbosi verticis arce
> Divisos Celtis late prospectat Iberos,
> Alquæ æterna tenet magna discrimia terris."
> *Silius Italicus*, iii. lin. 417.

When thus restricted, the chain terminates on the W. near Fuenterrabia, between lat. 42° 10' and 43° 30' N., and long. 3° 30' E. and 2° W., its length being about 270 m., its average breadth about 38 m., and the area about 1100 sq. m. This great range may be considered as consisting of two parts, both having the same direction, though not in one continued line, the point of division being formed by the Val d'Aran, near the head waters of the Garonne; of these the W. part is more southerly than that to the E. The steep escarpment is on the side of Spain; the slope towards France being much more gradual, owing to the elevation of the level districts about the Adour and Garonne. It may be remarked, also, that the French valleys ascend the crest of the chain by easy steps, more or less lofty, while the opposite side presents a continued succession of rugged chasms, abrupt precipices, and huge masses of naked rock. The chain at each extremity declines towards the sea; but the fall on the E. is much more sudden than on the W.: the elevation only 50 m. from the Mediterranean being 8500 ft., whereas on the opposite side this altitude is not found nearer than 70 m. from the coast. The following table gives the position and elevation of the principal summits of the Pyrenees.

	Lat.	Long.	Height in feet.
Le Canigou	42° 31' N.	2° 25' E.	9,141
Pic Pedrous . . .	42 34 —	1 56 —	9,511
Pic de Serrere . .	42 38 —	1 90 —	9,546
Monicalm . . .	42 41 —	1 39 —	10,069
Pic des Estats . .	42 40 —	1 23 —	10,611
Caraboulos . . .	42 48 —	0 37 —	10,518
Troumouse . . .	42 42 —	0 15 —	10,196
* La Maladetta, or Pic de Netou	42 38 —	0 47 —	11,424
Pic Posets . . .	42 40 —	0 31 —	11,277
* Mont Perdu . . .			11,168
Pic de Cascade . .			10,745
Do. Vignemale . .	42 46 —	0 4 W.	11,001
Do. Snobe . . .	42 49 —	0 21 —	10,376
Do. Midi de Bigorre	42 56 —	0 12 —	9,541

The above measurements are those given by Bruguière (*Orographie de l'Europe*), from the extensive survey's made by Charpentier, Ramond, Bebont, Vidal, &c. The summits marked with an asterisk, which comprise the three highest points of the chain, are on its S. face. Glaciers are found, as in the Alps, on some of the higher elevations; but these masses of permanent ice are much less extensive, the point of perpetual congelation being at a height of 10,000 ft., or about 2000 ft. higher than in the Alps, a difference owing chiefly to the vicinity of the sea on either side. The valleys of the Pyrenees are numerous, and of singular conformation; for, whereas the depressions of the Alpine system run from 40 to 70 m. nearly in the direction of the chain, all the great valleys of the Pyrenees are transversal, taking their origin at a col in the crest, and running N. and S. almost at right angles with the main ridge. The largest valleys are found in the central Pyrenees, the principal being those of the Garonne and Lavedan, each of which is from 40 to 50 m. in length. These transverse valleys are commonly the beds of rivers, or rather torrents: which, in the wider and less elevated parts, take a slow and serpentine course, but in the defiles become rapid and impetuous torrents, often interrupted by cascades, and sometimes by a series of falls. Lakes are frequently enclosed in the basins formed in the higher ridges of the valleys, and several of those in the central Pyrenees, which are sheltered from the sun, are perpetually covered with ice. Some valleys, also, instead of running in a deep and narrow defile, or a series of little basins, more or less extensive, rising by

degrees to the height of the central ridge, present at their origin a single basin, surrounded on three sides by a lofty wall of rock, and opening by a narrow gulley into the vale below; and these natural amphitheatres, or *cirques*, as they are termed (the principal of which is that of Gavarnia, in the valley of Barèges), constitute the grandest and most distinctive features of the Pyrenees. About 50 passes are formed through the valleys now noticed; but by far the greater number are practicable only for the peasantry, and *contrabandistes*, who are found in all parts of the Pyrenees. There are only five good carriage roads over the chain:— 1. the Col de Pertus, the extreme E. pass between Perpignan and Jonquera, practicable at all seasons, and by all kinds of vehicles; 2. the Pass of Puymorens, leading from the valley of Seine to that of the Arriège (6300 ft. high); 3. the Port de Caufrane, between Pau and Saragossa (6713 ft.); 4. the Port de Roncevaux, between St. Jean and Pampeluna (5771 ft.); and 5. the Pass of Bidassoa, along the high road connecting Burgos and Vittoria with Bayonne. The first and last of these passes are those most generally used by travellers between Spain and France. The two highest passes are the Port d'Or (9843 ft.), and the Brèche de Roland (9856 ft.); but they are practicable only during about two months of the year.

The geology of the Pyrenees is still very imperfectly known, notwithstanding the researches of Charpentier, Palassou, and other naturalists. The extent of primitive rock is smaller than in the Alps; but its arrangement is very peculiar; not in isolated masses, bursting here and there through the transition and secondary formations, but in a band or zone running in the direction of the chain, but only occasionally falling in with the crest or central chain; the granite in the W. section is on the S., and in that to the E. on the N. side of the main ridge. The primitive formation is extremely simple, consisting of three rocks only, granite, micaceous schist, and primitive limestone, which, together, form a pretty continuous band, stretching three fourths across the isthmus. The transition rocks, comprising the great bulk of the mountain system, are arranged in vertical bands flanking the primitive formation, and consist of argillaceous schist, schistose and common grauwacké, with blue limestone: these strata occur mostly on the N. side, the beds S. of the primitive formation being chiefly secondary rocks, as red sandstone, Alpine and Jura limestone, &c. The oolite and chalk formations are found in the lower parts of the chain on either side. The existence of volcanic action is proved by the trap and other similar formations, interspersed in different parts of the chain; by the peculiar manner in which many of the strata are upheaved; by the frequency of earthquakes on both sides of the range, and by the abundance of thermal springs, especially in the valley on the French side of the Pyrenees. The most celebrated springs are those of Bagnères de Bigorre and Barèges, Bagnères de Luchon and St. Sauveur; all of which are visited, during the summer months, by persons labouring under rheumatism and chronic disorders, as well as by others in search of pleasure and picturesque scenery.

Iron, copper, zinc, and lead are found in the Pyrenees; but, with the exception of iron, these ores are not wrought, though it be a well-known fact that great mineral riches were extracted from these mountains by the Carthaginians and Romans. Indeed, there are the remains of 300 or 400 deserted mines in different parts of the Pyrenees, some of which are said to be very rich. The veins of marble are numerous and valuable, one of them, a white marble, being equal, in purity and closeness of texture, to that of Carrara.

The flora of the Pyrenees comprises the rhododendron, the alpine rose, and a large variety of plants common to high elevations; the principal forest trees being the box, fir, pine, and, in lower situations, the deciduous oak. The fauna comprises the *izzard*, a species of chamois, the wolf, and the bear; with a variety of birds, many of them are migratory.

The Pyrenees give rise to numerous rivers. Those on the N. side comprise the Adour and its tributaries, the Pau, Oléron, Saison, Nive, &c.; the Garonne, with the Gers, Arriège, and others of its affluents, all sending their waters into the Bay of Biscay; and the Aude, the Tet, and the Tech, falling into the Mediterranean. The rivers descending from the S. slope are mostly tributaries of the Ebro, the principal being the Arragon, Gallego, Cinca, and Segre, all of which have numerous branches; the other rivers of this slope are the Ter and the Lobregat, the latter flowing into the sea close to Barcelona.

With respect to the scenery of the Pyrenees, as compared with that of the Alps, Mr. Inglis observes, that each has its peculiar charms, but that scenes of savage sublimity are more frequently met with in Switzerland than in the Pyrenees; the N. lat. of the former, as well as the greater size of its rivers and lakes, adding to its features of wild grandeur. The Pyrenees, however, appear to exceed the Alps in elevation, owing to the much lower level of the valleys, as compared with those of Switzerland, some of which are nearly 4000 ft. high, while those of S.W. France scarcely exceed 2000 ft. The presence of oaks, also, in the forests, clothing the sides of the mountains, gives a beauty to the Pyrenees which is wholly absent on the fir-covered steeps of the Alps. But, however worthy of the traveller's notice, these valleys will probably never become the frequent resort of the tourist, because access to them is extremely difficult, and the accommodation for travellers worse than in different; while Switzerland, accessible in several directions by good roads, is visited not only on its own account, but because it is the high road to Italy. (*Inglis's Switz., France, and Pyrenees*, p. 249-5.)

The mountaineers of the Pyrenees are shepherds, and small proprietors both of land and cattle; but owing to the deficiency of nutritious vegetation, their condition as grazers is far inferior to that of their brothers of the Alps. They are patient and industrious, though seldom raised above want; besides which, they are in every thing, but especially in food and clothing, more simple and primitive than the Alpine mountaineers, than whom, however, they are unquestionably handsomer and more vigorous. Their food usually consists of rye or barley bread and paste, made from Indian corn, with, occasionally, salted kid's flesh and pork. Mr. Inglis speaks favourably of their character for kindness and generosity; adding also, that "crime of every description is rare in the Pyrenees, theft very unfrequent, and murder altogether unknown." (P. 249.) On these mountains is found the extraordinary race of people, called *Cagots*, distinguished by their sallow and unhealthy countenances, stupid expression, want of vigour, relaxed appearance, imperfect articulation, disposition to goîtres, and inferior share of mental capacity. They live in the most retired valleys, secluded from and studiously avoiding intercourse with the rest of the inhabitants, by whom they are looked upon as a degraded race: and though they are not persecuted, enslaved, and debarred the privileges of religion as formerly, they are "still a separate family, still outcasts; a people having evidently no kindred with those who live around them. but the remnant of a different and more ancient family." The origin of this curious race is hidden in obscurity: but M. Ramond and others conceive them to be the miserable remains of the Germanic conquerors of the Spanish peninsula.

The Pyrenees, which seem to have been known to the Greeks under the name of Πυρήνη, are connected with many important historical events. Hannibal crossed them on his way to Italy, at the beginning of the second Punic war, most probably by the pass of Pertus, near the E. end of the chain. Julius Cæsar also traversed them with his army, when marching into Spain against Pompey. Charlemagne carried his victorious arms over these mountains, and added Spain to the empire of the Franks. Edward the Black Prince led his army over one of the western passes, when fighting in defence of Peter the Cruel against Henry of Trastamare; and these mountains have obtained a more recent celebrity from having been the scene of several obstinate struggles between the French and English at the close of the Peninsular war, the result of which set in a striking point of view the great military talents of the Duke of Wellington. (*Les Pyrénées, par Chausenque*, p. 9-43; *Palassou, Histoire Nat. des Pyrénées*, 2 vols., passim; *Encycl. Métrop.*, art. *Pyrenees, by the Rev. G. C. Renouard*; *Malte-Brun, Inglis. &c.*)

PYRENEES (BASSES), a frontier dep. of France, reg. S.W., formerly comprised in Gascony, Bearn, and Navarre; between lat. 43° 47' and 43° 35' N., long. 0° and 1° 48' W., having N. Landes, E. Hautes-Pyrenees, S. and W. Spain, and N.W. the bay of Biscay, on which it has a coast-line of about 25 m. Area, 749,490 hect. Pop. in 1836, 428,400. The Pyrenees bound this dep. on the S.; their highest point in this part of the range. the Pic du Midi, being 9546 ft. in height. Their ramifications cover the greater part of the dep., which is traversed by the rivers Gave de Pau, Oleron, Bidoussa, Niève. &c., tributaries of the Adour, and flowing N.W. Small lakes, mineral springs, &c., are very numerous. Except on the mountains, the climate is temperate and healthy; the soil in the lower parts of the country is very fertile, while the mountain sides are covered with fine pastures and forests. which maintain great numbers of cattle and hogs. In 1834, 156,223 hectares of the surface was said to be arable, 66,254 in artificial meadows, 23,175 in vineyards, 130,172 in woods, and 340,732 occupied with natural pastures, heaths, wastes, &c. (*French Official Tables.*) "Between Betharam and Pau. the country is beautiful. We are traversing the rich vales of Bearn, every inch of land is cultivated, and the road is a constant succession of villages and houses. The principal products of this country are fruit, wine, and Indian corn, all of which grow a great perfection. It is in this district that the prunes so much prized in England are grown and prepared; and very description of fruit that is produced in the lower parts

of Bearn is excellent of its kind. Here too, we find the vine, not as it is found in the other parts of France, an insignificant shrub, covering the acclivities, and possessing not much greater beauty than a potato-field; but trained from tree to tree, as in some parts of Italy and in the Tyrol. This district, excepting the valleys of the Pyrenees, is certainly the most beautiful part of France." (*Inglis's Switzerland, &c.*, p. 308.) Agriculture is, however, in a very backward state: by what would seem to be a singular contradiction. the sides of the hills are cultivated, while the plains, which, it may be presumed, would be much more productive, are left waste or in pasture; and the corn is insufficient for the home supply. Maize and wheat are the grains principally cultivated; flax and hemp are, also, raised in large quantities. The produce of wine is estimated at about 300,000 hectolitres a year, of which about one third part is consumed by the inhabitants; some growths, particularly those of Jurançon, near Pau, are of a superior quality. In 1836, there were estimated to be nearly 117,000 head of cattle in the dep. and 464,000 sheep, the produce of wool being supposed to amount to about 1,000,000 kilogr. The breed of horses has been greatly improved by the stud at Pau: a great many mules are bred for export into Spain. The hogs furnish the hams so well known under the name of *jambons de Bayonne*. In 1835, of 94,893 properties subject to the *Contribution foncière*, 47,058 were assessed at less than 5 fr., while 14 only were assessed at 1000 fr. and upwards. Copper, iron, sulphur, cobalt, slate, marble, and granite, are among the mineral products; and mining industry is carried on upon a pretty extensive scale. The manufactures comprise woollen and linen stuffs and yarn, printed handkerchiefs (called *mouchoirs de Bearn*), leather, hardware, earthenware, paper, chocolate, brandy, &c.; and the dep. furnishes supplies of planks. with cordage, &c., for the dock-yard at Bayonne. The value of the cattle, wine, hams, salted geese, &c., exported to foreign countries. especially contraband to Spain, is estimated at 4,000,000 fr. a year, and that of the exports to other parts of France at about 3,000,000 fr. Basses-Pyrenees is divided into five arronds.; chief towns, Pau, the cap., Bayonne, Mauleon, Oleron, and Orthez. It sends five mems. to the chamber of deputies. Registered electors in 1838-9, 1106. Total public revenue in 1831-2, £7,856,744. (*Hugo*, art. *Basses-Pyrénées; Dict. Géog.; French Official Tables; Inglis.*)

PYRENEES (HAUTES), a frontier dep. of France. reg. S.W., formerly included in Gascony, between lat. 42° 42' and 43° 35' N., and long. 20' W. and 35' E., having N. Gers, E. Haute Garonne, W. Basses-Pyrénées, and S. Spain, the ridge of the Pyrenees forming the line of demarcation between the two kingdoms. Area, 452,790 hectares. Pop., in 1836, 233,031. Within the limits of this dep. are some of the most remarkable places of resort and objects of curiosity in the Pyrenees, as the baths of Bareges, Bagnères, and Cauterets, the valley of Gavarine, *Broche de Roland*, &c. By far the greater part of the surface is covered with ramifications of the Pyrenees, among which the Gave de Pau, Gers, Adour, &c., take their rise. Small lakes are very numerous in the mountains. In 1834, of the surface of the dep., 34,539 hectares were estimated to be arable, 44,378 in meadows, 15,382 in vineyards, 84,611 in woods, and 173,579 in heaths, wastes, &c. There is a larger proportion of good soil in this than in the neighbouring dep. of the *Basses-Pyrénées*, though the produce of corn (chiefly maize and wheat), be still insufficient for the inhabitants. Property is much subdivided: most of the peasants are proprietors of the soil they cultivate; and the greater number of the other proprietors are engaged in the cultivation of their own lands. In 1835, of 78,713 properties subject to the *contribution foncière*, no fewer than 44,031 were assessed at less than 5 fr., and 12,965 at from 5 to 10 fr., while only eight were assessed at 1000 fr. or upwards. (*French Official Tables.*) Inglis describes a family in the mountains whose condition might be taken as a fair sample of that of the poorer mountaineers of the Pyrenees. "The property of a peasant," says he. "consisted of two cows and three goats. A small meadow in the neighbourhood of the hut was fertilized, and allotted to rye, and about a rood of land was laid out in potatoes and cabbages. The peasant and his family consumed the whole produce of the animals. Meat of no kind ever entered the cabin; but the lake, one and a half league distant, occasionally supplied a few fish. A kind of cheese, like some of the poorer Scotch cheeses, was made from the goats' milk; and the sale of this to the lower orders at Cauterets was the only source of the money necessary for the purchase of clothes and whatever else was not produced by cows and goats." (*Switzerland, the Pyrenees, &c.*, p. 280, 281.) The dep. produces about 270,000 hectolitres a year of inferior wine, about half of which is exported or converted into brandy. A good many cattle are reared, and the horses bred in the plain of Tarbes are extensively purchased for the service of the light cavalry. Mules are bred for export into Spain. The produce of wool is reckoned at about 370,000 kilogr. a year; and a good deal of butter is made of the milk of the sheep. The produce of honey and wax is also considerable. Minerals and manufactures, though of little importance, are not quite valueless, there being good marble quarries, and some fabrics of woollen, linen, and cotton stuffs, with tanneries, distilleries, &c. The dep. is divided into three arronds.; chief towns, Tarbes (the cap.), Argeles, and Bagneres de Bigorre. It sends three mems. to the chamber of deputies. Registered electors in 1836, 39,545. Total public revenue, in 1831, 3,118,056 fr. (*Hugo*, art. *Pyrenées-Hautes*; *Dict. Géog.*; *French Official Tables.*)

PYRENÉES-ORIENTALES, a marit. and frontier dep. of France, reg. 8., consisting principally of the old prov. of Roussillon, with a portion of Languedoc, E., having N. Ande, W. Ariege, S. Spain, the ridge of the Pyrenees forming the line of demarcation between the two kingdoms, and E. the Mediterranean. Shape nearly triangular. Area, 411,693 hectares. Pop., in 1836, 157,052. The Pyrenees in this dep. are less lofty than in the greater part of the rest of their course; still, however, the Canigou, their highest peak, attains an elevation of 9140 ft. The dep., though in great part mountainous, comprises a large extent of plain country. The principal rivers are the Tet, Tech, and Agly, but none of these are navigable. In 1834, 92.554 hectares of the surface were supposed to be arable, 9796 in artificial pastures (probably meadows), 58,442 in vineyards. 43,877 in woods, and 188,407 in heaths, wastes, &c. The arable lands are of two classes,—wet and dry. The irrigable, or wet, are always under culture; in the rotation to which they are subjected they usually yield three crops in two years; one of which is wheat, and the others trefoil, or beans, maize, potatoes, hemp, or flax, when they are kept in grass for an equal period. The dry, or non-irrigable, lands, are alternately under wheat or rye, or in fallow. But though agriculture has been better conducted than in most of the adjacent deps., the corn continues to be separated from the grain by treading out with horses, as in Aude. The total produce of corn, chiefly wheat, maize, and rye, was estimated, in 1835, at 442,160 hectolitres a year, a quantity obviously much below the demand of the population. The dep. also furnishes in ordinary years about 300,000 hectol. of wine, and 12,000 of oil. The *vins ordinaires* are highly coloured and heavy, and are but little drunk in a pure state in other parts of France, but are extensively employed to give colour and body to the weaker wines of other deps. The red wines of Bagnols, and the white wines of Rivesaltes are, however, of a very superior description. The former are *pleins de corps et de spiritueux, avec de la moelle, du velouté, et un fort bon goût. En vieillissant ils acquièrent de la finesse du bouquet, et un couleur de l'or.* (*Jullien*, p. 282.) The white Muscat wine of Rivesaltes is said by the same distinguished authority to be decidedly the best *vin de liqueur* produced in France. *Il est plein de finesse, de feu, et de parfum, il embaume la bouche et la laisse toujours fraîche; et il est bien constant que ce vin est l'un des meilleurs de l'univers, lorsqu'il provient d'une bonne année et qu'il a vieilli.* (p. 285.)

The horses of the department have been very much improved by the royal stud at Perpignan and otherwise; sheep and goats are, however, the most valuable domestic animals; the former yield annually about 50,000 kilogr. of wool for exportation, after supplying the home consumption. The mulberry has been introduced, but this department ranks *last* among those in which silk is grown. About 300,000 kilogr. a year of cork are collected. Property is very much subdivided. Of 54,906 properties subject to the *contrib. foncière*, in 1835, 27,018 were assessed at less than 5 francs, and 9454 at from 5 to 10 francs; only 32 properties were assessed at 1000 francs and upwards. Iron, copper, bismuth, and lead are raised, but in no great quantities. Many of the inhabitants of the coast are occupied in the tunny and anchovy fisheries; and there are manufactures of coarse woollen stuffs and hosiery, with iron forges, tanneries, paper-mills, distilleries, &c. The department is divided into three arronds.; chief towns, Perpignan (the cap.), Céret, and Prades. It sends three members to the chamber of deputies. Registered electors, in 1836, 39,849. Total public revenue, in 1831, 3,169,498 francs. Roussillon belonged successively to the kings of Majorca and of Arragon, till Louis XI. took it from the latter. It was restored by Charles VIII., and remained attached to the Spanish monarchy till 1640, when it was finally annexed to France. (*Hugo*, art. *Pyrenées-Orientales*; *Dict. Geog.*; *French Official Tables.*)

Q.

QUEBEC, a city and seaport of Canada, of which, and of the British possessions in N. America, it is the capital, on the N.W. bank of the St. Lawrence, at the point where

it is joined by the St. Charles, about 340 m. from the mouth of the former, lat. 46° 48' 49" N., long. 71° 10' 43" W. Pop., in 1831. 27,562, but now probably above 30,000. It occupies the extremity of a ridge, terminating in the angle formed by the junction of the two rivers, in the point called cape Diamond, rising to the height of nearly 340 ft. above the St. Lawrence. The cape is surmounted by the citadel, and the town extends from it, principally in a N.E. direction, down to the water's edge. It is divided, from the difference of elevation, into the upper and lower towns. The old town, which lies wholly without the walls, partly at the foot of cape Diamond and round to the St. Charles, has narrow and dirty, and, in parts, steep streets; "the most crowded parts of the old town of Edinburgh not being more irregular or confined than the lower town of Quebec." (*Macgregor.*) The ascent from the lower to the upper town, which crosses the line of the fortifications, is by a winding street, and by flights of steps. The streets in the latter, though rather narrow, are generally clean, and tolerably well paved, or macadamised. Both towns are wholly built of stone; and the public buildings, and most of the houses in the upper town are roofed with tin plates, the glitter of which, in the sun, has a brilliant and striking effect, but is destructive of every thing that Europeans have been accustomed to call grand and venerable. In the lower town the houses are mostly covered with shingles.

Quebec is very strongly fortified, and may, in fact, be called the Gibraltar of America. The citadel over cape Diamond includes an area of about 40 acres; and is formidable alike from its position and its works, constructed on a gigantic scale, and on the most approved principles. The line of the fortifications, which stretches nearly across the peninsula on the W., and runs along a ridge between the upper and lower towns, is intersected by five gates, and has an inner circuit of about 2½ miles. Beyond the ramparts, on the W., are the extensive suburbs of St. Roch, St. John, and St. Louis.

The public buildings are substantial, rather than elegant. The Château St. Louis, the residence of the governor-general, a large, plain building, on a height overhanging the river, was burned down early in 1835. The Roman Catholic cathedral of Notre Dame, the Protestant cathedral, with sundry other Catholic and Protestant churches; the old episcopal palace, now the seat of the Canadian legislature; the quadrangular edifice formerly the college of the Jesuits, but now a barrack; in the upper town, with the Quebec bank; the Exchange reading-room, well furnished with American, British, and other newspapers and periodical publications; and the government warehouses in the lower town, comprise the principal public buildings.

There are three nunneries, one of which, the Hôtel de Dieu, is a very valuable hospital. The nuns are here, in fact, a most useful class of persons, acting as nurses to the sick admitted within these establishments, and as instructresses of young girls.

Among the establishments for educational purposes, the first place is due to the French college. It has a principal, and professors of theology, rhetoric, and mathematics, with five regents for the Latin and Greek classes. Here is, also, a royal grammar school, a classical academy, a national school, and many French and English private schools. A Royal-Institution for the advancement of learning within the province, and a Literary and Historical Society, respectively enjoy the patronage of the government and of the principal inhabitants. A Mechanics' Institute was established in 1830; and it has numerous benevolent associations. The city public library, though not very extensive, is said to be well selected, and to contain a great variety of standard works. The garrison, also, has a good library. Several newspapers are published in the city.

Though not a manufacturing town, Quebec has various distilleries, breweries, with tobacco, soap, and candle works; and numbers of fine ships have been launched from its yards. It has two or three banks, and a savings'-bank. The markets are well stocked with most sorts of produce, except good fish, which is rather scarce and dear.

The climate, though on the whole good and healthy, is in extremes. In summer the heat is equal to that of Naples, while the cold of winter is not inferior to that of Moscow. This inequality occasions a corresponding difference in the modes of life during the different seasons in the year. In winter travelling is carried on by means of sledges and carrioles, in the same way as in Russia. The first view of Quebec in sailing up the St. Lawrence is striking in the extreme; and Mr. Macgregor speaks in high terms of the magnificent prospect from the citadel on cape Diamond.

The majority of the population being of French extraction, the French language, which is still spoken in some of the best circles with great propriety, and the Roman Catholic religion, predominate. Society is here more polished and refined than in any other town of British America; and the higher provincial gentry of French descent are distinguished by the courteousness and urbanity of their manner. But, in consequence, perhaps, of the seductive example of the mimic court established among them, all classes are much given to show, and generally indulge in expenses beyond their means. Hence Quebec is very expensive; and owing to the jealousies that exist, and the violence of provincial politics, society is split into different parties. Great attention is, also, paid to etiquette; and those admitted to the governor-general's parties rarely associate with those who do not enjoy that honour.

Vessels of the very largest burden arrive at Quebec. Its harbour or basin, between the city and the island of New-Orleans, is of great extent, having in general about 28 fathoms water, the tide rising from 16 to 18 ft. at neaps, and from 25 to 30 ft. at springs. Ships lie alongside the wharfs along the St. Lawrence. There are extensive flats between the lower town and the St. Charles, where, if it were deemed of importance, wet docks might be easily constructed.

The trade of the city is very extensive. It engrosses almost the entire trade of the prov. with the mother country, the West Indies, &c.; and is annually resorted to by vast numbers of immigrants, who partly settle in Canada, but who mostly re-emigrate to the United States. It has a regular intercourse, by means of steamers, with Montreal, and other ports higher up the St. Lawrence, and with Halifax, and other ports on the Atlantic. Still, however, it must

not be forgotten, that in so far as the United Kingdom is concerned, the trade with Canada and Quebec is wholly forced and factitious, and is not a source of profit, but the reverse. It is in fact, as has been already shown (i. 534.) almost wholly a consequence of the high discriminating duties laid in our ports on timber from the N. of Europe, by which we exclude a cheap and good to make room for a comparatively bad and dear article. But for this preposterous arrangement, the trade between Great Britain and Quebec would be extremely unimportant. Thus, in 1837, of the total value of the imports from Canada, amounting in all to £908,792, timber made £651,796! the only other article of any importance being ashes. The principal articles of import into Canada consist of corn, cottons, woollens, silk, and other manufactured goods; glass ware, spirits and wines, iron and hardware, sugar and tea, &c. The total value of the imports into Lower Canada (nine tenths of which went to Quebec), in 1837, amounted to £1,602,353.

It is material, however, to bear in mind, that little more than a third part of these imports were paid for by the exports; they were, in fact, principally paid for by the treasury at home, and are to be regarded as the means sent out by England to pay the troops and meet the other heavy expenses she has to incur in the preservation of this most unprofitable colony. We subjoin some returns relative to the trade and navigation of Quebec during the seven years ending with 1839.

Account of the Arrivals of Ships in Quebec, specifying the Amount of Tonnage, the number of Seamen employed, and Number of Immigrants; with the Number of Vessels built in Quebec, the Tonnage, and Number of Men employed in building the same; in each Year from 1833 to 1839, both inclusive:

	1833.	1834.	1335.	1836.	1837.	1838.	1839.
Arrivals	1,005	1,122	1,129	1,185	1,032	1,195	1,175
Tonnage	271,148	395,863	393,305	353,505	396,185	343,782	363,844
Seamen	12,384	13,546	13,756	14,400	12,844	13,413	15,727
Immigrants	22,062	30,217	30,217	27,513	21,855	2,950	7,413*
Ships built	13	11	18	3	18	16	17
Tonnage	3,952	4,601	8,482	749	9,230	8,448	9,312
Men employed	715	608	990	165	990	840	935

* The small number of immigrants in this and the previous year is ascribable to the then disturbed state of the province. The numbers have since become as large as formerly.

Account of the Ships arrived at Quebec and Montreal in 1838; specifying the Countries whence they arrived, and the Number of Ships, with the Tonnage and Crews from each.

	Vessels.	Tons.	Men.
Great Britain	726	262,130	10,575
Ireland	137	44,111	1,871
British North American Colonies	144	19,702	918
British West Indies	25	3,276	167
France	19	6,297	201
United States	13	5,279	201
Hamburg	9	2,694	98
Portugal	3	870	34
Spain	3	761	31
Saxony	2	856	38
Sicily	2	540	22
Cuba	2	284	10
Gibraltar	1	277	12
Prussia	1	298	10
Algerria	1	463	10
Africa	1	149	7
River La Plata	1	253	15
Total 1838	**1,091**	**347,574**	**14,380**

Clearances in 1838:

	Vessels.	Tons.	Men.
Great Britain	729	262,367	11,343
Ireland	139	60,492	2,502
British North American Colonies	198	8,161	495
British West Indies	16	2,089	117
River La Plata	1	343	17

Quebec, as every one knows, was taken from the French in 1759. A British army, under General Wolfe, having effected a landing near the city, attacked and defeated the French army under Montcalm, on the heights of Abraham, to the W. of the town. Wolfe fell in the moment of victory; and Montcalm, who was also mortally wounded in the action, expired soon after. The French, panic-struck by the loss of the battle and the death of their commander-in-chief, surrendered the city before even a single battery had been opened against it. A monument was erected, under the patronage of Lord Dalhousie, in the gardens of the chateau, inscribed with the "Immortal memory of Wolfe and Montcalm."

QUEDLINBURG, a town of Prussian Saxony, reg. Magdeburg, circ. Aschersleben, on the Bode, a tributary of the Saale, 7 m. S.S.E. Halberstadt, and 33 m. S.W. Magdeburg. Pop., in 1838, 12,903. It is well built, and is surrounded by irrrted walls, pierced by four gates. On an eminence immediately above the town is an old castle, now falling into decay, but which has a good library, and is in part convert-

ed into a school. It was formerly the residence of the abbesses of Quedlinburg, who were princesses of the empire, and had a seat on the Rhenish bench of bishops. Many of these abbesses are buried in the Stiftskirche, or church of the ancient abbey; in which are also the tombs of Henry the Fowler, his empress, the beautiful countess Königsmark, mother of Marshal Saxe, &c. Quedlinburg has several hospitals, public schools, and various charities; with manufactures of woollen stuffs, distilleries, sugar refineries, &c. The rearing of cattle and hogs employs many of the inhabitants. Klopstock, author of the "Messiah," was a native of Quedlinburg, having been born here on the 2d of July, 1733: since his death a monument has been erected to his memory in the Brühl garden near the town. Quedlinburg was formerly a free imperial city, and has been frequently the residence of the German emperors and the seat of ecclesiastical councils. (Berghaus, Allg. Länder, &c., iv. 654; Murray's Handbook for N. Germany.)

QUEEN ANNE, county, Md. Situated in the E. part of the state, on the E. side of Chesapeake bay, and contains 400 sq. m. Bounded N.W. by Chester river; Kent island in Chesapeake bay belongs to it. It contained in 1840, 9036 neat cattle, 12,881 sheep, 12,590 swine; and produced 113,411 bushels of wheat, 35,767 of rye, 501,274 of Indian corn, 117,765 of oats, 15,975 of potatoes. It had eight stores, five grist-mills, one tannery; six academies, 96 students; 12 schools, 364 scholars. Pop.: whites, 6138; slaves, 3960; free coloured, 2541; total, 12,633.

QUEENSBOROUGH, a bor., seaport, and par. of England, county Kent, lake Scray, on the E. coast of the Isle of Sheppey, at the junction of the channel of the Swale with the estuary of the Medway, 2 m. S. Sheerness, and 37 m. E. by S. London. Area of parish, 380 acres. Pop. of do., in 1831, 786. The town, which is poor and mean, consists principally of a main street, having the guildhall, with a jail underneath, near its centre. The church, an ancient structure, has a tower at its W. end; and there is, also, a chapel for Independents. A charity-school for the education of the sons of the freemen used to be supported by the corporation and the parliamentary representatives for the borough, and there are some minor-schools, and a sunday-school. The inhabitants are almost wholly engaged in the breeding and supplying of oysters for the London market.

Inconsiderable as it has long been, Queensborough enjoyed the privilege of sending two members to the House of Commons from the æra of Elizabeth down to the passing of the Reform Act, when it was most properly disfranchised. It was reckoned too unimportant to be included in the provisions of the Municipal Reform Act. A fine old castle in the vicinity of the town was demolished by order of parliament in 1650.

QUEENSBURY, p. t., Warren co., N. Y.— 57 m. N. Albany, 429 W. Bounded S. by Hudson river, and contains the village of Glen's Falls. It contains a Methodist and an Episcopal church, 25 stores, one fulling-mill, one grist-mill, nine saw-mills, one powder-mill; one academy, 44 students; 22 schools, 918 schools. Pop. 3789.

QUEEN'S COUNTY, an inland co. of Ireland, prov. Leinster, having N. King's co., E. Kildare and Carlow, and a detached portion of King's co., S. Kilkenny, and W. Tipperary. Area 396,810 acres, of which 60,975 consist of unimproved bog and waste lands. Surface generally flat; and soil, except where bog occurs, for the most part very fertile. Estates mostly large; but many of them are let on perpetual leases; the head lessees on these estates forming the middle class of gentry. These, however, have mostly relet their farms, generally in smaller divisions, to inferior tenants; and these again have subdivided them to others; so that many of the occupancies are extremely small, and held by persons too poor to be able to make any improvement. But where farms have been let on terminable leases, they are larger, and on these various improvements have been introduced, both as respects the rotation of crops, the implements of husbandry, and the stocks. Dairying is carried on to some extent, and a good deal of cheese is made for the Dublin market. Average rent of land 14s. an acre. Coal and limestone are found in this co.; but the former is not wrought. Principal rivers, Barrow and Nore. It is divided into eight baronies, and 50 parishes; and sends three members to the House of Commons, being two for the county, and one for the borough of Portarlington. Registered electors for the county, in 1839-40, 2347. In 1831, this county had 93,783 inhabited houses, 95,403 families, and 145,851 inhabitants, of whom 72,469 were males, and 73,382 females.

QUEEN'S, county, N. Y., the central county of Long Island, and contains 396 sq. m. It has the East river and Long Island sound on the N., and the Atlantic ocean on the S. There is a lighthouse on Sand's point on the sound. It contained in 1840, 14,181 neat cattle, 26,477 sheep, 21,518 swine; and produced 97,741 bushels of wheat, 105,399 of rye, 336,401 of Indian corn, 64,027 of buckwheat, 2603 of barley, 348,447 of oats, 214,121 of potatoes. It had 83 stores, 13 lumber-yards, seven fulling mills, seven woollen factories, seven flouring-mills, 41 grist-mills, 17 saw-mills, three paper-mills, one rope-walk, five tanneries, one distillery, one glass-cutting works, four printing-offices, three weekly newspapers; one college, 36 students; eight academies. 429 students; 7d schools, 3670 scholars. Pop. 30,324. Capital, North Hempstead.

QUEENSFERRY (SOUTH), a parl. bor. and seaport of Scotland, co. Linlithgow, on the S. shore of the frith of Forth, nearly opposite to N. Queensferry, 8 m. W. by N. Edinburgh. It is a poor decayed place, which was originally founded for, and is still principally dependant on, the ferry, which has long been established at this point, across the Forth. Pop., in 1831, 684. It consists chiefly of one street, running parallel to the frith. The only public buildings are the parish church, a dissenting chapel, and the town-hall: it has a soap manufactory and a brewery. New-hall, immediately to the E. of the town, is now the principal ferry station across the Forth; though, under certain circumstances, the ferry-men prefer the pier of the borough.

North Queensferry, on the opposite side of the frith, which is here less than 2 m. across, is still more inconsiderable. The principal ferry business across the Forth is now, in fact, carried on between Newhaven and Brunt-island, and Newhaven and Kirkcaldy. Dr. Wilkie, author of the "Epigoniad," was a native of S. Queensferry, having been born there in 1721. Queensferry unites with Dunfermline, Inverkeithing, Culross, and Stirling, in sending one member to the House of Commons. Registered electors in this bor., in 1839-40, 49. Hopetoun House, the splendid mansion of the Earl of Hopetoun, is situated a little W. from S. Queensferry. It occupies a commanding position, and has a noble view.

QUENTIN, ST. (an. *Augusta Veromanduorum*), a manufacturing town of France, dep. Aisne, cap. arrond., on the Somme and on the canal of St. Quentin, 24 m. N.W. Laon; lat. 49° 50' 51" N., long. 3° 17' 38' E. Pop., in 1836 (exc. comm.), 19,899. It was formerly a place of strength; but since the time of Louis XIV. its ramparts have given place to suburbs and public walks, and a fine public promenade extends on the E. side of the town, along the banks of the canal. St. Quentin is tolerably well built; its principal streets being wide, its new quarters handsome, and many of its houses modern. In the centre of the town is a large square, in which is the town-hall, a curiously ornamented Gothic edifice. The principal church, or cathedral, is a vast and majestic pile. 416 feet in length internally, and its nave 212 feet in height. The hospitals, the public library, with 14,000 volumes, the courthouse, belfry, theatre, and

concert-hall, comprise the other buildings worth notice. It has tribunals of original jurisdiction and commerce, a chamber of manufactures, a council *des prud'hommes*, &c., a communal college, drawing and commercial schools, courses of practical geometry and mechanics, and a society of arts and sciences.

St. Quentin was formerly the centre of an extensive manufacture of linen fabrics and yarn. This branch of industry has now almost disappeared; but its place has been supplied by the cotton manufacture. In 1834, within an area of 12 leagues round St. Quentin, embracing its arrond. with parts of the deps. Somme, Du Nord, and Pas de Calais, it was estimated that nearly 75,000 hands were employed in weaving, bleaching, and spinning cotton: besides ma..y more employed in subordinate departments. The principal articles are striped and spotted muslins, yarn, &c., and the town and its neighbourhood have about 700 bobbinet frames. The cotton spinners, whose number is about 4000, reside principally in the town; the weavers live in the villages and surrounding country, where most of them are petty proprietors, occupied in agricultural labours for three or four months of the year. But it is obvious that goods produced by semi-manufacturers of this sort could never withstand, for a moment, the free competition of goods produced in factories supplied with the best machinery, and with a proper distribution and division of labour, and that they must depend wholly on the artificial market produced by the exclusion of cheaper articles. At St. Quentin the weaving of 1200 and 1400 checked muslins of a coarse description, 39 inches wide, and sold for 9d. per ell of 44 inches, is paid at the rate of 10d. a day: a still coarser description is sold for 6½d. or 7d. the ell, and the weaver gets 8d. a day. For finer goods the weavers' wages vary from 1s. to 1s. 4d. a day: these prices are wholesale, with 5 or 8 per cent. discount, ready money. The wages are gross, the winding only being paid by the manufacturer. At an average, the weavers in the country gain, children from 3d. to 6d., women from 7d. to 10d., and men from 1s. 3d. to 3s. a day. In the cotton-mills in the town, the average wages of a male adult may be from 6½ to 7 fr. a week. The workmen on the bobbinet frames get from 1 fr. 50 c. to 1 fr. 75 c. a day; and from 40 to 50 Mechlin frames have been set up, for the produce of which a considerable demand has arisen, and on which the wages amount to 6 fr. and 7 fr. a day. The workpeople in the town, of all ages, work 13 hours a day. In the country the hours are not regulated: but they work nearly as much as in the town, especially in winter. (*Symons, in Handloom Weavers' Rep.*, p. 128.)

The cotton-mills of St. Quentin are by no means so extensive as those of the dep. Haut-Rhin: few employ more than 900 hands. Children are employed here at an earlier age than in the cotton factories of Alsace; but the workpeople of all ages appear to enjoy much better health and more comforts in St. Quentin than either at Mulhausen or Lille. "The working classes make no complaints: they seem all pretty well supplied with the necessaries, and many of the comforts of life. They are all well clothed, and have plenty of vegetables to eat; little meat is consumed, or desired by them." (*Symon's Rep.*)

Table linens, leather, soap, and sulphuric acid, are also produced here; and the commerce of the town with the adjacent parts of France, Belgium, and Germany, is much facilitated by good roads and by the canal of St. Quentin. The latter connects the inland navigation of France with that of the Netherlands, by forming a communication between the Oise, the Somme, and the Scheldt. It is remarkable chiefly for the tunnels cut through the high ground, about 4 m. N. St. Quentin. The first of these is 100 feet below the surface, 24 feet is width, the same in height, and 3 m. in length. The second tunnel is on a still larger scale, being 3 m. in length, and 200 feet below the surface. Daylight is admitted, at certain distances, by openings carried to the surface; and the tunnel being cut through a chalk rock, the sides are not built. It was finished in 1810. (*Hugo*, art. *Aisne*; *Villermé, Tabl. des. Ouvr.*; *Handloom Weavers' Rep.*)

St. Quentin, as previously stated, was formerly strongly fortified, and was regarded as one of the bulwarks of France on the N. In 1557, in the earlier part of the struggles between Philip II. and France, the army of the former, under the Duke of Savoy, having threatened to attack the town, defended by the famous Admiral Coligny, and a weak garrison, the constable Montmorency advanced with a considerable army to its relief, and succeeded in throwing some reinforcements into it. On his retreat, however, he was overtaken by the Spaniards, when a general action ensued, which ended in the total defeat of the French, who lost all their artillery and baggage, with about 7000 men killed and prisoners, including many persons of distinction. The town soon afterwards fell into the hands of the Spaniards. The battle having been fought on the 10th of August, St. Lawrence's day, the vast palace of the Escurial, built by

Philip II., was dedicated to the saint in commemoration of the victory. (*Watson's Philip II.*, vol. i., 70, &c.)

Pierre la Ramée, better known by his Latin name of *Ramus*, one of the earliest and ablest opponents of the scholastic system of philosophy, and the Marquis Condorcet, were natives of the vicinity of St. Quentin, though not, as is frequently stated, of the town itself.

QUERETARO, a city of Mexico, cap. of the state of its own name, in a rich and fertile valley, 110 m. N.E. Mexico, and 60 m. E.S.E. Guanaxuato; lat. 20° 36' 30" N., long. 100° 10' 15" W. Pop. at least 40,000. (*Ward.*) It is a well-built city, with three large squares, many handsome public and private edifices, and the usual excess of convents and churches. The Franciscan monastery is spacious, and surrounded with extensive gardens ; and the convent of Santa Clara is an immense building, inhabited by 250 females, including many young ladies sent thither for their education. The streets have side pavements, laid with flags of porphyry : the city is well supplied with water, brought to it by an aqueduct about 10 m. in length, carried across the valley upon 60 arches.

It is divided into five parishes ; four in the body of the town, and one in the suburbs, separated from the rest by a little stream. " We were much struck with the busy look of Queretaro, which has quite the air of a manufacturing town. More than half the houses contain shops, and the population is engaged either in small trades or in the wool manufactories, which are still very numerous. These are divided into two classes, *obrages* and *trapiches*. The first comprises the establishments that employ from 10 to 30 looms ; the last those in which only one or two are in activity. In both coarse cloths, of different patterns and sizes, are manufactured ; part of which is retailed upon the spot in the great Plaza, where a market is held every evening by torchlight, and part sent to the capital or other great towns of the confederation. The demand for these manufactures has decreased very much since the ports were opened to European imports : indeed, the woollen trade is now principally kept up by a government contract for supplying the army with clothing. The price paid for scarlet, green, and yellow cloths of the very coarsest texture, varies from 24 reals (12s.) to 15 reals (7s. 6d.) per vara : and there is no doubt that they might be obtained of a better quality at a much lower price from abroad. The wool is brought principally from the northern states. San Luis de Potosí, and Zacatecas ; its price fluctuates from 16 to 24 reals the arroba of 25 lbs., including carriage ; but the wool most esteemed is the produce of the state itself. It acquires its value, not from any superiority in the breed of the Queretaro sheep, but from the circumstance of the flocks being so much smaller than those of the north that they can be better attended to, fed in richer pastures, and kept more clear from thorns, which deteriorate the fleece. This wool sells for 2½ dollars (or 20 reals) the arroba." (*Ward's Mexico*, i., 183, 184.) The manufactures of this city are conducted on the same nefarious system that prevails elsewhere in Mexico (*see* this vol., p. 352), of inveigling the Indians into debt by the voluntary advance of money, and then shutting them up in the factories, under prison discipline, with criminals of all kinds, till they have liquidated the debt by their labour, a circumstance which every pains is taken to prevent, and which rarely occurs. (*Poinsett's Notes on Mexico,* p. 186, &c.)

QUIMPER-CORENTIN, a town and river port of France, dep. Finisterre, of which it is the cap., at the junction of the Air with the Odet, about 11 m. above where the latter falls into the bay of Benaudet, 115 m. W. by S. Rennes ; lat. 47° 5' 29" N., long. 4° 6' 15" W. Pop., in 1836, 9715. It stands at the declivity of a hill, and is divided into the old and new town. The former, surrounded by walls flanked with towers, is ill built ; but in the new town there are some good streets and houses. The cathedral, a handsome Gothic edifice of the 15th century, and other churches ; the military hospital ; the college, a large structure formerly belonging to the Jesuits ; the theatre, and some public baths, are the principal public buildings. The river is navigable as far as the town for vessels of 200 tons burden, those of greater size anchoring opposite its embouchure, in the bay of Benaudet. It has manufactures of earthenware, and building docks ; its inhabitants, also, engage in the pilchard fishery, and carry on a pretty brisk trade in provisions. Quimper is a bishop's see, and derived its present name from its first bishop, in the 5th century ; previously to which it was called *Corisopitum*. It was sacked by Charles of Blois in 1345.

Among the distinguished men, of whom Quimper has to boast, may be mentioned Hardouin, the commentator of Pliny, celebrated alike for his learning and his paradoxes ; and Freron, the most voluminous, but, at the same time, the most valuable of the French critics of last century. (*Hugo, Finisterre ; Dict. Géog., &c.*)

QUINCY, p. t., Norfolk co., Mass., 9 m. S. by E. Boston,

449 W. It contains large tracts of salt meadow. Three miles back from the shore is an elevated range, rising in some places 600 feet above the level of the sea, which contains an inexhaustible quantity of the celebrated Quincy granite, which is extensively exported. The Merchants' Exchange in the city of New-York is built of it. A railroad extends 3 m. from the quarry to tide water on Neponset river, constructed in 1836, at an expense of $30,000, the first railroad built in the United States. It has some vessels, employed chiefly in the fisheries. It contains four churches, two Congregational, an Episcopal, and a Universalist ; 15 stores, four lumber-yards, two tanneries, one printing-office, one weekly newspaper ; eight academies, 137 students ; six schools, 708 scholars. Pop. 3486.

QUINCY, p. t., Franklin co., Pa., 57 m. S.W. Harrisburg, 83 W. It has seven stores, one furnace, three forges, six flouring-mills, two grist-mills, eight saw-mills, two tanneries, three distilleries ; eight schools, 258 scholars. Pop. 2503.

QUINCY, p. v., capital of Adams co., Ill., 104 m. S. Springfield, 884 W. Situated on a bluff on the E. side of Mississippi river, 125 m. by water, above the mouth of Illinois river. It contains a brick courthouse, which cost $90,000, on a fine public square, four churches, a Congregational, Episcopal, Methodist, and Baptist ; 25 stores, a United States land office, a large steam flouring and saw-mill, 200 dwellings, and about 1500 inhabitants. There are 300 steamboat arrivals in a year, and flour and pork are exported to the amount of $100,000 annually.

QUITO, a celebrated city of S. America, cap. of the republic of Equator (*Ecuador*), in a ravine on the E. side of Pichincha (a volcanic mountain, which, at no very distant period, was in a state of activity), above 9500 feet above the sea, 160 m. N.N.E. Guayaquil, and 460 m. S.W. Bogota. Lat. 0° 13' 27" S., long. 78° 10' 15" W. Pop., variously estimated at from 40,000 to 70,000. Quito is, on the whole, one of the best built cities in the New World. It has four broad, straight, and well-paved streets, and three large and some smaller squares, in which are the principal public buildings, and the best private residences. The streets which run N. and S. are pretty level, but those which cross them ascend the skirts of the Pichincha on the one hand, and descend on the other towards a small river, over which is a small bridge ; and from this unevenness of the ground some of them are so steep as to be impracticable for carriages, besides being, for the most part, narrow, crooked, unpaved, and almost impassable after heavy rains. So numerous, also, are the crevices in the mountain, that, in the suburbs particularly, several of the houses have been raised on arches. The houses, which are large and convenient, are mostly built of unburnt brick, cemented with a species of mortar, used by the ancient Peruvians, which soon becomes extremely hard. On account of earthquakes, they are seldom more than one story in height, exclusive of the ground floor, or *rez-de-chaussée*. They are flat roofed, and have usually a balcony towards the street. Generally speaking, they are indifferently furnished, and deficient in cleanliness. The city has an abundant supply of indifferent water, obtained from several streams, which flow through its conduits. The principal square has, in its centre, a fine brass fountain ; and on its sides are the cathedral, the bishop's palace, the town hall, and the palace of the president. The last is a gloomy-looking building, with a terrace in front, ascended by two flight of steps. In it are the halls of the *audiencia*, treasury, and archives, the president's apartments, with the offices of the public secretaries, and the jail. The cathedral, a plain building, with a steeple at one corner, is much less handsome than several of the other churches. Quito has numerous convents. The ex-Jesuits' college has a beautiful front, with Corinthian columns, finely sculptured by native artists. The interior of this edifice is very rich, and when visited by Stevenson it had a library, said to comprise 20,000 volumes, including several rare works. A part of the edifice has been converted into halls for the university, and another part into barracks. Ulloa speaks in high terms of the Franciscan convent, which, he says, from its size, fine proportions, and beauty, might be classed with the best edifices in Europe. Previously to the revolution, the churches and convents were richly furnished with silver ornaments, plate, paintings, &c.; but a part of this wealth has, we believe, been since turned to more useful purposes. It has a workhouse and orphan asylum on a large scale, and said to be well conducted, a hospital, &c.

Quito ranks pretty high, at least among Spanish American cities, as a place of education. It had two universities before the time of Charles III., but they were then united into one. Besides this university, which still exists, there are several colleges under the guidance of the different religious orders. Ulloa states, that young men of distinction usually studied philosophy, theology, and jurisprudence. " They succeed pretty well," he adds, " in the exact sciences, but are extremely ignorant in all that respects politics, history, and those philosophical studies that contribute most to expand and enlarge the mind." There can be no doubt that this is a perfectly accurate statement; but we apprehend the students of Salamanca in Old Spain were, at the epoch of Ulloa's visit to Quito, but little more advanced than those belonging to the latter. It were absurd to suppose that colleges conducted by priests, under an arbitrary government, should supply any sound instruction, either in politics or philosophy.

The city was made a bishop's see in 1545, and is the residence of the president, and the seat of all the superior courts and offices of the republic.

Coarse cottons, and woollen cloth, baizes, flannels, ponchos, stockings, &c., are made in Quito, which is also highly celebrated for its confectionary; but its chief exports consist of the corn, and other agricultural products of the province. These, with some of its manufactures, are sent by way of Guayaquil to Central America, in exchange for indigo, iron, and steel; and to Peru, in return for brandy, wine, and oil, and for gold, silver, and other metals, Ecuador not being very rich in mineral wealth. The foreign imports comprise all kinds of European manufactured goods, with iron, steel, and some other raw materials. "The European manufactures most in demand are English broad cloths, kerseymeres, coloured broad flannels, calicoes, plain and printed dimities, muslins, stockings, velveteens; Irish linens, in imitation of German *platillas*; all kinds of hardware and cutlery, and foreign silk velvets, satins, &c., as well as English ribbons and silks. As for the Lima market, the articles should be of a good quality, and of the newest fashion; the more this point is attended to, the better the market will be found." (*Stevenson's S. America*, ii., 319.) The markets of Quito are abundantly provided with beef, mutton, pork, and poultry, both cheap and good; vegetables and fruits of all kinds, bread, &c. The consumption of cheese is said to be very great; its cost price being estimated at from 80,000 to 90,000 dollars a year, and many of the neighbouring farmers are principally engaged in its manufacture.

The inhabitants of Quito do not differ materially from those of other Spanish American cities; bull fights, masquerades, dancing, music, and religious processions being their principal amusements. When Ulloa visited the city, indolence was, and, we suspect, still is, the distinguishing characteristic, if we may so speak, of all classes. This is, in great part, no doubt, to be ascribed to the mildness of the climate, and the ease with which the ordinary necessaries of life may be produced. The city enjoys, as it were, a perpetual spring. Vegetation never ceases at any period of the year; but from December to March violent storms of rain and lightning almost daily occur in the afternoon. Earthquakes are, also, frequent; and one of those visitations that occurred in 1797 is said to have destroyed in the province above 40,000 persons, and to have had a permanent influence over the climate. A plain, about four leagues N.E. from the city, was made choice of by the French and Spanish astronomers, in 1736, for measuring a degree of the meridian; and an inscription on a marble tablet on the wall of the ex-Jesuits' church in Quito commemorates the event, and the labours of the commission; but the most enduring memorial of that great undertaking is to be found in the *Historical Voyages* of Ulloa, one of the best works of its kind that has ever been published. Quito was founded by Sebastian Benalcazar in 1534, and incorporated as a city by Charles V. in 1541. (*Ulloa*, liv. v., caps. 4, 5; *Stevenson's S. America*, ii., 279-325; *Geog. Account of Columbia, &c.*)

R.

RAAB (Hung. *Győr*, or *Nagy Győr*, an. *Jaurinum, Arabo,* or *Arabonia*), a royal free town of Hungary, cap. the co. Raab, at the confluence of the river of the same name with the Danube, 22 m. W.S.W. Comorn, and 39 m. S.E. Presburg. Population about 13,500. It stands in a low and marshy plain, and is rather unhealthy. Like Vienna, the city is separated from the suburbs by its old fortifications, and an open glacis, planted with trees, and forming public walks. Some of the streets are regularly built; and, besides three or four churches and a cathedral, the city has the bishop's palace, the public buildings belonging to the county and the corporation, and some handsome residences belonging to the Esterhazy and Zichy families. (*Bright's Travels,* 298.) It has, also, a royal academy of law and philosophy; and Roman Catholic, Greek, and Lutheran high schools. In the centre of the city is an immense Capuchin convent, its roof surmounted with two very high and conspicuous spires. There are various other conventual establishments, two work-houses, two barracks, a theatre, arsenal, &c. Raab is the seat of a larger trade in

corn than any other city in Hungary, and is a great depôt for the trade of Pesth with Germany and Italy. (*Berghaus*) It has several large annual fairs, its commerce depending, for the most part, on its favourable position on two navigable rivers. It was a strong post under the Romans, and has been generally kept in a defensive state by the Hungarian kings; but it was twice taken by the Turks, and, in 1809, an Austrian force was routed by the French under its walls. (*Oesterr. Nat. Encyc.*; *Berghaus*; *Bright's Travels in Lower Hungary*; *Walsh's Constantinople, &c.*)

RABUN, county, Ga. Situated in the N.E. part of the state, and contains 330 sq. m. Bounded S.E. by Chatoga river, S.W. by Tarroree river, which unite in the S. post of the county, to form Tugaloo river. It contained in 1840, 2790 neat cattle, 2094 sheep, 8526 swine; and produced 1452 bushels of wheat, 1635 of rye, 62,921 of Indian corn, 13,151 of oats, 4734 of potatoes, 3298 pounds of tobacco. It had three stores, 23 grist-mills, five saw-mills, one tannery, 13 distilleries; four schools, 131 scholars. Pop.: whites, 1828; slaves, 84; total, 1912. Capital, Clayton.

RACCONIGI, or RACONIGI, a town of the Sardinian dom., div. Coni, prov. Saluzzo, cap. mand., near the Macra, and 21 m. S. Turin. Pop., in 1838, inc. com., 10,162. It is walled, and tolerably well built: it has several good churches; but its chief ornament is a noble castle and park, belonging to the Prince of Carignano. Among the works of art in the castle are some pictures (if they may be so called), beautifully executed in silk. Silk weaving and spinning are the principal branches of industry in and round this town. (*Dict. Geog., &c.*)

RACCOON, t., Gallia co., O., 94 m. S.E. Columbus. Organized in 1806. Watered by Raccoon creek. It has five stores, three grist-mills, three saw-mills, two tanneries. Pop. 1610.

RACINE, county, Wis. Situated in the S.E. part of the territory, and contains 610 sq. m. Bounded E. by lake Michigan. Watered by Fox, Root, and head branches of Des Plaines rivers. It had in 1840, 4306 neat cattle, 349 sheep, 6459 swine; and produced 36,099 bushels of wheat, 36,169 of Indian corn, 3451 of barley, 1795 of oats, 41,523 of potatoes, 6051 pounds of sugar. It had two commission houses in foreign trade, 19 retail stores, two grist-mills, 13 saw-mills; two academies, 65 students; 29 schools, 607 scholars. Pop. 3475. Capital, Racine.

RADNOR, an internal co. of England, S. Wales, having N. the cos. of Montgomery and Salop. E. Hereford, S. Brecnock, from which it is separated by the Wye, and W. Cardigan. It is of a triangular shape, and comprises 271,80 acres. With the exception of some low and comparatively fertile tracts along the borders of Hereford and Salop, and in the valley of the Wye, the larger portion, by far of this county is wild, mountainous, and dreary. It is stated, in Davies' *Survey of S. Wales,* published in 1815, that about two thirds the surface consisted of waste land, nearly moor, but partly also bog! Several divisions and some extensive enclosures have, however, been effected in the interim; so that the extent of waste or common land, though still very great, has been materially diminished. At present the wastes are mostly depastured by sheep; and this county rears, in proportion to its size, more sheep than any other in the principality. They are mostly of a small hardy breed. Numerous encroachments have, from time to time, been made on the wastes or commons, by persons who had property adjoining, and by cottagers, who erected huts on their borders, and gradually extended their gardens, till they sometimes included acres of land. A good many of the manors on which these encroachments were made formerly belonged to the crown; and having been lately sold, the purchasers have attempted to oust those who had made these encroachments. But though, no doubt, the act was illegal, still under the circumstances it was beneficial rather than otherwise; and having been permitted in the first instance, their rights should have been protected. The farms in the low grounds vary from 20 to 200 acres: they are mostly held at will, or from year to year; and as there are no conditions to enforce a proper system of management, agriculture is in the most backward state, there being no proper rotation of crops, and the land being frequently foul and out of order. Many of the low farms have a portion of moor or common pasture attached to them. The cattle in the low grounds are principally of the Hereford breed: there are some extensive meadows, and irrigation is well understood. There are some rather large estates; but there are, also, many of an inferior size. Average rent of land, in 1810, 6s. 3d. an acre. The minerals and manufactures of the county are of no importance. Radnor is divided into 6 hundreds and 50 parishes; and sends two members to the House of Commons; being one for the county, and one for New Radnor and its contributory boroughs. Registered electors for the county, in 1839-40, 2034. In 1831 this county had 4437 inhabited houses, 4872 families, and 24,641 inhabitants, of whom 12,453 were males, and 12,198 females. Sum contributed to the relief of the poor, in 1834–32, £8836. Annual value of real property, in 1815, £103,378. The borough of New-Radnor, referred to above, is of great extent, embracing an area of nearly 30 m. in circumference. In 1831, it had a population of 2501 ; but the town itself has not more than 400 inhabitants, and would not be worth notice but for its being the head of the parliamentary borough. PRESTEIGN (which see) is the only town in the county of any importance.

RAGUSA, a town of Sicily, prov. Syracuse, co.. Modica, near the W. bank of the river of its own name, about 14 m. above the embouchure of the latter on the S. coast of Sicily, and 30 m. W.S.W. Syracuse. Pop., in 1831, 21,466 The inhabitants, who are said to be active and industrious, have considerable cloth and silk manufactures, and a rather extensive trade in the corn, oil, wine and cheese of the surrounding territory. It is near the site of the anc. *Hybla Herea* ; but being seldom visited by travellers, little is known of its state. (*Rampoldi, Corografa.*)

RAOUSA (Slav. *Dubrovnik*, Turk. *Paprovnik*, an. *Rausium*), a seaport town of the Austrian dominions, prov. Dalmatia, cap. circ. of its own name, on a peninsula in the Adriatic, 37 m. W.N.W. Cattaro ; lat. 42° 36' 30" N., long. 18° 11' 55" E. Pop., in 1837, 3039. It is walled, and has two good harbours, one to the N.W., and the other to the S.E. ; which, as well as the town, are protected by several modern forts. Its streets are narrow, except one, the Corso, which intersects it from N. to S. ; its houses are well built, in the Italian style, but many of them are now unoccupied. It has a cathedral, a Greek church, a Piarist gymnasium, high school, military hospital, (once the Jesuits' college), lazaretto, and theatre. It is the see of a Roman Catholic vicar-bishop, and the seat of the superior judicial courts for the circuit : it has manufactures of silk, leather, and rosoglio. Though Ragusa has greatly declined from her former importance, she still has a considerable coasting trade ; and in 1838, 236 vessels belonged to the port, 30 of which traded to the Levant and the western ports of the Mediterranean. Ragusa was founded in the seventh century by some fugitives from Epidaurus in Illyria, when that city was destroyed by the Slavi. It continued to be a republic under the successive protection of the Greeks, Venetians, and Turks, till 1806, when it was taken by Napoleon, who erected it into a dukedom, which he conferred on Marshal Marmont. On the fall of the former. Ragusa was given to Austria. (*Berghaus, Allgemeine Länder, &c.,* iv., 968 ; *Austria, Nat. Encyc., &c.*)

RAHWAY, p. t., Essex county, N. J., 9 m. S.W. Newark, 30 m. N.E. Trenton, 205 W. Drained by Rahway river, and its tributaries. The village is partly in Woodbridge t., Middlesex county, and partly in Rahway t., Essex county, on both sides of Rahway river, at the head of tide-water, 5 m. from its mouth. The village is in several detached parts, and contains six churches, a Presbyterian, Methodist, Baptist, two Friends, and an African, the Athenian academy, with an edifice 68 ft. long, and 36 ft. wide, and 2 stories high, the Rahway female institute, with a neat and commodious building, a bank, an insurance office, 30 stores, a large saw-mill, about 400 dwellings, and 2000 inhabitants. The railroad from Jersey city to Philadelphia passes through the place. The town contains 10 stores, 14 flouring-mills, four grist-mills, five saw-mills, one paper-mill, one tannery, one distillery ; one academy, 41 students ; seven schools, 344 scholars. Pop. 2533.

RAISIN, r., Mich., rises in Wheatland t., Hillsdale co., and flowing in a winding course, 130 m. by the course of the river, it enters lake Erie, 2½ m. below Monroe. It is one of the most important streams in Michigan, affording extensive water-power. It flows by several important villages.

RAJAHMUNDRY, a distr. of the Madras presid. in British India. (*See* CIRCARS, NORTHERN.)

RAJAHMUNDRY, a town of Hindostan, cap. of the above distr., on the Godavery, about 50 m. from its mouth, and 55 m. N.E. Masulipatan. It stretches along the river for some distance, and has an old fort, several mosques, a fine bazaar, &c. It is the seat of the district court, and a station for two companies of sepoys. (*Madras New Almanac, &c.*)

RAJEMAHAL (*The Royal Residence*), a town of British India, presid. Bengal, prov. Bhaugulpore, on the Ganges, 6 m. N.W. Moorshedabad. Under Acbar, and Sultan Shujah, the brother of Aurungzebe, it was the capital of Bengal ; and, though much decayed, it is still estimated to have 30,000 inhabitants. It consists principally of one long street of stone or mud houses, generally with two stories ; about a dozen market places, scattered over a wide extent of ground ; a few tombs and mosques, and the ruins of a spacious palace. The inhabitants have some traffic with the hill people of the district, but their main source of profit derived from their supplying necessaries to travellers on the Ganges. (*Hamilton's E. I. Gaz. ; Mod. Trav.*)

RAJESHAYE, a distr. of British India, presid. and prov. Bengal, principally between the 24th and 25th degs of N lat., and the 88th and 90th of E. long., having N. Dinajepoor, and Rungpoor ; E. Myemunsing, and Dacca Jelalpore ; W. Purneah ; and S. the Ganges, separating it from Jessore, Nuddea, and Moorshedabad. Area, 3950 sq. m. Pop., in 1822, 1,087,155 ; about two thirds Hindoos, and the rest Mohammedans. The whole surface is so low that from the beginning of July to the end of November it is nearly submerged by the inundations. Towards the E. it is thickly wooded. Few solid edifices, and no fortresses exist in this district ; but it comprises many populous commercial villages, and the towns of Nattore, Bauleau, and Hurrial. Land revenue, in 1827–30, 14,64,299 rupees. (*Parl. Rep., &c.*)

RAJPOOTANA, the largest prov. of Hindostan, towards its N.W. quarter, between 24° and 21° N. lat., and 70° and 77° E. long. ; having W. and N. Moultan and Lahore, E. Delhi and Agra, S. Malwah and Gujrat, and S.W. Sinde. Its length, N. and S., is estimated by Hamilton at 350 m., and its average breadth at 200 m., giving it an area of about 70,000 sq. m. It comprises a large extent of sandy desert, but in the S. and E. are many fertile tracts. This province, is wholly subsidiary to the British, consists of a number of feudal principalities, the chief of which are Judpoor, Jesselmere, Jeypore, Odeypoor, and Bicanere, which see. For farther and full particulars respecting this portion of Hindostan, the reader may consult the copious work on Rajasthan by Colonel Tod.

RALEIGH, city, capital of Wake county, N. C., and of the state, in in 35° 47' N. lat., and 78° 48' W. long., 123 m. N.W. Newbern, 60 m. N. Fayetteville, 164 m. S.S.W. Richmond, Va., 288 W. The population in 1820, was 2074 ; in 1840, 2224. It is pleasantly situated 6 m. W. of Neuse river and 27 m. N.W. Smithfield, the nearest ordinary point of navigation ; though boats sometimes ascend the Neuse, to within 8 m. of Raleigh. It is laid out with great regularity, having at its centre Union-square of 10 acres, from which extend four broad streets 99 ft. wide, dividing it into four quarters. In the centre of each of these quarters are four other squares of four acres each, and the streets which intersect the quarters are 66 ft. wide. The statehouse is a splendid granite edifice, in the centre of Union-square, 160 ft. long, and 90 ft. wide, surrounded by massive granite columns, 5¼ ft. in diameter and 30 ft. high, after the model of the Parthenon at Athens. It has a fine dome, and spacious and handsome legislative halls, and cost $500,000. The other public buildings are a courthouse, jail, governor's house, secretary of state's office, a theatre, market, two banks, five churches, an Episcopal, Presbyterian, Methodist, Baptist and Roman Catholic, and about 400 dwellings. According to the census of 1840, there were 43 stores, capital $191,200, four printing-offices, two binderies, five weekly, and one semi-weekly newspapers ; four academies, 95 students ; two schools, 33 scholars. The former state-house, containing a beautiful marble statue of Washington, by Canova, was burned in 1831, and this fine work of art was destroyed. A railroad extends 85 m. from this place to Gaston on Roanoke river, whence it communicates with a railroad leading to Petersburg, Va.

RALLS, county, Mo. Situated toward the N.E. part of the state, and contains 470 sq. m. Drained by Salt river, and its branches. It contained in 1840, 5440 neat cattle, 6344 sheep, 17,245 swine ; and produced 23,954 bushels of wheat, 1524 of rye, 240,225 of Indian corn, 34,148 of oats, 9366 of potatoes, 79,119 pounds of tobacco, 10,984 of sugar. It had nine stores, one flouring-mill, 15 grist mills, 11 saw-mills, one oil-mill, four tanneries, seven distilleries ; one academy, 38 students ; nine schools, 243 scholars. Pop.- whites, 4450 ; slaves, 1209 ; free coloured, 11 ; total, 5670. Capital, New-London.

RAMAPO, t., Rockland co., N. Y., 8 m. W. New-city, 132 m. S. Albany. Drained by Ramapo and Saddle rivers. The New-York and Erie railroad passes through it. It has eight stores, two fulling mills, two woollen factories, one cotton factory with 516 spindles, five grist-mills, two saw-mills ; eight schools, 315 scholars. Pop. 3222.

RAMGHUR, a distr. of British India, up the S. part of the prov. Bahar, between the 23d and 25th degs. of N. lat., and the 83d and 87th of E. long. ; having N. Shahabad, Bahar, and Bhaugulpore ; E. Beerbhoom, and the Jungle Mehals ; and S. and W. the territory ceded by the rajah of Berar. Area estimated at 22,430 sq. m. Pop. in 1832, 2,252,985. A large proportion of this district is rocky and unproductive, or covered with wood. Iron, coal, lead, and antimony exist in the hilly region of the S., but from want of capital and enterprise among the inhabitants, few, if any, mines are wrought. Many of the zemindars have very extensive possessions, and are, in a great measure, independent of British authority. Slavery, though of a very mild character, is universally prevalent. There are many old brick forts in Ramghur, affording protection to hordes of banditti, and other refractory persons; but few other durable buildings. This district has always been notorious for crime; and hitherto, notwithstanding its great extent, has been of little value to the British Indian government. The land revenue in 1828, amounted to only 161,293 rupees. (*Parl. Reps.; Hamilton's E. I. Gaz.*)

RAMILLIES, an inconsiderable village of Belgium, prov. S. Brabant, 18 m. S.S.E. Louvain. This village is famous in history from its being the scene of the great victory gained on the 23d of May, 1706, by the allied forces, under the Duke of Marlborough, over the French, under Marshal Villeroi. The French army amounted to about 80,000 men, being superior in number to that of the allies. The latter, however, owing, as is alleged, to the bad dispositions and incapacity of Villeroi, gained an easy as well as a complete victory. The French lost about 8000 men, killed and wounded, and nearly 7000 prisoners, including 600 officers, with all their artillery and baggage. The loss of the allies did not amount to 3000 men. The Duke of Marlborough, whose gallantry was as conspicuous as his great talents as a general, had a horse shot under him in the action; and the head of Colonel Brienfield, who was assisting his grace to remount, was carried off by a cannon-ball! (*Smollett's Continuation of Hume, cap. 9; Voltaire, Siècle de Louis XIV, cap. 20.*)

RAMPOOR, a large town of British India, prov. Delhi; on the Cosilla, a tributary of Ganges, 112 m. E. by N. Delhi. It has two brick palaces, a lofty mosque in the principal street, the magnificent mausoleum of a former chief, and some good houses; but the greater part of the town consists of sun-burnt brick houses, with thatched or tiled roofs. Its territory is exceedingly productive, and well cultivated.

Another town, called Rampoor, situated on the Sutleje, about 160 m. E. by N. Umritsir, is a favourite place of Hindoo pilgrimage, and an *entrepôt* for the commerce of Hindostan with the countries beyond the Himalaya, on which account it has a large yearly fair. (*Hamilton's E. I. Gaz.*)

RAMREE, a town of British India, prov. Aracan, cap. of the prov., and on the island of the same name, about 115 m. S. by E. Aracan. Pop. estimated, in 1835, at 8000, and increasing. It stands at the head of a creek, in which there is pretty good anchorage, and has a bazaar, supplied, though meagerly, with British goods. Its inhabitants, who are a fine athletic race, carry on a brisk trade with Chittagong, Sandoway, Bassein, and Calcutta. (*Pemberton; Rep. on the E. Frontier, p. 93, &c.*)

RAMSAY, a town and par. of England. co. Huntingdon, hund. Huntingstone. The par., which extends into the co. Cambridge, and has an area of no less than 17,660 acres, had, in 1831, a pop. of 3006, of whom about 1600 may belong to the town. The latter, about 10 m. N.N.E. Huntingdon, consists principally of one long street, running E. and W. with another branching off to the N. Houses mostly brick. The church, a fine old edifice, has a nave, chancel, and aisles, with an embattled tower at the W. end. The town formerly belonged to the Cromwell family, several of whom are buried in the church, but there are no monuments to their memory. It has a free school, founded and endowed in 1663, which educates about 70 boys; and a charity school, which educates about 50 girls, was founded towards the beginning of last century by John Dryden, Esq., a relation of the great poet. The town suffered severely from the fire in 1731. Ramsay is within the great level of the fens; and in its vicinity are several shallow lakes, or meres, that called the Whittlesea, about 4 m. N.W. from the town, being the largest in the kingdom. A magnificent Benedictine abbey, founded here in 969, acquired in the sequel, great wealth and celebrity. At the dissolution, its revenues amounted to £1983 a year. A ruined gateway is all that now remains of this once famous fabric. Market day Saturday; fair, 3d July, for pedlery.

RAMSAY, a town of the Isle of Man, which see.

RAMSGATE, a seaport, watering-place, and parish of England, co. Kent, on the E. coast of the isle of Thanet, 15 m. E. by N. Canterbury, and 65 m. E.S.E. London. Area of par., 960 acres. Pop. in 1841, 10,909. The town, which, till the beginning of the last century, was little more than a small fishing village, has risen to its present importance, partly in consequence of the construction of its artificial harbour, the largest of its kind in England, but principally from its having become a favourite summer resort of visiters from the metropolis. Its older portions, irregularly built, with narrow streets and mean-looking houses, occupied chiefly by the trades-people, lie in a flat opening towards the sea by a narrow gulley, (here called *gate*, whence the name of the town,) between two very steep cliffs: the latter, which are ascended by stone steps, are now covered with good looking, though, in general, not very substantial houses, laid out in terraces, crescents, &c. Building is still going forward on a pretty extensive scale. The town is well

lighted with gas. Till 1838 the supply of water was insufficient, and constituted no small item of expense to the inhabitants of the houses on the cliffs, but an ample supply was then introduced from the neighbouring village of St. Lawrence. A market-house stands at the intersection of the principal streets of the old town, opposite to which is the principal public library and town bank. Facing the sea are several hotels (in one of which are the assembly-rooms, a public library and well-constructed baths on the W. cliff, with the Clarence and other baths, bazaars, and a coffee-house, in the lower part of the town, abutting on the west basin. The church, recently erected, a large and handsome building, in the perpendicular style, with an octagonal tower, provides accommodation for about 2000 persons: the living (created by act of parl. in 1827) is a vicarage of the annual value of £400, patron the archbishop of Canterbury. A chapel-of-ease, till lately the only episcopal place of worship in the town, is supported chiefly by voluntary subscriptions and pew rents; besides which, there are places of worship for Wesleyan Methodists, Independents, Baptists, and Unitarians, and a Jews' synagogue. The town has four Sunday schools, attended by about 800 children, and there are national Lancastrian and infant schools, furnishing instruction to nearly 500 young persons of both sexes.

The most striking feature of the town, however, is its harbour, one of the most magnificent works of its kind in the kingdom, formed by double piers, built under the superintendence, successively, of Smeaton and Rennie. It was commenced, at the public expense, in 1750, with the intention of forming an asylum for vessels in the Downs, which might otherwise be driven on the Goodwin sands. The E. pier extends, in a curvilinear direction, upward of 260 yds. into the sea, its total length, including its angles, amounting to 900 ft., that of the W. pier being about 1500 ft. The width of the entrance is 240 ft., and the harbour area 46 acres.

The general breadth of the piers is 26 ft., including a strong parapet on the outer sides next the sea. They are constructed chiefly of Portland and Purbeck stone. After the piers were nearly finished, the deposition of sand and mud within the harbour being so great as to threaten ruin to the entire project, Smeaton recommended the construction of a basin within the harbour, to be filled at high water and let out again at ebb tide, so that any deposit might be carried off by the artificial current. This was accordingly done, and has been found to answer the purpose; and in 1787, an advanced pier was undertaken to facilitate the entrance of shipping in tempestuous weather. About the same period a dry dock was excavated, and storehouses erected. There is a lighthouse on the head of the W. pier, whence a clear red light is displayed at night, when there are 10 ft. water in the mouth of the harbour; this same being given, during the day, by a red flag from the head. A wet dock has lately been constructed near the basin for the repair of vessels. Still, however, it must be admitted that though no cost has been spared to render this harbour as useful as possible to the shipping in the adjacent dangerous part of the channel, it is, owing to the want of water, very defective, and, as it can be entered only at certain times of the tide. It is far from realising the expectations that were formed of its being a good refuge harbour. It is under the superintendence of an incorporated company of trustees, who appoint the chief and deputy harbour-master and other officers. It was made a royal port in honour of a visit from George IV. in 1821; and an obelisk near the pier-gates has since been erected to commemorate that most unimportant event.

Ramsgate had formerly an extensive commerce with the Baltic, but this has long declined. It has still some trade with France and Holland, chiefly in the importation of eggs, fruit, provisions, &c.; the gross customs revenue of the port (which includes Margate and Broadstairs), amounted in 1840, to £7068. Its coasting trade is pretty extensive, chiefly with London, Newcastle, and Sunderland. A considerable fishery of turbots and soles is carried on off the coast by boats from the W. ports of England; but only a small share of it is taken by the inhabitants of Ramsgate. Indeed the town displays little activity beyond what is caused by the influx of visiters during the summer season, and for whose accommodation steam-packets ply daily to and from the metropolis. The smoothness of the sands E. of the harbour, and the clearness of the water, make the beach particularly well adapted for bathing, and it constitutes a favourite resort for visiters, hundreds of whom may here be seen in the height of the season strolling about, lounging on chairs, and enjoying the sea-breeze. The pier furnishes another delightful promenade, and on the W. cliff, which is ascended from the inner basin by a flight of stone stairs, is a fine broad walk, extending westward towards Pegwell bay. Two well-conducted bath establishments are replete with accommodation for invalids, and libraries, news-rooms (one of which is in a marquee on the sands), bazaars, concerts, &c., furnish ample means of

RANDOLPH.   RANGOON.

occupation and amusement. Ramsgate is a member of the Cinque Port of Sandwich, and is governed by a deputy, appointed by the mayor of that borough; but judicial affairs are regulated by a local magistracy under a local act, passed in 1812. Markets on Wednesday and Saturday, but daily during summer, and exceedingly well supplied.

RANDOLPH, county, Va. Situated toward the N.W. part of the state, and contains 2660 sq. m. Drained by Cheat river and its branches, and by branches of the E. fork of Monongahela river. It contained in 1840, 10,166 neat cattle, 13,818 sheep, 9349 swine; and produced 27,212 bushels of wheat, 6861 of rye, 151,009 of Indian corn, 5039 of buckwheat, 86,784 of oats, 29,732 of potatoes, 7280 pounds of tobacco, 81,240 of sugar. It had 10 stores, 42 grist-mills, 25 saw-mills, five tanneries, 10 distilleries; five schools, 108 scholars. Pop.: whites, 5799; slaves, 216; free coloured, 193; total, 6208. Capital, Beverly.

RANDOLPH, county, N. C. Situated a little W. of the centre of the state, and contains 900 sq. m. Drained by Deep creek, and by branches of Yadkin river. It had in 1840, 10,111 neat cattle, 13,962 sheep, 25,529 swine; and produced 78,095 bushels of wheat, 1932 of rye, 295,829 of Indian corn, 47,671 of oats, 9607 of potatoes, 89,790 pounds of tobacco, 81,533 of cotton. It had 23 stores, two cotton factories with 1156 spindles, 24 flouring-mills, 27 grist-mills, 23 saw mills, 16 tanneries, 40 distilleries, two potteries, one printing-office, one weekly newspaper; two academies, 65 students; two schools, 32 scholars. Pop.: whites, 11,107; slaves, 1407; free coloured, 361; total, 12,875. Capital, Ashboro'.

RANDOLPH, county, Ga. Situated toward the S.W. part of the state, and contains 690 sq. m. Bounded W. by Chattahoochee river, and drained by its tributaries, and by branches of Flint river. It contained in 1840, 4534 neat cattle, 327 sheep, 5947 swine; and produced 1376 bushels of wheat, 63,740 of Indian corn, 2301 of oats, 6513 of potatoes, 327,067 pounds of cotton. It had four stores, two grist-mills, one saw-mill; three schools, 97 scholars. Pop.: whites, 5386; slaves, 2679; free coloured, 11; total, 8276. Capital, Cuthbert.

RANDOLPH, county, Ala. Situated in the E. part of the state, and contains 875 sq. m. Watered by Tallapoosa river and its branches. It had in 1840, 38,324 neat cattle, 3356 sheep, 19,740 swine; and produced 13,496 bushels of wheat, 912,997 of Indian corn, 5031 of oats, 21,541 of potatoes, 2239 pounds of tobacco, 240,540 of cotton. It had 11 grist-mills, five saw-mills; three schools, 302 scholars. Pop.: whites, 4446; slaves, 596; free coloured, 1; total, 4973. Capital, McDonald.

RANDOLPH, county, Ia. Situated toward the E. part of the state, and contains 440 sq. m. Drained by the W. fork of White river and by Mississinewa river, and their branches. It contained in 1840, 8564 neat cattle, 11,444 sheep, 27,341 swine; and produced 63,639 bushels of wheat, 2626 of rye, 401,291 of Indian corn, 115,846 of oats, 25,893 of potatoes, 8325 pounds of tobacco, 167,782 of sugar. It had 17 stores, two flouring-mills, 17 grist-mills, 20 saw-mills, eight tanneries, one distillery; 23 schools, 610 scholars. Pop. 10,684. Capital, Winchester.

RANDOLPH, county, Ill. Situated in the S.W. part of the state, and contains 546 sq. m. Drained by Kaskaskia river and its branches. Bounded S.W. by Mississippi river. It contained in 1840, 16,847 neat cattle, 7669 sheep, 25,338 swine; and produced 56,792 bushels of wheat, 1042 of rye, 101,342 of Indian corn, 76,051 of oats, 18,177 of potatoes, 1,174 pounds of tobacco, 5776 of cotton, 719 of sugar. It had 10 commission houses in foreign trade, 32 retail stores, one fulling-mill, one woollen factory, six flouring-mills, eight grist-mills, 11 saw-mills, six tanneries, four distilleries, two printing-offices, two weekly newspapers; one college, 9 students; one academy, 25 students; 14 schools, 403 scholars. Pop. 7944. Capital, Kaskaskia.

RANDOLPH, county, Mo. Situated N. of the centre of the state, and contains 450 sq. m. Watered by the E. fork of Chariton river and its branches. It had in 1840, 9349 neat cattle, 9443 sheep, 24,443 swine; and produced 13,693 bushels of wheat, 371,875 of Indian corn, 77,172 of oats, 13,187 potatoes, 841,304 pounds of tobacco, 9067 of sugar. It had 20 stores, 15 grist-mills, eight saw-mills, four tanneries, distilleries; 25 schools, 655 scholars. Pop.: whites, 4917; slaves, 1437; free coloured, 12; total, 7198. Capital, Huntsville.

RANDOLPH, county, Ark. Situated toward the N.E. part the state, and contains 690 sq. m. Drained by branches Big Black river. It contained in 1840, 4174 neat cattle, 46 sheep, 12,454 swine; and produced 5843 bushels of neat, 86,170 of Indian corn, 9940 of oats, 5761 of potatoes, 50 pounds of tobacco, 8156 of cotton. It had six stores, one woollen factory, one grist-mills, one saw-mill, one nery, eight distilleries; 25 schools, 655 scholars. Pop.: whites, 1973; slaves, 216; free coloured, 7; total, 2196. Capital, Pocahontas.

RANDOLPH, p. t., Orange co., Vt., 23 m. S. Montpelier, 499 W. Drained by branches of White river, which afford water-power. It contains three churches, a Congregational, Methodist, and Universalist, seven stores, one fulling-mill, one woollen factory, three grist-mills, 10 saw-mills, one oil-mill, four tanneries; one academy, 181 students; 21 schools, 915 scholars. Pop. 2678.

RANDOLPH, p. t., Norfolk co., Mass., 16 m. S. Boston, 444 W. It contains five churches, two Congregational and three Baptist, a bank, 20 stores, one grist-mill, four saw-mills; one academy, 90 students; 11 schools, 890 scholars. Pop. 3213.

RANDOLPH, p. t., Cattaraugus co., N. Y., 313 m. W. by S. Albany, 336 W. Watered by Alleghany river and its tributaries. It contains a church, five stores, one grist-mill, 13 saw-mills, two tanneries; seven schools, 246 scholars. Pop. 1283.

RANDOLPH, t., Morris co., N. J., 7 m. N.W. Morristown. It has Schooley's mountain in its N. part. Drained by Den's branch of Rockaway river. Magnetic iron ore is abundantly found. It has seven stores, one fulling-mill, one woollen factory, one furnace, one forge, one grist-mill, eight saw-mills, two tanneries, seven distilleries; seven schools, 367 scholars. Pop. 1801.

RANDOLPH, p. t., Portage co., O., 139 m. N.E. Columbus, 315 W. It has eight schools, 435 scholars. Pop. 1649.

RANGOON, a town, river-port, and the chief, and indeed almost the only, entrepôt for the foreign trade of the Birman empire, prov. Pegu, on the E. and principal branch of the Irawaddi (called the Rangoon river); about 26 m. from the sea, 50 m. S.S.W. Pegu, and 90 m. W. by N. Martaban; lat. 16° 42' N., long. 96° 20' E. Pop., according to a census in 1826, about 18,000, which is perhaps not far from its present (1841) amount, though it has been stated to amount to 40,000, and even 50,000! The town and suburbs extend lengthwise along the bank of the river for about 1 m. by ½ in depth; but the houses are very unequally scattered over the space. The fort, or rather wooden stockade, which constitutes the town properly so called, is an irregular oblong, entered by eight gates and several sallyports. (Crawfurd's Embassy, ii., 51.) According to Mr. Malcolm, who visited Rangoon in 1835, "the city is spread upon part of a vast meadow, but little above high tides, and at this season (May) resembling a neglected swamp. The fortifications are of no avail against modern modes of attack. They consist of merely a row of wooden timbers set in the ground, rising to the height of about 18 feet, with a narrow platform running round inside for musketeers, and a few cannon lying at the gateways in a useless condition. A dozen foreigners have brick tenements, very shabby: there are also four or five small brick places of worship, for foreigners, and a miserable custom-house. Besides these, it is a city of bamboo huts, in appearance as paltry as possible. The eaves of the houses generally descend to within 6 or 8 feet of the ground; very few being of more than one story, or having any other covering than thatch. Cellars are unknown, and all the houses are raised two or three feet above the ground, for coolness and ventilation. As the floors are of split bamboo, all dirt falls through; and what is not picked up by crows, dogs, fowls, &c., is occasionally swept out and burned. The streets are narrow, and paved with half-burned bricks, which, as wheel-carriages are not allowed in the city, are in tolerable repair. There is neither wharf nor quay: in four or five places are wooden stairs, at which small boats may land passengers; but even these do not extend to within 20 feet of low-water mark. Vessels lie in the stream, and discharge into boats, from which the packages, slung to a bamboo, are lugged on men's shoulders to the custom-house." (Travels in S. E. Asia, i., 76, 77.)

The river opposite Rangoon is about 600 yards across; and the town is accessible to ships of 1200 tons burden. The navigation, though somewhat intricate, is safe and practicable with the aid of ordinary native pilots. At neaps the tide rises and falls about 18 feet, and at springs, from 25 to 30 feet. Rangoon presents many facilities for ship-building. The banks of the river are so flat and soft, as to render docks almost unnecessary; and there is nearly a complete water communication with the teak forests of Pegu, by far the most abundant in India. Ship-building has, in fact, been carried on at Rangoon since 1786; and in the 38 years which preceded our capture of the town, 111 square-rigged vessels of European construction had been built, the total burden of which amounted to upward of 35,000 tons. Several were of from 800 to 1000 tons. But Malcolm says that this branch of business is now almost annihilated.

The markets of Rangoon are well and cheaply supplied with many sorts of provisions, as rice, poultry, fish, &c. The foreign commerce of the town is still considerable, though greatly crippled by enormous port-charges and absolute prohibitions against exporting rice or the precious

671

metals. Specie is, indeed, exported, but only clandestinely. The trade of the empire seaward is principally with the ports of Chittagong, Dacca, Calcutta, Madras, Masulipatam, and Penang, and occasionally with the Persian and Arabian gulfs. No direct trade has yet been carried on between Birmah and any European country. The articles exported to foreign countries from Rangoon are teak wood, catechu, stick lac, bees' wax, elephants' teeth, raw cotton, orpiment, gold and silver, ganne, and ponies, which are much esteemed at Madras. By far the most important of these commodities is teak timber; the quantity of this wood annually exported is said to be equal to 7500 full-sized trees. The principal imports are cotton piece goods from India and Britain, British woollens, iron, steel, quicksilver, copper, cordage, borax, sulphur, gunpowder, saltpetre, fire-arms, coarse porcelain, English glass ware, opium, tobacco, cocoa and areca nuts, sugar, and spirits. Of these by far the most important is cotton piece goods. In 1826–27, the exports and imports of Rangoon were estimated each at £300,000. (*Crawfurd, &c., in Commerc. Dict.*)

About 2 m. N.N.W. Rangoon, is the celebrated Shoe-Dagon temple, which, though rather smaller than that of Shoe-Madoo, at Pegu (which see), is much more highly ornamented. The two principal roads leading to it are lined on either side with numerous pagodas, some of considerable size. The great temple, similar to that of Pegu, stands on a planted terrace, raised upon a rocky eminence, and reached by an ascent of 100 stone steps. The area of this terrace is about two acres; the temple at its base is 310 feet in diameter, and 338 feet in height, surmounted by a *tee*, or umbrella of open iron-work, 45 feet in height, and richly gilt. "The golden temple of the idol may challenge competition, in point of beauty, with any other of its class in India. The building is composed entirely of teak-wood, and indefatigable pains are displayed in the profusion of rich carved work which adorns it. The whole is one mass of the richest gilding, with the exception of the three roofs, which have a silvery appearance. A plank of a deep red colour separates the gold and silver, which has a happy effect in relieving them. All round the principal pagoda are smaller temples, richly gilt, and furnished with images of Gaudma, whose unmeaning smile meets you in every direction; and the sight of which, accompanied by the constant tinkling of the innumerable bells hung on the top of each pagoda, combines, with the stillness and deserted appearance of the place, to produce an impression on the mind not speedily to be effaced." (*Mod. Trav.*, xi., 194, 125.) This temple, having long enjoyed a higher reputation than any other in the Birmese dominion, is resorted to by numerous devotees; and near it live 150 families, called "slaves of the pagoda," to whose care it is entrusted. (*Malcolm's S. E. Asia*, i., 76–82; *Crawfurd's Embassy to Ava*, ii., 51–57; *Mod. Trav.*, xi.)

RANKIN, county, Miss. Situated a little S. of the centre of the state, and contains 800 sq. m. Bounded W. by Pearl river, by branches of which it is drained. It contained in 1840, 4379 neat cattle, 1282 sheep, 25,797 swine; and produced 176,520 bushels of Indian corn, 9412 of oats, 25,125 of potatoes, 18,920 pounds of rice, 1,356,400 of cotton. It had eight stores. Pop.: whites, 9777; slaves, 1851; free coloured, 3; total, 4631. Capital, Brandon.

RAPHOE, t., Lancaster co., Pa., 12 m. N.W. Lancaster. Bounded S.E. by Great Chiques creek, and W. by Little Chiques creek, which unite at the S.W. angle of the t. It contains 12 stores, three fulling-mills, two furnaces, two flouring-mills, eight grist-mills, five saw-mills, two tanneries, four distilleries; 16 schools, 688 scholars. Pop. 3557.

RAPIDES, parish, La. Situated near the centre of the state, and contains 600 sq. m. Watered by Red river and its branches. It contained in 1840, 19,706 neat cattle, 4350 sheep, 31,065 swine; and produced 456,850 bushels of Indian corn, 3043 of oats, 44,752 of potatoes, 16,527,810 pounds of cotton. It had 25 stores, nine saw-mills, two printing-offices, two weekly newspapers; one college; 99 students; four academies, 136 students. Pop.: whites, 3243; slaves, 10,511; free coloured, 378; total, 14,132. Capital, Alexandria.

RAPPAHANNOCK, r., Va., rises in the Blue Ridge, and flowing S.E. 130 m., it enters Chesapeake bay, by a large æstuary. 25 m. below the mouth of the Potomac. It is navigable, for vessels requiring 10 feet of water, 110 m. to Fredericksburg.

RAPPAHANNOCK, county, Va. Situated toward the N.E. part of the state, and contains 100 sq. m. Drained by head branches of Rappahannock river. It contained in 1840, 9257 neat cattle, 13,195 sheep, 18,480 swine; and produced 179,850 bushels of wheat, 309,950 of Indian corn, 4784 of buckwheat, 93,794 of oats, 24,274 of potatoes, 4900 pounds of tobacco. It had 13 stores, 20 flouring-mills, 38 grist-mills, 36 saw-mills, eight tanneries; three academies, 49 students; 15 schools, 453 scholars. Pop.: whites, 5307; slaves, 3663; free coloured, 287; total, 9257. Capital, Washington.

RARITAN, r., N. J., is formed of two branches which unite in Somerset co., and enters Raritan bay at Amboy. Sloops ascend the river 17 m. to New-Brunswick, and 5 feet water, and steamboats run daily from that place to New-York city.

RARITAN, t., Hunterdon co., N. J., 20 m. N. Trenton. Drained by branches of Raritan river. It contains nine stores, four grist mills, four saw-mills, one oil-mill, one tannery, one distillery, one pottery, two printing-offices, two weekly newspapers; 15 schools, 351 scholars. Pop. 2510.

RASTADT, a town of W. Germany, grand duchy of Baden, circ. Middle Rhine, of which it is the cap., on the Murg, a tributary of the Rhine, 13 m. S.W. Carlsruhe. Pop. 5650. It is walled, and has wide streets, several Roman Catholic and Lutheran churches, a gymnasium, lyceum, various other schools, and a palace, the residence of the last margraves of Baden, in which is a curious collection of Turkish trophies. Rastadt is the seat of the superior judicial court for the circ. It has manufactures of tobacco, chicory, carriages, and machinery and instruments of various kinds; but is principally noted for two congresses held in it; the first, in 1714, when a treaty was concluded between Marshal Villars and Prince Eugene; and the second, in 1798 99, which terminated abruptly in the unexplained assassination of two of the French envoys. (*Berghaus; Dict. Géog.*)

RATHKEALE, an inland town of Ireland, prov. Munster, co. Limerick, on the Deel, 17 m. W.S.W. Limerick. Pop., in 1831, 4972. It principally consists of one street, nearly 1 m. in length, which has many good houses and shops. A parish church, a Roman Catholic chapel, with an independent and a Methodist meeting-house, a fever-hospital, dispensary, courthouse, and bridewell. Several families of German palatines are settled in the town. General sessions are held four times a year; petty sessions on Thursdays. It is a constabulary station, and has a considerable retail trade. Markets on Thursdays; fairs on 7th Feb., 4th April, 1st and 17th June, 5th Aug., 18th Sept., and 18th Nov. Postoffice revenue, in 1830, £376; in 1834, £508. A branch of the national bank was opened here in 1835.

RATISBON (Germ. *Regensburg*, an. *Castrum Regnum*, afterwards *Augusta Tiberii*), a city of S. Germany, capital of the Upper Palatinate, in Bavaria, on the Danube, across which it communicates with its suburb, Stadt-am-Hof, by a bridge 1000 German feet in length, 64 m. E.N.E. Munich. Lat. of St. Emmeran's Tower 49° 0′ 59″ N.; long. 19° 3 43″ E. Pop. about 22,000, two thirds being Roman Catholics. It is one of the oldest towns in Germany, and has a proportionally antique appearance. Its streets are narrow and irregular; all its houses, though lofty, are altogether old-fashioned and inconvenient. Many have tall battlemented towers, loopholes for musketry, &c.; and several of the large residences are several ornamented with heraldic bearings. The cathedral is a fine Gothic edifice, begun in the 13th century, but the greater part appears to have been constructed in the 15th: its interior was formerly crowded with a number of extraneous ornaments; but these have been removed. In it are a few fine works of art, including a *bas relief* on the tomb of the Primate Dalberg, designed by Canova. Two older cathedrals adjoin this edifice; one now used as a baptistry, is supposed to date from the 10th or 11th century, and the other is of a still earlier date.

Near the cathedral is also a massive square tower, probably a remnant of an ancient Roman fortress. The church of St. Emmeran has some very curious monuments of high antiquity, but it is now half in ruins; and St. Emmeran's abbey, a large though not a fine building, has been converted into the palace of the prince of Tours and Taxis. The establishment to which the latter belonged was founded long before the time of Charlemagne, by whom it was enlarged; and, at the time of its dissolution, it is said to have been 1200 years old. The Scotch Benedictine convent, a monastic institution in Ratisbon, the small revenues of which has prevented its secularization, supports two monks and five young Scotch ecclesiastical students. It has a very curious church, supposed to date from the 10th or 11th century. The town-hall, a gloomy and irregular structure, is interesting as being the seat of the German diet from 1663 to 1806. The apartments formerly occupied by the diet present little that is remarkable; but beneath the office are some dungeons, in which are preserved the rack, and other machinery of torture, formerly in use. See *Murray's Handbook for S. Germany, p. 66.*)

The old bishop's palace, in which the emperor Maximilian II. died, is now a brewery. The ramparts of Ratisbon are no longer of use as a means of defence, but serve as public walks. Within the city is a monument to the great astronomer Kepler, who expired here in 1630. Ratisbon has a royal academy, Roman Catholic lyceum, Lutheran gymnasium, theological seminary, observatory, public library with 20,000 volumes, school of design, museum of mechamatical and philosophical instruments, botanic society, &c. Since it has ceased to be the seat of the German diet, it has been mostly shorn of its importance; but a good many vessels for the navigation of the Danube are built here; and it has several breweries, distilleries, tanneries, and iron-works. It formerly enjoyed the exclusive right of the navigation of the Danube, upwards to Ulm, and downwards to Vienna. This privilege is extinct, but it still has a large share of the traffic on the river. At Donaustadt, on the Danube, about 6 m. distant, is the *Valhalla*, a fine Doric marble temple, built by the present king of Bavaria, for the reception of statues and busts of the distinguished men of Germany. This edifice, commenced in 1830, was to be completed in the course of 1841.

Ratisbon was the capital of the dukes of Bavaria, till their duchy was overturned by Charlemagne. It was afterwards a free imperial city, governed by a count of the empire. In 1809, some severe fighting took place before it between a part of the grand French army, under Napoleon, and the Austrians, to the disadvantage of the latter, who were forced to retire towards Bohemia. (*Berghaus, Allg. Länder. &c.*, iv., 135; *Dict. Geog., &c.*)

RAVENNA, a city of the Papal states, cap. leg. of the same name, on the Montone, amid extensive but fertile marshes, 43 m. E.S.E. Bologna, 84 m. N.W. Ancona, and 4½ m. from the Adriatic; lat. 44° 25' 16" N., long. 12° 12' 11" E. Pop. about 16,000. It is chiefly deserving of notice for its architectural remains. Having been the capital of Italy during the last years of the W. empire, and successively the residence of Honorius, Valentinian, Odoacer, Theodoric, and the succeeding Gothic monarchs, it presents many interesting specimens of the architecture of that period, few of which are elsewhere to be found. The empress Placidia, from 422 to 450, and Theodoric, from 492 to 526, embellished it with the best edifices the times were capable of producing; and many of these exist in singularly good preservation. The church of San Vittore is said to date as far back as the early part of the 4th century; but, according to Mr. Woods, what remains of it, even if the date be accurate, is a mere barn, without character. The earliest perfect church is that of Santa Agata Maggiore, completed in the early part of the 5th century, having granite columns, rich marbles, &c., apparently taken from some more ancient edifice. San Giovanni del Sagra was built by Placidia, anno 435; San Francesco apparently about the same date; St. Apollinaris Nuovo, a foundation of Theodoric, and having mosaics of that period; St. Apollinaris, at Chiassi, built by Justinian, on the ruins of an ancient temple of Apollo; and Spirito Santo, also of the 6th century. In all these churches the general plan and style are nearly the same. They consist of three naves divided by columns, supporting arches; the middle nave terminating in a semicircular recess, covered with mosaics. The roof is of timber, and not concealed. No very distinct marks of specific difference are observable in the workmanship, between the structures of the 4th and 6th centuries, except in the ornamental parts; the capitals and mouldings of the later period are much more fanciful. The cathedral of Ravenna was originally founded towards the end of the 4th century; but the present building is modern, and has some frescoes by Guido, bas-reliefs, and rich altars. Near it is the baptistry, an octagonal building, probably of the same date as the ancient cathedral, and all-over covered with mosaics, attributed to an archbishop who lived about 430. The baptism of Christ is represented on the dome in mosaic, "and the river-god seems also to be introduced into the composition." The church of San Vitale, another octagonal structure, dates from the first half of the 6th century. Here, also, is a little church, built by Placidia, whose tomb it contains, with those of Honorius, Constantius, Valentinian III., &c. Without the city is the mausoleum of Theodoric, constructed by his daughter Amalsunta.

But by far the most interesting of all the structures to be seen at Ravenna is the tomb of Dante, the immortal author of the *Divina Commedia*, who expired here, in exile on the 14th of September, 1321.

> "Happier Ravenna! on thy hoary shore,
> Fortress of falling empire! honour'd sleeps
> The immortal exile."
> 　　　　　　　(*Childe Harold*, iv., 59.)

was buried in the church of the Franciscan monastery, a handsome tomb erected by his protector Guido da Polenta, restored by Bernardo Bembo in 1483, again restored by Cardinal Corsi in 1692, and replaced by a more magnificent sepulchre in 1780, at the expense of Cardinal Luigi Valenti. The Florentines repeatedly demanded the ashes of the mighty dead, but these demands were uniformly met by flat and firm denial.

The *Porta Aurea*, a triumphal arch at the W. entrance of the city, and a few remains, said to be portions of Theodoric's palace, and having, according to Woods, some similarity to those of the palace of Diocletian, at Spalatro, complete the principal remaining antiquities. The city has arts, which, however, are of little service as means of

57

defence. Some of its squares are neat, and ornamented with statues of popes, &c.; and the streets are mostly wide and regular, but dirty; and the houses are old-fashioned, and gloomy. It has a few silk manufactures, and a large annual fair; but, like Padua, it is very dull, and fitter for study than for active business. A monastery has been appropriated to a public library, containing from 30,000 to 40,000 volumes; and a museum, with a few objects of natural history, antiquities, casts, and paintings. Few of the churches are rich in paintings; but there are some good private galleries. Byron praises Ravenna for its climate, and says, he found much education and liberality of thinking among the higher classes. It is an archbishopric, and the residence of a Papal legate.

Ravenna was originally founded by a colony of Thessalians, most probably on the sea-shore, but in the days of Strabo, it was, owing to accumulation of mud, surrounded by marshes. (Lib. v., p. 148.) Hence, says Silius Italicus,

"Lenta paludosae perscindunt stagna Ravenna."
(Lib. viii., v. 602.)

Being difficult of approach, and well fortified, its advantages as a stronghold and a naval station were perceived by Augustus, who constructed a new harbour, about 3 m. from the old town, which he connected with the Po and the old city by a canal, and with the continent by a causeway. Ravenna henceforward became the principal station of the Adriatic fleet, and the new and old cities were nearly joined by intermediate buildings. But the same cause, the accumulation of mud and other matters, brought down by the Po and other rivers, that had destroyed the port of the ancient city, in no very long time destroyed that constructed by Augustus: it is now, in fact, about 4½ m. from the sea, and so early as the fifth or sixth century of the Christian æra "the port of Augustus was converted into pleasant orchards; and a lonely grove of pines covered the ground where the Roman fleet once rode at anchor!" (*Gibbon*, v. 209, 8vo. ed.) But this very circumstance, though it lessened the naval importance, increased the strength of the new city, which, from the beginning of the fifth to the middle of the eighth century, was considered as the seat of government, and the capital of Italy. At length the Greek ex-archate being overthrown, and its territory was given by Pepin to the Holy See in 773. Subsequently Ravenna successively belonged to its own lords, the Bolognese, and the Venetians, till 1509, when it reverted to the pope. In 1512 the French, under the Duke de Nemours, gained a signal victory over the Spaniards, at a short distance from Ravenna; an event commemorated by an obelisk erected on the field of battle.

RAVENNA, p. t., capital of Portage co., O., 34 m. S.E. Cleveland, 140 m. N.E. Columbus, 319 W. Watered by a branch of Mahoning river, which flows to the Ohio, and by a branch of Cuyahoga river, which flows to lake Erie. The village, which was made the capital of the county in 1808, when the site and surrounding country were a wilderness, is a neat and flourishing place, laid out with streets which cross each other at right angles. State-street, running E. and W. through the centre, is 100 feet wide; the other streets are 60 feet wide. It has a handsome courthouse of brick, 68 by 45 feet, a jail of stone, 60 by 40 feet, five churches, a Congregational, Methodist, Baptist, a Disciples, and a Universalist, an academy, eight stores, about 100 mechanics of various descriptions, occupying 50 mechanic shops, 125 dwellings, and about 800 inhabitants. The Pennsylvania and Ohio canal passes through the S. part of the village, and affords water-power, and great facilities for business. There are in the t. nine schools, 371 scholars. Pop. 1542.

RAY, county, Mo. Situated toward the N.W. part of the state, and contains 570 sq. m. Bounded S. by Missouri river. Watered by Crooked river. It contained in 1840, 7925 neat cattle, 6702 sheep, 25,909 swine; and produced 15,984 bushels of wheat, 271,000 of Indian corn, 65,793 of oats, 45,600 of potatoes, 61,700 pounds of tobacco. It had 15 stores, eight grist-mills, seven saw-mills, two tanneries, four distilleries; 10 schools, 206 scholars. Pop.: whites, 5714; slaves, 834; free coloured, five; total, 6553. Capital, Richmond.

RAYMOND, p. t., Cumberland co. Me., 58 m. S.W. Augusta, 561 W. Bounded S.W. by Sebago pond. Incorporated in 1803. It has three grist-mills, four saw-mills; 19 schools, 488 scholars. Pop. 2032.

RAYMOND, p. v., capital of Hinds co., Miss., 18 m. W. by S. Jackson, 1028 W. It contains a courthouse, jail, a number of stores and neat dwellings, and about 500 inhabitants. Nett proceeds of the postoffice in 1841, $671.

RAYNHAM, t., Bristol co., Mass. First settled in 1650. The first forge in the U. States was erected here in 1652. It contains three churches, a Congregational, Unitarian, and Baptist, two stores, one furnace, one forge, five grist-mills, six saw-mills; seven schools, 358 scholars. Pop. 1329.

READFIELD, p. t., Kennebec co., Me., 12 m. W. Augusta, 604 W. A large pond in its S. part lies partly in Winthrop. It contains the Maine Wesleyan seminary, founded

in 1692, seven stores, one fulling-mill. one woollen factory, one grist-mill, two saw mills, one oil-mill, three tanneries; two academies, 390 students; 12 schools, 700 scholars. Pop. 2037.

READING, a parl. and mun. bor. and market town of England, hund. its own name, co. Berks, of which it is the cap., on the Kennett. close to its junction with the Thames, and on the line of the Great Bath-road and of the Great Western railway, 38 m. W. London. Area of parl. and mun. bor., which are coextensive, 2080 acres. Pop., in 1841, 19,512. "The town is of considerable size, and apparent prosperity, the inhabitants having increased rapidly within the last 15 years. A new high road into the heart of the town, recently completed, is an important improvement. There are many excellent houses, and the main streets are spacious, containing very good shops, and being well lighted with gas." (Bound. Rep.) The town-hall, completed in 1788, is a substantial building, well adapted for municipal purposes. The borough jail is small and inconvenient; and the house of correction, though large, is reported to be "deficient in various important requisites, badly ventilated, insecure, and without punishment cells." (Prison Inspectors' 5th Rep.) Reading has three parish churches. St. Lawrence is a large structure, with a fine tower of flint and stone, chiefly in the perpendicular style; the other churches are of a mixed architecture; and St. Mary's is remarkable for a handsome tesselated tower. Two of the livings are in the gift of the lord chancellor, and St. Lawrence belongs to St. John's college, Oxford. There is a chapel-of-ease and a licensed episcopal chapel in the parish of St. Mary's, and a district church in that of St. Giles; and there are places of worship for Roman Catholics, Baptists, Independents, Wesleyan Methodists, and the Society of Friends. The town has numerous Sunday schools, furnishing religious instruction to nearly 2000 children, and there are several subscription day-schools, attended by about 800 children. The grammar school, originally founded in the reign of Henry VII., and endowed with property belonging to the decayed hospital of St. John, received great additions from Archbishop Laud, and Sir Thomas White conferred on it two fellowships at St. John's college, Oxford. It enjoyed a high reputation as a classical seminary, while the Rev. Dr. Valpy was head master, and was then attended by 190 boys, partly day scholars and partly boarders; since then, however, it has greatly declined; but it continues to be a respectable school, and is improving. The school is open to all boys, whether natives or residents of Reading; but none are admitted free. (Char. Comm. 22d Rep.) A blue-coat school, founded in 1646, has an endowment producing an average income of £850 a year, and furnishes clothing, instruction, and an apprentice-fee to 40 boys (elected by the corporation): a girl's green-coat school, in St. Mary's parish, is supported by subscription. Reading has a county hospital, opened in 1839, and numerous almshouses and money-charities, few towns in England having so large an amount of property held in trust for charitable purposes. A dispensary, eye-infirmary, and savings' bank have been established within the last few years. The Reading institution has an excellent library and news-room, and it has also a subscription news-room, mechanics' institute, small theatre, and two weekly newspapers are published.

The trade of Reading is very considerable, but more of a general than specific character. The manufacture of woollens was formerly pretty extensively carried on, chiefly in a large range of buildings known as the Oracle; but the business is now wholly abandoned, and the buildings are occupied by riband and silk weavers, of whom there is a considerable number in and about the town. Floor-cloth and sail-cloth are also produced, and there are several large breweries. The trade of the town has been greatly facilitated by the Kennet and Avon canal and the Thames navigation, and it will be still more facilitated by the opening of the Great Western railway, which has here a principal station.

Reading, "which claims to be a borough by prescription," was incorporated by Henry III., and has received many subsequent charters. Under the Municipal reform Act. the borough is divided into three wards, the municipal officers being a mayor and five other aldermen, with 18 councillors. Corporation revenue in 1839, £3400, exclusive of £93 accruing from the sale of property. Quarter and petty sessions are held under a recorder, besides which there is a civil court for debts under £10. The Lent assizes and winter quarter sessions for the county are held here, besides weekly petty sessions. Reading has sent two members to the House of Commons since the 23 Edward I., the right of voting down to the Reform Act being in the inhabitants paying scot and lot. The electoral limits were left unchanged by the Boundary Act, and in 1839-40 it had 1141 registered electors. Reading is likewise one of the polling places at elections for the county. Markets on Wednesday and Saturday; on the latter for corn. Fairs, February 2d, May 1st, 674

July 25th, and September 21st; the last being for cattle and cheese, and extremely well attended.

Reading is of great antiquity, though its origin is unknown. In 1263 Henry III. held a parliament here, and another was adjourned thither in 1453. In the great civil war it was successively occupied by the forces of parliament and of the king. Archbishop Laud was the son of a clothier in this town, where he first saw the light in 1573. In his prosperity he did not forget the place of his birth, to which he was a liberal benefactor.

READING, p. t., Windsor co., Vt., 61 m. S. Montpelier. 62 W. Drained by Mill river, and branches of Queechy and Black rivers. It contains three churches, a Congregational, Methodist, and Universalist, five stores, four fulling-mills, two woollen factories, two grist-mills, seven saw-mills, two tanneries; 15 schools, 403 scholars. Pop. 1363.

READING, p. t., Middlesex co., Mass., 12 m. N. Boston. 432 W. Incorporated in 1644. Watered by Ipswich river. It contains four churches, two Congregational, a Unitarian, and a Baptist, 12 stores, four forges, four grist-mills, three saw-mills; one academy, 71 students; nine schools, 49 scholars. Pop. 2193.

READING, p. t., Fairfield co., Conn., 15 m. N.W. Bridgeport, 79 m. S.W. Hartford, 284 W. Watered by Saugatuck and Norwalk rivers. It has four churches, an Episcopal, Methodist, Congregational, and Baptist, six stores, five grist-mills, seven saw-mills; one academy, nine students; nine schools, 259 scholars. Pop. 1674.

READING, p. t., Steuben co., N. Y., 21 m. E. Bath, 194 m. W. by S. Albany, 305 W. Bounded E. by Seneca lake. It contains two churches, a Presbyterian and Methodist, five stores, eight saw-mills, two tanneries; 13 schools, 364 scholars. Pop. 1541.

READING, p. b., capital of Berks co., Pa., 57 m. N.W. Philadelphia, 52 m. E. Harrisburg, 145 W. Situated on the E. bank of Schuylkill river, 1 m. below the mouth of Tulpehocken creek. The streets are spacious and straight, crossing each other at right angles, chiefly covered with a hard white gravel from sandstone, superior to macadamised streets. It contains a courthouse, a jail, three banks, an academy, three public libraries, one of which is in German, a female seminary, 12 churches, a Lutheran, German Reformed, Episcopal, Presbyterian, Methodist, Baptist, Friends, Universalist, Roman Catholic, and three African. The courthouse is on elevated ground, on a square 280 feet long and 220 feet deep, and has a portico of sandstone in front, presenting an imposing appearance, and commanding an extensive prospect: it cost $50,000. The Lutheran church is a large and handsome structure, capable of containing 1500 persons, and has a steeple 201 feet high; said to be the highest in the state. The German Reformed church, erected in 1832, of brick, has a steeple 151 feet high. There are two market-houses, in which market is held twice a week. Seven weekly newspapers are published, three of which are in the German language. Fifty-five thousand dozen wool hats have been manufactured in a year, for the southern and western markets. There is a rolling-mill and nail factory, capable of rolling 3500 tons of bar iron, and manufacturing 1500 tons of nails annually. They are of excellency of anthracite coal. White wines, of an excellent quality, are made and sold to the amount of ..... a year. There are two flouring-mills in the borough, one of which is said to manufacture 8000 barrels annually. The place is supplied with spring water from a spring of great mount, conducted into a reservoir near the head of Penn street, and distributed in iron pipes through the various streets; and it has sufficient head to render it very useful in extinguishing fires. The Schuylkill and Union canals meet here, and the Philadelphia and Reading railroad passes through the place. There are two bridges across the Schuylkill, one of which was built in 1816, and cost $28,000. The borough had, in 1840, 23 stores, capital $161,000, three lumber-yards, capital $90,000, one forge, two grist-mills, three tanneries, one distillery, two breweries, one pottery, 1365 dwellings, and 8410 inhabitants. It was laid out in 1748, by Thomas and Richard Penn, proprietors and governors of the province.

READINGTON, p. t., Hunterdon co., N. J., 5 m. N.E. Flemington. Bounded S.W., S., and S.E. by the South branch of Raritan river. Drained by Rockaway creek and its branches. It has nine stores, six grist-mills, five saw-mills, one oil-mill, three tanneries, three distilleries; 13 schools, 306 scholars. Pop. 2273.

RED BANK, p. t., Clarion co., Pa., 190 m. W.N.W. Harrisburg, 256 W. Drained by Red Bank, Mahoning, and Beaver creeks. Alum is found in the S.W. angle of the t. It contains five stores, five grist-mills, eight saw-mills, one distillery, one pottery; 24 schools, 836 scholars. Pop. 3973.

RED HOOK, p. b., Dutchess co., N. Y., 22 m. N. Poughkeepsie, 49 m. S. Albany, 321 W. Bounded W. by Hudson river, on which it has two landings, called Lower Red Hook and Upper Red Hook, at each of which is a small village

Dutch Reformed churches, 11 stores, seven
e saw-mills; one academy, 33 students; five
iolars. Pop. 2699.

ind Ark., is the first large river which enters
above its mouth, and rises in the base of the
ocky mountains, near Santa Fe, in Mexico.
if it forms, for a considerable distance, the
en the United States and Texas. A greater
rse is through fertile prairies of a reddish
ea its colour to the waters, and to this cir-
wes its name. Its shores are covered with
and with grapevines, which produce a deli-
n its banks is the favourite range of the buf-
wild animals peculiar to the western prai-
) m. above Natchitoches commences the Raft,
t of a swampy expansion of the river, to the
30 m. The river divides into a great number
many of which are shallow; and these chan-
a obstructed by fallen trees, brought down by
upper parts. Between these masses the river
ees completely covered by the raft, on which
d willows in the alluvial soil collected on it;
erally the river is visible between the masses
n some places it has even been said that it
ed on horseback, but not without danger. At
ise, this raft has been so far removed that
hough not without difficulty, have passed
f this raft were effectually removed, the river
vigable for 400 m., and the fertile soil on its
invite settlers. Below the raft, and 4 m. above
. the whole volume of the river is again united,
des into a number of channels, and fills many
lakes which lie parallel to it. In its lower
n channel of the river is narrower than above
its depth is increased.

i, a market town and par. of England, co. Corn-
'enrith, on an acclivity on the high road be-
and St. Ives, 8 m. W.S.W. the former, and 11
:ter. Area of par., 3770 acres. Pop., in 1831,
town consists principally of one long street,
side of the hill. It has been greatly extended
d of late years, having fully doubled its popu-
1801; and is pretty well built, paved, and light-
Among the more recent buildings are a hand-
granite tower, with an illuminated clock in its
and the savings' bank, a neat edifice with a
. front. The parish church, at the foot of the
.l, half a mile S.W. the town, was rebuilt about
living is a rectory, in the patronage of Lord de
:, of the gross annual value of £501. A chapel
the pointed style, has, however, been erected
io the town, partly by a grant from the parlia-
mmissioners, and partly by subscription; the
:uracy, in the patronage of the rector. There
eting-houses for Baptists, Methodists, and Qua-
mmar school, erected by voluntary contribution
th various other private and Sunday schools, a
a reading-room, and a small theatre.
owes its importance wholly to the extensive cop-
r mines in its vicinity. The Consolidated and
les, between it and Gwennap, are the most ex-
any in Cornwall. In 1836 they employed in all
luals, and 21 steam-engines, some of which were
atest power. One of the shafts in these mines
than 300 fathoms deep, the temperature at the
ng from 96° to 99° Fahr.‡ (See De la Beche's
'Cornwall, p. 601; Statistics of the British Em-
I.) The ore is conveyed by railways to Deveron,
belonging to Falmouth harbour, and to Portreath,
., on the N. side of the peninsula. Redruth has
eral trade. Markets on Wednesday and Friday,
t is one of the largest corn markets in the W. of
Fairs, May 2d, August 3d, and October 12th,
cattle. Tehiddy-park, the seat of Lord de Dun-
is about 2 m. from the town.

EA (an. Arabicus Sinus, or Rubrum Mare), an
inland sea between Africa and Arabia, connected
Indian ocean by the strait of Bab-el-Mandeb, lies
n. across, between lat. 12° 40' and 30° N., and
and 44° E. Extreme length, 1420 m.; do. breadth,
average breadth, 135 m.; estimated area, 185,000
t runs in a pretty direct course from N.N.W. to
ing divided at its N. extremity into two arms, the
nd most westerly of which is the gulf of Suez, at
of which is the port of the same name, which see.
er branch, which runs N.N.E., having a length of
is called the gulf of Akaba, which see. On the
ory between these gulfs is the mountain group of
dnai, including mounts Sinai and Horeb, famous in
n history. The gulf of Suez is more than 180 m. in
ind its breadth may average about 22 m. The strait
al, the entrance to this gulf, is nearly 16 m. across.

The present charts of the Red sea were formed in 1829, by
order of the E. India company. under the superintendence
of Captains Elwin and Moresby; but a considerable section
of its S. part had been surveyed in 1801 by Sir H. Popham
under Lord Valencia. The more recent, however, is the
only complete survey that has been made of this important
gulf.

The Red sea, though, generally speaking, of great depth
(probably averaging 100 fathoms), is in parts studded with
rocky islets and hidden coral banks, which extend far into
the channel, and sometimes impede the course of vessels.
The islands are scattered pretty abundantly in all parts of
the sea. Several occur near the entrance of the two northern gulfs; but by far the greater number are found at its
S.W. extremity, nearly opposite Massouah, this group being
denominated the Dhalak archipelago. Farther S. are several other islands, one of which (called Djibel-Teer) comprises an active volcano, rising nearly 1900 feet above the
sea. (See Moresby's Sailing Directions for the Red Sea,
p. 26.) The island of Cameran lies S.W. Loheia; and
nearly opposite Hodeida, in lat. 15° N., is the group of the
Zebayer islands. The coral reefs of this sea are more numerous and extensive than in any other body of water of
equal extent. They extend most commonly in long strips
parallel to, and about a quarter of a mile from, shore, with
which they are in many cases united: the deep water runs
close up to their edge; but the banks themselves are seldom
more than 5 feet below the surface. Among the reefs unconnected with the shore, several are at some distance from
it, a channel intervening of sufficient depth to admit of navigation by small vessels, and having good anchorage in stormy
weather. These reefs are more numerous on the E. than
on the W. coast; but the Dhalak archipelago is, perhaps,
more extensively intersected with them than any other part
of the sea. There are also many isolated reefs; but they
present few obstacles, owing to the transparency of the
water, which renders them easily discoverable. No surf is
ever observed on them, how boisterous soever the weather,
a circumstance attributed to the porous nature of the coral
on the outer edge of the reef. (Wellsted's Arabia, ii., 246.)
After all, however, these reefs offer no considerable obstacle
to ships, and the shelter which they afford in some cases
facilitates navigation without decreasing its speed.

The existence of the islands and reefs just described has
led to a division of the Red sea into a central and two lateral channels. The central channel, between the outer extremities of the reefs extending from either shore, is very
deep throughout, and in some parts no bottom has been
found, even at a depth of 250 fathoms. The breadth of
this channel in the parallel of Djidda is 110 m.; but farther
S. it diminishes to little more than 40 m., and continues
gradually decreasing down to the strait of Bab-el-Mandeb.
The inner channel, on the Arabian side, is formed partly
by broken reefs and sunken rocks, partly by islands and
long-continued reefs, a large part also being open to the sea:
its average width is from 2 to 3 m., and the depth is very
considerable, though the anchorage is so insecure as to afford little protection from the sea. The inner channel, on
the opposite coast, is similarly bound to that now mentioned; but it is much narrower, and soundings can be obtained
only in a few places: the harbours, of which there are
seven or eight, furnish tolerably good anchorage, but it is
very unsafe to anchor on the rocky shelves projecting from
the reefs. Both these lateral channels are connected with
the open sea, by cross channels, some of which, especially.
N. of lat. 17°, are of great width and depth.

The winds of the Red sea are not uniform in its different
parts and channels. With respect to the central channel,
it may be observed that the N.E. monsoon here becomes a
S.E. wind of considerable force, decreasing, however, as it
passes northward, and disappearing about lat. 18°, where it
is replaced by the N. breezes that prevail during the warm
season. The S. winds commence in October, and subside
at the end of May, when they are succeeded by N.W. winds,
which usually bring with them thick, hazy weather, especially on the Arabian side. In the lateral channels, N.
winds, inclining to land and sea breezes, are more or less
prevalent, land-squalls being very common both in April
and May.

The currents of the Red sea appear to be entirely governed by the winds; for it is observed that with S. breezes
they set northward, and with N. winds to the S. It is probable, also, that they increase according to the strength of
either, as little or no current is perceptible during the prevalence of light, variable breezes, just before the monsoon.
N. of Djidda, however, both the winds and currents are very
variable throughout the year; but here even the latter depend on the former, and a strong S. wind will cause a current of 20 or 30, and even 40 m. a day. Tides have been
observed in a few parts of the coast; but the rise and fall
are not sufficient to allow of the conclusion that this sea is
subject to lunar influence. The idea that the waters of the

Red sea were formerly at a higher level than the Mediterranean, and that the surface has been depressed by the constancy of currents flowing towards the Indian ocean, is wholly chimerical: the fact is, that from May to October, during the prevalence of N. winds, the water is 2 feet lower than the average level; whereas from December to February, when the currents run in an opposite direction, the water collects at the N. part, which accordingly becomes unusually elevated. (*Geog. Journ.*, vi., 82; *Wellsted's Arabia*, ii., 300, and *Append.*; *Sailing Directions*, p. 211–219, &c.)

The country about the Red sea is more or less mountainous, though the high lands seldom abut directly on its coasts: indeed, this sea may be described as the lower part of a valley bounded eastward by the table-land of Arabia, and westward by a range of mountains rising from 4000 to 6000 feet above the sea. Between the high grounds and the shore, however, a level district intervenes, of some extent, and considerable fertility; but it is almost uncultivated, the inhabitants living in idleness, and contenting themselves with the wretched pittance that may be obtained by pasturage, and the plunder of the Hadj season. Nothing, indeed, can be conceived more wretched and degraded than the condition of the people dwelling in the villages bordering the Red sea.

The principal harbours on the E. side of this sea are Mocha, Hodeida, Loheia, and Djidda, the first and last being by far the most important: Suez, Cosseir, Suakim, and Massouah are the chief places on the African side; and a pretty active communication is kept up between the inhabitants of the opposite shores, more particularly as the sea is crossed by all the African pilgrims on their way to Mecca and Medina. The vessels employed in transporting the pilgrims and their merchandise may amount to about 400, each averaging 130 tons; and the pilgrims from Africa alone are said to average 20,000 annually. (*Geog. Journal*, vi., 89.) Grain, also, and slaves, are large articles of trade between the two shores: the grain is shipped at Cosseir, wholly on account of Mehemet Ali, pacha of Egypt; the slaves are brought almost exclusively from Suakin and Massouah. Mocha being the only port whence goods are exported, the Red sea has not, till within the last few years, been much frequented by foreign vessels, except those belonging to pilgrims from Persia and India, with the merchandise of both countries; wheat, tobacco, dates, and Persian carpets being brought by the former, and rice, sugar, muslins and other fabrics, indigo, spices, and handsome young females, by the latter. Coffee, frankincense, and gums are sent in exchange from Mocha; but the returns from the other ports are in cash, pearls, &c. (*Geog. Journal*, vi., 91.) Within the last few years, however, the navigation and commerce of this sea has greatly increased; and steamers have been regularly established between Suez and India, by means of which, and of the steamers from Alexandria to Marseilles, Bombay is now brought within less than 60 days' distance of England! The gales in the N. part of the Red sea offer some impediments to steamers; but we learn from Wellsted and other Indian officers, who have engaged in the survey and packet-service, that there are no obstacles which experience and perseverance may not overcome. *Dépôts* for coal, &c., are already established in different parts, and the steam-packets now run with almost undeviating regularity. The beneficial consequences of this communication, too, will, in all probability, not be confined to England, Egypt, and India; for it is scarcely to be supposed that the Red sea can become the high road from Britain to her Indian possessions, without a portion of the commerce carried on between the eastern and western worlds again reverting to this its ancient and most direct channel, and without the diffusion of industry and civilization among the barbarous hordes that now inhabit its coasts.

The Red sea is first mentioned in sacred history in connexion with the miraculous passage of the Israelites across the gulf of Suez. (*Exod.* xiv. 21.) In the time of Solomon two ports, Elath and Eziongeber, were established on the gulf of Akaba; and the Phœnicians seem to have carried on a large trade on this sea, though, probably, they had no direct communication with India. The early Greek writers, including Herodotus, seem to have had very vague notions respecting the Red sea; for the 'Ἐρυθρά Θάλασσα apparently comprises, in their estimation, the whole extent of coast from the Indus to the coast of Africa. During the flourishing period of the Persian empire, the Persian gulf was the medium through which Europe and Western Asia received the wealth of the east; but under the successors of Alexander, especially the Ptolemies, who exerted themselves to promote the riches of this sea, it became an important channel of intercourse between Europe and India and the east. This intercourse continued with little interruption, though not to the same extent at all times, till the discovery of the passage round the cape of Good Hope, by which commerce

was diverted into a wholly different channel. The flow, however, seems now to have arrived when the Red sea is again to recover a portion, if not the whole, of its ancient importance as a great commercial highway.

Great discrepancy of opinion has prevailed respecting the origin of the name. According to Pliny (*Hist. Nat.* lib vi. cap. 23) and Quintius Curtius (lib. viii. cap. 9), its Greek name, signifying red, is derived from Erythros, a king of the adjoining country. But the more probable opinion seems to be that it is derived from the great abundance of coral found in it. (*Wellsted's Arabia*, vol. ii. passim; *Moresby's Sailing Directions*, pp. 200–293; *Geog. Journ.* vols. vi. and vii., &c.)

REGGIO, (an. *Rhegium Julii*), the most S. city and seaport of Italy, in the Neapolitan dom., prov. Calabria Ultra I., of which it is the cap., on the E. side of the strait of Messina, 8 m. S.E. Messina, and 78 m. S.W. Catanzaro, lat. 38° 7′ N., long. 15° 55′ E. Pop. 8000. As a city, it is inferior to its opposite neighbour Messina; but it has a fine situation in the midst of orange groves, is well supplied with excellent water, and its climate is said to be the best in all the continental dom. of Naples. The former town was destroyed by the earthquake of 1783, since which Reggio has but slowly recovered its prosperity. It has been laid out on a regular plan, which, when completed, will render it a handsome city. A wide road, called the *Marina*, extends along the sea shore, parallel with which the principal street runs through the centre of the town. The houses are in general good; and as it stands on a gentle declivity, it is well drained. It has a collegiate and many other churches, several convents, a royal college, hospital, foundling asylum, and a handsome theatre. It is surrounded with walls, outside which are several suburbs. Its ancient fort no longer exists. Reggio is the seat of an archbishop, and of a civil and criminal tribunal. It has manufactures of gloves, stockings, &c., of silk, and produces some articles from the filaments of the *Pinna marina*; which, with oil and fruit, are its chief exports.

The ancient Rhegium was one of the most celebrated and flourishing cities of *Magna Græcia*. It was founded nearly 700 years B.C., by a party of Chalcidians, Zanclæans, and other Greek colonists; and was for 200 years the cap. of one of the principal republics of S. Italy. The government was subject to the same mutations as that of the other Greek cities, being sometimes under a democracy, but more frequently under an oligarchy, or a single tyrant. It was besieged by the elder Dionysius, tyrant of Syracuse, who having succeeded in cutting off all communication between the sea on the one hand and the country on the other, reduced the inhabitants to such distress for want of food, that a bushel of wheat is said to have been sold for 5 minas, or, according to the usual method of computing, about £15 12s. 6d. At last, after sustaining the most dreadful privations, they were obliged to surrender, when most of those who survived were sent as slaves to Syracuse. It, however, again recovered some portion of its former importance, and succeeded in repelling an attack of Hannibal. Augustus established a colony in the city. It produced several distinguished followers of Pythagoras, some historians of celebrity, and some distinguished sculptors. It suffered in antiquity, as well as in more modern times, from earthquakes. (*Mitford's Greece*, v. 452, ed. 1838; *Ancient Universal History*, viii. 29, 8vo. ed., &c.)

REGGIO (an. *Regium Lepidi*), a city of N. Italy, cap. of a distr. of the duchy of Modena, between the Tessone and Crostolo, tributaries of the Po, 16 m. W.N.W. Modena, and 15 m. S.E. by E. Parma. Pop. about 18,000. It is defended by ramparts and a citadel, and is a well-built and rather handsome, though miserably dull town. Its streets are regular, and bordered with arcades. It has some handsome churches, numerous convents, a town hall, theatre, hospital, public library, and a library and museum of antiquities, collected by Spallanzani. No antiquities remain, except a statue in the principal square, traditionally said to represent Brennus, the Gallic leader. Reggio has manufactures of silk and linen fabrics, horn, wooden, and ivory articles, &c., with some trade in cattle and agricultural produce, and a large fair, which lasts during the entire month of May. It is supposed to have been founded by M. Æmilius Lepidus, who constructed the famous Æmilian way. It was here that the elder Brutus was slain by order of Pompey. Napoleon erected Reggio into a dukedom for Marshal Oudinot. Among other distinguished individuals to whom this town has given birth, may be mentioned Ariosto, perhaps the greatest of Italian poets, born here on the 8th of Sept., 1474, and the naturalist Spallanzani: its vicinity has also the honour of being the birth place of the great painter, Antonio Allegri, surnamed Correggio. (*Cramer's Anc. Italy.* i. 65; *Rampoldi; Dict. Géog.*, &c.)

REHOBOTH, p. t. Bristol co., Mass. 40 S. by W. Boston, 413 W. Palmer's river affords water-power. It contains five churches, two Baptist, a Congregational, a Reformed

Methodist, and if Christian, five stores, two cotton-factories with 1840 spindles, four grist-mills, four saw-mills; 15 schools, 394 scholars. Pop. 9169.

REICHENBERG, a town of Bohemia, and, next to Prague, the largest and most flourishing in that kingdom, circ. Buntlau, in a well wooded part of the Lausnitz mountains, 27¼ m. N.N.W. Gitschin, and 56 m. N.E. Prague. Pop. 11,500. It has three parish churches, two castles, a fine school-house, a new theatre, and the largest brewery in Bohemia. It has manufactures of woollen cloth and yarn, cottons and linens, with numerous dyeing-houses, &c. (*Aust. Nat. Encyc.; Berghaus.*)

REIGATE, a parl. bor., market-town, and par. of England, co. Surrey, hund. its own name, in the valley of Holmsdale, at the foot of a range of chalk hills, traversing the co. from E. to W., 16 m. E. Guildford, and 18¾ m. S.S.W. London. Area of parl. bor., which is co-extensive with the par., 5000 acres. Pop., in 1831, 3307. The town, which consists of a principal thoroughfare running E. and W., crowed at one end by another at right angles to it, is "small, but remarkably neat, with a greater number than usual of gentlemen's houses attached to it." (*Bound. Rep.*) The church, in the fields a little E. the town, is a large structure, in the perpendicular style, with an embattled stone tower, the rest of the building being of grey limestone: in the interior are many handsome monuments, and in an attached building is a public library. The living is a vicarage of the annual value of £418, in private patronage. The Wesleyan Methodists, Baptists, and the Society of Friends, have their respective places of worship, with attached Sunday-schools; there is, also, a large national school for children of both sexes, and a small grammar-school. The market-house and town-hall is a small brick building of no pretensions to beauty; and near it is a clock-house, occasionally used as a prison. A castle once stood on an eminence N. from the town; but the only parts now remaining are the moat, and a curious excavated chamber, once used either as a prison or storehouse. The priory, the property of Earl Somers, an elegant modern mansion at the S. end of the town, built on the site of an Augustine monastery, has an attached park of 70 acres. Reigate formerly carried on a pretty considerable trade in oatmeal, no fewer than 20 mills being employed, previously to the American war, in its manufacture; but this business has since so much declined, that only one mill remains. At present it depends principally on its being on the line of the principal thoroughfare between Brighton and London; an advantage which, however, it will probably lose, now that the Brighton railway, which passes about 1½ m. W. from the town, is completed. Fuller's earth and white sand are found in considerable quantities in the neighbourhood, which produces an abundance of medicinal and other plants.

Reigate is a bor. by prescription, its governing officer being the bailiff of the lord of the manor. It sent two members to the House of Commons, from the 23 Edw. I., down to the Reform Act, the right of election being in the burgage holders, of whom, in 1831, there were only eight, wholly under the influence of Earl Somers, the lord of the manor. The Reform Act deprived the bor. of one of its members, and extended the electoral limits so as to comprise the entire par. Reg. electors, in 1839-40, 198. It is a polling place at elections for E. Surrey. Petty sessions for the hund. and the spring quarter sessions for Surrey are held here by the co. magistrates. Markets for corn, &c., on Tuesday; and a cattle market on the first Tuesday in each month. Fairs, Whit-Monday, Sept. 14, and Dec. 9.

REMI, or REMY (ST.), a town of France, dép. Bouches-lu-Rhone, cap. cant., in a valley 15 m. N.E. Arles. Pop., in 1836, ex. com., 3938. It was formerly surrounded with a double line of ramparts; but these have been levelled, and their place is occupied by a fine circular promenade. Streets, narrow and irregular; but there are many good ouses. The town-hall, new par. church, and a lunatic asylum are the principal public buildings. St. Remi is chiefly remarkable for its Roman antiquities, about 1 m. from the town, and supposed to belong to the ancient Glanum. They consist of two edifices; one an arch somewhat similar to the central portion of that at Orange, but much mutilated; the other a beautiful Corinthian monument, square at its base, but circular above, appearing from an inscription to have been erected by Sextus L. Marcus to his parents, whose statues are in the circular portion of the structure. This relic of antiquity, an engraving of which is given in *Wood's Letters*, i. 159, appears to be in good preservation.

The Abbé Expilly, one of the most laborious and useful geographical writers of the last century, was a native of Remi, where he first saw the light in 1719. His principal work, entitled *Dictionnaire Géographique, Historique, Politique des Gaules et de la France*, in six tomes folio, incomplete, reaching only to the letter S.: it contains a great fund of information.

REMSEN, p. t., Oneida co., N. Y., 18 N. Utica, 109 W.N.W. Albany, 405 W. Watered by Black river. It contains two churches, a Baptist and Presbyterian, six stores, one grist-mill, seven saw-mills; 13 schools, 600 scholars. Pop. 1638.

RENAIX, or RONSE, a town of Belgium, prov. E. Flanders, arrond. Oudenarde, cap. canton, 20 m. S.S.W. Ghent. Pop. in 1836, 12,390. (*Hennebling.*) It is a fine situation, and is embellished with several public fountains: its old castle is now in ruins, but it has some good public buildings. It has manufactures of cotton, woollen, and linen stuffs, hats, beer, and chocolate; with a considerable trade in linens, a monthly and two weekly markets, and two large annual fairs. (*Vandermaelen, Dict. Fland. Orientale.*)

RENFREW, a small maritime co. of Scotland, having N. and W. the river and frith of Clyde, S. Ayrshire, and E. Lanarkshire. Area, 145,280 acres, of which about a half may be arable. There is a large extent of hilly, moorish ground, in the W. parts of the county, and along the confines of Ayrshire; but from Port Glasgow, eastwards along the Clyde it is comparatively flat. Soil very various; being in parts thin and sandy, while in others it consists of a deep, loamy, fertile clay; and the country being in general well enclosed with hedges, and ornamented with gentlemen's seats and plantations, has a rich appearance. Tillage husbandry is still in a rather back state, and neither the rotation of crops nor the management of the land is so well understood as might have been expected. "Much, however, has been done in the way of improvement during the last 40 years; enclosures have become general; new roads have been constructed; the land has been drained, and limed, and the rotation of crops improved." (*New Statistical Account of Scotland*, art. *Lochkinnock, co. Renfrew.*) Farm-houses and offices have, also, been greatly improved. On the whole, however, the county from the humidity of the climate and the nature of the soil, is better adapted for grazing and dairying (the latter of these is extensively followed) than for tillage. There are several large estates, but property is, notwithstanding, a good deal subdivided. Farms of a medium size. Average rent of land, in 1810, 17s. 7d. an acre. There are valuable coal-mines near Paisley and the eastern parts of the county, and limestone and freestone are very generally diffused. Paisley is the principal seat of the Scotch shawl manufacture; and, next to Glasgow, of the Scotch cotton manufacture; manufacturing industry is also extensively prosecuted at Pollockshaws, Neilston, and other places. Greenock and Port Glasgow, which are both in this county, are considerable sea-ports. Principal rivers, White-Cart, Black-Cart, and Gryfe. The county is divided into 16 parishes, and sends three members to the House of Commons, being one for the county, and one each for Paisley and Greenock. Renfrew and Port Glasgow are associated with other boroughs in the return of a member. Registered electors for the county, in 1839-40, 2289. In 1831, this county had 11,153 inhabited houses, 28,904 families, and 133,443 inhabitants, of whom 61,154 were males, and 72,289 females. Valued rent £69,172 Scotch: annual value of real property in 1835, £265,534.

RENFREW, a parl. bor. of Scotland, co. Renfrew, of which it is the cap. within about 4 m. of the S. bank of the Clyde, 5 m. W. Glasgow. Pop. 2002. It consists merely of a single street, from which several lanes issue. The only public buildings are the parish church, town-hall, and jail. There is no dissenting chapel. There are no native manufactures in the town; but about 250 looms are employed on account of Glasgow manufacturers. It has also a distillery, and a considerable dairy establishment. The borough was connected with the Clyde by a canal in 1786, but it has been allowed to go into disrepair. A quay was constructed on the Clyde opposite to the town in 1835; and a railway, 3½ m. in length, connects the borough with Paisley. On the whole, however, it has but little trade or enterprise. The royal family of Stewart, so called from their office, had their original residence near this town. It joins with Port Glasgow, Dumbarton, Rutherglen, and Kilmarnock, in sending one member to the House of Commons. Registered electors in this borough, in 1839-40, 91; councillors, 19; corporation revenue, £1500.

RENNES (an. *Condate*, afterwards *Redones*), a city of France, dep. Ille-et-Vilaine, of which it is the cap., in a plain, at the confluence of the Ille and Vilaine, 61 m. N. by W. Nantes; lat. 48° 6' 50" N., long. 1° 40' 47" W. Pop. in 1836, ex. com., 29,909. The Vilaine divides Rennes into an upper and lower town. The former, which is the largest, is regularly built, and handsome; the lower town is quite the contrary. The houses in both are, however, of a dull grey stone, which gives the city a sombre appearance.

Rennes has several tolerable squares, as that of the *Palais de Justice*, in which was formerly a bronze equestrian statue of Louis XIV.; and the *Place aux Armes*, planted with lime trees, and forming a favourite promenade. But it has neither public fountains nor covered market-places; and, for a city of its size, there are few public buildings worth notice. Though not fortified, it has several gates, on one of which is an almost effaced Latin inscription in honour of the emperor Gordian. Opposite this gate is the cathedral, a heavy-looking edifice, with a front flanked by two square towers, and ornamented with five rows of columns of different orders. Several of the other churches are in much better taste. The town-hall is one of the best public edifices. It has been rebuilt, together with a large part of the city, since a destructive fire in 1720, and comprises a large saloon, used for public fêtes, the halls of various judicial courts, a public library of above 30,000 volumes, schools of design and architecture, and the apartments occupied by the mayor. The hotel appropriated to the use of the royal court, tribunal of commerce, law-school, &c., is a building in the Tuscan order, having some good paintings and arabesques.

Rennes has several hospitals, a house of correction, in which various manufactures are conducted, an arsenal, artillery forges, and various military schools. It is the seat of a bishop, whose diocese comprises the department Ille-et-Vilaine; it is also the seat of the royal court for the five departments of Brittany, the capital of the 13th military division, and has a chamber of manufactures, a faculty of law, a secondary school of medicine, two seminaries, a tolerable museum of painting, and some other scientific establishments. In the immediate vicinity are several good promenades; the principal, the Thabor, formerly a garden of the Benedictines, on a height above the city, has a statue of Duguesclin.

Though favourably situated for trade, Rennes has few manufactures: the principal are those of sail-cloth, for the navy, &c., fishing-nets, and twine. It has, however, a considerable traffic in linens, butter, cider, and provisions, which is much facilitated by the canal of Ille and Rance, and that between Nantes and Brest. It has 12 annual fairs.

Rennes was the capital of Brittany from the 9th century to the revolution. It has produced numerous distinguished men, among whom may be specified the famous constable Duguesclin, born in the castle of Motte-Broon, in the immediate vicinity, in 1314; La Bletterie, the author of the lives of Julian and Jovian; Ginguéné, the author of the *History of Italian Literature*, &c. (*Hugo*, art. *Ille-et-Vilaine*; *Guide du Voyageur*; *Dict. Géog.*)

RENSELLAER, county, N. Y. Situated in the E. part of the state, and contains 696 sq. m. Organized in 1791. Bounded W. by Hudson river. Watered by Hoosic river, Poestenkill, and Wynantskill. The Albany and West Stockbridge railroad, which has a branch to Troy, passes through it. It connects at Troy with railroads to Schenectady and Saratoga. It contained in 1840, 32,174 neat cattle, 134,864 sheep, 27,916 swine; and produced 21,454 bushels of wheat, 247,703 of rye, 329,193 of Indian corn, 54,767 of buckwheat, 9488 of barley, 810,333 of oats, 759,346 of potatoes, 3500 pounds of hops, 30,546 of sugar. It had 44 commercial and 13 commission houses in foreign trade, 408 retail stores, 19 lumber-yards, four furnaces, eight forges, 16 fulling mills, 13 woollen factories, 21 cotton factories with 54,035 spindles, 18 flouring-mills, 37 grist-mills, 106 saw-mills, eight oil-mills, four paper-mills, two powder-mills, three rope-walks, 22 tanneries, one distillery, four breweries, one glass factory, six printing-offices, two daily, four weekly, and one semi-weekly newspapers, and one periodical; 12 academies, 546 students; 239 schools, 11,512 scholars. Pop. 60,259. Capital, Troy.

RENSELLAERVILLE, p. t., Albany co., N. Y., 26 m. W. Albany, 369 W. Watered by Cataskill and Foxes creeks. It contains five churches, a Presbyterian, Episcopal, Methodist, Baptist and Christian, 22 stores, one fulling-mill, one woollen factory, one furnace, three grist-mills. 11 saw-mills, nine tanneries; one academy, 55 students; 20 schools, 1188 scholars. Pop. 3705.

REPTON, or REPINGTON, a par. and village of England, co. Derby, hund. Repton, on a small affluent of the Trent, 6 m. S.W. Derby, Area of par. with Bradby chapelry, 6440 acres. Pop., in 1831, 3083. The town consists principally of a street of scattered houses, about one mile in length; its inhabitants being chiefly agricultural. The parish church is a spacious structure, with an elegant spire, and several handsome monuments. The living, a perpetual curacy, in the gift of Sir G. Crewe, is worth £123 a year. A free-school, well endowed in 1566, is held in the remains of a priory of Black Canons, established in 1172.

REQUENA (an. *Loretum*), a town of Spain, in New-Castile, prov. Cuenca, on the Magro (a tributary of the Jucar), 43 m. W.N.W. Valencia. Pop., according to Miñano, 10,963. It is surrounded with walls, and commanded by an old castle, built on an eminence: the streets are tolerably straight, lined with well-built houses, and, as in most

Spanish towns, there is a spacious *plaza*, in the centre of which is an ornamental fountain. Three parish churches and a college are the only public buildings. The chief employment of the inhabitants is in weaving ribands and silk goods. There has been a great increase of activity since the restoration of tranquillity; and Requeña now furnishes a large supply of these articles to Madrid, Seville, and Cadiz. A fair is held annually in September. The neighbourhood, a portion of which is irrigated, is well cultivated, and furnishes corn, wine, fruit, saffron, and large quantities of silk. During the war of the succession, in 1706, the castle was taken by the English, but retaken the following year by the French under the duke of Orleans. (*Miñano; Dict. Géog., &c.*)

RETFORD (EAST and WEST), a parl. and mark. town of England, co. Nottingham, hund. Bassetlaw, on the Idle, a tributary of the Trent, 26 m. N.N.E. Nottingham, and 129 m. N.N.W. London. Area of the former parliamentary borough, which was coextensive with the parish of East Retford, 130 acres; its population, in 1831, being 2491. The modern municipal borough, however, comprises, together with the parish of East Retford, parts of those of West Retford, Clareborough, and Ordsall, and may have a population of between 4000 and 5000. The town consists principally of several thoroughfares, which meet in a common centre, and are united by several cross streets. To the N.E. are the hamlets of Moorgate and Spital Hill; and to the S. South Retford and Thrumpton, to which last houses extend from East Retford, so as to form almost one continued line of buildings. West Retford is divided from East Retford by the Idle; the bridge over which being in the line of the streets of both towns, they seem to be, what, in fact, is substantially the case, a single town. The houses of East and West Retford have a respectable appearance, and the streets are lighted with gas. (*Mun. Corp. Rep.*) East Retford church is a spacious edifice of different dates, with a lofty square tower. The church of West Retford is small, but has a handsome spire. The living of East Retford is a vicarage, value £160 a year: patron, Sir R. Sutton: that of West Retford is a rectory, in the gift of the corporation of East Retford, value £264 a year. (*Eccl. Rev. Rep.*) In the suburb of Moorgate is a new chapel of ease, in the Gothic style: there are in the town places of worship for Baptists, Independents, Wesleyans, &c. The town-hall is a convenient building, surmounted by a cupola, and having beneath a good market-place. There are two alms-houses, one for old men, the other for women; a free grammar-school, founded by the ward VI., of which the municipal authorities are trustees, a national school, established in 1813, various minor charities, a news-room, and a small theatre. There is no manufactory of any kind in the town, which depends entirely for its support upon the retail trade carried on with the inhabitants of the surrounding agricultural districts. (*Mun. Corp. Rep.*) Malting was formerly extensively carried on, and hat-making and the manufacture of worsted were introduced; but these branches have nearly ceased. The Idle is not navigable at Retford; but it communicates with the Trent by the Chesterfield canal, which passes S. of the town, and is carried over the Idle by an aqueduct. East Retford is governed by four aldermen and 12 councillors, and had formerly petty sessions, &c.; but the Municipal Reform Act orders that it shall not have a commission of the peace unless upon petition or grant. Previously to this act its police was reported to be very inefficient, and the administration of the corporate and other revenues had been very unsatisfactory. (*Mun. Rep., Append.*) It is said to have been a borough by prescription: it received many charters from Henry III. and subsequent sovereigns, down to James I. East Retford returned two members to the House of Commons in the 9th Edward II., and continued to enjoy this privilege down to 1830. But it having been proved that gross bribery had been practised at the election that then took place, it was proposed by one party to incorporate the adjoining hundred of Bassetlaw with the borough, and by another to transfer the franchise to Manchester, Birmingham, or some other of the great towns that were then unrepresented. After lengthened discussion, the first-mentioned plan was agreed to, so that the parliamentary borough of East Retford is now identical with the hundred of Bassetlaw. Registered electors, in 1839–40, 3785. Markets at East Retford on Saturdays; fairs, March 23 and Oct. 2, for horses, cattle and cheese. (*Parl. Reports, &c.*)

RETHEL, a town of France, dep. Ardennes, cap. arrond., on a steep declivity beside the Aisne, here crossed by a wooden bridge, 24 m. S.W. Mézières. Pop. in 1836, ci. com. 6,771. It is pretty well laid out, and is improving: but the houses are still mostly of wood, and there is no remarkable public building.

Rethel was formerly the cap. of the co. of Champagne, and was fortified. It has a court of original jurisdiction, a communal college, society of agriculture, theatre, several hospitals, and two prisons. It is chiefly noted for its woollen manufactures, which are estimated to employ from 1400 to 1800 hands, mostly spinners. As in most small manufacturing towns, the workpeople are generally employed at their homes. The prices of labour are somewhat lower than in Rheims; but provisions, &c., are cheaper, and, on the whole, the woollen spinners of Rethel are in a better condition than those in that city. Its trade has been considerably augmented by the opening of the canal of Ardennes; besides woollen manufactures, the town has many iron forges, breweries, and tanneries. (*Hugo*, art. *Ardennes; Villerms, Tableau des Ouvriers*, i. 248–252.)

REUS, a town of Spain, in Catalonia, 9 m. W. Tarragona, and 54 m. W. by S. Barcelona. Pop., according to Miñano, 24,600. It stands on a plain gently sloping towards the coast, and comprises several streets lined with good houses, with numerous churches, hospitals, barracks, an orphan asylum, theatre, and handsome public fountains. The inhabs. are extensively employed in the manufacture of silk and cotton fabrics, hats, soap, &c.; besides which, there are large dye-houses, bleaching grounds, tanneries, spirit distilleries, glass-houses, &c.; in fact, Reus may be regarded as one of the most important manufacturing towns of Catalonia, and the numerous handsome houses in its neighbourhood sufficiently indicate the industry and prosperity of its inhabitants. The town is connected by a canal with the port of Salou, whence its products are exported in exchange for rice, flour, codfish, anchovies, &c. Its weekly market is one of the largest in Spain, and is frequently attended by upwards of 10,000 persons. The town was founded in the middle of the 12th century; but its present importance is wholly consequent to the establishment of silk and cotton manufactories at the close of the last century. (*Miñano; Laborde, Itinéraire de l'Espagne; Dict. Géog., &c.*)

REUSS, a territory of central Germany, forming two independ. principalities, between lat. 50° 20′ and 51° N., and long. 11° 45′ and 12° 15′ E., having S. Bavaria, E. Saxony, and N. and W. the territories of Prussia, Coburg, Gotha, and Weimar, the last dividing it into two unequal portions. Area, 391 sq. m. Pop., in 1838, 101,800. Surface generally hilly; in the N. it is watered by the Elster, in the S. by the Saale. Tillage is less an occupation of the inhabs. than the rearing of cattle and sheep. The most fertile tract adjoins the town of Gera. Woods comparatively extensive, and one of the chief sources of national wealth. Woollen, linen, and cotton fabrics are produced; mining is little followed, and the only metallic works are a few iron forges. The inhabitants are almost all Lutherans. The territory of the elder branch of Reuss consists of the lordships of Greiz and Burgh, having an area of 144 sq. m., and a population in 1836, of 32,100. Greiz is the chief town, and seat of the superior judicial court whence appeal lies to the tribunal of Jena. The public revenue amounts to about 80,000 dollars a year; the contingent to the army of the confed. being 223 men. The younger branch of Reuss has a territory of 447 sq. m., composed of the lordships of Schleiz, Lobenstein-Ebersdorf, and Gera, with a population of 69,700; chief town, and seat of government, Schleiz. Appeal from the courts of this principality lies also to Jena. Public revenue, about 235,000 dollars; contingent to army, 522 men. Each branch has a separate vote in the full diet of the German Confederation, and, together with Hohenzollern, Lippe, Liechtenstein, and Waldeck, the 16th place, and a vote in the committee. (*Berghaus, Allg. Länder, &c.*, iv. 393–394; *Almanach de Gotha.*)

REUTLINGEN, a town of Wirtemberg, circ. Schwarzwald (Black Forest), of which it is the cap., on the Eschatz a tributary of the Neckar, 19 m. S. Stuttgard. Pop. in 1838, 11,500. Reutlingen presents a contrast to many other old imperial cities, having mostly regular streets, and well-built though antiquated houses. It is fortified, and has several suburbs. One of its churches has a tower, said by Berghaus, to be 380 ft. in height; the town-hall, lyceum, a well-endowed hospital, and orphan asylum, are the other most conspicuous edifices.

It has manufactures of leather, lace, net for women's caps, of the annual value of 100,000 florins; clocks and watches, &c.; with dyeing and bleaching factories, and printing establishments, whence many pirated editions of German works have issued.

Reutlingen was the first town in Swabia which embraced the Reformation. (*Memminger Beschr. von Wurtemberg; Berghaus, &c.*)

REVEL, a gov. of European Russia. (See ESTHONIA, 376.)

REVEL (Esth. *Tallin*, Russ. *Kolyvan*), a seaport town of Russia in Europe, cap. of the above gov.; on a small bay on the S. side of the gulf of Finland, 200 m. W.S.W. Petersburg; lat. 59° 26′ 33″ N., long. 24° 44′ 30″ E. Pop. about 1,000. The city proper, included within the ramparts, is small; and although it has many good brick houses, its streets are narrow and irregular. There are several Luthe-

ran, a Roman Catholic, and some Greek churches, all stone edifices; and various charitable and educational establishments, the latter including a gymnasium, episcopal seminary, and a school (*pension*) for nobles. The castle, a modern edifice, is appropriated to the provincial authorities: the municipal officers, who are elected by the city, reside in the town hall. The admiralty is the principal remaining public building. The suburbs, consisting mostly of wooden houses, cover a large extent of ground along the shore. Revel is much resorted to as a watering place, and has some good warm baths, a theatre, several clubs or *casinos*, and three or four public libraries, one of which, the property of the city, is said by Possart to comprise 10,000 vols. This town is one of the stations for the Russian fleet, and has a harbour defended by several batteries. This port, which was materially improved in 1820, is deeper than that of Cronstadt, though more difficult of entrance. The roadstead, formed by some islands, is well sheltered; the long duration of the frost is the principal drawback on Revel as a naval station, though that is a disadvantage which it shares in common with the other Russian ports in the Baltic. Though not connected with the interior by any navigable river, Revel has a considerable trade. Its principal exports are corn, spirits, hemp, flax, timber, and other Baltic produce; the imports consist of colonial produce, herrings from Holland and Norway, salt, cheese, wine, tobacco, fruits, dye stuffs, cotton yarn, stuffs, and other manufactured goods, &c. A portion of the customs' revenue is enjoyed by the town.

Revel was founded by the Danes in 1218, and afterwards sold by them to the Knights of the Teutonic Order. In 1561 it came into the possession of the Swedes, but was taken from them by the Russians in 1710. Near it is the Katharinenthal Palace, built by Peter the Great; the gardens of which are a favourite public promenade. (*Schnitzler; Possart, &c.*)

REYNOLDSBURG, p. v. capital of Humphreys county Tenn., 69 m. W. Nashville, 783 W. Situated on the E. side of Tennessee r., on a beautiful plain rising gradually from the river to the height of 15 feet above it, and entirely above high water mark. Laid out in 1812, and is 12 m. below the mouth of Duck river. It contains a courthouse, jail, 8 stores, and about 200 inhabitants.

RHEA, county, Tenn. Situated toward the S.E. part of the state, and contains 440 sq. m. Bounded E. by Tennessee r., and drained by its tributaries. It contained in 1840, 4156 neat cattle, 2843 sheep, 13,366 swine; and produced 19,648 bushels of wheat, 3851 of rye, 248,477 of Indian corn, 44,760 of oats, 5693 of potatoes, 5305 pounds of tobacco, 8741 of cotton. It had six stores, one forge, seven grist-mills, four saw-mills, seven tanneries, 13 distilleries. Pop.: whites, 3580; slaves, 377; free coloured, 28: total, 3985. Capital, Washington.

RHEIMS or REIMS (anc. *Durocortorum*, post. *Remi*) a celebrated city of France, dép. Marne, of which, though not the capital, yet it is by far the largest town, cap. arrond., in a plain near the Vesle, a tributary of the Aisne, 27 m. N.N.W. Châlons, and 95 m. E. by N. Paris; lat. 49° 14′ 41″ N. long. 4° 2′ 47″ E. Pop. in 1836, 38,359. It is surrounded by ramparts faced with stone, which, being planted with trees, form agreeable public walks; and there are other promenades in the immediate neighbourhood. It is about a league in circuit, and is entered by six gates, one of which, the *Porte Neuve*, a triumphal arch, with handsome doors of open iron-work, was raised in honour of Louis XVI. at his coronation. Two principal thoroughfares, which meet in the *Place Royale*, divide the city into four unequal portions. It is tolerably well laid out, its streets being wide, straight, and generally clean: it has several good squares; but its houses are small, having mostly only two floors, and are constructed with monotonous uniformity. Waterworks, beyond the walls, distribute the waters of the Vesle through the town, but, according to Hugo, in insufficient quantity; and there is not one handsome public fountain. It has, however, some remarkable public edifices.

The cathedral, one of the largest and most magnificent in Europe, is that in which the coronation of the French kings has taken place, with few exceptions, from the era of Philip Augustus. It was chiefly constructed between 1212 and 1242. It is 479 ft. in length, 99 ft. in breadth, and 144 ft. in height. It has a noble front, flanked with two square towers, 392½ ft. in height. Of the three grand entrances on this side, the central is 90½ ft., and those on either side 29½ ft., in width; above the former is a beautiful circular window. The whole front is ornamented with nearly 550 statues, and a great number of columns and bas-reliefs; and similar decorations abound in every part of the exterior. In one of the towers is a bell weighing 23,600 lbs. This cathedral has some fine stained glass, tapestry, marble pavements, &c.; a very fine organ; the "washing the feet," a *chef d'œuvre* of Poussin; various curiosities, including the font said to have been used in the baptism of Clovis, and the tomb of Jovinus, a citizen of Rheims, who arrived at the dignity of Roman consul, *anno* 366. The last, an admirable specimen of ancient art, was transferred from the church of St. Nicaise, destroyed during the phrenzy of the revolution.

Rheims, which was a place of great consideration under the Romans, had various other antiquities, but they were mostly destroyed or carried off during the period in question. The most remarkable ancient monument now existing is the *Porte de Mars*, one of the city gates; a triple archway, ornamented with eight Corinthian columns, and numerous bas-reliefs, though these are now greatly defaced. This arch appears to have been erected in honour of Cæsar and Augustus, when Agrippa was governor of Gaul. Without the walls are the traces of an amphitheatre.

The church of St. Remi, constructed in 1041, is considered the oldest in the city. It occupies almost as much ground as the cathedral; and though much less lofty and ornamented, is in a similar style of architecture. It contains the remains of the ancient and curious mausoleum of St. Remi. The town-hall, an edifice of the time of Louis XIII., with whose statue its front is ornamented, the new prison, *maison rouge*, which has some historical interest, theatre, several hospitals, &c., are among the other chief objects of notice. In the centre of the *Place Royale* is a bronze statue of Louis XV., surrounded with allegorical figures, erected in 1818, to replace a similar statue destroyed in 1793. Rheims is the seat of an archbishop, whose suffragans are the bishops of Amiens, Beauvais, Chalons, and Soissons, and of courts of assize, original jurisdiction, and commerce, a council *des prud'hommes*, chamber of manufactures and arts, a royal college, &c. It has a public library of 24,000 printed vols. and 1,000 MSS., a botanic garden, schools of mutual instruction, a *mont de piété*, savings' bank, &c.

Rheims is the centre of the manufacture of woollen stuffs, which extends over nearly the whole dép. of Marne, and the adjacent déps. of Aisne and Ardennes. In 1834, these manufactures were estimated to occupy 50,000 hands, of whom 12,000 were settled in Rheims. About 3000 of the latter were weavers, one tenth part of whom wrought at jacquard-looms; 1500 were employed in spinning yarn, 4000 in fulling, washing, and otherwise preparing the fabrics made. For the last 25 years the woollen manufacture of Rheims has made a considerable progress in most of its branches. Wages are good. Those of workmen vary, according to the work, from 1¼ up to 4¾ fr. a day; women from 75 c. to 1¾ fr.; and children from 50 to 75 c. But the average is estimated to be, for 1st class workmen, 2½ fr. to 3 fr.; 2d class, 1½ fr. to 2 fr.: and women 75 c. to 1 fr. 25 c. daily. They usually work for about 12 hours a day; those living in the city being mostly employed in the workshops of the manufacturers. In general the workpeople are well clothed and well fed; but they are said to be improvident, and few save money. The vice of drunkenness is very prevalent; and morals are as bad or worse, perhaps, than in most manufacturing towns of France. From 1825 to 1835, the proportion of illegitimate to total births amounted to about 1 in 4. (*Villermé*, *Tableau des Ouvriers*, i. 216–247.) Rheims also produces soap, candles, biscuits, and gingerbread; and has breweries, tanneries, leather factories, &c. It is a principal *dépôt* for the wines of Champagne, large quantities of which are stored up in cellars, similar to those of Epernay, which see. Besides its trade in woollen manufactures and wines, it has a considerable trade in cotton stuffs, flour, and other agricultural products, &c. Under the Romans, *Durocortarum* was the cap. of Belgica II., and was distinguished as a seat of letters and philosophy. It became a bishopric before the irruption of the Franks, and received many privileges from the Merovingian kings. In 1359, Rheims successfully resisted the arms of Edward III. In 1547, a university was founded in it, which lasted till the revolution, when it was suppressed. In 1814, it was taken by the Russians, who were soon after expelled by Napoleon, with great loss.

Among the great men of whom Rheims has to boast, the most distinguished by far is Colbert, minister of finance during the most splendid period of the reign of Louis XIV., born here on the 29th of August, 1619. It has also given birth to the Abbe de la Pluche, the historian Vely, &c. (*Hugo*, art. *Marne*; *Villermé*, *Tabl. des Ouvriers*; *D'Anville*, *Notice de la Gaule*, 280, &c.)

RHIN (BAS, or LOWER-RHINE), a frontier dép. of France, in the E. part of the kingdom, which, with Haut-Rhin, formerly constituted the prov. of Alsace, chiefly between the 48th and 49th degs. of N. lat., and the 7th and 8th E. long.; having N. Rhenish Bavaria, and the dép. of Moselle, W. Meurthe and Vosges, S. Haut-Rhin, and E. the Rhine, separating it from the Grand Duchy of Baden. Area, 464,781 hectares. Pop., in 1836, 540,213. The W. part of the dép. is covered by the Vosges mountains, and their ramifications. The average elevation is 2000 to 2500 ft.; but the Hochfield rises to 4460 ft., and the Schneeberg to 2850 ft. The surface declines towards the E. Principal rivers, the Ill, with its numerous tributaries, the Moder, Zorn, Sarre, &c. In 1834, the arable lands were estimated at 180,920 hectares, meadows 56,024 ditto, vineyards 13,123 ditto, orchards 5924 ditto, and woods 117,754 ditto. The marshes in the E., and the stony tracts of the W., are unsuitable for agriculture; and though the middle of the dép. be fertile and well cultivated, the produce of corn is not sufficient for home consumption. More potatoes are grown than in any of the neighbouring déps.; nearly 4,745,000 hectolitres having been produced in 1835. A good deal of tobacco is raised, and bought by the government. The annual produce of wine may be estimated at about [illegible] hectolitres: it is of a medium quality, but, speaking generally, is inferior to that of the Haut-Rhin. About a half of the produce is consumed in the departments, the rest being sent, partly to other departments and partly to Baden, &c. Hops, wood, hemp, onions, and other vegetables, are articles of extensive cultivation. The plough is usually drawn by horses. Horned cattle are pretty numerous, but sheep are scarce. Poultry is extensively reared near Strasburg; particularly geese, the livers of which serve to make the *pâtes de fois gras*, for which that city is so celebrated. In 1835, of 231,956 properties subject to the *contrib. foncière*, 141,962 were assessed at less than 5 fr., and only 83 at 1000 fr. and upward. Iron mines are wrought; lead, antimony, cobalt, coal, and bitumen are met with; and salt is made from springs, in the N. and W. The dép. has manufactures of cotton yarn, muslins, woollen, and linen cloths, leather, saddlery, carriages, glass wares, &c. There are many iron-forges; and fire-arms are manufactured at Mutzig and Klingenthal. Strasburg is the seat of an extensive general commerce and transit trade. Bas-Rhin is divided into four arronds.; chief towns, Strasburg the cap., Saverne, Schlestadt, and Wessembourg. It sends 6 members to the chamber of deputies. Registered electors (1838–39), 1782. Total public revenue (1831), 14,336,474 francs. The majority of the population is Protestant.

RHIN (HAUT, or UPPER RHINE), a frontier dép. of France, in the E. part of the kingdom, formerly comprised in the prov. of Alsace; between lat. 47° 27′ and 48° 18′ N., and long. 6° 45′ and 7° 35′ E., having N. Bas-Rhin, W. Vosges and Haute-Saône, S. Doubs, and the Swiss cantons of Neufchatel, Berne, and Basle, and E. the Rhine, separating it from the territories of Baden. Area, 406,032 hectares. Pop. in 1836, 424,258. In the W. are the chains of the Vosges and Jura: one summit of the former, the *ballon d'Alsace*, rises to 4608 ft. above the sea. The rest of the surface is mostly plain. Except the Rhine, the Ill is the only navigable river; but the canal between the Rhine and Rhone intersects the dép. in its whole length. As in Bas-Rhin, the most fertile portion of the surface is in the centre of the dép., where agriculture is in a pretty advanced state. In 1835, the arable lands were estimated to comprise 155,571 hectares; meadows, 52,566 do.; vineyards, 11,141 do.; orchards 5819 do.; and woods, 113,215 do. Principal crops, wheat and barley; 2,400,000 hectolitres of potatoes were said to be raised in 1835. The produce of wine is estimated at above 400,000 hectol. a year. Some of the white wines, especially those of Guebwiller, Riquevir, Thann, &c., are highly esteemed. The *vins gentils* are extensively purchased by the merchants of Cologne and Frankfort, who mix them with the wines of the Rhine, to which they impart strength and vivacity. They keep for more than twenty years, improve as they grow older, and sustain no injury from travelling. The dép. also produces excellent beer. In 1835, of 174,015 properties subject to the *contribution foncière*, 100,859 were assessed at less than 5 fr., and 26,810 at from 5 to 10 fr. There are various iron and other mines, and good building stone and other minerals are met with in various parts. "Hand-looms are scattered over nearly the whole of the déps. of the Haut and Bas-Rhin; the articles produced are chiefly muslins and fine calicoes. the wages of the hand-loom weavers vary; for the most skilled class, from 6 to 9 fr. per week: for the second-rate class, from 4½ to 6 fr.; and for women and children, forming the third or lowest class, from 2 to 4½ fr. These are nett wages, the manufacturers either beaming the warps and furnishing the looms in factories, or paying equivalently." (*Symon's Report*, p. 119.) Villermé states that in 1834, the cotton manufactures of the Haut-Rhin employed 100,000 individuals, or nearly a fourth part of the entire population and, since that period, he affirms the number has still further increased: but a large proportion of these persons prosecute the cotton trade as a subsidiary employment only, carrying it on in their cottages when they are not necessarily engaged in the culture of their little patches of ground. Mulhausen, Thann, Guebwiller, Soultz, and Sainte Marie aux Mines, are the chief seats of the cotton manufacture. "The homes of the weaving classes (in these towns) are, for the most part, dirty and comfortless, and evince every symptom of bad management and poverty. Even those who have children in the cotton-mills do not keep up any appearance of comfort. The Alsatian weavers have generally speaking a sufficiency of food, though in all other respects they are badly off. In the mountains of the Vosges, the peasantry are worse off still: the looms found there are chiefly on the system of the 'customer' weavers of Scotland, but there are some who weave for manufacturers at very low wages." (Symon's Reports, p. 120.)

Villermé gives a deplorable account of the abject poverty of a great proportion of the cotton weavers of the Haut-Rhin. many of whom come from Switzerland and beyond the Rhine. (Tableau des Ouvriers, i., 24–30, &c.) This dep. has also manufactures of hardware, clocks, and watches, with various iron and steel forges.

It is divided into three arronds.: chief towns, Colmar, the cap.. Altkirch, and Belfort. It sends five members to the chamber of deputies. Number of electors (1836-9), 596. Total public revenue (1831), 9,338,247 francs. Unlike Bas-Rhin, the majority of the population in this dep. are Roman Catholics, but Protestants are numerous.

Alsace, which formed a part of the kingdoms of Austrasia and Lorraine, afterward belonged to the German empire ill 1298, when it became mostly independent. It subsequently belonged to Austria; but was finally annexed to France by Louis XIV., in 1697. (Hugo, art. Haut-Rhin; Villermé; Parl. Papers; French Official Tables, &c.)

RHINE, a large river of W. Europe, rising on the N. side of the Alps, flowing through Switzerland and Germany, and falling into the North sea or German ocean, between l. 46° 30' and 52° N., and long. 3° 40' and 9° 50' E. length, measured along the stream, 950 m.: area of basin, including tributaries, 83,298 sq. m., or about 1-43d part of Europe. It originates in two principal streams, which are their sources on the N. side of the Pennine Alps: the principal of these, called by the Germans Vorder-Rhine, is formed by the junction of two small streams flowing from the N. side of mount St. Gothard, at an elevation of 6581 at above the sea, lat. 42° 32' N., long. 8° 53' E., only a few miles from the source of the Rhone. Hence the main stream, which soon becomes enlarged by the affluence of numerous brooks and mountain torrents, takes at first a E. direction through the magnificent and stupendous range of the Rheinwald, enclosed on both sides by almost perpendicular rocks, rising 3000 feet above the river, and clothed to their very summits with stately firs. At the lower end of this valley, and only a few miles above Chur, near Reichenau, the river is joined by its E. branch, the Hinter-Rhine, which rises on the side of the Moschelhorn or Peissberg, near the pass of St. Bernard. At Chur the river deflects N., maintaining that general direction through a fertile and romantic valley, abounding with vineyards, as far as the lake of Constance, into which it pours its waters, at a level at this point being 1255 feet above the sea, or 3 feet below the source. The river, leaving this beautiful lake at its W. end, near the town of Constance, enters smaller expanse of water, called the Unter or Zeller-See, which is the island of Reichenau, and thence, narrowing its channel, runs W. to Schaffhausen; 3 m. below which stream, pent between lofty rocks, and divided by craggy islets, falls over a ledge of rocks 76 feet in height, forming one of the most celebrated European cataracts. See (AFFHAUSEN.) The channel, from this point to Basle, is extremely tortuous, winding through lofty rocks, which confine the waters within a narrow compass, and consequently increase the rapidity of the current. At Zuzach, 11 m. above the confluence of the Aar, occurs a second town, below which, however, the natives venture their loaded boats, except during the spring-floods. The river maintains its direction through a rocky valley, interrupted with many cragns rising above the stream, as far as Basle, to it is crossed by a wooden bridge, 600 feet in length, on stone abutments. Here also commences the navigation of the river, its level at this point being 827 feet above German ocean.

Basle, indeed, seems to be the proper point of division between the upper and lower Rhine; for the navigation of this town is so interrupted by falls and rocks as to be scarcely of any importance, whereas from hence to the sea the boats pass at almost all seasons of the year. At or near Basle a pretty constant N. course, the Rhine forms the boundary between France and the grand duchy Baden, and afterward between Baden and Rhenish Bavaria. The cities and towns in this part of its course being Breisach, on the E. bank, at the confluence of the Neckar, Strasburg, Speier, Oppenheim, and Mayence, on the right bank. At the last of these towns, at the junction of the Mayn, the stream takes a sudden turn W. to Bingen, on which the river is bank, from which point the course of the river is uniformly N.W. to the delta at its mouth. Coblentz, at the confluence of the Moselle. Bonn, Cologne, and Cléves, are the chief towns on the W. bank; those on and near the E. side comprising Wiesbaden, Dusseldorf, Wesel, and Emmerich, of inferior size. The delta of the Rhine is

the largest in Europe, not even excepting that of the Wolga. It extends, with its ramifications, 110 m. along the coast from the E. shore of the Zuyder-Zee to the S. branch of the Maas; and the distance from the apex, about 10 m. below Emmerich, being 72 m., the total area of the country comprised within its limits is 4150 sq. m. When the river divides, the left or S. arm takes the name of Waal; and the other retaining that of the Rhine, is connected, a little farther N., by an artificial canal with the Yssel. Still lower down the Rhine takes the name of the Leck, in order to distinguish it from the old Rhine, now sanded up, which passed by Utrecht and Leyden to the sea at Catwyk. The Rhine has at present three mouths. About two thirds of its waters flow to the sea by the Waal, the remainder being carried partly to the Zuyder-Zee by the Yssel, and partly to the ocean by the Leck and Maas, on which is the great Rhenish port of Rotterdam. These branches, however, are so interlaced with natural and artificial channels, and there are so many lagoons, marshes, &c., in this district, that a map becomes indispensable to any detailed description. (Lyell's Geol. ii., 59–59.)

The entrance to the Maas, leading to Rotterdam, lies in lat. 51° 56' N., and is commonly called the Briel-gat or channel: it has a bar across, on which there are 7½ feet water at neap-tides. Spring-tides rise here from 11 to 12 feet, and neaps 6 or 7 feet. (Norie's Sailing Directions for the North Sea, p. 153.)

The breadth and depth of the Rhine in different parts of its course have been pretty accurately determined by repeated observations. At Chur, in the Grisons, it is about 250 feet in width, or about as broad as the Thames at Richmond; at Schaffhausen the width is estimated at 370 feet. and at Basle about 550 feet. The breadth of the stream between Basle and Strasburg is much increased by the numerous islands that interrupt the current; but from the last-mentioned place to Speiers the width varies from 1000 to 1900 feet. The size of the Rhine thence downward to Coblentz gradually increases to near 9000 feet in width; but near that point it enters a mountainous defile, and becomes much narrower, widening again at Cologne, where it is 1400 feet across; and lower down to its mouth, in the principal navigable branches, it has a breadth exceeding 2000 feet. The depth of the channel from Basle to Strasburg averages about 19 feet; and below the latter town the river is navigable by large steamers and vessels of considerable tonnage. July is the season when the river is fullest, and it then rises about 12 feet above its average height. May and October are the seasons of low water. The descent of the river, and consequently the rapidity of the current, are extremely variable in different parts; but the mean inclination from Strasburg is estimated at about 1-3 feet per mile; and the current may average somewhat more than 3 miles an hour, though below Cologne it is greatly decreased, the Rhine there becoming comparatively a sluggish river.

The affluents of the Rhine are very numerous; but the chief tributaries belong to the portion below Basle. The only navigable affluent above that point is the Aar, a stream which drains the greater part of Switzerland, and brings down nearly an equal volume of water with the Upper Rhine itself. It rises in the great mass of St. Gothard, passes through a narrow valley, in which is the cataract of Aarfall, 150 feet in height, afterward enters the lake of Brienz, and thence, after passing through the lake of Thun (1875 feet above the sea) runs past Berne, and enlarged by different tributaries, bringing down the waters of lakes Lucerne, Wallenstadt, Zurich, &c., joins the main river at Coblentz (47° 26' N., and 8° 15' E.), with a wide and powerful current. Its chief affluents are the Reuss and the Limmath; the former rising on the N.W. side of mount St. Gothard, receives the waters of lakes Luzerne and Zug, while the latter rises in the Oberland Alps, and runs through the lake of Zurich, both joining the Aar on its E. bank, about a mile from each other; both are navigable except during the spring floods, but owing to the rapidity of the current boats ascend empty.

Below Basle the Rhine is joined by many large tributaries, the most important of which enter from the E. or right bank. The first of these is the Neckar, which falls into the main river at Mannheim, about 214 m. below Basle. It rises in the Black Forest, in about lat. 48° N., and 8° 30' E.. and has a very tortuous course, first N.E., subsequently N.N.W., and lastly W., of more than 180 miles. It is navigable for large barges up to Heilbronn, and for small craft as far as Stuttgard. The Mayn, which enters the Rhine at Mayence, or Maintz, about 90 miles below Frankfort, which is on its N. bank, is a most important tributary. Large river barges, vessels of 100 tons burden, ascend as high as Kitzingen, which is 165 miles from the mouth: its entire course is estimated at 320 miles. (See MAYN.) The Lhan joins the Rhine about 48 miles below Mayence; it is about 140 miles long, but is not navigable above Limburg, 24 miles from the mouth. The Ruhr and the Lippe, are two other considerable tributaries on the E. side; both navigable for about 100 miles. The principal affluent on the W. bank is the Moselle, which rises on the W. side of the Vosges, at an elevation of 2356 ft., and after flowing past Nancy, Metz, Tréves (near which it is joined by the Sarre) enters the main river at Coblentz, after a course of 280 miles. The confluence of the Meurthe, 160 miles below the source, marks the extent of the navigation. The Meuse, or Maas, rises in the department of the Upper Marne, on one of the W. offsets of the Vosges, and running with a tortuous course, somewhat like the letter S., past St. Mihiel, Verdun, Sedan, Mézières, Namur, Liege, and Maestricht, joins the Waal, or principal stream of the delta, near Gorkum, below which, as already observed, the main stream assumes the name of Maas down to its mouth.

The geology of the valley of the Rhine has been rather extensively investigated by Boué, Von Buch, Brongniart, and other French naturalists. The bed of the Upper Rhine, from its source to the Chur, is formed of primitive rocks, chiefly gneiss and porphyritic granite; but at this point, grauwacké, blue limestone, and old red sandstone become the prevailing rocks, as far as the lake of Constance, where they are succeeded by tertiary formations, probably of more recent date than the gypseous strata of the Paris basin. Secondary and tertiary rocks line the river from Basle to the Neckar; but from this point to Bingen, below the confluence of the Mayn, granite, gneiss, and mica-schist form the substance of the high crags that line the river on both sides. Lower down the cliffs are composed of secondary limestone, with superimposed strata of new red sandstone, and in some parts volcanic-rocks are found curiously interspersed with the inferior chalk strata. The coal formations are found in the upper parts of the Ruhr and Lippe, this part of Rhenish Prussia furnishing the chief supply of that mineral for the purposes of steam-navigation. Below Dusseldorf the tertiary rocks are replaced by diluvial and alluvial formations, which form the subsoil of the delta. Geologists, however, are of opinion that the numerous islands in this intricate delta have been formed not so much by deposits brought down by the river, as by the inroads which the sea is continually making on this coast. (*Lyell's Geol.* ii. 53; and *Encyclopédie Méthod.*, art. *Rhin.*)

The scenery of the Rhine has been justly admired by travellers. Wildness and rude grandeur characterise it in the defiles above the lake of Constance, and the country from the Unter-See westward, as far as Rheinfelden, 6 miles above Basle, is almost equally romantic. But at this point the character of the scenery changes, and the river, formerly an inconsiderable feature in the landscape, becomes a broad and majestic stream, flowing as far as Manheim through a rich open valley, from 30 to 50 miles in breadth. The banks there begin to be more bold and rocky; but the scenery most generally admired is chiefly between Mayence and Coblentz. "The Rhine here pursues a meandering course pent between lofty and craggy mountains, and resembles rather a succession of lakes than a river. Here, indeed,

> 'The river nobly foams and flows,
> The charm of this enchanted ground,
> And all its thousand turns disclose
> Some fresher beauty varying round.'
>
> (*Childe Harold.*)

"These mountains, however, are after all only mountains in miniature. They have often, indeed the steepness, rudeness, and overhanging ridges of the mountains bordering the Rhone; but as compared to them in size, they are but molehills. The groves on the hill sides are few and far between; but there is no grove without a church spire rising in the midst, and overtopping the trees. Frequently a daring and fantastic castle, crowned by an ancient castle, frowns over the river, or rises majestically from the brow of the steep; but the woods, unlike those of the Rhone, look like plantations, and the vines obtrude an unceasing idea of the artificial." (*Leitch Ritchie's Travelling Sketches on the Rhine*, p. 79.)

The Rhine, with its various affluents, comprises a navigation of about 1500 m., and, in a commercial point of view, is perhaps the most important river in Europe, owing to the numerous states to which it affords a water conveyance. The following table exhibits the extent of the states, or portions of the states, included within the valley of this great river:

Switzerland	12,400 sq.m.	Prussia	31,159 sq.m.
France	15,000	Other Ger. States	9,042
Austria	860	Belgium	4,100
Baden	4,644	Netherlands	6,620
Bavaria	10,000		
Wirtemburg	5,390		83,996

The navigation of the Rhine has always been of considerable importance, but since the employment of steamers, and the abolition of the tolls, and other political obstacles to its free use, its importance as a channel of navigation and im-

portance has been immeasurably increased. Vessels of large burden ascend the river to Cologne, and Strasburg is reached by those of 80 or 90 tons. Recently, however, steamers have been regularly established on the Upper Rhine, between Strasburg and Basle, and an astonishing increase has taken place in the number of passengers, and the quantity of goods passing up and down the river. In 1830 for example, the number of passengers conveyed along the Rhine, in the Prussian territories, amounted to only 22,62 whereas, in 1836, they amounted to 145,961, and now may exceed 900,000! The increase in the quantity of merchandise conveyed along this great international highway has been equally great; and its importance will, no doubt, be still more rapidly increased by the opening of railways, and improved methods of communication with the great cities on its banks.

Besides the goods conveyed up and down the Rhine in steamers and sailing vessels, immense quantities of timber are sent down in the form of rafts. The smaller rafts are the Upper Rhine, and the smaller affluents of the river used formerly to rendezvous at Namedy, near Andernach, where they were consolidated into rafts of a larger size, that were sent down the river to Dordrecht, where they were generally broken up, and the timber sold and forwarded to its final destination. Of late years, however, the plan of constructing very large rafts has begun to fall into disuse though, as they are still sometimes met with, the reader may not be displeased to have them described.

"A little below Andernach the little village of Namedy appears on the left bank, under a wooded mountain. The Rhine here forms a bay, where the pilots are accustomed to unite together the small rafts of timber floated down the tributary rivers into the Rhine, and to construct enormous floats, which are navigated to Dordrecht, and sold. These machines have the appearance of a floating village, composed of 12 or 15 little wooden huts, on a platform of oak and deal timber. They are frequently 800 or 900 ft. in length, and 60 or 70 ft. in breadth. The rowers and workmen sometimes amount to 700 or 800, superintended by pilots, and a proprietor, whose habitation is superior in size and elegance to the rest. The raft is composed of several layers of logs placed one on the other, and bound together: a large raft draws not less than 6 or 7 ft. of water. Several smaller ones are attached to it, by way of protection, besides a string of boats loaded with anchors and cables, and used for the purpose of sounding the river and going on shore. The domestic economy of an East Indiaman or an English man-of-war is hardly more complete. Poultry, pigs, and other animals, are to be found on board; and several butchers are attached to the suite. A well supplied boiler is at work night and day in the kitchen; the dinner hour is announced by a basket stuck on a pole, at which signal the pilot gives the word of command, and the workmen run from all quarters to receive their messes. The consumption of provisions in the voyage to Holland is almost incredible; sometimes amounting to 40,000 or 50,000 pounds of bread; 15,000 or 20,000 of fresh, besides a quantity of salted meats; and butter, vegetables, &c., in proportion. The expenses are so great, that a capital of three or four hundred florins is considered necessary to undertake a raft. Their navigation is a matter of considerable skill, owing to the strength of the current, the rocks, and shallows of the river; and some years ago the secret was thought to be monopolised by a boatman of Rüdesheim and his sons." (*Autumn on the Rhine.*)

The *Rhenus*, or Rhine (Gr. Ῥῆνος), became first known to the Romans by the conquest of Julius Cæsar, who caused it twice to punish the Germans. It is thus described by him: *Rhenus oritur ex Lepontiis qui Alpes incolunt, et longo spatio per fines Nantuatium, Helvetiorum, Sequanorum, Mediomatricorum, Tribucorum, Trevirorum citatus fertur; et ubi oceano appropinquat, in plures diffluit partes multis ingentibusque insulis effectis quarum pars magna a feris barbarisque nationibus incolitur (ex quibus sunt qui piscibus atque ovis avium vivere existimantur), multisque capitibus in oceanum influit.* (Bell. Gall. iv. 10.) Ancient writers, though agreed with respect to its general source, differ respecting the number of mouths by which it falls into the ocean. Tacitus (Ann. ii. 6) speaks of two only, one of which, probably the modern Waal, he terms *Vahalis*, and the other *Rhenus*. Pliny and Ptolemy, however, say that there were three mouths, the most northerly of which, called Flevum, was supposed to have been formed by a channel dug by Drusus, to connect the Rhine with the Ivala, which is, most probably, identical with the Yssel, as Pomponius Mela (iii. 2) assures us that it fell into the lake Flevo, or modern Zuyder-Zee. No doubt, however, the channels of the river in the delta must have shifted, both prior and subsequently to the accounts given by the classical writers; besides which, the inroads made by the sea on the coast of Holland render it impossible to ascertain what may have been the exact number of its mouths at any very remote period. (*Encyclopédie Méthod.*, art. *Rhin.* Bru-

f Europe; Conversations Lexicon; Inglis's . &c.)

)VINCE OF), a prov. containing the S. por-
sso-Rhenish states; having H. Holland, E.
-stphalia, Nassau, and other German states,
'V. the latter, Belgium, and Holland. It
l° and 52° N. lat., and 6° and 9½° E. long.
q. m. Pop., in 1837, 2,433,230, of whom
atholics, and 564,728 Protestants. Principal
Jhapelle, Cologne, Coblentz, Dusseldorf, El-
s, Bonn, &c. It is divided into five regen-
again into 59 circles. Principal rivers, Rhine,
s almost the whole extent of the prov. Mo-
err, &c. Surface very various. In E. part
he Moselle consists principally of volcanic
i a chain of the same sort (Eyfel Gebirge)
a province between Malmedy and Coblentz.
volcanic rocks being particularly suitable for
the vine, it is very extensively cultivated:
the vines of the Rhine and the Moselle being
about 700,000 eimers a year, though but little
Exclusive of wine, the principal raw products
, flax and hemp, timber, tobacco, &c. Manu-
: Lower Rhine are both extensive and flour-
e towns of Aix-la-Chapelle, Eupen, Malmedy,
., along the Netherlands frontier, and in El-
nen, and others to the E. of the Rhine, im-
nery is to be met with; and the manufacture
n cloths, cassimeres, fine and coarse cottons,
rosecuted with great vigour and success. The
.d hardware manufactures are also important

an Official Account of the principal business
in the Rhine Province, in 1837, and of the
tablishments, and the number of work-people
:ach :—

Manufactures.	No. of Estab.	Workmen.
silk wares	60	5,190
and other woollen stuffs	101	8,029
for woollen yarn spinning	48	1,807
inning and weaving	43	4,144
.	1	302
.ing-mills	7	153
ing .	7	603
wares .	15	395
.ts .	38	1,406
e wares .	4	63
boilers	16	730
lucts	9	63
earthen wares .	3	199
ting and gilding	9	113
.s .	78	58
sirits	12	44
.s	7	104
.	1	90
.	1	9
.	2	684
.	30	63
.	3	41
.	6	693
.oaches	3	49
.	1	6
.ools .	1	38
Total	**389**	**26,145**

ECK, p. t., Dutchess co., N.Y., 55 m. S. Albany,
ounded W. by Hudson river. If contains two
i Dutch Reformed and Methodist; nine stores,
g-mill, two grist-mills, three saw-mills, one pa-
our schools, 27 scholars. Pop. 2650. At the
amboats from New-York to Albany stop and re-
ngers daily. Freight barges are towed twice a
:w-York.
ISLAND, still retaining under its new constitutional name of *Rhode Island and Providence Plan.*
the smallest state in its territory in the Union;
ated between 41° 22' and 42° 3' N. lat., and be-
6' and 71° 38' W. long., and between 5° 23' and 5°
.from W. It is about 49 m. long, and 29 m. broad,
1360 square miles, of which Narragansett bay la-
; and the whole state contains 870,400 acres.
.ded into five counties, which, with their popula-
0, were as follows:

.ience	58,073
.ort	16,874
.t	6,476
.	13,083
.ington	14,324

pulation in 1790 was 58,825; in 1800, 69,122; in
31; in 1820, 83,059; in 1830, 97,212; in 1840,
Of these, 51,302 were white males: 54,225 were
ales; 1413 were coloured males; 1825 were col-
ales. Employed in agriculture, 16,617; in com-

merce, 1348; in manufactures and trades, 21,271; in navi-'
gating the ocean, 1717; in the learned professions, 457. It
is the only state in the Union in which the number employed in manufactures and trades exceeds those employed in
agriculture.

Newport and Providence are the principal seats of government, and are considered the capitals, though the legislature sometimes regularly meets at South Kingston, Bristol, and East Greenwich.

About one third part of the state in the west and north-
west, is hilly and rocky; but there is nothing which properly deserves the name of a mountain or high hill in the
state. On Narragansett bay and the Atlantic ocean, and
in some places on the streams, the surface is level. In the
western and northwestern parts, the soil is thin, and has no
great fertility. But near Narragansett bay, and on the islands in it, the soil has great fertility. It is generally distinguished for the excellence of its cattle and sheep, butter and
cheese; and produces Indian corn, rye, barley, oats, and, in
some places, wheat; though it is better adapted to grazing
than to grain. Grass, fruits, and culinary vegetables are
produced in great perfection. Shell and other fish are
found in the rivers and bays, of an excellent quality.

There were in the state in 1840, 8094 horses and mules,
36,891 neat cattle, 90,146 sheep, 20,659 swine. There were
produced 3096 bushels of wheat, 34,321 of rye, 450,498 of Indian corn, 2970 of buckwheat, 66,490 of barley, 171,517 of
oats, 911,973 of potatoes, 183,820 pounds of wool, 363 tons
of hemp or flax. The products of the dairy amounted to
$223,239; of the orchard, to $32,098; of lumber, to $44,455.

The exports of the state consist chiefly of flaxseed, lumber, horses, cattle, beef, pork, fish, poultry, onions, butter,
cheese, barley, and especially cotton goods. In proportion
to its population, it is the greatest manufacturing state in the
Union, and especially it led the way in the manufacture of
cotton goods. Its shipping is also extensive. The exports
for the year ending September 30th, 1841, was $278,465;
and the imports were $339,592. The tonnage entered was
25,195 tons, and the tonnage cleared was 21,698 tons.

The climate is healthy, and on the islands more mild than
in other parts of New-England. The heat of summer is
tempered by the sea-breezes, and the severity of winter is
moderated by the proximity of the ocean. Newport is celebrated as a delightful summer residence, and is much resorted to, particularly from the southern states.

The rivers, though not large, furnish many fine mill-seats,
which have been extensively used for manufacturing purposes. The principal are Pawtucket, Providence, Pawtaxet,
Pawcatuck, and Wood rivers. Narragansett bay is a fine
body of water, extending 30 miles into the state, and containing a number of beautiful and fertile islands. Among
them are Rhode Island, which gives name to the state, 15
m. long, and on an average 3½ m. broad. Connanicut is 8
m. long, and 1 broad; Prudence is 6 m. long; and Block
island, 10 m. out in the Atlantic, is 8 m. long, and from 2
to 4 broad. It constitutes the township of New-Shoreham,
and contains a lighthouse. The entrance of Narragansett
bay is between point Judith on the W., and Seakonnet
rocks on the E., and it has many good harbours. The harbour of Newport, on the S.W. part of Rhode Island, is one
of the best in the United States, being spacious, safe, easily
accessible, and with a sufficient depth of water for the largest vessels. Providence, at the head of Narragansett bay,
35 m. above point Judith, has a good harbour, and in population, commerce, and wealth, is the second city in New-
England. Bristol on the E. side of the bay, 15 m. N. of
Newport, has a safe and commodious harbour. Pawtucket,
4 m. N. of Providence and Pawtaxet, 10 m. S. of Providence,
are distinguished as manufacturing villages.

There were in the state in 1840, 44 commercial and 57
commission houses in foreign trade, with a capital of
$2,043,507; 930 retail dry goods and other stores, with a
capital of $9,810,125; 58 persons engaged in internal trans-
portation, with 83 butchers, packers, &c., employed a capital of $71,050; 969 persons engaged in the lumber trade,
employed a capital of $254,900; 1160 persons employed in
the fisheries, with a capital of $1,077,157.

The manufactures of this state, small as it is, are deserving of particular notice. According to the census of 1840,
home-made or family goods were produced to the amount
of $51,180; 41 woollen factories, with 45 fulling-mills, em-
ployed 961 persons, and produced goods to the amount of
$642,172, with a capital of $685,350; 209 cotton factories,
with 518,817 spindles, employed 12.086 persons, producing
articles to the amount of $7,116,792, with a capital of
$7,396,000; 27 persons produced 1000 tons of anthracite coal,
with a capital of $6000; five furnaces produced 4198 tons
of cast iron, with a capital of $22,250; two paper-mills pro-
duced articles to the amount of $25,000, and other manu-
factories of paper produced to the amount of $9800, the
whole employing 15 persons, and a capital of $45,000; hats
and caps were manufactured to the amount of $92,465, and

straw bonnets to the amount of $86,106, the whole employing 411 persons, and a capital of $56,427; 27 tanneries employed 89 persons, and a capital of $72,000; 44 saddleries and other manufactories of leather produced to the amount of $182,110, with a capital of $70,695; 43 persons produced granite and marble to the amount of $36,202; 113 persons produced bricks and lime to the amount of $68,000; 534 persons produced machinery to the amount of $437,100; 164 persons produced hardware and cutlery to the amount of $138,790; 179 persons manufactured the precious metals to the amount of $283,500; 57 persons produced 1,237,050 pounds of soap, 157,250 pounds of tallow candles, and 264,500 pounds of spermaceti or wax candles, with a capital of $252,628; 161 persons manufactured carriages and wagons to the amount of $78,811, with a capital of $36,661; various mills produced articles to the amount of $23,663, employing 166 persons, and and a capital of $152,310; nine rope-walks employed 45 persons, produced cordage to the amount of $49,700, with a capital of $38,300; vessels were built to the amount of $41,500; 195 persons manufactured furniture to the amount of $121,131, with a capital of $63,300; four distilleries produced 885,000 gallons of distilled spirits, and three breweries produced 59,600 gallons of beer, with a capital of $139.000; six brick and 292 wooden houses were built, employing 887 persons, and cost $379,010; 16 printing-offices, eight binderies, two daily, four semi-weekly, and 10 weekly newspapers, and two periodicals, employed 122 persons, and a capital of $35,700. The total amount of capital employed in manufactures was $10,606,136.

Brown university at Providence, was founded at Warwick in 1764, and permanently located in Providence, in 1770, is the only college in the state, and is a flourishing institution. The president, and a majority of the trustees, are required to be of the Baptist denomination. The common schools of this state, formerly less attended to than in the other New-England states, have latterly received much attention, and are improving. In 1843 there was expended for instruction in the state $42,944. The state has a permanent school fund amounting to over $50,000. The sum of $25,000 annually is paid from the state treasury to the school committees of the several towns for the support of the public schools. In 1840 there were in Brown university, and in a high school which partakes of the nature of a college, 394 students. There were 52 academies or grammar schools with 3664 students, 434 common and primary schools, with 17,355 scholars.

The principal religious denominations are Baptists, Congregationalists, Episcopalians, and Methodists. In 1836, the Baptists had 90 congregations, and 18 ministers, besides nine other Baptist churches of a different description; the Congregationalists had 16 ministers and 16 congregations, with 2100 communicants; the Episcopalians had 16 congregations, 18 ministers and 1655 communicants; the Methodists had 10 ministers. Besides these, there are some Friends, Unitarians, Universalists, Christians, and Roman Catholics. There is a state prison at Providence, completed in 1838.

In the commencement of 1840, there were 62 banks in the state, with an aggregate capital of $9,880,500, and a circulation of $1,719,230. Although the banks are so numerous, averaging two to a township, though located in the principal places, yet they have generally proved their credit unimpaired. Rhode Island has no public debt.

Until recently, the only constitution of the state was the charter granted by King Charles II. in 1663. Within about 20 years, four attempts have been made, under the sanction of the general assembly, to form a constitution for this state, all of which attempts, except the last, failed. In March, 1840, the Rhode Island Suffrage association was formed, having in view the extension of the right of suffrage to all the native white male citizens of the United States resident in the state. In May, 1841, as a mass meeting held at Newport, this association took measures for calling a convention of the people, without regard to the existing government; and a convention was chosen by a part of the people in several towns, which assembled at Providence, October 9th, 1841. This convention proceeded to form a constitution for the state, and submitting it to a portion of the people, they adopted it, and the convention proclaimed this constitution to be the supreme law of the land. But the whole proceeding was repudiated by the existing government, who, in the mean time, had taken measures to assemble a convention on the first Monday in November, 1841, who formed a constitution, and submitted it to the people for their ratification; and there were for the constitution 16,702, against it 8013; majority in favour of it 8689. On the 4th of May, 1842, the charter government was organized, as usual, at Newport, the state being in too great commotion to organize, under the constitution recently adopted. The Suffrage party, having elected a governor and a legislature, under the so called people's constitution, organized May 3d, 1841, a government, under the protection of an armed force at Providence. May 18th, 1842, an attempt was made by an armed force, commanded by the governor under the "people's constitution," to capture the state's arsenal in Providence, and a civil war seemed to be commencing. In the third week of June, 1842, the general assembly passed an act, providing for another convention to form a constitution, to be held in September, and to be composed of delegates chosen by persons who had resided three years in the state, without regard to property, taxation, or military service as a qualification. In the last week of June, 1842, another desperate effort was made to overthrow, by force of arms, the regular government of Rhode Island; but the insurgents, with their governor at their head, were quickly defeated and dispersed at Chepacket, by the forces of the state. The convention provided for by the act of the general assembly in June, assembled at Newport in September, 1842. The constitution formed by this convention was submitted to the people on the 21st, 22nd, and 23rd of November, in 1842, and by a vote of 7032 for it, and 59 against it, it was adopted; the "Suffrage party" formally protesting against it, but inconsistently voting under it. Thus a threatened civil war has been avoided, without the interference of the power of the United States' government, and a check has been given to irregular attempts to alter the constitution of the government of a state; and things seem likely to settle down into a tranquil condition.

The executive power of the state is vested in a governor, elected annually by the people. In case of the inability of the governor to serve through death, impeachment, or absence from the state, his place is supplied by the lieutenant governor, elected also annually by the people; or in case of the inability of both, the office shall be filled by the president of the senate. The senate consists of the lieutenant-governor, and one member from each town or city in the state, and is presided over by the governor, or in his absence by the lieutenant-governor, who has no vote excepting on an equal division of the members. The secretary of state is also secretary of the senate, and in the absence of the governor and lieutenant-governor, presides over the body and a president is chosen. The house of representatives consists of not to exceed 72 members. Each town or city is entitled to at least one member, provided that no town or city shall elect more than one sixth of the entire number. The present ratio of representation is one member to every 1530 inhabitants, and each fraction exceeding one half is entitled to one representative. The present number of members is sixty-nine. The judicial power of the state is vested in one supreme court, and such inferior courts as the general assembly may, from time to time, ordain. The judges of the supreme court are elected by the general assembly in grand committee, and may be dismissed by a majority of all the members elect of each house separately. Every person, a citizen of the United States, of the age of 21 years, who has resided, and has had a home in the state for one year, and in the town or city where he offers his vote for six months next preceding the election, and real estate in said town or city to the amount of 134 dollars, or renting for seven dollars above all incumbrances whatsoever, is entitled to the right of suffrage; or if his property in another town or city in the state than that in which he resides, he must produce a certificate to that effect from the clerk of the city or town in which it lies, dated within ten days previous to the election, and showing that the deed, if there be any, has been recorded for 90 days. The name of the voter must have been legally enrolled for one year next previous to the election in the town or city where he offers his vote, and he must have paid a town or city tax of one dollar, and have been enrolled in the militia, and have performed military duty. A resident at any garrison or naval station in the state does not give a legal residence. The general assembly holds two sessions annually, one at Newport, on the last Monday of October, the other biennially at South Kingstown, and in the intermediate years, at Bristol and East Greenwich, alternately. An adjournment of the October session is held annually at Providence.

This state has several important works of internal improvement. The Blackstone canal connects Providence and Worcester, and lies partly in this state. The same is true of the Providence and Boston railroad. With this is connected a daily line of steamboats to the city of New-York. The Providence and Stonington railroad lies principally in this state, and is connected by a line of steamboats to New-York city. When the Long Island railroad shall have been completed through the island, which is expected to be done by July, 1844, this, with the Stonington and Providence railroad, and the Providence and Boston railroad, will constitute the most direct and speedy communication between New-York and Boston.

Rhode Island was first settled by Roger Williams in 1636, who located in Providence, and gave it its name, in memory of having found in it a refuge from religious persecution.

d from Salem, Massachusetts, on account
opinions, for which he held that men were
to their Maker; while his opposers thought
i measures were opposed to the peace of
iuted by them. In 1638 he purchased the
nicus and Miantonomoh, two chiefs of the
Ilians. In 1638, William Coddington and
persecuted for their religious tenets, follow-
Providence, and by his advice they pur-
adians the island of Aquetnec, now called
nd Coddington became their judge or chief
i also granted universal toleration to all
In 1643 Williams went to England as the
settlements, and by the assistance of Sir
tained a patent from the parliament, then
me power, by which the towns of Provi-
., and Portsmouth, were authorized to es-
vernment for themselves. And in 1647 all
sembled at Portsmouth enacted a code of
lished a civil government. An assembly
to consist of six representatives from each
executive and judicial power was vested in
four assistants, to which there was an ap-
courts established in each town for the trial
s. Some difficulties having arisen in regard
Williams and Coddington, upon an application
aries II., a charter was granted, incorporating
nd Providence plantations. It was provided
should be molested or called in question for
i in matters of religion, and the supreme
ited in a governor, deputy governor, ten as-
presentatives of the several towns chosen by
This is the venerable charter under which
d and prospered. until recently. In 1730, the
abitants was 18,000, and in 1761 it was 40,000.
ore an honourable part in the revolutionary
eral Greene, one of her citizens, was inferior,
ficer, only to Washington. She was the last
states, who adopted the federal constitution.
d in May 29th, 1790, by a majority of two
ecent constitution went into operation May

i famous island of the Mediterranean, off the
Asia Minor (an. *Lycia* and *Caria*), 10 m. S.
the city of Rhodes, at the N.E. extremity of
log in lat. 36° 26' 53" N., long. 28° 19' 51" E.
i m. in length, N.E. and S.W., and is, where
ut 18 m. across. Pop., according to Savary,
on rather more than a third are Greeks; but
t does not exceed 20,000. Probably it may be
about 25,000, a number that would have been
antiquity to have peopled one of its three cities.
ountains runs lengthwise from one end of the
other; their highest summit, mount Artemira
s, on which was a temple of Jupiter), com-
oble view of the island and of the adjacent
ia Minor. In antiquity this mountain chain
with dense forests of pine, whence the Rhod-
plies of timber for their fleets, and in modern
supplied considerable quantities for the dock-
istantinople. Speaking generally, the soil in
rts is dry and sandy; but it has some fine val-
well watered by the numerous streams that de-
he mountains. In antiquity it was famous for

*rμου pregnant with the fertile reed
X plants, and herbs, and fruits, and foodful grain
ch verdant hill unnumbered flocks dress feed ,
Unnumbered men possess each flowery plain.
            Pindâr, by West, Olympic Odes, No. 7.*

to the insecurity and extortion of which the in-
ve been long the victims, its agriculture is in the
ied state, many of its finest fields being allowed
:, and the island not producing corn sufficient
ecanly population. Its wheat is still, however,
quality; but according to the statement of Mr.
vine, which he calls " tolerable, red, and sharp,"
adly degenerated from that mentioned by Virgil
for the feasts of the gods !

*e age is, Dis of mensis accepta secund's
sonoria, Rhodia, et tumid», bumaste, racemis.
                              Georg . lib. ii., lin. 101.*

luces oil, oranges, citrons, and other fruits ; and,
grinding despotism by which it is weighed down,
ndace, in profusion, most necessaries and luxur-
ble is quarried in several parts of the island.
ate of Rhodes (*clarum Rhodon*, Hor.) is probably
n the Mediterranean. " It is," says Dr. Clarke,
lightful spot; and its gardens are filled with deli-
Here, as in Cos, every gale is scent-d with the
rful fragrance, wafted from groves of orange and
s. Numberless aromatic herbs exhale, at the

same time, such profuse odour, that the whole atmosphere
seems to be impregnated with a spicy perfume. The present inhabitants of the island confirm the ancient history of
its climate; maintaining that hardly a day passes throughout the year in which the sun is not visible." *Travels*, iii.,
276, 8vo. ed.) The heat, which otherwise would have been
oppressive, is tempered by the wind, which blows with
little variation from the N. and N.W.

The only beasts of burden used in the island are mules
and donkeys, there being no camels, and but few horses,
and those only belonging to the richer Turks. Partridges
are very abundant. Various species of excellent fish, with
coral and sponges, are found in the adjacent sea.

The city of Rhodes is situated, as already stated, at the
N.E. extremity of the island. It is built amphitheatrewise,
on ground rising gently from the water's edge; and is
strongly fortified, having a moated castle of great size and
strength, and being surrounded by walls, flanked with
towers. These works were constructed by its former masters, the knights of St. John; and Mr. Turner says, that,
though he had known nothing of the history of the island,
he should have perceived that its fortifications were the
work of the same master hand that had constructed those
of Malta. (ii., 11.) The town consists generally of narrow
winding lanes and mean houses. When in possession of the
knights it had many substantial stone houses, some of
which, as well as the public buildings, were ornamented
with the arms of the knights in *alto relievo*; but the greater
number of these houses are now in a state of ruin, and
such as have been rebuilt in their stead are mere wooden
fabrics. Contrary to what might have been expected, the
best streets in the city are in the quarter inhabited by the
Jews. The Greeks occupy a distinct quarter, behind and
S.E. from the city properly so called. On the land side the
city is surrounded by a burying-ground ; and beyond it are
the suburbs, consisting of detached and finely situated
houses, surrounded by gardens, many of which, however,
are said to be unoccupied. The ancient palace of the grand
master is now the residence of the pacha ; the large and
fine church of St. John is the principal mosque, and the
great hospital has been converted into a public granary. It
has two harbours ; the smallest, a fine basin, with a narrow
entrance, is protected on all sides from the wind ; but the
Turks having allowed filth and sand to accumulate in the
entrance, it can now be used only by the smaller class of
vessels : the other harbour is much larger, and has deep
water, but is safe only during westerly winds, those from
the N.E. throwing in a heavy sea ; on this account large
vessels prefer anchoring in the roads in 21 fathoms water,
from its being more convenient for getting out to sea, in the
event of the wind setting in strong from the N.E. A lighthouse is erected on a mole between the two harbours.
Several ships for the Turkish navy have been built at
Rhodes; but the trade of the town is quite inconsiderable.
There are some, though but few, remains of antiquity in the
city ; the barbarism of its Saracenic and Turkish conquerors, and the recurrence of destructive earthquakes, having
destroyed most memorials of its former splendour.

Historical Notice.—Rhodes was early distinguished by its
wealth, its naval power, the wisdom of its laws and institutions, and its superiority in art and science. Tlepolemus,
a prince of Rhodes, distinguished himself at the siege of
Troy ; and the island could then boast of two then famous
cities of Lindus, Jalysus, and Camirus. The city of Rhodes
is much less ancient, having been founded during the Pelop-
ponesian war. But its advantageous situation, and the excellence of its harbour, soon gave it a decided superiority
over the other towns of the island, many of whose inhabitants withdrew to it; and it was, in fact, one of the best
built and most magnificent cities of the ancient world. It
had been constructed with the greatest regularity ; its streets
being wide and straight, and the houses in each being of
the same height and built on the same model. Pliny calls
it *civitas libera et pulcherrima* (*Hist. Nat.*, lib. v., cap, 31) ;
and Strabo, who had seen Rome, Alexandria, and other
great cities of the ancient world, gives the preference to
Rhodes. "The beauty," says he, "of its harbours, streets,
and walls, and the magnificence of its monuments, render
it so much superior to all other cities as to admit of no
comparison." (Lib. xiv.) Its temples, especially those
dedicated to Bacchus, Diana, Isis, &c., were celebrated alike
for the magnificence of the building, and the statues and
paintings with which they were enriched. In the noble ode
already referred to, written about 500 years B.C., Pliny alludes as follows to the excellence of the Rhodians in statuary :

*Thence in all arts the sons of Rhodes excel ;
  Though b u' their forming hands the chisel guide ;
The in each street the breathing marbles tell,
  The stranger's wonder, and the city's pride.*

The most famous of the works of art in Rhodes were two
pictures, of the most transcendant merit, by Protogenes, the

contemporary and rival of Apelles (*Strabo*, lib. xiv. ; *Pliny, Hist. Nat.*, lib. xxxv., cap. 10), and the colossus, the work of Chares of Lindus, deservedly reckoned one of the wonders of the world. This magnificent brazen statue, erected in honour of Apollo, the tutelary deity of Rhodes, is said to have been 70 cubits (about 105 feet) in height, and of the most admirable proportions. It was set up about *anno* 278 B.C., and was thrown down by an earthquake 56 years thereafter; and it is a curious fact that it lay where it fell for nearly 890 years, or till A.D. 667, when the island, having been taken by the Saracens, they broke the statue to pieces, and sold the brass. Blaise de Vigénere, a writer of the 16th century, stated, for the first time, that the colossus stood with a foot on each side the entrance to the port, and that the largest vessels, under full sail, passed between its legs. This story, which carries absurdity on its face, and for which there is not a shadow of authority in any ancient writer, having been adopted by Rollin, has thence found its way into most modern works. (*Pliny*, lib. xxxiv., cap. 7 ; *Rollin, Hist. Ancienne*, iv., 137, 4to ed. ; *Savary's Letters on Greece*, English translation, 63 ; *Biographie Universelle*, art. *Chares.*) Exclusive of this matchless work, Rhodes had 105 colossal statues ; each of which might, according to Pliny (*loc cit.*), have sufficed to illustrate the town.

The wealth of the Rhodians was derived partly from the fertile soil and advantageous situation of their island, but more from their extensive commerce and commercial navigation, and the wisdom of their laws, especially those having reference to maritime affairs. Such, indeed, was the estimation in which the latter were held, that the rule of the Rhodian law *de jactu* was expressly embodied in the Digest (lib. xix., tit. 2), and has been thence adopted into all modern codes. Indeed, the fair presumption seems to be that most of the regulations as to maritime affairs included in the civil law were derived from the same source.

Rhodes was also famous for its science and literature. Æschines, on his retirement from Athens, opened a school of rhetoric in this city ; and towards the termination of the Roman republic, and under the early emperors, Rhodes was held, as a school of eloquence, literature, and philosophy, to be little, if at all, inferior even to Athens; and these, combined with the genial temperature of the climate, and the luxurious refinement of the capital city, made it be resorted to by some of the most illustrious individuals of whom Rome has to boast, including, among others, Pompey and Cicero. Julius Cæsar, too, had set out to study at Rhodes, and was only prevented by being captured on his voyage by pirates. (*Suetonius*, lib. i., cap. 4.) Tiberius resided for about 7 years in the island. It seems also to have been a favourite retreat of those Romans who wished to withdraw from the factions and turmoil of Rome. (*Cicero, Epist. ad Fam.*, lib. ii., epist. 28.)

The government of Rhodes, which, like that of most other Greek cities, was originally monarchical, was subsequently changed into a democracy, and thence into an aristocracy ; under which it enjoyed a degree of tranquillity and prosperity to which most Grecian cities were strangers. It was taken by Mausolus, king of Caria, but recovered its independence under his widow, the famous Artemisia. From this period Rhodes continued to enjoy profound peace, till it was attacked by Demetrius, the son of Antigonus, one of Alexander's successors. The siege of the city of Rhodes by Demetrius is one of the most celebrated in ancient history ; but all the science and efforts of Demetrius were defeated by the bravery and resolution of the Rhodians, and he was compelled to raise the siege, *anno* 303 B.C., after it had continued about a year. The expense of the colossus was mostly defrayed from the sums received by the Rhodians for the machines and other engines used by Demetrius in the siege, and which he gave up to them. It may be worth while to notice the fact, mentioned by Hume, that this siege affords the only example to be found in antiquity of the establishment of a cartel for the exchange of prisoners. (See *Essay on the Populousness of Ancient Nations.*)

The Rhodians were subsequently ranked among the steadiest of the allies of Rome ; they repulsed Mithridates, who made an attack on their city, and continued to enjoy their liberty till the reign of Vespasian, when Rhodes was made a Roman province. The island was overrun by the Saracens ; but having been recovered by the eastern emperors, it was presented, in 1308, by the Emperor Emanoel, to the knights of St. John of Jerusalem, who held it till 1522, when, after a desperate resistance, it was taken by the Turks, under whose barbarous brutalising despotism it has sunk to the miserable condition in which we now find it. "Beauty and mildness of climate, fertility of soil, variety and abundance of necessary and agreeable articles, a situation favourable for useful enterprise and a prosperous trade, nothing is wanting to this fine island, except to be delivered from the Turks, who possess, in so eminent a degree, the fatal talent of converting the most happy abode into a spot to be shunned and dreaded." *Sonnini's Trav-*

els *in Greece and Turkey*, i., 166, Eng. trans. ; see also, exclusive of the works already referred to, *Anc. Universal History*, viii., 137–212, 8vo. edit. ; *Savary's Greece, Letters* XII. and XIII. ; *Voyage d'Anacharsis*, cap. 73 ; *Schweinzing's Treatise on the Maritime Laws of Rhodes*, passim, &c.)

RHODEZ or RODEZ (an. *Segodunum*), a town of France, dep. Aveyron, of which it is the cap., on a hill beside the Aveyron, 163 m. E. by S. Bordeaux ; lat. 44° 21′ 8′ N., long. 2° 34′ 22′ E. Pop. in 1836, ex. com. 9138. Like most other very old towns, it is ill built : streets steep, narrow, dirty, and dark from the projection of the upper stories. The neighbourhood is, however, agreeable : and the town, which lies rather an imposing aspect from without, is closely surrounded with gardens and planted promenades. Almost the only edifice worth notice is the cathedral, a Gothic building, constructed between the 13th and 16th centuries. Its fine tower, which, it is said, may be seen at a distance of nearly 50 m., is 268½ feet high. square for two thirds its height, then octagonal, ornamented with delicate tracery, surmounted with a small cupola and a colossal statue of the Virgin ; it has minarets at each corner, on the summits of which are figures of the four evangelists. This church is rich in arabesques, and has a fine organ. The bishop's palace, prefecture, royal college, formerly belonging to the Jesuits, seminary, public library, with 16,000 volumes, town-hall, hospital, convent of Cordeliers, a building of the 14th century, &c., are the other chief public edifices.

Rhodez is a bishop's see, and has courts of primary jurisdiction and commerce, a chamber of manufactures, schools of drawing, and for deaf and dumb persons ; manufactures of coarse woollens for clothing troops, hats, wax candles, and playing cards ; and some trade in cheese, wool, corn, linens, &c. It has four annual fairs. In its vicinity many mules are bred. Rhodez was annexed to the dom. of France by Henry IV. (*Hugo*, art. *Aveyron*, &c.)

RHONE (an. *Rhodanus*), a celebrated river of S.W. Europe, rising in the Pennine Alps, traversing portions of Switzerland and France, and falling into the Mediterranean, between lat. 43° 15′ and 48° 15′ N., and long. 4° and 6° 22′ E. Length, 590 m. ; estimated area of basin, 37,380 sq. m. The highest source of this river is on the W. side of the great mass of St. Gothard, between the Furca, Galenstock, and Grimsel, at an elevation of 5780 feet above the sea ; but it scarcely assumes the form of a river till its junction with three or four other streams at the foot of the glacier of its own name, a beautiful fan-shaped cluster of ice, the lower edge of which is 5470 feet above the sea. Its course through the Valais is W.S.W. as far as Martigny, about 66 m. from its source, the height of the river at this point being 1523 feet above the sea ; but here the stream assumes a N.N.W. direction for about 24 m., entering the lake of Geneva at a level of 1236 feet above the sea, bringing with it a deposit of mud which has partially filled all the upper part of the lake. (*Lyell's Geology*, i., 333, 334.) The Rhone, on leaving this lake at the town of Geneva, where it is crossed by two bridges, is soon afterwards joined by the Arve, and then enters a rocky defile between the Alps and Jura chain, taking a S.W. direction for about 22 m., as far as the gorge called the *Perte du Rhône*, where its waters are hidden by limestone rocks nearly meeting over the stream. Its course thence is nearly due S. for 40 m., as far as St. Genix, at which point the river is about 645 feet above the sea. Here, however, it takes a sudden turn to the W.N.W., which direction it maintains with few exceptions to its junction with the Saone at Lyons, the average fall from the lake of Geneva to this city (a distance of about 150 m. along the stream) being estimated at somewhat more than 6 feet per mile. The *Perte du Rhône* is thus described by Mr. Bakewell: "The river, before arriving at *la Perte*, runs in a narrow bed, cut in soft clay strata reposing on a bed of calcareous stratum ; but on reaching this stratum, the waters have excavated a deep tunnel in it, into which they fall with considerable force ; the rocks on each side approaching so close, that before the space was widened by the Sardinian government to prevent smuggling, a man might have strode across, and sees the Rhone pass at a great depth between his feet. This tunnel is divided half-way down by projecting ledges of rock into an upper and lower channel. In winter and early in spring the river runs below these ledges, and is nearly concealed ; in one part, also, masses of rock have fallen down, and entirely covered the lower bed of the river for about 60 yards. This part may be traversed when the river is low ; but in summer, during the melting of the Alpine snows, it is much enlarged and flows over the intervening rock." The Rhone was in this state at the time of Mr. Inglis's visit ; and to this circumstance may be traced the disappointment felt by that traveller. (*Bakewell's Travels*, iii., 265, 266 ; *Inglis's Switzerland*, p. 175.) The Rhone leaves the hilly country a few miles E. of Lyons, where its deep, transparent blue, and very rapid waters are joined from the N. by the sluggish and muddy stream of the Saone: indeed, so marked is the difference between these rivers, that for many miles below Lyons, they flow side by side, the E. portion being clear and blue, the other of a muddy yellow colour. The course of the Rhone close to the city is from N.N.E. to S.S.W.: it has a medium breadth of about 650 feet, and is so liable to inundations that embankments have been formed to protect the town and its suburbs. From Lyons the united stream holds a course nearly due S. to the Mediterranean, receiving numerous streams both from the W. and E., but chiefly the latter: the Isère, a considerable river, rising on Mount Cenis, joins it between Tournon and Valence; and near Avignon (where the Rhone is 117 feet above the sea), it is joined by the Durance, a swift and turbid stream, which collects the waters from the western face of the maritime Alps.

The Rhone enters the Mediterranean by four mouths: the first separation occurs at Arles, where two branches are formed; one called the great Rhone, running S.E., the other known as the little Rhone, pursuing a S.W. course, and both together enclosing the alluvial island of Camargue, which has an area of about 1900 sq. m. Each of these again bifurcates a short distance above the mouth; but the E. channels are those only which admit of safe navigation. The Rhone, which has a very rapid course Rhodanus ferox), brings down a whitish sediment, discolouring the Mediterranean to a distance of 6 or 7 m., and here is every reason to suppose that there has been a constant, though slow, advance of the base of the delta during the last 18 centuries: indeed, Mese (an. Mesua Collis), stated by Pomp. Mela to be almost an island, is now far inland, and Notre Dame des Ports, a harbour in the 9th century, is now a league from the sea. The confluence of the Rhone with the currents of the Mediterranean forms bars cross the mouths of the river, and by these means considerable spaces become divided off from the sea, and subsequently from the river also, when it shifts its channels of flux. Some of these lagoons being subject to the occasional ingress of the river when flooded, and of the sea during storms, are alternately fresh and salt. Others, after being led from the sea, become more salt by evaporation, and are, in fact, natural salterns. The sea, opposite the mouth of the Rhone, deepens gradually from 4 to 40 fathoms within a distance of 6 or 7 m., the bottom being characterised by very curious alternations of marine and freshwater shells. [Lyell's Geology, i., 341–345.)

Among the tributaries of the Rhone, by far the most important is the Saone (an. Arar), which rises on the Vosges, that mass of high land which gives origin also to the Murthe, Moselle, and Meuse: the source is at Viomenil, 10 feet above the sea. Its course is tortuous, though generally S.S.W. as far as Chalons; 18 m. above which it receives on its E. bank the waters of the Doubs. This river is at an elevation of 2630 feet above the sea, in one of the longitudinal valleys of the Jura mountains; and after turning N.N.E. for about 60 m., is suddenly deflected southward by mount Terrible, whence its course is S.W., past Besancon, to its confluence with the main river. The general direction of the Saone from Chalons is S. by W., past Macon, Trevoux, &c.; and the average fall from the former is to its junction with the Rhone at Lyons scarcely exceeds 1 ft. per mile; and the channel being extremely tortuous, it has a very slow current. (Infinit incredibile lenitus ut occultis, in utram partem fluat, judicare non possit. Amm. lib. i., cap. 10.) Its waters are charged with marl, and the débris of Jura limestone. The Isère and Durance have already been mentioned. They are extremely rapid, charged with deposit from the secondary strata forming the main ridge of the Alps. The only considerable tributary is the Ardèche, rising in the Cevennes, not far from the Loire.

Owing to the rapidity of its current, the Rhone was formerly of comparatively little use in navigation, the principal trade being carried on by vessels down the stream, only from Lyons to Beaucaire, the boats that ascended ever being usually dragged up by horses. But since the introduction of steamers, a vast impulse has been given to navigation; the Rhone, Saone, and their greater affluents have become important channels of intercourse; and basins are now improving more rapidly than any other of France. We extract the following details from the interesting work of Mr. Maclaren, who sailed down the Saone and Rhone, from Chalons to Marseilles:

"I left Chalons at six A.M., and reached Lyons by the boat at two, distant about 85 m. The fare was only 3 francs. From Lyons to Avignon the fare is 30 francs, distance about 170 m.; time about 13 or 14 hours per boat. The steamboats are English built, and worked by English engineers. They are about 100 feet long, 25 and are neat and commodious, though by no means rapid. They are very flat in the bottom, drawing only 24 inches of water. The Rhone is full of sandbanks,

which, owing to the rapidity of the current, are continually shifting, and vessels drawing a greater depth than 2 feet cannot navigate the river with safety. The English engineer told me that his steamboat went about 9 m. an hour in still water; the current adds about 5 m. going down, and deducts as much going up; so that she moves at the rate of 14 m. the one way, and 4 the other.

"The Rhone passes through one of the most beautiful, picturesque, and delightful regions in the world. It is one continued vineyard, skirted and sheltered by mountains from 500 to 2000 feet in height, presenting every variety of form and aspect—now round and smooth—now rugged and peaked—now bare and sterile—now clothed with vines or mulberries, or cut into terraces, and carefully cultivated to their summits. Mount Pilatre, and others of the high Cevennes, on the W., and the Alps on the E., capped with snow, appear through openings in the lateral chains at intervals. The valley is often contracted to a space of one mile; again it spreads out in the form of a rich plain, to a breadth of 10 or 12 m. Twenty times the river appears closed in by the hills, and you are puzzled to conjecture where it escapes, till a bend in its course clears up the mystery. To the natural beauties of the country are added those which it derives from the industry and taste of its inhabitants. There is an almost unbroken line of large towns, villages, hamlets, cottages, and neat villas, along each side of the river, and not merely on its banks, but reaching back to the mountains. The glittering white walls of these buildings, surrounded by rich, well cultivated lands, give vivacity to the scene, and fill the mind with images of peace, abundance, security, and contentment. From Chalons to Marseilles, the marks of progress meet the eye everywhere. The whole district, 270 m. in length, is advancing with giant strides. At every step, in descending the river, we meet with houses or manufactories building, quays forming on the river, bridges erecting, roads or towing paths, or embankments making. Two facts will show that this is no exaggeration. It is only eight years since steamboats were introduced; and there are now six plying between Chalons and Lyons, and ten between Lyons and the sea, though the coal they use is brought from England, and costs about 50s. a ton. It is but ten years since suspension-bridges were heard of in the district; yet in the short intervening space, the industrious inhabitants of these districts have erected fourteen suspension-bridges over the Saone, and twelve over the Rhone. I doubt if there be as many at this moment in all England. And let it not be supposed that the bridges are paltry, or rude and imperfect works. They are light and elegant; the arches are often of great span, and the pathway is either level, or slightly and uniformly curved. Sometimes they have two arches, but in general three. The chains are sometimes single, but more commonly triple or quadruple, the suspension rods always single. The columns on the piers are sometimes slender obelisks of stone, sometimes thin tall slabs of cast iron. Taking them altogether, they are the lightest and most handsome structures I ever saw, and show great boldness, as well as skill and taste, in the engineer. But the fact on which I wish to fix attention is the enterprise, and the wealth which these works bespeak. The 26 bridges must have cost at least £200,000; and this sum has been raised by the public spirit of the district, and expended on one species of improvement, while many others were in progress." (Notes on France, &c., p. 32–37.)

The basin of the Rhone is connected by canals with the other principal rivers of France. The canal of the Rhone and Rhine connects the Doubs with the Ill, an affluent of the Rhine; the Canal du Centre unites the Saone to the Loire, and the canal of Burgundy connects the Saone with the Yonne, a navigable branch of the Seine; so that in this way the four principal rivers of France are all united. The navigation of the mouth of the Rhone is considerably improved by the canal of Arles, which runs close to the Great Rhone, and by the canal of Beaucaire, which leaves the river near the town of that name, and not only runs into the Mediterranean, but is joined by branch canals with the Canal du Midi, connecting the Garonne and the bay of Biscay with the Mediterranean. The Rhone is also connected with the Loire by a railway, which passes through the important manufacturing town of St. Etienne. (Bakewell's Travels, ii., 260–275; Lyell's Geol., i., 340–345; Encycl. Méthodique, art. Rhône.)

RHONE, a dep. of France, and next to that of the Seine, the smallest in the kingdom, though from its containing the city of Lyons, it is among the most populous, reg. S., between lat. 45° 28' and 46° 18' N., and long. 4° 20' and 4° 55' E., having N. Saone-et-Loire, W. and S. Loire, and E. Ain and Isère, from which it is principally separated by the rivers Rhone and Saone. Length, N. to S., 55 m.; average breadth, about 20 m. Area, 279,081 hectares. Pop., in 1836, 434,429. Surface mostly mountainous, being covered with ramifications of the Cevennes. The dep. is well watered; but there are no navigable rivers, except the Rhone and Saone. In 1834 it was estimated that it had 143,120 hectares arable land, 36,399 ditto meadows, 30,552 vineyards, and 34,466 of woods. The produce of corn is far below the internal demand. The produce of wine, the chief source of agricultural wealth, is supposed to exceed 450,000 hectolitres a year. The wines produced in that portion of this dep. called the Beaujolais belong to the class of Macon wines. Of the other growths, probably the best is the Côte-Rôtie, a red wine raised near Ampuis. It is an excellent wine, possessing *du corps, du spiritueux, de la finesse, une sève et un parfum tres-agréable.* (Jullien, p. 184.) It requires to be kept in the cask for five or six years before bottling. The white wines of Condrieux are also very good. The forests produce fine chestnuts; but the quantity of timber they yield is insufficient for the demand of the important manufacturing districts round Lyons and St. Etienne. Fortunately coal is abundant. Neither horned cattle, nor sheep, are very numerous; but a great many goats are kept, and in some places they have been crossed with the breed of Thibet. From the milk of the goats on Mont d'Or, a cheese is made, which fetches a high price. The silk worm is reared in some places, but the culture of the mulberry appears to be diminishing. Rhone is rich in minerals. Besides coal, iron, copperas, argentiferous lead, barytes, manganese, fine marbles, &c., are obtained; and in this department are the two most productive copper-mines in France, those of Chessy and St. Bel.

The principal manufacture is that of silk stuffs, which is carried on upon a most extensive scale. It however, has been fully noticed under the art. LYONS, to which the reader is referred. It has also manufactures of muslins at Tarare, linen and cotton thread at Thizy and other towns, and of hardware, jewellery, glass, paper, paper hangings, chemical products, &c. In 1835, of 81,044 properties subject to the *contrib. foncière*, 31,884 were assessed at less than 5 francs, and 12,358 at from 5 to 10 francs; but 212 were assessed at upwards of 1000 francs, which, taking into account the small extent of the department, exhibits a much greater proportion than ordinary of the larger class of properties.

Rhone is divided into two arronds.; chief towns Lyons and Villefranche. It sends five members to the chamber of deputies. Number of electors (1836–39), 4951. Total public revenue (1831), 25,900,703 francs; expenditure, 18,801,371 francs. (See Hugo, art. *Rhone*; *French Official Tables.*)

RHONE, BOUCHES DU. *See* BOUCHES DU RHONE.

RIAZAN, a gov. of European Russia, between lat. 53° and 55° 40' N., and long. 38° 30' and 41° 15' E.; having N. Vladimir, E. and S. Tambof, and W. Tula and Moscow. Area, 14,968 sq. m. Pop., in 1838, 1,341,700. Surface generally flat. The Oka, running from W. to E., divides it into two unequal portions of very different aspect. The country S. of that river is the more elevated; the air wholesome, and the soil fertile: in the N., on the contrary, the country is generally low, marshy, and covered with woods, or destitute of culture. More corn is grown than is required for home consumption; the average produce being between 5,000,000 and 6,000,000 chetwerts a year. The forests, which are very extensive, cover above a third part of the surface: those belonging to the crown comprise about 428,000 deciatines, or 1,284,000 acres. Hops, tobacco, and garden vegetables are, in some districts, raised in large quantities. The proprietors of the pasture lands let them to graziers belonging to the Ukraine, who bring thither large herds. The breed of horses is good; the Russian government has a *dépôt d'Etalons* at Shopine. Bees are supposed to produce about 250,000 roubles a year. There are a few iron mines and stone quarries. Manufactures have made some progress. Those of glass and hardware occupy the first rank; and there are others of woollen, cotton, and linen fabrics, cordage, potash, soap, &c., with dyeing establishments, tanneries, and distilleries. A portion of the manufactured goods is sent to Moscow, and by way of the Oka, down the Wolga; but the principal exports are the raw products of the government, consisting of corn, cattle, honey, lard, iron, timber and wooden articles. The population is principally Russian; but partly of the Tartar stock. Riazan is subdivided into 12 districts; chief towns, Riazan the capital, Zaraisk, and Kacimof. Education is very backward. In 1832, the pupils at schools and other seminaries amounted to only 1 in 934 of the population. There was then, also, but one printing-press in the government.

RIAZAN, a town of European Russia, cap. of the above gov., on the Trouberge, a tributary of the Oka, 110 m. S.E. Moscow: lat. 54° 37' 41" N., long. 39° 15' 54" E. Pop., about 9000. It consists of two distinct portions; an irregular fortress, with an earthen rampart, including numerous churches, the episcopal palace, formerly the residence of the princes of Riazan, the consistory, &c.; and the town-proper, in which are also numerous churches, with a fine

edifice for the government-offices, several convents, a seminary and public library, hospital, &c. The town has greatly increased in size and importance within the last 30 years; but most of the houses are still of wood, and planks occupy the place of pavement in the streets. Riazan is the seat of a military governor, with authority over the governments of Riazan and Tambof, and of the chief judicial courts of its government. It has a gymnasium, to which a society of arts was attached in 1820; a school of drawing and architecture, founded in 1824, schools for the children of official persons, &c.; and several of the principal manufactures in the government. The old town of Riazan, destroyed by the Tartars in 1568, is distant about 33 m. S.E. (*Schnitzler, La Russie; Possart Russland, &c.*)

RIBEAUVILLE (Germ. *Rappoltzweiler*), a town of France, dep. Haut-Rhin, cap. cant., at the foot of the Vosges, 7 m. N. Colmar. Pop., in 1836, ex. com., 6361. Above it are the remains of the old castle of Ribeaupierre; and in the immediate neighbourhood are some other ruined fortresses, and the principal remains of the ancient wall, called the *Heidenmauer* ("wall of the Pagans") erected at a remote period along the top of the most E. range of the Vosges. It has manufactures of calicoes and cotton handkerchiefs.

RICHFIELD, p. t., Otsego co., N. Y., 13 m. N.W. Cooperstown, 72 m. W. Albany, 376 W. Bounded S.E. by Canderaga lake, into which its streams flow. It contains sulphur springs, and has five stores, one fulling-mill, two grist-mills, four saw-mills, four tanneries, 13 schools, 487 scholars. Pop. 1680.

RICHFIELD, p. t., Summit co., O., 134 m. N.E. Columbus, 347 W. It contains three churches, a Congregational, Methodist, and Baptist, one academy, 96 students; 9 schools; 353 scholars. Pop. 1108.

RICHFIELD. t., Huron co., O. It has 13 schools, 471 scholars. Pop. 1599.

RICHLAND, district, S. C. Situated in the central part of the state, and contains 630 sq. m. Bounded E. by Wateree river, S.W. by Broad and Congaree rivers. It contained in 1840, 3878 neat cattle, 3022 sheep, 14,789 swine, and produced 506,585 bushels of wheat, 65,854 of rye, 844,511 of Indian corn, 34,737 of buckwheat, 9337 of barley, 329,169 of oats, 965,784 of potatoes. 1070 pounds of tobacco, 4,123 of sugar. It had 67 stores, nine fulling-mills, eight woollen factories, eight flouring mills, 74 grist-mills, 168 saw-mills, two oil-mills, 25 tanneries, five potteries, two printing-offices, one weekly newspaper; one academy, 52 students; 90 schools, 6635 scholars. Pop.: whites, 5208; slaves, 10,664; free coloured, 407; total, 16,387. Capital, Columbia.

RICHLAND, county, O. Situated toward the N. part of the state, and contains 900 sq. m. Drained by branches of Mohiccan creek and of Olentangy river. It contained in 1840, 34,930 neat cattle, 79,786 sheep, 51,448 swine, and produced 506,585 bushels of wheat, 65,854 of rye, 844,511 of Indian corn, 34,737 of buckwheat, 9337 of barley, 329,169 of oats, 965,784 of potatoes. 1070 pounds of tobacco, 4,123 of sugar. It had 67 stores, nine fulling-mills, eight woollen factories, eight flouring mills, 74 grist-mills, 168 saw-mills, two oil-mills, 25 tanneries, five potteries, two printing-offices, one weekly newspaper; one academy, 52 students; 90 schools, 6635 scholars. Pop. 44,532. Capital, Mansfield.

RICHLAND, p. t., semi-capital of Oswego co., N. Y., 155 m. W.N.W. Albany, 385 W. Bounded W. by lake Ontario. Pulaski village contains the courthouse and jail. It is a building of brick. Watered by Salmon river. It contains three churches, a Presbyterian, Methodist, and Baptist. 17 stores, two fulling-mills, one woollen factory, three grist-mills, 15 saw-mills, one oil-mill. three tanneries; one academy, 64 students; 37 schools, 866 scholars. Pop. 4850.

RICHLAND, t., Venango co., Pa. It has four stores, one fulling mill. three grist-mills, four saw-mills, two tanneries. Pop. 1385.

RICHLAND, t., Bucks co., Pa., 37 m. N.W. Philadelphia. Watered by Tohickon creek and its branches. It contains a Friends' church, five stores, two grist-mills, four saw-mills, two tanneries, three potteries; three schools, 85 scholars. Pop. 1781.

RICHLAND, L., Belmont co., O. It contains St. Clairsville, the capital of the county, which see. The National road passes through it from E. to W. It has two academies, 42 students; 12 schools, 294 scholars. Pop. 3748.

RICHLAND, t., Clinton co., O. It has three stores, two grist-mills, one saw-mill; 10 schools, 551 scholars. Pop. 1385.

RICHLAND, t., Fairfield co., O. Pop. 1982.

RICHLAND, t., Guernsey co., O., 95 m. E. by N. Columbus. Pop. 1777.

RICHMOND, a parl. and mun. bor., market town, and par. of England, the cap. of a district called Richmondshire, having a separate jurisdiction. W. div. wap. Gilling N. Riding, co. York, on the declivity of a hill rising from the Swale, which half encircles the town, and is crossed here by a stone bridge, 11 m. S.W. Darlington, and 41 m. N.N.W

York. Area of parl. bor., which comprises the pars. of Richmond and Aske. 5690 acres. Pop. of town, in 1831, 4000; ditto, in 1841, 4765. The town, which is most picturesquely situated, commands, from many points, very fine views of the Swale, its bold rocky banks, and the well-wooded country around; and its appearance is made more imposing by the ruins of its castle and keep, built on a rock above the river. The streets are irregularly laid out; but a very fine broad avenue leads from the Darlington and Northallerton roads into an extensive market-place, surrounded by the principal shops, town-hall, chapel-of-ease, and one of the largest hotels in the N. of England (comprising the assembly-rooms). In the centre of the market-square is a column, or cross, under which was a reservoir for water, brought by pipes from the neighbouring hamlet of Aislebeck; but the supply being deficient, a much larger quantity has recently been brought from Coalgate, and a new and larger reservoir has been constructed for its reception. The houses are mostly built of a reddish sandstone; and the town, being well paved, lighted with gas, and kept remarkably clean, has a very neat appearance. The church, built on a slope facing the river, is principally in the perpendicular style, with a fine pinnacled tower; but some portions are clearly of an older date. The living is a rectory in the gift of the lord chancellor; but the perpetual curacy of the chapel-of-ease has, till lately, been in the patronage of the corporation. The Wesleyans and Baptists have places of worship, and there is a large Roman Catholic chapel with an attached school. The town has a national school, attended by about 200 boys and girls, an infant school, and Sunday schools. A free grammar school, founded in 9 Elizabeth, is well endowed with property under the management of the corporation, who appoint its head master. All natives, and the sons of residents within the borough, are admitted gratis. The number of day-scholars, however, seldom exceeds 20; besides whom the master is allowed to take boarders. The school recently attained considerable distinction, having been till lately under the superintendance of the eminent scholar, the Rev. James Tate, whose eldest son is now (1841) head master. Another free school, for commercial purposes, is under the control of the corporation: in whose hands, also, are several charity estates including endowments for almshouses. See *Char. Comm., 7th Report.*) A scientific society and mechanics' institute have attached libraries, and there is a savings' bank. " There are no manufactures of importance at Richmond. Its market is a very considerable one for corn; but the quantity brought thither has recently decreased, and the prospects of the town are not improving. the failure of the lead-mines (about 12 m. higher up the river), the supply of corn from Ireland, and the facilities given by the railroad between Liverpool and Manchester, contribute to its decline." (*Mun. Corp. Rep.*) It is, however, inhabited by many wealthy families, who, with the neighbouring gentry, cause a pretty extensive retail trade. The training of race-horses is also a considerable business; and races are held annually in October, about a mile from the town, on a high moor, which has a commodious grand stand. About 1½ m. N.N.W. is Aske hall, a seat belonging to the Earl of Zetland; and near Catteick is Brough hall, the residence of Sir W. Lawson. Richmond, which received its first royal charter in 2 Hen. III., and was incorporated in the 19 Eliz., is governed, under the Municipal Reform Act, by a mayor, three other aldermen, and 12 councillors. Corporation revenue in 1840, 560l. Quarter and petty sessions for the borough are under a recorder, and petty sessions for the wap. of Gilling take place on alternate weeks. A civil court for the recovery of debts under £100, another for the dist. of Richmondshire, and a court leet for the manor, are occasionally; and there is a monthly ecclesiastical court under the Archdeacon of Richmond. The borough returned two members to the House of Commons since 27 Eliz., the right of election down to the Reform Act in the holders of burgage tenures. The Boundary enlarged the limits of the borough by the addition of parish of Easeby; and in 1839-40 it had 289 registered electors. It is also one of the polling-places at elections for the riding of the county. Market on Saturday; three chartered and other fairs.

history of Richmond is closely associated with that castle, founded by Alan, the first earl of Richmond, having received from William the Conqueror the forfeited estates of the Earl of Mercia, built the castle and the town of Richmond, to protect his family and property. Under the Norman monarchs, the title and earldom were possessed by different families, allied to the royal; and in the contest between the houses of York and Lancaster, they also several times changed possession, till at length they were vested in the crown by the attainder of Henry, Earl of Richmond, to the throne, under the name of Henry VII. Since this epoch, the castle has

been allowed to fall into partial decay. It still, however, bears marks of its former grandeur and importance. The keep tower, of which the walls are nearly entire, is a Norman structure, about 100 feet in height, the walls being 11 feet thick; the lower story is supported by a vast column in the middle, from which spring circular arches, closing the top. The floors of the two upper rooms have fallen in; but a modern wooden staircase leads to an older flight in the walls, by which the visiter may reach its summit, which commands an extensive and beautiful view of the surrounding-country. The ruins of several other parts of the castle still remain, and latterly they have been partially repaired. In the S.E. corner of the area is a ruinous tower, in which is a dungeon, 13 or 14 feet deep. The ground covered by the castle comprises nearly six acres: it belongs to the Duke of Richmond and Lennox, on whose ancestors it was conferred, with the title of duke, by Charles II. Here are also the ruins of a monastery of Grey Friars, the steeple of which is a remarkably fine specimen of the perpendicular style; and at Easeby are extensive and highly interesting remains of an abbey, built in the 19th century, for Premonstratensian monks: the buildings are chiefly in the early English style; but the windows both of the chapel and refectory have some very elegant tracery, well worth examination. (*Rickman's Arch.*, p. 276.)

RICHMOND (the Tivoli of London), a town and par. of England, upper div. hund. Kingston, co. Surrey, at the bottom, and on the slope and summit of an eminence rising from the S. bank of the Thames (crossed here by a handsome stone bridge of five arches), 10 m. W. by S. London. Area of par. 1230 acres; pop., in 1831, 7243. The principal street extends the whole length of the town, running along the ridge on its W. side to the summit of the hill, and being, in the lower parts, parallel to the river. The other streets are of very inferior importance. The houses in the lower part of the town are old-fashioned, and by no means large; but on the hill and outskirts are many handsome mansions occupied by families of distinction. The houses in the terrace and the Star and Garter hotel, at the summit of the hill, command very extensive and noble prospects of the Thames and its rich valley, Windsor castle being distinctly seen in the distance.

> " Here let us sweep
> The boundless landscape. * * *
> Heavens ! what a goodly prospect spreads around,
> Of hills and dales, and woods and lawns, and spires,
> And glittering towns and gilded streams, till all
> The stretching landscape into smoke decays."
>   *Thomson's Summer.*

The church is a respectable structure, with a low embattled tower, and, till 1568, was a chapel-of-ease to Kingston, to the rector of which parish the patronage still belongs. It has been much enlarged and repaired within the last century; and contains the remains of Thomson, the poet of the *Seasons*, who died here on the 27th of Aug. 1748, of Dr. Moore, author of *Zaluco*, and of *Views of Society in France, Italy, &c.*, Gilbert Wakefield, the scholar, Kean, the actor, &c. A new district church has also been recently erected in the pointed style. There is a Roman Catholic chapel, and the Independents, Baptists, Wesleyan, and Calvinist Methodists, have places of worship, to most of which Sunday schools are attached. A charity school, founded here in 1713, and subsequently endowed, furnishes gratuitous instruction to about 250 poor children, 60 of whom receive clothing. There are also three sets of almshouses, with pretty considerable endowments, and numerous money charities. A theatre is open during summer, and there are several excellent hotels, that on the hill, already alluded to, being a very extensive establishment in a magnificent situation.

Richmond is infinitely more a place of pleasure than of business, though it has a considerable retail trade for the supply of the resident families and visiters. It is a principal resort of visiters from London during the summer months, conveyed thither partly by coaches and partly by steamers and row boats; on the whole, however, the influx of visiters to Richmond, at least of the middle and lower classes, has materially diminished since the establishment of steamers on the river to Greenwich, Woolwich, Gravesend, &c.

Richmond park, the principal entrance to which, is at the W. end of the terrace contiguous to the Star and Garter hotel, formed by Charles I., comprises an area of 2253 acres, being about 8 m. in circumference. It consists mostly of poor soil; but has a great variety of surface, is well stocked with deer, and perfectly open to the public. The great lodge, which stands on rising ground, commanding a fine prospect, was built by Sir R. Walpole, ranger during the reigns of George I. and II., at an expense of £14,000. Here, also, is a new or stone lodge, built for a hunting seat by George I., and given by George III. to Lord Sidmouth, by whom it is still (1841) occupied.

What is called the *Old Park* extends along the Thames

from Kew to Richmond, and includes the royal gardens and pleasure grounds attached to Kew palace (see Kew). This park is, however, inaccessible to the public, except on certain days; and is considered as belonging rather to Kew than to Richmond

The *Green*, at the bottom of the town, forms a parallelogram almost as large as Lincoln's Inn Fields; it is used in summer for playing matches at cricket, bowls, &c.: on one side of the green is a handsome public walk.

Richmond (formerly called *Sheen*), was for centuries the site of a royal palace; but at what period it was erected is uncertain, though it became a fixed residence of royalty as early as the 14th century. Henry V. rebuilt it in a magnificent style. Henry VII. held a tournament here, in 1492, soon after which, the building having been destroyed by fire, a new palace was erected by that monarch, who gave the manor its present name, being that of his own title, previously to his accession to the crown: he died here in 1509. But its principal distinction consists in the fact, that when the emperor Charles V. visited England in 1522, he was lodged in this palace. Queen Elizabeth was imprisoned in it by her sister Mary, and it afterward became one of her favourite palaces, and here she died, on the 24th March, 1603. During the commonwealth, the palace was sold by the parliamentary commissioners for £10,783; and the whole appears to have been then dismantled and demolished, nothing now remaining except a few of the out-offices, its site being occupied by several modern mansions held on lease from the crown. On the N. side of the palace once stood a monastery, founded by Henry V., for Carthusian monks, the revenues of which, at the dissolution, were estimated at £953. A Franciscan convent, founded here in 1499, by Henry VII, was suppressed with the other in 1534.

The custom of Borough English, by which, in the event of the father's dying intestate, lands descend to the youngest son, or, in default of heirs male, to the youngest daughter, prevails in the manor of Richmond.

RICHMOND, county, N.Y., situated in the S. part of the state, and contains the whole of Staten Island, having an area of 53 sq. m. It is divided into four townships, and has a communication many times a day by steamboats, to New-York. It contained in 1840, 2517 neat cattle, 136 sheep, 3180 swine; and produced 18,989 bushels of wheat, 8865 of rye, 36,347 of Indian corn, 4538 of buckwheat, 5819 of barley, 33,729 of oats, 47,712 of potatoes. It had 49 stores; capital invested in the fisheries $36,000; one dyeing and printing works, 14 flouring-mills, eight grist-mills, one saw-mill, one printing-office, one weekly newspaper; one academy, 25 students; 14 schools, 604 scholars. Pop. 10,965. Capital, Richmond.

RICHMOND, county, Va., situated in the E. part of the state, and contains 200 sq. m. Bounded S.W. by Rappahannoc river. It contained in 1840, 6996 neat cattle, 4736 sheep, 14,329 swine; and produced 43,234 bushels of wheat, 1014 of rye, 231,493 of Indian corn, 25,990 of oats, 5434 of potatoes, 5738 pounds of tobacco, 9252 of cotton. It had 12 stores, eight grist-mills, three saw-mills; nine schools, 170 scholars. Pop.: whites, 3092; slaves, 2363; free coloured, 510; total, 5965. Capital, Richmond, C.H.

RICHMOND, county, N.C., situated in the S. part of the state, and contains 540 sq. m. Bounded W. by Yadkin river, N.E. by Lumber river. It had in 1840, 10,653 neat cattle, 7996 sheep, 18,841 swine; and produced 20,347 bushels of wheat, 1649 of rye, 247,169 of Indian corn, 12,242 of oats, 31,171 of potatoes, 4715 pounds of tobacco, 1,794,154 of cotton. It had eight stores, one cotton factory with 1600 spindles, 13 flouring-mills, 40 grist-mills, 15 saw-mills; six academies, 219 students; eight schools, 233 scholars. Pop.: whites, 4693; slaves, 3880; free coloured, 336; total, 8909. Capital, Rockingham.

RICHMOND, county, Ga., situated in the E. part of the state, and contains 384 sq. m. Bounded N.E. by Savannah river, S.W. by Briar creek. The railroad from Augusta to Decatur crosses its N. part. It contained in 1840, 5449 neat cattle, 758 sheep, 10,797 swine; and produced 1670 bushels of wheat, 183,015 of Indian corn, 1912 of oats, 28,079 of potatoes, 692,275 pounds of cotton. It had 12 commission houses in foreign trade, 265 retail stores, two furnaces, two cotton factories with 2000 spindles, one flouring-mill, 15 grist-mills, 19 saw-mills, two printing-offices, two daily, four weekly, and two semi-weekly newspapers, and two periodicals; eight academies, 294 students; four schools, 36 scholars. Pop.: whites, 5650; slaves, 6096; free coloured, 186; total, 11,932. Capital, Augusta.

RICHMOND, city, port of entry, capital of Henrico county, and of the state of Virginia, is in 37° 32' 17" N. lat., and 77° 27' 28" W. long. from Greenwich, and in 0° 26' 4" W. long. from Washington.' It is 23 m. N. from Petersburg, and 117 W. from W.' The population in 1800 was 5737; in 1810, 9785; in 1820, 12,067; in 1830, 16,060; in 1840, 20,153, of whom 7509 were slaves.

It is pleasantly situated on the N. side of James river, immediately below the falls, at the head of tide water, directly opposite to Manchester, with which it is connected by two bridges. Manchester may be regarded as a suburb of Richmond. Occupying a site between the upper and the lower country, it is remarkably healthy, being free from the peculiar diseases of both, and the deaths annually do not exceed one in 85 of its population. The place presents a variety of surface, and affords from many of its parts highly picturesque views. The streets generally cross each other at right angles, and are mostly 65 feet wide; but some are narrower, and others are wider. The part of the city which is built on is about three miles long, and three fourths of a mile wide; but as laid out, it contains about 3 square miles, much of which is not built upon. The city is divided into two unequal parts by Shockoe creek, an active stream, which enters James river about 800 feet east of Mayo's bridge. Shockoe hill, in the W. part of the city, and on which are beautiful and airy residences, and Richmond hill, stand opposite to each other, with the creek between them, and near the eastern extremity of the city is Church hill, a commanding eminence. The buildings are spread over these elevated grounds, and the valley between them, declining to the river. The city contains about 1200 dwellings, more than a thousand of which are of brick, covered with slate, and the remainder are of wood. The buildings are chiefly plain, with more regard to convenience than to architectural ornament, though there are exceptions to this remark. Shockoe hill, which is a favourite place of residence, is an elevated plain; and near its brow is Capitol-square, a beautiful public ground, containing about nine acres, surrounded by a handsome iron railing, and ornamented with gravelled walks, and shaded with a variety of forest and other trees. In the centre of this square stands the capitol or statehouse, in a conspicuous and commanding situation, having a portico in front, with an establishment, supported by lofty Ionic columns, of fine proportions, and an imposing appearance. It was constructed after a model brought by Mr. Jefferson from Nismes in France. In an open hall in the centre of the building within, is a marble statue of Washington, by Houdon, a French artist. This statue was erected in 1788, during the lifetime of Washington; and on its pedestal is the following inscription, from the pen of Mr. Madison. "The General Assembly of the Commonwealth of Virginia have caused this statue to be erected, as a monument of affection and gratitude to George Washington, who, uniting to the endowments of the hero the virtues of the patriot, and exerting both in establishing the liberties of his country, has rendered his name dear to his fellow-citizens, and gives the world an example of true glory." In a niche in the wall of the room, is a bust of La Fayette. The City hall, which is contiguous to the state house, is a costly and elegant edifice of Grecian architecture, having a portico at each end, with four Doric columns. It contains accommodations for the city courts, the common council-room, and several offices. The penitentiary is situated in the western suburbs of the city, and is an immense building, surrounding a hollow square, about 300 feet long from north to south, and 110 feet broad from east to west. The connected grounds belonging to it consist of several acres, which are enclosed. There is a county and a city jail; and an armory, with an edifice 325 feet long, and 260 feet wide. The almshouse is in the northern suburb of the city, and has a spacious edifice, well adapted to its purpose. There is a female orphan asylum, supported in part by funds of the corporation, and partly by private liberality, which is a highly useful institution. There is a public school for the education of poor children of both sexes, founded in 1816, with a commodious edifice, which is under the superintendence of trustees appointed by the city council, and is sustained by an annual appropriation from the literary fund of the state, and an appropriation from the city treasury. There are three market-houses in the city. Among the public buildings of the place is the Virginia Historical and Philosophical society, which has its seat in Richmond, though its members are scattered over the state. It was founded in 1831, and has been incorporated. There is a theatre in Richmond, which, however, is not extensively patronized. A great blow was given to theatrical exhibitions by the destruction by fire, during a performance, of the old theatre of Richmond, which occupied the site of the present Monumental Episcopal church, which has the following inscription on a monument on its west side. "In memory of the awful calamity that, by the providence of God, fell on this city, on the night of the 26th of December, in the year of Christ 1811, whereby the sudden and dreadful conflagration of the Richmond theatre, many citizens of different ages, and of both sexes, distinguished for talents and for virtues, respected and beloved, perished in the flames, and in one short moment public joy and private happiness were exchanged into universal lamentation, this monument to

erected, and the adjoining church dedicated to the worship of Almighty God, that in all future times, the remembrance of this mournful event on the spot where it happened, and where the remains of the sufferers are deposited in one urn, may be united with acts of penitence and devotion. Above sixty killed and many others maimed." This event deeply affected the whole country. Among those who perished, was George William Smith, Esq., governor of the state. The city water works are among the most splendid and useful establishments of the city. They were commenced in 1830, and cost about $120,000. Two forcing pumps, moved by water-power, raise 400,000 gallons of water each, from James river, in 24 hours, into three reservoirs, containing each 1,000,000 gallons, whence it is distributed over the city in pipes, along which are fire hydrants, at convenient distances.

Richmond contains 16 churches, three Episcopal, two Presbyterian, three Methodist, three Baptist, a Campbellite, Unitarian, Friends, Roman Catholic, and a Jews' synagogue. Some of the churches have large and elegant edifices. There are three banks, two of which occupy different apartments in the same fine building; and there are two insurance offices.

Richmond is about 150 miles from the mouth of James river, by the course of the channel, and between 50 and 60 m. above City Point. Vessels requiring 14 feet of water come to the bar, about 6 m. below the city; above that those requiring 10 feet of water come to its wharves, at ordinary tide. The channel of the river is winding, and the distance of the place from the ocean forms some impediment to navigation. The tide rises at Richmond about 4 feet. Four steamboats are employed in towing vessels to and from City Point. There are owned in Richmond two ships, two barques, three brigs, and three schooners; and as fewer than 100 vessels visit the port, in the course of the year. A line of five schooners sails once a week to Petersburg; and a line of five schooners once a week to New-York. Three steamboats form a line for passengers to Norfolk; and two steam packets form a line to Baltimore. The exports from Richmond consist principally of flour, tobacco, and coal; and the total amount of its cargoes to foreign countries and coast-wise, amount to about 000,000, annually. The tonnage of the port in 1840, was 6911 tons. A canal extends from Richmond 116 miles to Lynchburg, sometimes occupying the bed of the river, and is in progress beyond that place. At Richmond it is designed to enter a basin at the head of tidewater, which serves the shipping. A railroad extends from Fredericksburg through Richmond to Petersburg, and thence to Weldon Roanoke river, where it connects with other southern railroads. At Richmond it crosses James river on an raised bridge erected for the purpose. But the commercial facilities of Richmond, are exceeded by its manufacturing advantages. In the falls in James river, extending about 6 miles, it has an immense water-power; and this greatly, but is but partially improved. It has extensive flouring-mills, and iron works, with which are connected rolling and slitting-mills, and nail factories, and it is a very extensive cotton factory. According to the census of 1840, there were 17 commercial and 29 commission houses in foreign trade, with a capital of $3,063,000; retail dry goods and other stores, with a capital of 3,456; three lumber-yards, with a capital of $94,000; furnaces produced 600 tons of cast iron, and eight rolling-mills, &c., produced 1700 tons of bar iron, using 4100 tons of fuel, the whole employing 280 persons, and a capital of $317,900; 100 persons produced machinery to the amount of $198,000; nine persons manufactured the precious metals to the amount of $10,800; one factory with 5810 spindles, produced to the amount 900, employing 269 persons, with a capital of $175,000; persons manufactured tobacco to the amount of $693,340, capital of $492,250; 17 persons manufactured hats as to the amount of $15,000, with a capital of $3000; saddleries and other manufactories of leather produced amount of $6000, with a capital of $3800; 13 persons manufactured 182,000 pounds of soap, and 95,000 pounds of candles, with a capital of $13,000; 96 persons manufactured furniture to the amount of $81,000, with a capital of $39,000; 79 persons manufactured carriages and to the amount of $45,000, with a capital of $60,500; persons produced confectionary to the amount of with a capital of $2500; 21 flouring-mills produced barrels of flour, and with two grist-mills and three mills, employing 69 persons, produced to the amount of with a capital of $461,000; 155 persons produced wild liquor to the amount of $55,000; one paper mill, using 50 persons, produced to the amount of $55,000; were built to the amount of $5000 · eight printing-one bindery, two daily, six weekly, and two semiweekly newspapers, and one periodical employed 79 persons. capital of $48,700; 43 brick houses and five

wooden houses were built by 233 persons, and cost $304,500; the total amount of capital employed in manufactures was $1,372,950. There were one college, 63 students; 13 academies, 733 students; 14 schools, 503 scholars.

Richmond is governed by a mayor who is elected by the city councils, a recorder, and eleven aldermen. The recorder and aldermen are chosen from 27 individuals elected by the people, and the remaining 15 of these individuals, constitute the city council.

Richmond was founded by an act of the state legislature in 1742; and the seat of government was removed to this place from Williamsburg in 1780. In 1787, it contained about 300 houses. The canal around the falls in James river, which added so much to the commercial advantages, was completed in 1794. Its extension to Lynchburg since 1825, with other facilities of communication with other places, has added greatly to these advantages.

RICHMOND, p. t., Lincoln co., Me., 15 m. S. by W. Augusta, 585 W. Bounded E. by Kennebec river. It contains 12 stores, one grist-mill, one saw-mill; one academy, 50 students; nine schools, 472 scholars. Pop. 1804.

RICHMOND, p. t., Cheshire co., N. H., 61 m. S.W. Concord, 424 W. Chartered in 1752. Drained by branches of Ashuelot river. It has 15 schools, 418 scholars. Population 1165.

RICHMOND, p. t., Chittenden co., Vt., 13 m. S.E. Burlington, 27 m. W.N.W. Montpelier, 513 W. Incorporated in 1794. Watered by Onion or Winooski and Huntington rivers, which afford water-power. It contains a church with 16 sides, and a steeple rising from the centre, occupied by various denominations, three stores, one fulling-mill, one woollen factory, six saw-mills, two tanneries; eight schools, 210 scholars. Pop. 1054.

RICHMOND, p. t., Berkshire co., Mass., 142 m. W. Boston, 369 W. Drained by a branch of Housatonic river. It has four stores, three saw-mills; one academy, nine students; five schools, 183 scholars. Pop. 1097.

RICHMOND, t., Washington co., R. I., 30 m. S.S.W. Providence. Bounded W. by Wood river, S. by Charles river, branches of which afford water-power. It contains two fulling-mills, three woollen factories, six cotton factories, with 7036 spindles, one grist-mill : two schools, 50 scholars. Pop. 1361.

RICHMOND, t., Ontario co., N. Y., 14 m. W. Canandaigua, 209 m. W. Albany. Honeoye lake lies partly in its S. part, by the outlet of which it is watered. It contains an Episcopal and a Presbyterian church, six stores, one fulling-mill, one woollen factory, two flouring-mills, one grist-mill, three saw-mills, one oil-mill, three tanneries, one distillery; one academy, 35 students; 14 schools, 641 scholars. Pop. 1927.

RICHMOND, p. v., capital of Richmond co., N. Y., 159 m. S. by W. Albany, 237 W. Situated partly in three townships, near the centre of Staten Island, and contains a courthouse, jail, two churches, a Presbyterian and Episcopal, three stores, 30 dwellings, and about 200 inhabitants.

RICHMOND, t., Berks co., Pa. Watered by Moselem creek, which affords water-power. It contains two churches, common to Lutherans and Presbyterians, five stores, one furnace, four grist-mills, three saw-mills. Pop. 1997.

RICHMOND CITY, p. v., Lake co., O., 177 m. N.E. Columbus, 351 W. Situated on the W. side of Grand river, 1½ m. from its entrance into lake Erie. Steamboats and lake vessels come to its wharves.

RICKMANSWORTH, a market town and par. of England, co. Herts., hund. Cashio, on the Gade, 18 m. N.E. London. Area of par., 9740 acres. Pop. in 1831, 4574. The parish church has been rebuilt within these few years, with the exception of the embattled tower, which belonged to the old church. The living, a vicarage, in the patronage of the Bishop of London, is worth about £510 a year. It has an endowed national school, and other minor and Sunday schools, and two almshouses. It is governed by two constables and two head boroughs. In its vicinity are several streams, on which considerable flour and paper-mills have been erected. Its trade is facilitated by the Grand Junction canal, which passes close by the town. Moor-park, a seat of the Marquis of Westminster, in the vicinity, was once the residence of Cardinal Wolsey; but it has since been rebuilt in the modern style; the grove, belonging to Lord Clarendon, and other seats, are also in the immediate neighbourhood.

RIDGEBURY, p. t., Bradford co., Pa., 18 m. N.W. Towanda. Drained by Bentley and South creeks, tributaries of Tioga river. It has seven schools, 260 scholars. Pop. 1914.

RIDGEFIELD, Fairfield co., Ct., 31 m. W. by N. New-Haven, 81 m. S.W. Hartford, 281 W. Drained by Saugatuck and Norwalk rivers, and the E. branch of Croton river. It contains three churches, a Congregational, Episcopal, and Methodist, 12 stores, one furnace, one fulling-mill, two grist-mills, four saw-mills; three academies, 46 students; 14 schools, 392 scholars. Pop. 2474.

RIDGEWAY, p. t., Orleans co., N. Y., 10 m. W. Albion, 262 m. W. by N. Albany. 400 W. Drained by Oak Orchard creek, which has a fall of about 30 feet, affording good water-power. The Erie canal passes through it. It contains four churches, a Baptist, Methodist, Episcopal, and Presbyterian, 90 stores, two fulling-mills, two woollen factories, one furnace, three flouring-mills, two grist-mills, 11 saw-mills, two tanneries, one brewery, two printing-offices, two weekly newspapers: one academy, 97 students; 21 schools, 989 scholars. · Pop. 3554.

RIDGEWAY, t., Bradford co., Pa. Drained by Clarion river and Toby's and Kersey's creeks. It has three stores, one grist-mill, 10 saw-mills; seven schools, 260 scholars. Pop. 1214.

RIGA, an important city and river port of European Russia, cap. of Livonia, on the Dwina or Dana, about 9 m. from its embouchure, in the gulf of Riga, lat. 56° 57′ 12″ N., long. 24° 0′ 4″ E. Pop. in 1835, including the garrison of 10,000 men, 67,338. About two thirds of the resident population are Lutherans, the rest consisting of members of the Russo-Greek church, Catholics, &c.

Riga is strongly fortified. It consists of the town, properly so called, and the suburbs; the former being entirely inclosed by the fortifications. Streets in the town narrow, crooked, and houses generally brick: in the suburbs, which are much more extensive, the streets are broad and regular, and the houses mostly of wood. One of the suburbs lies on the left bank of the river, the communication with it being maintained by a floating bridge about 2400 feet in length.

Among the public buildings are the cathedral, consecrated in 1211, and rebuilt in 1547; the church of St. Peter, built in 1406, with a tower 440 feet in height, being the most elevated in the empire, and commanding a fine view of the city and adjacent country; the castle, the seat of the chancellery, and of the general and civil governors; hall of the provincial states; town-house; exchange; arsenal, &c. A magnificent column, surmounted by a colossal bronze Victory, was erected in 1817, by the mercantile body, in honour of the emperor Alexander and the Russian army.

Among the literary establishments are a gymnasium, a lyceum, a school of navigation, and various elementary schools, a public library, an observatory, a society of Lettonian literature, &c. Manufactures of no great importance, though, of late, materially improved. In 1835 there were in the city 25 different manufacturing establishments, employing in all 1397 workmen, of which those of cotton, cloth, and rugs, were the most important. There are also various sugar-houses, tobacco manufactories, breweries, &c.

Owing to her situation on a large navigable river, Riga is the entrepôt of an extensive country; and is, in respect of foreign commerce, the next town in the Russian dominions to Petersburg. Corn used to be the principal article of export, but it is now far surpassed by flax and flax-seed, the exports of which have increased very rapidly. The other great articles of export are hemp and hemp-seed, timber, including masts and deals, hides, tallow, coarse linen, and canvass, &c. The total value of the exports in 1836 amounted to 47,735,684 roubles, of which flax constituted 23,829,920 roubles, flax-seed 8,731,763 ditto, and timber 3.563,763 ditto. In 1835, the exports of hemp amounted to 855,422 pods, but this is rather above the average. The imports consist principally of sugar, and other colonial products, dye-stuffs, wines, cotton, cotton-stuffs and cotton-yarn, woollens, salt, herrings, &c. Their total value in 1836 was 15,662,675, being less than a third part of the exports. The largest part by far of the trade is carried on with England. There is a bar at the mouth of the river which has usually from 12 to 13 feet water; and it is customary for vessels drawing more than this to load and unload the whole or a part of their cargoes at Boldema, a small part outside the bar. The entrance to the river, at Dunamunde, is guarded by a fort, where a ship is also the custom-house. The ships arriving at Riga vary from 1000 to 1500 a year. In 1836, the arrivals were 1102, and the clearances 1127, about a third part of which were English. The customs duties during the same year exceeded 8,000,000 roubles. If we may depend upon the official accounts, Riga has increased very rapidly. Under Catherine II., its population did not exceed 20,000; in 1824 it amounted to 29,906; and in 1835, as already stated, to 57,338, exclusive of the garrison. It has occasionally suffered considerably from inundations. (See the official returns published by the Russian government; Schnitzler, La Russie, &c., p. 568; the art. DWINA, in this dictionary. &c.)

RIGA, p. t., Monroe co., N. Y., 241 m. W by N. Albany, 262 W. Drained by Black cr., a tributary of Genessee r. The Tonawanda railroad passes through it. It has three stores, four fulling-mills, one woollen factory, one furnace, two grist-mills, six saw-mills, two tanneries, one distillery; two academies, 65 students; 14 schools, 774 scholars. Pop. 1984.

RINDGE, p. t., Cheshire co., N. H., 67 m. S.S.W. Concord, 435 W. Watered by branches of Contoocook and Miliers rivers. Incorporated in 1768. Iron ore is found, and a paint resembling Spanish brown. It contains two churches, a Congregational and Methodist, four stores, three grist-mills, seven saw-mills; 13 schools, 383 scholars, Pop. 1161.

RIMINI, (an. Ariminum), a city of Italy, Papal states, leg. Forli, between the rivers Marecchia (an. Ariminus), and Ansa (an. Aprusa), within about two miles of the embouchure of the former in the Adriatic, 28½ m. E.S.E. Forli, and about the same distance S.E. Ravenna. Lat. 44° ? N.; long. 12° 34″ 35″ E. Pop. about 12,500; but including the suburbs, and immediately contiguous villages, the pop. amounts to 27,000. (Rampoldi.) It is walled, and entered by several gates. A long street traverses it, terminating on the N. at the Porta di San Giuliano, whence emerges the Æmilian way, leading to Piacenza; and on the S. at the Porta Romana, where ends the Flaminian way, conducting thither from Rome. The town is well built, having some good streets, and handsome marble palaces; but, like most other Italian cities, it has a dull, melancholy, and deserted appearance. The cathedral, said to have been built on the ruins of a temple of Castor and Pollux, was restored and altered by Alberti in the 15th century, and has a very elegant exterior, to which, however, the interior offers a lamentable contrast. In it are some tombs of the Malatesta family, once lords of Rimini. According to Eustace, this building was converted by the French into a military hospital. The church of San Giuliano has a fine altar-piece by Paul Veronese, and several good paintings by Guido and other masters; in that of St. Augustine is a ceiling handsomely painted in fresco. It has a handsome town-house, theatre, a bowling green, and a good fish market, and in the Gambalunga palace is an extensive library, liberally thrown open to the public. The castle built here by Sigismund I. is falling into decay. The principal square is embellished with a fountain, and a statue of Paul V.; and in another of the open spaces is a small platform, or pedestal of a column, which, according to an apocryphal tradition, was the suggestum on which Cæsar harangued his troops after passing the Rubicon! At the S. entrance of the city is a triumphal arch, raised in honour of Augustus, and in most respects worthy of admiration; but Eustace says, that it was surmounted in the middle ages by a Gothic battlement, by which it is still disfigured. Both rivers are crossed by solid bridges; that over the Marecchia appears, from the inscriptions, to have been commenced by Augustus, and finished by Tiberius. It is of marble, 220 ft. in length (Rampoldi), and has five arches. Critics differ as to its merit as a work of art; but being still in excellent preservation, there can be no difference of opinion as to the solidity of its construction. Without the walls are the traces of an amphitheatre, and other remains of antiquity. The port of Rimini, on the Adriatic, at the mouth of the Marecchia, is so much obstructed by sand, as to admit small vessels only; but the fragments of marble facing piers, &c., attest that it was formerly of considerable importance.

Rimini has manufactures of silk, glass, and earthenware and the surrounding country being very fruitful, it has a considerable trade in corn, and furnishes quantities of this to the neighbouring towns.

In antiquity Ariminum was of far more importance than at present. It was the first considerable town on the E. coast of the peninsula, after crossing the Rubicon,; the boundary between Cisalpine Gaul and Italy, property so called,) and was regarded as one of the bulwarks of the latter. A Roman colony was planted in it anno U.C. 485, and at a subsequent period it received another colony. It was occupied by Cæsar immediately after he had crossed the Rubicon, and was honoured with the especial patronage of Augustus. It was given by Pepin, with the rest of Romagna, to the Holy See, to which, with few interruptions, it has since belonged. (Rampoldi, Corografia dell' Italia; Woods's Letters of an Architect; Cramer's Ancient Italy. i. 255, &c.)

RIO DE JANEIRO, or simply RIO, a city and seaport of Brazil, of which it is the cap., and the largest and most important commercial city of S. America, on the W. side of one of the finest bays in the world. 80 m. W. Cape Frio, lat. 22° 54″ 15′ S., long. 43° 13″ 56′ W. Pop. probably about 200,000, of whom about a half are whites, and the rest mostly negro slaves. The city, which is in the shape of a parallelogram, is situated on level ground, at the foot of hills, and has a fine appearance from the bay. The older portion, or that adjoining the sea, is divided on the W. from what may be called the new town by a large open space, the Campo da Honra. The style of architecture is in general mean, resembling that of the older parts of Lisbon; and though great improvements have been effected since the emigration of the court of Portugal to Rio in 1807, a great deal remains to be done before it be entitled to mak even with a second-rate European town. The streets, which are mostly straight, and intersect each other at right angles, are pretty generally furnished with *trottoirs*, and paved with blocks of granite. But though many of them have been widened of late years, and otherwise improved, they are still, for the most part, narrow and dirty, with a water-course in the centre, the usual receptacle of the filth from the houses. (*Three Years in the Pacific, by an American*, i. 97.) The houses, which are mostly of granite, or of granite and wood, are seldom more than two stories in height, rough, or whitewashed, with red tile roofs. They are narrow, but deep in proportion to the height; the lower story is commonly occupied by the shop or workhouse, and, in the houses of people of distinction, by the stable and coach-house: the second story (and third, if there be one) comprising the family apartments. The mildness of the climate, which is here a perpetual spring, rendering artificial heat unnecessary, there are no fire-places except in the kitchens, and, consequently, very few chimnies, which, to a stranger from Europe, gives the city a bald and, as it were, truncated appearance. The windows in the second story generally open upon iron verandas, the *jalousies* having been removed by order of government. In the outskirts of the town the streets are unpaved, and the houses of only one floor, low, mean, and dirty, with doors and windows of lattice work, opening outwards to the annoyance of the passengers. The rents of houses are scarcely as high as in London. Inside of the houses it is usual for all the apartments on the same floor to communicate above the partitions, which do not extend to the ceiling. This, though it destroys privacy, is advantageous, by allowing that free circulation of air so important in hot climates.

There are about 40 churches in the city; but none of them can be called fine buildings, or are worth the notice of travellers from Europe. The cathedral or church of Nossa Senhora da Gloria, on a lofty hill, on the S. side of the city, is a conspicuous object from a distance, and especially from the bay. There are several conventual establishments; *misericordia*, with an attached hospital; a foundling hospital; an institution where female orphans, born of white married parents, are educated and portioned off in marriage, with various other charitable institutions.

Water is conveyed into the city from a neighbouring lofty hill (2268 ft.), called the Corcovado, by a magnificent aqueduct, constructed in 1740. The water is thence conveyed to public fountains in different parts of the city; and a good many persons earn a livelihood by carrying water from these fountains to private families. Nothing, however, would contribute so much to the cleanliness and health of the city as an increase in the supply of water, and its distribution by pipes to private houses. The town is very indifferently lighted. There are but few inns and hotels, and those mostly very inferior; they are wholly for the accommodation of strangers, being rarely visited by the townspeople.

The royal palace forms two sides of a *large*, or oblong space, opening to the bay near the principal landing-place. It consists partly of the old palace of the viceroys, and partly of a convent formerly belonging to the Carmelites, and is wholly destitute of architectural beauty. Among the other public buildings may be specified a new and handsome theatre, the exchange, the old college of the Jesuits, and the episcopal palace and royal villa of Christovao, in the environs.

Neither education nor literature is here in a very flourishing state. There are, however, sundry, lyceums and grammar schools, and a great many private licensed academies. Among the principal educational institutions may ve specified a school of surgery, military and commercial academies, an academy of arts, a national museum, &c. n the vicinity is a very good botanic garden, comprising bout 4 acres, including, among other exotics, numerous specimens of the tea plant. Near the imperial palace is the ablic library, containing from 50,000 to 70,000 vols. Several daily and weekly newspapers issue from the Rio press; at they are said to be, without exception, the most worthless publications of their class any where to be met with. he police and health of the city have been materially improved within the last dozen years.

The market-place consists of a collection of filthy booths r the sale of vegetables, which are at once abundant, excellent and cheap. The farina of the *Jatropha Manihot* is re the grand substitute for bread stuffs, and is extensively ed by all classes, forming, with bananas, the principal od of the negroes. The yam supplies the place of the a toe; and the finest oranges are sold at from 10 to 25 its per 100. Butchers' meat is sold in shops, scattered up down the city. The beef, though wholly without fat, tender. Pork is good; mutton bad and dear. Fish undant and good. The slave-market, described by Basil l and others, no longer exists; but the clandestine importation of slaves is still carried on to a very considerable extent. (*Three Years in the Pacific*, i. 103.) Slaves, however, are, speaking generally, very indulgently treated. "We may," says a Rio circular, "assure our philanthropic friends abroad, that, as regards the slaves, they are here better fed, better clothed, are on the whole better treated, laugh more and weep less, than the major part of the labouring poor in Europe."

Commerce.—The trade of Rio is very extensive, and has increased rapidly during the last few years. It is now by far the greatest mart for the export of coffee. The shipments of this important article which, in 1830, amounted to 396,785 bags, have increased with such unexampled rapidity that, in 1839-40, they amounted to 1,095,346 bags, that is (taking the bag at 154 lbs.), to 168,683,384 lbs., or 75,305 tons! being nearly equal to all the exports of coffee from all the other ports in the world! Sugar is also an important article of export from Rio, though latterly it has been decreasing, and does not now exceed 10,000 cases (15 cwt. each); the exports of sugar from Santos are, however, increasing, and amounting, in 1830-40, to 624,750 arrobas. (*See* SANTOS.) The other great articles of export from Rio are hides, rice, tobacco, rum, tapioca, ipecacuanha, manioc flour and other inferior articles. The export of cotton has almost entirely ceased; and that of gold, diamonds, &c., is mostly clandestine, and too inconsiderable to be worth notice. We subjoin

An Account of the Quantities and Values (in Reis) of the the Principal Articles of Brazilian Produce, exported from Rio, in 1840.

	Reis.	Reis.
Coffee, 6,955,960 arrobas, at 2600	18,906,396	
Expenses, including duty and commission, 17¼ per cent	3,319,390	21,615,094
Sugar, 575,000 arrobas, at 2100	1,207,500	
Expenses, as above, 15 per cent.	905,620	2 113,120
Hides, 184,266, at 6300	1,181,628	
Horns, 100,000, at 4500 per 100	4,500	
Tanned half-hides, 17,500, at 2500	43,750	
Rice, bags, 17,305, at 9000	161,145	
Tobacco, arrobas, 100,000, at 4600	460,000	
Rum, pipes, 3110, at 65,000	202,150	
Tapioca, barrels, 250, at 9000	2,250	
Ipecacuanha, lbs., 20,000, at 500	10,000	
Jacaranda, manioc flour, and various articles	200,00	
	2,234,384	
Expenses, as above, 12½ per cent.	279,350	2,514,184,
Total value of exports during the year 1940		26,242,398*

* Being at the medium exchange of the year, equal to about 3,400,000*l*.

The aggregate value of the exports, in the undermentioned years, have been—

	Reis.			Reis.
1836	 18,711,824		1839	 23,362,396
1837	 15,362,842		1840	 26,242,398
1838	 20,455,965			

The principal article of import consists of cotton goods, the value of which amounts to full one third of the total value of the imports. Next to cottons are woollens, linen, and silk manufactures, wines, jewellery, and iron mongery; flour, meat, fish, butter and other articles of provision; spirits, salt, earthenware, paper, and a host of other articles. Of the total value of the imports in 1838-9, estimated at 29,450,698 rs. that of the cotton goods, which were almost wholly supplied by Great Britain, amounted to 10,555,704 ! We subjoin

An Account of the Value of the Imports into Rio during each of the Three Years ending with 1838-9, specifying the value of those furnished by each Country :—

Countries	1836-1837.	1837-1838.	1838-1839.
	Rs.	Rs.	Rs.
G. Britain and her possessions	13,345,787	9,609,846	10,593,504
France	3,941,145	2,434,160	4,311,363
United States N. A.	1,051,474	1,667,863	1,799,687
Hamburg and Bremen	2,037,508	1,001,875	2,296,317
Boenos air. R. Plata	1,090,264	1,590,816	1,377,287
Portugal and her possessions	1,674,839	1,658,395	2,002,366
Spain	357,649	692,499	785,413
Italy	470,574	985,900	470,005
Parts of the Pacific	582,044	216,567	9,584
Baltic ports	166,689	155,049	356,355
Fisheries	130,206	794,751	160
Holland and Belgium	110,267	115,303	168 243
Austria	55,469	68,851	2,471
Cape of Good Hope	91,011	38,568	5,339
Sundry places	12,419		19,433
Coastwise duties paid	495,922	129,680	682,920
Duty, during seaport	168,568	307,122	37,286
Totals	25,890,105	21,316,225	24,456,988

The customs' duties at Rio, in 1840, amounted to 1,999,822 reis on imports, and to 1,920,406 rs. on exports. During the same year 858 ships arrived at Rio from foreign ports; and 812 sailed, of which 519 were laden with Brazilian produce, and 230 in ballast. On the 1st of June, 1841, there were in the port 21 British, 27 Portuguese, 12 American, &c. ships. The arrivals coastwise, in 1840, amounted to 1489. (*From Fresse's Rio Circular, and Private Returns from Rio.*)

The currency of Rio, and of Brazil generally, is in a very vitiated state. The par of exchange, when the silver currency was maintained, was 67½d., and the current rate was usually higher; but for some years past, owing to the introduction of paper and copper, the exchange has fallen, so that its average rate in Rio, in 1840, was 30¾d.

Harbour.—As already stated, the harbour of Rio is one of the finest in the world. Its entrance is marked by a remarkable hill, in the form of a sugar-loaf, 960 ft. in height, close to its W. side, while on the opposite side of the bay, at the distance of about 1¼ m., is the fort of Santa Cruz, on which is a lighthouse. There is, also, a lighthouse, having the lantern elevated about 300 ft. above the sea level, on *Ilha Raza* (Flat island), about 10 m. S. from the mouth of the harbour. Ships may enter either by night or day, there being no obstruction or danger of any kind. The water in the bay is sufficient to float the largest ships of war; and it is extensive enough to accommodate all the navies of all countries in the world.

RIPLEY, county, Ia. Situated in the S.E. part of the state, and contains 400 sq. m. Drained by Laugherty's and Graham's creeks. It contained in 1840, 6300 neat cattle, 5433 sheep, 6388 swine; and produced 22,173 bushels of wheat 115,893 of Indian corn 158,413 of oats, 9191 of potatoes, 5051 pounds of tobacco, 19,886 of sugar., It had 28 stores, three flouring-mills, five grist-mills, 13 saw-mills, seven tanneries, one distillery; one academy, 20 students; two schools, 70 scholars. Pop. 1039. Capital, Versailles.

RIPLEY, county Mo. Situated toward the S.E. part of the state, and contains 1080 sq. m. Drained by branches of Big Black, and by Current rivers. It contained in 1840, 4405 neat cattle, 2769 sheep, 12,694 swine; and produced 7442 bushels of wheat, 161,303 of Indian corn, 5791 of oats, 3539 of potatoes, 19,529 pounds of tobacco, 7315 of cotton, 5920 of sugar. It had six stores, 20 grist-mills, 14 saw-mills, eight distilleries; five schools, 60 scholars. Pop.: whites, 2777; slaves, 77; free coloured, 2; total, 2856. Capital Van Buren.

RIPLEY, p. t., Chautauque co., N. Y., 10 W. Mayville, 350 m. W. by S. Albany, 358 W. Bounded N. by lake Erie. It contains one store, one fulling-mill, three grist-mills, eight saw-mills, four tanneries; two academies, 35 students, 17 schools, 747 scholars. Pop. 3197.

RIPLEY, p. v., Union t., Brown co., O., 113 m. S.W. Columbus. 474 W. Situated on the N. side of Ohio river immediately below the mouth of Red Oak creek, 56 m. above Cincinnati. It contains two churches, a Presbyterian and Methodist, 12 stores, one iron foundry, one engine factory, two steam saw-mills, one steam woollen factory, one oil-mill, two carding machines, a steamboat yard, 100 dwellings and 800 inhabitants. Several steamboats are owned at this place, and much flour, pork, &c., are shipped to New-Orleans.

RIOM (an. *Ricomagus*), a town of France, dép. Puy-de-Dôme, cap. arrond., on a hill, 8 m. N. by E. Clermont. Pop., in 1836, ex. com., 11,050. It is well built; but the houses are mostly in an antiquated style, and being wholly constructed of Volvic lava, with which it is also paved, it has a singularly sombre appearance. It has several handsome public fountains, and its churches, hospitals, the various public offices, and the sub-prefecture, are good buildings. One of its promenades is ornamented with a statue of Desaix.

Riom is the seat of a royal court for the déps. Puy-de-Dôme, Allier, Cantal, and Haut-Loire; and of tribunals of primary jurisdiction and commerce. It produces cotton and linen cloths, candles, leather, and brandy; and has a large trade in agricultural produce. It was formerly the cap. of Auvergne, and the residence of its dukes, some remains of whose castle still exist; and it continues to rank as the second town within the ancient limits of that province. Among the distinguished individuals belonging to Riom may be specified Gregory of Tours, and the learned Jesuits J. and A. Sirmond. (*See Hugo.* art. *Puy-de-Dôme; Dict. Géog., &c.*)

RIPON, a city, parl. and mun. bor. and market town of England. N. riding co. York, lower div. wap. Claro, on the Ure, close to its confluence with the Skill, both of which are here crossed by stone bridges, 22 m. N.W. York. Population of parl. bor. (which comprises the township of Ripon with part of the district of Boudgate), in 1831, 5700. "The appearance of Ripon is that of a very respectable and wealthy country town; it is increasing in population, and, though it has no manufactures itself, there is an extensive manufactory of saddle-trees in Bondgate. The neighbouring country is rich and well wooded, containing many objects of local attraction and interest; and this may, perhaps, account for the fact, that of late years many respectable families have been induced to settle here. The town has,

therefore, increased in importance as a residence for persons of that class." (*Parl. Bound. Report.*) The streets are irregularly laid out, but clean, and well lighted with gas, many of them meeting in the market-place, a spacious square surrounded with good shops, &c., having in its centre an obelisk 90 ft. in height, surmounted by the city arms, and on its S. side a particularly handsome town-hall, comprising courts, offices, and a handsome room for balls and public meetings. Ripon minster, erected in 1836 into a cathedral, and built in the 14th and 15th centuries, is said to be one of the best proportioned churches in England. It is a cruciform structure, with a tower rising from the intersection of the nave and transepts, besides two others (each 110 ft. in height) at the W. end. Its length, from E. to W. is 266 ft.; breadth of choir and aisles, 67 ft.; and length of transepts, 132 ft. "This venerable edifice," says Mr. Britton, "contains various parts worthy of attention, particularly its W. front, a very fine specimen of bold early English, and, except the battlements and pinnacles, without alteration. A part of the church is Norman, and a great portion of the transepts but little later.' The choir is partly decorated; and S. of it is a Norman crypt, above which are some Norman buildings used as vestries. At the E. end of the choir, which is sadly disfigured by heavy modern galleries, is a decorated E. window of five lights with very elegant tracery. The nave is very light, exhibiting some extremely fine composition; and there is a considerable quantity of good screen-work, both in wood and stone." The chapter comprises a dean, subdean, and six canons, who divide among them a net revenue of £833. Ripon was erected into a bishopric in 1836, the diocese comprising most of the populous parts of the W. riding, with the liberty of Richmondshire in the N. riding. The annual revenue of the see is £4500, chiefly derived from the surplus fund formed out of the deductions from the revenues of the larger sees. Trinity church, built and endowed in 1828 at a cost of above £13,000, is a Gothic cruciform structure, with lancet windows, and has accommodation for upwards of 1000 persons. The Wesleyan Methodists have two places of worship, and there is one each for Independents and Primitive Methodists. The town has several Sunday-schools, besides which there are national schools both for boys and girls, and an endowed blue-coat school. A grammar-school, founded in 1555, has an endowment producing about £370 a year; is free for Latin, Greek, and English grammar, to the sons of all residents, besides whom the master may receive boarders and pay-scholars. The town has several alms-houses, money-charities, &c.; and a dispensary furnishes relief to the sick poor. A mechanics' institute, subscription library, and news-room, are comprised in "the public rooms," a large and rather handsome building, erected by subscription, having attached gardens and pleasure grounds. A theatre was built in 1792; but it has been converted into a depôt and riding-school for the York hussar yeomanry cavalry.

Ripon was formerly celebrated for its manufacture of spurs, and the expression, " as true steel as Ripon rowels," was once proverbial for men of integrity and intrepidity. It had also a considerable manufacture of woollen-cloths; but both these branches of industry have long ceased to exist. Linen-weaving, malting, and tanning are pursued to some extent; and, as seen above, saddle-trees and saddlery are made in pretty large quantities. It is a large staple for wool, bought up here by the clothiers of Leeds, Halifax, &c.; and it has an excellent corn-market. In 1767, the Ure navigation was brought up to the town, which is thus rendered accessible by barges of 30 tons. The Knaresborough and Claro banking company and Wensleydale banking company have branches here, and, in addition to these, there are two private banks and a savings' bank.

Ripon is a borough by prescription, but received a charter from James I. Under the Municipal Reform Act it is governed by a mayor, three aldermen, and 12 councillors. Corporation revenue, in 1839, £401, exclusive of £389 accruing from the sale of property. The borough has a commission of the peace under a recorder, and a separate minor court is held by the dean and chapter, having jurisdiction over the district, called "canon fee." Ripon has sent two members to the House of Commons since the reign of Edward VI., the right of election being previously in the Reform Act, in the holders of burgage tenures, all of which had, for many years, been in the possession of a single individual. The limits of the parliamentary borough, as fixed by the Boundary Act, comprise the entire township of Ripon with a part of the township of Almonderbury and Bondgate. Registered electors, in 1839-40, 383. Large markets on Thursday, abounding with all kinds of agricultural produce, and six yearly fairs.

Ripon obviously derives its name from its position on *rypt* of the river Ure; but whether it be of Roman or Saxon foundation is wholly unknown. A monastery, founded here A.D. 661, attained considerable celebrity. It was, however

604

RIVE-DE-GIER.

ever, destroyed by the Scotch in the reign of Edward II., and was not afterwards restored.

RIVE-DE-GIER, a rapidly increasing town of France, dép. Loire, exp. cant., on the Gier, a tributary of the Rhone, at the head of the canal of Givors, and on the railway from St. Etienne to Lyons, 12 miles N.E. the former, and 30 m. S.W. the latter. Population, in 1836, ex. com., 9040; but it is now (1841) probably much higher. It was formerly fortified, but its works have been destroyed, and it is new principally remarkable for its commercial activity, sharing largely in the growing prosperity of the country along the Rhone (which see), and the districts round Lyons and St. Etienne. It has extensive manufactures of glass wares; and its coal-mines furnish a large proportion of the coal required in the hardware factories of St. Etienne. It has also hardware manufactures of its own. The officers of the Givors canal company, and the noble reservoir belonging to that canal, are the most remarkable public works at Rive-de-Gier. (*Hugo, &c.*)

RIVERHEAD, p. t., capital of Suffolk co. N.Y., 90 m. E. New-York, 225 m. E.S.E. Albany, 305 m. W. The post-office is called Suffolk C.H. Bounded N. by Long Island sound. Watered by Peconic river, flowing into little Peconic bay. Small vessels come to the mouth of the creek, 2½ m. from the courthouse at Riverhead village. The village contains a courthouse, jail, three churches, a Congregational, Methodist, and New Jerusalem, an academy, 70 dwellings, and about 500 inhabitants. The courthouse has stood over 100 years. Wood is extensively exported to New-York and is loaded into the larger vessels at Jamesport, 5½ m. below Riverhead village. The town contains 17 stores, one fulling mill, one woollen-factory, three grist-mills, one saw-mill, two tanneries; 14 schools, 651 scholars. Pop. 2449.

ROANE, county, Tenn. Situated centrally toward the E. part of the state, and contains 600 sq. m. Watered by Tennessee and Clinch rivers. It contained in 1840, 11,140 neat cattle, 10,970 sheep, 43,034 swine; and produced 71,380 bushels of wheat, 639,084 of Indian corn, 148,962 of oats, 16,064 of potatoes, 9695 pounds of tobacco, 24,151 of cotton. It had 12 stores, one furnace, four forges, one cotton factory, with 504 spindles, 15 flouring-mills, 37 grist-mills, 22 saw-mills, one oil-mill, eight tanneries, 18 distilleries; one academy, 30 students; four schools, 86 scholars. Pop.; whites, 9590; slaves, 1296; free coloured, 60; total, 10,948. Capital, Kingston.

ROANOKE, r., Va. and N.C., is formed by the junction of Staunton and Dan rivers, near the S. boundary of Va., and flows into the head of Albemarle sound. Its navigation is unobstructed to Halifax near the foot of the Great Falls, 75 miles, for vessels of 45 tons. At the falls the river descends 100 ft. in twelve miles. A canal, 12 miles long, extends around these falls, opening a navigation for bateaux to the junction of Staunton and Dan rivers, both of which are navigable for some distance for boats of 5 tons.

ROANOKE, county, Va. Situated S.W. of the centre of the state, and contains 370 sq. m. Drained by branches of Staunton river. It had in 1840, 5337 neat cattle, 5087 sheep, 10,681 swine; and produced 140,506 bushels of wheat, 13,538 of rye, 181,534 of Indian corn, 98,246 of oats, 6303 of potatoes, 509,373 pounds of tobacco. It had 13 stores, one fulling-mill, two woollen-factories, eight flouring-mills, 12 grist-mills, 16 saw-mills, one oil-mill, two tanneries, three distilleries, one printing-office, one daily and one weekly newspaper; two academies, 104 students; five schools, 98 scholars. Pop.; whites, 3845; slaves, 1553; free-coloured, 101; total, 5499. Capital, Roanoke C.H.

ROANNA (supposed to be the *Rodumna* of Ptolemy), a town of France, dép. Loire, cap. arrond., on the Loire, here crossed by a new bridge, 39 m. N. Montbrison. Population, in 1836, ex. com., 9234. It is pretty, well built, open, straggling town, with a good quay, a large hospital, a handsome theatre, and a communal college, having a good library and cabinet of natural history, &c. It is the seat of a tribunal of original jurisdiction, and a chamber of manufactures; and is a depôt for the manufactures of Lyons and the S.E. of France, passing to the central and N.W. deps. It also manufactures muslins, calicoes, and woollen fabrics : and has some trade in corn, wine, flour, timber, and charcoal. Various remains of antiquities have been frequently discovered in and near the town. (*Hugo*, art. *Loire; Dict. Géog.*)

ROBERTSON, county, Tenn. Situated in the N. part of the state, and contains 350 sq. m. Drained by Red river and its branches. It contained in 1840, 9736 neat cattle, 12,015 sheep, 43,032 swine; and produced 85,068 bushels of wheat, 7416 of rye, 43,993 of Indian corn, 97,289 of oats, 89290 of potatoes, 1,168,833 pounds of tobacco, 12,394 of cotton. It had 10 stores, one cotton factory with 212 spindles, six flouring-mills, 30 grist-mills, 22 saw-mills, five tanneries, 19 distilleries, one printing-office, one weekly newspaper, and seven periodicals. Pop.; whites, 9977; slaves, 3790; free coloured 34; total, 13,801. Capital, Springfield.

ROCHDALE.

ROBESON, county, N.C. Situated in the S. part of the state, and contains 700 sq. m. Drained by Lumber river. It had in 1840, 14,830 neat cattle, 12,513 sheep, 29,374 swine; and produced 2646 bushels of wheat, 248,195 of Indian corn, 1379 of oats, 54,632 of potatoes, 547,505 pounds of cotton. It had six stores, 56 grist-mills, 13 saw-mills; two academies, 61 students; 1126 scholars in schools. Pop.: whites, 6062; slaves, 2885; free coloured, 1222; total, 10,370. Capital, Lumberton.

ROBESON, p. t., Berks co., Pa. Bounded N.E. by Schuylkill r. It contains a Presbyterian and Lutheran church; six stores, one furnace, seven forges, four flouring-mills, five grist-mills, eight saw-mills, three tanneries. Pop. 2018.

ROBINSON, t., Alleghany co., Pa., 6 m. N.W. Petersburg. Bounded N. by Ohio river, E. by Chartier's creek. It has one store, one fulling-mill, two woollen-factories, five flouring-mills, five grist-mills, two saw-mills; nine schools, 315 scholars. Pop. 1694.

ROCHDALE, a parl. bor., market town, and par. of England, middle div., hund. Salford, co. Lancaster, on the Roch, tributary of the Irwell, 11 m. N.N.E. Manchester, and 27 m. E.N.E. Liverpool. Population of town and parliamentary borough (which comprises all within the limits of a circle described with the radius of ¾ m. from the town-hall), in 1831, 25,000 : population or parish (which extends over 58,680 acres), 74,427. The town consists of several streets, greatly improved of late years : the principal thoroughfare, in which are the market-house and town-hall, being wide and lined with brick houses. The streets generally are well paved, lighted with gas, and well supplied with water from four reservoirs in the township of Wardleworth. Here are two assembly-rooms, a small theatre, a commodious jail, and workhouses. The parish church, a structure of Norman and early English architecture, with a square pinnacled tower, stands on an eminence, to which there is an ascent from the lower part of the town by a flight of 126 steps ; the living is a vicarage, of the annual value of £1720 in the gift of the Archbishop of Canterbury. The town has also two episcopal chapels, and within the parish are six others, chiefly in the patronage of the vicar. Here is a Roman Catholic chapel, and 12 places of worship for different denominations of dissenters, among whom Methodists, Independents, and Baptists are the prevailing bodies. Each church and chapel has a Sunday-school, attended by great numbers of children. The town has a large national school, another well-endowed establishment, called the *Moss school*, for the gratuitous education of 40 boys and 20 girls ; a free grammar-school, founded in 1564, and attended by about 45 boys (who, owing to the smallness of the endowment, pay a fee of £6 a year) ; an endowed girls' school, and numerous money charities. A dispensary and ladies charity furnish relief to the sick poor ; besides which, there are bible, tract, and other religious associations, a literary society, a horticultural society, and a savings' bank.

Rochdale is a principal seat of the woollen and cotton manufactures, especially the former: the woollen articles produced here consist principally of baizes, flannels, and kerseys ; and those of cotton of strong calicoes, fustians, &c. In 1839, there were in the parish (exclusive of Saddleworth township) 56 woollen-mills, 95 cotton, and one flax-mill, propelled by an aggregate steam power of 2969 horses, employing, in all, 9745 hands, of which 8129 belonged to the cotton-mills. There are also numerous hand-loom weavers, though power-looms have been extensively introduced within the last few years. Hat-making is pursued on a pretty considerable scale, and numerous hands are employed in the town and neighbourhood in making machinery. The Manchester and Liverpool District Banking Company have a branch here, and there are two private banks. Coal and stone abound in the neighbourhood, and the parish comprises 19 collieries. Iron ore has been found in considerable quantities in Butterworth township, and flags are quarried at Spotland. Rochdale has means of communication with Manchester and Liverpool westward, and with Halifax, Leeds, &c., eastward, both by canal and railway. The Rochdale canal, which passes near the town, uniting westward with the Duke of Bridgwater's canal, and eastward with the Calder and Ribble navigation, 33 miles in length, cost upwards of £600,000, and is supplied, at its summit-level, by large reservoirs, one of which covers 130 acres. It was opened in 1804, and the daily traffic may amount to 1400 tons. The Manchester and Leeds railway passes through the district; and in Calderbrook township is a tunnel 2860 yards long, chiefly cut through the solid rock, 80 ft. below the surface. The total length of the line is 50½ m., and the whole is now open.

Notwithstanding its population and manufacturing importance, Rochdale had no representative in the legislature till the passing of the Reform Act, which conferred on it the important privilege of sending one member to the House of Commons, the electoral limits being defined, as above stated, in the local act of 6 Geo. IV. c. 101. Registered

y
result

695

electors, in 1839–40, 965. It is also a polling-place for the S. division of Lancashire. The town is under the jurisdiction of the county magistrates; but a manor-court is held every three weeks for the recovery of debts under 40s. Markets on Monday and Saturday; fairs, May 14, Whit-Tuesday, and Nov. 7, for horses and cattle. (*Baines's Hist. and Geog. of Lancashire; Butterworth's Stat. of Lancashire; Parl. Reports, &c.*)

ROCHEFORT, a maritime town, and the third naval port of France, dép. Charente-Inférieure, on the Charente, about 12 m. (by water) from its mouth, opposite the isle of Oleron, and 10 m. S.S.E. La Rochelle; lat. 45° 56' 10" N., long. 0° 57' 34" W. Pop., in 1836, ex. com., 13,948; but said to have been, in 1841, upwards of 17,000. (*Ports, &c. of France*, 160.) It is situated at the extremity of an extensive plain, and is shaped like a bow, the arc formed by the ramparts, and the chord by the river. The town, which is wholly of modern date, having been founded under Louis XIV., is well laid out and built, though the houses want elevation. Some of the principal streets are planted with double rows of poplars; and in the centre of the town is the Place d'Armes, a large and regular square, planted, and ornamented with a fountain; which, with other fountains, provide the inhabitants with an abundant supply of river water. The port is capacious, and deep enough to receive vessels of the largest size, having 20 ft. water at low ebb, and more than 40 ft. at high tide. The mercantile harbour, separated from the *port militaire*, admits vessels of from 800 to 900 tons. The naval yard is entered by the *Port du Soleil*, a handsome gateway constructed in 1828; on either side of which are lodges for the guard, the agents for the surveillance of the port and officers of the customs. It comprises building docks for ships of from 60 to 120 guns; sawing, brass and copper mills impelled by steam; a sail-loft, model workshop, a *bagne*, or prison, capable of accommodating 1000 convicts; a rope-house, in which cables upwards of 400 yards in length are made, and a naval and military arsenal, biscuit manufactory, and stores for materials of every kind necessary in the fitting out of ships of war. The cables and ship biscuit made at this port are admitted to be the best in France. The naval hospital without the town comprises nine separate buildings, furnishing accommodation for 1200 patients. The residence of the naval commandant is a fine building, surrounded by gardens, which are open to the public. Rochefort is the seat of a maritime prefecture, and tribunals of original jurisdiction and commerce; and has schools of naval gunnery, hydrography (2d class), medicine, drawing, music, &c.; a society of arts and literature, Bible society, public library, &c. It has a few vinegar factories, and sugar refineries; but ship-building is by far the most important branch of industry, and the ship-builders of the mercantile port construct handsome vessels for the coasting trade and cod fisheries. The little trade otherwise enjoyed by Rochefort is principally in corn, wines, salt, and brandy. The town was formerly very unhealthy; but it has been, in this respect, greatly improved by the drainage of the adjacent country. (*Ports and Arsenals of France*, 150–173; *Guide du Voyageur en France; Dict. Géog.*)

ROCHELLE (LA), a town and seaport of France, dép. Charente-Inférieure, of which it is the cap., on the Atlantic, 76 m. S. by E. Nantes, and 93 m. N.N.W. Bordeaux; lat. 46° 9' 21" N., long. 1° 9' 40" W. Pop., in 1836, 14,857, "Rochelle has an admirable commercial position. The town forms, as it were, the bottom of a small gulf, which serves as an *avant port*. It is defended by two handsome towers, which, whether La Rochelle be approached by land or water, are seen at an immense distance. Opposite the town, at the extremity of the roadstead, are the isles of Ré and Oleron. The harbour is safe and commodious; it is protected by a strong jetty, and is capable of receiving vessels of 400 or 500 tons' burden. There has recently been created a dock or *arrière port*, where vessels are careened. The town itself is clean and well built." (*Ports, &c. of France.*) The streets are wide and straight, and have footpavements, mostly under arcades, on which the houses are built. Few of the private buildings are lofty or of much size; but the town has, notwithstanding, a striking appearance. The fortifications, constructed by Vauban, consist of ramparts, with 19 bastions and eight lunettes, the whole enclosed with a moat and a covered way. The town is entered by seven gates, one of which, the *Porte d'Horloge*, is a handsome structure, apparently of the 16th century. There are several good squares; and without the walls are the promenades called the Mail and the *Champ de Mars*. The cathedral, town-hall, courts of justice, hospital, orphan asylum, exchange, and a good bathing establishment, are the principal public buildings, though several more are worth notice. La Rochelle is the cap. of the 19th military division of France, which comprises the déps. Charente-Inférieure, Deux Sèvres, and Vendée; the seat of a bishop, of courts of primary jurisdiction and commerce, of a royal

academy of arts and sciences, &c., and the residence of several foreign consuls. It has a seminary and communal college, schools of navigation and design, a public library of 20,000 vols., botanic garden, &c.; several prisons, a mint, and a royal arsenal and foundry. Its trade is extensive, not only in wines and brandies, but in wood, iron, salt, cheese, butter, oil, sardines, and colonial produce.

La Rochelle appears to owe its origin to a castle constructed here to check the incursions of the Normans. It was for some time in the possession of the English, from whom it was taken by the French in 1224. During the religious wars, and especially after the massacre of St. Bartholomew, it was a stronghold of the Protestants. It was invested by the Catholic forces in 1572, and withstood a long siege, terminated by a treaty. "The numerous infractions of that treaty in the reign of Louis XIII., and under the ministry of Richelieu, led to a second siege which commenced in August, 1627, and which was as violent, and longer and more decisive than the former. The king, the Duke of Orleans, Marshal Bassompière, and all the most renowned generals of the time, were present at the siege. The circumvallation extended for three leagues round the town; but the sea being open, English vessels poured in provisions and ammunition. After six months of heroic resistance, the entrance to the harbour by an immense dyke, extending 1500 metres into the sea, and of which the remains are still visible at low water. The result was soon fatally apparent. Famine quickly decimated the ranks of the besieged; and after a siege of 14 months and 18 days, La Rochelle was compelled to capitulate. Richelieu made a triumphal entry into the city; the fortifications were demolished, and the Protestants deprived of their last place of refuge." (*Ports and Arsenals of France; Guide du Voyageur, &c.: Dict. Géog.*)

ROCHESTER, a city, parl. and mun. bor., and market-town of England, co. Kent, lathe Aylesford, at the W. end of and adjoining Chatham on the S. bank of the Medway, crossed here by a handsome stone bridge of 11 arches, 25 m. W. by N. Canterbury, and 28 m. E. by S. London. Pop. of parl. bor. (which includes with the old bor. additional portions of the par. of Stroud), in 1831, 12,658. The bridge over the Medway connects the town with Stroud, on the opposite bank of the river; so that the three towns of Chatham, Rochester, and Stroud form, as it were, a continuous street, upwards of 2 m. in length, along the road from London to Dover. The houses in Rochester generally have a somewhat antiquated appearance; and among them are several built chiefly of timber, with projecting gables and stories. The town is partially paved, and lighted with gas. Fortifications were erected for its protection in 1088; but fort Pitt is now used as a military hospital, and fort Clarence has become a lunatic asylum for soldiers. The town-hall, a spacious brick edifice, fronted by Doric columns, is open underneath, and above is a large hall with several portraits of public characters: at the back is a small bar. jail. On the site of the former guildhall is the clock-house, a neat building erected in 1706, at the expense of the celebrated admiral Sir Cloudesly Shovel. A theatre is occasionally opened; there are also assembly rooms and baths, and a record room at the end of the bridge. At the S.W. angle of the town, rising abruptly from the river, is the castle, anciently a strong fortress, but which has, for a lengthened period, been in a state of decay. The walls, which are of Kentish ragstone, enclose a quadrangular area of nearly 300 sq. ft., and, with their towers, are now in ruins. The walls of the keep, however, at the S.E. angle of the court, are in good preservation: it rises about 104 ft. from the ground, and has turrets at the angles rising 19 ft. above the rest of the buildings. "The style of the whole building is Norman, and it presents a fine specimen of the mode adopted at the date of its erection to enable a very small number within the castle successfully to resist a much greater number of besiegers; for which purpose, indeed, the access, the various successive gates, and other defences, are admirably calculated. The masonry in the interior is very good, especially that of the well, which is in one of the walls, and was accessible from several floors of the castle." (*Rickman's Arch.* p. 187.)

A little E. of the castle is the cathedral, originally founded by Ethelbert, about A.D. 600; but the present building was principally erected in the 12th century, from the plans of Bishop Gundulph, one of the first church architects of his day. It is a cruciform structure, with a central tower, of modern erection, rising from the intersection of the nave and transepts with another, now unroofed, called Gundulph's tower, on the N. side, close to the E. transept. According to Britton, the dimensions of the church are as follows: total inside length, 306 ft., of which 156 ft. belong to the choir; breadth of nave and aisles, 66 ft.; length of the great and small transepts, 122 ft. and 90 ft. respectively; extent of W. front, 81 ft.

"The exterior appearance of this cathedral is not very imposing, and the exterior walls of the nave are either much decayed or covered by modern repairs. The other parts of the church are surrounded by buildings, so that little more than one portion can be seen at a time. The W. front is a fine specimen of Norman enrichment, but has a very large inserted perpendicular W. window. The nave has Norman piers and arches, except those next the cross, which, with most of the E. portions of the church, are early English. There are other Norman portions on the other side, which appear to be the remains of the cloisters and other monastic adjuncts. The crypt is very spacious, extending under the whole choir: its character is early English; but a portion under the N. aisle may be considered almost Norman. There are a few monuments; but they are more remarkable for singularity than beauty. The whole cathedral, except the nave and S. aisle of the chancel, is adorned with early English groining, and, as at Canterbury, the floor of the choir is considerably raised above that of the nave." (*Britton*, quoted in *Rickman's Arch.* p. 184.)

The interior has very recently been repaired in excellent taste, and Rochester cathedral is one of the best specimens of the early English style in the kingdom. The ecclesiastical corporation comprises a dean and six prebendaries, who divided among them, in 1834, a nett revenue of £5106, and have the patronage of 20 benefices; there are, also four minor canons. The see of Rochester was worth, at an average of three years ending with 1831, only £1489 a year; but it either has been or is to be increased to £4000 a year. The town has two parish churches; one of which, St. Margaret's, is of very mixed architecture, and the other, St. Nicholas, in the perpendicular style: the living of the former is in the gift of the dean and chapter, and of the latter in that of the bishop. The Wesleyan Methodists, Independents, Unitarians, and the Society of Friends, have places of worship; and the Sunday schools within the bor. are attended by above 800 children. A grammar-school, founded in 1542, having six exhibitions at Oxford and Cambridge, is supported by the dean and chapter; besides which there is an endowed mathematical school, established in 1701. Two national schools give instruction to about 500 children; and there are two or three almshouses, with smaller money-charities.

"Rochester, like Chatham, was in a great degree dependent on the expenditure produced by the naval and military establishments at Chatham during the war, and the reduction of that expenditure has been severely felt. The greater part of the population consists of retail tradespeople. There is no manufactory in or near the town; but a considerable quantity of coal and other articles is imported for the supply of the country inward as far as Tunbridge; and there is a large export trade in hops." (*Municipal Report.*)

The great customs' duty received here in 1840 amounted to £58,496, being nearly double the amount received in the preceding year; by far the largest portion of this sum is, however, derived from the trade of Chatham. The town is said, in the *Municipal Boundary Report*, to be declining in prosperity; "the poor-rates have increased; many of the houses are uninhabited, and there is less expenditure and luxury among the upper classes than in former times."

The bor. of Rochester, first incorporated in the reign of Henry II., is divided, under the Municipal Reform Act, into three wards, and is governed by a mayor, five aldermen, and 18 councillors. Corporation revenue, in 1839, £3427. It has a commission of the peace under a recorder, and petty sessions are held twice a week; besides which, there is a court for the recovery of small debts, with two or three other minor local courts.

Rochester has returned two members to the House of Commons since 26 Edward I., the right of election down to the Reform Act being in the freemen not receiving alms. The electoral limits were enlarged, as above mentioned, by the Boundary Act. Reg. electors in 1839-40, 1194. Corn markets on Tuesday, and for provisions generally on Friday; fairs, disused.

The only event of any importance connected with the modern history of Rochester is the descent made by the Dutch in 1667. *See* CHATHAM.

ROCHESTER, city, capital of Monroe co., N. Y., is situated 43° 9' 17" N. lat., and 77° 51' W. long., 73 E. Buffalo, 141 Utica, 220 W. by N. Albany, 87 E.N.E. Niagara falls, N.W. Canandaigua, 309 W. The population in 1820, 1502; in 1830, 9269; in 1840, 20,191. It is situated on both sides of Genesee river, 7 m. S. of lake Ontario. It was laid out as a village in 1812; incorporated as a village in 1817; and as a city in 1834; and the city limits cover an area of 4324 acres. It is handsomely laid out, with considerable, though not entire regularity. The parts of the city connected by three bridges, and the river is also crossed in the middle of the city by the splendid aqueduct of the canal. The streets are spacious, with a width varying

from 60 to 80 feet, well paved in the centre, with convenient sidewalks; and there are several public squares, which are enclosed. Buffalo-street, which is broad and straight, runs through the centre of the city, crossing the river on a bridge; and on the E. side is called Main-street. The city is handsomely built, chiefly with brick, and many of the stores and dwellings are elegant. Many of the houses have fine gardens, ornamented with fruit trees and shrubbery. Some of the churches and public buildings are handsome structures. The public buildings and institutions are a courthouse, jail, county clerk's office, two market-houses, six banking-houses, besides a bank for savings, a mutual insurance company, 23 churches, four Presbyterian, three Episcopal, two Methodist, two Baptist, one Free-will Baptist, two Friends, a Reformed Presbyterian, an Evangelical Lutheran, a Universalist, a Free Bethel, a Free Congregational, two Roman Catholic, and two African, a museum, a Collegiate institute, two female seminaries, two orphan asylums, a mechanics' literary association, a young mens' association, an atheneum, and an apprentices library.

Rochester owes its great advantages and rapid growth especially to the vast water-power created by the falls in Genesee river, which amount to 268 ft. within the bounds of the city, in which are three successive perpendicular falls of 96, 20, and 105 ft., besides rapids. On these rapids and falls are many large flouring-mills, not surpassed in the world, and numerous other hydraulic works. It is estimated that, independent of the capital invested in the construction of these mills, it requires $2,000,000 annually to keep them in operation, and that they produce annually, to the amount of $3,500,000. Rochester is finely situated for commerce. Vessels come up the Genesee river from lake Ontario to Carthage, 2½ m. below the centre of the city, where steamboats arrive and depart daily, and to which there is a railroad from the city. The river is boatable above the place for 45 miles, to Mount Morris. The Erie canal passes centrally through the city, giving it access on the E. to Albany, and thence by Hudson river to New-York city; and W. to Buffalo, and thence to the upper lakes, and the various parts connected with them by canals. The Genesee valley canal, when completed, will connect it with Olean on the Alleghany river, and thence to Pittsburg. The chain of railroads from Boston to Buffalo passes through it, and gives it a ready access to both these places, and to all others on the route. These facilities for transportation have completed the advantages derived from its immense water power, and the eminently rich agricultural region by which it is surrounded; which if it does not yield a sufficiency of wheat for the supply of its mills, it can be readily obtained from the great wheat-growing state of Ohio, and the other fertile states of the west.

According to the census of 1840, there were in the city one commission and one commercial house in foreign trade, with a capital of $15,100; 256 retail dry goods and other stores, with a capital of $1,238,898; two lumber yards, with a capital of $30,000; 464 men engaged in internal transportation, with 71 butchers, packers, &c., employed a capital of $156,000; 53 persons produced machinery to the amount of $48,000; 25 persons manufactured hardware and cutlery to the amount of $9000; 10 persons manufactured 250 small arms; 14 persons manufactured the precious metals to the amount of $6800; 75 persons manufactured various metals to the amount of $95,900; 116 persons produced granite and marble to the amount of $57,000, with a capital of $7600; four persons manufactured granite and marble to the amount of $5000; 49 persons made bricks and lime to the amount of $14,015; four fulling-mills and four woollen factories employed 69 persons, producing to the amount of $50,000, with a capital of $58,616; one cotton factory with 3000 spindles, employed 80 persons, produced to the amount of $40,000, with a capital of $80,000; 58 persons manufactured tobacco to the amount of $73,000, with a capital of $16,000; hats and caps were manufactured to the amount of $44,900, and straw bonnets to the amount of $1600, the whole employing 196 persons, and a capital of $22,625; three tanneries produced 3760 sides of sole leather, and 5300 sides of upper leather, employing 165 persons, and a capital of $129,500; saddlery, and other manufactures of leather, produced to the amount of $946,500, with a capital of $50,795; 11 persons produced soap and candles to the amount of $33,500; three distilleries produced 195,000 gallons of distilled spirits, and three breweries 204,960 gallons of beer, the whole employing 37 persons, and a capital of $80,300; 21 persons produced drugs and paints to the amount of $42,000, and turpentine and varnish to the amount of $450, with a capital of $45,500; two persons produced glass to the amount of $3001, with a capital of $1000; one pottery, employing five persons, produced to the amount of $3500, with a capital of $1500; 16 persons produced confectionary to the amount of $22,700, with a capital of $9750; two paper-mills, employing 37 persons, produced to the amount of $35,000, with a capital of $23,500; one rope-walk, employing six persons, produced cordage to the amount of $7000, with a capital of $5500; four persons manufactured musical instruments to the amount of $5000, with a capital of $5000; 84 persons manufactured carriages and wagons to the amount of $70,800, with a capital of $35,900; 29 flouring-mills produced 311,665 barrels of flour, and with eight saw-mills and one oil-mill, employed 256 persons, producing to the amount of $1,841,975, with a capital of $945,600; vessels were built to the amount of $74,900; 284 persons manufactured furniture to the amount of $41,700, with a capital of $113,400; nine printing-offices, one bindery, four daily, five weekly newspapers, and two periodicals, employed 82 persons, and a capital of $32,500; 61 brick or stone, and 66 wooden houses were built, employing 418 persons, and cost $401,970. The total capital employed in manufactures was $1,963,017.

Such is Rochester; and the United States, the land of wonders in this respect, with the exception of Lowell, Mass., does not probably furnish another example of an equally rapid rise. Its early founders never anticipated such a growth. In 1812 there were only two framed dwellings on the spot, each containing a single room; and a few years before, when a proposal was made in the Legislature of the state to build here a bridge across the Genesee river, it was strongly opposed, and a member remarked, " that it was a God-forsaken place, inhabited only by muskrats, visited only by straggling trappers, through which neither man nor beast could gallop without fear of starvation or fever and ague." The act to build the bridge passed, but was reprobated by many, as an extravagant waste of the public money, on such " an outlandish and unfrequented spot." But, like the opposers of the Erie canal, they probably many of them lived to acknowledge their error.

The city, which contains over 2000 dwellings, is divided into five wards, and is governed by a mayor, a recorder, a marshal, and 10 aldermen, two from each ward. These constitute the common council, who appoint a city clerk, city attorney, and a treasurer.

ROCHESTER, p. t., semi-capital of Strafford co., N. H., 46 E. Concord, 505 W. Salmon falls river bounds it on the N.E., and Cocheco river passes through it, both affording water-power. It contains three churches, a Congregational, Methodist and Free-will Baptist, 11 stores, four fulling-mills, three woollen-factories, two grist-mills, four saw-mills, two oil-mills, two tanneries; 17 schools, 788 scholars. Pop. 3431.

ROCHESTER, p. t., Windsor co., Vt., 43 S.S.W. Montpelier, 482 W. Drained by White river, and by a branch which affords water-power. Chartered in 1781. It contains a Congregational and Methodist church, three stores, two fulling-mills, one grist-mill, seven saw-mills; 13 schools, 469 scholars. Pop. 1396.

ROCHESTER, p. t., Plymouth co., Mass., 54 S.S.E. Boston, 441 W. Bounded S.E. by Buzzard's bay. Watered by Mattapoiset and Sippican rivers. It contains five churches, three Congregational, a Baptist, and Universalist, 17 stores, two grist-mills, five saw-mills; one academy, 100 students, 23 schools, 1009 scholars. Pop. 3864. Mattapoiset village is on a branch of Buzzard's bay, has a good harbour, and is engaged in ship-building, the whale fishery, and the manufacture of salt.

ROCHESTER, t., Ulster co., N. Y., 76 S.W. Kingston, 74 S. by W. Albany. Watered by Rondout creek, and its tributaries. The Delaware and Hudson canal passes through it. It contains a Dutch Reformed church, nine stores, two fulling-mills, one woollen factory, seven grist-mills, 11 saw-mills, three tanneries; 10 schools, 503 scholars. Pop. 5974.

ROCK, r., Wis. and Ill., rises S.W. of Winnebago lake, and flows S. into Ill., and thence S.W., and enters Mississippi river, 3 m. below Rock Island. In Wisconsin it receives Catfish river, which passes through the "Four Lakes."

ROCKAWAY, beach, Hempstead t., Queens co., N. Y. Situated on the S. shore of Long Island, between Jamaica bay and the Atlantic ocean, and has become a fashionable watering-place. Here is the Marine Pavilion, a splendid boarding house, and a pleasant and healthy resort in the summer season. The accommodations are extensive, and of the finest kind, and nowhere can the sea air and sea bathing be enjoyed in greater perfection.

ROCKBRIDGE, co., Va. Situated in the central part of the state, and contains 680 sq. m. Drained by North river and its branches, which flow into James river. It derives its name from the celebrated Natural bridge, for a description of which, see Natural Bridge. It contained in 1840, 13,432 neat cattle, 19,821 sheep, 26,321 swine; and produced 263,756 bushels of wheat, 69,566 of rye, 504,646 of Indian corn, 4329 of buckwheat, 249,018 of oats, 35,579 of potatoes, 293,755 pounds of tobacco. It had 98 stores, three furnaces, three forges, 32 flouring-mills, 13 grist-mills, 56 saw-mills, two oil-mills, 14 tanneries, 36 distilleries, two potteries, two printing-offices, two weekly newspapers; one college, 63 students; four academies, 152 students; 28 schools, 746 scholars. Pop.: whites, 10,446; slaves, 3519; free coloured, 396; total, 14,284. Capital, Lexington.

ROCKCASTLE, co., Ky. Situated a little S.E. of the centre of the state, and contains 330 sq. m. Bounded S.E. by Rockcastle river. Drained by Dick's river. It had in 1840, 2863 neat cattle, 3703 sheep, 5613 swine; and produced 8472 bushels of wheat, 66,382 of Indian corn, 29,325 of oats, 3149 of potatoes, 4725 pounds of sugar. It had two stores, five grist-mills, four saw-mills, three tanneries. Pop.: whites, 3093; slaves, 377; free coloured, nine; total, 3003. Capital, Mount Vernon.

ROCKINGHAM, co., N. H. Situated in the S.E. part of the state, and contains 695 sq. m. Watered by Lamprey, Exeter and Beaver rivers. Bounded S.E. by the Atlantic for 18 m., constituting the whole seacoast of the state. It contained in 1840, 29,764 neat cattle, 38,389 sheep, 11,673 swine; and produced 20,309 bushels of wheat, 28,933 of rye, 904,504 of Indian corn, 3036 of buckwheat, 98,929 of barley, 83,177 of oats, 805,367 of potatoes. It had 94 commercial and commission houses, in foreign trade, 323 retail stores, six lumber yards, capital invested in the fisheries, $59,580, five furnaces, 16 fulling-mills, four woollen factories, six cotton factories, with 19,580 spindles, two flouring-mills, 89 grist-mills, 106 saw-mills, one oil-mill, one paper-mill, one powder-mill, 43 tanneries, two distilleries, four potteries, seven printing-offices, five weekly newspapers, 19 academies, 1015 students, 243 schools, 10,618 scholars. Pop. 45,771. Capital, Portsmouth.

ROCKINGHAM, co., Va. Situated a little N. of the centre of the state, and contains 833 sq. m. Watered by Shenandoah river and its tributaries. It contained in 1840, 28,122 neat cattle, 23,956 sheep, 38,765 swine; and produced 375,187 bushels of wheat, 90,886 of rye, 470,584 of Indian corn, 948,418 of oats, 41,186 of potatoes, 37,189 pounds of tobacco, 3330 of sugar. It had 80 stores, one furnace, nine fulling-mills, 35 flouring-mills, 16 grist-mills, 48 saw-mills, four oil-mills, 21 tanneries, 15 distilleries, three potteries, one printing-office, one weekly newspaper; one academy, 45 students, 96 schools, 799 scholars. Pop.: whites, 14,944 slaves, 1899; free coloured, 501; total, 17,344. Capital, Harrisonburg.

ROCKINGHAM, co., N. C. Situated in the N. part of the state, and contains 475 sq. m. Drained by branches of Dan and Dan rivers. It contained in 1840, 9735 neat cattle, 6237 sheep, 30,574 swine; and produced 61,013 bushels of wheat, 4805 of rye, 431,085 of Indian corn, 140,429 of oats, 9953 of potatoes, 1,777,905 pounds of tobacco, 23,947 of cotton. It had 18 stores, one furnace, one cotton factory with 864 spindles, 13 flouring-mills, 36 grist-mills, 25 saw-mills, two oil-mills, five tanneries, eight distilleries; 11 academies, 241 students; 10 schools, 211 scholars. Pop.: whites, 8185 slaves, 4572, free coloured, 275; total, 13,032. Capital, Wentworth.

ROCKINGHAM, p. t., Windham co., Vt., 93 S. Montpelier, 450 W. Watered by William's and Saxton's rivers, which afford water-power. Bounded E. by Connecticut river. It contains three churches, two Congregational and an Episcopal, 12 stores, seven fulling-mills, four woollen factories, five grist-mills, seven saw-mills, two paper-mills, two tanneries, one printing-office, two binderies, one weekly newspaper and one periodical; 17 schools, 703 scholars. Pop. 3188.

ROCKINGHAM, p. v., capital of Richmond co., N. C. 135 S.W. Raleigh, 400 W. Situated 5 m. E. of Pedee river, and contains a courthouse, jail, several stores, and about 200 inhabitants.

ROCK ISLAND, in Mississippi river, at the foot of Rock river rapids, and is 3½ m. long, and three fourths of a mile wide. The shores are of perpendicular rock, 50 ft. above high water. It contains fort Armstrong on its S. part.

ROCK ISLAND CITY, p. v., capital of Rock Island co., Ill., 150 N.N.W. Springfield, 875 W. Situated between Rock river and the Mississippi, has been laid out on an extensive scale, and contains in Stephenson village a courthouse which cost $98,000, a jail, 15 stores, 725 dwellings, among which is a splendid hotel, and about 700 inhabitants. The rapids extend 15 m. from the mouth of Rock river, in which distance they descend about 22 ft.; ledges of rocks extend across the river, which in low water form a serious impediment to navigation. It has here in contemplation to cut a canal from the Mississippi to Rock river, by which an immense water-power would be gained in the latter.

ROCKLAND, co., N. Y. Situated in the S.E. part of the state, and contains 179 sq. m. Bounded E. by Hudson river. It contains magnetic iron ore, and variegated marble. The New-York and Erie railroad, which commences at Piermont, crosses the county. It contained in 1840, 6235 neat cattle, 17,399 sheep, 11,511 swine; and produced 3539 bushels of wheat, 25,140 of rye, 41,136 of Indian corn, 36,111 of buckwheat, 47,055 of oats, 48,117 of potatoes. It had 53 stores, five lumber yards, one furnace, three forges, four fulling-mills, four woollen factories, three cotton factories with 1043 spindles, one dyeing and printing works, 20 grist-mills, 11 saw-mills, one paper-mill, five tanneries, two distilleries; 28 schools, 1099 scholars. Pop. 11,975. Capital, New City village, in Clarkson t.

ROCKLAND, p. t., Venango co., Pa., 212 N.W. Harrisburg, 284 W. It has three stores, three furnaces, one fulling-mill, five grist-mills, five saw-mills; three schools, 101 scholars. Pop. 1178.

ROCKLAND, t., Bucks co., Pa. Drained by Maxatawny and Sacony creeks. It has two stores, one furnace, one forge, three grist-mills, four saw mills, one paper-mill, two tanneries, one distillery. Pop. 1649.

ROCKPORT, p. t., Essex co., Mass., 32 N.E. Boston, 464 W. Situated on the N. part of cape Ann, and has a harbour for small vessels. It contains three churches, a Congregational, Baptist, and Universalist, 13 stores; six schools, 592 scholars. Pop. 2650.

ROCKPORT, p. t., Cuyahoga co., O., 132 N.E. Columbus, 367 W. Bounded N. by lake Erie. Watered by Rocky river. It contains two grist-mills, two saw-mills, two tanneries; eight schools, 516 scholars. Pop. 1235.

ROCKPORT, p. v., capital of Spencer co., Ia., 162 S.S.W. Indianapolis, 689 W. Situated on a high bluff on the N.W. bank of Ohio river, and contains a courthouse, jail, a tannery, a ship yard and about 300 inhabitants.

ROCKVILLE, p. v., capital of Montgomery co., Md., 36 W. Annapolis, 16 W. Situated on the head waters of Watts branch of Potomac river, and contains a courthouse, jail, county offices, several stores, 60 dwellings, and about 350 inhabitants.

ROCKY MOUNT, p. v., capital of Franklin co., Va., 179 W. by S. Richmond, 254 W. It contains a courthouse, jail, two stores, one furnace, in the vicinity, which employs 100 labourers, and about 200 inhabitants.

ROCKY MOUNTAINS, a very extensive mountain system of N. America, (See vol. i. p. 78.)

ROCROY, a town of France, dép. Ardennes, cap. arrond., in an extensive plain, near the Belgian frontier, 15 m. N. by W. Mézières. Pop., in 1836, 3692. It is surrounded by a rampart strengthened with bastions and demi-lunes; is the seat of a tribunal of original jurisdiction, and has a military hospital, a society of agriculture, and some hardware manufactures.

Rocroy having been besieged by a Spanish army in 1643, the Prince of Condé, then Duke d'Enghein, and only 21 years of age, advanced to its relief, with an army inferior in numbers and in the quality of the troops. But the extraordinary talent and brilliant courage of the prince more than made up for his inexperience and the inferiority of his force. The French gained a complete victory (19th May, 1643). The Spanish infantry, which had hitherto been invincible, was cut to pieces; and the French arms acquired a superiority which they preserved for more than sixty years, or till the battle of Blenheim. The humanity of Condé, henceforth called "le Grand," was as conspicuous on this occasion as his talents and his courage. (Voltaire, Siècle de Louis XIV., cap. 3.)

RODMAN, p. t., Jefferson co., N. Y., 11 S. Watertown, 63 N.W. Albany, 408 W. It contains three churches, a Methodist, Baptist, and Presbyterian, four stores, two fulling-mills, one woollen factory, two grist-mills; 21 schools, 60 scholars. Pop. 1702.

ROMANS, a town of France, dép. Drôme, in a fine plain, on the Isère, by which it is separated from the Bourg-de-Péage, on the opposite side of the river, the communication between the two being maintained by a fine bridge, 12 m. .E. Valence, and 35 m. W S.W. Grenoble. Pop., including Bourg-de-Péage, in 1836, 11,245. Romans, which is surrounded by an old wall flanked with towers, is said by Argo to be a handsome well-built town, and has some fine promenades. The parish church is the only remaining portion of the famous monastery founded here by St. Bernard, 837, to which the town owes its origin. Romans has a théâtre, a pensionnat, or school for the gratuitous education of young ladies, a tribunal of commerce, &c., with manufactures of silk and woollen stuffs, hosiery, gloves, &c. It is also an active trade in the produce of the dep., including silk, wool, wine, oil, truffles, &c., which is much favoured by the junction of the Isère with the Rhone. (ge, art. Drôme; Dict. Géographique.)

The famous Baron Lally, who, having distinguished himself at the battle of Fontenoy, was subsequently sent as commander-in-chief of the French forces to India, was a native of Romans, where he first saw the light, on the 15th January, 1702. Being of a violent, irritable temper, and involved in the greatest difficulties, Lally got embroiled with every body. After the fall of Pondicherry, in 1761, he returned to France, where, pursued by the hatreds he had excited in India, he was, after a lengthened imprisonment, tried by the parliament of Paris, and condemned to be decapitated. This unjust sentence was carried into effect

on the 9th May, 1766, (See, for a calm, dispassionate statement of the case of Lally, the Siècle de Louis XV., by Voltaire, cap. 34: the article on Lally in the Biographie Universelle is by a partizan.)

ROME, the most celebrated of European cities (Lux orbis terrarum, si ars omnium gentium, Cicero, Catil. 4. cap. 6.), famous alike in ancient and modern history; first, as the metropolis of the most powerful nation of antiquity, and afterwards, as the ecclesiastical capital of Christendom, and the residence of the pope, on both sides the Tiber, but principally on its E. bank, about 16 m. from its mouth, 115 m. N.W. Naples, and 145 S.S.E. Florence; lat. 41° 53' 54" N., long. 12° 29' 47" E. The pop., which in 1837, amounted to 155,552, had decreased, in 1838, to 148,903, owing to the influence of cholera and other causes. (Serristori Statistica, p. 94.)

Modern Rome, which interests alike by its classical associations, its antiquities, its churches, and its works of art, is surrounded by walls (mostly occupying the site of those constructed by the emperor Aurelian) in the form of an irregular polygon, about 14 m. in circ., the longest diameter being from the Porta del Popolo, N.W., to the Porta S. Sebastian, S.E., about 3 m. in length. The city has 16 gates, three or four of which, however, have been walled up: of these, the principal are the Porta del Popolo, on the road to Florence and Ancona; the Porta Pia, on the road to Tivoli; the Porta Maggiore, leading to Palestrina; and the Porta S. Giovanni, leading S.E. to Albano and Naples. But not more than a third part of the enclosed area is covered with buildings, the rest consisting of ruins, gardens, and fields, with some churches, convents, and other scattered habitations. The older part of the ancient city, where the principal ruins are found, is about ½ m. S. from the modern city; but it is needless to add, that the former, under the emperors, was much more extensive than the modern town, inasmuch as, besides the space within the walls, it had very extensive suburbs. The ground occupied by the city is mostly low, being only from 35 to 45 ft. above the level of the sea. Exclusive, however, of the low grounds, several low hills, or eminences, are comprised within the limits of the city; and in consequence of ancient Rome being popularly said to stand upon seven hills, it was sometimes called urx septicollis. The following measurements exhibit the height of the hills now referred to:—

	Eng. Ft.		Eng. Ft.
Capital, at the W. angle of the Tarpeian rock	151	Esquiline hill	218
		Quirinal hill at the	
Do. at N. end	160	Pope's palace	156
Palatine hill	170	Ancient pavement of	
Aventine —	156	the Forum	39
Celian —	168	Do. do. of Trajan's Forum	43
Pincian —	218		

The Palatine hill, the nucleus of the city, comprises a space of about 40 acres, and appears originally to have had precipitous edges; but excepting this, and the Capitoline hill, comprising about 16 acres, the rest have an easy ascent, and are, in fact, mere eminences. No doubt, however, owing to the accumulation of rubbish in the valleys, and the lowering of the hills by rains, the digging of foundations, &c., the elevations of the latter must have been much more striking in antiquity than at present. The seven hills, properly so called, on the E. bank of the Tiber, were included in the city so early as the reign of Servius Tullius; but at a later period the Mons Pincius (Pincian hill), to the N.E., and of the Mons Vaticanus, and the Mons Janiculus, on the W. side of the Tiber, were enclosed within the city walls. The hills consist chiefly of volcanic tufa intermingled with thin beds of travertine, making good building stone, as well as of silicious sand, and a few layers of pumice and scoriæ; while the low ground which has been raised several feet above its ancient level, apparently by the deposits left by frequent floods, is covered with calcareous sands, marls, clays, or silt.

Rome is divided into 14 districts, or Rioni, 11 of which are in the portion now inhabited; but the more popular, and for general purposes more intelligible, distribution of the city is into a central portion on the low ground E. of the Tiber, the ancient Campus Martius, a division to the E. of the latter, on the Pincian, Quirinal, and Viminal hills, and a third division on the W. side of the river. The first of these quarters is the chief seat of bustle and trade: it is intersected by the Corso, and has some other good streets. On the N. W. side of the Capitol is the meanest and dirtiest part of the city, chiefly inhabited by Jews, and many degrees worse than Monmouth-street or Houndsditch in London. The upper part of the city on the Pincian and Quirinal hills is less densely covered with houses, and chiefly comprises palaces and villas, churches, convents, and gardens; it is intersected near the pope's palace on the Quirinal hill by two fine avenues, crossing each other at right angles, and having four fountains at the point of intersection. Between the Quirinal and Esquiline hills are several streets inhabited by the lower orders, and extending to the Via Vaccina, a rather broad thoroughfare leading to the Roman forum. Beyond, N., E., and S., are gardens and fields, studded here and there with villas, churches, and ruins, including the baths of Diocletian and Titus, and the Colosseum. That portion of Rome which lies on the right or W. bank of the river, consists of two parts: the Vatican, northward, in which are comprised the castle of St. Angelo and St. Peters, the glory not merely of the eternal city but of Christendom; and the Trastevere or ancient Janiculum, southward, which has the street, called from its length (nearly 1 m.), Via Longara, a botanic garden, and the Villa Corsini.

Rome, meaning the modern city (as we shall afterwards notice the ruins), is dull, dirty, and, with a few striking exceptions, meanly built: "Indeed," says Mr. Forsyth, "whichever road you take, your attention will be divided between magnificence and filth." (*Italy*, p. 124.) Most of the streets are narrow, crooked, and badly lighted with melancholy-looking *reverbères*, or else by lamps hung over the statues of the Virgin, which seem to be indispensable to all considerable houses. From this general censure, however, must be excepted the Corso, already alluded to, extending in a straight line, more than a mile, from the Porta del Popolo to the foot of the Capitoline hill, and which, in the greater part of its extent, is 30 ft. wide, with *trottoirs* at the sides, which, however, are so narrow, and so often encroached on, as to be, in fact, of little service: it is lined with many handsome palaces, some of which may vie in grandeur with the noblest in Italy. Two other streets, the *Strada di Ripetta*, and the *Strada del Babiuno*, diverging from the open space before his holiness' palace, are tolerably straight, and tolerably wide. These three streets are crossed by a line connecting the bridge of St. Angelo with the Pincian hill, and forming one of the most bustling thoroughfares in Rome. With these exceptions, however, the streets are mean looking, and the houses have a shabby, dilapidated appearance, wholly unworthy so celebrated a city. The private houses are usually from three to five stories in height, built of tufa or brick, and plastered over. With these are intermixed many huge old buildings (*palazzi*), contrasting most strikingly with the surrounding slight, mean tenements, by their bulk, height, and air of antique grandeur. In consequence, however, of the decay of the families to which they belong, many of them are now turned into ecclesiastical colleges or hotels, or let to foreign ambassadors or noblemen; and in those which have escaped this fate, the lower story is sometimes let for shops, and sometimes retained for stables, coach-houses, and servants' rooms. The second story is generally a picture gallery, consisting of a suite of rooms all opening into each other, and richly adorned with marble and painted ceilings. The owner of the building occupies the third story, or, if there be a fourth, the third and fourth, throwing open his galleries to artists and all who choose to give two or three *pauli* to his servants. "The great size of many of the palaces, and the abundance as well as bold projection of the ornaments, produce, indeed, a general impression of magnificence; but if we can get space enough in front to examine the parts distinctly, we often turn away dissatisfied with the absurdity and disproportion they exhibit. Generally speaking, these buildings exhibit great simplicity of design, usually presenting to the street one simple continued line of surface, rarely decorated either with columns or pilasters. Ornaments round the windows are never omitted, but are generally too large, and in bad taste. The stories are generally divided by horizontal mouldings along the front, and great space is left between the ranges of windows. The whole is crowned by a large and rich cornice." Such, according to Woods (*Letters of an Architect*, i. 438), is the general external appearance of the best among these *palazzi*, of which Rome comprises in all upwards of 300, many, of course, being very inferior both in size and architectural beauty.

Rome has not a single square; and of the piazzas or paved areas, the Piazza del Popolo, the Piazza Navona, and that in front of St. Peter's, are the only three that deserve notice. They are adorned with obelisks, statues, and fountains; but the first and last being at the extremities of the town, are lost as places for walking in or meeting friends. The fine promenade on the Pincian hill, E. of the Piazza del Popolo, is planted with trees, and commands an extensive view; but it is of small extent, and is shut at sunset. Without the walls, however, on the N.E. side of the city, is the Villa Borghese, the grounds of which, nearly 3 m. in circ., planted after the fashion of an English park, and ornamented with statues, fountains, &c., are open to the public, and constitute the favourite resort of all classes, whether on foot or in carriages. Rome, like London and Naples, is destitute of promenades sheltered from the wea-

ther, and well lighted at night, a convenience which Paris has in perfection; and another defect is the absence of elegant shops, cafés, and restaurants, that form so striking a feature in the French capital. In the number and grandeur of its public fountains, however, and in the quality of its water, Rome surpasses every city. Almost every public *piazza* has its fountain, and almost every fountain has some peculiarity in its size, form, or situation, to attract attention. The finest of these, and perhaps the most magnificent in the world, is the fountain of Trevi, which has a white marble basin in a vast enclosure paved with large slabs of the same material. It represents a palace of Neptune placed on a rough, broken rock, and adorned with Corinthian pillars; in the centre of the building, under a rich arch, stands Neptune in his car drawn by sea-horses; and water runs down in torrents from the rock, making, as it were, a sea at its base. In the summer the waters overflow their usual limits, fill the whole marble basin, and rise to a level with the square, which after sunset is a favourite lounge of the inhabitants. The fountain in the Piazza Navona has an obelisk in its centre, surrounded by Tritons and Naiads, seated on rocks, and spouting forth water in magnificent *jets d'eau*. The *Fontana felice*, on the Viminal hill, discharges itself under an Ionic arcade through a rock, which a figure of Moses is striking with his rod. Other figures surround the prophet, and below are spouting lions hanging over the basin, as if eager to slake their thirst. These and the other fountains of the city are supplied from these only out of the numerous aqueducts which attest the luxuriance of the ancient city; and yet such is the quantity they convey, and so pure are the sources whence it is derived, that no city can boast of such a profusion of clear and salubrious water, poured forth spontaneously in *jets d'eau* without the aid of expensive machinery, as at Versailles, St. Cloud, and Chatsworth.

But, after all, water at Rome would seem to be provided rather for show than for use. Forsyth says that when he visited the city, the fountain of Trevi was inaccessible from filth! (P. 124.) There is, generally, in fact, a great deficiency of cleanliness, and a limited consumption of water, in all cities supplied by fountains, and wherever, indeed, it is not conveyed by pipes into private houses, and placed, as it were, under the hand of the occupiers.

The great glory of modern Rome consists in the magnificence of her churches, or rather in the matchless structure of St. Peter's.

"Thou, of temples old, or altars new,
Standest alone—with nothing like to thee—
Worthiest of God, the holy and the true.
Since Zion's desolation, when that He
Forsook, his former city, what could be,
Of earthly structure in his honour piled,
Of a sublimer aspect? Majesty,
Power, Glory, Strength, and Beauty, all are aisled
In this eternal ark of worship undefil'd."

This magnificent fabric occupies the site of an older structure raised by Constantine the Great. Pope Julius II. laid the first stone of the new building on the 12th of April, 1506, having selected the famous Bramante for his architect. But the latter dying in 1514, other architects were employed to carry on the work, till, in 1546, it was fortunately committed to the illustrious Michael Angelo, who nearly completed the dome, and a large portion of the building. After Michael Angelo's death, in 1563, the work was prosecuted under other architects, till its completion in 1614. The colonnade by Bernini was added in 1665-67. "St. Peter's," says Mr. Maclaren, "unlike many other celebrated edifices, surpasses expectation. The front is too low, and has some other defects; but the vestibule is admirable, and the interior is solemn, grand, rich, harmonious, beyond any thing that I had conceived. It is unquestionably the noblest building ever reared by human hands, the only work of art, as Madame de Staël observes, which produces an impression of grandeur akin to that which we receive from the works of nature. So vast are its dimensions, that colossal statues and monumental groups of figures are stowed away in its aisles and recesses without impairing the unity and simplicity of the plan, as they do in the St. Paul's of London. The interior of the dome (which is 140 ft. in diameter), as well as a considerable portion of the other surface, is covered with pictures, all of which, however, are, with one exception, of mosaic. The eye forms most erroneous estimates respecting the height of the different parts of St. Peter's, and most visitors are on this account disappointed by first impressions. The splendid bronze *Baldacchino*, or canopy, immediately under the dome and over the high altar, close also to the supposed tomb of St. Peter, is about 190 ft. high, though in appearance only 30. The chair of St. Peter, too, behind the high altar, appears from a distance as if raised only a few steps from the ground, whereas it is placed on an elevation 70 ft. above the floor. The pen seen in the hand of the prophet in one of the lower compartments of the dome might be supposed to be 12 or 18 inches in length, whereas its real length is 6 ft. The visiter has no adequate conception of the magnitude of the dome until he arrives at the roof (the passages of which are so contrived that one may ascend on horseback), when he finds it rising before him like a mountain. The view from the external gallery round the lantern is exceedingly fine and extensive, embracing the Campagna from the sea to the Apennines; besides which, on looking at the roof of the church, its 10 cupolas give it somewhat the appearance of a town, so astonishing is its size. The depth of the floor, as seen from the interior of the lantern, appears lessened from 400 ft. to 100 ft.; but it may be discovered that the eye is deceived, as the promenaders below appear only as tiny infants. When we stand in the interior gallery of the cupola corresponding with the whispering gallery of St. Paul's in London, and look at the mosaics* on the concave, we are surprised to find them composed of square bits of coloured stone, half an inch broad, clumsily put together, often with intervals between them; yet, seen from below, they might pass for oil paintings. We find a pictured face of an angel, close to us nearly a yard broad; but, when we look across the gallery, a similar face, and really of equal proportions, seems to be of the natural size. The lights in this splendid edifice are finely tempered, well distributed, and kept in admirable order. The profusion, also, of rare and beautiful marble, introduced in every part, together with the gilded roof, the statues, monuments, mosaic ceilings and pictures, forms a display of brilliant and unexampled magnificence, which requires weeks and almost years to contemplate." (*Maclaren*, 177, 178; *Burton*, ii. 131–140.)

The form of the church, as designed by Bramante, was that of a Latin cross; but this was changed by Michael Angelo to a Greek cross, which has the advantage of exhibiting the whole structure at one *coup-d'œil*. Unfortunately, however, the plans of the latter were afterwards departed from by Carlo Maderné. In the caustic, and, perhaps, unjust language of Forsyth, " a wretched plasterer came down from Como to break the sacred unity of the master-idea, and him we must execrate for the Latin cross, the aisles, the mean-looking attics, and the low, ugly front." (*Italy*, p. 179.) The latter, however, is 396 ft. in length, and 150 ft. in height; and with whatever defects it and other parts of the structure may be charged, still there can be no question that "St. Peter's is beyond all comparison the most magnificent temple ever raised by mortal hands to the worship of the Supreme Being. It is a spectacle that never tires: you may visit it every day, and always find something new to admire. Then its temperature is delightful—as pleasantly cool in summer, as it is comfortably warm in winter. The fact is, that the walls are so thick that the air is not affected by that without, so that, like a well-built cellar, it enjoys an equability of temperature all the year round." (*Matthew's Diary of an Invalid*, p. 86.)

Of the many august ceremonies performed in this magnificent temple, the most imposing is that of the *Tenebræ*, on the night of Good Friday, when the hundred lamps that burn over the tomb of St. Peter are extinguished, and a stupendous cross of light appears suspended from the dome between the altar and the nave, shedding over the whole edifice a soft lustre, delightful to the eye, and forming, with the objects animate and inanimate on which it sheds its light, a scene singularly striking, by a happy mixture of tranquillity and animation, of darkness and light, of simplicity and majesty; a scene, indeed, far more sublime and more deeply impressive than the illumination of the external dome on the night of St. Peter's day (June 29th). (*Eustace*, ii. 164–192.)

The dimensions of St. Peter's have been variously given by different authorities, and perhaps exactness is not attainable ; but the following measurements are adopted by Gwilt, and may, perhaps, be depended on. The dimensions of St. Paul's, of Milan cathedral, and St. Sophia's at Constantinople, are from Woods, Eustace, and Dallaway.

Dimensions.	St. Peter's.	St. Paul's.	Milan Cath.	St. Sophia.
	Ft.	Ft.	Ft.	Ft.
Extreme inside length .	607	510	442	269
Length of transepts .	445	292	284	243
Height from cross to floor . .	438	308	256	
Width of nave . .	107		177	
Total area, including outside walls .	227,000	84,000		

Comparing the Roman church with the British cathedral, which, though *longo intervallo*, may well claim to be the second in the world, the floor of St. Peter's covers nearly

* The mosaic work is made not far from the church, in a building, part of which is used for the office of the Inquisition. The pieces employed in making these pictures are square pieces of a vitrified substance, called *smalti*, composed of glass, lead, and tin; and there are about 15,000 different shades of colour. Marbles were first used, but the glare from their polished surface was found to destroy the effect of the picture. (*Burton*, ii. 140.)

five English acres (nearly the size of the Coloseum), while that of St. Paul's occupies only two acres; and the actual bulk or entire contents of the former, as compared to the latter, are as four to one. And, taking into account the number and splendour of the decorations of St. Peter's, we need not wonder that it is supposed to have cost, with its monuments, gilding, and embellishments, from 12 to 16 *millions sterling*, whereas the cost of St. Paul's did not exceed £750,000 ! " In the interior of these two noble buildings the difference is scarcely less striking than between one of our old barn-like meeting-houses and the most elegant of our modern episcopal churches ; but, as regards the exterior, all admit, that in symmetry, purity of design, and true architectural beauty, the English is superior to the Roman temple." (*Maclaren's Notes*, 177–180.)

It is to be regretted that the situation of this cathedral of Christendom has been remarkably ill chosen. " No building," says Woods, " of great consequence was ever so badly placed. There is no distant point of view in which this church gives the impression of great magnificence, or from which it has the appearance of being such an immense building as it really is. This is owing to its situation in a hollow between the Janiculine and Vatican hills, which are connected by a neck behind it : so that on three sides it is surrounded by slopes rising almost immediately from it to about the height of the nave ; and even in front, notwithstanding the large space before it, the building seems encumbered with houses, which occupy a slip extending towards the river. From the bridge of St. Angelo little is seen but the dome itself ; and even when a glimpse is at length caught of its front and of the circular colonnade by which it is approached, it appears much nearer than it is ; and the magnificent avenue, the *chef-d'œuvre* of Bernini, seems a finer object than the edifice to which it leads." (*Wood's Letters of an Architect*, i. 368.) This colonnade, which consists of two opposite semi-ellipses, forming each a broad covered passage leading to the front of the cathedral, comprises 256 Doric columns of travertine 40 ft. in height, arranged in four rows, and surmounted by 192 statues of saints. In the central space, between the colonnades, flanked by two fountains, is an obelisk, consisting of a single block of granite about 80 ft. in height, surmounted by a cross, the total altitude being 136 ft. ; it was brought from Egypt to Rome by Caligula, and formerly stood in the circus of Nero, having been removed to its present situation by Pope Sixtus V.

" All churches," says Forsyth, even of the patriarchal class (those called *basilicas*), stand at an awful distance from the majesty of St. Peters." That of St. John Lateran however (at the S.S.E. extremity of Rome, far from the modern buildings, and upwards of ½ m. from the Coliseum), is entitled to the second place, and in former times was superior to St. Peter's. The chapter of the Lateran even now takes precedence, and the popes are always crowned here. This church was built on the site of an older structure in the beginning of the 14th century. Its front, consisting of a magnificent colonnade, is certainly impressive, notwithstanding its numerous faults. There are 5 entrances ; that in the centre having a bronze door taken from the Temple of Peace in the Forum ; and on the top of the façade are 15 statues of our Saviour, and various saints. The interior is divided into 5 aisles, and is the pillars of the nave are colossal statues of the Twelve Apostles ; this church comprises also a chapel of the Corsini family, in the form of a Greek cross with a central dome, gorgeously decorated with marbles, gilding, and pictures, and said to be one of the richest in Rome. Adjoining this church is a palace, which, after having been for many centuries the residence of the popes, was converted, in 1693, into a hospital for the poor ; and at no great distance is the Scala Santa, a building celebrated for containing a staircase of 28 white marble steps, alleged to have belonged to the palace of Pilate at Jerusalem, and which orthodox Roman Catholics esteem a meritorious act of piety to ascend on their knees ; indeed, so great is the number of devotees, that, with a view to the preservation of the steps, they have been covered with planks of wood. At the top is a painting of our Saviour at the age of twelve, begun by St. Luke, but finished by miraculous agency ! (*Burton's Rome*, ii. 177.) The church of St. Paul's, outside the walls, one of the handsomest in Rome, and in many respects superior to that of St. John Lateran, was burned down in 1823, and is now in course, though very slowly, of being rebuilt. The basilica of St. Maria Maggiore is a very fine and large edifice ; but the profusion of its ornaments takes from the unity of the main design, and the narrow brick tower, rising above the whole, is in very bad taste. The interior has three aisles, the central one being lined by 36 Ionic pillars of white marble, which have a very beautiful effect : they are undoubtedly ancient, and may have belonged to the temple of Juno Lucina, that formerly stood here. A chapel in this church, belonging to the Borghese family, deserves notice for the richness of its decorations. The church of *St. Pietro in vincoli*, originally erected about anno 490, but since wholly rebuilt, presents to the eye a noble hall, supported by 20 Doric columns of Parian marble, open on all sides, adorned with some beautiful tombs, and terminating in a semi-circle behind the marble. But it is principally remarkable for its containing the tomb of Julius II., illustrated by the noble statue of Moses by Michael Angelo.* The other churches are so numerous that it would be an almost endless task to describe them; many of them command admiration from the splendour of their decorations and the articles of *virtù* which they contain; but perhaps the finest and most worthy of attention from the stranger are those of *Santa Agnese* in the Piazza Navona, *San Carlo al Corso*, *San Ignazio*, the *Santi Apostoli*, the *Chiesa di Gesù*, and the *Chiesa Nuova*; all of which are abundantly rich in marbles, gilding, &c., though not always disposed in the best taste.

The *Vatican*, the most ancient and by far the most celebrated of the papal palaces, is a mass of buildings erected at various times by different popes, said to cover a space about 1200 ft. in breadth, and to comprise 4000 apartments. "The effect," however, says Burton, "is any thing but pleasing; from no point of view does it present any extent of front or magnificence of design; while its proximity to St. Peter's interferes most unfortunately with the view of that building." (ii. 255.) The interior consists of a suite of galleries of small breadth, which, if placed in a continuous line, would extend two miles in length. It contains a countless multitude of inscriptions, statues, busts, relievos, urns, sarcophagi, and vases, to say nothing of its literary and numismatic treasures, its books, MSS., drawings, the number of which the visitor can only guess at by counting the presses that conceal them from his sight. Taken altogether, it is by far the richest museum in Europe, and the precious objects it contains are magnificently lodged; for when the church was rich, she patronised the arts liberally both by buying and building, and even now the posthumous benevolence of popes and cardinals occasionally expends itself in erecting a new gallery or embellishing an old one. (*Maclaren*, p. 174.) The collection of sculptures is beyond all comparison the largest and most valuable in Europe, comprising, among other great works, the unequalled group of Laocoon and his sons, which even Michael Angelo despaired of being able to restore, the celebrated Apollo Belvedere (found at Antium, near the close of the 15th century), the well known group of the Nile and his offspring, the Belvedere Torso of Hercules and Hebe, a noble statue of Adonis, and another of Marcellus with an excellent bust of Pius VII. by Canova. The library of the Vatican is alleged to comprise about 80,000 printed books and 35,000 MSS.; but, in point of fact, its literary riches are unknown, the catalogues having never been completed. There is reason, however, to think that its collection of ecclesiastical MSS. immeasurably surpasses any other in Europe; but it is very deficient in works of modern literature; and its value can be fully appreciated only by the churchman and the antiquary. The picture-gallery, which is by no means extensive, is, as compared with the collections now noticed, quite of modern creation. The frescoes of Raphael and Michael Angelo, indeed, have long adorned the walls of the Vatican; and the works of the former occupy three open galleries, called the *Loggie di Raffaelle*, which go round three sides of a square court; but the oil-paintings have been collected wholly by Pius VI. and subsequent pontiffs, including the present pope, to whom we are indebted for the completion of the gallery. It comprises several of the *grand* productions of the Italian schools, including among others the "Transfiguration," by Raphael, usually considered his *chef-d'œuvre*, and the noblest work of art in the world; the "Madonna di Foligno" by the same master, the "Communion of St. Girolamo" by Domenichino, the "Martyrdom of St. Sebastian," perhaps Titian's very best work, the "Madonna of Monte Luca" by Giulio Romano, &c. The Vatican has two chapels, the most celebrated of which, known as the Sistine chapel, was built by order of pope Sixtus IV.; its walls and ceiling were covered during the pontificates of Julius II. and Paul III. with frescoes from the masterly hand of Michael Angelo. Behind the altar is the magnificent painting of the Last Judgment, the theme of so much eulogy and so much criticism; and on the ceiling are represented the Creation, the Deluge, and other scriptural subjects, the windows being adorned, somewhat inconsistently, with full-length figures of Prophets and Sibyls It is a very large and lofty oblong room, with scarcely any of the furniture of a chapel: it is used on few occasions, except during the Holy week, when the *Tenebre* and the "overwhelming" *Miserere* of Allegri are sung by the papal choir. In this chapel

also the cardinals meet in conclave for the election of a new pope. The *Sala Regia*, a hall of great size and good proportions, connects the above-mentioned chapel with another dedicated to St. Paul, which, like the first, is painted in *fresco* by Michael Angelo, and used only on great festivals. The Vatican is now seldom inhabited by the pope, except during the grand festival of Easter, the present abode of the pontiffs being on the Quirinal-hill (now called Monte Cavallo, from the two horses on its summit, taken from the baths of Constantine.) This palace, which was begun by Paul III., presents two long fronts, plain and unadorned. Like those of most of the other palaces, the court within being upward of 300 ft. in length by 165 ft. in width, three sides being surrounded by porticoes, and the fourth having a double row of arcades surmounted by a clock tower. The grand staircase on the side to the right of the gateway conducts to the papal apartments, and the chapel, all of which are on a grand scale, and adorned with frescoes, especially those by Guido, which ornament a small private chapel. The adjoining gardens are spacious, refreshed by several fountains, and shaded by groves of laurel, pine, ilex, and poplar; but little attention being paid to them, they have a shabby, neglected appearance, made only more apparent by their great size. In front of the palace stand an Egyptian obelisk, flanked on either side by the statues of the horses which, as already stated, give the hill its present name. The Lateran Palace, near the church of St. John Lateran, has three lofty fronts of great extent and simplicity: a few apartments are reserved for the pope, when he comes to perform service at St. John's; but the rest is used as a hospital for 250 orphans.

Among the *public buildings* of modern Rome, the *Campidoglio*, or modern capitol, deserves notice, as being one of the best architectural works of Michael Angelo. The road to it is by a labyrinth of narrow dirty streets, leading from the Corso to two flights of steps, at the foot of one of which are two basaltic lions. At the top are colossal equestrian statues of Castor and Pollox, on a line with which stand several other statues and trophies. Opposite the steps is the senator's palace, the two other sides being occupied by the palazzo de Conservatori and the Museo Capitolino, the garden of which overhangs the Tarpeian rock.

"Fittest goal of treason's here,
The promontory whence the traitor's leap
Cured all ambition."

But, owing to the accumulation of soil at the bottom, the leap might now be taken without any very extraordinary risk.

The ancient buildings, to be noticed in the sequel, are nearly all gone; but there are many statues, and one in particular, a bronze equestrian figure of M. Aurelius, occupying the centre of the Piazza del Campidoglio, demands attention, not only on account of its beauty, but its acknowledged antiquity. In fact, quite enough is still furnished both by nature in the commanding position of the hill, and by art in the various architectural embellishments formed principally of ancient materials, to call up in the mind of the classical student those bye-gone days when decrees issued from the capitol fraught with the destinies of a subject world.

"Ages and realms are crowded in this space
This mountain, whose obliterated glories
The pyramid of empires pinnacled.
Hither the kingdoms and the nations came
In supplicating crowds to learn their doom."

Childe Harold.

The Capitoline Museum comprises a few and not very valuable paintings by the old masters; but on the other hand the collection of statues and marbles includes some of the most precious relics of ancient art, among which may be mentioned the "Dying Gladiator," the misnamed "Antinous," and the splendid group of "Cupid and Psyche." (*Conder's Italy*, iii. 347: *Maclaren*, p. 168–169.) The only other government buildings requiring notice are the new postoffice, in the Piazza Colonna, near the Corso, and the castle or citadel of St. Angelo: the central tower of the latter was built by Hadrian (thence called *Moles Hadriani*) for a mausoleum, and was unquestionably the most superb sepulchral monument ever raised in Rome. It began to be used as a fortress when the city was attacked by the Goths: its defences were strengthened by various pontiffs, the last and greatest improvement having been made by Urban VIII., who completed the *fossi* and bastions towards the meadows. It is now used as a state prison and house of correction, but it is better known to foreigners as the place whence are discharged the magnificent fireworks of the Easter festival.

Rome, as previously seen, comprises a great number of palaces, with the general plan of which the reader is already acquainted; but a few deserve more particular notice. The *Colonna* palace, in the square of its own name, fronting the Corso, is entered by a noble painted staircase leading to a gallery which, in point of size and architecture, is the finest in Rome: the roof of the saloon is supported by

* The magnificent ode or sonnet of Zappi, inspired by the contemplation of this statue may be seen, with a spirited translation, in *Roscoe's Life of Leo X.*, iv. 320, 3d edit.; it is also given with a different translation in *Burton's Rome*, ii. 234.

polished columns of *giallo antico*, and the ceiling displays the battle of Lepanto, the event of which raised a colonna to the honours of a Roman triumph. The collection of pictures, however, is about the worst of any in Rome. (*Williams's Letters*, ii. 80.) The *Borghese* palace is also a very splendid building, "remarkable for its extent, its porticoes, its granite columns, and its long suite of apartments, being still more distinguished by the well-supported magnificence that pervades every part, and gives the whole mansion, from the ground floor to the attic, an appearance of neatness, order, and opulence." These are the words of Eustace; and, according to latter writers, this palace is at present equally well kept up. The collection, which, contrary to the usual rule, is on the ground floor, occupies nine large rooms, and ranks among the first in Rome. The *Doria* palace, in the Corso, has three vast fronts, and comprises a spacious court, surrounded by colonnades; the gallery is particularly rich in paintings of the Italian and other schools, including many landscapes by Gaspar Poussin and Claude; in fact, nowhere else in Rome is there so varied a collection, or one so well suited for the purposes of study. (*Rome in the 19th century*, iii. 6-14.) The *Barberini* palace, one of the grandest buildings in modern Rome, built from the united designs of Maderno, Bernini, and Borromini, consists of a projecting centre, surmounted by a square lantern and two smaller wings: it comprises, also, besides statues and paintings, a noble library, freely thrown open to the public. The *Farnese* palace, a noble structure, to erect which the Colosseum and theatre of Marcellus were despoiled of their choicest ornaments, had formerly a magnificent and, in some respects, unrivalled collection of ancient sculpture, paintings, and books. But the king of the Two Sicilies having succeeded to the rich inheritance of this illustrious family, the collection has been carried to Naples. On the ceiling of a gallery belonging to the *Rospigliosi* palace, on the Monte Cavallo, is the famous Aurora, the *chef-d'œuvre* of Guido, which the beautiful engraving by Morghen must have made familiar to many of our readers. The *Bracciano* and *Corsini* palaces, and others too numerous to be mentioned, deserve attention for their architectural merits and the treasures of art with which they are furnished.

Rome, besides its palaces, has numerous *villas*, both within and without the present walls, built chiefly by a few cardinals, whose riches, taste, learning, and leisure, conspired to create these beautiful retreats. The Villa Borghese has already been mentioned as including pleasure-grounds that form the favourite resort of the modern Romans, and the mansion, with its collection of pictures and marbles, is equally beautiful. The Villa Albani, outside the Porta Salara, is of exquisite design, planned by Cardinal Albani, one of the profoundest antiquaries of modern times: it was here that, under his patronage, Winkelman pursued those studies that enabled him to write his history of ancient art. The collection, once far more extensive, is said to be extremely choice; in fact, it does not contain a single mediocre piece. The villas Aldobrandini, Pamfili, Santi, and Ludovisi, are all, more or less, adorned with frescoes and ancient statuary; besides being well situated and surrounded with fine gardens. The magnificent Medici villa, on the Pincian hill, is now appropriated to the use of the French Academy.

The Tiber, including its windings, runs for three miles through Rome. The greatest breadth of the stream within the walls is only about 400 ft., and the smallest 200 ft., the average being somewhat less than one third the breadth the Thames at London Bridge, and considerably less than that of the Clyde at Glasgow. It is deeper, however, than the Clyde, and has certainly a larger volume of water.

> " Vorticibus rapidi et multâ flavus arenâ
> In mare præruperat." *Virg. Æn. vii. 31.*

It well merits the epithet of *flavus*, as it is not only discoloured, but loaded with *yellow* mud beyond almost any other river; and this is not the consequence of accidental floods, for its waters are scarcely ever clear, and hence, no doubt, its ancient name *Albula*. Its banks are low and bare, constituting for the most part of crumbling soil, without verdure, and at many places without even protecting walls; they are not ornamented by a single promenade or decent street. (*Maclaren*, p. 150.) There are only two places where there is a sort of quay, or landing-place, one called *Ripetta*, on the E. bank, above the bridge of St. Angelo, where boats from the inland provinces land wine, provisions, &c., and the other at the S. end of the city, on the site bank, called the *Porto di Ripa Grande*, where sea-going vessels la nd their cargoes: and where there is a line of warehouses, and a custom-house. Three bridges cross the river within Rome; that most northward is the Ponte Angelo (anc. *Pons Ælius*), built by Hadrian, and restored in its present form by Clement IX.: it is a structure of great beauty, having a balustrade, on the top of which are several hideous-looking figures of angels. The next,

proceeding southward, is the Ponte Sisto (anc. *Pons Janiculensis*), built by either Trajan or Antinonus Pius, and rebuilt by Sixtus IV., in the 15th century. About half a mile lower down is the island of San Bartolomeo, the ancient *Insula Tiberina*, of oblong shape, about 1000 ft. in length, and 300 ft. in breadth, united with the E. bank by the Ponte San Bartolomeo, (anc. *Pons Cestius*), and with the W. bank and the district of Trastevere by the Ponte di quattro Capi, so called from a head of Janus Quadrifrons that formerly stood there; it is the Pons Fabricius of antiquity, and was constructed *anno* 61, B.C. Within Rome, also, are the remains of three ancient bridges, the principal being the *Pons Triumphalis*, so called from the circumstance of the triumphal processions crossing it on their way to the capitol: it is now entirely destroyed, but the piers of it may be distinguished by the agitation of the water a little below the bridge of St. Angelo. About half of the Pons Palatinus, the most ancient bridge of Rome, is still standing, a few yards below the island of San Bartolomeo, and a continuation has been made of wood for the accommodation of foot passengers. The Pons Sublicius or Æmilius, the most southerly of the old Roman bridges, has long disappeared. The Tiber is now, as it was in Virgil's time, subject to very high and also frequent floods, the water sometimes rising as high as the Piazza di Spagna, and very frequently laying under water all the streets in the busy quarter near the river. On the Porto di Ripetta are two pillars which mark the height of the different floods for some centuries past; and it appears from it that they have all happened between the months of November and February. The frequency of these inundations gave rise to various projects for preventing them, and Aurelian caused the banks to be raised and its channel cleared, though with what success we are not informed. The vast accumulation of soil, however, by which the surface of modern Rome is raised so many feet above the ancient city, must undoubtedly, make it less liable to suffer from floods now than formerly. (*Burton*, i., 236.)

The *classical monuments* of Rome are very numerous, their interest depending on their beauty, grandeur, and singularity, their intrinsic merits, or on the events and personages historically associated with them. Those who expect gratification from the first source will, in many instances be disappointed, as the greater number present little to please the eye or gratify the taste. The *aqueducts*, for example, those astonishing efforts of human industry, which stretch across the Campagna in various directions, exhibit their real greatness only to the understanding. To the eye, these works (of which there seem to have been 14, coming from 9 different sources,) present merely a series of naked brick arches, scarcely larger than a house-door in span, or higher than a park-wall, and without any sort of ornament. Near the mountains, and in crossing valleys, they may be lofty; but in the vicinity of the city they are low and tame; three only now remain in a state fit for use, viz. the Aqua vergine, Aqua felici (anc. the *Claudius aqueduct*), and the Aqua Sabatina, which supplies the Janiculum. The Roman *roads*, also, solidly built of large stones, may be called great works for their expense and utility, but they have no external attractions. The same remark applies to the *Cloaca*, of which a false idea is conveyed by calling them *sewers*. They were rather drains made to carry off the stagnant water of the pestilential marshes, which occupied much of the low ground near the Tiber, and the spaces between the Aventine, Palatine, and Capitoline hills. They were constructed at a very early period (according to some, in the time of the kings), for the obvious reason, that the marshes separated the first inhabited parts of the city from each other, and their desiccation became indispensable. The height and width of the *Cloaca maxima* are equal, each measuring 13¼ ft.: a view of it may be obtained at its mouth, where it flows into the Tiber a little below the Ponte Rotto, another portion being visible near the Arch of Janus Quadrifrons.

The *baths*, as they now exist, are an assemblage of naked, half dilapidated brick walls, which surprise by their huge size and the extent of ground they cover. We know of the former existence of *eight* thermæ, erected by different emperors; and the carcasses of three remain in considerable masses, those of Titus, Caracalla, and Diocletian, the ruins of the first two of which are in vineyards, a great part of the last having been transformed by Michael Angelo into the church and monastery of Santa Maria degli Angeli. Each consists of a labyrinth of apartments, the uses of some of which antiquaries have scarcely been able to conjecture. Caracalla's baths cover an area of 28 English acres, a space nearly three times the size of Lincoln's-inn-fields, in London. "But we must keep in mind the multifarious nature of these establishments, which included not only baths, with their caldaria, frigidaria, sudatoria, &c., but porticoes and gardens, libraries, lecture and conversation rooms for the philosophers, academies and halls for declamation, gym-

703

ROME.

nasia for the 'fancy,' theatres for the gay, temples for the devout, and, most probably, wine shops or places of refreshment for all; in fact, they were less baths than 'places of universal recreation.' The Roman citizen left his house early, and only returned at night to his bed, spending the day chiefly in the forum, the courts of justice, or at the baths. The latter establishments seem to have combined the uses of our coffee-houses, reading-rooms, libraries, lecture-rooms, and theatres, as well as baths. We may call them, indeed, national *club-rooms*, supplied with every species of accommodation then in vogue, and open to the whole free citizens of Rome. It is this circumstance which gives them their interest. We cannot tread these ghastly chambers, where no sound now falls on the ear save the echo of our own steps, without thinking of the animated throng that once peopled them, the crowd of Roman citizens of all grades and classes, from the emperor to the mendicant who received his monthly dole of wheat from the public granaries, the foreigners from every clime, princes, tax-gatherers, hostages, petitioners, litigants, soldiers, parasites, who came to the seat of empire for business or pleasure. All these must have resorted to the baths, some for ablution, exercise, or amusement, some to read in the libraries, some to listen to the philosophers, some to talk of the news and hear bulletins read from the armies, announcing battles on the Rhine or Euphrates, or insurrections in Spain or Gaul." (*Maclaren's Notes*, p. 142.)

Among the numerous temples that once adorned the ancient capital of the world, the Pantheon and the temples of Vesta, Peace, Fortuna virilis, and Bacchus, present extensive and very interesting remains; but incomparably finer than all the rest is the Pantheon (in the ancient *Campus Martius*), which, though stripped of its external ornaments, and disfigured by two modern belfries, erected by Bernini, is entire within. This exquisite temple was built, as Pliny tells us (*Hist. Nat.*, lib. xxxvi., cap. 15), and as the inscription on the portico testifies, by Agrippa, the friend and general of Augustus, who dedicated it to Jupiter the Avenger, *Jovi ultori*. It is a perfect circle, 180 feet in diameter.

> "Relic of nobler days and nobler arts
> Despoiled, yet perfect, with thy circle spreads
> A holiness appealing to all hearts—
> To art a model "

Its beauty consists in its admirable proportions; and its portico, 110 feet in length by 44 feet in depth, supported by 16 Corinthian columns of white marble, has a most majestic appearance. "The portal is more than faultless; it is positively the most sublime result that was ever produced by so little architecture." (*Forsyth.*) The dome is of great extent, and has a central aperture, from which the building receives its entire light. The Pantheon has been stripped of everything that could be taken away, in order to furnish materials for the embellishment of St. Peter's. It is now made the receptacle of monuments to those who have deserved well of their country, and contributed to sustain the reputation of Italy.

The great wonder of ancient Rome, however, is the Flavian, amphitheatre, now the Coliseum, or more properly Colosseum[*] unquestionably the most august ruin in the world, and by far the largest amphitheatre of which we have any knowledge.

> "Omnis Cæsareo cedat labor amphitheatro,
> Unum præ cunctis fama loquatur opus."
> *Martial de Spect.*

It consists of a vast ellipse, the length of the longest diameter measured from the outside of the outer wall being about 690, and that of the shortest 513 feet, so that it covers about 5½ English acres of ground! The longest diameter of the arena has been variously given at from 287 to 300 feet, and the shortest at from 180 to 190 feet; the space between the arena and the outer wall (from 160 to 167 feet) being occupied by the walls, corridors, and seats that rose tier above tier from the wall round the arena nearly to the top of the outer wall. The latter, which is about 179 feet in height, consists of three rows of vaulted arches rising one above another, exclusive of which, it had, when perfect, upper works of wood. This colossal amphitheatre is said to have had seats for 87,000 spectators, and standing room for 20,000 more! There is really, therefore, but little of exaggeration in the statement of Addison that the amphitheatre

> On its public shows unpeopled Rome,
> And held, uncrowded, nations in its womb.

The arena was sufficiently extensive for the exhibition, on the grandest possible scale, of the bloody sports that delighted the ancient Romans; and here hundreds, and even thousands, of gladiators and of wild beasts have frequently contended at once—

> " Butcher'd to make a Roman holiday!"

[*] The amphitheatre of Vespasian and Titus was, in the lower ages, called *Colosseum*, or *Coloseum*, from the vastness of the building, and no', as has been sometimes stated, from a colossal statue of Nero, supposed to have stood near it. (*Lumisden's Antiquities of Rome*, a valuable work, p. 329.)

704

This magnificent ruin has been much damaged by earthquakes, lightning, and the destroying influence of time; but it has suffered incomparably more from the injuries inflicted upon it by the successive masters of Rome. In the 13th century it was occupied as a fortress; and in the course of the next century it became, what it long continued to be, a common quarry, whence materials were taken to build a large portion of the modern city. (*Hobhouse's Illustrations of Childe Harold*, p. 363–366.) In consequence of these lengthened devastations, " not a single step is now remaining of all the seats of stone which rose in regular succession from the arena; but the wall which surrounded it, to prevent the escape of the wild beasts, is nearly entire. The interior presents a most complete scene of destruction. By means of broken staircases, we may climb up a considerable height, and almost be lost in the labyrinth of ruins. It is from such a view of these remains that the best idea of their vastness is formed; and if viewed by moonlight, when the shattered fragments of stone, and the shrubs which grow upon them, are seen at a distance in alternations of light and shade, the mind receives mingled impressions of gratification and melancholy which, perhaps, no other prospect in the world could produce." (*Burton's Rome*, i. ii. 55.) At length, however, an end was put to the spoliation of this most splendid relict of imperial Rome. "Benedict XIV. consecrated the spot which persecution and fable had stained with the blood of so many Christian martyrs." (*Gibbon*, xii., 455.) And subsequent pontiffs have repaired and strengthened portions of the existing ruin.

"The pile speaks powerfully to the imagination, through the bloody rites once celebrated in it. It was the scene of those savage fights of gladiators, those combats of wild beasts which were unknown among other ancient nations, and have fixed a brand of infamy on the Roman name. The interior has been carefully cleared out, and the boundaries of the arena which was so often soaked with blood are distinctly seen. When we stand among the broken arches of this vast edifice, now the abode of bats and owls, silent as the grave, and with not a single building near it, our thoughts are irresistibly carried back to the thousands of all ranks and both sexes who once filled its ample benches, to the roars and yells of the wild animals lacerating each other, the shrieks of the slaughtered human beings, and the shouts of the blood-thirsty multitude now applauding the blow which took away a fellow-creature's life, and now calling out for fresh victims. These cruel exhibitions most characteristic of Rome, both republican and imperial. The Greeks, wherever they established their power, carried with them the elevated sentiments and graceful mirth of the stage; and you may trace the dominion of the Romans over the ancient world, by the amphitheatres built for the gratification of their ferocity." (*Maclaren's Notes*, 152–3.)

The Romans, always fond of shows and games, were especially attached to those of the Circus.

> Duas tantum res anxius optat,
> Panem et Circenses.
> *Juvenal, Sat* 10, v. 81.

There are said to have been at one time no fewer than 13 circuses in the city and its environs. The principal of which were the circus Maximus, circus Agonalis, and the circuses of Nero and Caracalla. Of the circus Maximus (which Ammianus Marcellinus describes as being at once " the temple, the dwelling-house, the public meeting, and all the hopes of the ancient Romans"), there are now no remains; but Pliny informs us that it was capable of accommodating 260,000 spectators; Juvenal, using, perhaps, a poet's licence, goes the length of saying,

> " Totam hodie Romam Circus capit."

The form of the circus Agonalis (supposed to have been built by the emperor Severus) may be traced in the piazza Navona; and even the round end is not lost: it is about 750 ft. in length; and the races held here during the carnival forcibly remind us of the uses to which it was formerly devoted. The circus of Caracalla, outside the gate of San Sebastian, has its walls still entire, though the seats have fallen in, leaving a kind of terrace along the whole length of the walls. It appears to have been 1678 ft. in length, 435 ft. in width, and to have been capable of accommodating 20,000 spectators. Of the other Roman ruins (except those of the forum and capitol), the two columns of Trajan and Antonine, and the three triumphal arches of Titus, Constantine, and Severus, principally deserve attention, from their beauty and the taste displayed in their execution. The column of Trajan, erected to commemorate that emperor's successes in Dacia, is 115 ft. 10 in. in height, not including the statue of St. Peter, which Sixtus V. had the bad taste to substitute for that of its illustrious founder. A spiral staircase leads to the balustrade at the top, and the exterior is adorned with sculptures in basso relievo, spirally arranged round the column, representing the victories and achievements of the emperor. Napoleon's pillar at Paris is a good imitation of that of Trajan. The pillar of Antonine (or, more correctly, of M. Aurelius, for he erected it), in the piazza Colonna, is 122½ ft. high, and is now surmounted by a colossal statue of St. Paul: the bas-reliefs, similarly arranged to those on the other column, are not nearly so well executed, and the whole is much injured and defaced. The arch of Titus, built to commemorate his victories over the Jews, consisting of a single arch, was adorned with eight marble columns of the Composite order, and had its interior covered with sculptures, representing the emperor's triumph after the capture of Jerusalem; but it is in a state of great dilapidation, many of its rich decorations having been carried off to embellish the Farnese and other palaces: latterly, however, some attempts have been made towards its restoration. Till the time of Sixtus IV., the bas-reliefs were not visible, so much had the soil accumulated and buried the arch; but that pontiff ordered it to be excavated; and there is now a clear passage under it at the level of the ancient pavement, and, perhaps, on the pavement itself. The arch of Constantine, at the foot of the Palatine hill, near the Colosseum, is the most noble, because the best preserved structure of the kind in Rome; but it is indebted for its chief beauty to the spoliation of the arch of Trajan, which stood in the forum of that emperor, and which the senate, with equal barbarism and servility, stripped of its bas-reliefs and other rich materials, employing them to ornament the new structure. It consists of a large arch between two of smaller size, having on each side four fluted Corinthian columns of giallo-antico supporting the figures of eight Dacian captive warriors. It is covered with the bas-reliefs taken from the arch of Trajan, and with others of later date and of very inferior execution. The soil which had accumulated round this arch was excavated in 1804, when part of the *Via Triumphalis* was brought to light. The grass-grown platform at the top was once, probably, occupied by the victor in his triumphal car; but this has disappeared. We have already stated, that the paltry gateway in front of Buckingham palace is a wretched miniature imitation of this arch. (*Lumisden's Ancient Rome,* 397.)

The Capitoline hill, "that rock of triumph, that high place where Rome embraced her heroes," naturally kindles a feeling of enthusiasm; but of the topography of the ancient capitol we really know next to nothing. "Four temples, 15 chapels, three altars, the great rock, a fortress, a library, an athenæum, an area covered with statues, the enrollment-office, all these are to be arranged on a space 400 yards in length, and 200 in breadth; and of these, the last only can with precision be assigned to the double row of vaults crowded with salt, where the inscription of Catulus was discovered. The citadel may be believed to have extended along the whole side of the hill." (*Hobhouse's Illustrations of Childe Harold.*) But however little be known of the precise position of these ancient buildings, here was situated the *domus de canna straminibusque,* which passed for the house of Romulus, and was preserved with religious care till the time of the emperors; here the Roman people celebrated their most sacred rites, and kept their treasures, archives, trophies, records, Sibylline books, and other valued relics; and here 300 conquerors, in the space of 1000 years, deposited their spoils and consummated their glories, by the grand spectacle of a triumph. (*Maclaren,* p. 161.) The Roman forum, however, is, perhaps, the most melancholy object within the walls of "the eternal city." Its former grandeur is utterly annihilated; the ground has been applied to other purposes, and even the exact position of its various parts is much disputed, though it be probable that excavations, judiciously conducted, would set the question at rest. The forum, as described by Bunsen, the Prussian envoy, who took great pains on its investigation, appears to have been of no great size (about two acres), and to have owed much of its magnificence to the temples, basilicæ, curiæ, and other buildings that surrounded it. Indeed it was in consequence of inadequate size, that Julius Cæsar built a new one. Augustus, Trajan, and other emperors, followed his example, and Rome had ultimately a multitude of forums. If though the ancient forum Romanum and present Campo Vaccino be so desolate that we might apply to it Virgil's description of its appearance before the arrival of Trojan settlers. (*Æn.* viii., 360:)

"Passim armenta videres
Romanoque foro et lautis magiro carinis."

classical scholar turns with more pleasure to the pages of Byron:

"The Forum, where the immortal accents glow,
And still the eloquent air breathes, burns with Cicero!
The field of freedom, faction, fame, and blood;
Where a proud people's passions were exhaled
From the first hour of empire in the bud
To that when farther worlds to conquer failed."

forums of Augustus, Trajan, and Nerva, were laid out near the foot of the Capitoline hill, close to, though separated by buildings from the forum Romanum; and southward rises the Palatine hill, on which was built that mere village or collection of huts of which the masters of the world, in the days of their grandeur, loved to speak as "the cradle of their empire, the acorn, whence sprung the mighty oak that overshadowed the world." Cicero had a house here, and the brick ruins are still extant of the golden palace of Nero; but at present this spot, which once lodged the whole Roman people, is occupied by a single villa, surrounded by vineyards and gardens. All the more conspicuous monuments above described belong to the imperial times, for scarcely a shred remains which can be referred with certainty to the ages of the republic. The principal exceptions are the Tullian prison, comprising two dungeons, perfectly dark, and built with huge blocks of stone, answering in all respects, to the striking description given of it by Sallust; (*Bel Catalin.,* cap. 55); there are also two ancient tombs (one of which belonged to the Cornelian family, and contained the bones of the Scipios); and the Cloaca Maxima already mentioned; but these structures have little beauty, and derive their interest almost exclusively from classical associations.

Population.—It is extremely difficult to arrive at any just conclusion with respect either to the population of Rome, or of any other of the great cities of antiquity. Generally it has been exceedingly exaggerated. The great actions of the Romans, the vast extent of their empire, and the magnificence and splendour of their capital, the original seat of their power, seem naturally enough to lead to the conclusion that its population must have been immense. The strong national spirit of the Roman writers led even the most cautious among them to magnify the power and importance of the eternal city, which were exaggerated beyond all bounds by orators and poets, anxious to gain the favour of the public by flattering their prejudices, and exalting their power and greatness. The statements, too, of the classical writers as to the population of Rome and other great towns, are not only in themselves very vague, but, being extremely liable to mistakes in copying, have, no doubt, in many instances been magnified by copyists and others, always prone to exaggerate what is really great, and of which they have no distinct knowledge. And in addition to this, all inquiries into the population of Rome, Athens, and other ancient cities, are rendered peculiarly difficult from the circumstance of the returns of the censuses, and the statements in the classical authorities founded on them, usually or always referring to such free citizens only as were capable of bearing arms, without including children or slaves, though the latter formed in most instances a large, if not the largest portion, of the population. Our limits will not, however, permit of our entering into any detailed examination of the various statements that have been put forth with respect to the population of Rome. The exaggerations of Vossius, Lipsius, Châteaubriant, and others, who give to imperial Rome 14, 5, and 3 millions of inhabitants, are too absurd to deserve notice. Hume, who in his masterly *Essays on the Populousness of Ancient Nations,* has discussed the question of the population of Rome with his usual learning and good sense, arrives at the conclusion that Rome, when at the height of her greatness, might have been about as populous as London in 1760: in other words, that she might then have had from 700,000 to 800,000 inhabitants. Gibbon estimated the population at 1,200,000 (v., 286, 8vo. ed.); but it would appear the more moderate estimate of Hume is the more accurate, though the probability be that even it is beyond the mark. It appears from the very learned and elaborate researches of M. Dureau de la Malle (*Economie Publique des Romains,* liv. ii., cap. 10) that the area of Rome, included within the walls of Aurelian, which have been traced and laid down with the utmost precision, amounts to very near 1396½ hectares, that is to about three fifths the area of Paris: and the fair presumption is from the numerous forums and other open spaces in Rome, the number of the public buildings, and the great magnitude of many of the private residences, that its population, as compared with that of Paris, would be in a still less proportion. To the population within the walls has, however, to be added that of the suburbs, the amount of which is the subject of elaborate inquiry by the same learned critic. On the whole, he concludes, apparently on good grounds, that the population of imperial Rome, including its suburbs, in its most flourishing period, may be fairly estimated, allowing for troops and strangers, at between 560,000 and 570,000 (i., 403). And although we would not be understood as agreeing with M. Dureau de la Malle in all his statements, we have no doubt of their general accuracy, and of his estimate coming very near the mark. And, in fact, how small soever it may appear when contrasted with the statements that have been long current as to its vast magnitude, a population of 600,000 is really immense for a city like Rome, without either manufactures or trades, and the inhabitants of which chiefly depended for subsistence on the gratuitous distribution of the corn supplied by the conquered provinces.

During the troubles that devastated Italy, and especially Rome, from the 5th to the 13th century, the population of the city rapidly declined, and did not exceed that of a third or fourth rate town of the present day. But from the 14th century it began again to increase; and in the "golden days" of Leo X., it is supposed to have amounted to about 85,000. Towards the middle of the 17th century it was estimated at 90,000. In 1709, the inhabitants amounted to 138,568; in 1740, to 146,080; in 1765, to 161,899. (*Gibbon*, xii., 429) But owing to the influence of the French occupation, they had fallen off, in 1821, to 146,000, exclusive of Jews. It has, however, again increased; and at the last census, in 1838, it had 148,903 inhabitants. Owing partly to the unhealthiness of a portion of the town, and partly to the celibacy of large classes of the population, the deaths uniformly exceed the births; so that were it not for the influx of inhabitants from all parts of Italy, and indeed, of Europe, the population would rapidly decline. The decrease in 1837 was chiefly attributable to the ravages of the cholera.

The *inhabitants*, generally, are of a very mixed race; and it would be absurd to suppose, after so many changes, that they possess any considerable portion of ancient Roman blood. The men of the working and middling classes are generally stout and good-looking; though what are called Roman faces seem to be rarer than in England. The women, though good looking when young, soon become coarse; and, being large-boned, have a haggard appearance on losing their plumpness in old age. The men wear hats with crowns like a sugar-loaf, very wide cloaks wrapping round and round like a Scotch plaid; pieces of cloth tied about the legs with cords, instead of stockings, and sandals in lieu of shoes. The women generally wear a scarlet spencer with sleeves; and, for a head-dress, a piece of white linen, thickened on the crown by numerous folds, and with the end hanging down behind to the shoulders. Want of cleanliness is a common vice. The streets, public places, houses, and persons of the bulk of the population would all be improved by scrubbing, washing, and combing. Some of the most interesting objects are inaccessible from the accumulation of filth; and the appearance of the monks is absolutely disgusting; they are not redolent of holiness, but of dirt and vermin.

The cardinals and bishops being (under the pope) the rulers of the country, constitute the court-party, and claim the highest rank, after whom come the lay-nobility, subsisting on the revenues of their estates. The priesthood forms a very numerous portion of the inhabitants; for, besides 37 cardinals and bishops, it appears, according to the census of 1836, that there were then in the city 3490 priests and monks, besides 1384 nuns. The civil nobility, with a few exceptions, are few in number, poor, and without power or influence. The lawyers, who are divided into four classes (corresponding nearly with king's counsellors, barristers, attorneys, and notaries), form a pretty extensive section. After them rank the artists, a very numerous body, with a good deal of influence in society; and next to these are the *mercanti di Campagna*, a wealthy class, who farm extensively, and have warehouses at Rome for the sale of their produce. Rome has about 5500 shops; but their owners, with some few exceptions, rank below the classes above described. The foreigners, a mixed multitude, among whom the English and Russians are the most numerous, and, generally speaking, the best informed, constitute a class of themselves; enjoying also, in consequence of their wealth, many peculiar privileges. The police exercise no inquisitorial powers; and foreigners may live as they please without attracting attention, and do, with impunity, what would not be permitted to natives. To this circumstance, as well as to the fascinations of antiquity and modern art, we may attribute the visits of foreigners; for, of the English at least, a large proportion are led by motives very different from a love either of the fine arts or classic lore. (*Maclaren*, p. 145.) With respect to morals, it is admitted on all hands that they are extremely lax. The common people are intelligent, and obliging, but passionate; and, on the slightest provocation, strike at each other with knives. Revenge and jealousy often lead, among the lower orders, to assassinations; rendered more frequent by the almost perfect impunity with which they may be committed. The statements as to conjugal infidelity are, perhaps, exaggerated; though the circumstances, under which society is placed, the swarms of priests, monks, and others, having no excitement but that of intrigue, leave no doubt as to the prevalence of licentiousness, and the general corruption of morals. The modern Romans are prone to falsehood. "They never speak truth," says Mr. Maclaren, "at the expense of their own interest; and in the courts it is asserted that any quantity of false evidence may be got for money. Cheating, in all its forms, is practised by high and low; and provided it be cleverly done, and successful, they feel a pride in telling it. The judges and functionaries of all kinds have the reputation of being very corrupt. The higher classes are slaves to their vanity, and their indolent pleasures; the lower to the most abject superstition. This character, however, chiefly belongs to the past or passing generation. A large proportion of the young Romans in the middle and upper classes are described as liberal, gentlemanly, and honourable; but they, and indeed the educated classes generally of all ages, are Deists. They speak with contempt of the mummeries and pious frauds they daily witness, but go once a year to confession in order to avoid scandal. The Romans have, however, their redeeming qualities; they are very sober, social in their habits, fond of their children, and obliging to strangers. There is no town, perhaps, where foreigners feel so much at ease. They may dress as they please, live as they please, and indulge in all their personal tastes and eccentricities, without being annoyed, or even stared at. In private lodging-houses strangers often meet with much genuine and gratuitous kindness. Many of their vices may be ascribed to the operation of a bad political system on minds naturally acute and active; for falsehood, hypocrisy, and craft are the natural fruits of a government which crushes liberty of thought." (*Maclaren's Notes*, p. 88.)

The manners of the upper classes are indicative of extreme indolence. They rise late, and are never to be seen until four in the afternoon, when they take a drive up and down the Corso, which, narrow as it is, may be termed the Hyde park of Rome; after which, they resort to soirées in private houses, for the theatres are open only during the carnival. To walk in Rome is quite unfashionable; and a carriage of some kind or other is quite indispensable, even to those of the noblesse or gentry whose limited means deny them a comfortable meal. Dancing, conversation, and cards are the chief evening amusements; dinner-parties are almost unknown, and suppers are only given on great occasions. In the month of May, all the inhabitants that can afford it, go to the country for two months, and again in October for the same period, the air of the Campagna being then purified by the rains of April and September. On these occasions, they hire a house or lodging in one of the pretty towns 10 m. or 15 m. from Rome; and their principal amusement during their villegiatura consists in fishing and bird-catching; the chase, in any of its noble forms, being little followed.

The public amusements consist of theatrical representations, concerts, and religious ceremonies, with occasional frolics at the carnival and other festive seasons. There are three theatres, two of which, the opera seria and opera buffa, are open during a great part of the year; but the performances are of a very mediocre description; the concerts have little to recommend them, and among the people at large, music forms but a small part of their enjoyment, though a few wandering harpers (*cantastorie*) may sometimes be found trying to inspire them with the love of sweet sounds. An amphitheatre (*corvea*) for bull-fights, tumbling, horse-riding, &c., has been formed out of an ancient mausoleum of Augustus, and when open is a favourite resort. The carnival would require some space for its description in detail: it may be sufficient here to observe, that in its license and intrigue, its unbridled mirth, and its levelling of rank; nay, even in the scenes of its celebration, it bears an obvious resemblance to the Roman Saturnalia; but it approaches, perhaps, more closely to the fast of Cybele, when, according to Livy (xxiv., c. 14), the richest draperies were hung from the windows, masquerading took place in the streets, and every one, disguising himself as he pleased, walked about the city in jest and buffoonery. If the historian had informed us in addition that one of the principal amusements was a promiscuous pelting of sugar-plums or chalk-stones, he would have furnished us with a precise picture of the modern carnival. Religious festivals are very frequent, but occur oftener between Advent and St. Peter's day than at other seasons. The pope celebrates mass and confers his public benediction in St. Peter's on Christmas day, Easter day, Whit-Sunday, and other festivals, on which occasions the solemnities are unusually grand, and attract immense crowds to the church.

The exhibition of the illuminated cross has been already mentioned. The illumination of the exterior of the church is also very imposing: the appearance of this immense building, with its dome, lantern, and cross, all lighted with large paper lanterns, has a most striking and magnificent effect, which, however, is much heightened, when at a given signal thousands of globes and stars of vivid fire, suddenly ignited, as if self-kindled, blaze in a moment into one dazzling flood of light, all over that vast structure. Immediately after the above display, on the night of St. Peter's day, follows the Girandola, an exhibition of fireworks, from St. Angelo, which is generally admitted to be superior to any other of the kind in the world. These festivals cost the Papal treasury about 15,000 crowns a year.

We have already noticed (PAPAL STATES, *ante*, 462) the wretched state of literature and education in modern Rome. It has, indeed, a university, a college, and numerous public schools; but they either afford no instruction in the higher branches of literature and philosophy, or such only as is of the worst possible description. All foreign publications, that might lead to expand and enlighten the public mind, are rigidly excluded; all native works must be submitted to the revision of the licensers; and the only literary pursuits that meet with any encouragement are those having reference to antiquity and the fine arts; and even they feel the paralysis that affects the other and nobler branches of study. " Rome, once the mistress of the world, the seat of arts, empire, and glory, now lies sunk in sloth, ignorance, and poverty, enslaved to the most cruel as well as to the most contemptible of tyrants, superstition and religious imposture." (*Middleton's Cicero*, i., 494, 4to ed.)

Rome has numerous *charitable institutions*, the total annual revenue of which amounts to between 800,000 and 900,000 dollars, half of which comes from the Papal treasury, the rest being supplied by endowments or voluntary contributions. But however large be the number of these establishments at Rome, " a great proportion of them are of doubtful, ill-directed, and even pernicious charity. Not to speak of the foundling hospitals, or those which offer a premium to idleness and thoughtlessness; there are 13 societies for giving dowries to girls on marriage, and pecuniary gifts on taking the veil; and of 1400 women married here in a year, 1000 avail themselves of these societies." There is also much private alms-giving, especially by the pope, who thus spends about 35,000 crowns a year. The consequence of this indiscriminate charity is seen in the mendicity, squalor, wretchedness, idleness, and want that meets you at every step in the streets of Rome. There are in the city 21 establishments for the diseased, insane, and convalescent, of which eight are public and 11 private hospitals, accommodating, on the whole, about 4000 patients; the average mortality is about seven per cent. There are also eight foundling hospitals, in which are nearly 4000 children of both sexes. In fact, Rome is one of the great recipients for abandoned children, brought thither from remote provinces, and even from Naples. The mortality in these hospitals is absolutely frightful, upward of 72 per cent.

An account of the Papal government, its judicial system, &c., has been already given, with these statistical details, in the article PAPAL STATES. The city is governed by an ecclesiastical governor, and a council (*sacra consulta*), appointed by the pope; and though there be a *senator*, or civil governor, he enjoys only the name without its authority; and the title has, for many years, been conferred exclusively on a native of another Italian state, as it has not been bought safe to entrust it to a Roman. The police of the city consists of about 4000 carabineers, somewhat similar to the gens d'armes; but the inefficiency of this body, which i said to be even more imbecile than the old town guard of Edinburgh, is proved by the frequent robberies and assassinations committed with almost total impunity.

Rome, though the chief manufacturing city of the Papal ates, has no manufacture deserving much notice. The incipal are silk and woollen goods, especially velvets, ocades for the clergy, and the more expensive kinds of k goods. Hats of very good quality are made here to the lue of about 900,000 crowns a year. The manufacture mosaics and jewellery of an extremely varied character, cuples a great number of hands, and many also are employed in making casts, or imitations of antique models, &c. ather, and prepared skins, gloves, parchment, strings for sical instruments, glue, glass bottles, are among the other cles manufactured in the city; but they are of no great ortance; and, with the exception of works connected h the fine arts, all the manufactures are conducted in the most clumsy manner. The hospital of St. Michael has privilege of furnishing cloth for the apostolic palaces the pontifical troops. None but national wool is employ ed in the manufacture; the spinning is done by hand, fly by women in the prisons, the warping is effected by manual labour, and it is made a boast that no machinery is employed where the work can be done without. The establishment has 25 looms, employs 850 persons, produces about 77,500 yards a year of the most costly, it the best, cloth produced in Europe. Manufactures ome kind or other are carried on also, chiefly by hand-ur, in 12 conservatories, containing about 600 inmates. e has an insurance company, a public bank, besides private banking-houses, a savings' bank, and a *monte* *ta*, the last of which had, in 1836, a circulating capital 0,000 crowns.

A great discrepancy of opinion has prevailed with respect to the climate of Rome. The fact, however, seems to be, that wherever the houses are few, and the ground is mainly covered with gardens, fields, or ruins, malaria is felt during the summer months, though not in the same degree as in the open country outside the walls. Now, this is the condition of the greater part of ancient Rome, of all the districts E. and S. the Quirinal and Capitol; so that five of the seven hills are either wholly or partially unhealthy. The upper part of the Pincian hill, the road toward the Porta Pia, and the space between the baths of Diocletian and the Porta San Lorenzo are also considered unhealthy, and there are districts of the same character hardly inhabited, having a convent here and there, the rest being laid out in gardens, vineyards, &c. West of the Tiber, the district of Lungara is unhealthy. The more densely peopled parts, on the contrary, are sufficiently healthy; and it may be said, with truth, that *modern Rome*, which extends from the Quirinal and the Capitol to the banks of the Tiber, is generally free from malaria. There are unhealthy seasons in Rome, as in most other cities, and in particular years epidemic fevers prevail to a fearful extent in the dirty and densely peopled districts; but these have no connexion with malaria, being attributable rather to the absence of sewerage, and the filthy habits of the lower orders. The temperature of the city is generally mild and genial. Frosts are not frequent, and though snow falls occasionally, it seldom lies on the ground more than a single day. The *tramontana*, however, a piercingly cold N. wind, sometimes blows for days together. Rains are frequent and heavy in November and December; but fogs are rare. In summer the heat is often oppressive, especially during the prevalence of the *sirocco*. In summer, the hour after sunset is considered the most unwholesome period of the day, and then people generally avoid exposure to the air. (*For. Quart. Rev.*, xxi., 38, &c.)

The *history* of Rome, which includes, for many centuries, that of all the countries washed by the Mediterranean, and, at a later period, that of the western Christian church, is far too extensive to allow of any considerable details in a work of this nature. Its foundation is hidden in the obscurity of an age respecting which few records remained in the time of its historians; and the investigations of Beaufort and Niebuhr have thrown much doubt on its early traditional history. Chronologists, however, are pretty well agreed in assigning its foundation to Romulus, its æra, according to Varro, being 753 years B.C. According to the account of Livy, the founder was succeeded by six other monarchs; and the constitution during the kingly period was an *elective* monarchy, with a king, senate, and popular assembly, the king being, at the same time, chief magistrate, high priest, and commander of the army; though, in point of fact, as his election depended on the voice of the comitia, the " people" were the real source of power. The senate originally consisted of 100 members, to whom, in course of time, others were added. The comitia comprised the burghers only, and the decrees of the senate required their approval before they became law. The Romans during this period being successful in war, added considerably to their previously confined territory. The public and private vices of Tarquinius Superbus led (anno 510 B.C.) to the abolition of kingly government, and the establishment of the republic, under two consuls, annually chosen, originally from the patricians only, but afterward from either patricians or plebeians. The temporary ascendancy of the patrician party effected the institution (B.C. 500) of the dictatorship, by which, on extraordinary emergencies, the whole power of the state was committed to a single individual, who might act with despotical authority. In the sequel, after many delays, and much opposition, officers called tribunes were appointed by the people, who had a *veto* on the proceedings of the senate. The constitution was thus founded on the principles of a distribution of power between the aristocracy and the commonalty; and in this state it remained without any considerable change to the end of the Punic wars, the empire of Rome being in the meanwhile extended over Italy, Sicily, Corsica, and Sardinia, the N. coast of Africa, and part of Spain. Amid these successes the distinction of patricians and plebeians seemed to have disappeared; but the unequal distribution of the public lands, or of those conquered by the arms of the republic, led to new, protracted, and bloody struggles between the patricians, who had appropriated to themselves the lion's share of these lands, and the plebeians, who sought to bring about their more equitable division. This occasioned the introduction by the latter of an AGRARIAN LAW; not, however, meaning by this, as is commonly understood, a law to interfere with, or to effect, an equal distribution of private property, but merely a law to limit the extent of the public lands held by individuals, and to subject them to a real, and not a nominal, rent. (*See Niebuhr*, ii., *passim*.)

The history of the intestine troubles of Rome during the

long protracted contests respecting this law, and the extension of the franchise to all Italian subjects and allies of Rome (the latter of which led to the Social war), would lead us into details quite unsuited to the nature of this work. It is sufficient here to state that the principle of representation not being adopted in the Roman constitution, it could not long survive, after the extension of the franchise to the Italians in general. The deliberations of the city assemblies were henceforth liable to be controlled by an influx of citizens from a distance, and full scope was given for the exercise of all sorts of corruption and intimidation. The soldiers, too, after they had carried their victorious arms beyond the boundaries of Italy, gradually ceased to pay their accustomed deference to the orders from Rome, and began to regard themselves rather as the servants of the generals by whom they were commanded, and to whom they looked for advancement, than of the republic. In consequence, the whole power of the state came to be engrossed by the great military leaders; and Marius and Sylla, Pompey and Cæsar, Marc Antony and Augustus, were successively masters of the Roman world. The battle of Actium (anno B.C. 30) threw the whole power of the state into the hands of Augustus, and the public, weary with intestine wars and revolutions, were glad to enjoy tranquillity under his supremacy. The *imperator*, who had previously been merely the commander in chief, now began to concentrate all the powers of the state in his own person. He became, in effect, perpetual dictator, and held the sovereign power free from all constitutional responsibility. The senate, indeed, continued to exist under the emperors, and the prætors or judges retained their names; but the decrees of the former were recommended, or rather dictated, by the emperors, and the *edicta* of the latter were superseded by summary decrees called *constitutiones principum*. In this state the government of Rome remained about 400 years. The succession depended partly on the will of the reigning emperor, who sometimes appointed his successor, either by adoption, or by giving him the title of Cæsar. In the event of no successor being named by the previous emperor, the right of election devolved on the senate; but it was frequently usurped by the army and by the Prætorian guards; and sometimes rival emperors were chosen by the senate and the army, or by different armies, the pretensions of the candidates being decided in the field. Under such circumstances, and considering the degraded state of the Roman people, enervated by indolence, and corrupted by largesses, immunity from taxes, and indulgence in public shows, it may well excite surprise that the empire did not sooner fall to pieces.

Some speculative inquirers have classed the circumstance of the imperial dignity being elective among the causes that contributed to its decline; whereas it really appears to have been almost the only principle that enabled it to survive so long. In a government like that of Rome, where everything had to be transacted directly by the emperor, a hereditary monarchy, which supposes the occurrence of minorities, was out of the question. And how unworthy soever the means by which some of the emperors arrived at the imperial dignity, not a few of them owed it to their superior ability. Nerva, Trajan, the two Antonines, Severus, Aurelian, and other able princes, gave new vigour to the tottering fabric, and prolonged its existence.

At the close of the fourth century the Roman dominions, which still extended from Britain on the W. to the Euphrates on the E., were divided between Honorius and Arcadius. At this time, too, the barbarians, sensible of the growing weakness of the Romans, began to harass the empire with incessant hostilities, and one country after another was lost, till at length Italy itself was invaded by the Huns, and shortly afterward by the Heruli, whose general, Odoacer (A.D. 476), dethroned the impotent Romulus Augustulus, assumed the title of *rex*, and fixed his residence in Ravenna. Thus fell the greatest empire of the world, exactly 1229 years after its supposed foundation by Romulus. Odoacer gave way to Theodoric, king of the Ostrogoths, and during his reign Rome and all Italy enjoyed a period of peace and prosperity. But the calm was only temporary; Belisarius, the general of Justinian, and Totila the Ostrogoth, successively took Rome, which was stripped of some of its most splendid monuments, at the same time that its inhabitants were reduced to a state of wretchedness which they had not before experienced. After having become a province, or *exarchate*, of the eastern empire, Rome passed, in 774, under the dominion of the Franks, who retained it till the deposition of Charles le Gros, in 887, after which the possession of Rome and Italy became, during more than three centuries, the subject of contention between the emperors of Germany, the numerous states into which Italy had been parcelled, and the bishops of Rome, who with the title of pope assumed a right to temporal power. Nicolas III. at length obtained from Rodolph of Hapsburg, in 1278, the grant of an independent territory, called the States of

the Church; and thus began the sovereignty of the popes, which, with some interruptions, has continued to the present day. (Farther details respecting this part of Roman history will be found under the article PAPAL STATES.)

ROME, p. t., semi-capital of Oneida co., N. Y., 15 m. S.W. Utica, 107 m. N.W. Albany, 301 W. Drained by Mohawk river and Wood creek, the latter flowing into Oneida lake. These streams were connected by a canal 2 m. long, before the construction of the Erie canal, which was purchased when the Erie canal was made. The village is near the Erie canal, between Mohawk river and Wood creek, occupies the site of old fort Stanwix, of revolutionary memory, and is handsomely laid out with streets 100 feet wide, crossing each other at right angles, and has two public squares. It contains a courthouse, jail, county offices, a U. States arsenal, magazine, and workshops, a bank, six churches, two Presbyterian, an Episcopal, Methodist, Baptist, and Roman Catholic, all of which have handsome brick edifices, two private seminaries, 250 dwellings, 16 of brick or stone, and the rest of wood, and 2500 inhabitants. The Black river canal here unites with the Erie canal, and the Syracuse and Utica railroad passes through it. There are in the t., mostly in the village, 33 stores, two fulling-mills, one woollen factory, one cotton factory with 969 spindles, one furnace, two grist-mills, 15 saw-mills, three tanneries, one brewery, two potteries, one printing-office, one weekly newspaper; three academies, 307 students; 21 schools, 1990 scholars. Pop. 5680. Next to Utica, it is the most important town in the county.

ROME, p. v., capital of Floyd co., Ga., 161 m. N.W. Milledgeville, 762 W. Situated at the junction of Etowah and Oostanalau rivers, to form Coosa river. It contains a courthouse, jail, two churches, a Methodist and Baptist, an academy, three stores, 30 dwellings, and about 300 inhabitants.

ROME, p. v., capital of Perry co., Ia., 143 m. S. by W. Indianapolis, 653 W. Situated on the N. bank of Ohio river. It contains a fine brick courthouse, a stone jail, two churches, a flourishing public school, 40 dwellings, and about 250 inhabitants.

ROMFORD, a market town and par. of Essex, in the liberty of Havering-atte-Bower, on the high road from London to Norwich, 13 m. E.N.E. London, and 16 m. S.W. Chelmsford. Area of par., 3340 acres. Pop., in 1831, 4864. The town consists principally of a long wide street along the high road, having near its centre the market-house and town-hall, in which are held the petty sessions for the liberty. The church is an ancient structure, consisting of a nave, chancel, and N. aisle, with a tower at the W. end. The living is a curacy subordinate to that of Hornchurch, in the patronage of New college, Oxford, value £54, besides which the curate receives an annual stipend of £36. (*Ecclesiast. Report.*) The Wesleyan Methodists and Baptists have their respective places of worship, with attached Sunday schools; besides which there is a national school, partly endowed and partly supported by subscription. The town has also several almshouses and benefactions for the poor. At a little distance from Romford are cavalry barracks, erected in 1795, but now disused. The inhabitants are chiefly retail dealers or persons employed in market gardening and agriculture. The town derives, however, its principal advantage from its situation on the London road, and more recently from being one of the stations on the Eastern Counties railway. It is also one of the polling places for the S. division of Essex. Markets, especially for calves, well attended, on Mondays, Tuesdays, and Wednesdays; fairs, June 24th, for cattle and horses.

ROMNEY (NEW), a cinque-port, decayed bor. market town and par. of England, co. Kent, lathe Shepway, partly in lib. Romney marsh, and partly in hund St. Martin Pountney, 19 m. S.W. Dover, and 58 m. S.E. London. Area of bor. and par., 2290 acres. Pop., in 1831, 953. The town, which arose out of the ruins of Old Romney, was formerly a comparatively flourishing condition, being a considerable seaport; but the haven has for many years been completely filled up. It consists at present of a broad principal street crossed by one of inferior size, in which is the town-hall. Houses chiefly of brick, the market-house and town-hall being modern erections. The church is a spacious structure, consisting of a nave, aisles, and chancel, partly Norman and partly in the pointed style, with a large and curious tower at the W. end. The living is a vicarage in the patronage of All-souls' college, Oxford; of the nett value of £161 a year. The Wesleyan Methodists have a small chapel, and there are two Sunday schools, besides a free school and almshouses. The inhabitants, with a few exceptions, are employed in grazing cattle on Romney marsh, a rich tract of land, extending about 7 m. N. and W. from the town, and comprising about 47,000 acres. This tract is defended from the encroachments of the sea by an immense embankment called Dymchurch-wall, along which is a good road for carriage: this sea-wall is kept in repair by a rate levied on the proprietors of the marsh. The sheep depastured here furnish long combing wool.

New Romney is a borough by prescription, and returned two members to the House of Commons from the reign of Edward III. down to the Reform Act, by which it was disfranchised. It was not considered of sufficient importance to be included in the provisions of the Municipal Reform Act; but it still retains certain privileges reserved in that Act for the Cinque-ports. Old Romney, 2 m. W. the town, has now only a few houses surrounding the church. Markets on Thursday; fair, August 26, for pedlery, &c.

BONNEY, p. v., capital of Hampshire co., Va., 188 m. N.N.W. Richmond, 116 W. Situated on the E. side of the S. branch of Potomac river. It contains a courthouse, jail, several stores, and about 400 inhabitants. Nett proceeds of the postoffice in 1841, $412.

ROMORANTIN, a town of France, dep. Loir-et-Cher, cap. arrond. on the Seudre (a tributary of the Loire), where it receives the Morantin, 24 m. S.E. Blois. Pop., in 1836, ex. com., 6503. It was formerly the capital of Sologne, and was embellished by Francis I. It has an old castle, a spacious prison, a theatre, courts of original jurisdiction and commerce, and some manufactures of woollen stuffs and yarn. Romorantin was taken by Edward the Black Prince in 1356. Cannon appear to have been used in the siege; but this, though one of the earliest, is not, as has been alleged, by any means the first occasion on which they were so employed. But it is better known in history, by giving its name to the edict of 1650, drawn up by the chancellor l'Hôpital, which gave to bishops, and took away from the parliaments the power to try cases of heresy. It is said that the chancellor consented to this edict only to avoid a still greater evil, the establishment of the inquisition. (Henault Abrégée, Anno 1560; Hugo, &c.)

ROMSEY or RUMSEY, a mun. bor., market town, and par. of England, co. Hants, hund. King's Sombourn, the town being situated on the Test, a tributary of the Anton, on the Andover canal, 6 m. N.W. Southampton. Area of par., divided into Romsey-Extra and Infra, 9310 acres. Pop. in 1831, 5432. The town, which consists chiefly of a long and wide street, crossed by another at right angles, covers a considerable extent of ground. It has an audit-house, with a market-place beneath, and an old town-hall, in which petty sessions are held; but by far the most remarkable public building is the parish church. This interesting edifice is almost the only remaining portion of an abbey said to have been founded here by Edward the Elder. The present structure appears, however, to date from the beginning of the 12th century, and it is one of the most complete Anglo-Norman monuments in the kingdom. "It is a cross church, with a low massive tower; the general exterior appearance is Norman, of very good character, and much of it unaltered. The W. end is early English, very plain outside, and its details accommodated to the Norman part; but the inside of this W. portion is a very fine specimen of the early English, rich rather by composition than minute ornament. The central portion and the transepts, with the sides of the chancel, are Norman, showing various singularities and mixtures of pointed and round arches." (Rickman, Goth. Arch., p. 176.) This church has a fine high altar, much good tracery, stained glass, &c.; and a curious peculiarity is, that a large fruit bearing apple tree grows from its roof. The living, a vicarage in the gift of the dean and chapter of Winchester, is worth £365 a year. The Presbyterians have a meeting-house in Romsey; and it has an almshouse, a charity school for thirty boys, a free school, &c. The corporation are trustees for several charities which, with the affairs of the borough generally, appear to have been well managed. (Mun. Corp. Appendix.)

The borough was first chartered by James I.; its corporation consists of a mayor, six aldermen, and 12 capital burgesses. Romsey-Infra is the only corporate town; Romsey-Extra being without the jurisdiction of the former. Corporation revenue, £98 1s. 3d.; corporation debt, £1300.

The sole importance of Romsey is derived from its supplying goods by retail to a large agricultural district. "The borough is increasing in population, but the trade is decreasing. Formerly, many extensive manufactures of paper and other articles were carried on at Romsey, where are is abundant and excellent water-power; but the introduction of steam has greatly diminished the trade of Romsey." (Mun. Corp. Appendix.) Immediately adjacent the town is Broadlands, the seat of Viscount Palmerston. Markets on Thursday; fairs on Easter-Monday, August 26, and November 8, for cattle, sheep, and horses, (Parl. Rets. &c.)

ROMULUS, p. t., Seneca co., N. Y., 183 m. W. Albany, 9 W. Bounded E. by Cayuga lake, W. by Seneca lake. Contains a Presbyterian church, one store; eight schools, scholars. Pop, 2235.

RONDA, a city of Spain, in Andalusia, prov. Granada, on Guadiaro, 40 m. W. by N. Malaga, and 48 m. N.N.E.

Gibraltar. Pop., according to Miñano, 18,678. Its situation is peculiar, being built on lofty rocks beetling over the river, across which, at an elevation of 200 feet above the surface, are thrown two bridges, one of which consists of a single arch, 110 feet in span, and surmounted by another bridge of three arches, at a much greater elevation. A third bridge crosses the stream somewhat above the town. The river is wholly unnavigable; and several cascades are formed close to the city. One portion, called the Old City, overhangs the S. Cliff, and is encircled by an old embattled wall, built by the Saracens, and flanked by extensive outworks, while the more widely spread buildings on the opposite bank bear the name of El Mercadillo, or New Town. Within the fortifications stands the royal palace of Abon-Melic, the Moor, now a vast heap of ruins. The only entrance to the city is through a succession of gates, leading to a long and narrow, but tolerably straight street, running N. and S. for about ¾ m. to the upper or new bridge. This street is lined with handsome shops; and from it lead off both right and left numerous alleys, communicating with little courts and crooked passages, all of which, however, are lined with remarkably good houses. In fact, says Capt. Scott, "this labyrinth is the Mayfair of Ronda, the aristocratic location of all the Hidalguia of the province, who, proud of the little patch of land won by the swords of their forefathers from the accursed Moslems, would as soon think of denying the infallibility of the pope as of taking up their abode among the mercantile inhabitants of the mushroom suburb, which, however, is, beyond all comparison, the most agreeable place of residence." (Scott's Ronda and Granada, i., 103.) The principal streets of the new town are wide and tolerably straight: it contains some fine open plazas; and although the houses are thus more exposed to the sun, they enjoy a freer circulation of air. The absence of an enclosing wall tends, also, in point of coolness, to give the Mercadillo an advantage over the city. It is nearly as difficult of approach, however, and as incapable of expansion as the walled city itself, for cliffs bound it on three sides, leaving the access free only on its N. side The city has few public buildings, except its churches, which are numerous, and gaudily fitted up; but they have neither paintings nor statuary of any merit. The new town comprises a small, but commodious theatre, the stables of the Real Maestranza (or corporation of nobility for breeding horses), and the Plaza de los Toros, a circular covered building of stone, one of the handsomest in Spain, and capable of accommodating 10,000 spectators.

The inhabitants of Ronda are principally employed in agricultural and horticultural pursuits, though there are several manufactories of coarse woollen cloths and hats, two or three tanneries, and numerous water-mills. It is a place, also, of considerable commerce; its secluded, and, at the same time, central situation, making it a convenient dépôt for smuggled goods, in which, indeed, the present trade of Spain mostly consists. A very large fair is annually held here in May for cattle, horses, sheep, and general goods: it collects an astonishing concourse of people from all parts of the country, and offers an excellent opportunity for observing the costumes and shades of character peculiar to the inhabitants of the different provinces. "The society of Ronda is particularly good, consisting of some of the most ancient Andalusian families; who, with all the polish of the first circles, are exempted from the demoralising vices which distinguish the fashionable inhabitants of Madrid and other large cities. On the whole," continues Captain Scott, "I scarcely know a place where a few weeks can be more agreeably spent." (i., 118.) The climate of Ronda, also, is very delightful; neither oppressively hot nor disagreeably cold: and it is considered so favourable to longevity, that it has become a common saying—En Ronda los hombres de ochenta años son pollones ("at Ronda even men at eighty are chickens").

The neighbourhood is not only extremely picturesque, but produces an abundance of wine, oil, and corn, as well as the fruits and vegetables peculiar to a more N. climate. Cattle graze in large herds on the plains, and the hills abound with many varieties of game, including deer and wild boars. About 3 m. S.E. of Ronda is the singular mountain, called Cresta de Gallo, consisting of two parallel ridges joined at the bottom, one red, the other white: both of them possess mineral riches, which, under a better system of national economy, might, probably, be turned to good account.

Ronda has been supposed, though, perhaps, with little foundation, to owe its origin to the Romans. Nothing certain, however, is known respecting it prior to the domination of the Moors, who made it one of their principal strongholds. In 1831 it became the court residence of Abon-Melic, son of the emperor of Fez, who erected the castle and fortifications. It was finally taken from the Moors by Ferdinand of Castile in 1485. (Scott's Ronda and Granada, i., 99-129; Mod. Trav.; Miñano, &c.)

RONDOUT, creek, N. Y., rises in Sullivan co., and flows into Hudson river in Ulster co., 1 m. below Rondout village. The Wallkill, a main branch, enters it on the S. side, in Esopus. It affords good water-power. The Delaware and Hudson canal passes along the valley of this creek, and there is a lighthouse at its entrance into Hudson river. Rondout village is the depot of the coal of the Delaware and Hudson Canal Co.

ROSCOMMON, an inland co. of Ireland, prov. Connaught, having N. Leitrim and Sligo, E. and S.E. Longford, Westmeath, and King's county, from which it is separated by the Shannon, S.W. and W. Galway, from which it is separated by the Suck, and Mayo. Area 609,405 acres, of which 131,063 are bog and mountain, and 24,787 water. There are some mountainous tracts in the N. parts of the county, and elsewhere; but, speaking generally, its surface is nearly flat, exhibiting, for the most part, either green fields or bogs. Substratum principally limestone. Pastures most luxuriant. Stone fences, so common in Scotland, are, in Ireland, nearly peculiar to this county. Estates very large; many of them, however, are let on perpetual leases, the holders of some of which form an intermediate class between the great proprietors and the occupiers. A large proportion of land in pasture; but latterly tillage has been rapidly extending. Several improvements have been introduced both in the plan of husbandry and in the instruments employed in carrying it on. "But the general system of agriculture, excepting on lands held by wealthy individuals, still remains (1832) in a very imperfect state; and the smaller farms are cultivated in a manner at once slovenly and wasteful." (*Weld's Survey of Roscommon*, p. 654.) Tillage farms generally small. Oats and potatoes principal crops; but wheat is now rather extensively cultivated. A good deal of work is done by the *loy*, a species of spade. Cattle, long-horned; sheep, long-woolled; both breeds good; few dairies. Average rent of land, 13s. an acre. Some new cottages, on a few estates, are neat and comfortable, but the great majority continue to be as bad as possible. The same may be said of the farm buildings.

There are veins of coal and ironstone in the N. parts of the county, to the W. of lough Allen. These had been occasionally wrought to some extent for a considerable period, but in general to the heavy loss of those by whom the works were carried on. It was, however, contended that this happened from the want of capital, or want of skill on the part of those employed; and the most exaggerated and delusive accounts were, at the same time, published of the value of the mines. At length, during the memorable year 1825, three companies were formed for working the coal and iron mines at Arigna and other places in this county. One of these, after examining the ground, prudently declined proceeding any farther: the energies of another were paralysed by the fraud, jobbing, and mismanagement of some of its directors and agents; and the third (the Irish Mining Company), an enterprising and well conducted association, ultimately abandoned the undertaking, their collieries having proved, if not absolutely worthless, not worth the cost of working them. (*Weld's Survey*, p. 33, 77, 682, &c.) The linen manufacture was at one time pretty extensively diffused over the county, but it has latterly very much fallen off. Being washed through its whole extent by the Shannon, few Irish counties have greater facilities than Roscommon for the easy and convenient disposal of their products. It is divided into six baronies and 56 parishes, and sends two members to the House of Commons, both for the county. Registered electors, in 1839–40, 2092. Principal towns, Roscommon and Elphin. In 1831, this county had 41,369 inhabited houses, 44,259 families, and 249,613 individuals, of whom 123,031 were males, and 126,582 females.

ROSCOMMON, an inland town of Ireland, prov. Connaught, cap. of the above co., 6 m. W. from lough Ree, and 78 m. W. by N. Dublin. Pop., in 1831, 3306. It has a parish church, a Roman Catholic chapel, a public school, a market-house, a cavalry barrack, an extensive modern county court-house and jail, and an infirmary. Races are held annually in the vicinity. Under a charter of James I., in 1612, the corporation, which consisted of a sovereign, 12 burgesses, and a commonalty, returned two members to the Irish House of Commons till the union, when it was disfranchised. It principally occupies the S. slope of a gently rising hill; but it is straggling, ill-built, and poor, its hovels stretching along the principal lines of road by which it is approached. Latterly, however, it has been somewhat improved. In summer, it suffers from a deficiency of water. A manor court holds pleas for debts to the amount of £10. The county assizes are held here; as are general sessions twice a year, and petty sessions every Monday. It is a constabulary station, and has manufactures of coarse woollens, linens, and brown pottery, for the supply of the immediate neighbourhood. The trade in corn is increasing. Markets on Saturdays; fairs on Whit-Monday and 5th December. Postoffice revenue, in 1830, £800; in 1836, £966. Branches

of the Agricultural and National banks were opened here in 1836. The ruins of an old Dominican abbey, founded in 1257, has a fine monument of its founder, one of the O'Connor family. Here, also, are the remains of a fine old castle, built by the English soon after their entry into this part of the country. The town, and a considerable contiguous estate, which has hitherto been much neglected, is the property of the Earl of Essex. (*Statistical Survey; Priest's Guide.*)

ROSCREA, a town of Ireland, prov. Munster, near the N.W. extremity of the co. Tipperary, finely situated between the Slieebh-Bloom and Devils-bit ranges of hills, on a branch of the lesser Brosna, 40 m. N.E. Limerick. Pop. in 1831 5512. It is of great antiquity, having been made the seat of a bishopric in the 6th century, united to Killaloe in the 12th. Some remains of the old cathedral may still be seen in the W. front of the parish church; it has also a fine stone cross, a pillar tower, an old castle built by the Ormonde family, and the ruins of a Franciscan monastery. The town is regularly built. Among the public buildings are the parish church, a Roman Catholic chapel, meeting-houses for Quakers and Methodists, a school on the foundation of Erasmus Smith, a fever hospital, cholera hospital, and dispensary, market-house, bridewell, and barrack. A manor court, which holds pleas, to the extent of £10 Irish, sits monthly; petty sessions are held on Mondays. It is a constabulary station. It manufactures coarse woollens, has several flour-mills, two tanyards, two breweries, and a distillery, and carries on a very extensive trade in grain: its retail trade is also considerable, the surrounding country having a more than usual number of resident gentry. Markets on Mondays and Thursdays; fairs on 25th March, 7th May, 21st June, 8th August, 9th October, and 29th November. Postoffice revenue in 1830, £814, in 1836, £973. Branches of the Agricultural and National banks were opened in 1835.

ROSE, p. t., Wayne co., N. Y., 7 m. N. Lyons, 179 m. W. by N. Albany, 359 W. Drained by small streams flowing into lake Ontario. It has two stores, one fulling-mill, one grist-mill, seven saw-mills; seven schools, 573 scholars. Pop. 2038.

ROSS, t., Jefferson co., Pa. It has two stores, one fulling-mill, one woollen-factory, five grist-mills, 17 saw-mills; four schools, 100 scholars. Pop. 1421.

ROSS, p. t., Carroll co., O., 128 m. E.N.E. Columbus, 395 W. It has one school, 20 scholars. Pop. 1503.

ROSETTA, or ROSSETTA (Arab. *Rashed*), a town and seaport of Lower Egypt, on the W. branch of the Nile (the anc. *Bolbitine* mouth), near its embouchure, 36 m. E.N.E. Alexandria. Down to a late period, it was one of the most important commercial towns in the country, and had a population estimated to amount to about 25,000; but since the opening of the Mahmoudieh canal from Alexandria to the Nile, Rosetta has sunk into comparative insignificance; its population has dwindled down to about 4000, and the principal traffic consists in the removal to Alexandria of the bricks, and other materials, of its buildings! It was principally constructed of red brick, plastered over and white-washed. As elsewhere in the east, the streets are narrow, and the upper stories project, so as frequently to meet. Upon the whole, however, Rosetta is neater than many oriental towns, and its situation in the midst of date, lemon, and orange groves, is distinguished for beauty. The inhabitants are principally occupied in the rice, cotton, sal-ammoniac, and leather factories established by the pacha, at a wretchedly low rate of wages. There are now no resident merchants in Rosetta, and its shipping, which was formerly considerable, is reduced to a few boats. The port, though tolerably secure within, is difficult of entrance, there being a shifting bar at the mouth of the river, which can only be passed with safety during favourable winds, and at certain times of tide. Rosetta is famed for the supposed salubrity of its air, which attracts visitors thither during the summer and autumn months. It was founded by one of the caliphs about 870, near the site of the ancient *Bolbitinum*, but has no antiquities of its own. Here, however, was discovered the famous *trilingual* tablet, called the "Rosetta stone," to which we are mainly indebted for the discoveries of Young and Champollion. (*Private Information.*)

ROSS AND CROMARTY, two cos. of Scotland, in the Highlands, forming together a maritime district of great extent, stretching quite across the island, and including Lewis, in the Hebrides. These counties, though in some respects distinct, are united under one sheriff, and Cromarty being a small county, consisting of several detached portions, most of which are wholly surrounded by parts of Ross, they may be most conveniently noticed under one head. They are bounded N. by the co. of Sutherland, E. by the firths of Dornoch and Moray, S. by Inverness, and W. by the Atlantic. Area, 1,904,000 acres, of which 1,532,300 are mainland, and 373,200 islands; the freshwater lakes cover a space of 44,800 acres on the mainland, and of 12,899 in the islands. The E. parts of the province, consisting of the districts called the Black Isle, or the Peninsula, between the Beauly and Moray friths, the frith of Cromarty ; and Easter Ross, or the Peninsula, between the friths of Cromarty and Dornoch, are comparatively flat and fertile. Easter Ross has a considerable extent of clayey loam, and of light sandy soil. The soil of the Black Isle is very various ; much of it is poor, but the cultivated portion consists principally of clayey loam, good black mould, and sandy loam. In Strathpeffer, and the country round Dingwall, the soil is clayey ; but with these exceptions the rest of the county is wild, dreary, rugged, and mountainous, interspersed with lakes, and narrow glens, that afford pasture for sheep and black cattle. Estates, for the most part, very large ; but there are several that are not of much value. Farms of all sizes ; but the number of small occupancies, though still very considerable, is much diminished. Native breed of cattle hardy, compact, and well suited to the country ; but in the W. parts of the county, the Skye and Argyleshire breeds, or one closely allied to them, is most prevalent. Cattle were formerly much more abundant than at present. Sheep-farming has, for many years past, engrossed almost the whole attention of the principal farmers and improvers ; so that, besides a decrease in the number, it is also said that the breed of cattle has deteriorated. This, however, has been denied by others : and, at all events, the baneful practice of overstocking is no longer carried to anything like the extent to which it was formerly practised in this, as well as in other Highland counties. At no very distant period, oxen were extensively employed in field-labour ; but they are no longer used for that purpose. All sorts of improvements, both in breeding and cropping, have been tried by the principal proprietors, and by many intelligent and enterprising sheep farmers that have immigrated thither from the S. Most part of Easter-Ross, great part of the Black Isle, with the country round Dingwall, and along the N.W. shore of the Inner frith of Cromarty, now ranks with the finest districts of Scotland. It is traversed in every direction by excellent roads, is well fenced, and has a more than usual number of seats and plantations. Agriculture has been wonderfully improved ; and the crops of wheat and turnips are at present nowise inferior to those in the more S. cos. But exclusive of these districts, a great extent of mountainous country is still occupied by the old Highland tenantry. These are a brave and hardy race ; but poor, and without either enterprise or industry. They occupy the straths or valleys between the mountains, and along the banks of the rivers, which in some places are so thickly tenanted, that there is a family for every Scotch acre of arable land ! On this they raise oats, bear, or bigg (a species of barley), and potatoes ; frequently cultivating the ground with a crooked spade (caschrom) instead of a plough. The mode of ploughing, which was formerly general over the whole country, and which is still practised by the smaller tenants, is barbarous in the extreme. The smaller tenants uniformly possess a considerable extent of grazing ground, which is commonly contiguous to, but sometimes at a considerable distance from, their arable possessions. Their huts are for the most part wretched ; few of them having either chimneys or windows ; they prefer, indeed, living in the midst of smoke and filth ; and in winter the cattle are generally housed under the same roof with the family. Except for a few months, when sowing or reaping their crops, preparing and saving their fuel, &c., the greater part of their time is spent in the pursuit of game, in fishing, or in idleness.

Previously to the reduction of the duties on whiskey, in 1823, illicit distillation was very prevalent, and is still carried on, though to a comparatively small extent. "It cannot be said with truth that the class of people of which the great majority of the population consists enjoy the comforts of life in even a moderate degree. Poorly fed, scantily clothed, and miserably lodged, theirs is a life of penury and toil ; exposed to the temptations of idleness, without its ease, and to the slavery of labour, without its rewards, they drag out a wretched existence suffering under the continual fear of impending want, and uncheered by any prospect of amendment in their condition. (Art. *Parish of Glenshiel ; New Statistical Account of Scotland*, No. 12, p. 200.) Under these circumstances, no reasonable person can doubt that the measures adopted by many landlords during the last half century for consolidating the small possessions held by the native tenants, and introducing farmers possessed of capital and skill, have been, in a public point of view, eminently beneficial. In some instances the change may have been hastily effected ; but we hesitate not to say that it has, on the whole, been highly advantageous to the peasantry themselves. Having been obliged to repair to villages or to emigrate, they have also been obliged to lay aside their slothful habits ; so that, in point of fact, not only the wealth and industry, but even the population of the country, as gained materially by the introduction and extension of

that sheep-farming system that has been the theme of so much ignorant vituperation. In proof of this, we may observe that the population of this district amounted, according to the enumeration of Dr. Webster, in 1755, to 47,466. In 1800, it had increased to 55,877 ; and notwithstanding the increase of sheep-farming, and the prevalence of emigration in the interim, it amounted, in 1831, to 74,820. Minerals and manufactures of no importance. Average rent of land, including the islands, in 1810, 1s. 1d. an acre. Principal rivers.—Conon, Orin, and Beauly. This district is divided into 33 parishes, and sends one member to the House of Commons ; the boroughs of Dingwall, Tain, and Cromarty being associated with others in the return of a representative. Registered electors for the county, in 1839-40, 713. In 1839, Ross and Cromarty had 15,039 inhabited houses, 16,187 families, and 74,820 inhabitants, of whom 34,927 were males, and 39,893 females. Valued rent, £87,940 Scotch : annual value of real property, in 1815, £121,577.

ROSS, a town and parish of England, county Hereford, hund. Graytree, on the Wye. 15 m. W. by S. Gloucester. Area of parish, 3540 acres. Population of do. in 1831, 3078, of which the borough had about 2000. The latter is finely situated on an eminence above the river ; but its streets are steep, rough, and narrow. The church, in a conspicuous situation, has a lofty spire, and is partly in the perpendicular style ; but, according to Rickman, it has been injured by alterations and repairs. The living a rectory and vicarage, vested in the patronage of the bishop of Hereford, is one of the best in the county, being worth £1294 a year, nett income. There is a market-house, two charity schools, and an almshouse. In Camden's time Ross was celebrated for its cutlery and its cider ; the former, however, has entirely disappeared, but it continues to be distinguished by the excellence of the cider made in its vicinity. It is governed by a serjeant and four constables. Market on Thursday.

The "Man of Ross," immortalized by Pope (*Moral Essays*, iii. line. 250), was a Mr. John Kyrle, a native of the town, where he died in 1724, at the age of 84. The splendid eulogium of the poet did not really go beyond Kyrle's merits. He expended his time and income in promoting objects of public utility and benevolence, by which the town continues to be benefited. His portrait is still preserved in his house, now an inn, near the entrance of the road from Gloucester.

Ross, county, O. Situated toward the S. part of the state, and contains 650 sq. m. Watered by Scioto river and its tributaries. It contained in 1840, 20,578 neat cattle, 30,382 sheep, 48,678 swine ; and produced 390,790 bushels of wheat, 3580 of rye, 1,899,390 of Indian corn, 1505 of barley, 199,459 of oats, 27,070 of potatoes, 29,870 pounds of sugar. It had 101 stores, three bloomeries, three fulling-mills, four woollen-factories, 12 flouring-mills, 19 grist-mills, 29 saw-mills, one oil-mill, one paper-mill, 18 tanneries, 11 distilleries, two printing-offices, two binderies, two weekly newspapers ; two academies, 85 students. Pop. 27,460. Capital, Chillicothe.

Ross, t., Alleghany co., Pa., 4 m. N. Pittsburg. Bounded S. E. by Alleghany r., S.W. by Ohio r. It contains four gristmills, five saw-mills ; six schools, 330 scholars. Pop. 1675.

ROSSIE, p.t., St. Lawrence county, N.Y., 25 m. S.W. Canton, 183 N.W. Albany, 452 m. W. Watered by Oswegatchie and Indian rivers. It abounds with iron and lead ore. It contains three stores, three furnaces, three gristmills, four saw-mills, two tanneries ; five schools, 217

compartments, and finely sculptured. The capitals of the pillars are enriched with foliage, and a variety of figures, most elaborately and minutely cut.

The Earls of Orkney and Roslin were interred in a vault below the floor of the chapel · and it is a curious fact, that down almost to the æra of the revolution they were buried, not in coffins, but in complete suits of armour. This circumstance has been alluded to by Scott, in his fine ballad of "Rosabelle," in the "Lay of the Last Minstrel."

The glen in which the Esk runs from Roslin to Laswade is mostly narrow; has in parts high, precipitous, rocky, and well-wooded banks, and is celebrated for its romantic scenery. A little below Roslin, on the opposite side of the river, is Hawthornden, the seat of Drummond, the contemporary and friend of Ben Jonson, and one of the best poets of his age. The house stands on a brink of a precipice, overhanging the river, and, with the estate, is now in the possession of the descendants of the poet. Below the house are caves, or extensive apartments cut in the sandstone rock.

Roslin is much resorted to in summer by parties from Edinburgh; and most strangers who visit the latter, contrive also to visit Roslin. (*Pinnant's Scotland*, iii. 254, ed. 1790; *Chamber's Gaz. of Scotland, &c.*)

ROSSBACH, a village of Prussian Saxony, 16 m. S. Halle, celebrated in modern history for its being the scene of the great victory gained on the 5th of November, 1757, by Frederick the Great, with little loss to his own forces, over the French and Imperialists.

ROSTOCK, a commercial city and seaport of N. Germany, being the largest town, though not the capital of the grand duchy of Mecklenburg-Schwerin, on the Warnow, 9 miles above its mouth in the Baltic, and 40 m. N.E. Schwerin, lat. 54° N., long. 12° 19' E. Population, in 1837, 18,067. It is surrounded with old walls, and divided into three parts; the old, middle, and new towns. It has several suburbs, which, with the city, are built in an old-fashioned style. It has been frequently the residence of the grand dukes, and has a ducal palace, numerous churches, a convent, two hospitals, a town-hall, theatre, &c. The church of St. Mary's is interesting from its having a monument in honour of Grotius, the illustrious author of the treatise *De Jure Belli et Pacis*, one of the greatest men of modern times, who expired here on the 28th of August, 1645, far from his family and friends, an exile from his ungrateful country. In one of the squares, thence called *Blucher's Platz*, is a statue of the celebrated Prussian general, Blucher, who was a native of the town. The university of Rostock, one of the oldest in Germany, was founded in 1419. It has four professors of theology, five of jurisprudence, five of medicine, and ten of philosophy; besides three extraordinary professors, and ten private teachers, having attached to it a library of 45,000 printed volumes, theological and other auxiliary schools, an anatomical theatre, laboratory, botanic garden, and various scientific collections. It is, however, but poorly attended, having, in 1836, only 86 pupils. Rostock has a society of natural history, and other learned associations, and a commercial institute. It is one of the most active manufacturing towns of N. Germany, having numerous woollen factories, breweries, and distilleries, vinegar and soap works, &c. Rostock has a pretty extensive trade. The exports consist chiefly of good red wheat, barley, peas, rapeseed, and a few oats; with wool, rags of a very superior quality, oil cake, rape oil, bones, flax, horses, cattle, provisions, &c. The average export of all kinds of grain may be taken at from about 115,000 to 130,000 qrs. a year; and the total value of all sorts of exports may be estimated at about £980,000. The imports consist of colonial products, spices, wine, manufactured goods, &c. There belong to the port nearly 200 vessels of from 150 to 250 tons, which trade with most European nations, the United States, and Brazil. The outport of Rostock is Warnemunde, at the mouth of the Warnow. The depth of water at the latter varies from 10½ ft. to 12½ ft.; but when the W. pier, now in process of construction, shall have been completed, it is expected that the depth of water will be from 12 ft. to 14 ft. The depth of water in the river from Warnemunde up to Rostock is usually from 6 ft. to 9 ft.; so that vessels drawing more than this, must be lightened to get up to the latter. The commercial weights here are the same as at Hamburg. The duties are extremely moderate; on most imported articles they amount, in privileged vessels, to only 3 per cent, and in other ships to 4½ per cent., *ad val.* An export duty of about 5*d.* per quarter is charged on corn, and of about 4*s.* 8*d.* per hhd. on wine; wool is not subject to any export duty.

Rostock having been formerly one of the Hans towns, had, for a lengthened period, and till lately, some exclusive privileges. Its vessels bore the flag, not of the grand duchy, but of the city of Rostock; and it had its own separate jurisdiction, independent of the rest of Mecklenburg; appeals from its tribunals being carried to the central court at Lubeck. But it has now only a court of secondary jurisdiction, with appeal to the tribunal at Parchim; and its other privileges, have been either curtailed or abolished. (*Berghaus; Allg. Länder, &c.; Stein; Commerc. Dict.*)

ROSTOFF, a town of European Russia, on the Don, about 22 miles above where it falls into the sea of Asoff.

Population, about 8000. This and the contiguous town of Nakhitchevan are the principal *entrepôts* of the trade of the vast countries traversed by the Don. The inhabitants of the latter are the more commercial, but Rostoff would seem to enjoy the special favour of the government, the *dépôts* of provisions for the army, the fortresses of the Caucasus, and of the eastern coast of the Black sea, being established in it. The fort St. Dimitri, near the town, is a *dépôt* for the munitions of war required by the above-mentioned places. During the proper season there is a great deal of bustle and activity both here and at Nakhitchevan.

ROSSTRAVER, p. t., Westmoreland co., Pa., 17 m. S.W. Greensburg, 191 m. W. by S. Harrisburg, 217 m. W. Bounded E. by Youghiogheny river., W. by Monongahela river. It has three stores, one fulling-mill, one woollen factory, five flouring-mills, nine saw-mills, two tanneries, two distilleries, one pottery; six schools, 238 scholars. Pop. 1880.

ROTHENBURG, a town of Bavaria, circ. Middle Franconia, cap. distr., near the Tauber, 40 m. W. Nuremberg. Population 5700. It is beautifully situated, but irregularly laid out, and has few edifices worth notice. It was anciently a free city of the empire, and is surrounded by old walls flanked with numerous towers. It has manufactures of woollen stuffs, and an active trade in corn and cattle. (*Berghaus, &c.*)

ROTHERHAM, a manufacturing town, parish, and township of England, W. riding county York, liberty of Hallamshire, wap. Strafforth and Tickhill, on the Don, crossed here by a handsome stone bridge, and close to its junction with the Rother, (whence the name,) 6 m. E.N.E. Sheffield, and 149 miles N.N.E. London. Area of parish (comprising eight townships), 12,810; do. of township 2149 acres. Population of town, in 1831, 4083. The town, partly in a valley, and partly on the sides of two steep hills, has several steep, narrow, and irregular, streets, lined with indifferently built stone houses. Recently, however, great improvements have been made, the streets have been widened, new houses built, and gas generally introduced. The courthouse, jail, market-house, and public library, are handsome modern buildings. The church (chiefly built by Archbishop Rotherham, in the 15th century, and by his son-dered collegiate) is a large cruciform structure of perpendicular architecture, with a central tower and spire fully enriched with pannels, canopies, and crockets: "On the whole," says Mr. Britton, "This is one of the finest parish churches in the N. of England, and deserves the most attentive examination, both as to its composition and merit of its details." (*Arch.* p. 274.) The living, a vicarage in the gift of lord Howard of Effingham, is worth £178 a year. There is an episcopal chapel in the township of Tinsley; handsome churches have recently been erected at Greasborough and Thorpe, and a church is in course of construction at Kimberworth. The town has a Roman Catholic chapel, and places of worship for Wesleyan and Primitive Methodists, Baptists, and Unitarians; besides which, there is at Marsborough, on the other side the river, an Independent chapel, with an attached academy for the education of young men intended for the ministry belonging to that class of dissenters. There are seven Sunday-schools, a Lancasterian school for 400 children of both sexes, two endowed charity schools, and a grammar school, founded in 1584, slenderly endowed, but conferring certain advantages on its pupils at the English universities. A dispensary was established in 1806; and there are almshouses for poor unmarried females. An ecclesiastical college, founded here at the close of the 13th century, was suppressed by Edward VI.; a part of the building is now used as an inn.

Rotherham, from its position in the middle of a district abounding with iron and coal, has long been distinguished for its manufactures of cast-iron; and during the American and French wars it almost exclusively supplied the navy with cannon. The iron-work for Sunderland, Southwark, and other bridges, was also cast here. The manufacture is still carried on with great vigour, new establishments have sprung up; and at present fenders, engine-work, and every variety of hardware, are manufactured on an extensive scale. Gas, starch, naphtha, and soap are also produced; a flax-mill employs about 100 hands, and there are two extensive porter-breweries. The town enjoys great facilities for the transport of its manufactured produce. The Don was made navigable to Tinsley, above the town, in 1720, and is accessible to Rotherham for vessels of 30 tons. A railway to Sheffield was opened in 1838, and the distance between the two towns (5½ m.) is performed in about a quarter of an hour. The North Midland railway, also, passes close to the town, and the Marsborough station is, in point of traffic, one of the most important on the line. The distance by railway to London is 171 miles, and the journey is accomplished in about nine hours. Large markets for corn and cattle on Monday; fairs, Whit-Monday and Dec. 11. The town has no regularly constituted municipal authority; but a body chosen by the inhabitants, called "the feoffees of the common land of Rotherham," employ the proceeds of certain rents for the improvement of the town. The midsummer & arter sessions for the W. riding, and weekly petty sessions, are held here by the county magistrates; besides which, there is a court of requests for the recovery of small debts. It is, also, the chief place of a poor-law union.

ROTHESAY, a royal borough and seaport of Scotland, county Bute, of which it is the capital, at the head of a beautiful bay on the E. side of the island, 30 m. W. by S. Glasgow. Population, in 1837, 4994. Port Bannatyne, a favourite sea-bathing residence, is 2½ m. N.W. Rothesay. Being protected on the land side by surrounding hills, and towards the sea by the opposite coast of Argyle, only 3 m. distant. Rothesay has a very mild climate, and is much resorted to by sea-bathers, as well as by invalids. Exclusive of its castle, the principal public buildings are the town-hall and county buildings (under one roof), in the castellated form, with an elegant tower; two places of worship (one of them Gaelic), connected with the established church; and chapels belonging respectively to the Associate Synod, Cameronians, Independents, and Episcopalians. The parish church is situated ½ m., and Mount Stuart, the seat of the marquis of Bute, 3½ m. S. from the town. The means of education are ample: there are six public libraries in the town, and two reading rooms; with various friendly societies and charitable associations. A savings' bank was established in 1821.

The cotton manufacture has been introduced; and two cotton mills employ 455 hands. But the fisheries may be said to be the most important, as well as the oldest, branch of business carried on. The salmon fishery obtains to a limited extent, as also that of haddocks, whitings, and soles. But the herring fishery is more extensive than all these together. The fishery, however, is not carried on in the bay of Rothesay; or if so, only in a small degree; it centres chiefly in the kyles (straits) of Bute, and the adjacent saltwater lakes; but it is principally carried on with Rothesay capital. In the year ending 5th of April, 1840, 17,119 barrels of herrings were cured at Rothesay. About 60 vessels of 3000 tons, exclusive of steamers, belong to the port. There are two small building yards, and three branch banks.

The castle of Rothesay, a noble ruin, is of great antiquity. It was at one time a favourite royal residence, and Robert III. expired in it in 1405. John, earl of Bute, the favourite of George III., and Matthew Stewart, the mathematician, father of Dugald Stewart, were both natives of this borough. Previously to 1832, it joined with other boroughs in sending one member to the House of Commons; but it is now merged in the county representation.

ROTTERDAM, a celebrated commercial city of Holland, being, in point of population and importance, the second in the kingdom, prov. S. Holland, cap. arrond., on the N. bank of the Maas, where it is joined by the Rotte, whence its name, 17 m. (direct distance) from its mouth, and 35 m. S. S. W. Amsterdam; lat. 51° 55' 19" N., long. 4° 29' 14" E. Population, on the 1st of January, 1840, 78,098, having increased about 10,000 during the previous ten years. It is of a triangular shape, the base of the triangle extending along the river. It is surrounded by a moat, and, like every other Dutch town, is intersected by numerous canals, only one public thoroughfare, the High street, being without a canal in its centre. This street, which runs in a direct line E. and W. through the city, is somewhat raised above the rest, being built upon the dam by which the Maas is prevented from inundating the country behind the town. Being the principal seat of retail commerce, it is lined with shops throughout. Rotterdam has a striking appearance from the river.

The fine quay, called the Boompjes, from the rows of trees with which it is planted, extends along the river for nearly a mile: behind the trees is a line of well-built houses three or five stories in height, mostly of dark-coloured brick, all having an aspect of much grandeur. The quay being crowded with shipping, this part of the city is the great seat of business, and presents a scene of incessant activity. In penetrating through the town from the Boompjes, we come to street after street, each having a canal in the middle, lined with trees on both sides, and exhibiting a mixture of fifty gable fronts of houses, trees, and masts of shipping. These canals, or havens, stretch lengthwise and crosswise, into the meshes of a net, through the city; and at every street interval is perceived a drawbridge of white painted wood, constructed with ponderous balancing beams overhead, and raised by means of chains, for the passage of vessels. The ground beneath the trees is paved with small yellow bricks, and is chiefly occupied as quays for the landing of goods. The space from the trees to near the houses is used in the usual coarse manner for carts and carriages, and here the foot passengers are generally obliged to walk,

for small outshot buildings, flights of steps to doorways, and such like interruptions prevent any regular thoroughfare on the narrow brick trottoirs close by the houses. The havens are in a few places protected by chains from the streets, so that there is constant liability to accidents, particularly at night, when the darkness is but poorly relieved by oil lamps dangling, Parisian fashion, from the ropes stretched betwixt the trees and the houses. Latterly a portion of Rotterdam has been lighted with gas; but according to a parsimonious plan, the lamps are not lighted when the moon is expected to shine." (Chambers' Tour, p. 5, 6.)

The houses of Rotterdam are generally on a large scale, and lofty; and in many of the streets they are really elegant. Mr. Chambers speaks as follows of a large class of residences. "Each house may be considered the castle of a merchant, who both resides with his family, and carries on the whole of his commercial transactions within the same set of premises. The front part of the building exhibits an elegant door of lofty proportions, 15 or 20 ft. high, for instance, at the head of a flight of steps. On getting a glimpse into the interior, you see a lobby paved with pure white marble, and a stair of the same material leading to the story above, which consists of a suite of lofty rooms, and is the main place of residence of the family.

"Some of the rooms are finished in a style of great elegance, with rich figured cornices and roofs, silk draperies to the windows, smooth oak floors, and the walls most likely painted as an entire picture or landscape, in oil, by an artist of eminence. Near to the door of the house is a coachhouse door, which, on being thrown open from the street, discloses a wide paved thoroughfare leading to an inner court, the buildings round which are devoted to the whole warehousing department of the merchant. The bulk of the edifices of this great trading city are of the kind I describe." The ordinary houses are badly heated, and are in other respects not very comfortable; but the poorest house in the city is as clean as scrubbing and washing can make it, both inside and out; in this respect, indeed, the Dutch have no equals. Among the minor peculiarities which strike an Englishman, are the small mirrors affixed outside to almost all the first-floor windows, which are placed so as to show the inmates whatever may be going on in the street below. The want of good potable water is a great inconvenience.

Rotterdam has few public buildings of interest. The town-hall, a large modern structure in the Grecian style, has a noble council room, with rooms for the city library and philosophical apparatus; but, though superior to the generality of such edifices, it is much inferior to that of Amsterdam. The exchange, on the contrary, is a finer building than that of the capital; it is rectangular, with a court in the centre, surrounded with arcades, supported by 30 pillars, each of a single block. The Schieland palace, occupied in 1811 by Napoleon and Maria Louisa, is perhaps the finest of the public edifices. The custom-house and the former East India house on the Boompjes present nothing remarkable. There are from 15 to 20 churches, including one for English Episcopalians, and one for Scotch Presbyterians. The principal is the church of St. Lawrence, formerly the Roman Catholic cathedral. It is a large Gothic brick building, with a lofty square tower, and dates as far back as 1412; but nearly all its ancient ornaments were swept away at the Reformation. In it are the monuments of the celebrated Dutch admirals, De Witt, Kortnaer, and Brakel, each of which has an appropriate inscription. It has also a splendid organ, said to be superior in size to the great organ of Haarlem, generally considered the largest in Europe. This is 150 ft. in height, mounted upon a colonnade about 50 ft. in elevation, and has, according to Barrow, 5500 pipes, the largest being 32 ft. in height, and 16 in. in diameter. In the market-place is a bronze statue of Erasmus, the most illustrious by far of the natives of Rotterdam. The house where this great scholar, "the glory of the priesthood and the shame," first saw the light, on the 28th Oct., 1467, still exist; but (quantum mutatus!) it has been degraded into a gin-shop! Rotterdam has a naval dock-yard; but it is on a small scale, and contains little worth notice. In the neighbourhood of the city are many places of entertainment, as tea-gardens &c.; and in the town are several clubs, where English as well as the continental newspapers are taken in. Rotterdam is the seat of the marine department for the Maas, of the superior judicial courts for the province, and of a tribunal of commerce; the cap. of the 9th militia district of the kingdom; and the residence of a military commandant, a director of police, and numerous foreign consuls. It has a society of arts and experimental philosophy, founded in 1767, branches of the societies of Public Good and the fine arts, a college, a Latin school; many superior and intermediate and poor schools, in the whole of which, according to Chambers, about 8000 children are instructed; and various private academies. The central prison of Holland for juvenile offenders is at Rotterdam; it has also various workhouses and charitable institutions, and a ma-

ings' bank, paying interest at 4 per cent. There are manufactures of tobacco, refined sugar, needles, and pins, glass wares, corks, dyeing and chemical products, spirits. &c.; large markets are also held weekly for corn, flax, hemp, and other agricultural produce; and the annual fair of Rotterdam is the largest in Holland. (*De Cloet*.) Rotterdam is more advantageously situated in a commercial point of view than Amsterdam, or any other Dutch town. She is easily accessible from sea by the largest class of merchantmen; and from her position on the principal embouchure of the Rhine, as well as of the Maese, she is the grand emporium of the foreign trade of the countries which they traverse. The imports and exports are similar to those of Amsterdam, which see. The white Zealand and Rhenish wheat shipped here is of a superior quality; and it is the best market for madder, geneva, cheese, &c. The imports, in 1840, comprised, among other things, 20,115 tons coffee; 20,057 tons sugar; and 5029 chests indigo, mostly from Java; with 13,305 bales cotton, 17,411 hhds. tobacco, 21,012 slabs banca tin, &c. During the same year there entered the port 1871 ships of the burden of 329,584 tons. Rotterdam has a regular intercourse by means of steam-packets with London and other great over-sea ports, and with Dusseldorff, Cologne, Maestricht, and other ports on the Rhine, Maese, &c. The railway from Amsterdam to Haarlem is to be extended to Rotterdam.

Besides Erasmus, the great painter Adrian Vander-Werf, was a native of Rotterdam.

ROTTERDAM, p. t., Schenectady co., N. Y., 22 m. W.N.W. Albany, 390 W. Bounded N.E. by Mohawk r. It contains two Dutch Reformed churches, four stores, two grist-mills, three saw-mills, one oil-mill; 10 schools, 367 scholars. Pop. 2284.

ROUBAIX, a manufacturing town of France, dép. du Nord. arrond. Lille, cap. canton, on the canal of Roubaix 7 m. N.E. Lille. Pop., in 1836, ex. com., 13,426, or with com., 19,455. Like most Flemish towns, it is well built. It formerly laboured under a want of water; but of late an adequate supply has been obtained by means of Artesian wells. After Lille, Roubaix is one of the chief towns in the dép. for the manufacture of cotton goods; it has been estimated that, in the town, and immediately adjacent country, about 30,000 hands are alternately employed in the manufacture of cotton and woollen goods; the latter from about June to September, and the former during the remainder of the year. Mr. Symonds says, "of the various weaving and spinning districts of this manufacturing dép., there is none more prosperous than Roubaix, which has increased prodigiously within the last 10 or 15 years. The articles chiefly manufactured are Thibets, waistcoat-pieces, and thick cotton goods. There are about 12,000 looms in this district, of which half are Jacquard looms. The latter are principally in the factories, the manufacturers finding it necessary to preserve the privacy of their patterns. A weaver will on cotton goods, earn at an average 30 sous (15*d*.) a day; on the second class work, from 30 to 40 sous (15*d*. to 20*d*.); and on the Jacquard loom, from 2 fr. to 5 fr. per day, the average being about 3 fr.: these are gross wages. Nett earnings will be, weekly, about 12 fr. or 13 fr. for the Jacquard weaver, and from 6 fr. to 10 fr. for the plain weavers. Women and children obtain plenty of employment at the mills. Wages had risen by above 15*d*. in the last few years. I inspected a woollen spinning factory; the spinners were earning from 20 fr. to 24 fr. per week, nett, as a matter of course; and it is by no means uncommon for them to earn 30 fr nett. The girls earned about 7 fr. 30 cents, and the children 3 fr. per week. The rooms were all high, clean, and well ventilated; and the machinery, built by an engineer at Rheims, though not of the newest description, was extremely good. The yarn spun was fine and even, and superior to what I have sometimes seen in Scottish mills." (*Handloom Weaver's Rep.*, p. 129, 130.) The working population of Roubaix is increasing by continual immigrations from Belgium. Some of the labourers live in the town; but the greater number reside in the neighbouring villages and hamlets, coming daily to and from the factories, in which they work from 14 to 15 hours a day. They live mostly on meat, soup, potatoes, and beer; using butchers' meat four days a week. Symons says, that the morals of the working classes here are decidedly worse than in most other districts; but, according to Villermé, who, in this respect, is a better authority, the work-people of Roubaix and Turcoing are, whether as regards morals, cleanliness, clothing, lodging, food, or health, decidedly superior to those of Lisle. Drunkenness is here, and, indeed, every where else throughout French Flanders, a prevalent vice; but, in other respects, the conduct of the work-people seems to be good; and they have established numerous societies for their mutual support and assistance. (*Villermé*; *Tableau des Ouvriers*; *Hugo*; *Dict. Géog.*)

ROUEN (an. *Rothomagus*), one of the principal cities of France, and the great seat of its cotton manufacture, dép.

Seine-Inférieure, of which it is the cap., on the Seine, 44 m. (direct distance) from its mouth, and 67 m. N.W. Paris lat. 49° 26′ 27″ N., long. 1° 5′ 59″ E. Pop. in 1836, 91,907, but if the inhabs. of all its suburbs be included, the pop. will amount to upwards of 100,000. (*Villermé*, &c.) The city, which stands in a fine and fruitful country, is most admirably situated on a navigable river, by which it communicates with the capital on the one hand, and with the flourishing seaport of Havre on the other; and it is surrounded by a verdant and delightful country. Its numerous spires and towers, and the vessels that throng its quays give it a very imposing external appearance, to which its interior presents in most parts a striking contrast. Generally it is ill built. Streets mostly narrow, crooked, and filthy; houses principally of wood, or rather of lath and plaster, though in the W. and newer quarters of the city some are built of more solid materials, and have even considerable elegance. It is oval, or rather lozenge shaped, and was for a lengthened period pretty strongly fortified; but its ramparts are now demolished, and their place is occupied by a series of boulevards, which separate the city proper from the *faubourgs* Cauchois, Beauvreuil, Bonsecours, Martinville, &c.

The Seine here crossed by a bridge of boats, and one of stone, divides it from its large suburb of St. Sever. The boulevards, which are planted with trees, like those of Paris, and the fine broad quays and *cours*, which extend along the banks of the river, are the favourite and almost the only public promenades; the *squares* or open spaces are shabby and irregular, and except the *Place Royale*, near the centre of the city, are all of insignificant size. Some, however, are ornamented with public fountains, with which Rouen is well furnished: the *Fontaine de Lisieux* is a curious piece of antique sculpture, representing Mount Parnassus, with figures of Apollo, Pegasus, &c. In the square of La Pucelle, an indifferent statue of Joan of Arc is erected on the spot where that heroine suffered martyrdom in 1431.

The central parts of the city are the chief seat of general commerce; the upper classes principally reside in the faubourg Cauchoise, and the N. suburbs; while the lower quarters at the E. end of the town, and the faubourg St. Sever, are almost wholly inhabited by the manufacturing classes.

By far the most celebrated and striking public edifice is the cathedral, one of the noblest religious structures in France, or even in Europe. It was constructed principally between the 13th and 16th centuries inclusive: entire length, 434 ft.; breadth 103 ft.; length of transept, 178 ft. height of nave, 89½ ft.* Its richly ornamented front has three fine portals, over the central of which is a square tower, and spire of iron work, reaching to a height of 686 ft.; flanked by two lofty but dissimilar towers, the *Tour Romain* and *Georges d'Amboise*. The former, which dates from a period long anterior to the rest of the building, is in a simple and unadorned style; but the latter, built at the end of the 15th century, is much admired for the beauty of its architecture. It is ornamented with numerous sculptures; and before the revolution contained an enormous bell, which with many others belonging to this cathedral, was then sent to the cannon foundry. The interior of the edifice is lighted by 130 windows, many of which are ornamented with stained glass; and contains a vast number of tombs, including that of Richard I. Cœur de Lion of England, many dukes of Normandy, and 17 archbishops of Rouen; and the fine mausoleum of the two cardinals d'Amboise. The latter is very perfect; but many of the other monuments were much mutilated in the religious and revolutionary wars.

The church of St. Ouen in the Place Royal belonged to the oldest conventual establishment in Normandy, and occupies a larger extent of ground than the cathedral. It is an admirable specimen of the pointed Gothic; its fine octagonal tower rising from the centre of the building is 213 ft. in height. The town-hall adjoining this church was originally a portion of the conventual edifice; and, besides various public offices, is appropriated to the museum and public library, with about 80,000 volumes. Several of the other churches in Rouen well deserve notice, and some are of high antiquity. The city clock is placed in a square Gothic tower, erected in 1839, in the High-street. The *Palais de Justice*, with a noble saloon, was built for the parliament of Normandy, at the end of the 15th century. The numerous public halls of Rouen, for the exhibition and sale of different articles, are well adapted to their destination. They occupy three sides of a square, the centre of which forms an open exchange. A special apartment is devoted to every different kind of goods; the cotton cloth hall, where the most important branch of traffic is conducted, is 230 ft. in

---

* The dimensions are given by Hugo, and the *Guide du Voyageur*, in French ft., which are here converted into English ft.

length, by 53 ft. 4 in. in breadth. Every Friday, from 6 A.M. till noon, these halls display great commercial activity. There is another exchange adjacent to the quay. The exchequer-office, *chambre des comptes*, barracks, the *théâtre*, a spacious general prison, another prison for accused but untried persons, prefecture, archbishop's palace, mint, custom-house, college, two theatres, *hôtel-dieu*, the general infirmary, which, according to Hugo, has commonly 3268 inmates, Protestant church, and the remains of an old fortress, are among the other principal public buildings. Rouen has some private houses worth notice, especially those in which Fontenelle and Pierre Corneille were born; others in which are some curious works of art, &c. It is the seat of a royal court for the deps. Seine-Inf. and Eure; of tribunals of primary jurisdiction and commerce; a Chamber of commerce, and council *des prud' hommes*, royal and university academies, a royal college, and a mint established in the 9th century; the cap. of the 15th military division of France; the seat of an archbishop, whose diocese comprises the dep. Seine-Inférieure, and whose suffragans are the bishops of Bayeux, Coutances, Evreux, and Seez; the seat of a Protestant consistory; and the residence of many foreign consuls. It has a central society of agriculture; societies of public emulation, commerce, agriculture, medicine, &c.; a Bible society; schools of design and navigation; a botanic garden; savings' bank, and various charitable institutions.

Rouen is so eminent for its cotton manufactures that it has acquired the title of the French Manchester, and checked printed cotton cloths for women's dresses, are commonly known in France by the name of *rouenneries*. It was anciently celebrated for its linen fabrics, the manufacture and dyeing of which appear to have been carried on in it in the earliest times of the French monarchy. But so late as the middle of the last century, the workmen employed at Rouen were nearly all foreigners, Germans, Dutch, or Swiss; who remained in France only during a part of the year, returning to spend the remaining months in their native countries: and less than 50 years ago, the cotton yarn employed in the manufactures was wholly spun by hand. At present, however, both water and steam power are largely employed. The whole region round Rouen shares more or less in its branches of manufacture. Mr. Symons says there are 920 spinning mills in Normandy and the Seine, and 920,000 spindles. (*Hand-loom Weavers' Report.*) The prefect of the Seine-Inférieure, in 1835, estimated the weavers of cotton and woollen goods in that dep. to amount to about 130,000, four fifths of whom were resident in Rouen and its immediate neighbourhood: and Villermé (1840) states that 50,000 persons, men, women, and children, or about half the entire population of the city and suburbs, are engaged in the cotton manufacture. (*Tableau des Ouvriers*, i. 136-9.)

"The manufacturers of Rouen (says Mr. Symons) pride themselves greatly on the superiority of their products; and it is but justice to say that I have seldom seen printed cottons so good in colour and texture at 7d. per ell, as those of Rouen at that price. The goods produced by the Norman looms are in direct competition in third markets with those from the W. of Scotland. The wages of Norman weavers are, if anything, lower than in Scotland, but provisions are at least 20 per cent. cheaper. So that no great difference is perceptible in the condition of the two classes; and were I to be forced to choose whether I would be a pullicate weaver in Scotland or in Normandy, I think I should be sorely puzzled which to select, or rather which to consider the greater infliction." (*Hand-loom Rep.* p. 125.) The nett wages obtained by country weavers working on their own account in pullicates, thick calicoes, &c., are about 1 fr. a day, or, a week, and this may be taken as above, rather than below, the average. Children and women are both occupied in weaving at proportionate earnings. In the spinning mills, spinners get from 15 to 30 fr. a week, working from 50 to 84 hours; women and girls employed at the carding and drawing operations gain from 7 fr. to 10 fr. The power-loom weavers earn from 12 to 16 fr., and day labourers from 10 to 12 fr. a week. (*Hand-loom Rep.*)

In some branches, and especially that of spinning, wages diminished during the ten years preceding 1834; since which they have been somewhat augmented. According to Villermé, the working classes of Rouen are, upon the whole, in a much less depressed condition than those of Lille, and their health is also much better. Drunkenness is in both towns the prevailing vice among the lower classes; but it seems to be less prevalent at Rouen. The woollen manufactures of the city are unimportant; their chief seat in this dep. being at Elbeuf (which see). Broad cloths, velvets, hardware, superior earthenware, chemical products, and confectionary, for which Rouen is famous, are the other principal products. Vessels of 200 tons ascend to the city, which carries on a considerable trade with countries both in the N. and S. of Europe, the Levant,

America, and the other maritime deps. of France; the greater part, however, of its foreign commerce is carried on through the intervention of Havre.

Rouen was of sufficient importance in the third century to be created a bishop's see; it afterward became successively the capital of the kingdom of Neustria, and of the duchy of Normandy. Prince Arthur of Brittany having been put to death in Rouen, in 1203, by John, king of England, Philip Augustus besieged and took the city in the year following. It was retaken by Henry V. of England in 1417, and retained by the English till 1449, when it was finally annexed to the French crown. The Reformation made great progress here; and the city suffered much in consequence of religious feuds. But fewer individuals fell victims to the Massacre of St. Bartholomew and the phrensy of the revolution in this than in most other large French cities.

Rouen has given birth to some of the most illustrious individuals of whom France has to boast; among whom may be specified Pierre Corneille, deservedly surnamed *le grand*, one of the greatest modern dramatists, born here on the 6th of June, 1606; his brother, Thomas Corneille; Fontenelle, the academician, born here in 1657; Bochart, the famous oriental scholar; Daniel the historian; Brumoy, author of the *Théâtre des Grecs*, &c. (*Hugo*, art. *Rouen*; *Villermé, Tableau des Ouvriers*, i., 135-160; *Hand-loom Weavers' Rep.*)

ROULERS, a town of Belgium, prov. W. Flanders, cap. canton on the Mandelbecke, a tributary of the Lys, 26 m. W. by S. Ghent. Pop., in 1836, 9982. It has a high school, manufacture of linen fabrics, leather, soap, oil, &c., and some trade in butter of a superior quality. (*Hœuschling; Dict. Géog.*)

ROUSE'S POINT, p. v., Champlain t., N. Y., 158 m. N. Albany, 560 W. Situated on the W. side of lake Champlain, and has a small village containing a Methodist church and about 125 inhabitants, with a steamboat landing. A little N. of the village, the United States government, shortly after the last war, built a strong fortification, in the form of a round tower, pierced for several tiers of guns, on which was expended over $400,000. On accurately taking the latitude with the fine instruments used in that survey, it was found to be within the Canada line, and was abandoned. It is nearly in ruins, but is now, by the terms of the late treaty, within the United States, and is the most favourable site for a fortification which will command the entrance to lake Champlain. A fortification here would be of great importance in case of a war.

ROVEREDO (Germ. *Rovereit*), a town of the Tyrol, on the frontiers of Austrian Italy, cap. circ. of its own name, on the Leno, near its junction with the Adige, 13 m. S. by W. Trent. Pop. 7300. It is well built, many of its edifices being of marble. The most remarkable building is the castle, on a height commanding the town. It has superior civil, criminal, and commercial tribunals for the circ., a gymnasium and high school, and an English conventual establishment. In and round the town are numerous silk mills, which employ, according to the Austrian Encyclopedia, 8000 hands. It has also tobacco and leather factories. (*Oester. Nat. Encyc.; Berg.; Rampoldi, Corografia, &c.*)

ROVIGNO, a seaport town of Austrian Italy, circ. Istria, on the Adriatic, 39 m. S.S.W. Trieste. Pop. 9600. It has numerous fine churches and other public edifices, high and female schools, and two hospitals. Its principal church is built after the model of St. Mark's at Venice. It is the seat of civil, criminal, and commercial tribunals; has two harbours, one of which is tolerably secure; and carries on a considerable trade in wine, olives, timber, anchovies, and tunny. Its inhabitants are principally seafaring people, or engaged in the fisheries; they are, however, partly occupied in ship building and making cables. Near the town are some quarries of superior marble. (*Rampoldi; Oester. Nat. Encyc.; Berghaus.*)

ROVIGO (an. *Rhodigium*), a town of Austrian Italy, prov. Venice, cap. deleg., on the Adigetto, a branch of the Adige, in the swampy tract of the Polesin, 36 m. S.W. Venice, and 17 m, N.E. Ferrara. Pop., in 1837, 5669. It is fortified in the old style, with walls flanked with towers, a ditch, and a citadel; and is entered by six gates. According to Rampoldi, it is, rather a handsome town; but the *Austrian Encyc.* says, it is ill-built, and unhealthy. It has numerous churches, a seminary, and a hospital; two orphan asylums, a large and fine theatre, an academy of sciences and arts, and various superior public schools. The library of the Count Silvestri, comprising 36,000 volumes, is open to the public. It is the seat of the superior courts for the deleg., and the residence of the delegate, and the bishop of Adria. Though much decayed, it has a considerable trade in corn, a large fair from Oct. 20 to 28, and three weekly markets. General Savary was created by Napoleon, Duke of Rovigo. (*Rampoldi; Berghaus; Austr. Nat. Encyc., &c.*)

ROWAN, county, N. C. Situated centrally in the W. part of the state, and contains 375 sq. m. Drained by Yadkin river. It had in 1840, four schools, 98 scholars. Pop.: whites, 8646; slaves, 3365; free coloured, 98; total, 12,109. Capital, Salisbury.

ROWLEY, p. t., Essex co., Mass., 30 m. N. by E. Boston, 470 W. Watered by Rowley river, which affords water-power, and has at its mouth a harbour for small vessels. The first fulling-mill in the United States was erected here. It contains two churches, a Congregational and Baptist, three stores, two grist mills, two saw mills, four tanneries; one academy, 30 students; three schools, 218 scholars. Pop. 1903.

ROXBORO', t., Philadelphia co., Pa., 7½ m. N.W. Philadelphia. It contains the village of Manyunk on Schuylkill river, where a canal connects the pool at Fairmount with that of Flat Rock. The dam presents a beautiful cascade, which adds much to the picturesque scenery. Steatite or soapstone is found in the N. part of the township. It has 24 stores, one forge, 11 fulling-mills, four woollen factories, eight cotton factories, with 17,940 spindles, eight flouring-mills, two grist-mills, two saw-mills, three paper-mills; 12 schools, 746 scholars. Pop. 5797.

BOXBORO', p. v., capital of Person co., N. C., 54 m. N.N.W. Raleigh, 250 W. It contains a courthouse, jail, two stores, and about 100 inhabitants.

ROXBURGH, an inland and frontier co. of Scotland, having N. the co. Berwick and a small portion of Mid-Lothian, E. and S. Northumberland and Cumberland, and W. Dumfries and Selkirk. Area, 457,920 acres, of which nearly a half is occasionally under the plough. It is partly intersected and partly bounded by the Tweed; and is traversed from its S.W. border, where it has its source, N.E. to Kelso, by the Teviot, whence it is sometimes called Teviot-dale. It has every variety of surface and soil. The low arable lands in the valleys of the Tweed and Teviot, consist principally of light turnip soil. The mountainous or pastoral district is principally in the S.W. parts of the county, along the Dumfries, Cumberland, and Northumberland border. The hills, however, like the Cheviots, to which they are contiguous, are mostly smooth, dry, and well covered with good sheep pasture. Property mostly in large estates; but there are several of the smaller class of proprietors. Farms generally large; and some farmers frequently hold three or more farms. Arable husbandry is as well understood and practised in the lower parts of this county as in the most improved parts of the empire. It is also celebrated for having been the theatre where some of the principal improvements in modern farming were first introduced, and where others were first successfully practised in Scotland. Mr. Dawson, the great improver of Scotch husbandry, occupied the farm of Frogden, near Kelso, in this county; and in it, soon after 1760, he set to work the first plough drawn by two horses, driven by the first ploughman, that was ever seen in Scotland: And if Mr. Dawson was not the first to set the example of raising turnips, he was the first practical farmer by whom they were profitably cultivated on a large scale. (Survey of Roxburgh, p. 66, 90.) Farmers for dressing corn were also made and used in this county before they were seen in any other part of Scotland. (Ibid. p. 59.) Large quantities of wheat are now produced. Cattle, a mixed breed. Sheep, principally Cheviots. Within the last 30 years many important improvements have been effected in this district. A large extent of land that was entirely pastoral now bears luxuriant crops: bone manure has been introduced; agricultural management has been materially amended; a good deal of waste land has been planted; farm-houses and buildings have, in numerous instances, been rebuilt on approved plans; thrashing-machines have been erected on most considerable farms; and the habits and accommodations of the people have been materially improved. (New Statistical Account of Scotland, Roxburgh, p. 23, 33, 40, 123, &c.) There are some very productive orchards in the neighbourhood of Jedburgh, Melrose, and Kelso. Average rent of land, in 1810, 10s. an acre. Marl is found in vast quantities in several parts of the county, and it is also well supplied with limestone and freestone. Various branches of the woollen manufacture have been introduced, and are prosecuted with considerable vigour at Hawick and Wilton; and in a lesser degree at Jedburgh, Melrose, and Kelso. The parish of Kirk Yetholm, in this county, is celebrated as being the residence of the largest colony of gypsies in Scotland. Roxburg contains 31 parishes; and returns one member to the House of Commons for the county; and the borough of Jedburgh joins with other burghs in returning a member. Registered electors for the county, in 1839 40, 2227. Principal towns, Jedburgh, Kelso, Hawick, and Melrose. In 1831, this county had 6733 inhabited houses, 6930 families, and 43,663 inhabitants, of whom 20,761 were males, and 22,902 females. Valued rent, £314,663 Scotch. Annual value of real property, in 1815, £254,180.

ROXBURY, p. t., Norfolk co., Mass., 2 m. S. Boston, 436 W. It is connected to Boston by a neck of land, which constitutes a broad avenue, and may be regarded as a continuation of Washington-street, Boston. It contains the pond, 4 miles from Boston, from which the city is supplied with water, by four trains of cast iron pipes, the aggregate length of which, with the distributing pipes, is 68 miles. Incorporated in 1630. There are many elegant country-seats in the township. It contains 10 churches, five Congregational, one Unitarian, one Episcopal, one Baptist, and two Universalist, two banks, 83 stores, four lumber-yards, five grist-mills, four saw-mills, five tanneries, two printing offices, two binderies, one weekly newspaper; 14 academies, 350 students; 20 schools, 661 scholars. Pop. 589.

ROXBURY, p. t., Litchfield co., Ct., 53 m. W.S.W. Hartford, 314 W. Watered by Shepaug river. It contains two churches, Congregational and an Episcopal, two saws, two grist-mills, three saw-mills; seven schools, 221 scholars. Pop. 971.

ROXBURY, p. t., Delaware co., N. Y., 90 m. E. Delaware, 65 m. S.W. Albany, 352 W. Drained by Papacton branch of Delaware river. It has five stores, one fulling-mill, five grist-mills, 18 saw-mills, four tanneries; 18 schools, 900 scholars. Pop. 3012.

ROXBURY, t., Morris co., N. J., 14 m. N.W. Morristown. Schooley's mountain covers a great portion of its area. On its summit is Budd's pond, 2 m. long and 1 broad, which has its outlet into the S. branch of Raritan river. It has nine stores, one forge, two fulling-mills, one woollen factory, two grist-mills, two saw-mills, four tanneries, one distillery, one pottery; 14 schools, 587 scholars. Pop. 2221.

ROYALSTON, p. t., Worcester co., Mass., 74 m. W. Boston, 416 W. Incorporated in 1765. Miller's river and its branches afford good water-power. It has three saw, seven fulling-mills, two woollen factories, two grist-mills, seven saw-mills, two tanneries; 12 schools, 533 scholars. Pop. 1667.

ROYALTON, p. t., Windsor co., Vt., 33 m. S. Montpelier, 468 W. Watered by White river. It contains three churches, a Congregational, Episcopal, and Methodist, four stores, two fulling-mills, two woollen factories, one grist mill, six saw-mills, two tanneries; one academy, 30 students; 16 schools, 567 scholars. Pop. 1917.

ROYALTON, p. t., Niagara co., N. Y., 8 m. E. Lockport, 271 m. W. Albany, 406 W. Bounded S. by Tonawanda creek. At Gasport on the Erie canal, are inflammable springs. It has 11 stores, two grist-mills, seven saw-mills; 16 schools, 744 scholars. Pop. 2549.

ROYALTON, t., Cuyahoga co., O., 126 m. N.E. Columbus. It had one store: 15 schools, 454 scholars. Pop. 1052.

RUBICON. See ITALY, ante, p. 66.

RUDDLE, t., Independence co., Ark. It has eight mills, three grist-mills, two saw-mills, one tannery, two distilleries, one printing-office, two weekly newspapers; one academy, 55 students; two schools, 45 scholars. Pop. 1283.

RUDGELEY, a market town and par. of England, co. Stafford, and E. div. of hund. Cuttleslone, on the S. bank of the Trent, crossed here by a fine aqueduct of the Grand Trunk canal, 8 m. E.S.E. Stafford, and 129 m. N.W. London. Area of par., 7190 acres. Pop., in 1831, 3765. The town is well built, comprising many good houses, and a few that may even lay claim to elegance. The parish church, an ancient structure, with a handsome tower of its W. end, has recently been enlarged and almost rebuilt. The living is a vicarage in the gift of the dean and chapter of Litchfield. There are places of worship for several classes of dissenters, and several Sunday schools. A free grammar school was founded here by Queen Elizabeth; an endowed school furnishes clothing and instruction to 28 boys; besides which there is a national and infant school. Hopkins's almshouses afford relief to aged poor women, and there are several money charities. The principal manufacture of Rudgeley is that of hats and felts; but other articles are made here. A stream, which runs through the town to the Trent, turns several colour and corn-mills, and there are some iron-forges. At Brereton, within this parish, are extensive collieries, employing from 500 to 600 men. The town derives considerable advantages from its position on the great line of canal communication between the N. and S. counties. Its government is vested in two constables, chosen by the inhabitants. Markets on Tuesday; fairs April 17, June 5, and October 21, for horses and cattle.

RUGBY (an. Rocheberie, or Rokeby), a market town and par. of England, co. Warwick, hund. Knightlow, on the Avon, 98 m. E.S.E. Birmingham, and 75 m. N.W. London. Area of par. 2190 acres: pop., in 1831, 2501. The town, on an eminence S. of the river, consists of three streets, one of which, leading to the church, is broad, and lined with modern brick houses: indeed, great improvements have been made within the last few years, and the advantages derived by the town from its proximity to the Birmingham railway seem likely still further to promote its prosperity: in the older part of the town, however, there are many houses of plaster and timber, denoting the former poverty of the place. The church is an ancient building, possessing little architectural interest, with a square embattled tower, having a turret at its S.E. angle: the living is a rectory, of the annual value of £510, in the gift of Earl Craven. There is also a district church, of very recent erection. The Wesleyan Methodists and Baptists have places of worship; and there are several Sunday schools, a charity school for 30 children of each sex, with almshouses and other charities. The chief importance of Rugby, however, is derived from its great public school, to which the talent of several of its recent masters and the richness of its endowments have given a well-merited celebrity. It was originally a simple grammar school, founded in 1557, by Lawrence Sheriffe, citizen of London, a native of the neighbourhood, for the benefit of the town and neighbourhood of Rugby. Any person who has resided during two years in, or within 10 m. of the town, may send his sons to be educated free of expense: but if the parent reside out of the town, his son must then lodge at one of the boarding-houses of the school, paying the same rate for his board as those not on the foundation. The number of boys on the *foundation* is unlimited; but the masters may not receive more than 260 boys not on the foundation. The number of scholars may at present average about 280, about 60 of whom are on the foundation. The school property consists of land within the parish, and of about 8 acres of land, called the Conduit Close, in the neighbourhood of Lamb's Conduit-street, London; the value of which has so greatly increased, in consequence of the buildings erected upon it, that the annual revenues of the school, which, at the middle of last century, were under £120 a year, now exceed £5000. The management is vested in 8 trustees; and the school is under a head master and eight classical masters, with subordinate teachers of writing, French, &c. The study of classical literature is carried quite as high as elsewhere, and the success of the boys at examinations for scholarships at Oxford and Cambridge, shows that their knowledge is of a substantial character; but at the same time, history, both sacred and profane, modern as well as ancient, physical and political geography, arithmetic and mathematics (as far as conic sections), and French, constitute integral parts of the course of instruction. An annual examination is held at Christmas, and the names of the boys that distinguish themselves are published in a class paper. The school has 14 exhibitions, established by the founder. Three exhibitioners are elected every year by the trustees on the report of examiners sent from the universities of Oxford and Cambridge. These exhibitions are of the value of £60 a year, and may be held for seven years *during residence* in any college at their university. There are likewise six scholarships, of annual value of £25 each, supported by subscription: one scholar is elected every year, and his age must not exceed 14 years at the time of his examination. The scholarship is tenable for six years, if the boy holding it remains so long at Rugby. The ancient buildings of this great seminary consisted formerly of a master's house, and two or three school-rooms, all of very limited size and shabby exterior, totally inadequate to the wants of the increasing establishment. In 1809, however, the erection of a large handsome pile of buildings was commenced on the site of the old school-house at the S. extremity of the town. The edifice is of white brick, dressed with stone at the angles, windows, and cornices, the whole being of our architecture. The principal front is 230 feet in length, and the schools are entered by a turretted gateway up the street and leading to the principal court, a fine 90 feet in length, by 75 feet in breadth, having cloisters on three of its sides. The buildings on the S. side comprise the dining-hall of the head-master's boarders and school-rooms; on the W. side is the great school, and on the N. side are schools for the French and writing classes. The apartments of the head master are roomy and commodious, communicating also with the several dormitories running round the quadrangle over the school-rooms. The school-chapel is a detached building, in later pointed style, the interior being fitted up with and handsomely carved seats; the ceiling is decorated with paintings, and near the altar is the statue of Dr. —, a late head master, by Chantry. Rugby has no manufactures, and the inhabitants of the vicinity are principally agricultural. The trade of the town, however, has greatly increased by the opening of the Birmingham railway, which has one of its stations here; and it is now an important *entrepôt* between the surrounding country and metropolis. The Oxford canal passes, also, within a short distance of the town, connecting it with the principal navigation of England. On an eminence N.E. of the town are some slight remains of a castle supposed by

Dugdale to have been erected in the reign of Stephen. Large markets on Saturday for corn and provisions; a great horse fair Nov. 22, and 12 other fairs. (*Cher. Comm. 29th Report; Journal of Education,* vii., 234–249; *Railway Handbook, &c.*)

RUGEN, an island in the Baltic, belonging to Prussia, opposite to Stralsund, and separated from Pomerania, by a strait varying from 1½ to 2 m. in width. It is of an exceedingly irregular shape, being deeply indented by bays and arms of the sea. Area, 361 sq. m. Pop. about 20,000. It is very fertile, and sends annually large quantities of corn, &c., to Stralsund. Rugen differs much in appearance from the mainland part of Pomerania, its coasts consisting mostly of high, precipitous, chalky cliffs. It is well wooded; and being intercepted by ravines, as well as deep narrow bays, its scenery is highly picturesque. This circumstance, and its facilities for sea-bathing, render it a favourite resort in summer. The inhabitants are primitive in their habits and manners, industrious, and frugal. They are principally of the reformed religion; and their language is a patois of low German intermixed with Swedish. The fishing in the adjoining seas and bays is very productive. Unfortunately the island has no good harbour, and its coasts are very dangerous. A lighthouse, having the lantern elevated 197 feet above the level of the sea, has been erected on the most northerly promontory of the island, in lat. 53° 41' 12" N., long. 13° 57' 27" E. Bergen, the capital, situated in the centre of the island, has 2790 inhabitants. After being long in possession of Sweden, Rugen became part of the Prussian dominions in 1815.*

RUGGLES, p. t., Huron co., O., 91 m. N. by E. Columbus, 380 W. Watered by head branches of Vermillion river. It has two saw-mills; six schools, 266 scholars. Pop. 1945.

RUMFORD, p. t., Oxford co., Me., 52 m. W.N.W. Augusta, 614 W. Watered by Androscoggin river and its tributaries, which afford good water-power. Incorporated in 1800. It has six stores, one grist-mill, three saw-mills, two tanneries; 12 schools, 622 scholars. Pop. 1444.

RUMNEY, p. t., Grafton co., N.H., 47 m. N. by W. Concord, 525 W. Drained by Baker's river. First settled in 1765. It contains two churches, a Baptist and Christian, three stores, one grist-mill, eight saw-mills, three tanneries; 10 schools, 364 scholars. Pop. 1110.

RUNGPOOR (*Rangepore*), a district of British India, presid. and prov. Bengal, between lat. 25° and 27° N., and long. 88° and 91° E.; having E. Assam, S. Mymensing and Dinajpoor, W. the latter and Purneah, and N. Sikkim and Bootan. Area, 7856 sq. m. Pop. in 1822, 1,340,350. It is wholly on the N. side of the Ganges, and is intersected by the Brahmaputra and Teesta. The climate is not so hot as in most other parts of Bengal; the soil is inferior to that of the Dinagepore district. Tobacco is the staple product. Wheat is also a considerable crop; barley, oats, maize, cotton, and indigo, are little grown. A good deal of cotton-thread is, notwithstanding, spun in the district, the material being imported by way of Moorshedabad, &c. Total land revenue, in 1829-30, 1,063,178 rupees. Hindoos and Mohammedans are supposed to be nearly equal in point of numbers. Chief towns Rungpoor, the cap.; Mungulhaut, Chilmary, and Goalparah. Rungpoor, in lat. 25° 43' N., long. 89° 22' E., has been estimated to have a population of from 15,000 to 20,000.' (For copious information respecting this district, and that of Purneah, the reader is referred to the surveys by Dr. Buchanan Hamilton, in *Martin's British India,* vol. iii.; *Parl. Reports, &c.*)

RUPERT, p. t., Bennington co., Vt., 94 m. S.W. Montpelier, 494 W. Pawlet river and a branch of Battenkill afford water-power. It has three stores; nine schools, 346 scholars. Pop. 1086.

RUPPIN (NEW), a town of the Prussian dom., prov. Brandenburg, gov. Potsdam, cap. circ., on the lake of its own name, 37 m. N.W. Berlin. Pop., in 1837, 7995. It is well built, and has a council-house, high school, hospital, central prison, barracks, and a large covered military exercising ground; with manufactures of woollen goods, gloves, and leather. Its trade is greatly facilitated by the Ruppin canal between the Havel and the Rhin, forming a link in the communication between the Elbe and the Oder. (*Berghaus, &c.*)

RUREMONDE, or ROERMOND, a town of Belgium, prov. Limburg, cap. arrond., on the Meuse, where it is joined by the Roer, 37 m. N.W. Maestricht. It was dismantled by Joseph II.; but is still surrounded by ramparts, and is the residence of a military commandant. It is well built; is the seat of a court of primary jurisdiction, and a college; and has manufactures of woollen stuffs, and considerable trade.

* It is very probable that Rugen was at one time joined to the mainland; but there is no foundation whatever for the statement that it was separated from it by a violent storm in 1309.

717

RUSCOMB MANOR, t., Berks co., Pa. It has one store, one grist-mill, two tanneries. Pop. 1189.

RUSH, county, Ia. Situated toward the S.E. part of the state, and contains 400 sq. m. Drained by Big and Little Blue river, and Big and Little Flatrock creeks. It contained in 1840, 15,163 neat cattle, 29,091 sheep, 48,458 swine; and produced 99,712 bushels of wheat, 5375 of rye, 1,156,707 of Indian corn, 294,064 of oats, 19,039 of potatoes, 12,939 pounds of tobacco, 414,155 of sugar. It had 36 stores, 18 flouring-mills, 17 grist-mills, four tanneries, eight distilleries, one brewery, two printing-offices, two weekly newspapers; 48 schools, 1149 scholars. Pop. 16,454. Capital, Rushville.

RUSH, p. t., Monroe co., N. Y., 12 m. S. Rochester, 218 m. W. Albany, 363 W. Drained by Honeoye creek, a tributary of Genesee river. It contains a church, three stores, one fulling-mill, two tanneries; 15 schools, 631 scholars. Pop. 1999.

RUSH, t., Champaign co., O. It has two stores, three saw-mills; six schools, 179 scholars. Pop. 1236.

RUSH CREEK, t., Fairfield co., O. Pop. 2424.

RUSHFORD, p. t., Alleghany co., N. Y., 272 m. W. by S. Albany, 331 W. Drained by Cold creek, a tributary of Genesee river. It has two churches, eight stores, two fulling-mills, one woollen factory, one furnace, two grist-mills, seven saw-mills; nine schools, 273 scholars. Pop. 1512.

RUSHVILLE, p. v., capital of Rush co., Ia., 40 m. E.S.E. Indianapolis, 538 W. Situated on the W. side of Big Flat Rock creek, and contains a courthouse, jail, three churches, a Baptist, Methodist, and Presbyterian, all of brick, four stores, 80 dwellings, and about 550 inhabitants. Pop. of the township 2146.

RUSHVILLE, p. v., capital of Schuyler co., Ill., 60 m. W.N.W. Springfield, 838 W. It contains a courthouse of brick, two stores high, a jail, four churches, a Presbyterian, Methodist, Episcopal, and Campbellite, 12 stores, besides groceries, 300 dwellings, and about 1200 inhabitants.

RUSSELL, county, Va. Situated in the S.W. part of the state, and contains 1370 sq. m. Watered by Clinch river, and by head branches of the W. fork of Sandy river. It contained in 1840, 14,404 neat cattle, 15,431 sheep, 26,945 swine; and produced 59,400 bushels of wheat, 7898 of rye, 204,253 of Indian corn, 142,277 of oats, 20,658 of potatoes, 106,616 pounds of sugar. It had 10 stores, 25 grist-mills, 10 saw-mills, two oil-mills, eight tanneries, 100 distilleries; two schools, 41 scholars. Pop.: whites, 7159; slaves, 700; free coloured, 96; total, 7878. Capital, Lebanon.

RUSSELL, county, Ala. Situated in the E. toward the S. part of the state, and contains 865 sq. m. Bounded E. by Chattahoochee river, and drained by its tributaries. It contained in 1840, 17,375 neat cattle, 1806 sheep, 27,901 swine; and produced 11,479 bushels of wheat, 387,534 of Indian corn, 5341 of oats, 37,913 of potatoes, 1,112,719 pounds of cotton. It had 13 stores, 12 grist-mills, 11 saw-mills; three academies, 111 students; 14 schools, 325 scholars. Pop.: whites, 6944; slaves, 7906; free coloured, 3; total, 13,513. Capital, McDonald.

RUSSELL, county, Ky. Situated toward the S. and E. part of the state, and contains 920 sq. m. Watered by Cumberland river and its tributaries. It contained in 1840, 3698 neat cattle, 3809 sheep, 11,704 swine; and produced 8676 bushels of wheat, 144,119 of Indian corn, 96,300 of oats, 4594 of potatoes, 16,374 pounds of rice, 780,505 of tobacco, 5773 of cotton, 4409 of sugar. It had three stores, one cotton factory, with 168 spindles, eight grist-mills, three saw-mills, one oil-mill, two tanneries, 26 distilleries: two schools, 50 scholars. Pop.: whites, 3698; slaves, 406; free coloured, 4; total, 4938. Capital, Jamestown.

RUSSELL, p. t., St. Lawrence co., N. Y., 192 m. N.N.W. Albany, 476 W. Drained by West Canada creek. It has one store, five saw-mills; 13 schools, 371 scholars. Pop. 1373.

RUSSELL, t., Putnam co., Ia. It has four stores, six grist-mills, seven saw-mills, one tannery, five distilleries, one pottery: eight schools, 205 scholars. Pop. 1503.

RUSSELLVILLE, p. v., capital of Franklin co., Ala., 111 m. N. by W. Tuscaloosa, 796 W. It contains a courthouse, jail, several stores, about 30 dwellings, and 200 inhabitants. Nett proceeds of the postoffice in 1841, $382.

RUSSELVILLE, p. v., capital of Logan co., Ky., 172 m. S.W. Frankfort, 696 W. Situated between Green and Cumberland rivers, 36 m. from each. It contains a courthouse, jail, two churches, an academy, several stores, 175 dwellings, and 1196 inhabitants.

RUSSIA, the most extensive, and one of the most powerful empires, either of ancient or modern times. It comprises the whole northern portion of the eastern hemisphere, from the frontiers of Posen and the gulf of Bothnia on the W., to the Pacific ocean and Behring's straits on the E., or from the 18th to the 190th deg. of E. long., being a distance, on the 60th deg. of lat., of nearly 6000 mi. Its extent, from N. to S., though less vast, is still very great, stretching from

718

the 38th to the 70th, and in some parts to the 78th deg. of N. lat., exhibiting an average breadth of about 1200 m. And, exclusive of this, Russia claims a very large tract in the N.W. part of America; and is mistress of Nova Zembla, and some other large islands in the Arctic ocean, of the Aleutian islands off Kamtchatska, and of the Aland isles, &c., in the Baltic. Her superficial extent has not been determined with anything like accuracy. It was estimated by Hassel at 372,935 geog. sq. m., viz., Russia in Europe, including Finland, 72,809 sq. m.; Russia in Asia, 275,767 do.; and Russia in America, 24,000 do.: and Schnitzler, a his *Statistique Générale*, has adopted this estimate. The latest, however, and probably the most accurate estimate of the extent of this vast empire is that given by M. Kaypen, of the Academy of Sciences of Petersburg, in the employment of government. According to this gentleman, the area of its different great divisions is as follows:

	Geog. sq. m.
Russia in Europe, including the portions of the governments of Perm, Orenbourg, and Viatka, that extend into Asia	84,855
Northern Asiatic Russia, or Siberia	259,290
Southern Asiatic Russia, or Transcaucasian provinces	3,381
Grand duchy of Finland	6,495
Kingdom of Poland	2,351
American colonies	17,390
Grand total	**384,390**[*]

The reader may, perhaps, acquire a better idea of the vast extent of the Russian empire, when he is told that it includes nearly one seventh of the terrestrial part of the globe, and about one twenty-seventh part of the entire surface. But by far the greatest proportion of this prodigious superficies is almost uninhabited, and seems to be destined to perpetual sterility; a consequence partly of the extreme rigour of the climate in the provinces contiguous to the Arctic ocean, and partly of almost all the great rivers by which they are traversed having their embouchure in that ocean, and being, therefore, inaccessible either for the whole or the greater part of the year.

Face of the Country. Mountains.—Russia is, in general, level, and comprises some of the most extensive plains in the world. That part of the empire which is in the eastern hemisphere is naturally parcelled into the two great divisions of European and Asiatic Russia, by the Oural mountains, which stretch in a N.N.E. direction from the Caspian sea to the Arctic ocean; forming, through the greater part of their course, the boundary between Europe and Asia. The highest points in this chain have an elevation of about 6500 ft. above the level of the Caspian. In all the vast country, extending on the W. side of the central chain to the confines of Poland and Moldavia, there is hardly a single hill. The Valdai hills, or elevated grounds, between Novgorod and Twer, where the Wolga, the Don, and the Dniepr have their sources, are nowhere more than about 1200 ft. above the level of the sea, the country exhibiting a waving surface, and without any considerable elevations. There is nothing, in fact, save the forest, to break or interrupt the course of the wind, in all the immense space interposed between the Oural and the Carpathian mountains. The only great chain of mountains in western Russia is that of Caucasus, between the Euxine and Caspian seas, and this is almost at the southern extremity of the empire. Siberia, or Asiatic Russia, consists principally of a vast plain, slightly inclining to the N. Towards the S. and E., however, it is in parts mountainous, being separated from Mongolia and Manchouria by high and little explored ridges, in which the great rivers that flow through it to the Arctic ocean have their sources.

The most distinguishing feature in the appearance of Russia is her vast forests. Schnitzler, who estimates the surface of European Russia at about 400 millions of declatines, supposes that 156 millions are occupied by forests. They are so very prevalent in the governments of Novgorod and Twer, between Petersburg and Moscow, that it has been said a squirrel might travel from the one city to the other without ever touching the ground. The forest of Volkonski, at the source of the Wolga, is the most extensive of any in Europe. In the government of Perm, on both sides of the Oural mountains, containing 15 millions of declatines, no fewer than 11 millions are covered by forests! The forests of Asiatic Russia are, also, of vast size. In extensive districts, however, the surface is quite free from wood. This is particularly the case in the vast *steppes* or plains in the governments of Astrakhan and Omsk, which in many parts, indeed, are a mere sandy desert.

Rivers and Lakes.—The rivers of Russia are usually divided into five groups or systems, corresponding to the seas in which they have their embouchure, viz., the Arctic ocean, the Baltic, the Black sea, the Caspian, and the Pacific ocean. The first division is by far the largest. It comprises, in Europe, the Dwina, Mezen, and Petchora; while in Asia it includes, among a host of others, the Obi, Jenisei, and Lena, three of the largest rivers of Asia. All these rivers run from S. to N.; and the last three have a course of from 2000 to 2500 m. The rivers which fall into the Baltic, though of far greater importance in an economical point of view, are of very inferior magnitude. The principal are the Neva, which has Petersburg at its mouth, the Duna, and the Niemen. The rivers which fall into the Black sea equal those falling into the Baltic in commercial importance, and far exceed them in length of course and volume of water. Among others are the Dniestr, Dniepr, Bug, Don, and Kuban. The basin of the Caspian has, however, to boast of the largest and most important of the rivers of Russia, the Wolga. This great river has its sources in the government of Tver, about 180 m. S. by E. from Petersburg: including sinuosities, its course is about 600 leagues, while that of the Danube is only about 450! See WOLGA.) It is of vast consequence to the internal navigation of the empire. The Caspian sea, also, receives the Oural and the Emba.

Owing to the flatness of the country through which they flow, and the vast rapidity of their course, the rivers of Russia are but little interrupted by cataracts, flow with a tranquil stream, and afford great facilities to internal navigation. The severity of the climate no doubt prevents, during a considerable portion of the year, all intercourse by water; and, as already stated, renders the rivers falling into the Arctic ocean of comparatively little value. Lack, however, the frost, which interrupts navigation, affords the greatest facilities to land travelling. (see post.) The lakes as well as the rivers, of Russia are upon a gigantic scale. The lake of Baikal, in the government of Irkutsk, in Asiatic Russia, is one of the most extensive in the world. In European Russia, the lakes of Ladoga, Onega, Peipus, Ilmen, and Bielo-Onero, are also of great extent, particularly the first. The duchy of Finland is almost everywhere interspersed with lakes, and they are very abundant in other provinces, particularly in that of ....

Soil and Climate.—These, it is obvious, must differ exceedingly in so vast a country. Some provinces mostly consist of sandy barren plains, or vast morasses. But the most valuable portion of the empire, or that included between the Baltic, the gulf of Finland, and the Wolga, on N. and E.; the Black sea on the S., and Austria, Poland, Prussia, &c., on the W., has, speaking generally, a black mould, of great depth, mostly on a sandy bottom, easily wrought, and very fertile. In some places it lines to sand or gravel; in many, from the want of drainage, it's peaty or boggy: in Livonia, and parts of Livonia, it is clayey, but it nowhere inclines to chalk. ... following statements by Plescheyéf, whose accuracy well known, contain all the information with respect to soil and productiveness of the country that seems to be tired in a work of this description. "Russia," he says, divided into two great parts by the Oural mountains, which form an uninterrupted barrier through its whole width, and separate Siberia or Asiatic from European ....

That part of Russia which lies on this side of the Oural mountains, presents an immense plain declining westward by an easy descent. This plain, from its vast extent, a great variety of climates, soils, and products. Its western part, which sensibly declines towards the White Frozen seas, is covered with forests, marshy, and but fit for cultivation. The other, and more southerly on of this vast plain, includes the whole district along Volga, as far as the steppes or deserts between the ... and the sea of Azof, and constitutes the finest of Russia: generally it has a fertile soil, the arable meadow land preponderating over the woods and ....

... part of the country which extends towards Vo... Tambof, Penza, and Simbirsk, as far as the deserts, remarkable for the superior quality of every kind and other produce. It has everywhere an excellent, consisting of black earth, strongly impregnated saltpetre. But the tract which commences between of Azof and the Caspian, and extends near the of the latter, and between the Wolga and Oural, as the Emba, is little better than a desert, being level, barren, and full of salt lakes.

... country lying on the other side of the Oural mountains known by the name of Siberia, is generally a flat vast extent, declining imperceptibly towards the ocean, and rising thence by equally imperceptible towards its southern border, where at last it is lost immense mountain ranges which separate the Russian and Chinese empires. It is unnecessary to notice in

detail the different great divisions of this vast territory. As we proceed farther E., the climate in the same latitude becomes more and more severe, the summers shorter, and the frosts more intense. In general it may be stated, that the more southerly portion of Siberia, or that between the S. frontier of the empire and the 57th or 60th deg. of lat., as far E. as the river Lena, has, for the most part, a fertile soil; and that, notwithstanding the severity of the climate, it produces most kinds of grain. But, owing to the increase of cold and the nature of the soil, the more northerly portion of the region now noticed, or that extending from the 57th or 60th deg. of lat. to the Frozen ocean, and the whole country E. of the Lena, from the frontier of Manchooria northwards, is wholly, or almost wholly unfit either for cultivation, or for the grazing of cattle. In the E. a portion of this vast tract is mountainous, but it mostly consists of immense levels, full of swamps and bogs, covered with moss, which would be totally impassable were it not that the ice, which never thaws deeper than a few inches, gives a firm underfooting." (*Eng. Trans.*, p. 5.)

A country which, like Russia, extends from N. to S. through almost 40 degrees of latitude, might be supposed to have almost every climate; and this is, in some measure, the case. "When spring commences in one division of this vast empire, another experiences all the rigours of winter. Here the parched camel traverses arid burning deserts; there the rein-deer courses over heaps of snow, under which he finds a scanty supply of moss. The Samoide sleeps in his cabin, where the days are short and cloudy, while the Kirghisian feeds his flock under a clear serene sky. This variety of the products and diversity in the manner of living, gives Russia advantages not enjoyed by any other European country. She possesses, in the greatest abundance, all the most necessary articles, and the greater number of those which are reckoned luxuries; and she either furnishes or may procure all the products of different climates." (*Storch, Tab. de la Russie*, i., 4.)

But notwithstanding the heats that usually prevail during summer, especially in the southern provinces, cold, speaking generally, predominates very decidedly in Russia. With the exception, indeed, of the Crimea and the transcaucasian provinces, no part of Russia can be said to be generally hot; and even in them the frost in winter is often very severe. The climate of Russia is, in fact, proverbial for its severity; and this increases not only as we advance towards the N., but also as we advance towards the E.; the cold being decidedly greater in Siberia than in the same latitudes in European Russia, a difference which is also sufficiently perceptible in the provinces on the E. and W. sides of the latter. This, no doubt, is owing to various causes; but principally, perhaps, to the greater cultivation of the western provinces and their proximity to the Baltic; and to the vast extent of frozen sea and land traversed by the winds from the N.E. Beyond the 65th degree of lat. the ground is covered with snow and ice for about nine months in the year; and during the other three months ice is always found at a little distance below the surface. Corn crops cannot be depended upon in European Russia beyond the 62d degree of lat.; and the great agricultural provinces lie to the S. of the 58th deg. The fruits of temperate climates are seldom met with beyond the 52d deg. At Petersburgh, in lat. 59° 56', the mean maximum of cold is about 34°, and the mean maximum of heat 23°, Reaumur. The Neva is commonly frozen over before the end of November, and the ice never breaks up before the end of March. At an average of ten years it is calculated that there are annually at Petersburgh 97 bright days, 104 rain, 73 snow, and 93 unsettled. At Moscow, in lat. 55° 45½, the cold is more severe than at Stockholm in lat. 59° 20½. At Astrakhan in lat. 46° 21', nearly the same as that of Lyons, the Wolga is sometimes frozen over so as to bear loaded waggons. The sea of Azof is usually frozen over from November to the beginning of April. In Siberia, as already stated, the cold is much more severe than in the provinces to the W. of the Oural mountains. The breaking up of the ice on the Lena does not take place before the beginning of May.

But this severe cold is not unhealthy, and is much less inconvenient than might be supposed. While the frost lasts the air is pure and bracing, and its severity is guarded against by warm clothing, and by having the houses properly constructed and heated. At Petersburg and Moscow the winter is, in fact, the finest season. The inhabitants seem to revive at its approach. Sledge-roads over the snow render travelling commodious and agreeable; and a winter journey in a moderate frost by moonlight is a high enjoyment. The Russian peasants care only for warm covering for their legs and feet. At Petersburg, in a frost of 25° Reaumur, it is common to see women standing for hours together washing their linen through holes dug in the ice over the Neva!

Spring can hardly be said to have any place in the Russian calendar. The transition from frost to fine weather is usually very rapid. In a brief period after the snow and ice have disappeared, the fields and trees are clothed in the livery of summer, and vegetation makes an extraordinary progress. At Petersburg the summer is as mild and agreeable as in the S. of France; but there and in all the N. provinces it is very variable. As we advance towards the S. it becomes steadier, and the heats increase. At Astrak-

han the mercury in the thermometer sometimes rises to 103¼° Fah.; and in the transcaucasian provinces it runs still higher. The autumn, or the period of transition from summer to winter, is the most unpleasant season in Russia. The sky is generally cloudy, and rains and storms are very prevalent.

The Crimea, from its high S. lat., and its being washed by the Euxine, has the most agreeable climate in the empire.

NAMES, Area, and Population of the different Great Divisions and Governments, included in the Russian Empire in 1838:—

Governments.	Area in Geog. sq. m.	Pop. in 1838.	Governments.	Area in Geog. sq. m.	Pop. in 1838.
Northern Provinces:			Tchernigoff	80?	1,390,00
Archangel	15,212	230,000	Pultawa	1,002	1,621,00
Oinnetz	2,354	230,900	Kharkoff	1,346	1,374,00
Vologda	6,880	747,500	Voroneje	1,254	1,567,90
Great Russia:			Don Cossacks	2,550	640,30
Petersburg	710	585,000	*New Russia:*		
Novgorod	2,070	825,400	Ekaterinoslaf	1,186	791,00
Pskof	1,045	705,300	Kherson	1,680	763,00
Smolensk	954	1,061,300	Taurida	2,040	590,30
Moscow	550	1,249,700	Bessarabia	794	728,00
Twer	1,122	1,297,000	*Wolga and Caspian Provinces:*		
Yaroslaf	807	916,500	Kasan	1,104	1,930,00
Kostroma	1,438	1,058,700	Penza	674	999,00
Nijni Novgorod	878	1,071,100	Simbirsk	1,141	1,199,00
Vladimir	831	1,133,200	Saratof	3,472	1,664,00
Riazan	707	1,241,700	Astrakhan	2,699	254,50
Tambof	1,152	1,591,700	Caucasus, &c.	1,893	478,90
Tula	529	1,113,500	*Ooral Provinces:*		
Kaluga	541	914,900	Orenbourg	6,535	1,771,00
Orel	755	1,360,300	Perm	3,721	1,448,00
Koursk	794	1,527,300	Viatka	2,497	1,521,00
Baltic Provinces:			*Siberia:*		
Esthonia	315	282,200	Tobolsk	86,818	814,00
Livonia	896	740,100	Tomsk*		1,677,00
Courland	475	503,000	Irkutsk		597,30
White Russia:			Yakutsk		162,00
Witepsk	778	717,700	Kamtchatka	162,800	4,300
Mohilef	824	846,600	Okhotsk		7,00
Minsk	1,983	1,034,800	Yeniseisk		295,00
Lithuania:			*Transcaucasian Provinces:*		
Wilna	1,161	1,315,800	Georgia, &c.	3,381	2,890,00
Grodno	570	791,700	Finland	6,405	1,387,145
Bialystock	169	251,000	Kingdom of Poland.	2,987	4,398,90
Little Russia:			Russian America	17,580	61,853
Volhynia	1,073	1,314,100	Totals	364,368	58,673,300
Podolia	578	1,548,900			
Kief	798	1,459,800			

N.B. The 364,368 Geographical sq. m. of 15 to a deg. are equal to 7,785,000 Eng. sq. m. of 69·15 to a deg.

* The pop. of the old prov. of Omsk, suppressed in 1838, is wholly included in this table in that of Tomsk; in point of fact, however, it was ranged in part to Tobolsk, and in part only to Tomsk.

Classified ACCOUNT of the Population of the Russian Empire in 1838, according to the Official Statement published by the Minister of Finance:—

	Males.	Females.		Males.	Females.
Russian priests	59,331		Greeks of Nijny, gunmakers		
Deacons and sacristans	63,178		of Tula, &c.	10,882	10,980
Male children of priests, deacons, and sacristans	138,548		Citizens of Bessarabia	57,905	56,175
Total	254,057	249,748	*Inhabitants of Villages.*		
			Peasants (that is, slaves), the private property of the emperor and the imperial family, peasants annexed to the crown, &c.	10,441,300	11,882,00
Priests of the United Greek and Roman church	7,893	7,318			
Catholic priests	9,497				
Armenian priests	474	343	Peasants the property of nobles	11,403,700	11,985,000
Lutheran do.	1,003	955			
Reformed church	51	37	*Wandering Tribes.*		
Mohammedan Mollahs	7,850	6,071	Calmucks, Circassians, and Mohammedans of the Caucasus	945,715	262,00
Lamas (Tartar)	150				
Nobles.			*Territory beyond the Caucasus.*	(Nearly.)	(Nearly)
Hereditary	284,731	253,489	Georgia, Armenia, Mongolia, &c.	699,147	699,130
By virtue of service, &c., with their sons	78,929	74,273	Poland	2,077,311	2,110,911
Petty officers who have left the army, and are employed in the civil service, &c.	187,047	237,443	Finland	663,636	725,424
			Russian colonies in America	30,761	30,761
Foreigners of all classes	29,114	15,215	Total of population	98,896,923	30,237,343
Military colonies	950,898	961,467	Grand total of both sexes	59,133,566	
Inhabitants of Towns.			In this table, however, and in that given above, the private soldiers of the army and navy, with their wives and children, are not included; so that the sum total, in round numbers, may be estimated at sixty-one millions. In addition to which must be reckoned the inhabitants of the mountains between the Black sea and the Caspian, amounting to 1,445,000. There are also some wandering tribes of Cossacks and others, whom it is impossible to number.		
Merchants	131,347	190,714			
Shopkeepers, artizans, &c.	1,339,434	1,433,969			
Citizens in the eastern provinces	7,535	6,966			

Divisions, Population, &c.—The divisions of the Russian empire have differed materially at different periods. Peter the Great made some important changes in the distribution that had existed previously to his epoch. The whole, however, was remodelled and placed on a new footing by Catherine II. in 1775. She divided the entire empire into three great regions: those of the north, middle, and south. Each of these regions was subdivided into governments, of which there were at first 42, and at the end of her reign 50. Paul made some ill-advised changes on this distribution, which were set aside on the accession of Alexander. The existing divisions were mostly fixed by the latter in 1822, nearly on the basis laid down by Catherine. The empire, exclusive of the kingdom of Poland, is divided into governments, exclusive of certain territories called provinces, or *oblasts*, not formed into governments. Some of these divisions, particularly those in Asiatic Russia, are of vast extent; but neither their boundaries nor their population are well ascertained. But, as the best that can be had, we subjoin M. Kœppen's table of the area and population of the different provinces included in the empire. [See preceding page.]

Animal and Vegetable Products.—The animals of Russia include those commonly met with in the arctic circle, and in temperate climates, as well as some of those belonging more peculiarly to the intertropical regions. Exclusive of horses, oxen, sheep, &c., rein-deer and camels are both met with. The latter is employed in travelling through the deserts in some of the S. provinces, while the former constitutes the principal wealth of the Samoiedes, Tunguslans, Ostiaks, Tchouktchia, and other tribes inhabiting the extreme N. parts of the empire. The dog is common everywhere; and in parts of Siberia, where there are neither horses nor rein-deer, is of the very greatest utility; and besides being employed for draught and burden, is used as food. Bears are abundant: beavers and other fur-bearing animals are also common; and in many provinces the chase forms a principal part of the occupation of the people. The rivers and lakes swarm with fish.

All sorts of corn succeed in Russia; though, as already stated, the crops cannot be depended upon above the 62d degree, nor is cultivation attempted in any part of the empire beyond the 65th degree. Fruits of all sorts are abundant in the S. provinces. The vine is cultivated in the vicinity of Astrakhan, in the Crimea, and in the Caucasian provinces; but the wine made from it is of a very low quality. Though meadows are not abundant in Russia, the pastures in most parts are excellent. The forests will be afterwards noticed.

Minerals, Working of Metals.—The mines of Russia are of considerable importance and value. The principal are situated in the Oural and Altai mountains, and in the vicinity of Nertschink, in Siberia. Gold and platina are both found in considerable quantities in the Ouralian mountains, especially in the mines in the neighbourhood of Catherineburg; the former being also obtained from the Krilof mine, &c. in the government of Tobolsk, and from various stream works in the Oural. According to the official returns, the gold and platina obtained from mines belonging to the crown, and to individuals in the Oural, in 1836, were:

Gold:	Poods.	Lbs.	Platina:	Poods.	Lbs.
Crown mines	129	17	Crown mines	20	17
Private	134	21	Private	117	25
	263	38		138	42

In 1837, the same mines produced 309 poods 23 lbs. gold, and 118 poods 28 lbs. platina; and exclusive of the above, the Siberian mines produced in the same year 160 poods of gold, of which 30 poods were extracted from silver. On the whole, therefore, the produce of gold in Russia, in 1837, must have exceeded 470 poods; that is, taking the pood at 36 lbs., and the price of gold at £4 per oz., it must have amounted to £1,089,880!

Silver is chiefly found in the Altai mountains of Kholiva-Voskvessenski, and in the Siberian mountains contiguous to Nertschink. The annual produce of the silver mines exceeds 3000 poods, or 108,000 lbs., worth, at 5s. 2d. an ounce, 34,600. And if we add to the value of the gold and silver that of the platina, the aggregate value of the three will not be estimated at less than £1,500,000 a year.

Copper is found in the government of Olonets, and in the Oural and Altai mountains. It is produced to the extent of about 210,000 poods a year. About 40,000 poods of lead are obtained from the mines of Nertschink and Kholivano. The mines furnish a supply more than sufficient for the uses of the empire. The works in the Oural mountains are said to employ above 50,000 labourers; and iron is procured in the Altai, Caucasus, the Valdai hills, &c. The total product of iron is said by Schnitzler to amount to millions poods; and may now, perhaps, be taken at

from 10 to 12 millions poods, or from 160,000 to 180,000 tons. Wood only is used in the iron-works and foundries of Russia. Mercury, antimony, zinc, cobalt, &c., are found in Siberia.

Cast-iron articles are prepared at most mines where there are forges. There is an important cannon manufactory at Petrozavolsk, in the government of Olonets, which was brought to a high state of perfection by an Englishman of the name of Gascoigne. But the principal hardware manufactories are carried on at Tula, in the government of that name. (See TULA.) A great variety of articles of cutlery are produced; and the royal manufactory of fire-arms is very extensive, employing, it is said, about 7000 male and 9600 female workers. (*Schnitzler, La Russie, la Pologne, &c.*, p. 315.) Very different statements have been made as to the quality of the arms produced at Tula. Those of Dr. Clarke were speedily confuted by the exploits of the Russian forces; and no doubt the manufacture has been since materially improved. At present we are assured, on undoubted authority, that though the arms produced at Tula be inferior in point of finish to those made in England, they are of very good quality.

Russia is abundantly supplied with mines of salt and brine springs; but as most of them are at a great distance from the Baltic and western provinces, there is a large importation of salt from England and Austria. The salt mines and brine springs in the government of Taurida (the Crimea), are the most extensive, and furnish annually about 16 millions poods (nearly 266,000 tons): those of Perm, which annually supply about seven millions poods, are the next in importance : the mines of Iletsk, in the government of Orenbourg, furnish about 300,000 poods, exclusive of what is furnished by the salt lakes, and additional supplies are obtained from various other mines and springs.

Agriculture.—Landed property in Russia is generally divided into estates belonging either to the crown or the nobility. Some of those belonging to the latter are very extensive; but, owing to the compulsory division of estates among the children of a family on the death of the father,[*] this is not generally the case; and the too great subdivision of the land is, in fact, one of the evils with which Russia is threatened. The peasants occupying the crown estates are in a state of *prædial*, and those occupying the estates of the nobility are generally in a state of *absolute*, slavery. The value of a Russian estate formerly depended more on the number of labourers or slaves belonging to it, and which may be either sold, or let out by the proprietor, than on its extent, or the quality of the soil : but since the increase of population this, in many districts, is no longer the case ; and the proprietor is sometimes burdened with the charge of supporting and maintaining labourers, and paying the tax on them to government, for whose services he has little or no use. Different proprietors adopt different methods in the management of their estates. In the principal part of Great Russia, comprising all the central provinces of the monarchy, the system is very simple. The proprietors rarely farm any considerable portion of their estates, or interfere with the mode in which they are cultivated. They usually content themselves with distributing them among the peasantry ; their revenue consisting in the produce of an *obrok*, or capitation tax, imposed on each male peasant, by way of rent. In the Baltic provinces, on the other hand, the owners usually retain a quarter, or some less proportion, of their estates in their own possession, managing them either directly, or by the intervention of stewards or agents. The peasants on such estates are allowed cottages, having pieces of ground attached, generally, perhaps, about 15 acres, sufficient for their support ; their obrox or tax being paid by labour, or *services*, performed on the lands held by the proprietor, or let by him to others. In the Ukraine and other S. provinces, the peasants are partly free ; and these pay for the lands they occupy, sometimes money, but more frequently produce and labour rents. There is also a class of small proprietors who cultivate their own lands, but who have not the privilege of holding slaves.

It may seem at first sight that, provided its amount were moderate, the system of parcelling out land to occupiers charged with an obrok or capitation tax, could not be fairly objected to. And when such is the case, and the landlord allows the occupiers to reap the fruit of increased economy and exertion, this is no doubt the preferable mode of dealing with slaves. But the vice of the system is that, except on the crown estates, everything depends on the will of the proprietor, who, though occasionally enlightened and liberal, is too frequently ignorant of his real interests, careless, and embarrassed. The occupier has, in fact, no security whatever. If he improve his farm, or have about him the appearance of getting rich, the chances are that his obrok will be im-

---

[*] One seventh of a man's landed property goes, on his death, to his widow for ever ; one fourteenth goes to each daughter ; and the residue is equally divided among the sons. (*Venables' Russia, 132*.)

mediately increased; or that a portion will be taken from his farm, and given to another family. Even the predial slaves on the crown estates, from whom only a certain amount of labour or of corvées can be legally demanded, are frequently exposed to the extortion of those to whom the lands are let, or of the officers employed to superintend them. Hence, in the generality of instances, the peasants invariably follow a routine system; they avoid labour by which they are not to benefit; and it is seldom, except on the lands managed by the proprietors themselves, or on those occupied by the few free tenants to be found in the country, that any considerable improvements are ever attempted. When the population of an estate parcelled among peasants is increased, a new distribution of the land has to be made; and, if unoccupied land cannot be assigned to the new families, they are provided for by diminishing the shares apportioned among the old ones. On such occasions, if the occupiers complain that their obrok or tax is too heavy, it is usual to grant those on whom the proprietor can depend, license to leave the estate; and though they still continue slaves, many avail themselves of this permission, and migrate to towns, where they exercise some trade or profession, from the profits of which they frequently pay a much higher obrok than was imposed upon them in the country, and accumulate sums with which to purchase their freedom.

It is evident from these statements, and from the influence of the compulsory division of estates in preventing improvements by the proprietors, that agriculture in Russia must be at a very low ebb. But though this is in general condition, it differs materially in different provinces; and some estates, even in the most backward provinces, have been greatly improved. In Livonia, and the provinces bordering on the Baltic, and also in parts of the Ukraine, the husbandry is very superior, and the implements quite equal to the best that are to be met with in most parts of Germany. But, with the exception of a few estates, it is quite otherwise in the rest of the empire. The plough is there a wretched implement drawn by one horse, and calculated rather to scratch than to turn up the soil. The harrow is made of wood, and rollers and hoeing machines are entirely unknown. Were it not that the soil is generally light, friable, and very easily wrought, it would be impossible to cultivate it by such means. But these suffice to make it produce more than enough for the wants of the inhabitants. There is not, indeed, as Mr. Loudon has truly stated, another country in Europe where corn crops may be raised at so little expense of labour as in Russia.

Exclusive of the forests, and the sandy deserts of the south, vast tracts in the northern parts of the empire are, and always must be, unsusceptible of cultivation. Taking the whole surface of European Russia at 402 millions of deciatines, M. Schnitzler supposes that the cultivated land does not exceed 61¼ millions, and the meadows (chiefly in Livonia and Cornland) six millions deciatines. (*Essai Statistique de la Russie*, p. 34.) The products vary, of course, with the differences of soil and climate. All sorts of corn are raised; but rye being the common food of the peasantry, it is produced in much greater quantities than any other sort of grain. Next to rye is oats; and the value of (the crops of these two, taken together, is supposed to be more than double the value of the crops of wheat, barley, and of every other kind of corn. Orel, Kasan, Nijni Novgorod, Penza, Tambof, and Koursk, are the most productive provinces; and it is in them that the greatest quantity of wheat is raised. According to the official returns, which, however, in these matters are little to be depended upon, there were sown in European Russia, in the latter part of 1834, 20,349,149 chetwerts winter corn (rye and wheat); and in the following spring, 30,396,046 chetwerts spring corn; the return to the former being, at an average, 4¼ for 1, and that of the latter nearly 4 for 1. But, in the best districts, wherever the land is moderately well cultivated, the produce is much greater; and is, indeed, nowise inferior to that of the most favoured countries. Corn, in Russia, is very frequently kiln-dried in the sheaf, before it is either stacked or thrashed. Flax and hemp are very extensively cultivated; and, besides what is made use of at home, are very largely exported. Potatoes succeed almost everywhere; and this, also, is the case with hops. Tobacco is confined to the S. provinces, where it is an important article. It deserves to be mentioned, to the credit of the government, that it has latterly been exerting itself in the most efficient manner, for the improvement of agriculture. Professorships of agriculture have been established in the different universities; and an institution to which a model farm is attached has been established near Mohilew for educating 120 pupils, so as to fit them to act as stewards or managers of large estates.

Horses are very abundant in Russia. Speaking generally, they are coarse and ill-shaped, but hardy and active. In the southern provinces, however, whence the cavalry horses are brought, the breed is very superior. The khans or

chiefs of the nomadic tribes occasionally possess as many as 10,000 horses.

M. Storch states that there is no country in Europe where so many cattle are reared as in Russia, and none where they are taken so little care of. (*Tableau de la Russie*, i., 155.) Exclusive of the numerous herds, which constitute a principal part of the wealth of the pastoral and nomadic tribes, every peasant has a few head, and even the beggar has a cow or a goat! The ordinary Russian ox is small, lean, and bony; but those of the Ukraine, Podolia, Volynia, and some other provinces, are large, and of a very fine breed. Many thousand head are annually sent from the Ukraine to Petersburg and other Russian towns, and also to Silesia and Germany. Tallow is at present, and has been for some years, the most important article of export from Russia: the increase of the exports of this article from Odessa within these few years has been quite extraordinary. The wool of the common Russian sheep is hard and coarse; but latterly considerable efforts have been made to improve the breed by importing fine wooled sheep from Germany; and wool, notwithstanding the increase of factories at home, is becoming an important article of export. Thus, during the period from 1814 to 1834, the annual average export of wool amounted to only 1,384,068 lbs., whereas, in 1838, it amounted to very near 13,000,000 lbs., valued at £559,000. Hogs are everywhere abundant, and, in the northern provinces especially, furnish a principal part of the food of the people, while their bristles are an article of export. Goats are also abundant.

The rearing and management of bees is more attended to in Russia than in any other European country, and is, in fact, the principal occupation of several tribes. The wild bees, however, greatly exceed those that are domesticated. Their culture is principally attended to in the provinces of Kasan and Ourfa. Individuals among the Baschkirs possess 100 hives in their gardens, and upwards of 1000 in the forests! (*Storch, Tableau*, ii., 349.) Honey is very extensively used in many parts instead of sugar. The export of wax is very considerable. In 1834, it amounted to 22,949 poods, or 800,928 lbs.

Manufactures—are not generally in an advanced state. Since the reign of Peter the Great, their improvement and extension have, it is true, been favourite objects with the government; and heavy duties and prohibitions have, in consequence, been imposed on such foreign articles as it was supposed might interfere with similar articles of native growth. This however, was a very erroneous policy. The slavery of the peasantry is an all but invincible obstacle, in so far at least as they are concerned, to the formation of those habits of industry, perseverance, and invention, necessary to insure success in manufactures; while the thinness of the population, the variety of natural products, and the fertility of the soil, all concur in pointing out agriculture, including under that term mines and fisheries, as the natural and most advantageous employment that can be carried on upon a great scale, till civilization is more generally diffused. In fact, considering the peculiar circumstances under which Russia has been placed, and the deficiency of her capital, the wonder rather ought to be that she has made so great a progress as she has done in manufacturing industry, than that she should still be so backward. Among the peasantry generally, there is little or no subdivision of labour. Each family commonly supplies itself with all the clothing and furniture which it requires. Sometimes, however, a person superadds some particular employment to his ordinary avocations; and sometimes this principle is carried further, and the inhabitants of entire villages devote themselves to some particular trade. It is only in pretty large towns that the division of labour, such as it exists among us, is carried to any considerable extent; and even there it is a common complaint that the native products, though showy, are seldom substantial or good. "*Si le génie de l'invention lui manque,*" says Schnitzler, "*le Russe possède en revanche, au plus haut dégré, la facilité d'imiter ce que d'autres ont produit; et un esprit vif et exercé lui fait deviner les procédés qu'il faut suivre. Mais, pressé de gagner, et manquant de persévérance, il travaille à la hâte, plutôt pour l'apparence que pour la durée, et les produits de toute espèce restent toujours inférieurs au-dessous des ouvrages exécutées par des mains étrangères.*" (*Tableau de la Russie*, p. 135.) The versatility of the Russian peasant is astonishing. He is truly a Jack of all trades, and will turn his hand to whatever may be required. "He will plough to-day, weave to-morrow, help to build a house the third day, and the fourth, if he master need an extra coachman, he will mount the box, and drive four horses abreast as if it were his daily occupation. None of these operations, except, perhaps, the last, will be as well performed as in a country where the division of labour is more thoroughly understood. They will all, however, be sufficiently well done to '*serve the turn*,' a favourite phrase in Russia. The people are very ingenious, but perseverance is wanting; and though they carry many arts to a high degree of excellence, they generally stop short of perfection; and it will be very long before their products can come into competition, for finish, durability, or cheapness, with English goods." (*Venables' Russia*, p. 141.)

In certain departments, however, Russia is not merely equal but even superior to other countries. Her leather is excellent; and for some purposes, such as bookbinding, is decidedly superior to any other material. The process followed in the preparation of this important article has been often described; and foreigners have frequently engaged in the business in Russia, with the view of making themselves acquainted with the details, that they might undertake it at home. But whether it be owing to something in the bark or the water, or to some other undiscovered cause, none of the attempts to produce Russia leather in foreign countries have ever succeeded, and Russia continues to enjoy a monopoly of this valuable product, and to export it in large quantities. The sail-cloth, cordage and canvass, tick, felt, mats, pot-ashes, soap, candles, caviar, isinglass, spirits, and some other articles produced in Russia, are quite as good, or better, than those of any other country.

So late as 1788, almost all the cloth required for the clothing of the army was imported from abroad; but it is now wholly manufactured at home. Cloth of a superior quality is also made at Moscow and its vicinity, at Jamborg, near Petersburg, Sarepta, and other places; but, generally speaking, it is both inferior to what might be imported and dearer. Linen is principally manufactured in Vladimir, Kostroma, Moscow, and Kaluga; and sail-cloth and cordage in Archangel, Orel, &c. The silk manufacture of Moscow is extensive and thriving, and it is carried on to a less extent in other towns. Numerous establishments for the spinning of cotton have been recently founded. Generally, however, they supply only the coarser descriptions of yarn, the finer sorts being almost wholly imported from England. The cotton manufacture has recently made a rapid progress. It is principally carried on in the government of Vladimir; Chouia and Ivanova being its chief seats. In 1826 they had 15,619 looms, and employed 94,257 workpeople. During the same year there were sold in the province 5,610,600 lbs. of foreign cotton yarn, and 660,000 lbs. produced by Russian spinners. (*Schnitzler, La Russie, &c.*, p. 105.) The increase has been very great in the interval; for, it appears from the official returns, that there were in Vladimir, in 1839, 315 factories, which gave employment to 83,655 workpeople. The glass manufacture has, also, made a very rapid progress. Single plates are made at the Petersburg glass-works, that cost £800 each. The glass-works of the brothers Mainoff, in Tula, and elsewhere, are deservedly celebrated. The home consumption of glass is great, and is rapidly increasing. The manufacture of snuff and cigars, punch, and soap has, also, very rapidly increased. Paper, coarse and fine earthenware, jewellery, &c., are produced at Moscow, Petersburg, and other places.

Industry of all sorts has made an astonishing progress since the peace of 1815. In 1812 there were, in the entire empire, 2338 manufacturing establishments, employing 1,083 workpeople: whereas, in 1836, the former amounted to 6015, and the latter to 308,573, of whom about one in were free labourers. During the next three years progress was still greater, there being, in 1839, 1855 manufactories, employing 412,931 workpeople, exclusive of those engaged in mines, smelting-houses, furnaces, &c. In 1827, the total value of the manufactured produce of Russia was estimated at 500,574,397 roubles, and it is now (1841) probably not under 650 millions. Among the different manufacturing establishments, in 1839, 616 were appropriated to the production of woollens, 327 to that of silk, 467 to that of cotton, 367 to that of linen.

The progress of industry is strikingly evinced by the fact while, in 1830, there were only two steam-engines in government of Moscow, it reckoned about 100 in 1830! same government had, in 1839, 1058 factories, and 14 workpeople.

With the exception of the formidable restraints originating in the slavery of the peasantry, industry is quite free in Russia. There are no internal monopolies, save those of salt, spirits, and playing cards. There is nothing in the laws, or corporations, to check competition; and all who are free may exercise any art or profession, either in town or country, as may be most agreeable to themselves. Slaves have obtained a passport, or license from their owners, leave their estates, are, in this respect, in the same ison. Since 1826, lectures have been instituted in all Russian universities, for the instruction of manufactures or handicraft-tradesmen in mechanics, chemistry, &c., the applications of science to practice.

Commerce.—The commerce of Russia is already, notwithstanding the paralyzing influence of the prohibitive sys-

tem, very extensive; and will, no doubt, continue to increase with the growing wealth and population of the empire, and according as more liberal principles may be expected to prevail. The principal articles of export are tallow, which is more largely exported from this than from any other country; grain, particularly wheat; hemp and flax; timber, potashes, bristles, linseed and hempseed, linseed and hempseed oils, furs, leather; fox, hare, and squirrel skins; canvass and coarse linen, cordage, wool, caviar, wax, isinglass, tar, &c. The principal imports are sugar, especially from the Havannah; cotton stuffs and yarn, the latter being by far the most important article sent from Great Britain to Russia; coffee, but not in large quantities; indigo, and other dye-stuffs; woollens, oils, spices, wine, salt, tea, lead, tin, coal, fine linen from Holland and Silesia, &c.

The principal trading ports are Petersburg and Riga, on the Baltic, but particularly the former; Archangel, on the White sea; Odessa, on the Black sea; Taganrog, on the sea of Azof; and Astrakhan and Baku, on the Caspian sea. Moscow is the principal entrepôt of the interior commerce of the empire. The trade with China is mostly carried on through Kiacta; and the fair of Nijni Novgorod is celebrated all over Europe. There are also very large fairs at Irbit, Kharkoff, Koursk, and other towns. (*See* these towns.) The value of the goods offered for sale at the different great fairs, in 1838, is said to have amounted to 353,894,722 roubles. In 1839 the total value of the exports from the empire to foreign countries was estimated at 339,009,358 roubles (£12,853,000); of which three from Petersburg amounted to nearly a half, or to 138,044,000 roubles. At an average of the six years ending with 1838, the real value of the articles of British produce and manufacture exported from this country to Russia amounted to £1,686,391 a year, cotton-twist constituting three fourths of their total value.

Account of the Quantities of the principal Articles of Russian Produce and Manufacture imported into the United Kingdom from Russia in 1839.

Articles.		Quantities.	Articles.		Quantities.
Bristles	lbs.	1,272,164	Plain linen	pieces	13,948
Wheat	qrs.	371,693	Rhubarb	lbs.	55,051
Barley	—	76,138	Flax seed and		
Oats	—	816,923	linseed	bush.	3,987,486
Rye	—	14,030	Rape seed	—	55,871
Flax and tow	cwts.	705,708	Calf. &c. skins		
Hemp undressed	—	731,012	untanned	cwts.	16,900
Hides, untanned	—	9,131	Deals, &c. gt. hands.		15,316
Tallow	—	1,215,161	Battens. &c.	—	4,861
Tar	lasts	10,418	Masts, yards, &c. No.		3,404
Sheep's wool	lbs.	7,966,594	" " " ½ lds.		1,252
Linen yarn	cwts.	2,187	Fir and oak tim-		
Plain linen	ells	881,418	ber.	—	14,317

Water Communications.—The great road from Petersburg to Moscow is justly said by Lord Londonderry to be a most magnificent public work. It is nearly 390 m. in length, quite level, about double the width of the Great Northroad in England, and is macadamised throughout, and kept along the whole line in the most perfect repair. (*Tour,* i., 144.) But, with the exception of this and of a few other principal lines, there is a great want of good roads in Russia. This, however, is productive of less inconvenience than might be expected, from the circumstance of the frost rendering the worst roads fit for sledge travelling for a considerable period of the year; and from the number of navigable rivers and the extension that has been given to their navigation by the construction of numerous canals. By these means a water communication has been effected between the great navigable river the Wolga, which has its embouchure in the Caspian sea, and Petersburg and Archangel: the Wolga has also been united with the Don, which falls into the sea of Azof. The Pripet, an affluent of the Dniepr, which falls into the Black sea, has been connected with the Bug, an affluent of the Vistula, while the latter has been connected with the Niemen.

Few countries, in fact, have so extensive a command of internal navigation. Goods put on board in St. Petersburg may be conveyed to Astrakhan, a distance of above 1400 m., or to any port on the Caspian, and *vice versa*, without once being landed. The iron and furs of Siberia and the teas of China are received at Petersburg in the same way; but owing to the great distance of these countries, and the short period of the year during which the rivers and canals are navigable, they sometimes take three years in their transit!! Immense quantities of goods are conveyed during winter upon the ice in sledges, to the different ports, and to the nearest *pristans*, or places in the interior, where barks are built for river or canal navigation. They are put on board in anticipation of the period of sailing, that the barks may be ready to take advantage of the high water, by floating down with the current as soon as the snow and ice begin to melt. The cargoes carried up the river into the interior during summer are principally conveyed to their ultimate destinations by the sledge roads during winter. The conveyance by the latter is generally the most expeditious; and it, as well as the internal conveyance by water, is performed at a very moderate expense.

The barks that come from the interior are mostly of a very rude construction, flat-bottomed, and seldom drawing more than 20 or 30 inches water. When they arrive at their destination, they are sold or broken up for fire-wood. Those that leave the ports for the interior are of a superior description, and are comparatively few in number; the commodities imported being, at an average, of much greater value, relatively to their bulk and weight, than those that are exported.

As illustrative of the importance of the inland navigation of the empire, we may mention that, in 1839, no fewer than 46,850 boats and 17,469 rafts arrived from the interior at the different great ports and emporiums of the Russian empire, the goods so conveyed being worth 536,931,730 roubles. Of these, 22,842 boats and 784 rafts, value 196,974,904 roubles, arrived at Petersburg; 1362 boats and 1235 rafts, value 15,981,500, at Archangel; 1965 boats and 1373 rafts, value 32,437,378, at Riga, and so on.

Accounts in Russia are kept in roubles and kopeks: but the rouble is of two kinds, very different in value. The silver rouble is worth from 3s. 2d. to nearly 4s. English, varying according to the distance from the capital. The paper rouble, worth about 10d. English, and usually considered equivalent to a franc, is the basis of all mercantile calculations, and is divided into 100 kopeks. The kopek, worth consequently about a centime, is a copper coin. The only gold coin is the demi-imperial, value about 20 francs. Since 1828, platina coins, worth about £1 sterling, have been struck; but they are not yet of any practical importance. The Russian pound is rather larger than the avoirdupois pound; the last = 13·8 quarters; the chetwort, the measure for corn, = 5·75 English bushels; the deciatine, land measure = about 2·7 acres; the verst of 104·5 to a Geog. degree = 1167 yards, 3 versts being about equivalent to 2 English miles.

Native and Foreign Merchants, &c.—Every Russian carrying on trade must be a burgher, and have his name registered in the burgher's book; he thus acquires an unlimited freedom of trade. All whose names are in the burgher's books are either townsmen who have property within the city, or members of a guild. There are three guilds. Those belonging to the first report themselves to possess from 10,000 to 50,000 roubles; these may follow foreign trade, are not liable to corporal punishment, and may drive about the city in a carriage drawn by two horses. Those belonging to the second guild declare themselves possessed of from 3000 to 10,000 roubles: they are confined to inland trade. A capital of from 1000 to 5000 roubles entitles its owner to admission into the third guild, which comprises shopkeepers and petty dealers. The rates paid by the members of these guilds amount to 1½ per cent. upon their declared capital, the statement of which is left to the conscience of every individual. Burghers are not obliged to serve in the army, but may provide substitutes, or pay a fine. The *guests*, or foreign merchants who enrol themselves in the city register, on account of their commercial affairs, enjoy privileges nearly similar to those enjoyed by the members of the first guild.

None but native Russians are allowed to engage in the internal trade of the country; and hence a foreigner, who imports goods into Russia, must sell them to Russians only, and at the port where they arrive. A few foreigners, indeed, settled in Russia, and having connections with the natives, trade with the interior; but it is contrary to law, and the goods are liable to be seized. In 1836, the merchants of the three guilds amounted, in all, to 128,854.

The merchants engaged in foreign trade are mostly foreigners, of whom the English are the principal. The peculiar privileges formerly enjoyed by the latter are now nearly obsolete, and their rights, in common with those of other foreigners, are merely those of guests. The English factory at Petersburg is, at present, little else than a society formed of some of the principal English merchants, several of whom, however, do not belong to it: its power extends to little else than the management of certain funds under its control.

Owing to the scarcity of capital in Russia, goods, the produce of the country, are frequently paid in advance, and foreign goods are most commonly sold upon credit. From the month of November to the shipping season in May, the Russians who trade in flax, hemp, tallow, bristles, iron, &c., either come themselves to Petersburg or Riga, or employ agents to sell their goods to foreigners, to be delivered, according to agreement, in May, June, July, or August. The payments are made according to the circumstances of the sellers and buyers; sometimes the buyer pays the whole amount, in the winter months, for the

goods which are to be delivered in the summer or autumn; and sometimes he pays a part on concluding the contract, and the residue on delivery of the goods. The manufacturers and dealers in linen usually come to Petersburg in March, and sell their goods for ready money.

Foreign goods were formerly almost entirely sold at a twelvemonth's credit, and some at a still longer term; but of late years several articles, as coffee and sugar, are sold for ready money: still, however, the great bulk of foreign goods for the supply of the interior is sold on credit. Most of the Russians who buy goods on credit of foreigners, for the use of the interior, have no other connection or trade with Petersburg than merely coming there once or twice a year to make purchases; which, having accomplished, they set off with the goods, and the foreigner neither sees nor hears of them again till the bills become due. It is obvious, from this statement, that experience and security are nowhere more requisite in a merchant than here. He has nothing, in fact, but his own knowledge of the native dealers to depend upon: and it is highly creditable to the Russians, that foreigners do not hesitate to trust them with immense sums on such guaranty.

Government.—In Russia all power emanates from the sovereign, whose authority is uncontrolled, except by the respect he may yield to established customs, to the privileges of certain classes, and the prejudices of the people. The will of the monarch has no legal limits, so that he may be said to be absolute. The act of election of 1613, which conferred the crown on the house of Romanof, recognises the unlimited power of the sovereign. The empress Catherine and the emperor Alexander laboured to give order, simplicity, and regularity to the administration, and to reduce it to a system, so that it might be as independent as possible of the caprices of the sovereign. Alexander, indeed, proclaimed in 1811 that the law was in Russia superior to the sovereign, and gave to the senate the right of remonstrating against any *ukase* (as an Imperial decree is called) they thought contrary thereto. This, no doubt, seeing the way in which the senate is composed, is a very feeble check on the despotic power of the emperor. But it may well be doubted, whether, in the actual state of Russia, the present form of government be not better adapted to its wants than any that could be substituted in its stead. It is sufficiently clear, as well from general principles as from what has actually occurred, that Russian princes cannot safely follow a course of conduct generally disliked by the nation. On the other hand, however, the extent and unity of the sovereign power is the best security for the progress of civilisation, and for the improvement and well-being of the mass of the people. The latter being, for the most part, slaves, without property, intelligence, or influence, would be tyrannised over in an incomparably greater extent than at present, had the nobles any share in the government, or were they able to control its proceedings. What Poland was Russia would be, were the nobility or superior classes participating in the sovereign power. But the interests of the autocrat and those of the mass of the people are generally identical. Under his protecting ægis civilisation is daily extending, and a class of free labourers is gradually growing up. The emperor is afraid of the nobles, not a few of whom are supposed to be tinctured with liberal opinions; but he has nothing to fear from their slaves. Hence the despotic power of the former over the latter has been materially reduced; very great changes for the better have been made in the condition of the peasantry on the crown estates; and the government has gone steadily on, with quite as much rapidity as circumstances would warrant, endeavouring to improve and advance the servile portion, that is, the great mass, of its subjects, and to pave the way for their ultimate emancipation. An enlightened despotism is, in fact, the most suitable government for such a country as Russia. A representative constitution would merely put additional power into the hands of a comparatively small class, and would be as little adapted to the wants of such a country as an absolute government would be to England.

The monarch is the central point of the administration: his decisions are law. Every thing emanates from him in the first instance, and every thing is referred to him in the last. The labour he has to undergo is great, and requires incessant activity.

The public business is transacted, under the emperor, by different boards, councils, or colleges, which have each separate, but sometimes not easily distinguished functions. The *Imperial Council of the Emperor* was established, on its present footing, in 1810. It consists of a president and an indefinite number of members, of which the ministers always make a part. It is divided into five departments of legislation, war, civil and religious affairs, finance, and the affairs of Poland; and has the superintendence of all matters connected with the internal administration of the empire. The second college, or sénate, was founded by Peter the Great in 1711, and is reckoned the most important body in the state. It has various functions, partly of a deliberative and partly of an executive character, set forth in a ukase of 20th Sept., 1802. It is the high court of justice for the empire, and controls all the inferior tribunals. The members are nominated by the emperor; at present their number is about 100, and each receives a salary of 7000 roubles a year. The senate is divided into eight committees or sections, of which five sit at Petersburg, and three at Moscow. Each committee is authorized to decide in the last resort upon certain descriptions of cases, brought either immediately before it, or by appeal from the inferior courts. In a few cases, however, parties dissatisfied with its decisions may petition the emperor. The senators are mostly persons of high rank, or who fill high stations; but a lawyer of eminence presides over each department, who represents the emperor, and without whose signature its decisions would have no force. In the plenum, or general meeting of the sections, the minister of justice takes the chair, as high procurator for his majesty. Besides its superintendence over the court of law, the senate examines into the state of the public revenue and expenditure, and has power to inquire into public abuses, to appoint to a great variety of offices, and to make remonstrances to the emperor. Monthly reports of its proceedings are published in the gazette.

The third college consists of the *Holy Synod*, and to it is committed the superintendence of the religious affairs of the empire. It is composed of the principal dignitaries of the church. All its decisions run in the emperor's name, and have no force till approved by him.

The fourth college consists of the *Committee of Ministers*, of whom there are eleven, viz., the ministers of the imperial household, of war, finance, justice, interior, public instruction, imperial domains, postoffice, roads and public buildings, and the vice-chancellor and comptroller-general. The ministers have frequently colleagues who supply their place when they are either sick or absent. They communicate directly with the emperor, or with his *Chancellerie particulière*, in whose hands all the executive authority is centred.

The local administration differs in different provinces; government having always allowed conquered or annexed countries to preserve their own laws and institutions, except in so far as they were hostile to the general constitution of the empire. Finland, for example, has a special form of government; and the provinces wrested from Sweden by Peter the Great, Courland, and those formerly belonging to Poland, have peculiar institutions and privileges, which, however, have latterly been much modified. But, despite these exceptions, the form of the provincial government is, notwithstanding, sufficiently uniform.

The empire is divided into general governments, or vice-royalties, governments, and districts. There are, at present, 14 of the first, 50 or 51 of the second, and above 300 of the last. There are also, as already stated, extensive districts which, from the thinness of the population, or otherwise, are not organized into regular governments, which are called provinces, or *oblasts*. The viceroys, or general-governors, are the representatives of the emperor; and as such command the forces, and have the supreme control and direction of all affairs, whether civil or military. All the functionaries within their jurisdiction are subordinate to, and make their reports to them. They sanction or suspend the judgments of the courts, &c. A civil governor, representing the general governor, assisted by a council of regency, to which all measures must be submitted, is established in each government or province. In case of dissent, the opinion of the governor is provisionally adopted till the pleasure of the emperor with respect to the matter is ascertained. A vice-governor is appointed to fill the place of the civil-governor when the latter is absent or unwell. There are also, in every government, a council of finance under the presidency of the vice-governor, who manages the crown estates, and superintends the collection of the revenue; a college of general provision, which has the direction and inspection of all charitable foundations, prisons, workhouses, schools for the instruction of the poor, &c.; and a college of medicine, which attends to all matters connected with the public health, appoints district physicians, inspects pharmacopeias, &c. The districts have each their local functionaries. The towns have a municipal body, elected once every three years by the different classes into which the population is divided. And each town has, also, according to its importance, a commandant or bailiff, appointed by the crown, who has charge of the police, of the public buildings and magazines, and who executes sentences, pursues criminals, &c.

The Russian judicial system is complicated, and not easily understood, except by natives. There are civil and criminal courts in every circle and a supreme court of

justice, divided into civil and criminal sections, is established in every government. Cases decided in the inferior courts may be appealed to it. Its sentence is final in all criminal cases, and in all civil matters relating to sums under 500 roubles. Those involving property to a greater amount may be carried before the senate.

It is a curious fact that, notwithstanding the despotical nature of the government, all the provincial tribunals consist partly of elective functionaries. Thus, the superior court for a circle consists of a judge and secretary, and of two assessors chosen annually by the nobles, and two by the peasants: and the superior court of justice for a government, which is divided into a civil and criminal chamber, consists of a president, secretary, and four assessors for each chamber, two of the assessors being chosen by the nobility, and two by the burghers. It is, in fact, a principle in Russia that a portion of the judges in every court should belong to the same class as the party whose interests are under discussion, and be elected for that purpose by his compeers. In the case of the nobility and burghers, this is a most valuable privilege; but in the case of the peasantry, who stand most in need of protection, this privilege is quite illusory; their slavery and ignorance making them utterly incapable of profiting by it.

Previously to the reign of the Empress Catherine II., the judges, particularly in the inferior courts, were wretchedly paid. That princess increased their salaries; but they are still far too low. And seeing that the judges are removable at pleasure, and owe their situation to favour rather than to merit, we need not wonder that the greatest abuses continue to exist in the administration of justice. The proceedings are dilatory in the extreme. The prohibition against taking fees from suitors is rarely complied with; and in most tribunals it is affirmed, that if justice cannot be altogether defeated, it may at least be indefinitely postponed, by dint of money. These abuses have, however, been in part, at least, obviated by the publication, between 1826 and 1833, by the Legislative Commission, of an extensive digest (*Sweed Zakonow*, "Body of Law") of all the laws then in force relative to the rights of citizens and the administration of public justice. This publication has greatly simplified the law; and it is of vast importance from its being, as it were, a charter of rights which may be appealed to on all future occasions, and which it will be very difficult for any succeeding sovereign to abridge. But it would, notwithstanding, be idle to expect any very material improvement in the ordinary administration of justice, until the influence of public opinion, and of a comparatively free press, neither of which has at present any existence in Russia, be brought to bear on the administration of justice, and of public affairs generally. The latter, in fact, is the only security against abuse on which any reliance can safely be placed. Wherever judges are exempted from the control of public opinion, and the antmadversion of the press, they are most commonly the obsequious instruments of government, and seldom scruple to commit injustice when they believe it will be acceptable to their superiors.

There is in Russia, particularly in the great towns, a very efficient system of police. The officers are empowered to discharge various functions besides those which come more peculiarly within their province; such as the decision of differences between masters and servants, &c. Crime is not frequent in Russia; and property is as well protected in it as in any other country. Houses being generally built of wood, fires in great towns are apt to be very destructive; and the most effectual precautions are taken to prevent their occurrence. All strangers arriving in Russia must produce their passports at the police office, and notify their arrival in the public papers.

Punishments.—Capital punishments are rare in Russia, treason being the only crime visited with death. In cases of murder, fire raising, and other capital offences, the criminal, after receiving a certain number of lashes from the knout (a heavy thong whip), under the infliction of which he sometimes expires, is condemned for life to forced labour in the mines of Siberia. This part of the legislation of Russia has been the theme of much, though, as it appears to us, of little deserved eulogy. We agree with Mr. Coxe, that the fear of death is to most men the most efficient check on the commission of crime; and though it were conceded that Russian malefactors undergo a much severer punishment, still, as people generally know nothing of it, it makes no impression on them, and has little or no influence in deterring others from committing similar offences. (See *Coxe's Travels in Russia, &c.*, iii., 116.) The nostrils of criminals used also to be slit, and their face branded with a red-hot iron previously to their banishment to Siberia; but this needless aggravation of punishment was put an end to by the Emperor Alexander.

Torture was formerly universal in Russia, and was inflicted at the discretion of the superior justices in all parts of the country, by whom, as was to be expected, the power was often shamefully abused. Russia is indebted to the Empress Catherine for the abolition of this atrocious practice. And it is a singular fact, and worthy of the attention of those who are so fond of recommending immediate changes, that the prejudice of the Russians, in regard to the necessity of torture, was so deeply rooted that Catherine had to proceed with great caution in bringing about its abolition, which was effected rather by indirect than by direct means. (*Coxe, ubi supra.*)

According to an official return there were, on the 1st of Jan., 1835, 97,121 criminals in Siberia, of whom 93,264 were females. Of the convicts, about 10,000 are condemned to forced labour in the mines, and otherwise ; the others being mostly employed in agriculture. The greatest criminals, or (according to the interpretation of the Russian government) those sent thither for political offences, are mostly confined in N.E. Siberia, the climate of which is especially severe. The desertions amount to about 2000 a year.

Division of the People into Classes.—The people of Russia are divided into four classes, viz., 1. nobles, 2. clergy, 3. burghers, merchants, and other farmers, and 4. the peasants, or slaves.

1. *Nobles.*—Previously to the reign of Peter the Great, the Russian nobility consisted principally of the descendants of the ancient petty princes of the country, or of lords possessed of vast estates. They were in the exclusive possession of all situations of trust and emolument, to which they succeeded according to their rank. Peter, who early saw the disadvantage of this state of things, and the necessity of undermining the influence of the nobles, most of whom were violently opposed to his projects for the regeneration of the country, had recourse, in furtherance of his plans, to the scheme of creating a new order of nobility. In this view he divided all the civil and military functionaries in the service of the state into 14 classes, enacting, at the same time, that the eight highest classes should confer on the individuals in them the distinction of hereditary nobility ; that some of the other classes should confer the distinction of personal nobility, or of nobility for life ; and that those enrolled in the others should be deemed gentlemen, or *bien nées*. Some modifications were made in this arrangement by the Empress Catherine II. ; but it is still maintained nearly as it was contrived by Peter the Great.

The creation of a new nobility founded on merit, or on services rendered to the state, was, no doubt, a material improvement at the time. By illustrating many new families, it has served to lessen the influence of the old nobility, and to liberalise the order, at the same time that it has opened a prospect to all enterprising individuals of rising to the highest dignities. On the whole, however, it would seem that the system, having served its purpose, might now be advantageously abandoned. "*En grossissant,*" says M. Schnitzler, "*à l'infinis, le corps de la noblesse, ne nuit-il pas à cette institution peut-être nécessaire ; et en dépouillant le tiers état de tout ce qu'il a des citoyens distingués, n' affaiblit-il pas la considération dont il serait juste et utile d'entourer la classe laborieuse ? N'enlevait-il pas à jamais aux arts et à l'industrie des hommes capables, qui auraient contribué à leurs succès, s'ils n'avaient pas du à soutenir un rôle nouveau, et qu'ils ne craignissent de déroger à un rang auquel des longs efforts et les travaux de leurs pères les ont enfin élevées, en se livrant à des occupations qu'on regarde comme roturières ?*" (*Statistique Générale, p. 244.*)

According to the official accounts, the order of the nobility comprised, in 1836, 691,355 individuals, of whom 538,160 enjoyed hereditary and the others personal dignities. In Russia, properly so called, the nobles are not numerous; but they abound in Podolia, Volhynia, and other provinces acquired from Poland, and especially in Poland itself, which, in 1837, had 283,420 nobles ! Few, however, of the latter possess estates, and many of them are in a very destitute condition. In the Polish provinces, and in Courland, Livonia, and Esthonia, none but nobles can inherit landed property; but this is not the case in Russia Proper, though, with the exception of the crown estates, they are, in fact, almost the sole proprietors. The titles of prince, count, and baron have superseded those formerly in use. In the government of Tula, there are said to be more than 100 families having the dignity of prince. All the members of noble families are noble, and have the same title as the head of the family. On the death of a noble person, his estate is divided, according to a fixed scale (see *ante, p. 721*), among his children of both sexes. Nobles are exempted from all personal charges, and from the obligation to serve in the army, but they are obliged to furnish recruits according to the number of their vassals. Nobles are also exempted from corporeal punishment ; have leave to distil all the spirits required for the consumption of their establishments ; may engage in manufactures or trade ; have a right to all the minerals on their estates, &c. Precedence is determined, in Russia, by military rank ; and an ensign would

take the *pas* of a nobleman not enrolled in the army, or occupying some situation giving military rank. (*Schnitzler, Essai d'une Statistique, &c., p. 112, &c.*)

The property of a noble who has been condemned is not confiscated by the state, but goes to his family. The nobles also elect various local magistrates, assessors, &c., and deliberate at their meetings on various matters connected with the local administration. There is also in every government a committee of nobles to watch over the interests of the body, and to take care of the establishments that belong to it ; and every circle has a committee of nobles who manage the estates and affairs of nobles who are under age. These privileges, which are obviously of considerable importance, were embodied and set forth in a ukase in 1789 ; and another ukase of the Emperor Alexander prohibits all government functionaries from interfering with the election of the assessors, and other functionaries chosen by the nobles.

It is not easy to form a fair estimate of the character of the Russian nobles. Generally speaking, their education is more superficial than solid ; but many are, nevertheless, highly accomplished ; they are all well acquainted with French, and numbers with the English and German languages ; those who have travelled being distinguished by the superior polish and elegance of their manners. They are universally hospitable ; and most of them affect, and many relish, the society of literary men and artists. That they are more sensual, more given to ostentatious display, and less distinguished by a gentlemanly bearing towards their inferiors, than the higher classes in England or France is, no doubt, true. But the representations of their manners and conduct, given by Clarke, Lyall, and other travellers, of their caste, are, notwithstanding, mere vulgar caricatures, which, though they may perhaps apply to a few individuals, are generally quite as wide of the truth as M. Pillet's accounts of our fair countrywomen. Considering, indeed, that the Russian nobility have no exciting political occupation, that in most parts of the empire there is so middle class, and that the occupiers of their estates are not freemen but slaves, the wonder is, not that their tastes and habits should be, in some respects, barbarous, but that they should have made so great an advance as they have done since the reign of Peter the Great, and that they should be so intelligent and refined as they are found to be.

The Russian nobles, like those of England and other countries in feudal times, are in the habit of keeping great numbers of vassals in their houses as servants. The number of such retainers in some great families exceeds all belief, amounting sometimes to above 500 ! They receive only a trifling pittance as wages, but that is quite enough for their wants, as they are fed and clothed by their masters. Several Russian noblemen have recently distinguished themselves by their attention to their estates, and by the efforts they have made to introduce the improved processes and implements in use in more advanced countries. In some instances they have brought land stewards and labourers from England. Latterly, also, many of the principal nobles, have become extensive manufacturers, and some of the greatest manufacturing establishments of the empire are, at present, in their hands. They are driven, in fact, to adopt this course by the circumstances under which they are placed. All agricultural, and most out-of-door employments being suspended during winter, the noblemen, who must provide for the subsistence of their slaves, whether the latter be employed or not, naturally endeavour to avail themselves of their services during the interruption of husbandry pursuits, by setting on foot some species of manufacture. The latter, indeed, is frequently carried on only during winter, the peasants being employed in agriculture during the rest of the year. When, however, a nobleman establishes a manufacture on a large scale, and keeps it constantly at work, the peasants are usually put on the footing of hired labourers, and instead of getting an allotment of land, are paid for their work, and left to supply themselves with necessaries. Some manufactures conducted in this way have been eminently successful : though it be hardly necessary to add, that if they be of the higher class, or require any peculiar skill, economy, or attention, they are not of a kind that can be successfully carried on by the agents of noblemen ; and that the moment the protection afforded by oppressive custom-house duties, under which they have grown up, is withdrawn, they will straight way fall to the ground. (See *Venables' Russia, &c., p. 148.*)

Mr. Coxe and Dr. Pinkerton, among the best and most trustworthy of the English travellers who have visited Russia, speak very favourably of the Russian nobility. The former says, that though they have adopted the delicacies of French cookery, they neither affect to despise their native dishes, nor squeamishly reject the solid joints which characterise an English repast. The pleasant as well as the choicest viands are collected from the most distant quarters. At the tables of opulent persons in Petersburg may be seen sterile from the Wolga, veal from Archangel, mutton from Astrakhan, beef from the Ukraine, and pheasants from Hungary and Bohemia. The common wines are claret, Burgundy, and champagne; and English beer and porter may be had in perfection and abundance. It is usual to take a whet before dinner; but the stories engrafted on this practice, of the prevalence of inebriety among the higher classes are wholly without foundation. In this respect their habits have undergone a total change since the days of Peter the Great; and they are now remarkable for sobriety. The peasantry, however, often indulge to excess in their positions. (*Coxe*, ii. 151.)

The lengthened stay of the Russian armies in the western and more civilised European states, after the defeat of Napoleon's invasion, made a large number of the nobles, and of the more intelligent classes (which in Russia consist of the military officers), familiarly acquainted with a more advanced state of society, and a better form of civil polity. This circumstance, also, gave an increased stimulus to the desire for travelling that previously prevailed among the nobility, many of whom withdrew to France, England, and Italy. It is not to be denied, that the influence of these concurring circumstances has since, on various occasions, made itself sensibly felt in Russia; and that the government has sometimes had reason to believe that a considerable portion of the nobility, and even some of the most distinguished regiments, would not be displeased to see some limit set to the powers of the czar. To counteract this feeling, various obstacles have latterly been opposed to the emigration of Russian nobles and to their residence abroad: the most vigilant measures have also been adopted to hinder the employment of foreign tutors and governesses; and to prevent the introduction of foreign works not approved by the censor. It remains to be seen whether these measures will be effectual to maintain the present order of things; but, at all events, it is sufficiently clear, from what has been already seen, that, under existing circumstances, such a revolution in Russia as should materially modify the power of the czar, would not be for the advantage of the bulk of the people, but the reverse.

2. *Clergy.*—This body will be more fully noticed under the head religion. It comprises, in all, about 274,000 individuals, of whom about 254,000 belong to the established church. Including the wives of the priests, it is supposed that about 540,000 persons may belong to this class. They are exempted from all direct taxes, and from corporal punishment, and may acquire all sorts of fixed property.

3. *Merchants, Burghers, &c.*—This comprises the class intermediate between nobles and peasants, and is thus alluded to by the empress Catherine in her instructions for a new code of laws:— "This class, composed of freemen, belong neither to the class of nobles nor to that of peasants. All those who, being neither gentlemen nor peasants, follow the arts and sciences, navigation, commerce, or exercise trades, are to be ranked in this class. In it should be placed all those who, born of plebeian parents, shall have been brought up in schools or places of education, religious or otherwise, founded by us or by our predecessors. Also the children of officers and of the secretaries to the chancery," &c. Merchants and traders belong to this class; and they, as already stated, are distributed into their guilds according to the amount of capital they respectively possess (see *ante*), and enjoy various privileges on their paying a certain per centage on their declared capital. The burghers, or second division of this class, possess many privileges superior to the peasants; but they are distinguished from the merchants by being subject to the capitation tax and to enrolment in the army and navy. The Germans and other free colonists established in different parts of the empire, and the free cultivators and tenants found in certain districts, belong to this intermediate class. This class comprises about 3,000,000 of individuals.

4. *Peasants.*—Unhappily, however, the far largest portion of the people of Russia are slaves belonging either to the crown or to individuals; above 21 millions being the property of the former, and 23 millions of the latter. Count Scheremetief is proprietor of above 110,000 slaves, and the numbers of those belonging to some of the other great landholders are but little inferior. The nobles are obliged to pay a tax to government (at the rate generally of about four roubles per male), and to furnish recruits for the army according to the slave population of their estates. The time and labour of the slaves belonging to private individuals are absolutely at the disposal of their masters, who may seize whatever property they may happen to acquire. The most common practice is for the latter to impose on their peasants a obrock or capitation tax, which may amount, at an average for those resident in the country, to from 35 to 45 roubles per *mala*, young and old; but those who have received licenses to reside in towns, or who have learned any profession, or have been successful, are charged far greater sums, sometimes even as much as 1000 roubles a year or upwards!

Others, instead of an obrock, perform task-work; others, again, deliver a certain portion of their produce; and from some all these are demanded. Run-away slaves are punished by imprisonment and hard labour.

Besides having power to dispose of his time and labour, the master may inflict corporeal punishment on his slave; but he is forbidden by law (which, however is often evaded) from treating him with any great cruelty, and he is guilty of a capital offence if death arise from his chastisement within 24 hours. When one class may exercise such power over another, very great abuses cannot fail to exist. The insecurity, too, under which the peasants are placed, is necessarily fatal to their industry. Oppression and ill-treatment are now, however, a good deal less common than formerly; and it is certainly true that the condition of the boors is by no means so bad as might, *à priori*, be concluded, and that, as respects their command over the necessaries of life, they are in a much better situation than the peasantry of Ireland. Those on the estates of humane and enlightened landlords are in decidedly comfortable circumstances; while they mostly all have sufficient supplies of the articles they consider necessary to existence. Some licensed slaves have accumulated very large fortunes. One of this class of persons is mentioned as having 4000 labourers in his employment; and another planned and built the finest church in Petersburg.

The peasants are of a sound constitution, stout and firmly built, and generally of a middle stature. They live in wooden cottages, formed of whole trees piled upon each other, and built together in villages, the gables to the road. Sometimes they consist of two stories, but more frequently only of one. They are heated by stoves, and though dirty, are not uncomfortable nor ill suited to the climate. Their furniture consists generally of wooden articles, with a pan or two. Beds are little used, the family generally sleeping on the ground, on benches, or on the stove.

The dress of the peasant consists of a long coarse drugget coat, fastened by a belt round the waist, but in winter they wear a sheep-skin with the woolly side inwards. Their trowsers are of coarse linen; instead of stockings, woollen or flannel cloth is wrapped round the legs, and boots or shoes of matted linden bark are frequently substituted for those of leather. The neck, even in winter, is bare,[*] and the head is covered by a peaked round hat or cap. (*Voyage de deux Français dans le Nord de l'Europe*, tom. iv. p. 323.)

The Russian peasant considers himself well fed if he have rye-bread, which is the staple article of food throughout the empire, and sour cabbage soup, with a lump of fat, or hog's-lard, boiled in it, by way of relish. He uses butchers' meat on holidays, and at other times eggs, salt-fish, bacon, lard, and mushrooms, which, at the proper season, are extremely abundant, onions, &c. His favourite dish is a hodge-podge of salt or fresh meat, groats, and rye-flour, highly seasoned with onions and garlic. Salted cucumbers are a constant dish at the peasant's table all the year round. These and salted cabbages form an important article of national commerce. They are brought in large vats from the southern provinces, where the climate favours their production, to Moscow, Petersburg, and other large towns, and here they are constantly on sale in the public markets. The preparation, in autumn, of a sufficient supply of these pickled vegetables forming, in every family, an important part of domestic economy. This dependence of the Russian peasant on vegetable diet is, no doubt a consequence of the extraordinary number of fasts and fast-days, of which he is a careful observer, and which are multiplied to such an absurd extreme, that it is said there are only from 60 to 70 days in the year on which it is permitted to use butchers' meat. *Quas* a fermented liquor, made by pouring boiling water on rye or barley-meal, is the common beverage of the peasant. But he is also very fond of mead, and still more so of corn brandy, and other spirituous liquors. The consumption of the latter is immense, exceeding 80,000,000 gallons a year, and furnishing annually a large revenue to government. The use of tea is becoming more and more extended. A substitute for it, called *izbitzn*, consisting of herbs, honey, &c., boiled together, is also extensively used by the peasantry.

The peasants are exceedingly superstitious. A vessel of holy water hangs from the ceiling of every room, and a lamp lighted on particular occasions. Every house is provided with a sacred corner, supplied with one or more pictures of their tutelary saints, coursely daubed on wood, frequently resembling rather a Calmuck idol than a human head; but sometimes they are of a better quality, and neatly framed: to these they pay the highest marks of veneration. All the members of the family, the moment they rise in the morning, and before they retire to sleep in the evening, never omit their adoration to the saints: they cross

* This, according to the intelligent author of the *Voyage de deux Français*, is a decisive criterion by which to distinguish the genuine Russian.

727

themselves during several minutes, upon the sides and forehead, bow very low, and sometimes even prostrate themselves on the ground. Every person, also, on entering the room, pays his obeisance to these objects of worship previously to his addressing himself to the family.

The Russian peasantry have the vices incident to their situation. With a great capacity of endurance, and the most extraordinary talent for instruction, they have but little active vigour or steadiness of purpose. In accosting a person of consequence, or from whom they expect any favour or advantage, they prostrate themselves, touch the ground with their hands, and kiss the fringe of his garments. Their insecure position makes them anxious to enjoy the present moment; and their masters being obliged to provide for their support when they become old and infirm, they have little motive to providence or forethought. When they accumulate money, they most frequently bury it in the ground; a practice common to all countries where property is insecure.

The peasantry belonging to the richer nobles are, speaking generally, in all respects, much better off than those who belong to the class of poor and petty nobles; and, unfortunately for the peasants, the number of the latter is rapidly increasing by the subdivision of estates in every successive generation, and the constant augmentation of the nobility, from the influx into its ranks of individuals in the civil and military service. In 1822, however, the evil resulting to the peasantry from this state of things was partially obviated by restricting the right to purchase slaves to nobles possessed of a certain amount of property.

Previously to the reign of Peter the Great, it was customary for the Russians, of all ranks, to marry their children very early, even before the age of puberty. Though restrained by Peter and Catherine II., this custom of early marriage still prevails, and is said to be fraught with many pernicious consequences. A ukase issued in 1801, prohibits priests from solemnizing marriages, unless the man be 18 and the woman 16 years old.

The use of the vapour bath is universal in Russia, not being reckoned a luxury but a necessary; and public baths are met with in all parts of the country. They are resorted to by the peasantry, at least, once a week. Though the baths are highly heated, the bathers not unfrequently run out and in summer plunge into cold water, or, if it be winter, roll themselves in the snow! This sudden alternation of temperature is not found to be injurious to health. But, notwithstanding the frequent use of the bath, the boors are very deficient in cleanliness. (See *Coxe*, 5th ed., i. 269; *Schnitzler, Essai, d'une Statistique Générales*, passim; *Voyage de deux Français au Nord de l'Europe*, iv. 318-332; *Pinkerton's Russia*, pp. 69-80, &c.; *Foreign Communications on the Poor Laws*, p. 330, &c.)

Army.—The military power of Russia having been ridiculously exaggerated by some, and as ridiculously depreciated by others, deserves to be inquired into with some attention. The Strelitzes, the first regularly organized corps of infantry in the Russian service, seems to have had their origin about the middle of the 16th century; and continued till their suppression by Peter the Great, to constitute the principal strength of the army. They enjoyed various privileges; were always about the person of the emperor; and by their licentiousness and insubordination, as well as bravery, bore a close resemblance to the Pretorian bands of antiquity, and the Janissaries of the Ottoman Porte. The abolition of this formidable corps, and the reconstruction of the army on a plan similar to that followed in the more civilised countries of Europe, was undoubtedly one of the greatest services rendered by Peter the Great. At his death, the regular army amounted to about 110,000, exclusive of the imperial guard; and the success with which attended his contest with the Swedes showed that this army was a match for the best troops that could then be opposed to it.

Under Catherine II., the army was greatly augmented and improved. This able and ambitious princess augmented the pay of the troops and officers, and gave them a new, more commodious, and elegant uniform than that formerly in use. She formed the Cossacks into a light cavalry, which, after being successfully opposed to the Spahis of the Turks, has since distinguished itself in the great contests of more modern times. During the latter part of the reign of Catherine the regular army amounted to about 250,000 men; and little was wanting to place it on a level with that of the surrounding powers, save the better organisation of the commissariat department, and the choice of better educated and more skilful native officers.

It is, however, to the emperors Alexander and Nicholas that the Russian army is indebted for the more efficient organization, discipline, and power by which it is now distinguished. The momentous struggles in which the former was engaged called forth all the military resources of the empire; many abuses were rectified, and improvements introduced; and the armies of Alexander were at length enabled to contend successfully with those of the greatest capacity of the age. Under the present emperor, the discipline and organization of the army have been still further improved; and it is, at present, in a comparatively high state of efficiency.

The Russian army has been newly organized, by a ukase of the 9th August, 1833. Down to that period, two large armies were maintained; but those were then consolidated, and the staff of one of them reduced. The army is now divided into six *corps d'armé* of infantry, each corps into three divisions, each division into two brigades, each brigade into two regiments, and each regiment into six battalions of 1000 men each; four battalions take the field, and two remain as a reserve; so that each regiment, in fact, is 6000 strong, and each *corps d'armé* is 48,000 strong. To each of these six corps there is attached one division of light horse, in two brigades (Hussars and Huhlans) of two regiments each. Each regiment of horse consists of nine squadrons of 160 horse each, eight of which take the field, and one remains as a reserve; so that a regiment in the field has 1280, and a division 5120 horses. Besides this, each corps has a division of artillery united to it, consisting of three brigades, with four batteries each of eight guns, and one brigade of horse artillery, one reserved battery, one park of artillery, and three sapper battalions, together 6000 men. Thus a complete *corps d'armé* is 60,000 men strong, with 120 pieces of artillery; and the whole active army 360,000 men, and 720 pieces of artillery. Then comes the *corps of guards*, in three divisions of infantry, three divisions of cavalry, and one division of artillery, with 120 guns; then the grenadier corps, equal in strength to an infantry corps; both together 114,000 men, and 346 guns. There are two reserved corps of cavalry, each of two divisions (Hulhans and Cuirassiers) each division of two brigades of two regiments each, with two brigades of horse artillery, and a dragoon corps in two brigades of two regiments each, and one brigade of horse artillery: altogether this cavalry corps consists of 30,000 men and horses. Finally, there are two independent *corps d'armé*, of the Caucasus and Siberia, the first equal in strength to a whole corps, the latter to half a one; so that the entire strength of the regular army in time of peace amounts to 594,000 men; which, at present, is equal to one per cent. of the population, as the Asiatics, and particularly those tribes who serve as irregular troops, are not included in this account. (*Supp. to the Conversations Lexicon of Gegenwart*, Engl. trans. p. 196.)

But exclusive of the above, the troops not of the line, or those forming the irregular army, constitute a very formidable force. They consist, 1st, of upwards of 50,000 men in garrison in different parts of the interior, and along the frontier; 2d, of above 100,000 veterans, mostly employed for the same purpose; and 3d, of the irregular Cossack cavalry, and the colonized regiments.

The really effective force of the Russian army may, therefore, be reckoned at about 600,000 men; but from this various deductions must be made. Two out of the six battalions of the different regiments are almost always absent, constituting, in fact, *dépôts*, being employed in public works, in the conveyance and training of recruits, &c. It is believed, too, that the forces actually embodied rarely come up to the returns; the officers being driven, by the inadequacy of their allowances, to add to their means of subsistence by keeping up vacancies in the ranks. And if, in addition to these various causes of deduction, we bear in mind that Russia must always, in the event of her being engaged in foreign war, keep a large amount of forces at home to guard the frontiers, and to maintain tranquillity in Poland and other disaffected provinces, it will be seen that, at present, she would have considerable difficulty in marching 200,000, or even 150,000 men across the frontier. When Napoleon invaded Russia in 1812, the entire force brought to oppose him scarcely exceeded 260,000 men. During the last war with Turkey, the invading army did not exceed 120,000. Even if the troops really at the disposal of government corresponded with those in the official returns, Russia would find great difficulty in sending large armies into distant countries, and maintaining them when there. Her finances are far from being in a flourishing state; and owing to the abuses that prevail in her commissariat department, a great expenditure is incurred, at the same time that the troops are often very ill provided with the most indispensable necessaries.

At home, however, Russia is all but invulnerable. The severity of the climate renders it next to impossible for an invading army to maintain any permanent footing in the country; whilst the nature of the ground, without roads, and intersected with forests, rivers, and marshes, opposes the greatest obstacles to the advance of an invading force, and still more to its retreat. Even though the army of Napoleon had not had to contend with the rigours of an unusually early winter, the result of his expedition could not have been materially different. He could not possibly have maintained himself during the winter at Moscow. Sooner or later he must have retreated; and a retreat through such a country, and in presence of a powerful enemy, ready to take every opportunity of attacking, could not fail to be most disastrous.

The troops of the imperial guard are a very fine body of men. Generally, the Russian soldiers are, in respect of bodily vigour, inferior, perhaps, to those of England. They have no enthusiasm; and in respect of activity and intelligence, are very far below those of England, France, and Prussia. On the other hand, however, they possess, in the greatest perfection, the two first qualities of a soldier; the most unflinching courage, and the most implicit obedience. Subjected from birth to a master whose will is their law, the habit of prompt and absolute obedience becomes, as it were, a part of themselves. Regardless of dangers or difficulties, they will attempt whatever they are ordered; and will accomplish all that the most undaunted resolution and perseverance can effect. They also endure, without a murmur, the greatest hardships and privations, and support themselves in situations where others would starve. The Cossacks, Baschkirs, and other irregular cavalry, are very useful troops, and are well calculated either to improve a victory or to cover a retreat. Contrary to what might have been expected, the artillery is the department in which the Russians have made the greatest advances; but it is said to be in excess as compared with the other descriptions of force. Were the officers as intelligent and skilful as the soldiers are brave and docile, the Russian army would be most formidable. But this is far from being the case. Latterly, however, great efforts have been made to improve the education of the officers, and, exclusive of the establishments for that purpose mentioned below, a military academy was opened at Petersburg in 1832, where officers not above the degree of captain are instructed in military service; and in 1837 a school for 400 cadets for the artillery and engineers was opened at Woronesch. The pay of the officers, though still miserably low, has been increased; and the present emperor has endeavoured to excite the martial spirit of the people, and to make the service popular, by inflicting grand military spectacles. Some of these have been on a gigantic scale. At the grand military and religious festival in commemoration of the battle of Borodino, in 1839, no fewer than 120,000 troops were present! Marshal Marmont has spoken in high terms of the efficiency and discipline of the Russian forces he reviewed in the S. provinces in 1834.

Recruiting.—The army is recruited from the classes of peasants and artisans, every individual belonging to them being liable to compulsory service, provided he be of the proper age and stature. The levies are ordinarily in the proportion of 1 or 2 to every 500 males; but during war the portion is at least as 2 or 3 to 500, and sometimes as much 4, or even 5, to 500. This last proportion, however, or that of 5 to 500 males, may be taken as the maximum levy, and is rarely exceeded. The number of recruits to be furnished by the empire in general, and by each district in particular, is fixed according to the results of the preceding ones. The nobles nominate such of their serfs as they wish to complete their quotas, the only conditions being that they should have a good constitution, and be of the requisite size, and not less than 18 nor more than 40 years of age; and, as idle, ill-disposed individuals are sure to be included in preference for recruits, those who are averse to the service endeavour to distinguish themselves by industry and good conduct. The minimum standard height of infantry is not less than 1 mètre 594 millimètres; and of cavalry, 1 mètre 689 millimètres. The recruits are first sent to the recruiting establishments, and thence forwarded in corps to which they are assigned. Nobles, magistrates, clergymen, and students are exempted from the service. Merchants and traders enrolled in the different guilds are also exempted: the levies furnished by the Cossacks are regulated by particular treaties; and many half savage races are excused, partly on account of their diminutive size, and partly because of their great aversion to a military life. Generally, it is found that a levy of 2 on every 500 produces a supply of about 90,000 or 100,000 men. Substitutes are allowed, and may be effected by mutual consent, provided the noble be informed, and do not oppose.

A ukase of the 3 (15) Sept. 1827, fixes the period of service at 20 years in the imperial guard, and 22 in the other corps. Every individual, with his family, if he have one, becomes free the moment he is enrolled in the ranks of the army. In case of desertion, he is again enslaved; but desertion is exceedingly rare in Russia.

The guard is recruited from the grenadiers; the latter serve in the infantry of the line and light chasseurs.

Promotion.—After two years' service a soldier may become a sub-officer. The sub-officer who has served twelve years obtains of right the rank of sub-lieutenant or ensign.

Children of soldiers are educated at public military schools, or at schools belonging to the regiment: those who pass their examination with credit become sub-officers. There are at Petersburg schools for pages, engineers, officers of artillery, sub-officers of the guard, &c.; the rank of ensign being given to pages who have gone through a certain course, and to gentlemen cadets who have been two years in the service. But the principal establishment for the education of officers is that of the *Corps des Cadets* at Petersburg, founded in 1731. It has about 700 pupils, the sons of noble parents, that is, of those who have attained to the rank of captain in the civil or military service. The pupils are divided into five classes, and on leaving school become ensigns in regiments of the line. This school has materially contributed to diffuse information among the inferior nobility, and to supply the army with able officers. There are also schools for cadets at Moscow, Woronesch, Polotsk, Tula, Tamboff, and other towns. The pupils leave after a fixed time, and are ranked as ensigns.

During peace promotion depends upon seniority, from the rank of ensign to that of colonel: during war it is determined indifferently, by gallantry, selection, and seniority.

Pay, &c.—The Russian army is supported at very little expense. Exclusive of their pay, the higher class of officers receive considerable allowances, as mess-money, &c.; and they generally contrive to eke out their emoluments in various indirect ways. The pay of the subalterns is the most inadequate; and it is hardly possible for any one to serve as a subaltern in the cavalry, especially in the cavalry of the guard, unless he have private resources. Officers are allowed, according to their rank, one or more servants (*deutschik*), maintained by government, but equipped at the expense of their masters. They are taken from among the recruits, the least suitable for active service. The pay of a common Russian soldier does not exceed 25s. a year! and various deductions are made even from this miserable pittance. He gets a new uniform each year; and is allowed, in addition, three barrels of flour, 24 lbs. of salt, and a certain quantity of rye or oatmeal. On fête days the soldiers of the guard receive a certain allowance of butchers' meat, but this is very rarely tasted by their fellows. At home the soldier is paid in paper; but when he crosses the frontier he is paid in silver roubles; and as *one* of the latter is equivalent to *four* of the former, his pay, when abroad, is, of course, augmented in the same proportion. This may, perhaps, have been partly intended as a stimulus to the soldier to undertake offensive operations; but, besides having this effect, it was absolutely necessary, to enable him to subsist among foreigners without robbing. The cavalry horses are very good; and, fodder being very cheap, they are well kept.

Soldiers leaving the army on the expiration of their compulsory service are entitled to a small pension; and those who have been maimed or wounded are received and supported in some of the hospitals established in that view in different parts of the country. Soldiers who continue in the army after their term of compulsory service has expired acquire several advantages. They receive, exclusive of the retiring pension to which they are entitled, double pay; and after five years' voluntary service they are entitled to a retiring pension equal to three times their original full pay.

The inadequate pay of the officers and men is the grand evil in relation to the Russian army. It compels all classes to resort to underhand methods of making money; and hence the jobbing and corruption of the first, and the thievish habits of the latter. Government is plundered in every possible way; and while the army loses in strength and efficiency, it may be questioned whether it would not be more advantageous, even in a pecuniary point of view, for government to increase the pay of the officers and troops, so as to raise them above the necessity of indulging in practices injurious to the service, of the existence of which it is well aware, but at which, as matters now stand, it is obliged to wink.

Capital punishments are at all times rare in the Russian army, and are never inflicted except during war. During peace culprits are uniformly condemned to transportation to Siberia, and to forced labour in the mines. Corporal punishments may be ordered by the commanding officers of regiments, but they cannot be carried beyond twenty strokes of the knout, and are not to be inflicted except for very grave offences. Soldiers who continue in the army after their full period of compulsory service is exhausted, cannot be corporeally punished except by command of a council of war.

Military Colonies.—Exclusive of her ordinary forces, Russia has a considerable force of military colonists. These are a sort of agricultural soldiers established by a ukase issued in 1818, agreeably to the suggestion of General Araktchieff. The object was to create a military force at the least possible expense, by ingrafting military service upon the agricultural labours of the peasants. For this purpose certain districts belonging to the crown were selected in the environs of the lake Ilmen, in the government of Novgorod, and in some of the southern governments, the territory of which was distributed among the peasantry, at the rate of about 15 deciatines, or 45 acres of arable land to each head of a family, villages on an improved and uniform plan being at the same time erected for their accommodation. The stock and implements necessary for the cultivation of this land are furnished to the colonist by the crown, and he is charged with its cultivation, with contributing to the common magazine of the village, keeping up the roads, &c.; the surplus produce, after these outgoings and the provision for his family are deducted, being at his disposal. A soldier is assigned to each colonist, to be maintained by the latter; but the soldier is, in return, obliged, when not absent or engaged in duty, to assist the colonist in the labours of his farm. The colonists, as well as the soldiery, are deprived of their beards, and wear uniform, everything within the colony being subject to military regulation: there is no restraint on the marriage of the soldiers; and their male children, and those of the colonists, are all bred up to be soldiers. The girls are educated in separate schools; and, though there be no regulation to that effect, are generally married to the young men belonging to the colonies. Exclusive of the principal soldiers already alluded to, there is in every cottage a substitute or supplementary soldier, generally a son of the colonist, who is bound to take the place of the principal soldier in the event of his death or sickness, so that the regiments distributed among the colonies can never want their full complement of men.

The colonies may contain in all from 50,000 to 60,000 troops· but, notwithstanding what Marshal Marmont has stated to the contrary, the experiment is now generally regarded as a failure. It will always, indeed, be found to be impossible successfully to combine the business of agriculture with the military service. The soldiers get attached to their farms and families, and become unwilling to leave them and impatient of military restraint. Few, indeed, at all familiar with such subjects, will be surprised to learn that considerable discontent has, at different times, prevailed among the colonies in question. A dangerous mutiny, in which several officers lost their slaves, was not suppressed except by the presence of the emperor, who discovered on this occasion his usual courage and decision. Indeed, the general opinion is, that the military colonies will be gradually abandoned.

Navy.—Russia has a very considerable naval force, the fleet comprising about 50 ships of the line, 25 frigates, 10 steam ships, and about 600 smaller vessels. She is indebted for her naval power, as she is for her ascendancy by land, her civilization, and, indeed, everything else, to the creative genius of Peter the Great. Previously to his accession, Russia had no seaport, other than Archangel, and did not possess a single gun-boat. As soon, however, as Peter had acquired a footing on the Baltic, he set about creating a navy; and the better to qualify himself for the task of its construction, he visited Holland, where he not only made himself acquainted with the principles of naval architecture, but with the practical business of a ship's carpenter, by working himself at this employment! The monarchs since Peter, and especially Catherine II. and the present emperor, have all exerted themselves to increase and improve the fleet; and it is now, perhaps, in as high a state of efficiency as it is likely to attain.

The truth, however, is, that though the naval force of Russia be sufficient to give her an overwhelming influence in the three inland seas connected with her empire, or in the Baltic, the Black sea, and the Caspian, it is not in the nature of things that she should ever be able to cope with the maritime powers situated on the ocean. The Baltic, in fact, affords no proper field for the training and exercise of a fleet. Besides being limited in extent, it is frozen over for half the year; during all which time the ships have to be laid up; and the crews, being on shore, cannot possibly attain that skill in seamanship, and dexterity in manoeuvring, that is attained by sailors constantly afloat. And, by a singular contradiction, instead of attempting to obviate this state of things, and sending squadrons into the open seas, and keeping them afloat all year, it is a rule of the Russian service, that every third year the seamen, if so we may call them, shall not go to sea at all! Nothing, therefore, can be more idle and unfounded than the statements so frequently put forth as to the danger to be apprehended from the increasing naval power of Russia. Such dangers are wholly imaginary. The physical circumstances under which she is placed must always prevent her from becoming a great naval power. She is superior, by far, to any other power on the Baltic, the Black sea, or the Caspian; but there her ascendancy naturally stops; and any attempt on her part to construct fleets to cope with the maritime powers, properly so called, would be a most absurd and improvident waste of the national resources.

Russian ships, both in the Baltic and Black sea, last but a very short time, and, consequently, are very expensive.

The great naval stations are Cronstadt, in the gulf of Finland, and Sevastopol, in the Black sea.

Education in Russia is at a very low ebb. There have for more than a century been schools in all the great towns but these are but few in number; and the rural population is too much dispersed, even if it were not enslaved, and tied down to routine practices, to allow it to reap much benefit from country schools. But, notwithstanding the difficulties in its way, education is making progress, and has been much improved and extended within the present century. It has always been, and continues to be, an object of great solicitude with the government. A plan for a national system of instruction was laid down in a ukase of the emperor Alexander, issued in 1803; which, though it has undergone various modifications, contains the outline of the system that is still followed. The empire is divided, in respect of education, into a certain number of districts, each of which has, or is intended to have, a university, with a certain number of lyceums (at which the young men intended to fill civil offices are mostly instructed), gymnasiums, high schools, and elementary schools, varying according to its extent and population. At present the districts are those of Petersburg, Moscow, Kharkof, Kasan, Dorpat, Kieff, Helsingfors, Odessa, White Russia, the Transcaucasian provinces, and Siberia; but of these only the first seven have universities. A curator, or inspector, is placed at the head of each district, who is in constant communication with the minister of public instruction. The subjects and courses of study, the examinations to be gone through, the fees to be paid, &c., are all fixed by government. The sum placed annually at the disposal of the minister of public instruction amounts to above three millions roubles.

The report of the minister of public instruction to the emperor, at the close of 1835, gives various details as to the then state of the higher departments of public instruction, and has partly supplied us with the following information:

1. *University of Petersburg.*—This university, founded in 1819, had, in 1835, 73 professors and subordinate functionaries, and 385 students. The six governments dependent upon it had eight gymnasiums.

2. *University of Moscow.*—At the above epoch this university, founded in 1755, had 90 professors or functionaries, and 611 pupils. In its library there were 45,000 volumes. It had nine governments within its jurisdiction, and in these there were, in 1835, a lyceum, an institution for nobles, nine gymnasiums, five free schools, 73 central schools, and 138 parish schools. The surveillance of the system is committed to the care of an inspector and five sub-inspectors. A school has since been founded in Moscow for the gratuitous education of the sons of 50 decayed merchants.

3. *The university of Kharkof,* founded in 1803, had 51 professors, &c., and 315 pupils. The eight governments under its jurisdiction had, in December, 1835, seven gymnasiums.

4. *The university of Kasan,* founded in 1804, had, in 1836, 76 professors, &c., and 176 pupils. The nine governments under its jurisdiction had each a gymnasium. The Arabic, Persian, Turkish, and Mongolian languages are taught in this university.

5. *The university of Dorpat,* founded in 1632, one of the most celebrated in Russia, had, in 1835, 74 professors, &c. and 563 pupils. The university library had 38,000 volumes, and there is attached to the institution a botanical garden and a museum. The three governments under its jurisdiction had, in 1835, four gymnasiums, and 156 public schools with 3750 pupils.

6. *The university of Kieff,* called St. Wladimir, was founded in 1834, being intended to supply the place of that of Wilna, suppressed after the late Polish insurrection. It had, in 1836, 88 professors and subordinate functionaries, and 903 pupils. It is well endowed, its revenues having amounted, in 1835, to 250,000 roubles. There are four governments under the jurisdiction of this university, in which were one lyceum, seven gymnasiums, and 85 free schools with 461 pupils. Government provides for the education of 50 pupils at this university, of whom 26 are trained to be teachers, and 24 are to be instructed in the law, so as to be able then to fill judicial and other civil offices in the old Polish provinces.

7. *The university district of White Russia* includes five governments and a province. It has as yet no university but it had, in 1835, 12 gymnasiums and a high school.

8. *The university of Helsingfors,* in Finland, founded in 1827, instead of that of Abo, destroyed by fire in the course of that year. In 1839 it had about 40 professors, &c., and 425 pupils. There are in the grand duchy of Finland about 390 establishments for educational purposes.

9. *University district of Odessa.*—In this, as in the district of White Russia, no university has as yet been established. But there is a lyceum in Odessa; and in the three governments and provinces subjected to its jurisdiction there were, in 1835, five gymnasiums, and 13 high schools.

10. In the *Transcaucasian district* there is a gymnasium, a free school, and 12 central schools.

11. The number of schools in Siberia is not specified; but there is said to be a great want of teachers there, as there is in most other divisions of the empire.

Besides the above there are various schools founded for particular objects, and not coming directly under the control of the minister of public instruction. Among others may be specified the military schools in Petersburg, Moscow, and other towns; schools for the special use of the nobility; the schools of the surgico-medical academies of Petersburg, Moscow, &c.; schools founded and endowed by individuals, and those founded by, and placed under, the control of the clergy.

The latter, or the theological schools, intended principally for the instruction of the sons of the clergy, are among the most ancient and important of any in Russia. They consist of four principal academies at Kieff, Moscow, Petersburg, and Kasan, which give instruction in the higher branches, and confer the degrees of A.M. and M.D.: of 36 diocesan schools; and of between 350 and 400 district schools, at which considerable numbers of the inferior classes are instructed; and of a still greater number of parish schools. The total number of scholars in these four descriptions of schools may vary from 65,000 to 75,000. Though the sons of the clergy generally follow the profession of their fathers, this is not always the case. Occasionally they enter the civil service, and some of the most celebrated statesmen, historians, poets, &c., of Russia have sprung from this class. (*Dr. Pinkerton's Russia*, p. 251; *Schmitzler, &c.*)

Elementary instruction is in the most depressed state. According to the ukase of 1802, a grammar school should be established in every district, and an elementary school should be established in every parish, or at least in every two parishes, according to the population. But these regulations have, in very many instances, not been complied with; and when it is considered that the advantages of education are but little appreciated by the peasantry, and that it is frequently discouraged by the nobility, to whom it occasions some expense, it will not appear surprising that such should be the case. In despite, however, of every obstacle, education has made and is making a very considerable progress. This is seen from the following statement of the schools in existence in 1804, 1824, and 1835:

	1804		1824		1835	
		Pupils.		Pupils.		Pupils.
schools under the Minister of Public Instruction	498	36,481	1,411	109,699	1,661	95,707
Military schools	15	29,000	177	102,595	152	179,580
Ecclesiastical do.	100	15,000	544	36,000	601	67,484
special do.	13	31,775	4	41,300	1,502	157,954
Totals	647	108,56	2,118	263,444	3,966	560,575

Of the total schools existing in 1835, 2841 were maintained at the expense of government. Among the scholars, no fewer than 239,311 enjoyed bursaries, or were educated gratis. The total expense incurred that year by government for school purposes amounted to 28,734,141 roubles. In 1835 the number of schools has been considerably increased: and if we add to the pupils at school those receiving instruction at home, it will be seen that education made a rapid progress in Russia; and that though very backward, especially in the country districts, it is far more generally diffused than is generally supposed.

Since the epoch of the Polish insurrection, the government of Russia has discovered considerable jealousy with respect to education. In consequence all Russian subjects have been forbidden from studying at any foreign university. A strict surveillance is exercised over all descriptions of schools; no private schools are in future to be opened without permission from the proper authorities, and all masters and mistresses of such seminaries must be native Russi; and it is farther ordered, that no one shall be a teacher in a private family without being accredited by a university, and having a certificate of capacity and good conduct. The sciences principally taught in the universities are the history, literature, geography, and statistics of Russia.

Lectures on politics or political economy would be esteemed dangerous, and are forbidden. The object of these regulations is manifest; and we are not surprised that some of should have been adopted. But, whatever may be the case with the higher branches, the government has had to perceive that the diffusion of elementary instruction, including the principles of the useful arts, would not to shake the stability of the existing order of things, it would do more than anything else to raise the country from the state of ignorance and debasement in which we now find them, and to developed the resources of the country. Great numbers of new schools have been opened within the last half dozen years; and lectures on agriculture, and the application of science to art, have been established in the different universities.

The fact is, that a taste for instruction and reading is beginning to be widely diffused among the town population. Many new works, some of them of distinguished merit, annually appear; and many foreign works are translated into Russian. Numerous literary and scientific journals issue from the presses of Petersburg, Moscow, Riga, Odessa, &c.

The journals of a political cast being subjected to a severe censorship, afford no means of judging of the prevalent political feelings of the mass of their readers. All works and journals imported from abroad must also be submitted, under heavy penalties, to the inspection of the censors. This jealousy of whatever might tend to expand the minds of the people, and to make them acquainted with their rights and duties, is the grand obstacle to the civilisation of the higher classes and burghers. But, notwithstanding the censorship, about 350,000 volumes of foreign works, principally French and German, are annually imported.

Professors and teachers in universities and other seminaries obtain rank, and rise according to merit and seniority; and when unfit for the active duties of their station, they receive a small pension. But their regular salaries, like those of almost all other Russian functionaries, are totally inadequate to support their rank and station in society. A professor has only about £80 a year, and his assistant £32; a professor of languages has £34, and each of the students, supported by the crown, £8 a year! The institutions of Moscow and Tsarskoi-Zelo are the most distinguished of those intended for the instruction of nobles. In the first there are about 300 boarders, paid for by their parents; the rest attending as day-scholars. The first class pay £38 a year, and the second £24. (*Pinkerton's Russia*, p. 256.)

The pedagogical institution of Petersburg is one of the most important and valuable in the empire. It is exclusively appropriated to the education and training of school-masters. In 1835 it had 45 functionaries and 144 pupils.

Young men belonging to great families used formerly to be, for the most part, educated at home by foreign tutors, of whom not a few were ignorant, unprincipled, and servile. But an edict of the emperor Alexander contributed to subvert this practice, by excluding all young men, not educated at a public seminary, from the higher class of public employments; and, as already seen, it has been still farther discouraged by the present emperor.

Races.—The Russian empire embraces at present an immense variety of different races; but the great bulk of the nation, or the Russians properly so called, with the Poles, and also the Bulgarians and Servians, belong to the great Slavonic family. The Slavonians are most generally supposed to be the descendants of the *Sarmatæ* of antiquity; but, though probable, this is by no means certain. At all events, they are radically distinct from the Goths, on the one hand, and from the Tartars and other Eastern nations, on the other. There is no foundation whatever for the common opinion that they were denominated Slavonians from their being originally slaves. On the contrary, when first known to history, the Slavonians were as free as the Goths. The practice of slavery was gradually introduced; and in Russia it was not completed till the beginning of last century. (See Dr. Pinkerton's *Russia*, p. 276, and the authorities there referred to.) The individuals of Slavonic origin at present within the limits of the Russian empire are supposed to amount to about 46,000,000, being about three fourths of the entire population.

The next principal race is that of the Ouralians or Finns, inhabiting the grand duchy of Finland, Esthonia, Lapland, and several districts in the north of the empire. The Finnish population is believed to amount, in all, to above 3,000,000 individuals. The Letto-Lithuanian race, amounting to nearly 2,000,000, is principally found in Lithuania and the W. provinces. There are, also, about 2,000,000 Tartars; 2,000,000 Georgians, Armenians, &c. The Germans settled in various parts of the empire may be taken at about 450,000; and there are, besides, 1,064,000 Jews, with Samoyedes, Mongolians, Kamchatkadales, Americans, &c.

Language.—Those who are aware of the various races and the numbers of different people included within the Russian empire, need not be surprised that about *forty* distinct languages are in use, having attached to them an immense number of different dialects. The individuals belonging to the Slavonic race have two languages—the Russian and the Polish, both derived from the ancient Slavonic. This mother-tongue, augmented and modified by the influence of Christianity, which introduced into it a number of Greek words, and by the dominion of the Tartars, by whom it was loaded with Turkish and Mongolian terms, was gradually formed into the Russian. The primitive idiom continued, however, to be employed in the liturgy and the sciences till the reign of Peter the Great, when the Russian gained that ascendancy in religion and science it had already gained in conversation. The extraordinary advances that were then made in civilization occasioned the introduction of an immense number of new words. At length the language became tolerably well fixed. The alphabet, which consisted originally of 45, has been reduced to 37 letters, some of them borrowed from the Greek and others from the Latin. Some characters are, however, quite unlike those of any other language, and can hardly be pronounced by any save Slavonians. The grammatical forms are not well defined, and the conjugations are exceedingly irregular. Otherwise the language is rich, sonorous, flexible, natural, and elegant. The variety of its terminations is very remarkable. There is very little *patois* in Russia; the language of the country differing but little from that of the towns. There are, however, three principal dialects; 1st, the *Great-Russian*, the pure or cultivated language of the nation, spoken in Moscow, and all the central parts of European Russian; 2d, the *Malo-Russian*, or language of the S.E. parts of Russia in Europe; and 3d, the *White-Russian*, or dialect spoken in Lithuania. Volhynia, &c. It is a curious fact, that the first grammar of the Russian language appeared at Oxford in 1696. The best grammar is that of Dobrowski, published at Vienna in 1822. The Russian academy has published a dictionary of the language in 6 vols., 4to., 1806-1822.

Literature.—Russia has had several distinguished natural philosophers and mathematicians, but they have been chiefly foreigners (Germans principally) resident in the country. At present the literature of Russia occupies a respectable place in that of Europe. The introduction of Christianity was marked by the growth of a taste for letters among the ancient Slavonians; but the only remains of that early literature are some fragments of chivalrous poetry, and the annals of the monk Nestor. The Tartar invasion arrested the progress of literature, and Russia fell back into the abyss of barbarism, whence she did not begin to emerge till after the accession of the house of Romanoff. The attempts of the restorers of literature were at first confined to some feeble dramatic performances; and towards the close of the 17th and the beginning of the 18th centuries to miserable imitations of French and other foreign works. In the course of the 18th century, however, Lomonosoff created, by his precepts and his example, a national literature. Soumarokoff carried the drama to a high degree of perfection, and since then a crowd of writers have distinguished themselves in all departments, from the *epopeia* down to eclogue and fable; and the national literature continues to flourish with undiminished vigour. The History of Russia, by Karamzin, now in course of publication, is a work of great merit. Numerous journals or periodical publications, in different languages, devoted to politics, literature, and science, appear in different parts of the empire; but so long as these are subjected to a severe censorship, and as the government looks with jealousy on anything approaching to the expression of a free opinion, the political and philosophical works of Russian writers can be but little deserving of attention. The first Russian press was set up at Kieff in 1551. Previously to 1800, there had not been printed above 1000 works in Russia; in 1807, the number of such works was about 4000; in 1821, they amounted to 13,249, and at present (1840) to more than double this number, about a third part being translations from the French and other foreign languages.

Russia has some splendid libraries and museums. The imperial library at Petersburg contains about 400,000 vols. and 17,000 manuscripts; and the Romantzow museum contains a large collection of national antiquities and of every kind of curiosities.

Religion.—Most religions are to be found in the ancient continent have their adherents in Russia. The court, however, and the great body of the nation profess the Russo-Greek Christian faith, denominated by its votaries the orthodox or true Catholic faith. The points in which it principally differs from the Roman Catholic faith, are, its denying the spiritual supremacy of the pope, its prohibiting the celibacy of the clergy, and its authorising all individuals to read and study the Scriptures in their vernacular tongue. The prohibition of celibacy is carried to such an extent, that no priest can perform any spiritual function before he is married nor after he becomes a widower; and as he is not allowed to remarry, the death of his wife and the cessation of his functions as a priest (unless he be specially allowed by the bishop to continue them) are necessarily identical! The priests may, however, on the death of their wives, enter into a convent, and enjoy the barren privilege of becoming eligible to be dignitaries of the church. Pictures of saints are admitted into the Russo-Greek churches and houses; but all statues, bas-reliefs, &c., are rigidly excluded. There are several fasts, of which that of Easter, which

continues for *seven* weeks, is the longest and strictest. Divine service is performed in the native tongue; and singing in churches is unaccompanied by any sort of instrumental music. The total population professing the Russo-Greek faith may be estimated at about 36,000,000. No country in Europe possesses such a number of fine churches as Russia. The meanest village is generally furnished with a temple ornamented with gilt domes and spires. These edifices are nearly all in the Grecian style of architecture, substantially built of brick, plastered and painted with much taste, forming a striking contrast to the huts or izbas of the peasantry by which they are surrounded.

There are in all Russia nearly 500 cathedrals and about 29,000 churches attached to the established faith, the latter employing about 70,000 secular or parochial clergy. There are also about 550 convents, of which 450 are for men and 70 for women. Adjoining to each church, or near it, there is always a *kolokolnia* or belfry, commonly of great height, and provided with large bells, which are tolled several times during every service, and on holydays not ringing the whole day. The Russians are passionately fond of the sound of bells, and larger and finer ones are nowhere to be found; every church has in its steeple four or five of different sizes; and in many this number is doubled and even trebled. (*Pinkerton's Russia*, p. 298, &c.)

The Russian church was long subordinate to that of the Eastern empire, its metropolitan being nominated by the patriarch of Constantinople. But after the capture of the latter city by the Turks in 1453, the Russian clergy appointed their own metropolitan. This practice continued till the reign of Peter the Great, who declared himself the head of the Russo-Greek church, appointing, at the same time, a synod for the management of its affairs. The clergy are either secular or regular—the former consisting of the parochial clergy, and the latter of the higher dignitaries, monks, &c. The hierarchy is composed of bishops, archbishops, and metropolitans. There are in all thirty-eight dioceses.

In Russia, as in most other countries, the piety, or superstition of individuals, had conferred great wealth on the church, particularly on the monasteries. This having occasioned many abuses and irregularities, afforded a pretext, of which Peter the Great availed himself, not only to suppress various monasteries, but to deprive the church of the greater part of its wealth. In the reign of Catherine II. the degradation of the clergy was completed by the appropriation of the whole immoveable property of the church to the use of the state, pensions being assigned, in its stead, to the different functionaries to whom it had belonged. But with the exception of a few livings in Petersburg, Moscow, and other principal cities, the stipends of the clergy, even when increased by the offerings of the people, and by the perquisites on occasion of births, marriages, funerals, &c., are quite inadequate to provide for their comfortable subsistence. The total number of established clergy, of all ranks and orders, may be taken at about 254,000; and the sum allowed as stipends by government is so very small, that they are almost wholly dependent on their flocks. The revenue even of the senior metropolitan, the highest dignitary in the hierarchy, did not recently exceed £800 or £700 a year; an archimandrite, or abbot, the class next below a bishop, had not generally more than from £40 to £50 a year! (*Pinkerton's Russia*, p. 96.) Mr. Coxe says, that "besides the surplice fee, which is the poorest benefice amounts to £4 a year, and in the most profitable to but £20, they have only a wooden house, scarcely superior to that of the meanest among their parishioners, and a small portion of land, which they generally cultivate with their own hands; while the highest dignity to which they can ever attain, so long as they continue married, is that of a prototype of a cathedral, whose income scarcely exceeds £90 a year." (*Travels in Russia, &c.*, ii. 142.)

When such is the depressed condition, we need not wonder at the low state of learning and want of refinement among the great bulk of the Russian parochial clergy. Coxe tells us, that when he was in Russia, many of the parish priests were so miserably ignorant as to be unable to read, even in their own language, the gospel they were commissioned to preach! But, though still very far behind, such gross ignorance is now much less common than formerly. The duties of the Russian clergy of all orders are very laborious. Dr. Pinkerton, whose authority is unquestionable, says, that we find in the family circles of the secular or parochial clergy, a degree of culture and good manners peculiar to themselves. This description of clergymen wear long beards, and form, in fact, like the priests of old, a kind of distinct class, or caste. None but the sons of clergymen are educated for the church; nor is there one instance in a thousand of any one belonging to any other class entering the ranks of the secular clergy. The regular, or dignified clergy, on the contrary, though often the sons of priests, not unfrequently receive recruits from among the nobles and other classes; and all the higher stations in the church continue to be filled up from their ranks. (*Pinkerton's Russia*, p. 250.) Orders, and other marks of distinction are conferred on the Russian clergy; and at present a bishop is little thought of unless he be decorated with the star and riband of some order of knighthood.

The Russo-Greek church has, from an early period, had its schisms and dissenters. The latter are said to be split into about 70 sects. They are classed under the common denomination of *raskolniks*. The ritual, or service of the Russian church is contained in *twenty volumes folio*, in the Slavonic tongue!

With the exception of the restraints laid on the Jews, who are excluded from Russia Proper, almost all religions may be freely professed anywhere in the empire. No member of the Russo-Greek church is, however, permitted to renounce his religion; and when a marriage takes place between one of its members and a person belonging to another faith, the children must all be brought up in the established faith. Catholics are very numerous in the Polish provinces; there are, also, large numbers of Lutherans, chiefly in the Baltic provinces, with Mohammedans, Jews, worshippers of the Grand Lama, Feticists, &c.

Finances.—Owing to the low state of civilization in most parts of the Russian empire, and the want of manufactures and great towns, the public revenue is by no means so great as might be supposed from the vast extent of the empire, and the magnitude of the pop. In consequence, however, of the cheapness of most necessary articles in Russia, and the small rates of pay of the soldiers and other public functionaries, her limited revenue goes a great way, and she is able to meet outgoings that elsewhere could not be met with less than twice or three times the sum.

Most topics connected with the public revenue and expenditure are involved in a mystery which it is not always easible to penetrate. The former is derived from a few sources, consisting of,

1. The capitation tax, charged on all the male boors belonging to individuals, and also on some descriptions of semen. At an average it may be estimated at about roubles a head; and, estimating those subjected to it at 000,000, it will give 40,000,000 roubles.

2. The *Obrok*, or rent paid by all male boors on the own estates. Estimating this tax at 10 roubles, and we subjected to it at 9,000,000, it will give a sum of 900,000.

3. The tax of 1½ per cent. on the declared capital of the merchants. The amount of this tax may be taken at 10,000.

4. The customs' duties in 1839 produced very near 00,000 gross revenue. The expenses of collection amount to about 7½ per cent. of this sum.

The spirit duties produce a very large sum. In Russia properly so called, government reserves to itself a monopoly of distillation; but in the rest of the empire the issue of the distilleries, which any one may construct, is subjected to an excise duty. The consumption of spirits very large; and the revenue thence arising may be estimated at 100,000,000. Among their other privileges the clergy may distil all the spirits required for their establishments free of duty.

The salt mines and brine springs are monopolized by government, which sells their produce at the rate of a ruble per pood. This monopoly is supposed to produce about 10,000,000 a year.

The crown mines, and the duties payable by the proprietors of private mines, produce together about 16,000,000. The seignorage on coin may be taken at 8,000,000.

Stamps, licenses, &c., and the tax laid on the sale of revenable property, may be estimated at 7,000,000.

Miscellaneous items, such as the sums paid by the serfs to be exempted from furnishing recruits for the army, the rent of crown property let on lease, the profits of manufactures, &c., may be taken at 8,000,000.

Hence it may be concluded that the public revenue of the Russian empire amounts, in all, to 379,000,000 or 20,000,000 roubles a year, that is, to about £15,836,000.

The taxes are partly farmed, and partly collected by government officers. There is, as already stated, in every government, a council charged with the administration of things relative to the finances.

Information with respect to the expenditure is even less accurate than that respecting the income of Russia. In time of peace, however, they are understood to be nearly but during war, or on extraordinary occasions, large an increase of expenditure, the ordinary revenue is insufficient to meet the outgoings, and it is usual both to raise the rate of taxation and to resort to loans. The cost of the army and navy (the latter being about one fourth part of the former) amounts to more than revenue. The next great items are the interest and fund on account of the public debt; the civil list,

internal administration, public works, &c.; the diplomatic service, and various other items.

According to the report of the minister, M. Kankrin, the public debt amounted, in 1837, to 956,333,574 roubles.

Historical Sketch.—The ancients had very little acquaintance with the vast countries included in the empire of Russia. The monarchy is usually regarded as having been founded by Rurik about anno 862, his dominions, and those of his immediate successors, comprising Novgorod, Kieff, and the surrounding country. In 980-1015 Vladimir introduced Christianity, and founded several cities and schools. But, from this period down to 1237, when the country was overrun by the Tartars, Russia, with few exceptions, was the theatre of civil war. In 1328 the seat of government was transferred to Moscow; and in 1481 the Tartars were finally expelled. In 1613 the house of Romanoff, whence his present majesty is descended, was raised to the throne; and from this period the empire acquired strength and consistency. Under Alexis Mikhailovitch (1645-1676) White Russia and Little Russia were conquered from the Poles, and the Cossacks of the Ukraine acknowledged the supremacy of the Czar, various internal improvements were effected, and the power of Russia began to be felt and feared by all her neighbours. At length, in 1696, Peter the Great ascended the throne, and the destinies of Russia and of the northern world were immediately changed. This prince, who has, probably, a better claim than any other that ever existed to the epithets of "great" and of "father of his country," gave to the arms of Russia a decided preponderance in the north of Europe; he also gave her a fleet; conquered large provinces on the Baltic; laid the foundations of the noble city which bears his name; and introduced among his people the arts, the literature, the customs, and, to some extent also, the laws and institutions of the more civilized European nations. The difficulties he had to encounter in his projects for remodelling and civilizing his dominions were of the most formidable description; and could not have been overcome by any one possessed of less authority, or of a less stern decided character.

From this period Russia has progressively advanced in power and civilization. Under Catherine II. (1762-1796), a princess of extraordinary talent, Russia acquired a vast accession of power by her acquisitions in Poland and on the Black sea, where she has now the same ascendancy as in Baltic. The history of Russia, during the present century, is known to everybody. The attempt of Napoleon to dictate a peace to the emperor Alexander, in the ancient capital of the Czars, led to the overthrow of his colossal power, and gave a vast accession of influence and consideration to Russia; which has been maintained and extended under the present emperor.

It would be idle to speculate upon the permanency of the present order of things in Russia. A great deal, in such an empire, depends on the personal character of the sovereign. The present occupier of the throne has every quality—good sense, undaunted courage, great decision, and the utmost vigilance and activity—required in the ruler of such a country. But should the government fall into less able and skilful hands, it is not improbable that Russia may become the theatre of revolution and change, for which, at present, she certainly is not fitted.

The following table gives a view of the extent of the Russian dominions at different epochs:

	Germ. sq. m.
In 1535, at the accession of Joha the Terrible, his dominions comprised	37,900
— 1585, at his death	144,000
— 1613, at the accession of Michael Romanoff	148,000
— 1645, at his death	258,000
— 1725, at the death of Peter the Great	280,000
— 1741, at the accession of Elizabeth	325,000
— 1796, at the death of Catherine II.	336,000
And at present (1841)	364,000

Tables similar to this have been the theme of much silly declamation about the grasping, insatiable ambition of Russia. No doubt her rulers have the same desire to extend her territories as those of France, England, or any other power; but certainly they are not, in this respect, at all peculiar. In point of fact, however, by far the greater part of the territorial acquisitions of Russia have consisted of mere deserts, or of countries occupied by roving barbarians, and are worth little or nothing. Her really valuable acquisitions have been confined to those on the side of Poland and the Black sea. Her conquests in this direction have added materially to her power; and it is but fair to add, that they have also added materially to the well-being and civilization of the inhabitants.

RUSSIA, t., Loraine co., O. It has one store, one gristmill, one saw-mill, one printing-office, one bindery, one semi-weekly newspaper; the Oberlin Collegiate Institute with 498 students, three schools, 101 scholars. Pop. 1302.

RUSSIA, p. t., Herkimer co., N. Y., 16 m. N. Herkimer,

93 m. N.W. Albany, 408 W. Watered by West Canada creek and its branches. It contains a church common to Baptists and Presbyterians, eight stores, two lumber-yards, three fulling-mills, two cotton factories with 1004 spindles, two forges, four grist-mills, 11 saw-mills, four tanneries; 14 schools, 619 scholars. Pop. 2298.

RUSTCHUK, or RUTZSCHUK, a fortified city of Turkey-in-Europe, prov. Bulgaria, cap. Sanjak, on the Danube, 56 m. E. by N. Nicopolis, and 62 m. N.W. Shumla. Pop. variously estimated, but may probably be about 30,000. It is built on a steep bank, up which the streets ascend from the river. It is surrounded on three sides by walls, in the manner of Turkish fortifications; but towards the river it is partly open. At its N.E. extremity is a ruined citadel, on an abrupt height above the Danube. "The description already given of Belgrade applies, with very little modification, to this Bulgarian city, except that Rustchuk is not in such a state of dilapidation, and the Turks here appear more civilized than the Servians: they have schools for their boys; and several of their houses are furnished with glass windows. The comfort of fountains, simple as they are in exterior, and the luxury of coffee-houses, are not unknown to the Bulgarians; on the contrary, the one and the other abound in Rustchuk." (*Elliot's Trav.*, i. 180.) The streets are narrow, and gloomy; on either side they present only dead walls; and as in all the rest of Bulgaria and in Roumelia, each of the larger houses is a fortress in itself. The governor's palace, some of the mosques, and some public baths are the only edifices worth notice. Many of the buildings are white-washed, and their tall chimneys are visible at a great distance. Woollen, silk, and cotton stuffs are made here; and there are many Turkish, Greek, and Armenian merchants in the town, who carry on considerable trade with Vienna, Wallachia, &c., in cloth, corn, and indigo. Mr. Quin observed in the bazars of Rustchuk, "a rich display of military saddles and bridles, belts, and cartouche-boxes, gayly ornamented; of Persian carpets, Broussa silks, ataghans, pistols, pipes, umbrellas, Greek caps, scarlet jackets, yellow slippers, gold-headed canes, fine cloths, woollen and cotton stockings, and every article of grocery, fruits, vegetables, meat, fowl, fish, hardware, and jewellery. The floors of these shops were usually elevated above the level of the street, and the owners and their assistants sat inside upon the floors, some working as tailors, some as saddlers, and artisans of the ordinary trades." (*Voyage down the Danube*, 314–15.) In 1812, the Russians took and burned the citadel and a part of the town; and, in 1829, they entered the town after little opposition. (*Elliot's Trav.*, i. 178–182; *Quin's Voyage*, 288–315; *Slade's Germany and Russia*, 187–189.)

RUTHERFORD, county, N.C. Situated in the S.W. part of the state, and contains 1025 sq. m. Drained by head branches of Broad river. It contained in 1840, 14,086 neat cattle, 17,784 sheep, 22,059 swine; and produced 76,663 bushels of wheat, 6030 of rye, 1,090,388 of Indian corn, 1000 of barley, 57,955 of oats, 16,509 of potatoes, 4653 pounds of tobacco, 250,325 of cotton. It had 11 stores, two flouring-mills, 53 grist-mills, 16 saw-mills, two oil-mills, five tanneries, 15 distilleries, one printing-office; eight schools, 149 scholars. Pop.: whites, 15,875; slaves, 2901; free-coloured, 195; total, 18,902. Capital, Rutherfordton.

RUTHERFORD, county, Tenn. Situated near the centre of the state, and contains 540 sq. m. Drained by Stone's creek and its branches, flowing into Cumberland river. It contained in 1840, 25,295 neat cattle, 35,625 sheep, 81,430 swine; and produced 147,500 bushels of wheat, 10,000 of rye, 3,291,000 of Indian corn, 222,400 of oats, 114,666 of potatoes, 1,069,000 pounds of tobacco, 2,690,000 of cotton, 1200 of sugar. It had 90 stores, five woollen factories, one cotton factory with 612 spindles, five flouring-mills, 30 grist-mills, 25 saw-mills, 12 tanneries, 20 distilleries, two printing-offices, two weekly newspapers; five academies, 244 students, 24 schools, 633 scholars. Pop.: whites, 15,043; slaves 9072; free coloured, 166; total, 24,280. Capital, Murfreesboro'.

RUTHERFORDTON, p. v., capital of Rutherford co., N.C., 216 W. by S. Raleigh, 453 W. It contains a court-house, jail, an academy, several stores, and about 300 inhabitants.

RUTHERGLEN (pronounced Ruglen), a royal and parl. bor. of Scotland, co. Lanark, on the left bank of the Clyde, 2½ m. S.E. of Glasgow. Pop., in 1831, 4741. It consists of one leading street, straight and well paved, nearly ½ m. in length, 112 ft. broad, and of the parallel lane called the Back Row. From both sides of the main street, which lies in a direction nearly E. and W., go off a few cross lanes. There are no public buildings except the parish church, a *quoad sacra* place of worship connected with the establishment, a chapel belonging to the Relief, and the town-hall. A small cotton-mill employs about 80 hands; and there are two print-fields in the vicinity, a Turkey red dye work, and a chemical work. About 500 hand-loom muslin weavers are employed by Glasgow manufacturers.

Rutherglen was created a royal bor. in 1126, at which time it was of more importance than Glasgow, the latter being included within its municipal boundaries. But in 1226, Alexander II. granted a charter to Walter, bishop of Glasgow, relieving his town from certain servitudes previously due to Rutherglen. (*Mun. Corp. Rep.*, 1835, part i. p. 371.) Previously to the passing of the Reform Bill in 1832, Rutherglen, unimportant as it has been, enjoyed the same parliamentary privileges as Glasgow; being united with it and two other towns in sending one member to the House of C. Glasgow has since that period had two representatives for itself; while Rutherglen joins with Kilmarnock and three other bors. in choosing a mem. Registered voters, in 1840, 1170. Municipal councillors, 18; corporation revenue, £700. (*New Stat. Acc. of Scotland, sc. Lanarkshire*, p. 373–398; *Ure's Hist. of Rutherglen*, 1793.)

RUTHIN (Welsh *Rhudd-ddyn*, or *Rhuthyn*, the "Redfortress"), a parl. and mun. bor. market-town, and par. of N. Wales, co. Denbigh, hund. Ruthin, in the vale of Clwyd, 17 m. W.S.W. Chester, and 45 m. N.W. Shrewsbury. Pop. of mun. and parl. bor., in 1831, 3,376. The town, situated on rising ground, chiefly E. of the Clwyd, consists of a principal avenue, entered by several other inferior streets. At the summit is the market-place, in which is the town-hall, a substantial building, erected in 1663, but recently repaired. The county hall, a fine modern stone structure, forms with the jail an interesting feature in the town. The jail, recently enlarged by the addition of a building for female prisoners, has suitable arrangements for the classification of the inmates. The church, originally conventual, belonging to a community of Bonhommes, is an ancient structure of mixed architecture; the tower, S. and W. fronts, being comparatively modern, and much inferior to the rest of the building. The interior is elegantly fitted up with stalls, tabernacle-work, &c., and the roof is of carved oak, pannelled, and richly sculptured. It was made collegiate in 1310 by John de Grey, who formed an establishment for regular canons, and endowed it with valuable lands and numerous privileges. A part of the cloisters has been converted into a residence for the warden of Christ's hospital, founded here by Dr. Gabriel Goodman, for the support of 12 decayed housekeepers. The warden and pensioners are impropriators of the great tithes of Ruthin and Llan-Rhydd; and the warden, who is appointed by the dean and chapter of Westminster, is the vicar of both parishes, with an income of £263 a year. The free grammar-school is endowed with a moiety of the tithes of Llan-Elidan, and is under the superintendence of the warden, though the appointment of the head-master rests with Jesus college, Oxford: the school ranks as one of the best in N. Wales, and is attended by about 50 boys. A national school is established here for 40 children of each sex; and there are Sunday-schools attached to the chapels of the Independents, Wesleyan and Calvinist Methodists. No particular trade or manufacture is carried on in the town, exclusive of what is necessary for the accommodation of the inhabs., who are principally employed in agriculture; and no change of any consequence appears to have taken place in it for several years, neither are any causes in operation or likely to operate towards the extension of the town beyond its municipal limits. (*Mun. Bound. Rep.*)

The corporation of Ruthin consists, according to the Mun. Reform Act, of a mayor, three aldermen, and 12 councillors. Corp. rev., in 1839, £145. The bor. unites with Holt, Wrexham, and Denbigh in sending one mem. to the H. of C., the right of election down to the Reform Act having been vested in the resident burgesses. The parl. limits, according to the Boundary Act, include with Ruthin, parts of the pars. of Llan-Fwrog, Llan-Rhydd, and Llan-Fair-Dyffryn-Clwyd. Reg. electors, in 1839–40, for the united bors., 941. Ruthin is likewise one of the polling places at elections for the co., and the assize town; besides which the quarter sessions are held alternately here and at Denbigh. Markets well supplied with corn, on Monday, and a provision market on Saturday. Fairs, March 20, Friday before Whit Sunday, Aug. 6, Sept. 30, and Nov. 19.

Ruthin, according to the Welsh historians, is of high antiquity; but we have no authentic information respecting it prior to the reign of Edw. I., who built here a magnificent castle, overlooking the Clwyd, on its W. bank, which he presented, in 1281, to Reginald de Grey. The ruins have been restored with admirable taste by the present proprietor. The manor of Ruthin belongs to the Lady Grey de Ruthin, daughter of the 19th baron of that name, who has been a great benefactress to the town.

RUTLAND, an inland co. of England, surrounded by Lincoln, Leicester, and Northampton. It is the smallest of the English cos., containing only 95,360 acres, of which about 9000 are arable, meadow, and pasture. Surface gene-

734

; and the soil almost everywhere loa-
e W. part of the county, in which is the
celebrated by Drayton, is under grass, and
llage. It is particularly celebrated for its
d sheep. Estates and farms of various
ent of land, in 1810, 20s. 9½d. an acre.
d runs along its S.E. border, from Rock-
amford; and there is a canal from the river
he principal town. It is divided into five
arishes; and returns two members to the
s, both for the county. Registered electors,
n 1831 Rutland had 3935 inhabited houses,
19,385 inhabitants, of whom 9721 were
females. Sum expended for the relief
9–40, £7946. Annual value of real prop-
3,216: profits of trades and professions in

ty, Vt. Situated toward the S. part of
ains 958 sq. m. Watered by Otter creek,
eechee, and Pawlet rivers. The Green
uses through its E. part. It contains iron
marble. There were, in 1840, 40,023 neat
eep, 15,563 swine; and were produced
wheat, 38,013 of rye, 154,792 of Indian
kwheat, 154,119 of oats, 642,106 of pota-
of silk cocoons, 396,804 of sugar. It had
aces, five forges, 11 fulling-mills, eight
ne cotton factory with 1076 spindles, one
st mills, 103 saw-mills, two oil-mills, one
neries, two printing-offices, one weekly
academies, 362 students; 949 schools,
Pop. 30,699. Capital, Rutland.
capital of Rutland co., Vt., 62 m. S.W.
. Drained by Otter creek and its branch-
ch of Castleton river. It contains five
ngregational, an Episcopal, a Methodist,
ores, one woollen factory, two grist-mills,
e paper-mill, one tannery, one printing-
one weekly newspaper; 16 schools, 963
8. The principal village has an elevated
ains a courthouse, jail, two churches, 12
office, about 100 dwellings, many of them
out 600 inhabitants.
Worcester co., Mass., 55 m. W. by N.
ncorporated in 1722. A branch of Ware
-power. It is on the dividing ridge be-
f Connecticut and Merrimac rivers. It
ational church, a town-house; one acad-
13 schools, 712 scholars. Pop. 1960.
Jefferson co., N. Y., 6 m. E. Watertown,
eany, 422 W. Drained by Sandy creek
it has the remains of an Indian fortifica-
four stores, two fulling-mills, one grist-
mills, three tanneries; 11 schools, 253
0.
seaport, and watering-place of England,
he Isle of Wight, par. of Newchurch, and
liberty, 5½ m. E.N.E. Newport. Pop.
1, 4929, of which about 3800 belonged to
The latter has a peculiarly handsome
ortsmouth, opposite to which it lies, at a
m.; its white houses, interspersed with
ations, being ranged in successive rows
steep acclivity, rising directly from the
i hills, and surrounded with cultivated
of a principal street, running upward
d intersected by others, some of which
e detached residences, well adapted for
iers who flock thither during summer.
s-of-ease, one of which is a handsome
places of worship for Wesleyan Metho-
. Independents, with attached Sunday-
ndowed charity school; it has also a
embly-rooms, libraries, baths, &c., and
The accommodations for summer visiters
mproved within the last few years, since
to favour; and a pier, constructed on
¼ m. into the sea, making the town ac-
at all times of the tide. Steamboats run
r in the day, during summer, between
, Cowes, and Southampton; and the
uth-western railway has occasioned a
tern to this and its rival watering-place.
air is salubrious, and in the environs are
walks commanding fine land and sea
hing place, Ryde is inferior to Southsea,
ast of Hampshire, or to Cowes, on ac-
wness of the water for a lengthened dis-
It has no manufactures, and its trade
ined to the supply of visiters and the

l mun. bor., cinque port, market town and

par. of England, co. Sussex, hund. Godstow, rape Hastings,
on the Rother, about 3 m. from its mouth, 36 m. E. by N.
Brighton, and 53 m. S.S.E. London. Pop. of parl. borough
(which comprises, with the par. of Rye, that of Winchelsea and six others, with a portion of the par. of Brede), in
1831, 7700. The town, which stands on the edge of an extensive tract of marsh land, running along the coast as
far as Hythe, consists of several regular and well-formed streets, lined with old, but respectable-looking houses,
many of which command fine views of the channel and
surrounding country. In the centre of the town is the market-house and town-hall, with an old structure called the
Ypres tower, occasionally used as a jail. A public library is supported by subscription, and there is a small theatre. The church is a large cruciform structure, with a
central tower, partly of Norman and partly of early English
architecture: the aisles of the choir have fine lancet windows, and there is a large and fine perpendicular E. window. The living is a vicarage, in the gift of the earl of
Burlington. The Wesleyan Methodists, Independents, and
Baptists have their respective places of worship, with attached Sunday-schools; and it has a small endowed grammar-school, an English school, with several almshouses
and other charities. "There is no manufactory in the
town, which depends upon being the port and market
through which the surrounding rich agricultural district is
supplied. Some years since it is said to have been in a declining condition; the reverse, however, is now the case,
and it appears to be, indeed, in an eminently thriving state.
This is mainly attributable to the improvements lately
made in the harbour and in the navigation of the river,
which have made the town accessible to vessels of 200
tons. By means of the Rother and its branches, it supplies
the surrounding country to a distance of 8 m. with coals
and other articles; and there is a project in contemplation
to extend the navigation as far as Robert's bridge, a distance of 15 m. in a straight line. This, if carried into execution, will be a source of considerable advantage to the
town, which will then become the port through which the
country, as far as Battle, will be supplied." About 90 vessels, of the aggregate burden of nearly 5000 tons, belong to
the port; and in 1840 the gross customs' tons duties amounted to £6790. The exports are chiefly wool, oak timber, and
bark. A considerable quantity of hops is raised in the
neighbourhood, for the drying of which large quantities of
Welsh coal are annually imported. (Mun. Bound. and
Corp. Rep.)
Rye is a borough by prescription, and is governed, under the
Municipal Reform Act, by a mayor, three other aldermen, and
12 councillors, styled "the mayor, jurats, and communality
of the ancient town of Rye." Corp. revenue in 1839, £499,
exclusive of £295 accruing from the sale of property. Courts
of session are held under a recorder; and there is a court
for the recovery of debts under 40s. Rye returned two members to the House of Commons from the 42 Edw. III. down to
the Reform Act, which deprived it of one of its members,
and at the same time enlarged the electoral limits, so as to
comprise with the old borough the rest of the parish of Rye,
the town and parish of Winchelsea, with six other entire
parishes, and a small portion of the parish of Brede. Registered electors in 1839–40, 594. Markets on Wednesday and
Saturday; fairs, Whit-Monday and August 10th.
Rye is very ancient, but its early history is little known.
It appears to have been an original cinque port, and is
mentioned as a member of these ports in a charter granted
by Henry III. In the reign of Edward III. the town was
so considerable that it sent nine armed vessels to the royal
fleet when the king undertook the invasion of France. In
the next reign, A.D. 1377, it was plundered and burnt by
the French, but it soon recovered its consequence. The
rise of other ports on the same coast during the last century, and the filling up of its port, occasioned a considerable
decrease of its importance, which, however, as above seen,
it has now a fair prospect of in some measure recovering.
RYE, p. t., Rockingham co., N. H., 6½ m. S. Portsmouth,
42 m. S.E. Concord, 479 W. Bounded S.E. by the Atlantic.
The harbour admits small vessels engaged in the fisheries.
It has three churches, a Congregational, Methodist, and
Christian, three stores, five grist-mills, three saw-mills;
one academy, 25 students; four schools, 490 scholars.
Pop. 1205.
RYE, p. t., Westchester co., N. Y., 27 m. E.N.E. New-York
city, 146 m. S. Albany, 251 W. Bounded S.E. by Long Island
sound, E. by Byram river. It contains three churches, a
Methodist, Episcopal, and Presbyterian, five stores; three
academies, 54 students; six schools, 193 scholars. Pop.
1803.
RYEGATE, p. t., Caledonia co., Vt., 43 m. E. by S. Montpelier, 520 W. Bounded E. by Connecticut river. Watered by Wells river, which affords water-power. At Canoe
falls, in Connecticut river, is a dam, which affords good
mill seats. It contains two churches, an Associate Re-

formed Presbyterian and a Reformed Presbyterian, one store, one grist-mill, five saw-mills, two tanneries; nine schools, 459 scholars. Pop. 1293. First settled in 1774, by emigrants from Scotland, from whom most of the inhabitants are descended.

S.

SABINE r., La. and Texas, rises in 32° 30′ N. lat., in Texas, and, flowing south-eastwardly, it enters Louisiana; and, from the southern part of Caddo parish, it forms the boundary between Louisiana and Texas. It is about 300 m. long, and enters the gulf of Mexico through Sabine lake, which is 30 m. long, and from 1 to 8 m. wide.

SACKETT'S HARBOUR, p. v., Houndsfield t., Jefferson co., N. Y., 174 m. N.W. Albany, 415 W. Situated on the S.W. side of Black River bay, near the foot of lake Ontario. First settled in 1802, incorporated in 1821. During the last war with Great Britain, it was a military and naval station of the Americans, and has the best harbour on the lake. It contains three churches, an Episcopal, Methodist, and Presbyterian, a ship-yard, extensive U. States barracks, a bank, 22 stores, two furnaces, one steam-engine factory, two grist-mills, three saw-mills, one large tannery, one distillery, 300 dwellings, many of them of limestone, and about 2000 inhabitants. On a point S. of the village is a light-house. A canal brought from Black river, near Watertown, affords good water-power. Tonnage in 1840, 3637 tons.

SACO r., N. H. and Me., rises in the White mountains, within a few rods of the source of Ammonoosuc river, which flows W. to Connecticut river, while Saco river flows E. through the Notch in the White mountains, and finally enters the Atlantic, between Saco and Biddeford. It has four principal falls in Maine. of 70, 20, 32, and 42 feet respectively, affording in the whole an immense water-power. Great quantities of pine timber grow on its banks.

SACO, p. t., and port of entry, York co., Me., 15 m. S.W. Portland, 65 m. S.S.W. Augusta, 100 m. N.N.E. Boston, 530 W. Bounded S. by Saco river, which affords extensive water-power, by a fall of 42 feet, within a short distance. It contains 38 stores, three cotton factories with 17,760 spindles, two grist-mills, two saw-mills, three tanneries, two printing-offices, two weekly newspapers; five academies, 246 students; 21 schools, 800 scholars. Pop. 3358. The village contains five churches. a Congregational, Episcopal, Methodist, Baptist, and Universalist, a bank, an academy, numerous stores and mills, many handsome dwellings, and considerable trade, particularly in lumber. Tonnage of the port in 1840, 3358.

SADSBURY, p. t., Chester co., Pa., 37 m. N.W. Philadelphia, 59 m. E.S.E. Harrisburg, 129 W. The postoffice location is called Sadsburyville. Drained by Octarara creek and the W. branch of Brandywine river. It contains 10 stores, three fulling-mills, two woollen factories, two cotton factories with 2217 spindles, two forges, one furnace, one flouring-mill, five grist-mills, four saw-mills; eight schools, 390 scholars. Pop. 2292.

SADSBURY, t., Lancaster co., Pa., 16 m. S.W. Lancaster. It has three stores, three forges, five flouring-mills, three grist-mills, four tanneries; eight schools, 250 scholars. Pop. 2093.

SADSBURY, t., Crawford co., Pa. It has nine stores, three grist-mills, two saw-mills, one tannery, one distillery. Pop. 2441.

SAFFI, AZAFFI, or ASFI (an. Safia), a city and seaport of Morocco, prov. Abda, on the Atlantic, near cape Cantin, and 95 m. N.W. Morocco. Pop. estimated at 12,000, including about 3000 Jews. (Gråberg af Hemso.) It is built in a sterile ravine between two hills; being very hot in summer, and disagreeable in winter, " as the waters from the neighbouring mountains, occasioned by the rains, discharge themselves through the main street into the ocean, deluging the lower apartments of the houses." (Jackson.) It has thick and high walls, and a neat palace, formerly the occasional residence of the emperor's sons; a little way N. of the town is a small fort. Its roadstead is safe in summer, but in winter, when the winds blow from the S. or S.W., vessels are obliged to run to sea; which they have been known to do several times in the course of a month, while taking in their cargoes.

Saffi was formerly an emporium of the European trade with Morocco, but its commerce has declined with the rise of Mogadore; on the foundation of which, in 1760, the emperor ordered the removal thither of all the merchants of Saffi. The Moorish and Bedouin inhabitants are fanatical and intolerant, and said to be hostile to Europeans, with whom, however, they have now little intercourse. In the environs are many Mohammedan sanctuaries. This city is supposed to have been founded by the Carthaginians. In modern times, it belonged to the Portuguese, from 1508 till

1641. (Gråberg af Hemso, Specchio di Marocco; Jackson's Morocco, &c.)

SAFFRON-WALDEN (anc. Saffron-weald-den. " the woody hill abounding with saffron"), a municipal bor., market town, and par. of England, co. Essex, and hund. Uttlesford, 23 m. N.N.W. Chelmsford, and 37 m. S.N.E. London. Pop. of mun. bor., in 1831, 4350. The town is a valley close to a tongue of high land, surmounted by the church, comprises several good streets, and a spacious market-place, with a neat town hall. Many of the houses are good, and the place generally bears an appearance of neatness and comfort. The church, which stands so high as completely to overtop the town, is an elegant structure in the perpendicular style, with an embattled tower at its W. end; the living, a vicarage in the gift of Lord Braybrooke, is worth £237 a year. The Independents, Baptist, Unitarians, and Friends, have each places of worship. A free grammar-school, endowed by Edward VI., has an exhibition at Queen's college, Cambridge; but the education is now conducted on the national plan, and connected with it is a girls' school, the whole furnishing instruction to about 250 children of both sexes. There are likewise several Sunday-schools, some almshouses lately re-built, and various other money charities. Saffron-Walden has several large malting establishments, and enjoys a good retail trade for the supply of the numerous wealthy families in its vicinity; besides which it has well attended markets for corn, cattle, and provisions. The neighbourhood is extremely productive, and well cultivated; but the growth of saffron (whence the name of the town) has been abandoned. The borough was incorporated by Edward VI., in 1549, and is governed, under the Municipal Reform Act, by a mayor, three aldermen, and 12 councillors. Corporation revenue. in 1839, £147 19s. Petty sessions for the hundred are held every alternate week, and there is a court for the recovery of debts under £10. Saffron-Walden is also one of the polling places at elections for the N. division of Essex, and the principal town of a poor law union, embracing 21 parishes. Markets on Saturday; fairs, for horses and cattle, Saturday before Midlent Sunday, and November 1st.

SAGHALIEN. See Japan.

SAG HARBOUR, p. v., port of entry, East Hampton and Southampton townships, Suffolk co., N. Y., 110 m. E. New-York, 260 m. S.S.E. Albany, 340 W. Situated toward the E. end of Long Island, on a branch of Gardiner's bay and has a good harbour. It contains four churches, a Presbyterian, Methodist, Roman Catholic, and an African, numerous stores, 450 dwellings, and about 3250 inhabitants. The inhabitants are largely engaged in the whale fishery, having a considerable number of vessels and a large amount of capital embarked in it. The tonnage of the port in 1840 was 20,405 tons. The collection district, however, embraces the bays and harbours between Mootauk point and Oyster Pond point. Salt is made by the evaporation of sea water.

SAGINAW, river and bay, Mich. The river is one of the largest in the peninsula, and is formed by the union of Cap, Flint, Shiawasse, and Tittabawassee rivers. 25 m. in a direct line from its entrance into Saginaw bay. Its length from the head of some of its branches to its mouth is about 130 m. The bay into which it enters is 60 m. long and 18 wide, setting up from lake Huron. It is navigable for vessels of the largest class, and has some of the best harbours on the lake. At the entrance of the river there is a bar with about 5 or 6 feet of water; but above that it has a depth of 25 or 30 feet.

SAGINAW, county, Mich. Situated toward the N.E. part of the settled portion of the peninsula, and contains 1681 sq. m. Saginaw bay bounds it on the N.E. Watered by Saginaw river and its tributaries. It contained in 1840 1086 neat cattle, 1462 swine; and produced 4725 bushels of wheat, 9837 of Indian corn, 9841 of oats, 16,929 of potatoes, 12,229 pounds of sugar. It had five stores, one grist-mill, six saw-mills. Pop. 892. Capital, Saginaw.

SAGINAW, p. t., capital of Saginaw co., Mich. 97 m. N by W. Detroit, 621 W. The village is on the W. bank of Saginaw river, 23 m. from its mouth. The ground on which it stands is elevated 30 feet above the level of the river. It contains a courthouse, a U. States landoffice, a deputy collector's office, three stores, two warehouses, and two steam saw-mills. A steamboat and several vessels belong to the place. The t. has five stores, one grist-mill, five saw-mills. Pop. 837.

SAHARA, or the great central desert of Africa. See Africa, i., 29.)

SAID (an. Sidon), a town and seaport of Asiatic Turkey, celebrated in remote antiquity as one of the greatest emporiums of the Mediterranean, and as being the parent city of Tyre. Its modern representative is seated on the N. end of a cape extending into the Mediterranean. 22 m. N.N.E. Tyre, 90 m. S.S.W. Beyrout, and 55 m. W. Damascus. lat. 33° 34′ 5″ N., long. 34° 22′ 40″ E. Pop. estimated by at from 4000 to 5000 only, of whom more Christians of the Greek church. Olivier says, e compared with Beyroot as to its population d its streets are equally irregular and filthy ; on is more pleasant and advantageous, its ter built, and it has *khans* (hotels or private considerable extent, and which would not ity of Europe. Most of these have their own ependent of those distributed through the Aula supplies Said with water, which is it by a channel, kept in ill repair, for a dis-than a league." (*Voy. dans l' Emp.,* Oth. iv. te streets in the upper town, which stand ivity, are narrow and gloomy, being arched places, like those of Jerusalem. The lower vn is more cheerful. The bazaars are well :ially with leather goods. (*Robinson's Syria*) Most of the houses have gardens attached. de the town is defended by a high wall, now as well as by a fortress, on a hill to the S., said constructed by Louis IX. of France. There osques, both within and without the town-:nerally supposed that the ancient city was rther inland and, like many other maritime iity, at some distance from the sea-shore, on was situated. The non-existence of any an-1 notice at Said seems to favour this opinion ; :e called Old Sidon, at about the distance above aces of ancient walls and other buildings are ile.·

however, be no doubt that the harbour on the present town was the port of the ancient e modern town. Here is a quay formed of ls, in parts of which a tessellated pavement narbles, with representations of animals, fes-ill exists in tolerable preservation. Many is are also wrought into the walls, and others to a bridge of several arches, which runs dland to a castle built in the sea by Fahr-ed-ated emir of the Druses, in the 16th centu-r, aiming to render himself independent and only fortified Said, but, in order to make its :ssible to Turkish galleys, choked it up by :s filled with the *débris* of ancient buildings. gave a severe blow to the commerce of the lstead being so insecure that scarcely a fish-:an lie in it in safety. (*Bowring's Report.*) ie end of the last century, the French mer-had a considerable trade with Marseilles, to ported cotton, silk, and woollen goods, fruits, mony, galls, soda, and wax. At present the rces of the inhabitants are derived from dye-he manufacture of leather and silk goods. mentioned in Gen. x. 15, 19 ; and appears to importance at a very early period, since it in Joshua as the "great Zidon," (ch. xi. 8 ; ie division of Palestine it was allotted to the ' Asher, but we learn from Judges i. 31, that into the actual possession of that tribe. Its ere anciently eminent in ship-building, and d by Solomon in the construction of the being, among the Jews, none who had imber like unto the Sidonians." Pliny states famous for its glass manufactures. In its portance, it appears, however, to have been by Tyre, and afterwards generally followed f that city. In the middle ages, during the of Jerusalem, it was a lordship and an epis-r the Tyrian primacy. The crusaders, who recovered it from the Saracens in 1150 ; but bliged to surrender it to the latter in 1289. urhood is the convent of Mar-Elias, the the late eccentric Lady Hester Stanhope. *rta and Palestine,* i. 368, 373 ; *Olivier, Voy.* ch. ii. ; *Bowring's Report on Syria ; Mod.* &c.)

i city, river-port, and the chief emporium of Anam, prov. Tsiampa, of which it is the 'er of its own name, about 50 m. above its 0° 47' N. long. 107° 5' E. Pop. loosely esti-)0 among whom are many Chinese. It con-istinct towns, about 3 m. distant from each the intermediate road is lined all the way g houses. Pingeh, which has the citadel, of the governor, is on the W. bank of the 'iver ; whereas Sai-gon proper, the main seat s on a smaller river, which however, is nav-l-sized boats. The two towns are nearly of e streets, which are regular, and intersect right angles, are in some instances of great these, or along canals, many of which trav-he houses are disposed in straight lines, and

close to each other. They are built chiefly of mud, en-closed in bamboo frames, and plastered ; sometimes they are tiled, but more frequently thatched with palm-leaves or rice-straw : they have seldom more than one story ; but some of the better sort are surrounded with an open court-yard, with a gate towards the street. In Sai-gon proper, some of the streets are paved with flags ; and quays of stone and brickwork extend nearly a mile along the river. The citadel in Pingeh, constructed by a French engineer in 1790, has not been completed. It is a quadri-lateral fortress with earth ramparts, bastions, horn works a regular glacis, esplanade, and dry ditch. When Mr. Crawfurd visited the city, no guns were mounted on it, though there were several French cannon in the arsenal ; and at present it is not capable of a regular defence. The interior, which is neatly laid out, has barracks, officers' quarters, and the governor's residence. In the N. E. part of the city, on the banks of a deep creek, are the naval yard and arsenal, one of the most complete establishments in the empire, where many good junks and war-boats have been built. The rice magazines, the naval arsenal, and the royal palace, a brick edifice, are the other principal public buildings. Sai-gon communicates with the Cam-boja river by a canal 23 m. in length, about 80 ft. in width, and 12 ft. in depth. Its foreign trade, which is principally with China and Siam, is much less considerable than that of Bankok. The principal articles exposed for sale in the shops are Chinese earthenware, silks, paper, tea, &c. A few glass bottles and some broadcloths may be met with, but scarcely any other European goods. The markets are well supplied with poultry, hogs, oxen, &c. ; and the flesh of other animals less suited to a European taste, as dogs and alligators. Fruit is in great abundance, and the variety and excellence of the fish can hardly be surpassed. Ac-cording to Mr. Crawfurd, Sai-gon is far preferable, as a place of residence, to Bankok. Its vicinity is well culti-vated with rice, areca palm, &c. (*Crawfurd's Embassy to Cochin China,* i., 317-348 ; *White's Voyage, &c.*)

ST. ALBANS, p. t., Somerset county, Me., 50 m. N.N.E. Augusta, 645 W. Drained by Sebasticook river. Incorpor-ated in 1813. It contains four stores, two fulling-mills, two grist-mills, four saw-mills ; one academy, 50 students ; 13 schools, 651 scholars. Pop. 1564.

ST. ALBANS, p. t., capital of Franklin county, Vt., 25 m. N. Burlington, 63 N.W. by N. Montpelier, 537 W. Bound-ed W. by lake Champlain. It contains 20 stores, four saw-mills, two tanneries, two printing-offices, two binderies, two weekly newspapers ; one academy, 80 students ; 14 schools, 315 scholars. Pop. 2702. The village, situated 3 m. E. of the lake shore, is on elevated ground, and has a public square in the centre. It contains a courthouse, jail, three churches, a Congregational, Episcopal and Methodist, an academy, 16 stores, various mechanic shops, 100 dwell-ings, and about 700 inhabitants. At a bay on the lake shore, 3 m. W. of the village, is a landing, from which a steamboat plies to Burlington, by the way of Plattsburg and port Kent, and where there are a church, a store-house, and several dwellings.

ST. AUGUSTINE, city, port of entry, and capital of St. Johns co., Flor., 200 m. E. by S. Tallahassee, 880 W. It is situated 2 m. back from the shore of the Atlantic, on the S. side of a peninsula, connected to the main land by a narrow isthmus, and protected from the swell of the ocean by Anastasia island, which is so low as not to obstruct the sea breezes, or a view of the ocean. The site is not over 12 ft. higher than the ocean, and the soil is rich in calca-reous and vegetable deposites. The town is embosomed in orange trees. The climate is equal to any in the world ; frosts are experienced only one or two months in the year, and many winters pass entirely without them. The air is equal to that of Italy or the south of France. In the sum-mer season it is tempered daily by sea breezes, and the land breezes render the evenings cool and pleasant. It is a fa-vourite resort of invalids from the N. The city is in the form of a parallelogram, fronting on the E. on Matanzas sound, which spreads half a mile in width between it and Anastasia island, forming a harbour sufficiently capacious to contain a large fleet in perfect safety. The city is one mile long and three fourths of a mile wide, though not more than half of this extent is compactly built. The principal streets cross each other at right angles, but they are narrow, and many of them very crooked. The old houses are generally two stories high, built of shell stone, and the lower floors are covered by a composition which resembles stone, and in some instances the upper floors and the roof were formerly of the same material. A fine large square opens from Matanzas river into the eastern part of the place, on the W. side of which a neat courthouse has been erected of stone, which contains also the public of-fices. On the N. side stands a splendid Roman Catholic church.· On the S. side are several elegant residences, and Trinity church, a neat Gothic edifice. In front of the harbour is a fine market place. The intervals around the square are filled with dwellings, surrounded by orange groves, which give it a rural, rather than a commercial appearance. At the mouth of the harbour is a bar with not more than 9 ft. of water at low tide, though within it has 18 or 20 ft. It is completely commanded by fort Marion at the N. end of the city, directly opposite to the entrance. It is in the form of a trapezium, with bastions at the corners, and made bomb-proof. It is calculated to contain 1000 men, and formerly mounted 70 pieces of heavy ordnance; but they are at present dismounted, and it is now occupied as an arsenal and a jail for criminals. From the fort a sea wall is erecting, 7 ft. high and 5 ft. thick, which will extend through the whole front of the city, to check the inroads of the tide, which will cost $50,000. The city contains four churches, a Roman Catholic, Episcopal, Methodist and Presbyterian, a U. States' land office, 90 stores and groceries, various mechanic shops, 400 dwellings, and 2450 inhabitants. In the S. part of the city, fronting the Matanzas, are extensive barracks, rebuilt, after having been destroyed by fire, at an expense, to the U. States government, of $25,000. One half of the population of the city are from the United States; the remainder are Spaniards, French, and other nations, who are rapidly becoming American in their character. One half of the inhabitants are of the Roman Catholic religion. On the N. point of Anastasia island is a lighthouse. A packet plies regularly to Charleston. The city is governed by a mayor and four aldermen, chosen from any part of the place. *

SAINT BERNARD, parish, La. Situated in the S.E. part of the state, and contains 150 sq. m. It has lake Borgne on the N.E., and, for some distance below New-Orleans, includes both banks of the Mississippi. Bounded E. by the gulf of Mexico. It contained in 1840, 1862 neat cattle, 1154 sheep, 389 swine; and produced 65,150 bushels of Indian corn, 24,185 of potatoes, 34,600 pounds of rice, 4,308,000 of sugar. It had nine stores, two saw-mills, one distillery; one college, 21 students. Pop.: whites, 1035; slaves, 2137; free coloured, 65; total, 3237.

SAINT CHARLES, parish, La. Situated in the S.E. part of the state, and contains 512 sq. m. Bounded N.E. by lake Pontchartrain. Watered by Mississippi river. It has Allemande lake and bayou on the W., and Washa lake on the S. It contained in 1840, 3075 neat cattle, 3223 sheep, 1090 swine; and produced 207,000 bushels of Indian corn, 1500 of oats, 1500 of potatoes, 800,000 pounds of rice, 10,000,000 of sugar. It had two saw-mills; two schools, 23 scholars. Pop.: whites, 874; slaves, 3792, free coloured, 104; total, 4700.

SAINT CHARLES, county, Mo. Situated in the E. part of the state, and contains 470 sq. m. Bounded S.E. by Missouri river, N.E. by Mississippi river, N. by Cuivre river. It contained in 1840, 947 neat cattle, 4606 sheep, 19,324 swine; and produced 54,144 bushels of wheat, 271,251 of Indian corn, 5920 of barley, 48,705 of oats, 25,853 of potatoes, 301,686 pounds of tobacco. It had 23 stores, one flouring-mill, 15 grist-mills, seven saw-mills, six tanneries, seven distilleries, one printing-office, one weekly newspaper; one college, 104 students; two academies, 57 students; 13 schools, 391 scholars. Pop.: whites, 6286; slaves, 1597; free coloured, 28; total, 7911. Capital, St. Charles.

ST. CHARLES, p. v., capital of St. Charles co., Mo., 20 m. N.W. St. Louis, 110 m. E. Jefferson city, 829 W. Situated on the N. bank of Missouri river, on the first elevated land on the river above its mouth. It has a rocky shore, and the alluvial land commences at the lower end of the place. It has five streets parallel with the river, each about a mile and a half long. It is the principal crossing place over the Missouri from St. Louis to the N.W. part of the state, both for travel and merchandise. It contains a courthouse of brick, a jail of stone, a market house of brick, three churches, a Presbyterian, Methodist and Roman Catholic, a nunnery with a female school attached, 10 stores, two steam-mills. 130 dwellings, and 1457 inhabitants. It is the seat of St. Charles college, under the direction of the Methodists, founded in 1839, which has a president and four professors or other instructers, and 85 students. The commencement is on the last week in August. It has two academies, 57 students. The situation of St. Charles is pleasant and healthy, has good water, and a convenient landing place on the river, and is well situated for trade. There is stone coal in the vicinity. Pop. of the town 2918.

ST. CLAIR, lake, Mich., is situated between lakes Huron and Erie, and is 24 m. long, 30 m. wide, 90 m. in circumference, and 20 ft. deep. It receives Clinton river from Michigan, and Thames river and others from Canada, and receives by St. Clair strait, the waters of lake Huron and the other connected lakes, N. and W. of it.

SAINT CLAIR, county, Ala. Situated toward the N.E.

part of the state, and contains 840 sq. m. Bounded S.E. by Coosa river. Drained by Canoe and Broken Arrow creeks. It contained in 1840, 7905, neat cattle, 2940 sheep, 1972 swine; and produced 21,370 bushels of wheat, 236,605 of Indian corn, 21,000 of oats, 2190 of potatoes, 368,291 pounds of cotton. It had six stores, one flouring-mill, eight grist-mills, five saw-mills, four tanneries, 11 distilleries; 19 schools, 398 scholars. Pop.: whites, 4305; slaves, 225; free coloured, 8; total, 3538. Capital, Asheville.

SAINT CLAIR, county, Mich. Situated in the E. part of the state, and contains 930 sq. m. Bounded E. by St. Clair river. Watered by Black, Pine and Belle rivers. It contained in 1840, 3101 neat cattle, 1075 sheep, 3029 swine; and produced 10,836 bushels of wheat, 11,453 of Indian corn, 1230 of buckwheat, 12,641 of oats, 40,657 of potatoes. It had 16 stores, one flouring-mill, four grist-mills, 15 saw-mills, six tanneries, one printing-office, one weekly newspaper; 13 schools, 226 scholars. Pop. 4008. Capital, St. Clair.

SAINT CLAIR, county, Ill. Situated toward the S.W. part of the state, and contains 648 sq. m. Bounded W. by Mississippi river. Watered by Kaskaskia river and Silver and Indian creeks. It contained in 1840, 23,954 neat cattle, 9723 sheep, 34,955 swine; and produced 146,394 bushels of wheat, 1539 of rye, 630,025 of Indian corn, 12,387 of barley, 102,872 of oats, 24,134 of potatoes, 976 pounds of tobacco. It had two commercial and one commission houses in foreign trade, 36 retail stores, one woollen factory, three flouring-mills, 11 grist-mills, 14 saw-mills, one oil-mill, two tanneries, three distilleries, one printing-office, one weekly newspaper; one college, 195 students; two academies, 98 students; 16 schools, 505 scholars. Pop. 13,631. Capital, Belleville.

SAINT CLAIR, county, Mo. Situated in the S.W. part of the state, and contains 800 sq. m. Watered by Osage river and its branches. Capital, Osceola.

SAINT CLAIR, p. t., Bedford co., Pa. 113 m. W. Harrisburg, 139 W. It has one store, one fulling-mill, one flouring-mill, one saw-mill, one tannery, one distillery, two schools, 73 scholars. Pop. 1488.

ST. CLAIR, p. t., capital of St. Clair county, Mich., 45 m. N.E. Detroit, 579 W. Bounded E. by St. Clair strait or river. Watered by Pine and Belle rivers. The village is situated on the S. side of Pine river, at its entrance into St. Clair strait, on the site of old St. Clair fort. It contains a courthouse, jail, and a number of stores and dwellings, and has one grist-mill, three saw-mills; two schools, 48 scholars. Pop. 413.

ST. CLAIR, t., Butler co., O. Bounded S.E. by Great Miami river. Watered by St. Clair and Four Mile creeks. It contains Rossville village opposite to Hamilton, and has two schools, 41 scholars. Pop. 1174.

ST. CLAIR, t., Columbiana co., O. 11 m. S.E. New-Lisbon. Watered by Little Beaver river. The line of the Sandy and Beaver canal passes through it. It has two schools, 51 scholars. Pop. 1739.

ST. CROIX river, Me., constitutes the boundary between the United States and the British province of New-Brunswick. Its whole length is 130 m. It passes through Grand lake, and at its head has a monument, which designates the permanent boundary, from which the line proceeds due north. It has many falls, affording great water-power, and enters Passamaquoddy bay. It is navigable 12 m. for large vessels to Calais, where it is crossed by a bridge.

SAINTES (an. Mediolanum, aft. Santones), a town of France, dep. Charente-Inférieure, cap. arrond., on the Charente, here crossed by a stone bridge, 39 m. S.E. La Rochelle. Pop. in 1836, ex. com., 7893. Its situation is good, and it is entered from the S. by a finely planted promenade: but it is ill laid out, and is, for the most part, badly built. It has, however, some remarkable public buildings and antiquities. The cathedral was founded by Charlemagne; and the tower, with the principal entrance, is said to have belonged to the original edifice: but the rest of the structure dates principally from the 16th century. The church of St. Eutrope has a fine steeple, constructed under Louis XI. A celebrated abbey was founded here in 1043, into which Eleanor of Guienne retired after her divorce from Louis-le-Jeune; its buildings are now converted into cavalry barracks. The sub-prefecture, formerly the bishop's palace; the hospital, originally the seminary; the Protestant church, hall of justice, public library, and theatre, comprise the other chief public buildings. It has cabinets of natural history and antiquities, a communal college, and departmental accuracy grounds; and manufactures of hosiery, earthen-ware, dyeing-houses, tanneries, &c. Saintes is the centre of a district furnishing the best Cognac brandy; in which, and in corn and wool, it has a large trade.

Under the Romans Mediolanum was one of the chief cities of Aquitaine. Some Roman baths exist on the banks of the river; and without the walls are the remains of an amphitheatre, almost as extensive, though not nearly in such good preservation, as that of Nimes: adjacent to the

triumphal arch dedicated to Tiberius Drusus and
s (*Guide du Voyageur*); and the ruins of an
circus, &c., are still traceable. Saintes was the
he department Charente-Inferieure, from 1790
Dict. Géog.)

FRANCIS river, Mo. and Ark., rises in St.
Mo., and flowing S., enters Ark. and receives
r river, a large branch which rises in Mo., and
y parallel to it. It enters the Mississippi in
anty, and is said to be navigable in high water
n. Though it passes through many lakes and
waters are remarkably clear, and abound with

...ANCIS, county, Mo. Situated toward the S.E.
state, and contains 425 sq. m. Drained by St.
d sources of Big and Establishment rivers. A
portion of the celebrated Iron mountain is in
. It contained in 1840, 3740 neat cattle, 3548
l swine; and produced 21,903 bushels of wheat,
dian corn, 31,273 of oats, 441 of potatoes, 13,410
nbacco. It had 11 stores, four grist-mills, two
ive tanneries, one distillery; 11 schools, 286
'op.: whites, 2694; slaves, 501; free coloured,
ll. Capital, Farmington.

...ANCIS, county, Ark. Situated in the E. part of
d contains 1080 sq. m. Drained by Cache and
rs. It contained in 1840, 5598 neat cattle, 559
swine; and produced 1436 bushels of wheat,
Indian corn, 3242 of oats, 13,790 of potatoes,
of tobacco, 53,339 of cotton. It had seven
flouring-mill, six grist-mills, two saw-mills;
la, 70 scholars. Pop.: whites, 2122; slaves,
loured, 2; total, 2499. Capital, Franklin.

NCISVILLE, p. v., capital of West Feliciana
5 m. N.W. New-Orleans, 1170 W. Situated on
he E. bank of Mississippi river, 1 m. from it.
ou Sara it communicates with the Mississippi,
topping-place for descending boats, and export
ntity of corn. It has a court-house, 15 stores,
abitants.

...ENEVIEVE, county, Mo. Situated in the E.
S. part of the state, and contains 400 sq. m.
E. by Mississippi river. Watered by Au Vase
shment rivers. It has copper, lead and glac
it springs. It contained in 1840, 4329 neat cat-
sep, 10,591 swine; and produced 28,976 bush-
it. 203,754 of Indian corn, 18,351 of oats, 2590
81,855 pounds of tobacco, 1175 of sugar. It
mission houses in foreign trade, 14 retail stores,
mill, two grist-mills, 13 saw-mills, two tanne-
ademies, 60 students; four schools, 92 scholars.
a, 2583; slaves, 548; free coloured, 37; total,
al, St. Genevieve.

EVIEVE, p. v., capital of St. Genevieve co.,
l. by E. St. Louis, 190 m. E.S.E. Jefferson city,
uated on the W. bank of Mississippi river, 1
m it, on Gabourie creek, a small stream which
is boatable. It contains a Roman Catholic
academy with 40 students, and about 1000 in-
In front of the village was about 4000 acres of
, which was cultivated in common; but the
nts of the river have carried much of it away.
xtensive trade in lead. The inhabitants are
ench descent. Pop. of the town 1607.

RGE, p. t., Lincoln co., Me., 48 m. S.E. Augus-
Situated on a peninsula, bounded E. and S. by
; W. by Muscongus bay. It has some ship
considerable shipping employed in the coast-
l the fisheries. It contains 10 stores, one grist-
ools, 904 scholars. Pop. 2094.

RGE'S, p., hundred, New-Castle co., Del., 31
, 195 W. It has 10 stores, three grist-mills, two
hree schools, 92 scholars. Pop. 3127.

...NA, an island of the S. Atlantic ocean which
us in all time to come as being the scene of Na-
risonment and death. It belongs to Great Brit-
tuated 800 m. S.E. from Ascension island, and
from the coast of Benguela, in S. Africa, lat. 15°
5° 46' W. Length, 10½ m.; breadth, 6½ m.
acres. Pop., about 5000, of whom nearly a half
It appears from a distance like the summit of
arine mountain, rising abruptly from the shore
d almost perpendicular cliffs, varying from 300
height, diversified in a few places with deep
es, descending to the sea, and forming difficult
rs for the fishermen. In the largest of these,
N.W. is James' Town, the capital and port of
The interior is a plateau, about 1506 ft. above
ded into two unequal parts by a ridge of mount-
bout 500 ft. above the plain, and 2000 ft. above
highest summit of this ridge, called Diana's
ft. in height. The geological formation of the

island consists almost entirely of basalt, over which in some parts are strata of limestone mingled with tufa and other igneous substances, proving it to have once been the seat of volcanic action. A deep crater-like dell, however, called the Devil's Punch-bowl, is the only feature at all resembling an extinct volcano. The climate is mild, and little variable, the thermometer ranging between 57° and 70° Fahr. in winter (June, July, and Aug.), and between 68° and 74° in summer (Jan., Feb., and March). Rain is common, especially in Feb. and July; and the frequency of cloudy days tends greatly to moderate the heat. The influence of the climate over the constitution of residents in the island is, however, said to be most disastrous. According to O'Meara, it is "extremely unhealthy, hepatitis and dysentery prevailing to an extent and with a severity seldom paralleled, so that very few persons pass their 45th year." (*Voice from St. H.*, ii. 436.) About a fifth part of the entire surface is covered with soil, which, though not deep, is rich and productive. Luxuriant pastures are found on the high lands; agriculture, however, is but little attended to; and the inhabitants mostly depend on foreign supplies for their support. The valleys are chiefly laid out in garden ground; and here may be seen near each other, and all flourishing alike, the mangoe, banana, tamarind, and sugar-cane of the tropics, the orange, citron, grape, fig, and olive of S. Europe, and the apple, gooseberry, and currant of a still colder region. The yam and all the European vegetables are abundant, three crops of potatoes being often raised from the same ground within the year. The tops and sides of the hills are covered with the cabbage tree, dog-wood tree, and gum-wood; and the oak, also, has been introduced. The *Palma Christi* and common blackberry are so luxuriant as to be eradicated with difficulty. Cattle and sheep are not numerous, the latter being barely sufficient to supply the wants of the shipping. Goats graze in immense numbers on the high grounds. Neither birds nor beasts of prey, if we except rats, are found; but the latter are numerous and destructive, and there are many varieties of troublesome and poisonous insects. Game is abundant, and the coast furnishes excellent fish. Whales and turtles are often seen near the shore, especially in January and April.

St. Helena, of which the East India Company were "the lords proprietors" till 1834, is now under a military governor, assisted by a civil secretary; and the laws are administered by the chief judge. It is still used as a place of refreshment for vessels sailing northward on the Atlantic, those proceeding southward not being able to make the island. Its commerce is trifling; the value of the imports, principally consisting of stores for shipping, not exceeding from £40,000 to £50,000 a year.

James' Town, the port and the residence of the authorities, is the only town. The anchorage is good in 12 fathoms water, and the port is well protected from the winds. The town is entered by an arched gateway, within which is a spacious parade lined with official residences. The church, a handsome building, fronts the gateway; and close by it a street branches off into the inner parts of the town. The shops are plentifully supplied with English and Asiatic products, but the prices are invariably very high. The principal inhabitants reside on the higher and cooler parts of the island, and visit James' Town only on Sundays, or when the want of supplies obliges them to come to its shops and market. One of the handsomest of these villas is Plantation House, a mansion of considerable elegance, belonging to the governor, situated in the midst of extensive grounds, adorned with a variety of fine trees and shrubs.

Longwood, the residence of Napoleon, stands on the plateau, in the middle of an extensive park. When first occupied by the ex-emperor, it was of very limited dimensions; but some additions were subsequently made to it. After Napoleon's death, the house was for some time uninhabited. Latterly, it has been converted into a kind of farming establishment; and very recently, the room in which the conqueror of Austerlitz breathed his last was occupied as a cart-house and stable!

St. Helena was discovered on the 21st of May (St. Helena's day), 1502, by Juan de Noya, a Portuguese; but no establishment was formed nor was the island inhabited, till the Dutch became its masters in the middle of the 16th century. Captain Munden, in 1673, took it from the Dutch; and it was soon afterwards granted by Charles II. to the English East India Company, who, with the exception of the period of Napoleon's imprisonment, held the proprietorship down to 1834, when it was restored to the English government. (*St. Helena Gazette*; *Major Beatson's Letters on the Agr. of St. Helena*; *O'Meara's Voice from St. Helena*.)

Such is a brief notice of this very unimportant island; to which, indeed, we should not have thought it worth while to allude, but for the fact of its having been selected by Great Britain and the other allied powers as the place to which they banished Napoleon. It is unnecessary to perticularise the circumstances connected with this memorable event. Suffice it to say, that after the battle of Waterloo, and his second abdication, Napoleon, having retreated to Rochefort, addressed, on the 13th of July, 1815, the following letter to the Prince Regent, afterwards George IV.

" *Altesse Royale.*

" *En butte aux factions qui divisent mon pays, et à l'inimitié des plus grandes puissances de l'Europe, j'ai terminé ma carrière politique, et je viens, comme Thémistocle, m'asseoir au foyer du peuple Britannique. Je me mets sous la protection de ses lois, que je réclame de V. A. R., comme du plus puissant, du plus constant, et du plus généreux de mes ennemis.* " NAPOLEON."

But though Napoleon was more than entitled to compare himself to Themistocles, he erred widely in supposing that the Prince Regent was another Artaxerxes. It was, no doubt, indispensable for the tranquility of Europe and the world, and for the advancement of the projects of the allied sovereigns, that Napoleon should be placed under restraint. But a sense of what was due to themselves, and still more of what was due to the extraordinary individual the chance of war had placed in their power, should have secured him the best asylum and the most generous treatment consistent with perfect security. It is needless to contrast what they should have done, with what they actually did. The sending of Napoleon to St. Helena, and his treatment while there, constitute the most discreditable chapter in the history of modern royalty. It is painful to have to reflect that the government of such a country as England should have taken a conspicuous part in this unworthy treatment of a fallen foe. Every reader of ancient history heartily execrates the vindictive malignity with which the Romans pursued Hannibal. And yet, in comparing his treatment with that of Napoleon, it should be borne in mind that the Carthaginian hero never relaxed in his hostility to his ancient and hereditary enemies; he did not place himself in their hands, nor did he appeal to their generosity; and the probability is, that if they had got him into their power, they would have despatched him at once, and not have sent their illustrious captive to some miserable islet, to embitter and insult the few remaining years of his memorable life.

Napoleon arrived at St. Helena on the 13th of October, 1815, and there he expired on the 5th of May, 1821. His remains, after having been deposited for 19 years in a humble grave, near his prison-house, were, in 1840, conveyed with great pomp and ceremony to France, where, agreeably to the wish expressed in his last will, they now repose in the Hôtel des Invalides, in Paris.

SAINT HELENA, parish, La. Situated in the S.E. part of the state, and contains 1700 sq. m. Bounded W. by Anute river. Drained by Ticfah river and its branches. It contained in 1840, 5590 neat cattle, 1513 sheep, 16,300 swine; and produced 102,930 bushels of Indian corn, 4978 of oats, 38,762 of potatoes, 35,060 pounds of rice, 925,176 of cotton. It had 11 stores, 11 grist-mills, eight saw-mills, two tanneries; one academy, 37 students; four schools, 86 scholars. Pop. whites, 1945; slaves, 1573; free coloured, 7; total, 3525. Capital, Greensburg.

SAINT JAMES, par., La. Situated in the S.E. part of the state, and contains 250 sq. m. It lies on both sides of Mississippi river. It contained in 1840, 4702 neat cattle. 5107 sheep, 3990 swine; and produced 155.790 bushels of Indian corn, 1,039,950 pounds of cotton, 15,137,000 of sugar. It had 14 stores, two saw-mills; two colleges, 505 students; three schools, 48 scholars. Pop.: whites, 2762; slaves, 5711; free coloured, 75; total, 8548. Capital, Bringiers.

SAINT JOHN BAPTISTE, par., La. Situated in the S.E. part of the state, and contains 260 sq. m. Bounded N.E. by lake Pontchartrain. Mississippi river passes through it, along the margin of which is most of the land capable of cultivation. It contained in 1840, 2690 neat cattle, 3000 sheep. 950 swine; and produced 208,830 bushels of Indian corn, 112,000 pounds of rice, 11,000,000 of sugar. It had two saw-mills; eight schools, 81 scholars. Pop.: whites, 2141; slaves 3444; free coloured, 191; total, 5776. Capital, Bonnet Carre.

ST. JOHNS, river, Me., and New-Brunswick, rises in Somerset co., Me., near the head-waters of the Chaudiere in Canada, and the Penobscot, Me. Below the entrance of St. Francis river it forms the N.E. boundary of Me., until it crosses into New-Brunswick, where it pursues a S. and S.E. course until its entrance into the bay of Fundy. It is navigable for vessels of 50 tons, 80 m. from its mouth, and for boats 200 m., and its whole length is about 350 m. The free navigation of this river in New-Brunswick, secured to the United States, by the late treaty with Great Britain, will be of great importance to the N. part of Maine, by affording an outlet to the immense quantity of lumber which is found near its borders.

ST. JOHNS river, Flor. Its sources have not been ex-

plored, but are probably in extensive grass meadows, extending from E. to W. from 3 to 12 m., and from 20 to 30 north and south, and probably within 20 m. of the coast. It often spreads from 3 to 5 m. in width, but at other places is not one fourth of a mile wide. It is exceedingly crooked; and though in a straight line it may not be more than 150 m. from its source to its mouth, by the course of the river it is more than twice that distance. Vessels drawing 8 ft. of water enter lake George and Dun's lake, 120 m. from its mouth. At the entrance of the river there is a bar with 12 ft. of water, and here it is only a mile wide. There is a lighthouse on the S. side of the river at its entrance. Its general course is N. nearly parallel to the coast, but in its N. part, it curves suddenly to the east, and it enters the Atlantic.

SAINT JOHNS, county, Flor. Situated in the E. part of the peninsula, and contains 1450 sq. m. Watered by St. Johns and North river, and Matanzas sound. It produces oranges, citrons and lemons. Live oak grows extensively on St. Johns river. It contained in 1840, 663 neat cattle, 534 swine; and produced 3175 bushels of Indian corn, 1195 of potatoes, 25,000 pounds of sugar. It had 45 stores, two printing-offices, two weekly newspapers, three academies, 113 students; 4 schools, 77 scholars. Pop.: whites, 1685; slaves, 888; free coloured, 121; total, 9694. Capital, St. Augustine.

ST. JOHNSBURY, p. t., Caledonia county, Vt., 7 m. E. Danville, 37 m. N.E. Montpelier, 546 W. Chartered in 1786, first settled in 1786. Pasumpsic river and its tributaries, afford good water power. It contains three Congregational churches, six stores, three fulling-mills, one furnace, five grist-mills, seven saw-mills, one oil-mill, three tanneries, two potteries, one printing-office, one weekly newspaper; one academy, 25 students; 15 schools, 662 scholars. Pop. 1887.

ST. JOHNSVILLE, p. t., Montgomery co., N. Y., 61 m. N.W. Albany, 394 W. Bounded S. by Mohawk river. Drained by east Canada and Zimmerman's creeks. It contains a Dutch Reformed church, three stores, one grist-mill, three saw-mills; four schools, 113 scholars. Pop. 1893.

SAINT JOSEPH'S river, Mich., rises in the N.E. part of Hillsdale county, and flowing E. with a broad southerly bend by which it enters Ia., it turns to the N.W. and it enters lake Michigan. It is 250 m. long, and in length and volume of water, it is the second river in the peninsula. It is navigable for keel boats 130 m. to Lockport, and has a good harbour at its mouth, protected by a pier, which is sufficient for any number of vessels required by the lake navigation. There is a sand bar at its mouth with 6 ft. of water. It receives many tributaries, affords much water-power, and flows through a fertile territory and is in some places, heavily timbered on its borders.

SAINT JOSEPH'A, county, Mich. Situated in the S. part of the state, and contains 528 sq. m. Organized in 1829. Watered by St. Joseph's river, and its branches. It contained in 1840, 7865 neat cattle, 3986 sheep, 13,664 swine; and produced 131,451 bushels of wheat, 9436 of rye, 148,944 of Indian corn, 2997 of buckwheat, 11,323 of barley, 119,125 of oats, 66,396 of potatoes, 20,930 pounds of sugar. It had two fulling-mills, seven flouring-mills, 11 saw-mills, one tannery, two distilleries; 37 schools, 819 scholars. Pop. 7068. Capital, Centreville.

SAINT JOSEPH's, county, Ia. Situated in the N. part of the state, and contains 468 sq. m. Organized in 1830. Watered by St. Joseph's and Kankakee rivers. It contained in 1840, 6551 neat cattle, 3666 sheep, 14,289 swine and produced 102,690 bushels of wheat, 197,439 of Indian corn, 1920 of buckwheat, 133,647 of oats, 40,887 of potatoes, 1412 pounds of tobacco, 72,018 of sugar. It had one commission house in foreign trade, 20 retail stores, two flouring mills, two grist-mills, 14 saw-mills, five tanneries, two distilleries, one pottery, one printing-office, one weekly newspaper; one academy, 38 students; 22 schools, 898 scholars. Pop. 6425. Capital, South Bend.

ST. JOSEPH, p. t., Berrien co., Mich., 195 m. W. by S. Detroit, 652 W. Watered by St. Joseph and Pagaw rivers which here unite. Bounded W. by lake Michigan. The village, which is the capital of the county, is situated on the S. side of St. Joseph river, at its entrance into the lake. It is one of the most important places on the W. side of the peninsula. It contains a courthouse, jail, an Episcopal church, 19 stores, four large forwarding and commission houses, a bank, two steam saw-mills, a printing-office, issuing a weekly newspaper, and about 500 inhabitants. There have been appropriated, by the United States, $3500 for the improvement of the harbour. It has 2680 ft. of wharf, and admits vessels requiring 6½ ft. of water. Three steamboats are owned here, but few other vessels. A bridge here across St. Joseph river cost $15,000. It has two schools, 43 scholars. It commands a pleasant view of lake Michigan.

SAINT LANDRE, parish, La. Situated toward the

t of the state, and contains 2000 sq. m. Bounded
Atchafalaya river. W. by Bayou Nex Pique. Drain-
the and Vermillion rivers. It contained in 1840,
at cattle, 20,236 sheep, 20,841 swine; and pro-
4,508 bushels of Indian corn, 74,593 of potatoes,
ads of rice, 3500 of tobacco, 21,437,190 of cotton,
f sugar. It had 37 stores, three flouring-mills, 100
t, eight saw-mills, four tanneries, two printing-
ro weekly newspapers; one college, nine students;
lemies, 120 students; seven schools, 85 scholars.
lies, 7179; slaves, 7129; free coloured, 995; to-
l. Capital, Opelousas.

AWRENCE, river, United States and Canada,
t outlet of the five great lakes or inland seas of N.
and after a course of more than 2000 m., flows
gulf of St. Lawrence, receiving the drainage of a
of over 500,000 sq. m. With the middle of the
ough which it passes, it forms the boundary be-
we United States and Canada, until it intersects
degree of N. lat.; or rather what was settled,
somewhat erroneously, and marked as such before
y of 1783. On accurately taking the latitude of
y the most improved instruments of the present
a true line of 45°; but, by the late treaty it has
ablished as the true line. The river has different
different parts of its course. From lake Ontario
ath, is the St. Lawrence proper, though above
, it is sometimes called Iroquois river. Between
tario and Erie, it is called Niagara river; between
e and St. Clair, Detroit river; between lakes St.
l Huron, St. Clair river; between lakes Huron
rior, St. Mary's river. It is navigable for ships of
400 m. to Quebec; and for ships of 600 tons to
, a distance above Quebec of 180 m. From Que-
ontreal it has an average breadth of 2 m. The
of the gulf of St. Lawrence at its mouth, from
der to Mingan settlement in Labrador, is 150 m.
es many tributaries, the most important of which
side is Chambly or Richelieu river, the outlet of
amplain; and on the N. side, Ottawa or Grand
uch enters it a little above Montreal. For a more
ccount, see LAURENCE (ST.)

.WRENCE, county, N. Y. Situated toward the
t of the state, and contains 2717 sq. m. Bounded
St. Lawrence river. Watered by St. Regis, Rack-
Oswegatchie, and Indian rivers and their tributa-
ed ore is extensively found in the neighbourhood
. Iron ore and marble are also abundant. It
l in 1840, 61,455 neat cattle, 125,891 sheep, 41,889
nd produced 278,007 bushels of wheat, 23,571 of
984 of Indian corn, 34,312 of buckwheat, 94,018 of
14,009 of oats, 1,412,272 of potatoes, 848,132 pounds
It had two commission houses in foreign trade,
s, 16 furnaces, six forges, 26 fulling-mills, four
factories, two flouring-mills, 45 grist-mills, 103
t, one paper-mill, 27 tanneries, one distillery, two
two printing-offices, two weekly newspapers;
lemies, 761 students; 357 schools, 13,539 scholars.
16. Capital, Canton.

' LOUIS, county, Mo. Situated in the E. part of
, and contains 550 sq. m. Bounded E. by Missis-
r, N.W. by Missouri river. Watered by Mara-
r. It contained in 1840, 23,877 neat cattle, 8730
441 swine; and produced 59,177 bushels of wheat,
ye, 477,879 of Indian corn, 1908 of buckwheat,
oats, 90,966 of potatoes, 197,045 pounds of tobac-
d one commercial and 24 commission houses in
ade, 214 retail stores, 17 lumber-yards, two flour-
14 grist-mills, '13 saw-mills, one oil-mill, six tan-
ar distilleries, 22 printing-offices, six daily, seven
nd five semi-weekly newspapers; two colleges,
its; 14 academies, 602 students; 28 schools, 1534
Pop.: whites, 30,505; slaves, 4616; free colour-
otal, 35,979. Capital, St. Louis.

'UIS, city, capital of St. Louis co., Mo., is in 38°
lat., and 90° 15' 30'' W. long. from Greenwich,
l' 15'' W. long. from W. It is 90 m. by water be-
south of the Missouri; 1064 m. above the mouth
io; 1149 m. above New-Orleans; about 1100 m.
falls of St. Anthony; 190 m. E. Jefferson city;
The population in 1810, was 1600; in 1820, 4598;
194; in 1840, 16,469, of whom 1531 were slaves.
in commerce, 845; in manufactures and trades,
avigating rivers, &c., 893; in the learned profes-
It is the commercial metropolis of the state, and
rly the seat of government. The site is elevated
above the floods of the Mississippi, and is protect-
sem by a limestone bank, which extends nearly
advantage rarely enjoyed on the Mississippi,
generally bounded by high perpendicular rocks,
alluvial soil. This spot has an abrupt acclivity

from the river to the first bottom, and a gradual one from
it to the second bottom. The first bank presents a view of
the river, being elevated 20 ft. above the highest water;
the second bank is 40 ft. higher than the first, and affords
a fine view of the city, river, and surrounding country, and
contains the finest residences. The place was originally
laid out on the first bank, and consisted of three narrow
streets running parallel with the river. Fortifications were
erected on the second bank, as a defence against the savages. Soon after the American emigration commenced,
four additional streets were laid out back of the first, on
the second bottom, which is a beautiful plain, and these
streets are wide and airy. There are eight principal streets
parallel to the river, crossed by over 20, running from the
river, and crossing them at right angles. There are extensive tracts laid out N. and S. of this, called North St.
Louis, and S. St. Louis, which will probably be soon built
upon. The whole length of the place extends in a right
line 5¼ m., and by the curve of the river, 6¼ m. Its breadth
may ultimately extend 6 m. back from the river, but is at
present one third of that distance. The thickly settled
parts are confined within much narrower limits, and extend a mile and a half along the river, with half that
breadth. Front-street is open on the side toward the river,
and on the other side is a range of warehouses, four stories high, built of limestone, which have a very commanding appearance, and are the seat of a heavy business. In
First-street the wholesale and retail dry goods stores are
located; and in the streets immediately back of this are
the artisans and tradesmen. It contains many neat, and
some elegant buildings. The more recent houses are built
of brick, of an excellent quality made in the immediate vicinity; some are of stone quarried on the spot, and are
generally whitewashed. Many of the residences, particularly in the back parts of the place, have spacious and
beautiful gardens attached to them.

Among the public buildings of the city the city hall is a
splendid edifice of brick, the basement of which is occupied as a market, at the foot of Market-street, on a square
reserved for the purpose. There is another market in the
N. part of the city. The courthouse is in the centre of a
public square near the middle of the city; the Presbyterian church occupies an eligible site on the high ground of
the city; and is a large and well-finished building, surrounded with ornamental trees, which are carefully pruned.
The Unitarians have a large church of tasteful architecture. The Roman Catholic cathedral is a large and splendid edifice, 136 ft. long, 58 ft. wide, the walls of which are
40 ft. high, above which the tower of the steeple rises 90
ft. square to the height of 40 ft.; surmounted by an octagon
spire covered with tin, crowned with a brass gilt ball 5 ft.
in diameter, above which is a gilt brass cross 10 ft. high.
The lightning rod is 17 ft. higher than the cross. In the
steeple is a peal of six bells, the three largest of which
weigh from 1600 to 2600 pounds each. The front of the
building is of polished freestone, with a portico of four
massive Doric columns. The interior is splendidly finished and furnished, and contains several elegant paintings
of celebrated masters. The city contains 14 churches, two
Episcopal, two Methodist, two Presbyterian, an Associate
Reformed, a German Lutheran, a Baptist, a Unitarian, two
Roman Catholic, an African Methodist, and an African
Baptist. There is a United States land office, a theatre
and concert hall. There are several literary and benevolent institutions in the city. The St. Louis university is
under the direction of the Roman Catholics, was founded
in 1832, has a president, 12 professors or other instructers,
146 students, and 7900 vols. in its libraries. The commencement is on the third Thursday in August. It has a spacious building 4 m. N. of the city, which cost $30,000,
which has extensive grounds around it. Its medical department is within the city, and has a building with a hall
for lectures, a chemical laboratory, &c., which will accommodate 400 students. The Western academy of sciences
has an extensive museum of natural history, mineralogy,
&c. There is also a museum, containing many Indian antiquities, fossil remains, and other curiosities. The convent of the Sacred Heart is an institution of nuns, for conducting female education. The Protestant ladies conduct
an orphan asylum; and there is a Roman Catholic orphan
asylum, conducted by the sisters of charity, who also minister in the St. Louis hospital.

The city has a bank and two insurance companies. The
United States arsenal is in the southern limits of the city.
Jefferson United States barracks are on the bank of the
Mississippi, 10 m. below the city, and can accommodate
about 700 men.

The city is supplied with water raised by steam power from
the Mississippi to a reservoir on an elevated ancient mound,
whence it is distributed in iron pipes over the city. A company is also formed for lighting the city with gas. The
country for 15 m. W. of the city consists of a very fertile

prairie, Chonteau's pond in the upper part of the city is a beautiful sheet of water, fed almost entirely by pure springs, and has an outlet into Mississippi river.

The city is admirably situated for commerce, and already surpasses in its trade every other place on the river, N. of New-Orleans. The Mississippi and Illinois to the N.; the Ohio and its tributaries to the S.E.; and the Missouri to the W. afford it a ready access to a vast extent of country; while to the S. the Mississippi furnishes an outlet to the ocean for its accumulated productions. It is the principal depot of the American fur company, who have a large establishment with 1000 men in their employ. A vast amount of furs is here collected; and 10,000 dried buffalo tongues have been brought in, in a single year. Numerous steamboats ply from this place in various directions; and several cross the Mississippi as ferry boats. The landing is good at South St. Louis, on a bold rocky shore, where there is sufficient depth of water for boats of the largest class. The steamboat arrivals, in a single year, have amounted to 800, with a tonnage of over 100,000. The tonnage of the port in 1840, was 11,259 tons.

There were in the city in 1840, one commercial and 94 commission houses in foreign trade, with a capital of $717,000; 214 retail dry goods and other stores, with a capital of $3,875,050; 17 lumber-yards, with a capital of $287,529; 40 persons employed in internal transportation, together with 37 butchers, packers, &c., employed a capital of $141,500; furs, skins, &c., exported, were valued at $306,300; 167 persons manufactured machinery to the amount of $166,807; 13 persons manufactured 305 small arms; nine persons manufactured the precious metals to the amount of $5050; 65 persons manufactured various metals to the amount of $54,000; 60 persons produced granite and marble to the amount of $30,000; 249 persons produced bricks and lime to the amount of $122,500; 13 persons manufactured tobacco to the amount of $3550, with a capital of $9250; 28 persons manufactured hats and caps to the amount of $77,600, with a capital of $12,000; two tanneries employed 14 persons, and produced 8000 sides of upper leather, with a capital of $54,500; 12 manufactures of leather, as saddleries, &c., produced to the amount of $116,600, with a capital of $54,850; 15 persons produced 133,000 pounds of soap, and 243,000 pounds of tallow candles, with a capital of $16,700; one distillery produced 30,000 gallons of distilled spirits, and six breweries 370,700 gallons of beer, the whole employing 38 persons, and a capital of $48,800; eight persons produced paints and drugs to the amount of $13,500, with a capital of $7000; one rope-walk, employing three persons produced cordage to the amount of $5000, with a capital of $10,000; 78 persons manufactured carriages and wagons to the amount of $54,500, with a capital of $35,250; two flouring-mills produced 13,656 barrels of flour, and with six saw-mills, and one oil-mill, produced to the amount of $185,600, with a capital of $106,500; 22 printing-offices, six daily, seven weekly, and five semi-weekly newspapers, employed 82 persons, and a capital of $49,650; 210 brick or stone, and 130 wooden houses were built, employing 397 persons, and cost $761,980. The total amount of capital employed in manufactures, was $674,250. There were in the city 10 academies or grammar schools with 577 students, and seven common or primary schools, with 713 scholars.

St. Louis was first settled in 1664 by a company of merchants, to whom M. D'Abbadie, the director general of Louisiana, had given an exclusive grant for the commerce with the Indian nations on the Missouri. The company built a large house and four stores here; and in 1770, there were 40 private houses, and as many families, and a small French garrison. In 1780, an expedition was fitted out at Michilimacinac, consisting of 140 British and 1500 Indians for the capture of St. Louis and other places on the W. side of the Mississippi, which was successfully repelled by the aid of an American force under Gen. George Rogers Clark, who proceeded from their encampment on the opposite side of the river. In May 1821 the place contained 651 dwellings, 232 of which were of brick, or stone and 419 of wood, and the population was 5600. No inland town in the world is more favourably situated for commerce than St. Louis. It is the natural depot of the vast and fertile regions watered by the Missouri, the upper Mississippi, the Illinois and their numerous tributaries, it is rapidly increasing in population and importance, and is destined to be one of the principal cities of the west.

SAINT MARKS, p. v., port of entry, Leon co., Flor., 20 m. S. Tallahassee, 816 W. Situated on St. Marks river at the junction of Wakully river to form Appalachee river. It is the port of Tallahassee, with which it is connected by a railroad and is favourably situated for business. It is 6 m. from the entrance of the river where is a lighthouse, and the river has 8 ft. of water to this place.

SAINT MARTINS, parish, La. Situated toward the S.

part of the state, and contains 850 sq. m. Bounded N.I and E. by Atchafalaya river. Watered by Teche river. It has Chetimaches lake and Grand island in its S. part. It contained in 1840, 26,930 neat cattle, 5054 sheep, 29 swine; and produced 96,920 bushels of Indian corn, 1,783,263 pounds of cotton, 2,474,700 of sugar. It had 2 stores, two lumber-yards, two tanneries, one printing-office, one academy, 26 students; six schools, 110 scholars. Pop. whites, 3549; slaves, 4641; free coloured, 484; total, 8674 Capital, St. Martinsville.

ST. MARTINSVILLE, p. v., capital of St. Martinsville par., La., 178 m. W. New-Orleans, 1221 W. Situated on the W. side of Teche river, and is built chiefly on one street, on the elevated bank of the river. It contains a courthouse, jail, a Roman Catholic church, and about 60 inhabitants.

SAINT MARY'S, county, Md. Situated in the S.W. part of the state, and contains 300 sq. m. Bounded N.I. by Patuxent river, S.W. by Potomac river, E. by Chesapeake bay. It contained in 1840, 18,073 neat cattle, 11,206 sheep, 17,390 swine; and produced 62,279 bushels of wheat, 1563 of rye, 255,955 of Indian corn, 61,862 of oats, 11,762 of potatoes, 2,672,052 pounds of tobacco. It had 39 stores, one cotton factory with 294 spindles, four flouring-mills, 31 grist-mills, one printing-office, one weekly newspaper; four academies, 54 students; 19 schools, 351 scholars. Pop. whites, 6870; slaves, 5761; free coloured, 1362; total, 13,224. Capital, Leonardtown.

SAINT MARY'S, parish, La. Situated in the S. part of the state, and contains 870 sq. m. Watered by Teche river. It contained in 1840, 16,806 neat cattle, 6211 sheep, 6403 swine; and produced 133,410 bushels of Indian corn, 90,017 of potatoes, 4741 pounds of rice, 1,436,080 of cotton, 13,991,000 of sugar. It had 13 stores, one grist-mill, two saw-mills, two tanneries; eight schools, 110 scholars. Pop.: whites, 2266; slaves, 6586; free coloured, 288; total, 8950. Capital, Franklin.

ST. MARY'S, p. v., port of entry, Camden co., Ga., 322 m. S.S.E. Milledgeville, 802 W. Situated on St. Mary's river, 7 m. from its mouth. It has a safe harbour, and vessels requiring 21 ft. of water come to its wharves. It contains a Presbyterian church, an academy, several stores, 80 dwellings, and about 600 inhabitants. It is the most southerly town in the state, and as the country back of it is not well settled, it has not as extensive trade.

ST. MAWE'S, a market town and seaport of England, co. Cornwall, hund. Powder, par. St. Just-in-Roseland, on the E. shore of Falmouth harbour, 2 m. E.N.E. Falmouth. Area of par. St. Just, 3550 acres. Pop., in 1831, 5598. The town consists of one irregularly-built street, at the foot of a hill facing the sea, the inhabitants being principally fishermen and pilots. The harbour of St. Mawe's is a coast belonging to that of Falmouth, the entrance to which is defended by St. Mawe's castle, built in the time of Henry VIII. It is governed by a portreeve chosen at an annual court-leet. This inconsiderable place sent two members to the House of Commons from 1562 till the passing of the Reform Act, by which it was disfranchised. Markets on Fridays.

ST. STEPHENS, p. v., Washington co., Ala., 149 m. S. by W. Tuscaloosa, 981 W. Situated on the W. bank of Tombigbee river, at the head of schooner navigation, 120 m. above Mobile. It has a considerable number of dwellings, chiefly of none, but has the aspect of decay. It contains a United States land office, several stores, and about 150 inhabitants.

SAINT TAMMANY, parish, La. Situated in the E. part of the state, and contains 972 sq. m. Bounded S. by lake Pontchartrain, E. by Pearl river, W. by Tangipahoa river. Watered by Chifuncte river. It contained in 1840, 25,000 neat cattle, 1250 sheep, 18,500 swine; and produced 12,150 bushels of Indian corn, 4000 of oats, 18,300 of potatoes, 967,250 pounds of rice, 140,600 of cotton, 389,000 of sugar. It had 25 stores, 18 grist-mills, 21 saw-mills, one printing-office, one weekly newspaper; two academies, 15 students; one school, 15 scholars. Pop.: whites, 2373; slaves, 1940; free coloured, 385; total, 4598. Capital, Covington.

ST. THOMAS, one of the Virgin Islands in the West Indies, belonging to the Danes, in about lat. 18° 22' N., long. 65° W., 36 m. E. Porto Rico. Its area is estimated at 37 sq. m., and its population at about 7000, of whom 300 are whites, 1500 free blacks, and the remainder slaves. Surface mountainous, and the island generally less fertile than St. Croix. Droughts and violent hurricanes are frequently experienced. Sugar and cotton are the principal produce. St. Thomas has long been, and still continues to be, one of the principal emporiums in the West Indies. It owes this distinction partly to its convenient situation, partly to its spacious and safe harbour at St. Thomas on the S. side of the island, and partly and principally to the moderation of the import duties, which vary from 1 to 1½ per cent. St.

In consequence, become as it were a *dépôt* for
the neighbouring islands; goods being sent
rehoused till opportunity offers for conveying
final destination. The great articles of un-
manufactured goods, principally from Eng-
ly, also, from other countries of Europe, with
aber, &c., from the United States. We sub-

of the Import Trade of St. Thomas in 1840.

	Vessels entered.	Tonnage.	First cost Value of Imports.
			Dollars.
Stain . .	42	8,280	2,100,000
	36	5,944	640,000
	7	640	23,000
	9	1,290	53,000
h and Altona	32	5,390	980,000
t . .	15	2,865	41,000
	9	1,432	190,000
	5	306	18,000
of America			
R. America	217	38,679	968,000
hia . .	346	55,192	4,367,000

year the colonial arrivals were as follows;

	Vessels.	Tonnage.
ada and New Grenada .	55	4,612
Islands . .	600	9,923
ditto .	55	2,311
h ditto .	277	11,361
ditto .	381	13,637
ditto .	29	3,148
ditto .	15	369
sh ditto .	43	1,313
Totals . .	1,566	48,894

*y a great number of vessels neither landing nor load-
rch, being in that case free from port-charges.*

. Franklin co., Pa., 7 m. W. by S. Cham-
. S.W. Harrisburg, 98 W. Drained by Black
ains two churches, four stores, one furnace,
.ctory, five grist-mills, seven saw-mills, one
tanneries, two distilleries ; 11 schools, 364
. 1795.

NT. *See* VINCENT (ST.)

CA, (an. *Salmantica*), a celebrated city of
f Leon, and prov. of its own name, on the
of the Douro), crossed here by a handsome
27 arches, 92 m. S. by W. Leon, and 119 m.
id. Pop. according to Miñano, 13,918, not
lergy and university. It stands on three
d is surrounded by walls: streets generally
and crooked, extremely dirty, and with a
choly aspect. There are numerous public
tains ; but the only one worth notice is the
fine square, each side of which is 293 ft. in
led by houses of three stories, all of equal
t symmetry, with iron balconies, surmount-
alustrade : the lower part is open, forming
ning all round the square. Bull-fights are
n June. Among the public edifices, by far
cathedral, a gothic building with a super-
on the exterior ; it is 378 ft. in length, and
h, the height of the nave being 130 ft. In
ome good paintings, and a fine organ with
The chapter comprises a bishop, 10 dig-
i canons. The city has also 25 parish
rw of them merit description. The church
however, which was formerly attached to
vent, may almost vie in splendour of dec-
cathedral itself. Salamanca has, for ma-
n celebrated as the seat of a university,
as buildings belonging to the various col-
a principal feature in the city : indeed, so
amiards of the collegiate edifices of Sala-
somewhat pompously termed it *Roma la
e*). The university was founded in 1239,
revenue, in 1826, of 1,900,000 reals, or
though this has since been much dimin-
rression of the convents, and the applica-
is to secular purposes. In the 15th and
e university of Salamanca was attended
15,000 students; but its former glory is
question whether it have now 300 pupils.
it all regretted, for what advantage could
om the prelections of Spanish churchmen
ture, the philosophy of Aristotle and Tho-
man jurisprudence, and the canon law ?
niversity is a large library, furnished with
iks, and piles of scholastic divinity.
were four public and 25 private colleges
the university ; but many of these have
ly, and the buildings are little better than
id, but few of them retain any traces of

their former magnificence, their most valuable effects having been carried off during the Peninsular war, and the monastic libraries were burnt by wholesale in 1836. Before
the suppression of the monasteries, Salamanca had 580
clergymen ; and before the coffers of the churches and
convents were emptied to supply the wants of the state,
ample provision was made for the support of imposture and
idleness. Every street swarmed with vagabonds, not
merely those who were proper objects of compassion, but
those also who, if compelled to work, would have been
found abundantly able to maintain themselves. Mendicancy still prevails ; but the want of public support must
eventually compel the mendicants to apply themselves to
industrious callings.

The manufactures of Salamanca are inconsiderable,
comprising some fabrics of broad-brimmed hats (*sombre-
ros*) several tanneries, two or three establishments for
weaving woollen cloths, and a few others for making
starch, glue, earthenware, &c., besides a pretty large manufacture of shoes. A weekly market is held here, and an
annual fair in September. The suburbs abound with well
planted walks ; the *huerta*, or irrigated tract near the river,
is planted with fruit trees ; corn and leguminous plants
abound throughout the neighbourhood (the husbandry of
which, according to Townsend, is superior to that in most
parts of Spain), and the hills, clothed with oak trees, are
depastured by oxen, sheep, and goats, celebrated for the
delicate flavour of their meat.

Salamanca, though mentioned by the classical writers
under the name of *Salmantica*, appears to have been of
little importance under the Romans, though a Roman road
and some other monuments are still extant. Salamanca is
celebrated in the history of the late Peninsular war for
the victory gained in its vicinity on the 22d July, 1812, by
the Anglo-Portuguese army, under the Duke of Wellington, over the French under Marshal Marmont. The struggle was most severe, but the British were successful at all
points. (*Twiss's Spain and Portugal*, p. 58–63 ; *Townsend's Spain*, ii. 72–86 ; *Miñano ; Dict. Geog. ; Encycl.
Metrop. ; Napier's Peninsular War*, iv. 168.)

SALANKEMENT. a small village of the Austrian dominions, prov. Slavonia, on the Danube, nearly opposite
to the embouchure of the Theiss, 23 m. E. by S. Neusatz. It
deserves notice from its having been the spot where the first
decisive check was given to the progress of the Turks. A
powerful army of the latter, commanded by the justly celebrated Vizier Kiuperli, was encountered here on the 19th
of August 1691, by the Imperialists under Prince Louis of Baden. After an obstinate and well contested action, without
any decisive advantage to either party, Kiuperli fell ; when
the Turks, panic-struck by his loss, were totally defeated,
leaving above 20,000 men on the field of battle. The loss
of the Imperialists did not exceed 3000 men. (*La Croix,
Abrégé de l' Histoire Ottomane,* anno 1691.)

SALEM, co., N. J. Situated in the S.W. part of the state,
and contains 320 sq. m. Bounded N.W. and S.W. by Delaware r. and bay. Watered by Salem and Alloways creeks.
It contained in 1840, 11,277 neat cattle, 16,486 sheep, 26,329
swine : and produced 94,484 bushels of wheat, 36,995 of
rye, 371,384 of Indian corn, 305,691 of oats, 70, 644 of potatoes. It had 48 stores, three woollen factories, 17 grist-
mills, twelve saw-mills, four tanneries, one pottery, one printing-office, two weekly newspapers ; two academies, 71 students ; 43 schools, 1646 scholars. Pop. 16,024. Capital,
Salem.

SALEM, p. t., Rockingham co., N. H., 27 S.S.E. Concord,
458 W. Spiggot r. and its branches afford water-power.
It contains three churches, a Congregational and two Methodist, four stores, two fulling-mills, two woollen factories,
five grist-mills, three saw-mills ; 10 schools, 467 scholars.
Pop. 1408.

SALEM, city, port of entry, and one of the capitals of Essex co., Mass. is situated in 42° 31' 19" N. lat., and 70° 53'
57" W. long. from Greenwich, and 6° 7' 27" E. long. from
W. It is 14 N.N.E. Boston, 24 S. Newburyport, 454 W.
The population, in 1790, was 7921 ; in 1800, 9457 ; in 1810,
12,613 ; in 1820, 12,731 ; in 1830, 13,886 ; in 1840, 15,082.
Employed in commerce, 987 ; in manufactures and trades,
1189 ; in navigating the ocean 1201 ; in the learned professions, 58.

It is chiefly situated on a peninsula or tongue of land,
formed by two inlets or arms of the sea, called North and
South rivers. A handsome bridge over North or Bass
river connects the place with Beverly, was built in 1788, and
is 1481 ft. in length. The lowest bridge, on the S. side was
built in 1808. There are handsome settlements at North and
South Salem, on the opposite sides of the two rivers, connected with the town by these bridges, which are free. A
north-east arm of the peninsula, towards Beverly bridge,
is considerably settled ; and an easterly arm, called the
Neck, is occupied chiefly by the almshouse, hospital, and
two forts which are situated here. The almshouse has a

considerable farm attached to it. The compact part of the city is more than a mile and a half long, and three-fourths of a mile wide, extending across the peninsula. The city is handsomely laid out, though not with entire regularity, with spacious streets, crossing each other nearly at right angles. Essex-street occupies the height of land, and extends quite through the city, not more than from 20 to 25 ft. above high-water, nearly east and west, but its course is winding. Federal and Bridge streets are broad and straight. Chesnut-street is considered the handsomest in the place, though it is not the most public. The place contains 134 streets. In the eastern part of the city is an elegant public square or common, containing about 8½ acres, enclosed by a neat railing, laid out with gravelled walks, and bordered by a large number of elms. The city is well-built, partly of brick or stone, but largely also of wood; and some of the houses are elegant, particularly in the vicinity of the common. The streets are generally well paved and lighted, and the city is supplied with pure spring water from the neighbourhood, in pipes. Salem was long the second place in New-England in wealth, commerce, and population; but Providence and Lowell now exceed it in population, and New-Bedford in shipping. In early times it even disputed with Boston the right to be considered the capital. In proportion to its size, it is now one of the most wealthy places in the United States. The wharves, 31 in number, are along South river, which forms the principal harbour, which is so shallow, that vessels requiring more than 12 ft. of water must load and unload at a distance from the wharves, yet the harbour is accessible to ships of war, which find a safe anchorage, as was proved by the frigate Constitution, which here took refuge during the last war, when pursued by a superior force. The commerce of Salem is extensive, and her ships visit almost all parts of the world. From 1784 to 1812, and later, her citizens extensively prosecuted, and were greatly enriched by the East India trade, which was more extensively prosecuted here than in any other place in the United States. In 1818, there were 53 vessels employed in this trade, belonging to Salem, with an aggregate tonnage of 14,272 tons; but this branch of its trade, though still carried on, is less extensive than formerly. The number of vessels engaged in foreign commerce is about 100, and 13 in the whaling business, besides the vessels employed in the coasting trade and the fisheries. The tonnage of the district in 1840, was 37,021 tons. There are many islands in the harbour, most of which are small and rocky. Winter island lies on the north side of the entrance, and contains 36 acres. Fort Pickering is situated on its eastern point. The light-house is on Baker's island, which contains 55 acres. By the construction of a dam across North river, something after the form of the mill-dam at Boston, an important water-power has been produced, and several tide-mills have been erected. Among the public buildings are a courthouse, jail, an almshouse, a hospital, a market-house, a custom-house, an East India marine museum, with many curiosities from remote parts of the world, and a lyceum. There are eight banks with an aggregate capital of $2,330,000; three insurance companies with a total capital of $976,000, besides two mutual insurance companies, without a definite capital. There is also a bank for savings. There are two public libraries, an athenæum containing 9000 vols., and a mechanics' library containing 9000 vols., besides some others. There are 18 churches, four Congregational, four Unitarian, two Baptist, a Methodist, an Independent, Friends, Christian, Universalist, Roman Catholic, and a free church. Among the public institutions is a society, formed in 1801, of those who had as captains or supercargoes doubled the Cape of Good Hope or Cape Horn, for the relief, when necessary, of the families of its members, and for advancing the knowledge necessary for the East India trade. Strangers introduced by a member have free access to its splendid museum. There are various other charitable and benevolent societies. Harmony Grove cemetery, W. of the city, contains 35 acres of ground beautifully laid out, and ornamented with trees, and is a highly attractive spot.

According to the census of 1840, there were 45 commercial houses engaged in foreign trade, capital not given; 80 retail stores, with a capital of $430,000; the fisheries produced 7500 quintals, of smoked or dried fish, 10 barrels of pickled fish, 22,800 gallons of spermaceti oil. 968,500 gallons of whale or other fish oil, the whole employing 300 persons, and a capital of $900,000; four persons manufactured machinery to the amount of $3000; 20 persons manufactured the precious metals to the amount of $50,000; 76 persons manufactured various metals to the amount of $60,000; six persons manufactured granite and marble to the amount of $5000; 15 persons produced brick and lime to the amount of $7000; 100 men manufactured tobacco to the amount of $40,000, with a capital of 30,000; 16 persons produced hats and caps to the amount of $7000, with a capital of $5000; 17 tanneries produced 20,000 sides of sole lea-

744

ther, and 40,000 sides of upper leather, the whole employing 100 persons, with a capital of $75,000; 50 other manufactories of leather, as saddleries, etc., produced to the amount of $430,000, with a capital of $250,000; 15 persons produced 65,000 pounds of soap, 10,000 pounds of tallow candles, and 100,000 pounds of wax or spermaceti candles, with a capital of $65,000; four distilleries produced 380,000 gallons of distilled spirits, employing four persons, and a capital of $35,000; 20 persons produced drugs and paints to the amount of $50,000, with a capital of $140,000; four sidewalks, employing eight men, produced cordage to the amount of $103,000, with a capital of $33,000; six persons produced musical instruments to the amount of $4000, with a capital of $3000; ten persons produced carriages and wagons to the amount of $8000, with a capital of $8000; two grist-mills and two saw-mills, employing ten persons, produced to the amount of $107,000, with a capital of $50,000; 60 persons manufactured furniture to the amount of $50,000, with a capital of $40,000; three printing-offices, three binderies, two weekly, and two semi-weekly or weekly newspapers, employed 20 persons, and a capital of $9000; 10 wooden houses were built, employing 60 persons, and cost $25,000. The total capital employed in manufactures was $1,430,000. There was one academy, with 32 students; 77 schools, 2965 scholars.

Salem was chartered as a city in 1836. It is governed by a mayor, six aldermen, and 24 common-council men.

Next to Plymouth, Salem is the oldest place in the state, having been settled by John Endicott, at the head of a company of emigrants, in 1628. Its Indian name was Naumkeag or Naumkeek, by which it was designated in early times. They were strongly reinforced next year, by the arrival of 11 ships, and 1500 persons, part of whom went to Boston, Charlestown, Dorchester, and other towns. This fleet brought out the charter, under which John Winthrop had been chosen governor, and Thomas Dudley deputy governor. These settling at Boston, the seat of government was transferred to that town. In 1678 Salem contained 85 dwelling-houses and 300 rateable polls. Marblehead was taken from Salem in 1649; Beverly in 1668; Danvers in 1757. The celebrated delusion denominated "the Salem witchcraft," commenced in that part of the town which is now Danvers; but the court, the trials, and the executions were here; 19 persons were put to death in this and some of the neighbouring towns, including Boston and Charlestown. The place of execution in Salem is still denominated Gallows Hill, from which new peaceful spot, the finest view of the city is obtained. At length the severity was condemned, and the occasion of it was pronounced a delusion. These colonists undoubtedly acted exceedingly, but their error belonged to the age. The English laws at that time recognised witchcraft as a capital offence, and they received the sanction of that learned and upright jurist, Sir Matthew Hale.

Salem was distinguished as an early supporter of the American Revolution. On the closing of the port of Boston, the General Court was removed to this place; and General Gage hoped that the inhabitants would be conciliated by the expectation of increasing prosperity. But by a unanimous vote in a public town-meeting, they gave him to understand that they disdained to flourish on the ruin of Boston, and they retained an attempt of the British to seize some military stores, by taking up a draw-bridge in the town, before the battle of Lexington. During the Revolutionary war, Salem furnished 60 privateers, manned by 4000 men, which were eminently bold and successful.

The railroad from Salem to Boston, 14 miles, was opened in 1836, and is passed in 50 minutes; and Salem is thus brought into a near connection with the commercial capital of New-England.

SALEM, p. t., semi-capital of Washington co., N. Y. 45 m. N.N.E. Albany, 416 W. Bounded S. by Battenkill r. Drained by Black and White creeks, to which afford water-power. It contains a courthouse, jail, two churches, a Presbyterian and an Associate Reformed, 11 stores, one fulling-mill, one woollen factory, five grist-mills, five saw-mills, one printing-office, one weekly newspaper; one academy, 60 students; 22 schools, 731 scholars. Pop. 2855.

SALEM, p. t., capital of Salem co., N. J., 64 m. S. Trenton, 175 W. Watered by Salem cr. and its tributaries. The village is on Salem creek, 2½ m. from its mouth in Delaware bay, and contains a fine brick courthouse 40 by 40 feet, fire-proof brick county offices, a stone jail, a market house, a bank, an academy, and about 250 dwellings. The town contains 17 stores, two lumber yards, capital $23,000, two tanneries one pottery, one printing office, two weekly newspapers; two academies, 71 students; five schools, 130 scholars. Pop. 2007.

SALEM, p. t., Mercer co., Pa. 240, m. W.N.W. Harrisburg. 296 W. Watered by Little Shenango and Crooked creeks. It has one fulling-mill, two grist-mills, 13 saw-mills; 13 schools, 509 scholars. Pop. 1980.

", Stokes co., N C., 109 m. W. by N. Raleigh, a Moravian village, situated on a small branch er. It is built chiefly on one extended street, trees. The houses are chiefly of brick. It oravian church, a female academy of celebri- s four brick buildings, four stories high, on a quare, a branch of the bank of Wilmington, cotton factory, a paper-mill, a printing-office, 0) inhabitants.

v., capital of Crittenden co., Ky., 249 m. kfort, 781 W. It contains a courthouse, jail, one school, 25 scholars. Pop. 233.

, Columbiana co., O., 167 m. E.N.E. Columbus, as 10 schools, 324 scholars. Pop. 1900.

'hampaign co., O. It has two flouring-mills; 7 scholars. Pop. 1402.

lighland co., O. It has six schools, 233 schol- d.

edferson co., O. It has nine stores, one grist- neries; six schools, 375 scholars. Pop. 2044.

shelby co., O. It has three stores, one wool- e grist-mill, three saw-mills, two tanneries; l scholars. Pop. 1158.

'.. capital of Washington co., Ia., 93 m. S. In- i W. Situated on the head waters of Blue ontains a courthouse, jail, market house, a ary of brick, two churches, a Methodist and 12 stores, a grist-mill, oil-mill, and cotton fac- d by steam, and various other mills and man- o printing-offices and about 1500 inhabitants. istrict of British India, presid. Madras, be- h and 13th degs. of N. lat., and 77¼ and 80 ; having N. E. and E., N., and S., Arcot; ichinopoly; S.W. and W. Coimbatoor, from arated by the Cavery; and N.W. the Mysore a, 6518 sq. m. Pop. in 1836-7, 905,190. Its is above the E. ghauts; and its climate is ng, which makes it be much frequented by lida. It comprises the Barramahl districts, and forming its N. portion. Except the Ca- ir. it has few rivers and no lakes. In 1836, cres, 1,118,730 were estimated to be under ive of 661,500 ft for cultivation, and 2,391,600 re barren and mountainous. About three land is assessed under the ryotwar, and the er the zemindar system. Rather more than ation is supposed to be actively employed in: laize, rice, and a little cotton are grown, and s of teak, sandal, and black woods grow on the principal exports are cloth, ghee, tama- , jaghetry, oil seeds, and iron. Iron ore is and good steel is made. Cloth is, however, amodity, and is manufactured for export to and America. The chief imports are areca black pepper. Total revenue of the district, ,563 rup.; of which sum, 1,625,594 rup. were land-tax. Salem, the chief town, and resi- 3ritish authorities, stands in about lat. 11° O 13' E. It has some trade in cotton cloths, (*Madras New Almanac; Hamilton, E. I.*

considerable town in Sicily, intend. Trapani, hill 90 m. E. Marsala. Pop., in 1831, 12,162. ated, but has a most abject appearance. Its indolent, and the town has no trade, being, ussell (*Trav.*, p. 64), distinguished only by tition. Salemi occupies the site of the an- Sir R. C. Hoare supposes that it derives its from a Saracen chief of the 9th century. - *in Sicily*, il. 85.)

(an. *Salernum*), a celebrated city and sea- in the Neapolitan dominions, prov. Princi- vhich it is the capital; at the foot of a hill ore of the gulf of Salerno, 17 m. S.S.W. 8 m. S.E. by E. Naples, with both of which ected by good roads. Lat. 40° 44' N.; long. he modern city with 16,000 inhabitants, is ell built; and the narrow and dirty streets, errat levels from the immediate edge of way up the mountain, gives its interior a ince, and afford but inconvenient residen- tuation is most happy, and a marine, or ch skirts its whole length along the shore, the French, and contributes to render its e sea extremely imposing. There is no broken mole affording protection to the only, offers the semblance of such an ae- (*Craven's Tour, &c.*, pp. 369. 370.)

square has a good public fountain, and is several Gothic edifices, including the inten- theatre, and the cathedral. The houses le streets paved with lava. Ancient Gothic

walls, in tolerable repair, enclose the city; and on the hill above, amid the principal remains of the ancient Salernum, is a ruined citadel. The cathedral, a heavy Gothic struc- ture, is the most interesting of its public edifices. It was erected by the Normans on the site of an ancient building, and is dedicated to St. Matthew who is said to be buried within its walls. "The *atrium*, or court before it, is spa cious, and surrounded by a portico of antique columns of porphyry, granite, &c. (said to have been brought from Præstum 23 m. S.S.E.. by Robert Guiscard); upon which the Normans constructed a range of brick arches, bent more after the Saracenic than the Gothic or Grecian manner; these support a regular set of apartments. In the centre is a basin of granite, 15 ft. in diameter, constantly filled by a fountain of excellent water. Many excellent sepulchres are placed in the colonnade, and the church contains also some monuments of remarkable personage, as Roger and Wil- liam, dukes of Apulia, Margaret of Durazzo, and the rest- less pontiff Gregory VII., who died of chagrin at Salerno, in 1085. On each side the entrance of the choir is a pulpit raised upon pillars. Their pannels are formed by rich mo- saic of many colours, disposed in knots and stars. The choir is inlaid with square and oval plates of verde-an- tique, porphyry and serpentine; the great altar is decorated in the same barbarous, but splendid manner." (*Swinburne's Tour*, ii. 119.) Though the cathedral has but few paint- ings, it has, luckily, an ample supply of miraculous ima- ges! There are numerous other churches, one of which is said, but on doubtful authority, to be the burial-place of John of Procida, a native of Salerno, celebrated as the principal contriver of the conspiracy against the French in Sicily, which terminated on the 30th of March, 1282, in the massacre known by the name of the "Sicilian Vespers." Salerno has two hospitals, a workhouse, three government pawn-banks, a seminary, a royal lyceum, and a university. To the last belongs a school of medicine, which was once among the most famous in Europe, but which has, for a lengthened period, lost its pre-eminence. But the lyceum in this town is said to be superior to most others in the kingdom.

Salerno is an archbishop's see, the residence of the pro- vincial intendent, and the seat of a superior criminal court, and of a civil tribunal. Previously to the period when Na- ples attained to a decided lead among the cities of S. Italy, Salerno carried on a considerable commerce by sea; that, however, has now wholly disappeared, though it continues to possess a pretty extensive inland trade, and has two large annual fairs.

Its climate is mild, but it is unhealthy from the proximity of marshes and rice-grounds, the culture of which occupies many of the inhabitants. It is doubtful whether the an- cient Salernum was contiguous to, or at some distance from the sea; but, on the whole, the probability seems to be that it did adjoin the sea, or that it was within such a short distance of it as to justify its being reckoned among mar- itime towns. (*Cellarii, Geographia Antiqua*, i. 860.) After the fall of the Roman empire, Salerno became the capital of a flourishing republic, the sovereignty of which was contested by the Greeks, Saracens, Lombards, and Nor- mans; the latter of whom obtained possession of the city in 1076. Having been mostly burned down by the empe- ror Henry VI., it subsequently became a feudal possession of the Colonna, Orsini, and Sanseverini families, till it was re-annexed to the royal domains by the emperor Charles V. (*Rampoldi; Craven; Swinburne; Eustace, Classical Tour*, iii. 55.)

SALINA, p. t., Onondaga co., N. Y., 133 m. W. by N. Albany, 350 W. Drained by Onondaga creek, flowing into Onondaga lake, lying in the t. It contains the villages of Syracuse, Salina, Liverpool, and Geddes; and contains the richest salt-springs in the United States. It has in its sev- eral villages, eight churches, two Presbyterian, two Epis- copal, two Methodist, a Baptist, and Roman Catholic, about 130 stores and groceries; three academies, 150 stu- dents, 21 schools, 1848 scholars. The salt-springs are owned by the state, which receives a duty of six cents on a bushel from the manufacturers, which on 2,623,305 bush- els manufactured in 1840, amounts to $157,338. Pop. 11,012, being the most populous township in the state.

SALINE, co., Mo. Situated a little N.W. of the centre of the state, and contains 829 sq. m. Bounded N. and E. by Missouri r. Watered by La Mine r. and its branches. It contained in 1840, 7,176 neat cattle, 5798 sheep, 2475 swine; and produced 24,677 bushels of wheat. 304,095 of Indian corn, 36,763 of oats, 11,858 of potatoes, 17,250 pounds of tobacco. It had two commission houses in foreign trade, 10 retail stores, two flouring-mills, 19 grist-mills, 9 saw- mills, six distilleries, 14 schools, 303 scholars. Pop.; whites, 3635; slaves, 1615; free coloured, 8; total 5,258. Capital, Jonesboro'.

SALINE, co., Ark. Situated near the centre of the state, and contains 720 sq. m. Bounded S.W. by Saline r.

It contained in 1840, 1530 neat cattle, 273 sheep, 3880 swine; and produced 508 bushels of wheat, 9500 of Indian corn, 1690 of oats, 4200 of potatoes, 940 pounds of tobacco, 64.500 of cotton. It had three stores, five grist-mills, one saw-mill, one tannery, one distillery; one school, 36 scholars. Pop.: whites, 1662; slaves 399; total, 2,061. Capital, Benton.

SALINS, a town of France, dep. Jura, cap. cant., in a narrow valley on the Furieuse, 26 m. N.E. Lons-le-Saulnier. Pop., in 1836, ex. com., 6185. It is walled, and commanded by two forts on adjacent heights. Its principal street is paved, and lined with substantial houses. There are several churches, a spacious college, a public library, good barracks, a theatre, hospital, and prison: these buildings are nearly all new, the town having been almost wholly destroyed by fire in 1825. Salins has several iron-forges, stone works, and brandy distilleries; but its name and principal importance are derived from its brine springs, which were wrought in the time of the Romans. They occupy a large space in the middle of the town, inclosed by turreted walls. And according to Hugo, the produce amounts to 140,000 cwts. of salt a year; in addition to which a considerable quantity is made at Arc, about 4 leagues distant, to which an aqueduct conducts a portion of the water of the Salins springs. (Hugo; Dict. Géog.)

SALISBURY, or NEW SARUM, a city, parl. bor., and market town of England, co. Wilts., of which it is the cap., hund. Underditch, on the Avon. here crossed by three stone bridges (one of which has 10 arches), 21 m. W. Winchester, and 80 m. W.S.W. London. Pop. of parl. bor., which includes, with the city, the extra-parochial district of the Cathedral Close, and parts of parishes Fisherton and Milford, in 1831, 11,672. It is built with great regularity, having six principal streets running from N. to S., crossed by the same number, intersecting them at nearly right angles. The houses generally are large and respectable, some, also laying claim to considerable architectural elegance. It is well paved, lighted with gas, and kept remarkably clean by means of brooks running down the middle of the streets. The market-place, a large open square, on its W. side, has, at its S.E. angle, the council house, a brick structure, with a Doric portico, erected at the close of the last century at the expense of the Earl of Radnor, and since greatly enlarged. The interior is divided into court-rooms, offices, &c., for the business of the assizes, quarter-sessions, and corporation, and it has several good modern portraits. On the opposite side of the market-square is a curious old hexagonal-shaped building, having a conical roof supported by pillars; it is called the Poultry or Butter-cross, and was probably built in the reign of Edward III. In another part of the same square is the public library and reading-room, founded in 1819, and supported by subscription. The most striking feature in Salisbury, however, is its cathedral, which stands in the large open space called the Liberty of the Close, on the S. side of the city. The situation is remarkably good; the precinct or close is kept in the best order, and comprises some very fine trees which, as well as those in the palace-grounds, serve to embellish the views of the cathedral. The W. front, the N. side, and the E. end of the latter are all open, and may be seen from peculiarly favourable distances: indeed, the N.E. view is perhaps the best general view of a cathedral to be had in England, and displays the various portions of this interesting edifice to the best advantage. Salisbury cathedral has the advantage of being built in one style, the early English, and on a uniform and well-arranged plan. The centre tower and spire (the entire height of which is estimated at 404 ft.) are of later date, but admirably accommodated to the style of the building. The plan is that of a complete cathedral, having spacious cloisters, an octangular chapter-house, and a tower for a library and muniment room : there are two transepts, each of which has an aisle eastward, and the nave has a large N. porch. The extreme length of the church (including the Ladye chapel) is estimated at 474 ft.; breadth of nave and aisles, 78 ft.; height of nave 30 ft.; and width, including the great transepts, 210 ft. Modern alterations have taken away the altar-screen, and thrown the Ladye chapel open to the choir: the organ-screen, also, as well as a large portion of the tabernacle-work in the choir, is of modern construction. The E. window is filled with a beautiful painting on glass of the Resurrection, from the designs of Sir Joshua Reynolds; another window exhibits a painting on glass of the Elevation of the Serpent in the Wilderness; there are other painted windows, and in various parts of the church are several ancient monuments, some of which are extremely curious. The W. front is a beautifully enriched specimen of the pointed architecture, peculiar to this church: the angles are terminated by tolerably massive square towers, surmounted by spires, and pinnacles; and over the grand central entrance is a series of canopied arches beneath the great W. window, which is formed in

three divisions. The exterior of the church is enriched with a number of recesses situated in tiers at different heights all round the building. Many of the statues still remain, and it is supposed that originally there must have been at least 200. The cloisters are remarkably magnificent, forming an exact square, each side of which is 82 ft. in length. The cathedral library is built over the E side of the cloisters, and adjoining them, in the same direction, is the chapter-house. "On the whole," says R. Rickman, "the cathedral presents an object for architectural study hardly equalled by any in the kingdom : the purity of its style and the various modes of adapting that style to the purposes required, deserve the most minute attention." (Rickman's Gothic Architecture, p. 257, 258.)

Within the close, formerly surrounded by a wall and entered by several ancient gates, deserving admiration, are the residence of the bishop, dean, canons, &c. The deanery-house is opposite the W. front of the cathedral; and at a little distance S.E., surrounded by gardens, is the bishop's palace, a very irregular building, in different styles of architecture, having been enlarged and repaired at various periods, from the middle of the 15th century down to a recent period. The gardens are on a large scale, comprising an area of several acres, well planted with fine, large old trees. The episcopal see was removed from Old Sarum to this bury under the authority of a papal bull, in 1217, about which time the cathedral was founded, the expense of its erection, exclusive of the chapter-house, tower, and spire being estimated at 40,000 marks, or £26,667, an enormous sum in those days. The chapter comprises, besides the bishop, a dean, precentor, chancellor, and six canons residentiary, dividing among them a nett revenue of £2808 annually, and having residences and separate revenues, with the patronage of 18 benefices. There are likewise 31 prebendaries, besides choral vicars, &c. At an average of the three years ending with 1831, the revenues of the bishopric amounted to £3939 a year; but it either has been, or is to be augmented to £5000 a year. The bishop has also a large portion of the cathedral-patronage, besides that of 18 benefices; his diocese extends over the whole of Wiltshire and a portion of Shropshire. Salisbury has three other churches, one of which, St. Edmund's, is in the gift of the bishop, and that of St. Thomas's in the patronage of the dean and chapter. St. Edmund's is a perpendicular structure, with large windows and good tracery, the chancel having been modernized. The tower fell down, and was rebuilt in the 17th century. St. Thomas's is a large perpendicular church of good composition, with its tower standing on the S. side of the S. aisle: it has a nave and chancel, with aisles and a clerestory. St. Martin's is a large church in the early English style, with some more recent parts. The church at Fisherton is small and of mixed architecture. The Roman Catholics have a handsome chapel; and there are places of worship for Independents, Baptists, Wesleyan Methodists, and Unitarians, with attached Sunday-schools. A grammar-school is attached to the cathedral for the instruction of the choristers, and there is another in the city, founded by Queen Elizabeth, in the patronage of the corporation. The bishop supports a good school; and there is also an orphan school; and the different parishes have their respective, national, infant schools, &c. There are several charities, among which may be mentioned that of Bishop Le Poor, near Harnham bridge; Trinity hospital, founded in the reign of Richard II., for 12 aged matrons; and Bishop Ward's college for clergymen's widows; with several other almshouses, money charities, &c. An infirmary, founded near Fisherton bridge in 1766, is liberally supported by subscription; a mendicity society has been established with considerable success, and there are various minor benevolent institutions, bible, and tract societies, &c.

A county-jail has been recently erected in Fisherton, and there is a small, but neat theatre, little patronised, with assembly and concert rooms. A weekly newspaper, called the "Salisbury and Winchester Journal," has a pretty extensive circulation. Races are annually held near the town in August. "The town of Salisbury cannot be considered as increasing, or in a state of improvement, having very little trade. An extensive woollen manufacture was formerly carried on here; but it is now confined to a single factory; and cutlery, for which this town was once famous, is now brought for sale from Birmingham." (Parl. and Municipal Bound. Report.) It has hitherto derived some benefit from its position on one of the great roads to Devonshire; but the opening of the Western railway has deprived it of this advantage.

Salisbury received its first charter from Henry III., which was afterwards renewed by several monarchs. According to the Municipal Reform Act, it is divided into three wards, and governed by a mayor, six aldermen, and 18 councillors. Quarter and petty sessions are held under a recorder; besides which the assizes and quarter sessions for the county are held here. A court for the recovery of debts to any

d monthly by the bishop's bailiff, who also
et for the lord of the manor. Salisbury has
members in the House of Commons since the
d I., the right of election down to the Reform
'd in the corporation, consisting, in 1831, of
l'he limits of the parl. bor. were enlarged by
Act, so as to include, with the old bor., the
and certain parts of Fisherton and Milford
mentioned. Registered electors, in 1839–40,
is also the chief election town for the S. di-
hire. Markets on Tuesday and Saturday,
fairs on alternate Tuesdays. Fairs, Tues-
; Tuesday after March 25; Whit Monday
lay for horses, and Oct. 29 for butter and

s its foundation to the removal of the ec-
lishment from the once important but now
ld Sarum, the Roman station of *Sorbiodu-*
N. from the modern city. The quarrels
ps of Henry II. and Roger Le Poor, the tur-
that day, induced the latter to establish
ore peaceful and advantageous situation:
around the cathedral, and by the influence
and clergy, soon became an important
ther fell to decay, and was ultimately de-
abitants. Henry III. granted the city a
it to the same privileges as Winchester,
ensive local powers on the bishop of the
were occasionally held here during the
turies. The city became celebrated, after
es I., for the abortive attempt of the roy-
nel Wyndham, to proclaim Charles II.
eries existed here prior to the Reforma-
: no extant remains of these foundations.
aguished individuals to whom Salisbury
iy be specified James Harris, the author
and of other learned and ingenious phi-
iphysical treatises, born here in 1709.
clever, but not very learned, deistical
entury, was also a native of this city.
ry stretches the vast tract of downs and
bury plain; and about 6 m. N. of the city
monument of Stonehenge, which see.
ports; Rickman's Architecture, &c.)
Merrimac co., N. H., 16 N.N.W. C^{on.}
arsarge mountain lies in its W. part.
iter river. Chartered in 1768. It con-
, a Congregational and Baptist, three
iill, two grist-mills, two saw-mills; one
s; 10 schools, 394 scholars. Pop. 1389.
Addison co., Vt., 69 S.W. Montpelier,
lake Dunmore, the largest pond in Vt.,
fourths of a mile wide, the outlet of
ter river, affords water-power. Water-
'er. It has two stores, one fulling-mill,
one forge, one glass-factory, one grist-
0 schools, 440 scholars. Pop. 942.
ssex co., Mass., 42 N.N.E. Boston, 477
the Atlantic, S. by Merrimac river, W.
contains six churches, two Congrega-
Baptist, Christian and Universalist, 23
s, three woollen factories, three grist-
. three tanneries, one printing-office,
per; 11 schools, 658 scholars. Pop.

itchfield co., Ct., 53 W. Hartford, 337
1790, incorporated in 1740. Bounded
r, which between this town and Ca-
ilar fall of 60 ft. affording great water-
il ponds, the outlets of which afford
ilns three churches, a Congregational,
list, seven stores, three furnaces, 10
s, four saw-mills, two tanneries; one
13 schools, 695 scholars. Pop. 2562.
rkimer co., N. Y., 73 W.N.W. Alba-
y East Canada creek, and tributaries
. It contains five churches, a Pres-
it, and two Universalist, seven stores,
forge, four grist-mills, 11 saw-mills.
tanneries, one printing-office; 14
Pop. 1859.
ncaster co., Pa., 16 E. Lancaster, 51
l W. Drained by Pecquea creek,
ower. It contains nine stores, one
, two grist-mills, one saw-mill, one
10 schools, 390 scholars. Pop. 3059.
i co., Pa. It has three stores, three
mills; one school, 40 scholars. Pop.

tal of Rowan co., N. C., 118 W. Ra-
on a branch of Yadkin river. It
il, a church, an academy, and about

---

700 inhabitants. Here is an ancient stone wall, laid in cement, from 12 to 14 ft. high and 22 inches thick, which
reaches to within a foot of the surface; and the length of
what has been discovered is about 300 ft. The object, time
and purpose of its construction are uncertain. A smaller
similar work has been discovered, at the distance of 6 m.

SALMON, r., N. Y. rises in Lewis co., and enters lake
Ontario in Mexico bay. Twelve miles from its mouth is
a remarkable cataract of 107 ft. perpendicular, where the
river is 250 ft. wide at high water. Above the cataract the
rocky banks are 80 or 90 ft. high and below it 200 ft. high.
Many salmon are taken in the river below the falls, whence
it derives its name. It is navigable for schooners, 1½ m. from
the lake, and boatable in high water, to the falls.

SALON, a town of France, dép. Bouches du Rhone, cap.
cant., in a fertile plain within about 3 m. of the canal *de Cra-
ponne*, and 29 m. N.N.W. Marseilles. Pop., in 1836, 4446. It
is divided into an old and a new town, separated from each
other by a planted boulevard. It has, according to Hugo,
an air of opulence, of which many larger towns are destitute: its streets are regular, and it has many good houses
and public buildings; including the church built by the
Templars, the parish church, with several curious sculptures, town-hall, &c. On a rocky height, at the extremity
of the town, is an old castle, now converted into a house of
correction (or, according to Hugo, a barrack). It has manufactures of silk twist, hats, soap, and olive-oil; and a brisk
general trade. The remains of a temple in honour of Tiberius have been discovered here. (*Hugo, Dict. Géog., &c.*)

SALONICA (an. *Thessalonica*), a celebrated city and
seaport of European Turkey, cap. sandjak of its own name,
at the N.E. extremity of the gulf of same name, 185 m.
N.N.W. Athens. lat. 40° 20′ 47″ N., long. 22° 57′ 13″ E.
Its population was estimated by Mr. Walpole at about
42,000, by M. Beaujour at 60,000, and by later, though, perhaps, less cautious authorities, at 70,000, of whom about
three fifths are Turks, and the rest chiefly Jews and Franks,
with a few Greeks. Its appearance, when approached
from the gulf, is very imposing, as it is seen from a great
distance, placed on the acclivity of a steep hill, amid cypress trees and shrubs, surrounded by lofty white-washed
walls ascending in a triangular form from the sea, and surmounted by a fortress with seven towers. The domes and
minarets of numerous mosques rise from among the other
buildings, and, being surrounded with cypresses, give an air
of splendour to its exterior.

The circ. of the city walls probably exceeds 5 m.; but,
according to Dr. Clarke, " a great part of the space within
is void." (vol. 443, 8vo. ed.) Its interior presents the same
irregularity, and many of the deformities common to Turkish towns; but, on the whole, as respects cleanliness and
internal comfort, it may contrast favourably with most other
places in Turkey of large size and population. " The houses
are generally built of unburned bricks, and are, for the most
part, little better than so many hovels." (*Clarke.*) Those
of the principal inhabitants, Greeks and Turks, have here,
as in Yanuin, small areas connected with them, generally
occupied by a few trees. The bazaars, at the lower end of
the town, are very extensive, forming several long, but narrow streets shaded either by trellises with vines, or by projecting wooden sheds, with branches of trees thrown across.
The dealers are principally Greeks and Jews; and the shops
are well filled with manufactured goods and colonial produce; but in jewellery, shawls, and the richer articles of
oriental dress, they appear inferior. Some of the mosques
are worth notice from their size and antiquity, especially
two which were formerly Greek churches. Another remarkable edifice, called the Rotunda, after having successively served as a heathen temple and Christian church, has
been converted into a mosque: it has evidently been built
on the model of the Pantheon at Rome. The cupola is
adorned with mosaic work, appearing like eight frontis-
pieces of fine buildings, and in the dome is a circular aperture, as in that of the Pantheon. A fourth mosque has
been formed out of a fine temple of the Thermæa Venus.
This was originally a perfect parallelogram 70 ft. in length,
and 38 ft. in width, supported on either side by 12 columns
of the Ionic order, of the most exquisite proportions. The
Greeks spoiled this beautiful building by endeavouring to
make it cruciform; but the six columns of the *pronaos* remain; and M. Beaujour says, that if the Gothic disfigure-
ments were stripped off, the original edifice would be found

Its original height appears to have been 43 ft.; but the lower part, to the depth of 27 ft., is below the present surface: the span of the arch is 14 ft., and the masonry is of squared white marble blocks, having inscriptions and appropriate bas-reliefs. In the middle of the city is a magnificent ruin, called *Incantadas* by the Spanish Jews resident here, supposed to have been the *propylæum*, or entrance to a circus, consisting of five Corinthian columns, supporting an entablature on which are several figures much defaced, as large as life, and still exhibiting the traces of a master's hand. The castle, which forms a large area, separated by a rampart from the city, has lofty and well-built walls; and, at its highest point, stands the fortress surmounted, like that of Constantinople, by seven towers, called by the Turks *Yedi-koule*, and by the Greeks 'Εντάπυργοι. These towers, however, though occupying the site of the old acropolis, are comparatively modern, having been built by the Venetians.

Salonica, during that period of the late war, when the anti-commercial system of Napoleon was at its height, became an important depôt for British goods, whence they were conveyed to Germany, Russia, and other parts of Europe. They were transported on pack-horses, by long and laborious journeys, into the centre of Europe, through Bosnia into Austria, and through Bulgaria into different parts of Hungary, Transylvania, &c.; the time occupied in travelling from Salonica to Vienna being about 35 days. At all times, however, it has had a considerable trade, which, of late years, has rather increased, particularly as regards the importation of British cotton manufactures. The exports principally consist of silk and tobacco, wool, raw cotton, wheat, and other species of grain, linseed and hempseed, timber, nuts from Mount Athos, &c.; the entire value in 1837 being estimated at £136,614, while the imports for the same year (consisting principally of British manufactures and colonial produce), were valued at £90,810. The import-trade of the rayahs, most of whom are Jews, who purchase by firmans the same commercial privileges as the Franks, is carried on by credits on Vienna, few of the importers having any capital; and orders to England are commonly paid for by drafts on Vienna, where the charge for credit is at the rate of 4 per cent. per annum. Goods are generally sold in the interior at an advance of 10 or 14 per cent. on the invoice cost, leaving a profit varying from 4 to 8 per cent. for the importer. We subjoin

An Account of the number of ships, with their Tonnage, and the Value of their Cargoes, that entered and left Salonica in 1837, specifying the countries to which the Vessels belonged

Countries	Entered.			Cleared.		
	Vessels.	Tons.	Value of Cargoes	Vessels.	Tons.	Value of Cargoes
			£.			£.
British	5	846	17,764	5	846	30
Maltese	3	230	420	3	593	1,305
Ionian	4	183	1,817	4	183	370
French	6	1,124	5,173	6	1,097	20,114
Austrian	16	2,928	9,086	13	2,629	6,492
Russian	9	250	462	1	120	262
Sardinian	13	3,069	6,340	10	2,478	9,300
Greek	243	14,544	23,326	235	11,081	46,962
Turkish	29	6,454	77,169	29	4,766	41,711
Total	329	125,492	90,809	306	23,273	136,614

The intercourse with England was a few years ago principally carried on through Malta by Maltese or Greek vessels: but the trade is now almost exclusively carried on in English bottoms. The ordinary import and export duties are those common to foreign trade in Turkey, viz. 3 per cent. *ad valorem*, being farmed by the Porte under certain restrictions to the pacha, or governor of the city. Salonica has no port; but there is excellent anchorage in the roads opposite the town, which, from the configuration of the gulf, are nearly landlocked. Accounts are kept in piastres of 40 paras, or 120 aspers, and the coins are similar to those of Constantinople (which see). The weights and measures are the same as those of Smyrna, except that the *kislot*, or corn measure of Salonica, =3·78 kizlox of Smyrna.

The commercial classes consist chiefly of Jews and Franks, the Greek population having greatly diminished since the war of independence. Salonica, however, is a metropolitan see, with eight attached bishoprics, and there are numerous Greek churches. The Jews form an important section of the population; they are chiefly of Spanish descent, and obtain a livelihood by commerce and retail trade in the bazaars, those of the lower orders being employed as porters on the quays or in similar offices. The Franks, most of whom reside in the lower part of the city, consist almost exclusively of French and Germans, who have establishments for the management of the transit trade. The situation is said to be unhealthy, especially in autumn, owing to the vicinity of the marshes at the head of the gulf: intermittent fevers are then exceedingly com-

mon, as well as chronic visceral complaints, the result of repeated attacks of those diseases.

Thessalonica was at first an inconsiderable town under the name of *Therma*, by which it was known to Herodotus, Thucydides, and Æschines. Xerxes stayed here some days with his army (*Herod.* vii. 128), and it was occupied for a short time by the Athenians during the Peloponnesian war. According to Strabo (lib. vii.), Cassander changed its name to that of his wife Thessalonica, the daughter of Philip and sister of Alexander the Great. After the conquest of Macedonia by the Romans it was made the capital of the second of the four districts into which that country was divided. It was the residence of Cicero during a part of the time he continued an exile. Valerius raised it to the rank of a colony; and it had an amphitheatre, a hippodrome, and numerous splendid public buildings. It is also extremely interesting from its connection with the early history of Christianity; having been visited by St. Paul, who made many converts, to whom he addressed the Epistles to the Thessalonians. (*Holland's Travels; Clarke's Travels,* vii. 441–478; *Morning Chronicle for July 14, 1838; Cellarii Geog. Antiqua,* i. 1044.)

SALOP, or SHROPSHIRE, an inland co. of England, having N. Denbigh, a detached portion of Flint, and Cheshire, E. Stafford, S. Worcester, Hereford, and Radnor, and W. Montgomery. Area, 859,999 acres, of which about 790,000 are supposed to be arable, meadow, and pasture. Aspect much diversified. No part of the surface is perfectly flat; but the great plain of Salop or Shrewsbury is comparatively level. It extends, lengthwise, from Whitchurch, on the confines of Cheshire, S. to Church Stretton, a distance of about 30 mi.; and from Oswestry, on the confines of Denbigh, to Colebrook Dale on the E. about 22 mi. The Wrekin hill rises out of this extensive plain on its E. side. The S., or rather the S.W. parts of the county contain several ranges of flattish square-shaped hills, divided by beautiful valleys. Soil various, but generally fertile. In the N. it consists of a red sandy loam, like that of Cheshire: in the S., a mixture of clay and loam is most prevalent; and in the W. there is a good deal of gravelly light soil. The harvest is said to be a fortnight earlier on the E. than on the W. side of the county; a difference depending partly on soil, on the greater elevation of the ground on the W. side, partly, also, on difference of soil. Salop is principally under tillage; but, in the S. and W. breeding and dairy are carried on to a considerable extent. A great deal of cheese, sold under the name of Cheshire, but not the genuine article, is made in this county. The wool of the hilly tracts used to be of a peculiarly fine quality, but has deteriorated during the present century in consequence of the efforts of the farmers to increase the size of the animal and the weight of the fleece. The total stock of sheep in the county is supposed to exceed 420,000: producing annually above 7000 packs of wool. Hops are produced on the borders of Hereford. Property variously divided: some estates being very large, while there are many of very moderate degree of size. On the borders of Wales, farms are small, many not exceeding 30 acres; but on the E. side of the county, in the vicinity of Shiffnal, Wellington, Newport, &c., they vary from 100 to 500 acres or more.

The district of Clun forest, in the S.W. part of the county, is divided into small freehold properties, varying in value from £5 to £150 a year, the majority being of the middle class. Their occupiers, who, in most cases, are the owners, employ few labourers, the principal part of the work on their farms being executed by themselves and their families. Leases less common now than formerly, and farms generally held from year to year. Agriculture improving; but, owing to the want of leases of sufficient length, and with proper conditions as to management, continues to be very defective. The succession of crops taken in succession, has been materially diminished since 1890; but two wheat crops still not unfrequently follow each other. These remarks do not, however, apply to the district on the E. side of the county mentioned above, where the farms are large; for there the farmers are active and enterprising, and agriculture highly improved. Turnips extensively cultivated, and for the most part it follows few oats grown. Cattle of mixed breed, and rather inferior. Pork and bacon much used by the people. Large flocks of turkeys raised by some farmers. Average rent of land some places much wanted. Average rent of land is about 17s. 2d. an acre. Principal mineral products, coal, lead, limestone, and freestone. With the exception of Wales, Staffordshire, and Lanarkshire, more than the quantity of iron is made in this county than in any other county of the empire; the Salop furnaces produced above 73,000 tons of iron in 1828, and 100,000 tons in 1839. The furnaces are principally at brook Dale, between Wellington and Witley, &c. china ware, and a very superior species of pottery, are at Coalport on the Severn and its vicinity: pipes, baths,

oesley; carpets at Bridgnorth; gloves at
ome branches of the flannel manufacture
a Shrewsbury and its neighbourhood; but
t portion of the flannel sold in its markets
Menoneth and Denbigh. The Severn,
navigable at Poole, county Montgomery,
unty in a S.E. direction, dividing it into two
portions; and it is besides intersected by
anals. Roads formerly very bad, but now
proved, though still susceptible of much
alop is divided into 15 hundreds, or districts
at denomination, and 216 parishes. It re-
s to the House of Commons, viz., four for
two each for the boroughs of Shrewsbury,
Wenlock, and one for Ludlow. Registered
county, in 1839-40, 8815, of whom 5039
and 3776 for the S. division of the county.
id 47,293 inhabited houses, and 239,014 in-
iom 119,357 were males, and 119,657 fe-
ended for the relief of the poor, in 1839-40,
il value of real property, in 1815, £1,063,709.
and professions, in ditto, £879,933.
in island on the W. coast of Hindostan,
, immediately N. of Bombay island, with
rcted by a narrow causeway. Length 18
rage breadth of about 13 m. Pop. loosely
0, about one fifth of whom may be Portu-
There are two towns on the island, Tan-
ader; the first being neat and flourishing,
rt, several churches, and a considerable
nt. The more remote interior parts of the
ted by wild tribes, having no intercourse
of the coast; but who, being occupied as
coal, bring it down to particular spots,
ried away by dealers in the article, who
ice a payment, settled by custom, of rice,
tools.

arkable objects of Salsette are the cavery, among the most remarkable Buddhic
lia. They are of various sizes and forms,
different elevations over both sides of a
ging to a range of hills which divides the
early equal parts. The largest and most
bears a great resemblance to that of Car-
, 553), and was converted by the Portu-
ch. It is entered through a fine and lofty
a little to the left hand of which is a de-
pillar, surmounted by three lions seated
n either side of the portico is a colossal
early 20 ft. in height. The screen which
tibule from the temple has in its centre a
which are three windows in a semicircu-
re, it is covered with carved figures. The
is 91½ ft. in length, and 38 in breadth,
surrounded on every side but that of the
colonnade of octagonal pillars. Of these,
nearest the entrance have carved bases
rest are not finished in this manner. In
end is a dome-shaped rock, the *daggoe* of
traditionally said to have once supported
umbrella. The roof, like that of the Car-
micircular arch, supported by slender ribs
he various other caves in the hill are
fed, and attached to many are deep and
ns. There are other cave-temples in the
ages of Mompezier and Ambowlee; and
e ruins of a very handsome Portuguese
t monastery. (*Forbes; Lord Valentia;
i. Trev.*)

nf S. America, cap. of the prov. of its own
La Plata, on the high road from Buenos
20 m. N.W. the former city; lat. 24° 30'
/' W. Pop. from 8000 to 9000. "Upon
a neat appearance, and boasts of a cathe-
urches. It is, however, badly situated in
lley, through which flow the rivers Arias
aries of the Salado); the latter of which
abandoned its ancient bed, and seems to
tant period, to burst over the low marshy
ich the city stands." (*Parish's Buenos
s air is unhealthy, but its vicinity abounds
cattle, &c., in which, and in salt, wine,
the city has an active trade. It was
hilip de Lerma in 1582.
ecayed bor., market-town, and par. chap-
o. Cornwall, S. div., hund. of East, 17 m.
a, and 4 m. N.W. Plymouth. Pop. of
It stands on a steep rock, near the Ta-
the principal street runs at right angles,
one above another to the hill-top, on
chapel and town-hall. The latter is sup-
he open space beneath being used for a

market. Streets narrow and ill-built; the houses being for
the most part little better than cottages, though chiefly of
stone from the rock on which the town stands. The chapel
is small; and the living is a curacy subordinate to the vicarage of St. Stephen, value £45 a year. There are also
two places of worship for dissenters, with attached Sunday
schools, and there is a small free school. Saltash, which
appears to have been formerly of more importance than at
present, is principally inhabited by fishermen, or persons
connected with the docks and shipping of Devonport; and
in summer is a favourite resort for holiday-people from
Plymouth and the surrounding neighbourhood. It is likewise one of the chief entrances into Cornwall from Devonshire, and is approached by a ferry over the Tamar, the
revenues of which belong to the corporation. Saltash was
made a free borough in the reign of Henry III., and returned two members to the House of Commons from the reign
of Edward VI. down to the Reform Act, by which it was
disfranchised. It was considered of too little importance
to be included in the provisions of the Municipal Reform
Act. Markets on Saturday: fairs, Feb. 2, July 25, and the
Tuesdays before each quarter-day.

SALTCOATS, a seaport town of Scotland, co. Ayr, partly in the par. Ardrossan, and partly in that of Stevenston,
24 m. S.W. Glasgow, and about 1 m. S. Ardrossan. Pop.
in 1836, about 4000. It has some good houses; but, on the
whole, is indifferently and irregularly built, and mean looking. It has a town-house, with a handsome spire, clock,
and bell. Its name is derived from the salt-works, established in the town for the production of salt, by the evaporation of sea water; but since the repeal of the duty on
salt, they have been nearly abandoned. Magnesia, however, still continues to be produced to some extent. The
principal dependence of the inhabitants is on the weaving
and sewing of muslins, for the Glasgow manufacturers.
There may, in all, be about 500 looms so employed; principally on lappets, gauzes, shawls, trimmings, silks, &c.
About 30,000 tons of coal are annually shipped here for
Belfast, Dublin, &c. A good deal of ship building was formerly carried on, but latterly it has declined. Two congregations belong to the United Associate Synod, and one to
the Relief; and there is a Gaelic chapel. It has a sub
scription library, a parochial school, a free school, managed
by a committee of ladies, and other schools, a savings'
bank, and some friendly societies, a branch of the Ayrshire
Banking Company, &c. The harbour is very defective,
and in this respect it labours under great disadvantages as
compared with Ardrossan. (*New Statist. Acc. of Scotland,
Ayrshire, parishes Ardrossan and Stevenston.*)

SALT CREEK, t., Pickaway co., O. It has six stores,
one fulling-mill, two saw-mills, four tanneries, two distil
leries; five schools, 153 scholars. Pop. 1814.

SALT CREEK, t., Wayne co., O. It has three stores, two
grist-mills, three saw-mills, one oil-mill, two tanneries; one
school. 82 scholars. Pop. 1461.

SALT LICK. t., Fayette co., Pa. It has one store, one
furnace, one fulling mill, four grist-mills, 15 saw-mills; five
schools, 143 scholars. Pop. 1911.

SALT river, Ky., rises in Mercer co., and flowing N.
into Anderson co., and turning W. through Spencer and
Bullitt counties, receiving the Rolling fork, a large branch
from the S., it enters Ohio river at West point. It flows
through some of the most fertile portions of the state.

SALUZZO (Fr. *Saluces*), a city of the Sardinian dom.,
div. Coni. cap. prov., at the foot of the Alps, on an affluent
of the Po, 30 m. S.S.W. Turin. Pop. in 1838, 14,496. It
consists of two portions, one on the summit and declivity,
and the other at the foot of a hill. The upper town is
walled, tolerably well built, and has a castle, which was,
for three centuries, the residence of the marquises of Saluzzo; one of whom, between 1478 and 1480, constructed
the gallery through the Col de Viso. (*Murray's Handbook for Piedmont,* 344.) The lower town is the more
populous, and continues on the increase. The cathedral, a
handsome building, is in a suburb. Saluzzo has several
convents, an intendency, a court of primary jurisdiction, and
a royal college. It is a bishop's see. Its chief manufactures
comprise silk, leather, hats, and hardware: and it has some
trade in wine, corn, and cattle. Under the French, Saluzzo
was the cap. dep. Stura. (*Rampoldi, Cowg., &c.*)

SALZBURG, a city of Upper Austria, cap. of the circ.,
as it formerly was of an archbishopric of the same name,
on the Salzach, a tributary of the Inn, 67 m. S.W. Linz,
and 70 m. E.S.E. Munich. Lat. 47° 48' 10'' N.; long. 13°
1' 23'' E. Pop. in 1834 about 12,000. having decreased considerably since Salzburg ceased to be the cap. of an indep
territory. The Salzach, which here flows impetuously be
tween two masses of rock, divides the city into two portions, connected by a stone bridge, 370 ft. in length. It is
walled, and entered by eight gates; and on a lofty point,
commanding the town and adjacent country, is the *Hoen-
salzberg*, formerly the feudal citadel and residence of the

prince-archbishops, but now used as a barrack. On the opposite side of the river is the *Capucinerberg*, a similar height, surmounted by the Capuchin convent. Owing to the number of its churches, the profusion of marble statues, and flat-roofed houses, Salzburg has the aspect of an Italian city. According to Turnbull, one of the principal hotels, which, he says, forms a fair specimen of the general style of building, "is a heavy pile of masonry, five stories in height, and composed of walls, floorings, and internal divisions, from the basement to the roof, entirely of stone." Generally speaking, the city is dull and gloomy, and its streets narrow, irregular, and grass-grown. The cathedral, constructed in the 17th century, on the plan of St. Peter's at Rome, is large and imposing, and has numerous monuments, sculptures, paintings, and other works of art. It has a fine façade of white marble, occupying the whole side of a public square, with three entrances, flanked by two rows of marble statues. Several of the other churches are highly gilt, and decorated. Monasteries are numerous, but the number of their inmates has been much reduced. In the church of the Benedictine convent is the tomb of Michael Haydn, the musical composer, who, as well as Mozart, was a native of Salzburg. The Mirabel palace is a handsome modern edifice. One of the greatest curiosities in the city is a gateway or tunnel, 430 ft. in length, cut through the solid rock (*Spencer*), though for what purpose we are not informed.

Salzburg has a military and three civil hospitals, several charitable institutions, a government pawn-bank, and a prison. It had formerly a university; but this is now reduced to a lyceum of two faculties, medicine and jurisprudence, with a library of 30,000 volumes, and probably of 120 MSS., some of the 8th and 9th centuries, a botanic garden, zoological museum, &c. In the Benedictine convent is another extensive library, with collections of coins, &c. It has, also, a gymnasium, Ursuline female school, a spacious public cemetery, a public museum, and a theatre. Salzburg is still the residence of an archbishop, who has five suffragans; and is the seat of the superior courts for the circle, &c. It has manufactures of cotton yarn, leather, starch, gunpowder, iron wire, and files, and some transit trade, though this has very much diminished. It is well and cheaply supplied with provisions, but the prevalence of *goitre* is a drawback to its advantages. All travellers agree that it is hardly possible to exaggerate the romantic beauty of the scenery of the neighbourhood.

Salzburg is supposed to occupy the site of the ancient *Juvavia*, destroyed by Attila in 446. In 803, Charlemagne and the ambassadors of Nicephorus, emperor of the East, met in this town to settle the boundaries of their respective empires. In the 13th century the city became the capital of a territory, governed by its archbishops till 1802, when it was secularized. (*Cxtr. Nat. Encyclop.; Berghaus; Turnbull's Austria*, i., 130–143; *Spencer; Germany and the Germans*, ii., 309–312.)

SAMARANG, a town and seaport of Java, on its N. coast, the cap. of a prov., near the mouth of the river of its own name, 240 m. E.S.E. Batavia; lat. 6° 56' S., long. 110° 27' E. Pop. supposed to be at least 20,000, including many Chinese and some Europeans. It is tolerably well built, and is fortified with ramparts and a wet ditch, capable of resisting a native force. It has many good houses, a large church, town-hall, and hospital, a military school, theatre, and observatory. Before it is a deep morass, and it communicates with the sea only by two raised causeways and the river: it is, however, less unhealthy than the lower parts of Batavia. Provisions are cheap; and near the town are many country houses. "The river, or rather creek, is very shallow, and cannot be entered by loaded boats at low water. The roads are also exceedingly insecure; the town owes its importance, therefore, solely to the industry of the natives in the adjoining districts, who raise large quantities of coffee, pepper, and rice. Many ship-loads of the latter are annually exported to China, and to different countries in the archipelago." (*Earl's Eastern Seas*, p. 49.) Samarang is the seat of one of the three civil and criminal courts, and courts martial in the island, and the residence of a governor with extensive authority.

SAMARCAND, a city of indep. Tartary, in Bokhara, on the Sogd, or Zer-Afchan, 120 m. E. Bokhara, lat. 39° 30' N., long. 66° 50' 15" E. Pop., according to Sir A. Burnes, about 10,000. The out-works are said to be about 30 m. in circ., enclosing gardens, parks, fields, and extensive suburbs : the inner wall surrounding the city is of earth, and has four gates. Samarcand has the appearance of having been magnificently built, but it is now in a decayed condition, and gardens, fields, and plantations occupy the place of its numerous streets and mosques. There were formerly upwards of 200 mosques, many of white marble ; but most of these have become mere ruins. Of the 40 *medressas*, or Mohammedan colleges, only three are perfect, one of them forming the observatory of the celebrated Ulug

Beg being extremely handsome, ornamented with bronze and enamelled bricks. Another college, called *Sherdar*, is likewise of very beautiful architecture. The tomb of the famous Timour Bec, or Tamerlane, and his family, still remains ; and the ashes of the emperor rest beneath a mighty dome, the walls of which are superbly adorned with jasper and agate.

Samarcand has several bazaars, and three large khans; but its commercial importance is all but extinguished, Bokhara having been for many years the great entrepôt of its caravan traders, as well as the modern capital of the country. The ancient city, however, is still regarded with high veneration by the people, and till a king of Bokhara has entered Samarcand to his rule, he is not viewed as a legitimate sovereign : indeed, its possession becomes the first object on the demise of one ruler and the accession of another. Paper, made of silk, is said to have been early manufactured at Samarcand, but ordinary paper is now supplied from Russia. The situation of the city has been deservedly praised by Asiatics, since it stands near low hills, in a country elsewhere plain and level. The climate is dry and healthy; good water is supplied from a great number of fountains, communicating by pipes with the river, and the neighbourhood furnishes abundance of fruit, and other supplies for the market.

Samarcand, which was taken in 1220 by Jenghiz-khan from the Sultan Mahomet, became under Timour the capital of one of the largest empires in the world, and the centre of Asiatic learning and civilization, at the same time that it rose to high distinction on account of its extensive commerce with all parts of Asia.

It was reunited to Bokhara by Abdullah at the close of the 16th century, since which it has gradually fallen to its present rank as a mere provincial town ; and we may now search in vain for its ancient palaces, the beauty of which is eulogised by the Arab historians. (*Burnes's Bokhara*, i. 270–272; *Hagemeister sur l'Asie Occidentale; Ritter's Asien*, vol. v.; *Dict. Géog.*)

SAMBOR, a town of Austrian Poland, cap. circ. of same name, on the Dniestr, 44 m. S.W. Lemberg. Pop. nearly 10,000. (*Berghaus.*) It is tolerably well built, and has several Rom. Catholic and United Greek churches, a hospital, a criminal tribunal, mining court, salt ascendancy, gymnasium, &c. Its inhabitants are employed partly in the manufacture and bleaching of linens, and partly in making salt. Rhubarb is cultivated in the neighbourhood. (*Cestor. Nat. Encyc.*)

SAMOS, a famous island of the Ægean sea, now belonging to Turkey, off the W. coast of Asia Minor, from which it is separated by the narrow strait called the Little Boghaz, only 2 m. across. It has on the N. the gulf of Scala Nova, is about 30 m. in length, E. and W., by about 8 or 9 in mean breadth ; mount Kerki, on its W. extremity (m. Cataboto), from its collecting clouds and generating thunder), being in lat. 37° 43' 88" N., long. 26° 38' 21" E. The pop. was estimated by Tournefort at 12,000 ; Mr. Tennent, later, though inferior authority, estimates it at 60,000 but this, we have little doubt, was decidedly beyond the mark, even at the time when it was framed ; and since the revolution in Greece there has been a good deal of emigration from the island. A chain of mountains runs from one extremity of the island to the other : most of these are covered, as in antiquity, with forests of oaks and other trees, though in parts they are precipitous and bare. Tournefort says that they consist principally of white marble. It has several pretty extensive valleys, especially on its S. coast, which, being well watered by streams from the hills, produce, even with the most deficient culture, excellent crops of wheat and other grain, with olives, figs, oranges, and other fruits, wine, silk, cotton, &c. In antiquity it was celebrated for its extraordinary fertility ; it was then, also, cultivated with the utmost care, and the walls still exist which were built to form the sides of the mountain terraces, and to facilitate their culture. (*Tournefort*, i., 407.) It still continues to be the most productive island of the archipelago. It annually exports considerable quantities of corn ; from 25,000 to 30,000 cantars grapes, and about 15,000 barrels raisins. The only thing which Strabo did not admire in Samos was its wine (*lib. xiv.*) ; but Tournefort says that when properly made, its muscadel wine is very superior. Perhaps it was not produced in antiquity. Oil and valonia are also considerable articles of export. Wolves and other wild animals occasionally commit ravages among the oxen and sheep ; poultry are excellent, and partridges exceedingly abundant. Exclusive of marble, it is said to furnish iron, lead, and even the precious metals.

Having voluntarily surrendered to the Turks, this island has been less harshly treated by them than most others in the archipelago. It, however, zealously espoused the cause of the other Greeks during the revolutionary struggle, and, though it was assigned to the sultan by the treaty which recognised the independence of Greece, the inhabitants refused at first to submit to his officers. Previously to this event, the government of the island was substantially vested in three primates, chosen by the inhabitants. But this, we may believe Mr. Turner, by whom the higher order of unians are called "the most unprincipled miscreants in existence" (iii. 110), would seem to have been no great one. Besides being oppressed by the agents of the sultan and the primates, the island has, also, been fleeced by a swarm of Caloyers, Papas, and other Greek priests, whose very claim to live at the public expense is, that they are able to repeat mass from memory. A considerable sum is allotted to the bishop of Nicaria for his important service of reading, once a year, the water and cattle of the inhabitants! (*Tournefort*, i. 408.)

The present capital of the island, called Khora, or Megalis, is on its S. side, about 2 m. from the sea, on the low extremities of a mountain, on the ascent of which the site of the ancient city was situated. Though not withsome good houses, it is a miserable town, having stony, unpaved, and hardly passable streets. Vathi, on the one of the island, is larger than Khora, and has an exit harbour; but it, also, is a wretched place, with streets from 6 to 8 feet in width, execrably paved and steep. (*er, ill.*, 107.)

is the present state of an island that, in antiquity, one of the most famous in the Ægean sea. Samos attained to great distinction. She was one of the powerful of the states belonging to the Ionian confederacy; and was able, by means of her fleets, to maintain her independence after Cræsus and Cyrus had reduced the rest of Ionia, on the continent. The city of Samos, the shore of the island, was extensive, and populous, fortified, and adorned with many noble public works. Among the other great works executed by the Herodotus specifies a tunnel, which they had carried through the mountain, to convey a supply of water to an immense mole, constructed for the security of the harbour (of which the remains still exist), about 120 feet, and advanced in a curved line about the sea; and the largest temple of which he had any knowledge. (*Herod.*, lib. iii., cap. 60.) the people to which the venerable historian alludes was sacred to Juno, and stood a little to the W. of the city of mbrasus. The island, indeed, was especially sacred to Juno, and was supposed to have been the place and where she espoused Jupiter. Hence, says Virgil speaking of Carthage, where the goddess had also a temple,

hæc Juno fertur terris magis omnibus unam posthabita coluisse Samo. —*Æneidos, i., lin. 19.*

of the goddess in this temple was very ancient, been the work of Smilis, a contemporary of Dædalus; among other statues in and near the temple, of Jupiter, Minerva, and Hercules, by Myron, most celebrated sculptors of antiquity. Mark carried off these statues to Rome; but Augustus the Minerva and Hercules be returned to Samos, that of Jupiter.

Games, instituted in honour of Juno, called *Hpa*, were celebrated here with extraordinary Like other great temples, that of Juno was well who implored the protection of the goddess notices the arrival of deputies from praying that the *petustum Argli jus* might (*Ænæal, iv.*, cap. 14.) The subsequent historic edifice is but little known. It has, however, fice to the ravages of time, or of barbarism, when visited by Tournefort, more than a one of two columns were all that remaining.

tings Samos was famous in antiquity for was everywhere in great request; and touring it is even said to have been distinguished. of Samos experienced the mutations ornaments of most Greek states. Originally so were superseded by a mixed government, sometimes to democracy, and sometimes to occasionally it was subject to tyrants. Of celebrated is Polycrates, who attained the 6th century B.C. His object seems in the government partly by force, partly reducing the inhabitants, and partly by schemes of foreign conquest. (*Mitford's* at a period subsequent to the death of inveigled and crucified by the satrap province, the Samians were attacked by Pericles; who, after an obstinate months' duration, succeeded in reducing somewhat later period it received a s. During the contest between Mark

Antony and Augustus, Samos was, for a while, the head quarters of the former and of Cleopatra, who kept court here with more than regal magnificence. After Augustus had become the master of the Roman world, he passed a winter in this island, which he restored to its freedom, and at the same time conferred on it other marks of his favour. It afterwards became subject to the Greek emperors; and finally, in the 16th century, to the Turks, under whose brutalising sway it has been reduced to the miserable state in which we now find it.

Of the many illustrious individuals that Samos has produced, Pythagoras is by far the most distinguished. The era of his birth is not quite ascertained, but it appears to have occurred about 580 years B.C. He early visited Egypt and other ancient seats of learning; but, on his return from his travels, being, as is said, dissatisfied with Polycrates, he emigrated to Magna Grecia, and founded at Crotona a school of philosophy, that speedily attained to the highest celebrity. Samos also gave birth to Rhœcus, said by Herodotus to have been the architect of the temple of Juno, to Theodorus the sculptor, and a host of others. Anacreon was among the distinguished guests invited by Polycrates to Samos.

The narrow strait between Samos and the mainland is famous in ancient history for the great victory gained in it and the adjacent promontory of Mycale, over the fleet and army of Xerxes, on the same day that the forces he had left in Greece, under Mardonius, were destroyed at Platæa. (Exclusive of the authors already referred to, see *Ancient Universal History*, viii., 252–254, and xiii., 510, 8vo ed.; *Anacharsis*, cap. 74; *Fosciolati Lexicon*, voce Samus, &c.)

SAMPSON, county, N. C. Situated a little S.E. of the centre of the state, and contains 800 sq. m. Bounded W. by South river. Drained by Black river and its branches. It contained in 1840, 6735 neat cattle, 4432 sheep, 20,562 swine; and produced 3872 bushels of wheat, 1061 of rye, 130,951 of Indian corn, 2196 of oats, 96,127 of potatoes, 29,974 pounds of rice, 342,309 of cotton. It had eight stores, 16 grist-mills, 10 saw-mills; one academy, 35 students; 13 schools, 275 scholars. Pop., whites, 7475; slaves, 4425; free coloured, 257; total, 12,157. Capital, Clinton.

SANA, a city of Arabia, the cap. of Yemen, and the residence of the imâm, in a valley from 6 to 9 m. in breadth, and 4000 feet above the level of the ocean, near the head of the Shab river, and about 150 m. N.N.E. Mocha. Pop., according to the estimate of Mr. Cruttenden, in 1836, 40,000. The city is walled; as is also its suburb of Bir-el-Azab, which was open in Niebuhr's time. The city and suburb, together, are said to be 5¼ m. in circ. The walls are mounted with cannon, but these are in a very bad condition. At both the E. and W. extremities of the city is a castle, having each a palace of the imâm. The streets of Sana are narrow, though broader than those of Mocha, and some other Arabian cities. A handsome stone bridge is thrown across the principal street, down which a stream of water runs in wet weather. Houses principally built of brick, with open holes for windows, closed when necessary by wooden shutters; but some, belonging to the higher classes, have glass windows, beautifully stained. The palaces are built of hewn stone, plastered over with grey-coloured mortar. All the private residences in Sana appear to be furnished with fountains. There are about 20 mosques, very elaborately adorned, many having their domes gilt, especially those in which are the tombs of the imâms. The public baths are both numerous and good: they are on the same plan as those of Egypt, and "a favourite resort of the merchants, who meet here to discuss the state of trade, and the news of the day over their cup of keshr, and their never-failing hookah."

A part of the city is appropriated to the Jews, who amount to about 3000. Each pays about a dollar a year for permission to reside; and a sheikh is appointed, who is responsible for the regular payment of this impost, and of the heavy taxes laid upon their vineyards, gardens, &c. The Jews subsist chiefly by the sale of silver ornaments, gunpowder, and spirituous liquors, and many by working as common artisans, such as shoemakers, &c. There are also many Hindoos among the population, who, like the Jews, are obliged to conceal as much as possible the property they possess, for fear of exaction. The Mohammedan merchants are generally wealthy, and live in good style. The principal trade of Sana is in coffee, the city being in the heart of the coffee country of Yemen. The article is brought into the market in December and January; and considerable quantities of it are retained in the warehouses. It is, however, little used for home consumption, the favourite beverage being *keshr*, an infusion of the husk. The coffee-husk accordingly fetches here the higher price of the two, from 4 to 12 dollars per 100 lbs. being paid for it. Very fine silk goods, spices, sugar, &c., are exposed for sale in the bazaars. The imports are principally piece goods and Persian tobacco; with dates, and a great quantity of thread, or rather twist for weaving. Glass is in great request, and is principally supplied from Egypt. The import duties at Sana are so slight as to be almost nominal.

The climate is too dry to be healthy; rain seldom falls, and famine appears to be a frequent result. Some inscriptions, supposed to be in the ancient Himyari character, have been discovered here (see *Geog. Journal*, viii., 287), but travellers have hitherto found few, or no other antiquities. The greater part of the fortifications, and an aqueduct now ruined, are said to have been the work of the Turks, who held the sovereignty of the country till about two centuries ago. (*Cruttenden, in Geog. Journal,* viii.; *Niebuhr, Voyage en Arabie, &c.*)

SANDBORNTON, p. t., Belknap co., N. H., 21 m. N. Concord, 502 W. Bounded E. by Great bay of Winnipiseogee lake, W. partly by Pemigewasset river. Incorporated in 1770. It contains five churches, a Congregational, Methodist, Baptist, Free-will Baptist, and Christian; five stores, three fulling-mills, one woollen factory, one cotton factory with 700 spindles, seven grist-mills, 10 saw-mills, four tanneries, one printing-office, two binderies; one academy, 148 students; 23 schools, 1037 scholars. Pop. 2745.

SANDERSVILLE, p. v., capital of Washington co., Ga., 28 m. S. E. Milledgeville, 658 W. It contains a courthouse, jail, seven stores, 25 dwellings, and about 125 inhabitants.

SANDFORD, p. t., Broome co., N. Y., 20 m. E. Binghampton, 120 m. S.W. Albany, 304 W. Drained by Ocquaga creek, flowing into Delaware river. It contains one store, two grist-mills, nine saw-mills; nine schools, 369 scholars. Pop. 1173.

SANDISFIELD, p. t., Berkshire co., Mass., 126 m. W. by S. Boston, 361 W. Incorporated in 1726; first settled in 1750. Drained by a branch of Farmington river. It contains two churches, a Congregational and Baptist, eight stores, one fulling-mill, one woollen factory, two grist-mills, six saw-mills, three tanneries; 14 schools, 382 scholars. Pop. 1464.

SANDLAKE, p. t., Rensselaer co., N. Y., 17 m. E. Albany, 384 W. Watered by Poestenkill and Wynantskill creeks. It contains five churches, a Presbyterian, two Methodist, a Baptist, and a Lutheran; 11 stores, one cotton factory with 1000 spindles, one glass factory, one flouring-mill, four grist-mills, 27 saw-mills, two tanneries; 21 schools, 1258 scholars. Pop. 4303.

SANDUSKY, river, O., rises in Richland and Crawford counties, and after a course of 90 m., enters Sandusky bay.

SANDUSKY, county, O. Situated toward the N.W. part of the state, and contains 320 sq. m. Drained by Sandusky river and its branches. It contained in 1840, 10,513 neat cattle, 8452 sheep, 33,914 swine; and produced 119,122 bushels of wheat, 6890 of rye, 142,628 of Indian corn, 7732 of buckwheat, 1962 of barley, 77,073 of oats, 78,403 of potatoes, 83,722 pounds of sugar. It had two commission houses in foreign trade, 14 stores, three flouring-mills, eight grist-mills, 26 saw-mills, one oil-mill, three tanneries, one distillery, one pottery, two printing-offices, two weekly newspapers; 56 schools, 5159 scholars. Pop. 10,182. Capital, Lower Sandusky.

SANDUSKY CITY, p. v., port of entry and cap. of Erie co., O., 60 m. W. Cleveland, 110 m. N. Columbus, 414 W. Situated on the S. side of Sandusky bay, fronting the opening into lake Erie, three miles distant, of which it has a delightful view. The entire village is based on an inexhaustible quarry of the finest building stone, of which many of its edifices have been erected. Excepting for three winter months, its wharves are thronged with steamboats and other lake vessels, whose arrival and departure enliven the view. It contains four splendid churches, an academy of stone, three stories high, 26 stores besides groceries and provision houses, a ship-yard where steamboats and other vessels are built, 300 dwellings, and about 2000 inhabitants.

SANDWICH, a cinque-port, mun. and parl. bor. of England, co. Kent, lathe St. Augustine, hund. Eastry, on the Stour, about 2 m. from its mouth, and 65 m. E. by S. London. Area of town and port, 1960 acres. Pop., in 1831, 3136. The parliamentary borough however comprises, with the foregoing, the parishes of Deal and Walmer, and the extra-parochial hamlet of St. Bartholomew, having an aggregate area of 3810 acres, and a population in 1831, of 12,235. Sandwich is divided into the three parishes of St. Mary, St. Peter, and St. Clements. It is washed on the N.E. by the river Stour, and surrounded on every other side by a dyke, the remains of its old fortification. It is irregularly built, and has a more ancient appearance than, perhaps, any other town in the county. The streets are well paved, and lighted, and the inhabitants are supplied with excellent water from the river, and from a spring which rises near the Eastry, and is brought to the town by a canal, 3 m. in length. St. Clement's church is a spacious building, with a massive tower of Norman architecture rising from four semi-circular arches in the centre of the building, and supported on strong piers. In some parts it is curiously ornamented. The living is a vicarage, with a nett income of £310 a year. St. Mary's is also a vicarage, worth £117 a year nett. Both the foregoing parishes are in the gift of the archdeacon of Canterbury. St. Peter's is a rectory, in the gift of the crown and the corporation of Sandwich alternately, worth £144 a year nett. There are places of worship for Independents and Wesleyans; two hospitals, one founded in the 12th century, and accommodating 16 residents, who must be freemen; the guildhall, built in 1579, and a new house of correction, comprise most of the remaining public buildings. The free grammar school of Sandwich was founded in the reign of Elizabeth, and received considerable endowments in lands in 1563. Its governors are the mayor and corporation: it has four scholarships in Lincoln college, Oxford, of which two are in the appointment of the governors of the school, and two in that of the rector and fellows of the college; and four in Caius college, Cambridge, nominated in a similar manner. It has, also, a national school and other charities. The town has been for several years in a depressed and declining state. It has no manufactures, and its trade is trifling, consisting principally in the importation of coal for the use of the town and neighbouring country. The scheme of straightening the course of the Stour to the sea so as to form a canal, has been abandoned for want of capital and enterprise. (*Mun. Bound. Rep.*)

Sandwich was first incorporated by Edward III. Its corporation consists of four aldermen and 72 councillors, styled the mayor, jurats, and commonalty of the town and port. Their jurisdiction extends over Ramsgate, Sarr, and Walmer, and did formerly over Deal, which are all members of this cinque-port. Sandwich has, however, no commission of the peace except upon petition or grant. It has sent two members, usually styled barons, to the House of Commons since the 42d of Edward III. Previously to the Reform Act, the right of voting was in the freemen, resident and non-resident, the freedom being acquired by birth, gift, marriage, apprenticeship, ownership, and residence. Registered electors in the new parl. bor., in 1839-40, 977; corporation revenue, in 1839, £1339.

Sandwich, formerly called Lundenwick, appears to have risen into consequence, on the decline of Richborough, the an. *Rhutupium*, about the 6th century. It was long a place of considerable trade, and continued, till a comparatively late period, to be a kind of out-port to London, many goods being conveyed by land to and from the capital. Market, Wednesday and Saturday; fair, December 4, for clothing, &c. (*Mun. and Bound. Reports, &c.*)

SANDWICH, p. t., Carroll co., N. H., 49 m. N. Concord, 530 W. Bounded S.W. by Squam lake. Watered by Bearcamp river. Chartered in 1763. It contains five churches, a Congregational, two Methodist, and two Free-will Baptist; seven stores, two fulling-mills, five grist-mills, eight saw-mills, one oil-mill, four tanneries; one academy, 34 students; 20 schools, 785 scholars. Pop. 2525.

SANDWICH, p. t., Barnstable co., Mass., 59 m. S.S.E. Boston, 459 W. Bounded N.E. by Cape Cod bay, W. by Buzzards bay. A ship canal has been proposed to connect these bays, which would be 5 m. long, through level ground, and would save the tedious and somewhat dangerous navigation around cape Cod. The village contains four churches, a Congregational, Unitarian, Methodist, and Roman Catholic. In other parts of the town are six churches, four Methodist, a Congregational, and Friends. It contains 17 stores, one glass factory, seven grist-mills, one saw-mill; three academies, 180 students; 31 schools, 1130 scholars. Pop. 3719.

SANDWICH ISLANDS. (*See POLYNESIA.*)

SANDY, river, Va. and Ky., rises in Logan and Tazewell counties, Va., and flowing N.W. and N. for nearly 200 m., it enters the Ohio at Cattlesburg. For a distance of more than 70 m., it forms, by the Tug fork and main river, the boundary between Virginia and Kentucky. It is boatable for 500 m. from its mouth.

SANDY CREEK, p. t., Oswego co., N. Y., 161 m. N. by N.W. Albany, 391 W. Bounded W. by lake Ontario. Drained by Little Sandy creek. It has two churches, a Presbyterian and Methodist, four stores, one fulling-mill, one furnace, one grist-mill, five saw-mills; one academy, 45 students; 12 schools, 483 scholars. Pop. 2428.

SANDY HILL, p. v., semi capital of Washington co., N. Y., 53 m. N. Albany, 423 W. Situated on the E. bank of Hudson river, immediately above Baker's falls. The streets form a triangular figure, leaving an open area of that form in the centre. It contains a courthouse, jail, three churches, a Presbyterian, Methodist, and Roman Catholic, eight stores and groceries, various mills and manufactories, 150 dwellings, and about 1000 inhabitants. The river here affords great water-power, which is but partially improved. A navigable feeder of the Champlain canal, which is 1 m. distant, passes through the place.

SANDY HOOK, a sandy beach, Middletown t., Monmouth—

mooth co., N. J., is 6 m. long, and from half a mile to a mile wide, enclosing Sandy Hook bay, 7 m. long and 6 m. wide, forming the outer harbour of New-York. It has a light-house on its N. point.

SANDY LAKE, p. t., Mercer co., Pa., 228 m. N.N.W. Harrisburg, 263 W. Drained by Sandy creek, flowing into Sandy lake in its N.W. part. It contains three stores, one furnace, three fulling-mills, four grist-mills, nine saw-mills, one oil-mill; eight schools, 184 scholars. Pop. 1366.

SANDYSTON, p. t., Sussex co., N. J., 87 m. N. Trenton, 256 W. Bounded W. by Delaware river. It has one store, three grist-mills, three saw-mills; nine schools, 279 scholars. Pop. 1209.

SANFORD, p. t., York co., Me., 83 m. S.W. Augusta, 12 W. Incorporated in 1768. Watered by Mousam river. t has five stores, one fulling-mill, two woollen factories; 0 schools, 1052 scholars. Pop.2233.

SANGAMON, river, Ill., rises in Vermillion co., and by a winding course S.W. W., and N.W., it flows into Illinois ver. It is navigable for steamboats of 100 tons burthen to etersburg, and in high water, one has proceeded to within m. of Springfield. It flows through a highly fertile untry.

SANGAMON, county, Ill. Situated near the centre of the ate, and contains 900 sq. m. Sangamon river and its butaries afford extensive water-power. Organized in 21. It is one of the most fertile counties of the state. It ntained in 1840, 20,878 neat cattle, 18,233 sheep, 69,016 ine; and produced 74,529 bushels of wheat, 8533 of rye, 90,160 of Indian corn, 2359 of barley, 193,873 of oats, 532 of potatoes, 9500 pounds of tobacco. It had 50 res, five flouring-mills, 93 grist-mills, 40 saw-mills, 21 mills, nine tanneries, four distilleries, four potteries, re printing-offices, four weekly newspapers; three academies, 193 students; 90 schools, 844 scholars. Pop. 14,716. ital, Springfield.

ANGERFIELD, p. t., Oneida co., N. Y., 15 m. S. Utica, m. W.N.W. Albany, 372 W. Watered by Chenango r and Oriskany creek. It contains three churches, two sbyterian and a Baptist, six stores, two fulling-mills, one ace, one grist-mill, eight saw-mills, one tannery, two lleries; two academies, 150 students; 12 schools, 662 dars. Pop. 3251.

ANGERVILLE, p. t., Piscatiquis co., Me., 77 m. N.W. usta, 672 W. Incorporated in 1814. Watered by a I branch of Sebasticook river. It has four tanneries, 11 ols. 510 scholars. Pop. 1197.

NILAC, county, Mich. Situated on the N.E. part of ettled portion of the peninsula, and contains 730 sq. m. ded E. by lake Huron. Watered by Cass and Black . It is unorganized. The eastern part contains ex-t pine timber.

NQUHAR, a royal and parl. bor. of Scotland, co. ries, in the valley of the Nith, and near the left bank at river on the road from Dumfries to Ayr, 26 m. V. the former. Pop., in 1831, 1527. It consists prin-/ of a main street along the line of the high road. It town-hall, with a tower and clock, a handsome parish 1, built in 1823, two chapels in connection with the ated Secession Church, and a chapel for Anabaptists; parochial and other schools, a subscription library, gs' bank, &c. The inhabitants are principally de-t on the weaving of cottons, and on the sewing and dery of muslins for the Glasgow manufacturers. is an extensive carpet manufacture at Crawick Mill, m. from the town.

uhar seems to have derived its origin from its fine le, now in ruins. This, which formerly belonged to s of Sanquhar, having been purchased in 1630 by an · of the last duke of Queensberry, became, on the of the latter, with other vast possessions in Dum-e, the property of the family of Buccleugh. It was a royal borough in 1598, and is united with Dum-nan, Lochmaben, and Kirkcudbright in sending one to the House of Commons. Registered electors, 66. Corporation revenue, in 1839-40, £73. Coun-7. (New Statistical Account of Scotland, art. San-and Official Returns.)

A CRUZ. (See Teneriffe.)

ANDER (an. Portus Blendium), a city and seaport . coast of Spain, cap. prov. of its own name, on the / the bay of the same name, running into the bay . 50 m. W.N.W. Bilbao. Pop., according to Mi-716. It is built on the slope of a hill, and has ets lined with tolerably respectable houses, the public edifices being the cathedral, two parish and three hospitals. Few of these, however, iy architectural merit; and Captain Cook, an in-a veller, states, that "Santander is almost the only pain of similar magnitude, where no artist in any t has left a memorial of his skill." (Sketches in .) It is a thriving town, however, with a con-

iderable number of new houses, very unusual in Spain; and it is the chief seaport of Old Castile, it having been the principal object of the government for some years back to make it one of the principal marts for the supply of Madrid. It has a large trade with Cuba, to which it sends the wheat of Castile, large mills being erected in the neighbourhood for converting it into flour previous to embarkation. The exportation of wool is at present shared with Bilboa; but when the roads are completed, it will have the superiority over that port from its greater proximity to the wool-bearing districts. There are iron-mines in the neighbouring mountains; but, owing to the disturbed state of the country for some years past, they have been little wrought. The astilleros, or building establishment of the marine, formerly much employed, is now almost in ruins; and the forests of the Montana, which once supplied Spain with nearly all the timber for the navy, are now seldom used, except for the supply of fuel. The harbour of Santander is large, well sheltered, easily accessible, and sufficiently deep for all trading vessels. The vicinity produces an abundance of wheat and other grains, fruits of several varieties, and large quantities of cattle; the coast also swarming with salmon and other kinds of fish. (Cook's Sketches, i., 75-78; Minano, &c.)

SANTAREM (an. Praesidium Julium), a river-port and town of Portugal, formerly the residence of the court, prov. Estremadura, cap. Comarca, on the Tagus, 45 m. N.N.E. Lisbon. Pop. estimated at 8000. It is built on a hill, and consists of three separate parts; the Maravilla on the summit, the Ribera on the E. declivity, and the Alfange on the W. and S., descending to the river's bank, and commanded by the fortress of Alcazaba. Only a few portions of its old walls remain. It is well built, and has some good public edifices; but these are much neglected, and several have almost fallen to ruin since the removal of the court to Lisbon, in the 15th century. Besides numerous churches and convents, Santarem has several hospitals and asylums, and two Latin schools; and it is the seat of the Patriarchal seminary, the highest ecclesiastical establishment in the kingdom. Its environs are fertile, and well cultivated, and it has an active trade with Lisbon. (Dict. Geog., &c.)

SANTEE, river, S. C., is one of the largest rivers of the state, and is formed by the union of Congaree and Wateree rivers, about 25 m. S.E. Columbia. Both these branches rise, by different names, in N. C., where the main branch of the Wateree is called Catawba r., and of the Congaree, Broad r. It enters the Atlantic by two mouths, about 30 miles below Georgetown. It is navigable for boats in high water, by the Wateree and Catawba branch, 300 m. to Morgantown, N. C. Steamboats ascend the Congaree branch to Columbia. A canal 22 m. long connects Santee with Cooper r., by which the produce of a large section of this state, and of a part of N. C. finds a market at Charleston.

SANTORINI (an. Thera), an island of the Ægean sea, belonging to the S. Cyclades, 65 m. N.N.E. from the nearest point of Crete, Mount St. Elias, the highest point of the island, being in lat. 36° 20' 45" N., long. 25° 22' 8' E. Pop. 12,000. This island is shaped like a crescent, or rather horse-shoe, the concave side to the W., forming a bay, sheltered by the islands Therasia, Aspronisi, &c. The island has a dismal appearance from the sea, consisting wholly of black volcanic rocks, without wood, rivers, or rivulets; but it has, notwithstanding, some very fertile districts, the decomposed volcanic rocks and ashes supplying a fruitful soil, which being carefully cultivated, produces corn, cotton, and large supplies of wine. The inhab. have no water, other than that which they collect in cisterns; and the calcined rock, being of a light consistency, the houses are rather excavated in it than built. Pyrgos in the centre of the island, near the seat of the ancient Thera, and Scaros, on the coast of the bay, are the only towns of any consequence. The inhabitants are very industrious; and have sustained little other inconvenience from the Turkish dominion except that of paying the tribute due to the Porte.

It was the general opinion of the ancients, that this island, and others in its vicinity, had been thrown up from the bottom of the sea; and Pliny says that this event occurred in the 4th year of the 135th Olympiad. (Hist. Nat., lib. ii. cap. 87.) No doubt, however, this date is erroneous; as it appears from Herodotus, that the island was inhabited 1550 years B. C., or 1312 years before the epoch assigned by Pliny for its appearance. (Herod., lib. iv. caps. 147, 148, and 151.) Probably, unless the date given by Pliny be vitiated, he may have referred to some eruption that had occasioned an enlargement of the island. The convulsions of which it was anciently the theatre, have not been suspended in more modern times; a new island having been thrown up near its coast in 1573, and another in 1707, each being preceded by a violent volcanic eruption. In remote antiquity it was called Calliste, or the beautiful, an epithet that never could have been applied to it, had its appearance then been at all like that which it now exhibits. The ruins of its ancient city, Thera, on the hill now called St. Elias, evince its extent and magnificence. (*Tournefort*, i., 268, &c.; *Savvini's Greece and Turkey*, 237, &c. English trans.; *Herodote par Larcher*; *Tab. Georg.*, art. *Thera*, &c.)

SANTOS, a town and seaport of Brazil, prov. St. Paul, in a low and unhealthy situation on the N. side of the island of St. Vincent, 35 m. S.S.E. St. Paul, lat. 23° 56′ 15″ S., long. 46° 0′ 15″ W. "Santos is a place of considerable trade, being the storehouse of the great captaincy of S. Paulo, and the resort of many vessels trading to the Rio de la Plata. It is tolerably well built; and its pop., consisting chiefly of merchants, shopkeepers, and artificers, amounts to 6000 or 7000." (*Mawe*, p. 83.) The pop. has, however, increased materially since the publication of Mawe's work. Several rivulets flowing from the mountains unite in one great river a little above the town. The port is formed by the continent and the island St. Amaro. There are two entrances, but that of the S. is alone navigable by large vessels; the other, which is formed by the river Bertioga, being fit only for small craft. The harbour admits ships of large burden, which are sheltered from all winds except those from the S.S.W. round to the S.E. A pilot is not absolutely necessary on entering. "In advancing into the river Santos, you will have 10, 9, 8, and 7 fathoms water, until you near the bar, upon which there are only from 4½ to 5 fathoms: the entrance is narrow, but the starboard side is much the boldest, and has 19 fathoms water close to the shore. After passing the first *barra-grande*, the water deepens to 15 and 16 fathoms within 1½ fathoms of the shore. The best anchorage will be abreast nearly of the centre of the town, in 7 fathoms, on a bottom of mud. Provisions are abundant, and good water may be obtained by sending a boat about 7 m. farther up the river." (*Blunt's Sailing Directions*.) The part called the Narrows is defended by two forts.

Though the commerce of Santos will not bear to be compared with that of Rio or Bahia, it is very considerable. Sugar is the great article of export; and the shipments of it have latterly been increasing. The imports are similar to those of Rio, which see. Being, as it were, the port of St. Paul's, an extensive intercourse is carried on with the latter. We subjoin

An account of the exports of Sugar from Santos during the three Seasons ending with 1839–40.

Shipped for	1839–1840.	1838–1839.	1837–1838.	1836–1837.
	Arrobas.	*Arrobas.*	*Arrobas.*	*Arrobas.*
Europe	176,000	130,000	12,000	51,000
River Plate	165,500	74,000	108,380	208,000
Ports of Brazil	190,350	38,500	45,000	142,000
Valparaiso	162,000	104,000	66,500	140,000
United States	100,000	50,000	15,000	7,000
Total	624,750	397,100	249,000	530,000

SAONE (HATTE), a dep. of France, reg. E.; between lat. 47° 15′ and 48° N., and long. 5° 35′ and 7° E., having N. the dep. Vosges, E. Haut-Rhin, S. Doubs, and W. Cote-d'Or and Haute-Marne. Length, N.E. and S.W. about 70 m., breadth varying from 25 to 40 m. Area, 530,990 hectares. Pop., in 1836, 338,910. In the N. and E. are the Vosges mountains, and their ramifications. The general slope is to the S.W., in which direction the Saône traverses the dep. throughout its centre. The Oignon forms its S.E. boundary. There is a considerable extent of rich soil. In 1834, the arable lands were supposed to comprise 256,103 hectares, meadows 58,993 do., vineyards 11,769 do., and woods 154,938 do. Agriculture has made some progress within the present century; but it is still very backward. The produce of corn, pulse, &c., exceeds the demand for home consumption. Wheat, oats, and barley, are the principal crops. In 1835, according to the official returns, nearly 1,890,000 hectol. of grain of all kinds were harvested, besides about 1,000,000 do. potatoes. The vineyards form a principal source of wealth. The produce may be estimated at about 350,000 hectol. a year. The wines of Ray, Charicy, Navenne, Quincy, Gy, and Champlitte-le-Chateau, are the best; they have a fine colour, body *as bon goud*, and may be kept for a long time. *On peut les considérer comme de bons vins d'ordinaires de troisième qualité.* (Jullien, p. 135.)

Near the Vosges, large quantities of cherries are grown for the manufacture of *kirschwasser*. Timber is an important product: and the annual produce of wool is estimated at 130,000 kilogr. In 1835, of 129,312 properties subject to the *contrib. foncière*, 67,294 were assessed at less than 5 fr., and 20,455 at from 5 to 10 fr.; while only 91 were assessed at 1000 fr. and upwards. In minerals this dep. is one of the richest in France. Its iron forges employ about 5000 hands; and bar iron, iron plates, and wire, steel, and various iron goods are made to the annual value of

754

14,000,000 fr. Glass and earthenware, cotton goods, paper, and hats, are also produced: the exports are, however, mostly confined to agricultural products, and iron goods. Haute-Saône is subdivided into three arronds.: chief towns, Vesont, the cap., Gray, and Lure. It sends four *mems.* to the chamber of deputies. Registered electors, in 1838–39, 1052. Total public revenue (1831), 7861,653 fr. According to Hugo, "*Les habitans de la Haute Saône ayant été namé mêlés avec les conquérants francs ou Bourguignons qui ont d'autres parties de la France, représentent assez exactement par leur extérieur, l'ancienne race Gauloise ou la peuple Gallo-romain.*" (*Hugo*, art. *Haute Saône*; *French Official Tables*.)

SAONE-ET-LOIRE, a dep. of France, reg. E. principally between the 46th and 47th degs. of N. lat. and long. 3° 40′ and 5° 30 E., having N. Cote d'Or, E. Jura and Ain, S. Rhone and Loire, and W. Allier and Nievre. Area, 856,472 hectares. Pop., in 1836, 594,180. The E. and W. parts of the dep. are level; the centre is mountainous, the mountains dividing the basins of the Loire and the Saône. These two rivers are, however, united in this dep. by the canal du Centre. Nearly half the surface consists of a rich and fertile soil. In 1834, the arable lands were supposed to comprise 456,393 hectares, meadows 138,665 do., vineyards 37,936 do., and woods 156,694 do. The produce of corn exceeds what is required for the consumption of the dep.: in 1835, 2,556,000 hectolitres, principally wheat and rye, are said to have been harvested, besides 1,629,000 hectolitres potatoes; which last form the staple food of the inhabs. of the mountains. Some of the vineyards in this dep., especially those in the arrond. of Chalons-sur-Saône, produce wine that ranks in the first class of Burgundy. The wines produced in the other districts are known in commerce by the name of *vins de Mâcon*. They are excellent as *vins ordinaires*, but cannot be compared with the first-rate growths. Jullien says that their proper place is in the second class of burgundies, immediately after the finest growths of Beaune; and that they are *en general assez spiritueux, quelquefois trop fumeux, et toujours agréables.* (p. 121.) The produce of wine is estimated at about 500,000 hectol. The arrond. Charolles has some fine pastures; and, in 1830, the dep. was supposed to possess 228,848 head of cattle, and 405,000 sheep, being a much larger stock than in any of the neighbouring deps. A great number of hogs are reared. In 1835, of 129,312 properties subject to the *contrib. foncier*, 79,957 were assessed at less that 5 fr., 96,906 at from 5 to 10 fr., and 22,247 at from 16 to 90 fr.; and 989 at 1000 fr. and upwards. Coal, iron, manganese, and marble are raised; the glass and iron works and potteries are important. The commerce of the dep. centres principally in Chalons-sur-Saône. This dep. is divided into five arronds.; chief towns, Mâcon, the cap., Autun, Charolles, Chalons, and Louhans. It sends 7 mems. to the chamber of deputies. Registered electors, 1838–39, 3,943. (*Dict. Géog.*; *French Official Tables.*)

SARAGOSSA, ZARAGOZA (an. *Caesarea Augusta*), a city of Spain, kingdom of Aragon, prov. of its own name, in a fine plain on the Ebro, crossed here by two bridges, 52 m. S.E. Pampeluna, 156 m. W. by N. Barcelona, and 175 m. E.N.E. Madrid; lat. 41° 47′ N., long. 42° 43″ W. Pop., according to Miñano, 43,440. The limits of the town are marked by a wall partly of turf and partly of stone; and there are eight principal and two smaller gates. It is divided into four quarters and two suburbs, comprising 27 wards of 900 long, narrow, ill-paved, and dirty streets; indeed there is only one wide street in the whole city, viz. the Cosso, which sweeps round the outside of the town on the land side, connecting the market-place and the Ebro. (*Cook's Sketches in Spain*, i. 366.) The houses, generally speaking, are of brick, and three stories high; but few of them have any pretensions to architectural display. The town has an immense number of churches, two of which are cathedrals, thus characterised by Mr Townsend: "That called *El Aseo* is vast, gloomy, and magnificent, exciting devotion, inspiring awe, and inclining the worshipper to fall prostrate and adore in silence the God who seems to veil his glory; the other, called *El Pilar*, being spacious, lofty, light, elegant, and cheerful, inspires hope, confidence, complacency, and makes the soul impatient to express its gratitude for benefits received." *Spain*, i. 205.) This church, however, was nearly destroyed during the siege in 1808–9; and several of the other churches and convents were then also destroyed. The chapter of the united cathedrals comprises an archbishop, dean, 14 dignitaries, and 30 canons. Among the numerous other churches, 16 of which are parochial, that of Santa Engracia is worth notice on account of its valuable paintings, sculpture, &c.; and the conventual church of St. Domingo, in the place of the same name, is remarkable for a fine altar-piece and mausoleum of white marble. There are five *hospicios*, or public almshouses, one of which, the *Casa de Misericordia*, has accommodation for 700 men and ged persons of both sexes, and another affords a refuge for pwards of 1000 orphans and foundlings. The exchange. rar the *Puerta del Angel*, is an antique-looking, square uilding, ornamented with busts of the kings of Aragon, enclosing a spacious hall supported by 50 Doric columns; congruous to which is the sessions-hall of the *ayuntamiento*. here are two sets of barracks, and in the suburbs are several extensive and well-planted walks. A little W. of e city is the fortress of Alja-feria, so called from its under, the Moorish king Ben-Aljafe, who made it his lace. A university was founded here on the expulsion the Moors, in 1118, but was not incorporated till 1474: it as well attended at the close of the last century, but is w comparatively deserted. Among the other establishnts may be mentioned a royal economic society, with rfessors of chemistry and agriculture, botany, rural ecomy, &c.; a royal academy of the fine arts, a public library, i a *monte de piedad*. The manufacturing industry of ragossa, once very considerable, has all but fallen to de-r; the only manufactures, at present, being those of .net woollen cloths, parchment, shoes, and leather. The rn enjoys also considerable advantages for commerce, ing to its position in the midst of a fertile country, and the canal of Aragon, which runs from near Tudela to tago: its trade, however, is confined chiefly to the sport of grain to Tortosa in exchange for articles of be consumption.

On the whole, Saragossa may be said to be on the de. like all the provincial caps., many of the old famil-having gone to hide their poverty at Madrid; and many nificent houses, on a scale not exceeded any where in n, are now let out in tenements. Provisions of all , corn, wine, oil, mutton, game, and vegetables are .p, abundant, and excellent, this being probably the country for living in Spain. The people, generally, i vil and polished, as in all the old cities; but the lowes es have a bad reputation, and assassinations are said to mmon. The peasants of the environs wear a Moorish me, like those of Valencia, and in manners they are and more ferocious-looking than almost any other ntry of the Peninsula." (*Cook's Sketches*, i. 111.) e climate is temperate and healthy, though somewhat : the neighbourhood produced good crops of wheat, r, and maize, kidney beans and other vegetables, wine, ixits, and silk. The neighbouring hills depasture great ers of sheep, chiefly belonging to the *Ganaderos* or -grasers of Saragossa, an old and highly privileged ztion. (*Townsend*, i. 205–212; *Cook's Sketches*, i. 108– *Theodore Trav.*; *Miñano, &c.*)

gossa is very ancient, being said to have been founded Phœnicians or Carthaginians. It was greatly en-by Julius Cæsar, who made it the head quarters of teran legion; and Augustus gave it the name of :an Augusta, with the privileges of a free colony. Of man buildings, however, which, according to Strabo, a ve been numerous and handsome, there are scarcely -ttifices. Towards the close of the 5th century it was ,y the Goths, who were expelled in 712 by the Sara-zed at length, in 1017, it was made the capital of a Moorish state. A century afterwards it was be-a ard taken by Alphonso of Aragon; and it was sub-ly united to the kingdom of Castile. But it is prin-known in modern history from the obstinate re-made by its inhabs., under Palafox, in 1808–9 to mch, commanded successively by Marshals Mortier nex. The siege lasted, with some slight intermis-xzm July 15, 1808, to Feb. 21, 1809; when, after a bout 6000 men killed in battle, and above 30,000 ren, and children carried off by hunger, pestilence, fanatical excesses that raged in the unfortunate u rendered to the French. Colonel Napier's ac-txis famous siege has stripped it of more than half nze with which it was early invested in this "The "heroic" Palafox "for more than a month the surrender never came forth of a vaulted r hich was impervious to shells, and in which, much reason to believe, that he and others, of ived in a state of sensuality, forming a disgust-zt to the wretchedness that surrounded them." i. 49. 3d edit.) In obstinacy, fanaticism, and sa-ty. the Saragossans seem to have borne a striking to the Jews besieged by Titus. The loss of in the siege did not exceed 4000 men.

K, a town of European Russia, gov. and dist. b th sides the Saranga near the Insar, 70 m. N. Pop. 8750. Most of its houses are of wood; ever, two cathedrals, nearly a dozen other convent, various manufacturing establishments, nnual fair. (See, also, PENZA.)

C, r., rises in Saranac lake, in Franklin co., roe of Racket r., and flows N E. until, in Platts-W. and enters Cumberland bay of Lake Champlain, at the v. of Plattsburg. It affords good water-power in its course, and has a considerable fall af the village of Plattsburg, where are mills. On the banks of this river, the British and American land forces were engaged, in the battle of Plattsburg; the British occupying its N. side, and the American the S. side, the former not being able permanently to cross the r. A canal might be constructed along this river and Racket river, which would connect Lake Champlain with Lake Ontario, through St. Lawrence r.

SARANAC, p. t., Clinton co., N. Y., 15 m. W. Plattsburg, 181 m. N. Albany, 630 W. Watered by Saranac river. It abounds with iron ore, and has a Methodist church, five stores, one forge, one glass factory, one grist-mill, nine saw-mills; five schools, 100 scholars. Pop. 1402.

SARATOF, an extensive government of European Russia, between the 48th and 53d degs. N. lat., and the 42d and 50th E. long., having N. the govs. of Penza and Simbirsk, E. that of Orenbourg, S.E. and S. Astrakhan, and W. Tambof, Voronejе, and the country of the Don Cossacks. Length, and greatest breadth, about 350 m. each. Area estimated at about 73,600 sq. m. Pop., in 1838, 1,564,400. The Wolga intersects it from N. to S., dividing it into two portions of nearly equal size, but differing considerably in general character. The E. division is a wide steppe, destitute of wood, and covered in many parts with salt lakes, from one of which 10 million poods of salt are said to be annually obtained. The W. division is in part hilly, and though stony towards the S., has some tolerably fertile tracts in the N., where agriculture is the chief occupation of the inhabitants. Rye, wheat, oats, millet, and peas, are raised, and in ordinary years the produce, after supplying the demand for home consumption, leaves a considerable quantity for exportation. Potatoes, flax, and hemp, are also produced, and the cultivation of tobacco, hops, and wood, has been introduced by German and other colonists. The climate, in some situations, is sufficiently mild for the culture of the melon, grape, and mulberry. The principal forest trees are oaks, poplars, Siberian acacias, and firs. The woods are mostly in the N.W.; and those belonging to the crown are estimated at about 418,500 desiatines; but the supply of timber is not adequate to the home demand. The rearing of live stock is conducted on a large scale; and the more wealthy proprietors are endeavouring to improve the breed of sheep, by the introduction of Merino flocks. The rearing of bees and of silkworms is on the increase. The fisheries in the Wolga furnish large supplies of fish, both for home consumption and exportation. Next to salt, mill-stones and a little iron are the chief mineral products.

The population is very mixed, including Tartars and Kirghizes, and on the Wolga are numerous colonies, founded principally by German and other immigrants from W. Europe; originally attracted thither by grants of land, and privileges conferred by the empress Catharine, in 1763. In 1811, their numbers amounted to about 55,000; and in 1838, they are said to have increased to nearly 118,000. The colonists are free, and in most respects subject only to their own jurisdiction. They conduct the most important manufactures of the government, which consist of linen, cotton, and woollen fabrics, hosiery, iron ware, leather, and earthenware. There are numerous flour-mills and distilleries. This government is favourably situated for commerce: it communicates by the Wolga with the Nijni-Novgorod and the Caspian sea, and by the Medveditza and Don, with the sea of Azof. The Tartars have a large trade in sheep-skins, and the Kalmucks in horses of a very fleet though weak breed. About 5000 merchants, trading in corn, salt, fish, caviar, cattle, tobacco, and fruits, had a few years since an aggregate capital of 11,175,000 roubles. Sarniof is divided into 10 districts; chief towns Saratof, Volsk, and Tzaritzyne. The population are mostly divided among the Greek, Protestant, and Mohammedan religions. Education, except in the schools of the colonists, and of the capital town, is at a very low ebb; and in 1830 there was but one printing-press in the government.

SARATOF, a town of Russia in Europe, cap. of the above gov., on the Wolga, 335 m. S.S.E. Nijni-Novgorod, and 360 m. N.N.W. Astrakhan. Lat. 51° 31′ 34″ N., long. 46° E. The pop. (including military), according to the official accounts, exceeds 35,000; but this is believed to be beyond the mark. It consists of an upper and lower town; but, though founded so late as 1665, it is neither regularly laid out nor well built. It has some good and even handsome stone residences; but most of its houses are of wood, and it has frequently been in great part destroyed by fire. There are about a dozen Greek-Russian churches, some convents, a Protestant and a Roman Catholic church, a mosque, and a *gostinõe-duor*, or bazaar, a large stone building for the warehousing, exhibition, and sale of merchandise. Since 1833, a new and handsome archbishop's palace has been constructed; and there are several hospitals, a gymnasium, and an ecclesiastical seminary, established in 1828, and having at present (1841) about 500 students. The inhabitants manufacture cotton fabrics, cotton and silk stockings, clocks and watches, leather, wax lights, tallow, vinegar, beer, &c. Saratof, which is intermediate between Astrakhan, on one hand, and Moscow and Nijni-Novgorod on others, has an extensive trade, its exports being principally corn, salt fish, hides, cattle, and native manufactured goods; and its imports, tea, coffee, sugar, iron, glass, and earthenware, woollen, silk, and cotton stuffs, peltry, &c. It has three large annual fairs. (*Schnitzler, La Russie; Possart, Das Kaiserth, Russland,* p. 591–616.)

SARATOGA, county, N.Y., situated towards the E. part of the state, and contains 800 sq. m. It is celebrated for its mineral waters, and contains marl and bog iron ore. Drained by Sacanda river, and Fish and Kayaderosseras creeks. Railroads from Troy and Schenectady form a junction at Ballston, and proceed to Saratoga. It contained in 1840, 4810 neat cattle, 96,656 sheep, 51,601 swine; and produced 72,001 bushels of wheat, 169,950 of rye, 328,631 of Indian corn, 85,974 of buckwheat, 496,089 of oats, 1,019,639 of potatoes, 5767 pounds of hops, 100 of silk cocoons, 26,910 of sugar. It had 166 stores, 12 lumber-yards, two furnaces, 33 fulling-mills, 21 woollen factories, one cotton factory with 9098 spindles, four flouring-mills, 37 grist-mills, 130 saw-mills, one oil-mill, two paper-mills, 27 tanneries, two distilleries, one pottery, one printing-office, one weekly newspaper; 10 academies, 527 students; 164 schools, 6319 scholars. Pop. 40,553 Capital, Ballston Spa.

SARATOGA, lake, Saratoga co., N.Y., 4 m. S.E. Saratoga, 5 m. E. Ballston Spa, is a beautiful sheet of water, 9 m. long and 2 wide, well stored with fish, and much resorted to by visiters to Saratoga springs.

SARATOGA, t., Saratoga co., N.Y., 6 m. Ballston Spa, 36 m. N. Albany. Bounded E. by Hudson river, S.W. by Saratoga lake, from which Fish creek flows into Hudson river. It contains eight stores, two fulling-mills, two woollen factories, three grist-mills, three saw-mills; one academy, 56 students; seven schools, 454 scholars. Pop. 9694.

SARATOGA SPRINGS, t., Saratoga co., N.Y. Drained by small streams flowing into Kayaderosseras creek, which passes through its S. part. It contains the village of Saratoga Springs (which see), and contains 42 stores, one fulling mill, one woollen factory, two grist-mills, two saw-mills; four academies, 173 students; five schools, 253 scholars. Pop. 3384.

SARATOGA SPRINGS, p. v., Saratoga co., N.Y., 36 m. N. Albany, 181 m. N. New-York, 406 W. This is the most celebrated watering-place in the United States, and one of the most celebrated in the world. The village is built on five streets running N.N.E. and S.S.W., crossed by as many others, chiefly at right angles. Broadway, the principal street, is 140 feet wide, and a mile long. There are a number of splendid boarding houses, the principal of which are the United States, Union hall, the Congress hall, the Pavilion, and 70 others of less celebrity. The four principal houses contain about 1000 rooms. The principal springs are the Congress, Washington, Putnam's, Pavilion, Iodine, Hamilton, Flat Rock; and one mile E., the Ten Springs, and others, in the village, making 18 in the whole. There are four extensive bathing houses. The number of visiters at one time are, when full, about 3000, and through the season in the whole, about 35,000. The village was incorporated in 1826, and contains six churches, a Presbyterian, Methodist, Baptist, Episcopal, Universalist, and Roman Catholic; two academies and two female seminaries; 35 stores, two public libraries and reading-rooms, three printing-offices, issuing two daily and one weekly newspapers; 400 dwellings, and 2500 inhabitants. The springs differ slightly, some in their properties. Congress spring, much more resorted to than any other, and much more of which is sent abroad in bottles, is situated in the S. part of the village, and according to the analysis of Dr. Chilton in N.Y., May 1, 1843, it contained the following ingredients: one gallon of 231 cubic inches had, chloride of sodium, 363.829 grains; carbonate of soda, 7.900; carbonate of lime, 86.143; carbonate of magnesia, 78.621; carbonate of iron, .841; sulphate of soda, .651; iodide of sodium and bromide of potassium, 5.920; silicia, .472; alumina, .321; total, 543.998 grains. Carbonic acid, 284.65; atmospheric air, 5.41; making 290.06 inches of gaseous contents. In the spring of 1842, this fountain was cleansed and renovated by putting down a new curb, extending down to the rock from which the water issues, which increased its gaseous and saline qualities, and restored it to its original strength. The Iodine spring has a less quantity of iron, and the Pavilion spring more carbonic acid than any other. The latter is brought from an orifice in a rock, 40 feet underground, and well tubed. The following is an analysis of it by Dr. Thomas in May, 1842. Chloride of sodium, 296.69 grains; carbonate of magnesia, 69.50; carbonate of lime, 60.94; carbonate of soda, 4.70; oxide of iron, 3.10; iodide of sodium and bromide of potassium, 2.75; silicia, .62; alumina, .25; total of solid contents, 361.74 grains.

756

Carbonic acid, 480.01; atmospheric air, 8.00 cubic inches of gaseous contents. Next to the Congress spring, this is next extensively resorted to, and much of it is sent abroad in bottles. Myriads of small globules are thrown by its pa nearly a foot from its surface, and it is very lively and sparkling.

The visiters are from all parts of the United States, sometimes from the distance of over 3000 miles, and frequent from foreign countries. The arrivals are sometimes 150 in the course of a week, and the larger houses are then overflowing in the height of the watering season, which is chiefly during the month of August; but they do not receive visiters during a considerable part of the year. Some of the smaller houses are kept open for visiters during the whole of the year, when those who are found here are chiefly invalids.

The waters are useful in many cases of disease; but to other persons whose system wants tone and recruiting, they are the best of all remedies, and few persons who are in the diseased drink of them, even for a short season, without experiencing a sensible benefit. From one to three pints are generally taken before breakfast, and the appetite is strengthened, and the whole system is invigorated. These who drink of the water from bottles carried abroad, find it eminently beneficial. A single dealer in the city of New-York sells several thousand dozens of bottles annually, and there are several other places in the city at which it is sold, though in smaller quantities.

SARDINIA (Ital. *Sardegna*, Fr. *Sardaigne*, an. *Ichnusa*, from its resemblance to the print of a foot, post *Sardiniae* an island of S. Europe, and next to Sicily, which it nearly equals in size, the largest in the Mediterranean. It lies principally between the 39th and 41st degs. of N. lat., and the 8th and 10th of E. long., being separated from Corsica on the N. by the strait of Bonifacio. It is of an oblong form; length, N. and S., about 160 m.; average breadth, about 60 m.; area, with its dependent islands, 9300 sq. m. Pop., in 1838, 524,633.

Sardinia differs from Corsica in being more diversified, more fertile, and richer in minerals. A large proportion of the surface is hilly or mountainous. The principal mountain chains extend from N. to S., at no great distance from the E. coast; but in various parts of the island there are ranges of considerable length stretching in an opposite direction. The general elevation of the mountains is from 1000 to 3000 ft.; the peak of Limbarra, however, is 3388 ft. and that of Gennargentu, in the chain of that name (the *sani Montes* of antiquity), 5276 ft. in height, on which enables the people of Aritzo to trade in snow for the consumption of the capital. (*Smyth,* p. 67.)

There are many extensive plains, the principal being those of Oristano and Sassari in the N., that watered by the Tirsi in the centre, and the *Campidano,* between Oristano and Cagliari, in the S. The Tirsi, Flumendosa, Coghinas, Mannu, &c., flowing through these plains, are considerable rivers: the minor *campi* are watered by numerous small streams. Around the coasts are many lagoons, and some considerable bays, as those of Cagliari, Oristano, Sassari, Orisel, &c.

The mountain-chains of Sardinia and Corsica have a similar formation, being composed of granite, schist, primitive limestone, &c. Through the centre of Sardinia, from N. to S., extends a remarkable tertiary formation of a calcareous nature; and various volcanic products are scattered over this formation, while the traces of extinct craters are visible in many parts of the island. Earthquakes, however, are rare; nor are storms frequent, though the climate is proverbially variable as to temperature. According to Capt. Smyth, the mean temperature of the year, at the level of the sea, may be taken at 61.7 Fahr., and the mean height of the barometer at 29.69.

Extensive districts are very unhealthy, and in antiquity the island was celebrated alike for the excellence of its air, and the badness of its air. *Sardinia furtile, et non pari coeli malioris; atque ut fecunda, ita pene pestilens.* Mela, lib. ii. cap. 7.) "The intemperie, as the fever is here called, appears to be somewhat different from the *malaria* of Italy and Sicily; for though equally if not more acrimonious in effect, it does not always produce the swelled bodies and sallow skins which are the symptoms of the latter. Both diseases usually commence when the summer heat, assisted by light showers, decomposes the pure grass from the low grounds, and continues till the latter end of November, when heavy rains have precipitated the miasma, and purified the air. But they differ, as malaria is generally supposed to be weak, if not unless imbibed during sleep; whereas the intemperie is worst at night, is pernicious at all times." (*Smyth,* p. 89.) The chief source of insalubrity appears to be in the exhalations from the numerous marshes and stagnant pools of the plains, and might, therefore, it seems fairly concluded, be greatly abated by a proper system drainage. Fire is said to be a powerful antidote against the evil; and the lords of Oristano were formerly accustomed to light large fires round the town, which had the effect either of rarefying or destroying the mephitic vapours.

Notwithstanding her extent, the richness of her soil, her position in the centre of the Mediterranean, and her convenient harbours, Sardinia has been strangely neglected, not only by her own governments, but by the European powers generally; and has remained, down to our own times, in a semi-barbarous state. A long series of wars and revolutions, followed by the establishment of the feudal system in its most vexatious and oppressive form; the fact of her having been for a lengthened period a dependency of Spain, and, if that were possible, worse governed even than the dominant country; the division of the island into immense estates, most of which were acquired by Spanish grandees; the want of leases, and the restrictions on industry, have paralysed the industry of the inhabitants, and sunk them to the lowest point in the scale of civilization. Since 1750, however, improvements of various kinds have been slowly, but gradually gaining ground; and, within the last few years, several important and substantial reforms have been introduced, that will, it is to be hoped, conspire to free this fine island from the abyss into which it has been cast by bad laws and bad government.

Besides that portion of the island occupied by lakes and marshes, there are large sandy or stony districts, called *techie*, which, comprise, in the aggregate, more than one third part of the island; a similar extent may be assigned to forests and pastures; the remaining portion of the surface being laid out in corn-fields, vineyards, olive-grounds, orchards, gardens, &c. About one fifth part of the cultivated land is supposed to be allotted to the growth of corn, which, even under the present system of agriculture, is said give a return of 7 or 8 for 1; and, in some favoured districts, the average is said by Smyth to amount from 16 to for 1. Of the capacity of the island for producing the most luxuriant crops of corn, there can, indeed, be no manner doubt. In antiquity, Sardinia was reckoned, along with Sicily, a granary of Rome. "*Siciliam et Sardiniam benignissimas urbis nostrae nutrices.*" (Val. Max., lib. vi., cap. 6.)

Utraque frugiferis est insula nobilis arvis.
Nec plus Hesperiam longinquis messibus ulla,
Nec Romana magis compleveat horrea terra.

Lucan, lib., lin. 66.

But the unfavourable political and municipal regulations under which the island has latterly been placed, have gone to neutralise the advantages it owes to nature. The culturists of Sardinia principally consist of two great classes—those who cultivate small farms on the *métayer* principle, and those who work on the estates of others, getting in most instances, a patch of land for their support, and working it at such times as they are not employed on the lands. Both classes are excessively poor. The tenements under which the former class hold are seldom more than a year; the landlord furnishing the seed as as the land, and receiving half the produce. Those occupy land for which they are obliged to pay a rent services, or other feudal services, are, if possible, still off; having usually to borrow the seed either from landlord, or from the *Monti Frumentarii* established for purpose, and having also to defray the tithes and a other burdens. Another disadvantage, under which uses labour, is the want of houses on their farms: the must live together in villages, and have frequently to make a journey of several miles in going to and coming their farms. Probably, also, this may, in some degree, not for the frequent change of occupancies by the Sardinian peasants; though, as Marmora has truly observed, circumstances be rarely improved by such changes. belonging to a canton or commune are frequently used on a kind of partnership system, being divided into free portions: one of these, called *vidazzone*, comprises lands that are in cultivation, and which are distributed among certain individuals, while the other two are occupied in common as pasture. But, as a new rotation takes place every year, it is plain that no individual can take any interest in the improvement of the soil; sort of tenure becomes, in fact, the most effectual that be devised for the extinction of industry. Latterly, the government has been making efforts to promote the formation of inclosures and the division of the land, which, though opposed by the prejudices of the people, made some progress. (*Marmora, Voyage en Sardinia.* v., cap. 1.)

These, however, are not perhaps the greatest disappointments to agriculture. As if to annihilate the possibility of the peasantry emerging from their depressed condition, to oblige them to confine their industry to the supply their indispensable wants, it has been enacted that shall be exported if its price exceed 30 reals the and a heavy duty is laid on all that is exported, as a substitute for a general land-tax. Most other articles of export have been loaded with similar duties; and it would really seem that every device that ignorance and short-sighted rapacity could suggest had been practiced to reduce this "benignant nurse" of imperial Rome to a state of poverty and destitution.

Happily, however, the bounty of nature has proved an overmatch for the perverse ingenuity of man; and such is the fertility of this fine island, that, notwithstanding the influence of the duty now referred to, and the wretched system of agriculture, it exports in good years about 400,000 *starelli*, or 500,000 bushels of wheat, 200,000 st. barley, 8000 ditto maize, 100,000 ditto beans, 200,000 ditto peas, and 1000 ditto lentils. The culture of the vine is gradually increasing in importance, and about 3500 Catalan pipes are exported, chiefly from Alghero and Ogliastro. Olive oil, owing to the little care taken in its preparation, and its consequent bad quality, has hitherto been but little exported; but it is susceptible of an indefinite increase, and might be made an important article. Tobacco is a royal monopoly, and brings about seven millions livres a year into the public treasury. Flax, linseed, saffron, hemp, and barilla, are grown to some extent: silk is produced only in limited quantities, but its produce might, no doubt, be vastly increased; some cotton is produced, and also small quantities of madder, which last grows wild in the island. The mountains are clothed with forests of oak, beech, chestnut, and other timber; but from the want of roads, these are nearly useless. The agricultural implements and processes are excessively rude. The Sardinian plough, the counterpart of that described by Virgil, does little more than scratch the ground. It is without a coulter, and is very frequently wholly constructed of wood. Oxen only are used in ploughing and other field labour. Most of the garden grounds are wrought by the hoe, the spade and mattock being unknown, except in the Piedmontese labourers on the new roads. The corn is left in the fields till it be thrashed, an operation effected by the primitive practice of treading with horses and oxen.

We are glad, however, to have to state, that within the last few years some very important changes for the better have been introduced into the island, and that some of the worst of the abuses previously noticed have been obviated. In 1836, in pursuance of inquiries previously commenced, feudal jurisdictions were completely abolished; and since then the feudal system has been wholly subverted. And if, as is to be hoped, government follow up the enlightened course of policy on which it has entered, by giving freedom to commerce, the probability is, that the island will, at no very remote period, recover a large share of its ancient prosperity. According to a law passed in 1839, all lands were declared to be the property of individuals, communes, or the crown; the latter becoming the possessor of all waste lands, or those to which neither private parties nor communes could show any title. Lands which had been cultivated or applied to use, whether enclosed or not, were assigned in perpetuity to the occupiers, undisturbed possession being held to confer a sufficient right to the property in the absence of any other title: those whose interests were at all affected by the new changes received compensation in money or lands, or by an assignment of public funded property. The king substituted himself in the place of the barons: he took all the feudal rents into his own hands; and their value being estimated at 20 years' purchase, public securities to the amount, bearing five per cent. interest, were made over to the nobles in exchange for the privilege of which they had been deprived. All kinds of vassalage were, at the same time, made redeemable; and courts of law placed under the direct control of the state were substituted in the place of the feudal jurisdictions where the barons were at once suitors and judges! It is impossible to overrate the importance of these changes; and there cannot be a doubt that they will have the greatest and most beneficial influence. (*Von Raumer, Italy, &c.*, i., 295–301.)

The greater number of the oxen, horses, and other live-stock, wander wild over the island, bearing the mark of their owners, and browzing in the woods in winter, there being no wolves. They are generally, as might be expected, very inferior; but considerable pains are taken in the breeding of some descriptions of horses, and horse-races are a prevalent amusement. It is singular, notwithstanding the badness of the roads, that mules should be unknown. The Sardinian sheep is said, by Marmora, to be remarkable only for its degeneracy: its wool is of a very low quality, and is worth little, except in the island. Cheese, made of the milk of sheep and goats, is extensively exported; but this is a result, not of the goodness of the milk, but of its extensive supply, arising from the great number of these animals, there being about 800,000 sheep, and 530,000 goats. (*Marmora, p. 444.*)

The *mouflon* (the *Ophion of Pliny, Hist. Nat.*, lib. xxxviii., cap. 9) whence some naturalists suppose the sheep to be derived, is a native of Sardinia. "It is a ruminating animal, frequenting only the highest and most secluded woods; where, from its timidity and fleetness, it is with difficulty shot. The form of the ears, head, legs, and hoof, identify the mouffion with the sheep, though in. size it is rather larger, and is, moreover, clothed with hair instead of wool. The horns are neither full nor deciduous, but hollow, and precisely similar to those of the ram, while the bleat is the same: it propagates also very readily with the sheep, the mixed produce being the 'umbro.' Though so shy in its wild state, the mouffion soon accommodates itself to domestic habits." (*Smyth*, p. 120.) Deer, wild boars, and a variety of game, abound in the forests; and the skins of about 60,000 rabbits and hares, from 4000 to 5000 foxes, and 9000 martins, are annually exported, besides 5000 cantars of *cursucci*, or dried skins, for making glue.

Though various improvements have been effected of late years, it is still true that the interior of this island exhibits, at this moment, a degree of barbarism which can with difficulty be believed to exist in Europe. The shepherds, and others who occupy the mountainous parts of the island, are in the habit of wearing only coverings of tanned leather, or of shaggy goat or sheep skins. They are constantly armed to protect themselves from banditti; roaming with their flocks over the uninhabited tracks, enjoying a bare subsistence, and acquainted with no laws but those of their own formation. They sometimes sow small patches of wheat and barley round their temporary dwellings; but they subsist chiefly on fruit, game, and the produce of their flocks, each family constituting, as it were, a patriarchal association. Though this part of the population be inoffensive, the number of banditti in the mountains formerly rendered it unsafe for any one, whether a foreigner or Sardinian, to venture far into the interior without an escort; and the farmers in the plains have been accustomed to rely for protection from the depredations of their highland neighbours on a long established corps, called the *barancelli*. This is an armed association, chosen annually in the village districts; the members of which are bound to make restitution for all thefts, provided they receive immediate intimation of the robbery. Their remuneration arises from an annual sum subscribed by every landholder. An attempt was made by the government, in 1819, to disband this force, but it was unsuccessful; and, on the whole, the *barancelli* are well adapted to the condition of the country.

The banditti that have long infested and still continue to infest parts of this island, owe their origin to a variety of causes, among which, no doubt, may be included the influence of the feudal system, and the opportunities afforded by the state of the country, full of natural fastnesses, without roads, and without an efficient system of police, for their carrying on their depredations with impunity. Latterly, however, some stringent measures have been adopted for their suppression. The privileges of sanctuaries have been in most instances abolished. Roads have been made into districts that were previously inaccessible; the right to wear arms has been restricted; and these measures, combined with the abolition of the feudal system, and the establishment of royal courts for the speedy and more equal distribution of justice, will, probably in no very lengthened period, go far to suppress the robberies and assassinations which have so long disgraced the island.

The houses of the peasantry are most wretched, consisting usually of only one story without windows; or, if there be windows, they are not glazed. A whole family frequently dwells in a single room in which kids, chickens, and dogs seek indiscriminate accommodation with the naked children; while as ass is usually employed turning a corn-mill (*mola asinaria*) in the corner! The centre of the room has a square hole in the clay floor, for the fire, but there is no outlet for the smoke, except accidental holes in the roof or door. A few small low chairs, with an equally low table, constitute the usual movables. Earthenware not being common, the ordinary substitute is an oblong wooden dish. More flesh is used than in Sicily, but less polenta. Omelettes of curds, &c., and raw vegetables, are favourite articles of diet.

The towns and villages are mostly large and well situated; but with unpaved, narrow streets, mean houses, and a want of every convenience. Immense dunghills, the collection of ages, disfigure the principal entrances. In the N. half of the island the villages are constructed of free-stone or granite; but most of the country houses in the S. are built with sun-dried bricks made of mud and straw. In the towns some pretty good mansions are met with, though they are ill fitted up, and their *atria* generally as dirty as those of the ancients in the days of Juvenal. (*Present State of Sardinia*, p. 164, &c.)

The fish on the coasts and in the harbours of Sardinia are mostly caught by foreigners; Sicilians, Neapolitans, Tuscans, Genoese, &c. Anchovies and pilchards have become rare; but in 1838 upward of 17,000 tunnies were taken,

besides several thousand liberated from the nets, from the fishermen not having the means of curing them. The lagoons of Oristano, Cagliari, &c., abound with fine mullet, bream, and eels. From 200 to 300 boats arrive every year from Naples and Genoa, to the coral fisheries on the coast, each boat collecting, at an average, coral to the value of about 1500 dollars.

Sardinia has ores of silver, copper, lead, and iron, which, if wrought, would, it is believed, be among the most valuable of her resources. But the code of regulations for the working of the mines proves an effectual obstacle to all mining speculations. It obliges speculators to work their mines under the direction of government engineers, or officers of the royal corps of miners, who are to be consulted and furnished with plans on the erection of smelting mills, &c. Speculators are also prohibited from exposing for sale, or exporting, the produce of their mines, without permission from the intendant-general, and are, besides, to keep a journal of the daily produce of their mines, the same to be exhibited in a separate statement monthly at the intendant's office of the cap. district. (*Macgregor's Report*, 73.) The consequence is, that mining, and the quarrying of the fine porphyry, basalt, marble, &c., of the island, is almost wholly neglected.

Salt is a royal monopoly, and affords a considerable revenue. Until recently, Sweden drew almost all her supplies of this article from Sardinia; and it continues to be exported in considerable quantities. It is obtained by solar evaporation, principally near Cagliari. The expenses incurred by government in producing salt do not exceed one reals the salm; whereas the continental subjects of the crown buy many thousand salms at 30 dollars or reals each.

Except the royal gunpowder, salt, and tobacco manufactories, a few for cotton, woollen, and silk goods, and some coarse pottery and glass works, Sardinia has no manufacturing establishments, except such as are employed in preparing raw produce for sale. Very little skill is shown by any of the artisans; and watches, clocks, and even coarse cutlery, are *all* imported. The want of roads has hitherto proved a serious obstacle to manufactures, as well as to every other branch of industry. A good road, practicable for wheel carriages, has, however, been formed within the last few years from Cagliari to Sassari, and cross roads are being carried from it to some of the most considerable places in the island. But previously scarcely any roads were passable for travellers, except on horseback or on asses, the *lettiga* of Sicily being unknown. A cart for luggage was, indeed, used; but this vehicle was a mere ladder mounted on solid wheels fixed to the axletree, and stuck round the edge with triangular nails, being, according to Captain Smyth, a ruder machine than any he had seen in Spain, Greece, or Calabria.

The commerce of Sardinia has long been stationary, so that a statement of its trade in 1834 will be nearly applicable to the present time. In that year the value of the principal articles of export and import were estimated as follows:

Imports.		Exports.	
Articles.	Value	Articles.	
	Fr.		Fr.
Timber, &c. . . .	250,272	Brandy, wines, liqueurs, &c. .	
Hosiery	195,105	Grain, &c. . .	
Hemp and cordage .	133,025	Skins and leather	
Cotton goods . .	1,924,662	Fish, cured and fresh	
Drugs and spices .	698,298	Salt	
Woollen goods . .	667,766	Meats, hams, &c.	
Hardware and metals	808,339	Cattle . . .	
Skins and leather .	266,317	Timber, &c. .	
Linen fabrics, &c. .	216,409	Tobacco . .	
Corn, &c. . . .	148,514	Various . .	
Various	1,174,383		
Totals . .	4,968,110		

The custom duties received in the above year amounted to 997,932 fr. (*Marmora*, 460.)

Amounts (when not estimated, as above, in francs) are kept in lire, soldi, and denari: the lira of 20 soldi and 12 denari = about 1s. 6d.; the real of 5 soldi = 4½d.; the real of 10 reali = 3s. 9d. The Sardinian lb. of 12 oz. = 14 oz. 3 dr. avoird.; the rubbo = 25 lbs.; the moggio or staro of corn) = about 1 bushel 1 peck. The palmo = 9¼ English inches; the starello or maggio (of land) of Cartan = 1 roods 27 poles 19 yards; of Sassari = 1 rood 36 poles 24 yards.

Sardinia is governed by a viceroy, whose commission generally lasts for three years, and who is the chief of the civil and judiciary administrations, and the commander of the forces both by land and sea. The island is subdivided into two grand divisions, those of Cagliari and Sassari. 10 provinces, 32 districts, and 368 communes. The seven cities, or principal towns, are under the administration of the magistrati municipal bodies, each composed of a members. Each commune has a council of three, five or seven members, presided over by a sindaco. The Udienza Reale, created in 1661, and reformed in 1823, is the highest tribunal in the island. It is composed of 13 judges and two presidents, and is divided into three chambers, two civil and one criminal; and has at its head the regent, the first functionary in the island after the viceroy. Besides its functions as a supreme judiciary court, it participates in the legislative power, the decree of the viceroy published with the concurrence of the Udienza Reale, having the force of laws. Sassari has a tribunal resembling the Udienza Reale of Cagliari, to which appeals may be made from its decisions. In the two cities last named are tribunals of commerce. In the provinces justice is administered by prefects, whose decisions are final in civil causes to the amount of 10 scudi, and who have primary jurisdiction in criminal cases. The *curia*, or district tribunals, have a very limited jurisdiction.

Sardinia has a parliament, called the *Stamenti*, consisting of three chambers: the ecclesiastical, selected from the prelates; the military chamber, comprising all the nobles, 30 years of age, with or without fiefs; and the royal chamber, composed of the deputies of the towns and communes under the *capo giurato* of the capital. The stamenti are convoked and holden during the king's pleasure, but meet only on extraordinary occasions. Each section holds its sittings apart; and, after separately discussing the matter under debate, they communicate by deputies. The deliberations of the ecclesiastical body, respecting donations, must be submitted to the pope for his approval, before passing into a law. The supreme council of Sardinia has its seat in Turin: it is composed of a president and five councillor senators, and is similar to the ancient supreme council of Aragon. Beyond this tribunal there is no appeal; and it gives its opinion in all state affairs transmitted to it from the government of the island. (*Report on Sicily and Sardinia*, p. 77.)

The laws of Sardinia are partly comprised in the code titled the *Carta de Logu*, promulgated in 1395, and said to be drawn up, considering the period when it was issued, with great discretion and good sense. It has, however, been materially modified by the successive acts of the stamenti, the edicts of the different sovereigns of the house of Savoy, and the *pragoni* or decrees of the viceroys. In consequence of the numerous and, in many instances, conflicting enactments that thus been issued, the law has become exceedingly obscure. This encourages litigation, of the island is, in fact, to use the forcible expression of a rumor, *une mine inepuisable pour la chicane* (391). Recourse is had to the courts to determine the most trifling questions; and unfortunately the means of legal redress are once tedious, expensive, and uncertain. "The country fees are extremely poor; and venality is so common, that sentences are just and equitable only when the government has a criminal matter in hand. This is one of the leading evils of the administrations that have so stigmatised the island. It is an acknowledged difficult task to work a reform in detail; for if a magistrate prove himself more than ally active in his office, he is sure to receive the vengeance of adverse partisans; and the effect of the whole system and practice is a melancholy want of security both in persons and property." (*Smyth*, p. 135.)

nothing, in fact, would do so much to put down crime and to restore security and good order to the island, as the simplification of the law, and the nomination of superior and responsible judges. And now that the government has begun the work by the suppression of the feudal system, it may be hoped that it will apply its energies to a reform of abuses in the judicial system.

religion, other than the Roman Catholic, is tolerated in the island; and the secluded position and ignorance of the people have prevented the growth of any heresy, so that in this respect Sardinia boasts a higher purity than Rome herself.

The island is divided into three archbishoprics; those of Cagliari, Sassari, and Oristano; and eight bishoprics. These are rigorously exacted. The revenues of the church is estimated by Serristori at 960,000 Italian lire, of which the secular clergy, amounting to between 1800 and receive about 264,000. According to Marmora (p. there are 90 conventual establishments for monks, of there there are about 1130; and 14 similar establishments for nuns.

Public education in Sardinia has, of late years, been considerably improved. "There are now normal schools in of the 10 provinces, attended by about 6650 pupils. There are, besides, secondary schools in the two principal towns Cagliari and Sassari, which are frequented by about students. The university of Cagliari reckons about 365 students, and that of Sassari 225. The course of studies is into theology, jurisprudence, philosophy, medicine, surgery. By an ordinance of the late king, in 1823, village or commune has now a gratuitous school for

reading, writing, arithmetic, religious instruction, and the elements of agriculture. The effect of the diffusion of instruction among the people, aided by a better system of administration and police, is already visible in the decrease of crimes, especially of murders, which, from the frightful amount of 150 yearly, in a population of about half a million, had been reduced in 1828 to 90." (*Journ. of Educ.*, vol. iii., p. 23.)

A voluntary regiment of 1400 chasseurs is levied and maintained in Sardinia, for the service of that island. Respecting the rest of the armed force, &c., the reader is referred to SARDINIA, KINGDOM OF.

The finance department is managed by an intendent-general in Cagliari, and vice-intendents in Sassari, and a sub-intendent in each of the other provinces. The public receipts in 1824 amounted to 2,837,302 fr.; the direct taxes amounting to 1,088,250 fr. The paper money in circulation, in 1806, amounted to 3,840,000 fr.; but, in 1825, it had all been withdrawn, except 480,000 fr.

According to Captain Smyth, there is a striking resemblance between the Sards and Greeks. "It is impossible," he says, "for any one who has travelled in Greece, not to be struck with the similarity which, in many points, exists between the Sards and the Greeks. Not only are their arms, music, dances, dresses, and manners in close resemblance, but many of their words and superstitions are exactly the same; so that the opportunities I have had of comparing the two nations would lead me to infer the partial identity of their origin. The Sards are of a middle stature, and well shaped, with dark eyes and coarse black hair; except in the mountains, where fresh complexions and blue eyes are met with. They have strong intellectual faculties, though uncultivated, and an enthusiastic attachment to their country. They are active when excited, but extremely indolent in general. Their good qualities are counterbalanced by cunning, dissimilation, and an insatiable thirst for revenge." (*Smyth's Sardinia*, p. 141–192, &c.)

Though vassals in Sardinia could change their lord and residence at will, the degrading services and tenures of feudalism were in full vigour in most parts of the island down to its abolition in 1839. The dependence of a peasant on his lord commenced when he was deemed capable of earning his bread; and an annual tribute, either in money or kind, was exacted from all above the age of 18; and this in addition to the usual imposts on lands and stock; the contributions demanded for prisons, robberies, arson, and exemptions from the *reedia*, or one day's personal labour, as well as from other dominical services. These feudal burdens, with tithes, taxes payable to the king, alms, as they are called, to mendicant monks, and other grinding extortions, amounted, in many instances, to nearly 70 per cent. of the earnings of the peasant. And if to this amount of taxation be added the vicious customs that have prevailed in the letting of land, unintelligible laws, and venal judges, need we wonder at the poverty and semi-barbarism of the peasants, and that revenge has become, in their estimation, a sacred duty.

The Sards are enthusiastically fond of poetry, but the other fine arts have met with no encouragement; and there is not a native painter, sculptor, or engraver, of any eminence in the island; and the press being under a rigid censorship, the current literature, if so it may be called, is beneath contempt. The language of Sardinia is that dialect of the Italian which preserves the greatest portion of Latin.

We have little authentic information respecting the history of this island previously to its conquest by the Carthaginians, from whom it was taken by the Romans in the third Punic war. On the fall of the Western Empire it was successively possessed by the Vandals, the Goths, the emperors of the east, and the Moors; from whom it was taken, in 1022, by the Genoese and Pisans. It continued to be a subject of contention between these rival nations till 1295, when it was taken possession of by the kings of Aragon, and it remained attached to the Spanish monarchy till 1714, when, by the peace of Utrecht, it was ceded to Austria. In 1720 the latter exchanged it for Sicily with Victor Amadeus of Savoy. Previously to the French revolution, the Sardinian government is said to have been desirous to sell the island to the empress of Russia for £1,000,000 sterling; but the scheme was defeated by the interference of France and Spain. (*Young's Travels*, ii., 256.) It was unsuccessfully attacked by the French in 1793; and on the seizure by the latter of the continental portion of the Sardinian dominions, Cagliari became the residence of the royal family. Recently, as already seen, measures which promise to be of the utmost importance to the island, have received the sanction of the government. (For further information, see *Marmora, Voyage en Sardaigne*, Paris, 1826, an elaborate and valuable work; *Smyth's Present State of Sardinia*, in great part derived from the former; *Parl. Report on Sicily and Sardinia*; *Diet. Geog.*; *Serristori, Statist. d'Italia, &c.*)

SARDINIA (KINGDOM OF, Ital. *Stati Sardi*), a state

of S. Europe, comprising the whole of N. Italy W. of the Tessino, including the territory of Piedmont, Genoa, and Nice, and the adjacent duchy of Savoy on the W. side of the Alps, with the island of Sardinia in the Mediterranean, nearly the whole of these dominions being included between the 39th and 46th degrees of N. latitude, and the 5th and 10th of E. longitude. Its divisions, area, population, &c., are as follows:

Divisions.	Area in sq. m.	Pop. census of 1838	Pop. to sq. m.
Savoy	4,279	564,157	132·1
Turin	3,195	871,910	274·3
Coni	2,712	556,151	205·7
Alexandria	2,035	595,963	292
Novara	3,092	842,728	261·6
Nice	1,619	230,718	142·5
Aosta	1,204	76,110	63·3
Genoa (and Capreja)	2,105	674,385	320·6
Island of Sardinia, &c.	9,241	524,633	57
Totals	29,102	4,930,365	169

The insular portion of this kingdom being described under the previous article, we have now to deal only with the continental portion, lying between lat. 43° 40' and 46° 24' N., and long. 5° 33' and 10° 5' E., having W. France, N. Switzerland, E. Austrian Italy and Parma, and S. the Mediterranean. The Alps separate this territory into three great divisions: Piedmont in the centre, forming the upper part of the valley of the Po; Savoy, a rugged and mountainous region on the W. and N.W.; and the province of Genoa, Nice, &c., in the S., between the Maritime Alps and Apennines, and the sea.

The most valuable portion of the kingdom is the plain of Piedmont, extending from the foot of the Alps to that of the Apennines on the S., and to the Tessino on the E. The soil is everywhere a rich, sandy loam, with little appearance of clay, and of great fertility. Owing to the heat of the climate in summer, water is here the great desideratum; and advantage has accordingly been taken of the numerous streams that pour down from the mountains, which are distributed with infinite skill all over the low grounds. Nowhere, indeed, is the art of irrigation carried to greater perfection than in that part of the great plain of the Po included in Piedmont. "Water is here measured with as much accuracy as wine. An hour per week is sold, and the fee simple of the water is attended to with the same solicitude as that of the land." (Young, ii., 160.) The irrigated lands being under the influence of a southern sun, produce the most luxuriant crops.

Lands in Piedmont are mostly enclosed, generally by ditches, but, in many parts, with hedges also, which in some districts equal those in the best English counties. The crops, however, are generally divided by lines of fruit trees of different kinds, intermixed with mulberry trees, poplars, and oaks; and that the benefit of these trees may not be limited to the shade they produce, they support vines. Speaking generally, farms in Piedmont are small, and are usually held on the métayer system, the landlord receiving half the produce, and paying the taxes and repairing the buildings. On the whole, however, the agriculture of this part of the Sardinian dominions bears so close a resemblance to that of the rest of Lombardy, that we beg to refer the reader for farther particulars to the article ITALY (AUSTRIAN) in this volume.

Few countries have so large a disposable produce as Piedmont. It has an immense number of cities and towns; and yet the Riviera of Genoa, Nice, and the country as far as Toulon, are supplied with corn and cattle from its superabundant produce. The produce of maize is considerable: and it constitutes the principal support of the country population, who make use of it under a variety of forms. The most usual course of husbandry consists of what might be called in England a four-shift, the first year being maize, the second wheat, the third clover or fallow, and the fourth wheat. It is customary to mix French beans and hemp along with the maize. Wheat is sown on narrow ridges, and is earthed over by the plough; which in Piedmont is an implement of a better kind than in most parts of Italy. Wheat harvest takes place in the beginning of July; it is thrashed by means of cylinders drawn by horses over the straw, which is turned up by forks. According to Arthur Young (ii., 908), the common produce of the wheat crops in Piedmont does not exceed six times the seed; which, considering the quality of the soil, he is justified in calling "miserable;" but the better crops yield between 10 and 11 seeds, or even more; and with a better rotation, and more care, this might be made the average produce of the plain.

To the corn crops must be added those of hemp, which is sometimes considerable, and silk, for which Piedmont is famous, with wine, vegetables, and fruit; the produce of the farm-yard, and the profit of rearing and fattening stock. The latter branches are the principal sources of profit from cattle; for though the latter be very abundant in Piedmont, the farmers have not learned, like those of the Milanese, to

760

derive much advantage from their milk. A farm of medium size was estimated by Chateauvieux to consist of about 60 acres, one fourth being in pasture. Such a farm would support a family of eight or nine persons, maintain upwards of 20 head of live stock, produce silk to the value of £25 a year, and more wine than was required for home consumption. The crops of maize and French beans go far to maintain the labourers, so that nearly the whole crop of corn may be carried to market, as well as a considerable quantity of the inferior articles. (Chateauvieux, 30, 31.)

Savoy, which is remarkable for the grandeur and beauty of its scenery, though a poor country, produces enough to supply the wants of its inhabitants. The peasants are all, or mostly all proprietors. The plough is of use only in the valleys; on the high grounds the peasants break up the soil with the pickaxe and spade, and, to improve it, carry up mould and manure in baskets from the valleys. Small reservoirs are prepared near the tops of the hills and mountains, from which water is let out at pleasure in spring and summer: while, to prevent the earth from being washed down the declivity, small stone walls are erected, so that, by dint of skill and industry, cultivation is extended over tracts which would otherwise be a continued surface of naked rock.

Wheat, oats, barley, rye, and hemp, are the principal grains cultivated: in some favourable situations the vine is grown; and the white wine of Montmelian is especially esteemed. The walnut is the olive of Savoy, supplying the inhabitants with oil, not only for home consumption, but also for exportation to France and Geneva. The kernels are crushed by a mill, into a paste; which is pressed to extract the oil, and afterwards dried in cakes called pain amer eaten by children and poor people. A good many cattle and sheep are reared, and driven for sale into Piedmont and Lombardy. The butter and cheese of Savoy are of good quality, and are important products. Many mules and horses are bred for the transit trade between France, Germany, and Italy.

The olive, though but little grown in the other parts of the kingdom, is the chief article of culture S. of the Apennines. The land in the Genoese territory is generally hilly and rocky; but has mostly a S. aspect, suitable for the olive and the vine. The cultivated land is supposed to comprise about one fourth part of the surface. The land here is divided into very small farms, those near the towns comprising only about 5 acres, and those in the interior about twice as much. Only a small proportion of the land is cultivated by proprietors: it is usually let on leases of 3, 5, 7 or 9 years, but never more; the rent of cultivated near Genoa is very high.* In the greater part of the Genoese territory the rent is paid in cash or produce, as wine, oil grain, &c. rated at a fixed price; but in the provinces of Novi and Levante the rent is paid, as is usual in the rest of the kingdom, on the métayer principle; the landlord furnishing the land and one third the seed, and receiving two thirds the produce. Wheat and maize are generally sown alternately on the same land; and good land is said to yield usually from four to six for one, or double that quantity when tilled with the spade, as is customary in some parts. Each farm of four or five acres supports a family. Labourers get from £3 to £5 a year, with board and lodging. Their usual diet consists of Indian corn, chestnuts, potatoes, beans, and fruit; making little or no use of butchers' meat. Women work in the fields, and tend the cows, in addition to spinning, weaving, and other domestic work, in which they are very industrious. The occupiers of farms are not in a prosperous condition; and we may add, that they never will be in such condition while farms are so very small, and held under such a tenure. Peasant, however, are more numerous in the towns than in the country. The government makes no provision for the support of the poor; but there are various private charities for their maintenance. (Parl. Rep. on Agriculture, 1836.)

The mineral riches of this country are little explored, but iron of good quality, lead, copper, sulphur, manganese, and cobalt abound in the mountains of Piedmont and Savoy. The mines of Pesey, in the Tarentaise, formerly yielded from 30,000 to 40,000 cwts. of lead, and about 4000 marcs silver a year. Alabaster, fine marble, serpentine, slate, &c., are plentiful. Salt is found both in mines and in springs; and near Moutiers are government salt-works, said to produce 3,600,000 lbs. a year. There are some forges, and other iron-works; but the principal manufactures consist of silk stuffs, velvets, and stockings, mostly consumed in Italy. Coarse woollen and linen goods are made in several provinces; and coarse stuffs for the rural population in Savoy. Sail-cloth, cables,

---

* It is stated, in a report from Genoa, in the Parl. paper, No. 94, Sess. 1839, that the average rent of cultivated land near Genoa varies from 20l. to 50l. an acre! that wood and pasture land, in a good situation, fetches 20l. an acre of rent; and that the rent of "rocky land," where sheep and goats are said to feed, is 2l. an acre! We apprehend that this is pretty much on a level with the statement in another Parl. paper, that the government of Tuscany, in Russia, produces annually 36,000,000 quarters wheat! (See Tuscany.)

house furniture, paper, white lead, glass, earthenware, optical and surgical instruments, jewellery, and works of art and viris are among the articles made at Genoa, Nice, and other principal towns ; and there are numerous brandy and liqueur distilleries, and tanneries.

Trade.—It is difficult to form any fair estimate of the trade of the Sardinian dominions ; a large proportion of the imports not being destined for the supply of the country, but merely for transit to France, Switzerland, Germany, &c., and no distinction being made between these portions. In fact, if we may depend upon the following returns of the trade of the kingdom in 1835, it would appear that, with the exception of raw silk, olive oil, and woollen fabrics, the imports of all the great articles exceeded the exports. The articles of exports not included among those imported, consist of rice, paper, and vermicelli. It is abundantly certain, however, that the exportation of raw and silk stuffs, including velvets, exceeds very considerably the imports of such articles retained for home use, as is the case with oil, flax, and various other articles.

Account of the Value of the Principal Articles imported into and exported from the Sardinian States by Sea, in 1835.

Imports.		Exports.	
	Francs.		Francs.
Corn . . .	9,767,000	Corn . . .	2,644,500
Cocoa . .	827,800	Coffee . .	1,478,000
Coffee . .	2,877,400	Hemp . .	771,100
Hemp . .	1,247,100	Cotton wool .	941,600
Cotton wool .	3,211,100	Cotton yarn .	671,700
Cotton yarn .	540,000	Cheese . .	1,142,700
Iron . . .	1,437,500	Olive oil . .	15,504,800
Olive oil . .	15,244,600	Indigo . .	979,500
Indigo . .	1,497,300	Paper . . .	1,824,500
Woollens . .	5,237,500	Pulse or vermicelli .	559,100
Iron . . .	5,796,200	Hides . . .	3,665,200
Salt fish . .	3,284,600	Salt fish . .	1,748,500
Raw silk . .	2,705,400	Rice . . .	2,405,000
Sugar . . .	12,610,300	Raw silk . .	3,495,700
Tobacco . .	4,354,800	Sugar . . .	4,924,500
Silk stuffs . .	15,214,500	Tobacco . .	1,111,400
Woollens do. .	4,960,100	Cotton stuffs .	18,441,000
Flax and hemp do.	1,305,100	Woollens do. .	5,566,600
Silk do. . .	4,794,700	Flax and hemp do.	1,150,000
Wines . . .	1,951,300	Silk do. . .	2,657,200
Other articles .	24,673,300	Other articles .	45,920,900
Total .	120,765,100	Total .	95,665,900

Of the above imports, those from Great Britain amounted 12,695,800 fr. ; France, 22,135,700 fr. ; Naples and Sicily, 171,500 fr. ; Austria, 8,290,100 fr. ; Brazil, 5,416,100, &c. The exports those to France were the largest, amounting 27,719,100 fr. ; those to Austria amounted to 14,935,300 fr. to Naples and Sicily, 9,954,500 ; and those to Great ain only 1,350,000 fr. During the same year there entered the different ports of the Sardinian states 3927 ships, the aggregate burden of 968,100 tons, of which 87 ships the burden of 15,068 tons were from the United Kingdom. Genoa, Nice, and Cagliari, especially the first (which see), the principal seats of the foreign trade of the kingdom. In 1835, the import and export trade of the arrondissements of Genoa, Nice, Cagliari, and Port Maurice, were as ws :

	Imports.	Exports.
	Frs.	Frs.
Genoa . . .	96,196,700	63,015,100
Nice . . .	29,446,100	19,498,600
Cagliari . .	4,061,400	2,471,800
Port Maurice .	48,900	910,700
Totals .	129,765,100	95,665,900

ne of our readers may, perhaps, be disposed to con-, seeing the wide discrepancy between the value of imports and exports in the above accounts, that one or of them must be wholly erroneous. But though we do not certainly to guarantee their authenticity in question does not prove their inaccuracy, inasmuch as the countries that received the goods at second hand through the Sardinian dominions, may have them otherwise than by the export of produce through transit duties on goods passing through the Sardinian have been abolished. The duties on consumption in the interior of the kingdom are moderate. The imports of are principally from the Black sea, but partly also from

government is a monarchy, hereditary in the male and though the regal authority be, in some degree, described by a supreme council in the island of Sardinia is absolute in the continental portion of the kingdom it has, however, been a constant object of the government to restrain the extravagant privileges of the nobility, and corporations, and to enlarge the rights and liberties of the bulk of its subjects. Hence, says Count , "Piedmont was the first country which, in 1729, destroyed nearly the whole system of feudal authority and

personal service, leaving scarcely any but honorary privileges in force. It then also limited the right of primogeniture and of entail, and consequently gave greater scope to the free cultivation of the soil ; and diminished the powers of the clergy, more particularly that of investing land in mortmain. These constituzioni were revised and confirmed in 1770." (*Hist. of Pol. Econ.*, p. 232.) The king is assisted in his administration by five ministers, or secretaries of state, for the interior, War, Marine, Finance, Justice, and Foreign Affairs ; and by a council of state, consisting of a president, 14 ordinary, and an unlimited number of extraordinary members. In each province the whole power of government is in the hands of an intendant, who, like all other functionaries, is appointed by the king; who, of course, may, also, dismiss him at pleasure. Intendants of an inferior grade are appointed for districts and towns, who manage all the public business of their respective localities, though every town has also a magistracy, varying in its number of members according to circumstances, which regulate its municipal and private affairs. In each district is a judge, with authority in civil causes, to the amount of 300 lire ; but from whose decision appeal may be made, when the amount exceeds 100 lire. Each of the 37 provinces has a tribunal, with a president, from two to six councillors, a government advocate, &c. In 1838, a new code of laws for the Sardinian states was adopted, which, though certainly an improvement on the heterogeneous code it displaced, exhibits some glaring defects. Among others, the use of torture is retained in certain cases ; the most arbitrary means are used to extend the Roman Catholic religion ; and Jews are subject to the most illiberal restrictions. Duelling is punished with death, even though neither combatant should be killed or wounded. The whole population is Roman Catholic, except about 7000 Jews, 21,000 Waldenses in the N.W. part of Piedmont, and a few Protestants in Genoa and elsewhere. Public instruction is less diffused than in Austrian Italy, and is in general of a very inferior kind. It is wholly under the direction of a central board in Turin, entitled the *Magistrato di Riforma*, which has under it a council *di riforma* in each province. The principal university at Turin has four faculties, and about 1280 students. Genoa, also, has a university, and there are secondary universities at Chambery, Asti, Mondovi, Nice, Novara, Saluzzo, and Vercelli, with 97 royal colleges in the larger, and 34 communal colleges in the smaller towns. The professors may be either clergymen or laymen, but where the candidates are otherwise equal, clergymen are always preferred. Foreign books pay an oppressively high duty ; and the censorship, being, at the same time, rigid in the extreme, all but insuperable obstacles are opposed to the dissemination of sound information as to politics and political economy. But, despite this jealousy, it is due to the Sardinian government to state that it has of late discovered a very enlightened spirit, and done much for the improvement of its subjects ; and schools and colleges, as well as most branches of the public service, have been materially ameliorated. Various new and wholesome laws have been enacted for the regulation of commerce, roads, weights, and measures, sanitary police, vaccination, prisons, forests, the game laws, &c. A good many roads, canals, and bridges, have been constructed, and hospitals, museums, baths, and public establishments of all kinds have been founded ; the harbours have been improved ; lighthouses and barracks built, and the army has been thoroughly reorganised.

The Piedmontese infantry is composed of two classes of soldiers, viz., the permanent and the contingent (Provinciali). The former serve eight years in the standing army, and at the end of that term receive their discharge from all future service. The latter serve nominally 16 years, during eight of which they are considered as forming part of the standing army, and the remainder they belong to the reserve. They are consequently divided into 16 annual contingents, eight for the first period, and eight for the second. Each soldier has to perform active service for 14 months after his enrolment, on completing which he may return home, but is subject to be called upon during the remainder of the first term of eight years to re-enter temporarily the ranks, for the annual formation of a camp of manoeuvres, or upon any occasion of state emergency. The Sardinian army on the peace establishment consists of about 23,000 men ; but in time of war it may be raised to 132,600 men, of whom 67,900 are infantry. 50,000 depôt and reserve ditto, 6000 cavalry, and 6300 artillery men. Perhaps no country in Europe has a better organized army, or a finer soldiery, in proportion to its extent. The subaltern and field officers are, in general, a fine body of young men, well educated in the duties of their rank. But the general officers are not supposed to possess the experience adequate to the proper exercise of their commands, inasmuch as they mostly owe their rank to court intrigue and royal favour, and not to the value or length of their services. In this respect, however,

Sardinia is not singular. Napoleon and Frederick the Great selected their officers, because they were aware of their merit, and cared for nothing else. But ordinary sovereigns can do nothing of the sort; and court favour and parliamentary influence, without regard to merit, must, speaking generally, be at all times omnipotent in the disposal of places in the army, as in everything else.

The Public Revenue and Expenditure for 1839 were estimated as follows:

Revenue.		Expenditure.	
	Liri.		Liri.
Customs and taxes on consumption	42,500,000	Royal household., &c.	4,000,000
Direct taxes, &c.	27,000,000	Justice	4,300,000
Foreign and post-office	2,500,000	Foreign affairs	3,000,000
Interior (inc. mines)	500,000	Interior	7,400,000
Coinage, &c.	800,000	War office	28,100,000
Treasury fees, interest, &c.	908,000	Artillery	2,900,000
Naval department	300,000	Navy	2,100,000
		Finance	6,105,000
		Customs	3,500,000
		The queen-dowager	3,0,000
		The Prince of Carignan	150,000
		Public debt	8,682,000
Total	73,800,000	Total	74,474,000

According to the above statement the expenditure would appear to exceed the income; but such is not really the case, the above being merely the estimated amount of the revenue, which, for several years past, has been considerably under its real amount. In 1837 there was a surplus of 2,300,000 lire over the expenditure. The only exemptions from the land-tax are in favour of the royal palaces, domains, and manufactories; the residences and gardens of the clergy, churches, and churchyards. The public debt is about 145,000,000 frs., bearing interest at 4 and 5 per cent. The credit of the state is high; a consequence partly of the progressive reduction of the debt, and partly of the punctual payment of the interest. (Von Raumer's Italy, i., 536–574.)

Savoy was the nucleus of this monarchy. It was governed, as early as the 10th century, by its own counts, whose descendants acquired Nice in 1396, and Piedmont in 1418. The sovereigns of Savoy and Piedmont were long celebrated for their ability and the skill with which they preserved and extended their limited dominions, notwithstanding the difficulty of their position in the immediate vicinity of the great European powers.

This territory was recognised as a separate kingdom at the peace of Utrecht, in 1713. Sicily was then added to the Piedmontese dominion; but, in 1790, it was exchanged for the island of Sardinia. Genoa and its territory, Monaco, &c., were annexed to the Sardinian crown at the peace of 1815. (Serristori, Statistica d'Italia, Von Raumer's Italy, &c., i., 906, 300; Châteauvieux, Young, &c.)

SARDINIA, p. t., Erie co., N. Y., 28 m. S. E. Buffalo, 976 m. W. Albany, 343 m. W. Bounded th. by Cattaraugus cr. Drained by head branches of Cazenove cr. It contains a Baptist church, one store, one fulling-mill, one furnace, two grist-mills, two saw-mills, two tanneries; 13 schools, 446 scholars. Pop. 1743.

SAREPTA, a town of European Russia, near the frontiers of the government of Saratoff, on the Sarpa, near its confluence with the Wolga. Pop. above 2600. This town was founded in 1765, by a colony of Herrnhutters in Moravia: it is well built, neat, clean, and fortified, so as to be secure from the predatory incursions of the contiguous nomadic tribes. Its inhabitants are distinguished by their industry: they manufacture linens: silk and cotton stuffs, with stockings and caps, in great request all over the empire. They also raise and manufacture tobacco, distil spirits, &c.

SARGUEMINES, a town of France, dép. Moselle, cap. arrond., on the Sarre, 41 m. E. by N. Metz. Pop. in 1836. 4113. This town, under the name of Guemond, was formerly one of the strongest in Lorraine; but no portion remains of its ancient fortifications except a dismantled citadel, now appropriated to the gendarmerie. The sub-prefecture, hall of justice, and college, occupy the buildings of a Capuchin convent, founded in 1721. There are some spacious prisons. Sarreguemines has manufactures of cotton thread, forks, spoons, and earthenware of a superior quality; and is the entrepôt for the papier mâché snuff boxes made in the surrounding villages, and of which it is said to export 200,000 dozen a year. (Dict. Géog.; Guide du Voyageur.)

SARI, a very ancient city of Persia, prov. Mazanderan, of which it is the cap., about 18 m. from the S. shore of the Caspian, and 115 m. N. E. Teheran. Previously to 1836 it is said to have had from 30,000 to 40,000 inhabitants, who carried on a brisk trade with Astrakhan, and the interior of Persia (see Fraser's Trav. on the Caspian, p. 14); but about that time it was nearly depopulated by the plague. "Sari is surrounded by a ditch and a mud wall, flanked by pentagonal brick towers. The gateways have fallen down, and roads have been broken through them in every direction. The appearance of the town differs essentially from that of

any other in Persia S. of Elburz. The houses are built of burnt brick, and neatly tiled; some of the streets are well paved; and, although the marks of ruin are everywhere visible, Sari has something of the appearance of an English village, or small market town." (Major & Arey Tra. in Geogr. Journal, viii. 104.) Sari is frequently mentioned by the poet Ferdousi. Its vicinity is flat, woody, and well watered. (Kinnier's Persian Empire, 163.)

SARK, or SERCQ, one of the islands belonging to Great Britain in the English Channel, lat. 49° 28′ N., long. 2° 24′ W., intermediate between Guernsey and Jersey, seven m. E. the former, and nine m. N. W. the latter; length, and greatest breadth, about two m. each. Pop. in 1831, 542. It is divided into two portions, Great and Little Sark, united by a narrow neck of land. It differs little from the adjacent islands in its physical features. The soil is sandy and produces most kinds of grain and vegetables. A good many fish and sea-fowl are taken round its coasts. The inhabitants make cheeses, and knit stockings, gloves, Guernsey jackets, &c.; which they send to the ports in the W. of England, in exchange for colonial and manufactured goods.

SARNO, a town of the Neapolitan dom., prov. Principato-Citra, cap. cant., at the head of the river Sarno (an. Sarnus), 11½ m. N. W. Salerno. Pop. estimated at 12,600. It is a well built and flourishing town, having a handsome cathedral, several convents, an old castle, belonging to the Barberini family, a seminary, hospital, some sulphureous baths, and manufactures of paper and copper wares.

Sarno is celebrated in history for the desperate battle fought in its vicinity, anno 553, between the troops of Justinian under Narses, and the Goths under their king Teias. The entire defeat of the latter, and the death of their monarch, terminated the Gothic kingdom and power in Italy. (Gibb., vii. 391; Crasen's Tour in Abruzzi, &c., ii. 185–188.)

SARTHE, a dep. of France, reg. N. W., between lat. 47° 35′ and 48° 40′ N., and long. 0° 25′ W. and 0° 39′ E.; having N. Orne, E. Eure-et-Loir and Loir-et-Cher, S. Indre-et-Loire and Maine-et-Loire, and W. Mayenne. It is of a compact shape, 60 m. in length, N. and S., by about the same in breadth, E. and W. Area, 621,600 hectares. Pop., in 1831, 457,372. Surface generally level, except in the N. W., where there are a few hills. The principal rivers are the Sarthe, with its tributaries the Vegre, Husne, and Loir. The former rises near Mortagne, in the dep. Orne, and runs with a very tortuous course, S. and S.W., to the vicinity of Alençon near which it receives the Loir, and unites with the Huisne to form the Maine, after an entire course of nearly 160 m. In the 15th century it was navigable to Le Mans, but its navigation is now difficult for some distance below that city. Besides Le Mans, Alençon stands on the Sarthe. The soil of this dep. is very various; in some parts there are rich lands, but poor sandy tracts predominate, especially in the S. E. In 1834, the arable lands were supposed to comprise 393,456 hectares, meadows 58,120 ditto, vineyards 10,081 ditto, orchards 19,479 ditto, and woods 68,328 ditto. Wheat, barley, and rye are the principal corn crops; and are sufficient, along with potatoes, of which at least 1,160,000 hectol. were raised in 1835, for the consumption of the population. The produce of wine is not enough for the consumption; but about 220,000 hectol. of cider and perry are said to be annually manufactured. Live stock pretty abundant and good. Bees are but little reared; and the wax, to which Le Mans has a considerable trade, comes mostly from the neighbouring departments. In 1835, of 131,791 properties subject to the contrib. foncière, 46,057 were assessed at below 5 fr., 22,949 at from 5 to 10 fr., and 22,032 at from 10 to 20 fr. Hardware, paper, woollen fabrics, leather, wax-candles, sail-cloth, glass and earthenware, soap, and other articles of necessity, rather than of luxury, are the goods principally manufactured in Sarthe. The two forges, &c., produce annually about 1,000,000 kilogr. of good iron. The étamines of Maine formerly enjoyed a great celebrity, but other fabrics have superseded them; so that St. Calais, and other towns where they were chiefly made, have fallen into decay. Sarthe is divided into four arrondissements chief towns, Le Mans the capital, La Flèche, Mamers and St. Calais. It sends seven members to the chamber of deputies. Registered electors, in 1838–39, 2345. Total public revenue, in 1831, 10,538,907 fr. In this department the females among the peasantry, and even in the classes above them, appear to enjoy but little consideration. "Si la maîtresse (la fermière) accouche, on demande: 'Est-ce un gas' Quand le contraire arrive, on dit: 'Ouen, ce n'est qu'ene creïature' (une fille); et, en effet, un homme a ici quatre et cinq fois autant de valeur qu'une femme. Telle forte et robuste servante, propre à tous les ouvrages, ne gagne que 30 fr. et sa nourriture par an, tandis qu'un laboureur est payé de 150 à 200 fr. pour l'année." (Hugo, art. Sarthe; Dict. Géog.; French Official Tables.

SARUM (OLD), an ancient, and now totally ruined city and bor. of England, co. Wilts., on a hill, two m. N. Salisbury, or New Sarum, which see. It was the Sorbiodunum of the Romans; and, being surrounded by walls and defended by a castle, became a place of considerable consequence under the Saxons. Under William the Conqueror. the bishop of Shireburn and Sunning removed his see thither; and such was its importance. that parliaments were held in it under subsequent Norman kings. But it always laboured under various inconveniences, the principal of which was the total want of water; and in consequence of this, and of disputes between the crown and the church as to the possession of the castle, the inhabitants began gradually to remove to the more convenient situation of New Sarum, or Salisbury; and the seat of the bishopric being translated to the latter, in the reign of Henry III., old Sarum fell into a state of total decay, and was almost wholly deserted in the early part of the reign of Henry VII. For a lengthened period there has been hardly even any vestiges of its ruins. (Camden, Gibson's ed., p. 114.)

Old Sarum sent two members to the House of Commons in the reign of Edward III.; and notwithstanding its total decay, the proprietor of the burgage tenures in the borough, or of the land on which it once stood, was permitted to exercise this important privilege in its name down to the passing of the Reform Act, when it was disfranchised. Not having a single house or inhabitant, Old Sarum afforded the most perfect example of a nomination borough. The property several times changed hands; and though the estate was of little intrinsic value, the privilege it possessed of manufacturing two law-makers for the British empire, made it sell for a very large sum. It may well excite astonishment that such an outrage on the principle of representation should have been permitted to exist for so lengthened a period.

SARUN, a district of Hindostan, presid. Bengal, prov. Bahar, and one of the richest and most prosperous in British India, between lat. 25° 30' and 27° 30' N., and the 84th and 86th degs. of W. long.; having W. Goruckpoor, S. Ghazepoor, Shahabad, and Patna; E. Tirhoot, and N. Nepaul. Length, N. to S., about 110 m.; breadth, varying from 25 to 90 m. Area, 5760 sq. m. Pop., in 1822, 1,464,075. It is well watered; the Ganges forms its entire S. boundary, and the Gunduck intersects it near its centre; it supplies in abundance all the principal products of the east, besides good timber for ship-building, masts, spars, &c. There is little jungle or waste land; cattle, though not numerous, are of good quality. Manufactures few; the principal is that of saltpetre. a great deal of which is produced in this district. The Mohammedans form but a small portion of the entire population. Chief towns, Chuprah, Bettiah, and Maisey. Total land revenue, in 1829–30, 15,00,563 rupees. (Hamilton's E. I. Gaz.: Parl. Reports.)

SASSARI, a city of Sardinia, cap. of its N. division, in the W. part of the island, on the Turritano, about 10 m. from mouth at Porto Torres, in the gulf of Sassari, 58 m. N. by W. Oristano, and 100 m. N.N.W. Cagliari. Lat. 39° 20' N.; long. 8° 35' 20" E. Pop., in 1838, including its commune and port of Torres, 24,408. It is surrounded by a wall, strengthened by square towers, with five gates and a citadel, the latter being now used merely as a barrack. It has good main street; and is surrounded by public walks, shaded by trees. Sassari has numerous churches, convents, nunneries, a Tridentine seminary, and a public hospital. The cathedral, a massive structure, has a disproportionately high spire, and a very elaborate façade; but its interior is clean and airy, and it has several good sculptures, including a monument by Canova. The university is established in its former Jesuits' college. The palace of the governor is an extensive edifice, and the public buildings in general well adapted for their intended purposes. There are some pretty good inns and coffee houses, and the shops are very equal, if not in some instances superior, to those of Cagliari. (Smyth's Sardinia, p. 267, 268.) It is the seat of an archbishop, of a tribunal of secondary jurisdiction, an appeal to the Audienza Reale of the island, and of a tribunal of commerce; and is the residence of the vice-intendant and vice-treasurer of Sardinia, and of a military governor. It has a considerable trade in tobacco, oil, corn, &c.

Porto Torres (an. Turris), its port, 10 m. distant, can only accommodate small vessels; ships of large size being obliged to anchor in the roads nearly one mile outside, where, however, the anchorage is pretty good. Sassari rose on the decline of Turris, during the insecurity of the middle ages. Again it appears to be better conducted in its vicinity than any other part of the island. Immediately without its walls is the fountain of Rosello, an abundant source of water, embellished with much architectural ornament. (Smyth's Present State of Sardinia; Official Reports, &c.)

SASBACH or Saltzbach, a village of the Grand Duchy Baden, bailiwick of Achern, at the foot of the mountain Black Forest on an affluent of the Acher, 17 m. E.N. of Freiburg. This village, which has about 1000 inhabitants has acquired a high degree of historical interest from

the famous Marshal Turenne having been killed in its vicinity, by a random shot, on the 27th of July, 1675. The circumstances attending the death of this great general have been detailed by Voltaire (Siècle de Louis XIV., cap. 12), and other distinguished writers. His remains, deposited by order of Louis XIV. in the royal burying-place in the abbey of St. Denis, escaped, at the era of the revolution, the fanatical violence that scattered the dust of so many kings. At length, after various vicissitudes, they were deposited in the church of the Invalids, by order of the still more illustrious captain now entombed within the same sacred precincts. A monument, in honour of Turenne, erected in 1781 on the place where he fell, was repaired in 1801 by Moreau, and was reconstructed of granite in 1829. The funeral orations in honour of Turenne, by Flechier and Mascaron, are chefs-d'œuvre.

SATALIEH, or ADALIA. See ADALIA.

SATTARAH, a considerable town and fortress of Hindostan, prov. Bejapoor, about 60 m. S.S.E. Poonah. Lat. 17° 42' N.; long. 74° 12' E. The fort stands on a scarped hill; at the foot of which is the town, built partly of stone, and partly of mud or unburnt bricks, but comprising no edifice of note, if we except a new palace built within the last 20 years. The fort, though naturally strong, was taken by Sevajee from the Bejapoor sovereign in 1673, by Aurungzebe in 1699, and by the British in 1818. The British cantonments are about 2 m. to the E. Sattarah was, under Sevajee and his immediate successors, the cap. of the Mahratta empire. The Sattarah rajahs, however, had been reduced to the condition of rois fainéants by their ministers, the Peishwas, for a lengthened period previously to 1818, when the British vested the rajah with a limited authority over a portion of the dominion of his ancestors. The Sattarah territories since then have comprised an area of about 8000 sq. m., with a population of perhaps, 1½ millions. In his evidence before the committee of the House of Commons, in 1830, Colonel Briggs stated, "The administration of the government of Sattarah may be deemed a good specimen of the management of a native government. The country is divided into districts, each yielding from a lac to 1½ lac of rupees, containing from 150 to 200, and even 300, villages. Over this district is an officer, called a soubahdar. That district is then subdivided among a great number of junior officers, each having from 6 or 8 to 20 villages under his charge. The whole civil and judicial business is conducted through those officers. The annual assessment of the land is fixed with reference to the sum yielded in former years; the assessment varying every year with the quantity of land cultivated. The revenue is always paid in money." In 1828, it amounted to about 15,60,000 rupees. This state was subsidiary to the British till about 1839, when the rajah, having been detected in a conspiracy against the British authorities, was dethroned, his dominions having since that time been amalgamated with and administered under the Bombay Presidency. (Hamilton's Gazetteer: Parl. Reports.)

SAUGERTIES, Ulster co ' _ , 10 m. N. Kingston, 45 m. S. by W. Albany, 331 m W. Bounded E. by Hudson r. Catskill mt. extends into its W. part. Watered by Platte-kill and Esopus creeks. It contains six churches, a Presbyterian, Dutch Reformed, Episcopal, Methodist, Baptist, and Roman Catholic, 23 stores, one woollen-factory, 12 bloomeries, four grist-mills, seven saw-mills, one paper-mill, three tanneries; 17 schools, 1109 scholars. Pop. 6216.

SAUGUS, p. t., Essex co., Mass., 11 m. N. Boston, 451 m. W. Taken from Lynn in 1815. It contains two churches, a Congregational and Methodist, three stores, one fulling-mill, one woollen-factory, one grist-mill; five schools, 375 scholars. Pop. 1098.

SAULT DE ST. MARIE, p. v., capital of Chippewa co., Mich., (commonly written Sault St. Mary, and pronounced Soo St. Mary,) 400 m. N. Detroit, 921 m. W. It occupies the site of an old French fort, on the S. side of St. Mary's

man Catholic, three stores, one of which belongs to the American Fur Company, who have a trading house; and fort Bradey with a garrison. Connected with the Baptist church is a missionary school for Indian children, to which $1000 annually are contributed by the United States government. The Methodists also have a missionary school, and there is a school in the fort for the children of the officers

portance, as settlements are making on its southern shore, which contain great mineral treasures, particularly copper. The mail arrives at this place weekly in summer, and once in six weeks in winter. The navigation extends from the first of May, to the 20th of November. The thermometer often sinks in winter to 25° or 30° below zero of Farenheit, and the mercury sometimes congeals. The permanent population is estimated at 800, though it is sometimes greatly increased by the influx of fur traders and Indians. Great quantities of white fish and trout are caught here, of a superior quality, and are extensively exported.

SAVANNAH, river, Ga., forms the N.E. boundary of the state, separating it from S. C. It is formed by the union of Tugaloo and Seneca rivers, the latter from S. C., which unite near the S.E. corner of Franklin co., 100 m. above Augusta. Thence it flows S.E. and enters the Atlantic through Tybee sound, in 32° N. lat. Tybee island at its mouth is 5 m. long and 3 broad, and contains a lighthouse on its N.E. point, 80 feet high. Large vessels come to *Five fathom hole*, 3 miles below Savannah, and 14 miles from the ocean, and large brigs come to the wharves in Savannah. Steamboats go to Augusta, 127 m. by land and 340 by water from its mouth; and pole boats go 150 miles above Augusta. The tide flows up the river but 55 miles.

SAVANNAH, p.t., Wayne co., N.Y., 11 m. E. Lyons, 168 m. W. Albany, 348 m. W. The S.E. part is covered by the Montezuma marshes. Crusoe lake, 1¼ m. in circumference, has its outlet into Seneca river. It has one store, one saw-mill; eight schools, 511 scholars. Pop. 1718.

SAVANNAH, city, port of entry, capital of Chatham co., and the largest city in Georgia, is in 32° 4′ 56″ N. lat., and 81° 8′ 18″ W. long., and 4° 5′ 54″ W. long. from W. It is 120 m. S.S.E. Augusta, 190 m. E.S.E. Macon, 90 m. W.S.W. Charleston, S. C., 158 m. E.S.E. Milledgeville, 662 W. The population in 1810, was 5195; in 1820, 7523; in 1830, 7776; in 1840, 11,214, of whom 4694 were slaves. Employed in commerce, 604; in manufactures and trades, 707; in navigating the ocean, 201; do. rivers and canals, 40; in the learned professions, 131.

The city is built on the S. side of Savannah river, on a sandy plain, elevated 40 feet above the level of the river, 17 m. from the ocean. The entrance over the bar from the sea does not change, and being full a mile wide, and having from 18 to 21 feet of water at low tide, the harbour is one of the finest on the southern coast of the United States. The entrance is so plain and direct that any navigator, on his first visit, having made the lighthouse on Tybee island, can, by means of the Coast Pilot, even during a heavy gale of wind, run into good anchorage, without risk or difficulty. The facility of entrance is equalled by no port of the United States S. of Newport, R. I. The navigation of the river was obstructed during the revolutionary war, to protect the city from the British; and these impediments and their effects have not yet been entirely removed. The government of the United States undertook their removal, and have still an officer in charge of the work. Vessels requiring 13 feet water load at the wharves of the city; and those requiring 15 or 16 feet of water, four or five miles below, at good anchorage. The plain on which the city is built is nearly a perfect level, and extends along the river a mile E. and W., and continues for several miles S., increasing in width back from the river. The city is regularly and beautifully laid out with streets and lanes running east and west; and crossed at right angles with streets only, running north and south. Between every other street are public squares, which in the whole are 18 in number, generally railed in, surrounded and interspersed with trees of different kinds, and form small parks, covered with grass, which give the city during the spring and summer months a cool, airy, and rural appearance. The streets generally are lined on either side with trees, and some have single, and others double rows of trees running through their centres, the latter forming perfect arcades, and serving at all times for delightful and shady walks. The houses are built with wood, brick, and stone, many of them are of fine architecture, and of beautiful proportions, and cover, at the present time, a space a mile in length, and three quarters of a mile in width. It contains 13 churches, four Baptist, three of which are African, an Independent Presbyterian, a Presbyterian, two Episcopal, a Methodist, a Lutheran, a Unitarian, a Mariners' church, a Roman Catholic, and a Jews' Synagogue. The Independent Presbyterian church is constructed of a light coloured granite, and cost $100,000. The other public buildings are a courthouse, an exchange, arsenal, a guard-house, jail, United States barracks, an academy, theatre, female asylum, widows' asylum, a poorhouse and hospital, a market-house, and several fine banking houses. There are many benevolent societies which have no buildings, also, a public library, and a historical society. Some of the private dwellings, many of which are of brick, and some smooth cast to resemble marble and other stones, are tasteful and elegant. The warehouses are numerous, generally lining the wharves, and built of brick or stone, most of them three or four stories high. The business is generally done in an upper story, entered from the top of the bluff, while the lower stores serve to receive the merchandise directly from the ships. There are two iron foundries, four steam saw mills, three marine railways, and three steam rice-mills. The population, according to a census taken in the summer of 1836, amounted to 12,758 permanent inhabitants; to which should be added a transient winter population of 3000, according to the computation of the officer who took the census, making in the whole 15,758. Little business being done in the summer, during which it was formerly considered unhealthy, many of the inhabitants have acquired the habit, during that season, of residing which accounts for the great increase of the __ __ during the winter season. But since the rebuilding of the city after the fire of 1820, the improvement in __ dwellings, the greater attention to the cleanliness of the streets and lanes, the organization of a board of health, the substitution of a dry for a wet cultivation of rice in the vicinity, to effect which the inhabitants contributed to the planters the sum of $70,000, and a favourable change of the habits of the people, and of the climate of the country generally, have unitedly made Savannah to rank among the healthy places of the United States. In the opinion of the most eminent physicians, the winter is now considered as the most unhealthy season.

The exports of Savannah during the commercial year 1843, were 285,754 bales of cotton, 25,038 tierces of rice, 7,500,000 feet of lumber, 5,175,000 shingles, and 64,362 staves. The direct foreign imports amounted only to $279,896 in value. The registered, licensed, and enrolled tonnage of the port, in 1840, was 17,930 tons. Savannah is at present the centre of commerce for a large extent of country, which, from her peculiar location, must be very much enlarged. Her immediate neighbourhood to Florida, which, from the shallowness of her harbours, can never accommodate large shipping; the dangerous navigation around the cape of Florida; and the facility of passing round the Alleghany mountains, which terminate in Georgia, by a railroad, will eventually cause her to assume a commanding position in the south. Already a railroad is completed 199 miles to Macon, in the centre of the state, which is to be continued 101 m. farther west, 60 of which are already finished, where it will join the State, or Atlantic and Western railroad, which is designed to run 136 m. still farther northwest to Tennessee river. These works, which will probably be finished by the close of 1845, will, by a direct route of 427 miles, open to Savannah the commerce of the west. The State railroad will extend by a branch to Rome, on the Coosa river, which is navigable 100 m. to its falls. There is also in contemplation a branch of the Monroe railroad to the Chattahoochee river, at a point leading directly to a union with the Montgomery railroad in Alabama, as a continuation of the great route of southern and western travel. There is likewise in progress, and nearly finished, a railroad from the Flint to the Oakmulgee rivers, which may be extended 40 m. to the Chattahoochee, to draw the products of western Georgia and a part of Alabama to Savannah. These things show the prospect of the extension of her trade by land conveyance. A canal extends from Savannah, 16 m. to Ogeechee river. There are on the water already two companies employed with six steamboats, four of which are of iron, with 30 towboats of 150 tons burthen each, running to Augusta, 250 m. up the Savannah river. There are also two companies who have 11 steamboats and 42 towboats, running through an inland coast navigation to Darien, and up the Altamaha and Oakmulgee rivers, 69 m. to Macon. There are also running through an inland coast navigation, two steamboats to and from Florida; four steamboats ply to Charleston; two by the inland passage, and two by the outside passage. The boats run steadily during the winter season, and whenever the state of the river permits, during the summer, with occasional assistance from other boats. There are three lines of brigs, with six vessels in each, making 18, in the whole, to the city of New-York, one vessel of which sails every two days from each place, occupying on an average seven days on the passage between the two places.

According to the census of 1840, there were in Savannah two commercial and 50 commission houses in foreign trade, with a capital of $943,500; 191 retail dry goods and other stores, with a capital of $653,190; eight lumber-yards, with capital of $49,900; 15 persons produced machinery to the amount of $7300; 65 pounds of reeled silk, worth $250, were produced by three men and two women, with a capital of $500; five manufactories of leather, as saddleries, &c., produced to the amount of $9400, with a capital of $19,000; five persons manufactured 15,000 pounds of soap, and 41,000 pounds of tallow candles, with a capital of $6000; 25 persons manufactured drugs and paints to the amount of $35,000, with a capital of $35,800; seven persons manufactured confectionary to the amount of $600, with a capital of $550; four persons manufactured carriages and wagons to the amount of $1700, with a capital of $700; 14 persons manufactured furniture to the amount of $10,300, with a capital of $15,000;

four printing-offices, two binderies, three daily, three weekly, and three semi-weekly newspapers, employed 22 persons, and a capital of $22,000; three brick or stone, and 43 wooden houses were built, employing 192 persons, and cost $138,100. The total capital employed in manufactures was $105,460. There were three academies, 385 students; seven schools, 470 scholars.

The city was founded by General James Oglethorpe in 1733, and was incorporated in 1761 by a charter, which was amended and enlarged in 1787. In December, 1778, Savannah was taken by the British under Colonel Campbell, which they did not evacuate until 1782. On the 10th of January, 1820, 463 buildings were burned, occasioning a loss to the amount of $4,000,000; but it has been since rebuilt, chiefly with brick, with additional convenience and beauty.

SAVANNAH, p. v., cap. of Hardin co., Tenn., 49 m. N.W. Florence, Ala., 131 m. S.W. by W. Nashville, 807 m. W. Situated on the E. bank of Tennessee river. It is considered one of the healthiest places on the river above the Muscle shoals, and contains a courthouse, jail, a church, several stores, various mechanic shops, and 250 inhabitants.

SAVANNAH, p.v., capital of Carroll county, Ill. 202 m. N. Springfield, 872 m. W. Situated on the E. side of Mississippi r., and contains a courthouse jail, and a number of stores and dwellings. Nett proceeds of the postoffice in 1841, $126.

SAVE (Germ. Sau, an. Savus), a river of the Austrian empire, and one of the principal tributaries of the Danube. It rises towards the N. extremity of Carniola, in about lat. 46° 30′ N., long. 14° E., and runs at first S.E. through the government of Laybach and Croatia, to about lat. 45° 15′, long. 17°. It thence has more of an E. direction, forming the boundary line between the Austrian prov. of Slavonia, on the N., and Turkish Croatia, Bosnia, and Servia on the S., till it enters the Danube at Belgrade, after a course or 590 m. (Rerghaus.) Its chief affluents are the Kulpa, Unna, Verbas, Bosna, and Drina. Though not very rapid, its inundations are often very destructive. Being navigable as far as the mouth of the Kulpa, for vessels of from 150 to 200 tons, it is a good deal used for commercial purposes. Few towns of any consequence are, however, situated on its banks, the principal being Brod and Krainburg; Laybach, Agram, Petrinia, and Posega are, however, at no great distance, and some of them are seated on its tributaries. (Berhaus, Allg. Länder; Oest. Nat. Encyc.)

SAVERNE, (an. Tabernæ), a town of France, dép. Bas-Rhin. cap. arrond., on the Zorn, a tributary of the Rhine, 19 . N.W. Strasburg. Pop. in 1836, ex. comm., 5118. Though nely situated, it is but indifferently built, and has no remarkable edifice, except an old palace, formerly belonging the bishops of Strasbourg, but now used for the police barracks and prison. The town has manufactures of woolen cloths, hosiery, hardware, &c., with some trade in timber, floated down from the Vosges by the Zorn.

SAUGAR, or SAUGOR, a large town of Hindostan, prov. alwa, in the ceded districts on the Nerbuddah, lat. 23° 48′ . long. 78° 47′ E.: taken by the British in 1818.

Saugor is also the name of an island of the Sunderbunds, the mouth of the Hooghly, about 60 m. S.S.W. Calcutta. railway to connect it with that city was projected a few ars ago, and is now probably completed.

SAUMUR, a town of France, dép. Maine-et-Loire, cap. und., on the Loire, 28 m. S.E. Angers. Pop., in 1836, ex. m., 11,576. The Loire here forms several islands, and is sed by five or six bridges, one of which, a stone bridge 12 arches, 284 yards in length, long considered as one of finest in France, connects the town with its suburb of Croix' Verte. Saumur is built partly at the foot, and tly on the declivity, of a hill crowned by a citadel. Its ner portion is tolerably well laid out, and has a handsome ay and terrace facing the river; but the upper town is irular, and the streets inconveniently steep. The castle, ich appears to have been constructed at different periods ween the 11th and 13th centuries, was the occasional idence of the kings of Sicily and the dukes of the house Valois : for some time previously to the revolution it s a state prison; it now serves as an arsenal. There several churches worth notice; one of which, curious n its antiquity, is supposed to have been constructed in 5th or 6th century; and another, Notre Dame des Ariers, is remarkable for its beauty; having a fine dome ported on Corinthian columns, an altar-piece by Philip Champagne, &c. The barracks are among the best of description of edifices in France; they are four stories eight, and can accommodate 1200 men. The town-. public library, public baths, and theatre, are the other cipal buildings. Not far from the town is a famous rischool. It has manufactures of linen cloths, handkerchfs, necklaces, copper and iron wares, leather, saltpetre, , with a brisk trade in provisions, and four large annual . Inglis says he should "greatly prefer Saumur as a dence to Angers; it is more airy and lively, the country

quite as beautiful, and provisions are a shade cheaper." (Switzerland, &c. p. 347.)

Saumur was taken in 1026 by Fulk of Anjou, and, after many vicissitudes, was annexed to the French crown in 1570. A Protestant academy, founded here by the famous Duplessis Mornay, the friend of Henry IV., governor of the town for a lengthened period, was dissolved by Louis XIV. in 1684. (Hugo, art. Maine-et-Loire; Guide du Voyageur, Dict. Geog.)

SAVONA, a town and seaport of N. Italy, in the Sardinian States, div. Genoa, cap. prov. on the Mediterranean, 20 m. S.W. Genoa. Pop. (1838), with comm., 16,911. It has ramparts, which, however, are of no great strength, and many good public and private buildings; but its streets are narrow, winding, and badly paved. It had formerly two harbours, the best of which was filled by the Genoese in 1525, from jealousy of its capabilities: the other, formed by a mole projecting E. into the sea, is small, and is rather difficult of approach, from the accumulation of sand and mud near its mouth. Savona is a bishop's see, and the seat of judicial and commercial tribunals; it has manufactures of silk goods, iron and earthenware, and exports oranges and lemons, grown in its vicinity. Savona was the birthplace of Popes Sixtus IV. and Julius II., and is said to have been for some time the residence of Columbus. Pope Pius VII. was also detained in it in 1810-11, by order of Napoleon. (Dict. Geog. &c.)

SAVOY. See SARDINIA (kingdom of).

SAXONY (KINGDOM OF), a secondary state of central Europe, and of Eastern Germany, principally between lat. 50° 10′ and 51° 30′ N., and the 12th and 15th degs. of E. long.; having W. the Indep. Saxon principalities; N. Prussian Saxony and Brandenburg, and S. Bohemia. It is of a triangular shape. Length, E. to W., about 140 m.; the greatest breadth nearly 90 m. The area, pop., subdivisions, &c., are as follow:

Circles.	Area in sq. m.	Pop. 1837.	Pop. to sq. m.
Dresden	1,870	420,217	232
Leipsic	1,338	367,753	274 9
Zwickau	1,796	564,707	327-6
Budisin	985	292,913	272-4
Total	5,759	1,635,190*	284 1

The Erzgebirge (ore mountains), and the Riesengebirge (giant mountains), extend along almost the whole of the S. and S.E. frontier, but they nowhere rise to 4000 ft. of elevation. Their declivity is more gradual and undulating on the Saxon than on the Bohemian side; so that they cover the greater part of the country with their ramifications, rendering it either mountainous or hilly. There is, however, a very considerable extent of level ground, extending from the foot of the hilly tract, or from Colditz, Meissen, and Bautzen, northward, all along the frontier of Prussian Saxony. The country to the S.E. of Dresden, where the Elbe forces its way through the mountain chain, has been called the "Saxon Switzerland." It is about 30 m. in length by 24 in breadth, diversified, and highly picturesque; but its name is likely to convey a wrong impression of its scenery, its highest summit, the Schneeberg, being only 2150 ft. in height. The spurs given off by the Erzgebirge to the N. enclose the valleys of the Elbe, the two Muldas, the Zochoppau, Elster, Pleisse, &c., all of which flow to N.W., and, except the first, which is navigable throughout the whole extent of the kingdom, rise in Saxony.

Saxony has a milder climate than most parts of continental Europe in the same latitude; the mean temperature of the year is about 47° Fahr.; that of the winter quarter being 34°, and of the summer 59°, at an average of the entire country, which has a mean elevation of about 1100 ft. above the sea. Landed properties are rather of limited size; but in all the rural districts the people appear to be contented and, on the whole, comfortable: pauperism is rare. According to Mr. Gleig, "There is, perhaps, no country in the world where more is made of the land than Saxony. Every spot of earth which seems capable of giving a return is cultivated; and the meadows are mowed twice or thrice in the course of each summer. You never met with such a thing as a common or waste, while the forests are all guarded with a strictness proportionate to their value. As farmers, I should say that the Saxons were more clean and industrious than skilful. The fields are always well cleared of weeds, and in their crops they have a succession: but the favourite grain is rye; and either because it does not require to be pampered, or that manure is a scarce article with them, they do not seem disposed to fatten the soil too frequently."† (Germany, &c., i., 237–238.) Rye, wheat, and

* This is exclusive of the troops, which, when added, raise the total population to 1,652,114
† It is obvious from this extract, that Mr. Gleig knows nothing of agriculture; but his testimony as to the appearance of the country is, notwithstanding, of some value.

785

barley are scarcely grown, except in the low country; in the mountain region they are met with only in the valleys; oats and potatoes being there the chief crops. Pease, vetches, millet, teasel, flax, oil seeds, tobacco, and garden vegetables, are pretty generally cultivated; and artificial grasses are nearly universal. But notwithstanding the improvement of agriculture, and the industry of the people, considerable quantities of corn have to be imported. A great deal of fruit is grown; and between 7000 and 8000 morgen of land are occupied with vineyards. In 1835, upwards of 53,000 eimers of wine were made, some of very tolerable quality. The forests which occupy about 1-4th part of the entire surface, consist of fir, pine, beech, oak, elm, maple, larch, &c. Upwards of 1-3d part of these woods belongs to the crown, yielding an annual revenue of 2,000,000 dollars; and nearly 10,000 individuals are engaged in wood-cutting.

Saxony is celebrated for her breeds of sheep, which are among the finest in Europe. The late king, when elector of Saxony, introduced the breed of Merino sheep into his dominion, and exerted himself to promote the growth of this valuable race of animals, with such success, that they are now found to succeed better in central Europe than in Spain; and, notwithstanding the rapidly increasing importations from Australia, the greater portion of the immense quantity of wool that we import still continues to be brought from Saxony and other German states. According to Berghaus, there are only 696,000 sheep in the kingdom; but we apprehend that there must be some signal error in this statement, and that their number must be very materially greater. Indeed, they are stated, in *Horschelman's Stein* (ii., 475), to exceed 2,000,000. We may farther state that, in 1837, Prussian Saxony had above 2,000,000 sheep; and though it be more extensive than the kingdom of Saxony, we believe it is not so well stocked as the latter. The best wool is produced on the sheep-walks of the Saxon Switzerland. The cattle of Saxony, the number of which exceeds 550,000, are also of a superior description; but the butter is usually indifferent, while, to increase its weight it is frequently overloaded with salt. But such as it is, the demand for it is universal. "Never," says Mr. Strang, "did I witness so much butter daily consumed, as I have seen since I entered this kingdom. Here, in short, bread and butter is the order of the day at all hours. It is the perpetual family staple, and essential as a make-weight at every meal. You find it with equal propriety at breakfast, at lunch, at dinner, and at supper. A larder in Saxony may well be called the *buttery!*" (*Germany, &c.,* i., 140.) Horses are not so extensively reared as other live stock, and hogs are not numerous. The game laws are very rigidly enforced, all sorts of birds being included in their enactments; and rights of fishing, &c., appear to be preserved with the most scrupulous tenacity.

Mining is one of the principal occupations of the inhabitants. Few parts of Europe equal the Erzgebirge in the variety and extent of their mineral riches. The basis of these mountains is granite, covered by gneiss, mica, and clay slate in succession, between which are other strata containing metallic ores. Upwards of 500 mines are wrought, which are said to employ about 11,000 workmen; and between 50,000 and 60,000 persons, or about 1-30th part of the entire population, derive their subsistence from mining industry and the manufacture of metallic products. The total annual value of the metals obtained is estimated by Berghaus, at 1,760,000 dollars; the silver producing nearly 930,000, and the iron and iron-wares 400,000 dollars.

Lead, bismuth, arsenic, antimony, cobalt, and manganese are the other principal metals. Freiberg (which see, i. 947) is the centre of the silver mining district. The neighbourhood of Meissen yields the fine porcelain clay, of which the "Dresden China" is made. About ½ million scheffel coal are annually produced. Salt is scarce, since the salt mines, formerly included in the Saxon dominion, were separated from them in 1815, and this important necessary is mostly imported from Prussia. Serpentine marble, and fine building stone, are abundant; as are various gems, including the topaz, jasper, agate, tourmaline, &c.

The most important branch of manufacturing industry in Saxony is that of cotton. Its extension has been attributed to the nearly contemporaneous introduction of the potatoe, called by German writers the "manna of the mountains," and which has enabled the Saxon weavers to obtain a sufficiency of food at exceedingly low wages. In 1830, there were in Saxony 86 spinning establishments, having 361,202 spindles, employing 5380 adults, and 2443 children: of these establishments, only 3 were wrought by steam. In 1837, there were 124 spinning establishments, with 490,395 spindles. (*Bowring's Rep. on the Prussian. Comm. League,* p. 35.)

Most descriptions of cotton fabrics are now produced, and many new factories have been established in Chemnitz, Zwickau, Auderan, Freyburg, &c. Great efforts are making to improve the construction of machinery; and joint stock companies for the purpose have been established near

Chemnitz and Dresden. Coal has within these few years been found in the neighbourhood of Dresden; and it is said that the mines are becoming productive, and promise a good supply. Cotton-printing works are on the increase, and have been much improved within the last few years. Although the Saxon prints, in general, are not equal to the best English in beauty of pattern or brightness and fastness of colouring, they are said to make up for these deficiencies by the cheaper rates at which they can be produced. (*Kryser, in the Handloom Weavers' Report,* part II., 510, 211.)

The only article, however, in which the Saxons come into competition with us in the American and other foreign markets is cotton hosiery, particularly the inferior descriptions. Dr. Bowring says that he had seen some stockings intended for the American markets which were sold at the rate of 3s. 4d. a dozen! The number of stocking frames amounted, in 1815, to about 9,000; in 1831, to 14,000; and in 1836, to 20,000. The number of master-workmen, in 1834, was 7,165; hired workmen, 3444; apprentices, 2852; in all, 13,461 persons. The number of frames in work was 13,361, and the average weekly return per frame was 1 doll. 4 gr., =3s. 4d. sterling. There is no branch of industry which seems more appropriate to Saxony than this. It requires only a small outlay of capital for the stocking-maker; his wooden frame is not expensive; the cost of his stock of cotton twist is small; and by associating agricultural with manufacturing industry, he supplies himself from his own little farm with the principal necessaries. If we may depend on the statements of Bowring and others, it would appear that the stocking-weavers of Saxony are in a state of progressively increasing prosperity. Most of them are independent labourers, buying for themselves the raw material and selling their manufactured stockings to a number of small collectors, who furnish the Chemnitz or the Leipsic markets. (*Bowring's Report,* p. 36.)

We confess, however, that we are not a little sceptical as to several of these statements, and have very great doubts whether persons in the condition of the Saxon stocking-weavers can ever come into competition with those of England in the production of any but the commonest description of goods. In illustration of what is now stated, we may mention, that during the year ended the 30th September, 1839, hosiery of the value of 412,410 dollars was imported from the Hanse Towns into the United States, of which, probably, the principal portion was from Saxony; and that, with yarn of the value of 22,310 dolls., made up the whole of the imports of cotton goods into the United States from these ports in the course of that year! (*Offic. Rep. Cong.*)

The manufacture of linen in Saxony is also of considerable importance. The weaving business employed, in 1833, from 12,000 to 13,000 looms; of which about 950, employing about 3000 persons as weavers, loom-builders, pattern-drawers, &c., were appropriated to the weaving of damasks and table-linen. The weekly earnings of the damask weaver vary from about 2s. 3d., up to 7s. 6d.; but for white linens, the utmost amount of a man's wages, per week, would be from 2s. to 2s. 4d. sterling. The spinning of flax employs numerous hands; but notwithstanding about 38,000 cwt. of yarn is annually imported from Silesia, Bohemia, &c., and latterly there have been considerable importations from England. (*Handloom Rep.,* p. 511.)

In 1837 there were in Saxony 196 establishments for the spinning of woollen yarn, with 71,866 spindles, and eighteen establishments for combed yarn, with 20,766 spindles; making together 101,861 spindles, being an increase of 31·1 per cent. since 1834, and of 127 per cent. since 1831. The progress of the Saxon cloth manufacture in the three years from 1834 to 1837 is stated in an official report to have been greater than in the 30 years preceding! Great improvements have been introduced, not only in the fabric, but in the finish of woollen goods, particularly by the production from the Netherlands of a new steam brushing machine. In 1839 the woollen manufacture employed from 3,000 to 4,000 looms, which produced about 160,000 pieces of cloth. Much attention has been paid of late to the manufacture of machinery; though it be still far behind what is met with in the manufacturing districts of Great Britain. The Jacquard loom is gradually being introduced, and there are schools of manufacture at Dresden, Chemnitz, Planen, &c. Finis and figured silks, of very fair quality, are made at Annaberg, Penig, and Frankenberg. The government is very desirous of promoting the culture of silk, and some establishments for the propagation of the worms exist at Dresden and Leipsic; but in such a climate they can hardly be expected to have much success. Wooden wares are made in the country, of the estimated value of about 300,000 dollars a year (*Berghaus*); and Saxony supplies furniture of every description, musical instruments, &c., to a great part of Germany. Porcelain, and modern antiques, are articles made in large quantities, particularly at Meissen. The china produced here formerly enjoyed a very high reputation throughout Europe, but owing to the extraordinary improvements

ments made in the manufacture in this and other countries, Meissen china has declined considerably from its ancient celebrity. Almost every article of use or luxury is made in Saxony; the chief deficiency is in paper, of which the Saxon manufactures do not produce nearly enough for the immense consumption of the presses of Leipsic and Dresden.

The extension of the cotton and woollen manufactures of Saxony, since 1833, is wholly, or almost wholly, ascribable to the circumstance of her having then joined the Prussian commercial league. This opened a widely extended market for her products among the German states, from the greater number of which they had previously been either wholly excluded, or admitted only clandestinely, and under great difficulties. Saxony, in fact, has derived the greatest advantage from the league, much more, in proportion to her extent and population, than Prussia. Little or no cloth of Saxon manufacture has hitherto found its way to the United States.

The extensive commercial relations of Saxony owe their origin to the enlightened policy of Frederick Augustus, the Elector, afterwards king of Saxony, who, at a time when protecting and prohibitory tariffs surrounded his states, adopted a liberal commercial system, and converted Saxony, and especially Leipsic, into one of the most important marts, not merely for the supply of central and northern Europe, but part even of Asia with all sorts of manufactured produce. The fairs at Leipsic were for a lengthened period the great sources whence Russia, as far as the borders of China, Poland, the provinces on the Danube, and many parts of the Turkish and Persian dominions, were supplied with manufactures; and though they have latterly declined, they still continue to be much resorted to. Leipsic has been for a lengthened period the centre of the book trade of Germany, being, London and Paris only excepted, the greatest literary emporium in the world. The value of the works sold at the Easter fair of late years has been estimated at about 3,000,000 dollars; and as many as 600 booksellers are said to have been assembled at some of these fairs from all parts of Germany to dispose of their publications and adjust their accounts! (*See* LEIPSIC.)

The Convention dollar, coined in Saxony,=32 groschen, or 4s. 1½d. Engl.: the Saxon dollar of 24 gr. is an imaginary coin. 100 Dresden *scheffels* are equivalent to 195 of Berlin. The other measures, &c., are comparatively unimportant, and the coins of Prussia are generally current.

Government.—Saxony is an hereditary and limited monarchy, having a senate or upper house, and a house of representatives. Previously to 1830, it had states: but these had comparatively little power; and the imposition of taxes and other public burdens, and the regulations of the public expenditure, mostly depended on the pleasure of the king. But the French revolution of 1830, was speedily followed, in Saxony, by some very important political changes. The king was obliged to associate his nephew, the present sovereign, with him in the government, a representative system was organized on a new principle, and the system incident to the feudal system were suppressed.

The senate consists of 32 members, and the house-of representatives of 300. Of the senators some take their seats by virtue of their offices, as the Roman Cath. bishop, the king's principal chaplain, a Protestant bishop, the dean of the university, and one or two great officers of state. Of the remainder, some sit in the right of their estates, and transmit their privileges, as senators, with their estates to their children; the others are elected for each parliament by a constituency of their own order. The qualification for a seat in the senate is the same that is required to entitle a man to vote in the election of a senator, viz: a landed estate worth 6000 dollars a year. The lower house is elected by a species of household suffrage, or by such heads of families as contribute in any way to the public burdens. The candidate must possess property to the amount of 1000 a year. Senators receive 7 dollars a day during the sitting of parliament, deputies 3 dollars. But owing to the high qualification required for deputies, and other causes, the new constitution is not so popular as it might have been expected. "Though," Mr. Gleig, all "money-bills must needs, as among ourselves, originate in the ch. of dep., the minister has but to coax his plan through an agent, and he is sure to carry it. Meanwhile the constituency, especially in the rural districts, complain loudly that their parliament is worse than useless; that it saddles them with burdens which, under the system, they were not required to bear; that now they 350 sovereigns to maintain instead of one! The rigid guardianship still exercised over the press keeps the Saxon peasant profoundly ignorant of what is passing in their parliament; though as to speaking, such a thing is unknown in the chamber of deputies. In the upper house the case is what different; for it is composed of men who do occasionally speak." (*Gleig's Germany* in 1837, i. 182-194.) With all its defects the newly organized representative system is of great importance to the kingdom; and it

will, no doubt, acquire greater influence and independence according as experience makes the deputies better acquainted with their rights and duties.

Perhaps, however, the inroad made on the feudal system, which was maintained in the rural districts nearly to the fullest extent, was the most advantageous of the various changes effected in 1830.

Hereditary jurisdictions were abolished, except in very rare cases; and the more oppressive privileges enjoyed by the lords were, at the same time suppressed. All towns now elect their own municipalities, and are governed by laws of their own making; while the rural districts, being divided into departments, each of which has its own magistrates, whom the people not only choose, but may, also, in case of malversation, degrade from office. The municipal officers, also, though elected by the citizens for life, are liable, on conviction of incapacity or unfair dealing, to be degraded. Their powers are very considerable in reference both to person and property, for they regulate the police, hear and determine civil causes, and both fix the amount of local rates to be levied on the citizens, and determine how the produce shall be expended. In the election of the magistrates, every rate-payer has a voice. They are all salaried officers. It is necessary, however, to enable any one to be appointed to the magisterial office, that he should have received a legal education, and be possessed of some small portion of land, and of the house in which he resides.

There are civil and criminal courts in the capital of each circle, and a high court of appeal in Dresden, in which latter all capital cases are tried. Executions take place by decapitation. There are special military tribunals, a superior fiscal court, university court at Leipsic, mining tribunal at Freiberg, patrimonial tribunals, &c. The reigning family is Roman Catholic, but there are not more than 30,000 Roman Catholics in the kingdom, the great bulk of the population being Lutherans. Literature and the fine arts have flourished more in Saxony than in any other part of Germany; and there is scarcely any country in Europe where primary instruction is so widely diffused: the number of the individuals attending the schools, and other seminaries, is said so be as high as one in six of the population. The university of Leipsic is the principal seminary.

Every male inhabitant 20 years of age is, with certain exceptions, obliged to serve in the army for six years in time of peace, and for three years subsequently in the reserve corps. The armed force is pretty extensive; it consists of about 13,000 men, of whom 10,000 are privates under arms; besides the reserve corps of 3000 more. This kingdom holds the fourth rank among the German states, having four votes in the full diet and one in committee, and furnishes a contingent of 12,000 men to the army of the confed. Its public revenue from 1840 to 1842 has been fixed at 5,500,297 doll., and its expenditure at 5,494,755 doll. At the end of 1836 the public debt amounted to 10,926,457 doll.

The Saxons are among the best specimens of the old Teutonic race. In person, manners, &c., travellers have remarked that they bear a striking resemblance to the English agricultural population. The Saxon royal family is said to be descended from Witchind, sovereign of this territory in the time of Charlemagne.

Saxony was created an electorate in 1422, which title it retained till 1806, when Napoleon erected it into a kingdom. During the late war the king of Saxony was, from the battle of Jena downwards, a firm ally of Napoleon, who made extensive additions to his dominions; and he did not abandon the fortunes of his benefactor till after the battle of Leipsic had compelled the French to evacuate Germany. This conduct led to the dismemberment of the kingdom by the treaty of Vienna in 1815: some of its most valuable provinces were then assigned to Prussia; and, but for the opposition of Austria, it is probable that Saxony would then have ceased to exist as a separate state. (*Berghaus, Allg. Länder, &c.; Stein; Strang's Germany and the Germans; Gleig's Germany, &c.; passim.*)

SAXONY, a prov. of the Prussian states, consisting of the territories dismembered from the kingdom of Saxony in 1815, with the Saxon states formerly belonging to Prussia, has on the N.E. and E. Brandenburg, S. the kingdom of Saxony and the Thuringian states, and on the W. Hesse, Brunswick, and Hanover. It is of a very irregular outline, has several *enclaves*, and includes within its frontiers the independent principalities of Anhalt, Sondershausen, &c. Area. 9765 sq. m. Pop., in 1837, 1,539,353, of whom 1,437,333 are Protestants and 97,932 Catholics. It is divided into three regencies, and these again into 41 circles. Principal towns Magdeburg, Halle, Erfurth, Merseburg, Naumburg, Burg, &c. The Hartz mountains lie on the W. frontier of the province; but, with this exception, there are no hills of any considerable magnitude. Principal rivers, the Elbe and its affluents, the Saale, Mulda, Unstrut, &c. Soil in parts sandy and unproductive, but in general loamy and fertile. The plain of Magdeburg is reckoned about the best land in Prussia, and is very well cultivated. Principal products, wheat and wheat and other sorts of corn, flax and hemp, excellent wool, tobacco, &c. The vine is cultivated in the neighbourhood of Merssburg and some other places. Productive mines of coal, iron, rock-salt, &c. are wrought in different parts of the province. The stock of sheep exceeds 2,000,000 head, and wool, which has been vastly improved by crossing with merinos and other fine-woolled breeds, has become a staple product. Manufactures important and valuable, consisting of fine woollens, linens, earthenware and porcelain, hardware, &c.

SAYBROOK, p. t., Middlesex co., Ct., 42 m. S.S.E. Hartford, 334 W. Bounded S. by Long Island sound, E. by Connecticut river, which here enters the sound with a bar at its mouth of 12 feet water at the highest tides. The harbour is on a cove which sets up from Connecticut river, which is here seldom completely frozen in winter. The shad fishery in the spring is very valuable, and white fish are largely taken for manure. The village contains a Congregational and an Episcopal church. The borough of Essex, 7 m. above the mouth of the river, contains four churches, a Congregational, Episcopal, Methodist, and Baptist; an academy, 10 stores, an extensive rope-walk, and about 1000 inhabitants, and has considerable ship-building and a number of vessels employed in the coasting trade. A ferry crosses Connecticut river, leading from the village at the centre to Lymes. The township was first settled in 1635, and was early of great importance. Yale college was for some time located here, during which 15 commencements were attended. The township contains 16 stores, two lumber-yards, four grist-mills, four saw-mills; two academies, 66 students; 13 schools, 578 scholars. Pop. 3417.

SCARBOROUGH, a seaport, parl. and mun. bor., market town, and par. of England, N. riding co. York, and E. div. of the wap. of Pickering Lythe, on a rocky slope, rising from an extensive bay, 35 m. N.E. York. Area of parl. bor. (which includes, with the old bor. and par., the extra parochial district of the castle), 2160 acres. Pop., in 1831, 8760; in 1841, 10,060. It has a very striking appearance from the sea, from which it rises amphitheatrewise to a considerable height. It is well built; the streets in the upper part of the town are spacious and well paved; and the houses generally have a handsome appearance. It is also extending S.W. towards Falsgrave, and southward along the shore. The principal public buildings are the town-hall, trinity-house, news room, assembly-rooms, a neat and well-conducted theatre, a sea-bathing infirmary, five bathing establishments, and two public libraries. But the handsomest and most classical building belonging to the town is the museum, a rotunda 37½ ft. in diameter, by 50 ft. in height, in the Roman Doric style. It is constructed f the Kelloway limestone, a fine building material, presented from his extensive quarries at Hackness, by Sir J. V. B. Johnstone, Bart., M.P. for the borough; and though of recent erection, has a valuable collection of specimens illustrative of the geology and natural history of the N. riding. A fine iron bridge of four arches, supported on massive stone piers, 70 ft. in height, has been thrown across a ravine to connect the higher town with the spa, ½ m. to the S. This handsome structure, which cost £9000, raised by subscription, was completed in 1828. The parish church, which was given by Richard I. to the abbey of Citeaux, in Burgundy, stands on an eminence not far from the ruins of the castle: it was formerly much larger than at present, but the part now used is commodiously fitted up for divine worship. Christ church, built in 1828, of stone furnished by Sir John Johnstone, in the early English style, has accommodation for 1300 persons. A chapel of ease was erected in 1840, in the lower part of the town, near the quay side, by voluntary subscription, principally for the accommodation of the poor: it contains nearly 560 sittings, of which about three fourths are entirely free. There are places of worship, also, for Roman Catholics, Wesleyan, primitive, and association Methodists, Independents, Baptists, and the Society of Friends, to most of which, as well as to all the churches, are attached well-attended Sunday schools. A grammar school, founded in the 9th century, is but slenderly endowed, but there are several good subscription schools, including two on the National and one on the Lancastrian plan. The Amicable society also clothes and educates between 70 and 80 boys and girls. A seaman's hospital is under the government of the Trinity house, and there are almshouses, and several other benevolent, as well as religious institutions.

The harbour, which is easy of access, is protected by a handsome stone pier, of modern erection; but, unfortunately, it labours under a deficiency of water, having only from 4 ft. to 5 ft. at low ebb springs, and from 8 ft. to 9 ft. at low ebb neaps; but from the first quarter flood to last quarter ebb, vessels drawing 8 ft. water may enter the harbour with safety. A small foreign, and pretty considerable coasting trade is carried on. In 1840 there belonged to the

port 205 ships, of the aggregate burden of 32,867 tons. During the same year the gross customs' duties amounted to £1887 7s. 2d.; there has, however, owing to the privilege of bonding having been conferred on the port, been a remarkable increase this year (1841) in the amount of the duties; the receipts for the 11 months ending 30th Nov. being £3721 12s. 3d. A great deal of fish is brought in here, and forwarded to the populous districts in the West Riding; and of late years several persons have embarked in the herring fishery, which is becoming an important and profitable source of employment to the fishermen. From 40 to 50 yawls belong at present to Scarborough and Filey. In addition to the numerous small boats used for fishing ashore; and it is not unusual for 150 or 200 boats to enter the harbour, during the season, at the same tide with herrings. As an encouragement to the fisheries, the corporation remit the tithe of fish to which they are entitled; and a society has been formed to raise an honorary fund to meet the casual losses of nets, lines, and tackle, of such provident fishermen as become subscribing members; and thus ensuring to them, at a slight charge, an advantageous protection.

"Since the peace, ship-building, which formerly contributed in no slight degree to the prosperity of the town, has greatly declined, and now has almost ceased; now, from the shallowness of its artificial harbour, is it every likely to become a very important trading place. It is supported by, the resort of strangers to it for the purposes of sea-bathing and amusement; and these are principally of the middle classes, and from the manufacturing districts of Lancashire, Durham, and the W. riding of Yorkshire." (*Parl. Bound. Report.*) Scarborough is, however, frequented not only for the purpose of sea-bathing, but on account of its two mineral spring; which have lately been analysed by Richard Phillips, Esq., F.R.S. We subjoin the result of his analysis of a gallon of water from each spring.

Asiatic gas	North Spring.	South Spring.
Chloride of sodium (common salt)	3-5 cubic inches	7.5 cubic inches
Crystallised sulphate of magnesia	86-64 grains	39-63 grains
Crystallised sulphate of lime	160-68 —	385-16 —
Bicarbonate of lime	104-00 —	116-75 —
Bicarbonate of protoxide of iron	69-26 —	47-40 —
Total contents	1-94 —	1-91 —
Specific gravity of the water	365-42 —	615 36 —
	1-0825 —	1-2945

Temperature 49° with very little variation.

It is probable that the spas may, from the growing reputation of the town, and their being so conveniently connected with it by the bridge above alluded to, again acquire some portion of that celebrity which they formerly enjoyed. The recent erection of a commodious saloon, in the castellated style, with embattled towers, the architectural beauty of the wells, the massive sea-wall, forming at once a secure protection to the spas and a delightful promenade, especially at high water, combined with the newly-laid out ornamental walks and grounds, have materially increased the natural attractions of Scarborough as a watering-place. These improvements have been effected at an expense of upwards of £8000.

Scarborough, which received its first charter from Henry II., in 1252, is divided under the Municipal Reform Act into two wards, the government being vested in a mayor, five aldermen, and 18 councillors. Corporation revenue, in 1839, £3990. Quarter sessions are held under a recorder, and petty sessions are held weekly both for the borough and North riding. The borough has sent two members to the House of Commons since 23 Edw. I., the right of election down to the Reform Act being in the common council of the borough, a body comprising 44 individuals. The Boundary Act included with the old borough the extra-parochial precinct of the castle. Registered electors in 1841-42, 539. It is one of the polling-places at elections for the N. riding, and the chief town of a poor-law union, comprising 33 parishes. Markets on Thursday and Saturday; cattle fairs, Holy Thursday and Nov. 22.

N. of Scarborough, on a bold craggy eminence commanding a very extensive sea view, stand the ruins of a castle built in the reign of Stephen, on which Piers de Gaveston, the minion of Edward II., fled for refuge from the vengeance of the exasperated barons. The castle, after sustaining two sieges from the parliamentary troops, was dismantled at the close of the civil wars: and though a portion of it was repaired in 1745, and barracks have been subsequently built in its immediate vicinity, it is principally in ruins. The remains of the keep consist of a square tower nearly 100 ft. in height: the entire surface included within the outer walls comprises nearly 19 acres. A strong gateway still remains, with portions of the circular towers occurring at intervals in the line of the fortifications. It was, in fact, previously to the invention of artillery, one of the principal strongholds in the kingdom. (*Baines's Gaz. of Yorkshire; Granville's Spas; Parl. and Mun. Bound. Reports;* and *private information.*)

SCARBOROUGH, p. t., Cumberland co., Me., 10 m. S.W. Portland, 59 m. S.W. Augusta, 536 W. Bounded S.E. by the Atlantic. It has four stores, four grist-mills, two tanneries; 14 schools, 854 scholars. Pop. 2172.

SCAGHTICOKE, p. t., Rensselaer co., N. Y., 20 m. N. Albany, 390 W. Bounded W. by Hudson river. Watered by Hoosic river, and Tomhanic creek. It contains two churches, a Presbyterian and Methodist; 10 stores, one fulling-mill, four cotton factories with 5807 spindles, two grist-mills, three saw-mills, two powder-mills; 15 schools, 839 scholars. Pop. 3389.

SCHAFFHAUSEN, the most N. canton of Switzerland, and, after Zug and Geneva, the smallest in the confederation. It is between lat. $47°\ 40'$ and $47°\ 50'$ N., and long. $8°\ 25'$ and $8°\ 55'$ E., being separated by the Rhine from the cantons of Zurich and Thurgeau, while, on all other sides, it is surrounded by the territory of the grand duchy of Baden. Area, 116 sq. m. Pop., in 1837, 31,125. Surface undulating, its loftiest hill, the Raadenberg, in the N., rising only to about 1200 ft. above the Rhine. The soil is generally calcareous, but fertile; and the climate is among the mildest in Switzerland. It is an agricultural rather than a manufacturing canton; and, according to Picot (*Statist.* 363), its agriculture has greatly improved within the last half century. Formerly, indeed, the supply of corn was quite insufficient for home consumption; whereas, in good seasons, considerable quantities are now exported. Artificial pastures have also been materially increased, as well as the number of cattle; fruits are abundant, particularly cherries, from which a good deal of *kirschwasser* is made; and the produce of timber is amply sufficient for the wants of the inhabitants. (*Picot.*) There are nearly 5000 arpents of vineyards, which furnish the principal article of export, wine being sent to St. Gall and Appenzell, the Black forest, and other neighbouring districts; but of late years the competition of the wines of Baden, &c., and the duties imposed on the Schaffhausen wines in Germany, have crippled the trade.

One of the principal branches of industry in Schaffhausen is the conveyance of goods through the canton, which is greatly facilitated by the navigation of the Rhine. Salt, from Wirtemberg, timber, &c., are conveyed through Schaffhausen to Switzerland; but the circumstance of this confederation not being comprised within the Prussian customs union is injurious to the transit and export trade of the canton; and in consequence a desire to join the German league has long prevailed in Schaffhausen and some other cantons, though not in the majority. (*Bowring's Rep. on Switzerland*) The manufacturing establishments comprise a few cotton and hardware factories. Accounts are kept in florins, of 60 *kreutzers*=20d. English. The foot is the same as that in Zurich, the lb. a little larger. This canton is divided into 64 districts. The male inhabitants, of full age, and not bankrupts, paupers, or suffering penal sentence, choose the legislative body. The latter, grand council, consists of 74 members, 24 of whom form the petty council, which is intrusted with most part the executive power. The grand council meets in January every year, and is presided over by a burgomaster, who is named annually. The population is wholly Protestant. Education appears to be well attended to.

Schaffhausen was not included in ancient Helvetia, and the inhab. resemble their Swabian neighbours rather than Swiss. It was admitted into the Confederation in 1501.

SCHAFFHAUSEN, (originally *Schiff hausen*, or *Ship-house*), town of Switzerland, and the cap. of the above canton, on the Rhine, 25 m. W. by N. Constance, and 49 m. E.N.E. do. Pop. about 7500. It is walled, and defended by the *Unoth*, an old citadel supposed to be of Roman origin, but which is now furnished with extensive bomb-proof casemates. Streets ill paved; and the buildings are remarkable for their quaint and antique architecture: many are ornamented in front with stucco, carved or fresco work. The *Minster*, founded in 1052, is a massive edifice in the round arched style, with numerous monuments in its cloisters. An ordinary bridge across the Rhine replaces that unique marvel of art consisting of one arch 364 ft. in length, designed by the French under Marshal Oudinot in 1799. Schaffhausen has a gymnasium, a college with nine professors, a high female school, and an excellent library. The *Minster* comprises the books that belonged to the celebrated historian Müller, by far the most illustrious of the sons of Schaffhausen, where he first saw the light, on the 1 of January, 1752. The town is a principal *dépôt* for goods passing between Switzerland and Germany, and France, and Zurich; consisting of silk, cotton, and woollen raw cotton, colonial produce, Nuremberg manufac-

tures, Swiss cheeses, &c. Schaffhausen is supposed to have originated about the 8th or 9th century; it was subjected by Austria in 1330, but has been independent since 1415.

The celebrated falls of Schaffhausen are situated about a league S.S.W. from the town, where the Rhine breaks through a ramification of the Black-forest mountains. The height of these falls, which, in some respects, are the grandest in Europe, varies, according to the season, from 50 to 75 ft., being greatest in June and July, when the river is swollen by the melting of the snow on the mountains. The stream, which, immediately above the fall, is about 300 ft. in width, precipitates itself over a ledge of limestone; four rocks projecting from which divide it in its descent into five portions. The greatest body of water falls between the first of these rocks and the castle of Laufen, on the S.E. bank of the river; from which, according to Murray, the best view of the falls is obtained.. "It is not," says Mr. Spencer, "the height of the fall but the immense body of water broken into spray in the most picturesque manner over the rocks, that constitutes the great beauty of the cataract. In other respects, it cannot bear the slightest comparison with either those of Terni or the Staubbach." (*Germany and the Germans,* ii., 61.)

In 1790, Lord Montagu, a young British nobleman of great promise, was drowned in a rash attempt to descend these falls; and, by a curious coincidence, his death occurred nearly at the same time that his noble seat, Cowdrey house, near Midhurst, was burned down. (*Ebel, Manuel du Voyageur; Picot, Statistique; Bowring, &c., on Switzerland, &c.*)

SCHAUMBURG-LIPPE (PRINCIPALITY OF), one of the minor states of N.W. Germany, principally between lat. $52°\ 10'$ and $52°\ 30'$ N., and about long. $9°\ $ E., surrounded by the territories of Hesse-Schaumburg, Hanover, and Prussian Westphalia, exclusive of some detached lordships enclosed in the territory of Lippe-Detmold. Area, 217 sq. m. Pop., in 1838, 27,600 mostly Lutherans. It is hilly towards its S. extremity, but flat in the N., where the lake called the Steinhuder Meer occupies about 11,000 morgen. The productive portion of the surface comprises about 74,000 morgen, besides nearly 34,000 morgen of forest-land, chiefly in the W. The soil is in general superior to that of Lippe-Detmold, and agriculture and cattle-breeding are more advanced. The inhabitants of both principalities employ their intervals from rural labour in spinning flax and weaving linens. Coal is raised in the S. to the value of about 30,000 dollars a year; and forms, with corn, wool, timber, and linen goods, a principal article of export.' The constitution, which dates from 1816, is a limited monarchy; the powers of the prince being similar to those of the sovereign of Great Britain; the *landstande*, or parliament, consisting of all the noble landed proprietors, with four deputies for towns, and six representatives of the peasantry. Appeal lies from the decisions of the courts of this principality to the superior court of Wolfenbuttle. Public instruction as in Lippe Detmold, is well attended to. Public revenue, about 137,000 dollars. There is no public debt. Schaumburg-Lippe has one vote in the full diet of the Germanic Confederation, and with Lippe-Detmold, Hohenzollern, Reuss, Waldeck, and Liechtenstein the sixteenth place, with one vote in the committee. Its contingent to the army of the Confederation amounts to 240 men. (*Berg., &c.*)

SCHELDT, (Fr. *Escaut*), a river of France and Belgium, which rises in the dép. Aisne, near St. Quentin, and runs mostly in a N.N.E. direction, through the dép. du Nord, and the provs. of Hainault, E. Flanders, &c., to Antwerp, after which it turns N.N.W. and, dividing into the E. and W. Scheldt, which enclose the islands of Beveland and Walcheren, enters the North sea in about the same lat. as the Thames. Its entire length is estimated at about 200 m., its breadth at Dendermond is about 650 ft., at Antwerp, 1700 ft.; and the width of its mouth varies from $2\frac{1}{4}$ to $3\frac{1}{2}$ leagues. It is navigable from Valenciennes. Its principal tributaries are the Scarpe, Lys, and Durme, on its W., and the Dender and Rupel on its E. side. St. Quentin, Cambray, Valenciennes Tournay, Oudenarde, Ghent and Antwerp, are on its banks. Its current is slow; and in the lower part of its course, where it runs through a completely flat country, its banks are fenced by dykes to prevent inundation. It is connected by the canal of St. Quentin and other canals with the Somme, Seine, and Loire, and with the principal rivers and cities of Belgium, in its neighbourhood. During the commercial ascendancy of Antwerp the Scheldt enjoyed a larger share of traffic than any other European river; but its importance in this respect, though still considerable, has since greatly declined. "There was nothing," says Barrow, "on this noble river, in our progress upwards, that conveyed any impression of an active or extensive commerce. In sailing up or down the Thames, or on approaching London within 4 or 5 m., the multitude of shipping affords indications not to be mistaken of the commercial wealth and prosperity of London. But the Scheldt, when we ascended it. was a vacant river; we neither met nor overtook a single sail; and, with the exception of two or three American ships, and some 10 or 12 small vessels, mostly brigs, there was little appearance of trade along the common quay of Antwerp." (*Tour in Holland*, p. 11.) This, however, was before the revolution of 1830 had made Antwerp once more the commercial emporium of Belgium; and in the interval, the Scheldt has certainly regained some portion of its former consideration. (*See* ANTWERP.)

SCHELESTADT, a fortified town of France, dép. Bas-Rhin. cap. arrond., on the Ill, a tributary of the Rhine, 26 m. S.S.W. Strasburg. Pop., in 1836, ex. com., 9353. It was fortified by Vauban, and is naturally strong from its being in a great measure surrounded by marshes. It has a hospital, prison, communal college, theatre, manufactures of cotton and linen fabrics, iron wire, soap, and earthenware, for which last it was famous as long ago as the 13th century, with breweries, distilleries, &c. It is supposed to have been the ancient *Elsebus*, destroyed by Attila, where Charlemagne and his successors had afterwards a palace. The Swedes took it in 1632, but restored it to the French two years afterwards. (*Hugo*, art. *Bas-Rhin*; *Dict. Géog.*)

SCHEMNITZ (Hun. *Selmecz-Banya*), a famous mining town of Hungary, co. Honth, in a mountainous distr., on the Schemnitz, a tributary of the Gran, 46 m. N. by E. Gran. Pop., in 1837, with its suburbs, 17,036. The town is entered by an old and strong gateway, which conducts to a long, narrow, steep street, wretchedly paved, and so hemmed in by sloping hills that there is scarcely room for a row of houses on either side. At the end of this street is a mountain amphitheatre, the *proscenium* of which is occupied by the churches and other large buildings; while the hill sides are covered with the white cottages of the miners embosomed among trees. (*Paget's Hungary*, i. 327.) The town has many good-looking houses, with shops and inns; but its fine old ruined castle appears to be the only edifice of much interest.

The mines of Schemnitz, which extend under the town, and have been wrought for several centuries, furnish considerable quantities of silver, whence gold is again extracted. The ores vary greatly in productiveness; but, speaking generally, the mines have not been very profitable. There are six principal veins or courses, each from 10 to 20 fathoms in thickness, running nearly E. and W. almost parallel to, and at the distance of from 60 to 300 or 400 fathoms from each other, and connected by various small branches. In these extensive courses there are 12 royal mines, besides a number belonging to private individuals, who are obliged to dispose of all the ore they obtain to the royal smelting works at a fixed rate. The whole of these mines communicate with the emperor Francis's adit or level, at the depth of nearly 200 fathoms. (*Bright's Travels*, p. 149.) At a still greater depth is the adit of Joseph II., a magnificent work, 12 mining ft. in height by 10 ft. in breadth, extending from Schemnitz to the valley of Gran, a distance of nearly 10 Eng. m. This adit, which is still unfinished, will carry off the water from mines which cannot now be wrought, and is so constructed that it may be used either as a canal or a railway. It has been already no fewer than 40 years in progress, and it is supposed that, when complete, it will have cost, at least. £400,000. Dr. Clarke, who descended into the mines of Schemnitz, says, "All the imperial mines are connected with each other; offering, in their whole extent, a subterraneous passage which reaches to the astonishing length of 3000 fathoms, nearly 3½ m.! The sight of the interior of the Paquerstohin (one of the mines) convinced us that there are no mines in the world like those of Hungary. How wretched, in comparison, appear the mines of Cornwall and Wales, where it is sometimes necessary to creep upon the hands and knees, wet through, over all sorts of rubbish, to get from one shaft to another. The inside of a Hungarian mine may be compared to the interior arrangement of one of our best frigates, where space has been so husbanded, and cleanliness so strictly maintained, that nothing is seen out of its place, and there is room enough for every operation." (*Travels*, viii. 303, 8vo ed.)

Dr. Clarke should, however, have added that the mines of Cornwall and Wales are wrought by private individuals for the sake of profit only, whereas the imperial mines which he visited are wrought at the expense of government, to which profit is a secondary consideration. Hence the greater outlay on the latter; hence also, in part at least, their comparative unproductiveness, and the bad and costly manner in which, according to Mr. Paget, all the Austrian mining establishments are conducted.

The ore, besides silver and gold, contains lead, and sometimes iron, copper, zinc, or arsenic. In consequence of the want of wood and water but little ore is smelted on the spot, being principally sent to Neusohl or Kremnitz. About 20,000 miners are employed in the Schemnitz district.

Mr. Paget states that the officers and workmen are all very indifferently paid; and hence there is not unfrequent-

ly a good deal of embezzlement. The workmen, who are paid by the piece, are not permitted to earn more than about 3s. a week: it would seem, indeed, rather to be the object of the government to keep up the mines for the employment they afford, than for the inconsiderable profit made by them.

A school of mining, in imitation of that at Freiberg, was established at Schemnitz in 1760, which has five professors and about 200 students, who are all educated free of cost, several of them being farther furnished with an annual donation of from £90 to £30, to assist in their maintenance. It is believed, however, that this school is, in respect of science and practical knowledge, far behind that of Freiberg, and most other mining schools. It has a pretty good library, but its collection of minerals is very inferior. (*Paget's Travels in Hungary*, i.)

SCHENECTADY, co., N. Y. Situated towards the E. part of the state, and contains 200 sq. m. Watered by Mohawk river, along which are extensive and fertile flats. The railroads from Albany to Utica, from Troy to Schenectady, and from Schenectady to Saratoga springs, pass through it. It contained in 1840, 10,808 neat cattle, 18,094 sheep, 13,063 swine; and produced 13,113 bushels of wheat, 32,06... of rye, 62,597 of Indian corn, 41,388 of buckwheat, 103,55... of barley, 216,968 of oats, 240,535 of potatoes, 4423 pounds of sugar. It had 45 stores, two tanneries, seven fulling-mills, two woollen factories, one cotton-factory with 200 spindles, one flouring-mill, eight grist-mills, 23 saw-mills, one oil-mill, four tanneries, one brewery, one printing-office, one weekly newspaper; one college, 230 students; three academies, 83 students; 57 schools, 1272 scholars. Pop. 17,387. Capital, Schenectady.

SCHENECTADY, city, and capital of Schenectady co., N. Y., 16 N.W. Albany, 15 S.W. Ballston Spa, 304 W. Situated on the S. bank of Mohawk river, and is a mile long by half a mile broad, laid out with considerable regularity on 19 streets, not always crossing each other at right angles. It contains a courthouse, jail and county offices in one neat building, a market, female academy, a lyceum, two banks, besides a savings bank, an insurance company, nine churches, a Dutch Reformed, Presbyterian, Episcopal, Methodist, Baptist, Cameronian, Universalist, Roman Catholic, and an African, 60 stores and groceries, one cotton-factory, two flouring-mills, two furnaces, one brewery, three tanneries, and various mechanical establishments, 1089 dwellings, and 6784 inhabitants. It is the seat of Union college, founded in 1795, which has a president and 10 professors or other instructors, 2125 alumni, of whom 308 have been ministers of the gospel, 213 students, and 13,000 vols. in its libraries. The commencement is on the fourth Wednesday in July. Its philosophical and other apparatus is very complete. The buildings, of which the two principal ones are 600 ft. apart, are each 200 ft. long and four stories high, on elevated ground east of the city, in a commanding situation. A centre building and other edifices are necessary to complete the original plan. Attached to the colleges are 239 acres of land, a part of which is designed to be laid out in grounds and walks. The reputation of the institution is deservedly high.

SCHIEDAM, a town and port of S. Holland. cap. cant., on the Schie, a tributary of the Maas. 3 m. W. Rotterdam, and 1 m. N. from the Maas. Pop., in 1837, 11,815. It is well built in the usual style of Dutch towns, and has numerous churches, an exchange, a Latin school, a chamber of commerce and manufactures, and a branch of the Society of Public Good. "It is conspicuous both by the smoke which issues from the chimneys of its distilleries and the vast number of windmills by which it is surrounded. The whole horizon, in fact, in the direction of Schiedam, seems animated with life and bustle. Schiedam is the chief seat of the manufacture of Dutch gin, or Hollands. The quantity of that spirit produced here annually is very great, there being in the town as many as 160 distilleries, while many thousands of pigs are supported by the refuse of the malt employed in the manufacture. The gin of Schiedam is strong, but mild in flavour, and is usually sold in Holland for 9d. a bottle, or 4s. 6d. a gallon; the price of the gin on its importation into England being increased by freight and duties to about 28s. or 30s." (*Chambers's Tour in Holland*, i.) The duty on Hollands, amounting to 2s. 6d. a gall., is, in fact, one of the most objectionable in our tariff. The entries of Hollands for home consumption, which formerly amounted to about 200,000 galls., were reduced, in 1840, to 15,992 galls.! We doubt, however, whether the consumption be really diminished. The exorbitancy of the duty has not taken away the taste for Hollands, but it has thrown the trade into the hands of the smuggler, to the injury alike of the legitimate trader and the revenue. Schiedam has rope-walks, building-docks, and a small though convenient port on the Schie. It sends one deputy to the states of the prov.

SCHOHARIE, co., N. Y. Situated toward the E. part

of the state, and contains 691 sq. m. Organized in 1795.
Watered by Schoharie creek and its tributaries. It contains
water limestone, bog iron ore and sulphur springs. The
atter at Sharon are becoming much celebrated. It contained in 1840, 37,533 neat cattle, 71,258 sheep, 31,865
wine ; and produced 73,671 bushels of wheat, 129,349 of
re. 67,800 of Indian corn, 80,609 of buckwheat, 217,478 of
arley, 47,953 of oats, 600,396 of potatoes, 133,706 pounds
f sugar. It had 81 stores, 30 fulling-mills, one woollen-
actory, 36 grist-mills, 160 saw-mills, one paper-mill, 32 tan-
eries, one brewery, one printing-office, one weekly news-
aper ; five academies, 306 students, 199 schools, 9294
cholars. Pop. 32,358. Capital, Schoharie.

SCHOODAC. t., Rensselaer co., N. Y., 14 S. Troy, 7 S.
E. Albany, 392 W. Bounded W. by Hudson river, and
ained by its small tributaries. It contains a Dutch Re-
rmed church, 13 stores, three fulling-mills, two grist-
ills, five saw-mills ; 23 schools, 1133 scholars. Pop. 4125.

SCHOOLEY'S MOUNTAIN, forms a part of a granite
ain, which extends N.E. and S.W. across the state of
w-Jersey. It crosses the N.W. part of Morris co., and
308 ft. above its base, and 1100 ft. above tidewater. Near
top in Washington t., Morris co., 56 N. Trenton, 218 W.
a mineral spring of some celebrity. It has a temperature
56° of Farenheit, discharges 30 gallons an hour, and con-
is muriate of soda, muriate of lime, muriate of magnesia,
phate of lime, carbonate of magnesia, etier, and carbonic
d oxyde of iron. The carbonic acid which the water
tains, is altogether in a state of combination. It is a
lybeate of the purest kind, and beneficial in chronic dis-
se and general debility. The pure air and romantic
ery of this region, with its elevated position, render it a
and pleasant place of summer resort. There are three
l-kept hotels, besides several private boarding houses in
vicinity. It has generally a considerable number of
ors in the summer season.

CHROON, t., Essex co., N. Y., 25 N. Albany, 25 S.S.W.
abeth. Watered by Schroon river, and contains Schroon
and other small lakes. It has two stores, two fulling-
, two forges, two grist-mills, 36 saw-mills, two tanne-
16 schools, 490 scholars. It supplies much valuable
er. Pop. 1669.

HUYLER, co., Ill. Situated in the W. part of the
and contains 360 sq. m. Drained chiefly by Crooked
and its branches. It contained in 1840, 6760 neat cat-
345 sheep, 17,888 swine ; and produced 36,800 bushels
eat. 226,033 of Indian corn, 37,998 of oats, 23,545 of
es, 1494 pounds of tobacco. It had one commission
in foreign trade, 19 stores, nine grist-mills, 17 saw-
seven tanneries, two distilleries, one printing-office,
aily newspaper ; 14 schools, 545 scholars. Pop. 6972.
l. Rushville.

IUYLKILL, r., Pa., rises by three principal branches
uylkill co., and pursuing a S.E. course, it enters Dela-
river, 7 m. below Philadelphia. It is about 140 m.
and receives Tulpehocken creek from the W., and
men creek from the E., besides many smaller tributa-
To the wharves in western Philadelphia, where
s a depth of water 13 or 14 ft. at common tides, ves-
from 300 to 400 tons come. By means of canals and
is navigable from Fairmount dam above Philadel-
108 m. to Port Carbon in Schuylkill co., and on it an
se amount of coal is transported. The Union canal
the Schuylkill river near Reading, and following
ocken creek, it passes over to Swatard creek, and
tes at Middletown on Susquehanna river ; and is
ing, thus completing the line of canal from Philadel-
Pittsburgh, excepting the railroad of 36½ m. across
egheny mountains.

YLKILL, co., Pa. Situated toward the E. part of the
d contains 660 sq. m. Drained by Schuylkill river
branches. It is celebrated for immense quantities
acite coal which it contains, and exports. It con-
1840, 6565 neat cattle, 6107 sheep, 7864 swine ;
luced 23,744 bushels of wheat, 85,858 of rye, 45,971
corn, 26,731 of buckwheat, 75,810 of oats, 102,267
es. It had 99 stores, produced 431,370 tons of an-
coal, had four furnaces, three forges, five flouring-
grist-mills, 189 saw-mills, two powder-mills, 11
one distillery, five breweries, four printing-offices,
ery, four weekly newspapers ; three academies,
nts ; 31 schools, 880 scholars. Pop. 29,053. Capi-
gsburg.

KILL, p. t., Chester co., Pa., 78 m. S.S.E. Harris-
W. Bounded N.E. by Schuylkill river. Water
ench and Stoney creeks. It has six schools, 439
Pop. 2079.

KILL. t., Schuylkill co., Pa. It abounds with an-
wit, and has four stores, two grist-mills, nine saw-
ree schools, 150 scholars. Pop. 1334.

LKILL HAVEN. p. v., Manheim t., Schuylkill
8 N.E. Harrisburg, 168 W. Situated immediately

below the junction of the W. branch, with the main branch
of Schuylkill river. The Schuylkill canal passes through
it, and the West Branch railroad extends from it to the
coal mines, at the foot of Broad mountain. It contains 150
dwellings and 990 inhabitants, mostly connected with the
coal trade.

SCHWABACH, a town of Bavaria, circ. Middle-Franco-
nia, 9 m. S.S.W. Nuremberg. Pop. 7600. (Berghaus.) It
is walled, and pretty well built, having several Protestant
churches, a synagogue, a mint, hospital, &c. It is the sent
of various manufactures, the principal being that of pins ;
but there are others of hosiery, hats, gold and silver lace,
tobacco, paper, printing types, and Jews' harps. It owes
its distinction as a manufacturing town to the influx of
emigrants from France, after the revocation of the edict of
Nantes.

SCHWARTZBURG-RUDOLSTADT, a principality of
central Germany, between lat. 50° 30' and 51 N., and about
11 W. long., inclosed by the territories of Saxe-Weimar, Cobourg, Meiningen, and Hildburghausen. Area, 405 sq. m.
Pop. in 1837, 65,600, mostly Lutherans. It comprises a portion of the N. declivity of the Thuringian forest mountains,
and is watered by the Schwartza, Ilm, and Saale. It does
not yield sufficient corn for home consumption : timber and
salt are its principal products. Iron, and a few other metals, are found ; and woollen cloths, earthenware, glass, and
other kinds of goods are manufactured. Since 1816, the
government has been a limited monarchy ; the representative body consisting of five deputies of the nobility, five of
the citizens, and five of the rural pop. (Berghaus ; but
Horschelmann says there are six of each.) The deputies
are elected every six years. The parliament has the control of the public funds, and no new law can be adopted
without its consent. The principal judicial courts are at
Rudolstadt and Frankenhausen ; from which appeal lies to
the superior tribunal of Zerbst, in Anhalt-Dessau. Public
revenue about $290,000. The public debt has been in rapid
process of diminution since 1821, and in 1836 it amounted
to only $80,152. This principality furnishes 539 men to the
army of the German confederation. Chief towns, Rudolstadt the cap. on the Saale, with 4500 inhabs. ; and Frankenhausen with 5000 inhabs., and a considerable trade in
corn and wool.

SCHWARTZBURG-SONDERSHAUSEN, a principality
of central Germany, between lat. 51° 12' and 51° 26' N.,
and about long. 11° E., inclosed by territories belonging to
Prussia on every side except the W., where it joins a detached district of Saxe-Gotha. Area, 358 sq. m. Pop., in
1834, 54,060, mostly Lutherans. Surface undulating, and
traversed by several affluents of the Unstrui, flowing in a
W. direction. The lower part of the country yields more
corn than is required for home consumption ; the higher
portion has extensive forests, and timber and potash are
among its principal products. Iron is found ; and forges and
hardware factories are the principal manufacturing establishments though some woollen and linen goods are woven. The government is an unlimited monarchy. Appeal
may be made from the judicial courts to the superior court
of Zerbst in Anhalt-Dessau. Public revenue, in 1836,
99,936 dollars ; expenditure, 94,411 ditto. Public debt, about
79,000 ditto. Contingent to the army of the Confederation,
451 men. Chief towns, Sondershausen, the cap., on the
Wipper, with 3800 inhabitants ; and Arnstadt (which see).
This principality, like the preceding, has one vote in the
full diet of the German Confederation, and shares the 15th
place and one vote in the committee with Oldenburg, and
the Anhalt principalities. (Berghaus, &c.)

SCHWEIDNITZ, a town of Prussian Silesia, cap. circ.
finely situated at the Riesengebirge mountains, on the Wei-
stritz, in a fertile and beautiful country, 30 m. S.E. Breslau.
Pop. about 9500. It is well built, and strongly fortified, the
fortifications which had been dismantled by order of Napoleon in 1807, having been re-constructed since the peace on
an improved plan, and rendered more formidable than ever.
Its castle, formerly the residence of the Piast dukes, is now
a workhouse. It has a magnificent Roman Catholic church,
a fine town-house, a gymnasium, a house of correction, and
the usual government offices of the cap. of a circ. : with
manufactures of woollens, cottons, linens, &c. Near it is
the castle of Fürstenstein, a fine antique feudal edifice,
bought by the late king of Prussia.

SCHWERIN. See MECKLENBURG SCHWERIN.

SCHWYTZ, or SCHWEITZ (CANTON OF), one of the
four Forest Cantons of Switzerland, which gave its name
to the Confederation, in the central part of which it lies,
between lat. 46° 50' and 47° 20' N., and long. 8° 30'' and 9°
E., having N. and N E. the canton of Zurich and St. Gall,
E. Glarus, S. Uri and Unterwalden, and W. Zug and Lucerne. Area, 338·3 sq. m. Pop., in 1837, 40,650. Nearly
the whole surface is mountainous : the Rosstock rises about
8200 ft., and the Righi about 6150 ft. above the sea. (Bra-
guiers Orographie.) The Rossberg, the fall of a portion

of which in 1806 had most destructive effects, is partly in this canton and partly in that of Zug. The Sihl and the Muotta are the principal rivers: the former falls into the lake of Zurich, which forms most part of the N. boundary of the canton, and the latter into the lake of Lucerne, which limits the canton on the S.W. The Linth canal, between the lakes of Wallenstadt and Zurich, runs along its N.E. extremity. The soil and climate are more favourable to cattle-breeding than to agriculture, which is so much neglected, that in some valleys, according to Picot, the plough and flail are unknown! On the other hand, however, the inhabitants are distinguished by their superior treatment of live stock; the cattle of Schwytz are accounted among the best in Switzerland; and upwards of 20,000 head are annually sent from the S. side of the Alps to depasture on the mountains during summer. Near Kussnacht, on the lake of Lucerne, the vine is grown; and apples, which produce cider, are tolerably abundant. The forests are extensive, and the supply of turf is all but inexhaustible: cotton thread, and this in very small quantity, is almost the only article of manufacture. The principal exports are cattle, cheese (sent mostly across mount St. Gothard), and timber. The transit trade is of little importance; and on account of the badness of the roads is mostly confined to the lakes and the navigable parts of the rivers.

The government is a pure democracy, the sovereign power residing in the people at large. The male population above 16 years of age form the general assembly of the canton, which meets every two years, on the first Monday in May, at Schwytz, to appoint, by a show of hands, the landamman, and other supreme officers, the deputies to the diet; and to deliberate on alliances, declarations of war, treatises of peace, &c. A council of high functionaries and 270 ordinary members, assembles usually twice a year, to prepare instructions for the deputies, and hear their reports; and another council of 90 members is entrusted with the general executive power. The canton is divided into six districts, each of which has its own council and tribunal of primary jurisdiction, the decisions of which are final in cases not above the amount of 200 florins. The chief tribunal sits in Schwytz, and is composed of members, two thirds of whom belong to the districts of the cap., and the rest to the other districts of the canton. The inhabitants are exclusively Roman Catholics, subordinate to the bishop of Chur. Public education is more backward in this than in most other cantons; and it has no public library. At 16, every male inhabitant is enrolled in the militia, and Schwytz furnishes a contingent of 602 men to the army of the Confederation, in which it holds the fourth place immediately after the three directorial cantons. Schwytz, the cap. of the canton, at the foot of mount Mythen, 26 m. S.S.E. Zurich, has about 3300 inhabitants. (*Picot, Statistique de la Suisse; Ebel, Manuel, &c.; Inglis's Switzerland, passim.*)

SCIACCA, (an. *Therma Selinntina*,) a town and seaport of Sicily. intend. Girgenti, on the S. coast, nearly 30 m. S.E. the ruins of Selinuntum, and 30 m. N.W. Girgenti. Pop., in 1831, 12.668. The town, situated on the declivity of an eminence rising from the bay, is surrounded by an irregular wall, in tolerable repair, with bastions towards the sea, and the castle of Luna at its E. angle. At a distance it has a respectable appearance; but, notwithstanding its large churches, convents, and magazines, it appears to be poverty-stricken and wretched. Smyth says that it has not recovered from the influence of the long-continued and painful feuds between the families of Luna and Perollo, in the 16th century. (*Swinburne's Travels,* ii. 949, &c.)

Some of the famous hot springs, whence the city had its ancient name, are a little without the walls towards the E. But the steam-baths, the construction of which was ascribed, in antiquity, to Dædalus, and now called the stufe of St. Calogero, are on the summit of an isolated mountain, about 3 m. N.E. of the town, and correspond exactly with the description of Diodorus Siculus. They continue, as of old, to be frequented by patients, and consist of several sudorific grottoes, or caverns, the outer one of which has seats excavated in the rock. (*Hoare's Classical Tour,* ii., 86; *Smyth's Sicily,* p. 218.)

Sciacca is one of the principal ports on the S. coast of the island, for the exportation of corn, and the rock upon which the town stands is, in numerous places, hollowed out into *caricatori*, or corn cellars. In summer ships anchor at about 1 m. off town, in from 7 to 12 fathoms, on a bottom of sand and clay; but being exposed to every wind from the S.E. round to the W., it is not resorted to in winter, except by boats and flat-bottomed craft. (*Smyth's Appendix.*)

Agathocles, tyrant of Syracuse, famous alike for his great talents, perfidy, and cruelty, was a native of Sciacca, where he was born *anno* 359 B.C. He was of low origin. his father being a potter banished from Reggio, his native city. (*Biographie Universelle.*) Fazelli, the historian of Sicily, was also a native of this town.

778

SCILLY ISLANDS, a group of islands, belonging to England, lying about 30 m. W. by S. from the Land's End, supposed by some to be the *Cassiterides,* or the islands of the ancients. There are supposed to be, in all, about 139 islands and rocks, but there are only about half a dozen of any importance. St. Mary's, the largest, is said to contain about 1640 acres:- the entire area of the group, as given in the population returns, is only 5570 acres. In 1831 they had a population of 2465. From their situation they necessarily have a mild equable temperature; and though few be common, the islands are very healthy. They produce good barley, rye, and oats, the latter being principally of the variety called *pillas,* or *avena nuda.* Potatoes are extensively cultivated. Horses and cattle small; sheep numerous, and of good quality; sea-fowl are found in great numbers, and partridges are, also, abundant. There are shrubs, but few or no trees. The inhabitants used to make great quantities of kelp; they are also expert fishers, and act as pilots to such ships as require their services. As already stated, the islands are generally supposed to be the *Cassiterides,* or tin islands of the ancients. But it is most probable that the W. extremity of Cornwall was included under this term; and, at all events, there is now no tract of tin, nor, indeed, of mines of any sort, in any of the islands. Heugh-town, the capital of the islands, and their only town, is situated on the W. side of St. Mary's. It has a pier and a custom-house, and is a place of some consequence, being defended by a fort, called the Star castle, with a small garrison.

Persons accused of felonies are sent to Cornwall to be tried at the county assizes; but all minor offences and civil suits are tried by a court consisting of twelve of the principal inhabitants, delegated by the proprietor of the islands under the duchy of Cornwall, of which they form a part. This court sits once a month at Heugh-town, St. Mary's, for the trial of cases. Vacancies in it are usually filled up by election; but it may be dissolved and a fresh appointment made by the proprietor.

The position of the Scilly islands renders them of very considerable importance in navigation. Lying at the point of junction, as it were, of the English and St. George's channels, ships passing from the one to the other, should the wind be unfavourable, often take shelter under these islands: it is sometimes, also, convenient for vessels to take shelter among them, rather than beat about at sea in bad weather; and a strong gale from the E. usually brings a numerous vessels. There are four principal sounds or roads among the islands, exclusive of smaller inlets. Of the roads, St. Mary's, between the islands of St. Mary and St. Agnes, is the best; but it is generally the safest place in entering it, or any of the other roads, to make use of pilots. The latter are always in readiness to offer their services.

A lighthouse of the first class was erected on St. Agnes island, the most southerly of the group, in 1680, the lantern of which is elevated 138 ft. above high water mark. It is, according to the ordnance survey, in lat. 49° 53 37' N., long. 6° 19' 23' W.

But, notwithstanding the warning given by this light these islands have been the scene of numerous shipwrecks. The most distressing of these catastrophes took place on the night of the 23d of October, 1707, when the fleet from the Mediterranean, under the gallant Sir Cloudesley Shovel, got foul of the islands; the ship bearing the flag of the admiral and two other line-of-battle ships struck upon the rocks near the lighthouse, and were totally lost, with every soul on board. Some of the other ships were in extreme danger. It is not exactly known how the accident arose. The night was dark, but there was very little wind, otherwise the whole fleet must have been destroyed. It is probable that the light had been mistaken for another. The body of the admiral was cast ashore, and buried in St. Mary's; but it was soon after removed to Westminster Abbey where a monument, creditable to the liberality, though not to the art of the nation, was erected to his memory. See *Borlase's Account of the Scilly Islands,* 4to., Oxford, 1756 Cornwall, in *Lyson's Magna Britanniæ; Burnet's Hist. of His Own Times,* iv., 205, ed. 1753; *Smollett's England,* anno 1707, &c.)

SCIO (an. *Chios*), a celebrated and beautiful island of the Ægean sea, belonging to the Turks, about 3 m. W. from cape Blanco, in Asia Minor; Chio, its chief town, on an E. side. 53 m. W. Smyrna, being in lat. 38° 22 37' N., long. 26° 9' E. It is about 32 m. in length N. and S. and where broadest, about 16 m. across. Though for the most part mountainous and rugged, it has a considerable extent of level and gently sloping ground. Its climate is mild and delightful; and it has numerous fine springs and rivulets. Dr. Clarke says it is the "paradise of modern Greece more productive than any other island, and yielding to none in grandeur." (III., 236 8vo ed.) In antiquity and in modern times, down to the late dreadful catastrophe, was cultivated with the greatest care and assiduity. Owing to the limited extent of the arable land, and the greater unsuitableness of the soil for other crops, the principal part of the corn required for the use of the inhabitants has always been brought from the ports on the Black sea and other marts. The staple articles of produce are silk, mastic, figs, lemons and oranges, wine, oil, cotton, almonds, &c. Its mineral wealth has been but little explored, but it contains abundance of marble, jasper, and a kind of green earth, resembling verdigris.

The wines of Chios, especially those produced in the district of *Arousia*, were among the most esteemed of any in the ancient world. They have been celebrated by Virgil (*Ecl.* v., lin. 72) ; and Horace asks

"Quo Chium pretio cadum
Mercemur ?"

According to Pliny, Chian wine was served up by Julius Cæsar at his most splendid entertainments ; and it is thought worthy of notice, that Hortensius left a very large stock of this famous beverage to his heir. (*Hist. Nat.*, lib. xiv., cap. 14, 15.) The wine of the island still preserves some portion of its ancient celebrity ; but the produce is scanty, and it is said to be injured by transportation.

Mastic, the method of gathering which is fully described by Tournefort (i., 376), is the most esteemed of the modern products of the island, being in great request among the Turkish ladies. All the mastic trees are supposed to be the property of the grand seignior, or rather of the sultana mother, of whom this island is the peculiar demesne. But formerly the trees were left, with the island itself to the inhab., with but little interference on the part of the Turks, on addition of their annually furnishing a certain quantity of mastic to the Cadi for the use of the imperial seraglio, and paying a moderate capitation tax. And it is to the comparative exemption it has thus enjoyed from Turkish despotism and rapacity, that the sprightly vivacity of its inhabitants, and their greater industry, enterprise, and prosperity, are to be ascribed. Besides its chief city, the island had, previously to its late calamity, several considerable towns and numerous villages. The population which was very dense, has been variously estimated at from 80,000 to 100,000 ; of whom from 30,000 to 35,000 belonged to the capital.

The latter on the E. coast of the island, constructed by the Genoese, along the sea shore at the foot of the mountains (an. *Pelinæi Montes*), on the slope of which stood the ancient city, is the cleanest, handsomest, and most desirable town, as a residence, in the Levant. It is well built, extended, previously to 1822, with its gardens and villas for about 4 m. along the sea. Its houses are commodious, and its shops and warehouses were then well furnished ; it had numerous Greek and Roman Catholic churches, schools, and even a college. The silk manufacture carried on upon a large scale, and the velvets, damasks, and other silk goods of Scio, were highly esteemed. It had for many years the principal *entrepôt* of the archipelago, and carried on an extensive and flourishing commerce. Some say that the shallowness of its harbour, which is suitable only for the smaller class of vessels, was the drawback on its prosperity ; but this defect was to a great extent compensated by the excellence of its roadstead, which affords secure anchorage, and every facility for getting to sea. (*Thornsford*, i., letter 9 ; *Semini's Travels in Greece and Turkey*, cap. 27 ; *Walpole's Journey*, quoted by Clarke ; *Carne's Letters from the East, &c.*) Unhappily, however, these statements apply to the past rather than to the present state of this fine island. In 1822, during the progress of the revolutionary struggle in Continental Greece, a force landed in the town, and a part of the inhabitants, who had hitherto pursued a strict neutrality, having joined them, they attacked and took the citadel, defended by a small Turkish garrison, which they put to the sword. A strong Turkish force having landed immediately after, took the most desperate revenge for the outrage that had been committed. The island was given up to indiscriminate pillage and massacre. The inhabitants, taken by surprise, and enervated by long peace and prosperity, offered no effectual resistance to their murderous enemies. It is said that above 30,000 individuals were put to the sword ; that as many more, principally women and children, were carried off and sold as slaves ; and that the capital was converted into a heap of ruins, and every village in the island laid waste ! These statements are most likely a good deal exaggerated ; but still there cannot be a shadow of a doubt that the visitation was of the most severe and tremendous description. Whether Scio be destined to recover from this wholesale butchery is problematical ; but it were absurd to expect that it should so long as it remains in the deadly grasp of the debased barbarians who have perpetrated such atrocities. If the principal inhabitants as were fortunate enough to escape being massacred, immediately fled from the island ;

and that commerce which had been its principal support has been transferred to Syra, Napoli, and other places. (*Carne's Letters from the East ; Tract on the Greek Revolution*, by the Rev. T. S. Hughes, Lond., 1822, &c.)

In antiquity, Chios gave birth to many distinguished individuals ; among whom may be specified Ion, the tragic poet, Theopompus, the historian, Theocritus, the sophist, and Metrodorus, the physician and philosopher. But Chios aspires to a still higher honour, that of being the native country of the first and greatest of poets.

" The blind old man of Chio's rocky isle,"

of whom Velleius Paterculus has justly as well as forcibly said—*quod neque ante illum, quem ille imitaretur ; neque post illum, qui eum imitari posset, inventus est.* (Lib. i., cap. 5.) And it is admitted by the ablest critics that, of all the cities that contended for the honour of having been the birth-place of Homer, the claims of Chios and Smyrna were apparently the best founded.

The Chians were for some time, in possession of the empire of the sea. They are said to have been the first who traded in slaves ; and the oracle, informed of the fact, declared that it had drawn upon them the anger of heaven ; one, says Bartholomi, of the noblest, but at the same time, least regarded answers the gods have communicated to man. The Chians took a prominent part in the great revolt of the Ionian cities against the Persians, by whom they were afterwards reduced, and punished with great severity. At a subsequent period we sometimes find them on the side of the Athenians, and sometimes on that of the Lacedæmonians.

" Moderate in prosperity, blameless towards their neighbours, and using their increasing wealth and power for no purpose of ambition, but directing their politics merely to secure the happiness they enjoyed," the Chians were among the most respectable of the Greek states. (*Mitford*, iii., 316, 8vo ed.)

They became the allies of Rome during the wars with Mithridates. After innumerable vicissitudes Scio came, in the middle ages, into the possession of the Genoese, who, as already stated, built its capital. It was taken by the Turks in the 16th century. (Exclusive of the authorities referred to above, see *Ancient Universal History*, viii., 296, &c., 8vo ed. ; *Cellarii Geographia Antiqua*, ii., 19 ; *Voyage d'Anacharsis*, cap. 72, &c.)

SCIOTO. river, O., is the second in magnitude of the rivers which have their whole course in the state. It rises in Hardin county, and flowing E. and then S. it enters the Ohio river at Portsmouth. It is 180 m. long, and receives the Olentangy or Whetstone river at Columbus, a large branch from the N. It is boatable in high water as far as the Little Scioto in Marion county. The Ohio canal strikes this river about 10 m. below Columbus, and follows it to its mouth. It flows through a highly fertile country.

SCIOTO. county, O. Situated in the S. part of the state, and contains 600 sq. m. Bounded S. by Ohio river. Watered by Scioto river and its branches. The Ohio canal passes through it, and enters the Ohio river at Portsmouth. It contained in 1840, 7547 neat cattle, 9509 sheep, 11,077 swine ; and produced 44,370 bushels of wheat, 332,360 of Indian corn, 1000 of barley, 81,577 of oats, 14,960 of potatoes, 7049 of sugar. It had five commercial and four commission houses in foreign trade, 44 retail stores, four flouring mills, 15 grist-mills, 15 saw-mills, two tanneries ; 42 schools, 1943 scholars. Pop. 11,192. Capital, Portsmouth.

SCIPIO, p. t., Cayuga co., N. Y., 8 m. S. Auburn, 164 m. W. Albany, 329 W. Bounded E. by Owasco lake. Drained by Salmon river. It contains two Baptist churches, five stores, one fulling-mill, two grist-mills, two saw-mills ; one academy, 36 students ; 18 schools, 828 scholars. Pop. 2925.

SCIPIO, t., Seneca co., O. It has three stores, one grist-mill, two saw-mills ; eight schools, 197 scholars. Population 1556.

SCITUATE, p. t., Plymouth co., Mass., 26 m. S.E. by S. Boston, 454 W. Bounded N.E. by Massachusetts bay, S.E. by North river. The harbour is small and of difficult access, and has a number of coasting and fishing vessels, with a village of about 30 dwellings. The town has seven churches, two Unitarian, a Congregational, two Methodist, a Baptist, and a Universalist. 16 stores, nine grist-mills, nine saw-mills ; 21 schools, 1025 scholars. Pop. 3886.

SCITUATE, t., Providence co., R. I., 12 m. W. Providence. Watered by the N. branch of Pawtuxet river. It contains three churches, two Congregational and a Baptist, a bank, 15 stores, one fulling-mill, one woollen factory, 11 cotton factories with 19,634 spindles, six grist-mills, six saw-mills, two academies, 62 students ; 19 schools, 864 scholars. Pop. 4090.

SCOTLAND, one of the secondary European kingdoms, comprising the northern and smaller portion of the island of Great Britain, and forming one of the three great divisions of the united kingdom of Great Britain and Ireland, between lat. 54° 38' and 58° 40' 30" N., and long. 1° 46' 30", and 6° 3' 30" W., or, including the Hebrides, 7° 44' W. It is surrounded by the ocean on all sides, except on the S., where it is separated from England by the Solway frith, the Cheviot hills, and the Tweed. Its greatest length, N. to S., from Dunnet Head to the Mull of Galloway, may be estimated at about 260 m.: its breadth is very unequal; varying from 32 m., between Alloa on the frith of Forth and Dumbarton on the Clyde, to 140 m. between Buchanness point in Aberdeenshire and Rowanmoon point in Inverness-shire. Its subdivisions, area, and population, are as follow:

Counties.	Total Area in sq. m.	Area of Land in Acres.	Area of Lakes, &c in Acres.	Pop. in 1831.	Pop. in 1841.	Acres of Land to an individual.
Aberdeen	1,970	1,254,400	6,439	177,651	192,383	9·525
Argyll, ex. Islands	2,390	1,408,000	38,400	66,289	63,788	22·076
Ayr	1,045	664,960	3,840	145,055	164,522	4·041
Banff	647	412,800	1,980	48,604	50,076	8·242
Berwick	442	282,880		34,048	34,427	8·276
Caithness	697	439,680	6,400	34,529	36,197	12·145
Clackmannan	48	30,720		14,729	19,116	1·607
Cromarty	266	163,840	6,400	11,299	11,362	14·419
Dumbarton	259	145,920	19,840	33,211	44,205	3·294
Dumfries	1,253	801,920	6,400	73,770	72,825	11·011
Edinburgh	354	226,560		219,345	225,623	1·084
Elgin	496	302,720	4,480	34,231	34,204	8·659
Fife	470	298,880	1,990	128,839	140,310	2·130
Forfar (Angus)	892	568,320	2,560	139,606	170,400	3·335
Haddington	272	174,080		36,145	35,781	4·865
Inverness, ex. Islands	3,036	1,858,560	84,480	54,059	55,531	33·468
Kincardine	382	243,900	1,980	31,431	33,052	7·358
Kinross	79	46,080	4,480	9,072	8,763	5·258
Kirkcudbright	834	525,760	8,000	40,590	41,099	12·792
Lanark	945	604,880	1,920	316,819	427,113	1·416
Linlithgow	120	76,800		23,291	26,648	2·860
Nairn	196	124,800	1,900	9,354	9,219	13·538
Peebles	319	204,180		10,578	10,580	19·406
Perth	2,638	1,656,320	32,000	142,894	138,151	11·929
Renfrew	227	144,000	1,920	133,443	154,735	930
Ross, ex. Islands	2,199	1,394,160	38,400	48,990	58,603	28·167
Roxburgh	715	457,600	390	43,663	46,083	9·947
Selkirk	265	168,320	960	6,833	7,969	21·089
Stirling	502	312,960	8,390	72,621	82,179	3·898
Sutherland	1,801	1,122,560	30,080	25,518	24,666	45·510
Wigtown	450	288,960	4,800	36,258	39,179	7·375
	26,014	16,332,800	316,160	2,202,854	2,451,668	
Average of acres to an individual in the mainland .						6·651
Islands.						
Bute, &c.	161	103,040	2,560	14,151	15,695	6·585
Isles of { Argyll	929	594,560	13,440	34,501	33,332	17·688
Isles of { Inverness	1,150	736,000	37,760	40,728	42,084	17·489
Isles of { Ross	560	358,400	12,800	14,541	17,015	21·063
Orkney Isles	425	272,000	9,600	28,847	30,441	8·935
Shetland Isles	855	547,200	16,000	29,392	30,355*	18·026
Totals	29,600	18,944,000	408,320	2,365,114	2,690,610	
Average of acres to an individual in the islands .						15·436
Average of acres to an individual in the kingdom .						7·028

* The various totals of the above counties include 4715 persons in barracks, and 1784 persons in vessels, &c., in harbours.

Scotland is extremely irregular in its surface and outline, and, compared with England, may be said to be sterile, rugged, and mountainous. This is so much the case, that, estimating the whole extent of the country, inclusive of islands and lakes, at 19,000,000 acres, perhaps not more than 6,000,000, or, at most, a third part, are arable; while the surface of England and Wales, amounting to 37,000,000 acres, comprises at least 29,000,000 acres, or more than three fourths, of arable land. With the exception, indeed, of a few rich alluvial tracts, there are no extensive vales in Scotland; its surface, even where least mountainous, being generally varied with hill and dale. It is divided by the frith of Clyde, loch Lomond, and the Grampians, into the two grand divisions of the Highlands and Lowlands; the former comprising the N., and the latter the S. part of the country. The Highlands again are divided into two unequal parts by the remarkable narrow and deep valley through which the Caledonian canal has been constructed. The arable lands in the Highlands are mostly confined to the E. parts of Ross and Cromarty, a slip along the S. side of the Moray frith, and the E. parts of Aberdeenshire. With these exceptions, the far greater part of the Highlands consists of mountains, moors, and morasses; and in various parts, especially along the W. coast, they are extremely bleak and barren. In Caithness there is a considerable extent of low ground, but it is mostly moor. The lowland division of the country comprises, also, a large extent of mountainous districts; but the mountains are not so lofty nor so bleak and rugged as in the Highlands, and there is a much greater extent of low fertile land.

The mountains of Scotland run generally in chains from S.W. to N.E., though there are many detached groups not following this distribution. They are frequently rocky, bare and precipitous; though mostly covered with heath. The principal and most celebrated chain is that of the Grampians, which comprises nearly all of the highest of the Scottish mountains, except Ben Nevis. This may girdle extends across the island from the arms of the sea, called loch Etive and loch Fyne in Argyllshire. E. by N. to Stonehaven on the E. coast, and Echt in Aberdeenshire, forming, as already stated, in the greater part of its course, the line of demarcation between the Highlands and Lowlands, and separating the waters which flow into the Forth, Tay, and South Esk, from those which join the Spean, Spey, and Dee. Its most elevated summits are near the head of the Northern Dee. Ben Macdha, 11 m. N.W. Braemar, lat. 57° 6' N., long. 3° 37' W., 4390 ft. above the level of the sea, is at once the culminating point of the Grampians, and the highest of the British mountains, being 20 ft. higher than Ben Nevis which was long considered as the highest of the Scotch mountains, and 819 feet higher than Snowdon in Wales. The other principal summits in the Grampian chain are Cairntoul, 4245 ft.; Cairngorm, 4095 ft.; Ben Lawers, 3945 ft.; Ben Avon, 3987 ft. height, &c. Ben Lomond, in Stirlingshire, 3195 ft. in height on the E. side of loch Lomond, the best known of all the Scottish mountains, also belongs to this chain. The Grampians are distinguished by their sterility, and desolate aspect; their sides, in many places, exhibiting vast perpendicular ledges of rock. The principal passes through this chain are those of Aberfoyle, Leal, Glenshie, and Killin, &c. &c.

Ben Nevis, alluded to above, lies to the N.W. of the chain, in about 56° 49' 30" N. lat., long. 5° W., being separated from the Grampians by the moor of Rannoch. It rises to an elevation of 4370 ft. above the sea. Its summit, which commands a magnificent view, extending from the

SCOTLAND.

Paps of Jura to Cuchullin in Skye, Cairngorm, Ben Macdhu, &c., is during the greater part of the year covered with snow. From Ben Nevis N. to loch Broom, several mountains rise to nearly 4000 ft. in height; and the country is so thinly inhabited that frequently for many miles not a house is visible. But from loch Broom to cape Wrath the surface diminishes considerably in elevation, and though bleak in the extreme, is, for some distance from the W. coast inland, not more than about 1000 ft. above the sea.

In the Lowlands, the Sidlaw and Ochill hills, which run parallel to the Grampians, nowhere rise to 2500 ft. Indeed, Broadlaw, on the N. border of Dumfriesshire, the highest mountain in the S. of Scotland is only 2741 ft. above the sea. The more elevated tracts in the Lowlands, including the mountains of Roxburgh, Dumfries, Peebles, Selkirk, and Lanark shires, are frequently smooth, and covered with a fine sward, affording good pasturage for sheep.

Though the valleys and level tracts in Scotland be few and of limited extent, as compared with those of England, some of them are extremely fertile, and they are mostly well cultivated. The carse of Stirling and Falkirk, on the banks of the Forth; that of Strathearn and Gowrie, on the Tay; and the merse of Berwickshire, all low alluvial tracts, are not inferior in point of productiveness to any lands in the empire. Teviotdale, or the low lands along the Teviot; Tynedale, or the low lands along the Tyne, in E. Lothian; the How of Fife, or the low ground along the Eden, in Fifeshire; and Strathmore, or the low grounds between the Grampian mountains and the Ochill hills, consist, for the most part, of rich loamy soil, and are extremely well farmed. It should also be observed that, the general inequality of the surface makes the lower parts of the country appear to be much less fruitful than they really are: the hollows between the small eminences being often extremely fertile, and the eminences themselves, even when they are unsusceptible of tillage, frequently furnish excellent pasture. This is particularly the case in the S.W. counties: large tracts of lands in Galloway and Dumfriesshire that let from 20s. to 30s. an acre and upwards, would appear to one not well acquainted with the country and its capabilities, worth little or nothing. A good deal of level, but generally high-lying land, especially in the Highlands, and in some parts also of the Lowlands, consists of moors; having for the most part a clay subsoil, covered with peat-earth, or moss impregnated with water, not unlike the bogs of Ireland. Many of these moors are of very considerable extent; the largest, probably, as well as the most desolate and worthless, is the moor of Rannoch, to the S. of Ben Nevis, comprised in the shires of Argyle, Perth, and Inverness.

Rivers.—Scotland has numerous rivers, several of which are of considerable size. They differ from those of England in being more precipitous, rapid, interrupted by cataracts, and subject to sudden overflowings. Except the Tyde, the others mostly disembogue on the E. coast. The Tweed, which rises on the confines of Dumfriesshire and Lanarkshire, falls into the North sea at Berwick, after a course of about 100 m., only a small portion of which is navigable. Proceeding northwards, the next river of any considerable magnitude is the Forth, which rises on the E. side of Ben Lomond, and has in general an easterly, but very tortuous course to Kincardine, where it unites with its salt estuary, or rather arm of the sea, the frith of Forth, the *Bodotria* of Tacitus. It receives on its N. side, the Teith and Allan, and from the S., the Devon, &c.: Aberfoyle, Stirling, and Alloa are on its banks. The Forth is held for some considerable distance from its source; but during the greater part of its course it runs through a flat country with many windings: vessels of 300 tons ascend the Forth as far as Alloa, and those of 70 tons ascend to Stirling. It is connected with the Clyde by the great canal from Grangemouth to Dunglass. The Tay is the largest of Scotch rivers, and is supposed to carry more water to the sea than the Thames, or any other river in Great Britain. (*See* TAY.) The N. and S. Esks, Dee, Don, Spey, Findhorn, all discharge themselves on the E. coast; in the N. division of the Highlands, are the Nairn, Ness, Beauly, &c. The Spey is one of the largest rivers in Scotland, and certainly the most rapid. It rises in loch Spey, and pursues mostly a N.E. course to the Moray frith, which it enters after a course of about 90 m. It receives no large tributary, but innumerable mountain torrents, in consequence of which it is subject to frequent and destructive inundations. The Clyde, the *Glotta* of Tacitus, though not the largest, is decidedly the most important of which river in a commercial point of view, Lanark, Hamilton, and Glasgow being situated on its banks. It rises in the highest part of the Lowlands, close to the sources of Tweed and Annan; and runs at first N., but afterwards generally N.W., to the frith of Clyde, with which unites 7 or 8 m. below Glasgow, after a course of between 70 and 80 m. It receives from the S. the Douglas, Nethan,

Avon, Cart, &c., and from the N. the Kelvin, Leven, &c. The Clyde has been rendered navigable, for vessels of 400 tons, as far as Glasgow. (*See* CLYDE and GLASGOW.) The Southern Dee, Nith and Annan, into the Solway frith, are the only other streams it is necessary to notice.

The *lochs*, or fresh-water lakes of Scotland, are numerous, and highly distinguished for their picturesque scenery. loch Lomond (which see *ante*, p. 185) is the largest lake in Great Britain; being about 94 m. in length, and from 7 m. to 7¼ m. across in the broadest part. It is estimated to cover about 25,000 acres. Lochs Awe, Ness, Maree, Tay, Shin, &c., in the shires of Argyle, Ross, Perth, and Sutherland are among the other principal lakes. Most of these are long, narrow, and deep, filling up the bottoms of the valleys between the mountains. They abound with trout, perch, pike, &c.; but loch Leven, in Kinross-shire, is the only lake that yields any revenue to its proprietors.

The coasts of Scotland are mostly bold and rocky; and, on the W. side in particular, they are very much indented by arms of the sea, termed friths and lochs, that extend far inland, and, for the most part, carry deep water to their very head. These friths and inlets are of considerable importance in a commercial point of view, especially as few of the rivers are navigable to any great distance inland. On the E. coast are the friths of Forth and Tay, which, especially the first, are of great importance, as affording facilities of communication to the richest districts of the country; N. of the latter, on the same coast, are the friths of Moray, Dornoch, Cromarty, &c.; on the W. coast, the frith of Clyde, and lochs Broom, Torridon, Linnhe, Fyne, &c., deeply indent the country. The harbours of Leith, Grangemouth, Queensferry, Bruntisland, &c., are in the frith of Forth, and those of Dundee and Perth in the frith of Tay. Between the Tay and Buchan Ness are the harbours of Montrose, Aberdeen and Peterhead: the frith of Cromarty, N. of Buchan Ness point, is unquestionably the best asylum for shipping on the E. side of Great Britain, and one of the finest, indeed, that is anywhere to be met with. Between the latter and Duncansby head are the small harbours of Wick, St. Clair's bay, &c. From cape Wrath to the Clyde, the narrow arms of the sea, though deep and secure, are little frequented. The ports of Greenock and Glasgow are the principal in the frith of Clyde, and enjoy an extensive trade; but Lamlash bay, on the E. side of the Isle of Arran, is the best harbour in the neighbourhood. There are some pretty good harbours on the coasts of Wigtown and Kidcudbright shires. The principal headlands are St. Abb's head, Fife Ness, Peterhead, Tarbet Ness, and Duncansby head, on the E.; Dunnet head and cape Wrath, on the N.; Ru-Rea, Ardnamurchan, and the Mulls of Oe and Cantire, on the W.; and the Mull of Galloway and Burrowhead on the S. coast.

There are few or no islands off the E., but many of large size lie contiguous to and off the W. coast. These are mostly included under the Hebrides (which see). The islands of Orkney and Shetland (which see) lie off the N. coast of Scotland; the Orkneys, the nearest, being separated from the mainland by the Pentland frith, 6 m. across.

General Aspect of the Country.—The finest parts of the low country of Scotland usually want the rich luxuriance of an English landscape. Within the last sixty or seventy years a great deal has, no doubt, been done in the way of raising plantations; and the strictures of Dr. Johnson, as to the deficiency of wood, would at present be quite inapplicable, however just they may have been when dictated. In Scotland, however, plantations are not spread generally over the country, but are mostly congregated in the neighbourhood of gentlemen's seats, while in many large tracts they are wholly wanting. In most parts, too, we look in vain for those hedgerow trees that give so much of a woody appearance to the southern part of the island. Generally, also, the enclosures are a good deal larger than in England; and the fences being either stone walls (dykes) or hedges, that occupy only a small space of ground, having little of the breadth and roughness of those of England, the country, however well farmed, seems to an Englishman deficient in vegetation and verdure, and cold and comfortless. On the other hand, however, the succession of

from 300 to 400 ft. above the sea-level, the mean temperature of the year is 47·8°, which may be taken as that of the inland parts generally in the S. of Scotland, the mean of the coldest month being 38·3°, and the warmest 59·4°. A great deal of rain falls in Scotland, but very unequally ; for on the E. coast it ranges from 22 to 28 in. ; whereas on the W. coast and in the Hebrides, it ranges from 30 to 44 in. The average fall of rain in Edinburgh is about 23½, and in Glasgow about 29·65 in. Excess of humidity, and the occurrence of heavy rains in August, September, and October, and of cold, piercing E. winds, especially along the E. coast, in the months of April, May, and the first half of June, are the great drawbacks on the climate of Scotland. It is rare, indeed, that the crops suffer from heat or drought ; but they frequently suffer from wet and from violent winds, especially in the W. part of the country. The climate is however highly salubrious, and favourable alike to longevity, and to the development of the physical and mental powers.

Scotland is supposed to possess about 3230 indigenous plants, of which 870 are dicotyledonous, 290 monocotyledonous, and 2080 cryptogamic. Most of the forest trees of England are met with. In the Highlands are several extensive forests of pine (*Pinus sylvestris*), covering the valleys, and ascending to an elevation of 2500 feet up the mountains. Apples, pears, plums, peaches, apricots, &c., ripen in the open air as far N. as Iverness, and in warm sheltered situations to the N. extremity of the kingdom. The sea-weed, which grows in great profusion round the coasts, used to be extensively manufactured into kelp, and the business, though much diminished, in consequence of the preparation of *soude factice*, (artificial soda), is still carried on to a considerable extent.

The *wild animals* of Scotland are mostly the same as those of England, including the stag, wild roe, hare, rabbit, fox, badger, otter, wild cat, hedge-hog, &c. ; though some of these are becoming scarce. The wolf and beaver, formerly natives of the country, have been long extinct ; and the only existing remains of the *urus*, or native breed of cattle, are restricted to a few preserved in the duke of Hamilton's park, near Hamilton. One of the domestic animals peculiar to Scotland is the *colley*, or true shepherd's dog, and many specimens of the unmixed breed are extant. Several species of eagles, and other birds of prey, and aquatic birds in great numbers, inhabit the rugged coasts, and the pheasant, ptarmigan, black cock, grouse, and partridge abound inland. The noble species of game called the capercailzie, or cock of the wood (*Tetras Urogallus*), was formerly abundant in Scotland ; but it appears to have been exterminated about 1760. Within these few years, however, it has been reintroduced by the marquis of Breadalbane, the earl of Fife, and other extensive forest proprietors ; and there can be no doubt that, if properly protected, it will succeed very well ; but it is too obvious and too tempting a mark for the poacher to maintain itself. Scotland has also most of the English singing-birds, except the nightingale, which is rarely, indeed, found N. of the Trent. The fish are similar to those of England : the rivers teem with the finest salmon, trout, &c. ; and the salmon fisheries of the Tay, Tweed, Forth, Spey, &c., are highly valuable.

Geology.—A line drawn in a N.E. direction from the mouth of the Clyde to Stonehaven, on the E. coast, separates the two principal geological regions. The first, to the N. of that line is mostly composed of primary rocks, granite, gneiss, mica-slate, covered at the foot of the mountain chains with beds of conglomerate and red sandstone ; whereas the second to the S. of the above line, is the region of transition formations, in which rocks of that kind mostly prevail, overlain in various parts by trap, red sandstone, and coal beds : granite is, however, largely developed in the S.W. part of the kingdom, in the stewartry of Kirkcudbright, at Criffel, and in the Cairnsmuir range. Little if any coal exists in the primary division of the country ; few metals are discovered there, and its most important mineral products are building-stone and roofing-slate. Some lead mines are, however, wrought at Strontian in Argyleshire ; and in Inverness-shire plumbago of inferior quality has been found, imbedded between laminæ of mica slate. None of the secondary calcareous formations, so extensively prevalent in England, have been found to exist in Scotland, nor any tertiary formations.

Coal and Iron.—The great coal district of Scotland may be considered as bounded on the N. by a line drawn from the mouth of the Tay, to the N. extremity of the isle of Arran, and on the S. by another line drawn from St. Abb's head to Girvan, in Ayrshire. These limits comprise a band of country, in which are several large coal fields detached from each other, the most valuable extending along the banks of the Forth, with a breadth of from 10 to 12 m. on either side of the river. The Edinburgh coal-field, to the S. and E. of that city, occupies an area of 80 sq m. ; and from Bathgate the coal deposits extend W. to Glasgow and Paisley, and have, in fact, been the principal

cause of the wonderful progress made by the former in manufactures, wealth, and population. There are several small detached coal-fields in Ayrshire and some of the other of the S. counties. Iron is of frequent occurrence in the coal districts, especially in Lanarkshire, where the ores are of the very best quality ; and the iron trade of that county, and of Scotland generally, has latterly increased with an extampled rapidity, and is now of the greatest importance. So much so is this the case, that, while in June, 1835, there were in Scotland only 29 furnaces in blast, estimated to produce 75,000 tons of iron, it appears from the following statement, obtained from private sources, on which every reliance may be placed, that in June, 1840, there were 61 furnaces in blast in Scotland, producing at the rate of 239,500 tons a year.

Works.	Blast Furnaces in blast	Tons Pig-iron per Week
Lanarkshire :		
Clyde	4	36
Govan	5	50
Calder	4	34
Dundyvan	5	50
Garnherrie	3	30
Summerlee	3	30
Carnbroe	4	36
Monkland	3	24
Wilsontown	1	8
Coltness	2	20
Castlehill	2	20
Shotts	2	18
Ayrshire :		
Muirkirk	3	170
Gartsherrie	1	?
Renfrewshire :		
Hawkhill	1	10
Stirlingshire :		
Carron	4	40
Clackmannanshire :		
Devon	1	8

" Abstract Account of Pig-iron made in Scotland in 1840 :

	Furnaces	Tons per Week	Tons per Annum
Lanarkshire	38	4,190	204,880
Ayrshire	9	230	11,960
Renfrewshire	1	20	1,040
Stirlingshire	4	240	12,480
Clackmannanshire	1	20	1,040
Total	61	4,700	239,500

" The whole of the above iron is made by means of heated air with the exception of one furnace at Carron.

" It will be observed that the weekly quantity is multiplied by 50 only, the average stoppages being equal to about two weeks.

" There are several new furnaces building, so that by next summer (1841) there will probably be ten additional to these.

" The bar-iron trade of Scotland is in its infancy. Dundyvan works have just been completed, and are now in full operation, making . . . . . 200 tons per week.

Govan is not yet completed ; but by the end of the year will also make . . 200

Monkland is in a similar situation as Govan, ... 200

Muirkirk makes about 40

Total . . 640

Equal to 32,000 tons a year of rolled bars, boiler plates, sheets and rods."

Gold has been occasionally found in the streams near the lead hills in Lanarkshire, and elsewhere ; and silver has been met with in various places : but the precious metals are not so abundant as to defray the expense of working them. Next to iron, lead and copper are the most valuable metals. The mines of Wanlockhead and Leadhills, on the borders of Lanarkshire and Dumfriesshire, furnish annually about 700 tons of lead, but the produce varies considerably. Small quantities of cobalt, bismuth, manganese, &c., are met with. Scotland produces marble in great variety, and of very superior quality, slates, excellent building stone, and many species of gems. Brick is but little used in building in Scotland ; the houses being everywhere almost of stone.

Fisheries.—The salmon fishery is the most important of what may be called the domestic fisheries ; and since 1788 London has received the greater part by far of her supply of salmon from Scotland, considerable quantities having also sent to Liverpool. The fish are brought up, even in the hottest weather, quite fresh, being packed in pounded ice. Previously to the introduction of this plan, salmon used to be consumed principally in the country where it was taken ; and, in some parts of Scotland, domestic servants used to stipulate that they should not be obliged to dine on salmon more than three or four times a week. The salmon fisheries seem to have attained their maximum value toward the end of last war, when the fisheries in the Tweed were let for from £15,000 to £18,000 a year ! and those of the Tay, Dee, Spey, &c., were proportionally valuable. But the value of the Scotch salmon fisheries has, speaking generally, declined greatly of late years; in consequence partly and principally of a diminished supply of fish in the rivers, but, in some degree, also, from the greater facility of the communication between London and Liverpool, and the consequent importation of Irish salmon into the London markets. We have been fortunate enough to obtain from a source on which every reliance may be placed, the following

Account of the Quantity of Salmon packed in Ice imported into London, from Scotland, during each of the eight Years ending with the 14th of Oct., 1841, and of the wholesale Price of the same.

Years ending Oct. 14.	Weight of Fish.	Average Price, about.	Total Value.
	Lbs.	*d.*	*L.*
1834	3,432,800	9½ per lb.	135,960
1835	4,740,960	9	177,800
1836	2,751,840	10½	120,400
1837	3,617,600	10	150,750
1838	2,396,800	10½	104,160
1839	1,830,680	11	83,880
1840	1,687,840	11	77,860
1841	3,192,672	8¾	116,400

This, it will be observed, is independent of the pickled salmon brought from Scotland, the quantity and value of which varies as much as that of the fresh salmon. But we are well assured that, at an average of the last eight years, its value has not exceeded £12,000 a year. At an average, the retail price of salmon in London may be taken at from 50 to 75 per cent. above the wholesale price.

We may remark by the way that as by far the largest portion of the salmon made use of in London comes from Scotland, the above statement shows that its consumption in the metropolis is not nearly so great as is generally supposed. In fact, it is little used, except by the more opulent classes; and nothing that is not generally used by the middle classes, or by them and the lower, is ever of much importance.

The herring fishery has latterly been prosecuted with considerable success, and is now becoming an important branch of industry. Down to a recent period, it had been attempted to bolster up this department by granting bounties on the fitting out of vessels for the fishery, on the herrings taken and exported, and so forth. But, notwithstanding the very large sums expended in this way, their influence was found to be injurious rather than otherwise; partly by the meddling they occasioned on the part of government, and partly by the temptation the bounty afforded to small farmers and others, nowise acquainted with the business, to engage in it to the injury of the regular fishermen. At length, however, the abolition of the duty on salt, and the growing conviction of the inexpediency of the bounty, led first to its gradual, and ultimately to its total, abolition in 1830; since which, we are happy to have to state, the business has not only been carried on upon a solid foundation, but has progressively increased. This is obvious from the following

Account of the total Quantity of Herrings cured, branded, and exported, in the undermentioned Years.

Years ending April 5.	Total Quantity of Herrings cured.			Total Quantity of Herrings branded.	Total Quantity of Herrings exported.		
	Gutted.	Ungutted.	Total.		Gutted.	Ungutted.	Total.
	Barrels.	*Barrels.*	*Barrels.*	*Barrels*	*Barrels.*	*Barrels.*	*Barrels.*
1811	65,430	26,397½	91,827½	55,662½	18,880	19,253	38,133
1815	105,372½	54,767	160,139½	83,376	68,938	72,367½	141,305½
1821	347,190½	35,301	382,491½	309,700½	244,096	9,420	253,516
1825	303,397	44,268½	347,665½	270,844½	201,882½	134	202,016½
1830	280,933½	48,623½	329,557	218,418½	177,776	3,878½	181,654½
1831	371,096	68,274½	439,370½	237,085	260,976	3,937	264,903
1832	313,113½	49,547	362,660½	157,830½	214,820½	2,679½	217,499½
1833	353,684½	63,279½	416,964½	168,259½	218,492½	2,253	220,684½
1834	382,677½	68,853½	451,531½	178,000½	269,133½	2,960	272,093½
1835	217,242½	60,074½	277,317	85,079½	156,225½	2,580	158,805½
1836	399,324	98,290½	497,614½	192,317	270,846½	2,547	273,393½
1837	290,169	107,660½	397,829½	114,192	187,238½	2,027	189,265½
1838	382,400	125,374½	507,774½	141,532	229,160½	5,997	235,158
1839	382,229	173,330½	555,559½	153,659½	233,690½	6,040	239,730½
1840	405,379½	138,565½	543,945	152,231	250,554	1,968	252,522

N. B.—In the six years ending 5th April, 1816, the bounty on herrings cured gutted was 2s. per barrel, while there was a bounty at the same time of 2s. per barrel, payable by the excise on the exportation of herrings, whether cured gutted or ungutted, but which ceased on the 1st of June, 1815. In the years ending 5th April, 1826, the bounty on herrings cured gutted was 4s. per barrel (about 25 per cent. on the wholesale price). In the four succeeding years, the bounty was reduced 1s. per barrel each year, till the 5th April, 1830, when it ceased altogether, and has not since been renewed.

The cod and white-fish fishery is also extensively carried on, about 160,800 cwts. having been cured in the year ending the 5th of April, 1840. The large share taken by Scotland in the northern whale fishery is evinced by the statement, that, in 1834, of 76 vessels engaged in the trade, 41, of the aggregate burden of 13,242 tons, were from Scotland, and brought home 4912 tons of oil and 257 tons of whalebone. This department of industry has since, however, rapidly declined, not from any diminution of skill or enterprise, but from the increasing risks and unprofitable nature of the business. Fortunately the loss of the whale fishery has been more than compensated by the extension of the herring fishery, and the success that has attended it.

Races of Inhabitants.—It is generally allowed that the first immigrants into Scotland, like those into England, whence, perhaps, they originally came, belonged to the great Celtic family; and Mr. Chalmers and others have endeavoured to prove that the population continued to be wholly Celtic till it was alloyed, first by Roman, and subsequently by Gothic, invaders. (*Caledonia*, vol. i., p. 496, &c.) This opinion does not seem very tenable. Tacitus expressly affirms that the Caledonians, or inhabitants of Scotland, were of Germanic or Gothic origin. "*Rutilæ Caledoniam habitantium comæ, magni artus, Germanicam originem asseverant.*" (*Vit. Agricolæ*, cap. ii.) Agricola, however, from whom Tacitus derived his information, knew little or next to nothing of the country N. of the Grampians; and as there is every reason to think that Berwickshire, the Lothians, Fife, and other parts of the low country on the E. coast of Scotland, were, like the same tracts in England, early occupied by Belgic, or other Gothic colonists from the opposite continental coast, it seems most probable that Tacitus, in ascribing to the Caledonians a Germanic origin, had these only in view. The fair presumption is, that in the northern, as in the more southern part of the island, the old Celtic inhabitants maintained their ascendancy in all the mountainous, and, comparatively, inaccessible districts; and this reasonable presumption is corroborated by various circumstances.

In the third century, the terms Picts and Pictland began to be substituted for Caledonians and Caledonia. It is pretty generally believed that these terms apply to the same people, and the same country. It seems, indeed, to be perfectly clear, that the Picts were descended from the Scythians, or Goths;* and if we be right in our statement, as to the origin of the Caledonians, it follows that, if the Picts were not identical with them, they belonged, at all events, to a congenerous race.

About the period of the withdrawal of the Romans from Britain, a tribe called *Scoti*, or *Atacotti* (the *Dalriads* of the venerable Bede), began to be distinguished as a leading tribe in Ireland; and seems, indeed, to have given its name to the island, which, for some centuries, was called *Scotia*. (*Pinkerton's Geography*, i., 137, ed. 1811.) It would seem, that, previously to the 11th century, a colony of the Scoti from Ireland had established themselves in the West Highlands; and this colony, in no very long time, gave its name, first to the Highlands, whence it was subsequently extended, on its being united under one government, to all that part of the island N. of the Tweed and the Solway frith. The Scoti ceased to be heard of in Ireland not long after they had obtained a footing in the Highlands, and the ancient names of that island were revived.

Everything connected with the history of the Scoti is involved in impenetrable obscurity. But it is agreed, that, whatever may have been their remote origin, they had, when they settled in the Western Highlands, the language and habits of the Irish Celts, or Gael, a congenerous race with the Highland Celts, and speaking, in fact, the same language. But the Scoti-colonists had a written language, which the old occupiers of the country had not; and they were also decidedly superior to the latter in knowledge and

* See Pinkerton's chapter on the Origin of the Picts in his *Inquiry into the Early History of Scotland*.

civilization. (*Pinkerton on the Early History of Scotland*, ii., 160.) These circumstances sufficiently account for the ascendancy they acquired, and for their being able to give their name to the Highlands, without having recourse to the hypothesis, for which there is not a tittle of evidence, of their having conquered the country.

After the Romans withdrew from Britain, some Gothic or Saxon tribes, following the example set by those who had previously settled in the more southerly parts of the island, established themselves, during the sixth century of our æra, between the Tweed and the frith of Forth. (*Turner's Anglo-Saxons*, 5th ed., i., 209.) These new immigrants were afterwards followed by others, at the same time that they drew recruits from their brethren established in England; and Mr. Chalmers supposes that, their power being thus progressively augmented, they gradually acquired a complete ascendancy in all the southern parts of the kingdom, and communicated to it their language and manners. (*Caledonia*, ii., 7.) This, however, would have been an extremely difficult task; but if, which seems abundantly certain, we conclude, with Pinkerton, that the Picts, who were in possession of all the low country in the sixth century, were congenerous with the Saxons, by whom it was then invaded, the two races would readily amalgamate, and the early prevalence of the Scandinavian or Gothic tongue in the lowlands is rationally and satisfactorily accounted for.

Towards the end of the eighth century, a fresh colony from Ireland established itself in the district now known by the name of Galloway, in the S.W. part of Scotland. But though these colonists succeeded in giving a name to the country, they were not sufficiently numerous to introduce their language into common use.[*] And for several countries, long indeed before the inhabitants had any considerable intercourse with other parts of the kingdom, the finest tongue has been in as universal use in Galloway as in any part of the lowlands of Scotland.

Exclusive of the Celts, Goths or Picts, Romans, Scoto-Irish, and Saxons, colonies of Danes and Norwegians established themselves in Caithness, and other parts of the mainland, as well as in Orkney and Shetland, and parts of the Western Isles. Generally, however, it may be said, notwithstanding the late great influx of Irish settlers into Glasgow, Paisley, and other large towns, that at present the inhabitants of the Lowlands of Scotland are principally of Saxon, while those of the Highlands, with the exception of Caithness, are almost entirely of Celtic extraction.

Population.—We have few data to guide us in estimating the amount of the population previously to the period of the union, in 1707, at which time Scotland is supposed to have had about 1,050,000 inhabitants. In 1755, the population was ascertained by Dr. Webster to amount to 1,265,000, and since 1801 its progress has been as follows:—

Counties	1801	Inc. per cent	1811	Inc. per cent	1821	Inc. per cent	1831	Inc. per cent	1841		
									Males	Females	Total
Aberdeen	123,099	10	135,075	15	155,387	14	177,651	8·2	89,398	102,755	192,203
Argyle ‡ . . .	71,859	19	85,585	14	97,316	14	100,973	·	47,864	49,491	97,149
Ayr	84,306	23	103,954	23	127,299	14	145,055	13·4	78,976	85,552	164,528
Banff	35,807	2	36,668	19	43,561	12	48,604	3·	23,425	26,451	50,076
Berwick . . .	30,621	1	30,779	8	33,385	2	34,048	1·1	16,527	17,920	34,447
Bute	11,791	2	12,033	15	13,797	3	14,151	10·9	7,108	8,587	15,695
Caithness . . .	22,609	4	23,419	29	30,238	14	34,529	4·8	16,093	19,594	36,197
Clackmannan . .	10,858	11	12,010	10	13,263	11	14,729	29·7	9,331	9,785	19,116
Dumbarton . .	20,710	17	24,189	19	27,317	22	33,211	33·3	23,505	21,798	44,588
Dumfries ‡ . .	54,597	15	62,960	13	70,878	4	73,770	·	34,097	38,728	72,825
Edinburgh . .	122,954	21	148,607	29	191,514	15	219,345	9·8	106,769	122,914	225,683
Moray	26,705	5	28,108	11	31,162	10	34,231	2·2	16,071	18,892	34,964
Fife	93,743	8	101,272	13	114,556	19	128,839	8·9	66,735	74,575	140,310
Forfar . . .	99,127	8	107,264	6	113,430	23	139,606	22·	79,934	91,146	170,689
Haddington ‡ .	29,986	4	31,164	12	35,127	3	36,145	·	17,383	18,288	35,701
Inverness . .	74,292	5	78,336	15	90,157	5	94,797	3·	45,508	50,100	97,615
Kincardine . .	26,349	4	27,439	6	29,118	8	31,431	5·1	15,504	17,345	33,049
Kinross ‡ . .	6,725	8	7,245	7	7,762	17	9,072	·	4,194	4,540	8,743
Kirkcudbright .	29,211	15	33,084	15	38,903	14	40,590	1·2	18,638	22,981	41,669
Lanark . . .	146,699	31	191,752	27	244,387	30	316,819	34·8	208,289	218,744	427,113
Linlithgow . .	17,884	9	19,451	17	22,685	3	23,291	15·2	13,766	13,669	26,840
Nairn ‡ . . .	8,257	·	8,251	9	9,006	4	9,354	·	4,239	4,959	8,541
Orkney, Shetl. .	46,844	·	46,153	15	53,124	10	58,239	4·3	27,343	33,925	61,268
Peebles ‡ . .	8,735	14	9,935	1	10,046	5	10,578	·	5,192	5,395	10,589
Perth ‡ . . .	126,366	7	135,093	3	139,050	3	142,894	·	63,339	78,812	138,152
Renfrew . . .	78,056	19	92,596	21	119,175	19	133,443	15·9	73,795	82,306	155,205
Ross, Cromarty .	35,342	10	63,853	13	68,898	9	74,820	5·5	36,461	42,119	78,685
Roxburgh . .	33,682	11	37,230	10	40,892	7	43,663	5·3	21,580	24,873	46,463
Selkirk . . .	5,070	16	5,889	13	6,637	2	6,833	16·9	3,272	4,017	7,289
Stirling . . .	50,825	14	58,174	12	65,376	11	72,621	13·1	41,070	41,900	82,773
Sutherland ‡ .	23,117	2	23,029	·	23,840	7	25,518	·	11,307	13,208	24,519
Wigtown . .	22,918	7	26,891	23	33,940	9	36,258	8·	18,058	18,982	37,071
Totals	1,599,068	14	1,805,688	16	2,093,456	13	2,365,114	10·6	1,241,376	1,378,334	2,620,603

Note.—The returns for 1841 include 4745 persons resident in barracks in various counties, and 1775 persons on board ships in harbour.

This increase, though rapid, is less than the increase of the population of England during the same period; and it is, also, much less than the increase in Ireland from 1801 to 1831. This, however, is a favourable system, for there are good grounds for thinking that the wealth of Scotland had increased more rapidly during the above period than that of either of the other great divisions of the empire; and if so, it is plain, inasmuch as her inhabitants have not increased so fast, that their condition must have been proportionally improved; and, in point of fact, they are, speaking generally, in more prosperous circumstances at present than at any former period. Owing to the sterility of the soil, and other causes, the population of Scotland is much less dense than that of the sister kingdoms. The increase of population has been chiefly in the great towns: during the 10 years ending with 1820, the population of the country at large increased 16 per cent., and during the 10 years preceding 1830, 13 per cent.; while the population of the large towns increased respectively 26½ and 20½ per cent. during the same periods. The population in some of the counties has latterly rather declined, in consequence of the consolidation of farms, and the extension of sheep walks.

Agriculture.—Scotland, from being about the middle of last century one of the worst cultivated countries of Europe, is now one of the best. At this moment, indeed, the agriculture of the best farmed counties of Scotland is

certainly equal, and is by many deemed superior, to that of Northumberland, Lincoln, and Norfolk, the best farmed counties of England. The proximate causes of this extraordinary progress must be sought for in the rapid growth of manufactures and commerce, and consequently of large towns and the proportionally great demand for agricultural produce since the peace of Paris in 1763, and especially since the close of the American war. Fortunately, too, the influence of these favourable circumstances was not counteracted by any vicious institutions, or by anything unfavourable in the mode of letting and occupying land. Next to the open field system, the wonderful improvement of Scotch husbandry may be ascribed to the prevalence of leases of reasonable length, usually 19 years, and which generally embody clauses to prevent the exhaustion of the soil; the absence of tithes and in most instances, of poor rates, and of all oppressive

[*] "Galloway, in the Latin writers of the middle ages, Gallia or Gallovidia, so called by the Irish who overran these districts and who themselves, in their native language, Gael." (*Chalmers' Pref. to Caledonia*; also the *Appendix to Symson's Galloway*, p. 114.)

[†] This rather has been, in all respects, most injurious to the people of Scotland and England, where it has taken place to so capital that without conferring any corresponding advantage on the Irish. (See *Hist. of Great Britain*, 2d edition, i., 307.)

[‡] The population of the counties marked thus, has decreased between 1831 and 1841.

as; the prohibition of subletting, and the in-
the lease by the heir-at-law; the introduction
aing into the Highlands, and the great improve-
n the construction of all sorts of farming im-
'he general use of thrashing machines, many
impelled by steam, and of ploughs with two
, by the ploughmen, constitute, in fact, one of
ashing characteristics of Scotch, as compared
, agriculture.

,pperty in Scotland, as compared with its extent
in fewer hands than in England, there being
more than 8000 proprietors in the whole coun-
.oat subdivided in the counties of Fife, Mid-
afrew, and Kirkcudbright, but even in these
 any large estates; and in most other parts of,
he greater portion by far of the land is distrib-
y large estates, many of which are held under
trict and perpetual entail. It might be supposed
of tenure would be prejudicial to agriculture,
ly, its effect is not found to be nearly so great as
been, a priori, anticipated. This results from
ance of the courts of law having decided that
, let lands belonging to an entailed estate, either
ually long period, or by fines (Scottice gras-
at, in truth, there is little or no difference be-
ditions under which entailed and unentailed
ccupied; and as the proprietors of the former
npowered to burden the estates, proportionally
e, with sums laid out on necessary improve-
uildings, they are, speaking generally, in quite
r, and as productive as the others.

of all sizes; varying from 50 to 500 acres and
the improved tillage districts, and from 500 to
d upwards in the hilly and mountainous tracts.
few of the sequestered glens of the Highlands,
he improved systems have not been introduced,
of the land is nowhere carried to such an extent
udicial to agriculture; and, in most parts of the
ms have been gradually consolidating and in-
ize since the American war. At an average of
, arable farms may vary from 150 to 300 acres;
farms from 500 to 2000 acres.

he class of the American war, the farm build-
, parts of Scotland were mean and inadequate in
. In the Lothians they were commonly ranged
ving the dwelling-house in the middle, with a
end, and cattle-houses at the other. In other
were frequently huddled together without any
r. The walls were always low, in most instan-
e and clay, the roof being invariably thatch.
Il was universally opposite to the door; and so
t in wet weather it was no easy matter to get
ise with dry feet. The change that has taken
se respects during the last half century has been
omplete. In none but the least accessible and
red districts are any of the old houses now to be
Perhaps, indeed, the other extreme has not been
avoided; buildings having, not unfrequently,
d that seem to be both larger and more expensive
ary. The offices are mostly constructed in the
square. In some instances, the dwelling-house
of its sides; but in the better class of farms it is
some distance from the offices. It is generally
high, and is well, and sometimes handsomely,
Both houses and offices are almost always slated.
e of buildings is uniformly defrayed by the land-
be tenant, for whose accommodation they are in
stance erected, sometimes pays a percentage upon
laid out upon them. Sometimes, also, the ten-
take to carry the materials used in building.
:es in many parts of Scotland consist of dry stone
uch, though destitute of beauty, make, when
ilt, a capital fence. This species of enclosure
ave originated in the S. W. Kirkcudbright had
vere early subdivided with excellent stone dykes,
ow celebrated all over the kingdom by the name
vay dykes."* They are of very various heights,
s of goodness; but the best are built double to a
ght, when they are capped with broad flat stones
a little on each side, over which others are usu-
ngle; but sometimes those laid over the cap-stone
o lock firmly together. The best dykes vary from
in height; and, where they have been carefully
,ell built, and constructed of good stones, they
ust excellent and very durable fence. Examples
re of their standing for 60 or 70 years without re-
most any repair. In a few instances they have
d at above 100 years of age, in a state of perfect
on! But unless they be of superior material and

an excellent account of the Galloway dykes, with judicious re-
r construction, in Smith's Survey of Galloway, p 90—93.

workmanship, they seldom last more than from 25 to 30
years. Most of the dry stone walls now to be met with all
over Scotland have been built, sometimes with more sometimes with less success, after the Galloway model.

In respect of farming implements, Scotland has very much
the advantage over England. The improved Scotch plough
is every where met with in the agricultural districts, and is
uniformly drawn by two horses, driven by the ploughman.
Iron harrows are common. Thrashing machines are introduced far more extensively than in England; and there is
hardly, indeed, a considerable farm in any part of the country without one. The Scotch labourers have never been so
absurd as to attempt to advance their interests by destroying those valuable engines.

Scotland may be divided into three agricultural districts :—
Of these the first, or most southerly, extending from the English border to the rivers Forth and Clyde, contains a large
extent of mountainous and pasture land. But extensive
tracts in Berwickshire and the Lothians, on the E. coast,
are naturally fertile, and are farmed with a degree of skill,
economy, and success, unequalled almost in any other part
of the empire. There are also large tracts of fertile and
well-farmed land in Lanark, Renfrew, Ayrshire, Galloway,
and Dumfries; but the climate on the W. coast is not so
favourable, and agriculture is not so far advanced on that
side the island as on the E. The second agricultural division stretches from the Forth and Clyde to the great chain
of lakes connected by the Caledonian canal, that runs from
Inverness to the island of Mull. The mountains in this division are on a grander scale than in the southern division;
and the proportion of waste land much greater. It, however, contains some of the finest land in the empire. The
carse of Gowrie, stretching from Perth to Dundee, consists
of the richest alluvial soil; but its agriculture, though good,
is not equal to that of some other districts. Strathern, lying to the W. of Perth, is also very fertile. Most part of the
extensive county of Fife is arable, and is, in general, very
highly improved. There are also very large tracts of fine
land in Forfarshire and Angus, and smaller portions in
Aberdeenshire and Moray. The third division of Scotland,
or that which embraces the country lying to the N. of the
Caledonian canal and the lakes, is, with the exception of
the eastern parts of Ross-shire, and a few patches besides,
wild and mountainous. Black cattle, sheep, and wool, are
its principal products.

Except in the S.E. counties, oats are grown in far greater
quantities than any other kind of grain; and, from more
attention being paid to their culture, or the greater suitableness of the climate, or both, the produce is greater than in
England, varying from 25 and 30 up to 70 bushels an acre,
and even more. Oatmeal, which, till a late period, formed,
in cakes and porridge, the principal part of the food of the
great bulk of the people, is still in very extensive demand;
but, latterly, the use of wheaten bread has become very
general in the rural districts, as well as in the towns. Turnips and potatoes are successfully cultivated throughout
most part of the Lowlands. The raising of the former is,
perhaps, nowhere so well conducted as in E. Lothian and
Berwickshire. Potatoes have of late been grown in large
quantities in some of the E. cos. for the London market;
and they form an important article of food in all parts of the
country. The practice of taking two white crops in succession has been almost wholly abandoned in the Lowlands.

Dairy husbandry is mostly pursued in the shires of Ayr
and Renfrew, in the former of which Dunlop cheese is
made; but it is also introduced into Wigtown and other
counties. Cows of the genuine Ayrshire breed are admirable milkers, and the average quantity of butter produced
by each has been estimated at upwards of 250 lbs. a year.
The Galloway, Fifeshire, and Highland breeds are the best
for fattening, and yield, especially the first and last excellent beef. The Galloway cattle are mostly sent up half fed,
to be fattened in Norfolk and Suffolk for the London markets. Three principal breeds of sheep are reared in Scotland: the dun-faced, or Scandinavian breed, said to have
been imported into Scotland from Denmark, of which a few
are found in the counties N. of the frith of Tay; the black-
faced, or heath breed, very widely diffused, and very hardy;
and the Cheviots, the famous breed native to the Cheviot
hills. The latter are now found to thrive in districts that
were formerly supposed to be suitable only for the black-
faced breed, and have already, to a considerable extent superseded the latter; the carcase and fleece being both much
more valuable.

Quantity and Value of Agricultural Produce.—These are
subjects as to which there is but little information on which
it would be safe to place much reliance. In the statistical
tables in the *General Report of Scotland* (III. Append. p.
5), the arable land is estimated at 5,043,450 English acres :
of these the proportion in grass is estimated at 2,489,725;
leaving 2,553,725 in tillage, which is supposed to be distributed as follows :

	Acres.		Acres.
Wheat	140,193	Potatoes	80,000
Barley	280,193	Turnips	407,125
Oats	1,280,352	Flax	16,500
Rye	500	Gardens	32,000
Beans and Peas	118,000	Fallow	218,050

But a large extent of waste land has been brought under cultivation during the last 20 years; and we are also satisfied, from the greatly increased consumption of wheaten bread in Scotland, and other circumstances, that the quantity of land assigned to the growth of wheat has increased both absolutely and relatively. In our view of the matter, the distribution of the land in tillage would be more correct, were it made as follows:

	Acres.		Acres.
Wheat	280,000	Turnips	350,000
Barley	280,023	Flax	15,000
Oats	1,275,000	Gardens	32,000
Beans and Peas	100,000	Fallows	150,000
Potatoes	120,000	Total	2,253,000

Assuming this distribution to be nearly correct, the quantity and value of the crops may be estimated as follows:

Crops.	Acres.	Produce per Acre.	Total Produce.	Price per Quarter.	Value.
				s.	L.
Wheat	220,000	3 qrs.	660,000	50	1,650,000
Barley	280,000	3½	980,000	30	1,470,000
Oats	1,275,000	4½	5,737,500	25	7,171,875
Beans and Peas	100,000				
Potatoes	120,000	5l. 5s.			5,220,000
Turnips	350,000	8l.			128,00
Flax	16,000				
Gardens	32,000	13l.			416,000
Total					13,355,875

The average value per acre of the pasture land of Scotland is estimated in the General Report at £3, which we incline to think is not far from the mark: and on this hypothesis, the produce of 2,480,795 acres of pasture may be taken at about £5,000,000 a year. But to this has to be added the produce of about 14,000,000 acres of mountain pasture, wood-land, and waste land; which, taken at an average value of 2s. an acre, is £1,400,000. Hence the total annual value of the land produced in Scotland, will be :

Value of crops and gardens	£13,355,875
—pasture land	5,800,000
—uncultivated land and wood	1,400,000
Total	19,755,875

We believe, for reasons similar to those stated in reference to England (L. 842), that the rent of land in Scotland does not differ very materially at this moment from the rental of 1810, which amounted to £4,851,404. There have certainly been considerable variations in particular districts ; but from all that we have been able to learn, we incline to think that the fall in some quarters has been fully balanced by an equal rise in others ; and that, on the whole, the aggregate rent of the country at the two periods is pretty nearly equal.

It is difficult to decide as to the portion of the gross rent which should be set apart as the rent of the 14,000,000 acres of uncultivated land ; but there are pretty good grounds for thinking that it does not exceed £850,000, leaving £4,000,000 as the nett rent of the arable portion, being at the rate of about 16s. an acre. Though there are some considerable exceptions, there can be no manner of doubt that, speaking generally, the arable land of England is superior to that of Scotland ; but in consequence of the greater skill and economy of the farmers in the latter, and of the advantage they enjoy in the possession of leases, and the absence of tithes and poor-rates, they are able to pay decidedly higher rents for lands of equal fertility. The rent of corn land in Scotland varies from 7s. to £3 an acre, and occasionally even amounts to £5 and £6. The best pasture land rarely fetches more than £3 per acre ; and that which is of a medium quality may vary from 10s. to 25s.

Rise of rent.—Rent has increased much more rapidly in Scotland than in England. This is ascribable partly to the extremely backward and depressed state of Scotch agriculture till after the peace of Paris, in 1763, and partly to the extraordinary advance it has made since the close of the American war. The entire rental of the kingdom is not supposed to have exceeded £1,000,000 or £1,200,000 in 1770. In 1764 it is believed to have rather exceeded £2,000,000 ; and since then it has been more than doubled ! So rapid an increase of rent is probably unmatched in any old settled country, and indicates an astonishing degree of improvement.

Condition of Occupiers and Labourers.—We are happy also to have to state, that the wealth of the farmers, and the comfort and well-being of the agricultural labourers,

have increased in quite as great a proportion as the rents of the landlords. We have already noticed the extraordinary improvement that has taken place in farm-houses and offices since the close of the American war, and especially during the present century ; and the same improvement is everywhere visible in farming stock and implements ; in the furniture and other accommodations of the farm-houses ; and in the dress and mode of life of their occupants. We have, indeed, no hesitation in affirming, that no old settled country, of which we have any authentic accounts, ever made half the progress in civilization and the accumulation of wealth, that Scotland has done since 1763, and especially since 1787.

Some very great improvements have been introduced into agriculture within these few years, and are now rapidly spreading over the country. Among these may be specified furrow-draining, subsoil ploughing, and the use of bone manure. These improvements have already changed the face of various extensive districts. Farms that were formerly wet, late, and suitable only for oats, are now, by the aid of furrow-draining, made thoroughly dry, early, and suitable for any species of crop. The introduction of bone-manure has occasioned a great extension of the turnip culture. The facilities afforded by steam navigation for the conveyance of fat cattle and sheep to the great markets of London and Liverpool, as well as to those of Edinburgh and Glasgow, have also been of vast importance, and have enabled the remotest districts to come into successful competition with those that are most favourably situated. In consequence, agriculture is at this moment in a rapid state of advancement ; and the fair presumption seems to be that, great as has been its improvement during the last half century, it will be equal or greater during that on which we are now entering.

The only thing that seems at all likely to defeat this presumption, or to retard the future progress of agriculture, is the fact of the Reform Act having conferred the elective franchise on all occupiers of land worth £50 a year and upwards. This has been in every respect a most pernicious, ill-advised measure. Formerly the landlords rarely interfered as to the politics of their tenants ; and provided they paid them their rents, and managed their lands according to the stipulations in their leases, they might be of any political or religious party they pleased. But now it is altogether different. The landlords, desirous, like other people, of extending their political influence, endeavoured to control, or rather command, the suffrages of their tenants, and to multiply the dependent voters on their estates. In furtherance of these objects, they have not scrupled, in many instances, to resort to intimidation, and to adopt vindictive measures against such of their tenants as have voted contrary to their wishes. This, however, though the most prominent at the time, is but the least evil resulting from the new state of things. It has already led in many instances to a change in the mode of letting land ; and there is but too much reason to fear that it may, in the end, subvert that system of giving leases for 19 or 20 years certain that has been a main cause of the astonishing improvement of agriculture. It has also occasioned, in many instances, a subdivision of farms for the mere purpose of creating votes ; and these cannot, indeed, be a question that, however well intended, the conferring the elective franchise on the tenants has been one of the greatest blows ever struck at their independence, and at the prosperity of agriculture. For is there anything in this but what might have been, and in fact was, anticipated from the measure. Tenants, as such, whether they held farms worth £50 or £500 a year, nor about the very last description of persons to whom the franchise should be conferred. Very many of them are indebted to, and dependent, to a greater or less extent, on their landlords ; and the few who are independent are so because they have accumulated property, and would, in consequence, have been entitled to the franchise, had it been conferred, as it should have been, on those only who possessed a certain amount of realised property. If that be, as it unquestionably is, the best system of voting that keeps the greatest number of independent electors to the poll, and keeps back the greatest number of those that are dependent, the giving the franchise to the tenants and occupiers of land must be about the very worst system, for they are, of all classes, that which is most dependent, and the most at the mercy of others.

The condition of the agricultural labourer has, also, as already stated, been vastly improved. With the exception of those districts in the Highlands and isles, luckily few in number, into which improvements have not yet made their way, the cottages of the peasantry have been mostly rebuilt during the last half century ; and though still, in most instances, without the rustic beauty and neatness that so frequently distinguish English cottages, they are far from uncomfortable. In most parts of the country each of the farm-labourers as are married, and have families, receive the greater part of the wages in specific quantities of farm

SCOTLAND.

h do not vary with the variations of price; so
e not so well off as the manufacturing work-
trade is brisk and prices low, neither are they
ffer like the others, when there is little de-
>ur, and prices are high: on the whole, they
? generally, be said to be in decidedly comfort-
.ances. The unmarried servants frequently
arm-house. They are almost all excellent
all of them are able to read and write: and
i unfrequently emerge from obscurity, and at-
:tion.*

ires.—For a lengthened period after the union
i, Scotland made little or no progress in manu-
d It was not till after the peace of Paris, in
e public enterprise began to be turned into this
il, and that a rapid extension took place of most
itry. A considerable depression ensued towards
d after the termination of the American war.
iot of any very lengthened duration; and the
of the cotton trade having been laid about this
ufactures have continued, from 1787 down to
time, progressively to gain ground in Scotland,
een prosecuted with equal skill, industry, and
ccording to the report of Mr. Stuart, factory in-
839, there were in that year the following fac-
otland:

Mills or Factories.	Wrought by Steam.	Wrought by Water.	Persons employed.
	148	75	55,575
153	37	116	40,582
251	108	91	17,897
6	5	.	762
otal	676	395	280	114,861

Of whom 86,387 were upwards of 18 years of age.

ere were in Scotland 17,721 power-looms, of
31 were employed in weaving cotton fabrics.
been a very material increase of power looms in
il; and exclusive of them there were, in the 8.
Scotland, in 1838, about 51,060 hand-looms for
rica, employing, as was estimated, about half that
families. The cotton manufacture, which prin-
itres in Glasgow and Paisley (which see), is of
rely recent introduction, the first steam-engine
n factory having been constructed so late as 1792.
roollen manufacture has been of long standing,
formerly much more widely diffused than at pre-
ing been, in fact, with that of linen, a domestic
ire, and pursued in every cottage. It was the
practice of the peasantry, and occupiers of land,
t home, the greater part of their own wool, as a
employment, and to send the yarn to be made into
th in the nearest village.

s still a class in Scotland called *customer weavers*,
over the country, but now principally confined to
anda, employed by private families to weave yarn
e fabrics for domestic use. Most of these are also
nl labourers, weaving only in the intervals of their
avocations; they earn from 1s. to 2s. 6d. a day,
ugher than 1s. 9d. But except these, and persons
watering-places, and on parts of the E. coast, where
ige in fishing or boat-letting for a part of the year,
bulk of the weavers of Scotland subsist entirely
on, and engage in no other pursuits. (*Symons in
m Weaver's Rep.*, p. 5.)

sent, and for some time past, this class has been in
ressed circumstances. Owing to the facility with
he business of weaving may be learned, and the
dependence it confers on the weaver, it has always
ivourite employment; and, consequently, except in
f great prosperity, the wages of weavers have gen-
en rather low. Of late years, however, the intro-
of power-looms has gone far to supersede to a great
he business of the regular hand loom weaver, es-
of those engaged in the manufacture of cottons;
i fair presumption seems to be, that in no very
ied period the business of the hand-loom weavers
totally annihilated. But though there can be no
iat, in a public point of view, this change will be
ve of great advantage, it involves, in the meantime,
s of hand-loom weavers in the greatest difficulties;
point of fact, nine tenths of that manufacturing dis-
which we have recently (1841) heard so much has
ed from hand-loom weavers thrown out of employ-
the competition of power-looms, or forced to labour
merest pittance. That such persons are proper ob-
public sympathy, none can doubt; and everything

*rder who may wish to become more particularly acquainted with
rdinary changes that have taken place in the agriculture and mode
he people of Scotland since 1760, would do well to consult *Robert-
al Recollections.* It is at once an authentic, interesting, and in-
work.

that is practicable should be done to facilitate their employment in other businesses, or their emigration to the colonies. Till such time as this transference has been effected,
there will be the same unvarying, and generally unfounded
tale of manufacturing distress.

It appears from the statements given in Burns' account
of the cotton manufacture for 1840, that the total quantity
of cotton spun in England during that year amounted to
308,804,959 lbs., and that spun in Scotland to 37,878,892
lbs. Now, taking the entire value of the manufacture at
£35,000,000, and supposing it to be proportioned in both
countries to the quantity of yarn spun in each, that would
give about £3,500,000 for the value of the portion belonging
to Scotland. We believe, however, that the value of the cotton goods annually produced in Scotland may be estimated
at about £5,000,000; for a large proportion of the Scotch
manufactured articles consists of the finer descriptions of
muslins and other superior and costly fabrics, which makes
their aggregate value exceed what might be inferred from
the comparative amount of yarn produced in Scotland.

Compared with the woollen manufacture of England,
that of Scotland is inconsiderable. Flannels, blankets,
shawls, plaids, stockings and stocking yarn, tartans, carpets,
druggets, &c., are produced to a considerable extent at Galashiels, Hawick, Jedburgh, and Aberdeen. Some of the finer
descriptions of cloth are made at Aberdeen and in its vicinity, and some of its woollen mills and factories are on a
large scale. Kilmarnock is the seat of a very extensive and
flourishing carpet manufacture. The power-loom having
hitherto been but little adopted in the woollen manufacture,
the weavers employed in this department get good wages,
are well clothed and lodged, and in all respects exhibit a
marked contrast with the hand-loom weavers engaged in
the cotton manufacture.

The linen manufacture of Scotland is of very considerable value and importance: Dundee is its chief seat; and
the statements previously given (*see* art. DUNDEE i. 786)
show that its increase since 1811 has been quite extraordinary. Osnaburghs, sheeting, cotton bagging, sail-cloth, dowlas, and other coarse goods, are the articles principally made
in Dundee and in Kirkcaldy, Arbroath, Forfar, Montrose,
Aberdeen, and other seats of the manufacture in the E. of
Scotland. The finer descriptions of linen fabrics, as dama-
asks, diapers, shirting, &c., are principally produced in
Dunfermline, and its immediate vicinity. Power-looms have
not hitherto been extensively introduced into this department, but they are gradually gaining ground. At present it
employs about 17,000 hand-looms in summer, and from
22,000 to 23,000 in winter. The linen weavers occupy an
intermediate position between the woollen weavers on the
one hand, and the cotton on the other. The silk manufacture is of little importance: its chief seat is Paisley (which
see, *antè*, p. 449.)

The iron-works at Carron, near Falkirk, established in
1760, were for a lengthened period the most extensive in
the kingdom; but they are now far surpassed by those of
Gartsherrie, Calder, Clyde, &c., in Scotland, and by many
in England. A good deal of Scotch ironmongery, comprising anchors, bolts, axles, mill and engine-work, &c., is exported to the colonies and foreign countries. The manufacture of machinery is an important branch of industry.
Coach-making is carried on in all the large towns, and ship-
building in many of the ports. Glass wares, chemical products, soap, candles, and starch, are among the other principal manufactured goods.

ACCOUNT of the Quantities of British Spirits entered for
Consumption in Scotland, with the Rates of Duty on the
same, and the Amount of Duty in each Year since 1820.

Years.	Rate per Gallon.	No. of Imp. Gallons.	Nett Revenue.
			L. s. d.
1821	5s. 6d. per wine gall.	2,229,435	727,680 19 7
1822	Do. do.	2,079,556	691,136 6 6
1823	Do., after Oct. 20, 2s.	2,229,728	586,854 17 8
1824	Do.	4,350,301	520,624 12 4
1825	Do.	5,981,650	642,945 11 1
1826	2s. 10d. per imp. gall.	5,988,789	662,383 4 0
1827	Do.	4,152,199	672,441 6 6
1828	Do.	5,716,190	809,559 6 7
1829	Do.	5,777,330	818,448 0 0
1830	3s. 10d., &c., and 3s. 4d. do.	6,007,631	839,968 6 0
1831	3s. 4d. do.	5,700,689	950,041 4 3
1832	Do.	5,407,097	901,142 16 8
1833	Do.	5,948,599	985,051 3 3
1834	Do.	6,045,042	1,007,707 3 4
1835	Do.	6,013,932	1,002,322 0 0
1836	Do.	6,620,426	1,103,471 0 0
1837	Do.	6,124,035	
1838	Do.	6,259,711	
1839	Do.	6,188,582	1,031,213 0 0
1840	Do. and 3s. 6d.	6,180,136	1,00-,049 3 4

The favourite beverage of the people of Scotland has, for
a lengthened period, been whiskey, a spirit generally distilled
from malt or raw grain. Owing to the excess of the duties

with which this spirit has occasionally been charged. Its smuggling has sometimes been carried to a great extent; but since 1826, when the duties were reduced from 5s. 10d. to 9s. 10d. and 3s. 4d. per imperial gallon, clandestine distillation and smuggling have been comparatively rare.

Commerce.—Having little industry, and being thinly peopled, Scotland had formerly a very limited foreign trade. The exports consisted of wool, skins, hides, and other raw materials, exchanged for corn, wines, spices, &c. Even so late as the era of Cromwell her mercantile marine comprised only 93 vessels, of the aggregate burden of 2734 tons, and 18 barks. During the reign of Charles II. the trade of the country, especially that with Holland and the countries round the Baltic, began to increase. It was not, however, till after the completion of the union, in 1707, when the trade to the American and W. Indian colonies was, for the first time, open to the enterprise and activity of the Scotch, that the commercial energies of the nation began to be awakened. But for a while the merchants of Glasgow, who first embarked in the trade to America, carried it on by means of vessels belonging to English ports; and it was not till 1718 that a ship built in Scotland (in the Clyde), the property of Scotch owners, sailed for the American colonies.

(*See* art. GLASGOW, i. 995.) The establishment and rapid extension of manufactures in Scotland, since 1763, has necessarily occasioned a corresponding increase of commerce, and the mercantile marine of Scotland is now very considerable indeed. In 1839 it comprised 3143 vessels of the aggregate burden of 300,636 tons. Such, however, has been its increase in the interval, that on the 31st of December, 1840, it comprised 3479 vessels, of the aggregate burden of 429,204 tons, manned by 28,429 seamen! The commerce of Scotland too is greater, in fact, than appears from the Customs' returns; inasmuch as a good many articles of foreign produce imported at second hand through the English ports, and articles of native produce exported in the same way, do not appear in the list of imports or exports. At present the principal articles of export consist of cotton and linen stuffs, cotton and linen yarn; wool, iron, and hardware all goods, coals, spirits, and beer; black cattle; herrings, salmon, and other salted and fresh fish; stationery, &c. The great articles of import consist of tea, sugar, coffee, and other colonial products; raw cotton, flax and hemp, tobacco, raw silk, wine, dye stuffs, &c. The great emporiums are Glasgow (inc. Greenock and Port Glasgow), Leith, Aberdeen, Dundee, and Montrose. We subjoin an

Account of the Gross and Nett Customs' Duties collected at the different Scotch Ports in 1839 and 1840.

Ports.	Gross Receipts.						Nett Receipts.					
	1839.			1840.			1839.			1840.		
	L.	s.	d.	L.	s.	d.	L.	s.	d.	L.	s.	d.
Aberdeen	71,831	14	2	80,018	3	9	63,035	5	1	71,163	18	1
Ayr	1,006	4	5	1,897	1	7				853	7	11
Banff	909	18	11	1,337	7	8	1,634	5	8			
Borrowstoness	4,894	13	7	4,879	14	8	3,956	11	9	3,363	16	1
Campbeltown	696	11	2	483	12	10						
Dumfries	9,986	10	11	9,107	4	3	5,861	9	5	5,465	8	7
Dundee	92,502	14	2	63,346	8	9	83,782	15	5	54,454	8	2
Glasgow	466,974	19	2	479,363	19	9	446,939	17	11	453,396	19	10
Grangemouth	38,230	19	1	31,216	9	11	35,349	11	5	26,539	4	4
Greenock	315,064	4	6	341,647	19	2	303,637	3	7	369,965	10	10
Inverness	6,965	13	1	6,171	3	3	1,864	9	10	2,639	13	11
Irvine	3,794	18	5	3,592	9	3	1,063	8	4	1,527	7	3
Kirkaldy	6,594	12	0	4,307	0	0	2,738	4	2	71	5	2
Kirkwall	1,190	17	4	671	2	11						
Leith	573,685	13	7	602,999	6	3	455,829	17	8	465,941	4	10
Lerwick	603	11	7	707	16	6						
Montrose	33,135	5	11	33,483	16	10	27,446	6	11	27,797	6	7
Perth, from 5th July, 1840				10,766	9	7				10,353	5	5
Port Glasgow	68,045	13	10	84,369	7	3	62,681	0	11	79,462	9	2
Stornoway	925	2	7	6446	2	5						
Stranraer	300	14	8	587	4	3						
Wick	2,082	11	7	1,140	19	1	263	19	9			
Totals	1,699,131	17	8	1,755,781	12	11	1,486,366	17	10	1,530,665	18	2

In proof of the extraordinary progress made by Scotland, we may state that at the epoch of the union, in 1707, the revenue amounted to only £110,694. In 1788 it had increased to £1,099,148; in 1813, when the income-tax was at its height, it amounted to £4,204,097; and in 1839, notwithstanding the repeal of the income-tax, and of a great number of other taxes, the gross revenue of Scotland amounted to £5,254,694, and the nett revenue to £4,959,460, a rate of increase wholly unexampled in, either of the other divisions of the United Kingdom. We subjoin an

Account of the Gross and Nett Produce of the Revenue of Scotland in 1840.

	Gross Produce.	Nett Produce.
	L.	L.
Customs	1,755,782	1,735,279*
Excise	2,559,968	2,396,491
Stamps	559,850	555,402
Taxes	249,822	248,551
Postoffice	26,300	23,699
Totals	5,151,777	4,959,528

Currency.—The currency of Scotland has, for a lengthened period, principally consisted of the notes of the different banking companies. These, for the most part, are joint stock associations, with numerous bodies of partners, and have been managed with great skill and discretion.

Very few bankruptcies have occurred among the Scotch banks; and they have, no doubt, contributed materially to forward the improvement of the country by the facilities they have afforded to industrious and deserving individuals of obtaining loans: and still more by the practice, which has long been acted upon, of taking very small sums in deposite, and allowing interest upon them at about 1 per cent. below the market rate at the time. This has brought, as it were, a number of substantial and well-organized savings' banks within reach of all classes; and by furnishing every facility for the safe and profitable custody of the smallest and largest sums, has powerfully stimulated the desire to save and amass. The deposits in the Scotch banks are sup-

* Deducting drawbacks and repayments.

posed to amount, at this moment, to about 27 millions sterling; and we believe it may safely be affirmed, that, but for the facilities that have been afforded for the accumulation of the smaller class of sums, at least £15,000,000 of this amount never would have existed. The amount of the notes of the Scotch banks in circulation may be taken, at an average, at about 3½ millions.

Roads.—With the exception of the military roads, constructed in the Highlands after the suppression of the rebellion, in 1745, the roads of Scotland were, speaking generally, in the most execrable state down to the American war. But such, and so great has been the improvement in the interval, that they are now quite equal to the best roads in England, and are not, indeed, surpassed by any in Europe. They are laid down on the most approved principles; and, notwithstanding the rugged nature of the country, it is but seldom that horses in a carriage may not be driven along at a smart trot. The facility with which excellent materials for their construction may almost every where be obtained has materially contributed to forward their formation. The roads within what is called the Highland districts have been partly constructed by means of advances from government, and the public money has very rarely been so profitably expended.

Within the present century several railways have been projected, some of which are already completed, and others in a state of great forwardness. They are mostly in the manufacturing districts, of which Glasgow and Dundee are the centres. The principal, however, will be that between Edinburgh and Glasgow, about 46 m. in length. It was begun in 1838, and is expected to be completed in the course of the present year (1841), or early in the next, at an expense of about £1,200,000.

Various projects are now on foot for connecting Edinburgh or Glasgow with some of the great lines of railway trading from London to the N. of England; and the presumption is that this junction will be effected at no distant period. In the mean time, however, an individual may, by means of the railway to Preston or Lancaster, and the small reaches from the latter to Glasgow or Edinburgh, or the steamers from Fleetwood, near Preston, to Ardrossan, reach Glasgow or Edinburgh from the metropolis in about twenty-four hours! A journey which, so late as the era of the Americans war, took about a week for its completion!

Canals.—Of the Scotch canals, the most important by far is that called the Great canal, uniting the Friths of Forth and Clyde, and, consequently, forming an easy water communication between the E. and W. coasts of the island. Including its branch to port Dundas, in the vicinity of Glasgow, it is about 38 m. in length; the medium width at the surface is 36 ft., and at the bottom 27 ft.; average depth, from 9 ft. to 10 ft.; summit-level, 156 ft. above the sea; it has 39 locks. This important work was begun in 1768, but was not finally completed till 1790. It has been as profitable to the shareholders as it is advantageous to the public; the dividend on the original stock having been latterly about 28 per cent.

The Union canal joins the Forth and Clyde canal near Falkirk, and stretches thence to Edinburgh, being about 31½ m. in length. It was completed in 1822, but has been, in all respects, a most unprofitable undertaking. Hitherto, the proprietors have not received any dividend, and their prospects, we understand, are but little, if anything improved.

There are other canals in the vicinity of Glasgow: the Crinan canal stretches across the Mull of Cantire; and there is also a canal in Aberdeenshire. But the greatest work of this class in Scotland, or, perhaps, in the empire, is the Caledonian canal. It stretches quite across the island, through the centre of the Highlands, N.E. and S.W., from the Beauly frith on the E. coast to loch Linnhe on the W. It is chiefly formed by the chain of lakes, including loch Ness, loch Oich, and loch Lochy, which occupy the bottom of the remarkable glen or depression through which it is carried. Its total length, including the lakes, is 60½ m.; but the artificial or excavated part is only 21½ m. At the summit it is only 94½ ft. above the level of the W. ocean. It is mostly constructed upon a very grand scale, being intended to be 20 ft. deep, 50 ft. wide at bottom, and 122 ft. at top; the locks are 20 ft. deep, 172 in length, and 40 ft. in width; and had it been wholly executed as was originally intended, frigates of 32 guns, and merchant ships of 60 tons' burden, might have passed through it. It was opened in 1822, being executed entirely at the expense of government, from the designs and under the superintendence of Thomas Telford, Esq. The entire cost has exceeded a million sterling. There can, however, be no doubt that it was projected without due consideration, and may be regarded as about the least advantageous public work that has been undertaken for a lengthened period. During the year 1836-37, the total revenue of the canal, arising from tonnage dues and all other resources, amounted to only £3279, while the ordinary expenditure during the same year amounted to £3079; and it has increased little, if anything, in the interum. But this is not all. Owing to a wish to lessen the expense, and to hasten the opening of the canal, parts of it were not excavated to nearly their proper depth, while others were executed in a hurried and inefficient manner. Hence the canal does not really admit vessels of above 250 or 300 tons burden; and owing to the want of steam-tugs on the lakes, these are frequently delayed in making their passage across for a lengthened period. During 1837 and 1838 the works sustained considerable damage; and so large a sum will now be required for the repair and for the purchase of steam-tugs, &c., that it has been gravely debated whether it would not be better only to abandon and fill up the canal! There is naturally, however, a great disinclination to desert a work which, how inexpedient soever originally, has executed at an enormous expense; and various schemes have been suggested for relieving the public from the expense of keeping it up without involving its destruction. Among others, it has been proposed to assign it to a joint company, on their agreeing to complete the works, keep them in repair, and an act authorising such transfer passed in 1840. Hitherto, however, it has not been able to dispose of it in this way; so that, at this moment (December, 1841), it is not easy to conjecture what is come of this undertaking.

Constitution of Scotland has been, from the earliest, what is called a limited monarchy. Originally the great, or great council of the nation, consisted of the nobles, and the principal ecclesiastics. Burgesses, representatives for the towns, were admitted into the parliament by Robert Bruce, in 1326; and in 1427 lesser barons, or freeholders in the different counties, authorised to send representatives: but so little was the privilege valued, that it was hardly exercised for 160 or till the reign of James VI., when the freeholders compelled to send representatives.

and, however, derived little or no benefit from her tenant. The nobility and clergy sat and voted in the chamber with the representatives of the lesser barons

and of the towns; so that, even if the latter had been more powerful and independent than they were, they could have made no effectual opposition to any measure patronised by the nobility and clergy. In point of fact, however, the representatives of the counties were mere nominees of the great lords; and the towns having neither wealth, population or importance, the representatives were necessarily as impotent as themselves. The nobility and clergy were, in truth, for a lengthened period, everything, and the people nothing.

Even had it been, in other respects, better constituted, the institution of the Lords of the Articles would have rendered the parliament of Scotland good for nothing as a check on the sovereign. This was a committee, consisting of a few members, chosen either directly or indirectly by the crown, to which all matters to be brought before parliament had, in the first instance, to be referred; and which had power to reject such as it disapproved of, and to modify and alter the others in any way it thought proper. This committee had, therefore, a negative before debate; and the whole duty of parliament was confined to meeting for a day or tow at the end of the session, to confirm the proceedings of the Lords of the Articles! With such an instrument at his command, we need not wonder at the preponderating influence possessed by the sovereigns in the Scotch parliaments; and had their ability to carry laws into effect been, in any respect, equal to the facility with which they could get them passed, the kings of Scotland would have been the most, instead of the least, powerful of European princes.

The committee of the Lords of the Articles was suppressed at the revolution; but owing to the defects in the mode of choosing representatives, the constitution of parliament was but little improved by its suppression; and down to the passing of the Reform Act, in 1832, Scotland had the shadow merely without the substance—the disadvantages without any of the advantages—of a representative government. Happily, however, its representative system is now placed nearly on the same footing as that of England. (For further details, *see* vol. i p. 464 and p. 651.)

According to the articles of union in 1707, the peerage of Scotland is represented in the House of Lords of the United Kingdom by 16 peers, chosen by the whole body of Scotch peers at the commencement of each parliament: it was then also arranged that the counties of Scotland should be represented in the House of Commons by 30, and the boroughs by 15 members. This arrangement was continued till 1832, when the borough representation received an addition of eight members, the members for the counties continuing as before.

Courts of Law.—The court of sessions, which was constituted by an act of the Scottish parliament in 1537, is the highest civil court of Scotland, having jurisdiction in all civil questions of whatever nature. It was intended to supply the place of the previously existing courts, and more especially of a judicial committee of parliament called the "lords of session," whence the name of the court and the titles of the judges. Originally it consisted of seven laymen and eight churchmen, including the president. In 1640, however, an act was passed, providing for the exclusion of churchmen from the court; and, though repealed in 1661, the principle laid down in it has ever since been acted upon. Other important improvements were introduced at different periods, particularly after the revolution, when right of appeal from the court to parliament was, for the first time recognised. At the union power was given to all individuals who considered themselves aggrieved by judgments of the court of sessions to appeal to the house of lords; and it is a curious fact, that, at this moment, and for a lengthened period, the principal judicial business of the House of Lords has consisted in hearing and deciding Scotch appeals. Originally, and down to 1808, the whole 15 judges sat together in one court; but in that year an act was passed dividing the court into two chambers, the lord president presiding in the first division of seven judges, and the lord justice-clerk in the second, of six; the two remaining judges trying cases in the first instance, or, as it is technically termed, sitting as lords-ordinary. Since then the number of judges has been reduced to 13; four belonging to each of the divisions, and five acting as lords-ordinary, or sitting as single judges. The judges were at first chosen by the Scotch parliament; but since 1554 they have been appointed by the crown. They are indifferently styled lords of session, or senators of the college of justice, which last embraces the whole body of barristers (advocates), and attorneys or solicitors who practise before the court. They must be 25 years of age; and. by the treaty of union, no person can be named to the office unless he have served as an advocate or principal clerk of session for five years, or as a writer to the signet for ten. The salaries of the ordinary judges have recently been raised to £3000 a year each; those of the lord justice clerk, and lord president, being respectively, £4000 and £4500.

SCOTLAND.

At its outset the court of session was intended to serve as a standing or perpetual jury for the trial of cases; the introduction of petty juries into the trial of civil cases in Scotland being only of very recent date, as well as of limited application. It was, in fact, unknown till 1815, when a special or jury court was instituted, for the trial of cases involving questions as to the value of property, the amount of damages, or the determination of some fact. But in 1830 this court was suppressed; and the court of session now avails itself of the assistance of petty juries in the trial of the above description of cases.

The high court of justiciary was remodelled, and placed nearly on its present footing, in 1672. It consists of five judges of the court of session, specially commissioned by the sovereign, together with the justice-general and justice-clerk; the former, or, in his absence, the latter, being president. In 1836, the office of lord justice-general was conjoined with that of lord-president of the court of session. The jurisdiction of this court extends to all criminal cases, except those of high treason, which are tried by a special commission, in the English form, on the finding of a grand jury, which is not used in other cases in Scotland. The judgments of the court of justiciary are final, no appeal lying from them to the House of Lords. Circuit or assize courts are held twice a year, by the judges of this court, in the principal towns of Scotland, two judges usually going on each circuit; and an additional circuit-court is held at Glasgow during the Christmas holidays. The circuit-courts have power to decide in appeals from inferior courts, where the subject in dispute does not exceed £25. Cases brought before the justiciary court are tried by petty juries of 15 persons, who decide by plurality of votes, not being compelled, as in England, to give unanimous verdicts. It is not going too far to say, that, down to a very recent period, this was, in as far as respects political cases, one of the most corrupt, and worthless tribunals in Europe. Owing, as has been previously stated (see vol. i. 466.), to the mode in which juries were selected, it was always in the power of the lord advocate, or public prosecutor, to get a jury appointed favourable to his own views; and the judges, having been appointed by the crown, and looking to it, most probably, for farther advancement for themselves or their families, were, with few exceptions, obsequious tools. Hence, in Scotland, to be prosecuted for a political offence was, for a lengthened period, nearly equivalent to being condemned. Luckily, however, this disgraceful state of things has been thoroughly reformed. Juries in Scotland are now fairly selected; the accused has the same right of peremptory challenge as in England; so that, however disposed, the judges can no longer dictate verdicts. The old court of exchequer, commission of teinds, admiralty and consistory courts, are now merged in the court of session.

The inferior courts of law are those of the boroughs, justices of the peace, and sheriffs. The first are called "bailie" courts, from being presided over by a bailie or alderman, with, in some cases, the assistance of a legal assessor. Their civil jurisdiction within the borough depends on circumstances, being sometimes nearly equivalent to that of sheriffs in counties; but their criminal jurisdiction extends only to petty riots and common police offences. The justices of peace decide without appeal in actions where the demand does not exceed £5 besides costs. They commit criminals and hold petty sessions, at which two are a quorum, and quarter sessions; but have in no instance the power of transportation.

The sheriff courts are very important, and transact most of the county business. Each county has a principal sheriff, called a sheriff-depute, from his being deputed or appointed by the crown; who, in addition to duties similar to those devolving upon English sheriffs, has a very extensive civil jurisdiction. He holds office ad vitam aut culpam, his salary varying from £300 to £600 a year, according to the supposed onerousness of his duties. In the counties of Edinburgh and Lanark, the residence of the sheriff-depute is enforced; but in the other counties he is rarely resident, his presence not being necessary, except on particular occasions. Sheriff-deputes are, in fact, usually practising lawyers in Edinburgh, and the ordinary business of the county is devolved on the sheriff substitutes, or deputies of the principal, who are always resident. In extensive counties there are usually several sheriff-substitutes. This very useful class of judges must be chosen from advocates, writers to the signet, solicitors of the supreme courts, or solicitors of three years' standing before a sheriff court: and, though nominated by the sheriff depute, they cannot be displaced without the concurrence of the lord president and lord justice-clerk. At present their salaries, which were raised in 1840, vary from £300 to £550 a year, exclusive of fees. The sheriff, or his substitute, holds small-debt courts for the decision of questions of debts and costs to the amount of £8 4s. 8d., in which the pleadings are all vivâ voce, the expense small, not exceeding 2s. 6d. or 3s., and the judgment final. In his ordinary courts, however, the authority of which extends to all personal actions without limit of amount, the pleadings are mostly in writing. Until a comparatively late period the sheriff exercised a criminal jurisdiction, in some cases to capital punishment; but his powers, in this respect, are now greatly abridged. He still occasionally tries criminal cases with a jury, but the sentence may be appealed from to the court of justiciary. No sentence, except for petty offences, involving fine, imprisonment, or, at most, banishment from the county or borough, can be pronounced by any legal authority in Scotland without a jury; nor can any person be now imprisoned for any debt under £8 4s. 8d.

Religious Establishments.—The Roman Catholic religion and the jurisdiction of the pope, were abolished in Scotland in 1560: a confession of faith, on Calvinistic principles, drawn up by the celebrated John Knox, was then agreed to, and the Protestant religion was established in an act of the legislature. Knox, having studied under Cardan at Geneva, introduced the Genevese or Presbyterian form of church government; but, though organised, it did not receive the sanction of the legislature till 1592. After the accession of James VI. to the throne of England, he endeavoured, notwithstanding the strenuous opposition of the great bulk of the nation, to re-establish episcopacy; and a struggle was carried on between the abettors of episcopacy and presbytery, who alternately prevailed, according as the court or popular party happened to have the ascendancy, till the revolution, when presbytery was definitely established.

Some, however, of the parishes are collegiate, or have two clergymen; and latterly some of the more extensive parishes have been divided, and assistant or quoad sacra ministers have been appointed to them. These are called quoad sacra or quoad spiritualia ministers, because they are not entitled to participate in the civil endowments belonging to the parish, and are wholly supported by a sum granted annually by the sovereign. These quoad sacra clergymen, of whom there are 41, were admitted, by an act of the general assembly of 1833, members of presbyteries and other church courts. But, though acquiesced in, the legality of this act is very questionable.

At present, and since the reign of Queen Anne, (1712,) the privilege of appointing clergymen to parishes has been vested in the crown or in private patrons, with the proviso that they must be selected from among those who have gone through the course of study prescribed by the church, and been examined and licensed as preachers by some presbytery. The right of patronage has long, however, been exceedingly unpopular. Its enforcement, in despite of public opinion, occasioned the great secession from the church in 1741; and latterly it would seem to have become more unpopular than ever. The general assembly, by a measure called the veto act, passed in 1834, gave the congregations belonging to parishes a right to reject a presentee, if he was not acceptable to them: but the exercise of the privilege conferred by this act has been found ineffectual by the court of session and the House of Lords; and for some time back the church of Scotland has been greatly agitated by this question. It were much to be wished that it were satisfactorily disposed of. No doubt there are enormous difficulties in the way, but they are far from being insuperable; and we are clear that either the privilege of selecting their clergymen should be given to parishes, or that they should be authorised peremptorily to reject any presentee not acceptable to them. The latter, perhaps, would be the least exceptionable mode of disposing of the question, and it might at once be effected by giving the veto act of the general assembly the force of law. This privilege is, in fact, of the very essence of presbytery. It is certainly a popular institution; and it is idle to suppose that those who are conscientiously attached to it should ever approve a system of absolute patronage. That control over the election of clergymen, for which the majority of the clergy and people of Scotland are now contending, is not only right and proper in itself, but is in keeping with the other institutions of the country. All magistrates of boroughs, members of parliament, and other functionaries, are now chosen by popular election; and we have yet to learn why a different practice should be followed in the case of clergymen and that they, whether acceptable or not, should be thrust upon the public. Such a system is sure, in the end, to destroy itself. To keep it up can serve no purpose, unless it be to lessen the utility of the church, to occasion agitation, and to add to the number of dissenters.

The Scotch church is a perfect democracy, all the members being equal, none of them having any power or pre-eminence of any kind over another. There is in each parish a parochial tribunal, called a kirk session, consisting of the minister, who is always resident, and a greater or smaller number of individuals, of whom, however, there must always be two, selected as elders. The principal duty of the elder is to superintend the affairs of the poor, and to assist in visiting the sick. The session interferes in certain cases of scandal, calls parties before it, and inflicts ecclesiastical punalties. But parties who consider themselves aggrieved, may appeal from the decision of the kirk session to the presbytery in which it is situated, the next highest tribunal in the church.

A presbytery consists of the clergymen of an indefinite number of contiguous parishes, and of an elder from each kirk session. It has cognizance of all ecclesiastical matters within its limits, examines, licenses, or rejects preachers, or candidates for the ministry, reviews the decisions of kirk sessions, &c. Originally presbyteries met every week, but now, in general, only once a month. Appeal may be made from their decisions to the synods.

A synod consists of the clergymen of an indefinite number of contiguous presbyteries, with an elder from each of the different kirk sessions. This court, which usually meets twice a year, reviews the decisions of presbyteries; but its own decisions may be reviewed by the general assembly, the highest ecclesiastical authority in the kingdom. We subjoin

A STATEMENT, exhibiting an Account of the Number of Synods, Presbyteries, Parishes, and Clergymen, belonging to the Church of Scotland:

Synods.	Presby- teries	Par. ishes.	Clergy- men.
Synod of Lothian and Tweedale .	7	120	128
Merse and Teviotdale .	8	56	66
Dumfries .	5	56	66
Galloway .	3	37	37
Glasgow and Ayr .	8	157	159
Argyle .	5	55	57
Perth and Stirling .	5	69	90
Fife .	4	67	72
Angus and Mearns .	6	81	87
Aberdeen .	8	104	109
Moray .	7	56	58
Ross .	3	27	27
Caithness and Sutherland .	3	24	25
Glenelg .	5	41	41
Orkney .	3	90	21
Shetland .	2	14	14
Total number . 16 Synods	90	1,023	1,050

The GENERAL ASSEMBLY, which consists partly of clerical and partly of lay members, chosen by the different parishes, boroughs, and universities, comprises 386 members, as follows:

Eighty Presbyteries send, ministers	218
elders	94
City of Edinburgh, a minister	2
Sixty-five other royal burghs	65
University of Edinburgh,	
University of Glasgow,	one minister or
University of St. Andrew's,	one elder each ... 5
Marischal College, Aberdeen,	
King's College, Aberdeen	
Churches in India, a minister and an elder	2
Total number of members	386

The general assembly meets annually in May, and sits 10 days; but it has power to appoint a commission, with powers equal to its own, to take up and consider any matters it may have left undecided. The assembly is honoured during its sittings with the presence of a nobleman, the representative of the sovereign, with the title of high commissioner. He is, merely, however, a state appendage, and cannot interfere in any way with its proceedings. All matters brought before the assembly are decided, after debate, by a vote. Party sometimes runs as high in the assembly as in the House of Commons; and its discussions are frequently as acrimonious as eloquent. The stipends of the Scotch clergy are principally derived from the wreck of the tithes and other property that belonged to the Rom. Cath. church, which was principally seized upon at the reformation by the nobility and gentry. The court of session, as commissioners of teinds (tithes), have power from time to time to augment the livings of clergymen as may not be already in the receipt of the whole disposable tithes of their respective parishes. But in many parishes the tithes have been wholly taken up or exhausted; and in 1812 an act was passed to raise, at the public expense, the incomes of such clergymen as had less than £150 a year, exclusive of glebes and houses, to that amount. At this moment (1841) the average income of the clergy of the church of Scotland, exclusive of the ministers of quoad sacra parishes, amounts to about £300 a year, over and above their glebes and houses, the average value of which may be estimated at about £30 a year.

The dissenters from the church consist principally of the members of the United Associate Synod, Burgher Synod, Relief Synod. The secession took place in 1741, in consequence, as already stated, of the enforcement of the patronage. There are some other bodies of dissenters; and of late years, owing to the influx of Irish into Glasgow, the Roman Catholics have received a great accession of numbers. The doctrines of the church of Eng-

land have also become fashionable among a good many of the higher classes. On the whole, however, the number of the dissenters has been very greatly exaggerated; and at this moment (1841) we do not suppose that, altogether, they exceed 600,000, or, at the outside, 650,000.

Public Instruction.—An endowed school has been for a lengthened period established in every parish in Scotland. The landlords are bound to build the school-house, and a house for the residence of the master, and to pay him a salary, which at present varies from £25 13s. 9d. to £34 4s. 5d. The power of nominating and appointing schoolmasters is vested in the landlords and minister of the parish. It is usually expected that a Scotch parochial teacher, besides being of unexceptionable character, should be able to instruct his pupils in the reading of English, the arts of writing and arithmetic, and the more useful branches of practical mathematics, and be possessed of such classical attainments as qualify him for teaching Latin and the rudiments of Greek.

Exclusive of the statutory allowance, schoolmasters receive fees from their pupils, according to a scale regulated by the landlords and clergymen. They are in general very moderate, varying, for the different branches of education, from 1s. 6d. to 7s. 6d., and, in a few instances, 10s. a quarter. An efficient system of education has, in consequence, been brought within the reach of all classes, and has been productive of the greatest advantage.

In the largest country parishes there are often subsidiary schools, established by the landlords, the masters of which are allowed a portion of the statutory salary; and in all the more considerable towns by far the greater proportion of the children are educated at non-parochial schools. Speaking generally, classical instruction is not carried to the same extent, in any of the Scotch schools, that it is carried in Eton, Harrow, and the higher class of English schools; but, on the whole, they furnish an extremely good and useful education.

Several returns have been published of the number and description of the various Scotch schools, and of the number of young persons of both sexes by whom they are attended. But these returns are all very incomplete, little to be depended upon, and, in fact, little better than worthless. At an average, we believe, it may be estimated that from an eighth to a tenth part of the inhab. of Scotland are at schools or academies of one description or another.

The higher branches of education are taught in the universities of Edinburgh, Glasgow, Aberdeen, and St. Andrew's (which see). Each of these universities has faculties of literature, philosophy, law, and medicine. No religious test is required from the students; the latter do not live in college halls, as in the English universities, and are not subject to any college jurisdiction when beyond the walls of the university.

Edinburgh University was long celebrated as a medical school, and still, indeed, preserves some considerable portion of its ancient celebrity. Probably, on the whole, the instruction afforded by the Scotch universities, though in many respects defective, is as good as can reasonably be expected, so long as the vicious practice is followed of allowing the professors to judge of the qualifications of candidates for degrees and literary honours; that is, of the merits of their own pupils, and, by consequence, of the solidity and efficiency of their own plan of instruction. The Scotch universities labour under a great disadvantage, from the want of a superannuation fund. The professors, having no resource on which to fall back in the event of their getting into ill health, are frequently obliged to cling to their chairs, as a means of subsistence, long after they have been disabled, by sickness and otherwise, from a proper discharge of their duties.

Language and Literature.—It has long been a prevalent opinion that the English and Scotch languages are merely different dialects derived from the same common source; and there are very good grounds for thinking that this opinion is correct. There is, however, no reason for concluding, as some have done, that the Scotch is merely a corrupt dialect of the English. It is quite as ancient as the latter; and both, in fact, are dialects derived from the same original tongue. The Gothic occupants of the lower parts of England and Scotland did not all come from the same parts of the continent, so that there would most probably be at the epoch of their immigration considerable differences in their dialects: and while these, on the one hand, would be diminished by their intercourse with each other in their new settlements, they would, on the other hand, be, in some instances, likely to be increased by their intercourse with the Celts and ancient inhabitants of the country, and with new comers; till, at last, they would be moulded into new dialects of the same common language. One of the most ancient existing specimens of the Scotch language is a poem on the death of Alexander III. In 1286, written soon after the event. Barbour, a contemporary of Chaucer, is also a distinguished writer; and the language was farther improved and perfected by James I., Dunbar, and Douglas, Bishop of Dunkeld, translator of Virgil, &c. But since the union of the crowns, and especially since the union of the kingdoms in 1707, the Scotch has been gradually giving way to the English, and it is now used only by the vulgar and illiterate. The poems of Burns, and some of Scott's novels, have tended, in no ordinary degree, to perpetuate and popularize the Scotch language; but notwithstanding their powerful influence, the fair presumption seems to be, that, at no very remote period, it will wholly cease to be a spoken language. In fact, it may be said to be almost extinct already, for, though most persons use Scotch words in their ordinary conversation, there are now very few indeed, if any, even among the lowest classes, who speak Scotch without a large intermixture of English.

The Erse or Gaelic, the language generally spoken by the lower classes in the Highlands, is, no doubt, the language of the ancient occupants of the country. It differs but little from the Irish; so little, that after a short intercourse, persons speaking Irish and those speaking Erse have no difficulty in understanding each other. But it is a curious fact that the Irish and Erse are wholly unintelligible to the Welsh. The Manks, or native language of the Isle of Man, is a dialect of the Erse or Irish.

It would be useless in a work of this kind to enter upon any statements with respect to the literature of Scotland. It is sufficient to say that there are but few departments in which it cannot boast of writers of the highest excellence. Speaking generally, the literature of Scotland, as contrasted with that of England, may, perhaps, be said to be less learned, less practical, and less playful; and to be more metaphysical, scholastic, and sustained. It would not, perhaps, be very difficult to specify the causes which have occasioned this difference in the literature of the two divisions of the island; but, owing to the more intimate connexion that now subsists between the people of both, it is probable that it will gradually become less perceptible.

Down to a comparatively recent period the Scotch newspaper press was alike inefficient and degraded. This was not a consequence of any indifference on the part of the public to newspaper discussions, but was wholly ascribable to the corrupt state of the court of justiciary. Full scope was given to the panegyrists of the government of the day, how worthless soever it might be; but any one who happened to question its merits, or who ventured to espouse or recommend any doctrine or theory not approved by the lord advocate for the time being, was, in fact, at the mercy of the latter, and might be banished almost at pleasure. The servility of the judges, and the facility of packing juries, afforded the agents of government a ready means of crushing any obnoxious writer; and, in fact, it may be said that Scotland had no newspaper press, or none worthy of the name, till after the close of last war. During the year ending the 5th January, 1841, 73 newspapers were published in Scotland, to which 4,751,043 stamps were issued.

The establishment of the *Edinburgh Review*, in 1809, is an important epoch, not only in Scotch, but in European literature: it effected a total and most advantageous change in the previous style of criticism and periodical writing.

The Provision for the Poor that has long existed in Scotland, originated, like that of England, in attempts to repress mendicity. The acts of the Scottish parliament in reference to this subject bear, in many respects, a close similarity to the English acts, and are partly, indeed, copied from the latter. They differ, however, from the English laws in this important respect, that they make provision only for the maimed and impotent poor, and not for those who are able-bodied; and it is now generally laid down by the best authorities that the latter have no title to claim relief as matter of right. The administration of the poor laws in Scotland differs extremely from their administration in England; and to this, more than to any difference in the laws themselves, the wide discrepancy that now exists between the provision for the poor in the two countries is mainly to be ascribed. In Scotland the administration of the poor's funds is usually vested in the kirk sessions; but, at the same time, any landlord may call the session of any parish in which he has property to account for their administration of those funds, and may attend and vote at their meetings in reference to such matters. No justice of the peace, sheriff, or other inferior judge, is permitted to interfere with the proceedings of the kirk sessions and landlords in their conduct of the affairs of the poor; and as the members of the sessions consist, for the most part, at least in country parishes, either of landlords or farmers, holding under leases of considerable length, they have always been anxious to keep the charge for the poor within the narrowest limits.

So economically, indeed, have the poor's funds been administered, and so anxious have the administrators been to allow none but really necessitous persons to participate

in them, and to keep the allowances as low as possible, that assessments for the support of the poor have not been introduced into more than about a third part of the parishes of Scotland, and that in the others the poor are supported by collections made at church doors, and other voluntary contributions. "The Scotch," to use the words of a Report by the General Assembly in 1820, "have uniformly proceeded on the principle, that every individual is bound to provide for himself by his own labour, so long as he is able to do so; and that his parish is only to make up that portion of the necessaries of life which he cannot earn or obtain by other lawful means. Even in cases of extreme poverty, the relations and neighbours of the paupers have a pride in providing for their necessities either in whole or in part. This circumstance will account for the small number of paupers in some very populous parishes; and serves at the same time to explain a fact which is obvious in so many returns in the country districts, that the sums given to paupers appear to be so disproportioned to what their real necessities require. A small sum given to aid their other resources affords them the relief which is necessary; and it would be both against the true interests and the usual habits of the people were a more ample provision made for them by their parishes."

Latterly, however, an opinion has been gaining ground that, under the existing system, economy is carried to an excess; and that the really necessitous poor are not adequately provided for. We incline to think that this opinion is perfectly well founded; and the statements in Dr. Alison's valuable publications, and those previously laid before the reader in articles EDINBURGH and GLASGOW in this work, exhibit a state of things that calls loudly for alteration and amendment. Economy is not the only thing to be attended to in the administration of the affairs of the poor. No doubt it is a most important consideration; but the claims of humanity are not to be trampled under foot for its sake, as they certainly have been in many parts of Scotland. We should be sorry, however, to see any serious inroad made on the existing system; and, perhaps, the objections now made to it might be obviated by giving paupers a right of appeal to the sheriffs, and instructing the latter to see that adequate provision is made for the really necessitous. We subjoin, from the Report by the General Assembly of 1838, the following statement, exhibiting an account of the parishes, &c. (See next page.)

Historical Sketch.—The early history of Scotland is at once obscure and uninteresting. The country was long one of the most barbarous in Europe; and though Kenneth II. (*anno* 838) is said to have united the extensive territories from the Tyne N. to the Pentland frith into one kingdom, it is abundantly certain that various extensive districts were in a great measure independent of the crown for several centuries after this period. In consequence of their early holding Northumberland, Cumberland, and other lands in England, the kings of Scotland were accustomed to appear in the English court to perform homage for these possessions; in the same way as the English monarchs were themselves accustomed to perform homage to the kings of France for Normandy and other provinces held by them in that kingdom. (*Stuart's Public Law of Scotland*, vol. viii.) On the extinction of the direct line of the Scottish kings in 1290, by the death of Margaret of Norway, John Baliol and Robert Bruce, descendants of the Scottish king David I., appeared as competitors for the crown. The pretensions of both were supported by powerful parties; and, to avoid a civil war, it was agreed to refer the matter to the amicable decision of Edward I., king of England. This able and politic prince availed himself of the opportunity to advance the principle, for which the homage that had been performed by the Scottish princes for their English possessions afforded a colourable pretext, that the kings of England were the paramount sovereigns or liege lords of Scotland, and that the competitors for the crown should do homage to him as such. This was consented to; and Edward, finding Baliol most suitable to his views, decided in his favour. The latter, however, being less submissive than was expected, was speedily set aside by Edward, who attempted to seize the kingdom on pretence of its having escheated to him through the rebellion of his vassal.

The nation, however, was not to be so transferred. Sir William Wallace raised the standard of independence; and in the sequel, the famous Robert Bruce, grandson of the competitor of Baliol, appeared in the field; and after unparalleled exertions, continued through a series of years, the great victory of Bannockburn (1314) secured the independence of Scotland, and established the conqueror and his family on the throne.

The only daughter of Robert Bruce having married the lord high steward, Robert, the issue of that marriage, and the first of the family of Stuart who arrived at the royal dignity, succeeded to the crown on the death of David II. in 1371. From this period, the history of Scotland is compar-

SCOTLAND.

Account of the Parishes, and of the Number and Description of Persons, in Scotland, that participated in the Poor's Funds, in 1838; the Amount of such Funds, and the Sources whence they were derived; the Mode and Expense of Management, &c.

I. Parishes and Population.	Not assessed.	Voluntarily assessed.	Legally assessed.	Total.
Parishes	517	126	236	879
Pop. per census of 1831	872,696	305,654	1,137,646	2,315,996
II. Number of Poor, Amount of Relief afforded.				
Poor on permanent roll:				
Number of persons	24,379	6,598	26,998	57,989
Rate per cent. of these to the pop. .	2·79	2·16	2·37	2·50
	L. s. d.	L. s. d.	L. s. d.	L. s. d.
Sums distributed to them . .	24,546 0 11	12,890 0 8	73,984 18 7	111,721 0 2
Rate of relief per annum { Highest general	3 10 7	5 3 9·3	6 1 0·10	4 9 0·4
to each individual { Lowest general	1 3 10·3	1 13 6·3	1 19 5·4	1 8 10·6
{ Average	1 0 4·7	1 19 1·3	2 14 9·6	1 18 6·6
Rate per head of the burden to the pop.	0 0 8·10	0 0 10·1	0 1 3·7	0 0 11·2
Lunatics:				
Number of persons	211 2·3	186 2·3	712 2·3	1,111
Rate per cent. to the pop. . . .	·024	·06	·062	·048
	L. s. d.	L. s. d.	L. s. d.	L. s. d.
Amount applied to their maintenance, &c.	2,131 19 9	2,992 19 9	7,359 18 4	11,784 17 10
Average rate for each lunatic . .	10 1 9·2	13 6 1·5	10 6 8	10 12 4
Occasional poor:				
Number of persons	6,969	2,494	11,645	20,368
Rate per cent. to the pop. . .	·71	81	1·02	·87
	L. s. d.	L. s. d.	L. s. d.	L. s. d.
Amount distributed to them . .	4,903 12 11	1,676 7 9	9,103 8 0	14,983 8 8
Average rate of relief to each individual	0 13 0·5	0 13 0·6	0 15 7·8	0 14 8
Total poor:				
Number of persons	38,509	23,73	39,356	79,490
Rate per cent. to the pop. . .	3·53	3·03	3·44	3·42
	L. s. d.	L. s. d.	L. s. d.	L. s. d.
Amount applied, including cost of education	31,730 10 0	17,038 15 4	91,726 16 0	140,495 1 4
Average rate to each pauper . .	·1 0 7·1	1 16 9	2 6 8·9	1 15 4·10
Rate per head of the burden to the pop.	0 0 8·6	0 1 1·4	0 1 7·4	0 1 2·6
III. Funds for Support of the Poor.				
Legal assessments			77,230 19 0	77,230 19 0
Rate per head on the pop. . .			0 1 4·3	0 0 8
Church-door collections . . .	17,816 15 0	6,384 0 4	14,099 14 10	38,300 10 2
Rate per head on the pop. . .	0 0 4·11	0 0 5	0 0 2·11	0 0 3·11
Other voluntary contributions . .	6,604 14 5	9,164 8 9	3,207 7 0	18,976 10 2
Rate per head on the pop. . .	0 0 1·10	0 0 7·2	0 0 0·8	0 0 1·1
Session funds	10,570 5 10	4,275 10 6	5,758 16 6	22,604 12 10
Total funds	34,991 15 3	19,823 19 7	100,305 17 4	155,121 12 2
IV. Management.				
Hired agents { For levying assessment		48	282	330
{ For managing the poor .	66	23	113	202
{ Total . .	66	71	395	532
Gratuitous agents { Members of session	3,191	829	2,015	6,035
{ Other persons .	242	194	1,071	1,507
{ Total . .	3,433	1,023	3,086	7,542
V. Expense of Management.	L. s. d.	L. s. d.	L. s. d.	L. s. d.
Average annual expense of levying assessment		189 2 1	3,930 14 11	4,119 17 0
Do. do. managing the poor	293 15 3	98 4 11	2,576 8 1	2,968 8 3
Do. of litigation { as to assessment			573 9 1	573 9 1
{ as to settlement of paupers, and claims for relief	40 4 0	44 14 8	262 8 1	347 6 9
Total expense	333 19 3	332 1 8	7,344 0 2	8,009 1 1
Rate per pauper	0 0 2·7	0 0 8·10	0 3 8·11	0 2 0·3
Rate per head on pop. . . .	0 0 0·1	0 0 0·3	0 0 1·5	About 1·13

ely well known; and the continued and extraordinary fortune that attended the lengthened series of princes of House of Stuart has invested it with more than ordinary rest.

The principles of the reformers were early introduced Scotland, and were eagerly adopted by the great bulk the nobility and people. The Protestant religion obtained ascendancy in 1560, shortly before the return of the beautiful, but ill-fated Queen Mary from France, where she had been sent to be educated. At this period the royal authority was at a very low ebb; the most violent contentions railed among the nobility; and it would have required reign of no ordinary ability and energy of character to act the government under such difficult circumstances. We need not, therefore, be surprised at the failure of Mary, though not without good talents, was wholly inexperienced, at the same time that she had the misfortune to have strongly imbued with anti-Protestant prejudices, and the violence of her passions made her sacrifice her reputation and innocence, and the well-being of the men, to their gratification.

Having been deposed in 1567, Mary was succeeded by her James VI., then a minor. The latter succeeded, on the death of Queen Elizabeth, in 1603, to the crown of England, when the two British kingdoms were happily united one sovereign. (See vol. i., 859, &c.)

On the accession of the House of Stuart to the union of the crowns, a period of about 230 years, Scotland, speaking generally, was in a most unsettled, turbulent state. The feudal system had been early introduced into the country; and the great estates and influence enjoyed by several of the nobles enabled them to rival the sovereign in power and importance, and sometimes to despise his orders and insult his person. In France, England, and other countries, the sovereigns, by enfranchising the inhabitants of the great towns, and attaching them to their interests, succeeded, through their assistance, after a lengthened struggle, in abating the pride and independence of the barons, and reducing them to obedience. But the kings of Scotland had no such support on which to fall back: there was not, in fact, a single great town in the kingdom; and they had nothing to trust to but the supplies of men and money they could draw from the crown estates, and from the contributions of such of their vassals as happened to be at the time in their interest, or they could coerce. In consequence of these and other concurring causes, the power of the Scottish kings was circumscribed within the narrowest limits; the civil broils in which they were almost always engaged, were, in most instances, fomented and abetted by the government of England; and, a few short intervals excepted, the country was involved in continuous anarchy and confusion. (See the admirable introductory chapter to Robertson, and the Histories of Pinkerton, Tytler, &c., passim.)

The union of the crowns in 1603 introduced a great

change for the better into the state of domestic affairs in Scotland. The barons could no longer look to England for open or underhand support in their contests with their sovereigns; while, at the same time, the power of the latter was vastly increased by their being able to employ the resources of a far more civilized, populous, and powerful monarchy in their disputes with their ancient subjects. Hence, though Scotland laboured under various grievances, resulting principally from the unreasonable hostility of the sovereigns to the Presbyterian form of church government, to which the great majority of the public were enthusiastically attached, she gained prodigiously in tranquillity and good order subsequently to 1603.

The union of the kingdoms in 1707 was, as it were, the necessary result and completion of the union of the crowns. Though excessively unpopular at the time, and opposed by many of the best Scotch patriots, it has been of vast advantage to Scotland, as well as to the empire generally.

The consequences of the rebellion of 1745 were also, notwithstanding the unnecessary and disgusting cruelty exhibited in its suppression, advantageous. It extinguished for ever the long-cherished hopes of the Jacobites; and it stimulated government to take effectual measures for abating the barbarism that prevailed in the Highlands, and for the introduction of a more efficient administration of justice into all parts of the country. In this view the old feudal hereditary jurisdictions were abolished, and sheriffs nominated and paid by the crown were appointed in their stead; and this most salutary measure being accompanied by the construction of military roads, that were carried into the remotest districts, the empire of the law was fully established; disorders of all sorts were promptly repressed; and at length the public energies were happily turned into those departments of industry and enterprise in which they have achieved such astonishing results.

SCOTT, county, Va., situated towards the S.W. part of the state, and contains 624 sq. m. Drained by the N. fork of Holston river and by Clinch river. It contained in 1840, 9520 neat cattle, 13,532 sheep, 24,095 swine; and produced 39,534 bushels of wheat, 2729 of rye, 294,705 of Indian corn, 111,849 of oats, 16,648 of potatoes, 7339 pounds of tobacco, 59,547 of sugar. It had eight stores, 10 flouring-mills, 28 grist-mills, 13 saw-mills, four tanneries, 63 distilleries; eight schools, 206 scholars. Pop.: whites, 6911; slaves, 344; free coloured, 48; total, 7303. Capital, Estillville.

SCOTT, county, Miss., situated a little S.E. of the centre of the state, and contains 576 sq. m. Drained by branches of Pearl river, and by head branches of Leaf river. It contained in 1840, 5808 neat cattle, 607 sheep, 7533 swine; and produced 1255 bushels of wheat, 50,564 of Indian corn, 1453 of oats, 8020 of potatoes, 6290 pounds of rice, 1430 of tobacco, 376,156 of cotton. It had seven stores, three grist-mills, two saw-mills, six tanneries. Pop.: whites, 1189; slaves, 462; free-coloured, 2; total, 1653. Capital, Hillsboro'.

SCOTT, county, Ky., situated towards the N. part of the state, and contains 252 sq. m. Drained by North Elkhorn, and Eagle creeks, find their branches. It contained in 1840, 13,723 neat cattle, 22,380 sheep, 37,596 swine; and produced 105,428 bushels of wheat, 62,940 of rye, 999,470 of Indian corn, 167,990 of oats, 20,869 of potatoes, 2300 pounds of sugar. It had 31 stores, two woollen factories, 16 flouring-mills, 19 grist-mills, 90 saw-mills, one paper-mill, four tanneries, one printing-office, one semi-weekly newspaper, one college, 114 students; two academies, 160 students; 26 schools, 719 scholars. Pop.: whites, 8290; slaves, 5339; free coloured, 109; total, 13,668. Capital, Georgetown.

SCOTT, county, Ia., situated in the S.E. part of the state, and contains 200 sq. m. Drained by Graham's fork of the E. branch of White river and its branches. It contained in 1840, 2589 neat cattle, 3647 sheep, 4603 swine; and produced 15,229 bushels of wheat, 52,253 of Indian corn, 31,254 of oats, 3146 of potatoes, 6615 pounds of tobacco, 3795 of sugar. It had seven stores. eight grist-mills, four saw-mills; two schools, 55 scholars. Pop. 4942. Capital, Lexington.

SCOTT, county, Ill., situated in the W. part of the state, and contains 240 sq. m. Bounded W. by Illinois river, and drained by its small tributaries. It contained in 1840, 8354 neat cattle, 6409 sheep, 17,518 swine; and produced 41,925 bushels of wheat, 424,000 of Indian corn, 51,435 of oats, 15,461 of potatoes. It had one commission house in foreign trade, 30 retail stores, seven flouring-mills, 12 grist mills, 17 saw-mills, four tanneries, two distilleries, two potteries; one academy, 60 students; 16 schools, 415 scholars. Pop. 6125. Capital, Winchester.

SCOTT, county, Mo., situated in the S.E. part of the state, and contains 936 sq. m. Bounded N.E. and E. by Mississippi river, W. by Whitewater river. It contained in 1840, 3140 neat cattle, 1738 sheep, 26,909 swine; and produced 10,111 bushels of wheat, 398,090 of Indian corn, 13,174 of oats, 7718 of potatoes, 161,150 pounds of tobacco, 4466 of cotton. It had 15 stores, 19 grist-mills, five saw-mills, one

tannery, three distilleries, three potteries; six schools 124 scholars. Pop.: whites, 5028; slaves, 928; free coloured, 18; total, 5974. Capital, Benton.

SCOTT, county, Iowa, situated in the E. part of the territory, and contains 540 sq. m. Bounded S.E. by Mississippi river, N. by Wabesipinica river. It contained in 1840, 823 neat cattle, 273 sheep, 5643 swine; and produced 3,738 bushels of wheat, 46,057 of Indian corn, 11,738 of oats, 30,009 of potatoes. It had 16 stores, two flouring-mills, three grist-mills, nine saw-mills, one distillery, one pottery, one printing-office, one weekly newspaper; one academy, 25 students; five schools, 122 scholars. Pop. 2145. Capital, Davenport.

SCOTT, county, Ark., situated in the W. part of the state, and contains 950 sq. m. Drained by a branch of Arkansas river. It contained in 1840, 3616 neat cattle, 391 sheep, 10,272 swine; and produced 1546 bushels of wheat, 39,115 of Indian corn, 2996 of oats, 7309 of potatoes, 13,199 pounds of cotton. It had two grist-mills, one saw-mill, one distillery; five schools, 103 scholars. Pop.: whites, 1542 slaves, 131; free coloured, 21; total, 1694. Capital, Boonville.

SCOTT, p. t., Cortland co., N. Y., 149 m. W. Albany, 35 W. Drained by branches of Toughnioga river, and the inlet of Skeneateles lake. It has two stores, two grist-mills, nine saw-mills, one oil-mill; nine schools, 447 scholars. Pop. 1332.

SCOTTSVILLE, p. v., capital of Powhattan co., Va. 22 m. W. Richmond, 150 W. Situated on elevated ground on the dividing ridge between Appomattox and James rivers, and contains a courthouse, jail, several stores, and mechanic shops. 30 dwellings, and about 200 inhabitants.

SCRIBA, p. t., Oswego co., N. Y. 167 m. N.W. Albany, 377 W. It contains part of the village of Oswego. Bounded N.W. by lake Ontario, W. by Oswego river. Old fort Oswego was erected near the mouth of Oswego river, in 1727, the remains of which are still found. It has two stores, three grist-mills, one saw-mill, one tannery; 17 schools, 1047 scholars. Pop. 4051.

SCRIVEN, county, Ga., situated towards the S.E. part of the state, and contains 748 sq. m. Bounded N.E. by Savannah river, S.W. by Ogeechee river. Watered by Brier creek. It contained in 1840, 9538 neat cattle, 278 sheep, 8004 swine; and produced 1806 bushels of wheat, 1866 of rye, 147,305 of Indian corn, 1380 of oats, 11,598 of potatoes, 858,309 pounds of cotton, 1890 of sugar. It had six stores, seven grist-mills, six saw-mills; four schools, 65 scholars. Pop.: whites, 2162; slaves, 2622; free coloured, 9; total, 4794. Capital, Jacksonboro'.

SCUTARI (an. Chrysopolis), a town of Asiatic Turkey, being, however, in fact, a suburb of Constantinople, opposite the latter, on the other side of the channel of Constantinople, on the Bithynian shore, about 1½ m. E. Seraglio point. Its pop. has been variously estimated at from 35,000 to 60,000. It is built on the declivity of several hills, and has a very picturesque appearance from the opposite shore; its interior is similar to that of the Turkish capital, and it is built in the same style. A palace of the sultan, with extensive gardens, barracks constructed by the late sultan Mahmoud, several handsome mosques, a noted college of howling dervishes, several large cemeteries, with public baths, bazaars, &c., are the principal edifices and public establishments. It is a rendezvous for the merchants and caravans from Armenia and Persia trading to Constantinople, and is the first station for the assembling of the Turkish troops in Asia.

Scutari is very ancient, and is said to owe its importance to the circumstance of the treasury of the Persians having been established in it when they attempted the conquest of Greece. Its vicinity was memorable for the decisive victory obtained by Constantine the Great over his rival Licinius. (Cramer's Asia Minor, i. 191; Turner, Clarke, Elliott, &c., passim.)

SCUTARI, a fortified town of European Turkey, and the cap. of a pachalic in Albania, S. of the Lake Scutari, in Labeatis Palus) at the confluence of the Bojana and Drinassi, about 16 m. from the embouchure of the former in the gulf of Drino, in the Adriatic, 45 m. S.S. Cattaro. Its pop. has been estimated at 20,000. It has a pretty strong citadel, on an isolated rock; with various mosques, and Greek and Rom. Cath. churches. It stands on rising ground, and is built in a very straggling manner. A considerable active trade in timber is carried on by vessels which ascend the Bojano to the lake; the inhabitants also manufacture cotton fabrics and arms, and build small vessels. Scutari is supposed to occupy the site of the ancient Scodra, the cap. of the Illyrian king Gentius; and of which subsequently appears to have become a Roman colony. It is still a place of some importance; and when Sir J. Hobhouse travelled in Albania, the power of its chief was the only counterpoise to that of Ali Pasha. (Hobhouse's Albania, 448; Cramer's Anc. Greece, i. 40, 41, . Horschelmann.)

SCYLLA and CHARYBDIS: The former is a famous

it and town of S. Italy, at the N. entrance to the narrow
it separating Italy from Sicily ; and the latter is an equal-
famous whirlpool in the strait near the Sicilian coast.
rlla is a bold rocky headland, about 200 ft. in height,
jecting into the sea, and hollowed at the base into cav-
s by the action of the waves. It is surmounted by a cas-
in lat. 38° 14′ 15″ N., long. 15° 44′ E. There is a sandy
/ on each side the rock ; and the town of Scylla, built
icipally on the steep acclivities of the ridge, stretches
vn to the shore on either side. It has about 5,000 inhabits, said to be expert fishermen, seamen and divers.

This little town suffered tremendously from the earth-
ke that devastated Calabria in 1783. A large portion of
inhabitants, with the prince at their head, fled to the
ch, believing it to be least exposed to danger. But they
not been long there when an adjoining cliff fell into the
: and the waves, driven back by its fall, rushed forward
in with such tremendous fury as to rise high upon the
re, sweeping along with them in their recoil 2,475 per-
i, not one of whom escaped alive ! (Swinburne, ii. 419.)
he rock of Scylla is exactly 6,047 Eng. yards, or nearly
i. from the opposite point of Faro, at the N. E. extrem-
if Sicily. The whirlpool of Charybdis is not, however,
site to Scylla, but within the strait outside the tongue
nd enclosing the harbour of Messina. From its prom-
t position at the mouth of the straits, Scylla is exposed
ie full action of the sea making a loud noise
ie caverns it has hollowed in the rock, which, of course,
uch increased in storms. Charybdis seems to be formed
ie main current passing through the straits from the N.,
g thrown over to the Sicilian shore by the point of Pez-
ind meeting the lateral current running in an opposite
tion. It is from 70 to 90 fathoms deep, circling in quick
» ; and Captain Smyth says that small craft are some-
i endangered by it, and that he had seen a 74-gun ship
ried round on its surface." (Memoir of Sicily, 123.)
notwithstanding the action of the contrary currents,
:he formidable appearance of Charybdis, there is no
r any real danger in navigating the straits, provided
aution be exercised. Although, however, it be quite
ue that Homer, in depicting the terrific dangers en-
cred by Ulysses in this famous strait (see Pope's Odys-
b. xxi. lin. 87, &c.) has made a very liberal use of the
e allowed to a poet, still it is abundantly certain that
it have been much more dangerous in antiquity than
sent. It was a generally received opinion among the
its that Italy and Sicily were once united, and that
uad been torn asunder by some great convulsion of na-
Pliny says, "Sicilia, quondam Bruttio agro cohœrens,
terfuso mari avulsa." (Lib. iii. cap. 8.; see also Si-
alicus, lib. xiv. lin. 2.; Virgil, &c.) But whether
ere so or not, it is plain that the action of the current
eriod of more than 3,000 years must have materially
id and deepened the strait, and worn down those
prominences that render such narrow channels pe-
/ dangerous. The configuration of the strait has al-
doubt, been materially altered in the interval by the
akes so prevalent in this region ; so that we are by no
entitled to ascribe the statements of the ancients, in
o its dangers, solely to their ignorance of navigation
love of the marvellous.

ts in passing through the straits, in order to avoid
within the vortex of Charybdis, sometimes run upon
which gave rise to the famous proverbial expres-
'incidit in Scyllam cupiens vitare Charybdim ;" appli-
those who, to avoid a less, run into a greater danger.

ROOK, Rockingham co., N. H., 17 m. S. S. W.
uth, 47 m. S. E. Concord, 480 W. Watered by
irown's, and Walton's rivers. Chartered in 1768.
ns bog iron ore on some of its streams, has much
at building, and is finely situated for the fisheries.
mploy many of the people. It has five stores, 2
s, 2 saw-mills ; 1 academy, 25 students ; 4 schools,
ars. Pop. 1392.

L., Pike co., O. Bounded W. and N. by Scioto
'atered by Beaver creek. It contains eight stores,
g-mill, two grist-mills, four saw-mills ; 10 schools,
irs. Pop. 1635.

Y, county, Ark. Situated in the N. part of the
i contains 850 sq. m. Watered by a branch of
'er. It contained in 1840, 1,391 neat cattle, 650
69 swine ; and produced 1,071 bushels of wheat,
ndian corn, 1,742 of oats, 1,380 of potatoes, 8,826
tobacco, 3,878 of cotton. It had two stores, five
Pop.: whites, 933; slaves, 3; total, 936. Capi-
on..

MONT, p. t., Waldo co., Me., 38 m. E. Augusta,
Vatered by St. George's river. It has two stores,
-mill, two grist-mills, six saw-mills, two tanne-
schools, 460 scholars. Pop. 1374.

), lake or pond, Cumberland co., Me., is 12 m.
ne wide, and has several islands, the largest of

which contains 700 acres. It discharges its waters by Pre.
sumpscot river into Casco bay. The Cumberland and Ox.
ford canal extends from tide water near Portland to Sebago
pond or lake, 20¼ m. in length. It has 26 locks, and was
completed in 1829. By means of a lock constructed in Son.
go river, the navigation is continued into Brandy and Long
ponds, making in the whole a water communication natu.
ral and artificial of 50 miles. The total cost of the work
was $250,000.

SEBASTIAN (ST.), a fortified frontier city and seaport
of Spain, cap. prov. Guipuscon, in Biscay, at the extremity
of a low and sandy tongue of land, projecting into the bay
of Biscay, 10 m. W. by S. Fuenterrabia, and 40 m. E.N.E.
Bilbao. Pop., according to Miñano, 9,720, exclusive of its
garrison ; but, including this, and its suburbs, the pop. may
be taken at 13,000. Having been in a great measure rebuilt
since 1813, it is now one of the neatest and most regularly
constructed towns in the peninsula, presenting a favourable
contrast to most other Spanish cities. It is defended on the
E. and W. by strong walls, washed by the sea; on the N.
by the castle of Mota, on mount Urgullo, a rugged cone
near 400 ft. in height; and on the S. it is shut off by advanced military works, and by the little river Urumea, from
its suburb of St. Catharine. The castle is well supplied with
water, and is a fortress quite independent of the city, with
which it communicates by two routes, both defended by various batteries. St. Sebastian has some handsome squares,
several churches and convents, a civil and military hospital, &c.; its streets are clean, and it is abundantly supplied
with water, though not of the best quality. The Urumea
is crossed by a stone bridge of eight arches. Though secure and well defended, the harbour is difficult of entrance ;
it is formed by a mole, and is of small size. St. Sebastian
has always been a place of considerable trade, and was the
seat of the Philippine Company. It is the port whence
Pampelufia, Vittoria, Logrono, &c., obtain most part of their
supplies of colonial and other foreign goods ; and at which
the greater part of the French and English manufactures
destined for Madrid and other towns in the interior, are imported. Its exports are chiefly iron and wool. It had formerly some large cordage factories and tanneries, but these
have mostly fallen into decay. It is the residence of a military governor, two justices, and the seat of a sub-delega-
tion of police ; a lottery department, a tribunal of commerce,
&c. From its being, as it were, one of the keys of Spain,
its possession has always been an object of great importance
in the contests between the French and Spaniards. The
former took it in 1719, 1794, and 1808 ; and held it from the
last-mentioned epoch till 1813, when it was taken by the
British forces under Sir Thomas Graham, now Lord Lyne-
doch. The latter were repulsed in their first attempt to carry it by storm in July ; but they succeeded, though not without an enormous loss, in the second assault on the 31st of
August. A fire having broken out in the town during the
assault, it was all but destroyed. We regret to have to
state that, notwithstanding every exertion on the part of the
officers, the most horrible excesses were committed by the
victorious soldiers. (Miñano, Dic. de España, &c.; Napier's Peninsular War, vi. 65—197.)

SEBEC, p. t., Piscataquis co., Me., 96 m. N.N.E. Augusta, 693 W. Watered by Sebec river, issuing from Sebec
pond or lake, which lies W. of it, and extends into it. This
lake, situated mostly in Foxcroft and Bowerbank, extending into Sebec, is 10 m. long, and on an average 1 m.
wide, and enters Piscataquis river. The town was incorporated in 1812, and has three stores, one fulling-mill, one
woollen factory, one grist-mill, one saw-mill ; 12 schools,
258 scholars. Pop. 1,116.

SEDAN, a fortified town of France, on its N. E. frontier,
dép Ardennes, cap. arrond., on the Meuse, 19½ m. E.S.E.
Mezières, lat. 49° 42′ 59″ N., long. 4° 57′ 51″ E. Pop., in
1836, ex. comm., 12,906. Though an important frontier
town, the Dict. Geog. says that its works, some of which
were constructed by Vauban have latterly been neglected ;
and its citadel, at its S.E. extremity, has been converted
into an arsenal. It stands on very uneven ground, and is
separated into two unequal parts by the Meuse, here crossed by a stone bridge. It is well built ; the streets, which
are wide and clean, are ornamented with numerous fountains ; the houses are mostly of stone, roofed with slate, and
in the environs are various public walks.

Sedan has excellent cavalry and other barracks, a military hospital for 500 patients ; other military establishments ;
a Calvanist, and several Roman Catholic churches, a public library, communal college, a handsome theatre, &c. It
is the seat of a tribunal of primary jurisdiction, a chamber
of manufactures, &c. The water is said to have a tendency to produce goitre.

Sedan has been long celebrated for its woollen manufactures, consisting principally of fine black cloths, and casi-
meres. In 1836-37, from 11,000 to 12,000 work-people were
employed in the woollen manufactures of the town and its

vicinity, of whom from 3,000 to 4,000 belonged to the town; from 2,000 to 2,500, belonging to the neighbouring villages, went to work daily within the town; and the remainder, consisting principally of weavers, inhabited the country for a distance of from three to four leagues round. The last, who also occupy small patches of land, work at the loom in their own cottages; whereas those who live in and near the town are mostly employed in large manufacturing establishments. In these, they work nominally 10, and really from 14 to 15 hours a day. But, notwithstanding these long hours, the work-people of Sedan are decidedly better fed, clothed, and lodged, than those of most other manufacturing towns of France: men get from about 2 to 3½ fr. a day; women from 1 to 2 fr., and children from 10 sous to 1 fr. (*Villermé*, i. 284.)

The greatest harmony subsists between the work-people and the manufacturers. Drunkenness is comparatively rare, though in other respects their habits might be a good deal improved. Instruction is much more extensively diffused among the work-people than at Rheims, and, speaking generally, they have all the signs of good health; circumstances chiefly consequent on the non-introduction of children into the factories at too early an age. Villermé farther adds, that their education and morals are both in a state of improvement. (*Tableau des Ouvriers*, i. 253–279.)

Hosiery, leather, arms, and hardware, are also produced at Sedan; and it has numerous dying-houses, with an extensive trade in drugs, &c. Previously to the revolution Sedan was the cap. of a principality, which had often changed hands in the middle ages, but which was finally exchanged with Louis XIV., for some other fiefs by the Turenne family. One of the greatest of the French generals, the famous Marshall Turenne, was a native of this town, in the citadel of which he first saw the light on the 16th of September, 1611. His statue, in bronze, ornaments the principal square. (*Hugo*, art. *Ardennes; Dict. Geog., &c.*)

SEDGEWICK, L. Hancock co., Me., 85 m. E. Augusta, 681 W. Situated on a peninsula, having Blue Hill bay on the E., and a strait which separates it from Deer Island on the S. It has a good harbour, and a considerable number of vessels, employed in the coasting trade and the fisheries. Ship-building is a considerable business. It has seven stores, one fulling-mill, three saw-mills, three tanneries; 15 schools, 896 scholars. Pop. 1,922.

SEEKONK, p. t., Bristol co., Mass., 46 m. S. S. W. Boston, 404 W. Bounded W. by Seekonk or Providence river. This river affords water power. It contains two churches, a Congregational and Baptist, five stores, one fulling-mill, three cotton factories with 6,010 spindles, six gristmills, 12 saw-mills; one academy, four students, 13 schools, 343 scholars. Pop. 1,996.

SEGORBE (an. *Segobriga*), a city of Spain, cap. distr. of its own name in Valencia, near the Palancia, 18 m. N. W. Murviedro. Pop. between 6,000 and 7,000. Streets wide, and most of its houses well built: it has several squares, numerous public and private fountains, a cathedral in which are some good paintings, several convents, a prison, workhouse, and other public edifices. Its inhabitants are occupied in the manufacture of starch, earthenware, and paper, the distillation of brandy, and the quarrying marble in the vicinity. Two large fairs are annually held here. Various Roman antiquities have been found within the city. (*Miñano; Fisher's Picture of Valencia, &c.*)

SEGOVIA (an. *Secuvia*), a city of Spain, Old Castile, cap. of the prov. of its own name, 48 m. N.N.W. Madrid. Pop., according to Miñano, 12,880. It is built on two hills and the intervening valley, the unevenness of the site giving it a wild look. Most of the streets are crooked and dirty, the houses also are ill built, and chiefly of wood. The public buildings comprise 18 churches, including the cathedral, five hospitals, a mint, a college for cadets in the old castle or Alcazar, and military barracks. The cathedral, which is described by Swinburne as one of the handsomest churches in Spain, has a tower 330 ft. high, and exhibits a mixture of the Gothic and Arabian styles, nearly resembling that in the great church at Salamanca. The interior is characterised by a simplicity rarely seen in Spanish churches, the effect of which is infinitely superior to that of the gildings and ornaments elsewhere observable. The Alcazar is in great preservation, occupying a commanding situation on a rock rising above the open country. Towards the town is a large court before the great outward tower, formerly used as a prison, but now as a college of cadets. The rest of the buildings form an antique palace, once the favorite residence of Ferdinand and Isabella: it comprises several magnificent halls, with gilt ceilings; and along the cornice of the grand saloon are 52 wooden statues of the kings of Spain seated in state. The military college was re-modelled on the formation of the present constitution, and the instruction given by the professors embraces most branches of knowledge connected with military science. The great glory of Segovia, however, is its aqueduct, supposed to have

been built in the time of Trajan, and certainly one of the most perfect specimens of Roman architecture in Spain. Swinburne (ii. 248.) says, that "it is not only an immense monument for its solidity and good masonry, which has withstood the violence of barbarians and the inclemency of the seasons during so many ages, but a wonderfully beautiful and light in its design. It consists of 161 arches in two ranges; that nearest the ground comprising 118, of which 43 are surmounted by an equal number of others: the whole is built of square stones, without mortar; and at the top is a channel, once hollow, but now filled up, only 8 ft. wide, and without a parapet. The total length of the aqueduct is 750 yards, and its height in crossing the valley (immediately close to the *Plaza del Azoguejo*, where two of the arches cross the street) is estimated by Twiss at 102 ft., though other travellers say only 94 or 95 ft. (*Comp. Twiss, ii. with Townshend, ii. 118., and Inglis, i. 284.*) Swinburne considers it superior in elegance even to the Pont du Gard, near Nismes; but, in point of fact, the two differ considerably; the latter having three instead of two rows of arches, and the extreme height being 143 ft.

The mint of Segovia, the most ancient in Spain, is situated at the bottom of the city, on the small river Eresma, the water of which turns its machinery; for many years, however, its operations have been confined to the coinage of maravedis, quartos, and other copper pieces. Segovia is said to be a decayed city, and most hands in Spain remain accounts of the former flourishing state of its woollen manufactures; but Capmany has shown that, if not wholly unfounded, these accounts are, at all events, very greatly exaggerated; and, that when most flourishing, the number of looms in Segovia did not exceed 300 (*Questiones Criticas, p. 37.*), which, perhaps, is about their present number. It also, produces paper, earthenware and glass. A fair held here in June is much frequented. In the neighbourhood are mines of lead and copper, as well as quarries of black marble.

The early history of Segovia is somewhat obscure; but, like most other cities of Castile, it belonged successively to the Romans, Goths, and Moors, from the last of whom it was taken at the beginning of the 15th century. During the Peninsular war the town was occupied by the French from 1808 to 1814. (*Townsend, ii. 130—134; Inglis, i. 281—87; Miñano.*)

SEINE, the smallest, but most populous, wealthy, and important dép. of France, being that in which the capital is situated. It extends between lat. 48° 43' and 49° 2' N., and long. 2° 10' and 2° 35' E., being entirely surrounded by the dép. Seine-et-Oise. It is of a nearly circular shape, about 15 m. in diameter. Area, 47,549 hectares. Pop. in 1836, 1,106,891. The Seine traverses this dép. in its centre, with a general direction from N.W. to S.E., and receives the Marne within its limits. There are a few hills, but none of much elevation. Mont. Valerien does not rise to 450 ft. above the level of the Seine, and Montmartre is only 344 ft. in height. The soil is chiefly calcareous, the dép. forming the centre of the remarkable tertiary region called the Paris basin. (*See Paris anté, p. 486, and PLANCE, vol. i. 850.*) But the chalk is covered with a bed of argillaceous mould of considerable thickness; and the manure supplied by the capital renders the dép. very productive. The arable lands are estimated at 29,298 hectares; meadows, 1543 ditto; vineyards, 2784 ditto; and orchards, gardens, &c., 3502 ditto. Corn is not extensively grown; the little that is produced is but indifferent. It furnishes very superior peaches, and other fruits; and there are numerous market gardens for kitchen vegetables, &c. A good many cattle, and other live stock, are fattened for the Paris markets, and there are some flocks of superior sheep. In 1836 66,897 properties subject to the contrib. foncière. C.57 were assessed at less than 5 fr.; while 19,323 were rated at from 100 to 330 fr., and 9006 at 1600 fr. and upwards. The manufacturing industry centres in Paris, which see. It is divided into 3 arronds., and sends 14 members to the Ch. of Dep. Registered electors, in 1836-39, 16,571. Total public rev., in 1831, 95,392,189 fr.; expenditure, 15,363,521 fr. (*Hugo, art. Seine, &c.*)

SEINE (an. *Sequana*), a river of France; and though by no means the largest, yet one of the most important in the kingdom, being that on which the capital is built. It rises in the dép. and mountain-chain of the Côte d'Or, by which it is separated from the basin of the Loire, about lat. 47° 30' N., long. 5° E., 20 m. N.W. Dijon. It flows generally in a N.W. direction, between the basins of the Marne and long. 0° 10' E., nearly opposite Newhaven in Sussex. Its entire course, in consequence of its numerous windings, is estimated at 500 m., for nearly 330 of which it is navigable. Its source is about 1600 ft. above the level of the sea. For more than three fourths of its descent...

the first 100 m. of its course, for at Troyes it is not more than 440 ft. above the sea; and at Paris its mean elevation above the latter is only 51 ft., and at Rouen 26 ft. (*Brugnere Orographie.*) From its not rising in mountains of any great elevation. it is neither subject to serious inundations, nor has it a rapid current; and the latter circumstance, together with its gentle rate of descent in the lower parts of its course, renders it highly suitable for navigation. The chief obstacles to its utility are the shifting sand-banks in its estuary, and some shallows between Quillebœuf and Rouen. At Paris the Seine is from 300 to 380 ft. in width; at its mouth it is 7 m. in width. It is here subject at the return of every tide to a phenomenon termed the *barre*, similar to the *mascaret* in the Dordogne, the *bore* in the Solway Frith, Ganges, &c. This consists of a wave of great magnitude with an almost perpendicular front, impelled inwards from the sea with much violence as high as Jumieges, and sometimes even as far as Rouen. It gives notice of its approach by a noise which, according to Hugo, is heard for forty minutes beforehand, but it is, notwithstanding, frequently productive of damage to shipping. The tide in the Seine is usually perceptible as high as Rouen, to which city the river is navigable for vessels of 200 tons. Respecting the trade of the Seine, *see* the articles HAVRE and PARIS.

The Seine receives several considerable tributaries; as the Aube, Marne, and Oise, from the N. E., and the Yonne, Eure, and Rille, from the S. and W. Besides Paris, several large and flourishing commercial cities and towns are seated on the Seine; as Rouen, Elbœuf, Troyes, Melun, and Montereau, with Chatillon, Bar, Nogent, Corbeil, St. Germains, and Honfleur; and at its mouth is Havre, which, Marseilles excepted, is the first commercial port of France.

The Seine and its tributaries are connected by the canals of Briare, Orleans, and Nivernais, with the Loire; by that of St. Quentin with the Somme and Scheldt; by that of Ardennes with the Meuse, and by that of Burgundy with the Loire. The canal of Ourcq (*see* PARIS, p. 472) also communicates with it. The banks of the Seine below Paris have been much praised for their beauty. "I reached," says Mr. Maclaren, " Rouen by the diligence from Paris in 11¼ hours. The road passes along the valley of the Seine, which is extremely beautiful, but deficient in variety. It is 'from 2 to 10 m. in breadth, and is bounded on both sides, not by hills, but by *plateaus*, or table-lands, of a very uniform elevation. The lands are carefully cultivated, but enclosures are rare, and the cottages small and mean. I went from Rouen to Havre by the steamboat in eight hours." *Notes in France and Italy*, pp. 195, 196.)

SEINE-INFERIEURE, a maritime dep. of France, reg. i., formerly comprised for the most part in the prov. of Normandy, having E. the deps. Somme and Oise, S. Eure and Calvados, and W. and N. the British channel. Area, 602,912 hectares. Pop., in 1836, 693,683. The S. boundary consists mostly of the Seine and its estuaries. There are some hill chains, but none of much consequence. Coasts generally abrupt, presenting a succession of calcareous cliffs. Climate moist, and colder than on the opposite coast of England. Soil generally calcareous or sandy; but the arronds. of Havre and Yvetot there are some very fertile tracts, consisting of a fine light clay. In 1834, the surface is said to have been distributed as follows, viz.: 8,016 hectares arable, 28,024 do. meadows, 61,173 do. orchards and gardens, and 68,844 do woods. Agriculture more advanced in this than in most other deps. Ploughs a superior kind have latterly been introduced; the efficacy of manures is well understood; and fallows have been to a considerable extent superseded by the introduction of an improved rotation. Near Havre, on the large arms only, a few acres are in fallow: of the arable land, but one third part may be in wheat; one third in oats, barley, and rye; one sixth in clover; one twelfth in peas and vetches, and one twelfth in flax; and these are about proportions throughout most part of the dep. In 1835, 134,071 properties subject to the *contrib. foncière*, 36,910 were assessed at less than 5 fr., while 2413 were assessed from 300 to 500 fr., 1599 at from 500 to 1000 fr., and 754 1000 fr. and upwards; so that the proportion of large parties is greater in this than in any other dep. of France; to this its superior agriculture is, in a great measure, to be ascribed. The large proprietors seldom or never farm their own lands. Some farms run from 200 to 300 acres, but in general they are much smaller: they are almost always let for a term of 9 years, at a rent varying from 30s. to 36s. per hectare. which is always paid in money. According to the official tables, 5,313,000 hectolitres of corn were harvested in 1835, exclusive of 582,000 hect. potatoes, the cultivation of which is increasing on the small farms. A good deal of cider is made, and most of the peasants' gardens are surrounded by small orchards. Cattle, horses, and sheep are all of good breeds, and are, in fact, among the best in France: the produce of wool is estimated at about 650,000 kilogs. a year. Wages are high. Farm ser-

vants always live with their masters; and ploughmen receive from £8 to £12 sterling a year; women from £3 to £6, generally in money. Their food consists of bread, vegetables of all kinds, soup, eggs, cheese, &c., with cider for drink; and butchers' meat once or twice a week. The occupiers of farms are prosperous, though not rich. The farmer pays the land, house, window, and personal taxes, and is frequently called on for the support of paupers. The latter are numerous, and as no legal provision is made for them, they are wholly dependent upon voluntary contributions. (*Parl. Rep. on Agriculture*, 1834.) This dep. which ranks third in France with respect to population, is inferior to none in manufacturing industry. Rouen (which see) is at the head of the cotton manufacturing towns; and Elbœuf is one of the chief seats of the French woollen manufacture. Manufactures of most other descriptions are carried on; and the dep. has, through Havre, &c., a most extensive trade with England, America, and most parts of the world; and by the Seine, with Paris and the interior of France. It is divided into five arronds.; chief towns, Rouen, the cap., Dieppe, Havre, Neufchâtel, and Yvetot. It sends 11 members to the Chamber of Deputies. Registered electors, in 1836–39, 7599. Total public revenue (1831), 60,179,981 fr.; expenditure, 15,185,772 fr. (*Hugo*, art. *Seine Inf.; French Official Tables, &c.*)

SEINE-ET-MARNE, a dep. of France, reg. N., mostly between the 48th and 49th deg. of N. lat., and long. 2° 30' and 3° 30' E.; having N. the deps. Oise and Aisne; E., Aube and Marne; S., Yonne and Loiret; and W., the last-named and Seine-et-Oise. Area, 663,662 hectares. Pop., in 1836, 323,863. Surface, undulating; its slope being from E. to W. The highest hills are in the S. The Seine, here joined by the Yonne, traverses the S., and the Marne the N. part of the dep. The Ourcq, Loing, and Grand Morin are the other principal rivers. The geological formations are mostly calcareous, overlain in many parts by a deep layer of vegetable soil. This, indeed, is one of the finest agricultural deps. in France, and has a good deal of rich land. In 1834, it was said to comprise 367,864 hectares arable land, 33,393 ditto meadows, 16,979 ditto vineyards, 6607 ditto orchards and gardens, and 70,963 ditto woods. It has a large surplus of corn, principally of wheat and oats, for exportation. Potatoes are also pretty extensively grown. The produce of wine may amount to nearly 600,000 hectols.; but it is mostly of low quality, and is principally used for home-consumption only. Cider is also produced.

In 1819, according to Mr. Jacob, this dep. was better cultivated than those more in the E.; the soil is well adapted for turnips, small patches of which are occasionally met with.

Meaux is finely situated in the midst of rich natural pastures, which fatten great numbers of cattle, and the dairy husbandry is also carried on to some extent. The breed of sheep has been much improved by crossing with the merino breed, and the total annual produce of wool is estimated at 1,200,000 kilogs. Wax and honey are important articles. In 1835, of 173,606 properties subject to the *contrib. foncière*, 85,862 were assessed at less than 5 fr., 26,290 at from 5 to 10 fr., and 709 at 1000 fr. and upwards. No mines are wrought, but a good many hands are employed in quarrying paving and other stone. Manufactures principally of cotton and linen fabrics, hardware and cutlery, earthenware, leather, and paper. This dep. is divided into five arronds.; chief towns, Melun, the cap., Coulommiers, Fontainebleau, Meaux, and Provins. It sends five members to the chamber of deputies. Registered electors, in 1836–39, 2781. Total public revenue, in 1831, 19,888,754 fr.; expenditure, 7,577,351 fr. (*Hugo*, art. *Seine-et-Marne; Jacob's Tour, &c.*, pp. 442, 443.)

SEINE-ET-OISE (formerly *Isle-de-France*), a dep. of France, reg. N., principally between lat. 48° and 49° N., and long. 1° 30' and 2° 30' E; having N. the dep. of Oise, E. Seine-et-Marne, S. Loiret, W. Eure and Eure-et-Loire; it encloses the metropolitan dep. of the Seine. Area, 560,537 hectares. Pop., in 1836, 448,180. It has no hill 400 feet in height. The Seine traverses this dep. from N.W. to S.E., receiving the Oise and Essonne within its limits. As it belongs to the great tertiary basin of Paris, the soil of the dep. is principally calcareous; a large portion, however, is sandy; and it is not particularly fertile, except in the neighbourhood of Paris, where it is liberally manured. In 1834, according to the official returns, the arable lands comprised 367,741 hect.; meadows, 20,091 do.; vineyards, 16,711 do.; orchards, 7680 do., and woods 77,213 do. Principal corn crops, oats and wheat. The annual produce of wine is estimated at about 700,600 hectol., but it is of very indifferent quality: cyder is also produced, to the extent of about 100,000 hect. a year. The culture of figs, cherries, strawberries, and other fruits and vegetables for the Paris markets, is an important branch of industry. A good many sheep are bred and cows for their milk, which is sent to Paris; but few other kinds of live stock are reared.

In 1835, of 217,344 properties subject to the *contrib. fonciers*, 104,305 were assessed at less than 5 fr., and 508 at 1000 fr. and upwards. The proximity of Paris has given rise to a great variety of manufactures. Yarn and stuffs of all kinds, paper, hair fabrics, leather, earthenware, beet-root sugar, chemical products, &c., are among the principal goods manufactured. The dep. has a very extensive general trade. It is divided into six arronds.; chief towns, Versailles, the cap., Mantes, Pontoise, Rambouillet, Etampes, and Corbeil. It sends seven members to the Chamber of Deputies. Registered electors, in 1838-39, 3400. Total public revenue, in 1831, 23,036,434 fr.; expenditure, 12,803,713 fr. (*Hugo; French Official Tables, &c.*)

SELBY, a market town, river port, and par. of England, W. Riding, co. York, chiefly in Barkston-Ash wapentake, on the Ouse, 11 m. S. by E. York. Area of parish 3180 acres. Pop., in 1831, 4600. The town is pretty well built, paved, and lighted, and has latterly been much improved. A handsome Gothic market-cross, the parish church, and the town-hall, a neat brick edifice built in 1825, are the principal public buildings. The church is a portion (almost the only one remaining) of Selby Abbey, founded by William the Conqueror in 1069, in which Henry I. was born. It is a large and magnificent cross church, of mixed Norman and early English architecture. The choir is a most beautiful specimen of decorated work; the E. end is particularly fine, with very beautiful windows, and octagonal turrets, having rich pinnacles. It has some very superior stone screen-work, and ancient stained glass. (*Rickman's Gothic Architect.*) The living, worth £97 a year, is a perpetual curacy, in the diocese of York, in private patronage. Selby has meeting-houses for Friends, Independents, Calvinists, Wesleyans, Unitarians, and Roman Catholics; a grammar-school founded by Edward VI., a hospital for seven poor widows, &c. The Ouse, which is here crossed by a moveable timber bridge, is navigable to Selby for vessels of considerable burden; and it carries on an extensive intercourse by water with Goole and Hull. It is also connected by railways with Hull and Leeds; and a branch custom-house being established here, it has become a considerable *entrepôt*. It has also manufactures of sailcloth, leather, and iron goods, and slips for building river craft. Petty sessions for the wapentake are held here, and courts leet and baron twice a year by the lord of the manor. Market-day, Monday; fairs, Easter Tuesday, Monday after June 22, and October 10, for cattle, wool, linen, tin and copper wares, &c.

SELKIRK, an inland co. of Scotland, being one of the smallest, and the least populous in that part of the United Kingdom; having N. Mid-Lothian, E. Roxburg, S. Dumfries, and W. Peebles. Area, 169,280 acres, of which not more than 1-10th is supposed to be arable. This county is, in most respects, similar to that of Peebles, and the statements as to the one will apply, with little modification, to the other. The greater part of the surface is mountainous; but the hills are green and smooth to the summits, and afford excellent sheep pasture. The county is watered by the Tweed, and its two tributaries, the Ettrick and Yarrow; there is some excellent arable land in the valleys traversed by these rivers, but the extent is inconsiderable.

Selkirk has fully participated in the wonderful improvements that have been made during the last half century in most parts of Scotland. Its agriculture, breeds of cattle and sheep (now wholly Cheviot), roads, buildings, food and clothes of the inhabitants, &c., have all been signally improved. Average rent of land, in 1810, 4s. 8d. an acre. The woollen manufacture is carried on with spirit and success at Galashiels. The county sends one member to the House of Commons. Registered electors, in 1839-40, 612. Selkirk is divided into seven parishes, and had, in 1841, 1446 inhabited houses, and 7980 inhabitants, of whom 3973 were males, and 4017 females. Valued rent £80,308 Scotch. Annual value of real property in 1815, £43,584.

SELKIRK, a market town and royal bor. of Scotland, cap. of the above county, on the W. side of a range of mountains, about 1½ m. from the right bank of the Ettrick, and 33 m. S.E. by S. Edinburgh, on the road leading from the latter to Carlisle. Pop., in 1841, 1800. It consists chiefly of one wide, irregular street, which, at the market place, expands into a triangular open space. The only public buildings are the town hall, with a spire 110 ft. in height; a jail, the parish church, and a chapel belonging to the United Associate Synod. Besides schools, it has a mechanics' institute, three subscription libraries, and a reading-room. Mungo Park, the African traveller, was born within a mile of the town, and a monument is at present (1841) about to be erected to his memory. Abbotsford, the seat of Sir Walter Scott, on the right bank of the Tweed, is within 4 m. The town has no manufactures; but on the neigh-

bouring banks of the Ettrick are woollen-mills, for the making of hosiery, tweeds, blankets, and similar stuffs. In remote times Selkirk was distinguished for its manufacture of shoes: hence the expression "souters" (shoemakers) of Selkirk" was, and still is used as denoting the whole inhabitants. But shoe-making is not now carried on to any considerable extent. Poor-rates were introduced in 1841: the present annual assessment is about £300.

Selkirk was in ancient times a royal residence; the *Forest*, as the county was once called, being a favourite hunting scene of the Scotch monarchs. Its history is intimately connected with the border wars. A standard, taken from the English at the battle of Flodden, by the "souters of Selkirk," is still preserved. The battle of Philiphaugh (1645), in which the Marquis of Montrose was signally defeated by General Leslie, was fought within ½ m. of the town. Since the Reform Act the borough electors have been added to those of the county. (*New Stat. Acc. of Scotland, § Selkirkshire, 1-10; and Jeffrey's Guide to the Borders.*)

SELMA, p. v., Dallas co., Ala., 83 m., S.S.E. Tuscaloosa, 844 W. Situated on the N. bank of Alabama r. It contains several stores, and about 1000 inhabitants. Nett proceeds of the postoffice in 1841, $12.50. A railroad has been projected N. to Tennessee r. at Gunter's landing; about 30 m. have been graded, but little else has been done on it. It is abandoned for the present.

SELMA, p. v., Jefferson co., Mo., 108 m. E. by S. Jefferson city, 818 W. Situated on the W. bank of Mississippi r., 35 m. below St. Louis, and 5 below Herculaneum. It has a good landing, a large mercantile house, a shot factory, and considerable trade in lead.

SEMLIN, a frontier town of the Austrian empire, in Slavonia, on the Danube, 3 m. N.W. Belgrade, and 40 m. S.E. by E. Peterwardein. Pop. about 9200; a motley collection of Slavonians, Germans, Greeks, Servians, Croats, Gypsies, Jews, &c. It consists of an inner town and a suburb: it is not fortified, but only surrounded with a stockade. It has some good houses and churches, but its streets are mostly unpaved, mean, and dirty. At its N. extremity is the ruined castle of the famous John Huniades: it stands on a commanding height, having on its sides the huts of the Gypsey quarter. Semlin has a large quarantine establishment, at which travellers entering from Turkey are usually detained for from 10 to 40 days. The hospital, a high female school, and a German theatre, are the other principal public establishments in the town, which is the residence of a Greek *protopapas*, and the chief *entrepôt* of the trade between Austria and Turkey. Its principal imports from the latter are raw cotton and cotton twist, honey, saffron, hare and rabbit skins, pipe-bowls, &c.; its exports thence woollen stuffs, earthen and glass wares, and other manufactured goods. (*Oesterr. Nat. Encyc.; Berghaus, Burgess, Quin, &c.*)

SEMPACH, a small town of Switzerland, canton Lucerne, on the E. bank of the lake of same name, 7 m. N.W. Lucerne, famous in Swiss history for the victory gained in its vicinity on the 9th of July, 1386, by a Swiss force of about 1400 men, over 4000 Austrians, commanded by the archduke Leopold II. The Swiss historians ascribe their success in this battle to the patriotism and devotion of a knight of Unterwalden, who, grasping a number of the spears of the Austrian pikemen in his hands, showed his countrymen, at the expense of his own life, how they might make their way into the enemy's phalanx. But whatever truth there may be in this story, we believe that the easy and complete victory of the Swiss was principally owing to the less poetical fact of the archduke having been killed at the beginning of the action, and to the panic his death produced in his army. Besides the duke, about 2000 Austrian troops fell in the battle and pursuit: while the loss of the Swiss is said not to have exceeded 200 men. (*L'Art de Vérifier les Dates, Partie Moderne, xvii., 95, 8vo ed.*)

SEMPRONIUS, p. t., Cayuga co., N.Y., 16 m. S.E. Auburn, 164 m. W. Albany, 233 W. A part of Skeneateles lake lies on its N.E. border. Its small streams flow into Skeneateles and Owasco lakes. It has two stores, one grist-mill, nine saw-mills, two tanneries; nine schools, 377 scholars. Pop. 1304.

SENAAR. *See* NUBIA.

SENECA, lake, N.Y., is one of the most beautiful of the lakes of western New-York, is 40 m. long, and from 2 to 4 m. wide, is very deep, and never frozen entirely over. The depth, 12 m. from its outlet, has been found to be 632 feet. Its outlet affords great water-power at Waterloo, 6 m., and Seneca falls, 10 m. from the lake. Crooked lake outlet enters this lake from the W. at Dresden, 12 m. above Geneva. Steamboats run regularly on this lake from Geneva near its foot, to Jefferson at its head, 40 m. A canal extends from it to the Erie canal, and another canal connects it with Crooked lake. Its shores are highly picturesque and beautiful, and the delightful village of Geneva on its N.W. part, ornaments and is ornamented by it. Its outlet enters the N. part of Cayuga lake.

SENECA, r., N.Y., is formed by the outlets of Cayuga, Canandaigua, Owasco, Skeneateles, and Onondaga lakes. After receiving the outlet of Oneida lake, it becomes Oswego river, and flows into lake Ontario at Oswego village.

SENECA, county, N.Y., situated in the central part of the state, and contains 308 sq. m. Bounded E. by Cayuga lake and Seneca river. Bounded W. chiefly by Seneca lake, and watered by its outlet, which affords great water-power. Clyde river enters Seneca river in its N. part. It contained in 1840, 21,222 neat cattle, 63,824 sheep, 25,981 swine; and produced 398,505 bushels of wheat, 5526 of rye, 177,795 of Indian corn, 19,798 of buckwheat, 11,147 of barley, 232,446 of oats, 199,387 of potatoes, 1903 pounds of hops, 25,845 of sugar. It had 55 stores, one furnace, seven fulling-mills, two woollen factories, one cotton factory with 2500 spindles, 11 flouring-mills, 10 grist-mills, 25 saw-mills, five oil-mills, one paper-mill, nine tanneries, four distilleries, one brewery, four potteries, four printing-offices, three weekly, and one semi-weekly newspapers; three academies, 215 students; 15 schools, 4260 scholars. Pop. 24,874. Capitals, Ovid and Waterloo.

SENECA, county, O., situated in the N. part of the state, and contains 540 sq. m. Drained by Sandusky river and its tributaries. It contained in 1840, 18,035 neat cattle, 21,027 sheep, 34,256 swine; and produced 346,194 bushels of wheat, 9,923 of rye, 306,051 of Indian corn, 9510 of buckwheat, 74,736 of oats, 89,755 of potatoes, 1911 pounds of tobacco, 99,706 of sugar. It had 36 stores, one flouring-mill, 14 grist-mills, 34 saw-mills, six tanneries, four distilleries, two breweries, three potteries, two printing-offices, two weekly newspapers; 89 schools, 2389 scholars. Pop. 18,128. Capital, Tiffin.

SENECA, t., Ontario co., N.Y., 15 m. E. Canandaigua, 179 f. Albany. Bounded E. by Seneca lake. Drained by Flint creek. It contains the important village of Geneva in N.E. part, and has 48 stores, three fulling-mills, three grist-mills, 10 saw-mills, four tanneries, two breweries, two printing-offices, one weekly newspaper, and one periodical; one college, 196 students; six academies, 393 students; 21 schools, 1275 scholars. Pop. 7073.

SENECA FALLS, p. t., Seneca co., N.Y., 4 m. E. Waterloo, 166 m. W. Albany, 342 W. Taken from Junius in 1825. Watered by Seneca river. The village was incorporated in 1831, is situated on the outlet of Seneca lake, which is here a fall of 47 feet, over four dams. It contains five churches, a Presbyterian, Episcopal, Methodist, Baptist, and Roman Catholic; numerous stores, mills, and manufactures, 400 dwellings, and about 3000 inhabitants. There is in the township 28 stores, three fulling mills, one cotton factory with 2500 spindles, seven flouring-mills, one grist-mill, three saw-mills, one oil-mill, one paper-mill, one tannery, two distilleries, one brewery, two potteries, two printing-offices, two weekly newspapers; one academy, 104 students; 14 schools, 743 scholars. Pop. 4281.

SENEFFE, a village of Belgium, prov. Hainault, 6 m. S. Nivelles, famous from its vicinity having been the scene of one of the most sanguinary conflicts of modern times. Here, on the 11th of August, 1674, a French army, under the famous Prince of Condé, attacked the rear-guard of the Confederates, commanded by the Prince of Orange, toward William III., and gained a considerable advantage. But, not satisfied with this, Condé imprudently attacked the main body of the Confederates, who had taken a very strong position; on which, notwithstanding the most astonishing efforts, he could make no impression. The loss on both sides was nearly equal; and such was the slaughter, that above 20,000 men were left on the field of battle. In fact, to use the words of Voltaire, "La grande ombre bataille de Senef ne fut qu'un carnage." (Siécle de Louis XIV., cap. 12.) Both armies withdrew next day, neither attempting to molest the other. This was the last battle fought by the Prince of Condé. A well-known feat, ascribed to the prince in reference to this conflict, is of very doubtful authenticity.

SENEGAL, a large river of W. Africa, which, till the time of Delisle and D'Anville, was considered identical with the Niger of the ancients, but which is now ascertained to be wholly unconnected either with the Quorra, generally supposed to be the Niger, or with the Nilotic system of the African continent. Its sources have not been explored; but Mungo Park ascertained that they were separate from the basin of the Niger and Quorra by the Mandingo race. The Ba-fing (black-water), regarded as the nearn, rises in this mountain region, in about lat. 10° N. long. 11° W. Its course is generally N.W. to near lat. 14° and long. 17°, when it turns W., and falls into the Atlantic a little below the French settlement of St. Louis, after a course estimated at 1000 m. Its chief affluents are, the Kokoro on the right, and the Falémé on the

left, both of which join it in the upper half of its course. Timbo stands near the head of the Ba-fing: on its banks are the French forts of Faf, Dagana, Podhor, Bakel, St. Joseph, and Mussala. The early course of this river and its tributaries is through a broken country, diversified by rugged and precipitous hills, and intersected by numerous streams, the sands of which are copiously impregnated with gold dust. At Felloo, from 400 to 450 m. from the sea, it forms a cataract, up to which it is navigable all the year for flat-bottomed boats. After passing Galam, the Senegal rolls over a level plain, with a very gentle current; and after passing Podhor, a French station about 60 leagues from its mouth, the level is so complete, that Adanson does not conceive the total fall of the river from that station to the sea to be more than 2¼ feet. The tide is perceptible in the river for upwards of 60 leagues inland. The Senegal, in this part of its course, is bordered by vast forests, obstructed by thick underwood, and filled with numberless species of wild beasts and birds. At about 35 m. (direct distance) from the ocean, the Senegal divides into two arms, which enclose a delta. The principal or E. arm is deep enough to be navigable for the largest ships, but is obstructed by a bar at its mouth, which cannot be crossed, except during the inundations, by ships drawing more than from 10 to 12 feet of water. Vessels under this draught may, however, always navigate the river as far as Podhor; and in the rainy season vessels of from 130 to 150 tons ascend to Galam. Like the Nile, the Senegal annually overflows and fertilizes the adjacent country; and in July, when the inundation begins, some French vessels sail up as far as the river is navigable, trading with the natives for gum and other products. A fair, lasting fifteen days, is held annually at fort St. Joseph. After this, as soon as the waters begin to subside, the vessels return, spending only about a fortnight in the downward journey, but consuming nearly three months in their upward voyage.

The Senegal forms a part of the line of demarcation between two regions widely differing in every respect. To the N. within a few miles of its banks, is the great desert of Sahara, with here and there a few Moors; while to the S. are the fertile regions of Nigritia, inhabited by negroes. (Ritter's Africa, Fr. trans., ii., 20–27; Hugo, art. Senegal; Dict. Géog., &c.)

SENEGAL, a name derived from the above river, given to some small French colonial establishments on the W. coast of Africa, comprising several islands, and small portions of the African continent, between the Senegal and Gambia rivers. It is divided into two arronds., the N. consisting of the isles of St. Louis, Bavaghé, Safal, and Gheber, near the mouth of the Senegal, with some few establishments on the banks of that river, and trading stations along the coast between Capes de Verd and Blanco; and the S. arrond., comprising the island of Goree, Albreda, on the bank of the Gambia, and the other stations S. of Cape de Verd. The total pop. of these dependencies amounted, in 1836, to 18,040; of whom about 18,000 were Mohammedans and blacks: two thirds of the population inhabited St. Louis and its arronds.

This part of the African coast is nearly destitute of good harbours; those of St. Louis and Goree are the best. The soil of the isles and continental shore is sandy, but improves in quality farther inland, where it is covered, S. of the Senegal, with dense forests, and the most luxuriant vegetation. The climate, though not so very pestiferous as that of Sierra Leone, is extremely bad. The heat of summer is most relaxing and oppressive, especially during E. winds, though the thermometer does not stand extremely high. The wet season, which lasts, with S.W. winds, from June till Oct., is particularly fatal to Europeans, who are attacked with dysenteries, liver complaints, and various kinds of fevers. The mineral products are few. There are traces of iron, but little ore is wrought. Basalt, &c., are found at Goree, but scarcely any stone elsewhere; and at St. Louis the most solid buildings are only of brick. Gold is procured from the countries towards the head of the Senegal, but the efforts of the French to form settlements there have hitherto proved abortive. Near the mouth of the Senegal are some salt-pans; and in some parts of the interior natron effloresces on the soil. The vegetable products are the most varied and abundant. They include the gigantic baobab (Adansonia digitata), palms, mimosas, and gum trees of numerous kinds, Senegal ebony, and other valuable timber; with cotton, indigo, coffee, arnatto, olives, hemp, and other fibrous plants; cassia, sweet potatoes, millet, maize, &c. Among the wild animals are the elephant, lion, hippopotamus, wild boar, buffalo, tiger-cat, great numbers of deer, game of all species, and an immense variety of birds and reptiles. Oxen, buffaloes, horses, asses, &c., are used for domestic service, as in Europe; and goats, sheep, and hogs are reared. Several kinds of artificial grasses are grown, but the culture of products for food, or exportation, is pursued only to a very small extent, Senegal being a trading *entrepôt* rather than an agricultural colony. Few of the colonists are employed in manufactures, except in the working of iron, and shipbuilding. The making of bricks, lime, and salt employs a few hands; the negroes weave such clothes as they require, but other manufactured articles are obtained from Europe. We subjoin an

Account of the Quantities and Value of the principal Articles exported from the Senegal Colonies in 1836.

Articles.		Quantities.	Values.
Raw hides	. . . kilogr.	327,726	396,481 f.
Wax	. . . —	45,134	30,389
Elephants' teeth	. . . —	12,248	85,423
Gum Senegal	. . . —	1,791,510	2,544,694
Cabuet-woods	. . . —	40,509	14,178
Gold	. . . grammes	59,925	150,675
Specie	. . . —		74,809

The total value of the exports, including that of goods reexported, amounted to 4,051,263 fr.: the imports to 6,961,894 fr., the principal being linen and cotton fabrics, slops and apparel, brandy, liqueurs, wines, and other provisions.

Senegal is governed by a superior naval officer, who resides at St. Louis: Goree is the seat of a lieutenant-governor. There appears to be neither a representative assembly nor a colonial council. A court of primary jurisdiction sits at St. Louis; from the decisions of which, appeal lies to a court composed of the governor, the other chief functionaries, and certain principal inhabitants of the colony. The European force in Senegal consists of half a battalion of marines, a company of marine artillery, and a company of sappers, &c., altogether amounting to about 370 men. The French established themselves here in 1637, but no settlement of much importance was made till the formation of the Senegal company in 1664. The English took Senegal in 1758, but it was retaken by the French in 1779: it was again held by the English from a period shortly after the French revolution till the peace of 1814. (*Hugo*, art. *Senegal*; *Etats des Colonies Françaises*.)

SENLIS (an. *Augustomagus*, post *Sylvanectae*), a town of France, dép. Oise, cap. arrond., on the Nonette, a tributary of the Marne, 29 m. S.E. by E. Beauvais. Pop., in 1836, 5016. It stands on the declivity of a hill, and consists of the town proper, and three suburbs. The town is surrounded with thick walls, parts of which are supposed to be the remains of those constructed by the Romans. It is tolerably well built; but the streets are mostly narrow and crooked, and it has few public buildings worth notice. The cathedral, however, has a handsome spire, 225 ft. in height. Chicory, starch, and cotton thread are the principal manufactures. The town was of importance in the middle ages: under the Carlovingians it had the right of coinage; and in 1180 Philip Augustus espoused Elizabeth of Hainault at Senlis. (*Hugo*, art. *Oise*, &c.)

SENNET, p. t., Cayuga co., N. Y., 158 N. by W. Albany, 338 W. Drained by Broad cr. Taken from Brutus in 1807. It contains two churches, a Presbyterian and Baptist, one store, three grist-mills, one tannery, one distillery; 17 schools, 785 scholars. Pop. 3060.

SENS (an. *Agedincum*, post *Senones*), a town of France, dép. Yonne, cap. arrond., on the Yonne, 30 m. S.E. Auxerre. Pop. in 1836, ex. com., 9029. It is surrounded with decayed walls, attributed to the Romans, and various Roman antiquities exist in and round the town. It is said by some authorities, but not by others, to be well laid out and well built; and is kept clean by streamlets, which traverse its principal thoroughfares. It has a fine Gothic cathedral of the same proportions as Notre Dame, in Paris, though of less size. In it is the splendid marble mausoleum of the dauphin, son of Louis XV., and father of Louis XVI., Louis XVIII., and Charles X., a *chef-d'œuvre* of Coustou. In the chapter-house is a painting of the death of Thomas-à-Becket, who took refuge at Sens about 1166. The communal college is a large building, with a museum of antiquities, and a public library of above 6000 vols. Sens has a seminary, some public baths, a handsome theatre, a court of primary jurisdiction, &c.; manufactures of serge, druggets, wax candles, and glue, with breweries and distilleries; and an active trade in agricultural produce, timber, oak bark, leather, bricks, &c. Under Valens it was made the cap. of the 4th Lyonnaise; and it became an archbishopric on the establishment of Christianity in the empire. Several councils were held here in the middle ages, including that in 1140, at which Abelard was condemned for heresy. (*Hugo*, art. *Yonne*; *Guide du Voyageur*, &c.)

SERAMPORE, one of the Danish settlements in Hindostan, prov. Bengal, consisting of a town on the Hooghly, about 12 m. above Calcutta, and immediately opposite Barrackpoor. Pop. about 15,000. (*Malcom*.) It extends for 1 m. along the river, and is without fortifications, having only a small battery for saluting. "Serampore is a handsome place, kept beautifully clean, and looking more like a European town than Calcutta, or any of the neighbouring

794

cantonments. Since the Copenhagen rupture (and more especially since it ceased to be an asylum for debtors from Calcutta), it has grievously declined, and its revenues scarcely meet the current expenses. Many persons of different nations, who like a cheaper residence than Calcutta, take houses here." (*Mod. Trav.*, ix., 110.) It has long been the head quarters of the Protestant missions in India; and has a large and handsome college for the instruction of native youths, an extensive missionary printing establishment, &c. It was here that the Scriptures were translated into various Indian dialects, under the superintendence of Dr. Carey and others. (*Malcom's S. E. Asia*, ii., 41; *Hamilton's E. I. Gaz.*)

SERES, a large town of Turkey in Europe, in Macedonia, cap. of a beylik; on a declivity a little N. of the lake Takinos, and 44 m. N.E. Salonika. Its pop. is probably between 25,000 and 30,000; but this part of Europe is so seldom visited by travellers, that we have little accurate information respecting it. Seres is surrounded by a wall flanked with towers, and commanded by a citadel. It is said to be well built, the houses being interspersed with gardens: it has some spacious *khans*, numerous mosques, churches, and fountains, and several public baths; with linen and cotton manufactures, and an active trade in cotton, grown in large quantities in its vicinity. Seres states that 70,000 bales of cotton are annually exported from Seres, 30,000 of which go to Vienna. (*Herschkman's Stein*, &c.)

SERINGAPATAM (*Sri Renga-Patan*, "Vishnu's city"), a decayed town and fortress of India, S. of the Krishna, which, under Hyder Ali and Tippoo, was the capital of Mysore. It stands at the W. angle of an island in the Cavery, about 4 m. in length by 1½ m. in breadth, and is about 250 m. W.S.W. Madras. Lat. 12° 25' N., long. 76° 45' E. The fortress, constructed by Tippoo, is an immense mass of building, but in several respects injudiciously planned. It was, however, when invested by our troops, strongly fenced with six redoubts, and other strong outworks. All a capital, the town was but mean. It has one good bazaar, and a broad road under the ramparts, but the other streets have a very indifferent appearance; the houses also are shabby, and the public buildings few. Hamilton speaks of an arsenal, a gun-carriage manufactory, &c.; but it is probable that these, as well as the other military establishments, have been since removed to Mysore. On an eminence in the centre of the island is a large and handsome suburb, in which is the mausoleum of Hyder Ali and Tippoo: and across the Cavery, near the city, is a native bridge of granite, remarkable for its size and solidity.

Seringapatam was besieged by the English on three different occasions: the first two sieges took place in 1792 and 1799; at the latter, Tippoo purchased a peace by ceding half his dominions, and paying three crores and 30 lacs of rupees to the British and their allies. Another war, however, broke out in 1799; and on the 4th of May, in the same year, Seringapatam was stormed by the British and the Nizam's forces, under General Harris. On that occasion Tippoo was killed, with the greater part of his garrison, amounting to 8000 men, and the dominions of the last formidable enemy of the British in the Indian peninsula were added to our Indian empire. (*Mod. Trav.*, viii.; *Hamilton*, *E. I. Gazetteer*, &c.)

SERVAN (ST.), a town and seaport of France, dép. Ille-et-Vilaine, on the Rance, immediately behind St. Malo, of which town it may be considered the continental suburb, though comprised in a distinct commune. Pop. in 1836, with comm., 9948. St. Servan is well built, and has a good harbour for merchant vessels, divided into two parts by the *Solidor*, an isolated tower about 60 ft. in height. The dockyard, which derives its name from this town, has few slips, three of which are appropriated to the construction of frigates; and during the war many frigates were built here. The naval establishments at St. Servan are considerable, and a floating dock, to connect the port with that of St. Malo, is now (1841) in rapid progress towards completion. (*Ports, &c., of France*, 290, 291.) St. Servan has manufactures of sail-cloth, cordage, ship-biscuit, &c., and is the general *entrepôt* for the trade of St. Malo. Among its inhabitants are many English families, attracted thither by the cheapness of living, and the beauty of the neighbourhood. (*Guide du Voyageur*; *Hugo*; *Frès. h/.*.)

SERVIA (an. *Mœsia-Superior*, with part of *Illyricum*), one of the principalities on the Danube, nominally included in the dom. of Turkey-in-Europe, but in a great measure independent of the Porte. It extends between the 43d and 45th degs. of N. lat., and the 19th and 23d of E. long.: having N. the Hungarian prov. of Slavonia and the Banat, from which it is separated by the Save and Danube; E. Wallachia and Bulgaria, from the first of which it is also separated by the Danube; S. Macedonia, the Dushkan forming the boundary line in this direction; and W. Bosnia, from which it is divided by the Ibar and the Drin. Greatest length, N. to S., about 130 m.; breadth varying from 30 to 160 m. Area roughly estimated at 20,000 sq. m., and the pop. at about 1,000,000, mostly Christians of the Greek church. The greater part of the country is covered with mountains, those in the W. being ramifications of the Dinaric Alps, and in the S. and E. branches from the Balkhan. There are, however, some tolerably extensive plains, particularly in the N. and along the course of the Moravia. This river, which, after those above named, is the principal in Servia, nearly traverses the country from S. to N. The climate is remarkably variable, and much colder in winter than would be inferred from the latitude, the Danube and the Save being often thickly frozen over. The heats of summer are proportionally intense: the autumn is the most agreeable season, but ague is very prevalent then and in spring. The soil is almost everywhere fertile, though to a great extent uncultivated. Every species of grain common in Europe is raised, except rice. Maize is the principal, but much more wheat is produced than formerly, and maize bread is not now generally made use of by the inhabitants of Belgrade and other large towns. Owing to the inland situation of the country, and the want of markets, the price of corn is usually very low.

The vine is pretty generally grown; but, from defects of culture, the grapes of the same vineyard usually differ greatly in quality, and being all used promiscuously in the making of wine, it is, for the most part very indifferent. In the district of Belgrade, however, superior red wine approaching to claret has been made, though to no great extent. In fact, but little wine is drunk in Servia; a spirituous liquor, distilled from plums, called *slivovitza* or *rakia*, sold at about ½d. a quart, being used in its stead. Hemp, flax, tobacco, and cotton are cultivated, but only in small quantities. The pasture-grounds are extensive and good, though little can be said in favour of the breeds of cattle and sheep. Both are meagre and impoverished; and the former, though universally employed, with buffaloes, for draught, are not very numerous. The horses, also, are poor and diminutive, though latterly Prince Milosch has made considerable efforts to improve the breed. Hogs are by far the most valuable and favourite stock. No peasant's family is without these animals. They overspread the country in vast herds, being branded with the proprietor's name, and turned loose in the forests, where they feed on acorns, except in winter, when they are scantily fed at home on maize, and other dry provender. Hogs constitute the principal export from Servia: about 290,000 are said to be annually sent to the Austrian dominions, where they pay a considerable import duty, having also paid an export duty on leaving the Servian frontier. The wool of its Servian sheep is very inferior; but about 80,000 lamb and goat skins are annually disposed of to Austrian merchants.

The forests, which overspread a large proportion of the country, might, if they could be turned to good account, be made, under judicious management, an almost inexhaustible source of wealth. Oak, extremely well adapted for ship-building, ash, pine, &c., are the principal trees, and *valonea* is produced in great plenty. But, as if the natural difficulties in the way of its exportation were not enough, government has prohibited the felling of oak timber! and the forests in many places are so thick as to be all but impassable, and are at the same time encumbered with putrescent vegetation. The collection of leeches, which abound in the marshy districts, has been carried to some extent of late years. They are disposed of to French merchants settled in Belgrade and Semlin, who forward them by way of the Danube, &c., to Paris, which they reach in 12 or 14 days: but this, which promised to become a business of considerable importance, has been monopolised by government. Iron, copper, lead, quicksilver, and coal are found in Servia: but few or no mines are wrought, partly from a want of capital and enterprise, but partly also, it is alleged, from a wish on the part of government to conceal such tempting sources of national wealth, to avoid exciting any desire in its neighbours to possess themselves of the principalities.

Until a more extensive commerce take place on the Danube, or a free communication of some kind be established between the upper Save and the Austrian ports on the Adriatic, the great natural resources of Servia must continue all but unavailable. Her produce being similar to that of the S. provinces of Austria and Russia, these states row obstacles in the way of her commerce; at the same time that the adjacent Turkish provinces, have no need of her staples. The want of roads is also a great drawback the prosperity of all the provinces in this remote part Europe: the only high road in Servia is that which leads from Belgrade to Adrianople.

Servia, however, is less inconvenienced than most of the antiguous provinces by the want of roads, their deficiency being, in part at least, compensated by the easy access to navigable rivers by which she is almost surrounded. A more liberal and bolder commercial policy on the part of Austria would do much to develope the resources and advance the civilization of the Servians; and by attaching them to her interests, it would seem, also, to be the safest in a political point of view.

The Servians belong to the widely-spread Slavonian stock, with which most part of E. Europe is peopled. Their language is the most refined of the southern Slavonian dialects, and their poetry ranks high among that of the E. European nations. (See *Bowring's Specimens of the Popular Poetry of the Servians.*) In their manners and customs the Servians differ little from the other Slavonic tribes in their vicinity (*see* TURKEY, &c.): they are in general almost equally uncivilised, backward in the arts, &c., ignorant and superstitious; though in some of the larger towns some degree of advance has of late been perceptible.

Servia is divided into six provinces and 13 districts; chief towns, Belgrade, the capital, Semendria, Nissa, Jogodina, Kragujewasz, and Poschega. In the middle ages it formed an independent kingdom, the dominion of which extended over parts of Bulgaria, Bosnia, Albania, &c.: it was conquered by the Turks in 1365. The Turks still garrison Belgrade, which is the residence of a pacha; but nothing is left them beyond this military occupation, an acknowledgment of the supremacy of the Porte, and a small yearly tribute to the sultan. The internal government is wholly in the hands of the Servians. Early in the present century a successful revolt took place, headed by Czerny-George, a native chief, who, in 1806, took Belgrade from the Turks, and continued to govern the country till the peace of 1814, when it again submitted to the Turks, and Czerny-George took refuge in Russia. A new revolt, under Milosch Obrenowitsch, in 1815, was equally successful, and Milosch has till lately held the reins of government. He established a representative assembly, a council of ministry, &c. In 1835, a general poll-tax, to meet the state expenses, and various other financial plans, were adopted. But whether it were owing, as has been alleged, to Russian influence, or to other circumstances, Milosch was obliged to resign the government, and retire to his estates in Wallachia in 1839, since which period he has been succeeded by his second son, Prince Michael. Servia has a small standing military force of about 1750, men, 1500 infantry, 200 cavalry, and 50 artillerymen; but all males capable of bearing arms are enrolled in the militia, and a force of 40,000 men may be collected on an emergency. (*Priv. Report on Servia*; *Quin's Voyage on the Danube*, p. 225–234; *Paget, Burgess*, &c. passim.)

SETUBAL, or ST. UBES, a city and seaport of Portugal, prov. Estremadura, cap. Comarca, on the N. side of the bay of its own name, which receives the Sadao at its S.E. extremity, about 18 m. S.E. Lisbon; lat. 38° 28' 54" N., long. 8° 53' 32" W. Pop. according to Miñano, 15,000. It extends for about ¾ m. along the beach, consisting mostly of two or three parallel narrow streets, crossed by others, and some squares, in one of which is a handsome public fountain. It is enclosed by walls partly in ruins, and defended by the castle of St. Philip and a few other detached forts. It has several convents and hospitals, Latin schools, and courts of justice, broad quays, and a convenient harbour for merchantmen. Its environs, which are very picturesque and fertile, produce large quantities of muscadel and white wines; which, with oranges, lemons, and salt, are its principal articles of export. The exports of salt from St. Ubes have long been of very considerable importance; and furnish, indeed, the greater part of the demand of Sweden, and various other countries. Being preferable for certain purposes to the salt of this country, we import from 250,000 to 350,000 bushels a year. The pilchard fishery employs a good many hands, and a large fair is held annually from the 25th to the 29th of July. Near it is the famous convent of Arrabida, to which pilgrimages are performed. The ancient *Cetobriga* is supposed to have stood on the opposite shore of the bay, where various remains of antiquity have been found; one of which, a Corinthian pillar surmounted by a crucifix, stands in the square of the city. After the expulsion of the Moors, Setubal was all but deserted, till it was repeopled under Alonzo Henriquez and his son Sancho. It was fortified during the war of independence in the 17th century. (*Miñano.*) It suffered severely from the earthquake so disastrous to Lisbon in 1755. (*Dict. Geog.*; *Southey, &c. in Med. Tree., xix.*)

SEVASTOPOL, or AKTIAR, a town and seaport of European Russia, on the W. coast of the Crimea, lat. 44° 36' N., long. 33° 30' E. Pop. fluctuating, consisting principally of the garrison, and of the various individuals connected with the fleet. Sevastopol stands on a creek, on the S. side of one of the finest bays in the world, the *Ktenus* of Strabo. It stretches E. into the country about 5 m., with a breadth, where greatest, of about a mile; it has, till within a short distance of the bottom, near Inkerman, from six to eight fathoms water. There are in the cove on which the town is built five fathoms water close in shore. The bottom is clay and mud, and it is quite free from rocks and shoals. The bay is defended by strong forts on both sides the entrance. Merchantmen are excluded from Sevastopol, and it has become the principal station of the Russian fleet in the Black sea, for which it is incomparably better fitted than either Kherzon or Nicolaeff. Streets wide and regular, intersecting each other at right angles; houses extremely good, and built in the modern Italian style; principal edifices, admiralty, arsenal, hospital, barracks of the garrison, marine barracks, &c. The calcareous rocks at Inkerman (town of caverns), in the vicinity of Sevastopol, have been cut into the most extraordinary caverns, or rather into chapels, monasteries, cells, &c., " which, by their multiplicity and intricacy, astonish and confound the beholder." (*Clarke*, ii., 203; see also p. 272. where there is a good plan of the bay of Sevastopol; *Lyall*, i., 290, &c.)

SEVENOAKS, a market town and parish of Kent, lathe Sutton-at-Hone, hund. Codsheath, on a ridge of hills near the Darent, and on the road from London to Rye, 21 m. S.S.E. London. Area of par., with the liberties of Riverhead and Weald, 6790 acres. Pop., in 1841, 4827. The town consists principally of two wide streets, in one of which is the market-house. Many of the houses are large, and inhabited by opulent families. The parish church is spacious and handsome, and is a conspicuous object for several miles round. The livings are a rectory and a vicarage in the gift of the Curteis family; the former worth £305 and the latter £629 a year, nett: two curacies are established in the parish, one at Riverhead and the other at Weald chapel, respectively worth £45 and £100 a year. There are meeting-houses for Baptists and Wesleyans; an hospital for aged persons, and a free grammar school, both founded and endowed by Sir W. Sevenoke, in 1418. The latter was farther endowed by Queen Elizabeth, whose name it bears; and it has now an annual income of about £1000, with seven exhibitions, five of scholarships in any college of either university, and two in Jesus' college, Cambridge. In another school, founded in 1675, about 300 poor children are instructed on the national system. Near the town is Knowle or Knoll, the magnificent seat of the dukes of Dorset: it is at present occupied by the Countess Amherst, but has belonged, with little intermission, to the Sackville family, since the time of Elizabeth. It is a large, fine, castellated edifice. The interior, which is nobly furnished, has various pictures by celebrated masters, and other splendid works of art. Sevenoaks has no manufactures: there were formerly some silk mills in the vicinity, but they no longer exist. The town is governed by a warden, a bailiff, and four assistants, chosen at an annual court leet. Petty sessions are held at the Crown hotel on the last Saturday in every month; and a court of requests is held on the first Friday in each month. Sevenoaks is the head of a parochial union. Markets on Saturday; fairs, July 10 and Oct. 12, for hogs and poultry; and the third Tuesday in every month for cattle.

SEVERN, a river of England, being inferior only to the Thames in magnitude, and perhaps. also, in importance. It has its source in a small lake on the eastern side of Plinlimmon mountain, in Montgomeryshire. At its outset it is called the Hafren, the name by which, through its whole course. it was known to the Britons. It flows first towards the S.E., and afterwards turns to the N.E., as it approaches Newton, where it takes the name of Severn. Hence, through the vale of Montgomery, its course is almost due N., till, entering the great plain of Salop, beyond Welshpool, it turns abruptly to the S.E.; and pursuing the same direction, it almost encircles Shrewsbury. Flowing through Colebrook dale, and passing Bridgenorth, it follows a southerly course as it leaves Salop, and enters Worcestershire at Bewdley. Being now become a broad and deep river, crowded with barges, it rolls through a pleasant country in a tranquil stream, passing the city of Worcester, and traversing the vales of Evesham and Gloucester. In the latter it divides into two channels, one of which washes the walls of Gloucester: but, being again united, it forms a great tidal river. Its course from Gloucester to Nass point is tortuous; from the latter it flows S.W., till it assumes the name of the Bristol channel, expanding and insensibly losing itself in the Atlantic ocean.

The Severn, particularly below Gloucester, has frequently overflowed its banks, and occasioned much damage to the surrounding country. It is remarkable for its tide, which rushes in with a head 4 or 5 ft. high, and a loud noise. This, no doubt, arises from the wide expanse of the waters of the Atlantic in the Bristol channel being gradually narrowed, till at length they are forced violently up the river. Outside the Bristol channel, spring tides rise from 22 to 24 and 26 ft.; but in King's road, at the mouth of the lower Avon, they rise to the height of 48 ft., and sometimes more; and at Chepstow the rise is 60 ft. (*Norie's Sailing Directions for the Bristol Channel*, 98.) The opposition which the current from the sea meets with from

the adverse current of the river occasions that damming and grinding of the waves known by the name of bgrs or eagrs.

The Severn is navigable from Flatholm lighthouse. where it loses itself in the Bristol channel, to Welshpool, a distance of about 178 m.; and its navigation is continued by the Montgomery canal to Newton. If its, consequently, of the highest importance as a channel of internal communication; its capacity in this respect being materially increased by its numerous large tributary streams, and by the canals and railroads that join it. By means of the latter, it commands a large share of the commerce of Birmingham, and of the various trading towns of Staffordshire, Warwickshire, &c., and is united with the Thames, the Trent, and the Mersey. From Welshpool to the sea it has a gradual fall of 225 ft. (*Priestley on Inland Navigation, &c.*, 596.)

The navigation of the Severn from Nass point to Gloucester is both tedious and difficult. To obviate this inconvenience, a canal on a large scale has been dug from Berkley Pill to Gloucester. It is 18½ m. in length, from 70 to 90 ft. in width, and from 15 to 18 ft. in depth; and may consequently be navigated by vessels of 350 tons. There is a basin at each end for the accommodation of shipping. This canal, which was opened in 1827, has become the channel of an extensive commerce; and Gloucester is now rising fast in importance as a trading and shipping town. (*See* GLOUCESTER.)

The barges which navigate the Severn are about 128 ft. in length, from 19 to 20 in breadth, and five in depth. They carry above 100 tons. The trows are from 60 to 70 ft. long, 20 broad, and five deep, carrying 75 tons. They carry a square-sail, and have a mainmast and topmast.

Of the tributaries of the Severn, the most important are the Teme, the upper and lower Avon, the Wye, and the Usk.

SEVERNDROOG, or SAVENDROOG, a strong hill fortress of Hindostan, in the Mysore territory, 50 m. W. by S. Bangalore. Though it be impossible to invest this place closely, it was, nevertheless stormed and taken without the loss of a single man, by the British, under Lord Cornwallis, in 1791.

SEVIER, county, Tenn. Situated in the E. part of the state, and contains about 600 sq. m. Bounded S.E. by Smoky mountain. Watered by French Broad, and Little Pigeon rivers. It contained in 1840, 6,307 neat cattle, 3276 sheep, 19,896 swine; and produced 48,038 bushels of wheat, 1648 of rye, 339,113 of Indian corn, 50,689 of oats, 18,425 of potatoes, 4110 pounds of tobacco, 18,552 of cotton, 33⅓ of sugar. It had 10 stores, one furnace, one forge, five flouring-mills, 15 grist-mills, 13 saw-mills, four tanneries, 17 distilleries, one pottery; one academy, 23 students: 20 schools, 513 scholars. Pop.: whites, 6048; slaves, 354; free coloured, 40; total, 6442. Capital, Sevierville.

SEVIER, county, Arkansas. Situated toward the S.W. part of the state, and contains 2500 sq. m. Bounded S by Red river. Watered by North Little and Saline rivers and their branches. It contained in 1840, 7676 neat cattle, 997 sheep, 16,091 swine; and produced 1297 bushels of wheat, 150,730 of Indian corn, 8404 of oats, 12,730 of potatoes, 468 pounds of rice, 3477 of tobacco, 705,304 of cotton. It had 10 stores, six grist-mills, two saw-mills, one distillery; seven schools, 168 scholars. Pop. : whites, 3076; slaves, 327; free coloured, 7; total, 2810. Capital, Paraclifta.

SEVIERVILLE, p. v., cap. of Sevier co., Tenn., 213 m. E. by S. Nashville, 494 W. Situated on the S. bank of Little Pigeon river, just above the junction of the S. fork. It contains a courthouse, jail, a church, an academy, three stores, various mechanic shops, and about 175 inhabitants.

SEVILLE, a celebrated city of Spain, famous " for oranges and women" (*Byron*), the cap. of Andalusia, and of the prov. of its own name, in a wide and fruitful plain on the Guadalquivir, crossed here by a bridge of boats, 62 m. N.E. Cadiz, and 212 m. S.S.W. Madrid. Lat. 37° 27′ 30″ N., long. 5° 47′ 47″ W. Pop. acc. to Miñano, 91,380. It has numerous suburbs, but the city proper is about 4 m. in circuit; enclosed by a line of circumvallation 13 m. in circ. The ancient suburb of Triana is on the right bank; but with this exception, Seville lies wholly on the E. side of the river. "The streets, with a few exceptions, are narrow and crooked, some of them being so constructed that one may touch both walls at the same time. Few are wide enough for carriages; and many through which coaches pass, show, by the deep furrows in the walls, that one have touched, and often both at the same time." (*Townsend's Spain*, ii., 315.) The street or place called the Alameda, in the centre of the town, planted with elm trees, is, however, very magnificent; being 600 yards in length by 150 in width, decorated with three fountains, and with statues of Hercules and Julius Cæsar. And since Mr. Townsend's time various improvements have been introduced; old streets have been repaved, obstructions and irregularities removed, and numerous modern wide streets built in straight lines with regular and handsome houses. On the whole, however, Seville has all the peculiarities of a Moorish town, and furnishes a good specimen of the architecture of the Moors in their streets and houses, the former of which, narrow, close, and dirty, appear in strange contrast with the extensive and airy mansions that open on them, neatly whitewashed, and studded with numerous windows, each having its cool-looking, green Venetian shutters. The *Pasco* and the *Delicias* are the principal public walks, and perhaps in point of rural beauty are superior to any in Spain. The former is here what the *Prado* is in Madrid; and in it the population may be studied to the best advantage. Among the public buildings are 31 churches, including the cathedral, numerous large edifices formerly conventual, but many of which have lately been turned into manufactories; an exchange, guildhall, 10 hospitals, one of which is military, an asylum for decayed priests, eight sets of barracks, seven prisons, and two theatres.

The cathedral, built in the 14th and 15th centuries, occupies the site of a Moorish mosque; but it seems highly probable that it was a Christian church prior to the Mohammedan conquest. It has five naves, but no dome or central tower. It is, according to Townsend, 420 ft. in length by 263 ft. in breadth: the height from the floor to the roof being estimated at 126 ft. "The cathedral," says Mr. Swinburne, "is more cried up than I think it deserves; it is by no means equal to York minster for lightness, elegance, and Gothic delicacy. The clustered pillars are too thick, the aisles too narrow, and the choir, by being placed in the centre, spoils the whole *coup d'œil*, and renders the rest of the church little better than a heap of long passages. The ornamental parts are but clumsy imitations of the models left by the Moors. Not one of the great entrances or porches is finished; and to disfigure the whole pile, a long range of buildings in the modern style has been added to the old part." (ii., 23.) The only remaining parts of the mosque are the Giralda, or belfry, and the great gate of the cloisters, the latter of which is a fine specimen of the best style of Moorish architecture.

The most admired feature in the whole cathedral, however, is the *Giralda* (weathercock), a brick tower, 258 ft. in height, and exactly square, each side being 50 ft. in breadth. This is surmounted by four smaller towers, which are crowned by a small cupola, the whole terminating in the *giralda*, which gives its name to the tower, a colossal bronze statue of Faith, bearing a flag and palm-branch, 14 . in height, and of great weight, but so delicately poised as to turn with the slightest variation of the wind. The weight from the ground to the top of the statue is said by ónano to be 364 Spanish feet. The ascent to the top of the great tower is by an inclined plane, so gradual that one may ride up without inconvenience; the view from the summit is superb, extending over the entire plain. In point of riches, this cathedral ranked second only to that of Toledo. It had, also, with the church of the Capuchins, and a chapel of the hospital *de la Caridad* some noble pictures by Murillo; but, though some of these have been reserved, others have become, either by purchase or the right of conquest, the property partly of private individuals, and partly of Marshal Soult and other French generals.

Some of the Castilian monarchs are buried within the cathedral; but, as Bourgoing has observed, these tombs are no emotions compared to those excited by the sight of the slab, in front of the choir, which once covered the remains of COLUMBUS. It is inscribed, *A Castilla y Aragon Mundo dio Colon*—To Castile and Aragon Colon (Colon) gave another world. A sublime and not too abstruse an epitaph. We may add, that the remains of Columbus, after reposing here for about 30 years, were carried across the Atlantic, and deposited in the cathedral of Domingo: but, in 1795, on the capture of St. Domingo by the blacks, the ashes of the illustrious dead were again disturbed by the whites, and carried to the Havana, where they are now deposited. It is worthy of notice that the organ belonging to the cathedral was begun in 1580, by bequest of 20,000 volumes left for the purpose by Hero, one of Columbus's sons.

the large organ, which is considerably larger than that of Haerlem, has altogether 5300 pipes, with 110 stops. "Nothing," says Inglis, "can exceed the majesty of the tones awakened by this organ, and at times, the effect is almost too overpowering for human senses." (*Inglis*, ii., I.; *Cook's Sketches*, i., 139, and ii., 93-96.)

Immediately under the Giralda, occupying one side of a square, is the archbishop's palace, with a handsome and opposite to it is the *Lonja*, or exchange, a quadrular edifice, with a central *patio*, comprising apartments, some of which are still used by the merchants, in the greater part has been converted into an *Archivio papeles de Indias*, or repository for American archives; the voluminous records here preserved being as tidily placed and ticketed as if Spain still continued to

give law to her *ci-devant* transatlantic possessions! The floors are laid in chequered marble, and remarkably handsome. (*Scott's Ronda and Granada*, ii., 105.) A little removed from the Lonja is the Alcazar, a royal palace and gardens, said to have been constructed in imitation of the Alhambra, principally by Peter the Cruel and Charles V. Swinburne correctly terms it, "a pasticcio of Saracenic, conventual, and Grecian architecture." The exterior has a miserable appearance; but the first court, after entering the gate, has a grand effect. It is 93 ft. in length, by 69 ft. in breadth, flagged with marble, and surrounded with a colonnade of white marble Corinthian pillars, of handsome proportions and well executed, the walls behind being covered with grotesque designs in the Moorish taste. Next to the Court of Lions, in the Alhambra, this court is perhaps the best piece of Arabic building in Spain for execution and delicacy of design, though the ornaments of the palace in Seville be much inferior to those of that in Granada. The Alcazar comprises a suite of 78 successive apartments, having carved ceilings, with walls, like those of the Alhambra, with well-preserved arabesques. By far the most splendid, however, is the Hall of Ambassadors, a splendid apartment, adorned with designs in stucco, and with a floor of variegated marble. Within the Alcazar are many fine paintings, by Murillo, Velasquez, Luis de Vargas, and other Spanish masters, with a few specimens of the Italian school; but several of the best pictures have, within the last few years, been removed to the public gallery at Madrid. A considerable portion of the palace is now let out in lodging-houses, and to private individuals; the portion reserved for the sovereign comprising only a small section of the entire pile. The gardens, which are of small extent, are laid out according to the Moorish taste, in formal alleys, with chipped myrtle hedges and trees, cut to resemble warriors armed with clubs. The walks in some parts are laid with tiles, through which *jets-d'eau* are made to flow, which, by turning a screw, suddenly water not only the garden, but its unwary visiters. The *Casa Pilata*, another of the sights of Seville, is a private house, said to have been built on the exact model of that of the Roman governor of Jerusalem! Within the city, also, are many structures of Roman origin, which still show traces of their former magnificence. The octagon tower, or *Torre d'Oro*, was probably built by one of the Cæsars. The *Caños de Carmona*, a Roman aqueduct of 410 arches, still conveys water to the city from Alcala; and the gates, especially that of Triana, are very magnificent, though of equally ancient origin. Most of the other objects worth notice are without the walls. The first in order, is the *Plaza de los Toros*, or circus for bull-fights, half wood and half stone, and capable of accommodating 14,000 spectators.

The next remarkable object is the royal tobacco manufactory, a huge edifice, 440 ft. in length by 280 ft. in breadth, so strongly built and guarded by walls and ditches, as to appear like a fort or citadel, raised to overawe the citizens. It employs about 400 hands, of which more than half are engaged in making cigars. But, despite all the precautions of government, fully nine tenths of the cigars made use of are said to be clandestinely imported. (*Scott*, ii., 110.) The cannon-foundry is, on the whole, a creditable national institution, though not at present in any great activity. Among the other public establishments may be specified the cavalry barracks, royal saltpetre manufactory, military hospital, &c. The market-place is large and admirably suited to its purpose, the buildings being arranged in streets, an open space surrounding the whole, with gates and ornamental fountains. In the suburb of Triana is a separate market for the supply of the *gitanos*, or gypsies, its chief inhabitants.

The arrangement of the streets is very different from that observable in most other Spanish towns, and is mainly the effect of the hot climate. To a similar cause may be traced the internal arrangement of the houses. They are built almost universally in the form of a square, with a spacious courtyard, or *patio*, frequently paved with marble, and surrounded by piazzas opening on the apartments of the ground-floor; the exterior as well as every other part of the house being kept carefully whitewashed, the massive green wooden blinds of the windows being kept closely shut during the day. In addition to this, the rooms, which are usually paved with tiles, are furnished with ponderous window-shutters half a foot thick, kept shut till the sun is off the windows, when they are partially opened to admit the breeze. Hence the houses are so very dark that visiters at first with difficulty distinguish the inmates. The climate may also be said to divide the houses into two distinct parts. During the winter months (commencing in October and ending with April) the family inhabits the upper parts of the house, which are then thickly matted, and the rooms artificially heated by brasiers of charcoal; but when the hot weather sets in, these apartments are shut up, and a general move is made to the ground-floor, which, being considerably cooler, and opening on the *patio*, renders the heat more endurable "It is a pretty sight, indeed," says Sir A. C. Brooke (*Travels in Spain and Morocco*, i., 45). "to saunter during the delicious moonlight evenings of summer along the fashionable streets of the city; and nothing can be more strikingly brilliant than the appearance of the houses and hotels of the nobility and wealthier classes. On looking through the trellissed iron door opening to the street, you perceive the entire *patio* brilliantly illuminated, well furnished, and with pictures suspended from the marble columns of the arches. An awning forms a sufficient roofing by night as well as by day, and converts the space below into a spacious and lofty saloon, in the centre of which different *jets d'eau* spout forth from a marble fountain, both cooling the air and watering a variety of sweet, odoriferous plants, scattered around in flower-pots. Here the young ladies of the family may be seen enjoying the coolness of the evening, engaged in work, amusing themselves with music and singing, and receiving the visits of their friends." These summer habits are truly Moorish; and even in trifles glimpses of them become easily visible, as, for instance, in the contempt of chairs, for which mats and low stools are pretty generally substituted by all classes.

The aspect of the population of Seville differs greatly from that of Madrid. Even in the upper ranks, there is something in the ladies of an eastern appearance: they are more frequently veiled, their cheeks seem tinged with a hue of Moorish blood, and, along with the fire of a Castilian eye, there is mingled a shade of Oriental softness. Among the lower orders of the women, also, as among the Moors, may be remarked an extravagant and tasteless profusion of gaudy ornaments, immense ear-rings and bracelets, numerous rings, &c.; and the dress of the Andalusian peasant is even more grotesque and ornamented than that of the women, his jacket and waistcoat being almost always trimmed with gold or silver, and every article of his dress covered with silk cords and buttons. Another striking difference between Madrid and Seville is in the great mass of ragged, wretched-looking people in the latter, in consequence mainly of the heat of the climate, which renders labour a disagreeable exertion, especially in a country where subsistence is so easily procured. Let a small loaf of bread be given to one of these sons of idleness, he makes a hole in it, begs a little oil, not worth refusing, which he pours in, and soaking his bread as he eats it, he is set up for the day; and if he succeeds in getting a two-quarto piece, he may procure as many grapes as his heart can desire. What incitement has such a one to be busy? The upper and middle ranks of Seville live more luxuriantly, but not better than those of Madrid; for the luxuries of the former, their iced waters, lemonade, and pomegranates, their cool patios, fountains and baths, are necessary to health and comfort. But even in his ordinary diet, the Andalusian has the advantage over the Castilian; for though it be true that, like the inhabitants of the northern provinces, he dines on the eternal *puchero*, its ingredients are better in Andalusia than in Castile, the pigs being fed on the ilex-nuts and the vegetables of South Spain, being perhaps the finest in the world. The difference between Andalusia and Castile is still further observable from the state of society in the two provinces. The terrulin of Seville is quite different from that of Madrid, the former being at any rate more animated, if not more intellectual, and the dullness helped out with cards, dancing, forfeits, and other amusements, independent of mere chit-chat and *persiflage*. Balls and suppers are reserved for great occasions; but certainly substantial entertainments are more general than in the capital, perhaps because wealth is more generally diffused. Morals are at a very low ebb.

> The feast, the song, the revel here abounds:
> Strange modes of merriment the hours consume,
> Nor bleed these patriots with their country's wounds;
> Nor here War's clarion, but Love's rebeck sounds:
> Here Folly still has weavers inthralls;
> And young eyed Lewdness walks her midnight rounds:
> Girt with the silent crimes of capitals,
> Still to the last kind Vice clings to the tottering walls.
>
> *Childe Harold*, i., st. 46.

In Seville it is almost a derision to a married woman to have no *cortejo*, and a jest against a señorita not to have her *amante*. Indeed, the gallantries of the latter are not unfrequently carried quite as far as the intrigues of the former. (*Inglis*, ii. 39—41.)

But with all this corruption, the course of society runs smooth: jealousy appears not to disturb the *ménage*, the parties living together with all the outward show of mutual esteem, and inflicting the history of their private bickerings only on their most intimate friends. The amusements of the middle and higher classes consist of the daily promenading on the *Paseo* or *Alameda* (the Hyde Park or Regent-street of London); theatrical entertainments, of which they are passionately fond, and no mean judges; and the tertulia, which are so arranged as to succeed each other in the

arrangements of the day. The lower classes are fond of dancing; but of music they have little knowledge, in nothing can well be more disagreeable than their cray *guitars*. (*Scott's Ronda*, ii. 123—124; and *Inglis*, ubi supra.)

Seville, as a place of residence for a stranger who can only for sensual gratifications, is perhaps preferable to any other Spanish city. It is said that there is not a day throughout the year in which the sun does not shine on Seville. Winter is scarcely felt; and if the heats of summer be oppressive, as they truly are during the prevalence of the *solano*, the streets, houses, and economy of life are admirably adapted to lessen their influence. The surrounding country, with its orange and lemon groves, acacias, and other flowering trees and shrubs, is all that one can desire, and of many varieties and choice flavour may be had almost for nothing, and every necessary of life may be procured in abundance, and at very moderate rates. Game, fruit, and vegetables are excellent; and the bread (brought to market from the neighbouring village of *Alcala des Panaderos*) is said to be the best in Spain. Meat is reasonable, but of rather indifferent quality.

Seville has several establishments for the promotion of learning, science, and general education; but of these few, if any, can be considered as very efficient. Its university, founded in 1502, is in the most backward state possible. The other scholastic establishments comprise a school of medicine, two mathematical schools, a college of agriculture, and an academy of the fine arts, besides the ancient, though decaying, school of St. Elmo for navigation and gunnery. Seville has also several societies for the promotion of different branches of literature and science: but they exercise little influence, owing to the general want of sound elementary education.

If we might believe the stories which book-maker after book-maker has repeated, *nauseas ad nauseam*, of the former flourishing state of manufactures in Seville, we should certainly conclude that in the 15th and 16th centuries it was decidedly superior, as a manufacturing town, to what Manchester now is! It is said, for example, to have had in 1519, 16,000 silk looms, and 130,000 persons employed in the various branches of the silk manufacture! Such a statement carries absurdity on its face; and Capmany has shown conclusively, from the letter of the Venetian ambassador, Navagero, who visited Seville in 1525, and otherwise, that it is doubtful whether it had then a dozen silk looms; and that, instead of being a city with some 20,000 or 600,000 inhabitants, as must have been the case had it had 130,000 engaged in the silk trade only, there is no means to think that it was then larger or more populous than at present. (*Questiones Criticas*, p. 87, &c.) It is true that at a subsequent period the silk manufacture attained to considerable importance in Seville, there being, in 1688, about 3000 looms engaged in the business. The manufacture has since undergone many vicissitudes; but in the earlier part of the present century it employed about 500 looms. Owing, however, to the loss of the colonial markets, and still more to the harassed state of the country for many years back, the number of looms is at present reduced to from 300 to 600. Coarse woollen cloths are made in considerable quantities, but they are both much inferior to, and much dearer than similar English fabrics. There are several large tanneries, manufactories of hats, combs, and earthenware, &c.; but, as in the rest of Spain, the processes are so clumsy, that, speaking generally, all manufactured articles are of inferior quality. The tobacco manufactory, iron foundry, and saltpetre establishment, have been already mentioned as government monopolies. The trade of Seville rose to considerable importance after the discovery of America, in consequence of its being vested with the monopoly of the commerce between Spain and the New World. This advantage, however, was soon lost, from the difficulty of navigating the Guadalquivir with large vessels; and the trade was transferred to Cadiz. The river, at certain times of the year, is accessible as far as Seville for ships of 160 tons; but generally speaking, all vessels drawing more than 160 tons are obliged to load and unload 6 m. below the city. Some efforts, however, have lately been made for the improvement of the navigation. The exports comprise wool, gum, leather, oil, silk, and fruit, particularly oranges. The trade in oranges is carried on principally with England, to which about 40 cargoes are sent every year, consisting of about 5000 chests, 1-10th part of which are bitter, and the rest sweet oranges: the chief part of the export takes place in Nov. and Dec. The imports comprise various manufactured from England (many of which, however, are contraband), hides, hemp, and flax, from the Baltic; iron from Bilbao, and colonial produce from Cuba and Porto Rico. A considerable coasting trade is carried on with Cadiz, Malaga, Barcelona,

---

* Among others, these statements are given as if they came sanctioned by Laborde (*Itineraire*, iii. 324. ed 1809); and by Moreau de Jonnès, in his *Statistique d'Espagne*, 148. The latter indeed, is a wretched performance, utterly unworthy of confidence.

and other ports of Spain; and there is daily steam communication with St. Lucar and Cadiz.

Seville stands on the site of the *Hispalis* of the Romans. It opened its gates to the Moors in 711, soon after their invasion of Spain, and continued in their possession above five centuries, being the seat first of a regal, afterwards of an aristocratical government. It was taken by the Christians in 1247, after one of the most obstinate sieges mentioned in Spanish history; but since then it has seldom been the scene of military exploits. It is known in diplomatic history by a treaty concluded in it in 1729, by Spain, England, France, and Holland. In the autumn of 1800, it was visited by the pestilential fever which caused such mortality at Cadiz, and it is said that between the 12th August and 1st November of that year, it lost nearly a fourth part of its inhabitants, half the sufferers being, however, Gitanos or gypsies, inhabiting the suburb of Triana. On the invasion of Spain by Napoleon, in 1808, Seville asserted the national independence, and received the junta when driven from Madrid. It however surrendered to the French, on the 1st February 1810, and remained in their hands till the 27th August, 1812, when they left it, in consequence of their defeat at Salamanca.

Seville has given birth to several distinguished individuals, among whom have been included in antiquity the emperors Adrian, Trajan, and Theodosius. There can, however, be little or no doubt that these illustrious individuals were all natives of Italica, a Roman city, a few miles N.E. from Hispalis. Among the more remarkable individuals of whom Seville has to boast in modern times may be specified Las Casas, bishop of Chiapa, the defender of the Indians; Antonio de Ulloa, the traveller and economist; Lopez Rueda, the father of Spanish comedy, &c. The famous navigator, Magellan, or Magelhaens, sailed from Seville on the 20th of September, 1519, on the expedition in which he recovered the straits that bear his name.

SEVRES, a small town of France, dep. Seine-et-Oise, on the Seine, about midway between Paris and Versailles, being 5 m. N.W. the latter city. Pop. about 4000. It has been long famous for its royal manufactory of porcelain, or *vres China*; which, for elegance of design and excellence of quality, is equal, if not superior, to any made in Europe. large museum is established here, in which are collected specimens of most kinds of earthenware manufactured in ance and other countries. There is a warehouse in the e Rivoli, in Paris, for the sale of Sevres china. The arries whence the clay used in the manufacture of the celain has been obtained form extensive wine vaults. e Seine is crossed here by a handsome stone bridge.

EVRES (DEUX-), a dep. of France, reg. W., principally ween the 46th and 47th degs. N. lat., and 0° and 1° W. g.; having N. Maine-et-Loire, E. Vienne, S. Charente Charente Inférieure, and W. Vendée. Area, 607,350 ares. Pop., in 1836, 294,850. A hill chain, running a S.E. to N.W., divides the dep. into two portions, very ke each other in their general aspect, the southern being nearly flat, and the northern very much diversified. cipal rivers, the two Sevres (or *Niortaise* and *Nantr*), whence the name of the dep.: one discharges itself the Atlantic in Vendée, the latter falling into the Loire. rge proportion of the soil is stony, but there are some tracts. In 1834, the arable lands were estimated at 155 hectares; meadows, 74,953 do.; vineyards, 20,893 orchards, &c., 9675 do.; and woods, 36,090 do. Shallakes, pools, &c., occupy at least 10,000 hectares. Agture is generally very backward, being, in most parts, guished by an obstinate attachment to old methods; ore corn is raised than is required for home consump.

Flax, hemp, various fruits, &c., and about 350,000 l. of wine are annually produced. The quality of the is, with few exceptions, very inferior, and about half odnce is made into brandy. The annual produce of is estimated at 400,000 kilogrammes. Fat cattle, hogs, y, timber, brandy, and vinegar are the principal ex of the department. In 1835, of 127,942 properties subthe *contrib. foncière*, 69,394 were assessed at less than and 17,811 at from 5 to 10 fr. Minerals unimportant. manufacturing industry of the department is of little quence. It is divided into four arronds.; chief towns, the cap.), Pressuire, Melle, and Parthenay. It sends embers to the chamber of deputies. Registered electors in 1831, 5,747,475 *fugts*, art. *Deux-Sevres; French Official Tables.*)

VARD, t., Schoharie county, N.Y., 15 m. W. Scho-47 m. W. Albany. Drained by Cobbleskill cr. It stores, three grist-mills, nine saw-mills, three tan-13 schools, 600 scholars. Pop. 2088.

ICKLY, t., Westmoreland co., Pa. Drained by ly cr. It has five stores, one fulling-mill, four flour-la. six saw-mills, two paper-mills, two tanneries, three ries. Pop. 1572.

FTESBURY, a parl, and mun. bor. and market

town of England, co. Dorset, partly in Sixpenny Handley hund., and partly in Alcester liberty, on the border of Wilts, 29¼ m. N.E. Dorchester, and 95 m. S.W. London Previously to the Reform Act, the mun. and parl. boundaries of the bor., which were co-extensive, comprised only portions of the parishes of the Holy Trinity, St. Peter, and St. James: with a population, in 1831, of 2749. But since then the municipal limits have been enlarged, so as to include the whole of those parishes, with that part of Motcomb in which Enamore Green and Long Cross are situated; and the parliamentary boundary comprises the entire parishes of Cann, St. Rumbald, Melbury and Compton Abbas, Stower Provost, East Stower, Todbere, St. Margaret's Marsh, Motcomb, Donhead (in Wilts), and the chapelry of Hartgrove, making a total area of 20,910 acres. Population, in 1831, 8969.

The town is situated on the top, and extends nearly to the verge of a high narrow hill. It is healthy, though, from its situation, the air is often bleak and piercing. Though irregular, it is well built, a large proportion of the houses being constructed of freestone quarried in the neighbourhood. It formerly had anciently 12 churches, besides several chantries, a celebrated monastery, a hospital, &c. It has now but three churches, the principal of which, St. Peter's, is of great antiquity, and has some elegance, though much disfigured by modern alterations. Holy Trinity is joined with St. Peter's, St. Martin's, and St. Lawrence, in a rectory worth £166 a year. St. James's is also a rectory, worth £236 a year. Both livings are in the diocese of Bristol, and the gift of the Earl of Shaftesbury. (*Eccl. Rev. Rep.*) In the spacious and well-planted churchyard of Holy Trinity is inclosed a considerable portion of the wall of Shaftesbury Abbey, being all that remains of that once famous edifice. It is said to have been erected by the wife of Edmund, great grandson of king Alfred, for Benedictine nuns. (*Camden*, Gibson's ed. i. 60.) It was afterwards called St. Edward's Abbey, from Edward the Martyr, who was murdered at Corfe castle, having been buried in it. After the churches, the principal public buildings comprise a handsome town-hall, recently built by the Marquis of Westminster, at an expense of £3000, and meeting-houses for Friends, Independents, Wesleyans, &c. A free school, for 20 poor boys, was founded in 1719; and there are almshouses for both men and women. From its elevated position, Shaftesbury labours under a deficiency of water, which is conveyed up the hill in carts or on horseback, its supply affording employment to a number of persons. The town had formerly a manufacture of shirt buttons, which employed many women and children; but it has now ceased, and it has few outward signs of prosperity, though it is said that its condition has latterly improved. (*Boundary Rep. &c.*)

Shaftesbury is mentioned as a borough in Domesday Book; but its only existing municipal charter is that of James I., confirmed by Charles II. It sent two members to the House of Commons from the reign of Edward I. down to the passing of the Reform Act, which deprived it of one of its members, and at the same time increased its boundaries as already stated. The election for members was formerly vested in inhabitants paying scot and lot. Registered electors, in 1839-40, 491. Since the Municipal Reform Act it has been governed by a mayor, 3 other aldermen, and 12 councillors. Corporation revenue, in 1840, £180. No courts are held within the borough. Market-day, Saturday; fairs, Palm Saturday, June 24. November 23, for all kinds of cattle.

Shaftesbury is supposed to be on or near the site of an ancient British town called *Caersepton*; but it was of little importance till the foundation of its monastery, and has latterly depended principally on its political privileges. It gives the title of earl to the noble family of Ashley Cooper. (*Parl. Munic. and Boundary Reports, Appendices, &c.*)

SHAFTESBURY, p. t., Bennington co. Vt., 110 m. S. Montpeller, 414 m. W. Watered by branches of Walloomsac and Battenkill rivers. It contains three churches, two Baptist, and a Methodist, two stores, one fulling-mill, one grist-mill, four saw-mills, two tanneries; one academy, 20 students, 10 schools, 305 scholars. Pop. 1885.

SHAHABAD, a district of British India, presid. Bengal, prov. Bahar, between the districts of Patna, Bahar, and Ramghur, on the E. and S., and Benares, Ghazepoor, and Sarun, on the W. and N. Area, 4650 sq. m. Pop. (1822), 908,850, nearly all Hindoos. The Ganges bounds it N., the Sone W., and the Carumrassa E. It is very fertile, its staples being opium, tobacco, cotton, sugar, indigo, and hemp; it is celebrated for the excellence of its roads, a distinction mainly owing to a salutary reservation in the original land settlement with the Zemindars of a certain annual sum to keep them in repair. Total land revenue (1829-30), 14,60,483 rupees.

SHAHJEHANPOOR, a district of British India, prov. Delhi, having N.E. Nepaul, E. Oude, S. the latter and the district of Furruckabad, and W. Saiswan, Bareilly, and Pillibheet. Area, 1690 sq. m. Pop. uncertain. Total public revenue in 1829-30, 11,94,136 rupees. Its capital town of the same name. 175 m. S.E. Delhi, is reported to be more wealthy and nearly as populous as the latter city; so that it may probably have 50,000 inhabs. (*Hamilton*.) There are three other towns, named Shahjehanpoor, in Malwah, Lahore, and N. Hindostan.

SHALERSVILLE, p. t., Portage co., O., 5 m. N. Ravenna, 144 m. N.E. Columbus, 330 W. Watered by Cuyahoga r. It has seven schools, 265 scholars. Pop. 1281.

SHAMOKIN, p. t., Northumberland co., Pa. 76 m. N. Harrisburg. 189 W. Watered by Shamokin creek. It contains four churches, six stores, four grist-mills, three saw-mills, three tanneries, one distillery; six schools, 205 scholars. Pop. 1083.

SHANDAKEN, p. t., Ulster co., N.Y., 83 m. S.S.W. Albany, 342 W. Drained by Esopus creek and Nevisink r. It lies on the range of the Catskill mountains, and contains two stores, two grist-mills, 17 saw-mills, four tanneries, seven schools, 301 scholars. Pop. 1455.

SHANNON, a river of Ireland, being by far the largest and most important in that island, and hardly indeed inferior, if it be not superior, to any in the United Kingdom. It has, in many respects, particularly in its nearly insulating an extensive province, in the direction of its course, the length of its navigation, and the magnitude of its æstuary, a striking resemblance to the Severn. Its source is generally traced to the base of Cuilcagh mountain, in the N.W. part of Cavan. After running a few miles, it falls into lough Allen, about 10 m. in length, and from 4 to 5 m. broad; its course thence to Limerick being S., with a small inclination to the W.; issuing from lough Allen, it passes Leitrim, Carrick, Tarmonbury, &c., entering lough Ree, at Lanesborough. This, which is a very irregularly-shaped extensive sheet of water, is about 17 m. in length. Leaving it, the river, now greatly augmented, passes Athlone, and then winds by Shannon Bridge and Banagher to Portumna, near which it expands into lough Derg, a narrow lake, 23 m. in length, with deep bays and inlets. Escaping from the S. extremity of this lake, it flows on to Limerick. Here having met the tide, it takes a W.S.W. direction; and, gradually expanding into a noble æstuary, unites with the Atlantic, between Kerryhead and Loop-head, about 70 m. lower down.

From the head of lough Allen to its mouth, the Shannon has a course of about 214 m.; viz., lough Allen, 10 m.; lough Allen to lough Ree, 43 m.; lough Ree, 17 m.; lough Ree to lough Derg, 36 m.; lough Derg, 23 m.; lough Derg to to Limerick, 15 m.; and thence to the river's mouth, 70 m. Loop-head and Kerry-head are about 8 m. apart.

But the distance to which it has been rendered navigable is the most extraordinary circumstance connected with the Shannon. In this respect, indeed, it is superior to the Thames, Severn, Trent, or any English river. If lough Allen be (as it is considered by some) reckoned its source, it is navigable to its very head; but, tracing its origin to the base of Cuilcagh mountain, there are only 6 or 7 m., out of its entire course of about 230 m., that may not be navigated! It is unnecessary to insist on the value of a river of this sort flowing through the very centre of Ireland, insulating the great province of Connaught, and "washing the shores" of 10 out of the 32 counties which the island occupies.

Unluckily, however, the navigation of the Shannon, like that of most other rivers not of very great depth, is, in certain places and at certain seasons, a good deal obstructed. It may be navigated, with no very serious difficulty, from the sea to Limerick by ships of 400 tons burden. But immediately above the city, and in some other places, its course is impeded by rocks and rapids, and large sums have been expended in improving those parts of the navigation, partly by making lateral cuts, and partly by deepening the bed of the river. The level of lough Allen is about 144 ft. above high water-mark at Limerick, the ascent being in a great measure overcome by one double lock and twenty single locks, placed in those situations where lateral cuts have been made to avoid the rapids. These cuts are from 13 to 14 ft. wide at bottom, having the usual slopes, and are calculated for a depth of water varying from 4 to 7 ft. in ordinary seasons. Still, however, it must be admitted that considering its paramount importance, the navigation of the Shannon is by no means in a satisfactory state. In dry seasons it is impeded by shallows, on which there are sometimes only from 2 to 3 ft. water; and during floods the channel of the river, owing to its frequent expanding into extensive lakes, and the lowness of its banks, is not easily discovered. Had it been an English river, these difficulties would have been overcome long ago; and the money expended upon it might, had it been properly and effectually applied, have sufficed to obviate them. But the works have not unfrequently been very unskilfully and insufficiently executed. It is now, however, under much better management; but it will require a considerable additional expenditure to put the works into proper order, and to ensure at all times, what is so very essential, a safe and easy navigation. The introduction of steam tugs and steam vessels on the loughs of

800

the Shannon has been of infinite service; without them, indeed, it never could have been turned to much account.

The Suck, the principal tributary of the Shannon, rises in Roscommon. Its course is S., inclining to the E. dividing the counties of Roscommon and Galway, by Castlerea, Athleague, and Ballinasloe, till it unites with the Shannon at Shannon Bridge. On its E. side the Shannon receives the Inny, the upper and lower Brosna, Mulkerna, Maig, Fergus, &c. The last two are navigable to a considerable distance.

The importance of the Shannon, as a commercial river, has been materially increased by its junction with the Grand and Royal canals from Dublin. Though defective both in their plan and execution, and made at an immense expense, still it is not to be denied that they are, particularly the Grand canal, of great public utility. In connection with the Shannon, they have opened a communication by water across the island, so that persons living in its centre may send their produce, at a moderate expense, to Dublin or Limerick, as they find most advantageous. This laying open of new and almost boundless markets has given a stimulus to the improvement of the central parts of Ireland, of which it is not easy to overrate the influence, and which will, no doubt, be as permanent as it is powerful.

From its situation at the head of the æstuary of the Shannon, in a country naturally of the most exuberant ferality, 70 m. from the sea, Limerick is the principal emporium of the W. of Ireland; and its commerce is both extensive and rapidly increasing. The value of the produce, such as corn, flour, bacon and pork, butter, beef, &c., shipped from the port in 1820, amounted to £525,625; and in 1835, when prices were very low, to £726,000. In ordinary years it may now (1842) be taken at £1,200,000.

The badness of the accommodation for shipping is, however, a heavy drawback upon the trade of Limerick. At low water ships are obliged to lie aground; and as the bottom consists of hard, rugged, limestone rock, vessels of considerable burden, and those that are sharp built, are liable to be seriously injured by grounding.

But, as already stated (see art. LIMERICK), measures are now in progress to obviate these inconveniences, by constructing a floating harbour in the bed of the river opposite the city. (*Statistical Acc. of the British Empire.* i. 330.)

SHANNON, co., Mo. Situated in the S. part of the state, and contains 2400 sq. m. Drained by Current river and its branches, and by Big Black river. Organized since the last census. Capital, Shannon C.H.

SHAPLEIGH, p. t., York co., Maine, 89 m. S.W. Augusta, 522 W. Incorporated in 1785. It has several ponds, one of which gives rise to Mousom river, and another to Salmon Falls river. It has four stores, one fulling-mill, one woollen factory, three grist-mills, eight saw-mills, two tanneries; 634 scholars in schools. Pop. 1510.

SHARON, p. t., Windsor co., Vt., 40 m. S. Montpelier, 495 m. W. Watered by White r., which affords water power. It contains a Congregational church, and some Methodists and Baptists, two stores, one fulling-mill, one grist-mill, eight saw-mills, one paper-mill, two tanneries; 15 schools, 407 scholars. Pop. 1571.

SHARON, p. t., Norfolk co., Mass., 17 m. S.S.W. Boston, 423 m. W. Watered by Neponset r., issuing from Massapoag pond. It contains three churches, two Congregational and a Baptist, three stores, one cotton factory with 1152 spindles; five schools, 270 scholars. Pop. 1075.

SHARON, p. t., Litchfield co., Ct., 48 m. W. Hartford, 322 m. W. Incorporated 1739. Bounded E. by Housatonic r., across which are several bridges connecting it with Cornwall. It has a pleasant village at the centre, built chiefly on one extended street, containing three churches, a Congregational, Episcopal, and Methodist, an academy, and about 50 dwellings, some of them neat. The town has six stores, one cotton factory with 790 spindles, one furnace, one forge, two grist-mills, three saw-mills, two tanneries: 16 schools, 475 scholars. Pop. 2407.

SHARON, p. t., Schoharie co., N.Y., 43 m. W. Albany, 338 W. It contains a celebrated mineral spring, strongly resembling the White Sulphur springs of Va., highly efficacious in rheumatic, cutaneous, and dyspeptic complaints. A gallon of water by the analysis of Dr. Chilton, of New-York, contains sulphate of magnesia 42·40 grains, sulphate of lime 111·02, chloride of sodium 2·34, chloride of magnesium 7·42, hydrosulphuret of sodium, hydrosulphuret of calcium, and vegetable extractive matter 2·29; total, 166·24 grains, and sulphuretted hydrogen gas, 16 cubic inches. A splendid hotel has been here erected for the accommodation of visitors, and the situation is airy, and the surrounding scenery highly picturesque. Where sulphur waters are required, they are unsurpassed in the country, and are increasingly frequented. The town contains six churches, a Methodist, Dutch Reformed, two Lutheran, a Baptist, and a Universalist, eight stores, three fulling-mills, two grist mills, seven saw-mills, three tanneries; 14 schools, 783 scholars. Pop. 2520.

SHAWANGUNK, p. t., Ulster county, N.Y., 94 m. S.W. Kingston, 87 m. S. Albany, 296 W. Drained by Shawangunk cr. and Wallkill r. Nine mammoth skeletons have been dug up in this and the adjoining towns, one of which is exhibited in Peale's museum at Philadelphia, and another in Europe. It contains six stores, one woollen-factory, four grist-mills, two saw-mills; 19 schools, 795 scholars. Pop. 3836.

SHAWNEETOWN, p. v., Gallatin co., Ill., 195 m. S.S.E Springfield, 782 W. Situated on the N.W. bank of Ohio river, 10 m. below the mouth of Wabash river, and 117 m. above the mouth of Ohio river. It derives it name from the Shawnee Indians, to whom it formerly belonged. The great United States salines are 12 m. back of the town. The bank of the Ohio has a gradual ascent, but is subject to inundation at extreme floods. It has 15 stores, a bank, a United States land-office, a printing-office issuing a weekly newspaper, and 862 inhabitants. It was laid out in 1814, and is one of the most commercial places in southern Illinois.

SHEBOYAN, co., Wis. Situated in the E. part of the territory, and contains 500 sq. m. Pop. 133. Capital, Sheboygan, at the entrance of Sheboygan river into lake Michigan.

SHEERNESS, a seaport and market-town of England, in he par. of Minster, lathe S. Cray, co. Kent, on a low tongue of land at the N.W. extremity of the Isle of Sheppy, at the confluence of the Thames and Medway, on the E. bank of he latter, 18¾ m. W.N.W. Canterbury, and 36½ m. E. by S. London. Pop., including parish, in 1831, 7983. The town, which owes its rise to the formation of the naval dockyard, is divided into three parts, called respectively, Sheerness-roper, Blue-town, and Miletown, the first two being enclosed by fortifications. During the last few years, also, and especially since the fire of 1817, which destroyed about 300 houses, the town has been much enlarged, as well as greatly improved, by the erection of good brick houses, and is for formation of several new streets, well paved, and lighted with gas. The town was formerly very ill supplied with water; but at the beginning of the present century a well as sunk by the board of ordnance to the depth of 360 ft., which supplies water, not only to the town and garrison, but the shipping in the Medway. A pier with a causeway runs down from the town to low-water mark; and facing the river and sea is a wharf of considerable extent. Several old ships of war, also, have been stationed on the ore as breakwaters; formerly they used to serve as dwellings for many of the poorer townspeople, but few of them now inhabited. The parish church is at Minster, but a handsome district church has been erected in the Gothic le; and attached to the garrison is a chapel, the appointment to which is with the board of admiralty. The Baptists, independents, Wesleyan Methodists, Unitarians, and non Catholics, have their respective places of worship, there is a Jews' synagogue. Sunday-schools, are attached to the town church, and to several of the chapels: an infant-school is attended by about 200 children; a proprietary school has upwards of 40 boys, and there is a small endowed charity-school. The trade of Sheerness arises chiefly from the dockyard, and other government establishments, and considerable shipments are made to London of corn and seeds produced on the island, and of oysters from the existing oyster-beds. Pyrites are collected from the falling cliffs for the copperas works in the neighbourhood; and many of the inhabitants make a living by picking up or dredging for septaria (an oxide of iron), used in making Parker's Roman cement. Sheerness has also become, to a certain extent, a resort of sea-bathers, for whose accommodation there are reading-rooms, baths, bathing-machines, &c. Steamers run daily to and from London in summer, besides passage-boats to and from Chatham, which is about 11 m. up the Medway. Markets on Saturday

dockyard, which covers an area of about 50 acres, enclosed by a substantial brick wall, has been greatly enlarged and improved during the last 25 years, at an expense of above £1,000,000 sterling. It has every convenience for building, repairs, and fitting out of ships. It comprises a dock or basin of about 3¼ acres, capable of accommodating 10 sail of the line, and in which they may take on board their stores, ammunition, and provisions, and be, in all respects, equipped ready for sea. Three dry docks, each capable for the accommodation of a line-of-battle ship, have been constructed on the E. side of the basin, and open to it. It has also very extensive storehouses, with mast-houses, mast-ponds, and slip, smithery, and artificers' workshops; if every description; with handsome residences for the commissioners, port admiral, and other officers of the establishment. The principal offices of the ordnance department were, a few years since, removed to Chatham, and have formerly occupied by them has been added to the dockyard. The wharf wall, on the S. side of the basin

in front of the mast house, is 300 ft., and that on the river front 60 ft. in width, lined on both sides with granite. Numerous convicts are employed in the dockyard and on the hulks, chiefly in the improvement and repairs of the former. Sheerness, which so late as the time of the commonwealth was a mere swamp, was fixed upon after the restoration as an important position for a fortress. The works, however, were still incomplete when the Dutch under De Ruyter, in 1667, took and destroyed the fortress and the shipping. (See CHATHAM.) The fortifications were afterwards constructed on a larger scale; numerous batteries of heavy artillery were planted on both banks of the river. The dockyard was begun early in the last century. The mutiny of the fleet at the Nore, in 1798, threatened the town and dockyard with destruction, which, however, was happily averted. (Encyc. Brit., new ed., art. Dockyards; Hasted's Kent, &c.)

SHEFFIELD, a parl. bor., market town, and par. of England, cap. of the district of Hallamshire, W. riding county York, upper div. of wap. Strafforth and Tickhill, at the confluence of the Don and Sheaf, the former of which is crossed by three and the latter by two bridges, 30 m. S. Leeds, and 140 m. N. by W. London. Area of parl. borough and parish, which are co-extensive, 22,830 acres; population of parish, in 1801, 45,755, in 1831, 91,692, and in 1841, 110,891. Population of Sheffield township in 1831, 59,011; in 1841, 67,967. The town, originally confined to the slope of a hill rising S.E. from the Don, is now nearly 2 m. in length by 1 m. in breadth, and occupies the bottom and sides of several low hills, rising in various directions both from the Don and Sheaf, the whole being well paved and flagged, lighted with gas, and abundantly supplied with water. The older streets are steep, narrow, and singularly irregular; but the more modern streets are wide and straight, lined with good brick houses, and many of the shops are but little inferior to those of the metropolis. The smoke, however, proceeding from the numerous steam-engines, forges, and factories, gives the town a dingy, mean appearance, contrasting strangely with the extreme beauty of the surrounding country, embellished as it is, in every direction, by the numerous villas of the opulent bankers, merchants, and manufacturers of Sheffield. The market-place occupies a wide open space in the High-street, and near it are the parish church and principal inns: it is of modern construction, and comprises large shambles, and other accommodations. The corn-exchange, also, is a handsome modern building, comprising excellent accommodation for those frequenting the markets. The cutlers' hall in Church-street, belonging to the ancient corporation of cutlers, is a handsome stone building, with rooms for the transacting of corporate business, public meetings, dinners, &c. The parish comprises 12 churches, seven of which are in the town, and the rest in the rural townships, the patronage of all, except two, belonging to the vicar, though the appointment of the four assistant clergy men at the parish church belongs to the 12 burgesses, or chief inhabitants of Sheffield. The mother church of the Holy Trinity is a noble Gothic structure, about 240 feet in length by 130 feet in breadth, and from its centre rises a tower surmounted with a lofty spire, of handsome proportions: the part now used for divine service, which excludes the ancient chancel, was rebuilt in 1800, and is fitted up in a solid and handsome manner, with accommodation for upwards of 2000 persons: in the chancel are some curious old monuments, and a fine bust of the late vicar by Chantrey. St. Paul's, in Norfolk-street (erected by subscription in 1720), is a rather heavy Greek structure, with a tower surmounted by a dome, are in a similar style, and equally handsome, both having been erected, like St. George's, at the expense of the parliamentary commissioners. St. John's is a still more modern building. Besides the churches, which have accommodation for upwards of 15,000 persons, there are 20 places of worship for different denominations of dissenters. Six of these belong to the Wesleyan Methodists, and are among the largest as well as the most ornamental buildings of the town; the Independents have five large places of worship, and there is a handsome Roman Catholic chapel. Most of the places of worship have attached burial grounds; but the custom of interring the dead in the town will probably grow into disuse, from the opening of an extensive cemetery in the S.W. suburbs. Connected with the various places of worship are numerous Sunday-schools, furnishing religious instruction to about 7000 children, fully 3090 of whom are taught by the Wesleyan Methodists, here a numerous and intelligent body. Three national schools are attended by 500 boys and 650 girls; four Lancastrian schools (one of which is supported by the Wesleyans) by 750 boys and 400 girls, and seven infant-schools by upwards of 1000 children of both sexes. The out-townships have also numerous Sunday-schools, with eight national and three infant-schools. A grammar-school, founded in the reign of James I., has an endowment of about £140 a year, three fifths of which are paid to the head-master and two fifths to the usher, both of whom receive entrance-fees, and other extra payments, from the pupils. The management of the school, and the appointment of the masters, is vested in the vicar and 12 burgesses of Sheffield. A charity-school, established in 1796, provides clothing, board, and instruction, with an apprentice fee for 90 boys, and a similar establishment for 70 girls was formed in 1786. A collegiate school, founded a few years ago on a joint-stock principle, is well attended; and more recently, the Wesleyan body have established a proprietary school, in which 300 boys are boarded and liberally educated, partly with the view of providing for the better elementary instruction of the future ministers of that denomination. Among the many charities belonging to the parish of Sheffield, the principal is Lord Shrewsbury's hospital, for 18 men and the same number of women: the buildings, which have been erected on a new site, consist of a centre and wings, in the later English style. Hollis's hospital, a similar establishment founded in 1703, is endowed with funds for the support of 16 widows of cutlers, and a small charity-school. There are numerous minor charities, apprentice funds, &c., in the town, and each out-township has its separate charities. (See *Char. Com. 17th and 32d Reports.*)

The general infirmary, which stands about ¼ m. N. from the town, is a handsome stone building, with semi-circular wings and a central portico, its interior comprising many large and airy wards, with accommodation for about 200 in-patients. Adjoining, but distinct from the infirmary, is a large building, now almost completed, to contain fever wards. The medical and domestic arrangements are complete; and, on the whole, it is one of the best regulated provincial hospitals in the kingdom. It was opened in 1797, having cost above £90,000, raised by subscription. It has, also, several dispensaries, lying-in charities, Dorcas societies, provident institutions, a large auxiliary bible society, and various religious associations connected both with the established church and the several bodies of dissenters. A theatre was built in 1762, with attached assembly-rooms; and there is a handsome music-hall, in which concerts are frequently given.

A public subscription library in the music-hall is supported by 200 subscribers; the library attached to the mechanics' institute contains about 3000 vols.; the literary and philosophical society (established in 1822), has a good collection of minerals, fossils, plants, &c., with apparatus for experiments; and the botanical society has a garden comprising 18 acres, tastefully laid out, and a glass conservatory, 300 ft. in length, filled with rare exotic plants. There are likewise two well-supported news rooms, one of which, in High-street, has a handsome frontage, and a lofty elegant saloon. The chief commercial buildings are the post-office, excise-office and assay-office, erected in 1773, soon after the rise of the silver-plating trade. Sheffield has also two private banks and a savings' bank, besides five joint stock banking companies; and three weekly newspapers.

Nothing is known of the early history of Sheffield, or of the origin of that business for which it is now so famous. But it had attained to eminence in the making of knives so early as the 13th century; for Chaucer, contemporary with Edward III., mentions, in his *Reve's Tale*, the Sheffield "thwytel," or whittle, in such a way as shows it was then in common use. It does not appear ever to have lost the reputation for cutlery it had thus early acquired. In 1575, the earl of Shrewsbury, lord of the manor of Sheffield, sent to his friend lord Burleigh "a case of Hallamshire whittels, beinge such fruites as his pore cuntrey affordeth *with fame throughout the realme.*" (*Metal Manufactures, Cabinet Cyclop.*, ii. 5.) In 1624, a corporation was formed for the "good order and government of the makers of knives, scissors, shears, sickles, and other cutlery-wares in Hallamshire," the government being vested in a master, two wardens, six searchers, and 24 assistants; consisting of freemen only. The principal object in the formation of this corporation seems to have been the regulation of the marks or other devices which every individual was to strike or impress on the goods he made for sale. But these regulations can hardly be said to be any longer in operation.

The corporation continued on the footing fixed in 1624, till 1814, when an act was passed, permitting all persons indiscriminately, without their being freemen, or having served an apprenticeship, or obtained a mark from the corporation for their goods, to carry on business anywhere within the district of Hallamshire. This liberal and judicious measure has been of great service to the town, by inducing men of talent and enterprise, from all parts of the country, to settle in it, where their competition and industry have had the best effects.

For several centuries the manufactures of Sheffield were confined almost entirely to the making of sheath-knives, scissors, sickles, and scythes. About the beginning of the 17th century, a common tobacco-box and the Jew's harp were added to the list of manufactured articles; but it was not till about forty years after, that the manufacture of clasp-knives, razors, and files, for which it is now so famous, was introduced. It has been remarked, that for about a century after this period, the manufacturers discovered more of industry and perseverance, than of enterprise or ingenuity in the conduct of their business. About 1750 they began, for the first time, to carry on a direct trade with the continent. The manufacture of plated goods was soon after commenced; and from that period down to the present time Sheffield has made an astonishing progress in the career of industry; and in many branches of the hardware manufacture has no superior, and in some no rival. (*Rees's Cyclop.*)

Like Birmingham, Sheffield was most probably indebted to her situation for her early application to the hardware business. Coal and iron are found in her immediate vicinity. The Don, on which she is built, and four smaller rivers which flow into the Don near the town, supply her with power to work mills for forging, cutting, and preparing the iron and steel used in her manufactures; and in this respect she has an advantage over Birmingham. The river was made navigable to within about 3 m. of the town, so early as 1751; and a lateral canal has since prolonged the navigation to the town.

Cutlery, as it was the earliest, so it is still the largest and most important branch of industry. The principal articles are table-knives and forks, pen and pocket-knives of every variety and description, scissors, razors, surgical, mathematical, and optical instruments, scythes, sickles, saws, with all sorts of carpenters' tools, and so forth. The most beautiful and highly finished articles of cutlery exhibited in the shops of the metropolis, though stamped with the vendor's name, are mostly made in Sheffield, and the cutlery of the town is deservedly held in the highest estimation in all parts of the world.

With the exception of plated saddlery ware, almost all the other descriptions of plated goods made at Sheffield are reckoned superior to those made at Birmingham or anywhere else. Some of the best plated articles have silver edges, and, when used with ordinary care, last for a very long time, and can with difficulty be distinguished from silver. An extensive manufacture of articles of Britannia metal is carried on.

Mr. Jacobs says that, in as far as he was able to ascertain by inquiries of platers, the owners of flatting-mills, and of the manufacturers of plated goods, he was disposed to estimate the silver used for plating in Birmingham and Sheffield, including with it that used at Walsall and other places, by saddlers' ironmongers, at about 730,000 oz. a year; and deducting from the 150,000 oz. used at Birmingham, and 100,000 oz. for that used at Walsall and other minor towns, there will remain 550,000 oz. for the consumption of Sheffield.

Sheffield produces few articles in copper and brass, and no toys; but in lieu of these she has some peculiar and important businesses. The conversion of iron into steel is carried on to a far greater extent here than in any other part of the empire; and most of the steel used at Birmingham and other places is prepared at Sheffield. The works of the Messrs. Sanderson for the converting and preparing of steel by tilting, rolling, &c., are the most extensive and celebrated in the world.

The manufacture of files is one of the staple trades of Sheffield. Files are used in immense quantities at home, and are largely exported. Any one who has ever seen the process of file-cutting would be likely to conclude that it was an operation which might be successfully performed by machinery, and a great variety of contrivances have been set on foot in that view. Hitherto, however, none of them has completely succeeded; so that the best files continue now, as heretofore, to be cut by the hand.

Few comparatively, of the Sheffield manufacturers have large capitals, and the business is not so generally carried on in workshops and factories, as at Birmingham. A person worth a few shillings may commence business on his own account as a cutler; and in this class, individuals are not unfrequently journeymen one year and masters another, and conversely.

In 1833 some of the staple trades carried on in the town, and the individuals engaged in them, were specified as follows:—Table-knives and forks, 3880 hands; pen and pocket-knives, 2050; razors, 754; scissors, 690; files, 1749; saws, 363; edge-tools, 703; stove-grates, fenders, &c., 1539; white-metal, 645; silver-plated goods, 500; making in all 13,430. (*Report on Manufactures, &c.*, 1833, *Min. of Evidence*, p 175.) But the hands engaged in the conversion of iron into steel, and in several other important trades, are not included in this specification. Wages in Sheffield vary at present (1841) from about 12s. to 60s. a week. The labour in some of the departments is very severe, and in others great skill is required. Grinders, particularly those who do not use water in the operation, inhale the finer particles of stone and steel, and are usually short lived. Many efforts have been made to obviate this, as well as to lessen the risk of accidents in the grinding mills; but the employment continues to be more than usually unhealthy and dangerous; and as much skill is required in grinding the finer descriptions of knives and razors, wages, being influenced by both circumstances, are generally high.

It was stated before the British Association that there are in Sheffield 56 workmen's clubs (most of which are unions), and their aggregate revenues may probably amount to £36,000 a year, divisible among about 9000 members. Many hands are employed in grinding spectacle-glasses, most of which, indeed, come from Sheffield. The showrooms and manufactories of the leading houses are freely opened to all respectable strangers, and afford abundant proofs of the ingenuity that has raised the town to its present importance. The workmen of Sheffield have been accused of a tendency to riot and insubordination; and no doubt several destructive riots have taken place during the present century, which have required the interference of the military for their suppression; but these have all originated in extreme distress, or in some temporary and accidental cause, and, speaking generally, the inhabitants are distinguished by their orderly, good conduct. Few of the inhabitants live in cellars, like the poorer classes in Liverpool and Manchester, but occupy comfortable tenements, being, in this respect, much better off than those in most other large manufacturing towns. Sheffield enjoys the advantage of a direct canal communication, eastward to Hull, and by a very circuitous route westward to Manchester, Liverpool, &c. The Don was made navigable to Tinsley in 1751; other canals have subsequently been cut for the transmission of heavy goods, and the canal-basin of Sheffield is accessible to vessels of 50 tons; but the principal improvement has taken place within the last few years, by the opening of the railway to Rotherham, and its more recent connection with the N. Midland railway, by means of which, and other connected lines, a speedy communication is established with London, York, Leeds, Birmingham, Manchester, &c. The facility of transit will also be greatly increased on the completion of the Sheffield, Ashton, and Manchester railway, which will bring the last mentioned town within an hour and a half's distance of Sheffield.

This important manufacturing town had no voice in the legislature till the Reform Act, by which the parish was raised a parl. bor., with the privilege of sending two members to the House of Commons. Registered electors in 1839–40, 4451. The county magistrates administer justice, and the lighting and watching is conducted by the police commissioners under the authority of a local act; but there is no proper municipal corporation, and an application made to the privy-council in 1836 for such was rejected, owing to opposition of some of the most respectable inhabitants. A more recent application is still under consideration. The steer cutler is, at present, the returning officer at elections for the borough. Sheffield is also one of the polling-places for elections for the W. Riding. The county magistrates exercise jurisdiction within the town, and the police, regulated similarly to that of Manchester, Liverpool, &c., is very active. The parish of Sheffield constitutes, with its outtownships, a poor-law union, the expenditure of which for pauper maintenance amounted, in 1839, to £14,637. Markets on Tuesdays and Saturdays; fairs, Trinity-Tuesday for horses and cattle, and on Nov. 28 for cheese.

Sheffield is of great antiquity, and there can be but little doubt that close to or near it was once a considerable Roman station. A town existed here under the Saxons; and the reigns of the Plantagenets it was considered of sufficient importance to be defended by a strong castle. Mary Queen of Scots was confined for nearly 14 years the town, a country seat near the town, belonging to the Earl Shrewsbury, the owner, also of the castle. The latter seized in the civil wars by Sir John Gell, one of the parliamentary generals, and was demolished, by order of parliament, in 1646, there being now no remains except of the foundations. Its site, however, is still called Castle Hill. (*Hunter's History of Hallamshire; Statistical Account of the British Empire; Private Information.*)

SHEFFIELD, p. t., Berkshire co., Mass., 20 m. S. Lenox, E. Hudson, N. Y., 138 W.S.W. Boston, 349 W. Watered by Housatonic river and its tributaries. Chartered in 1733 and is the oldest town in the co. It contains marble of a good quality, and has a Congregational church, and

some Baptists and Methodists, eight stores, two fulling-mills, one grist-mill, eight saw-mills, three tanneries, three distilleries; one academy, 160 students, 13 schools, 618 scholars. Pop. 2322.

SHELBURNE, p. t., Chittenden co., Vt., 35 m. S. Burlington, 48 W. by N. Montpelier, 505 W. Bounded W. by lake Champlain, from which Shelburne bay sets up in a S. direction for four miles, and is a beautiful body of water. Laplatte river affords water-power, and enters the head of Shelburne bay. It contains three churches, a Congregational, Episcopal, and Methodist, one store, one fulling-mill, one grist-mill, one saw-mill, two tanneries; 10 schools, 207 scholars. Pop. 1098.

SHELBY, co., Ala. Situated near the centre of the state, and contains 950 sq. m. Bounded E. by Coosa r. Drained by Cahawba r. and its branches. It contained in 1840, 5894 neat cattle, 1944 sheep, 14,630 swine; and produced 17,093 bushels of wheat, 179,650 of Indian corn, 9946 of oats, 7283 of potatoes, 1197 pounds of tobacco, 510,382 of cotton. It had six stores, two forges, two flouring-mills, five saw-mills, two distilleries; six schools, 45 scholars. Pop.: whites, 4494; slaves, 1616; free coloured, 2; total, 6112. Capital, Columbiana.

SHELBY, co., Tenn. Situated in the S.W. corner of the state, and contains 600 sq. m. Bounded W. by Mississippi r. Drained by Loosahatchy, Wolf, and Nonconna rivers. It contained in 1840, 13,545 neat cattle, 4732 sheep, 39,963 swine; and produced 12,919 bushels of wheat, 551,790 of Indian corn, 23,452 of oats, 36,562 of potatoes, 36,870 pounds of tobacco, 1,006,050 of cotton. It had 27 commission houses in foreign trade, 30 retail stores, nine grist-mills, four saw-mills, two printing-offices, two weekly, and two semi-weekly newspapers; nine academies, 245 students; seven schools, 234 scholars. Pop.: whites, 7605; slaves, 7043; free coloured. 73; total, 14,721. Capital, Raleigh.

SHELBY, co., Ky. Situated in the N. towards the E. part of the state, and contains 442 sq. m. Drained by Brashear's cr. and its branches, flowing into Salt r. It contained in 1840, 18,747 neat cattle, 20,360 sheep, 73,011 swine; and produced 172,791 bushels of wheat, 49,449 of rye, 1,349,900 of Indian corn, 197,680 of oats, 17,306 of potatoes, 947,560 pounds of tobacco. It had 35 stores, two flouring-mills, 46 grist-mills, 21 saw-mills, five tanneries, nine distilleries; six academies, 387 students; 11 schools, 301 scholars. Pop.: whites, 11,256; slaves, 6355, free coloured, 157; total, 17,768. Capital, Shelbyville.

SHELBY, co., O. Situated in the W. part of the state, and contains 418 sq. m. Organized in 1819. Drained by Miami river and its branches. The Miami canal passes through its W. part, with a side cut to Sidney. It contained in 1840, 10,509 neat cattle, 10,143 sheep, 18,941 swine; and produced 78,091 bushels of wheat, 5498 of rye, 253,492 of Indian corn, 1795 of buckwheat, 131,010 of oats, 29,918 of potatoes, 79,830 pounds of sugar. It had 23 stores, one fulling-mill, three woollen factories, seven flouring-mills, seven grist-mills, 22 saw-mills, one oil-mill, eight tanneries, one distillery, two printing-offices, one weekly newspaper; 21 schools, 598 scholars. Pop. 12,154. Capital, Sidney.

SHELBY, co., Ia. Situated S.E. of the centre of the state, and contains 432 sq. m. Drained by Big and Little Blue rivers, and Sugar cr. It contained in 1840, 11,339 neat cattle, 13,452 sheep, 39,618 swine; and produced 61,611 bushels of wheat, 2775 of rye, 779,101 of Indian corn, 85,725 of oats, 16,017 of potatoes, 116,254 pounds of tobacco, 47,561 of sugar. It had 23 stores, one fulling-mill, one flouring mill, 19 grist-mills, 19 saw-mills, one oil-mill, eight tanneries, 12 distilleries; 11 schools, 249 scholars. Pop. 12,005. Capital, Shelbyville.

SHELBY, co., Ill. Situated a little S.E. of the centre of the state, and contains 1080 sq. m. Watered by Kaskaskia r. and its tributaries. It contained in 1840, 8423 neat cattle. 6693 sheep, 20,362 swine; and produced 18,595 bushels of wheat, 385,220 of Indian corn, 74,302 of oats, 6720 of potatoes. It had 16 stores, two flouring-mills, 18 grist-mills, 13 saw-mills, three tanneries, four distilleries; 13 schools, 437 scholars. Pop. 6559. Capital, Shelbyville.

SHELBY, co., Mo. Situated towards the N.E. part of the state, and contains 432 sq. m. Drained by Salt, and by North Two rivers, which affords water-power. It contained in 1840, 6376 neat cattle, 9850 sheep, 11,799 swine; and produced 5195 bushels of wheat, 134,970 of Indian corn, 14,775 of oats, 8117 of potatoes, 34,896 pounds of tobacco. It had six stores, five grist-mills, nine saw-mills, one distillery; 10 schools, 256 scholars. Pop.: whites, 2587; slaves, 458; free coloured, 11; total, 3056. Capital, Shelbyville.

SHELBY, p. t., Orleans co., N. Y., 19 m. S. Albion, 262 W. by N. Albany, 395 W. Drained by Oak Orchard cr. It contains two churches, a Methodist and Universalist, three stores, two fulling-mills, one grist-mill, three saw-mills; one academy, 83 students; 19 schools, 950 scholars. Pop. 3643.

SHELBYVILLE, p. v., capital of Bedford co., Tenn. 50 m. S.E. Nashville, 682 W. Founded in 1809, incorporated in 1819. Situated on the N.E. bank of Duck r., on moderately elevated and uneven ground, surrounded by cedar groves. It contains a courthouse, jail, a printing-office, two churches, 13 stores, various mechanic shops, and about 800 inhabitants.

SHELBYVILLE, p. v., cap. of Shelby co., Ky., 23 W. by N. Frankfort, 565 W. Situated on Brashear's cr., 12 m. above its entrance into Salt r., and contains a courthouse, jail, a bank, two churches, an academy, a printing-office, 180 dwellings, and 1335 inhabitants.

SHELBYVILLE, p. v., capital of Shelby co., Ill., 60 m. S.E. Springfield, 724 W. Situated on the W. bank of Kaskaskia r., and contains a brick courthouse, 40 ft. square, two stories high, with a cupola, a jail, nine stores and about 50 dwellings. It contains a large sulphur spring.

SHELDON, p. t., Franklin co., Vt., 62 m. N. Montpelier, 546 W. Watered by Missisque r. Its tributary, Black r., affords water-power. It contains three churches, an Episcopal, Congregational, and Methodist, five stores, one grist-mill, four saw-mills, two tanneries; nine schools, 226 scholars. Pop. 1734.

SHELDON, p. t., Wyoming co., N. Y. 265 m. W. Albany, 263 W. Drained by Tonawanda and Seneca creeks. It contains two churches, a Presbyterian and an Episcopal, six stores, one fulling-mill, two grist-mills, 12 saw-mills, three tanneries; one academy, 100 students; 18 schools, 772 scholars. Pop. 2453.

SHELTER ISLAND, t., Suffolk co., N. Y., 20 m. E. Riverhead, 245 S.S.E. Albany. It comprises the whole of Shelter Island, six m. long and four broad, lying off the E. end of Long Island, between Gardiner's and Great Peconic bays. A ship channel passes entirely around the island, and its headlands afford picturesque views. A ferry, 120 rods long, connects it to Southold. It contains a Presbyterian church, one store, one grist-mill; one school, 90 scholars. Pop. 379.

SHENANDOAH, r., Va., rises by two branches, of which the eastern or principal branch rises in Augusta co., and flowing along the W. side of the Blue ridge, receives the N. branch in the S. part of Frederic co., whence the united stream flows N.E. to its junction with the Potomac, shortly before its passage through the Blue ridge, at Harper's Ferry, where it presents a grand and picturesque view.

SHENANDOAH, co., Va. Situated towards the N. part of the state, and contains 410 sq. m. Watered by the N. fork of Shenandoah r. It contained in 1840, 10,582 neat cattle, 12,345 sheep, 16,494 swine; and produced 164,275 bushels of wheat, 32,357 of rye, 298,649 of Indian corn, 105,090 of oats, 34,980 of potatoes. It had 31 stores, three furnaces, three forges, two fulling-mills, four woollen factories, 28 flouring-mills, nine grist-mills, 20 saw-mills, one oil-mill, 19 tanneries, two distilleries, five potteries, one printing-office, one weekly newspaper: two academies, 108 students; eight schools. 247 scholars. Pop.: whites, 10,320 : slaves, 1033; free coloured, 255; total, 11,618. Capital, Woodstock.

SHEPTON-MALLET, a market town and par. of England, co. Somerset, hund. Whitstone, on a branch of the Brue. surrounded by several small hills, about 5 m. E.S.E. Wells. Area of par., 3770 acres. Pop, in 1841, 5332. The town, which comprises a number of narrow streets and lanes, has been latterly much improved by the construction of a new bridge and opening new roads, &c.; near its centre is a curious market-cross, erected in 1500. The church, in the early English style, is a spacious cruciform structure, with a tower and spire at the W. end. The living is a rectory, in the gift alternately of the crown and Mr. Wickham; its gross value is £890 a year; but the curate's salary of £150, and other demands, reduce the nett income to £533. (Eccl. Rev. Rep.) Here is the county bridewell, a large and conspicuous edifice; and here also petty sessions are held. There are places of worship for Baptists, Independents, Wesleyans, and R. Catholics; a convent of Visitation nuns, the only one of that order in the kingdom; an almshouse, founded in 1699; and a free school established in 1813. The town had formerly a flourishing manufacture of woollen goods, but this branch of industry is now much fallen off. Markets, Tuesdays and Fridays; fairs, Easter-Monday, June 18, and Aug. 8, for cattle and cheese.

SHERBORNE, or SHERBOURN, a market town and par. of England, co. Dorset, hund. Sherborne; on the Ivel, which divides the town into two parts, Sherborne and Castleton, 16¼ m. N. by W. Dorchester, and 110 m. W.S.W. London. Area of par., 4900 acres. Pop., in 1831, 4075. Sherborne is finely situated, partly on the acclivity of a hill, and partly in the fertile vale of Blackmore. It is compactly built; its principal streets, running E. and W., are crossed by smaller streets in a contrary direction. It was made a bishop's see in the 8th century, and continued such till 1075, when the see was removed to Salisbury, and Sherborne cathedral became an abbey church. Portions

of the abbey, including the refectory, &c., still remain. It was in great part destroyed by fire in the time of Henry VI. The church, however, chiefly rebuilt after that event, still exists, and is the modern parish church. It is a building of very different dates, but mostly in the perpendicular style; the S. porch is Norman. The groining is generally rich and good, and in the interior are several ancient monuments. (Rickman's Gothic Architecture.) The tower is upwards of 150 feet in height; in it are six bells, the largest of which, weighing upwards of three tons, was presented by Cardinal Wolsey. The living is a vicarage in the gift of the crown, worth £260 a year. Castleton (so called from the remains of an ancient castle, held by the royalists in the wars of Charles I.) is a separate living, a perpetual curacy, in the gift of Earl Digby, worth £81 a year.

The Wesleyans, Independents, Friends, &c., have meeting-houses. The buildings of the free grammar-school, founded by Edward VI., adjoin the church, and are built round three courts, two of which are used as play-grounds. They comprise a good house for the master, formed of the ancient lady chapels, with upper and lower school-rooms, dining-hall, library, numerous dormitories, &c. This school should have, according to its charter, 20 governors, and has an annual income, from rents, &c., of near £980. The head master has a salary of £125 a year, with a house, and a fee of £1 1s. from each boy on entrance, and the like sum annually; the under master has £100 a year and a house: both masters must be university graduates, and members of the church of England, attendance on the service of which is compulsory on the pupils. Subordinate masters are also engaged; some for teaching writing and arithmetic, which are paid for separately. The course of instruction is principally modelled on the Eton system, though some deviations from it have latterly been introduced. In 1836, there were 43 boys on the foundation, and 80 private pupils of the masters. This school has four exhibitions at the universities of £60 a year each, which may be granted for four years to pupils who have already kept four years on the foundation; but these exhibitions have not always been filled up. Sherborne has as almshouse founded by Henry VI., which had in 1135, 24 aged inmates, and an income of £666 a year. There are numerous other charities, including Lord Digby's school for girls, annual and Lancastrian schools, &c.; and the parish authorities have the privilege of keeping three boys at Christ's Hospital, London, on the produce of lands left for that purpose in 1670. (Charity Commissioners' Thirtieth Report, pp. 104–132.)

In the immediate neighbourhood is Sherborne castle, the seat of Earl Digby, built by Sir W. Raleigh. The mansion is in the form of the letter H; the body four stories in height, having hexangular towers at the four angles, which are united with as many wings. It has some antique tapestry, and five paintings. The park comprises 340 acres, and some of the finest oaks in the county. A bridge of three arches over the Ivel leads to the house. Pope was a frequent visiter at Sherborne castle; and on a monument in the church is inscribed his beautiful epitaph in memory of his young friends, the Hon. Robert Digby and his sister Mary.

Sherborne has some silk and woollen fabrics; but these, as well as other branches of industry formerly carried on in the town, have greatly decayed. It is within the jurisdiction of the county magistrates. Assizes were regularly held here till the reign of Edw. IV., but have since been only occasional. General quarter sessions are held here on the Tuesday after Easter. Though not a modern port, &c., Sherborne sent members to the House of Commons in the reign of Edw. III. Market-days, Tuesday, Thursday, and Saturday; fairs, May 29, July 18 and 26 (nominally), and Oct. 14, chiefly for cattle and pedlery.

SHERBURNE, p. t. Chenango co., N.Y., 103 m. W. Albany, 347 W. Watered by Chenango r. and its tributaries. It contains four churches, a Presbyterian, Episcopal, Methodist, and Baptist, 13 stores, two fulling-mills, three grist-mills, eight saw-mills; one academy, 108 students; 19 schools, 830 scholars. Pop. 2791.

SHERIDAN, p. t., Chautauque co., N. Y., 324 W. by S. Albany, 351 W. Drained by Walnut cr., flowing into lake Erie, which has its bounds it on the N.W. It has one store, 10 saw-mills, two tanneries; 12 schools, 591 scholars. Pop. 1883.

SHERMAN, p. t., Chautauque co., N. Y., 357 W. by S. Albany, 346 W. Drained by French cr. It has two stores, one furnace, one fulling-mill, two grist-mills, five saw-mills; six schools, 163 scholars. Pop. 1099.

SHETUCKET, r., Conn., rises in Windham and Tolland counties by two head streams called Willimantic and Natchaug rivers, in the S. part of Lisbon, between Norwich and Preston. It receives Quinnebaug r. At Norwich city it is joined by Yantic r., and below that, to its mouth, it is called Thames r., and is navigable for vessels.

SHIAWASSEE, co., Mich. Situated in the central part of the state, and contains 544 sq. m. Watered by Shiawassee. Looking-Glass, and Maple rivers. Organized in 1837. It had in 1840, 2143 neat cattle, 375 sheep, 2807 swine; and produced 19,584 bushels of wheat, 13,772 of Indian corn, 10,937 of oats, 23,007 of potatoes, 25,933 pounds of sugar. It had six stores, one flouring-mill, eight saw-mills; 11 schools, 260 scholars. Pop. 2103. Capital, Corunna.

SHETLAND ISLANDS. See ORKNEY AND SHETLAND.

SHIELDS (NORTH). See TYNEMOUTH.

SHIELDS (SOUTH), a parl. bor., market town, seaport, and township of England, co. Durham, E. div. of Chester ward, par. Jarrow, on the S. bank of the Tyne, near its mouth, and directly opposite to North Shields, about 8 m. below Newcastle, and 16 m. N.N.E. Durham. Area of parl. bor., 1760 acres. Pop., in 1831, 18,756, exclusive of 2516 absent seamen. This, and its sister town on the opposite bank of the river, may be regarded in some measure as the outports of Newcastle, their population and importance having grown up with the increasing magnitude of the coal trade and commerce of the latter. Its lower part consists principally of a narrow, crooked and inconvenient street, extending for nearly 2 m. along the river; but the streets in the upper part of the town are wider and better built, and lighted with gas. The principal edifices and institutions are the town-hall, also used as an exchange, a neat building in the centre of a spacious market-place; a theatre, a scientific and mechanics' institution, charity-school, dispensary, and the various places of worship. The church, dedicated to St. Hilda, is of considerable antiquity, and has been frequently repaired and modernised. The living is a curacy, in the gift of the dean and chapter of Durham, worth £330 a year nett. There are chapels for various dissenting sects, to most of which are attached Sunday-schools, and various charities, and benevolent societies. In the town-hall petty sessions are held twice a month, besides courts leet and baron by the dean and chapter of Durham, as lords of the manor. Although the appearance of South Shields has little to recommend it, and its buildings are far from imposing, yet it is a place of much importance. The river Tyne is here about two thirds the width of the Thames below London bridge; and the vessels which belong to or rendezvous at N. and S. Shields are disposed in tiers on each side, as in the port of London. The town is rapidly increasing; a considerable quantity of ground is marked out for building in the E. and S. directions, and no doubt can be entertained that if land, upon a freehold tenure, could be procured, the rate of increase would be much more rapid, and the scale of the buildings greater. The whole of the chapelry is the land of the dean and chapter of Durham. South Shields had formerly many salt-pans, and an extensive manufacture of salt; but this has been abandoned, and ship-building is now the staple business of the town, and is very extensively carried on, 12 ships, of the burden of 2596 tons, having been built in 1840. It has also very extensive glass-works, a pottery, a coal mine (which may be said to be in the town), and manufactures of soda, alum, &c. (Boundary Report.) Still, however, the main dependence of the town is on the coal trade of the river. Indeed, most of the large colliers belonging to Newcastle load at South Shields, the coal being brought down the river in lighters, or keels; and it is said that as many as 500 vessels are frequently seen lying together in the haven. The port is included, with that of N. Shields, in the port of Newcastle, but there belonged to it specially, in 1840, 354 ships. The town communicates by the Brandling Junction railway with the railway from Newcastle to Carlisle. The Reform Act conferred on South Shields the privilege of sending one member to the House of Commons. The parl. bor. comprises the townships of South Shields and Westoe. Registered electors, in 1839-40, 686.

Mr. Greathead, of this town, invented the life-boat, the first being built here by subscription in 1790. Markets on Wednesdays; fairs, last Wednesday in April, and first in May, last in Oct., and first in Nov.

SHIPPENSBURG, p. b. Cumberland co., Pa., 21 m. W. Carlisle, 34 m. S.W. Harrisburg. 101 W. Situated on Swan's run, a branch of Conedogwinit creek, which affords water-power. Incorporated in 1819, and contains four churches, a Presbyterian, Associate Reformed, Lutheran, and Methodist, 18 stores, three tanneries, one distillery, one eatery, one printing-office, one weekly newspaper; six schools, 350 scholars. Pop. 1473.

SHIRAZ, the second city of Persia, prov. Fars or Persia proper, formerly the cap. of the empire, in a valley 115 m. E. Bushire, and 220 m. S S.E. Ispahan; lat. 29° 36' N., long. 52° 44' E. Pop. variously estimated at between 20,000 and 40,000. Shiraz has always been celebrated for the beauty and fertility of its neighbourhood, which has been warmly eulogised by the poet Hafiz, a native of the city. It is surrounded by gardens, and had, till lately, an imposing

aspect from a distance; but this is said to be no longer the case, its domes and minarets having been levelled with the ground by the earthquake of 1824. Morier makes Shiraz nearly 4 m. in circ. but says that 1-3d part of its buildings to the S E. are in ruins. It is surrounded with pretty high walls, flanked with round bastions, and has six gates, each flanked with two towers. "On entering the city, the houses, which are in general small, together with the narrow filthy streets, give the stranger but a mean idea of the second city of the empire. The great bazaar, or market-place, built by Kerim Khan, forms, however, a distinguished exception to this general remark. It is about ½ m. in length, made of yellow burnt brick, and arched at the top, having numerous skylights, which, with its doors and windows, always admit sufficient light and air, whilst the sun and rain are completely excluded. The ark, or citadel, in which the begler-beg of Fars resides, is a fortified square of 80 yards. The royal palace within is far from being an elegant structure; and some pillars, its greatest ornament, were removed by Aga Mahomed Khan, to adorn his palace at Teheran." (Kinneir's Persian Empire, 62, 63.)

Shiraz seems to be rapidly hastening to decay: and most of its public structures, once very numerous, are already in a ruined or neglected state. The principal mosque is a very large edifice, having been the palace of Atabeg Shah, its founder. When visited by Morier, it had 15 considerable mosques, besides many others of inferior note, 11 madressehs, or colleges, 14 bazaars, 13 caravanserais, and 26 hummums, or baths, &c. The principal college has upwards of 100 rooms; but it, as well as most of the others, is now nearly abandoned by students. Within the walls of the city are numerous Mussulman tombs. The climate was formerly distinguished for salubrity, but it has been materially changed for the worse in this respect since the last earthquake. The heat of summer is excessive, rising sometimes to 110° Fahr. in the shade. (Morier, ii. 97-113.) The water of Shiraz owing to the neglect of the city authorities, is also very bad.

About ½ m. from the town is the tomb of Hafiz, the Anacreon of Persia. It stands within a quadrangular inclosure and consists of a block of marble, on which two of the poet' odes are sculptured, with the date of his death. His works are not, as has been stated, chained to the tomb, but a copy of them is kept in an adjacent chamber. Adjoining are the stream of Rocknabad, and the bower of Mosselia, so celebrated in the verses of the poet; and the former consisting merely of a small brook of clear water, not more than 2 ft. wide; while of the bower not a shrub remains and its site is only marked by the ruins of an ancient tower. The celebrated garden of Jehan Nama, near the tomb of Hafiz, is a walled enclosure about 200 yards square, laid out in walks bordered with cypress trees, and watered by a variety of marble canals and artificial cascades. This, and many other gardens in the neighbourhood, are ordinary places of resort, where the citizens chat, smoke, and drink coffee. The tomb of the poet Saadi is also in the vicinity of Shiraz with various conventual buildings for devotion, &c.

Shiraz is celebrated for its wine. The principal vineyards are situated at the foot of the mountains to the N.W. of the town, where the soil is rocky, and the exposure extremely favourable. It would appear, however, that the culture of the vine has degenerated; and whatever may have been the case formerly, little care is now taken in the preparation of the wine. It is of very various qualities; but the best of the white varieties is certainly inferior to good Madeira, and the best varieties of the red (ruby wine of Hafiz) are not unlike tent, and seem to have but slender claims to the praises that have been lavished upon them. (Henderson on Wines, 365.) The produce of wine may amount, in all, to from 80,000 to 100,000 galls., of which from 10,000 to 15,000 galls. may be exported to India, Bagdad, Bussroah, &c. The commerce of the city is still rather extensive; it is principally with Bushire, Yezd, Ispahan, and the cities in the N.W. of Persia. From Bushire, the chief imports are spices, Chinese and Indian goods of all kinds, iron, lead, quicksilver, glass wares, woollen cloths, muslins, linens, arms, ammunition, cutlery, and other European manufactures. These goods, with salt from the neighbouring lakes, are sent to Ispahan, Teheran, Yezd, &c., in exchange for the manufactures of those cities and other products. The exports to Bushire consist principally of wine, rose water, and attar of roses; assafœtida, dried fruits, silk, goats' hair, Caramanian wool, saffron, drugs, horses, orpiment, madder, and tobacco. The trade between Shiraz and Bushire employs above 2000 mules. (Rich's Journey to Persepolis, p. 263.)

Shiraz has no vestiges of remote antiquity, and was probably not founded till after the propagation of Mohammedanism. It had become a populous town in the 10th century, and soon afterwards it was surrounded with walls. Its greatest benefactor, however, appears to have been Kerim Khan, who reigned in the latter half of the 18th century; since his death it has gradually declined. (*Kinneir's Persian Empire; Waring, Alexander, Morier, Porter, Ousely, &c., in Mod. Trav., xii.; Rich's Journey to Persepolis, &c.*)

SHOREHAM (NEW), a market town, seaport, and par. of England, co. Sussex, rape Bramber, hund. Fishergate, at the mouth of the Adur, here crossed by a long wooden bridge, about 1 m. from the English Channel, and 6 m. W. Brighton, with which town it has been since 1840 connected by a railroad. Area of par., 170 acres. Pop., in 1831, 1503. New Shoreham appears to have risen on the decay of Old Shoreham, now an insignificant village about 1 m. distant, but formerly a piece of some importance. The town is built in a singular manner; and near its centre is the market-house, supported on Doric pillars. The parish church is the remaining portion of a large cross church, of which nearly all the nave has been destroyed; it has various portions of fine late Norman gradually running into early English forms and details. (*Rickman's Gothic Architecture*, p. 248.) The interior is remarkable for elegance and richness. The living is a vicarage, worth £127 a year, in the gift of Magdalen college, Oxford. Here are meeting-houses for Independents and Wesleyans; with a national school, &c.

The town stands on the Brighton and Chichester turnpike road, which now crosses a suspension-bridge, recently built by the Duke of Norfolk over the Adur. Shoreham has only a tide-harbour, but it is the best on this part of the coast; and having 18 ft. water at spring tides, it is sometimes frequented by ships of considerable burden. Ship-building is the principal business, and vessels of 700 tons have been built here. It has, also, a considerable general trade, the gross customs' revenue collected here in 1840 having amounted to £22,146. It is governed by two constables, chosen annually at the court-leet of the lord of the manor. It was a borough by prescription, and sent two members to the House of Commons from 1298 down to 1771, when the electors, having been convicted of gross corruption, the rape of Bramber was incorporated with the borough. The latter, however, was also disfranchised by the Reform Act. Market on Saturdays. Fair, July 25., for pedlery, &c.

SHOREHAM, p. t., Addison co., Vt., 73 m. S.W. Montpelier, 469 W. Bounded W. by lake Champlain, here half a mile wide, and has a ferry to Ticonderoga, N. Y. Lemonfair river affords water-power. It contains three churches, a Congregational, Baptist and Universalist, four stores, one fulling-mill, two grist-mills, three saw-mills, three tanneries; one academy, 35 students; 14 schools, 420 scholars. Pop. 1674.

SHREWSBURY, a parl. and mun. bor. and market town of England, co. Salop, of which it is the cap.: nearly in the centre of the co., in a peninsula formed by the Severn, on two gentle declivities, 50 m. S by E. Liverpool, 136 m. N.W. London. Pop., in 1841, 21,525. It is chiefly separated from the river by garden and meadow ground, skirted by a range of genteel houses, and its exterior appearance is from many points striking and majestic. The streets, as in most ancient towns, are irregular, and many of the houses have an antique appearance, presenting gables and over-hanging stories to the road; but various improvements have been made of late years in the thoroughfares, especially in lighting, flagging, &c. The river is here crossed by two handsome stone bridges, built by subscription, called respectively the English and Welsh bridges: the former, completed in 1774, at a cost of £16,000, is 410 ft. in length, and consists of seven semicircular arches; the other, or Welsh bridge, finished in 1795, at a cost of £8000, is 266 ft. in length, and has five arches. Adjoining the latter is a quay and warehouses.

Among the chief public buildings are the royal free grammar school (see below); the town and county hall from a design by Sir R. Smirke, cost £12,000; it is a handsome building, and affords excellent accommodation for the assize business; the market-house, built in the reign of Elizabeth, and unequalled in point of ornamental decoration by any similar structure in the kingdom; a spacious and elegant music hall, with news rooms, &c.; the county jail and bridewell for the town, near the castle, built on Howard's plan in 1793, at an expense of £30,000; the Doric column, at the entrance to the town from London, in honour of Lord Hill, 116 ft. in height, surmounted by a colossal statue of his lordship, cost £5973, raised by a subscription; a neat infirmary, 170 ft. in length (established in 1745), and rebuilt in 1830 at a cost of £18,736; the house of industry, on the E. bank of the Severn, for the poor of the six parishes of Shrewsbury, finished for a founding hospital in 1765, cost £12,000; the theatre, rebuilt in 1834, on the site of the royal residence of the princess of Powysland; a butter and cheese hall in Castleforegate, and a new savings' bank.

Shrewsbury has nine churches, most of which are embellished with rare and beautiful specimens of stained

glass. The new church of St. Chad is a handsome modern structure, formed by the intersection of two circles, with a tower and portico attached; the smaller of the circles being occupied by a grand staircase, and the larger one, 100 ft. in diameter, being the body, chancel, &c. of the building. St. Mary's, a cross church, of Norman and early English architecture, has a spacious chancel and chantry chapels, and a fine tower surmounted by a spire, one of the loftiest in the kingdom. The abbey church, the W. portion of a Benedictine monastery, founded by Roger de Montgomery, first earl of Shrewsbury, in 1803, displays many curious features of Norman architecture, combined with the earlier pointed style: the great W. window of the tower is only equalled by that of York cathedral; the aisles contain several fine old monuments, and opposite the S. entrance is an elegant octagonal stone pulpit: the interior forms a beautiful oriel, the roof being vaulted on eight delicate ribs: it formerly stood in the refectory. St. Giles's is a small but handsome edifice, built in the early part of the 12th century. St. Alkmund's was rebuilt in 1795, in the modern Gothic style, with the exception of the tower and spire, 184 ft. in height, which are singularly elegant. St. Julian's is a plain oblong building of brick, rebuilt in 1748; the tower belonged to the old church. St. George's, St. Michael's, and Trinity churches, have been erected during the last 16 years. The first is of freestone, and cruciform, in the lancet, or early pointed style. The two latter are of brick, in the Doric style, affording ample accommodation in free-sittings. Some of the parishes extend into detached parts of the adjacent country, where there are four chapels of ease belonging to St. Mary's, and one to St. Chad's.

Besides the churches, there are places of worship for Roman Catholics, Wesleyan Methodists, Independents, Baptists, and Unitarians, with attached Sunday-schools; and a meeting-house for the Society of Friends. A large national school founded in 1776, is attended by about 300 children of both sexes, and nearly an equal number of boys are instructed in a Lancastrian school. Allatt's charity school provides clothing, instruction, and an apprentice-fee, for 20 children of each sex. Bowdler's school, in the parish of St. Julian's, was established in 1724 for a similar purpose; and in the suburb of Frankwell, across the Welsh bridge, is a hospital, founded in 1734, which, besides supporting 12 aged people of each sex, furnishes instruction to 25 boys and 25 girls of Frankwell. There are several almshouses belonging to different foundations, and attached to particular parishes.

The most distinguished public charity of Shrewsbury, however, is the free grammar-school, founded and endowed by Edward VI., but greatly enlarged by Queen Elizabeth. This school, prior to the close of the last century, had, owing to certain defects in the original rules and ordinances, fallen to decay; but in 1798 an act was passed "for the better government and regulation of the free grammar-school of Edward VI.," by which the management of the school was invested in the bishop of Lichfield and Coventry and 13 trustees or governors, one of whom is the mayor for the time being. At the same time, the number of masters on the foundation was reduced from four to two, and their appointment was vested solely in St. John's college, Cambridge. The income arising from the endowment is about £3600 a year, besides which it confers several advantages on its alumni at both universities, viz. four exhibitions of £70, and four of £15 each, at St. John's, Cambridge; four of £60 each at Christ Church, Oxford; one of £100, and three of £35 each, either at Oxford or Cambridge; three contingent exhibitions; and six scholarships, with one bye-fellowship at Magdalen college, Cambridge. Prizes are annually awarded to merit; and to the best scholar proceeding to college is given the Sidney gold medal, having on its obverse the bust of Sir Philip Sidney, who, with his friend Fulke Grevill, afterwards Lord Brook, the poet, was educated in Shrewsbury school. During the present century the school has attained high celebrity, from the learning and talents of its master and (afterwards) visitor, Dr. Butler, late bishop of Lichfield and Coventry. In consequence of the success of its pupils in competitions for prizes, exhibitions and other honours, at the universities, the sons of the gentry have been sent here from all parts of the kingdom, and the establishment has, for many years past, comprised, independently of the free scholars (who must be sons of burgesses), many pupils paying handsome sums to the masters for board, lodging, and instruction: in fact, few public schools in England are superior to that of Shrewsbury. The school-house, erected in 1630, on the site of a more ancient wooden building, is a lofty structure of freestone, forming two sides of a court, the third side of which is formed by the library and chapel: a court is entered by a gateway, having columns on each side, with a Greek inscription over the arch. Two large houses belonging to the masters, contiguous to the schools, comprise every accommodation for boarders; and there are large play-grounds in front and at the back of the schools.

The town has four weekly newspapers; a literary and philosophical society; a mechanics' institute, and a public library with nearly 6000 vols.; the assembly rooms and theatre are well attended during winter; and races are annually held in the neighbourhood. On the S. side of the town is one of the most celebrated promenades in the kingdom called the Quarry. It is formed in meadow ground gradually sloping to the river, along which extends a graceful avenue of lofty lime trees, 540 yards in length.

Shrewsbury was formerly of considerable importance as a mart for flannels from Welshpool, Newtown, &c.; but this branch of trade is nearly extinct. It has, however, a large factory for spinning flax, with some smaller factories, and a large iron foundry, the whole furnishing employment to several hundred persons. The prosperity of the town, however, does not depend solely on its trade, as it is a favourite place of resort for persons of small income, or who have retired from business. The Severn, which even here is celebrated for its salmon, is navigable as far as Shrewsbury by vessels of from 30 to 80 tons, and a canal to Wombridge opens a communication with the coal districts of Staffordshire. (*Mun. and Parl. Bound. Rep.*) The banking establishments of Shrewsbury comprise various private banks and joint-stock banking companies, besides a savings' bank. The vicinity being a good barley country, the malting business is carried on to a considerable extent.

Shrewsbury which has received many royal charters especially from Richard I. and Charles I., is divided into five wards, and is governed by a mayor and five other aldermen, with 30 councillors. Corporation revenue, in 1839, £1900. Quarter and petty sessions for the borough are held here under a recorder, and there is a court of requests for the recovery of small debts. The lent and summer assizes are held here, as well as the quarter-sessions for the county. Shrewsbury has sent two members to the House of Commons since the reign of Edward I., the right of election down to the Reform Act being in burgesses paying scot and lot, and not receiving alms or charity. The electoral limits were enlarged by the Boundary Act, so as to include one entire parish, and parts of two others, with the whole borough. Registered electors, in 1839–40, 1865. Markets on Tuesday and Saturday: Fairs on the second Tuesday Wednesday in each month, for cattle, horses, cheese, butter.

Shrewsbury is supposed to have been built after the Roman station *Uriconium* had been destroyed in the 5th century. William the Conqueror gave the town and surrounding country to Roger de Montgomery, one of his followers, who built here a strong baronial castle, the keep of which remains, being converted into a modern dwelling-house. 1062 the castle and property was forfeited to the crown. Shrewsbury, from its situation close to Wales, was the scene of many border frays between the Welsh and English; and in 1277, Edward I. had his quarters here. On 21st of July, 1403, a desperate battle was fought near the town, between the royal army, commanded by Henry IV., and that of the rebel Earl of Northumberland, under command of the famous Lord Percy, surnamed Hotspur: the death of the latter, by an unknown hand, decided the victory in the king's favour: the loss on both sides was immense. During the wars of the Roses, Edward IV., after defeat and death of his father, Richard Duke of York, levied an army among the towns-people, with which he defeated the opposite faction at Mortimer's Cross. In the war between Charles I. and the parliament, the inhabitants they espoused the cause of the former; but in 1664 the town yielded to the parliamentary troops under Col. Milton, and the fortifications were destroyed. Dr. Taylor, the learned editor of "Demosthenes," and the author of "Elements of the Civil Law," was the son of a barber of this town, where he first saw the light in 1703. It was, also, the birth-place of Dr. Burney, the author of the "General History of Music." The surrounding country is picturesque and highly cultivated, on the plain extending every way about 13 m., beyond which are lofty ranges of hills. About 22 m. from the town is Boscobel House, where the Penderell family concealed Charles II. after his defeat at battle of Worcester. (For further particulars the reader is referred to *Owen and Blakeway's History of Shrewsbury*; *Owen's Memorials of Shrewsbury*; *Corp. Reports*; and *vale Information*.)

SHREWSBURY, p. t., Rutland co. Vt., 73 m. S.S.E Montpelier, 466 W. It contains Shrewsbury peak of the Green mountains, 4100 ft. above tide water. Watered by branches Otter-creek river, and contains a Congregational church, stores, two saw-mills, 10 schools; 436 scholars. Pop.

SHREWSBURY, p°t., Worcester co., Mass., 37 m. W. Boston, 404 W. Long pond, four miles long and from 40 to rods wide, lies between this town and Worcester. pond is the principal feeder of the Blackstone canal. contains a Congregational and Baptist church, two stores,

one grist-mill, two saw-mills; seven schools, 275 scholars. Pop. 1481.

SHREWSBURY, p. t. Monmouth co., N. J. 152 m. E. Trenton, 218 W. Drained by Nevisink, Shrewsbury, and Shark rivers. Bounded E. by the Atlantic, along which the shore is high, without marsh. Long Branch in this town is a great resort in the summer season, for sea air and bathing. It has fine boarding houses for the accommodation of visiters. It contains two churches, a Presbyterian and an Episcopal, 34 stores, one furnace, four grist-mills, three saw-mills, one tannery, one distillery; eight schools, 405 scholars. Pop. 5017.

SHREWSBURY. p. t., York co., Pa., 38 m. S. Harrisburg, 76 W. Watered by branches of Codorus cr. The borough in the S. part of the town contains two stores, one tannery; one school, 65 scholars. Pop. 340. The town exclusive of the borough has two stores, three woollen factories, seven grist-mills, seven saw-mills, one paper-mill, two tanneries, 10 distilleries; three schools, 40 scholars. Pop. 1398.

SHROPSHIRE. *See SALOP.*

SHUMLA (the an. *Marcinopolis?*), a city and strong military position of Turkey in Europe, being, in fact, one of the keys of Constantinople, on the N. declivity of the Balkhan (an. *Mons Hæmus*), on the great road from Constantinople to Rustchuk, 63 m. S.E. the latter, and 290 N.N.W. the former. Pop., according to Boué, 20,000. In a military point of view, Shumla is to be regarded as a vast entrenched camp. It occupies the declivity of a gorge in the mountains, which encloses it on three sides, like a horse-shoe; and on the fourth side, which descends into the plain, it is protected by a small hill, on which is a strong redoubt. The space occupied by the town is about 3 m. in length, by 2 m. in breadth. In the last century it had pretty strong walls; but these have been all but destroyed, and it is now merely surrounded by a ditch. It is, however, defended by some outworks and by a citadel, which has been greatly enlarged and strengthened since 1836. Its real defence consists, however, in the strength of its position: the plain to the N. of the town, on which the attacking army must encamp, is exceedingly unhealthy; and the surrounding mountains being steep, separated by deep rocky ravines, and covered with thick brushwood, which affords excellent cover for troops, Major Keppel, and other military authorities, appear to think that, if well defended, it would be all but impregnable. The Russians attempted to take Shumla, in 1774, 1810, and lastly in 1828; but they failed on every occasion. Its principal defect, in a military point of view, is the great number of troops required for its effectual defence; and the fact, as shown by the Russians in 1829, that it may be turned.

Shumla is intersected by a rivulet, and is divided into the upper and lower towns. The former is principally inhabited by Turks; it has fine new barracks, numerous mosques, covered with tin and copper, and, which is unique in Turkey, a town clock which strikes the hours, with a bell, &c., introduced by a pacha, who had been in Russia. The lower town, in which the Jews and Christians reside, is unhealthy, from, as is said, the influence of the adjacent marshes, but more probably from the filth of all sorts thrown into the rivulet which flows through the town. The tinmen and braziers of Shumla are the best in Turkey, and supply Constantinople with their wares. It has also some manufactures of silk and leather; and readymade clothes are manufactured in large quantities for sale to the merchants of the capital. It is the residence of a pacha and a Greek archbishop. (See *Keppel's Journey across the Balkhan, &c.*, i. 351, &c.; *Boué, Turquie d'Europe*, ii. 236, &c.)

SHUSTER, a city of Persia, prov. Khuzistan, on the Karoon (anc. *Eulæus?*), 165 m. S.W. by W. Ispahan, and 50 m. E.S.E. Shus, with which city it has been disputed, though, as generally supposed, unsuccessfully, the distinction of representing the an. *Susa.** Lat. 39° N., long. 48° 50' E. It was formerly the capital of the province; but having been depopulated by the plague in 1832, Dexphoul is now the capital. It may still, however, have 15,000 inhabitants. (*Rawlinson.*) According to Col. Chesney "Shuster is about the size of Shiraz, and contains from 10,000 to 12,000 houses, all on the left bank of the Karoon. The town spreads E. from the river in a semicircular form, covering undulating ground, surrounded in its whole circuit by a wall of unburned bricks, and washed by an artificial canal on one side, and the Karoon on the opposite." The houses are principally of stone. The canals, dykes, &c., about the town are extensive, but ill kept. Water is conveyed to all parts of the city by petty aqueducts. The ruins of a castle exist on a height near the remains of an ancient

---

* Major Rawlinson contends, that neither of these cities is the representative of *Shuster*, the anc. cap. of Persia, which, he says, is to be sought for in the ruined city of *Susan*, also on the Eulæus, but in the E.N.E. of Shus and Shuster. (See his researches in the *Geog. Journal*, ix., 69–93.)

bridge, carried away by a flood in 1832. There are no remains at Shuster that show it existed prior to the Sassanian dynasty (*Rawlinson*); but on the opposite bank of the Karoon there are numerous chambers excavated in the rock, and N. of the city walls are the traces of a much more ancient town, which appears to have extended on both banks of the river, being in this respect different from the ancient Susa. (*Caesary, Rawlinson, &c. in Geog. Journ.* iii. and ix.) The inhabitants formerly manufactured large quantities of woollen stuffs, which they exported to Bussorah, in return for Indian commodities brought from thence. (*Kinneir's Pers. Empire*, p. 97.)

SIAM (called by the Birmese *Yoodra* or *Yuthia*), an extensive country of India-beyond-the Brahmapuira, comprising, with its dependent states, most of the central and S. parts of that peninsula; extending between the 6th and 20th degrees of lat., and the 96th and 105th of E. long.; having N. the Laos country, E. the empire of Anam, W. the Birmese empire, the British provinces of Tenasserim, and the Indian ocean, and S. the gulf of Siam, which it encloses on three sides. Its area has been very variously stated, but according to Crawfurd, it amounts to 190,000 sq. m. Its population has been estimated, though on very vague and unsatisfactory data, by Crawford, at 2,790,500, and by Malcolm at 3,000,000; of whom, probably, 1,500,000 are native Siamese, 800,000 Shans, 250,000 Malays, Peguans, &c., and 450,000 Chinese settlers; but other authorities estimate the population at from 6,000,000 to 8,000,000.

Physical Geography.—The central part of this kingdom consists of the fertile valley of the Me-nam, one of the principal rivers of S.E. Asia; and the province of Chantibon, on the E. side of the gulf of Siam, is also very fruitful; but, with these exceptions, most part of the country, in so far at least as it is known to Europeans, is mountainous and rugged. The mountain chain which traverses the Malay peninsula separates Siam-proper on the W. from the valley of the Than-lweng or Saluen river, sometimes rises to the elevation of 5000 ft.; and a similar chain (the height of which, however, has not been ascertained), shuts it off on the E. from its Cambojan province of Battabang. The only navigable rivers of any consequence are the Menam, the Me-koo or river of Camboja, and the Than-lweng. The last two belong only partially to Siam, and are noticed in vol. i. of this work (p. 95, and 372). The Me-nam or Mei-nam (mother of waters) runs, on the contrary, through the heart of Siam, the principal towns of which are situated on its banks. According to native accounts, the Me-nam has its origin in the table-land of Yun-nan, whence it flows generally in a S. direction to the head of the gulf of Siam, entering the latter near lat. 13½°, and long. 101° E., after a course roughly estimated at 800 m. It is navigable for small boats as far as Changmai, or Zimmey, and large vessels ascend to Yuthia, the old capital of Siam. In its progress it encircles several islands; and at Bangkok, about 15 m. direct from the sea, it divides into three separate channels. Only the most easterly of these, or Pak-nam river, is navigable for large ships, the others being obstructed by shallow bars at their mouths; and even the Pak-nam branch has a bar 10 or 12 m. broad, with but 1½ fathom water at low tide; so that, even when lightened, vessels entering or leaving the river not unfrequently get aground, though, the bottom being soft mud, they sustain no injury. (*Malcolm*, ii., 130.) The Me-nam, its numerous tributaries, and the other rivers of Siam, annually overflow the country in July and the succeeding months.

The Climate, except in the marshes left after the inundation, is usually salubrious, though the small-pox and cholera sometimes make great ravages. At Bangkok the mean temperature of the year is about 83° Fahr.; the heat is, however, not of an oppressive character, and the annual range of the thermometer is stated to be only about 13 degs. What is called the cool season lasts from Nov. to Feb.; March, April, and May constitute the hot season; and the wet season continues during the rest of the year.

Natural Products, &c.—Iron is found in the mountain ridges on either side the valley of the Me-nam; there are also mines of tin, copper, and lead; and the precious metals are procured in small quantities. But the mineral products of the country are but little known or explored.

Siam is, perhaps, the cheapest country in the world for rice, which is commonly under 2s. and often costs only 1s. per cwt. This is ascribable principally, no doubt, to the natural richness of the soil, and the fact of its being annually overflowed by the Me-nam, or Nile of Siam. *L'inondation annuelle fait à Siam la sureté et l'abondance de la récolte de riz, et rend ce royaume le nourricier de plusieurs autres.* (*La Loubere, Voyage à Siam*, i., 53.) No doubt, however, the extraordinary cheapness of rice is in part, also, owing to the lowness of the land tax, which, as estimated by Mr. Crawfurd, is for rice land only about 1s. 3d. per English acre, or from one sixth to one tenth part of the produce, instead of from one fourth to a half, as in British India.

Here, as in most eastern countries, government is supposed to be the principal proprietor of the soil, but the tenants who pay the land tax run but little chance of being ejected; it is said, however, that gardens, orchards, and houses, are viewed as the private property of the occupants. The Chinese are at once the principal cultivators and traders in every branch of industry.

Besides rice, Siam yields nearly all the most valuable products of the east, and, under an intelligent government, might furnish vast quantities for exportation. Sugar, pepper, tobacco, the finest fruits, are principal articles of culture; and the forests, which cover a large proportion of the surface, produce teak, sandal, sapan, rose, eagle, and a variety of other variegated and perfumed woods, with numerous gums, &c. The teak is said by Hamilton to be of the same quality as that of Birmah; it is floated down 300 m. from the interior to the capital, and is there almost wholly employed in the construction of native junks, very little being exported. Iron, copper, tin, lead, and gold, are the principal mineral products; the gold is obtained by washings, the tin mostly from the tributary Malay territories. The wild animals are similar to those of Hindostan and the adjacent Ultra-Gangetic countries: the elephant is most abundant, and is extensively employed. A very rare, or white variety of the elephant, is sometimes found here, and is held in the highest estimation. Indeed, one of the titles of the Siamese monarch is, "lord of white elephants," several of which are maintained as state appendages at the royal court. "He who discovers one of these animals is regarded as the most fortunate of mortals. The event is of that importance that it may be said to constitute an era in the annals of the nation. The fortunate discoverer is rewarded with a crown of silver, and with a grant of land equal in extent to the space of country over which the cry of the elephant may be heard. He and his family, to the third generation, are exempted from all sorts of servitude, and their land from taxation." (*Finlayson*, 154.) The rhinoceros is more plentiful in this than in most other countries; tigers, though inferior in strength to those of Bengal, are also common.

Races of Inhabitants.—The Siamese appear to be of the same stock with the Laos Shans, to whose country their traditions point as their original seat. They are characterised by a broad forehead, a hairy scalp descending so low as to cover, in some instances, the whole of the temples; the lower jaw is long, and remarkably full under the zygoma, so as to give a square appearance to the countenance. Eyes small and oblique; lips thick; mouth large; beard scanty; hair coarse, lank, and uniformly black, but that of the chin is softer and of a lighter colour than is usual among the Ultra-Gangetic nations, height toward among the upper ranks by a bright yellow wash. Limbs thick, short, and stout; trunk square; they have a strong tendency to obesity; average height of men about 5 ft. 3 inch. They possess, says Mr Finlayson, from whose valuable work we have taken these particulars, the frame without the energy of London porters. (*Mission to Siam and Hue, 224.*) Travellers agree in representing the Siamese as crafty, mean, ignorant, conceited, servile, and rapacious. Indolence, as might be expected, is also one of their prominent traits. They have, however, some redeeming qualities: being exceedingly attached to their children, reverential to parents, temperate, and, except on great provocation, gentle in their manners. The upper classes, however, are offensively coarse, manifesting a total disregard for the feelings of others, and an unbounded arrogance. The Laos, or Shans, tributary to Siam, inhabit principally the N. part of the country, where they are divided into several tribes (*see* Laos). The Chinese settlers are mostly immigrants from the provinces of Canton and Fokien, and the island of Hainan. They resort to Siam unaccompanied by their families, intermarry with the Siamese, and adopt their form of religion, with most of their habits. (*Geog. Journ.*, ii., 292.) Each male above the age of 20 pays a capitation tax. The greater number of them are employed as traders, or, if within the tributary Malay states, in working gold and tin. At Bangkok there are a good many Cochin-Chinese and some settlers from Hindostan, most of whom are Mohammedans. The Portuguese Christians, or their descendants, of whom there may be about 2000, are engaged in commercial pursuits, as interpreters, &c., and are mostly in indigent circumstances.

Arts, &c.—The Siamese have made very small progress in the useful arts; nor, under existing circumstances, can it well be otherwise. All mechanics who evince any skill are immediately seized upon, and made retainers of the king or of some person in authority, who employs them for life in some useless service of vanity or ostentation. Hence there is, as it were, a premium on barbarism, and labour is dear and difficult to procure. The ordinary mechanics are, in fact, usually natives of China or Cochin-China. In no one useful art have the Siamese ever attained distinction.

They make no fabric that can bear to be compared with the cottons of Hindostan, the silks of Birmah, or the porcelain of China. Even in the fabrication of jewellery, a proficiency in which has been often remarked among ruder people, they exhibit little skill; and, in fact, their gold and silver trinkets, plate, and articles of zinc, tin, and brass, are all imported from China, or obtained from the Chinese settlers. It is through the ingenuity of the latter that the iron ore with which the country abounds has been of late years rendered available. At present a good deal of malleable iron is produced, and at Bangkok there are several extensive manufactories of cast-iron vessels; but these are wholly conducted by Chinese. The latter have also introduced the culture of sugar, now become a staple product, and have created a taste for commerce and the means of carrying it on. The cutlery and tools in use among the Siamese are of the rudest and simplest description; and, though the people fabricate arms, they have acquired no skill in the art, and fire-arms have always been supplied by Europeans. Even the coarse brown pottery in use is mostly made by Peguans. The art of dyeing is on the lowest scale, though the country abounds in the necessary materials; and printing silks and cottons is not practised by the Siamese in any shape or form. (Crawfurd, ii., 20–22.) Some Europeans have proposed to introduce steam-engines, saw-mills, cannon-foundries, the culture of indigo and coffee, &c.; but they have been always treated with scorn and violence, and their offers have been uniformly rejected, with the exception of those of a Frenchman, who, in the end, however, had to leave the country in disgrace, after having commenced the construction of an engine for rifling guns! (Gutzlaff, in Geog. Journ.)
Architecture is in the same low state as the other arts. The habitations in the alluvial grounds are all raised on piles, as in the rest of India beyond the Brahmaputra, though on the higher lands piles are dispensed with. But their houses are nearly all of the same fragile materials, among which the bamboo and Nipa palm-leaf are the principal; and it is only in the capital or in the other towns that any are to be seen constructed of brick and mortar, and roofed with tiles. The temples, though surrounded with brick enclosures, consist chiefly of timber-work; and, though laboriously carved, gilt, and otherwise adorned, exhibit no taste. Edifices for public convenience and utility seem to have no existence; and neither piety, superstition, vanity, nor interest, seems to have led the rulers of this country to construct bridges, wells, tanks, or caravanserais. The bridges, even at the capital, consist only of planks, and where do we observe any attempt to construct an arch. The absence of public roads is no less remarkable. There are but two of any consequence in the kingdom; one from Bangkok to Yuthia, and another from Chantibon to Tung.
In the N. and on the Malay isthmus elephants are used to convey merchandise across the narrow mountain ways; but these animals are prohibited, except to a favoured individuals, in most of the towns, and even at Bangkok wheel-carriages are unknown. But internal navigation is so extensive, cheap, and commodious, that in the central part of the country it supersedes the necessity for roads.
Commerce.—The foreign trade of Siam is conducted chiefly with China, Anam, Java, Singapore, and the other British ports within the straits of Malacca, with an occasional commerce with Bombay and Surat, England and America. The most important branch by far of the foreign trade is with China. This is estimated to employ at least 300 junks yearly, many of which are of 500 or 600, and some not less than 1000 tons. They are all of Chinese built, though partly constructed in Siam; some are owned by Siamese merchants, but many more by Chinamen residing in Bangkok, and the crews are never Siamese. These vessels make but one voyage a year; going in one monsoon, and returning in the other. Most of them arrive at Bangkok in October and January, but they continue to come from the distant provinces till April, and sail from the Me-nam in June and July. Numerous small vessels keep up a constant intercourse with the coasts of the gulf of Siam, the neighbouring islands; and two or three Siamese ships built on the European model, trade to Singapore. An artificial canal, kept in good order, connects the Me-nam with the Camboja river; but the trade by it, as well as the Cochin-Chinese sea-going vessels, has been depressed of late years, owing to hostilities between Siam and Anam.
Bangkok is the great emporium of trade; and, according to Malcolm, it has the largest commerce next to Canton, of any city not peopled by Europeans, or their descendants. The imports into Siam from China consist of hardware and porcelain, spelter, quicksilver, tea, lacksoy, fruits, raw silk, crapes, satins, and other silk fabrics, nankeens, shoes, fans, umbrellas, writing paper, incense and Chinese immigrants. From the Malay archipelago and the countries to the westward, the chief imports are British and Indian piece goods, arms and ammunition from Europe, woollen cloths, a little glass-ware, and commodities suited for the Chinese market, as pepper, tin, dragon's blood, rattans, *biche-de-mer*, esculent swallows' nests, and Malay camphor. Besides these articles, the principal exports to China and elsewhere are sugar, cardamoms, eagle, sapan and rose woods, mangrove bark, cotton, ivory, stick lac, areca nuts, salt fish, the hides and skins of oxen, buffaloes, elephants, rhinoceroses, deer, tigers, leopards, otteys, &c.; buffalo, ox, deer, and rhinoceros horns; bones, sinews, feathers, &c. Malcolm, who visited the country in 1837, speaks of the Chinese market as very common, and on the increase in Siam. But, in 1839, the importation of opium was prohibited by the government, conformably, as is said, to the wish of the Chinese government. Since that period but little trade in that article has been carried on with Siam. The trade in several of the most valuable products is of a royal monopoly; but the trade in sugar and pepper, the two great staples of the country, is free: the exports of sugar being estimated at 10,000 tons, and those of pepper at 3500 tons. In 1838, 16,196 cwt. of the former, and 23,759 lbs. of the latter, were imported from Siam into the United Kingdom.
Gold and copper are not used as money in Siam: the only coin is of silver, being merely a small bar turned in at the ends, and stamped on one side. Cowries are the ordinary medium of traffic, and 12,800 go to the tical, which is estimated to be worth 2s. 6d. The ordinary weights are the picul and catty; the former is the same as the Chinese, and divided into 50 catties of 2½ lbs. each. The Siamese fathom is about 6 ft. 6 in.: the sen, a land measure of 20 fathoms square.
The Government is an absolute monarchy, the sovereign being every thing, and the people nothing. The manners of the court, and the etiquette observed, seem to be nearly the same in the present day as they are described by the earliest European travellers; and are of a more servile description than those, perhaps, of any other court in the east. The king, one of whose titles is "the God Boodh," is supposed by his subjects to be a deity, and is reverenced as such; an immeasurable distance being supposed to exist between him and the highest of the nobility. "The people," says Mr. Finlayson, "are governed by opinion, absurd and unjust, not by reason, by sense, or by kindness. The most degrading and brutal tyranny is mistaken for well-meaning patriarchal kindness; and the oppression of the multitude, and the grinding of the many, is regarded as the will of the Deity. No man either wishes for, or aspires to, freedom of thought or action; and tyranny has cast its roots so deep, that change would seem hopeless." (*Finlayson,* 156.) Next to the sovereign, the nobility absorb most of the legislative and executive power; there being, except in some cases of appeal, no establishments exclusively for judicial purposes. The Siamese have, indeed, written laws; but it frequently happens that the king, on his accession, publishes a new edition, with his own interpolations, though neither the original code nor the changes introduced appear to be much regarded by the administrators of the laws. The same chiefs who are charged with the military, civil, and revenue administration, are the only judges and magistrates. According to the laws of inheritance, a man must leave his property to his family in preference to strangers; but no claim of primogeniture is recognised, the children usually sharing equally. The nature of the marriage contract is much the same as in other eastern countries, polygamy being permitted, and divorces obtained without difficulty. A breach of the marriage vow is not visited with so severe a penalty as in Anam, but is usually expiable by a pecuniary fine. The penal code of Siam bears a strong analogy to that of China, especially in the liberal and indiscriminate use of the bamboo for the punishment of all minor offences. For crimes of magnitude, the punishments, as in Birmah, are of the most savage description; torture may also be applied to extort evidence. They have, also, the same sort of ordeals for determining the guilt or innocence of accused parties that were common in Europe during the middle ages. And La Loubère and Crawfurd mention that, in the event of goods being stolen, should suspicion fall on different parties, it is customary to administer emetics to them all; in which case, the person with the weakest stomach, or who vomits first, is held to be the culprit! (*La Loubère,* i., 334; *Crawfurd, &c.,* passim.)
Armed Force.—Every male inhabitant, from the age of 21 upwards, is obliged to serve the state for four months a year. The following individuals are, however, excepted: members of the priesthood, the Chinese settlers, who pay a commutation tax, slaves, public functionaries, the fathers of three sons liable to service, and those who purchase exemption by a fine of from six to eight ticals a month, or by furnishing a slave or some other person not subject to the conscription, as a substitute. There is no standing

3 E   809

army. The king has for a good many years made large annual purchases of muskets; and Mr. Malcolm estimates that he now possesses upwards of 80,000 stand of arms, besides a considerable stock of cannon. The principal force of the Siamese consists in their elephants; but when contrasted with Europeans, their army may be said to be contemptible. At Bangkok, there is a numerous navy of war-junks, galleys, &c., built on the Chinese model, and mounting heavy guns, manned by Chinese and other foreigners.

The *Public Revenue* is estimated by Crawford at about £3,145,000 sterling a year; of which sum, the poll-tax and fines for non-service in the army may, perhaps, produce £2,500,000; the land-tax. £287,000; tax on fruit trees, &c., £65,000; on pepper, £50,000; on spirits and gambling, about £57,000 each; the customs, £33,000, &c. But exclusive of the taxes paid in money or produce, the people are subjected to *corvées*, and other oppressive burdens. The collectors receive no salary, being remunerated by a tithe of the revenue realised; an arrangement which generates a variety of abuses. The receipts and expenditure are said nearly to balance each other; and there is seldom any large sum in the public treasury.

Religion.—The worship of Boodh is nearly universal in the countries lying E. of Hindostan; but the Buddhism of S. is very different from that of N. Asia.* Gaudama is worshipped in Siam under the name of Somona Codom. Every male Siamese must enter the priesthood once in his life, though he may quit it again at pleasure. The *talapoins*, or priests, live together in monasteries, sometimes containing several hundred individuals, endowed by the government or by wealthy persons. The Papal church has maintained missions in Siam for nearly 200 years; but, according to Malcolm, there are only about 2200 Roman Catholics in the country, including 800 Anamese, and several descendants of Portuguese. Neither do the American Baptist and other missions appear to have made many proselytes. (*Malcolm's S. E. Asia*, ii., 155–164.)

Manners, Language, &c.—In the articles ANAM and BIRMAH, in this Dict., various details have been given respecting the manners and customs of Ultra-Gangetic nations, most of which apply to the inhabitants of Siam; though the latter are decidedly lower in civilisation than either the Anamese or Birmans. They are less gross, however, in their eating than the former; and women are not so much degraded among them as among the latter. They are also more generally acquainted with reading and writing than the Birmese. Both sexes dress nearly alike, and wear fewer clothes than almost any other semi-civilized people of the east; a cotton garment reaching downwards from the loins, with sometimes a scarf across the upper part of the body, usually completes the Siamese costume. Jewellery and trinkets are little used, but the teeth are always stained black. They are nearly as much addicted to gambling and cockfighting as their Malay neighbours; they are also very fond of theatrical entertainments and music, in which last they display considerable skill. Their language is radically monosyllabic, and cognate with those of the Laos Shans and Anamese; but many words have been introduced into it from the Cambojan, a polysyllabic language, and the *Pali* or sacred tongue; which last the common dialect imitates in the form of its written characters.

As in other Asiatic countries, slavery is common, and some chiefs have hundreds or even thousands of slaves. Some of the conquered districts have been almost depopulated, to bring their inhabitants to Siam; and at all times an active slave trade is carried on along the Birman frontier. Persons are sold into slavery for debt; and men may sell their wives and children at pleasure. A common custom is for the master not to support his slave, but to allow the latter to work for himself for two or three months, to supply necessaries for the rest of the year. Children inherit their parents' bondage.

History.—The Siamese are said to possess records which go back for 1000 years; but little in their accounts possesses any interest till 1511, when the first intercourse of Europeans with Siam took place. The Portuguese and Dutch had traded with the Siamese for the best part of a century, when the first British ship went up the Me-nam in 1612. In 1683, Constantine Phalcon, a Cephalonian Greek, had found means to get himself raised to the dignity of foreign minister of Siam, and soon afterwards opened a communication with France. Louis XIV. sent an envoy (the celebrated M. de la Loubère, to whom we are indebted for an excellent description of the country) to Siam in 1685. The French were, however, expelled the country in 1690; since which time numerous wars, either aggressive or de-

fensive, with the surrounding states, have been the most conspicuous events of Siamese history. (*La Loubère, Voyage de Siam*, 2 tomes, Paris, 1691; *Crawfurd's Embassy to Siam and Cochin China*, one of the best works on the Ultra-Gangetic nations; *Malcolm's S. E. Asia*, ii., 130–164; *Finlayson's Miss. to Siam*, &c., *Gaz. &c.*)

SIBERIA, a vast territory of N. Asia, belonging to Russia, which *see*; and *see*, also, the article ASIA, i. 142.

SIBKIM, or SIKKIM, a state of N. Hindostan, tributary to the British, between the 26th and 27th degs. N. lat., and about the 88th E. long., having N. Thibet, E. Bootan, W. Nepaul, and S. the Bengal territory. Area, about 460 sq. m. Pop. estimated about 166,000. From its situation on the S. slope of the Himalaya, its geography, products, &c., are nearly similar to those of Bootan and Nepaul, to which articles we beg to refer for its general description. It was placed under British protection in 1816. (*Hamilton's E. I. Gaz. &c.*)

SICILY (an. *Sicilia*), the largest, finest, most important, most fruitful, and most celebrated island of the Mediterranean, constituting that portion of the kingdom of Naples entitled the *Dominii al di la di Faro*, lies between lat. 36° 38', and 38° 18' N., and long. 12° 20', and 15° 40' E. It is separated from the S. extremity of Italy by the narrow strait of Messina, only 2 m. across; and from cape Bon in Africa by a channel 85 m. in width. It is of a triangular shape, and was hence, in antiquity, sometimes called *Triquetra*, but more commonly *Trinacria*, from its name nating in the three promontories of Boeo (an. *Lilybæum*, Passaro (an. *Pachynum*), and Faro (an. *Pelorum*.) It seems to have derived its usual name of Sicilia from the Sicani or Siculi, its earliest inhabitants. Length, E. and W. about 215 m.; greatest breadth 150 m. Area, pop., subdivisions, &c., as follow:

Intendencies (Valli)	Area in sq. m.	Pop. in 1831.	Pop. to sq. m.
Palermo	1,717	475,574	276
Messina	1,478	317,489	212
Catania	1,768	348,952	196
Syracuse	1,528	228,646	144
Caltanisetta	1,532	188,389	123
Girgenti	1,921	325,516	166
Trapani	1,047	171,287	163
Totals	10,826	1,945,346	180

The Neptunian or Madonian chain of mountains, stretches from the straits of Messina, at the N.E. extremity of the island, along its N. coast to cape Boeo at its W. extremity. Some of its summits are of considerable altitude. It gives off several spurs to the S., which with their ramifications cover a considerable portion of the surface. But exclusive of these, there are some mountains which are quite detached from and unconnected with any system. The principal of these is Etna, the most celebrated of European mountains, near the E. coast of the island, and by far the loftiest in Sicily, being not less than 10,872 ft. above the sea. (See ETNA.) There are some extensive plains: the principal is that of Catania, at the foot of Etna; the next in point of size being those of Milazzo, Terra Nova, Syracuse, and that extending along the S.W. shore for about 100 m. E. of Trapani. The rivers, though generally insignificant in point of size, are mostly celebrated in classic history or poetry. The principal is the Salso (an. *Himera*), which, as well as the Platani, Belici, &c., discharges itself on the S.W. coast. The Giaretta (an. *Simetus*) waters the plain of Catania. A great number of small brooks and torrents disembogue on the N. coast; but none of the rivers is navigable, or otherwise available for the purposes of trade. The only lake worth notice is that of Biveri, or Lentini, in the plain of Catania.

Except in some low and marshy tracts, the air of Sicily is generally salubrious, and the climate, though rather hot, is, for the most part, delightful. Cold weather is sometimes experienced, but the severity of the winter is never such as to affect the verdure of the country. Ice and snow are never seen except on Etna, and the highest summits of the Madonian chain; but the summer heats, especially during the prevalence of the sirocco, or S.E. wind, are often very oppressive. The range of the thermometer throughout the year at the level of the sea is from about 36° to 113° Fah.; its mean height being estimated by Smyth at 63°3°, and that of the barometer at 29·80. "Whilst the sun is in the northern signs, the sky, although it seldom assumes the deep blue tint of the tropics, is, nevertheless, beautifully clear and serene; but after the autumnal equinox the winds become boisterous, and the atmosphere heavy and dense; the dews and fogs increase, particularly on the coasts, and the rain falls in frequent and heavy showers." (*Smyth, p. e.*) Sicily has, on various occasions, been subject to destructive earthquakes, which usually take place towards the end of winter.

The primary rocks in the mountains are principally gra-

, quartz, and mica. These are overlaid in many parts limestone rocks, and most of the lower hill ranges are arreous, abounding with metallic ores. The soil, though y various, is almost everywhere endowed with the great-fertility, and has been famous alike in ancient and dern times for its extraordinary productiveness. Sicily s, in fact, the principal granary (*horreum*) of Rome. It ald by Livy to have been *Populosque Romanus, pace ac s, fidissimus annona subsidium.* (Lib. xxvii., cap. 5.) I the third oration of Cicero against Verres, or that en-d *De Frumento*, affords in every part the most conclu-proofs of the fertility of this fine island, and of the it importance of the supplies of corn which it furnish-o Rome. In some of the valleys the soil consists of a loam, from 20 to 30 ft. in depth! The decomposed anic products scattered over the surface of large por-s of the country are also extremely fruitful, being suit-alike for the production of corn, wine and oil. Even resent, despite the wretched system of agriculture, rith says that the usual produce is from 10 to 16 times, in favourable seasons, and on the best lands, 98 times seed! Immense beds of sulphur are found in the cen-and S. parts of the island.

he vegetable products of Sicily embrace numerous trop-as well as European plants. The surface has been di-d, according to its elevation, into the following five re-s, each distinguished by its vegetation.

Region	Height Ft. Ft.	Products
Sub-tropical .	— to 800	Papyrus, sugar-cane, date and dwarf palms, olives, agrumi, &c.
Evergreen . .	800 — 2,000	Similar to those of Apennines, at same elevation.
Oak and chestnut	2,000 — 4,000	(Mountains not so thickly wooded as in Italy.)
Beech wood	4,000 — 6,000	Maize, wheat, &c., to 4,500 feet.
Upper reg. . .	6,000 & above	Birch, juniper, &c. (See *Etna*)

lly was believed, in antiquity, to have been the native try of corn. (*Diod. Siculus*, lib. v.) Homer says of rly inhabitants:

> "Untaught to plant, to turn the glebe, and sow,
> They all their products to free nature owe.
> The soil untill'd a ready harvest yields.
> With wheat and barley waves the golden fields;
> Spontaneous wines from waggdry clusters pour,
> And Jove descends in each prolific shower."
> *Pope's Odyssey*, lib. xi., lin. 121.

iculture, also, is said to have originated in the island the auspices of Ceres. But (*quantum mutatus!*) are now few, if any, countries in Europe in which rt is in so degraded a state. There seems every rea-think, from the number and magnitude of its cities, ther circumstances, that its population in antiquity have been much larger than in modern times. In-the fair presumption seems to be that it must then amounted to at least from 3½ to 4 millions. And yet, hstanding this greater density of population, it was, eady seen, able to export vast quantities of corn to It does not, however, appear very difficult to ac-for this melancholy change. After the overthrow Roman power, Sicily was occupied successively by eeks, Saracens, Normans, and French, till at length ne a dependency first of the crown of Spain, and ecently of that of Naples. It is to this dependence, the introduction of the feudal system by the Nor-hat its backward state is principally to be ascribed. altiplied abuses which grew up in Spain under the Ferdinand and his successors of the Austrian line, ed with equal insurance in Sicily, and have proved destructive of the industry and civilization of its ints than of those of Spain. The Neapolitan re-s been equally pernicious; and misgovernment, the of the feudal system, insecurity, and unequal and taxes, have here, as everywhere else, paralysed and impoverished the people.

he grand curse of Sicilian, as of Sardinian indus-probably be found in the oppressive restrictions re been laid on the exportation of corn. Down to riod no corn could be exported without leave being from the *Real Patrimonio*, a body that pretended to ccount of the crops, and which determined whether-were to be any exportation; and in the event of allowed, it issued, or rather sold, licenses to a few individuals, authorising them to export certain quantities! Even had Sicily been ten times ductive than she really is, it is quite impossible culture could have flourished under such dis-ents. Luckily, however, these oppressive re-have recently been abolished, and there are no obstacles to the free exportation of corn. Oppres-

sive taxes, the want of leases of a reasonable length, and of practicable roads, are at present, perhaps, the greatest obstacles to agriculture.

The property of the island was valued in 1811, when the English garrison and fleet occasioned a great demand, and high prices for produce of all kinds, and this valuation has been continued to this day, as the basis on which the land and house tax (*fondiaria*) is levied. A rate of 7½ per cent. on this valuation was first charged, which was subsequently raised to 12½ per cent., at which it is now fixed. Owing, however, as is stated, to the fall in the price of agricultural produce since 1811, this tax is alleged by Mr. Macgregor to be more than equivalent to a duty of 25 per cent. on the produce of the soil taken at its present (1840) value, and to be a very great obstacle to improvement. We believe, however, that its influence in this respect, though considerable, has been much overrated; and that the backward state of Sicily is principally owing to other and different causes.

Though there be in Sicily a very considerable number of small proprietors, by far the greater part of land belongs to the crown, the church, and the nobility, some of whom have very extensive and valuable estates. Down to a recent period these were held under a system of strict entail, and their occupiers, as well as those on the estates of the crown and church, usually held under triennial leases, and were in a state of feudal bondage, and subject to numerous exactions on the part of their lords. Under such circumstances, even though there had been neither restrictions on exportation nor a land-tax, the depressed condition of the peasantry, and the low state of agriculture, need not be wondered at.

But we are glad to have to state that the dawn of a better day seems to have arisen, and that several important changes have lately been introduced. We have already noticed the removal of the restrictions on exportation; and in 1812 and 1838 laws were passed for the abolition of the feudal system, and the complete emancipation of the peasantry. And, notwithstanding the poverty and ignorance of the latter will hinder them from immediately profiting, to the extent that might be anticipated, from the passing of this law, it cannot fail, in the end, to be productive of the best effects. It was also enacted in 1819, that in future, on the death of any individual possessed of an estate in land, and having more than one son, the half only of the estate shall descend to the eldest son, and that the other half shall be divided in equal shares among the other children. This law, which appears to have been framed on the model of that which regulates succession to property in France, will probably have nearly similar effects. In both cases tries the abuses of entails might have been obviated without running into the opposite extreme, and establishing a system that can hardly fail, in the end, to occasion the too great division of landed property.

The arable lands in Sicily have been roughly estimated to comprise about 3,700,000 acres; vineyards, 115,000 do.; vegetable and fruit gardens, 260,000 do.; woods and olive plantations, 1,125,000 do.; the remainder of the surface being mostly waste. Ploughs and most other agricultural implements appear to have undergone no improvement for centuries; and "a bunch of brambles, drawn by an ox, supplies the place of a harrow." (*Smyth.*) The magnificent crops which are occasionally met with are wholly ascribable to the fertility of the soil, and not to the labour or skill of the cultivators. In fact, such is the carelessness of the husbandmen, or such the difficulties under which they have laboured, that, according to Mr. Macgregor, nearly a third part of the pop. have to depend for support on the fruit of the cactus (Indian fig, or *figuisine*), found in the greatest profusion in most parts of the country. (*Report*, 46.)

We suspect, however, that there is a great deal of exaggeration in this statement; and, at all events, it is sufficiently certain that, independently of the changes already alluded to, there has of late years been a decided increase of the means of subsistence. The growth of pop. which increased from 1,660,267, in 1798, to 1,943,366, in 1831, being an augmentation of 283,099, is an evidence of this; and has, no doubt, been at once a cause and a consequence of the improvements which though slowly, and almost imperceptibly, have begun to make their way into the island. At present (1842) the pop. is probably not under 2,100,000.

Exclusive of wheat and barley, hemp, flax and cotton are raised with but little labour. The culture of the last is said to be extending of late years, especially in the neighbourhood of Mazzara. It is mostly short-stapled, and but little is exported, and that only to Naples and Trieste. It is probable, however, that by attention to its culture, and the introduction of improved varieties, its quality might be improved, and it become an article of some importance. The sugar-cane was formerly a staple product of the S. shore of Sicily. But, owing to the introduction of cheaper sugar from the W. Indies and Brazil, the culture of the cane is now restricted to some small plantations near Avola, and will, probably, at no distant period, be wholly abandoned.

The district round Marsala is the principal seat of the wine culture; and, thanks to the exertions of some English capitalists established in that city, the production of wine is become an important branch of industry, and it forms a principal article of export. (*See* MARSALA.) Vines are generally treated as in France; being cut low, and not festooned along other trees, as in S. Italy. But, except in the English establishments, little care is in general bestowed on the vintage: the wine press is a very rude machine, and in some districts it is altogether wanting; the process of crushing the grapes being performed in large vats, by the treading of bare-footed peasants. Along the N. coast, the mountain slopes and valleys are almost wholly covered with olive groves; though elsewhere they are rare, and do not furnish sufficient oil for the inhabitants. But for the imperfections in the mode of its preparation, the oil of Sicily would be excellent. The olives, however, are permitted to hang on the tree till they come off with shaking, or beating with light canes: and they are then kept in vats till they get quite black, so that the oil becomes pungent and rancid, and, though fit for the lamp, is totally unfit for the table. It is only near the capital, and in a few other places, that a more improved process is followed. Lemons and oranges, which grow luxuriantly, are of excellent quality, well adapted for long voyages, and, when intended for exportation, are collected with more care than any other agricultural product. They are largely exported, and are, altogether, highly important. Almonds, pistachios, dates, madder, the barilla-plant, hazel-nuts, the *Ricinus palma*, or castor-oil plant, saffron, tobacco, &c., might all be raised in any quantity; but their culture is, for the most part, neglected, or ill-conducted. The mulberry is grown in the vicinity of Messina, and in the N. E. part of the island; but the produce of silk does not exceed 400,000 lbs. a year. The manna ash is grown near the capital, and manna not being monopolised by the government in Sicily as in Naples, it might be a profitable article of trade, if there were any public enterprise. Liquorice is found growing wild in several parts of the island, and considerable quantities of juice are exported. The culture of shumac is more attended to, and it forms a principal article of export. Potatoes, which have been introduced during the present century, are become a principal article in the diet of the peasantry. The farm labourers, who are very badly lodged, receive, according to M. Simond, from 3 to 4 *carlini*, or from 1s. to 1s. 4d. a day. Besides potatoes, their food consists mostly of maize polenta, onions, garlic, salt fish, cheese from sheep's milk, oil, and beans, which last are a staple in the food of both men and cattle. The peasantry sometimes eat salted pork, but rarely any other kind of flesh. (*Simond, Galt, Blaquière, &c.*, passim.)

The want of improved means of communication is one of the greatest drawbacks on agriculture; except in the vicinity of Palermo and other great towns, there are all but unknown; and the only mode of travelling is by means of the *lettiga*, a kind of fly without wheels, holding two persons, and carried like a sedan chair by two mules, one before and the other behind. Happily, however, government appears to have at length become alive to the evils arising from the want of roads, and 1½ per cent. of the land tax is henceforth to be applied to their construction: and permission has also been given to raise a loan of 1,000,000 dollars, at 5½ per cent., for the same purpose. (*Raumer's Italy*, ii.)

Formerly there were only certain ports from which corn could be exported; a limitation which gave rise to the establishment at these ports of public magazines, or *caricatori*, where the corn may be deposited till an opportunity occurs of shipping it off. Provided it be of good quality, and be brought in immediately after harvest, or at farthest, in August, it is warehoused free of expense; what it gains in bulk after that period (about five per cent.) being sufficient to defray all expenses. The receipt of the *caricator*, or keeper of the magazine, is negotiable like a bill of exchange, and is the object of speculative purchases on the exchange at Palermo, Messina, &c., according to the expected rise or fall in the price of corn. The depositor of a quantity sells it in such portions as he pleases, the whole being faithfully accounted for. The public magazines, in some parts of the island are either excavations into calcareous rocks, or holes in the ground shaped like a bottle, walled up, and made water-proof, containing each about 300 salme of corn, or about 2,250 Engl. bushels. The neck of the bottle is hermetically closed with a stone fastened with gypsum. Corn may be thus preserved for an indefinite length of time; at least it has been found in perfectly good order after the lapse of a century. (*Simond*, p. 540.; *Swinburne*, ii. 405.)

The rearing of live stock occupies even less attention than tillage. In general, the horses, mules, and asses of Sicily are small and ill made; the mules of Modica and the asses of the Pantellarian breed being exceptions. The Tunis, or reddish-brown, and long horned breed of cattle, are large, strong, and well-formed, and there is a good breed of goats. But the sheep, excepting a few merino flocks, are very inferior, and their wool is used only in the coarse manufactures of the country. Hogs are of the worst possible breed. Forests, owing to waste and mismanagement, have nearly disappeared, except on the flanks of Etna, and the S. mountains. Staves for wine casks, and ship timber, are mostly imported from other countries, and even firewood is scarce.

The fisheries are chiefly conducted by corporations of fishermen, or monied individuals. That at Palermo employs, during the season, from 900 to 1,000 boats, and 3,500 fishermen; and the produce is valued at from £20,000 to £25,000 a year. The fishermen of Palermo belong to two corporations, each of which has a physician, surgeon, chaplain, and other officers, who are paid from a fund raised by a subscription from each member, of about 3 per cent. on his share of the produce. This fund is also applied to the relief of members, and other general purposes. Tunnies, or fish principally caught on the Sicilian coasts, are taken as in other parts of the kingdom of Naples (which see). The valuable fish, which was in great request in antiquity, as well as in modern times, is of large dimensions, being generally from 4 to 8 ft. in length, with a nearly equal girth. Its flesh is highly nutritious. The shoals of many enter the Mediterranean early in the year. The *tonnare*, or fishing establishments on the Sicilian coasts, are more extensive and valuable than those in any other part of the Mediterranean. The nets belonging to the one in the bay of Palermo are so very strong as to be able to arrest the progress of a ship when under sail. The fishery of the sword-fish is confined chiefly to the straits of Messina, and the anchovy and pilchard fisheries to Sciulania. Lentini has some trade in *botargo*, made of the roe of the mullet. The coral fishery, near Bona in Africa, is principally frequented by fishermen from Trapani, at which city the coral is polished, and brought for exportation to Catania, Naples, Leghorn, &c.

The minerals of Sicily are important and valuable. Sulphur ranks first; it is found in great quantities imbedded in blue marl, or in gypsum and limestone, over most of the central and south parts of the island. The sulphur mines have been wrought for upwards of 300 years; but it is only since 1830 that any extraordinary quantity has been prepared for exportation. Subsequently to 1833, the trade with this country increased so much that the export of sulphur to the United Kingdom rose from 19,192 tons in the above year, to 36,654 tons in 1838. In this year, however, the Neapolitan government granted to a French company the monopoly of the sulphur, the production of which was to be limited to 600,000 quintals, to be supplied to the company by the proprietors of the mines at certain fixed prices, on condition of the latter paying to the government a bonus of 400,000 Neapolitan ducats a year! It is needless to dwell on the impolicy and absurdity of such a project. Instead of attempting to limit the export of sulphur, government should have given it every possible facility; and taking the export under a fine system, at only 1,000,000 quintals, it would have yielded, at the low duty of 2s. a cwt., on export, a larger sum than was to be paid by the company for their monopoly. Luckily, however, a firm remonstrance by England occasioned the suppression of the monopoly, and the sulphur trade is again restored to its former state. Some sulphur mines are wrought by English speculators with machinery brought from England, and workmen from Wales, Cornwall, and Scotland; but in most of the other mines, the processes are very rude, and, in melting a great part of the sulphur is allowed to escape in gas, to the destruction of the surrounding vegetation. Sicily furnishes saltpetre of excellent quality, in sufficient quantity for her own consumption, but from want of enterprise, none is produced for exportation. Rock salt, bitumen, gypsum, and marble of different kinds, are found in various places; and good salt is made at Trapani and other coast towns. There are also ores of copper, lead, mercury and iron; but very few of these are wrought. There are no iron foundries in the island; and iron and the goods are principally imported from England, lead from Spain, and steel from Germany.

In some of the principal cities there are a few manufactures of silk, woollen, cotton, and linen stuffs; the cotton and woollen yarn being imported from Naples, Saxony, &c. A successful attempt has of late been made at Trapani to spin low nos. of cotton twist by steam power: and some progress is making, both at Palermo and Messina, in the manufacture of ordinary printed muslins, and such like articles. At Palermo there are also oil-cloth and glass factories. But both glass and oil-cloth, with cotton and coarse woollen goods, India handkerchiefs, crapes, and earthenware, are principally supplied by England; fine woollens, printed cottons, and silk goods, porcelain, &c., come from France and Belgium; Germany and Holland send the principal part of the linen goods; paper and Swiss goods are ported from Genoa; and dye woods and colonial products me direct from America. The duties on most articles of import are so very high, that a large proportion of the goods consumed in the island, especially sugar and other colonial oducts, are smuggled. The following is an

ACCOUNT of the Quantities and Values of the principal Articles, the Growth and Produce of Sicily, exported from that Island in 1839:

Articles	Weight or Measure	Totals.	
		Quantities.	Value.
			£.
Argol and cream of tartar	Cwts.	10,086	12,019
Barilla	—	49,465	26,179
Brimstone	—	641,344	116,142
Cantharides	—	187	2,826
Cheese	—	2,656	2,381
Corn, grain, rice, and pulse	Quarts	6,371	11,694
Cotton wool	Cwts.	1,460	8,195
Essences	lbs.	41,118	9,415
Fish, salted	Cwts.	24,476	13,265
Fruits { Dried and pickled	Boxes	102,108	163,175
{ Oranges and lemons		5,60,434	119,737
Lemon juice	Gallons	294,357	6,315
Linseed	Quarts	32,492	55,112
Liquorice paste	Cwts.	14,163	23,073
Manna	—	5,490	53,740
Oils { Linseed	Gallons	18,142	1,705
{ Olive	—	692,479	98,869
Rags	Cwts.	29,554	18,940
Salt	Tons	22,055	15,361
Seeds	Cwts.	23,3 6	5,920
Sumac	—	286,042	253,367
Silk	No.	55,124	61,969
Shrub	No.	313,710	8,716
Wines and spirits	Gallons	4,601,517	156,315
Wool	Cwts.	60	750
Other articles	—		103,9 2
Total			1,340,493

uring the same year the value of the import into Sicily amounted, according to the official returns, to £568,996; the cipal articles being sugar and other colonial products; en stuffs, yarn and wool; woollens, silks, and linens; a, hardware, fish, &c. But, as already stated, the official returns afford no real test of the amount of the imports, value of which may be safely estimated at above 00,000. There belonged, in 1838, to the different Sicilian ports about 9,250 vessels, of the aggregate burden of 1 43,000 tons, employing above 25,000 men.

Accounts are kept in ducats = 3s. 5·2d. of 10 tarini; the = 4·1d., equivalent to 10 bajocchi, of 2 grani and 8 piccioli. The oncia of 30 tarini = 10s. 3d. The lb. = 7 lb.; the salma of wheat, &c. = 7½ Eng. bushels. The palm = 10 inches 3 lines Eng.; the braccio = 3 palmi; the canna = 8 palmi.

Government.—The feudal system was introduced into by Count Roger, soon after the expulsion of the Saracens, in 1072. He also established a representative assembly, or parliament, which subsisted notwithstanding the changes the island has undergone, down to our own This assembly consisted of 3 estates, or braccios, rst, or braccio ecclesiastico, comprised 66 prelates, abbots, and other clergymen; the second, or braccio militare, used 227 nobles, among whom were 38 princes, 27 and 37 marquises, but the larger portion of the nobility had no seat in the assembly: the third, or braccio demaniale comprised 43 representatives of as many free towns. The prince of Butero was hereditary president of the assembly (Swinburne, ii. 170.) It is obvious, from this statement, that the nobles and clergy had an overwhelming majority in this assembly: and while the possession of by far the greater portion of the landed property of the island made tantial and real equal to the numerical ascendancy of two classes, the establishment of majorats and ond the servitude of the peasantry, who were in the absolute state of dependency on their lords, interested her in the support of abuses that opposed insurmountable obstacles to the public prosperity. No wonder, therefore, that the Sicilian parliament should have failed in producing the advantages we are accustomed to ascribe to such ones; and that it should, in fact, have become a bulwark for the defence and protection of the most oppressive abuses and privileges.

The crown was quite as anxious as the burghers to limit the privileges of the braccio militare, provided that could without extending the privileges of the people in a rational point of view. But not daring openly to attack so powerful a body, it fell upon the device, worthy of the and imbecile government of old Spain, of ruining the industry of the country, by laying restrictions on the exportation of its produce, that it might, in this way improve the barons! (Brydone, p. 350, edit. 1806.) This id system, if so we may call it, was acted upon during the whole of last century, and Sicily was a prey to every species of abuse. At length, in 1812, a new constitution was established, under the auspices of Lord William Ben-

tinck, commander of the British forces in the island. Under this constitution, which was formed on the model of that of England, the legislative power was vested in the king, and in an upper house consisting of barons and bishops, and a lower house elected by the people. Unluckily, however, Sicily was not in a condition suitable for the working of such a form of government. The upper house had everything to lose, the lower everything to gain; and though some members of the former saw the expediency or rather necessity of yielding up injurious privileges and making timely reforms, the far greater number were firmly opposed to all innovation. Under such circumstances no improvements could be effected; and the constitution becoming unpopular with all parties, the crown had little difficulty in effecting its abolition in 1816, and in establishing a nearly arbitrary system of government. Since then the administration of Sicily has been assimilated to that of Naples; and, as already seen, several important and, on the whole, highly beneficial changes have been introduced. (Reumer's Italy, ii. 288, &c.)

Trials are public, but not by jury. Until very lately, the police service was conducted, as in Sardinia, by a number of companies, having each at their head a captain, who chose his men at pleasure, and who was bound to make compensation for all robberies and thefts committed in the day-time on the high roads within his district. The companies were paid out of the public treasury; but since 1837 they have been abolished, and their place supplied by gendarmeric. Sicily is not subject to any forced levy of soldiers for the Neapolitan army.

The Rom. Catholic is the established religion, but others are tolerated. There are about 58,000 Greeks in the island, chiefly living in the Piana dei Greci, near Palermo, and a few thousand Jews. There are three archbishoprics, those of Palermo, Messina, and Monreale; 10 bishoprics, and priests in all the communes. The church is chiefly maintained by revenues derived from landed estates. There were, in 1832, no fewer than 658 monasteries; and the monks, whose subsistence forms a heavy tax on the industry of the labouring class, amounted, at the same time, to 7,591. (Giornale di Statistica. Palermo, 1836, No. I. 111.) Education is almost wholly in the hands of the clergy, but it is better conducted than in the rest of the Neapolitan dom.; and if the quality of the instruction were at all commensurate with the number of schools, the people of Sicily would, in this respect, be fully on a level with most other countries. Palermo and Catania have flourishing universities, both of which have had many distinguished individuals among their professors; there are colleges and academies in 21 towns, and primary and secondary schools in each commune. In these popular schools, besides reading, writing, and arithmetic, the pupils are taught linear drawing and the geography of Sicily. In the prov. of Catania the method of mutual instruction has been adopted. There are several Jesuits' schools, three Episcopal academies for divinity students, and boarding-schools for the nobility, &c., at Palermo. Females are usually educated in convents till they be 18 or 20 years of age. Some scientific journals are published, especially at Catania, a city distinguished for the superior education and morals of its inhabitants. But it would be nugatory to expect that there can be anything like an efficient system of superior education, or any literature or philosophy deserving the name, in a country where the press is subject to the most severe censorship; and where all foreign works that might tend to expand the minds of the people, and to make them acquainted with their rights and duties, and with the elements of national prosperity, are as rigorously excluded as if they were fraught with pestilence. Sicily has numerous hospitals and other public institutions, but they are said to be generally ill-conducted. In most large towns there is a monte-di-pieta, or government pawn-bank.

Besides the fondiaria, or land tax of 12½ per cent., the public revenue is derived principally from a tax of 13 tari and 19 grains per salma on the grinding of corn, a duty of 4 gr. per rotolo on the meat consumed in the provincial capitals, with customs duties, and duties on shipping, stamp and registration duties, tobacco duty, the lottery, the post-office, and a duty on the salaries of all persons in official situations. The whole may amount to about 1,900,000 once, that is, to about £950,000 a year gross revenue. We subjoin a copy of that part of the budget for Sicily for 1835, which gives a view of the revenue of the island, and of the sources whence it is derived. [See top of next page.]

Each intendancy is under the control of an intendant or prefect, with a council and secretary, and each district under a sub-intendant, council, and secretary. The board of police for the island, which sat at Palermo, has been dissolved, and the intendants and subintendants now communicate directly with the ministry. Each community is under a syndic elected by the inhabitants from among their number. In each commune and every quarter of the

Sources of Revenue.	Amount.	
	Partial in ounces. 1 ounce=10s.	Total in ounces. 1 ounce=10s.
Ordinary Revenue of Landed Property.		
Land-tax 12½ per cent. . .		458,419 0
General Administration of indirect Duties.		
On grinding corn, 15 tarís and 12 grains per salma . . .	640,800 0	
Custom and navigation duties .	399,496 20	
Duty on bolette for the consultive board of trade . . .	775 0	
Corn stores	1,705 0	
Stamp on playing cards . .	604 0	
Duty on tobacco . . .	10,000 0	
Ditto on books imported from foreign countries . . .	400 0	
Stamp duty on national manufactures	232 0	
Duty on weights and measures for the Porto-franco of Messina .	1,932 0	
Sundries	40 0	
	992,564 20	992,254 20
Branches depending on the general Administration of Rami Dritti diversi.		
Tax on merchants and bankers, exclusive of foreign merchants at Messina	11,889 0	
Duties on registrations, sott of judiciary expenses which cannot be recovered . . .	72,788 19	
5 and 25 per cent. on pensions .	7,051 27	
Duty of 4 grains per rotolo on meat consumed in the provincial capitals	36,880 27	
From shops and markets (wine shops in the military stations and barracks excepted) . .	1,772 8	
Physican Inspector . .	1,475 5	
Cruciata for bulls . .	19,284 11	
Stamp duty on gold and silver manufactures . . .	587 3	
Fees on the royal exequatur .	714 26	
Quit rent on the salt ponds leased to the college of Trapani .	676 6	
Curtain rents . . .	7,472 19	
Uncertain duties . .	1,514 26	
	160,729 17	160,729 17
Uncertain Revenue.— Particular Administrations.		
Royal Lottery . . .	177,494 0	
Postoffice . . .	17,564 7	
Fire arms and shooting licences	6,685 5	
2½ per cent. on salaries of government officers, for pensions to widows and retired persons .	6,309 17	
10 per cent. on the salaries of all persons holding situations under government . . .	46,591 27	
Stoppage of six months' pay on promotions and appointments .	3,800 0	
Savings on account of vacant employments . . .	16,000 0	
Extraordinary revenue, not included in the collector's statements . . .	4,760 12	
From the superintendence of roads and forests . . .	3,946 0	
	72,563 26	72,563 26
Total ordinary revenue .		1,746,310 19
Extraordinary Revenue.		
Arrears of credit down to 1835 .	20,000 0	
Arrears from 1836 to 1839 .	45,000 0	
	65,000 0	65,000 0
Second deduction of 10 per cent. on material expenses . .		1,585 4
Total extraordinary revenue		64,886 4
Recapitulation of receipts :		
Ordinary revenue . .		1,746,310 19
Extraordinary ditto . .		64,886 4
Presumed increase in the coffers of collectors of land tax in the divide of Reggia, duty on tobacco, postoffice, and antiquated credit . . .		62,276 0
Total		1,883,504 23

principal cities, there is a *conciliatore* nominated by the king on the recommendation of the inhabitants, who gives summary decisions in disputed matters not exceeding the value of six ducats ; a judge for each *circondario* resides in every principal town, and each intendancy has a civil tribunal with a president, three judges, an attorney-general, and a chancellor, and a superior criminal tribunal. The superior courts in the intendencies of Palermo, Catania, and Messina are at once civil and criminal tribunals, and have six judges each. That at Palermo has the supreme jurisdiction throughout the island. There are, in fact, no fewer than 250 judges among a population of two millions! The 150 judges of circondario receive from £36 to £84 each, the civil judges about £150, criminal do. from £210 to £250, and the judges of the supreme court from about £400 to

£600 a year ; but this excessive multiplication of courts and judges is a nuisance rather than anything else. With few exceptions, too, the judges and other legal functionaries are said to be notoriously corrupt. But the principal trouble in the way of the proper administration of justice consists in the obscurity and contradiction of the laws. This affords great facilities for, and temptations to litigation, and the country is, in consequence, overrun with swarms of law, pettifogging attornies. There can, indeed, be no doubt that one of the greatest improvements of which the island is susceptible would be the simplification of the law, and the dismissal of more than half the present judges, giving their salaries to those that were still retained. "The prisons," says Mr. Macgregor, "especially those for political offenders, seem to claim a large share of the attention of government ; not, however, for the comfort of the prisoners, but for their security. The state and criminal prisons, on the island of Maritimo, contains, perhaps, the most horrid and strongest dungeons in the world. The prisoners are lowered down several hundred feet from the rocky height above, and are seldom, if ever, heard of afterwards." (*Report,* p. 45.)

Inhabitants.—The Sicilians are of middle stature, well-made, with dark eyes, and coarse black hair ; their features are better than their complexions ; and they attain maturity and begin to decline earlier than the inhabitants of more northern regions. They are cheerful, inquisitive, and forceful, with a redundance of unmeaning compliments ; showing they are not so deficient in natural talents, as in their due cultivation. Their delivery is vehement, rapid, full of action, and their gesticulation violent ; the latter is so significant as almost to possess the powers of speech ; and animates them with a peculiar vivacity, bordering, however, rather on conceit than wit, on farce than humour. The upper classes are incorrigibly indolent, and fond to excess of titles, and such like marks of distinction. Here, in fact, every house is a palace, every handicraft a profession, every respectable person at least an excellency, and every consul-boy is charged with an embassy! This love of ostentation is so inveterate, that the poorer nobility and gentry are penurious to an extreme in their domestic arrangements, and almost starve themselves to be able to appear abroad in the evening with a mean and poverty-stricken equipage. Notwithstanding the energies of the peasantry be impaired by the mildness of the climate, and the multiplied occasions of which they are the victims, they may be said, as compared with the upper classes, to be industrious ; they are also sober ; but passionate, ignorant, credulous, and superstitious. They are, however, bigots, rather than fanatics ; and are civil and kind to such hereditical strangers as may be thrown in their way. There is a great want of keeping and of comfort, even in the best houses ; and in them, and, indeed, everywhere, there is a disgusting want of cleanliness. (*Smyth,* p. 26–66 ; *Stolberg,* iv. 339, &c.)

Sicily early became the seat of many flourishing Greek colonies, of which Syracuse and Agrigentum were the most celebrated. At a subsequent period it was the scene of an obstinate and lengthened contest between the Carthaginians and Romans, and became the first and most valuable acquisition made by the latter beyond the limits of Italy. After the fall of the Western Empire, it was successively held by the Vandals, the Goths, and the Greek emperors, till 827, when it was overrun by the Saracens. In 1072 it was taken by the Normans, who, as already seen, established the feudal system, and kept possession of the island till the establishment of the Swabian dynasty, in 1194. In 1265 Charles of Anjou became master of Sicily ; but the massacre planned by John of Procida, known by the name of the "Sicilian Vespers," 29th March, 1282, put an end to the sway of the Angevines. It soon after became a dependency of Spain, and was governed by Spanish viceroys till 1706, when a popular revolution annexed it to Savoy. By the peace of Utrecht, in 1711, it was ceded to Victor Amadeus of Savoy, who, in 1720, was compelled by the emperor Charles VI. to exchange it for Sardinia. In 1735 the Austrians were driven out by the Spaniards, and the infant Don Carlos was then crowned king of the two Sicilies. While the continental dominion of Naples were held by Napoleon, Palermo was the residence of the court, the island being defended by an English fleet and garrison. An insurrection that broke out in 1821 was speedily suppressed by the Austrians. (*Smyth's Sicily ; Serristori Statistica ; Macgregor's Report on Sicily ; Rosmer's Italy ; Swinburne, Blaquière, Hughes, Simond, &c., passim.*)

SIDMOUTH, a seaport, market town, and par. of England, co. Devon, hund. E. Budleigh, on the Sid, at its mouth, in the English channel. 13½ m. E.S.E. Exeter. Area of par. 1970 acres. Pop., in 1831, 3126. It is situated between two steep ranges of hills, which enclose it on every side except the S., where it is open to the sea. From its sheltered situation, freedom from fogs, and the beauty of its surrounding scenery, Sidmouth has long been a favourite watering place, and was frequently visited by George III., his queen, and court. The port was formerly of some consequence; but being choked up by sand, it is now accessible only by small vessels. The sands, however, have contributed to its popularity as a bathing place; and it has a handsome promenade on the beach, warm baths, good assembly, reading, and billiard rooms, and the other establishments usual at such places for the accommodation of visitors. Attached to the town are the suburbs of Western Town and the Marsh. The parish church, an ancient building, belonged, in the 13th century, to St. Michael's monastery in Normandy; it contains a monument to the celebrated Dr. Currie of Liverpool, the first and best biographer of Burns. The living, a vicarage, in the patronage of the present incumbent, is worth £481 a year. There are Baptist, Independent, and Unitarian chapels, a national school, several charities for the relief of the poor, a theatre, &c. Petty sessions for this and the neighbouring parishes are held monthly at Sidmouth; and courts-leet and baron, at which the peace-officers, &c., are chosen, are held annually. Sidmouth gives the title of viscount to the Addington family. Markets on Saturdays. Fairs, Easter Monday and Tuesday, and third Monday in September, for cattle. (*Parl. Rerts, &c.*)

SIDNEY, p. t., Kennebec co., Me., 9 m. N. Augusta, 604 -. Bounded E. by Kennebec river, W. by Snow's pond, a large and beautiful sheet of water, the outlet of which was through Waterville to Kennebec river. Incorporated 1792. It has two stores, one fulling-mill, two grist-mills, eight saw-mills, two tanneries; 18 schools, 633 scholars. p. 2190.

SIDNEY, p. t., Delaware co., N. Y., 93 m. S.W. Albany, 1 W. Drained by Ouleout and other small creeks flowing into Susquehanna river. It has a Presbyterian and Baptist church, one store, one fulling-mill, one woollen factory, two grist-mills, six saw-mills; 11 schools, 462 scholars. 1732.

SIDNEY, p. v., Clinton t., capital of Shelby co., O., 79 m. by N. Columbus, 473 W. Situated on the E. branch Miami river, and contains a brick courthouse 44 feet are, a jail, 11 stores, 90 dwellings, and 713 inhabitants.

SIENA, or SIENNA (an. *Sena Julia*), a city of central Italy, Grand Duchy of Tuscany, cap. prov. of its own name, three small hills, between two tributaries of the Ombro, 30 m. S. by E. Florence. Pop., in 1836, 18,975. The labouring country is rugged and naked; but the city itself is embosomed in trees, and entered by a fine avenue, which gives it an agreeable and imposing appearance from without. Its streets, however, are narrow, steep, and uneven; and its houses, though often dignified with the title palaces, are built of brick, and are nowise remarkable for architecture. The inhabitants are active, intelligent, and industrious.

Though fallen from its former rank as a republican with 150,000 inhabitants (?), to that of a provincial town, with the melancholy title of the capital of the Maremma, Sienna exhibits no signs of decay, but has, on the contrary, every appearance of active industry; with scarcely any beggars; the streets are well paved and very clean; shops numerous and well supplied; the people well-fed, and the women remarkably graceful and good-looking." (*Simond, Tour, &c.*, 571.) The principal public building is the cathedral, a vast, and, on the whole, magnificent Gothic edifice, founded in the 13th century, though wholly of one date, and built in alternate courses of black and white marble. It has been severely, and, we venture, justly, censured by Forsyth for its incongruousness, want of taste, and barbarous emblematical devices. Over others supporting the nave is a series of the heads of Popes, and the pavement is a kind of mosaic-work, much which is very beautiful. The sacristy is adorned with a work of Pope Pius II., partly painted by Raphael at a very early age, and partly from his designs; and in the same room a most beautiful antique group of the three graces. (*see Letters*, i., 315.) Under this building is a subterranean church, which, if the cathedral, as is affirmed, stand on site of a temple of Minerva, is, most probably, of remote antiquity. The churches of St. Dominico and St. Agnes, the hospital, city hall, and theatre, are worth seeing. The city hall is in the great piazza, a sloping semicircular space, laid out in walks, ornamented with statues, forming the principal lounge of the inhabitants. The facing the main street, has an esplanade and rampart planted with trees, which also form favourite public walks. The antiquities include a Roman gate, the remains of ancient walls, &c., and it has a fountain celebrated in the *Inferno*, and several good public and private galleries of paintings.

Sienna has some reputation as a seat of learning. Its university, founded in 1230, has a library of 25,000 volumes, formerly 60 professors. Its importance has greatly decayed; but it is still celebrated as a school of medicine,

and may have about 300 pupils. It has, also, an ecclesiastical and several other seminaries (including some for *pugilistic science* (for which Sienna was renowned in the 13th century), and various academies, and learned societies, among which last are the *Rozzi*, and *Intronati*, considered the oldest establishments of their kind in Europe. The Sienese pique themselves on speaking the Tuscan language in its greatest purity.

Sienna, which is an archbishop's see, has about 40 woollen factories, besides manufactories of hats, paper, leather, &c. Its chief trade is, however, in corn and other agricultural produce, and the marble of its vicinity. Augustus sent thither a Roman colony, previously to which this city appears to have been insignificant, though boasting of very high antiquity. In the middle ages it was, like Pisa, Florence, &c., the capital of a republic, constantly at war with its neighbours; and it was generally flourishing and independent, till Philip II. of Spain took and conferred it, with its territory, on Cosmo I. of Florence. The French took it in 1808; and previously to 1814, it was the capital of the dep. of Ombrone. No fewer than seven popes, including Pius II., Gregory VII., and Alexander III., with Socinus, and other eminent individuals, have been natives of Sienna. (*Simond; Forsyth; Bowring's Rep.; Conder's Italy, iii., 60-60, &c.*)

SIERRA LEONE, a colonial establishment of Great Britain, on the W. coast of Africa, consisting of a peninsula, about 25 m. in length, N. and S., washed by the Atlantic on the N.W. and S., and partly bounded on the E. by a bay formed by the Sierra Leone river. Free Town, the capital, on the N. shore of the peninsula, is in lat. 8° 29' 40" N.; long. 13° 15' W. It had, in 1839, a population of nearly 42,000, all black or coloured, except about 100 Europeans. The peninsula consists principally of a range of conical mountains, from 2000 to 3000 feet in height, surrounded by a belt of level ground, from 1 to 3 m, in breadth. This is probably the most unhealthy situation in which Europeans have ever attempted to establish a settlement. The principal characteristic of the climate is its extreme humidity. The enormous quantity of 314 inches of rain appear to have fallen at Sierra Leone during three months of 1836, and more fell in two successive days, the 22d and 23d of August, than in Britain throughout the entire year! There are two seasons, the wet and the dry. At Sierra Leone and the Isles de Loss, the former extends from May to November, and at the Gambia from June to September or October, and is always ushered in and carried off by tornadoes. Nothing can exceed the gloominess of the weather during this period; the hills are wrapped in impenetrable fogs, and the rain falls in such torrents as to preclude all exercise and amusement out of doors. At this period fevers of nearly every type, dysenteries, liver complaints, and other diseases, usually make their appearance; though they are so prevalent at all times that they can scarcely be said to belong exclusively to any season. During the 18 years, from 1819 to 1836, every white soldier in the colony was, at an average, three times a year under hospital treatment, and nearly half the force perished annually; and in 1825 and 1826, when the mortality was at its height, three fourths of the troops were cut off! To accept a situation in Sierra Leone has, in fact, been little else than a species of suicide. Nor is the destructive influence of the climate confined to the whites only; the blacks, though in a less degree, are affected by it, and often die in great numbers. (See *Tulloch, Report on the Health of the Troops, &c.*)

Objects of the Colony.—This colony was founded in 1787, partly as a commercial establishment, but more from mistaken and ill-considered notions of humanity. Being intended to consist principally of free blacks, who were to be instructed in the Christian religion, and in the arts of Europe, it was supposed it would become, as it were, a focus whence civilization might be diffused among the surrounding tribes. About 1200 free negroes who, having joined the royal standard in the American war, were obliged, at the termination of that contest, to take refuge in Nova Scotia, were conveyed thither in 1792. To these were afterwards added the Maroons from Jamaica: and, since the legal abolition of the slave trade, the negroes taken in the captured vessels, and liberated by the mixed commission courts, have been carried to the colony. But the efforts made to introduce order and industrious habits, and to lay the foundations of civilization amongst the blacks, though prosecuted at an enormous expense of blood and treasure, have, we regret to add, been signally unsuccessful. And this, after all, is the only result that could have been rationally anticipated. The laziness of the blacks has been loudly complained of, but without reason. Men are not industrious without a motive; and most of those motives that stimulate all classes in colder climates to engage in laborious employments, are unknown to the indolent inhabitants of this burning region, where clothing is of little importance, and all but dispensed with, where sufficient supplies of food may be obtained with comparatively little exertion, and where more than half the necessaries and conveniences of Europeans would be positive incumbrances. And had it been otherwise, what progress could a colony be expected to make, into which there are annually imported thousands of liberated negroes, who, if not wholly incapable of civilization, are, at all events, in the lowest stage of barbarism? The hopelessness of making any beneficial change in the character and condition of the blacks, by keeping up this most pestilential establishment, is now so very apparent, that it may be hoped it will be speedily abandoned. Latterly, indeed, some of the liberated Africans have been carried to the West Indies, where they may be of some use, which is not the case here.

Commercially considered, Sierra-Leone appears to quite as little advantage as in other points of view. The country round the settlement consists of a vast and all but impenetrable forest, only small patches of which have been cleared and cultivated. The principal articles of export consist of teak and cam wood, with ivory, palm oil, hides, gums, and a few other articles; but their value is inconsiderable, amounting to not more than from £60,000 to £100,000 a year. The great article of export from the coast of Africa is palm oil, and of this more than fifty times as much is exported from the coast to the S. of the Rio Volta, several hundred miles from Sierra Leone, as from the latter. We doubt, indeed, whether the commerce with the western coast of Africa will ever be of much importance. The condition of the natives would require to be very much changed, of which there is not the slightest prospect, before they can become considerable consumers of European manufactures. It is singular, that speculative persons in this country should be bent on prosecuting, without regard to expense, a trade with barbarous uncivilized hordes, while they contribute to the neglect or suppression of the incomparably more extensive and beneficial intercourse we might carry on with the opulent and civilized nations in our immediate vicinity. We are bold to say that the equalization of the duties on Canadian and Baltic timber, and the abolition of the existing restraints on the trade with France, would do ten times more to extend our commerce than the discovery of 50 navigable rivers, and the possession of as many forts or factories on the African coast.

The government of Sierra Leone is vested in a lieutenant-governor, assisted by a legislative council of five official members. The chief justice presides in the supreme court of law, held alternately in the course of the year at the different stations under his command; and there are mixed commission courts for the adjudication of vessels taken in the slave-trade. The colony is subdivided into six districts and about 16 parishes, in each of which are one or more schools on the Lancastrian or the national system. Total military force (1841) 306 men, besides militia, &c. Total public revenue of Sierra Leone and Gambia in 1839, £30,000; expenditure, £106,050. In fact this wretched dependency has cost the British public several millions, independently of the enormous sacrifice of life. St. George, or Free Town, the cap., is on the S. side of the estuary of the Sierra-Leone river, being surrounded on all other sides by an amphitheatre of mountains about 1 m. distant. The town and par. had a population in 1838, of 13,523. Its houses are mostly of wood, and disposed in broad and regular streets; around it are the country-houses of the white residents. The drainage of the town has of late been materially improved, but without much perceptible effect on its salubrity.

The villages dispersed in different parts of the colony consist of huts, built of wood and thatched with straw, so light that they are easily moved from place to place. The woods and mountains are infested with a great variety of wild animals, and the rivers swarm with alligators. Insects are so numerous and offensive as to be really a plague.

The settlements on the Gambia consist of St. Mary's Island, at the mouth of the river; area, 5 sq. m.; and M'Carthy's island, about 330 m. up the river. Aggregate pop., in 1839, 4495, of whom only 49 were whites. The trade of this settlement, though inconsiderable, is of more importance than that of Sierra-Leone, the value of the exports, in 1839-40, having amounted to £124,669.

The Cape Coast command, S.E. of Sierra-Leone, consists of the stations of Cape Coast castle (which see), Dixcove, Annamaboe, Accra, &c., along the Ashantee coast. The climate is nearly as insalubrious as that of Sierra-Leone; but since 1828, British troops have not been sent to die in these settlements, the defence of which has been entrusted to a native force of about 200 men. The trade of this part of the coast is of considerably more importance than that of Sierra-Leone, the value of the exports, chiefly palm-oil, dye-woods, Guinea grains, gold dust, and ivory, having amounted, in 1840, to £325,008. An active trade in palm-oil appears to have been growing up of late at the river Bonny. (*Tulloch's Reports*; *Report on the Colonies of W. Africa*, 1841, &c.)

816

SIGMARINGEN (HOHENZOLLERN-), one of the minor indep. principalities of Germany, which, with Hohenzollern-Hechingen, lies mostly between lat. 48° and 8° 3′ N., long. 8° 35′ and 9° 25′ E., inclosed by Würtemberg, on all sides except the S., where it touches the territory of Baden. H.-Sigmaringen is separated into two portions by H.-Hechingen: its S. portion is watered by the Danube, and the N. by the Neckar, &c. It has an area of 22 sq. m., with a pop., in 1838, of 41,900, nearly all Roman Catholics. Except in the S. the soil is generally poor; still, however, rather more corn is grown than is required for home consumption. Potatoes, flax, and fruits, are also grown, but the chief wealth of the state consists of its timber, cattle, and hogs. The rural population is partly occupied in manufacturing cotton and linen cloths, and hardware. Since 1833 it has had a constitutional government; but Berghaus does not say how the representative assembly is constituted. Appeal from the high court at Sigmaringen lies to the superior court at Stuttgart. The revenues of the prince amount to about 300,000 florins a year, partly derived from his estates in Holland. The contingent to the army of the Germ. Confed. consists of 356 men. The cap., Sigmaringen, on the Danube, has a population of 1400. There is no larger town.

Hohenzollern-Hechingen has an area of 138 sq. m., with in 1838, a pop. of 19,900. It is mostly mountainous, and cattle breeding forms the chief occupation of the inhabitants. It has a constitutional assembly formed of two representatives for the town of Hechingen, and 10 for the rural districts. Appeal from its superior law court presided over by the prince, may be made to the high court of Stuttgart. Total public revenue, 120,000 flor. Contingent to the army of the Germ. Confed. 145 men. Hechingen, the cap., on the Starzel, has 3000 inhabitants, several fine buildings, and some woolen manufactures. A little to the S. in the castle of Hohenzollern. Each of these principalities has a vote in the full diet of the Germ. Confed.; and, with Reuss, Lippe, Waldeck, and Liechtenstein, they have the 16th place and one vote in the committee.

SILESIA, an important and valuable prov. of the Prussian dominions, having N. the prov. of Posen, E. Poland and Cracow, S. Austrian Silesia, Moravia, and Bohemia, and W. Saxony and Brandenburg. It lies between 49° 48′ and 52° N. lat., and 14° 25′ and 18° 12′ E. long., and has an oblong shape, extending N.W. and S.E. about 94 m. Area, 15,711 sq. m. Pop., in 1837, 2,645,166: of whom 1,396,607 are Protestants, 1,226,533 Catholics, and 23,530 Jews. The proportion of Protestants to Catholics has increased considerably since the Prussian conquest. Principal towns, Breslau, Liegnitz, Glogau, Gorlitz, Neisse, Glatz, Oppeln. &c. It is divided into three regencies, and these again into 57 circles. Surface rugged and mountainous along its S. and S.W. frontier, but in other parts it is either flat, or but slightly undulating; this is particularly the case on the E. side of the Oder. The river now mentioned traverses the whole length of Silesia; and, being navigable for barges almost to the extreme S. limits of the province, it forms a valuable channel of communication. The other great rivers, the affluents of the Oder, are the two Neisses and the Bober on the S., and the Malapane and the Bartsch on the N. Soil very various, being in many parts loamy and highly productive, and in parts marshy and sandy. Great part of the regency of Oppeln is covered by vast forests. Principal products, corn, flax, and hemp, produced in very large quantities: the stock of sheep amounts to about 2,600,000 head, wool, of a very superior quality, now forms, next to linen, the principal article of export from the province; among the other products are beet-root sugar, timber, madder, tobacco, silk in small quantities, &c. Silesia is rich in mineral products. Coal is found in many parts, particularly in the vicinity of Schweidnitz, and Neisse, Glatz, &c. There are also valuable mines of iron, lead, zinc, copper, &c.

Manufactures are important and valuable. Linen is the principal product; but for some years past is been declining, the cotton manufacture having grown up in the interval to a considerable state of advancement. The woollens manufactured are generally coarse, but they occupy a considerable number of hands. No accounts have been published, on which any reliance can be placed, of the products and values of the different manufactures established in Silesia. The condition of the inhabitants of this province has been vastly improved, both as respects their command over the necessaries and conveniences of life, and their intelligence, since they became subjects of Prussia.

An intelligent observer, speaking of Silesia, observes "In a country where linen is a staple commodity, the majority of the men are weavers, which trade they often exercise in conjunction with their employment as agriculturists; and the women, without exception, are spinners of flax, for we frequently see the better classes pursuing their thread-making occupation, not only in the saloon, but in the promenade, and the lower orders in their huts and on the high

/

while their heads are heavily laden with provi-
market; but instead of the wheel they use the
b. I was informed, the prime cause of the
silence of the Silesian linen, as the thread is by
rendered more soft, round, and less inclined to
re is, however, a wide difference between the
of Silesia descended from German colonists and
Sclavonians, particularly those who people the
he frontiers of Poland. The former are indus-
ly, and manufacturing; while the latter are de-
torance, mendicity, and superstition; they also
sir neighbours the Poles, not only in their lan-
h is a species of Polish patois, but in their sheep-
and greasy kappetas, neither of which are ever
contaminate soap and water: another point of
their inordinate attachment to bodka, and a
tion for Madonnas, saints, and crucifixes. But,
distinctive trait of manners more characterizes
her humiliating mode of acknowledging a kind-
expression of gratitude being the servile 'Upa-
' (I fall at your feet); which is no figure of
they will literally throw themselves down and
et for the trifing donation of a few halfpence.
is the state to which feudal vassalage and su-
ave reduced this people!" (*Germany and the
ol. i.*)

1A (Turk. *Dristra*), a city of European Turkey,
rin, cap. sanjiak, on the Danube. 63 m. E.N.E.
Pop. estimated at 20,000. (*Stern, &c.*) Silistria
military post, and resisted the attempts of the
1828, though they took it after a nine months'
9. When visited by Mr. Elliott, in 1837, it was
ds, having been one of the pledges given by the
he payment of the sums due to Russia. New
hed the whole bank of the river, and an exten-
opposite the town was covered with provender
horses; while 8000 regular troops garrisoned the
r a strong lunette were posted a sufficient num-
nons to form a *post volant*, requiring only three
united so as to form a complete military com-
between the two banks. (*Travels in Turkey*,
136.) Since then Silistria has been restored to
The town is ill laid out, and many of the houses
is. The citadel, several mosques and public
a large bonding warehouse and custom-house,
cf buildings. The inhabitants exchange timber
with the Wallachians for salt and hemp; but
is not of much consequence. Near the city are
s of some fortifications thrown up during the By-
pire.

CREEK, t., Greene co., O. It contains the p. v.
wn. 10 m. E.S.E. Xenia. 64 m. W.S.W. Colum-
. The t. has two saw-mills, two tanneries, three
two schools, 52 scholars. Pop. 2447.

SPRING, t., Cumberland co., Pa., 7 N.E. Carl-
ed by Conedogwinit creek and its tributaries.
three stores, five flouring mills, four saw-mills,
y. two distilleries. Pop. 1938.

3K, a government of European Russia, on both
Volga, having N. the government of Kasan, E.
S. Saratoff, and W. Penza and Nijegorod. Area,
n. Pop. in 1838, 1,200,000. It consists of a gen-
ting plain, having a black, and generally very
Besides the Wolga, it is watered by the Sura
ffluents of the former. Climate in extremes, the
ing hot, and the winter cold. The Wolga is un-
en over for about five months. Rye, wheat, and
, are raised in quantities more than sufficient for
nption. Hemp is largely cultivated, with flax,
poppies, &c. Except among the Kuimucks, the
cattle is not much attended to. In the N. for-
indant. Distilleries numerous; and besides the
is manufactured by the peasants, there are estab-
for the manufacture of cloth, coarse linen and
ad coverlets, with glass-works, soap-works, can-
&c. (*Schnitzler, La Russie, &c.*, p. 676, &c.)
t, the capital of the above government, on the
an isthmus between it and the Swiaga, lat. 54°
, long. 49° 22' 15" E. Pop. above 13,000. It
ly on an eminence, which commands a fine view,
on a plain. Streets broad and straight; houses
wood, but neat and commodious inside. There
was churches, which, with one exception, are all
ad two convents. The town is in a fertile coun-
endes large quantities of corn, exports the pro-
r fisheries on the Wolga. It is a good deal re-
r the surrounding nobility.

ROPOL, or AKMETCHET, a town of European
the Crimea, of which it is the capital, 40 m. N.E.
. Pop. 4900. It stands in a fine but not very
nation on the river Salghir, and consists of two
new built by the Russians, in the European style,

the other, old and occupied by the Tartars. The streets in
the former are wide and regular; and it contains the government offices, and a cathedral, said by Dr. Lyall to be by
far the handsomest ecclesiastical edifice he had seen in Russia. (I. 142.) Within the last few years some improvements have been made in the Tartar part of the town, but
the streets continue to be narrow, crooked, and filthy, and
it has a mean, miserable appearance.

The celebrated traveller and naturalist Pallas lived for
15 years in this town. It was his own wish to emigrate
thither, and to enable him to gratify it, the Empress Catherine II. made him a present of an estate in the best part of
the peninsula. But being cut off from the society he had
enjoyed in Petersburg, and exposed to family annoyances,*
Pallas became dissatisfied with the country and with the
climate he had so highly panegyrized. Having sold his estate, he left Simpheropol in disgust in 1811, and returned,
after an absence of 42 years, to his native city Berlin, where
he died in the course of the same year. (*Biographie Universelle*, art. *Pallas*.)

SIMPLON, a celebrated pass over the Alps, where a
magnificent road was constructed by order of Napoleon, establishing an easy carriage communication between Geneva and Berne, in Switzerland, and Milan. (*See* art. ALPS,
vol. i. 67.)

SIMPSON, co., Miss. Situated a little S. of the centre of
the state, and contains 550 sq. m. Bounded W. by Pearl
river, and drained by its small tributaries. It contained in
1840, 7693 neat cattle, 1120 sheep, 13,371 swine; and produced 2131 bushels of wheat, 132,066 of Indian corn, 2677
of oats, 18,136 of potatoes, 19,365 pounds of rice. 679,915 of
cotton. It had nine grist-mills, seven saw-mills. Pop.:
whites, 2473; slaves, 907: total, 3380. Capital, Westville.

SIMPSON, co., Ky. Situated in the S. part of the state,
and contains 288 sq. m. Drained by Drake's creek and its
branches, which flow into Big Barren river, and by branches of Red river, a tributary of Cumberland river. It contained in 1840, 4948 neat cattle, 8065 sheep, 14,134 swine;
and produced 45,590 bushels of wheat, 1486 of rye, 281,756
of Indian corn, 106,333 of oats, 8198 of potatoes, 706,131
pounds of tobacco, 24,941 of cotton, 1776 of sugar. It had
seven stores, six cotton-factories with 244 spindles, three
flouring-mills, seven grist-mills, two saw-mills, two tanneries, two distilleries. Pop.: whites, 5004; slaves, 1493; free
coloured, 40; total, 6537. Capital, Franklin.

SIMSBURY, p. t., Hartford co., Ct., 13 m. N.N.W. Hartford. 349 W. Watered by Farmington river. In the N.E. extremity, on the W. side of Farmington river, is the flourishing manufacturing village of Tariffville, where is a carpet-
factory producing 132,000 yards of carpet annually. The
Farmington c nal, extending from New-Haven to Northampton, Mass., passes through the town. It contains three
churches, a Congregational, Episcopal, and Methodist, five
stores, one woollen-factory, employing 95 persons, producing
articles to the amount of $50,000; with a capital of $100,000,
one grist-mill, two saw-mills, two tanneries, two distilleries; 13 schools, 376 scholars. Pop. 1805.

SINAI (MOUNT), a mountain of Arabia, near the gulf
of Suez, or upper part of the Red sea, famous for its connection with some of the most memorable events of sacred
history. It is generally supposed to be identical with the
mountain called by the Arabs *Djibel Mousa*, or Mountain
of Moses, or simply *El Tor*, the Mountain, in the peninsula
between the gulfs of Suez and Akaba, in about lat. 28° 25'
N., long. 34° 10' E. The group of mountains to which Sinai belongs, and which also includes mount Horeb, mount
St. Catherine, and other remarkable summits, is surrounded
on all sides by deserts occupied only by tribes of Bedouins,
or wandering Arabs. The mountains are penetrated by
deep chasms, edged by bare perpendicular ledges of rock;
and the whole has a singularly wild and sterile appearance.

The convent of St. Catherine, founded by the Emperor
Justinian, in a valley on the slope of the mountain, is the
halting place whence pilgrims set out to ascend to the summit. Being exposed to the attacks of the Arabs, it looks
more like a fortress than a convent. It is an irregular quadrangular edifice, surrounded by high and solid walls and
covers a considerable extent of ground. To prevent being
surprised by their troublesome neighbours, the entrance gate,
which is rarely opened, is built up; and on ordinary occasions all access to the convent is by an entrance about 30 ft.
from the ground, to which travellers, provisions, &c., are
raised in a basket made fast to a rope, pulled up by a wind-
lass. The interior of the convent presents little remarkable,
all the apartments and chapels being built of rough stone,
without symmetry or order, communicating by crooked and
dark passages. The church of the Transfiguration alone

possesses any pretensions to magnificence. It is 80 ft. in length, and 53 in breadth, paved with marble, adorned with a variety of figures. The event to which it relates is represented in mosaic. But the grand treasure of this church, and that which is supposed, by zealous Catholics, to confer on it peculiar sanctity and importance, is the possession of the relics of St. Catherine, borne by angels to the neighbouring mountain, which still bears her name, and subsequently collected and deposited in a marble sarcophagus in this building! The skeleton of the hand, covered with rings and jewels, is the only portion of the remains of the saint that is exhibited to her faithful votaries.

Mount Sinai, as every one knows, is almost as famous in the sacred history of the Mohammedans as of the Jews; and it is a curious fact, that there is a Mohammedan mosque within the precincts of this convent. It has also an excellent garden at a little distance, which is reached by a subterraneous passage, secured by iron gates. It produces fruits, plants, and vegetables, in the utmost profusion. The climate is temperate, in consequence of the elevation; and snow even falls in winter.

The ascent to the mountain, which lies through a ravine to the S.W., commences close to the convent. It is steep, but the labour of ascending has been greatly facilitated by rude steps cut in the rock. At the height of about 500 ft. from the convent is a spring of fresh and cold water, covered by a rock, which protects it from the sun and rain. After ascending a little higher, the traveller gains the summit of mount Horeb, which forms to use the expression of Laborde, a kind of breast, from which Sinai rises. "Continuing our route from this halting-place by a path, still more rugged and steep than before, we arrived in about 45 minutes at the summit of Sinai, the apex of a peak not more than 50 yards across at its widest part." (*Wellsted*, ii. 95.)

The height of mount Sinai has been variously estimated, but according to observations taken by Mr. Wellsted, it may be estimated at about 7500 ft. above the level of the sea, and about 2500 ft. above the convent of St. Catherine.

On the summit of the mountain is a dilapidated church, which tradition represents as founded on the spot where, amid thunder and lightning, and the smoke of the agitated mountain, Moses received the Decalogue from the hands of the Almighty. (Exodus, cap. xx.) Truth, however, is seldom unaccompanied with error; and but a few yards distant from the church are the ruins of a mosque; this mountain, by a singular coincidence, being hallowed alike in the estimation of Jews, Christians, and Mohammedans.

"It seems," says Sir Frederick Henniker, "to a person on the summit of Sinal, as if the whole of Arabia Petræa had once been an ocean of lava, and that, while its waves were literally running mountains high, it had suddenly been commanded to stand still." Mount Sinai itself, Mount St. Catherine, which is still higher, and the adjacent mountains, rise in sharp, conical, granite peaks; and from their steep and shattered sides huge masses have been thrown down. The prospect from the summit of Sinai is most extensive: the gulf of Akaba, on the one hand, and that of Suez on the other, with mount Agrib, on the Egyptian coast, are distinctly visible. Barrenness and desolation are, however, its grand characteristics. "No villages and castles, as in Europe, here animate the picture; no forests, lakes, or falls of water, break the silence and monotony of the scene. All has the appearance of a vast and desolate wilderness, either gray, darkly brown, or wholly black." (*Wellsted*, ii. 98.) But it is the associations connected with the mountain, and the astonishing events of which it is believed to have been the theatre, that inspire those feelings of awe and veneration felt by all who have either beheld or ascended mount Sinai.

Considerable doubts have, however, been entertained whether the mountain now described be really the mount Sinai of the Pentateuch. It might be expected that the summit of the mountain should exhibit some traces of the stupendous phenomena that are said to have accompanied the manifestation of the Divine presence. But, according to Burckhardt, neither Sinai, nor any of the adjoining summits, exhibits any traces of volcanic action. It is supposed by some that the *Djibbel Katerin*, or Mount St. Catherine, has the best title to be regarded as the true Sinai.

There are really, however, no means by which to arrive at any satisfactory conclusions on the subject. All that can with confidence be stated (for modern legends and traditions go for nothing), is that mount Sinai must be somewhere in this vicinity; and that though the hypothesis, that the *Djibbel Mousa* and the Sinai of the Bible be identical, be not free from difficulties, it is as much so, perhaps, as any other that has been advanced in its stead. (*Calmet, Dictionnaire de la Bible, art. Sinai; Shaw's Travels*, p. 316, 4to ed.; *Laborde's Arabia Petræa*, p. 242; *Wellsted's Arabia*, ii. 90, &c.; *Burckhardt's Arabia, &c.*)

SINDE, an extensive country of N. W. India, between

Hindostan and Beloochistan, comprising the lower country and delta of the Indus; extending between lat. 23° and 29° N., and long. 67° and 71° E.; having N. the Punjab and Bahawulpoor territories, E. Rajpootana, S. the Ran of Cutch and the Indian ocean, and W. Beloochistan and Cutch-Gundava. Its length, N. to S., is about 350 n., its breadth is very variable; but its entire area may be about 100,000 sq. m.; and Burnes estimates the population at about 1,000,000. Mr. Elphinstone dwells on the striking resemblance between this country and Egypt. "One description might serve for both. A smooth and fertile plain is bounded on one side by mountains, and on the other by a desert. It is divided by a large river, which forms a delta as it approaches the sea, and annually inundates and enriches the country near its banks. The climate of both is hot and dry, and rain is of rare occurrence in either country." But here the similarity ends. Egypt has populous towns, numerous canals, and abundant harvests; while in Sinde, such is the barbarism of its country and government, that miserable villages are even few, and agriculture languishes equally with commerce; the policy of the ameers being to keep the land in a state of nature, that their territories may not attract the cupidity of the surrounding tribes. (*Burnes*, i. 244.) The various products of Sinde differ little from those of the rest of India. Rice, indigo, tobacco, and sugar-cane are among the principal; vines, figs, pomegranates, and even apples, are successfully raised at Tatta; and wheat, barley, and the common Indian grains, are grown to great perfection in Lower Sinde. There are vast herds of horned cattle and sheep, which are generally larger than those of Hindostan. Camels and buffaloes are numerous. Game is very plentiful, though wood is scarce. Salt and saltpetre effervesce almost everywhere on the soil; and in the hands of an intelligent government, Sinde might become a rich country.

The main exports are salt, rice, ghee, hides, saltpetre, cotton, oil, shark's fins, bark for tanning, &c.; with an-fætida, and other gums; Cashmere shawls, saffron, horses, leather, musk, alum, and various drugs and gums from the countries on the N. and W. The principal imports from India are metals, ivory, tea, tutenague, and other China wares, chintzes, broadcloths, arms, and other Indian and European manufactures; but particularly opium, in transit from Malwah to Bombay. From Persia and Arabia the Sindians also obtain silks, swords, carpets, dates, rosewater, coffee, &c. (*Pottinger*.) But nearly all the trade centres in Kurachee on the Belooch border; there is less commerce anywhere else, even on the Indus. Merchants, in prosecuting their journey to Candahar from the sea, or the Indian desert, quit the disturbed and unsafe Sindian territories with all despatch. The only encouragement the chiefs give to any trade is to that in opium, and on that they levy the enormous duty of 250 rupees the camel load. The revenue from this article was estimated not long since to amount to seven lacs rupees a year.

The government of Sinde is a military despotism. The country is divided among a number of unruly chieftains, called ameers, whose aggregate revenue has been estimated at 30 lacs of rupees (£300,000) a year. The chief towns are Tatta, Hyderabad, Kurachee, Khyrpoor, Shikarpoor, Larkhana, &c. The Sindians are of a middle size, and darker than most of the inhabitants of Hindostan. Most of them are Mohammedans of the Shiah sect.

There is little to praise in their character; they have nearly all the worst vices of an enslaved people. They are, however, brave in the field, and, unlike other Asiatics, pride themselves on being foot soldiers, preferring the sword to the matchlock; but from want of discipline, their army is a mere rabble, and could oppose no effectual resistance to regular troops. (*Burnes's Bokhara, &c.*, i. 238-259; *Pottinger's Beloochistan; Geog. Journ.* vii., &c.)

SINGAPORE, a settlement belonging to Great Britain, in S.E. Asia, consisting of a small island at the S. extremity of the Malay peninsula, loc. the town of the same name; the latter being in lat. 1° 17′ N., long. 103° 51′ E. The island is of an elliptical form; greatest length E. and W., about 27 m.; average breadth 11 m. Area, computed at 275 sq. m. Pop., in 1836, 29,984; of whom 13,749 were Chinese settlers, and 9638 Malays. The island is separated from the main land by a strait, which, though scarcely ½ m. in breadth in its narrowest part, was the route usually followed by ships between India and China in the early stages of European navigation. But the grand commercial highway between the E. and W. portions of maritime Asia now passes along the S. side of Singapore, between it and a chain of desert islands about 9 m. distant; the safest and most convenient track being so near to Singapore that ships, in passing and repassing, approach close to the roads. The town is wholly undersized for the rapid rise and growing importance to be attached to the strait. This has rendered it not merely a convenient entrepôt for the trade between the western world and India, on the one hand, and China on the other : but also for that between the former and the eastern archipelago, the Philippines, &c.

The surface of the island is low and undulating in some parts, rising into rounded hills, covered with jungle ; though none of these is 400 ft. in height. It consists principally of laterite resting on sandstone ; its N. and E. portions are, however of granite. It abounds with iron ore ; but this is the only metallic product hitherto discovered ; though, from the great abundance of tin in the neighbouring countries, it probably exists here also. The climate is hot, with but little variation of temperature: the thermometer usually ranges from 71° to 89° Fah ; in 1835 the range lay wholly between 77° and 87°. (*Malcom*.) The total annual fall of rain is about 100 inches : the monsoons are little felt ; but the island is kept in a perpetual state of verdure by frequent showers. According to Mr. Crawfurd, the summits of the hills are generally sterile ; but on their slopes and in the intervening valleys there is occasionally a good deal of fertile soil. Gambier or catechu, and fine tropical fruits and vegetables, are grown in tolerable quantities. Nutmegs, coffee, and pepper have succeeded ; and the produce of the Chinese pepper gardens for 1836 was estimated at 10,000 piculs. Cloves have wholly failed, and the settlement depends for rice on Java, Bengal, and Sumatra, and for pigs, poultry, and cattle, on Malacca. Down to 1837, when they were in part remodelled, the regulations as to land were great obstructions to the clearing, cultivation, and prosperity of the island.

The absence of the elephant and tiger, and other formidable wild animals, and of the swarms of insects common in warm climates, are circumstances favourable alike to agriculture and the comfort of the inhabitants. The only quadrupeds are some small species of deer, the otter, porcupine, and a few others ; but it has a great variety of birds and reptiles. Tripang, and *agar-agar*, a delicate fernlike sea-weed, are furnished in great abundance by the neighbouring coral reefs and shoals.

A few manufactures, including that of pearl sago, agricultural implements, arms, &c., are carried on principally by the Chinese. But, as already stated, the entire importance of Singapore consists in its being an emporium—an *entrepôt*, as it were, for the commerce of the adjacent countries, and of that between eastern and western Asia, &c. We subjoin an

Account of the Quantity and Value of the Principal Articles of Asiatic Produce shipped at Singapore during the Year ending April 30, 1836.

Exports.		Quantities.	Value in Spanish dollars
Beche-de-mer	piculs	4,198	71,773
Birds' nests	—	277	168,438
Coffee	—	18 641	136,497
Cassia	—	5,676	40,449
Ebony	—	14,283	45,955
Gambier	—	37,396	64,670
Gold-dust	bunkals	17,507	823,422
Pepper	piculs	86,144	236,982
Rattans	—	39,980	79,194
Raw silk	—		154 694
Sago	—	29,514	71,609
Tortoiseshell	—	256	144,405
Sugar	—	29 370	162,407
Tin	—	27,025	410,506

In the same year 539 ships, of the aggregate burden of 3,053 tons, entered the port.

The chief imports are cotton and woollen goods, iron, and spelter from Great Britain ; opium, Indian piece-goods, and canvass bags from Calcutta, &c. ; ebony and cloves for the China market, from the Mauritius ; Banca, tin, ice, and spices, from the Dutch settlements ; raw silk, tin, tea, camphor, and nankeens from China, for Europe ; hier-of-pearl, sugar, rice, oil, bullion, and some Chinese ice, from the Philippines ; nearly the same articles from n and Cochin China ; and rice, oil, sapan wood, tortoiseshell, birds and feathers, camphor, spices, antimony ore, ammin, catechu, eagle wood, &c., from the various islands of the E. archipelago. The British piece-goods and lens go mostly to Manilla, China, Rhio, Siam, Borneo, Celebes ; cotton twist to the same countries ; arms to e and Rhio ; glass and iron wares to Manilla, China, Java, &c.

The port regulations are upon as liberal and convenient a footing as possible. The attempts hitherto made to impose duties on imports and exports have been successfully ed ; and there are no anchorage, harbour, lighthouse or fees of any description. A register is, however, of all exports and imports ; and to enable this to be reported must be made to the master attendant by the rs of vessels, and invoices delivered to the superintendent of imports and exports. The principal merchants gents are Englishmen ; but some, also, are Chinese, comprise the bulk of the shopkeepers, with by far the

most valuable part of the labouring population. The European merchants transact business on their own account ; but the principal part of their employment consists in acting as agents for houses in London, Liverpool, Amsterdam, Batavia, Canton, and the capitals of British India. The language of commercial intercourse is generally Malay. Merchants' accounts are kept in Spanish dollars, divided into 100 parts. The principal weights in use are the picul = 133⅓ lbs. ; the coyan of 40 piculs ; the bunkal (for gold-dust) = 832 gr. troy ; the bag of rice = 2 Bengal maunds, &c. We subjoin a

Comparative Statement of the Value of the Exports to the under-mentioned Countries during the Years 1837-38, and 1838-39.

Countries.	Value of Exports in Spanish dollars.	
	1837-38	1838-39
Total Great Britain	970,574	1,114,410
Foreign Europe	70,118	116,729
United States	115,152	175,924
Mauritius, Cape, and N. S. Wales	82,550	77,167
Calcutta	1,115,759	960,092
Madras and Coromandel	128,618	172 23
Bombay	253,695	223,519
Arabia	81,057	53,469
Manilla	444,735	201,558
Ceylon	8,756	2,811
China	1,169,760	132,696
Java	589,709	417,73
Rhio	113,591	133 751
Siam	225,342	234,324
Cochin China	93,971	54,441
Sumatra	203,459	224,709
E. Coast Peninsula	311,228	773,467
W. do.	25,496	23,777
Celebes and other islands	517,123	770,948
Borneo	253,371	322,734
Bali	116,516	141,540
Other places	106,313	100,707
Total Spanish dollars	7,315,590	6,379,704

Singapore, which is said to have been the earliest place settled by the Malays emigrating from Sumatra, and an ancient seat of considerable trade, was purchased of the sultan of Johore by the E. I. Company, in 1819. It was then an inconsiderable village ; but Sir Stamford Raffles, who recommended the purchase, clearly apprehended the advantages of its situation for a commercial *entrepôt*, and the importance of its occupation. It was placed at first under a resident, but had no organized government for several years afterwards. Mr. Crawfurd, author of the able work on the eastern archipelago, was governor of Singapore from 1823 to 1826. The governor is now assisted by a council of several salaried officers, and a recorder's court has been established. The military force consists of a wing of the Madras native regiment, and a small detachment of artillery. The public revenue, derived from an excise on the consumption of pork, opium, and home-made spirits, government rents, dues, fines, &c., amounted at an average of the three years from 1833 to 1836, to 264,973 Spanish dollars a year ; and the expenditure for the same period to 243,294 ditto. (*Newbold*.)

The town of Singapore is, as already seen, situated on the S. side of the island, on both banks of the rivulet or salt creek of its own name, stretching thence E. for about ¼ m. to another small creek of the same kind. Its central port is occupied with the dwellings of the merchants and the military cantonments ; the Malay quarter is at the E., and the principal Chinese and commercial quarter at the W. extremity, on the right bank of the rivulet, crossed by a wooden bridge. The streets are in general regularly laid out, and the houses superior to those of Penang, though the best are only of brick. On a hill N. of the town is the government house: the other principal buildings are the court-house, jail, new custom-house, missionary chapel, Armenian church, and the Singapore Institution, founded by Sir S. Raffles, for the cultivation of the languages of China, Siam, and the islands of the Malay archipelago. At present it has an English, a Malay, and a Tamul school, and about 70 pupils : it receives a small pension from government, but is principally dependent on subscription. Ships lie in the roads, or outer harbour, at from 1 to 2 m. from town. The assistance of a number of convenient lighters, which are always in readiness, enables ships to load or unload with scarcely any interruption throughout the year; and the creek being accessible to the lighters for ¼ m. inland, the goods are taken in and discharged at convenient quays before the principal warehouses.

On a small island, about 60 m. S.E. Singapore, is the Dutch settlement of Rhio, the seat of a Dutch resident, with about 34 000 inhabitants. It was originally settled in 1785, and colonized a second time by the Dutch in 1818; but its trade has been almost wholly superseded by that of Singapore. (*Newbold's British Settlem. in Malacca, &c.*, i. 286-398 ; *Crawfurd's Embassy to Siam*, ii 345-405 ; *Malcom's S.E. Asia ; Singapore Free Press, &c.*)

SING SING, p. v., Mount Pleasant t., Westchester co., N.Y., 116 m. S. Albany. 33 m. N. New-York, 258 W. Situated on the E. bank of Hudson r., partly in a deep ravine, and partly on a hill rising gradually to the height of 200 ft. above the level of the river. It was incorporated in 1813, and contains four churches, a Presbyterian, Episcopal, Methodist, and Baptist, Mount Pleasant academy, Mount Pleasant female seminary, many stores and mechanic shops, 250 dwellings, and about 2500 inhabitants. The Croton aqueduct crosses Sing Sing creek by an elevated aqueduct bridge, of substantial stone masonry. Half a mile S. of the village is the Sing Sing state prison. the buildings of which are constructed of marble quarried ou the spot, and were erected by the prisoners. In May 1828 the convicts were removed from the old state prison in the city of New-York, and in 1829 the first building was completed. This edifice is now 480 ft. long, 44 ft. wide, and five stories high, containing 1000 cells for the prisoners. In front and rear are various workshops, with the keeper's house, a chapel, kitchen, hospital, and storehouses. These as well as the principal edifice, are of rough dressed marble. A separate building has been constructed exclusively for female convicts, with well furnished apartments for the matrons. Attached to the prison are 130 acres of land, which contain a vast quantity of marble. which is extensively wrought by the convicts, and exported.

SINIGAGLIA, or SENEGAGLIA, (an. *Sena Gallica*), a town of central Italy, Papal States, leg. Urbino, on the Miss. about ⅓ m. from its mouth, in the Adriatic. 17 m. W.N.W. Ancona. Pop. 7000. It is regularly, though not strongly, fortified with a mound and bastions, and the gates are handsome. Its cathedral, of the Corinthian order, in the form of a Greek cross, has some good paintings, but nothing else very remarkable. The streets are broad, and the town has a neat appearance ; but it is indifferently supplied with water, and is said not to be very healthy.

Sinigaglia is the seat of the greatest of the Italian fairs. The fair commences on the 14th of July, and should terminate on the last day of that month ; but it usually continues five or six days longer. The duties on goods brought to the fair are extremely moderate, and everything is done to promote the convenience of those frequenting it. All sorts of cotton and woollen goods, lace, iron and steel, hardware, jewellery, brandy and liqueurs, raw and refined sugar, dried fish, cacao, coffee, spices, &c., are brought thither by the English, French, Austrians, Americans, Swiss, &c. These are exchanged for the various raw and manufactured products of Italy and the Levant; consisting, among others, of raw, thrown, and wrought silk, oil, fruits, cheese, alum, soda, sumach, sulphur, &c. The value of the imports at some recent fairs has been estimated at about £2,000,000. Accounts are kept in scudi of 90 soldi ; the scudo = 4s. 4d. very nearly ; 100 lbs. Sinigaglia = 73¼ lbs. avoirdupois. The ell, or braccio, measures 25·33 English inches.

The port belonging to the town, at the river's mouth, is fit only for small vessels. Sinigaglia is a bishop's see. According to Polybius, it was colonized by the Romans, A.U.C. 471. Having espoused the cause of Marius, it was taken and sacked by Pompey. (*Commercial Dict.* ; *Cramer's Anc. Italy.* i., 258.)

SISTOW, or SISTOVA, a town of European Turkey, prov. Bulgaria, on the Danube, which is here more than three quarters of a mile broad, 36 m. W.S.W. Rustchuk. Pop., estimated in Horschelmans Stein, at only 10,000 ; but by other authorities at 20,000 and upwards, including many Armenian and Greek merchants. It occupies a large extent of ground, surrounded by a palisade and a dry ditch. Mr. Quin says, "Sistow is beautifully situated. A range of well-wooded hills commence a league or two to the W., and extends a considerable way along the right bank of the Danube. The town, rising at the water's edge, winds its way up the undulations of the eminences. After ascending for a while, the houses are lost ; then they appear higher up, and the whole is protected by a citadel, which crowns the summit. (*Steam Voyage on the Danube,* i., 288.) Sistow has some trade in leather and cotton. It was here, in 1791, that a treaty of peace was signed between Austria and Turkey, after the latter had lost Rimnik, Ismail, &c., to the Russians. (*Slade's Germany and Russia, &c.*)

SITTINGBOURNE, a town and par. of England, co. Kent, lathe Scray, hund. Milton, on the road from London to Canterbury. 15 m. W. by N. the latter. Area of par., 1980 acres. Pop., in 1831, 2189. It consists chiefly of one wide street, running along the high road, and has several good inns. The parish church, a spacious building, has been greatly renewed since 1762, when it was destroyed by fire. The living, worth £212 a year, is in the gift of the archbishop of Canterbury. In 1508, Queen Elizabeth incorporated Sittingbourne under a mayor and jurat, with the privilege of sending two members to the House of Commons, and of holding a weekly market and fair ; but these privileges seem never to have been exercised, except as respects the fairs, which are held on Whit Monday, Tuesday, and Wednesday, and October 10. The markets are held once a month.

SIVACHE, or PUTRID SEA, the *Palus Putris* of the ancients, a lagoon on the E. side of the Crimea. On the N. it communicates with the sea of Azoff by the narrow strait of Yenitchi, being everywhere else separated from it by a narrow, low, sandy tongue of land, stretching N.N.W. from Arabat in the Crimea to opposite Yenitchi, a length of nearly 70 m. The lagoon is shallow, and its W. side, forming the E. shore of the Crimea, is extremely irregular. When the wind blows from the E., the water of the sea of Azoff is forced. through the strait of Yenitchi, and covers the whole surface of the lagoon ; but at other times it exhibits a large extent of mud, the exhalations from which are, in summer, exceedingly unhealthy. The Salghir, the principal river of the Crimea. falls into this lagoon.

SKENEATELES, p. t., Onondaga co., N.Y., 147 m. W. Albany, 340 W. Skeneateles lake, 15 m. long, and from a half to three fourths of a mile wide, lies partly in this township, by the outlet of which it is watered. It contains four churches, a Presbyterian, Methodist, Episcopal, and Baptist ; 11 stores, six fulling-mills. three woollen factories, six flouring-mills, three saw-mills, three tanneries, two distilleries, one brewery, two printing-offices, two weekly newspapers, and one periodical ; one academy, 60 students ; 20 schools, 907 scholars. Pop. 3981.

SKIBBEREEN, a town of Ireland. prov. Munster, in the most southerly portion of the co. Cork, on the Ilen, which is navigable from Baltimore to within half a mile of the town, 40 m. S.W. Cork. Pop. in 1831, 4436. It is a brisk, thriving town, and has a considerable retail trade. It has a parish church, a Roman Catholic chapel, a Methodist meeting house, several schools, a dispensary, market-house, barrack, courthouse, and bridewell. Petty sessions are held on Wednesdays: and it is a constabulary and coast-guard station. It has several large flour-mills and a brewery. The exports, which principally consist of corn. meal, flour, and provisions, are mostly shipped from Baltimore, a little lower down the river, where vessels of 200 tons load and unload. Markets on Wednesdays and Saturdays: fairs, May 14, July 10, August 2, October 12, and December 11 and 23. Postoffice revenue, in 1830, £507 ; in 1834, £633. A branch of the Agricultural bank was opened in 1835.

SKIPPACK, p. t., Montgomery co., Pa., 86 m. E. Harrisburg, 165 W. Drained by Skippack creek. It contains four stores, one printing-office, one weekly newspaper. Pop. 1485.

SKIPTON, a market town and parish of England, W. riding co. York, wapentake Staincliff and Ewcross, E. div. in the mountainous district of Craven, and on the Aire, 34 m. W. York. Area of par., 26,760 acres. Pop., in 1831, 6183. The town. consisting principally of one spacious street, is built wholly of stone from the neighbouring hills. The parish church has some monuments of the Clifford family. The living, a vicarage in the gift of the dean and chapter of Christ-church, is worth £185 a year. The Friends, Independents, and Wesleyans, have meeting-houses ; and there is a flourishing grammar school, founded in 1598, the pupils of which are eligible to the exhibitions of Lady E. Hastings at Queen's college, Oxford, and to two exhibitions in Christ college. Cambridge. The town has also another endowed, or clerk's school, a national school, &c.

Near the church is Skipton castle, the seat of the Earl of Thanet, supposed to have been originally built soon after the conquest. Though not well placed for a fortress, it was of some consequence in that capacity during the wars of Charles I. It was dismantled in 1648, but is now a splendidly fitted up noble residence. Skipton is governed by a constable elected annually at the manorial court-leet; and the general quarter sessions for the W. riding are held in its town-hall. It has some paper and cotton works, and a considerable trade in corn. sheep, and cattle, much facilitated by the Leeds and Liverpool canal, which passes close by the town. Market day, Saturday. Fairs, March 25; Palm Sunday eve, Easter eve, and three Tuesdays next after Easter, for cattle, horses, and sheep ; Whitsun eve, August 5. November 20 and 23, for horses, woollen, and linen cloths, mercery and pedlery.

SKOWHEGAN, p. t., Somerset co., Me., 38 m. N. Augusta, 631 W. Bounded S. by Kennebec river, which here has a considerable fall, affording great water power. A bridge crosses it to Bloomfield. It has nine stores, two fulling-mills, one woollen factory, one grist-mill, two saw-mills ; 11 schools, 612 scholars. Pop. 1584.

SKYE. one of the Hebrides, which see.

SLAVONIA (Hungr. *Tot Orszag*), a prov. of the Austrian empire, usually regarded as forming a part of Hungary, and chiefly included within its military frontier ; between the Drave and Danube on the N. and E. and the Save on the S., dividing it from Servia, Bosnia, and Turkish Croatia, and having Austrian Croatia on the W.

Area, estimated at about 3600 sq. m.; pop. 315,000, principally Slavonians of the Greek church; but partly, also, German colonists, gypsies, and Jews. A branch of the Carnic Alps, almost wholly of calcareous formation, runs E. and W. through Slavonia; but these mountains are of no great elevation, and a large part of the surface is flat. The plains are very fertile, though frequently unhealthy from the presence of extensive marshes along the large rivers. Wheat, rye, oats, barley, and most other grains, are produced in abundance, besides flax, hemp, tobacco, madder, &c. A good deal of strong wine is made, especially in Sirmia. Fruit is extremely plentiful, and there are vast orchards of plums, whence the favourite liquor *sliwowitza* is distilled. The hill ranges are covered with forests, consisting of excellent oak, &c. The breeding of live stock, particularly of hogs, which feed at large in the forests, is extensively pursued. The silkworm succeeds, but it is not much reared. The province is said to possess iron, salt, sulphur, and perhaps coal; but marble only is raised in any quantity. Manufactures, mostly domestic, excepting some of glass and earthenware, potash, &c. The trade partly consists in the exportation of the raw produce of the province, and partly in the transit of products. The principal articles of export are cattle and hogs, which go to the most distant provinces of the Austrian empire; hides, skins, rye, wheat, honey and wax, galls, timber, &c. Slavonia is divided into three counties, and four military districts: chief towns, Esseck, the cap., Peterwardein, and Posega.

Slavonia and the Banat comprise the most important portion of Hungarian *military frontier*, the system of defence organised in which deserves some notice. "The object has been to maintain, at the least possible cost, along the whole Turkish frontier of Hungary, a force which, in peace, might be employed for the purposes of quarantine and customs, and in war serve as a portion of the standing army. This has been effected so perfectly, that in peace nearly 40,000 men do duty along 800 m. of frontier; and they not only feed and clothe themselves, but pay heavy taxes in money besides, and perform, also, a considerable quantity of labour without pay. The land acquired by government along the whole of this district is held as fiefs on the tenure of military and civil service, from 36 to 50 acres constituting a fief. Each of these is bound to furnish, and to maintain and clothe, according to its size, one or more men-at-arms. The fiefs are given to families composed of several members, of which the eldest is the *house-father*, and who, with the *house-mother*, has the direction of the farm, the care of the house, and the right to control the whole family. The fiefs cannot be sold: the land is cultivated for the common good of all the members of a family; and the profit, if any remain after the taxes and other expenses are defrayed, is divided among them. In most cases, many married couples, with their children, sometimes to the number of 50 individuals, live under the same roof, cultivate the same land, eat at the same table, and obey the same father. The border family has to do civil service for the state, as in the repair of post-roads and bridges, draining of swamps, &c., one day per annum for every English acre, and eight days a year for the village. The border's chief tax, besides the furnishing the uniform (government supplying the arms, books, &c.), is the land-tax, amounting, for an entire fief, to from 2s. to 30s. a year. In time of peace, the man at-arms repairs to his military station for seven days at a time, where the family provides him with food. Besides this, he has the duty of transporting letters, as well as the money and baggage of his regiment, and of performing exercise. For the ordinary service, the number of men on duty amounts to 4180. In times of disturbance on the Turkish side, or when the plague is drawing near, it is increased to 6800, and in times of still greater danger, to 10,000. In time of war, the borderer must form part of the regular army, and march out of the country if required. The regular disposable force amounts to 34,800; but if the reserve and the *landwehr* be called out, to 100,000. If driven to the last extremity, they can muster to the amount of 200,000 men. (*Capt.*) By means of alarm-fires and bells, this immense force may be summoned together through the whole extent of the frontier in the space of four hours! The borderers divided into seven regiments. Every regiment receives orders ultimately from the council of war at Vienna. The Hungarian diet has no control over the levy and supply of these troops: and the schools, the language of the service, and many of the laws in the military border, are exclusively German." (*Paget's Hungary, &c.,* ii., 93–103; *plovics, Gemälde von Ungarn; Oesterr. Nat. Encyc.*)

LEAFORD (NEW), a market town and par. of Eng-l, co. Lincoln, wapent. Flaxwell in Kesteven, on the ., 16 m. S.S.E. Lincoln. Area of par., 3160 acres. Pop., 331, 2587. The town consists principally of three streets, has a prosperous appearance. The church, founded in , is interesting for its architecture. The tower, at the

W. end, is much the oldest part, and is early English; but surmounted by a spire, 144 ft. in height, of a later date. The aisles, transept, &c., are in the decorated, and the chancel and other parts in the perpendicular style. The whole of the details of this church, in all its styles, are very good. (*Rickman's Gothic Archit.*) In the chancel are several monuments to the Carr family, one of whom founded and endowed a free school in the town, and a hospital for 12 poor men. The school was closed from 1816 to 1825, during which period the schoolhouse was rebuilt; in the latter year it was reopened, when it had an income of £189 a year. There are several other charities, particularly Alvey's school, for 40 children, which has received many endowments since that of its founder. [*33d Rep. on Charities*, part iv.] Formerly the bishop of Lincoln had a palace at New Sleaford, but it no longer exists. A new Gothic sessions house has lately been built, in which petty sessions are held; and various other improvements have been effected in the town The living, a vicarage worth £170 a year, is in the gift of the prebendary of Lincoln cathedral. Market days, Mondays; fairs. Plow, Easter, and Whit Monday, for horses, cattle, and sheep; August 12, for provisions; and October 20, for cattle and sheep. (*Parl. Reports, &c.*)

SLESWICK (Germ. *Schleswig*), a duchy belonging to Denmark, comprising the S. part of the peninsula of Jutland, between lat. 54° 13' and 55° 30' N., and long. 8° 40' and 10° 10' E., having Jutland Proper to the N., S. Holstein, from which it is separated by the Eyder; E. the Baltic, and W. the North sea. Area, including the *adjacent* islands, 3450 sq. m. Pop., in 1835, 338,192. Surface low, and generally flat, being in parts varied only by a few undulating hills. Almost the whole of its western coast is either below or elevated very little above the sea, being defended from its irruptions (from which, however, it has frequently suffered much) by immense dykes and sluices. The country so protected consists principally of very rich marsh land, affording pasturage for large herds of very superior cattle, as well as great numbers of fine horses. In the interior the soil is sandy, interspersed with heaths, and not very productive, but on the eastern side it is fertile. There are no minerals of importance. The produce of corn, consisting principally of rye and barley, is sufficient for home consumption: and flax, hemp, and potatoes, are also grown. But the raising of cattle and horses forms the staple employment; and these, with butter and cheese, form the principal articles of export. The fishery is carried on to some extent. The deficiency of timber for fuel is compensated by the abundance of turf. The country is mostly open; but it is in parts enclosed with quickset hedges, and the farmhouses are neat, and have a comfortable appearance. The manufactures, which are unimportant, consist mostly of linen, hempen, and woollen fabrics, made in the peasants' cottages. Lace is produced at Tondern, and there are a few paper, tile, and other factories. Flensborg is the principal place of trade: Sleswick, Kiel, and Tonningen, are the other chief towns. This duchy preserves several of its ancient laws and institutions; feudal vassalage, however, was abolished in 1805.

SLESWICK, a seaport town of Denmark, cap. of the above duchy, at the bottom of the narrow gulf, or arm of the sea, called the Sley, 31 m. from its mouth, and 70 m. N.N.W. Hamburgh; lat. 54° 31' 15" N., long. 9° 34' 45" E. Pop., according to Horschelmanns Stein, 11,000. Though irregularly built, its brick houses, neatness, and manner of building, make it look like a Dutch town. It has three churches, including the cathedral, with several monuments, and a remarkable altar-screen; several hospitals, a deaf and dumb asylum, schools for the poor, a patriotic union, and other societies, a nunnery, a savings' bank, with manufactures of lace, woollen stuffs, earthenware, &c. Its commerce has been a good deal increased since the improvements in the navigation of the Sley; still, however, it is accessible only by the smaller class of vessels. It was formerly a member of the Hanseatic league, and a town of some note as early as the ninth century. In its immediate vicinity is the castle of Gottorp, formerly the residence of the dukes of Holstein-Gottorp, and now occupied by the governor of the duchies of Sleswick and Holstein. (*Stein's Handbuch der Geog., &c.*)

SLIGO, a marit. co. of Ireland, prov. Connaught, having N. the Atlantic ocean, E. the co. Leitrim, S.E. Roscommon, and S.W. and W. Mayo. Area, 434,887 acres, of which 168,711 are mountain and bog, and 8980 water. Surface much diversified; but though it has a considerable extent of level, rich land, it is, speaking generally, mountainous, rough, and boggy. There are a few pretty large estates; but a considerable portion of the county is divided among small proprietors. The statements as to the mode of occupying land, its management, and the condition of the inhabitants, given under the notice of the co. LEITRIM (which see), may be applied with little or no modification to this county. The great increase within the last few years in the amount of the exports from the town of Sligo, show that there must have been a corresponding extension of cultivation in this county and the contiguous portions of Leitrim. But unhappily the extension, and even improvement of tillage in Ireland, is not always accompanied by any corresponding improvement in the condition of the occupiers, which is here extremely bad. The *con-acre* system (see *ante*, p. 34) has made much progress in this county; the competition for land is extreme; and the occupier of any overrented patch that may choose to part with it, never fails to get a considerable sum as "tenant's right." Average rent of land, 10s. 8d. an acre. It has neither minerals nor manufactures of any importance. Principal rivers, Gavoge, Arrow, Awinmore, &c. It is divided into six baronies and 30 parishes, and returns three members to the House of Commons, two being for the county, and one for the borough of Sligo, the only town of any importance in the county. Registered electors for the county, in 1839–40, 1574. In 1831, Sligo had 29,588 inhabited houses, 32,903 families, and 171,765 inhabitants, of whom 83,730 were males, and 88,035 females.

SLIGO, a parl. bor. and seaport town, on the W. coast of Ireland, prov. Connaught, cap. of the above co., at the bottom of Sligo bay, and at the mouth of the river Garvoge, 107 m. N.W. Dublin; lat. 54° 22′ N., long. 8° 22′ W. Pop. in 1821, 14,151; in 1831, 15,152. The town, which is of considerable importance, and increasing both in wealth and population, is intersected by the Garvoge, which has its source in lough Gill, distant about 3 m. The larger portion, which is on the S. side of the river, is connected with that on the N. by two bridges. The streets are irregularly laid out, and those in the older parts of the town are narrow, dirty, badly paved, and badly lighted. Of late years, however, several new markets, warehouses, and lines of streets, have been erected; and it has a good deal of the bustle and appearance of a place of trade. Water is supplied by public pumps. The town has a library, two news-rooms, a small theatre, and a cavalry barrack. The ecclesiastical buildings comprise the parish churches of St. John and Calry; a large Roman Catholic chapel, a Dominican convent, and places of worship for Presbyterians, Independents, and Wesleyan Methodists. The ruins of Sligo monastery deserve notice for their architecture and for a monument of O'Connor Sligo, who died in 1693. There are two parish schools, a school in connexion with the board of national education, one on the foundation of Erasmus Smith, and one under the Incorporated society. The county infirmary, fever hospital, and dispensary, and a mendicity association, are within the town. Lunatics are sent to the district asylum for Connaught, at Ballinasloe.

The borough, which was chartered by James I. in 1614, is divided under the Irish Municipal Reform Act, 3 and 4 Victoria, cap. 108, into three wards, and is governed by a provost, five other aldermen, and 18 councillors. Sligo returned two members to the Irish House of Commons, and since the union it has sent one member to the Imperial House of Commons.

The boundary of the parliamentary borough extends one Irish mile from the market cross in every direction. Registered electors, in 1839–40, 1131. The boundary of the municipal borough is, however, of more limited dimensions. A board of commissioners, appointed under local acts, superintends the police of the town, and the improvement and regulation of the quays and harbour, with power to impose rates for those purposes. Under their management, the port has been a good deal improved. An extensive new quay and warehouses have been erected outside the bar; and though rather difficult of access, the port is now very tolerable. There are about 12 ft. water close to the quay, so that vessels of 250 and 300 tons come up to the town. The assizes and general sessions of the peace for the county are held here; the latter four times in the year, and petty sessions every Thursday. The courthouse, though a modern structure, is too small for the convenient despatch of business. The county jail, a large and well constructed building on the polygonal plan, is furnished with a tread-mill.

The linen trade, which was formerly carried on with some spirit, is now nearly extinct. The town has several flour-mills, a distillery, and four breweries. The markets for corn and butter, on Tuesday and Saturday, are held in buildings erected for the purpose. There is a valuable salmon fishery close to the town.

Sligo is the *entrepôt* of an extensive country, and has, in consequence, a considerable and increasing trade. The exports consist almost wholly of agricultural produce: thus, in 1835, the total estimated value of the exports amounted to £369,490; of which the value of corn, meal, and flour made £185,414, and that of provisions (beef, pork, &c.) £181,836. The imports consist of colonial products, English goods, tobacco, wine, coal, salt, &c., for the use of the town and the country dependent upon it. Gross customs' revenue in 1840, £35,680. Postoffice revenue in 1830, £2303;

in 1836, £9860. Branches of the bank of Ireland, its provincial bank, National and Agricultural ditto, are established in the town, which has also two newspapers.

The intercourse between the port and lough Gill, and its surrounding country, by the Garvoge, is ... by a bar thrown across the latter for the use of the large ... This obstruction is much complained of, and will, ..., at no distant period be removed. The county is in its capacity is highly picturesque, and has many fine seats. The population is comparatively respectable and well off. (*Boundary and Railway Reports, &c.*)

SLIPPERY ROCK, t. Butler co., Pa., 15 m. N.W. Butler, 22 m. W.N.W. Harrisburg, 263 W. Drained by Slippery Rock creek and its tributary, Wolf creek. It has three stores, three furnaces, four grist-mills, seven sawmills, four tanneries, one pottery; 10 schools, 472 scholars. Pop. 1507.

SLIPPERY ROCK, t. Beaver co., Pa. It has two stores, seven grist-mills, eight saw-mills, one oil-mill, two nurseries, one distillery; three schools, 161 scholars. Pop. 1528.

SLIPPERY ROCK, t. Mercer co., Pa. It has three stores, one fulling-mill, six grist-mills, six saw-mills, two tanneries, 11 schools, 385 scholars. Pop. 2066.

SMETHPORT, p. v., cap. of McKean co., Pa., 196 m. N. W. Harrisburg, 279 W. Situated at the confluence of Stanton and Potatoe creeks. It contains a courthouse and county offices, all of brick, a jail of stone, an academy endowed with a fund producing $2000 annually, two stores, one grist-mill, one saw-mill, one fulling-mill, 3 dwellings, and about 300 inhabitants.

SMITH, county, Tenn. Situated in the N. part of the state, and contains 590 sq. m. Watered by Cumberland river and its tributaries. It contained in 1840, 16,161 neat cattle, 20,944 sheep, 73,172 swine; and produced 81,430 bushels of wheat, 45,82 of rye, 933,348 of Indian corn, 117,528 of oats, 12,779 of potatoes, 3,017,013 pounds of tobacco, 61,945 of cotton. It had 14 stores, 53 grist-mills, 47 saw-mills, 31 tanneries, 24 distilleries, one pottery; one college, 22 students; four academies, 87 students; 15 schools, 389 scholars. Pop.: whites, 16,037; slaves, 4289; free coloured, 164; total, 21,179. Capital, Carthage.

SMITH, county, Miss. Situated a little S.E. of the centre of the state, and contains 520 sq. m. Drained by Leaf river. It contained in 1840, 9994 neat cattle, 741 sheep, 10,413 swine; and produced 1379 bushels of wheat, 66,403 of Indian corn, 557 of oats, 8740 of potatoes, 9819 pounds of rice, 442,503 of cotton. It had 28 tanneries; four schools, 70 scholars. Pop.: whites, 1543; slaves, 419; total, 1962. Capital, Raleigh.

SMITH, t. Belmont co., O. Pop. 1956.

SMITH, t. Washington co., Pa. It has four stores, four grist-mills, four saw-mills; nine schools, 450 scholars. Pop. 1345.

SMITH, t. Columbiana co., O. Pop. 1457.

SMITHFIELD, p. t., Providence co., R. I., 16 m. N. Providence, 413 W. Watered by Blackstone river. Woonsocket village, at the falls, is partly in this township, and has extensive manufactures. It contains 23 stores, 10 fulling-mills, one woollen factory. 31 cotton factories with 88,208 spindles, 13 grist-mills, 19 saw-mills, one paper mill, one academy, 40 students; 45 schools, 1541 scholars. Pop. 9534.

SMITHFIELD, t. Madison co., N. Y., 106 m. W. Albany. Drained by Cazenovia and Cowassalon creeks. It has four stores, one fulling-mill, two grist-mills, seven saw-mills, two tanneries; 12 schools, 535 scholars. Pop. 1099.

SMITHFIELD, t. Bradford co., Pa. It has four stores, one fulling-mill, one grist mill, 11 saw-mills, three tanneries, one distillery; 11 schools, 420 scholars. Pop. 1637.

SMITHFIELD, p. v., cap. of Isle of Wight co., Va., 62 m. S.E. Richmond, 298 W. Situated on a bay or creek of James river, 15 m. above Hampton roads. It contains a courthouse, jail, three churches, an Episcopal, Methodist, and Baptist; 19 stores, 300 dwellings, and about 1000 inhabitants. It has considerable trade, particularly in hams.

SMITHFIELD, p. v., cap. of Johnson co., N.C., 27 m. S.E. Raleigh, 305 W. Situated on the E. side of Neuse river, and contains a courthouse, jail, several stores, and about 200 inhabitants.

SMITHTOWN, p. t., Suffolk co., N. Y., 36 m. W. Riverhead, 191 m. S.S.E. Albany, 46 m. E. New York, 271 W. Bounded N. by Long Island sound. Nesaquake river affords water-power. It contains a Presbyterian church, 10 stores, one fulling-mill, one woollen factory; seven schools, 309 scholars. Pop. 1939.

SMITHVILLE, t. Chenango co., N. Y., 15 m. W. Norwich, 130 W. Albany. Drained by Genegantslett creek and other tributaries of Chenango river, which bounds it on the S.E., along which passes the Chenango canal. It contains five churches, two Methodist, a Congregational, and two Baptist; three stores, one grist-mill, 15 saw-mills, 3 schools, 667 scholars. Pop. 1762.

SMITHVILLE. p. v., cap. of Brunswick co., N. C., 173 m. S. Raleigh, 390 W. Situated on the W. side of cape Fear river, 3 m. above its mouth. It has a safe and commodious harbour, and contains a courthouse, jail, a hospital, a Methodist church, three stores, 60 dwellings, and about 350 inhabitants. It is regarded as one of the most healthy places in the state.

SMITHVILLE. p. v., cap. of De Kalb co., Tenn., 61 m. E. Nashville. 620 W. Situated on the W. side of Caney fork of Cumberland river, and contains a courthouse, jail, a Methodist church, several stores, and about 250 inhabitants.

SMOLENSKO, a government of European Russia, between the 53d and 57th degs. of N. lat. and the 30th and 6th E. long., having N. the governments Pskof and Tver, E. Moscow and Kaluga, W. Witepsk and Moghilef, and S. Orlof and Tchernigof. Area estimated at 20,220 sq. m.; pop. in 1838, 1,064,200. Surface mostly an undulating plain, some parts marshy; in the N. is a more elevated plateau, which the Dniepr and several other rivers have their source. The soil is generally fertile, and more corn, principally rye, is grown than is required for home consumption. Hemp, flax, tobacco, and hops, are cultivated. Cattle breeding is less attended to; but a good many hogs are reared. The forests are very extensive, and are, in fact, the chief sources of wealth. Game is very plentiful; and bees are reared almost everywhere. Iron, copper, and salt are found. Manufactures few, being nearly confined to leather, glass, wares, pitch, &c.; with sawing works, distilleries, and a few carpet factories in the capital. The raw produce of the government is exported, in large quantities, Riga, Wilna, and Moscow. It is divided into 12 circles. SMOLENSKO, a town of European Russia, cap. of the above gov., on the Dniepr, 230 m. W. by S. Moscow, lat. 54° 47′ 11″ N., long. 32° 3′ E. Pop. nearly 10,000. It is situated on both sides the river, and is surrounded by a massive wall flanked with towers. It appears to advantage at a distance, but is in reality a poor town, the houses being mostly of only one story, and built of wood; but since it was burned by the Russians, previously to their evacuating it on the advance of the French in 1812, it has been partially rebuilt of stone and brick.

Smolensko has three cathedrals, in one of which is a bell weighing 350 cwts.; 16 Greek churches, three convents, a theran and a Roman Catholic church, a seminary, gymnasium, a military school for nobles, several hospitals, and a carpet, hat, soap, and leather factories. In 1838, the erection of an iron pyramid was commenced here, intended commemorate the resistance made by the town to the French in 1812.

Smolensko is of considerable antiquity. It has suffered serious vicissitudes, but has always been a town of some consequence. (*Schnitzler, La Russie; Possart, Das Reich Russland.*)

SMYRNA, an ancient and celebrated city and seaport of Minor, the greatest emporium of W. Asia. on the W. of the Meles, a stream which, though of small dimensions, has acquired an immortality of renown, at the bottom of the gulf of its own name (an. *Hermius Sinus*), lat. 38° 25′ N., long. 27° 6′ 45″ E. Its pop. may be estimated from 120,000 to 150,000, more than half being Turks, and the rest Greeks, Jews, Armenians, Franks, &c. It is surrounded, at some distance, by an amphitheatre of lofty mountains, which leave it open only towards the sea; and is directly adjoining the city, on the W., is the ancient *Pagus*, which commands a noble view. This eminence is now called the Castle hill, from a citadel erected the summit, in the 13th century, by the Emperor John Comnenus. "A triangular plain, spread at the foot of this eminence along the shore, and the slopes of the hill itself, comprise the site of Smyrna. One side extends along the shore from N. to E. for about 3¼ m. The Turks occupy the upper part of the city, their streets hanging down the slopes of the hill; the Armenians are in the centre; the Jews have three different places around both; and the Franks themselves in the flat ground and close to the shore, following S.E. is a plain filled with gardens; and every part of the city is interspersed with shady trees." (*Burgess, Greece and the Levant,* ii., 67, 68.)

The view of Smyrna from the bay, rising amphitheatre-wise from the water's edge, backed by the hill crowned by its old castle, is grand and impressive. Unfortunately however, its interior has all the odious features common to Turkish towns. "If a first view be calculated to give a favourable impression, this is not confirmed by an examination of the interior of the city. The Frank quarter is long, paved, and narrow; in addition to which, it is rendered almost impassable by long strings of camels and horses carrying huge bales of cotton. The houses (except those of the consuls and principal merchants, which are large and commodious) are miserably built; the sides are often of planks, and when of bricks, the walls are

---

too thin to keep out cold and damp. Neither windows nor doors are made to shut close; and if locks appear on the latter, it is too much to expect that they should be serviceable. There is a great lack of accommodation for travellers. The only inn in the town contains but a single decent room; and the noise of revelry is incessant. Besides this, there are three boarding-houses, but furnished lodgings are not to be procured, nor can furniture be hired for a few weeks or months. The apparatus commonly used for supplying warmth to the body in cold weather is a brazier placed under the table, which is covered by a large cloth held by each member of the family circle up to the chin, to prevent the heat from escaping. Grates and stoves have of late years been introduced, but they are still rare, and to be seen only in Frank dwellings. The shops are little dark rooms, but tolerably supplied with European articles. The bazaars, with their long covered rows of stalls, built with sundry precautions against fire, whose ravages are awfully common, are secured by iron gates closed at night. As to the rest, Turkish towns in general offer little variety, and the description already given of Constantinople applies to Smyrna, except as regards the finer buildings, greater extent, and gaudy exterior of the capital." (*Elliot's Travels,* ii., 34–36.)

The principal buildings of Smyrna are the bazaar and *bezestein*, or market-place; the vizier-khan, constructed of the marble ruins of the ancient theatre; the palace of the *mutsellim*, or governor, and the various mosques, churches, hospitals, &c. There is a large public hospital in the N.E. part of the Frank quarter, supported by the Greeks, Franks, and other Christians, which ranks high in Turkey for its school of medicine. Its buildings comprise a laboratory and three sets of wards around a courtyard shaded by rows of trees. The castle on mount Pagus is very extensive, and occupies the site of the ancient acropolis. This fortress has been frequently repaired by the Turks, and accordingly presents an incongruous intermixture of architecture; but it is now mostly deserted, and in ruins, though a few old cannons be still mounted on its walls. Within are some vaults and cisterns, supposed to be coeval with its foundation; and a large but abandoned mosque, formerly a church dedicated to St. Polycarp, bishop of Smyrna, who is supposed, though on no very good authority, to have suffered martyrdom near the same spot.

Smyrna was the seat of one of the seven apocalyptic churches. (Rev. ii., 9.) And, according to Mr. Elliot, "there is not one of these churches within whose precincts the trumpet of the gospel now gives so distinct and certain a sound. While Mohammed is acknowledged in 20 mosques, and Jews assemble in several synagogues, the faith of the Messiah is taught in an Armenian, five Greek, and two Roman Catholic churches, and in two Protestant chapels, one connected with the English, the other with the Dutch consulate." (*Trav.,* ii., 45, 46.) The Armenians have a large academy at Smyrna.

Being surrounded, as already stated, by an amphitheatre of mountains, which concentrate the rays of the sun and interrupt the breeze, the heats at Smyrna, from June to the middle of September, are usually intense; and if the *imbat,* or sea-breeze, fail, the inhabitants are almost suffocated. This great heat, and want of ventilation, joined to the filthy and crowded state of the streets and houses, and the want of any efficient precautions on the part of the authorities, seldom fails to generate the most destructive diseases; among which, the plague not unfrequently makes its appearance, and commits dreadful ravages. At such periods all commercial and social intercourse immediately cease; and the Frank inhabitants retire to and shut themselves up in their country houses in the surrounding villages. The Turks, who are firm predestinarians, have hitherto taken few or no precautions to counteract the progress of the infection, or to guard against it; but it is stated by late travellers that some change is now beginning to take place in this respect, and that the principal Turkish inhabitants are slowly adopting some of the devices by which Europeans attempt to ward off the malady. (*Chandler's Asia Minor,* chap. 19; *Burgess,* ii., 70.)

Port and Commerce.—Though frequently overthrown by earthquakes, and laid waste by hostile incursions, the excellence of her port, and her advantageous situation for commerce, has always made Smyrna be rebuilt; and she still continues to be a great city, while Ephesus, Miletus, and other celebrated emporiums of the same coast, have, from the filling up of their harbours, been long since reduced to total ruin. The gulf of Smyrna, the entrance of which is between the island of Mytilene on the N., and cape Carabourun on the S., is deep and angular, the distance, following a ship's course from the entrance to the city, being about 11 nautical leagues. There is excellent anchorage in most parts of the gulf, merely avoiding the shoals on its N. side. Ships of large burden usually anchor abreast of the city in from 5 to 7 fathoms; but the water is so deep that they may lie close alongside the quays. The *imbat*, or sea-breeze, blows from morning till evening during the hot months, and is always waited for by ships going up to the city; and there being no obstructions in the way, the services of pilots are not required. In the night a land breeze generally blows from the city out to sea.

The principal articles of import consist of grain, furs, iron, butter, &c., from Odessa and Taganrog; and of cotton stuffs and twist, silk and woollen goods, coffee, sugar, cochineal and dye woods, iron, tin, and tin plates, rum, brandy, paper, cheese, glass, wine, &c., from Great Britain, France, Italy, the United States, &c. Coffee is, perhaps, the most considerable article: it comes principally from America and England, and the annual consumption is estimated at 3,000,000 okes.

The exports consist principally of raw silk and cotton, the former produced about Brusa, and sent chiefly to England; fruits, particularly raisins, and figs; opium, which goes chiefly to America and Holland; rhubarb, and a variety of drugs and gums; olive oil, madder roots, Turkey carpets, valloneа, sponge, galls, wax, copper, hare-skins, from 350,000 to 400,000 annually; goats' wool, safflower, &c. Burgess, who visited Smyrna in July, 1834, says, "British commerce, if I may judge from the present state of the harbour, is by far the most considerable. There are now 30 vessels under cargo, of which more than a half are bound for Great Britain and Ireland. The English appear to engross the commerce of fruit, the French of cotton; the Dutch trade is entirely fallen off." (ii., 80-85.) The principal foundation of the commerce of Smyrna consists in the caravans, composed principally of Armenians, that arrive from Persia and the east, at fixed periods, which are nearly identical with those of the arrival and departure of most of the foreign ships frequenting the port. They carry to and convey away the greater portion of the exports and imports. Bargains are principally effected by Jew brokers, many of whom have amassed considerable fortunes. Upon an average, the selling charges, generally, may be calculated at about 12 per cent., and the charges on purchasing at about eight per cent. The money, weights, and measures are mostly the same as at Constantinople. Accounts are kept in piastres of 40 paras. The value of the piastre fluctuates according to the exchange; but, in 1834, Burgess received 97¾ for £1 sterling, and the Spanish dollar, at the same time, was worth 21¼ piastres. The oke, which is the principal weight = 12 lbs. 13¾ oz. avoirdupois; the quintal = 127·48 lbs. avoirdupois. (*Com. Dict.*)

Smyrna is well supplied with provisions of all kinds. Fish, including red mullet, oysters, &c., are very plentiful in the bay; and game of all kinds, mutton, the flesh of wild boars, &c., are good and abundant. Whey and clotted cream are used in great quantities. Sweet lemons, oranges, citrons, water-melons, figs, and grapes, are grown in great perfection in the environs, particularly at Menomen, and the other villages on the opposite side of the gulf; whence boats carrying fruits and other provisions are continually passing to Smyrna. Mr. Burgess gives the prices of certain articles as follows: eggs, 4 to 6 paras each; milk 30 ditto per oke; a small loaf 6 or 8 ditto; melons 20 paras the oke, grapes 30 ditto; mutton 2 piastres 10 paras the oke; beef little more than half that price, &c.

Most travellers speak of the agreeable society met with in Smyrna; and the Greeks of the city have begun to adopt the manners and costumes of western Europe.

Historical Notices.—The accounts of the foundation and early history of Smyrna are obscure and somewhat contradictory. The most probable seems to be that it was founded by a colony from Ephesus. (*Strabo*, lib. xiv.) After undergoing various vicissitudes, it was destroyed by Alyattes, king of Lydia, the inhabitants being dispersed among the surrounding villages. At the distance of about 400 years, a project for reconstructing the city would appear to have been entertained by Alexander the Great; but, if so, it was not carried into effect by that conqueror, but by Antigonus and Lysimachus. The city built by them was not, however, on the site of the old city, which stood on the flat shore on the other side of the Meles, about 2¼ m. N.E. from the modern city. The admirable port and other advantages enjoyed by the newly-built city rendered it in a short time one of the most populous, wealthy, and handsomest of the Asiatic cities. "It is," says Strabo, "the finest city of Asia. Part of it is built on a hill; but the finest edifices are on the plain not far from the sea, over against the temple of Cybele. The streets are the most beautiful that can be: straight, wide, and paved with freestone. It has many stately buildings, magnificent porticoes, majestic temples (including a *Homerium*, or temple in honour of Homer), a public library, and a convenient harbour, which may be shut at pleasure." (Lib. xiv., *sub init.*) Under the Romans, Smyrna enjoyed the greatest consideration, and Marcus Aurelius rebuilt the city, after it had been almost destroyed by an earthquake. It was much frequented by the Sophists; and, along with Ephesus, became renowned as a school of oratory and science.

In more modern times it has undergone innumerable calamities, from which, as already stated, nothing but its admirable situation for commerce could have enabled it to recover. It was taken and given up to military execution by the famous Tamerlane, or Timour Bec, in 1402, and finally came into the possession of the Turks in 1424.

Smyrna is one of the numerous cities that contended for the honour of being the birthplace of Homer; and Chios perhaps, excepted, she would seem to have the best claim to this proud distinction. "*Homerum Smyrnæi suum esse confirmant; itaque etiam delubrum ejus in oppido dedicarunt.*" (*Cicero pro Archia, cap.* 8.) From being born on the banks of the Meles, which washed the wall of the ancient as well as of the modern city, Homer is sometimes called *Melesigenes.*

"Blind Melesigenes thence Homer call'd,
Whose poem Phœbus challeng'd for his own."

Thence, also, Tibullus (lib. iv., eleg. i., v. 200) calls his poems *Meleteas Chartas*; and Pausanias says (*Achaic.*, cap. 5) that a cave was pointed out as the source of the Meles, where they were said to have been composed.

Owing to the influence of earthquakes, and the still more destructive attacks of barbarians, Smyrna has but few considerable remains of antiquity. They consist principally of portions of the old walls, especially along the castle-hill, with some vestiges of the theatre and stadium. Many pedestals, statues, inscriptions, and medals have been and are still discovered in digging; and perhaps no place has contributed more than this to enrich the collections and cabinets of Europe.

It has been supposed by Chandler and others that the mud and other detritus brought down by the Kodus (an. *Hermus*), which has its embouchure on the N. side of the gulf, will, in the end, fill up the channel; and by depriving the city of its port, effectually consummate its ruin. But though this effect may ultimately be brought about, it is abundantly certain, comparing the banks at the river's mouth with the space that has to be filled up, that a lengthened series of ages must previously elapse.

Besides the authorities already referred to, see *Ancient Universal History*, vii., 410, 8vo. ed., &c.

SMYRNA, p. t., Chenango co., N. Y., 103 m. W. Albany, 351 W. Drained by tributaries of Chenango river, which afford water-power. It contains seven churches, two Methodist, a Baptist, a Congregational, a Union, and two Friends; five stores, three fulling-mills, one woollen factory, two grist-mills, 12 saw-mills, one oil-mill, two tanneries, 15 schools, 703 scholars. Pop. 2246.

SMYTHE, county, Va. Situated in the S.W. part of the state, and contains 480 sq. m. Drained by head branches of Holston river and New river. It contained in 1840, 9157 neat cattle, 10,857 sheep, 16,392 swine; and produced 51,390 bushels of wheat, 6811 of rye, 221,431 of Indian corn, 410 of buckwheat, 178,330 of oats, 33,829 of potatoes, 54,948 pounds of sugar. It had 11 stores, two forges, 21 grist-mills, 15 saw-mills, 10 tanneries, six distilleries, one pottery; three academies, 100 students; seven schools, 198 scholars. Pop.: whites, 5539; slaves, 838; free coloured, 145; total, 6522. Capital, Marion.

SNOWDON, a mountain of N. Wales, in Caernarvonshire, being at once the highest in the range of which it forms a part, and in S. Britain. The mountain, which is about 10 m. S.E. from Caernarvon, terminates in various peaks; the particular peak, the Wyddva (conspicuous), to which the name Snowdon is more particularly applicable, and which scarcely out-tops several of the surrounding summits, is 3571 ft. above the level of the sea. The W. side of the mountain is very precipitous, and is composed partly of pentagonal basaltic columns. The view from the summit is very extensive. "I saw from it," says Pennant, "the county of Chester, the high hills of Yorkshire, part of the N. of England, Scotland, and Ireland, a plain view of the isle of Man; and that of Anglesey lay extended like a map before me, with every rill visible." (*Tour in Wales*, ii., 377, ed. 1810.)

SNOW HILL, p. v., port of entry, capital of Worcester co., Md., 115 m. S.E. Annapolis, 155 W. Situated on the E. side of Pocomoke river. It contains a courthouse, jail, an academy, five churches, a Presbyterian, Episcopal, Methodist Episcopal, Methodist Protestant, and African Methodist, nine stores, 100 dwellings, and about 800 inhabitants. Tonnage in 1840, 7640.

SOCIETY ISLANDS. *See* POLYNESIA.

SOCOTRA (an. *Dioscoridis Insula*), an island in the Indian ocean, about 230 m. from the S. coast of Arabia, and 120 m. E. cape Gardafui, in Africa; its chief town being in lat. 12° 39′ 2″ N., long. 54° 6′ 29″ E. It is of an elongated shape. Area estimated by Mr. Wellsted at 1000 sq. m. Pop. probably 4000 or 5000, principally Bedouins, with some

ths, African slaves, and descendants of Portu-
e S. coast of Socotra preserves a convex and
roken line, but on the N. it is indented with
and harbours. The interior may be described
ig of mountains, nearly surrounded by a low
in 2 to 4 m. in width, extending from their base
The mountains are highest towards the N.E.
island, where their granite peaks rise to about
sewhere they average nearly 2000 ft. in height,
mostly of a compact cream coloured primitive
The island is not well-watered; but the E. is,
pect, better than the W. portion. The climate
ppear to be particularly salubrious, though it be
erate than in the adjacent continent. Among
tural products, the most important is aloes (*Aloë
Socotrina*), for which the island has been fa-
the earliest period. This plant is found grow-
eously and in great abundance on the sides and
the limestone mountains, at an elevation of
3000 ft. above the level of the plains. Its leaves
l at any period, and after being placed in a skin,
i suffered to exude from them. In this state
ostly shipped for Muscat. Formerly the parts
id producing the aloe were farmed out to differ-
unls, the produce being taken at a low fixed
e Sultan. At present, any one collects the aloe
o chooses to take the trouble, and nothing is
he Sultan's account. The quantity exported of
ried very much: in 1833, it amounted to 83 skins,
; and the best sold for one rupee the Bengal
/ one lb. English); while of the more indiffer-
ers might be procured for a dollar. (*Wellsted.*)
food is the article next in importance: it is the
a leguminous tree, the *Pterocarpus draco*, which
he mountains. Tamarinds, tobacco, and dates
as food), are also grown. Agriculture is in an
/ low state; a species of millet being the only
ated, and it is little used unless a failure of milk
se experienced. The animals are camels, sheep,
i, asses, and civet cats. Sheep and goats are
ge flocks in every part of the island: they are
f inferior kinds, while the cattle, on the contra-
small, are very superior, and appear to be of
an variety. The trade is principally with Mus-
e dates and other provisions are chiefly import-
ding to Arrian, the inhabitants of this island
ntiquity, subject "to the kings of the incense-
r Southern Arabia. At present Socotra belongs
an of Kissoen, but his supremacy is little more
nl, the government being chiefly delegated to
principal inhabitants, who again exercises lit-
ty, except over the Bedouin, or native popula-
es exposed for sale were dates, grain, tobacco,
tribute to the Sultan barely amounts to 200 dol-
The population is wholly Mohammedan. The
unveiled, and are partly occupied in tending
partly in making glue, and carding, spinning,
ig wool.

r town of any consequence is Tamarida, on the
, in the centre of a bay which affords tolerable
Having been ruined by the Wahabees, in
sists of only about 150 straggling and dilapidated
t more than one third of which were inhabited
the town had then but two shops, where the
es exposed for sale were dates, grain, tobacco,
(*Wellsted in Geog. Journal*, v. 129–229.)

p. t., Wayne co., N. Y., 12 m. N. Lyons. 195 m.
Albany, 371 W. Bounded N. by lake Ontario,
dus bay. Drained by Salmon creek. It contains
ies, a Presbyterian, Methodist, two Baptist, and a
st, eight stores, three fulling-mills, four grist-
w-mills, four tanneries, one distillery; 22 schools,
ars. Pop. 4472.

, a market town and par. of England, co. Cam-
nd. Staplos, near the borders of Suffolk, 6 m.
. Area of par., 13,420 acres. Pop., in 1831, 3667.
which is irregularly built, covers a good deal of
the parish church is of various dates, one por-
: late Norman. The living, a rectory, worth
ir, is in the patronage of the present incumbent,
holds the rectory of Soham-Earl, worth £515 a
ecl. Rev. Rep., 1831.) Soham has numerous
specially Bishop Laney's, for apprenticing child-
e parish, with an annual revenue of near £400;
arity school, and several almshouses. The fen
hich once covered the adjacent country, has been
d cultivated, and supplies most part of the dairy
or which Soham is celebrated. Market days,
Fair, May 7, for horses and cattle. (*Parlia-
report.*)

NS (an. *Noviodunum*, post. *Augusta Suessio-
rtified town of France, dep. Aisne, cap. arrond.;
ne, here crossed by a handsome stone bridge, 17

m. S.W. Laon. Pop. in 1836, 7864. Its fortifications consist merely of bastioned ramparts. It is well built, and
clean, the houses being mostly of stone, roofed with slate.
There are many curious and venerable public buildings;
including the cathedral, founded in the 12th century, with
an altar-piece by Rubens, representing the "Adoration of
the Shepherds;" the remains of an abbey founded in the
11th century; the castle, on the site of that which was
the residence of various Merovingian kings, &c. The
college, hospital, house of correction, public library with
18,000 vols. and theatre, deserve notice; and in the vicinity are the ruins of St. Medard's abbey, founded in 557, in
which Pepin, Carloman, &c., were crowned, and Louis-le-
Débonnaire was confined by his sons. Soissons is a bishop's see, and has a court of primary jurisdiction, a communal college, two seminaries, a society of arts and sciences,
manufactures of coarse woollens, hosiery, and earthenware,
and a considerable trade in agricultural produce.

Soissons has been the theatre of various historical
events. It was here, in 486, that Clovis finally extinguished the last remains of the Western Empire, by his victory over the Roman general Syagrius. The town then
became the capital of the Franks, and afterwards of a
kingdom of its own name in the 6th and 7th centuries. It
was frequently besieged and taken in the middle ages, and
was the scene of some severe fighting between the French
and allies in 1814. (*Hugo. art. Aisne; Dict. Geog, &c.*)

SOLESBURY, t., Bucks co., Pa. Bounded N.E. by Delaware river. It contains Ingham's spring, clear and cold
in summer, which never freezes in winter, and with a fall
of 18 or 20 ft., would drive two pairs of millstones through
the year. It has five good mill seats between its head and
Delaware river, 3 m. distant. It contains four stores, two-
flouring-mills, seven grist-mills, six saw-mills, one paper-
mill; eight schools, 381 scholars. Pop. 2038.

SOLON, p. t., Somerset co., Me., 18 m. N. Norridgewock,
52 m. N. Augusta, 647 W. Bounded W. by Kennebec river. It contains four stores, one fulling-mill, two grist-mills,
one saw-mill; eight schools, 507 scholars. Pop. 1139.

SOLON, p. t., Cortland co., N. Y., 12 m. E. Cortland. 136
m. W. Albany, 323 W. Drained by Ostelic creek and its
tributaries. It has one store, one grist-mill,-three saw-mills;
20 schools, 747 scholars. Pop. 2311.

SOLOTHURN (French *Soleure*), a canton of Switzerland, in the N.W. part of the Confederation, between lat.
47° and 47° 30', and the 7th and 8th degs. of E. long.;
having N. Basie, E. and S.E. Aargau and Lucerne, and on
its other sides the canton of Berne. Area 255 sq. m. Pop.,
in 1837, 63,196, mostly all Roman Catholics. Though of a
very irregular shape, it may be divided into two nearly
equal portions; the N.W. covered with ranges of the Jura
mountains, and the S.E. comprised in the valleys of the
Aar and Emmen. Some of the summits in the former rise
to about 4000 ft. above the level of the sea; but though
rugged, this part of the canton has a large extent of fine
upland pastures. In the other, or lower portion of the
canton, the ground is fertile and well cultivated; so that,
on the whole, Solothurn is regarded as one of the most
productive portions of Switzerland. More corn is grown
than is required for the consumption of the inhabitants;
the vine does not succeed, but a good deal of fruit is notwithstanding grown for exportation. The rearing of live
stock is here, however, as in most other Swiss cantons,
the chief branch of rural industry. In 1838, it was estimated to have about 28,000 horned cattle, 14,000 sheep,
and 16,400 hogs: the latter are fed in the woods, which
are tolerably extensive. The cattle are esteemed among
the best in Switzerland; they are of a peculiar large-tailed
breed, and with horses, cheese, cherry brandy, fire-wood,
and marble, constitute the principal articles of export.
Only a few hands are employed in mining; and the manufacturing establishments are mostly confined to a few iron
works, stocking and cotton looms, paper-mills, tanneries,
and printing houses. The currency, weights, and many
usages of this canton, are similar to those of Berne.

Under the constitution, as modified in 1831, the greater
council consists of 109 members, 96 of whom are chosen
by the towns and the 10 rural districts into which Solothurn
is divided, and the remaining 13 by the members already
elected. The lesser or executive council, composed of 17
members, is chosen with its president, or *avoyer*, from
among the greater council. The assembly meets twice a
year for 15 days, during which period each member receives
three francs a day. The town of Solothurn, and each of
the districts has a court of primary jurisdiction for civil
causes; but all criminal cases, as well as the final jurisdiction in civil suits, belong to a central tribunal of 14 members, presided over by the avoyer of the state assembly.
Every male inhabitant above the age of 16 is liable to military service: the contingent to the army of the confederacy is 600 men. Total public revenue in 1835, 408,184 francs;
expenditure, 400,599 francs, of which 18,000 francs was

contributed to the general Swiss treasury. There is no town worthy of notice, except the capital, Olten, Dornek, &c., being mere villages.

SOLOTHURN, or SOLEURE (an. *Solodurum*), a town of Switzerland, and the cap. of the above canton, on the Aar, near the foot of the Jura mountains, and 18 m. N. by E. Berne. Pop., in 1835, 4254. The river divides it into two unequal parts, which communicate by two wooden bridges. It was surrounded, in the 17th century, by cumbrous fortifications, the removal of a part of which was ordered by the cantonal council in 1835. It is tolerably well built, and has several conspicuous public edifices; including, among others, the cathedral of St. Urse, one of the best churches in Switzerland, with several other churches and convents, a town hall, a square clock-tower in the market place, the arsenal, with an extensive and curious collection of armour, a museum, government house, with some good sculptures, hospital, house of correction, barracks, theatre, &c. It has, also, a gymnasium, a botanic garden, and a public library, said by Ebel to comprise 10,000 volumes. On the whole, however, the town is dull, having few manufactures, and but little trade. The Polish patriot and general, Kosciusko, resided here during the last two years of his life, which terminated on the 16th of Oct., 1817. His remains were carried to the cathedral of Cracow, where they repose beside those of the famous John Sobieski. (*Stralmeier, Der Kanton Solothurn; Picot, Statist. de la Suisse; Ebel.*)

SOMERS, p. t., Tolland co., Ct., 23 m. N.E. Hartford, 359 W. Watered by Scantic river. It contains three churches, a Congregational, Methodist, and Universalist, six stores, one fulling-mill, one wollen factory; 10 schools, 446 scholars. Pop. 1621.

SOMERS, p. t., Westchester co., N. Y., 90 m. N. White Plains, 119 m. S. Albany, 274 W. Watered by Croton river. It contains two churches, two stores; four schools, 69 scholars. Pop. 2082.

SOMERSET, a maritime co. of England; having N. and N.W. the Bristol channel, the Severn, and Gloucestershire, E. Wiltshire, S. Dorset and Devon, and W. the latter. Area, 1,052,800 acres, of which about 900,000 are supposed to be arable, meadow, and pasture. With, perhaps, the single exception of Yorkshire, there is no county in England that has a greater variety of surface and soil than Somerset. In some places, particularly in its W. and N.E. divisions, it is hilly, and even mountainous; in its middle part, between the Ax and Parry, there are very extensive tracts of marsh land, which, in some places, are of extraordinary fertility: in other places again, there are extensive moors of which Exmoor, at the W. extremity of the county, is the principal. But, exclusive of these, the county contains a large extent of land equally adapted for tillage and pasturage. The vale of Taunton is one of the richest and most beautiful tracts in the kingdom. Tillage husbandry is neither extensively carried on, nor in the most approved manner. The land is not injured by over-cropping, but it is not properly wrought, and is frequently foul, and out of order. Principal crops, wheat, oats, barley, and beans. In the southern and interior parts, the rotation is, 1. fallow, 2. wheat, 3. beans or seeds, 4. oats; in the E. part of the county it is generally, 1. fallow, 2. wheat, 3. oats or barley, 4. seeds. Potatoes are pretty extensively grown; but turnips are not cultivated to any extent in any part of the county. Lime is frequently used on the arable land; and, with the exception of dung, is the only manure that is employed. Drilling but little practised; beans mostly planted by the dibble. Labourers said to be indolent; and that they seem, when employed, as if they would rather stand still than exert themselves. (*Kennedy and Granger*, i., 308.) A large proportion of the county is in grass, the dairy and fattening systems being both extensively carried on. Cattle, principally of the Devonshire breed; but a great variety of other breeds are met with. The celebrated Cheddar cheese is so called from a village of that name on the W. side of the Mendip hills; but it is now principally made in the marshes round Glastonbury. Bridgewater cheese is made from the marshes between that town and Cross. The stock of sheep is the county is supposed to amount to about 500,000 head, partly long, and partly short-woolled; producing, in all, about 16,500 packs a year. Large quantities of excellent cider are made in various parts, but particularly in the vale of Taunton. The woods and woodlands are supposed in all to cover from 20,000 to 23,000 acres; and it is distinguished by the stately growth of its hedgerow timber. Property very variously divided; some large estates, but a good deal of land occupied by yeomen who farm their own estates. Farms of various sizes, but the majority small. Leases, when granted, are usually for 8 or 12 years: in many instances, however, farmers hold at will. Average rent of land, 25s. 9d. an acre, being, with the exception of Leicester and Middlesex, the highest in the kingdom. Mineral products nu-

merous and valuable, consisting principally of coal, lead, calamine, fuller's earth, limestone, freestone, &c. Coal, however, to the lead mines having been nearly exhausted, or becoming more difficult to work, the produce of lead is now quite inconsiderable. The woollen manufacture used to be extensively carried on at Taunton; but is now given place to the silk trade, introduced in 1772 and at present prosecuted on a pretty large scale. The woollen manufacture is still, however, carried on at Frome, Shepton Mallet, Wellington, and some other places in the county; but it has long been in a declining state. Gloves amply manufactured at Yeovil. Principal rivers, Lower Avon, Ax, Brue, Parret, and Exe. The Parret is navigable from Langport to its mouth. Taunton and Bridgewater are united by a canal; and there are canals in other parts of the county. Somersetshire is divided into 40 hundreds and seven liberties, and contains 473 parishes. It returns 3 members to the House of Commons; viz., four for the county, two each for the cities of Bath and Wells, two each for the boroughs of Bridgewater and Taunton, and one for Frome. Registered electors for the county, in 1839-40, 18,783, of which 9759 belonged to the E., and 9024 to the W. division. In 1841, Somerset had 81,637 inhabited houses, and 436,002 inhabitants, of whom 200,421 were males, and 236,581 females. Sum expended for the relief of the poor, in 1839-40, £156,099. Annual value of real property, in 1815, £2,308,792. Profits of trade and professions in ditto, £1,329,966.

SOMERSET, county, Me. Situated in the N.W. part of the state, and contains 3600 sq. m. Watered by Kennebec river, flowing from Umbagog lake on its E. border. It contained in 1840, 37,366 neat cattle, 66,063 sheep, 11,039 swine; and produced 132,487 bushels of wheat, 16,601 of rye, 71,134 of Indian corn, 18,372 of barley, 167,369 of oats, 1,083,718 of potatoes, 94,121 pounds of sugar. It had 77 stores, 13 fulling-mills, one woollen factory, 44 grist-mills, 75 saw-mills, one oil-mill, 20 tanneries, two potteries, one printing-office, one weekly newspaper; four academies, 136 students; 333 schools, 13,179 scholars. Pop. 33,912. Capital, Norridgewock.

SOMERSET, county, N. J. Situated a little N. of the centre of the state, and contains 275 sq. m. Watered by Raritan river and its branches. It contained in 1840, 16,773 neat cattle, 16,754 sheep, 14,096 swine; and produced 40,780 bushels of wheat, 134,793 of rye, 573,949 of Indian corn, 52,060 of buckwheat, 1846 of barley, 323,945 of oats, 76,845 of potatoes. It had 64 stores, six lumber-yards, two fulling mills, six woollen factories, one flouring-mill, 29 grist-mills, 30 saw-mills, one oil mill, 14 tanneries, 14 other tanneries, one pottery, three printing-offices, one bindery, two weekly newspapers: one college, 115 students, two academy, 56 students; 48 schools, 1313 scholars. Pop. 17,455. Capital, Somerville.

SOMERSET, county, Pa. Situated toward the S.W. part of the state, and contains 1000 sq. m. Drained by branches of Conemaugh river and by Castleman's river, a branch of Youghiheny river. It contained in 1840, 31,498 neat cattle, 36,900 sheep, 25,790 swine; and produced 138,183 bushels of wheat, 169,550 of rye, 32,940 of Indian corn, 46,650 of buckwheat, 691,900 of oats, 134,498 of potatoes, 299,800 pounds of sugar. It had 46 stores, 13 fulling mills, two woollen factories, two flouring mills, 64 grist-mills, 141 saw-mills, four oil-mills, 26 tanneries, 47 distilleries, one brewery, nine potteries, two printing-offices, two weekly newspapers; 21 schools, 476 scholars. Pop. 19,650. Capital, Somerset.

SOMERSET, county, Md. Situated in the S.E. part of the state, and contains 500 sq. m. Bounded N.W. by Nanticoke river, S.E. by Pocomoke river. Watered by Wicomico and Manokin rivers. It contained in 1840, 11,234 neat cattle, 11,558 sheep, 20,010 swine; and produced 34,713 bushels of wheat, 426,102 of Indian corn, 125,887 of oats, 33,039 of potatoes. It had 79 stores, three lumber-yards, 26 grist-mills, 19 saw-mills, five tanneries: one college, 29 students, two academies, 96 students; 23 schools, 614 scholars. Pop.: whites, 11,485; slaves, 5377; free colored, 2646, total, 19,508. Capital, Princess Ann.

SOMERSET, p. t., Bristol co., Mass., 44 m. S. Boston, 49 W. Bounded E. by Taunton river, which is navigable to this place for vessels of considerable burthen. It has four churches, two Baptist, a Methodist and Friends, six stores; eight schools, 143 scholars. Pop. 1605.

SOMERSET, p. t., Niagara co., N. Y., 14 m. N.E. Lockport 276 m. W. by N. Albany, 419 W. Bounded N. by lake Ontario, and drained by small streams flowing into it. It has two stores, two saw-mills; six schools, 319 scholars. Pop. 1742.

SOMERSET, p. b., Somerset t., capital of Somerset co., Pa., 40 m. W. Harrisburg. 166 W. Incorporated 1804. It contains a courthouse, jail, an academy, three churches, a German Reformed, Methodist and Lutheran, seven stores, one fulling-mill, three tanneries, one brewery, two grist-

r-offices, three weekly newspapers; one
t. Pop. 638. The town, exclusive of the
two stores, 11 grist-mills, 12 saw-mills,
one school, 30 scholars. Pop. 2711.

TH, t., Strafford co., N. H., 45 m. E. Con-
N.E. and E. by Salmon Falls river. Great
the river contains extensive cotton and
, and has 2500 inhabitants. Vessels of
hin a mile of it. The town contains six
st, Free-will Baptist, Methodist, Christian,
nd Universalist, 30 stores, 12 fulling mills,
ry, four cotton factories, with 40,012 spin-
ls ; one academy, 90 students ; 13 schools,
). 3283.

, p. v., capital of Somerset co., N. J., 31
W. Situated a mile N. of Raritan river,
ourthouse, jail, a Dutch Reformed and
, an academy, a female seminary, seven
over 100 dwellings, many of them neat,
ts. The Elizabethtown and Somerville
stern termination here.

v., capital of Fayette county, Tenn.
e S. of Loosahatchie river, and contains
a male and female academy, a bank,
d 750 inhabitants. Somerville college
here, and the building is in progress.
market town and par. of England, co.
ground beside the Carey, 11 m. S.S.W.
r., 6030 acres. Pop., in 1831, 1786. The
ome small streets, with houses mostly
; and has a town-hall, in which petty
d one of the county jails. The church,
e, has an eight-sided embattled tower,
The living, a vicarage, worth £259 a
f the Earl of Ilchester. Somerton has
and an almshouse for eight poor wom-
o-ed to have been a Roman station,
ttion respecting it till the Heptarchy,
siderable fortified town, and the resi-
er kings of Wessex. Hence it abounds
t, including parts of the ancient walls,
he castle in which John king of France
sequently to his capture at the battle of

f France, reg. N., comprised mostly in
ardy, between lat. 49° 37' and 50° 20'
nd 3° 10' E. ; having N. Pas-de-Calais,
d W. Seine-Inférieure and the English
247 hectares. Pop., in 1836, 545,024,
owards the N.W., which direction is
pal rivers, consisting of the Somme,
o two nearly equal parts, the Authie,
; N. boundary, and the Bresle, bound-
The Somme rises at Fonsomme in
erally N.W. to the English channel,
ttle below St. Valery, nearly opposite
ry from 3 to 4 m. wide, after a course
principal affluents are the Avre and
Iam, Peronne, Amiens, and Abbeville
le Somme is navigable for about half
igation is interrupted by shoals. The
Somme (Canal de Picardie), 96 m. in
t Abbeville, and connects this river
department has generally a naked ashing to the official tables, it comprised
ole land, 15,432 do. meadows, 20,550
d 51.207 do. woods. Agriculture is
in most French departments; le cul-
t et dégagé de préjugés. (Hugo.)
an is required for home consumption;
as estimated at nearly 5,758,000 hec-
est, maslin, and rye. About 200,000
nd 100,000 do. of beer, are produced
sheep are numerous, and the produce
about 780,000 kilog. a year. In 1835,
sbject to the contrib. foncière, 193,745
than five francs. Mineral products
e ; but the department is distinguish-
res. Woollen, cotton, and silken
s, are made at Amiens and Abbe-
thread, oil, and leather at Peronne.
Escarbotin is the seat of some of the
ories in the kingdom; and machine-
ot sugar, are produced in considera-
e is divided into five arrondissements:
e capital, Abbeville, Doullens, Mont-
sends seven members to the cham-
stered electors, 1838-39, 3971. To-
31, 16,184,848 francs. (Hugo, art.
i Tables.)

l, a group off the E. archipelago, 4th
ttending from the N.E. part of Bor-

neo to Mindanao, the most S. of the Philippine islands, between the 4th and 7th degs. of N. lat., and the 120th and
123d of E. long.; having S. the sea of Celebes, and N. the
Sooloo sea. They consist of about 60 islands, taking their
name from Sooloo, one of the largest, about the middle of
the group. They produce rice, sweet potatoes, yams, and
many of the finest fruits of the east; but sago is the principal food of the inhabitants. Pearls, mother-of-pearl, and
cowries, are among their most valuable products. Formerly the inhabitants carried on a large trade with Japan;
at present their commerce is chiefly with the adjacent
islands of Celebes, Mindanao, Borneo, &c., and a few junks
that come yearly from China. The Sooloos are mostly
Mohammedans, and live under a sultan, whose power is,
however, much limited by a kind of feudal aristocracy.
They are distinguished for their piracies, and their continual hostility to the Spaniards of the Philippines; in 1775
they destroyed an establishment formed by the East India
Company on the neighbouring island of Balambagan.
(Hamilton, E. I. Gaz.)

SOPHIA, or TRIADITZA, a city of European Turkey,
prov. Bulgaria, near its W. extremity, in a fine plain, on
the Bogana, a tributary of the Isker, 93 m. S.S.E. Widin,
and 155 m. S.W. Rustchuk. Hoeschelmann (Stein), the
Dict. Geog., &c., agree in rating its population as high as
from 45,000 to 50,000; and it is said to have 30 mosques,
and 10 churches. Mr. Burgess says, "Sophia, although
one of the meanest cities I ever saw, must still be considered as the capital of Bulgaria, and as holding a high
rank among the cities of European Turkey. The situation appeared to me the most unfavourable that could
have been chosen for a city; sunk in a hollow, it is constantly liable to be inundated; and without canals to carry off the superabundant waters of the Isker, the plain is
almost lost to the labour of the agriculturist. The habitations are all made of baked mud: and I scarcely saw
one which ought to be qualified with any other appellation
than that of hovel." (Greece, &c., ii., 209.) Sophia is,
however, the residence of the begler-beg of Roumelia, and
of Greek and Roman Catholic archbishops. It has manufactures of woollen and silk stuffs, leather, and tobacco, and
an extensive general trade. There are some warm baths.
This city, founded by Justinian, was built, it is said, on the
ruins of the ancient Sardica. (Burgess's Greece, and the
Levant, i., Dict. Geog., &c.)

SORA, a city of the Neapolitan dom., Terra di Lavora,
cap. distr, on the Liris, 50 m. N.N.W. Capua. Pop 10,000.
(Rampoldi.) "The episcopal town of Sora, retaining its
ancient name and situation, is about 3 m. distant from Isola,
along an excellent road, which terminates with the valley itself at its gates. Here the Liris, flowing from a glen
of narrower dimensions, but considerable length, forms a
bend round the city, and is crossed by two bridges. The
place is consequently in a flat but not unpleasant position;
one whole flank being watered by the river, and the hinder
extremity resting against an insulated rocky hill, on which
are seen the ruins of its Gothic castle, and those of its still
more ancient walls. The dwellings are large, the streets
wide and well paved, and the pop. apparently easy and industrious." (Craven's Excursion in the Abruzzi, &c. i.
112.)

After its cathedral, in the front of which are a number of
inscriptions and fragments of sculptures, the principal buildings are four other churches, several convents, hospitals,
and seminaries, a showy modern gateway, &c. The adjacent country is both fertile and well cultivated.

Sora was of Volscian origin, but became, at an early period, attached to the party of the Samnites; though subdued and colonized by the Romans, it repeatedly threw off
their yoke, and vindicated its ancient freedom. Juvenal
enumerates Sora among the country towns in which an individual, tired of the bustle and dissipation of Rome, might
find a comfortable residence:

Si potes avelli Circensibus, optima Sora
Aut Fabrateriæ domus aut Frusinone paratur.
Sat. iii., 223.

Sora was the birth-place of the Cardinal Baronius.

SORIA (an. Numantia), a city of Spain in Old Castile,
cap. prov. of its own name on, and at no great distance
from, the source of the Douro, here crossed by a fine stone
bridge, 113 m. N.E. Madrid. Pop. about 5500. It is enclosed by old walls, and has numerous churches and convents, a hospital, Jesuits' college, prison, &c. On the E. it
is commanded by an old fortress, now partly in ruins; and
on the S. is a considerable suburb. It is dirty and ill-built,
with a few silk fabrics, and some trade in wool.

Adjacent to the town, on the N., are the ruins of the
famous city of Numantia, destroyed by the Romans, anno
139 B.C. No people ever discovered greater bravery, than
the inhabs. of this small state. Numantia, quantum Car-
thaginos, Capua, Corinthi, opibus inferior, its virtutis nom-

tus at honore par omnibus, summumque, si viros æstimes, Hispania decus. (*Florus*, lib. ii. cap. 18.) The conduct of the Romans in this contest was distinguished alike by perfidy and vindictive malignity. The Numantines having defeated Pompey, grandfather of Pompey the Great, who had besieged their city, he concluded a treaty with them. But the Romans having, on various pretexts, broken this treaty, sent a powerful army against the Numantines under the consul Mancinus. The latter, however, being even more unsuccessful than Pompey, was obliged, to save himself and his army from total destruction, to conclude a new treaty with his successful adversaries; who stipulated for nothing but that they should retain their independence, and be reckoned among the friends and allies of Rome.

Tiberius Gracchus then quæstor in the consular army, was a principal party to this treaty, the observance of which was sworn to by all the chief officers of the Roman army. But though the Numantines spared by this treaty the lives of 10,000 Roman soldiers that were in their power, and stipulated for nothing that a generous or high minded people could, under any circumstances, have refused to concede, the senate and people of Rome were base enough to annul the treaty; and sent Scipio Africanus, who had destroyed Carthage, to wage a war of extermination against the Numantines! Scipio who knew the bravery of those he had to contend with, did not attempt to carry the city by storm; but having surrounded it by strong lines of circumvallation, left famine to effect its reduction. Notwithstanding their inferior numbers, the Numantines made the most astonishing efforts to break through and destroy the works of the Romans; but having been repulsed, they were reduced to the most dreadful extremities. It is uncertain how the final catastrophe of this noble city was consummated; whether, as Florus affirms (lib. ii. cap. 18.), the Numantines set it on fire, and perished in the flames; or whether, as Appian states (*In Iber.* p. 311.), having surrendered, the small remnants of its inhabitants that were found alive were sold as slaves. One thing only is certain, that the struggle reflects the highest credit upon the Numantines, and the most indelible disgrace on the Romans. It is due to the character of Tiberius Gracchus to state, that he reprobated in the strongest manner the perfidy and bad faith of his countrymen in refusing to ratify the treaty with the Numantines. (Besides the authorities already referred to, see *Ancient Universal History*, xii. 400, 8vo ed.)

SORRENTO (an. *Surrentum*), a city and seaport of the Neapolitan dom., prov. Naples, on the S. side of the bay of Naples, 18 m. S.E. that capital. Pop. about 8000. (*Rampoldi.*) It is well built and clean; and has been celebrated in antiquity, as well in modern times, for the beauty of its situation, and the mildness of its climate, being hence called by Horace *Surrentum Amœnum.* (Epist. ii. 18, lin. 52.) It is the seat of an archbishopric; and besides the cathedral, has several churches, numerous convents, a hospital, seminary, college, school of navigation, and some silk manufactures. It was supposed in antiquity to have been the seat of the Sirens. (*Plin.* hb. iv. cap. 5.) But it derives its principal illustration from its having been the birth-place of Torquato Tasso, the greatest of Italian, and, perhaps, of modern poets, born here on the 11th of March. 1544. "Among the many respectable houses termed palaces, which adorn Sorrento, that in which the author of the *Gerusalemme Liberata* was born, naturally excites the greatest interest; it is placed on the cliff rising immediately from the sea, and offers some pretensions to elegance of architecture, but probably retains in its outward form no remains of its ancient appearance. Sorrento is a place of high antiquity; and the various inscriptions, bas-reliefs, &c., found in it at different periods, are collected under an archway in the town, which thus forms a kind of open museum, accessible to every visiter. The ancient walls and towers can scarcely be referred to the Lombards, who erected this territory into a small independent principality. Oil, milk, meat, and game, are all excellent in their different kinds at Sorrento; while its veal is, by some, reputed the best in Europe. The capital is supplied from it with many of these articles, by boats plying at regular times of the day." (*Craven*, 304–306.) The beautiful bay of Sorrento, 3 m. wide, is surrounded by a semi-circular range of wooded hills, between which and the sea is a rich plain, the *Piano di Sorrento*, in which are many villages and detached houses.

It was on the hills bounding the plain *Colles Surrentini witiferi*, that the famous wine was produced, which, in antiquity, vied with the Falernian and Cæcuban.

Surrentina bibes? nec Martbian picta, nec surum
Seme: debent callous hæc tibi vina suos.
                              Mart. Epig, hb. xiii. 110.

It was a powerful wine, and did not arrive at perfection till it had been kept above 20 years. Owing to the want of care, the wine now produced from Surrentine grapes is among the poorest in Italy. Near Sorrento are the remains

of the villa of Pollius, described by Statius. *Sn Insburne's Travels.* i. 98.)

SOUND. *See* ELSINEUR.

SOURABAYA, a considerable town of Java, long one of the three principal seaports of that island on the N. coast of which it is situated, about 100 m. E. Samang. It stands about 1½ m. from the strait of Madura, on both sides a river, said to be navigable by boats for 18 a. kms the sea, and deep enough at its mouth to receive vessels of 250 tons. The town itself is of small extent, but it has extensive suburbs, and round it are a number of handsome villas its vicinity, though low, being less unhealthy than that of Batavia. Mr. Earl says, that it is also much more gay and lively than the latter, and well supplied with provisions of all kinds. Sourabaya seems to have the only secure harbour on the N. coast of Java, and the only one in which the shipping is well defended by the batteries on the shore. Its chief entrance is commanded by a strong fort on a small island 9 m. from the town. Several English merchants, agents to houses at Batavia, are settled here, though Europeans are upon the whole few. There are numerous Arabs and their descendants. According to Earl, more ships are built at Sourabaya than at any other port of the E. Archipelago; when he visited the place, in 1836, the great ship-builder was an Englishman, employing from 200 to 900 workmen, all of whom lived in a large village adjacent his dockyard. (See *Earl's Eastern Seas*, 47, 72, Hamilton, &c.)

SOUTH AMBOY, t., Middlesex co., N. J. Bounded E. by Raritan r., N.E. by Raritan bay. Drained by South r. and its tributaries. The Camden and Amboy railroad passes through it. The village is at the head of Raritan bay, at the mouth of Raritan r., 1 m. S. of Perth Amboy. The railroad here connects with steamboats to the city of New York. It has a good harbour, and contains three stores, a large stone-ware factory supplied with excellent clay in the vicinity, a grist-mill, saw-mill, paper-mill, six schools, 198 scholars. Pop. of the t. 1825.

SOUTHAMPTON, a parl. and bor., seaport, and market town of England, being a county of itself in Hampshire, at the embouchure of the Itchen, in an inlet of the sea, called Southampton-water, 12 m. S. by E. Winchester, and 70 m. W.S.W. London. Lat. 50° 53′ 59″ N.; long. 1° 24′ W. Pop. of the parl. bor., which is co-extensive with the county of the town, in 1831, 19,324; in 1841, 28,652. The approach to the town from the London road, through a fine avenue of trees and a well-built suburb, is exceedingly striking. The principal entrance is through Bargate, one of the ancient gates, which also divides the town into two parts, called respectively Above-bar and Below-bar. The High-street, below bar, which is more than ½ m. in length, leads directly to the quay, for the improvement of which the water-gate was removed about 45 years ago. The ancient part of the town was formerly encased with walls about 1½ m. in circuit, of which there are considerable remains on the W. side of the town, and two old gates in addit ion to Bargate. Many smaller streets branch both E. and W. from the principal avenue, and buildings are rapidly increasing, the space occupied by streets and houses now exceeding 3½ m. in circuit. The old town occupies nearly the whole of the parishes of St. John, St. Lawrence, Holyrood, St. Michael, and All-Saints-intra. The parishes of St. Mary, and All-Saints-extra are extensive, and were till of late years principally agricultural. The town, however, now extends into both of them: and in the latter the new buildings consist principally either of handsome town-houses or detached villa residences. A new and has been formed, extending E. from the High-street to the steam-ferry over the Itchen; several new streets are now ready in progress, and docks are about to be formed on the river, near the ferry-pier. The parishes of St. Lawrence and Holyrood, through which the High-street passes, contain the dwellings of the most respectable and principal tradespeople: in the latter are the market-place, market-house, custom-house, several of the principal banks, the town quay: indeed, in a commercial point of view, these two parishes comprise the most important part of the town." (*Mun. Bound. Report.*) The whole town is well paved, lighted with gas, and is exceedingly clean: the inhabitants have hitherto been supplied with water from several springs in the neighbourhood, but from the increase of population, and the deficiency of water of late years, the commissioners of water-works have entered into a contract for the construction of an artesian well on Southampton common, to supply 40,000 cubic feet per day, which is now (January, 1842,) nearly finished, the workmen having reached a depth of 552 ft., and entered the chalk stratum. The old reservoirs on the common will be supplied from this well in dry seasons. Besides the buildings devoted to commerce and other purposes, the town has a theatre and assembly-rooms. The military orphan asylum for girls, established by the late Duke of York, and occupying the den-

barracks, has been removed, and the buildings are now
prised to the establishment for the trigonometrical
r. recently removed thither from the Tower of Lon-
The handsome suite of baths on the beach h ve been
rted into a dock house and offices for the Southamp-
ock company; but there are very convenient baths
er parts of the town. The old Saxon castle, repair-
Richard II. in the view of protecting the harbour,
lied down some years ago, and a private chapel, in
the Church of England service is performed, built
s site.
own has five parish churches, three of which are
ift of the lord chancellor, one in that of the bishop
chester, and another in the patronage of Queen's
Oxford. Holyrood church is an ancient edifice,
wer and spire, the portico being the site, before the
Act. for the hustings at elections for the borough.
ael's is a large structure in the Norman style, with
ine tower and spire between the nave and chan-
Saints is a Grecian building, with a turret, sur-
by six Corinthian pillars on a square pedestal. St.
ivo, is a modern structure, and its extensive buri-
has hitherto served as a general cemetery for the
t owing to its extremely crowded state, the requi-
s have been given, and an act will be applied for
m of parliament for the formation of a cemetery
s on Southampton common. The Roman Cath-
ependents, Baptists, Wesleyan-Methodists, and
e each places of worship; and there is a Friends'
use. There are two chapels, one proprietary,
der trustees in connection with the establish-
well attended Sunday-schools, with several day-
pported wholly or in part by endowment. The
hool, founded by Edward VI., has a small en-
he premises have recently been rebuilt, and fur-
moodition for about 40 boys, boarding with the
r. A hospital, called the Domus Dei, founded
of Henry III., provides lodging, clothing, and a
nd to four aged people of e ch sex. Among
he charities, is one left, in 1760, by the will of
aunton, which, besides providing for the in-
10 boys, furnishes a stipend of £10 a year for
ns, and gives rewards to deserving fem le wer-
also a female penitentiary, public dispensary,
ospital, a royal humane society, several ben-
and a school of industry for 50 girls, founded,
zh the influence of Queen Adelaide. There
igious societies, a literary society, a polytech-
with 300 members, an infirmary, and several
od subscription libraries. A regatta takes
mmer on Southampton-water, under the di-
ronage of the Southampton yacht club, re-
aed, and races are held in autumn on the
of the town. Two newspapers, also, are
Saturday.
water affords good anchorage; and ships
load and unload alongside the town quay,
the custom-house. A new pier of wood
projects about 400 yards from the shore,
ient landing-place for passengers from
well as a promenade for the inhabitants
t has a carriage-drive to its extremity.
ensive scale, are now (1842) in the course
ted, which will contribute materially to
acilities now enjoyed by the port.
and Oriental steam navigation company
le of their vessels from this port to Alex-
n; and, more recently, the Royal West-
company have started their steam ships
o, and have purchased a large tract of
of the Itchen for constructing repairing
ing dock.
on an inlet of the sea, stretching N.W.
mooth and the isle of Wight at least 17
ry, and which has been prolonged by
on to Winchester, 12 m. inland. South-
rium of an extensive district, and conse-
etty extensive trade. In 1840 there be-
197 vessels, of the aggregate burden of
ring the same year the gross customs'
£77,943. It may be farther mentioned,
increasing importance of the port, that,
f the aggregate burden of 51,307 tons,
h cargoes from foreign ports; whereas,
an 555 ships, of the aggregate burden
l inwards with foreign cargoes!
r position with respect to the opposite
hampton has been for a lengthened pe-
iation for travellers to and from the
respect, however, her importance has
d since the opening of the south-west-
th the town has been brought within a

three hours' journey of the metropolis. This great undertaking, in which above £1,000,000 have been expended, has
gone far to make Southampton an outport, as it were, of
London. In fact, she is now become the principal station
of the steamers for Havre, Dieppe, and other French ports,
as well as of those for Lisbon, the Mediterranean, &c. By
setting off from Southampton, the difficult navigation from
the N. foreland and round by Dover and Beachy head is
avoided; and ships are enabled to proceed on their voyages
with comparatively little chance of being delayed by adverse winds. Vast numbers of tourists are also brought
down by the railway, attracted by the mildness of the climate, and the beauty of the scenery of the New forest, the
isle of Wight (to which there are steamers every hour), &c.
The completion of the railway, now in progress of construction, from Havre by Rouen to Paris, will confer still
greater importance on Southampton.

Southampton has various private banking establishments,
two joint-stock banks, and a savings' bank. Markets on
Tuesday, Thursday, and Saturday for provisions, and on
Friday for corn, well attended. Fairs, May 6 and 7, and
on Trinity Monday and Tuesday.

Southampton was first incorporated in the reign of Henry
I. Under the Municipal Reform Act it is divided into five
wards, with 10 aldermen and 30 councillors, from which 40
members are chosen the mayor, sheriff, and two bailiffs.
The mayor and bailiffs are the returning officers of the borough. Corporation revenue in 1839, £3059. Quarter and
petty sessions are held under a recorder; and there is a
court for the recovery of debts to any amount. The borough h is returned two members to the House of Commons
since the 23d Edw. I., the right of election down to the Reform act being vested in the inhabitants paying scot and
lot. The electoral limits were left unchanged by the Boundary act. Registered electors, in 1839–40, 1463. It is also
the election town for the southern division of the county of
Hants.

Southampton is said to have arisen out of the neighbouring Roman station Clausentum, E. the Itchen, which was
succeeded by the Saxon Hantune, on the site of the present
town. The castle, as already stated, was much enlarged
by Richard II., who also strengthened the fortifications about
the town and harbour. Henry V. set sail from this port in
August, 1415, at the head of the troops which, on the 25th
of October following, gained the great victory of Agincourt.
The inhabitants were actively engaged in the wars of York
and Lancaster, in which the latter party was defeated with
great loss. Its celebrity as a watering-place dates from the
middle of last century, when baths were erected, a chalybeate spring was discovered, and great additions were made
by the formation of new streets and terraces, the laying out
of public gardens, &c. The shores of Southampton-water,
being richly clothed with wood and studded with villas,
afford a succession of finely-diversified scenery, set off by
the ruins of Netley abbey, about 2 m. S.E. from the town.
This structure was founded in 1229, by Henry III., for Cistercian monks. The refectory and kitchen are in tolerable
preservation; and there are some fine remains of the abbey church, which was cruciform, and had at its E. end
a noble window. The whole is embosomed in wood; and
near it is a modern tower, used as a tea-house, on the foundations of a fort erected by Henry VIII., commanding a
fine view of Southampton-water. (Mun. and Parl. Bound.
Reports, and Private Information.)

SOUTHAMPTON, p. t., Hampshire co., Mass., 9 m. S.W.
Northampton, 102 m. W. by S. Boston, 372 W. Marshan
r. affords water-power. The Hampshire and Hampden canal passes through it. It contains a Congregational church,
four stores, one fulling-mill, two grist-mills, six saw mills;
one academy, 40 students; seven schools, 370 scholars.
Pop. 1157.

SOUTHAMPTON, p. t., Suffolk co., N. Y., 250 m. S.S.E.
Albany, 330 W. It is a large township in the E. part of
Long Island, extending from Great and Little Peconic bays
on the N. to the Atlantic on the S. It has four churches,
two Presbyterian, and two Methodist, 53 stores, two woollen factories, 10 grist-mills, three saw-mills, one pottery,
two printing-offices, two weekly, and one semi-weekly
newspaper: two ac demies, 90 students; 31 schools, 1234
scholars. Pop. 6205. It contains the village of Sag Harbour, which see. . .

SOUTHAMPTON, co., Va. Situated in the S.E. part of the
state, and contains 648 sq. m. Bounded E. by Blackwater
r., S.W. by Meherrin r. Watered by Nottaway r. It contained in 1840, 10 003 neat cattle, 7967 sheep 43.663 swine;
and produced 9730 bushels of wheat, 3032 of rye, 55,1875 of
Indian corn, 71,312 of oats, 88,036 of potatoes, 1080 pounds
of rice, 25,346 of tobacco, 851,315 of cotton. It had 20
stores. 24 grist-mills; two academies, 40 students; 22
schools, 400 scholars. Pop.: whites, 6171; slaves, 6555;
free coloured, 1799; total, 14.525. Capital, Jerusalem.

SOUTHAMPTON, t., Bucks co., Pa., 12 m. S.E. Doylestown.

Watered by Poquessing and Pennypack creeks. It contains a Friends' church, four stores; two schools, 68 scholars. Pop. 1256.

SOUTHAMPTON, t., Cumberland co.. Pa., 18 m. S.W. Carlisle. It has three stores, two furnaces, three flouring-mills, seven saw-mills; six schools, 240 scholars. Pop. 1484.

SOUTHAMPTON, t., Franklin co., Pa., 13 m. N.N.E. Chambersburg. Bounded N.W. by Conedogwinnit cr. It contains two stores, two fulling-mills, one woollen factory, two furnaces, one flouring-mill, seven grist-mills, seven saw-mills ; eight schools, 335 scholars. Pop. 1703.

SOUTHAMPTON. t., Bedford co., Pa. It has one store, one fulling-mill, one flouring-mill, one saw-mill ; three schools, 125 scholars. Pop. 1543.

SOUTH BERWICK, p. t., York co., Me., 95 m. S.W. Augusta, 500 W. Bounded S.W. by Salmon Falls r., in which the Great Falls afford extensive water-power, where is a flourishing manufacturing village. The t. contains 16 stores, one fulling-mill, three woollen factories, one cotton factory with 6912 spindles, two grist-mills, five saw-mills, three tanneries ; one academy, 73 students ; 14 schools, 871 scholars. Pop. 2314.

SOUTHBOROUGH, p. t., Worcester co., Mass., 26 m. W. Boston, 414 W. Watered by a branch of Sudbury r. Incorporated in 1727. It contains three churches, a Congregational, Unitarian, and Baptist, three stores, one fulling-mill, one woollen factory, three grist-mills, three saw-mills ; six schools, 337 scholars. Pop. 1145.

SOUTHBRIDGE, p. t., Worcester co., Mass., 61 m. W.S.W. Boston, 381 W. Quinnebaug r. affords water-power. It contains three churches, a Congregational, Methodist, and Baptist, 10 stores, one woollen factory, eight cotton factories, with 14,660 spindles, one grist-mill, six saw-mills ; one academy, 25 students ; seven schools, 368 scholars. Pop. 2031.

SOUTH BRISTOL, p. t., Ontario co., N. Y., 12 m. S. Canandaigua, 513 m. W. Albany, 349 W. Bounded S.E. by Canandaigua lake. Watered by Mud cr. It contains a Presbyterian church, two stores, one fulling-mill, one grist-mill, 12 saw-mills ; 14 schools, 631 scholars. Pop. 1375.

SOUTH BRUNSWICK, t., Middlesex co., N. J., 12 m. S.W. New-Brunswick. Bounded W. by Millstone r. It has eight stores, four grist-mills, five saw-mills, nine distilleries ; 12 schools, 360 scholars. Pop. 2797.

SOUTHBURY, p. t., New-Haven co., Conn., 20 m. N.W. New-Haven, 54 m. S.W. Hartford, 304 W. Pamperaug r. affords water-power. It contains a Congregational and Methodist church, nine stores, three fulling-mills one woollen factory, four grist-mills, 10 saw-mills, three distilleries ; two academies, 61 students ; 10 schools, 298 scholars. Pop. 1542.

SOUTH CAROLINA, one of the southern United States, is situated between 32° 2′ and 35° 10′ N. lat., and between 78° 24′ and 83° 30′ W. long., and between 1° 45′ and 6° 15′ W. long. from W. It has an average length of about 200 m., and a breadth of 160 m., containing 30,213 sq. m., or 19,336,390 acres. It is bounded N. by North Carolina ; S.E. by the Atlantic ocean ; S.W. by Georgia, from which it is separated by Savannah river. The population in 1790, was 249,073 ; in 1800, 345,501 ; in 1810, 415 115 ; in 1820, 502,741 ; in 1830, 581,185 ; in 1840, 594,398, of whom 327,038 were slaves, or something more than one half of the population. Of the free population 130,496 were white males ; 123,588 were white females ; 3864 were coloured males ; 4412 were coloured females. Employed in agriculture, 198,363 ; in commerce, 1958 ; in manufactures and trades, 10,325 ; in navigating the ocean, 381 ; do. rivers, canals, &c., 348 ; in the learned professions 1481.

The state is divided into 29 districts, which, with their population in 1840, were as follows :—

Districts.	Total Pop.	Districts.	Total Pop
Abbeville	29,351	Lancaster	9,907
Anderson	18,493	Laurens	21,584
Barnwell	21,471	Lexington	12,111
Beaufort	35 794	Marion	13,932
Charleston	82,661	Marlborough	8,408
Chester	17,747	Newberry	18,350
Chesterfield	8,574	Orangeburg	18,519
Colleton	25,548	Pickens	14,356
Darlington	14,822	Richland	16,397
Edgefield	32,8 2	Spartanburg	23,839
Fairfield	20,165	Sumter	27,492
Georgetown	18,274	Union	18,936
Greenville	17,839	Williamsburg	10,327
Horry	5,755	York	18,918
Kershaw	12,281	Total	591,398

Anderson and Pickens districts, constituting the district of Pendleton, are divided for judicial purposes only. The state is also divided into parishes for the more equal distribution of the political power of the state. They were originally made for the purpose of church government, but are now continued for political convenience.

Columbia, on the E. bank of Congaree river, immediately below the junction of Broad and Saluda rivers, form it, is the seat of government.

Carolina presents a great variety of soil and surface. Along the seaboard and for 40 miles into the interior, the face of the country is flat and unpromising ; covered with extensive tracts of pine barren, swamp, and marsh, or open meadow without wood ; comprising the most fertile and the most sterile extremes of soil. Ascending toward the centre of the state, the country rises into a tract of moderate elevation. Advancing still further in a north-westerly direction, it becomes mountainous, and very picturesque. The first section, which is generally called the lower country, includes the sea-islands, famous for producing the finest kind of cotton, called the sea-island cotton, which bears a higher price than the other kind. The tide lands are equally celebrated for their valuable crops of rice. The high lands of this region are generally poor : interspersed with strips of great fertility. The climate is moist, very changeable, and during the summer and autumnal months, extremely unhealthy. The region which lies between the tide lands and the granite or mountain ridges, is called the middle country, less healthy in summer than the latter, but much more so than the former. In winter and spring, it may be regarded as much more healthy than either. It is well irrigated by rivers and water-courses. It possesses, amidst long and barren tracts of swamp and forest, many fine spots for culture, and produces, in considerable abundance, the kind of cotton which is called upland, or short staple. In addition to cotton and rice, which are chiefly produced in the lower and middle sections, the state produces Indian corn and potatoes; wheat, pease, rye, and oats; tobacco, indigo, lumber, tar, pitch, and turpentine, oils, and silks. The fruits which mature and flourish are figs, apricots, cherries, nectarines, peaches, pears, melons, and pomegranates. Oranges are uninjured in ordinary winters. Among the metals are found gold, iron, and lead; plumbago, pyrites, and asbestos; granite, oil and soap-stones.

According to the census of 1840, there were in the state 129,921 horses or mules, 573,808 neat cattle, 232,981 sheep, 878,532 swine, poultry valued at $266,364. There was produced 968,354 bushels of wheat, 44,736 of rye, 14,722,965 of Indian corn, 3967 of barley, 1,486,208 of oats, 2,682,313 of potatoes, 299,170 pounds of wool 69,390,869 of rice, 61,710,274 of cotton, 51,519 of tobacco, 38,689 of sugar, 15,857 of wax, 9080 of silk cocoons, 94,616 tons of hay. The products of the dairy were valued at $577,819 ; of the orchard, at $52,275 ; of lumber, at $537,694.

Charleston, at the confluence of Ashley and Cooper rivers, is much the largest and most commercial place in the state. Georgetown, at the head of Winyaw bay, 13 miles from the ocean, admits only small vessels. Beaufort, on Port Royal island, has much the best harbour in the state, but is not a place of much trade. Columbia, the capital of the state, on the E. side of Congaree river, is a flourishing place, as is Georgetown, on Winyaw bay, Camden, on Great Pedee river, and Hamburg, on Savannah river, opposite to Augusta.

Great Pedee river, 450 miles long, rises in North Carolina, and runs through the eastern part of the state. It is navigable for sloops 130 miles. The Santee, formed by the junction of the Congaree and Wateree, rises in North Carolina, and has steamboat navigation about 200 miles. The Saluda river is a branch of the Congaree. The Edisto is navigable for boats 100 miles. The Savannah, common to this state and Georgia, washes the whole S.W. border of the state, and is navigable for steamboats 250 miles, and for pole-boats 150 miles further. Among the smaller rivers are Ashley, Cooper, and Coosahee.

The most important literary institution in the state is the college of South Carolina, founded in 1804. There is a theological seminary connected with the institution. In both departments there were in 1843, 216 students. Charleston college was founded in 1795, and has about 50 students. The medical institution in Charleston has eight professors, and 158 students. There were in the state in 1840, 117 academies or grammar schools, with 4326 students, and 566 common or primary schools, with 12,520 scholars. There were 20,615 free white persons over 20 years of age, who could neither read nor write.

The Methodists, Baptists, and Presbyterians are the most numerous religious denominations. At the commencement of 1836, the Methodists had 37,303 communicants ; the Baptists had 314 churches, 226 ministers, and 36,638 communicants ; the Presbyterians had 90 churches and 28 ministers ; the Episcopalians had 50 churches, one bishop, and 43 ministers ; the Lutherans had in 1842, 34 ministers, 34 congregations, and 1667 communicants ; and there were a number of congregations of Roman Catholics, Universalists, Friends, Universalists, and Jews.

At the commencement of 1846, there were 14 banks

in the state, with an aggregate capital of
and a circulation of $4,459,404. The state
use of 1840, amounted to $3,764,734, and in
ed to nearly $5,000,000.

s of the state, for the year ending Sept. 30th,
d to $8,043,284; and the imports to $1,567,431.
entered was 55,690 tons, and the tonnage
5 tons. There were in 1840, 41 commercial
sion houses engaged in foreign trade, with a
568,050; 1253 retail stores, with a capital of
57 persons employed in the lumber trade.
1 of $100,000; 125 persons employed in in-
riation, who with 46 butchers, packers. &c.,
apital of $112,900; 53 persons employed in
with a capital of $1617.
of home-made or family manufactures, was
e were three woollen factories, employing
oducing articles to the amount of $1000, with
100; 15 cotton factories, with 16,355 spindles,
persons, producing articles to the amount of
a capital of $617,450; four furnaces, pro-
as of cast iron, and nine forges, producing
iron, the whole employing 248 persons, and
13,300; five smelting houses employing 69
ing gold to the amount of $37,418, with a
00; one paper-mill, employing 30 persons,
ie amount of $90,800, with a capital of
ms produced hats and caps to the amount
nneries employed 981 persons and a capi-
; 243 other manufactories of leather, as
produced articles to the amount of $100,472,
$45,662; eight potteries employed 49 per-
articles to the amount of $19,300, with a
50; 137 persons produced machinery to the
561; 36 persons produced hardware and
amount of $13,465; 420 persons produced
agons to the amount of $189,270, with a
90; 164 flouring mills produced 58,458 bar-
ed, with other mills, employed 2122 per-
articles to the amount of $1,301,678, with
35,804; 1281 persons produced bricks and
int of $193,608. with a capital of $72,445;
manufactured 586,327 pounds of soap, and
tallow candles; 251 distilleries produced
of distilled spirits, employing 219 persons,
$14,342; vessels were built to the amount
persons manufactured furniture to the
55, with a capital of $133,500; 111 brick
and 1594 wooden houses were built, em-
and $1,527,576; 16 printing-offi-
es, three daily, 12 weekly, and two semi-
ers, and four periodicals, employed 164
pital of $131,300. The total amount of
in manufactures was $3,216,970.
tution of South Carolina was formed in
ich was formed in the Union. The pres-
as ratified at Columbia, June 3d, 1790.
ected for two years by the joint vote of
legislature, and is ineligible for the next
iust be 30 years of age. have resided in
ears, and possess, within the state, pro-
unt of 1500 pounds sterling, above his
nant governor is elected at the same
manner, and with similar qualifications;
death, removal, resignation, or absence
ischarges the duties of the office. The
15 members, elected for four years by the
the number is elected biennially. A
a free white citizen of the age of 20
and must have been a citizen and resi-
r the five years next preceding his elec-
freehold estate in the district for which
e value of 300 pounds, clear of debt.
ithin the district, he must possess with-
00 pounds clear of debt. The house of
sists of 125 members, elected biennially
member must be a free citizen of the
ears of age or upwards, and have been
c for at least three years next previous
s, or a real estate of 150 pounds ster-
If a non-resident, he must possess
00 pounds sterling, clear of debt. The
or courts are elected by the joint ballot
The courts consist of 10 judges, and
r :—of law, equity, and appeal. The
inion to all the districts; the latter is
ion and Columbia. The secretary of
l surveyor-general are elected in the
for the same eriod as the governor.
ints of the state, who have paid taxes
g the election, and who have resided

for six months in the county where they offer their vote,
are entitled to the right of suffrage.

South Carolina has several important works of internal
improvement. The Santee canal, extending 22 miles from
Charleston harbour to Santee river, was finished in 1802,
and cost $650,667. Through this canal, and the improve-
ment of the Santee and Congaree rivers, a steamboat communication has been opened from Charleston to Columbia.
Winyaw canal extends from Winyaw bay 7½ miles to
Kinlock creek, a branch of Santee river. The navigation
of Catawba river has been improved by five short canals,
with an aggregate length of 11½ miles. The South Carolina railroad extends 135½ miles from Charleston to Hamburg. This road was commenced in 1830, and completed
in 1834, and cost $1,750,000. It has since been sold to the
Louisville, Cincinnati, and Charleston railroad company
for $2,400,000, paid for in the stock of the latter company.
The entire length of this road, from Charleston to Cincinnati, will be, when completed,718 miles. The Branchville
and Columbia railroad extends from Branchville, on the S.
Carolina railroad, 66 miles to Columbia, and is to form a
part of the Charleston, Louisville, and Cincinnati railroad.

No permanent settlement was made in South Carolina
by Europeans until 1670, when a small body of English
emigrants, under William Sayle, arrived at Port Royal
Island. From this place they removed in 1679 to the present site of Charleston, at a tongue of land then called
Oyster Point, at the mouth of Ashley and Cooper rivers.
In 1682, the province was divided into three counties. In
1706, the French and Spaniards made a combined attack
upon Charleston and were defeated. In 1715, the war with
the Yemassee Indians threatened the destruction of the colony; but the Indians were defeated by Governor Craven.
In 1720, the proprietary government was thrown off, and
that of the crown established. In 1740, Charleston was
half destroyed by fire; and in 1741, indigo was first planted. In 1769, South Carolina was divided into seven precincts, viz: Charleston, Georgetown, Beaufort, Orangeburg, Cheraw, Camden, and Ninety-Six. In 1775, the importation of British goods was prohibited, and the first
military force raised for the defence of the colony against
the crown. In 1776, the British were defeated in an
attack on fort Moultrie, at the entrance of Charleston
harbour. In 1780, Charleston was besieged by Sir Henry Clinton, and was taken at the end of six weeks. Several important battles were fought in South Carolina
during the Revolutionary war, the most important of
which was that of Eutaw Springs, in 1781, which in effect
terminated the war in this state. In 1785, the British evacuated Charleston. In 1785, the first Methodist church was
established; and in 1791, the first Roman Catholic church
was founded. In 1794, cotton was first exported. In 1822,
an insurrection among the negroes of Charleston was discovered and defeated. In 1832, counter-proclamations
were issued by President Jackson and Governor Haynes of
S. C., on the subject of nullification, originating in the tariff, but the subject passed off without ultimate collision.
In convention, May 23d, 1788, this state adopted the constitution of the United States; yeas, 149, nays, 73; majority, 76; and hitherto this constitution has gained increased
firmness and stability by every obstacle which it has encountered. Esto perpetua, is responded by every state in
the Union.

SOUTHEAST, p. t. Putnam co., N. Y., 107 m. S. Albany, 291 W. Drained by Croton r. It contains eight stores,
two fulling-mills, six grist-mills, seven saw-mills, one paper-
mill; 13 schools, 543 scholars. Pop. 1910.

SOUTHFIELD, t. Richmond co., N. Y., 2 m. E. Richmond, 155 m. S. Albany. Situated on Staten Island,
bounded S.E. by Raritan bay, E. by the Narrows. It contains fort Tompkins and fort Richmond, on its E. border,
and has seven stores, one flouring-mill, two grist-mills, one
printing-office, one weekly newspaper. Pop. 1619.

SOUTH HADLEY, t. Hampshire co., Mass., 88 m.
W. Boston. 377 W. Bounded W. by Connecticut r., which
has here a fall of 50 ft. in 80 rods. Here the first canal on
Connecticut r. was constructed around these falls, 2 m. long,
with five locks; and for the distance of 300 ft. is cut 40 ft.
deep through solid rock. A dam 1100 ft. long is thrown
across Connecticut river, which affords great water-power,
but partially improved. It contains a Congregational and
Methodist church, Mount Holyoke female seminary, eight
stores, two fulling-mills, two woollen factories, one grist-
mill, three saw-mills, three paper-mills, three tanneries;
one academy, 119 students; six schools, 476 scholars. Pop.
1458.

SOUTH HUNTINGTON, t. Westmoreland co., Pa.,
It has nine stores, one fulling-mill, seven flouring-mills, 15
saw-mills, three tanneries, seven distilleries; one school,
117 scholars. Pop. 2793.

SOUTHINGTON, p. t. Hartford. co., Conn., 18 m. S.W.
Hartford, 291 W. Quinnipiac r. affords water-power. In-

corporaten in 1779. The Farmington canal passes through it. It contains three churches, a Congregational, Episcopal, and Baptist, an academy, five stores, three grist-mills, five saw-mills; nine schools, 457 scholars. Pop. 1887.

SOUTH KINGSTON, t. capital of Washington. co., R. I., 30 m. S.S.W. Providence. Bounded S. by the Atlantic. E. by Narr. ganset bay, and is the largest in territory of any town in the state. It has a large number of ponds, one of which contains 3000 or 4000 acres. Near the centre is North Kingston village, containing a courthouse, a jail, a bank, a Congregational church, and about 30 dwellings. In the S.E. part is another village, containing an Episcopal church. There is also in the t. a Baptist, and two Friends' churches. It contains 27 stores, 11 fulling-mills, 10 woollen factories, one cotton factory with 1000 spindles, eight grist-mills, four saw-mills, five tanneries, one printing-office; one academy, 30 students; 18 schools, 455 scholars. Pop. 3717.

SOUTH MIDDLETON, t., Cumberland co., Pa. Drained by Yellow Breeches cr. Letart's spring, on its N. border, gives motion to two mills. It has five stores, one furnace, one forge, one fulling-mill, three woollen factories, four flouring-mills, seven saw-mills, one paper-mill; one school, 60 scholars. Pop. 2055.

SOUTH MOLTON (or MOULTON), a munic. bor., market town, and par. of England, N.E. part co. Devon, hund. S. Molton, on an eminence near the confines of Exmoor, 12 m. E. by S. Barnstaple. Area of par., 6160 acres. Pop. in 1831, upwards of 4000. The limits for the municipal borough and parish are co-extensive. The town consists chiefly of three streets, diverging from a spacious marketplace. Many of the houses are good; streets well paved; the footpaths have been flagged, at a considerable expense, by the corporation. It is well lighted, and the public walks are kept remarkably clean. (Mun. Corp. Rep. Appendix.) The parish church has some good monuments, and a large organ. The living, a perpetual curacy, worth £157 a year, is in the gift of the dean and canons of Windsor. Here are free and charity schools; a guildhall, in which petty sessions are held every three weeks; and a jail, rebuilt a few years since. The population is partly manufacturing, and partly agricultural; the manufactures are chiefly of coarse woollen cloth, but that of lace has recently been introduced, and the trade of the town is said to be increasing. This borough sent members to the House of Commons in the 30th of Edward I.; but it does not appear to have subsequently exercised the privilege. It was chartered by Elizabeth and Charles II.; and is governed by a mayor, three other aldermen, and 12 councillors, the latter being elected for life among the resident inhabitants. The corporation, which is trustee for several charities, has an income of £985 a year, out of which the mayor receives an annual income of £50. About 3 m. N.E. from the town is Castle Hill, the seat of Earl Fortescue. (Municipal Corp. Rep. and Appendix.)

SOUTHOLD, p. t., Suffolk co., N. Y., 15 m. E. Riverhead, 241 m. S.E. Albany, 321 W. Bounded N. by Long Island sound, and by Great Peconic and Gardiner's bays on the S. It contains Fisher's, Gull, Plum, and Robbins's islands. It contains three churches, a Presbyterian, Methodist, and Universalist, 28 stores; 23 schools, 1071 scholars. Pup. 3907.

SOUTHPORT, p. t., Chemung co., N. Y., 4 m. S. Elmira, 197 m. W.S.W. Albany, 277 W. Bounded N. by Chemung r. It has three stores, two fulling-mills; 11 schools, 553 scholars. Pop. 2101.

SOUTH READING, p. t., Middlesex co., Mass., 11 m. N. Boston, 451 W. It contains two churches, a Congregational and Baptist, ten stores; one academy, 75 students; seven schools, 400 scholars. Pop. 1517.

SOUTHWARK, district of, a suburb of Philadelphia, and is a mile and a quarter long, and of the same breadth. It is governed by 15 commissioners, five of whom are elected annually, for the term of three years. It contains 780 acres of level alluvial land, and about 5300 dwelling houses, many of them handsome and commodious, but a large portion of them frame or brick buildings, two stories high. Most of the streets are paved and lighted, and have a watch. It is supplied with water by the Schuylkill water-works. The United States navy yard, and several ship and boat yards are on Delaware r. It has a lofty brick shot tower, and contained in 1840, five commercial and commission houses in foreign trade, capital $81,000; 259 ret'il stores, capital $362,109; nine lumber yards, capital, $243,000; two dyeing and printing works, one tannery, seven distilleries, two breweries, two potteries, one sugar refinery, four rope-walks, two printing-offices, one weekly, and one semi-weekly newspapers; six academies, 1050 students; 40 schools, 2070 scholars. Pop. 27,584. See PHILADELPHIA.

SOUTH WHITEHALL, p. t., Lehigh co., Pa., 87 m. E.N.E. Harrisburg, 160 W. Watered by Jordan and Cedar creeks. It contains a Lutheran church, four stores, five

grist-mills, two saw-mills, two tanneries; 11 schools, 335 scholars. Pop. 2290.

SOUTHWICK, p. t., Hampden co., Mass., 16 m W.S.W. Springfield, 103 m. W. by S. Boston, 359 W. Watred by a branch of Westfield r. It contains three churches, a Congregational, Baptist and Episcopal, two stores, two saw-mills, two powder-mills; one academy, 45 students, seven schools, 185 scholars. Pop. 1214.

SOUTHWELL, See LONDON.

SOUTHWELL, a market town and par. of England, co. Nottingham, Southwell and Scrooby liberty, on an eminence near the Greet, 13 m. N.E. Nottingham. Area of par. 639 acres. Pop., in 1831, 3385. The town is neat, well built, and well paved. It has a convenient suite of assembly-rooms, a theatre, &c.; but its principal building is the minster, or parish church. This, which is a large and magnificent edifice, is said to be, in part at least, as old as the time of Harold. Its extreme length is 306 ft., its breadth 50 ft., and the length of the transept 121 ft. The W. front has two lofty square towers, divided into seven stories. There is a low massive centre tower, and a chapter house on the N. side. The nave and transepts are Norman, the parts E. of the centre early English, and the chapter-house early decorated. There are some perpendicular insertions, particularly a very large W. window. Within the church are the monuments of five archbishops of York. The chapter-house, which is light and graceful, has 16 prebendal stalls. "The early English portions, which consist of the choir, its aisles, and small E. transepts, forms one of the best examples of this style in the kingdom. The whole of this church deserves the study due to a cathedral; and though it be not so varied in its style as some edifices, it claims attention for its purity and good preservation." (Rickman's Gothic Architecture, p. 221.) The ruins of an ancient palace of the archbishops of York, the favourite summer retreat of Cardinal Wolsey, stands in the park, and a part is now appropriated as a sessions house for the liberty. The general bridewell for the county is at Southwell, which has also meeting-houses for Wesleyans and Baptists, and a free school, with two scholarships at St. John's college, Cambridge. What little trade the town possesses is chiefly in malt, hops, and tan. The living of Southwell is a vicarage, worth £144 a year, in the gift of the prebendary of Southwell. (Excl. Rev. Rep.) Market-days, Saturday; fairs, Whit-Monday and October 21.

SOUTHWOLD, a seaport mun. bor., market town, and par. of England, hund. Blything, on an eminence on the E. coast of the co. Suffolk, about one m. N. from the mouth of the Blythe, 30 m. N.E. Ipswich, and 94 m. N.W. London. Pop., in 1831, 2079. "It consists principally of a long row of houses commencing near the bridge over the Bess creek, and extending S.W. to the edge of the cliff, besides numerous other houses with gardens, lying N. and S. of the principal street. Near the sea there are several good houses, but the others are chiefly of an inferior description. The town is not lighted nor watched; but the footpaths have been partly flagged by subscription, and the inhabitants are well supplied with water from several excellent springs." The guildhall is a handsome stone building, and a new jail was built in 1819. On the cliffs are two batteries, one of which has a parapet and six eighteen-pounders, but the other has only two guns. The church, built in the middle of the 15th century, is a fine edifice in the later English style, with a lofty tower and steeple of freestone intermingled with flint of various colours. The S. porch is extremely elegant; and above the clerestory roof is a light, open lantern. The interior is highly ornamented with gilding and carved work; and on the whole this is one of the finest churches in the county. The Independents, Baptists, and Wesleyan Methodists have each places of worship; the town, also, has three Sunday-schools and a national school. "The retail trade of Southwold is trifling; but a somewhat more important traffic is carried on in the importation of coal and the exportation of salt (which is made here), and herrings, and malt. The principal business, however is connected with the influx of visiters, who resort to the town as a watering-place during the summer season," and for whose accommodation there are baths, reading-rooms, a grand promenade, &c. (Mun. Bound. Rep.) The entrance to the haven is by the river, and vessels trading to the port land their goods at Black sheet quay, about 1 m. S.S.W the town. The superintendence of the haven is vested in commissioners; and it has been much improved by the erection of two piers at the mouth of the river, which has been made navigable to Halesworth, 8 m. W by N. the town. The borough of Southwold, which was incorporated in the reign of Henry VII., is governed under the Munic pal Reform Act by a mayor and three aldermen, with 12 councillors. A court of record is established for the recovery of small debts, and there is a court of admiralty for the regulation of the port, which is subordinate to that of Yarmouth. Markets on Thursday; fairs, Trinity-Monday, and August 24.

Southwold, or Sole Bay, E. of the town, is celebrated as the scene of the great naval engagement which took place on the 28th of May, 1672, between the combined English and French fleets under the Duke of York (afterwards James II.) and Marshall d'Estrées and a Dutch fleet under the famous De Ruyter. The action was most obstinately contested; the loss on both sides being very great and nearly equal. The Earl of Sandwich, who behaved with the greatest gallantry, was blown up along with his ship. The French suffered but little, in consequence, as is supposed, of their having received secret instructions to spare their ships. (*Yuns*, cap. 65.)

SPA, a town and watering-place of Belgium. prov. Liege, the borders of Rhenish Prussia, 14 m. S.E. Liege. It consists of "a cluster of neat white houses, thrown into the form of two or three irregular streets and open promenades, a whole embowered amidst trees and gardens, and overlooking on the N. and E. by a woody mountain range." (*Chambers's Tour on the Cont.*) It was at one period a place of great resort, and so highly distinguished for its mineral waters, that "Spa" became a common name for mineral springs and bathing-places wherever found. One of its greatest patrons was Peter the Great of Russia, who frequently visited it, and built the pump-room over the main spring, the only edifice the place having any architectural pretensions. It is now, however, greatly fallen off. The tide of fashion has set in favour of Carlsbad and Wisbaden; and it is not supposed to have more than about 3000 resident inhabitants. The Pouhon, or main spring, is a strong and active chalybeate, impregnated with carbonic acid gas, which gives it vivacity, and fits it for being preserved and sent in bottles to all parts of the world. There are several similar springs at from 2 to 5 from the town, at all of which there are pump-rooms, to some baths are attached. Spa has two libraries with supplied reading-rooms, a theatre, card-rooms, &c., a parish church, and a Capuchin convent; and during the season an English church is opened. The hotels and boarding-houses are very respectable, and are abundantly supplied with provisions. A manufacture of painted wood-boxes, &c., carried on in the town, employs a good many hands. (*Chambers's Tour in Ed. Journal, Murray's handb.; Cont's Belgium, &c.*)

AFFORD, p. t., Onondaga co., N. Y., 18 m. W. Syracuse, 154 W. by N. Albany, 334 W. Bounded S.W. by Onteles lake, N.E. by Otisco lake. It has five stores, a fulling-mill, two grist-mills, eight saw-mills, two tanneries; 10 schools, 700 scholars. Pop. 1873.

AIN (an. *España*, Span. *Hispania*. Fr. *Espagne*), an extensive and once powerful kingdom of S. Europe, occupying the E. and largest portion of its S.W. peninsula; between lat. 36° 5' and 43° 36' N., and long. 3° 20' E., and 9° W.; having N.E. France, from which it is separated by the Pyrenees; N. the Bay of Biscay; W. Portugal and the Atlantic; and S. and E. the straits of Gibraltar and the Mediterranean. Greatest length, E. to W., about 650 m.; greatest breadth, 550 m. The names, area, and population of its ancient and modern subdivisions have been stated as follows:
(See top of next column.)

The shape of Spain resembles that of a very irregular fan, the longest side of which faces the N. The coast on the whole, pretty regular, without those great and sudden indentations that characterise the shores of other countries, though an exception may be made as regards the coast of Gallicia, which is fringed with bays and islands, the principal among the former being the Bays of Arzos, Pontevedra, and Vigo; and among the latter capes Estaca, Ortegal, and Finisterre. The other capes are principally on the coast of the Mediterranean. Tarifa abuts on the strait of Gibraltar; and further capes Gata, Palos, La Nao, and Creux, the last being the extreme E. point of the peninsula. The surface is much diversified, and intersected with mountains; the whole may be described as a table land of considerable elevation, Madrid, the capital, being 2173 ft. above the sea, which is the average height of the towns in the interior. The chains of mountains, are pretty clearly defined, running from E. to W. through the peninsula. 1. The the PYRENEES (which see) not only divides France from Spain, but runs in a continuous chain parallel to, and at a distance from the N. shore upwards of 600 m. as far as cape Finisterre. The E. division is known as the Pyrenees properly so called, the W. portion consisting of Cantabrian mountains; the highest point in the former is the Netore on mount Maladetta (11,424 ft.), and in the latter the Peña de Peñaranda, S.W. of Oviedo (11,031). 2. A range extends W.S.W. from the Ebro, near Tudela, through Old and New Castile; Leon and Estremadura, and finally running S.S.W., through Portugal, to cape Roca, in Spain: the culminating point is the Sierra de Grados at the S.W. angle of Old Castile; but the average does not exceed 4500 ft. 3. A chain branching from that last mentioned divides the basin of the Ta-

Provinces.	Area in sq. m.	Population.	Totals.	
			Sq. m.	Population.
1. New Castile.				
Madrid	1,315	349,126		
Guadalazara	1,946	159,044		
Toledo	5,774	276,962		
Cuenca	11,201	234,582		
Ciudad Real	7,543	277,789	30,582	1,317,492
2. Old Castile.	30,892	1,317,492		
Burgos	}	224,607		
Logroño	} 7,474	147,218		
Santander	}	166,720		
Oviedo	3,998	434,835		
Soria	4,078	115,619		
Segovia	3,466	134,951		
Avila	2,560	137,903		
Leon	5,994	267,458		
Palencia	1,723	148,491		
Valladolid	3,279	184,647		
Salamanca	5,026	210,314		
Zamora	3,562	158,425	41,585	2,296,181
3. Galicia.	72,447	3,643,673		
Coruña	}	435,670		
Lugo	} 15,897	357,272		
Orense	}	319,088		
Pontevedra	}	360,002	15,897	1,471,288
4. Estremadura.	89,344	5,191,855		
Badajos	} 14,829	316,822		
Caceres	}	231,368	14,829	548,390
5. Andalusia.	102,673	5,989,675		
Seville	}	367,376		
Huelva	} 8,989	133,470		
Cadiz	}	324,298		
Jaen	4,451	299,918		
Cordova	4,159	315,459	17,599	1,407,354
6. Granada.	120,272	7,077,589		
Granada	}	376,974		
Almeria	} 9,682	234,789		
Malaga	}	338,442	9,682	950,305
7. Valencia.	129,894	8,027,754		
Valencia	}	451,595		
Alicant	} 7,683	318,444		
Castellon-de-la-Plana	}	190,059		
Murcia	} 7,877	390,694		
Albacete	}	229,713	15,560	1,430,805
8. Catalonia.	145,454	9,439,342		
Barcelona	}	449,479		
Tarragona	} 12,180	233,477		
Lerida	}	151,929		
Gerona	}	214,150	12,180	1,041,432
9. Aragon.	157,634	10,480,764		
Zaragoza	}	304,823		
Huesca	} 14,728	214,974		
Teruel	}	214,989	14,728	734,686
	172,362	11,724,449		
10. Navarre.	2,450	281,794	2,450	281,798
	174,810	11,456,177		
11. Guipuscoa.				
Alava	1,082	67,523		
Biscay	1,367	111,436		
Guipuscoa	622	104,491	2,971	288,450
	177,781	11,738,627		
12. Balearic Islands.				
Palma	1,757	229,197		
Canary Islands	3,220	159,960	4,977	429,147
	182,758	12,168,774	182,758	12,168,774

gus from that of the Guadiana; the central portion, S. of Toledo, called the Sierra de Guadalupe, attains a height of 5110. 4. A range, called the Sierra Morena, runs along the S. border of La Mancha, in New Castile, which, though not continuous and of no great extent, forms the water-shed

(See Essay on the Phys. Geog. of Spain, in *Laborde, Itinéraire d'Espagne*, vol. v. last ed. *Bruguière, Orographie de l'Europe; Berghaus, Erdbeschreibung*, p. 316—318, *Antillon*, p. 226—270. The altitudes are given exclusively from Bruguière.)

The mountain-chains now described regulate the course of the principal rivers, some of which are of great extent, and have numerous tributaries. Immediately S. of the Pyrenees is the Ebro, which, rising on the Asturian range, near

Reynosa, runs S.E. through a succession of narrow valleys, receiving its chief tributaries from the S. face of the Pyrenees, and flows into the Mediterranean about 20 m. below Tortosa: its entire length somewhat exceeds 400 m., and the area of its basin is estimated by Berghaus (*Erdbeschreibung*, p. 237) at 25,800 sq. m. Among the other rivers flowing into the Mediterranean are the Guadalaviar and Jucar, falling into the bay of Valencia, and the Segura in Murcia; the rest are unimportant. Five large rivers run westward into the Atlantic ocean. The most N. of these is the Minho, which rises in the Asturian mountains, and running first S. and then S.W., enters the sea a little below Caminha, after a course, including its windings, of about 150 m. The Douro rises in the mountains of Old Castile, a few miles N. of Soria, and takes a generally W. course, by Aranda, Tordesillas, and Zamora, as far as Miranda, where, turning southward, it forms a portion of the boundary of Portugal, through which it flows westward into the sea close to Oporto: its length is estimated at 500 m., and the country drained by itself and tributaries somewhat exceed 34,000 sq. m. The Tagus has its source in the Sierra de Albarracin, in Arragon, whence it flows W.S.W. by Aranjuez, Toledo, Talavera, and Alcantara, to the confines of Portugal: it then turns S.S.W., and after expanding into a fine estuary, enters the Atlantic a little below Lisbon, built on its N. bank. The Tagus has numerous important tributaries, the chief of which are the Henares, Alberche, Alagon, and Zatas, the last being in Portugal: the extent of its basin is estimated at 29,000 sq. m. The Guadiana, rising in La Mancha, runs first N.W., then W. as far as Badajoz, where it curves southward, and enters the sea at Ayamonte, after a course of 420 m.; it has several pretty large tributaries, and drains an area of about 25,600 sq. m. The Guadalquivir, which, with its tributaries, drains a large portion of Andalusia, rises in the Sierra de Alcaraz, and taking a S.S.W. direction by Andujar, Villafranca, Cordova, and Seville, turns southward, and after crossing a low, unhealthy swamp, enters the Atlantic at San Lucar, after a course of 330 m.: its largest affluent is the Genil, and the area of the entire basin is nearly 18,000 sq. m. But, with the exception of those portions of the Douro and Tagus within the limits of Portugal, these rivers, notwithstanding their length, offer few advantages for navigation, owing to the rocks, shallows, and falls with which they are encumbered.

The Ebro has, however, been made navigable to a considerable extent by means of the canal of Aragon; and the channel of the Tagus is also in course of being improved, so as to make it accessible for boats as far as Aranjuez. Vessels of 100 tons ascend the Guadalquivir, within about 8 m. of Seville. The rivers on the N. side of Spain are comparatively insignificant, owing to the closeness of the Asturian mountains to the sea: of these, the Bidassoa, forms the dividing line between France and Spain. There are no lakes of any considerable size, though in the Pyrenees and other chains there are several small mountain-lakes. Swamps and morasses, however, are both numerous and extensive: the principal being the Gallocante, in Aragon, the Nave, near Palencia, and the Lagunes of Palomeras and Caldera.

A central band of granite and mica-schist stretches along the Pyrenees from the Mediterranean to the bay of Biscay, flanked successively by beds of secondary and cretaceous formations; the primary rocks, however, are by no means so extensive as in the Alps, and do not extend westward beyond the Bidassoa, all the mountains of Asturias and Galicia being of the sandstone and carboniferous limestone that form the lower parts of the main chain. The lofty range that divides the two Castiles, and forms the watershed between the Douro and Tagus, consists chiefly of granite and other primary rocks, which pass eastward under the sandstone, forming the lofty uplands of Soria, in Old Castile; it is flanked on both sides by sandstone and limestone; but in New Castile are extensive beds composed of the *débris* of primitive rocks associated with marls and gypsum, the marly subsoil being remarkable for the fertility of the surface, whereas the gypseous districts are remarkable for their barrenness and dismal appearance, such as is exhibited in the neighbourhood of Madrid. The Sierra Morena also exhibits a large proportion of primary rocks, partly covered by secondary and other rocks, with the nature of which we are but little acquainted. The Sierra Nevada is a mass of mica-slate and serpentine, flanked northward by secondary and more recent rocks, containing some of the richest marbles of Spain, many of which adorn the churches of Granada and Seville; the S. side forming the deep valleys of the Alpujarras, is principally of secondary limestone resting on slate (highly metalliferous), greenstone, and blue limestone. The limestone strata of the Sierra de Gador are remarkable for lead mines, which are extremely rich, and sufficient to supply the whole of Europe with this mineral for many centuries to come. With respect, indeed, to the mineral wealth of Spain, there can be little doubt that

it is very great, and by no means exhausted by the workings of the ancients. There are valuable copper-mines in the Sierra Morena and the Alpujarras, and near Teruel in Aragon. Quicksilver is found at Almaden, on the N. side of the Sierra Morena, and graphite occurs near Malaga, in the neighbourhood of Malaga. Graphite is found also in the high Pyrenees, scarcely inferior to that of Cumberland, but wholly unwrought. Sulphur occurs in several parts of Valencia, and saltpetre, alum, and salt are found in many parts of the country. Iron and coal abound in the mountains of Biscay and Asturias, and are wrought to a considerable extent; besides which there are extensive iron mines, with smelting-houses, &c., at Marbella, near Malaga, and in the Sierra Morena, near Pedroso. Coal occurs also in large seams throughout Aragon and Catalonia, as well as on the Guadalquivir near Seville; and traces of the same mineral have been discovered near Malaga.

"The *soil* of the Peninsula exhibits great diversities. The central region consists for the most part of arid, unsheltered plains either of sand or gypsum, intersected with lofty mountains, which reflect with intolerable fierceness the scorching heat of summer, and sharpen into more intense keenness the intense cold of winter. The lower region of the coast, sloping gradually towards the sea, is broken into an alternation of mountains and valleys, producing the most agreeable variety, and presenting a pleasant contrast to the bleak and barren sameness which characterises the central region. It is every where fertile, or may be rendered so by irrigation." (*Foreign Quarterly Review*, No. IX. p. 152, written by the author of this article.) The alluvial soil of Old Castile is tolerably productive, even without irrigation; New Castile has every variety, from the gypseous marl composing the poor soil about Madrid, to the red marl of Guadalaxara and the limestone of Arganda. The valleys of the Sierra Morena, and the whole of Estremadura, have a soil formed of detritus from primary rocks, and cannot be excelled in beauty and natural fertility. The soil of Andalusia is chiefly of marl and clay interspersed with red sandstone marls, and it is by irrigation only that it can be made productive. The Vega of Malaga, however, is naturally of surprising fertility, owing partly to the long establishment of irrigation, but partly, also, to the fact of its being in a great measure alluvial. Valencia has a poor ungrateful soil, yielding crops only by forced cultivation and the use of water. In Catalonia and Aragon the detritus of limestone is found alternating with fine red marls and waste tracts of gypseous marls, similar to those near Madrid. On the whole, the valleys of the Sierra Morena and the Alcarria, the provinces of Toledo and Guadalaxara, the Vega of Malaga, and the country between Gibraltar and Cadiz, would probably repay the labours of agriculture better than other parts of Spain." (*Cook's Sketches in Spain*, ii. 338, 339, in which are many valuable remarks on its geology and national economy.)

The *climate* of Spain is greatly diversified, being modified by the physical conformation of the country. The temperature of the air always varying less near the coast than in the interior, is much more equable in the maritime than the other provs. On the N. and W. coast westerly winds prevail, and, being loaded with moisture from the Atlantic, discharge abundant rains in winter and spring. The coast of the Mediterranean has a calmer atmosphere, with a prevalence of E. winds, and a temperature generally rising above 57° Fahr., and seldom descending so low as 39°. Winter, indeed, is almost unknown on a coast sheltered by the elevated land of the interior, and warmed by the rays of a cloudless sun; while the heat of summer is very great, and would be all but intolerable, were it not lessened by the sea breeze, which lasts during the greater part of the day. On the plateau of Castile, the mean height, of which, according to Bruguière, is about 1960 ft. above the sea, heat accumulates much more slowly, and it is only during the month of July that the temperature ascends as high as 79° Fahr. In August, the mornings and evenings begin to be cold; and in winter the severity of its climate forms a very striking contrast with the heats of summer. Except in the N. provs., the climate of Spain is every where remarkable for dryness: a freedom from rain, and a cloudless sky bring advantages that may generally be counted on; but the dryness occasionally becomes so excessive that the crops are dried up, vegetation destroyed, and men and animals die miserably of thirst. (*For. Quart. R.*, ix. 153.) Two kinds of winds are very troublesome in Spain. The *galego* is N. and N.W. wind, which comes down from Galicia, is very cold and piercing; causing, besides other diseases, painful affections of the eyes, often ending in blindness, which is very common in all the more elevated districts. This ophthalmia, however, is attributed by some writers to the vast quantities of minute nitrous particles blown up from the waste lands, and held in suspension by the wind. (*Ponze, Anales de Midi*, p. 5-6.) The S. prov. are visited by the *solano*, which, like the *sirocco* of Italy, relaxes the system, s giddiness, inflammation, and even death, extreme and sudden variations, the climate of ateau is far from healthy. The Madrid chollangerous, and often fatal to strangers; besides is a general tendency to pulmonary consumper diseases of the lungs. Scrofulous diseases e common than in Russia, and epilepsy is by ... The yellow fever, which often ravages the s created much discussion among physicians, om treat it as epidemic, while others are of it is brought to Spain by infection. Insanity · or less in all parts, but especially in the provs. the Mediterranean. (*Faure*, p. 69–84; *and* t. 156–158.)

il products of Spain are rich and various, and ily be made the source of vast wealth. By a idence Spain itself was to the ancient what its ssessions have been to the modern world, the cee of the supply of the precious metals. It is loubtful, however, notwithstanding the numete to the contrary, whether the Carthaginians ver discovered any mines of gold and silver in more probable opinion seems to be, that the ally obtained from washings; and that the silvas by far the more abundant and important extracted from the lead, which was then for the sake of the silver, in vast quantities. *eografia*, 149.) The mine of Guadalcanal, ding to Cook (ii. 73), is the only one of silver vrought, was, with that of Cazalla and others, rg after Spain had been evacuated by the Rollon, loc. cit.)

aluable of the existing Spanish mines are those inada; and the supplies obtained from them du20 years have been so large that they have he abandonment of several less productive er countries, and a considerable fall in the price quicksilver mines of Almaden, in La Mancha, emely productive, and supply, indeed, most nicksilver imported into this country, and large r the New World. Exclusive of innumerable there are mines of rock salt at Migranilla, in and the mountain of Cardona: in Catalonia, V. Monserrat is a vast and solid mass of pure The iron trade will be afterwards referred to; intimony, and other minerals are found in vathe country, with every variety of marble, and ilding stone. There can, indeed, be no doubt, government capable of developing the nationthe mineral wealth of Spain would be found if not superior, to that of most other countries. *and Animal Products.*—The wheat of Spain, ry various qualities, is generally excellent, and aid to be the best in Europe. In some districts grown is insufficient for the consumption, the ing made up from the surplus produce of other importation, though, owing to the badness of d the consequent difficulty and cost of carriage, a great difference in the prices of corn in marent distance from each other. Wine is raised throughout the country; and the coast districts ota, Malaga, Benicarlo, and Alicant, furnish ies for exportation. The wine of the interior, om exported, in consequence of the bad roads of transport, is sometimes of good quality; and le Peñas, in La Mancha, in particular, a dry red tained a high reputation for its superior flavour '. Grapes are also exported, both in a fresh ste. Among the other productions of the soil ley, maize, rice, oil, sugar, hemp, flax, *esparto* ton, saffron, barilla, honey, and silk, with all in vegetables, and some even of those of the ons.

of the S. are lemons, bitter and sweet oranges, es, dates, olives, almonds, and pistachio nuts; s, cherries, peaches, and chestnuts are grown ovs. Immense quantities of hazel nuts are exCatalonia, and the fruit of the carob-tree is ling cattle. On the Pyrenees, Asturian mounern Morena, &c., are luxuriant forests; but, on Spain has less timber than any other extensive urope; a circumstance owing, not to any inapsoil for the growth of forest trees, but to an ini inexplicable prejudice of the people against i are mercilessly cut down or destroyed before any considerable size. Indeed, so universal is ity in the central provs., that the most rigorous e necessary to preserve the avenues of Aranjuez n destruction; and all statutes for the encourplanting have signally failed of their object. . *Review,* ix. 154, 155.) Spain has eight varie; among which are the evergreen oak, or *Quer-*

cus *ballota*, which has edible acorns, in taste resembling chestnuts; the cork oak (*Q. suber*) and the cochineal oak (*Q. coccifera*), on which is found the false cochineal, yielding a fine crimson dye. The true *Quercus robur*, however, which furnishes the best materials for ship-building, scarcely exists, except in the N. provs. (*Cook's Sketches*, ii. 242–255.) Among the other forest trees may be enumerated tamarisks, pines, beeches, chestnut trees, nut trees, firs, poplars, and the sumach (*Rhus coriaria*), the bark of which is used for tanning.

Among the animal products of Spain, the horse is entitled to particular notice. The Arabs, when in possession of the country, stocked it with their finest breeds; and though the race has degenerated, it still shows many of the points by which it was originally distinguished. In beauty, grace, and ducility, the horses of Andalusia are said to be superior to those of England; but it may be doubted whether they are equal to the same amount of labour. In fact, the number of good horses is rapidly decreasing in Spain, chiefly owing to the preference given to mules for domestic and agricultural purposes· the importation of horses to improve the breed, and the exportation of colts are alike forbidden, and " the number of horses bred at present is quite inconsiderable, notwithstanding the decrees, premiums and encouragements of every kind that have been offered by government. The celebrated breed of the sovereigns of Spain at Cordova is nearly extinct; in the Serrania de Ronda (once the Cleveland of Spain) only miserable animals, called *serranos*, are now reared; the wealthiest Andalusian nobles have only two or three indifferent saddle-horses, and there is scarcely a horse in the whole country fit for the draught of artillery." (*Cook*, ii. 50–61.) Great numbers of mules are bred in Old Castile, being sent to come to their full size in the rich pastures of Estremadura, whence they are supplied to the rest of Spain. The asses are very different animals from those seen in England, being of a large size, carefully bred, and in strength, docility, and sure-footedness, nearly equal to the mules. Cattle are small, and not of fine appearance. The bull of Andalusia is found wild in the Sierra Morena. Hogs are bred in vast numbers, and those which feed on acorns are celebrated for the delicacy of their meat, which is, perhaps, unequalled. Sheep, however, are the favorite stock of Spain, and are everywhere raised in considerable numbers (see *post*); nor are there wanting wild animals, such as wolves, lynxes, wild cats, wild boars, and foxes. The bear, which used to be common two centuries ago, is now found only in the Pyrenees. Monkeys are met with in the Sierra de Ronda, besides which there are various reptiles, as chameleons, lizards (some 3 ft. in length), vipers, and snakes. Among the birds may be noticed several species of vultures, falcons, owls, ravens, magpies, Cornish choughs, partridges, quails, bustards, and plovers. (*Cook's Sketches*, i. 58–62, 254–287; *and For. Q. Review,* ix. 156.)

Agriculture.—" No country in Europe," says Laborde, " is so generally fertile as Spain, or has equal advantages at all seasons of the year. Spain, after its conquest by the Romans, became the granary of the Roman empire. Under the Goths vast canals and sluices were formed for irrigating the land, and the amount of corn then raised was sufficient not only for the home supply, but also, to a considerable extent, for exportation. Agriculture under the Moors was in a still more flourishing state; for when they invaded the country they carried with them their methods of husbandry, broke up the uncultivated lands, augmented the number of plantations, carried the art of irrigation to a degree now scarcely attainable, introduced the culture of rice, and greatly improved the breed of horses; in fact, every kind of production was increased under their improving hands; and the era of their expulsion designates the epoch of the decline of agriculture. The Spaniards, thus deprived of the assistance of the Moors, were compelled to till the land themselves; but for such pursuits they possessed neither talents, activity, nor patient industry. Hence the whole system fell into a state of languor, from which it has, owing to several causes, never recovered." (*Laborde, vol. iv.*)

We have quoted this passage, because it states, in a few words, what has long been the popular opinion in regard to the ancient as compared with the modern state of Spain. We believe, however, that it is wholly erroneous. The fertility of the country has been greatly exaggerated; and we much doubt whether her agriculture was ever in so advanced a state as at this moment. A great portion of Spain is, owing to the heat of the climate and the want of water, wholly unfit for husbandry; and she has in consequence of the frequency of droughts, been at all times subject to the most destructive famines. Owing to the numerous ridges of mountains by which she is intersected, her internal commerce has always laboured under the greatest difficulties; and there is no evidence that her artificial communications, that is, her roads, canals, bridges, &c., were at any former period in a more improved state than that in which we now find them. Owing to vicious institutions, bad government, and other causes, Spain has, for a lengthened period, continued stationary, or made little progress, while other nations have advanced with giant steps in the career of improvement; but there is no real foundation for the prevalent notion of her having been comparatively well cultivated, rich, and industrious previously to the expulsion of the Moors, or in the reigns of Ferdinand and Isabella and Charles V. Capmany, in his *Questiones Criticas* (cap. i.) has proved, beyond all controversy, that there were in the 15th and 16th centuries the same complaints of the wretched state of agriculture, of the idleness of the Spaniards, of their contempt for industry and the useful arts, and their dependence on foreigners, that are still made against them. It is needless to say, that without tranquillity and good order there can be nothing like a flourishing agriculture. But at the very time that it is said to have been most flourishing, that is, previously to and during the reign of Ferdinand and Isabella, the Spanish historians represent the country as a prey to rapine, outrage, murder, and every sort of violence and disorder. Indeed, so early as the 13th century, the principal cities of Aragon and Castile had formed an association, called the *Santa Hermandad* (Holy Brotherhood), for their mutual protection against the robbers and plunderers with which the country was infested; and during the reign of Ferdinand and Isabella this institution was still further extended. (*Robertson's Charles V.*, vol. i., note 36.) And if these facts were not enough to demonstrate the entire worthlessness of the statements as to the flourishing state of agriculture in Spain previously to the expulsion of the Moors, the organization of the laws respecting the *mesta* (migratory flocks) would sufficiently evince the truth of what has now been advanced: for had the country not been at the time in a half occupied semi-barbarous state, every one must see that the oppressive privileges conferred on the owners of the sheep never could have been carried into effect, or had any practical existence.

Having thus briefly disposed of the apocryphal statements as to the former flourishing state of agriculture in Spain, we have now to inquire into its present state, and the circumstances to which its long-continued depression are principally to be ascribed.

With the exception of a few districts which have peculiar facilities for irrigation, agriculture is in the most backward state imaginable. "Great part of the land is not tilled, and that which is tilled is executed in so careless and slovenly a manner as to produce a starved crop of corn in spots where they might command the most abundant harvests. The corn is usually choked up with stones, filth, and weeds of every kind." (*Clarke's Letters*, p. 285.) Generally speaking, tillage farms are small, and rents low; but owing to the exorbitant taxes, and other expenses wholly exclusive of rent, the farmers are wretchedly poor, and when they require money are obliged to obtain it at exorbitant interest, by mortgaging their crops. The system of letting land is very various, money rent being taken in some parts, while in others the rent consists of a stipulated quantity of the produce, and in others the *métayer* system prevails. Generally, however, large estates are not let out in farms, but are managed by agents, who, for the most part, are totally ignorant of the business of agriculture, and whose great object is to squeeze out of the land all that it can be made to produce by the most compendious processes. Farm houses are rarely seen, except along the E. coast. The farmers live in huts of the meanest construction, crowded together in villages, so that farm buildings, often so expensive in other countries, cost almost nothing. Spring-corn is generally sown on the ground before it has been turned up, and is still covered with the winter weeds; and is then ploughed down, or rather scratched in with a miserable instrument, and left to nature. Owing to the dryness of the climate, this is a less ruinous system than might have been supposed, for when the heat sets in the corn ripens, while the weeds perish. When ripe the corn is gathered in the field, and after being thrashed or trampled out by mules and asses, is left in heaps on the ground till it be sold. The corn speculators of Castile preserve grain in *silos*, or subterraneous caves, sometimes for five or six years, or till a market opens for it. Public granaries, or *positos*, are, also, established in most districts, where corn may be warehoused till it can be disposed of. The implements of husbandry are of the rudest description; it is not uncommon in the S. to see men returning from plough seated on a mule, to the sides of which their whole apparatus is tied; the use of fanners is unknown, except in the neighbourhood of seaport towns, to which they have been imported from England; corn is winnowed by throwing it up in the air, and it is more frequently ground by hand, than by either wind or water mills. (*Cooke's Sketches*, ii. 40–42.) Land is not supposed to yield to the proprietors more than 1½, or at most 2 per cent., for, when the tenant has paid tithe, *primicias*, and other taxes, little more remains than half the produce,

to pay both rent and labour. It is exceedingly difficult to estimate the rent of land by the English acre, from the great uncertainty and irregularity of the measures. The term *fanegada* is used to indicate the extent of land on which a *fanega* of wheat may be sown, an extent which varies in every village: this quantity of land, whatever it may be, lets, according to circumstances, at from 12s. to 34s., the average value of a fanega of wheat being about 3s. 6d. Vines and olive-lands are measured by the *aranzada*, an equally vague standard. The wages of farm labourers average about 13d. *per diem*; or, if boarded with their masters, from 6½d. to 10d. But, though tillage has been greatly extended during the present century, it is still true that in most parts of Spain no improvement has been made during the last 150 years; but it is otherwise in Biscay, Navarre, and Aragon. In the first hoe and spade husbandry pretty generally prevails, and every inch of arable ground near the roads seems to be carefully cultivated. The wheat raised in Biscay perhaps exceeds the consumption of the district, and considerable crops are also raised of rye, maize, barley, and oats. In Leon, Castile, and Andalusia, agriculture, which is in the most degraded state, is confined to the growth of wheat.

The most careful cultivation is found in the *huertas*, or irrigated lands of Granada, Murcia, and Valencia. These tracts, indeed, are considered as the gardens of Spain, and abound not only with every variety of fruits, but all kinds of vegetables and plants, useful either as food or materials for manufactures. The mild red pepper raised in the *huerta* of Murcia is celebrated all over Spain, and forms a considerable article of trade with the interior. Rice is the chief product of Valencia. The sugar-cane of Granada and Valencia is as good as that of the West Indies; but it is cultivated at much greater expense, and its growth has, in consequence, been almost wholly abandoned. Considerable quantities of corn are raise in different parts along the S.E. coast. Mulberry trees are carefully cultivated in the E. provinces; those of Murcia and Valencia are white, those of Granada black. In the cultivation of vines pains are not used; but the cuttings are planted, and, not being permitted to attain any great height, gradually form thick and very stout stocks. Espaliers also are numerous, especially in Andalusia, and the grapes on these vines attain an extraordinary size, the bunches often weighing from 12 to 14 lbs. The rich level lands produce the largest quantities of wine, but here, as elsewhere, that raised on gravelly soils on the hilly slopes is the best. The quality of the wine varies greatly in different districts; but it may be said with truth that, except the wines of Xeres, Rota, Malaga, Alicant, and Benicarlo, which are intended for exportation, few of the Spanish wines are equal even to those of third-rate quality in France. Being very generally kept in skins, smeared with pitch, they acquire an *olor de bota*, or peculiar taste, and a flavour not disliked by the natives, but very disagreeable to foreigners.

It appears, from the official returns published in 1803, which, though not to be altogether depended upon, are sufficiently accurate to give a just notion of the general state of the country, that, taking the extent of Spain at between 18,900 and 19,000 square leagues, the surface was distributed nearly as follows:

	Leagues.
Cultivated lands and fallows	4,200
Pastures and commons	11,400
Forests and copses	1,300
Mountains and rivers	1,300

It is certain, from the increase of population, and the nearly total cessation of importation since 1842, that the proportion of cultivated lands must have increased considerably in the interval, though at this moment they are still but more, perhaps, than about a fourth part of the entire surface. The Pyrenees, the hilly parts of Biscay and the Asturias, the vast plains of Andalusia, the two Castiles, Estremadura, and Leon, are almost wholly in pasture; and some parts the traveller may journey for many miles without seeing either a house or an individual. In point of fact, however, half the pastures really consist of heaths, and neglected tracts covered with thyme and other wild herbs, that at present are next to worthless. There are few of the irrigated meadows, and hay is seldom or never prepared for fodder. Indeed, notwithstanding the preference given to pasturage, and the privileges that have been long enjoyed by the migratory flocks, it is not supposed that the stock of sheep in Spain at this moment (1841) exceeds 12,000,000 or, at most, 14,000,000; whereas, in England, which has not one fourth part of the pasture and waste land that belongs to Spain, the stock of sheep is certainly not under 25,000,000, while that of horses and cattle is proportionally great.

The Spaniards distinguish their sheep into the *ordinary*, or those who remain in the same place during the year;

ad the migratory, or those who move from place to place, be latter, or *transhumantes*, consisting chiefly of the menos, or fine-woolled breeds, are depastured during winter i the vast plains of Andalusia, Castile, Leon, and Estremadura; and are driven in summer to the nearest mountains. These migratory flocks are collected for their journeys in large bodies of 10,000 and upward, called *mestas*, eir peregrinations being regulated by a peculiar code of ws, and by immemorial custom. It is obvious that this ignatory system has originated in natural causes; and at, in fact, it is an important branch of the rural economy Spain. In winter, when the mountains are covered with ow, the plains are in the greatest verdure and beauty; d in summer, again, when the herbage of the plains is ithered and burnt up by the heat and drought, the pastures the Sierras, and other mountain tracts, are in a state of mparative luxuriance. Nothing, therefore, can be more tural than this shifting of the flocks; it is for the mutual erest of the occupiers of the hills, and those of the plains, d no doubt has prevailed in Spain from the remotest antiquity, and will necessarily continue to prevail.

The laws and customs, however, under which the migrans of the flocks are conducted have been, for a lengthened period, singularly inexpedient and oppressive. It appears it, about the middle of the 14th century, the depopulation large tracts of country by a pestilence gave a consideraextension to pasturage; and enabled the proprietors of migratory flocks to usurp certain privileges, which they re since succeeded in maintaining. (*Townsend*, ii. 61.) us they are not only allowed to drive them over village tures and commons, but the proprietors of such cultivated lands as lie in their path are obliged to leave for them 'ide path, and, which is still worse, no new enclosures be made in the line of their migrations, nor can any i that has once been in pasture be again cultivated till as been offered to the mesta at a certain rate! In conuence of the perverse arrangements, disputes, which frently terminate in bloodshed and murder, are perpetually ng place between the herdsmen and those through whose is the flocks have to pass. On the whole, however, we ine to think that the mischiefs said to be entailed on in by the laws and customs in question have been a good exaggerated. As already seen, the migration of the ts is essential in Spanish rural economy; and it does appear, were government to set resolutely about the er, that any insuperable difficulty would have to be untered in defining and fixing the roads to be taken by locks, and in otherwise regulating their migrations, so prevent them from being injurious to third parties. may be worth mentioning, that Capmany ascribes the great improvement in the wool of Spain to the introon of a flock of sheep from England in 1394, being a on of the dowry brought by Catherine, daughter of the of Lancaster, to her husband, the eldest son of the of Castile. (See *Questiones Criticas*, 9; *and Memorias rricas sobre la Marina, Comercio, &c., de Barcelona*, 36.)

e low state of agriculture in Spain may be ascribed r to physical and partly to moral causes. At the head e former must be placed the heat of the climate, and ridity of the soil. Most part of the rivers with which ountry is intersected run in deep beds, and are but available, except in a few favoured localities, for purof irrigation. Probably, however, moral have had nore influence than physical causes in retarding the ass of agriculture in the peninsula. At the head of rmer must be placed the vast extent of the lands beg to the nobility, clergy, and corporations. Mr. Townmentions, that the estates of three great lords—the of Osuna, Alba, and Medina Celi—cover nearly the of the immense province of Andalusia, and several other provinces are hardly less extensive. (ii. 238.) vast possessions are uniformly held under strict enand, speaking generally, are all managed by stewards, s only to remit money to their masters, who are atly in embarrassed circumstances. The younger es of the great families, though they inherit all their nherit little or none of their wealth. They are, for et part, exceedingly ill educated; and when not emin government service, pass their days in a state of dependence. It is singular, notwithstanding their se possessions, that the Spanish grandees have little aste for a country life, or for the improvement of tates; and the fact is, that from the one end of the ia to the other there is no such thing as a fine coun-. The great estates belonging to the corporations, is, are held in common; and, in consequence, are or almost wholly, in pasture. Luckily, however, e estates that belonged to the church have been ted during the late revolutions; and their sale and will materially increase the number of smaller proand give a powerful stimulus to improvement;

and a stop has also been put to the practice of entailing. The interruption given to labour, by the immense number of religious festivals and saints' days, has also been exceedingly injurious to agriculture and all sorts of industry.

The Spanish character is also unsuitable to success in agriculture and manufactures. During the prolonged struggle with the Moors, a taste for daring adventures, and for an irregular predatory mode of life, was widely diffused throughout the nation; and the discovery and conquest of America, which occurred nearly at the same time that the power of the Moors was annihilated by the conquest of Granada, afforded a new and boundless field for the exercise of the peculiar taste and talents formed in the Moorish wars. In addition to the means thus afforded of arriving at wealth and distinction by a more compendious and less laborious, though more hazardous, route than that of sober industry, those honorary distinctions of which the Spaniards are extremely fond were conferred only on those who followed the profession of arms, and who could show that their ancestors had not degraded themselves by engaging in the debasing pursuits of agriculture, manufactures, and commerce! And while the higher and more aspiring classes were thus led to regard the useful arts with contempt and disdain, the multiplication of convents and such like establishments afforded the means of keeping a vast number of individuals in pampered idleness. We need not, therefore, wonder at the repeated complaints that have been made by native and foreign writers of the pride and laziness of the Spaniards. (See *Capmany, Questiones Criticas*, p. 46-49, &c.) What else could be expected in a country where agriculture and the useful arts have, for a lengthened period, been looked upon as mean and sordid, and below the attention of a gentleman? The heat of the climate was enough, of itself, to enervate the inhabitants, and to render them indolent; and when we add to this the powerful influence of the causes now shortly glanced at, with the want of leases and roads, and the oppressiveness of taxation, need we wonder at the backward state of agriculture and of the other useful arts in Spain?

There are several societies in Spain, assuming the title of "Friends of the Country," for the encouragement of agriculture and the arts; most of them were founded in the reign of Charles III., and were warmly patronised by Campomanes, the most enlightened minister of whom Spain has to boast, and by Count Florida Blanca. Hitherto, however, they seem to have rendered but little service, if we except that of Madrid, to whose exertions the famous Memoir of Jovellanos (*Informe de la ley Agraria*) is principally to be ascribed. The reader will find this memoir in an English dress in the 4th volume of the translation of the *Itinéraire* of Laborde.

Manufactures.—It might have been expected, from the abundance of wool and silk in Spain, and her extensive colonies in America, that her manufactures would be in a comparatively flourishing state. This, however, is not, nor has it ever been, the case. Capmany, and other able writers, have shown that the statements as to the flourishing state of manufactures in Spain, in the 14th and 15th centuries, have no better foundation than those respecting the flourishing state of agriculture, and the magnitude of the population, at the same period. Some of the circumstances that have contributed to depress agriculture have also contributed to depress manufactures; but they have also been affected by others of a peculiar description, among which may be specified the oppressive influence of the *alcabala*, and other taxes (see *post*), corporation privileges, monopolies on the part of government, and the want of competition and emulation through the exclusion (in as far as practicable) of foreign manufactured goods. Catalonia, Biscay, and Valencia are the most industrious provinces, and in them manufactures are most advanced. Those of silk and cotton, especially the first, are carried on to a considerable extent in Barcelona, Valencia, and other towns; but, though the fabrics be excellent, the colours are wretched. The blonde mantillas of Almagro, in La Mancha, are, perhaps, the best of the Spanish manufactured articles. The finest broadcloths are made at Alcoy, in Valencia; and coarse cloths (*paño pardo*) are extensively manufactured in Catalonia, and in various districts throughout the country. But, with the exception of silks, all the woven fabrics produced in Spain, whether woollens, cottons, or linens, are at once badly finished and enormously dear: even the coarse, hardspun *mantas*, that serve the muleteers for cloaks and blankets, fetch prices that would astonish the peasantry of England and France. In the N. provinces tanning is the most important branch of manufacture, furnishing the principal supply of leather for the interior: the business was introduced by, and is chiefly in the hands of, refugee Basques, from the French side of the Pyrenees. The few tan-works of Andalusia are mostly in the hands of Englishmen. The manufacture of paper and hats has been established with some success; and there are numerous potteries, though the products be principally of coarse quality. In Valencia and Catalonia, however, finer articles are made; but even there the art is only in its infancy. There is a royal porcelain manufactory at Madrid, on the plan of that of Sèvres, occasioning, like its prototype, a constant loss. Soap is made on a somewhat extensive scale in various parts of Spain, that of the best quality being exported. In Biscay the production and manufacture of iron has been for many years conducted with considerable activity; and it is probable that the depression occasioned by the late civil war, of which Biscay was the principal seat, has already been removed. It is impossible, however, that the iron trade of Biscay, how abundant soever the ore, can rise to any great importance; since wood fuel is scarce, and coal, being at a considerable distance, and the roads extremely bad, is little used; while English coal, which might be procured at about one third the price, is strictly prohibited. Still, however, to some extent, at least, in almost every village of the province, the iron-ware manufacture is carried on. Horse-shoes and nails, coarse locks, guns, and bedsteads are the leading articles with which the Biscay manufacturers supply the interior: large copper utensils are also made on a considerable scale in this part of Spain. Muskets, pistols, and sabres are manufactured by the government in Valencia; and several minor establishments exist at Saragossa, Barcelona, Malaga, Cadiz, and Seville. The sabre manufacture of Toledo, once so highly celebrated for its finely tempered sword-blades, has all but fallen to decay, only 50 workmen being employed at the time of Inglis's visit. (*Spain*, i., 312.) The manufactures of saltpetre and gunpowder, brass cannon, tobacco, porcelain, tapestry, and mirrors are conducted exclusively by government; the supply is very limited, the prices of the articles produced extravagantly high, and, excepting tobacco, they are all productive of loss. In some parts mats, shoes, and other articles are extensively made of the esparto rush.

Commerce.—It is the peculiar misfortune of Spain that every part of her political system has been alike vicious and objectionable. Had her commercial policy been liberal, it would, in some degree, have compensated for the defects in the distribution of property and political power, and would, no doubt, have given a powerful stimulus to industry. But, unluckily, it has been in perfect harmony with her other institutions, and is, in all respects, worthy of the favourite seat and stronghold of the Inquisition. From the reign of Ferdinand and Isabella down to the present time, the policy, if so we may call it, of the Spanish government has been wholly anti-commercial. Their grand object has been to exclude foreign manufactures from the peninsula, and to preserve a monopoly of its markets, as well as of those in the colonies, to the home manufacturers. It is, however, almost needless to say, that their efforts to bring about this result have been signally unsuccessful. The oppressive taxes imposed on the manufacturers, the multiplication of fasts and holydays, the government monopolies, the badness of the roads and other means of communication, made it impossible for the Spanish manufacturers, even if they had evinced greater enterprise and industry than they have done, to produce manufactured articles as cheap as the English, the French, and others less unfavourably situated. Under such circumstances, the prohibition of certain descriptions of commodities, and the oppressive duties laid on others, could have no effect except to suppress the legitimate commerce of the country, and to throw it wholly, or almost wholly, into the hands of smugglers. Any one who takes up a map of Spain must be satisfied at a glance that it would be impossible, even for an army of customs' officers, to prevent her being deluged with smuggled products, provided they were materially cheaper than her native products; for, besides her extensive sea frontier, they may be introduced by way of France and Portugal, and also through the Basque provinces, which have distinct laws, and enjoy an exemption from the commercial code inflicted on the rest of the kingdom. We need not, therefore, be surprised that every effort to prevent the clandestine introduction of foreign productions should have completely failed. The severities occasionally inflicted on the smugglers, instead of abating, seem really to have increased the evil. The contraband trade has long been a favourite occupation, and has been eagerly followed by the adventurous, the necessitous, and the desperate. It is believed that for nearly three centuries from 100,000 to 150,000 individuals have been pretty constantly engaged in this occupation; that is, they have been engaged in trampling on the laws, obstructing their officers, and committing acts of violence and blood. When Mr. Townsend travelled in Spain, the country was a prey to the disorders occasioned by this wretched system. But it was to no purpose that the experience of two centuries, and the writings of various able men, had conclusively demonstrated its destructive influence. The government and the people, thanks to the influence of the Inquisition and the clergy, were so ignorant

and infatuated as to shut their eyes to its effects, and to resist every attempt to modify it, or to render it less noxious to the public interests.

And, strange to say, notwithstanding all the ravages Spain has undergone in the interval, her old anti-commercial policy still continues to maintain its ascendancy. At this moment a half or more of the entire trade of the country is in the hands of the *contrabandistas*. The tariff is divided into 15 classes; and in addition to hundreds of prohibitions, and duties varying from 50 to 200 per cent. on many articles of the first importance, the numerous forms to be observed at the custom-house, and the delay in entering any article, constitute of themselves a considerable premium on smuggling. It is stated that at present about 3000 actions are annually instituted against contraband duties and others engaged in illicit trade, which involve in the ruin of a vast number of families; at the same time that the courts of law are filled with perjury, and the country with bloody conflicts. And yet these atrocities secure no one object government has in view; foreign goods may be bought, though at an enhanced cost, in every market in Spain; the home manufactures are in the lowest state of degradation; industry of all kinds is paralysed; and the customs' revenue does not exceed a fourth part certainly of what it would amount to under a liberal commercial code! The truth is, that a thorough reform in her commercial policy is absolutely indispensable to give Spain even a chance of being regenerated. So long as the present tariff is maintained, so long will she be a theatre of sanguinary contests: without industry, with civilization,—a reproach and a disgrace to Europe.

As a specimen of the way in which the tariff is framed, we may mention that, notwithstanding the vast importance of a cheap and abundant supply of iron to agriculture and manufactures, wrought-iron articles are wholly excluded, at the same time that a duty of 900 per cent. is imposed on iron in bars (*en bruto*)! This, of course, is done to protect the iron-masters and founders of the Basque provinces! in other words, a trifling advantage is conferred on one individual in 100, at the expense and to the serious injury of the other 99!

Inasmuch as the returns published by the Spanish authorities of the value of the import and export trade of the country include only the commodities that pass through the custom-house, it is obvious, from the previous statements, that they are really good for nothing. According to the *Blanca Mercantil* for 1836, the aggregate value of the goods imported at 11 of the principal ports, during the course of that year, amounted to £1,615,000; whereas the value of the exports from the same ports, during the same year, is said to amount to £3,113,868. It is, however, abundantly certain that the value of the imports must have exceeded that of the exports; and the reason why they are apparently so much below the latter is to be found in the fact, that, being more generally subject to prohibitions and heavy duties, the trade in them is thrown, to a greater extent, into the hands of the smugglers. In illustration of what has now been stated, we may mention that the value of the exports to Great Britain, in 1838, is estimated, in the *Balanca Mercantil*, at £1,397,470; while that of the imports from Great Britain is set down at only £149,465. In point of fact, however, we send about £400,000 worth of produce direct to Spain, exclusive of the farther and more considerable amount smuggled through Portugal, Gibraltar, &c. We incline to think that, allowing for smuggling, the import and export trade of Spain may each be estimated at about £4,000,000 sterling, or, perhaps, a little more. And considering the vast, and, as it were, unexplored resources of the country, and the infinite variety of desirable products she could supply to others, we have no doubt, that under a liberal commercial system, her commerce would speedily be trebled, and that, at no very distant period, it would be increased in a tenfold proportion.

The great articles of export from Spain consist (exclusive of silk manufactures) of raw products. Of these, wine, wool, fruits of various kinds, lead, quicksilver, brandy, barilla, olive oil, raw silk, wheat, &c., are the most important, and are almost all susceptible of an indefinite increase.

The great articles of import are colonial products, obtained principally from Cuba, Porto Rico, &c.; cottons and cotton wool; linens, and hemp, and flax; woollens; salted fish, hardware, glass, and earthenware; timber, rice, hides, butter, and cheese, &c.

The importance of the trade that Spain formerly carried on with her vast possessions in the New World, was, at all times, much exaggerated; and she, in truth, was little better than an agent in the business, the greater part of the

---

* Moreau de Jonnes, in his *Statistique de l'Espagne*, gives the official returns of the imports and exports, without assigning any doubt as to their correctness. It is the same with most other statements in that work; which, in fact, is good for little, unless it be to bring statistical extravagance into discredit.

goods sent in Spanish bottoms to the colonies being, in reality the property of foreign merchants. Spain, notwithstanding the emancipation of Mexico and South America, has still some very valuable colonies; and if nothing else can, the astonishing progress made by Cuba and Porto Rico, since the abolition of the prohibitive system, should satisfy her of its ruinous tendency.

Owing to the badness of the roads, and their unfitness for carriages, the principal carriers of merchandise are the *arrieros*, or muleteers, who traverse the country in all directions along beaten tracks, many of which are accessible only to them. They form a large portion of the provincial population, and, on the whole, have a good character for honesty to their employers, though they are nearly all, more or less, engaged in smuggling transactions. The extent of his traffic may be estimated from the fact, that about three fourths of the entire inland traffic in corn is carried on by their means. Recently, however, wagons have begun to be introduced on all the practicable roads, and should the latter be improved, the business of the *arrieros* will proportionally fall off. We subjoin, in illustration of the export trade of Spain, an

ACCOUNT of the Quantities of the principal Articles of Merchandise imported into the United Kingdom from Spain and the Balearic Islands in 1836 and 1838.

Articles.		1836.	1838.
Barilla and alkali	cwts.	3×,624	2×,7×4
Cochineal	lbs.	7,544	
Coffee	—	115,993	
Figs	cwts.	2,557	5,953
Lead	tons	1,217	1,510
Lemons and oranges	chests	27,951	31,077
Liquorice juice	cwts.	1,297	394
Oil, or oil	galls.	57,548	138,167
Quicksilver	lbs.	1,430 3×0	2,253,002
Raisins	cwts.	192,119	180,009
Raw silk	lbs.	4,467	901
Thrown do.	,,	1,261	1,924
Lamb skins (undressed)	No.	10,611	
Brandy	galls.	19,560	220,527
Sheep's wool	lbs.	1,814,×77	2,109,524
Spanish wines	galls.	3,312,×54	4,052,707
Wheat	qrs.	421	17,741
Peruvian bark	lbs.		1,717

Currency.—The trade of a banker, as it is understood in eat Britain, is unknown in Spain; but there is notwithstanding, an extensive circulation of inland bills of exchange. I merchants in good credit call themselves bankers, do nking business, and have agents and connections in the ferent towns, to facilitate their operations; but there is, twithstanding, considerable difficulty in remitting money m place to place, and a different rate of exchange frequently exists between towns only a few leagues distant. ordinary transactions there are no substitute for cash, I a good deal of trouble and inconvenience is experienced counting, examining, and weighing the coins.

Accounts are kept in *reales de vellon*, of which about 90 equivalent to £1. The pistole is worth 16s. 9d. British rency. The money in circulation consists of gold and er coins of very various value, and of copper. Dollars rarely seen, especially in the N. and near the sea-coast, consequence of the premium they bear in France, to ich they are smuggled in large quantities, notwithstanding the penalties consequent on their exportation. Travellers are allowed to carry out of the country a sum for their enses, amounting to £90 in gold only; and should they ound, on examination, to have more, the whole may be ed. Oil is sold by the *arroba mina*, 100 of which are al to 335 English wine-gallons: four arrobas are equal quintal, or 102 English pounds. The *cahis*, or measure corn, is divided into 12 fanegas, five of the latter being al to one quarter. One hundred Spanish *varas*, or I's, are equal to 92.5 English yards, and a Spanish *legua* ains 5000 varas. The traveller tries in vain to find a whereby he may compare the Spanish land measures the English acre; and, with respect both to weights measures generally, they vary greatly in different proes.

Roads and Canals.—Spain is singularly destitute of roads, other means, for the speedy and easy transport of ellers and products from place to place. The king's ways (*caminos reales*), the only roads worthy of the e, extend only between the more important places: of them, as, for instance, those between the capital Pampeluna, Saragossa, Badajoz, Seville, and Granada, generally speaking, kept in good repair; and the great from Pampeluna, by Vittoria, Burgos, and Valladolid, adrid, is stated by recent travellers to be scarcely in to the second-rate roads of England. The great road lencia, though probably somewhat improved since the ution, was previously in a wretched state, and so ive is that running by Tarragona and Albacete into na. In Catalonia, the roads are comparatively nuus and good; stage-coaches, also, run between the

more important towns. The road from Saragossa to Barcelona is in good repair, and a diligence has been established on it since 1831. The roads of Biscay and Navarre, also (owing to their being placed under a provincial government), are more numerous, better constructed, and more carefully managed, than in the rest of the country. The great complaint with respect to these roads, whether originally made with road-metal, or formed partly by tracks of carts, and afterward improved, is, that they are seldom provided with bridges over the numerous torrents pervading the country, and scarcely ever kept in proper repair. The mountain-roads are mere paths worn by the feet of the mules during a long series of years. The revenue applicable to the construction and repair of roads is derived partly from local taxes, levied by postage-dues, and duties on articles of consumption; and partly, also, on tolls levied at intervals of 10 or 12 English miles. The tolls appear to be light; but the government is said to derive from them a greater revenue than it expends on the roads. The *caminos reales* are under the superintendence of a board under the presidency of the board of finance, and the government funds are available only for these roads, the rest being left either to the chance sums levied on travellers, or to the wheels of carts and the feet of mules! With respect to the sums of money employed in repairing roads in Spain, it may suffice to state, that although that country be more than three times the size of England, and naturally more difficult, the outlay on roads in it is hardly one twentieth part the sum expended on those in England. The diligences on the principal roads are decidedly better than might have been expected; and travel at a rate of 6 or 7 miles an hour; they are far, however, from being safe conveyances, and instances are every now and then occurring of their being stopped and the passengers robbed.

No country in Europe is worse provided with canals than Spain, though, looking at the map merely, one would suppose that in none were there greater facilities for their construction. But the imbecility of the government, the ignorance of the people, the porousness of the soil, and heat of the climate, oppose very serious obstacles to their formation. Still, however, some advances have been made, and the government of Isabella II. may, in this respect, be advantageously contrasted with that of Charles II. During the reign of the latter, a company of Dutch contractors offered to render the Mancanares navigable from Madrid to where it falls into the Tagus, and the latter from that point to Lisbon, provided they were allowed to levy a duty for a certain number of years on the goods conveyed by this channel. The council of Castile took the proposal into their serious consideration, and, after maturely weighing it, decided, "That if it had pleased God that these two rivers should have been navigable, he would not have wanted human assistance to have made them such; but, as he had not done it, it is plain he did not think it proper that it should be done. To attempt it, therefore, would be to violate the decrees of his providence, and to mend the imperfections which he designedly left in his works!" (*Clarke's Letters on Spain*, p. 284.)

But such undertakings are no longer looked upon as sinful, and several have been projected, and a few completed, since the accession of the Bourbon dynasty to the throne of Spain.

The canal of the Ebro, from Tudela to Santiago, 41 m. below Saragossa, was chiefly executed in the reigns of Charles III. and IV., under the administration of Count Florida Blanca; and though of insufficient depth for navigation on any large scale, it is made available during nearly its whole extent for barges of small draught, besides being extremely useful for the irrigation of the surrounding country.

The most important project of this kind at present on foot is the canal of Castile, intended to open a communication between the vast and fertile plains of Old Castile and Leon, and the N. sea, and to afford an outlet for their surplus produce. It has been constructed from Segovia on the S., past Valladolid and Palencia to Aguilar del Campo: a branch runs westward to the Rio Seco, and another is in course of construction to Burgos. The inefficiency of the engineering processes, the difficulty of procuring good labourers, and the nature of the soil, have presented serious obstacles to the undertaking, which has now been in progress, with certain intermissions, since 1753, and has not yet been completed! A large portion is now, however, available for navigation; and the advantages that would result from its completion are so great, that it may be expected that a vigorous effort will speedily be made to have it terminated. The navigation of the Tagus has, also, engaged the attention of different Spanish sovereigns; and at the close of the 16th century the river is said to have been made navigable for barges from Toledo to its mouth; but, if so, it was subsequently rendered useless through neglect, and it is only within the last few years that a company has undertaken to make it navigable from Aranjuez (23 m. above Toledo) down to Lisbon. The long-pending question respecting the right of the Spaniards to navigate the lower part of the river has at length been settled, and there is now no reason why small steamers should not ascend as high as Toledo into the richest part of New Castile, thus establishing a valuable trading connection between the inhabitants of the interior and of the greatest emporium on the coast of the peninsula.

The Guadalquivir was once navigable for flat-bottomed boats up to Cordova, but Seville is the highest point reached at present. Many projects have been set on foot for improving the river by deepening the channel; but the great and sudden floods to which it is subject must operate as a bar to its successful navigation; nor, even if the part above Seville were considerably improved, is it at all probable that the bars and sand-banks of the marshy district known as the *Marisma* could be so far removed as to make the river accessible by sea-borne vessels even as far as Seville.

Population.—We have already noticed the exaggerated and unfounded statements, with respect to the former flourishing state of agriculture, manufactures, and commerce in Spain. Inasmuch, however, as the population of a country is mostly dependent on its agriculture and manufactures, it follows that the same facts and reasonings which show that their extent and prosperity in the 14th and 15th centuries had been greatly exaggerated, go far, also, to show that this must have been the case with regard to the accounts of the comparatively dense population of Spain at the era alluded to. Down to the 15th century, or to the junction of the crowns of Castile and Aragon by the marriage of Ferdinand and Isabella, Spain was divided into a number of states, between which the most violent animosities subsisted, and most parts of the peninsula were a prey to violence and every species of disorder. It would be contradictory and absurd to suppose that a country placed under such circumstances could be densely peopled; and Capmany has conclusively shown that there are really no grounds whatever for thinking that Spain had been at any time more populous than at the period (1807) when his acute and learned work, the *Questiones Criticas*, was published.

No doubt, however, the population of Spain declined considerably during the disastrous reigns of Philip III., Philip IV., and Charles II. This decline has been ascribed, in great part at least, to the expulsion of the Moors in the reign of Philip III. But, though it be impossible too strongly to condemn this measure, and that of the expulsion of the Jews during the reign of Ferdinand and Isabella, and the detestable fanaticism in which these acts originated, we are, notwithstanding, inclined to think that their influence has been a good deal over-rated. The numbers expelled were magnified far beyond the truth; and it is farther obvious that the vacuum, such as it was, created by their expulsion, would, in ordinary circumstances, have afforded a new field for the employment of those who continued in the country, and have acted, in fact, as a stimulus to population. Neither are we inclined to lay any stress on the statements of those who contend that Spain was depopulated by the emigrations to America. These were far too inconsiderable to have any such result. No one pretends that the emigrations to America depopulated England; and yet they were quite as extensive as those from Spain. Indeed, Ulloa, Ustariz, and Campomanes have conclusively shown that the emigrations from the peninsula had, in no degree, the effect ascribed to them.

The truth is, that the decline of population between the demise of Philip II. and the termination of the war of the succession, was a consequence of continued and systematic misgovernment, rather than of any particular acts of oppression. The rapacity, intolerance, and influence of the Spanish clergy, the *falo de se* character of her financial system, the destructive contests in which she was engaged, the weakness of her sovereigns, and, in a word, her wretched internal policy, prostrated her energies, paralysed her industry, and not only prevented her recovering from, but aggravated, the wounds inflicted by the bigotry of her rulers. But, under the milder, more intelligent, and equitable administration of the Bourbons, some improvements took place; and the population having increased gradually, though slowly, during the whole course of last century, is now certainly as great as at any former period of her history, and perhaps greater. "*No per eso puede decirse, que la España haya estado mas poblada que al presente, ni en tiempo de los Romanos, ni en el siglo 16; en cuya época suponen algunos escritores arbitrariamente que llegó a 20 ó 21 millones el número de sus habitantes. Por el contrario, todos los datos mas exactos, y las combinaciones mas racionales, persuaden, que no hubo entónces sobre la superficie de España mas de los 10½ millones de almas, a que el ultimo censo reduce su poblacion actual.*" (Antillon, *Geografía*, 147.) In 1787 the population amounted to 10,268,150, or

perhaps 10½ millions, as it is believed, on apparently good grounds, that the official returns were below the mark; and since then it has increased nearly two millions.

Under the existing constitution, there is to be a deputy for every 50,000 inhabitants; and according to the decree issued on the 3d of Aug. 1837, appointing the deputies for the different provinces, the total population of the kingdom amounted to 12,168,000 (see Table at the beginning of this article). We believe, however, that no very great dependence can be placed on this return, though, probably, in the result it is not far from accurate.

Religion.—Spain has long been, and still is, the favourite seat of the Roman Catholic religion, the country in which it has been maintained in the greatest purity, and to the exclusion of every other. The Inquisition was introduced, or, at all events, was vested with a vast increase of power, in the reign of Ferdinand and Isabella; and that formidable tribunal ultimately succeeded, by dint of the stake and the rack, and such like atrocious means, in exterminating heresy, or, in other words, all difference of opinion as to religious matters in Spain; and it was, also, mainly instrumental in prevailing on its weak and bigoted sovereigns to banish the Moors. According to Llorente, no fewer than 13,000 individuals, accused of heresy, were publicly burned by the different tribunals of Castile and Aragon; and 191,413, accused of the same offence, suffered other punishments in the brief space between the establishment of the modern Inquisition in 1481 and 1518, only two years after the death of Ferdinand; and since then the number of its victims has been incomparably greater! (*Prescott's Ferdinand and Isabella*, iii., 465.) Probably this statement may be exaggerated; but it is, notwithstanding, abundantly certain, that all other persecutions of which we have any authentic accounts have been mild compared with those inflicted by this blood-thirsty tribunal. In point of fact, however, the mischievous influence of the Inquisition did not consist so much in its judicial murders, and other atrocities perpetrated in the sacred name of religion, as in its deadly influence over the mind and feelings of the nation. It was, as every one knows, quite as hostile to all sorts, of political and philosophical knowledge as to heresy in religion; it was, in fact, the deadly foe of everything like free inquiry; and while the importation of most useful works from foreign countries was a capital offence, nothing could be printed at home unless it were approved by the inquisitors. Under such circumstances, need we wonder at the ignorance of the Spanish people, their bigotry, intolerance, and the profound veneration they so long displayed for whatever is most worthy of contempt?

The numbers and wealth of the clergy and monastic orders were such as might be expected in a country where the Inquisition was triumphant, and where to commit a murder was a less offence than to insinuate a doubt as to the "real presence!" According to an official statement drawn up in 1812, it appears that the clergy were then in possession of about one fourth part of the landed property of the kingdom, exclusive of tithes and other casual sources of income, producing in all a total gross revenue of about eleven millions sterling a year! The revenues of some of the dignified ecclesiastics were quite immense: the archbishopric of Toledo is said to have been worth from £65,000 to £80,000 a year.

According to the official returns of the census of 1788, the ecclesiastics of all descriptions, including 61,617 monks, 32,500 nuns, and 2795 inquisitors, amounted to 168,425 individuals. (*Townsend*, ii., 213.) And it appears, from the official returns published in the *Correo Literario* of Madrid in 1833, that, notwithstanding the attacks made upon the ecclesiastical state during the French war and subsequently, it then comprised 175,574 individuals, of whom 61,727 were monks, and 24,007 nuns.

Happily, however, a very great change for the better has been effected in the interim. A decree, passed on the 2d of July, 1835, suppressed all conventual establishments with not more than 12 inmates; and the example thus wisely set was followed up by the decree of the 5th of March, 1836, which entirely suppressed all conventual establishments, and religio-military orders. The monks who were thus turned out of their old haunts were to receive small stipends; and it is to be regretted that, owing to the difficulties in which the country has since been involved, these stipends have been very irregularly paid. But the inconveniences thence arising affect only a few individuals, whose claims on the public sympathy were of the slenderest description; whereas the measure in which they originate cannot fail to be productive of great national advantage, and is, in fact, one of the most beneficial results of the late changes.

The whole of the vast property formerly belonging to the church has been confiscated for the use of the state, and a considerable portion of it has been already sold. According to the constitution the nation undertakes to support the public worship and clergy of the established church; but, owing to the intestine commotions that have prevailed in the country, and its financial difficulties, this condition has not been effectively carried out; and not a few of the clergy are, at present, but little removed from a state of indigence.

It should further be borne in mind that, during the last half century, and especially since the commencement of the late struggle with France, the bigotry of the Spaniards, especially of the inhabitants of towns, and the influence of the priests, have materially declined. And, by a necessary, though unfortunate, consequence, the abuses and vices of the clergy have reacted against religion itself; and, at this moment, most intelligent persons in Spain, though making an outward profession of religion, entertain a profound contempt for the mummeries enjoined by the clergy, and are mostly, indeed, decided sceptics. According, however, as the church is purified, and ceases to be identified with everything most deserving of reprobation, religion will, no doubt, recover its proper influence, and will cease to be degraded in the public estimation by the intolerance, extortion, and immorality of its professors.

Government.—At the period of the union of the crowns of Castile and Aragon, by the marriage of Ferdinand and Isabella, each of the kingdoms had representative assemblies, or cortes, that shared in the legislative authority, and enjoyed very extensive privileges. Unluckily, however, though the crowns were united, by the marriage now referred to, the kingdoms were not; each continued to preserve its own laws and institutions; and their mutual jalousies enabled the sovereigns to employ the one against the other, and ultimately to crush the liberties of both. This result was greatly facilitated by the extensive conquests of the Spanish sovereigns. In the reign of Ferdinand and Isabella, Granada, Navarre, and Naples, were subjected to the Spanish crown; so that the princes became, in a great measure, independent of the constitutional control of the cortes of their hereditary states. Under Charles V., who possessed, in right of his father, all the dominions of the house of Austria, and under whom all but boundless territories were acquired in the New World, the preponderance of the external dominions of the crown was greatly increased; and the defeat of the forces of the rebellious Castilians, under Padilla, in his reign, and the execution of the Justiza of Aragon in that of Philip II., completed the extinction of all constitutional control on the acts of the sovereign; at the same time that the Inquisition having controlled, and next eradicated, all energy and independence of mind, the nation gradually sunk into a state of torpor and stupid indifference.

This state of things continued, with slight interruptions, till the invasion of Spain by Napoleon; when the mortified pride of the nation made them take arms in defence of its independence, and of the rights of the worthless princes who had abdicated the crown. It is useless to enter into any details as to the events that followed. The novel circumstances under which the nation was now placed made it necessary to convoke her ancient cortes, and in the end a constitution was formed on a representative basis. This constitution was, however, abolished by Ferdinand the moment he was set at liberty by the French, in 1814; and from this period down to 1820, the ungrateful and vindictive bigot exerted himself to effect the ruin of those to whom he was mainly indebted for the crown, and even so far as to restore the Inquisition that had been supported by the French. But the army, and a large portion of the nation, disgusted at these measures, broke out in revolt, and Ferdinand was compelled to accept the constitution of 1812. Owing, however, to the influence of the clergy, and the ignorance of a large proportion of the people, the constitution was by no means acceptable to very classes; and the French having entered Spain with a powerful army, under pretence of restoring order, the constitutionalists were everywhere defeated, and Ferdinand was once more restored to absolute power.

Ferdinand having expired in 1833, his infant daughter, Isabella II., was proclaimed queen, in virtue of a law entitling her to the crown in preference to her uncle, Don Carlos, the heir of the crown under the Salic law, which previously obtained in Spain. This led, as every one knows, to an obstinate civil war, which fortunately terminated in the total defeat of Don Carlos and his claims.

In 1834, after the demise of Ferdinand, Christina, the regent and mother of Isabella, proclaimed a character of the Spanish nation, called the *Estatuto Real*; but it gave little or no satisfaction to the liberal or constitutional party whom the pretensions of her daughter were supported; and the queen was obliged to issue a decree, pledging her to adopt the constitution of 1812, with such modifications as the cortes might agree to. And this constitution having in consequence been subjected to a careful and its revision, a new constitution, which is now (1842)

in force, was promulgated at Madrid on the 16th of June, 1837. As this is a document of much importance, and has been, and no doubt will be, much referred to, the reader may not, perhaps, be disinclined to have it fully laid before him:

CONSTITUTION OF THE SPANISH MONARCHY, PROCLAIMED IN MADRID ON THE 16TH OF JUNE, 1837.

DONNA ISABELLA THE SECOND, by the grace of God and the Spanish monarchy, Queen of Spain; and in her royal name, and during her minority, the Queen Dowager, her mother, Donna Maria Christina de Borbon, Regent of the empire; to all those to whom these presents may come. Be it known, That the Cortes-general have decreed and approved, and that We in due form have accepted the same, as follows: It being the will of the nation to revise, in virtue of its sovereignty, the political constitution promulgated in Cadiz on the 19th day of March, 1812, the Cortes-general, assembled for this purpose, decree and approve the following Constitution of the Monarchy of Spain:

"Art. 1. Spaniards are, 1st. All persons born in the Spanish dominions. 2d. The children of Spaniards, though born out of Spain. 3d. Strangers who have obtained letters of naturalization. 4th. Those persons who, without letters of naturalization, obtain a right of settlement in any part of the monarchy of Spain. The rights of Spanish citizenship are forfeited by naturalization in a foreign country, and by accepting employment under any other government without the permission of the sovereign of Spain.

"Art. 2. All Spaniards may print and publish their thoughts freely, without a previous censorship, but subject to the laws. The determination of offences by the press belongs exclusively to juries empannelled for that purpose.

"Art. 3. Every Spaniard has the right of petitioning in writing to the cortes and king, as the laws prescribe.

"Art. 4. The same code of laws shall govern in all parts of the monarchy, and in them shall be recognised by all Spaniards but one *right* in common trials of a civil or criminal nature.

"Art. 5. All Spaniards are eligible to public offices, according to merit and capacity.

"Art. 6. Every Spaniard is obliged to defend the country with arms in his hands whenever he may be called upon to do so by the law, and to contribute according to his abilities to the expenses of the state.

"Art. 7. No Spaniard can be detained, imprisoned, or taken from his family, nor his house entered, excepting in those cases and according to the forms determined by the laws.

"Art. 8. If the security of the state require, in extraordinary circumstances, the temporary suspension, in whole or in part, of the Spanish monarchy, the provisions of the preceding article are to be determined by the law.

"Art. 9. No Spaniard can be prosecuted or sentenced, except by a judge or competent tribunal, in conformity with laws enacted anterior to the commission of the offence, and in the manner laid by them.

"Art. 10. The confiscation of property is abolished, and no Spaniard is to be deprived of his property, except in cases justified by public utility, and with a previous indemnification of losses sustained.

"Art. 11. The nation is obliged to maintain the public worship and ministers of the Catholic religion professed by the Spaniards.

"*Of the Cortes.*—Art. 12. The power of enacting laws resides in the Cortes, in conjunction with the king.

"Art. 13. The Cortes is composed of two co-legislative bodies equal in powers—the Senate and the Congress of Deputies.

"*Of the Senate.*—Art. 14. The number of senators shall be equal to three fifths of the whole number of the deputies.

"Art. 15. The senators are appointed by the king, from a triple list proposed by the electors of each province who elect the deputies.

"Art. 16. To each province belongs the right of proposing a number of senators proportional to its population; but each is to return one senator at least.

"Art. 17. To be a senator it is necessary to be a Spaniard, to be forty years of age, and to be possessed of the income and other qualifications defined in the electoral law.

"Art. 18. All Spaniards possessed of these qualifications may be proposed for the office of senator in any of the provinces of the monarchy.

"Art. 19. Each time that there is a general election of deputies, whether in consequence of their term of office having expired, or of a dissolution of the congress, the third part of the senate, in the order of seniority, is to be renewed, those going out being re-eligible.

"Art. 20. The sons of the king and of the immediate heir to the throne are senators of right at the age of twenty-five years.

" *Of the Congress of Deputies.*—Art. 21. Each province shall appoint one deputy at least for every 50,000 souls of the population.

" Art. 22. The deputies are elected by the direct method, and may be re-elected indefinitely.

" Art. 23. To be a deputy it is necessary to be a Spaniard, in the secular state, to have completed the 25th year, and to possess all the other qualifications prescribed by the electoral law.

" Art. 24. Every Spaniard possessing these qualifications may be named a deputy for any of the provinces.

" Art. 25. The deputies shall be appointed for three years.

" *Of the Meeting and Faculties of the Cortes.*—Art. 26. The cortes are to assemble each year. It is the right of the king to convoke them, to suspend and close their meetings, and dissolve the cortes ; but under the obligation, in the latter case, of convoking and re-assembling another cortes within the period of three months.

" Art. 27. If the king should omit to convoke the cortes on the 1st of December in any one year, the cortes are notwithstanding to assemble precisely on that day ; and in case of the conclusion of the term of the congress holding office happening to occur in that year, a general election for the nomination of deputies is to commence on the first Sunday of the month of October.

" Art. 28. On the demise of the crown, or on the king being incapacitated to govern through any cause, the extraordinary cortes are immediately to assemble.

" Art. 29. Each of the co-legislative bodies is to form rules for its own internal regulation, and to scrutinize the legality of the election, and the qualifications of the individuals who compose them.

" Art. 30. The congress of the deputies is to name its president, vice-president, and secretaries.

" Art. 31. In each legislature the king shall appoint from among the members of the senate the president and vicepresident of that body, the latter appointing its own secretaries.

" Art. 32. The king shall open and conclude the sittings of the cortes in person, or by his ministers.

" Art. 33. One of the legislative bodies cannot be convoked for business without the other being assembled at the same time, except in the case in which the senate sits in judgment on the king's ministers.

" Art. 34. The legislative bodies are not to deliberate in conjunction, nor in the presence of the king.

" Art. 35. The sessions of the senate and of the congress shall be public, and only in cases requiring reserve can private sittings be held.

" Art. 36. The king and each of the co-legislative bodies possess the right of originating laws.

" Art. 37. Laws relating to taxes and public credit shall be presented first to the congress of deputies ; and if altered in the senate contrary to the form in which they have been approved by the congress, they are to receive the royal sanction in the form definitely decided on by the deputies.

" Art. 38. The resolutions of each. of the legislative bodies are to be determined by an absolute plurality of votes ; but in the enactment of laws the presence of more than half the number of each of these bodies is necessary.

" Art. 39. If one of the co-legislative bodies should reject any project of law submitted to them, or if the king should refuse it his sanction, such project of law is not to be submitted anew in that legislature.

" Art. 40. Besides the legislative powers which the cortes exercise in conjunction with the king, the following faculties belong to them : First, to receive from the king, the immediate successor to the throne, from the regency or regent of the empire, the oath to observe the constitution and the laws. Second, to resolve any doubt that may arise of fact or of right with respect to the order of succession to the crown. Third, to elect the regent, or appoint the regency of the empire, and to name the tutor of the sovereign while a minor, when the constitution deems it necessary. Fourth, to render effective the responsibility from the ministers of the crown, who are to be impeached by the deputies, and judged by the senators.

" Art. 41. The senators and deputies are irresponsible and inviolable for opinions expressed and votes given by them in the discharge of their duties.

" Art. 42. Senators and deputies are not to be proceeded against or arrested during the sessions without the permission of the legislative body to which they may belong. If not taken in the act of committing flagrant crime ; but in this case, and in those in which they are prosecuted or arrested while the cortes are closed, they are to give immediate information to their respective co-legislative bodies for their cognizance.

" Art. 43. Deputies and senators who receive from the government or from the royal family pension, employment which may not be an instance of promotion from a lower to a higher office of the same kind, commission with sundry honours or titles, are subject to re-election.

" *Of the King.*—Art. 44. The person of the king is sacred and inviolable, and is not subject to responsibility. His ministers are responsible.

" Art. 45. The power of executing the laws resides in the king, and his authority extends to all matters that conduce to the preservation of public order in the interior, and to the security of the state abroad, in conformity with the provisions of the constitution and the laws.

" Art. 46. The king sanctions and promulgates the laws.

" Art. 47. Besides the prerogatives granted to the crown by the constitution, he possesses the following : 1. To issue decrees, regulations, and instructions, which may be conducive to the execution of the laws. 2. To provide that justice be promptly and efficiently dispensed throughout the empire. 3. To pardon criminals, according to the provisions of the law. 4. To declare war and make peace, afterward giving an account and documents to the cortes. 5. To dispose of the military forces of the country, distributing them as may be most convenient. 6. To conduct diplomatic and commercial relations with other states. 7. To provide for the coinage of money, on which is to be impressed his bust and name. 8. To decree the application of the funds destined for each branch of the public administration. 9. To appoint public officers, and to confer honours and distinctions on all classes, in conformity with the law. 10. To name and dismiss his ministers without restriction.

" Art. 48. The king is obliged to be authorized by a special law : 1. To alienate, to grant, or exchange any part of the Spanish territory. 2. To admit foreign troops into the kingdom. 3. For the ratification of treaties of offensive alliance, of special treaties of commerce, and those which stipulate to give assistance to any foreign power. 4. To absent himself from the kingdom. 5. To contract matrimony, and to permit those who may be called to the throne to enter into that state. 6. To abdicate the throne in favour of his immediate successor.

" Art. 49. The income of the king and royal family is to be settled by the cortes at the commencement of each reign.

" *Of the Succession to the Crown.*—Art. 50. Donna Isabel II. de Borbon is the legitimate queen of Spain.

" Art. 51. The succession to the throne of Spain shall be in the regular order of primogeniture and representation, always preferring the anterior to the posterior line of succession ; in the same line also preferring the nearer degree of kindred to the more remote ; in the same degree, the male to the female line of descent ; and in the same sex, the eldest to the younger branches of the family.

" Art. 52. The line of the descendants of Donna Isabel II. de Borbon becoming extinct, her sister and her natural by the father's side, male as well as female, and their legitimate descendants, shall succeed, if not excluded specially by law.

" Art. 53. If the lines of succession pointed out become extinct, the cortes shall name the sovereign as may be best for the interests of the nation.

" Art. 54. The cortes shall exclude from the succession such persons as are incompetent to govern, or who have done anything which should cause them to deserve forfeiting their rights to the throne.

" Art. 55. During the reign of a female her husband is to take no part whatever in the government of the kingdom.

" *Of the Minority of the Sovereign, and of the Regency*—Art. 56. The sovereign is to be considered a minor until he is fourteen years of age.

" Art. 57. When the king is incapacitated from exercising his functions, or the crown is vacant in consequence of the minority of the immediate successor, the cortes shall appoint a regency for the government of the kingdom, consisting of one, three, or five persons.

" Art. 58. Until the cortes shall appoint the regency the kingdom shall be governed provisionally by the father or mother of the king, and in their absence by a council of the ministers.

" Art. 59. The regency shall exercise all the functions of royalty, in whose name shall be published all the acts of the government.

" Art. 60. The guardian of the king, while a minor, shall be the person appointed in the will of the deceased sovereign, always providing that such guardian shall be a Spaniard by birth ; if the deceased king should not have appointed such guardian of the successor, a must to the throne, then the father or mother, being in the state of widowhood, shall be the guardian. In the absence thereof, the cortes shall appoint ; but the offices of guardian and regent cannot be discharged by the same individual, except in the case of a father or mother.

" *Of the Ministers.*—Art. 61. All commands or dispositions issued by the sovereign shall be signed by the respective ministers; and no public functionary is to execute such orders if not thus signed.

"Art. 62. The ministers may be senators or deputies, and take part in the discussions of the two legislative bodies; but they are permitted to vote in that body only to which they belong.

"*Of the Judicial Power.*—Art. 63. To the tribunals and judges alone belong the power to apply the laws in civil and criminal cases, and without exercising any other functions than those of judges and of ministers of justice.

"Art. 64. The laws are to determine the nature of the tribunals and judgments which are to exist, the organization of each, its faculties, the mode of proceeding, and the qualifications of the officers belonging to them.

"Art. 65. Judgments in criminal cases to be public, and in the form prescribed by the laws.

"Art. 66. No magistrate or judge can be deposed, for a shorter or longer period, except by a written sentence, nor suspended from the discharge of the duties of his office, except by a judicial act, or in consequence of the order of the king, after he, on sufficient grounds, commands him to be tried by a proper tribunal.

"Art. 67. Judges are personally responsible for all infractions of the law committed by them.

"Art. 68. Justice is to be administered in the name of he king.

"*Of the Provincial Deputations and Corporations.*—Art. 9. In each province there shall be a provincial deputation, composed of a number of persons specified by law, and appointed by the same electors who return the deputies to the cortes.

"Art. 70. For the internal government of towns a corporation shall exist in each, to be elected by the inhabitants to whom this right pertains by law.

"Art. 71. The law shall determine the organization and duties of the provincial deputations and the corporations.

"*Of Taxes.*—Art. 72. Each year the government shall submit an estimate of the expenses of the state for the following year, and a schedule of the contributions and of the means of raising them; and in little manner the accounts of the collection and disbursement of the public revenues, for the examination and approval of the cortes.

"Art. 73. No tax or contribution is to be imposed or collected which has not been authorized by the law of the estimates or other special authority.

"Art. 74. A similar authorization is necessary to dispose of the property of the state, and for raising money by loans on the public credit.

"Art. 75. The public debt shall be under the especial protection of the nation.

"*Of the National Military Forces.*—Art. 76. The cortes, the proportion of the king, shall each year determine the number of the permanent military forces by sea and land.

"Art. 77. In each province there shall be corps of national militia, whose organization and duties are to be decided by a particular law; and the king may, in case of necessity, dispose of these forces within their respective provinces, but not out of them without the express authorization of the cortes.

"ADDITIONAL ARTICLES.—Art. 1. The laws shall define the time and manner in which judgments by juries for every class of offences are to be established.

Art. 2. The provinces beyond the seas shall be governed by special laws."

The business of the Spanish government is carried on by the regent (at present, 1842, General Espartero, duc de Victoria), by ministers of the interior, foreign affairs, war, marine, and finance. The kingdom is now, as lately seen, divided into 47 provinces, exclusive of the Baleario and Canary islands. The constitution declares a provincial deputation, or cortes shall be elected in every province, for the superintendence of its internal affairs; and that corporations are to be established in the different towns. Such corporations have, indeed, long existed in Spain; and every *pueblo* in the kingdom, great and small, has, for a long series of years, had its *ayuntamiento*, or corporation, which manages the common property (often very large) of the pueblo; applots and levies the taxes required for public and private purposes, &c.; and otherwise has considerable power and influence.

The laws of Spain, previously to the late revolution, and a greater number of those now in force, are embodied in codes known by the titles of *Fuero juzgo*, *Leyes de las Partidas*, *Ordenamiento Réal*, *Fuero Réal*, and *Novis-Recopilacion*. The first of these is, in the main, an enactment of the Theodosian code, originally published by :, son of Euric, one of the Gothic conquerors of the nation, and successively augmented by the addition of laws. The Ordenamiento Real contains the code of established by Ferdinand and Isabella. The Leyes Siete Partidas is a compound of Gothic, Roman, and canon law. The Fuero Real (a mixture of Roman and Gothic law) was compiled at Huesca, in 1248, for the use of the kingdom of Aragon; and the Novisimo Recopilacion is a digested collection of edicts issued by the kings of Spain, and enjoys the highest authority. It cannot be surprising that, with so many different and often conflicting codes, the general system of jurisprudence should be extremely defective. But the administration of the laws is incomparably worse than the laws themselves, being slow, complicated, and protracted to a ruinous degree. There are endless appeals from one jurisdiction to another, and the whole machinery of the courts is adapted to screen the venality of the judges, and to afford a rich harvest to the *escribanos*, or attorneys, the only medium of communication between the client and the judge. "Justice of no kind has any existence; there is the most lamentable insecurity of person and property; redress is never certain, because both judgment and execution of the laws are left to men so inadequately paid, that they must depend for their subsistence upon bribery. Nothing is so difficult as to bring a man to trial who has anything in his purse, except to bring him to execution; this, indeed, is in most parts impossible, as money will always buy indemnity. Everything in Spain connected with the following out the laws is in the hands of the *escribanos* : these are the friends of all bad men; for, whatever be the action a man may commit or meditate, he has only to confide in the *escribano*, and pay for his protection." (*Inglis's Spain* in 1830, i. 251.)

The wretched defects in the administration of justice were, in some degree, obviated, in so far, at least, as petty cases were concerned, by the adoption of a sort of arbitration system. Individuals called *alcaldes*, annually chosen, according to the different privileges of the different towns, boroughs, and villages in which they reside, decide the cases brought before them, like Sancho Panza in the island of Barataria, according to their own sense of what is right and equitable. But, with the exception of this rude and defective tribunal, every other part of the Spanish judicial system is a tissue of the most scandalous abuses. There is, in fact, but little security for property, and still less for life.

A person robbed or assaulted may prosecute, but, if unsuccessful, is bound to pay all the expenses; and is, indeed, forced to lodge a sum of money with the *alcalde pedaneo* (sheriff's officer), before any steps be taken in the business. In cases of murder and assassination, the last of which is horribly frequent, witnesses are afraid to come forward, as it very frequently happens that they are imprisoned until they establish their innocence. But even when, braving all these dangers, individuals boldly denounce a crime, there are at least *five* chances to one that the culprit escapes from prison, or compounds his felony with the judges; and in that case, the accusers have everything to dread from the vengeance of the criminal. The banditti are numerous and powerful; not only overawing those among whom they live, but keeping all the petty officers in their pay; so that they are enabled to prosecute their murderous career with comparative impunity. Even when a robber or assassin has been convicted and sentenced, there is no certainty that punishment will follow. Thus, from execrable laws, still more execrably administered, it is commonly said in Spain, that not one crime in ten is ever brought before the courts; and, though this be probably overstated, if we take the proportion at one in four or five, we shall certainly be within the mark. Under such circumstances, need we wonder at the dreadful frequency of the most atrocious crimes?

Language, Literature, and Education.—It seems probable that the Cantabrian was the most ancient language of Spain, of which remnants are supposed by some still to exist in the modern Basque, spoken by the Biscayans and other inhabitants of the districts bordering on the Pyrenees. The old language of the Peninsula must, no doubt, have been considerably alloyed by the admixture of Phœnician words and phrases during the Carthaginian dominion; and when the Romans conquered Spain, they introduced their language, which, for several centuries, was the principal medium of communication of all except those living in the most remote districts. The Visigoths, who followed the Romans in the possession of the peninsula, introduced the *lingua Romana*, a mixture of the Latin and German languages; but the Latin, though corrupt, still continued to be spoken in many parts. Again, when the Moors overran the country, expelled the Visigoths, and established their own power, they brought with them the Arabic language, already highly cultivated, and well adapted for poetry; and this, in turn, became the general language of the country. Thus, out of numerous elements was gradually formed a new language—the Spanish; and though numerous dialects necessarily arose in the different petty kingdoms into which the country was split, that of Castile became at length the classical language of Spain. Its basis ia Latin; and many of the ancient inflexions, as well as words, are still preserved. There are also a large number of Teutonic words; but the admixture of Arabic, though very considerable, is less than in the Portuguese. Force of expression, depth of sound, and mellifluous cadence, are the peculiar characteristics of the Spanish; which, however, has a guttural accent, derived probably from its Teutonic origin. The abundance of vowels and liquids makes the language harmonious, when spoken by native Castilians: it is essentially poetical, and poetry may be considered as the germ of the national literature. It is a curious fact, that there is very little patois among the Castilians, and that the language is spoken by the lower classes with remarkable purity and precision.

The rise of Spanish literature cannot be traced further back than the middle of the 12th century, for the songs of the Troubadours belong to a period antecedent to the settlement of the language. The ballads composed in honour of Rodrigo Diaz de Vivar, called el Campeador, or more popularly the Cid, are among the earliest specimens of Spanish writing, and display at once great independence of thought, and felicity of expression. No doubt, however, the Moorish ballads, or those written to celebrate the chivalrous contests between Christian and Moslem knights, that preceded and accompanied the fall of Granada, form the most striking and distinctive part of the national literature of Spain. "The Moorish wars had always afforded abundant theme of interest for the Castilian muse; but it was not till the fall of the capital that the very fountains of song were broken up, and those beautiful ballads were produced, which seem like the rays of departed glory lingering round the ruins of Granada. They present a most remarkable combination of, not merely the exterior form, but the noble spirit of European chivalry, with the gorgeousness and effeminate luxury of the east. They are brief, seizing single situations of the highest poetic interest, and striking the eye of the reader with a brilliancy of execution so artless in appearance withal as to seem rather the effect of accident than study. We are transported to the gay seat of Moorish power, and witness the animating bustle, its pomp, and its revelry, prolonged to the last hour of its existence." (Prescott's Ferdinand and Isabella, ii. 200.) But it was, perhaps, hardly necessary to say so much about the Spanish ballads, as the admirable translations of Mr. Lockhart have made their spirit, at least, familiar to most readers.

The honour of being the first to introduce regular dramatic writing into Spain has been ascribed to Torres de Naharro, in the early part of the 16th century. He was followed by Lopez de Vega, born at Madrid in 1562, at once the most original, most unequal, and most voluminous of the peninsular dramatists. Calderon, born in 1600, carried the Spanish drama to its highest perfection. Like his great precursor, Lopez de Vega, his plays are most unequal, the finest scenes being mixed up with the most revolting barbarism and extravagance. The astonishing fecundity of these writers may in some degree account for, though it cannot excuse, the defects and inconsistencies in their dramas. The published works (which do not, however, embrace nearly all his pieces) of Lopez de Vega, consist of 25 vols. 4to., each containing 10 or 12 plays; and 127 dramas are ascribed to Calderon, besides a still greater number of vaudevilles, interludes, &c. The Spanish drama, however, has long fallen into decay. The humiliation of the country during the disastrous reigns of Philip IV. and Charles II., and the deadening influence of the Inquisition, were little favourable to its culture; and after the accession of the Bourbon dynasty to the throne, French criticism and taste obtained an ascendancy, while the troubles in which Spain has been more recently involved have stifled all poetical talent. Some endeavours, indeed, have been made to revive the national drama; but they have signally failed, and no modern name connected with this branch of literature deserves notice, except perhaps that of Martinez de la Rosa, the author of the Viuda de Padilla.

The Araucana of Ercilla, born in 1525, is the only poem that Spain has produced that has any pretensions to be classed among epics.

Chivalrous romance was early and assiduously cultivated in Spain. Happily, however, the inimitable satire of Cervantes destroyed at once and for ever the whole race of knights errant. His Don Quixote, however, still continues to interest all classes of readers by its exhaustless wit, the truth of its delineations, and its practical good sense. It has been rendered into almost all languages; and, how defactive soever the translation, it never fails to amuse and instruct.

But, with the exception of this unique and admirable work, Spanish works are but little known in foreign countries; and in most departments, indeed, the literature of Spain is poor in the extreme. And how could it be otherwise? In 1502 the censorship of the press was established;

and the power of carrying it into effect was very much entrusted to the Inquisition. " Il s'est établi dans Madrid," says Beaumarchais, with quite as much of truth as of wit, " un système de liberté sur la vente des productions, qui s'étend même à celles de la presse; et que, pourvu que je ne parle en mes écrits ni de l'autorité, ni de culte, ni de la politique, ni de la morale, ni des gens en place, ni des corps en credit, ni de l'Opéra, ni des autres spectacles, ni de personne qui tienne a quelque chose, je puis tout imprimer librement, sous l'inspection de deux ou trois censeurs." (Marriage de Figaro, acte v.) Under such circumstances, it would be contradictory and absurd to expect that the Spanish writers should have distinguished themselves in philosophical research, original discussion, or in any pursuit requiring freedom of inquiry. Spain has a few respectable, but no great or eminent authors.

Even so late as 1830, a respectable, and, which is not singular, an intelligent priest stated to Mr. Inglis "that nothing was so difficult as to obtain a license to publish a book, even though it contained no allusion to politics; and the better the book," said he, " the more difficult it is to obtain a license, and the more dangerous to publish; because government does not wish to encourage writing or even thinking upon any subject; and the publication of a good book sets men a-thinking." (Inglis, i. 212.)

Since 1830, however, a great change for the better has taken place; the censorship of the press has been suppressed, and the influence and authority of the clergy greatly diminished. Hopes, may, therefore, be reasonably entertained that literature will again revive; but no sudden development of the mental resources of the nation need be expected, and many years must elapse before literature acquire any material influence.

It would be to no purpose to take up the reader's time by entering into any lengthened details with respect to the state of education in Spain. It has been wholly in the hands of the clergy and the Inquisition; and, instead of contributing to expand and enlighten the mind, it has been so contrived as to prevent every sound principle, and imbue it with the grossest and most unworthy prejudices. The education of the Spaniards has been, in fact, the complement of their religion; and, like it, has been fitted and intended to enslave and debase the people, and to make them believe that it was alike their duty and their interest to submit, without murmur and without inquiry, to whatever their temporal and spiritual rulers might direct. Nothing, of course, like moral or political philosophy was taught in any seminary of Spain; there was scarcely that little progress made in the study of the learned languages, and science was in the most abject state. A thorough reform of the educational establishments of the country is indispensable; and it were better, indeed, that they should be wholly shut up than that they should continue on anything like their ancient footing. Efforts have been already made to introduce an improved system of elementary instruction; but, unfortunately, it has been necessary, from the want of other qualified parties, to intrust the superintendence and management of the schools to the parochial clergy, and they, speaking generally, are decidedly hostile to the new order of things, which they not unnaturally regard as the last degree inimical to their power and interests.

Taxes.—The taxation of Spain has been quite in harmony with her institutions, and, like them, seems to have been intended to obstruct as much good, and inflict as much evil on the country as possible. Probably, indeed, the alcabala is the most objectionable tax that has ever been imposed. It originated in 1341, and consisted at first of 10 on all commodities, whether raw or manufactured, as often as they were sold, and raised always according to the selling price! Such a monstrous impost was of itself sufficient to annihilate industry. It effectually hindered the manufactures from making any progress in Castile and the other provinces subject to its destructive influence. And Uztariz, Ulloa, and Campomanes, Spanish authors of the highest credit, agree in opinion with Mr. Townsend, that it is to their exemption from this odious tax that the comparatively flourishing state of industry in Catalonia and Valencia is mostly to be ascribed.

The alcabala, however, no longer exists in its original form; but, even in its amended shape, it is in the last degree objectionable. It is now converted into a local one, or octrois (derechos de puertas), paid on bringing produce into towns and villages.

In some instances the rate of duty has been diminished, but in the larger number it has been greatly increased. The same duties are not imposed in different towns; almost every one having a peculiar rate for itself. In general it is fixed by the ayuntamiento, or council of the town. Most foreign manufactured goods pay about 30 per cent. ad valorem, some as much as 60 per cent., exclusive of the customs duties charged at the frontier. Wine pays a duty which in many places is equal to 100 per cent. upon its value; and oil from 50 to 75 per cent. The duties press severely on every class, but chiefly on the poor, and are the subject of universal complaint. The *ayuntamiento* receives a certain per centage upon the amount collected at the gate of the town for local objects; the rest goes to the public treasury. In small villages it is levied only on necessaries, as meat, fowls, eggs, oil, corn, &c.

In towns which are not walled, or have no gates, a tax called the equivalente is levied, that is, the inhabitants are assessed in such a sum as the alcabala would probably produce, calculating from the revenue afforded by other towns of the same size. The alcaide and ayuntamiento are responsible for its payment, and they divide the sum very arbitrarily among the householders, according to their estimate (influenced of course, by every sort of partiality) of the products each ought to consume.

Next to the alcabala, tithe is the most oppressive tax in Spain, and the most complained of. Formerly it was exacted with the greatest rigour, and from articles (among others, from *los malos ungrros*, *de lo que pasan con su cuerpo*—Ley de Partida, 3d tit. 20. part 2), which might have been considered beyond the reach of even clerical rapacity. Arguelles estimated the value of the tithe at 600 millions of reals; and, according to official documents, it amounted in 1808, to 690 millions. In 1829, however, it was estimated, by a commission appointed to inquire into the subject, at only 335,694,000 reals; but it is believed that this estimate was as much below the truth as the former was probably above it. The clergy, however, did at no time receive the whole of this immense income; and, in consequence of the events of the last few years, but a very small ortion of it now finds its way into their coffers.

Among other taxes are the *Frutos civiles*, six per cent. on the produce of all rented lands; but this has recently been converted into a direct tax on land, trade, and manufactures; *Medias Annatas*, being the first half year's rent of entailed lands on the accession of the heir; *Lanzas*, a imposition tax in lieu of the troops which the nobles and mayorazos were formerly bound to furnish; *Subsidio del comercio*, a tax of £100,000, levied upon the merchants. The finance minister fixes the proportion which each intendancy must pay. The intendants then fix the proportion payable by each town; and the *ayuntimientos* assess the individuals. It is so unequally divided, that merchants of the first class pay £40 in some towns, and only £20 in others. It is generally believed that the amount really received, under pretence of this tax, is double the nominal amount; there being no means of checking the misconduct of the officers. There are, also, taxes on houses, on patents, cock-fights, the lottery, &c.; with the enormous customs duties already referred to.

The bulk of the taxes were formerly divided into two great classes, and the division is not yet wholly abandoned, the *rentas generales* and the *rentas provinciales*. The former are collected throughout all Spain, with the exception of Biscay. They include the revenue derived from the customs, the stamp duties, customs, &c., together with royal monopolies of salt, tobacco, and gunpowder. The *rentas provinciales* are collected only in the provinces belonging to the crown of Castile, and do not, therefore, affect Navarre, Catalonia, Aragon, or Valencia, which are peculiar and less burdensome taxes.

The tax denominated *paja y utensilio* is appropriated to support of the army. Previously to the reign of Philip the inhabitants of the different provinces in which the troops were quartered, were obliged to furnish them gratuitously, with various articles of subsistence. But, in 1719, this was changed into a money contribution, at certain special rates for officers and men. Nothing, however, can be more unjust and unequal than this tax, for it imposes a heavy burden on the places where troops are stationed, which other parts of the country, though perhaps larger, are entirely exempted.

The revenue derived from the tobacco and salt monopolies very considerable. Every one may buy any quantity of tobacco he chooses, provided he buy it in the government *cos*; but salt is the subject of assessment. The intendant fixes the consumption which should take place in each town under his charge, and the total quantities are delivered to the alcaides, who fix the specific quantities for each individual must pay, whether he use it or not. land-owners, farmers, &c. are charged with a quantity portioned to the number of individuals in their employ; and of the cattle and sheep which they possess.

The provinces of Navarre and Biscay, and some others, make by an annual contribution. which does not amount above £150,000, an exemption from a number of petty...

The *cruzada*, or bull granting permission to eat on Fridays, and four days every week during Lent, is general demand all over the kingdom, and is supposed to bring from £300,000 to £500,000.

Such is a rough outline of the taxation that has existed in Spain down to a very recent period, but which, it is highly probable, will now undergo very essential modifications, particularly the customs duties, which are of themselves sufficient to retain the country in the barbarism into which it has been plunged. We are unable to lay before the reader any authentic account of the revenue of Spain, but its gross amount may, perhaps, be estimated at about 850 millions reals, or £9,400,000 sterling. This, however, is little better than a guess; for no really trustworthy account of the revenue of Spain has been, nor, we believe, could be published for these many years back. The *ayuntamientos* of the different *pueblos*, into which the kingdom is divided, are mostly disinclined to render a fair account of their proceedings; and as they have great influence, a lengthened period will probably elapse, supposing the present government to maintain its footing, before any accurate account of the different items of revenue be made public. In the meantime it is abundantly notorious that the expenses of Spain very much exceed her income; and that she is, at this moment, wholly without the means of meeting her engagements. According to some recent semi-official statements, the total debt of Spain amounted in October, 1841, to 14,160,968,047 reals, or £157,344,080, of which sum the internal debt amounted to 8,318,985,279 reals, and the external debt of every description, to 5,841,982,768 reals, or to very near £65,000,000. A large amount of this debt is due to the English; and the interest on it has not been paid for a lengthened period.

Bad, however, as is the condition of Spain in a financial point of view, it is not desperate. The bankrupt state in which we find her is the fruit of centuries of despotism and vicious legislation, and of lengthened civil contests. But a liberal government has been organized; and, provided it be able to maintain itself, the institutions that have entailed so much ruin on the country will gradually be amended, and her vast resources, and the talents of her people, be again developed.

The present government of Spain has manifested a laudable anxiety to make good the engagements of the country; and, if it be not overthrown, the fair presumption seems to be that it will effect its object. The value of the unsold national property belonging to the state was estimated by Señor Mendizabel, on the 22d of May, 1840, at nearly £96,000,000, and as sales were effected in the course of the following year to the extent of about £16,500,000, its value may now (Dec. 1841) amount to about £79,500,000;[*] the whole of which is, according to existing arrangements, to be applied to the payment of the public debt. We are aware that these statements are not entitled to any great degree of confidence, partly from the difficulty of obtaining any accurate account of the public property, and partly from the degree in which its value is liable to be affected by the supposed stability of the government, the quantity of property thrown upon the market, the conditions of sale, &c. On the whole, however, the estimate now referred to would seem to be moderate. And, supposing that the government maintains itself, and is able to carry into effect the liberal policy to which it has given its sanction, the regeneration of Spanish finance and of the kingdom will be secured. A thorough reform of the customs laws would alone put 300,000,000 of reals, or above £3,000,000 sterling a year into the public treasury, exclusive of the other and still more important advantages that would result from it.

Army.—The regular army may amount to about 60,000 men; and there is, besides, an extensively organized militia. But, though abstemious, brave, and patient under fatigue and privations, the Spaniard wants that steady perseverance and intrepidity essential to the military character. The troops are, also, for the most part, badly disciplined, badly appointed, and will not bear to be compared, in any respect, with those of England, France, Germany, or Russia. The cavalry are particularly ill equipped; and the engineering department is such as might be expected from the state of science in the country. But, how deficient soever, the troops are, at all events, superior to the officers; and it is a curious and not easily explained fact, that, during the whole of the late war with France, not a single individual attained to distinction as a commander, or evinced any superior ability in the art of war, or in the conduct of considerable masses. Some of them were good guerilla leaders, but nothing more.

The Spanish navy, once so formidable, is now sunk to a state of almost total decay.

Races, Character, &c.—There are *four* distinct races in Spain: 1st, the Spaniards, who form the bulk of the population; 2dly, the *Basques* (about 500,000), descended from the ancient Cantabrians, and living in Navarre and the Basque provinces; 3dly, the *Morescoes*, descendants of the Moors, about 60,000 of whom still reside in Granada and the *Alpujarras*; and lastly, the *Gitanos*, or gypsies, a race (comprising about 50 000) spread all over the peninsula, but especially on the S.E. coasts; not strolling from place to place, as in England, but generally pursuing fixed occupations in the towns.* The Spaniards are middle-sized, thin, with well-proportioned limbs, dark hair, black piercing eyes, overshadowed by thick eyebrows, sharp features, and sallow complexions. The women are generally of middle or low stature, but gracefully formed, with almost aquiline noses, full, dark, expressive eyes, dark hair, and complexions varying from the flesh tint of N. Europe to the light olive of the Moors.

The character of the Spaniards has been very variously drawn; but, though it differs materially in different provinces, its discriminating features are not to be mistaken. Though commonly slow, cautious, and deliberate, they become, when their passions are roused, rash, violent, and precipitate in the extreme. Though formal, they are courteous in their bearing, and though grave, polite. The pride of the Spaniards is proverbial, and they entertain the most overweening opinion of themselves and their country. Though friendly, they are easily offended, vindictive, and more inclined to revenge real or fancied insults, than to remember favours. They are fond to excess of show and ostentation; and will endure the greatest privations at home to make a display in public. Their vicious institutions and their climate have made them in the last degree indolent and procrastinating. They are infinitely less jealous now than formerly; and their bigotry has become passive rather than active. They have ceased, in fact, to care much about religion; and are satisfied if they observe the fasts and unmeaning mummeries which it enjoins. Their ignorance often makes them attached to what is most ruinous to themselves; and those who think to gain their favour by denouncing some flagrant abuse, frequently find, to their surprise, that it is the object of popular attachment. They are temperate in eating and drinking, though it may be doubted whether this be not more the consequence of necessity than of choice. Morals are, speaking generally, even more corrupt in Spain than in Italy; and Mr. Townsend mentions, that at the period of his visit, in 1790, the priests divided with the officers the duty of *cortejos* to the ladies!

We subjoin, from the works of two of the most intelligent Englishmen who have ever visited Spain, Mr. Swinburne and Col. Napier, author of the classical work on the *Peninsula War*, the following notices of the Spanish character:—

"The listless indolence, equally dear to the uncivilized savage and to the degenerated slave of despotism, is no where more indulged than in Spain; thousands of men in all parts of the realm are seen to pass their whole day wrapped up in a cloak, standing in rows against a wall, or dozing under a tree. In total want of every excitement to action, the springs of their intellectual faculties forget to play, their views grow confined within the wretched sphere of mere existence, and they scarce seem to hope or foresee any thing better than their present state of vegetation; they feel little or no concern for the welfare or glory of a country, where the surface of the earth is engrossed by a few overgrown families, who seldom bestow a thought on the condition of their vassals. The poor Spaniard does not work, unless urged by irresistible want, because he perceives no advantage accrue from industry. As his food and raiment are purchased at a small expense, he spends no more time in labour than is absolutely necessary for securing the scanty provision his abstemiousness requires. I have heard a peasant refuse to run an errand, because he had that morning earned as much already as would last him the day without putting himself to any further trouble. Yet I am convinced that this laziness is not essentially inherent in the Spanish composition, for it is impossible without seeing them, to conceive with what eagerness they pursue any favourite scheme, with what violence their passions work upon them, and what vigour and exertion of powers they display when awakened by a bull-fight, or the more constant agitation of gaming,—a vice to which they are superlatively addicted. Were it again possible, by an intelligent spirited administration, to set before their eyes, in a clear and forcible manner proper incitements to activity and industry, the Spaniards might yet be roused from their lethargy, and led to riches and reputation; but I confess the task is so difficult that I look upon it rather as an Utopian idea than as a revolution ever likely to take place.

"Their soldiers are brave and patient of hardships; wherever their officers lead them, they will follow without flinching, though it be up to the mouth of a battery of cannon; but unless the example be given them by their commander, not a step will they advance. Most of the Span-

lards are hardy, and, when once engaged, go through difficulties without murmuring, bear the inclemency of the season with firmness, and support fatigue with strong perseverance. They sleep every night in their beds on the ground, are sparing in diet, perhaps more from a want of habitual indigence than from any aversion to luxury, whenever they can riot in the plenty of another man's table, they will gormandize to excess, and not content with stuffing their fill, will carry off whatever they can stuff into their pockets. I have more than once been a witness to the bringing of a supper by the numerous beaux and admirers which the ladies lead after them in triumph wherever they are invited. They are fond of spices, and scarce eat any thing without saffron, pimento, or garlic; they delight in what tastes strong of the pitched skin, and of oil that has a rank smell and taste; indeed, the same oil feeds their lamp, swims in their pottage, and dresses their salad, is also the lighted lamp is often handed down to the table, that each man may take the quantity he chooses. Much tobacco is used by them in smoking and chewing. All these hot dry kinds of food, co-operating with the parching qualities of the atmosphere, are assigned as causes of the spare make of the common people in Spain, where the priests and the innkeepers are almost the only well-fed, portly figures to be met with.

"The Spanish is by no means a naturally serious melancholy nation: misery and discontent have cast a gloom over them, increased, no doubt, by the long habit of distrust and terror inspired by the Inquisition; yet every village still resounds with the music of voices and guitars, and their fairs and Sunday wakes are remarkably noisy and riotous. They talk louder and argue with more vehemence than even the French or Italians, and gesticulate with equal if not superior eagerness. In Catalonia the young men are expert at ball; and every village has its *pelota*, or ground for playing at fives; but in the south of Spain I never perceived that the inhabitants used any particular exercise. I am told that in the island of Majorca they still wield the sling, for which their ancestors, the Baleares, were so much renowned.

"Like most people of southern climates, they are fiery in their persons, and overrun with vermin." (*Travels*, i. 104.)

"The Spanish character," says Col. Napier, "is distinguished by inordinate pride and arrogance. Dilatory and improvident, the individual as well as the mass, all possess an absurd confidence that every thing is practicable which their heated imaginations suggest; once excited, they can see no difficulty in the execution of a project, and the obstacles they encounter are attributed to treachery. Kind and warm in his attachments, but bitter in his anger, the Spaniard is patient under privations, firm in bodily suffering, prone to sudden passion, vindictive, bloody, remembers insult longer than injury, and cruel in his revenge. There is not upon the face of the earth a people so attractive in the friendly intercourse of society. Their majestic language, fine persons, and becoming dress, their lively imaginations, the inexpressible beauty of their women, and the air of romance which they throw over every action, and appeal to every feeling, all combine to delude the senses and impose upon the judgment. As companions, they are, incomparably, the most agreeable of mankind; but danger and disappointment attend the man who, confiding in their promises and energy, ventures upon a difficult enterprise. 'Never do to-day what you can put off till to-morrow,' is the common proverb in Spain, and rigidly followed." (i. 38, &c.)

In Spain there is a good deal of aristocratic pride, and the distinction of ranks is much attended to. The *hidalgos*, or gentry, claim to be descended from those Spaniards who, on the subjugation of the rest of the country by the Moors, found an asylum in the fastnesses of the northern provinces, whence they again gradually spread their conquering arms over the whole country. Besides the *hidalgos de sangre*, or by descent, there are also *hidalgos de privilegio*, or by office, conferred on them by the sovereign; but of these there are comparatively few. According to the official returns, there were in the kingdom, in 1787, 480,589 hidalgos, of whom no fewer than 491,040 belonged to the Asturias, Biscay, Burgos, Galicia, and Leon. Many of the hidalgos are extremely poor, and they are all inordinately proud.

The grandees of Spain who are the real nobility of the country, are the descendants of those who, in consideration of their eminent services, acquired the privilege of speaking in the cortes covered in presence of the king. A man may be a *titulado*, that is, he may enjoy the title of duke, marquis, or count, without being noble; and in Spain, in fact, such titles are of little more consequence than that of baronet in this country.

The hidalgos formerly enjoyed various privileges, but these have now been very much curtailed. Among others, they could entail lands or establish majorats; these, however, are now abolished.

The Manners and customs of the inhabitants very greatly

parts of Spain, and are much influenced by
the articles MADRID and SEVILLE.) The diet
of the and higher classes consists of chocolate for
with mutton, beef, and pork, especially the latter,
various ways, and accompanied by cabbage,
Spanish beans), onions, and large peas called
the olla, or cocido, is a favourite dish; and the
orizos) of Castile are said to be about the best
Wine is used only in small quantities, and the
common use are seldom much stronger than the
wines of France. The siesta, or repose during
the day, is customary to all classes throughout
in one to four o'clock, in Madrid and most other
shops are either shut, or a curtain drawn before
the shutters of every window are closed, and
sectable person is to be seen in the street. But
t the siesta is over, all is again instinct with
life. Exercise is usually taken in the evening,
the entire population is abroad. Tertulias, or
ties, are very frequent in the great towns. The
the frequented. Bull-fights, though discounte-
overnment, are in Spain what the circus was
only, the national pastime, favourite resort, and
ment of all classes. Though by no means entitled
t as musicians, the Spaniards have considerable
te; and all orders are passionately fond of
national dances being the bolero and fandango;
a graceful easy movement, the latter a dance of
nore licentious character, seldom seen in good

er classes live on wretched fare, rarely eating
ish only occasionally, except on the coast. The
ourers fare somewhat better, the chief articles
g bread, soup, garlic, bacon, and garbanzos, with
aliments of wine and oil. Notwithstanding the
of the convents, mendicity is still exceedingly
and perhaps the only remedy for this inveterate
the abolition of all endowments for paupers, and
hment of some provision for the poor, as in Eng-
alms and broad-brimmed hats are very generally
a men; and the mantilla and fan are in universal
females.

several and all other travellers in Spain have
iptions of bull-fights. We extract the following
om Mr. Inglis's work, not because it is the best,
but it is one of the most recent:

all-fight is the national game of Spain, and the
Spaniards for this spectacle is almost beyond be-
lay, in Madrid, is always, during the season of
hts, a kind of holiday; everybody looks forward
yment of the afternoon, and all the conversation
toros. Frequency of repetition makes no differ-
true amateur of the bull-fight; he is never weary
1 times he finds leisure and money to dedicate to
te pastime. The spectacle is generally announced,
the of his majesty, to begin at four o'clock; and,
n, all the avenues leading towards the gate of
in commotion; the Calle de Alcala, in particu-
hout its whole immense extent, is filled with a
vd, of all ranks and conditions, pouring towards
A considerable number of carriages are also seen,
oyal carriages; but these arrive later; and there
nany hack cabriolets, their usual burden being a
ad two girls dressed in their holiday clothes, for
o way of showing gallantry so much approved,
e lower orders, as treating to a bull-fight; and
is carried so far as to include a drive in a red and
riolet, the peasant need sigh no longer.

been able to secure a place in one of the best
te spectacle was most imposing: the whole amphi-
hd to contain 17,000 persons, was filled in every
d and round, and from the ground to the ceiling,
he imagination back to antiquity, and to the butch-
Roman holiday. The arena is about 230 feet in
this is surrounded by a strong wooden fence,
feet in height, the upper half retiring about a foot,
ave in the middle of the fence a stepping-place,
the men may be able, in time of danger, to throw
es out of the arena. Behind this fence there is an
e about 9 feet wide, extending all the way round,
a retreat; and where, also, the men in reserve
iling, in case their companions should be killed or

Behind this space is another higher and stronger
inding the amphitheatre, for the spectators: from
e the seats decline backward, rising to the outer
d above these there are boxes, which are all roofed,
of course, open in front. The best places in the
it about 4s.; the best in the amphitheatre below,
6d.; the commonest place, next to the arena, costs
in the centre of the W. side is the king's box;
tered here and there, are the private boxes of the
and amateurs. In the boxes I saw as many wom-

en as men; and in the lower parts the female spectators
were also sufficiently numerous: all wore mantillas; and
in the lower parts of the amphitheatre, which were exposed
to the sun, every spectator, whether man or woman, carried
a large circular paper fan, made for the occasion, and sold
by men who walk round the arena before the fight begins,
raising among the spectators their long poles, with fans suspended, and a little bag fixed here and there, into which the
purchaser drops his four quartos (1½d.).

"The people now began to show their impatience, and
shouts of el toro were heard in a hundred quarters; and
soon after a flourish of trumpets and drums announced that
the spectacle was about to commence. This created total
silence, and the motion of the fans was, for a moment, suspended: first entered the chief magistrate of the city on
horseback, preceded by two alguacils, or constables, and
followed by a troop of cavalry, who immediately cleared the
arena of every one who had no business there; next, an
official entered on foot, who read an ordonnance of the
king, commanding the fight, and requiring order to be kept;
and these preliminaries having been gone through, the
magistrate and cavalry retired, leaving the arena to the
two picadors, who entered at the same moment. These are
mounted on horseback, each holding a long lance or pike,
and are the first antagonists the bull has to encounter;
they stationed themselves on different sides of the area,
about 20 yards from the door at which the bull enters; and
at a new flourish of trumpets, the gate flew open, and the
bull rushed into the area: this produced a deafening shout,
and then total silence. The bulls differ very widely in
courage and character; some are rash—some cool and la-
trepid—some wary and cautious—some cowardly—some,
immediately upon perceiving the horse and his rider, rush
upon them; others run bellowing round the arena—some
make towards one or other of the chulos, who, at the same
moment that the bull appears, leap into the arena with coloured cloaks upon their arms; others stop, after having advanced a little way into the arena, look on every side, and
seem uncertain what to do. The blood of the bull is generally first spilt: he almost invariably makes the first attack,
advancing at a quick trot upon the picador, who generally
receives him upon his pike, wounding him somewhere about
the shoulder. Sometimes the bull, feeling himself wounded, retires to meditate a different plan of attack; but a good
bull is not turned back by a wound—he presses on upon his
enemy, even if, in doing so, the lance be buried deeper in
his flesh. Attached to the mane of the bull is a crimson
riband, which it is the great object of the picador to seize,
that he may present to his mistress this important trophy of
his prowess. I have frequently seen the riband torn off at
the moment that the bull closed upon the picador.

"The first bull that entered the arena was deficient both
in courage and cunning: the second was a fierce bull of
Navarre, from which province the best are understood to
come: he paused only for a moment after entering the
arena, and then instantly rushed upon the nearest picador,
who wounded him in the neck; but the bull, disregarding
this, thrust his head under the horse's belly, and threw
both him and his rider upon the ground: the horse ran a
little way, but, encumbered with trappings, he fell; and
the bull, disregarding for a moment the fallen picador, pursued the horse, and, pushing at him, broke the girths, and
disengaged the animal, which, finding itself at liberty, galloped round the arena, a dreadful spectacle, covered with
gore, and its entrails trailing upon the ground. The bull
now engaged the chulos: these young men show great dexterity, and sometimes considerable courage, in the running
fight, or rather play, in which they engage the bull, flapping
their cloaks in his face, running zigzag when pressed, and
throwing down the garments to arrest his progress a moment,
and then vaulting over the fence, an example which is sometimes followed by the disappointed animal. But this kind
of warfare by the bull of Navarre seemed to consider child's
play; and leaving these cloaked antagonists, he made furi-
ously at the other picador, dexterously evading the lance,
and burying his horns in the horse's breast: the horse and
his rider extricated themselves, and galloped away; but
suddenly the horse dropped down, the wound having proved
mortal. The bull, victorious over both enemies, stood in
the centre of the arena, ready to engage another; but the
spectators, anxious to see the prowess of the bull directed
against another set of antagonists, expressed their desire by
a monotonous clapping of hands and beating of sticks, a
demonstration of their will perfectly understood, and always
attended to.

"The banderilleros then entered: their business is to
throw darts into the neck of the bull; and, in order to do
this, they are obliged to approach with great caution, and to
be ready for a precipitate retreat; because it sometimes happens that the bull, irritated by the dart, disregards the cloak
which the banderillero throws down to cover his retreat,
and closely pursues the aggressor. I saw one banderillero so

closely pursued, that he saved himself only by leaping over the bull's neck. The danger, however, is scarcely so great as it appears to the spectator to be; because the bull makes the charge with his eyes shut. The danger of the picador who is thrown upon the ground is much greater; because, having made the charge, the bull then opens his eyes, and the life of the picador is only saved by the address of the chulos, who divert the attention of the victor. Generally the banderilleros do not make their appearance until the bull appears, by his movements, to decline the combat with the picadors, which he shows by scraping the ground with his feet, and retiring. If the bull show little spirit, and the spectators wish that he should be goaded into courage, the cry is 'fuego,' and then the banderilleros are armed with darts, containing a kind of squib, which explodes while it sticks in the animal's neck.

"When the people are tired of the banderilleros, and wish to have a fresh bull, they signify their impatience in the usual way, and the signal is then given for the matador, whose duty it is to kill the bull. The matador is in full court dress, and carries a scarlet cloak over his arm, and a sword in his hand: the former he presents to the bull; and when the bull rushes forward, he steps aside and plunges the sword in the animal's neck; at least so he ought to do, but the service is a dangerous one, and the matador is frequently killed. Sometimes it is impossible for the matador to engage upon equal terms a very wary bull, which is not much exhausted. This was the case with the sixth bull which I saw turned out: it was an Andalusian bull, and was both wary and powerful. Many times the matador attempted to engage him, but without success; he was constantly upon the watch; always disregarding the cloak, and turning quick round upon the matador, who was frequently in imminent danger. At length the people were tired of this lengthened combat, and seeing no prospect of it ending, called for the semi-luna, an instrument with which a person skulks behind, and cuts the ham-strings of the animal: this the bull avoided a long while, always turning quickly round; and even after this cruel operation was performed, he was still a dangerous antagonist, fighting upon his knees, and even pursuing the matador. The moment the bull falls he is struck with a small stiletto, which pierces the cerebellum; folding doors, opposite to those by which the bull enters, are thrown open, and three mules, richly caparisoned and adorned with flags, gallop in; the dead bull is attached by a hook to a chain, and the mules gallop out, trailing the bull behind them: this is the work of a moment—the doors close—there is a new flourish of trumpets, and another bull rushes upon the arena.

"And how do the Spaniards conduct themselves during all these scenes? The intense interest which they feel in this game is visible throughout, and often loudly expressed; an astounding shout always accompanies a critical moment: whether it be the bull or the man who is in danger, their joy is excessive; but their greatest sympathy is given to the feats of the bull. If the picador receives the bull gallantly, and forces him to retreat: or, if the matador courageously faces and wounds the bull, they applaud those acts of science and valour; but if the bull overthrow the horse and his rider, or if the matador miss his aim, and the bull seems ready to gore him, their delight knows no bounds. And it is certainly a fine spectacle to see the thousands of spectators rise simultaneously, as they always do when the interest is intense: the greatest and most crowded theatre in Europe presents nothing half so imposing as this. But how barbarous, how brutal, is the whole exhibition! Could an English audience witness the scenes that are repeated every week in Madrid? A universal burst of 'shame!' would follow the spectacle of a horse, gored and bleeding, and actually treading upon his own entrails, while he gallops round the arena; even the appearance of the goaded bull could not be borne—panting, covered with wounds and blood, lacerated by darts, and yet brave and resolute to the end.

"The spectacle continued two hours and a half; and during that time there were seven bulls killed, and six horses. When the last bull was despatched, the people immediately rushed into the arena, and the carcass was dragged out amid the most deafening shouts." (i., 148.)

Historical Notice.—After being in part occupied by the Carthaginians, Spain became the prey of the Romans, by whom she was finally subdued in the reign of Augustus. After enjoying a lengthened period of tranquillity and prosperity under the sway of the Romans, Spain was invaded, in the beginning of the 5th century, by the Vandals, and other Gothic tribes; and in the next century the Visigoths acquired the ascendancy, and established their supremacy in every part of Spain. The latter, however, were not long permitted peaceably to enjoy this fine and fertile country. In 711 a powerful Arabian force crossed the strait of Gibraltar, and having defeated the Visigoths, and killed Roderick their king, in a great battle near Xeres de la Frontera, in

Andalusia, they speedily overran the whole country, driving the remains of the Visigoths into the fastnesses of the North, where they did not think it worth their while to know them. But the Saracens having been signally defeated by Charles Martel in France, and their fiery zeal having subsided, the Christians began to descend from the mountains of the Asturias, and gradually recovered portions of the several countries. The kingdom of Leon was founded under Alphonso I., about the middle of the 8th century; and from that period, notwithstanding the superior civilisation, learning, and splendour of the Saracenic sovereigns, the Christian power was progressively increased at the expense of that of the Mohammedans.

The provinces that were wrested from the Moors were not formed into one, but into several independent states, of which Castile and Aragon gradually absorbed most of the others. In the 15th century these two leading states were united by the marriage of Ferdinand of Aragon, with Isabella of Castile; and having conquered Granada, the last possession of the Moors, in 1492, and subsequently annexed all that part of Navarre to the S. of the Pyrenees, the whole of Spain was united under the same government; and Naples being at the same time conquered, and America discovered, Ferdinand, besides being one of the ablest princes of his day, became the most powerful.

Ferdinand was succeeded by his grandson, Charles I., known in history as Charles V., emperor of Germany; who added, by his father's side, the archduchy of Austria and the Low Countries to the vast inheritance of Spain and the Indies, now augmented by the conquests of Mexico and Peru. Charles, the most illustrious by far of the Spanish sovereigns, was succeeded in his Spanish dominions and in the Low Countries by his eldest son, Philip II., the husband of Queen Mary, of England, who, having conquered Portugal, in 1580, reduced the entire peninsula into one kingdom. The conquest of Portugal may be said to mark the culminating point of the Spanish monarchy. The tyranny and intolerance of Philip had already, indeed, raised a rebellion in the Low Countries; which, after a struggle unexampled for duration, for the sacrifices it entailed on the weaker party, and for its beneficial consequences, terminated in the independence of the Seven United Provinces. The power of Spain now began rapidly to decline. The seeds of that decay had, however, been profusely scattered in the reign of Ferdinand and Isabella. The establishment of the Inquisition and of the censorship of the press, and the attacks made on the ancient rights and liberties of the nation, paralysed its energies; and the unsuccessful rebellion of the commons of Castile, under Charles V., and the brutal and ferocious bigotry of Philip II., extinguished every spark of civil and religious liberty, and subjected the country to the vilest of all despotisms, that which principally depends for support on intolerance, superstitious zeal, and religious quackery.

Under such a government Spain either continued stationary or retrograded, while the surrounding nations made rapid advances in the career of civilization. Her sovereigns were as imbecile as the country; and on the death of Charles II., the last prince of the Austrian line, the monarchy was dismembered; and it was the arms of Louis XIV., and the talents of the Duke of Berwick, that secured the will of the deceased monarch or the wishes of the Spaniards, that placed a Bourbon dynasty on the throne.

The new dynasty was less intolerant than that to which it succeeded; and some reforms were introduced during the course of last century. These, however, were of comparatively slight importance; and it was clear that the abuses under which the country laboured were so deeply seated, and so entwined with every existing institution, and with the habits and prejudices of the people, that they could not be eradicated, nor even materially abated, otherwise than by a revolution.

This was brought about by the weakness of the Spanish sovereigns and the immeasurable ambition of Napoleon. Not satisfied with a considerable subsidy from Spain, Napoleon wished to reduce it to the state of a province of France; and in this view he procured the abdication of the imbecile monarch, Charles IV., and proceeded to seat his brother Joseph on the vacant throne. The opposition made by the Spaniards to this transfer, and the important results to which it led, are well known to every body, and need not be here alluded to. We are not, however, of the number of those who are inclined to ascribe any great merit to the Spaniards for their opposition to Napoleon, who, but for the interference of England, would have had no great difficulty in suppressing the insurrection. It is idle to talk of the patriotism of the Spaniards. The government, for which they took up arms, was nothing but a tissue of abuses, and their "beloved" sovereigns the merest drivellers. But the pride of the nation was hurt; and the priests, who knew that the ascendancy of the French would be a death-blow to them, did not fail to set them in the most odious light, and used every possible means to make them the objects of fanatical hatred. But, as already stated, it was the arms of England, and the extraordinary talents of her great general, and not the stupid fanaticism of the Spaniards, that repulsed the French beyond the Pyrenees. It is unnecessary to recapitulate the events that followed the restoration of Ferdinand. After years of civil war and bloodshed, a liberal government has happily been established. Whether it be destined to last, time only can show; but if it be, there can be no doubt that the abuses which still infest the country will be gradually exterminated, her gigantic resources developed, and the well-being of her people, and her power and importance in the scale of nations, vastly increased.

SPALATRO, a city and seaport of Dalmatia, on the Adriatic, opposite the island of Brazza, lat. 43° 30′ 12″ N., long. 16° 26′ 33″ E. Pop. 7500. It is surrounded by ruined walls, is the seat of an archbishop, has a cathedral and several other churches, a lazaretto, several convents, a gymnasium, normal school, &c., with barracks, and a military hospital. It has both an outer and inner harbour, the former affording secure anchorage to vessels of any burden. It has rather a considerable trade, consisting principally in the exportation of the produce of the surrounding country, as well as of products brought from Bosnia, including cattle, horses, figs, rosoglio, wax, &c. There are thermal springs in the immediate vicinity of the town.

Spalatro, however, would hardly have been worth mentioning in a work of this kind, but for its containing the ruins of the magnificent palace built by Diocletian. The emperor belonged to Salona, a now ruined city about 3 m. N.E. from Spalatro; and being warmly attached to his native country, he retired thither to spend the remainder of his days, after his abdication of the imperial purple, A.D. 305. From the vastness of the palace, it is all but certain that he had begun its erection long previously to his abdication. Though, most probably, in the contemplation of that extraordinary event. The situation seems to have been most judiciously chosen. "The soil is dry and fertile, the air pure and wholesome; and, though extremely hot during the summer months, this country seldom feels those sultry and noxious winds to which the coasts of Istria, and some parts of Italy, are exposed. The views from the palace are no less uitful than the soil and climate are inviting. Towards W. lies the fertile shore that stretches along the Adriatic; in which a number of small islands are scattered in a manner as to give this part of the sea the appearance of a great lake. On the N. side lies the bay which led to the ancient city of Salona; and the country beyond it appearing in sight, forms a proper contrast to that more extensive prospect of water which the Adriatic presents both to S. and the E. Towards the N., the view is terminated by high and irregular mountains, situated at a proper distance, and, in many places, covered with villages, woods, vineyards." (*Adam's Ruins of Spalatro*, p. 6.) The want of good water, its only defect, was obviated by the construction of an aqueduct, which conveyed an abundant supply from Salona.

The palace was in the form of a quadrangle, flanked by towers. Its longer sides, including the towers, were each about - in length, and its shorter 592 ft., so that it covered in nearly 9½ English acres! It was constructed of a beautiful freestone, but little inferior to marble. "Four streets, intersecting each other at right angles, divided the several parts of this great edifice, and the approach to the principal apartment was from a very stately entrance, still denominated the golden gate. The approach was terminated by a peristyle of granite columns; on one side of which we enter the square temple of Æsculapius (now the church of St. John the Baptist), and on the other the octagon temple of Jupiter (now the cathedral). The buildings were all high from the top, and appear to have consisted only of masonry." (*Gibbon*, cap. 13.)

Diocletian expired in this splendid retreat, A.D. 313. The emperor Constantine Porphyrogenitus, who could only see that it was in a neglected and decaying state, affirms that no description could convey a proper idea of its grandeur. Now, though the town of Spalatro has been principally built out of its ruins, its prodigious remains give a vivid idea of the wealth and magnificence of the Roman emperor. (*Gibbon ubi supra.*)

SPALDING, a market town and par. of England, co. Lincoln, part. Elloe, div. Holland, on the Welland, 23 m. S. Lincoln. Area of par., 12,070 acres. Pop., in 1831, though within the fens, the town is well drained, its clean and well paved, and the houses have a nice entrance. It consists of four principal streets, and various market-place, at one end of which is the t, erected at the expense of a private gentleman, part being let in shops for the benefit of the poor. One open space is the house of correction for the part of Holland, a brick building of four stories, constant an expence of £16,000 but said to be in several

respects defective. (See *Inspect. of Prisons*, 3d. Rep.) The parish church, originally erected in the 13th century, but rebuilt, with some additions, during the 15th, is a light structure in the perpendicular style, with a fine tower and crocketed spire, and a handsome porch. The living, a perpetual curacy, in the gift of a body of trustees, is worth £250 a year nett. (*Ecclesiast. Rev. Report.*) There are places of worship for Independents, Baptists, Wesleyans, Friends, &c., and many charities; including Queen Elizabeth's grammar school, which had, in 1835, an income of £199 15s., common, blue-coat, and other schools, and an aggregate of endowments for the benefit of the poor, termed the "Town Husbands," amounting to about £450 a year. (*Thirty Second Rep. on Charities*, Part IV.) There are assembly and card-rooms in the town-hall, a subscription library, a literary club, &c.; and formerly an antiquarian society was established at Spalding, of which Sir Isaac Newton, Sir Hans Sloane, Dr. Stukely, and other eminent persons, were members. The Welland is navigable thus far for sloops of 30 tons burden, which come up to the centre of the town, and land or take in cargoes, at the doors of the warehouses. Spalding, which is a member of the port of Boston, has a considerable trade in corn, coal, and Yorkshire and Norfolk wool; and its weekly market is the largest in the county for the fat cattle reared in the adjacent marsh lands. Most part of the neighbourhood is appropriated to grazing.

Spalding has long been the principal seat of the law-courts for the division of Holland. At present, quarter and petty weekly sessions, courts baron and leet, a court of requests, and a court of sewers, are held in the town-hall. Market-days, Tuesdays. Fairs, April 27th, June 30, Aug. 28, Sept. 25, and Wed. before Dec. 6, chiefly for cattle, horses, hemp, and flax.

SPANDAU, a strongly fortified town of Prussia, prov. Brandenburg, gov. Potsdam, at the junction of the Havel with the Spree, 7 m. W. Berlin. Pop., in 1837, ex. garrison, &c., 6753. "Spandau, in the time of the great Frederick, was, and still continues to be, the state-prison of Prussia. It may at any moment be covered from a hostile approach by letting out the waters of the rivers which meet under its walls. Being filled with troops, it has more the air of an enormous barrack than of a town; and is, in point of fact, so regarded. Both its citadel and penitentiary are deserving of notice; the former, on account of its position on an island of the Spree, the latter, because it is said to be managed with exceeding skill." (*Gleig's Germany*, &c., i. 56.) The citadel is a regular square with four ramparts, 40 ft. in height, and good casements: the penitentiary was formerly the residence of the electors of Brandenburg, and now has, says Berghaus, 750 inmates, many of whom are criminals sent from the capital. The principal streets are clean, airy, and spacious, in spite of the disproportionate height of the houses. The church of St. Nicholas, constructed in the 16th century, has a great number of monuments. Spandau is the seat of a civil tribunal, and a forest-board; and has some manufactures of woollen and linen cloths, tobacco-pipes, and earthenware; with breweries, distilleries, &c. It was the scene of Baron Trenck's captivity. It was taken by the Swedes in 1631, and the French in 1806. (*Berghaus; Stein, &c.*)

SPANISH TOWN (or *Santiago de la Vega*), the cap. and seat of the legislature of the island of Jamaica, co. Middlesex, on the river Coire, about 6 m. from the sea, and 11 m. W.N.W. Kingston. It is the official residence of the governor, and the commander-in-chief, and the seat of the court of chancery, the supreme court of judicature, &c.; but has otherwise very little importance, and a population of at most only 5000. It is very dirty, badly drained, and the inhabitants are at all times very subject to febrile diseases. (*Tulloch's Report on the W. Indies*, p. 58.)

SPARTA, p. t., Livingston co., N.Y., 241 m. W. Albany, 336 W. Drained by Canaseraga creek. Dansville branch of Genesee Valley canal passes through it. It contains four churches, a Presbyterian, Lutheran, Methodist, and Episcopal; 25 stores, four fulling-mills, one woollen factory, four furnaces. seven grist-mills, 16 saw-mills, one oil-mill, four paper-mills, four tanneries, one distillery, one brewery, one printing-office; one academy, 12 students; 29 schools, 1829 scholars. Pop. 5841.

SPARTA, p. v., capital of Hancock co., Ga., 93 m. N.E. Milledgeville. 4488 W. It contains a courthouse, jail, a Methodist church, a female academy of a high order, 10 stores, 50 dwellings, and about 600 inhabitants, nearly half of whom are coloured.

SPARTA, p. v., capital of White co., Tenn., 83 m. E. by S. Nashville. 598 W. Situated at the foot of Cumberland mountains, on a branch of Caney fork of Cumberland river. It contains a courthouse, jail, a church, an academy; five stores, various mechanic shops, and about 350 inhabitants.

SPARTANBURG, district, S.C., situated in the N. part of the state, and contains 1650 sq. m. Bounded N.E. by Broad river, S.W. by Ennoree river. Drained by Pacolet and Tygar rivers and their branches. It contained in 1840, 20,496 neat cattle, 11,104 sheep, 31,251 swine; and produced 58,670 bushels of wheat, 792,751 of Indian corn, 67,815 of oats, 2529 of potatoes, 1,595,303 pounds of cotton. It had 39 stores, two furnaces, three forges, four cotton factories with 2207 spindles, six flouring-mills, 56 grist-mills, 41 saw-mills, eight tanneries, 37 distilleries, two potteries; five academies, 179 students; 17 schools, 375 scholars. Pop.: whites, 17,994; slaves, 5687; free coloured, 58; total, 23,669. Capital, Spartanburg.

SPARTANBURG, C.H., p. v., capital of Spartanburg dist., S.C., 96 m. N.W. Columbia, 471 W. It has a pleasant and healthy location, and contains a courthouse, jail, and about 1000 inhabitants.

SPENCER, county, Ky., situated a little N. of the centre of the state, and contains 280 sq. m. Drained by the E. fork of Salt river and its branches. It contained in 1840, 5613 neat cattle, 8495 sheep, 23,497 swine; and produced 70,786 bushels of wheat, 14,083 of rye, 394,765 of Indian corn, 71,722 of oats, 3908 of potatoes, 49,900 pounds of tobacco. It had 14 stores, 16 grist-mills, nine saw-mills, one tannery, 21 distilleries; two academies, 118 students; nine schools, 275 scholars. Pop.: whites, 4650; slaves, 1911; free coloured, 20; total, 6581. Capital, Fayetteville.

SPENCER, county, Ia., situated in the S. part of the state, and contains 400 sq. m. Bounded S. by Ohio river, E. by Anderson's creek, W. by Little Pigeon creek. It contained in 1840, 5673 neat cattle, 5713 sheep, 17,943 swine; and produced 18,339 bushels of wheat, 53,548 of Indian corn, 17,719 of oats, 12,796 of potatoes, 96,407 pounds of tobacco, 2706 of sugar. It had 15 stores, 12 grist-mills, five saw-mills, four tanneries, one printing-office, one weekly newspaper; 19 schools, 300 scholars. Pop. 6305. Capital, Rockport.

SPENCER, p. t., Worcester co., Mass., 53 m. W. Boston, 395 W. Incorporated from Leicester in 1753. Watered by branches of Chickapee river. It contains three churches, a Congregational, Baptist, and Universalist; five stores, six grist-mills, nine saw-mills; 11 schools, 598 scholars. Pop. 2004.

SPENCER, p. t., Tioga co., N.Y., 15 m. W. Oswego, 179 m. W.S.W. Albany, 280 W. Drained by Catatonog creek and its branches. It contains a Methodist and Presbyterian church, seven stores, three grist-mills, 20 saw-mills, two tanneries: 429 scholars in schools. Pop. 1532.

SPENCER, t., Guernsey co., O., situated on the head-waters of Will's creek. Pop. 1669.

SPENCER, p. v., capital of Owen co., Ia., 50 m. S.W. Indianapolis. Situated on the N.W. bank of the W. fork of White river. It contains a courthouse, jail, two churches, a Methodist and Baptist; four stores, an oil-mill, rope-walk, copper and tin factory, and other mechanic shops, and about 375 inhabitants.

SPEY, a river of Scotland, in the Highlands. It has its source in loch Spey, within about 6 m. of the head of loch Lochy, and thence pursues a N.E. course through Badenoch and Strathspey to Fochabers, below which it falls into the Moray frith. It receives innumerable mountain streams, but no very important tributary. Following its windings, the course of the Spey is about 96 m.; but it is only about 75 m. in a direct line from its source to its mouth. It drains about 1300 sq. m. of country, and, besides being one of the largest, is admitted to be the most rapid of Scotch rivers. Being fed wholly by mountain torrents, it is very liable to sudden and destructive inundations. It flows through what is the best wooded portion of the Highlands. The Duke of Richmond (proprietor of the Gordon estates) has several valuable salmon fisheries on this river.

SPEZZIA (Ital. Spezia), a town and seaport of N. Italy, Sardinian dom., div. Genoa, prov. Levante, of which it is the capital, at the extremity of the gulf of its own name, 50 m. E.S.E. Genoa. Pop., in 1838, inc. comm., 9796. It is finely situated, is tolerably well built, and has an excellent harbour. Napoleon, aware of the advantages of its position, is said to have intended making it a naval station and arsenal; and since he drew attention to its importance, its commerce has improved. The gulf of Spezia (anc. Portus Lunæ), is about 3½ m. in length, by an average breadth of half as much. It exhibits in one part the phenomenon of a powerful spring of fresh water, which bubbles up from the bottom, and preserves its purity, unmixed by the surrounding salt water, nearly to the surface. (Conder's Italy, i. 267.)

SPILSBY, a market town and par. of England, co. Lincoln, soke Bolingbroke East, in Lindsey, on an eminence near the Limb, 10 m. from the sea, and 27 m. S.E. Lincoln. Area of par., 2340 acres. Pop., in 1831, 1384. The town consists of four streets, diverging from a central square, which forms the market-place. The latter is ornamented on its E. side by the market-cross, a plain octagonal shaft, with a quadrangular base, elevated on five steps; and on the W. by the town-hall, built in 1764. The parish church is an irregular structure, consisting of two aisles.

with a handsome embattled tower at the W. end, said to have been built in the reign of Henry VII. In the interior are several antique monuments. The living, a rectory, in the gift of Lord Willoughby d'Eresby, is worth £109 a year. Near the town is a new residence, house and prison for the division of Lindsey, occupying about ten acres of ground, with a Doric portico in front, constructed at a cost of £25,000. Spilsby being the principal town in the S. part of Lindsey, is the seat of the general quarter sessions, and of petty sessions. It has several charities, particularly a free school, founded by Lord Willoughby in 1611, and which has now an income of £98, and instructs 44 poor children of the parish, besides whom there are about 50 pay-scholars. (Charity Rep., 39, pt. iv.) Market-days, Mondays; fairs, Monday before and after Whitsuntide, annually, and 3d Monday in July.

SPIRES (Germ. Speier; anc. Noviomagus), a city of W. Germany, in Rhenish Bavaria, of which it is the cap. on the Rhine, where it is joined by the Speyer, 164 m. N.E. Landau; lat. 49° 19' N., long. 8° 26' 16" E. Pop., in 1833, 528, of whom about 3000 were Roman Catholics. This is supposed to be one of the most ancient, as it long was one of the chief cities of Germany. In the 14th century it is stated to have had 97,000 inhabitants; and in the 16th and 17th centuries it was the seat of the imperial chamber (Reichskammergut), or superior court of appeal for the German empire; and previously to 1668, it had five suburbs, enclosed within ramparts, 13 gates, and 64 towers provided with artillery. But in that year it was taken and all but destroyed by the troops of Louis XIV.; and though rebuilt about ten years afterward, it has never attained its previous prosperity. It still occupies a large extent of ground, but its walls, which are entered by five gates, enclose numerous open spaces. The cathedral, which withstood the attempts of the French wholly to destroy it, is the most remarkable building. It was founded and completed in the 11th century, on the site, as is alleged, of a Roman temple of Venus; and it is perhaps the most unquestionable edifice existing in the round arched style. Nine German emperors, and many other celebrated personages, have been buried in it, but their tombs were ransacked and mutilated by the French in 1689 and 1794.

Since 1819, however, the Bavarian government has done much to repair the interior of the cathedral, and the duke of Nassau has erected a splendid modern monument to his ancestor the Emperor Adolph. Spires has numerous Roman Catholic and two Lutheran churches, a gymnasium, an orphan asylum, house of correction, forest school, botanic garden, and a hall of antiquities, in which many curiosities found in the province are deposited. The outer walls are still standing of an old palace, in which no fewer than 49 diets have been held. Of these, the most celebrated by far was that held in 1549, on the subject of the religious disputes that then agitated the empire. On this occasion the majority, consisting of the party attached to the church of Rome, agreed to a resolution by which all changes in the doctrine and discipline of the established church, not previously approved by a general council, were declared to be unlawful and of no effect. The minority, including the princes and others attached to the doctrine of the Reformers, presented, on the 19th of April, 1529, a protest against the above resolution; and from this circumstance they acquired the name of Protestants, which has since become the distinguishing term for those who have renounced the communion of the church of Rome, how much soever they may differ among themselves. (Menkenius, iv., 73, ed. 1728.) Noviomagus was included by the Romans in Germania Prima. It was the winter quarters of Cæsar, by whom it was fortified, as a check on the incursions of the neighbouring Allemanni. Several Roman, Frank, Saxon, and Suabian emperors embellished and made it their residence; and Henry V. of Germany gave the citizens of Spires a monopoly of the transit trade of the Rhine, and other valuable privileges. During the French ascendancy Spires was the capital of the department of Mont-Tonnerre. (Schreiber, Guide du Rhin; Berghaus, Allg. Länder, &c.)

SPITZBERGEN (formerly called E. Greenland), a group of islands in the Arctic ocean, being the most N. land hitherto discovered, between the 76th and 81st degs. of N. lat., and the 9th and 23d of E. long., about £80 m. N.N.W. the North cape, and nearly midway between Greenland and Nova Zembla. There are four principal islands, Spitzbergen Proper, N.E. and S.E. islands, and Prince Charles's island to the W. of the others; besides many islets and rocks. Their united area does not appear to be equal to that of Iceland. They rise in many places into mountains, of from 1000 to 2000 ft. in height, the peaks of which are covered with snow, coeval, perhaps, with their creation. The coasts are iron-bound, presenting only a few inhospitable harbours; the best of these is Smeerenberg, on the W. coast, where the Dutch had once a considerable establishment. The surface is, for the most part, destitute of any vegetable or animal products; but there are a few bears and foxes, which live upon fish, &c. Spitzbergen was formerly a principal station of the whale-fishers; but the whales have for a considerable period been nearly extirpated in the surrounding seas, and it is now but little visited. It was originally discovered by Sir Hugh Willoughby in 1533, and was first visited by the Dutch in 1595. Its shores were principally surveyed by Captain Phipps, in 1773.

SPOLETO (an. *Spoletium*), a city of central Italy, Papal States, cap. deleg., and formerly of the duchy of its name, on the slope of an isolated rocky hill, 33 m. S.E. Perugia, and 60 m. N. Rome. Pop. 8000. Forsyth says that it is neanly built, with steep and dirty streets, and that is interesting only from its remains of antiquity; but Eustace and others represent it in a less unfavourable light. It is connected, across a deep ravine, with a neighbouring hill, by means of a stupendous aqueduct, serving both as a conduit and a bridge, raised upon a lofty range of 10 pointed arches, which, though repaired in modern times, is no doubt of Roman origin. The cathedral, said to have been built in the time of the Lombards, is of a very mixed style, having a front of five Gothic arches supported by Grecian pillars, while, internally, it is in the form of a Latin cross, with a noble range of Corinthian columns. It has some showy decorations: near it is a handsome fountain. The citadel, massive stone fortress, built by Theodoric, and repaired by Narses, stands on a height overlooking the town. The temple of Concord has been converted into a church; but, addition to it, Spoleto has two arches, a bridge, the ruins of a theatre, and several other Roman remains: on one of the gates is an inscription importing that Hannibal's troops were repulsed in an attack on the town, after the battle of Thrasymene. It has a few insignificant manufactures of woollen stuffs and hats. Under the French it was the capital of the department Thrasymène. It was of incomparably greater importance in antiquity than in modern times, and is reckoned by Florus among the *municipia Italiæ splendissima* (lib. iii., cap. 21). (*Woods; Eustace; Forsyth.*)

SPOTTSYLVANIA, county, Va., situated in the E. part the state, and contains 408 sq. m. Bounded N. by Rappahannock river, S.W. by North Anna river. Drained by and branches of Mattapony river. It contains the city of Fredericksburg, and had in 1840, 7971 neat cattle, 7670 sheep, 12,455 swine; and produced 58,450 bushels of wheat, 349 of Indian corn, 101,774 of oats, 9787 of potatoes, 147 pounds of tobacco, 4454 of cotton. It had 22 stores, furnace, one fulling-mill, one woollen factory, two flour-mills, 37 grist-mills, 15 saw-mills, four tanneries, two tanning-offices, four semi-weekly newspapers; 15 academies, 415 students; 20 schools, 243 scholars. Pop.: whites, ; slaves, 7590; free coloured, 785; total, 15,161. Cap. Spottsylvania C.H.

SPRINGFIELD, p. t., Sullivan co., N. H., 35 m. N.W. Concord, 492 W. It occupies the height of land between mac and Connecticut rivers, the head waters of Black flowing into the former, and of Sugar river into the r. Chartered in 1769. It contains a Congregational ch, one store, one grist-mill, three saw-mills; seven ols, 430 scholars. Pop. 1232.

SPRINGFIELD, p. t., Windsor co., Vt., 77 m. S. Montpelier, N. Bounded E. by Connecticut river. Watered by river, which affords good water-power. It contains churches, a Congregational, Methodist, Baptist, and common to Episcopalians and Universalists; eight stores, furnace, six fulling-mills, two woollen factories, one factory with 650 spindles, three grist-mills, eight mills, one oil-mill, one paper-mill, three tanneries; 19 s, 904 scholars. Pop. 2625.

SPRINGFIELD, p. t., capital of Hampden co., Mass., 24 m. Hartford, Ct., 91 m. W. Boston, 383 W. Pop., in 1830, in 1840, 10,985. Incorporated in 1645. It is one of most important inland towns in New-England. Bounded by Connecticut river. Watered by Chicopee and vers. On the river are rich alluvial meadows, exzly fertile, and back of them the ground rises to a arable elevation, and terminates in a plain of moderate uity extending E. to Wilbraham. The principal village built chiefly on Main-street, spacious, and over two a length, which is well built, and contains some elegances. This street runs parallel with the river, and E. of it, on rising ground, are handsome sites for gs. overlooking the village. It contains a court-jail, six churches, two Congregational, a Unitarian, cal, Methodist, and Baptist, and one of the most exarsenals of construction in the United States. A 1234 feet long, connects it with West Springfield. mory is pleasantly situated on elevated ground, half E. of the village. The buildings are arranged on a nare, and consist of one brick edifice, 240 by 32 feet, ries high, occupied by lock-filers, stockers, and fin brick forging shop, 150 by 32 feet; a brick build- y 32 feet, two stories high, the second story form-

ing a spacious hall devoted to religious worship; a brick building, 100 feet by 40, and two stories high, used as a depository of arms; and numerous smaller shops and stores, and 17 dwelling-houses. The water-works are situated on Mill river, about 1 m. S. of the arsenal, on three different sizes called the Upper, Middle, and Lower water-shops, the whole comprising five workshops, 98 forges, 10 triphammers, 18 water-wheels, exhibiting a great assemblage of water-works. In 1837 the public lands and buildings were valued at $210,000; machinery at $56,000; one hundred and seventy thousand muskets on hand, $9,040,000; muskets manufactured during the year ending April 1, $154,000; ordnance and stock on hand, $80,000; and 260 men employed, which have since been increased. The water-power owned by the United States would admit of a great extension of these works. On Chickapee river, 4½ m. N. of Springfield village, is the flourishing manufacturing village of Chickapee, which contains three churches, a Baptist, Methodist, and Congregational, large cotton and other manufactories, and nearly 2000 inhabitants. Four miles from Springfield, on Chickapee river, is the recent manufacturing village of Cabotsville, which contains two churches, a Universalist and Baptist, and about 2000 inhabitants. The railroad from Boston to Albany passes through the principal village of Springfield. A steamboat plies regularly to Hartford, Ct., and a railroad is nearly completed, connecting the Hartford and New-Haven railroad with the Boston and Albany railroad, at Springfield. According to the United States census of 1840, there were in Springfield 68 stores, with a capital of $250,000; two lumber-yards, with a capital of $6000; seven persons produced machinery to the amount of $120,000; 30 persons produced hardware and cutlery to the amount of $25,000; 280 persons produced 30 cannon and 14,000 small arms; 25 persons manufactured various metals to the amount of $50,000; 14 persons produced granite or marble to the amount of $10,000, with a capital of $15,000; 35 persons produced bricks and lime to the amount of $16,000; eight cotton factories, with 43,700 spindles, employed 1700 persons, producing articles to the amount of $975,000, with a capital of $1,650,000; 20 persons produced hats and caps to the amount of $4000, and straw bonnets to the amount of $3000, with a capital of $2000; three tanneries produced 1500 hides of sole leather and 25,000 sides of upper leather, employing 71 persons, and a capital of $30,000; other manufactories of leather, as saddleries, &c., produced to the amount of $55,000, with a capital of $10,000; six persons produced 50 tons of soap and 30 tons of tallow candles, with a capital of $10,000; two breweries employed eight men, producing 48,000 gallons of beer, with a capital of $28,000; eight persons produced confectionary to the amount of $50,000, with a capital of $6000; four paper-mills, employing 80 persons, produced to the amount of $300,000, with a capital of $290,000; 30 persons produced carriages and wagons to the amount of $30,000, with a capital of $12,000; three grist-mills and three saw-mills employed six persons, and a capital of $18,000; 25 persons manufactured furniture to the amount of $10,000, with a capital of $5000; seven printing-offices, two binderies, and four weekly newspapers, employed 20 persons, and a capital of $4500; 31 brick or stone, and 22 wooden houses were built, employing 85 persons, and cost $94,000. The total amount of capital employed in manufactures was $2,631,500. There were three academies with 140 students; 36 schools, 1512 scholars.

SPRINGFIELD, p. t., Otsego co., N.Y., 8 m. N. Cooperstown, 61 m. W. Albany, 379 W. The N. end of Otsego lake extends to the middle of the township. Bounded W. by Canderraga lake. It contains four churches, a Presbyterian, Associate Reformed, Methodist, and Baptist; five stores, four fulling-mills, one woollen factory, one furnace, four grist-mills, 13 saw mills, three tanneries; 16 schools, 800 scholars. Pop. 2382.

SPRINGFIELD, p. t., Essex co., N. J., 8 m. W. Newark, 52 m. N.E. Trenton, 217 W. Bounded E. by Rahway river, and drained by its tributaries. It is distinguished for its paper manufactories, and contains two churches, a Presbyterian and a Methodist; seven stores, three grist-mills, two saw mills, 11 paper-mills; seven schools, 453 scholars. The village on Rahway river contains 200 dwellings, and about 1200 inhabitants. Pop. of the township 1651.

SPRINGFIELD, t., Burlington co., N. J., 5 m. N.E. mount Holly. It is inhabited mostly by Friends. Drained by Assiscuk creek, and by branches of Rancocus creek. It contains three Friends' churches, three stores; seven schools, 411 scholars. Pop. 1632.

SPRINGFIELD, p. t., Bradford co., Pa., 154 m. N. by E. Harrisburg, 261 W. Drained by Bentley's creek, and branches of Sugar creek. It contains one fulling-mill, one grist-mill, seven saw-mills; 12 schools, 413 scholars. Pop. 1487.

SPRINGFIELD, t., Bucks co., Pa., 15 m. N. Doylestown, 40 m. N. Philadelphia. Drained by Durham creek, and by branches of Tohickon creek. It contains a church, six stores, one fulling-mill, two woollen factories, seven grist-mills, six saw-mills, one oil-mill, two tanneries. Pop. 9072.

SPRINGFIELD, t., Mercer co., Pa. Drained by Neshannock creek. It contains four stores, six grist-mills, eight saw-mills, two tanneries; 13 schools, 476 scholars. Pop. 1804.

SPRINGFIELD, t., Erie co., Pa. Drained by Crooked, Elk, and Raccoon creeks. It contains three stores, one fulling-mill, two grist-mills, seven saw-mills; one school, 18 scholars. Pop. 2344.

SPRINGFIELD, t., York co., Pa. It has four stores, four grist-mills, four saw-mills, two oil-mills, two tanneries, 16 distilleries. Pop. 1207.

SPRINGFIELD, p. v., capital of Robertson co., Tenn., 26 m. N. by W. Nashville, 708 W. Situated on a commanding elevation, one mile S. of the Sulphur fork of Red river. and contains a courthouse, jail, four stores, various mechanic shops, and about 350 inhabitants.

SPRINGFIELD, p. v., capital of Washington co., Ky., 68 m. S.S.W. Frankfort, 590 W. It contains a courthouse, jail, a bank, and 598 inhabitants.

SPRINGFIELD, p. v., Springfield t., capital of Clarke co., O., 43 m. W. Columbus, 436 W. Situated on the S. side of the E. fork of Mad river. The National road passes through it. It contains a handsome courthouse, county clerk's office, a jail, a male and a female academy, four churches, a Presbyterian, Methodist, Reformed Methodist, and an Associate Reformed; 30 stores, a paper-mill, grist-mill, fulling-mill, carding-machine, a brewery, a distillery, a printing-office, issuing a weekly newspaper, 400 dwellings, and 2062 inhabitants. Pop. of the township, exclusive of the village, 2349.

SPRINGFIELD, p. v., capital of Sangamon co., and of the state of Illinois, 105 m. N. by E. St. Louis, Mo., 780 W. Situated near the centre of the state, 4 m. S. of Sangamon river, on the border of a beautiful and extensive prairie. It was laid out in 1822, and in 1823 it contained 30 families, chiefly inhabiting log cabins. It became the capital of the state in 1840, and has had a rapid growth. It contains a statehouse, for the erection of which $50,000 were appropriated, a courthouse and market-house, on a public square, a jail, a United States land-office, three academies, six churches, two Presbyterian, an Episcopal, Methodist, Baptist, and a Reformed Baptist; 34 stores, one iron foundry, four carding-machines, three printing offices, and three weekly newspapers. Pop. 2579.

SPRING GARDEN, t., York co., Pa. Bounded W. and N.W. by Codorus creek. It contains one store, two fulling-mills, one woollen-factory, five grist-mills, one saw-mill, two oil-mills, one paper-mill, two tanneries, six distilleries, one pottery; one school, 22 scholars. Pop. 1819.

SPRING GARDEN, p. v., Philadelphia co., Pa., 100 m. E.S.E. Harrisburg, 140 W. It is in fact a constituent part of the city of Philadelphia, N.E. of the city proper. It is governed by 13 commissioners, elected for three years. It contains the Fairmount water-works, the Eastern penitentiary, the House of Refuge, the city hospital, an extensive floor-cloth factory, five commission-houses in foreign trade, capital, $35,000; 106 retail stores, capital $234,650; 13 lumber-yards, capital $271,000; four woollen factories, four cotton factories with 7802 spindles, five dyeing and printing works, one flouring-mill, one grist-mill, one paper-mill, one rope-walk, three tanneries, one distillery, three breweries, one pottery; eight academies, 702 students; 29 schools, 7037 scholars. Pop. 27,849. *See* PHILADELPHIA.

SPRINGPORT, t., Cayuga co., N. Y., 9 m. S.W. Auburn, 165 m. W. Albany. Bounded W. by Cayuga lake, into which its streams flow. It contains seven stores, one fulling-mill, one woollen factory, three grist-mills, three saw-mills, two tanneries; 10 schools, 566 scholars. Pop. 3890.

SPRINGWATER, p. t., Livingston co. N. Y., 16 m. S.E. Geneseo, 226 m. W. Albany, 341 W. Drained by the inlet of Hemlock lake, which affords water-power. It contains three stores, two fulling-mills, three grist-mills, 21 saw-mills, three tanneries; 17 schools, 888 scholars. Pop. 2632.

SQUAM, lake and river, N. H. This lake lies between Sandwich, Holderness, Moultonborough, and Centre Harbour townships, and is 6 m. long, and, at its greatest width, 3 m. wide, containing a number of islands. It is a beautiful sheet of water, surrounded by picturesque scenery. The outlet is Squam river, which flows into Pemigewasset river.

STADE, a town of Hanover, cap. district of its own name, on the Schwinge, near its mouth, in the Elbe, 20 m. W.N.W. Hamburg. Pop., with its suburbs and garrison, about 5700. On the bank of the Elbe, adjacent to the town, is the castle of Brunshausen, near which a vessel is stationed to receive the toll exacted by the Hanoverian government on all vessels passing up the Elbe. Stade has three Lutheran churches, a gymnasium, a cavalry school, a central workhouse, &c. Its inhabitants are engaged in manufactures of flannel, hosiery, &c., and have some share in the transit trade on the Elbe.

Since 1736 English vessels have been allowed to sail up to Hamburg without stopping to pay the duties at Brunshausen; they have, however, to be paid at Stade, before the vessel can be cleared out. The duties are very heavy on certain descriptions of goods, and are, in fact, a great obstruction to trade. It certainly is not a little surprising that the different nations of Europe, and especially the English should have submitted, for so long a period, to the exaction of a toll on what is really one of the most important commercial channels in Europe. If it be impossible otherwise to get rid of the nuisance, it would be good policy at once to buy an exemption from the duty.

STAFFA, a small island of Scotland, belonging to the Hebrides, famous for its basaltic columns and caverns, off the W. coast of the island of Mull, 9 m. N.N.E. Iona. It is of an oval shape, about 1½ m. in circumference, consisting of an uneven table-land, resting on cliffs of variable height, the highest being about 144 ft. above the sea. The cliffs and the caves by which they are perforated, consist mostly of basaltic columns, resting on conglomerate trap or tufa. The columns are partly perpendicular, partly oblique or horizontal, and partly bent. The average diameter of the columns is about 2 ft.; but they sometimes extend to 3 and 4 ft. They are generally pentagonal and hexagonal; they sometimes, however, have seven or nine sides, but are rarely triangular or rhomboidal. They are not so exquisitely united, nor are their angles as sharp as those of the Giant's Causeway.

Except on the N.E. shore of the island, at the landing place, it is almost everywhere surrounded by cliffs hollowed with caverns. But the W. side being exposed to the full swell of the Atlantic, and beat by a heavy surge, has been comparatively little explored, and the principal caverns of which we have any certain information are on its E. side. Of these the most celebrated by far is Fingal's Cave, visited and described by Sir Joseph Banks, Dr. M'Culloch, and Sir Walter Scott. The height of the cave, as given by M'Culloch, from the surface of the water, at mean tide, to the centre of the ceiling or arch, is about 66 ft., the height of its sides 36 ft., and its depth 227 ft.* Its sides are formed by ranges of nearly perpendicular columns: a deep channel-like fissure, parallel to the sides, extends along the whole length of the ceiling, which is formed of the bottoms of columns whitened by the infiltration of carbonate of lime into their interstices. The sea never entirely ebbs from the cave, the inmost recesses of which may be discovered from without. In moderate weather boats sail up to its farthest extremity.

"It would be no less presumptuous than vain," says Dr. M'Culloch, "to attempt a description of the picturesque effect of that to which the pencil itself is inadequate. But if this cave were even destitute of that order and symmetry, that richness arising from multiplicity of parts combined with greatness of dimension and simplicity of style, which it possesses; still, the prolonged length, the twilight gloom half concealing the playful and varying effect of a reflected light, the echo of the measured surge as it rises and falls, the transparent green of the water, and the profound and fairy solitude of the whole scene, could not fail strongly to impress a mind gifted with any sense of beauty in art or in nature." (*Geology of the W. Islands.*)

But the noblest description of this magnificent cave is that given by the great minstrel:

"— That wondrous dome,
Where, as to shame the temples deck'd
By skill of earthly architect,
Nature herself, it seem'd, would raise
A minster to her Maker's praise!
Not for a meaner use ascend
Her columns, or her arches bend;
Nor of a theme less solemn tells
That mighty surge that ebbs and swells,
And still, between each awful pause,
From the high vault an answer draws,
In varied tone prolong'd and high,
That mocks the organ's melody.
Lord of the Isles, cant. iv. s. 8.

In a note on this splendid passage, the author says, 'I would be unpardonable to detain the reader upon a wonder so often described, and yet so incapable of being understood by description. This palace of Neptune is even greater upon a second than the first view. The ... unius which form the sides of the cave, ... strength of the tide which rolls its deep and heavy swell up to the extremity of the vault, the variety of the tints formed by white, crimson, and yellow stalactites or petrifactions, which occupy the vacancies between the bases of the broken pillars which form the roof, and intermix them with a rich, curious, and variegated chasing, occupying each interstice; the corresponding variety below water, where the ocean rolls over a dark red or violet-coloured rock, from which, as from a base, the basaltic columns arise; the tremendous noise of the swelling tide, mingling with the

* These measures differ in some respects, but are materially from those given by Sir Joseph Banks.

oned echoes of the vault, are circumstances elsewhere unaralleled."

Mackinnon's Cave and the Boat Cave, though inferior to hat now described, are also magnificent caverns.

Staffa was first made known to the public by the interesting account of it given by Sir Joseph Banks, by whom it as visited in 1772. (*Pennant's Tour in Scotland*, ii. 300, d. 1790.) It is now, during summer, frequently visited by eamers. It is uninhabited, and has not even a house or ut in which any one can take shelter during a storm.

STAFFORD, a central co. of England, having N. Chese, E. Derby and Warwick, S. Worcester, and W. Salop. rea, 737,760 acres. Aspect very various. The N. part, or at portion of the co. lying to the N. of a line drawn from ttoxeter, on the confines of Derbyshire, to Newcastle-under-Lyne, consists principally of moorlands. The hills, in ese parts of this district, rise to an elevation of about 1000 above the level of the sea; sometimes consisting of vast aps of gravel, and sometimes of huge cliffs, having immense masses of rock scattered round their bases. With e exception of some beautiful valleys, the whole of this strict is sterile, cold, and dreary. The soil in many places peat; but in some parts, particularly between the rivers ve and Churnet, it is of a superior quality, and produces d herbage. The middle and S. parts of the county are eeably diversified with hills, level lands in pasture and n, plantations, and gentlemen's seats; but, in its extreme angle, the iron-works are its most prominent feature. e valley of the Trent is particularly fertile and beautiful. anock heath, immediately to the W. of Rudgely, consed, in 1813, above 28,000 acres; and is by far the largest the remaining tracts of waste land in the county. Mr. estimated the cultivated land, including parks, at 600,000 s, of which, he supposed, 100,000 may be meadow and ture, and 500,000 arable. The latter he distributes as ws: viz. 200,000 acres of clay loam, or more friable ed loam; 200,000 acres of gravelly or sandy loam, or r mixed, including calcareous soils; and the remainder, 00,000 acres, of light, sandy, gravelly, or other substances, p. 13.) The air is sharp and cold; and, in the W. particularly, there is a great deal of rain. Stafford is a mining and manufacturing than an agricultural country husbandry, though not so far advanced as it might s, of late, very considerably improved. Wheat, oats, s, and barley are the principal crops. The usual rotation the clay land is, 1. fallow; 2. wheat; 3. beans; 4. ; 5. oats. Various important improvements have been ted within the last twenty years, particularly on the se of the Duke of Sutherland, who has expended large on drainage, on the building of new, commodious, and lent farm-houses, and on other substantial improves. The cattle of Staffordshire are principally of the horned breed; but, within no very distant period, they been extensively crossed with short-horns; and the s of some of the principal breeders consist, at present, ly of the latter. Dairy husbandry is extensively practcheese is the principal product, and it is but little into that of Cheshire and Derby. The sheep stock is esed at about 187,000, and the produce of wool at about xcks. Property in estates, varying from £10,000 a own to 40s.: farms of all sizes, from 25 to 500 acres, e smaller class is decreasing: leases frequently grant t the greater number of farms held at will. Average land, in 1810, 20s. an acre. Coal, iron, and lime are in the greatest abundance in most parts of the county is particularly famous for its potteries and iron-foun-

The chief seat of the former is in a district denom-*The Potteries*, between Newcastle-under-Lyne and -on-the-Moors, in which there are several very conle towns and villages, mostly supported by the busi-The neighbourhood affords abundance of fine clay al; but the finest clays are principally brought from k, in Dorsetshire, soapstone from Cornwall, and flints e chalk pits near Gravesend, and from Wales and . The iron-works are principally situated in the S. f the county, in the vicinity of Walsall, Wednesbury, &c. Their increase has been quite extraordinary. , there were, in all, 123 furnaces, estimated to pro-out 212,000 tons a year, being about double the pro-1820. And since then the increase has been but litior; for, though the number of furnaces has not been d, they have been rendered so much more powerful ient, that their produce in 1839 has been estimated 367,000 tons! a quantity very decidedly exceeding was produced in all England besides! In this reideed, Staffordshire is exceeded only by South which has, within these few years, taken the lead ghly important branch of national industry. (*Serithe Iron Trade*, 292.) The manufacture of locks, ge tools, bridles, spurs, and an infinity of other e articles, is prosecuted upon a very large scale at ampton, Bilston, and Walsall, and their vicinity.

Soho, the famous establishment of Messrs. Bolton and Watt, where there is the greatest manufactory of steam-engines in the world, is situated within this county, on its extreme southern border. Glass is also made on the confines of Worcestershire. Hats, shoes, and boots, are prepared at Stafford for exportation, as well as home consumption; and cotton-mills have been erected at Rochester and other places. Principal rivers, Trent, Dove, and Stour. The Trent and Mersey canal passes through the county, dividing it into two pretty equal parts; and it is intersected by an immense number of other canals, and more recently by various railways. It is divided into 5 hundreds, exclusive of the city of Lichfield and the boroughs of Stafford and Newcastle-under-Lyne, and 145 parishes. It returns 17 members to the House of Commons; viz. 4 for the county; 2 each for the city of Lichfield and the boroughs of Newcastle-under-Lyne, Stafford, Tamworth, Wolverhampton, and Stoke-upon-Trent; and 1 for Walsall. Registered electors for the county in 1839–40, 18,489; of whom 10,020 were for the N., and 8469 for the S. division. In 1841, Stafford had 97,576 inhabited houses, and 510,206 inhabitants, of whom 258,729 were males and 251,477 females. Sum contributed for the relief of the poor, in 1838–39, £82,971. Annual value of real property in 1815, £1,200,385; profits of trades and professions in ditto, £516,721.

STAFFORD, a parl. and mun. bor. and market town of England, hund. Pirehill, in the above co., of which it is the cap., on the Sow, crossed here by a neat stone bridge, 23 m. N.N.W. Birmingham. Pop. of parl. bor., which includes, with the old bor., a portion of the par. of Castle-church, in 1841, 9152. It is generally well-built, the houses, several of which are handsome, being of brick and slate; and is paved and lighted with gas, under the provisions of a local act: a good supply of water is procured from several public pumps. The principal street runs N.N.W. from the bridge: and near its centre is the market-square, in which is the county-hall, a large modern building of stone, comprising several handsome apartments, besides an assembly-room, a grand jury room, courts for the assizes and sessions, mayor's office, and other apartments. The county jail is also a modern structure of extensive dimensions, and well arranged, both for the health and classification of prisoners, 200 of whom may be accommodated in separate cells. Stafford has two parishes, St. Mary's and St. Chad's, now consolidated and in the patronage of the lord chancellor. St. Mary's church is a large cruciform structure, in the early English style, from the centre of which rises a lofty octagonal tower; about £9000 having been raised by voluntary contributions for the repair and restoration of this venerable structure. Christchurch and St. Paul's have been recently erected; the latter is a good specimen of Gothic architecture. St. Chad's is a Norman structure, with more recent English additions. There is a small but very handsome Roman Catholic chapel; and the Wesleyan Methodists, Independents, and the Society of Friends have each places of worship, mostly with attached Sunday-schools. The grammar school, on ancient foundation, was much enlarged by Edward VI.; the income from the endowment exceeds £370 a year, two thirds of which are paid to the head master, and the remainder to the usher. It is open to all boys of the town; but the number of those on the foundation seldom exceeds 20. The appointment of the masters is vested in the corporation,subject to the approval of the bishop of Lichfield and Coventry. A diocesan national school is established here, and a Lancastrian school is supported by subscription. The institution for the relief of the orphans and widows of the clergy within the archdeaconry of Stafford is not only liberally supported by subscription, but has an income of £2400 a year, arising from funded property. Superannuated or infirm clergymen, also, are eligible for the benefits of this charity. The county infirmary, in the Foregate, has accommodation for 190 in-patients, and relieves a much larger number of patients at their own dwellings. It has a respectable medical staff. The county lunatic asylum, established in 1818, receives patients not only from the county itself, but from the kingdom generally, though the former are received on lower terms than the others. This establishment is supported both by subscription and funded property; it is admirably conducted, and may justly rank among the principal asylums in the kingdom.

The buildings comprise accommodation for 170 patients, and the gardens cover an area of several acres. There is an alms-house; but it is only poorly endowed, and fast failing to decay. The manufacture of shoes is the principal employment of the inhabitants, and several manufacturers employ 150 hands: a good workman can earn from 20s. to 30s. a week, and there is a steady demand for labour. The tanning of leather is no longer carried on to any extent. Stafford is noted, in common with the neighbourhood, for the excellent quality of its ale.

The Manchester and Birmingham railway passes close to the town, where it has a principal station. It has been, and no doubt will continue to be, of great advantage to the town.

It has two weekly newspapers, and two banks. Stafford was incorporated in the reign of John. It is divided, under the Municipal Reform Act, into two wards, its officers being a mayor, five aldermen, and 18 councillors.

The assizes and quarter sessions for the county are held here. The borough has returned two members to the House of Commons since the 23d Edward I., the right of election down to the Reform Act being in the resident freemen. The Boundary Act included a part of the parish of Castlechurch with the old borough. Regis. electors, in 1839-40, 1265. The custom of borough-English, by which lands descend to the younger son, to the exclusion of his elder brothers, prevails within the town and liberties.

Stafford is, also, the election town for the N. division of the county, as well as the principal seat of a poor-law union comprising 21 parishes. Markets on Saturday: fairs, April 5, May 14, June 25, October 3, and December 5, chiefly for horses and cattle. There is also a fortnightly cattle market.

About a mile and a half S.W. the town, on the site of a very old castle, demolished at the close of the parliamentary wars, Sir George Jerningham, now Baron Stafford, has built a massive structure after the design of the old fabric. (*Private Information.*)

STAFFORD, t., Tolland co., Ct., 6 m. N.E. Tolland, 24 m. N.E. Hartford, 73 m. W.S.W. Boston. Watered by Willimantic river and its branches. It contains bog-iron ore, which is extensively manufactured into hollow ware. It contains three churches, a Congregational, Methodist, and Universalist, eight stores, one fulling-mill, five woollen factories, four furnaces, four forges, two grist-mills, four saw mills, two tanneries; 19 schools, 652 scholars. Pop. 2469. It contains Stafford mineral springs. There are two springs of different qualities; one contains a solution of iron, sustained by carbonic acid gas, a portion of marine salt, and what is called natron or native alkali. This is a highly efficacious chalybeate spring. The other contains a large portion of hydrogen gas, of sulphur, and a small portion of iron. A spacious hotel has been erected for the accommodation of visitors, and the country around is peculiarly romantic.

STAFFORD, p. t., Genesee co., N. Y., 243 m. W. by N. Albany, 377 m. W. Drained by Black cr. It contains three churches, a Presbyterian, Baptist, and one common to Christians and Universalists, four stores, three fulling-mills, one woollen factory, five grist-mills, eight saw-mills; one academy, 99 students; 13 schools, 773 scholars. Pop. 2361.

STAFFORD, t., Monmouth co., N. J., 38 m. S. Freehold. It has nine stores, one grist-mill, four saw-mills; seven schools. 215 scholars. Pop. 2149.

STAFFORD, county, Va. Situated in the E. part of the state, and contains 335 sq. m. Bounded S.W. by Rappahannock river, N.E. by Aquia creek, E. by Potomac river. It contained in 1840, 5357 neat cattle, 5195 sheep, 9068 swine; and produced 30,516 bushels of wheat, 4281 of rye, 212,183 of Indian corn, 68,166 of oats, 11,548 of potatoes, 34,031 pounds of tobacco, 760,287 of cotton. It had seven stores, 16 grist-mills, two saw-mills; 11 schools, 195 scholars. Pop. whites, 4489; slaves, 3596; free coloured, 360; total, 8454. Capital, Falmouth.

STAMFORD, a parl. and mun. bor. and market town of England, S.W. extremity co. Lincoln, on the Welland (crossed here by a stone bridge of five arches), 38 m. S. Lincoln, and 80 m. N. by W. London. Pop. of parl. bor., which includes Stamford-baron with the old bor., a small portion of the par. of St. Martin, on the opposite side of the river, and in co. Northampton, in 1841, 7760. It is well built, principally of stone, partly paved, well lighted with gas, and supplied with water from Wothorpe, about 1. m. distant; but the streets are so irregularly laid out as scarcely to admit of description. An act of parliament was, however, passed in 1841 for correcting some of these defects, and for the better cleansing and paving the town; and the commissioners for executing the act seem resolved that its powers shall be put into full operation. Mr. Newcomb, proprietor of the *Stamford Mercury*, has lately improved the N. entrance of the town, called Scotgate, by the erection of numerous houses in an admirable style of architecture; but the fact of the land N. from the river being common, and that in the parish of Stamford-baron, consisting of the park and demesne of Burghley, preclude the probability, at present, of any great increase. (*Mun. Bound. Report.*) The town-hall, rebuilt in 1776, a large and fine edifice, comprises a sessions-room, jail, and muniment-office; the town has also, a small well-arranged theatre, and assembly-rooms. The other public buildings are the churches, of which only 6 remain out of 14. St. Mary's, considered the mother-church, was built about the end of the 13th century, and is chiefly in the later English style, having a very fine tower and spire. All Saints is an extremely handsome building, with a tower and octangular crocketed spire. The livings are all in the gift of

the Marquis of Exeter, except St. John's, of which Richard Newcomb, Esq. is joint patron. The Wesleyan Methodists, Independents, and Roman Catholics, have each places of worship, and there are numerous Sunday-schools. The grammar-school, founded in 1548, is well endowed, the net income of the master being nearly £700 a year; but the school has for many years past been of little service to the inhabitants. A blue-coat school affords clothing and instruction to about 150 boys: the petty-school was founded in 1604, and a girl's national school was established in 1813. The endowed charities are numerous and valuable: several hospitals, or alms-houses, have been founded at different times; besides which, there are several considerable bequests for the relief of the aged poor. A handsome infirmary has recently been built near the town, and furnishes accommodation for about 30 in-patients and 150 out-patients. There are no manufactures; but a considerable business is carried on in malting, and in a retail trade with the neighbourhood. The Welland is navigable for barges from hence to the sea. In the town are excellent hot and cold baths; and races are held in March and July, on Wittering heath, in the vicinity. It has two weekly newspapers, and three banks. Markets on Monday and Friday, the latter for corn. Fairs, Mid-lent Monday, Monday before May 12, and November 8.

STAMFORD (an. *Staan-forde*, meaning the paved ford) was incorporated in the reign of Edward IV. It is divided, under the Municipal Reform Act, into two wards, its officers being a mayor and 5 aldermen, with 18 councillors. Corp. revenue, in 1839, £4107 12s. 2d. Quarter and petty sessions are held under a recorder, and there is a court for the recovery of debts under £40. The custom of borough-English, by which landed property descends to the youngest son, to the exclusion of his elder brothers, prevails here; but there is only one copyhold house in the town. Stamford has, with some intermissions, sent two members to the House of Commons since the reign of Edward I., the right of election down to the Reform Act being in the resident freemen and inhabitants paying scot and lot. The electoral limits were enlarged by the Boundary Act so as to include a portion of Stamford-baron S. the river. Registered electors in 1839-40, 679. Stamford is likewise the principal town of a poor-law union comprising 37 parishes. The remains of conventual buildings, which are found abundant in different parts of the town, shows that it was formerly of some importance as an ecclesiastical settlement; and in the reign of Edward III. it became, for a brief period, the seat of a university which, however, soon fell in decay.

Within a short distance of the town, on the E. side, is Burghley-house, the magnificent seat of the Marquis of Exeter, one of the finest Elizabethan residences in the kingdom. (*Bound. and Mun. Reports, &c.*)

STAMFORD, p. t., Fairfield co., Ct., 77 m. S.W. Hartford, 263 m. W. Bounded S. by Long Island sound, on which is a good harbour, at the mouth of Mill river. Drained by Mill and Mianus rivers. It contains 17 stores, one furnace, one forge, one tannery, one printing office, one weekly newspaper; 14 schools, 451 scholars. Pop. 3598.

STAMFORD, p. t., Delaware co., N.Y., 61 m. W.S.W. Albany, 357 m. W. Drained by head branches of Delaware r. It contains three churches, a Methodist, Episcopal, and Union, 11 stores, one fulling-mill, one woollen factory, two grist-mills, four saw mills, two tanneries, 14 schools, 6 scholars. Pop. 1686.

STANDISH, p. t., Cumberland co., Me., 62 m. S.S.W. Augusta, 529 m. W. Bounded N. by Sebago lake, incorporated in 1785. It has eight stores, one grist-mill, six saw mills; 17 schools, 483 scholars. Pop. 2798.

STANFORD, t., Duchess co., N.Y., 16 m. N.E. Poughkeepsie, 72 m. S. by E. Albany. Drained by Wappinger's creek. It has six stores, one grist-mill, one saw mill, one paper-mills; eight schools. 269 scholars. Pop. 2074.

STANFORD, p. v., cap. of Lincoln co., Ky., 32 m. S.S.E. Frankfort, 563 m. W. Situated on a small branch of Dick's river, and contains a courthouse, a jail, a seminary, and 263 inhabitants.

STANHOPE, p. v., on the S. boundary of Byram t. Sussex co., N. J., 60 m. N. Trenton, 230 W. Situated on the contcong r. and on the Morris canal. It contains two stores, three forges, and about 30 dwellings. The river affords good water, having, by leading it out of its bed, a fall of 30 ft.; and the canal, by an inclined plane, overcomes an elevation of 76 ft.

STAPLETON, v., Southfield t., Richmond co., N.Y. Situated on the E. side of Staten Island, 2 m. N. of the Narrows. It contains a Methodist church and a Dutch Reformed chapel attached to the seamen's retreat, one store, a printing office, 50 dwellings, and about 400 inhabitants. The Seamen's Retreat is a hospital for the support of sick or disabled seamen, supported by a tax levied by law of $150 on each master of a vessel, and one dollar for each mariner on a foreign voyage, and 25 cents for each seaman for each voyage coastwise, which entitles each individual to the benefits of the hospital when sick or disabled. It has a fine building which cost $115,000, attached to which are 37 acres of land which cost $10,000.

STARGARD, a town of the Prussian states, prov. Pomerania, on the Ihna, in a very fertile country, 21 m. E. by S. Stettin. Pop, 8500. The cupola of St. Mary's church is supposed to be one of the most elevated in Germany. It has a college or gymnasium, founded by a private citizen in 1631, a school of arts, &c., with distilleries and different branches of manufacture.

STARK, co., O. Situated towards the N.E. part of the state, and contains 650 sq. m. Drained by Tuscarawas r. and its branches. It contained in 1840, 25,690 neat cattle, 51,119 sheep, 39,240 swine ; and produced 753,977 bushels of wheat, 52,751 of rye, 273,925 of Indian corn, 13,196 of buckwheat, 5043 of barley, 504,651 of oats, 156,673 of potatoes, 96,688 pounds of sugar. It had nearly 32 bushels of edible grains, exclusive of potatoes, to every individual of its population. It had 22 commission houses in foreign trade, 105 retail stores, three furnaces, one forge, 11 fulling-mills, two woollen-factories, five flouring-mills, 34 grist-mills, 99 saw-mills. two oil-mills, 26 tanneries, 96 distilleries, 11 breweries, 11 potteries, five printing-offices, four weekly newspapers; 33 schools, 800 scholars. Pop. 34,603. Capital, Canton.

STARK, county, Ia. Situated towards the N.W. part of the state, and contains 432 sq. m. Drained by Kankakee and Yellow rivers. There are extensive marshes on the Kankakee river. It is unorganized. Pop. 149.

STARK, county, Ill. Situated N.W. of the centre of the state, and contains 288 sq. m. Drained by Spoon river. It contained in 1840, 1461 neat cattle, 1392 sheep, 5269 swine ; and produced 18,876 bushels of wheat, 61,655 of Indian corn, 16,850 of oats, 8112 of potatoes, 590 pounds of tobacco, 3130 of sugar. It had three stores, three grist-mills, seven saw-mills; eight schools, 214 scholars. Pop. 1573. Capital, Toulon.

STARK, t., Herkimer co., N.Y., 12 m. S.E. Herkimer, 69 m. N.N.W. Albany. Drained by Otsquaga creek. It contains a Dutch Reformed church, two fulling-mills, one cotton factory with 2424 spindles, one furnace, two grist-mills, eight saw-mills, two tanneries; eight schools, 385 scholars. Pop. 1766.

STARKS, p. t., Somerset co., Me., 40 m. N.N.W. Augusta, 633 W. Bounded N.E. by Kennebec river. Watered by Sandy river, which here enters the Kennebec. Incorporated in 1795. It contains one store, one grist-mill, three saw-mills ; 17 schools, 693 scholars. Pop. 1559.

STARKEY, p. t., Yates co., N.Y., 10 m. N.E. Penn Yan, 8 m. W. Albany, 315 W. Bounded E. by Seneca lake, and drained by small streams flowing into it, of which Big cream has a fall of 140 ft. in the course of half a mile. It contains a Presbyterian and a Methodist church, 12 stores, or fulling-mills, one woollen factory, one furnace, two curing-mills, four grist-mills, 12 saw-mills, two tanneries ; schools, 1832 scholars. Pop. 2426.

STARKSBOROUGH, t., Addison co., Vt., 42 m. W. by Montpelier, 498 W. Chartered in 1780. Watered by Huntington river and Lewis creek. The religious denominations are the Congregationalists, Methodists, Friends, and we will Baptists. The Friends have a church. It has one re, one forge, one fulling-mill, one grist-mill, three saw-ls ; 14 schools, 488 scholars. Pop. 1263.

STATEN ISLAND, N.Y., centrally distant 11 m. from city of New-York, with which it is connected by sever team ferry boats, plying many times daily to different ding places on the island. It constitutes the county of hmond, which see. The view from the summit of lamond hill, 307 ft. above the level of the ocean, embraces the harbour of New-York, with its islands, its fortifications and its shipping, the city, and the surrounding country Long Island and in New Jersey, is exceedingly various picturesque. The island is 14 m. long, and its greatest idth 8 m. wide, and at a mean width 5 m. It was organized as a county in 1683. On a hill near the Narrows is a graph, communicating with another on the back part of exchange in New-York. The island is divided into four iships, all of which were organized in 1788. At New hton, at the N. end of the island, is a neat village with splendid boarding-houses, much resorted to in the summer season, 6 m. from New-York.

TAUNTON, p. v., cap. of Augusta co., Va., 116 m. W. . Richmond, 156 m. W. Situated on Lewis cr., one of head branches of Shenandoah r., and is one of the old-illages of middle Virginia. The land rises from the cr., he streets are regularly laid out, but narrow. It contains a handsome courthouse, a jail, three churches, an only, numerous stores, 250 dwellings, and about 2500 itants. The western lunatic asylum is located here, has a handsome and commodious edifice. Well prected of the postoffice in 1841 $1506.

EPHENSON, county, Ill. Situated in the N. part of ate, and contains 500 sq. m. Watered by Pekatonica

river and its branches. It contained in 1840, 3108 neat cattle, 741 sheep, 9693 swine ; and produced 41,903 bushels of wheat, 96,974 of Indian corn, 1101 of buckwheat, 3714 of barley, 52,218 of oats, 45,138 of potatoes, 2091 pounds of tobacco, 6543 of sugar. It had seven stores, five grist-mills, nine saw-mills ; 10 schools, 177 scholars. Pop. 2200. Capital, Freeport.

STEPHENTOWN, p. t., Rensellaer co., N. Y., 98 m. E. S.E. Albany, 376 W. Watered by Kinderhook river. It contains three churches, a Presbyterian and two Baptist, six stores, one fulling-mill, three woollen factories, one cotton factory with 1648 spindles, four grist-mills, 10 saw-mills ; 13 schools, 915 scholars. Pop. 2753.

STERLING, p. t., Worcester co., Mass., 41 m. Boston, 411 W. Watered by Still river. It contains three churches, a Congregational, Baptist, and Universalist, three stores, three grist-mills, seven saw-mills, two tanneries; 11 schools, 485 scholars. Pop. 1647.

STERLING, p. t., Windham co., Ct., 49 m. E. by S. Hartford, 376 W. Watered by Moosup river. It contains two churches, a Baptist and Congregational, three stores, four cotton factories with 5400 spindles, four grist-mills, five sawmills, one oil-mill ; seven schools, 259 scholars. Pop. 1099.

STERLING, p. t., Cayuga co., N. Y., 179 m. W. Albany, 362 W. Bounded N.W. by lake Ontario. Watered by Sodus creek entering little Sodus bay. It contains two stores, one fulling-mill, three grist-mills, seven saw-mills, two tanneries ; 13 schools, 773 scholars. Pop. 2533.

STETTIN, an important town and river port of the Prussian states, cap. Pomerania, and of a reg. of the same name, on the left bank of the Oder, about 36 m. above where it unites with the Baltic, lat. 53° 23' 20" N., long. 14° 33' E. Pop., in 1838, 31,100. (Berghaus.) It communicates by a bridge with a suburb on the right side of the river, and is very strongly fortified. It is well built, and is the most ancient as well as the principal town of Pomerania. Principal edifices, the royal castle, governor's house, mint, exchange, arsenal, and theatre. It has several churches, of which the principal, St. Mary's, was founded in 1263. The warehouses, belonging to the salt company are the most extensive of any in Prussia. The royal square is ornamented by a statue of Frederick the Great. It is the residence of the provincial authorities : and has a court of appeal for the circle, a gymnasium (illustre) founded in 1543, an observatory, a seminary for the training of schoolmasters, a public library, and various other literary institutions. Ship and boat building, and the forging of anchors, are extensively carried on ; there are also distilleries, with a considerable variety of manufactures. Stettin is the seat of an extensive and growing commerce, and is now, indeed, the principal port of importation in Prussia. She owes this distinction mainly to her situation. The Oder, which flows through the centre of the Prussian dominions, is navigable for barges as far as Ratibor, near the extreme southern boundary of Prussian Silesia, and is united by means of canals with the Vistula, Elbe, Spree, &c. Stettin is, consequently, the principal emporium of some very extensive and flourishing countries ; and is not only the port of Frankfort-on-the-Oder, Breslaw, &c., but also of Berlin ; being, next to Dantzic, the principal port in the Prussian dominion. Hence, at the proper seasons, its wharfs are crowded with lighters that bring down the produce of the different countries traversed by the river, and carry back colonial products and other articles of foreign growth and manufacture. Vessels of considerable burden, or those drawing above 7 or 8 ft. water, load and unload by means of lighters at the mouth of the river, at Swinemunde, the outport of Stettin, on the E. coast of the isle of Usedom. (See SWINEMUNDE.) There is a great wool fair in the month of June each year. A bank, similar to that of Berlin, is also established here, with an insurance company, &c. The principal articles of export consist of linens, corn, wool, timber, and staves, zinc, manganese, bones, oil cake, bottles, &c. The imports consist of sugar, coffee, and other colonial products ; wine, indigo, and other dye stuffs ; cotton stuffs, yarn, and raw cotton ; herrings, hardware, oil, tallow, coal, salt, &c. Stettin has considerable manufactures of woollen stuffs, hosiery, leather, sail-cloth, tobacco, &c., and the most extensive sugar refinery in Prussia. In 1834 there belonged to the port 200 vessels of the burden of 49,892 tons, being about one third part of all the ships belonging to Prussia. During the same year there entered the port 1744 ships, of the aggregate burden of 170,348 tons.

STEUBEN, county, N. Y. Situated in the S. towards the W. part of the state, and contains 1400 sq. m. Watered by Conhocton and Tioga rivers and their branches. The Corning and Blossburg railroad meets the New-York and Erie railroad in this county. Crooked and Seneca lakes lie partly in this county. It contains iron ore and some mineral springs, and had in 1840, 43,476 neat cattle, 148,133 sheep, 34,309 swine ; and produced 390,275 bushels of wheat, 13,929 of rye, 102,974 of Indian corn, 80,311 of buckwheat,

23,543 of barley, 387,980 of oats, 580,959 of potatoes, 341,948 pounds of sugar. It had 101 stores, 19 fulling-mills, three woollen factories, four flouring-mills, 36 grist-mills, 207 saw-mills, three oil-mills, one paper-mill, 31 tanneries, two distilleries, three printing offices, three weekly newspapers; four academies, 394 students; 339 schools, 15,318 scholars. Pop. 46,138. Capital, Bath.

STEUBEN, county, Ia. Situated in the N.E. corner of the state, and contains 225 sq. m. Drained by Pigeon river, and Crooked and Fish creeks. Organized in 1837. It contained in 1840, 2031 neat cattle, 873 sheep, 2702 swine; and produced 22,149 bushels of wheat, 31,906 of Indian corn, 26,940 of oats, 15,866 of potatoes, 1177 pounds of tobacco, 16,043 of sugar. It had six stores, three grist-mills, eight saw-mills, four tanneries, one distillery, one pottery; 17 schools, 429 scholars. Pop. 2578.

STEUBEN, p. t., Oneida co., N. Y., 101 m. W.N.W. Albany, 407 W. Drained by Cincinnati creek, and by tributaries of Mohawk river. It has one store, five saw-mills, two tanneries; 20 schools, 539 scholars. Pop. 1903.

STEUBENVILLE, p. t., capital of Jefferson co., O., 141 m. E.N.E. Columbus, 38 m. W. by S. Pittsburg, Pa., 264 m. W. Bounded E. by Ohio r. Watered by Cross cr. The village was laid out in 1798, and incorporated in 1805. It is situated on the W. bank of Ohio r., and is regularly laid out with streets crossing each other at right angles. It contains a courthouse, jail, an elegant town-house, with a market-house in the lower story, a bank, an academy, and six churches. It is governed by a president, recorder, and seven trustees. Most of the following statistics of the census of the town in 1840 belong to the village. It had 26 stores with a capital of $52,400; three lumber yards, capital $18,000; one furnace produced 130 tons of cast iron; 10 persons produced 180,000 bushels of bituminous coal, with a capital of $1800 dollars; six persons produced granite or marble to the amount of $3000, with a capital of $300; one person produced machinery to the amount of $1000; five persons manufactured 108 small arms; one person manufactured various metals to the amount of $1000; 12 persons produced bricks and lime to the amount of $6000; five woollen factories employed 200 persons, producing to the amount of $137,000, with a capital of $138,000; one cotton factory with 10,294 spindles, employed 36 persons, and a capital of $15,000; one person produced 218 pounds of reeled silk, worth $250; three persons manufactured tobacco to the amount of $3000, with a capital of $1800; six persons manufactured hats and caps to the amount of $3000, and straw bonnets to the amount of $500, with a capital of $1800; two tanneries produced 2800 sides of sole leather, 3800 sides of upper leather, employing 10 persons, and a capital of $16,000; seven other manufactories of leather as saddleries, &c., produced to the amount of $7000, with a capital of $4300; two persons produced 1000 pounds of soap, and 500 pounds of tallow candles, with a capital of $300; two breweries employing two persons produced 16,000 gallons of beer, with a capital of $600; one paper-mill, employing 60 persons, produced to the amount of $60,000, with a capital of $30,600; one rope-walk, employing two persons, produced cordage to the amount of $400, with a capital of $100; 10 persons produced carriages and wagons to the amount of $9000, with a capital of $2500; four flouring-mills produced 15,000 barrels of flour, and with one saw-mill produced to the amount of $45,000, employing eight persons, and a capital of $30,000; 15 persons manufactured furniture to the amount of $6000, with a capital of $2000; two printing-offices, one bindery, and two weekly newspapers, employed nine persons, and a capital of $4000. The total amount of capital employed in manufactures was $251,050. There was one college, with 208 students, one academy, 30 students, five schools, 591 scholars. Pop., in 1830, 2937; in 1840, 5203.

STEYNING, a market town and par. of England, county Sussex, hund. Bramber, hund. Steyning; area of par. 3290 acres. Pop., in 1831, 1436. The town, on the Adur, 5 m. from the English Channel, and 11½ m. S. Horsham; consists of four indifferently built streets. It has a very curious Norman church, with a great variety of excellent and very elaborate detail. (Rickman's Gothic Archit.) The living, a vicarage worth £308 a year, is in the gift of the Duke of Norfolk. Brotherhood-hall, an old edifice of the time probably of Henry VIII., is appropriated to a free school, founded in 1614, for the classical education of 10 boys. This town has never been of any considerable importance, and, in fact, would not have been worth mentioning in a work of this kind, but for the circumstance of its having sent two members to the House of Commons, from the reign of Edward II. to the passing of the Reform Act, by which it was disfranchised. The franchise was nominally vested in the inhabitant householders paying scot and lot; but was really in the hands of the Duke of Norfolk, the proprietor of the borough. It is a polling place for the W. division of the county, and has petty sessions.

STEWART, county, Tenn. Situated in the N. toward the W. part of the state, and contains 575 sq. m. Bounded W. by Tennessee river, watered by Cumberland river, and drained by small tributaries of both. It contains in 1840, 9285 neat cattle, 6999 sheep, 32,731 swine; and produced 27,974 bushels of wheat, 443,470 of Indian corn, 64,63 of oats, 735? of potatoes, 993,495 pounds of tobacco. It had three furnaces, thee forges, four tanneries; eight schools 2.5 scholars. Pop.: whites, 6317; slaves, 2117; free-coloured, 153; total, 8587. Capital, Dover.

STEWART, co., Ga. Situated in the W. toward the S. part of the state, and contains 682 sq. m. Bounded W by Chattahoochee river, and drained by its tributaries, and by Kitchafoona river, a tributary of Flint river. It contained in 1840, 20,766 neat cattle, 2596 sheep, 34,632 swine; and produced 13,114 bushels of wheat, 399,540 of Indian corn, 21,265 of oats, 46,340 of potatoes, 4,981,937 pounds of cotton. It had two academies, 162 students, 10 schools, 315 scholars. Pop.; whites, 8174; slaves, 4741; free coloured, 18; total 12,933. Capital, Lumpkin.

STILLWATER, p. t., Saratoga co., N. Y., 29 m. N. Albany, 392 W. Bounded E. by Hudson river, N.W. by Saratoga lake, S. by Anthony's kill. In the N. part of this t. Burgoyne was captured by General Gates, October 17th 1777, which nearly decided the question of American independence. It has 11 stores, one cotton factory with 2888 spindles, three grist-mills, four saw-mills; two academies, 173 students, eight schools, 389 scholars. Pop. 2733.

STILLWATER, p. t., Sussex co., N.J., 77 m. N. Trenton, 234 W. Drained by Paulinskill creek. It has three stores, four grist-mills, three saw-mills, two tanneries; 12 schools, 300 scholars. Pop. 1476.

STIRLING, a central and marit. co. of Scotland, having N. the co. Perth, E. the frith of Forth and Linlithgow, S. the latter, Lanark and Dumbarton, and W. the latter and loch Lomond. Area 321,280 acres, including 6388 acres water. Surface extremely diversified, consisting partly of high mountains; partly of extensive moors, bogs, &c.; and partly of very rich alluvial carse lands. Ben Lomond, the most celebrated and best known of the Highland mountains, in the N.W. part of the co., immediately above loch Lomond, has an altitude of 3191 ft. The Fintry, Campsie, and Lennox hills lie in the middle and S. parts of the county; the surface, from Denny N.W. to loch Lomond, is in most places very bleak and sterile. The low alluvial or carse lands, which are extremely productive, lie on both sides the Forth but principally on its S. bank, extending from Falkirk to above Stirling. They are supposed to comprise in all, from 35,000 to 45,000 acres. They consist principally of a bluish clay, intermixed with sand. In the W. parishes clay soil predominates; and, as it rests on a bottom of hard ferruginous clay (till), it is cold and wet. In some places along the rivers the soil is light and gravelly. In the high moors it is mossy; and in the lower grounds there are considerable peat bogs. Several large estates; but property a good deal divided. Farms in the lower districts vary from 20 to 300 acres; but, in the hilly and mountainous districts, they are much larger. Agriculture very various; but generally well suited to the situation and climate. Drainage has recently been practised on a very extensive scale. In the carse, wheat, beans, barley, and clover, but particularly the first two, are the principal crops. On the lighter lands turnips are largely cultivated; oats being the prevailing crop on all the poorer high lands. Potatoes generally cultivated. Sheep mostly of the black-faced Linton breed; but Cheviots have been largely introduced. Besides the cattle bred in the county, which are not remarkable for their goodness, great numbers of highland cattle are annually purchased for feeding at the Falkirk trysts. These are the greatest fairs, or markets for cattle of any in Scotland. They are held on the 2d Tuesday of August, September and October; the last being the largest. Cattle in all sorts of condition are brought to them from all parts of Scotland, but principally from the north; as are also sheep and horses. At an average, it is supposed that about 80,000 cattle, 55,000 sheep, and 5000 horses are annually disposed of at these trysts. Estimating the cattle to be worth £7 each, the sheep 11s., and the horses £10, their entire value will be nearly £850,000. (Youatt on Cattle, &c., p. 121.)

Stirlingshire is said to have about 13,000 acres of natural wood, and above 10,000 acres of plantations. The E. parts of the co. have a finely diversified appearance: and the view from Stirling castle is, perhaps, unequalled by any other in Britain. (See next art.) Coal abundant: and there are large supplies of ironstone, freestone, &c. Average rent of land in 1810, 11s. 4d. an acre. Extensive works have long been established at Carron, for the smelting of iron, and the manufacture of all sorts of cast-iron goods, whether for civil or warlike purposes. (See CARRON.) Exclusive of distilleries, some branches of manufacture, on a pretty considerable scale, are carried on at St. Ninian's, Stirling, Falkirk, and other towns. Principal river, Forth; to which are tributary the Carron, Bannockburn, and other small streams. Stirling has 25 parishes, and returns one member to the House of Commons. Registered electors for the county in 1839-40, 2223. The borough of Stirling unites with the boroughs of Inverkeithing, Dunfermline, Queensferry, and Culross; and the borough of Falkirk with those of Linlithgow, Lanark, Airdrie, and Hamilton, in returning representatives. In 1841 Stirling had 15,837 inhabited houses, and 82,179 inhabitants, of whom 41,070 were males, and 41,109 females. Valued rent, £108,509 Scotch. Annual value of real property, in 1815, £218,761.

STIRLING, a royal and parl. bor., river port, market town, and fortress of Scotland, cap. of the above county, on the Forth, 30 m. W.N.W. Edinburgh, and 22 m. N.E. Glasgow. Pop. of parl. bor. which includes the suburb of St. Ninian's, in 1841, 10,745. The situation of Stirling is magnificent in the extreme. It is built on the S.E. declivity of a hill, which, rising from one of the most fertile plains in Scotland, terminates on the N. and W., immediately over the river, in an abrupt basaltic rock, about 300 ft. in height, surmounted by a fine old castle. It consequently bears a striking resemblance to Edinburgh; but its situation is finer, more central, and in all respects more suitable for the capital of Scotland. The view from the castle, if not unrivalled, is at all events, certainly unsurpassed by any other in the empire. It combines all that can give variety, interest, and grandeur to a prospect. To the E. it extends over the richest valley in Scotland, as far as Edinburgh, commanding all the windings of the Forth; to the W. is the fertile strath of Monteith, the view in this direction being bounded by Ben Lomond; on the N. it is bounded by the range of the Ochill hills: and immediately to the S. is the field of Bannockburn, the Marathon of Bruce and of Scotland.

"The principal street, which extends from the castle down the ridge of the hill, with narrow cross streets branching from it down the declivity on each side, is open and spacious; and the houses, though many of them bear marks of antiquity, are generally lofty and comfortable. A new ridge over the Forth (finished in 1835) has been begun, and streets are proposed to be built from it on the low ground the E. of the town. Several neat houses have lately been erected in the neighbourhood. Many respectable families have been induced to settle in Stirling, in consequence the cheapness of living, the beauty of the surrounding country, and the society which the town affords." (Parl. Rep.)

The principal building is the castle. A fortress is said to have been erected on its site by the Romans, and there can be no doubt of its great antiquity. Its inaccessible situation in the centre of the kingdom, at the point where the Forth first comes fordable, renders it, as it were, the key of the lowlands on the one hand, and of the Highlands on the other. No wonder, therefore, that Stirling early became a place of great importance, and that it was for a lengthened period a favourite royal residence, and the seat of the legislature. Previously to the invention of artillery the castle was a place of great strength; but, notwithstanding the additions made to the works in more modern times, it could oppose any effectual resistance to an army properly plied with artillery. It is a quadrangular building, with open area in the centre, and besides other structures, includes the old royal palace, principally built by James V., the parliament house; but these venerable structures are, by a scandalous outrage on taste and national associations, been degraded into barracks for common soldiers! It is stipulated in the articles of union with England, that king castle shall be always garrisoned, and kept in repair. Among the public buildings in the town are the town-house, jail, Cowan's hospital, founded in 1639, and richly endowed; the athenæum, with a spire 120 ft. in height, a for the circuit and sheriff courts, &c. The old church, venerable Gothic edifice, a portion of which formed part the Franciscan monastery, founded in 1494, has long been used into two places of worship; James VI., when a child, was crowned in it, on the 29th of July, 1567, the coronation sermon being preached by the famous reformer, Knox. Three churches belong to the secession; and the Reformed Presbytery, Independents, Episcopalians, Wesleyan Methodists, and Roman Catholics, have each a church. There are altogether four churches belonging the establishment. Some of the houses formerly occupied by the principal Scotch nobles, are still met with in different parts of the town. Religious dissent prevails in Stirling, perhaps to a greater extent than in any other town in Scotland. Ebenezer Erskine, one of the ministers of Stirling, was (along with his brother Ralph, one of the ministers of Dunfermline) a founder of the secession, or Associate Synod, about 1740. This town has been long eminent for schools, particularly those for classical literature. It has also a mechanics' institute, and several public libraries. Exclusive of Cowan's hospital, noticed above, founded by a citizen of that name, Stirling has two other well en-

dowed hospitals, exclusive of the interest of £4000 left in mortmain for behalf of the poor. But "where the carcass is, there will the eagles be gathered together;" and, notwithstanding the ample provision in question, pauperism is quite as prevalent in this as in most other Scotch towns.

The chief manufacture is that of tartans, tartan shawls, carpets, and yarns, which is carried on to a large extent; and the dyeing of yarns, home-made cloths, and silks is also carried on. Cotton goods are manufactured, though to no great extent, with ropes, malt, leather, soap, and candles. The town has extensive markets, the Corn Exchange being one of the finest in Scotland; it has also a considerable coasting and retail trade. About 100 vessels are said to be engaged in the trade on the Forth up to Stirling, and steam-packets, which in summer are crowded with passengers, ply daily between the town and Granton Pier, near Edinburgh. (Bound. Rep.) The town has a branch bank of the Bank of Scotland, and four other branch banks, and two weekly newspapers.

Stirling received its first charter from Alexander I., in 1190: it is now governed by a provost, three ballies, and 18 councillors. Corporation revenue, about £3000 a year. The borough unites with Dunfermline, Culross, Inverkeithing, and S. Queensferry, in sending one member to the House of Commons. Registered electors, in 1839-40, 497.

Stirling and its immediate vicinity has been the scene of some of the most memorable events in Scotch history. In 1297 Wallace defeated a formidable English army close to the town; and the victory of Bruce at Bannockburn, in 1314, secured the independence of Scotland. James II. was born in the castle; and there, in 1452, he basely murdered, with his own hand, the Earl Douglas, whom he had inveigled thither by the grant of a safe conduct, and the assurances of friendship! Stirling was also the birth-place of James V., and it continued to be his favourite residence. Here, also, James VI. resided with his tutor, the celebrated George Buchanan (Scotorum sui seculi facile princeps), till he was 13 years of age; and here, as already stated, he was crowned in 1567.

The abbey of Cambuskenneth, one of the richest and most magnificent in Scotland, stood on the banks of the Forth, a short way from Stirling. Its ruins are still very considerable. (Private Information.)

STOCKBRIDGE, a market town and par. of England, co. Hants, hund. King's Somborn, on the Test, a tributary of the Anton, and on the Andover canal, 8 m. W.N.W. Winchester. Area of par., 1290 acres. Pop. in 1831, 851. The town comprises only a single street, in which are numerous inns. Its principal support being derived from its situation on the high road between London and Exeter. It has a town-hall, a neat edifice, erected in 1810, by the present Marquis of Westminster. From the first of Elizabeth down to the passing of the Reform Act, by which it was disfranchised, this petty place, or rather its proprietors, enjoyed the privilege of sending two mems. to the House of C.

STOCKBRIDGE, p. t., Windsor Co., Vt., 49 m. S. by W. Montpelier, 475 W. White river passes through its N. part, and in one place is compressed into a channel only a few ft. wide, affording good water-power. Watered by Tweed river, a branch of White river. Steatite or soap stone is found. Chartered in 1761. It contains two churches, a Congregational and one other, and has some Methodists, Baptists, and Universalists, two stores, two fulling-mills, one woollen factory, two grist-mills, seven saw-mills; 10 schools, 437 scholars. Pop. 1418.

STOCKBRIDGE, p. t., Berkshire co., Mass., 133 m. W. Boston, 363 W. Watered by Housatonic river. Marble and limestone are extensively found. A tribe of Indians resided here on a reservation of six miles square, granted to them by the state, from 1736 until their removal to Oneida, N. Y., in 1788; and much was done to civilize, instruct and Christianize them by the labours of Rev. John Sergeant and Rev. Jonathan Edwards, and others. They were the fast friends of the whites, and exceedingly useful to them in the French and Revolutionary wars. They have since removed to Green Bay of lake Michigan, on a tract purchased for them. The village of Stockbridge is one of the pleasantest in Western Massachusetts, situated on the N. side of Housatonic river, on a plain elevated considerably above the level of the river, and contains two churches, a Congregational and Baptist, a bank, an academy, several stores, and over 40 dwellings, many of them neat. There are in the town a Congregational church besides the above, eight stores, two cotton factories, with 3800 spindles, one furnace. one grist-mill, six saw-mills, two tanneries; 12 schools, 398 scholars. Pop. 1992.

STOCKBRIDGE, p. t., Madison co., N. Y., 5 m. N.E. Morrisville, 113 m. W. by N. Albany, 366 W. Drained by Oneida creek. It has a Presbyterian and a Methodist church, four stores, one woollen factory, one flouring-mill, two grist-mills, seven saw-mills, two tanneries; one academy, 33 students, 13 schools, 747 scholars. Pop. 2340.

STOCKHOLM.

STOCKHOLM, a celebrated city of the N. of Europe, the cap. of Sweden, at the junction of the Lake Mœlar with the Baltic, 440 m. W. by S. Petersburg, lat. 59° 20′ 31′′ N., long. 17° 54′ E. Pop., in 1839, 63,885. 'It is very strikingly situated, partly on a number of islands, at the entrance of the lake, and partly on the mainland, upon both sides of the strait, covering altogether an area of about 4½ sq. m. The view of the city, when approached from the Baltic, commands the palace, the principal bridge, and other prominent objects, and is extremely grand and imposing. "On the islands, and more particularly on those called *Stockholm* (island of the Castle), *Riddarholm* (Knight's island), and *Helge Aud's holm* (island of the Holy Spirit), all so near each other that they are united by 13 short bridges of stone, and others of wood, stand the king's palace, the great cathedral, the bank, the hall of the diet, and most of the more conspicuous ornaments of the city ; but the larger portion of the private houses are built on the mainland, which on the N. side, called the Nörrmalm, or N. suburb, slopes gradually backwards from the shore ; but on the S. side or the Södermalm, rises in bold abrupt cliffs, where the white houses nestle beautifully among shading trees. The streets on the mainland are in general pretty regular, though not very wide ; but many of those on the islands are as narrow and crooked as a lover of irregularity could desire. On the three principal islands most of the houses are of stone, but in the N. and S. suburbs the greater part are of stuccoed brick, painted white, yellow, or faint blue. In the remote suburbs, wood is still the only material employed. In most of the houses the stairs and lobbies are of a dark blue stone, with such a slippery surface that the stranger is exposed to many a tumble before he gets accustomed to them. Most of the great thoroughfares are tolerably well paved, but they are almost wholly destitute of footpaths. In all parts of the city it has been necessary from the nature of the ground, to build upon piles. There are no fortifications of any kind round Stockholm." (*Bremner*, i. 343.)

Except some churches, few buildings of importance are situated in the N. suburb. It contains, however, the two principal squares of the city ; one of which, the "King's Garden," bordered with large shady trees, has a good statue of Charles XIII. ; while the other, called the square of Gustavus Adolphus (one side of which is formed by the opera house, in which Gustavus III. was assassinated in 1792), has a well-executed statue of the hero whose name it bears. From this square a very handsome bridge opens a communication with the principal island and the royal palace ; and, as all heavy goods are here carried by water, it forms the most fashionable and agreeable lounge in Stockholm. The line of this bridge crosses Helge-Aud's island, cutting off a small portion of the latter, which, being fitted up as a garden, is the resort of the best society of the capital.

The Södermalm, or S. suburb, is connected with the city by a bridge, underneath which are sluices for drawing off the waters of the lake. It has two fine churches, and a statue of Charles XII. on the military parade.

Though Stockholm has numerous statues, it has but few public edifices of an ornamental character. The palace, however, an immense quadrangular edifice, begun during the reign of Charles XII., has a majestic appearance from whatever point it may be viewed. Its N. and S. faces being prolonged eastward, towards the sea, inclose between them a flower-garden. The lower part is of polished granite ; the upper, of brick covered with cement. It contains a museum of antiquities and sculpture, with several good works by Swedish artists ; a picture gallery, in which are a few fine paintings by Raphael, Teniers, Paul Potter, Ostade, &c. ; some other superior paintings in the queen's apartment ; the royal library, with 40,000 vols. (*Bremner*), in which a copy of every book printed in Sweden must be deposited ; the repository of national archives, and a chapel, very splendidly fitted up. Broad and massive quays, which surround the palace on its E. and N. sides towards the principal harbour, effectually prevent the noise and confusion of the shipping, which at one point is very dense, from reaching the royal apartments. On one of the quays, within view of the palace windows, is a statue of Gustavus III.; and at the other end of the avenue leading from this point, a small, but handsome obelisk, has been raised to his memory. The king's stables, on an adjacent island, form a very extensive quadrangular building, having accommodation for 120 horses. The churches, externally, are in general devoid of architectural merit; but the cathedral or St. Nicholas' church, adjoining the palace, in which the kings of Sweden are crowned, is imposing from its mass, and has some paintings and sculptures of merit, and a remarkable piece of carving, representing St. George and the dragon. But the most interesting church is the Riddarholm, in which the kings of Sweden are buried, and in which are preserved many national military trophies. The spire of this church, destroyed by lightning in 1835, has since been replaced. There are also Ger-

man, Finnish, and Scotch Presbyterian churches : a synagogue, &c. The Riddarhuset, or Hall of the Diet, erected in the time of Christina, is in a peculiar style of architecture, and has its roof ornamented with statues. The hall in which the diet assembles is of moderate size. Its walls are hung with the armorial bearings of the principal Swedish families, and its seats are subdivided into four distinct compartments, without, however, any difference as to the mode of their fittings. The president's chair, a fine specimen of Dutch workmanship, is at the upper end of the hall ; the nobles' seats being on the right, those of the clergy on the left, and those of the town and country deputies in front. In the intervals of the diets the hall is sometimes used as a concert-room. The town-house is an old-fashioned building, with four wings. The military hospital, on one of the more remote islands, is among the best establishments of its kind in Europe. The hospital, founded by Gustavus III., though spacious, is not so well conducted.

There are several well-ordered prisons, and public charities of various descriptions ; including a foundling hospital, to which many children are carried, and which, in fact, is one of the principal sources of the prodigacy for which the city is distinguished. In the country parts of Sweden, the proportion of legitimate to illegitimate children is about 10 to 1 ; in the towns, as 6½ to 1 ; whereas in Stockholm, it is under 2½ to 1 ! The pop. of Stockholm has long been nearly stationary, or has but slowly increased. The mortality is very heavy, the deaths exceeding the births by above 100 a year ; a result ascribable partly to the mortality occasioned by the foundling hospital, and partly to the prevalence of dram-drinking, and the poverty and dirty lodgings of a considerable portion of the lower classes.

There are several public parks in the neighbourhood of Stockholm. The most celebrated is the *Dyurgarten* (deer garden), to the E. of the city. From its great extent, and romantic character, Bremner says it "is, without exception, the finest public park in Europe. The rugged peninsula of which it occupies the greater part, is so finely varied with rocks and trees, that art, which must do everything in the parks of other great capitals, has here only not to injure nature. The margin of the peninsula is covered with old-fashioned eating-houses, &c. Within this confined circle runs the beautiful carriage drive, lined with mansions often of classical design, Swiss cottages, Italian mansions, &c. Among these are placed coffee-houses, equestrian theatres, and dancing-rooms, while the space between them and the road is occupied with flower-plots and shrubbries, through which rustic seats are scattered. In this park is the bust of Bellman, a lyric poet of great excellence, in the time of Gustavus III. The anniversary of this poet's birth is kept with great rejoicings by all classes, but especially the bacchanalian Club, whose members, headed by the king himself, come out in festive array, to parade round the park, which is very appropriately decorated with green and vine-leaves." (*Bremner's Excurs.*, i. 308, 361.) Not far from this popular monument is *Rosendahl*, a favourite summer residence of the king. This, like all the other houses at the park, is a *portable wooden* edifice, which species of structure is very common all round Stockholm. Drottningholm, Haga, &c., at different distances from the capital, are other royal summer residences. Near Rosendahl Park is an immense porphyry vase, 9 ft. in height, and 18 ft. in diameter, which cost about £10,000! Opposite Haga is the new burial ground, a spacious cemetery, in which are some fine monuments. Carlberg and Ulriksdal were formerly royal palaces ; but the former has been converted into a military academy, and the latter into a military hospital.

Stockholm is the principal emporium of Sweden. The entrance to the harbour from the Baltic is intricate, and should not be attempted without a pilot ; but the water is deep, and the harbour itself capacious and excellent, the largest vessels lying close to the quays. The principal exported ports are iron, timber, and deals. The shipping engaged in foreign trade has materially declined of late years : it amounted, in 1830, to only 130 ships, of the aggregate burden of 22,470 tons. (*See Bremner*.)

Stockholm is also the principal manufacturing town of the kingdom, having woollen cloth, cotton, linen, silk, glass and earthenware factories, iron-works, including a foundry for the construction of steam-engines, &c. conducted by a Mr. Owen from Glasgow, sugar refineries, breweries, &c. "The stranger who has lately come from Copenhagen is greatly struck with the contrast between its streets and those of Stockholm. The silence and order of the former capital are here replaced by a degree of turmoil and activity, indicative at once of greater wealth and greater industry. The number of showy or well-filled shops is not very great, nor are they much frequented ; but in all the more central streets, especially near the palace, gay equipages are in constant motion ; and the principal quay, except at the resting hour of noon, exhibits all the bustle of a thriving seaport.' (*Bremser*, i. 357.) The more important branches of commerce are generally carried on by natives of the city; settlers from the provinces being comparatively few, except such as are engaged in the more laborious kinds of occupations. Foreigners, consisting principally of Germans and Englishmen, with about 600 Jews, are but few in number. Considering, indeed, that the king is a native of France, and the proximity of the city to Russia, the small number of Frenchmen and Russians may well excite surprise.

A few years since it was a common complaint that there were no good hotels, and that lodgings were both scarce and dear in the city. But several inns have been recently built, and, according to Bremner, the Stockholm hotels may now be considered fully equal to those of second-rate towns in other parts of the continent. They do not supply diners, which may, however, be had at the eating-houses, where a pretty good dinner may cost about 2s. "Even at the most fashionable places in the park," says Bremner, "the charge, as compared with England, is extremely moderate. We happened to see the bill for a dinner that would not have disgraced the Star and Garter. Fish, fruit, and every delicacy of the season, were served in profusion. The wines, also, were first-rate, and included port (here a dear wine), sherry, claret, Rudelsheimer, champagne, and punch, supplied to 24 guests; yet the whole charge, including attendance, was only 250 rix-dollars, or about 12s. a head."

During the summer season nearly all persons in tolerable circumstances spend the greater part of their time in the environs. At all the public places visiters are waited on by women; and a stranger is surprised to see many employments entrusted to men in other countries here undertaken by females. The ferry-boats, for instance, are almost all rowed by Dalecarlian females, in their peculiar native costume; though, if we may rely on Mr. Murray's account, their beauty is not very likely to be much injured by this masculine employment.

Stockholm is not the seat of a university, but it has several distinguished academies, including the academy of sciences, established about 1740, with an admirable museum of zoology; the Swedish academy, founded by Gustavus III.; the academy of painting and sculpture, which has produced Sergel, Fogelberg, Bystrom, &c.; and that of literature, a college of medicine, schools of navigation, drawing, &c., with societies of agriculture, commerce, and philosophy. There are also several clubs and reading-rooms, on much the same plan as those of London, and various newspapers.

Stockholm appears to have been founded by Birger, regent of Sweden, in the 13th century. It became the residence of the Swedish sovereign soon after Birger's death, it was not recognized as the capital till the 17th century, previously to which, Upsala had been the seat of the art. (*Stockholm and its Environs*, Stockholm, 1839; Bremner, *Excurs. in Denmark, Sweden, &c.*, ii., 340–405; Murray's *Handbook for the N. of Europe; Coxe, &c.*)

STOCKHOLM, p. t., St. Lawrence county, N. Y., 223 m. N.W. Albany, 507 W. Drained by St. Regis river and tributaries. It contains a Presbyterian church, three res, two fulling-mills, one woollen factory, two gristls, five saw-mills, one tannery, one pottery; 11 schools, scholars. Pop. 2995.

STOCKPORT, a parl. bor., and manufacturing town of gland, co. Chester, on the borders of Lancashire, 5 m. . Manchester, and 10 m. N. Macclesfield. The parl. mun. bor. comprises the township of Stockport, with t of those of Brinnington and Heaton Norris, and the alets of Brinksway and Edgely in the adjacent parish 'headle; and had, in 1831, a population of about 42,000*, in 1841, of 50,495. The town-proper, which is supd to occupy the site of a Roman military station, is t on an abrupt hill beside the Mersey, which, sweeping id its E. and N. boundary, is here joined by the Tame. n the bank of the river the houses rise in successive round the sides of the hill, from the base to the summe having apartments excavated in the sandstone ; and the numerous extensive factories elevated e each other, and spreading over the town, give it, cially when lighted up at night, a striking appearance. most ancient part of the town surrounds the church market-place on the top of the hill, whence various ts diverge in different directions. The principal street, d the Underbank, follows the direction of the old Ro road S. to Buxton. Three bridges across the Mersey ect the town-proper with its suburbs of Portwood and n-Norris.

twood, in the township of Brinnington, is large, populand of considerable manufacturing and commercial ance. To the W. of Stockport numerous streets,

houses and factories, cover the greater part of the hamlets of Brinksway and Edgeley. Heaton-Norris, which is situated in Lancashire, communicates with the better part of Stockport, by a new line of road, made within the last 15 years, and a noble bridge of 11 arches across the valley and the river. The arch over the river, built of hard white stone, has a span of above 90 ft., and an elevation of 40 ft. above the water. The arches on the Cheshire side are carried across several streets, leaving thoroughfares underneath. Stockport is well paved, and lighted with gas; and there is an ample supply of water, at a reasonable rate. The parish church, the chief public edifice, appears to have been erected in the 14th century, but has been much patched up in later times. The chancel had a fine decorated E. window; but this has been removed. At the W. end of the church is a lofty square tower, crowned with a pierced parapet and pinnacles; and in the interior are several ancient monuments. The living, a rectory of the clear annual value of £1882, is in the gift of Lady Vernon. There are two chapels of ease, St. Peter's and St. Thomas's, both perpetual curacies, the former worth £220, in private patronage; and the latter, worth £110 a year, and in the gift of the rector of Stockport. St. Thomas's, an elegant building in the Grecian style, was erected in 1825, by a parliamentary grant, at an expense of £14,555. There are places of worship for Independents, Methodists, Friends, Unitarians, Roman Catholics, &c.; and a good and handsome subscription news-room. A covered market is not much frequented; but the large open market in the centre of the town is well supplied and well attended. (*Mun. Corp. Rep.*) A free grammar school, founded in 1487, is under the government of the Goldsmiths' Company of London. It gives gratis instruction to 150 boys, from 6 to 14 years of age, sons of inhabitants of Stockport, &c., in the ordinary branches of education. Their nomination rests with three visiters appointed by the Goldsmiths' Court of Assistants. The master has a salary of £210, and the usher of £105 a year. Some handsome buildings for this foundation have been lately erected, at a cost of about £4000. A large national school was established at Stockport in 1805; and handsome school-houses, &c., were built, at an expense of £10,000. A great number of children of both sexes are educated here; and without the town this establishment has several branch schools. Most of the religious denominations have their own Sunday schools. There are almshouses, for six poor men, founded in 1683; and various other charities for the relief of the poor. (*31st Rep. on Charities.*) Two weekly newspapers are published in the town, and it has a joint-stock bank and two branches of other proprietary banks, exclusive of a savings' bank.

Formerly, the winding and throwing of silk were the principal branches of industry in Stockport; but these have declined in favour of the cotton manufacture, which now occupies the greater part of the population. Several large factories have been constructed of late years. There are also several silk-mills in full activity, the rivers affording an ample supply of water. (*Mun. Rep.*) The importance of Stockport as a manufacturing town is, however, chiefly owing to its abundant supply of coal, obtained from Poynton and the districts on the line of the Manchester and Ashton canal, with which it communicates by a branch canal. In 1839, there were in the parish 23 cotton-mills, wrought by 1309 steam-engines, and employing upwards of 6900 work-people. (*Mills and Factories Rep.*) The weaving of calico has spread itself over all the neighbouring villages; and calico-printing is carried on to a great extent, there being many large dye-houses in the vicinity. Fine woollen cloths, hats, &c., are also manufactured; and the construction of machinery is an important department. The municipal borough is divided, under the Municipal Reform Act, into six wards, and is governed by 14 aldermen and 42 councillors. The ancient charter of incorporation is of uncertain date. The office of mayor was, till a late period, mostly honorary; the town now has a commission of the peace. Corporation revenue in 1840, £1519. The Reform Act conferred on this borough, for the first time, the important privilege of returning two members to the House of Commons. Registered electors, 1839–40, 1270. By far the most interesting object in the vicinity of Stockport is the stupendous viaduct of the Manchester and Birmingham railway over the Mersey, erected at a cost of £100,000. This town was a military post of some consequence previously to the Conquest; but as it is not mentioned in Domesday Book, it had probably been destroyed at that epoch. No remains now exist of its old castle. In the civil war, Stockport was first garrisoned by the parliament; then taken by Prince Rupert; but finally retaken by the parliamentary troops, who retained it till the termination of the contest. (*Bound. and Mun. Rep., &c.*)

STOCKPORT. p. t., Columbia co., N. Y., 5 m. N. Hudson, 24 m. S. Albany, 346 W. Bounded W. by Hudson river.

he *Municipal Corporation Report* says, approximatively, 43,000.

Drained by Kinderhook creek. Organized in 1833. It contains a Presbyterian and a Methodist church, 11 stores, three fulling-mills, two woollen factories, three cotton factories, with 7968 spindles, one fleuring-mill, one grist-mill, one paper-mill; one academy, 92 students; five schools, 356 scholars. Pop. 1815.

STOCKTON, p. t., Chautauque county, N. Y., 7 m. E. Mayville, 331 m. W. by S. Albany, 338 W. Drained by Bear creek. It contains a Presbyterian church, three stores, one fulling-mill, one grist-mill, seven saw-mills, three tanneries; 14 schools, 761 scholars. Pop. 2078.

STOCKTON-ON-TEES, a town, seaport, mun. bor., and par. of England, being next to Newcastle and Sunderland, the principal port, in the kingdom for the shipment of coal, county Durham, ward Stockton, on the Tees, near its mouth, 17½ m. S.E. Durham. Area of parish, which comprises the townships of Stockton, Preston-on-Tees, and Hartburn, 4190 acres. Pop., in 1831, 7991. But nearly all this population is agglomerated within the limits of the municipal borough, which lies wholly within the township of Stockton. The town, one of the cleanest and handsomest in the N. of England, consists of a straight and wide main street, about ¼ m. in length, running from N. to S.; in which are many good houses, built chiefly of brick, though a few are of stone, taken from the old castle. From this street, smaller ones branch off on the E. towards the river; while on the W. a great many new houses and streets have been recently built. In the N.E. part of the town is a spacious square, lately enclosed and planted, in which are some good buildings. About the middle of the high street is the town hall, a commodious square edifice, with court, assembly, and other public rooms; but partly occupied as a hotel; and near it is a handsome Doric column, on the place formerly occupied by an open cross. Near the S. end of the town is a handsome stone bridge, with five elliptical arches, erected by subscription, between 1764 and 1769, at an expense of £8000. The tolls of the ferry over the Tees were previously the property of the bishop of Durham, to whom a considerable annuity was made payable by the shareholders; but the whole debt having been paid off, the bridge became toll free in 1820. A little farther S. the Tees is crossed by a suspension-bridge, forming part of the Middleborough branch of the Darlington and Stockton railway. The Port Clarence railway terminates on the Tees, a little N.E. of Stockton. The town is watched by an efficient police, and well lighted with gas.

The parish church is a neat brick edifice, with a tower 80 ft. in height at its W. end. The living, a vicarage, worth £947 a year nett, is in the gift of the bishop of Durham. There are places of worship for Baptists, Friends, Independents, Methodists, Unitarians, and Roman Catholics, several having Sunday schools attached; a mechanics' institute and library, a subscription-library, and a neat theatre. A charity school was founded here by subscription in 1721, and a school for girls in 1803; and Stockton, with the adjacent parish of Norton, as a scholarship at Brazennose college, Oxon. Some almshouses, established in 1682, were rebuilt in 1816, and afford accommodation to 36 poor persons. It has a dispensary, a savings' bank, and many benefit societies.

The only manufacture is that of sail cloth; for which there were, in 1837, three considerable establishments, one employing nearly 400 hands. The two railroads which pass to the town employ a good many hands; and Stockton is said to be in a prosperous state. (Mun. Corp. Rep.) New coal mines of large extent have recently been discovered in the neighbourhood, and the trade of the port has increased considerably of late years. In 1839, there were shipped coastwise from Stockton 1,308,778 tons coal, exclusive of 86,669 tons shipped for foreign ports. In 1840, Stockton supplied the metropolis with 1835 cargoes of coal, amounting to 495,369 tons. Linen and worsted yarn, and lead, &c., are also shipped in considerable quantities. Gross amount of customs' duties collected at the port, in 1840, £97,235. The port dues belong to the bishop of Durham, as lord of the manor, but are leased to the corporation at a nominal rent. Stockton is supposed to have been incorporated about the 13th century. The municipal borough is now divided into two wards, and is governed by a mayor, five other aldermen, and 18 councillors. Corporation revenue, in 1840, £1201. The borough has, under the Municipal Reform Act, a commission of the peace; besides weekly petty sessions and courts baron eight times a year, for debts not above 40s. There is, however, no jail, but only a lock-up house in the borough. Stockton is of considerable antiquity, and was long the occasional residence of the bishops of Durham. Its castle was demolished by order of parliament in 1647. (Mun. Corp. Rep. and Appendix, iii., &c.)

STODDARD, county, Mo. Situated in the S.E. part of the state, and contains 900 sq. m. Bounded W. by St.

Francis river. It contained in 1840, 4980 neat cattle, 394 sheep, 20,356 swine; and produced 7506 bushels of wheat, 255,973 of Indian corn, 7005 of oats, 6569 of potatoes, 935 pounds of cotton. It had 12 stores, seven grist-mills, two distilleries; three schools, 80 scholars. Pop. whites, 3082; slaves, 71; free coloured, 1; total, 3153. Capital, Bloomfield.

STODDARD, p. t., Cheshire co., N. H., 44 m. S.W. Concord, 448 W. It occupies the height of land between Connecticut and Merrimac rivers, and contains a Congregational and a Universalist church, two stores, one grist mill, four saw-mills; 13 schools, 295 scholars. Pop. 1808.

STOKES, county, N. C. Situated in the N. toward the W. part of the state, and contains 836 sq. m. Drained by Yadkin river, and a branch of Dan river. It contained in 1840, 11,583 neat cattle, 11,963 sheep, 31,817 swine; and produced 74,989 bushels of wheat, 8347 of rye, 423,673 of Indian corn, 107,756 of oats, 16,129 of potatoes, 349,383 pounds of tobacco, 56,481 of cotton. It had 22 stores, one furnace, six forges, one woollen factory, one cotton factory with 2000 spindles, five flouring mills, 46 grist-mills, 38 saw-mills, 10 oil-mills, eight tanneries, 146 distilleries, three potteries; two academies, 225 students; 13 schools, 317 scholars. Pop.: whites, 13,418; slaves, 2993; free coloured, 765; total, 16,965. Capital, Germantown.

STOKE-UPON-TRENT, a parl. bor. township, and par. of England, co. Stafford, hund. Pirehill, on the Trent, 9 m. E. Newcastle-under-Line, and 15 m. N. by W. Stafford. Area of par., 10,490 acres. Pop., in 1831, 37,220. The parliamentary borough comprises the most populous portions of the parish, including the townships, or rather towns, of Hanley, Shelton, Lane End, Fenton, &c., with those of Burslem, Tunstall, and Ruston Grange, and the hamlet of Sneyd, in adjacent parishes, being, in fact, co-extensive with the district termed the Potteries. It has an entire area of about 16,000 acres, and a population in 1831 of nearly 51,000, and in 1841 of about 70,000. Stoke-upon-Trent is lighted with gas, and well supplied with water. The old parish church being much decayed, a new and handsome church was erected in 1829, partly by subscription among the inhabitants. In it has been placed a bust of the great founder of the earthenware manufacture, the celebrated Josiah Wedgwood, who died in 1795, and who was admirable alike for the improvements he introduced into the fabric of the ware, and for the classical elegance and purity of his designs. The living, a very valuable rectory, being worth £9717 a year nett, is in the gift of J. Tomkinson, Esq. It has, also, chapels for various dissenting sects, and a large national school. The Trent and Mersey canal passes through Stoke parish, and on its banks are numerous wharfs for shipping the earthenware, which is the great, and, indeed, almost the only product of the district. The contributary townships of Burslem and Hanley have been already noticed. (I., 487, 960.) The towns now included within the district called the Potteries have almost all risen to importance since about 1760, when Wedgwood commenced his career. The Reform Act erected in 1832 the districts specified above into a parliamentary borough, and gave it the privilege of sending two members to the House of Commons. Registered electors, in 1839-40, 1693. The smallness of this number, as compared with the population, is a consequence of the low value of houses in the district. "The cheapness of building materials, and the abundance of building ground, render a house rated at somewhat below £10 sufficient for the wants not only of the higher order of mechanics, but even of many other classes. This low rate of house-rent does not arise from a depression of trade and wages: no place that we have visited appears in more full employment, more prosperous, or more steadily advancing in improvement, than this important district." (Bound. Report.)

The towns and villages comprised in the Potteries, or in the parliamentary borough of Stoke-upon-Trent, are so near each other, that their limits are not easily defined, and to a stranger the entire district has the appearance of a large straggling town. A very large proportion of the population is engaged in, and a still greater is dependent on, the pottery manufacture. With the exception of the clay used in gilding, most of the materials employed are worth very little; so that the value of the finished articles as well as their exquisite beauty, and adaptation to every purpose of utility and ornament, is mainly ascribable to the skill and labour expended upon them. The wives and children of the workmen are usually employed; but though they work together in factories, yet, as they reside in separate cottages, the manufacture partakes largely of the domestic character. The wages of a whole family amount to a very considerable sum. "The workmen process of glazing, so injurious to the health of these employed, has been rendered nearly free from its deleterious effects by the substitution of borate acid for lead, which was formerly wholly used, but now only in the proportion 'of eight per cent. The people employed in that branch were formerly not admissible into clubs, and were considered as degraded objects from the insalubrity of their employment; but they are now gladly received into benefit societies. (*First Factory Report*, b. ii., p. 78.) At present (1841) the Potteries are rather in a depressed condition; but, speaking generally, the workpeople have a healthy comfortable appearance. The Sunday schools in the district are extremely well attended. (*Statistics of the Brit. Emp.*, i., 712, 3d ed.)

STONE, a market town and par. of England, co. Stafford, hund. Pirehill, on the Trent, and on the high-road from London to Liverpool, 20 m. N.W. Lichfield. Area of par. 20,030 acres. Pop., in 1831, 7808. The town consists principally of two streets crossing each other, and is pretty well built. The parish church is a modern structure, with a low square tower: the living, a perpetual curacy, worth £214 a year, is in the gift of the crown. Here are chapels for Independents, Wesleyans, and Roman Catholics; with a free school founded in 1558, and other charitable endowments. The Trent and Mersey canal passes through the town, the inhabitants of which are principally engaged in the manufacture of shoes. Petty sessions are held here every fortnight. Stone is supposed to owe its origin to a monastery founded in 670, afterwards made subservient to that of Kenilworth. Market days, Tuesdays: four cattle fairs are held yearly. Meaford, in Stone parish, was the birth-place of Sir John Jervis, afterwards Earl St. Vincent, from his great victory over the Spanish fleet, off cape St. Vincent, on the 14th Feb., 1797.

STONEHAM, p. t., Middlesex co., Mass., 10 m. N. Boston, 450 W. It contained in 1840, 29 manufactories of leather, chiefly of shoes, produced to the amount of $115,166, with a capital of $84,750. It has five stores, one printing-office, one weekly newspaper; six schools, 251 scholars. Pop. 1017.

STONEHAVEN, a bor. of barony, seaport, and market town of Scotland, co. Kincardine, of which it is the cap., on the German ocean, at the point where two small rivers, the Curron and Cowie, fall into a small bay, flanked on both sides by lofty rocks, 14 m. S. by W. Aberdeen. Pop., in 1831, 3050. It consists of two parts: the old town, on the S. side of the Carron, is irregularly and badly built; but the new town, on the N. bank of the river, on the estate of Mr. Barclay of Ury, which has rapidly grown up, is comparatively well built and handsome: it consists of two parallel streets and cross streets, with a large square in the centre, and is far superior, in wealth and population, to the other. The two towns are connected by a handsome stone bridge. The parish churches of Dunottar and Fetteresso are in the immediate vicinity of the town, which has also two chapels, belonging respectively to the Episcopalians and Seceders. Exclusive of other seminaries, the town has a free school for the education of 60 poor children. The harbour, which is a natural basin, has been a good deal improved by the erection of piers, and affords a safe refuge for the smaller class of vessels.

The inhabitants engage to some extent in the herring and haddock fisheries, having, in 1839–40, had 172 boats employed in this department. It has a flax and a woollen mill, but neither is of considerable size, and two branch banks. The trade of the town is inconsiderable.

Dunottar castle, about 2 m. S. from the town, stands on lofty peninsulated rock, projecting into the sea, being separated from the mainland by a vast chasm or natural fosse. The summit of the rock, which is mostly occupied by the ruins of the castle, comprises about 1¼ acre. This castle was, for a lengthened period, the property and residence of the noble family of Keith, Earls Marischal. It was forfeited and dismantled after the rebellion of 1715, on the attainder of its noble proprietor. Owing to its position, was formerly a place of considerable strength, and has been repeatedly besieged. (*Municipal Boundary Report; Brewster's Scotland*, iii., 150.)

STONEHENGE, the same given to a gigantic ruin, consisting of vast stones, partly upright and partly fallen, in Salisbury plain, co. Wilts, England, 2 m. W. Amesbury, and 7 m. N. Salisbury. Though its present appearance be that of a confused mass, justifying, in some degree, Camden's epithet of *insana substructio*, it is seen, on a little examination, that its original form, which may be easily traced, was circular. When perfect, it had consisted of two outer concentric circles of stones, with two inner groups of stones. The outer circle, the diameter of which is 100 appears to have originally consisted of 30 upright stones, which 17 are still standing. Their average height is about 14 ft., and their sides 7 ft. by 3 ft. Each of these upright stones has tenons on its upper end, on which were fixed horizontal stones or imposts, with mortices to correspond with the tenons; and these imposts being connected together, formed a continuous circular architrave all round the fabric. The inner circle, 8 ft. 3 in. within the outer cir-

cle, consists of smaller stones, more irregularly shaped than those in the outer circle, and without imposts. Only eight stones of this circle are now standing, but there are remains of 12 others on the ground. Within the inner circle are two groups of stones, having between them a large flat stone called the *altar*. Some of these interior stones are of vast size, and have imposts similar to those of the outer circle. According to what appears to be the most accurate calculation, Stonehenge, when entire, must have comprised, in all, 129 or 130 stones. They consist mostly of a fine, white, compact sandstone, closely resembling, or, rather, identical with, the grey-weathers and other detached masses of stone scattered over the surface of the downs in the vicinity of Avebury and Marlborough.

This gigantic structure is surrounded by what must originally have been a deep trench, about 30 ft. in breadth; and connected with it are an avenue and *cursus*. The former, a narrow road of raised earth, extends in a direct line from what is supposed to have been the grand entrance to the structure, a distance of 594 yards, when it divides into two branches, one leading to a row of barrows, and the other to the *cursus*, an artificially-formed flat tract of ground. The latter, ½ m. N.E. from Stonehenge, is bounded by parallel banks and ditches, measuring 3036 yards in length, by 110 yards in breadth.

Such is a brief notice of this stupendous monument, and of its principal appendages. Similar remains are found at Avebury, in the vicinity, and various places in Brittany, the Orkney islands, &c. Conjecture has exhausted itself in vain, though frequent ingenious attempts have been made to explain the origin and use of this wonderful fabric, and others of its class. The most common opinion is that it was raised by the anc. Britons for a druidical temple. We have elsewhere (*see* AVEBURY, i., 255) stated our reasons for believing the statement of its having been connected with the worship of the druids as altogether unfounded; and there is no evidence to show that the ancient Britons raised, or could raise, so extraordinary a structure. In truth, we know nothing of this and the other monuments of the same kind, beyond the fact of their existence. They belong to a period of which all records have irretrievably perished; and it is extremely improbable that the veil by which their origin, and the purpose of their founders, is now hid, should ever be drawn aside. Inigo Jones, the learned Dr. Stukely, Dr. Smith, Sir R. C. Hoare, Gough, in his edition of Camden, &c., have given descriptions of Stonehenge. A good account of the ruins, with a view of the different theories as to the origin and purpose of the structure, may, also, be found in *Rees's Cyclopædia*.

STONEHOUSE. *See* PLYMOUTH.

STONELICK, t., Clermont co., O. Watered by Stonelick fork of Little Miami river. It has two stores, one grist-mill, one saw-mill; four schools, 160 scholars. Pop. 1477.

STONINGTON, p. t., New-London co., Ct., 12 m. E. New-London, 60 m. S.E. Hartford, 355 W. Watered by Mystic and Pawcatuck rivers, which afford water-power. The principal village or borough is on a rocky point of land which projects half a mile into the E. end of Long Island sound, and has a good harbour, protected by a breakwater constructed by the United States government at an expense of $50,000. It contains two churches, a Congregational and Baptist; two academies, a bank, 150 dwellings, and over 1000 inhabitants. It has a considerable number of vessels employed in the whaling and sealing business. A railroad connects this place with Providence, which, with the Long Island railroad, now about completed, will constitute the most direct route from New-York to Boston. The township contains 18 stores, two lumber-yards, one fulling-mill, four woollen factories, four grist-mills, one tannery; two academies, 103 students; 15 schools, 807 scholars. Pop. 3698.

STORNOWAY, a bor. of barony, seaport, and market town of Scotland, on the E. side of the island of Lewis, one of the Hebrides (which see), on a fine bay, 36 m. W. by N. from the nearest point of Cromarty, on the main land, lat. 58° 13' N., long. 6° 16¼' W. Pop. of the town and immediately contiguous villages about 3000. This, which is the most considerable town in the Western islands, has grown up, within no very long period, from a paltry hamlet of about a dozen houses, in consequence of its favourable situation for carrying on the herring and cod and ling fisheries, especially the latter. Though not regularly built, the houses are substantial, and slated, and there are some good shops. The harbour is formed by a pier; and the bay, which is spacious and has deep water, is formed by two low headlands and an island. Seaforth lodge, the occasional residence of Mackenzie of Seaforth, proprietor of the island of Lewis, is in the immediate vicinity of the town. There is a parish church, town-house, and custom-house,

people are Roman Catholics, there is not a single Protestant dissenter. The means of education, which formerly were very deficient, have been increased materially of late years. Gaelic is the language generally spoken throughout the island: in Stornoway, however, it is giving way to English, and divine service is now performed one part of the day in Gaelic, and the other in English.

Small packets, supported by government, ply weekly between Stornoway and Pollew, on the coast of Ross; and in summer Stornoway is sometimes visited by steamers from Glasgow.

The herring fishery has latterly declined; but during the year ending the 5th of April, 1840, 15,810 cwts., or 790 tons, cod and ling were cured at Stornoway and Barra: it may be worth from £12 to £15 a ton. During the same year 574 boats, manned by 3336 men and boys, were fitted out from the same places for the fishery. The town has a branch bank, and a rope manufactory.

With the exception of a small district immediately around Stornoway, the island of Lewis is in the most backward state imaginable, and the inhabitants poor and wretched in the extreme; nor, owing to the embarrassment of the present proprietor, is there much prospect of any speedy change for the better. (*New Statistical Account of Scotland, Ross, and Cromarty*, 115, 140; *Anderson's Guide to the Highlands*, 482, &c.)

STOUGHTON, p. t., Norfolk co., Mass., 90 m. S. Boston, 496 W. Drained by head waters of Neponset river. It contains four churches, a Congregational, Methodist, Baptist, and Universalist; 11 stores, one fulling-mill, one woollen factory, three cotton factories with 580 spindles, two saw-mills; 10 schools, 692 scholars. Pop. 2142.

STOURBRIDGE, a market town of England, co. Worcester, hund. Halfshire, par. Old Swinford, on the Stour, here crossed by a stone bridge, 18 m. N.N.E. Worcester. Pop. of township in 1831. 6148. Though irregularly built, the houses are pretty good: it has a handsome market-house, a theatre, a subscription library, &c. The episcopal chapel, erected by subscription in 1742, is beyond the jurisdiction of the bishop: the living, a curacy, in the gift of the inhabitant householders, is worth £134 a year. There are places of worship for various sects of dissenters, and a well endowed free school, founded by Edw. VI., in which Dr. Johnson received the rudiments of his education. Stourbridge has a national school, and a great number of benevolent and benefit associations. It is governed by a bailiff, town-clerk, &c.: and has petty sessions, and a 40s. court of requests. It has manufactures of glass and earthenware, and hardware: the iron trade of the town and neighbourhood is considerable, and most part of the iron-work used in the construction of the custom-house and new postoffice, London, came from Stourbridge. The town communicates, by a branch canal, with the Staffordshire and Worcestershire canal, by which great numbers of bricks are sent hence to the metropolis and elsewhere. Markets on Fridays. Fairs, Jan. 8 and Mar. 29, for horses and cattle; Sept. 8, for cattle and sheep.

STOURPORT, a market town of England, co. Worcester, hund. Halfshire, par. Kidderminster, at the confluence of the Severn and Stour, 3½ m. S.S.W. Kidderminster, with the pop. of which town and par. its own is returned. It is wholly of modern date, owing its origin to the Staffordshire and Worcestershire canal, which joins the Severn on its S. side. It is well built, principally of brick, and is partially paved, and lighted with gas. The Severn is here crossed by an iron bridge, the central arch of which has a span of 150 ft., rising to 30 ft. above the surface of the river. Stourport has an extensive transit trade, being, in fact, one of the principal *entrepôts* between the E. and W. parts of the kingdom. Markets on Wednesdays; fairs, Easter Monday, Sept. 15, Dec. 18, chiefly for hops and cattle; also a meeting every Thursday for hops.

STOW, p. t., Lamoille co., Vt., 22 m. N.N.W. Montpelier, 535 W. Drained by Waterbury river and its branches, a tributary of Onion or Winooski river. It contains three churches, a Congregational, Methodist, and one common to Universalists and Christians; four stores, two fulling-mills, one woollen factory, three grist-mills, six saw-mills, two tanneries; 12 schools, 447 scholars. Pop. 1371.

STOW, p. t., Middlesex co., Mass., 25 m. W. by N. Boston, 427 W. Assabet river affords water-power. It contains two Congregational churches, five stores, two woollen factories, four grist-mills, four saw-mills; two academies, 50 students; five schools, 300 scholars. Pop. 1230.

STOW, p. t., Summit co., O., 131 m. N.E. Columbus, 399 W. Watered by Cuyahoga river. It contains four stores, one flouring-mill, three saw-mills; one academy, 37 students; 13 schools, 387 scholars. Pop. 1533.

STOW-MARKET, a market town and par. of England, co. Suffolk, hund. Stow, on the Gipping, a tributary of the Orwell, and on the road and half way between Ipswich and Bury. Area of par., 1940 acres. Pop. in 1831, 2672.

The town consists principally of a main street, with many good houses, and has a bustling and thriving appearance. The parish church is large and handsome, part of it being in the decorated and part in the perpendicular style between): it has a tower and slender spire of considerable height, a peal of eight bells, &c. The living, with that of Stow-Upland, adjacent, a vicarage worth £201 a year, is in the gift of — Wilcox, Esq. (*Eccl. Rev. Rep.*, The Baptists, Methodists, &c., have meeting-houses, and there are various schools and benevolent societies. Stow-market is connected by a navigable canal with Ipswich, and has a brisk trade in malt and barley, with manufactures of cotton ade and sacking. Near it is the house of industry for the hundred, a handsome edifice on an eminence, erected at an expense of £12,000. It has petty sessions, a manorial court, &c., and is a polling place for the W. division of the county. Markets on Thursdays; fairs, three times a year.

STOYTOWN, p. b., Quemahoning t., Somerset co., Pa., 131 m. W. by S. Harrisburg, 157 W. Situated on Stony creek, and contains a German Reformed church, three stores, six grist-mills, 15 saw-mills, two tanneries, two distilleries; four schools, 100 scholars. Pop. 257.

STRABANE, an inland town of Ireland, W. side county Tyrone, prov. Ulster, on the Mourne, near its confluence with the Finn or Foyle, 1 m. E. Lifford, and 14 m. S.S.W. Londonderry. Pop. in 1836, 5147. It is built on the skirts of the Marquis of Abercorn, in a fine valley enclosed by lofty mountains; and has a good linen market, an extensive retail trade, and a considerable yard in the export of grain and provisions, by way of Londonderry. The older parts of the town, along the river, are low, with narrow, dirty streets and mean houses; but in the newer parts there are some comparatively good streets, shops, and houses. It has a parish church, a Roman Catholic chapel, two Presbyterian and two Methodist meeting-houses, a fever hospital and dispensary, a market-house and a session-house, and a bridewell. It is connected by a bridge with the suburb on the left bank of the river. Under the Municipal Reform Act (3 and 4 Victoria, cap. 108), it has a corporation entitled the sovereign, free burgesses, and commonalty. Previously to the union it returned two members to the Irish House of Commons, but was then disfranchised. It has a parochial school, a Lancasterian school, and some other schools. The trade of the town is much facilitated by a canal, about 4 m. in length, from it to where the Foyle becomes navigable for barges of 40 tons. (See *Railway Report*, &c.)

Quarter sessions are held in April and October, petty sessions on alternate Tuesdays, and a manor court, with jurisdiction to the amount of £3, is held once a month. Adjoining the town is a valuable salmon fishery, the produce of which is mostly sent, either fresh, packed in ice, or pickled, to the British markets. Post-office revenue in 1829, £1158; in 1836, £1958. Branches of the Agricultural, Provincial, and Belfast banks were opened in 1838. "I saw little or nothing of rags in Strabane: there was a respectable look about the people and everything they did" (*McGilpie's Ireland*, ii., 167.) Persons coming to attend the assizes for the county Donegal, held at Lifford, usually take up their residence in Strabane.

STRABANE, t., Adams co., Pa., 6 m. N.E. Gettysburg. Drained by Conewago creek and its branches. It has three stores, one flouring-mill, one grist-mill, two saw-mills, two tanneries, one distillery. Pop. 1370.

STRAFFORD, county, N.H. Situated in the S.E. part of the state, and contains 350 sq. m. Bounded by the common falls or Piscataqua river, which separates it from Maine. Drained by Lamprey, Cocheco, and Isinglass rivers. It contained in 1840, 57,393 neat cattle, 53,554 sheep, 26,717 swine; and produced 99,898 bushels of wheat, 33,855 of rye, 250,497 of Indian corn, 9290 of buckwheat, 32,055 of barley, 101,578 of oats, 1,988,409 of potatoes, 168,139 pounds of sugar. It had 947 stores, capital $402,121; 37 fulling-mills, 19 woollen factories, 17 cotton factories with spindles, three furnaces, 86 grist-mills, 115 saw-mills, four oil-mills, 44 tanneries, one distillery, one pottery, five printing-offices, four binderies, three weekly newspapers. Total capital in manufactures $2,674,874. One college, 65 students; 12 academies, 1330 students; 485 schools, 30683 scholars. Pop. 61,127. Capital, Dover.

STRAFFORD, p. t., Strafford co., N.H., 25 m. E.N.E. Concord, 506 W. Watered by Isinglass river, issuing from Bow pond, which is 650 rods long and 400 rods wide. It contains four churches, two Freewill Baptist, a Methodist, and Christian; seven stores, one saw-mill, one academy, 85 students; 18 schools, 776 scholars. Pop. 1800.

STRAFFORD, p. t., Orange co., Vt., 34 m. S.S.E. Montpelier, 509 W. Watered by a branch of Onnipcompanoosuc river, which affords water-power. On one point of iron, from which large quantities of copperas are manufactured, is found in its S.E. corner. It contains four churches, one Congregational, and three Union or Free; three stores, one

falling-mills, two grist-mills, nine saw-mills; 12 schools, 690 scholars. Pop. 1762.

STRALSUND, a strongly fortified town of the Prussian states, prov. Pomerania, cap. reg. and circ. of the same name, on the narrow strait separating the isle of Rugen from the continent, lat. 54° 19' 26'' N., long. 13° 7' 20'' E. Pop. (1834), 14,713. It was founded in 1209. Streets narrow and dirty, houses ill built; but it has a fine arsenal, and some good public buildings. It is encompassed on the land side by lakes and marshes, so that it can only be approached by bridges. Its fortifications, which had been dismantled, have been renovated and greatly improved since 1815, so that it is now one of the strongest places in the monarchy. It has a gymnasium, two public libraries, and an orphan asylum, with breweries, distilleries, and various manufactures; and carries on a considerable commerce, exporting corn, timber, beer, linens, &c. Its port, though small, is convenient and safe, but it labours under a deficiency of water. Close to the town the depth does not exceed 7 feet, at a little distance it increases to 10 feet, and in the offing there is 13 feet. In 1834 there belonged to Stralsund 73 ships, of the burden of 6861 lasts.

STRANRAER, a parl. and royal bor. and seaport of Scotland, co. Wigtown, on flat ground, on the inner or S. shore of the inlet of the sea called loch Ryan, on the high road from Dumfries to Portpatrick, 21 m. W. by N. Wigtown. Pop. of royal bor. in 1841, 3439; but the parl. bor., which includes some contiguous suburbs, had then a pop. of 4589. It consists of three streets parallel to the shore, united by several cross streets. The principal streets are neat and pretty well built; but there are a great many wretched hovels in the lanes and outskirts of the town, mostly occupied by Irish immigrants. It has a town-hall, jail, a church belonging to the establishment, with various dissenting chapels, one parochial and seven unendowed schools, two very good subscription libraries, and a good public reading-room. It is the centre of a considerable retail trade, but has no manufactures, except some hand-loom weaving, on account of the Glasgow manufacturers. It has three branch banks, a savings' bank, and a weekly newspaper. Loch Ryan is a fine basin. Opposite to a place called Cairn, on its E. shore, there is good anchorage, and water sufficient to float the largest ships. Stranraer harbour dries at low water; but it would not be difficult, by carrying out the pier to a greater distance be to the loch, to make it accessible at all times of the tide. In 1840, 36 vessels belonged to the port, of the aggregate burden of 2039 tons. Its customs revenue is inconsiderable, and inadequate to defray the cost of the establishment. Steamers regularly ply between the port and Glasgow and Belfast. A large proportion of the population are Irish, mostly in an abject state of poverty. Stranraer was made a royal borough in 1617. It has 18 councillors. Corporation revenue in 1840, £267. It unites with Wigtown, Whithorn, and New Galloway, in sending one member to the House of Commons. Registered electors in 1839-40, 230. (New Statistical Account of Scotland, No. XXI., 94; Official Returns; and Private Information.)

STRASBOURG (an. Argentoratum), a fortified city of France, on its E. frontier, dep. Bas-Rhin, of which it is the cap., on the Ill, within a short distance of the Rhine, to which its glacis extends, and across which it communicates with Kehl by a bridge principally of boats, about 100 m. S.W. Mentz, and 250 m. E. by S. Paris; lat. 48° 34' 56'', long. 7° 44' 51'' E. Pop. in 1836, 50,239 (many of whom are Protestants and Jews), exc. the garrison, generally mounting to 6000 men. The city is of a triangular form, enclosed by a bastioned line of ramparts strengthened by numerous outworks, entered by seven gates, and has on its side a strong pentagonal citadel, built by Vauban. By means of sluices constructed under Louis XV., the adjacent country may be laid under water; and several additional defences having been constructed since the peace, it is now one of the most important fortresses and arsenals in France, and has the largest dépôt of artillery. Strasbourg is agreeably situated, and generally well laid out: its streets are mostly narrow, with lofty houses; but it has several rather large and regular squares. Though for a lengthened period united to France, it still has all the outward appearance of a German town, with which the costume and language of inhabitants correspond. The Ill and its branches intersect the city in all directions, and are crossed by numerous wooden bridges. Without the walls are several suburbs. By far the most remarkable public edifice is the minster, cathedral, one of the noblest structures of its kind. It is said to have been originally founded by Clovis, in 504; but Charlemagne constructed the choir, the only part that survived the destruction of the old cathedral by lightning in 1007. The modern building was begun in 1015, but not finished till the 15th century. The entire length of the interior is 378 ft.; breadth, 140 ft.; height from the pavement to the roof of the nave, 76¼ ft. The W. or grand entrance

has, on its N. side, a spire of the extraordinary height of 437½ Paris, or 468½ English ft. (Schreiber); being, if the dimensions be accurate, about 7 ft. higher than St. Peter's in Rome, and about 5 ft. higher than the great pyramid of Cheops. It is of open work, and combines with the most perfect solidity extraordinary lightness and elegance. The view from the top of this spire is one of the most extensive and finest that can be imagined; it is, however, enjoyed by few only. The ascent to the top of the tower may, indeed, be accomplished without much difficulty, and the view from it is superb; but the ascent thence to the lantern requires very powerful nerves, and, in fact, ordinary visiters are not permitted to attempt it. The erection of this famous spire was commenced in 1276, by Erwin de Steinbach, and continued by his son, his daughter Sabina having also contributed some fine sculptures to the principal portal. It was finished in 1439, under the direction of Schulz, an architect of Cologne. Beside the grand portal are equestrian statues of Clovis, Dagobert, Rodolph of Hapsburg, and Louis XIV.; and over its centre is a marigold-shaped window of stained glass, 51 ft. in diameter. The interior has a fine stone carved pulpit, with numerous monuments, statues, &c. The famous astronomical clock, constructed by Isaac Habrecht, which indicated the days of the month, the places of the sun and moon, and other celestial phenomena, has been standing still for a considerable number of years. Napoleon is said to have contemplated the repair of this curious piece of mechanism. In the church of St. Thomas is the magnificent monument in honour of Marshal Saxe, the

of St. William, the Château Royal, the large public library, said to contain (though we do not place implicit confidence in the statement) 135,000 volumes, the new synagogue, the theatre (with a handsome front of six Ionic columns), the prefecture and other public residences, the arsenal, barracks, cannon foundry, and other military establishments, and various public schools, are deserving of notice. There are several hospitals and asylums, civil and military prisons, an exchange, corn, fish, and other markets, various assurance companies, a royal dépôt d'étalons, a botanic garden, &c. The environs are well cultivated, and Strasbourg has many good public walks; the principal of which is the Contades, without the city, laid out by the marshal of that name in 1764.

Strasbourg is a bishop's see, the seat of a court of primary jurisdiction, council of prud'hommes, and a chamber of commerce; of a university academy, royal college, mint, boards of forest economy, roads, and bridges, and of customs; the Lutheran consistory for the S. of France, faculties of law, medicine, sciences, &c., Roman Catholic and Protestant seminaries, and societies of agriculture, literature, and arts. The museum of the academy is very rich in the natural products of Alsace. The city is divided into four cantons, in each of which is a justice of the peace and a police commissary. It has an extensive royal manufactory of snuff, for which Strasbourg is famous; with considerable manufactures of woollen, linen, and cotton stuffs, leather, hats, paper, playing cards, earthenware, shell articles, printing types, chemical products, &c., exclusive of dye-houses, breweries, printing establishments, and sugar refineries; the pâtés de foies gras of Strasbourg have obtained to high gastronomical celebrity. The trade of Strasbourg is very extensive, its situation on the Rhine having rendered it a great frontier entrepôt.

Strasbourg is very ancient, and most probably, indeed existed previously to the Romans. It assumed the name of Strateburgum in the 5th century. On the first partition of the Frankish territory it was included in the kingdom of Austrasia, and on the second in Lorraine. In the 10th century it belonged to the German emperors, and subsequently became a free city of the empire, which it continued to be till 1681, when it was taken possession of by Louis XIV., and finally annexed to France. Pierre Schœffer, who contests with Gutenberg the honour of being the inventor of printing, and Generals Kellermann and Kleber, are among the distinguished natives of Strasbourg. The latter is buried in the cathedral, and a monument has been erected to his memory on the artillery parade. (Hugo, art. Bas-Rhin; Schreiber, Guide du Rhin; Guide du Voy. en France; Murray's Handbook, &c.)

STRASBURG, p. t., Lancaster co., Pa., 46 m. E.S.E. Harrisburg, 190 W. Bounded N. and W. by Pequea cr. Drained by Little Beaver cr. It contains a church, 10 stores, 10 flouring mills, 8 grist-mills, 5 saw mills, 4 tanneries, 5 distilleries; 8 schools, 390 scholars. Pop. 4155.

STRATFORD-UPON-AVON, a town, mun. bor. and parish of England, co. Warwick, hund. Barlichway, celebrated as the birth-place of Shakspeare, 7 m. S.W. Warwick. Area of par., 6860 acres. Pop., in 1831, 5171. The town is finely situated on a gentle acclivity rising from the W. bank of the Avon, which here expands to a breadth of about 130 yards, and is crossed by a bridge of 14 arches, built in the reign of Henry VII., but repaired and widened in 1814. In the older parts the houses, though intermixed with others of more modern date, have an antique appearance; several new streets, have, however, been constructed of late years, and the corporation has distinguished itself by the aid it has given to improvements of all sorts. (*Municipal Bound. Report*.) It has a large, handsome, cruciform church, with a square embattled tower, surmounted by a lofty spire; the transepts, tower, and some parts of the nave, are early English; the rest of the building is mostly a mixture of decorated and perpendicular. (*Rickman*.) It has several fine old monuments. Of these the most interesting by far is that of Shakspeare, on the N. wall of the chancel. It is constructed partly of marble, and partly of stone; consisting of a half-length bust of the poet, with a cushion before him, placed under an ornamental canopy, between two columns, supporting an entablature. Under the bust are the following lines:

Judicio Pylium, genio Socratem, arte Maronem,
Terra tegit, populus mæret, Olympus habet.

These are followed by six lines in English verse; and on a flat stone, which covers the grave, is an entreaty not to disturb the dust "enclosed heare," and an imprecation against such as might profane the ashes of the mighty dead. The living of this church, a vicarage, worth £239 a year, is in the gift of the Earl of Plymouth. Here is also a chapel, that once belonged to the guild of the "Holy Cross," suppressed at the reformation: it is of the age of Henry VII., in the perpendicular style, and has several curious fresco paintings on its walls. Attached to this building is a hall for the brethren of the guild, since used for the meetings of the corporation; alms-houses for 24 poor persons of both sexes, and a free grammar-school for children, natives of the borough. The modern town-hall, a building of the Tuscan order, erected in 1768, has a hall 60 ft. in length by 30 ft. in breadth. Having been dedicated, at the jubilee in 1769, to the memory of Shakspeare, it is thence called the Shakspeare Hall. It is decorated with pictures, by Wilson and Gainsborough, of the great poet and Garrick; and outside the building is a statue of the poet, which, with the pictures, was presented by Garrick. Here are national, Lancastrian, and other schools, two public libraries, a neat theatre, &c. The *Mun. Corp. Report* says that some of the charities in the town are highly spoken of. The town is governed by a mayor, 3 aldermen and 12 councillors, but has no commission of the peace; and even its court of record has fallen into disuse. Corporate revenue, in 1840, £2194. The only manufacture carried on belongs to one of the departments of button-making, and it is by no means extensive. Markets on Fridays: fairs six times a year, for cattle, corn, provisions, &c. The Avon is navigable by barges from the Severn to Stratford, where it unites with the Stratford canal, which is itself connected with the Worcester and Birmingham canal.

Little, unfortunately, is known of the life of the illustrious poet to whom Stratford owes all her celebrity. He first saw the light on (as is supposed) the 23d of April, 1564. Having married in 1582, he soon after went to London, where he produced the greater part of his immortal works; and having returned to Stratford to spend the evening of his days, died there in 1616, on the 23d of April, the anniversary of his birth. This brief notice comprises nearly *all* the authentic information we possess regarding the greatest of dramatic poets, notwithstanding his death occurred little more than two centuries ago! "No letter of his writing; no record of his conversation; no character of him, drawn with any fulness by a contemporary, can be produced." (*Hallam*.) The house in which the great poet was born, in Henley-street, is still standing; and is the resort of all visitors to the town. It has, however, been converted into two houses, and otherwise much altered. The house in which Shakspeare passed the latter years of his life was, to the disgrace of those concerned, demolished in 1759; when the famous mulberry-tree he is said to have planted in its garden was also cut down! (*Whaler's Stratford-on-Avon; Municipal Boundary Report, &c.*)

STRATFORD, p. t., Fairfield co., Ct., 50 m. S.W. Hartford, 287 m. W. Bounded E. by Housatonic r. It contains five churches, a congregational, Episcopal, Methodist, Baptist, and a Universalist, 12 stores, one flouring-mill, two gristmills, one saw-mill; one academy, 36 students; nine schools, 393 scholars. Pop. 1808.

STRATFORD (STONY), a market town and parish of England, co. Bucks, hund. Newport, on the Ouse, which is here the boundary of the co., and is crossed by a stone bridge, 4½ m. N.E. Buckingham. Area of par., 70 acres. Pop., in 1831, 1053. It is built on the line of the ancient Watling-street, and is supposed to occupy the site of the *Lactodorum* of the Romans. The houses are mostly of freestone,

extending for about one m. on either side the road. The parish church was rebuilt in the Gothic style, in 1777; the living, a perpetual curacy, worth £130 a year, is in the gift of the bishop of Lincoln. There are meeting-houses for various dissenters, national and Sunday-schools, a nursery for apprenticing children, &c. At one time in this town the person of the young king Edward V. was seized; and Grey and Vaughan arrested by Richard duke of Gloucester. The only manufacture is that of lace; but the inhabitants have some trade in corn. Markets, which are well supplied with provisions, are held on Fridays: fairs, 21st August, and Fridays before 10th October, for hiring servants; and 5th November for cattle.

STRATSBOROUGH, p. t., Portage co., O., 130 m. N.E. Columbus, 330 m. W. Watered by Cuyahoga r. and Tuscarawas cr. It contains two churches, a Presbyterian and Methodist, eight schools, 189 scholars. Pop. 983.

STRONG, p. t., Franklin co., Me., 43 m. N.W. Augusta, 632 m. W. Watered by Sandy r. It has 6 schools 63 scholars. Pop. 1109.

STRONGSVILLE, p. t., Cuyahoga co., O., 131 m. N.N.E. Columbus, 361 W. It contains three churches, one of which is a Congregational, one of the finest in the co., seven stores, one fulling-mill, one woollen factory, one grist-mill, three saw-mills; 11 schools, 363 scholars. Pop. 1167.

STROUD, a parl. bor., market town and par. of England, co. Gloucester, hund. Bisley, on the Slade or Stroud-water, near its junction with the Frome, 9 m. S. Gloucester. Area of par., 3990 acres. Pop., in 1831, 8607; in 1841, 8680. The parl. bor., however, is not confined to the town, but includes the whole clothing district, of which it may be regarded as the centre, comprising about 14 parishes. The general characteristics of the district are "the situation of the mills on streams in deep ravines; the scattered and irregular manner in which the houses are built on the hill side; and the contrast between the high land (in many cases either wood or common, with few inhabitants) and the valleys studded with houses and thickly peopled." (*Parl. Bound. Report.*)

Stroud stands on the side of a hill, and consists chiefly of a long street, crossed by another at its base: the houses are good, and the streets well paved and lighted. The parish church, a large edifice, has a tower with an octangular steeple at its W. end. The living, a perpetual curacy, worth £132 a year, is in the gift of the bishop of Gloucester. There are places of worship for Baptists, Independents, Wesleyans, &c.: and several charities for educating poor children, and giving relief to the poor.

Stroud, and the district of which it is the centre, owes much of its prosperity to the Stroud water, which is not only made available for the working of machinery, but is said to be peculiarly adapted to the dyeing of scarlet and other colours. The clothing trade has, in consequence, extended itself principally along the banks of the river, on which there are numerous fulling-mills, &c. In 1821 there were 14 mills in Stroud parish, about half impelled by water and half by steam, which employed together 1299 work-people. (*Mills and Factories Rep.*) The prosperity of the town and district depends, of course, upon the state of the clothing trade, and partakes of the fluctuations incident to the latter. Power-looms have been to be introduced into the manufacture, but hitherto they have not made much progress. The Thames and Severn canal, which passes close to the town, is of great advantage to its trade.

The Reform Act conferred on Stroud, and its adjacent district, as specified above, the important privilege of sending two members to the House of Commons. Registered electors, in 1839-40, 1902. Petty sessions for the hund. are held in Stroud. Markets, on Fridays, fairs, May 10 and August 21, for cattle sheep and hogs.

STROUDSBURGH, p. t., Stroud t., capital of Monroe co., Pa., 3 m. N.W. Delaware Water Gap, 114 m. N.E. by E. Harrisburg, 217 W. Situated at the junction of Brodhead's and Pocono's creeks, and contains a court-house, jail, two churches, a Methodist and Friends, six stores, one flouring-mill, two printing-offices, two weekly newspapers, 10 schools, 460 scholars. Pop. 467.

STUHLWEISSENBURG, a town of Hungary, cap. of a co. of its own name, 18 m. N.E. lake Balaton, and 35 m. S.W. Buda. Pop., with its two suburbs, 20,000. (*Berghaus.*) "Though formerly a Roman town, and a scene of frequent occurrence in Hungarian history, it contains nothing remarkable. The palace of the bishop, and some of the houses connected with it, are handsome; but the streets are badly paved, and the whole town disagreeably placed in the centre of a huge bog." (*Paget's Hungary, i. 527.*) It was, for a lengthened period, the residence of the sovereigns of this country, and has a royal mausoleum in which 14 of them are buried. It has a gymnasium and other Roman Catholic schools, a military academy, a Magyar theatre, &c., and manufactures of woollen cloth, flannel, soap, and leather. (*Austr. Nat. Encyclop., &c.*)

STURBRIDGE, pt. t., Worcester co., Mass., 61 m. W.S.W. Boston, 379 m. W. Watered by Quinnebaug r., which affords water-power. It contains two churches, a Congregational and Baptist, five stores, one fulling-mill, five cotton factories with 11,612 spindles, three grist-mills, eight saw-mills, two tanneries; 13 schools, 567 scholars. Pop. 2005.

STUTTGARD, a city of S. Germany, cap. of the kingdom of Wirtemburg; on the Nesen, a small tributary of the Neckar, about 1¼ m. from its embouchure in that river, 38 m. S.E. by E. Carlsruhe, and 120 m. N.W. Munich. Lat. (of the gymnasium) 48° 46' 32" N.; long. 9° 10' 45" E. Pop., including some suburban villages, about 38,500. (Berghaus.) It is situated in an amphitheatre of fertile hills; and having been, for the most part, laid out during the present century, is one of the cleanest and handsomest towns of Germany.

"The approach of Stuttgard is exceedingly pretty. The road passes through an avenue of lofty poplars, and you enter the broad and handsome Neckar Strasse, without encountering any of those disagreeable or vulgar appearances that frequently distinguish the suburbs of a capital. The ty may be said to stand in the centre of a garden: on every side it is surrounded by vineyards and orchards. In the ancient part of the town, the streets are narrow and crooked; but in the modern, they are broad and handsome. The court and the military, with the other necessary concomitants of a capital, give it rather a lively air: and there is usually a great bustle in the streets." (Strang's Germany in 1831, ii. 394.) Other travellers say that it is very dull, presenting little to interest a stranger; and that, like most of the capitals of petty states, it has a parade of importance, which it is really not entitled. It has also been evidently a mistake to plant the city on the dirty and stagnant Nesen instead of the Neckar, a fine navigable stream, which it have opened for it a considerable traffic to and from Rhine. And from being placed in a deep hollow, it is hotter, according to Spencer, enveloped in mists and fogs, and in summer it is unhealthy from malaria. (Germany of the Germans, ii. 337.)

The new royal palace has the advantage, if such it be, of being situated both in town and country; opening on one side to a fine park, and on the other into a spacious square, adorned with trees and fronting the Königs Strasse, or high-street, the finest in the city. The palace is an imposing freestone edifice, begun in 1746, and completed by the king. It has a centre and two projecting wings; the whole forming, like Buckingham palace, three sides of a square. The parapets are decorated with handsome statues, the roof immediately above the grand entrance is surmounted, we cannot say adorned, with a large gilt crown, &c. The same bad taste does not prevail in the interior, though even there ostentation and costliness are here visible. (Autumn near the Rhine, p. 316.) There are a vast number of apartments, and several are hung with splendid tapestries, from the Gobelins at Paris, and from Napoleon to his ally the late king. Besides these the palace contains many good Flemish paintings and sculptures by Danekker and Canova. In the same square in which the new palace is situated are the opera and the theatre. The former, now occupied by officers of the court or government, has the aspect of a fortress; and behind it is a Gothic church, in which are the tombs of the dukes of Wirtemberg. The theatre is a wooden building; but it has usually a good company.

Other buildings are worthy of notice; as the palatier members of the royal family, the Ständehaus, or of the parliament of Wirtemberg, to the debates sessions of which the public are always admitted; the town-house, chancery court, new barracks, post-office, city schools, large hospital, workhouse, royal studs, crowning the palace, with an extensive stud of fine royal stables, and a riding school, &c. Stuttgard contains theran churches, a Calvinist and a Roman Catholic, and a synagogue. The public library, open daily, and from 3 to 5, a very large and valuable collection, comprises from 170,000 to 180,000 volumes, including significant collection of Bibles. The museum of tory comprises a remarkable collection of fossils in annstadt. There are royal cabinets of medals, models, maps, charts, &c., and many private collections. Without having the pretensions of the Munich or Dresden, to be considered the seat of the arts, Stuttgard has been distinguished as the birth place of several of the most eminent German literati, as Schiller, Danekker, Menzel, long the editor of the Kunst-Blatt, Baron Cotta, the famous publisher, of Danekker's finest works are in this city, and wrote his Robbers. Stuttgard is an industrious though unfavourably situated for trade. Cotta's establishment is one of the most extensive on the continent, it extends to printing and book-binding, the weaving

73

of woollen and cotton goods, and the making optical, mathematical, and musical instruments, are the chief branches of manufacturing industry. Some agreeable effervescing wine is made on the surrounding hills; and about 3 m. N.E. the city is Kannstadt, resorted to by the citizens and others as a favourite watering-place.

Stuttgard suffered severely in the wars of the 16th and 17th centuries: but it escaped with little loss during the last war, though repeatedly occupied by the armies on both sides. (Berghaus; Memminger, Beschr. von Wurtemberg; Strang; Spencer; Murray's Hand-Book for S. Germ., &c.)

STUYVESANT, p. t. Columbia co., N. Y., 12 m. N. Hudson, 17 m. S. Albany, 351 W. Bounded W. by Hudson r., on which is a good landing, from which steamboats and barges ply to New York. It contains a Dutch Reformed church, seven stores, one fulling-mill, one woollen factory, one cotton factory with 1600 spindles, one saw-mill · five schools, 380 scholars. Pop. 1779.

SUDBURY, a parl. and mun. bor. of England, principally in the co. Suffolk, but partly also in Essex, on the Stour, here crossed by a stone bridge, 18 m. W. Ipswich, and 50 m. N.E. London. The parl. bor. includes the three pars. of St. Peter, St. Gregory, and All Saints, and some small extra-parochial districts on the Suffolk side of the river, with the hamlet of Ballingden-cum-Brunden in Essex: its area may amount, in all, to about 2100 acres, and it had, in 1831, a population of about 5500, of which 4677 belonged to the old borough. "The town of Sudbury lies towards the S. extremity of the borough. It is a neat, clean, and well-built town. Under the provisions of a local act, it is lighted, the footpaths flagged, and the roads kept in repair: it is also watched occasionally. The town has been much improved of late years internally, but the buildings have not extended beyond its former limits, and, indeed, this cannot occur; for the lands immediately surrounding the town, which are adapted for this purpose, are subject to a right of common pasture during part of the year. It is said, that if this right could be abolished, the buildings would probably increase." (Mun. Corp. Rep.) Sudbury has three churches, All Saints, St. Gregory, and St. Peter; mostly in the perpendicular style, but some of the tracery, and other parts, have been much mutilated. (Rickman's Gothic Arc.) The living of All Saints, a vicarage, worth £119 a year, is in the gift of —— Sperling, Esq.: those of St. Gregory and St. Peter, curacies, worth £160 a year, are in the gift of Sir L. Maclean. Except a large independent chapel and a Baptist chapel, there are few or no other buildings worth notice; and no remains exist of the Benedictine priory, founded in the reign of Henry II. The grammar-school, founded in 1491, has an income of about £100 a year: and there are various other charities, including a national school, at which about 150 children attend; but the education of the poorer classes is said to be very defective.

Sudbury was one of the towns in which the immigrant Flemish woollen manufactures were established by Edward III. Its woollen trade has, however, been for many years nearly discontinued, though, luckily, it has been replaced by that of silk. In 1838 there were estimated to be about 600 looms in the town; of which 270 were wrought by men, 250 by women and girls, and 80 by boys. They were then, however, little more than half employed. Mantels, lutes, and gros-de-Naples are the articles principally manufactured. Jacquard looms are not very generally introduced. The Sudbury weavers are said to be less expert in making fine goods than the weavers in Spitalfields; but the wages are nearly the same in both. The average per week, after deducting expenses, was, in 1838, for velvets and satins, 19s.; for figured goods, 19s.; and for gros-de-Naples, about 7s. 6d. In addition to silk weaving, there is at Sudbury a manufacture of buntings, which employs about 200 looms in the town. These are wrought by women and children, or old men unfit for silk weaving. The wages obtained at this employment average from 3s. 2d. to 4s. 6d. a week. (Handloom Weavers' Rep. ii.)

Sudbury is governed by a mayor, three other aldermen, and 12 councillors; and has a commission of the peace, a weekly court of record, &c. Corporation revenue, in 1840, £317 19s. The parliamentary and municipal boroughs are co-extensive.

Sudbury has sent two members to the House of Commons from the early part of the reign of Elizabeth. The boundary act increased its limits, as already noticed; but it is said, whether justly or not we do not undertake to decide, to be now, as formerly, distinguished by the venality of a large portion of the constituent body. Registered electors, in 1839-40, 594. Markets on Saturdays. Fairs, March 12, July 10, and September 4, for earthenware, glass, and toys. (Parl. Rep.)

Gainsborough, the eminent artist, worthy, as Sir Joshua Reynolds has stated (Fourteenth Discourse), to rank among the first painters of the English school, was a native of Sudbury, where he first saw the light in 1727. Sudbury was also the birth-place of Dr. Enfield, whose *Compendium of Brucker's History of Philosophy* is drawn up with great skill and judgment, and is, in fact, the best work on the subject in the English language.

SUDBURY, p. t., Middlesex co., Mass., 20 m. W. by N. Boston, 421 W. Watered by Sudbury r., a tributary of Concord r., which latter bounds it on the E. It contains three churches, two Congregational and a Methodist, five stores, four grist-mills, two saw-mills, one paper-mill; six schools. 367 scholars. Pop. 1422.

SUEZ, a town and seaport of Egypt, near the N. extremity of the gulf of its own name, which is also the N.W. angle of the Red sea, 76 m. E. Cairo; lat. 29° 59' 10" N.; long. 30° 15' 5" E. Pop. recently estimated at only 1500; but this must be independent of the numerous pilgrims and merchants, who are continually passing through the town, Suez being on the main route between Cairo and Mecca, and on that by which the commerce of Egypt with the countries to the eastward is carried on. The head of the gulf on which it stands has always been the seat of a considerable transit trade, and the ancient cities of Arsinoe and Kolzum stood in the neighbourhood; but Suez is a comparatively modern as well as a very mean town. Turner says, "Take it for all in all, Suez is indisputably the most miserable place I have seen in the Levant. Its only gate is to the N.W.; three cannon are mounted near it, and there are eight more on the banks of the sea. In its present state, 50 men could take it with ease. Suez produces nothing, being on all sides surrounded by the desert. The clothes, and even the provisions, of the inhabitants are all brought from Cairo, to the last loaf. Frequent caravans come from Jaffa and Jerusalem, bringing oil, tobacco, and soap." *Turner's Levant*, ii. 414, 415.) It suffered much from the French, by whom it was in great part destroyed; and it now consists merely of sun-dried brick houses and unpaved streets, with about a dozen mosques, a Greek church, custom-house, &c.; the whole surrounded by a ruined wall and some entrenchments thrown up by the French. It is wholly destitute of water, which is brought to it by the Arabs from wells several miles distant, and, besides being high-priced, is of a nauseous description. The port is accessible only by boats of from 30 to 60 tons. Steamers and other vessels belonging to the E. I. Company moor outside a sand bar at a distance of 2 m. from the town. But since the establishment of what has been called the over-land route to India, Suez has become a place of considerable importance, and is now the residence of an agent for the E. I. Company, and of several commercial agents.

The gulf of Suez, which at low water is in many parts so shallow as to be fordable, is memorable in Sacred History as the scene of the submersion of Pharaoh and his host. The isthmus of Suez, connecting Asia and Africa, is a sandy waste, between 70 and 80 m. across. Near Suez may still be seen the vestiges of the canal cut by Pharaoh Necho and Ptolemy Philadelphus, to unite the Red sea with the Nile. (*Turner's Levant; Dict. Géog.; Private Information, &c.*)

SUFFIELD, p. t., Hartford co., Ct., 17 m. N. Hartford, 352 W. Bounded E. by Connecticut river. It contains four churches, two Congregational and two Baptist, the "Connecticut Literary Institution" founded by the Baptists in 1835, which has an edifice 72 feet long, 34 feet wide, and 4 stories high, containing 24 rooms for students. It has also 16 acres of land connected, and a house for the steward, and it has usually over 100 students. The township was chartered in 1670, and until 1752 was under the jurisdiction of Massachusetts. It contains eight stores, two grist-mills, nine saw-mills, two paper-mills; one academy, 144 students; 11 schools, 611 scholars. Pop. 2669.

SUFFOLK, a marit. co. of England, having N. the co. Norfolk, E. the German ocean, S. Essex, and W. Cambridge. Area, 969,600 acres, of which about 820,000 are supposed to be arable, meadow, and pasture. Surface generally flat. Soil various; that of the middle and most extensive district consists principally of a strong loam, on a clay-marl bottom. The district, bounded by the rivers Stour, Orwell, and Brett, S. from Burstall, is a very rich loam, of extraordinary fertility. The maritime district, lying along the E. coast, consists of sandy loam and sand, which in some places is covered with heath. The soil in the N.W. parts is comparatively poor, consisting partly of sand and partly of peat. On the whole, Suffolk is not inferior, in respect of natural fertility, to any county in the kingdom. The climate is dry; but frosts are severe, and in spring the N.E. winds are sharp and prevalent. Tillage husbandry is prosecuted with great skill, spirit, and success. Ploughing, in every part of the county, is performed, as in Scotland, by a pair of horses driven by the ploughman, and is extremely well executed. Fallowing is uniformly practised on the heavy lands. These, also, are particularly adapted for the growth of beans, which, as well as peas, are extensively cultivated. Turnips not so extensively

866

grown as in Norfolk, being principally raised on the borders of Cambridgeshire. On clover leys when it is very generally planted by the dibble; but, when the land will stand all sorts of grain, as well as turnips, are drilled. But of the land in beans, peas, tares, &c., is now drilled without any ploughing, being merely scarified and scuffled so as to be rendered fine enough for the drill to work. The usual rotation in the turnip lands is, 1st, turnips; 2d, barley; 3d, seeds; 4th, wheat: on the heavy lands, 1st, fallow; 2d, wheat; 3d, seeds or beans; 4th, wheat: when the 3d is seeds, beans or oats come in 5th. Hemp has been cultivated for a lengthened period, and is reckoned of the finest quality. Carrots are a good deal grown; and hops are raised in the vicinity of Stowmarket. Suffolk is famous for its breeds of horses, cattle, and hogs. The horses are called punches and are, as the term implies, short and compact, being well fitted for regular farm work. The cattle have sprung from the Galloways, many of which are fastened in the county. Like their progenitors, they are polled, and rather medium sized. They are better suited than the Galloways for the dairy, being excellent milkers. The produce of butter is not, however, supposed to be in proportion to the milk, though it be also very considerable. Arthur Young supposed that about 40,000 firkins of butter were annually sent from Suffolk to London. (*Survey*, 2d ed., p. 204) ; and, as the management of cows is now better understood, and more attended to, the quantity may, at present, be increased to 50,000 or 55,000 firkins. Stock of sheep about 300,000. Warrens were formerly numerous in the sandy district, but they are now much diminished. Property much divided; a good deal in the hands of respectable yeomen, who cultivate their own estates. Farms generally large, but many small. They are usually let on leases for 7 or 14 years. Tenants mostly restrained from exceeding their corn crops to a fallow ; but, in other respects, they are left pretty much at liberty. Farm buildings indifferent. Cottages generally bad. Average rent of land, 14s. 2d. an acre. Minerals of no importance. This county was formerly celebrated for its manufactures, particularly those of woollen, but they are now much decayed. Mixed silks and worsted stuffs are still, however, manufactured at Sudbury, Colchester, and other places. Gun-flints are made in large quantities at Brandon. There is a considerable manufacture of stays at Ipswich.

Suffolk is well watered, being intersected by the rivers Orwell, Deben, Ald, Blythe, and Lech. It is separated from Essex by the Stour. Suffolk is divided into 21 hundreds, and 510 parishes. It returns 11 members to the House of Commons; viz., four for the county, two each for the boroughs of Bury St. Edmunds, Ipswich, and Sudbury : and one for the borough of Eye and contiguous parishes. Registered electors for the county, in 1839–40, 11,086, viz. 6404 for the E , and 5091 for the W. div. In 1801, Suffolk had 64,081 inhabited houses, and 315,189 inhabitants of whom 154,167 were males, and 161,022 females. Sum expended for the relief of the poor in 1839-39, £145,671. Annual value of real property in 1815, £1,157,365. Profits of trades and professions in ditto, £453,485.

SUFFOLK, county, Mass. It is in the E. part of the state, and contains 116 sq. m., being the smallest in territory in the state; but as it includes the city of Boston, it is the most important. It consists of the city of Boston and the small township of Chelsea, with the islands belonging to each. It contained in 1840, 703 neat cattle: 6 sheep, 73 swine ; and produced 130 bushels of wheat, 129 of rye, 2908 of Indian corn, 1134 of barley, 279 of oats, 2,190 of potatoes. It had 2202 houses in foreign trade, capital $11,695,000 ; 583 retail stores, capital $4,013,229, 21 lumber yards, capital $284,010 ; capital invested in the fisheries $35,100 ; six furnaces, one cotton factory, with 1188 spindles, four grist mills, three glass factories, three tanneries, 17 distilleries, two breweries, one pottery, 25 printing offices, 28 binderies, seven daily, 11 weekly, and seven semi-weekly newspapers, and seven periodicals; 14 academies, 300 students; 146 schools, 14,877 scholars. Pop. 95,773. Capital, Boston.

SUFFOLK, county, N. Y. It comprehends the E. part of Long Island, with Shelter, Gardiner's and Fisher's islands adjacent. The Long Island railroad extends to Greenport, in the N.E. part of the county. Salt is manufactured by the evaporation of sea-water. It contained in 1840 23,836 neat cattle, 46,751 sheep, 20,534 swine ; and produced 882,7 bushels of wheat, 79,693 of rye, 355,314 of Indian corn, 49,707 of buckwheat, 9469 of barley, 258,328 of oats, 179,159 of potatoes, 1834 pounds of hops. It had 169 stores, 39 lumber-yards, capital invested in the fisheries $846,500, five fulling-mills, eight woollen factories, two cotton factories, with 1038 spindles, one flouring-mill, 51 grist-mills, 94 saw-mills, three paper-mills, three printing offices, four weekly and one semi-weekly newspapers ; six academies, 338 students ; 164 schools, 7673 scholars. Pop. 32,469. Capital Riverhead.

FFOLK. p. v., capital of Nansemond co., Va., 28 m.
¼. by W. Norfolk, 85 m. S.E. Richmond, 218 W. Situ-
on the E. side of Nansemond river, which is navigable
its place for vessels of 250 tons. It contains a court-
ie. jail, four churches. an Episcopal. two Methodists,
a Baptist, 20 stores, 300 dwellings, and about 1500 in-
uants. The railroad from Portsmouth, opposite to Nor-
to Weldon. N. C., passes through the place.
'GAR CREEK, t., Green co., O. It has three stores,
flouring-mills, five grist-mills, one saw-mill, six distil-
. Pop. 2369.
GAR CREEK, t., Stark co., O. It has two schools, 44
ars. Pop. 1862.
DAR CREEK, t., Wayne co., O. It has seven stores,
flouring-mills, nine saw-mills. Pop. 2293.
LLIVAN, county, N. H. Situated toward the S.W.
f the state, and contains 530 sq. m. Bounded W. by
ecticut river. Watered by Sugar river and its branches.
ee lake, 9 m. long, and 1¼ broad, lies on its E. border.
tained in 1840, 22,628 neat cattle, 88,296 sheep, 10,917
; and produced 26,573 bushels of wheat, 29,033 of
6,686 of Indian corn, 16,394 of buckwheat, 9085 of
, 156,438 of oats, 505,376 of potatoes, 142,541 pounds
ar. It had 44 stores, one furnace, 11 fulling-mills, six
in factories, one cotton factory, with 350 spindles, 27
nills, 80 saw-mills, two paper-mills, 18 tanneries, three
g-offices, two weekly newspapers; two academies,
dents; 97 schools, 7053 scholars. Pop. 20,340. Capi-
lwport.

IVAN, county, N. Y. Situated in the S.E. part of
te, and contains 919 sq. m. Bounded S.W. by Dela-
iver. Watered by Mongoup and Nevisink rivers,
hes of Delaware river. . The Delaware and Hudson
nd the New-York and Erie railroad pass through it.
lined in 1840, 18,057 neat cattle, 19,476 sheep, 10,047
and produced 8793 bushels of wheat, 66,090 of rye,
f Indian corn, 51,942 of buckwheat, 196,229 of oats,
of potatoes, 45,359 pounds of sugar. It had 65
0 lumber-yards, one furnace, four smelting houses,
ng 100,000 pounds of lead, three fulling mills, 29
ls, 174 saw-mills, 19 tanneries, three distilleries,
ting-office, one weekly newspaper; one academy,
nts; 101 schools, 3894 scholars. Pop. 15,629. Capi-
icello.

IVAN, county, Tenn. Situated in the N.E. part of
t, and contains 520 sq. m. Drained by Holston
d its branches. It contained in 1840, 12,368 neat
4,057 sheep, 41,087 swine; and produced 99,061
of wheat, 5662 of rye, 390,025 of Indian corn, 1272
, 179,896 of oats, 23,016 of potatoes, 5360 pounds of
It had 13 stores, one furnace, four forges, one
tory, with 512 spindles, 18 flouring-mills, six grist-
saw-mills, one paper-mill, one powder-mill, eight
, three distilleries; three academies, 130 students;
s, 700 scholars. Pop.: whites, 9504; slaves, 1037;
ired, 195; total, 10,736. Capital, Blountville.
AN, county, Ia. Situated in the W. toward the S.
e state, and contains 430 sq. m. Bounded W. by
iver, and drained by its small tributaries. It con-
1840, 12,442 neat cattle, 11,886 sheep, 33,915 swine;
ced 44,458 bushels of wheat, 2614 of rye, 538,543
corn, 86,700 of oats, 17,907 of potatoes, 20,446
tobacco, 85,625 of sugar. It had 15 stores, 23
, 22 saw-mills, four tanneries, four distilleries,
'y; one academy, 45 students; 20 schools, 620
Pop. 8315. Capital, Benton.
.n, p. t., Madison co., N. Y., 122 m. W. N. Al-
W. Bounded N. by Oneida lake. Drained by
a and Chittenango creeks. It contains four
a Presbyterian, two Methodist, and a Dutch Re-
i stores, two fulling-mills, one woollen factory,
e, one flouring-mill, four grist-mills, 18 saw-mills,
il, two tanneries, one distillery, one printing-
weekly newspaper; one academy, 40 students;
1240 scholars. Pop. 4390.
x, p. t., Tioga co., Pa., 146 m. N. by W. Harris-
N. Drained by head branches of Tioga river,
stores, one fulling-mill, one grist-mill, nine saw-
school, 45 scholars. Pop. 1378.
RA, the most W. island of the E. archipelago,
Borneo, the largest in the E. seas, between lat.
0 S., and the 95th and 105th degs. of E. long.,
n the N.E. from the Malay peninsula by the
alacca, and on the S.E. from Java by the straits
aving E. the sea of Java, and surrounded on
ther sides by the Indian ocean. Its direction is
to S.E., nearly parallel to the Malay peninsula,
ded by the equator into two nearly equal por-
of an elongated shape, about 1050 m. in length,
area variously estimated at from 122,000 to
n. Its population is wholly unknown; but it
nstanding, been estimated at about 2,000,000.

Various mountain chains run through the island longitudi-
nally, sometimes in treble or four-fold ranges, rising under
the equator to 14,000 or 15,000 feet in height, and always
much nearer to the W. than to the E. coast. The E. half
of the island is, in fact, almost wholly low, flat, and intersected by numerous rivers. Some of these, as the Palembang, Jambi, Indragiri, and Siak rivers, are of considerable
size, but they have been hitherto little explored by Europeans. The W. side of the island is also well supplied
with water; and in the interior are numerous small lakes.
The climate, even in the plains, is not so hot as might be
expected in a country situated in the midst of the torrid
zone. The thermometer at mid-day generally fluctuates
between 82° and 85° Fahr., though it sometimes rises to
87° or 88°, at sunrise it is usually as low as 70°. Dense
fogs, thunder storms, and water-spouts, off the coasts, are
very frequent. The soil on the W. side of the island.
which is that best known to Europeans, is a stiff reddish
clay, and a great proportion of the surface, especially towards the S., is an impervious forest. Gold dust, which is
very abundant, copper, iron, sulphur in the numerous volcanic districts, saltpetre, coal of indifferent quality, and
naphtha appear to be the chief mineral products. The
copper ore is very rich, but, owing to the indolence of the
inhabitants, the mines are little wrought, except over a
very limited district. Gold dust is, however, an article of
considerable traffic, and is brought by merchants from the
interior to the sea-coast. where it is bartered for iron, iron
tools, and the piece goods of the E. Indies and Europe.
Among the vegetable products, the most important is pepper, the average produce of which may amount, at present,
(1842) to about 30,000,000 lbs. a year, of which about
21,000,000 are furnished by the W., and 9,000,000 by the E.
coast. This supply amounts, in fact, to more than all the
pepper supplied by all the other countries in the world.
Nearly the whole of this extensive trade is in the hands of
foreigners, especially Dutch, English, and Americans. But
little Sumatra pepper goes to China.

After the capture of the Moluccas by the British, in 1796,
the nutmeg and clove were introduced at Bencoolen; and,
in 1825, their culture had so far succeeded, that the produce, in that year, was estimated at 89,000 lbs. of nutmegs,
34,000 lbs. cloves, and 22,000 lbs. of mace; but their quality
being very inferior, as compared with the products of Amboyna and the Banda isles, the culture has made little
farther progress. Camphor is one of the most valuable
kinds of produce, and the Sumatran camphor sells in China
for 12 times the price of that of Japan. It is the concrete
juice of the Dryobalanops Camphora, and a tree of the
order Guttiferæ (whereas the Japan camphor is derived
from a species of laurel). It grows only in the N. part of
Sumatra, not being found S. of the line, or beyond the 3d
deg. of N. lat. It is a stout tree, its trunk sometimes measuring 6 or 7 feet in diameter. The same tree produces,
under different circumstances, camphor, oil, or pitch, which
are found in cavities of the trunk, not extending the whole
length of the tree, but in small portions of 1 and 1½ feet
long at certain distances. "The method of discovering the
camphor is, by making a deep incision with a Malay axe,
till the camphor is seen: bundreds of trees may be thus
mutilated before the sought-for tree is discovered: when
attained, it is felled and cut in junks, a fathom long, which
are again split, and the camphor is found in the heart, occupying a space of the thickness of a man's arm. The
produce of a middling-sized tree is about eight Chinese catties, or nearly 11 lbs.; and of a large one, double the quantity." (Roxburgh in Asiat. Researches, xii.) Rice is the
principal species of grain. It is of very different varieties,
comprised in the two great classes of upland and lowland,
the former being considered the best. The land destined
for its culture is chosen at the approach of the dry season;
and as fresh ground is frequently cleared for the purpose,
the wanton destruction of fine timber is immense. The
rice is sown at the beginning of the rains, and ripens in
about five months from that time. The same spot of low
ground is, for the most part, used without intermission for
several years, the degree of culture bestowed by turning up
the soil, and the overflowing water, preserving its fertility.
Fallows occur occasionally; but as occupancy in most parts
of Sumatra gives the right of property in the land, they are
not very frequent.

The sawahs, or fields adapted for rice in low and wet
situations, are prepared by turning into them a number of
buffaloes; or in parts where it is less permanently moist,
the soil is turned up either with a wooden instrument between a hoe and a pickaxe, or with a plough. While the
sawahs are in preparation, a small, adjacent, and convenient spot of good soil is chosen, in which the seed grain is
sown as thick as it can well lie on the ground; and after
having grown to the height of several inches, it is taken up,
in showery weather, and transplanted to the sawah, where
holes are made, four or five inches asunder, to receive the

plants. To the minute care thus bestowed upon the latter, Marsden attributes the large proportion of produce obtained, which, he says, averages 100, and is sometimes as high even as 140 fold ! (*Marsden's Sumatra*, p. 77.) A singular method is adopted for separating the grain from the ear. The bunches of paddy are spread on mats, and the Sumatrans rub out the grain under their feet, supporting themselves for the more easy performance of this labour by holding with their hands a bamboo, placed horizontally over their heads. As an article of trade, Sumatran rice seems to be of a more perishable nature than that of most countries, the upland rice not being expected to keep longer than 12 months, and the lowland showing signs of decay after six. Sago is common in Sumatra, and is used occasionally as food, though not an article of general use. Millet is cultivated, but in no great quantity. The cocoa-nut, betel, bamboo, sugar-cane, various palms, and an abundance of tropical fruits, are indigenous. The sugar-cane is cultivated not for the manufacture of sugar, but for the sake of chewing the juicy reed; and hemp, instead of being used for the supply of materials for cordage, furnishes an intoxicating preparation. Turmeric, ginger, cassia, indigo, coffee, caoutchouc, rattans, many scented woods, and in the N. benzoin, are among the other principal kinds of produce.

Buffaloes are the most important live stock; the ox does not appear to be naturalized. The breed of horses is small, but well-made and hardy; sheep also are small. The hog and goat are both domestic and wild. Elephants, and many species of deer, abound; and tigers of a large and powerful species, the rhinoceros, hippopotamus, orangoutang, bears, &c., are met with, besides other animals in great variety. Around the shores extensive coral islands are continually forming; and coral is one of the principal articles of export, the other exports being pepper, rice, camphor, and other native products. The imports are chiefly Indian piece goods, salt, silks, and opium, from Hindostan; coarse porcelain, iron pans, gold thread, and many small articles, from China; striped cottons, spices, krises, and other weapons, from Java, Celebes, and the rest of the archipelago; metals, hardware, cutlery, and broadcloths from Europe.

According to native traditions, Sumatra and the adjacent islands have been the original seat of the Malay race; the type of which is certainly there met with in its greatest perfection. Except the Achinese, inhabiting the N. extremity of the island, and whose commixture with the Moors of Western India has distinguished them from the other tribes, the Sumatrans, according to Marsden, may generally be described as follows: "They are rather below the middle stature; their bulk is in proportion; their limbs are for the most part slight, but well shaped, and particularly small at the wrist and ankles. Upon the whole they are gracefully formed, and I scarcely recollect ever to have seen one deformed person. The women, however, have the preposterous custom of flattening the noses and compressing the heads of children newly born while the scull is yet cartilaginous. They likewise pull out the ears of infants to make them stand at an angle from the head. Their eyes are uniformly dark and clear, and among some, especially the southern women, bear a strong resemblance to those of the Chinese. Their hair is strong, and of a shining black, the improvement of both which qualities it probably owes, in a great measure, to the early and constant use of cocoa-nut oil. The men are beardless, and have chins so remarkably smooth, that, were it not for the priests displaying a little tuft, we should be apt to conclude that nature had refused them this token of manhood. But the boys, as they approach to the age of puberty, rub their chins, upper lips, &c., with quick lime, and the few hairs which afterward appear are plucked out from time to time with tweezers, which they always carry about them for that purpose. Their complexion is properly yellow, wanting the red singe that constitutes a tawny or copper colour. They are, in general, lighter than the half-breed of the rest of India; those of the superior class, and particularly their women of rank, approaching to a great degree of fairness." (p. 44–46.) No negro or other distinct race appears to occupy the mountainous regions, as in other parts of the archipelago; and the personal difference between the Malays of the coast and the country inhabitants is so little marked, that it requires some experience to distinguish the two. (*Marsden*, p. 49.)

The original clothing of the Sumatrans is the same with that found by navigators among the inhabitants, of the South sea islands; consisting of the inner bark of a tree, beaten out to the degree of fineness required, some fabrics being nearly equal in softness to the most delicate kid-skin. The dress of the men comprise short drawers, a close waistcoat, with filigree buttons, a robe extending sometimes to the knees, a party-coloured scarf, a sash in which the *kris* or dagger is stuck, and a small turban, or umbrella-hat,

while that of the women consists of a bodice, a petticoat, reaching from the bosom to the feet, a robe with sleeves, and sometimes a gauze veil, &c., with various figured ornaments. Both sexes file, blacken, and otherwise deface their teeth; and the great men sometimes set them a gilding, by casing the under row with a plate of that metal.

The villages are always on the banks of some river or lake, and consist of houses built chiefly of bamboo, and on posts, as in other countries of S.E. Asia. They are, however, much superior to those constructed among many other Ultra-Gangetic nations. Their furniture is very simple consisting, in the best kind of houses, chiefly of mats of a fine texture, serving for beds, some low tables, coarse earthenware, brass waiters, and iron pans. In eating, as the knives, spoons, nor any substitutes for them are used. The diet of the Sumatrans is mostly vegetable, but they eat the flesh of buffaloes, goats, fowls, &c., curried or otherwise dressed. In a few species of manufacture the Sumatrans exhibit great skill; such as in working gold and silver filligree. This art, described by Marsden (p. 173, 187, is conducted with the rudest tools. The wire-drawing instrument is made of a piece of iron-hoop; "an old hammer head, stuck in a block, serves for an anvil; and I have seen a pair of compasses composed of two old nails tied together at one end. In general, they use no bellows but blow the fire with their mouths, through a joint of bamboo; and, if the quantity of metal to be melted is considerable, three or four persons sit round their furnace, which is an old broken *kwali*, or iron pot, and blow together." (p. 179). Yet the manufactured material is celebrated for its delicacy and beauty, not only throughout the E. but in Europe. They weave silk and cotton cloths for home consumption and some of their work is very fine, and the patterns pretty, tily fancied. Different kinds of earthenware, brass, and fire-arms are made; and it is said that formerly cannon were cast at Achin. Little skill is, however, commonly shown in forging iron, or in carpenter's work. The Sumatrans are wholly strangers to painting and drawing; their carvings are always grotesque; and their proficiency in the exact sciences is very limited. Medicine is in the lowest state, being entrusted to old people, who, in a great measure, depend on charms and talismans for the cure of diseases! The Sumatrans are fond of music, and have many musical instruments, though most part of these have been borrowed from the Chinese. Their poetry is by no means contemptible, and is much favoured in point of harmony by the Malay language, the smoothness and sweetness of which have gained for it the appellation of the Italian of the East. They write in the Arabic character.

Sumatra is divided among a number of minor states, the principal being Menankabowe, Achin, Siak, Palembang, and Lampong. The European settlements, Bencoolen, Padang, &c., are on the S.W. coast, trenching on the country of the Rejangs. Among this people, the inhabitants of the villages are under the jurisdiction of magisterial headmen, or *dupatis*, who meet at stated periods and places in an assembly at which the *pangeras*, or feudal superior, presides. These *pangeras* claim despotic sway, but, as the *dupatis*, have, in fact, little more than a patriarchal and judicial power; they levy no tax, nor seem to have any revenue, other than accrues from their determination of cases referred to them in appeal from the decisions of the *dupatis*. And in the immediate neighbourhood of the more powerful states, the *pangeras* seem to acknowledge a kind of vassalage to the sovereigns of the latter. The Rejangs are said to be totally without religious worship of any kind, though not destitute of a belief in superior beings. A large proportion of the inhabitants of Sumatra are, however, Mohammedans.

Menankabowe, occupying the centre of the island, appears to have been always the leading state; but the meagre records of its history exist. Sumatra was first visited by the Portuguese in 1509, by the Dutch in 1598, and by the English two years afterwards. The latter contrived to establish factories and form settlements in the east during the 17th century; but principally in 1685-86. These settlements were retained by the British till 1825, when they were ceded to the Dutch in exchange for Malacca, &c. (*Marsden's Sumatra*, the most copious and by far the best work on the subject; and to which we refer for further details.)

SUMMIT, county, O. Situated toward the N.E. part of the state, and contains 432 sq. m. Watered by Cuyahoga and Little Cuyahoga rivers, which afford extensive water power. The Ohio canal passes through the middle of it from N. to S. It contained, in 1840, 22,762 neat cattle, 41,054 sheep, 25,301 swine; and produced 317,403 bushels of wheat, 23,070 of rye, 209,600 of Indian corn, 9854 of buckwheat, 5475 of barley, 188,463 of oats, 137,433 of potatoes, 107,998 pounds of sugar. It had six commercial houses in foreign trade, 53 stores, six furnaces, one forge, nineteen fulling-mills, 11 woollen factories, eight flouring

, 26 grist-mills, 50 saw-mills, two oil-mills, two paper-
, one powder-mill, sixteen tanneries, eight distilleries,
cewery, five potteries, four printing-offices, one bindery,
weekly newspapers; one college, 112 students, six
uries, 217 students, 114 schools, 3860 scholars. Pop.
). Capital, Akron.

urr, p. t., Schoharie co., N. Y., 16 m. S.W. Scho-
32 m. W. Albany, 378 W. Drained by Charlotte r.
choharie cr. It has four stores, two fulling-mills,
grist-mills, six saw-mills; eighteen schools, 692
rs. Pop. 9010.

UNER, county, Tenn. Situated in the N. part of the
and contains 625 sq. m. Bounded S. by Cumberland
and drained by its small tributaries. It contained,
), 18,965 neat cattle, 28, 801 sheep, 71,356 swine; and
ced 194,484 bushels of wheat, 12,780 of rye, 1,800,935
an corn, 498,248 of oats, 351,083 of potatoes, 2,615,100
s of tobacco, 161,379 of cotton, 1500 of sugar. It had
es, 62 grist-mills, 27 saw-mills, nine tanneries, 18 dis-
s, one brewery, two printing-offices, two weekly
apers; five academies, 336 students, 15 schools, 473
rs. Pop.: whites, 14,891; slaves, 7286; free,coloured,
tal, 22,445. Capital, Gallatin.

ver, p. t. Oxford co., Me., 8 m. N. Paris, 40 m. W.
ta, 605 W. Incorporated in 1798. Drained by
es of Androscoggin r. It has three grist-mills, three
lls, eight tanneries; nine schools, 486 scholars.
69.

PTER, district, S. C. Situated a little S.E. of the
of the state, and contains 1240 sq. m. Bounded W.
by Santee and Wateree r., N.E. by Lynches creek.
d by Black r. and its branches. It contained, in
,961 neat cattle, 8655 sheep, 47,260 swine; and pro-
125 bushels of wheat, 1536 of rye, 681,977 of Indian
230 of oats, 963,711 of potatoes, 368,505 pounds of
,719 of cotton. It had 35 stores, 50 grist-mills, 30
ls, five tanneries; seven academies, 145 students,
ls, 603 scholars. Pop.: whites, 8644; slaves, 18,875;
ured, 373; total, 29 937. Capital, Sumpterville.

ter, county, Ala. Situated in the W. part of the
d contains 1200 sq. m. Bounded N.E. and E. by
ee r., by the W. fork of which and smaller tribu-
is drained. It has three academies, 150 students,
ls, 382 scholars. Pop.: whites, 13,901; slaves,
free coloured, 116; total, 29,937. Capital, Liv-

ter, county, Ga. Situated toward the S.W. part of
, and contains 675 sq. m. Bounded E. by Flint r.,
hes of which it is drained. It contained, in 1840,
at cattle, 2352 sheep, 23,472 swine; and produced
ushels of wheat, 1800 of rye, 231,870 of Indian
0 of oats, 23,400 of potatoes, 740 pounds of tobacco,
f cotton. It had 13 stores, seven grist-mills, 10
s; eight schools, 234 scholars. Pop.: whites,
ves, 1643; free coloured, 1; total, 5759. Capital,

TERVILLE, p. v., capital of Sumpter district, S. C.,
Columbia, 449 W. It contains a handsome court-
jail, three churches, 10 stores, and about 400 in-

RY, p. b., Augusta t., capital of Northumberland
58 m. N. by E. Harrisburg, 168 W. Pleasantly
n a plain on the E. side of Susquehanna r., 2 m.
confluence of the W. and N. branches at Nor-
und, below the Shamokin dam, 2783 feet long,
r the passage of the Pennsylvania canal. It has
ourt-house and county offices, a stone jail, three
a Presbyterian, Methodist, and German Reform-
stores, three tanneries, one distillery, one pottery,
ng-office, one weekly newspaper; four schools,
rs, 250 dwellings, and 1108 inhabitants. A mile
borough, a bridge 1825 ft. long, crosses Shamokin
l the N. branch of the Susquehanna, and con-
place with Northumberland. It cost $90,000.

RY, p. v., port of entry, Liberty co., Ga., 40 m.
Savannah, 212 m. S.S.E. Milledgeville, 792 W.
n the S. side of Medway r., at the head of St.
s sound. It contains a Baptist church, an aca-
e stores, 30 dwellings, and about 200 inhabitants.
ur is safe and commodious, and the situation is
nd healthy, and is resorted to by the planters
sickly season. In early times it was a rival of

RLAND, a parl. and mun. bor., sea-port, and
tland, being, next to Newcastle, and Stockton,
t port in the kingdom for the shipment of coal,
n., ward Easington, on the Wear, close to its
he North sea, 13 m. N.E. Durham, with which
onnected by a railway, and 245 m. N.N.W. Lon-
of light house) 54° 55' 12" N., long. 1° 21' 16"
of Sunderland parish 120 acres. Population of
l1, 17,060. But the parliamentary borough com-

prises, along with Sunderland, the townships of Bishop-
Wearmouth and B.-W. Pans on the S. side of the Wear,
and those of Monk-Wearmouth, M.-W. Shore, and Southwick, on its N. side; the whole including an area of 5095
acres, and a population, in 1831, of 40,735. The portions
of the town on the N. and S. sides of the river are connected by a magnificent iron bridge (see post.)

"Sunderland and Bishop-Wearmouth, at no distant time,
were two distinct towns, at a considerable distance from
each other. All the houses in Sunderland appear to be of
considerable age; but in Bishop-Wearmouth the intervening space has been gradually curtailed, and at last filled up
by buildings, so that at present the two form only one
town. With the exception of one street, in which there
are some respectable houses and shops, Sunderland presents the appearance of one mass of small houses crowded
together, with interstices of narrow lanes rather than
streets. The population in them is so dense as to give the
appearance of unhealthiness as well as absence of cleanliness. This is not the case in Bishop-Wearmouth; in the
new part of the town that adjoins Sunderland, there are
some good streets, and excellent houses, and it is in this
part that the higher classes of inhabitants reside. This
town is increasing rapidly; several new streets have been
recently built, and others are in course of building. Bishop-
Wearmouth Pans is a small district running along the bank
of the river from the parish of Sunderland to nearly as far
as the bridge. Its population is very dense; it contains
some glass manufactories and iron works for the manufacture of such articles as are required for the shipping. Monk-
Wearmouth Shore is a large township immediately opposite Sunderland, and part of Bishop-Wearmouth, and has a
dense population, with but few houses or inhabitants of the
higher classes. Adjoining Monk-Wearmouth Shore, on the
W., and extending for some distance along the river, is
Southwick. In it are some coal pits and a railway; the
greater part of the township, however, is agricultural, and
contains very little trading population. Monk-Wearmouth
lies to the N. of Monk-Wearmouth Shore, and does not
come down upon the river. Its population is almost entirely
connected with the trade of the port. The only carriage
communication between the two sides of the river is by the
bridge, on passing which, from the right to the left bank, a
toll is taken. There are, however, several ferries for foot
passengers." (Mun. Corp. Rep.)

The cast-iron bridge, now alluded to, over the Wear, is
the most remarkable object in this part of the county. It
was constructed between 1793 and 1796, at an expense of
about £33,400, and consists of one magnificent arch 236 ft.
in span, elevated in the centre above 100 ft. above high-
water mark, so that large ships sail under it by merely
lowering their top-masts. In 1816, this bridge was disposed
of by a lottery of 6000 tickets, at a price of £30,000.

The parish church of Sunderland is large and handsome;
its E. end is particularly elegant, the altar being placed in
a circular recess under a dome. The living, a rectory, is
worth £264 a year. St. John's chapel-of-ease, built in
1669, is a perpetual curacy, worth £122. Bishop-Wearmouth church has an ancient chancel, and an early decorated E. window, but the rest is modern. The living is a
very valuable rectory, being worth £2899 a year. A new
chapel-of-ease was built by parliamentary grant in 1827.
Monk-Wearmouth church has had nearly all its ancient
features obliterated by modern alterations, though it still
possesses a rude Norman tower, &c. (Rickman's Gothic
Architect.) The living is a perpetual curacy, of the clear
annual value of £225. The foregoing livings are all in the
gift of the Bishop of Durham, except the last, which is in
that of Sir H. Williamson. There are numerous places of
worship for Dissenters in the town and vicinity, including a
synagogue. The exchange, a neat edifice in the High-
street, erected in 1814, at a cost of £8000, comprises commercial, news, and court rooms, an auction mart, &c. The
theatre, assembly rooms, barracks, custom-house, and excise office are among the chief public buildings. Bathing
machines are kept on the sands, E. of the town; and on
the town moor, and at Hendon, S.E. of the town, baths
have been established, Sunderland being a good deal resorted to in the bathing season. It has a subscription library, a mechanics' institute, at which lectures are delivered, and several other literary institutions. A school, founded and endowed in 1778, educates and clothes 36 poor girls,
and it has, also, national, friends', and various inferior
schools. A large infirmary occupies a new building raised
in 1822; and an almshouse for 10 widows or daughters of
master mariners, was founded and endowed in 1820. There
are numerous other almshouses and charities of different
kinds, and four freemason's lodges in the town and neighbourhood. The town is lighted with gas, and well supplied with water.

The port, immediately within the river's mouth, is formed by two grand piers, each about 450 yards in length,

which project one from the S. and one from the N. side of the river into the German ocean. At the extremity of the N. pier is a lighthouse, having the lantern elevated 73 feet above the sea at high water : there is also a harbour light on the S. pier, which shows during ebb and ½ flood. At springs there is from 15 to 17 ft. water over the bar, and at neaps from 10 to 12 ft.: the channel is close by the N. pier head. To prevent the crowding of shipping in the river, a large basin or dock has been constructed on its N. side.

The staple businesses of the town are the building of ships and the shipment of coal. The former is carried on to a greater extent here than anywhere else in the kingdom. In 1840, for example, there were built in Sunderland no fewer than 302 ships, of the aggregate burden of 87,023 tons. During the same year there belonged to the port 918 ships, of the aggregate burden of 188,769 tons, manned by 8078 seamen, being a greater amount of shipping than belongs to any other port of the United Kingdom, London and Liverpool only excepted. It would, indeed, appear, from the parliamentary returns, that Newcastle has a greater amount of shipping; but this results from the shipping belonging to N. and S. Shields being included in that of Newcastle, of which it makes fully a half. In 1839, 913,960 tons of coal were shipped coastwise from this port, and 370,620 tons were exported to foreign countries. In 1840, Sunderland supplied the metropolis with 2474 cargoes, amounting to 730,148 tons of coal. Sail-cloth, chain cables, glass and earthenware, are also extensively manufactured in the town; and these, with lime, grindstones, wrought marble, &c., constitute, next to coal, the principal articles of export. The gross customs duties amounted, in 1840, to £119,681. The fisheries are important; and the adjacent village of Deptford, on the Wear, has a large rope-factory wrought by steam. The Sunderland Joint Stock banking company was established here in 1836; and it has also a branch of the Newcastle, Shields, and Sunderland bank, with several private establishments. Two weekly newspapers are published in the town.

Sunderland was created a borough by Bishop Pudsey, at the end of the 12th century. Under the Municipal Reform Act, it is divided into seven wards, and is governed by a mayor, 13 other aldermen, and 42 councillors. Though it has long been of considerable importance, Sunderland had no voice in the legislature till the Reform Act conferred on it the important privilege of sending two members to the House of Commons. Registered electors, in 1839–40, 1057. Corporation revenue, in 1840, £4846. It has a commission of the peace; and weekly sessions are held, besides courts leet and baronial by the bishop of Durham. Market-day, Saturday; and for cattle every other Tuesday. Fairs, May 11 and 12, Oct. 12 and 13, for cattle, &c. This was the first town in England attacked by cholera in 1832. (*Parl. Reports, and original returns obtained for this work.*)

SUPERIOR, lake, the largest of the five great lakes of N. America, and supposed to be the largest body of fresh water in the world. It is about 380 miles long, and 130 wide in its widest part, and 1200 or 1500 miles in circumference, 900 ft. deep, and 641 ft. above the level of the ocean. Its waters are remarkably clear and transparent, and abound with fish, particularly trout, white fish, and sturgeon. The two former are extensively taken and exported. The trout generally weigh about 12 pounds, but in some instances exceed 50 pounds. This lake is subject to as violent storms as the Atlantic, and the navigation is equally dangerous. The coast is generally elevated, rocky, and in some parts mountainous. It contains a number of islands in its N. and N.E. parts, the largest of which is Isle Royal, 100 m. long and 40 m. broad. The land on the borders of the lake is generally sterile. It receives more than 30 rivers, and discharges its waters into lake Huron by the straits of St. Mary, where are rapids which entirely obstruct the navigation. The vessels which navigate lake Superior never leave that lake. A ship canal is in contemplation around the falls or rapids of St. Mary, which is practicable at a moderate expense. The pictured rocks on the shore toward the E. end, are a great curiosity. They form a perpendicular wall, 300 feet high, extending about 12 miles. They have numerous projections and indentations, with vast caverns, which receive the waves with a tremendous roar. At one place a considerable stream is thrown from them into the lake by a grand cascade, 70 feet high, and is projected so far that boats pass dry between it and the rocky shore. The Doric rock or arch appears like a work of art, consisting of an isolated mass of sandstone with four pillars, supporting a stratum or entablature of stone covered with soil, and giving support to a handsome growth of spruce and fir trees, some of which are 50 or 60 feet high. Copper abounds in the country S. of lake Superior, and may hereafter add to the importance of the navigation of this great lake.

SURAT, a large town of Hindostan, cap. prov. of Gujrat, and of the British distr. of its own name, under the presid.

Bombay, on the Taptee, about 20 m. above its mouth, in the gulf of Cambay, and 198 m. N. by E. Bombay, in 21° 11′ N., long. 73° 7′ E. The population was formerly estimated, in 1818, at 157,000. It is about 6 m. in cir. encompassed like a bow, the chord, formed by the Taptee, having at its centre, a small castle garrisoned by a few sepoys and Europeans. On other sides the town is surrounded with a wall flanked with semi-circular towers. Within the wall are some good European houses, formerly occupied by the French, but now the residences of the English officers: the houses within the town are very mean, consisting only of timber frames filled up with brick, their upper stories projecting over each other. The streets, too, are narrow and irregular. Surat has an English church, and an English school, with numerous Hindoo schools; a large European cemetery, containing the tomb of Sir C. ... endon, one of the earliest governors of British India, &c.; but the most remarkable building or institution is a hospital for sick animals, similar to that at Baroach. During the last century this hospital contained "horses, asses, oxen, sheep, goats, monkeys, poultry, pigeons, and a variety of birds; also an aged tortoise, which was known to have been there 75 years. The most extraordinary was that was that appropriated for rats, mice, bugs, and other noxious vermin, for whom suitable food was provided." Hamilton, E. I. Gaz.]

Surat had formerly a large trade in all kinds of eastern produce; but this has greatly declined, and its exports consist at present (1842) principally of cotton wool, which is sent in large quantities to Bombay. Most part of the old manufactures of Surat, except lungees and shawls, for which there is little demand, have been superseded by those of Great Britain, and the greater number of the native merchants have become poor. Among the traders however, are numerous Parsees, the descendants of those expelled from Persia by the Mohammedans, who have the reputation of being wealthy. Vessels of 20 or 40 tons may come up to Surat; but those of greater size must be about 15 m. lower down the river; and few boats larger than the ketches in the E. I. Company's service ever ascend so high as the town.

Surat is the residence of a British collector, judge, military commandant, &c., and is the seat of a board of civil tons, a circuit court, and of the Sudder Adawut or chief tribunal for the entire presidency of Bombay. It is supposed to be one of the most ancient cities of Hindostan being mentioned in some of the earliest records. The English factory, founded here in 1615, was the first mercantile establishment of the E. I. Company in the Mogul dominions, and continued to be the chief British station in India till Bombay became the seat of supreme authority, in 1687. (*Hamilton; Mod. Trav.*, x. 144–149, &c.)

SURINAM. *See* GUIANA (DUTCH).

SURREY, a co. of England, which, though inland enjoys, from its being skirted on the N. by the Thames, all of the advantages of a maritime county. It has to the N. Middlesex, and a small part of Bucks, from both of which it is separated by the Thames: on the E. it is bounded by Kent, on the S. by Sussex, and on the W. by Hampshire and Berks. It comprises all that portion of the metropolis to the S. of the Thames, and is thus, in fact, a metropolitan county. Area, 485,120 acres, of which about 400,000 are arable, meadow, and pasture. With the exception of the Weald, the surface consists of alternate hill and dale. Some of the hills rise to a pretty considerable height, affording highly diversified and beautiful prospects. The soil comprises every variety, from the richest loam to the poorest moor. There are three portions, the soils of which are particularly well defined; viz. 1st, the Weald, occupying all the S. part of the county, from Crowhurst to Farnmere; 2dly, the sandy loam district, lying between the Weald and the downs; and, 3dly, the downs, or chalk land, occupying the whole E. side of the county from Croydon to Tilsey, but gradually decreasing as we advance W., till at Farnham, on the border of Hants, it is reduced to a narrow strip. To the N.E. of the downs, between the ... and the Thames, there is a great variety of soil, partly consisting of strong dark clay, partly of sandy loam, &c. In the N.W. and S.W. parts of the county, but especially in the former, there are very extensive tracts of heath and rich lands ground; and smaller tracts of the same kind are met with in various other places. On the whole, however, the county may be said to be of an average degree of fertility. Climate good; and, owing to the variety of surface, the abundance of wood, and its contiguity to the metropolis, it is one of the most desirable counties in England for a residence. A large proportion of Surrey is in tillage; but agriculture, speaking generally, is in a decidedly backward state, and two, or even more, white crops will not unfrequently follow in succession. On the rich, friable, consequently loams between Croydon and Epsom, six quarters of wheat an acre are not unfrequently reaped; and on the

h andy loams near Godalming, 5 quarters is no uncom-
on crop; but on the poorer soils, and in the Weald, the
duce seldom exceeds from 2½ to 3 quarters. The turnip
lture was introduced into Surrey sooner than into any
er English county; but even at present turnips are but
dom drilled: their management is but imperfectly under-
od, and tares are generally preferred by the Surrey
mers to any other species of green crop. Turnwrist
ughs are used in many quarters, but the swing plough
most common. It is drawn by three, four, or even five
ny horses yoked in line! Lime is extensively used as
nure; and the application of salt for that purpose is daily
oming more general. Hops are raised in considerable
ntities; and those grown in the neighbourhood of Farn-
n are preferred to most others. Peppermint, lavender,
mwood, chamomile, &c., are raised in the physic gar-
s about Mitcham; and a considerable extent of land at
't oea, and other places along the banks of the Thames
ppropriated to the production of asparagus and other
tables for the London market. There is no peculiar
d of cattle in Surrey, but the short horns and the Sussex
d are the most prevalent. A considerable number of
p are kept, principally on the down-land. Large num-
of hogs are fed; they consist principally of the Berk-
e and Chinese varieties. The Dorking breed of fowls
high estimation; they are large, handsome, perfectly
c, and are distinguished by having six claws to each
and at other places, it is crossed by bridges. Port Deposite,
There are no very large estates in Surrey. Farms
f sizes; the largest are on the down lands, and the
lest in the Weald; but, at an average of the county,
ize of farms is not supposed to exceed 170 acres. They
ommonly held under leases for 7, 14, or 21 years; but
... ous customs that prevail as to entry defeat the ad-
iges that might otherwise have resulted from this
e. (See *Kennedy and Grainger, Letting of Land,* I.
In the Weald the farm-houses are mean and ruin-
but they are better in other places. Cottages good,
equently ornamented with vines and flowers. Aver-
ent of land, in 1810, 15s. 2¼d. an acre. There is a
deal of valuable timber and coppice wood in Surrey,
ularly in the Weald. Large quantities of fuller's
are dug up in various places; and there are also ex-
t quarries of freestone and limestone. Except in so
it is connected with the metropolis, Surrey has very
manufactures; and those of little or no importance.
es the Thames, it is watered by the Wey, the Mole,
e Wandle; and it is traversed by the Surrey and
n canals, and by the Brighton and S.W. railways.
ike ronds good; but cross ronds, particularly in the
, very indifferent. It contains 13 hundreds exclusive
boroughs of Southwark and Lambeth, and the town
dford; and is divided into 142 parishes. It returns
nbers to the House of Commons; viz., four for the
, two for the borough of Guildford, one for the bo-
of Reigate, two for Lambeth, and two for the borough
thwark. Registered electors for the county, in 1839-
7, of whom 6146 were for the E., and 3621 for the
stion. In 1841, Surrey had 55,375 inhabited houses,
3,613 inhabitants; of whom 278,186 were males, and
females. Sum expended for the relief of the poor,
—39, £164,227. Annual value of real property, in
1,549,702. Profits of trades and professions in ditto,
533.

RY, co., Va. Situated toward the S.E. part of the
nd contains 324 sq. m. Bounded N.E. by James r.,
V Blackwater r. It contained in 1840, 4136 neat cat-
5 sheep, 12,593 swine; and produced 9217 bushels
at, 185,040 of Indian corn, 35,900 of oats, 33,559 of
, 4692 pounds of tobacco, 63,954 of cotton. It had
s, one cotton factory with 240 spindles, one flouring-
grist-mills, 166 distilleries; eight schools, 186 schol-
op.: whites, 2557; slaves, 2853; free coloured.
tal, 6840. Capital, Surry C. H.

v, co.. N. C. Situated in the N. towards the W.
he state, and contains 726 sq. m. Drained by Yad-
nd its branches. It contained in 1840, 12,609 neat
2,128 sheep, 38,168 swine; and produced 48,804
of wheat, 20,542 of rye, 484,976 of Indian corn,
f oats, 23,866 of potatoes; 83,070 pounds of tobac-
8 of cotton. It had 19 stores, 10 forges, one cotton
with 400 spindles, 52 grist-mills, 27 saw-mills, one
nill, 15 tanneries, 225 distilleries, three potteries;
ls, 335 scholars. Pop.: whites, 13,093; slaves,
e coloured, 208; total, 15,079. Capital, Rockford.
UEHANNA, river, the largest river in Pa., and
he larger of the United States, is formed by two
nches. The eastern branch rises in Otsego co.,
f its sources constitutes the outlet of Otsego lake.
nnngo, a considerable river from the N. unites
nt Binghamton. Proceeding W. and S.W., it
ito Pennsylvania, and seven miles below the line,
es Tioga or Chemung r., from the state of N. Y.

It then flows S.E. and S.W., until, at Northumberland, it
receives the western branch, its largest tributary, which
rises in Cambria co., in the Appalachian valley, and has a
course of about 200 miles. Its course is then S. to the
junction of the Juniata r. from the W., which has a course
of about 200 miles. Its direction is then S.E. until its entrance into Chesapeake bay, in the N.E. corner of Maryland. It is navigable for sloops five miles from its mouth
to Port Deposite, at the head of tide water. Above that,
for nearly 50 miles, it is obstructed by a continual succession of rapids, which render the ascending navigation in
boats impossible, and even the descending navigation in
boats at all times difficult and hazardous. In time of high
water, for a short season, a great amount of lumber and
produce descends the river in rafts, arks, &c., which are
sold, with their freight, at their place of destination. In the
upper parts of the river, the obstructions to boat navigation are not great, and might, without great difficulty, be
overcome. In respect to its capacity for navigation, and as
a medium of commerce, this great river is of less importance than many others which drain a less extent of surface, and carry a less volume of water. At the union of the
Tioga river with the main stream, it is 600 ft. wide; at
Wilkesbarre it is 700 ft. wide; at Nescopeck and Berwick it
is 1226 ft. wide; at Northumberland, 1825 ft. wide; at Harris.
burg, 2676 ft. wide; at Columbia 5690 ft. wide. At these,
and at other places, it is crossed by bridges. Port Deposite,
in Cecil co., Md., is about a mile below the lowest bridge
across the river; and at this place the produce not sold
above, is accumulated, and much of it sent by sloops to
Baltimore. Canals are extensively constructed along its
borders, for which it affords great facilities. Its whole
length, to its remote sources, is about 450 miles.

SUSQUEHANNA, co., Pa. Situated in the N.E. part of the
state, and contains 825 sq. m. Drained by branches of
Susquehanna r. It contained in 1840, 29,275 neat cattle,
72,157 sheep, 16,840 swine; and produced 60,828 bushels of
wheat, 17,382 of rye, 83,650 of Indian corn, 44,222 of buckwheat, 1890 of barley, 237,185 of oats, 362,218 of potatoes,
252,137 pounds of sugar. It had 53 stores, eight fulling-
mills, three woollen factories, 24 grist-mills, 90 saw-mills,
one oil-mill, one paper-mill, 18 tanneries, one distillery,
three printing offices, one bindery, one weekly newspaper;
five academies, 104 students; 173 schools, 5074 scholars.
Pop. 21,195. Capital, Montrose.

SUSQUEHANNA, t., Dauphin co., Pa. Watered by Paxton
cr. Harrisburg lies partly in this t. It has, exclusive of Harrisburg, one store one fulling-mill, one flouring-mill, one grist-
mill, three saw-mills; two schools, 33 scholars. Pop. 1452.

SUSSEX, a marit. co., of England, on its S. coast, having N. and N.E., Surrey and Kent; S. and S.E. the English
channel; and W. Hants. Area, 938,240 acres. Surface and
soil, very various. A ridge of chalk hills to which (though
in strictness applicable only to a part) the term South
Downs is usually applied, runs through the co. from South
Harting and Miland Chapel, on the confines of Hants, to
Beechy-head, where it terminates in high precipitous cliffs.
Their N. declivity is rather steep, but that on the S. is gently sloping. The soil on the South Downs is generally a
light hazelly mould on a substratum of loose chalk. On
the S. side of this range, along the const from Emsworth,
gradually decreasing to near Brighton, there is a considerable extent of fine. rich, loamy land. To the N. of the S.
Downs is the extensive tract called the Weald of Sussex,
uniting on the E. with the Weald of Kent, and stretching as
far W. as Petworth. The soil of the Weald is similar to that
of the Weald of Kent; being, for the most part, a stiff,
tenacious clay, with occasional sandy and gravelly patches
intermixed. It is thickly covered with oak wood; and,
when viewed from the South Downs, appears like an immense forest. In the E. parts of the co., in what is called
Pevensey level, and near Winchelsea, are considerable
tracts of very fine, deep marsh land. Climate, mild, dry,
and early. A large extent of Sussex is under the plough;
but husbandry is in a backward condition; and Messrs.
Kennedy and Grainger truly state that no very material improvement need be expected till those pernicious habits
with respect to the letting and entry to farms that prevail
here, as well as in Kent and Surrey, be totally changed.
(*Tenancy of Land,* i. 337.) Crops principally cultivated.
wheat oats, and barley; and on all the light lands, turnips
are extensively grown. Great quantities of hops are raised, particularly in the eastern parts of the country; there
being in 1838, 11,517 acres under this crop. Sussex is deservedly celebrated for its breeds of cattle and sheep; each
being about the very best of its kind. The oxen are of a
deep red colour, and have tapering turned-up horns; they
fatten easily, produce excellent beef, and are not inferior to
any other breed in field labour. The greater part of the
tillage in the Weald is performed by ox-teams. The native
cattle do not, however, answer for the dairy. The peculiar
breed of sheep belonging to the co. is called the South

Down, from its being found in the greatest perfection on the South Down chalk hills. The breed is now widely diffused; but owing to the extension of tillage on the Downs, and the increase in the size of the animal and the weight of the fleece, neither the mutton nor the wool is supposed to be so good as formerly. Total stock of sheep estimated at nearly 900,000.

Sussex has been celebrated from the remotest period for the abundance and excellence of its timber; and in these respects it continues to be decidedly superior to every other English county. Oak is the principal timber of the Weald; but in other parts beech is the most prevalent. To the abundance of wood is principally to be ascribed the circumstance of Sussex being formerly distinguished for the number of its iron-works; but since pit-coal began to be generally employed in the smelting and refining of iron, these have been wholly abandoned, as well as those that were formerly established in Kent. Property much divided. Average size of farms in the Weald 100 acres; in the Downs, from 1200 to 2000 acres. A great proportion of the farms held by tenants at will; and owing to injurious customs as to entry, a large part of the capital of the tenant is swallowed up in the useless payments he is compelled to make; so that much of the land is very insufficiently stocked. Offices invariably thatched and weather-boarded. Average rent of land, in 1810, 11s. 8¼d. an acre. Manufactures of little importance. Ironstone, fuller's earth, limestone, and sandstone are all met with. The rivers are of no great magnitude. The principal is the Arun. It communicates by a canal with Langport harbour on the W., and with the Wye and the Thames on the N. In the Weald there are several ponds in which freshwater fish are fed for the London markets. Sussex is divided into six rapes, and these again into 65 hundreds, and contains 310 parishes. It returns 18 members to the House of Commons; viz. four for the county; two for the city of Chichester; two each for the boroughs of Brighton, Lewes, Hastings, and Shoreham; and one each for Arundel, Horsham, Midhurst, and Rye and Winchelsea. Registered electors for the co., in 1839–40, 8902, of whom 5316 were for the E., and 3586 for the W. division of the co. In 1841 Sussex had 54,066 inhabited houses, and 152,198 females. Sum expended on the relief of the poor, in 1838–39, £142,410. Annual value of real property, in 1815, £919,350. Profits of trade and professions in do., £372,059.

SUSSEX, co., N. J. Situated in the N. part of the state, and contains 415 sq. m. Drained by Wallkill, Pequannock, Pequest, and Paulenskill creeks. Hopatung pond, which supplies the summit level of Morris canal, lies on its S.E. border. It contained in 1840, 26,346 neat cattle, 23,963 sheep, 30,236 swine; and produced 64,400 bushels of wheat, 228,316 of rye, 319,618 of Indian corn, 147,067 of buckwheat, 217,196 of oats, 201,090 of potatoes. It had 56 stores, four lumber yards, five furnaces, 13 forges, five fulling-mills, 47 grist-mills, 57 saw-mills, 15 tanneries, 24 distilleries, two printing-offices, two weekly newspapers; five academies, 172 students; 121 schools, 3369 scholars. Pop. 21,770. Capital, Newton.

SUSSEX, co., Del. Situated in the S. part of the state, and contains 860 sq. m. Bounded E. by Delaware r., bay, and the Atlantic. Drained by Nanticoke and Marshy Hope rivers, which flow into Chesapeake bay. Mispillion cr. bounds it on the N., and with other small creeks, flows into Delaware bay. Indian r. enters the Atlantic through Rehoboth bay. It contained in 1840, 18,956 neat cattle, 17,996 sheep, 33,054 swine; and produced 30,000 bushels of wheat, 8644 of rye, 872,817 of Indian corn, 48,189 of potatoes. It had 111 stores, 54 grist-mills, 78 saw-mills, 10 tanneries, two printing offices; six academies, 220 students; 46 schools, 1309 scholars. Pop. 25,093. Capital, Georgetown.

SUSSEX, co., Va. Situated in the S.E. part of the state, and contains 465 sq. m. Drained by Nottaway r. and its branches. It contained in 1840, 8831 neat cattle, 6030 sheep, 23 957 swine; and produced 18,777 bushels of wheat, 404,793 of Indian corn, 103,916 of oats, 34,815 of potatoes, 175,672 pounds of tobacco, 477,043 of cotton. It had 18 stores, 21 grist-mills, one tannery; seven academies 165 students; 10 schools, 196 scholars. Pop.: whites, 3584; slaves, 6384; free coloured, 811; total, 11,229. Capital, Sussex C. H.

SUTHERLAND, a marit. co. of Scotland, occupying the N.W. angle of the country, has on the N. and W. the Atlantic, E. the co. Caithness and the Moray frith, and S. the frith of Dornoch, Ross, and Cromarty. It contains 1,152,640 acres, of which 30,080 are under water. The aspect of the country is wild, bleak, and in many parts, savage. The E. shore has a small fringe of good arable land; but the rest of the surface is rugged and mountainous, being, however, interspersed with various narrow straths, or glens, and some considerable lakes and morasses.

Sutherland, like the other highland counties was till recently occupied by native tenants, similar in their respects to those of Ross. These, however, have, for the greater part, been recently removed either to villages on the coast, or have emigrated; and the lands have been divided into extensive sheep-farms, furnished with excellent houses and offices. The native breed of cattle is small, but when crossed by those of Argyle and Skye, it is said to be equal to any that the highlands can produce. Galloways and other varieties, have also been introduced. Owing to the extraordinary extension of sheep-farming in this county, the stock of cattle has been diminished in a still greater degree than in Ross; but sheep being much better adapted to the country, the change has been, both locally as regards public point of view, highly advantageous; that having been, through its means, coupled with a very extensive drainage, rendered considerably productive land were formerly almost useless. Cheviots are kept here in remarkable well in almost all parts of Sutherland. About 40,000 sheep, and 180,000 fleeces are said to be annually sent to the S. from this county. (Anderson's Highlands, p. 12.) Four-fifths of the county belongs to the Duke of Sutherland, who has expended vast sums in the formation of roads and loans, the building of bridges, piers, farmhouses, and villages, and other expensive and substantial improvements. Since 1811, above 100 m. of road have been made in the county by the parliamentary commissioners, and above 350 by individual exertion and statute labour! The fringe of arable land along the E. coast has been divided into moderate-sized farms, well inclosed and drained, and presenting as good a specimen of the improved turnip husbandry, as is to be found in any part of the island. Nowhere, indeed, in Scotland have improvements been attempted on a greater scale, or prosecuted with more zeal, skill, and success, than in this remote county. Whether the change should not have been more gradually introduced, we shall not undertake to say: but there cannot be a doubt, that the character and habits of the people, as well as the rural economy of the district, have been signally improved.

Sutherland has three great deer forests; and ptarmigan, grouse, and blackcock, alpine hares, &c., are abundant. Limestone and freestone are met with. Average rent of land in 1810, 6d. an acre. The herring fishery is carried on with spirit and success, both on the E. and W. coasts, but principally from Helmsdale. Principal ports Oyckel, Fleet, Bosa, and Helmsdale. It contains 13 parishes, but no considerable town, Dornoch, the largest, not having a population of 600! It returns one member to the House of Commons. Registered electors in 1837–40, 172. Dornoch unites with Kirkwall, Wick, Tain, Dingwall, and Cromarty, in returning a member. Valued rent £59,480 Scotch. Annual value of real property, in 1816, £23,756. SUTTON, p. t., Merrimac co., N. H. 36 m. W.N.W. Concord, 480 W. Kearsarge mt. lies on its E. border. Branches of Warner r. afford water-power. Chartered in 1768. It contains four churches, two Baptist, a Free-will Baptist, and a Universalist, four stores, one grist mill, five saw-mills, two tanneries; 13 schools, 337 scholars. Pop. 1232. SUTTON, p. t., Caledonia co., Vt. 53 m. N.N.E. Montpelier, 564 W. Watered by branches of Passumpsic r. A pond of 200 acres flows N. into Barton r. It has a Baptist church, two stores, one fulling mill, one grist-mill, five sawmills; 11 schools, 379 scholars. Pop. 1063. SUTTON, p. t., Worcester co., Mass. 44 m. W. by S. Boston, 400 W. Watered by Blackstone r., along which passes the Blackstone canal. Soapstone and granite are found. It contains five churches, two Congregational, two Baptist, and an Episcopal; six stores, one fulling mill, one woollen factory, four cotton factories with 6096 spindles; 12 schools, 762 scholars. Pop. 2270.

SUTTON-COLDFIELD, a market town and par. of England, co. Warwick, hund. Hemlingford, on the road from Birmingham to Lichfield, 6 m., N.N.E. Birmingham. Area of par., 13,030 acres. Pop. in 1831, 3684. The town, on an acclivity, in a bleak situation, consists principally of one long street. Houses good, and the inhabitants are supplied with water. The parish church, an older establishment, belongs to the diocese of the 13th century, has a statue of Vesey, Bishop of Exeter in the time of Henry VIII., a native and a great benefactor of the town. A flourishing free school, founded by Vesey, and national schools, almshouses, and several other charitable endowments, exist at Sutton. The inhabitants are principally employed in the manufacture of Birmingham goods. The town, which is of great antiquity, was chartered by Henry VIII., under a warden, aldermen, and two justices. The corporation had various privileges, which have since become void, though some are, however, still held quarterly. Markets on Mondays; fairs, Trinity Monday and Nov. 8, for sheep and cattle. SUWANEE, r., Flor., rises in Okefinokee swamp, a Ga. It receives the Allapahaw from the N., and the Withlacoochee from the N.W., and enters the gulf of Mexico by several mouths, which have not more than 5 ft. of water, though above the bar 15 ft. of water may be had for 35 m. It flows over a rocky bed; and in one place has a natural bridge for about 3 m., the subterraneous passage of which is unable to receive the water in high floods, when the stream passes over it.

SWAFFHAM. a market town and par. of England. co. Norfolk, hund. South Greenhoe, 25 m. W. Norwich. Area of par. 8130 acres. Pop., in 1831, 3245. The town is finely situated on rising ground, and is well built. The parish church is a spacious and handsome structure, chiefly in the perpendicular style, with a lofty nave, the vaulted roof of which is richly adorned with figures carved in Irish oak; it has a well proportioned tower, with enriched embrazures and pinnacles, some handsome pillars and monuments, and curious inscriptions. The living of Swaffham, with Threxton vicarage and rectory, worth £738 a year, is in the gift of the Bishop of Norwich. The Friends. Baptists, Wesleyans, &c., have meeting-houses. The county bridewell and a beautiful market-cross, surmounted by a figure of Ceres, erected by Lord Orford, in 1783; a public assembly-room, and a neat theatre, are the other principal public edifices. Swaffham has a free grammar, and a national school, and various almshouses, &c. Quarter-sessions for the county are held here at midsummer, besides annual courts leet and baron, and weekly petty sessions. Markets principally for butter, on Saturdays; fairs, May 12th, July 21st, and Nov. 3d, for cattle, sheep, and toys.

SWANSEA, a parl. bor., seaport, and par. of South Wales, co. Glamorgan, hund. Swansea, on the W. bank of the Tawe, at its mouth in the Bristol channel, 34 m. W.N. W. Cardiff, lat. 51° 37' 12" N., long. 3° 55' 30" W. Area of par., 2261 acres. The parl. bor., however, includes also the par. of St. John Lansamlet and the hamlets of Morriston, Clas-Lower, &c., on both sides of the river; having a total area of about 5000 acres, with a population, in 1801, of 6631; in 1831. of about 18,880; and in 1841, of 21,239. The compact portion of the town is about 1 m. in length, N. to S., by somewhat more than ½ m. in average breadth, and consists of three or four parallel lines of thoroughfares, crossed by numerous others. It is generally clean, and pretty well built, and has been of some reputation as a watering-place. It has an excellent market, attended by all the neighbouring district, with a handsome courthouse, a which the assizes, quarter and petty sessions are held, an infirmary, assembly-rooms, theatre, royal institution for literary and scientific purposes, with a good library and museum; mechanics' institution, two reading or newsrooms, savings' bank, poor-house, house of correction, a Orcas and benevolent societies, a society for prosecuting felons, and a branch of the bank of England. The town is paved, lighted with gas, and well supplied with water, and has a small police. (Mun. Corp. Rep.) The parish church is comparatively a modern edifice, with a square tower; the living, a vicarage, worth £291 a year, is in the gift of society. St. John's, a perpetual curacy in the gift of Sir John Morris, Bt., is worth £87 a year. There are, also, a synagogue, Roman Catholic, and numerous other dissenting chapels; and on an elevated site near the centre of the town is Swansea castle, founded in 1099, now partly converted into barracks and stores. A free school was founded in the town in 1682; but like many other institutions formerly established at Swansea, it has become nearly extinct. (See Charities' Rep. 32, iii.) There are, however, several national and Lancastrian, and numerous private schools. Swansea is highly prosperous and increasing. It owes its importance principally to its collieries, and extensive works for the smelting of copper and other ores established in its neighbourhood. The latter are on a very great scale; and, in fact, by far the largest part of the copper ore produced in Ireland, Cornwall, and other parts of the United Kingdom, as well as in Cuba, &c., is brought here for smelting. Swansea has also a very extensive trade in the shipping of coal, having exported 460,201 tons coastwise in 1840, exclusive of 15,676 sent to foreign parts. It has also two large potteries. Tawe at its mouth expands so as to form a harbour, which is protected seaward by two handsome piers; vessels of considerable burden enter at high water, but at ebb the harbour nearly dries. But it is intended to obviate inconvenience by constructing a floating harbour, and also intended to unite the town with the hamlet of St. has by a bridge, &c. A lighthouse is erected on the pier, and the Mumbles' light is about 4 m. S.W. from fort. A canal goes from Swansea to Henoyadd in Brecknockshire; and two canals on the opposite bank of river communicate, one with the harbour of Neath. There is a road to the Mumbles and Oystermouth westward, by which coals are taken out, and lime and limestone brought and tram-roads also connect the different works, and

the canals and wharfs. Exclusive of coal culm and copper ore, iron ore, limestone, clay, rotten stone, tin plates, and timber, are brought to Swansea for its own consumption, or for exportation inwards or outwards. Gross customs' revenue, in 1840, £2935. It has numerous manufactories. The municipal and parl. boroughs are co-extensive. The former is divided into two wards, and is governed by a mayor, six aldermen, and 18 councillors. Its earliest charter extant dates from temp. Jno. There was no criminal court within the borough till 1835, but now it has a commission of the peace; and a court-baron for causes under 40s., and a court of pleas for those above that amount, are held every third Monday. Corp. revenue, in 1839. £3450. Swansea was formerly a contributory bor. to Cardiff, the right of voting having been in the burgesses by birth, marriage, or gift, resident or non-resident. It is now joined with Aberavon, Kenfig, Loughor, and Neath, in sending one member to the House of Commons. Registered electors in 1839-40, for Swansea, 681; for the entire district, 1247. It is also a polling-place for the county. Markets, Wednesday and Saturday. Fairs, second Saturday in May, July 2, Aug. 16, Oct. 8, and the two following Saturdays. (Bound. and Mun. Corp. Rep., and App.; Charity Rep. &c.)

SWANTON, p. t., Franklin co., Vt., 71 m. N.W. Montpelier, 545 W. Bounded W. by lake Champlain. Watered by Missisque river, which has a fall of 90 feet, 6 m. from its mouth, affording extensive water-power. Variegated marble is found and extensively exported. It covers over 300 acres to an unknown depth, but is generally obtained at a depth of from 2 to 8 feet. It contains two churches, a Congregational and Baptist, eight stores, one grist-mill, five saw-mills, two tanneries; 16 schools, 945 scholars. Pop. 2312. Missisque r. is navigable for canal boats. 6 m. to the v.

SWANZEY, p. t., Bristol co., Mass., 48 m. S. by W. Boston, 414 W. An arm of Mount Hope bay sets up into the t., on which is a small v. Incorporated in 1667. King Philip's Indian war commenced in this t., June 20th. 1675. It contains three Baptist churches, nine stores, one grist-mill, one saw-mill, two paper-mills; seven schools, 137 scholars. Pop. 1484.

SWANZEY, p. t., Cheshire co., N.H., 53 m. S.W. Concord, 429 W. Watered by Ashuelot r. It contains iron ore. Chartered by Mass. in 1733, by N. H. in 1753. It contains three churches, a Congregationalist, Baptist, and Universalist, four stores, one fulling-mill, one woollen factory, one cotton factory with 680 spindles, four grist-mills, 14 saw-mills; 12 schools, 577 scholars. Pop. 1755.

SWATARA, t., Lebanon co., Pa., 9 m. N. Lancaster. Drained by Swatara cr. and its branches. It has six stores, one fulling-mill, four grist-mills, four saw-mills, two tanneries, one distillery, one pottery; two schools, 60 scholars. Pop. 1506.

SWEDEN (Sverige), a kingdom of northern Europe, comprising, with Norway and Lapland, the whole of the Scandinavian peninsula, of which it forms the eastern, southern, and most important portion; between lat. 55° 20' and 69° N., and long. 11° 18' 30" and 24° 13' E., having N.E. Russian Finland, from which it is separated by the Tornea and one of its affluents; E. and S. the gulf of Bothnia and the Baltic, S.W. the Sound, Kattegat, and Skagerrack; and W. and N. Norway, from which it is for the most part divided by the great mountain chain of Scandinavia. Length N. to S. 950 m.; average breadth about 190 m. Area about 170,700 sq. m. Sweden is divided into three principal regions, Gotaland (Gothia) in the S., extending to about lat. 58° 45' N.; Sweden proper occupying the centre as far as lat. 60° 40' N.; and Norland (by far the largest portion), comprising the remainder. It was formerly divided into 18, but is now divided into 24 läns, or governments, as follows:

Regions and Läns	Area in English sq. m.	Pop. in 1839.	Regions and Läns	Area in English sq. m.	Pop. in 1839.
Norland:			*Gatäland:*		
Pitea . .	33,060	46,172	Juneeping	4,414	148,698
Umea . .	29,4 5	54,256	Kronoar	4,243	179,300
Hernosand .	9,-16	85,242	Halmstadt	1,846	94,832
Gstersund .	19,618	43,517	Weibj		
G-fle . .	7,842	109,381	(Gottland)	1,962	42,589
Sweden Prop.:			Weksio	3,796	119,309
Fablun .	12,272	141,204	Christianat		
Carlstadt	6,957	192,479	Kraalf	2,439	162,809
Orebro .	3,270	121,313	Carlscrona	1,137	93,249
Westeras	2,445	92,411	Malmo	1,456	218,074
Upsal .	2,0-2	85, 93	The läns		
Stockholm	2,916	195,427	Woner,		
Nykoping	2,512	113,733	Wetter,	3,642	
Gotaland:			Maelar,		
Linkoping	4,270	200,888	Hjelmar,		
Mariestadt	5,323	179 449	&c.		
Gottenburg	1,908	164,583			
Wernersburg	5,015	215,698	Total .	170,7-5	3,100,772

Topography, Mountains, &c.—The Scandinavian peninsula rises gradually from the W. coast of the Baltic until it reaches its greatest elevation in the great mountain chain, usually called the Scandinavian Alps, or Doffrine hills, dividing Sweden and Norway. This chain extends from the Sylt-fjell in about 63º N. lat. and 13º E. long. to the N. cape, in the general direction of N.N.E. and S.S.W. It differs from the Alps and Pyrenees in not being a continued chain of summits, but a succession of large *plateaux* from 20 m. to 30 m. across, from which the culminating points project. The Sylt-fjell, the loftiest point on the Swedish frontier, is 6552 ft. above the level of the sea. The other principal peaks belonging to the same chain are the Sulitelma 6342 ft., and the Saulo 5095 ft. in height. The Helags, within the frontier, has an elevation of 6100 ft.

Speaking generally, Sweden may be said to be a flat country. There are, indeed, some ranges of high grounds and detached hills, but, on the whole, it is wonderfully level. This is so strikingly the case, that all the way from Gottenburg to Stockholm, by the Orebro road, there is not a single hill or declivity till within a few miles of the capital. (*Thomson's Travels*, p. 393.)

According to Forsell, 1-12th part of the surface of Sweden is 1900 ft., more than 2-5ths 760 ft., and 7-10ths 285 ft. above the level of the Baltic. The remainder, consisting chiefly of the coasts, is of less elevation. These are, for the most part, fenced by numerous rocks and islets. The islands of Gothland and Œland, in the Baltic, belong to Sweden: they are situated opposite the S.E. shores of the kingdom, and Œland is separated from the main land by a narrow strait, which in one part (opposite Kalmar) is only about 4 m. across.

The S. provinces consist chiefly of vast sandy plains interspersed with small lakes and hills, which are sometimes bleak and barren; but elsewhere clothed with woods. The central region contains extensive plateaux of table land covered with forests. The N. part of the kingdom is diversified with mountains, deep valleys, and glens, alternating with sandy wastes and vast forests.

Rivers.—Sweden is extremely well watered. Through its N. and central parts, 12 large rivers flow into the gulf of Bothnia. The Tornea, which has the longest course, runs almost due S. for about 290 m.; but the largest is the Angerman, 230 m. in length, so deep that ships of 600 tons load at Nyland, about 70 m. from the sea. Next to these are the Umea, with a course of 350 m., and the Windel, 235 m. in length. The general direction of the rivers falling into the Baltic is N.W. to S.E. Few of them are of any considerable size, and notwithstanding the generally flat country through which they flow, their navigation is much impeded by rocks and numerous cataracts, and is rendered perilous during the inundations occasioned by the melting of the snows. Some of them increase 18 or 20 ft. in height so rapidly as to carry away large trees, and even to detach immense blocks of granite from the mountains; still, however, the inundations occasion but little damage, owing to the number of lakes, which serve as so many basins for the reception of the surplus water. There are, in fact, upwards of 80 considerable lakes, occupying in the aggregate a very large surface. The principal of these is the Wener, the largest lake in Europe, after that of Ladoga, between lat. 58º 22' and 59º 25' N., and long. 12º 20' and 14º 12' E., above 90 m. in length, by 56 m. in its greatest breadth, 147 ft. above the level of the sea. It receives many streams, the only outlet for its waters being a channel about 200 yards in width, immediately below which is the celebrated cataract of Trolhoüa. Though in parts very deep, a great portion of this lake is so shallow as to render its navigation difficult and dangerous. The lake next in size is the Wetter, 86 m. in length, by 16 m. in its greatest breadth. It is about 25 m. S.E. the Wener, and 295 ft. above the level of the sea. In some places it is 70 fathoms deep: it is often agitated by sudden and violent storms. The Mœlar lake is an inlet of the sea, extending westward from Stockholm, near its entrance from the Baltic, about 70 m., with a breadth varying from 2 to 20 m. It is deep and clear, contains some hundred islands, and is regularly navigated from April to November. The Hjelmar, a lake lying to the S.W. of the Mœlar, to which it is united by a canal, is 35 m. in length, varying to 15 m. in width.

Climate.—For five or six months of the year the surface of the N. parts of the country, from the summits of the mountains, to the bottoms of the valleys, is covered with ice and snow. The rivers and lakes are also frozen from October to April. In the central parts, the winter seldom lasts more than three or four months; and in the S. and W. parts, the climate is very similar to that of the N. of Germany. In the N. division a great degree of heat is experienced during a short period of the year. The transition from winter to summer is there, also, very rapid, often occurring within the space of a few days.

On the whole, however, the climate of Sweden is much milder than might be expected from its high N. lat. The winter is not so cold as in countries in the same lat. further to the E., at the same time that the mean temperature of

the summer is but little inferior. According to Dr Twemson, the mean annual temperature at Petersburg is 7 [?] Fahr.; whereas at Stockholm it is 41·9º. The advantage on the side of Stockholm is chiefly in the six winter months, the mean temperature of these being in it 29 F; and at Petersburgh only 21·9º. During the six summer months the advantage on the side of Stockholm amounts to only 5 and in July and August the temperature is 1º hotter at Petersburg than in Stockholm. The winter is considered by the inhabitants as peculiarly pleasant. The roads are always dry; and as the winds are seldom violent, travelling is both rapid and agreeable, the traveller dressed to himself from the cold by warm clothing. The great defect of the climate is the occurrence of frosts in Aug. and Sep., by which the crops are often injured. (*Thomson's Travels*, p. 405.) Near Tornea, at midsummer, the sun is visible during the whole night. The longest day at that town is 21½, and the shortest 2¾ hours. At Stockholm the longest day is 18½, and the shortest nearly 6 hours in length.

Geology and Minerals.—In point of structure, the whole of Sweden may, with few exceptions, be considered primitive. Granite and gneiss are the predominant rocks; but the former is more extensively diffused than the second, which is found chiefly on the shores of the Baltic, and in the S. part of the kingdom. A remarkable geological feature, which Sweden has in common with some parts of N. Germany and Denmark, is the presence of a vast number of enormous erratic blocks of granite scattered over its surface, especially in the central and S. provinces. In the S. they are collected in long spits, or tongues, resting upon the plains, which are quite unconnected with them, more in the N. they are scattered indiscriminately, and so profusely, that scarcely an acre of land is without one or more heaps of them. They seldom exceed 30 or 40 ft. in height above the surface, and form many islands in the lakes, as well as heaps on the plains. (*Laing, Sweden*, p. 41, 42.) These are mountains of secondary formation in Jemtland, Narrica, E. and W. Gothia, and in the islands of Gothland and Œland. Shelly limestone, chalk, &c., are met with in Scania. Deposits of oceanic shells are found in the country near Uddevalla; but at Stockholm, Upsala, Hernsand, and at other places on the E. side of the peninsula, the shells discovered are of the kinds belonging to the Baltic, without any mixture of the oceanic. Sweden is rich in mineral products. Among these are iron, the best in Europe, copper, cobalt, zinc, lead, antimony, gold and silver, alum, nitre, sulphur, with porphyry, marble, alabaster, limestone, millstone, whetstone, asbestos, potters' earth, &c. But the only metals that occur in any considerable quantity are iron, copper, and lead; iron being the most abundant and least the scarcest of the three. There is a remarkable deficiency of the more valuable products found in secondary formations, as coal and salt. The former, indeed, has been discovered and wrought, near Helsingborg, in the S. of the kingdom, but it is of very inferior quality: there are no salt beds nor brine springs, and the waters of the Baltic not being sufficiently impregnated with salt, it is wholly imported.

Vegetable Products.—The forests of Sweden are estimated to occupy about 98,000 sq. m., or four sevenths of the whole surface of the country. Those of the N. region consist of birch, pines, firs, &c., which in the central parts, are mixed with ash, willow, linden, and maple; and in the S. with oak, beech, apple-elm, &c. Few beeches are N. of lat. 57º; oaks are found as far N. as Sundsvall; linden is found as far N. as latitude 61º, the hazel upon the cherry and ash as 63º, and the general limit of the oak and pine woods is lat. 69º 30'. The small dwarf birch, pen, mountain-ash, and dwarf grey alder, are found in the N. as 70º, but only in the valleys and sheltered mountains. The walnut and mulberry are almost entirely confined to Gotaland; the chestnut is very rare. The forests were formerly much neglected; and there is now in many countries districts a great deficiency of timber. Indeed, a considerable proportion of the firewood required for the consumption of Stockholm, is brought from Finland. Lately, however, a great deal more attention has been paid to the forests. Those belonging to the state have been placed under the care of a special institution, and very extensive plantations of oaks, firs, &c., have been made. The power of private proprietors to cut down timber was formerly limited, but this restriction no longer exists. In the interior of the country, however, and in such parts as have no facilities of water-carriage, or otherwise, for the conveyance of timber to the seaports, and are distant from mines, there is but little hope that the forests will ever become an object of considerable attention. Pears, apples, and plums of all kinds grow in the open air in the S.; but the grape, fig, apricot, and peach, do not ripen except in hothouses. All kinds of melons are grown, currants up to lat. 60º 30. and gooseberries every where, even as far N. as lat. 70º. The soil is suitable for all kinds of pulse crops. Asparagus is

ires hot beds in lat. 60° 30', cabbages cease to come to itunty in lat. 64°, carrots and parsnips grow to lat. 66° 20' , turnips and potatoes nearly to lat. 70°. The yellow et root is produced spontaneously ; the red is cultivated. close sward of common grass is rarely seen ; but docks, stles, ragweed, and such roots as infest the land in more countries are seldom observed, even by the road side, or the most neglected spots.

Animals.—The most common wild animals are the wolf, r. fox, elk, reindeer, roebuck, glutton, ermine, and a spe- i of lynx. The wild boar is now found only in the isle of 'and. Whales and sea-calves are occasionally found in Baltic and Gulf of Bothnia ; and the porpoise (*Delphi-Phocæna*, Linn.) commits great ravages among the fish nuse seas. There are few hares, but abundance of other ls of game. The cock of the wood, or capppercailzie (*Te-Urogallus*), formerly met with in Scotland, and recent-eintroduced into that part of the United Kingdom, is non in the Stockholm markets, whence it is sometimes ght to London ; though inferior in flavour to grouse, it uch larger, sometimes weighing from 14 to 16 lbs., and together a very fine species. Partridges are very plen- as are woodcocks and web-footed wild fowl. Eagles alcons inhabit the cliffs ; the wild swan and eider are ed for their down ; and the eggs of the latter are highly med. The seas surrounding Sweden abound with fish ; ding sturgeon, cod, lamprey, rays, soles, turbot, pil- lk. herrings, and the stremming, a small species of her- w hich has been latterly very abundant on the E. Swe- oasts. Excellent mackerel and oysters are found in gat. The rivers and lakes are well supplied with sal- pike, perch, trout, eels, and numerous fish of the genus ni. The pike, perch, barbel, and crayfish, are found Baltic, as well as in the lakes and rivers. Many of h of this sea appear to be of a mixed character, be- oceanic fish, and those of fresh water.

iculture.—The soil of Sweden, though mostly thin or. has been greatly improved by the industry of the tnnts. The coast land is usually bare of soil, the rock appearing every where. The flat alluvial lands lake Wener, and in the basins and valleys connect- it, consist of a harsh crystalline sand, impregnated in, and not very productive ; but on the N. side of the the neighbourhood of Carlstad, the soil is of a su- lescription. In the country lying between the lake and the Baltic, there are some very fertile tracts ; nd this lake, and in the district round Carlstad, es- the latter, agriculture has made great advances ; is are well cultivated in large farms, and the coun- nibles some parts of the interior of England, except lands under cultivation are not fenced by hedges, a ooden palings. Of the 170,715 sq. m. forming the of the country,

	Square Miles.
ble lands occupy	3.490
idows and common pasturage	7.385
uitivated forest and mountain land	137.000
es and marshes	22,055

gricultural products consist chiefly of rye, barley, -lin (a mixture of barley and oats), wheat, potatoes, emp, flax, and almost all the fruits and legumes urope. In the S., rye is the most cultivated ; in the y, the culture of the latter increasing in proportion oceed farther towards the pole ; but the grains of are generally less nutritious than those of the S. of nd are more difficult to preserve. Wheat suc- far N. as 63°, but does not ripen in W. Bothnia. oin ripen N. of lat. 63° 20', but barley is grown al- ie limits of the pine woods, in lat. 69° 30'. Hops ated up to 62°, tobacco to 62° 30', and flax to near- Buckwheat, madder, and woad are grown in Sca- iones parts of the S. the produce is equal to that st cultivated lands in England and France. In canin a return of seven for one is obtained ; but in Sweden the proportion does not exceed four for e uncertainty of the climate and the chances of ts, are the greatest obstacles with which the agri- ins to contend ; and some singular devices are re- to counteract their effects. In Jemtland, for ex- i people pile up large quantities of wood along the the small patches of land sown with corn, the e wind should blow from the N. or N.E. in the if August, they may set them on fire to protect from the frosts! It is usual, also, in the S. parts ntry, to prevent the crop from being injured by in the ear, to draw ropes across the heads of the shake off the dew before sun-rise, which, but for then be frozen.

ig to the official returns for 1837, the annual pro- in and potatoes, after deducting the seed, amount-

	Swed. Barrels.
Wheat	244,619
Rye	2,378,366
Barley	1,800,909
Oats	1,532,046
Maslin	774,678
Pease	299,109
Potatoes	4.113,442
	11,043,161

equivalent to about 44,000,000 Eng. bushels.

In the N. potatoes supply the deficiency of corn, and are preferred to all other kinds of food. Tobacco is cultivated near Stockholm, but not to any extent. After that of Hol- land, the flax produced in Sweden is probably the best in Europe. Hemp is at present not much grown ; but the gov- ernment is endeavouring to extend its culture.

The whole arable surface of Sweden is divided into 66.441½ *hemmans* of land. The word hemman signifies merely an estate, or homestead, and gives no idea of the value or extent of the land, some being incomparably lar- ger and more valuable than others. It is, in fact, a fiscal division, for the purpose of levying the land tax according to ancient assessments. Originally, however, the hemmans belonged, for the most part, to single proprietors ; but they are now generally divided into 3, 4, 8, 16, or more parts, and it is rare for a family to possess a hemman entire. Of the 66.441½ hemmans, 771½ belong to towns, 50,000 to private in- dividuals, 359 to the crown, 373 to academies and universi ties, 201 to colleges and schools, 289 to the church, 204 to hospitals and asylums, 183 to military schools, 31 to sailors, and 4,045 to the army. According to circumstances, the lands are subject to a different amount of taxation : of the estates belonging to the nobles, 3462 are wholly exempted from all public burdens ; and 17,929 estates, partly belong- ing to them and partly to other privileged parties, enjoy a partial exemption from taxation.

The estates that originally belonged to the nobles, but which, since 1810, may be indifferently held by nobles or commoners, are exempted from the land-tax, and also from the obligation to furnish a soldier for the army, the nobles themselves having been originally bound to personal ser- vice in the army. This inequality in the rate of taxation is practically, however, notwithstanding the statements of Mr. Laing to the contrary, of no real importance, the land tax was fixed at a certain amount of produce centuries ago. It can no longer, therefore, be fairly regarded as a burden on the land, the value of which really depends on its nett revenue, after this fixed charge has been deducted. There is no injustice in the circumstance of certain lands in En- gland being subject to tithe, while others are not ; and it is quite as idle to talk about the injustice of the unequal distribution of the Swedish land-tax.

There is, however, in Sweden, an assessment of 5 per cent, laid on the nett annual value of *all* estates. But this, though apparently an equal, is in fact, a very unequal and impolitic tax ; inasmuch as it makes no distinction between the income derived from the rent of land properly so called, and that which is really derived from the capital laid out on the land, and as it operates as an obstacle to improvement. The occupiers of crown-lands in Sweden have long had, and still have, leave to constitute themselves the absolute proprietors of such lands on their paying a sum equal to six years' value of the land-tax laid on the land. It is not, therefore the amount of the burdens falling on the land in Sweden, which, despite the statements to the contrary, are really very moderate, but the influence of the 5 per cent. as- sessment in discouraging improvements, and still more, the minute subdivision of the hemmans, occasioned by the con- tinued division and subdivision of heritages, in consequence of the law of equal partition among the children of a fam- ily, that are the principal obstacles to improvement. Pro- perty is, in many instances, divided into such minute propor- tions as to be wholly unsusceptible of a proper system of cultivation ; and the occupiers are often in the poorest cir- cumstances. There are parcels of land of not more than 40 yards sq., and a Dalecarlian peasant sometimes sells his landed property for 2 or 3 rix-dollars (3s. 6d. to 5s.), the re- gistration of the sale costing as much as the estate !

In some extensive districts there are not, at an average, above 14 acres of arable land to a farm ; and in the district of Carlstad, where farms are largest, and agriculture most advanced, the average extent of arable land in each farm may be taken at about 72 acres. At an average of the en- tire kingdom, the arable land may be estimated at about 28 acres per farm. (*Thompson's Travels in Sweden*, p. 426.)

But, notwithstanding these disadvantages, and those that originate in its backward climate and not very fertile soil, agriculture has made a very material progress in Sweden since 1815. This is partly ascribable to the encouragement afforded by government, and to the establishment of model farms, some of which are managed by agriculturists from

Great Britain. One of these, in the vicinity of Linkopping, on the S. border of the lake Wener, consists of about 15000 acres, of which about 500 are under the plough. The manager of this farm had, in 1839, twenty young gentlemen boarded with him in the house, and 20 scholars of the working class. Government allows him 10,000 dollars a year towards defraying his expenses, and the establishment is said to be alike successful and advantageous. Mr. Stevens, an experienced Scotch agriculturist, who visits Sweden every summer, and assists landed proprietors in laying out their estates, and putting them under an improved system of management, says that "Of late years an enthusiasm has sprung up for the improvement of agriculture among all classes of people not to be equalled in any other continental country. This has been owing in a great measure to the exertions of the agricultural societies established within the provinces, and the great interest the landed proprietors now take in the improvement and management of their estates. English and German works on agriculture are studied; improved agricultural implements from Great Britain and other countries are introduced; and in many parts Scotchmen and Germans are seen directing the plough or conducting the operations of the field." (Bremner, ii. 218.)

The best evidence, however, of the improvement and extension of agriculture is to be found in the fact that, previously to 1820, there was generally a large importation of corn into Sweden from Dantzic and other parts, whereas that importation has now, in ordinary years, wholly ceased, and there is, on the contrary, a considerable exportation. No doubt a good deal of this improvement is to be ascribed to the more extensive culture of the potato. But, independent of this, the improvement in cultivation generally has been most striking. Mr. Stevens says, " that the necessity under which the population formerly laboured, in the N. parts of the country, of having every now and then to use the bark of trees as a substitute for meal in the making of bread, has been in a great measure obviated; and that both the public and private magazines are completely filled with corn."

Houses in the country in Sweden are mostly constructed of wood; and are roofed with timber, turf, and straw. Gentlemen's houses, however, and houses in towns, are usually covered with tiles. Recently, thick coarse paper prepared with tar has been used for roofing, and is said to answer very well. Slates are very scarce; and Dr. Thomson states that he only saw two houses in the kingdom roofed with slate. (p. 399.)

It is estimated by Forsell, that seven ninths of the whole population are employed in agriculture: peasants, proprietors of the soil they cultivate, have been reckoned at 147,974; those who live on land not their own at 1,688,717; husbandry labourers, holding houses and lands under proprietors, at 470,091; and servants, living in the house with their employers, at 277,466. Masters and mistresses are authorized, by an old law, to inflict summary corporal chastisement on their servants, with no other limit than that they do not kill or maim; and Mr. Laing, founding on this fact, states that servants in Sweden are little better than slaves. But it might have occurred to him that laws here, as elsewhere, become obsolete; and we have been assured, by those who are thoroughly acquainted with the country, that the corporal chastisement of servants is quite as rare in Sweden as in England, and that they are treated with great kindness. The poverty of the soil, and short duration of summer, require a great number of hands during the season for agricultural employments; but during the remainder of the year, they are comparatively idle. Since 1830, the price of agricultural labour has been about 8d. or 1s. a day in the S. and centre of Sweden; but in the N. it costs 1s. 4d. per day. Labour is generally cheaper in Sweden than in Norway, from there being a greater number of the agricultural classes who are destitute of property. According to Laing, the condition of the middle and lower classes, in Sweden, is much less prosperous than that of the same classes in Norway. In this respect, however, as in many others, he is a suspicious authority; as there can be no doubt that he has represented the condition of the Norwegians under much too favourable, and that of the Swedes under much too unfavourable colours. Indeed, the Swedish peasant appears, even from Laing's account, to be far from being in a distressed situation. He says that, "compared with the cotter, or labourer, in Scotland, the Swedish peasant is better provided with physical comforts: he is far better lodged, better fed, and has access to fuel and food generally better."(277.) Rent is, most commonly, some proportion of the produce, and is usually paid in kind, there being but few districts in which it is paid in money. Labourers are frequently paid by getting a piece of land, which they cultivate for themselves, working on the proprietor's domain certain days in the week.

Mr. Coxe, one of the best and most trustworthy of travellers, gives the following details with respect to the condition of the Swedish peasantry. " I had frequent opportunities of observing the customs, manners, and food of the peasants. On entering a cottage, I usually found all the family employed in carding flax, spinning thread and in weaving coarse linen, or cloth. The peasants are excellent contrivers, and apply the coarsest materials to some useful purpose; they twist ropes from hogs' bristles, horses' manes, and bark of trees, and use eel skins for bridles. Their food principally consists of salted flesh and fish, eggs, milk, and hard bread. At Michaelmas they usually kill their cattle, and salt them for the ensuing winter and spring. Twice a year they bake bread, in large round cakes, which are strung on files of sticks, suspended from the ceilings of the cottages: this bread is so hard as to be occasionally broken with a hatchet, but is not unpleasant. The peasants use beer for common drink, and are much addicted to malt spirits. In the districts towards the W. coasts, and at no great distance inland, tea and coffee are not unusually found in the cottages, which are procured in great plenty, and at a cheap rate from Gottenburg.

" The peasants are well clad in strong cloth of their own weaving. Their cottages, though built with wood, and only of one story, are comfortable and commodious. The room in which the family sleep, is provided with ranges of beds in tiers (if I may so express myself) one above the other: on the wooden testers of the beds in which the women lie, are placed others for the reception of the men, to which they ascend by ladders. To a person who has just quitted Germany, and been accustomed to tolerable inns, the Swedish cottages may, perhaps, appear miserable hovels; but to me, who had been long used to places of far inferior accommodation in Russia, they seemed comfortable places of reception. The traveller is able to procure many conveniences, and particularly a separate room from that inhabited by the family, which could seldom be obtained in the Polish and Russian villages. During my course through those two countries, a bed was a phenomenon which seldom occurred, excepting in the large towns, and even then, not always completely equipped; but the poorest huts of Sweden were never deficient in this article of comfort: an evident proof that the Swedish peasants are more civilized than those of Poland and Russia." (Coxe's Letters, iv., 277-279.)

According to the official returns, Sweden had, in 1837, 385,000 horses; 1,557,976 head of horned cattle; 1,412,649 do. sheep; and 513,692 do. hogs. In general, all kinds of domestic animals are inferior. The horses are everywhere small. There is a fine breed in the isle of Öland, not more than 3 or 4 ft. high: these, however, are rapidly decreasing. In the S. provinces, the number of horses, as compared with the population, is much greater than in France, or even in England; there being, it is said, in Scania, 243 horses to every 1000 inhabitants! As we proceed N., the number of horses diminishes; and in Swedish Lapland they disappear altogether, their place being supplied by rein-deer, of which some proprietors possess 1000 head. In Lapland, the rein-deer and dog are the only domestic animals. Swedish black cattle are also small: the best are those of E. Gothia and Dalecarlia; in summer they are driven to the mountains, where chalets, similar to those of Switzerland, are constructed. The sheep-folds are well kept, and government has endeavoured to improve the breeds by crosses with those of Spain, France, England, and Saxony. Sheep are not reared N. of lat. 63°; goats thrive as far as lat. 65°. (Journal de Travaux Statistiques, &c., p. 131-143; Laing, p. 266-272.)

Fisheries, form a very considerable branch of industry. The herring fishery on the W. and S. coast commenced in 1740, about which time herrings began to appear in large shoals on the coasts. The quantities annually taken increased until 1798, since which they have decreased; the place of the herring being now supplied by the sprat, a fish about the size of the sprat, but of much finer flavour. From 1790 to 1796, the towns of Gottenburg, Bangelf, and Marstrand disposed of 1,972,214 barrels salt herrings, and 261,971 hhds. fish-oil, which fetched together £480,000, about three fourths being sold to foreigners. But since 1805, the average produce of the fishery has not exceeded 2000 barrels, the herring having, in a great measure, abandoned the coasts. The streaming is cured like the herring, and is often eaten raw out of the pickle: it is extensively used in Finland and the N. of Russia, and forms a favourite dish even with people of condition. The principal streaming fisheries are on the coasts of the gulfs of Finland and Bothnia. The principal salmon fishery is at Dyefors, on the Klarely, a river which falls into the lake Wener. The salmon fisheries of Norrkoping, Gefle, and Hernosand, are also very productive. A company in London employs two packet boats, with wells in the bottom, in trading to Gottenburg for lobsters, which are bought there for 3½d. or 5d. each.

Mines.—The mines of Sweden, though inconsiderable as compared with those of this country, are a considerable source of national wealth. They are principally situated in the central provinces, which have no fewer than 261 out of the 586 mines said to exist in the kingdom. Swedish iron is of very superior quality, and that of the Danemora mines is especially well fitted for conversion into steel; but owing to injudicious restrictions and the want of coal, the production in Sweden is not supposed (including what is licensed and what is made for home consumption without a license) to exceed 85,000 or 90,000 tons bar iron, of which about 70,000 are exported. In 1839 we imported 17,049 tons of Swedish iron. The copper mines produce, in all, only about 750 tons a year; the metal is not so good as that of England, and is impregnated with iron. Fahlun, the chief mine, has long been in a declining state, the number of workmen at present employed not exceeding 500. The works of this mine are conducted entirely by water-power, and are remarkable for their completeness; connected with them is also a manufactory of sulphuric acid. The smelting furnaces and iron works are licensed to produce certain quantities, some licenses being as low as 50 tons, and others as high as 400 or 500 tons; and some fine bar iron works have licenses for 1000 tons each. These licenses are granted by the college of mines which has a control over all iron works and mining operations. The iron masters make annual returns of their manufacture, which must not exceed the privileged or licensed quantity, on pain of the overplus being confiscated. The college has established courts of mines in every district, with supervising officers of various ranks. All iron sent to a port of shipment must be landed at the public weigh-house, the superintendent of which is a delegate of the college; so that it is impossible for an iron master to send more iron to market than his license authorises. It is true that sales are made to inland consumers at the forges, of which no returns are made out, and in so far the licenses are exceeded; but it is not supposed that the quantity so disposed of exceeds a few thousand tons a year. Every furnace and forge pays a certain annual duty to the crown. Its amount is fixed by the college when the license is granted; and care is taken not to grant the license to any

one unless he have the command of forests equal to the required supply of charcoal, without encroaching on the supply of this material, required for the existing forges in the neighbourhood. As the supply of pig-iron is limited to the quantity licensed to be made, the college, in granting new licenses to bar-iron works, always takes into consideration how far this may be done without creating a scarcity of pig-iron. Hence, the erection of new forges depends —1st, on having a supply of charcoal, without encroaching on the forests which supply your neighbours; and 2d, on the quantity of pig-iron which the college knows to be disposable. The courts of the mines decide all disputes that arise among the iron masters regarding the exceeding of their licenses, encroachments, &c.; an appeal to the college lying from their decision, and ultimately to the king in council, or to the supreme court of the kingdom.

It is needless to dwell on the impolicy of such regulations. No doubt it is quite right for government to interfere to prevent the waste and destruction of the forests; but having done this, it should abstain from all other interference, and leave every one at liberty to produce as much iron as he may think proper. Mines of any importance are usually held by a society of shareholders. Some of them are only worked occasionally; and, as the labour is performed by peasants, who live ostensibly by husbandry, it is impossible to form any correct estimate of the numbers engaged in mining industry.

Manufactures.—For many ages, Sweden had none of any importance; the Hanseatic towns took away its raw materials, and re-exported them manufactured to the country: the other manufactures were then, as they still are in great part, domestic. But about the middle of the 17th century, various manufactures, including those of glass, starch, brass, pins, silk fabrics, leather, soap, steel, and iron articles, besides printing-presses and a sugar refinery, were established: the workers in these establishments were mostly from Germany and the Low Countries. In 1771, there were in Sweden 885 manufactories; in 1824, 1177; and in 1831, 1184, employing 12,143 hands, exclusive of miners. We subjoin an official

ACCOUNT of the Number of Factories, Looms, and Workmen, in each Department of Manufacturing Industry in Sweden, in 1838 and 1839, and of the Value of the Produce in each.

	1838.				1839.			
	Factories.	Looms.	Workmen.	Value.	Factories.	Looms.	Workmen.	Value.
				Rixd. Banco.				*Rixd. Banco.*
Cotton and linen weaving .	42	636	805	426,581	44	761	949	465,580
Riband ditto	10	69	107	45,152	11	71	103	45,194
Cloth ditto	108	558	3,455	3,863,439	114	685	3,642	4,045,989
Stuffs ditto	7	22	29	20,405	5	19	22	19,968
Silk ditto	16	355	527	467,495	16	372	571	494,431
Silk spinning . . .	11	.	26	27,600	12	.	21	28,000
Canvass and sailcloth .	10	233	496	248,659	10	239	595	250,912
Dyeing . . .	304	.	922	400,769	317	.	980	437,324
Glassworks . .	15	.	547	366,227	13	.	615	354,600
China or earthenware .	2	.	364	155,142	2	.	375	176,292
Perfumery . . .	14	.	15	22,706	14	.	15	24,906
Paper mills . . .	89	.	1,333	756,878	87	.	1,241	803,494
Soapworks . . .	15	.	34	127,845	17	.	37	114,054
Sugar refinery . .	28	.	397	2,489,256	25	.	458	2,625,763
Morocco leather . .	4	.	14	25,464	5	.	13	15,920
Tobacco manufactories .	87	.	790	1,018,528	81	.	765	1,003,036
Watch ditto . .	149	.	230	36,622	143	.	246	37,760
Leather curriers . .	255	.	687	678,076	258	.	675	599,728
Oil manufactories . .	47	.	98	148,567	48	.	93	138,905
Wax candle ditto . .	6	.	21	60,005	7	.	19	56,425
Woollen and cotton spinning mills . .	9	.	763	745,047	9	.	810	879,968
Rope manufactures . .	17	.	82	64,612	22	.	93	108,339
Porter brewery . .	1	.	116	184,479	1	.	86	175,437
Machine makers . .	16	.	252	98,299	19	.	290	120,342
Vinegar distillers . .	19	.	38	22,109	19	.	35	22,494
Calico printers . .	11	.	104	42,179	9	.	120	76,094
Sundry minor manufactories	812	114	2,029	547,928	789	110	1,983	546,634
Total . . .	**2,104**	**1,987**	**14,311**	**13,090,089**	**2,097**	**2,177**	**14,861**	**13,597,809**

Sweden has but few facilities for the formation of great manufacturing establishments; but, owing to the long winter nights, during which most out-of-door occupations are necessarily suspended, she has great facilities for the carrying on of domestic manufactures; and, in point of fact, the Swedish peasantry not only supply themselves with most descriptions of agricultural implements and household furniture, but with nearly all the coarse woollen, linen, and cotton goods, required for their ordinary use. No foreign or factory-made goods, however cheap, can supersede or materially interfere with this domestic manufacture; for, as the people would otherwise be idle, its products may literally be said to cost them nothing. Several factories have, however, been established in Sweden for the production of the finer descriptions of woven fabrics, some of which have had considerable success. The

government of Elfsborg is the grand seat of the domestic manufacture of cotton in Sweden. A factory recently erected in this government, driven by water, has 150 power-looms employed in the production of cambrics and shirtings; and it is at present (1842) in the course of being greatly enlarged.

Of the woollen manufactures, that of cloth is the principal. It is well made, chiefly of the wool produced in the country; but being principally intended for domestic use, and the cultivators making most of what they require, the sale is but small. Norrkoping and Stockholm are the towns in which the largest quantities are made. Foreign cloths are prohibited; but the contraband trade is extensive. The manufacture of other woollen stuffs is confined to flannels, serges and bombazines, which were formerly prohibited, but are now imported in considerable quantities.

Handkerchiefs, which form the usual head-dress of the women, form the principal produce of the silk manufacture, though taffetas, gros de Naples, levantines, and ribands, are also produced. The manufacture has been a good deal improved by the introduction of Jacquard looms from France. The manufacture of sail-cloth is increasing. The principal glass factory is at Bromeo, in Westrogothia. Eskelstuna is the principal seat of the hardware and cutlery business, being a sort of miniature Sheffield: fire-arms are made in it at a factory established by government. The quality of Swedish paper has latterly been much improved, and the quantity so much increased that considerable supplies are now sent to Denmark and Germany. The distillation of corn brandy has been constantly increasing since the reign of Gustavus III. In 1772, government, in order, as is supposed, effectually to suppress drunkenness, prohibited distillation; but, as might have been foreseen, the increase of smuggling and clandestine distillation rendered the prohibition useless, and made it be withdrawn. The Swedes are great consumers of ardent spirits. Every proprietor and occupier of land has a right to distil spirits; the size of the still and the amount of the duty depending on the value of the property. Mr. Stevens states that, in 1829, there were 167,764 stills going, which were calculated to make, within the year, about 30,000,000 gallons worth as many rix-dollars, and paying a duty of 434,396 dollars, a little more than a farthing a gallon! As the distiller is not bound to use any particular kind of grain or malt, a considerable quantity of potatoes is used in distillation. We understand that, in this respect, but little change has taken place during the last ten years; but, taking the consumption at only 25,000,000 gallons, it gives, taking the population at three millions, an average annual supply of 8¼ gallons to every individual, young and old, being about three times the average consumption of the people of Scotland. This is not a new habit in Sweden. "Le Suédois est sobre, sur tous les points, a l'exception de l'eau de vie. Cette funeste habitude commence dès l'enfance, et doit être regardée comme une des causes de la dépopulation de la Suède." (Voyage de Deux Français dans le Nord de l'Europe, ii., 422.) A porter brewery is established near Gottenburg, but the demand for its produce is very limited, not exceeding 5000 hhds. a year. Excepting oak timber and hemp, Sweden possesses every material necessary for the construction of ships. Saltpetre, potash, and tar, are among secondary articles of manufacture. There are two establishments for the instruction of persons intended for trade or manufacture, one at Stockholm, and the other at Gottenburg. A school for mining is established at Fahlun. Schools, where gratuitous instruction is given in naviga-

tion, have been established in five of the principal ports, and no individual can be appointed master or mate of a merchantman without passing an examination in one of these schools, and receiving a certificate of his ability properly to discharge the duties of such situations.

ACCOUNT, showing the Quantities of Cotton Goods made in Sweden during the Ten Years ending with 1840, distinguishing those manufactured in regular Factories from those made by the Peasantry, in the Government of Elfsborg:

Years.	Cotton goods manufactured in regular Factories.	Cotton goods made by the Peasantry of the government of Elfsborg.	Total of Cotton Goods produced.
1831	577,099	1,277,749	1,704,795
1832	638,944	2,846,574	3,480,545
1833	945,751	2,506,262	3,452,527
1834	945,192	2,456,875	3,402,071
1835	997,374	2,575,075	3,572,371
1836	915,673	3,109,004	4,025,865
1837	880,093	4,811,563	4,691,766
1838	1,166,863	2,880,544	4,047,247
1839	1,310,249	1,954,528	3,265,776
1840	1,290,822	4,098,562	5,389,354

Observations.—1. The Swedish aln is equal to about two-thirds the English yard.

2. Beside the goods measured by the yard, a considerable quantity of cotton goods, raised by the piece, is annually made, such as shawls, handkerchiefs, waistcoats, &c.

3. By the tariff of 1836, which took effect on the 1st January, 1837, several articles of cotton previously prohibited were allowed to be imported.

4. The tariff of 1835, which came into effect on the 1st January, 1837, gave additional facilities for importation.

Trade.—The trade of Sweden, which, from the situation of the country, must necessarily be of limited extent, has been reduced below even its natural bounds by the policy of the government in endeavouring to bolster up manufactures. Latterly, however, this system has been relaxed; and the trade and industry of the country have both experienced the beneficial influence of the more liberal policy that has been adopted. The exports consist almost wholly of raw produce, of which iron and timber, especially the former, are by far the most important articles next to these are copper, alum, corn, tar, cobalt, &c. The imports principally comprise sugar, coffee, and other colonial products; salt, wines, silk, and wool; cotton, cotton twist, and cotton stuffs; hemp, hides and skins, oil, &c. The foreign trade is principally carried on with Great Britain, the United States, Holland, Hamburg, Denmark, &c. It principally centres in Stockholm and Gottenburg. We subjoin an

ACCOUNT, exhibiting the Official Value of the Exports from, and Imports into, Sweden, and the Amount of Customs' Revenue for the Ten Years ending with 1840.

Years.	Value of Exports.	Value of Imports.	Total Value of Exports and Imports.	Customs' Revenue.		Total
				On Exports.	On Imports.	
	Rix-dollar banco.	Rix-dollar banco.	Rix-dollar banco.	Rix-dollar banco.	Rix-dollar banco.	
1831	13,505,000	12,303,000	25,493,000	428,426	1,717,845	
1832	14,647,000	13,755,000	28,404,000	423,139	2,172,660	
1833	16,963,000	13,886,000	30,569,000	451,004	2,865,408	
1834	15,882,000	14,527,000	30,709,000	425,474	2,906,300	
1835	18,580,000	15,562,000	34,147,000	512,971	2,611,062	
1836	18,834,000	15,537,000	34,371,000	427,761	2,599,152	
1837	17,451,000	16,451,000	33,909,000	328,890	2,903,374	
1838	22,160,000	19,499,000	41,650,000	480,573	3,277,053	
1839	21,018,000	19,363,000	40,381,000	503,345	3,026,916	
1840	20,434,000	18,308,000	38,872,000	722,226	3,855,903	

Observations.—1L. sterling is equal to 12 rix-dollars Swedish banco. The revenue on exports is almost entirely derived from the duty on iron; duty was, in 1840, reduced to half its former amount. The export duty on wool is to cease in 1842.

Roads, Posting, &c.—The main roads to and from Stockholm are generally excellent and well kept; but the cross roads are comparatively neglected. A landholder is bound to keep in good repair that part of the public road which passes through his possessions; but it is needless to say that it is very difficult to enforce this regulation. The system of posting, though affording every facility for the traveller, is onerous on and injurious to the agriculturists. On all the principal routes, post-stations are established every 7 or 10 m. a part, to which the farmers and peasants of the district are compelled to furnish horses and a driver to the next post-station, at a very low rate, for any traveller who may require them. The station-master has the privilege of being the only innkeeper out of the towns; but he also is obliged to keep horses to perform the same duties as those of the farmers on certain days of the week. Severe penalties, and even corporal punishments, are inflicted on the peasantry for any default in the fulfilment of this duty. The rate of hire paid for each horse is equivalent to about 1d. per Eng. mile. (Laing, p. 203–205, &c.)

Canals.—The formation of a system of internal navigation that should connect the Cattegat and the Baltic, has long engaged the attention, and occupied the efforts, of the

people and government of Sweden. Various motives conspired to make them embark in this undertaking. The sound, and other channels leading to the Baltic, being commanded by the Danes, they were able, when at war with the Swedes, greatly to annoy the latter, by cutting off all communication by sea between the E. and W. parts of the kingdom. And hence, in the view partly of obviating this annoyance, and partly of facilitating the conveyance of iron, timber, and other bulky produce from the interior to the coast, it was determined to attempt forming an internal navigation, by means of the river Gotha and the lakes Wener, Wetter, &c., from Gottenburg on the Baltic. The first and most difficult part of this enterprise was the perfecting of the communication from Gottenburg to the lake Wener. The Gotha, which flows from the latter to the former, is navigable throughout far the greater part of its course, for vessels of considerable burden; but, besides other obstacles less difficult to overcome, the navigation at the point called Trollhatta is interrupted by a series of cataracts, about 112 ft. in height. Owing to the rapidity of the river, and the manner in which granite rocks over which it flows, and by the perpendicular banks of which it is bounded, the attempt to cut a channel

nal, and still more to render it directly navigable, presented the most formidable obstacles. But, undismayed by these, on which it is, indeed, most probable he had not efficiently reflected, Polhem, a native engineer, undertook, about the middle of last century, the Herculean task of constructing locks in the channel of the river, and rendering it navigable! Whether, however, it were owing to the insuperable obstacles opposed to such a plan, to the defective execution, or deficient strength of the works, they were wholly swept away, after being considerably advanced, and after vast sums had been expended upon them. From this period, down to 1793, the undertaking was abandoned; but in that year the plan was proposed, which could have been adopted at first, of cutting a lateral canal through the solid rock, about 1½ m. from the river. This new enterprise was begun under the auspices of a company incorporated in 1794, and was successfully completed in 1800. The canal is about 3 m. in length, and has about 6½ ft. water. It has eight sluices, and admits vessels above 100 tons. In one part, it is cut through the solid rock to the depth of 72 ft. The expense was a good deal less than might have been expected, being only about 1,000. The lake Wener, the navigation of which was opened with Gottenburg, is, as already seen, very safe, and is encircled by some of the richest of the Swedish provinces, which now possess the advantage of a convenient and ready outlet for their products.

As soon as the Trollhetta canal had been completed, there could be no room for doubt as to the practicability of extending the navigation to Söderköping. In furtherance of this object the lake Wener has been joined to the lake Wetter by the Gotha canal, which admits vessels of the same size as that of Trollhetta; and the prolongation of navigation to the Baltic from the Wetter, partly by two lakes of equal magnitude with the above, and partly by art, is now completed. The entire undertaking is called Gotha navigation, and deservedly ranks among the first of the kind in Europe. Besides the above, the canal of Arboga unites the lake Hielmar to the lake Maeler; and, since 1819, a canal has been constructed from the latter to the Baltic at Söderstelge. The canal of Strömsholm, so called from its passing near the castle of that name, has effected a navigable communication between the province of Dalecarlia and the lake Maelar, &c. (For further details, see *Coxe*, iv. 253–266, and v. 56–66; *Thomson's Travels*, p. 35, &c.)

Currency, &c.—The currency consists almost wholly of paper, and though, since 1835, bank notes may be freely exchanged for paper, there is little or no demand for the coin. The *rix-dollar banco*, in which all mercantile transactions are carried on, is worth about 20d. sterling; the *gold dollar*, used as the medium of exchange in ordinary transactions, being worth two thirds the former, or 1s. Rix-dollars banco, are exchanged for rix-dollars specie, at the rate of 2⅔ the former for one of the latter; the rix-dollars are divided into 48 skillings. From this circulation vary from eight skillings to 500 dollars banco. As the prejudice in favour of paper money, that, in small towns and remote districts, coins, excepting of copper, to a small value, are often refused as payment.

Measures of Length.—The Swedish foot =11·684 Eng. The ain = 2 feet; the fathom = equal 3 ells; the rod =

Public Finances, &c.—In 1840, the budget of revenue and expenditure was fixed for that and the following years, to the next meeting of the diet, at 10,742,588 rix-dollars banco, or £895,240 sterling. The amount is derived as:

	Rix-dollars banco.
tax and other perpetual revenues	4,566,380
ums, stamps, and other taxes, voted by every diet	6,176,500
Total	10,742,880

The land-tax and other perpetual taxes being fixed, constitute, in fact, a portion of the property of the state belonging to the state, and cannot justly be regarded as taxes. Hence it follows, that the total sum levied in the shape of taxes for public purposes, exclusive of the maintenance of the indelta soldiers (see *post*), and other local amounts to only 6,176,500 rix-dollars banco, or £ sterling; so that, despite the statements of Laing and others to the contrary, there cannot be a doubt that Sweden is at this moment (1842) the most lightly taxed country of Europe.

Government is a monarchy, hereditary in the male line as a representative diet, one of the most ancient states. The king must be a Lutheran, and his person inviolable. He is assisted by a state council, composed of members, including the ministers of justice, foreign affairs, marine, interior, finance, and public worship,

and three councillors. The army and all foreign relations are under the immediate control of the king; but he cannot decide on any matter touching any other branch of government, without the concurrence of the council. He nominates to all appointments, both military and civil; concludes foreign treaties, declares war, and makes peace; and has right to preside in the supreme court, and to grant pardons. The princes of the blood-royal are excluded from all civil employments. The different apartments of justice, war, marine, mines, commerce, &c., are called *colleges*.

The diet, or representative assembly, consists of four separate chambers, consisting respectively of deputies from the nobility, clergy, burghers, and peasants or cultivators; the latter class having acquired the privilege of sending representatives towards the end of the 15th century. Since 1840, the proprietors of iron works have obtained the privilege of sending three deputies to the chamber of burghers to watch over their interests. The king nominates the presidents or speakers of the chambers of nobles, burghers, and peasants; the archbishop of Upsal being president, *ex officio*, of the chamber of clergy. The diet is convened every *five* years, and usually sits for three or four months, but occasionally, as in 1840–41, for a much longer period. The head of every noble family is, by law, a member of the chamber of nobles; but notwithstanding that the nobility includes in all about 13,500 individuals, it is but seldom that the chamber of nobles is attended by above 500 individuals. The clergy have 80 deputies, the burghers 85, and the peasants generally from 140 to 150, chosen by the arrondissements; the deputies for the clergy, burghers, and peasants, receive salaries during the sitting of the diet from their constituents. No new tax or impost can be established without the concurrence of the diet; nor can any modification of the constitution be legally effected without the concurrence of all the chambers composing the diet.

The four chambers deliberate and vote separately, but all questions must, previously to their decision in the chambers, be referred to standing committees chosen at the commencement of the diet, consisting of an equal number of members from each order. In constitutional questions, which cannot be decided in the same diet in which they are raised, the unanimous consent of the four orders is required, but in other matters the decision of three orders is valid. When two orders are opposed to two, the subject, according to its nature, is either dropped, or referred to the decision of a special committee, composed of 30 members of each order. Differences on minor points are adjusted by the committee, to which the matter was originally referred.

In most cases the decrees of the diet must be submitted to the king, who has an absolute veto; and it is a curious circumstance, peculiar to what M. de Pradt called the *semi-constitutional* government of Sweden, that frequently the king has refused his sanction to the resolutions of the diet, and the diet has negatived the proposals of the king, without occasioning a change of ministry, or exciting any deep feeling of animosity on either side. The king used his privilege of the veto to a great extent in negativing measures agreed to by the diet of 1840–41. This anomaly is increased by the absolute legislative power which the constitution confers on the king in all matters of internal administration and police, in regard to which the diet merely presents addresses and petitions expressive of their views and wishes.

" Previously to the diet held immediately subsequently to the revolution of 1809, the nobility enjoyed several valuable privileges and fiscal immunities. These, however, they then wisely surrendered, stipulating only for the general freedom of trade, externally and internally—a stipulation which has not hitherto been fully carried out. The division of the diet into separate chambers, representing particular orders of the state, is, therefore, less objectionable now than formerly, though it be still necessarily productive of considerable inconvenience.

The chamber of clergy, though said by Mr. Laing to be the most enlightened and independent order of the diet, have interests to support that are in many respects peculiar, and which may sometimes, perhaps, be opposed to those of the public, at the same time that they are mostly all more or less dependent on the crown.

The chamber of burghers consists of representatives of the guilds, trades, and corporations of the different towns. These, as every body knows, are possessed of certain franchises and immunities which go to obstruct competition, and, consequently, to enhance the cost of the articles furnished by the privileged class.

The deputies of the peasants represent by far the greater number of the people, though not the greater portion of the property of the country; and are, themselves, seldom in such circumstances as to enable them to act a really independent part. And hence, in consequence partly of the constitution of the diet, which opposes the greatest obstacle to all organic

879

SWEDEN.

changes, how expedient soever, and partly to apparent rather than real inequalities in the privileges of the different orders, a good deal of discontent prevails in Sweden. It is, indeed, hardly possible that the present complicated and vicious system should be able to maintain its ground much longer; and the best way to avoid the recurrence of another of those revolutions so frequent in Swedish history, will be to adopt measures furobvia ting the defects inherent in the existing political organization of the country, and for making the diet a representation, not of class interests, but of the intelligence and property of the kingdom. Still, however, there can be no doubt, notwithstanding the defects in its constitution, that the country has made a rapid progress during the last twenty years, and that there is every prospect of this progress being continued.

Justice.—The 24 *läns* are subdivided into 117 *føgderier*, or districts, each comprising one or more *hørødes*, or cantons. At the head of each län is placed a governor, charged with the civil and military jurisdiction, and the receipt of the revenue. Every canton is under the superintendence of a länsman, who is the executive officer of the administration, and subordinate to a *kronofogde*, or kind of sub-prefect, with authority over four or five cantons. There are 264 courts of original jurisdiction, or hæred courts, with a judge presiding over each. These courts sit three times a year, and 12 peasants are elected by the peasantry of each hæredø, who serve as jurymen for two years. There are three royal judicial courts; viz. at Stockholm, for the upper or N. provinces: at Jonkopping, for the middle, and at Christianstad, in Scania, for the S. provinces. The first has, subordinate to it, seven inferior tribunals, the second six, and the third three. These are the highest courts of appeal from the inferior tribunals, and have alone jurisdiction in all criminal cases affecting life or property, as well as in all affairs above the value of £4. The decisions of these courts are subject to the review of the supreme court of justice, composed of 12 councillors, and presided over by the minister of justice. There are *lagmans' courts*, to which appeal is first made from the inferior tribunals, but they are generally considered superfluous establishments. Questions of divorce are brought before the ecclesiastical courts. Although the forms of justice be little complicated, the decision of the courts are often long delayed. A new civil and criminal code is at present in course of preparation.

State of Crime, &c.—Sweden being almost wholly an agricultural country, with but few manufactures, and only one large town, and having, also, a constitutional government, and a widely diffused system of public instruction, it might be expected, *à priori*, that it would exhibit a high state of moral feeling, and a remarkable paucity of crime. Such, however, we regret to say, is far from being the case; and though there can be no doubt that the representations as to the depravity and immorality of the Swedes, given in Mr. Laing's work, are far too highly coloured, still it must be confessed that crime and immorality prevail to an extent not easily accounted for. According to the official returns published in the *State Gazette*, the number of persons prosecuted for criminal offences before all the Swedish courts, in the year 1835, was 26.273 ; of whom 21.262 were convicted, 4915 acquitted, and 96 remained under examination. In 1835 the total population of Sweden was 2,983,144 individuals. In this year, therefore, 1 person of every 114 of the whole nation had been accused, and 1 in every 140 persons convicted, of some criminal offence. By the same official returns it appears, that in the five years from 1830 to 1834 inclusive, 1 person in every 49 of the inhabitants of the towns, and 1 in every 176 of the rural population, had on an average, been punished each year for criminal offences. In 1836, the number of persons tried for criminal offences, in all the courts of the kingdom, was 26,925 ; of whom 22,291 were condemned, 3688 acquitted, and 945 under trial or committal. The criminal lists of this year are stated to be unusually light; yet they give a result of 1 person in every 112½ of the whole population accused, and 1 in about every 134 convicted, of criminal offences ; and, taking the population of the towns and the rural population separately, 1 person in every 46 individuals of the former, and 1 in every 174 individuals of the latter, have been criminally convicted within the year 1836." (*Laing's Sweden*, p. 109, 110.)

This certainly is an immense amount of crime ; but, when examined, it is found to be far less than it appears to be. In Sweden, the police interferes with every thing ; and offences of the most venial kind, and which, in fact, cannot with any propriety be called crimes, are, notwithstanding, punished as such, and appear in the list of criminal cases. Thus, if the peasantry neglect to repair the roads of the parishes to which they belong, or if they neglect or delay to bring horses, when required, to the posting stations, they are subjected to fine and imprisonment, and included in the list of criminals. In towns, in like manner, the neglect of sweeping chimneys, repairing and cleaning streets, &c., are reck-

oned criminal offences ; as are drunkenness, adultery, abuse of a parent, or of a wife or husband, and so on. Hence, it is obviously necessary, in order to the [...] of any thing like a fair comparison between crime in Sweden and other countries, to strike off from the list of crimes all that numerous class guilty only of petty offences not noticed except in Sweden, and to include those only that are included in the criminal returns of the countries with which it is compared.

But, notwithstanding this deduction, the extent of crime in Sweden, though nothing like what it appears to be at first sight, is still unusually great. This is shown by the following specification of the criminal offences committed in Sweden in 1837, 1838, and 1839, omitting all of minor importance, or that can in any way be regarded as cases of police :—

	1837.	1838.	1839.
Blasphemy	1	4	1
Murder . by violence .	25	24	23
— (by poison, &c.)	4	7	4
— (with knife, &c.)	0	1	1
— (child) .	13	17	11
— (abortion)	8	15	15
Arson . . .	3	3	5
Burglary and highway robbery	6	8	9
Sacrilege . .	3	13	15
Abominable offence .	11	6	1
Incest, &c. .	12	9	17
Perjury . . .	11	17	4
Forgery .	130	157	130
Rape . .	8	9	1
Total . .	235	299	284

In addition to the above, there were in 1837, 2456 cases of theft ; in 1838, 3290 ; and in 1839, 2814 do. But there are no means of distinguishing in these cases between petty or police cases and those of a graver description. (Art. on *Sweden*, Foreign Quarterly Review, No. 55, p. 169.)

The frequency of forgery in so poor a country is a consequence of the general use of paper money, which supplies at once the greatest temptations to and facilities for the commission of the crime. On the whole, however, it is very difficult to give any satisfactory explanation of the causes of the many crimes, some of them of a very atrocious kind that take place in Sweden. A considerable influence has been ascribed, and with justice, to the obligations imposed on peasants, without any regular occupation or trade, of finding sureties, or bondsmen, to guarantee their payment of the taxes due to government, or else of their being subjected to imprisonment. Great stress has also been laid on the wretched and crowded state of the prisons and houses of correction, in which, as already seen, great numbers of persons who have committed the most trifling offences, are shut up with the greatest criminals. The prevalence of dram-drinking is, no doubt, also a most prolific source of crime ; as is the interruption given to the labours of the peasantry, and the idle habits generated by the posting system ; and, more probably than any other, the increase of poverty arising out of the continued subdivision of the land, and the want of an efficient system for the support of the unemployed and necessitous poor.

The proportion of illegitimate to legitimate births is higher in Sweden than in most other countries. Mr. Laing has given some statements illustrative of what he considers to be the cause of this state of things. We doubt, however, whether they be entitled to much weight ; and, in point of fact, a good many of the unions which give rise to illegitimate births, are really equivalent to a species of marriage though without the sanction of the law. A good deal, also, is to be ascribed to the establishment of foundling hospitals in the capital and other large towns. But, whatever may be its causes, this demoralization is not, as Mr. Laing seems to have supposed, of recent origin. " Il y a beaucoup de libertinage dans les grandes villes ; il commence quelquefois avant l'age de 12 ans, et il est poussé à l'excès jusqu'à l'âge de 30. Alors les jeunes personnes deviennent sages, c'est à dire qu'elles n'ont plus d'un amant ; et après quelques années, elles se marient, ort avantageusement pour l'ordinaire : les hommes ne font nulle attention à la vie antérieure." Voyage de Deux Français, ii. 422.)

It is only, however, by attending to the statistical returns of committals and convictions, that a traveller in Sweden is made to suspect the existence of any considerable amount of immorality. " Whatever," says Mr. Laing, " may be the want of morals in this country, there is no want of manners. You see no blackguardism, no brutality, no revolting behaviour. You may travel through the country, and come to the conclusion that the people are among the most virtuous in Europe." Mr. Laing further tells us, that though he travelled slowly across the country, stopping every night or ten miles at the houses at which the people are supplied with spirits, he only saw one party of peasants a little tipsy, but by no means drunk. We suspect, however, that Mr. Laing's optics had been somewhat defective, or that he had

896

been too squeamish to look into the places or join the parties where intoxication was most likely to prevail. At all events, other and later travellers tell a very different story. There cannot, in fact, be a doubt of the frequency of drunkenness, notwithstanding it is liable to be punished by a fine of about 5s., or six days' hard labour in the house of correction !

Army.—The Swedish army comprises three different kinds of troops ; viz. enlisted soldiers, always on pay and duty, *indelta* soldiers, and the conscription, or local militia. The numbers of the two first are seen by the following table :—

Enlisted Troops.			Indelta Force.		
Horse Guards	.	1,000	Cavalry	. .	7,000
Artillery	.	5,416	Infantry	. .	94,800
Infantry	.	2,930			
Total		9,346	Total	. .	59,846

The militia is roughly estimated at about 95,000 men. The indelta system, which is peculiar to Sweden, originated with Gustavus Adolphus, was permanently established by Charles XI., and has continued, with some trifling modifications, in full operation to the present day. "To understand it fully, it must be borne in mind that the whole of Sweden is divided into military districts or provinces, each of which is bound to contribute a certain number of men to this branch of the national force. Each holder of as much crown land as forms a *hemman* is bound to provide a man, to whom he assigns a croft of land, with a cottage, cowhouse, and barn, and an annual money allowance of about £1 8s., one suit of rough clothes, and two pair of shoes. The croft is cultivated by the soldier himself while at home; but during his absence on service with the army at the annual reviews, or on any government employment, it is cultivated by the landholder for behoof of the family. When the soldier dies, his widow and children transfer the house, &c., to his successor, whom the landholder, under a considerable penalty, is bound to provide within three months. To furnish a cavalry soldier with his horse, &c., two or three hemmans are united ; but both in regard to cavalry and infantry, the provinces are divided in such a manner that the colonel of each regiment shall have his farm (also provided in the way just explained) as nearly as possible in the centre of the regiment ; a captain in the centre of his company ; and so down, through the lowest non-commissioned officers. The farms occupied by officers are large and valuable. The landholders are bound to transport the men, with their baggage, to the annual reviews, and to allow them so much a day for their expenses. Government furnishes the uniforms, and in time of war gives the men higher pay, which is afterwards raised from the landholders. In time of peace these soldiers are turned to excellent account, by employing them on roads and other public works; and when not required for these purposes, they are bound to labour for the respective landowners, at the current rate of daily wages. The number of officers in his corps, as indeed in the whole Swedish army, is unusually small, there being only one officer to about every 40 men, while in France and Austria, there is an officer to very 12." (*Bremner*, 408.) Sundays are the usual days f inspection. Mr. Bremner and other travellers speak in he highest terms of the appearance of the Swedish roops.

The militia consist entirely of foot soldiers, provided with lothing, arms, &c., by the government. The artillery train composed of about 230 pieces of various calibre. The lief arsenals are at Stockholm, Gottenburg, and Christianstad. The principal fortresses are, Wanas, on the lake Zetter ; Waxholm, near Stockholm ; Carlscrona, and Christianstad. In the island of Gothland, where there are no nds fit for the maintenance of the troops, all the male inhabitants between the ages of 20 and 50 may be called on to ke arms in defence of the island, if attacked.

The annual expense to the country of an indelta regiment 1200 men, amounts to about £8500 sterling. The whole st of the army and fortresses. exclusive of the maintenance of the indelta troops, is fixed in the budget of 1840 at 06,110 rix-doll. banco, equivalent to about £342,000 a ar.

Navy.—The naval force of Sweden consisted in 1840 of ships of the line, 8 frigates, 8 brigs, &c., and 247 gunits, 3 royal yachts, and sundry steam-vessels. The permanent seamen at command of the government may amount about 8000 men. They are maintained in the same way the indelta troops, by assignments of lands. Together h conscripts, the whole naval force may be augmented to ut 24,000 men. The Swedes are excellent sailors, and ectally skilful in the management of small craft. The f naval stations are Carlscrona, Stockholm, and Gottung. The annual expense of the navy amounts to 1,430 r. doll. banco, or about £110,000.

he Religion of the state and of nearly all the inhabit-

ants, is the Lutheran ; there being only about 2000 Catholics, and under 1000 Jews. There is one archbishopric, that of Upsala ; and 11 bishoprics. The functions of public worship are exercised by about 3080 ecclesiastics. The higher order of the clergy are nominated by the king from lists presented by each diocese : the election of curates and others of the inferior orders, is left to the people at large. The revenues of the clergy generally are derived from church lands, &c.: the bishops receive, in addition, a tithe on corn, and one from the inhabitants of the four or five parishes surrounding the episcopal residence. The revenue of the archbishop of Upsala does not exceed £800 a year. The richest bishopric, that of Linkopping, is worth about £560 a year. The bishop of Hernosand has scarcely £940 a year. The clergy are an important body.

All sects are tolerated in Sweden, but with this important restriction, that Lutherans only can be advanced to any employment under the state. According to Laing, the people, generally speaking, are extremely superstitious.

The churches are generally well kept, and great attention is paid to the outward forms and ceremonials of religion. Much more liberality is shown towards Jews in Sweden than in Norway; and there are synagogues at Stockholm, Gottenburg, Norkopping, and Carlscrona. A dissenting sect called *Läsere*, or readers, has lately become very numerous in Lapland and the N. parts of the country. " In Sweden, generally, all kinds of amusements begin the moment that public worship is over—in the country, dancing and drinking; in the capital and large towns, theatres, equestrian exhibitions, rope-dancing, balls, &c. In fact, the Swedes appear to regard the sabbath as terminated with the service of the day; but to atone for shortening it so much, they commence its observance, at least in the rural parishes, at six o'clock on the Saturday evening. As soon as that hour strikes, all week-day labour ceases, the whole family clean themselves, and the devotions of the evening are begun." (*Bremner's Excursions in the North of Europe*, ii. 229.)

Public Instruction.—Elementary instruction is in a very advanced state in Sweden. Every adult person must give proof of ability to read the scriptures before be can exercise any act of majority ; and notwithstanding the dispersion of the population, it is said that there is not one individual in 1000 of the adult population unable to read. Parents in the humblest circumstances are all able to give instruction in reading and writing to their children. No qualification is required in a teacher by the local authorities other than good character, it being left to the public to decide as to the capacity of the teacher, and the merits of his modes of instruction.

There are two universities, viz. those of Upsal and Lund, at either of which the instruction is of a very superior description. Subordinate to these are the gymnasia or provincial high-schools, in which are taught the branches of education necessary for the students before entering the universities. In 1830 there were 3060 establishments for elementary instruction, attended by 11,195 pupils, of whom 572 belonged to the class of nobles, 1410 to the clergy, 3499 to public functionaries, 2999 to burghers, and 2815 to the peasantry. The budget for 1842 and subsequent years includes the sum of £82,407, appropriated for the use of the ecclesiastical department, universities, schools, &c.

An academy for perfecting the Swedish language was founded by Gustavus III. in 1786, and a royal academy of sciences originally established by Linnæus. There are special schools for the military and naval service, and others of history, antiquities, *belles-lettres*, &c.

Public Press.—The press is free by law, every man being responsible for what he publishes. In 1812, however, a temporary power for the seizure of periodical publications was granted by the diet, and has been since continued, notwithstanding the efforts to obtain its abolition. Upwards of 80 newspapers are published in the kingdom, 19 of which are issued in Stockholm. Several, however, consist almost entirely of advertisements, which, as they pay no tax, are very numerous, and are employed as the means of transacting business among even the lower orders. In all offences of the press the same jury officiates both as grand and petty jury, there being no revision or appeal from the first decision of the court.

Arts, &c.—The arts and sciences have been successfully cultivated in Sweden. Antiquities formed the first objects of national research ; but their study was superseded in the time of Linnæus and Scheele by that of natural history and chemistry. The reign of Gustavus III. was the most flourishing period of the arts and literature. The Swedes annually import from £6000 to £7000 worth of foreign books, mostly French, English, and German. Among distinguished authors and men of science, Sweden has produced Linnæus, Tycho-Brahe, Scheele, Bergman, Puffendorf, Berzelius, &c. The taste of the ancient Scandinavians for music appears in the present day to have descended to only the higher and middle classes. At Stockholm there is an opera, which, as well as the theatre at Gottenburg, is regularly open during a part of the year.

Races.—With the exception of a few Finns and Laplanders, in the more northerly parts of the kingdom, the inhabitants of Sweden are wholly of Gothic descent. The Finns, however, are supposed to have at one time occupied the whole country, and to have been driven to the forests and fastnesses of the north by an irruption of Goths some centuries before our era. And, whatever truth there may be in this theory, it is, at all events certain, that as no irruption of any other tribe has taken place into Sweden since the supposed Gothic invasion, the blood of the Goths must be found there in a state of comparative purity. The description of the Germans given by Tacitus might, indeed, be applied to the Swedes of the central and southern parts of the kingdom, who are a tall, robust, fine race of men, with fair complexions, light hair, and blue eyes. But to the N. of 62½° or 63° lat., these characteristics begin to disappear; light hair becoming uncommon, and the complexion being frequently brown and even tawny.

During the disastrous period from 1800 to 1810 there was a progressive diminution of the pop.; but since the accession of his present majesty, a great change for the better has taken place. The pop., which amounted to 2,584,690 in 1800, had increased to 3,109,772 in 1839, being an increase of 595,000. And the produce both of corn and potatoes having increased still more rapidly, it follows that the pop. is not only increased in amount, but, which is of more importance, has acquired a proportionally greater command over necessaries and conveniences.

Historical Notice.—The early history of Sweden is obscure, and has little interest, at least to foreigners. The Swedes being discontented with their king, Albert of Mecklenburg, who had been raised to the throne, in 1365, Margaret, Queen of Denmark, styled the Semiramis of the North, a princess of extraordinary talent, availed herself of the opportunity to establish her authority in Sweden. In this object she was completely successful; and by the famous treaty of Calmar, concluded in 1397, the three kingdoms of Denmark, Sweden, and Norway were united under the sway of Margaret. But the Swedes speedily became dissatisfied with this union; and the cruel and tyrannical proceedings of Christian II. excited a rebellion that terminated in the emancipation of Sweden. The famous Gustavus Vasa led the Swedes in their struggle for independence. He hoisted the standard of revolt in 1590: and having entered Stockholm in triumph in 1523, was raised by the unanimous suffrages of his fellow citizens to the throne. Gustavus, who subsequently introduced the Protestant religion, died in 1560, in the 70th year of his age, and the 40th of his reign. "Equally great as a legislator, a warrior, and a politician, he distinguished himself in every station; whether we consider his cool intrepidity and enterprising spirit, his honest integrity, and political foresight, his talents for legislation, his attachment to letters, and encouragement of learning, his affability, and his solid and enlightened piety. These great qualities, set off by a graceful and majestic person, and heightened by the most commanding eloquence, drew general esteem and admiration; and it may be justly said of him, that the most arbitrary monarch never exercised a more unbounded sway over his vassals, than Gustavus possessed from the voluntary affection of his freeborn subjects." (*Coxe*, iv. 158.)

Eric, the son and immediate successor of Gustavus, manifested symptoms of that insanity, which, unhappily, has since been exhibited on more than one occasion by the princes of the house of Vasa. Gustavus Adolphus, grandson of Gustavus Vasa, ascended the throne in 1611. Under this great prince, who was at once an enlightened ruler and the greatest general of his time, the glory and power of Sweden attained to a maximum. At the outset of his reign he was involved in hostilities with the Russians, the Poles, and the Danes, which he terminated with the most triumphant success, having acquired Ingria and Carelia from the Russians, Livonia from the Poles, with sundry valuable territories from the Danes. These successes, and his reputation for ability and disinterestedness, naturally made him the leader of the Protestant party, in the struggle they had to wage against the power and ambition of the house of Austria. And though his glorious and successful career was prematurely terminated by his death at the battle of Lutzen, in 1632, his exertions were mainly instrumental in bringing about that freedom of religious worship, and that equal distribution of power, established by the treaty of Westphalia. The success that had attended the arms of Sweden under Gustavus, continued to attend them under his daughter Christina, who abdicated the throne in 1564, and his other successors down to Charles XII., who became king in the year 1697.

This extraordinary individual, celebrated alike for his successful exploits and his reverses, well nigh consummated the ruin of Sweden. Inflexible in his resolutions, which

898

were inspired by an ambition that was closely allied to madness, the success that attended his early campaigns made him regard everything as possible, and precipitated him into the most extravagant projects. But the battle of Poltowa (which see) put an end to his career of conquest; reduced him to the condition of a fugitive; and gave Russia a lasting ascendancy over Sweden.[*]

Charles XI. and Charles XII. enjoyed a nearly absolute authority; but the calamities entailed on the country by the folly, or rather insanity of the latter, led, on the accession of his sister Ulrica Eleonora to the crown, to the enacting of stipulations, by which the royal authority was very materially circumscribed. It was, however, again enlarged in the year 1772.

Gustavus the III. having been assassinated in 1792, was succeeded by Gustavus IV., then a minor. As soon as this prince had been declared major he embroiled himself in hostilities with France, from which Sweden certainly had nothing to fear. He next engaged in a quixotic contest with Russia; and when the latter had overrun Finland and was threatening an attack on Stockholm, he had the unparalleled folly to reject the assistance of 10,000 English troops who had arrived at Gottenburg! Under these circumstances the dethronement of the king became indispensable to the safety of the state; and this was effected by a bloodless revolution in 1809, when his uncle, who took the title of Charles XIII., was raised to the throne, prince Christian of Holstein-Augustenborg being, at the same time, declared crown prince and successor. On the premature death of the latter Marshal Bernadotte, prince of Ponte Corvo, was elected successor to the crown by a diet held at Orebro in 1810; and having accepted the honour he soon after arrived in Sweden, of which he became king on the death of Charles XIII., in 1818.

There can be no question that the revolution which brought Marshal Bernadotte to Sweden has been of vast advantage to that kingdom. The taint of insanity in the princes of the house of Vasa, even had it been less obvious than in the cases of Charles XII. and Gustavus IV., was quite sufficient to justify a change of dynasty. And if great services, a mild, equitable, and enlightened system of government, and an unblemished private character, be any title to the esteem and affection of a people, few princes have a better claim than Charles-John to the esteem and regard of their subjects. (Exclusive of the works referred to in the course of this article, we have derived much valuable information from private parties in Sweden.)

SWEDEN, p. t., Monroe co., N. Y., 227 m. W. by N. Albany, 386 W. Drained by Salmon creek. It contains the village of Brockport, on the Erie canal in its N. part, which has about 350 dwellings, and 2000 inhabitants. The town contains three churches, a Presbyterian, Methodist, and Baptist, one store, two saw-mills; 10 schools, 694 scholars. Pop. 1834.

SWEET SPRINGS, p. v., Monroe co., Va., 200 W. Richmond, 256 W. This is one of the oldest watering places in Virginia. The water resembles the Bristol Hot Wells, England. It has accommodations for 300 persons. The waters are slightly acidulous in taste, and sparkle like champaign, and are efficacious in many complaints. The temperature is about 70 of Farenheit. It sends forth a large stream filling two plunging baths, and turns a mill, 20 yards from its source.

SWINEMUNDE, a town of Prussia, in Pomerania, on the E. coast of the island of Usedom, on the middle mouth of the Oder, or rather of the lagoon, or haff, which receives it previously to its falling into the sea, lat. 53° 55′ N., long. 11° 15′ 15″ E. Pop. 3700. It is the outport of Stettin, all vessels destined for the latter, that draw more than 7 or 8 ft. water, being obliged to load and unload by means of lighters at Swinemunde. Formerly there were not more than 7 ft. water over the bar at the river's mouth; but it it has recently been so much improved by dredging, the construction of piers, &c., that vessels drawing from 19 to 21 ft. water, come to the quays of Swinemunde, and its port is now the very best on the whole S. coast of the Baltic. In 1840 there arrived at the port 1744 vessels of the aggregate burden of 170,348 tons, the value of the imports for the same year being estimated at £1,228,900. (See STETTIN.)

SWITZERLAND (an. *Helvetia*, including part of *Rhætia*), an inland and mountainous country of central Europe, having Germany on the N. and E., Italy on the S., and France on the W. It lies principally between the 46th and 48th degrees of N. lat., and the 6th and 11th of E. long. Its greatest length N. and W. is 210 m.; greatest breadth N. and S. 140 m. It is a republic formed by the union of 22 confederated states, or cantons, the area, pop., &c., of which are as follow:

---

[*] The life of Charles XII., by Voltaire, is one of the most interesting pieces of biography ever published. See also the admirable character of Charles XII. by Dr. Johnson, *Vanity of Human Wishes*, v. 147.

Cantons.	Area in sq. m.	Population at end of 1837.						
		Citizens of canton.	Citizens of other cantons.	Foreigners.	Total pop.	Pop. to sq. m.	R. Catholics.	Protestants.
urich	665·3	217,219	7,091	6,366	231,576	338·0	about 2,000	about 230,240
erne	2,561·5	366,681	16,029	5,203	407,913	159·3	41,000	356,860
ucerne	587·4	120,512	3,383	626	124,521	219·1	124,468	53
:hwytz	420·8	12,948	537	34	13,519	32·2	13,519	
ri	339·3	39,326	1,128	196	40,650	190·0	40,650	
iterwalden (Upper)	{ 262·8 {	11,857	506	11	12,368	} 86·1 {	22,571	
(Lower)		9,804	388	11	10,203			
arus	279·8	28,217	821	310	29,348	105·1	3,800	25,548
ig	85·4	14,193	1,019	110	15,322	180·0	15,322	
iburg	563·9	83,234	6,010	1,901	91,145	161·8	82,745	8,400
lothurn (Soleure)	254·6	59,214	3,274	708	63,196	248·8	63,196	
ale (city)	{ 184·6 {	10,611	8,481	5,229	24,321	} 355·5 {	6,000	46,500
(canton)		35,900	3,952	1,161	41,103			
iaffhausen	119·7	29,492	1,409	254	31,125	261·5	600	31,125
penzell (Out. Rhodes)	{ 159·8 {	38,701	1,898	481	41,080	} 334·6 {	9,796	40,080
(Int. Rhodes)		9,671	89	36	9,796			
Gall	747·7	144,350	11,139	3,355	158,853	212·6	99,300	58,400
sons	2,968·0	79,601	2,967	1,938	84,506	28·5	24,000	62,000
rgau (Argovia)	502·4	174,992	5,965	1,798	182,755	363·5	87,500	79,800
urgau (Thurgovia)	268·3	78,160	4,463	1,501	84,124	313·9	18,500	72,191
ino (Tessin)	1,034·7	110,445	299	3,179	113,923	110·1	113,923	
id	1,181·9	164,686	14,931	3,985	183,582	155·4	3,000	180,882
ais	1,561·6	73,673	778	1,347	75,798	43·6	75,798	
ifchatel	280·2	40,868	14,534	3,914	58,616	209·3	2,400	54,400
ieva	91·3	38,156	8,677	11,835	58,668	644·6	17,000	41,668
Totals	15,233·0	2,012,580	190,682	54,767	2,188,000	143·7	847,086	1,325,926

ysical Geography.—Simond has not inaptly remarked—
" some idea may be formed of the Helvetic geography
imparing the country to a large town, of which the
ys are the streets, and the mountains groups of con-
is houses." (*Travels in Switzerland,* i. 141.) Indeed,
r the larger portion of Switzerland consists of moun-
comprising many of the highest summits of the Alps.
r is, however, a considerable extent of flat ground in
. W., in the cantons of Friburg, Berne, and Solothurn.
eneral distribution of the great Alpine chains in the
l E. parts of Switzerland has been already noticed in
ucle ALPS, in this *Dict.,* l. 68-70, and need be only
indicated here. Two great parallel chains, enclos-
e Valais, extend between mount Blanc, in Savoy,
he S.W. boundary of Switzerland, and Mount St.
rd. To the most southerly of these chains, called
nnine Alps, belong mount Rosa, 15,150 ft., and mount
, or the Mütterhorn, 14,836 ft. in height. (*Saussure.*)
N. chain, or the Bernese Alps, belong the Finster-
n, 14,035 ft., the Monch, 13,497 ft., the Jungfrau,
ft. in height, &c. (*Tralles.*) E. of mount St. Go-
which may be considered the central point of the
Alps, the Rhætian Alps stretch through the Grisons;
in the N., other chains cover with their ramifications
art of the four Forest cantons (Lucerne, Schwytz,
alden, and Uri). Among the loftiest summits of the
n Alps are the Dödiberg, 11,765 ft.; Muschelhorn,
ft. in height. (*Tralles.*) The Alps of the Forest
 have several summits, as the Gallenstock, Suosten-
.c. the height of which is but little inferior. Most
receding chains have a general direction from S.W.
But the direction of the main ranges throughout
of Switzerland is generally towards the N. or N.W.,
also, corresponds with the general slope of the coun-

e W., however, beyond the lakes of Neufchâtel
'nne, the slope of the surface is towards the N.E.
uintain-system of this part of Switzerland is that
ura (see *ante,* 89.); a system composed of several
ranges of mountains, inclosing very long and narrow
but nowhere rising to 6000 ft. in height. (*Brugu-*
ographie del' Europe; Picot, Ebel, &c.)
rent rivers Rhine, Rhone, Inn, Ticino, and Doubs
names), have their sources in Switzerland; after
he chief river is the Aar. The Aar (see the article)
he foot of the Finsteraarhorn, and runs at first E.,
wards N.W. through the lakes of Brienz and Thun,
9 m. W. Berne, when it turns N.E., and finally
the Rhine, near Klingenau, after a course of about
This river, which drains by far the greater part of
and, receives on the right the Emmen, Wigger,
iss, Limmat, &c.; and on the left the Simmen,
nd Thiele. Unterseen, Thun, Berne, Solothurn,
iu are on its banks. The Thur and Birs, tributa-
ie Rhine, are the only other streams that deserve

rland has a greater number of lakes than any other
ountry of equal extent in Europe, excepting, per-
· grand duchy of Finland. All these lakes are
, and remarkable for the depth and purity of their
ad their great variety of fish. The following is a

STATEMENT showing the probable Area, Height of Surface
above the Sea Level, and greatest ascertained Depth of
the principal Swiss Lakes.

Lakes	Area in sq. m.	Height above Sea.	Greatest Depth.
		ft.	*ft.*
Geneva, or Leman	246	1,290	1,012
Constance	220	1,205	784
Neufchatel	90	1,330	453
Lucerne	44	1,340	909
Zurich		1,342	640
Thun		1,806	756
Brienz		1,852	600
Zug		1,365	1,296?
Bienne		1,418	400
Wallenstadt		1,335	309
Sempach		1,738	

The lakes Maggiore and Lugano are partly, also, in
Switzerland. A notice of most of these lakes will be found
in this work under their several heads, or those of the different cantons in which they are situated.
Switzerland is almost wholly composed of primary and
sedimentary rocks; volcanic formations are rare. The geological constitution of the mountain chains has been already
noticed. (ALPS, JURA, &c.) The central portion of the
Alps consists of granite, gneiss, porphyry, and other primary
rocks, inclosed successively by transition and secondary formations: the Jura is wholly of a remarkable limestone formation. But the region between the Alps and the Jura is
occupied with a peculiar formation of green sandstone,
called *molasse,* or *nagelfluh,* alternating occasionally with
limestone and grauwacke, which extends throughout all
the lower parts of Switzerland into S. Germany. This deposit has been classed with those of a tertiary kind, and
Brongniart and other geologists suspected it to be of a
date posterior to the formation of the Paris basin. For particulars respecting the complicated geology of Switzerland,
the reader may consult the works of Saussure, Humboldt,
Brongniart, Lyell, &c.
The mineral riches of the mountains are but little known
or explored; a few iron mines in the Jura being the only
ones that deserve notice. There are numerous mineral
springs, many of which are resorted to medicinally; and
those at Bex, and others in the canton of Basle, furnish considerable quantities of salt. A few insignificant coal beds
have been met with; but the remaining mineral products of
any value are mostly confined to slate, marble, gypsum,
granite, and other kinds of building stone.
The climate is not only dependent on elevation, but on
the influence exercised by the glaciers in cooling the atmosphere, the openings and exposure of the valleys, &c. But,
on the whole, Switzerland is a much colder country than
its lat. and situation in Europe would appear to warrant.
At Berne, the mean annual temp. is about 45° Fah.; at
Basle, 49°; and at Geneva (1200 ft. above the sea), 46½°
Fah. The climate in the Alpine regions is believed by some
to have become colder in recent times; since the line of
perpetual snow (which here varies from about 9300 to 9800 ft.
above the sea) has certainly descended lower, as compared
with a former period; the glaciers have increased in number; and many tracts are now bare, which were formerly
covered with forests and pasture-grounds.

SWITZERLAND.

The vegetable products of nearly all the different zones of continental Europe are found in Switzerland. The Valais, which has the widest range of vegetation among the Swiss cantons, produces, without culture, nearly 2000 species of plants, exclusive of 1000 *cryptogamia*. In respect of its vegetable products, the country may be classed into seven distinct regions or zones, according to its elevation, as follows:—

Regions.			Productions.
	Ft.	*Ft*	
Lower Region .	—	to 2,100	Limit of the vine. In lower parts of Tessin and Valais, the fig, pomegranate, &c.
Forest do.	. 2,100	— 3,500	Limit of the elm. Buckwheat and maize to 3500 feet. Chesnut ceases at 3000 feet.
Beech do.	. 3,500	— 5,300	Flax, hemp, and barley flourish at 4000 feet, about which Italian poplar, ash, and wild cherry cease.
Pine and Fir do.	5,300	— 6,800	Neither potatoes, apples, or pears grown.
Lower Alpine do.	6,800	— 8,500	Limit of trees of every kind. Includes some good pasture land.
Upper do.	. 8,500	— snow line.	Only shrubs and Alpine plants.
Snow region, above line of perpetual snow.			*Saxifraga oppositifolia*, gentians, chrysanthemums, &c.

There are, no doubt, various exceptions to this table, consequent on difference of lat., position, &c.; but it may be considered as applying to the country generally. (*Wahlenberg, De Veget. in Helv.; Kasthofer, Voyage, &c.; Nots in Foreign Quarterly Review.* i. 910–913.)

Among the wild animals of Switzerland are the bear, wolf, lynx, wild boar, chamois, ibex, deer, and game of all kinds, the marmot, ermine, &c. The chamois is growing scarce. The remarkable variety of the spaniel, so useful, and the breed of which is preserved with such care at the *hospice* of St. Bernard, is of Spanish descent, and frequently attains the height of 2 ft. and the length of 6 ft. The birds of prey comprise numerous species of eagles and vultures, one of which latter, the *lammergeyer* (lamb-destroyer), is said to be the largest native bird of Europe. Salmon, trout, carp, &c. inhabit the lakes. There is only one venemous serpent, the *Coluber berus;* but the insect tribes are more numerous than we might be led to suppose from the rugged and elevated nature of the country.

Property, Agriculture, &c.—Switzerland is a country of small proprietors. An estate of 150 or 200 acres, belonging to an individual, worth perhaps from £90 to £100 a year, would be considered large every where, except in the canton of Tessin, or the Emmenthal, in Berne, and a few other districts, where local customs exist to prevent the too great division of property. Except in certain of these districts, the property of individuals is at their death divided in equal shares among their children, without respect to sex or seniority. In certain cantons, however, as Glarus, landed property cannot be left to any one not a direct descendant, and, falling such heirs, it becomes the property of the government. Indeed, several of the cantons and governments, as that of Berne, and the greater number of the towns, possess a very great extent of landed property. But this is generally apportioned in small lots to the different parties having right to it, or is depastured in common. Switzerland, in fact, is almost wholly a pastoral country : little corn is produced, and the crops are scanty and precarious. Cattle, sheep, and goats constitute the chief riches and dependence of the inhabitants. There are, generally speaking, no farmers ; each proprietor farming his own small portion of land, and the mountainous tracts belonging to the different communities being, as already stated, depastured in common. No foreigners can become possessors of land, nor can native Jews in several of the cantons.

Switzerland has been estimated to comprise 2,250,000 *morgen* arable land, 900,000 do. land in artificial pastures, 190,000 do. vineyards, and 2,400,000 do. forest. (*Neigebaur's Schuts. Allg. Erdkunde,* xxi. 51.) It is only in the canton of Thurgau that corn is produced in any considerable quantity, and even there the home growth does not exceed two thirds the required supply. In Uri no corn is raised ; and in certain parts of the Bernese Oberland wheat is treated as an exotic, and trained carefully over twigs! Rye, oats, and barley are principally cultivated ; maize, however, is grown in some parts in considerable quantities. Beans, lentils, potatoes, turnips, pumpkins, flax, hemp, woad, madder, poppies, and tobacco are also grown, but to an insignificant extent. Vines flourish in several of the cantons ; as on the shores of the lake of Geneva, Vaud, the Valais, Neufchâtel, Aargau, &c. The canton of Neufchâtel has been estimated to produce, at an average, about 700,000 gallons, 400,000 of which, at least, are sold in the neighbouring cantons. The manufacture of sparkling wine, in imitation of champagne, has of late years been on the increase in Neufchâtel, and from 130,000 to 140,000 bottles are now annually exported. Along the banks of the lake of Constance and in the cantons on the Rhine, apple, pear, and cherry orchards are numerous ; and cider, perry, kirschwasser, &c. are made in large quantities.

" Vineyard husbandry," says Mr. Laing, " is altogether a garden cultivation, in which manual labour, unassisted by animal power, scarcely even by the simplest mechanical contrivance, does every operation ; and this gives the character to *all* their husbandry : hand labour is applied to all crops, such as potatoes, Indian corn, and even common grain crops, more extensively, both in digging and cleaning the land, than with us. It is not uncommon to find agricultural villages without a horse ; and all cultivation done by the hand, especially where the main article of husbandry is either dairy produce or that of the vineyard." (*Notes of a Traveller,* p. 355.)

Cows, goats, and sheep, as already stated, constitute the principal wealth of the Swiss, the inhabitants of the manufacturing towns excepted ; or, to discriminate more accurately, the goats, in a great measure, support the poorer class, while the cows supply the cheeses, from which the richer derive their limited wealth. The Swiss peasant is extremely fond of his cow ; and to pass the winter without a cow to care for would be to him extremely irksome. The cantons of Glarus, the Grisons, Appenzell, Berne, Tessin, and the Valais, are those most distinguished for the extent and excellence of their pastures. With little exception, all the land not covered with forests, in the cantons of Schwytz and Uri, is used for the pasturage of cattle. The Alpine pastures are estimated, not by their extent, but by the number of cows they will maintain ; in the lower Alps about three acres, and in the upper from 10 to 15 acres, being the usual average allowed to each. In several of the W. cantons, these pastures are mostly private property ; in the E. they commonly belong to the cantons, being apportioned among the different parishes, each having its *alp*, or common pasture for its cows. Each inhabitant is entitled to a share of this pasture from June to October. Few individuals, however, have such a number of cows as would repay the labour of attending them in summer on the mountains, properties being in general so small as rarely to be able to maintain above five or six cows in winter, and usually, indeed, not more than half that number. The practice, therefore, is for parishes to hire herdsmen and assistants to take care of the cows in summer when on the mountains, and to make the cheese. The owners of the cows get credit daily for the quantity of milk furnished by their cows ; and the produce of the sale of cheese at the end of the season, the expenses being deducted, is divided among them in proportion to the total quantity of milk furnished by each. (*Laing's Notes of a Traveller,* 351.) When let, the mountain pastures are rented from the middle of May to the middle of September, the cattle being kept in the lowlands during the remaining eight months of the year. The term of the lease on which they are let rarely exceeds a summer. Six or eight goats, or about four calves, sheep, or hogs, are deemed as in feeding, equivalent to a cow ; but a horse is reckoned equal to five or six cows, because he roots up the grass. The Swiss cows are very handsome animals, and so valuable that, even in Switzerland, they fetch about £36 each. They yield more milk than those of Lombardy, where they are in great demand. In some parts of Switzerland, with 40 cows, a cheese of 45 lbs. may be made daily ; and in the vicinity of Althorf, they make in the course of 100 days, from the 20th of June, two cheese daily of 25 lbs. each, from the milk of 18 cows. Cheese appears to have been an important article of export from Switzerland from a remote period. Many varieties are made ; the most celebrated of which are those of Schabzieger (see GLARUS), and of Neufchâtel and Gruyère (which see). About 24,000 cwt. Gruyère cheese is said to be annually exported ; and from the middle of July to October, about 300 horses are employed in transporting Swiss cheeses over mount Gries.

The total number of cattle in Switzerland has been vaguely estimated at 800,000, of which 300,000 are cows. (*Schuts.*) They are principally of two distinct breeds : one of large size, with branching horns, mostly inhabiting the lower parts of the country ; and another called the *Oberlander*, a small and inferior species, confined chiefly to the Alps. The best cattle are those of the Simmenthal, the district of Saanen, and the cantons of Fribourg and Solothurn ; the last being especially remarkable for the excellence of its oxen. Cows, as well as oxen, are employed for the plough. The horses, though not handsome, are strong and spirited, and well adapted for cavalry and artillery service, for which they are exported to France and elsewhere. Asses and mules are bred in the S. cantons, where they are mostly used for the conveyance of passengers and merchandise. The stock of sheep is estimated at half a million, and of goats at about the same number. There are two varieties of sheep, one native covered with a coarse ite wool; and the other a Flemish breed, with fine wool dun and yellowish colour. But sheep are not a favourstock, and it is only in a few places that tho improvest of their fleeces, by crossing with merinos, has been ll attended to. flogs are of a large but coarse breed, are principally kept in the Forest cantons.

he urgent necessity of collecting fodder for the supof the cattle during winter makes the collection of s for hay a work of paramount importance. Hence, rever it is found it is carefully collected; and the peashaving crampons on their shoes, to prevent them slipgather hay in places inaccessible to cattle! Grass, hree inches high, is sometimes cut three times a year; n the valleys, the fields are shaven as close as a bowlreen, and all the inequalities clipt as with a pair of ors. In Switzerland, as in Norway, the art of mowing s to be carried to its highest perfection, and no where much skill and attention displayed in harvesting corn nay. But arable and meadow cultivation are both in kward state, owing principally to the pertinacity with n the people reject innovations, and cling to old and tive methods of husbandry. On arable lands fallows place every 4th or 5th year; and the culture of tur'or cattle feeding is unknown. Great attention is, rer, paid to the collection of both solid and liquid re, but they are said not to be very judiciously eni ; and the neglect of irrigation, which might almost where be easily effected, and the rudeness of agrial implements, especially ploughs, are obvious. (*Kas- Voyage dans les Petits Cantons et dans les Alpes nnes.*)

wages of agricultural labour are low; but, on the the rural population may be considered well off. mons says, that "it would require 30s. a week in d, in the neighbourhood of any country town, to aan, his wife, and three children (two of whom shall ve 15 years of age), in the same condition, and in all al respects, on a footing with the average of Swiss peasants having the same family." (*Rep. on Swiss om Weavers.*) We incline, however, to think that a very exaggerated statement; and from all that learn, the agricultural labourers in most parts of d and Scotland have no reason to envy the condithose of Switzerland. The diet of the Swiss conore of porridge than is general in England, and nore milk and cheese is consumed. In other rewine and cider being substituted for beer, the cataf articles of food is much the same among the ry in both countries. The houses inhabited by the ipulation are mostly of wood, but capacious, and d generally with all the articles required for daily he herdsmen who tend the cattle in the Alpine are lodged in *chålets*, or rude log huts formed of iks of pines, and having rarely any furniture, exnecessary dairy utensils.

bis hunting, fishing in the lakes, and boat building, some of the inhabitants in the intervals of agriclustry, but to no great extent. A great number of nigrate to foreign countries, where they act as vaembark in various trades, especially those of cons and bakers; always returning, however, to spend their gains in their native land. The Swiss have for centuries the *condottiers* of Europe; and have een ready to barter their blood and bravery, or nto the military service of any sovereign or repub'hose to hire their services, and to support any wever unprincipled or unjust! There were formerous Swiss regiments in the service of France n; and they are still extensively employed by the vaples, the pope, and the king of Sardinia. A fact, will do anything for money, and nothing t. Hence the proverb, *point d'argent, point de* The foreign mercenaries are extremely economisuch of them as survive return home with all have been able to amass.

ictures.—Notwithstanding the geographical diss of Switzerland, the inhabitants have carried iches of manufacturing industry to a considera- of advancement Various parts of the country ed, been noted since the 13th century for their ires; and despite the competition of this and atries, they are now more extensively carried on

It is impossible, however, from their being ally domestic, to estimate the number of persons or the annual value of the produce of any of manufactures. Most part of the agricultural inire almost wholly occupied during winter at the other branches of manufacturing industry; and ngage in them during the evenings throughout nd when their little patch of land does not reattention. In the districts devoted to handir from one to four looms are usually to be found

in a cottage; the weavers being furnished with the warp, woof, &c., by the manufacturers, to whom they return the woven goods. A line drawn through Switzerland in a N.N.E. direction, across the canton of Fribung, and through the Seanen, pretty accurately divides the German from the French population, each portion of which presents in its manufactures some distinguishing characteristic of its origin. In the French cantons the manufacture of watches, musical boxes, jewellery, &c., are most extensively carried on; while cotton and silk fabrics are the principal employment of the inhabitants of the E. and N.E. cantons. In the Grisons, and the Italian cantons S. of the Swiss Alps, there are few, if any, manufactures. The principal manufacturing cantons of German Switzerland are Appenzell (outer Rhodes), St. Gall, Thurgau, Zurich, Aargau, and Basle; and in the French part of the country those of Geneva and Neufchâtel. Appenzell and St. Gall are the principal seats of the cotton trade, which Mr. Symons has roughly estimated to employ between 600,000 and 700,000 spindles. Raw cotton is imported from England, France, Holland, and Trieste; cotton twist of the higher numbers being almost wholly brought from England. Cotton printing is conducted to some extent in Neufchâtel, where the quality of the water, and perhaps other physical circumstances, conspire to produce brilliant and beautiful colours, in which, indeed, consist the whole excellence of the Swiss goods. The silk manufacture is mostly conducted in Zurich and Basle. The raw silk is drawn from various foreign states, but chiefly from Lombardy, from which country also four fifths of the organzine are brought. The watch-making trade in Geneva and Neufchâtel is of very great importance. (See the articles.) France furnishes to Switzerland about 50,000 movements (*ébauches*) annually, and receives all her fine works and watches from the Swiss manufacturers. The watch-making business in France is in reality of no great importance, most of the artisans being employed in what is called the *repassage* of the works originally produced in Switzerland. (*Bowring's Rep.*, p. 12.) Nearly 190,000 watches are made annually in the elevated regions of Neufchâtel, and many more, besides jewellery, in the canton of Geneva, a large portion being smuggled into France. Linen fabrics, damasks, &c., rivalling those of Belgium, are made in Berne, in which canton, however, there are few manufactures of consequence except those of linen. In addition to the above, woollen cloths, paper, leather, straw plait, iron goods, &c., are made in various places, but many of these have declined in favour of those above specified.

It is easy to see that the foundation of Swiss manufactures is laid in the peculiar distribution of property in the country, and the necessities of the inhabitants. Most families have a small patch of land; but as its cultivation does not occupy half their time, and is besides unable to afford them more than a scanty supply of the most indispensable necessaries, they naturally endeavour to eke out their limited means by engaging in weaving and such like employments. And inasmuch as all they make in these employments is so much clear gain, so much added to the fund on which they must otherwise subsist, it is plain they can afford to work at the lowest possible rate of wages; and, in point of fact, their eulogists, Symons and Bowring, admit that their wages are reduced to the smallest pittance.

The Swiss, from their situation in the centre of Europe, are obliged to pay an enhanced price for their cotton and yarn; so that their whole advantage consists in their being able to reduce wages to next to nothing without being driven from the business. We believe, however, that even this resource will not be found to afford them adequate protection; and that they will be compelled, by the competition of the power-looms of this and other more favourably situated countries, to relinquish all but the finer and more difficult sorts of weaving.

The influence of the circumstances now alluded to has been increased by the wise and liberal policy followed by the government. Switzerland is a country in which the great principles of free labour at home, and free intercourse with foreigners, have been fully carried into practice. No restrictions exist upon the pursuit of any branch of trade. "Industry has been left to itself. Wealth has not been diverted, by legislative interference, from its own natural tendencies. There has been no foolish struggle encouraged by the government between the protected monopoly of the few, and the unprotected interests of the many. Two millions of men have made, under every disadvantage, the experiment of free trade as a system. The consumer has been allowed to go to the cheapest market, the producer to the dearest; and activity is every where visible alike in the trading and agricultural districts. One element only is wanting to make Switzerland the most prosperous of manufacturing nations. Capital is rapidly increasing, by the action of unrestricted, unfettered, unprotected industry."

SWITZERLAND.

(*Bowring's Rep.*, 3-4.) The general prosperity is also favoured by other extraneous circumstances: land is, for the most part, released from tithes and taxes, and the people subjected to very trifling fiscal burthens. In many of the cantons there is no national debt; and some of them, indeed, nearly discharge the expenses of their government out of the interest of the capital accumulated from the surplus revenues of previous years. The establishment of the Prussian Customs' Union occasioned at first serious apprehensions among the Swiss; and some of the N. cantons, as Schaffhausen and Thurgau, were anxious to join it. But a commission, appointed by the confederation in 1833, decided that such a step was altogether unadvisable; and there seems to be no great probability that Switzerland will join the League.

Several of the cantons derive a considerable portion of their revenues from a *droit de péage*, or duty, generally less than 1*d.* a cwt., on goods at the turnpikes on the various cantonal frontiers. A federal duty of from one to two *batzen* per cwt. is paid at the frontier of the republic on goods imported from foreign countries; but goods exported from Switzerland pay no dues, and from the absence of *ad valorem* duties no estimate can be made of the value either of the imports or exports of the Confederation. (*See* GENEVA.) Silk, cotton goods, lace, watches and jewellery, straw plait, cattle and cheese, wine from Vaud and Neufchâtel, liqueurs, herbs, &c., constitute the chief articles of export. In 1834, according to the French accounts, the value of the wine, oil, madder, brandy, salt, fruits, colonial produce, silk, woollen, and cotton manufactures, &c., exported from France to Switzerland, amounted to nearly 30,000,000 francs; for which Switzerland sent back horned cattle, cheese, ribands, linens, thread, and other produce, to the amount of 13,700,000 francs, the balance being, no doubt, liquidated, by the clandestine exportation of watches, jewellery, &c. Wheat is brought principally from South Germany; salt (about 500,000 cwt. a year), leather, hemp, flax, chicory, tobacco, and oils, come principally from Germany, but partly from France; raw cotton, cotton twist, cotton cloths for printing, hardware, iron and other metals, fancy wares, colonial produce, drugs, dyes, &c., from England, and partly also from the Netherlands; woollen stuffs from Belgium, Swabia, and Saxony; silk from Piedmont and Lombardy, &c. Switzerland enjoys a large share of the transit trade between Germany and Italy, Austria, and France. The principal lines are from the lake of Constance across the country to Geneva; from Schaffhausen and Basle to Geneva; but especially from Basle, through Lucerne, mount St. Gothard, to Milan, Genoa, &c. Another line passes from Basle to Zurich, through the Grisons, and across the Splugen, to Lombardy and Trieste. The roads, which are maintained by the cantonal governments, are every where, almost, in good order; but from the rugged nature of the country, carriage is costly as compared with that in the neighbouring states. The carriage of a ton weight of goods over a distance of 30 m. frequently costs 25*s.*

Hardly a country in Europe has so complicated a currency, or set of weights and measures, as Switzerland, nearly each canton having its own, which differs more or less from the rest. The Swiss franc of 10 *batzen* = 1½ French francs, and nearly 1*s.* 3*d.* English. German money is, however, common in the E. and N., and the coinage of Milan in the S., while French money is almost universally met with.

Government.—The 22 cantons are united on equal terms in a confederation for mutual defence; but in most other respects each has its own independent internal administration. The government is wholly republican in every canton, except Neufchâtel, in which the king of Prussia exercises the right of sovereignty. Before 1831, when important reforms took place in the Swiss constitutions, the cantons were divided into aristocratic and democratic; but at present the government is more or less democratic in all. In Uri, Schwytz, Unterwalden, Glarus, Zug, and Appenzell, the functions of legislation and sovereignty are vested in the *lands gemeinde*, or general assembly of the inhabitants, in which every citizen of full age, without any property qualification, has a vote. In the other cantons the legislative power is delegated to the *landrath*, or council of representatives, elected in the primitive or general assembly of the inhabitants, the elective suffrage in which is universal, or nearly so. And in some of these cantons, as St. Gall and Basle (country), the primitive assemblies have a vote on the decisions of their grand council in all matters of an organic character; while in others the people at large have a right to revise their constitution in primitive assembly after a certain number of years. The grand council, where it exists, elects the petty council, or executive power, at the head of which is the burgomaster or *avoyer* of the canton; in other cantons the *landamman*, or *landshauptmann*, is chosen by the general assembly. The

general diet of Switzerland is composed of deputies from all the cantons, from two to three being sent by each, though each canton has but one vote. The deputies vote according to the instructions received by them from their several governments. The diet declares war, concludes peace, contracts foreign alliances, nominates its diplomatic representatives, determines the amount of military force, and governs the expenditure of the finances of the confederation. It meets annually in the chief town of the directorial canton, on the first Monday of July, or at any other time, on the demand of any five of the cantons, being presided over by the burgomaster of the canton in which it meets. Its place of meeting is Berne, Zurich, and Lucerne, every two years alternately. While not sitting, its powers are confided to the *vorort* (grand council of the directorial canton for the time being), to which, in case of matters of importance pending, the cantons add six additional members. In the case of sudden danger from within or without, the canton affected has the right to demand assistance from any adjacent canton, but must also give immediate notice to the vorort, which convokes the diet, and the latter determines whether the expenses of such assistance shall fall upon the canton requiring it, or upon the general body. A federal chancery, composed of a chancellor and a secretary of state, is established in the directorial canton by the diet. The revenue of the confederation consists chiefly of a money contingent of about 548,000 Swiss francs, contributed by the different cantons proportionally to their military contingent, and of the tolls on imports collected by the frontier cantons, and accounted for by them to the diet. The money-contingent of the different cantons varies from 5 to 12 Swiss francs for every man of their military contingent. The number of men each canton furnishes to the federal army varies according to its amount of population and resources. The total armed force of the confederation, according to the scale adopted in 1848, amounts to 64,000 men, viz. 51,800 infantry, 5800 artillery, 5700 cavalry, 700 engineers. Switzerland has, however, no standing army in the strict sense of the word. It is only a militia force, in which every male Swiss must serve for a certain number of years, holding himself at all times ready if called on for cantonal or federal service. Every two years a federal camp is formed for exercise; and at Thun, in the canton of Berne, a school for the instruction of officers is held for two months each year.

The judicial power of the confederacy is very limited: the diet cannot in any way act as a court of justice, even for the purpose of mediating differences between the several cantons, the decision of which is always referred to special arbitration. If the arbitrators cannot agree, the diet may indeed appoint an umpire; but from his decision no appeal lies to the diet. Judicial tribunals, independent of the executive, are established in all the cantons. They are of three grades, and vested equally with civil and criminal jurisdiction. In cases of rebellion, a federal tribunal may be established by the diet, though the offenders may also be tried by their respective cantonal courts. Each canton has its own code of laws, which are, in general, similar to those prevalent in Germany. In some cantons, as Fribourg, Schwytz, &c., the Caroline, or penal code of Charles V., is in force, or was in 1836; but in the interval some new arrangements have been made. Down to that period trials were not public, and juries did not exist. The prisons in most parts of the country are in a bad state; but in Berne, and especially in the French cantons, improved and benevolent systems of discipline have been adopted, and at Geneva the panoptic penitentiary system of Bentham has been introduced.

Religion.—Besides the Catholic and Protestant population, the respective numbers of which are given in the table at the beginning of this article, there are about 600 Anabaptists and 1800 Jews. The latter enjoy no political rights. Many very bigoted provisions are in force with respect to religion in the Roman Catholic cantons. In Basle city, indeed, the Protestants retaliate, no Catholic being able to acquire in it the right of citizenship. But in Valais it is peremptory that all the children shall be brought up in the Catholic faith; and in Uri, Schwytz, and Unterwalden, any native who should marry a Protestant would be deprived of all the rights of citizenship, and banished the canton! In general, according to the religious compact of Aarau, no Protestant minister is permitted to preach in a Catholic canton, and *vice versâ*; though in parishes in which only one church exists, in Glarus, St. Gall, and other cantons, that edifice is used at different times by both Protestants and Catholics. The Catholic are much more numerous than the Protestant clergy, comprising altogether about 6000 individuals (regular and secular), the incomes of many of whom are very considerable. There are four Roman Catholic dioceses: Chur and St. Gall, Basle[a], Lausanne, and Fribourg.

_____
a The titular bishop of Basle resides at Solothurn, and of Lausanne at Fribourg.

mnae, and Sion; the bishops of which are suffragans of the archbishop of Milan. Tessin is in the diocese of the bishop of Como. The government of the Protestant church is considered a branch of the department of public instruction, and as such belongs to the magistrates in the various cantons.

The Swiss Protestant church, as everybody knows, was originally Calvinistic in principle, and is Presbyterian in its form. But the zeal by which the Swiss Protestants were formerly distinguished appears, if we may depend on the statements of Mr. Laing and others, to have wholly evaporated; and it is a singular and not easily explained fact, that, in the Protestant cantons, religion is, at present, less cared for, and has less influence, than anywhere else in Europe. The people are not infidels; but are wholly indifferent to, and, in fact, careless about religion! Mr. Laing has endeavoured, though not, as we think, with any great access, to account for this apathy to religious truths on the part of the Swiss. (*Notes of a Traveller.*) No doubt it is the result of a variety of causes; and is principally, perhaps, to be ascribed to something defective in the system under which the clergy are appointed, and in their training. It is right, however, to state that, notwithstanding the neglect of religion, the Swiss Protestants are eminently moral in their habits; and, though mercenary, are honest and upright in their dealings.

Public education is very widely diffused in the cantons Zurich and Aargau; for, if we may rely on a parliamentary report of 1837, the pupils in their public schools 1832 were to their whole population as one to five. In and and Neufchâtel the proportion was about as one to ; and in Switzerland at large, in 1834, as one to nine; being consequently, in respect of the attendance at school, before Great Britain, the Austrian empire, Belgium, and France. Parents must give their children some sort of education, from the age of five to that of eight years; or their neglect may be punished by fines, and, in some cases, even by imprisonment. (See *Symons's Rep., &c.*) The obstinate refusal of parents to send their children to school however, a rare case; because no child becomes able to exercise the rights of citizenship, or is taken into service of the kind, without having first received the sacrament, which is administered to those only who have attained a certain degree of instruction. In every district there are primary schools, in which the elements of education, geography, history, singing, &c., are taught; and secondary schools for youths of from 12 to 15, in which instruction is given in ancient and modern languages, geometry, natural history, the fine arts, music, calligraphy, &c. In both these the rich and the poor are educated together, the poor being admitted gratuitously. There are normal schools in several of the cantons for the instruction of schoolmasters; who are subsequently paid, by the cantons, salaries varying usually from £10 to £50 a year. Sunday-schools exist in several cantons, and Lancastrian schools at Geneva and Vaud. (*Journal of Education*, vol. iii.) There are superior gymnasia in all the chief towns. Basle has a university, which was formerly much frequented; since 1832 universities have been established in Berne and Zurich. The ordinary expenses of a student at Berne, including living, &c., may, perhaps, be covered by from 600 to 800 Swiss francs a year: the expense at Zurich is rather In the principal towns there are good libraries, and literary associations; and between 20 and 30 newspapers, also magazines, &c., are published in Switzerland, some which are above mediocrity, though the former are said sadly deficient and incorrect in their foreign news and local politics.

Every parish or community is obliged to support its own who become chargeable on their own commune. Only those having the rights of citizenship have a right to missionary support; the privilege not being extended to others, though born in the commune. In most instances, communes have poor-funds administered independently of cantonal government; but if these are not found sufficient, a poor-rate is levied. This rate is always limited; in Zurich, no more than about 2½d. a year from each individual. The number of poor appears to be on the decrease; and it is only in Uri, Tessin, Valais, and one or other cantons, that pauperism is at all common.

One peculiar feature in the condition of the Swiss population, the great charm of Switzerland, next to its natural ', is the air of well-being, the neatness, the sense of property imprinted on the people, their dwellings, their land. They have a kind of Robinson Crusoe in about their houses and little properties; they are daily building, repairing, altering, or improving something about their tenements. The spirit of the proprietor to be mistaken in all that one sees in Switzerland. cottages, for instance, are adorned with long texts of scripture painted on or burnt into the wood in front the door; others, especially in the Simmenthal and Haslethal, with the pedigree of the builder and owner. These show, sometimes, that the property has been held for 200 years by the same family. The modern taste of the proprietor shows itself in new windows, or additions to the old original picturesque dwelling, which, with its immense projecting roof, sheltering or shading all these successive little additions, looks like a hen sitting with a brood of chickens under her wings.

"None of the women are exempt from field-work, not even in the families of very substantial peasant proprietors, whose houses are furnished as well as any country houses with us. All work as regularly as the poorest male individual. The land, however, being their own, they have a choice of work, and the hard work is generally done by the men. The felling and bringing home wood for fuel; the mowing grass generally, but not always; the carrying out manure on their back; the handling horses and cows, digging, and such heavy labour, is man's work: the binding the vine to the pole with a straw, which is done three times in the course of its growth; the making the hay, the pruning the vine, twitching off the superfluous leaves and tendrils, these lighter, yet necessary jobs to be done about vineyards or orchards, form the women's work. But females, both in France and Switzerland, appear to have a far more important *rôle* in the family, among the lower and middle classes, than with us. The female, although not exempt from out-door work, and even hard work, undertakes the thinking and managing department in the family affairs, and the husband is but the executive officer. The female is, in fact, very remarkably superior in manners, habits, tact, and intelligence to the husband, in almost every family of the middle or lower classes in Switzerland. One is surprised to see the wife of such good, even genteel, manners and sound sense, and altogether such a superior person to her station, and the husband very often a mere lout. The hen is the better bird all over Switzerland." (*Laing's Notes*, p. 336.)

If we divide the people of Switzerland according to their language, nearly 1,300,000 speak a German dialect, 450,000 French, and about 123,000 a corrupt Italian: in a large part of the Grisons, the Romansch tongue, bearing a very close analogy to the ancient Latin, is spoken in several dialects. The distinctions of language are the principal among the Swiss: there are few physical differences in the inhabitants of the different parts of the country, except that the natives of the mountainous parts are the more muscular and active. The Swiss are unquestionably a brave people, devoted to their home and their freedom, for the maintenance of which they have often made great sacrifices and exertions. The situation in which they are placed, their scanty means of subsistence, the necessity of husbanding their resources, and the difficulty of increasing them, have made them sober, industrious, and economical; but also, we must say, mean and mercenary. There is nothing they will not undertake, how degrading soever, provided they think they can make money by it. To attain the rank of valet in the family of some foreign nobleman seems the summit of their ambition. Though attached to liberty themselves, 2s. 6d. or 3s. a day will make them flock to the banners of its most inveterate enemies. In this respect, indeed, they have no predilections, and the emperor of Russia and the president of the United States may equally command their services.

"Man and steel, the soldier and his sword,"

continue to be the most marketable of Swiss products. Though attached to their country, they have no relish for its magnificent natural beauties; and though an honest, laborious, prudent, and, on the whole, respectable people, they have little that is amiable or attractive in their character.

After the conquest of Helvetia by Julius Cæsar, the Romans founded in it several flourishing cities, as Aventicum, &c., which were afterward destroyed by the barbarians. On the decline of the Roman empire, it successively formed a part of the kingdom of Burgundy and the dominions of the Merovingian and Carlovingian kings; while the E. part of Switzerland became first subject to the Allemanni, and subsequently it was wholly included in the German empire under Conrad II. in 1037.

The house of Hapsburg had, from an early period, the supremacy over all the E. part of Switzerland; and it preserved its ascendancy till about 1307, when Uri, Schwytz, and Unterwalden, entered into a confederacy for mutual aid against Austria, which compact was confirmed after the defeat of Leopold Duke of Austria, at the battle of Morgarten, in 1315. From 1332 to 1353, Lucerne, Zurich, Glarus, Zug, and Berne, joined the Confederation. Aargau was conquered from Austria in 1415; the abbey and town of St. Gall joined the other cantons in 1451–54; Thurgau was taken in 1460; Fribourg and Solothurn admitted in 1481; the Grisons in 1497; Basle and Schaffhausen in 1501, and Appenzell in 1513. About this time Tessin was conquered from the Milanese; and Vaud was taken from Savoy, by the Bernese, in 1560. The remaining cantons were not finally united to the Confederation till the time of Napoleon; and the present compact, by which all are placed on a perfect equality, only dates from the peace of 1814.

The principal authorities consulted for this article have been *Picot, Statistique de la Suisse*; *Ebel, Manuel du Voyageur, &c.*; *Simon. Voyage en Suisse*; *Hoffman; Inglis's Switzerland*; *Dict. Geog.*; *Bowring and Simone's Reports, &c.*, passim.

SWITZERLAND, county, Ia. Situated in the S.E. part of the state, and contains 216 sq. m. Bounded S. by Ohio river, and drained by small creeks flowing into it. The vine is successfully cultivated. It contained in 1840, 6737 neat cattle, 9672 sheep, 11,254 swine; and produced 76,889 bushels of wheat, 269,385 of Indian corn, 79,531 of oats, 35,375 of potatoes, 55,167 pounds of tobacco, 3148 of sugar. It had 29 stores, one flouring-mill, eight grist-mills, 15 saw-mills, five tanneries, one printing-office, one weekly newspaper; 22 schools, 210 scholars. Pop. 9920. Capital, Vevay.

SYCAMORE, t., Hamilton co., O. Watered by Mill creek and its branches. It contains several villages. Pop. 2207.

SYDNEY, a town of E. Australia, the cap. of the British colony of New South Wales, on a cove on the S. side of the magnificent bay, or inlet of the sea, called Port Jackson, about 7 m. from its mouth; lat. 33° 55' S., long. 150° 10' E. Pop., in 1841, about 30,000. We subjoin the following details respecting the census of that year:

Born in the colony	7,000
Arrived free	17,332
Other free	3,356
Ticket-of-leave holders	207
Convicts in government employ	1,018
Ditto in private service	1,060
Total souls in Sydney	29,973

In the District of Sydney.

Males	20,733
Females	14,744
Grand total in town and district of Sydney	35,507

Sexes.

Married males	6,082
Ditto females	5,891
Married	11,973
Single males	14,651
Ditto females	8,883
Single	23,534

Of the population there belong to the

Church of England	19,903
Church of Scotland	3,565
Wesleyans	937
Other Dissenters	973
Roman Catholics	9,552
Jews	476
Mohammedans and Pagans	1,101

The town stands principally on two hilly necks of land, bounding Sydney Cove on the E. and W., and on the intervening flat ground for nearly 2 m. inland, and would appear from the extent it covers to contain a much larger population than really belongs to it; but the houses in many parts are not more than one story in height, and are generally surrounded by gardens. In the older part of Sydney, termed 'the Rocks,' the streets are comparatively irregular, for, owing to a want of attention at first, they were laid out, and the houses built, according to the views of individuals, without any fixed or regular plan. But latterly this defect has been to a considerable degree remedied in the old streets: and the new ones are systematically laid out. On the left side of the cove many handsome houses rise in successive terraces. The E. peninsula is almost wholly occupied by the government domain, a circumstance

which is rather to be regretted, since the water, being deeper there than on the W. side of the cove, it is better adapted for the erection of warehouses, &c. It has been proposed, however, to construct a public wharf along the E. side of the cove. The government house is an edifice, built, at different periods, by successive governors, and having in front a fine plantation of English oaks and Cape pines, the walk round which forms the favourite promenade of the citizens. The other chief public buildings are the great barracks, occupying one side of the principal square; the convict hospital, a large stone building with verandahs to both stories; a smaller military hospital, convict barracks, the courthouse, police and commissariat offices, custom-house, new jail, &c. Sydney has two English churches, a Roman Catholic chapel, a handsome Gothic building, with chapels for Presbyterians, Wesleyans, &c. Among its educational institutions, are the Australian college and Sydney college, which furnish superior instruction in classics, mathematics, and English literature; a normal institution, with Episcopalian, Presbyterian, and Roman Catholic schools. There are, also, numerous boarding-schools, and other private seminaries, some of which are said to be very well conducted. Several newspapers are published in the town; and many works have been published, the printing and plates in some of which would do no discredit to the London press. It has also a museum, and botanical garden. "The shops in the town are frequently laid out with great taste; they are not, as in America, 'stores,' where every article may be bought under the same roof; but each trade or business has its own distinct warehouse. House rent is high at Sydney, as may be inferred from the fact that building land has been recently sold, in George-street, at the rate of £20,000 an acre! Several commercial establishments are of considerable size; auction-rooms have been lately built by one individual at a cost of £5000, and £30,000 has been expended on one distillery. Large sums have been spent also in erecting steam-engines, mills, &c.; a good theatre has been built on speculation, and the hotels and inns are numerous and excellent." (*Martin's British Colonies*, 424.) The town is partially supplied with water by means of a newly-constructed tunnel, nearly 2½ m. in length.

Sydney is admirably adapted for the capital of a great trading colony. Port Jackson is one of the finest natural basins in the world. It stretches about 15 m. into the country, and has numerous creeks and bays; the anchorage is everywhere excellent, and ships are protected from every wind. The entrance to this fine bay is between two gigantic cliffs, not quite 2 m. apart. On the most northerly, in lat. 33° 51' 30" S., long. 151° 16' 30" E., a lighthouse has been erected, the lantern of which is elevated 67 feet above the ground, and about 345 feet above sea. It is navigable for ships of any burden to the distance of 15 m. from its entrance, or 7 m. above Sydney, up what is called the Parramatta river. Ships come close up to the wharves and stores of the town, their cargoes being hoisted from the ship's hold into the warehouses. Sydney is consequently the emporium of all the settlements in this part of Australia, and has a very extensive trade.

Wool is the great article of export, and next to it are whale oil and whalebone, the produce of the southern whale fishery. Timber is also exported in considerable quantities; but the exports of other articles are inconsiderable.

The colony being much more suitable for sheep pasture than for tillage, but little corn is raised. In consequence grain and provisions form very important articles of import from Van Diemen's Land and other places. Considering the character of a large proportion of the population, one need not be surprised at the circumstance of drunkenness being a prevalent vice, and consequently, that spirits and wines are largely imported. The other great articles of importation are manufactured goods and apparel of all sorts, hardware, earthenware, saddlery, and carriages, &c., from England.

We subjoin some statements illustrative of the trade of Sydney for some years past.

Account of the Value of the Imports into New South Wales, nine-tenths being into Sydney, during each of the Three Years ending with 1840.

Articles Imported.	1838.	1839.	1840.
Liquors of all kinds, or spirits, wine, ale, beer, cider, and perry	£106,516	£290,734	£338,691
Grain, provisions, and other edibles, including sugar, tea, &c.	250,173	470,317	382,149
Other articles of consumption, as salt, soap, candles, tobacco, &c.	71,817	108,069	169,682
Forage	4,503	22,396	4,551
Wearing apparel, clothing, and bedding	481,804	365,855	276,209
Articles for domestic or personal use, as furniture, carriages, plate, &c.	112,239	108,645	192,509
Ditto for intellectual purposes, as books, prints, instruments, printing materials, &c.	54,811	42,946	38,602
Ditto for use in Agriculture, manufactures, trade, and commerce	203,049	948,190	452,596
Coin	38,280	24,445	6,467
Totals	£1,383,750	£1,786,391	£2,482,652

The value of the imports into Sydney has, for a lengthened period, uniformly exceeded the value of the exports,
the balance being met by the expenditure on account of
government, the disbursements of ships visiting the port,
&c. But during the three years ending with 1840, there has
been a great deal of overtrading at Sydney, and the value
of the imports having very greatly exceeded the means of
payment, and the real wants of the colony, much temporary embarrassment and distress have followed.

Among other articles, the imports of spirits, in 1838,
mounted to 1,151,583 galls. ; in 1839, to 1,744,474 do.; and
1840 to 2,260,774 do. The total quantity of wheat im-
orted during the above three years amounted to 539,773
bshels.

During the year 1840, the value of the exports from Syd-
ey amounted to £1,251,544; of which wool and other co-
nial articles amounted to £569,172; the produce of the
sheries to £265,099; and British and foreign goods re-ex-
rted, to £423,452. We subjoin

STATEMENT of the Quantities of Wool, Whale Oil, and
Whalebone exported from N. S. Wales during each of
the five years ending with 1840.

Years.	Wool.	Sperm Whale Oil.	Black Whale Oil.	Whalebone.
	Lbs.	Tuns.	Tuns.	Tuns.
1836	3,611,118	1,052	1,148	79
1837	4,273,715	2,569	1,506	73
1838	5,426,993	1,591½	3,055	174
1839	6,597,961	1,779	1,229	135
1840	7,565,980	1,264	4,298	250

n 1840, wool was valued for export at 15d. per lb. ; sperm
ale oil at £85 per tun, black whale oil at £18 per do.,
whalebone at £100 per ton.

anking has been for a lengthened period one of the prin-
il businesses carried on in Sydney. Four or five joint-
k associations have been established in the town for
ing on banking business, exclusive of several joint-
k banking associations in London which have branches
ydney. The oldest of the native establishments, the
t of New South Wales, was founded in 1816. The
s have frequently divided at the rate of 20 and 28 per
. per annum of profit. Recently, however, there has
a great over-issue of paper, which has been at once a
e and a consequence of the late over-trading. It has
estimated that the paper afloat in Sydney, in March,
and the bills discounted, amounted together to above
00,000 ; whereas, in March 1837, they did not together
d £1,000,000 ! Sydney has also a savings bank, with
rous joint-stock associations for conducting the busi-
of insurance, and steam, gas, and auction companies
been established.

e aggregate number of convicts that arrived in New
Wales from 1828 to 1836, inclusive, was 30,090, and
e emigrants during the same period, 10,634.

e British settlements in New South Wales were ori-
y intended to serve as penal establishments, to which
cts might be transported, and employed in public and
e works, and are still used for that purpose. The first
with convicts arrived in January, 1788, at Botany
but it having been found quite unsuitable for a colony,
ttlement was removed to Sydney. The progress of
lony has been much more rapid than might have been
ated, considering its immense distance from Europe,
eral inferiority of the soil, the prevalence of droughts,
e large amount of convict pop. Its progress has been
nly owing to the boundless extent of its unoccupied
their suitableness for sheep pasture, and the unpre-
ed and extraordinary increase in the stock of sheep,
the exports of wool. In illustration of this increase
y state, that in 1822, the exports of wool from the
amounted to only 152,890 lbs ; in 1825 they had in-
l to 411,600 lbs. ; in 1830 to 899,750 lbs. ; and in 1840
prodigious amount of 7,668,960 lbs. ; and the proba-
that the increase in time to come will be equally

the circumstance of the great majority of the con-
d other emigrants being males, a great disproportion
nays existed between the sexes in the colony. Gov-
t of late years endeavoured to lessen this dispropor-
sending out considerable numbers of young unmar-
ales free of expense. The most conflicting accounts
en received as to the conduct of these females on
iding, and the influence of their immigration on the
It appears to be sufficiently established that the
s, in many respects, fallen short of the anticipations
omoters ; but on the whole, there can be no doubt
measure has been advantageous, and that it has
ot only to increase the population, but to improve
ls of the colony.

he plan was acted upon of selling lands in the co-
ernment has been in the habit of expending the

75

whole, or a large proportion of the revenue derived from
such sale in bounties for the encouragement of immigration.
According to a proclamation issued in the colony in 1840, a
bounty of £38 is granted towards defraying the expense of
every agricultural labourer, carpenter, bricklayer, &c., arriving in the colony with his wife, provided neither be above
40 years of age. Bounties are also paid on the children of
such persons; and on unmarried males and females arriving in the colony. In consequence of this encouragement
a very powerful stimulus has been given to immigration.
In 1840, no fewer than 14,392 persons left the different ports
of the United kingdom for Australia, of whom 7648 were destined for Sydney; and in 1841 the emigration was still greater. (*Sydney Almanac for* 1841; *Private Information with
respect to Sydney; Parl. Returns as to Emigration, &c.*)

SYLHET, a distr. of British India, presid. Bengal, beyond
the Brahmaputra, and chiefly between the 24th and 25th
degs. of lat. and the 91st and 93d of E. long., having N. the
territory of the Cosseahs, and Jynteah, E. Cachar, S. Tipperah, and W. the distr. of Myennusing and Dacca. Area,
3538 sq. m. Pop. in 1822, 1,083,790, it being one of the most
densely peopled portions of the British dominions in the
east. Its borders are mountainous, and on the E. and S.
the mountains rise to the height of about 6000 ft. ; but its
central part, which is flat and intersected by the Barah, and
a great many other rivers tributary to the Brahmaputra, is
covered with rice-fields, &c. Cotton and sugar are raised
in considerable quantities ; and Sylhet produces the finest
oranges and limes throughout British India : they are grown
in extensive plantations, or rather forests, and continue to a
great extent. Chunam, wax, aloe wood, wild silk, and elephants, are among the other chief products; and coal of a
very fair quality has been somewhat recently discovered.
Boat-building is pursued pretty extensively, and Sylhet
shields are articles much prized by the natives of Hindostan. The land is, in general, very much divided : land revenue in 1829-30, 303,516 rupees. Mohammedans are very
numerous in this district. Sylhet, its chief town, and the
residence of the principal authorities, is on the Soormah, in
lat. 24° 53' N., long. 91° 40' E. (*Hamilton; Pemberton, on
the E. Frontier.*)

SYRA (an. *Scyros*), an island belonging to Greece, in the
group called the N. Cyclades, the port of Syra. on the E.
side of the island, 15 m. W. from the greater Delos, being
in lat. 37° 26' 30" N., long., 24° 55" E. It is about 10 m. in
length, N. and S., and 5 m. in breadth. Though rugged and
not very fruitful, it is well cultivated : and the pop., which
in 1825 was not supposed to exceed 4,500, is now (1842) estimated at above 25,000 ! It is indebted for this extraordinary increase of pop. to the convenience and excellence of
its port and its central situation, which have made it a considerable commercial entrepôt. Most part of the trade that
formerly centred at Scio is now carried on here; and the
island has not only received numerous immigrants from
that island, but also from many other parts of Greece. Great
Britain and most European powers have consuls in Syra;
and it also is the principal seat of the Protestant missionaries to the Levant. The town, which is in great part old,
has several new streets and houses, and has an appearance
of great bustle and animation. Pherycides, one of the most
celebrated of the ancient Greek philosophers, the disciple
of Pittacus, and the master of Pythagoras, was a native of
this island. We subjoin an

ACCOUNT exhibiting the Number of Ships, their Tonnage,
and the Value of their Cargoes, that arrived at Syra in
1839, specifying also the different Countries to which
they respectively belonged.

Countries.	Vessels.	Tonnage.	Value of Cargoes.
British . . .	61	10,392	184,963
Greek . . .	292	65,179	188,110
Ottoman . . .	291	8,069	35,257
Russian . . .	75	16,736	19,755
Austrian . . .	85	12,597	19,776
steamers	51	7,943	31,403
French . . .	23	3,309	5,210
Ionian . . .	77	6,399	2,508
Sardinian . .	39	4,407	5,543
Belgian . .	1	100	1,416
Tuscan . .	1	193	1,773
Neopolitan .	4	1,062	15
Hanoverian .	1	900	2,328
Jerusalem .	3	290	1,312
Total . .	1,864	136,443	464,146

SYRACUSE (an. *Syracusa*), a famous city of Sicily, cap.
of an intendency, district, and cant., on the E. coast of the
island, 31 m. S.S.E. Catania ; lat. 37° 2' 58" N., long. 15°
16' 10" E. Pop. in 1831, 16,805; but said to have been considerably reduced during the interval from the influence of
cholera and other causes. The modern city is wholly confined to that small portion of the site of the ancient city included in the island of Ortygia, separated from the main-

land by a fossé, and projecting S. in the shape of a narrow peninsula, inclosing between it and the mainland the noble basin called the Great Harbour, which its security, and the facility of its access, render one of the best ports in the Mediterranean. (*Smyth's Sicily*, 163.) Outside the peninsula is the Little Harbour (an. *Trogilus*). Syracuse is pretty strongly fortified, being defended by a bastioned wall, and other works. The port is protected by the castle of Maniaces, near the S. extremity of the peninsula. The modern city has little except its ancient renown, its noble harbour, and the extreme beauty of its situation, to recommend it. The temple of Minerva has been converted into the cathedral; but the portico and front having been destroyed by an earthquake, are modern, and in bad taste. It has several other churches, with numerous convents, a seminary for the clergy, a college for general studies, a hospital, a lazaretto, extensive barracks, a museum, and a public library. There are some remains of the temple of Diana, but they are unimportant. The famous fountain of Arethusa (*see* ARETHUSA), the great glory of ancient Syracuse, is now defiled by the admixture of the sea, and is degraded into a sort of washing-tub for the poorer classes of town's-women. The commerce of the city, the principal source of its wealth in antiquity, is also quite inconsiderable; its exports consisting only of trifling quantities of oil, corn, fruits, hemp, saltpetre, &c. "Its streets are narrow and dirty; its nobles poor; its lower orders ignorant, superstitious, idle, and addicted to festivals. Much of its fertile land is become a pestilential marsh; and that commerce which once filled the finest port in Europe with the vessels of Italy, Rhodes, Alexandria, Carthage, and every other maritime power, is now confined to a petty coasting-trade. Such is modern Syracuse! Yet the sky which canopies it is still brilliant and serene; the golden grain is still ready to spring almost spontaneously from its fields; the azure waves still bent against its walls to send its navies over the main; nature is still prompt to pour forth her bounties with a prodigal hand: but man, alas! is changed; his liberty is lost; and with that the genius of a nation rises, sinks, and is extinguished." (*Hughes's Greece, &c.*, i. 55, 8vo edit.)

The ancient Syracuse was founded by a colony from Corinth, about *anno* 735 B.C. Its advantageous situation, and the commercial enterprising spirit of its inhabitants, speedily raised it to the highest distinction. Cicero calls it the greatest and most beautiful of Greek cities: *Urbem Syracusas maximam esse Græcarum urbium, pulcherrimam omnium, sæpe audistis.* (In Verrem, lib. iv. cap. 52.) As soon as it had outgrown the limits of the original city, which, like the modern, was confined within the island of Ortygia, it began to extend towards the N., covering, when in its zenith, a large triangular space; which, rising precipitously from the sea on the one hand, and the plains to the W., on the other, admitted of being easily fortified. This new city terminated on the N. in the hill of Epipolæ, which, however, was not included within it till the time of the elder Dionysius, who constructed at that point the fortress of Hexapylon, the vast ruins of which still attest its former strength and importance. The city was defended partly by lines of rocks, and partly by strong walls. Its circuit is estimated by Strabo at 180 stadia. or about 20 Eng. m.; and supposing that the sinuosities of the walls were followed, this statement is probably not very wide of the mark. Among the advantages of the situation chosen for the site of the new city, was its inexhaustible supplies of fine freestone; which, though soft, and easily wrought in the quarry, became, by exposure to the air, sufficiently hard.

The space included within the walls of the new or N. city comprised, 1. the quarter of Acradina, the largest and most populous of the whole, adjoining the island of Ortygia, having E. the sea; it contained the temple of Olympian Jupiter, the forum, the prytaneum, &c.; 2. the quarter called Tyche, from its temple to Fortune (Τύχη), lying N.W from Acradina: and 3. the quarter called Neapolis, or the New City, from its being the last built: this, which was the most westerly portion of the city, and was bounded in part by the Great port, contained a spacious theatre, cut in the rock, upon the slope of a hill; and two temples, one dedicated to Ceres, and one to Libera or Proserpine. (*Cicero, ubi supra.*)

Among the existing remains of Syracuse, the most extraordinary, perhaps, are the *latomiæ*, or prisons. These are immense excavations cut in the solid rock to a great depth, with steep overhanging sides, whence all egress is impossible. They appear originally to have been quarries (whence their name), and to have been subsequently formed into prisons. They have been forcibly and admirably described by Cicero:— *Latomias Syracusanas omnes audistis, plerique nostis; opus est ingens magnificum regum ac tyrannorum; totum est ex saxo in mirandam altitudinem depressum, et multorum operis penitus exciso. Nihil tam clausum ad exitus, nihil tam septum undique, nihil tam tutum ad custodias, nec fieri nec cogitari potest.* (In Verrem, ult. cap. 27.) In the

N.W. angle of the *latomia* of the Neapolis is the famous cavern called the ear of Dionysius. It runs into the heart of the hill, in the form of the letter S, the sides being dilated quite smooth, had the roof gradually narrowing to a point, along which runs a groove, which collected, as is supposed, the sounds of the voices of the prisoners. It derives its name from the popular belief that Dionysius was accustomed to incarcerate in it those he supposed inimical to his authority; and that, by applying his ear to one end of the groove, and listening to their conversation, he ascertained whether his suspicions were well-founded. There appears, however, to be little or no foundation for this story; though, from the care bestowed on its construction, it was evidently have been intended for some special purpose.

The *latomiæ* on the hill of Epipolæ were selected as the place of confinement for the miserable remains of the vast armament fitted out by Athens for the reduction of Syracuse. About 7000 men are said to have been shut up in this prison, exposed alternately to the beams of a vertical sun, rendered more intolerable by its reflection from the surrounding rocks, and to the chills of the evenings, with insufficient supplies of food, and without any means of preserving cleanliness, or even of escaping from the contact of the sick and dead. Every hardship was accumulated on the heads of the unhappy sufferers; till at length, after an interval of above two months, most part of those that survived were brought forth, to be sold as slaves. (*Thucyd.*, lib. vii. *ad finem.*) The *latomiæ* were also used by Verres for the imprisonment, not merely of Syracusans, but of Roman citizens.

The catacombs, in the Acradina, are of vast extent, and may be truly called a city of the dead. They consist of a principal and several smaller streets, all excavated in the rock, with deep contiguous recesses on each side, containing cells for the reception of the dead. Various theories have been formed as to the era of the formation of these vast subterranean excavations, which, no doubt, belong to a very remote antiquity.

On the whole, however, considering the great extent of the city, and the number and magnificence of its public buildings, the continental portion of Syracuse, with the exception of the *latomiæ* and catacombs, and some remains of the walls, and of aqueducts, has very few monuments of antiquity of which to boast. Swinburne (ii. 334) and Hughes (i. 82) express their astonishment at the almost total disappearance of all vestiges of the great public and other buildings, with which the city was once filled. This, however, is not inexplicable: the sea has undermined a portion of the walls of Acradina; and the perishable nature of the stone of which the city was built, added to the influence of earthquakes, the ravages of war and of barbarians, and the accumulation of rubbish, have made Syracuse, like Carthage, *non agnoscenda propriis ruinis*. Among the ruins of some baths, excavated in 1810, was found the torso of a Venus, worthy of the best age of the art, and now the pride of the museum.

Various estimates have been formed of the population of Syracuse when in the summit of its prosperity. These, however, are mostly all exaggerated, and entitled to but little attention. Thucydides says that it was nowise inferior to Athens (lib. vii. p. 503); and that it was a very large and splendid city is a fact of which there can be no manner of doubt; but owing to the great extent of the open spaces and public buildings within its walls, its population could not be in any degree proportioned to what would be contained in a modern city of the same size. Probably it may have amounted to 200,000; or, at most, 250,000; though, if anything, we suspect that this estimate is beyond the mark. Syracuse appears at first to have been under a republican government; but it subsequently became subject to kings, or tyrants, of whom Gelon and Hieron were among the earliest and most celebrated, the triumph of the latter in the chariot-race at the Olympic games having been the subject of one of Pindar's noble odes. But Thrasybulus, the younger brother of the latter, having been expelled the city, the republican form of government was restored.

The Syracusans having been involved, during the course of the Peloponnesian war, in contests with other cities of Sicily, the Athenians sent a fleet to the assistance of the latter, and from less to more, Athens became so much mixed up with Sicilian affairs, that she determined to bring them to a satisfactory conclusion, by undertaking the conquest of Syracuse itself. The greatest exertions were made to effect this grand object; the zeal of the public was supported by the zeal of private individuals; and the armament fitted out by Athens for the reduction of Syracuse is universally admitted to have been the greatest and most splendid ever sent forth by any Greek state. The events of this contest, which fixed the attention of all Greece, have been described by Thucydides, and form the most interesting portion (lib. vi. and vii.) of his history. It is sufficient here to state, that the failure of the expedition was as complete as the hopes of success had been sanguine. Alcibiades, who had assisted in planning the expedition, and whose genius might have conducted it to a successful issue, having been unwisely removed from the command, was succeeded by Nicias, an able general, but one who had been hostile, from the outset, to the project, and who, though brave and experienced, wanted the ability and decision required for the conduct of such an enterprise. After various vicissitudes, the besiegers and besieged changed places. The defeat of the Athenian fleet, which had been cooped up in the great harbour, in an attempt to force a passage through their enemies, may be said to have terminated the expedition, and with it the glory and empire of Athens. "In hoc portu," says Cicero, speaking of the great harbour, "Atheniensium nobilitatis, imperii, gloria naufragium factum existimatur." (In Verrem, v. cap. 37.)

A few years after the defeat of the Athenians, which occurred anno 413 B.C., the supreme direction of affairs at Syracuse was usurped by Dionysius the Elder, whose character presents a singular compound of greatness and meanness, generosity and cruelty. Dionysius the Younger, who succeeded his father, was finally expelled from Sicily by Timoleon; who, having demolished the citadel constructed by the elder Dionysius, and his magnificent tomb, restored the Syracusans to their freedom, and having vanquished their enemies, retired into private life.

They did not, however, long preserve the liberty given them by Timoleon. In the course of a few years, Agathocles attained to the supreme authority. After his death, the city underwent various revolutions, being sometimes the ally of the Carthaginians and sometimes of the Romans, the end it was subjugated, though not without a vigorous resistance, by the latter.

The siege of Syracuse by the Romans under Marcellus is one of the most celebrated in history. It withstood, for a gthened period, all the efforts of the Roman general, who had to contend, not only against the natural strength and fortifications of the place, but also against the extraordinary engines and wonderful machines of Archimedes, the great geometer, and one of the greatest geniuses of antiquity. length, however, the Romans gained possession of the, anno 200 B.C., partly by stratagem, and partly by the chery of one of the Syracusan leaders. Archimedes unfortunately lost his life in the confusion that followed the ng of the city. (Livius, lib. xxv. cap. 23-31.)

nder the Romans, Syracuse continued to be a great and ortant city. It was taken by the Saracens in 878, and n up to military execution. But, notwithstanding this many subsequent calamities, it continued to be of considerable importance till 1693, when it was laid in ruins, and t part of its ancient monuments destroyed, by the dreadful earthquake of that year.

addition to Archimedes, Syracuse has to boast of having given birth to Theocritus, the first and greatest of pastoral poets, and to Moschus. (In addition to the works referred to above, see Sir R. C. Hoare's Classical Tour, ll. 175; History of Syracuse, in Ancient Universal History, vols. vii. and viii., 8vo ed.; Plutarch's Life of Marcellus, &c.)

SYRACUSE, p. v., Salina L, capital of Onondaga co., N. Y., a. W. by N. Albany, 348 W. Situated on the Erie canal the junction of the Oswego canal. Incorporated in and has been raised to its present importance by the, particularly the Erie canal. It contains a courthouse, jail, clerk's office, an academy, two banks, six churches. Episcopal, a Presbyterian, Methodist, Baptist, and two, 100 stores, two flouring-mills, three furnaces and machine-shops, and numerous other mills and manufactories, dwellings, and 6500 inhabitants. The village and town are celebrated for the manufacture of salt, made from springs which abound here. Fine salt is made by solution by heat, and coarse salt by solar evaporation. the salt works in this vicinity there were produced in 2,864,634 bushels, of which 524,461 bushels were in use. Salt is also manufactured in Geddes, Liverpool, ulina, villages in the vicinity. The salt works are an ant source of revenue to the state, which receives six a bushel on all the salt made. The great line of n railroads from Albany to Buffalo, passes through ce. This, with the junction of two important canals, great facilities for business, and has given to this rapid growth.

IA and PALESTINE or JUDEA, two of the most ted regions of the E. hemisphere: the former in the ancient Phœnicia; and the latter is sometimes he Holy Land, from its being the theatre of most at events recorded in sacred history. These famous es have, for many centuries, ceased to be independ have for a lengthened period formed a portion of Turkey: they extend principally between the 31st h degs. N. lat., and the 34th and 41st E. long., having pachalics of Itchli and Marash, in Asia Minor; d E. the Euphrates; S.E. and S. the Arabian des-

ert; and W. the Mediterranean. Previously to the subjugation of the country by Mehemet Ali, it was divided into the four pachalics of Aleppo, Tripoli, Acre, and Damascus. Palestine, or the Holy Land, comprised in the two latter pachalics, forms the S. portion of the region, being about 200 m. in length by 80 m. in its extreme breadth. The entire length of Syria and Palestine may be about 450 m. N. and S.; its breadth varying from 100 to 280 m. Its area has been loosely estimated at 48,000 sq. m.; and its population at from 1¼ to 2, and even 3 millions; but probably 1,500,000 may not be far from the mark. (Bowring's Report on Syria.) Colonel Campbell, a few years since, estimated it at 1,864,000; of whom 997,000 were supposed to be Mussulmans (Turks, Arabs, &c.), 22,000 Ansarians, or Bedouins, 17,000 Metualis and Yezidis, 46,000 Druses, 230,000 Maronites and Christians of the Romish church, 345,000 Christians of the Greek church, and 175,000 Jews. The practice of polygamy among several of these tribes, aided by the extensive conscriptions of Mehemet Ali, is said to have produced a great excess of females over males. (Bowring's Physical Geog., p. 4, 5.)

The W. or coast portion of Syria is mountainous, while the more inland portion, or that to the E. of the Orontes and Damascus is mostly flat. The mountains run mostly N. and S., parallel to the Mediterranean. The principal chain, in different parts of its course, is termed Alma-Dagh, (an. Mons Amanus,) Jeb-el-Anseyry, and Lebanon; it runs at an average distance of about 24 m. from the sea, from the range of Taurus, in Asia Minor, as far S. as the vicinity of Tyre, where it terminates. The chain of Anti-Libanus detaches itself from the foregoing in about lat. 34°, and running S. parallel to, and at no great distance from the latter, encloses the famous valley anciently called Cœle-Syria. About where Libanus terminates, Anti-Libanus divides into two chains, enclosing the valley of the Jordan, the Dead sea, &c., continuing subsequently to bound the valley of El Ghor to the head of the gulf of Akaba. Both Libanus and Anti Libanus give out numerous lateral spurs, some of the former extending so as to project, like mount Carmel, in bold headlands on the coast. These subordinate ranges, with the W. declivity of Libanus, and the E. declivity of Anti-Libanus, are by far the most fertile portions of the mountain system. Cœle-Syria, though fertile itself, is enclosed between precipitous and barren heights. The mountains which surround the Dead sea, and those to the W. of the Jordan, are arid, stony, and full of precipices and caverns, and have a melancholy, desolate appearance, harmonising well with that of the desert by which they are bounded on the E. For further details respecting the mountain region of Syria, see Lebanon, said, p. 155. Palestine consists principally of rugged hills and narrow valleys. It has, however, some fertile plains of considerable size W. of the mountains; as that of Esdraelon (an. Megiddo), 30 m. in length, by 16 m. in breadth; that of Sharon, famous in antiquity for its roses (supposed to be the flower of a species of cistus, with which it is covered); but by far the most extensive and valuable plain is that of Haouran, E. of the Jordan, and of mount Gilead, and trenching on the 'Desert.' "By the great Syrian desert, however, we are not," says Mr. Addison, "to understand a bare wide waste of sand, like the great African desert. The term must be considered to mean destitute of settled inhabitants, towns, villages, and houses, and peopled only by roving pastoral tribes. Instead of sand, the uninhabited district beyond Damascus consists of a fine black soil covered with long, burnt-up, rank grass and herbs; and inhabited by antelopes, wild asses, and wild boars, which search out the thinly scattered spots where water is to be found. The same description of country, we are told, continues the whole way to Palmyra. In summer the soil is parched and cracked into innumerable fissures by the burning rays of the sun, and the herbage and vegetation are all killed; but having previously come to maturity, and scattered their seed upon the ground, no sooner do the winter rains commence than the dry grass is beaten down and rotted, and the seeds, moistened by the abundant rains, sprout up with astonishing luxuriance. In summer, the Bedouins are obliged to congregate in the vicinity of pools and wells; but in winter they spread themselves over the wide surface of the desert, and make long journeys with their flocks and herds." (Damascus and Palmyra, ii. 216, 217.)

The principal rivers of Syria are the Euphrates, Jordan, and Orontes, severally noticed in this work. The coast line is watered by numerous small streams, falling into the Mediterranean, which contribute greatly to fertilize the land: but of these none are navigable. The largest and most remarkable lake is that of Asphaltites, or Dead sea (which see, l. 736.) The next in size is that of Tiberias, or Gennesareth, the theatre of some most remarkable miracles. (Luke and Matt. viii.; Matt. xiv. 25.) It is about 16 m. in length, from 5 to 6 m. in breadth, and is traversed throughout its centre by the Jordan, of which, in fact, it may be regarded as an expansion. On its E. side it is confined by bold, barren, and precipitous mountains; but elsewhere its shores are generally level. According to Dr. Clarke, "It is longer and finer than any of our Cumberland or Westmoreland lakes, though it be perhaps inferior to lake Lomond, in Scotland. It does not possess the vastness of the lake of Geneva, although it much resembles it in certain points of view. In picturesque beauty it comes nearest to the lake of Lucarno in Italy, although it be destitute of anything similar to the islands by which that majestic piece of water is adorned. It is inferior in magnitude, and perhaps in the height of its surrounding mountains, to the lake Asphaltites, but its broad and extended surface, added to the impression under which every Christian pilgrim approaches it, gives to it a character of unparalleled dignity." (Travels, iv. 216.) Its unbroken margin, and the total absence of wood on its shores, without a boat or vessel to be seen throughout its whole extent, give it a melancholy, monotonous appearance.

Several combats took place on this lake between the Jews and Romans, and its banks were formerly the seat of several flourishing cities. Of these, however, Tabaria, the miserable representative of the ancient *Tiberias*, is almost the only existing relic. The lake of Gennesareth, like all other inland seas, is subject to squalls and sudden gusts of wind, that render its navigation rather dangerous. The *Bahr-el-Margi*, near Damascus, and the lakes of Horus, and of *Agi Dengis*, near Antioch, are the only others worth notice. The coast of Syria and Palestine presents a nearly straight line, extending through six degrees of lat., being but little indented by arms or inlets of the sea, the principal being the bays of Scanderoon and Antioch; and though it was in antiquity the seat of a great maritime people, it has very few good harbours; the best are those of Scanderoon and Acre. The former, however, is inconveniently placed, quite at the N. extremity of the country, and is besides very unhealthy. The harbours of Tyre, Sidon, &c., so famous in antiquity, are now, for the most part, filled with sand, or otherwise choked up.

Geology and Minerals.—Of these we have no authentic information. The prevalent rock is limestone, abounding in fossil remains, and hollowed into numerous caverns. The higher parts of the Libanus ranges seem, however, to consist of graüwacke, slate, and other transition rocks, and the rocky mountains skirting the Dead sea, of granite, gneiss, dolomite, &c. Antioch is situated in a great tertiary basin, everywhere broken, however, by serpentine and diallage rocks. (*Geog. Journ.* vii. 420.) "The whole of the Hauran," says Elliott, "is covered with a species of blue stone, very hard, yet porous, and of which all the mill-stones of Syria are made." Volcanic matters cover a considerable extent of country, and the traces of extinct volcanoes are met with in many places. There are occasional indications of coal; but, except building stone, salt, with which a great part of the soil is highly impregnated, and asphaltum, from the vicinity of the Dead sea, are almost the only mineral products of much value.

Climate.—Owing to the great differences of elevation and exposure, the greatest dissimilarity prevails with respect to temperature. On the whole, however, the country may be said to have two climates; one very hot, which is that of the coast and the interior plains, such as those of Balbec, Antioch, Tripoli, Acre, Gaza, Hacuran, &c.; and the other, or that of the mountains, at least at a certain height, temperate, and similar to that of France. (*Volney*, i. 314, Eng. trans.) In most parts, the occurrence of the rainy season, as well as the quantity of rain which falls, are very variable. The winter in the plains is so moderate, that the orange, date, banana, and other delicate trees, flourish in the open air; and it appears equally extraordinary and picturesque to the European at Tripoli to behold under his windows, in the month of January, orange-trees loaded with flowers and fruit, while the lofty summits of Lebanon are seen covered with ice and snow. But in the more northerly parts of the country, and to the E. of the mountains, the winter is more rigorous, without the summits being less hot. This is occasioned by the E. plains being high above the level of the sea, exposed to the parching blasts of the E. and N.E. winds, and screened by the mountains from the humid winds from the W. and S.W. that sweep over the Mediterranean. At Aleppo winter commences about the middle of December, and usually lasts for six weeks or two months. The frosts, however, are seldom of any considerable intensity; snow rarely lies above a day: narcissus' are in flower during the whole of this season; and hyacinths and violets make their appearance before it is over. Spring commences in February, and is extremely pleasant, having no defect but its short duration. Early-in May corn is nearly ripe; and by the end of that month the heats commence, and the country begins to assume a parched and barren aspect. From this period to the middle or end of September no rain ever falls; and the inhabitants sleep exposed on their terraces, without danger from damps or other noxious influences. At Aleppo an interval of between 20 and 30 days usually occurs between the first and second rains: during

which period the weather is serene and extremely brightful; and if the rains have been at all heavy, though but of a few hours' duration, the country assumes a new face. After the second autumnal rains the weather becomes variable, and winter approaches by degrees. The vernal are heavier than the autumnal rains; and, like the latter, are often accompanied with thunder. The trees frequently retain their leaves, till the beginning of December. The heats of summer are usually tempered by westerly breezes; but when during this season the *sessid* occurs, that is, when the winds blow from the Arabian and Persian deserts, or from the E. inclining to the S., the heat becomes suffocating and excessive; and the inhabitants have no resource but to shut themselves closely up in their houses. Luckily, however, these winds are not of very frequent occurrence; and sometimes they do not occur once in a summer. Shocks of earthquakes are common; and, in 1822, Aleppo and several other towns were nearly destroyed by one of these visitations. (*Volney*, i. 315; *Russel's Aleppo*, 10–14, &c.)

It is clear, therefore, as Volney has stated, that "Syria unites a great variety of climates, and collects within a narrow compass pleasures and productions which nature has elsewhere dispersed at great distances of time and place. With us, for instance, seasons are separated by months; there, we may say, they are only separated by hours. If in Saïde or Tripoli we are incommoded by the heats of July, in six hours we are in the neighbouring mountains, in the temperature of March: or, on the other hand, if chilled by the frosts of December, at Besharrai, a day's journey brings us back to the coast, amid the flowers of May. The Arabian poets have therefore said, that 'the Sannin (Lebanon) bears winter upon his head, spring upon his shoulders, and autumn in his bosom, while summer lies sleeping at his feet.' I have myself experienced the truth of this figurative observation, during the eight months I resided at the monastery of Mar-Hanna, seven leagues from Beyrout. At the end of February I left at Tripoli a variety of vegetables which were in perfection, and many flowers in full bloom On my arrival at Antoura I found the plants only beginning to shoot; and at Mar-Hanna everything was covered with snow. It had not entirely left the Sannin till the end of April; and already, in the valley it overlooks, roses had begun to bud. The early figs were past at Beyrout when they were first gathered with us, and the silk-worms were in cod before our mulberry-trees were half stripped.

"To this advantage, which perpetuates enjoyments by their succession, Syria adds another, that of multiplying them by the variety of her productions. Were nature assisted by art, those of the most distant countries might be produced within the space of twenty leagues. Even at present, despite the barbarism of a government inimical to all industry and improvement, we are astonished at the variety this province affords." (l. 317.)

It is true that Syria and Palestine are sometimes visited by the plague; but this is a consequence of slothfulness and the want of proper care and precautions. Dysenteries, leprosies, &c., are also frequent; but, on the whole, the country is highly salubrious; and has no peculiar disease, except the pimple or ulcer of Aleppo. (See ALEPPO.)

Products and Resources, &c.—The beauty, fertility, and various products of Syria made her be regarded as one of the finest and most fruitful of countries; and her superiority in these respects has been extolled by the best modern travellers. It seems unnecessary, therefore, to dwell on what is so generally admitted. But the question as to the fertility of Palestine has given rise to some conflicting statements, and as the subject possesses peculiar interest, we shall notice it at some little length.

In the sacred writings, the fertility of the Holy Land is described in the most striking manner. Moses calls it "a land that floweth with milk and honey; a land of brooks and waters, of fountains and depths, that spring out of the valleys and hills; a land of wheat and barley; of vines, figs, and pomegranates; of oil, olives, and honey; a land where there is no lack or scarcity of any thing; whose stones (or rocks) are iron; and out of whose mountains brass may be dug up." (Deuteronomy, viii. 7 &c.)

It may, perhaps, be permitted to suppose, that as Moses wished to reconcile the Jews to the territory on which they were about to enter, and to extinguish any lurking desire in their part to return to the "fleshpots" of Egypt he would represent the "promised "land under the most favourable colours. On the whole, however, it would seem, despite the statements that have been made to the contrary, that his description is substantially correct. It is strikingly confirmed by Tacitus, who says, speaking of Palestine, *Rari imbres, UBER SOLUM. Exuberant fruges, nostrum ad morem; præterque eas balsamum et palmæ.* (*Hist.* lib. v. cap. 5.) It is true that Strabo, in his 16th book, speaks in very contemptuous terms of the country round Jerusalem; but he was by no means so well acquainted with Palestine either with the history of the Jews or with Judea; and in-

sides, even though the accuracy of his statement as to the country to which he has referred were admitted, that would not authorise any inference to be drawn unfavourable to the general fertility of Palestine. In antiquity Judea was very carefully cultivated; and notwithstanding the great density of its population, it is said when in the zenith of its prosperity, under Solomon, to have exported considerable quantities of corn. (1 Kings, v. 11.) The declivities of the hills were formed into terraces, of which the vestiges still remain (*Maundrell*, p. 66., ed. 1740), and were covered with plantations of figs, vines, and olives. It was, as Tacitus has stated, particularly celebrated for its palm-trees, which, in fact, were the emblem of the country; and the aromatic plants that grew in the uncultivated parts furnished the wild bees with the honey which they stored in the hollows of the rocks and trees. Indeed Maundrell, whose accuracy is unquestionable, states that he perceived in many such places " a smell of honey and wax as strong as if one had been in an aviary." (*In loc. cit.*) We cannot, however, form any fair estimate of the state of the country in antiquity from the condition in which we find it at the present time, seeing it has groaned for centuries under the yoke of barbarous tyrants, and been subjected to every species of tyranny and oppression. "The Holy Land," says Dr. Shaw, " were it as well peopled and cultivated as in former times, would still be more fruitful than the very best part of the coast of Syria and Phœnice, for the soil is generally much richer, and, all things considered, yields a more preferable crop. Thus the cotton that is gathered in the plains of Ramah, Esdraelon, and Zabulon, is in greater esteem than what is cultivated near Sidon and Tripoli. Neither is it possible for pulse, wheat, or any sort of grain, to be more excellent than what is sold at Jerusalem. The barrenness, or scarcity rather, which some authors may either ignorantly or maliciously complain of, doth not proceed from the incapacity or natural unfruitfulness of the country, but from the want of inhabitants, and the great aversion there is to industry and labour in those few who possess it. There are, besides, such perpetual discords and depredations among the petty princes who share this fine country, that allowing it were better peopled, yet there would be small encouragement to sow, when it was uncertain who should gather in the harvest. Otherwise the land is a good land, and still capable of affording its neighbours the like supplies of corn and oil which it was known to have done in the time of Solomon." (*Trav.*, p. 336, 4to ed.)

At a more recent period Dr. Clarke said of the Holy Land, "The delightful plain of Zabulon appeared everywhere covered with spontaneous vegetation, flourishing in the wildest exuberance. The scenery is to the full as delightful as in the rich vales upon the S. of the Crimea: it reminded us of the finest parts of Kent and Surrey. The soil, although stony, is exceedingly rich. We found the valleys W. of Jerusalem covered with plentiful crops of tobacco, wheat, barley, Indian millet, melons, vines, pumpkins, and cucumbers." (*Trav.* iv. 423.)

These statements are more than sufficient to attest the natural riches and fertility of this famous region. As an agricultural or corn-growing country it is, indeed, far inferior to Egypt and many other states: but the variety of its surface and products, the salubrity of the climate, and the productiveness of its cultivated lands, would make it were it possessed by an industrious, well governed people, a most desirable country.

Recently, however, the condition of Syria and Palestine has been changed materially for the worse, as compared even with what it was at the epoch of Clarke's visit. The destructive contests of which it has been the theatre, the consequent destruction of property, the interruption of all sorts of industry, and the emigrations occasioned by the conscriptions of Mehemet Ali, have reduced Syria and Palestine to a state of depression to which they had never previously sunk, and from which there is but little hope of their recovering under their present rulers.

During the ascendancy of the Egyptians Mehemet Ali attempted to introduce the same compulsory or forcing system into Syria and Palestine that he has introduced into Egypt (which *see*). In this view the principal officers of the government and the army, and the more opulent inhabitants, were compelled to undertake the task of restoring ruined villages, and the culture of their lands. Government intended, by means of the increased cultivation of wheat and barley, to render Syria independent of supplies from without; and, if possible, to obtain a surplus for exportation. In good harvests, indeed, Syria, particularly its N. part, previously produced sufficient corn for its own consumption; and had the measures undertaken by the Egyptian government been persevered in, there can be little doubt that there would have been a great increase of produce. But in 1837 the influence of the new system was paralysed by a drought; and the events that speedily followed overturned at once the power and the projects of the Egyptians.

The landed property of Syria, as of the rest of the Turkish empire, is supposed to belong to the sultan, as the vicegerent of God and the prophet; and the principle, that it did so, in fact, was acted upon at the conquest of the country, by the caliph Omar, in the 7th, and by the Turks under Selim I., in the 16th century. At present, however, this assumed property of the sultan is a mere legal fiction. Soon after Selim's conquest, the ruinous effects of the general confiscation became so apparent that measures were taken for giving the occupiers a right of property in the land on paying a small quit-rent. Land may now be classed under three heads; that belonging to the sultan and government; *vacuo* or entailed lands; and real property, belonging to the proprietors, and descending by inheritance. The lands and property belonging to the sultan and government are those escheating in different ways; such as lands abandoned in consequence of non-cultivation during three years, lands left by the extinction of families, lands confiscated, &c. Entailed property, called *vacuo et harameus*, consists of that settled by private individuals for the maintenance of public caravanserais, fountains, and charitable institutions; and of that vested in the hands of the clergy for behoof of certain pastles, and their heirs or nearest kin. Some lands are settled on the eldest heir in perpetuity, and cannot be sold, though they may be exchanged. According to the rule of the Ottoman law, Franks and other foreigners cannot hold land in the Turkish dominions; but, in fact they do hold it, by means of long leases or otherwise, which make it little less secure than freeholds. In the succession to property, the sons inherit twice as much as the daughters.

In Lebanon, almost every male inhabitant is a small proprietor; and in the neighbourhood of Beyrout there are a great number of land holders who, for the most part cultivate the white mulberry tree. Large proprietors are few, except among the emirs of mount Lebanon, some of whom have extensive estates, which they either cultivate on their own account, or let out to farming tenants. (*Bowr. Rep. on Syria*, p. 101, 102; *La Syria jusqu'en* 1840, p. 75–83, 148, 149.)

The *miri*, or land-tax, is not assessed in Syria by any invariable rule, or according to any admeasurement of the land. A government is assessed in a certain amount, which is apportioned among the different villages according to their greater or less amount of population, or more or less extent of land; and the peasants themselves apportion the payments each has to make. In the cultivation of all kinds of produce, except silk, the landed proprietor supplies the peasantry with seed, and a certain sum of money to buy oxen, cattle and implements of husbandry; and receives 10, 15, or 20 per cent. of the produce, according as the ground is more or less taxed. The remainder is divided into two equal parts, one of which the proprietor takes, and the other is for the peasants. These last are obliged to repay the money advanced to them, but not the seed. (*Colonel Campbell's Report*.)

The old Roman plough, drawn by bullocks, is that almost universally employed. Wheat, barley, maize, millet, lentils, sesamum, &c., are grown principally in the plain of the Haouran, which has always been considered the granary of Syria. It is inhabited by Turks, Druses, and stationary Arabs, and is visited in spring and summer by several Bedouin tribes. Burckhardt computes the resident population at from 50,000 to 60,000. The fertility of the soil depends entirely upon the water with which it is supplied; and the harvest is, therefore, in proportion to the abundance of the winter rains, and the extent of artificial irrigation. Lands which cannot be irrigated usually lie fallow every other year; though a part is sometimes sown in spring with sesamum, cucumbers, melons, and pulse. Where an abundance of water may be obtained from neighbouring springs, the soil is sown with lentils, peas, sesamum, &c., after the grain harvest. In middling years, wheat is said to yield 25 times the seed; and the produce of barley is said sometimes to average 50 and even 80 times the seed: though these statements are usually much exaggerated, and but little to be depended on. The first harvest is that of horse-beans, at the end of April: vast tracts are sown with these, to serve as food for cows, sheep, and camels. Next comes the barley, and, towards the end of May, the wheat harvest. The wealth of a cultivator is estimated by the number of *fedhans*, or yokes of oxen, he employs. The owner of two or three is estimated rich, and he will probably possess, besides, two camels, a mare or gelding, or a couple of asses, and forty or fifty sheep. Taxes are very heavy in the Haouran. There is, first, the *miri*, paid to the pacha, and which is levied on the fedhans, the amount depending on the sum at which the whole village is rated in the pacha's books, and which must be paid so long as the village is inhabited, be the number of fedhans employed few or many. Next is the obligation to supply the troops, &c., with provender; and the third and heaviest contribution paid by the villagers is the *khone*, or tribute (identical with the *black-meal* of the Scotch,) claimed by the Bedouins, in return for their protection, or rather forbearance. Each village pays *khens* to the sheikh of a tribe. who is then bound to protect the inhabitants, and pays a tribute of from 30 and 40 to 400 piastres to the pacha for this privilege. Lastly come the unlimited contributions exacted by the pachas. The receipt of the *miri* of the whole pachalic, which may amount to £250,000, is in the hands of Jew bankers, who not only get about 5 per cent., but contrive to extort something further on their own account. Families in the Haouran are constantly moving from one place to another. In the first year of their new settlement the sheikh acts with moderation towards them; but his exactions becoming insupportable, they migrate to some other place, where they have heard that their brethren are better treated; they soon find, however, that the same system prevails over the whole country. (*Mod. Trav.*, iii. 80–83.) In addition to all these exactions, the crops in the Haouras are sometimes destroyed by mice, though not so often as in the neighbourhood of Horus, and Hamah. But the worst enemies of the agriculturist are the clouds of locusts which sometimes devastate the country, devouring every " green thing." They are not, however, an unmitigated nuisance, having been used for food from time immemorial, and are said to be both wholesome and palatable.

The most careful cultivation in Syria is exhibited on the slopes of mount Lebanon and other inaccessible districts, where the inhabitants enjoy a comparative exemption from the exactions of their Turkish masters. " Stimulated by their sense of security, they have, by dint of art and labour, compelled a rocky soil to become fertile. Sometimes, to profit by the water, they conduct it, by a thousand windings, along the declivities, or stop it, by forming dams in the valleys; while in other places, they prop up ground, ready to crumble away, by walls and terraces. Almost all these mountains, thus laboured, present the appearance of a flight of stairs, or an amphitheatre, each step of which is a row of vines or mulberry-trees. I have reckoned 120 of these gradations on the same declivity, from the bottom to the top of the mountain. So powerful is the influence of even the feeblest ray of liberty and security." (*Volney*, i. 300.)

The mulberry-tree flourishes on the coast and through the more fertile parts of the Lebanon range, and a little more attention to the culture of silk would render it in a few years the principal article of export. The mulberry-plants are set in rows 6 or 8 ft. apart; they are cut off at a considerable height, and suffered to retain only the fresh twigs. The arrangement generally made with the peasantry is to allow them one fourth part of the silk for taking care of the worms, and reeling it off the cocoons. The land owner provides the leaves, which are gathered by the peasants. He also erects the sheds in which the cocoons are kept, which are simple reed enclosures, without any roof. The quantity of silk annually produced on mount Lebanon is estimated at about 1900 cantars, or 240,000 okes, fetching from 190 to 195 piastres the oke, of which about two thirds are exported. About 400 cantars is considered an abundant crop in the Tripoli district. Aleppo receives about 250 cantars from Antioch, and other quarters. Its chief consumption there is in the manufacture of the cotton and silk goods used for upper garments by the wealthy inhabitants; but it is also sent into all parts of the Turkish empire; and, in 1836, 100 cantars were sent to Genoa, France, and England. The average annual produce of cotton in the vicinity of Acre, Jaffa, Nablous, and other places in the S., is estimated at from 30,000 to 35,000 cantars, worth about £350,000. In the N. the crops are exposed to great vicissitudes. The quality of the cotton is sometimes good, but more commonly inferior. The export is chiefly to Smyrna, and other parts of Turkey. Not more than from 1000 to 2000 cantars reach W. Europe, the quantity that comes to England being very trifling. The oil harvest is very precarious. From 8000 to 10,000 cantars may be about the average consumption in Aleppo, half of which is produced in the neighbourhood. The average produce around Damascus is estimated at from 4800 to 5000 cantars. The oil has of late years been considerably improved, and its quantity augmented by the introduction of oil-presses from France. Wine might become an important article both of consumption and export; and at some of the convents of Lebanon (where the vine is suffered to trail on the ground) a very good wine, called *vino d'oro*, is met with. Madder and indigo grow wild; and the former, as well as the sugar-cane, has been partially cultivated. Ibrahim Pacha introduced the cochineal insect into Syria with every prospect of success; for the cactus, on which it feeds, grows there to an immense size, and forms, in fact, most of the hedges in the country. The dates of Syria are not equal in quality to those of Egypt or Nubia; but the date palm is so abundant that, as already stated, it was anciently the symbol of Judea; and it is probable that Phœnicia was so called from the abundance of this plant (*φοῖνιξ*). Tobacco is grown in almost every part of Syria,

its consumption being universal both by males and females. The best is found in the districts of Aleppo, Latakia, Tripoli, and mount Lebanon, large exports taking place from Latakia and Tripoli to Egypt and elsewhere. The total produce is estimated by Col. Campbell at 10,700 cantars a year. Scammony, the juice of a species of convolvulus, which grows in N. Syria, is a valuable article of export, and that from Aleppo is esteemed the best in the markets of Europe. But it is rarely obtained pure, the confectioner first adulterating it with flour or starch, to give it colour and consistency; and with myrrh, to give it a bitter, aromatic taste. It is then sold to the Jew dealers, who further adulterate it in the same manner, mixing four or five rottoli of starch with one rottoli of scammony, in which state it is sent to England at a price of from 250 to 300 piastres per rottolo. From 1200 to 1500 loads of hemp are produced in the Damascus district; but it is not an article of export. From 200 to 250 cantars of bees' wax are annually collected in the Aleppo district, nearly half of which is sent to Europe.

The forests of N. Syria have lately supplied large quantities of timber; the arsenals and dockyards of Egypt have been principally furnished from this source. From 70,000 to 80,000 trees, large and small, or about 14,000 tons of timber, principally, pine, oak, and beech, were shipped in 1835 for Alexandria.

The Holy Land in antiquity was eminently distinguished for its abundance of cattle, including sheep, goats, camels, and asses; and though much diminished in numbers, these animals still constitute a principal part of the wealth of the occupiers. No very large or formidable wild animals exist at present in Palestine; the fallow deer, gazelle, wild goat, jackall, fox, and porcupine, are the principal. There are, however, numerous birds, including two species of vultures, great quantities of game, and wild fowl; and a great variety of reptiles is met with. (*Mod. Trav.* i. 15, 16.)

Conflicting statements have been put forth with respect to the actual condition of the peasantry. According to Mr. Consul Moore, " the fellah, or peasant, in Syria, earns little more than a bare subsistence." But Dr. Bowring, on the other hand, states that " the condition of the labouring classes is, comparatively with that of those in England, easy and good. They feed on mutton at three piastres per oke, several times a week; bread daily; sometimes rice pillaus, and always bulgur pillaus (a preparation of wheat boiled and bruised, or half ground); the pillaus are made either with butter, olive, or sesame oil; cheese, eggs, olives, various dried fruits, and an abundance of vegetables, such as roots, turnips, and radishes, preserved in brine or vinegar, and cucumbers and capsicums in vinegar, for winter use. Their clothing is not especially coarse; the fine cloaks permit them to wear light cotton and other similar apparel; and in the short winter they are generally well covered. Their lodging is good; generally each family has a separate house, or a set of rooms. Lodging generally in Syria is cheap, comparatively with most other countries. The Mussulmans have few holydays; the Christians have a great many, and their amusements are much of the same sort as the Mussulmans, if any thing, less sober; but, on the whole, none of the classes of the population can be said with habitual indolency. But it is rare that any of the working classes can lay by some adequate to enable them to pass the decline of life without labouring. In Syria a great portion of the labour is done by females; they are constantly seen carrying heavy burdens, fetching water, &c., they bring home timber and brushwood from the forests, and assist much in the cultivation of the fields." (*Rep. on Syria*, 49, 50.) Field labour near Beyrout is paid at from 5 to 6½ piastres (1s. to 1s. 3d.); and artisans, as masons, carpenters, &c., get 14 or 15 piastres (2s. 10d. to 3s. 0d.) a day. The yearly expenditure of one of the labouring classes may average from £12 to £16. (*Ibid.*, p. 51.)

The fisheries are unimportant, except that of sponge, which is obtained along all the N. half of the coast; and in a good season, about 3500 okes are gathered, which are principally sent to Smyrna, Rhodes, Marseilles, &c.

Few of the *manufactures* for which Syria was renowned, survive at present. In Damascus there are about 4000 looms for silk and cotton stuffs, each producing for a five pieces a week, worth from 80 to 95 piastres each. In Aleppo, nearly 6000 looms of the same description were at work in 1829; but the number in 1838 had dwindled to 1900, the consumption of rich stuffs having fallen off in favour of cotton goods, for which British twist is employed, and which occupy about 500 looms in that city. About 300 looms are also said to be occupied in the manufacture of the gold and silver thread stuffs, and the total product of the looms of Aleppo is estimated at £250,000 sterling a year. (*Bowring's Rep.*, p. 84.) In good old years, from 7000 to 8000 cantars of soap are made in Aleppo; and perhaps 18,000 cantars at Damascus, Jerusalem, Nablous, and other parts of the country; it is not, however, exported to any great distance. Coarse woollens, glass, earthenware, and leather are among the other chief goods manufactured. Horus, Hamah, and Beyrout, are the other principal manufacturing towns. The ancient art of dyeing in purple is lost at Tyre, and Damascus blades have no longer their former reputation. In Palestine, a considerable trade is carried on in the manufacture of crosses, beads, rosaries, and such like trumpery.

Commerce.—In remote antiquity, Sidon and Tyre were the principal emporiums of the world: they were succeeded by Damascus, Antioch, Joppa, &c.; and in later times by Palmyra, whose grandeur was mainly owing to her situation on the great route of traffic between E. Asia and Europe and W. Asia. But for a lengthened period the commerce of Syria has been comparatively inconsiderable. The internal trade of the country is greatly impeded by the want of good roads; those that exist being mostly mere mule or camel tracks. But, notwithstanding, gum arabic, tragacanth, asafœtida, opium, &c., are brought from the surrounding countries; galls and barilla from beyond the Euphrates; saffron from Persia and Natolia; hare, fox, and jackal skins; yellow berries and goats' hair from Asia Minor; and these, with cotton, goats' and sheeps' wool, silk, tobacco, and other kinds of raw produce, previously specified, form the principal exports. The imports consist chiefly of colonial produce and European manufactures; coffee (W. India), from France, Italy, and England; sugar, from France and Great Britain; pepper, spices, rice, dyeing drugs, copperas, cotton manufactures, cambrics, shirtings, nankeens, imitation shawls, and cotton twist, for the most part from England; iron, tin, sal-ammoniac, woollen cloths, from France and Belgium; silks from France, glass wares from Bohemia, by way of Trieste, &c., are the most important. The caravans from Bagdad, Mosul, and Erzeroum, to Mecca, pass through Syria, bringing galls, indigo, Mocha coffee, buffalo-skins, tombac, gum, Cashmere shawls, and a few Indian manufactures, in return for European manufactures and cochineal; and constant caravans travel between Aleppo and Aintab, &c., bringing oil, grain, and leather for the use of the former, which is by far the most important *dépôt* in the interior of Syria. The progressive increase of the foreign trade after the Egyptian conquest was obvious; and chiefly in favour of the British, at the expense of the French and Sardinian. In 1839, Syria and Palestine received from Great Britain 10,976,500 yards of cotton fabrics, worth £195,770, and 777,135 lbs. cotton twist, with earthenware, steel, wrought and unwrought, cochineal, indigo, sugar, pepper, pimento, &c., the value of the whole amounting to £251,509. (*Board of Trade Reports*, 1841.)

The following is an account of the values of the imports into, and the exports from, Syria in 1836, specifying those from and to each country:—

Imports.		Exports.	
	Piastres.		*Piastres.*
From Austria	1,941,500	To Austria	957,700
Egypt	14,034,000	Egypt	12,090,000
France	6,542,000	France	6,495,000
Great Britain	7,261,800	Great Britain	550,000
Greece	144,400	Greece	345,630
Sardinia	3,700	Sardinia	—
Tuscany	9,021,000	Tuscany	2,133,520
Turkey	8,241,400	Turkey	4,671,300
Total	46,210,500	Total	29,390,380

The balance was principally paid in specie, or in European bills of exchange. The slave trade is not carried on to any great extent in Syria.

Government, &c.—The immediate influence of the conquest of Mehemet Ali was exhibited not only in the increase of commerce, but in a better system of police, and a better administration of justice, an increase in the value of land and labour, an increase of cultivation, and a greater religious toleration. But the *fayahs*, and working classes generally, though better protected, were more burdened and impoverished. They were forced to labour for sums far below the ordinary rate of wages; their camels, cattle, &c., were continually seized for the service of the government, and their property and resources, of whatever kind, were subjected to fresh exactions. According to Mr. Werry, nothing was done to improve the means of communication in Syria during the Egyptian ascendancy; few public works having been undertaken, except extensive barracks in the large towns. Neither did the government make public education so much an object of its care as in Egypt. The forced cultivation introduced by Ibrahim Pacha enriched only the government, not the subject. But the short period during which Syria was held by Mehemet Ali, and the uncertainty of his tenure, was sufficient to hinder him, however much disposed, from undertaking or effecting any considerable reforms or changes; though, if we may judge from what he has done in Egypt, his changes, had he been allowed to introduce them, would hardly have been improvements. On the whole, however, we incline to think that, whatever might have been the influence of his government, it would, at all events, have been preferable to the worn-out, extravagant despotism of the Porte.

Under the Egyptians, Syria was divided into six districts; those of Aleppo, Damascus, Jaffa, Tripoli, Saida (Sidon), and Adana. Every town had a *mutsellim*, or head police magistrate; and in all having a population of above 2000 persons, *Sciori divans*, or town councils, were established by Ibrahim Pacha. These bodies consisted of from a dozen to 20 of the chief inhabitants, without distinction of religion, and acted as a civil and commercial court, the decisions of which were subject to appeal to the divans of Aleppo or Damascus, and finally to the supreme government at Cairo. These courts greatly circumscribed the duties of the *cadi* sent annually from Constantinople to make the judicial tour of Syria. Justice was remarkable for its promptitude and severity. Murder, burglary, highway robbery, and other capital crimes, are, however, comparatively rare in Syria. Europeans are subject to the jurisdiction of their own consulates.

No law exists making provision for the poor, though there are many private Mussulmen endowments; and the other religious sects mostly support and relieve their own sick, paupers, &c. In every parish, or mosque district, there are Mohammedan primary schools; and Jewish, Christian, and other primary schools, are established wherever those sects prevail. But the instruction in these is mostly limited to that derived from religious books; and there is no native school in Syria where a more advanced education is given than in reading and writing, with the exception of the Greek college at Beyrout, where geography is studied from books printed at the Protestant press. The American missionaries have a superior college in that town, and various other schools in the country.

The *army* maintained by Ibrahim Pacha in Syria was greatly over-rated in amount. Mr. Moore estimates the total number of men at from 45,000 to 50,000; 35,000 infantry, 4400 artillery, and from 6000 to 8000 cavalry, independent of the irregular Bedouin troops. The forces in Syria were, for the most part, Egyptian, the Syrian conscript being usually sent to Egypt, and replaced by others from thence. The discipline introduced into the Syrian army was copied from that of the French. Ibrahim Pacha had the address to disarm all the civil population, and attempted to settle some of the Bedouin tribes on the skirts of the desert. But subsequent events appear to have restored their arms to the former; and the latter, most probably, prowl lawlessly over the country, as heretofore.

The *public revenue* in 1836, derived from the land, house, cattle, capitation, toleration, &c., taxes, government rents, fines, customs, and *octroi* duties, &c , was supposed to be equivalent to about £606,000; but the expenditure in the same year was estimated at nearly £1,900,000, the deficiency having been made up by contributions from Egypt.

The *inhabitants* of Syria comprise a mixture of different races, consisting partly of the posterity of those who occupied the country when it was overrun by the Arabs, that is, of the Greeks of the lower empire; partly of the posterity of the Arab conquerors of the country; and partly Turks, or Ottomans. And these, again, have been intermixed with each other, with the crusaders, who invaded and held a portion of the country for a considerable period, and with the wandering Bedouins, Kurds, &c. But, how different soever their origin, these races have, in the course of time, become equally naturalized to the country. The inhabitants are generally of a middling stature; those belonging to the southern are more swarthy than those belonging to the northern plains; and these, again, than those belonging to the mountains. On Lebanon, indeed, and in the mountainous districts generally, the complexion does not differ materially from that of the inhabitants of the S. of France. Arabic is the general language of the country; and Volney affirms, in opposition to the statements of Niebuhr, that neither Syriac nor modern Greek is any where in common use.

But notwithstanding the family or national resemblance by which the Syrians are now distinguished, they are distributed into different classes or tribes, all differing from each other in more or fewer particulars. Of these tribes, one of the most famous is that of the Druses, occupying the S. parts of Lebanon and Anti-Lebanon, parts of the Haouran, &c. They are supposed to be of Arabic origin, and to be disciples of a Mohammedan heretic of the 10th century. Their religion, notwithstanding the late researches of Mr. Jowett, Mr. Robinson, and others, still continues involved in a good deal of mystery. According to Volney, they appear to have a contempt for all that the Mohammedans hold most sacred; for, he says, they neither practise circumcision nor prayer, nor fasting, nor observe festivals nor prohibitions; and that they drink wine, eat pork, and allow of marriage between brothers and sisters, though not between fathers and children. They have an emir of their own, and enjoy a rude independence, to which, no doubt, their "openness, sincerity, and engaging manners" (*Clarke*, iv. 306) are mainly to be ascribed. They are divided into two great classes, the learned, or initiated (*akout*) and unlearned, uninitiated (*djaheis*). The former, who enjoy various privileges, are distinguished by their white turbans. Robinson says that "the uninitiated perform no religious rite whatever, unless when circumstances oblige them to assume the appearance of musulmen." (ii. 11.) They are eminently tolerant; and live on good terms with both Christians and Mohammedans. Mr. Elliot is not very favourable to the Druses. "Outwardly," says he, "they are moral in their deportment; but it is doubted whether similar decorum prevails behind the scenes. Though polygamy be permitted, yet few have more than one wife, who, however, may be divorced at pleasure. They are extremely hospitable; yet, where no breach of hospitality is involved, the rights of blood and friendship are unhesitatingly sacrificed to interest. They have little personal, but much public pride. The women are distinguished by an appendage as strange, unmeaning, and hideous, as female fancy ever devised. Other nations may laugh at the long trains of the ladies of England, the infantine shoes of China, or the monstrous nose-rings of India; but the *tantour* of Lebanon surpasses them all. It is a plated silver, or gilt tube, resembling a straight horn, 18 inches long, and standing out like a unicorn's, at an angle of 45° from the centre of the forehead, or from one side of the head; it is fastened by means of a spring, balanced by three heavy tassels hanging down the back, and covered with a white transparent veil." (*Elliot's Trav.*, ii.)

The *Maronites* are a Christian sect, principally inhabiting the country about Lebanon, adjacent to the Druses. They originated in the 6th century, and profess themselves to be followers of the monk Maron, whence their name. They effected a union with the church of Rome, from which they had never differed very widely, about 1215. They are divided into the two classes of sheiks or chiefs, and common people, and have a spiritual head, with the title of patriarch of Antioch. They are all husbandmen; property is sacred among them; and, on the whole, they bear a good character. Like the Druses, they have succeeded in maintaining their independence, paying merely a moderate tribute to the pachas. The Metualis, another tribe, are Syrian Mohammedans, of the Shiite, or Persian sect. The Ansarians, Yezidi, Samaritans, &c., have complicated religious systems, partly Mohammedan and partly idolatrous; but for accounts of these and the other Syrian tribes, we beg to refer the reader to *Volney*, which still continues to be the best work on this interesting region.

The *ancient history* of Palestine, is familiar to every reader of the sacred writings. Under Solomon, it became a rich and powerful kingdom; and after undergoing various vicissitudes, it finally became tributary to the Romans. At the period of the advent of the Messiah, it was divided into five provinces, Judea, Samaria, Galilee, Perea, and Idumea. We have already (art. JERUSALEM) noticed the conquest and destruction of that city by Titus, and the final dispersion of the Jews. In more modern times the Holy Land became the seat of a violent struggle. A singular combination of credulity and superstition gave birth to the crusades; and for some centuries the recovery of the Holy Land, and especially of the holy sepulchre, was sufficient to precipitate hundreds upon hundreds of thousands of blood-thirsty fanatics upon the east. At length, after oceans of blood had been spent, the victories of Saladin put an end to this deplorable phrenzy. In 1516, the country was taken by the Turks. Very recently, or in 1832, Ibrahim, son of Mehemet Ali, pacha of Egypt, undertook and speedily effected the conquest of Syria and Palestine. It is doubtful, as already seen, whether they would have gained any thing by the change; but it would be very difficult, indeed, to show that they could have lost any thing. The great European powers, however, with the exception of France, determined not to permit Syria to be dismaided from the Porte; and, in 1840, a British fleet, after a short but tremendous cannonade, took Acre, and Ibrahim was compelled to agree to evacuate the province. It does not, however, seem that the states by whom this revolution was effected took any step whatever to ensure the better government of the country in future; to obviate any one of the grievances by which it has been oppressed, or to make any stipulation of any kind in favour of the inhabitants.

In the art. TYRE, the reader will find some notice of the history of Phœnicia. (See *Rolands Palestina; Bowring, &c., Rep. on Syria; Castillo, La Syrie sous Mehemet Ali; Marment, Voyage, &c.*, ii., iii.; *Russell's Aleppo; Volney, passim; Elliot, Wilde, Robinson, Addison, &c.; Mod. Trav.*, i., ii., iii., &c.)

SZEGEDIN, a royal free town of Hungary, co. Csongrad, of which it is the cap., on the Theiss, where it is joined by the Maros, 60 m. W.N.W. Arad, and 100 m. S.E. Pesth. Pop. 32,300. (*Berghaus.*) It consists of the town-proper, tolerably well built, and chiefly inhabited by Germans; the fortress, the residence of a commandant and garrison, con-
8.6

nected with the town by two bridges; the upper and lower suburb, and the corn market. It has a house of a-------- (which, according to the *Aust. Nat. Encyc.*, is the o--- in Hungary), a lyceum, gymnasium, Piarist college, m----- school, &c.; and is the see of a Greek protopope. P--- says it is one of the most disagreeable towns in Hungary it has, however, a good deal of trade, chiefly in corn, s--- soap, and tobacco, with several soap and other f------ It also supplies some of the best river craft in the kingdom (*Berghaus; Paget's Hungary, &c.*, i. 1481.)

T.

TABRIZ, or TAURIS, a city of Persia, prov. Adverbijan, of which it is the cap., in a large and f--- plain, on a small river which falls into lake Urumea, 390 m. W.S.W Tebern, lat. 38° 10' N., long. 46° 37' E. Its population was estimated by Chardin at upwards of half a million; but it has declined so much in the interval, that it is now probably under 30,000; and it is said by Mr. Kinneir to be one of the most wretched cities he had seen in Persia. (*Persian Empire*, p. 151.) Being surrounded by a forest of orchards, it appears from the high ground above it to be of immense extent; and a modern traveller considers the circuit of the gardens of Tabriz to measure not less than 30 miles. (*Rennison in Geog. Journal*, z. 3.) But the town itself, which is nearly in the centre of this area, is only about 2¼ m. in circuit; it is surrounded with a brick wall and towers, and is entered by seven gates. It has few public buildings of note, the principal is the citadel of Ali Shah, a part of which is now converted into an arsenal, where many European artisans are or have been employed. A considerable portion of the population live in the suburbs, which straggle over the area of the ancient city, and are built of its ruins. Tabriz is said by D'Anville to represent the ancient Gazaca where Cyrus deposited the treasures of Cresus; and it was afterwards taken by Heraclius; and it has been also supposed to be identical with the Γάζαρα of Ptolemy. But according to other authorities, it was built under Harun a Raschid, of whom it was certainly a favourite residence. Its trade, which was formerly extensive, has greatly declined. Few cities have suffered so much from the ravages of war and earthquakes. Its climate is praised by the natives for its salubrity; but the changes of temperature are extremely great and sudden, and in winter the cold is so intense, that "many instances have occurred of individuals accidentally excluded from the city by arriving after the gates were shut, being found frozen to death in the morning." (*Mod. Trav.*, xiii. 253–258; *Kinneir's Persian Empire*, &c.)

TADCASTER, a market town and par. of England, W Riding, co. York, partly in the liberty of St. Peter of York, and partly in Barkstone-Ash wapent.. on the Wharf, here crossed by a stone bridge; 10 m. W.S.W. York. Area of par., 6100 acres. Pop. of ditto, in 1831, 2655. The town is well built. The church, which is handsome, is the perpendicular style, has a fine tower. (*Rickman's Gothic Architecture.*) The living, a vicarage in the gift of Lord Egremont, is worth £240 a year. It has chapels for Methodists and other dissenters, Jesus' hospital for four poor men, a free grammar school, founded in the time of Queen Elizabeth, Sunday schools, for some of which spacious buildings have been erected, &c. There are no manufactures, but a good of retail trade. Markets on Thursdays. Fairs on Wednesday in April, May, and October, for sheep and cattle.

TAGANROG, a fortified seaport town of Russia in Europe, on the N. shore of the N.E. angle of the sea of Azoff, denominated the gulf of the Don, about 10 m. from the mouth of that river; lat. 47° 13' 45" N., long. 3° 30' E. Pop. about 18,000. The foundations of Taganrog were laid by Peter the Great in 1698; but it afterwards fell into the possession of the Turks; and it was not till the reign of Catharine II. that it became of any considerable importance. It has ten churches, of which three are built of stone; a gymnasium, a poor's hospital, &c. It was assisted by its illustrious founder to replace Azoff, the ancient emporium of the Don, the port of which had become unfit for inaccessible; and its whole consequence is derived from this circumstance, or from its being the entrepôt of the commerce of the vast countries traversed by that great river. The exports consist principally of corn, particularly wheat, iron and hardware from Tula; with cordage, hemp and sail-cloth, copper, tallow, wool, leather, furs, wax, ashes, caviar, isinglass, &c. The imports consist principally of wines, fruit, drysalteries, cotton and woollen goods, spices, dye-stuffs, tobacco, sugar, coffee, &c. The largest portion by far of the trade is carried on with Constantinople, Smyrna, and other Turkish ports; but a good deal is also carried on with the Italian and other foreign ports; and there is an tensive coasting trade with Odessa and other Russian ports. In 1836 the total value of the exports from Taganrog to foreign ports amounted to 7,422,377 roubles, and that of the imports to 7,864,118 do., the value of the exports the Russian ports amounting, during the same year to 969,925 roubles, and that of the imports to 1,829,233 do.

Seeing that Taganrog was built to obviate the difficulties it had to be encountered, by vessels entering the Don, rough the shallowness of the water, it might have been pposed that care would be taken to place it in a position which it should be, in as far as possible, free from this defect. This important consideration seems, however, to have en in a great measure overlooked. The gulf of the Don is dom navigable by vessels drawing more than from 8 to 9 water, and even these cannot approach within less than out 700 yards of the town. They are principally loaded carts, drawn each by a single horse, the expenses being very considerable that it costs from 190 to 150 copecks to p a chetwert of wheat

To obviate these inconveniences it has been proposed to to Kertsch, on the E. coast of the strait of Yenikali, a St for the produce of the sea of Azoff; and while the er would be much easier of access to foreign ships, the ters that at present bring down the products of the n of the Don from Nakhitchevan and Rostoff to Taganwould be able to bring them direct to Kertsch, where might be landed and shipped with much greater facility and less expense. In 1836, 761 vessels arrived at, and sailed from, Taganrog; but owing to the shallowness e water they were chiefly of small burden. With the ption of a few foreign houses, the merchants are mostly ther Greeks, or of Greek origin, and are not wealthy. e emperor Alexander, whose reign will always form a orable and brilliant era in the history of Russia, expired ganrog on the 19th of November, 1825. (For further parrs see Schnitzler, La Russie, p. 717; Hagemeister on the of the Black sea, p. 31, &c.; and Russian Offic. Acc.)

GHKANNIC, p. t., Columbia co., N.Y., 12 m. E. Hudl m. S. Albany, 347 W. Drained by Copake cr. It ro stores, two fulling-mills, one woollen factory. two ng-mills, two grist-mills, two saw-mills; six schools, llars. Pop. 1674.

HKANNIC. mts. N.Y., a range in the E. part of the state, extends 50 m. near the line of the state, commencee Rensellaer co., and extending through Columbia and es counties. Its highest summits are in Sheffield, 3000 feet above tide water. They divide the waters flow into Hudson r. from those which flow into Long sound.

US (Span. Tejo, Portug. Tojo), the principal and elebrated river of the Span. peninsula, through the of which it flows from E. to W., between the basins Ebro and Douro on the N. and the Guadiana on the S. is source in the Sierra Albarracin. on the borders of on and New Castile, about lat. 40° 25' N., long. 1° 35' n. W. Teruel, and only 90 m. from the Mediterranean. It runs N.W , but after having been joined by the Mocourse is generally W. or S.W., through New Castile remadura, in Spain ; and in Portugal between Beira mtejo, and through Estremadura to the Atlantic, t enters after expanding into a wide estuary, about ow Lisbon. Its entire length may be estimated at 00 m., about three fourths of which are in Spain. ipal tributaries are the Jarama, Alberche, Alagon, ere from the N., and the Rio del Monte, Salor, Sora, n the S. Aranjuez, Toledo, Talavera, Almaraz, e, Abrantes, Punheto, Santarem, and Lisbon, are nks: At its entrance into Portugal, the Tagus is s in width, and at Punheto upwards of 300 yards. labem, it expands into a wide basin, from 2 m. to ess; but opposite that city its breadth contracts to 2 m. The Tagus has been celebrated, both in and in modern times, for its picturesque beauty: however, can be more incorrect than these poetical ns. It flows in fact, for the most part, through an try, bare of wood, and uncultivated; its banks are steep, its current impetuous, and its waters turbumuddy. It was famous in antiquity for its golden grus aurifferas arenis celebratur. (Plin. Hist. Nat., p. 22; see. also, Silius Italicus, lib. vii., v., 755; etamorph., lib. ii. v. 251, &c.) At present, however few particles of gold are ever found in the sands ir ; and though they may have been more abundquity, the fair presumption seems to be, that it is or its celebrity, in this respect, rather to the yel- of its sands than to its gold.

the Tagus above Lisbon has not been of much l importance, though it be navigable as high as Attempts have, however, been made to render le from Toledo, and even Aranjuez; and no id the present liberal government maintain its Spain, this will be effected. Inglis says, that in

the winter of 1829, the passage downwards to the sea was successfully undertaken by a boat from Toledo. But, he adds, " this could not have been done at any other season ; because, in dry weather, the water is in many places almost wholly diverted from its natural channel, for the use of the mills that have been erected upon its banks." (Juglio's Spain, i. 297; Miñano Diccionario, &c.; Cellarii Geographia Antiqua, i. 68, &c.)

TAIN, a royal and parl. bor. of Scotland, co. Ross, on rising ground, near the S. shore of the frith of Dornoch, and near the mouth of the river Tain, 94 m. N.N.E. Inverness. Pop. of parl. bor. in 1841, 1857. " Tain has improved of late years, and may be considered in a thriving condition. Many new houses have been built, and a good deal of ground has been fenced. It possesses a good academy (founded in 1809), which has attracted a number of families to the town for the education of their children ; but it has little or no manufacture, and the sand bars on the coast deprive it of any advantage it might have derived from its maritime situation. A new entrance into the town, from the S., is contemplated." (Parl. Bound. Report.) It has a handsome par. church, erected in 1815, a grammar school, and three branch banks. It is associated with Cromarty, Dingwall, Dornoch, and Kirkwall, in sending one member to the House of Commons. Registered electors, in 1839-40, 86. It is governed by a provost, two bailies and 12 councillors. Corporation revenue, in 1840, £206.

TALAVERA DE LA REYNA, a city of Spain, New Castile, prov. Toledo, on the Tagus, 42 m. W. by N. Toledo, and 65 m. S.W. Madrid. Pop. about 8000. It stands in a large and fertile plain, and is divided into two parts by the river, which is here crossed by a stone bridge of 35 arches, and 530 yards in length. The remnants of old fortifications exist, but the town cannot, at present, be said to be fortified. " The modern town is very irregularly built, with low houses, and narrow and ill-paved streets ; it has eight parish churches, eight monasteries, and five nunneries, which appear to have nothing about them remarkable, (except, perhaps, the collegiate church, in which Miñano says there is a good picture of the Assumption). There are two handsome alamedas, or public promenades, which are but little frequented, the inhabitants being as generally sunk in apathy and sloth as in the days of their townsman Mariano." (Mod. Trav. xix. 221.) Talavera has an economical society, schools of Latin, philosophy, theology, &c., and had formerly some manufactures of silk stuffs, and porcelain. Its markets are tolerably well supplied with provisions. It is supposed to represent the Talabrica of the Romans: it was taken by the Moors in 714, and various Moorish remains are still to be seen in the city and its neighbourhood. After many vicissitudes it was destroyed by the Moors in 1109, but was speedily rebuilt. It afterwards became an apanage of the queens of Spain, whence its name.

In modern times it has been rendered famous by the obstinate battle fought in its neighbourhood, on the 27th and 28th of July, 1809, between the British and Spanish forces under Sir Arthur Wellesley (now Duke of Wellington), and the French under Joseph Buonaparte, assisted by Marshals Jourdan and Victor. The French, who commenced the attack, were repulsed at all points. The slaughter was great, and nearly equal on both sides. (Mariano; Napier's Peninsular War, ii. 393, &c.)

TALBOT, county, Md. Situated toward the E. part of the state on the Eastern shore of Chesapeake bay, and contains 250 sq. m. Bounded E. and S. by Choptank r., W. by Chesapeake bay, N.W. by Wye r. Several bays set up from the Chesapeake, the chief of which are Tread haven and St. Michaels. It contained in 1840, 8850 neat cattle, 9640 sheep, 14.296 swine: and produced 222,822 bushels of wheat, 4494 of rye, 517,239 of Indian corn, 40,151 of oats, 96,294 of potatoes. It had one commission house in foreign trade, 59 retail stores, six grist-mills, one saw-mill, two tanneries ; two printing-offices, one weekly newspaper; four academies, 78 students; 21 schools, 870 scholars. Pop.: whites, 6083; slaves, 3687; free coloured, 2340 ; total 12,090. Capital, Easton.

TALBOT, county, Ga. Situated S.E. of the centre of the state, and contains 400 sq. m. Bounded N.E. by Flint r. It contained in 1840, 15,822 neat cattle, 6342 sheep, 46,095 swine : and produced 71,743 bushels of wheat, 1705 of rye, 671,430 of Indian corn, 33,850 of oats, 16,755 of potatoes, 6,772,393 pounds of cotton. It had 34 stores, four flouring-mills, 38 grist-mills, 34 saw-mills, five tanneries ; two academies, 101 students; 17 schools, 610 scholars. Pop. : whites, 8861; slaves, 6746; free coloured, 90; total, 15,697. Capital, Talbotton.

TALBOTTON, p. v., capital of Talbot co., Ga., 92 m. W.S.W. Milledgeville, 798 W. It contains a courthouse, jail, several stores, two churches, a Methodist and Baptist, two academies, and about 800 inhabitants. Nett proceeds of the postoffice in 1841, $861.

TALLADEGA, county, Ala. Situated toward the E. part of the state, and contains 1200 sq. m. Bounded W. by

Coosa r. Drained by Talladega r. It contained in 1840, 16,696 neat cattle, 3825 sheep, 39,803 swine; and produced 41,106 bushels of wheat, 610,537 of Indian corn, 96,218 of oats, 24,005 of potatoes, 2495 pounds of rice, 6026 of tobacco, 2,610,191 of cotton. It had 17 stores, 15 grist-mills, 11 saw-mills, three tanneries, one distillery, one printing-office, one weekly newspaper; six academies, 189 students; 15 schools, 358 scholars. Pop.: whites, 7663; slaves, 4896: free coloured, 26; total, 12,587. Capital, Talladega.

TALLIAFERO, county, Ga. Situated in the N.E. part of the state, and contains 130 sq. m. Drained by branches of Little and Ogeechee rivers. It contained in 1840, 5487 neat cattle, 3527 sheep, 12,215 swine; and produced 16,627 bushels of wheat, 165,411 of Indian corn, 21,336 of oats, 5136 of potatoes, 1,486,406 pounds of cotton. It had 10 stores, nine grist-mills, six saw-mills; six schools, 201 scholars. Pop.: whites, 2295; slaves, 2856; free coloured, 39; total, 5190. Capital, Crawfordsville.

TALLAHASSEE, city, capital of Leon county and of the territory of Florida, 20 m. N. St. Marks, 292 m. W.S.W. St. Augustine, 896 W. The situation is elevated, and a fine mill-stream, issuing from several springs, flows on its E. border, with a fall of 15 or 16 feet, but soon sinks into a cleft of limestone rock, a frequent occurrence in this country. It is regularly laid out, containing a number of public squares, and has a statehouse, a courthouse, jail, market-house, an academy, three churches, an Episcopal, Methodist, and Presbyterian, a bank, 30 stores, three printing-offices, three weekly newspapers, 400 dwellings, and 1616 inhabitants, which in the winter season at times amount to 2500. A railroad to St. Marks, which may be regarded as its port, greatly facilitates business.

TALLAHATCHEE, county, Miss. Situated toward the N.W. part of the state, and contains 1188 sq. m. Drained by Tallahatchee river and its branches. It contained in 1840, 4737 neat cattle, 890 sheep, 10,259 swine; and produced 1887 bushels of wheat, 136,760 of Indian corn, 7046 of oats, 11,230 of potatoes, 1,596,965 pounds of cotton. It had five grist-mills, two saw-mills, three schools, 48 scholars. Pop.: whites, 1362; slaves, 1591; free coloured, 2; total, 2985. Capital, Charleston.

TALLAPOOSA, county, Ala. Situated toward the E. part of the state, and contains 910 sq. m. Watered by Talla-poosa river and its branches. It contained in 1840, 11,991 neat cattle, 1025 sheep, 13,738 swine; and produced 11,683 bushels of wheat, 159,580 of Indian corn, 2903 of oats, 10,484 of potatoes, 1550 pounds of tobacco, 217,609 of cotton. It had 12 stores, eight flouring-mills, 19 grist-mills, 13 saw-mills, three tanneries, three distilleries; one academy; 10 schools, 948 scholars. Pop.: whites, 4494; slaves, 9013; free coloured, 7; total, 6444. Capital, Dadeville.

TALLMADGE, p. t., Summit co., O., 128 m. N.E. Columbus, 332 W. Watered by Cuyahoga and Little Cuyahoga rivers. It contains the village of Middlebury, and a part of that of Cuyahoga Falls. In the N.W. part is an extensive body of bituminous coal, which is mined and considerably exported. It contains two Presbyterian churches, eight stores, two furnaces, one distillery, three saw-mills, one oil-mill, two paper-mills, one brewery, one printing-office, one weekly newspaper; one academy, 35 students; nine schools, 273 scholars. Pop. 2134.

TAMBOFF, a central government of European Russia, principally between the 52d and 55th degs. of N. lat., and the 40th and 43d of E. long., having N. Vladimir and Nijni-Novgorod, E. Pensa and Sarato, S. Voronege, and W. chiefly the latter and Riazan. Its length N. to S. is about 350 m., breadth varying from 100 to 250 m. Area, 24,420 sq. m. Pop. in 1838, 1,591,700. Surface flat, except in a few parts, where it is slightly undulating. Principal rivers the Tsna and Mocksha, tributaries of the Oka, flowing N.; and the Vorona, a tributary of the Don, flowing S. In the N. the soil is sandy and marshy; a large proportion of the country, principally the marshes, being covered with forests: in the E., or steppe, so called from its being bare of wood, the soil consists principally of a black mould, and is comparatively fertile. The government comprises in all 5,912,229 declatines (a declatine = 2·7 Eng. acre), which, according to the notice published in 1833, by M. Korsakoff, formerly vice governor of Tamboff, were distributed as follows:

	Declatines.
Cultivated and cultivable lands	2,936,177
Meadows and pasture grounds	1,513,388
Forests of the crown . . 153,788 }	1,035,441
Do. individuals . . 881,675 }	
Towns and villages, with their dependencies	93,197
Communal properties	796,749
Roads, marshes, and waste lands	315,270
Total	5,912,222

Corn is the principal product; but according to the official accounts, the crops are extremely variable, and scarcities frequently occur. The crop of 1809, for example, was es-

timated at 9,294,897 chetwerts, and that of 1821 at only 5,923,796 chet. (a chetwert = 6 bushels nearly). In 1822 an abundant year, 800,000 chetwerts were exported to Moscow and Petersburg. We should not have thought it worth while to be so particular in these notices, but for the fact of its having been stated, in a consular return from Petersburg, that in 1835 no fewer than 26,000,000 quarters of corn were produced in this government! That any public officer should have made such a statement, and that the foreign office should have printed it without note or comment of any kind, is, if not a singular, at all events rather a remarkable circumstance. It may be affirmed, with the utmost confidence, that the entire produce of the government never amounted to 12 million quarters, either in 1835 or any other year. We have already seen that the cultivated and cultivable lands in the government amount to about 2,936,00 declatines, or to very near 6,000,000 acres. Now supposing this land to be all in cultivation (which it is not), the invariable practice here is to take two crops and then fallow, so that a third part of the land is constantly waste, leaving 4 millions of acres under crop. If each acre produced, at an average, three quarters, the total produce would be 12 million quarters; but when we bear in mind the backward state of agriculture, and the poverty of the soil in the E. half of the government where the return is said not to exceed 3 or 4 times the seed, it will be immediately seen that if we estimate the entire produce in ordinary years at 6,000,000 quarters, we shall most probably be a good deal beyond rather than within the mark. It is to be observed, too, that by far the larger portion of the produce is rye, wheat not constituting above one fifth part of the total produce, and we may farther observe, that it would cost from 15s. to 20s. to convey a quarter of this wheat either to Petersburg or to a ship in the roads of Taganrog, in the sea of Azoff. The peasantry are well treated, and in good circumstances. Hemp is extensively grown; the value of the quantity exported amounting, according to Schnitzler, to 1,000,000 roubles a year.

The forests along the Mocksha supply a good deal of timber for ship and boat building; and the inhabitants are there principally wood-cutters, carpenters, coopers, or pitch and tar makers. Cattle, principally brought from the steppes of the Don, the Wolga, and the Caucasus are numerous, and are extensively fattened for the neighbouring governments, and for Moscow and Petersburg. In 1832 the stock of sheep was estimated at 1,000,000 head, and 670,000 hogs. The horses belonging to the gentry are good, and have been much improved by the stud kept by the Oztoff family; but the horses of the peasantry are wretched. The manufacture of woollen cloth is carried on to a considerable extent. Peter the Great established an extensive cloth manufactory, for the service of government, at the village of Bonden. This, however, was burnt down in 1836; but having been since rebuilt on a great scale, it now (1842) gives employment to about 9000 males, and 1150 females: the consumption of wool is stated at 50,000 poods a year; and besides furnishing 440,000, arschines of cloth annually for the army, it produces other goods, worth 1,500,000 roubles. It has also numerous forges, distilleries, tallow factories, salts, of which a very considerable part belongs to Count Kanakrine, &c. Principal towns, Tamboff, Morchansk, Chausk, Jelazno, Lipetsk, &c. (See Schnitzler, La Russie, p. 36; Possart das Kaiserth Russland, p. 563; Venables' Russia, p. 333, &c.)

TAMBOFF, a town of Russia in Europe, cap. of the above gov., about the centre of which it is situated, on the Tsna, 385 m. S.E. Moscow; lat. 52° 43′ 17″ N., long. 41° 43′ 17″ E. Pop. in 1830, according to an official statement, 21,142. The town, which is about 2 m. in length, by 1 m. in breadth, was originally founded and fortified in 1636 as a defence against the incursions of the Nogai Tartars. The houses are principally of wood: but there are various stone churches, a large monastery, gymnasium, civil hospital, a military orphan asylum, &c. In the school of cadets at Tamboff, about 100 pupils, sons of nobles, are instructed in French, German, military exercises, &c.; and they most intelligent are afterwards sent to the corps de cadets at Petersburg. A high school for young ladies was founded in 1834, and there are various other schools. Manufactures of woollen cloth, alum, vitriol, &c., are established; and the town has active general trade.

TAME, a river of England. See TEAMES.

TAMWORTH, a parl. and munic. bor., market town, and par. of England, principally in the co. Stafford, but partly also in Warwickshire, being divided into two parts by the Tame, where it is joined by the Anker, 64 m. S.E. Lichfield. Area of Parl. bor., which the Boundary Act made co-extensive with the par., 19,990 acres. Pop. of do. in 1831, 7189. The town is well built, and a handsome bridge is thrown across each of the rivers. According to the Munic. Bound. Rep., the pop. of both the bor. and par. appeared, at the time (1837), to be stationary, or decreasing rather than otherwise. This may be attributed in some measure to the decay of the staple manufacture of the place, and to the breaking up of the large establishment of the late Sir R. Peel. The town was not then lighted; but the inhabitants were about to assess themselves under the general act for that purpose. The gas works were nearly completed, and it is now well lighted. It is paved at the expense of the corporation. The parish church, dedicated to St. Editha, is supposed to occupy the site of a very ancient nunnery. "It is a large, handsome edifice, with a fine tower, and a crypt under part of the church. Some portions are of decorated date, and some perpendicular, and both good; some of the windows have had very fine tracery. In the tower is a curious double staircase." (*Rickman's Goth. Arch.*) Numerous monuments adorn the interior of this church. The living, a perpetual curacy, worth £170 a year, is in the gift of — Ressington, Esq. There are various dissenting chapels, a hospital for 14 poor men and women, founded and endowed by Thomas Guy, the founder of the famous hospital in Southwark which bears his name; a grammar school, which received endowments both from Edward VI. and Elizabeth. Very recently a free-school has been established by the present Sir Robert Peel, and there are other schools. Boys from the grammar school are eligible to a scholarship in Catherine Hall, Cambridge, and a native of the town to a fellowship in St. John's college, Cambridge. On an artificial height, near the town, is Tamworth castle, a seat of the Townshend family. This castle, though now much modernised, is of great antiquity, having, according to some authorities, been founded by Ethelfleda, daughter of Alfred. It was conferred, with the town of Tamworth, by William the Conqueror, on Robert de Marmion, Lord of Fontenay in Normandy, the exploits of one whose supposed descendants have been immortalised in the best of Scott's poems.

Calico-printing, and the manufacture of superfine narrow woollen cloths, were the chief branches of industry at Tamworth; but of late years they have materially declined.

Tamworth, which is on the line of the Roman Watling Street, was a place of much consequence, and the favourite residence of the Mercian kings during the heptarchy. It appears to have been a borough by prescription; but was re-incorporated by Elizabeth. The municipal borough, which is much less extensive than the parliamentary borough, is governed by four aldermen and 12 councillors; it has no commission of the peace, though quarter sessions for civil causes are held. Corporation revenue, in 1840, £377 7s. 8d. The commissioners of inquiry into the municipal affairs of the different boroughs, speak highly of the past government of Tamworth. "The governing body is wholly self-elected; it does not appear, however, that the power thus vested in the body has been in any respect abused. Neither does it appear that the corporation, either as regards the appointment of members to the body corporate, or the exercise of the elective franchise, have been subject to the operation of any undue local influence. The absence of all complaint leads to the conclusion that the objects of municipal government have been satisfactorily attained in this borough; that the governing body has been judiciously selected, justice well administered, and the revenues carefully applied to public purposes." Tamworth has sent two members to the House of Commons once the 5th of Elizabeth. Previously to the Reform Act, the right of voting was in the inhabitants paying scot and lot. Registered electors, in 1836–40, 501. The present distinguished parliamentary leader, Sir Robert Peel, Bart., has presented Tamworth for a lengthened period, and has been one of its greatest benefactors. His seat, Drayton manor, is about 1 m. S. the town. (*Mun. Corp. and Bound. Rep. Appendix, &c.*)

TAMWORTH, p. t., Carroll co., N. H., 55 m. N. Concord, 9 W. Watered by Bearcamp, Conway, and Swift rivers. Partered in 1766, first settled in 1771. It has five churches — a Congregational, two Methodist, and two Free-will baptist, three stores, one fulling-mill, seven saw-mills; 19 schools, 672 scholars. Pop. 1717.

TANEY, county, Mo. Situated in the S., toward the W. of the state, and contains 1496 sq. m. Drained by white river and its branches. It contained, in 1840, 8133 cattle, 2424 sheep, 11,496 swine; and produced 6996 bushels of wheat, 182,102 of Indian corn, 5577 of oats, 3844 potatoes, 17,205 pounds of tobacco, 10,650 of cotton, 5000 sugar. It had three stores, 12 grist-mills, six saw-mills, one distilleries. Pop.: whites, 2212; slaves, 40; free coloured, 12; total, 2264. Capital, Forsyth.

TANJORE, a distr. of Hindostan, presid. Madras, and one of the most valuable in British India. ranking in point of cultivation and productiveness next to Burdwan in Bengal. It is principally between lat. 10° and 11° 30', and the 79th and 80th degrs. of E. long., having N. the district of Trichinopoly, W. Madura, and the ocean on the S. and E. Area, 3376 sq. m. Pop., in 1831, 1,198,730. About half the prov-

ince is a flat alluvial delta formed and completely irrigated by numerous branches of the Coleroon, which constitutes the N. boundary. This delta is justly considered the granary of the Madras territories; almost the whole of it is cultivated with rice, which is here produced in larger quantities and with more certainty than in any other district on the E. coast. The average gross produce in rice (not paddy) yearly is estimated at 58,046 *garce.* (*Madras New Almanac,* 1839.) The rest of the district S. of the delta is on a considerably higher level; its surface is undulating, and it comprises many varieties of soil.

Tanjore was formerly assessed under the ryot-war system, but this was afterward abandoned for the village settlement, under which last both the revenue and cultivation have increased very considerably. (*Cotton, in Rep.* 1832.) The district is famous for its excellent roads.

The population is for the most part Hindoo, and chiefly agricultural; but there are some manufactures of cotton and silk stuffs, of copper utensils at Comboconum, Manar-goods, &c. These, however, have declined greatly of late years, owing to the importation of cheaper English goods, though some manufactured articles are still exported with the agricultural produce to Bengal, Achin, Tranquebar, and the adjacent districts. The imports are iron, saltpetre, dry grain, oil, glue, wax, tamarinds, &c., from Coimbatore, Salem, Trichinopoly, &c. The trade of Tanjore is very considerable: in 1837–38 the value of the exports was estimated at 2,619,259 rup. Total public revenue in the same year, 4,738,607 rup.; expenditure, 398,677 rup.; so that in a pecuniary point of view it is a very valuable possession.

Tanjore was never permanently conquered by the Mohammedans, and Hindoo institutions and edifices have been preserved in it in much purity and perfection. In almost every village there is a pagoda, with a lofty gateway of massive, though not elegant, architecture, in which sundry Brahmins are maintained; and on all the great roads leading to these places are choultries for the accommodation of pilgrims. The district has been noted for the prevalence of *suttees.* In antiquity it constituted the principality of Chola, whence the whole coast afterward acquired the name of Coromandel. It was conquered by the Mahrattas in 1675; but we came quietly into its possession in 1799, on condition of allowing the rajah a lac of star pagodas and a fifth part of its nett revenue annually. (*Madras Almanac for* 1838; *Hamilton; and Parl. Reps.*)

TANJORE, a large city of Hindostan, cap. of the above distr., in a plain S. of the Coleroon, and 170 m. S.E. Madras. Lat. 10° 45' N., long. 79° 19' E. Its population is not stated, but is probably from 35,000 to 40,000. It is said to be nearly 6 m. in circumference, and consists of two separate portions, both fortified; one comprising the palace and other public buildings; and the other a celebrated pagoda, perhaps the finest specimen in India of a pyramidical temple. Its grand tower is 199 feet in height, and is remarkable for its simplicity. In a covered area in this temple is a bull carved in black granite, 16 feet in length by 19½ feet in height, deemed one of the best works of Hindoo art. Close to the temple stands an English church. According to Hamilton, Tanjore appears to have been pretty strongly fortified; and the city is more regularly built, and has a larger proportion of solid and ornamental edifices, than any other native town S. of the Krishna. In the palace is a group of the late rajah and his tutor, the Danish missionary Schwartz, executed at the desire of the rajah by Flaxman. The British residency is outside the walls, to the S. Tanjore was unsuccessfully besieged by the British in 1749, and the French in 1758; but was taken by the former in 1773. (*Hamilton; Heber, in Mod. Trav., &c.*)

TAORMINA (an. *Tauromium,* or *Tauromenium*), a town of Sicily, prov. Messina, cap. cant., on a high, craggy mountain, on the E. coast of the island, about half way between Messina and Catania, being 30 m. S.W. the former, and 31 m. N.E. the latter. Pop., in 1831, 3029. Travellers speak in the highest terms of the surrounding scenery. "Were I," says Swinburne, "to name a place that possesses every grand and beauteous qualification for forming a picture—a place on which I should wish to employ the powers of a Salvator or a Poussin, Taormina should be the object of my choice. Everything belonging to it is in a large, sublime style." It is fortified by an irregular wall and lines, constructed by the Saracens, surmounted by an old Saracenic castle and more modern works; and above all, on the summit of a tabled cliff, is the inconsiderable town and military post of Mola. Though Taormina has an immoderate proportion of convents and large buildings, it is ill built and dirty, and, notwithstanding its elevated situation, it is said to be but indifferently healthy. (*Smyth.*) On a fountain, in the main street, part of the statue of a centaur, with the addition of a copper nimbus, is held to represent St. Pancras, a native of the town, and its protector!

Taormina has some splendid remains of antiquity. Its

theatre, which is most probably of Greek origin, is the object of universal admiration. It is of very ample dimensions, being capable of accommodating no fewer than 40,000 spectators (*Smyth*), and is wonderfully well preserved. It is principally excavated in the slope of the mountain, its seats being hewn out of the rock: the *proscenium* and parts connected with the stage are built of brick, and are nearly entire; the space allotted to the orchestra is also preserved, as well as the dressing-rooms of the actors. Its greatest breadth (measured on the plan in *Russell's Sicily*) is about 360 feet; its extreme length, 300 feet; and it is so admirably contrived that, even now, the slightest noise, as the tearing of a piece of paper on the stage, is distinctly heard in the most distant part of the theatre. The seats command the most superb views of mount Etna, Aci Reale, Catania, and, it is said, of the country even as far as Syracuse. "The spot," says Sir R. C. Hoare, "seems to have been created for a public edifice: behind and before are steep precipices, which leave just room sufficient to place this most noble and magnificent structure. I visited it frequently, and never left it without regret." (ib., 193.) In addition to the theatre, Taormina has an entire side of a naumachia, upward of 350 feet in length, with the remains of the aqueduct and the reservoirs that supplied it with water; and in every direction round the town are sepulchres, cenotaphs, tesselated pavements, remains of remarkable edifices, &c., attesting its ancient wealth and magnificence. The churches have little remarkable, though that of St. Pancras appeared to Sir R. C. Hoare to be of Grecian origin, and probably, he says, the oldest building in the town. The Dominican convent has a large court, surrounded with columns of fine brown and white marble. The inhabitants have some trade in wines and hemp, the former being, it is said, of superior quality, though very inferior to what they must have been in antiquity, when they occupied a high place at the Roman banquets.

Tauromenium is of uncertain origin: it was taken by Dionysius the Elder, in the 94th Olympiad, or about anno 403 B.C. A Roman colony was settled in it by Julius Cæsar. The ancient city was ruined by the Saracens in 963; since which it has never recovered any considerable portion of its ancient importance. (*Smyth's Sicily*, p. 128, 129; *Russell*, p. 241–248; *Sir R. C. Hoare, Class. Tour*, p. 194–200; *Swinburne; Brydone*.)

TAPPAHANNOÇ, p. v., port of entry, and capital of Essex co., Va., 50 m. N.E. by E. Richmond, 112 W. Situated on the S.W. side of Rappahannock river, 50 m. from its mouth in Chesapeake bay, in a low and unhealthy location. It contains a courthouse, jail, a church free to all denominations, a female seminary, four stores, 50 dwellings, and about 500 inhabitants. A ferry here crosses Rappahannock river. Large vessels come to the place, and all the shipping belonging to the towns on the river is entered at the custom-house at this place. Tonnage in 1842, 5682.

TAR r., N. C., rises in Person co., and, running southeastwardly about 180 m., it enters Pamlico sound. Vessels requiring 9 feet of water navigate it to Washington, and small steamboats to Tarboro'.

TARANTO (an. *Tarentum*), a famous city and seaport of S. Italy, kingd. of Naples, cap. of the prov. Otranto, anciently one of the wealthiest and most celebrated cities of Magna Grecia, near the N. extremity of the gulf of Taranto, 42 m. W.S.W. Brindisi, and 160 m. E.S.E. Naples. Lat. 40° 28' N., long. 17° 33' E. Population estimated by Burgess and others at 18,000. It stands on what was formerly an isthmus, but is now an island, separating the gulf, or outer sea, from an inner bay, called the Little sea (*Mare Piccolo*), 15 or 16 m. in circumference. At its N. extremity is the old channel, leading to the Mare Piccolo, crossed by a bridge about 160 yards in length, over which an aqueduct is brought, conveying water to the city from the mountains of Mutina, about 15 m. distant. The channel on the S. side of the town is artificial, having been originally opened by Ferdinand I., and deepened by Philip II.: it also is crossed by a bridge about 50 yards in length. In antiquity the citadel occupied the site of the modern city.

The harbour of Taranto is excellent, and might, with little difficulty, be made all but perfect. In antiquity the Mare Piccolo, or inner bay, was the principal rendezvous of the Tarantine ships, where they lay perfectly secure from hostile attacks, and as safe in other respects as if they had been in dock. The entrance to the inner bay is now, however, so choked up with rubbish, that it is accessible only to small boats; but it might be easily cleared out, and the basin rendered as useful as ever. Adjoining the town, the Mare Piccolo has from 4 to 6 fathoms water. The present or outer harbour is at once extensive and safe. There are 4 fathoms water close to the town; and the bay, which is capacious, is protected by the islands of St. Peter and St. Paul.

Its situation is striking and singular, and despite the change in its fortunes, Swinburne says its appearance is

replete with wonderful beauties. The ancient city extended along the shores both of the gulf and the Mare Piccolo, and the walls which ran from the one to the other had the base of the triangular space which it covered. Of its magnificent buildings, which included a temple of Neptune, the guardian deity of the city, scarcely any vestige is left, except the outlines of an amphitheatre, some substructures of apparently a Roman work, and an immense mass composed of fragments of pottery. "The shape of the modern city has been, with some justice, assimilated to a ship, being wide in the centre, and tapering at each end. The principal street runs from one of its extremities to the other in a waving line; and narrow and tortuous communications lead to two other parallel streets; one of which extends along the waters of the outward gulf, but considerably above their level, and is defended from their fury by a parapet wall and projecting battery. Here the best houses are situated. The Marina, on the contrary, which borders the inner bay, or Mare Piccolo, is scarcely raised above its surface; and nothing can present a stronger contrast than the crowded, filthy, but lively appearance of the last, opposed to the quiet, clean, but deserted aspect of the former. The Marina is inhabited entirely by fishermen and their families, who constitute more than half the population of the place." (*Craven, Tour*, p. 176.) The cathedral, dedicated to San Cataldo, a native of Ireland, is richly adorned within, and has a silver statue of its patron saint, some ancient columns, inlaid work in *pietra dura*, &c. Taranto has a formidable looking castle commanding its harbour, numerous convents, a diocesan seminary, two hospitals, an orphan asylum, and manufactures of linen and cotton fabrics, velvets, &c. The great articles of commerce are oil and shellfish. The taste for the latter is said to prevail at Taranto in a greater degree than in any other part of the kingdom; and when Craven visited the city, the governor drew a revenue from this branch of trade of about 34,000 ducats a year. Swinburne speaks of having partaken at a single meal of 15 different sorts of shell-fish. (*Trav.*, i., 27.) The neighbourhood was anciently famous for the murex and purpura, but these have been superseded by muscles, oysters, &c., which are reared in immense numbers in the Mare Piccolo.

Tarentum was either originally founded, or, as is most probable, occupied by a colony from Sparta, about anno 700 B.C. The colonists, influenced, no doubt, by the advantageous situation of their new country for a seat of commerce and commercial navigation, became, in no very lengthened period, distinguished for their proficiency in these departments of industry, and their city is admitted to have been the greatest emporium of S. Italy, or Magna Grecia. *Tarentus Lacedæmoniorum opus, Calabriæ quondam, et Apuliæ, totiusque Lucaniæ caput, cum maritimis et maris portuque nobilis, tam mirabilis situ: quippe ut ipsis Adriatici maris faucibus posita, in omnes terras, Istriam, Illyricum, Epirum, Achæam, Africam, Siciliam vela dimittit.* (Florus, lib. i., cap. 18.) Polybius also has stated the commercial advantages enjoyed by Tarentum. (Lib. x., Frag. 1.)

The history of this great city is very imperfectly known. Her government, like that of most other Greek states, was different at different periods, being sometimes administered by kings or tyrants, and sometimes by the people. She was distinguished not only by her wealth and commerce, but by the splendour of her public buildings and works of art. She also became a favourite seat of literature and science; and the followers of Pythagoras, though proscribed in other parts of Italy, found here a safe asylum. The famous philosopher Archytas, a disciple of Pythagoras, was repeatedly placed, by the suffrages of his fellow-citizens, at the head of the government; and showed, by his judicious conduct in civil affairs, and as leader of the armies of the republic, that he was no less eminent as an administrator and a general, than as a moralist, a mechanist, and a geometer. (*Bruckeri, Hist. Philosoph.*, i., 1118; see also *Horace's Ode to Archytas*, lib. i., ode 28.)

The refinement produced by the accumulation of wealth and the culture of literature and the fine arts, has been supposed by most ancient writers, and by their copyists in modern times, to have had a most injurious influence over the martial virtues of the Tarentines, and to have occasioned an all but universal degeneracy and corruption of manners. There does not, however, appear to be any real ground for such imputations. When the progress of Roman conquest and universal dominion brought her armies and fleets into the territories and seas adjoining Tarentum, the latter did not seek to purchase a treacherous truce, by submitting to the dictates of the Roman generals. On the contrary, she made every effort to maintain her independence; and, as she knew that her own forces were inadequate for such a struggle, she wisely sought assistance from others; and it was at her instigation that Pyrrhus invaded Italy. After the departure of Pyrrhus, Tarentum attached herself to the rty of Hannibal; and it was not owing to any deficiency of bravery, but to treachery, that Fabius ultimately obtained possession of the city.

The conduct of the Romans on this was consistent with their behaviour on every similar occasion. The city was delivered up to military execution; and such of the inhabitants, amounting to about 30,000, as had escaped the massacre, were sold for slaves. (*Livius*, lib. xxvii., caps. 13- *Plutarch's Life of Fabius, &c.*)

Tarentum never fully recovered from this dreadful blow; though, notwithstanding the preference shown by the Romans for Brundusium, she had again become, in Strabo's time, a considerable city. A little to the N.E. of Tarentum, on the Galesus, were situated the fertile valley and ridge union, the beauties of which have been described in glowing terms by Horace. (Lib. ii., od. 6.) In addition to the authorities referred to above, see also *Ancient Universal History*, xii., 146, and 208, 800, id.)

TARARE, a manufacturing town of the S. of France, Rhone, cap. cant., in a narrow valley on the road from to Lyons, 20 m. N.W. the latter. Pop., in 1836, 5990; sc. comm., 7769. It is the centre of a manufacture of and figured muslins, which, within a circle of from 20 leagues of mountainous country, employs wholly part at least 50,000 hands, about 20,000 being adult loom weavers, from 15,000 to 16,000 women and children employed subsidiary to these, from 4000 to 5000 employed as agents or otherwise by the manufacturers, the rest, ly females, being occupied in embroidering or figuring plain goods. Most of the weavers, &c., work at their homes, and the manufacturers do not generally carry business on a large scale, or employ many hands. There is a factory, however (that of the Messrs. Macculloch Glasgow), in which from 180 to 200 hands are employed; goods are produced which are said to be fully equal to those of Glasgow, though we are not disposed to place it confidence in this statement. "A portion of the workers in the country get from 75 cents. to 1½ fr. a day; of the town from 2 fr. to 3 fr. 50 c. and sometimes 3 according to their skill. The manufacturer furnishes the reed and the upper mounting, all the rest being at pense of the workmen. Those who are in the town all the year round, whereas those in the country do have for more than seven months a year, the remaining months being occupied in agricultural employment. *Loom Weavers' Rep.*, i., 124.)

Tarare weavers are pretty well lodged, fed, and clothed Villermé relates, *Je ne connais aucune fabrique où les ticserands m'aient paru avoir des mœurs et des habitudes meilleures, aucune ville manufacturière qui offert moins d'ivrognes et moins de libertins que Tarare* (Tableau des Ouvriers, &c., i., 188.)

TASCON, a town of France, dep. Bouches-du-Rhone, cant., on the Rhone, opposite Beaucaire, with which it communicates by a new suspension-bridge, 13 m. E. by S. Pop., in 1836, 9290. It is surrounded with walls by towers, and is commanded by a castle on a rock changing the river, built in the 13th century, and formerly the residence of the counts of Provence. The streets are and regular, and one of the principal is lined with The parish church, a fine Gothic edifice of the century, has a richly sculptured entrance, and a subterranean chapel, in which is a marble statue of St. Martha. It has a public library, a theatre, town-hall, court house, two hospitals, barracks, &c.; and in the neighbourhood is a very extensive nursery called the *Pepinière elle*. It has also manufactures of silk and woollen some trade in boat-building, and in wine, brandy, (*Hugo*, art. *Bouches-du-Rhone; Guide du Voyageur.*)

TES (an. *Bigorra*, post *Turvia*), a town of France, hautes Pyrénées, of which it is the cap., in a fine plain, verdour, here crossed by a stone bridge of six arches, by S. Pau. Pop., in 1836, 12,500. It is one of the it and cleanest towns in the S. of France. Its constructed chiefly of marble, stone, or brick, and with slate, have usually gardens attached, of considerable size. A wide main street, containing numerous cafés, runs through the centre of the town, which divided into three nearly equal portions, by two open spaces; one, the *Place de Maubourguet*, being with trees, and forming a favourite promenade, streets crossing the main thoroughfare are almost wide and regular, and nearly all lead into suburbs, there are five, surrounding the town on every quarter of Tarbes is well supplied with water river. There are few public edifices worth notice most so is the prefecture, formerly the bishop's building of different dates, but with an imposing effect. The cathedral is on the site, and, it is said, of a portion of, the ancient fortress of *Bigorra*; it is ly adorned with some columns of Italian *breccia*.

76

The old castle of the counts of Tarbes now serves for the prison. The college and theatre are handsome. Tarbes is a bishop's see, the seat of tribunals of primary jurisdiction and commerce, a forest board, &c., and has schools of design and architecture, a royal *depôt d'étalons*, and some manufactures, principally of copper, iron, and other metals. It is the great commercial *entrepôt* for the country immediately N. of the Pyrenees, and has a large market, once a fortnight, frequented by individuals from a distance of even 20 leagues round. "Here," says Inglis, "for the first time, one perceives a slight approximation to the usages of the country that lies beyond the majestic barrier of the Pyrenees. This is visible in the dress of the women, who no longer cover their heads with bonnets, hats, caps, or handkerchiefs, but with scarlet squares of woollen stuff, trimmed with black, which they throw over the head and shoulders, something in the form of the Spanish mantilla, &c." (*Inglis's Pyrenees, &c.*, 298.) Tarbes is on the direct road to Bagnères de Bigorre (which see), and to Baréges, distant only about 22 m. S. (*Hugo*, art. *Hautes Pyrénées; Dict. Geog. &c.*)

TARBORO', p. v., capital of Edgecombe co., N. C., 76 m. E. by N. Raleigh, 209 W. Situated on the W. side of Tar river, at the head of steamboat navigation, 85 m. above its mouth in Pamlico sound. It contains a courthouse, jail, an academy, a bank, and about 600 inhabitants.

TARN, a dep. of France, reg. S., formerly inc. in Languedoc; principally between lat. 43° 30' and 44° 10' and long. 1° 30' and 3° E., having N. and N.E., Aveyron; S.E. and S., Herault and Aude; S.W., Haute Garonne; and N.W., Tarn-et-Garonne. Area, 573,977 hectares. Pop., in 1836, 335,844. This dep. is enclosed by mountain-ranges on the N.E. and S.; it slopes to the W., in which direction its rivers, the chief of which are the Tarn, Agout, and Viour, have their courses. The Tarn rises in mount Lozere, and flows in a general S.W. direction, through the deps. Lozere, Aveyron, and Tarn, to about 15 m. from Toulouse, where it turns N.W., and ultimately falls into the Garonne, 22 m. above Agen. Its principal affluents are the Aveyron, on the right or N. side, and the Agout on the left. Florac, Milhau, Alby, Montauban, and Moissac are on its banks. It is navigable for about 90 m. from its embouchure. In 1834, the arable lands in this dep, were estimated to comprise 395,610 hectares; meadows, 41,848 do.; vineyards, 31,943 do.; woods, 80,291 do.; and heaths and wastes, 61,439 do. (*Offic. Tables.*) With the exception of the mountain tracts, the soil, speaking generally, is extremely good; and the valleys are not inferior in fertility to any in France. Agriculture, however, is in a very backward state; and the rotation of crops can hardly be said to be introduced. But the supply of corn, notwithstanding, exceeds what is required for home consumption. It produces from 400,000 to 450,000 hectolitres of wine, of which that of Gaillac, partly red and partly white, is the best. The former has *une couleur tres foncée beaucoup de corps, du spiritueux, et un bon gout*. (*Jullien*.) It is improved by a sea voyage; the white wines have similar qualities. Before the introduction of indigo into commerce, a good deal of woad was raised in the dep., and it is still cultivated round Alby. Cattle, of a good breed, are rather numerous; and the produce of wool is estimated at 150,000 kilogr. a year. In 1835, of 94,479 properties subject to the *contrib. foncière*, 49,613 were assessed at less than 5 fr., and only 107 at from 1000 fr. upwards. There are mines of iron, copper, coal, and marble. Near Alby is a very extensive work for the conversion of iron into steel. The manufacture of cotton and woollen fabrics and yarn, of which Castres is the centre, employs about 15,000 hands. Silk furniture stuffs are made at Lavaur. Morocco leather, paper, cords, glass, copper wares, files, &c., in various parts. Tarn is divided into four arronds.: chief towns, Alby, the cap., Gaillac, Castres, and Levaur. It sends five members to the chamber of deputies. Registered electors, in 1838-9, 2461. Public revenue, in 1831, 6,433,455; public expenditure, in do. 4,513,064. The land-land tax, or *contribution foncière*, is said to be very oppressive in this dep. This dep. is one of the chief seats of Protestantism in France, and was a principal scene of the crusades against the Albigenses. (*Hugo*, art. *Tarn; Dict. Geog.; French Official Tables.*)

TARN-ET-GARONNE, a dep. of France, reg. S., in about lat. 44° and principally between the 1st and 2d degrees of E. long.; having N. the dep. of Lot. E. Aveyron, S.E. and S. Tarn and Haute Garonne, and W. Gers and Lot-et-Garonne. Area, 366,976 hectares. Pop., in 1836, 242,250. Surface generally undulating; in the S. and E. there are, however, some hill ranges of considerable height, the sources of a number of small streams. Principal rivers, Garonne, Tarn, and Aveyron, all flowing through the S. half of the dep. In 1834 arable lands were estimated to comprise 229,294 hectares, the meadows 17,346 do., vineyards 36,703 do., and woods 45,387 do. A greater proportion of wheat is raised in this than in any other of the S. deps. of France;

the estimated produce in 1835 having been 1,190,000 hectol. Rye, maize, and oats are also grown; and the total produce of grain considerably exceeds the home demand. The produce of wine amounts to about 450 hectol. a year, of which about 250,000 are exported, and converted into eau de vie. The finest are the red wines of the arrond. of Castel-Sarrazin, the best of which have a fine colour, du spiritueux, et un bon gout. (Jullien, 244), but the greater portion are inferior. Prunes, flax, hemp, and oil-seeds are among the other principal articles of culture. Irrigation is not well understood; and the produce of hay is small. Live stock are, in consequence, less numerous than in the adjacent deps.; the quality of the wool is inferior. Hogs and poultry are extensively fattened; the former for export to Spain. In 1835, of 85,711 properties subject to the contrib. foncière, 32,712 were assessed at less than 5 fr., and only 95 at 1000 fr. or upwards. The want of capital is a formidable obstacle to the progress of manufacturing industry. Some coarse woollen stuffs, with stockings and other fabrics of silk are made at Montauban; serges, linen cloths, and woollen yarn in the arrond. of Castel-Sarrazin: and there are some considerable tanneries, paper and flour-mills, &c. Tarn-et-Garonne was made a dep. by Napoleon, on account of the importance of its capital, Montauban. It is divided into three arronds., and sends four members to the chamber of deputies. Registered electors, in 1838-9, 2125. Total public revenue, in 1831, 6,108,515 fr. (Hugo, art. Tarn-et-Garonne; Official Tables, &c.)

TARRAGONA, (an. Tarraco), a city and seaport of Spain, in Catalonia, cap. prov. of its own name, at the mouth of the Francoli, in the Mediterranean, 45 m. W.S.W. Barcelona. Pop., according to Miñano, 17,000. This once famous city is now contracted to a space which covers only a small portion of its ancient limits, and is ill built and dirty. A large and broad street, with some handsome edifices has, however, been laid out within the present century. Its fortifications consist, besides the town walls, of two castles, and several batteries to protect the harbour; but the height on which the city stands is commanded by mount Olivo. The river Francoli, adjacent to the city is crossed by a narrow bridge of six arches, and the town is entered by six gates. Townsend says, "Tarragona, of all the cities in Spain, would give the most agreeable employment to the antiquary. Here he would admire the remains of an amphitheatre, of a circus, of the palace of Augustus, of temples, and of an extensive aqueduct, with fortifications, which although of a more recent date, are ancient." (Journey, &c., iii. 311.) But, according to Inglis, in 1833, the remains of the amphitheatre were then little more than visible. Near Tarragona is the building called the tomb of the Scipios, in which the father and uncle of S. Africanus, who were killed in battle with the Carthaginians, are said, though on no good authority, to have been buried. It is about 19 ft. sq. and 28 ft. in height, resembling the tomb of Theron at Girgenti. In the front, facing the sea, are statues of two warriors in a mourning posture, roughly cut out of the stones of the sepulchre, and much worn by the sea air. The inscription is so much defaced that it can hardly be deciphered. The cathedral of Tarragona is worth a visit, particularly the court and cloisters, which are surrounded with numerous pillars.

The archbishopric is one of the most ancient in Spain, having existed in the 7th century. It has several convents, an hospital, a seminary, academies of design and naval architecture, other superior schools, a theatre, &c. Tarragona is the chief exporting port of Catalonia. Its export consists of nuts, almonds, wines, and brandy. The nuts sent to the English market are known by the name of Barcelona nuts; but they are neither grown near, nor exported from Barcelona. They are grown more in the interior of the prov., and are all exported from Tarragona. The average export of nuts to England is from 25,000 to 30,000 bags (four to a ton) a year. The export of almonds is about 12,000 bags. From 5000 to 5500 pipes of wine are exported from Tarragona to Rio Janeiro, Guernsey, Jersey, &c.; and about 400 pipes of brandy are exported chiefly for Cette and Cadiz, from which places it finds its way into the wine butts of Bordeaux and Xeres. Cork wood and cork bark are also exported from Tarragona." (Inglis's Spain in 1830, ii. 277.)

Pliny says that Tarraco was founded by the Scipios, who planted a colony in it (lib. iii. cap. 3); but most probably it had been founded previously, and was only increased by the Scipios. It was the seat of a principal tribunal, and was, in fact, not merely the capital of Hispania Citerior, or Tarraconensis, but of Spain, under the Romans. Augustus resided in it for a short period, and Hadrian enlarged its port and erected a mole. It was taken by the Goths in 467, and by the Moors in 714, from whom it was retaken by Alfonso of Aragon in 1220. It was several times the place of meeting of the states of Catalonia. In 1705 it was captured by the English, who at first intended to retain it as a military post, but afterwards abandoned it for Gibraltar. In 1811 it was taken and sacked by the French under Suchet. Orosius the historian is said to have been a native of Tarraco, though the fact has been disputed. (Miñano; Townsend; Inglis; Mod. Trav., &c.; Cellarii Geog. Antiqu., i. 140.)

TARSUS, a celebrated city of antiquity, and still a town of some importance, in Asia Minor, pachalic of Itchil, cap. sanjak, on the Cydnus, about 12 m. from the Mediterranean, and 89 m. W.N.W. Scanderoon. Lat. 36° 46' 20" N.; long. 34° 45' 45" E. Its permanent pop. is estimated at about 7000. (Bowring's Rep.); but during winter a great many Turkish, Greek, and Armenian families flock into the town. The modern town does not cover one fourth part of the area occupied by the city under the Romans, and few vestiges remain of its former magnificence. The remains of a theatre, and of a spacious circular building, an ancient gateway, and beyond the walls a singular and solid structure, 132 paces in length, by about 60 in breadth, are among the principal. Some traces are perceptible of the more ancient walls, but those now inclosing the town are not supposed by Kinneir to be of an earlier date than the time of Haroun Al Raschid, in the 8th or 9th century; and the castle is said to have been built by Bajazet. The houses seldom exceed one story in height; they are terrace roofed, and the greater part are constructed with hewn stone, furnished by the more ancient edifices. There are two public baths, a number of mosques, several caravanserais, a small church, &c. The plain around Tarsus is very fertile, and cultivated by Greeks, chiefly for corn and cotton, which last is a principal article of export, the others being wool, beeswax, gall-nuts, copper, goats' hair, and skins, ox and buffalo hides, and hair sacks. The river Cydnus is now navigable only by very small boats, and the greatest part of the produce exported is shipped at Mersia, a port, or roadstead about four hours' journey W., at which there is said to be good anchorage all the year round. The value of the imports and exports may amount to about £100,000 a year each. (Bowring, Rep. on Syria; Geog. Journ., ii. 506, &c.; Kinneir, &c.)

Nothing is known of the origin of Tarsus; but it is abundantly certain that it was very ancient, and that it had either been originally founded by Greeks, or had subsequently received a Grecian colony. It was the metropolis of Cilicia, and was captured by both Cyrus and Alexander. It continued to flourish under the successors of the latter, and under the Romans. Strabo says it was very populous and powerful; and he further adds, that its schools of philosophy, literature, and science were superior even to those of Athens and Alexandria (lib. xiv.); and though this be obviously an extravagant eulogy, there can be no question that it was a most distinguished seat of learning. St. Paul, the apostle of the Gentiles, was a native of Tarsus, where he was born in the second year of the Christian æra, and where he acquired a competent knowledge of Greek literature before he went to study the law of Moses at Jerusalem. To ingratiate themselves with Julius Cæsar, the inhabitants changed the name of the city to Juliopolis; and it is plain, from the statement of St. Paul (Acts xxiii. 28), that some of them, if not all, ranked as Roman citizens. Tarsus produced several other distinguished individuals; among whom may be specified Antipater, the stoic; Athenodorus, the philosopher, and friend of Augustus, &c. (Cellarii Geographia Antiqua, ii. 254.; Ancient Universal History, vi. 12. 8vo ed., &c.)

TARTARY, TAHTARY, or TURKESTAN, a very extensive region of Central Asia, partly comprised in the Chinese empire, and partly distributed among the states of Bokhara, Budukshan, Khiva, Kokan, the Kirghis Steppe, which see.

TATNALL, co., Ga., situated toward the S.E. part of the state, and contains 1000 sq. m. Bounded S.W. by Altamaha river, N.E. by Cannoochee river. Watered by Great Ohoopee river. It contained, in 1840, 2877 neat cattle, 492 sheep, 2107 swine; and produced 596 bushels of wheat, 9259 of Indian corn, 4201 of potatoes, 2587 pounds of cotton, 1865 of sugar. It had one store, four grist-mills, three saw-mills. Pop.: whites, 1878; slaves, 841; free coloured, 5; total, 2794. Capital, Reidsville.

TATTA, a town of N.W. Hindostan, and one of the principal in Sinde, near the Indus, about 130 m. above its mouth, and 55 m. S.W. Hyderabad. Lat. 24° 44' N., long. 68° 27' E. Pop. estimated at less than 15,000, by Burnes, who describes it as "an open town, built on rising ground in a low valley. The houses are formed of wood and wickerwork, plastered over with earth; they are lofty, with flat roofs, but very confined, and resemble square towers. Some of the better sort have a base of brickwork, but stone has been used only in the foundation of one or two mosques. A spacious brick mosque, built by Shah Jehan, still remains, but it is crumbling to decay; and there is little else to remind one of its former greatness. The commercial prosperity passed away with the empire of Sinde.

"Of the weavers of *lorsgeee* (silk and cotton fabrics), for which it was once so famous, 125 families only remain; and there are not 40 merchants in the city. Such has been the decay of Tatta, so populous in the days of Nadir Shah. The country in its vicinity lies neglected, and but a small portion of it is brought under tillage." (*Bokhara, &c.*, i. 27.) Tatta has been supposed to represent the *Pattala* of the ancients, and with some reason, since at this point the Indus, as stated by Arrian (lib. vi.), divides itself into two branches; but no conclusive evidence has been elicited on this point.

TAVISTOCK, a parl. bor., market town, and par. of England, co. Devon, hund. Tavistock, on the Tavy, 13 m. N. Plymouth. Area of par., which is identical with the parl. bor., 11,600 acres. Pop. in 1831, 5602. The town is on the N.W. bank of the river, here crossed by three bridges, and from which the ground rises, by a steep acclivity on both sides, to the height of several hundred feet. A very contracted valley from the N. is also occupied by houses closely packed together. The parts of the town built on higher ground to the N., or overlooking the more expanded valley to the W., are of more modern date. The streets, in many parts, are irregular and indifferently paved; but the houses are good, and the town generally is pleasant. Tavistock appears to have owed its origin to an abbey of black friars, founded here by an earl of Devon, in 961. At the dissolution of the monasteries, this abbey, along with the lordship of the town, was given by the king to John Lord Russell, the ancestor of the present Duke of Bedford. Some remains of the monastic edifice still exist; the former refectory is now used as an assembly room, and near one of the bridges is a large, handsome, arched and pinnacled gateway, apparently of the time of Henry VI. The principal remains of Tavistock abbey are in the perpendicular style. The par. church has four aisles, a chancel, a tower at the W. end, and in its interior are several good monuments. The living, a vicarage, worth £398 a year, is in the gift of the Duke of Bedford. There are meeting-houses for Wesleyans, Independents, Unitarians, Friends, &c., a large and convenient workhouse, a national school, chiefly supported by the Duke of Bedford, some small educational endowments, almshouses, and other public charities. Tavistock was one of the four stannary towns in the co., and is governed by a portreeve, chosen yearly at the lord's court, who is also the returning officer of the borough. It has sent two members to the House of Commons since the 23d of Edward I.; the right of voting, down to the Reform Act, having been in freeholders of inheritance to possession inhabiting within the bor. The population is chiefly agricultural, though some serges and coarse linens are made, and mining and the working of iron occupy a few hands. Markets on Fridays. Fairs, Jan. 17, May 6, Sept. 9, Oct. 10, and Dec. 11, for cattle. Tavistock gives the title of marquis to the Duke of Bedford. Sir Francis Drake, the famous navigator and naval commander, belonged to the immediate vicinity of Tavistock, where he was born in or near 1545. (*Boundary Report, &c.*)

TAUNTON, a parl. bor. and market town of England, co. Somerset, W. division, hund. Taunton Dean: on the Tone, here crossed by a stone bridge of two arches, 37 m. S.W. Bristol. Area of parl. bor. about 1450 acres. Pop. in 1831, 12,148. Taunton is one of the principal towns in the co.; the main streets are spacious, well paved, and lighted with gas. They run mostly from E. to W., and from N. to S., the town being about 1 m. in length, and nearly as much in width. The houses are generally good, and have frequently extensive outlets and gardens; the appearance of the town indicates a prosperous, respectable community. There are, however, several lanes and courts (popularly called *colleges*) branching from the main street, which were formerly filled with inhabitants but little above the condition of paupers, who had been drawn into these close and unwholesome recesses to be within the limits of the bor., and to exercise the franchise extended to every inhabitant housekeeper. (*Bound. Report.*) The most striking public edifice is St. Mary Magdalen's church, in an open space in the heart of the town. It is 98 ft. in length by 86 ft. in greatest breadth. Its nave is divided into five aisles by four rows of clustered columns, supporting bluntly-pointed arches; and at its W. extremity is an elegant quadrilateral tower, with a pinnacle at either corner, their entire height being 150 ft. This church is richly decorated both without and within, and has numerous monuments, a fine organ, &c. Much of its decoration is said to be due to Henry VII., in return for the strenuous support of the Lancastrian cause by Taunton; but the tower and other parts of the edifice seem to have been erected somewhere about the end of the 4th century. Rickman (*Gothic Architecture*) says that his church is a very fine example of the enriched perpendicular style. St. James's, the old conventual church of Taunton priory, is a plain but strong and well-finished building. The living a perpetual curacy worth £254 a year, is in the gift

of Sir T. C. Lethbridge. There are chapels for Independents, Wesleyans Baptists, Unitarians, Friends, Roman Catholics, &c.; the last named is a handsome building of the Ionic order. There is also a Franciscan convent, occupying what was originally intended for a general hospital. At the W. ent. of the town is the castle, built in the time of Henry I. on the site of another fortress, built about 700 by Ina, king of the West Saxons. This edifice comprises the hall, in which assizes for the co. are held in Lent, general quarter sessions at Michaelmas, and a court of requests weekly. The market-house is a large brick edifice, comprising the corn-market, town-hall, assembly-rooms, &c.; and beside it is a handsome building in the Ionic order, the lower part of which is a fish and poultry market, and the upper a library and reading room, museum, &c. The Taunton and Somerset institution, established in 1823, has a spacious reading and news-room, and a valuable, though not extensive, library. There is a neat theatre in the town. The numerous charities of Taunton include the grammar-school, founded in 1522, and having a small endowment; a school for 80 boys and 50 girls, supported by voluntary contribution; several almshouses; the Taunton and Somerset hospital, with accommodation for 26 patients; an eye infirmary, a lying-in charity, &c. The town has a weekly newspaper.

Taunton was one of the first towns in England in which the woollen manufacture was established; but the woollen trade of the town has greatly declined, and the industry of the inhabitants is now chiefly exercised in manufactures of silkstuffs, as crapes, sarsenets, &c., and of lace. The town derived considerable advantage from the construction of the Taunton and Bridgewater canal, by which a good deal of Welch coal is now brought to it in return for the agricultural produce of the vicinity. The trade of Taunton has also experienced renewed activity since the opening of the Great Western railway, as far as Bridgewater. Taunton was formerly a municipal borough, but in consequence of neglect in filling up the vacancies in the corporation, it lost its charter in 1792. It is a parliamentary borough by prescription, and appears to have sent two members to the House of Commons as early as the 23d Edward I. Previously to the Reform Act, which confirmed its privilege of sending two members, the right of voting was in potwallopers not receiving alms. Registered electors 1839-40, 1010. The returning officers are the bailiffs, chosen at an annual court-leet. Taunton, though not alluded to in the "Itinerary of Antoninus," was, in all probability, known to the Romans, as a great number of Imperial coins have been found in and near it. In the time of the Heptarchy it was a place of considerable note. In the civil war, it sided with the Parliament, and in 1645 its castle sustained, with success, a long siege against the royal forces under Lord Goring. Markets, Wednesday and Saturday. Fairs, June 17, and July 7 to 10, chiefly for cattle and horses. (*Parl. Reps.; Priv. Inf.*)

TAUNTON, p. t., semi-capital of Bristol co., Mass., 33 m. S. Boston, 20 m. E. by N. Providence, 4920 W. Pop. in 1830, 6042; in 1840, 7645. Situated on Taunton river, which is navigable to this place, 20 m. from its mouth in Mount Hope bay, for vessels of 50 tons. Watered by Oance and Rumford rivers, tributaries of Taunton river, which afford good water-power. The village contains in the centre, a handsome public square, neatly enclosed, called Taunton Green, on one side of which stands the courthouse, having a portico with four Ionic columns. It has also a jail, a townhouse, nine churches, four Congregational, two Baptist, an Episcopal, a Methodist, and a Roman Catholic. It has three banks, with an aggregate capital of $450,000, and two insurance companies. Iron works were established here in 1652, and nails are now manufactured to a large amount. A branch of the Boston and Providence railroad extends to this place. Thirty coasting vessels belong to the port. There were in the t. in 1840, 62 stores, with a capital of $139,800; four lumber-yards, capital $19,000; two furnaces and two forges; hardware and cutlery was manufactured to the amount of $50,000; one fulling-mill, six cotton factories with 19,956 spindles, one dyeing and printing works, 12 saw-mills, one paper-mill, one tannery, one pottery, three printing-offices, two binderies, two weekly newspapers; two academies, 179 students; 43 schools, 1960 scholars.

TAURIDA, a government in the S. of European Russia, consisting partly of the peninsula of the Crimea, and partly of a tract on the mainland, lying between the Dniepr, the Black sea, the sea of Azoff, and the Berda. Area, 43,948 sq. m. Pop. in 1838, 520,000. The mainland part of the government, which, though the least interesting, is the most extensive, consists almost entirely of vast, and in many parts sterile plains, denominated the Steppe of the Nogais, from the Tartar tribes, by which it is principally occupied. "These," says Dr. Clarke, "are a very different people from the Tartars of the Crimea; they are distinguished by a more diminutive form, and by the dark copper colour of their complexion which is sometimes almost black. They bear a remarkable resemblance to the Laplanders, although their dress and manner have a more savage character." (ii., p. 318, 8vo ed.) Above 17,000 Germans are colonized to the E. of the river Molotchna. (For farther particulars, see art. CRIMEA in this work.)

TAY, a river of Scotland, being the largest of the Scotch, and, in respect of the quantity of water it conveys to the sea, it is the greatest even of the British rivers. It rises in the high mountainous country a little to the N. of loch Lomond, and, flowing N.E. by Killin, expands into the beautiful long narrow lake called loch Tay. Issuing thence, its course is N. and E. to Logierait, S. to Dunkeld, E. to Kinclaven, S., inclining a little to the W., to Perth; N.E. to the point of Rhind; then north-easterly, past Dundee, till it falls into the sea between Tentsmoor Point and Buttonness. From Rhind Point to Dundee the channel of the river expands into an estuary called the frith of Tay. From its source to Dunkeld the Tay flows with a rapid current; partly through a very wild, and partly through a highly picturesque, romantic country. Its subsequent course as far as Perth is through a comparatively fruitful country; and from the latter to the sea, it flows through the richest and finest valley in Scotland.

From Buttonness to Dundee the river is navigable for ships of 500 tons' burden; and at high water, vessels of above 100 tons' burden reach Perth, 20 m. above Dundee. Two lighthouses have been erected on Buttonness, to mark the entrance to the river. The bar at its mouth has 2¼ fathoms water over it. Dundee, the port of the Tay, has wet docks and a pier harbour; the latter dries at low ebb; but at high water springs it has a depth of 14 or 15 feet, and at neaps of 9 or 10 feet. Large ships anchor in the channel of the river. The mouth and channel of the Tay are a good deal encumbered with sand-banks; and its navigation is rather difficult, partly on that account, and partly from the strength of the tides.

Among the more remarkable of the tributaries of the Tay, may be mentioned the Lyon, which joins it near Fortingal. The Tummel has its sources in the moor of Rannoch, and flowing through the loch of that name, is joined, near the pass of Killiecrankie, by the Gary, from the confines of loch Ericht. The united river falls into the Tay at Logierait. Near Kinclaven the Tay receives the united waters of the Airdle, the Isla, and other rivers flowing S. from the mountains on the confines of Aberdeenshire. At Rhind Point it receives its important tributary, the Earn, flowing E. from loch Earn. The basin of the Tay comprises a space of about 2400 sq. m.; and Mr. Smeaton ascertained that it carries to the sea more water than even the Thames. Its course from its source to Buttonness is estimated at about 110 m. It is the finest salmon river in Great Britain; its fisheries let for a large sum; the fish being mostly conveyed, packed in ice, to London.

TAYLORSVILLE, p. v., capital of Patrick co., Va. The postoffice is called Patrick C.H., 296 W.S.W. Richmond, 201 W. Situated on Mayo river, and contains a courthouse, jail, two stores, one flouring-mill, two tobacco factories, one tannery, 40 dwellings, and about 300 inhabitants.

TAYLORSVILLE, p. v., capital of Spencer co., Ky., 35 m. S.E. Louisville, 30 m. W.S.W. Frankfort, 581 W. Situated on the N. side of Salt river, at the junction of Brashear's creek. It contains a courthouse, jail, a church, and 366 inhabitants.

TAZEWELL, county, Va. Situated in the S.W. part of the state, and contains 1600 sq. m. Drained by Tug fork of Sandy river, and head waters of Holston and Clinch rivers. It contained in 1840, 10,460 neat cattle, 11,170 sheep, 14,656 swine; and produced 33,688 bushels of wheat, 13,965 of rye, 149,973 of Indian corn, 126,432 of oats, 15,596 of potatoes, 43,064 pounds of sugar. It had six stores, 14 grist-mills, eight saw-mills, four tanneries, 32 distilleries. Pop.: whites, 5466; slaves, 786; free coloured, 38; total, 6290. Capital, Jeffersonville.

TAZEWELL, county, Ill. Situated a little N. of the centre of the state, and contains 1130 sq. m. Bounded N.W. by Illinois river. Drained by Mackinaw river and branches of Sangamon river. It contained in 1840, 7785, neat cattle, 5422 sheep, 19,973 swine; and produced 184,983 bushels of wheat, 1844 of rye, 423,751 of Indian corn, 73,630 of oats, 40,378 of potatoes, 4940 pounds of sugar. It had 14 stores, two flouring-mills, five grist-mills, 19 saw-mills, one tannery, three distilleries, two potteries; four academies, 191 students, 10 schools, 290 scholars. Pop. 7221. Capital, Tremont.

TAZEWELL, p. v., capital of Claiborne co., Tenn., 221 m. E. by N. Nashville, 473 W. Situated on Russel's creek, a branch of Powell's river. It contains a courthouse, jail, two churches, a Presbyterian and Methodist, seven stores, various mechanic shops, and about 600 inhabitants.

TCHERNIGOFF, a government of European Russia, to the E. of the Dnieper, and between the government of Smolensk on the N., and that of Poltava on the S. The

estimates of the area differ very widely; perhaps it may be taken at about 19,000 sq. m. Pop., in 1838, 1,300,089. Surface flat; soil fertile; climate dry, healthy, and mild. Principal river Dniester, which bounds it on the W., and Desna, by which it is intersected. All sorts of corn are raised, but principally rye, barley, and oats. Produce of the harvest estimated at about 4,000,000 chetwerts. Flax and hemp, tobacco, hops, &c., are also cultivated. Oxen, of a very fine breed, are raised and fattened to a great size. Horses small, hardy, and active. There are some pretty extensive forests. Free cultivators are common in this and the other governments of Little Russia. Manufacturing industry, though still very backward, has made much progress during the present century. Spirits largely consumed, and there are numerous distilleries. Commerce considerable: the exports consist principally of cattle, tallow, hides, &c., spirits, honey and wax, potash, hemp-seed, &c. (Schnitzler, La Russie, &c., p. 459.)

TECUMSEH, p. t., Lenawee co., Mich., 57 m. S.W. Detroit, 511 W. Watered by Raisin river. It has three commercial houses in foreign trade, 12 retail stores, two flouring-mills, two grist-mills, two saw-mills, one tannery, one distillery, one printing-office, one weekly newspaper; one college, 20 students; 12 schools, 803 scholars. Pop. 2300.

TEFLIS, or TIFLIS, a city of W. Asia, the cap. of Georgia, and of all the Caucasian and Trans-Caucasian provinces of Russia; on the Kur (an. Cyrus), 230 m. E. by N. Trebizond. Lat. 41° 30' 30" N., long. 42° 1' 30" E. Pop. probably about 30,000, most of whom are Armenians, with some Mussulman families. "Teflis occupies the right bank of the Kur, in a contracted valley formed by irregular mountains, parallel with the stream on the side of the city, and hills coming down in a point quite to the water's edge on the other. A circular fort covers this point, and together with a small suburb is united to the city by a bridge of a single wooden arch, thrown over the river; while the ruined walls of an old citadel crown the top, and extend down the side of a part of the opposite mountain. The old and native part of the city is built upon the truly Oriental plan of irregular narrow lanes, and still more irregular and diminutive houses, thrown together in all the endless combinations of accident. Here and there European men, aided by Russian power, has worked out a passable road for carriages, or built a decent house, overlooking and putting to shame all its mud-walled and dirty neighbours. A line of bazaars, too, extending along the river, and branching out into several streets, together with much bustle and business, display some neatness and taste, and is connected with two or three tolerable caravanserais. Several old and substantial churches, displaying their belfries and cupolas in different parts, complete the prominent features of this part of the city. In the N. or Russian quarter, clean palaces, government offices, and private houses, lining broad streets and open squares, have a decidedly European aspect, and exhibit in their pillared fronts something of that taste for showy architecture which the edifices of their capital have taught the Russians to admire. Teflis has the appearance of an excessively busy and populous place. Its streets present not only a crowded, but a stirring many Oriental cities, a lively scene. Every person seems hurried by business. Nor is the variety of costumes, representing different nations and tongues, the least noticeable feature of the scene." (Smith and Dwight's Missionary Researches, 121–194.)

The Armenian cathedral is a large and somewhat striking edifice; there are two mosques, and among the other places of worship is a German Protestant chapel. It has also a French and a German hotel; but they are represented as being, in most respects, the reverse of what they should be. House-rent is high; but otherwise living is not expensive. Teflis has many remarkable sulphureous hot springs, their temperature varying from 100° to 112° Fahr.; and to these, it is supposed by some, the city owes its name. Over some of these the Russian government has erected the covered baths, a plain edifice, but which, by being kept in good order, differs widely from all the other bath establishments in the city, and realizes a handsome revenue. Teflis is very favourably situated for trade, and its commerce is pretty extensive, having greatly increased during the period of Russian occupation. Almost all the trade is, however, in the hands of Armenians; in 1830, scarcely half a dozen mercantile houses existed belonging to any other foreigner, and only one European consul (a Frenchman) resided here. In 1830, the Russians founded a school in Teflis, which has since been erected into a gymnasium; and there are several other schools. Georgia in general, and its capital in particular, has long been celebrated for the beauty of its women; and, according to the missionaries referred to above, "this has not been over-rated, for we have never entered a city so large a proportion of whose females were handsome in form, features, or complexion, as Teflis."

Teflis does not boast a very high antiquity. It is said to

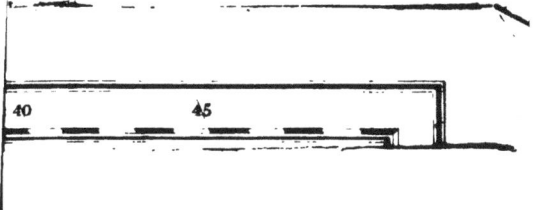

40   45

a remarkable ress
dress and manner
314, 8vo ed.) Al
E. of the river I
art. CRIMEA in th
  TAY, a river o
and, in respect of
sea, it is the grou
the high mounts
Lomond, and, fl
beautiful long i
thence, its cours
E. to Kinclaven,
N.E. to the point,
dee, till it falls in
Buttonness. Fre
the river expands
From its source t
rent; partly thro
ly picturesque, ro
far as Perth is t
and from the lat
and finest valley
  From Buttons
ships of 500 tons
above 100 tons' I
Two lighthouses
the entrance to
fathoms water o
wet docks and a
but at high wate
at neaps of 9 or I
of the river. 1
good deal encum
rather difficult, g
strength of the i
  Among the m
may be mention
The Tummel ha
flowing through
pass of Killiec
loch Ericht. 1
gierait. Near K
of the Airdle,
the mountains
Rhind Point it
flowing E. from
prises a space
certained that I
the Thames. ;
estimated at ab
Great Britain;
being mostly o
  TAYLORS
postoffice is ea
201 W. Situat
jail, two stores
tannery, 40 dw
  TAYLORSVIL
S.E. Louisville
on the N. side
creek. It co
habitants.
  TAZEWEL
of the state, a
of Sandy riv
rivers. It o
sheep, 14,656
12,965 of rye,
of potatoes, 4
grist-mills, ei
Pop.: whites
6990. Capits
  TAZEWELL
of the state,
Illinois river,
Sangamon ri
5422 sheep, I
wheat, 1844
40,272 of po
two flouring
nery, three s
students, 10
Tremont.
  TAZEWEL
E. by N. Ne
branch of I
two church
various mea
  TCHER
the E. of I
Smolensk I
92

have been built in 400 by Vachtang, the founder of a dynasty which ruled from the Euxine to the Caspian. It was taken by Jenghiz Khan; by the Turks in 1576; sacked by Aga Mehemti Khan in 1795; and fell to the Russians in 1801. It suffered greatly from the ravages of the cholera in 1830. (*Smith and Dwight's Miss. Researches; Lyell's Trav. in Russia; Med. Trav., xvii.; Dict. Geog., &c.*)

TEHERAN, or TEHRAUN, the modern cap. of Persia, prov. Irak-Ajemi, near the foot of mount Elborz, which divide that prov. from Mazanderan, 211 m. N. Ispahan; lat. 35° 40' N., long. 51° 22' 50" E. Pop., during the residence of the court, estimated at from 80,000 to 70,000, or upwards. It is about 5 m. in circuit, and is enclosed by a strong earthen wall, flanked with numerous towers, surrounded by a glacis, outside which is a large dry ditch. The appearance of the city from a distance is picturesque; but it has few public edifices worth notice; and notwithstanding it has many good shops and bazaars, it is said by Morier to have a "mudlike" look within, its houses, like those of other Persian towns, being constructed of sun-dried bricks, while many of its streets are wretchedly paved. The *Ark*, or citadel, comprises, besides the royal residence and harem, quarters for the guards, the record chamber, treasury, hall for receiving ambassadors, and other public offices, 10 baths, two or three gardens, reservoirs, &c. The grand saloon in the palace is said to be very magnificent: the throne is a platform of pure white marble, raised a few steps from the ground, and carpeted with shawls and cloth of gold; and the whole interior of the department is profusely decorated with carving, gilding, arabesque painting, and looking-glass, the last material being interwoven with all the other ornaments, from the vaulted roof to the floor. The mosques, colleges, and caravanserais, though not very numerous, are in good repair. Much less than a century go, the present metropolis of Persia would hardly have been considered of sufficient importance for the capital of a province. It first became the metropolis towards the end of the last century, under Aga Mahomed Khan, who seems to have selected it for that dignity partly on account of its good position in a military point of view, and partly from a vicinity to the hereditary possessions of his family. Its greatest drawback is its unhealthiness from damp, which, with the excessive heats in summer, oblige the sovereign and his court to remove at that season, and encamp in pavilions and tents on the plains of Sultaneea, or Oujan; at which period the resident population of Teheran is reduced perhaps 10,000. The environs of Teheran are not unpleasant, the plain both to the E. and W. being covered with villages, and abounding in grain. To the N. of the town is a handsome palace, which its situation and the fine gardens that surround it make a delightful residence.

A short distance S. from Teheran are the ruins of the city *Rhé*, generally supposed to be identical with the ancient *Rhages*, the capital of the Parthian kings, where Alexander halted for five days in his pursuit of Darius. The ruins cover a great extent of ground, having in their centre a modern village, with a noble mosque and mausoleum—an oasis in the midst of the surrounding desert. (*Kinneir's Persia*, 118; *Fraser's Persia*, 61; *Morier, &c., Med. Trav.*) It should, however, be mentioned that Major Rawlinson and others contend that the ruins now traced are not those of Rhages, but of an Arabian city, called Rhei; and that the ruins of Rhages are to be found Kalah-Erig, 30 m. E. Teheran. (*Geog. Journal*, x., &c.*)

EIGNMOUTH, a market town and seaport of England, Devon, hund. Exminster, at the mouth of the Teign, in English channel, 12 m. S. Exeter. It is intersected by Teign; the communication between its two divisions, 1 of which constitutes a parish, being kept up by a long stone bridge across the river, having a drawbridge at the extremity for the accommodation of vessels. Area both parishes, 1280 acres. Pop. in 1831, 4688. W. Teignmouth, or the portion on the W. side the river, is irregularly built, and ill paved; but E. Teignmouth is beautifully situated, and having been of late much improved, is now of the most favourite watering-places in the S.W. part England. The church of W. Teignmouth is a modern singular structure; the living being a curacy worth £80 per, under the vicarage of Bishop's Teignton. In this of the town there are independent and Calvinistic ing-houses, a national school, a quay on the river, and kyard, in which sloops of war and vessels of 200 tons have been built. E. Teignmouth church is mostly modern edifice: the living, a perpetual curacy, worth a year, is in the gift of the vicar of Dawlish; and also, are a Baptist chapel, an endowed school for 13 children, some good inns, a theatre, reading-rooms, their establishments usual in a watering-place. Teignmouth is governed by a portreeve, chosen annually at the leet of the lord of the manor. Many of its inhabitants employed in the coast fishery, and others in the supply

of goods to visiters: a good deal of the Haytor granite is also shipped from this port; but the bar at the mouth of the river renders the harbour accessible only to coasting vessels. Teignmouth is of high antiquity, and is said to be the place at which the Danes first landed in 787. It gives the title of baron to the Shore family.

TELLICHERRY, a town of British India, presid. Madras, and one of the principal seaports in the district of Malabar; on the W. coast of Hindostan, 42 m. N.N.W. Calicut; lat. 11° 45' N.; long. 75° 23' E. Population uncertain. This was the chief trading settlement of the British on the Malabar coast previously to 1800, when the E. I. Company's warehouses were transferred to Mahe, about 6 m. to the S.E. The most wealthy natives, however, still reside at this town; which a few years since continued to be the mart for the best sandal wood from above the Ghauts, and cardamoms from Wynaad. (*Hamilton's E. I. Gaz.*)

TELFAIR, county, Ga. Centrally situated toward the S. part of the state, and contains 950 sq. m. Watered by Ocmulgee river and its branches, and by head waters of Santilla river. It contained in 1840, 12,939 neat cattle, 2578 sheep, 1015 swine; and produced 3921 bushels of wheat, 43,192 of Indian corn, 1136 of oats, 18,939 of potatoes, 3450 pounds of rice, 80,780 of cotton, 22,510 of sugar. It had eight stores, seven grist-mills, 13 saw-mills, one tannery; seven schools, 65 scholars. Pop.: whites, 2001; slaves, 757; free coloured, 5; total, 2763. Capital, Jacksonville.

TEMESWAR, a royal, free, and fortified town of Hungary-beyond-the-Theiss, cap. co. of its own name, in a marshy plain, on the Alt Bega river, 72 m. N.N.E. Belgrade; lat. 45° 49' 27" N., long. 2° 14' 17" E. Pop., with its four suburbs, about 13,000, including numerous Germans, Greeks, Wallachs, and Jews. Mr. Paget says, "Temeswar, the capital of the Banat, and the winter residence of the rich Banatians, is one of the prettiest towns I know anywhere. It has two handsome squares, and a number of very fine buildings. The county-hall, the palace of the Bishop of Csanad, the residence of the commander, and the townhouse, are all remarkable for their size and appearance." Temeswar was taken from the Turks in 1716 by Prince Eugene, who laid out and strongly fortified the modern town, which is now one of the principal fortresses of the Austrian monarchy. It has a Roman Catholic and a Greek cathedral, a synagogue, seminary, Piarist gymnasium, arsenal, military school, some barracks, and various other military establishments, and is the seat of the principal civil establishments and authorities of the Banat. Good water is raised by machinery for the supply of the town. It has manufactures of silk and woollen stuffs, paper, tobacco, oil, &c.; and an extensive trade in these articles and in the transit of agricultural produce. Its inhabitants are said to be generally opulent. Its commerce has been considerably facilitated by the excavation of the Bega canal, about 72 m. in length, which, passing by the town, unites it with the navigable portion of the Bega, and, consequently, with the Theiss and the Danube. This canal has also been advantageous from its assisting in drying the marshes by which the town is surrounded: though in summer it is still rather unhealthy. Temeswar is supposed by D'Anville to represent the an. *Tabiscus*, to which Ovid was banished. It was taken by the Turks, under Polyman, in 1551, who held it till 1716. (*Paget's Hungary; Berghaus; Austr. Nat. Encyc.*)

TEMPE, a famous valley and defile in the N.E. part of Thessaly, stretching from near Baba to the gulf of Salonica, from 6 to 8 m. in length, between Olympus on the N., and Ossa on the S. It is traversed by the Selambria (an. *Peneus*), and is, in parts, so very narrow that there is merely room for a military road along side the river. In some respects the defile bears a striking resemblance to the pass of Killiecrankie in Scotland; but the scenery is incomparably more magnificent. The appearance of the chasm, and the traditions current in antiquity, leave little doubt that the rocks had been rent asunder by some tremendous convulsion of nature, which opened a passage for the waters that must previously have deluged the greater part of Thessaly. In some parts it is grand in the extreme. The precipices consist of naked perpendicular rocks, rising to a prodigious height; so that the spectator can scarce behold them from below without giddiness. Livy's description, therefore, in addition to its intrinsic grandeur, has all the majesty of truth: *Rupes utrinque ita abscissa sunt, ut despici vix sine vertigine quadam simul oculorum animique possit: Terret et sonitus et altitudo per medium vallem fluentis Penei amnis.* (Lib. xliv., cap. 6; *Clarke*, vii., 370.)

The character of this gorge or defile is evidently that of wildness and savage grandeur, and does not, therefore, harmonize with the descriptions the poets have given of the *Zephyris agitata Tempe* (Hor., Od. iii., v. 24), and the *viridis Tempe* (Catulus, Carm. lxiii., v. 285). No doubt, however, their descriptions apply not to the pass itself, but to a vale at the mouth of the pass next the sea, "which, in situation, extent, and beauty, amply satisfies whatever the poets have said of Tempe." (*Cramer's Ancient Greece,* i., 378.)

TEMPLEMORE, an inland town of Ireland, prov. Munster, co. Tipperary, about 1½ m. W. from the Suir, and near the E. foot of the Devil's Bit mountains, 74 m. S.W. Dublin. Pop., in 1831, 2936. This is a neat town, in a comparatively rich and improved part of the country. It has a handsome parish church, a Roman Catholic chapel, a Methodist meeting-house, a school on the foundation of Erasmus Smith, a fever hospital and dispensary, a market-house, a bridewell, and large infantry barracks. Petty sessions are held on Wednesdays; fairs on Jan. 30, March 30, May 17, June 28, July 30, Sept. 3, Oct. 21, and Dec. 7. It is a constabulary station. Postoffice revenue in 1830, £827, in 1836, £778.

TEMPLETON, p. t., Worcester co., Mass., 61 m. W.N.W. Boston, 411 W. Watered by branches of Miller's and Chickapee rivers. It contains two churches, a Congregational and Unitarian, four stores, eight fulling-mills, two woollen factories, one furnace, two grist-mills, 16 saw-mills, four tanneries; ten schools, 480 scholars. Pop. 1776.

TENASSERIM PROVINCES, the name given to a long and comparatively narrow strip of territory in India-beyond-the-Ganges, belonging to Great Britain, comprised within the Bengal presidency, with which, however, it has no natural connexion. It consists principally of the provinces taken from the Birmese in 1825–26, or of the W. or coast districts of Siam, comprising Martaban, or Amherst, Ye, Tavoy, and Mergui, with its archipelago, extending between 11° and 19° N. lat., and about 98° and 99° E. long., having N. the independent Shan country, E. and S. Siam, and W. the Birmese empire and the Indian ocean. Area loosely estimated at 39,500 sq. m. The population being said to amount to only 85,000, though, notwithstanding its thinness, we believe that this is greatly below the mark, and that 160,000 would now be a nearer approximation. These provinces are shut off from Siam by one of the great mountain ranges, which branching from the table-land of Yunnan, traverse the Ultra-Gangetic peninsula in a S. direction. This chain rises, in this part of its course, to the height sometimes of 5000 ft., and is everywhere clothed with dense forests. Immediately along the coast the country is an alluvial flat covered with mangroves; and more inland, wherever it has been cleared, is found to be very fertile. It is extremely well watered; the great river, Than-lweng, or Thanluen, divides the province of Martaban from Birmah, and there are a variety of minor rivers, some of which are navigable to a considerable distance.

In the N. provinces, the year may be divided into the rainy and dry seasons, each of about six months' duration; but the latter resolves itself into the cold and hot seasons. These last are not very distinctly marked; but the coldest months are December and January, when the thermometer is sometimes, in the morning, down to 55°, but in the hottest part of the day ranges between 70° and 80°. The hot season immediately succeeds the cold, and continues until the rain begins to fall in April or May. The rains then commence, and continue until September or October; and although during a part of this time the sun be in its zenith, yet the almost incessant fall of rain renders this season the most refreshing part of the year. The annual fall of rain, during these six months, is about 200 inches. (*Mason, in the Pinang Gazette.*) In the S., where both the climate and products in many respects differ, it rains at least every fortnight throughout the year. The Tenasserim provinces are, upon the whole, much more healthy than many parts of India. As in other tropical climates, the most frequent diseases, common to both natives and foreigners, are, dysentery, with intermittent, remittent, and bilious fevers. They seldom prove fatal to the natives, and usually do so to Europeans only when the constitution has been impaired from other causes. The staple product is rice, of which a dozen different varieties are cultivated. Upland rice is grown on the hill sides by the Karean tribes, and much rice by the Birmese and other settlers. In the lowlands, in the beginning of April the farmer weeds his ground, and about the first week in June, when the rains commence, he hires a drove of buffaloes, if he have none of his own, and drives them about in a compact body over the wet field until the whole soil has been sufficiently worked to receive the seed. The principal harvest is in December, when the grain is reaped by a sickle, somewhat like the English. The ears are trodden out by buffaloes, and the rice is husk-ed and bruised by means of a wooden mortar, or by a hand-mill, formed of two grooved logs of wood, set upright and fitted into each other; a kind of machinery superior to that in use for the like purpose among other nations in a similar stage of civilization. Indian corn is not much raised, but sweet potatoes, yams, beans, onions and cucumbers are common. Tobacco is cultivated all over Tenasserim, as every one smokes. "from the child of three years of age to

the decrepid grandsire; from the governor's wife to the rice beater. (*Low's Hist. of Tenas.*) The sugar cane is also pretty general, though no marketable sugar is made. Cotton, hemp, indigo, pepper, &c., are only partially cultivated. Cardamoms, long pepper, catechu, and cocoa nuts, are gathered wild. Fruits are numerous, and the finest produce teak, sapan, aloe, and many other valuable woods, with bamboos, rattans, and many balsamic and medicinal plants. Iron ore is found in Ye and Tavoy, tin is very abundant in the S. provinces, but has not been seen in the N. Like gold, which is also widely diffused, it is obtained chiefly by washings. Trenches are dug, leading into the creeks, down which rapid streams run in the rainy season, and wash down the metallic particles. The workmen goes into the water, with a wooden dish in the form of an inverted cone, and having filled it with sand and pebbles, whirls it round on the surface of the water, by which action the lighter materials fly out, and leave the heavier down in the vortex of the inverted cone, consisting of a teaspoonful, or upwards, of tin and sand. Without further cleaning it goes to the smelter, and produces, I am told, from 50 to 75 per cent. of pure metal. Although all persons, Birmans or Kareans, are at liberty to procure the metal without any interference from government, yet few engage in the work; from which we may fairly infer that the returns are not remarkably profitable. (*Mason.*) Coal has been discovered in Mergui. Salt is made in numerous parts along the coast, and large quantities of saltpetre have been obtained from the bats' dung, collected in immense limestone caves in different parts of the country. The number of elephants inhabiting these provinces is supposed by Dr. Helfer to be proportionally greater than in any other part of India. The Birmese settlers hunt the elephants, and carve many kinds of articles from their ivory. Rhinoceros' horns are an article of trade, but the valuable skin of that animal is not met with in commerce, and, indeed, the trade in all kinds of hides, which might be made very profitable, has hitherto been wholly neglected. (See *Helfer's Rep.* 79-3.) Wax, honey, tortoise-shell, and edible birds' nests, are the principal commercial produce derived from the animal kingdom. The insect races are in great variety, and constitute one of the greatest pests of the country.

The manufactures are few; those of cloth and silks are the principal, but they have been, in a great degree, superseded of late years by the cotton goods imported from England and Hindostan. The weavers are almost exclusively woman, and there will hardly, perhaps, be found a loom throughout the provinces which has not a loom. The inhabitants of Tenasserim and Mergui carry on a brisk petty trade with the ports between those towns and Europe. They also occasionally visit Pinang, the Nicobar islands, Achin, Chittagong, and Dacca, exchanging their own produce for betel nut, raw and wrought silks, white muslins, earthenware, woollens, petroleum, cutlery, Chinese umbrellas, a little opium, ambergris, &c. The merchants of Tavoy and the S.; their exports are of much the same kind; their imports are cotton, tobacco, petroleum, Piece goods, hose &c, iron in bars, European and Bengal articles, &c. Mergui is the province but situated for commerce. Besides its trade coastwise, are Maulmain, ante, p. 294), a great deal of internal petty traffic is carried on by boats of from 3 to 30 tons burden; and caravans arrive occasionally from the confines of China, bringing lac, drugs, swords, manufactured cottons and silks, raw silk, candied sugar, earth nuts, black beans, ivory, and horns; and taking back salt, spices, cotton, quicksilver, assafoetida, borax, chintzes, piece goods, broadcloth, and various European articles. Capt. Low says, that the average of a late general estimate, from authentic documents, for one year, makes the value of the imports amount to 650,000 rupees, and that of the exports to only 173,880 rupees, though we doubt whether any such discrepancy can, in fact, exist between them. The weights and measures, as well as the usages and habits of these provinces, are mostly Birmese; the present inhabitants being of Birman extraction, though, according to tradition, the earliest inhabitants of the country were Siamese. Maulmain is the capital and residence of the governor and chief British authorities; subordinate officers are resident in the towns of Ye, Tavoy, and Mergui. The public revenue, derived from an assessment on grain of 20 per cent., one of 25 per cent. on other landed produce, taxes on gaming, opium, arrack, betel, &c., birds' nests, fines, capitation tax, &c., has been estimated at 237,000 rupees a year. (*Low's Hist. of Tenasserim in Asiat. Journal; Helfer's Report, &c.*)

Malcolm, the American missionary, has pointed out some of the advantages derived by these provinces from their connexion with England. "English influence," says he, "in a variety of ways has improved their condition. It has abolished the border wars, which kept the people and their neighbours continually wretched. Various other improvements are perceptible. Coin is getting introduced instead of masses of lead and silver; manufactures are improving; implements of better construction are used; justice is better administered; life is secure; property is sacred; religion is free; taxes, though heavy, are more equitably imposed, and courts of justice are pure generally. Formerly men were deterred from gathering round them comforts superior to their neighbours, or building better houses, for fear of exactions. Now, being secure in their earnings, the newly built houses are much improved in size, materials, and workmanship. The presiding officer in each province sits as magistrate on certain days every week; and before him every citizen, male or female, without the intervention of lawyers, may plead his cause, and have immediate redress. Everywhere in British Birmah the people praise English justice." *Malcolm's S. E. Asia*, 173–4.) Indeed, notwithstanding the present thinness of the population, emigration from Birmah into the Tenasserim provinces has been going on ever since we have held possession of the latter. Such, in truth, is the destructive influence of the Birmese government, that during the time it possessed Tenasserim, it had all but converted it into a desert.

TENBY (Welch, *Dynbych-y-Pysgod*), a parl. and munic. bor., market town, and seaport of Wales, co. Pembroke, hund. Narbeth, on the summit of a promontory on the W. side of Carmarthen bay, 9 m. E. Pembroke. Area of the in-liberty of the parish of St. Mary, which is co-extensive with the borough, 329 acres. Population, in 1831, 1942. The town consists principally of one long and broad street, lined with good houses, and pretty well paved. It is partly surrounded with walls, and is further defended by some batteries on the shore; its castle, supposed to have been built by the Flemings, by whom this part of the country was formerly occupied, is in a state of decay. The church of St. Mary is a spacious structure 146 ft. in length and 83 in breadth, with an elegant spire 152 ft. in height, supposed to be the loftiest in Wales, and painted white to render it a conspicuous mark for seamen. The roof of the nave is supported by arcades, having fluted pillars, and the ceiling is formed of carved wood ornamented with several figures, armorial bearings, &c. In the interior are several monuments. The living, a rectory and vicarage, worth £317 a year, is in the gift of the crown. (*Eccl. Rev. Rep.*) "An ancient edifice within a few feet of the W. entrance is now used as a school. A flight of steps on the outside leads to the apartment near which is a small arch in the style of Henry VII., and two others occur in an old wall opposite. These are supposed to have formed the principal entrances to St. Mary's college, once a convent of Carmelite friars, founded in 1399. The remains of St. John's chapel are situated in a marshy spot 200 or 300 yards from the town. Those of St. Julian's stand upon the pier, near its extremity. Opposite the town are some wild masses of rock, forming the islands of St. Catherine, and more distant are those of St. Margaret and Caldy. Eastward stretch the Norton sands bounded by grand and high cliffs; here are several bathing machines. Round the S. and W. sides of the town are the white or whit-sands, presenting a romantic and agreeable walk, 2 m. in length, to Glitar." (*Nicholson's Cambrian Guide*, p. 608.) The town-hall, courthouse, new market and slaughter houses, public baths, assembly-rooms, theatre, reading-rooms, bowling-green, several good hotels, &c., are the other most conspicuous objects at Tenby. The town was formerly a place of much commercial importance; and after the settlement of the Flemings here, it exported considerable quantities of woollen cloths. At present its trade is inconsiderable.

Tenby is a creek of the port of Milford. Nine or ten vessels from Plymouth, Brixholme, &c., make it their station during the fishing season, and supply the Bristol market. Its oysters, which are of superior quality, are sent pickled to London, Liverpool, and other places. Tenby is now principally distinguished as a watering-place, for which it is singularly well adapted, by the great beauty of its situation, and the protection from rough weather which it receives from the surrounding head-lands. It has grown rapidly, especially in good houses; most of which have been built within the last 20 years. A great many are occupied as lodging-houses during the season, and left empty for the rest of the year. Property of this sort appears likely to increase fast and steadily: houses are continually springing up in the many agreeable situations which are found unoccupied. (*Boundary Report*.) The public baths are both extensive and elegant; they comprise numerous bath and dressing rooms, warm and vapour baths, bed-rooms for invalids, a handsome promenade room, &c.; and are approached by an excellent carriage road. The water of their large reservoirs is changed every tide. The bathing-machines are the property of the innkeepers: the terms 1s. each time, and 6d. the guide. (*Cambr. Guide*, p. 604.)

Tenby is governed by a mayor, three other aldermen, and 12 councillors. Its earliest charters appear to have been granted about the time of Edward III., by the earls of Pembroke, but the earliest extant is of Richard III. Previously to the Municipal Reform Act, there were nearly 400 burgesses in the corporation, but their functions were merely nominal: and the whole management of the borough rested with the common council, which consisted of about 40 members. The town has only a few small endowments for charitable purposes. Courts of petty sessions weekly, a manor court, &c., are held; but the only jail consists of two large cells, formerly the old garrison dungeon. Prisoners are rarely confined here: when the term of imprisonment exceeds a month, they are sent to the jail at Haverfordwest. Corporation revenue, in 1840, £1104. The Reform Act did not

or 15 years of age, but retains its flavour and strength for a much longer period. The town, on the N.E. side of the island, is defended by a small fort. On the N. the port is protected by a pier, and it has pretty good anchorage. In antiquity it was a sort of *depôt* for the produce destined for Constantinople; and Justinian erected in it a large warehouse, the ruins of which are still extant, where vessels loaded with corn from Alexandria discharged their cargoes when they happened to be prevented, as was frequently the case, by contrary winds, from making a passage through the Hellespont to the capital. (*Tournefort*, i. 397.) Tenedos, according to Strabo (lib. xiii.), had a temple dedicated to Apollo; but it is principally known from its having been mentioned in connexion with Troy by Homer:

"Thou source of light! whom *Tenedos adores*;"
Pope's Iliad, i. lin. 55.

and by Virgil. According to the latter, it was the place to which the Grecian fleet made their feigned retreat before the sack of Troy:

"Est in conspectu (Troiæ) Tenedos notissima fama
Insula, dives opum, Priami dum regna manebant."
Æneid, ii. lin. 21.

TENERIFFE, or TEYDE (PEAK OF), a famous conical and volcanic mountain in the centre of the island of Teneriffe, of which its basis occupies the greater portion (See CANARY ISLANDS), rising, according to Borda, to the height of 11,434 French, or 19,172 English feet above the level of the sea. According to Von Buch the Peak consists of an enormous dome of trachyte, covered with layers of basalt. The summit of the cone (*El Piton*, the sugar loaf) is terminated by a crater, surrounded, as it were, by a circular wall or parapet. Humboldt, who descended to the bottom of the crater, found the heat was perceptible only in a few crevices, which gave vent to aqueous vapours, with a peculiar buzzing noise. (*Personal Narrative*, i. 173; *Engl. Trans.*)

The ascent of the Peak is distinguished by a difference of vegetation somewhat similar to that which is observed on the ascent of Etna (which see). Above the lower and more fertile tracts near the sea, where date trees, plantains, sugar-canes, Indian figs, vines, and olives, flourish in profusion, rises what is called the region of laurels. These are fed by a vast number of springs, that rise up amid a turf covered with perpetual verdure. Extensive plantations of chestnuts occur in the lower part, above which rise four species of laurel, and an oak resembling that of Thibet. The underwood in the lower part consists of arborescent heath, and in the upper part of ferns. Above this commences a vast forest of fir and pine trees which characterize the colder regions of the earth. Succeeding to this is a vast plain, like a sea of sand, covered with the dust of pumice stone, which continually fills the air. It is embellished with tufts of the beautiful shrub called the *retama* (*Spartium nubigenum*, Aiton), growing to the height of 9 ft., and loaded with odoriferous flowers, which are said to communicate a peculiar excellence to the flesh of the goats that feed upon them. At the entrance of this plain the rich verdure of the island terminates, as well as all appearance of habitation; and the traveller ascends aftewards through a complete solitude. Above this sandy plain are the *Malpays*, a name which the Spaniards apply to grounds destitute of vegetable mould, and covered with loose and broken fragments of lava. The ascent here is steep, and extremely fatiguing, as the blocks of lava roll from beneath the feet, and often leave deep hollows. At the extremity of the Malpays is a small plain called the Rambleta, from the centre of which the Piton, or conical summit rises to the height of about 2350 feet. Here are found those spiracles which are called by the natives the Nostrils of the Peak, consisting of crevices whence issue watery and hot vapours. The ascent of the Piton is steep, and rendered difficult by the loose ashes with which it is covered. At the top there is scarcely room to stand, and the crater, as already stated, is enclosed by a steep wall.

The view from the top of the Peak, though characterized by peculiar beauty, falls far short of the magnificent prospect from the summit of Etna. The cultivated and wooded parts of the island are, however, seen in close proximity, and the steep and naked declivities of the upper parts of the mountain strikingly contrast with the smiling aspect of the country beneath. The transparent atmosphere enables the spectator to distinguish minute objects, such as houses, sails of vessels, and trunks of trees. Beyond the eye wanders on all sides over the vast expanse of the Atlantic, and commands the whole archipelago of the Canaries. It has been alleged that the view extends as far as cape Bojador, on the coast of Africa.

The summit of the Peak is a *solfatara*, or extinguished volcano, whence no eruption has taken place since its discovery by Europeans; but some eruptions have taken place from the sides of the mountain during the course of last century. In 1704, one occurred in the district of Guimar, which buried several valleys, and approached within a short distance of the port of Orotava. Two years after, in 1706, the lava, issuing forth in a different quarter, buried the town and port of Garachico, then the finest and most frequented in the island. Another eruption happened in June, 1798, not far from the summit of the Peak, but it was not productive of much damage.

Notwithstanding its proximity to the equator, and to the coast of Africa, the Piton, or cone, is covered with snow during several months of the winter, and snow is always found in the hollows not exposed to the sun's rays. A powerful heat is always felt on the ground at the summit of the cone, and Humboldt mentions that his hands and face, and those of his party, were frozen, while their boots were burnt by the heat of the soil on which they walked. (*Dictionnaire Géographique; Humboldt's Personal Narratives.* i., 147–194. *Eng. Trans.*; *Lyell's Geology*, ii., 138.)

TENNESSEE, river, runs chiefly in the state to which it gives name, and is the largest tributary of the Ohio, carrying nearly as much water as the main stream. Its whole length is nearly 1200 m., which is longer than the course of the Ohio below Pittsburg, though, reckoned to its remote source in the Alleghany river, the Ohio is considerably the longest. Its principal head branches are Holston and Clinch rivers. Clinch river receives Powell's river; and Holston river receives French Broad river. The Clinch and Holston rivers rise in the Alleghany mountains in Virginia, and unite at Kingston, Roane county, Tennessee. The united stream then pursues a course generally S.W. to near the line of Georgia whence it pursues a winding course westwardly, until it turns to the S. and enters Alabama. It then makes a bold sweep to the S.W. and W. and again enters Tennessee at the N.W. corner of Alabama, whence it pursues a northerly course across Tennessee, which it leaves in the N.W. corner of Stewart county, and crossing Kentucky it enters Ohio river 11½ m. below the mouth of Cumberland river; and 47½ m. above the mouth of the Ohio. It receives, through its whole course, many tributaries, none of which, excepting those already specified, are of very great extent. The country which it drains contains about 41,000 sq. m. There are no perpendicular falls, and few rapids which obstruct navigation, through its whole course. It is susceptible of navigation by boats for 1000 m., 130 of which lie in the state of Alabama. It is navigable excepting at low water, for steamboats, from its mouth to Florence, Alabama, a distance of 259 m., at the Muscle shoals. A canal has been made around these shoals, 36 m. in length; and above the navigation is unobstructed for 250 m. to the suck or whirl in Marion county, Tennessee, where the river passes through the Cumberland mountains. Just as the river enters the mountains, a great rock projects from the northern shore, which causes a sudden bend in the river. The water is thrown with great violence against the southern shore, whence it rebounds around the point of the rock, and produces the whirl. The river here is compressed to the width of 70 ft., and boats ascend and descend the whirl with great difficulty. Immediately above the whirl are the trembling shoals, which at a low stage of the water are a great impediment to navigation, and above that the

river is rapid, and chiefly favourable to a descending navigation.

TENNESSEE, one of the western United States is situated between 35° and 36° 40′ N. lat., and between 81° 45′ and 90° 15′ W. long., and between 3° 35′ 36′′ and 12° 17′ 36′′ W. long. from W. It is bounded N. by Kentucky and Virginia; E. by North Carolina; S. by Georgia, Alabama and Mississippi; and W. by Mississippi river, which separates it from Arkansas and Missouri. Its medial length is about 430 m., and its medial breadth 104 m., containing 45,080 sq. m. or 28,800,000 acres. The population in 1790, was 35,691; in 1800, 105,602; in 1810, 261,727; in 1820, 422,813; in 1830, 681,904; in 1840, 829,210, of whom 183,059 were slaves. Of the free population 325,434 were white males; 315,193 were females; 2796 were coloured males, 2796 were coloured females. Employed in agriculture, 227,739; in commerce, 2817; in manufactures and trades, 17,815; in navigating the ocean, 55; do. rivers and canals, 382; in the learned professions, 2042.

The state is divided in 79 counties, which, with their population in 1840, were as follows:

Counties.		Total Pop.	Counties.		Total Pop.
Eastern District.			Jackson		12,821
Anderson	.	5,658	Lawrence		7,121
Bledsoe	.	5,976	Lincoln		21,493
Blount	.	11,745	Marshall		14,558
Bradley	.	7,385	Maury		28,125
Campbell	.	6,149	Montgomery		14,982
Carter	.	5,372	Overton		8,570
Claiborne	.	5,574	Robertson		12,501
Cocke	.	6,888	Rutherford		24,238
Granger	.	10,572	Smith		21,179
Greene	.	16,076	Sumner		22,445
Hamilton	.	8,175	Stewart		9,367
Hawkins	.	15,685	Warren		10,926
Jefferson	.	12,076	Wayne		7,705
Johnson	.	2,689	White		10,747
Knox	.	15,485	Williamson		27,006
Marion	.	6,070	Wilson		24,460
McMinn	.	13,719			
Meigs	.	4,794	Total .		411,749
Monroe	.	12,046	*Western District.*		
Morgan	.	2,660	Benton		4,722
Polk	.	3,570	Carroll		11,382
Rhea	.	3,985	Dyer		4,484
Roane	.	10,543	Fayette		21,182
Sevier	.	6,443	Gibson		13,409
Sullivan	.	10,736	Hardeman		14,300
Washington	.	11,751	Hardin		8,215
			Haywood		13,638
Total .		224,360	Henderson		11,425
Middle District.			Henry		14,385
Bedford	.	20,546	Lauderdale		3,485
Cannon	.	7,185	Madison		16,530
Coffee	.	8,184	McNairy		9,385
Davidson	.	30,509	Obion		4,354
De Kalb	.	5,982	Perry		7,438
Dickson	.	7,074	Shelby		14,721
Fentress	.	3,550	Tipton		6,286
Franklin	.	12,033	Weakley		9,634
Giles	.	21,494			
Hickman	.	8,618	Total .		189,221
Humphreys	.	5,195	Total of State		829,210

Nashville, on the S. bank of Cumberland river, is the seat of government. Cumberland mountains run through the state in the direction of N.E. and S.W., dividing into what has been called East and West Tennessee. Since the purchase in 1818 of all the territory between the Tennessee and Mississippi rivers, this portion of the state has been denominated the Western District; so that the whole territory is now divided into Eastern, Middle, and Western Tennessee. In eastern Tennessee the country is mountainous, having many branches of the Alleghanies, the most elevated of which are Cumberland and Laurel mountains. This district, although sufficiently fertile for ordinary farming, can never vie with the middle and western sections in profitable agriculture. With this exception, it yields most other vegetable productions of the states, and its vast waterpower fits it for a manufacturing region. Middle Tennessee is comparatively a level country, though it has some ranges of hills. A large portion of the lands are fertile and lie well for cultivation. Western Tennessee presents generally an undulating surface and a light and sandy soil, well adapted to the growth of cotton, which is here the great staple. The soil of the state is various, but generally fertile; and even the mountainous region in the E. part, though not generally very productive, has among its mountains, fertile valleys.

On the eastern boundary is a chain of mountains denominated in its different parts, Unka, Iron, Smoky and Bald mountains which constitute a continuous range. Some of the mountains of Tennessee are over 2000 ft. high, and they are generally wooded to their tops, though in some instances too rough for cultivation. Iron ore is found in nearly every county in Eastern and Middle Tennessee, and in many places it is wrought, furnishing iron equal in quality to any in the country. Fine beds of mineral coal are also found in this region. Bahr and other mill stone, grind, marble, rock crystals, gypsum, paints and dyestuffs, ochre

TENNESSEE.

and nitrous earths may be procured in any quantities in the mountainous districts, and there are numerous fine mineral springs. On the borders of Georgia and North Carolina some gold has been found, and a beautiful variegated marble near Nashville. All the forest trees common in the western country are found in this state. Juniper, red cedar, and pine exist in various places. Beautiful groves of pine abound in the mountains, and spirits of turpentine, tar, resin and lampblack are manufactured in considerable quantities. The mulberry tree flourishes, and silk may become a considerable article of production. The vine is cultivated and flourishes, and peach trees do well, if they escape the late frosts. Taken as a whole, cotton a Middle and Western Tennessee, is unquestionably the staple production; but tobacco, wheat and Indian corn are largely raised. Whiskey, hogs, horses, cattle, flour, gunpowder, saltpetre, coarse linen, poultry, bacon, lard, butter, apples, pork, tobacco, Irish and sweet potatoes, tar, turpentine, resin and lampblack, constitute the loading of boats for the southern market. In East Tennessee, cattle, horses and hogs are driven over the mountains to an eastern market.

There were in the state in 1840, 341,409 horses or mules; 2,851 neat cattle; 741,593 sheep; 2,926,607 swine; poultry valued at $606,969. There were produced 4,569,692 bushels of wheat; 304,390 of rye, 44,986,188 of Indian corn, 17,118 of buckwheat, 4800 of barley, 7,035,678 of oats, 04,570 of potatoes, 1,060,332 pounds of wool, 29,550,432 tobacco, 7977 of rice, 27,701,277 of cotton. 50,907 of wax, 7 of silk cocoons, 256,073 of sugar, 31,233 tons of hay, 4 of hemp and flax. The products of the dairy were valued at $472,141; of the orchard at $367,105; value of other produced $317,606; 3336 barrels of tar, pitch, &c., re made.

The climate is mild and generally healthy. The winter Tennessee resembles the spring in New-England. Snow soon falls to a greater depth than 10 inches, or lies longer than 10 days. Cumberland river has been frozen r but three or four times since the first settlement in country. On some low grounds in the western parts of state, the inhabitants are subject to bilious fevers and r and ague in the autumn.

The usual route to market is down the Cumberland and nessee rivers to the Ohio, and thence down the Missippi to New-Orleans. Foreign goods come extensively the coast through Pittsburg, and also from New-Orleans through the Mississippi and its tributary waters.

Tennessee river has its chief course in this state, though es in Virginia, runs extensively in Alabama, and enters the Ohio in Kentucky. It is 1200 m. long, is navigable for steamboats to Florence, 250 m. from its mouth, and boats more than 250 m. farther. Cumberland river in Kentucky, but has its course chiefly in Tennessee. navigable for steamboats 120 m. to Nashville, and for 300 m. farther. It enters the Ohio in Kentucky, 59 m. its mouth in the Mississippi. The Holston, Clinch, h Broad and Highwassee are branches of Tennessee Obion, Forked Deer, and Wolf rivers in the western f the state, flow into Mississippi river, and are navigable boats.

hville is much the largest and most commercial place state. Knoxville, on Holston river, is the largest n East Tennessee, and was formerly the seat of government; Murfreesboro was formerly also the seat of government, and is surrounded by a fertile country in Middle see. Memphis on Mississippi river, is a place of rable business. Clarksville, Franklin, Jonesboro', ester and Columbia are considerable places.

e were in the state in 1840, 13 commercial and 52 sign houses in foreign trade, with a capital of 600; 1032 retail dry goods and other stores, with a of $7,357,300; 1125 persons engaged in the lumber vith a capital of $6700; 31 persons engaged in transportation, who, with five butchers and packers, ed a capital of $96,811.

—made or family goods amounted to $2,886,661. were 26 woollen factories and four fulling-mills, employed 45 persons, producing articles to the of $14,290, with a capital of $25,600; 36 cotton i with 16,813 spindles, employed 1542 persons, and i articles to the amount of $325,719, with a capital 240; 34 furnaces produced 16,128 tons of cast iron, forges produced 9673 tons of bar iron, employing sons, and a capital of $1,514,736; four persons produced to the amount of $1500, with a capital of $400; lting houses for lead employed four persons, and a f $350; 21 persons produced 12,942 bushels of bi- s coal; five paper-mills produced articles to the of $46,000, and other manufactories of paper pro- the amount of $14,000, the whole employing 87 and a capital of $63,000; 177 persons produced caps to the amount of $104,949; 454 tanneries em-

ployed 909 persons, and a capital of $464,114; 374 other manufactories of leather, as saddleries, &c., produced articles to the amount of $350,050, with a capital of $154,540; 29 potteries employed 50 persons, produced to the amount of $51,600, with a capital of $7300; 142 persons manufactured hardware and cutlery to the amount of $57,170; 266 persons produced machinery to the amount of $257,704; 34 persons manufactured 564 small arms; 11 persons manufactured the precious metals to the amount of $28,400; 10 persons manufactured granite and marble to the amount of $5400; 417 persons produced bricks and lime to the amount of $170,371; 1426 distilleries produced 1,109,107 gallons of distilled spirits, and six breweries produced 1636 gallons of beer, the whole employing 1341 persons, and a capital of $218,182; 518 persons manufactured carriages and wagons to the amount of $219,897, with a capital of $60,878; 28 rope-walks employed 256 persons, producing cordage to the amount of $139,630 with a capital of $84,930; 202 persons manufactured furniture to the amount of $79,580, with a capital of $90,650; 255 flouring-mills produced 67,881 barrels of flour, and with other mills, employed 2100 persons, producing articles, to the amount of $1,020,064, with a capital of $1,310,195; 193 brick or stone houses, and 1098 wooden houses were built by 1467 persons, and cost $457,402; 41 printing-offices, five binderies, two daily, six semi-weekly, and 36 weekly newspapers, and 10 periodicals, employed 191 persons, and a capital of $112,500. The total amount of capital employed in manufactures in the state, was $3,731,560.

Greenville college, at Greenville in East Tennessee, was founded in 1794; Washington college, in Washington county, was founded in 1794; the university of Nashville, the principal literary institution in the state, was founded under the name of Cumberland college in 1806; the East Tennessee college at Knoxville, was founded in 1807; Jackson college near Columbia, was founded in 1830; the South Western Theological seminary, at Maryville was founded in 1821. The number of students in these institutions in 1840, was 369. There were in the state 152 academies with 5539 students, and 923 common and primary schools, with 25,090 scholars. There were in the state 58,531 white persons over 20 years of age, who could neither read nor write.

In 1836 the Methodists had 127 travelling preachers, and 34,966 communicants; the Baptists had 413 churches, 219 ministers, and 98,472 communicants; the Presbyterians had 190 churches, 90 ministers, and 10,000 communicants; the Episcopalians had a bishop and eight ministers. There were besides many Cumberland Presbyterians, and some Lutherans, Friends, Christians, and Roman Catholics.

The constitution of this state was formed in 1796, at Knoxville. This constitution was revised and amended, and ratified by the people, in March, 1835.

The governor is elected by the people biennially, and is not eligible more than six years in any term of eight years. He must have attained the age of 30 years, must be a citizen of the United States, and a citizen of the state for seven years next preceding his election. The senate consists of 25 members, elected by the people once in two years. Every senator must have attained the age of 30 years, must be a citizen of the United States, must have been an inhabitant of the state for three years, and of the district for which he is elected for one year, immediately preceding his election. The house of representatives consists of 75 members, elected at the same time and for the same period as the senators. Every representative must be a citizen of the United States, of the age of 25 years, must have been a citizen of the state for three years, and a resident in the county for which he is elected, for one year immediately preceding his election. All judges are elected by the joint vote of the general assembly. The judges of the supreme courts are elected for 12 years, and must have attained the age of 35 years. The judges of the inferior courts are elected for eight years, and must have attained the age of 30 years. The state attorneys are elected in the same manner, for the term of six years. The secretary of state is elected by the joint vote of the general assembly for four years; and the state treasurer, in like manner, for two years. Every free white male citizen of the United States, who has been a citizen of the county where he offers his vote for six months next preceding the day of election, enjoys the right of suffrage; and no person is disqualified from voting on account of his colour, who is, by the laws of the state, a competent witness in a court of justice against a white man. The legislature meets biennially at Nashville, on the first Monday of October.

The state owns the whole capital of the bank of Tennessee, to the amount of $2,000,000; and of the Union bank to the amount of $500,000; making a total of $2,500,000. The state debt amounts to $3,300,000. It owns about half of all the internal improvements of the state, and has uniformly a surplus in the treasury.

Some works of internal improvement have been completed, and greater have been projected. A railroad has been completed from Memphis on the Mississippi, 50 m. to La Grange in Lafayette county. Somerville branch extends from the main road at Moscow, 16 m. to Somerville. Among the contemplated railroads is the Hiwassee railroad from Knoxville, 98½ m. to the Georgia line, to unite with the Western and Atlantic railroad of Georgia. But this and other railroads contemplated are suspended for the present.

This state was originally included in the charter of North Carolina, given by Charles II., in 1664. In 1757 fort London was built, and garrisoned; and the Indians, to induce artisans to settle among them, made donations of land. Fort London was established on the N. side of Little Tennessee river, about 1 m. above the mouth of Tellico, in the centre of the Cherokee country. A war with that Indian nation having occurred, the garrison was besieged, and compelled to surrender for the want of provisions. By the terms of the capitulation they were to retire beyond the Blue Ridge; but after proceeding about 20 m., the Indians fell upon and massacred the whole number, excepting nine persons, amounting to between two and three hundred. This happened in the year 1760. In 1761 Colonel Grant marched against the Indians and subdued them, and compelled them to sue for peace. The only settlements which had been made in the vicinity of fort London were broken up by the war; but tranquillity having been restored, 15 or 20 persons formed themselves into a company and came to a place now called Carter's valley, in East Tennessee. In 1768 an exploring party came into the country from Virginia. The first permanent settlements were made in 1768 and 1769. The settlers were chiefly from North Carolina and Virginia. The settlements continued to increase until 1774 and 1775, when an extensive purchase of land was made from the Indians by Henderson and company, but not without warm opposition from the chief, who declaimed against the encroachments of the whites, without effect. In 1776 war with the Indians occurred. but after some fighting an arrangement was made by the states of North Carolina and Virginia, by which the boundaries of the territory, now state of Tennessee, were definitively settled. In 1779 Capt. James Robertson and others from East Tennessee crossed Cumberland mountain, and explored the country in the neighbourhood of Nashville, and planted corn that season on the ground where Nashville now stands. They all returned for their families excepting three, who remained to keep the buffaloes, which abounded in this region, out of the corn. In 1780 was fought the celebrated battle of King's mountain, by which the British, tories and Indians were defeated by the intrepid backwoodsmen. Toward the close of 1781 the Cherokees and Chickasaws sued for peace. This year North Carolina allowed to the settlers on the Cumberland the *rights of pre-emption*. Each head of a family, and every single man of 21 years of age, was allowed 640 acres of land, who had made actual settlements prior to June 1780. This year also, North Carolina established courts of equity in all the districts of the state. In 1783, in addition to bounty lands to the soldiers of the regular army, 25.000 acres were granted to Gen. Green for his revolutionary services. In 1784 North Carolina ceded the territory, now the state of Tennessee, to the United States, if they should accept of it within two years from the passage of the act, North Carolina retaining the jurisdiction until congress should make provision for a territorial government. In the mean time the inhabitants determined to organize on their own responsibility a territorial government. The assembly of North Carolina repealed the act ceding the western country to congress, and considerable confusion followed. In December 14th, 1784, a convention of deputies formed a constitution for the new state, denominated *Frankland*, and announced to North Carolina that they considered themselves independent of her. A part of the people adhering to North Carolina, two conflicting courts were exercising authority in the territory, and other conflicting acts were passed by the two legislatures. In 1789 the legislature of North Carolina authorized and required their senators in congress to execute deeds of conveyance for the territory of Tennessee, which they did on the 25th of February 1790. In May, 1790, congress passed a law for the government of the country south west of the Ohio, and William Blount was, by President Washington, appointed the first governor of the territory, who in October 1790, established his residence in East Tennessee. In 1791, the "Knoxville Gazette" was established at Rogersville. The population of the territory then amounted to 36,043 including 3417 slaves. A war with the Creeks and Cherokees took place in 1793 in which, after severe fighting, they were subdued. On the 19th of October the governor authorized an election of a colonial legislature by the people. The assembly met at Knoxville on the fourth Monday of February, 1794, and was regularly organized. The Creeks and Cherokees continuing their depredations, war

was made upon them, until despairing of preventing the settlement of the whites, in June, 1794, they applied for peace, which was concluded by a treaty ratified at Philadelphia on the 20th of July. In 1795, the territory was found to contain 77,262 inhabitants, which entitled them to become a state, a constitution was formed in February, 1796, and on the 6th of June, 1796, they were admitted to the Union. The citizens of this state bore an important part in the last war with Great Britain, particularly in the defence of New-Orleans.

TENTERDEN, a mun. bor., mar. town, and par. of England, co. Kent, lathe Scray, the mun. bor., comprising all the hund. of Tenterden, and the par. of same name, together with a part of the par. of Ebony; 43 m. S.E. London. Area of par. 8690 acres. Pop., in 1831, 3177. The town, which is finely situated on an eminence, environed by hop-grounds, is well built. It consists principally of a single street, continuous with which is the straggling village of Bird's isle to the N., and a little to the E. is the hamlet of Lys Green. The parish church, a spacious structure, consists of a nave, N. aisle and chancel, with a well-built lofty tower at the one end, on which are sculptured the arms of the monastery of St. Augustine, to which foundation this church was appropriated in 1259. The living of Tenterden, a vicarage, worth £177 a year, is in the gift of the dean and chapter of Canterbury. It has, also, chapels for Baptists, Wesleyans, and Unitarians. a free school founded in 1521, for six scholars, and which has been greatly enlarged by voluntary subscriptions, and several minor charities. Tenterden has no manufactures, and depends entirely on its retail trade with the rich agricultural country in which it is situated, and upon its proximity to Romsey marshes: which has made it be selected as a place of residence by persons engaged in the grazing of sheep and cattle on the marsh. There are also a few gentlemen of independent fortune living in and about the town. There seems but little probability of its becoming a place of much greater importance than it is at present. It is neither paved, lighted, nor watched. (*Mun. Bound. Rep.*) In the reign of Henry VI., Tenterden was incorporated and annexed to the town and port of Rye; but the earliest existing charter is that of Elizabeth. It is governed, under the Municipal Reform Act, by a mayor, three aldermen, and 12 councillors, who hold petty sessions weekly, and a court of record every fortnight. Corporation revenue, in 1840, £505. Market day, Friday; fair, first Monday in May, for horses, cattle and pedlery. (*Mun. Bound. and Corp. Rep. Append.*)

TEQUENDAMA (FALL OF), a celebrated cataract in the republic of New-Granada, Colombia, in the course of the Bogota river, a tributary of the Magdalena, a few miles W. from Bogota. A short distance above the fall the river is 140 ft. in breadth; but being forced into a narrower, though deep bed of only 40 ft. in width, it is precipitated at two bounds down a perpendicular rock, to the depth of 650 ft.; and even in very dry seasons, Humboldt says the volume of water still presents a side view of 98 sq. metres. "The cataract forms an assemblage of every thing that is sublimely picturesque in beautiful scenery. This fall is not, as has been commonly said, the loftiest in the world; but there scarcely exists a cataract which, from so lofty a height, precipitates so voluminous a mass of waters." (*Humboldt's Researches*, i., 17.) The body of water, when it first parts from its bed, forms a broad arch of glassy appearance; a little lower down it assumes a fleecy form; and ultimately, in its progress downwards, it shoots forth into millions of small tubular masses, which chase each other like sky-rockets. The noise which attends the fall is quite astounding; and dense clouds of vapour are sent up, which rise to a considerable height, and mingle with the atmosphere, forming in their ascent the most beautiful rainbows. The comparative smallness of the stream which runs off from the foot of the fall, proves that a large proportion of the water is lost by evaporation. (*Med. Tra., xxvii., 330.*) What gives the fall of Tequendama a remarkable appearance, is the great difference in the vegetation surrounding its different parts. At the summit the traveller "finds himself surrounded, not only with the aralia, begonia, and the yellow bark tree, but with oaks, elms, and other plants, the growth of which recalls to his mind the vegetation of Europe; when suddenly he descends, as from a terrace, and at his feet, a country producing the palm, the banana, and the sugar-cane." The true cause of this phenomenon has not been satisfactorily explained. The difference of altitude, about 175 metres, is, as Humboldt has stated, too inconsiderable to have much influence over the temperature of the air. (*Researches, p. 79, &c.*)

TERAMO (an. *Interamnia Prætutia*), a city of the Neapolitan dom., prov. Abruzzo Ultra. of which it is the cap., 947 ft. above the sea, in the angle formed by the Vezzola, where it joins the Tordino, 16 m. W. from the embouchure of the latter in the Adriatic, and 19 m. N.N.E. Monte Corno the highest summit of the Appennines. Pop., in 1830, 10,231. It was formerly surrounded by strong walls, but is now quite open. With one exception, its streets are narrow and dirty, and its houses, for the most part, mean-looking. In the outskirts, however, some of the houses are in better taste. The cathedral has been modernized. There are several convents, hospitals, asylums, &c. It has but few manufactures or industrial establishments; but it is he seat of the civil and criminal tribunals of the province, and has a royal college, a seminary, or establishment for he instruction of the clergy; and is the residence of several opulent families. Its vicinity is in general fertile, producing corn, wine, and oil in abundance: in the time of the Romans its wine was in high estimation:

Tum qua vitiferos demittat Prætutia pubes
Læta laboris agros.   Silius Italicus, lib. xv., v., 568.

ome buried arches, the vestiges of a theatre, baths, and ome other edifices, are the principal remains of the ancient city. (Del Rè Descrizione delle Due Sicilie, ii., 47, c.; Craven, Excurs. in the Abruzzi, i., 310.)

TERCERA. See Azores.

TERLIZZI, an inland town of S. Italy, kingd. of Naples, ov. Bari, cap. cant., on an elevated site, 18 m. W. Bari. p., said to amount to about 10,000. Notwithstanding its ze, it seems to possess little worth notice beyond the usu-superabundance of religious edifices, if we except a gal-y of pictures, comprising works by several of the great lian masters, in the mansion of a noble family.

TERMINI (an. Therma Himerenae, and simply Ther-), a maritime town of Sicily, on the N. coast of the and, intend. of Palermo, cap. district and canton, near mouth of the river of its own name, 24 m. E.S.E. Palmo, lat. 37° 57′ 28″; N. long. 14° 42′ E. Pop., in 1831, 942. It is finely situated on the declivity of a hill rising m the sea; and besides being surrounded by an old wall, arther defended, towards the sea, by a castle on a high k, commanding the town and port. The streets are, for most part, narrow and dirty; but it has some pretty d public buildings, among which are several churches convents, a royal college, two hospitals, a monté-di-a, an asylum for females, and convenient baths over hot springs, for which the city has been celebrated a the remotest epoch, and from which she has derived modern as well as her ancient name. The town is a catore, or shipping port, and exports (mostly coastwise 'alermo), corn, oil, shumac, dried fruits, manna, &c. sardine and anchovy fisheries are also actively carried The harbour, which is but indifferent, is open to the (Smyth's Sicily, p. 95; Rampoldi, &c.)

out 6 m. E. by S. from Termini, are the ruins of the nt Himera, near which Gelon, tyrant of Syracuse, to-defeated and destroyed an army of Carthaginians, to have comprised no fewer than 300,000 men, com-ied by Hamilcar, grandfather of Hannibal, who lost fe in the action. The Carthaginians were the allies of es, and their defeat is said by Herodotus (lib. viii., cap. to have happened on the same day that the Greeks d the victory of Salamis, anno 480 B.C. But though abundantly certain that Gelon gained a great victory Hamilcar, it is extremely improbable that the forces e latter amounted to half the number mentioned. Hannibal never had 100,800 men at any one time his command; and the probability is that 30,000 l be much nearer the number of Hamilcar's army 300,000. The ships which had conveyed the troops iera, are said to have been drawn up on the beach; 'aptain Smyth says that so great a fleet and army not possibly have been accommodated in the situation the battle is said to have been fought. In fact, very dependence is to be placed on the statements as to rs, in most ancient authors; they are almost invari-uch exaggerated.

a subsequent period Hannibal avenged his grand-s disaster by taking and utterly destroying Himera. of its citizens as escaped the massacre which took on this occasion, sought an asylum in Therma. (Ci-Verrem, ii., cap. 35.) Augustus raised the latter to k of a colony. Stesichorus, one of the most ancient lebrated of the Greek poets, was a native of Himera. NATE. See Molucca Islands.

NI (an. Interamna), a town of the Papal states, poleto, in a rich and fine valley, near the right bank Nera (an. Nar),

" Sulfurea Nar albus-aqua."
Æneid. lib. vii., lin. 514,

m. W. from the famous falls of the Vellino, and .N.E. Rome. Pop., circ. 9000. It is surrounded by and towers; but though it has wide streets, salome e buildings, and a charming situation, it is, on the poor and mean, retaining but few traces of its an-plendour. It has a cathedral with a superb altar,

a hospital, a monte-di-pièta, and some other charitable foundations, a handsome theatre, and a building, erected in 1827, for the reception of the waters of the Vellino for the public accommodation. Among the remains of antiquity are some vaults of an amphitheatre constructed under Tiberius, portions of temples of the Sun and Cybele transformed into churches, and the remains of public baths. The surrounding country is extremely productive, fœcundissimos Italiæ campos (Tacit. Annal. lib. i., cap. 79); and on the river are flour and oil-mills, tanneries, &c. (Rampoldi, &c.)

The historian Tacitus is said to have been a native of Interamna; but there is no evidence that such was really the case. The emperors Tacitus and Florianus are also said, but on no better grounds, to have belonged to it.

The falls of the Vellino, called the Cascata del Marmore, about 4 m. E. from Terni, are among the most striking objects of the kind that are any where to be met with. The total height of the fall, which is divided into three leaps, is probably (for there is the greatest discrepancy in the statements on the subject) from 650 to 750 ft. The water is conveyed to the fall in an artificial channel, more than 1 m. in length, originally dug by the consul Curius Dentatus, anno 274 B.C. (Cicero, Epist. ad Atticum, Epist. 15); but, the channel having been filled up by a deposition of calcareous matter, it was widened and deepened, and in part altered, in 1596, and again in 1785. Byron has appropriated some magnificent stanzas to a notice of these falls (Childe Harold, cant. iv., st. 69-72); and he adds in a note, " I saw the Cascata del Marmore of Terni twice at different periods; once from the summit of the precipice, and once from the valley below. The lower view is far to be preferred, if the traveller have time for one only; but in any point of view, either from above or below, it is worth all the cascades and torrents of Switzerland put together." (See Cellarii Geographia Antiqua, i., 751-778; Eustace, i., 327, oct. ed.; Wood's Letters, ii., 97, &c.)

TERRACINA (an. Anxur and Terracina), a seaport town of the Papal states, deleg. Frosinone, at the S. extremity of the Pontine marshes, close to the Neapolitan frontier, 59 m. S.S.E. Rome; lat. 41° 18′ 14″ N., long. 13° 33′ 37″ E. Pop. about 6000. This town, which is on the Appian way, and adjoining the embouchure of the canal for the draining of the marshes, stands partly on low ground, and partly on the declivity of a hill. With the exception of the portion along the shore, it is ill built; and, owing to the deleterious air of the contiguous marshes, it is unhealthy, and the inhabitants have a sickly appearance. On the hill is the cathedral, erected, as is supposed, on the site of the temple of Jupiter Anxurus; higher up are the ruins of the ancient Anxur—

" Impositum saxis late candentibus Anxur,"
Hor. Sat., lib. i., sat. 5.

And crowning the brow of the rock which overhangs the modern town are the ruins of the palace of Galba, repaired and. re-occupied by Theodoric, commanding a magnificent view of the Pontine marshes, Monte Circello, and of Gaeta and the Neapolitan shore, as far as Baiæ. Pope Pius VI. endeavoured, by improving the drainage of the marshes, and by clearing out and deepening the harbour, which had been completely filled up, to recover for Terracina some portion of its former importance. But his efforts have not had the anticipated success; and though the fishery be carried on to some extent from the port, it has little or no trade. In 1810, Napoleon suppressed the bishopric of Terracina.

Anxur, which was originally a town of the Volsci, subsequently became a Roman colony, and an important naval station. It was sacked by Alaric, and was occupied by the Arabs for about a century. It was also taken and sacked by a French force in 1798. (Rampoldi; Cellarii Geographia Antiqua, i., 812; Wood's Letters, ii., 176, &c.)

TERRANOVA (an. Gela), a seaport town of Sicily, prov. Caletanisetta, cap. district, on the S. shore of the island, near the mouth of the river of the same name, 18 m. E. Alicata; lat. 37° 4′ 30″ N., long. 14° 15′ E. Pop., in 1831, 9780. It is well situated on a bank near the sea, and has a fine palace, belonging to its proprietor, the Duke de Monteleon; but the streets are irregular and dirty, and its castle, churches, and convents appear to be neglected. It has a tolerably good hospital. Water is said by Swinburne to be scarce and dear. Coarse cloth is manufactured in the town; and having a caricatore, or shipping station, it has some trade in the exportation of corn, wine, sulphur, soda, &c. The cloth made in the town finds a good market at the commercial fair held in August. The anchorage at Terranova is opposite to, and about 1 m. from the shore, in from 7 to 11 fathoms. It is, like other places on the same coast, open to the southerly gales, which sometimes throw in a heavy sea. In 1839, 67 ships (35 of which were British), of the burden of 4981 tons, cleared out from the port. Smyth says that a party of strolling players has existed in this town for more than half a century, called the *Com-pagnia degli Uniti*, from their all sharing alike in the gains of the society.

Though the question be not free from difficulty, there seems every reason to think that Terranova, and not Alicata, occupies the site of the ancient Gela. It has some remains of antiquity, consisting of the foundations and mutilated fragments of a great temple, and of a Doric column. Gela was a Rhodian colony, and early attained to considerable distinction. But it is principally memorable for having given birth to Gelon, prince or tyrant of Syracuse, famous alike for his virtues, and for his great victory over the Carthaginians, commanded by Hamilcar, grandfather of Hannibal. Gela was subsequently destroyed by Phintias, tyrant of Agrigentum, and is included by Strabo among the uninhabited towns of the island. The modern town was founded by Frederick of Arragon, towards the close of the 13th century. (*Smyth*, 196 ; *Swinburne*, ii., 301, &c.)

TERRE BONNE, parish, La. Situated in the S. part of the state, and contains 1850 sq. m. Bounded N.E. by Bayou La Forcha, S. by the gulf of Mexico. Much of the soil is too low for cultivation, but fertile, adapted to the sugar cane. It contained in 1840, 4764 neat cattle, 1013 sheep, 5473 swine ; and produced 115,002 bushels of Indian corn, 26,283 of potatoes, 17,880 pounds of rice, 1,003,850 of cotton, 6,675,600 of sugar. It had one store ; two schools, 40 scholars. Pop.: whites, 2075 ; slaves, 2300 ; free coloured, 35 ; total, 4410. Capital, Williamsburg.

TERRE HAUTE, p. v., capital of Vigo co., Ia., 73 m. W. Indianapolis, 644 W. Situated on a beautiful high ground on the E. bank of Wabash river. The National road here crosses the river on a fine bridge. It contains a brick courthouse, a jail, two churches, a Congregational and Methodist, a bank, a market-house, a seminary, two steam mills, a brewery, five commission houses in foreign trade, 50 retail stores, a printing-office, and 2000 inhabitants. It commands the trade of an extensive and fertile country, and is flourishing.

TERUEL (an. *Turdeto*), a town of Spain, cap. prov. of its own name in Aragon, on a hill, at the foot of which flows the Guadalaviar, 75 m. N.W. Valencia. Pop. about 8000. It is walled, and tolerably well built. Being a bishop's see, it has numerous churches and convents; one of the latter, belonging to the Jesuits, being the largest edifice in the town. The cathedral, an extensive building, though its architecture be not wholly in good taste, is sumptuously adorned within, and has, or, at all events, had, many fine paintings. It has several fountains supplied with water by an ancient aqueduct. Its manufactures comprise woollen and linen fabrics, shoes, and earthenware, with fulling-mills, dyeing-houses, tanneries, &c. The vicinity is very fertile, and near it are some celebrated warm sulphur springs. Teruel is the residence of a military governor, and was a fortress of some consequence under the Moors, from whom it was taken by Alphonso II. in 1171. (*Miñano, &c.*)

TESCHEN, a town of Austrian Silesia, cap. circ. and duchy of same name, on the Olsa, a tributary of the Oder, 36 m. E.S.E. Troppau. Pop. about 7000. It is well built and has three suburbs, a ducal castle, several Roman Catholic churches, a Lutheran church, and gymnasia for both persuasions, that of the former possessing, it is said, a library of 12,000 vols. (*Berghaus*.) There are several other superior schools, and a military asylum. Teschen is the seat of the circle tribunal and other courts, and has manufactures of woollen cloths, cassimeres, leather, fire-arms, &c. Here was signed a treaty between Austria and Prussia, in 1779.

TESSIN, or TICINO, the most S. canton of Switzerland, between lat. 45° 50′ and 46° 37′ N., and long. 8° 25′ and 9° 19′ E., being separated by the main chain of the Alps from Uri and the Grisons on the N., while on other sides it is surrounded chiefly by the Austrian and Sardinian territories, the lakes Maggiore and Lugano forming parts of its S. frontier. Area estimated at 1034 sq. m. Pop. in 1837, 113,993, all Roman Catholics. Most part of this canton is either mountainous, or divided into numerous valleys by alpine ramifications: in the S., however, it sinks to the level of the plain of Lombardy. The Ticino, whence this canton derives its name, has its sources in mount St. Gothard, in the Valli Bedretto, Piora, Blegno, &c. Its course is generally southward, and after intersecting the canton near its centre, and traversing the Lago Maggiore in its entire length, it forms the boundary between Lombardy and Piedmont, falling into the Po at Pavia, after an entire course of about 100 m., about 60 of which are navigable. The climate of Tessin is mild ; and though its pastures be not so good, nor so well watered as those of the cantons N. of the Alps, its soil is generally very fertile. Agriculture is, however, extremely backward, partly from the ignorance and want of industry of the inhabitants, and partly from the too great subdivision of the surface into small properties, por-

tions of which at great distances from each other sometimes belong to the same proprietors. Wheat, rye, and maize are the principal grains raised ; a good deal of tobacco is cultivated. Wine is grown in many districts ; but, like the Italian wines, that of this canton will not keep for any considerable period. The silk of Tessin is of superior quality ; and a supply worth from 200,000 to 300,000 Swiss francs is sent annually into other parts of Switzerland. (*Picot*.) Most of the fruits common to Lombardy flourish here : the chestnut woods are extensive, and chestnut flour is largely consumed by the inhabitants. The canton abounds with timber, but much of it is useless from the want of roads and expense of carriage. About 2000 quintals a year of cheese are sent into Italy, and calves, sheep, and hogs are also exported. The chamois is a native of this canton. It sometimes breeds with the domestic goat, and the resulting progeny is greatly prized for its skin. There are scarcely any manufactures, and the trade of Tessin is chiefly in the conveyance of goods between Switzerland and Italy. A great many of the male natives of the canton emigrate to Milan, Venice, Trieste, Turin. Marseilles, and the adjacent countries, where they serve as confectioners, chocolate manufacturers, waiters in coffeehouses, &c. ; leaving the labours of the field and the care of the cattle to the women.

Tessin was merely a territory subordinate to Switzerland till 1815, when it was admitted into the confederation, in which it holds the eighteenth rank. Its government was materially altered in 1830 : when the grand council, which holds the sovereign and legislative power, was made to consist of 114 members, chosen in the different communes by all the citizens born in the canton twenty-five years of age, and who possess immoveable property to the value of 200 Swiss francs, or the usufruct of such property to the value of 300 francs. It chooses its own president, and meets each year by rotation in Bellanzona, Locarno, and Lugano. The executive body, or lesser council, consists of nine members, chosen by the greater council among its members. Equality before the laws, the freedom of the press, and the right of petition, are guaranteed. Tessin is subdivided into eight districts and 38 circles, and the lesser again into communes. In every commune there is a municipal council of from 3 to 11 members, with whom rests the direction of the local police. A justice of the peace sits in each circle ; in each district there is a court of primary jurisdiction ; and for the whole canton there is a supreme tribunal of 13 members. Criminal processes appear to be more common in this than in most other cantons of Switzerland. The public revenue, principally derived from salt and custom duties, in 1833-34, amounted to 637,339 Swiss livres : expenditure to 787,640 ditto. Public debt, in 1836, 5,041,450 ditto. Contingent to the army of the Swiss Confederation, 1804 men.

The inhabitants of Tessin are of middle stature, and generally square and strongly built ; though, on account, it is said, of their irregular mode of life, among other causes, they seldom attain a great age. In many respects they resemble their Italian neighbours, and their language is a dialect of the Italian. Among them have been several eminent painters, sculptors, and architects ; the latter including Domenico Fontana, who completed the dome of St. Peter's, and executed many other great works in Rome. But the bulk of the inhabitants are very backward in point of education, and some of the communal magistrates can neither read nor write. It has no council of public instruction, no literary association (except, perhaps, in Lugano), and scarcely a library. The habits of the people are dirty and depraved. According to Picot, " *L'éloge de la sobriété Italienne ne convient au canton du Tessin, ni sous le rapport de la boisson, ni sous celui de manger. Quoique donné naturellement les dispositions les plus heureuses, ils ne manquent d'amour du travail, d'industrie, et de ressources, en sorte qu'ils sont inférieurs aux autres peuples de la Suisse en moralité et en aisance, malgré tout ce que la nature a fait pour eux.*" (Picot, Statistique de la Suisse, p. 474-498 ; Ebel, &c.)

TETBURY, a market town and par. of England, co. Gloucester, hund. Longtree, near the source of the Avon, 16½ m. S.S.E. Gloucester. Area of par., 4980 acres. Pop., in 1831, 2550. The town consists of several streets, meeting in its centre, in which is a large market-house. It is well built, the houses being mostly of stone. The parish church is a handsome edifice : having, with the exception of the tower and spire, been rebuilt in 1781. It has chapels for Independents and Baptists, a grammar school, a well endowed Sunday school for all the poor children of the parish, an almshouse for eight poor persons, &c. The businesses of woolcombing and wool-stapling are carried on, but to no great extent. The supply of water used formerly to be very deficient ; but the deficiency has been obviated by the sinking of several deep wells. Markets on Wednesdays ; fairs, four times a year, for cows, cheese, cattle,

ambs. and horses. A fortified camp, probably of the an-
ient Britons, formerly existed here ; and Roman coins have
een frequently dug up in and near the town.

TETUAN, a town and seaport of Morocco, kingd. Fez,
rov. Hasbat, on the shore of the Mediterranean, 33 m. S.E.
'angier. Pop. said to amount to 16,000 ; of whom 9000
iay be Moors, 4200 Jews, 2000 blacks, and 800 Berbers.
Gräberg af Hemsö.) The town stands on the declivity
" a hill crowned with a square castle, the residence of the
overnor. It is of considerable extent, and its walls are
anked in different parts with square forts, on which a few
eces of ordnance are mounted. Cannon are also placed
the castle, and on a square tower at the mouth of the
'er forming the port ; but it could not oppose any effectual
sistance to a European force. The streets are narrow
d dirty, and as in Fez and other cities of Morocco, many
nearly covered in by the upper stories of the houses.
ie latter are frequently of two stories, and tolerably well
ilt and finished ; and there are several good mosques.
commercial importance Tetuan ranks next to Fez, from
ich place it receives the goods brought by the caravans
n Tunis, Algiers, Alexandria, Timbuctoo, &c. Wool,
o, and other provisions, wax, hides, cattle, leather, some
nufactured stuffs, and other African produce, are ex-
ted to Spain, France, and Italy, in return chiefly for
opean manufactures. The port of Marteen is about 2
'rom the sea, on a small river, the mouth of which is
' so choked up with sand as to admit only of the ence of small craft. The roadstead, formed by a high
t of land which runs out into the see sw. of the river,
eltered from W. winds, but during the prevalence of
e from the E. vessels must retire to some other place.
an was formerly the residence of several European
uls ; at present, however, no Europeans being allowed
ide in the town, their functions are performed by vice-
ils, who are mostly Jews. (*G. of Hemsö, Imp. del
ecco ; Mod. Trav. xx.*)

WKESBURY, a parl. and mun. bor., market town,
ar. of England, co. Gloucester, hund. Tewkesbury, on
von. near its confluence with the Severn, on the bor-
f Worcestershire, 9 m. N.E. Gloucester, and 90 m.
V. London. Area of the mod. parl. bor., which is
cal with the par., 1890 acres. Pop., in 1831, 5780 ; in
5861. It consists of two principal thoroughfares,
ig in the form of the letter Y, and from which many
'r streets branch out. The three principal streets are
und respectable, but the other streets are inferior, and
ncipally occupied by the poor and labouring popula-
(*Mun. Bound. Rep.*) The town is nearly insulated
" Mill Avon" (an ancient cut, by which the Avon
en nearly diverted from its original channel), and its
ries, the Carran and Swillgate brooks ; and it is in
uence compactly built. Many of its houses are
ne, and it is well paved and lighted with gas. The
'ncircling the town are crossed by several bridges ;
1826, an elegant cast-iron bridge, having an arch
in span, was thrown over the Severn, about ½ m.
ie town. The parish church, which formerly be-
o a flourishing Benedictine abbey, that grew out of
stery founded here in 715, is a large and noble struc-
ts length is 317 feet within the walls, and that of
sept 122 feet : the choir and side aisles are 70 feet
th, and the W. front 100 feet : the height from the
he roof is 120 feet ; and the height of the tower is
Mr. Rickman says of this church, that it " is one
ost curious and magnificent edifices in the county.
e is Norman, the piers are round and very lofty ;
tersection of the cross is a very fine Norman tower,
with arches both within and without, in several
The choir has a multangular east end, with addi-
iapels and a chapter-house, all of excellent deco-
tracter ; the windows of the aisle and transepts
decorated and some perpendicular. The W. win-
perpendicular, inserted into a very lofty Norman
reat depth, with shafts and mouldings. In the
of the choir are considerable remains of ancient
lass. There are some traces of the cloisters re-
n the S. side of the nave ; they were perpendicu-
ry rich. There are several portions of very good
irk and stalls. The abbey gate is standing, though
pidated." (*Gothic Architect.*) It has many fine
nents. The living of Tewkesbury, a vicarage in
the crown, is said in the Ecclesiastical Report to
nly £313 nett, but it is affirmed to be worth above
ie market-house, a handsome structure, has Doric
nd pilasters, supporting a pediment in front. The
tho lower part of which is used for the courts,
pper part as a council-hall and assembly-room,
d in 1788 by Sir William Codrington. The other
dings include various dissenting chapels, a thea-
ough jail, and house of industry. Tewkesbury
grammar school, founded in 1576 ; blue coat,

77

national, and Lancastrian schools, with almshouses, a dispensary, lying-in-charity, and several other benevolent establishments. The town formerly produced considerable
quantities of woollen cloth and a superior kind of mustard.
At present (1842) its principal manufactures consist of cotton hosiery. About 600 stocking-looms are now employed
in the borough. Wages vary from 3s. to 8s. a week, the
average being about 6s. About 30 persons are also employed in the bobbinet-lace trade, earning about 15s. a
week ; and 25 workmen are engaged in the manufacture
of nails. The carrying trade up the Severn and the corn-
market have declined since the improvements in the navigation at Gloucester and the construction of the railroad
between Stratford and Moreton ; but, on the whole, the
town is still in a thriving state.

Tewkesbury has returned two members to the House of
Commons since the 7th of James I. The right of voting,
down to the Reform Act, was in freemen and holders of
burgage tenements, of whom there were then 500. Registered electors, in 1839-40, 409. The municipal is co-extensive with the parliamentary borough. The town is governed by a mayor, four aldermen and 12 councillors. It
has a commission of the peace, a separate court of quarter
sessions, a court of record for debts not above £50, &c.
Corporation revenue, in 1840, £1306. Markets, Wednesday
and Saturday, for corn, cattle, poultry, &c. Fairs are held
nine times yearly.

In a field in the immediate vicinity of the town, still
called, from the circumstance, the " Bloody Meadow," was
fought, on the 4th of May, 1471, a decisive engagement between the Yorkists, under Edward IV., and the Lancastrians, under Queen Margaret and her son. The Lancastrians were totally defeated, with the loss of a great many
persons of distinction, and about 3000 soldiers left on the
field. Margaret and her son having been taken prisoners,
the latter was immediately assassinated. (*See Bound.
and Municipal Reports and Appendix ; Hume's England,
cap. 23.*)

TEWKSBURY, t., Hunterdon co., N. J., 14 m. N.E.
Flemington. Drained by Rockaway creek and its branches,
and by Lamington river. It contains three churches, a
Lutheran, Methodist, and Presbyterian, nine stores, four
grist-mills, five saw-mills, two tanneries ; eight schools, 274
scholars. Pop. 1944.

TEXAS, a new and independent republic of N. America,
between the United States and Mexico, extending from 26°
to 40° of N. lat., and from 94° to 108° W. long. It is separated from Mexico on the W. and S. by the Rio Grande, or
Bravo del Norte ; on the N. the Red river, and the Arkansas chiefly separate it from the W. territory of the United
States : on the E. the river Sabine divides it from Louisiana ;
and S.E. it borders the gulf of Mexico. Its area has been
estimated at 310,000 sq. m., of which rather more than one-
fourth part has been appropriated. Its population is differently stated at from 200,000 to 350,000, but may probably
amount to at least 300,000, chiefly Anglo-Americans.

The general aspect of the country is that of a vast inclined plane, gradually sloping from the mountains on the
W. eastward to the sea, and intersected by numerous
rivers, all having a S.E. direction. The territory is, however, naturally divided into three separate regions, which
in many respects differ from each other. The first or level
region extends along the coast, with a breadth inland varying from 100 m. where greatest in the centre, to 70 and 30
m., being most contracted towards the S.W. extremity.
The soil of this region is principally a rich alluvium, with
scarcely a stone, yet singularly free from stagnant swamps.
Broad woodlands fringe the banks of the rivers, between
which are extensive and rich pasture lands. The second
division, the largest of the three, is the undulating or rolling-
prairie region, which extends for 150 or 200 m. farther inland, its wide grassy tracts alternating with others that are
thickly timbered. These last are especially prevalent in
the E., though the bottoms and river valleys throughout
the whole region are well wooded. Limestone and sandstone form the common substrata of this region : the upper
soil consists of a rich friable loam, mixed indeed with sand,
but seldom to such an extent as to prevent the culture of
the most exhausting products. (*Iken.*) The third or mountainous region, situated principally in the W. and S.W.,
forms part of the great Sierra Madre, or Mexican Alps (see
AMERICA, vol. i., 78), but little explored and still unsettled.
At its remote extremity it consists of an elevated table land,
" where the prairies not unfrequently resemble the vast
steppes of Asia except in their superior fertility." The
mountain sides are clothed with forests of pine, oak, cedar,
and a great variety of trees and shrubs, and they inclose
extensive alluvial valleys, most of which are susceptible
of irrigation and culture, Mr. Kennedy says, that the sides
of the mountains, and even not a few of their summits, are
adapted to agriculture ; and, if we might depend on the
lately published statements, it would appear that there are

3 M   913

very few, if any, countries of the same extent which have so small a proportion of unproductive land.

After the rivers already named, the principal, proceeding from N. to S., are the Neches, Trinidad, Brazos de Dios, Colorado, Guadalupe, San Antonio, and Nueces. They all fall into the gulf of Mexico, or rather (except the Brazos de Dios) into its bays and lagoons. The latter bear a considerable resemblance to the haffs along the S. shore of the Baltic, except that they are upon a much larger scale: and the coast, as Humboldt has stated, presents everywhere formidable obstacles to navigation, in the long, low, narrow belts of land by which it is fenced, and which bound the lagoons, in the want of harbours for vessels drawing more than 12½ feet water, and in the bars at the mouths of the rivers; still, however, steam vessels have been able to enter and ascend the Sabine to a considerable distance. The Neches is navigable for small steamboats for upwards of 100 m.; Trinidad river for 300 or 400 m.; and the Brazos de Dios, for at least half that distance. The Rio Colorado, a river larger than the Thames between Chelsea and Richmond, is obstructed by a raft about 10 m. above its mouth, but measures are in progress for removing this, when it is anticipated that it will be navigable for small steamers as high as Austin, the capital of Texas, about 290 m. from its mouth. The San Antonio and Nueces are navigable only for comparatively short distances; but the Rio Grand del Norte, a noble stream, having an estimated course of 1800 m., will most probably, though in parts broken by rapids, become hereafter an important commercial channel. Galveston bay, into which the Trinidad flows, by far the finest on the coast, is about 35 m. in length N. and S., and from 12 to 18 m. E. and W. Its average depth is from 9 to 10 feet, but in the channel there are from 18 to 30 feet water.

The Texan year is divided into a wet and a dry season. The former lasts from December to March, during which N. and N.E. winds are most prevalent; the latter, from March to the end of November, during which the winds vary from the S.E. round to S.W., may be subdivided into the spring, summer, and autumn. From April to September the thermometer in different parts of the country has been found at a general average to range from 63° to 100°; average heat, 9 A.M., 73° F.; at noon 83°; 3 P.M., 77°. (Kennedy.) These great heats are, however, tampered by continual and strong breezes, which commence soon after sunrise, and continue till about 3 or 4 o'clock P.M., and the nights throughout the year are cool. From March to October little rain falls, though thunder storms frequently occur. During the rest of the year wet weather is prevalent; the rivers swell and inundate the country, and the roads are generally rendered impassable. Snow is seldom seen in winter, except on the mountains. The climate is said by its panegyrists to be decidedly more salubrious than that either of Louisiana or of the adjacent parts of Mexico; but it is very difficult to imagine any satisfactory reasons why such should be the case, and we confess we entertain some considerable doubts as to the accuracy of the statement. On the low alluvial sea coast, intermittent fevers are admitted to be prevalent in summer, though not, it is said, to an epidemic extent; and the yellow fever, it is further stated, rarely, if ever, occurs: indeed, Mr. Kennedy goes so far as to affirm that "nine tenths of the republic are considered healthier than the most healthy parts of the United States." (i., 74.)

But, after making every fair allowance for exaggeration, there can be no doubt that Texas is both a fine and a fertile country. The surface is in most parts covered with luxuriant native grass, comprising, with the common prairie grass, the gama, musquite, wild clover, wild rye, &c., and affording excellent pasturage. It has, also, an ample supply of timber, as well for use as for ornament. Live oak (Quercus sempervirens), so valuable for ship-building, is here more abundant and of better quality, perhaps, than in any other part of America. White, black, and post oak, ash, elm, hickory, musquite (acacia), walnut, sycamore, bois d'arc, so called from the Indians using it to make their bows, cypress, caoutchouc, &c., are among the common trees: and the mountainous parts in the S.E. abound with pine and cedar of fine quality. Among the natural curiosities of the country is the "Cross-timber" of N. Texas, a continuous series of forests, varying in width from 5 to 50 m., and extending in a direct line about the long. of 97° W. from the woody region at the sources of the Trinidad, northward to the Arkansas river. It appears at a distance like an immense wall of wood; and from the W., such is its linear regularity, that it looks as if it were planted by art. It forms the great boundary of the W. prairies.

Texas is amply supplied with fruits and garden products. The climate of the lowlands is too warm for the apple, but almost every other fruit of temperate climates comes to perfection. Peaches, melons, figs, oranges, lemons, pineapples, dates, olives, &c., may be grown in different localities with little cost.

Grapes are abundant; and being free from the "foxy" flavour common to the grapes of most parts of America, very tolerable wine has been made from them. Vanilla, indigo, sarsaparilla, and a large variety of dyeing and medicinal shrubs and plants are indigenous; and on all the river-bottoms is an undergrowth of cane, so thick as to be almost impervious. Along the water-courses also, and near the sea, the larger trees are sometimes wreathed with Spanish moss, which serves both for fodder and for the manufacture of cheap bedding, &c. The flora of Texas is particularly rich and copious.

Cotton is the great agricultural staple of the republic; and it is affirmed, and perhaps truly, that it is very decidedly superior, as a cotton-growing country, to the best districts in the United States; producing a greater quantity of cotton per acre, and of a longer and finer staple. The best of the long-stapled cotton is produced in the low alluvial soils, and the short-stapled on the rolling or undulating lands. According to Mr. Iken, whose statements, however, we do not presume to guarantee, the advantages of the cotton planter in Texas over the planter in the United States consist in the following particulars: "He has cheaper land, a larger crop, a better staple, an earlier season to plant, and therefore to pick; likewise a longer season for the latter precarious operation prior to the rains and frosts doing injury; by the superior facilities for raising stock, he can feed his labourers about 50 per cent. cheaper than in the United States; their clothing, owing to the lower tariff, will be far less expensive, and the more salubrious climate will make their lives a better purchase." (p. 63.) Superior cotton-growing lands yield, it is said, from 1½ to 2 bales of clean cotton per acre, worth, in the Galveston market, perhaps 4d. per lb. Its cultivation hitherto has been principally on the Brazos and Colorado, Red and Trinidad rivers, and Caney creek; but it is steadily on the advance, and, according to Mr. Iken, the crop of 1841 may be estimated at 50,000 bales, of which, however, a considerable portion came down the Red river, and was included in the exports from New-Orleans. Cotton planting begins in February, and picking in June. The latter employment is an easy and profitable occupation for women and children. Texas cotton has been for several years shipped direct to Liverpool in British bottoms.

The grains chiefly cultivated are maize and wheat. The average crop of the former, on good ground, is from 38 to 60 bushels per acre; but 75 bushels are said to be frequently obtained, and two crops may be gathered in the year, the first being usually planted in February, and the second late in June. (Kennedy.) A crop of excellent wheat has been cut in May, and the same land has yielded a heavy crop of maize in the ensuing October. The corn, we are further told, is reckoned worth from 1½ to 2 dollars per bushel at the farms, and from 2 to 3 dollars in the market. (Kennedy, i., 93.) But we incline to think that there must be some error about this statement: such prices are really higher than those of England, and appear to be wholly irreconcilable with the accounts of the wonderful productiveness of the soil. We have, in fact, little doubt that either one or the other statement, or, which is most probable, that both are erroneous. Rye, barley, oats, &c., are raised in the upper country, and rice near the river estuaries; but small quantities only of these grains have hitherto been raised. The sugar-cane is also said to attain to greater perfection than on the Mississippi; and Mr. Kennedy states that the produce on a small plantation, despite the waste arising from very imperfect machinery, has averaged about 3500 lbs. to the acre. (i., 92.) Tobacco will probably hereafter become an important staple. The mulberry grows vigorously, and the experiment of rearing silkworms has already, we are assured, been successful; and common and sweet potatoes are said, like every thing else in this fortunate land, to attain to perfection!

The rearing of live stock has, however, been long the principal and favourite occupation of the Texan settlers, and many of the prairies are covered with a valuable breed of oxen, which scarcely require, and certainly do not receive, much more care or attention than the prairie deer. It is usually estimated that 100 cows and calves, purchased for 1000 dollars, will in ten years have increased about 35 fold, thus numbering 3600, worth, at the same price, 36,000 dollars; though we may remark by the way, that, with such an enormous increase, it is not easy to see why the price should be the same. A profitable trade in cattle is opened with New-Orleans, &c., and the West India islands offer coffee, of which the Texans use large quantities, in exchange for cattle. Salt, for curing beef, is obtained every where near the coast; and the hide, horns, and tallow, shipped to Europe, Mr. Iken says, will alone pay more than the cost of the animal. The rearing of horses and mules is also pretty extensively pursued: sheep thrive on the upper lands, but require folding. Hogs are very profitable: and bees, which are produced in great numbers, might also be

tide productive, there being an extensive demand for wax in Mexico. Vast herds of buffaloes and wild horses wander over the prairies, and deer are every where abundant. Bears, cougar, panthers, peccaries, wolves, foxes, raccoons, &c., are common; and most of the planters are obliged to keep packs of large and powerful dogs to prevent the destruction of their stock. Most of the birds known in the United States are common to Texas, and the bays, &c., abound with fish of excellent quality, beds of good oysters, and other *testacea*. Alligators of 16 ft. in length are sometimes met with in the rivers, particularly Red river and its estuaries: turtles, tortoises, &c., in the estuaries. There are several venomous serpents, and, as in all other warm countries, moschetoes and other insect plagues are common. If Mr. Iken, notwithstanding his wish to represent every thing belonging to Texas in the most favourable light, admits that "the inclination for luxurious indolence, to which the climate predisposes," is a worse evil than either serpents or moschetoes; and "the settler," he adds, "will have much greater reason to be on his guard against this agreeable poison, than against that of the *anguis in herbâ!*"

After the statements previously made, the reader will not be surprised to learn that Texas is represented as a peculiarly eligible country for emigrants, particularly agriculturists. Mr. Iken, among other practical hints to emigrants, lays it down that the first objects of the settler should be to establish himself in a decidedly healthy situation, with a fertile soil, plenty of timber, and good water; and, truly, if the grant whose estate comprises these advantages be not satisfied with his lot, it would be hard to say where he would go to improve it. In purchasing lands, the price varies greatly; but good land, with an unquestionable title, may be obtained for from 3 to 5 dollars an acre, or even less per. Inferior capitalists will find the rearing of stock easiest and most profitable pursuit; and those of a still poorer class may, it is said, employ themselves profitably in rearing a garden in the neighbourhood of some town, which generally affords a ready market for garden produce. The formation of such gardens is looked upon by Mr. Iken as an object secondary only to the planting of fields." Agricultural labourers without capital, are said to find little difficulty in connecting themselves with farmers already established, on advantageous terms. "The modes of husbandry in Texas are of the most simple description. The object of the farmer, after building a small and temporary log-house, is to enclose a sufficient space of the open ground adjoining, by the erection of a rail fence. He then proceeds to break up the land with a light plough, which is usually drawn by oxen. A yoke of large oxen, broken, is from 30 to 60 dollars: a horse, for general agricultural purposes, about 30 dollars. The Texan farmers generally content themselves with one ploughing previously to sowing. Manuring is altogether dispensed with. The time for maize, cotton, and most other crops, is in February and March. A few hoeings to destroy weeds, to hand to earth up the young plants, is all that is required on the part of the farmer to bring them to perfection." *Texas,* 51, 52)

any parts of the rolling prairie region, coal of a superior quality, and iron ore have been found; and it has been supposed that beds of these valuable minerals extend over a great part of the country. Silver mines were wrought in Santa Fé, in the N.W., till the works were destroyed by the Comanche Indians. Nitre abounds in the soil, is obtained from numerous lakes and springs; and in several places. Granite, limestone, gypsum, &c., are abundant, except in the low alluvial region.

The Progress of Texas.—The reader will have already seen that we are disposed to entertain some considerable doubts with respect to the statements as to the extraordinary fertility of Texas. But admitting them to be true ever, still we should not be at all sanguine as to its progress, and should think that European emigrants would well to pause before they decide on establishing themselves in Texas. The soil is too fruitful, and the climate especially in the lower and more fertile parts of the country, decidedly too hot and relaxing to permit of any continued employment being vigorously prosecuted by free men.

Had the free importation of slaves into Texas permitted, its progress might, and most probably would have been as rapid as that of the southern states of the union. But the importation of slaves into Texas, from the United States, is prohibited; and if the demand in this quarter should fail, and the prohibition of their importation from other quarters be really enforced, its progress, we apprehend, be comparatively slow. It is a distinction and an absurdity to suppose that free settlers of a country should exhibit any considerable portion of energy and industry they would exhibit if they were located upon a less fertile soil, or were placed under a worse climate. Mexico and the United States may both be quoted in illustration of this principle: industry is in

the former at the lowest possible ebb (*see* Mexico) though it is needless to add that the incentives to labour are there quite as great as in Texas. And does any one suppose that the cultivation of cotton and sugar would have been carried to near its present extent in the southern United States, but for their all but unlimited command of slave labour? We, therefore, are well convinced that the future progress of Texas will depend principally upon the fact, whether it can or cannot derive ample supplies of slave labour. If it can (with or without the sanction of the law is in this respect of no importance), the fair presumption is, that it will make a rapid progress; whereas if it cannot, its progress will, most probably, be comparatively slow; and we should anticipate that, under such circumstances, its inhabitants would gradually fall into the same state of semi-barbarous indolence into which the Mexicans have already sunk.

The geographical position of Texas is eminently favourable to the growth and extension of commerce. Its rivers and the facilities which the country affords for the completion of railways, will enable the traders and agriculturists to forward their produce easily to the coast; whence it may be forwarded to the European markets, and to those of Cuba and the West Indies generally. In exchange for the cotton and other products sent to Great Britain, the Texans import British manufactured goods, not only for their own consumption, but partly, also, for the supply of the N. states of Mexico. Indeed, Santa Fé has been, since 1825, the great emporium of N. or New-Mexico; and in it the traders of that country meet those of the United States; the former purchasing the manufactures brought by the latter with peltry and bullion, so that a good deal of specie reaches the United States by this route. The annual amount of the trade at Santa Fé was estimated in 1834 at 2,000,000 dollars; and in 1841 Iken estimated it at 3,000,000 dollars. Most of the articles for this trade are purchased in Philadelphia, whence they are transported overland to Pittsburgh, where they are shipped for St. Louis, being thence conveyed in wagons to Santa Fé, which they reach after a journey of at least 4000 m. From St. Louis to Santa Fé, about 1900 m., the road is extremely bad, running through a country so infested by hostile Indians, that the United States government is obliged usually to send an escort of cavalry with the larger caravans: indeed, 300 dragoons were sent in 1839, for the protection of one body of traders: (*Kennedy.*) But Santa Fé is only 600 m. from the Texan coast, so that it may not unreasonably, perhaps, be anticipated, that eventually Galveston, and other Texan ports, will be the principal routes by which European goods will reach N. Mexico. As already stated, a good deal of Texan cotton wool is exported to Europe from New-Orleans, to which it is brought by way of the Red river, without its appearing to be the growth of Texas. The total export of cotton direct from the latter in 1840 amounted to 15,304 bales; in addition to which, hides, deer, otter, beaver, and other skins, cattle and other stock, and bullion from Mexico, are the principal articles of export. Money is very scarce in Texas: not one sale in ten is made for cash; and Mr. Iken is entitled to credit for recommending "shippers to be cautious as to the extent of their consignments, and to recollect that, however great the commercial prospects of the country may be, it is at present only a new market, whose own consumption must necessarily be limited, and the channels of whose interior or transit trade are as yet but very partially opened." (p. 66, 67.)

The duties on most manufactured goods imported into Texas are nominally 45 per cent.; but, being paid in a depreciated paper currency, do not actually exceed 10 per cent. Books, farming and other tools and utensils, wearing apparel, and French wines, are mostly imported duty free. The same rules and regulations are observed on the importation of goods into the United States are observed in Texas, consular certificates being, however, dispensed with. The currency, weights, &c., are similarly calculated, except that land measures and a few others are identical with those of Mexico. The principal Texan ports of entry are Galveston, Matagorda bay, and Aransas, to all which pilots are attached. Vessels of about 250 tons, or not drawing more than 10 or 11 ft. water, are those best suited to the trade. (*Iken's Texas, &c.*) The tonnage duties on merchant ships in the Texan ports are 60, and those on steamers 30 cents per ton.

Texas is an integral, and not, like the United States or Mexico, a federal republic. The president is elected for three years, and is not again eligible for a similar term. In other respects the constitution generally resembles that of the United States. The legislative power is vested in the congress, composed of a senate and a house of representatives. The latter body consists of not less than 40 nor more than 100 members, who are annually chosen by universal suffrage, and each of whom must be at least 25 years of age, and have resided in the county or district which he represents for the six months next preceding his election. The senators are annually chosen by districts as nearly equal in free population as practicable, and are not less than one third or more than half the number of representatives. One third of their number is replaced annually. Ministers of religion are ineligible to a seat in either house of congress. All members of the government are paid for their services: the president's salary is 10,000 dollars a year, and that of members of congress 8 dollars a day during session. A bill passed by a vote of two thirds of the members of the legislature becomes law without the sanction of the president; and congress reserves to itself the right of declaring war, levying imposts, contracting loans, &c. The executive power is in the hands of the president, assisted by the vice-president, both of whom must be full 35 years of age, and have been inhabitants of the republic for three years preceding their election. The different branches of public business are conducted by 12 committees of five members, each appointed by the legislature.

Texas is subdivided into three great departments: those of Bexar in the S., Brazos in the centre, and Nacogdoches in the N.; and at present into about 40 counties. It is farther divided into seven judicial districts, in each of which is a judge. The judges are appointed by congress, and hold office for four years. The salary of the chief justice is 5000 dollars, and of the district judge 3000 ditto. The supreme court consists of the chief justice of the republic, and the judges of each district, and sits at Austin on the second Monday in each year, continuing in session till all the business before it be disposed of. Assize courts, courts of probate, justices' courts, &c., are held in each county; the common law of England, together with the acts of congress, having been adopted as the law of the land.

The towns are incorporated on the plan of the United States, the principal being Austin, Galveston, Houston, Bexar, Goliad, Matagorda, and Santa Fé. The last has a population of about 7000; Houston and Galveston about 3000 each. Freedom of the press, and the free exercise of religion, are guaranteed. Slavery is permitted; and the number of negro slaves in 1840 was estimated at 11,300; since which there has been some increase. Most of these slaves have been brought from the United States by the planters from that country who have settled in Texas; and, as already stated, the importation of slaves from elsewhere is declared to be piracy punishable with death. Unreasonable or cruel treatment of a slave is punishable by a heavy fine. According to Iken, there are several schools and colleges in Texas; Sunday schools, bible, temperance, &c., societies are already numerous; 12 newspapers were regularly published in Texas in 1841, and few free inhabitants are unable to read and write. (p. 78, 79.)

The present military force is chiefly composed of volunteer troops and militia, whose services are sometimes required against the Comanches and other hostile Indian tribes of the N. and W. The navy consists of a sloop-of-war, two brigs, an armed steamer, several schooners, &c. The public revenue, derived from a land tax of ½ to 1 per cent. on its estimated value, custom duties, licences, &c., amounted in 1840 to 802,054 dollars, which is said to have considerably exceeded the expenditure. The public debt is stated by Kennedy to have been, according to the last accounts, 5,893,000 dollars.

Previously to 1690, Texas formed a remote and merely nominal part of the conquests of Cortez, inhabited almost wholly by predatory Indian tribes; but in that year the Spaniards, having driven out a colony of French who had established themselves at Matagorda, made their first permanent settlement at San Francisco. On the consummation of Mexican independence, Texas was constituted one of the federal states of Mexico, in conjunction with the adjacent state of Cohahila; a union very unpopular with the Texans, and which was productive of the first disagreement with the central government. The war of separation commenced towards the end of 1835, and on the 21st of April, 1836, the independence of Texas was finally secured by the defeat of the Mexican president, Santa Anna, at San Jacinto. (For farther particulars, we beg to refer the reader to Kennedy's Texas, 2 vols. 8vo, and the shorter compendium of Iken; from which works we have derived most of the statements in this article.)

TEXEL (THE), an island belonging to Holland, at the entrance of the Zuyder-Zee, off the point of the Helder, at the N. extremity of the prov. of North Holland, from which it is separated by the channel, about 2½ m. across, called Mars Diep, its most southerly point being in about lat. 53° 1' N., long. 4° 46' E. It forms a canton of the arrond. Alkmaer; length, N.E. to S.W., 13 m., and where broadest, nearly 6 m. in width. Pop. from 5000 to 6000. It is low, and in part marshy, but is defended from the irruptions of the sea, partly and principally, by a line of dunes or sand-banks, which extend along its W. coast, and partly by strong dykes. The district of Eyerland (country of eggs), so called from the vast number of eggs deposited by the

sea-fowl on its shores during the breeding season, was formerly a distinct island, having been united to the Texel by a dyke in 1630. The soil, which is extremely fertile, is mostly employed in the feeding of cattle and sheep, the number being of a peculiarly fine, long-wooled breed. The inhabitants, who occupy a town, Burg, in the centre of the island, and some villages, in addition to agriculture, engage in fishing, boat-building, &c., and act as pilots. There is an excellent roadstead on the E. coast of the island, which is the usual place of rendezvous for merchantmen from Amsterdam waiting for a favourable wind to leave the Zuyder-Zee. The number of sandbanks make the approach to the island difficult; and on the W. side it is all but inaccessible.

During a tremendous storm in February, 1825, the sea broke through the dykes by which the island is defended, and laid a large portion of its surface under water, destroying great numbers of cattle and sheep, and otherwise occasioning a heavy loss of property. The breach, however, was soon after repaired, and it is now supposed to be better protected than ever.

Several naval engagements have taken place off this island. Of these, the most celebrated was that between the Dutch fleet under the famous admiral the senior Tromp, and the English fleet under Monk, afterward Duke of Albemarle, on the 31st of July, 1653. The action was maintained with the utmost vigour on both sides, till the death of Tromp, who was shot through the heart by a musket-ball, decided it in favour of the English. (Busching, Geographie Universelle, xiv., 137, Fr. ed. of 1779; Dict. Géographique; Campbell's Lives of the Admirals, ii., 35, ed. 1785.)

THAME, or TAME, a market town and par. of England, co. Oxford, hund. Thame, on the Thame, a tributary of the Thames, 12 m. E. Oxford. Area of par., 5310 acres. Pop., in 1831, 2885. The town consists of three principal streets, uniting in a spacious market-place. It has also a market house, over which is the town-hall. The parish church, a large, well-built cruciform structure, comprises a nave, two aisles, a N. and E. transept, a chancel, and has a fine embattled tower, supported by four meagre pillars. The exterior is of noble proportions, and contains numerous monuments, but is ill laid out, and spoiled by irregular galleries, &c. The united living of Thame, Towersey, Tetsworth, and Sydenham (two vicarages and two curacies), worth £300 a year, is in the gift of — Long, Esq. Near the church are some remains of the prebendal house of Thame, now occupied by offices belonging to the parsonage farm; and in Thame park, about 1 m. S.E. the town, considerable portions of an ancient Cistercian monastery adjoin the mansion. In 1558, Lord Williams established a free school, "of noble dimensions," at Thame: it is open to all boys of the parish, and in trust of the warden and fellows of New college, Oxford, who nominate the master, subject to the approbation of the Earl of Abingdon. It had a high character during the 17th century, but is now much fallen off. Another free school, an almshouse for five poor people, and various annual donations to the poor, exist here. The population is mostly agricultural; lace-making by women and children is the only manufacture. The Thame, being navigable for barges, promotes the traffic of the town; and the market is well supplied with corn and cattle. Thame is supposed to have been a Roman station, and was of some consequence in the time of the Saxons, and during the civil wars of Charles I. The famous constitutional lawyer, Sir John Holt, Chief Justice of the King's Bench, was a native of this town, where he first saw the light in 1642. Market on Tuesday; fairs, Easter Tuesday and Old Michaelmas day, for cattle, &c. (Beauties of England, Wales, Oxfordshire, &c.)

THAMES, a river of England, being the largest in that part of the U. Kingdom, and, in a commercial point of view, the most important in the world. It rises in Gloucestershire, being formed by the junction of the Isis, Lech, Colne, and Churnet, rivulets which have their sources in the Cotswold hills. The first, which is the most important, rises on the borders of Wiltshire, a little to the S.W. of Cirencester. It flows E. by Cricklade; and, being augmented by the other streams, the combined river takes the name of Thames, and becomes navigable for barges at Lechlade, on the confines of Gloucestershire and Berkshire. Its course is thence N.E., till, being farther augmented by the Windrush and the Evenlode, from the borders of Gloucestershire, it turns a little to the N. of Wytham-house, to the S. After passing Oxford, it bends suddenly to the W. by Newnham park to Abingdon. Having again resumed its southerly direction, it is joined, a little below Dorchester, in Oxfordshire, by the d'hame.

This latter river has several sources, of which the most remote are in the central parts of Buckingham, near Kirsleigh and Wendon lodge. They unite at Thame in Oxfordshire, from which point, to where it joins the Thames, it is navigable.

It may here be proper to state that, according to the ommon opinion, the Thames obtained its name (said to be 'hame-isis, shortened to Thames) from the junction of the 'hame with the Isis, or with the river coming from Glou-estershire. Probably, however, this opinion, notwithstand-g its apparent accuracy, has no good foundation. At all vents, it appears to be abundantly certain that the river hich passes Lechlade, formed by the junction of the rivu-ts already referred to, has from a very remote period been lled the Thames; and that the name Isis, given to it by e literati of Oxford, is not mentioned in ancient charters by ancient historians, and is wholly unknown to the mmon people in the country through which it flows. *amden's Britannia*, Gibson's edition, i., 109; *Campbell's litical Survey*, i., 139.)

From Wallingford, a little below the influx of the Thame, river flows almost due S., till, passing Basildon park, it ns E. to Reading, where it is joined by the Kennet: it n flows N.E. to Great Marlow: thence S. to Maidenhead, l S.E. by Windsor and Staines, till it receives the Wey. course is then E., with many bold sweeps, to London; l flowing through the metropolis, and being augmented he Lea from Hertfordshire, and the Darent, it continues ourse E. till it unites with the sea at the Nore light, 45½ elow London bridge.

he distance from London bridge to Lechlade, where the mes becomes navigable, following the windings of the ., is 146½ m.; the total rise from low-water mark at the er to the latter being about 258 ft. This ascent is come by means of several locks, constructed at different ds, of which the first is at Teddington, 18¾ m. above lon bridge; this, consequently, is the limit to which ide flows. The low-water surface of the river, from ington lock to London bridge, falls about 16 ft. 9 in., out 10¾ in. in a mile at an average. The high-water at Teddington is about 1 ft. 6 in. above the high-water at the bridge; and the time of high water is about ours later. The average fall in the bed of the river, Teddington to London bridge, is about 1 ft. a mile; eadth of the river at London bridge is 692 ft.

ugh not a rapid, the Thames is by no means a slug-iver; it rolls forward with an equable and steady cur-nd is remarkable for the purity of its waters. It has dmirably described by Denham, in his *Cooper's Hill*:

> "Though deep, yet clear; though gentle, yet not dull;
> Strong without rage; without o'erflowing, full."

it is as a navigable and commercial river, having n on its banks, and bearing on its bosom numberless 'raught with the produce of every country and every , that the Thames is principally distinguished. Its of water is so great, that, as a shipping port, London peculiar advantages; even at ebb tide there is from 12 . water in the fair way of the river above Greenwich; e mean range at the extreme springs is about 22 ft. er, is, in fact, navigable as far as Deptford for ships burden; to Blackwall for those of 1400 tons; and to Katherine's docks, adjoining the Tower, for vessels tons. As already stated, it is navigable by barges to nfines of Gloucestershire; and the navigation is continued by canals through Cirencester and Stroud evern; but the usual water communication between , Bath, and Bristol, is by the Kennet, which unites e Thames at Reading. The conveyance of goods channel usually occupies about seven days; and rigation is besides exposed, particularly between and London, to much interruption from droughts, c. The whole course of the river, from its source ore, is reckoned at from 205 to 210 m.

emoval of the old London bridge has caused a con- change in the river above, and also, though in a ree, below the bridge. Owing to the contracted hrough which the water had to make its way at ridge, there was a fall of from 4 ft. 9 in. to 5 ft. at er; this fall is now reduced to about 2 in.; so that water line above the bridge is nearly 5 ft. lower at des than formerly. In consequence, a greatly in-body of tidal water flows up and down the river; meets with no obstruction, it flows with a decided-r velocity. The effect of this is to scour and deepen nel of the river; its influence, in this respect, being ensibly felt as far up as Putney bridge, 7¾ m. above bridge. The shores above the latter, that were foul and muddy, are now becoming clean shingle el, and, near low water, the beach is quite hard The shoals are also decreasing below the bridge; can be little doubt that the change will, at no dis-d, be felt from the Nore up to Teddington.

the removal of the old bridge, a barge, starting pool with the first of the flood, could not get far-Putney bridge without the assistance of oars. er similar circumstances, a barge now reaches Mortlake, 4 m. farther up, before using oars, and, with a little help, she may reach Richmond, and, taking horses there, may get to Teddington in a tide. The descent down the river has been equally facilitated; the mean velocities of the flood and ebb, between London bridge and West-minster bridge, are, flood, 3 m. an hour, extreme, 3¼; ebb, 3¼ m., extreme, 3¾.*

Of the tributaries of the Thames, the Kennet, Wey, Lea, and Darent only are navigable, and are, therefore, the only ones that we need notice.

> "The Kennet swift, for silver eels renowned,"

rises on Marlborough Downs, in Wiltshire, and, pursuing an easterly course, falls into the Thames at Reading. It has been made navigable as far as Newbury; whence the canal previously mentioned is carried, by Devizes and Brad-ford, to Bath and Bristol. The Wey falls into the Thames near Oatlands; it has its source in the eastern part of Hamp-shire, and has been rendered navigable from Godalming to the Thames, a distance of about 20 m. The first naviga-tion locks used in England are said to have been constructed on this river. The Lea rises in the chalk hills near Luton, in Bedfordshire; and, preserving a southerly course, falls into the Thames near the East India docks. It has been made navigable, by collateral cuts and otherwise, as far as Hertford. This navigation, which is of considerable impor-tance, began to command the attention of the legislature so early as 1425, in the reign of Henry VI. It has not yet, however, received all the improvement and extension of which it is capable. (*Priestly on Inland Navigation, &c.*, p. 411.) The Darent has its source near Westerham, in Kent; it falls into the Thames about 4 m. below Dartford, to which it is navigable. (See *Statistics of British Em-pire*, i., 30.)

Thames, r., Ct., is formed by the union of Shetucket and Yantic rivers at Norwich city, whence it flows S. 14 m. to Long Island sound at New-London. Steamboats and con-siderable vessels go up to Norwich city; and the harbour at New-London, 3 m. from its mouth, is the best in the state, and one of the finest in the United States. It is easy of access, spacious and safe, with sufficient depth of water for ships of the line. It is defended by fort Trumbull in New-London, and fort Griswold on Groton heights, on the opposite side of the river.

THANET (ISLE OF). *See* Kent.

THAXTED, a market town and par. of England, co. Es-sex, hund. Dunmow, on the Chelmer, near its source, 34 m. N.E. London. Area of parish, 5890 acres. Pop., in 1831, 2293. The town is irregularly built, and, excepting its church, has no public edifice worth notice; this is a large and fine structure, in the perpendicular style, its earliest ex-isting part probably dating from the middle of the 14th cen-tury. It is built cathedral-wise, with a transept between the nave and chancel: its internal length is 183 ft.; breadth, 87 ft.; and at its W. end is a tower, with a very rich crock-etted spire, 181 ft. in height. The whole fabric is embattled and supported by strong buttresses, terminated by canopied niches and pinnacles, curiously purfled. The N. porch is richly ornamented with sculpture, and the cornice and upper part charged with various figures. Above the en-trance are two escutcheons, one containing the arms of France and England, and the other those of the House of York; a part of the edifice having been constructed at the expense of Edward IV., the rest chiefly at that of the noble families of Clare and Mortimer. "The nave is curious, being not so wide as either of the aisles. Most of the but-tresses of the aisles are enriched with panneling, and have fine pinnacles. Some of the windows are square-headed; their tracery has been much mutilated. This church had, at one time, a considerable portion of fine stained glass, which has, however, long been gradually diminishing." (*Rickmann, Gothic Architecture.*) The living, a vicarage, in the gift of Lord Maynard, is worth £450 a year. Thax-ted has meeting-houses for several sects, a parish school for 50 children, an endowment of nearly £4000, by Lord Maynard, in 1698, for general charitable purposes, and many minor charities. It was a municipal borough till the reign of James II., when, on the corporate officers being served with a *quo warranto*, its privileges were dropped, and its former guildhall is now the workhouse. The town is of high antiquity, its church being mentioned in the time of Edward the Confessor. Markets on Fridays; fairs, Monday before Whitsunday and August 10, for horses, &c. (*Parl. Reports, &c.*)

THEBES, THEBÆ,† or DIOSPOLIS (the city of Jupi-ter), a once famous, but long ruined city of Upper Egypt, the capital of the kingdom of the Pharaohs when in the ze-nith of their power, and whose remains exceed in extent and magnificence all that the most lively imagination could

* We are indebted for these details to John Smeaton, Esq., engineer

† It has been supposed that the word *Thebæ* is derived from the Egyptian word *Thbaki* (the city).

figure to itself. The ruins are situated in about lat. 25° 43′ N., long. 32° 39′ E., in the narrow valley of the Nile, stretching about 7 m. along both banks of the river, and extending to the mountains on either side. One might suppose, seeing the vast magnitude of its public edifices, that its private buildings would be in a corresponding style of magnificence, but Diodorus tells us that the Egyptians were little solicitous in respect of the latter; and, at all events, all traces of private fabrics have disappeared; and temples, palaces, colossal statues, obelisks, and tombs, alone remain to attest the wealth and power of its inhabitants. Thebes was undoubtedly one of the most ancient, as well as one of the greatest and most splendid of cities. Its most flourishing period was probably from about *anno* 1700 to *anno* 700 B.C. Homer has alluded to her in terms which, but for the ruins, might have been deemed extravagant:

"Not all proud Thebes' unrivalled walls contain,
The world's great empress on the Egyptian plain
That spreads her conquest o'er a thousand states,
And pours her heroes through a hundred gates;
Two hundred horsemen and two hundred cars,
From each wide portal issuing to the wars."
Pope's Iliad, ix., lin. 500.

Modern travellers have not been able to find any distinct traces of walls round the ruins; and the opinion has prevailed from a very remote epoch, that Homer, in the passage now referred to, did not allude to gates in the city walls, but to the gates of the different temples, or, as Pomponius Mela supposes, to the palaces of great men (lib., i., cap. 9). Probably, however, the poet, by this expression, merely meant to convey a lively idea of the prodigious population and power of the city.

The seat of government had been removed from Thebes to Memphis (near Cairo), previously to the invasion and conquest of Egypt by the Persians under Cambyses. This event took place *anno* 525 B.C., when, according to Diodorus, the Persians plundered and set fire to Thebes. It appears, however, to have, in some degree, recovered from this disaster. But after the conquest of Egypt by the Greeks, their whole attention was directed to the improvement and embellishment of Alexandria, so that the cities in Upper Egypt, and especially Thebes, progressively declined in importance and population. Its fall was accelerated by its having revolted against Ptolemy Philopater, by whom it was subsequently reduced, and given up to military execution. In Strabo's time it was only partially inhabited. In the earlier ages of the Christian æra it was still of some little consequence; but for these many centuries it has been only inhabited by a few wretched Copts and Arabs, who, with bats and owls, occupy miserable hovels, mostly in the courts, and sometimes on the roofs of the ancient structures.

The principal ruins on the E. or Arabic side of the river are those of Carnac and Luxor, about 1½ m. apart. The first of these, which there can be no doubt is the temple of Ammon, the Jupiter of the Egyptians, is described by Diodorus as a vast structure, or rather collection of structures, the principal being erected on an artificial elevation. It has various entrances, the avenues to which have been flanked on each side with rows of sphinxes. The principal front to the Nile is of enormous magnitude, being 368 ft. in length by 148 ft. in height, with a doorway in the middle 64 ft. in height. Entering this superb gateway, and passing through a large court, we pass between two colossal statues through another propylon, entering by a flight of steps to a vast hall, the roof of which, consisting of enormous slabs of stone, has been supported by 134 huge columns. This gigantic hypostyle hall is about 338 ft. in width, by 170½ ft. in depth, so that its area comprises 57,629 sq. ft., being considerably more than 1½ acre, or more than five times the area of St. Martin's church, Trafalgar-square, London; and yet this magnificent hall does not occupy one seventh part of the space included within the walls of the temple! (*Egyptian Antiquities; Library of Entertaining Knowledge*, i. 89.) The entrance to what is supposed to be the adytum of this famous temple is marked by four noble obelisks, each 70 ft. in height, but of which three only are now standing. "The adytum consists of three apartments, entirely of granite. The principal room, which is in the centre, is 20 ft. long, 16 wide, and 13 high. Three blocks of granite form the roof, which is painted with clusters of gilt stars on a blue ground. The walls are likewise covered with painted sculptures of a character admirably adapted to the mysterious purposes mentioned by Herodotus, on the subject of the virgins who were introduced to the Theban Jupiter. (*Herod.* i. 182.) Beyond this are other porticoes and galleries, which have been continued to another propylon at the distance of 2000 ft. from that at the W. extremity of the temple." (*Hamilton's Egyptiaca.*)

The great temple is supposed to have had four grand entrances, one fronting each of the cardinal points. Deducting its porticoes or propyla, the length of this stupendous structure, measured on the plan of the French *savans*, is 1215 ft., and its least breadth 321 ft.; so that its area must be rather above nine acres! And "besides the great edifice, with its propyla, obelisks, and avenues of colossal sphinxes, it has magnificent temples to the N. and S., altogether forming an assemblage of remains such as, perhaps, no other spot on earth can offer." (*Egyptian Antiquities*, i. 94.)

Champollion says, with reference to the ruins of Carnac, "*Là m'apparut toute la magnificence Pharaonique, tout ce que les hommes ont imaginé et exécuté de plus grand. Tout ce que j'avais vu à Thèbes, tout ce que j'avais admiré avec enthousiasme sur la rive gauche, me parut misérable en comparaison des conceptions gigantesques dont j'étais entouré. Il suffira d'ajouter qu'aucun peuple, ancien ni moderne, n'a conçu l'art d'architecture sur une échelle aussi sublime, aussi large, aussi grandiose, que le firent les vieux Egyptiens: ils concevaient en hommes de 100 pieds de haut; et l'imagination qui, en Europe, s'élance bien au dessus de nos portiques, s'arrête et tombe impuissante au pied des 140 colonnes de la salle hypostyle de Karnac.*" (*Lettres Ecrites de l'Egypte, &c.*, 98.)

The palace of Luxor (*El kusr*, the ruins) about 1½ m. S. from Carnac, on the same side of the river, though inferior in size to the latter, is also a structure of vast dimensions. Its principal entrance facing the *N.* is most magnificent. On either side the doorway stood two obelisks, or monolithes, each formed out of a single block of red granite, 80 ft. in height, about 8 ft. square, and most beautifully sculptured. Recently, however, one of these obelisks has been taken down and conveyed at an immense expense to Paris, where it has been erected in the Place de la Concorde; but it is as little in unison with the objects among which it is now placed as a Pharaoh would be at the court of the Thuilleries, and it is to be regretted that it should have been separated from the venerable structure of which it formed so splendid an ornament. Between the obelisks and the propylon are two colossal statues, each measuring about 44 ft. from the ground. The entire palace is about 800 ft. in length, by about 200 ft. in breadth. It is in a very ruinous state, but though most part of the outer walls have been thrown down, the greater number of the columns in the interior are still standing. It is sadly encumbered with the hovels of the modern Copts and Arabs, and with the accumulated filth and rubbish of centuries. The victories of Sesostris are sculptured on the E. wing of the propylon and on other parts of the palace with infinite spirit, and the greatest amplitude of detail. "It was impossible," says Mr. Hamilton, "to view and reflect upon a picture so copious and detailed, without fancying that I here saw the original of many of Homer's battles, the portrait of some of the historical narratives of Herodotus, and one of the principal groundworks of the stories of Diodorus; and to complete the gratification, we felt that, had the artist been better acquainted with the rules and use of perspective, the performance might have done credit to the genius of a Michael Angelo or a Giulio Romano. Without personally inspecting this extraordinary edifice, it is impossible to have any adequate notion of its immense size, or of the prodigious masses of which it consists. In both these respects, and combined with them, in respect to the beauty and magnificence of its several parts, it is, I should imagine, unrivalled in the whole world." (*Egyptiaca*, 121.) This palace is supposed to have been founded by Amenophis Memnon, about *anno* 1650 B.C.

The ruins on the W. or Libyan side of the Nile are not less interesting than those on its E. side. About 2200 ft. from the river are two sitting colossi, each about 38 ft. in height, and seated on a pedestal of corresponding dimensions. The probability seems to be, that the most northerly of these colossi is the statue of Memnon, which has obtained an immortality of renown, from its being believed to have emitted a sound when it was first struck by the rays of the morning sun! (See vol. i. 830.) Champollion has, however, shown from the hieroglyphics on its back, that this famous statue really represents the Pharaoh Amenophis II., who reigned about *anno* 1680 B.C. These statues are supposed, by the same distinguished authority, to have decorated the façade of the principal front of the celebrated structure, the *Ameinophon* of the Egyptians, and Memnonium of the Greeks. But if such be really the case, the destruction of this building has been incomparably more complete than that of any one else of the famous structures belonging to the city; and it is now, indeed, next to impossible to form any thing even like a ground plan of the ruins. (*Lettres d'Egypte*, 307.)

Between Medinet-Abou and Kournak are the remains of a noble building, about 530 ft. in length and 200 ft. in breadth, supposed by some to be the tomb of Osymandias, described by Diodorus, but which has been more generally supposed to be the Memnonium. Champollion, however, has shown that neither of these suppositions is correct; and that it was built by, and had in fact been the residence of Rhamses the Great, or Sesostris, the most illustrious of the

Egyptian monarchs. The *Rhamesseion*, for such is its proper name, is very much dilapidated; but its immense and noble proportions, and the beauty of its sculptures, make it one of the most interesting, as well as magnificent of Theban structures. Between the propylon and the front of the palace, a distance of about 56 paces, are the fragments of a stupendous colossal statue of Rhamses the Great. It has been broken off at the waist, and the upper part is now prostrate on the ground. This enormous statue measures 63 ft. round the shoulders, and 13 ft. from the crown of the head to the top of the shoulders. The barbarian energy exerted in its destruction has been such, that nothing of the general expression of the face can now be discerned; and, as Mr. Hamilton has truly stated, "Next to the wonder excited by the boldness of the sculptor who made it, and the extraordinary powers of those by whom it was erected, the labour and exertions that must have been used for its destruction are the most astonishing." (P. 167.)

It would be to no purpose, even did our space permit, to attempt giving any account of the innumerable hieroglyphics, pictorial tablets, and bas-reliefs on the ruins of the Rhamesseion. They principally relate to the triumphs of its illustrious founder, and his adoration of the gods of his country. The author of *Scenes and Impressions in Egypt* alludes as follows to the representation of the victories of Sesostris:—"The hero, as compared with the rest of the figures, is of great size; he stands erect in his chariot; his horses on their speed—a high, cloud-pawing gallop; his arrow drawn to the head; the reins fastened round his loins: you have the flight of the vanquished; the headlong fillings of the horses and the chariot: you have the hurrying crowd of the soldiers on foot; a river; drowning; the succouring of warriors on the opposite bank; and, in a compartment beyond, you have a walled town; a storm; the assailants climbing ladders; the defenders on the parapet; the upheld shield; the down-thrust pike; a sad, but yet a stirring picture, bringing to your mind many a historic scene, like memorable and melancholy." (P. 95.)

The following, according to Champollion, is the dedication of the great hall of the palace, sculptured in the name of the founder, in beautiful hieroglyphics, upon the architraves of the left side:—

'Haroeris, all-powerful, the friend of truth, the lord of the upper and lower regions, the defender of Egypt, the instigator of countries; Horus, the resplendent possessor of palms, the greatest of conquerors, the king-lord of the world, sun, guardian of justice, approved by Phré, the son the Sun, the well-beloved of Ammon; RHAMSES, has used these structures to be erected in honour of his father Amon-Ra, king of the gods. He has caused to be conducted in good white sandstone the great hall of assembly, supported by large columns with capitals imitating fullblown flowers, and flanked by smaller pillars with capitals imitating a truncated bud of the lotus; and he has dedicated all to the Lord of Gods, for the celebration of his assemblies: this is what the king ever living has done." (*Lettres d'Égypte*, p. 273; we have used the translation given in the article on *Egypt* in the new ed. of the *Encyc. Britannica*.)

The tombs of the kings of Egypt in the valley or rather valley ravine of Biban-el-Moluk, to the S.W. of the ruins to the W. side of the river, are not less extraordinary than the structures previously noticed. They have been described as follows in the elaborate and learned article on *Egypt* referred to:*

"The site chosen for the royal necropolis appears to be eminently suited to its melancholy destination; for a valley or ravine, encased as it were by high precipitous rocks, or mountains in a state of decomposition, presenting large masses, occasioned either by the extreme heat or by interlinking down, and the backs of which are covered by bands or patches, as if they had been in part burned, a spot which, from its loneliness, desolation, and apparent barrenness, harmonizes well with our ideas as to the most fit locality for a place of tombs. No living animal, it is said, frequents this valley of the dead; even the fox, the hare, and the hyena, shun its mournful precincts; and its dull echoes are only awakened at intervals by the foot of some solitary antiquary, led by inquisitive curiosity to pry into the very secrets of the grave. The catacombs, or Syringes, as they are all constructed on nearly the same plan; yet no two of them are exactly alike; some are complete, others were never to have been finished, and they vary much in the depth to which they have been excavated. In general, the entrance is by the exterior opening of a passage 20 ft. wide, which descends gradually about 50 paces, then exhibiting whilst the descent becomes more rapid, and is continued for some distance farther. On either side of this passage is a horizontal gallery, on a level with the lowest part of the first descent: at the interior extremity there is a spacious and lofty apartment, in the centre of which is placed the royal tomb; and beyond this there are commonly other small chambers at the sides, whilst in some cases the principal passage is continued a long way into the rock. The royal tomb is for the most part a sarcophagus of red or grey granite, circular at the one end, and square at the other; but where there is no sarcophagus, a hole or grave is discovered, cut in the rock to the depth of from 6 to 30 ft., and which appears to have been covered by a granite lid. Almost all the lids, however, belonging to the graves excavated in the rock, have either been removed or broken. In those sepulchres which have been finished, the walls from one end to the other are all covered with sculptures and paintings, executed in the best style of ancient art; and owing to the unparalleled dryness of the atmosphere in Egypt, the colours, where they have not been purposely damaged, are as fresh as when first laid on. The labours of Belzoni in exploring these tombs, and the success with which they were rewarded, are well known. Strength and resolution as herculean and inflexible as his were required to overcome the suspicions of the Arabs, the want of mechanical aid, and the heat and closeness of the caverns; but his perseverance was amply recompensed by the discovery of six tombs in this hypogean city of the dead. The most remarkable of these, with all its galleries, is upwards of 300 ft. in length, and is called by Belzoni the tomb of Apis, from his having found the mummy of a bullock in one of its chambers. In another apartment was a magnificent sarcophagus of white alabaster, almost as transparent as crystal; and the whole excavation, sculptured and painted in the most finished style of art, was in the most perfect preservation. These catacombs, as already stated, were the sepulchres of the kings of the three Diospolitan dynasties; and accordingly, by means of the hieroglyphical inscriptions, Champollion discovered the tombs of six kings of the 18th dynasty; that of Amenophis-Memnon, the most ancient of all, in an isolated part of the valley towards the W.; and lastly, those of Rhamses-Meïamoun, and six other Pharaohs, his successors, belonging either to the 19th or 20th dynasty. No sort of order, either in regard to dynasty or succession, appears to have been observed in the choice of situations for the different royal tombs; on the contrary, each sovereign seems to have caused his own to be dug wherever he found a vein of stone adapted for the purpose of sepulture and the immensity of the projected excavation. The royal catacombs, however, which have been thoroughly completed and finished, are but few in number: these are, the tomb of Amenophis III., or Memnon, the decoration of which has been almost entirely destroyed; that of Rhamses-Meïamoun; and of Rhamses V.; probably also that of Rhamses the Great; and lastly, that of Queen Thaoïsi. All the others are incomplete. The tomb of the great Rhamses, or Sesostris, still exists, according to M. Champollion, and is the third on the right of the principal valley: but it has sustained greater injury than almost any other, and is filled nearly to the ceiling with rubbish."

Such is a very brief and imperfect notice of some of the more important ruins scattered over the site of this ancient capital of the Pharaohs—"*ceterum Thebarum magna vestigia.*" (*Taciti Annal.*, lib. ii. cap. 70.) Their vastness is such as almost to stagger belief; and the traveller who finds himself among these gigantic monuments of remote antiquity feels an almost overpowering sensation of astonishment and awe. It is extremely difficult to form any apparently satisfactory conclusions as to the means which the Theban monarchs must have put in motion to raise such stupendous edifices. Their extraordinary magnitude, the size and hardness of the blocks of stone (usually granite) of which they are built, and the countless numbers, depth, and nicety of the hieroglyphics and pictorial tablets with which they are profusely covered, must have occasioned the employment of an enormous quantity of labour, and an all but boundless expense. Most probably the work was principally executed by slaves, or by requisitions of compulsory labour furnished by subjugated countries; but, in whatever way it may have been effected, it must, especially when we consider the limited advance then made in mechanical science, have involved an outlay which only a very great revenue could have sufficed to meet.

It is impossible to form any just idea of what Thebes must have been in the days of her glory, previously to the Pharaohs leaving her palaces for those of Memphis, while her porticoes were crowded with merchants and merchandize, and before

"Relentless war had pour'd around her wall."

Thebes had little in common with most ancient and still less with most modern cities. She in fact was, as it were, the capital of a by-gone world, of which we know little or nothing save what may be learned and conjectured from her own monuments.

THEBES, a famous city of ancient Greece, the capital of

---

* This article was written by the late Dr. Brown of Edinburgh, and is a fine specimen of his great learning and research.

THEBES.

Bœotia. The modern town is, however, of comparatively limited dimensions, being confined to the eminence occupied by the Acropolis of the ancient city, and the pop. is supposed not to exceed 5000. It is the cap. of a prov. of the same name, and is situated in a fine plain 29 m. N.N.W. Athens, lat. 38° 22' 30'' N., long. 25° 45' 15'' E. When seen from a distance, the modern town still assumes the appearance of a considerable city. Prodigious ramparts and artificial mounds appear on its outside; it is surrounded by a deep fosse, and the traces of its old walls may yet be discovered. But the contrast between its external and internal appearance is most striking. Previously to the late revolution, the streets were narrow and dirty, the houses being either constructed of the ruins of ancient edifices, or mere wooden hovels. It had, however, some handsome mosques and minarets, with a bazaar shaded by gigantic plane trees, and extensive gardens; but these have been almost wholly destroyed during the late struggle. The town, however, is again beginning to be rebuilt, and Mr. Mure says that " the principal street is of considerable width, with some good new houses on each side, interspersed with the usual number of hovels, wooden sheds, ruins, and rubbish." (Tour in Greece, i. 260.) It retains very few traces of its ancient magnificence, and the sacred and public edifices mentioned by Pausanias and others have wholly disappeared. It is now, however, as of old, extremely well supplied with excellent water.

The ancient city of Thebes, or rather its citadel, is said to have been founded by Cadmus (and hence called Cadmeia), a Phœnician, or perhaps Egyptian adventurer, who introduced the knowledge of letters into Greece, anno 1549 B.C. (Larcher, Chronologie d'Herodote, p. 569.) Its walls were constructed at a later period by Amphion and Zethus, the former of whom is believed to have been the earliest of Greek musicians, and hence

" Dictus et Amphion, Thebanæ conditor arcis,
Saxa movere sono testudinis, et prece blanda
Ducere quo vellet."
Hor. Ars. Poet. lin. 394.

The city had seven gates; its circ. is variously stated at from 43 to 70 stadia, and its pop. might perhaps amount to about 50,000. It had many magnificent temples, theatres, gymnasiums, and other public edifices, adorned with noble statues, paintings and other works of art. Its government, like that of all other Greek cities, was fluctuating and various. Originally it was subject to kings or tyrants; and after the republican government had been established, the aristocratical and democratical parties alternately prevailed. Owing to her proximity to Athens, from which, of course, she had everything to fear, Thebes was for a lengthened period what may be called the natural enemy of Athens, and during the Peloponnesian war was the most efficient ally of Lacedæmon. But after the failure of the expedition against Syracuse had broken the power of Athens, and Thebes had no longer any fear of her hostility, dissensions began to spring up between her and Lacedæmon, and the Thebans, under their great leaders Pelopidas and Epaminondas, acquired a decided superiority over the latter, and became for a short while the leading Greek state.

After the battle of Chæronea, in which the Thebans bore a principal part, Philip placed a garrison in the citadel of Thebes; but, on his death, the Thebans rose in arms against his son, Alexander the Great. The latter, however, having taken the city by storm, anno 335 B.C., razed it to the foundations, the house that had been occupied by Pindar being alone excepted from the general destruction; such of the inhabitants, amounting, it is said, to 30,000, as had not been killed, being at the same time sold as slaves. (See Mitford's Greece, vii. 339, 8vo ed., and the authorities there quoted.)

But about twenty years after this catastrophe, the city was rebuilt by Cassander, when the Athenians, forgetting the ancient animosities that had subsisted between them and the Thebans, generously contributed towards the reconstruction of the walls. Subsequently the city underwent many vicissitudes. It appears to have suffered from the exactions of Sylla. Strabo calls it a poor village (lib. ix.); and Pausanias, who describes its temples and other remains, says, that, with the exception of the temples, the lower town was wholly destroyed. (Lib. ix. cap. 7.) The fertility of the surrounding plain, which produces corn, wine, and oil in the greatest abundance, and the excellence of the air and water, appear to have been the principal cause why Thebes has been able to survive so many disasters, and is still a considerable and increasing town.

Thebes is particularly famous in the early and heroic ages of Greek history. " Nec cedentes Athenis claritate, quæ cognominantur Baotia Thebæ, duorum numinum, Liberi atque Herculis, ut volunt, patria." (Plin. Hist. Nat. lib. iv. cap. 7.) The stories, also, of Laius, Jocasta and Œdipus, and their unfortunate progeny, and of the wars of the seven chiefs and their descendants, the Epigoni, against Thebes, have supplied topics of the deepest interest that have engaged the attention of the greatest poets of antiquity and of modern times.

The air of the Bœotian plain is less pure than that of Attica, and this circumstance was pretty generally believed in antiquity to be the cause of the dulness of the Thebans who, speaking generally, wanted the quickness, penetration and vivacity, that distinguished the Athenians. But the difference of character was probably owing rather to a difference in the education and institutions of the two people than to any difference of soil or climate. In respect of illustrious men, Thebes need not fear a comparison with any city of ancient or modern times. The names of Hesiod and Pindar, of Pelopidas and Epaminondas, are also sufficient to illustrate and ennoble a nation. It was, indeed, as already stated, the extraordinary talents and virtues of the latter that raised his country to a preponderating influence in the affairs of Greece. It deserves, also to be mentioned, to the honour of the Thebans, that the odious practice tolerated in other Greek states, of exposing children at their birth, was forbidden in Thebes. (In addition to the authorities already referred to. see Clarke's Travels, vii. 61, &c., 8vo ed.; Dodwell's Greece, i. cap. 9; Voyage d'Anacharsis, cap. 34, Ancient Universal History, vi. p. 180-200, &c.)

THEISS, (an. Tibiscus), a great river of Hungary, being the most important of the tributaries of the Danube, parallel to which it flows, in the lower part of its course through the great Hungarian plain. It has its sources in the Carpathian mountains, on the confines of the Bukowina, within a short distance of the sources of the Pruth, in about lat. 48½° N., long. 24½° E., being formed by the junction of two streams, the black and the white Theiss. Its course, which throughout is extremely tortuous, is first generally W.N.W. to Tokay, from which point it flows with innumerable windings, S.S.W. and S., till it enters the Danube, nearly opposite to Salankement, in lat. 45° 10' N., long. 20° 23' E., after a course of at least 500 m., taking only its more important windings into account, for the greater part of which it is navigable. It constitutes the line of separation between two of the four great divisions of Hungary, thence called Hungary-on-this-side- (or N. and W.), and Hungary-beyond- (or S. and E.; the Theiss. Its chief tributaries are the Bodrog, Schaya, with the Hernad and Zagyva on the right, and the Szamos, Körös, Maros, &c., with the other principal rivers of Transylvania, on the left. The area of its basin is estimated by Berghaus at upwards of 6000 sq. m. (Atl. Länder, &c., iv. 735.) The traffic on the Theiss is at present mostly confined to the conveyance downward of salt (from the co. Marmaros, in which it rises), and of timber in rafts. " Hitherto no steamboat has been established on the Theiss; but from the extreme richness of the surrounding country, the size and importance of many of the places on its banks, and above all from the exceedingly bad roads in its neighbourhood, there can be little doubt that the establishment of steam navigation will be undertaken before long. The depth, width, and force of the stream of the Theiss are as favourable as could be desired; but it is objected that the windings of the river require to be cut off by canals; and in some cases 30 or 40 m. would be saved by a cut of 3 or 4. Should the canal be formed between the Danube at Pesth and the Theiss at Szolnok, as is contemplated, this river will assume an importance far greater than is at present imagined. The slow muddy waters of the Theiss seem to suit the fish better than those of any other river in Hungary. It is said that, after an overflow, they have been left in such quantities as to be used for feeding the pigs, and manuring the ground. The sturgeon of the Theiss, though smaller than that of the Danube, is remarkable for its fatness and delicate flavour." (Paget's Hungary and Transylvania, i. 478, 479.)

THERMOPYLÆ (from θερμός, hot; and πύλα, a gate, or pass); a famous defile on the shore of the Maliac gulf, on the N.E. coast of Greece, near the mouth of the Recheck (an. Sperchius), between the steep precipices at the E. termination of mount Œta and the sea, in about lat. 38° 38' N., long. 22° 39' E. The defile is about 5 m. in length, and where narrowest, was not anciently, more than 60 paces across. In cujus valle ad Maliacum sinum evergente stat est non latius quam sexaginta passus. Hac una militaris via est, qua traduci exercitus, si non prohibeantur, possunt. Liv. lib. xxx. cap. 15.) At present the only practicable road through the strait is by a narrow causeway, on either side of which is an impassable morass, bounded on the one side by the mountains, and on the other by the sea. This pass is now, as in antiquity, the principal, and, indeed, almost the only road by which Greece can be entered from the N.E.; and as it may be defended by a comparatively small force, its occupation is of the utmost importance for the defence of the country. At the narrowest part of the pass are hot springs, a circumstance which, as seen above, has given the defile its peculiar name. (Clarke's Travels, vii. 314, &c., 8vo ed.)

It was in this pass, as every body knows, that, anno 480 B.C., the Spartan king Leonidas, with about 4000 Greeks, resisted for awhile the whole force of the Persian army invading Greece under Xerxes. After the Persians had succeeded in opening a passage by another route across the mountains. Leonidas, having dismissed almost all the other Greeks, devoted himself with 300 Spartans, in obedience to the laws, which forbade Spartans, under whatever disadvantage, to fly from an enemy, and, agreeably to the answer of the oracle, a sacrifice to insure the independence of his country. (*Herodotus*, lib. vii. cap. 210–228.) This event has given Thermopylæ all its interest, and will make it be held in "everlasting remembrance." After the final defeat of the Persians a magnificent monument, the ruins of which still remain, was erected in honour of Leonidas and his heroic companions. It had an inscription, said by Cicero, by whom it has been translated, to have been written by Simonides (*Tuscul.*, i. cap. 42.), and which has been rendered into English as follows :—

"To Lacedæmon's sons, O stranger, tell
That here, obedient to their laws, we fell !"

The ground near the Sperchius, on which the army of Xerxes was encamped during the attack on Thermopylæ, could not possibly have accommodated his troops had their numbers approached to any thing like those specified by Herodotus. But there cannot be so much as the shadow of a doubt that these are grossly, and, indeed, ludicrously, exaggerated. To suppose, as is stated by the venerable father of history, that the army which Xerxes led to Thermopylæ and his fleet comprised 5,283,220 troops, sailors, and male followers of all descriptions (*Herod.* lib. vii., cap. 187), exclusive of women, eunuchs, &c., is a palpable absurdity. ' It may be confidently affirmed that no such force ever was brought together, and that if it were it could neither be fed nor kept together for the shortest period. If we estimate the troops, seamen, and other followers of all kinds employed by Xerxes in this expedition at 500,000 individuals, we shall certainly be not within, but far beyond, the mark. The statements of Herodotus are founded merely on rumour, which is always sure to exaggerate that which is really great ; and the Greeks were particularly prone to magnify their exploits beyond all reasonable bounds.

THETFORD, a parl. and mun. bor. of England, chiefly in the co. Norfolk but partly in Suffolk ; being separated by the little Ouse into two unequal parts, connected by an iron bridge, constructed in 1829, at the intersection of the roads from Newmarket to Norwich, and from Rottesdale to Lynn ; 7 m. S.W. by W. Norwich, and 79 m. N.E. London. Area (parl. bor., which comprises the three parishes of St. Cuthbert, St. Mary, and St. Peter, 8270 acres. Pop., in 1831, 482. The town is straggling, and irregularly built, with little trade or manufactures ; but it does not appear to be decaying, and has a clean and respectable appearance. It is not paved, lighted, nor watched. It has increased of late years very little beyond its former limits. (*Bound. and fun. Corp. Rep.*) St. Peter's, called the "black church." from being constructed mostly of flint, was principally rebuilt in 1780 ; it is provided with buttresses, battlements, &c. The living is a rectory, worth £35 a year. St. Cuthbert's and St. Mary's (which last parish is in Suffolk), are both perpetual curacies, the former worth £50 and the latter £83 a year. All the livings are in the gift of the duke Norfolk. The guildhall is a fine old building, erected in the time of Charles II. The market-house, roofed with iron ; the jail, a large but ill-contrived building ; the bridewell, workhouse, several dissenting chapels, and a theatre, occasionally opened, are the other principal buildings. A hospital for two poor men and two women, and a free grammar school, were established in the reign of James I. ; it has, besides, almshouses founded in 1690, a national school, funds for apprenticing poor children, and many micharities. Thetford is governed by a mayor, three other aldermen, and 12 councillors : its earliest extant charter is of William III. It has no commission of the peace, but petty sessions and a court of record are held weekly. The corporation revenue is principally derived from the tolls on navigation from Thetford to White House ferry, under local ; gross amount in 1840, £638. Thetford has sent two members to the House of Commons since the reign of Edward VI. : previously to the Reform Act the right of voting was vested in the mayor, burgesses, and commonalty, limits of the parliamentary borough were not affected he Boundary Act. Registered electors, in 1839–40, 160. town is a polling place for the W. division of Norfolk. tford is generally supposed to occupy the site of the *magus* of the Romans. During the heptarchy it was the al of the E. Anglian kingdom, and on the E. side of the t are remains of intrenchments, &c., supposed to date that period. In the time of Canute a convent was ded in the town, some remains of which are still existing. The gateway, &c., of a priory, founded in 1104, and traces of a monastery, established at a later period, also be seen. In the reign of Edward III. it is said to

have had 24 principal streets, five market-places, 20 churches, eight monasteries, and six hospitals, beside other public foundations ; but these statements are of doubtful authenticity, and are most probably much exaggerated. It has been occasionally visited in more modern times by some of the British sovereigns, particularly James I., who had a hunting-seat in the neighbourhood. Among the natives of Thetford who have attracted notice, the most celebrated by far was Thomas Paine, author of the once famous but now forgotten pamphlets entitled "Common Sense," "Rights of Man," "Age of Reason," &c. Paine was born on the 29th of January, 1737 ; his father, who was a stay-maker in Thetford, belonged to the Quakers. Markets, Sat. ; fairs, May 14, Aug. 2 and 16, for sheep ; Sept. 25, for cattle, cheese, and toys.

THETFORD, p. t., Orange co., Vt., 43 m. S.S.E. Montpelier, 498 W. Bounded E. by Connecticut river. Watered by Ompompanoosuc river. It contains two churches, a Congregational and Baptist, six stores, one fulling-mill, one woollen factory, four grist-mills, 13 saw-mills, one oil-mill, two tanneries ; two academies, 176 students ; 17 schools, 708 scholars. Pop. 2065.

THIBET, or TIBET (native *Tiup'ho*, *Bhote*, and *Pud-koschim*, "snowy region of the north"), a very extensive region of central Asia, mostly comprised within the Chinese empire, between lat. 29° and 31° N., and long. 79° and 104° E., having N. Chinese Turkestan and the desert of Cobi ; E., the Chinese prov. of Szechuen ; S., Yun-nan ; N., Birmah, and the great Himalaya, separating it from Assam, Bootan, Sikkim, Nepaul, and the upper British provinces, ; and W., the Punjab territories N. of the Himalaya, Budukh-shan, the Beeloot Tagh mountains, &c. The W. part of this vast tract, called Little Thibet (including Ladakh, Lé, Baltee, &c.), appears, however to be independent of China. Its boundaries on every side but the S. being so uncertain, and our knowledge of the country so limited, it is impossible to form any thing like an accurate estimate either of its area or population. Thibet, though it does not include the highest summits of the Himalaya, comprises a large portion of the elevated table-land in the centre of the continent, with the sources of almost all the great rivers of S. Asia, including the Indus, Sutleje, Ganges, Brahmaputra, Irrawadi, Than-lweng, and Menam-kong or river of Camboja, as well as those of the great Chinese rivers, the Yang-tse-kiang, and Hoang-Ho. Its mountain-chains generally run parallel to the great Himalaya, of which Thibet is the N. slope ; but some are said to stretch in a N.E. direction to the frontiers of Koko-nor, and others extend from N. to S. between the valleys of the great rivers in the S.E. Thibet has numerous lakes ; the chief are the Tengkiri-nor, the largest, about 110 m. N.W. Lassa, and the lake Palte or Yamo-rouk, S. of the San-po river, which surrounds in the form of a ring a large island of a shape similar to its own. (*Chinese Repos., &c.*)

According to Mr. Turner, there is a very striking contrast in the face of the country in passing from Bootan to Thibet. " Bootan presents to the view mountains covered with perpetual verdure, and rich in forests of large and lofty trees, while not a slope or narrow slip of land between the ridges lies unimproved. Thibet, on the other hand, strikes a traveller at first sight as one of the least favoured countries under heaven, and appears to be in a great measure incapable of culture. It exhibits only low rocky hills, without any viable vegetation, or extensive arid plains, both of the most stern and stubborn aspect, promising full as little as they produce. Its climate is cold and bleak in the extreme, from the severe effects of which the inhabitants are obliged to seek refuge in sheltered valleys and hollows, or amidst the warmest aspects of the rocks. Yet the advantages that the one country possesses in fertility and in the richness of its forests and fruits, are amply counterbalanced in the other by its numerous flocks and invaluable mines. As one seems to possess the pabulum of vegetable, in the other we find the superabundance of animal life. The variety and quantity of wild-fowl, game, and beasts of prey, flocks, droves, and herds in Thibet, are astonishing. In Bootan, except domestic creatures, nothing of the sort is to be seen." (*Turner's Thibet, &c.*, p. 216.)

The same division of the seasons prevails here as in Bengal. The spring, from March to May, is marked by a variable atmosphere, heat, thunder-storms, and occasionally refreshing showers. From June to September is the damp season, when heavy and continued rains throughout most parts of the country swell the rivers, which bear off the surplus waters to augment the inundation of Bengal. From October to March a clear and uniform sky succeeds, seldom obscured either by fogs or clouds, and for the first three months of this season a degree of cold is felt, among the lofty mountain ranges of the S., far greater, perhaps, than is known to prevail in Europe.

This region is remarkable, at all seasons, for the dryness of the winds, and meat and fish are prepared for carriage to

THIBET.

any distance, and will keep to any season of the year, by being dried up by exposure to the frosty air. Vegetation is frequently dried to brittleness, and every plant may be rubbed between the fingers into dust. (*Turner*, p. 303.)

Goitre, syphilis, and smallpox appear to be the most severe diseases in Thibet, and, unfortunately, are very prevalent. Syphilis is said, by an English surgeon, to make a more rapid progress, and rage with more violence here than in any other country. (*Saunders*, p. 410.) Catarrhs and rheumatism are more frequent than in Bengal.

Of the geology of Thibet we have only scattered notices. Moorcroft found that the hills in part of Little Thibet consist apparently of clay-slate, fragments of granite, quartz, &c., being strewn upon their sides. (*Moorcroft's Travels*, i. 439.) The latter rocks, with primary limestone, talc, and similar formations, seem to enter largely into the mountain ranges, where they are often interspersed with beds of clay and sand, and occasionally of chalk. Tincal is obtained in inexhaustible quantities; rock salt is met with in many parts, and nitre effloresces abundantly on the surface of the soil. Gold is found in lumps and irregular veins, or in the form of dust in the rivers, and is frequently of great purity. There are mines of lead, silver, copper, and cinnabar, but few if any of iron, though chalybeate springs are very frequent. The difficulty of procuring fuel for smelting the less valuable ores proves an insuperable obstacle to success in mining: timber of all kinds is rare, and the dung of animals is the only substitute for fire-wood. The discovery of a coal mine would be an invaluable acquisition to Thibet.

The usual crops are, barley, coarse pease, and wheat. The first forms by far the largest proportion of the whole; wheat never enters into the food of the poorer classes, and rice is not cultivated. A brief notice of the agriculture and vegetable products of Little Thibet will be found under the article LADAKH (*ante*, p. 136.) Turnips and radishes are almost the only garden vegetables, and fruits are of little variety. For most vegetable products, and, indeed, medicinal plants, Thibet is dependent on Bootan, Nepaul, and the other countries S. of the Himalaya.

Among the useful animals of Thibet, sheep merit a distinguished rank. The flocks of these are numerous; and upon them the chief reliance of the inhabitants is placed. A peculiar variety which seems indigenous to the country, is of small size, with black heads and legs, and soft wool; their mutton, which is almost the only animal food used in Thibet, being said by Mr. Turner to be the finest in the world. The sheep are occasionally employed as beasts of burden, being laden with salt, grain, &c. They are the bearers of their own costs to the best marts, where the wool is usually made into a narrow cloth resembling frieze or thick coarse blanketing. The skins of both sheep and lambs are commonly cured with the wool on; and, in order to secure a silky softness of the fleece, the ewes are sometimes killed before their time of yeaning, when their skins bear a high price in China and all over Tartary. (*Turner*, p. 302.) The Thibet goat (*Capra hircus*), which affords the valuable material for the shawl manufacture, feeds, like the sheep, in large numbers together. These are perhaps the most beautiful among the whole tribe of goats. Their colours are various: black, white, of a faint bluish tinge, and of a shade something lighter than a fawn. They have straight horns, and are of a lower stature than the smallest sheep in England. The material used for the manufacture of shawls is of a light fine texture, and clothes the animal next the skin. A coarse covering of long hair grows above this, and preserves the softness of the interior coat. The creature is, no doubt, indebted for the warmth and fine quality of the latter to the nature of the climate and country it inhabits. On removing some of the goats to the hot atmosphere of Bengal, Turner says they quickly lost their beautiful clothing, and a cutaneous eruptive humour soon destroyed almost all their coat. He was also unsuccessful, after repeated trials, in attempting to acclimatise the animal in England. (*Turner's Thibet*, p. 356.) Recently, indeed, the Thibet goat has been naturalised in France; but it is quite certain, from the great difference of the climate, that the wool will, in no long time, lose all its distinguishing and most valuable qualities. The most valuable species of cattle is the yalk, or grunting ox (*Bos grunniens*), which is also indigenous to the country. Their cows supply an abundance of rich milk: they are very useful as beasts of burden, and throughout Hindostan their bushy tails are in great request as chowries to drive away flies, &c. For agricultural labour, small cattle, like those of Bengal, are chiefly employed. Most of the native animals of Thibet, as the hare, bharal (*Ovis ammon*), dog, &c., have long furry coats. Among the wild animals, perhaps the most curious is the musk deer, which delights in excessive cold. It is about the height of a moderately-sized hog, which it closely resembles in the figure of the body. It has a small head, a thick and round hind quarter, no tail, and extremely delicate limbs. The hair with which it is covered is prodigiously copious, and grows erect all over the

body, in some parts to between two and three inches in length, thin, flexible, and undulated. Its colour at the base is white, in the middle black, and brown at the points. The musk is a secretion formed in a little bag at the navel, and found only in the male. The musk deer, valuable for this product is deemed the property of the state, and is hunted only by permission of government. (*Turner*, p. 306.)

At the end of the last century, the valley of Jhansu in Thibet was particularly famous for the manufacture of woollen cloth, for which there was an extensive demand. The cloths, which were confined to two colours, garnet and white, seldom exceeded half a yard in breadth, and were woven very thick and close. A good deal of cloth is also said to be made at Lassa, great quantities of a red colour being annually exported into China. Moorcroft describes the process of weaving at Pili, in Little Thibet, as follows:—"The two ends of the warp are fastened together, and it is then stretched upon two rods, one fixed to the body of the weaver (who is invariably a woman) by a cord, which admits of the work being loosened or tightened at pleasure, and the other well fastened to some stones at a distance, equal to half the length of the cloth. The whole is close to the ground, on which the workwoman sits, but the portion close to her is slightly elevated by a third rod; loops, each including a thread, and received upon a small stick like a rattan, supply the place of a heddle: of these there are three sets, which draw up parts of the warp alternately as required. A large heavy reed, into which a thin bar of iron is inserted, is a substitute for the reed, and three or more heavy strokes are made with its armed edge upon every thread of the woof. The last instrument must be taken out after the insertion of each piece of yarn, and when placed perpendicularly, with its two edges separating the warp, abundance of room is given for the passage of the balls of worsted made use of without the covering of a shuttle. This part of the process is tedious, but the warp is prepared in a quick and simple way: several pegs are driven into the ground, so near each other that the whole may be reached without any material movement of the body; the yarn is fastened to one of them, and carried on round the others till a sufficient quantity has been wound; all are then taken out except three, which have their pieces supplied by rods, and the warp only requires spreading. Every woman knows how to weave, but only half their number may be considered as employed in the manufacture, for if a house contain two, one is usually busy in domestic affairs. 26,000 yards, 17 inches wide, may be fabricated annually in the Pili district, of which about half is exported. Farther N. a coarse loom is in use, not very unlike that common in Europe. Several varieties of cloth are manufactured; some thick and heavy, with a long nap, others fine. All the wool used is of a coarse kind, and in consequence the finer cloths have a hardness, something similar to that of camlet or plaid, to which they are little inferior. Very good sacking is also made of the hair and wool from the yalk. (*Moorcroft, Trav.*, p. 71–74.)

Thibet has, from time immemorial, been a country of considerable traffic; but here, as in Bootan, foreign trade is monopolised by the government, and a few of the first officers of state. The commerce is principally with China, the Chinese trade being carried on partly at Sin-sing, a garrison town on the W. frontier of China, and partly at Lassa, by caravans, which come there in October. These consist of 300 or 600 persons, who bring goods on cattle, mules, and sometimes horses, exchanging tea, silver bullion, brocades, fruits, &c., for fine and coarse woollen cloths, gold dust, and Bengal goods. The imports from China are large. Turner was informed, that in the territory of Teshoo Loomboo the tea to the value of £80,000 or £70,000 sterling was annually consumed, and Bootan is supplied with tea from Thibet. The other imports from China are tobacco, quicksilver, cinnabar, furs, porcelain, musical instruments, European cutlery, pearls, coral, &c. From Bootan and Bengal Thibet receives English broadcloths, piece goods, Allahabad cloth, kimcots, coarse sugar, tobacco, indigo, paper, rice, sandalwood, spices, gums, and otter skins. Many of these articles come through Nepaul, which receives all its Chinese imports through Thibet. The trade with Assam is very limited, but small quantities of rice, coarse silk, iron, stick lac, are imported; from Turkestan come horses and camels. From Ladakh E. Thibet receives dried fruits, shawls, gumboge, saffron, &c. The general returns of Thibet are in gold-dust, silver, tincal, musk, woollen cloths, goat and lamb skins, and rock salt; the goats' hair is almost all sent through Ladakh to Cashmere for the manufacture of the Cashmere shawls. (See vol. i., p. 567 of this *Dict.*)

Moorcroft says that in Little Thibet traffic is carried on chiefly by barter, and money is almost unknown. Salt, wool, turquoises, sheep, and goats, are imported from Chanthan, or Chinese Thibet, and are paid for with grain, woollen cloth, and horses. From Biahar and Kutu (or Pili) come iron cooking utensils, brass, copper, tobacco, rice, dried fruits, tea-cups, timber, amber, &c., paid for in a similar way. The iron and metal vessels of the S. are sent to Lé, in exchange for tea, coarse cloth, coral, and dyeing drugs. (Tree. ii., 71.) Further details respecting the trade of Little Thibet will be found in the art. LADAKH (anté p. 136).

The modes of conveyance in Thibet differ altogether from those of Bootan. In the latter all species of goods are carried on the shoulders of the people, chiefly the females; in Thibet, they are conveyed by the chowry cattle, horses, mules, and asses. The horses which are very docile, are not natives of Thibet, but mostly brought from Turkestan, after having been carefully emasculated to prevent their propagating their species.

China has been justly celebrated for her magnificent bridges and public works, but Thibet is far from sharing in this celebrity. Over one river, crossed by Mr. Turner, was constructed a long bridge, upon nine piers, of very rude structure. "The piers were composed of rough stones, without cement; but to hold them together, large trees, with their roots and branches, had been inserted; and some of them were vegetating. Slight beams of timber were laid from pier to pier; and upon them large flat stones were loosely placed, that tilted and rattled when trod upon: and this, I fear, is a specimen of their best bridges. Many were extremely dangerous to pass over." (P. 229.) The boats, also, used to cross the rivers are of a very rude kind; some are made chiefly of leather, consisting of a rude skeleton of wood, with thwarts and ribs, over which a bull's hide is stretched.

The country is politically divided into Wei and Tsang, or Hither and Farther Thibet. Wei is that division bordering on China, and having for its capital Lassa, or H'lassa, the residence of the Dalai Lama. It is divided into eight cantons, that of Lassa being the principal; and 30 feudal ownships, called teasers, which lie northward, contiguous o some similar townships in the country of Ko-ko-nor, Tsang, or Ulterior Thibet, is W. of the former, and extends V. from about long. 90° E. It is divided into seven cantons, its capital being Teshoo-Loomboo. These two provinces are under the direction of two ministers, sent from he imperial cabinet at Pekin; and of two high priests of Thibet, called Dalai Lama and Bantchin-Erdeni. The he ministerial residens govern both provinces conjointly, onsulting only with the Dalai Lama for the affairs of Hither, and with the Bantchin-Erdeni for those of Farther Thibet. All appointments to offices of government and titles nobility must be approved by the Chinese officers. But minor matters the residents do not interfere, leaving such fairs to the secular deputies of the high priests, called heba. The government of the 30 feudal townships in ither Thibet, and of the Tamuh or Dam Mongols inhabiting the N. frontier, is entirely in the hands of residents. Chinese Repository, i., 174.) Two officers, natives of the untry, are sent to each canton from Lassa, and relieved ery three years. The subordinate management of the mmunities is entrusted to two officers in each, the d'heba d vazir, the former appointed from Lassa, the latter a tive of the place, who, with the chief lama of the village, m a sort of local council, dependent on the provincial aurities; who again are obliged to refer to the capital for tructions in all extraordinary cases. (Moorcroft, i. 365.) Thibet is remarkable as being the central seat and headarters of Buddhism, where the Buddhic religion is prered in its greatest purity. The whole nation is divided two distinct and separate classes, those who carry on business of the world, and those who hold intercourse h heaven. No interference of the laity ever interrupts regulated duties of the clergy, nor do the latter ever ploy themselves in secular affairs. In this, and in the ence of castes, consists some of the most striking differes between the religion of Thibet and that of Hindostan, listinction of caste being utterly repudiated by the Buddfaith. The priests of Thibet are all called lamas; and Grand, or Dalai Lama, who resides at Lassa, is believed is adherents to be an incarnation of the Divinity in a man form. On the dissolution of this body, he is supd to reappear in the body of some infant, who subseatly passes through the term of his mortal existence all the honours of the Grand Lama. The mode of rtaining the identity of the new lama is described at th in Hamilton's E. I. Gazetteer, art. THIBET. The 100 Lamas, and others, are also supposed to be divine marions, occupying successively different bodies; and er (p. 335–336,) gives an amusing account of an interwith the former, who, although only 18 months old, ars to have conducted himself with astonishing dignid decorum! The Buddhists of Thibet have convents en and women, and their religious institutions present al striking coincidences with those of the Roman Catthurch.

e written laws of Thibet, which are said to be of high

antiquity, have in recent times been modified in accordance with those of China. Robbery or dacoity is usually punished by perpetual banishment; murder by death. Adultery is not classed among serious criminal offences; and strict chastity before marriage is not expected in the fair sex. In Thibet, as in Bootan and other countries of the Himalaya, the practice of polyandria is common; a female associating herself with all the brothers of a family, without restriction of age or numbers. The choice of the wife is the privilege of the elder brother.

The people of Thibet belong to the great Tartar family. Their physical appearance has been already noticed in this work (see ASIA, i. 183). They are said to be mild and humane, but their intellect is sluggish, and they have never exhibited the enterprise of their neighbours either to the N. or S. At Dras, in W. Thibet, Moorcroft found the population much addicted to pilfering; but he says that this is not the character of the people in general, especially of those who follow the faith of Buddha: the people of Dras are Mohammedans, and like those of Ladakh generally, have suffered much moral detriment from contact with the Cashmerians. (Tree., ii. 43.) In this part of Thibet the houses are built of pebbles, cemented with earth, having terraced roofs, without chimneys. Further E. the peasants' dwellings are mean structures, resembling brick-kilns in shape and size, and built of rough stones heaped upon each other without cement. The great scarcity of timber in Thibet prevents the higher class of inhabitants from boarding the floors of their rooms, which are accordingly of stone or marble. Bedsteads appear to be wholly unknown; the general custom is to spread on the floor, by way of a bed, a thick mattress, which serves for a seat by day. Both sexes dress chiefly in woollens, in which yellow and red are predominant colours, with upper garments of sheep, goat, or jackal skins, and high and thick boots, but the upper classes parity in silks, and in cloaks lined with sable or other furs. Their food principally consists of barley, variously prepared, with tea, spirits, beer, and mutton, which last they prefer raw. Their meals are taken at no stated times, but under the impulse of hunger. The business of the day usually begins by prayer; they then follow their peculiar avocations till evening, which is always spent in recreation, music and dancing being among their principal amusements. Mr. Turner found the priests acquainted with the signs of the zodiac, the satellites of Jupiter, Saturn's ring, &c.

The art of printing has also, from a very remote age, been practised in Thibet. But no improvements appear to have been made in this country in any branch of science known to the inhabitants. Their mode of printing has probably been derived from China; but they esteem the city of Benares as the traditional source of both their learning and religion. (Turner, p. 281.) There appears to have been, from the remotest time, a connexion between Thibet and India; and the uchen character, in which the sacred writings of this people are preserved, bears a strong resemblance to that of the Sanscrit. The umén, or ordinary character of business and correspondence, is distinct from the former.

Several remarkable customs prevail in Thibet. In every visit of ceremony a silk scarf, usually white, and with the mystic sentence Dom mane pee me oom interwoven at both ends, is invariably exchanged, and accompanies every letter sent, between people of every rank and station in life. "This usage," says Turner, "is observed in all the territory of the Deb Rajah (Bootan); it obtains throughout Thibet; it extends from Turkestan to the confines of the Great Desert; it is practised in China, and, I doubt not, reaches to the limits of Manchoo Tartary." Another custom, which the people share with the Parsees, is that of exposing the bodies of the dead among the laity to be devoured by carnivorous birds. The bodies of sovereign lamas after death are dried by exposure to the air, and preserved enshrined; those of inferior lamas are usually burnt, and their ashes enclosed in little metallic idols. Other corpses are committed to the rivers; but the inhumation of the dead is totally unknown.

Thibet appears to have had relations with the Chinese empire at a very early period; but it was governed by its own princes till about 1720, when the emperor Kang-he acquired its sovereignty. Still the greater share of power was left in the hands of the grand lama till the invasion of the Nepaul Gorkhas in 1790; when, on their expulsion by the Chinese, the present form of government was established, and strangers, formerly permitted to enter the country, were altogether excluded. (Note in Chinese Repository, i., 175; Turner's Embassy to Thibet; Moorcroft's Tree. in the Himalayan Provs., &c.)

THIELT, a town of Belgium, prov. W. Flanders, arrond. Bruges, cap. cant., a little S. of the railway between Bruges and Ghent, 13 m. S.S.E. the former. Pop. in 1836, including commune, 11,660. It has no government establishments or public buildings of any consequence; but it is a busy and flourishing town, with manufactures of leather, hats, soap, and lace, being indebted for its prosperity to its situation near a tributary of the Lys, which gives it the advantage of a considerable inland navigation. It unites with Poperingen in sending three members to the provincial states. Among the natives of Thielt was Oliver Ledain, the barber, and afterward the favourite, of Louis XI. This unworthy minion, who figures as an important personage in Scott's novel of Quentin Durward, did not escape the fate due to his deserts, having been hanged, in 1484, after the death of Louis.

THIERS, a town of France, dep. Puy de Dôme, cap. arrond., on the Durolle, 23 m. E.N.E. Clermont. Pop. in 1836, 6807. The appearance of the town is picturesque, being situated on the declivity of a hill, and tolerably well built; but its streets are narrow and steep, and its vicinity is so arid and bare that its inhabitants have been always obliged to depend mainly on their manufacturing industry. It has considerable fabrics of hardware and cutlery, and of woollens, paper, leather, &c. It is the seat of tribunals of primary jurisdiction and commerce, a chamber of manufactures, a council des prud'hommes, and a communal college. It owes its origin to a castle existing here in the earliest period of the French monarchy, and is supposed to derive its present name from Thierri, king of Metz, early in the 6th century. (Hugo; Dict. Géog., &c.)

THIONVILLE, a fortified town of France, dep. Moselle, cap. arrond., on the Moselle, 16 m. N. Metz. Pop. in 1836, 4901. It is, in general, well built; and, unlike most fortified towns, has broad streets. It is entered by three gates, and communicates with its citadel across the river by a wooden bridge. It has a handsome place d'armes, three sides of which are occupied by barracks, and the fourth by the cavalry stables, considered among the best in France. The new parish church, corn market, theatre, college, civil hospital, and the former mansion of the governor, now the sub-prefecture; the tribunal of primary jurisdiction, mayor's residence, and gendarmerie, with the botanic garden, are all deserving of notice. Hosiery, woollen cloths, candles, leather, liqueurs, and spirits are manufactured in the town and its vicinity. The kings of France, of the first and second races, frequently resided here. After the Carlovingians, Thionville successively belonged to the counts of Luxemburg, and to Burgundy, Austria, and Spain. It was repeatedly besieged and taken in the 16th and 17th centuries, but has belonged to France ever since it surrendered to the prince of Condé, in 1643. (Hugo, Guide du Voyageur; Dict. Géog.)

THIRSK, a parl. bor., market town, and par. of England, co. York, N. Riding, wapent. Birdforth, on the Codbeck, an affluent of the Swale (here crossed by two stone bridges), by which the town is divided into old and new Thirsk, 22½ m. N.W. York. The parliamentary borough comprises the townships of Thirsk, Sowerby, Carlton-Miniott, and Sand-Hutton, in the parish of Thirsk, with the adjacent townships of S.Kelvington and Bagby, including an area of 9810 acres. Population in 1821, 4672. Thirsk is a country town, situated in an agricultural neighbourhood: the old town is chiefly composed of one long street of rather an unpromising appearance, at the commencement of which is a small open space. In the new town, also, the houses are, for the most part, of an inferior class, and inhabited by small tradesmen. About quarter of a mile to the S. of Thirsk is Sowerby, an extremely neat country village. Many of the houses are of a superior class, and have an air of neatness and respectability. Sowerby consists of one long and very wide street; it has been selected as a residence by many persons, who, having retired from business, live upon the produce of the capital they have accumulated. (Bound. Rep.) The parish church is a large and handsome edifice, in the perpendicular style: it has a lofty W. tower, and is wholly of one design, with pierced battlements; the details are good, and the general appearance elegant. (Rickman's Gothic Architecture.) A part of this church is said to have been built out of the ruins of the castle, belonging to the Mowbray family, erected in the 10th and destroyed in the 19th century, on the site of which new Thirsk is partly built. The living of Thirsk, a perpetual curacy worth £143 a year, is in the gift of the archbishop of York. There are several places of worship for Dissenters, charity schools, a dispensary, &c. The principal employment is the manufacture of coarse linens and sacking. Thirsk appears to be in a stationary condition. It is governed by a bailiff, chosen annually, and sworn in at the court leet of the lord of the manor. The former borough comprised only a part of the old town of Thirsk. It sent two members to the House of Commons in the 23d Edward I., and again from the reign of Edward VI. down to the passing of the Reform Act; the right of election being in the owners of burgage tenements. The Reform Act deprived Thirsk of one of its members. Registered electors in 1839-40, 327. Markets on Mondays. Fairs, eight times a year for cattle, horses, sheep, and leather. (Parl. Reps., &c.)

THOMAS, county, Ga. Situated toward the S.W. part of the state, and contains 1089 sq. m. Drained by Ochlockonee river and its branches, and by branches of Withlacoochee river. It contained in 1840, 24,699 neat cattle, 288 sheep, 17,960 swine; and produced 1049 bushels of wheat, 149,736 of Indian corn, 334 of oats, 49,169 of potatoes, 22,65 pounds of rice, 1,565,262 of cotton, 52,445 of sugar. It had 17 stores, 13 grist-mills, nine saw-mills; two academies, 30 students; eight schools, 169 scholars. Pop.: whites, 3010; slaves, 2930; free coloured, 26; total, 6766. Capital, Thomasville.

THOMASTON, p. t., Lincoln co., Me., 42 m. S.E. Augusta, 619 W. Bounded E. by Penobscot bay, W. by St. George's river. It contains excellent limestone, from which 350,000 casks of lime are annually manufactured. It is estimated that there are $14,000,000 worth of limestone within 20 feet of the surface, and nearly half a million of dollars are annually realized from the sale of lime. The state prison is situated on the bank of St. George river, attached to which are 10 acres of ground. The buildings are of stone, surrounded by a high stone wall. The keeper's house is 30 by 40 feet, the hospital 23 by 48 feet, and there are 50 cells in the prison. The convicts are chiefly employed in cutting and preparing for exportation a fine blue granite, found on the river. Major-general Henry Knox, of the Revolutionary army, died here in 1807, and was buried, by his own request, in the middle of a thick cedar grove, on his own ground, near his beautiful dwelling. The t. contains 28 stores, two fulling-mills, three grist-mills, one saw-mill, one pottery, two printing-offices, two weekly newspapers; one college, 15 students; three academies, 166 students; 27 schools, 2423 scholars. Pop. 6227. The Baptist Theological Institute was founded here in 1837, has two professors, 23 students (1842-3), and 500 volumes in its library.

THOMASTOWN, an inland town of Ireland, prov. Leinster, co. Kilkenny, on the Nore, 18 m. N. Waterford. Pop. in 1831, 2831. The town, which consists mostly of mean houses, is principally occupied by agricultural labourers and small traders, generally in very depressed circumstances. The public buildings include the parish church, a Roman Catholic chapel, a courthouse, and a bridewell. Lighters of from 20 to 30 tons come up to the town, which has three large flour-mills, and exports considerable quantities of flour, corn, and provisions: it has also a tannery and two breweries. Thomastown sent two members to the Irish House of Commons, but was disfranchised at the Union. Quarter sessions are held in January, April, July, and October, and petty sessions every alternate week. It is a constabulary station. Markets on Mondays and Saturdays. Fairs, March 17, May 25, June 29, and September 15. Post-office revenue, in 1830, £264; in 1836, £605. A branch of the Agricultural Bank was opened in 1835. There are several resident landlords in the neighbourhood of Thomastown, and the agriculture and trade of the country has improved. But Inglis affirms that the condition of the people has, notwithstanding, visibly deteriorated of late years (i., p. 57); and this, we regret to say, how anomalous soever it may appear, is true of most country districts in Ireland.

THOMPSON, p. t., Windham co., Ct., 48 m. E.N.E. Hartford, 384 W. Quinebaug, French, and Five Mile rivers afford good water-power. It contains four churches, a Congregational, two Baptist, and a Methodist; 15 stores, seven grist-mills, two saw-mills; one academy, 30 students; 15 schools, 827 scholars. Pop. 3535.

THOMPSON, p. t., Geauga co., O., 190 m. N.E. Columbus, 349 W. It has one grist-mill, four saw-mills, eight schools, 536 scholars. Pop. 1022.

THOMPSON, t., Seneca co., O. It has two saw-mills, five schools, 180 scholars. Pop. 1404.

THORN, a fortified town of the kingdom and province of Prussia, reg. Marienwerder, cap. circ., on the Vistula, here crossed by a long wooden bridge, about 90 m. from its mouth, and 52 m. S.S.W. Marienwerder. Pop., in 1836, 7668. It consists of an old and a new town, separated by a wall and ditch. There are three Roman Catholic and two Protestant churches; several convents and asylums, a Lutheran gymnasium, &c. It is the seat of the courts for the circ., and has various manufactures, and a considerable trade. It is very strong, its fortifications having been greatly improved and augmented since 1815. It was founded by the first grand master of the Teutonic order, in 1231; and most part of its principal edifices are of old date. But its chief claim to notice is derived from its having been the birth-place of Copernicus, the discoverer, or rather restorer, of the true theory of the world, born on the 19th February, 1472. His great work, De Revolutionibus Orbium Cœlestium, in six books, was published at Nuremberg, in 1543, a few days before the death of its illustrious author, which took place on the 24th of May of the same year.

THORN, t., Perry co., O. It has eight stores, three tanneries, 10 schools, 370 scholars. Pop. 2097.

THORNBURY, a market town and par. of England, co.

Gloucester, hund. Thornbury, in the vale of Berkley, 11 m. N. by E. Bristol. Area of par., 11,580 acres. Pop., in 1831, 4375. The town consists principally of three streets, arranged in the form of the letter Y. The church is a handsome cruciform structure, with a lofty tower, ornamented with rich open-worked battlements and pinnacles. The living, a vicarage worth £500 a year nett, is in the gift of the dean and chapter of Christchurch, Oxford. There are two subordinate curacies in the parish, at Oldbury and Falfield. I: has also Baptist, Independent, Quaker, and Wesleyan meeting-houses; a grammar-school for boys, another free school for 36 children, almshouses, &c. But it is principally remarkable for the remains of a magnificent castle, begun by Stafford, duke of Buckingham, in 1511, but left in an unfinished state, when he suffered on the scaffold, in 1522. Its site is very commanding; Rickman says its style is the late perpendicular, with good details; and it is especially interesting from its affording some interesting specimens of the last gradation of castellated architecture. (Gothic Architect.) Thornbury was formerly a municipal borough, governed by a mayor and 12 aldermen; but no charter is extant, and the body being found useless, the corporation was abolished by the Municipal Reform Act. The clothing trade was formerly carried on pretty extensively, but it is now nearly extinct. Market-day, Saturday; fairs, Easter Monday, August 15, Monday before St. Thomas, and December 21, for cattle and pigs. (Municipal Report; Append., &c.)

THORNE, a market town and par. of England, co. York, V. Riding, wapent. Strafforth, &c., near the Don, and on the borders of Lincolnshire, in a low, flat, and mostly fertile, at marshy country, 23¼ m. S. by E. York. Area of par., 7,840 acres. Pop., in 1831, 3779. The town appears to be prosperous; it is tolerably well built, and the streets are paved. The parish church is a neat building, with a square tower and pinnacles. The living, a perpetual curacy, worth £117 a year, is in the gift of the heir of Sir H. Etherington. (Eccl. Rev. Rep.) There are several dissenting chapels, two free schools, &c. At a suburb called Kingman-hill, on the Don, about 1 m. from the town, reservoirs of considerable burden are built, and a brisk trade in coin and other goods is carried on at Thorne, which is greatly promoted by the Stainforth and Keadby canal. Market-day, Wednesday; fairs, Monday, Tuesday, and Wednesday, after both June 11 and October 11, for cattle, horses, and pedlery.

THORNTON, p. t., Grafton co., N. H., 52 m. N. by W. concord, 533 W. Watered by Pemigewasset river. Millbrook has a fall of 42 feet perpendicular. It has three arches, a Congregational, Free-will Baptist, and Universalist; three stores, one fulling-mill, one grist-mill, five mills; 12 schools, 435 scholars. Pop. 1045.

THOUSAND ISLANDS, N. Y., situated in St. Lawrence river, a little below lake Ontario. The number is said to exceed 1500 in the distance of 27 m. The largest are Grand Howe islands in Canada, and Carlton, Grindstone, and Isle islands in the United States, which latter belong to Jefferson co. The passage between some of them is narrow and intricate. It was a matter of some difficulty to fix the boundary between the United States and Canada at this point.

THRASYMENE (LAKE OF), or lake of Perugia (Lat. Thrasymenus), a famous lake of Central Italy, Papal States, deleg. Perugia, 10 m. W. the city of that name. It is of a circular shape, about 30 m. in circumference, has several small islands, and is rather shallow, the water not exceeding 24 feet; it is well stocked with fish. Its banks are covered with olive plantations. Being unfed by ramifications of the Appenines, it has no natural outlet for its waters; and in consequence, when it is swollen by rains, it was apt to overflow its banks, and deluge the surrounding country. To obviate this danger, a tunnel (or emissario), similar to that of lake Albano (see No.), above ½ m. in length, has been cut through a hill on the S.E. side of the lake, by which its surplus waters are conveyed to an affluent of the Tiber. This useful work, if not wholly constructed, was, at all events, repaired and perfected, by a lord of Perugia in the early part of the 15th century. (Rampoldi, Trasimene, Lago di.)
This lake is famous in history for the great victory gained on its banks by Hannibal over the Romans, anno 217 B.C. A battle appears, according to the statements of the best authorities, to have been fought in a valley near Passignano, on the E. shore of the lake, by which it is entered from the N. by a narrow defile, and is shut up on all sides by steep hills and mountains. Hannibal having entered this defile, posted his Numidian cavalry on either side, and in this position waited the arrival of the Romans, by whom he was imprudently followed. As the latter entered the valley at night-fall, and at break of day on beginning their march, they were assailed on all sides with tremendous fury. The disorder caused by this

unexpected attack was increased by the circumstance of a thick fog arising from the lake and concealing their enemies. But, notwithstanding they were thus, as it were, caught in a trap, the Romans displayed their accustomed bravery, and struggled, if not for victory, at least to sell their lives as dearly as possible. It is mentioned, as evincing the fury of the contest, and its all-absorbing interest, that a violent earthquake, which in great part overturned several towns of Italy, and otherwise committed great ravages, occurred during the heat of the fight without being noticed by any one! (Livy, lib. xxii., cap. 5.) In the end, however, the triumph of Hannibal was complete. The Romans left 15,000 men, including their consul, Flaminius, whose rashness had led them into the snare, dead on the field of battle; and, according to Polybius, they lost about the same number, taken prisoners. The loss of the Carthaginians did not exceed 1500 men. (See Livy, ubi supra, and the excellent account of Polybius, General History, lib. iii., cap. 8.)
In noticing the lake of Thrasymene, Byron has alluded to the incident of the earthquake as follows:

> "And such the storm of battle on this day,
> And such the frenzy, whose convulsion blinds
> To all save carnage, that, beneath the fray,
> An earthquake reel'd unheededly away!
> None felt stern Nature rocking at his feet,
> And yawning forth a grave for those who lay
> Upon their buckler for a winding-sheet;
> Such is the absorbing hate when warring nations meet!"
> Childe Harold, iv., 63.

THREE RIVERS, or TROIS RIVIERES, the third town of Lower Canada, cap. distr. of its own name, on the St. Lawrence, where it is joined by the St. Maurice, 66 m. S.W. Quebec, and 75 m. N.E. Montreal; lat. 46° 23' N., long. 72° 29' W. Pop., estimated by M'Gregor at between 4000 and 5000. It derives its name from two small islands at the mouth of the St. Maurice, which divide it into three channels, but the town is on the W. bank of that river. The situation is agreeable, though not the town itself, which is one of the oldest in Canada. It contains about 490 dwelling houses, mostly built of wood, a handsome courthouse, a strong jail, a decent looking Catholic and a Protestant church, an Ursuline convent, founded in 1677, &c. The river is deep near the town, and the steamers stop to take on board passengers and fuel. Here the courts of justice for the district are held; and here, at one period, a great share of the fur trade centred. Some furs are still brought down by the Indians, and purchased by the agents of the Hudson's Bay Company; and there are a few breweries, potash factories, inns, shops, a printing-office, &c. But its general trade has been mostly absorbed by Montreal and Quebec. (M'Gregor's British America, &c.)
Trois Rivières was of much more importance formerly than at present, having been originally the capital of Canada. (Encyc. America.) Its population is still principally French, and the names of its streets are all traceable to Paris. It sends two members to the house of assembly.

THURGAU, or THURGOVIA, a canton of Switzerland, in the N.E. part of the confed., between lat. 47° 20' and 47° 40' N., and long. 8° 40' and 9° 30' E.; having S. St. Gall, W. Zurich and Schaffhausen, and N. and E. the Rhine and the lake of Constance. Area estimated at 268 sq. m. Pop., in 1837, 84,124; this being, with the exception of Geneva, the most densely peopled of the Swiss cantons. Thurgau, though it cannot be called mountainous, has a very uneven surface, consisting of low hills interspersed with narrow valleys. The canton derives its name from the Thur, which traverses it about its centre; next to which, the principal rivers are the Murg and Sitter. Agriculture is the principal occupation of the inhabitants, and though the soil in certain parts requires a great deal of manure, Thurga may, on the whole, be considered one of the most fertile cantons of the confed. There are extensive vineyards over nearly half the canton, and the value of the produce of wine, in average years, is estimated at £80,000 sterling. The internal consumption may be about a fourth part of the whole quantity, the remaining three fourths being exported to St. Gall and Appenzell. Considerable quantities of fruit and cider are exported in the same directions, but the quantity sent to Germany has of late years greatly diminished. About 100,000 hectols. of wheat are raised annually, being about two thirds the consumption. Oats, hemp, flax, potatoes, and hops are the other articles chiefly raised; oats and brandy are among the exports to Appenzell. The breeding of cattle is unimportant; but a large proportion of the S. part of the canton consists of fine pasture-land, and lean cattle being imported and fattened, are subsequently exported to the neighbouring states. The possessors of landed property would seem to be in a very distressed state; for the president of the canton is said to have stated, in 1836, that its estates were mortgaged to neighbouring cantons to the amount of £980,000 sterling, on which an annual interest of £40,000 was paid. (Bowring's Report.) The wages of agricultural labour are from 4½d. to 7¼d. per day, the mater providing food, which consists of oat cakes, barley, and sometimes wheaten bread, potatoes, &c. Among the classes engaged in manufactures, the use of coffee is being introduced. The peasantry eat meat once or twice a week: their ordinary beverage is wine or cider, though beer is also brewed in the canton. The male peasantry generally dress in woollens, the females in light cotton stuffs.

The commercial interests of the canton are said to have been injured by the Prussian league; at all events the linen and cotton goods sent into Germany, are much less now than formerly. Nearly one third part of the inhabitants are more or less engaged in manufacturing labour, principally in weaving cotton and linen fabrics, and spinning flax. The manufacture of linen is, however, declining, and its annual value is not now supposed to exceed £8000 a year. From 3000 to 5000 looms are employed in weaving cotton goods, the chief dépôts for which are St. Gall and Zurich. Weavers' wages range from about 7d. to 1s. 8d. a day. There are some rather extensive establishments for cotton printing, the prints being sent chiefly to the Levant. Silks are manufactured for the French markets; and a good deal of the packing canvass sold in Basle, Zurich, &c., is made in Thurgau.

The government, revised in 1831, is democratic. The great council of 100 members, which has the sole legislative power, consists of the representatives of the 32 circles, into which the canton is divided, elected by all the citizens above 25 years of age who pay taxes on property to the value of 200 florins, and are not paupers, or otherwise disqualified. The great council assembles twice a year, for 15 days at a time, unless its sessions be prolonged on special accounts; it is wholly renewed every two years, half the members going out yearly. The executive duties are intrusted to a council of six members, who must be 30 years of age, and who hold office for six years. Two landammans are chosen annually, and preside for six months alternately in the great and little council. Each commune had its own council, composed of the syndic, or mayor, and four other members, and its police and petty civil tribunal. There are courts of original jurisdiction in each of the eight districts of the canton, and a supreme court of appeal, in Frauenfeld, the capital. In 1837, about 72,000 of the inhabitants were Protestants, and 18,500 Roman Catholics. Public education is very widely diffused. Public revenue, in 1834, 157,920 florins; expenditure, 107,990 do.

As early as the 5th century Thurgau was governed by its own counts. It afterwards passed to the dukes of Zaehringen, and the counts of Kyburg; and, in 1264, to the house of Hapsburg. In 1460, it was conquered from the latter by the confederated Swiss cantons, and governed by their bailiffs or prefects till 1798. It was then constituted a separate member of the confed., in which it now holds the 17th place, furnishing a contingent of 152 men to the confederate army. (Pupikofer, Der Kant. Thurgau; Bowring's Rep. on Switzerland; Picot Statist. de la Suisse; Ebel, &c.)

THURLES, an inland town of Ireland, prov. Munster, co. Tipperary, on the Suir, 77 m. S.W. Dublin. Pop., in 1831, 7054. The river intersects the town, the communication between its different parts being maintained by a bridge. The public buildings comprise a fine Roman Catholic chapel, which serves for the cathedral of the see of Cashel, a Roman Catholic college, two nunneries, with chapels annexed, a market-house, a courthouse and bridewell, and a barrack. Mr. Inglis speaks favourably of Thurles. "It cuts," says he, "some figure at a distance, owing to the new and very handsome Roman Catholic chapel, and the unfinished Roman Catholic college. The town stands on a wide, scantily wooded, uninteresting plain. It contains about 7000 inhabitants, and is tolerably prosperous; for, having no larger town nearer to it than 40 or 50 m., it supplies an extensive interior district, and is besides an important market for country produce. There are no fewer than 15 annual fairs and two weekly markets held at Thurles. I saw scarcely any beggars in the place, and the cabins in the outskirts were not of the worst kind. There are two nunneries; in one of which are 20 nuns and 50 boarders: there is also an excellent Roman Catholic school belonging to the Roman Catholic institution. I observed no shops to be let, and saw several houses in the course of being built. Land lets very high in the neighbourhood." (i., p. 107.)

General sessions are held twice a year; petty sessions on Saturdays; it is a constabulary station. Markets on Tuesdays and Saturdays. Fairs on Easter Monday, August 21, December 21, and first Monday of every month. Postoffice revenue, in 1830, £452; in 1836, £560. Branches of the National and Agricultural banks were opened in 1835 and 1836.

THURSO, a seaport town of Scotland, N. shore, co. Caithness, on the Pentland frith, at the bottom of Thurso bay, between Dwarrick Head on the E., and Holburn Head on the W., at the point where Thurso river (here crossed by a handsome bridge) falls into the bay, 8½ m. S.W. Dunnett

Head. Pop., in 1841, about 2500. It is irregularly built, and rather ill paved; but in the suburbs are some neat freestone houses, and the church, built, in 1832, at an expense of £6000, is a handsome structure; it has also a meeting-house for original seceders, an independent chapel, and several schools. A short way to the E. is Thurso castle, the seat of Sir George Sinclair, Bart., proprietor of the town. Though the most northerly post town of Great Britain, it has a daily communication by a mail coach with Inverness and the south, and it communicates by regular traders and steamers with Leith, Wick, &c. There is a harbour at the mouth of the river for the accommodation of vessels drawing 12 feet water; and ships of any burden may anchor on the W. side of the bay, in Scrabster roads, under cover of Holburn Head. At present (1842) about 15 vessels belong to the port. A little straw-plait is manufactured in the town; and it has also a rope-walk and some tanneries. The town has three magistrates, appointed by the Sinclair family, the superiors of the borough. On the whole, the town may be regarded as in a nearly stationary state; and its progress has been by no means commensurate with the sanguine expectations of its late proprietor, the well known Sir John Sinclair. It is obvious, indeed, that the poor and thinly peopled country in its vicinity, and the nature of the remote and dangerous frith on which it is placed, are all but insuperable obstacles to its ever becoming of any considerable importance. (New Statistical Account of Scotland, art. Thurso.)

TIBER (an. Tibris, Ital. Tevere), the most celebrated though not the largest river of Italy, rises in the Tuscan Apennines, about 5 m. N. Pieve San Stefano, and has a general S.S.E. course to within 20 m. from Rome, where it turns S.W., and enters the Mediterranean by two mouths, 17 m. below that city, after a course of about 150 m. It is said to have been anciently navigable for vessels of considerable burden as far as Rome, and for small boats to within a short distance of its source (Dion. Hal., lib. 44; Strab., v., 219); and it still continues to be navigable, in certain seasons, as far as the confluence of the Nera; 38 m. N.N.E. Rome; but its navigation is at all times difficult, especially at its embouchure, and in the vicinity of Rome, and requires continual attention. The entrance of the river from the sea, and its subsequent navigation, are, in fact, so troublesome, that the harbour of Ostia, at its mouth, was relinquished in antiquity for that of Centum-cellæ, now Civita Vecchia, which still continues to be the port of Rome, though it be considerably more than twice the distance of Ostia from the city, with which it is connected merely by a road. (See the articles CIVITA VECCHIA and OSTIA in this work.) Its principal tributaries are the Topino, Nera, and Teverone from the E., and the Nestore, Chiana, and Xepl, from the N. and W. By the Chiana, it communicates with the Arno. Besides Rome, Borgo San Sepolcro, Città del Castello, Fratte, Orte, Otricoli, Magliano, and Ostia are on its banks, and Perugia and Orvieto in its immediate vicinity. In antiquity the Tiber divided Etruria from Umbria, and the territories of the Latins and Sabines: at present it separates the delegs. of Spoleto and Rieti from the deleg. of Viterbo and the Comarca di Roma.

Notwithstanding its immortality of renown, its banks are not generally picturesque, and at first sight it generally disappoints strangers. It is muddy, and during the floods to which it is very subject, verifies the description of Horace:

"Vidimus flavum Tiberim, retortis
Litore Etrusco violenter undis;
Ire dejectum monumenta regis."   Od. i. 2.

But at other times it flows with a comparatively gentle current;

"Ad terram Hesperiam veniet, ubi Lydius arva,
inter opima virum, leni fluit agmine Tibris."
     Æneid, ii. v. 781.

It was anciently called Albula, and this name, as well as the epithet flavus, given it by Horace, and other writers, was no doubt derived from the yellowish hue of its waters, discoloured by the mud with which they are loaded. "Some travellers, measuring its mass of waters by its bulk of fame, and finding its appearance inferior to their preconceptions, have represented it as a petty and insignificant stream. However, though far inferior in breadth to all the great rivers, yet as it is generally, from a few miles above Rome to the sea, about 300 ft. wide upon an average,[*] it cannot, with justice, be considered a contemptible rill." (Eustace, ii. 296, 8vo ed.) And a much higher authority than Eustace, Mr. Maclaren, says, that though the Tiber at Rome be not so wide as the Clyde at Glasgow, "it is deeper, and has certainly a larger volume of water." (Notes of a Tour, 138.) "Above and below the city it runs through groves and gardens, and waters the villas and retreats of the richer Ro-

[*] Mr. Matthews (Diary, &c., p. 78,) says, at the Ponte S. Angelo, at Rome, the breadth is about 312 English feet. This is the narrowest part. At the Ponte Molle, the breadth increases; and 2 m. above Rome, the Tiber is nearly twice as broad as within the walls.

mans: but beyond *Ponte Molle* it rolls through a long tract of plains and hills, fertile and green, but uncultivated and deserted. Yet these very banks, now all silence and solitude, were once, like those of the Thames, covered with life, activity and rural beauty, lined with villages, and not unfrequently decorated with palaces. *Pluribus prope solus quam ceteri in omnibus terris amnes accolitur aspiciturque vallis.* (*Plin. Hist. Nat.* lib. iii. 5.) Below the city, when it has passed the *Villa Malliana*, it falls again into a wilderness." (*Eustace, ubi supra.*) It terminates in a marshy pestiferous tract, its two arms inclosing the *Isola Sacra* of the ancients. (See *ante*, ROME, p. 703; *Cramer's Anc. Italy*, i. 240; *Matthew's; Dict. Geog.,&c.*)

TICONDEROGA, p. t., Essex co., N. Y., 97 m. N. Albany, 469 W. Bounded E. by lake Champlain., It contains the ruins of old fort Ticonderoga, and a convenient steamboat landing in its S.E. part. The village is situated on the outlet of lake George, which affords great water-power. The view of lake Champlain from the ruins of the old fort, is highly picturesque. A ferry crosses the lake to Shoreham, Vt. The town contains two churches, a Congregational and Baptist, eight stores, one fulling-mill, one woollen factory, one furnace, one forge, two grist-mills, 21 saw mills; one academy, 50 students, 10 schools, 464 scholars. Pop. 2169.

TIERRA (vulg. TERRA) DEL FUEGO, " *The land of fire*," so called from its apparent volcanoes by its discoverer fagellan, or Magalhaens, a large island, or rather group of lands, lying of the S. extremity of South America, from which it is separated by the strait of Magellan. The group which extends between the 53d and 56th degs. of S. lat., and the 64th and 75th of W. long., consists of King Charles's, south Land, Navarin, Hoste, Clarence, and some other lands, cape Horn forming the most S. point. The E. part King Charles' South Land, is low, with plains like those Patagonia; but near its W. side it is traversed by mountain chains nearly 4000 ft. in height, covered with perpetual snow. Slate is abundant, but hornblende is said to be the prevailing rock here and in all the adjacent islands. (*King Geog. Journ.*, vol. l.) Lava and other volcanic products ve been found. The country, in many parts seems well oded, and Winter's bark (*Drimys winteri*), introduced a medicine in 1579, was discovered here. A kind of birch *titula antarctica*), with a stem from 30 to 46 inches in diameter, is one of the principal forest trees, and probably er trees may be found; but the interior has hitherto n very little explored. Guanacoes and foxes appear to be the most numerous wild animals. The Fuegians are a ruler race of savages, in nearly the lowest stage of barbarism. Their most striking physical peculiarities are a small low forehead, prominent brows, small eyes, wide ris, large mouth, thick lips, long black hair, and large r, as compared with the extremities. They go nearly ed, but smear over their bodies with various substances; in wigwams made of the trunks of trees, and subsist almost wholly on fish, seals, and testaces. They are occasionally cannibals, and have, in fact, no objection to any of food. They are not wholly ignorant of the arts, being acquainted with the use of fire, and availing themselves ws and arrows, and in the N. of the *bolas* of the Patagon. On the whole, however, they would appear to be ledly below many of the lower animals in respect of ort, and to be but little above them in sagacity and inon. Their language is said to present many affinities the Araucanian. (For further and numerous particulars ve refer the reader to the *Voyages of the Adventure* and *la*; *Cook*; *Weddell, &c.*)

FIN, p. v., Clinton t., capital of Seneca co., O., 86 m. lumbus, 423 W. Situated on the E. side of Sandusky contains a courthouse, jail, several stores, a printing-issuing a weekly newspaper, and about 600 inhabitants. IN, t., Adams co., O., 106 m. S. Columbus, 437 W. sains West Union v., the capital of the co., and has hools, 52 scholars. Pop. 1540.

RIS. *See* EUPHRATES.

BURG a town of Holland, prov. N. Brabant, cap. in the arrond. of Bois-le-Duc, near the Ley, 13¼ m. Bois-le-Duc. Pop. about 11,000. This is decidedly st built town in the prov., though, from lying out of eat road, it is little visited by travellers. It has three ies, a chapel, and a handsome castle; and has very ve fabrics of fine and coarse woollen cloths, cassimere &c. It sends three deputies to the provincial states.

3IT', a town of the Prussian states, prov. Prussia. reg. nnen, cap. circ. on the Niemen, or Memel, where it d by the Tilse, 60 m. N.E. Konigsberg. Pop., in 1838, It consists principally of a long and wide street, few good-looking houses. The Niemen, which is ue up to the town, is here crossed by a bridge of 1150 ft. in length. The exports consist of timber, amp, flax, provisions, wax, leather, &c., sent down ar in flat-bottomed boats, for shipment at Memel.

The cutting of the canal of Oginsky has, by uniting the Niemen with the Dniepr, effected a communication between the Baltic and the Black sea. It has an old castle, several churches, a royal gymnasium, hospital, and board of taxation, with manufactures of woollen cloth, hosiery, gloves, leather, hardware, &c.

This town is famous in diplomatic history for the treaty signed here on the 7th of July, 1807, by France, Russia, and Prussia. The conferences that led to this treaty were held between Napoleon and Alexander, who met, for the first time, with great pomp and ceremony, on a raft moored in the middle of the Niemen, on the 25th of June.

TIMOR (THE EAST), an island of the E. archipelago, 2d division (*Crawfurd*), principally belonging to the Dutch, between the 123d and 128th degrees of S. lat., and the 13th and 15th degs. of E. long., 100 m. S.E. Floris, and 360 m. from the N.W. coast of Australia. It extends obliquely from N.E. to S.W., its length being estimated at 250 m., and its average breadth at 35 m. Horschelmänn, in his edition of *Stein's Handbuch*, estimates its area at about 8900 sq. m., and its population at 800,000. The natives of the interior are Papuan negroes; the coasts are inhabited mostly by Malays, Chinese, Dutch, and Portuguese, the latter possessing the town of Dilli, on the N.E. side of the island. Surface mountainous, but without volcanoes. Its rivers are small; and the soil is, upon the whole, not particularly fertile. Sandal-wood and wax are the most valuable products; about 10,000 cwt. a year of the first were formerly exported to Java. The natives cultivate rice, maize, millet, yams, sweet potatoes, and cotton; rice and maize, with the sugar of the lontar palm, and sago, are their principal articles of food. Gold is found both in grains and large pieces; but the aborigines are said to have a strong aversion to search for it, and once massacred a party of Dutch, sent inland to collect the metal. The imports are rice, arrack, sugar, tea, coffee, betel nut, and Chinese, Indian, and European manufactures; the duties on the trade have been said to suffice for the keeping up of the Dutch establishments on the island. The Dutch fixed themselves at Coepang, on the S.W. coast, in 1630; but we learn, from recent accounts, that they have now all but abandoned Timor for Sandalwood island, about 200 m. more to the W., which abounds with fertile and grassy plains; and where the colonists are much less likely to be disturbed by the hostility of the natives. About 300 m. N.E. Timor is Timor Laut, an island 70 m. in length, by about 25 m. average breadth. Little, however, is known as to the state of this island. (*Hamilton's E. I. Gazetteer, &c.*)

TIMBUCTOO, or TOMBUCTOO, a town of central Africa, on the S. border of the great desert of Sahara, about 8 m. N. from the Joliba, or Niger, but near one of its arms or tributaries, in about lat. 17° 50' N., long. 3° 40' W. Stationary pop., probably 10,000 or 12,000. The existence of this city has been long ascertained: but as till recently it was only known to Europeans by vague reports and suspicious narratives, the most conflicting and contradictory reports have been made respecting it. Certainly, however, the *grestuas pro magnifico* has seldom been more strikingly exemplified than in this instance; the most exaggerated and unfounded statements having been put forth respecting its magnitude, commerce, and population. These, however, have now been completely dispelled, and Timbuctoo is ascertained to be a very poor town in a wretched country.

It is, in fact, situated amid burning and moving sands on the verge of a morass. It is of a triangular form, about 3 m. in circuit, and is surrounded by walls. The better sort of houses, built of bricks dried in the sun, are large, but not high, consisting entirely of a ground floor. Each house forms a square, containing two inner courts, round which are ranged the chambers, without windows or chimneys, and serving at once for magazines and bed-rooms. But within and without the town are many straw huts of a circular form, serving as lodgings for the poor and for slaves, who sell merchandise for their masters. The streets are said to be clean, and sufficiently wide to allow three horsemen to pass abreast. It has seven mosques, two of the largest of which have walls about 15 ft. in height, and are each surmounted by a tower. It is chiefly inhabited by negroes of the Kissour nation; but it is also the residence of a considerable number of Moors, who carry on the trade of the town, and who leave it as soon as they have accumulated a little property. The inhabitants are zealous Mohammedans.

The trade of Timbuctoo, though much exaggerated, is considerable, it being a station for the caravans between N. Africa and the Soudan, or Nigritia, and also a *dépôt* for their produce. Salt is, however, the staple merchandise of the place. This important article, which is wholly wanting in Soudan, is brought from the mines of Towdeyni, in the desert, about 335 m. N. from the town, being conveyed thither in the form of cakes on the backs of camels. In addition to salt, the caravans from the Barbary States bring dates, stuffs of European manufacture, with fire-arms, gunpowder, hardware, glass ware, coral, tobacco, paper, and other articles, which they exchange for slaves, gold-dust, ivory, ostrich-feathers, palm-oil, gums, &c. Owing to the sterility of the surrounding country, all the provisions required for the use of the town, have to be bought from Jenné, on the Niger about 300 m. S.S.W. Timbuctoo. These are conveyed by an arm of the river to Cabra, whence they are carried by camels, about 3 m. to the town. Jenné, according to Caillié, is a more important, richer, and more commercial town than Timbuctoo. The Touariks, a warlike and savage tribe, on the banks of the Niger, exact heavy duties on all the commodities imported into Cabra, and occasionally commit extensive outrages. Timbuctoo is governed by a negro governor or prince; who receives presents, but imposes no duties either on the inhabitants or the products brought to the town. The government is, in fact, patriarchal, and the prince is said, by Caillié, to be mild and just. The slaves, of whom there are great numbers, are well treated.

Timbuctoo is said to have been founded A.D. 1213, and to have soon after become the capital of a great Moorish monarchy (Walckaaer, Recherches Géographiques, p. 14); and since it ceased to enjoy this distinction, its trade, as well as its importance, is believed to have greatly declined. But it is not at all likely that a town in such a situation should ever have been the capital of any considerable state; and we are inclined to think that the accounts of its ancient have but little better foundation than those of its modern prosperity. We have elsewhere endeavoured to show, that the commerce of which Timbuctoo is one of the centres depends on natural causes; and that it will, most probably, continue to be carried on in time to come in the same way in which it has been carried on from the remotest period down to the present day. (See vol. i. 35.) Ritter, who has collected and discussed the different accounts of Timbuctoo published previously to that of Caillié, has exaggerated alike its importance and its trade, and has farther indulged in some rather fanciful speculations as to the increase of its trade. (Geography of Africa, Fr. trans., ii. 81–112.) It would, indeed, be easy to show that the barbarism of Africa depends on natural and permanent, and not on artificial or accidental circumstances; and though its commerce and civilization may, no doubt, be materially increased in the course of time, the fair presumption seems to be that, owing to the nature of the country and climate, the wants of the natives and their industry will always be much too limited to admit of their ever becoming extensive demanders of European products. (See Caillié's Travels to Timbuctoo, ii. 49–78, Eng. trans. This is not only the most recent but by far the most authentic and best account of the town that has ever been published. See also, Walckaaer, Recherches sur l'Afrique Septentrionale, passim, for an account of the progress of discovery in this part of Africa in modern times.)

TINIAN, one of the Ladrone islands, which see.

TINICUM, t. Bucks co., Pa., 37 m. N. by E. Philadelphia, 12 m. N.E. Doylestown. Bounded N.E. by Delaware river, Watered by Tinicum creek. The Delaware canal passes through it. It contains two churches, five stores, five grist-mills, five saw-mills, three tanneries; four schools, 94 scholars. Pop. 1770.

TINNEVELLY, a district of British India, presid. Madras, at the E. extremity of Hindostan, between lat. 8° and 10° N., and principally between long. 77° and 78° E., having N. the district of Madura; E. and S.E. the gulf of Manaar, separating it from Ceylon; and S.W. and W. Travancore, from which it is divided by a chain of mountains. Area, 5590 sq. m. Pop., in 1836–37, 807,964. "The face of the country is a little undulated, but the general appearance is that of an extensive plain interspersed with small hills. The lower parts are well supplied with tanks, and afford great quantities of rice. On the banks of the rivers are also numerous paddy fields. There are several flats that run to a great distance, on which abundance of cotton is produced; the higher grounds are well cultivated, and covered in the season with luxuriant crops of dry grain. S. of Palmacottah, towards the extremity of the peninsula, the country becomes sandy and bare, covered in general with Palmyra tops. Towards the E. coast, and all round to the S., there are several hills of red sand, with which the atmosphere is often darkened during the windy season. Close to the sea beach, all along from Tutucorin to cape Comorin, the small villages are inhabited by fishermen, who are all Christians, and several Roman Catholic churches are situated close to the sea." (Madras Almanac.)

Tinnevelly is intersected by many winding rivers, which are supplied with water by both monsoons. The climate of some parts is remarkable. In the N. it is similar to that of Madura, but in the mountains on its W. side are several openings or passes, which, while the rest of the country on the E. side of India is parched up with heat, admit the cool winds prevailing at that period on the Malabar coast. The

chief of these is the Arungole pass, near which is Kotalam, a place of great resort for Europeans, on account of its bracing climate.

Rice and cotton are the chief products of this district, the last, which is of the Bourbon variety, is grown of a superior quality. Many fruits, roots, and greens are cultivated, but some of the most common Carnatic products are neglected, and in unfavourable seasons rice is imported from Travancore. While Ceylon belonged to the Dutch, an attempt was made to establish spice plantations in Tinnevelly, and cinnamon, nutmegs, &c., were planted, but on our acquiring possession of Ceylon these attempts were abandoned. Tinnevelly is assessed under the village system, but for a lengthened period great irregularities attended both the assessment and collection of the revenue. (See Rep. on E. I. Affairs, 1832, App. Revenue, p. 479; Hamilton's E. I. Gaz.) Great improvements, have, however, been effected in these branches of late years, and in 1837–8, the total revenue of the district amounted to 24,15,105 rupees.

Tinnevelly is subdivided into 11 taloohs or circles. Its chief towns are, Tinnevelly, the capital and residence of the collector and judge, in about lat. 8° 46' N., long. 78° 7' E.; and Palmacottah, the head military station, about 5 m. to the E., where a good many long cloths, silks, muslins, &c., are made for exportation to Madras and elsewhere. Iron is forged and saltpetre obtained in many parts of the district. The inhabitants of Tinnevelly appear to live in greater comfort than those of the neighbouring districts, and their dwellings are mostly well constructed. Mohammedans are few, and the primitive Hindoo manners and customs are scarcely anywhere seen in greater purity. (Madras New Almanac; Parl. Reps., &c.)

TIOGA, county, N. Y. Situated in the S. part of the state, and contains 490 sq. m. Watered by Susquehanna river, down which lumber is extensively floated. The line of the New-York and Erie railroad passes through it, along the Susquehanna r. It contained, in 1840, 41,576 neat cattle, 43,230 sheep, 14,967 swine, and produced 167,392 bushels of wheat, 4967 of rye, 117,449 of Indian corn, 47,181 of buckwheat, 1058 of barley, 180,967 of oats, 289,193 of potatoes, 6876 pounds of hops, 116,760 of sugar. It has 73 stores, 13 fulling-mills, three woollen factories, 37 grist-mills, 259 saw-mills, two oil mills, 15 tanneries, two distilleries, two printing-offices, two weekly newspapers; one academy, 908 students, 6814 scholars in schools. Pop. 29,827. Capital Owego.

TIOGA, county, Pa. Situated in the N. part of the state, and contains 1200 sq. m. Drained by Tioga r. and branches of Pine cr. It contained, in 1840, 27,443 neat cattle, 25,430 sheep, 11,905 swine, and produced 98,623 bushels of wheat, 33,694 of rye, 66,610 of Indian corn, 22,874 of buckwheat, 110,526 of oats, 289,590 of potatoes, 181,064 pounds of sugar. It had 52 stores, four fulling-mills, two woollen factories, one flouring-mill, 26 grist-mills, 145 saw-mills, one oil-mill, 13 tanneries, two distilleries, two printing-offices, two weekly newspapers; one academy, 76 students; six schools. 2156 scholars. Pop. 15,498. Capital, Wellsborg.

TIOGA, p. t., Tioga co., N. Y. 176 m. W.S.W. Albany, 277 W. Bounded S.E. by Susquehanna river, and watered by its tributaries. It contains four stores, three-fulling-mills, one woollen factory, four grist-mills, 42 saw-mills; 9 scholars in schools. Pop. 9464.

TIPERAH, a district of British India, presid. Bengal, between lat. 23° and 24° 30' N., and the 91st and 93d degs. of E. long., having N. Sylhet, E. the Muneepoor territories, S. Chittagong and the sea, and W. the Brahmaputra, separating it from the district of Dacca. Area, 6830 sq. m. Pop., in 1822, 1,372,980. This district yields cotton, rice, and betel-nut of a very superior quality. Elephants of large size are found in the forests; and in the S. salt is manufactured. The coarse cotton goods, baftaes, commaes, &c., made here are durable and substantial, and were formerly exported in large quantities by the East India Company and by private merchants. The inhabitants are similar in most respects to those of the adjacent districts beyond the Brahmaputra, though the upper classes have adopted many Hindoo usages. In respect of public education Tiperah appears to be extremely backward. It was acquired by the British in 1765. Total land revenue (1899–30) 816,417 rupees. (Parl. Reports, &c.)

TIPPAH, county, Miss. Situated in the N. toward the E. part of the state, and contains 1008 sq. m. Drained by Hatchie, Muddy, Wolf, and Tippah creeks, and by head branches of Tallahatchee r. It contained, in 1840, 15,292 neat cattle, 5034 sheep, 36,904 swine; and produced 15,782 bushels of wheat, 395,160 of Indian corn, 9128 of oats, 79,045 of potatoes, 1,061,768 pounds of cotton. It had 17 stores, one flouring-mill, 18 grist-mills, 14 saw-mills, five reserves, two distilleries; one academy, 40 students; 11 schools, 367 scholars. Pop. whites, 7300; slaves, 2134; free coloured, 1; total, 9444. Capital, Ripley.

TIPPECANOE, county, Ia. Situated N.W. of the centre of the state, and contains 504 sq. m.   Watered by Wabash r. and its tributaries.   It contained, in 1840, 15,361 neat cattle, 18,197 sheep, 44,031 swine; and produced 137,943 bushels of wheat, 7586 of rye, 990,160 of Indian corn, 1842 of buckwheat, 158,353 of oats, 27,648 of potatoes, 67,110 pounds of sugar. It had 35 stores, two woollen factories, one cotton factory, with 200 spindles, two fulling-mills, 12 flouring-mills, six grist-mills, 25 saw-mills, one oil-mill, 10 tanneries, 14 distilleries, two breweries, two printing-offices, two weekly newspapers; two academies, 50 students; 24 schools, 589 scholars.   Pop. 13,724.   Capital La Fayette.

TIPPECANOE, p. t., Tippecanoe co., Ia., 82 m. N.W. Indianapolis, 640 W.   It has three schools, 79 scholars.   Pop. 1374.

TIPPERARY, an inland co. of Ireland, prov. Munster, having N. the Shannon, by which it is separated from Galway; E. King's county, Queen's county, and Kilkenny; S. Waterford; and W. Cork and Limerick.   Area, 1,013,173 acres, of which 182,147 are unimproved mountain and bog, and 11,328 water.   The mountainous districts are in the S.W.   Adjoining Waterford and Cork, in the S.E. angle, is Sliebhnaman mountain : and a chain of mountains runs across the county from Limerick to King's county.   The bog is mostly a portion of the great bog of Allen.   With these exceptions, Tipperary consists principally of extensive and fertile plains, with a calcareous subsoil, forming a rich land as is to be met with in any part of the empire.   Some very large estates, but many of a moderate size.   Tillage farms generally small, and mostly held under middle-men.   The con-acre system is very prevalent in some parts of this county, as it is indeed in most counties of Ireland, though it be carried to the greatest extent in Connaught.   By con-acre is meant a pernicious custom prevalent among the landlords and occupiers of the larger class of farms, of letting to the peasantry, or cottiers, small slips of land varying from a perch to half an acre, for a single season, to be planted with potatoes or cropped.   Old grass-land is frequently let out on this system ; and then it is usual to allow the surface to be pared and burned !   The rent of this land s enormous, running from £7 to £19 or £13 an acre !   Potatoes are invariably planted on con-acre land when it is broken up from grass ; and afterwards it is usual to take corn it successive crops of corn.   Wherever this practice xists, there cannot, of course, be the least improvement ; nd nothing but the extraordinary fertility of the soil could nable it to produce anything under so destructive a system.   But, despite the prevalence of con-acre, some considerable nprovements have been effected of late years, in the introduction of improved implements and improved stock, the extension of green crops, &c.   Grazing, however, was formerly, and still is, the principal employment in Tipperary he native Irish breed of long-horned cattle attain to a very large size, and are found in the greatest perfection in this unty.   There are also many fine flocks of long-woolled sheep.   Average nt of land, 17s. 8½d. an acre.   Unfortunately, the condi-on of the peasantry, instead of being improved with the provements that are admitted to have taken place in ag-ulture is, on the contrary, more depressed now than at v former period ; and, in consequence, they are extreme-turbulent, and agrarian outrages are probably more fro-ent in this than in any other Irish county.   The manu-ture of broadcloth was formerly carried on to some ex-t at Carrick, but is now wholly relinquished.   Tipperary has copper and lead mines, coal, slate, &c.   Exclusive he Shannon, the principal river is the Suir.   It contains aronies and 186 pars., and returns four members to the se of Commons ; two being for the county, and one each the bors. of Clonmel and Cashel.   Registered electors the co., in 1839-40. 4143.   In 1831, Tipperary had 60,264 abited houses; 66,309 families; and 409,563 inhabitants, whom 197,713 were males, and 204,850 females.

IPPERARY, an inland town of Ireland, prov. Munster, of the above co. near the Arra, an affluent of the Suir,). S.E. Limerick.   Pop., in 1831, 6972.   It is well paved cleaned under the superintendence of commissioners, is, for an Irish town, pretty well built.   "Tipperary," Inglis, "is most agreeably situated, in a fine undulating ng country, and within a few miles of a beautiful range ills, which divides the counties of Tipperary and Lim-.   Tipperary though inconsiderable in size to bear the ? of the county, is rather a flourishing town ; and is : a mercantile traveller would call 'a good little town.' e is no town westward nearer than Limerick, and there nsequently, a busy retail trade, the result of country s.   There is also a good weekly market, which makes rary the depôt of agricultural produce for a range of 15 m. round.   But, notwithstanding the better circum-es of the tradesmen, I found the condition of the la-ng classes little better than elsewhere.   Wages were 4d. a day, without diet, and there is nothing approach-constant employment for the population.   Enormous

rents, varying from £2 10s. to £4, are paid for wretched cabins in the suburbs." (I. p. 190.)   Tipperary has a parish church, a Roman Catholic chapel, a Methodist meeting-house, schools on the foundation of E. Smith, a market-house, a barrack, and a bridewell.   Petty sessions are held on Thursdays.   A chief police magistrate resides here.   Markets on Thursdays and Saturdays.   Fairs, April 5, June 24, Oct. 10, and Dec. 10.   Postoffice revenue, in 1830, £726 ; in 1836, £1066.   Branches of the National and Agricultural banks were opened in 1835.

TIPTON, co., Tenn.   Situated in the W. part of the state, and contains 415 sq. m.   Bounded W. by Mississippi river.   Watered by Hatchy river, and its branches, and by Forked Deer river.   It contained, in 1840, 9974 neat cattle, 1207 sheep, 21,851 swine ; and produced 12,819 bushels of wheat, 9493 of rye, 359,989 of Indian corn, 21,890 of oats, 96,935 of potatoes, 74,110 pounds of tobacco, 1,015,899 of cotton.   It had two commercial and two commission houses in foreign trade, three retail stores, two flouring-mills, two grist-mills, three saw-mills ; one academy, 55 students ; seven schools, 140 scholars.   Pop.: whites, 3637 ; slaves, 3128 ; free coloured, 31 ; total, 6800.   Capital, Covington.

TIRHOOT (native Tyrahhueti), a district of British India, presid. Bengal, prov. Bahar, between lat. 25° and 27° N., and long. 85° and 87° E. ; having N. Nepaul, E. Purneah, W. Sarun, and S. Bhangulpore and the Ganges, which separates it from the districts of Bahar and Patna.   Area, 7732 sq. m.   Pop., in 1822, 1,007,700.   The surface is undulating and well watered ; the climate is more healthy than that of the districts more to the S.   In the N. there are ex-

digo on the ground which had been previously given up to corn." (Evidence before the Lords, 1830.)   Tirhoot is one of the principal districts in India for the growth of indigo ; besides which, sugar, opium, tobacco, turmeric, ginger, rice, &c., are its chief vegetable products, and great quantities of saltpetre are procured from the soil.   It also supplies great numbers of cavalry and other horses.   Timber abounds in the N., but is of little utility from the absence of roads and the shallowness of the rivers.   Total land revenue, in 1829-30, 1,560,563 rupees.   Tirhoot appears to have formed an independent Hindoo principality till 1237.   It was annexed to the crown of Delhi in 1325, and acquired by the British in 1765.   (Hamilton's E. I. Gaz., &c.)

TIRLEMONT, (Flemish Thienen), a town of Belgium, prov. S. Brabant, arrond. Louvain, 11 m. S.E. from the city of that name.   Pop., in 1836, 7996.   Its extensive, but now dismantled walls, enclose a large extent of ground not built on, with a large square, in which is the ancient town-hall, church, &c.   It has manufactures of woollen cloths, flannel, and hosiery, oil, soap, earthenware, paper, and saddlery, with potteries, breweries, distilleries, &c.   It was formerly much more populous and thriving, having been one of the most important places in Brabant ; but being repeatedly taken and retaken by the Spaniards, French, and Dutch, in the 17th and 18th centuries, it suffered greatly in consequence.   It sends two deputies to the prov. states, and two others are sent by its districts.

TISBURY, t., Dukes co., Mass., 85 m. S.S.E. Boston, 23 m. S.E. New Bedford.   It occupies the central part of the island of Martha's Vineyard.   Holmes' Hole, in its N.E. part, has a good harbour, and a village which contains a Baptist and a Methodist church, and about 100 dwellings.   There are besides in the t. a Congregational and a Baptist church in the W. part.   It has six stores, one fulling-mill, one woollen factory, five grist-mills ; one academy, 40 students, six schools, 205 scholars.   Pop. 1520.

TISHAMINGO, co., Miss.   Situated in the N.E. corner of the state, and contains 1300 sq. m.   Bounded N.E. by Tennessee river.   Drained by Yellow and Tuscumbia creeks, and by head branches of Tombigbee river.   It contained, in 1840, 11,120 neat cattle, 2324 sheep, 22,371 swine, and produced 5130 bushels of wheat, 981,451 of Indian corn, 9293 of oats, 9108 of potatoes, 175,475 pounds of cotton.   It had 15 stores, six cotton factories, seven grist-mills, six oil-mills, three tanneries ; 14 schools, 301 scholars.   Pop.: whites, 5852 ; slaves, 828 ; free coloured, 1 ; total, 6681.   Capital, Jacinto.

TITCHFIELD, a market-town and par. of England, co. Hants, div. Portsdown, hund. Tichborne ; on the Titchfield river, near the mouth of Southampton Water, 7½ m. E.S.E. Southampton.   Area of par., 15,960 acres.   Pop., in 1831, 3712.   The town is small, but well built, and is the residence of many families of respectability.   A part of the church is said to have been built by William of Wykeham, in the 14th century, and other parts are still more ancient.

The living, a vicarage worth £966 a year, is in the gift of H. P. Delme, Esq. There is an Independent meeting-house and a charity school for 24 children. Near the Town are the remains of Titchfield House, in which Charles I. took refuge after his escape from Hampton Court, in 1647, built by the first earl of Southampton, on the site of a former Premonstratensian abbey; but the mansion is now nearly dilapidated, the entrance gateway and the stables being the only extant remains. Titchfield gives the title of marquis to the Bentinck family. Markets on Fridays. Fairs, March 9, May 14, Sept. 25, and Dec. 7.

TITICACA (LAKE OF). This lake, the largest and most elevated of the S. American continent, is partly comprised in the Bolivian republic, and partly in that of Peru: being enclosed by the Cordilleras S. of the table-land of Cuzco, and extending chiefly between lat. 15° and 17° S., and long. 69° and 71° W., about 135 m. S.E. Cuzco. Its outline is very irregular, being divided by a number of headlands into a main body, of an oblong form, and three subsidiary portions. Its area has been estimated by Mr. Pentland at 4000 sq. m., and its height above the ocean at 12,795 ft. ! It is said to be in many places nearly 500 ft. in depth. It contains many small mountainous islands, and the largest, at its S.E. extremity, the lake has received its name, which signifies "the Leaden Mountain." This island is 3 leagues in length by 1 in width, and about 1 mile from the shore. It is mostly uncultivated, but very fertile; and on it tradition places the first appearance of Manco Capac. (Pentland, in Geog. Journ., v. 89.) The island was consequently held in great veneration; a temple was erected on it by the succeeding incas, in which a great deal of wealth is said to have been accumulated; and this, it is alleged, was thrown into the lake by the Indians on the Spaniards becoming masters of the country. Mr. Pentland states that numerous Peruvian ruins are still found on the island.

The lake of Titicaca receives several rivers, its only visible outlet being the Desaguadero, which flows S., and is soon afterwards lost in the lake of Paria. Its waters, though not very potable, abound with fish, and sudden squalls and storms render its navigation rather dangerous. "The low banks of the lake are lined with rushes, which are here of great utility, being employed for almost as many purposes as the bamboo in the E. The huts of the poor are made of rushes; as also mats for the floor, and bed covers. The boats used on the lake are also made of rushes twisted together; the rudder and the mast only being of wood. These boats are frequently made with great taste and ingenuity; the larger venture to some distance from the banks of the lake, which, even in calm weather, is subject to a heavy swell." (Meyen, Reise um die Erde.; Mod. Trav., &c.)

TIVERTON, a parl. and mun. bor., town, and par. of England, co. Devon, hund. Tiverton, on the Exe, where it is joined by the Loman, 13 m. N. by E. Exeter, and 154 m. W.S.W. London. Area of par., which is identical with the parl. bor., about 20,000 acres. Pop., in 1831, 9766. The town is situated partially on the tongue of land between, and partly on the opposite banks of, the two rivers, each of which is crossed by a stone bridge. It is nearly 1 m. in length, N.E. to S.W., by about 5 fur. in its greatest breadth: it consists chiefly of several tolerably broad and well-paved streets, running N. and S. on both sides the Exe, and mostly joining Fore-street, the main thoroughfare, at right angles. The more narrow lanes and streets are B. of Fore-street.

Tiverton is lighted with gas, and watered by small streams from a branch of the Loman, called the Town leet. (Bound. Rep.) On an eminence between the rivers are the remains of the castle, a conspicuous object, occupying about an acre of ground. This fortress was erected in the reign of Stephen, and afterwards came into the possession of the Courtenays, earls of Devon: in the civil wars it was garrisoned by the royalists, but after a short siege was taken by Fairfax. The church, on an eminence near the castle, is reckoned the finest ecclesiastical edifice in the county, after Exeter cathedral. It has a tower 116 ft. in height, and is chiefly in the perpendicular style, though there are some remains of an earlier date, and an enriched Norman doorway. A chapel, built by a merchant named Greenway, in 1517, is a good example of the gorgeous style of ornament which then prevailed. The ceiling is coved, and has tracery and rich pendants; like many works of that time, the design is better than the execution. The church has some rich screenwork. (Rickman's Gothic Archit.) In the nave are some curious monumental inscriptions; and the pulpit, which was probably made about the time of Charles II., is ornamented with the arms of many Devonshire families painted in separate compartments. The altar-piece is a rich painting of Peter delivered from prison.

Tiverton consists of several ecclesiastical divisions. Clare, Pitt, and Tidcome quarters, or portions, are all rectories, in the joint patronage of the Earl of Harrowby, Sir W. Carew, Sir R. Vivian, and the Rev. W. Spurway; the nett value

of Clare rectory being £453, of Pitt with Cove £675, and of Tidcome £735 a year. A handsome modern chapel-of-ease has been erected, at which each of the portionists officiates in turn. There are various meeting houses, a spacious market-house built in 1830, a corn-market, town-hall, bridewell, assembly and subscription reading-rooms, a meat house, &c.

Tiverton has numerous charities. A free grammar-school, in an ancient and venerable stone edifice founded and endowed by a rich clothier of the town, named Blundell, in 1604, has now an income of nearly £700 a year. It furnishes instruction for 150 boys, and sends six students to either of the universities, and one to Balliol college, Oxford; and it has two exhibitions of £30 a year each, besides other scholarships. Another free school was founded by R. Chilcott, in 1611, and there are several minor schools of a similar kind. The almshouses for nine poor men, founded by Greenway in 1529, have an income of nearly £300, and the market trust money distributed to the poor by the corporation amounts to £336 a year. The aggregate income of the various charities of Tiverton is estimated by the charity commissioners at £2660 a year. (Digest of Charity Reps.) The manufacture of lace employs from 1300 to 1500 people. The town was at one period famous for its baizes, serges, plain cloths, kerseys, and other woollen goods, and even as late as 1612 was regarded as the head manufacturing town in the W. of England; but its manufacture received a severe blow from a most destructive fire, which occurred on the 5th of August, 1612, from which it never fully recovered, and the introduction of Norwich stuffs, in the middle of the last century, completed its decline. At present the woollen manufacture employs only a few hands. The town supplies an extensive rural district, but its trade is not said to be increasing. In 1834, a few small houses of £5 or £6 rent were the only class of buildings in progress in the town; and, on the whole, it may be said to be stationary. (Bound. Rep. and Mun. Corp. Append. 1.) Tiverton is now divided into three wards, and is governed by a mayor, five other aldermen, and 18 councillors. It has a commission of the peace, and a court of record for civil actions to the amount of £100 is held once a fortnight. A jail with six cells was built about 25 years since. The Boundary act made no change in the limits of the parl. and mun. bor., which, as already stated, are co-extensive with the par. Tiverton was incorporated by James I. who also conferred on it the privilege of sending two members to the House of Commons, the right of election, down to the Reform act, being vested in the corporation, consisting of 25 individuals. Reg. electors, in 1839-40, 495. Corp. revenue, in 1840, £610. Markets, Tuesdays and Saturdays, and a large cattle market on the second Tuesday in each month. Fairs, Tuesday after Whitsuntide and Sept. 29.

TIVERTON, p. t., Newport co., R. I., 13 m. N.E. Newport, 24 m. S.E. Providence, 420 W. Bounded W. by the E. passage of Narragansett bay, and by Mount Hope bay. A substantial stone bridge, 1000 ft. long, unites it with Newport. It is finely situated for navigation, and has a considerable number of vessels, employed chiefly in the fisheries. It contains a Baptist church, 14 stores, two woollen factories, six cotton factories with 1600 spindles, eight grist-mills, four saw-mills; 19 schools, 527 scholars. Pop. 3183.

TIVOLI (an. Tibur), a town of the Papal States, comm. of Rome, on a steep ridge, on the Teverone (an. Anio), 18 m. E. by N. Rome. Pop. 6000. In antiquity, Tibur was to Rome what Richmond is to London; but though in a magnificent and highly salubrious situation, the modern town is dirty and disagreeable, with narrow, steep, and ill-paved streets, and inferior houses. It has a cathedral and some other churches. Tibur or Tivoli, which is out of the most ancient cities of Italy, derives its entire interest from the classical associations connected with its ancient name, its scenery, and its remains of antiquity. The Teverone, come here to the edge of the cliffs that separate its valley from the Campagna, is precipitated downwards in a series of cascades, the beauty of which has been admired from the age of Horace down to the present time.

" Me nec tam patiens Lacedæmon,
Nec tam Laræ percussit campus opimæ,
Quam domus Albuneæ resonantis,
Et præceps Anio, et Tiburni lucus, et uda,
Mobilibus pomaria rivis."

Hor., lib. i. Od. vii., v. 10.

In modern times, the upper or principal fall was in a great degree, artificial, from the water having been precipitated over an embankment that had been formed to dam up the river. In 1826, however, one of those destructive floods that occasionally occurred in antiquity (Plin. Epist., lib. viii. 17), as well as in our own times, swept away the whole of this embankment, along with a church and some contiguous houses, so that the upper fall was nearly destroyed: not the same time that a branch of the river which ran through the town was dried up. But new channels have been since cut, by which, we believe, the river has been again precipitated down a lofty fall.

In the court-yard of an inn in the town, overhanging one of the cascades, is the classical ruin of a temple, supposed to be either that of the *Sibylla Tiburtina* or of *Vesta*, constructed in the reign of Augustus. This beautiful structure is a rotunda, 22 ft. 11 in. in diameter, surrounded by an open portico of composite columns. Though exposed to the weather without any roof or covering, it is better preserved than might have been expected. "It derives," says Eustace, "much intrinsic merit from its size and proportions, but it is not architectural merit alone which gives it its principal interest. Placed on the verge of a rocky bank, it is suspended over the *praceps Anio*, and the *domus resonantia* of the Naiads; Augustus and Mæcenas, Virgil and Horace, have reposed under its columns ; it has survived the empire, and even the language of its founders ; and after 1800 years of storms and tempests, of revolutions and barbarism, it still exhibits its fair-proportioned form to the eye of the traveller, and claims at once his applause and admiration." (*Classical Tour*, ii. 238, 8vo ed.)

It may be worth while mentioning that an English nobleman, the late Earl of Bristol, obtained permission from the authorities at Tivoli to take down and carry away this classical ruin, in the view of setting it up again in his park in England ! Luckily the desecration was prevented by the interference of the Papal government.

Near this temple are the remains of another, now forming a portion of the church of St. George, and an inn in the town is supposed to occupy the site of the temple of Hercules, whence Augustus borrowed the treasures collected by the piety of ages. But, besides these, little remains of the ancient Tibur. But though its temples and its theatres have crumbled into dust, its orchards, its gardens, and its cool recesses, bloom and flourish in unfading beauty. The declivities in its vicinity were anciently interspersed with splendid villas, the favourite residences of the refined and luxurious citizens of Rome. Among these may be enumerated the villas of Sallust, Mæcenas, Tibullus, Varus, Atticus, Cassius, Brutus, &c. The existing remains of what is supposed to have been the villa of Mæcenas sufficiently attest its ancient magnificence ; but probably the modern *villa d' Estense*, erected in the immediate vicinity of the ruins, in the 16th century, by a Cardinal d'Esté, exceeds in extent and grandeur that of the minister and favourite of the master of the Roman world ! Horace, who has over and over again expressed his admiration of Tibur, is supposed to have had a villa in its vicinity, and some ruins in a delightful situation are pointed out as those of his residence. But, notwithstanding the statement of Suetonius to the contrary, we are disposed, in this instance, to agree in opinion with Eustace, that the probabilities are, that the poet was not really master of a Tiburtine villa, and that all his allusions the gratifications he experienced in the groves and streams Tibur—

> "Circa nemus, avidique
> Tibur is ripss operon parvæ
> Carmina fiago,"
> *Hor.*, lib. iv. Od. ii. v. 35,

y be explained by his frequent visits to the villas of Mæcenas and his other friends.

Such is the mutability of human affairs, that two contents, which raise their white towers above the dark green de of the olive trees, are now the most striking structures in the neighbourhood of Tivoli ; and that monks loiter away their worthless existence under the shades where great and Horace elaborated their immortal works ! (Besides the works already referred to, see *Gell's Topography of the country round Rome*, art. *Tibur*; *Wood's Letters*, ii. *Matthews' Diary of an Invalid*, p. 226, &c.)

OBAGO, one of the W. I. islands belonging to Great ain ; in lat. about 11° 15' N., and long. 60° 40' W., 16 m. : Trinidad, and 82 m. S.E. Grenada. Area, 57,408 es. Pop. 13,700. "It is 32 m. in length and 12 in idth, on the N. extremity rugged and mountainous, and the sea appears like a mass of dark abrupt precipices. wards the S. and W. the ground descends into a succession of conical hills and ridges of no great elevation, which, they approach the sea, terminate in broken plains and lands. The E. districts is also mountainous. The soil the valleys is generally a rich dark mould, and is well red by numerous streams and rivulets. Cultivation g for the most part confined to a portion of the low near the sea on the S. side of the island, the greater of the interior is still in a state of nature, the high nds covered with forests, the deep ravines choked up vegetation, and the bottoms of the valleys, being very w and not possessing free drainage, generally of a wet by character. The climate and seasons here are much ame as at Trinidad, only rather more humid ; but we no measurement of the quantity of rain which falls ally. In some of the low grounds, excluded from the

influence of the breeze, the heat is described as being exceedingly oppressive, particularly at Scarborough, the capital, which lies at the foot of a hill on the S. side of the island. The troops enjoy the advantage of a more moderate temperature, being quartered in fort King George, on the summit of the hill above Scarborough, where the heat is modified by a constant breeze, and the mean temperature of the year does not exceed 79° Fah. On the average of the 20 years ending 1836, the mortality amounted to about 153 per thousand of the white, and 34 per thousand of the black troops, annually, the former being nearly double the rate which prevails throughout the whole windward and Leeward command. But as the climate has by no means affected the blacks in any corresponding degree, the deaths on the whole, are considerably less than at most of the other stations." (*Tulloch's Report on the Health of the Troops in the W. Indies*.) This island is beyond the range of the hurricanes; though Grenada, at so short a distance, is so subject to them as the rest of the Antilles.

Tobago produces almost every kind of plant that grows in the Antilles, besides many common to the adjacent parts of S. America. It was formerly supposed to have given its name to the narcotic plant tobacco, now so widely diffused ; but Humboldt has shown that there is no foundation whatever for this opinion, and that tobacco is a word of Mexican origin. (*Nouvelle Espagne*, iii. 50, 2d ed.) Indian and Guinea corn, pease, beans, figs, pineapples, and all kinds of tropical fruits, are grown, as well as potatoes, yams, carrots, turnips, onions, manioc, &c. Horses, cows, asses, sheep, deer, &c., probably introduced by the Dutch, have multiplied greatly, and wild hogs are very abundant.

The principal imports into the U. Kingdom from Tobago, during the three years ending with 1839, were

Articles and Quantities.			1837.	1838.	1839.
Sugar (unrefined)	.	cwt.	80,892	71,621	66,044
Rum	.	galls.	261,779	258,411	249,546
Molasses	.	do.	5,777	7,181	4,012
Cocoa	.	lb.	10,226		

The total value of the articles imported into Tobago from the U. Kingdom usually amounts to from £65,000 to £75,000 a year.

Tobago has its governor, council of nine members, and house of assembly of 16 members, whose powers are similar to those of Jamaica. It is divided into seven parishes. Number of pupils is 18 public schools, in 1839, 1942. Number of militia in do., 230 men. The sum awarded by government, in 1835, for the manumission of slaves in Tobago amounted to £234,064. This island, which was discovered by Columbus in 1498, was colonized first by the Dutch, and next by the Courlanders. It was ceded to Great Britain by France in 1763 ; but was retaken by the French in 1781, who retained possession of the island till 1793, since which it has belonged to England. (*Parl. Reports, &c.*)

TOBOLSK, a very large government of Asiatic Russia, comprising a large portion of the basin of the great river Obi, or the country between the 56th and 73d degrees of N. lat. and the 60th and 80th of E. long., having E. the gov. Yenisiesk, S. that of Tomsk and the Kirghis territ., W. the govt. of Orenburg, Perm, and Archangel, and N. the sea of Kara, gulf of Obi, &c. The area may amount to from 900,000 to 1,000,000 sq. m. ; and in 1838, the population was estimated at 685,000. Except on its S. and W. frontiers, it is almost everywhere level, or but a slightly waving plain, though varying greatly in point of fertility. From lat. 58° or 60° to lat. 65° or 66°, the country is generally occupied by vast forests of fir and birch; from the woody region N. to the Arctic ocean, the country, a low plain called the Tundra, is the most sterile imaginable, consisting of all but boundless moors and morasses, interspersed here and there with some stunted shrubs, and occupied by only a few Ostiak tribes, who subsist chiefly by fishing, and the chase of fur-bearing animals. Such is the severity of the climate, that this portion is usually covered with ice and snow for about nine months of the year: and during the other months, ice is always found at a little distance below the surface.

Immediately to the S. of the woody region, or between about lat. 56° or 60° on the N., and 54° or 55° on the S., is the agricultural portion of the government, including extensive tracts watered by the Irtish, a part of the Ishim, and the Tobol. Though not generally fertile, this district comprises some very productive tracts, and it has a considerable number of towns, though few of them be of any great size. Even in this part of the government, the climate is very severe; for, though the summer heats be sometimes oppressive, they are but of short duration, and the winters are long and excessively cold. Rye, oats, barley, and buckwheat are the principal crops. Between the agricultural district and the mountains separating the government from the country of the Kirghizes is the very extensive tract called the Steppe of Ishim, from its being in part traversed by that river and its affluents. Except along the river banks, it is mostly sterile; and in extensive tracts the soil is covered with a salt efflorescence.

Iron and copper are extensively raised in various parts of the Oural chain; and gold and silver are produced both there and in the Altai. At Catherinenburg, Kolyvan, and Barnaoul, are extensive forges; and soap and tallow-works, tanneries, mat-manufactories, &c., are found in different parts: but the commerce of the government is of more importance than its manufacturing industry. Except the clergy, and persons in the government employment, all the inhabitants are more or less engaged in traffic, exchanging their arable and other furs, cattle, cassia, fresh and dried fish, and game, with the Russian traders for corn, flour, hardware, &c. The merchants of Tobolsk, Toumen, and the principal towns in the S. and W., send every summer boats laden with flour and other provisions, by way of the Irtish and Obi, to Berezov, and other small towns in the N., which return with cargoes of fish, and with valuable furs, procured from the Ostiaks and other tribes. These furs are afterward partly sent, with soap, tallow, and hides, to the fair at Nijni Novgorod; partly to the Kirghiz, to be bartered for horses, cattle, and cotton goods, obtained through Bochara; and partly to Kiachta, on the Chinese frontier, where they are exchanged for tea, silk fabrics, and other Chinese products. The government, in common with the rest of Siberia, lies under the greatest disadvantages with respect to water communication; the frozen shores of its N. coast are inaccessible for the purposes of trade; and its rivers, though equal in magnitude to any belonging to the Asiatic continent, are covered with ice for the greater portion of the year. The chief mode of travelling and conveying goods throughout a great portion of the government is, as in the N. part of Europe, in sledges drawn by dogs or reindeer.

Mr. Bell and Captain Cochrane agree in representing the Tartar villages in the agricultural part of the government, as neat, clean, and comfortable. Their white plastered chimneys and ovens reminded the latter of his own country. The houses consist in general of one or two rooms. Near the hearth is an iron kettle, and at one end of the apartment a bench covered with mats or skins; on this all the family sit by day, and sleep by night. The walls are of wood and moss; a layer of moss between every two beams. A square hole is cut out for a window, and to supply the want of glass a piece of ice is often put in; two or three pieces will last the whole winter. They use no stoves, and have neither chairs nor stools. The furniture consists of a few earthenware utensils, and a set of portable appendages. The women never eat nor drink till the men have done, and then seldom in their presence. (*Mod. Trav.* xvii. 322.) Owing to the thinness of the population, and the immense distances between the different towns, education is very little diffused, and besides the schools in the capital (*see post*), there are, perhaps, hardly a dozen in the rest of the government. Except Tobolsk, the capital, and Toumen, there are no towns worth notice.

TOBOLSK, a city of Asiatic Russia, the cap. of W. Siberia, and of the gov. of its own name, and, indeed, of the whole of N. Asia; on the Irtish, close to its junction with the Tobol, lat. 58° 11′ 42″ N., long. 68° 6′ 13″ E. Pop., in 1835, 15,379. The town proper is built principally on the flat summit of a hill commanding an extensive view, and is surrounded by a strong brick wall with square towers and bastions. When approached from the W. it has a remarkably fine appearance, and it really contains some good and solid buildings, most of the government offices, and the residences of the Russian and German settlers, being within the walls. Along the banks of the river are suburbs, inclosed by a ditch and palisade, and inhabited mostly by Tartars. Tobolsk had, in 1835, 18 churches, and 1762 houses, of which only 25 were of stone. (*Journ. de St. Petersbourg.*) The streets, which cross each other at right angles, are mostly paved with wood. Among its public edifices, the most remarkable are, the cathedral, in the Byzantine style of architecture, with five cupolas, the archbishop's and governor's palaces, a monastery, and a large hospital. The climate in winter is very severe, so much so as sometimes to freeze mercury; and next to Yakutsk, Tobolsk is one of the coldest towns in Siberia: but the dress and houses of the inhabitants being fitted to resist its influence, it is not so disagreeable as might be supposed, and, in other respects, it is not an unpleasant residence. The rivers furnish an inexhaustible supply of fish, and provisions, fur, and game of all kinds are cheap and abundant; and shops, theatres, and places of public amusement are numerous. Being on the great road from Russia to China, it is well supplied with most European and Chinese goods; and French wines, English porter, and books of all kinds, are to be met with. Dobell says, "the society of Tobolsk may fairly stand a comparison with that of some of the best provincial towns in Russia." Among the inhabitants are many descendants of the Swedish officers, sent thither after the battle of Pul-

tawa, to whom Tobolsk is mainly indebted for its superior civilization. This city, which was founded in 1587, is the residence of the governor-general of W. Siberia, comprising the governments of Tobolsk and Tomsk: it has two ecclesiastical, and several Lancastrian schools, and various charitable institutions. No convicts or malefactors are sent thither from European Russia, although persons banished to Siberia for political offences are sometimes permitted to reside in Tobolsk. (*Mod. Trav.*, vol. xvii.; *Erman, Reise um die Erde; Dobell's Travels in Siberia; Journ. de St. Petersbourg.*)

TOBOYNE, t., Perry co., Pa., 20 m. W. Bloomfield. It contains five stores, two flouring-mills, two grist-mills, 11 saw-mills, three tanneries; 10 schools, 230 scholars. Pop. 1442.

TOBY, t., Clarion co., Pa., 15 m. N. Kittanning. 190 m. N.E. Harrisburg, 226 W. Drained by Licking, Cherryrun, Catfish and Red Bank creeks. Bounded N. by Clarion r. It contains four stores, one furnace, one fulling-mill, four grist-mills, four saw-mills, three tanneries, one distillery, one pottery: four schools, 22 scholars. Pop. 1829.

TODD, county, Ky. Situated in the S. toward the W. part of the state, and contains 612 sq. m. Drained by branches of Green river and by Elk fork of Red r., flowing into Cumberland r. It contained in 1840, 8250 neat cattle, 11,557 sheep, 23,405 swine; and produced 53,777 bushels of wheat, 9659 of rye, 633,960 of Indian corn, 9,861,665 of oats, 19,684 of potatoes, 3,123,698 pounds of tobacco, 9566 of cotton, 11,160 of sugar. It had 18 stores, seven flouring-mills, 25 grist-mills, 10 saw-mills, four tanneries, 18 distilleries, one printing-office; nine academies, 245 students; 16 schools, 340 scholars. Pop.: whites, 6670; slaves, 3579; free coloured, 42; total, 9991. Capital, Elkton.

TOCAT (an. *Berisa*), a town of Turkey in Asia Minor, pach. Sivas, on the Tosmulu, near its confluence with the Jesil Irmak (an. *Iris*), on the military road from Trebisond to Kharput, 25 m. S.S.E. Amasia, and 55 m. N.N.W. Sivas, lat. 40° 7′ N., long. 36° 30′ E. These would appear to have latterly declined rapidly in population and importance. Tournefort, by whom it was visited in the early part of last century, says it was then much larger than Erzroum; and he estimated its population at 29,000 Turkish families, with 4000 Armenian, and 300 or 400 Greek do., which on the most moderate hypothesis, would make an aggregate of above 120,000 inhabitants. We incline, however, to think, despite the high authority on which it is made, that this estimate must have been beyond the mark. At all events the population was estimated by Kinneir, in 1813, at only 60,000; and according to Mr. Brant, by whom it was visited in 1830, it was then reduced to a population of 6730 families (between 35,000 and 40,000 individuals), of which 5200 were Turkish, 1500 Armenian, 30 Roman Catholic, 50 Jewish, and 150 Greek. The position of the town is striking and singular, being built partly at the bottom, but principally, on the declivities of two steep hills, on the side of the narrow valley in which it is situated. Tournefort says that the greater number of the houses, which are mostly of wood, have two stories; that the streets are pretty well paved, and that the springs rising in the hills on which the town is built are so numerous, that each house has its peculiar fountain. (*Lettres*, ii. 432.) According to the same distinguished authority, it was, at the period of his visit, famous for its copper foundries, its Turkey leather, and its dye works; and was then also the centre of the commerce of Asia Minor. (*Lettres ubi supra.*) But Mr. Brant states that the manufactures and, still more, the trade of Tocat, have greatly diminished; that its numerous khans are empty, and that it has no symptoms of activity. Owing to its situation, the climate at certain seasons is excessively hot; and it is then, also, apt to be unhealthy. With the exception of the mosques, Armenian churches, and khans, it does not appear to have any building of consequence. The valley for about 3 m. above the town is occupied by gardens and vineyards. According to D'Anville, Tocat occupies the site of the ancient *Berisa*. (*Tournefort, ii. 431-438; Kinneir's Asia Minor, 556; Geographical Journal, vi. 219 &c.*)

TODMORDEN, a market town and chapelry of England, partly in the par. of Rochdale, co. Lancaster, and partly in that of Halifax, co. York, 17 m. N.N.E. Manchester. Pop. about 5000. The inhabitants are principally employed in the manufacture of fustian, dimity, velveteens, and other cotton goods with woollen fabrics similar to those manufactured at Halifax and Rochdale. The Rochdale canal, which passes by Todmorden, has greatly promoted its prosperity, which appears to have increased rapidly within the last 20 years. (*Parl. Reports, &c.*)

TOKAY, a town of Hungary, co. Zemplin, at the confluence of the Bodrog with the Theiss, 112 m. N.E. by E. Pesth; lat. 48° 7′ 12″ N., long. 21° 24′ 4″ E. Population 3500. It has a cathedral, a Lutheran, a Reformed, and a United Greek church, a convent of Piarists, and one of

Capuchins, and was formerly defended by a castle demolished in 1705.

Tokay derives its whole celebrity from its being the *entrepôt* for the sale of the famous sweet wine of the same name, made in the hilly tract called the Hegallya, or submontine district, extending 25 or 30 m. N.W. from the town. The Tokay is produced by allowing the grapes to become dead-ripe; the finest quality, or essence, being that which flows from the grapes, before they are trodden, by the mere pressure of their own weight: the next quality (*ausbruch*) is that which is obtained by treading the grapes, with the addition of a certain quantity of *must*, or juice derived from common grapes; the third and lowest quality (*maslas*) is that which is obtained by the application of a greater degree of pressure to the grapes, and the addition of a still larger quantity of *must*. When new, Tokay wines are of a brownish yellow muddy colour, which, when very old, changes to a greenish tint. The wine made in favourable seasons will keep for almost any length of time, and continues to improve with age. The best qualities are extremely rich and luscious, but cloying; and, unless very old, too sweet for palates accustomed to austerer wines. The finest and oldest varieties of Tokay fetch immense prices, as much as seven ducats a bottle having been paid for it! The best qualities are bought up for the imperial cellars; small quantities being sent as most acceptable presents to foreign princes and distinguished individuals. (*Aust. Stat. Exeme.; Henderson on Wine,* p. 298.)

TOLEDO (an. *Toletum*), a celebrated city of Spain, formerly its metropolis, in New Castile, cap. prov. of its own name, on the Tagus, 38 m. S.S.W. Madrid, lat. 39° 52′ 24″ . long. 4° 49′ W. Pop., which in the 15th century is said to have amounted, though this, no doubt, is a gross exaggeration, to 200,000, is now reduced to about 15,000. It stands on a rocky hill, nearly surrounded by the river, and is compassed by a wall flanked with about 150 small towers, built by the Moors. Internally Toledo is acknowledged to be ill built, poor, and mean; with narrow, steep, and dirty-paved streets. But Inglis says that, with the exception of Granada, its situation is the most striking of any in Spain. Its fine irregular line of buildings covers the summit and upper part of the hill, behind which, as approached on Madrid, the dark range of the Toledo mountains forms a majestic background. "Besides the numerous towers of convents, churches, and stupendous cathedral—the metropolitan church of Spain—the outline is broken by other buildings of a more grotesque or more massive form; while, here and there, the still greater irregularity of the outline points to ages too remote to have left to modern times any other legacy than their ruins." (*Spain in* 1830, i. 293.) Down to the recent changes by which, as already seen (*vol.* 840), the Spanish ecclesiastics have been stripped of a greater portion of their enormous wealth, the revenues of the archbishop and clergy of Toledo were immense; and the population of the city consisted principally, in fact, of priests and friars, and their dependants. The cathedral, founded in 587, is in the same style as those of Seville, Burgos, Milan, Siena, and Bologna. Swinburne says that it is to be compared with many we have in England; but as attracted the admiration of most other travellers; and he says that, excepting the cathedral of Seville, it is the finest and most magnificent of Gothic temples. (*Spain,* i. 304.) According to Twiss, it is internally 384 ft. (english) in length, 191 in breadth, and 107 ft. in height. It has a tower and spire, but the latter is said by Swinburne to be in the style of the Flemish and German spires, a cap of blue turrets piled one upon another. The roof is sustained by 85 columns, which divide the church into five aisles. The columns that run along the aisles are 45 ft. in circumference. There are 80 painted windows; and surrounding the choir and the high altar are 156 marble and porphyry pillars. (*Twiss's Trav.,* 183; *Inglis.*) Its interior is elegantly, as well as most magnificently adorned. The choir is covered with carvings representing the conquest of Granada, executed by Berruguete, a pupil of Michel Angelo, and Philip de Borgona; and among the paintings are (or at all events were) works by Rubens, Titian, Ludovico Greco, Vandyke, Guido, E. Caxes, Vincente Carduchio, Bassano, and other masters of the first celebrity. The ceiling of the sacristy is painted in fresco, by L. Giordano and has a picture of the Assumption, by Carlo Maratti. The pope and the king of Spain are always canons of this cathedral; and the revenue of its archbishop once amounted; it is believed, to little less than £100,000 a year! The silver, and jewels, the plunder of Mexico and Peru, served in the church, mostly escaped falling into the hands of the French. The archbishop carried away the most valuable articles to Cadiz; those that remained in places being redeemed for the comparatively trifling amount of 90 arrobas, or 2250 lbs., of silver. Townsend states that the treasures of the cathedral struck him with astonishment; and in 1830, Inglis was told that their total

value amounted to upwards of 40 millions of ducats, or £10,000,000 sterling! We have little doubt, however, that had the generally intelligent, but sometimes credulous traveller, inquired into the fact, he would have found that the relics, so precious in the estimation of the clergy, made up the far greater part of this enormous sum. We apprehend that any capitalist who should offer £1,000,000 for all the gold and silver that is at present, or that ever was in the cathedral, would make a very bad speculation.

The Alcazar, once the residence of the Moorish, and afterward of the Castilian sovereigns, is the other principal edifice in the city. It is a noble pile of three stories, surmounted by a balustrade, and forming a square of 256 feet, as measured by Twiss. (*Trav.,* 184.)[*] It is built chiefly in the Corinthian and Composite orders, of the dark stone with which the Escurial is built. The N. and S. fronts were erected in the time of Charles V., the former by Covarrubias and Vergara, and the latter by Juan de Herrera. When Toledo ceased to be the metropolis of Spain, the Alcazar was converted into a workhouse, and it was subsequently employed for a silk manufactory, established by the archbishop; but it is now untenanted, and so utterly neglected, that in one of its extensive vaults under ground Inglis encountered a party of wandering gypsies assembled round a huge fire.

Besides the cathedral, there are, or rather were, innumerable churches, monasteries, nunneries, and other religious buildings. Few, however, of these are worth notice. The Franciscan convent is, indeed, a fine edifice, and has a church built in the time of Ferdinand and Isabella. The late archbishop Lorenzano established a lunatic hospital at Toledo; built the modern edifice for the university, which in 1830 had more than 700 students, principally in jurisprudence; and founded a college for girls, each of whom is dowried, provided they do not go into a convent afterward. There are several other colleges, numerous hospitals, and asylums, a handsome town hall, two bridges over the Tagus, one originally a Roman work, bearing a Roman inscription, and terminated on one side by an arch with Corinthian columns; a mint, supposed to date from the time of the Romans, &c. There are a few pleasant promenades around the city, but the only public lounge in Toledo is the *Plaza Real;* and there, says Inglis, "at certain hours, particularly about two o'clock, it seems almost like a convent-hall of recreation, and a sacristy of a cathedral united; for canons, prebendaries, and curates, and 20 different orders of friars, are seen standing in groups, strolling under the piazzas, or seated upon benches, refreshing themselves with melons or grapes." But this square is half monopolized with blacksmiths' shops; and all the others are small, mean, and principally useful as market-places. The houses are mostly floored with brick, and are consequently dusty; and the Roman aqueducts being destroyed, water is sold about the streets, carried in small barrels on asses' backs. There is no public place of diversion of any kind: formerly there was a theatre, but it was suppressed by a royal order obtained through the head of the university. "Bull fights even are forbidden in this priest-ridden city; so that unless processions of saints, &c., are to be considered an amusement, the inhabitants have positively no resource but in the *tertulia.* Nowhere are Spanish customs seen more pure than in Toledo, and nowhere is the monotony of the *tertulia* more striking. The sole amusements are talking, or playing *basto* for a very low stake; and after a glass of *agua fresca,* the party separates. In Toledo, a certain circle agrees to form a *tertulia;* one house is selected where it is to be held, and the same individuals assemble at the same house, and at the same hour, every day throughout the year! This is Toledo society. No admixture of foreign, or even of modern innovation, is to be seen in Toledo. Men of all ranks wear the cloak; and the small round high-crowned Spanish hat is worn, not only by the peasantry, but almost universally by persons of all classes. Among the women no colours are to be seen; black is the universal dress, and scarcely any one enters a church unveiled." (*Inglis,* i., 294, 295–303.) We must add, however, that morals are said to be more correct there than in almost any other Spanish city.

Toledo has, from a remote period, been famous for its manufacture of sword blades. The royal sword manufactory, which is of great extent, and about 2 m. from the city, is close to the river which turns its machinery. It once employed many hundred hands; but, when visited by Inglis, in 1830, only 50 were employed, who finished about 8000 swords a year. They work by the piece, and make usually about 100 reals (20s.) per week; some of the most industrious 24 reals more. The art of tempering the steel had, for some time, declined, but it has since revived. "The flexibility and temper of the blades are surprising: there are two trials which each blade must undergo before it be

[*] Townsend states, "the quadrangle is 100 feet by 150 feet, and with the great staircase, the gallery, and the colonnade, has an air of elegant simplicity."

pronounced sound; the trial of flexibility, and the trial of temper. In the former, it is thrust against a plate on the wall and bent into an arc, at least three parts of a circle. In the second, it is struck edgeways upon a leaden table, with the whole force which can be given by a powerful man, holding it with both hands. The blades are polished upon a wheel of walnut wood." (*Inglis*, i., 312.) In addition to its sword manufactory, Toledo fabricates church ornaments, a few woollens, for hospital use, with paper, guitar strings, coarse glass, &c., and has some dyeing and fulling works. Its general trade is very insignificant; and a few years since there was no conveyance, nor even a continuous road, between it and Madrid.

The origin of Toledo is lost in obscurity. After having belonged to the Carthaginians it became a Roman colony. Few traces of Roman edifices, however, exist. except part of an amphitheatre, and some scattered remains of the Roman walls. In 467, it was taken by the Goths, and became the capital of their kingdom in Spain, till taken by the Moors in 714. Alphonso VI. and Rodrigo Diaz expelled the latter from Toledo in 1085; and, notwithstanding three vigorous sieges in the succeeding century, it has remained in the hands of the Spaniards ever since. Its decay dates from the removal of the court to Madrid, under Philip II. The celebrated Cardinal de Ximenes, regent of Spain during the minority of Charles V., was, for a lengthened period, archbishop of Toledo. (See *Dict. Géog.; Miñano, Géog. de España; Antillon; Swinburne; Twiss; Townsend; Inglis; Med. Trav.*, xix., 50-56; &c.)

TOLEDO, p. v., Port Lawrence t., Lucas co., O., 134 m. N.N.W. Columbus, 464 W. Situated on the W. side of Maume river, near its entrance into Maume bay. It contains six warehouses, 41 stores, two steam saw-mills cutting 16,000 feet per day, two iron founderies, two brick yards, which have manufactured 2,500,000 bricks in a year, 35 mechanic shops, two printing-offices, two weekly newspapers, and 9079 inhabitants. The number of steamboat arrivals is over 400 annually, and of schooners, about 250. Steamboats ply continually between Toledo and Detroit. The Wabash and Erie canal, now completed, will add to the business and importance of the place.

TOLLAND, county, Ct. Situated in the N. toward the E. part of the state, and contains 337 sq. m. Watered by Willimantic, Hop, and Scantic rivers. It contained in 1840, 17,485 neat cattle, 43,654 sheep, 7713 swine; and produced 7459 bushels of wheat, 50,709 of rye, 87,348 of Indian corn, 20,105 of buckwheat, 98,680 of oats, 281,090 of potatoes, 10,508 pounds of silk cocoons. It had 53 stores, six furnaces, two forges, 17 fulling-mills, 19 woollen factories, 11 cotton-factories, with 8622 spindles, two glass factories, 35 grist-mills, 74 saw-mills, three oil-mills, three paper-mills, 15 tanneries, nine distilleries; two academies, 53 students; 194 schools, 4880 scholars. Pop. 17,980. Capital, Tolland.

TOLLAND, p. t., capital of Tolland co., Ct., 18 m. E.N.E. Hartford, 354 W. Watered by Willimantic river, and by Hop river, issuing from Snipsic pond on its W. boundary, which is 2 m. long and 100 rods wide. The central village contains a courthouse, jail, a bank, three churches, a Congregational, Methodist, and Baptist, several stores, and 40 dwellings, some of them neat. The township contains six stores, three grist-mills, eight saw-mills; one academy, 25 students; 12 schools, 457 scholars. Pop. 1566.

TOLOSA (an. *Iturisa*), a town of Spain, in Biscay, prov. Guipuscoa, of which it is the cap., on the Oria and Arajes, 13 m. S. by W. St. Sebastian. Pop., according to Miñano, about 5000. It is placed in a narrow defile, surrounded by a pentagonal wall, flanked with towers, and entered by several gates. It is said to be handsome, and well built; the streets, which are furnished with footways, are clean, and lighted at night; and it is tolerably well supplied with water. Here are two parish churches, both fine buildings, two convents, a hospital, prison, post-house, a stone bridge across either river, with manufactures of arms, copper and earthen wares, woollen cloths, paper, hats, leather, &c.; three fourths of its inhabitants being artisans. (*Miñano.*) A large market is held on Saturdays. Tolosa is one of the 18 independent towns in which the provincial assembly of Guipuscoa is held, one of the four alternately the seat of the high judicial court of the province, and the place in which the provincial archives and military stores are kept. (*Miñano; Antillon; Inglis, &c.*)

TOMBIGBEE, r., Miss. and Ala., rises in the N.E. part of Mississippi, and joins Alabama river in Ala., to form Mobile river, 45 m. above Mobile bay. Its whole course by the stream is nearly 500 m. It is navigable for large vessels, nine months in the year, to St. Stephens, and for steamboats to Columbus, Miss.

TOMPKINS, county, N.Y. Situated a little S.W. of the centre of the state, and contains 380 sq. m. The S. part of Cayuga and Seneca lakes extend into the county, and through them it has access to the Erie canal, and through the lakes a passage by steamboats. Watered by

Fall and Six Mile creeks, and Cayuga inlet, which afford extensive water power. It contained in 1840, 29,929 neat cattle, 86,595 sheep, 23,772 swine; and produced 27,591 bushels of wheat, 4579 of rye, 216,515 of Indian corn, 71,122 of buckwheat, 9104 of barley, 268,695 of oats, 339,557 of potatoes, 2102 pounds of hops, 68,747 of sugar. It contains 95 stores, 11 lumber-yards, 19 fulling-mills, two woollen factories, one cotton factory, with 1572 spindles, eight flouring-mills, 30 grist-mills, 178 saw-mills, three oil-mills, two paper-mills, 27 tanneries, five distilleries, one brewery, one rope-walk, five printing-offices, three weekly newspapers; two academies, 337 students; 225 schools, 12,678 scholars. Pop. 37,948. Capital, Ithaca.

TOMPKINS, t., Delaware co., N.Y., 22 m. S.W. Delhi, 109 m. S.S.W. Albany. Watered by the W. branch of Delaware river. The line of the New-York and Erie railroad passes through its S. part. It contains three churches, a Presbyterian, Methodist, and Baptist, five stores, two fulling-mills, two grist-mills, 26 saw-mills, two tanneries; 13 schools, 697 scholars. Pop. 2635.

TOMKINSVILLE, p. v., Castleton t., Richmond co., N.Y., 6 m. S. New-York, 154 m. S. by W. Albany, 234 W. Situated on the N.E. part of Staten Island, overlooking the Quarantine ground, and contains three churches, a Dutch Reformed, Episcopal, and Methodist, 10 stores, 396 dwellings, and 1400 inhabitants. At the Quarantine ground are three hospitals, one of which is appropriated to pestilential diseases. Attached to the buildings are 30 acres of land, the whole surrounded by a substantial brick wall.

TOMSK, a town of Asiatic Russia, cap. of the gov. of same name, on the Tom, a tributary of the Obi, 656 m. E. by S. Tobolsk. Lat. 56° 29' 6" N., long. 85° 9' 51" E. " It has nearly 2000 houses, and from 8000 to 10,000 inhabitants. Here are workhouses for exiles; coarse cloth, leather, and soap manufactories; barracks, public magazines, military and other hospitals; an orphan house, a dispensary, &c. There are a number of handsome houses in Tomsk, but the town is irregularly built, except the part that occupies a hill overlooking the river Tom and the country round. Next to Krasnojarsk, Tomsk is the cheapest and most plentiful spot in Siberia." (*Dobell's Trav.*, ii., 140.) Its principal buildings are the cathedral and another church, the arsenal, treasury (in which are the magazines of furs collected as tribute), and two convents. The inhabitants carry on a brisk trade with the Calmucks and Ostiaks, in cattle, fish, &c.; and the town is an emporium for distilled spirits and Chinese goods. It was founded in 1604.

The government of which Tomsk is the capital is, with that of Tobolsk, under the authority of the governor-general of W. Siberia. Since 1838, it has comprised a portion of the former government of Omsk; and is supposed to have from 1,000,000 to 1,108,000 inhabitants. In 1838, no fewer than 255 pools 19 lbs. and 3133 zolot. gold were obtained from the different gold washings in this government. In its general features, it is very similar to the more southerly parts of the governments of Tobolsk, Yenisisk, &c. (which see).

TONAWANDA, p.t., Erie co., N.Y., 296 m. W. Albany, 392 W. It includes Grand Island in Niagara river. Drained by Tonawanda creek, which for the last 10 miles of its course it forms the Erie canal, and at its termination is a dam from which are locks into Niagara river. The village here has a good harbour for steamboats and other vessels. The canal is continued along the margin of the river to Buffalo. The township contains a church, 11 stores, four saw-mills, one brewery; four schools, 71 scholars. Pop. 1261.

TONNEINS, a town of France, dep. Lot-et-Garonne, cap. cant., on the Garonne, 20 m. N.W. Agen. Pop. in 1836, ex. comm., 4176; or inc. comm., 7086. It is one of the best situated and most active towns in the dep., having a considerable trade in cordage, hemp, and dried fruits, and a royal tobacco factory. It is clean and well built, and communicates with the opposite bank of the river by a new suspension bridge. The esplanade, a good square, on the site of an old castle destroyed in the religious wars, the town-hall, a neat theatre, and some public baths, are worthy notice.

TOPLITZ (or *Teplitz*), a town and watering-place of Bohemia, circ. Leitmeritz, and next to Carlsbad, the most popular place of resort of its kind in Germany. It is pleasantly situated on the Saubach, a small stream in a valley between the Erzgebirge and Mittlegebirge mountains, 6 m. N.W. Prague. Its resident population amounts to little more than 2700; but in the height of the season, in July and August, it is sometimes visited by 15,000 strangers. (*Murray's Handbook for S. Germ.*) More than 1-4th part of its houses are inns, and nearly all the rest are lodging-houses. The town is neat, and has been improved of late years by the addition of foot-pavements in the streets, and it is well lighted at night; but it has no buildings worthy of notice, except such as are connected with the baths.

The principal baths are distributed in four distinct buildings; the Steinbad, Fürstenbader, Fürstliche-Frauenzimmerbad, and the Herrnhaus, or mansion of Prince Clary, the proprietor of the town. All these are in the *Baude platz*, or bath square. The Steinbad includes three baths, for the gratuitous use of the public; one for the men, a second for the wives and daughters of citizens, and the third for the female peasantry, &c.: the first and last are under ground, and vaulted over, and may be compared to large inundated cellars. In the same house are some very comfortable private baths, supplied directly from the source. The Fürstenbad and Frauenzimmerbad comprise a number of superior private baths, the first for gentlemen, and the second for ladies. In the Herrnhaus, which was the usual residence of the late king of Prussia, when at Töplitz, there are many bathing apartments fitted up with great elegance; and attached to this mansion are some extensive and beautiful gardens, always open to the public, a theatre, &c. The baths in the Girderhaus also in the Baude-platz, the Jews' baths and others are supplied from the main springs. Without the town, and in the neighbouring hamlet of Schönau, are many baths of a lower temperature than in the town. In all there are about 90 private baths, which are in such constant requisition when Töplitz is full, that, by a strict regulation, no person is allowed the use of a bath and dressing-room for more than an hour at a time, for which from 10 to 20 kreutzers are usually paid. The springs are saline, with a dash of iron; the hottest, or *hauptquelle*, has a temperature of about 122 Fahr. It emerges from a porphyry rock, and so abundantly that its supply, per hour, has been estimated at 1,389,670 cubic feet of water. The waters of Töplitz are particularly esteemed in gout, and rheumatic affections, diseases of the joints, &c., requiring tonic treatment. The invalids of the Prussian, Russian, and Austrian armies are often sent here, and lodged in appropriate buildings. Of late years a pump-room has been established in the gardens of Prince Clary. The hotels and lodging-houses are good and cheap. Dr. Granville says, " The living at Töplitz is, beyond comparison, cheaper than in any other watering place I have visited. A dinner at a *table d'hôte* without wine will cost about 1s. 3d. Apartments may be hired at one of the best hotels, consisting of a bed-room and sitting-room, for not quite a guinea a week." (*Spas of Germany,* 336.) Bathing is the chief occupation of the morning. The dinner hour is one or two o'clock; the afternoon is commonly spent in excursions; the evening in the theatre or the *salons*; but " except on ball nights, and on the occasion of some great concert, the town is buried in dead silence by 10 o'clock." Public gaming is not allowed; but it is alleged that gambling is, notwithstanding, extensively carried on. Töplitz was the seat of a diplomatic congress in 1813, and again in 1835. (*Granville; Spencer; Gleig; Austrian Natural Encyclopædia; &c.*)

TOPSFIELD, p. t., Essex co., Mass., 25 m. N. by E. Boston, 465 W. Watered by Ipswich river. It contains two churches, a Congregational and Methodist, three stores, two grist-mills; one academy, 104 students; four schools, 411 scholars. Pop. 1050.

TOPSHAM, a market town and par. of England, co. Devon, hund. Wonford, on the Exe, where it is joined by the Clyst, 4 m. S.E. Exeter, of which it may be considered the port. Area of par., 1740 acres. Pop., in 1831, 3184. It consists of several good streets; the Strand, in particular, at its S. extremity, has many respectable residences, and is inhabited by some families of good fortune. The church is built on an eminence overlooking the river. The living, a perpetuate curacy, worth £227 a year, is in the gift of the dean and chapter of Exeter. The chief business of the town is ship-building, and most of the inhabitants are connected with shipping. The quay, which was originally constructed in the 14th century, is spacious and convenient; but from the corporation of Exeter (to which it belongs), neglecting to clean the channel of the river, vessels drawing more than 9 or 9½ feet are unable to come up to it. *Mun. Append. Rep. on Exeter.*) An active coasting trade is, however, carried on from Topsham. Markets on Saturdays; fair, first Wednesday in August.

TOPSHAM, p. t., capital with Wiscasset and Warren, of Lincoln co., Me., 30 m. S. by W. Augusta, 571 W. Bounded S. and E. by Androscoggin river, which affords great water-power, and furnishes great conveniences for navigation. It has considerable ship-building, and largely exports lumber. It is connected with Brunswick by a substantial bridge, and has a courthouse, nine stores, one fulling-mill, one grist-mill, 13 saw-mills; 12 schools, 642 scholars. Population, 1883.

TOPSHAM, p. t., Orange co., Vt., 23 m. S.E. Montpelier, 25 W. Chartered in 1761, first settled in 1781, organized 1790. Watered by head branches of Wait's river. It contains three churches, a Congregational, Baptist, and Freewill Baptist, a town house, occupied as a church, two

stores, two fulling-mills, three grist-mills, 10 saw-mills, three tanneries; 18 schools, 630 scholars. Pop. 1745.

TORBAY, a spacious bay of the English channel on the S.E. coast of Devonshire. It is of a semicircular shape, opening to the E., and nearly 4 m. across from Torquay or Rob's Nose its N. to Berry Point its S. boundary. Its shores at its mouth are on both sides formed by ramparts of rock, but between these, in the centre, at the bottom of the bay, the ground forms a vale, gently declining to the water's edge. Ships anchor all over the bay in 6, 7, 8, and 9 fathoms water. The ground is strong clay, and holds remarkably well. This spacious basin has frequently afforded shelter to the fleets of England, and is celebrated in history as the place where our great deliverer, William III., landed on the ever-memorable 5th of November, 1688.

TORGAU, a town of Prussian Saxony, reg. Merseburg, cap. circ. Torgau; on the Elbe, here crossed by a covered bridge, 65 m. S.E.W. Berlin. Pop. about 6500. It is pretty strongly fortified, is the seat of the principal courts, &c., for its circle; and has manufactures of woollen cloths and hosiery, leather, &c., with some trade in corn and timber. The vicinity of Torgau has been the scene of several conflicts. Of these, the most important took place on the 23d of November, 1760, when the Prussians, under Frederick the Great, forced, after a desperate resistance, the intrenched camp of the Austrian army, under Marshal Daun, and gained a decisive victory.

TORNEA, a town of the Russian dom., N.W. frontier of the Grand Duchy of Finland, on a peninsula in the river Tornea, where it falls into the gulf of Bothnia, lat. 65° 50' 50" N., long. 24° 12' 15" E. Pop. from 500 to 700. This little town, which was built by the Swedes, in 1602, consists of two principal streets of wooden houses. It has a considerable trade in the exportation of stock-fish, reindeer, skins, furs, iron, planks, tar, butter, pickled salmon, &c. The climate is very severe, though less so, perhaps, than might be expected from its high latitude. In June the sun is visible at midnight above the horizon.

Tornea is celebrated in the history of science for the visit made to it in 1736, by the French academicians Maupertuis, Clairaut, Monnier, and Camus, accompanied by the Swedish astronomer Celsius, with a view to the determination of the exact figure of the earth. The operations do not, however, appear to have been conducted with sufficient accuracy; and there is a discrepancy of about 300 toises between the length of the degree as determined by the academicians and that measured by the Swedish astronomer Svanberg in 1801. This town, along with the Grand Duchy of Finland, was ceded to Russia by Sweden, by the treaty of Fredericksheusen, in 1809. (*Clarke's Travels,* ix., 542, &c., 8vo ed.; *Schmitzler, La Russie, &c.,* p. 694; *Biographie Universelle,* art. *Maupertuis.*)

TORO, a town of Spain, in Leon, prov. Zamora, cap. intend.; on a hill, at the foot of which runs the Douro, crossed by a narrow stone bridge of 22 arches, 39 m. N.N.E. Salamanca. Pop. about 10,000. (*Miñano.*) It is enclosed by old and dilapidated walls, and entered by six gates. The streets, though broad, and on a declivity, are dirty, and the houses indifferent. It has a collegiate, and 18 par. churches. It formerly had 13 convents, with three hospitals, a ruined alcazar, or Moorish castle, a palace belonging to the dukes of Berwick, barracks, a prison, &c. The inhabitants are principally occupied in the growing and trading in wine, but they have also manufactures of coarse woollen and linen cloths, brandy, and leather. Toro is of great but uncertain antiquity. It is famous in history for the victory obtained in the vicinity in 1476, by Don Ferdinand of Arragon, over Alphonso V. of Portugal; and for the collection of laws framed in 1505, and inserted in the Spanish statutes under the name of *Leyes de Toro.* (*Miñano; Dict. Géog., &c.*)

TORONTO, formerly York, a town of Upper Canada, of which it is the cap., on the N. shore of lake Ontario, towards its W. extremity, in lat. 43° 22' N., long. 79° 20' W. Pop., in 1832, 12,571. It was founded by Governor Simcoe in 1794, and was burnt by the Americans in 1812. In 1831 it had only about 4000 inhabitants, its subsequent progress having been more rapid than that of any other town in Canada. It is now a handsome town, with spacious streets crossing each other at right angles; many of its buildings being of brick, to which timber is gradually giving place. The public edifices are well adapted to their purposes, though none of them has a very striking appearance: they comprise government, parliament, and court houses; a new college, consisting of five brick buildings, the central one being surrounded by an ornamental dome, government stores, a jail, bank, hospital, grammar school, Episcopal church, &c. The garrison is stationed about 1 m. W. of the town, where the entrance to the harbour is guarded by a battery and two blockhouses. Toronto harbour, or bay, is formed by a long and narrow peninsula, stretching out to the S.W. for about 6 m., and terminating in Gibraltar point, on which a lighthouse has been erected. The bay is nearly circular, and about 1½ m. across; it has a considerable depth of water, and affords extensive and safe anchoring ground. (*Stevenson's Civil Engineering, &c.*)

Toronto is the residence of the governor, the seat of the superior judicial courts, and the place where the parliament of Upper Canada assemble. A good military road leads from it to Cook's bay in lake Simcoe, 37 m. distant. (*Mac Gregor's British America*, vol. ii.)

Kingston, at the other extremity of lake Ontario, on its N. shore, about 140 m. E.N.E. Toronto, was the former capital of Upper Canada, and, though less central, has been considered by many as more eligible than Toronto for this distinction. It has an excellent harbour, where ships of the line may lie close to the shore; and is also the site of the principal naval dockyard in the colony. It covers a considerable extent of ground, and many of its houses are of stone; it has several good public buildings, and, though its population reached, in 1838, to 3977, it continues to be the principal entrepôt of the trade between the upper and lower province. It was founded in 1782, on the site of fort Frontenac.

TOROPETZ, a town of European Russia, gov. Pskof, on the Torops, 945 m. S. Petersburg. Pop. about 7500. It is entirely surrounded by lakes and rivulets, and communicates by the Torops with Riga, which renders it a place of some trade. It has 13 churches, including a cathedral, and two convents. A few of its houses are of brick or stone, but the major part are of wood, the streets also being paved with planks. On an island in the Torops is a dilapidated fort. This town, under the name of Krivitch, is mentioned as early as the introduction of Christianity by Vladimir, about 990. It was the capital of a republic, which lasted through the whole of the 12th century, but which in the 13th became subject to hereditary princes. Towards the end of the 15th century it belonged to the Poles, but it was retaken by the Russians in 1500. (*Schnitzler, &c.*)

TORRINGTON, a mun. bor., town, and par. of England, co. Devon, hund. Fremington, partly on the summit and partly on the declivity of an eminence on the E. bank of the Torridge, here crossed by two bridges, 5½ m. S.S.E. Bideford, and 30 m. N.W. Exeter. Area of par., 3640 acres. Pop., in 1831, 3093. It consists principally of two parallel lines of thoroughfares, nearly 1 m. in length, connected by several short streets. The parish church, which had been mostly blown up by an explosion of gunpowder in the civil wars, was rebuilt in 1651; and in 1830 a new tower and octagonal spire were erected. The living, a pe. rectual curacy, worth £162 a year, is in the gift of Christch. rch college, Oxford. It has a market place surrounded by good houses, a neat town-hall, places of worship for Baptists, Independents, Wesleyans, &c.; almshouses, a charity or blue-coat school for 39 boys, and various minor charities. Near the town is Stevenston, the seat of Lord Rolle, who cut a canal in 1823, which skirts the hamlet of Taddiport, on the opposite side of the Torridge. At this hamlet is a hospital for the poor of the parishes of both Great and Little Torrington. A bowling-green now occupies the site of a castle erected on an eminence S. of the town in the 14th century. Torrington is watched during the winter months, but is not lighted; and its police is said to be very inefficient. It has a very indifferent jail. The chief occupation of the industrious classes is the manufacture of gloves, which is not confined to the town, but gives employment to many families of the surrounding district. (*Mun. Corp. Rep.*) Torrington appears to have been first chartered by Philip and Mary: it is now governed by a mayor and three other aldermen, and 12 councillors, who hold petty sessions every three weeks. Other courts formerly held have gone into disuse. Corporation revenue in 1840, £371. Torrington sent members to the House of Commons down to the reign of Henry VII., when it appears to have lost or relinquished the privilege. At the Restoration the earldom of Torrington was conferred on General Monk; and it now gives the title of viscount to the Byng family. Markets on Saturdays; fairs, May 4th, July 3d, and October 10th, for cattle.

TORRINGTON, p. t., Litchfield co., Ct., 28 m. W. by N. Hartford, 335 W. Drained by Naugatuck river and its branches. It contains, in the S. part, the flourishing manufacturing village of Wolcottville. It has five churches, three Congregational, a Methodist, and Baptist, three stores, one woollen factory, four grist-mills, eight saw-mills; 14 schools, 416 scholars. Pop. 1707.

TORSHEK, or Torjok. a town of European Russia, gov. Tver, cap. distr., on the Tverza, 138 m. N.W. Moscow : lat. 57° 56′ N., long. 57° 55′ E. Pop. about 12,000. (*Possart.*) It was nearly burnt down in 1767, since which it has been rebuilt with considerable regularity and with rather wide streets, and, though its houses are still generally of wood, its public buildings are mostly of stone : the latter includes a cathedral and 20 other churches, two convents, a govern-

ment house, normal school, orphan asylum, &c. It is famous for a holy spring, which attracts pilgrims from all directions. Being on the high road from Petersburg to Moscow, and having also the best inn on this road, it is a place of considerable name, and has three large manufactories. Its principal manufacture is that of saffron, or coloured and prepared Russia leather. A large traffic is carried on in shoes, gloves, and various articles of this material, embroidered with gold and silver; but the traveller had better be on his guard against imposition, as the dealers not only ask three times as much for any article as it is worth and they will take, but also endeavour to substitute articles made of sheep skin for the genuine leather, which is a preparation of ox hides. (*Schnitzler, Possart, Murray's Handbook, &c.*)

TORTOLA, one of the Virgin Islands, in the W. Indies, belonging to Great Britain, lat. 18° 27′ N., long. 64° 34′ 45″ W., between St. John's and Virgin Gorda. It is about 12 m. in length by 3 or 4 in its greatest breadth. Pop. in 1838, 7731 ; of whom 5192 were blacks. "This island consists of a range of hills rising in some places to the height of 1600 feet, and encircling a spacious harbour, or basin ; they are, for the most part, barren, rocky, and precipitous, and there is but one valley of any extent throughout the island. The town of Tortola is on the W. side of the harbour, at the foot of these hills, which rise so close behind it that many of the houses are built within ss-mark, and consequently suffer from damp. The barrack and hospital for the troops are at the S.E. extremity of the town, and as they lie open to the trade winds, which blow across the harbour, they are not much incommoded by heat. But considerable sickness, particularly from fever, has been found to prevail among the troops at Tortola." (*Tulloch's Report*, p. 37.) In 1839, 5420 cwts. sugar, 7935 gallons rum, and 11,287 lbs. cotton were imported into the U. Kingdom from Tortola : the total value of the exports from the island in 1837 amounted to £94,796, and that of the imports to £10,498.

TORTONA (an. *Dertona*), a town of N. Italy, in the Sardinian states, div. Alessandria, cap. prov. of its own name, at the foot of a hill crowned by a ruined castle, 13 m. E. by S. Alessandria. Pop. in 1838, 10,821. It was a place of considerable strength till dismantled by the French in 1796. It is the see of a bishop, the seat of a court of primary jurisdiction, and has manufactures of silk, starch, &c.; and some trade in corn and wine. It appears from inscriptions to have been a Roman colony, under the name of *Julia*.

TORTOSA (an. *Dertosa*), a town of Spain, in Catalonia, on the N. bank of the Ebro, about 25 m. from its mouth, and 93 m. S.W. Barcelona. Pop., according to Minano, between 10,000 and 11,000. It is defended by several outworks, and is divided into the old and new towns, both of which are walled. In the old all built, and has but one public fountain. The cathedral is near the river, and under the protection of the castle. The front is Ionic, with massive pillars, some of single stones ; the choir is of Corinthian architecture : but the whole edifice is void of taste, and its interior is much overloaded with ornament. Townsend observed in the cloisters a chapel bearing indications of the most remote antiquity. The see of Tortosa is a bishopric, and was very rich. There are several parish churches, nine convents, a Latin school, hospital, public granary, &c.; but, next to the cathedral, the principal edifices are the bishop's palace, and the mansion of the Vall Cabra family. Tortosa is the residence of a military governor, the seat of an ecclesiastical court, &c., and has manufactures of earthenware, paper, and leather ; a considerable trade in corn and silk, and an active fishery and coasting trade. Within a league of the city are some quarries of valuable marble, known as Tortosa jasper. "The leaves, or plain of Tortosa," says Townsend, "is most delightful. Far as the eye can reach, you look down upon a plain covered with vines, olives, figs, pomegranates, apricots, mulberries, and all kinds of grain ; and through this fertile vale you trace the meanderings of the Ebro, which is here wide and navigable." (ibid., 305.) The town had the privileges of a Roman *municipium* conferred on it by Scipio. On one occasion, during the wars between the Spaniards and the Moors, the women of Tortosa distinguished themselves so much, that in 1170 the military order of La Hacha, or the "Flambeau," was instituted for them. They also enjoyed several privileges, most of which are now lost ; but it is said that in all matrimonial ceremonies they still maintain the right of precedence. (*Peyron in Mod. Trav.* xviii. ; *Townsend ; Minano.*)

TOTNESS, a parl. and mun. bor., town, and par. of England, co. Devon, hund. Coleridge, on the Dart, 9 m. from its mouth, and 20 m. S. by E. Exeter. Area of parl. bor. which comprises the whole parish of Totness with the manor of Bridgetown, 1411 acres. Pop. in 1831, 4158. The town, which is neat and clean, consists principally of one long street, that communicates at its E. extremity, by a bridge of three arches, with the suburb of Bridgetown, on the opposite side of the river. The town is finely situated, the main street gradually rising from the water till it reaches the site of the ancient castle, now a ruin, on an immense artificial mound, commanding an extensive view of the neighbouring country. It was anciently surrounded by a wall, and some of the gateways still remain. The houses are old fashioned, some of them having piazzas, and their upper stories frequently projecting beyond the lower. But, with the exception of a few on the Plymouth road, all the modern buildings are in the Bridgetown division; and it is here that a farther extension of the town may be looked for. (*Mun. Bound. Rep.*) The church is a spacious, handsome structure, in the perpendicular style, with a well-proportioned tower at the W. end, which has octagonal pinnacles and rich buttresses. In the chancel is a rich stone screen: it has also a stone pulpit, enriched with tracery and shields; but the altar-piece is Grecian, and does not correspond with the rest of the building. This church appears to have been rebuilt about 1432. The living, a vicarage worth £900 a year, is in the gift of the crown. There are meeting-houses for Independents, Wesleyans, and Unitarians, an old guild-hall and council-chamber, a small theatre, an assembly-room, &c.

Judhael de Totnais, to whom the manor was given by William the Conqueror, erected the castle and also a Benedictine priory, which, at the dissolution, had a revenue of £194 10s. a year. It has numerous foundations, among which is a grammar school, established in 1554, having an income of £70 a year. Totnes had formerly a thriving woollen trade; but at present it has no manufactures. Many of the inhabitants are employed in agriculture, some in fishing, and some in navigation, the Dart being navigable to the town for small vessels. Warehouses have been built at Bridgetown, on the E. bank of the river, and many houses are building in the neighbourhood. The town has some trade in the importation of coal and other articles, and may, on the whole, be said to be improving. (*Mun. Corp. Rep., Append.*) It was first incorporated by King John; and is now governed by a mayor, three other aldermen, and 8 councillors. It has no commission of the peace; but a court leet is held once a year, and petty sessions occasionally. Corporation revenue, in 1840, £390. The borough has sent two members to the House of Commons since the 1d of Edward I., the right of voting, down to the Reform ct, having been in the corporation and freemen. Registered electors, in 1839–40, 341.

Among other distinguished individuals, Totnes has given rth to Edward Lye, the learned author of the *Dictionarium izonico et Gothice Latinus*, 2 vols., folio, 1772, and to Dr. Kennicott, the Hebraist, to whom the learned world is indebted for a most elaborate and excellent edition of the Hebrew bible. In his younger days Kennicott was master the grammar school in the town.

TOUL, a town of France, dep. Meurthe, cap. arrond., on e Moselle, here crossed by a handsome stone bridge of ven arches, 13 m. W. Nancy. Pop. in 1836, ex. comm., 79. It is generally well built, and its streets have recently been macadamized. Its principal buildings are the cathedral, a fine Gothic structure of the 15th century; the vn-hall, formerly the bishop's palace; the cavalry barracks and magazines, civil hospital, corn-hall, college, abattoir, &c. Its manufactures comprise calicoes, muslins, ollens, hosiery, and earthenware. This town was anciently the capital of the *Leuci* conquered by Cæsar. It s ceded by Charles the Simple to the emperor Henry Fowler, and was not definitively annexed to France 1552. Baron Louis, Admiral de Rigny, and several distinguished generals have been among the natives of Toul. igo. &c.)

TOULON, a famous seaport town of France, being the 2nd, or, perhaps, since the conquest of Algiers, the first al port in the kingdom; dep. Var, of which, though not capital, it is by far the largest and most important town, he bottom of one of the finest harbours of the Mediterranean, 32 m. E.S.E. Marseilles, and 190 m. S.S.E. Lyons; 43° 7' 10'' N., loug. 5° 55' 40'' E. Pop. in 1836, ex. com., 18, and inc. com., 35,329; but in 1841 the population estimated at 45,000; though this, probably, includes garrison, and the *forçats* in the *bagne*. The town, ch is of an oval shape, the longest side facing the sea, es gracefully and majestically towards the N., extending her ramparts to the foot of a chain of high mountains, ching from the E. to the W. The position of the place ld be picturesque and beautiful, were there the least ure; but the rocks and mountains are arid, bare, and ly destitute of covering, or umbrage of any kind. The i is strongly fortified, being surrounded by a double part, and a large and deep ditch, defended to the E. and N. by hills covered with redoubts. Among the that of La Malgue (on a peninsula to the S.E.) is the

most remarkable, not only for its extent, but the solidity of its construction. Latterly works have been in progress for uniting the town to this fortress, and a solid rampart with fosses has been already raised." (*Ports, &c., of France,* 202–205.) Toulon is divided into the old and new town; both are tolerably well built, but the streets of the former are narrow and crowded, and all the squares, except one, are small and irregular. The new quarter, in which are most of the naval establishments, is much superior in point of appearance. The principal street, the *Rue de Lafayette,* which intersects the town in its whole extent, and is partially planted with trees, is the seat of the principal market, and is a scene of great bustle and activity. It terminates near the port in the *Place d'Armes,* a handsome square, planted with trees, one side of which is formed by the admiralty-house. The town-hall, facing the commercial port, with two colossal statues in front, by Puget, regarded as *chef d'œuvres;* the house occupied by that distinguished sculptor, the old cathedral, three other churches, the court-house, military arsenal, occupying an ancient convent, naval, military, civil, and foundling hospitals, and a handsome communal college, are the other chief public buildings. Though on an arid soil, Toulon is well supplied with water, by springs from the mountains; and several of its numerous public fountains are ornamented with statues, &c. (*Guide du Voyageur; Hugo, &c.*) The suburbs are not only increasing, but, from the rapid augmentation of the population, and importance of the place, of late years, it has been found necessary to add additional stories to the older houses. Since 1830, two new quarters have sprung up without the walls; one on the road north-eastward to Valette, and the other on the road westward to Ollioulles. The latter is filthy, fetid, and abominable. It goes by the name of Navarin, and is chiefly occupied by the Genoese labourers, who occupy the same place in this that is occupied by the Irish labourers in most English towns. Owing to its situation at the foot of high bare hills that intercept the winds from the N., and reflect the sun's rays, the climate in summer is extremely hot. (*Ports, &c., of France.*)

Toulon is the Brest of the Mediterranean, and may be looked upon as the Plymouth of France; though, since the construction of the breakwater in Plymouth sound, the latter is superior, perhaps, as a roadstead to the inner road of Toulon. Both the old and new harbours are artificial. The latter, formed by hollow and bomb-proof jetties, running off from the E. and W. sides of the town, is sufficiently extensive to accommodate 30 sail of the line, as many frigates, and an equal proportion of small craft. The entrance is shut by a boom, and it is never ruffled by any wind to occasion damage. The outer sides of the jetties present two large batteries, even with the water's edge. The entrance to the inner road, on which the harbour opens, is between the *Grosse tour* on the one side, and fort Eguilette on the other, about 620 fathoms apart: the road is a good deal encumbered with banks, and the anchoring-ground is, in part, foul and rocky; but in other places this is not the case, and altogether it is a very fine basin. The outer, or great roadstead, to the E. of the latter, bounded on the S. by the narrow peninsula terminating in cape Cepét, has deeper water and better anchorage than the inner, but it is open to E winds, which sometimes throw in a heavy sea. The lazaretto stands on a secure cove, La Veche, on the S. side of the outer road, with from 4 to 8 fathoms water.

The arsenal of Toulon is one of the finest in Europe. It occupies a surface of 354,140 sq. metres (87 acres), and has dry docks, and every accommodation for the construction, repair, and outfit of ships. In general, from 3000 to 4000 free workmen are employed within its walls; but in 1841, when unusual activity prevailed in all the French ports, there were between 5000 and 6000 labourers employed, exclusive of above 3500 criminals.

The rope house, constructed by Vauban, is 1120 feet in length and 54 in breadth. The docks, slips, sheds, masthouse, sail factory, magazines, &c., are on a grand scale, though, as a ship-building port, Toulon has hitherto been inferior to L'Orient and Rochefort. A new arsenal, meant as an appendage to the old, has been recently laid out. The depôt of oak timber is the largest in France. The *bagne*, instituted in 1682, is, from want of room on shore, established on board some hulks: it is occupied by criminals condemned to hard labour for 10 years and under. The cost of each criminal amounts to very near one franc a day.

The mercantile port, which is bordered by a fine quay, is shut off from the harbour, for men-of-war, by a line of dismasted vessels. (*Hugo.*) The imports consist chiefly of corn, flour, salt provisions, timber, &c., for the use of the naval establishments; and the exports of oil, capers, figs, raisins, almonds, oranges, and other fruits, with cloth, soap, &c., manufactured in the town. The trade of the port has hitherto been inconsiderable, but it has materially increased since the conquest of Algiers, and will probably continue to increase. In 1841 eight steamers were continually plying between Toulon and Africa, Corsica, Italy, and the east, and two small vessels to La Seyne. Toulon is the capital of an arrondisement and two cantons, and is the residence of a naval prefect, a commissary-general, and of numerous other government officers and foreign consuls; it has tribunals of primary jurisdiction and commerce, a board of customs, a college, schools of hydrography and marine artillery, courses of geometry and mathematics, a society of arts, and an excellent naval museum, public library, government, pawn, and savings' banks, a theatre, with a stationary company, public baths, &c.

Toulon appears to have existed in the time of the Romans. In more modern times it was occasionally attacked by African corsairs, and to defend it from these incursions, Louis XII. commenced the erection of the Grosse-tour at the entrance to the inner road, which was finished by Francis I. Henry IV. commenced the construction of the old port, now appropriated to merchant vessels, in 1594. But it is wholly indebted for its modern importance as a great naval port and a strong military position, to Louis XIV., who expended vast sums on its fortifications, and on the arsenal and harbour. It was unsuccessfully besieged by Prince Eugene in 1707. Having, in 1793, been delivered up by the royalists to the English and Spaniards, it was retaken by the republicans, after a siege in which Napoleon gave the first decided proofs of his extraordinary military talents. On evacuating the town the allies set fire to the magazines, and to the ships they were unable to carry off; the fortifications have since been thoroughly repaired, and several new works constructed, so that it is stronger now than ever, and if properly garrisoned would be all but impregnable. (*Ports and Arsenals of France*; Hugo, *Art. Var.*; *Guide des Voyageur en France*; *Dict. Geog.*)

TOULOUSE, or THOULOUSE (an. *Tolosa*), one of the principal and most ancient cities in the S. of France, dep. Haute Garonne, of which it is the cap., on the Garonne, at the junction of the canals of Languedoc and Brière with that river, 132 m. S.E. Bordeaux; lat. 43° 35' 46" N., long. 1° 26' 30" E. Pop., in 1836, ex. com., 66,615; or with com., 77,372. Inglis calls it a fine large flourishing place, situated in the midst of abundance, and containing many fine edifices, and remarkable objects (*Switzerland and the Pyrenees*, 219); but, according to Hugo, Toulouse, before the revolution of 1789 was *fort laide*, and is still far from being a fine town, notwithstanding all that has been done for its embellishment. It is very irregularly laid out; its streets, which are mostly narrow, crooked, ill-paved with rounded pebbles from the river, and dirty, form a complete labyrinth. Nearly all the buildings are of red brick cemented with bad mortar, which, being blackened by age, gives the town a gloomy appearance; the older houses, and those in the lower quarters, consist of sun-dried bricks, in frames of wood work, and are greatly dilapidated. But within the last 20 years the civic authorities have undertaken many improvements : *on enlargit, on redresse les principales rues ; on plâtre ce qu'on ne peut rebâtir ; on assainit ce qu'on ne peut embellir ; on déblais les anciennes places, on en forme de nouvelles ; on jette à bas des constructions religieuses, inutiles depuis la suppression des couvents ; on utilise des églises abandonnées.* (Hugo, art. *Haute Garonne*.) The shape of Toulouse is an irregular oval ; the city comprises an island in the Garonne, and on that side is bordered by good quays : on other sides it has been till lately inclosed by walls, flanked with large round towers. But these are gradually disappearing; and their place is being occupied by good houses, and regular streets. Inglis says that he had not seen any provincial town in France with such extensive suburbs as Toulouse. The city communicates with the suburb of St. Cyprian, across the Garonne, by a massive free-stone bridge of seven arches, built in the latter half of the 16th century, at the further end of which a modern triumphal arch has been erected.

The principal open space is the *Place du Capitole*, serving for the grand daily market, which, according to Inglis, is admirably supplied. The large quadrangle is ornamented at each of its four corners with a handsome fountain. The capitol, or town-hall, from which it derives its name, on its E. side, in the Ionic order, is nearly 130 yards in length. In it are several spacious halls ; one is ornamented with the busts of the most distinguished natives of the city ; and another, called the *Salle de Clemence Isaure*, has a marble statue of that distinguished lady, the great patroness of the "Floral Games." The theatre occupies one wing of the capitol ; but it is badly planned and decorated. The other sides of the square are chiefly occupied by hotels and cafés. Several of the other squares are ornamented with fountains, and planted like the public promenades on the banks of the canals, and the Garonne.

Several of the churches are worth a visit. The cathedral is planned on a magnificent scale, but unfinished: the nave, which is the oldest part of the building, probably dates

from about the 12th century. The most ancient ch. that of St. Servin, has been erected, according to Hug. on the site of a very celebrated temple of Apollo : it is a heavy Gothic building, part being said to date from the 9th century, though the greater portion is much more modern, particularly the choir. Among its ornaments, if so we say call them, Inglis noticed a bas-relief caricature of Calvin, as a hog in a pulpit preaching ! and that there might be no mistake in the matter, the words, *Calvin le porc prêchant*, were inscribed below ! This church is rich in relics presented by several popes, and other persons : it has a cupola supporting a lofty spire. The churches of La Daurade and Dalbiade deserve mention : in the former Clemence Isaure is supposed to have been buried, and on the grand altar are preserved the golden flowers presented to the successful poets at the floral games. Toulouse is said to have had, at one period, 80 churches, but many have been converted to other purposes. One serves for the museum, in the picture gallery of which are some productions of the best masters of Italy, with a much larger number of copies ; while in the cloisters attached is a fine collection of antiquities excavated near Martres, in 1827. The best modern building after the capitol, is the prefecture, formerly the archbishop's palace.

Toulouse was till recently ill supplied with water : but it is now amply provided with that great necessary, furnished to numerous public and private fountains from a handsome *château d'eau*, or reservoir. Among the other public buildings may be specified two large hospitals, the *Ponts-Jumeaux*, or double bridge over the two canals at their junction, the veterinary school, public slaughter-house, mint, new edifice for the royal court and tribunal of primary jurisdiction ; the public libraries, containing together about 60,000 vols., school of artillery, barracks, arsenal, polygons, gunpowder and other large mills, &c. Toulouse is the cap. of the 10th military div. of France, and an archbishop's see. Previously to the revolution it was the seat of one of the leading French universities ; and it has now a university academy, with faculties of law, sciences, literature, and theology : it has also a royal academy of sciences, &c., societies of medicine, painting, arts, and agriculture, some of which grant prizes to successful candidates, after the example set at the floral games of old. The *jardin des plantes* at Toulouse is the largest and finest in France after that of Paris. (*Guide du Voyageur, &c.*)

The floral games, previously alluded to, appear to have been instituted in the early part of the 14th century. They were originally held on the 1st of May ; and consisted of a trial of skill among the poets and *troubadours* of the vicinity, those who recited the best verses receiving the prize, which consisted of golden flowers. Clemence Isaure, the lady mentioned as the great patroness of these games, bequeathed, in 1540, the bulk of her fortune to the civic authorities, to be expended by them in *fêtes* and prizes at floral games, to be held annually in her house, on the 1st and 3d of May. These games were finally superseded by the creation of the academy of *belles lettres* in 1694, the directors of which gave prizes for the best papers.

Toulouse has manufactures of coarse woollen cloths, silks, gauzes, printed cottons (*indiennes*), cotton yarn, flax and steel wares, paper, wax-lights, musical strings, and vermicelli, with dyeing-houses, distilleries, a cannon foundry, and a royal tobacco manufactory. It has also a large trade in the produce of the surrounding country, Spanish wool, and colonial produce. Inglis says, "The neighbourhood of Toulouse will be found one of the cheapest places of residence in Europe. Within the city, every thing is about one fourth part dearer than in its immediate vicinity, owing to the *octrois*. But in the markets held in the neighbouring villages, meat is sold at 3d. and 3½d. per lb. ; fowls are not more than 10d. a pair ; a fine turkey costs but 2s. 6d. or 3s. ; eggs, fruit, and vegetables are remarkably abundant and cheap ; and wine does not exceed 1½d. per bottle. The country is thickly covered with country-houses ; and one of these, furnished, and suitable for a small family, and with an excellent garden, may be had for 400 fr., or £15 per annum."

Nothing is known of the origin of this city but that it is very ancient. It was the cap. of the Tectosages ; and having been taken by the Romans, anno 105 B.C., they afterwards embellished it with numerous splendid edifices : but owing to the combined influence of time and the attacks of the barbarians, these structures have been almost wholly destroyed, so that some vestiges of the amphitheatre, and of a few other buildings, are all that now remain to attest the wealth and power of its Roman masters. It was successively the cap. of the Visigothic kingdom of Gaul and Aquitaine, and was thenceforward governed by its own feudal counts till 1271, when it was annexed to the crown of France.

Toulouse is principally celebrated in recent times for the sanguinary conflict that took place in its vicinity on the 10th of April, 1814, between the allied army, under the Duke of Wellington and the French, under Marshal Soult. The allies were superior in point of numbers, but the advantage of position was on the side of the French. Notwithstanding a desperate resistance, the latter were driven from the Mont Rave, and obliged soon after to evacuate the city. The loss on both sides was very great, especially on that of the victors, who had 4650 men killed and wounded: the French loss amounted to about 3000 men. Unfortunately, this was a useless sacrifice, as Napoleon had already abdicated; but, though the contrary has been stated, it is certain that Marshal Soult was wholly ignorant of the circumstance. (*Napier's Peninsular War*, vi. 639, &c.)

The inhabitants of Toulouse appear, even in the estimation of their countrymen, to be endued with a very large share of that versatility which has been said, though perhaps without much foundation, to be a distinguishing trait in the French character. "Comme l'exécution n'est pas précédée chez le Tolousain d'un jugement réfléchi, il se livre malheureusement, avec trop de facilité, aux excès dont ensuite il a lieu de se repentir: ainsi, on le voit massacrer, en quelque sort avec joie, les Protestans lors de la St. Barthélemy, et durant les guerres du Calvinisme; le vertueux président Durantz, à l'époque de la Ligue; les royalistes pendant la Revolution, et les patriotes à la Restauration: il accueille avec transport Napoléon durant son règne, et plus tard, avec le même enthousiasme, Wellington et son armée; il assassine ensuite le général Ramel, qui veut le sauver de l'anarchie: ainsi, il est toujours le jouet de la versatilité de son caractère, et l'instrument le plus docile de toutes les scènes de barbarie auxquelles le poussent les hommes ardens qui veulent l'égarer." (*Dict. Géog.*)

Toulouse has given birth to many distinguished individuals, among whom may be specified Cujas, the greatest civilian of modern times, born here in 1520; Raymond, count of Toulouse, so celebrated in the first crusade: M. de Villele, late minister of France, &c. (*Martinière, Grand Dictionnaire*, art. *Toulouse*; *Hugo*, art. *Haute Garonne*; *Dict. Géog.*; *Guide du Voyageur*; *Inglis's Switzerland*.)

TOURNAY, (Flem. *Doornik*), a town of Belgium, prov. Hainault, cap. arrond. and of two cantons, on the Scheldt, close to the French frontier, 45 m. W.S.W. Brussels; lat. 0° 36′ 20″ N., long. 2° 23′ 17″ E. Pop. in 1836, 26,919. Tournay covers nearly as much ground as Lisle, though so much less populous. Its former fortifications were demolished by the emperor Joseph II., but since 1814 it has been surrounded anew with military works, and has a good citadel. The Scheldt, crossed here by several flying bridges, divides Tournay into two parts, called the old and new towns: the latter is well built, and has a fine quay along the river, which forms a favourite promenade; but, excepting its historical recollections, the former has little to render it interesting.

The cathedral, a fine Gothic building, with five towers and spires, supposed to have been a bishop's see as early as the 5th century, was formerly richly adorned, but suffered greatly from the French revolutionary phrenzy. The old abbey of St. Martin has been of late years levelled with the ground, to give place to the town-hall and public gardens. Few other buildings are worthy of notice; though there are several hospitals and asylums, including one for aged ecclesiastics, a court of primary jurisdiction, chamber of commerce, exchange, theatre, athenæum, academy of fine arts, episcopal seminary, many good private schools, a *mont-de-piété*, &c. Without the walls are several suburbs. Tournay is one of the most active manufacturing towns of the Netherlands, and must have been celebrated for its industry at a very early period, since it is mentioned in the *Notitia Imperii* of the 5th century as one of the fifteen towns in the empire having manufactures of woollen and linen yarn. At present, Vandermaelen states, that three fourths of its population are employed in its various manufactures, and that from 12,000 to 16,000 looms are supposed to be employed in its commune. (*Dict. de Hainaut*.) Hosiery, calicoes, and linen fabrics, camlets, yarn of various kinds, waistcoats, and all kinds of articles of dress, with carpets, for which Tournay is deservedly famous, are the principal articles of trade; but large quantities of earthenware, bronze goods and hardware, curaçao, and other liqueurs, are made; and there are numerous breweries, distil-refineries, tanneries, drying-houses, &c. Full-aged cotton-spinners get about 1 fr. 35 c. a day, working about 12 to 14 hours; but one fourth part of the hands employed are children. Weavers get from 1 fr. 25 c. to 1½ fr. In the royal carpet factory, the workmen obtain from 2 to 2½ fr. a day, though there also many children are employed, who labour at five years of age. (*Hand-loom Weavers' Report*.) Tournay has various charitable institutions for the benefit of sick workpeople, for pensioning the widows of weavers, &c. and children are not suffered to be too laboriously employed. (See *Vandermaelen's Dict*.) The Scheldt, which is navigable to the town for vessels of 150 tons, is the prin-

cipal channel for the conveyance to the town of coals, spices, dyeing materials, tobacco, deals, brandy, wines, and for the export of the manufactured goods, chalk, building stone, oil, &c., produced in the town and its vicinity. Tournay has probably experienced as many vicissitudes as any town in Europe. It is the *Civitas Nerviorum*, taken by Julius Cæsar, and has since belonged to an infinite number of masters, and been taken and retaken over and over again. (*Vandermaelen Dict. de Hainault; Dict. Geog.; De Cloet; Parl. Reps.*)

TOURS (an. *Cæsaromagus*), a city of France, dep. Indre-et-Loire, of which it is the cap.; it is surrounded by extensive and fertile plains, and is itself placed on the narrow tongue of land between the rivers Loire and Cher, close to the point of their confluence, 127 m. S.W. Paris. Lat. 47° 23′ 46″ N., long. 0° 41′ 38″ E. Pop., in 1836, 26,669. "Tours is well known as a favourite retreat of English absentees. Great part of the town is new; and the streets, several of which are spacious, and the houses clean, substantial, and many elegant, give it an air of ease, pleasure, and abundance, possessed by few other cities in France. The beauty of Tours has arisen since the Revolution, and has, indeed, sprung out of it, for great part of it has been rebuilt upon an improved plan." (*Inglis*, p. 352.) And, in fact, the older parts of the city still consist of narrow, crooked, and dirty streets, with mean and ill-built houses. It is surrounded by planted boulevards on the site of its ancient fortifications; it has 12 different entrances, and five suburbs. It communicates with the opposite bank of the Loire by one of the finest bridges in Europe, constructed chiefly between 1762 and 1777: it is of stone, level on the summit, 475 yards in length by 16 in breadth, and has 15 arches, each 26¼ yards in span. Over the Cher are two bridges, one of 17 and the other of 8 arches. From the bridge over the Loire a noble street, the *Rue Royale*, straight, spacious, bordered with footways, and lined with uniform buildings of freestone, intersects the town in its entire breadth, terminating on the S. in the *avenue de Grammont*, leading to the smaller bridge over the Cher. At the commencement of this street, close to the Loire, is a handsome square; in which are the town-hall and the departmental museum, new and symmetrical buildings, the latter containing upwards of 200 paintings. The cathedral is said to have been founded in the 4th century, burnt down in the 6th, rebuilt by Gregory of Tours, but again burnt down in the 12th century; after which its reconstruction proceeded so slowly that it was not completed till 1550. It has a noble front, flanked by two towers, built by Henry V. of England. Its interior, though not beautiful, is richly ornamented, and contains much stained glass, together with the mausoleum of the children of Charles VIII. A curious collection of MSS. is kept in this cathedral. The other churches are mostly small and gloomy, and possess little worth notice. The so-called tower of Charlemagne is the only remaining portion of the abbey of St. Martin, destroyed in 1797, of which the kings of France used to be the abbés. The archbishop's palace is one of the handsomest in the kingdom: the prefecture, courthouse, college, general hospital, exchange, theatre, barracks, prison, and a highly ornamented fountain in the market-place, are the other most conspicuous objects. This city is the seat of courts of primary jurisdiction and commerce, a chamber of commerce, council of prud'hommes, societies of agriculture, sciences, arts, &c.; and has courses of practical geometry, a library said to comprise 40,000 volumes, a cabinet of natural history, and botanic garden. It was here that Louis XI. established the silk manufacturers he invited out of Italy; and it was for a considerable period famous for its silks; but it has long been far surpassed in this department by Lyons, which has peculiar advantages for the prosecution of the silk manufacture. Next to that of silk, which is carried on to a considerable extent, the manufactures of Tours consist principally of woollen cloths, carpets, and woollen yarn; but they are not extensive; and the trade of the city is chiefly in the retail supply of its inhabitants and visiters.

"The promenades round Tours are charming: among these the elm-avenue is the most conspicuous; the quay is also pleasant. The environs of the city furnish the most agreeable walks; innumerable little paths lead in every direction through the fields, and among the knolls and copses. Tours, 20 years ago, was as cheap a residence as any place on the Loire; but a great advance in the prices of every thing, and particularly house-rent, has naturally followed the approbation of Tours by the English. Immediately after the war, a large house, with every convenience, and a garden of two or three acres, might have been had for £20 a year; but this sum may now be more than doubled. Provisions are still moderate in price, and wood is less expensive here than in most other parts of France." Besides the English, Tours is much resorted to by French gentry, who, though in independent circumstances, are not rich enough to afford the expense of living in the metropolis. The cas-

TOUMEN.

TRALEE.

tle of *Plessis les Tours*, built by Louis XI., where he principally resided, is about 1 m. from the city. It is constructed of brick; is embosomed in wood; and has a venerable appearance. In its chapel is a portrait of Louis, dressed in armour, taking off his helmet to the Virgin and Infant. (*Inglis*, p. 354.)

Among the eminent men to whom Tours has given birth, may be specified Cardinal Amboise, prime minister of Louis XII., and Rapin, the author of the much admired Latin poem, *De cultu Hortorum*, and of several critical publications. Rapin has not forgotten to celebrate the praises of his native city and the surrounding country:—

Adde omnem istam rivis, et fontibus oram,
Pratorum immensos tractus, et amoena serendum
Flumina, vitiferosque utroque ex litore colles:
Quid memorum variis opulentam mercibus urbem,
Et studia, et mores populi, quem serica texta
Tractantem, facit culti clementia blandum?
Adde umbris nemorum storum: et mollis temper
Gramina prata novo, et nusquam sine floribus hortos.
Lib. i., line 489, ed de Brotier, Paris, 1788.

Grecourt, the poet, was also a native of Tours, and St. Gregory, hence called Gregory of Tours, was for a lengthened period bishop of the see.

Tours was anciently the capital of the *Turones*, conquered by Cæsar, *anno* 55 B.C. In the 5th century, it became the capital of the 3d Lyonnaise. After many vicissitudes it fell into the hands of the Plantagenets; and formed a part of the English dominions, till 1204, when it was annexed to the French crown. (*Dict. Geog.; Hugo* art. *Indre-et-Loire; Guide du Voyageur.*)

TOUMEN, a considerable town of Asiatic Russia, gov. Tobolsk, on the Toura, 190 m. S.W. Tobolsk. Pop., in 1835, 9213. It is situated in a fertile tract, and its inhabitants are said by Cochrane to be both wealthy and hospitable; though, according to Dobell, little can be said in favour of their morals. (*Travels*, ii., 115.) In almost every house the manufacture of a coarse kind of carpeting sold all over Siberia is carried on; and its tanneries, which are more extensive than any others in the government, employ nearly 300 workmen, and produce goods to the value of more than 1,000,000 roubles a year. (*Journ. de St. Petersbourg*.) In the neighbourhood are extensive forests, and vast quantities of mats, with carriages, and various wooden articles, are made for exportation; besides which the town has a large trade in timber, tallow, hides, embroidery, vegetables grown in the vicinity, cattle, &c. It was the first town founded by the Russians in Siberia, having been built in 1586, on the site of a previous Tartar city called *Tsinghis-Tora*, or "the town of Genghis." (*Cochrane; Dobell; Erman, Reise um die Erde; Journ. de St. Petersbourg*.)

TOURNUS, a town of France, dep. Saône-et-Loire, cap. cant., on the Saone, here crossed by a bridge of five arches, between Macon and Chalons, 16 m. N.N.E. the former, and about the same distance S. the latter. Pop., in 1836, ex. comm. 4480. It stands on a declivity crowned by the remains of a Benedictine abbey, which formerly possessed extensive privileges. It is clean, well-built, and has some good public edifices. Its trade is principally in corn, wine, and building stones, sent down the Saone to Lyons. (*Hugo, &c.*)

TOWAMENSING, t., Northampton county, Pa., 87 m. E.N.E. Harrisburg, 194 W. Bounded W. by Lehigh river. It contains a Lutheran church, six stores, three lumber-yards, one fulling-mill, two furnaces, one forge, five grist-mills, eight saw-mills; five schools, 140 scholars. Population, 1847.

TOWANDA. p. t., capital of Bradford co., Pa., 134 m. N. by E. Harrisburg, 244 W. Bounded E. by Susquehanna river. Drained by Sugar creek. It contains bituminous coal. The borough is situated on the N. branch of Susquehanna river, and contains a courthouse, jail, 16 stores, one furnace, one tannery, one distillery, three printing-offices, three weekly newspapers; one academy, 80 students; one school, 63 scholars. Pop. 912. The town, exclusive of the borough, has one fulling-mill, one woollen factory, one furnace, two grist-mills, three saw-mills, one oil-mill; four schools, 227 scholars. Pop. 1009.

TOWCESTER, a market town and par. of England, co. Northampton, hund. Towcester, on the Tow, here crossed by three bridges, 8 m. S.W. Northampton. Area of par. 2790 acres. Pop., in 1831, 2671. The town consists principally of three streets, at the union of the roads from Stony Stratford, Northampton, and Daventry. It stands on the ancient Watling Street, and was probably a Roman station. The church is a neat edifice, supposed to date from the 11th century. The living, a vicarage, worth £317 a year, is in the gift of the bishop of Lichfield and Coventry. Here are chapels for various dissenting sects; a grammar-school, founded at the dissolution of the monasteries, and endowed with part of the revenues of a college dating from the reign of Henry VI.; several almshouses, &c. Markets on Tuesdays: fairs four times yearly.

940

TOWNSEND, p. t., Windham co., Vt., 117 m. S. Montpelier, 439 W. Watered by West river. It contains two churches, a Congregational and Baptist, the Leland comical and English school with a brick building 54 by 36 feet, and a large boarding-house for its students, four stores, one fulling-mill, three grist-mills, four saw-mills, two tanneries; one academy, 156 students; nine schools, 438 scholars. Pop. 1348.

TOWNSEND, p. t., Middlesex co., Mass., 41 m. N.W. Boston, 437 W. Drained by Squanticook river, a branch of Nashua river. It contains two churches, a Congregational and a Unitarian, six stores, one cotton factory with 256 spindles, one furnace, four grist-mills, nine saw-mills, two tanneries; one academy, 70 students; 11 schools, 655 scholars. Pop. 1892.

TRAFALGAR (cape), a promontory of the S.W. coast of Spain, prov. Andalusia, 25 m. N.W. Tarifa, in the strait of Gibraltar, of which, indeed, it forms the N.W. extremity, lat. 36° 10′ 13″ N., long. 6° 1′ W. This cape, which in antiquity was called the promontory of Juno (*Junonis promontorium*), is low, and terminates in two points, the most easterly of which is surmounted by a round tower.

Cape Trafalgar is famous in naval history for the great battle fought in its vicinity on the 21st of October, 1805, between the combined French and Spanish fleet, under Admirals Villeneuve and Gravina, and the English fleet under Lord Nelson. The former had 33 sail of the line and 7 large frigates, while the fleet of the latter only amounted to 27 sail of the line and 3 frigates; but the superior skill and gallantry of the British admiral, and of his officers and men, far more than compensated for the nominal inferiority of the English fleet, and secured for the country the greatest naval victory recorded in her annals. No fewer than 19 French and Spanish line of battle ships were captured, and 4 that had escaped from the action were subsequently taken by Sir Richard Strachan; the other vessels that escaped into Cadiz being, at the same time, mostly rendered unserviceable. Unfortunately this great and decisive victory was not acquired without a very heavy loss. Nelson, who was mortally wounded early in the action, lived only to be made aware of the destruction of the enemy's fleet.

TRAJANOPOLI (called by the Turks *Orikhove*), a town of European Turkey, prov. Roumelia, sanj. Gallipoli, on the Maritza, 45 m. S.S.W. Adrianople, lat. 41° 7′ 3″ N., long. 26° 18′ 15″ E. It is said to have a population of 15,000 inhabitants; is the see of a Greek archbishop, and has a considerable commerce; but it lies so much out of the routes usually resorted to by travellers, that we have but little information respecting it.

TRALEE, a parl. bor. and marit. town of Ireland, prov. Munster, co. Kerry, of which it is the cap., within about 3 m. of the head of Tralee bay, near the Ballymullen river, 55 m. N.W. Cork. Pop. of parl. bor., which includes an area of 534 acres, in 1831, 9562, and now (1848, probably 13,000. Tralee is a rapidly-increasing, thriving town. At the close of the late war it was little else than a congregation of cabins; but now, to use the words of Mr. Inglis, "it has streets that would not disgrace the best quarters of any city; and these not streets of business, which it also has, but streets containing gentlemen's houses, or, at all events, houses which no gentleman might be ashamed to live in. I have no hesitation in pronouncing Tralee altogether the most thriving town I have seen since leaving Clonmel; and in some respects it leaves Clonmel behind it. Its retail trade is extensive and improving, and many of the dealers are wealthy. As good shops are to be found in Tralee as in Cork; and the stock in many of them is very extensive. I was at Tralee on a market day, and I do not recollect to have seen a busier place. Independently of an extensive supply of country produce, there was a very abundant exhibition of all kinds of manufactured goods and apparel, and every shop in the town was crowded to the door. House rent is high here; higher, in fact, than in any English county town. There is a spa in the vicinity a good deal resorted to for its waters: its situation is pleasant, and a number of pretty country houses have been erected in the neighbourhood." (i., 252, 263.)

The public buildings and establishments comprise a handsome parish church, two large Roman Catholic chapels, a nunnery, to which an excellent school for girls is attached, several meeting-houses for dissenters, a new county courthouse, "which is in every way a handsome and commodious structure" (*Inglis*), a fever hospital, a lunatic asylum, a county infirmary, infantry barracks for 650 men, two breweries, a distillery, &c. The town stands on the estate of Sir Edward Denny, who has thrown open the pleasure-grounds, attached to the castle in its immediate vicinity, to the inhabitants. Sir E. Denny is also patron of the living of Tralee, worth above £400 a year. Exclusive of the girls' school, Tralee has a Catholic free-school, and other schools, two of which are subordinate to the Board of education in Dublin. It sent two members to the Irish House of Commons, and since the union it has sent one member to the Imperial House of Commons. Previously to the Reform Act the right of election was nominally vested in the old corporation, dissolved by the act 3 and 4 Victoria, cap. 108 ; but it was, to all intents and purposes, a nomination borough ; and it is stated by the Boundary Commissioners, " that it has generally happened that the people of the town and their parliamentary representative were total strangers to each other ; and it is a fact worth mentioning, that several intelligent and respectable persons, among whom was the provost himself, could not tell us the name of the present member !" Reg. electors, in 1839–40, 296.

The county assizes are held here ; and general sessions four times a year ; and petty sessions on Tuesdays. It is, also a constabulary and coastguard station. Markets on Tuesdays and Saturdays : fairs May 3, Aug. 4, Oct. 9, Nov. 7, and Dec. 13. Post-office revenue, in 1830, £1237 ; in 1836, £1511. The Provincial bank had a branch opened here in 1826, the bank of Ireland and the National bank in 1835, and the Agricultural in 1836.

The port is included in that of Limerick. Owing to the shallowness of the water in the river, barges of more than forty tons were, till lately, loaded and unloaded at Blennerville, about 2 m. S.W. from the town, while vessels of greater burden were compelled to load and unload by means of lighters, at the Samphire islands, in the bay about 6 m. W. from the town. In the view of obviating this inconvenience, a ship canal, 15 ft. deep, has been cut from the bay to a basin adjacent to the town, where vessels of 300 tons may now load and unload. If this canal have the success that has been anticipated, it will be of great service to the trade of Tralee. The value of the exports, which principally consist of corn and provisions, amounted, in 1835, to £49,315.

It is a singular and not easily-explained fact, that, notwithstanding all the proofs of prosperity found in Tralee and its vicinity, wages are extremely low, employment scarce, and the condition of the lower classes very much depressed.

TRANI, a seaport town of S. Italy, kingd. Naples, prov. Terra di Bari, cap. cant. on the Adriatic, 26 m. N.W. Bari, lat. 41° 17′ 5″ N., long. 16° 25′ 36″ E. Pop. 13,600. It is surrounded by a bastioned wall, with a fosse on the land side, and has a citadel, but is not a place of any strength. The streets, which are narrow and dirty, are flanked with ill-built houses, except round the port, where there are some private dwellings that would not disgrace the best parts of Rome. Its large cathedral, erected more than six centuries ago, is said by Swinburne to be in a very mean taste, with preposterous ornaments and clumsy pillars. Exclusive of the cathedral it has about 20 other churches, with six convents for monks (one of which, the monastery of St. Clare, is a magnificent structure), two nunneries, an orphan asylum, a large seminary, and a theatre, said to be inferior only to those of Naples. Trani is the seat of one of the great civil courts of the kingdom, of a superior criminal court, of a civil tribunal for the province, and is the residence of many old families. It labours under a great deficiency of spring water, so that the inhabitants are obliged principally to depend on rain water collected in cisterns. It exports corn, oil, sweet wine of good quality, figs, almonds, and other products of the vicinity. Some cotton stuffs are produced in the town, of cotton raised in the cant. The ramparts command a fine view both towards the interior and the sea.

The harbour, which is nearly encircled by the town, has naturally deep water, but owing to the accumulation of sand thrown in by the sea, and of the filth from the surrounding houses, it is so much filled up as to be accessible only to the smallest boats, while in summer the stench is intolerable. Of course, were it in the hands of a vigorous enterprising people, it would be very soon cleaned out. The few vessels that carry on the languishing trade of the town are obliged to anchor about 2 m. off shore, being laden by lighters.

In 1502 a contest took place under the walls of this town, between 11 French, and as many Spanish knights. The combatants fought till there remained only six Spanish and four French knights : the latter then alighted and defended themselves behind their horses, as behind a rampart, till might put an end to the contest. (Swinburne's Two Sicilies, i., 180 ; Orsera's Tour, p. 99 ; Rampoldi, &c.)

TRANQUEBAR, a town and seaport of Hindostan, belonging to Denmark, on the Coromandel coast, surrounded by the British district of Tanjore, between two arms of the Cavery, 140 m. S.S.W. Madras, lat. 11° 0′ 15″ N., long. 81° 54′ 30″ E. Pop. of the town and its small territory about 20,000. Tranquebar is surrounded by bastioned ramparts faced with masonry, and at its S.E. angle is the citadel of Dansburg, in which is an old castellated building, serving for the government offices, and having a lighthouse on its

highest point. The town is small but very neat and clean, there not being a native hut or other mean structure within its walls. The principal streets may be called handsome, the whitened houses being of two or three stories, with little Grecian porticoes of three or four pillars projecting into the street, and windowed generally with rattan lattices. The government house, two Protestant churches and a Portuguese Roman Catholic chapel are in the town, the religious missions at Tranquebar are said by Malcolm to have greatly declined of late. (S.E. Asia, ii., 68.) There is no harbour in the Cavery for vessels of a larger class than boats, which have accordingly to anchor outside the surf in the bay. It has, however, some traffic by sea with Bengal, the Malabar coast, the straits' settlements, Ceylon, &c. ; it has also manufactures of salt, and cotton goods. The revenues are derived from the government share of the rice cultivation, the sale of arrack, tobacco, fish, oil, &c., and the customs, but they are scanty. (E. I. Gazetteer, &c.)

TRANSYLVANIA (Germ. Siebenbürgen, Magy. and Slav. Erdeli, an. Dacia Mediterranea), the most E. prov. of the Austrian empire, comprised between the 45th and 48th degs. N. lat., and chiefly between the 22nd and 26th E. long., having Hungary on the N. and W., and on the E. and S. Moldavia and Wallachia, from which it is separated by the main chain of the Carpathians. It is of a square shape : greatest length and breadth about 140 m. each. Area estimated at 20,400 sq. m. Pop., in 1839, 2,064,900. (Berghaus, Allg. Länder, &c.) Most part of the surface is covered with ramifications of the Carpathian mountains, which rise in mount Bukhest, near Kronstadt, to nearly 8700 ft. in height : these, however, give place in the N. to the valley of the Szamos, in the centre to that of the Maros, and in the S. to that of the Aluta. All these rivers, of which the Maros is the principal, rise in Transylvania, and have, more or less, a W. course, the general slope of the country being towards the W. The first two are tributaries of the Theiss, the last joins the Danube in Wallachia : the banks of all, and particularly the Maros, are densely wooded (whence the modern name of the province), and possess considerable picturesque beauty. It is in general well watered, and in the S. are some extensive marshes. As the country at large is rather an elevated table land, the climate is cold, though in most parts healthy. The soil is of very various qualities ; the mountains are generally granitic or calcareous, but the plains and valleys are often very fertile, and, notwithstanding the backwardness of agriculture, a surplus of corn over the quantity required for home demand is generally produced. Wheat, barley, oats, rye, buckwheat, and maize, most kinds of pulse, potatoes, and garden vegetables, are cultivated ; wine is one of the leading products of the country ; in the orchards apples, pears, plums, apricots, almonds, mulberries, chestnuts, &c., are grown ; and tobacco, hemp, flax, saffron, and clover, are ordinary crops. The lands are, in general, held under a feudal tenure, as in Hungary, except in the Saxon-land, which division of the province is by far the best and most industriously cultivated.

Transylvania is divided principally among three distinct nations ; the Magyar, the Szekler, and the Saxon, each of which has a share in the government of the country. They inhabit different districts : the Magyars (with the Wallachs) occupy the whole W. and centre ; the Szeklers the E. and S.E. ; and the Saxons the greater part of the S., with a portion of the N.E. (Paget, ii., 360 and Map.) The first occupy at least three fifths of the entire principality, of which Clausenberg is the capital, and the Szeklers and Saxons about one fifth each ; Maros-Vasarhely being the chief town of the former, and Hermanstadt of the latter. With these races are intermixed a number of Poles, Gypsies, Jews, Armenians, &c. We subjoin a statement, which is probably not far from accurate, of the number of the different races of people inhabiting Transylvania, and of the numbers attached to the different religions :

Races of People		Religions	
Magyars	597,000	Roman Catholics,	
Wallachs, &c.	1,336,000	United Greek	480,000
Szeklers .	540,000	*Greek Catholics	990,000
Saxons .	200,000	Calvinists .	635,000
Poles .		Lutherans .	200,000
Gypsies .	40,000	Unitarians .	53,000
Jews, Greeks, Armenians, &c.	20,000	*Jews and all other sects .	632,000
Total	2,060,000	Total	2,000,000

The Magyars and Wallachs have been already described (ante, p. 1139, 1139, &c.). The Szeklers termed by the Latin writers of the lower Empire Siculi, are probably the descendants of a barbarian horde that had settled in the province during the decline of the Roman power. The

* The religions of the sects thus marked are tolerated only, whereas the others are established, by government.

Magyars, on entering the country in the 10th century, finding the Szeklers cognate with themselves in features, language, character, &c., left them in the undisturbed possession of the lands they had inherited, on condition of their guarding the Magyar frontier on that side. They were not even rendered tributary, and to this day the Szeklers hold themselves to be noble born, free, and equal. But in the lapse of centuries many changes have crept into their condition. "The richer and more powerful have gradually introduced on their own estates the system in operation in other parts of Transylvania, and the peasant and the seigneur are now found in the Szekler land as elsewhere. Titles, too, and patents of nobility have been freely scattered through the country; taxation, also, and the forcible introduction of the border system instead of the desultory service of former times, have made great changes. As almost all these changes, however, have been introduced without the consent of the people, and often by the employment of open force, they are still regarded as illegal by the Szeklers, who are consequently among the most discontented of any portion of the Transylvanians." (Paget, ii., 390, 391.)

The Saxons appear to owe their origin to a colony transplanted thither from the Rhine by one of the sovereigns of Hungary in the 12th century. They live under a count, or chief, who, like their clergy, is elected by themselves; and they enjoy freedom from tolls within their district, and other important privileges. "One of the fundamental laws of the Saxons is the equality of every individual of the Saxon nation. They have no nobles; no peasants. Not but that many of the Saxons have received letters of nobility, and deck themselves out in all its plumes; yet, as every true Saxon will tell you, that is only as Hungarian nobles, not as Saxons. Their municipal government was entirely in their own hands; every village chose its own officers, and managed its own affairs, without the interference of any higher power. But a few years ago, a great and completely arbitrary change was made in this institution; the effects of which have been to deprive the Saxon communities of the free exercise of their privileges, and to deliver them into the power of a corrupt bureaucracy, over which they have little or no control. The Saxons, however, are a slow people, and, though they have long complained, they have scarcely ever ventured to demand a restitution of their rights. Hitherto they have been among the most certain adherents of the crown: they have rarely joined the liberal party. They preserve, for the most part, the dress, language, habits, &c., their ancestors brought with them from Germany. For the rest, the Saxons are undoubtedly the most industrious, steady, and frugal of all the inhabitants of Transylvania; and they are consequently the best lodged, best clothed, and best instructed." (Ibid. ii., 424-433, &c.)

The peasants of Transylvania are in a more depressed condition, and much more ignorant, than those of Hungary. Among the greatest evils to which the Transylvanian peasant has to complain, is the want of any well-defined code of laws to which he may refer. The peasant land, too, has never been classed, as in Hungary, according to its powers of production; nor has the size of the peasant's portion, or fief, been accurately determined. The amount of labour, therefore, cannot be fairly and legally proportioned to the quantity and value of the land. Nor is the amount of labour itself better regulated. In some parts of the country it is common to require two days a week; in others, and more generally, three are demanded; and in some the landlord takes as much as he can possibly extract out of the half-starved serfs settled on his estates. It is rare that the peasant's cottage has more than two rooms, sometimes only one; his furniture is scanty and rude, his crockery coarse, and those little luxuries which in the Hungarian denote something beyond the indispensable are rarely seen in Transylvania. The ignorance of the Transylvanian peasant is often intense; and he is generally superstitious and deceitful: these qualities are most conspicuous in the Wallachs, but the Magyars are by no means free from them. Schools are extremely rare. The peasants belonging to the Greek church are undoubtedly the most ignorant; those of the Unitarian and Lutheran churches the best educated.

"We had remarked," says Mr. Paget, "throughout the Szekler-land generally, a better state of cultivation and greater signs of industry than in most other parts of Transylvania. But the Saxon-land, on the Aluta, appeared like a garden in comparison even with the former. The whole plain seemed alive with ploughs and harrows, and on every side teams were moving about, manure was spreading, and the seed was being scattered abroad with a busy hand. The most startling feature in the picture was the very active part taken by the women. Some were sowing corn, others using the fork and spade, others holding the plough, and others driving the team." (Paget, ii., 311-316, 495-498.)

Transylvania may hereafter rank high as a wine growing

country: it abounds with declivities of a rugged or volcanic soil. No less than 1-9th part of its present population is dependent on the culture of the vine; all the gentlemen, and even superior tradesmen, grow their own wines. The mode of making them is very ill understood; but there are several superior kinds of wine produced, mostly in the valleys of the Maros and its tributaries. They are in general white, well-flavoured, and full bodied. The highest price it is an ordinary year of the better sorts is about 2s. the cruse (15 bottles).

The rearing of horses and other live stock is one of the most important branches of national industry. In the Szekler mountains a small wiry horse, similar to the Welsh pony, appears to be indigenous; but, for improved breeds, no less than 60 celebrated studs are said to exist in this small territory, 20 of which have probably a greater or less infusion of English blood, the English breed and modes of treatment of horses having been introduced of late years. Buffaloes, scarce in Hungary, are common here. The sheep, which are long-woolled and curly-horned, are sent into Wallachia to graze in the winter. The oak, beech, &c. forests, which are estimated to cover nearly 3,940,000 acres, feed large quantities of hogs.

Her mineral produce is a principal source of the wealth of Transylvania. There are numerous gold mines in the country, and almost every stream and river is auriferous; the annual produce of gold is estimated at from 2000 to 2200 marcs, and of silver 5000 marcs. (Austr. Nat. Encyc.) The gold mines of Zalathna, in the basin of the Maros, are supposed to have been wrought ever since the time of the Romans, and those round Nagy Banya are certainly of that era. From the latter, and some other mines, the ore is sent off monthly to Kremnitz, to be smelted. Gold-washing in Transylvania is almost monopolised by the gypsies. Government grants a gypsy band the privilege of washing the sands of a certain brook, on condition of their paying a yearly rent, which is never less than three ducats of pure gold per head for every washer. A gypsy captain settles this matter with the government, and is answerable for the rest of the tribe, from whom he collects the whole of their earnings, which he re-divides among them, after paying the tribute. (Paget, ii., 364, 365.) Iron, lead, copper, antimony, arsenic, mercury in the form of cinnabar, &c., are also found in Transylvania; and the mines of Inzkrem are the richest in tellurium of any in Europe, and these in which this metal was first discovered. Marshal Marmont states that coal, of very good quality, is found in some parts; but it is not made use of. Salt is much more important rock-salt abounds at Maros and Szamos-Ujvar, &c., about 600,000 centners being annually produced, which, excepting about 30,000 centners consumed in the neighbourhood, is wholly exported to Hungary. The miners work from 3 to 11 A.M., and get about 10d. a day. The centner of salt is delivered at the pit's mouth for about the same sum, and sold in Transylvania at 3½ florins, or 7s. the centner. The greater part, however, is sent by the Maros to Szegedin at an expense of 10d. more each centner, and sold there at 7½ guilden, or 15s. the centner. The E. of Transylvania is supplied from mines in the Szekler-land, where Paget (ii., 399) says he saw an entire hill of salt. This hill, in consequence of the strict monopoly exercised by the government over the article, was surrounded by guards to prevent the peasants from stealing the salt! Alum, sulphur, saltpetre, sulphate of soda, and many crystals and inferior kind of gems, are found in the province.

Except those of woollen, cotton, and some other fabrics in Cronstadt, Hermanstadt, and other parts of the Saxon-land, few manufactures are carried on to any great extent. Woollen and linen stuffs, cotton fabrics, &c., hats, leather, shagreen, potash, earthenware, paper, and gunpowder are made in different places; the clothing of the peasants being generally of domestic manufacture. Some forges, breweries, and vinegar factories are scattered over the country; but woollen, silk, and linen fabrics, jewellery, hardware, glass wares, &c., are principally imported from abroad. In return for salt, corn, cattle, horses, hogs, hides, wax and honey, timber, metals, and other raw produce. The trade is mostly in the hands of the Greeks and Armenians; and, as yet, little facility is afforded for commerce with Hungary and Wallachia. Retail tradesmen, who sometimes have kept dealings with Pesth and Vienna, will give money on bills, or transmit considerable sums for a per centage; but there is not one regular banker in the whole country! (Paget, ii., 477.) The Maros and Szamos are navigable, and are the chief routes for the conveyance of goods. The roads and bridges are everywhere in the most wretched state, and, except in a few towns, inns are unknown.

Transylvania sends deputies to the Hungarian diet, but has also a diet of her own, composed as follows:— Every county and free town sends its members: the Magyars about 46, and the Szeklers and Saxons 16 each. The Catholic church sends two members, representatives of abbeys;

the Catholic and United Greek bishops claim each a seat also. Besides these are the *Regalists*. Some of these are nominated by the crown for life; others, as the lords-lieutenant, privy councillors, and secretaries, have seats in virtue of their office. The number of regalists is said to have been limited to 59 by Maria Theresa; but this regulation has been grossly infringed, the present number exceeding 200! A governor, aided by a privy council, secretaries, and others corresponding with the Transylvanian chancery at Vienna—in other words, acting under the direction of an Austrian minister—constitute the executive body, whilst the legislative is formed by a diet to be held every year. The appointment of the executive is vested jointly in the diet and the crown, the former nominating for every office three individuals from each of the received religions, from among which twelve the crown appoints one. Besides the *candidation* of the executive, the duties of the diet may be said to consist in the making and altering of laws for the internal government of the country; the voting supplies of troops; the levying, but not voting, the contribution; and the conferring the *indigenat*, or right of citizenship, upon strangers." (*Paget*, ii., 372–373.)

Magyar Transylvania is divided into 11 counties (three of which, in the N., have been lately annexed to Hungary); the Szekler-land into five, and the Saxon-land into nine, *stuhls*, besides some subordinate districts. The government of the Magyar counties and Szekler stuhls, and of the towns, is nearly the same as in Hungary: that of Saxon Transylvania has been already noticed. In the capital of each county and stuhl is a court of primary jurisdiction, subordinate to the Transylvanian chancery at Vienna.

A band or zone of country along the S. and E. frontier, with a population of about 140,000, forms the Transylvanian military frontier. Here are maintained two Wallach and two Szekler infantry border regiments, and one regiment of Szekler hussars. The inhabitants of this tract are subject to the Austrian military frontier laws. (*See* SLAVONIA, *ante*, p. 890, 891.)

The majority of the clergy, and particularly the Wallach priests of the Schismatic Greek church, are little superior to the peasantry in point of education. Those of the United Greek church are better educated, having a lyceum, gymnasium, and normal school at Balasfalva, and enjoy the same general privileges as the clergy of the Roman Catholic faith, which is that most favoured by the government, and entitled to the tithes in case of dispute. The great body of the Protestant clergy is also derived from the poorer classes of society; and its members, during the period of their education, are commonly maintained by the lord of the village to which they belong, till sent to college. Besides six gymnasia, the Calvinistic church has four superior colleges; one of which, that of Enyed, stands higher for general education than any other college in Transylvania, and has an annual revenue of £1000 from funds deposited in the Bank of England. (*Paget*, ii., 386.) The Lutherans have a college at Cronstadt, and seven gymnasia. The government of the Reformed churches in Transylvania is somewhat like that of the Presbyterian church of Scotland, and is described at length by Mr. Paget, ii., 487, 489.

The Unitarian is a established religion only in Transylvania; where it was introduced by the Polish queen of Apolya I. in the 16th century, and for some time continued to be the religion of the court. The Unitarians include all the Poles, with some of the Magyars and Szeklers, and reside chiefly in the Szekler-land, where they have about 100 churches; they have a college at Klausenbury and two monasia elsewhere.

" The habits of society in Transylvania, in many respects, differ little from those of England about the end of the last century. In some of the old-fashioned houses almost a triarchal simplicity is kept up. The houses of the richer nobles are large and roomy, and their establishments are conducted on a scale of some splendour. It is true that they are deficient in many things which we should consider absolute necessaries; but, on the other hand, they exhibit many luxuries which we should consider extravagant with ace their incomes. It is no uncommon thing, for instance, in one-storied house, with a thatched roof and no uncarpeted or, to be shown into a bedroom where all the washing apparatus and toilet is of solid silver. Bare whitewashed walls and rich Vienna furniture; a lady decked in jewels which might dazzle a court, and a handmaid without shoes or stockings; a carriage and four splendid horses, with a coachman whose skin peeps out between his waistcoat and expressibles—are some of the anomalies still to be found in Transylvania." (*Ibid.*, 318, &c.)

This principality had been connected with Hungary for many centuries previously to the conquest of that country by the Turks, after which it threw off its allegiance, and became a quasi-independent kingdom, alternately tributary to Turkey, under the influence of Austria, to which latter empire it was annexed by Joseph II. in 1699. Since this period

it has enjoyed comparative tranquillity. Of late years a vigorous opposition has grown up in the diet, and in 1836 government was obliged to recall the archduke Ferdinand, and to concede to the Transylvanians the election of the president of the diet, the free publication of debates, &c. (*Paget's Hungary and Transylvania*, 2 vols., the best work by far that yet has appeared on those countries; *Austrian Nat. Encyclop.*; *Berghaus: Marmont, Voy. en Hongrie*, &c., i., 105–132.)

TRAPANI (an. *Drepanum*, from ὄρεπανος, a scythe, the tongue of land on which it is built being curved in the form of that instrument), a seaport town of Sicily, cap. prov., dist. and cant. of same name, on a projecting point of land on the W. coast of the island, 46 m. W. Palermo, the lighthouse on Colombaria rock, at the mouth of the harbour, being in lat. 38° 2′, N. long. 12° 30′ 18″ E. Pop., in 1831, 24,735. It is a military post of the second class, being surrounded by a wall and bastions, with ravlins in good repair, and covered by a glacis. The harbour, on the S. side of the town, is protected by Sigia fort, at the extremity of the tongue of land on which the city is built, the fire of which is crossed by that of a battery on Colombaria rock. The castle, in the N. angle, though unworthy of the name, is the residence of the governor, and other military authorities. The streets are regular, and the town is commodious and pretty well built. The cathedral and senatorial-palace are fine edifices. It has many convents and nunneries, and nearly 40 churches, with two hospitals, a college, two seminaries, a well-conducted *monte-di-pieta*, and an oratorio. The church of San Lorenzo is said to be " a simple and majestic specimen of correct architecture." Despite the number of its priests and friars, its inhabitants are said by Captain Smyth to be industrious and enterprising, and to afford the best artisans and sailors of the island. It has produced excellent scholars, painters, and architects, and the art of engraving on gems, which had been lost during the dark ages, was here recovered, and brought to perfection by Mazarielli: indeed, the inhabitants are now principally distinguished as sculptors and carvers of coral, amber, wood, shells, rings, and alabaster. To the W. of the town is a well-designed but still unfinished promenade. The marina forms a good walk under the line wall. The harbour is said to have been much damaged by the great earthquake of 1542; but though small, it is secure, and might be easily enlarged. It has a tolerably good mole, on which is the pratique office, accessible to vessels of 300 tons, vessels of larger burden anchoring near the Colombaria, in 8 or 9 fathoms water, muddy bottom. Water is conveyed to the town by an aqueduct from the foot of mount San Giuliano (an. *Eryx*), a little to the N.E. of the town. See GIULIANO (SAN).

The trade of the town is very considerable. The *salinas*, a little to the S.E., are the most extensive of any in the island. The salt, which is of good quality, costs about 8s. a ton, and is largely exported. The Trapanese carry on the coral fishery on the coast of Africa to a considerable extent; and the cutting and polishing of coral is one of the principal branches of industry carried on in the town. Besides salt and coral, the exports comprise soda, alabaster, rough or cut into vases, statues, &c., and a variety of other articles. In 1839, the port was entered by 114 foreign vessels, of which 73 were Austrian, and only two English.

Excepting vestiges of the mole formed by Fabius to join Colombaria to the continent, two mutilated lions' heads, that grace a fountain, and some fragments of marble, there are no remains of antiquity here, though coins of Drepanum have been occasionally found. (*Smyth's Sicily*, p. 237, &c.; *Macgregor's Report on Sicily, &c.*)

Drepanum is very ancient. It is represented by Virgil as having been visited by Æneas, and as the place where Anchises breathed his last. (*Æneid*, iii., lin. 707.) It was early occupied by the Carthaginians; and from its advantageous position and excellent port, was considered by them as of the first importance. During their struggle with the Romans it was the scene of frequent contests. Of these the most celebrated was the great sea-fight, *anno* 237 B.C., between the Roman fleet under the consul Claudius Pulcher, and the Carthaginian fleet under Adherbal. The latter gained a complete and decisive victory, with comparatively little loss on their part. (*Polybius*, lib. i., cap. 4.)

TRAVANCORE, a state of Hindostan, subsidiary to the British, and forming the S. extremity of the Indian peninsula, between the 8th and 10th degrees of N. lat., and the 76th and 78th degrees of E. long., having E. the British districts, Tinnevelly and Dindigul, N. Cochin, and on other sides the Indian ocean. Length, N. to S., about 140 m.; breadth, 60 m. in the N., and gradually diminishing to 20 m. in the S. Area estimated at nearly 4000 sq. m., and population at somewhat less than 1,000,000. The surface, which is varied with hill and dale, rises in the E., into a mountain chain, covered with forest trees and jungle. It is well watered, and highly adapted, by its climate, &c., to the wet cultivation, and rice is grown in large quantities; besides which, cardamoms, cassia, ginger, turmeric, betel nut and cocoa nuts, are among the chief vegetable products. Tobacco is principally imported from Ceylon, and is a government monopoly, from which the rajah is stated to derive a revenue of 13 lacs of rupees a year. (*Colebrooke in Revenue Rep.*) Elephants, buffaloes, and large tigers inhabit the more remote parts, and ivory, bees' wax, and some other valuable products are among the exports. Indications of coal are said to be met with.

According to Colonel Munro (*Evidence*, 1832), the whole of the land belongs to the government, or to individuals; village institutions being rare, or rather wholly wanting in Travancore. The land is assessed on the ryot-war system, a fresh survey having been made under the native government, every 10 or 12 years. Though the country is poor, the inhabitants are said to be less impoverished than in many parts of the E. I. Company's territories, the land-tax being less heavy. Lands, the property of the government, are assessed according to the quantity of seed sown on them, and the rent, in general, amounts to less than half the produce: lands, the property of individuals, pay, in many cases, under five per cent. on the produce. The lowness of the land-tax was formerly compensated for, to the native government, by the monopolies of pepper, betel, cardamoms, and other valuable products, which the inhabitants were obliged to supply to the state at very low prices. Most of these monopolies were destroyed, and replaced by a more equitable system of taxation, while the country was under the administration of the British. Except, however, as respects the lightness of the land-tax, the native government of Travancore was most oppressive. There was a chain of officers from the *dewan* to the lowest inhabitant, exercising all the powers of government, military, judicial, civil, revenue, without any check or control whatever; and besides this, several of the subordinate classes, subject to a capitation-tax, were formed into companies of about 100 men each, under a separate officer, and obliged to perform all kinds of work for the benefit of the government at the pleasure of the revenue servants. Under the British, this chain of revenue service was abolished with the capitation-tax, and the various monopolies, except those of pepper and tobacco. At the same time, however, the land-tax was increased, a circumstance which in so far countervailed the other improvements. In 1814, the country reverted to its former authorities, and since then, the old order of things is said to have been in a great measure revived.

Travancore, being an integral portion of the ancient Malabar, the prevailing usages and customs are generally similar to those which prevail along the adjacent parts of the W. coast of Hindostan. The sovereignty of the country, honorary dignities, and even property, descend in the female line, as in Canara, &c. The ruling family is Hindoo, and the principal part of the pop. consists of Brahmins and Nairs; but there are also many Mopiays (Mohammedans) and it is estimated that in Travancore and Cochin, there are 100,000 Syrian Christians. In some communities Christian churches are considerably more numerous than pagodas or mosques. The Travancore rajah, about the middle of last century, subdued most of the smaller states in his neighbourhood, and extended his dominion to their present limits, but, in 1790, these would have fallen a prey to Tippoo Saib but for our intervention. The final subsidiary treaty with the British was entered into in 1809. Travancore furnishes to the Anglo-Indian army a contingent of three battalions of infantry, and eight lacs of rupees a year to the Indian treasury. Total revenue, in 1826-27, 40,42,645 rupees; expenditure, including subsidy, 37,68,392 rupees. Principal towns, Trivanderum, the cap., Anjengo and Quilon; Travancore, the former cap., is now in a state of decay. (*Parl. Reps., &c.*)

TREBIZOND (an. *Trapezus*), a city and seaport of Asia Minor, on the S.E. coast of the Black sea, 190 m. N.W. Erzeroum; lat. 40° 1' N., long. 30° 44' 59" E. Pop. variously estimated at from 15,000 to 30,000. The town is built on the slope of a hill declining to the sea, and backed by steep eminences rising behind. Its central portion is surrounded by a castellated and lofty wall; on either side of the walled portion is a deep ravine, filled with trees and gardens, both ravines being crossed by bridges. Overlooking the town is a citadel; but it is dilapidated and neglected, and commanded by neighbouring heights. The walls of the city are, however, sufficiently strong to serve as a defence against troops without artillery. The space included within the walls is of great extent, but it is principally filled with gardens and plantations. The houses, which are mostly built of stone and lime, and roofed with red tiles, have in general only a ground floor, and as each is environed by a garden, the town from the sea has the appearance of a forest, scarcely a house being visible. The walled city is solely inhabited by Mohammedans; the Christians live outside the walls (principally in the eastern suburb), where are also most of the bazaars and khans. Besides nearly 20 church-

944

es and chapels, still retained for the service of the Greeks and Armenians, almost all the mosques have formerly been Christian churches. The handsomest mosque is that of St. Sophia, 1 m. W. of the city. "It is of small dimensions, built of hewn stone, in the form of a cross, and divided into a nave and two aisles, lighted from a cupola, supported by four marble pillars. The principal entrance is adorned with four white marble Corinthian columns; the Roman eagle is conspicuous over the gate; below it are numbers of small reliefs, and a beautiful cornice runs around the exterior of the edifice." (*Kinneir's Asia Minor, p. ...*) Several of the other mosques and churches are in the same style; but the most curious edifice in the city is the ... a huge square structure with two small windows in each front, probably erected by the Genoese as a powder magazine. (*Ib.,* p. 341.) A high square tower, and the ... remains of many other buildings crown the eminences near the mosque of St. Sophia; but none of them, nor any other remains at Trebizond, are of an age anterior to the Christian era, and Mr. Hamilton regards all the existing ruins, called Genoese by the Turks and Greeks, as clearly Byzantine. (*Geog. Journ.,* vii. 62.)

Trebizond has two ports; one on either side of a small peninsula projecting from the town into the sea. That on the E. is the best sheltered, and is the place of anchorage for the largest ships. It is exposed to all but S. gales; but it does not appear that with ordinary precaution any danger need be apprehended. The ground from ¼ to ¾ m. E. from the point is clean, and holds extremely well. Ships moor with open hawser to the N. and a good hawser and stream anchor on shore as a stern-fast. At night the wind always comes off the land. After the autumnal equinox, the Turkish and European vessels resort to Platana, an open roadstead, about 7 m. to the W. of Trebizond. (*Com. Dict; Brant, in Geog. Journ.*) In antiquity, and in more modern times previously to the conquest of Constantinople by the Turks and the exclusion of all foreign vessels from the Black sea, Trebizond was the seat of an extensive trade. It is the natural emporium of all the countries to the S.E. of the Black sea, from Kars on the E., round by Derbshir to Amasia on the W. Previously to 1830, however, its trade had dwindled to the export of a few products of the country to Constantinople, the import of iron from Taganrog, and a traffic with Abassah, carried on in small craft, which carried away salt, sulphur, lead, and Turkish manufactures; bringing in return the raw productions of the Caucasus, slaves, &c. But the treaty of Adrianople, by opening the Black sea to European ships, restored the old channel of communication between Europe and India, Persia, &c., through Trebizond; and the Russian policy of 1831, by putting an end to the immunities enjoyed by the Russian ports S. of the Caucasus, has given Trebizond an importance it did not previously possess. Its principal articles of import are manufactured cottons, mostly from Great Britain, sugar, coffee, rum, salt, tin, wine, &c. More than half the articles imported are destined for Persia; and while, in 1836, only 5000 bales of European merchandise passed through Trebizond on their way to that country, in 1835, nearly 22,000 proceeded by the same track to the same destination. (*Brant.*) The exports to Europe consist of silk, sheep's wool, tobacco, carpets, shawls, galls, and drugs of various sorts, beeswood, nuts, &c., with some wax, honey, and beans to Constantinople; but all in comparatively trifling quantities. Rich veins of copper and lead exist in the neighbouring mountains, but they are badly, if at all, wrought; and as the export of timber and corn is no longer prohibited by the Turks, it is not so difficult as formerly to obtain return cargoes.

This city was originally founded by a colony from Sinope, but subsequently outstripped its parent city, and all its inferior ports along the coast, in wealth and importance. It was a flourishing emporium when it was reached by Xenophon and the ten thousand at the close of their memorable retreat. It continued to be an important city of the Greek empire till the subjugation of the latter by the Crusaders, when its duke of the Comneni family assumed the dignity of emperor. His dominion extended from Sinope to the Phasis, and his family reigned for more than 250 years, till Trebizond came into the possession of the Turks in 1461. (*Smith and Dwight, Researches, &c., 454-458; Geog. Journal, vi. vii.; Kinneir's Asia Minor; Com. Dict., &c.*)

TREGONY, a market town of England, co. Cornwall, hund. Powder, par. of Cuby, on the Fal, 16 m. S.W. Bodmin. Area of the par. of Cuby, 2410 acres. Pop. in 1831, 1282. This insignificant place would not have been worth notice in a work of this kind, but for the circumstance of its having sent two members to the House of Commons from 1509 down to the passing of the Reform Act, when it was most properly disfranchised. The right of election was in potwallopers residing within the bor. The pop. is almost exclusively agricultural.

TRENT (an. *Tridentum*), a town of the Tyrol, but within the natural limits of Italy, on the Adige, which is here crossed by a fine bridge, 14 m. N.N.E. Roveredo, lat. 46° 6' 36" N., long. 11° 3' 45" E. Pop. 13,000. It is seated in a small but beautiful valley; being, however, from its elevation, exceedingly cold in winter, and, from the reflection of the surrounding mountains, equally hot in summer. It is surrounded by a pretty high wall, is well built with houses in the Italian style, has well paved though irregular, narrow, and dirty, streets, and a square ornamented with a handsome fountain. The palace belonging to the old prince-bishops of Trent, in a corrupt Gothic style, is of large dimensions, has some good apartments, fine fresco paintings, rare marbles, and extensive gardens. It has also a cathedral and several other churches, in one of which, Santa Maria Maggiore, the famous council held its meetings. In 1805, the roof of this church fell in, and we have not learned whether the original picture representing one of the sittings of the council, with portraits of its more distinguished members, that belonged to this church, escaped being destroyed. (Rampoldi, iii. 1245.) It has also three convents for monks, and a nunnery, a large hospital, an orphan asylum, a lyceum with several professors, a gymnasium, &c.; with manufactures of silk and other fabrics. It is one of the seats of the transit trade between Germany and Italy; and exports wine, corn, tobacco, and iron, produced in the surrounding country.

This town, which is very ancient, became, in the middle ages, the cap. of a lordship under its bishops, by whom, in 1363, it was united to the Tyrol, in which it has since been comprised. The bishopric was secularised in 1803; but the bishop is still in the enjoyment of a handsome revenue. Under the French it was the cap. of the dep. of the Upper Adige, and is now the seat of the government of the circle of the same name. (Marcel de Serres, Voyage de Tyrol, ii. 362; Eustace, i. 102, 8vo ed., &c.)

But the celebrity of Trent is entirely owing to its having been selected as the place of meeting for the famous general council of the church, convoked by Pope Paul III., and which, after much procrastination, met for business on Dec. 13, 1545, and continued, though with several interruptions, through 25 sessions, till 1563, under three successive pontiffs. It consisted of dignitaries of the church, representatives of the different universities, and of ambassadors from the princes and states attached to the communion of Rome. It was intended to revise, fix, and declare the doctrines of the church, to remove the abuses that had crept into its government, and the conduct of its functionaries,—

Saccurrere lapsis
Legibus, et vorace revocare in pristina mores.

and, if possible, to restore peace and unity to the church. It may be said to have fully accomplished the first object, and in some degree, also, the second; but, as might easily have been foreseen, it wholly failed in the third object, or in the attempt to smooth the differences and allay the violent struggles and animosities that then divided and agitated the Christian world. The constitution of the council, indeed, and the commanding influence which the papal legates early acquired over its deliberations, deprived it of all pretence to the character of an impartial tribunal, and fully justified the Protestants in repudiating its authority and rejecting its decrees. The latter were subscribed by 255 legates, cardinals, archbishops, bishops, and other dignitaries, and have been generally admitted to contain, along with the creed of Pope Pius IV., a complete, authoritative, and well digested synopsis of the principles and doctrines of the Roman Catholic religion.

The intrigues of which this council was the theatre have been developed with singular talent by Sarpi, in his famous *History of the Council of Trent*. But as Sarpi was the implacable enemy of the court of Rome, and has dexterously endeavoured to show that its pretensions were almost always unfounded, and its advocates in the council almost always in the wrong, his conclusions, or, rather, the conclusions drawn by the reader from his statements, are not always to be depended on. The history of Sarpi, though an able and ingenious, cannot be said to be an honest or trustworthy work. Tiraboschi cautiously says of it, "*Io son ben lungi di sostinere, che gli si debba credere ciò ch' ei raconta, solo perchè egli il raconta !*" (Letteratura Italiana, viii. 231, ediz. Modena, 1793.)

TRENT, a river of England, being, next to the Thames and Severn, by far the most important stream in that part of the U. Kingdom, not only on account of the length of its course, but of the fertile districts through which it passes, the immense number of canals with which it communicates, and the considerable rivers it receives in its progress.

It has its source near the Cheshire border, in the moorlands of Staffordshire, about 4 m. N. from Burslem. At first its course is nearly S.E., when it makes a sudden turn by the E. to the N. near Burton-on-Trent. It afterwards divides Leicestershire from Derbyshire; and pursuing a

N.E. course, by Nottingham to Newark, it turns more and more to the N. After dividing Nottinghamshire and Lincolnshire, and passing Gainsborough, it enters Lincolnshire at West Stockwith; and flowing N., with a little inclination to the E., unites with the great estuary of the Humber, at a place called Trent falls. It may be navigated by vessels of 200 tons as far as Gainsborough, and by barges as far as Burton-on-Trent, a distance of about 117 m., having in this lengthened course a fall to low water mark of only 118 ft., or very near a foot per mile. From Burton-on-Trent to its source, the rise of the river is about 376 ft.; at least, the summit level of the Caldon canal, which passes near the head of the Trent, is 494 ft. above the sea. (Priestley's Map of Canals, &c.)

Of the subsidiary streams that fall into the Trent, the most considerable are the Blythe, Tarn, Dove, Derwent, and Soar; but of these it is only necessary to notice the last two. The Derwent rises in that part of Derbyshire called the High Peak, after passing Matlock, Cromford, and Derby, it has a circuitous course, from the latter to Wilden Ferry, where it unites with the Trent. It is navigable as far as Derby, about 13 m.; but it has been superseded, as a channel of communication, by the Derby canal. The Soar rises E. from Winckley, in Leicestershire, it flows through a rich grazing country, and more than half encompasses the ancient town of Leicester. After receiving the Wreke, its course is N., with a little inclination to the W., till, passing Loughborough, it falls into the Trent near Cavendish bridge. It is navigable to near Loughborough, a distance of about 7 miles.

The canals that communicate with the Trent are of the greatest importance; assisted by them, it affords an easy means of export for the manufactures of a large district of Lancashire; the salt of Cheshire; the produce of the potteries of Staffordshire; the coal of Derbyshire; and the agricultural products of Nottinghamshire, Leicestershire, and Lincolnshire. It also opens a communication with the sea by way of Lincoln and Boston; through which channels, as well as the Humber, the articles above enumerated are conveyed; and, in return, the interior of the country is supplied, either by Hull and Gainsborough, or Boston and Lincoln, with such commodities as are required by an immense population. (Priest. Treat. on Rivers, Canals, &c., p. 681.)

TRENTON, p. t., Oneida co., N. Y., 12 m. N. Utica, 96 m. W.N.W. Albany, 401 W. Drained by Nine Mile and West Canada creeks, on the latter of which are Trenton Falls, which are often visited by travellers as a curiosity. The creek falls 312 ft. by a succession of cascades, in the course of about 2 m., and the scenery is picturesque and beautiful, and when the water is high it is grand. The t. contains 10 stores, three fulling-mills, two woollen factories, one furnace, four grist-mills, 12 saw-mills, three tanneries, one distillery; one academy, 160 students, 23 schools, 951 scholars. Pop. 3178.

TRENTON, city, Mercer co., N. J., and the capital of the state, is in 40° 14' N. lat. and 74° 36' W. long., and 2° 30' E. long. from W. It is 10 m. S.W. Princeton, 26 m. S.W. New-Brunswick, 30 m. N.E. Philadelphia, 60 m. S.W. New-York, 166 W. The population in 1810 was 3003; in 1820, 3942; in 1830, 3925; in 1840, 4035. Of these 162 were em-

in a general description should be considered as belonging to it, and in the city and S. Trenton there are 1000 dwellings, and about 6000 inhabitants. At the foot of the falls or rapids a beautiful covered bridge crosses the Delaware, 1100 ft. long, resting on five arches, supported on stone piers, which is regarded as a fine specimen of this species of architecture. It is of sufficient width to allow of two carriage ways, one of which is appropriated to the accommodation of the railroad from New-York to Philadelphia. The public buildings and institutions of Trenton are a state house, 100 ft. long and 60 ft. wide, built of stone and stuccoed to resemble granite. It is beautifully situated on the Delaware, commanding a fine view of the river and the surrounding scenery. The house for the residence of the governor of the state is a plain and commodious edifice. There are three fire proof public offices, two banks, a public library, founded in 1750, a lyceum, and seven churches, a Presbyterian, Dutch Reformed, Episcopal, Methodist, two Friends, and an African Methodist. All these are in the city proper. In South Trenton there are a courthouse, a state prison, four churches a Methodist, Baptist, Reformed Baptist, and Roman Catholic.

The Delaware and Raritan canal, forming an inland navigation from Trenton to Brunswick, passes through the place, and is here entered by a navigable feeder, taken from the Delaware, 23 m. above the city. The canal is 42 m. long, 75 ft. wide, and 7 ft. deep, and capable of affording a passage for small sloops. It would have been of vast importance during the last war, when this route was an immense thoroughfare. It was finished in 1834, and cost $2,500,000. The canal crosses the Assunpink creek E. of the city, on a fine stone aqueduct. Above the falls, the Delaware is navigable for large boats as far as Easton, which adds much to the commercial advantages of Trenton. The New-Jersey railroad passes through the place. A company has been chartered with a capital of $200,000, for the purpose of taking the water from the river by means of a dam and raceway, and carrying it through and below the city, with outlets for mills, which will create an extensive water-power for manufacturing purposes. The Assunpink creek, which enters the Delaware immediately below the city, affords some water-power.

Trenton contained, in 1840, 50 retail stores with a capital of $196,300; four lumber yards, capital $49,000; two flouring-mills, two grist-mills, three saw-mills, three paper-mills with a capital of $30,000, three tanneries, one brewery, one pottery, one ropewalk, three printing-offices, two binderies, two weekly and one semi-weekly newspapers. Total capital employed in manufactures $247,800. There were four academies with 164 students; 10 schools, 314 scholars.

The town was first settled in 1720; incorporated as a city in 1799; and is governed by a mayor, recorder, three aldermen, and 13 assistants.

In December 25th, 1776, at a gloomy period of the Revolutionary war, Gen. Washington crossed the Delaware in the night, and suddenly attacked and captured 1000 Hessians of the British army, which greatly revived the spirit of the nation, and had an important influence on the final result of the war.

TREVES (Ger. Trier), a city of the Prussian dom.; prov. Rhine, cap. of a reg. of the same name, on the Moselle, near its confluence with the Saar, and near the frontier of Luxemburg, 60 m. S.W. Coblentz, lat. 49° 46' 37" N., long. 6° 38' 20" E. Pop., in 1838, 14,941. Streets broad and straight; and some of the public buildings are imposing. Among the latter may be specified the cathedral, remarkable for its altars and marble gallery; the church of St. Simeon, of great antiquity; the elector's palace, now turned into barracks; the bridge over the Moselle, 600 ft. in length, the piers of which are supposed to have been built in the 28th year of the Christian era; the gate of Mars (Porta Martis), of colossal dimensions and great antiquity, &c. Its ancient university was suppressed in 1794, but it has a college or seminary for the education of Catholic clergymen, a gymnasium, a collection of medals, and a public library, both of which belonged to the university; the latter comprises above 80,000 volumes, many of which are scarce and valuable; it has also several hospitals, and a theatre. It is the seat of the government, has a prov. council, a tribunal of appeal for the prov., a tribunal of commerce, &c., with manufactures of linen, woollen, and cotton stuffs. Boats for the navigation of the Rhine are built here; and it has a considerable trade in Moselle wine, &c.

Treves is, perhaps, the most ancient, and was long the most celebrated, of the German cities. A Roman colony was planted in it during the reign of Augustus; and thence it was called Augusta Trevirorum. From that period it became a place of great importance, and was reckoned one of the bulwarks of the empire on the side of Germany. Constantine the Great and several other emperors occasionally resided in Treves. Ammianus Marcellinus calls it Domicilium principum clarum (lib. 15, s. 27). Ausonius, in his poem De Claris Urbibus, celebrates its praises, and notices the extensive commerce it carried on by the Moselle. Besides the bridge, the Porta Martis, &c., other remains of buildings that still exist, and many coins and relics found in the town and its vicinity, attest the power and splendour of its Roman masters. Beyond its walls are the ruins of an amphitheatre, cut in the side of a hill, where Constantine is said to have exposed some thousand Gauls to be torn by wild beasts. Treves was successively laid waste by the Huns, Goths, Vandals, and Franks, and was as often rebuilt. It was for a lengthened period the cap. of the archbishopric or electorate of Cleves. Latterly the pop. has increased considerably; though there is little probability that it will ever recover its ancient fame and importance. (Cellarii. Notit. O-bis Antiqui. 1. 317; Schreiber, &c.)

TREVISO (an. Tarvisium), a town of Austrian Italy, prov. Venice, cap. deleg. of its own name; on the Sile, 16 m. N. by W. Venice. Pop., in 1837, 11,598. It is irregularly built, and is surrounded by old walls. Most of the streets are wide and well paved, with colonnades in front of the houses; and there are numerous palazzi and religious structures. The old cathedral of St. Peter, a fine though unfinished structure, has in it a painting by Titian, and others by P. Bordone (a native of Treviso), Domenichino, &c.

The Gothic church of S. Nicolo, the town house, and theatre, are good buildings; and in the monte di pietà the picture by Giorgione. Treviso is a bishop's see, and has an episcopal seminary, gymnasium, public library, house gallery, several hospitals, a castle, theatre, &c. The manufactory, afterwards removed to Padua, was first established here. A large hardware factory, an extensive pottery, and manufactory of silk twist and stuffs, woollen cloths, paper, and cutlery, with some trade in corn, wine, cattle, fruit, &c., occupy most part of the inhabitants. A large fair is held each year, from the 3d to the 18th of October. The town appears to have been a Roman municipium. Under the Lombards, it was the cap. of one of the two marches or margraviates which they established on the confines of their kingdom in Italy (Ancona having been the cap. of the other). Under the French it was the cap. of the dep. Tagliamento. Napoleon conferred the title of Duke of Treviso on Marshal Mortier. (Rampoldi; Austr. Nat. Encyc.)

TREVOUX (an. Trevae, or Trevaux), a town of France, dep. Ain, cap. arrond. on the declivity of a hill, on the Saône, 13 m. N. Lyons. Pop., in 1836, 229. It was formerly surrounded by walls and towers; and on the summit of the hill on which it is built, are the ruins of its old castle, commanding a most extensive view over the surrounding plain. It has an antiquated appearance, with narrow streets, and mean-looking houses. Having been formerly the cap. of the principality of Dombes, and the seat of a parliament, courts of justice, a mint, &c., it has still to boast of some considerable ancient edifices, including the hall in which the parliament met, the hall of the courts of justice, a hospital founded by Anne Marie Louise d'Orleans, a quay on the Saône, &c. It has, also, a tribunal of original jurisdiction; a cloth manufactory; a royal establishment for the refining and assay of gold and silver, and some trade in the products of the surrounding country. It is very ancient. The emperor Severus defeated, anno 197, his competitor Albinus under its walls.

Trevoux has attained to considerable distinction in literary history. Louis Auguste de Bourbon, prince of Dombes, endeavoured to make it a sort of literary capital, and, in this view, he established, in 1695, a considerable printing-office in the town, in which he also intended to found a college. And not long after, or, in 1701, the well known and very learned monthly publication, entitled the Journal de Trevoux, conducted by the Jesuits, began to issue from this press; where it continued to be printed till 1734, when it was transferred to Paris. Here, also, appeared, in 1704, the first edition of the Dictionnaire de Trevoux, in 3 vols. folio. There were several subsequent editions of this valuable work, most of which, however, were printed and published in Paris. Of these the last and best edition, in 1771, was so much enlarged as to comprise 8 vols. folio. (Biogr. art. Ain; Moreri, art. Trevoux.)

TRIANGLE, a p.-t., Broome co., N.Y. 156 m. W.S.W. Albany, 317 W. Bounded W. by Castle river, S.W. by Toughnioga river. It has four stores, two fulling mills, two grist-mills, five saw-mills, three tanneries. 13 schools, 221 scholars. Pop. 1692.

TRICALA or TRIKHALI (an. Tricca), a town of European Turkey, cap. of the prov. of same name, identical with the an. Thessaly, on the E. side of a mountain ridge, 2 m. N. from the Selymbria (an. Peneus), and 37 m. W. by S. Larissa. Pop. estimated at from 10,000 to 15,000, chiefly Turks. It is of considerable extent; and the houses being intermixed with gardens and trees, it appears to be built in a wood, and the lofty minarets of its mosques rising above the trees give it a picturesque appearance. At the height of 30 or 32 feet above the pavement a wooden trellis-work interwoven with vines is carried over the streets, completely shading the passengers below. The shops are clean, and tolerably well furnished, and their possessors, who are chiefly Greeks or Jews, have a respectable appearance. (Holland's Trav. p. 249.) According to Strabo, this city had a magnificent temple of Æsculapius; but no traces of this edifice are now known to be extant. On a hill above the town are the ruins of a castle apparently dating from the time of the Greek emperors, and commanding a fine view over the plains of Thessaly. These are depastured by numerous flocks of sheep, and also produce a good deal of corn, the manufacture of blankets, coarse woollens, and cotton stuffs, occupying many of the inhabitants of Tricala. Its trade is also pretty extensive, from its being on the principal road from Yanina to Constantinople, and commanding the only pass by which supplies of corn and other provisions are brought from Thessaly into Albania. The circumstance renders it important as a military post. (Hughes, Trav. in Albania, &c., 1, 139; Cramer's Anc. Greece.)

TRICHINOPOLY, a distr. of British India, presid. Madras, chiefly between lat. 10° 30' and 11° 20' N., and long. 78° 10' and 79° 30' E., having N. Salem and S. Arcot, E. Tanjore, S. the latter and Madura, and W. Salem and

Coimbatore. Area, 3169 sq. m. Pop., in 1836-37, 554,730. The Cavery runs from W. to E. through the country, irrigating a considerable extent of rice land. In addition to rice, sugar-cane, with tobacco and betel-leaf, are grown in the tracts watered by tanks and wells: in the dry lands the other usual products of the Carnatic are extensively cultivated, and there is good pasturage for sheep and cattle, which are numerous. The principal imports are glue, oil, tobacco, pepper, and areca nut; while the exports comprise cloth, indigo, saltpetre, and cotton. The principal manufactures are cloth, for domestic use, and indigo, with some subsidiary articles made in the town of Trichinopoly (which see). Total land revenue, in 1836-37, 1,428,853 rupees.

TRICHINOPOLY, a large fortified town of British India, presid. Madras, cap. of the above distr., on the Cavery, 186 m. S.W. Madras. Pop., exclusive of troops, estimated at 14,000. (*Madras Almanac.*) It is of an oblong form, nearly 1 m. in length, N. to S., by about ⅜ m. in breadth. Exclusive of some outworks, it is surrounded by a double wall and ditch, with a covert way and glacis. But its defences are now mostly in a ruinous state, except the citadel near its N. extremity, which, being situated on an elevated rock, commands any military operations carried on in any part of the vicinity. On this rock also stands a large and massive pagoda, and a pillared square building, with a statue of Hanuman, occupies the highest peak, while in the S. face of the rock is a small sculptured excavation in the style of some of the cave temples at Ellora. The jewellery made at Trichinopoly had formerly much celebrity; and Trichinopoly chains are still in request. Cotton cloths, ble linen, harness, &c., are made here; and the town is a emporium for a great variety of manufactures. It is ell adapted for a military station, as, besides being well supplied with different kinds of merchandise and artisans, e roads about it are so good as to admit, at every season, an easy communication with Madras, Vellore, and Mysore. Hence, also. diverge all the great roads leading to Tanjore, Madura, and Dindigul, the three chief stations in the S. part of India. (*Madras Almanac, &c.*)

TRIESTE (an. *Tergeste*), a town and principal seaport the Austrian empire, cap. gov. and circ. of its own name. prov. Illyria, on the Adriatic, near its N.E. extremity, 73 m. by N. Venice. Lat. 45° 38' 37" N., long. 13° 46' 27" E. Pop. in 1836, of the city only, 51,346. The population of the own and its district, comprising about 40 sq. m., amounted, in 1839, to 75,551, having increased to that amount from 323 in 1821! (*Bowring's Rep.; Bergbaus.*) Trieste is divided into the old town, the new town, or Theresienstadt, Josephstadt, and the Frauzenvorstadt (Francis' suburb): the old town stands at the foot and on the declivity of a steep hill crowned by the citadel: it has dark, narrow, winding, and frequently steep streets, with gloomy-looking houses, and is surrounded by the remains of ancient fortifications. The new town, immediately N.W. of the former, is built on level ground, partly taken from the sea, consists, on the contrary, of handsome streets, crossing each other at right angles, and lined with neat buildings. It is chiefly intersected by the canal cut by Maria Theresa, by means of which vessels drawing 9 or 10 feet water may load and unload at the doors of the warehouses. Between the new and old towns runs the *Corso*, the principal thoroughfare, broad but winding, furnished with good shops and coffee-houses, and opening successively into spacious and handsome squares. The principal of these is the *Piazza Granda*, with a fine public fountain, and the column statue of the Emperor Charles VI., to whom Trieste is principally indebted for its importance in modern times. In this square the great vegetable and fruit market is held, on one side of it is the *locanda granda*, or principal hotel, commanding a fine view of the harbour. The exchange, the finest building in the city, stands in another square, in which is a statue of Leopold I. Continual improvements appear to be taking place in and around Trieste; many new streets and promenades have been laid out and public walks planted with trees; new moles, and an antic hospital, the cost of erecting which has been completed at 800,000 florins, have very recently been contracted.

The cathedral, in the old town, is supposed to occupy the site of a temple of Jupiter. It is in the Byzantine style; its interior, like St. Mark's at Venice, is ornamented with mosaics; and many Roman inscriptions, carvings, &c., are built up in the walls. It contains the monument of Winckelmann the antiquary, author of the famous work, *Histoire de l'Art chez l'Antiquité*, assassinated here in 1768. There are five other Roman Catholic, two Protestant, and three Greek churches, a synagogue, and an English chapel. The finest of these edifices are the Greek churches, particularly that at the head of the great canal, with a magnificent marble altar, and an organ esteemed among the finest in Italy. The church of the Jesuits merits attention for its architecture and fine paintings, and the palace of the

governor is also an imposing structure: the handsomest residence is the house formerly belonging to a Greek merchant of the name of Careiotti, who, having begun business in Trieste almost as a pedlar, is said to have died worth £1,000,000 sterling! (*Spencer.*) The castle formerly constituted the main protection of the town and harbour, and is still maintained in a tolerable state of defence. The great theatre is spacious, and there are several minor theatres. Among other objects worthy of notice are the barracks, post-house, dockyard, lazarettos, one of which is among the most perfect establishments of its kind, and the terrace of the casino ornamented with several statues.

Trieste is in the S. what Hamburg is in the N., the great commercial entrepôt of Germany. A harbour, which, though rather limited in size, is easy of access and convenient, has been formed by the Theresian Mole, founded on a ledge of sunken rock, and projecting N.W. into the sea from the S. extremity of the old town. At its termination has been formed an irregular platform about 1100 feet in circumference, on which have been erected a fortress and lighthouse, with an intermittent light 106 feet above the sea. Another lighthouse, having the lantern 143 feet above the sea, has been erected on the point of Salvore, about 18 m. W. by S. Trieste. The port, with the mole, forms a crescent 1½ m. in length, being a continued quay, faced with hewn stones, and with stairs and jetties for the convenience of embarkation. On the N. side of the port is a dock or harbour, appropriated exclusively for vessels performing quarantine. It is walled round, and furnished with hotels, warehouses, and every sort of accommodation for passengers and goods. Ships under 300 tons burden lie close to the quays; those of greater size mooring in the roads in front of the city. The principal defects of the port are its limited size, and its being exposed to N.W. winds, which sometimes throw in a heavy sea. The gales, however, are seldom of long continuance; and the holding ground being good, when proper precautions are taken, no accident occurs.

Trieste being a free port, goods destined for its consumption, and that of the adjoining territory, pay no duties; but such as are taken into the interior for consumption pay, of course, the duties in the Austrian tariff. The transit duties and shipping charges are extremely reasonable.

The exports are very various, consisting partly of the raw and partly of the manufactured products of Austria Proper, Illyria, Dalmatia, Hungary, and Italy; with foreign articles imported and warehoused. Among the principal articles of raw produce may be specified, corn, chiefly wheat and maize, with rice, wine, oil, shumac, tobacco, wax, &c.; silk, silk rags and waste, hemp, wool, flax, linen rags, hides, furs, skins, &c.; the produce of the mines ranks as important item, consisting of quicksilver, cinnabar, iron, lead, copper, brass, litharge, alum, vitriol, &c.; the forests of Carniola furnish timber, for ship-building and other purposes, of excellent quality, and in great abundance, with staves, cork wood, box, hoops, &c.; marble also ranks under this head. Of manufactured articles, the most important are, thrown silk, silk stuffs, printed cottons from Austria and Switzerland, coarse and fine linens, and all sorts of leather; under this head are also ranked soap, Venetian treacle, liqueurs, &c., with jewellery, tools and utensils of all sorts, glass ware and mirrors, Venetian beads, refined sugar, and a host of other articles. Trieste is also a considerable *dépôt* for all sorts of produce from the Black sea, Turkey, and Egypt.

The principal articles of importation consist of sugar, coffee, dye-stuffs, cotton goods and cotton yarn, silks, oil, tin plates, salted fish, and a host of other articles. The value of the imports always exceeds that of the exports, occasioned in part by their being subsequently transshipped to other ports, and partly by there being an excess of exports as compared with imports from other parts of the of rosoglio, wax-lights, leather, soap, playing cards, musical instruments, &c., with dyeing-houses, sugar refineries, potteries, and distilleries. It communicates by *diligence* three times a week with Vienna; and by steam packets once a month with Smyrna and Constantinople, and continually with Venice and other towns on the Adriatic. It is better

withstanding the cheapness of most articles, owing to the absence of duties, Trieste is not a desirable residence for persons not engaged in business. Water is scarce and bad; the climate is in extremes; and the E.N.E. wind, known by the name of Bora, is very piercing. A mixture of all nations is met with here, and all the principal merchants and traders are foreigners. German is spoken by the authorities and in the public offices, but Italian is the prevailing language of the middle classes, while the lower speak a Slavonic dialect. We subjoin

An ACCOUNT of the Quantities of the principal Articles of Foreign Raw Produce imported into Trieste during each of the three years ending with 1840.

Articles.	1838.	1839.	1840.
Cassia lignea (chests)	1,700	2,251	2,175
— (mats)			
Cocoa (bags and barrels)	5,490	4,116	5,396
Coffee do. do.	140,820	120,147	196,176
— (casks)	5,739	6,585	3,657
Cotton, American and Brazil (bales)	23,278	24,460	51,968
Levant and Egyptian do.	34,638	60,084	32,931
Hides, dry, ox and cow (No.)	46,702	8,436	83,382
Indigo (chests)	581	11 582	566
— (serons)	49	21	43
Nankeens (pieces)	1,000	2,900	
Pepper (centners)	18,145	9,000	29,000
Pimento (bags)	1,811	2,743	1,150
Rum (casks)	1,596	2,105	1,797
Sugar, Cuba (boxes)	14,004	14,941	20,743
Brazil (chests)	28,580	29,236	26,125
— do (boxes, barrels, and bags)	14,609	24,821	23,290
West India, raw (casks)	1,821	3,524	1,793
Egyptian (packages)			
East India (bags and cases)	362	3,638	467
— refined (hogsheads)	14,694	12 701	13,418
Stockfish (vogs)	84 190	90,000	47,760

At an average, the value of the imports into Trieste may amount to from 4 to 4½ millions sterling.

We are indebted to the *Lloyd Austriaco* for the following returns respecting the navigation of Trieste during 1839:—

Arrivals.	Vessels.	Tons.
Sailing vessels engaged in foreign trade	1,854	228,855
Steam vessels (ditto)	84	7,749
Sailing boats engaged in the coasting trade	2,818	105,712
Steamboats (ditto)	203	46,880
Smaller craft (ditto)	7,714	179,238
Total	10,657	567,841

Clearances.	Vessels.	Tons.
Sailing vessels engaged in the foreign trade	1,807	236,353
Steam vessels (ditto)	84	7,748
Sailing boats engaged in the coasting trade	3,469	107,598
Steamboats (ditto)	203	46,490
Smaller craft (ditto)	6,196	143,992
Total	11,699	542,569

Traces of an amphitheatre and other Roman remains exist at Trieste. During the middle ages it was the capital of a small republic; but its history presents little remarkable till 1719, when Charles VI. made it a free port. The French took it in 1797, and again under Massena in 1805. (*Austr. Nat. Encyc.*; *Berghaus*; *Bowring's Report on the Lombardo-Venet. States*; *Turnbull's Austria*, i., 361–382; *Spencer's Germany*; *Commercial Dict.*, &c.)

TRIGG, county, Ky. Situated in the S. toward the W. part of the state, and contains 510 sq. m. Bounded W. by Tennessee river. Watered by Cumberland river and its branch Little river. It contained in 1840, 7993 neat cattle, 8134 sheep, 30,015 swine; and produced 37,172 bushels of wheat, 4975 of rye, 499,255 of Indian corn, 93,270 of oats, 9387 of potatoes, 1,879,537 pounds of tobacco, 21,361 of cotton, 4235 of sugar. It had 13 stores, seven grist-mills, six saw-mills, four tanneries; 11 schools, 965 scholars. Pop.: whites, 5614; slaves, 9059; free coloured, 50; total, 7716. Capital, Cadiz.

TRIM, an inland town of Ireland, prov. Leinster, co. Meath, of which it is the cap., on the Boyne, here crossed by a bridge, 25 m. N.W. by W. Dublin. Pop., in 1831, 3282. This is a very old town, having been given by Henry II., as part of the palatinate of Meath, to Hugh De Lacy. The latter constructed the castle, which, from its extent, strength, and elevated situation on the banks of the river, was at once the largest and most important of the numerous fortifications erected by the English within the limits of the Pale. The ruins sufficiently attest its ancient grandeur. On the other side of the river are the ruins of St. Mary's Abbey, an ancient and extensive edifice; and there are some other ecclesiastical remains. The town had, also, been surrounded by walls, considerable portions of which are still entire. Indeed it was anciently the occasional seat of the lords-lieutenant; and several parliaments have been held within its walls. It was taken, without opposition, by Cromwell, in 1649.

At present, however, notwithstanding it is the county town, Trim is of little importance. Its principal public building is the new county jail, an extensive structure on

the radiating plan. It has also, an ancient parish church, a Roman Catholic chapel, a dispensary, an infantry barrack, with a county infirmary, schools, &c. It returned two members to the Irish House of Commons till the Union, when it was disfranchised, and since then it has ceased to decline. The assizes for the county are held here, and general sessions twice a year, and petty sessions on alternate Saturdays. It is a constabulary station, and has a flour-mill, a brewery, and a tannery. Markets on four flour-mill, a brewery, and a tannery. Markets on four days; fairs March 27, May 6, Wednesday after Trinity Sunday, Oct. 1, and Nov. 16. Postoffice revenue in 1834, £364; in 1836, £377.

About 3 m. S. from the town, on the road leading to Summerhill, is Dangan, formerly the property of the Earl of Mornington, and memorable as the birth-place of the Duke of Wellington. The house is which the great general first saw the light has, however, been wholly pulled down; but a handsome pillar, surmounted by a statue of his Grace, has been erected in the town, in commemoration of his achievements, and of his connection with the vicinity.

TRIMBLE, county, Ky. Situated in the N. part of the state, and contains 150 sq. m. Bounded N. and W. by Ohio river. Drained by Little Kentucky river. It contained in 1840, 4887 neat cattle, 6373 sheep, 12,964 swine; and produced 27,571 bushels of wheat, 177,990 of Indian corn, 30,557 of oats, 6691 of potatoes, 383,140 pounds of tobacco. It had seven stores, two flouring-mills, 16 grist-mills, five saw-mills, three tanneries, three distilleries; eight schools, 265 scholars. Pop.: whites, 3787; slaves, 673; free coloured, 20; total, 4480. Capital, Bedford.

TRINCOMALEE, a marit. town, of Ceylon, on the N.E. coast, near the entrance to one of the finest bays in the world, about 159 m. N.E. Colombo. Lat. 8° 33' N., long. 81° 37' E. The town, which is but inconsiderable, is built at the foot of a rock, on which is the fort, on the isthmus of a narrow peninsula or tongue of land bounding the harbour on the E. It has but few European inhabitants, and, what is remarkable, few Singalese; the lower classes being principally Malabar Roman Catholics. The fortifications form a sweep of above 1 m. in length along the shore. Fort Frederick is a station for four companies of a European regiment, a company of royal engineers and artillery, and detachments of the Ceylon rifle corps. Fort Ostenburg, at the termination of a ridge of hills, about 3 m. S.W. This commands the entrance of the harbour, and the dockyard close beneath. It forms the head quarters of a detachment of artillery and a European company. The fortifications here were mostly constructed by the Portuguese: the Dutch did little or nothing for the improvement of the place while in their possession.

The harbour of Trincomalee was styled by Nelson "the finest harbour in the world." It is almost land-locked, and the water is so deep that it is all but practicable in most places to step from the shore on board large vessels moored alongside. During the N.E. monsoon, when all the other places on the Coromandel coast and in the bay of Bengal are obliged to put to sea, Trincomalee is their principal place of refuge, and a vessel from Madras can reach it in two days. The town, which may be considered as the military cap. of Ceylon, surrendered to the English in 1795. (*Bertolacci; Modern Trav., &c.*)

TRING, a market town and par. of England, co. Hertford hund. Dacorum, on the London and Birmingham railway, and on the road from London to Aylesbury, 32 m. N.W. London. Area of par. 7360 acres. Pop. in 1831, 3422. The town consists principally of two streets; it is pleasantly and built, the houses being mostly modern. The church is an embattled structure, with a massive tower and low spire at the W. end. The living, a perpetual curacy, worth £72 a year, is in the gift of the dean and chapter of Christ. There are meeting-houses for Baptists and Independents, a Lancastrian school, &c. The inhabitants are principally employed in the manufacture of straw plait, canvas, and a few silk fabrics. Markets on Fridays; fairs, Easter Monday, and Oct. 11. The railway here attains a height of 60 ft. above the level of the sea, being its highest or summit level. Tring park, in the vicinity, was built in the reign of Charles II.

TRINIDAD, an island of the West Indies, or Antilles, being the most southerly of the group called the Windward Islands, and next to Jamaica, the largest and most valuable of the islands belonging to Great Britain in this part of the world. It lies immediately off the N.E. coast of Columbia and the N. mouths of the Orinoco, between the 10th and 11th degs. of N. lat. and the 61st and 62nd degs. of W. long. N.W. extremity being only about 13 m. from Punta de la Pena, the extremity of the peninsula of Paria; and its S.W. cape being but 7 m. from the Delta of the Orinoco. On the W. Trinidad bounds the gulf of Paria, and on all other sides it is surrounded by the Atlantic. It is of a square or oblong form, with considerable projections at all its angles or ept the S.E. Length, N. to S., 50 m.; average breadth, exclusive of its projections), about 33 m. Area estimated, by the best authorities, at 1,287,680 acres, or about 2000 sq. m., though it has also been estimated at above 1,500,000 acres. Resident population, in 1838, 39,328, of whom 3601 were whites; but there are doubts as to the correctness of this return. The mountain chains run from W. to E., and may be regarded as continuations of the chains on the opposite coast of Venezuela, from which this island has most probably been detached by some convulsion of nature. Along the N. shore a bold range of mountains rises to the height of 3000 ft., broken into the most rugged and abrupt forms, and clothed to the summit with forest trees. Towards the S. extends a chain of hills of less elevation, and 'a more pastoral character, while the centre of the island occupied by a group of flat or round-topped hills, dividing as it were, into two extensive valleys, which are occasionally intersected by a succession of hill and dale. The whole island is well watered by numerous streams in every direction. The principal are on the W. coast: the Caroni, navigable for 6 leagues from its mouth; and on the E. the Ocupuche and Nariva, which last is said to be navigable for vessels of 250 tons to a league from its source. The N. and E. coasts are not well furnished with harbours; which unfortunate, as the winds blow from those quarters for three fourths of the year. But the W. coast has numerous bays and inlets; and the gulf of Paria is an extensive inland sea, in which ships of all sizes may ride securely, and anchor anywhere without the smallest risk, and in any convenient depth of water. (*Blunt's American Coast Pilot, p. L.*)

The greater part of the interior of this island is uncultivated, and, indeed, in a considerable degree unexplored. The grounds are in parts marshy, while the more elevated tions are, for the most part, covered with a dense vegetation of forest and underwood. The accounts best entitled to credit represent the island as being naturally extremely fertile. The soil is, in general, deep, stiff and tenacious; it is said that, if properly cultivated, it could alone supply sugar adequate for the consumption of England. It might be supposed that, in an island so extensive, mountainous, and covered with forests, the atmosphere would be generally overloaded with moisture. It does not, however, appear that the fall of rain is as great as in Guiana, the average being about 65 in. a year; and this is said to diminish as the progress of cultivation. The dry season commences in December, and ends in May; but it is a peculiar advantage of this island that it is exempted from those destructive droughts common to all the other W. India islands from Barbadoes to Cuba. During June and July showers are frequent; and in August, September, and October, the rain falls in torrents, often accompanied by violent storms. The weather generally moderates, and the rains become slight towards the end of October, and there is seldom rain fall after the beginning of December. The nights are really cool and pleasant. The mortality, during the thirty years ending with 1836, averaged about 10 per cent. for white, and 4 per cent. of the black troops a year. Fevers and dysenteries cut off most of the whites. (*Papers relating to the W. Indies, 1841–2, Trinidad; Tulloch's Rep. on Health of the Troops in the W. Indies, p. 17–19.*) It has been estimated that only about 1-30th part of the surface of this island is incapable of cultivation. The settlements of Trinidad are mostly confined to the N.W. A few places along the S.W. coast. It is stated in a report by a sub-committee of the planters of Trinidad, that, in 1,208,379 acres of land had been appropriated, of which 65 were in cultivation, and that 1,079,301, were then unappropriated, and belonged to the crown. (*Papers relating W. Indies, Trinidad, p. 103.*) Of the cultivated land, 1 acres were said to be under the sugar-cane, being divided into 184 estates, the capital invested in which is estimated in the same document at £2,000,000 sterling. The of the cultivated lands were occupied by cocoa plantations (6910 acres), coffee do. (1095), and provision and pasture grounds.

It is stated in the report now referred to, and the same has since been repeated in a tract published by Mr. —, that there is in Trinidad an extreme disproportion between the extent of granted land and fixed capital in the , and the amount of the labouring population, and that deficiency of the latter enables the labourers to dictate terms of contract and the rate of wages, to the grievous loss not only of the capitalists but also of the labourers selves. Some liberated Africans from Sierra Leone have been imported into Trinidad in the view of lessening it, but the supply has been too limited to have had effect; and the committee propose, as the only effective means by which it can be obviated, that leave should in to import free labourers from Africa into the colony. We apprehend, however, that it would be extremely difficult, or rather perhaps, we should say, impossible, to

prevent this license from being abused; and considering the efforts we have made to prevent similar practices on the part of others, (*Bandinel on the Slave Trade, p. 237*), it is, not easy to see how we could sanction them in our own case. There is greater reason to fear that the license in question would reproduce a considerable number, if not most of the evils connected with the slave-trade, and it is not, therefore, very likely that the power will be conceded.

The truth is, that the mischiefs complained of are not so much the effect of any positive scarcity in the supply of labour as of the circumstances under which the labourers are placed. They can provide for the comparatively few wants incident to such a climate with but little exertion, and having done this, is it to be expected they should do more? It is idle to suppose that blacks in the W. Indies, now that they are free, will make the same exertions they were compelled to make when they were slaves. The free inhabitants of all fertile tropical countries, are, we believe without a single exception, uniformly indolent. And to suppose it should be otherwise is equivalent to supposing there may be an exertion without a motive, an effect without a cause. If an hour's labour a day will fully supply a man's wants, it is absurd to suppose he should continue to labour three or four hours.

Cocoa is more extensively grown in Trinidad than in any of the other British Antilles, and is of superior quality. The cocoa-tree somewhat resembles the cherry-tree, and grows to about 15 feet in height. It flourishes most in the new soil on the bank of a river, delighting in shade, to procure which plantains or coral-bean trees (*madre del cacao*) are planted between every other row. The cocoa seeds are placed in small mounds, two seeds being sown together, and the weakest plant of the two afterwards destroyed; and the survivor is transplanted after attaining 15 or 18 inches in height. Until the age of five years, all the flowers are destroyed as they appear. The fruit grows in a pod, which, as it ripens, changes to a bluish red or lemon colour. The crop is gathered throughout the year, but principally in June and December. The ripe pods are broken or cut open, and the seeds extracted with a wooden spatula. They are afterwards spread out to dry in the sun on rush mats. When quite dry and hard, the nuts are lightly packed in boxes or bags, and kept dry for exportation. Coffee, indigo, tobacco, and cotton, come to perfection, though mostly grown only in small quantities. Here, also, are all the fruits and vegetables of the adjacent tropical climates, and the vines transplanted from France or Spain, are said to equal their parent stocks. The mountains, like those of the adjacent continent, consist chiefly of argillaceous and micaceous schist; milky quartz, ferruginous sand, pyrites, arsenic, alum, sulph copper, plumbago, sulphur, &c., are found; but the most abundant mineral is asphaltum, which may be supplied in any quantity. It is found in the greatest profusion in the lake Brea, or pitch lake; an area of about 150 acres in the N.W. side of the island, about 30 m. S. from Port Spain, and about 80 feet above the level of the sea. Though called a lake or lagoon, this depôt of pitch is for the most part quite solid, rent, however, by chasms, varying from 3 to 30 feet in width, but of no great depth, so that they are traversed without much difficulty. Here and there, wherever there is any soft, are clumps of stunted trees. The liquid part of the lake, on the side nearest the sea, is supposed to be about 3 acres in extent, and consists of fluid pitch of unknown depth, in a state of slow ebullition, and exhaling a strong bituminous and sulphurous odour. This vast pitchy cauldron must be approached with extreme caution. It has been attempted to apply the asphaltum brought from this lagoon to the same purposes as pitch and tar, but it is found to require so large an admixture of oil that it becomes too expensive. If it could be economically applied, Trinidad might furnish abundant supplies for the whole world. (*Trinidad Almanac for 1840, App. c. 4.*)

Exclusive of the pitch lake, Trinidad has several extinct volcanic craters, active mud volcanoes, and other evidences of volcanic agency. Slight shocks of earthquakes have also been occasionally felt, but happily the island appears to be exempted from the scourge of hurricanes.

Trinidad was greatly neglected by the Spaniards, and previously to 1783, when emigration to it was first actively promoted by them, no more cocoa, indigo, and other products were exported than sufficed to freight a small schooner two or three times a year to St. Eustatius. Since then the progress of cultivation has been comparatively rapid.

Trinidad, like St. Lucia and British Guiana, is governed by a governor and council, acting under the orders of the home government. The legislative council of the island consists of 12 members, 6 of whom are styled official, holding high offices, and 6 non-official, being selected from among the inhabitants; all are removable at the pleasure of the crown. The laws of the island are a mixture of those of Spain and England, and it is said that much mischief has been occasioned from the circumstance of the judges and other functionaries sent from England being ignorant of the former. The office of coroner does not exist here, nor trial by jury in the supreme criminal court. Every person about to leave the island must first give public notice of his intention, and obtain a pass from the governor.

The settled part of Trinidad is divided into 11 districts. The cap. and seat of government, Port Spain, which in 1839 had 11,663 inhabitants, is situated on the W. coast of the island, near the mouth of the river Caroni. It is one of the handsomest towns in the W. Indies, being built wholly of stone or brick, with wide and well kept streets, some of which are shaded with rows of noble trees. It has Protestant and Roman Catholic churches, a Presbyterian secession church, and a Methodist chapel. The stores and magazines are crowded with valuable merchandise, which,

however, is partly destined for the supply of Columbia. In the vicinity of the town are fort George, now hardly dismantled, and St. James' barracks. The latter are sad, in the *Trinidad Almanac*, " to be placed, on account of an infamous job, in one of the most pestilential spots in the island." (p. 98.) The harbour is good, and, in fact, so already stated, the entire gulf of Paria may be regarded as a magnificent harbour. Numerous government and other schools are established in Trinidad, which in 1839 were attended by about 950 pupils. The regular military force amounts to about 3400 men, including officers, but every freeman of the island is enrolled in the militia. Public revenue in 1835, £42,430; expenditure paid by the island, £38,297.

The following table exhibits an

Account of the Quantity and Value of the principal Articles exported from Trinidad in 1839.

	Sugar.	Cotton.	Coffee.	Cocoa.	Rum.	Molasses.
	Lbs.	Lbs.	Lbs.	Lbs.	Galls.	Galls.
To Great Britain . .	28,473,506	110,942	82,689	630,322	8,900	809,400
West Indies . .	556,204		93,136	112,610		1,380
North America . .	179,144	800	160	7,496	1,568	23,469
United States . .				59,840	180	
Other states . .		10,000		1,761,790		
Total . .	29,208,854	121,042	175,985	2,571,988	9,946	834,889

Value of the above {	Sugar . . £333,579	Cotton . . £2,823	Coffee . . £5,934	Cocoa . £35,925
	Rum . . 1,123	Molasses . 40,917	Sundries . . 4,167	Total 494,343 sterling

Trinidad was discovered by Columbus in 1498, and was taken possession of by the Spaniards in 1588, an event followed by the almost total extermination of the Indians. Raleigh visited it in 1595. The French took it in 1696, but soon afterwards restored it to the Spaniards, who held it till taken by the English under Abercrombie in 1797. (*Trinidad Almanac for* 1840 ; *Parl. Papers.*)

TRIPOLI, the most easterly of the Barbary states, the dominions of which, exclusive of Tripoli proper, comprise Barca and Fezzan, noticed in other parts of this work. Tripoli proper lies between lat. 29° and 33° N., and long. 10° and 20° E.; having E. Barca, W. Tunis, S. Fezzan and the desert, and N. the Mediterranean. It stretches along the North African coast about 300 m. E. and W. Its breadth inland varies greatly, owing to the frequent interruption of the desert; but its area has been estimated at nearly 100,000 sq. m., and its pop. at from 1½ to 2 millions, principally Moors and Berbers, with some Turks, Negroes, Jews, and Christians.

In antiquity, Tripoli proper was called the *Regio Syrtica*, from its lying between the *Syrtis Major*, now the gulf of Sidra, on the E., and the *Syrtis Minor*, now the gulf of Cabes, on the W. The former, or *Syrtis Major*, is a very extensive bay, extending from Bengazy on the E. to cape Mesurata on the W. about 280 m., having where greatest a breadth of 150 m. This gulf was reckoned in antiquity, next to the strait of Scylla and Charybdis, by far the most dangerous part of the Mediterranean, principally on account of the shallowness of its waters, which were said to be encumbered with quicksands, and partly also from the irregular action of its tides. " *Vorum importuosus atque atrox, et ob vadorum frequentium brevia, magisque etiam ob alternos motus pelagi adfluentis et refluentis infestus.*" (*Pomp. Mela,* lib. i. cap. 7.) The dangers of the Syrtis have also been frequently alluded to by the poets, who have given it the epithet of inhospitable:

" *Per inhospita Syrtis Litora, per calidas Libya sitientis arenas.*"

Lucan, lib. i. v. 367.

See also *Virgil, Æneid,* l. v. 110 : *Horace,* Od. l. 22, &c.

But though the navigation of the greater and lesser Syrtis, especially the former, be not free from danger, this has been greatly exaggerated by the ancients. During strong N. gales a very heavy sea is certainly thrown into the gulf, and the S. shore being low and sandy, a considerable portion of it is submerged, and the waters of the entire gulf have an extremely agitated and turbid appearance ; but in ordinary weather it may be navigated by middling-sized vessels with little or no difficulty. " The gulf of Sidra," says Captain Smyth, " has few or no dangers, excepting little heads of rocks scattered about different points, and the tides are insignificant. With the hand-lead going, a vessel may approach all parts ; but of what utility can it be to enter here, there being but one place in the whole gulf worthy to be called a port ? We could find anchorage for small vessels only at Bushaifa and Braiga, at the bottom of the gulf, and Gnarra island, Karhora, and Bengazy [which see], on the E. coast.

But it is obvious from this statement, and from the want of harbours and roadsteads, that when the vessels of the ancients, who had comparatively little skill in navigation, got embayed in this gulf during the prevalence of northerly gales, they must have been in an exceedingly perilous situation, and we need not therefore be surprised at the exaggerated terms in which they have described its terrors.

The coast-lands, except at the bottom of the gulf of Sidra, where the desert and sea are conterminous, are here, as in the rest of N. Africa, extremely fertile. These, however, seem to be the only valuable portions of the surface. The Atlas ranges approach nearer the sea here than in most other parts of Barbary, and immediately beyond them the country is a sandy arid desert. Tripoli has no river of any consequence, though a number of small rivulets descend from the mountains to the sea. Neither are there any lakes in the country, which, accordingly depends for its irrigation and consequent fertility almost solely upon the rains. These, when they occur, fall incessantly for several days and nights; they then cease suddenly, and not a drop more descends for several months together. The most severe famines are sometimes experienced from a continuance of drought ; but when this is not the case, the country appears to have lost none of its ancient productiveness. According to Mr. Blaquiere, " a more luxuriant tract than that in the immediate neighbourhood of the capital cannot be imagined. Country-houses, extensive pleasure-gardens, groves of orange-trees, and innumerable fountains, together with the incessant progress of vegetation, form an assemblage of rural beauty here which is rarely to be met with. The fairy scene does not, however, reach more than 5 m. inland ; when nothing but an immeasurable waste of sand is presented to the eye, and forms a striking contrast with the cultivated fields, to the edges of which it approaches. It should be observed, that a want of industry, and of proper encouragement from the government, are the only reasons why cultivation is not extended beyond its present limits. There is probably no country so highly favoured by nature as this in which respect to a rapid succession of the crops. The rains generally begin after gathering the dates, towards October, in the beginning of which month the Arabs plough and sow their grounds. In December and January the weather becomes dry and extremely pleasant, like our spring in England. In the beginning of April, the market before Tripoli is abundantly stocked with cattle, poultry, and vegetables of every kind. Towards June, almonds, figs, apples, pears, plums, peaches, nectarines, grapes, and melons are in season, and incredibly abundant. Cotton has been cultivated very successfully by various individuals; but, owing to a want of encouragement, does not form an article of export. Formerly, a quantity of raw silk formed one of the exports ; but its cultivation has latterly been neglected. Mulberry-trees, however, to be found near the capital in great numbers, so that silk may at any time become again a staple commodity of the country. The castor-tree (*Ricinus Palma Christi*) is found in the vicinity of Tanjoura, where a great deal of that oil is made annually, though it has not hitherto been exported in any great quantity." (*Blaquiere,* in *Modern Trav.* xx.) The exportation of corn, which otherwise would be very considerable, is prohibited except when carried on by the pacha for his exclusive benefit. Tobacco, saffron, madder, &c. are grown in small quantities ; some and galls are produced in the mountains, and the camel and horse are indigenous. But dates constitute the principal food of the inhabitants. The dates of Tripoli are finer than those of any other part of Barbary, and, besides its fruit, the date-tree yields a juice called *lagabbi,* which, when drunk fresh, is a very agreeable beverage, and, when suffered to ferment, forms an intoxicating fluid, "in great request among the Mohammedans, notwithstanding the prohibitions of the Koran." (*Della Cella*, 16.)

Each village is usually surrounded by plantations of date and olive trees, the surplus produce of which, with the straw mats, earthen jars, &c., made by the villagers, are partly exported, but are mostly disposed of to Bedouin traders. The vines along the coast yield grapes and raisins of the finest quality, and might be made to produce excellent wine. According to Della Cella, the neglect of such an advantage is less owing to the denunciations of the prophet than the exceeding sloth and ignorance of the people. Cattle, sheep, and poultry are reared in large numbers in some places; and as animal food is little consumed, they are principal objects of exportation. During the war, Malta drew large supplies of cattle and other live stock from Tripoli, and still imports most part of the cattle sent out of the country. Beef, though small, is very good, as is lamb; mutton is of inferior quality. A kind of wild cattle, the wild hog, antelopes, bustards, and several other wild animals useful to man, are met with in abundance. Large beds of rock salt exist in different parts of the country. On the coast, fish of every kind are most abundant; but, with the exception of a few boats employed from the capital, fishing does not form a part of public industry.

The natives of this regency manufacture carpets, bournooses, halks, and other woollen fabrics, camlets, mats of palm leaves, goats' hair sacks, Morocco leather of different kinds, earthen ware, prepared skins, and a few other articles. The manufacture of potash, like the exportation of salt, is a monopoly of the bey. The principal trade consists in the barter of European produce for those of the interior of Africa. From Tripoli, caravans go to Mourzouk, where a large fair is held in December and January, and to which the products of Bornou, Soekatoo, Houssa, Timbuctoo, &c., are brought. (See art. FEZZAN in this *Dict.*, i. 996.) The Fezzan merchants proceed in February and March to Tripoli, where they receive their goods for the fl. upon credit, paying by exchange one year for the goods purchased in the interior. They bring from the interior annually about 1500 negro slaves, 10,000 *metacali* (small parcels, each worth a Venetian sequin), of gold dust, 700 cwt. of natron, 1600 quintals of senna leaves, &c. The articles they take back are swords, pistols, mock pearls, brass, tin, coral, writing paper, cotton stuffs, &c.; and these articles, with provisions, colonial produce, timber, pitch, spirits, cochineal, indigo, damask, and other silk fabrics, spirits, looking glasses, toys, &c., constitute the principal imports from Europe. The exports from Tripoli by sea are wool of excellent quality, mantles, and other articles of dress, oil, senna, and other drugs, madder, barilla, hides, goat and sheep skins dressed, dates and other fruits, cattle, ostrich feathers, iron, gold dust, saffron, &c. (For further details see succeeding article.)

The government is in the hands of a bey, or pacha, who rules with despotic sway, and is chosen from among the Turkish officers resident in Tripoli, being confirmed in his authority by a firman from the Turkish sultan. He presides in the divan, and is assisted in his various duties by a bey-commander-in-chief; an aga commanding the Turkish soldiers; the kaya, or grand judge, who dispenses justice daily at the castle gate of the capital; the chief officers of the treasury and household; the *sheik-el-bled*, or head police magistrate; the *mufti*, or head of the priesthood; the *cadi*, or judge in matters respecting the Mohammedan faith, &c. The district governors seem to have powers equivalent to the bey, in their own districts; thus, the aga of Mesurata, besides his military attributes, unites in his own person all the judicial and legislative powers of the state, if we may so describe the functions exercised by an officer uncontrolled either by equity, reason, custom, or public opinion. (*Della Cella*, p. 45.) The revenues of the bey are derived from the tribute of the district governors, and the Arab tribes in the interior, taxes on the Jews and merchants, a tax of 10 per cent. on all land-produce, import and export duties, monopolies, presents, and exactions, fines for the mitigation of punishment, confiscations, &c.; their amount being estimated by Blaquiere at from £85,000 to £95,000 sterling a year; in addition to which a large portion of the necessaries for his use are procured by extortion from his subjects. His standing army is said to amount to 3000 men; but in time of war an army of 10,000 irregular cavalry and 40,000 foot may be raised by levies from the Arab tribes. The naval force is insignificant, consisting almost wholly of a few small vessels.

The character of the natives of Tripoli appears to be very indifferent. Captain Lyon says drunkenness is more common than in most towns in England. There are public wine-houses, at the doors of which the Moors sit and drink without any scruple; and the greater part of the better sort of people also are great drinkers. Blaquiere says he was unable to discover any good qualities to be contrasted with

the well known attributes of revenge, avarice, treachery, and deceit, which predominate alike in the prince and the peasant. There is probably no country on earth where the inhabitants are more inclined to be vicious. (*Blaquiere*, in *Med. Trav.* XX. 50.) And yet it is said such is the promptitude with which justice is administered, that crimes in Tripoli are less frequent than in European countries, and the people are more civilized than in most parts of Barbary. Intolerance towards Christians was formerly very strongly marked; but foreigners are now treated with respect, piracy and Christian slavery having been wholly abolished.

This territory contains some Roman antiquities, but they are much less frequent than in the adjacent territory of Barca. In the middle ages it generally shared in the fortunes of the rest of this portion of Africa. In 1592, Tripoli was given by the Emperor Charles V., who had become possessed of some authority over it, to the Knights of Rhodes; but these were driven from it by the Turks in 1551. Fezzan was rendered tributary about 1714; but the authority of the pacha, over either that country or Barca, appears to be little more than nominal, or at any rate very much disturbed.

TRIPOLI (an. Œa), a maritime city of N. Africa, cap. of the above regency, on a low rocky tongue of land, projecting into the Mediterranean : the castle being in lat. 38° 53′ 58″ N., long. 13° 10′ 58″ E. Pop. estimated at 25,000; 9000 of whom are Jews, residing in a suburb of their own. The town is much smaller than either Algiers or Tunis; it may be ¾ m. in length, by 5 furlongs in breadth; but its shape is very irregular. It is encompassed by high and thick walls, the original stone-work of which appears to have been very good; but they have been patched up in all directions with mud and fragments. A good many cannon are mounted on the ramparts, and Tripoli has some degree of strength as a fortress; it is entered by two gates, one to the E. and the other to the S. Viewed from the sea the town appears to be semicircular; and the extreme whiteness of its square flat buildings, covered with lime, which in this climate encounters the sun's fiercest rays, is very striking. The baths form clusters of large cupolas, to the number of eight or ten, crowded together in different parts of the town. The mosques are in general surrounded by plantations of Indian figs and date-trees, which at a distance, give the whole city a novel and pleasing aspect. Internally, however, it has narrow and irregular streets, and mean houses. The pacha's castle is at the E. end, within the walls, with a dock-yard adjoining. It is very ancient, and enclosed by a high strong wall; and the numerous buildings which have been added at different periods to its interior, to receive the junior branches of the royal family, have both deprived it of all symmetry, and increased it to a little irregular town. (*Mod. Trav.*) Tripoli is, in most respects, inferior to the capitals of the other Barbary regencies. Ali Bey, however, says, "In point of tranquillity and cleanliness, Tripoli might be taken as a model by some European towns in the Mediterranean. Though it possess neither the elegance nor the regularity of Valetta, you never hear of acts of violence being committed in the streets, and robberies are altogether unknown; the result of a well mounted police. Independent of a nightly patrol, there is a guard stationed in each street, who is responsible for whatever may occur in it. There is, besides, always a number of persons kept for the express purpose of sweeping the town. The caravanserais, mosques, and houses of the different consuls and higher classes, are usually built of stone, and regularly white-washed twice a year; the dwellings of the lower orders are of earth, small stones, and mortar. Tripoli has six mosques of the first rank, with minarets, and six smaller ones. The great mosque has a roof composed of small cupolas, supported by 16 elegant Doric columns of fine grey marble." There are three synagogues, one or two places of worship for Christians, several market-places, cafés, European hotels, &c. E. of the town, on a tract of rocky and elevated ground, is the site of the ancient cemetery, where several remains of antiquity have been discovered; and some portions of Roman tesselated pavements, fragments of columns, and entablatures, built up in modern walls, are met with in the city. The most striking relic of antiquity, however, is a magnificent triumphal arch, near the marine gate, at present used as a storehouse. Though half sunk in sand, its upper part is still in good preservation; and an inscription shows that it had been erected in 164, in honour of the emperors Aurelius Antoninus and L. Verus. It is built of huge blocks of marble, without cement, and has been ornamented with warlike trophies and other carvings in relief. The ceiling, also, is beautifully sculptured. (*Lyon, Trav. in N. Africa*, p. 18.)

The harbour of Tripoli, though not very spacious, is safe, and capable of accommodating a large fleet of merchant ships. Small frigates, whose draught of water does not exceed 18 ft., may also ride there in perfect safety. (*Bisquire, Letters,* i. 93.) It is formed by a long reef of rocks running out to the N.E., and by other reefs to the E. In the deepest part there are from 5 to 6 fathoms water. It is defended by the new Spanish and French forts, the reef and insulated rocks on the W., and by two other forts on the beach to the E. It is the principal *entrepôt* for the maritime trade for the regency. We subjoin an

Account of the principal Articles of Import and Export at Tripoli in 1836, with their estimated Value in Francs:

Imports.		Exports.	
Articles.	Value in fr.	Articles.	Value in fr.
Wheat and barley	541,531	Mantles	57,090
Beans and pease	106,855	Oil	9,383
Wool	82,257	Hides, &c.	10,813
Hardware	20,250	Saffron	4,070
Coffee	26,779	Seeza	7,903
Sugar	20,954	Wool	5,837
Cotton, manufact.	114,906	Cattle	3,334
Wines	22,166	Various	51,583
Various	472,035		
Total	1,360,336	Total	130,567

But it is obvious that this statement is but little to be depended upon; and that there can be in reality no such difference between the imports and exports. If the amount of the former be not overrated, we may be quite sure that that of the latter must exceed one million francs.

TRIPOLI, or TARABLUS, a town and seaport of Syria, cap. of the pachalic of its own name, on the Mediterranean, 130 m. S.W. Aleppo. Lat. 34° 26′ 22″ N.; long. 35° 51′ 32″ E. Pop. estimated at about 15,000. The town stands at the foot of a branch of mount Lebanon, on a small triangular plain at some little distance from the sea: the Marina, S.W. from the city, on a projecting point of land, is the place where merchant ships usually load and unload their cargoes. Tripoli is one of the neatest towns in Syria, and is surrounded by fine gardens; but its neighbourhood being marshy, its climate is frequently unhealthy. It is traversed by the small river Kadisha, which, however, is too shallow to be navigable even for boats. The houses are principally of stone, and many parts of the city bear traces of the architecture of the Crusaders, particularly some high Gothic arcades over certain streets; but there are no public buildings worth notice. Tripoli is commanded by an old castle on the heights behind, built during the crusades by the Count de Toulouse. The name *Tripoli* is, no doubt, derived from its being formerly divided into three separate towns at short distances from each other; and, in fact, El Mina, as the Marina is sometimes called, is a distinct town from Tripoli proper. Numerous granite columns and other ruins may still be seen along the shore. (*Pococke, Burckhardt, &c. in Mod. Trav.*) The port of Tripoli, to the N. of the Marina, and opposite the town, is merely a roadstead, sheltered on the W. by some rocky islets; and is safe only in fine weather. It is dangerous in winter, and particularly at the equinoxes, from the foulness of the bottom and the prevalence of strong gales. (*Purdy's Sailing Direct.*) It has, however, some trade, exporting silk, wool, cotton, and tobacco, with small quantities of oil, wax, cochineal, galls, and soap, manufactured in the town. There are numerous Greeks among the inhabitants, and a large share of the trade is in their hands. It is, also, the see of a Greek bishop, and the residence of several European consuls, (*Bowring's Rep. on Syria; Mod. Trav.*)

Tripoli was taken by the Crusaders, in 1108. It had previously been one of the most flourishing seats of oriental literature, and possessed a very large collection of Persian and Arabic works. It is said that 100 copyists were constantly kept employed copying manuscripts, and that the princes of Tripoli were in the habit of sending messengers into foreign countries to discover and purchase rare and valuable works. Unfortunately, however, this extensive and precious collection, amounting, it is said to 100,000 vols., was destroyed by the Crusaders, who displayed on this occasion the same fanatical zeal of which they have accused, though we believe unjustly, the Arabs in the case of the Alexandrian library. (*See* Alexandria.) A priest in the suite of Count Bertrand de St. Giles, having visited an apartment of the library in which were a number of duplicate copies of the Koran, reported that it contained none but the impious works of Mohammed, and that, consequently, it should be destroyed! And, as a matter of course, it was forthwith set on fire!

Balbi, despite his pretensions to superior criticism and sagacity, states, after Quatremere de Quincy, that this library contained no fewer than 3,000,000 vols.! (*Bibliothèques de Vienne,* p. 81.) Michaud most properly rejects this statement as incredible and absurd, and adopts in preference the reasonable account given by Novairi. (*Historie des Croisades,* ii. 43., ed. 1841.)

TRIPOLIZZA, or TRIPOLITZA, a town of the kingdom of Greece, cap. dep. Mantines, and under the Turks, the

cap. of the Morea, near the centre of which it is situated, 20 m. S. by W. Argos. It stands in a plain nearly 2000 ft. above the level of the sea, and before the late war was about 3 m. in circuit, and probably more populous than Athens; but we can form no estimate of its present population. It is of modern origin, and is supposed to owe its name to its having been principally constructed of the ruins of the three cities of Tegea, Mantinea, and Pallantium, the sites of all which are at no great distance. Previously to the Greek revolution it had some large and conspicuous buildings; but it suffered severely during its capture and sack by the Greeks in October 1821, and its ruin was completed when it was retaken by Ibrahim Pacha in 1825; indeed, an arched gateway is now said to be the only existing relic of the Turkish period. (*Mure's Journal,* ii. 217.) But it is again rising from its ruins, and is the seat of one of the first class treasuries in Greece and of other government establishments. It is said, however, that the government contemplates the removal of these to some other town; and if so, it is not very probable that Tripolizza should ever regain its former importance. (*Gell; Burgess' Greece, &c.,* i. 210.; *Strong; Greece as a Kingdom, &c.*)

TROAD (The), or site of the ancient city of Troy, and the scene of the battles described in the Iliad. The situation of this classical region has been pointed out with sufficient precision by Homer, and has been admitted, from the earliest antiquity, to comprise that portion of Asia Minor bounded by and immediately S. of the W. entrance to the Hellespont, opposite the island of Tenedos, having mount Ida on the E. and the gulf of Adramyti on the S. Here, no doubt, are the *campi ubi Troja fuit*:

" Hac ibat Simois ; hic est Sigeia tellus ;
  Hic steterat Priami regia celsa senis :
  Hic Æacides, illic tendebat Ulysses ;
  Hic laser admissos terruit Hector equos."
                                    Ovid. Epist. l. No. 33.

But, notwithstanding the immortality of renown that has been conferred on the " heaven-built " city, and the interest which the Troad has always excited, such and so great have been the changes brought about by the influence of war, the ravages of barbarians, and the lapse of ages, that it is now no easy matter to reconcile the descriptions of Homer with the existing appearances of the country.

The Troad has been examined by several learned travellers, including Chandler, Wood, Chevalier, Clarke, Hobhouse, Gell, and others. But as none of them had the means of making a proper topographical plan of the country, and as its appearance, and especially the magnitude and even number of the rivers, differ at different seasons of the year, we need not be surprised at their conflicting statements, even had they not been mostly under the influence of some preconceived theory. The system of Chevalier, which for awhile was pretty generally acquiesced in, was founded on the assumption that the Mendere, the principal river of the Troad, was the Simois, and that the small river to the S. of the latter, the Bournabashi or Kerki-jess, was the Scamander of Homer. This hypothesis is now, however, generally abandoned, and it is indeed surprising that it should ever have obtained currency.* Inasmuch, however, as it would be impossible to make minute details intelligible without the aid of a map, we shall merely observe that Major Rennel and Mr. Maclaren have all but demonstrated that the Mendere is identical with the Scamander of the Iliad; and the suggestion of Dr. Chandler, that the Thymbrius, (now the Dumbrek-sou,) a river to the N.E. of the Mendere, with which it unites before they fall into the sea, is the Simois of Homer, appears to be satisfactorily established by Mr. Maclaren. It is, in fact, the only river in the Troad, excepting the Mendere, that in any respect corresponds with the descriptions given in the Iliad of the Simois; and the plain between the Mendere and the Thymbrius is the only one of sufficient extent to allow of the battles described by the poet being fought.

Dr. Clarke has conclusively shown (lib. 133, 8vo ed.) that the ruins at *Palaio Callifat,* or *Isarlik,* are certainly those of the New Ilium of Strabo. They are situated on a rising ground about 3 m. from the sea, and about midway between the Mendere and Thymbrius. Here the learned traveller found not only the traces but the remains of an ancient citadel; and at the very moment of his visit the Turks were employed in raising vast blocks of marble from the foundations of this edifice, which exhibited the colossal and massive style of architecture peculiar to the early ages of Greek history. The ground around was covered with fragments of broken pottery, and medals have been discovered among the ruins.

In the time of Strabo, New Ilium, whose position is thus clearly identified, was believed by its inhabitants to occupy the identical site of the ancient city, and such had been the

* Chevalier's theory has been espoused by 'Gibbon' ; see his *Delen Riviera,* i. 97, &c ; but he adds nothing to Chevalier's statements, and does not seem even to have been aware of the existence of Mr. Maclaren's work.

belief uniformly entertained by them from the earliest period: Hence," says Tacitus, " *Homeru* antiquitatis gloria politissat. (Annal, lib. iv. cap. 55.) Strabo, however, places the old city considerably more to the E., but we agree with Mr. Maclaren in rejecting this statement, and in believing that the old and new city stood upon the same site. The fact is, that a city taken by an enemy, and given up to military execution, is never completely destroyed; the foundations, with portions of its walls and temples, are always sure to remain, and these with the ruins afford many facilities for the construction of a new city. There is no reason to think that the destruction of Troy was in any respect more complete than that of Thebes by Alexander the Great, and yet the latter was rebuilt in the course of 20 years. And it is further to be observed that the conqueror now named visited New Ilium, in the full conviction that it represented the ancient city, sacrificed to Minerva and the manes of Priam, conferred immunities on the inhabitants, and gave orders that the walls of the town should be rebuilt, which intention was carried into effect after his death by Lysimachus. It is childish to suppose that Alexander should have done this unless he had been satisfied of the identity of the old and new city; and neither Arrian, nor any one else of his historians, so much as insinuates a doubt upon the subject. It would be rash and unwarrantable to set aside such evidence on the sole authority of Demetrius of Scepsis, who has, in this instance, been followed by Strabo, more especially as it has been shown that the site of New Ilium corresponds incomparably better with the Ilium of Homer than any other site on the Troad.

Perhaps it may be said, that, before endeavouring to point out the situation of Troy, it might have been as well to inquire whether that city ever existed, and whether any such war as that of Troy was ever carried on. But such inquiries would be wholly misplaced in a work of this kind; and though it had been otherwise, they would be wholly superfluous. It is the mere wantonness of scepticism to call in question the existence of Troy. Even if there were nothing more, the Iliad, which obviously describes real and not fictitious events, would be conclusive of the question; and when we add the concurrent testimony of the most ancient and best Greek authors, including Hesiod, Herodotus, and Thucydides, and the traditions universally prevalent as to the event, we should be quite as much disposed to deny the existence of Nineveh, Babylon, or even Jerusalem, as of Troy.

Exclusive of Troy, the Troad contained, at a later period, some other cities, such as Sigeum, on the sea shore, at the mouth of the Hellespont, near the promontory of the same name, and adjoining the barrow or mound called the tomb of Achilles. It was founded posterior to the siege of Troy by an Æolian colony. It had, however, ceased to exist in the time of Strabo. But the town of Alexandria Troas, on the coast, about 17 m. S. from Sigeum, was by far the most important of the towns in the Troad built after the destruction of Troy. It was founded by Antigonus, one of Alexander's generals; and became, under the Romans, one of the most flourishing of their Asiatic colonies. (*Strabo*, lib. xiii.) It is twice mentioned in the "Acts of the Apostles," and was the scene of a miracle. (*Acts*, caps. xvi. and ix.) Its site, now called Eski Stamboul, is identified by the remains of walls and other buildings, including a theatre, gymnasium, a magnificent aqueduct, &c., that sufficiently attest its ancient magnificence. (See *Chevalier on the Plain of Troy, with notes by Dalzell,* 4to, 1791 ; *Rennell on the Topography of the Plain of Troy,* 4to, 1814 ; *Chandler's History of Ilium or Troy,* 4to, 1802 ; *Maclaren's Dissertation on the Plain of Troy,* 8vo, 1822, &c. In 1794, (most probably, for the date is not given,) Mr. Bryant published at Eaton his singular dissertation, to shew that no such city as Troy ever existed, and that the expedition against it never was undertaken. This dissertation was answered, and, as we think, completely refuted, by Mr. J. B. S. Morritt, in his vindication of Homer, 4to, 1798 ; a variety of other tracts have appeared on this curious question, but the above, with the replies of Bryant and Morritt, exhaust the subject.

TRONDHJEM (vulg. *Drontheim*), a town and seaport of Norway, cap. prov. of its own name, on the Nid (whence its ancient name *Nidrosia*) at its mouth, in the deep gulf called Trondhjem-fiord, 275 m. N.E. Bergen, lat. 63° 25' 50" N., long. 10° 23' 23" E. Pop., in 1835, 12,358. A great deal of expense has been incurred in its fortification; but, as it is commanded by heights beyond the Nid, which surrounds it on the S. and E., it is not really strong. The fortress of Munkholm, bristling with cannon, stands on a small island in the fiord opposite the city; but it is, perhaps, more serviceable as a prison than a military outwork.

Trondhjem is, on the whole, well built, though its houses are almost all of wood. The streets are spacious, with water cisterns at their intersections; and the town has a singular air of cleanness and comfort. The most remarkable edifice is the cathedral, built principally of stone, and founded early in the 11th century, though little of the original structure remains; that little, however, is enough to show that it had been originally one of the most magnificent ecclesiastical structures in Europe. (*Clarke's Travels,* x. 232.) Part of the architecture is Saxon, the rest Gothic, and round and pointed arches are frequently intermixed. The extreme length has been 346 ft., and its breadth 84 ft.; but the W. end, where was the grand entrance, had a chapel at each corner, making the breadth of that front 140 ft. (*Laing*, p. 69.) Only the transept and E. end of the building are now roofed in and used; the W. part serves for a timber yard. There are three other churches, all plain structures; a hospital for the aged and infirm; a workhouse; a public library and museum, in which are collections of books, minerals, natural history, and antiquities, better than might have been expected in so remote a part of Europe; a public grammar-school, Lancastrian schools, a theatre, and many hospitals and charitable institutions. The palace of the military commandant, though constructed wholly of wood was, in the opinion of Clarke, the first edifice of its kind in Scandinavia. There are no regular inns in Trondhjem, but several good boarding houses.

The town is governed by a corporation of twelve persons, elected from among the mercantile body; and is the seat of the superior courts, &c., for all the country N. of the Dovre-fjeld. The roadstead of Trondhjem is but indifferent, being unprotected both on the N. and W., and the bottom loose ground in 20 fathoms: the river does not admit vessels drawing more than 10 or 12 ft. water. Dried fish, tar, deals, and copper from Roraas, are the principal articles of export.

According to Dr. Clarke, Trondhjem-fjord never freezes; and the cold of winter, though severe, is not nearly so great as at Roraas, which lies more to the S. The surrounding country is studded with merchants' villas; and immediately beyond the town is an extensive suburb, reached by a good wooden bridge across the river.

Trondhjem is now connected by a continuous carriage-road with the Swedish capital; and in 1838 a government steamboat commenced running between Trondhjem and Hammerfest, calling off Tromsoe and other intermediate ports, from spring till September. English is understood and spoken by many of the inhabitants. Mr. Barrow says the manners and appearance of the upper classes nearly resemble our own ; and most of them are, in fact, more or less connected with mercantile houses in England, many having been educated in England, and others being accustomed to visit it once a year. (*Barrow's Excursions in the N. of Europe,* p. 345, 346.) The lower classes generally read and write; and among the opulent many are distinguished for their literary taste. (*Clarke, Preface to Scandinavia,* x. &c. ; *Laing ; Barrow, &c.,* passim.)

TROIS RIVIERES. See THREE RIVERS.

TROND (ST.) Flem. *St. Truyen,* a town of Belgium, prov. Limbourg, cap. cant., on a tributary of the Demer, 20 m. W. by S. Maestricht. Pop., in 1836, 8,490. It is supposed to owe its origin to a Benedictine abbey, founded here in 657: it was formerly fortified; but its works were dismantled in 1697. It has a considerable manufacture of fire-arms; and some trade in lace, leather, &c. A sanguinary action took place between the French and Austrians, in its vicinity, in 1793.

TROON, a seaport town of Ayrshire, on a point of land projecting into the sea, 6 m. N. by W. Ayr, and 8 m. S.W. Kilmarnock. Pop., in 1841, 2148. It is a neat, well-built town. The parish church is at Dundonald, about 4 m. distant ; but it has a chapel-of-ease and a chapel belonging to the United Secession church. Troon harbour on the N. side of the promontory on which the town is built, is the most accessible of any on the Ayrshire coast, has 16 feet water at low spring ebbs, and sufficient accommodation for a great number of ships. Its advantages had, however, been wholly neglected till it came into the possession of its present noble proprietor, the Duke of Portland, who has constructed a large dry or graving dock, not surpassed by any other in Scotland, for the building or repair of vessels of a large size ; a smaller do. ; and is now constructing a wet dock, that will accommodate 50 sail. It is also furnished with commodious warehouses, and a harbour-light. Through the exertions of his grace, the Troon has been united, by a railway, with Kilmarnock, by which it has been made, to some extent at least, the port of the latter. Great quantities of coal are raised in the vicinity, which, being brought to this port by the railway for shipment, are exported to the amount, in ordinary years, of about 108,000 tons. Ship-building employs from 100 to 200 hands ; and rope and sail-making, and the trades connected with them, are also carried on to some extent. About 4000 tons shipping belong to the port. Branches of the Ayr bank, and of the Glasgow Union bank, have been opened in the town. The Glasgow, Paisley, and Ayr railway, passes within less than a mile of Troon, with which it is about being connected by a branch railway. (*Private Information.*)

TROPEZ (ST.), a maritime town of France, dep. Var,

on a bay of the Mediterranean, 36 m. E. by N. Toulon. Pop., in 1836, 3637. The inscriptions, medals, &c., found here, prove that it occupies the site of *Heraclea*, an important maritime town of antiquity. It has a citadel, and towards the sea is defended by some old walls; its port is spacious and good, but is little frequented, except by fishermen, which comprise a large proportion of its inhabitants. St. Tropez is the seat of a tribunal of commerce, a council of *prud' hommes*, and a school of navigation.

TROPPAU, a town of the Austrian dominions, cap. Austrian Silesia, and of the principality and circle of its own own name; on the Oppa, a tributary of the Oder, 37 m. N.E. Olmutz. Pop., in 1837, including its suburbs, 12,536. It is well situated, is walled, and entered by four gates, and is well built. Its principal edifices comprise a castle, townhall, theatre, high school, and sundry churches. . It is the seat of courts for its province, circle, and duchy, a tribunal of commerce, a gymnasium, to which a flourishing museum was attached in 1814, and considerable manufactures of woollen and linen fabrics, with others of soap, leather, liqueurs, &c. Troppau was, from 20th Oct. to 20th Nov., 1820, the place of meeting of the diplomatic congress, which afterwards removed to Laybach. (*Oesterr. Nat. Encyc.; Borghaus.*)

TROUP, county, Ga. Situated in the W. part of the state, and contains 430 sq. m. Drained by Chattahoochee r. and several tributaries of it, the principal of which is Yellow Jacket creek. It contained in 1840, 13,730 neat cattle, 6325 sheep, 30,995 swine; and produced 68,525 bushels of wheat, 2357 of rye, 469,635 of Indian corn, 35,655 of oats, 23,369 of potatoes, 1746 pounds of tobacco, 2,996,049 of cotton. It had 47 stores, 19 flouring-mills, 25 grist-mills, 15 sawmills, five tanneries, 11 distilleries, one printing-office, one weekly newspaper; five academies, 318 students; 90 schools, 520 scholars. Pop.; whites, 8639; slaves, 7093; free coloured, 26; total, 15,733. Capital, La Grange.

TROUPSBURG, p. t., Steuben co., N. Y., 246 m. W. by S. Albany, 282 W. Drained by Cowanesque cr. It has two stores, one grist-mill; 10 schools, 275 scholars. Pop. 1171.

TROY, p. t., Waldo co., Me., 30 m. N.E. Augusta, 634 W. Drained by branches of Sebasticook r. Incorporated in 1812, received its present name in 1827. It has one fulling-mill, two grist-mills, three saw-mills; 13 schools, 609 scholars. Pop. 1375.

TROY, city, port of entry, and cap. of Rensselaer co., N. Y., is in 42° 44' N. lat., and 73° 40' W. long., and 3° 21' 24" E. long. from W. It is 6 m. N. Albany, 151 m. N. New-York, 376 W. The population in 1810 was 3895; in 1820, 5264; in 1830, 11,405; in 1840, 19,334. Of these, 9203 were white males, 9630 were white females, 243 were coloured males, 258 coloured females; 796 were employed in commerce, 2279 in manufactures and trades, 16 in navigating the ocean, 192 in navigating rivers, lakes, and canals, 215 in the learned professions. It is pleasantly situated on the E. bank of Hudson river, on an alluvial flat, raised somewhat above the level of the river, and bordered on the E. and N. by abrupt and considerably elevated hills. Mount Ida, as it is called, is directly in the rear of the improved part of the city, and mount Olympus is an isolated eminence, 120 feet high, in the N. part. The summits of both present commanding views of the city and of the surrounding country. The city extends 3 m. along the river, and back from it 1½ m. Poestenkill, which enters the Hudson immediately S. of the compact part of the city, and Wynantskill, which enters 2 m. farther S., are both fine mill-streams, and have a descent, by cataracts and cascades, of more than 270 feet within the bounds of the city, and of 400 feet within the distance of 4 m. The gorge in mount Ida through which the Poestenkill rushes is very narrow, and lined with perpendicular rocks, overhung with trees and various shrubbery, and the scenery is wild and picturesque. The Wynantskill has a wider valley, which admits of a road through it to the upper level of the country, but the scenery is rugged, precipitous, and picturesque. The city is laid out with great regularity. The main business street, called River-street, follows the course of the river, which here has a curve. The other streets are straight, and cross each other at right angles. There are 15 streets running N. and S., and most of them successively run into River-street; these are crossed at right angles by 19 other streets, running E. and W. They form blocks 400 by 280 feet, intersected by alleys. The streets are generally 60 feet wide, and are either paved or substantially gravelled, and have good sidewalks laid with flagstones or bricks, and handsomely shaded with trees, and publicly lighted. River-street has much of the bustle of business, but the streets back of this have a quiet and rural appearance, and present many desirable residences. The houses are mostly built of excellent brick, and some of them of marble, and are large, neat and commodious. The courthouse is a large marble building, with a fine Grecian front of the Doric order. There is a brick jail, and a county poorhouse, with a farm attached of 200 acres. The Rensselaer Institute is designed to give a scientific and practical education, and has fitted numbers for the practice of civil engineering. The Troy Female Institute sustains a high literary character, has educated many very respectable females, over 300 in number, of whom one tenth have been teachers, and it has furnished principals to the most distinguished female seminaries in all parts of the Union. It has about 90 teachers and 200 pupils. The Troy Academy is a flourishing institution. There is a lyceum of natural history, with a valuable library, and a cabinet of minerals and of natural history. The Young Men's Association has a large library, a cabinet, and a reading room. There are in the city two excellent market-houses. Several of the churches have an imposing appearance. St. Paul's (Episcopal) church, is a noble Gothic structure, which cost $50,000. There are 18 churches: four Presbyterian, one Associate Presbyterian, one Congregational, three Episcopal, two Methodist, two Baptist, one Universalist, two Roman Catholic, and two African, one of which is Presbyterian and the other Methodist. There are six banks, with an aggregate capital of $1,575,000; besides a bank for savings and two insurance companies. The city is supplied with water from a basin in the town of Lansingburg, elevated 72 feet above the level of the city, distributed through the streets in iron pipes, sufficient not only for the supply of families and of the public hydrants, but for many beautiful fountains. The force of the water alone, from the hose attached to the hydrants, is sufficient to throw the water over the highest houses, and in case of fire, nearly to supersede the use of fire-engines. The supply of water by the water-works is one and a half million of gallons daily, and the cost of the works was $190,000. About 1½ m. above the centre of the city a dam is thrown across Hudson river, which is also feet high, and more than 1100 feet long, which is furnished with a lock 114 feet long and 25 feet wide, by which sloops may pass to Lansingburg and Waterford, and boats may descend to Troy, after having passed through the locks of the Champlain canal at Waterford; or they may reach the city by the side cut in the junction canal at Watervliet, which communicates directly with Hudson river. This dam affords great water-power, and equally facilitates commerce and manufactures; and it will doubtless hereafter be more extensively employed to give motion to machinery than it is at present, as it gives control in Troy over the water of half of the Hudson, with a fall of about 11 feet. Troy is well situated for commerce. It is happily located at the head of natural sloop navigation on Hudson river, and on this account, before the construction of the dam and lock, it had great advantage over Lansingburgh. There are owned here sixty masted vessels, three large steam passage boats, two of which (the Empire and Troy) are among the largest and finest on the river. The Empire was built in 1842 and 1843, is 330 feet long, breadth of beam 26 feet, extreme width 69 feet, tonnage 1040 tons, and draws only 4 feet 10 inches of water, with wood and water on board for passage. The Troy was built in 1839 and 1840, is 294 feet long, has 26 feet breadth of beam, with an extreme width of 61 feet, has a tonnage of 721 tons, and draws only 4 feet of water, with wood and water on board for a passage. There are two smaller steam passage boats, and five steam tow boats, with 22 barges, which ply between Troy and New-York, transporting annually an immense amount of produce and merchandise. Numerous canal boats ply from Troy both on the Erie and Champlain canals. The Rensselaer and Saratoga railroad extends from Troy to Ballston Spa, a distance of 24 m., where it unites with the Schenectady and Saratoga railroad. A railroad is in operation from Troy to Schenectady, 21 m., which is to be extended from Troy to Greenbush, opposite to Albany, there to connect with the Albany and West Stockbridge railroad. There is a macadamized railroad completed from Troy to Bennington, Vermont, a distance of about 36 m. The northern turnpike, extending through Rensselaer and Washington counties, and thence into Vermont, is a very superior road, and brings an immense amount of produce to Troy as a market. The enterprising spirit and large amount of capital in Troy, which have produced many of these facilities for trade, have wisely improved them. According to the United States census of 1840, there were in Troy 61 commercial and 13 commission houses in foreign trade, with a capital of $2,274,621; 270 retail stores, with a capital of $944,963; 18 lumber-yards, with a capital of $296,600; 1138 persons occupied in internal transportation, with 70 teachers, packers, &c., employed a capital of $1,626,130; four furnaces produced 1520 tons of cast iron, and eight forges, &c., 37,000 tons of bar iron, the whole, including these employed in mining, employing 254 persons, and a capital of $279,000; 11 persons produced machinery to the amount of $17,000; 135 persons produced hardware and cutlery to the amount of $265,400; three persons manufactured 150 small arms; seven persons manufactured the precious metals to the amount of $10,000; 73 persons manufactured various metals to the amount of $69,209; 18 persons manufactured bricks and lime to the amount of $17,400; three fulling-mills and one woollen factory employed 36 persons, and produced to the amount of $30,000, with a capital of $28,000; seven cotton factories, with 35,500 spindles, and two dyeing and printing works, produced to the amount of $253,900, employing 880 persons and a capital of $352,150; 15 persons manufactured tobacco to the amount of $13,500, with a capital of $4250; 154 persons manufactured hats and caps to the amount of $117,393, and straw bonnets to the amount of $130, with a capital of $46,700; seven tanneries produced 5850 sides of sole leather and 13,900 of upper leather, employing 64 persons and a capital of $81,000; 40 other manufactories of leather, as saddleries, &c., produced to the amount of $231,050, with a capital of $463,150; 18 persons produced 215,000 pounds of soap, and 170,000 pounds of tallow candles, with a capital of $19,000; one distillery produced 900,000 gallons of distilled spirits, and three breweries 694,750 gallons of beer, the whole employing 41 persons and a capital of $110,000; three paper-mills produced to the amount of $80,000, and other manufactures of paper to the amount of $27,000, the whole employing 38 persons and a capital of $88,000; one pottery employing six persons produced to the amount of $7000, with a capital of $1500; one rope-walk employing 40 persons produced cordage to the amount of $30,000, with a capital of $40,000; 10 persons produced musical instruments to the amount of $10,000, with a capital of $6000; 131 persons produced carriages and wagons to the amount of $162,000, with a capital of $170,850; 13 flouring-mills produced 191,548 barrels of flour, and with two saw-mills produced to the amount of $1,171,966, the whole employing 50 persons and a capital of $374,000; vessels were built to the amount of $109,000; 83 persons manufactured furniture to the amount of $79,100, with a capital of $45,700; four printing-offices, two binderies, two daily, three weekly, and one semi-weekly newspapers, and one periodical, employed 48 persons and a capital of $23,900; 41 brick or stone and 91 wooden houses were built, employing 115 persons, and cost $199,430. The total amount of capital employed in manufactures was $2,493,135. The city contained 11 academies, 446 students; 40 schools, 1961 scholars.

West Troy, Watervliet t., on the W. side of Hudson river, though in a different county, is properly a suburb of Troy, with which it is connected by a bridge and two ferries. This flourishing village was incorporated in 1836, and contains eight churches, the Watervliet bank with a capital of $150,000, an extensive United States arsenal, 800 dwellings, and 5000 inhabitants. In the N. part of the village, the S. branch of Mohawk river unites with Hudson river, and here is a lateral canal, by which the Erie canal enters Hudson river; there is a second lateral canal 1 m. below, the waters of which pass through the grounds of the United States arsenal, comprising about 100 acres, inclosed on three sides by an iron fence, and in the rear by a stone wall; and an extensive water-power is obtained by the waste water of the canal for the use of the establishment. About 200 officers, soldiers, and workmen are attached to the arsenal, and arms and munitions of war to the amount of $100,000 annually are manufactured. In the yards of the arsenal are found cannon which were captured at Saratoga, Yorktown, and others cast in New-York and Philadelphia during the revolution. There were in 1840 in the township of Watervliet 94 stores, capital $100,853; four lumber-yards, capital $73,000; machinery was produced to the amount of 35,000; three fulling-mills and three woollen factories, with a capital of $95,000; two cotton factories, with 5160 spindles, employed a capital of $295,000; one brewery, capital $100,000; one paper-mill, two rope-walks, two flouring-mills, 10 grist-mills, 12 saw-mills, one printing-office, one weekly newspaper. Total capital in manufactures, $723,115. There were 13 schools, 1800 scholars. Pop. 10,141.

In 1787, on the ground on which Troy now stands, there were but four dwelling-houses, and the surface was covered with shrub oaks and pines. It was laid out by the name of Troy in 1796, and incorporated as a village in 1801. The city charter was granted in 1816. The city is divided into eight wards, and is governed by a mayor, recorder, eight aldermen, and five assistant aldermen.

TROY, p. t., Bradford co., Pa., 149 m. N. by E. Harrisburg, 256 W. Drained by Sugar cr., flowing E. into Susquehanna river. It has four stores, one fulling-mill, one furnace, one grist-mill, five saw-mills, two tanneries, one distillery, one printing-office, one weekly newspaper; one academy, 90 students; one school, 474 scholars. Pop. 1664.

TROY, p. v., cap. Obion co., Tenn., 104 m. W. by N. Nashville, 846 W. Situated on Davidson's cr., a N. branch of Obion river. Incorporated in 1831. It contains a court-house, jail, and about 250 inhabitants.

TROY, p. t., Oakland co., Mich., 22 m. N.W. Detroit, 546 W. Watered by head branches of Red r. It has three stores; nine schools, 662 scholars. Pop. 1482.

TROWBRIDGE, a market town and parish of England, co. Wilts., hund. Melksham, on a tributary of the Avon, the Were, which is here crossed by a stone bridge, 21 m. N.W. Salisbury. Area of par. 1960 acres. Pop., in 1831, 10,863; in 1841, 11,050. The principal street is spacious, but the others are generally narrow and inconvenient, and though some of the houses are good, the greater proportion are but indifferent. Many are, however, constructed of stone, and the town is paved, and lighted with gas. Trowbridge church, a large and striking edifice, consists of a nave, chancel, two side aisles, with attached chapels, in the windows of which is a good deal of stained glass, and lofty N. and S. porches. The two side aisles are separated from the nave by five pointed arches, springing from clustered columns, and are externally embattled and ornamented with crocketed pinnacles; at the W. extremity is a large tower with a tapering spire. The living, a rectory, worth £600 a year, is in the gift of the duke of Rutland. Trinity church at the W. end of the town, erected within these few years, is a fine building, with a considerable number of free sittings. It has also, meeting-houses for General and Particular Baptists, Presbyterians, Wesleyans, Independents, &c., (dissenters being very numerous in Trowbridge); a free school and an alms-house, and most of the different sects support Sunday-schools. The manufacture of woollen cloth was established in Trowbridge at an early period. Cassimeres, fancy kerseys, tweeds, &c., and other narrow woollens, are the principal products. In 1839, 19 woollen-mills were in full work in the parish, employing together 1278 hands, and there may be in the town and its immediately contiguous district about 1650 looms. (*Hand-loom and Factory Rep.*) But at present (1842) the trade is very much depressed. The Kennet and Avon canal passes about 1 m. N. of the town, placing it in communication with London on the one hand and Bristol on the other. Trowbridge is under the jurisdiction of the county magistrates, who hold petty sessions here monthly, and a court of requests, for debts not above £5, every three weeks. Markets, Tuesdays, Thursdays, and Saturdays; fairs, Aug. 5 to 7, for cattle, pedlers, millinery, &c. Crabbe, the celebrated poet, was for 18 years rector of this parish, where he died, Feb. 3d, 1833.

TROYES (an. *Tresæ* and *Augustobona Tricassium*), a town of France, dep. Aube, of which it is the cap., on the Seine, which partly surrounds it, and is partly diverted into its interior by numerous canals, for the supply of its various factories, 99 m. E.S.E. Paris. Lat. 48° 18′ 3″ N., long. 3° 4′ 49″ E. Pop., in 1836, 25,563. It is inclosed by an old wall in pretty good condition, and has several suburbs. The town is but ill built, most of its houses being constructed of timber; though some of its new quarters are clean and sufficiently well laid out. Before the revolution, Troyes comprised 28 parishes, but their number has since been greatly diminished. The cathedral is a fine Gothic structure, chiefly constructed on the site of a previous edifice, in the 13th century, though not finished till towards the end of the 16th. Its interior length is 374 ft.; breadth, 164 ft.; height of the vault, 96 ft., and of the cupola externally, 304 ft. There is a good deal of curious stained glass in this church, the figures representing the kings of France, counts of Champagne, bishops of Troyes, and other personages of the 13th century, in the peculiar costume of that period, and of the size of life. Some of the other churches deserve being visited. The town hall is an edifice of the 17th century, with a handsome front, and a hall in which are the marble busts of the most distinguished natives of Troyes. The museum contains collections of mineralogy, natural history, and paintings; and the public library is said by Hugo and others to comprise 55,000 printed vols., and nearly 5000 MSS. The hall in which these works are placed is about 180 feet in length, and 30 in width; and on its pannels are paintings by Gonthier, representing the principal achievements of Henry IV. The prefecture, bishop's palace, seminary, hospital, courthouse, public baths, and abattoir are the other most remarkable buildings. The environs are particularly beautiful. It is the seat of courts of primary jurisdiction and commerce, a chamber of commerce, council of *prud'hommes*; the residence of an inspector-general of navigation, &c. It has manufactures of cotton, hosiery, calico, woollen cloths, blankets, and cotton and woollen yarn; with bleaching establishments, for which the waters of the Seine is said to be most suitable; paper mills, &c. Troyes was formerly the capital of Champagne; and it was here that Henry V. of England espoused Catherine of France. In 1429, it was taken from the English by the French troops, under Joan of Arc. In the campaign of 1814, it was the head quarters of Napoleon. Among the distinguished individuals, natives of Troyes, may be specified Pope Urban IV., the sculptor Girardon, and the painter Mignard. (*Hugo, art. Aube, &c.*)

TRUMBULL, co., O. Situated in the N.E. part of the

state, and contains 875 sq. m. Drained by Mahoning river, and Musketoe, Shenango, Meander and Mill creeks. It was named in honour of Governor Trumbull of Connecticut, and is the oldest county on the Western Reserve. The Pennsylvania and Ohio canal passes through it. It contained in 1840, 48,835 neat cattle, 77,486 sheep, 41,970 swine; and produced 941,563 bushels of wheat, 24,858 of rye, 388,494 of Indian corn, 104,557 of buckwheat, 221,962 of oats, 254,984 of potatoes, 906,359 pounds of sugar. It had eight commission houses in foreign trade, 45 retail stores, eight fulling-mills, one woollen-factory, 44 grist-mills, 190 saw-mills, three oil-mills. 23 tanneries, four distilleries, one brewery, two printing-offices, two weekly newspapers; one academy, 179 students, 986 schools, 13,572 scholars. Pop. 38,167. Capital, Warren.

TRUMBULL, p. t., Fairfield co., Conn., 5 m. N. Bridgeport, 59 S.W. Hartford, 288 W. Incorporated from Stratford in 1796. Drained by Pequannock river, which flows into Bridgeport harbour. It contains four churches, two Episcopal, a Congregational and Methodist, three stores, one woollen-factory, three grist-mills, four saw-mills; five schools, 142 scholars. Pop. 1204.

TRURO, a parl. and mun. bor. and market-town of England, co. Cornwall, hund. Powder, on the Fal, 7 m. N. by E. Falmouth, and 230 m. W.S.W. London. Pop. of parl. bor., in 1831, 8921. This is decidedly the handsomest, and including its suburbs, the largest town in Cornwall. It owes its increase and prosperity partly to its being in the centre of an important mining district, and a principal stannary town, and partly to its situation on a river navigable thus far at high water by vessels of 100 tons. The streets are partially paved, and lighted with gas. The town and bor. are comprised in the three parishes of St. Mary, Kenwyn, and St. Clement's. St. Mary's church is built of Roborough stone, which, at first sight, seems like granite, in a rich and beautiful perpendicular style. The interior has, however, been modernized; and a modern steeple has been attached to the church. (Rickman's Goth. Archit.) The living, a rectory worth £135 a year, is in the gift of Lord Mount Edgecumbe. There are meeting-houses for Independents, Wesleyans, Baptists, Unitarians, Friends, Bryanites, and various other sects; and a grammar-school, founded in 1760, which has an income of £100 a year, with two exhibitions at Exeter college, Oxford. Sir H. Davy received his early education in this school. It has also an almshouse, and several minor charities, town and cottage halls, a county infirmary, barracks, theatre, workhouse, a small jail, a subscription library, and several societies for instruction or amusement. The exports of Truro consist chiefly of tin and copper; with some paper and carpeting made in the town. Truro is of high antiquity, and had formerly a castle built in the reign of Henry II., and some other ancient edifices; but these no longer exist. It is said to be a bor. by prescription; its earliest charter appearing to have been one granted by Reginald, Earl of Cornwall, at an uncertain date, but certainly before 1290. It is now divided into two wards, and is governed by a mayor, five other aldermen, and 18 councillors. It has a commission of the peace, a weekly court of record, and some minor courts. Corp. rev. in 1840, £1967. It has sent two members to the House of Commons since the 23 Edward I., the right of voting down to the Reform Act, having been in the mayor, and 25 capital burgesses. The old parl. bor. comprised only the central part of the town, and the adjacent part of Kenwyn parish; but the Boundary Act at least doubled its former extent. Registered electors, 1839–40, 644. Markets, Wednesday and Saturday, and a cattle-market the first Wednesday in every month; fairs, four times a year, principally for cattle. Foote, the comedian, was a native of Truro, where he first saw the light 27th Jan. 1721. (Parl. Boundary and Mun. Corp. Reports; and Appendix, &c.)

TRURO, p. t., Barnstable co., Mass., 112 m. S.E. Boston, 513 W. By water it is only 65 m. from Boston. It occupies the N. part of cape Cod. Bounded N.E. by the Atlantic, S.E. by Cape Cod bay. It has four churches, two Congregational, a Unitarian and Methodist, 16 stores, four grist-mills; one academy, 52 students, 12 schools, 567 scholars. Pop. 1920.

TRUXILLO, or TRUJILLO, a town of Spain, in Estremadura, prov. Caceres, cap. dep. on the Tozo, a tributary of the Tagus, and on the high road between Madrid and Lisbon, 134 m. S.W. by W. the former. Pop. about 4500 (Miñano.) It is divided into the city, the old town, and the citadel, which successively occupy the foot, acclivity, and summit of a hill facing the S. The city is the newest portion; it is well laid out, and has a fine square, and several handsome residences, most of the wealthy inhabitants having removed thither from the old town. In the square is a large mansion, once belonging to the family of Pizarro, a native of Truxillo; the front of which is ornamented with bas-reliefs, representing the conquest of Peru. The old town, surrounded by a wall, is ill-built and dirty; but the

castle, with a mixture of ancient and Saracenic architecture, is imposing; and the appearance of Truxillo at a distance is very prepossessing. It has the usual complement of churches and convents, with several hospitals, a college, post-houses, &c., and manufactures of leather and linen fabrics. The name of this town appears to be a corruption of Turris Julia. It is supposed to be the Castra Julia of Ptolemy; and several Roman antiquities have been discovered in the town and neighbourhood. (Miñano; Mod. Trav., xix.)

TRUXILLO, or TRUJILLO, a town of Colombia, in Venezuela, cap. prov. of its own name, in a mountainous valley, 150 m. S.E. Maracaybo. Pop. estimated at about 8000. It is said to have been one of the finest and most opulent cities of this part of America, previously to its being pillaged by the buccaneer Gramont, in 1678, when most of its inhabitants fled to Merida. The valley in which the city is built is so narrow as to admit nowhere of more than two parallel streets, and the houses are small and mean. There are Dominican and Franciscan convents, a college, several schools, &c. The climate is healthy: the adjacent lands produce sugar, cocoa, indigo, coffee, and wheat; the mutton is larger and finer than in any other part of the prov. The inhabitants make superior cheese and preserves; and are famed for cleaning and carding wool. Its trade is principally northward with Carora and Maracaybo. (Depons in Mod. Trav. xxvii. 319; Codazzi, Géog. de Venezuela.)

TRUXTON, p. t., Cortland co., N. Y., 22 m. N.E. Cortland, 130 m. W. Albany, 327 W. Drained by Toughnioga r. It has two churches, a Presbyterian and Baptist, five stores, one furnace, two fulling-mills, one woollen-factory, five grist-mills, 25 saw-mills, one oil-mill, three tanneries; 31 schools, 1905 scholars. Pop. 3658.

TSCHERKASK (NOVI or NEW), a town of European Russia, cap. of the country of the Don-Cossacks, on a hill adjoining a tributary of the Don, 398 m. S.S.E. Voronej; lat. 47° 94' 29" N., long. 390 36' E. Pop., according to Schnitzler, about 14,000, but, according to Possart, only 12,000. It is wholly modern, having been founded under the auspices of Platoff in 1805: its streets are regular and broad, but some years since they were neither paved nor lighted, and most of the buildings were of wood. A triumphal arch, of hewn stone, stands at either extremity of the main thoroughfare, and there is a large square in which Platoff had begun to build a residence for himself. The town is a bishop's see; and has a new cathedral, gymnasium, circle-school, hospital, arsenal, &c. It is the seat of all the government offices for the Don-Cossack country, which were removed thither in 1807, from Stara, or old Tscherkask on the Don, about 10 m. S. by E., in consequence of the inundations to which the latter was subject. The new cap. is favourably placed to avoid this evil, but it labours under a great drawback in being near no navigable river. Staro-Tscherkask, which formerly had 15,000 inhabitants, has now dwindled into insignificance. The new town has annually four large fairs; to one of which goods to the value of upwards of 2,000,000 roubles are frequently brought. (Possart Kaiserth. Russland; Schnitzler; Lyall, Trav. in Russia.)

TUAM, an inland city of Ireland, prov. Connaught, co. Galway, on a small river, 15 m. E. lough Corrib, and 145 m. W. by N. Dublin. Pop. in 1831, 6883. This town has greatly improved, and increased in pop. during the last 20 years. Several new buildings are in progress. The principal streets diverge from the market-place, in the centre of the town; and some of them have latterly been widened and much improved. Still, however, much squalid poverty is to be found in the town and its wretched outlets. It is of considerable importance in an ecclesiastical point of view, having been till recently the seat of a Protestant, as it still is of a Catholic archbishop. But, in 1839, on the demise of the late Protestant prelate, the see was reduced from an archbishopric to a bishopric, suffragan to Armagh. The Protestant cathedral is a small plain building; but the Rom. Cath. cathedral is a splendid structure, and is, indeed, by far the finest of the modern Rom. Cath. churches in Ireland. Here, also, is the Rom. Cath. college of St. Jarlath, founded in 1814, usually attended by about 140 pupils. It has also a nunnery, a diocesan school, and other public schools, a courthouse and jail, barracks, dispensary, &c. The town comprises the palace and handsome demesne of the Protestant bishop. Tuam sent two members to the Irish House of Commons down to the union, when it was disfranchised. General sessions are held twice a year, and petty sessions on Wednesdays: it is a constabulary station. The manufacture of coarse linens and leather is carried on to some extent; and it has a brewery and flour-mills, and a weekly newspaper. Markets on Wednesdays and Saturdays; fairs, May 10, July 4, Oct. 29, and Dec. 11. Postoffice revenue in 1830, £840: in 1836, £1072. Branches of the agricultural and national banks were opened in 1835 and 1836.

974

It is worthy of remark, that, notwithstanding Team has been for a lengthened period the seat of a Protestant archbishop, with a large revenue at his disposal, there were, in 1834, only 428 individuals in the parish that belonged to the Established Church, whereas the Catholic pop. amounted to no fewer than 14,511! The country round Tuam is flat, badly cultivated, and the peasantry poor and depressed in the extreme. (*Municipal Boundary Report ; Fraser's Guide to Ireland, &c.*)

TUBINGEN, a town of S. Germany, kingd. Wirtemburg, circ. Schwartzwald on the Neckar, 17 m. S.S.W. Stuttgard. Pop. in 1838, 7250. It is old and irregularly built; its principal edifice is the castle, formerly the stronghold of the Pfalzgraves of Tubingen, but now appropriated to the university of Wirtemburg. This university was founded in 1477; and the famous reformers, Melancthon and Rauchlin, were among its earliest professors; it has both a Roman Catholic and a Protestant theological faculty. In 1835, it was attended by 734 students; of whom 289 attended divinity, 82 law, 163 medicine and surgery, and 181 philosophy, &c. About 100 students were subjects of other German states. (*Journ. of Educ.* vol. ix.) This university has an observatory, botanic garden, cabinets of mineralogy, zoology, &c., and a library. The chief support of the inhabitants of Tubingen is derived from the retail supply of this and the other public schools; but they have also a few manufactures of woollens, gunpowder, &c. (*Berghaus; Memminger Beschreibung von Wirtemb.*)

TUCUMAN, the cap. of the state of the same name, in the S. American confederacy of La Plata, in a fertile plain on a tributary of the Medinas, and on the high road between Buenos Ayres and Potosi, about 315 m. N.N.W. Cordova; lat. 26° 49' S., long. 64° 55' W. Pop. estimated at 12,000. It has a cathedral, several convents, a Jesuit's college, &c; but the inhabitants generally, from their remote inland position, appear to have made little progress in science, or the arts of civilized life. Their principal trade is in oxen and mules. Tucuman was founded in 1665. It is memorable in history as the place at which the declaration of the independence of the Plata provs. was first promulgated, and where their first congress was held in 1816. (*Dict. Géog.*)

TUDELA, (an. *Tutela*), a city of Spain, prov. Navarre, in which it holds the second rank on the Ebro, where it is joined by the Queilos, and near the commencement of the great canal of Aragon, 50 m. N.W. Saragossa. Pop. 8150 (*Miñano*). The Ebro is here crossed by a noble bridge, of uncertain origin, 400 Spanish (or nearly as many English) yards in length, and having 17 arches. Tudela was formerly fortified, but nothing remains of its ancient walls, except the gates, or of its citadel more than one tower. Its streets are narrow, crooked, and dirty; its houses lofty, and mostly of brick: there are many private and some public fountains, and the remains, in several places, of baths, constructed by the Moors. Along the river are some shaded public walks. It has a cathedral, in which Blanche, of Castile, the queen of Peter the Cruel was buried, many other churches and convents, two hospitals, an orphan asylum, workhouse, prison, society of public good, Latin and medical schools, &c. Its inhabitants manufacture coarse woollens, hair fabrics, soap, tiles, bricks, and earthenware, and trade in oil, flour, and wine, esteemed the best in the prov. Tudela has two large annual fairs; one from 1st to 21st March, and the other from 23d July to 10th Aug. It has given birth to several distinguished characters, including, among others, the Jewish traveller of the 12 century, the rabbi Benjamin Ben Jonah, commonly called Benjamin of Tudela. On the 23d of Nov., 1808, a French army, under Marshal Lasnes, completely defeated, in the vicinity of this town, a greatly superior Spanish force under Castaños. The latter lost about 3000 men, killed, wounded, and prisoners, and were completely dispersed. (*Miñano; Napier's Peninsular War,* i. 406.)

TUFTENBOROUGH, p. t., Carroll co., N. H., 44 E. by N. Concord, 525 W. Bounded S.W. by Winnipiseogee ake, which presents a fine water prospect. It contains four churches, two Methodist, a Free-will Baptist, and Christian, three stores, one fulling-mill, two grist-mills, two æw-mills; 10 schools, 415 scholars. Pop. 1281.

TULA, a government of European Russia, principally between the 53d and 55th degs. of N. lat., and the 36th and 39th of E. long., having N. the government of Moscow, E. hat of Riazan, S. Orloff, and W. Kaluga. Length about 130 m.; average breadth, about 85 m. Area estimated at 1,200 sq. m. Pop., in 1838, 1,113,500, this being one of the nost populous of the Russian governments. It slopes generally to the N. and E., in which direction the Oka flows, orming its N.W. and N. boundary. The Don rises in this overnment. The surface is an undulating plain, and, though ot very fertile, it produces a good deal of corn, with beans, urnips, mustard, flax, hemp, tobacco, potatoes, and other egetables. The peasants, almost everywhere, have gardens in which they grow fruit, &c.; the climate being tol-

erably mild and healthy. In 1830, there were estimated to be about 360,000 head of cattle, 360,000 horses, and upwards of 1,000,000 sheep in Tula. Iron is abundant, and in the neighbourhood of the cap. local mines extend over an area of 16 sq. m.; but the metal is of inferior quality, and iron is one of the chief imports into the government. A bad sort of coal has also been met with; but wood and charcoal continue to be the principal fuel used in the forges and other factories. Forests cover about one sixth part of the surface. Dr. Lyall says, that S. of Tula, there is not so profuse a waste of timber in the construction of the peasants' houses as nearer Petersburg. Indeed, some of the houses are not built in the usual way, with trunks of trees morticed together at the corners, but consist of wattled wicker-work. The dwellings, or rather the huts of the peasants, which range along both sides of the road, are more paltry in their appearance, and more simple in their structure than those between the capitals. Indeed, they gradually become more miserable as we proceed S. till we come to regions where stone abounds. (*Tres.* i. 51-2. Except in the capital there are hardly any manufacturing establishments, other than tanneries, breweries, and distilleries, the last two being on a very extensive scale. The exports consist principally of corn, hemp, and flax, with cutlery, jewellery, hardware, &c., from Tula; the latter, with Brelef, being the chief seat of commerce. In this government is the canal of Ivanof, uniting the Oka with the Don, excavated by the Swedish prisoners in Russia early in the 18th century. Tula has been a separate government since 1796: it is divided into 12 districts; chief towns, Tula the cap., Brelef, Venef, and Odölef. Its inhabitants are nearly all Russians, with some German colonists. In respect of public instruction, Tula is subordinate to the university of Moscow; in 1835, 1476 children were attending the public schools. Public revenue in 1821, 9,319,172 roubles.

TULA, a town of European Russia, cap. of the above government, on both sides the Upa, 110 m. S. Moscow, lat. 54° 11' 25" N.; long. 37° 1' 34" E. Pop., including the government workmen, but exclusive of troops, about 35,000. (*Possart,* 527.) This town, the "Sheffield and Birmingham" of Russia, is one of the most interesting in the empire. Clarke says that, as seen from a distance, it has an imposing appearance. A very handsome church, with white columns, appears above the town, which occupies an extensive vale, and is filled with spires and domes. The entrances on both the N. and S. sides are through triumphal arches, made of wood painted to imitate marble. (*Tres.,* i. 237.) It is divided into several quarters, the communication between them being kept up by a number of wooden and stone bridges; and there are several suburbs. There are two convents and 26 churches in Tula, all of stone; but the edifices which chiefly attract the stranger's attention are the gun manufactory, the gymnasium for the government; Alexander's school, opened in 1802 for the education of youth, at the expense of the nobility; the foundling hospital, a branch of that of Moscow; the house of correction, prison, arsenal, theatre, *gostinoi door,* or building for the preservation and sale of merchandise, &c. The shops in the latter present more activity and industry than are usually met with in Russian towns, and some of the merchants are reputed rich. There is a continual mixture of wood and stone houses; but some streets are lined on both sides with stone edifices, many of which are massive and in good taste. (*Clarke, Lyall, &c.*)

The musket manufactory, though commenced at an earlier period, is indebted for its original importance to Peter the Great. It was remodelled and improved by Catherine II. in 1785; but its present excellence is mainly owing to Mr. Jones of Birmingham, invited into Russia in 1817. About 7000 men and 9600 women are employed in this factory, besides 3500 hands in subsidiary occupations. About 70,000 muskets and 50,000 swords are said to be annually made here, exclusive of great numbers of carbines, pistols, bayonets, pikes, &c. The metal employed comes wholly from Siberia, and is of excellent quality. The workmen in the gun factory enjoy peculiar immunities and privileges; they form a separate body, and have their judges selected from among themselves. They are divided into five trades, barrel-makers, lock-makers, stock-makers, furnishing-makers, and makers of small arms. The arms made at this factory have been ridiculously depreciated by some travellers, and as extravagantly extolled by others. The exploits of the Russian armies speedily showed the entire worthlessness of the statements made by Clarke as to the badness of the Tula muskets; and, in point of fact, though they want the neatness and finish of the muskets of Birmingham, they are of very good quality. Some, also, of the fire-arms and swords made here are very highly finished; but these are comparatively high priced. Among the other fabrics of Tula are mathematical and physical instruments, jewellery, and platina wares. with silk and hat fabrics, tanneries, &c. The town is the residence of a military governor, with authority extending over the governments of Tula, Tambof, Riasan, Oriof, Vornoeje, and sometimes Kaluga.

Ancient Tula, which existed in the 12th century, did not occupy the site of the modern town, though it was on the Upa, at no great distance. The present city was founded in 1509, by Vassili-Ivanovich, who fortified it with a stone and brick wall, &c. Its defences, however, were insufficient to prevent its being frequently plundered by the Tartars, it being on the high road to Moscow from the Crimea. It has often suffered severely from fire, the last visitation being in 1834. (*Schnitzler*; *Possart*; *Clarke's Trav.*, i.; *Lyall's Trav. in Russia*; *Modern Trov.*, xv.)

TULLAMORE, an inland town of Ireland, prov. Leinster, King's co., of which it is now the cap., on the Tullamore river, an affluent of the Broana, and on the line of the Grand canal, in the centre of the Bog of Allen, 49 m. W. by S. Dublin. Pop. in 1831, 6342. In consequence of its advantageous position on the Grand canal, this town, which in 1790 was an obscure village, has risen to be the principal town of the county. The streets are wide and regular, and the shops and private dwellings are most respectable. In addition to the ordinary passage-boats between Dublin and Ballinasloe and Limerick, which all touch here, it has a daily communication with Dublin, by means of the swift iron boats lately established on the canal. Large quantities of corn and other articles of provision are shipped here for the metropolis. In consequence of its increasing size and importance, the assizes and other county business were transferred thither in 1833, from Philipstown. The principal public buildings are the courthouse and jail, on the radiating plan, which stand contiguous, on a raised platform, at the W. end of the town. It has also a parish church, a large Roman Catholic chapel, a Quaker and two Methodist meeting-houses, some large public schools, a market-house, barracks, and infirmary. It is a constabulary station, has three breweries, and two distilleries; and large quantities of bricks are made in the vicinity. Markets on Tuesdays and Saturdays; fairs May 10, July 10, and Oct. 21. Postoffice revenue in 1830, £646; in 1836, £688. A branch of the bank of Ireland was opened here in 1836. Adjoining the town is Charleville forest, the seat of its noble proprietor, the earl of Charleville, to whose liberality and munificence the town is greatly indebted. The pleasure-grounds are open to the inhabitants.

TULLE, a town of France, dep. Correze, of which it is the cap., on the Correze, 72 m. S.W. Clermont. Pop., in 1836, ex. comm. 7285; or inc. comm., 9700. It stands partly on the steep declivities on either side the river, and partly on the narrow space of ground between. "It is small, and its buildings are old and unprepossessing; but it has a pleasant promenade on the river's bank, good quays, many bridges, a church in a semi-Gothic, semi-Carlovingian style, a well-planned court of justice, some large buildings appropriated to a royal manufactory of fire-arms, carried on upon a very extensive scale, a well kept hospital, gendarmerie barracks, a departmental prison, college, seminary, theatre, and public library of 9000 vols. Its inhabitants appear to have a decided taste for embellishment, and the town is expected thereby to manifest marked improvement in a few years. It has several mansions ornamented with Gothic and other sculptures, testifying the opulence of its ancient families. One house in particular, in the principal square, called the *Maison Sage*, and dating from the 14th century, has its front decorated with arabesques in good taste and of superior execution. The cemetery of Tulle is in a remarkable situation, on an isolated hill, commanding the town, on which also is a lofty square tower, supposed to have been built by the Romans, which has long served for a prison." (*Hugo, and Guide du Voyageur.*)

Tulle has tribunals of original jurisdiction and commerce, a school of geometry, &c., a society of agriculture, and manufactures of wax candles, oil, nails, and hardware, paper and leather; but it is a curious fact that though the linen fabric called *Tulle*, most probably derived its name from this town, it is no longer produced either here or in the neighbourhood. It has 12 fairs a year, one of which lasting the three first days of June, is a great mart for horses. The principal races within a circle of several departments are held near Tulle. This town is supposed to be not older than the 7th century; but about 3 m. northward are the ruins of Tintignac, probably the *Ratiastum* of Ptolemy, exhibiting traces of a large amphitheatre, and of other extensive edifices. (*Hugo, art. Cor.: Dict. Geog.*)

TULLY, p. t., Onondaga co., N. Y., 10 m. S. Syracuse, 197 m. W. by N. Albany, 328 W. Drained by Onondaga creek and Toughnlogn river. It contains two churches, a Presbyterian and Baptist; three stores, one fulling-mill, one woollen factory, three grist-mills, seven saw-mills; nine schools, 540 scholars. Pop. 1663.

TULPEHOCKEN, t. Berks co., Pa. Drained by Tulpehocken creek and its tributaries. It contains three church-

es, common to Lutherans and Presbyterians, two stores, one fulling-mill, one woollen factory, two forges, five grist-mills, six saw-mills, four tanneries, five distilleries. The Union canal passes through it. Pop. 1581.

TUNBRIDGE, or TONBRIDGE, a market town and par. of England, county Kent, laths Aylesford, hund. Tunbridge, on the Medway, 27 m. S.E. London. Area of par. ish, 14,730 acres. Pop., in 1831, 10,380. The town appears to have owed its origin to a strong fortress erected in the 11th century, of which the entrance gate, flanked by two round towers, and part of the keep still remain. It consists principally of one long, wide, and pretty well built street, paved, lighted, and very clean. The public buildings include the church, grammar school, town-hall, and market house. Several bridges cross the Medway, which is here divided into different streams, the principal being sides which, £366 is paid for master's salary, &c., leaving erected in 1775, from a design by Mr. Milne, architect of Blackfriars-bridge, London. The living, a valuable rectory, in the gift of — Deacon, Esq., is worth £763 a year, nett. There are several dissenting chapels. The grammar school, founded in 1554, by Sir Andrew Judd, a native of the town is under the government of the Skinner's Company, and has one exhibition to either university, of £18, two of £12, six of £10, and several to a less amount; besides which, £366 is paid for master's salary, &c., leaving a considerable annual surplus. At present there are about 45 scholars on the foundation. Holmes's school at Southborough, at which 50 children are taught the rudiments of instruction, has an income of £100 a year; and there are several minor establishments for education, besides various other endowments for the benefit of the poor, amounting altogether to upwards of £50 a year. (*Parl. Reps.*) It has a market on Friday, and four annual fairs. The grammar school has had some very distinguished masters, among whom may be specified the learned Vicesimus Knox, D.D., author of *Moral and Literary Essays*, a treatise on *Liberal Education*, and various other popular and exceedingly useful works. Dr. Knox succeeded his father as master of the school in 1778; and having held the situation for 33 years, or till 1812, he was in his turn succeeded by his son. The doctor died at Tunbridge in 1821.

The favourite watering-place, Tunbridge Wells, is partly in this parish and partly in those of Speldhurst and Frant in Sussex, being about 5 m. S. Tunbridge. It consists of several different divisions, as mounts Ephraim, Sion, and Pleasant, and the Wells, the pump and assembly-rooms, public parades, chapel of king Charles the Martyr, &c., being in the latter. The springs, which were first discovered in the reign of James I., soon attracted the notice of the fashionable world. Henrietta Maria, queen of Charles I., paid a visit to the wells; but there being at that period no houses nearer than Tunbridge, and those not particularly suitable for such a guest, her majesty and her suite lodged in tents pitched on Bishop's Down! The wells were also visited by Catherine, queen of Charles II., queen Anne, and other distinguished personages. The water is a chalybeate, with an excess of carbonic acid gas, very similar to that of the *Pouhon* spring at Spa in Belgium. Tunbridge Wells resembles Spa in some other particulars; as in its manufactures, toys, boxes, and turned wares being made here in great variety, and also in its being much less frequented now than formerly by the leaders of the *beau ton*. The season for taking the waters continues from May to November. There are races in August, which are tolerably well attended. The chapel at Tunbridge is situated at the junction of the three parishes of Speldhurst, Tunbridge, and Frant, and is partly in each. The air of this district is pure and salubrious, and is, perhaps, little less efficacious than the waters in removing complaints.

TUNBRIDGE, p. t., Orange co., Vt., 32 m. S.S.E. Montpelier, 495 W. Chartered in 1761. Watered by a branch of White river. It contains five churches, a Congregational, Methodist, Free-will Baptist, and a Union, two stores, one grist-mill, five saw-mills; one academy, 40 students; 20 schools, 540 scholars. Pop. 1811.

TUNICA, county, Miss. Situated in the N.W. part of the state and contains 600 sq. m. Bounded W. by Mississippi river. It contained in 1840, 1868 neat cattle, 77 sheep, 2069 swine; and produced 436 bushels of wheat, 24,236 of Indian corn, 1146 of potatoes, 222 pounds of cotton. It had three stores, two grist-mills, two saw-mills; one school, 68 scholars. Pop.: whites, 566; slaves, 251; free coloured, 4; total, 821. Capital, Peyton.

TUNIS (an. *Zeugitania* and *Bizacium*, the E. portion of the *Africa* of P. Mela, with part of *Getulia*), a kingdom or regency of N. Africa, a nominal dependency of the Turkish empire, principally between the 33d and 37d degrees of N. lat., and the 9th and 11th of E. long.; having 3 E. the regency of Tripoli, N.W. that of Algiers, S. and W. the desert, and N. and E. the Mediterranean. Length N. to S. about 400 m. Its area has been roughly calculated at 72,000 sq. m. The population has been very variously estimated; but, perhaps, it may be taken at about 2 or 2½ millions, of whom probably from 7000 to 10,000 may be Turks, about the same number Christians, 112,000 renegades, 100,000 Jews, and the remainder Arabs, Moors, and Berbers, the Arabs being the most numerous. This territory is traversed by several branches of the chain of Atlas, one of which separates it from the *Bilad-al-Jerid*, or "country of dates." The S. part of the regency is mostly a sandy waste, and some other parts are desert; but many tracts are of the highest fertility, particularly those watered by the Mejerdah. This river, the *Bagradas* of the ancients, is formed by the union of two streams, on the W. frontier of Tunis, and runs thenceforward generally N.E., entering the Mediterranean about lat. 37° N., long. 10° E., a few miles N. of the site of Carthage. According to Shaw, it is "equal to the Isis united with the Charwell." Flowing through a rich and fertile country, it becomes highly impregnated with salt:

> "Turbidus arentes lento pede sulcat arenas
> Bagradas." *Silius Ital.*, i, vi. 140.

The Mejerdah receives no large tributary, nor is there any other considerable river in the regency. In the S., about 40 m. inland, is the Sibhah, a remarkable tract 79 m. in length N.E. to S.W., portions of which formed the *Palus Lybiæ, P. Tritonis*, &c., of antiquity. In winter, it is covered with water to the depth of 2 or 3 ft., but at other times it is a dry plain, the surface being entirely covered with a salt incrustation. Sir G. Temple, who, in the dry season, spent seven hours in crossing the Sibhah, says that, on approaching it, "the grass and bushes become gradually scarcer; then follows a tract of sand, which, some way beyond, is in parts covered with a very thin layer of salt; this, as you advance, becomes thicker and more united; then we find it in a compact or unbroken mass or sheet, which can, however, be penetrated with a sword or other sharp instrument, and here I found it to be 11 inches in depth; and finally, in the centre, it becomes so hard, deep, and concentrated, as to baffle all attempts at breaking its surface, except with a pickaxe. The salt is considerably weaker than that of the sea, and is not adapted to preserving provisions, though its flavour is very agreeable." (ii., 160–163.) About the centre of the lake are the foundations of a circular tower, where caravans halt to feed their camels; and in several parts are elevated plateaus, forming islands in the rainy season, the largest of which, covered with a luxuriant vegetation of date palms, is the *Pala* of Herodotus. The Arabs have a tradition that this lake once communicated with the sea by means of a river, but no traces of such communication appear to exist at present. There are no other inland lakes of consequence, but several considerable arms of the sea, as the gulf of Biserta (an. *Sinus Hipponensis*), the lake of Tunis, &c. The coasts of Tunis are greatly indented by bays, those of Tunis, Hamamet, and the gulf of Gabes, or Lesser Syrtis, being the principal. The principal promontories and headlands are the Dakhul, a long tongue of land terminating in cape Bon (an. *Prom. Mercurii*), the scene of several events in the 5th book of the Æneid, capes Serra, Ras-el-abiad, or the white promontory (an. *P. Candidum*), Ras-Zibeeb (an. *P. Apollinis*), &c. The shores in the N. are frequently bold, but in the S. they are low and sandy.

The *geology* of this country has been little or not at all studied; nor have its mineral resources been turned to profit for many ages. Copper and lead were among the exports of the Carthaginians; and these metals, with silver, are still to be found in the mountains: there is also a quicksilver mine near Porto Farina, but mining is altogether neglected. The *climate* appears to be less hot than might have been supposed. Sir G. Temple says, the average heat in August and September at Tunis is 83° Fah.; and in the year of his visit the thermometer seldom rose to 90°, and never exceeded that limit. From the 19th December, 1832, to the 19th January, 1833, it averaged 55½°; the highest range during that period being 60°, and the lowest 52°. Rainy weather commences about the end of October, and continues, at intervals, till May. As early as January the surface is covered with fresh verdure; and, on the whole, the climate may be said to be healthy as well as pleasant. It is true that the plague is not unfrequent, and that its ravages have been supposed to be a principal cause of the depopulation that is admitted to have taken place during the last half century. But this is to be ascribed far more to the slothfulness of the inhabitants, and the want of precautions, than to anything unfavourable in the climate.

The *vegetation* is, for the most part, the same as in the adjacent regency of Algiers, and on the opposite shores of Sicily and S. Italy; the olive, pistachio, carob, with dates, onions, the lotus, &c., are common products. This region was, in antiquity, deservedly celebrated for its extraordinary fertility. It exported large quantities of corn to Rome,

of which it was one of the granaries. Pliny, in speaking of the extraordinary productiveness of the soil, assures us that a plant of wheat (*triticum*), sent from it to Augustus, had little short of 400 stalks; and another, sent to Nero, had 340! In antiquity, indeed, the common opinion was, that in this favoured region the labour of the husbandman was rewarded by the enormous increase of 100-fold! Hence, says Silius Italicus:

> "See sunt Byzacea cordi
> Rura magis, centum Cereri fruticantia culmis."
> Lib. ix., lin. 204.

And it would still seem to be endowed with the same wonderful productiveness. Sir G. Temple says, that "whilst halting in a field of young barley to feed our horses with its tempting crop, I counted on one plant 97 shoots or stalks; and this was not selected by me as being the largest, but as the nearest to where I was sitting." (*Excursions*, ii., 108.) In fact, there cannot be so much as the shadow of a doubt, that were Tunis subject to an intelligent government, it would, at no distant period, furnish large quantities of corn for exportation. At present, indeed, such is the undiminished fertility of the soil, that a surplus is raised for exportation notwithstanding the oppression and extortion to which the husbandman must submit. The government assessor goes into a field while the crop is in ear, and values it according to his caprice; taking care, however, to be always above, and never below the mark. The owner is then obliged to pay a tithe on this supposed value of his future crop; though, when harvest time has arrived, he finds, perhaps, that it does not exceed one fourth part of the sum at which it was estimated! The same is the case with olives, the principal resource of the country; and these, moreover, are not allowed to be gathered till an order to that effect has been received; and in consequence of the great delay which often takes place in sending it, the fruit frequently falls and rots on the ground, the owner not being permitted to pick it up; he is also obliged to send his olives, when they have finally been collected, to mills established by the bey, who derives therefrom a considerable profit. (*Temple*, i., 225, 226.) We need not, therefore, be surprised that agriculture should be almost wholly neglected and abandoned, no one venturing to cultivate more ground than is sufficient to supply his immediate wants, and to furnish the taxes to government. Wheat, barley, sorgho, maize, and millet are the grains principally raised; in the S., the date tree supplies the Arabs, not only with their principal nutriment, but, also with their fuel, and the materials for most of their domestic furniture. Cotton and indigo have been introduced into culture somewhat recently; in some parts saffron, white mulberry, opium, &c., are grown; and tobacco is pretty general. The sugar cane succeeds well, but no sugar is made. All the fruits of S. Europe, as pomegranates, oranges and lemons, figs, jujubes, &c., and the vines on the N. coast yield excellent raisins, most of which are dried for exportation; but apples and pears degenerate. Among other products of importance is hennah (*Alkennah Arabum*), so much used "as a dye for ladies' hands and horses' legs," and which is a chief article of trade at Gabes. This plant, where not annually cut, and kept low, grows to 10 or 12 ft. in height, putting out clusters of small flowers, having an odour of camphor. (*Shaw*, 114.) The dye is a bright orange, or tawny saffron. The leaves are picked twice a year, dried and powdered, and in this state sold in all the markets of the E. The powder, formed into a paste, is applied to the part required, and then bandaged round. The plant is cut level with the ground as soon as the leaves have been picked. The hennah, like the date palm, requires to be frequently watered, for which purpose the plantations are divided into squares, and enclosed by banks; a stream is then admitted into them, and allowed to flow for a certain time every week, generally an hour a day, and two hours during the night, each square being watered in turn. The expenses of watering are defrayed by the various occupants, in proportion to their number of squares. (*Semple; Shaw.*) This system of irrigation is noticed by Pliny. (I. lib. xviii., cap. 22.)

Horses, mules, camels, and oxen are used for field labour, and, with sheep, are the principal domestic animals. The breed of horses has deteriorated, in consequence of the government seizing for its use those that are most valuable. The cattle are small, but good, and many are sent to Malta. Some of the sheep are very fine, and all have the large fat tail which characterises the African breeds. The Merino breed is said to have been originally introduced into Spain from Barbary. (*Temple*, i., 227.) The lion, panther, jackal, wild boar, jerboa, genet, &c., are among the wild animals. Most of our readers are, no doubt, aware, that the banks of the Bagrada are celebrated in history for the stubborn resistance which an enormous serpent (120 feet in length), found on its banks, is said to have opposed to the progress of the Roman army under Regulus! (*Liv. Epit.*, lib. xviii.;

TUNIS.

Aulus Gellius, lib. vii., cap. 3.) But it is now generally acknowledged that many apocryphal statements have been mixed up with the history and fate of Regulus; and the *prælium grande atque acre* with the serpent does not certainly seem to be the most authentic part of the story. At all events, this gigantic brood of reptiles has now luckily disappeared; and Sir G. Temple says that the largest of those existing never exceed 12 feet in length. The locusts, which often visit the country in clouds, eating up "every green thing," are incomparably more destructive than the reptiles. Large quantities of fine coral are found round the coasts, which are visited in consequence by Sicilian and Neapolitan fishermen.

Manufactures are few: they comprise some silk, linen, and woollen fabrics, leather, &c.; but the principal are soap and the *beretti*, or red caps of Tunis, so well known throughout the Mediterranean. The principal soap-works are at Susa. The soap is of good quality, and the soft especially is much esteemed. Little is prepared on a speculative anticipation of a demand for exportation, but any quantity may be had by contracting for it a few months beforehand. The manufacture of skull-caps is said to have employed formerly more than 50,000 persons, and 3000 bales of Spanish wool were annually used. (*Macgill*, in *Mod. Tres.*) At present it is reduced to one third of this extent; spurious imitations of the Tunis caps having been made in Marseilles and Leghorn. The dye employed for the caps is the kermes chiefly; the process of dyeing is conducted mostly at Zaghwan, an inland town, the water of which gives great brilliancy and permanency to the colours. Previously to their being dyed, the caps are boiled for a whole day in alum-water. The woollen cloth made in the regency, principally at Jerba, is thin, resembling soft serge. Morocco leather is made in considerable quantities, and dyed skins are articles of extensive export. Though cochineal is used in dyeing, and the prickly pear grows most luxuriantly in Tunis, no attempts have yet been made to introduce the insect into the country.

Trade.—None of the Barbary states is so well situated as Tunis for an extensive commerce, particularly with Europe. Three caravans come annually from the interior of Africa, bringing slaves, senna, ostrich feathers, gold-dust, gum, ivory, &c.; cattle bear woollen cloths, muslin, linen and silk fabrics, red leather, spices, fire-arms, gunpowder, &c. Other caravans come from Constantinople with virgin wax, dried skins, cattle, and sheep, which they exchange for cloth, muslin, Tunis mantles, linen, raw and manufactured silks, colonial produce, essences, &c. According to a statement by the French consul, the value of the different articles of export from Tunis may be estimated as under:

Articles.	Value.
	Francs.
Oil	4,140,854
Grain and pulse	878,437
Pistachies and dates	27,250
Wool	1,165,451
Cattle	97,101
Sponges	210,304
Senna	80,017
Wax	45,140
Hides	240,770
Tunny fish	221,437
Gold-dust and elephants' teeth	400,000
Red caps	1,884,451
Soap	75,090
Burnouses, shawls, blankets, and various other articles	521,060
Total	9,465,436

The imports are woollen goods, from France and England; cottons and linens, from the latter and Germany; with coffee, spices, sugar, tin, lead, and iron, silk, wool, wine, &c. The government monopolises the trade in many articles; as tobacco, wax, wool, and provisions, which it farms out to various individuals.

The *Government* is in the hands of a *bey*, who rules with despotic power; he receives the caftan, with the dignity of a pacha of three tails, from the sultan, but is not otherwise in any way dependent upon Turkey. The divan is composed of 37 members, each of whom has a vote in council: but this body has only a nominal authority. The revenues of the bey have been estimated at 24,000,000 piastres, or upwards of £1,500,000 a year; though at present that derived from regular sources is supposed not to exceed one fourth part of this sum. Its principal sources are the customs, which are farmed every year to the highest bidder; the tithes upon the cultivation of olives, corn, and other products; the sale of permits for the exportation of necessaries and the importation of wines and spirits, usury taxes, the bey's domains, the sale of government offices, a poll-tax on the Jews, the traffic in slaves, and private mercantile speculations of the bey, occasional extortions from the wealthy, and the property of those who die without heirs, of which the exchequer takes forcible possession.

The armed force consisted, in 1832, of nearly 50,000 men; but of these 40,000 composed the contingents (chiefly cavalry) furnished by the different Arab tribes, and the standing army consisted of only about 6000 men. The regular infantry, a body of 2000 men, were originally organized by a French officer in 1831. There are 3000 Turkish infantry, 2000 spahis, or paid cavalry, 300 Mamelukes, forming the body-guard, and 16 pieces of artillery. The naval force now consists of only a corvette, a few brigs and schooners, and about 30 gun-boats; and Tunis is no longer formidable for piratical expeditions. By a treaty with France in 1830, piracy and Christian slavery were wholly abolished.

The people, their manners, customs, &c., are similar to those of Algiers, to which we beg to refer the reader. The Tunisians may, however, claim to be considered the most civilized and tolerant nation in Barbary: though in negotiations with them, as well as the neighbouring powers of N. Africa, Sir G. Temple thinks an attitude of firmness, and not of conciliation, should be adopted, the latter being always supposed to indicate fear and weakness.

This region, which in antiquity was the centre of the Carthaginian dominions, remained in the possession of the Romans from the destruction of Carthage to the beginning of the fifth century, when the Vandals settled themselves in Africa. In 690 it became subject to the caliphs, and after belonging to several successive dynasties, was conquered by Barbarossa in 1534. The emperor Charles V., in 1537, took Tunis, and restored the dethroned Muley Hassan; but in 1570 the country was taken anew by the Turks, and it has only regained independence by the gradual decline of their empire. (See *Shaw's Travels in Barbary*; *Temple's Excursions in the Mediterranean*, 2 vols.; *Mod. Trav.* xx.; *Private Information*.)

TUNIS (an. *Tunes*), a marit. city of N. Africa, cap. of the above regency, on the W. side of the gulf of Tunis, being separated from it by a large salt-water lake or lagoon, about 4 m. W. from the sea, and 3 m. S.W. from the ruins of the ancient Carthage; lat. 36° 47' 58'' N., long. 10° 11' E. Its population, which is greater, perhaps, than that of any other African city, Cairo excepted, has been estimated at from 100,000 to 150,000, of whom, perhaps, 30,000 are Jews and 2000 Christians. It stands on the edge of the lagoon, upon rising ground, backed on the W. by heights, which are crowned by the *kasbah*, or citadel. The town is enclosed by a wall of earth and stones, and a second wall surrounds its three suburbs, the outer wall being about 5 m. in circumference. Towards the N. it is defended by two castles, and other heights around it on the S. and E. are protected by detached forts; but notwithstanding the town is laid out on its defence, it is not a well fortified or strong town, and has been repeatedly taken. The houses, though of stone, are mean and poor, and the streets narrow, unpaved, and filthy: the bazaars, which are superior to those of Algiers, are vaulted overhead, and sometimes furnished with footways. There are a great number of mosques, several of which are handsome, and one was converted into a Catholic cathedral during the Spanish occupation. The palace built by the late bey, in which Queen Caroline lodged during her visit to Tunis, is a square edifice, magnificently decorated within. The rooms are paved with marble, and all open upon marble courts, with fountains in their centre. For about 10 feet from the floor the walls of the rooms are lined with glazed tiles, and above this with stucco-work peculiar to the Moors; while the ceilings are traced in different-coloured patterns, with much taste. The great hall of justice has never been finished, but different parts of the city are five large barracks, also built by the late bey; and a very extensive edifice of the same description, fitted to accommodate 4000 men, was, a few years since, in the course of being built. (*Temple*, i., 175.) In digging the foundations of this edifice, two sarcophagi were found, and an ancient cistern of great extent, and in good preservation. The citadel, though large, is in a ruinous state, having but one efficient battery: in it is a powder factory. Tunis has many gates, one of which called the *Bab-Kartajinah*, or Carthage-gate, has in its vicinity the Protestant burial-ground. It has also a Roman Catholic convent, church and chapel, a Greek church, an English consulate, and a theatre at which Italian operas and comedies are performed three or four times a week. About 1½ m. W. from the town is the Bardo, or summer palace of the bey. It resembles a little fortified town, with its ramparts, bastions, &c., and has a population of at least 4000 persons, employed in some way or other about the court.

The gulf of Tunis opens to the N., in the form of a horse-shoe; it is 16 m. deep, and has good anchorage all over at from 4 to 10 fathoms water. The N. and N.E. pass sometimes throw in a heavy sea, which, however, seldom occasions any damage. The port is at the Goletta, or channel, passing through the narrow belt of land separating the lagoon of Tunis from the sea. There is at all times about

960

15 feet water in the canal, and ships may use it on paying a fee of 3 dollars a day. It is not, however, much resorted to, all vessels of considerable burden loading and unloading from their moorings in the bay, by means of lighters. The Goletta is pretty strongly fortified, though commanded by a hill to the N. A harbour light, 40 feet in height, was erected at the entrance to the canal in 1819. A great number of boats are employed in conveying goods and passengers across the lagoon between the port and the city.

The lagoon of Tunis was formerly, as Procopius states, a deep port, with water sufficient to float large ships. But now, from its being the receptacle of the filth conveyed to it by the common sewers of the city, and other causes, its greatest depth does not exceed 6 or 7 feet; while round the shores it is comparatively shallow. An island in its centre, opposite the city, is defended by a fort. It does not receive any rivulet, and its loss by evaporation is supplied by a current which sets into it through the Goletta.

The average annual value of the exports from the regency of Tunis from 1830 to 1837 has been estimated at 7,527,000 piastres, and that of the imports at about 13¼ millions do. Subjoined is an

ACCOUNT of the Value of the Exports and Imports of Tunis in 1837, specifying the Value of those sent to each Country:

Countries.	Exports. Value in Piastres.	Imports. Value in Piastres.
France	2,134,091	1,941,229
Algiers and Bona .	437,728	98,925
Great Britain . .	685,174	1,289,286
Austria . . .	81,960	199,596
Sardinia . . .	916,271	613,418
Tuscany . . .	703,668	4,299,010
Turkey . . .	1,470,366	1,161,695
Barbary States .	66,630	83,146
Greece . . .	448,560	852,476
Naples and Sicily .	78,584	164,895
U. S. of America .	93,419	
Spain . . .		112,350
Total . .	7,063,686	10,982,407
Equal to .	£.289,630	£.380,140

But this, though it be the most considerable native trade of any city on the Barbary coast, is certainly not a tenth part of what it would amount to were the country subjected to an intelligent government, and its gigantic resources properly developed. Naval and military stores imported into Tunis pay no duty; other articles pay 3 per cent. ad valorem on a rated tariff. Accounts are kept in piastres (worth about 1s. 1d.) of 16 carobas, or 52 aspers each. The Tunis lb. of 16 oz. = 7773 grs.: the principal commercial weight is the cantaro of 100 lb. = 111 lb. avoird. The caffz, for corn, = 14½ imp. bushels; the mattar, for oil, = about 5 gallons. The pic varies from 18 to 26 in.

According to Strabo, Tunes existed before the foundation of Carthage. The chief events in its history are its numerous sieges and captures. Louis IX. of France died before walls in 1270; and it was taken by the emperor Charles V, who defeated Barbarossa under its walls, in 1535. On this occasion about 20,000 Christian slaves were freed from bondage; but, unfortunately, 30,000 Moslem inhabitants of the city were, at the same time, put to the sword, despite the efforts of the emperor to prevent it, by the victorious troops, while 10,000 more were carried away, and sold as slaves. (Robertson's Charles V., ii., 375, 4to ed.; Shaw's Travels; Temple's ditto; Comm. Dict.; and private information.)

TUNKHANNOCK, p. t., capital of Wyoming co., Pa., m. N.N.E. Harrisburg, 250 W. Bounded S.W. by Susquehanna river. Drained by Tunkhannock and Mishoppen cks. It contains a courthouse, jail, 12 stores, one fulling-l, three grist-mills, 11 saw-mills, one distillery; eight bols. 288 scholars. Pop. 1933.

URBUT, t., Northumberland co., Pa., 14 m. N. Sunbury, nded W., by the W. branch of Susquehanna river, and ned by its tributaries. It has five stores, three grist-s, two distilleries, one tannery; eight schools, 415 schol. Pop. 3872.

URCOING, or TOURCOING, a town of France, dep. Nord, cap. two cantons, immediately adjoining the Belgian frontier, 10 m. N.E. Lille. Pop., in 1836, 8749. It is fairly laid out, and pretty well built: the town hall in great square, two churches, a college, a charitable asylum, and the remains of an old feudal castle, are the most picuous objects. The inhabitants share in the manufactures common to Lisle and Roubaix, and their condition been noticed in the article on the latter (anté, p. 603—

"Tourcoing has fewer looms than Roubaix: the arts woven are chiefly coarse cotton goods and linens. weavers gain 9 or 10 francs gross per week; the other sing classes (adult males) get from 1½ to 2½ francs a The weavers are the most moral class in the town; attend to the education of their children, and seem

contented with their condition." (Symons in Hand-Loom Weavers' Report.)

TURIN (Ital. Torino, an. Augusta Taurinorum), a city of N. Italy, the cap. of the Sardinian dominions, prov. of same name, in Piedmont, near the Po, where it is joined by the Dora, 80 m. W.S.W. Milan. Lat. 45° 4′ 5″ N., long. 7° 40′ 15″ E. Civil pop. in 1838, of the city proper, 83,466; do. with suburbs included, 104,078; and including the canton, or immediately adjacent territory and the garrison, 123,892. The city is of an oval shape, and about 4 m. in circuit: it was formerly fortified, but is now an open town, standing in a rich, well watered, and well cultivated plain: it is approached by four fine roads shaded with forests trees; the surrounding hills being covered with handsome edifices, among which the church of La Superga is pre-eminent. The impressions which Turin produces on the traveller are very much governed by the circumstance of its being the first or the last city he visits in his progress through Italy. Mr. Woods, who had already seen the best productions of architecture, states that being built on a flat, Turin makes no show at a distance; the domes and towers are neither numerous nor lofty, and on looking down on the city from the neighbouring hills, the dingy red the roofs have a disagreeable appearance. (Letters of an Architect, ii., 422.) But Forsyth, a severe as well as an excellent judge, says that Turin is admired for the regularity of its plan, the cleanness of its streets, the symmetry of its squares, the splendour of its hotels, and the general elegance of its houses: and Simond adds, "Turin forms a perfect contrast with all the cities we have been accustomed to see in Italy: it is new, fresh, and regular, instead of antique and in decay; and the buildings, all alike, are collectively magnificent if not quite so in detail, the materials being only brick coated over in imitation of stone. A profusion of running water keeps the fine wide pavement clean. All round the town, ancient trees of luxuriant growth oppose their impenetrable shade to the intolerable heat of the sun, and the views of the Alps are magnificent." (Trav. p. 606.) On the whole it may be truly said, that, were it not for the taste for meretricious ornament which is offensively prevalent everywhere in Turin, it would be one of the very finest cities of Europe. It has indeed comparatively few modern works of art, and little to interest the antiquary; and there is scarcely anything to characterise it as an Italian city: to most travellers it has appeared rather like a new and handsome French town.

Except in the old town, which forms about one sixth part of the whole, the streets, which are bordered by houses four or five stories high, are straight and cross each other at right angles; and here, as in the new town of Edinburgh, and the N.W. and other parts of London, entire rows and streets of considerable extent are of precisely similar architecture. The royal palace stands in the centre of the town, in the Piazza Reale or di Castello a very large and elegant square, surrounded by many other public buildings, and having in its centre the former palace of the dukes of Savoy, a castellated mansion environed by a moat. The Strada del Po, a noble street ½ m. in length, leads to this square from the river, which is here crossed by a fine stone bridge of five arches, erected by the French; but which is said to be surpassed by a new bridge over the Dora recently completed. The Strada di Po, like the Strada Nuova and di Dora Grossa, the Piazza Reale, S. Carlo, &c., is embellished in its whole length with arcades over the footways, which give a most agreeable and imposing appearance to these parts of the city. The royal palace is little remarkable in its architecture, but it has some spacious and richly adorned apartments, and a good collection of paintings, including many of the Flemish school, and others by Titian, Guercino, Albani, Murillo, &c. In this edifice is the equestrian statue of Amadeus I., the figure in bronze, the horse in marble. Attached to the palace are gardens open to the public, the fashionable resort during day; the Rondo between the city and the Po, and Valentino grounds, being the favourite resort in the evening. The old palace of the dukes of Savoy is a singular building with four fronts of different architecture. According to Forsyth "three of these are hideous in themselves, and derive comparative ugliness from the beauty of the fourth. This last front, composed of one Corinthian peristyle raised on a plain basement, is the noblest elevation in Turin, where it holds the post of honour." (p. 412.) The private palaces would strike a stranger who had just crossed the Alps as very magnificent, but there are many in Italy equally large and in a much purer taste. That of Prince Carignano has a remarkable staircase by Guarini, who, along with Guvarra, has been the architect of most of the principal edifices in Turin.

The cathedral, a Gothic structure built about the end of the 15th century, has been praised for the richness of its appearance, particularly the W. front, which is ornamented with well executed bas-reliefs, &c. In it is the chapel of the

Santa Sinods, in which the winding-sheet of our Saviour is preserved with all the attention due to so important and so authentic a relic. This cathedral was formerly among the wealthiest churches in Italy; but its plate has been sold, and the produce applied for the most part to secular, indeed, but certainly quite as useful, purposes as those to which it had been previously appropriated. In fact, the useless riches of this cathedral, its images, vases, candlesticks, &c., defrayed the cost of erecting the bridge across the Po in this city, and embanking that river, as well as of improving the Tuileries, and building the Rue de Rivoli in the French capital. (*Conder's Italy*, i. 159.)

The church of *San Filippo*, *San Christina*, and many others, are richly adorned; but they all yield the palm to *La Superga*, situated on a hill about 5 m. from Turin. It was on this spot that the duke of Savoy (Victor Amadeus) and Prince Eugene met to concert their plans for the attack of the French, and the deliverance of the city in 1706; and the church was constructed by the duke as a monument of his gratitude to the God of battles for having given a signal victory to his arms. The edifice is not unworthy its origin. It is of a circular form, and surmounted by a dome. Eustace says, " All the columns in this building are of marble of different colours, and give the edifice an appearance unusually rich and stately. Instead of pictures, the altars are decorated with bas-reliefs; the pavement is of variegated marble; in short, all the different parts of the building, and even the details of execution, are on a scale of magnificence." (*Class. Tour*, iv. 100.) It must be admitted, however, that the building has some considerable defects, which have been pointed out by Mr. Woods.

The university of Turin was founded in 1405. It consists of five faculties, or colleges, consisting of theology, with four professors; law, with five do.; medicine, with six do.; surgery, with five do.; and the arts with 22. It is usually attended by about 1900 students, who board out in private families. (*Journ. of Educ.* xvi.; *Simond, &c.*) Its library is said by Rampoldi to have 60,000, and by Valery 112,000 vols.! Its buildings are extensive and well arranged: the court is surrounded with a double tier of porticoes, under which is a collection of ancient sculptures, bas-reliefs, &c., from the excavated Roman city of *Industria*, about 18 m. distant. In the museum of the academy of sciences is the valuable collection of Egyptian antiquities, formed by Drovetti, and purchased by the king of Sardinia for 400,000 fr. It comprises several colossal statues of Egyptian sovereigns, domestic and agricultural implements, MSS. and papyri, the famous Isiac table, &c. Under the same roof are museums of natural history, anatomy, and medals; and the royal library, comprising an extensive and valuable collection of historical and other works, including an extensive series of Bibles.

The citadel of Turin is a regular pentagon, planned by Urbino in the 16th century: it has extensive subterranean galleries, and is still of considerable strength. The gates of the city, which were cased with marble, were demolished by the French, and the ramparts dismantled and converted into public walks. But, to use the words of Rampoldi, *con tale destrusione divenne una città tutta nuova e d' un aspetto deliziosissimo*. The Valentino palace, and the other royal seats around the city, are now either deserted or appropriated to schools and museums; these with nine hospitals, two asylums, the colleges of the Jesuits and Ignorantelli, an arsenal, with a school of military engineering, a grand opera-house, ranking as the third theatre in Italy; two smaller theatres, the cemetery of the aristocracy, observatory, botanic garden, royal academy of painting, and *monte di pietà*, comprise most part of the other establishments worth notice. Turin is the residence of the king and seat of the royal senate. or superior court for the kingdom, and of a tribunal of primary jurisdiction, an archbishop's see, and has chambers of agriculture and commerce. Its manufactures consist chiefly of silk fabrics and twist; but it has others of woollen and cotton goods, hardware, arms, paper, glass, earthenware, liqueurs, &c.; and its printing business is pretty extensive. The shops and hotels of the city are good, but the supply of water is bad, and the prevalence of fogs render it rather an unpleasant residence in autumn and winter. The manners, habits, language, &c., of the inhabitants are more French than Italian.

Turin was made a military station by Julius Cæsar on his invasion of Gaul. In 312 Constantine gained in its vicinity a great victory over Maxentius. Charlemagne annexed this city to the Marquisate of Susa; and it came into the possession of the dukes of Savoy in 1039; and became their capital in 1251. (*Rampoldi.*) It was taken by Francis I. in 1536, and held for 26 years by the French, who again took it in 1640. But the most celebrated by far of the sieges of Turin took place in 1706, when it was invested by a powerful French army. Voltaire from the immense preparations made for this siege (*Siècle de Louis XIV.*, cap. 20.): but the incapacity and disagreement of the

French generals, and the talents of Prince Eugene and the duke of Savoy, secured for the latter an easy and complete victory. All the vast stores accumulated by the French fell into the hands of the conquerors, and the besieging army was wholly dispersed. Under the French ascendancy, from 1800 to 1814, Turin was the capital of the department of the Po. (*Rampoldi; Eustace*, iv. 60-102; *Forsyth: Woods*, ii. 422-6; *Ross's Letters from the N. of Italy: Conder's Italy*, i. 156-173, &c.)

TURIN, p. t., Lewis co., N. Y., 126 m. N.W. Albany, 435 W. Drained by tributaries of Black r. It contains two churches, a Presbyterian and Methodist, five stores, one flouring-mill, two grist-mills, nine saw-mills, one oil-mill, two tanneries; 16 schools, 571 scholars. Pop. 1704.

TURKEY, or the OTTOMAN EMPIRE, a very extensive country, partly in S.E. Europe, and partly in W. Asia, comprising some of the most celebrated, best admired, and naturally finest provinces of the continents to which they belong. The limits of the Turkish empire are not easily defined; inasmuch as it is usually represented as including several extensive countries, that are either substantially or virtually independent. Moldavia, Wallachia, and Servia, in European Turkey, are now connected with the Porte only by the slenderest ties; though, as some of their fortresses are garrisoned by Ottoman troops, and as they continue to pay tribute to the Porte, they may still, perhaps, be properly included within the wide range of the Turkish dominions. Egypt, however, and the other African territories that formerly belonged to the Porte, may now be considered completely dismembered; and, but for the interference of England and the other European powers, Syria and Palestine would have been annexed to the dominions of the pacha of Egypt.

European Turkey, including Wallachia, Moldavia, and Servia, comprises, with the exception of the new kingdom of Greece, almost the whole of the most easterly of the three great peninsulas of S. Europe, extending from 39 to 48½ deg. N. lat., and from 15½ to 29¼ deg. E. long. It is bounded on the N. by the Austrian empire, from which it is separated by the Save, the Danube, and the E. Carpathian mountains; on the N.E. it is separated from the Russian province of Bessarabia by the Pruth; on the E. it has the Black sea, the Bosphorus, the sea of Marmora, and the Hellespont; on the S., Greece; and on the W. the Mediterranean, the Adriatic, and the Austrian province of Dalmatia.

Very different estimates have been formed of the extent and population of this vast country, and neither is known with anything approaching to precision. Perhaps, however, we shall not be far wrong if we estimate its extent at 210,000 sq. m., and its population as from 14,000,000 to 15,000,000.

The Turkish dominions in Asia are of still greater extent than those in Europe, but their population is much less considerable. They embrace the whole peninsula of Asia Minor and the adjacent islands, the greater part of Armenia and Koordistan, with Syria and Palestine, Mesopotamia, and a portion of Arabia. In all, they may comprise an area of about 437,000 sq. m., with a population probably of about 10,000,000.

Physical Geography.—Bruguière includes the mountains of Turkey in Europe in the Alpine system. But according to other authorities, there are several different mountain systems in Turkey, having little connection with each other; and Boué states that most maps of the country err greatly in their representation of the direction, position, and height of the mountain chains. The high table-land anciently called *Mæsia Superior*, extending between Sophia and Pristina, and dividing the basin of the Morava on the N. from those of the Vardar, Strumas, &c., on the S. and of the Lower Danube on the E., forms the central nucleus of the Turkish mountains. From this centre branches run off northward, bounding Servia on the W. and E.; on the E. the Balkhan chain (aa. *Hæmus*) stretches in a nearly straight line from the sources of the Isker to the E. of Sophia, E. to the Black sea, a distance of about 250 m. dividing Bulgaria from Roumelia, and the waters that flow into the Lower Danube on the N. from those that flow into the Maritza on the S. The Despoto-Dagh (aa. *Rhodope*, and the mountain chains that run through Macedonia, branch off from the central nucleus on the S.; while on the W. it gives off various chains that unite with the true Alpine chains, which ramify through Croatia, Bosnia, Servia, govina, Montenegro, and Albania. Nearly in a direct line S. from Pristina runs a chain which divides Albania from Macedonia, and thence extending into Thessaly and Greece under the name of Pindus, separates the waters flowing into the Ægean from those flowing into the Adriatic and Ionian seas. The interposition of those mountain chains frequently renders the communication between contiguous provinces rare and difficult. But with the exception of a few heights, as mount Scardus, nearly 10,000 ft. in eleva-

TURKEY.

tion, and Scomius and Pindus, near Mezzovo (about 9000 ft.), the Turkish mountains seldom reach an altitude of 8000 ft.) Mount Dinara, whence the Dinaric Alps derive their name, is only 7458 ft. in height; the Albanian mountains are generally under 7700 ft.; mount Athos is 6778 ft. and mount Menikon (an. *Cercina*), the loftiest of the Balkhan chain, 8395 ft. in height. (*Bruguière, Orographie de l'Europe*.) The Balkhan has recently acquired a greater degree of interest than most of the other chains, from its being supposed to form an all but insurmountable barrier to an invading army. This, however, does not appear to be really the case. The W. portion of the Balkhan is seldom more than 4000 ft., and its more easterly portion, near the Black sea, not more than from 1800 to 2000 ft. in height, while it is traversed by half a dozen different passes, none of which is fortified. Hardly one of those appears, in fact, to present any very formidable obstacle to an invading army; and Major Keppel expresses his surprise that the Russians did not cross the Balkhan long before their last irruption into Turkey. (*Keppel, Journey across the Balkhan*, ii. 11.) We may further mention that there are more lines of communication for carriages across the Balkhan, between Thrace and Bulgaria, than between any of the other Turkish provinces. Indeed, there is only one road between Macedonia and Bosnia, and one between Macedonia and Servia; the last, though the only route by which the produce of Macedonia is conveyed to the N., being merely a mule track. There are three passes between Macedonia and Albania, but only one between Albania and Thessaly.

European Turkey has numerous narrow valleys, and some very extensive plains. By far the largest of the latter is that of Wallachia, Moldavia, and Bulgaria, traversed in its centre by the Lower Danube, and ranking at least as the third, if not the second, of the great plains of Europe. A considerable portion of Thrace, and some parts of Macedonia, are level, and Thessaly principally consists of a very fertile basin. Almost every part of the country is well watered; and, besides the Danube and Save (which last constitutes a great part of its N. boundary), Turkey has several rivers of very considerable size. Among those on the N. side of the great central plateau and its ramifications, affluents of the Save and Danube, are the Unna, Verbas, Bosna, Drin, Morava, Timok, Schyl, Isker, Aluta, Jalomnitza, Sereth, Pruth, &c. Among the rivers to the S. of the central plateau, the following may be specified, viz. the Maritza (an. *Hebrus*) has its sources in the N.W. angle of Roumelia, in the Balkhan and Despoto-Dagh mountains, and flows generally E. or S.E. to the centre of Thrace. Near Adrianople, where it receives the Tondja (the *Tænus* of Ptolemy), and thence S. or S.W. to the Ægean, which it enters close to the gulf of Enos, after a course of about 240 m. Its greatest width is about 3 furlongs. Adrianople, Philippolis, Demotica, Ipsala (an. *Cypsela*), &c., are on its banks, which, in many parts, are covered with forests of oak and elm. The Maritza is navigable from the time of the autumnal rains till May, as far as Adrianople, for boats of 200 tons; but during the summer months even craft ascend only as high as Demotica. (*Keppel*, i. 253.) The Kara-su (*Nestus*), Struma (*Strymon*), and Vardar (*Axius*), which traverse Macedonia in a S. E. direction, are all of considerable size, but generally shallow and unfit for navigation. The Selembria (*Peneus*) rises near Mezzovo, and drains the basin of Thessaly, falling into the gulf of Salonica at the mouth of the famous defile and vale of Tempe. The principal rivers flowing into the Adriatic are the Narenta, in Herzegovina, and the Drin and Vojutza (*Aous*), in Albania.

European Turkey has no lakes of any very great extent. The principal are those of Ochrida (*Palus Lychnitis*), about 20 m. in length by 8 m. in breadth, Scutari (*Palus Labeatis*) and Yanina, in Albania: there are numerous small lakes in Macedonia and Thessaly.

The physical geography of Asiatic Turkey requires but a brief notice, having been already treated of in the arts. NATOLIA, KURDISTAN, SYRIA, &c. in this work. Asia Minor consists chiefly of an extensive table-land, traversed by many parallel mountain ranges from W. to E., extending into Armenia and Kurdistan. This table-land appears generally to increase in height as we proceed eastward; mount Ida, overlooking the plain of Troy, being only about 5000 ft., while mount Bisutun, the culminating point of N. Kurdistan, is 12,000 ft. above the sea. From this lofty plateau several mountain ranges are given off to the S., inclosing the basins of the Euphrates, Tigris, Jordan, Orontes, &c., which, with the Halys (see NATOLIA, *said*, p. 426), Sangarius, Araxes, &c., are the principal rivers in this part of the empire. The largest lake is that of Van (which see); next which are the Dead sea and lake of Tiberias, in Palestine: many small lakes exist in Natolia. The N. part of Asiatic Turkey is mountainous, the surface declining towards the S., where it spreads out into extensive plains (an.

Chaldæa, *Mesopotamia, E. Syria, &c.*) of much natural fertility, but at present for the most part desert and uninhabited.

The coasts of Turkey, both in Europe and Asia, are in general bold and rocky. In many parts they present a long and tolerably uniform line with few gulfs or harbours of any magnitude. This is particularly the case with the coasts of the Black sea, Syria, and a part of Albania. But the shores of the Ægean and the adjacent seas are deeply indented with numerous bays and inlets, and present many good harbours, as those of Smyrna, Salonica, Constantinople, &c. Varna is the only good Turkish port on the Black sea. Durazzo (the an. *Dyrrachium*), on the Albanian shore, might easily be rendered an admirable port (see *Urquhart's Turkey, &c.*, p. 199); but at present there is not a single safe or convenient harbour along the whole W. coast of European Turkey.

The *Geology* of the two great portions of the Ottoman empire presents considerable differences. The great mountain chains of Europe consist of granite, gneiss, trachyte, syenite, serpentine, talc, mica, and clay-slate, and many other primary and transition rocks, inclosed between beds of sandstone or limestone; the latter being the most prevalent formation in the alpine ranges of the W. provinces and in Thrace. This latter province, with Bulgaria, &c., consists, in great part, of shelly limestone, marly clay, and other tertiary formations. Iron and other metallic ores are found in great abundance; but volcanic formations appear to be scarcer in Europe than in Asia. In Asia Minor, according to Mr. Brant, "the whole range of mountains, from sea to sea, is limestone. Volcanic rocks are frequently found, and granite rises up occasionally. The mountains abound in veins of copper and lead, the last being rich in silver. Mineral springs frequently occur; most of them hot." (*Geog. Journal*, vi., 188.)

As the country rises towards the E., granite and the other primary rocks become more prevalent. The lower basins of the Euphrates, the Danube, and other large rivers, are mostly alluvial.

Climate, and Natural Products.—In a region extending through nearly 20 degs. of lat. and more than 30 degs. of long., having every variety of elevation, exposure, soil, and subsoil, there must necessarily be the greatest variation of climate. The climate of European Turkey is much colder than that of the parts of Italy and Spain under the same latitudes, and is so very changeable that at Constantinople Fahrenheit's thermometer is said sometimes to fall 31° within an hour. In the Danubian provinces snow lies several feet deep, on the higher mountains for six months together; the thermometer frequently stands between 10° and zero, and in Moldavia it has been known to descend to 15° below zero. On the other hand, the summer heats are oppressive, and even in the N. the grape ripens by the end of July. The temperature and salubrity of Asiatic Turkey is almost equally variable with that of European Turkey. In the highlands of Armenia, even the plains are covered with snow as late as May; and the fine season, properly so called, does not comprise more than four months of the year, during which period both sowing and reaping are completed.

Asia Minor has but two seasons, the transition between them being scarcely perceptible. In winter, while the uplands are covered with snow, the lowland plains and valleys are visited by perpetual rains and N. winds. During summer there is scarcely any rain, but the soil is fertilised by heavy night-dews. Caramania suffers from very arid winds; and in the delta of the Euphrates and Tigris the barometer often rises to 40°. The climate of Syria, Mesopotamia, &c., has been already noticed in the arts. SYRIA, BAGDAD, and BUSSORAH.

The best indication of the relative temperature of different parts of Turkey is afforded by their vegetable products. In Croatia, Bosnia, and the adjoining provinces, the mountains are covered with forests of oak and elm; S. of the Balkhan the country is covered with forests of the sycamore, carob, and plane trees; gardens of roses, jasmine, and lilac; vineyards and orchards of nearly all kinds of fruit-trees; but is destitute of the olive, which, except in some particularly favourable situations, does not thrive N. of lat. 40°. The flora of Albania is similar to that of the opposite coast of Italy; and in Thessaly, the garden of European Turkey, oil, wine, cotton, tobacco, figs, citron, pomegranates, oranges, lemons, &c., grow to perfection. The same fruits, and other products, flourish in the more sheltered parts of Asia Minor, even on the shores of the Euxine: where, however, owing to the severity of the N. winds, among other causes, the forests seldom extend up the mountains above 5000 feet. In Armenia and Koordistan, the olive and orange ripen only in the warmer valleys, and we find on the high grounds much of the vegetation that prevails in the mountainous provinces on the Danube

963

and Save. S. of Taurus we enter an entirely new region, where the date palm, oriental plane, Babylonian willow, banana, pistachio, sugar-cane, and indigo, betoken a close approach to the vegetation of tropical climates. (*Geog. Journ.* x., 505–569, &c.)

The forests of European Turkey are infested by bears, wolves, jackals, &c.; to which, in parts of Asia, may be added. it is said, the lion and tiger. The gazelle, and deer of various kinds, hares, and other kinds of game, are very abundant. The great bare-necked vulture inhabits the ranges of Taurus, and the ostrich wanders over the sandy deserts of the South. The camel, a native of this region, is the chief beast of burden throughout the greater part of Asiatic Turkey. The other domestic animals will be mentioned hereafter.

Population.—We have already stated that the population, as well as the area of the Turkish empire, is very imperfectly known. Hence there is the greatest discrepancy in the estimates which have been formed of its amount, which vary from 6 or 7 to 21 or 22 millions for Turkey in Europe, and in about the same ratio for the Asiatic provinces. Lately, however, the Turkish government has directed some portion of its attention to statistical inquiries, and the population, though the accounts of it be still very vague, is better known now than formerly. In as far as respects Turkey in Europe, the estimates on which most reliance may be placed are those of Mr. Urquhart (*Turkey and its Resources*, p. 272), and of M. Boué (*Turquie d' Europe*, ii., 32.) We subjoin these estimates:

	Pop., according to Mr. Urquhart.		Pop., according to M. Boué.
Wallachians and Moldavians	1,500,000	Wallachia, 1839 .	2,404,887
Osmanlie	700,000	Moldavia, 1838 .	1,418,106
Greeks (Hellenic race and language, all Christians)	1,120,000	Servians	526,000
		Mussulmen in Servia	10,400
		Bosnian	700,000
Albanians (Skipetar race and language, two thirds Mohammedans)	1,600,000	Bosnian	200,000
		Herzegovinians	222,000
		Croats	160,000
		Montenegrins .	100,000
Tribes of Slavonic race and language, Bosnian, Turcomans, Pomac, one third Mohammedans; the rest (Servians, Bulgarians) Christians of the Greek and (Myrdites, Croatians) of the Latin church	6,000,000	Bulgarians . .	4,500,000
		Albanians . .	1,600,000
		Greeks . . .	980,000
		Zinzares (Wallachians of Pindus) .	200,000
		Turks	700,000
		Armenians . .	100,000
		Jews	250,000
		Gipsies . . .	150,000
		Franks, &c. . .	60,000
Vlachi Greek church	608,000	Total . .	14,577,593
Gipsies . . .	204,000	Or at most	16,570,000
Jews	260,000		
Armenians . .	100,000		
Franks, &c. . .	50,000		
Total	12,130,000		

The Turks or Osmanlis, who have, for about four centuries, been the dominant race, were originally of Scythian or Tartar extraction. We have already noticed their general characteristics as they are found in Asia at the present day (L. 178). But it is of importance to observe that even there the Turkish blood has been largely intermixed with the Mongolian and the Persian; and in Europe the higher class of Turks have generally furnished their *harems* with the finest women of Circassia and Georgia; while the inferior Turks have allied themselves with Servians, Albanians, Bulgarians, Greeks, &c. In consequence the original and distinguishing features of the race are now, in Europe at least, very much obliterated; and the Turkish, from being one of the ugliest of Asiatic nations, is become, speaking generally, one of the handsomest; though, from the peculiar mode in which the race is maintained, there is necessarily the greatest variety in their stature and appearance.

Turkish ladies have, in general very white delicate complexions, a consequence of their sedentary mode of life, and of their habit of veiling themselves when they take the air. Their mode of life, and their great addiction to the bath, render them rather disposed to *embonpoint*; but it is absurd to allege that this constitutes the *ne plus ultra* of a Mussulman's idea of beauty. Had such been the case, the Circassians and Georgians would not have constituted the pride of the harem. (*Boué*, i., 58.)

The national character and dispositions of the Turks have changed as well as their physical constitution, but in a far less degree. They are now, as of old, at once excessively proud and excessively sensual. Their pride is a consequence of their ignorance, and of the recollection of their former victories and conquests; and their sensuality is a consequence of the peculiar nature of the Mohammedan paradise, and of their wish to realise in this world

some portion of that felicity which is to be the *portion* of all true believers in the next. Other nations have *offered* to believe in the doctrine of predestination, but in this respect the Turks alone have given a practical effect to their speculative tenets; and their stationary state and *contempt* for the inventions and discoveries of other nations may be in no small degree, ascribed to their conviction of their *in*utility; from their belief that everything that *comes* is determined by an overruling Providence, against whose decisions it would be alike vain and *impious* to contend. Speaking generally, the Turk is true to his word; he is not prone to anger, nor liable to sudden gusts of passion; but when provoked, his fury has no limits; and he becomes brutal and ferocious in the extreme, involving the *innocent* and the guilty in one common ruin. His religion interdicts the use of wine; and though not always respected, this precept has, on the whole, a great and salutary influence.

Though capable, on emergencies, of great and vigorous exertion, laziness and apathy are distinguishing characteristics of the Turks. There is nothing in which they take so much delight as in reclining in the shade from sunrise to sunset, apparently in a state of total indifference, occasionally sipping coffee, and inhaling the fumes of tobacco. Whatever may be their object, they saunter through the streets with the same measured and monotonous step. They converse little, and the presumption is that their mind is as indolent as their body.

Perhaps no nation ever possessed so little talent for governing others as the Turks. They have never struck their roots, or acquired any solid footing, in the countries they have conquered. They are encamped in and occupy them; but they hold them by no tie other than the sword. They have never coalesced or associated with the original inhabitants; they look upon themselves as the *nation*, and the rest of the people, or those at least who have not embraced Mohammedanism, as an inferior and degraded race, which it is, if not a duty, at all events not a venial offence, to insult and trample upon. In this respect they differ widely from the Tartars who overran China, and, indeed, from every other people; and to this more than any thing else their weakness, and the wretched state of the countries subject to their dominion, is to be ascribed.

We have elsewhere noticed, under the names of the countries which they principally inhabit, the more important features in the constitution and character of the other great races inhabiting the Turkish empire; and to these we beg to refer the reader. (*See* ARABIA, ARMENIA, BULGARIA, GREECE, SERVIA, SYRIA, WALLACHIA, &c.)

Property.—There is, in many respects, a considerable similarity between the mode in which property has been distributed in Turkey, and that in which it was distributed in Europe during the middle ages. In both cases the object in view in this distribution was the establishment and support of a militia, who should be bound to repair, at their own expense, to the standard of the sovereign, and to follow him in his campaigns. *La gouvernement militaire est dans sa constitution fondamentale de tous les états Musulmans. Chaque individu s'y reconnait soldat : toujours il est prêt à prendre des armes et à marcher sous l'étandard du prophète. On doit enfin considérer la nation (Turquie) comme un grand corps d'armée dont le souverain est le généralissime.* (*D'Ohsson*, iv., 392.) Hence when the Turkish sovereigns made any new conquest, they were in the habit of dividing a portion of the territory into estates called, from their greater or lesser size, *saïmets* and *timariots*, which they assigned to the more deserving or most favoured of their followers. The latter, however, did not succeed to the hereditary or absolute property of these estates. On the contrary, they only held them during life or good behaviour; and whenever any vacancy occurred, whether by death or forfeiture, the sultan made a new appointment to the vacant fief; and it is affirmed that instances have been known of the same lordship having been held by eight different masters in the course of a single campaign! It is farther to be observed, that the rights of the peasantry (*repaïs* or cultivators) on these estates were carefully preserved; and, in point of fact, the new feudal lord, or lord of the manor (*spahi*), was merely entitled to demand from them, in full of rent, a tithe of the produce of their land and of the increase of their stock; and in consideration of this, he was not only bound to perform military service to the sultan, but also to protect the cultivators on his estate. When the Turkish institutions were in their vigour, and the exertions of the pachas and feudal lords were restrained by the vigilance of the sultan, this state of things contrasted most favourably with the rapine and anarchy that then prevailed in the greater part of Europe. 'I have seen,' says a contemporary writer, 'instances of Hungarian rustics set fire to their cottages, and fly with their wives and children, their cattle and instruments of labour, to the Turkish territories, where they were freed, besides the payment of the tenths, they would be subject to no imposts or vexations.' " (*Leunclavius in Turc. Imp. Stat.*)

According to the imperial survey ordered by Solyman the Magnificent, the number of *zaimets*, or estates, estimated at 500 acres of land, and upward, amounted to 3192, and the number of *timars*, or estates valued at from 300 to 500 acres of land, amounted to 50,160 ; the whole furnishing a revenue of nearly 4,000,000 rix dollars, appropriated to the maintenance of an army of about 150,000 men. (*Present State of Turkey*, L. 290, 291.) Olivier states that in his time it was computed that there were in the European part of the empire 914 *zaims*, and 8356 *timars* ; the number in Asia being nearly the same, and the whole furnishing a militia of above 60,000 men.

In 1818, it would appear that there were still 914 *zaims* in Europe, and in Asia 1479 ; the annual revenue from these amounting to from 25,000, 100,000 aspers each, which, at 100 aspers to a Turkish piastre, would give a yearly income of about £50 on an average from each ; but more recently the numbers of both have been still farther reduced.

Latterly, too, or since the disorganization of the empire, all sorts of abuses have crept into the management of the estates held by the spahis, or feudal lords. These have been oppressed by the pachas ; and they, in their turn, oppress the cultivators, increasing their demand for corvées or other services, and claiming and exacting, though illegally, a much greater portion of the produce than the tenth, to which they are legally entitled. And yet, despite their pillage of the cultivators, many spahis have, like the zemindars in Hindostan, been forced to abandon their estates ; and indeed, in many districts, especially in Asia Minor, owing partly to the illegal exactions of the lords, but still more to the arbitrary exactions of the pachas, the cultivators have wholly deserted the lands. The truth is, that, in most parts of Turkey, power makes law ; there is no real security, the rights of the people being trampled on at the pleasure of those in authority.

But it would be the greatest imaginable error to suppose that all, or that even the greater portion of the lands conquered by the Turks, were distributed in the way previously stated. The revenues of extensive tracts were appropriated to mosques, the great officers of state, the mother and mistresses of the sultan, the children of the imperial family, and the sultan himself ; and after these deductions, the residue, which still amounted to a very large proportion of the whole, was left, burdened with a tithe or land-tax of one tenth part of the produce, to the ancient proprietors. These, if Mohammedan, had the privilege of going to war ; others, whether Turks or Christians, that is, infidels, who, from choice or civil disability, devoted themselves to the arts of peace, and enjoyed their estates under the protection of the law, commuted their military service by the payment of a tribute instead.

It is commonly said that Turkey is a country in which there is no security for property ; and if by this be meant that it is exposed to illegal exactions of all kinds, partly by the feudal lords, and partly and principally by the pachas and their subordinate authorities, nothing can be more correct. But it is not true, speaking generally, to allege that in Turkey private property is not recognised by law, or that it may be seized at the pleasure of the sultan. This is the case, no doubt, with the property of persons in the public service, whose lives and fortunes must answer for their real or imputed misconduct. But all other sorts of property are respected in Turkey ; and even a pacha, or other public functionary, who has acquired property by the most objectionable means, may, if he please, easily place it beyond the grasp of the grand seignior. To accomplish this, he has merely to settle it on his family and direct heirs, leaving the reversionary interest in it to some mosque, which, on receiving a nominal quit-rent, takes charge of the property, which can no longer be either forfeited or affected by the crimes or misconduct of the original founder of the family or his heirs. Property so left is denominated *vacouf*. But this device, though quite effectual for the object in view, necessarily tends, in the end, to accumulate much too great a quantity of property in the hands of the church ; so that in obviating one abuse it occasions another.

If the Hatti-Scheriffs, or imperial decrees, issued within the last three or four years, were really carried into effect, they would effect a total and most beneficial change in the conditions under which property is held in Turkey ; practically, however, we believe they will have little effect, at least for many years to come.

Agriculture.—In Turkey the cultivators do not live dispersed over the country in hamlets, or in single farm-houses, but are congregated in villages, which, owing to the depopulation of most districts, are frequently at great distances from each other. These villages present a very striking picture of primeval manners, each family providing itself with most articles required for its consumption, while

their municipal affairs, or those in which the community have an interest, are conducted by their elders. The village communities in certain districts, especially of European Turkey, enjoy considerable powers ; and wherever this is the case, or where a tract of country happens to belong to a powerful individual, the cultivators, speaking generally, are comparatively prosperous. This, however, is the exception, oppression and a want of security being the usual consequences of Turkish ascendancy.

Turkey is not dependent upon any foreign country for the subsistence of its inhabitants ; it yields, on the contrary, corn and other produce, sufficient not only for the home demand but also for exportation. Ten times the produce might, however, be raised in these fine countries were a better policy adopted, and the inhabitants protected against vexatious exactions. The native rayahs or peasants, by whom cultivation is carried on, have generally little or no capital ; and as the tax on the crop has generally to be paid before the produce is gathered, they are in most cases obliged to borrow the money for this purpose at a ruinous rate of interest. Agriculture is accordingly in a very backward state throughout most parts of the empire. In Thrace, the rotation of crops is tolerably well understood ; but elsewhere in European Turkey cultivation is extremely depressed. Manuring is next to unknown, and in the mountainous parts, particularly in Servia and Albania, an immense waste of timber occurs, from the forests being burnt that the ground may be fertilized by their ashes. The ploughs (except perhaps in Wallachia and a few other provinces) are of the most wretched description, being seldom shod with iron, and fit only to scratch the surface of the earth : a bunch of thorns performs the functions of a harrow ; and the other farming implements, if so we may call them, are in general equally bad. Thrashing is performed, as in most eastern countries, by treading out the grain with cattle ; the straw being subsequently chopped by dragging over it a sort of heavy cylinder stuck with sharp flints. But the fertility of certain portions of the empire, as Thessaly, the valley of the Maritza, &c., is so great, that, despite the low state of husbandry, the average produce of corn is said to amount to from 15 to 30 times the seed.

On the whole, the cultivation of the soil appears to be better understood in Bulgaria than anywhere else. (*Boué, La Turq. d'Europe*, iii. 4.) Some notices of the agriculture of this and the other European provinces will be found under their separate heads in this Dict.

Maize is the principal species of grain cultivated in European Turkey, in the mountain-valleys as well as the plains, except in Bosnia, &c., where the climate is too cold. Wheat, rye, barley, oats, and buckwheat are also pretty generally cultivated ; and millet is grown in the more sheltered places. Rice is grown only along the banks of the Maritza and other marshy tracts in the S. provinces. The quantity of this grain produced in European Turkey being insufficient for the consumption, a portion of the required supply is imported from Egypt and Asia Minor. Great quantities of haricots, beans, cabbages, onions, melons, cucumbers, tomatas, capsicums, &c., are raised as articles of food ; but the potato is eaten only in Bosnia, Croatia, Herzegovina, Montenegro, and a few places in Servia and Albania : lentils, turnips, artichokes, asparagus, beet-root, and many other vegetables common among us, are almost unknown in Turkey. Though not usually drunk by the Mussulman (except those of Lower Albania), wine is grown in most provinces of Turkey in Europe, but Wallachia and Moldavia (which see), Bulgaria and Servia, are the principal wine countries.

Turkish wines are mostly red ; white wine is produced only in Wallachia, W. Bulgaria, and at Semendria, and a few other places. The best wines are very high coloured and somewhat similar to those of Cahers, and of Radicopani in the Papal states. These are grown chiefly in Macedonia, in the basins of Scutari and Prisren in Albania, Mostar in Herzegovina, on the hills along the Servian Morava, in Thrace, and in the vicinities of Lovdscha in Bulgaria, and Meteora in Thessaly. Certain growths in the S.W. of Macedonia, deserve particular mention. The inhabitants of those places possess, in fact, a valuable source of wealth, for there is no doubt that several of their wines might be advantageously exported. (*Boué*, ii. 251.) The want of proper cellars for storing the wines, and their rude preparation, detract greatly from their excellence. In Thessaly and Albania they are commonly spoiled, at least in the estimation of most foreigners, by the addition or absorption from the barrels, skins, &c., of resin or turpentine. In certain cautions, however, the use of such substances is not supposed to be necessary, and the wines there are accordingly very palatable. In Bosnia the vine is replaced by the plum, and the favourite beverage there is a liqueur made from its juice, called *slivovitza*. Peaches, apples, cherries, almonds, hazel nuts, &c., are grown, but grafting, and all other horticultural operations, are either unknown or much neglected. The olive flourishes in the S. prov. only, and most of the Turkish oil is grown in Asia. Flax, hemp, sesamum, saffron, cotton, tobacco, castor oil, and madder, are among the principal remaining articles of culture. The care of the forests does not seem to occupy much attention; nothing like a forest-board exists, and in many districts, formerly well wooded, wood for fuel is becoming scarce.[*] Live hedges are rare; where the fields are enclosed it is either with dry wood or stone walls, and where neither material is plentiful, as in E. Thrace, the fields are entirely open.

We have already given some notices of the condition of property and cultivation in Asiatic Turkey in the arts. NATOLIA, KURDISTAN, SYRIA, &c. In regard to Armenia, Mr. Brant says, "I had scarcely seen any place in this region approach to the state of apparent prosperity enjoyed by the inhabitants of the plain of Kharput. An Armenian farmer there had ten pair of draught oxen, and a few cows and sheep. The produce of his land was—wheat about 375 bushels, valued at £75 a year; cotton to the value of perhaps £28; grapes about 3300 lbs.; which, together with millet, lentils, and other sundries, made up the value of about £142 sterling. The millet, and fifty bushels of wheat, the grapes, and sundry produce were consumed by the farmer and his family. The cotton sold, after the lord of the soil had taken his rent, was about sufficient to pay the tax of 10 per cent on the whole produce to the pacha. I was not informed how much land he had in cultivation; there is no measure of land; it is estimated by the quantity of seed used in sowing, or the number of oxen necessary to plough it. They do not manure much, but allow the land to lie fallow every alternate year. Such is the general system of agriculture throughout Armenia." (*Geog. Journ.*, vi. 907–908.) The vine, and mulberry, tobacco, cotton, and oil, share the chief attention of the agriculturists in Asiatic Turkey, after the production of the grains, &c., necessary for food. The culture of silk is extensively carried on in several districts, but especially round Brusa in Asia Minor, where the mulberry-tree is kept cut short, and receives a good deal of careful attention.

In consequence, however, of the oppressions practised on the cultivators, vast tracts of land in Asia Minor are wholly deserted, or occupied only by the scanty population of a few wandering tribes. Nowhere, indeed, is the destructive influence of Turkish misgovernment so apparent as in the present state of this celebrated country, favoured alike by situation and climate, and which, in antiquity, was the seat of many noble cities, and powerful and refined nations. Industry and civilization have all but disappeared. "No care whatever," says Mr. Kinneir, "is taken to improve the land; nor can this be a matter of surprise, when we reflect that the farmer is liable to be turned out at a moment's warning, and is certain of being taxed or plundered in exact proportion to the yearly produce of his farm. It is not, indeed, uncommon, should there be a prospect of a plentiful harvest, for the crops upon the ground to be seized by the pacha at a low valuation, and then put up to the highest bidder. This system, so destructive of industry, may be traced to the ill-judged but favourite policy of the Porte in continually changing the governors of their provinces, lest by being settled for a considerable period in their governments, they should shake off their allegiance, as many have already done. The pacha, therefore, who, during the short time he remains in favour, has not only to feed the avarice of the imperial ministers, but also to accumulate an independency for himself before his retirement from office, is heedless of the interests of the farmer, or of those who are to succeed him, and only anxious to collect wealth. We consequently observe that those provinces where the chiefs maintain their independence, are invariable the richest, best peopled, and in every respect the most flourishing; since they find it their interest to encourage the cultivators of the land, who are continually deserting those parts of the country immediately governed by the sultan's officers, to place themselves under their protection. The prosperity of the provinces of Asia Minor is in this manner always fluctuating, according to the actions and dispositions of their respective rulers. Sometimes they are well peopled and cultivated (I speak comparatively), and at others waste and forsaken; whole villages emigrate from one district to another without much trouble or expense, since their houses are simple and of easy construction, and their articles of furniture so trifling as to be transported with facility on the backs of the cattle, which supply them with milk during the journey, and every where find abundance of pasture. The Greeks, called *Uroomi* by their Turkish lords, constitute a considerable portion of the peasantry in this part of

the empire, and are not, in my opinion, the fallen and dastardly race usually represented. The political or religious institutions of a state affect, without doubt, the character of a people, and this is no where more conspicuous than throughout those quarters of the globe where the blighting doctrines of Mahomet have been diffused. The unjust and cruel persecutions carried on by the Turks have damped the fiery spirit of the Greeks, and rendered distrust and deception absolutely necessary to the safety of their persons and property; whereas, under a more enlightened and less despotic government, the national character of that people would probably rise to the standard of the inhabitants in most of the civilized countries of Europe. To me they have always appeared as dispirited and broken-hearted; but at the same time ready to rise, if supported, and crush their vindictive rulers to the earth." (*Asia Minor, &c.*, p. 51.)

There are in Turkey great numbers of sheep and goats, the flesh of which constitutes the principal animal food of the inhabitants; but there are proportionally fewer cattle than in other countries of Europe, beef being seldom, and veal never eaten by the Turks. The sheep are nearly all of a small, thick-bodied breed, with a white fleece; marinas, large-tailed, or other improved breeds, are met with only as Servia, into which they were introduced by Prince Milosch, or in Bosnia. In Wallachia the sheep have tall spiral horns, and their wool is a principal source of wealth. The pastures there are fine and extensive, and large flocks are brought thither from Transylvania, to be depastured during winter. At the same season the sheep from the table-land of Cappadocia, &c., are driven into the plains of N. Syria; and many of the migratory Koord and Turkman tribes of Asia seek the pasture lands about Angora, the traders of which town supply their various wants, receiving in return the wool, skins, and other produce of their flocks, in which articles Angora has a very considerable trade. (*Geog. Journ.*, vi. 213.) The cattle along the banks of the Save and Danube appear to be a degenerate Hungarian breed. Those of the more S. prov. are different, being of medium size, and short-horned. Oxen are every where employed in field labour. Buffaloes are common, particularly in Bulgaria and Thrace. In Bosnia and other W. prov. some tolerably good cheeses, similar to Gruyère, are made; but the cheese of most parts of Turkey is in general too indifferent to suit our taste. In making cheese, the milk of ewes and goats is partially employed, but in general only is the absence of that of the cow and buffalo. Turks abhor the hog so much, that they do not generally permit its sale in the towns, where they form the chief proportion of the pop. and Boué says, "that the carcasses of hogs are only suffered to be brought to Constantinople at certain periods of the year, under an especial firman." (i. 508.) Nevertheless, they are reared in vast numbers in Servia, Bosnia, and other N. and W. prov., and, in fact, constitute the chief resource of the Servians. The Turks are good horsemen, and take pride in their horses. Little of that care is, however, bestowed on them that is common in W. Europe. They are fed only twice a-day, sometimes they are not put into stables, and are not groomed and trimmed as in Europe. The horse of European Turkey is generally of middle size, or rather below it, with a short neck, strong limbs, and a bay, chestnut, reddish brown, or white, seldom a grey colour. They are usually fed on barley; oats being used for horses only in the N.W. prov. The horses of Asiatic Turkey seem to be chiefly of Arabian descent. "The Montefik are an excellent race of horses, bred by a great tribe of that name on the banks of the Euphrates. In Armenia and Koordistan a prodigious number of fine animals might be procured at a cheap rate for the cavalry: the horses of Bagdad are large, and many of them show a great deal of blood; but those bred in the desert bordering on Damascus, are upon the whole the finest. I have heard of a poor Arab at Antioch in Syria, refusing 26,000 piastres for a mare of that breed. The only blood-horse I ever met with in Asia Minor was bred near Oascat in the plains of Cappadocia, and may be descended from that which was so much admired by the Romans." (*Kinneir's Asia Minor, &c.*, p. 508.) The ass is much used in Roumelia, S. Albania, &c. Mules are scarce in those prov., but very numerous in Montenegro, and other mountainous parts of Turkey.

There are mines of copper, argentiferous lead, iron, &c. in various parts of both European and Asiatic Turkey; and it is generally believed that several of the mountain chains which bound or intersect the Turkish prov., contain ores, not only of the useful, but of the precious metals. The Wallachian and Moldavian gipsies collect from the beds of the rivers pellets of gold mixed with a small quantity of silver, by means of which they are enabled to pay into the treasury the annual tribute of a drachm of gold imposed upon each man. But mining industry is not profitable in Turkey from various causes. "The ignorance of the people in the art of working the mines with economy is perhaps one cause of the neglect with which the Ottomans appear to treat this source of wealth; but the chief obstacle to exploration is the rapacity of government, which would seize upon the advantages of any new discovery, and subject the provincialists to the unrecompensed labour of opening the mines and extracting the ore." (*Thornton*, ii. 31, 32.) Asphaltum, nitre, salt in Wallachia, &c., and coal in Bulgaria, are among the mineral products of the empire, but are by no means raised to the extent that they would be under a liberal political system. Between Eski-shehr and Servi-Hissar, in Asia Minor, the substance called *maer-schaum*, so much used for German pipes, is found in large quantities. A specimen, procured by Mr. Ainsworth at this place, fresh from the mines, proves it to be a hydrated silicate of magnesia. It is a porous friable stone, almost entirely composed of small-grained vitreous or transparent felspar, decomposing and passing into a variety of porcelain earth. Great quantities of pipe-bowls are manufactured from this material, and sent to Constantinople for export into Germany, &c. (*Geog. Journ.*, x. 490, 491; *Clarke's Trav.*, &c.)

The *manufactures* of Turkey are more numerous, and display greater excellence, than might have been expected in a country so backward in the arts: indeed, her success in manufacturing industry is, upon the whole, greater than that of several countries ranking higher in civilization. Thornton, who though somewhat partial to the Turks, is, on the whole, an excellent authority, says, "I know not whether Europe can equal, but certainly it cannot surpass the Turks in several of their manufactures. The satins and silk stuffs, the velvets of Brusa and Aleppo, the serges and camlets of Angora, the crapes and gauzes of Salonica, the printed muslins of Constantinople, the carpets of Smyrna, and the silk, linen, and cotton stuffs of Cairo, Scio, Magnesia, Tokat, and Costambol, establish a favourable, but not an unfair criterion of their general skill and industry. The workmen of Constantinople, in the opinion of Spon, excelled those of France in many of the inferior trades. They still practise all that they found practised; but, from an indolence with respect to innovation, they have not introduced or encouraged several useful or elegant arts of later invention. They call in no foreign assistance to work their mines: from their own quarries their own labour extracts the marble, and more ordinary stone which is employed in their public buildings. Their marine architecture is by no means contemptible, and their barges and smaller boats are of the most graceful construction. Their foundry of brass cannon has been much admired, and their musket and pistol barrels, and particularly their sword blades (though the sword blades of Damascus are not so famous as formerly) are held in great estimation even by foreigners." (*Present State of Turkey*, i. 67, 68.) Their manufactures of Morocco and other leather, and of gold and silver lace, &c., deserve also to be mentioned with praise.

But if the Turks be more successful in the practice of some of the useful arts than is commonly supposed, they have made no progress in the fine arts, and are necessarily ignorant of the higher sciences. "Their buildings are rude incoherent copies, possessing neither the simplicity nor unity of original invention. Heavy in their proportions, they are imposing only from their bulk: the parts do not harmonize, nor are they subservient to one leading principle; the details are bad both in taste and execution; the decorations have no use, no meaning, no connection with the general design; there is nothing which indicates the conception of genius. The energies of the latter are chilled and repressed by the monotony of Turkish habits and the austerity of their customs. Their cities are not adorned with public monuments, whose object is to enliven or to embellish. The circus, the forum, the theatre, the pyramid, the obelisk, the column, the triumphal arch, are interdicted by their prejudices. The ceremonies of religion are their only public pleasures. Their temples, their baths, their fountains, and sepulchral monuments, are the only structures on which they bestow any ornament. Taste is rarely exerted in other edifices of public utility, *khans* and *besestins*, bridges and aqueducts. Sculpture in wood or in stucco, and the engraving of inscriptions on monuments or seals, are performed with neatness and admirable precision; and the ceilings and wainscoting of rooms, and the carved ornaments in the interior of Turkish houses, show dexterity and even taste. But their paintings, limited to landscape or architecture, have little merit either in design or execution; proportion is ill observed, and the rules of perspective are unknown. They reckon time by lunar revolutions, so that in the space of 33 years the Turkish months pass through every season. Their knowledge of geography does not extend beyond the frontiers of their empire. Their surgery is rude, from want of science, of skill, and of instruments." (*Ibid.* 69–77.)

The domestic manufacture of cotton stuffs in Turkey is pretty general; and Cannabich estimated the consumption of raw cotton in the Turkish empire at 90,000 bales a year,

10,000 of which are consumed in the fabrics of Thessaly alone (*Handb. der Geog.*), the best Turkish yarn being made in this province. Urquhart says, that in the S. provinces the poorest family requires 20 okes of uncleaned cotton, and 10 of wool, for its yearly consumption; and the manufacture of these materials occupies a large share of the peasant's indoor labour. Handkerchiefs, shirting, long-cloths, napkins, coarse cotton stuffs, and clothing in general, are the goods principally produced by their looms; and, according to Urquhart, 24,000,000 lbs. of cotton manufactures, worth £5,000,000 are made annually in European Turkey and Greece, (p. 150.) Very little dependence can, however, be placed on these statements; and there can be no doubt that the native manufacturers, who produce goods not for domestic consumption but for sale, have been involved in the greatest distress in consequence of the importation of English and other foreign goods. The manufacture of cotton yarn has been especially interfered with, and English cotton twist is now generally used for warp in such Turkish looms as are still at work, and is an article of increasing consumption.

The *commerce* of Turkey owes most part of its activity to the immunities and protection enjoyed by those engaged in it; which are not extended to individuals occupied in other avocations. The cultivator of the soil is ever a helpless prey to injustice and oppression, and the manufacturer has to bear his full share of the common insecurity; he is fixed to the spot, and cannot escape the grasp of the local governor. The raw material monopolized by a bey or ayan may be forced upon him at more than its fair value, and perhaps its quality may be inferior; fines may be imposed on him; he may be taken for forced labour, or troops may be quartered on his workshop. (*Urquhart*, p. 139.) It was not till 1837 that a firman was issued by the sultan allowing the free exportation of wheat to foreign countries. The Turkish gov. had previously been accustomed to prevent the exportation of grain from any part of the empire till Constantinople had been first abundantly supplied. In this view, the principal corn-growing provinces were obliged to furnish to the officers of the sultan a quantity of wheat equal to about a 19th part of the produce of their harvests. This contribution was called *istira*. The *istiragi* or collectors, on receiving the corn from the proprietor, paid him 20 paras for every kilo (about 60 lbs.). The total quantity of corn thus purchased for the supply of the capital amounted to about 1,000,000 kilos a year: this was sent by sea to Constantinople, and lodged in public granaries on the N. side of the harbour. As this stock was considered a resource against times of scarcity, it was not distributed till it began to be damaged, unless when it could be sold with considerable benefit. This frequently happened; for individuals were nor suffered to lay up their corn in magazines to resell in a similar manner; and Olivier estimated the yearly produce of this monopoly at 10,000 purses, or 5,000,000 piastres. After the treaty which opened the Black sea to the commerce of foreign nations, vessels with cargoes from the Russian ports were allowed the free passage of the Bosphorus and Hellespont; "a privilege," says Thornton, "so important, that I have known ships, which had surreptitiously loaded wheat, the produce of the Turkish provinces, sail to the Russian port of Odessa, and subject themselves to the delays and expenses of performing quarantine, paying the harbour fees and custom-house duties, for no other purpose than to obtain a certificate of their cargo being the produce of Russia, and thereby rescuing it from the vexations and extortions of the officers of the Turkish *miri*." (i. 256.) Other articles of provision, such as sheep, oxen, butter, cheese, wax, tallow, &c., used to be bought up in the same manner by the officers of government at their own price; but probably this system has now been in great part abolished; and except in the article of provisions, no restriction on commerce ever existed in Turkey. All foreign articles may be imported into the Turkish ports, without let or hindrance of any kind, on payment of an import duty of 3 per cent. *ad valorem*; and all articles of foreign and domestic growth or manufacture may be freely conveyed all over the empire. Her commercial system is, in fact, by far the best feature in the policy of Turkey.

The internal traffic of Turkey is greatly impeded by the badness or rather total deficiency of roads. Burgess says the sultan has not an inch of road in his dominions, which

from England, France, and Austria, supplies all the fairs throughout Roumelia and Bulgaria. (*Keppel's Journey across the Balkan*, p. 255.) Next to Smyrna, Aleppo is the chief seat of commerce in Asia. Caravans bring thither pearls, shawls, Indian and Chinese goods, from Bussorah and Bagdad; camels from Arabia; cotton stuffs and thread, Morocco leather, goats' hair, and galls, from the pachalics of Mosul, Diarbekir, Orfa, Aintab, &c.; furs, goats' hair, wax, gum ammoniac, &c., from Van, Erzeroum, and Kars; silk, copper, furs, and linens, from Asia Minor; silk, Mocha coffee, soap, scented woods, ambergris, drugs, and pearls, from Syria and Arabia; rice, coffee, and Egyptian produce, from Latakia; silk manufactures from Brusa and Damascus; European cotton and woollen stuffs, printed muslins, hardware, watches, wrought amber, and fur, from Smyrna and Constantinople. The principal articles of export are, sheep's wool, goat's hair, cattle, horses, hogs (from Servia, &c.), hides, hare skins, wheat, raw cotton and silk, tobacco, raisins, figs, almonds, mastic and other gums, gall-nuts, vallonea, leeches, honey, wax, saffron, madder, anise and linseed, turpentine, safflower, meerschaum pipes, whetstones, carpets, silk and cotton fabrics, leather, copper, and metallic wares, orpiment, &c., with Arabian, Persian, Indian, and Chinese goods. The principal imports are linen, woollen, cotton, and silk goods, colonial products and dye stuffs, hardware and earthenware, paper, furs, &c. The British trade in Manchester, Glasgow, Birmingham, and Sheffield manufacture, and other British produce has, however, been steadily increasing during the last 10 years. The following is an Account showing the Quantities of the Principal Articles imported into the United Kingdom from Turkey,* in 1838, 1839, and 1840.

Articles.		1838.	1839.	1840.
Corn (wheat)	qrs.	5,150	43,740	4,826
Figs	cwt.	12,946	14,825	17,363
Madder root	—	29,059	47,527	58,449
Oil (olive)	galls.	26,765	40,203	21,396
Opium	lbs.	80,654	177,451	50,745
Raisins	cwt.	23,942	22,856	44,233
Saltpetre	—	—	7,889	9,654
Flax and linseed	bush.	97,967	48,970	16,408
Silk (raw, &c.)	lbs.	476,755	791,905	785,189
Skins (lamb)	No.	242,565	159,763	163,354
Vallonea	cwt.	108,795	127,009	143,095
Wool (cotton)	lbs.	680,555	102,423	463,979
— (sheep's)	lbs.	761,018	1,183,532	655,884

The wealthier class of Turks are generally too apathetic and indolent for commercial pursuits, which they leave to the Greeks, Armenians, Arnaouts, and Jews. "The extreme simplicity of commerce, from the absence of all legislation on the subject, is visible in the establishment of a merchant: no books, save one of common entry, are kept; no credits are given; no bills discounted; no bonds, nor even receipts: the transactions are all for ready money; no fictitious capital is created; no risk or loss from bankruptcy to incur. A merchant, whose capital may exceed £20,000, will, very possibly, be without a clerk; and a small box, which he places on his carpet, and leans his elbow upon, encloses, at once, his bank and counting house." (*Urquhart*, 136.)

Accounts are kept in piastres of 40 paras of 3 aspers each; or in purses of 500 piastres. But the rate of exchange is very variable, on account of the continual deterioration of the coin. In 1816, the pound sterling was worth only 12 piastres; but in 1839, it was equivalent to 104, and in 1842, was worth nearly 190! The most common measures and weights are the oke = about 2¾ lbs., and the quintal of 44 okes. The arschine = 2 ft. Engl. Distance is commonly measured by the hour = about 3 m.

Government, &c.—Châteaubriand said of the Turkish government that it was an absolute despotism, tempered by regicide! In truth and reality, however, the government of Turkey is a species of theocracy. The grand seignior is supposed to be the lieutenant and vicegerent of the prophet, and consequently, also, in some degree, of the Deity himself. But though, at first sight, this may appear to confer all but unlimited powers on the sovereign; and though, in some respects, it certainly gives very great latitude to his actions, it at the same time subjects him to various restraints and limitations, which he dares not contemn or break through. His authority, in fact, is principally bottomed on the Koran; and were he to abandon its doctrines, and to act in the teeth of its precepts, or those deduced from it by eminent commentators, and sanctioned by custom, the foundations of his authority would be loosened, he would cease to be the lawful sovereign, and would be regarded as a usurper whom it is meritorious to dethrone. Hence, though absolute in some respects, the power of the grand seignior is, in others, in the last degree limited. He may put those engaged in his service to death at pleasure, but were he to interfere in any way with property left in trust to a mosque, or to outrage

the law by drinking wine in public, he would run a great risk of exciting discontent, and, if he persisted in such conduct, of being dethroned.

"The Turks," says an accurate observer, "learn very early that, if the prince be of right divine, he founds it on the Koran; that he is constituted such by the sacred code of laws, which, as a true believer, he has studied, and knew, before his accession to the throne, it would ever be his duty to observe; and that, consequently, he is as much bound and tied by all those laws as they themselves are.

"This is so explicitly and fully laid down in the Koran, that Mohammed thought it necessary to throw in rules of exception expressly for himself.

"Hence, when the people are notoriously aggrieved, their property or that of the church repeatedly violated; when the prince will riot in blood, or carry on an unsuccessful war; they appeal to law, pronounce him an infidel, a tyrant, unjust, incapable to govern; and, in consequence, depose, imprison, and destroy him." (*Porter's Observations on the Turks*, i. 109, 12mo ed.) And every one who has any knowledge, how slender soever, of Turkish history, is aware that this principle has not been inoperative; and that the Turks have, over and over again, exercised the right of resistance to what they looked upon as arbitrary power.

It may not, however, be out of place to mention that this dependence of the sultan on the Koran, though it limits, in some degree, his power to tyrannise over his subjects, opposes, at the same time, the most formidable obstacles to his attempts to introduce any organic changes, how expedient or necessary soever. The rights and social condition of the people, living in the Turkish empire, who have not embraced the religion of the conqueror, is supposed to be determined by the Koran. And hence the difficulty—without, as it were, overturning the very foundations on which the monarchy rests—of effecting any material changes in the situation of the dependent population. The Turks cannot, unless they abandon their own religion, amalgamate with them, or raise them to the same level as themselves; so that the nation must always consider of two distinct parts—the Turkish, or ruling portion; and the rayahs, or subjugated infidels, who exist upon sufferance, and who can never arrive at any situation of power or emolument. The character of the Mahommedan religion is, in truth, an all but insuperable obstacle to any thing like real reform. Though less intolerant than many others, it inculcates on the mind of its votaries the most exalted ideas of their own importance, and the most profound contempt for all sorts of unbelievers. There may, no doubt, be an imitation of European tactics, and an attempt to introduce something like the practices and institutions that prevail in European states; but it is impossible, so long as the religion of the prophet maintains its ascendancy, that they can have any considerable influence. Submission to their power has saved the unbelieving population of the country from death; but nothing short of their embracing the religion of the conquerors can effectually protect them from insult and contempt, and consequently, also, from extortion and tyranny.

The grand seignior is assisted in the government of the empire by a cabinet-council or *divan*, consisting of the principal ministers of the empire, and of the mufti or head of the law. Until very recently the sultans were in the habit of delegating the greater portion of their authority to the grand vizier (*vizier azem*), who became, as it were, vicegerent of the empire, being at the head of the civil government, and generalissimo of the military and naval forces. But of late years the powers of this high functionary have been very much curtailed. Indeed, the place was wholly abolished by the late, though it has been revived by the present emperor. The functions of the other ministers correspond with those of minister for foreign affairs (*reis effendi*), of the interior, commerce and finance (*tefterdar*), and of the commander-in-chief (*serraskier*), a grand admiral, &c. The court of Constantinople is generally known in other European countries by the title of the *Sublime Porte*, a designation derived from the *Bab Humayen*, or principal outer gate of the seraglio, whence the hati scherifs, or imperial edicts are usually issued.

The *sheik-ul-islam* (mufti), or head of the clergy and chief interpreter of the Koran and the canonical laws, is a very important functionary. He nominates to all the principal offices in church and law; and takes precedence of every other subject in the empire, even of the grand vizier. On most great occasions the sultan applies to the sheik-ul-islam for a *fetwa*, or legal opinion, to ascertain whether his intended course of action be in accordance with the Koran. But this is not indispensable, and has very rarely been refused. Latterly, too, the opinions of the mufti have become of less importance.

The mufti is always chosen from the *ulema*, a body comprising the clergy with the interpreters and administrators of the law. But, though they all study together, the lawyers and judges are quite distinct from the clergy; it being

* In the official accounts, the commerce of Turkey is not distinguished from that of Greece N. of the Isthmus; but the latter is quite inconsiderable.

left to every young man brought up in one of the colleges of the order to determine for himself, when he has attained a proper age and acquired a sufficient stock of learning, whether he will become a priest, or a doctor of law, or a judge: but it is to the latter, or the lawyers, that the title of ulema is more peculiarly appropriated.

Throughout Turkey, the ministers of religion are all subordinate to the civil authorities, who exercise over them the powers of diocesans. Magistrates may supersede and remove clergymen who misconduct themselves, or who are unequal to the proper discharge of the duties of their office. The magistrates themselves may also, whenever they judge proper, perform all the sacerdotal functions; and it is in virtue of this prerogative, joined to the influence which they derive from their judicial power and their riches, that they have so marked a pre-eminence, and so preponderant an authority, over the ministers of public worship. (*Thornton*, i. 198.)

The members of the ulema constitute a sort of aristocracy. They pay no taxes or public imposts, and, by a peculiar privilege, their property is hereditary in their families, and is not liable to arbitrary confiscations. Their persons are sacred; their blood may on no account be shed; nor can they be legally punished in any way but by imprisonment and exile. It is to be observed, however, that the power and dignity of the ulemas are not hereditary in individuals, but in the order. Formerly they held their offices for life; but about the end of the 17th century they were made removable at pleasure, like all other public functionaries. They now are appointed only for a year. Each individual, however, enjoys all the privileges of the order, independently of his holding any office, or exercising any public employment. There have been instances of muftis declining to obey the commands of the grand seignior, and of their remonstrating with him on the impropriety or illegality of his conduct; though, as the sultan makes the mufti, and can depose and exile him at pleasure, such conduct must necessarily be very rare, except when some formidable conspiracy is on foot, and when the powers of the sultan are consequently circumscribed. In the reign of Mustapha, the people put to death the mufti for having, as they alleged, misled the sultan. Cantemir says, that Murad IV. commanded a mufti to be pounded in a marble mortar, saying, that heads, whose dignity exempts them from the sword, ought to be struck with the pestle! but the fact is doubtful. (*Thornton*, i. 130.) Speaking generally, the influence of the mufti and ulema is uniformly opposed to all measures of reform; at least, to such as might be supposed to militate in any way against the peculiar doctrines and regulations enforced by the Koran.

Besides the ulema, there is a privileged order, limited to the descendants of Mahommed by his daughter Fatima. These are called *eomrs*, or *ameers*, have *syud* prefixed to their names, and are authorised to wear green turbans. Inasmuch, however, as they are very numerous, *eomrs*, like brahmins in India, are found in even the most abject ranks of life.

The government of the provinces is extremely rude, and is practically, indeed, little else than a tissue of abuses. European Turkey was formerly divided into the two great governments, or *ejalets*, of Roumelia and Bosnia; the former of which was subdivided into 16 sanjacks, or great governments, or pachaliks, and the latter into seven, besides some inferior governments. The power of the pachas within their respective districts is, in many respects, unlimited. They have under them *mussetims*, or sub-pachas, to whom they delegate a portion of their authority, and who watch over a certain extent of territory. Every pacha or governor, is supposed to represent the sovereign within the limits of his own jurisdiction, is invested with his authority, and exercises his prerogatives in all their plenitude. Nominally, however, continuous jurisdiction, or the determining differences between subjects, is left to the *cadi*, or judge, in conformity with the principles of mussulman government and the practice of the sultan.

Formerly the pachaliks (or rather *beylerbeyleks*, for such as the name given to the larger governments) were much more extensive than at present; and it not unfrequently happened that a pacha, at the head of a large government, having succeeded in getting his creatures made pachas of the surrounding governments, acquired such a degree of power as to be able to cast off his allegiance, and defy the sultan. Latterly, however, it has been the policy of the government to diminish the sanjacks, and so to lessen the danger of insurrection on the part of the pachas. The latter are appointed only for a single year; and the limits of the different pachaliks are being constantly changed. (*Bové*, 181, &c.)

All mussulmen, how humble soever their origin, are eligible to, and may fill, the highest offices in the state. In Turkey, birth confers no privilege, all true believers being equal in the eye of the law. But this sound principle is rendered of little or no value, or rather positively injurious,

by every thing being made to depend on the pleasure of the sultan. With the exception, indeed, of the law and the church, no previous study or preparation, nothing, in short, but the favour of the prince, which is most frequently obtained by the most unworthy acts, is required to elevate individuals from the very lowest to the very highest stations! And hence it is, that though individuals be sometimes found in Turkey admirably adapted for the situations they fill, these are very rare exceptions, incapacity and unfitness for their duties being the distinguishing characteristics of Turkish functionaries.

Till this vicious system be wholly abandoned, and individuals be appointed to important situations from other motives than the mere caprice of the sultan, it is nugatory to suppose there can be any substantial improvement. In this respect, however, little or no progress has hitherto been made. When Marshal Marmont visited Constantinople, towards the close of the late sultan's reign, who had been called, and not without reason, the Turkish reformer, a black eunuch was a general of brigade; and Achmet Pacha, who was then a general of cavalry, had been bred a shoemaker, and practised at a more recent period as a waterman in the harbour! And a short while subsequent to this the same Achmet Pacha was made *capitan pacha*, or high admiral of the fleet, of the duties of which station, it is hardly necessary to add, he knew no more than he did of the Principia of Newton. This, we apprehend, may be taken as a pretty fair specimen of Turkish reform. Marshal Marmont, who is a most intelligent and unexceptionable judge, says, that *aujourd'hui, comme autrefois, le faveur et le caprice du maitre sont les seuls titres pour occuper les emplois les plus importans. Les prétendus progrès en civilisation du gouvernement Turc n'ont pas encore commencé le principe qu'il faut apprendre ce que l'on veut savoir*, (ii. 58, 12mo ed.)

But if the rapid advancement of individuals from the lowest to the highest stations be so common in Turkey as hardly to excite attention, the sudden downfall and total ruin of the most exalted functionaries is no less common. In fact, a man who takes office in Turkey loses from that moment all feeling of security, and tacitly admits the right of the sultan to deprive him at pleasure of his office, his property, and his life. "All the officers of government owe," to use the words of Mr. Thornton, "their appointment to the sole favour of the sultan, without respect to birth, talent, services, or experience. They are deposed and punished without the liberty of complaint or remonstrance, and at their death the sultan inherits their property." (i. 162.)

Formerly the government of pachaliks and other important situations, if not bestowed by the sultan on some of his favourites, were regularly sold to the highest bidders, the leases being renewed annually, provided the pacha or other functionary remitted to Constantinople a sufficient douceur, or, if not, he was superseded by some less parsimonious competitor. And, when in office, the only criterion of an approved administrator was the magnitude of his douceurs, and the amount of tribute he remitted to the public treasury, no inquiry being ever made into the means by which this revenue was made. *Quocunque modo rem* was the brief and comprehensive maxim by which their conduct was regulated. "To rob those below him that he may bribe those above him, is the constant aim and sole object of each petty tyrant, through all the gradations of this baleful despotism." (*Modern Trav., Turkey*, p. 94.) The legitimate revenues of the pachas arise from the rents or produce of lands assigned for their maintenance, and from certain fixed imposts on the cities, towns, and villages of their pachalik. These, however, are in most instances the smallest portion of the revenue of the pachas. The far greater portion consists of illegal demands of all kinds, which the people have no means of resisting. M. Beaujour states that, during the time he resided in Salonica, the pacha enjoyed a revenue of about 140,000 piastres, derived from the rent of land, casualties, and other legitimate sources; and that, in addition to this, he made about 100,000 piastres more by *avanies*, or extortions! And yet this pacha was regarded as a man of singular justice and humanity! (*Tableau du Commerce de la Grèce*, i. 47.) Judge, then, what must be the state of a province governed by a covetous and rapacious pacha, which is the general character of these functionaries.

The flagrant abuses consequent on such a system have brought the Turkish empire to its present state of weakness and degradation! and the necessity of making some very decided changes in the administration has been long obvious. In consequence, a *hatti scheriff*, or imperial decree, was issued 3d November, 1839 (see *post*), which if it could be *bona fide* carried into effect, would go far to suppress most part of the existing abuses, and to introduce security and good order. But, unfortunately, the age of miracles is past, and nothing short of a miracle would suffice for the regeneration of the Turkish empire. Some of the grosser abuses may be suppressed; but, speaking generally, they

3 F

are too deeply seated, too much interwoven with the religion of the country and the constitution of society, to allow any one to suppose that they can be extirpated otherwise than by the agency of some tremendous revolution, that should overthrow everything that now exists. There is neither public virtue nor knowledge in Turkey sufficient to accomplish any considerable reforms. Corruption and venality are everywhere the order of the day; and M. Boué, who has very recently been in Turkey, and is well acquainted with the people and their institutions, affirms that now, as heretofore, the important places of pachas and cadis are sold to the highest bidder, *sont, pour ainsi dire, des enchères au plus offrant!* (Bi. 224.) And he farther affirms, that the most trifling as well as the most important affairs are all settled by the intervention of *douceurs!* A sovereign, with the absolute power, sagacity, and sternness of purpose of Peter the Great, might, perhaps, effect a substantial reform of the Turkish government; but to suppose that such a sovereign should be bred in the slothful luxury of the seraglio, is a contradiction and an absurdity.

In addition to the various sources of weakness and decay originating in vicious institutions and a bad system of government, may be added the imperfect subjugation of the countries comprised within the limits of the empire. The inhabitants of several districts, both of European and of Asiatic Turkey, enjoy, some almost a total and others a sort of semi-independence; forming so many asylums to which discontented and rebellious subjects from the adjoining provinces may retreat and form new schemes, and holding out the seductive and dangerous example of successful resistance. Exclusive of Servia, which is now only nominally under the Porte, there are numerous districts in Albania and Thessaly that are all but independent. Of these, the most important is the country called Myrditz, or the mountainous country occupied by the Myrdites on both sides the Drin. They can bring 10,000 men into the field, pay no taxes, and do not even allow a single Turk to remain within their boundaries! In many parts of Epirus there are similar independent communities. The extensive district of Montenegro, under the government of a bishop, is also substantially independent; as are several districts in other parts of the country. In Asiatic Turkey the Druses, Maronites, &c., in Syria, have succeeded in maintaining their independence; and many of the Turkman and other tribes found in Asia Minor are only nominally dependent on the Porte. (*Boué,* iii. 192-209, &c.) With such elements of disorder scattered over the whole face of the country, the only wonder is that anarchy and insubordination are not more widely diffused than we find them to be.

Mr. Urquhart devotes a large portion (caps. ii. iii. and iv.) of his work on Turkey to a description of its municipal establishments, of which he greatly overrates the importance. No doubt, however, they are the best part of the Turkish institutions. They form, as it were, so many little republics. Their authorities or elders, which are freely chosen by the inhabitants, assess and collect the poll, house, and land taxes; manage the municipal funds, arbitrate in petty matters, &c. These communities are modelled upon a plan similar in almost all respects, to the village system of Hindostan.

Justice.—The Ottoman empire is governed by a code of laws called *multeka,* founded on the precepts of the Koran, the oral laws of Mohammed, his traditions, usages and opinions, together with the sentences and decisions of the early caliphs, and the doctors of the first ages of Islamism. This code comprises a collection of laws relating to religious, civil, criminal, political, and military affairs; all equally respected as being theocratical, canonical, and immutable, though obligatory in different degrees, according to the authority which accompanies each precept. In some instances it imposes a duty of eternal obligation, as being a transcript of the divine will revealed to the prophet: in others it invites to an imitation of the prophet in his life and conduct. And though to slight the example be blamable, it does not entail upon the delinquent the imputation or penalty of guilt; while the decisions of doctors on questions that have arisen since the death of the prophet are of still inferior authority. When a matter occurs that has not been foreseen or provided for by the early promulgators of the law, the Sultan pronounces a decision; and his authority is absolute in all matters that do not interfere with the doctrines or practical duties of religion. The code *multeka* is, however, alone considered as paramount law: the decisions or decrees of the sultan (hatti scheriff), of which a compilation was made by Solyman the Magnificent, under the name of *canon nameh,* are considered as emanations from human authority, and, as such, are susceptible of modification, or even abolition, remaining in force only during the pleasure of the sultan or his successors. (*Thornton,* i. 107, &c.) The *adets* or provincial customs are also allowed considerable influence.

In all the districts and towns of the empire, justice is administered by judges (*cadis*), who are of five different ranks, according to the importance of the place in which they are established, each cadi being assisted by a deputy, or *naib.* Nothing can be more simple and expeditious than the form of proceeding in Turkish courts. Each party represents his case, unassisted by counsellors, advocates, or pleaders of any kind, and supports his statements by the production of evidence. The deposition of two competent witnesses is admitted as complete legal proof, in all cases whatever.

The promptitude of Turkish justice has been often praised; but though dilatoriness be, in this respect, highly blamable, we apprehend that it is a far less evil than the other extreme. In Turkey no ordinary legal authority can detain an untried man in prison more than three days; and in criminal cases the execution of sentences follows close upon the decision of the judge; but neither of these regulations appears to be advantageous; for, in the one case, sufficient time is not allowed to prepare either a defence or an accusation; and in the other, the immediate execution of the sentence prevents the power of appeal to a superior tribunal, and consequently takes away the only means of getting an unjust decision reversed, and, what is of more consequence, an unjust or ignorant judge exposed and degraded. In the greater number of civil cases appeals may be made from the *naib* to the *cadi,* from the latter to the *cadi-asker,* or judge of the province, and hence to the sultan. The latter, however, is rarely practised; and is effected only by presenting a petition for redress to his highness on his way to the mosque. Bastinado, fine, imprisonment, the galleys, and capital punishment, by hanging, drowning, beheading, or strangling, are the principal means of disposing of criminals. Death is sometimes awarded for what we should consider comparatively trivial crimes, as, for instance, unfair dealing on the part of tradesmen; though a butcher or baker convicted of short weight is more frequently nailed by the ear to the door of his shop.

Speaking generally, the administration of justice is in the most disgraceful state in Turkey. According to law, all the judgments of the pachas and of their deputies should be submitted to the cadi, and can only be legally carried into effect when approved by the latter. But, in practice, this salutary regulation is generally disregarded, and in most cases the sentences of the pachas are executed whether they be approved by the cadi or not. (*Boué,* iii. 351.) But the grand vice of Turkish justice consists in the venality of the judges and the toleration of perjury. "The monarch's despotism is not the greatest evil in Turkey: his subjects would, perhaps, bear that without much murmuring or great distress. The radical destruction of all security lies in the iniquitous administration of their laws, which are as impending sword in the hand of corruption, ever ready to cut off their lives and properties." (*Porter,* ii. l.) Mr. Thornton seems to think that Turks have rarely to complain of injustice, and that, speaking generally, the decisions of the judges, in cases where both parties are Mussulmen, are fair and impartial. We are assured, however, by those well acquainted with the fact, that this statement must be received with great modification, and that a rich or powerful Turk has, in most instances, little or no difficulty in obtaining a decision in his favour, however unjust his cause; and that as respects Christians and Jews, they have no chance in a litigation with a Turk, unless they succeed beforehand in securing the good offices of the judge. It is a principle of Turkish law that written testimony is of no avail when opposed to living witnesses; and hence every precaution should be taken to render the latter trustworthy. But, instead of this, the most detestable perjurers enjoy an all but total impunity, and carry on a lucrative as well as an infamous profession. False swearing is punished by leading the culprit through the streets seated on an ass, with his face turned to the animal's tail; and even this punishment, trifling as it is when imposed on such wretches, is rarely enforced. Magistrates are compelled to decide according to the evidence of notorious perjurers, unless they detect their falsehood at the moment. The subjects of foreign powers residing in Turkey are allowed, in virtue of treaties to that effect, to support their claims by written evidence. (*Thornton,* i. 196, &c.)

M. Boué, whose remarks on the administration of justice are as superficial as can well be imagined, is good enough to inform his readers that *si on se permet pas citer de justes témoins, on ne témoins subornés à prix d'argent, et même des juges qui se laissent gagner, la justice Turque ne ferait tout aussi bien ce nom que la nôtre!* (iii. 353.) True, and on the same principle we might say that if A. were not a thief, he would be as honest as B.

There is a considerable discrepancy in the accounts of the state of the police in Turkey, though most recent travellers say that it is extremely defective. No doubt, however, considering the absence inherent in every department of the administration, it is superior to what might have been expected. This is mainly ascribable to the regulation which makes every district of the country responsible for

TURKEY.

all the murders, robberies, and other crimes of violence committed within its bounds; and which consequently makes their repression the business of all the more respectable inhabitants.

Owing to the jealousy of the Turks of the invasion of their privacy, no writ of search can at any time be executed in the interior of the house of a Turk but in the presence of the IMAM; nor in that of a Christian, except accompanied by a priest, nor of a Jew unless a rabbi be present. The rooms occupied by the women, which are never entered, frequently shelter criminals.

Army.—The Turkish land forces may comprise, exclusive of artillery, about 100,000 infantry, of whom about 70,000 are regulars (*nizam*), and about 100,000 cavalry, of which by far the largest proportion formerly consisted of irregular troops, partly and principally furnished by the spahis, and other holders of estates, on condition of military service; but of late years the number of this description of troops has been greatly reduced. Previously to 1826, the janizaries formed the nucleus and main strength of the Turkish armies; and, though now destroyed, a short notice of that once famous militia, long the terror and scourge of Christendom, may not be unacceptable.

The most probable opinion seems to be that the janizaries were originally established by Amurath I., in 1362, and consisted at first of about 13,000 Christian captives, who were renewed by incorporating with them a fifth part of the prisoners of war. "But when the royal fifth of the captives were diminished by conquest, an inhuman tax of the fifth child, or of every fifth year, was rigorously levied on the Christian families. At the age of 12 or 14 years, the most robust youths were torn from their parents: their names were enrolled in a book; and from that moment they were clothed, taught, and maintained for the public service. According to the promise of their appearance, they were selected for the royal schools of Brusa, Pera, and Adrianople, entrusted to the care of the pachas, or dispersed in the houses of the Anatolian peasantry. It was the first care of their masters to instruct them in the Turkish language: their bodies were exercised by every labour that could fortify their strength; they learned to wrestle, to leap, to run, to shoot with the bow, and afterwards with the musket; till they were drafted into the chambers and companies of the janizaries, and severely trained in the military or monastic discipline of that order. The youths most conspicuous for birth, talents, and beauty, were admitted into the inferior class of the *agiamoglans*, or the more liberal rank of *ichoglans*, of whom the former were attached to the palace, and the latter to the person of the prince." (*Gibbon*, chap. 65.)

It is needless to add that the janizaries were taught to believe implicitly in the doctrines of Mohammedanism; and, having no relations or family ties to bind them to society, they regarded themselves not merely as the soldiers, but as the children of the sultan. They enjoyed, partly by the voluntary and partly by the forced concessions of their sovereigns, several valuable privileges and immunities. They formed the body guard of the sultan; they were stationed in the capital, and paid, not like the rest of the army, by assignments of land, but in money; and their captain was one of the most important officers in the public service.

But though formidable at first only to the enemies of the country, they gradually became hardly less formidable to their sovereigns. According as the severe discipline by which the Turkish armies had been originally distinguished was relaxed, and the sceptre fell into feebler hands, the janizaries became unruly, insolent, and overbearing. On various occasions they insulted the majesty of the throne, and, in 1623, they even proceeded to depose Osman II. Hence it had long been a favourite object with the sultans to endeavour to weaken the force and influence of the janizaries. But their efforts in this view had, until very recently, but little success. Selim III. having endeavoured to countervail the influence of the janizaries, by creating a regular army (*nizam*,) the former mutinied, and Selim lost his life in the commotions that ensued. But his successor, the late sultan, was more fortunate in his struggle with this unruly soldiery. In 1826 he issued a *hatti scheriff*, directing that the janizaries should be incorporated with the regular troops. The janizaries refused to obey this order; but the sultan having previously secured the co-operation of the mufti and of their agha, they were completely defeated; and such of them as escaped the conflict in which they were involved*, were deprived of their former insignia, and distributed among the new regiments of the line, so that there is now hardly a trace to be found of this once powerful force.

Previously to and since the destruction of the janizaries,

* The number of janizaries who fell in this conflict, or massacre, as it has been called, has been grossly exaggerated. Marshal Marmont says that not more than 600 lost their lives in Constantinople in the conflict, and in the punishments by which it was followed. A few were killed in other parts of the empire.

it has been a favourite object with the last and present sultans to organize and discipline their troops after the European fashion. But it does not appear that hitherto their efforts have been attended with much success; and it seems doubtful whether the troops have gained so much by the change in their discipline and tactics, as they have certainly lost in the decline of their enthusiasm and sense of nationality. They are now fully aware that they have become the pupils and copyists of those very infidels they were so long accustomed to hold in contempt; and that circumstance has made them lose that confidence in themselves, for the want of which it is very difficult, if not impossible, to compensate.

The regular troops are raised by a conscription among the Turkish part of the population. Inasmuch, however, as little or no attention is paid to the number of children in a family, or even to the health, size, or age of the conscripts, the conscription is found to be an intolerable hardship; and the recruits produced by it are of the most motley kind, youths of 15 and 17 being intermixed with men of 40, 50, and 60 years of age! The mortality among the conscripts is very heavy. According, however, to the hatti scheriff of 1839 (see post), the defects in the present recruiting system are to be obviated; and the period of service is to be reduced to four or five years.

The uniform of the regular troops is blue, faced with red. The household troops, or body guard of the sultan, comprise the *élite* of the army: their pay and appointments are better than those of the other troops. Marshal Marmont, whose intelligence and experience entitle his opinion on military matters to the greatest weight, makes the following statements with respect to the state of the Turkish troops:

"The lot of the Turkish soldiers is a very happy one. They are better fed than any other troops in Europe, having an abundance of provisions of excellent quality, and partaking of meat once, and of soup twice a day. Their magazines are filled with stores, and the regiments have large reserves. The pay of each soldier is 20 piastres per month*; the whole of which he receives, as there is a prohibition against withholding from him any part of that sum. In short, everything has been effected that could promote the welfare of the soldier.

"If no fault can be found on the score of the *matériel*, much is to be said against the *personnel* of this force. On the arrival of Achmet Pacha, we repaired to the exercising ground. Four battalions were in line; and, after inspecting them, they manœuvred before me. Nothing could be worse than this exhibition; indeed, these men ought not to be looked upon as troops, but merely as a mass of people, bearing the stamp of misery and humiliation; and they are evidently depressed by a knowledge of their own weakness. They all seem to have a willingness about them, but feel ashamed of their occupation; and, from the private to the colonel, not an individual among them has any conception of his duty. Moreover, the men are diminutive in stature, and wretched in appearance; many of them are too young for service; and we are led to inquire what is become of that noble Turkish people, the lofty, proud, majestic, handsome race of former days; for now we find no trace of them in the existing troops.

"I have endeavoured to discover why they have not hitherto succeeded better with the new system, and I thus account for the failure. The sultan was desirous of organizing troops according to the European mode; and his ambition was to form an army on the instant. He accordingly raised at once a great number of regiments; but the instructors, being generally individuals of an inferior station of life, without capacity or talent, who had been led to Constantinople by the circumstances which attend revolutions, were unfitted to accomplish the object in view.

"The new organization commenced simultaneously in all the corps; and the same description of persons were universally employed in endeavouring to carry it into effect. In none of the grades had any man confidence either in himself or in others; and no one, therefore, had a right to the command, which should always be derived from some superior claim. It is only as a consequence of such a principle that men are ever found disposed to yield obedience. In the troops of all the other powers of Europe, there are two admitted titles to precedence—birth and merit. The former has its basis on a higher social grade, which, by giving opportunities for better education, leads to the expansion of the mind; the latter, on the experience and information resulting from previous service. In Turkey there are no gradations in the social order, and the son of the water-carrier is on a par with the vizier's child, having often the same education. Hence, there is no admitted superiority in those invested with power; and the previous equality

* Strictly speaking, this is the pay of the guards only; the others receive 15 piastres per month. (*Soup*, iii., 230.)

971

indisposes others to obey authority obtained through mere caprice.

"As to the right derived from merit or experience, there can be none where all are novices.

"Such were the radical defects that prevailed in the formation of the Turkish army. The remedy would be to reduce things to their elements; and to re-commence by establishing in public opinion a respect for talent and capacity, in order to obtain that obedience and confidence in superiors, without which an army cannot exist: for it is such confidence that produces discipline and order, and creates the moral power requisite to give unity, compactness, and energy to the whole.

"If, instead of attempting to raise an army, as it were, by a mere decree, the sultan had been content with forming a single battalion, and had obtained the services of 30 or 40 really good officers, and a chief capable of comprehending the importance of his duties, it is probable that, in two years, he would have succeeded in producing a battalion to serve as a model for the rest; and this result once obtained, the sultan would have possessed the elements required. At the end of six months, or, at the utmost, of one year, by adding to the numbers of those first enrolled, and dividing the whole into two battalions, he might have formed a complete regiment; for the men of the first levy would, in the eyes of the recruits, have appeared as old and instructed soldiers. It is obvious that in ten years he would then have obtained an army. Whereas, according to the system followed, such a result is improbable, for an union of men like the present cannot be said to merit this title." (*Translation by Smith*, p. 61, &c.)

If the opinions of so eminent a judge as Marshal Marmont of the bad quality of the Turkish troops required any confirmation, it would be found in the history of their campaigns with the Egyptians. How superior soever in point of numbers, they never were able to make any head against the latter, till supported by European troops. The irregular troops consist principally of Kurds and Albanians, who, when properly commanded, make excellent troops.

The same corruption that infects the other departments of the Ottoman government, extends to that of the army. The pachas, commanders of regiments, and other functionaries, contrive to make large sums by keeping on their books a greater number of troops than they really have, and putting the pay and other emoluments drawn on their account into their pockets. At inspections, the place of the deficient troops is supplied by substitutes hired for the occasion, who disperse as soon as the inspecting officers have retired! (*Boué*, iii., 332.,

The attempts to reform the Turkish army would really, therefore, appear to have been one of the most abortive that has ever been made. The nationality, fanaticism, and confidence of the troops in themselves has been destroyed, and nothing but a miserable imitation of European tactics and discipline substituted in its stead. The ignorance of the officers, and the mischievous principle, if we may so call it, on which they are selected, were the first evils that should have been corrected; and, till they have been obviated, no other improvement can be of the smallest consequence. Marshal Marmont speaks very favourably of the school established at Constantinople for the instruction of the guards; but supposing it and other schools to be kept up, a lengthened period must elapse before they can have any material influence. Unless, indeed, the sultan should be able to avail himself, which probably the prejudices of his subjects will not permit, of the services of European officers in the command and organization of his army, we apprehend that it is not destined speedily to acquire any considerable degree of efficiency or strength.

Navy.—It is unnecessary to enter into any lengthened details with respect to the Turkish navy. At no time has Turkey been considerable as a naval power; and as the Turks have no taste for the sea, her best sailors have always been Greeks, Christian slaves, or renegades. In 1770 the Turkish fleet was destroyed by the Russians in the harbour of Tcheshmeh; and the defenceless state in which the coasts, and even the capital, were then placed, awakened the attention of government to the subject. Since that period many fine line of battle ships have been built in the Turkish ports, principally under the direction of Europeans; but, owing to the unskilfulness of the crews, and the all but total ignorance of the officers, most of whom have not been bred to the sea, and were appointed to be lieutenants and captains, can hardly distinguish between a rudder and a main-mast, the money laid out on the ships has been little better than thrown away. The battle of Navarino inflicted a severe blow on the Turkish navy, and soon after the accession of the present sultan, the capitan-pacha, or high-admiral, carried off the fleet to Alexandria, and delivered it up to Mehemet Ali! The latter, however, has since been obliged to return it. The emancipation of Greece, which formerly furnished

the best sailors to the Turkish fleet, has been a serious injury to the latter.

Houses and Mode of Life.—The houses of the Turks are built in contempt of all architectural rules. They are mostly only of one story, and are very rarely more than two stories in height, constructed of wood and sun-dried bricks, those of the better class being plastered and painted over on the outside. The windows when they open on a street or other exposed situation, are uniformly covered with lattice-work, which prevents the most inquisitive eyes from obtaining even a glimpse of what is going on within. But though mean and shabby on the outside, the houses of the more opulent Turks are often very sumptuously fitted up in the interior. The most convenient and magnificent apartments belong to the *harem*, or to the portion of the house appropriated to the exclusive use of the women; and this very frequently opens on a court having a fountain in the middle, and sometimes on a garden. The houses of the poorer classes are most uncomfortable, their windows being generally without glass, and their rooms without fire places. In winter they are usually heated by means of braziers, or pans of charcoal, which suffocate while they warm.

Lady Mary W. Montague has given a lively and accurate description of the houses of the higher class of Turks. "Every house," says her ladyship, "great and small, is divided into two distinct parts, which only join together by a narrow passage. The first house has a large court before it, and open galleries all round it, which is the very agreeable. This gallery leads to all the chambers, which are commonly large, and with two rows of windows, the first being of painted glass: they seldom build above two stories, each of which has galleries. The stairs are broad, and not often above thirty steps. This is the house belonging to the lord, and the adjoining one is called the harem, that is, the ladies' apartment (for the name of *seraglio* is peculiar to the grand seignior); it has also a gallery running round it towards the garden, to which all the windows are turned, and the same number of chambers as the other, but more gay and splendid, both in painting and furniture. The second row of windows is very low, with grates like those of convents; the rooms are all spread with Persian carpets, and raised at one end of them (my chambers are raised at both ends) about 2 ft. This is the sofa, which is laid with a richer sort of carpet, and all round it a sort of couch, raised half a foot, covered with rich silk, according to the fancy or magnificence of the owners; mine is of scarlet cloth, with a gold fringe; round about this are placed, standing against the wall, two rows of cushions, the first very large, and the next little ones; and here the Turks display their greatest magnificence. They are generally brocade, or embroidery of gold wire upon white satin; nothing can look more gay and splendid. There seats are also so convenient and easy, that I believe I shall never endure chairs as long as I live. The rooms are low, which I think no fault; and the ceiling is always of wood, generally inlaid or painted with flowers. They open in many places with folding doors, and serve for cabinets, I think, more conveniently than ours. Between the windows are little arches to set pots of perfume, or baskets of flowers. But what pleases me best is the fashion of having marble fountains in the lower part of the room, which throw up several spouts of water, giving at the same time an agreeable coolness, and a pleasant dashing sound, falling from one basin to another. Some of these are very magnificent. Each house has a bagnio, which consists generally in two or three little rooms, leaded on the top, paved with marble, with basins, cocks of water, and all conveniences for either hot or cold baths.

"You will, perhaps, be surprised at an account so different from what you have been entertained with by the common voyage-writers, who are very fond of speaking of what they don't know. It must be under a very particular character, or on some extraordinary occasion, that a Christian is admitted into the house of a man of quality; and their harems are always forbidden ground. Thus they can only speak of the outside, which makes no great appearance; and the women's apartments are always built backward, removed from sight, and have no other prospect than the gardens, which are enclosed with very high walls. They have none of our parterres in them; but they are planted with high trees, which give an agreeable shade, and, to my fancy, a pleasing view. In the midst of the garden is the *chiosk*, that is, a large room, commonly beautified with a fine fountain in the midst of it. It is raised nine or ten steps, and enclosed with gilded lattices, round which vines, jessamines, and honeysuckles, make a sort of green wall. Large trees are planted round this place, which is the scene of their greatest pleasures, and where the ladies spend most of their hours, employed by their music or embroidery. In the public gardens are public *chiosks*, where people go that are not so well accommodated at home, and drink their coffee, sherbet, &c."

Owing to the houses being mostly built of wood, fires are very frequent at Constantinople, and have sometimes been so very extensive as to threaten destruction to the entire city. The sultan generally attends in person to superintend the efforts made to suppress the fury of the flames. When rebuilt, little or no alteration is ever made in the form of the streets.

It should, however, be observed, as especially marking the character of the Turkish government, that these fires are not always accidental. Indeed, how singular soever the circumstance may appear, there can be no doubt that fires in Constantinople are made to perform the functions of petitions and public meetings in England! In fact, the city has been set on fire, over and over again, for a number of nights together, in order that the grand seignior may be made aware of the deep discontent of his subjects, and of their dissatisfaction with his measures or his favourites, or both. The frequency and continuance of the fires evince their origin; and they have seldom failed to produce a change in the measures of government, and the dismissal or execution of the unpopular favourites! (*Porter,* i., 160; *Thornton,* i., 187, &c.)

Public baths and *khans* are varieties of public buildings, that are found in most parts of Turkey. The use of the warm bath is universal among persons of both sexes, and all classes. Many of the public baths are handsome, and a few are really magnificent structures. They are mostly built of hewn stone, and comprise several apartments. "On entering one of these establishments, the visiter is conducted into a spacious and lofty hall, lighted from above: round the sides are high and broad benches, on which mattresses and cushions are arranged: here the bather undresses, wraps a napkin about his waist, and puts on a pair of wooden sandals before going into the bathing-room. The first chamber is but moderately warm, and is preparatory to the heat of the inner room, which is vaulted, and receives light from the dome. In the middle of the room is a marble estrade, elevated a few inches: on this the bather stretches himself at full length, and an attendant moulds or kneads the body with his hand for a considerable length of time. After this operation, the bather is conducted into one of the alcoves, or recesses, where there is a basin supplied by pipes with streams of hot and cold water; the body and limbs are thoroughly cleaned by means of friction with a horse-hair bag, and washed and rubbed with a lather of perfumed soap. Here the operation ends: the bather stays a few minutes in the middle chamber, and covers himself with dry cotton napkins: thus prepared he issues out into the hall, and lies down on his bed for about half an hour." (*Thornton,* ii., 202.)

The baths for ladies are similar, in most respects, to those for the other sex; but are more handsomely fitted up. Lady Mary W. Montague visited one of these baths at Adrianople, and has given an interesting account of it, and of the reception she met with from the Turkish ladies.

Khans are a description of public inns, or caravanserais, sometimes built by sultans and munificent individuals, for the public use and accommodation; and sometimes constructed, as in England, on speculation. They are of very various kinds. Exclusive of apartments for the use of travellers, and stables for their horses and camels, the larger khans have rooms in which the goods of merchants may be stored up. These are generally quadrangular structures, consisting of a series of apartments that open upon a terrace, which surrounds an inner court, and having stables in the back part of the building. The merchants store their goods in separate apartments, or in the rooms which they occupy; the muleteers, with their horses, encamp in the open air in the court, or retire to the stables; and the gateway, by which alone the court and rooms can be entered, being shut up at night, all are as safe as if they were in a fortress. In many towns these are the only taverns. Each khan has its *khanjy* (landlord), a *kakbia* (major-domo), a *khawajy* (coffee-maker), and an *oda-basher* (waiter), who attends to the commands of all the inmates. Sometimes the rooms are furnished, and sometimes not; and frequently, especially in Asiatic Turkey, the apartments are lighted by a window, having paper for glass, opening on the terrace, so that they are, for the most part, dark and gloomy. Food is sometimes, but not generally, furnished in these establishments, the usual method being to have it cooked abroad, and sent in. Coffee, however, is generally prepared in the establishment. (*Missionary Researches,* p. 67.)

Rice is the principal food of the lower orders, but the wealthier classes have a great variety of dishes. The breakfasts of the latter consist of fried eggs, cheese, honey, liban (coagulated milk), &c. The hour of dinner is very early. At entertainments the guests sit cross-legged on sofas or cushions round a low table. In the houses of persons of distinction, handsome ivory spoons (the use of gold or silver for such purpose not being permitted), and small pointed sticks, are laid beside each plate. The dishes are served

singly, and in rapid succession, sometimes to the number of 20 or 30: the guests help themselves, sometimes with their spoons, and sometimes with their fingers. Hashed lamb, poached eggs and lemons, stewed fowls, pigeons, &c.; pillaws, roasted meats, a whole lamb stuffed with rice, almonds, &c., are favourite dishes: they are all highly seasoned with salt and spices, and sometimes with onions and garlic. The dessert consists of sweetmeats, of which the Turks are exceedingly fond; with coffee, sherbet, fresh honey, grapes, figs, &c. During dinner, nothing is drank but water or lemonade. The supper is very similar to the dinner. (*Russell's Aleppo,* p. 105, &c.)

The month of Ramadan is observed as a fast; and from dawn till sun-set, during this month, the Turks neither eat nor smoke. But the moment the sun goes down, they eat a hearty meal; and the practice is, for the richer classes to keep the fast, if we may so call it, by sleeping at this season during the day, and sitting up eating and drinking during the greater part of the night! (*Russell,* p. 108.)

The national dress of the Turks is loose and flowing, that of the women, with the exception of the turban, differing but little from that of the men. The shape and colour of the turban serves to distinguish the different orders of the people, and the functions of public officers. Latterly, however, it has become fashionable to imitate the dress and manners of the other European nations; though the former is inconvenient in consequence of the numerous ablutions, the performance of which is enjoined by the Koran.

Every body knows that when females appear in the streets their faces are carefully veiled. And such is the privacy of the harem or women's apartments, that unless on very rare occasions, all males are excluded from them except the master of the family. "*Les plus proches parens, tels que les frères, les oncles, les beaux pères, n'y sont reçus qu'à certaines époques de l'année, c'est-à-dire, dans les deux fêtes de bayram, et à l'occasion des noces, des couches, et de la circoncision des enfans.*" (D'Ohsson, Tab. Générale, iv. 318.) Polygamy is authorized by the law of the prophet; but is a privilege not often resorted to. If a man marry a woman of equal rank, the marriage of any other wife is frequently guarded against by the marriage contract. In cases of polygamy, the wives are usually either slaves or women of an inferior rank to the husband.

There is a regular slave-market in Constantinople: but slaves in the east, and especially in Turkey, are far from being in the depressed condition we might suppose. The laws of Turkey protect the slave from ill-usage: and, in this respect, the customs of the country are in complete harmony with the laws. "The most docile slave rejects with indignation any order that is not personally given him by his master; and he feels himself placed immeasurably above the level of a free or hired servant. He is as a child of the house; and it is not unusual to see a Turk entertain so strong a predilection for a slave he has purchased, as to prefer him to his own son. He often overloads him with favours, gives him his confidence, and raises his position; and, when the master is powerful, he opens to his slave the path of honour and of public employment. If we seek for any confirmation of the truth of this assertion, let us look around the sultan, and observe who are the most distinguished men within his empire. Khosrew pacha, the old seraskier, the man who has governed and ruled all things in Constantinople, was a slave from the Caucasus, purchased by a Capudan pacha, whose protection has raised him to the highest offices. Halil-pacha, the son-in-law and most distinguished servant of the sultan, and to whom the brightest prospects are open, was a slave to the seraskier." (*Marmont,* Smith's trans., p. 25.)

The Turks are excellent horsemen, and throw the *djerid,* or lance with the greatest dexterity and force; but, excepting this exercise and that of wrestling, they indulge in no active exertion. "Their delight is to recline on soft verdure under the shade of trees, and to muse without fixing their attention, lulled by the trickling of a fountain or the murmuring of a rivulet, and inhaling through their pipe a gently inebriating vapour. Such pleasures, the highest which the rich can enjoy, are equally within the reach of the artizan or the peasant." (*Thornton,* p. 203.) They never dance themselves, but enjoy public dances, the performers in which, however, are reckoned infamous.

Turkish usages are, in truth, in almost all respects, quite the opposite of ours. "Our close and short dresses, calculated for promptitude of action, appear in their eyes to be wanting both in dignity and modesty. They reverence the beard as the symbol of manhood and the token of independence, but they practice the depilation of the body from motives of cleanliness. In performing their devotions, or on entering a dwelling, they take off their shoes. In inviting a person to approach them, they use what with us is considered a repulsive motion of the hand. In writing they trace the lines from right to left. The master of a house does the honour of the table by serving himself first from

TURKEY.

the dish: he drinks without noticing the company; and they wish him health when he has finished his draught. They lie down to sleep in their clothes. They affect a grave and sedate exterior: their amusements are all of the tranquil kind: they confound with folly the noisy expression of gaiety; their utterance is slow and deliberate; they even feel satisfaction in silence; they attach the idea of majesty to slowness of motion: they pass in repose all the moments of life which are not occupied with serious business: they retire early to rest; and they rise before the sun." (*Thornton,* ii. 186.)

Language, Literature, and Education.—The principal languages spoken in European Turkey are: 1. The *Turkish* and *Tartaric* languages, spoken by the Osmanlis, Tartars, and Yaruks (descendants of the Turcoman settlers in Macedonia). The Turkish language is very much intermixed with Arabic and Persian. It is expressive, soft and musical, and easy to speak, but not easily written. Its construction is artificial and laboured, and its transpositions are more remote from the natural order of ideas than the Latin or German. 2. The *Arabic,* the written language, used at court and in public worship. 3. The new *Greek* or *Romaic,* which consists of a great many dialects, and differs from the Hellenic, or ancient Greek, still in some measure preserved among the inhabitants of the Epirotic mountains, and in the valleys of the Cassiodorus (Suli), though greatly intermixed with foreign words and phrases. 4. The *Slavonic,* in several dialects, spoken by the Slavonians, Bosniacs' Croatians, and Bulgarians. 5. The *Armenian.* 6. The *Wlachian,* derived mainly from the Latin, but much intermixed with other languages. 7. The *Arnautic, Skipetarian,* or *Albanian,* is not, as was formerly supposed, a jargon formed of the admixture of a number of languages, but is a peculiar and distinct language, having regular grammatical forms and an essential character of its own. 8. The *Jewish,* i. e. Hebrew, intermixed with Spanish and Italian words. 9. The *Lingua Franca.*

The Turkish characters are, with some slight difference, the same as the Arabic and Persian, but they have a variety of hand-writings. The Arabic Kufi, in later times Meschi (literally the characters used in transcribing), is only used in copying the Koran, and other sacred works. The Diwani is the hand-writing used in business, letters, public documents, judicial proceedings, diplomatic affairs, official orders, passports, &c. It is written from the right to the left in an oblique direction, especially at the conclusion; all the letters are joined to each other, and twisted together, and the more they are so the more elegant is the writing considered. Dsheri is used in patents, diplomas, inscriptions on mosques, burial places, and other public edifices; its beauty consists in its oblique direction, upward and downward, and in the words being placed alternately above and below each other. The Talik writing was originally borrowed from the Persians, and properly signifies a flying or running hand. It is more pleasing to the eye than the other hands, and is a fine specimen of Oriental calligraphy. The Kirma, consisting of disjointed letters, is used in ledgers and registers. The Sulus (i. e. trebly thick) serves for title-pages, devices, and inscriptions upon coins, &c. There are many more varieties of hand-writings, which it is needless to specify. Instead of pens the Turks write with a reed (*Calam, Calamus*), which is cut like our pen, without a slit. The vowels, which, as in Arabic and Persian, are placed above and below the consonants, are generally omitted in writing (the Koran excepted), which renders the reading difficult to an unpractised eye. There are no marks of punctuation. As they are without tables, they usually write upon the left knee or hand, and instead of moving the hand, they move the paper in the process of writing.

The first printing press introduced into Turkey was established at Constantinople, in 1576, by Jews, who were, however, prohibited from publishing any Arabic and Turkish works. About 150 years afterwards, Ibrahim, a Hungarian renegade, succeeded in establishing a Turkish press; and it is worthy of remark, that in the hatti sheriff authorizing its introduction, the Sultan, Achmet III., felicitates himself that Providence had reserved so great a blessing to illustrate his reign, and to draw down upon his august person the benedictions of his subjects and of all Mussulmen, to the end of time. (*D'Ohs., Tab. Gén.* ii. 500.)

Down to 1742, 17 works in 23 volumes had issued from the press. From 1742 until 1755 it was not employed at all, and only at intervals until 1783, when it was attached to the newly-established school of engineers. In the beginning of the present century, it was transferred to Scutari, and attached to the military barracks in that place. The late sultan did much to advance its efficiency and extend its utility. The Turks, however, have a prejudice against printing, originating partly in an apprehension lest the Koran should be printed, which they would regard as the highest profanation, and partly in the opposition of the

vast numbers of scribes and copyists, which the guard use of the press would throw out of employment, and who, in consequence, take every opportunity to inflame the prejudices of their fellow subjects against it; but despite these difficulties, the art has made, and is continuing to make, some progress. A few years ago, a government newspaper was established, entitled, *Tables of Events,* which contains the different resolutions and orders of the divan. The *Moniteur Ottoman,* in the French language, also published at Constantinople, consists only of extracts from the former.

No sooner were the Turks converted to Islamism, that they began not only to study the Koran, but since the occasion of the Arabians, their superiors in civilization. It is a vulgar prejudice to suppose that the Koran discourages learning; on the contrary, "the ink of the learned and the blood of martyrs are," according to the prophet, "of equal value in the sight of heaven." (See the curious article on Kin (science), in the *Bibliothèque Orientale,* i. 698.)

Their favourite studies are law and theology. In the interpretation of the Koran and of the traditions, they follow the Arabian authorities, and most Turkish divines occupy their time with biographies of the prophet, and evidences and reasonings in favour of the Mohammedan religion: these with the innumerable commentaries on the Koran, form a mass of works which fill the greater part of their libraries. History poetry, and philosophy, however, are by no means neglected. Hammer, in his elaborate and valuable *History of the Ottoman Empire,* has consulted an immense number of Turkish historians; and in his *History of Ottoman Poetry* (the first volume of which was published at Pesth in 1836) he gives short sketches of the lives of 212 Turkish poets, with specimens of their works. At the same time, however, Hammer acknowledges that the Turks have no genius for original poetry, and that the whole of their poetry consists of translations from, and imitations of the Arabic and Persian poets, to whom they stand in nearly the same relation that the Roman poets did to the Greek.

Jurisprudence, a favourite pursuit of the Turks, is studied in the works and laws of the learned imams, sheiks and sultans, the traditional law of the prophet (Sunna). The most remarkable printed collections of Fetwas (decisions) are by Mufti Dahemalli, Abdubrahman, and Mustapha Kedssi: the work of the latter, published at Constantinople in 1882, contains several thousand fetwas of 20 muftis in the 18th century. In 1727, there issued from the press 34,000 fetwas by Abdubrahman, from 1645 to 1676, in 2 vols. folio; and in 1830, 5,600 by Ali Effendi, which with the collection of Abdulkerim Effendi, forms a work of high authority.

Turkish literature is particularly rich in collections of bon-mots, puns, proverbs, tales, anecdotes, and even novels; and they possess several encyclopaedias, and works upon the history of literature. The first volume of a bibliographical dictionary, in which are enumerated the titles of more than 30,000 different works in the Arabic, Persian, and Turkish languages, has been translated by Fluegel, and published by the "Oriental Translation Fund." The reign of Solyman the Magnificent may be considered as the Augustan age of Ottoman literature. This great prince was a liberal patron of the arts and sciences, and of literature and learned men.

Public schools are established in most considerable Turkish towns, and *medresses,* or colleges, with public libraries, are attached to the greater number of the principal mosques. But, owing to the total want of efficient masters, and if good elementary books, the instruction afforded by these establishments is of comparatively little value. In schools the pupils are taught to read and write the first elements of the Turkish language; the class-books being the Koran, and some commentaries upon it. In the medresses, which are the colleges or schools of the ulemas, the pupils are instructed in Arabic and Persian, and learn to decipher and write the different sorts of Turkish characters: instruction in a species of philosophy, logic, rhetoric, and morals founded on the Koran; and these, with theology, Turkish law, and a few notions on history and geography, complete the course of study.

"If," says Mr. Thornton, "we call the Turks an illiterate people, it is not because learning is universally neglected by individuals: for, on the contrary, the ulemas, or theological lawyers, undergo a long and laborious course of study; the Turkish gentlemen are all taught certain necessary, and even ornamental, parts of learning; and few children, at least in the capital, are left without some fixture of education. It must be acknowledged, however, that the objects of Turkish study, the rhetoric and logic, the philosophy and metaphysics of the dark ages, do in reality only remove men farther from real knowledge. The learning without which the researches of the soundest natural philosopher would be imperfect, are either entirely unknown in Turkey, or known only as childish playthings, to excite the

* "Geschichte des Osmanischen Reiches, 10 vols. 8vo. Pesth, 1822-34.

992

admiration of ignorance, or to gratify a vain curiosity. The telescope, the microscope, the electrical machine, and other aids to science, are unknown as to their real uses. Even the compass is not universally employed in their navy, nor are its common purposes thoroughly understood. And it may be truly said, that navigation, astronomy, geography, agriculture, chemistry, and all the arts which have been, as it were, created anew since the grand discoveries of the two last centuries, are either unknown, or practiced only according to a vicious and antiquated routine." (1. 29.)

The facts and détails given by M. Boué show that this statement is as applicable to the present period, as it was to that when it was written, nearly 40 years ago. At this moment (1842) the useful sciences are, without exception, in the most abject state. A school for medicine and surgery transferred, in 1839, to Galata Seraï, the school for the guards already alluded to, and a naval school, are the only establishments in which any attempt is made to supply really useful information. But even these are very far behind. The great deficiency, according to Boué, consists in the want of good elementary books; and he justly thinks that it would be of the greatest importance to get the best elementary works on the different branches of science translated into Turkish, either by native Turks, or by foreigners acquainted with the language. In 1839, the late sultan appointed a commission for the establishment of useful institutions; which among other projects, proposed to found scientific academies on an improved plan, at Constantinople, Smyrna, Adrianople, and other large towns. But nothing of the sort has hitherto been carried into effect.

It must, also, be borne in mind that Turkish schools are mostly attended by boys only. In Turkey education is not considered necessary to a girl; so that by far the greater number of women, knowing little or nothing themselves, can communicate nothing to their children.

Taxation.—The Turkish exchequer consists of two parts; the *miri*, or public, and the *hazné*, or sultan's private treasury. The former is derived from various sources, of which the principal are, 1st, the *haratch*, or poll-tax, imposed with very few exceptions, on all the males in the empire, not Mohammedans, between the ages of 7 and 60 years. Formerly the tax varied, under different circumstances, from 3 to 14 piastres per individual; but the value of money was greater then than now, and at present it varies, in different provinces and under different circumstances, from 10 to 60 piastres. Various districts compound for the poll-tax, the amount of which is then added to the land-tax. Mussulmen, though exempted from the poll-tax, pay the *averici*, amounting to five eighths of the haratch; 2d, the land-tax, of one tenth of the produce of the estates not subject to military service; 3d, taxes on movables, such as cattle, sheep, and goats, assessed taxes, &c.; 4th, customs' duties, and octrois; and 5th, the excise upon gunpowder, tobacco, salt, wine, &c. No authentic details have been given of the amount of the miri, and the estimates put forth by the best informed writers differ extremely. Probably, however, we shall not be far wrong if we estimate it at about £3,000,000 sterling a year.

The *hazné*, or private revenue of the sultan is derived partly from the imperial domains, or estates belonging to the crown; partly from the sums paid by the pachas and other dignitaries, on their accession to office, from presents from the same parties, on occasion of the beyram, and other public festivals, and from confiscations and inheritances; and partly from the contributions paid by the tributary provinces of Servia, Wallachia, &c. The female branches of the imperial family have their own especial revenues derived from lands appropriated to their use, or from peculiar taxes.

But in addition to the taxes now noticed, the subjects of the Porte are liable to be called upon, at the pleasure of the pachas, for contributions in kind for the maintenance of troops passing from one part of the country to another, for *corvées*, or requisitions of forced labour, and, in time of war, for forced loans, &c. In fact, the pachas and their satellites squeeze out of the people all that they possibly can, without inquiring or caring whether their demands be legal or otherwise. They know right well if they pay the stated amount of tribute, and secure by bribes and presents the good opinion of those in power, they may fleece the cultivators without let or hindrance. Since the time of Mohammed II. the revenues of Turkey have been farmed, or let to the highest bidder.

Exclusive, too, of the taxes on account of the general government, and the *avanias* of the pachas, the people have to provide for their local and municipal expenses, which are so very heavy as considerably to exceed the amount of the miri. (*Boué*, iii. 233, &c.)

The more intelligent Turks have long been sensible that the arbitrary power of the pachas, and especially their *avanias*, or extortions, was the most objectionable part of their political system, and would, if not redressed, terminate in universal poverty and disaffection. So early as 1660 the inconveniences of the existing order of things had become apparent, and it was then proposed to obviate them by commuting the different taxes on the rayas, or cultivators, and their contributions of forced labour and horses, for a single tax; which it should not, in any case, be permitted to exceed. (*Ham. Osmanische Geschite*, vi. 551.) Nothing, however, was done; and abuses of all sorts have continued to multiply according as the machine of government became relaxed down to the present day. The haïti scheriff previously alluded to as having been issued on the 3d of Nov., 1839, is principally directed to the obviating of these abuses. But, how indispensable soever, we apprehend it will be found to be wholly impracticable, so long as the present form of government is maintained, to remedy the abuses in question. All the agents of a despot are despots in their peculiar sphere; and though the sultan may be most anxious to suppress abuse, he is neither omnipresent nor omniscient, and it is to be feared that the same means that have hitherto maintained the pachas in power despite their oppressions, will be effectual for the same purpose in time to come. We should be happy to be able to think otherwise; but our firm conviction is that self reform is impossible in Turkey, and that the abuses of which she is the victim, will necessarily increase in number and virulence, till foreign force or domestic violence overthrow the religion and government, of which they are the bitter but legitimate fruits. We subjoin the Haïti Scheriff, or Imperial Decree, now alluded to. It is, at all events, a most remarkable document; and the fact of its having issued from a Mahommedan government is not the least singular circumstance in the history of the present times.

"Every one knows that in the beginning of the Ottoman Empire, the glorious precepts of the Koran and the laws of the empire were held as rules always revered, in consequence of which the empire increased in strength and greatness; and all its subjects, without exception, attained the highest degree of welfare and prosperity. Within the last 150 years a series of events and variety of causes have, from not abiding by the holy code of laws, and the regulations that arose from it, changed the welfare and strength into weakness and poverty. Thus it is that a nation loses all its stability by ceasing to observe its laws. These considerations have constantly presented themselves to our notice, and since the day of our accession to the throne, the public weal, the amelioration of the state of the provinces, and the relief of the people, have never ceased to occupy our thoughts. Bearing in mind the geographical position of the Ottoman empire, the fertility of its soil, the aptitude and intelligence of its population, it is evident that by bringing into operation efficacious means, we may obtain, by the assistance of God, the object we hope to insure, perhaps in the space of a few years. Thus, full of confidence in the Almighty, and relying on the intercession of our Prophet, we deem it necessary to seek, by new institutions, to procure to the state which compose the Ottoman empire the happiness of a good administration.

"These institutions should have three objects in view:—first, to guarantee to our subjects perfect security of life, honour, and property; secondly the regular levying and assessing of taxes; and thirdly, a regular system for the raising of troops, and fixing the time of their service.

"For, in truth, are not life and honour the most precious of all blessings? What man, however averse his disposition to violent means, can withhold having recourse to them, and thereby injure both the government and his country, when both his life and honour are in jeopardy? If, on the contrary, he enjoys in this respect full security, he will not stray from the paths of loyalty, and all his actions will tend to increase the prosperity of the government and his countrymen. If there be absence of security of property, every one remains callous to the voice of his prince and country. No one cares about the progress of the public good, absorbed as one remains with the insecurity of his own position. If on the other hand, the citizen looks upon his property as secure, of whatever nature it be, then, full of ardour for his interests, of which for his own contentment he endeavours to enlarge the sphere, thereby to extend that of his enjoyments, he feels every day in his heart the attachment for his prince and for his country grow stronger, as well as his devotedness to their cause. These sentiments in him become the source of the most praiseworthy actions.

"The assessment of regular and fixed taxes is a consideration of vital importance, since the state, having to provide for the defence of its territory, can only raise the means necessary for the maintenance of the army by contributions on the people. Although, thanks be to God, the inhabitants of this country have lately been freed from the curse of monopolies, formerly improperly looked upon as a source of revenue, a fatal practice still remains in force, although it cannot fail to give rise to the most disastrous consequences. —it is that of venal corruption, known under the name of

Htizam. According to this system of civil and financial practice, a district is abandoned to the arbitrary rule of one individual, but too often notorious for his rapacity, and the most cruel and most insatiable disposition; for, should this farmer of the revenue not be a virtuous man, he will have no other care but that tending to his own advantage.

"It becomes, then, necessary for every member of the Ottoman society to be taxed according to a fixed rate, in proportion to his means and circumstances, and that nothing further should be exacted from him, and that special laws should also fix and limit the expenses of our army and navy.

"Although we have already observed the defence of the country is a most important consideration, it becomes the duty of the inhabitants to supply soldiers to that object: it becomes essential to establish laws to regulate contingents which each district is to supply, according to the urgency of the moment and to reduce the time of the military service to four or five years, for it is at the same time doing an injustice, and inflicting a mortal blow on agriculture and industry, to take, without regard to the respective populations of each district, from one more, from other fewer men, than they can afford to provide, and it is also reducing the soldiers to despair, and contributing to the depopulation of the country, to retain them all their lives in the service. In short, without the different laws of which the necessity has been shown, there is neither strength, riches, happiness, nor tranquillity for the empire, and it has to expect these blessings as soon as these laws come into operation.

"It is therefore that in future the cause of every individual shall be tried publicly, according to our divine laws, after mature inquiry and examination; and till a regular sentence has been pronounced, no one shall have it in his power, either secretly or publicly, to put an individual to death, either by poison or by any other means.

"It is not permitted to attack the honour of any individual, unless before a court of justice.

"Every individual shall be allowed to be master of his own property, of whatsoever kind, and shall be allowed to dispose of it with full liberty, without any obstacle being offered by any one. For instance, the innocent heirs of a criminal shall not forfeit their right to his property, nor shall the property of a criminal be any longer confiscated.

"These imperial concessions extend to all our subjects, of whatever religion or sect they may be, and these advantages, they shall without exception, enjoy.

"Thus we grant full security to the inhabitants of our empire of life, honour, and property, as we are bound to do, according to the text of our holy law.

"As to the other subjects, they are subsequently to be regulated after the decision of the enlightened members of our Council of Justice, the members of which will be increased according to necessity, which is to meet on certain days, which we shall appoint. Our ministers and dignitaries of the empire will assemble to establish laws for the security of life and property, and the assessment of taxes, and every member of these assemblies shall be free to express his opinion and to give his advice.

"Laws concerning the regulation of the military service will be debated at the military council, which will hold its meetings at the palace of the seraskier.

"As soon as one law is settled, in order that it may be for ever valid, it shall be presented to us, and we shall honour it with our sanction, and to the head thereof we shall affix our imperial seal."

Since the above hatti scheriff was issued, the following statement has appeared in the Turkish Gazette:—

"The sultan, ever since his accession, has most ardently desired to signalize his reign by the re-establishment of the Ottoman power on the basis of the common well-being of his subjects. His efforts have on various occasions been crowned with the most signal success: but one fundamental reformation was requisite to crown his labours, and to assure to his people the benefits which he sought to confer upon them. The collection of the revenue has remained up to the present time laden with abuse, oppressive to the subjects, and detrimental to the state. Numerous firmans have been issued—inquiries have taken place; but the sultan, during his late journey through the provinces, having employed himself in examining into the state of the administration, has been convinced that no sensible improvement has been effected, and that more decisive measures are required.

"In order to proceed methodically in this reformation, his highness has ordered an extensive inquiry to be instituted, so as on the one hand to ascertain the amount of the contributions actually paid by each district. and, on the other, to ascertain the actual disbursements for the army, the marine, the arsenals, and the other military establishments.

"The council of the Porte has, therefore, been assembled in presence of the high functionaries of the state, to deliberate on the best means of carrying the intention of his highness into execution; and after a long debate, it has been resolved as follows:—
976

"'That a table shall be constructed, exhibiting the sums received—1st, for the treasury; 2d, for the Valis and Voivodes; 3d, for the expenses of travelling functionaries; the amount of contributions in kind to different departments, paid in saltpetre, corn, timber, &c.; 5th, the value of things to which certain towns and districts were liable, under the denomination of Angaria (corvée); 6th, the sums paid to local police, judges, &c.

"'That an exact statement or balance-sheet be prepared of the whole revenue, fixed and casual, of the state.

"Henceforward every tax unauthorised by the imperial canon shall be abolished.

"'The properties of the high functionaries of the state, whether military or civil, and the persons attached to their services, shall be equally assessed with those of the nation.

"'Every exemption from taxation, and every privilege through which the common burdens were avoided, shall cease.

"'The imposts shall be imposed with complete impartiality, at a rate of so much per thousand, which shall yearly be settled in the month of March, according to the new ordinance.

"'Each individual shall receive a ticket bearing the seal of the community, stating the amount of his contributions, and these sums shall be entered in the public register of each municipality.

"'Men of recognised probity and intelligence shall be commissioned at the public expense, to prosecute the necessary inquiries throughout the empire.

"'The above regulations shall immediately be carried into execution in the two provinces nearest to the capital, Broussa and Gallipoli, so that the effects and advantages of the change may be observed, and with the least possible delay extended to the remainder of the empire.

"'From the day of the execution of this order, the two provinces designated shall be exempt from the payment of the impost termed 'Ichtisah' (internal customs).

"'The confiscation of private property shall in no instance be allowed. The government shall in no case appropriate to itself the property of individuals, except on the death of persons who have no heirs.

"'The government will reserve to itself the right of previous liquidation in the case of a holder of government money dying without sufficient effects to cover his debts.'

"These regulations, fixed by the Council of the Porte, have been confirmed by the High Council, and sanctioned by the Imperial Firman.

"As these present institutions have for their object to cause the religion, government, nation, and empire to reflourish, we solemnly bind ourselves to do nothing in contravention to them. As a pledge of our promise, it is our determination, after having them deposited in the hall which contains the glorious mantle of the prophet, in presence of all the ulemas and dignitaries of the empire, to abide by these institutions in the name of God, and there order the ulemas and grandees of the empire to take the same solemn oath. After that, he who shall violate these institutions shall be liable, without any regard being paid to his rank, consideration, or credit, to corresponding punishment to his faults after once it has been made clear.

"A penal code shall be drawn out to this effect.

"As every functionary receives at present a suitable salary, and as the pay of those who are not yet sufficiently rewarded is to be subsequently increased, rigorous laws will be promulgated against the sale of patronage and places under government, which the divine law reprobates, and which is one of the principal causes of the downfall of the empire.

"The above resolutions being a complete renovation of ancient customs, this imperial decree shall be published at Constantinople and in all the provinces of our empire, and shall be communicated officially to all the ambassadors of friendly powers residing at Constantinople, in order that they may be witnesses to the granting of these institutions, which, if it may please God, are to endure for ever.

"May the Almighty God extend his protection to us! Let those who may presume to violate the present institutions be the object of divine maledictions, and be deprived of happiness now and for ever! Amen."

Historical Notice.—Othman, chief of the Oguzian Tartars, is generally accounted the founder of the Turkish empire. On his succeeding his father in 1288, his dominions were comparatively inconsiderable, being confined to the lordship of Siguta, in Bithynia, and a small tract of adjoining territory: but the talent of Othman, and the bravery and zeal of his followers, enabled him to add greatly to his paternal inheritance, and to bequeath the whole of Phrygia and Cappadocia to his son and successor. From this period the tide of Turkish conquest began to roll forward with a force that could not be checked by the feeble resources of the Greeks. In 1338, the Ottomans first obtained a footing in Europe. In 1362, Amurath, the grandson of Othman, in stituted the Janizaries—the first, and for a lengthened period the most powerful, numerous, and best-disciplined standing army established in modern times. The conquests of Timour threatened to subvert the Turkish power; but it soon recovered from the rude shocks it had sustained, and, in 1453, Mahomet II. entered Constantinople sword in hand, and established himself on the throne of Constantine and Justinian! But the undisturbed possession of all the counties from mount Amanus to the Danube did not satisfy the restless and insatiable ambition of the Turks. Selim, the grandson of Mahomet II. added Syria and Egypt to the dominions of his ancestors; and Solyman the Magnificent, the contemporary of the Emperor Charles V., and the most accomplished of all the Ottoman princes, conquered the greater part of Hungary, and in the east extended his sway to the Euphrates. At this period, the Turkish empire was, unquestionably, the most powerful in the world. "If you consider," says the historian Knolles, who wrote about two centuries since, "its beginning, its progress, and uninterrupted success, there is nothing in the world more admirable and strange; if the greatness and lustre thereof, nothing more magnificent and glorious; if the power and strength thereof, nothing more dreadful and dangerous; which, wondering at nothing but the beauty of itself, and drunk with the pleasant wine of perpetual felicity, holdeth all the rest of the world in scorn." Nor had this mighty power even then reached its greatest height. Solyman was succeeded by other able princes; and the Ottoman arms continued to maintain their ascendancy over those of Christendom until, in 1683, the famous John Sobieski, king of Poland, totally defeated the army employed in the siege of Vienna. This event marked the era of their decline. For awhile they continued to oppose the Austrians and Hungarians with doubtful fortune and various success; but the victories of prince Eugene gave a decisive superiority to the Christians. The crescent, instead of recovering its former lustre, has fallen like a star plucked from its place in heaven. Province after province has been dismembered from the empire: the Russians, now its most formidable enemies, have advanced to Adrianople, and, but for the mutual jealousies and animosities of the different Christian states, last century would have witnessed the final extinction of the Ottoman power.

When considered with attention, it does not seem difficult to discover the causes of these apparently anomalous results. The Turks, like their Tartar ancestors, are naturally a brave, patient, and hardy race. After their emigration from central Asia, they were long exposed to the greatest difficulties and privations. Pressed on all sides by Mongols, Turkmans, Saracens, and Greeks, they could not maintain their footing in Asia Minor without waging incessant hostilities with their neighbours: they were thus early inured to habits of pillage and blood. And, after they embraced the Mahommedan faith, they found in the law of the prophet not a license only, but a command to desolate the earth, and to propagate their religion and empire by violence. The peculiar tenets and doctrines of the Koran made a profound impression on the ferocious, ignorant, and superstitious minds of the Turks, who early became the most zealous apostles of a religion of which implicit faith and unconquerable energy are the vital principles. Their fanaticism knew no bounds. They literally believed that the sword was the key of heaven, and that to fall fighting in defence of the true faith was the most glorious of deaths, and was followed by the largest portion of eternal felicity. Firm and unshaken believers in the doctrine of predestination, assured that no caution could avert, and no dangers accelerate, their inevitable destiny, they met their enemies without fear or apprehension. Tribute, slavery, and death to unbelievers, were the glad tidings of the Arabian prophet; and have been loudly proclaimed by his followers over half the Old World. The Ottomans did not, like the crusaders, require an impulse from pontiffs or preachers to stimulate them to engage in the great work of conquest and conversion: the precept was in their law, the principle in their hearts, and the assurance of success in their swords. To such desperate energies, wielded by a succession of sultans distinguished for various and great ability, the Greeks had nothing to oppose but dispirited troops, and generals destitute alike of courage and capacity. From the age of Justinian the Eastern Empire had been gradually sinking. The emperors were alternately prodigal and avaricious, cruel, profligate, and imbecile; the people were a prey to all the evils of civil and ecclesiastical tyranny; their bodies were emasciated by fasting; and their intellectual powers stagnated in theological controversies, alike futile and unintelligible. The total defeat of Bajazet, the great grandson of Othman, afforded an opportunity which, had it been right-improved, might have enabled the Greeks to expel the Turks from Europe: but the Greeks were totally incapable of profiting either by this or any other event: and the schism of the west, and the factions and wars of France, England,

and Germany, deprived them of all foreign assistance, and enabled the Turks to repair their shattered fortunes, and again become the terror and the scourge of Christendom.

But the same cause to which the Turks principally owed their success—the intolerant bigotry and fanaticism of their religion, proved also the principal cause of their decline. It isolated them from the rest of Europe, and taught them to look down with contempt and aversion on the arts, sciences, and attainments of the infidel world. "There is," said they, "but one law, and that law forbids all communication with infidels." The more the surrounding nations have distinguished themselves by their advances in civilization and literature, the more determined have the mass of the Turkish people become to resist their example, to keep within the pale of their own faith, and to despise their progress. The fiery and impetuous zeal by which they were distinguished in the fourteenth and fifteenth centuries has long since subsided; but had it continued to burn with undiminished force, it could no longer have rendered them really formidable. The invention of gunpowder, and the various improvements that have been made in the art of war, have happily opposed an invincible obstacle to the success of multitudes without discipline, and of courage without skill. "Tant qu'il ne s'agissait que de reasembler et de retourner les drapeaux une multitude des soldats animés par la fanaticisme, l'avantage fut pour les Ottomans; mais cet avantage disparut quand la guerre appela le concours des sciences humaines, et que le génie, avec ses découvertes, devint la redoubtable auxiliaire de la valeur." (Michaud, Hist. des Croisades, v. 344, Ed. 1841.) That fanatical fervour, contempt of danger, and superiority of numbers, which formerly gave so decided a superiority to the Ottoman troops failed to enable them to withstand the science, cool deliberate courage, artillery, and tactics of the troops of Austria and of Russia. The Turks have degenerated both in their civil and military institutions, but their present weakness is to be ascribed more to their not keeping pace with the progress of their neighbours, than to their positive decline. Haughty, confiding, and illiterate, they have experienced all the fatal consequences of ignorance without once suspecting the cause. Resolved to employ no other means than force, they sank into despondency when it no longer availed; and having now almost abandoned the hope of recovery, they present, to their own astonishment and the mockery of Europe, the um-bra magni nominis—the mighty shadow of unreal power:—"We effected our conquests," said the mufti to the Baron de Tott, "without any aid from European tactics, and we do not now stand in need of them. Our defeats are not the effects of human force; they are the chastisement of our crimes; the decree of heaven has reached us, and nothing can avert the wrath of Omnipotence!"

The despotism of the sultans, and the vast extent of their power, have for a lengthened period contributed to accelerate the progress of decline. For awhile, however, it was otherwise. The perilous circumstances under which the Turks were originally placed, and the difficulties and dangers with which they had to struggle, obliged their chiefs to exert all their faculties. Having to rule over bold and fanatical subjects, to act as their generals in war and their legislators in peace, they were compelled to practise the military and peaceful virtues; to inspire confidence by superior knowledge and resolution; attachment by kindness; respect by dignity; emulation by discernment in the bestowing of rewards; and discipline and good order by a steady adherence to a uniform system. We do not say that nothing is to be ascribed to the personal character of the sovereign; but if we reflect, that, except in a single instance, a period of nine reigns, and of 264 years, is occupied from the elevation of Othman to that of Solyman, by a series of warlike and able princes (Gibbons, xii. 57), it must be allowed that something more than chance, that the necessities of the times had produced this long line of able monarchs.[*] No sooner, however, had the tide of Turkish conquest been stopped by the firm resistance of the Hungarians and Germans on the one side, and that of the Persians on the other, than the Ottoman monarchs began rapidly to degenerate. The evil was greatly aggravated by the regulation of Solyman the Magnificent, who, in order to hinder the rebellions and internal divisions that had sometimes occurred, established it as a principle, which has ever since been strictly adhered to, that none of the sultan's sons should be appointed to the command of armies or the government of provinces. This regulation had a fatal effect: instead of being educated, as formerly, in the council or the field, the heirs of royalty and of almost omnipotent power have since been brought up in the slothful luxury of the palace. Shut up constantly in their seraglios, ignorant of public affairs, benumbed by indolence, depraved by the flattery of women,

* Some of the foregoing statements have been borrowed from the notice of the increase and decline of the Turkish power, in the translation of Malte-Brun, written by the author of this article.

of eunuchs, and of slaves, their minds contracted with their enjoyments, their inclinations were vilified by their habits, and when they succeeded to the throne, their government became as vicious, corrupt, and worthless as themselves.

The vast extension of the Turkish empire was another cause of its decline. It multiplied the enemies, not the subjects of the state. To animate the various and discordant classes of people comprised within its widely-extended limits with the same spirit, and give them one common interest, would have required the adoption of a liberal and enlarged system of policy : but to act in this manner was utterly repugnant to the maxims of Ottoman legislators. The inhabitants of the conquered provinces who refused to embrace the religion of the prophet were branded with the title of infidels, and looked upon with aversion and contempt. To associate with such persons on any thing like a footing of equality, or to admit them to the enjoyment of political privileges, was out of the question. They existed only on sufferance ; and though their rights were legally defined, their proud and fanatical masters seldom hesitated about trampling them under foot, and subjecting them to every species of insult, extortion, and ill-treatment. Perhaps, however, it is true that the very weight of the tyranny to which the non-Mohammedan portion of the population has been subject, has, by subduing their energies and debilitating their minds to the level of slavery, tended to secure the tranquillity of the empire ! But, whether this be so or not, it has, at all events, ensured its depopulation, impoverishment, and degradation. Under this miserable government, palaces have been changed into cottages, cities into villages, and freemen into slaves. Sandys, who visited the Turkish empire early in the 17th century, when it was comparatively flourishing and vigorous, has described the unhappy state of the regions subject to its destructive despotism, with a truth and force of eloquence that will not speedily be surpassed : " These countries, once so glorious and famous for their happy estate, are now, through vice and ingratitude, become the most deplored spectacles of extreme misery: the wild beasts of mankinde having broken in upon them, and rooted out all civilitie ; and the pride of a sterne and barbarous tyrant possessing the thrones of ancient and just dominion ; who, ayming onely at the height of greatnesse and sensualitie, hath, in tract of time, reduced so great and goodly a part of the world to that lamentable distresse and servitude under which (to the astonishment of the understanding beholders) it now faints and groneth. Those rich lands at this present remaine waste and overgrowne with bushes, receptacles of wild beasts, of thieves, and murderers ; large territories dispeopled or thinly inhabited ; goodly cities made desolate ; sumptuous buildings become ruines ; glorious temples either subverted or prostituted to impietie ; true religion discountenanced and oppressed ; all nobilitie extinguished ; no light of learning permitted nor virtue cherished ; violence and rapine insulting over all, and leaving no security, save to an abject mind and unlookt on povertie." (Preface, ed. of 1637.)

Such is the government which the great powers of Christendom, including, we are sorry to say, England, profess themselves desirous to maintain in all its integrity ! We hardly, however, think that it is destined to a much longer endurance ; and, happily, into whatever hands it may fall, there cannot be so much as the shadow of a doubt but that the overthrow of the Turkish government and power will be productive of the greatest possible advantage to the interests of humanity.

TURKEYFOOT, p. t., Somerset co., Pa., 15 m. S.W. Bedford, 160 m. W.by S. Harrisburg, 178 W. Drained by Castleman's r. and Laurel hill cr., branches of Youghiogheny r. It has one fulling-mill, five grist-mills, six saw-mills, two tanneries, five distilleries ; two schools, 46 scholars. Pop. 1422.

TURNER, p. t., Oxford co., Me., 30 m. W.S.W. Augusta, 586 W. Bounded E. by Androscoggin r. Incorporated in 1786. It has 10 stores, two fulling-mills, five grist-mills, eight saw-mills, one oil-mill ; one academy, 80 students ; 17 schools, 1067 scholars. Pop. 2479.

TURNHOUT, a town of Belgium, prov. Antwerp, cap. arrond., in a wide healthy district, 25 m. E.N.E. Antwerp. Pop., in 1836, 12,909. (Hasseling). It is well built, and has manufactures of cutlery, linen cloths, lace, carpets, and oil, with bleaching and dyeing establishments, tanneries, brick and tile factories, &c. It sends three deputies to the states of the province.

TUSCALOOSA, county, Ala., Situated a little N.W. of the centre of the state, and contains 1350 sq. m. Drained by Black Warrior r. and its branches. It contained in 1840, 11,163 neat cattle, 3150 sheep, 21,160 swine ; and produced 17,956 bushels of wheat, 1359 of rye, 348,929 of Indian corn, 16,105 of oats, 16,863 of potatoes, 2,576,002 pounds of cotton. It had 31 stores, one flouring-mill, 24 grist-mills, 20 saw-mills, one oil-null, four tanneries, one distillery, five potteries ; four academies, 319 students. Pop. : whites, 9943 ; slaves, 6554 ; free coloured, 86 ; total, 16,583. Capital, Tuscaloosa.

TUSCALOOSA, city, cap. of Tuscaloosa co., Ala., and of the state, 160 m. S.S.W. Huntsville, 355 m. N. Mobile, by the river, 818 W. Situated on the S.E. side of Black Warrior or Tuscaloosa r., at the lower falls, on an elevated plain, at the head of steamboat navigation, 236 m. N. of Mobile by land. It is regularly laid out with spacious streets crossing each other at right angles. It contains a handsome statehouse in the centre of a public square, a courthouse, jail, an United States' land-office, four churches, a Presbyterian, Episcopal, Methodist, Baptist, a masonic hall, Washington and La Fayette academy, an athenæum for young ladies, a lyceum for boys, the Alabama institute, 15 or 20 stores, and 1940 inhabitants. About one mile E. of the state-house, stand the halls of the Alabama university. It was founded in 1828, has a president and seven professors, 74 alumni, 88 students, and 6000 vols. in its libraries. The commencement is on the Wednesday after the second Monday in December. The buildings are five in number, besides houses for the professors. In the centre is a rotunda, a large circular building three stories high. On the first floor is a large circular room for public occasions, the second is a gallery for spectators, and the third is occupied by the library. There are three edifices three stories high for students, and another for a laboratory and recitation rooms.

TUSCANY (GRAND DUCHY OF), (an. Etruria), a state of N. and central Italy, being the third in rank in that peninsula, principally between lat. 42° and 44½° N., and the 10th and 12th degs. of E. long.: the main body of the country has N.E. and S. the Papal States, from which it is mostly shut off by the Apennines, W. the Mediterranean, here called the Tyrrhene or Tuscan sea, and N.W. the duchies of Lucca and Modena ; but exclusive of the above, it has some detached portions of territory collectively termed Lunigiana, surrounded by the dominions of Lucca, Modena, Parma, and Sardinia. The greatest length of the compact portion of the Grand Duchy from N. to S. is 165 m. ; greatest breadth 135 m. The area and population are estimated as follows :—

Provinces.	Area in sq. m.	Pop. in 1836.	Chief Towns.
Florentine	2,168	671,862	Florence.
Pisano	1,716	295,482	Pisa.
Senese	462	139,051	Siena.
Aretino	1,243	239,416	Arezzo.
Grossetano	2,352	67,373	Grosseto.
Total	8,841	1,436,785	

The main chain of the Apennines, where it forms the boundary between Tuscany and the Papal dominions, lies wholly within the former. It is neither so lofty nor so rugged as in other parts of Italy, seldom rising to 4000 ft., though one summit, Boscolungo, attains to an elevation of 4178 ft. The principal passes across the Apennines in this part are those of Fumarolo, Pietra Mala, and Pontremoli, by which last, Tuscan Lunigiana communicates with Parma. The principal valley is that of the Arno, which comprises about one sixth part of the entire surface of the duchy. The Arno, one of the secondary rivers of Italy, rises in Monte Falterona, near Patro Vecchio ; and after running at first S.E., and then N.W., it flows generally W. to its mouth in the Mediterranean, 6 m. W. Pisa, after a course of about 130 miles. Its principal tributaries are the Chiana canal, the Sieve, Ombrone di Pistoja, &c. ; Florence, Empoli, and Pisa, are on its banks. It is usually navigable to Florence, but the navigation is bad, and in the early part of the 17th century a canal was formed from Pisa to Leghorn, avoiding the mouth of the river. The greater Ombrone (an. Umbro), drains the S. part of the country : the Serchio enters from Lucca ; and the Chiana and Tiber rise in Tuscany, as do nearly all the rivers of the N. part of the Papal dominions. There are no lakes of any consequence : those of Castiglione, now in process of filling up, Orbetello, Bientina, &c., are mere lagoons or arms of the sea. Few countries are better furnished with streams for irrigation, and the greater part of Tuscany is so fertile and diversified with hill and dale as to be one of the most pleasant regions of Italy. Mr. Maclaren says, " Florence lies in the centre of a magnificent basin. From the tower of the cathedral the eye ranges over a breadth of 30 m., which seems one vast grove, diversified with wooded hills of moderate height, and yet so thickly studded with glittering villages, hamlets, villas, and houses, that if Florence were a larger town, the whole country, back to the mountains, might be considered as its suburbs. The landscape has all the luxuriant beauty of the view from Richmond hill, with the grandeur of mountain scenery superadded. With the addition of a sheet of water, the scene would reach the perfection of rural beauty. The Arno is, indeed, much finer than the Tiber ; but it wants volume suitable to the grand scale of the scenery. It is but a

* The islands of Elba, Pianosa, Gorgona, Palmajola, Troja, and Meloria, in the Mediterranean, belong to the province of Pisa; and those of Giglio, Monte Cristo, Giannuti, &c., to the province of Grosseto.

978

thread in the rich valley through which it meanders."
(*Notes on Italy*, 185–6.)

Along the coast there is a succession of marshy plains, and all the S.W. part of Tuscany is occupied by the Maremme, an undulating and pestiferous tract of country, similar, in most respects, to the Campagna di Roma. The Maremme, which have long been the abode of desolation, were in remote antiquity among the richest and best peopled portions of Italy, and the seat of many of the chief cities of Etruria. For a lengthened period the drainage and improvement of this neglected tract has been a principal object of the reigning dynasty. The present grand duke, in particular, has prosecuted this great work with singular vigour; and from 1829 to 1832, he is said to have expended £200,000 sterling of his own property in hydraulic works, roads, bridges, buildings, &c., in the Maremme. (*Bowring's Report*, p. 49.)

The lake of Castiglione had, in 1836, been already half drained, as well as other stagnant pools; and the river Cornia has been diverted to fill up by its deposits the marshes of Piombino. Large tracts in the Maremme belong to an impoverished nobility, whose profit from the land is principally derived from wood-cutting, and the pasturage of sheep and goats, at from 10d. to 1s. 4d. a head during the winter season, or of buffaloes and other cattle. But the government has lately purchased up many of these estates, and frequently grants small tracts of land to those who agree to build houses and settle on them; or lets out farms at a low ground rent, which is easily raised by the tenant from the sale of the wood he clears from the estate. At Borghieri, the Count Gherardesca has introduced the cultivation of olives, vines, and mulberry trees to a great extent; and good crops of wheat are now grown in many places where a few years since a few wandering fishermen with difficulty obtained a scanty subsistence.

But it is not in the Maremme only that important drainages have been effected in Tuscany. The Grand Duke Leopold I. in 1788 began the canal which unites the Chiana with the upper Arno, by which a large extent of very fertile land in the Val di Chiana, formerly a pestiferous marsh, has been rendered productive and salubrious. The canal brings down a large quantity of river deposit, particularly during floods; and between 1816 and 1833, 3000 quadrati, or nearly 4 sq. m. of cultivable land, were acquired in this manner. (*Bowring, Rep.* 10, 47–51.)

The *Climate* of the Apennines is severe in winter, when the snow often lies for a month together. But in the valleys the winter is but imperfectly defined; the snow seldom lies for more than a day at a time; and vegetation is scarcely interrupted. The average temperature of the year at Florence is about 59° Fahr. (*Berghaus, Allg. Länder, &c.*) The *libeccio* and *scirocco* are occasionally prevalent, but the latter is much less annoying than at Naples. Putrid and intermittent fevers, with dropsy, scurvy, &c., as in the Campagna di Roma, are common in the marshes of Pisa and the Maremme, particularly in the autumn, when these tracts are deserted by almost all their inhabitants. But the other parts of the country are salubrious, and often favourable to longevity: in the province of Arezzo, during the decennial period, ending with 1835, the annual deaths sometimes did not exceed 1 in 40 of the population, and were never so high as 1 in 36. (*Bowring's Rep.*)

The *geological* rocks of the great Apennines, in this part f Italy, are chiefly serpentine, talc, mica, clay-slate, and quartz: gneiss, which is here absent, appears in the Maremme. The marble found near Seravezza is little inferior to that of Carrara. Tertiary deposits are frequent, consting of sandstone, marl, coarse limestone, chalk, and gpsum; and in the vale of the Arno, the country about enna, &c., they contain numerous fossil remains. A eat variety of chalybeate, sulphurous, and other springs exi in the Maremme and elsewhere, and one of the Apennos, the *Monte di Fo*, is a volcano scarcely yet extinct; ice, at intervals, it continues to send forth smoke, &c. any other places exhibit distinct traces of volcanic agen-. (*Hoffman's Europe; Stein; Schuts, &c.*)

Land and produce.—Of 6,138,993 *quadrati* of land comued in the grand duchy, according to a government sur-/. in 1834, 1,835,636 were estimated to be in pasture; 51.718 in forests; 997.672 arable; 844.285 in vineyards; ;.184 under vines and olives; 361,308 in chestnut woods, l the rest in meadow lands, buildings, roads, &c. The ne survey gives, in all, 192,067 separate properties, including from 50,000 to 60,000 farms; there are, in all, about .000 landed proprietors. Except in the Maremme, properes are mostly small; and the metayer system, by which produce of the soil is divided between the proprietor and cultivator, is generally prevalent. In this system the prietor supplies the capital, and the cultivator the labour utensils; the produce is equally divided between them, n as regards the profits arising from the sale of cattle. I cultivator is only obliged to supply the labour required

in ordinary cultivation. Should the proprietor desire to make new plantations, or reclaim waste lands, the expense falls wholly upon him, and he is obliged to pay the cultivator wages for extra work. The manure and seed is supplied at their joint expense; but the proprietor is obliged to furnish the cultivator with as much of the latter as is necessary for his support. On all occasions it is the proprietor who receives or disburses the moneys for the sale or purchase of cattle. The contract, which is not in writing, holds good only for a year, and the proprietor may at its termination discharge his cultivators; but in such a case he loses all claim upon them should their accounts be in arrear; and in general good tenants continue on the same estate from generation to generation.

The cultivators do not reside in villages, but in isolated houses, or cottages in the centre of their farms. They seldom, see their neighbours, except on holydays or at church. A good cultivator rarely goes to market; he neither buys nor sells; the great recommendation of the metayer system being that his land supplies the farmer with all he wants. The same field is frequently, at the same time, under different crops; and there is, in fact, but little division of labour either in husbandry or any thing else.

The coloni, or occupiers, seldom sink into absolute poverty; but, on the other hand, they as seldom rise to anything like wealth or comfort; and, though they may nearly all be above want, they enjoy extremely few of the luxuries and conveniences of life, other than those conferred on them by the bounty of nature. They are at once poor, ignorant, and incapable of vigorous and systematic exertion. "The metayer," says Sismondi, "lives from hand to mouth. He has rarely any corn in store, and still more rarely any oil or wine. He sells his oil when in the press, and his wine when in the vat. He has no provision of salt meat, butter, cheese, leguminous plants, &c. His kitchen utensils are of earthenware; and the furniture of his cottage consists only of a table, and some wooden chairs, one or two boxes, and a miserable bed, on which father, mother, and children sleep. Hence, as they possess nothing, they would in a bad year die of hunger, if they were not assisted by the proprietors, who rarely refuse making them a loan on security of the ensuing crop. Their debts are paid after the vintage, but before winter is over new ones are contracted. Of ten metayers, there is hardly one who owes nothing to his master. Such is their idleness, that a hired labourer will execute three times as much work in a day as a metayer." (*Tableau de l'Agriculture Toscane*, p. 212–216.)

"I had more than once occasion," says Bowring, "to see four generations inhabiting the same cottage: but the last had not added a particle of knowledge to the ignorance of the first: the same gross superstitions; the same prejudices against books; the same unwillingness to introduce any species of improvement; the same reference to ancestral usages. In innumerable cases, families have occupied the same farms for hundreds of years, without adding a farthing to their wealth or a fragment to their knowledge." The metayer system of agriculture is, however, not only the oldest, but the only one that is understood in Tuscany. At Pistoja, indeed, and some other places, the system of letting the land to the cultivator at a fixed rent has been partially adopted; but it has taken no deep root, and the mezzeria has been introducing itself also with the extension of cultivation into the Maremme. Any change to the existing mode of occupancy must, therefore, be remote.

Every species of cultivation, except that of wheat, vines, and olives, is neglected. Forests have been destroyed or badly managed; and the proper treatment of meadow lands is wholly unknown. All kinds of produce not suitable for direct consumption are but little esteemed. Mulberry trees, which answer admirably well, are not in favour; and the rearing of silkworms, though carried on more or less in the house of every cultivator, is reckoned of inferior importance, and is abandoned to the females. The produce of silk, therefore, is much less than it might be; and the same may be said of most articles of export.

The culture of the corn, and other articles required for home consumption is, however, conducted with the utmost care and attention. The hill sides are formed into a succession of terraces; and a small extent of land of very moderate fertility suffices for the support of a family of 10 or 15 individuals. "The Val d'Arno," says Mr. Maclaren, "is cultivated like a garden. Much of the land is in drills, about 2 ft. wide and 1 ft. deep, planted with maize, &c.; and about every 100 yards there is a neatly cut drain bordered by a row of poplars to serve as vine props. The rich bottom land is skirted with low hills, which are also carefully cultivated, and bear great numbers of olives. Though not a foot of land be wasted, and not a tree grows which has not been planted by human hands for use or ornament, the whole valley, from the bare hills on one side to those on the other, looks like a forest." (*Notes on Italy*, 185.)

But, as already seen, the appearance of the country furnishes no sure criterion of the condition of its inhabitants. The land is split into very small portions; and here, as in most other parts of Italy,

> "The poor inhabitant beholds in vain
> The redd'ning orange and the swelling grain;
> Joyless he sees the growing oils and wines,
> And in the myrtle's fragrant shade replnes;
> Starves in the midst of Nature's bounty curst,
> And in the laden vineyard dies for thirst."

M. de Chateauvieux states that the occupiers in the Val d'Arno are "never able to lay by any thing as a reserve against unfavourable years. On entering their houses we find a total want of all the conveniences of life, a table more than frugal, and a general appearance of privation." (Eng. Trans. p. 78.) It is true that, owing to the mildness of the climate, the temperance and contented disposition of the people, poverty is here productive of much less inconvenience than in England or less favourably situated countries. It appears, indeed, to have had but little influence over the increase of population. In 1801, for example, the pop. of Tuscany amounted to only 1,096,641, whereas, as already seen, it amounted in 1836, to 1,436,785.

Corn is the principal object of culture; but the quantity grown is inadequate to the consumption, and a good deal is imported at Leghorn from the Black sea and elsewhere. Maize, wheat, barley, and rye, are the grains principally raised. The husbandry is not uniformly the same, but, according to Chateauvieux, it is generally under a five years' rotation as follows:—1st, maize, French beans, peas, or other vegetables, manured; 2d, wheat; 3d, winter beans; 4th, wheat; 5th, natural clover, sown after wheat in the spring, and followed by *sorghe*, a large species of parsnip. The terrace cultivation absorbs a good deal of capital, time, and labour. Walls of turf, sod, and sometimes stone, are raised in succession along the sides of the hills, and support the soil brought down by the rains. The sunny sides of the hills thus terraced are chosen for vines; the shady sides for olive trees. The culture of olives and vines is not conducted on any scientific principle; but, from the care bestowed on their produce, the oil and wines of Tuscany hold a respectable rank. The best oil is raised near Pisa, Pescia, and Calci. The wines of Tuscany were formerly very celebrated: they include the Aleatico, and the red muscadel wine of Montepulciano, which last was, with excusable partiality, preferred by Redi to all other wines:

> "Montepulciano d'ogni vino è il re."
>
> Bacco in Toscana.

Two centuries ago the wines of Chianti were well known in England: those of Carmignano, Monte Catini, Ponte a Moriano, &c., have some reputation. The annual produce of silk is estimated at about 250,000 lbs.

Out of five years' crops in the Val d'Arno it will be seen that only one is for cattle, the rearing of which is but a subordinate branch of Tuscan husbandry. Except the herds and flocks belonging to the grand duke, and a few other large proprietors, in general only as many are kept as are indispensable for manuring the fields and other agricultural purposes. In 1832, the cattle, including buffaloes, were estimated at about 356,000, and the sheep at 600,000 head. The oxen are mostly of the Hungarian breed; they are not so much used for draught as buffaloes. From 60,000 to 70,000 sheep are pastured in the Maremme from Oct. to May, and in the Apennines the rest of the year. The wool of the Tuscan sheep is coarse, and little used for any but home manufactures; though during the short period of the French domination there was a considerable introduction of merino flocks. Cheese made of the ewes' milk is consumed in the country, but it is of very inferior quality. The horses of the grand duchy are among the worst in Europe, small, weak, and ill-shaped; and though some attempts have been made to improve the breed, we believe the success has been only very partial. On the other hand, however, the Tuscan asses are the strongest and finest in Italy. Goats, being easily fed, are kept in large numbers in most districts; and the *ricotta* cheeses are made of their milk. Pigs are very abundant in the provs. of Sienna and Grosseto, where they roam at large in the oak woods of the Maremme; and poultry and bees are pretty numerous.

It is a curious fact, that notwithstanding the houses of the peasants are so miserably ill-furnished, they are themselves of a very superior description. "In no country are the peasantry so well lodged. Probably half or more of their houses have been rebuilt within the last 60 years, and the remainder have been improved. It is reckoned, taking one house with another, they must have cost 1000 crowns or 5000 fr. each, and the average value of a farm is 2000 crowns. The living of the peasantry does not correspond to the luxury of their habitations: it is wholesome, though frugal. In most of the provs. bread is a mixture of rye, barley, and maize, with a little wheat; in some places, however, it is of pure wheat. Next to bread, beans form the principal nourishment of the cultivators. They drink but little wine; more

frequently aquarello or piquette. To eat fresh meat once a week is considered a luxury: and the poorest are satisfied with a piece of bacon. Salt fish is a good deal eaten. The import of salt cod into Tuscany exceeds 1,000,000 ft. a year" (*Bowring's Rep.* p. 44.) The day labourer gets on an average 10d. a day for 11 hours' work in the summer, and 8 hours in winter. An English gentleman farming an estate of 40 acres near Florence stated to Dr. Bowring, that the value of its gross produce amounted to about £215 a year, which after deducting the half share of the contadino, and the expenses either shared by him or falling solely upon the proprietor, with casual expenses, left him a nett profit of about £3 10s. per acre, or not quite 3¾ per cent. interest on his outlay. (See *Report*, p. 46, 47.)

Tuscany has a great variety of mines, which were formerly extensively wrought; but many of these have been abandoned. The principal are the iron mines of Elba (which see, i. 694.) Copper is obtained at Monte Catini, Montieri, &c.; cinnabar, lead, silver, &c., in different places. During a recent decennial period 413,800 lbs. a year of sulphur were produced at Pertia; and alum, nitre, various kinds of marble, &c., are found in different parts of the Maremme. But the most remarkable mineral product is borax, obtained from a collection of lagoons, unique in Europe, if not in the world, and spread over a surface of about 30 sq. m. at and near M. Cerboli, about 15 m. S. by W. Volterra. "As you approach the lagoons, the earth seems to pour out boiling water as if from volcanoes of various sizes, in a variety of soil, but principally of chalk and sand. The heat in the immediate vicinity is intolerable, and you are drenched by the vapour which impregnates the atmosphere with a strong and somewhat sulphurous smell. The ground, which burns and shakes beneath your feet, is covered with crystallizations of sulphur and other minerals. The vapours break forth violently in different parts of the mountain recesses: they only produce boracic acid, when they burst with a fierce explosion. In these spots artificial lagoons are formed by the introduction of the mountain streams. The hot vapour keeps the water perpetually boiling, and after it has received its impregnation during 24 hours at the most elevated lagoon, the contents are allowed to descend to the second lagoon, where a second impregnation takes place, and then the third, and so forth, till it reaches the lowest receptacle; and having thus passed through from six to eight lagoons it has gathered half per cent. of the boracic acid. It is then transferred to the reservoirs, and thence, after a few hours' rest, to the evaporating pans, where the hot vapour concentrates the strength of the acid, by passing under shallow leaden vessels from the boiling fountains above. There are from 10 to 90 pans, in each of which the concentration becomes greater at its descent, till it passes to the crystallizing vessel; from whence it is carried to the drying rooms, when after two or three hours it is ready to be packed for exportation. There are in all nine establishments. The whole amount produced varies from 7000 to 8000 lbs. of 12 oz. a day. The borax-lagoons have been brought into their present profitable action within a very few years. In 1833, only about 650,000 Tuscan lbs. were obtained; in 1836, 2,500,000 lbs. But the produce does not appear susceptible of much extension, as the whole of the water is now turned to account." (*Report*, p. 36-38.) The territory around Volterra is rich in salt springs, and the royal salt works employ about 90 labourers, and produce nearly 18,000,000 lbs. of salt a year, exclusive of the produce of Elba.

Among the *manufactures*, one of the principal is that of straw plait, so large an export from Leghorn. It is made with the straw of the beardless wheat, grown on soil the poorness of which renders the reed white, and is cut before it be quite ripe. Small patches of ground are chosen for its culture, on calcareous hills, and it is never manured. The seed is sown very thick; and Chateauvieux was assured that a crop of two acres would supply straw sufficient for the whole hat manufacture of Tuscany. (*Letters*, p. 73.) This manufacture employs a vast number of hands in Florence, Prato, and in all the districts from Florence to Pistoja on one side, and to Pisa on the other. It is estimated that between Florence and Prato only there are fifty manufactories; besides which, the females of nearly all the *contadini* families in the Val d'Arno employ themselves, more or less, in making straw hats. This branch of industry has latterly diminished; but when Chateauvieux wrote, it brought "an annual return of 3,000,000 fr. to the families of the country, for the men have no concern in it. Every young woman, for a few pence, purchases the straw she has occasion for; she exerts her talent to braid it as fine as possible; and sells, for her own profit, the hats she has made; the money which she thus earns at length forming her dower. Each individual in a family can earn from 30 to 40 sous (15d. to 20d.) a day in plaiting straw, while they can hire a poor woman from the Apennines for 3 or 10 sous (4d. or 5d. to do the domestic work." (*Ib.* p. 74, 75.)

From 3600 to 4000 looms in Florence are engaged in the silk manufacture; and some at Prato are engaged in the silk manufacture; and, in many parts of Tuscany, there are steam mills for the chain and tram. At Sienna, Prato, &c., are several factories for woollens, including *berretti* or red woollen caps, in imitation of those of Tunis. Pistoja has extensive iron works, hardware factories, and paper-mills; Prato, many copper foundries; and at Follonica nearly all the iron of Elba is smelted and wrought; carpets at Florence; leather and rope at Leghorn; marble and alabaster goods at Vol terra, &c.: glass, earthenware, hats, hempen and linen fabrics are the other articles principally manufactured. At Prato, which town may give a fair average, the wages of a working man, in most trades, may be about 2½ pauls, or 1s. a day; few earn more than 3 pauls, (15d.), or less than 2 pauls (10d.) The retail prices of food are, bread about ½d. per lb.; mutton and beef, 1½d. to 2½d. per lb.; oil, 3½d. per lb.; *vin-ordinaire*, 2½d. to 5d. for about three bottles, &c. (*Bowring's Report*.)

The exports are principally oil, charcoal, borax, straw plait and hats, cork, rags, potash, tanned hides, marble, coral, woollen caps for the Levant, timber, paper, soap, tartar, &c.: the imports, colonial produce, spices and dye stuffs, manufactured goods, hardware and earthenware, salted fish, &c. Leghorn (which see *anté*, 160) engrosses almost the whole foreign trade of the country. Accounts are kept in *lire* (= about 8d.) of 20 *soldi*, and 240 *denari* each. The paolo = 1½ lire; the pezza = 5½ lire. The Tuscan lb. = about 12 oz. troy. The staje, for corn, &c. = 3-4ths bushel: the moggio = 25 staje. The barile (of wine) = 10 imp. galls.; (of oil)= 7½ galls. The braccio = about 2 ft.: the seccata of 660 pertiche = about an acre.

The Government is an absolute monarchy, mildly exercised, being, in a practical point of view, not only the best in Italy, but one of the least exceptionable in Europe. The grand duke is assisted by a council of four ministers. Justice is usually administered by the syndics in the small towns and villages, from whose decision appeal may be made to the *vicario* of the canton. There are courts of primary jurisdiction in Florence, Leghorn, and Pistoja, from which appeal lies to the highest court, the *ruota* in Florence. Crime is rare, and is mostly confined to petty frauds and robberies. The civil legislation is far more defective than the criminal; and is often dilatory in the extreme. The total number of criminals and petty offenders convicted before the Tuscan courts in 1833 amounted to 1390. All the inhabitants, except a few Jews, Protestants, Greek communicants in Leghorn, &c., are Roman Catholics, being subordinate to the three archbishops of Florence, Pisa, and Sienna, and seventeen bishops. Education is almost exclusively in the hands of the ecclesiastics, but is notwithstanding better conducted than in most parts of Italy, especially the Papal states and the kingdom of Naples. There are universities in Pisa and Sienna; the former, in 1836, was attended by 543 students, and the latter by 254: and in Florence the famous schools of the Scolopj are attended by upwards of 1000 students. Numerous other superior schools and many learned societies exist, and no better evidence can be given of the increase of knowledge than the fact, that in 1814 there were but six printing presses in Florence, whereas in 1836 they amounted to 35. In 1835, about 21,300 children were attending the public schools. "Since 1830," says the *Journ. of Educ.* No. xvi., "much has been done to forward elementary education, though no general system be established. In the towns there are very good gratuitous schools, and Lancastrian schools have been established in various parts of the country. Holiday schools have also begun to be established." No. iii. of this *Journ.* has a very full account of the state of education, &c., in Tuscany.

Tuscany, as compared with its population and resources, has the smallest military force of any European state: it consists altogether of about 6000 men, of whom only 4500 are usually under arms. Military service is obligatory upon all classes, recruits being chosen by lot. The army is maintained on an economical footing, and officers are retained on half-pay till advanced in years, to avoid a multiplicity of pensions. The naval force is insignificant; there being only one ship of war to guard the port of Leghorn. (*Oudinot, Italie et ses Forces Milit.* p. 290–292.)

The roads, bridges, canals, &c., of Tuscany are well kept; the roads, in particular, are so good, that the conveyance of goods by the rivers and canals is comparatively little resorted to. Between Florence and Leghorn there is a daily post; between the other towns about two posts a week. Mendicity is prohibited by law; but there are numerous and richly endowed charitable institutions. Workhouses exist at Florence, Sienna, and Arezzo, supported partly by the state, and partly by voluntary contributions and the labour of the inmates. Pisa and Sienna have deaf and dumb institutions; and, together with Florence, Leghorn, and Pistoja, establishments at which orphans are taught different trades. Several institutions distribute food, bedding, cloth-

ing, and working tools to persons in want, and at Prato is one which makes annual grants for study, and furnishes loans for various purposes. Government pawn banks and savings banks are established in the principal towns; the latter paying 3½ per cent on deposits up to a certain amount. The public revenue, derived principally from the government rents, custom dues, stamps, salt monopoly, lottery, tobacco duties, patents for carrying arms, fees, fines, &c., amounted in 1830 to 25,104,898 Tuscan lire; and the expenditure to 23,078,029 lire. In every respect we may consider Tuscany as the most flourishing state of the Italian peninsula. At two distant epochs, in remote antiquity, and at the revival of science and commerce in the middle ages, Tuscany has been the seat of the highest degree of civilization in Europe. The Tuscans of the present day are admitted generally to excel the other inhabitants of Italy in their taste for the fine arts, and the polish of their manners; and the Italian, or, as it is sometimes called, *Lingua Toscana*, is here spoken in the greatest purity and perfection. Even within this limited territory, there are, however, no fewer than five different dialects—the Florentine, Senese, Pistojan, Pisan, and Aretine. The best Italian is said to be spoken in Sienna.

History.—Etruria was finally conquered by the Romans about *anno* 280 B.C. After the fall of the Western Empire, it successively belonged to the Goths and the Lombards, by the last of whom it was erected into a duchy. Charlemagne conquered Tuscany with the rest of the Lombard dominions; but under his feeble successors its marquises made their government hereditary and independent. The Tuscan territories were afterwards divided, in the 12th and 13th centuries, among the famous republics of Florence, Pisa, and Sienna; but these were re-united in 1531 into a duchy under Alexander de Medici, in whose family it continued till its extinction in 1737, when it fell into the hands of the house of Austria. In 1801, by the treaty of Luneville, Napoleon erected it into the kingdom of Etruria for the prince of Parma; but in 1808 it was incorporated with the French empire, and subdivided into the *depts.* Arno, Mediterranean, and Ombrone. Since 1814 it has reverted to Austria, and is now governed by one of the Austrian archdukes. (*Bowring's Rep. on the Statistics of Tuscany; Rampoldi; Sarristori, Statist. d'Italia; Sismondi, Agriculture Toscane; Chateauvieux, &c.*)

TUSCARAWAS, county, O. Situated centrally in the E. part of the state, and contains 655 sq. m. Drained by Tuscarawas river and its branches. First settled in 1803 or 1804 by emigrants from W. Pennsylvania and Virginia who were chiefly of German descent. It contained in 1840, 17,446 neat cattle, 32,798 sheep, 26,535 swine; and produced 332,098 bushels of wheat, 29,190 of rye, 965,973 of Indian corn, 36,278 of buckwheat, 1349 of barley, 263,285 of oats, 108,906 of potatoes, 15,000 pounds of tobacco, 100,409 of sugar. It had 22 commission houses in foreign trade, 57 retail stores, two fulling-mills, seven flouring-mills, 30 grist-mills, 71 saw-mills, two oil-mills, 17 tanneries, six distilleries, two potteries, one printing-office; 143 schools, 9866 scholars. Pop. 25,631. Capital, New Philadelphia.

TUSCUMBIA, p. v., Franklin co. Ala., 66 m. W. Huntsville. 195 m. N. by W. Tuscaloosa, 762 W. Situated 1 mile S. of Tennessee r. It contains three churches, a Presbyterian, Methodist and Campbellite, all of brick, two male and two female seminaries, an iron foundry and workshops connected with the railroad, 200 dwellings, and about 2000 inhabitants. An extraordinary spring flows from a large fissure in a limestone rock, which discharges 20,000 cubic feet of water per minute, and it flows a considerable stream 2½ m. to Tennessee r. The depth of the stream is 8 feet where it issues from the rock, and it supplies the inhabitants with water. The Tuscumbia, Courtland and Decatur railroad extends from Tuscumbia to Decatur on Tennessee r., 44 miles, avoiding the Muscle Shoals.

TUY (an. *Tuda ad Pinus*), a fortified town of Spain, in Galicia, cap. prov., on the Minho, which separates it from the Portuguese territory, 57 m. S. by W. Saint Jago. Pop. about 7000. (*Miñano.*) It stands on a height surrounded by several small rivulets, and has always been a fortress of some strength, and a key of Spain on this side. It is regularly laid out, and has well paved and clean streets, several squares and bridges, a cathedral, several hospitals and convents, a seminary, college, and two barracks. Its principal manufacture is of table linen, in which its inhabitants trade with Portugal; but it also produces hats, leather, liqueurs, &c. Its climate is rendered unhealthy by adjacent marshes.

TVER, a gov. of European Russia, between the 56th and 59th degs. of N. lat., and the 30th and 38th of E. long.; having N Novgorod, E. Jaroslavl and Vladimir, S. Moscow and Smolensko, and W. Pskof. Area estimated at 23,800 sq. m. Pop. in 1836, 1,397,980. The surface of this government is generally more elevated than that of other parts of European Russia; and several large rivers, as the Wolga,

Tvertza, Mologa, Mednevitza, &c., rise within its limits. In its W. part are several lakes. The Wolga has its scource in the lake of Selighur, and afterwards traverses the government in nearly its whole length from W. to E. The climate is severe, and the soil is but indifferently fertile. The harvests are precarious, and scarcely ever produce more than sufficient for home consumption. A good deal of hemp and flax, with beans, &c., are grown; but few kinds of fruit succeed. The forests are extensive, particularly in the N.; and about 319,000 deciatines of forest land belong to the crown. Manufactures of little consequence, but increasing: those of dyeing materials and spirituous liquors are the principal; and there are others of bricks, glass-ware, ropes, leather, woollen cloths, &c. This government is, however, distinguished for its commercial activity; and the capital of its merchants has been estimated at 17 million roubles. The trade centres mostly in Tver the capital and is facilitated by the Vischnij-Volotschok canal, which establishes a water communication between the Baltic and Caspian seas. The district of the government traversed by this canal is inhabited by a tribe of Carelians, and in the capital is a German colony; but the population is mostly Russians, of the Greek church. This government is divided into 12 districts; chief towns, Tver the capital, Torshok, Rjef and Bejetsk.

TVER, a town of European Russia, cap. of the above gov., on the Wolga, here crossed by a wooden bridge 550 ft. in length, where it is joined by the Tvertza and Tmaka, and on the high road between Moscow and Petersburg; 90 m. N.W. the former; lat. 56° 51′ 44″ N., long. 35° 57′ 36″ E. Pop. in 1830, about 20,000. In respect of the regularity of its streets and buildings, Tver ranks next to the two Russian capitals; but wants their bustle and animation. It is divided by the several rivers into the town proper, suburbs, and citadel. The last surrounded by a rampart of earth, comprises the governor's residence, an imperial palace, the cathedral, and seminary; and its numerous towers and cupolas give it, at a distance, an imposing appearance. The cathedral is a square edifice, with a lofty spire, surrounded by a guild copper dome, and surrounded, lower down, by four similar domes. The seminary, founded in 1727, for the instruction of 700 pupils in the sciences and ancient languages, is established in a convent built in the 13th century. There are numerous churches, government buildings, barracks, inns, a theatre, &c., and several public promenades, planted with trees. Tver owes its present regularity and beauty to a fire which almost totally destroyed it in 1763; after which the Empress Catherine ordered it to be rebuilt on a uniform plan. Some houses are of stone, but the greater part are of wood; and the paving is mostly of the same material. Captain Cochrane states that an impost is levied upon every horse that passes the gates, expressly to pave the streets. (*Mod. Trav.*, xvii. 114.)

Tver is a place of considerable trade, a large part of its population being merchants, or engaged in the navigation of the Wolga. It is an entrepôt for corn from the S. destined for Petersburg, and for goods conveyed overland to and from Riga. It is of considerable antiquity, having been the capital of a principality as early as the middle of the 13th century. It has frequently suffered from the plague, and been taken by both the Tartars and Poles; but it has remained, with little interruption, attached to the dominion of the Russians since 1490. (*Schnitzler; Possart; Mod. Trav., &c.*)

TWEED, one of the principal rivers of Scotland, forming, in the lower part of its course, the boundary between Scotland and England; has its sources on the E. side of Errickstane hill, about 6 m. from Moffat. Its course is first N.E. to Peebles; thence E. with a little inclination to the S. to Melrose; it next passes Coldstream and Kelso; and pursuing a N. easterly direction, falls into the sea at Berwick. The descent from the source of the Tweed to Peebles is 1000 ft., and thence to Berwick about 500 ft. more. (*New Statistical Account of Scotland*, p. 2.) The waters of the Tweed are particularly pure and limpid. The first part of its course is through a fine pastoral country, especially celebrated in Scottish song, and the latter, through one that is rich and well cultivated. Including windings, its length is reckoned at rather more than 100 m. Notwithstanding it conveys a large body of water to the sea, it is not navigable for any considerable distance. The salmon fisheries on the Tweed are of very considerable value and importance; being, in this respect, second only to those of the Tay. The fish is almost all conveyed packed in ice, to London.

Among its principal tributaries is the Etterick, which, flowing from the S. parts of Selkirkshire, joins it at the Eildon hills. A little lower down it receives the Gala, from Mid Lothian, and the Leader from the borders of East Lothian. The Teviot rises in Roxburghshire, on the confines of Dumfriesshire; and flowing N.E., and receiving several tributaries, it falls into the Tweed at Kelso. The Till rises in Northumberland, near Ingram, and, pursuing a

N. westerly course, falls into the Tweed at Tilmouth. Near Berwick, the Tweed receives the Adder, a considerable river, formed by the junction of the Blackadder and Whiteadder, having their sources in the Lammermoor hills. The basin of the Tweed is estimated at about 1870 sq. m.

TWENTY-FOUR PERGUNNAHS, a district of British India, presid. and prov. Bengal, between lat. 21° 38′ and 23° N., and long. 88° and 90° E., having N. Nuddea, E. Jessore, W. the districts of Calcutta, Hooghly, and Midnapore, from which it is divided by the Hooghly river, and S. the bay of Bengal. Area 3690 sq. m. Pop., in 1822, 599,595. Its surface is a dead flat, intersected by arms of the Ganges, and comprising many extensive jeels or marshes, with a considerable portion of the jungly tract known as the Sunderbunds. It has long been progressively increasing in productiveness and importance; but, like the adjacent district of Bachergunge, has been notorious for the prevalence of dacoity or gang-robbery. The Hindoos are reckoned in proportion to the Mohammedans as 3 to 1. Total land revenue, 1829–30, 1,139,888 rupees.

TWICKENHAM, a village and par. of England, co. Middlesex, hund. Isleworth, on the Thames, 10 m. W. by S. London, and 2 m. S.W. Richmond. Area of par., 2440 acres. Pop. of do. in 1831, 4571. The village consists of a street nearly parallel to the river, and of a number of detached villas. It would hardly, however, have been worth notice in a work of this kind, but for its having possessed the villa famous as the favourite residence of the most elegant, harmonious, and correct of English poets, where he composed many of his noblest works, and where he expired on the 30th of May, 1744. But to the disgrace as well of the country as of the parties more immediately concerned, "Pope's Villa" has been levelled with the ground! The structure now lives only in his immortal verses; and even his grotto,

> "Where, nobly-pensive, St. John sate and thought;
> Where British sighs from dying Wyndham stole,
> And the bright flame was shot through Marchmont's soul,"

has been suffered to go to ruin!

The church, a brick building, erected in 1714, contains the remains of the poet and of his parents. Pope himself raised a monument to the memory of the latter; and a monument to himself was raised, some years after his death, by his friends and literary legatee, Bishop Warburton. It is of grey marble, in the pyramidal form, and has a bust or medallion of the poet.

Among the existing villas in the vicinity of Twickenham, the most celebrated is that of Strawberry-hill, long the residence of Mr. Horace Walpole, by whom it was built, in a sort of trumpery Gothic style, and filled with a singular collection of rare, though mostly trifling articles. In the course of this year (1842), however, the collection has been sold by auction; and the villa itself, which is a very paltry affair, will probably soon share the fate of that of Pope. A national school, for the education of children of both sexes, was established in the village in 1809.

TWIGGS, county, Ga. Situated near the centre of the state, and contains 410 sq. m. Bounded S.W. by Ockmulgee r. It contained in 1840, 7401 neat cattle, 3813 sheep, 22,236 swine; and produced 14,743 bushels of wheat, 8549 of rye, 329,528 of Indian corn, 14,094 of oats, 16,325 of potatoes, 3,257,359 pounds of cotton. It had one commercial house in foreign trade, 10 retail stores, nine grist-mills; two saw-mills, one college, 29 students; three academies, 90 students. Pop.: whites, 4214; slaves, 4165; free coloured, 43; total, 8422. Capital, Marion.

TWINSBURG, p. t., Summit co., O., 142 m. N.E. Columbus, 300 W. Named from twin brothers who settled here in 1827. Drained by Tinker's creek, which flows into Cuyahoga r. It contains two churches, two stores, several mills; one academy, 70 students; eight schools, 385 scholars. Pop. 1039.

TYLER, county, Va. Situated in the N.W. part of the state, and contains 855 sq. m. Bounded N.W. by Ohio river, and drained by its tributaries, Middle Island, Fishing and other creeks. It contained in 1840, 6955 neat cattle, 11,047 sheep, 12,854 swine; and produced 52,730 bushels of wheat, 1375 of rye, 293,079 of Indian corn, 57,535 of oats, 34,349 of potatoes, 1116 pounds of tobacco, 46,995 of sugar. It had 13 stores, one flouring-mill, 14 grist-mills, 18 saw-mills, three tanneries, one distillery, one pottery; 13 schools, 436 scholars. Pop.: whites, 6864; slaves, 85; free coloured, 5, total, 6954. Capital, Middlebourn.

TYNE, an important river in the N. of England, is formed by the junction of two very considerable streams, the N. and S. Tyne. The latter rises on the borders of Durham and Cumberland, near Cross-Fell, one of the highest mountains in the great central range; and the former in the moorlands of Northumberland close to the Scottish border. They unite a short way from Hexham. After their junction, the river takes an easterly direction: and dividing Northumberland from Durham, and passing Newcastle, falls into the sea at Tynemouth, having the towns of N. and S. Shields close to its embouchure.

The Tyne is navigable for ships of from 300 to 400 tons burden, as far as Newcastle, and is navigated a few miles farther by keels, a peculiar description of craft employed to carry coal to the coal-ships. The banks of the Tyne at Newcastle are steep, and the ground rises on each side to a considerable height. Down to a comparatively late period the salmon fisheries in this river were of considerable value and importance. In 1761, no fewer than 260 fish were caught at one draught at Newburn; and in 1775, 275 were landed at one draught at the Low Lights, near the mouth of the river. The fisheries have, however, for several years past, been all but annihilated; a circumstance which has been variously accounted for, but which, perhaps, is most properly to be ascribed to the locks constructed at Bywell to improve the navigation of the river, preventing the ascent of the fish in the breeding season to the shallow streams in the upper parts of the river. For an account of the trade and shipping of this river the reader is referred to the articles NEWCASTLE, SOUTH SHIELDS, and TYNEMOUTH, in this work.

TYNEMOUTH and NORTH SHIELDS, a parl. bor., co. Northumberland, on the N. bank of the Tyne, at its mouth in the German ocean, immediately opposite South Shields, and 7 m. E.N.E. Newcastle. The parl. bor., consisting of the townships of Tynemouth, N. Shields, Chirton, Preston, and Cullercoats, had in 1831, a pop. of 25,301, and in 1841, of 25,808. The township of Tynemouth occupies its most E. angle, and at its S.W. extremity is the town of N. Shields. The township of Chirton stretches along the whole S.W. side of the par. adjoining N. Shields. Preston adjoins both that town and the township of Tynemouth; it is of small extent, but contains some excellent houses, and a large population for its small area, as compared with some of the other townships. At the N. extremity of the township of Tynemouth is that of Cullercoats, which contains the fishing town of that name. This township comprises only about 5 or 6 acres of land, the greater part of which is covered with buildings. (Bound. Rep.) The village of Tynemouth has been much enlarged of late years; it is in general well built, and during the summer season is much resorted to for bathing. Its most remarkable edifice is the castle, originally a priory erected in the 11th century upon a previous foundation; it stands on a lofty and rocky peninsula, and is approached from the W. by a gateway flanked by towers, the whole being enclosed by a wall which runs for the most part along the edge of the cliff, at the N.E. angle of which is a lighthouse. Great care is taken by government to preserve the remains of the edifice, which forms a sea-mark for ships approaching the harbour or navigating the coast. The ruins of the priory within consist of a turret, now serving as a barrack, other buildings converted into military magazines, &c. Mr. Rickman says, that these exhibit very fine specimens of monastic remains, and the parts now existing, which belonged to the E. end, and some other portions of the church, are of remarkably beautiful design. The style is early English, with considerable enrichment, and though the stone is much decayed, it shows great delicacy of execution. (Goth. Archit.) A monument is about to be here erected in honour of Lord Collingwood. This church was parochial till 1657, when a new church was built near N. Shields. The living of Tynemouth, a vicarage, worth £298 a year, is in the gift of the duke of Northumberland. Two other churches have recently been erected; one at the W. end of the parish, and the other at the village of Tynemouth, which are curacies in the appointment of the vicar.

North Shields has increased rapidly of late years in size and importance, along with the increasing trade of the Tyne. It has many good streets and squares, a good market place, gas and water works, a public library, scientific institution, neat theatre, Lancastrian and other schools, a sailors' relief society, meeting-houses for most of the principal dissenting sects, and a weekly newspaper. The Master Mariners' asylum, the site for which was granted by his grace the duke of Northumberland, is a neat stone edifice, recently built on the road leading from North Shields to Tynemouth. "The condition of the town of North Shields is certainly prosperous; it is progressively increasing in importance. The manufactories in this town are merely of those articles which are required by the ship-builder. Much building is in progress; and many improvements have been both commenced and agreed upon. New roads are to be made through the parish towards the W. and N.W.; and a railroad, 6¼ m. in length, now (1842) connects this town with Newcastle, passing through the township of Chirton. The town is chiefly extending itself on the W. and N.W. into Chirton and Preston townships, and in the directions of the town of Tynemouth. There is no doubt that in a few years the village of Chirton will be nearly united to North

Shields by a continuous street, and that a great portion of the township of Preston will be occupied by that town. Already several buildings of a superior class have been erected in that township in its immediate neighbourhood; and the whole of a small part of that of Tynemouth, adjoining both Preston and Chirton is either covered with new buildings, or marked out as their future site. It is stated that the town had taken the direction of Tyemouth and Preston, rather than Chirton, in consequence of the land of the latter township adjoining that town being in settlement." (Bound. Rep.) There is a bar at the river's mouth, but at high water it may be crossed by vessels of 500 tons, and those of 300 tons lies close to the quays. Ships, as explained in the article on SOUTH SHIELDS (which see), lie in tiers in the river, and were formerly loaded with the coal brought down the river in lighters; but of late years, staithes and drops having been erected, the intervention of lighters is in a great measure superseded. The town as well as that of S. shields is included in the port of Newcastle; but, in 1840, there belonged to it especially, 337 ships; and during the same year nine vessels were built in the port, of the burden of 808 tons. The entrance to the Tyne is defended by Clifford's fort, on its N.E. bank, near which is the low lighthouse: the high lighthouse being on the hill opposite Dockwray-square. Its dependence on Newcastle is much complained of, as it obliges all ships to clear out from the latter. The shipping of coal is the staple business of the port.

The town is under the jurisdiction of the county magistrates, but has a court for the recovery of debts under 40s., held at Easter and Michaelmas, by the steward of Tynemouth manor. It is lighted and watched by commissioners under a local act. The erection of a bridge over the river was formerly contemplated; but that project has been abandoned, and the communication between the towns of North and South Shields is maintained by means of a steam ferry. The Reform Act conferred on the borough of Tynemouth the privilege of returning one member to the House of Commons. Registered electors, in 1839–40, 764.

TYRE (Τύρος), the principal city of Phœnicia, and the most celebrated emporium of the ancient world, on the S.E. coast of the Mediterranean, where the inconsiderable town of Tsour, with 1500 inhabitants, now stands, lat. 33° 17' N., long. 35° 14' 35" E. The harbour of the modern town is choked up; and we have introduced this article merely that we might have the opportunity of laying before the reader some statements respecting the commerce, arts, and navigation of the Tyrians, the most distinguished mercantile people of antiquity.

Tyre was founded by a colony from Sidon, the most ancient of the Phœnician cities. The date of this event is not certainly known, but Larcher supposes it to have been 1600 years before the Christian æra. (Chronologie d'Hérodote, cap. ii. p. 131.) It is singular that while Homer mentions Sidon, he takes no notice of Tyre, whose glory speedily eclipsed that of the mother city; but this is no conclusive proof that the latter was not then a considerable emporium. The prophets Isaiah, Jeremiah, and Ezekiel, who flourished from 700 to 600 years before Christ, represent Tyre as a city of unrivalled wealth, whose "merchants were princes, and her traffickers the honourable of the earth." Originally, the city was built on the main land; but having been besieged for a lengthened period by the Babylonian monarch Nebuchadnezzar, the inhabitants conveyed themselves and their goods to an island at a little distance, where a new city was founded, which enjoyed an increased degree of celebrity and commercial prosperity. The old city was, on that account, entitled Palætyre, and the other simply Tyre. The new city continued to flourish, extending its colonies and its commerce on all sides, till it was attacked by Alexander the Great. The resistance made by the Tyrians to that conqueror showed that they had not been enervated by luxury, and that their martial virtues were nowise inferior to their commercial skill and enterprise. The overthrow of the Persian empire was a less difficult task than the capture of this single city, which was not effected till a mound had been carried from the mainland to the island on which it was built. The victor had not magnanimity to treat the vanquished as their heroic conduct deserved. In despite, however, of the cruelties inflicted on the city, she rose again to considerable eminence. But the foundation of Alexandria, by diverting the commerce that had formerly centered at Tyre into a new channel, gave her an irreparable blow; and she gradually declined till, consistently with the denunciation of the prophet, her palaces have been levelled with the dust, and she has become "a place for the spreading of nets in the midst of the sea."

Commerce, Colonies, &c., of Tyre.—Phœnicia was one of the smallest countries of antiquity. It occupied that part of the Syrian coast which stretches from Aradus (the modern Round) on the N., to a little below Tyre on the S., a distance of about 30 leagues. Its breadth was much less considerable, being for the most part bounded by mount Libanus to the E., and mount Carmel on the S. The surface of this narrow tract was generally rugged and mountainous: and the soil of the valleys, though moderately fertile, did not afford sufficient supplies of food to feed the population. Libanus and its dependent ridges were, however, covered with timber suitable for ship building; and besides Tyre and Sidon, Phœnicia possessed the ports of Tripoli, Byblos, Berytus, &c. In this situation, occupying a country unable to supply them with sufficient quantities of corn, hemmed in by mountains, and by powerful and warlike neighbours, on the one hand, and having on the other, the wide expanse of the Mediterranean, studded with islands, and surrounded by fertile countries, to invite the enterprise of her citizens, they were naturally led to engage in maritime and commercial adventures; and became the boldest and most experienced mariners, and the greatest discoverers, of ancient times.

From the remotest antiquity, a considerable trade seems to have been carried on between the eastern and western worlds. The spices, drugs, precious stones, and other valuable products of Arabia and India have always been highly esteemed in Europe, and have exchanged for the gold and silver, the tin, wines, &c., of the latter. At the first dawn of authentic history, we find Phœnicia the principal centre of this commerce. Her inhabitants are designated in the early sacred writings by the name of Canaanites,—a term which, in the language of the east, means merchants. The products of Arabia, India, Persia, &c., were originally conveyed to her by companies of travelling merchants, or caravans: which seem to have been constituted in the same way, and to have performed exactly the same part in the commerce of the east, in the days of Jacob, that they do at present. (Genesis xxxvii. 25, &c.) At a later period, however, in the reigns of David and Solomon, the Phœnicians, having formed an alliance with the Hebrews, acquired the ports of Elath and Eziongeber, at the N.E. extremity of the Red sea. Here they fitted out fleets, which traded with the ports on that sea, and probably with those of southern Arabia, the W. coast of India, and Ethiopia. The ships are said to have visited Ophir; and a great deal of erudition has been expended in attempting to determine the exact situation of that emporium or country. We, agree, however, with Heeren in thinking that it was not the name of any particular place; but that it was a sort of general designation given to the coasts of Arabia, India, and Africa, bordering on the Indian ocean; somewhat in the same loose way as we now use the terms East and West Indies. (See the chapter on the *Navigation and Commerce of the Phœnicians*, in the translation of Heeren's work.)

The distance of the Red sea from Tyre being very considerable, the conveyance of goods from the one to the other by land must have been tedious and expensive. To lessen this inconvenience, the Tyrians, shortly after they got possession of Elath and Eziongeber, seized upon Rhinocolura, the port in the Mediterranean nearest to the Red sea. The products of Arabia, India, &c., being carried thither by the most compendious route, were then put on board ships, and conveyed by a brief and easy voyage to Tyre. If we except the transit by Egypt, this was the shortest and most direct and for that reason, no doubt, the cheapest, channel by which the commerce between southern Asia and Europe could then be conducted. But it is not believed that the Phœnicians possessed any permanent footing on the Red sea after the death of Solomon. The want of it does not, however, seem to have sensibly affected their trade; and Tyre continued, till the foundation of Alexandria, to be the grand emporium for eastern products, with which it was supplied by caravans from Arabia, the bottom of the Persian gulf, and from Babylon, by way of Palmyra.

The commerce of the Phœnicians with the countries bordering on the Mediterranean was still more extensive and valuable. At an early period they established settlements in Cyprus and Rhodes. The former was a very valuable acquisition, from its proximity, the number of its ports, its fertility, and the variety of its vegetable and mineral productions. Having passed successively into Greece, Italy, and Sardinia, they proceeded to explore the southern shores of France and Spain, and the northern shores of Africa. They afterwards adventured upon the Atlantic, and were the first people whose flag was displayed beyond the Pillars of Hercules.[*]

Of the colonies of Tyre, Gades, now Cadiz, was one of the most ancient and important. It is supposed by M. de St. Croix to have originally been distinguished by the name of Tartessus or Tarshish, mentioned in the sacred writings (*De l' Etat et du Sort des Anciennes Colonies*, p. 14.) Heeren on the other hand, contends, as in the case of Ophir, that by Tarshish is to be understood the whole southern part of Spain, which was early occupied and settled by Phœnician colonists. (See also *Huet, Commerce des Anci-*

---

amnes, cap. 8.) At all events, however, it is certain that Cadiz early became the centre of a commerce that extended all along the coast of Europe as far as Britain, and perhaps the Baltic. There can be no doubt that by the Cassiterides or Tin islands, visited by the Phœnicians, is to be understood the Scilly islands and Cornwall. The navigation of the Phœnicians, probably also extended a considerable way along the western coast of Africa; of this, however, no details have reached us.

But of all the colonies founded by Tyre, Carthage has been by far the most celebrated. It was at first only a simple factory; but was materially increased by the arrival of a large body of colonists, forced by dissensions at home to leave their native land, about 883 years B.C. (St. Croix, p. 90.) Imbued with the enterprising mercantile spirit of their ancestors, the Carthaginians rose in a no very long period to the highest eminence as a naval and commercial state. The settlements founded by the Phœnicians in Africa, Spain, Sicily, &c., gradually fell into their hands; and after the destruction of Tyre by Alexander, Carthage engrossed a large share of the commerce of which it had previously been the centre. The history, commerce, and institutions of Carthage, and the misfortunes by which she was overwhelmed, have, however, been already noticed in this work (See CARTHAGE); and we shall only, therefore, observe, that commerce, instead of being, as some shallow theorists have imagined, the cause of her decline, was the real source of her power and greatness; the means by which she was enabled to wage a lengthened, doubtful, and desperate contest with Rome herself for the empire of the world.

The commerce and navigation of Tyre probably attained their maximum from 650 to 550 years B.C. At that period the Tyrians were the factors and merchants of the civilized world; and they enjoyed an undisputed pre-eminence in maritime affairs. The prophet Ezekiel (chap. xxvi., has described in magnificent terms the glory of Tyre, and has enumerated several of the most valuable productions found in her markets, and the countries whence they were brought. The fir trees of Senir (Hermon), the cedars of Lebanon, the oaks of Bashan (the country to the E. of Galilee), the ivory of the Indies, the fine linen of Egypt, and the purple and hyacinth of the isles of Elishah (Peloponnesus), are specified among the articles used for her ships. The inhabitants of Sidon, Arvad (Aradus), Gebel (Byblos) served her as mariners and carpenters. Gold, silver, lead, tin, iron, and vessels of brass; slaves, horses, mules, sheep, and goats; pearls, precious stones, and coral; wheat, balm, honey, oil, spices, and gums; wine, wool, and silk; are mentioned as being brought into the port of Tyre by sea, or to be made by land, from Syria, Arabia, Damascus, Greece, Turkish, and other places, the exact site of which it is difficult to determine.[†]

Such, according to the inspired writer, was Tyre, the "Queen of the waters," before she was besieged by Nebuchadnezzar. But, as has been already remarked, the result of that siege did not affect her trade, which was so successfully and advantageously carried on from the new city as from the old. Inasmuch, however, as Carthage soon after began to rival her as a maritime and mercantile state, this may, perhaps, be considered as the era of her greatest celebrity.

It would not be easy to over-rate the beneficial influence of that extensive commerce from which the Phœnicians derived such immense wealth. It inspired the people with whom they traded with new wants and desires, at the same time that it gave them the means of gratifying them. It everywhere gave fresh life to industry, and a new and powerful stimulus to invention. The rude uncivilized inhabitants of Greece, Spain, and northern Africa, acquired some knowledge of the arts and sciences practised by the Phœnicians; and the advantages of which they were found to be productive secured their gradual though slow advancement.

Nor were the Phœnicians celebrated only for their wealth, and the extent of their commerce and navigation. Their fame, and their right to be classed among those who have conferred the greatest benefits on mankind, rest on a still more unassailable foundation. Antiquity is unanimous in ascribing to them the invention and practice of all those arts, sciences, and contrivances that facilitate the prosecution of commercial undertakings. They are held to be the inventors of arithmetic, weights and measures, of money, of the art of keeping accounts, and, in short, of everything that belongs to the business of a counting-house. They were, also, famous for the invention of ship-building and navigation; for the discovery of glass; for their manufactures of fine linen and tapestry; for their skill in architec-

---

[*] Mons Calpe and Mons Abyla, the Gibraltar and Ceuta of modern times.

[†] There is, in Dr. Vincent's *Commerce and Navigation of the Ancients*, and in the *Indian Ocean* (vol. ii. pp. 644—651), an elaborate and able notice of most parts of that work; probably even necessary on this chapter of Ezekiel, in which most of the names of the things and places mentioned are satisfactorily explained. (See also *Heeren on the Phœnicians*, cap. iv.)

ture, and in the art of working metals and ivory: and still more for the incomparable splendour and beauty of their purple dye. (See the learned and invaluable work of the President de Goguet, *Sur L'Origine des Loix, &c.*, English trans. vol. i. p. 296, and vol. ii. p. 95–100; see also the chapter of Heeren on the *Manufactures and Land Commerce of the Phœnicians*.)

But the invention and dissemination of these highly useful arts form but a part of what the people of Europe owe to the Phœnicians. It is not possible to say in what degree the religion of the Greeks was borrowed from theirs; but that it was to a pretty large extent seems abundantly certain. Hercules, under the name of Melcarthus, was the tutelar deity of Tyre; and his expeditions along the shores of the Mediterranean, and to the straits connecting it with the ocean, seem to be merely a poetical representation of the progress of the Phœnician navigators, who introduced arts and civilisation, and established the worship of Hercules, wherever they went. The temple erected in honour of the god at Gades was long regarded with peculiar veneration.

The Greeks, were, however, indebted to the Phœnicians, not merely for the rudiments of civilisation, but for the great instrument of its future progress—the gift of letters! No fact in ancient history is better established than that a knowledge of alphabetic writing was first carried to Greece by Phœnician adventurers; and it may be safely affirmed, that this was the greatest boon any people ever received at the hands of another.

Before quitting this subject, we may briefly advert to the statement of Herodotus with respect to the circumnavigation of Africa by Phœnician sailors. The venerable father of history mentions, that a fleet fitted out by Necho, king of Egypt, but manned and commanded by Phœnicians, took its departure from a port on the Red sea, at so epoch which is believed to correspond with the year 604 before the Christian æra, and that, keeping always to the right, they doubled the southern promontory of Africa; and returned, after a voyage of three years, to Egypt, by the Pillars of Hercules. (*Herod.* lib. iv § 42.) Herodotus further mentions, that they related that, in sailing round Africa, they had the sun on their right hand, or to the north—a circumstance which he frankly acknowledges seemed incredible to him, but which, as every one is now aware, must have been the case if the voyage were actually performed.

Many learned and able writers, and particularly Gosselin (*Recherches sur la Géographie Systématique et Positive des Anciens*, tome i. p. 204–217), have treated this account as fabulous. But the objections of Gosselin have been successfully answered in an elaborate note by Larcher (*Herodote*, tome iii. p. 458–464, ed. 1802); and Major Rennel has sufficiently demonstrated the practicability of the voyage. (*Geography of Herodotus*, p. 682, &c.) Without entering upon this discussion, we may observe, that not one of those who question the authenticity of the account given by Herodotus presume to doubt that the Phœnicians braved the boisterous seas on the coast of Spain, Gaul, and Britain; and that they had, partially at least, explored the Indian ocean. But the ships and seamen that did this much, might, undoubtedly, under favourable circumstances, double the cape of Good Hope. The relation of Herodotus has, besides, such an appearance of good faith, and the circumstance which he doubts, of the navigators having the sun on the right, affords so strong a confirmation of its truth, that there really seems no reasonable grounds for doubting that the Phœnicians proceeded, by 2000 years, Vasco de Gama in his perilous enterprise. (We have borrowed this article from the *Commercial Dict.*)

TYRINGHAM, p. t., Berkshire co., Mass., 130 m. W. Boston, 369 W. Watered by branches of Housatonic river. It contains three churches, a Congregational, Baptist, and Methodist, and a society of *Shakers*, four stores, one fulling-mill, two grist-mills, seven saw-mills, one paper-mill; 15 schools, 407 scholars. Pop. 1477.

TYROL and VORARLBERG (an. *Rhætia*, with part of *Noricum*), a prov. of the Austrian empire, principally between the 46th and 48th degs. of N. lat., and the 10th and 13th of E. long.; having E. the archd. of Austria (Salzburg, &c.), and Carinthia; S., the Lombardo-Venetian kingdom; W., Switzerland (the Grisons, &c.) and the princip. of Liechtenstein; and N., Bavaria. Length, E. to W., about 220 m.; average breadth somewhat less than 100 m. Area estimated at about 11,000 sq. m. Pop., in 1838, 831,998. This country may be regarded as an extension of Switzerland towards the E. It is traversed in its whole extent by the main ridge of the Alps, which has here some of its loftiest summits, including, among others, mount Orteler 12,823 ft., and the Gross Gluckner 12,567 ft. above the level of the sea. This grand chain separates the waters that flow N. to the Rhine and the upper Danube from those that flow S. to the Po and the Adriatic, and the lower Danube. But, exclusive of this gigantic chain, an inferior chain from 50 to 60 m. S. of the latter divides the country into three por-

tions: the valley of the Inn, to the N. of the high Alps; that of the Drave, between the high Alps and the inferior chain; and the country to the S. of the latter drained by the Adige, Piave, and other rivers flowing into the Adriatic. The Vorarlberg, N.W. from the Tyrol, forms part of the basin of the Rhine, being drained by the Ill and Bregenz, and bounded on the N.W. by the lake of Constance. There are many small lakes in the country, but none are of any consequence: the Achensee, in the S. is the principal. The climate is very various. To the N. of the high Alps, and in the intermediate district, or valley of the Drave, it is very severe. Some very extensive mountain tracts are covered with immense glaciers and the accumulated snows of ages. The medium temperature of the year at Innsbruck is about 50°; at Botzen, or Bolzano, 57° Fahr. But the narrow valleys in the S., which unite with the plain of Lombardy, are very hot in summer; and frequently, indeed, experience the sirocco. In general the spring and summer are wet, and autumn is the most agreeable season.

The central chain of the Alps is composed chiefly of granite, flanked on either side with a zone of slate, overlapped by limestone: the accompanying ranges on both the N. and S. sides are mostly calcareous. Estimating the total extent of land in the prov. at about 7,000,000 acres, it has been distributed to the S. of Becker's *Handel's Lexicon*, Vienna, 1836, as follows: viz., cultivated or arable land, 536,520 acres: vineyards, 78,636 do.: meadows and gardens, 615,620 do.; commons, 929,503 do.; and forests, 2,767,490 do., making in all 4,990,873 acres; leaving consequently, above 2,000,000 acres of land occupied by inaccessible mountains, glaciers, and snow-tracts. The products and husbandry in the S. are much the same as in the N. parts of Lombardy. In other parts of the Tyrol, maize, wheat, and pulse are grown in the bottoms, and scanty crops of buckwheat, rye, and oats on the mountain sides; but the produce of corn is insufficient for the consumption. The Tyrol, is in fact, like the greater part of Switzerland, a pastoral country; the chief wealth of its inhabitants consisting in their cattle and other live stock. The cattle are kept in the valleys throughout the winter, but are in spring driven to the uplands, proceeding higher and higher as the lower meadows become exhausted and the upper divested of snow; and returning again in September. The meadows yielding the thickest grass are set aside for a hay crop. The hay, when cut, is carefully dried under cover, and stored up in sheds; but it is quite insufficient for the winter supply of the cattle, many of which have to be fed on maize stalks, ash leaves, &c. In the circle of Roveredo, and other parts of the country adjoining Italy a good many silk-worms are reared: and the annual average produce of silk is estimated at 3900 centners. The rearing of canary birds, though apparently an insignificant branch of industry, is pretty extensively carried on at Imst, and other places in the valley of the Inn; and, in fact, the Tyrol supplies most parts of Europe with these songsters. Among the wild animals are wolves, wild boars, and bears; the clefts of the rocks afford shelter to the marmots; and the chamois finds refuge on the highest summits, or in places secure from the approach of the hunter.

The precious metals and copper are met with; but they are of little importance. Iron and salt are abundant in certain districts; and though mining industry be in a rather backward state, are produced in considerable quantities. Silk is manufactured in the S.; next to which, iron ware, plates, nails, and other kinds of hardware are the principal products. Leather, linen fabrics, wooden articles (some of which are executed with great skill, and display much ingenuity), glass, paper, toys, and (at Innsbruck, Imst, &c.), some cotton goods are produced. But the principal exports are cattle, cheese, silk, iron, salt, wine, timber, tobacco, and other raw produce, in return for corn and most sorts of manufactured goods. The inhabitants are exceedingly industrious, ingenious, and inventive; but the poverty of the country obliges them, notwithstanding, to migrate in great numbers; and several thousands annually leave their homes for Swabia, Bavaria, Italy, and more distant countries, where they exercise various functions, and continue for a longer or shorter time till, by dint of economy, they have saved what they suppose will maintain them at home, when they immediately return. A considerable transit trade is carried on across the Tyrolese Alps, between Italy and the S. parts of Germany. The principal route for this is by the road over the Brenner pass, between Innsbruck and Brixen; and thence to Bolzano and Roveredo. At its highest point this road attains to an elevation of 4634 ft. At the W. extremity of the Tyrol is the famous military road over Monte Stelvio, rising to the prodigious height of 8960 ft. above the sea! being the highest elevation of any carriage road in Europe.

In the official registers of the Austrian empire, Tyrol bears only the appellation of county; but it has its own diet composed of four orders of members—the clergy, nobility, the deputies of the towns, and those of the peasantry, all of whom assemble in one chamber. No new tax can be imposed without the consent of this body; and when it is granted, the sovereign is bound to make an explicit acknowledgment that the states might have refused it had they chosen. In addition to the states, there exists a permanent deputation and tribunal, in which the peasantry are represented. The only imposts are a land tax, payable indiscriminately by all classes, and a charge on the higher classes, consisting of a per centage on pensions, tithes, and rents.

The Tyrol is divided into seven circles, their chief towns being Boizen, Schwatz, Imst, Bruneck, Trent, Roveredo, and Bregenz; in each of which is a court of justice. Innsbruck is the general cap. and the seat of the highest judicial tribunal. The population is almost wholly Roman Catholic, under the superintendence of ten bishops subordinate to the archbishop of Salzburg.

The character of the Tyrolese is said to contrast favourably with that of the Swiss. In the N. or German portion of the country they are neither so fawning nor mercenary as the latter; and in the S. they approach the Italian standard in their manners and disposition as well as their language. Though quite as attached to personal and national liberty as the Swiss, the Tyrolese have always been steadfast adherents of Austria; and, next to the archduchy, the Tyrol may be depended upon as the prov. most likely to remain firmly attached to the house of Hapsburg in the event of any future dismemberment of the empire. But the Austrians draw little or no disposable military force from the Tyrol. Its inhabitants form an irregular militia, and act with the greatest vigour and alacrity in the defence of their country; but their natural repugnance to a disciplined military life is so great, that all attempts to extend the conscription to this prov. have hitherto proved unsuccessful. Of late, indeed, a part of the regiment of *Jägers*, raised in the Tyrol for its own defence, has been removed into another prov.; but this measure appears to have produced much dissatisfaction.

The dress of the peasantry is peculiar. The principal finery of the men consists of a straw hat ornamented with ribands and nosegays: the dress of the women consists of a thick and short gown, stockings with cross stripes, and a cap tapering in the shape of a sugar-loaf. Music and dancing, rifle-shooting and athletic exercises, are the favourite amusements of the Tyrolese; in all which they excel.

From the fall of the Roman empire, this region ceased to be permanently united under one head till 1296; not long after which period, it passed by inheritance to the dukes of Austria, to whose descendants it has ever since belonged, with the exception of the period from 1806 to 1814. From 1804 to 1809 it belonged to Bavaria. The government of the Bavarians was, however, very disturbed; and the Tyrolese under Hofer maintained a doubtful contest with them and the French till 1810; when Hofer, having been taken and shot at Mantua, the Tyrol was governed by the French till 1814, when it reverted to Austria. (*Austrian Nat. Encyc.; Berghaus; Malte-Brun, &c.*)

TYRONE, an inland co. of Ireland, prov. Ulster, having N. Londonderry, E. lough Neagh and Armagh, S. Monaghan and Fermanagh, and W. Donegal. It contains 754,365 acres, of which 171,314 are unimproved mountain and bog, and 27,951 water, being a fraction of lough Neagh. Surface in many places, especially on the N. and W., rough and mountainous; but there is, notwithstanding, a large extent of fertile land. Property mostly in very large estates. Farms of various sizes : those in the mountainous districts large, and seldom much subdivided. Tillage farms small, and generally held under partnership leases; and it is almost needless to add that wherever this is the case agriculture is execrable. A great deal of work is done by the spade; and where ploughs are used, they are sometimes drawn by horses, bullocks, and milch cows, all yoked together! Potatoes and oats, the principal crops. Cattle and sheep very inferior. "Tenants may do what they will in regard to the management of their farms, if they only pay the rent." (*Poor Inquiry*, Append. F., p. 323.) Average rent of land, 14s. 6d. an acre. Habitations of the bulk of the people extremely mean : they live principally on oatmeal and potatoes, rarely tasting butchers' meat. Linen manufacture pretty generally diffused. A coal mine is wrought between Dungannon and Stewartston, but the coal is inferior. There is a good pottery near Dungannon. This is one of the counties in which illicit distillation was most prevalent. Principal rivers, Blackwater, Foyle, Ballinderry, with several others of inferior importance.

Tyrone contains four baronies, and 35 parishes; and returns three members to the House of Commons, being two for the county, and one for the borough of Dungannon. Registered electors for the county in 1839-40, 3594. In 1831 Tyrone had 54,063 inhabited houses, 57,035 families, and 304,468 inhabitants, of whom 149,410 were males, and 155,058 females.

TYRONE, p. t., Stenben co., N. Y., 302 m. W. by S. Albany, 310 W. It contains two churches, a Methodist and

Presbyterian, six stores, two fulling-mills, two grist-mills, six saw-mills; 13 schools, 660 scholars. Pop. 2132.

TYRONE, t., Perry co., Pa., 7 m. S.W. Bloomfield. Drained by Sherman's creek and its branches. It contains the county poorhouse, nine stores, two furnaces, two fulling-mills, one woollen factory, three flouring-mills, four grist-mills. 11 saw-mills, five tanneries, three distilleries, three potteries; 13 schools, 640 scholars. Pop. 2391.

TYRONE, t., Huntingdon co., Pa. It has four stores, two forges, one fulling-mill, four grist-mills, four saw-mills, three tanneries, one distillery; six schools, 331 scholars. Pop. 1288.

TYRREL, county, N. C. Situated in the E. part of the state, and contains 740 sq. m. Bounded N. by Albemarle sound, E. by the Atlantic, along which are Roanoke and several other islands belonging to it. Drained by Alligator river, which enters Albemarle sound by a broad estuary. It contained in 1840, 4123 neat cattle, 1947 sheep, 6549 swine; and produced 8083 bushels of wheat, 105,641 of Indian corn, 2517 of oats, 21,704 of potatoes, 90,335 pounds of rice, 22,963 of cotton. It had 22 stores, six grist-mills, one saw-mill, 69 tanneries, 16 distilleries. Pop. : whites, 3160 ; slaves, 1411; free coloured, 86; total, 4657. Capital, Columbia.

TYSDRUS or TYSDRA, an ancient and considerable, but now ruined, city of N. Africa, reg. of Tunis, at present represented by the inconsiderable village of El Jemme, 110 m. S. by E. Tunis, and 30 m. W. by S. the port of Mehedlah, or Africa. The walls of the ancient town may still be distinctly traced, and it comprises, besides the foundations of temples and other buildings, the mutilated fragments of columns, statues, &c. But the distinguishing feature of the place, and that which gives it all its present interest, is its superb amphitheatre. This noble ruin, the exterior of which is in a high state of preservation, is of vast size and magnificence, being 420 feet in length, by 368 feet in breadth, and 96 feet in height, so that it is inferior only, in respect of magnitude, to the Colosseum and the amphitheatre of Verona. It consisted originally of 64 arches, and four rows of columns of the Composite order, placed above each other. At each extremity was a grand entrance; but one of these, with an arch on either side, was destroyed, about a century ago, by one of the beys of Tunis, to prevent the ruin being occupied as a fortress by his rebellious subjects. The arena is nearly circular. There are no inscriptions by which we may learn the date or founders of this magnificent structure; but Dr. Shaw supposes, from its similarity to other structures of the same period, that it is of the age of the Antonines; and as the elder Gordian was proclaimed emperor in Tysdrus, he concludes that he may, probably out of gratitude to the citizens, have presented them with this grand structure for the celebration of those barbarous sports then so much in fashion. (*Travels.* p. 117, &c. ed.; see also *Temple's Excursions in Algiers and Tunis*, i., 138.)

U.

UDINE, a town of Austrian Italy, cap. deleg. Udine or Friuli, on the Roja, 60 m. N.E. Venice. Pop. in 1837, 19,282. (*Berghaus.*) It is fortified and well built, but, from its situation in a wide and level plain, its external appearance has nothing striking. Its streets are lined with arcades; and in the great square is a fine monument in commemoration of the treaty of Campo Formio (which village is in the immediate neighbourhood). Principal buildings, the cathedral, with some handsome marble columns and bas-reliefs, two parish churches, the chief guard-house, surmounted by a tower and two iron figures to strike the hours, the town hall, bishop's palace, a good opera-house, &c. The old castle, on elevated ground in the middle of the town, is now a prison. The French, during their occupation, constructed several public walks and otherwise embellished the town. Udine is the seat of the provincial assembly and superior courts, and has a lyceum, two gymnasia, a high school, episcopal seminary and library, a society of agriculture, several hospitals, asylums, &c. The inhabitants are principally engaged in the silk trade, but they also manufacture linen fabrics, leather, paper, and liquors. (*Aust. Nat. Encyc., &c.*)

UIST, NORTH AND SOUTH. See HEBRIDES.

ULEABORG, a town and seaport of Finland, cap. län, or district of its own name, on a peninsula at the mouth of the Ulea in the gulf of Bothnia, 66 m. S.S.E. Tornea. Pop. about 5000. It is regularly built, and is, after Abo, the principal commercial town of the province. Its harbour is, unfortunately, in great part choked up with sand. The chief exports are pitch, tar, fish, and salted butter. It was founded in 1605, and has frequently suffered from fire, by which it was nearly destroyed on May 25, 1822. (*Schnitzler, La Russie; Possart.*)

ULM, a frontier town of Würtemberg, cap. circ. Danube,

on the Danube where it begins to be navigable, 45 m. S.E. Stuttgard, and 44 m. W. by N. Augsburg. Pop. about 14,000. It has an antiquated appearance, and, though it has some traffic, a garrison, &c., it is dull. The cathedral, a fine Gothic building, has an unfinished tower, 337 German feet in height. The body of the building is 416 feet in length, 166 feet in breadth, the nave being 152 feet in height (Stern), so that it is larger than any other church in Germany. This edifice was erected, between 1377 and 1494, at the sole expense of the citizens of Ulm. It has some beautiful stained glass and carved work, and a tablet commemorating a showman's feat of the Emperor Maximilian, in 1492, who is said to have stood on the parapet of the tower, on one foot, balancing a coach-wheel with the other! Several other buildings are worthy of notice, as the town-hall, government and custom houses, corn-hall, and arsenal. Ulm has a gymnasium, a large and richly-endowed hospital, a female orphan asylum, and the house of correction for the circle. Tobacco pipe bowls, linen fabrics, leather, paper, vinegar, &c., are made by the inhabitants, many of whom also engage in horticulture, boat-building, the transit of goods, and the rearing of snails for export to Bavaria and Austria. Large quantities of Rhenish, Swiss, and other wines, are brought thither to be shipped down the Danube. Ulm was formerly strongly fortified, and a military post of importance. (*Borghaus; Memminger's Wirtemburg; Spencer's Germany.*)

In 1805, Ulm was the theatre of some most important military events. Austria, having declared war against France, pushed forward a strong army into Bavaria, under General Mack, who established his head quarters at Ulm. But Napoleon having succeeded by a series of masterly manoeuvres in cutting off Mack's communications with Austria, the latter was cooped up in the city with all that portion of his army, amounting to about 26,000 men, that had not already fallen into the hands of the French. Considering the strength of the place, and the numbers of the garrison, a vigorous resistance might have been anticipated; but, instead of this, Mack capitulated on the 17th of October, and delivered up the town, and his army as prisoners of war, without so much as firing a shot!

ULSTER, one of the provinces into which Ireland is divided, and the most northerly, comprising the counties of Donegal, Londonderry, Antrim, Tyrone, Fermanagh, Monaghan, Armagh, Down, and Cavan.

ULSTER, county, N. Y. Situated toward the S.E. part of the state, and contains 1096 sq. m. The Shawangunk and Catskill mountains cover a considerable portion of its surface. Watered by Rondout, Esopus, and Shawangunk creeks, which afford good water-power. The Delaware and Hudson canal passes in a N.E. direction, through the county, and terminates at Eddyville, on Rondout creek, 3 m. from Hudson river. Water limestone is found in large quantities on Rondout creek. It contained in 1840, 38,459 neat cattle, 50,840 sheep, 46,228 swine; and produced 57,877 bushels of wheat, 168,809 of rye, 161,805 of Indian corn, 305,087 of buckwheat, 293,133 of oats, 364,696 of potatoes, 28,945 pounds of sugar. It had four commission houses in foreign trade, 146 retail stores, 21 lumber-yards, one furnace, 12 forges, eight fulling-mills, 11 woollen factories, one cotton factory, with 2136 spindles, two flouring-mills, 42 grist-mills, 120 saw-mills, two paper-mills, one powder-mill, two glass factories, 33 tanneries, two potteries, three printing-offices, three weekly newspapers; one academy, 35 students; 165 schools, 6880 scholars. Pop. 45,822. Capital, Kingston.

ULVERSTONE, a market town and par. of England, hund. Furness, co. Lancaster, about 2 m. from the W. side of the embouchure of the Leven in Morecambe bay, on a tract apparently abandoned by the sea. 14 m. N.W. Lancaster. Area of par., 29,100 acres. Pop., in 1831, 7741. Of late years the town has greatly improved: it has now a neat theatre, two subscription libraries, news and assembly-rooms, &c.; and, besides the parish church, Trinity church has been erected within these few years. The livings of both are perpetual curacies, in the gift of — Braddyll, Esq.; that of the parish is worth £149, and that of Trinity church £143 a year. The parish church is a handsome modern structure, in the style that prevailed in the time of Henry VIII., with a good altar-piece of the Descent from the Cross, and an E. window of stained glass. There are also meeting houses for Dissenters, and some public schools of a minor kind. In 1795 a canal was cut from the river Leven, by which vessels of 400 tons reach a large basin, and load or unload close to the town. The inhabitants principally manufacture cotton goods, canvass hats, &c., and are occupied in conveying coastwise copper and iron ore, limestone, corn, and slates; the latter being exported in large quantities. Ulverstone belongs to the port of Lancaster. It is the seat of petty sessions for the hundred, manorial courts or the recovery of debts under 40s., &c.

ULYSSES, t., Tompkins co., N. Y., 6 m. N.W. Ithaca.

Bounded E. by Cayuga lake, and drained by its tributaries. Halsey creek has falls of 386 feet in the course of a mile, and a single cascade is 216 feet high, and sometimes 60 feet wide and 2 feet thick, affording great water-power. It contains three churches, a Presbyterian, Methodist, and Baptist, 15 stores, one fulling-mill, three saw-mills, two tanneries, one distillery; 14 schools, 865 scholars. Pop. 2976.

UMBAGOG, lake, is a beautiful sheet of water, on the boundary between N. H. and Me., 18 m. long, and 10 m. wide at its greatest width. Its outlet constitutes a considerable branch of Androscoggin river.

UNADILLA, p. t., Otsego co., N. Y., 94 m. W. by S. Albany, 334 W. Drained by Unadilla and Susquehanna rivers. It contains an Episcopal church, six stores, one fulling-mill, two grist-mills, two saw mills; 11 schools, 496 scholars. Pop. 2272.

UNDERHILL, p. t., Chittenden co., Vt. 15 m. N.E. Burlington, 56 m. N.W. Montpelier, 532 W. Chartered in 1763, first settled in 1786. Watered by head branches of Brown's river. It contains two churches, a Congregational and Methodist, two stores, 10 saw-mills; eight schools, 370 scholars. Pop. 1441.

UNION, county, Pa. Situated near the centre of the state, and contains 520 sq. m. Bounded E. by Susquehanna river and its W. branch. Drained by Penn's, Buffalo, White, Deer, Middle and West Mahaantango creeks. It contained in 1840, 14,665 neat cattle, 18,196 sheep, 16,578 swine; and produced 310,010 bushels of wheat, 135,367 of rye, 172,191 of Indian corn, 94,461 of buckwheat, 263,501 of oats, 107,570 of potatoes, 8000 pounds of tobacco. It had 51 stores, 13 fulling-mills, one woollen factory, two furnaces, one bloomery, 13 flouring-mills, 32 grist-mills, 75 saw-mills, six oil-mills, 24 tanneries, 18 distilleries, two breweries, eight potteries, six printing-offices, one weekly newspaper; two academies, 45 students; 49 schools, 2540 scholars. Pop. 22,787. Capital, New Berlin.

UNION, district, S. C. Situated in the N. toward the W. part of the state, and contains 340 sq. m. Bounded N.E. by Broad river, S.W. by Ennoree river. Drained by Tiger and Packolet rivers and their branches. It contained in 1840, 17,063 neat cattle, 7535 sheep, 32,215 swine; and produced 61,661 bushels of wheat, 1578 of rye, 757,319 of Indian corn, 63,405 of oats, 23,930 of potatoes, 1,612,594 pounds of cotton. It had 25 stores, two smelting houses, producing gold to the amount of $9375, five flouring-mills, 20 grist-mills, 12 saw-mills, seven tanneries, 15 distilleries; six academies, 240 students; 32 schools, 738 scholars. Pop.: whites, 10,485; slaves, 8354; free coloured, 97; total, 18,936. Capital, Unionville.

UNION, county, Ga. Situated in the N. part of the state, and contains 600 sq. m. Drained by Hiwassee river and its branches. It contained in 1840, 3051 neat cattle, 2593 sheep, 11,235 swine; and produced 7343 bushels of wheat, 2071 of rye, 129,340 of Indian corn, 8005 of oats, 8788 of potatoes, 6092 pounds of tobacco. It had two stores, 15 grist-mills, six saw-mills, one tannery, nine distilleries. Pop.: whites, 3065; slaves, 87; total, 3152. Capital, Blairsville.

UNION, parish, La. Situated in the N. part of the state, and contains 1200 sq. m. Bounded E. by Washita river, and drained by its tributaries. It had in 1840, six schools,

1738 scholars. Pop. 8422. Capital, Marysville.

UNION, county, Ia. Situated in the E. towrad the S. part of the state, and contains 234 sq. m. Drained by the E. fork of Whitewater river and its branches. It contained in 1840, 7945 neat cattle, 14,160 sheep, 23,271 swine; and produced 80,890 bushels of wheat, 3052 of rye, 512,340 of Indian corn, 136,097 of oats, 9312 of potatoes, 61,394 pounds of sugar. It had 20 stores, five fulling-mills, two woollen factories, one cotton factory, with 195 spindles, six flouring-mills, 12 grist-mills, 21 saw-mills, three oil-mills, six tanneries, 10 distilleries; 19 schools, 646 scholars. Pop. 8017. Capital, Liberty.

UNION, county, Ill. Situated toward the S. part of the state, and contains 380 sq. m. Bounded W. by Mississippi river. Drained by Clear creek and its branches. It contained in 1840, 6620 neat cattle, 4769 sheep, 22,968 swine; and produced 26,898 bushels of wheat, 310,559 of Indian corn, 52,913 of oats, 13,708 of potatoes, 3002 pounds of tobacco, 7464 of sugar. It had 15 stores, one flouring-mill, 20 grist-mills, three saw-mills, three tanneries, 15 distilleries, one pottery; one academy, 30 students; 12 schools, 387 scholars. Pop. 5594. Capital, Jonesboro'.

UNION, county, Ark. Situated in the S. part of the state, and contains 2600 sq. m. Drained by Washita river and its branches. It contained in 1840, 6695 neat cattle, 789 sheep, 9099 swine; and produced 1098 bushels of wheat, 111,365 of Indian corn, 1765 of oats, 10,021 of potatoes, 404,599 pounds of cotton. It had six stores, 20 grist-mills, two saw-mills; four schools, 64 scholars. Pop.: whites, 1081; slaves, 906; free coloured, 2; total, 9889. Capital, Union C.H.

UNION, p. t., Lincoln co., Me., 31 m. S.E. Augusta, 616 W. Drained by St. George and Muscongus rivers. Incorporated in 1786. It has six stores, three fulling-mills, four grist-mills, eight saw-mills, one paper-mill; 13 schools, 723 scholars. Pop. 1784.

UNION, p. t., Broome co., N. Y., 146 m. W.S.W. Albany, 268 W. Bounded S. by Susquehanna river. Drained by Nanticoke river and other tributaries of Susquehanna river. It has six stores, one fulling-mill, three grist-mills, 98 saw-mills, two tanneries; one academy, 65 students; 23 schools, 981 scholars. Pop. 3165.

UNION, p. t., Essex co., N. J., 6 m. S.W. Newark, 48 m. N.E. Trenton, 214 W. Watered by Elizabeth and Rahway rivers. It contains a Presbyterian church, two stores, two grist-mills, four saw-mills; one academy, 18 students; five schools, 183 scholars. Pop. 1462.

UNION, t., Berks co., Pa. Bounded N.E. by Schuylkill river. It has four stores, one furnace, two forges, one flouring-mill, two grist-mills, two saw-mills. Population, 1372.

UNION, t., Luzerne co., Pa. Bounded S.E. by the N. branch of Susquehanna river, and drained by its tributaries. The North Branch canal passes through it. It has two stores, two grist-mills, five saw-mills; eight schools, 285 scholars. Pop. 1242.

UNION, t., Fayette co., Pa. Drained by Redstone creek. It abounds with iron ore, and has two stores, one furnace, seven grist-mills, 14 saw-mills, two tanneries, one distillery; nine schools, 405 scholars. Pop. 2723.

UNION, p. v., capital of Monroe co., Va., 229 m. W. Richmond, 266 W. It contains a courthouse, jail, a Presbyterian and a Methodist church, four stores, two tanneries, and about 600 inhabitants.

UNION, t., Belmont co., O. Drained by Stillwater creek. The National road passes through it from E. to W. It has five schools, 134 scholars. Pop. 2196.

UNION, t., Brown co., O. Bounded S.W. by Ohio river. Drained by Red Oak and Eagle creeks. It has one college, 22 students; one academy, 45 students; four schools, 190 scholars. Pop. 3316.

UNION, t., Butler co., O. It has three schools, 98 scholars. Pop. 1828.

UNION, t., Clermont co., O. It has three stores, one flouring-mill, one saw-mill; three schools, 115 scholars. Pop. 1423.

UNION, t., Clinton co., O. It has eight stores, one fulling-mill, two woollen factories, one flouring-mill, three grist-mills, six saw-mills, one oil-mill, two printing-offices, two weekly newspapers; 15 schools, 1218 scholars. Pop. 2984. It contains Williamsburg village on Little Miami river, the former capital of the county.

UNION, t., Fayette co., O. It contains Washington village. It has 11 stores, one flouring-mill, one grist-mill, two saw-mills, two printing-offices, one weekly newspaper; 13 schools, 400 scholars. Pop. 1945.

UNION, t., Miami co., O. It has seven schools, 397 scholars. Pop. 1967.

UNION, t., Muskingum co., O. It has four churches; one college, 40 students; two schools, 58 scholars. Pop. 1625.

UNION, t., Warren co., O. It contains Deerfield village, and has seven schools, 435 scholars. Pop. 1617.

UNION, t., Shelby co., Ia. It has four grist-mills, three saw-mills, six distilleries. Pop. 1550.

UNION, t., Montgomery co., Ia. It has four schools, 90 scholars. Pop. 2943.

UNION, p. v., capital of Franklin co., Mo., 73 m. E. Jefferson city, 883 W. Situated on a branch of Maramec river, 7 m. S. Missouri river. It contains a courthouse, jail, four stores, and about 200 inhabitants.

UNION, p. v., capital of Union co., Ark., 206 m. S. Little Rock, 1251 W. Situated on the S.W. side of Washita.

river, and contains a courthouse, jail, several stores, and about 100 dwellings.

UNION TOWN, p. b., capital of Fayette co., Pa., 179 m. W. by S. Harrisburg, 195 W. Situated on Redstone creek. It contains a courthouse, jail, six churches, a Presbyterian, Episcopal, Methodist Episcopal, Methodist Protestant, Baptist, and an African Methodist, Madison college founded by the Methodists in 1825, chartered in 1827, 23 stores, one furnace, two grist-mills, three saw-mills, three tanneries, one distillery, one pottery, three printing-offices, three weekly newspapers, and two periodicals; one college, 119 students; five schools, 350 scholars. Pop. 1710.

UNIONVALE, t., Dutchess co., N. Y., 12 m. E. Poughkeepsie, 85 m. S. Albany. Drained by branches of Fishkill creek. It contains a Methodist church, six stores, one cotton factory, with 544 spindles, one flouring-mill, two grist-mills, two saw-mills; seven schools, 250 scholars. Pop. 1468.

UNION VILLAGE, v., Warren co., O., 4 m. W. Lebanon. It has been built and is inhabited by Shakers, who have a large church in the centre. It consists of six families, the largest of which contains 100 persons, each with an extensive house, surrounded by out-houses and workshops, and delightful gardens under fine cultivation, with excellent orchards, and the best breeds of domestic animals. They exhibit great neatness and skill in their mechanical productions.

UNITED STATES (THE) of America occupy the middle division of North America, constitute the most extensive...

The territory of the United States is divided into 26 states, and three territories, each of which has a separate government, and the District of Columbia, which is under the immediate jurisdiction of the United States Congress. The following is a list of the states, which are divided into the Northern or Eastern, the Middle, the Southern, and the Western states, with their population in 1840.

States and Territories.	Total Pop.	States and Territories.	Total Pop.
Northern or Eastern States.		Alabama	
Maine	501,793	Mississippi	
New Hampshire	284,574	Louisiana	
Vermont	291,948	*Western States.*	
Massachusetts	737,699	Ohio	
Rhode Island	108,830	Kentucky	
Connecticut	309,978	Tennessee	
Middle States.		Michigan	
New York	2,428,921	Indiana	
New Jersey	373,306	Illinois	
Pennsylvania	1,724,033	Missouri	
Delaware	78,085	Arkansas	
Maryland	470,019	*Territories.*	
Southern States.		District of Columbia	
Virginia	1,239,797	Florida	
North Carolina	753,419	Wisconsin	
South Carolina	594,398	Iowa	
Georgia	691,392	Total	

* The last three will probably be soon admitted into the Union.

Washington, on the Potomac river, in the District of Columbia, is the seat of government of the United States, having become such in the year 1800.

The country is intersected by two principal ranges of mountains, the Alleghany and the Rocky mountains. The Alleghany mountains in the E. run nearly parallel with the Atlantic coast, from Georgia, through Tennessee, Virginia, and Pennsylvania to New-York. The Rocky mountains in the W. run across the territory nearly parallel with the coast of the Pacific ocean, at the distance from it of several hundred miles. The Alleghany mountains run in separate and somewhat parallel ridges, with a breadth of from 60 to 190 m., and at a distance from the sea-coast of from 80 to 250 m., and terminate in the Catskill mountains E. of Hudson river, though some would extend them to the White mountains in New-Hampshire. The general height of the Alleghany mountains is nowhere above 2080 or 3000 feet above the level of the ocean, and not more than one half of that above the adjacent country. The highest peak in this range is Black mountain, in the W. part of North Carolina, which is 6476 feet high. Round Top, the highest peak in the Catskill mountains, is 3804 feet. The Rocky mountains may be regarded as a part of the great chain of the Cordilleras, and extend from Mexico to the 70° N. lat., running at an average distance of 600 m. from the Pacific ocean, with a general height of about 8000 or 9000 feet above the level of the sea, but not more than 5000 feet above the level of their base. Some of their elevated peaks rise to the height of 10,000 or 12,000 feet high. The Green mountains, a minor range, commence near New-Haven, Connecticut, and run through that state, Massachusetts and Vermont to the borders of Canada. Mansfield mountain, the highest peak in the chain, in a township of the same name in Vermont, is 4280 feet high. The White mountains in New-Hampshire are very elevated, the highest peak, mount Washington, is 6428 feet high. West of lake Champlain, in the state of New-York, are the Adirondack mountains, the highest peak of which, mount Marcy, is 5460 feet high.

West of the Rocky mountains the rivers generally flow W. to the Pacific, the principal of which is the Columbia. The rivers between the Alleghany and Rocky mountains are the Mississippi and its numerous tributaries, which flow into the gulf of Mexico, with the exception of a few of the smaller class which flow into the northern lakes. East of the Alleghany mountains the rivers flow into the Atlantic. The following are the principal rivers which flow into the Atlantic, with about their length in miles, Penobscot, 250; Kennebec, 200; Androscoggin, 170; Saco, 160; Merrimac, 200; Connecticut. 410; Hudson, 324; Delaware, 300; Susquehanna, 450; Potomac, 500; James, 500; Roanoke, 400; Cape Fear, 350; Pedee, 450; Santee, 450; Savannah, 500; Alatamaha, 400; St. Johns, 300. The following rivers flow into the gulf of Mexico: Appalachicola, 500; Alabama, 600; Tombigbee, 450; Mississippi, 3000. The following are tributaries of the Mississippi, Red, 1500; Arkansas, 2150; White, 1200; Missouri, before its junction, 3180; Kansas, 1100; Platte, 1600; Osage, 500; Yellowstone, 1100; Ohio, 1350; Illinois, 500; Des Moines, 800; Tennessee, 900; Cumberland, 600; Wabash, 500. The following rivers are W. of the Rocky mountains: Columbia, 1500; Multnomah, 200; Lewis's, 900; Clark's, 600. These are mostly include their remote sources. The two largest lakes that lie wholly in the United States are Michigan, 330 m. long and 60 m. broad, and Champlain, 120 m. long and 15 m. broad. But the great lakes Superior, 380 m. long and 130 m. broad; Huron, 240 by 150 m.; Erie, 240 by 60 m.; and Ontario, 190 by 55 m., are one half in the United States, the boundary between the United States and Canada passing through the middle of them. The gulf of Mexico on the S. of the United States, is a large branch of the Atlantic ocean, and receives the waters of the Mississippi valley. The principal bays are the Chesapeake, Delaware, Massachusetts and Penobscot. The principal capes are Ann, Cod, May, Henlopen, Charles, Henry, Hatteras, Lookout, Fear, Florida and Sable.

The soil of New-England is generally rocky and rough, better adapted to grazing than to grain. The valley of the

Connecticut, and some parts of Maine are exceptions to this remark. The low country on the Atlantic coast is generally light and sandy, not very fertile excepting on the margins of the rivers. The hilly country back of this is generally fertile. The soil generally in the valley of the Mississippi has great fertility. But toward the Rocky mountains it becomes sterile. The country beyond the Rocky mountains, with some exceptions, is but moderately fertile.

Beef, pork, butter, and cheese are the principal productions of the eastern states, though a great amount of wool is raised, and various grains for home consumption. Wheat is the staple of the middle states. In the northern portion of the southern states, wheat and tobacco are chiefly cultivated; and in the low country at the south, cotton, rice, and sugar are extensively raised. The western states are the granary of the United States, and indeed of the world, and it is scarcely possible to set bounds to the bread stuffs which they are capable of producing.

Among the mineral treasures of the United States, iron ore, coal and limestone are very extensive. The anthracite coal of Pennsylvania is inexhaustible, and the bituminous coal farther west is equally abundant. The lead region of Missouri, Illinois, Wisconsin, and Iowa, is probably the finest in the world. Gold is found to a considerable extent in Virginia, North Carolina, and Georgia, and marble and gypsum are very abundant.

The climate of the United States has great variety, extending as it does through more than 90 degrees of latitude, with a great variety in the elevation of its surface. In the northern part it is subject to great extremes of heat and cold, but is generally healthy. The Atlantic coast S. of New-Jersey and the borders of the gulf of Mexico, have an unhealthy climate from July to November. Back from the sea-coast, the elevated country is generally healthy, as are the western states, with the exception of some low and marshy portions.

The principal harbours of the United States proceeding from the N. to the S. are Portland, Portsmouth, Boston, Newport, New-London, New-York, Philadelphia, Baltimore, Norfolk, Charleston, Savannah, Mobile, and New-Orleans. Some of these harbours are unsurpassed in the world. The following are the most populous cities and towns in the United States, with their number of inhabitants in 1840: New-York, 312,710; Philadelphia, 228,691; Baltimore, 102,313; New Orleans, 102,193; Boston, 93,383; Cincinnati, 46,338; Brooklyn, 36,233; Albany, 33,721; Charleston, 29,261 (see Charleston); Washington, 23,364; Providence, 23,171; Louisville, 21,210; Pittsburg, 21,115; Lowell, 20,796; Rochester, 20,191; Richmond, 20,153; Troy, 19,334; Buffalo, 18,213; Newark, 17,290; St. Louis, 16,469; Portland, 15,218; Salem, 15,082.

In its commerce, the United States is the second country on the globe, being inferior only to Great Britain. In 1840, the capital invested in foreign trade by importing and commission merchants was $119,295,307; in domestic retail dry goods and other stores, $350,301,799; in the fisheries, $16,439,020. The registered tonnage of the United States for the year ending September 30th, 1842, was 975,354; the enrolled and licensed tonnage was 1,045,753; and of fishing vessels was 71,278; making a total of 2,092,390. Of the registered and enrolled tonnage there were employed in the whale fishery 151,612.

STATEMENT of the Commerce of each State and Territory, commencing on the 1st October, 1841, and ending on the 30th September, 1842.

| States and Territories | VALUE OF IMPORTS. | | | VALUE OF EXPORTS. | | | | | | | Total of domestic and foreign produce |
| | | | | Domestic Produce. | | | Foreign Produce | | | |
	In American vessels.	In foreign vessels.	Total.	In American vessels.	In foreign vessels.	Total.	In American vessels.	In foreign vessels.	Total.	
	Dollars.	Dollars.	Dollars.	Dollars.	Dollars.	Dollars.	Dollars.	Dollars.	Dollars.	Dollars
Maine	547,956	56,905	604,861	925,743	47,429	1,043,172	1,797	5,554	7,351	1,050,593
New-Hampshire	55,256	5,226	60,481	24,897	3,712	28,419		124	124	28,517
Vermont	200,468		200,468	550,293		550,393	7,216		7,219	557,509
Massachusetts	16,495,976	1,090,460	17,586,433	5,906,854	890,287	6,719,115	2,988,640	694,355	3,087,995	9,807,110
Rhode Island	280,362	5,364	225,692	329,962	4,625	323,457	25,250		25,250	348,696
Connecticut	349,580	6,127	325,707	531,313	1,279	532,592				532,592
New-York	51,883,665	699,680	57,825,404	16,580,510	4,158,476	20,739,082	4,788,522	1,089,920	6,337,492	27,076,775
New-Jersey	145		145	64,921		64,921	5,976		5,976	70,907
Pennsylvania	6,757,920	528,880	7,285,985	3,785,281	508,503	3,393,814	304,127	82,786	476,913	3,770,727
Delaware	1,212	1,345	3,557	50,369	4,708	55,863				55,986
Maryland	3,998,965	418,713	4,417,078	3,536,501	1,099,006	4,636,507	233,017	56,949	289,968	4,901,765
District of Columbia	23,984	5,122	29,066	398,961	177,989	494,920	1,394	921	2,455	501,673
Virginia	275,536	39,169	316,705	3,301,417	443,316	3,715,217	5,150		5,159	3,730,386
North Carolina	131,555	5,849	137,401	340,375	34,375	341,850				344,850
South Carolina	1,042,436	317,011	1,360,465	5,097,823	3,410,466	7,508,309	6,373	10,751	17,394	525,795
Georgia	230,525	111,330	341,784	3,861,824	1,687,347	4,529,151	130	976	1,106	4,800,257
Alabama	239,170	123,701	363,371	5,957,570	4,948,105	9,905,675				9,965,675
Mississippi										
Louisiana	6,179,987	1,854,683	8,066,690	21,508,399	5,910,108	27,427,423	882,801	254,656	998,737	28,404,149
Ohio	12,179	872	13,051	591,504	208,282	809,786				809,786
Kentucky	17,306		17,306							
Tennessee	5,687		5,687							
Michigan	79,392	929	80,754	282,329		282,329				282,329
Missouri	31,137		31,137							
Florida	164,412	13,568	178,980	23,383	9,223	32,606	2	778	775	33,384
Total	84,734,290	11,437,897	100,162,057	71,187,634	21,604,303	92,969,928	8,425,850	3,296,149	11,721,519	104,691,494

UNITED STATES.

STATISTICAL View of the Commerce of the United States, exhibiting the Value of Imports from, and Exports to each Foreign Country; also the Tonnage of American and Foreign Vessels arriving from, and departing to, each Foreign Country, during the Year ending on the 30th September, 1842.

Countries.	COMMERCE.				NAVIGATION.			
	Value of Imports.	Value of Exports.			American Tonnage.		Foreign Tonnage.	
		Domestic Produce.	Foreign Produce.	Total.	Entered into the United States.	Cleared from the U. States.	Entered the United States.	Cleared from the U. States.
Russia	$1,350,106	$316,096	$520,567	$836,593	8,068	5,691	1,597	1,530
Prussia	18,192	149,141	7,547	156,688	603			2,963
Sweden	890,934	238,948	105,970	344,918	3,304	1,311	13,391	4,440
Swedish West Indies	23,342	139,727	3,320	133,047	1,206	2,063	72	726
Denmark		70,766	27,819	98,585	453	785	231	917
Danish West Indies	584,391	791,628	157,960	949,088	21,698	26,749	1,334	781
Holland	1,067,438	3,226,238	396,968	3,623,296	94,502	33,349	2,988	12,884
Dutch East Indies	741,048	85,578	193,580	279,158	4,861	794		
Dutch West Indies	331,370	251,650	15,581	267,331	8,974	4,254	766	599
Dutch Guiana	74,764	101,055		101,055	3,900	5,454		
Belgium	619,588	1,434,038	176,646	1,610,684	19,132	19,949	7,840	19,673
Hanse Towns	2,974,019	3,814,994	749,519	4,564,513	14,195	16,779	46,988	54,666
England	33,446,499	36,681,808	2,932,140	39,613,948	307,943	285,679	141,989	130,854
Scotland	655,050	1,522,735	80,979	1,603,014	4,736	6,380	27,722	19,645
Ireland	102,700	49,968		49,968	3,309	631	28,787	
Gibraltar	12,268	486,937	115,981	589,998	3,997	19,115		1,726
Malta	7,300	11,644	8,981	19,906	581	726		
British East Indies	1,530,364	399,979	283,825	682,904	10,009	9,678	985	1,199
Mauritius						925	380	
Australia	28,693	59,651		58,651	1,905	1,787		
Cape of Good Hope	23,815							
British African ports					318	378		117
British West Indies	896,481	3,904,346	23,367	3,927,713	64,363	58,691	27,408	36,632
British Guiana	15,004	115,991	2,462	118,453	2,445	5,334	7,908	3,965
Honduras	202,808	197,339	36,648	163,987	5,371	3,678	274	17
British American colonies	1,762,001	5,950,143	240,166	6,190,309	334,634	382,215	28,580	67,668
Other British colonies					68			
France on the Atlantic	16,015,280	15,340,798	1,076,684	16,417,412	116,358	138,865	11,877	10,848
France on the Mediterranean	958,678	1,674,570	73,868	1,748,438	15,597	20,344	2,065	3,147
Bourbon								598
French African ports		3,899	80	3,979				502
French West Indies	199,160	495,397	23,609	519,006	13,396	28,769	4,139	1,390
French Guiana	50,172	44,063	1,030	45,093	1,998	1,519	382	477
Miquelon and French fisheries		4,938		4,938		2,665		203
Hayti	1,966,907	844,459	55,514	899,966	96,531	91,115	698	203
Spain on the Atlantic	79,735	333,392	1,900	334,489	11,946	11,636	698	1,202
Spain on the Mediterranean	1,065,640	921,898	16,578	938,476	16,587	5,319	2,800	73
Teneriffe and other Canaries	91,411	12,793	518	13,941	1,856	498	1,833	612
Manilla and Philippine islands	773,372	935,732	100,444	326,176	7,817	4,797	511	
Cuba	7,650,494	4,197,468	573,981	4,770,449	170,797	188,456	14,737	9,718
Other Spanish West Indies	2,517,001	610,813	19,718	630,531	56,635	29,585	1,304	1,006
Portugal	142,587	72,723	1,388	74,111	8,990	3,205	1,981	287
Madeira	146,182	43,054	1,939	44,984	1,944	9,253		
Fayal and other Azores	41,049	49,183	19,600	68,783	2,376	1,695	166	
Cape de Verd islands	17,866	103,557	11,529	115,086	446	3,919		
Portuguese African ports						198		
Italy	987,598	515,577	304,940	820,517	4,589	7,387	1,881	1,668
Sicily	539,419	237,261	195,797	433,658	18,369	1,979	4,167	3,663
Sardinia		40,208		40,908	314	1,153	552	770
Ionian islands	14,994				315			
Trieste	413,210	748,179	136,526	884,705	4,547	16,589	338	301
Turkey	370,948	125,521	76,515	202,036	4,957	1,615		
Morocco, &c.	4,779							
Texas	480,802	278,978	127,951	406,929	92,490	94,316	1,708	1,390
Mexico	1,905,696	969,371	564,862	1,534,233	13,481	15,918	1,806	1,080
Venezuela	1,544,342	499,380	166,832	666,212	12,987	9,742	9,706	3,811
New Grenada	176,216	57,363	46,361	103,724	1,637	1,615	744	146
Central America	194,994	46,849	23,617	69,466	3,931	698		146
Brazil	5,948,814	2,295,571	375,931	2,601,502	37,058	39,778	5,500	5,605
Argentine Republic	1,835,623	265,336	145,905	411,261	11,617	9,190	9,988	
Cisplatine Republic	581,918	201,999	67,968	980,967	6,104	14,215	936	619
Chili	831,039	1,270,941	368,735	1,639,676	3,072	7,609		
Peru	904,768				316			
South America generally		147,222	1,900	148,423		1,587		
China	4,934,645	737,509	706,888	1,444,397	12,125	7,230	303	300
Europe generally		19,290		19,390				
Asia generally	979,689	283,367	294,914	578,981	3,961	6,156		177
Africa generally	539,458	472,841	51,135	523,976	8,195	6,662	286	730
West Indies generally		205,913	1,790	207,703		16,680	71	
Atlantic ocean					9,892	9,886		
South seas	41,747	128,856	17,394	146,380	30,946	50,481		
Sandwich islands					799	518		
Northwest coast of America			2,370	2,370		982		
Uncertain places	10,144							
Total	100,162,087	92,969,996	11,721,538	104,691,534	1,510,111	1,536,451	728,775	765,452

ABSTRACT of the Tonnage of the principal Districts of the United States on the 30th June, 1843.

New-York city	496,965	Bath, Me.	Salem, Mass.
Boston, Mass.	202,509	Waldoborough, Me. 51,401	Belfast, Me.
New-Orleans, La.	144,408	Baltimore, Md. 50,434	Penobscot, Me.
Philadelphia, Penn.	104,349	Barnstable, Mass. 74,695	Sag Harbour, N. Y.
New-Bedford, Mass.	100,081	New-London, Conn. 41,650	Portsmouth, N. H.
Portland, Me.	56,272	Nantucket, Mass. 35,391	Charleston, S. C.
999		34,349	

UNITED STATES.

SUMMARY STATEMENT of the Value of the Exports, of the Growth, Produce, and Manufacture of the United States during the Year commencing on the 1st day of October, 1841, and ending on the 30th day of September, 1842.

Articles.	Value.			Articles.	Value.		
	Dollars.	Dollars.	Dollars.		Dollars.	Dollars.	Dollars.
THE SEA.				Hats		63,863	
Fisheries—				Saddlery		95,808	
Dried fish, or cod fisheries		587,788		Wax		103,626	
Pickled fish, or river fisheries, herring, shad, salmon mackerel				Spirits from grain		50,708	
		188,394		Beer, ale, porter, and cider		51,674	
Whale and other fish oil		1,315,411		Snuff and tobacco		525,490	
Spermaceti oil		228,114		Lard		682,428	
Whale oil		228,388		*Linseed oil and spirits of turpentine*		34,773	
Spermaceti candles		319,927		Cordage		20,457	
			2,823,010	*Iron—*			
THE FOREST.				Pig, bar, and nails		188,454	
Skins and furs		508,687		Castings		69,507	
Ginseng		65,702		All manufactures of		920,561	
Product of Wood—				Spirits from molasses		247,745	
Staves, shingles, boards, hewn timber	2,509,287			Sugar, refined		281,499	
Other lumber	253,981			Chocolate		3,094	
Masts and spars	37,780			Gunpowder		161,292	
Oak bark and other dye	111,082			Copper and brass		97,021	
All manufactures of wood	625,718			Medicinal drugs		188,315	
Naval Stores—				*Cotton, piece goods—*			4,458,671
Tar, pitch, rosin, and turpentine	743,329			Printed and coloured	695,040		
Ashes, pot and pearl	292,741			White	3,397,364		
		4,896,673		Twist, yarn, and thread	37,925		
AGRICULTURE.			5,515,292	All other manufactures of	256,261		
Product of Animals—				*Flax and Hemp—*			
Beef, tallow, hides, horned cattle	1,219,498			Bags, and all manufactures of		2,370,690	
Butter and cheese	208,185			Wearing apparel		1,080	
Pork (pickled), bacon, lard, live hogs	2,658,408			Combs and buttons		58,319	
Horses and mules	399,654			Brushes		34,714	
Sheep	36,992			Billiard tables and apparatus		1,986	
Vegetable Food—				Umbrellas and parasols		1,300	
Wheat	916,616			Leather and morocco skins not sold per lb.		5,388	
Flour	7,375,366			Printing presses and types		22,502	
Indian corn	545,150			Fire engines and apparatus		19,611	
Indian meal	617,517			Musical instruments		1,304	
Rye meal	124,396			Books and maps		16,255	
Rye, oats, and other small grains and pulse	175,092			Paper and stationery		44,546	
Biscuit, or ship bread	323,750			Paints and varnish		69,982	
Potatoes	58,844			Vinegar		27,370	
Apples	32,945			Earthen and stone ware		10,206	
Rice	1,907,387			*Manufactures of*		7,618	
		11,903,908		Glass		38,748	
Tobacco			16,475,494	Tin		5,898	
Cotton			9,540,785	Pewter and lead		16,789	
Other Agricultural Products			47,593,464	Marble and stone		13,921	
Flaxseed		34,891		Gold and silver, and gold leaf		1,323	
Hops		96,547		Gold and silver coin		1,170,754	
Brown sugar		8,998		Artificial flowers and jewellery		7,859	
Indigo		1,042		Molasses		19,040	
			81,670	Trunks		3,916	
MANUFACTURES.				Bricks and lime		5,726	
Soap and tallow candles		485,198		Domestic salt		39,064	
Leather, boots, and shoes		168,925		*Articles not enumerated—*			4,614,491
Household furniture		290,987		Manufactured		508,976	
Coaches and other carriages		48,509		Other articles		1,359,163	
							1,868,139
				Total			94,369,996

The United States are chiefly an agricultural people, to which they are led by the extent of their territory, and the fertility of the soil; and the agricultural resources of the nation are becoming yearly more and more developed. The following agricultural statistics are derived from the census of the United States for 1840. There were 4,335,699 horses and mules; 14,971,586 neat cattle; 19,311,374 sheep; 26,301,293 swine; poultry was raised to the value of $9,344,410. There were produced 84,823,272 bushels of wheat; 7,291,743 of buckwheat; 377,531,875 of Indian corn; 18,645,567 of rye; 4,161,504 of barley; 123,071,341 of oats; 108,298,060 of potatoes; 35,802,114 pounds of wool; 219,163,319 of tobacco; 80,841,422 of rice; 790,479,275 of cotton; 155,100,809 of sugar; 1,238,502 of hops; 628,303 of wax; 61,552 of silk cocoons; 10,248,108 tons of hay; 95,251 of hemp and flax. There were about 29½ bushels of edible grains, exclusive of potatoes, to every individual of its population. The products of the dairy were valued at $33,787,008; of the orchard at $7,256,904; of lumber at $12,943,507. And there were also made 124,734 gallons of wine.

The manufactures of the United States, though not equal to its agriculture and commerce, and of recent origin, have already risen to great respectability. A large amount of capital has been invested in them, machinery has been extensively introduced, and they supply a great amount of articles for home consumption, and already, considerable for exportation. No country in the world can compete with the United States in the article of coarse cotton goods, neither as to quality nor price. Cottons which in 1812 were worth two shillings a yard, can now be bought of a better quality for sixpence. And even in the finer quality of goods, great advancement has been made. It is only since the peace of 1815 that manufactures have made great progress; though they were commenced in Rhode Island many years before, and had made some progress. It was the policy of the British government before the revolution to discourage American manufactures, and thus to keep the country in a state of great dependence. But that time has gone by; and should events ever cut off a supply of British manufactures, the country would be able to do without them. Unless Great Britain and other countries shall consent in a fair way, to receive American breadstuffs in exchange for their manufactures, the Americans will be compelled to become their own manufacturers, and they will thus secure their substantial independence. Home-made or family goods were produced in 1840, to the amount of $29,023,380. There were 1240 cotton factories with 2,284,631 spindles, which employed 72,119 persons, produced articles to the amount of $46,350,453, with a capital of 51,109,359. Woollen manufactures employed 21,342 persons, producing goods to the amount of $20,696,999, with a capital of $15,765,124; paper-mills employed a capital of $4,745,939; hats and caps were manufactured to the amount of $8,704,342, and straw bonnets to the amount of $1,476,505; 20,018 persons were employed in tanneries, with a capital of $15,650,929; saddleries, and other manufactories of leather, employed a capital of $12,981,989; carriages and wagons employed 21,904 persons, and produced to the amount of $10,697,887, with a capital of $5,551,632; mills of various kinds employed 60,788 persons, produced to the amount of $76,545,946, with a capital of $65,858,470; vessels were built to the amount of $7,016,094; furniture was made by 18,003 persons, and employed a capital of $6,989,971. There were 1582 printing-offices, 447 binderies, 138 daily, 125 semi-weekly or tri-weekly, and 1141 weekly newspapers, and 227 periodicals, the whole employing 11,593 persons, and a capital of $5,873,815. Iron manufactures employed a capital of $20,432,131; glass manufactures employed a capital of $2,084,100, producing articles to the amount of $2,890,293. The anthracite coal employed a capital of $4,355,609; bituminous coal a capital of $1,268,862; and lead a capital of $1,246,756. The total amount of capital employed in manufactures, was $267,726,579. For a more particular account of the manufactures, see the articles on the respective states.

The revenue of the United States has arisen chiefly from customs on imports, and from the sale of the public lands. The government has rarely found it necessary to resort to direct taxation. By these means, the national government was enabled, Jan. 1st, 1837, not only to complete the payment of the public debt contracted during two wars with Great Britain; but after reserving $5,000,000, they were able to distribute to the states the sum of $37,468,859, which by an act of June 23d, 1836, was deposited with them according to the number of their electoral votes, liable to be recalled in case of necessity, but which will probably never be recalled. The great expense of the late Indian war in Florida, and the diminution of the customs arising from the stagnation of trade, have caused the expenses of the government to exceed the revenue; so that a small debt has been contracted, amounting on December 1st, 1842, to $20,590,355.

The public lands have recently been a great source of revenue. These lands have been ceded to the United States by the new states, or have been derived from the purchase of Louisiana in 1803, and of Florida in 1819. They are considered as belonging to the native tribes of Indians who inhabit them, until the title has been regularly extinguished by purchase and by treaty. When this is done, they are surveyed, and sold at a dollar 25 cents the acre, as the lowest price. This source of revenue is much less considerable than formerly. In 1836, it amounted to the large sum of $25,167,000, but it has now diminished to less than $5,000,000 annually. The law for the distribution of the proceeds among the states, has been repealed. The revenue of the United States for the year ending December 15th, 1842, amounted, with a small balance in the treasury at the commencement of the year, to $34,733,077; and the expenditure to $35,208,634, leaving a deficiency in the treasury on the 31st of December, 1842, of $575,557. The United States have 272,646,356 acres of public land surveyed and unsold, and much more which is not surveyed.

The mint of the United States was established at Philadelphia in 1793; and in 1838 branches were established at Charlotte, N. C., at Dahlonega, Ga., and at New-Orleans, La. At the mint in Philadelphia, the whole coinage from the commencement amounted to 255,087,171 pieces, with a value of $35,873,052; at the branch of Charlotte, 162,118 pieces, with a value of $666,030; at the branch of Dahlonega, 178,534 pieces, with a value of $897,638; at the

branch of New-Orleans, 14,179,656 pieces, with a value of $3,155,443; making a total of 289,697,479 pieces, with a total value of $90,592,163.

In June 30th, 1843, there were 13,814 postoffices in the United States; the amount of transportation during the previous year was 35,252,805 miles, at a cost of $2,975,310. The total expenditures for the service of the year ending at that time was $4,374,713; the gross amount of the revenue for the year was 4,296,925.

STATEMENT of the Revenue and Expenditure of the Postoffice Department for the fourteen years ending the 30th of June, 1842.

Year ending	Revenue.	Expenditure.
	Dollars.	Dollars.
June 30, 1829	1,707,418.42	1,782,132.57
— 1830	1,850,526.10	1,976,877.97
— 1831	1,577,511.54	1,936,122.42
— 1832	2,258,570.17	2,266,171.98
— 1833	2,617,011.88	2,930,414.47
— 1834	2,823,749.34	2,910,946.39
— 1835	2,993,345.76	2,757,360
— 1836	3,408,323.28	2,841,766.34
— 1837	4,100,945.43	3,288,319
— 1838	4,238,077.77	4,021,937.16
— 1839	4,477,514.04	4,624,717.42
— 1840	4,539,265.68	4,759,110.34
— 1841	4,379,317.78	4,467,379.39
— 1842	6,546,349.13	4,656,462.17

The following exhibits a general view of the regular army of the United States, according to the law of 1842.

Commissioned officers	712
Eight regiments of infantry, each composed of non-commissioned officers, musicians, and privates, 510	4080
Four regiments of artillery, each composed of non-commissioned officers, musicians, and privates, 660.	2640
Two regiments of dragoons, each composed of non-commissioned officers, and privates, 660 .	1320
Cadets	250
Total	9012

The principal reliance of the country for defence is on the militia of the several states, exhibited in the following table, according to the returns received at the office of the adjutant-general of the army, as stated in the Army Register for 1843.

General Abstract of the Militia Force of the United States, as stated in the Army Register for 1843.

States and Territories.	General Officers.	General Staff Officers.	Field Officers, &c.	Company Officers.	Total Commissioned Officers.	Non-commissioned Officers, Musicians, and Privates.	Aggregate.
Maine . .	97	89	567	1,846	2,599	49,823	45,322
N. Hampshire .	9	30	337	1,289	1,665	30,806	32,471
Massachusetts .	9	30	98	416	553	86,682	87,215
Vermont . .	12	40	215	906	1,173	26,362	27,536
Rhode Island .	5	35	99	276	415	14,540	14,955
Connecticut .	9	30	311	1,059	1,409	45,061	46,470
New-York . .	135	863	2,590	6,576	10,164	178,915	181,659
New-Jersey .	19	58	435	1,476	1,988	37,183	39,171
Pennsylvania .	55	183	948	4,070	5,256	246,709	251,965
Delaware . .	4	8	71	363	447	8,726	9,389
Maryland . .	28	68	544	1,703	2,307	44,467	46,984
Virginia . .	28	61	1,961	4,740	6,890	105,898	111,982
N. Carolina .	28	67	793	2,969	3,787	61,431	63,898
S. Carolina .	20	134	436	1,897	2,487	48,079	51,366
Georgia . .	36	96	746	2,212	3,808	54,580	57,322
Alabama . .	31	187	584	1,382	2,184	42,168	44,322
Louisiana . .	10	46	183	542	781	14,027	14,880
Mississippi . .	15	70	388	348	895	35,929	36,804
Tennessee . .	25	79	859	2,644	3,607	67,845	71,362
Kentucky . .	43	150	1,046	3,825	4,864	72,412	77,376
Ohio . . .							188,988
Indiana . .	31	110	566	2,154	2,861	51,859	53,947
Illinois . .							69,891
Missouri . .	45	213	658	1,692	2,608	57,081	59,689
Arkansas . .					157	1,871	2,028
Michigan . .	6	11	97	466	580	19,286	19,866
Florida Territory		1	9	33	43	784	827
Wisconsin Ter.	1	6	36	126	169	5,064	5,233
Iowa Territory							
D. Columbia .	1	3	94	69	96	1,153	1,249
	627	2,670	13,813	44,938	62,925	1,385,845	1,711,582

The U. States have navy yards at the following places: Portsmouth, Boston, New-York or Brooklyn, Philadelphia, Washington, Norfolk, and Pensacola. In July, 1843, there were 67 captains; 94 commanders; 334 lieutenants; 69 surgeons; 9 passed assistant surgeons; 69 assistant surgeons; 64 pursers; 22 chaplains; 133 passed midshipmen; 410 midshipmen; 31 masters; 7 master's mates; 23 professors of mathematics; 3 teachers of naval schools; 37 boatswains; 40 gunners; 36 carpenters; 35 sailmakers. Of the marine corps, there were 1 colonel commandant; 1 lieutenant colonel; 4 majors; 13 captains; 20 first lieutenants; and 20 second lieutenants. The pay of the officers of the navy is greater than that of a similar grade in the land forces.

The navy of the United States is the most efficient for its

size of any in the world. In July,'1843, it contained the following ships and other vessels of war, with the number of guns attached.

Name and Rate.	Guns.	Name and Rate.	Guns.
Ships of the Line.—10.		Levant	20
Franklin	74	Saratoga	20
Columbus	74	Ontario	18
Ohio	76	Marion	16
North Carolina	74	Decatur	16
Delaware	74	Preble	16
Alabama	74	Yorktown	16
Vermont	74	Dale	16
Virginia	74		
Pennsylvania	120	*Brigs.*—3.	
New-York	74	Dolphin	10
		Porpoise	10
Frigates, 1st Class.—14.		Pioneer	
Independence, Razee	54	Consort	
United States	44	Bainbridge	10
Constitution	44	Perry	10
Potomac	44	Somers	10
Brandywine	44	Truxtun	10
Hudson	44		
Santee	44	*Schooners.*—9.	
Cumberland	44	Grampus	10
Sabine	44	Shark	10
Savannah	44	Enterprise	10
Raritan	44	Boxer	10
Columbia	44	Experiment	4
St. Lawrence	44	Flirt	
Congress	44	Wave	
		Phenix	
Frigates, 2d Class.—2.		On-ka-hy-e,	
Constellation	36		
Macedonian	36	*Steamers.*—4.	
		Fulton	
Sloops of War.—17.		Poinsett	4
John Adams	20	Mississippi	*10
Boston	20	Union	
Vincennes	20	Princeton	
Warren	20		
Falmouth	20	*Store Ships.*—3.	
Fairfield	20	Relief	6
Vandalia	20	Erie	6
St Louis	20	Lexington	6
Cyane	20		
* Paixhan Guns.			

The constitution of the United States forbids the establishment of religion by law: but every person who does not interrupt the peace of society, is protected in the free exercise of his religion. The voluntary principle, as it is sometimes called, has been found to be more efficient than any legal enactment for the support of religious institutions. The union of church and state is generally regarded with aversion by the people, as that which is calculated to corrupt religion, and to afford no benefit to the state; while no people are more deeply sensible of the fact, that the stability, and even the existence, of a free government depends upon the prosperity of religious institutions, and their moral influence upon the principles and habits of the people. This is the great secret of the success of the American experiment, which has presented an example of a free government that has furnished great encouragement to the friends of freedom throughout the world, and already exerts a salutary influence on every civilized nation of the earth. The puritan fathers of New-England came to this country to found a religious and intelligent community, they planted the church and the schoolhouse side by side, and the influence of their character, and the principles and habits which they formed and cherished, are now felt to the remotest ends of the nation, and form the surest basis for political and religious freedom. In accordance with the

amounts to the sum of $5,500,000 annually, and that there is on an average, one evangelical Protestant minister to rather less than 1135 souls; and if we were to add the Roman Catholics, Unitarians, Universalists, and other preachers of that class, one preacher to every 800 souls. Thus it appears that the voluntary principle in the United States is exceedingly efficient in the support of religious institutions.

ACCOUNT of the Churches or Congregations, Ministers, Communicants, &c., belonging to the principal Religious Bodies in 1842.

Denominations.	Churches or Congregations.	Ministers.	Members or Communicants.	Population.
Baptists	7,896	4,741	573,702	
— Freewill	961	647	47,917	5,000,000
— Seventh Day	48	54	5,009	
— Six-Principle	16	10	2,117	
Catholics	512	562	.	1,300,000
Christians	1,000	800	150,000	300,000
Congregationalists	1,390	1,150	160,000	1,400,000
Dutch Reformed	197	192	22,515	450,000
Episcopalians	950	1,099	55,497	600,000
Friends	500	.	.	100,000
German Reformed	600	180	30,000	
Jews	.	.	.	15,000
Lutherans	750	267	62,266	540,000
Menonites	200	.	30,000	
Methodists	.	10,971	906,363	3,500,000
Moravians or United Brethren	94	33	5,745	12,000
Mormonites	.	.	12,000	12,000
New Jerusalem Church	27	33	.	5,000
Presbyterians	2,807	2,295	274,084	
— Cumberland	500	450	50,000	2,175,000
— Associate	183	87	16,000	
— Reformed	40	20	3,000	
— Associate Reformed	214	116	12,000	
Shakers	15	45	6,000	6,000
Tunkers	40	40	3,000	30,000
Unitarians	300	200	.	200,000
Universalists	653	317	.	600,000

The people of the United States, from the first settlement of the country, have been very attentive to the cause of popular education, and this cause is continually gaining a stronger hold on the community. It is recommended by all the governors of the states in their annual messages to their respective legislatures. Most of the older states have a respectable funds devoted to the support of common schools, and in the new states the general government have provided funds for the support of schools, by setting apart one 6th section in each township, containing each 1 sq. m., for the purposes of common education. The amount of land already set apart for educational purposes E. of the Mississippi is computed to amount to 8,000,000 of acres. The same spirit is also extending W. of the Mississippi, and has penetrated even to the Indian tribes; and the Choctaw nation has applied $18,000 per annum out of the moneys which

they receive from the United States to the support of schools. Knowledge and virtue are regarded as the main pillars of the republic. In less than 90 years from the landing of the pilgrims on the rock of Plymouth, Cambridge college was founded, and numerous similar institutions (perhaps too many) have been successively established from that day to the present time. The following are among the principal colleges and universities in the country, with the date of their establishment: Cambridge college, now Harvard university, in 1638; Yale college, at New-Haven, in 1700; Nassau-hall, or the college of New-Jersey, at Princeton, in 1746; Brown university, at Providence, in 1764; Dartmouth college, at Hanover, New-Hampshire, in 1769; the university of North Carolina, at Chapel Hill, in 1789; Bowdoin college, at Brunswick, Maine, in 1794; and among the more recent institutions, the university of Nashville, Tennessee, in 1806; the university of Virginia, at Charlottesville, in 1819; Amherst college, at Amherst, Massachusetts, in 1821; and many others. Perhaps if fewer institutions had been chartered, and they had been more liberally endowed, the beneficial results would have been greater, though the number educated would probably have been less.

According to the census of 1840, there were in the United States 173 colleges or universities, with 16,233 students; 3242 academies, with 164,159 students; 47,909 common and primary schools, with 1,845,244 scholars. In the above enumeration, theological and medical institutions, where they are separate from colleges, are ranked among universities and colleges. In the academies, the ancient and modern languages, grammar, history, logic, rhetoric, natural and moral philosophy, &c., are taught. The common schools are extensively provided with libraries, and appropriate apparatus for illustrating the sciences taught in them.

There are 38 theological seminaries, belonging to different denominations, designed to succeed a collegiate course, some of which are connected with colleges. The principal of them are the theological seminary at Andover, Massachusetts, Congregational; the theological seminary at Princeton, Presbyterian; the theological seminary at Auburn, Presbyterian; the theological seminary of the Episcopal church, New-York, Protestant Episcopal; the theological institution at Newtown, Massachusetts, Baptist; and the theological departments of Yale college and Harvard university.

There are eight law schools in different parts of the country. The earliest institution of this kind was founded in 1798 by the Hon. Tapping Reeve, and taught afterward by him in connexion with the Hon. James Gould, both judges of the supreme court of Connecticut. At this institution many of the principal civilians in the United States have been educated. It is now discontinued.

There are 28 medical schools, some of them connected with colleges. The principal are the medical departments of the university of Pennsylvania; of Harvard university; of Yale college; of Dartmouth college; of Transylvania university; of the university of Maryland at Baltimore; of the university of New-York; the college of surgeons and physicians, New-York; the Louisville medical institution; the Vermont academy of medicine at Castleton; the Berkshire medical school at Pittsfield, &c.

It is astonishing to reflect on the great works of internal improvement which have been recently undertaken and effected in the United States. Steamboats have rendered the great rivers, especially of the west, available for the purposes of commerce. More recently, canals and railroads have been constructed, and they have bound together the different parts of the confederacy by the strong ties of interest. The complete success of the Erie canal, one of the first and greatest works undertaken, has entirely silenced all opposition to the work, and proved a powerful stimulus to other works of the kind. (For an account of these works see the descriptions of the particular states.) With the true Anglo Saxon spirit, the United States are a great people for going ahead, and in works of acknowledged utility they will not ordinarily stop until they get against an insurmountable obstacle. During the ruinous credit system of 1836 and 1837 many splendid projects were formed for works beyond the present necessities and means of the country, and debts were contracted beyond the power of the states to discharge. If the plan had been to complete whatever was undertaken, and to commence no farther than could be completed, there would be nothing to regret. These works, so far as completed, have generally yielded a reasonable profit. To be whirled through the country by the aid of an iron horse, that knows no fatigue, at the rate of from 15 to 30 m. an hour, and with less personal fatigue in going 300 m. than would be involved in walking 3 m., is extremely pleasant; and to transport goods from New-York city to remote parts of Michigan, for instance, for a dollar for 100 pounds, is an immense advantage to trade. Individuals may at times, though not generally, have lost by these undertakings, but the country has been a great gainer. These works remain, with all their immense advantages, whatever may have been the result to those who projected and completed them. It is to be hoped that a blow has been given to the credit system from which it will not soon recover. The resources of the country are great, and it will at length recover from its embarrassments. To the disgrace of some of the states, the debts which have been contracted have been partially repudiated; but the doctrine is generally abhorred, and will not ultimately prevail. Foreign creditors, when they trust individual states, should recollect that they are not dealing with the general confederacy, who are not answerable for the debts of the states individually contracted, though it is believed that no state can maintain a respectable standing in the confederacy which does not ultimately and honestly pay its debts. Some great works which have been commenced are sure yet to be completed. The railroad

through Long Island, which, with connected railroads, will constitute the most direct and expeditious route from New-York to Boston, has recently been completed. The Erie railroad will eventually go through; it may be with a hard struggle. Some of the most important southern and western railroads are making progress amid all the embarrassments of the country, and their completion will add to the general prosperity. The banking system of the country is coming to a sounder state; and those banks which do not pay specie for their notes when presented have been extensively obliged to wind up their concerns. On the whole, the country is prosperous in its business, and giving steadily forward in all its great interests; and whatever abuse or mismanagement may exist finds a corrective in public sentiment and controlling circumstances.

The government of the United States must appear extremely complicated to a foreigner accustomed to the unity of a monarchical government, and from this circumstance its operations are often not well understood. The whole country is subject to the national or general government, consisting of three branches, the executive, legislative, and judicial. The executive power is lodged in the hands of the president, who is appointed for four years, by electors chosen for that purpose by the people, generally by districts. Each state is allowed as many electors as it has members of Congress. These are chosen differently in different states; but if by districts, each district chooses one elector, for the sole purpose of electing a president and vice-president. The latter presides over the senate; and in case of the death of the president, he immediately occupies his place. In this case his office, otherwise almost nominal, becomes exceedingly important. The person who has a majority of all the votes is president; and if no one has such a majority, the house of representatives chooses a president from three candidates having the greatest number of votes. In the election of president the votes are given by states. The vice president is chosen at the same time and in the same manner. The president appoints the secretary of state, and the ministers of the various departments of the administration, as the treasury, navy, war, and the post-master-general, and, directly or indirectly, he appoints to all the offices in the national government; in the more important ones, with the advice and consent of the senate. The national legislature consists of a senate and house of representatives. The senate is composed of two persons from each state in the Union, and chosen by the legislatures of the states respectively for the term of six years. One third of the senate is chosen each year. The house of representatives consists of member chosen for two years by districts. The number of representatives is proportioned to the number of the inhabitants, and the ratio has been fixed at one for every 70,680 inhabitants, three fifths of the slaves being omitted in the enumeration. The house of representatives represents the people, the senate represents the states.

No person can be elected president who is less than 35 years of age, who is not a native born citizen of the United States, or was not a citizen at the time of the adoption of the constitution of the United States, and who has not been a resident in the United States for 14 years. He is the commander-in-chief of the army and navy, and of the militia when in the actual service of the United States. With the advice and consent of the senate, he makes treaties, appoints ambassadors, judges of the supreme court, and other officers of the national government, whose appointment is not otherwise provided for by the constitution. He takes care that the laws be executed, and commissions all officers. He has power to grant reprieves and pardons for all offences against the United States, except in case of impeachments. In making treaties, the concurrence of two thirds of the senate is necessary.

A senator must be at least 35 years of age, and have been a citizen of the United States for nine years. It belongs to the senate to try all cases of impeachment of the president and vice-president.

A representative must be at least 25 years of age, and have been a citizen of the United States for seven years. Congress has power to lay and collect taxes; to provide for the common defence and general welfare; to borrow money; to regulate foreign and domestic commerce; to establish uniform laws of naturalization and bankruptcy; to coin money and regulate its value; to fix the standard of weights and measures; to establish postoffices and postronds; to grant patent and copy rights; to constitute tribunals inferior to the supreme court; to define and punish piracies and offences on the high seas, and against the laws of nations; to declare war, and grant letters of marque and reprisal; make rules respecting captures; raise and support armies; provide and maintain a navy; provide for the calling out of the militia, to execute the laws of the Union, suppress insurrections, and repel invasions; and to exercise exclusive jurisdiction over the District of Columbia. No

UNITED STATES.

member of Congress is allowed to hold any office under the government of the United States while he continues such. All bills for raising money must originate in the house of representatives.

The judicial power is vested in a supreme court, consisting at present of nine judges, appointed by the president, with the consent of the senate. They can be removed by impeachment before the senate, and they hold a yearly winter session in Washington, the capital of the United States. When not thus united, they hold circuit courts in different parts of the country. The whole country is divided, also, into districts, each having a judge appointed by the president, with the consent of the senate, for the decision of causes that fall within the cognizance of the United States' courts, and from whose decisions an appeal lies to the supreme court. This last court decides on the constitutionality of the laws passed by the national congress, and also by the several state legislatures, and on all questions between individual states, or between the United States and an individual state, and questions arising between a foreigner and either the United States or any other state. Although the supreme court have at times decided that the laws of particular states have been unconstitutional, and of course void, their decisions have never yet been permanently resisted. The judges hold their offices during good behaviour, and their salaries cannot be diminished during their continuance in office.

Each individual state has a government for the regulation of its local and internal concerns, consisting of a governor, senate, and house of representatives, to which are reserved all the powers not expressly granted by the constitution to the general government. Without this provision a republican government could not probably be maintained over a country so widely extended. The states, though bearing a close general resemblance, differ in many of the details of their constitutions, as in the term during which the governor holds office, and the extent of his power, the terms for which the senators and representatives are elected, and for which the judges of the several courts are appointed, their salaries, &c. The territory of all the states is divided into counties, having courts of justice attached to each, and officers for many local purposes, as maintaining the roads, providing for the poor, &c. In South Carolina and Louisiana, parishes answer to counties. In New-England, New-York, New-Jersey, Pennsylvania, and most of the states in the valley of the Mississippi, the counties are divided into townships, averaging 6 or 8 m. square, which form important civil districts and corporations: the inhabitants meet once a year or oftener for local purposes—the appointment of local officers and committees. In these primary assemblies the inhabitants acquire habits of transacting public business, fitting them for legislation and government in national and local affairs. In these assemblies the affairs of the American revolution were extensively discussed and matured, by which the cause of freedom was materially promoted. In Maryland, Virginia, the Carolinas, Kentucky, and Tennessee, the counties form the smallest territorial division. The larger towns are incorporated as cities and boroughs, which have municipal governments.

No country can hope successfully to maintain a government like that of the United States unless the inhabitants have been prepared and educated for freemen; and it is no slight preparation that is required. Without this, all attempts of the kind will be likely to result in anarchy or despotism.

The national existence of this country commenced July 4th, 1776, when the delegates from the states, in congress, assembled at Philadelphia, declared that "the United States are, and of right ought to be, free and independent;" but long and arduous was the struggle by which they made good the declaration. With a population of about 3,000,000, nothing but stout hearts, indomitable perseverance, and a devoted patriotism could have enabled them—unprovided with the means of carrying on a war with one of the most powerful nations of the earth—to conduct the contest to a successful issue. The world has never produced but one Washington. He learned the art of war in the best possible school, that of leading the colonial militia against the French, particularly at fort Du Quesne (now Pittsburgh), and displayed his youthful prowess in conducting the retreat of the shattered forces after the defeat of Braddock, which resulted from his neglect of the advice of Washington. The colonial militia were trained to the art of war in the protracted contest with the French in Canada, and the American officers often fought side by side with the British officers, to whom they were afterward opposed in mortal combat. The latter were well acquainted with the prowess of many of the officers of the revolution. A British officer once asked Gen. Putnam, immediately before the commencement of the revolution, whether he did not think that 5000 veteran British forces were able to overrun the country. The reply was characteristic. He said "that he would say nothing of the men, but the women would beat all their brains out with their ladies and broomsticks before they had got half through the country." It was happy that Washington was made commander-in-chief of the American forces Without the prudence, skill, and courage of this American Fabius, the contest would not probably have been conducted to a successful issue. Washington refused all compensation for his arduous services in the revolutionary war, excepting his expenses, of which he kept an accurate account.

The following were the principal battles of the revolution, with the commanders, and loss on each side. The war commenced with the battle of *Lexington*, April 19th, 1775; the American loss 84, British loss 245. *Bunker hill*, June 17th, 1775; American (Prescott) loss 453, British (Howe) 1054. *Flatbush*, August 12th, 1776; British (Howe) loss 400, American (Putnam and Sullivan) 2000. *White Plains*, October 28th, 1776; American (Washington) 300 or 400. British (Howe) 300 or 400. *Trenton*, December 25th, 1776; American (Washington) 9, British (Rahl) 1000. *Princeton*, January 3d, 1777; American (Washington) 100, British (Mawhood) 400. *Bennington*, August 16th, 1777; American (Stark) 100, British (Baum and Breymen) 600. *Brandywine*, September 11th, 1777; British (Howe) 508, American (Washington) 1200. *Stillwater*, October 17th, 1777; American (Gates) 350, British (Burgoyne) 600, 5732 men surrendered. *Monmouth*, June 25th, 1778; American (Washington) 230, British (Clinton) 400. *Rhode Island*, August 29th, 1778; American (Sullivan) 211, British (Pigott) 260. *Briar Creek*, March 30th, 1779; British (Prevost) 16, American (Ash) 300. *Stony Point*, July 15th, 1779; American (Wayne) 100, British (Johnson) 600. *Camden*, August 16th, 1780; British (Cornwallis) 375, American (Gates) 730. *Cowpens*, January 17th, 1781; American (Morgan) 72, British (Tarleton) 800. *Guilford Courthouse*, March 15th, 1781; American (Greene) 400, British (Cornwallis) 593. *Eutaw Springs*, September 8th, 1781; American (Greene) 555, British (Stewart) 1000. The war closed by the surrender, at Yorktown, of Cornwallis, October 19th, 1783, with 7073 British soldiers, to Washington. The whole amount of the expenses of the revolutionary war, estimated in specie, was $135,193,703.

The following table will show what proportion of the war was borne by the several states, to which the population, in round numbers, in 1790 is subjoined.

States.	Continentals.	Militia.	Pop. in 1790.
New Hampshire	12,497	2,093	141,000
Massachusetts (including Maine)	67,907	15,155	475,000
Rhode Island	5,908	4,284	69,000
Connecticut	31,939	7,792	235,000
New-York	17,781	3,304	319,000
New-Jersey	10,726	6,055	173,000
Pennsylvania	25,678	7,357	431,000
Delaware	2,386	367	51,000
Maryland	13,912	4,127	216,109
Virginia	26,678	5,620	454,000
N. Carolina	7,263		393,000
S. Carolina	6,417		133,000
Georgia	2,679		53,000
Total	231,971	56,163	3,044,000

It thus appears that Massachusetts and Connecticut, in proportion to their population, bore the largest share of the war. Vermont, not then admitted to the Union, bore her full proportion.

Provisional articles of peace, acknowledging the independence of the United States, were signed in Paris, November 30th, 1782, by John Adams, Benjamin Franklin, John Jay, and Henry Laurens on the part of the United States, and Mr. Fitzherbert and Mr. Oswald on the part of Great Britain. The definitive treaty was signed September 30th, 1783. The confederation of the states, which in time of the war had given to the resolves of Congress the force of law, now that the danger was passed, evinced that it was inadequate to the purposes of an efficient government to meet the claims against the United States, and provide for the public debt, to raise a revenue, and to harmonize the jarring interests of the states. The difficulties which attended the formation of a new government, though different in kind, were not less than those of achieving independence. But by a happy concurrence of circumstances, a constitution was at length formed and ratified, which has secured the unexampled prosperity and happiness of the people, and stands as an illustrious proof of the wisdom of the fathers of the revolution, and a model for other nations in the pursuit of freedom. On the second Monday of May, 1787, delegates from the several states assembled at Philadelphia, for the purpose of forming a constitution for the United States, and George Washington, who had led the American army, was called to preside over them. On the 17th of September, after a debate of four months, a constitution was adopted, and sent to the several states for their approval. It was provided that the ratification of nine states should be sufficient for its establishment. It was warmly debated by the state conventions, but finally adopted by them all. (For the vote in the several states, see the particular articles on the states.) In several states amendments were recommended. The people were entering on an untried experiment, and they proceeded with extreme circumspection and caution. From the moment the proceedings of the general convention at Philadelphia transpired, the public mind was exceedingly agitated, and suspended between hope and fear, until nine states had ratified the plan of the federal government. Indeed the anxiety continued until Virginia and New-York had acceded to the system. But this did not prevent the demonstration of their joy on the accession of each state. "On the ratification in Massachusetts, the citizens of Boston, in the elevation of their joy, formed a procession in honour of the happy event, which was novel, splendid, and magnificent. This example was afterward followed, and in some instances improved upon, in Baltimore, Charleston, Philadelphia, New-Haven, Portsmouth, and New-York, successively. Nothing could equal the beauty and grandeur of these exhibitions. A ship was mounted on wheels, and drawn through the streets; mechanics erected stages, and exhibited specimens of labour in their several occupations as they moved along the road; flags with emblems, descriptive of all the arts and of the federal union, were invented and displayed in honour of the government; multitudes in all ranks in life assembled to view the majestic scenes; while sobriety, joy, and harmony, marked the brilliant exhibitions by which the Americans celebrated the establishment of their empire." It was equal to a second declaration of Independence.

The constitution was finally ratified by Congress, July 14th, 1788. On the first Wednesday of January, 1789, electors of president and vice-president were appointed. The electors met on the first Wednesday of February, 1789, and General George Washington, "first in war, first in peace, and first in the hearts of his countrymen," was unanimously chosen president, and John Adams was chosen vice-president. Gen. Washington was inaugurated as first president on the 30th of April, 1789, in the open gallery of the old Federal hall in New-York, where the Customhouse now stands; and never did a heartier shout proceed from the assembled multitude of freemen than that which greeted the conclusion of the ceremony with, "Long live George Washington." And now, after an experiment of more than 54 years, after having seen how completely this constitution secures all the purposes of a good government, and at how cheap a rate, the fear and trembling which marked its commencement are exchanged for steadfast confidence and unbounded hope; it stands like a lighthouse on the shore of the sea of liberty to direct the political voyager in his perilous course to the port of freedom. The authority and the boundaries of the several departments of the government are becoming daily better settled and understood, and it is striking a deeper root in the sentiments of the people. Every fourth of July increasing proof of the attachment of the citizens to their excellent form of government, and affords increasing evidence of its stability and perpetuity; and the citizen cannot be found who would be willing to exchange it for any other government on the earth.

The following is a list of the presidents of the United States.

George Washington,	of Virginia,	from 1789 to 1797, 8 years.
John Adams,	" Mass.,	" 1797 " 1801, 4 "
Thomas Jefferson,	" Virginia,	" 1801 " 1809, 8 "
James Madison,	" Virginia,	" 1809 " 1817, 8 "
James Monroe,	" Virginia,	" 1817 " 1825, 8 "
John Quincy Adams,	" Mass.,	" 1825 " 1829, 4 "
Andrew Jackson,	" Tenn.,	" 1829 " 1837, 8 "
Martin Van Buren,	" N. Y.,	" 1837 " 1841, 4 "
William H. Harrison,	" Ohio,	died one month after his inauguration, 1841.
John Tyler,	" Virginia,	as vice-president succeeded, 1841.

General Washington died suddenly at his residence at Mount Vernon, December 14th, 1799, of an inflammation of the throat, aged 68 years; and the nation everywhere mourned for him as for a father, with a sincere and profound grief. She had but one such man to lose. Funeral orations were delivered in every considerable place in the country; and the respect of the world has added its sanction to the nation's tears.

In 1803 Louisiana was purchased of the French for $15,000,000; and Florida was ceded to the United States in 1821 by Spain, in compensation for spoliations on American commerce, for $5,000,000.

On the 4th of June, 1812, war was declared with Great Britain by the American Congress, by a vote in the House of Representatives of 79 to 49; and in the Senate by a vote

of 19 to 13. This war continued with varied success, and peace was concluded at Ghent, December 24th, 1814.

The original 13 states that adopted the constitution were New-Hampshire, Massachusetts, Rhode Island, Connecticut, New-York, New-Jersey, Pennsylvania, Delaware, Maryland, Virginia, North Carolina, South Carolina, Georgia. To these, 13 new states have been added: Vermont in 1791, Kentucky in 1792, Tennessee in 1796, Ohio in 1802, Louisiana in 1812, Indiana in 1816, Mississippi in 1817, Illinois in 1818, Alabama in 1819, Maine in 1820, Missouri in 1821, Arkansas in 1836, Michigan in 1836.

In the American part of this work, the statistics of townships are given, as containing the greatest amount of information which can be condensed in a small compass. The statistics of counties are given, particularly because in the states south of Pennsylvania and of the Ohio river the census of the United States was taken only by counties, with the exception of cities, and they will enable a person to assign the statistics, in all parts of the country, to their respective districts. The census of the United States has been in all cases strictly followed, unless it is otherwise specified. It has doubtless some errors, but as few as could be reasonably expected; and no person can minutely examine it, without regarding it with great respect. It has been taken by officers paid by the government, and sworn faithfully to perform their duty. The census and statistics of the several states have been so minutely given, that less particularity is necessary, than would be otherwise advisable, in the general account of the United States.

UNITY, p. t., Waldo county, Me., 34 m. N.E. Augusta, 629 W. Drained by Sebasticook river. Incorporated in 1804. It has two stores, two fulling-mills, four grist-mills, seven saw-mills, two tanneries; 13 schools, 584 scholars. Pop. 1457.

UNITY, p. t., Sullivan co., N. H., 45 m. W. by N. Concord, 471 W. Drained by Little Sugar, and Beaver Meadow rivers. Chartered in 1764. It contains three churches, a Methodist, Baptist and Universalist, two stores, one grist-mill, three saw-mills; one academy, 85 students; 11 schools, 452 scholars. Pop. 1238.

UNITY, t., Westmoreland co., Pa., 7 m. E. Greensburg. Drained by Crabtree and Big Sewickly creeks. It has four flouring-mills, four grist-mills, three tanneries, four distilleries; 11 schools, 495 scholars. Pop. 3002.

UNITY, p. t., Columbiana co., O., 146 m. N.E. by E. Columbus, 277 W. It has three schools, 175 scholars. Pop. 1896.

UNST. (See ORKNEY and SHETLAND Isles.)

UNTERWALDEN, one of the four forest cantons of Switzerland, near the centre of the confed. in which it holds the 6th rank; between lat. 46° 49' and 47° N. and long. 8° and 55' E., having W., Lucerne; N. the same cant., and the lake of Lucerne; E., Uri; and S. nese Oberland. Area, estimated at 262 sq. m. the Bernese. In 1837, 22,571, all Roman Catholics; and of whom 11,668 belonged to Upper, and 10,903 to Lower, Unterwalden. The territory consists principally of four valleys, inclosed by mountains of various heights, the loftiest of which, the Titlis, rises to nearly 11,000 ft. above sea. Two streams, called Aa, hardly deserving the name of rivers, flow into the lake of Lucerne; and there are several small lakes, and numerous cascades. The climate is temperate, particularly in the E., where various kinds of fruit are grown. The valleys and lower hills afford fine pasturage; which makes cattle breeding the chief occupation of the inhabitants. The cattle are small, but a good cow is estimated to yield a profit of from 50 to 100 florins a year to the owner; and about 10,000 head of cattle are annually depastured in the canton. The cheese of Unterwalden is reckoned inferior only to that of the Emmenthal; and a good deal is sent into Italy. In good years the value of the cheese exported from the valley of Engelberg amounts to about 48,000 florins. Agriculture is comparatively neglected; and corn, to the value of at least 190,000 florins, has to be imported from Lucerne. The vine does not succeed; and the place of wine is supplied by liquors made from different fruits. The forests are a chief source of wealth, a good deal of timber and fuel being exported. Firs and fir, and after these, beech, oak, and elm, are the principal trees. the extensive forest of Kernwald has formed the line of separation between the republics of Lower and N. and Upper and S. Unterwalden, ever since the 12th century. Mining and manufactures are insignificant; some linen thread is, however, spun in the valley of Engelberg.

In both parts of the canton, the constitution is wholly democratic. Upper Unterwalden consists of seven communes, all the male inhabitants of which attend in person of age meet in a general assembly, exercising the sole deliberative and legislative power, on the last Sunday in April, at the capital Sarnen. The executive body, consisting of 14 principal functionaries, chosen by the general assembly, and 65 other members appointed by the different parishes, ar-

ercises all the high judicial and other functions, except in case of capital punishment; when the *triple council*, an assembly composed partly of special delegates, must pass sentence. Lower Unterwalden consists of 13 communes. Its general assembly is similarly constituted; but its government, carried on at Stanz, the capital, is more complicated than that of the other part of the canton, being conducted by a great variety of councils and assemblies. Public education is every where very backward. Both parts of the canton have but one united voice in the Swiss diet; they contribute 382 men to the Swiss army, and 1907 francs a year to the federal treasury.

Unterwalden, with Uri and Schwytz, formed the nucleus of the Swiss confederation early in the 14th century; but little worth notice has occurred in its subsequent history, except that the inhabitants of Lower Unterwalden made a vigorous opposition to the French revolutionary troops in 1798, for which they suffered proportionally. (*Picot; Stat. de la Suisse; Ebel, &c.*)

UPPER, t., Cape May co., N. J., 13 m. N.E. Cape May courthouse. Bounded S.E. by the Atlantic, N. by Tuckahoe creek. It has two churches, an Episcopal and Baptist, four stores, one grist-mill, four saw-mills; five schools, 219 scholars. Pop. 1217.

UPPER ALTON, p. v., Madison co., Ill., 79 m. S. by W. Springfield, 806 W. Situated on elevated ground 2¼ m. back from Mississippi river, and E. of Alton. It contains three churches, a Presbyterian, Methodist and Baptist, five stores, one steam flouring and saw-mill, 320 dwellings, and 1002 inhabitants. It was laid out in 1816, and since 1827 it has had a rapid growth. (*See* ALTON.)

UPPER ALLOWAYS CREEK, t., Salem co., N. J., 7 m. S.E. Salem. Drained by Alloways and Stow creeks. It has four stores, one woollen factory, six grist-mills, seven saw-mills; eight schools, 400 scholars. Pop. 2235.

UPPER BERN, t., Berks co., Pa. Bounded E. by Schuylkill river, by a branch of which it is drained. It contains a church common to Lutherans and Presbyterians, two stores, three tanneries, four distilleries. Pop. 2906.

UPPER DARBY, t., Delaware co., Pa., 8 m. W. Philadelphia. Darby's and Cobb's creeks afford water-power. It has two stores, one woollen factory, two cotton factories with 6000 spindles, one flouring-mill, three grist-mills, four saw-mills, four paper-mills; four schools, 395 scholars. Pop. 1480.

UPPER DUBLIN, p. t., Montgomery co., Pa., 106 m. E. Harrisburg, 152 W. Drained by tributaries of Wissahiccon creek. It has six stores, one fulling-mill, two flouring-mills, four grist-mills, one saw-mill. Pop. 1322.

UPPER FREEHOLD, p. t., Monmouth co., N. J., 15 m. S.W. Freehold courthouse. Drained by Millstone river and Tom's and Crosswick's creeks. It has 23 stores, two grist-mills, one saw-mill, seven tanneries, nine distilleries; 18 schools, 1200 scholars. Pop. 5026.

UPPER HANOVER, p. t., Montgomery county, Pa., 35 m. N.W. Philadelphia, 82 m. E. Harrisburg, 178 W. Watered by Perkiomen creek and its branches. It contains four stores, seven grist-mills, five saw-mills, four oil mills; four schools, 181 scholars. Pop. 1467.

UPPER MACUNGY, t., Lehigh county, Pa. Drained by Lehigh creek. It has four stores, two grist-mills, two saw-mills, three tanneries; one school, 90 scholars. Pop. 1769.

UPPER MAHANTANGO, p. t., Schuylkill county, Pa., 65 m. N.E. Harrisburg, 175 W. Drained by two branches of Mahantango creek. It has five stores, eight grist-mills, eight saw-mills, one distillery. Pop. 1291.

UPPER MAKEFIELD, t., Bucks co., Pa., 25 m. N.N.E. Philadelphia. Bounded N. by Pidcock's creek. It has one store, one flouring-mill, one grist-mill, one saw-mill; four schools, 238 scholars. Pop. 1490.

UPPER MARLBORO', p. v., capital of Prince George's county, Md., 23 m. S.W. Annapolis, 17 W. Situated on a branch of Patuxent river, and contains a courthouse, jail, a church, and about 800 inhabitants.

UPPER MOUNT BETHEL, t., Northampton co., Pa. It has seven stores, one fulling-mill, one flouring-mill, five grist-mills, six saw-mills, two tanneries, two distilleries; one school, 22 scholars. Pop. 2643.

UPPER MERION, t., Montgomery co., Pa. Valley and Gulf creeks afford water-power. It has 10 stores, four woollen factories, seven flouring-mills, two grist-mills, six saw-mills; one academy, 30 students; six schools, 556 scholars. Pop. 2894.

UPPER MILFORD, t., Lehigh co., Pa. Drained by the N. branch of Perkiomen creek. It contains two churches, a Lutheran and a German Presbyterian, seven stores, one fulling-mill, seven grist-mills, six saw-mills, one oil-mill, two powder-mills, two tanneries, four distilleries; three schools, 120 scholars. Pop. 3081.

UPPER NAZARETH, t., Northampton co., Pa. Drained by branches of Manokissy creek. It contains Nazareth village, and has four stores, one fulling-mill, one grist-mill,

one tannery, one brewery; one academy, 86 students; three schools, 95 scholars. Pop. 1118.

UPPER OXFORD, t., Chester co., Pa. It has two stores, one woollen factory, one cotton factory with 1080 spindles, one flouring-mill, three grist-mills, six saw-mills, two paper-mills, one pottery; six schools, 157 scholars. Pop. 1277.

UPPER PAXTON, t., Dauphin co., Pa., 22 m. N. Harrisburg. Drained by Mahantango, and Great and Little Wiconisco creeks. Bounded W. by Susquehanna river. It has six stores, two woollen factories, four flouring-mills, one grist-mill, 10 saw-mills, one oil-mill, one tannery, one distillery, one pottery; nine schools, 800 scholars. Pop. 11,814.

UPPER PROVIDENCE, t., Montgomery co., Pa. Bounded S.W. by Schuylkill river. Drained by Perkiomen and Mingo creeks. It has six stores, two flouring-mills, five grist-mills, four saw-mills. Pop. 2244.

UPPER SAUCON, t., Lehigh co., Pa., 6 m. S.E. Northampton. Drained by Saucon creek, a branch of Lehigh river. It contains iron ore, and has seven stores, four grist-mills, eight saw-mills, one oil-mill, two tanneries; four schools, 130 scholars. Pop. 2072.

UPPER ST. CLAIR, t., Allegheny co., Pa. Drained by Chartier's creek. It has one fulling-mill, one woollen factory, four flouring-mills, two grist-mills, five saw-mills, two tanneries, one distillery; one academy, 12 students; 11 schools, 360 scholars. Pop. 2302.

UPPER SWATARA, t., Dauphin co., Pa. Drained by Swatara creek. It has one woollen factory, three flouring-ing-mills, one tannery, one distillery, one brewery, one pottery. Pop. 1935.

UPPER TULPEHOCKEN, t., Berks co., Pa. Drained by tributaries of Tulpehocken creek. The Union canal follows the latter stream along the S. boundary of the town. It contains a church common to Presbyterians and Lutherans, five stores, one forge, one woollen factory, five flouring-mills, six grist-mills, five saw-mills, two tanneries, one pottery. Pop. 2941.

UPSAL, or UPSALA, a city of Sweden, cap. prov. of same name, on the Sala, by which it is divided into two parts, 37 m. N. by W. Stockholm. Pop., 4500. "Upsala, built on a gentle height, and part of the adjoining plain in a very level and fertile country, is one of the most beautiful old-fashioned cities in Europe. The view on approaching it is very fine: an old red palace on the hill, occupied by the governor of the province, with the newly-finished university-buildings glancing white beside, and the grey towers of the cathedral rising calmly over both, give an air of grandeur to the place, as seen by the distant spectator, beyond that of any town we had yet visited in Sweden. If, on entering, this city loses somewhat of its dignity, still its broad, quiet streets, and its strange old houses, generally of brick covered with stucco, or of wood painted red and roofed with turf, impressed us at once as in excellent keeping with the character of this favoured retreat of science." (*Bremner*, ii., 299.)

Only a few of the inhabitants are engaged in manufactures, or in the little trade carried on by the river, on which a steam navigation is kept up with Stockholm. The greater number depend for support on the university, the principal in the kingdom. This establishment was founded by Steno Sture in 1478, and modelled on the university of Paris. It was warmly patronized by Gustavus Vasa, who was partly educated in it. At a subsequent period, however, it was transferred to Stockholm, but was again restored to Upsal by Charles IX. It has long enjoyed a very extensive celebrity, and is at present attended by from 1350 to 1450 pupils, though of these only from 800 to 900 may be resident at any one time. Thus in the winter session of 1837, the university had in all 1376 students, of whom 874 were resident. Of the entire number, 299 attended the theological, 305 the legal, 142 the medical, and 408 the philosophical classes. The students, like those of Scotland and Germany, lodge in the town. Their average expenditure may be estimated at about £30 for the session. As many as 150 students are maintained free of expense from endowments left by Gustavus Adolphus. The public lectures are all gratuitous, and but few private lectures are attended. The lectures commence in October, and continue till Christmas, when there is a six weeks' vacation, after which they are resumed, and continue till May. No fixed number of years' study is prescribed for a degree, the rule being, that a candidate may obtain it as soon as he can pass the required trials, which few attempt till they have studied in the university for some years. The public examinations are not very formidable; but those which have first to be gone through privately with each professor, are conducted with great strictness. Printed lists of the students and professors are published every year: the former are divided into eight *nations*, named from different Swedish provinces, each of which has an inspector, curator, librarian, &c. In the lists for a recent year, there apgear 4 theological, 3 judicial, 5 medical, and 14 philosophical professors, in all 96; besides 19 adjuncts, and 24 docents, with a music director, and masters for German, French, English, dancing, fencing, riding, &c. Most of the professors lecture in their own houses. None have a salary of more than £300 a year, and many have much less. The university revenues amount to about 130,000 thalers banco a year. A professor who has continued in office for the space of 30 years, is allowed to retire, with the title of Emeritus, and enjoys his salary during life. The new university is a handsome and spacious edifice, built of freestone, in the Florentine style of architecture. Most part of it has been devoted to two splendid halls, one on the first story for the principal library, and another on the second for oratorios, the conferring of academic degrees, &c. The ground floor is occupied by rooms appropriated to MSS.; but, until very lately, the library, comprising 100,000 vols., remained in the old university building. The greatest curiosity in this collection, is a manuscript of the four gospels, called, from its silver letters, the *Codex Argenteus*, supposed to be a copy of the Gothic translation made by Ulphilas, the apostle of the Goths, in the 4th century. Much controversy has existed among the learned as to the characters in which this famous codex is written; especially whether it be in those used by the Goths of Moesia, ancestors of the present Swedes, or in the Frankish idiom. At all events, however, there can be no doubt of its high antiquity; it is admitted on all hands not to be later than the 6th century, and may be of the 4th or 5th, while it certainly has the further advantage of having been translated from the original Greek. The codex was found in 1597, in the library of the Benedictine abbey of Werden, in Westphalia; and having found its way (whether honestly is doubtful) into the library of Isaac Vossius, was, on his death, purchased for the comparatively trifling sum of £250 by the Count de la Gardie, who presented it to the university. (*Coxe*, iv., 173, &c., 8vo ed.)

In the same edifice are preserved a beautiful cabinet, presented to Gustavus Adolphus by the city of Augsburg in 1632, and the large chest, sealed and left by Gustavus III., with instructions that it should not be opened till 50 years from the day of his death. The prescribed period expired during the past April (1842); but we have not learnt whether the chest has been opened. Here, also, is a botanic garden, and a museum, in which is a fine statue of Linnæus by Bystrom. This university has had many celebrated individuals among its professors, especially in the department of natural history; among others may be specified, Linnæus, Bergman, Afzelius, &c.

The cathedral is an imposing edifice, though built only of brick: in some respects its appearance is similar to that of Notre-Dame at Paris. It is 350 feet in extreme length, with a lofty nave and a magnificent altar. The Swedish monarchs were formerly crowned in this cathedral, and here, also, they are mostly interred. Among the numerous tombs the most interesting is that of Gustavus Vasa; distinguished in every station of life, and equally great as a legislator, a warrior, and a politician. Linnæus, also, is entombed in this cathedral, and the house is still shown which he occupied in the town.

Except the cathedral and university buildings, Upsal has no other remarkable edifice; but it has some public walks, on one of which a fine obelisk has been erected to the memory of Gustavus Adolphus. It is the metropolitan see of Sweden, and possesses an ecclesiastical school, a cosmographic, and other scientific societies.

About 2 m. N. is Old Upsal, a mere village of huts round an old church, the origin of which is lost in remote antiquity, but which certainly was once dedicated to the worship of Odin. Near it are some remarkable tumuli, and many curious antiquities have been discovered in the neighbourhood. (*Coxe*, iv.; *Bremner's Excursions*, ii.; *Dict. Géog.*; *Voyage de deux Français, &c.*)

UPSON, county, Ga. Situated a little S.W. of the centre of the state, and contains 285 sq. m. Bounded S.W. by Flint river, by branches of which it is drained. It contained in 1840, 5595 neat cattle, 1791 sheep, 17,846 swine; and produced 29,333 bushels of wheat, 221,340 of Indian corn 10,058 of oats, 8139 of potatoes, 6,477,334 pounds of cotton. It had 12 stores, two cotton factories, with 1790 spindles, four flouring-mills, 12 grist-mills, five saw-mills, seven distilleries; six academies, 303 students; seven schools, 171 scholars. Pop.: whites, 5536; slaves, 3868; free coloured, 4; total, 9408. Capital, Thomaston.

UPTON, p. t, Worcester co., Mass., 35 m. W.S.W. Boston, 410 W. Incorporated in 1735. Drained by West river, a tributary of Blackstone river. It contains three churches, a Congregational, Baptist, and Christian, three stores, one fulling-mill, one woollen factory, three grist-mills, six sawmills. Pop. 1466.

UPTON ON SEVERN, a market town and par. of England, co. Worcester, hund. Pershore, on the Severn, here

996

crossed by a stone bridge of 6 arches, 9 m. S. Worcester. Area of par., 3110 acres. Pop. of do., in 1831, 2343. The town is neatly built with well-paved streets. The church, rebuilt in 1758, is a handsome structure. The living, a valuable rectory, worth £979 a year, is in the patronage of the bishop of Worcester. The Baptists, and other sects, have also places of worship. A charity school for 18 girls, founded and endowed in 1718, to which a boys' school was attached in 1797, has been incorporated with two national schools supported by voluntary contribution. A subscription library has been founded. The river, which is navigable thus far for vessels of 100 tons, has a commodious wharf, and a good harbour for barges. It has no manufacture of any importance; but a considerable trade is carried on in corn, malt, coals, &c.; and a good deal of cider is brought here for shipment from Hereford and other parts. A manor court is held occasionally, and petty sessions once a fortnight. Upton is a place of great antiquity. During the civil wars it suffered mostly to the royal cause. Market-day, Thursday; fairs, four times a year.

URBANA, p. t., Steuben co., N. Y., 211 m. W. by S. Albany, 307 W. It contains the S. portion of Crooked lake, and is drained by small streams flowing into it. Hammondsport on the lake has a communication through it by water, to New-York city. It contains two churches, a Presbyterian and Episcopal, nine stores, two fulling-mills, one woollen factory, one flouring-mill, two grist-mills, seven saw-mills, two tanneries; two academies, 69 students; 12 schools, 650 scholars. Pop. 1694.

URBANA, p. v., seaport, capital of Middlesex co., Va., 86 m. E. by N. Richmond, 146 W. Situated on the S.W. side of Rappahannock river, 18 m. from its mouth. It contains a courthouse, jail, four stores, 20 dwellings, and about 175 inhabitants.

URBANNA, p. t., capital of Champaign co., O., 45 m. W. by N. Columbus, 439 W. The village contains a courthouse, jail, a Methodist church, a market-house, nine stores, a printing-office, 150 dwellings, and 1670 inhabitants. There are in the township 10 stores, one fulling-mill, one woollen factory, two tanneries, and, exclusive of the v., a pop. of 1293.

URBANA, p. v., capital of Champaign co., Ill., 52 m. E.N.E. Springfield, 691 W. Situated on the S. side of the west branch of Salt fork of Big Vermillion river. It contains a courthouse, jail, and about 30 dwellings.

URBINO (an. *Urbinum Hortense*), a fortified town of central Italy, in the Papal states, cap. leg., and formerly of the duchy of its own name, on a mountain near the Metauro, 90 m. S. W. Pesaro. Pop. of the town only about 7000, but, including the suburbs, 13,000. (*Rampoldi.*) Its fortifications are a good specimen of the military works of the 14th century. Its old castle now forms part of a Carmelite convent, in which and in the churches are several fine works by Raphael and other distinguished artists. Urbino has an ancient ducal palace, which, like the *Palazzo Albani*, is a large and fine building: a newly built cathedral, with a rich chapter and archbishopric; a university with about 200 students; a college, hospital, seminary, an association called by the singular title of *Academia Accurdiorum*, and a court of primary jurisdiction. But the legate and other chief authorities of the leg. of Urbino and Pesaro reside at the latter town. Urbino is celebrated for its hams, manufactured to the value of about 14,000 crowns a year (*Bowring's Rep.*). It has four annual fairs, and markets on Wednesday and Saturday.

Among other illustrious individuals, Urbino has to boast of having given birth to Raphael, hence called by the Italians Raffaello d'Urbino, unquestionably the greatest painter of modern times. He was born on Good Friday, 1483, and died at Rome in 1520, on the anniversary of his birth, at the early age of 37—a wonderfully brief space in which to have attained to such matchless proficiency, and to have completed so many great works. This was also the birthplace of Bramante, one of the greatest modern architects, a relative of Raphael; of the machinist Zabaglia, and other distinguished personages. (*Rampoldi; Cowder's Italy, &c.*)

URI, a canton of Switzerland, in the S.E. part of the confed., between the 46th and 47th degs. of N. lat., and 8° 30′ and 9° E. long., having N. Schwytz and the lake of Lucerne, E. Glarus and Grisons, S. Tessin, and W. the Valais, the Bernese Oberland, and Unterwalden. Area estimated at 335 sq. m. Pop. in 1837, 40,658, all Roman Catholics. The canton consists mostly of one principal valley, that of the Reuss, into which several others open laterally; and the whole are shut in, except at the N. and S. extremities of the canton, by lofty and generally impassable mountains. Among the mountains comprised in this canton is the celebrated St. Gothard, the passage across which is the principal route from Italy into E. Switzerland. The transit trade by this road was, and perhaps still is, the most profitable occupation of the inhabitants of Ca. Various other lofty and remarkable summits are included in the ranges surrounding the canton. The principal river is the Reuss, which rises in mount St. Gothard, and after a course usually N., through Uri, in its whole length, falls into the lake of Lucerne, near Altorf. "It is," says Inglis, "a remarkable river, not for the length of its course, or the volume of its waters, but for its extraordinary rapidity, far exceeding that of the Rhone; and the magnificent scenery on its banks. Its whole course is a succession of cataracts; and in the short space of four leagues, its inclination is no less than 2,500 ft.'" (*Inglis's Switzerland*, p. 100.) Besides the Lake of Uri (the S. extremity of that of Lucerne) there are numerous small alpine lakes; and the canton is a good deal visited by strangers, both on account of its sublime scenery and the historical interest attached to it as the land of Tell, said to have been a native of Bürglen, near Altorf.

The pastures of this canton are very superior, and these and their cattle constitute the entire wealth of the inhabitants. A good many cattle, tended by Bergamesque cowherds, are sent thither from Italy to feed during summer. The cheese of the canton is in high repute; but it is not extensively produced; and the stock of sheep, goats, hogs, &c., is not more than sufficient for the wants of the population. Agriculture is entirely neglected. "Throughout the whole of the upper part of the valley of the Reuss and in the vale of Ursern, not one stalk of any kind of grain, nor one cultivated vegetable of any kind, is to be seen. There is no doubt, however, that these might be successfully cultivated. The vale of Ursern is better sheltered than almost any part of the Engadine, and is less elevated than many other parts where rye is grown abundantly, where other grain, even wheat, is not a failure, and where all the hardier vegetables are plentiful. About 2 leagues lower down than Andernat, the valley widens. Cottages are sprinkled here and there, and now and then a hamlet; still, however, grass only was to be seen. I saw many warm stripes, and even little plains, along the river side, where wheat and vegetables could have been successfully cultivated; but I still continued to meet cars laden with flour and potatoes. As I descended still lower in the valley, the scenery became more varied and beautiful. Charming meadows lay by the river side, prettily diversified by clumps of walnut and pear trees; and cottages and hamlets thickly dotted the slopes. Altorf is surrounded by gardens and orchards; yet even here, where the climate is mild, and where the ordinary fruits come to great perfection, scarcely an ear of corn is to be seen." (*Inglis*, p. 1045.) This, however, does not originate in any want of industry, but in what is most probably a well founded conviction on the part of the inhabitants, that it is most advantageous for them to confine their attention to the breeding and depasturing of cattle, and to import their corn, flour, and other provisions, and such manufactured goods as they do not produce in their cottages. The devastations of the late war, and the construction by Napoleon of the route over the Simplon, by diverting a considerable portion of the travelling and trade carried on through this canton into a new channel, have been especially hostile to its interests. Several metallic ores are met with, and Picot (p. 237) enumerates a long list of valuable minerals found on mount St. Gothard; but mining industry is quite insignificant.

Uri is subdivided into two districts, Uri and Ursern; Altorf is the capital. The constitution is strictly democratic. The legislative power resides in the general assembly, composed of all the male population above 20 years of age, which assembles every year on the first Sunday in May, to choose the cantonal council of 44 members, to which is confided the direct executive power. There are several inferior councils for separate departments of service. Each district has its own assembly and tribunals; and in the capital is a tribunal of appeal, composed of 15 members, and presided over by the *landamman*. The communes are generally too poor to support public schools all the year round, and education is very backward; no libraries or literary societies worthy of notice exist in the canton. The inhabitants are under the ecclesiastical jurisdiction of the bishop of Chur: they are mostly of the German stock, though in the S. an Italian dialect is spoken. Contingent to the Swiss army, 236 men; do. of money to the federal treasury, 1184 francs a year. This canton is supposed to have derived its name from the *urus*, or wild bull, which formerly inhabited its valleys in great numbers. Uri was one of the three cantons which revolted from the German empire in 1307, and formed the nucleus of the Swiss confederacy. It was a principal theatre of war between the French and Austrians, in 1799–1800. (*Picot, Statist. de la Suisse, &c.*)

URUGUAY, or BANDA ORIENTAL, a republic of South America, between lat. 30° and 35° S., and long. 52° and 59° W.; having N. Brazil, E. and S. the Atlantic ocean and the estuary of the La Plata, and W. the Uruguay river (see *ante*, p. 611), by which it is separated from the Argentine or La Plata territories. It is of nearly circular form,

and is supposed to embrace an area of above 200,000 sq. m. while its population is not supposed, Indians included, to exceed 150,000, or at most 200,000. It is but very little known. The coast presents the aspect of a low flat plain, without wood of any kind, and as far as the eye can reach quite level in appearance with the water. Inland, however, and particularly in the N., the country is intersected by many hill-ranges alternating with valleys traversed by considerable affluents of the Uruguay. In this territory, the humidity of the soil, which is watered by numerous rivers, is corrected by the Pampero, a remarkably dry wind. The climate of Uruguay is proverbially healthy, and it is evident that the thinness of the population must arise from the mode of life followed by the settlers, or from political causes, and not from any deficient fertility of the soil, or other natural or necessary cause. It is divided into nine departments, and possesses three principal towns, Monte Video, La Colonia, and Maldonado, 15 small towns, and eight hamlets, without including estancias or farms, and ranchos or cottages. Monte Video, having a better port and as good a government, bids fair to become a city of greater trade and wealth than its opposite rival, Buenos Ayres. (For its trade and that of the republic generally,

S. America, 52–56.)

URUMEA, OORMIAH, or SHAHEE, a town and considerable lake of Persia, prov. Azerbijan. The town stands in a fine plain, watered by the Shar, 8 m. W. the lake, and 65 m. S.W. Tabriz. Its population has been roughly estimated at 12,000. It is fortified with a strong wall and deep ditch, which may be filled with water from the river. Though supposed to be the *Thebarma* of Strabo, it is said to possess no remains of antiquity worth notice: it has, however, been rarely, if at all, visited by recent European travellers.

The lake of Urumea or Shahee was visited by Major Rawlinson so late as 1838. "It extends above a degree of lat. in length, and is about 1-3d of that distance in extreme breadth. The greatest depth of water that is found in any part is four fathoms, the average is about two fathoms, but the shores shelve so gradually that this depth is rarely attained within 2 m. of the land. The specific gravity of the water, from the quantity of salt which it retains in solution, is so great, that a vessel of 100 tons burden has a draught of no more than 3 or 4 feet. This heaviness of the water prevents the lake from being much affected by storms, which, from its extreme shallowness, would otherwise render its navigation dangerous. A gale of wind raises the waves but a few feet; and as soon as the storm has passed, they subside. It is an old opinion that the waters of this lake are too salt to support animal life. No fish, certainly, are found in it, but the smaller class of zoophytes are met with in considerable numbers. The islands in the lake, until lately, were barren and uninhabited; but the largest has been recently colonized, and settlements on the rest have been proposed." (*Geog. Journ.*, x. 7, &c.)

On the E. side of the lake is the village of Shishewan, the residence of a Persian prince, Melik Kasim Mirza, a brother of the late shah, who has adopted in every particular European habits and pursuits. He has built a palace in the European style, near which he has established mulberry gardens for the culture of silk, a farm yard, a glass work, a pottery, looms for weaving cotton, silk, and worsted goods, and various other kinds of manufactures. He has also built a vessel of 100 tons, which he employs in trading on the lake; on which also, he contemplated the introduction of steam navigation. (For farther particulars, see *Rawlinson in Geog. Journ.* x., 5–9, &c.)

USHANT (Fr. *Ouessant*), the most W. of the islands off the coast of France, forming a portion of a group near the W. coast of Brittany, dep. Finisterre, in lat. 48° 28' 8" N., long. 5° 3' W., 26 m. W.N.W. Brest. Its area is about 2 sq. leagues, and its population rather exceeds 2000. It is difficult of access, but is tolerably fertile, affording pasture to a good many sheep and horses. It has a village, several hamlets, an old castle, and a small harbour frequented by fishing boats. Sir Edward Hawke totally defeated a French fleet, under Admiral Confians, off the coast of this island, in 1759. Owing to the violence of the weather, two of the British ships accidentally got ashore, and were lost. At a later period, on the 27th July, 1778, an indecisive action took place off the island between the English fleet under Admiral Keppel, who had been second in command in the former action, and that of the French under Count d'Orvilliers.

USK, a parl. and mun. bor., market-town, and par. of

England, hund. Usk, co. Monmouth, on the Usk; here crossed by a stone bridge of five arches, nearly in the centre of the co., 12 m. S.W. Monmouth. The modern bor., which is more extensive than the ancient, has an area of 410 acres, with a population, in 1831, of 1160. The town is neatly, but irregularly built, the houses being mostly detached and interspersed with gardens and orchards. It is indifferently paved; and down to a recent period was not lighted. The church, which belonged to an ancient priory, appears to have been erected in the Norman period. It was originally cruciform, but has been very much altered; the square embattled tower now at its E. end, seems to have been formerly in its centre; its interior contains little worth notice, except an inscription on a brass plate, which has not yet been satisfactorily explained. The living, a vicar-age, worth £250 a year, is in the gift of — Williams, Esq. Here, also, are chapels for Independents, Wesleyans, and Roman Catholics. Over the market-place is the town-hall, a handsome building, erected at the expense of the late duke of Beaufort; attached to which is a lock-up house for the custody of prisoners till they can be conveyed to the county jail. The educational establishments comprise a free grammar school, founded and endowed in 1624, to which a writing school has since been attached; a national school for both sexes, supported by voluntary contributions; and almshouses for 24 inmates. Near to the latter are the remains of the ancient priory previously alluded to.

"Usk is increasing not only from natural increase of population, but also from the beauty of its situation, attract-ing persons of independent property to build cottage villas in its vicinity. It has a small manufacture of Pont-y-pool japan ware, which employs four or five hands, and is the only one remaining in this part of the country, the trade having removed to Birmingham. It has no other trade peculiar to it, and very little of any kind." (Parl. Bound. Report.) Most part of the inhabitants are engaged in hus-bandry and salmon-fishing. The earliest charter of the borough dates from 1398; but it has not been governed for a long period by this or any other charter. The local authority is vested in a portreeve, chosen annually; a re-corder, two bailiffs, four constables, and an indefinite num-ber of burgesses. It is associated with Newport and Mon-mouth, in returning one member to the House of Commons. Total electors for the three boroughs, in 1839-40, 1304. Quarter sessions for the county are held here, as well as a court for the recovery of small debts, and a court leet twice a year.

Usk is a place of remote antiquity, and appears to have formerly been of much more importance than at present. On an eminence adjoining the town are the extensive re-mains of its ancient castle, formerly one of the most con-siderable structures of its kind in the country. It came through the Mortimers, earls of March, into the possession of the crown, and was the favourite residence of Richard duke of York, nephew of Henry VI., whose sons, Edward IV. and Richard III., were born within its walls. At a subsequent period it belonged to the earls of Pembroke; and is now the property of the duke of Beaufort.

Market-day, Thursday; fairs, four times a year, for wool, horses, cattle, and pediery; and once a month for cattle only. (Beauties of England and Wales; Parl. and Mun. Bound. Reps.)

USKUP or SCOPIA (an. Scopi), a considerable town of European Turkey, prov. Macedonia, cap. sandjlack, on the river of its own name, a tributary of the Vardar, 110 m. N.W. Salonica. Its pop. is estimated at about 10,000. It has a good many handsome mosques, Greek churches, &c.; but its streets, though wide, are filthy in the extreme. It is the see of a Greek archbishop, and has some manufac-tures of leather. A Turkish garrison is stationed in its old dilapidated castle.

USTIUG (VELIKI, or 'the Great'), a town of Russia, gov. Vologda, at the confluence of the Joug and Souchonia, tributaries of the Dwina, about 400 m. from the White sea, and 550 m. E. by N. Petersburg. This town, though in so remote and desolate a region, has 8000 inhabitants, and is the seat of a considerable trade between Europe and Asia. It has, according to Possart, three cathedrals and 29 parish churches, many of which are built of stone; and several private buildings, and a large exchange, are of the same material. It has numerous tallow, soap, candle, leather, and tile factories, with saw-mills, and some jewellery and silver works: but its trade is chiefly in corn, lard, linen, ship timber, and sail cloth. Its merchants, who trade with the Siberian towns generally, as far as Kiachta, were es-timated, in 1830, to possess an aggregate capital of 288,000 roubles; and the town revenues are estimated at 20,000 do. a year. A large annual fair is held here on the 8th July. (Possart, Kaisorth; Russland.)

UTICA, city, Oneida co., N. Y., is situated in 43° 7' N. lat., and 75° 13' W. long., 99 m. W.N.W. Albany, 140 m. E. Rochester, 202 m. E. by N. Buffalo, 76 m. E.S.E. Os-

wego, 388 W. Pop. in 1820, 2972; in 1830, 8323; in 1840, 12,782. The city is beautifully situated on the S. side of Mohawk river, on an inclined plane rising from the river so as to command, from its elevated parts, many fine pros-pects. The streets, which are numerous, are laid out with a good degree of regularity, generally, but not always crossing each other at right angles. It is well built, ex-tensively with brick, having many fine stores and large and elegant dwellings. The streets are neat and spacious, some of them 100 ft. wide, nearly paved, with convenient side walks. It contains a courthouse, offices for the clerks of the supreme and United States courts, 18 churches, three Presbyterian, a Dutch Reformed, two Episcopal, three Methodist, four Baptist, two Roman Catholic, a Universal-ist, a Friends, and an African, an Exchange building, a male and a female academy, the Utica library, a Mechan-ics' association which sustains popular lectures, an Appren-tices library, a museum, a Protestant, and a Roman Catho-lic orphan asylum, and various other charitable institu-tions, four banks with an aggregate capital of $1,265,000, besides a bank for savings, a mutual insurance company, and 1600 dwellings. The state lunatic asylum is located about a mile W. of the centre of the city, and has fine buildings, with a farm of 160 acres attached to it. The Erie canal passes through the city, and is here 70 ft. wide and 7 ft. deep, passed by a number of lofty and elegant bridges, and adds much to its business facilities. Cheaan-go canal, which extends to Broome county, here enters the Erie canal. The great Western railroad from Albany to Buffalo, passes through the place, consti-tuting it a great thoroughfare; and numerous stages on fine roads, proceed to various places N. and S. The coun-try around Utica is fertile, populous and rich, and it is the centre of an extensive trade. A culvert has been comple-ted within the city from the canal to the river, which cost $100,000. On the whole, Utica is one of the finest places in western New-York. It occupies the site of old fort Schuyler. In 1794 there were only three or four poor houses on the spot, and it dates its great prosperity from the completion of the Erie canal. There were in 1840, two commercial and three commission houses in foreign trade with a capital of $56,000; 188 retail stores with a capital of $1,678,595; three lumber-yards, capital $41,000; five furnaces produced 1059 tons of cast iron, employing 34 persons, and a capital of $59,000; 98 persons employed in internal transportation, together with 90 batchers, packers, &c., employed a capital of $12,980; 103 persons produced machinery to the amount of $166,555; six persons manu-factured 110 small arms; 12 persons manufactured the precious metals, and produced articles to the amount of $28,800; 13 persons manufactured various metals to the amount of $29,000; 21 persons produced bricks and lime to the amount of $9000; 19 persons manufactured tobacco to the amount of $20,000, with a capital of $4000; 7 per-sons manufactured hats and caps to the amount of $125,666, and straw bonnets to the amount of $9600, with a capital of $40,300; six tanneries produced 16,300 sides of sole leath-er, and 68,450 sides of upper leather, employing 28 persons, and a capital of $103,000; 10 other manufactories of leath-er, as saddleries, &c., produced $63,000, with a capital of $12,500; 11 persons produced 140,000 pounds of soap, and 115,000 pounds of tallow candles, with a capital of $12,980; two breweries employed 16 persons, producing 148,000 gal-lons of beer, with a capital of $26,000; seven persons man-ufactured paints and drugs to the amount of $18,000, with a capital of $2100; three persons produced confectionery to the amount of $10,000, with a capital of $5000; one pa-per-mill, employing four persons, produced to the amount of $6000, with a capital of $4000; three persons manufac-tured musical instruments to the amount of $1000, with a capital of $500; 86 persons manufactured carriages and wagons to the amount of $77,000, with a capital of $40,000; one flouring-mill produced 2000 barrels of flour, and with two grist-mills and two saw-mills, employed six persons, and a capital of $3000; 74 persons manufactured furniture to the amount of $100,395, with a capital of $29,110; six printing-office, three binderies, and six weekly newspapers, employed 72 persons, and a capital of $51,900; 62 brick or stone houses and 30 wooden houses were built, employing 208 persons, and cost $253,000. The total capital employ-ed in manufactures was $496,130. There were 14 acad-mies with 670 students, and 36 schools, 981 scholars.

UTRECHT (an. Ultra-Trajectum), a city of Holland, cap. prov. of its own name, on the old Rhine, by which it is divided into two parts, 20 m. S. by E. Amsterdam. Pop. 44,000. It is oval-shaped, and is one of the best situated towns in the kingdom, being built on an undulating tract of land, more elevated than the surrounding territory, and having, in consequence, a drier and purer atmosphere. The country immediately around is finely wooded, and well sprinkled with farm-houses and cottages. Fields of wheat and other grain are seen instead of unvarying green pas-res; and the novel spectacle presents itself of rivers and canals flowing below the general level of the country, and of above it on the tops of mounds, as elsewhere in Holland. The approach to Utrecht, from Amsterdam, is very ne, being through a long avenue of lime trees, which orms a favourite public walk. The city was formerly defended by lofty brick walls; but these are now broken and dismantled, and the old ramparts have been converted into bulevards. Since the separation of Holland and Belgium, however, some outworks have been thrown up.

Utrecht has an antique appearance, many of the houses being built in the Gothic style; as in other Dutch towns, the material for building is mostly brick. Mr. Jacob says, The streets are more regular, the houses more modernised, nd the squares more spacious, than in other towns of Holland. In some of the streets there are canals, or rather anches of the Rhine, for though they soon terminate in agnant canals, they have here some motion. The water so far below the level of the thoroughfares, that the wharfs on its sides have doors opening to a kind of caverns ander the streets, in which many of the poorer classes find abitations." (Tour in Germany, &c., p. 45.)

The fine public walk called the Mall, on the E. side of the town, about 2000 yards in length, is divided into alleys y rows of linden trees. The cathedral, formerly a fine lifice, has been so much dilapidated that only the choir, ansept, and tower remain: the last is wholly detached om the other parts, leaving room for a wide street on the ace formerly occupied by the nave. It must have been nginally larger than York minster, but less elegant, being osstly constructed of brick. The transepts are shut up, nd the only portion at present in use is the choir, fitted up n a plain manner for the Presbyterian service, though it as some fine monuments. The tower is a huge square ructure, 388 ft. in height; and from its summit the view ctends over a wide extent of country, comprising many ties, towns and villages.

Utrecht has a celebrated university, founded in 1636 at he expense of the city, which ranks next to that of Leyen: it has five faculties and 19 professors, and is attended y about 600 students. The university buildings have no atward show, but they comprise a valuable library, a retty good museum of natural history, and extensive collections in anatomy, pathology, &c.: especially one of beautifully-executed models in coloured wax. This city has so acquired distinction from the number and excellence ' its superior schools. It has a museum of national agriltural implements, established in a fine building, once ie residence of Louis Bonaparte, but this, according to r. Chambers, is an inferior collection. It has also a trimal of commerce, an academy, founded in 1778, which ives prizes for the best memoirs on scientific subjects, a anch of the society of public good, a mint, the machin- y of which is partly wrought by air-pressure, &c.

Utrecht, being in the centre of a populous agricultural strict, is more bustling than the small Dutch cities in neral. It has considerable manufactures of cloth and her woollen stuffs, velvets, linen fabrics, silk twist, fowlg-pieces, pins, &c., with bleaching-ground, sugar and salt fineries, brick and tile works, &c. It sends six deputies the provincial states. The famous act, called the union ' Utrecht, declaring the independence of the seven Unid Provinces, was signed here on the 29th of January, 79; and the treaties of Utrecht, which terminated the ar of the Spanish succession and gave peace to Europe, ere concluded here, in 1713 and 1714. Among other disnguished individuals, pope Adrian VI., the preceptor of harles V., was a native of this city. (Schreiber, Guide du him; Chambers' Tour in Holland; Dict. Géog., &c.)

UTRERA (an. Iliturgus Vericulum), a town of Spain, in ndelusia, prov. Seville, cap. distr. on the road from Madrid Cadiz, 14 m. S E. Seville. Pop., according to Miñano, ,050, but, according to Capt. Scott, 15,000, mostly agriclralists. "Utrera stands in the midst of a vast plain, at may be considered the first step from the marshes of ie Guadalquivir towards the Ronda mountains, 12 m. distnt to the E. A slight mound, that rises in the centre of e town, and is embraced by an extensive circuit of dipidated walls, doubtless offered the inducement to build town here; and these walls, some parts of which are ery lofty, and in a tolerably perfect state, appear to be Roan, though the castle and its immediate outworks are oorish. The town is large, and not walled in; the streets e wide and clean, and a plentiful stream, remarkable as ing the only running water within several miles, rises ar, and traverses, the place." (Scott's Ronda and Granla, ii., 141, 142.)

It has a spacious square, two parish churches (one of agular architecture), various convents and hospitals, a ood town-hall, prison, cavalry barracks, &c.; and near it a convent, resorted to by a great concourse of devotees uring a festival which lasts eight days from the 8th of

September. The bulls bred in the vicinity are the most ferocious of any in Andalusia; and a considerable trade is carried on here in cattle and horses; it has, also, some manufacture of hats, soap, starch, wax, and leather. In its vicinity are some productive salt springs. Utrera is an important military post, being at the divergence of several cross-roads. The French when advancing upon Cadiz in 1810, made strenuous efforts to reach it before the Spanish troops under the Duke of Albuquerque; but, being anticipated by the latter, Cadiz was prevented from falling into their hands. (Scott's Ronda, &c., ii.; Miñano.)

UTTOXETER, a market town and par. of England. co. Stafford, hund. Totmonslow, near the Dove, which is here crossed by a fine stone bridge of six arches, connecting the cos. of Stafford and Derby. Area of par. 8920 acres. Pop. of do. in 1831, 4864. The town, 17 m. W. by S. Derby, stands on a hill sloping towards the river, and consists principally of three streets, diverging from the market-place in its centre. With the exception of the tower and lofty spire, which are ancient, the church has been recently rebuilt: the living, a rectory, worth £135 a year, is in the gift of the dean and canons of Windsor. Here, also, are several dissenting chapels. It has a free school for 14 boys, founded in 1588 by Mr. Thomas Allen, a native of the town, celebrated by Selden, Camden, and others for the extent of his mathematical and antiquarian learning; a national school supported by subscription; almshouses for 12 inmates; a fund for the apprenticeship of poor children, &c. Petty sessions are held weekly by the county magistrates. The hardware manufacture is carried on to some extent, and there are numerous iron forges in the vicinity. Owing to the fertility of the surrounding country, especially the excellence of the pastures along the Dove, the market held here for agricultural produce, cattle, sheep, &c., is one of the best in the county. Its trade is facilitated by the Caldon canal, which joins the Trent and Mersey canal, coming within a short distance of the town. Market day, Wednesday. Fairs 10 times a year, chiefly for cattle, horses, and sheep.

UWCHLAN, p. t., Chester co., Pa., 30 m. S.S.W. Philadelphia, 69 m. E.S.E. Harrisburg, 129 W. Bounded S.W. by Brandywine creek, which affords water-power. It contains one church, one forge, one woollen factory, one flouring-mill, three grist-mills, four saw-mills, two paper-mills, two tanneries, one distillery, one pottery; seven schools, 210 scholars. Pop. 1565.

UXBRIDGE, a market town and chapelry of England, par. of Hillingdon, co. Middlesex, hund. Elthorne, border of Buckinghamshire, on the Colne and Grand Junction canal, over each of which it has a bridge, and on the high road from London to Oxford, 15 m. N. by W. the former. Pop. in 1831, 3043. It is situated on a slight eminence sloping to the river, and consists mostly of one long street, with a smaller one on the road branching off to Windsor. The main street is paved and lighted with gas; and many of the modern houses are handsome and substantially built. The market house, erected in 1789, is a brick building 140 ft. in length by 49 in width, supported on about 50 wooden columns. St. Margaret's chapel, built in 1447, is an irregular edifice of flint and brick in the pointed style, with a low square tower: the living, a perpetual curacy, worth £111 a year, is in private patronage. The Baptists, Friends, Independents, and Methodists have each places of worship. Two free schools, on the Lancastrian system, are held over the market-place, and are supported partly by voluntary contributions, and partly by annual subscription from the manor-funds, market-tolls, &c. A school of industry was founded in 1809, and there are several other charitable institutions for education and the relief of the poor. A reading room and public library has been established.

Uxbridge has the largest country corn-market in the kingdom. A great deal of excellent flour is made at the flour-mills in the town and its immediate vicinity, and its millers and mealmen are in general as opulent as they are respectable. Large quantities of malt are, also, produced in the town, the trade of which is greatly facilitated by the Grand Junction canal. It has also an extensive manufac-

for numerous hands. The municipal government is vested in the hands of two constables, four headboroughs, and other officers elected annually. Petty sessions for the town and several adjoining parishes are held every fortnight, and a court of requests for debts under 40s. once a month. The unsuccessful negotiation between Charles I. and the parliament in 1644 took place in an old brick building called the "Treaty House," which has latterly been converted into an inn. Markets, Thursday and Saturday; fairs, four times a year. (Beauties of England and Wales; Prior. Inf.)

UXBRIDGE, p. t., Worcester co., Mass., 38 m. S.W. Boston.

ton, 403 W. Watered by Munford and West rivers. Blackstone river and canal pass through it. It contains three churches, a Congregational, Unitarian and Friends, four stores, 12 fulling-mills, five woollen factories, three cotton factories with 5500 spindles, three grist-mills, seven saw-mills; two academies, 42 students; 13 schools, 549 scholars. Pop. 2004.

UZES, a town of France, dep. Gard, cap. arrond., on the Auzon. 12 m. N. by E. Nismes. Pop. in 1836, ex. comm., 5986. It is built on the declivity of a hill, at the foot of which rises the stream which was anciently conducted to Nismes by the *Pont du Gard*. It is old, ill built, and ill laid out. It was a bishopric in the time of the Visigoths, and the old episcopal palace, and the former residence of the dukes of Uzès, are the principal edifices: the last is a huge castle, inclosed by high walls flanked with round towers, and bearing a good deal of resemblance to the Bastile in Paris, destroyed in 1789. Uzès suffered much in the religious wars, when its bishop, chapter, and most part of its inhabitants, embraced Protestantism, and destroyed their cathedral. It has a court of primary jurisdiction, communal college, &c., and manufactures of silk hosiery, coarse woollens, pasteboard, &c. (*Hugo. Art. Gard*; *Guide du Voyageur en France.*)

V.

VAL-DE-PENAS, a town of Spain, in New-Castile, prov. La Mancha, part. Ciudad-Real, in a tolerably fertile plain. 112 m. S. by E. Madrid. Pop., according to Miñano, 10,248. It is well built; the mansion of the Marquis of Santa Cruz, and the warehouse of the royal tithes, being conspicuous among many other good edifices. This town would appear to be less overstocked with religious edifices than most others in Spain, for it is said to have only one parish church and one convent. It is, however, best known by the dry red wine produced in its neighbourhood, and hence called Val-de-Penas, which is in much request in Madrid, and approaches in quality to some of the stronger Bordeaux wines. The bread here is also of peculiar excellence. Some linens and soap are also manufactured, and woollen stuffs are sent to the town to be dyed. It has a large fair on the 7th of Aug. (*Miñano.*)

VALAIS (Germ. *Wallis*), a canton of Switzerland, in the S.E. part of the confed., between lat. 45° 50' and 46° 40' N., and long. 7° and 8° 25' E., having N. the Bernese Oberland, N.E. Uri and Tessin, E. and S. Piedmont, and W. Savoy and the canton of Vaud. Area estimated at 1600 sq. m. Pop. in 1837, 75,796, all Roman Catholics. This canton consists of the valley of the Upper Rhone, and may be described as "an immense trough," 70 m. in length, 1½ m. in depth, and 2 m. wide at the bottom. The mountains on each side are the highest in Europe; they form two walls of rock, rising from 10,000 to 14,000 feet above the Rhone: this valley may therefore be considered as the deepest in the known world. On the N. side are the Alps, to which belong the Finsteraarhorn, Jungfrau, Breithorn, and other enormous peaks; while the S. boundary is formed by the great chain from mount Blanc to St. Gothard, including the Cervin and M. Rosa. At the upper end of the valley these two ranges unite. Sixteen lateral valleys, some of considerable extent, open into the main valley of the Rhone; and where they join it, the width of the flat part of the valley is increased. Thirteen of these lateral valleys are inhabited. That part of the valley below Sion, formerly distinguished as the *Bas-Valais*, is a flat plain, swampy, and unhealthy, where the heat of summer is intense, and millions of mosquitoes, with intermittent fevers, cutaneous diseases, and cretinism in its worst forms, abound. Above Sion, in the *Haut-Valais*, the marshes disappear, and fine pasturage and vineyards indicate a more favoured region." (*Conder's Italy, &c.*, i., 280.) The Valais is remarkable as presenting within the smallest known area all the different climates and kinds of vegetable products met with between Italy and Iceland. At the foot of the Bernese alps the vine succeeds remarkably well, and very good wine is produced, though in its management the inhabitants be both unskilful and slovenly. Indian figs, almonds, chestnuts, pomegranates, grow with little or no culture along the banks of the Rhone, and corn of all kinds is produced at different elevations. The cattle of the Valais are inferior to those of most of the other Swiss cantons; but the rearing of stock is the principal branch of rural industry; the surplus produce in cattle, cheese, &c., being exported chiefly across the Simplon into Lombardy. The population is much poorer than that of the Bernese Oberland; which Mr. Bakewell attributes, though, perhaps, with less justice than he supposes, to two causes : first, that the land belongs to a few great proprietors, and the peasantry, being merely tenants, feel but little

interest in its improvement; and next, the prevalence of the Roman Catholic religion, with which there is usually found combined, in Switzerland at least, much ignorance and indolence. Iron, copper, lead, silver, cobalt, and small quantities of gold, &c., are met with: but mining industry is altogether neglected. Hempen cloths, woollen yarn, and woollen cloths, with a few other articles, are made, but these are almost the only manufactures. Next to cattle breeding, the transport of goods across the Simplon is the principal occupation of the inhabitants, for which purpose many mules are bred. The mineral waters of Brieg and Loueche are of some advantage to the canton, being resorted to by a good many invalids.

This canton is divided into 13 districts, called dizains; Sion (Germ. *Sitten*), on the Rhone, a town of 2500 inhabitants, a bishop's see, is the capital. The supreme power is in a diet, composed of four deputies from each dizain, chosen for two years by the dizain councils, which are elected by all the citizens above the age of 18 years. The presidents of the dizain are members of the diet by virtue of office, and the bishop of Sion has a voice in the diet equivalent to four votes. The diet meets annually on the first Mondays in May and November. The executive power is in a state council of five members elected by the diet, and who are all re-eligible, except the *grand-bailli* or president, who is eligible only after an interval of two years. Each commune has a court of primary jurisdiction; from which appeal lies to the district tribunals: the supreme court for the canton consists of 13 judges, nominated by the diet for two years, but always re-eligible. In respect of education, the Valais is behind most other parts of Switzerland. The public revenue, derived chiefly from salt and transit duties, is estimated at about 380,000 Swiss livres.

The inhabitants of the *Bas-Valais* are chiefly of the French, and those of the *Haut-Valais* of German descent. The Valais formed a part of the Burgundian, and afterwards the Frankish dominions; it next became subject to its own count-bishops; but in 1032, it was annexed to the German empire. It was allied, not associated, with the cantons that formed the Swiss confederacy in the 14th century. In 1798 it entered the confederacy as a canton; but in 1802 was detached from it to form a separate republic under the protection of France. In 1810, it was incorporated with the French empire as the dep. of the Simplon; and in 1815 it again became a canton of the confederacy, in which it holds the 20th rank, contributing 1280 men to its army, and 9886 livres to its treasury. (*Picot, Statist. de la Suisse*; Ebel; Bakewell; Conder's Italy, &c.)

VALDIVIA, a town and harbour of Chili, prov. of same name, of which it is the capital. The town, on the river Calcutta or Valdivia, about 16 m. from its mouth, is an insignificant village of wooden huts, and was, for the most part, ruined by the earthquake of Nov. 7, 1837; but the harbour is probably the finest, as it is one of the most strongly fortified, in the Pacific. It is in lat. 39° 53' 29" S., long. 77° 33' 24" W., and consists of an estuary, formed by the Valdivia and several smaller rivers, entered by a narrow strait, the shores of which are lined by numerous batteries, mounting, in all, nearly 130 pieces of cannon. Ships of the line ride here in perfect safety; the depth of water, in the centre of the bay, being from 6 to 7 fathoms, and close to the shore 5 fathoms. During their struggle for independence, this valuable station was captured, with a very inferior Chilian force, by Lord Cochrane, on the 3d of February, 1820. (*Miers's Chili: Voyage of the Adventure and Beagle.*)

VALENCE (an. *Julia Valentia*), a city of France, cap. of the dep. Drôme, on the Rhone, here crossed by a handsome suspension bridge. 59 m. S. Lyons. Pop. in 1836, 9390, or, inc. comm., 10,967. "Valence lies pleasantly on the left bank of the river, surrounded by a fertile country abounding in mulberry, almond, and other fruit trees. Opposite the town, a conical hill rises close to the Rhone; and about a mile beyond, a long range of vine-covered hills runs parallel with the river, producing the different species of St. Peray." (*Inglis's Switzerland, &c.*) The wine now alluded to, and which is not so much known in England as it deserves to be, is a dry white wine, characterised by great delicacy and sprightliness, and a flavour that partakes of the odour of the violet. When bottled in the spring following the vintage, St. Peray effervesces like Champagne. (*Henderson on Wines*, p. 176.) The town is enclosed by an old wall flanked with towers, and entered by several gates. It is irregularly laid out; but is pretty well built, and has latterly been much improved, both in its houses, a large proportion of which are of stone from the quarries of St. Peray, and in respect of cleanness. It has several public squares and promenades, as the *Champ de Mars*, planted with trees, the *places d'Orleans, aux Clercs, &c.* The cathedral or church of St. Apollinarius is a building said, by Mr. Wood, to be in a degraded Roman style. (*Letters, i., 108.*)

VALENCIA.

It has a large square tower of four stories, and in the interior is the tomb of Pope Pius VI., who died, at Valence, in 1799, with a bust of that pontiff by Canova, and some fine paintings by A. Caracci and other artists. Near this church is a private mausoleum, worth notice for its architecture. The barracks and citadel, the old residence of the governor, the courthouse, prison, public library with 15,000 volumes, and theatre, are among the principal buildings. Near the town is the polygon, a ground for the artillery-school practice.

Valence is a bishop's see, the seat of a court of primary jurisdiction, a communal college, chamber of manufactures, societies of agriculture, commerce, and arts, &c. It has two seminaries, a school of design, several asylums, a convent, a Lutheran church, a yard for building boats for the navigation of the river, manufactures of cotton goods, silk organzine, gloves, hosiery, leather, and earthenware, with marble works, lime and brick kilns, and sawing yards. It is a depôt for the wines, silk, and other produce of the S. of France, and has six fairs a year. It was anciently the capital of the Segalauni, and became a Roman colony under Vespasian; but it possesses no Roman antiquities. It formed successively part of the kingdoms of Burgundy and Arles, and of the dominions of the counts of Provence and Toulouse; it subsequently became the capital of the Valentinois, and was governed by its own feudal nobles till ceded to Louis XI., in 1449. From the 4th to the 13th century eight church councils were held in Valence. (*Hugo, Art. Drôme, Guide du Voyageur, &c.; Inglis; Woods; Duct. Géog.*)

VALENCIA, a kingdom or grand division of Spain, in the E. part of the peninsula, principally between the 38th and 41st deg. of N. lat., and 0° 35' E. and 1° 35' W. long., consisting of a long and comparatively narrow strip of country, extending along the Mediterranean, which bounds it on its whole length on the E.; and having inland from the N. round to the S., Catalonia, Aragon, New Castile, and Murcia. Area, 7683 sq. m. Pop. about 969,000. This is, upon the whole, one of the finest and most productive parts of the peninsula. A great proportion of the surface, particularly in the N. and W., is mountainous and rude; but the plain country, which stretches along the coast, and is watered by the Guadalaviar, Xucar, Mogra, Segura, &c., is a species of garden. All travellers coincide as to its extreme beauty, and superior cultivation, compared with the rest of Spain. Inglis, who has travelled over many of the finest parts of Europe, says, "The view of the plain is superb. Though not greener or more beautiful than the vale of Murcia, its immense extent and great populousness produce a more striking effect. The plain is probably little less than 30 m. long, and 20 wide; on three sides it is bounded by the mountains, and on the fourth by the sea; and throughout the whole of this vast extent, there is not an acre that does not produce its crop of grain, vegetables, or rice. The olive, mulberry, ilex, algarrob, orange tree, and palm, with all of which the plain is thickly dotted, give it the appearance of a union of garden and orchard; but the populousness of the plain is even more striking than its beauty and fertility. It forms altogether a prospect that, in richness and animation, cannot be equalled in any other country.

"The plain of Valencia produces every kind of crop congenial to the climate; two and three crops in the year are taken from it, and the greater part of the land returns 8 per cent. The rice crops are among the most valuable; they are chiefly produced in the territory of Albufera, surrounding the lake of the same name. This was the property first proposed to be granted to the Duke of Wellington, but the Cortes of Valencia objected to it, and the estates near Granada were substituted. The rice grounds produce only one crop in the year, but the return is from 8 to 10 per cent. The rice is put into the ground in June, and cut in September; water is then let in upon the ground, and when the stubble rots, the land is ploughed up, and no other manure is required. In Valencia and its neighbourhood, rice is in universal use by all classes, but the produce is much greater than the consumption of the plain; and the surplus is exported to the different ports of Andalusia. The whole produce is estimated at 12,000,000 arrobas, one half of which at least is exported; and the average price may be taken at about 3s. the arroba, or nearly 1½d. per lb. The other chief product is the white mulberry, once the source of great riches, through the silk factories of Valencia. The produce of silk from the plain is now computed at about 1,000,000 lbs. a year; by far the greater part is exported in its raw state, at an average price of from 8s. to 10s. per lb. The export of fruit from Valencia is large, particularly of raisins; these are of two kinds, the muscatel and an inferior raisin; but neither is equal to the raisins of Malaga. In 1829, 47,000 quintals of the best, and 42,000 of the inferior sort, were exported exclusively for the English market. The export of figs, oil, and wine from the provinces and different ports of Valencia, is also considerable; particularly the latter, called *Beni Carlo*, exported from the town of the same name. This wine is sent chiefly to Cette, from whence much of it finds its way, by the canal of Languedoc, to Bordeaux, to give body and colour to the clarets." (*Inglis, ii., 258–262.*) Barilla, sugar-cane, saffron, aniseed, &c., are also grown.

But, despite the abundance of the province, the peasantry, as is frequently the case in the most productive parts of the continent, appear to be in a very depressed condition. The *huerta* of Valencia, says Inglis, belongs, for the most part, to great proprietors. The duke of Medina Cœli has a revenue of 75,000 dollars a year from his estate there, and the families of Villa Hermosa and Benevento have almost as much; and, in fact, there are very few persons who cultivate their own land. And, though state taxes do not weigh down industry in this province, the Valencians are subject to heavy feudal services and seignorial demands, levied in kind on the produce of the soil, amounting to 1-7th, 1-6th, and, in some places, 1-4th of the crop! (*Mod. Trav.*) But these circumstances do not sufficiently account for the depressed condition of the peasantry; which is rather to be explained by the fewness of their wants, originating in the mildness of the climate, and their want of all desire to improve their situation.

No great number of cattle or horses are kept; and the sheep, though pretty numerous, yield wool of only a mediocre quality. A good many tunnies are taken on some parts of the coast by a method similar to that practised in the Neapolitan dominion. Mercury, copper, sulphur, arsenic, argentiferous lead, iron, coal, &c., exist in many places; but are procured only in small quantities. Salt from springs, marble, and potters' clay are the principal mineral products. Manufactures are unimportant; woollen and linen fabrics are indeed made in most of the towns, and silk goods in Valencia, Gandia, &c., but, at present at least, in very inconsiderable quantities. Cordage is made from the fibre of the esparto (*stipa tenacissima*), aloe, juncus, &c.; and tiles, similar to the Delft manufacture, soap, glass, paper, &c., are made in small quantities. An active internal traffic is kept up, the N. supplying the S. districts with timber, earthenware, linen and woollen stuffs, esparto, brandy, cattle, &c.; while the latter sends to the former corn, fish, Levant produce, silk, algarrobs, &c.; Valencia sends rice, silk, fruit, and fish to Aragon and Castile, for corn, wool, and cattle; and flax, hemp, silk, oil, rice, soap, &c., to Murcia and Granada. Its foreign commerce is chiefly with Italy, England, France, Holland, and S. America; from which countries, corn, salt, fish, ship-timber, pitch, tar, iron, fine linens, and other manufactured goods, are imported. Alicante, Valencia, Vinaroz, Murviedro, Benicarlo, and Guardamar, are the principal ports. (*Fischer's Picture of Valencia.*)

Very different opinions respecting the character of the population have been entertained by different travellers. "Upon the whole, the Valencians would appear to be an airy, lively, active, but effeminate people, very different in manner from the Castilians; while their character forms apparently a still stronger contrast to the savage heroism of the Catalonians and Aragonese. Their dialect, though much akin to the Catalonian, is said to differ from it in retaining more of the provençal." (*Mod. Trav., xviii., 167.*) This province was successively subject to the Carthaginians, Romans, and Visigoths, from whom it was taken by the Moors in 713. They held possession of it till 1238, when it was conquered by the Spaniards and annexed to Aragon. It afterwards formed a component part of the Spanish monarchy, but continued to preserve its representative body and its privileges, till the early part of the 18th century, when having, in the war of the Succession, taken part against the Bourbon dynasty, it was, on their establishment on the throne, deprived of its old constitution, and obliged to conform to the laws of Castile. (*Miñano; Antillon; Inglis's Spain; Swinburne; Townsend, Twiss, &c.*)

VALENCIA (an. *Valentia Edetanorum*), a city of Spain, cap. of the above kingdom, on the Guadalaviar, about 4 m. from its mouth, and 188 m. E.S.E. Madrid. Lat. 39° 28'

the whole length of the city. A fine view is obtained from any one of the bridges; the line of irregular buildings following the curve of the river, and the bridges, one beyond another, with the great Moorish gates, give it an air of much grandeur. Few cities, even in Spain, have, or rather had, so many religious edifices as Valencia: among these were reckoned no fewer than 27 convents for men and 92 for women, 16 churches, and 24 chapels and hermitages, amounting altogether to 86; so that we need not wonder that the streets abounded with friars and priests, and that the influence of the religious bodies was predominant! (*Inglis*, ii., 251.) The cathedral is of intermixed Greek and Gothic architecture, surmounted by a dome. It has numerous altars, a good deal of fine marble, some bas-reliefs, and paintings by some of the first Spanish masters. It is very rich in plate and relics. Some of the churches have domes, but the greater part tall, slender turrets, with all sorts of pinnacles and whimsical devices. In the multitude of sacred edifices, some may be found that excel in particular parts, or strike by the richness of their decorations; but all are overloaded with ornaments. (*Swinburne's Trav.*, i., 153.) In most, however, are fine paintings by Juanes, Espinosa, Ribalta, Ramirez, Victoria, a pupil of Carlo Maratti, and many other artists, all natives of Valencia. The famous Supper of Ribalta is in the Corpus Christi college; and the same subject by Juanes, a work reckoned among the finest pictures of Valencia, is in the church of St. Nicholas, which also possesses sundry other fine pictures.

The exchange, custom-house, the temple, a palace built for a military order, by Charles III., the archbishop's palace, college of Pius V., and several noble residences are worthy of notice; there are five hospitals, one of which is a large establishment for medical treatment of all kinds, several asylums, prisons, barracks, a theatre, &c. Inglis says, "Valencia is one of those cities, in which traces of Moorish dominion are the most visible; not in any splendid Alhambra or Alcazar, but in everyday sights and common objects. Gateways are occasionally seen sculptured in marble upon Moorish designs; stones over the doors, or underneath the windows, show by their chiselled marks their ancient fashioners. All the Moorish tokens also distinguishing the populations of Seville, Malaga, and San Felipe, are found in even greater distinctness in Valencia." (*Spain*, ii., 253.)

The university, founded in 1411, was formerly considered the best in Spain, particularly for the study of medicine. In 1830, it had nearly 2500 students, who were principally divided between law and philosophy. The professors, of whom there are about 70, are friars, except those who instruct in law: their salaries are from £50 to £130 a year. Lectures are delivered from October 11 to May 31. Education in the university is nearly gratuitous, and many of the students were in the habit of receiving portions of the food distributed daily from the convents. The university library has not more than 15,000 volumes, but its deficiencies are compensated by a good library in the archbishop's palace, with attached cabinets of antiquities, medals, &c., open for six hours daily. There are six other colleges, and many inferior academies; the royal academy of Saint Charles, for the instruction of students in the fine arts, is the only institution in the city not under the superintendence of the priests.

The manufactures of velvets, taffetas, flowered damasks, and other silk stuffs, are said, at the end of last century, to have employed upwards of 3000 looms; but if so, they have greatly declined in the interval. The existing manufactures comprise woollen fabrics, camlets, hats, table and other linen, gauzes, artificial flowers, leather, glass, paper, &c., with the "Valencia tiles," used for the flooring of houses in all the cities of the S. of Spain. These tiles are at once cool, and highly ornamental; but they are far from cheap, those of the best quality being much more expensive than an equal extent of the most sumptuous carpeting. A good workman employed in painting the tiles earns about a dollar a day.

The port of Valencia is at Grao, about 2 m. distant. It is connected with the city by a broad planted avenue, forming the favourite public promenade around the city. The harbour is suitable only for boats, and the roadstead is exposed to the S. and S.W. gales. The climate, though hot, is agreeable and healthy; and the city is a good deal resorted to by invalids. Society in Valencia appears to differ little from that in other S. Spanish towns. Many persons of rank and wealth reside here, but without any display. Travellers bear testimony to the agreeable vivacity, ready wit, freedom from affectation, and obliging disposition of the inhabitants. But Inglis says, that in Valencia, more than in any other city, he was struck with the absence of mental cultivation. But we apprehend the traveller must have been unlucky in his acquaintance. Mental culture may and most probably is at a low level; but the city which is the seat of perhaps the most flourishing university in the country, which has, also, produced many distinguished natives, and in which

1004

the first printing-press introduced into Spain was established, is not likely to be the zero of the intellectual scale.

Valencia was held by the Moors from 1715; but taken from them, in 1094, by the celebrated Cid, Ruiz de Diaz de Bivar. After his death it was governed by his widow Ximene, under whom it sustained successfully one siege against the Moors of Cordova, but ultimately capitulated to them in 1101. It was retaken by James I. of Aragon, in 1238, and peopled afterwards with Catalans and French settlers. It was taken by the French under Suchet, in 1812, and held by them till June, 1813. (*Mitana, &c., ut supra.*)

VALENCIA (NUEVA), a city of Colombia, republic of Venezuela, prov. Carabobo, in a plain about 3 m. W. of the lake of Valencia, and 18 m. S. by E. Puerto Cabello. Pop. estimated at 16,000. It covers a large extent of ground, most part of the houses having only a ground floor. The streets are very broad, and the market-place is of great size. The entrance to the town from the N. is by a good bridge of three arches, built of stone and brick; near which is the *glorieta*, a large circular space where the inhabitants meet in the evening for dancing and festivity. Humboldt says, "It is regretted, and perhaps justly, that Valencia has not become the capital of the country." Its situation near Puerto Cabello, with which it is said to communicate by a good road, gives it many advantages, and it is a place of brisk traffic. But it has the great disadvantage of being infested with white ants, whose excavations underground at certain seasons become very dangerous to the buildings of the city.

The lake of Valencia or Tacarigua is larger than the lake of Neufchâtel in Switzerland, but in its general form it has a nearer resemblance to that of Geneva, which is about the same height above the sea. The opposite banks of the lake also offer a similar contrast. Those on the S. are desert and almost uninhabited, and a screen of high mountains gives them a gloomy and monotonous appearance: the N. shore, on the contrary, is cheerful, pastoral, and set off with sugar, coffee, and cotton plantations. There are numerous islands in this lake, the waters of which are gradually diminishing. Its mean depth is from 12 to 15 fathoms; where deepest, it is not more than 40 fathoms. It abounds with fish, and is used for the purposes of commerce; but it is a singular fact, that for more than two centuries none of its navigators ever thought of using a sail! An English traveller, of no very distant period, says that a native of Nacay, settled in Valencia, had then first tried the experiment,—and of those who were assembled at the pulperia of La Cabrera." (*Mod. Trav.* xxviii. 178; *Humboldt, Pers. Narrative,* iv. ch. xvi; *Codazzi, Geog. de Venezuela; &c.*)

VALENCIENNES (an. *Valentiana*), a town of France, dép. du Nord, cap. arrond.; on the Scheldt, where it is joined by the Rhonelle, 26 m. S.S.E. Lille. Lat. 50° 21′ 47″ N., long. 3I° 31′ 55″ E. Pop., in 1836, ex. com., 16,879. It is a secondary fortress, and has a citadel constructed by Vauban. It is tolerably well built, but ill laid out; a part of it was much damaged by the severe bombardment it received in 1793, the marks of which are still visible. The Scheldt intersects the town from N. to S., dividing it into two unequal parts. There are several public buildings. The town-hall, built in 1612, is of mixed architecture, highly decorated, and contains some fine apartments; the second story is appropriated to a picture gallery, in which are some works by Rubens. Attached to it is a belfry, constructed in the 13th century, 180 ft. in height. The lower part of a handsome theatre serves for a corn-hall. The general hospital, founded in 1751, is one of the largest establishments of its kind. The public library has 18,000 vols., and the museum of natural history is rich in collections. There are military and founding hospitals, barracks, magazines, an arsenal, a college, an academy of the fine arts, founded in 1782, &c. Besides Valenciennes lace, a good deal of fine cambric, cotton yarn, hosiery, and blankets, iron plate and nails, starch, saltpetre, linseed oil, chicory, earthenware, and toys, are made in the town, which has also cotton-printing works, tanneries, distilleries, and salt refineries; and a considerable trade in wine, brandy, oil, soap, timber, and charcoal.

Valenciennes has sustained several sieges, the most memorable of which took place in 1793, when it was invested by an Anglo-Austrian army under the late Duke of York, to whom it surrendered at the end of six weeks, during which the besiegers sustained great loss. It was re-taken by the French from the Austrians in the ensuing year. Among the many distinguished natives of this town, have been—Froissart, the historian of chivalry, Watteau, the painter, and D'Argenson, the minister. (*Hugo, art. Nord; Guide du Voyageur, &c.*)

VALERY (ST.), usually called St. Valery-en-Caux, a town and seaport of France, dep. Seine-Inférieure, cap. cant.; on the British channel, 23½ m. N.W. by E. Rouen. Pop., in 1836, 3636. Its port, enclosed between two cliffs, is small, and not fit for the accommodation of vessels of considerable burden. St. Valery has manufactures of soda and cotton thread, with some trade in the exportation of Rouen fabrics, and in the importation of timber, dried fish, corn, &c. Many of its inhabitants are engaged in the cod, herring, and mackerel fisheries, and they enjoy the reputation of being good seamen. The country round is mostly in pasture.

St. Valery-sur-Somme is another seaport town, on the N. coast of France, dep. Somme, 11 m. N.W. Abbeville. Pop., in 1836, 3285. Its port is much larger than that belonging to the above town, and admits vessels of from 300 to 400 tons. It has large salt magazines, with manufactures of cordage and sail-cloth, and a brisk general traffic: of late years, an intercourse by steam has been occasionally kept up between this town and London. (*Hugo*, arts. *Seine-Inférieure and Somme; Dict. Géog. &c.*)

VALETTA (LA). *See* MALTA.

VALLADOLID (an. *Pintia*), a city of Spain, cap. prov. and dep. of its own name, in Old Castile, on the Pisuerga, here crossed by a stone bridge of 10 arches, where it is joined by its tributary the Esqueva, over which many small bridges are thrown, 73 m. S.W. Burgos, and 100 m. N.W. Madrid. Lat. 41° 45′ N., long. 4° 6′ W. Pop. 20,960 (*Miñano*). It stands in a plain bounded by limestone hills, and is surrounded by an old wall which encloses a large extent of ground. Swinburne says "Valladolid is a very large rambling city, full of edifices, which, during the reign of Philip III., who made it his constant residence, were the palaces of his great officers and nobility. Being abandoned by their owners, who have followed the court in all its different emigrations, they are fallen to decay, and exhibit a picture of the utmost desolation. The private houses are ill-built and ugly. The great square, some streets built upon porticoes, and many colleges and convents, are still grand, and denote something of former magnificence; but in general, Valladolid has the appearance of having been run up in a hurry to receive the court, and as if it had been meant to rebuild it afterwards, at leisure, of more durable materials than bad brick and mud, the composition of most of its present houses." (*Trav.* ii. 254, 255.) Upon passing the first gate from the Madrid road, the traveller enters the *campo grande*, a spacious area surrounded by 17 convents, the scene of repeated *auto de fees*, signalized by the burning of not a few heretics.

Valladolid has, or recently had, 16 parish churches, upwards of 40 convents, nine chapels, eight colleges, three hospitals, several asylums, barracks, &c., though, except some of the religious buildings, none deserve much notice. The cathedral, an unfinished edifice, built by Juan de Herrera, at the expense of Philip II., was intended to be one of the most sumptuous in Spain; but, according to Townsend and others, it is heavy and inelegant. The church and convent of St. Benito are handsome, but the church of St. Paul is probably most worth attention, from its general elegance, and the finish of its bas-reliefs and ornaments, which, after a lapse of 300 years, seem to have suffered little by their exposure to the weather. (*Townsend*, i. 365.) Laborde speaks of fine sculptures by Gregory Hernandes, &c., in this and several other churches in Valladolid; but whether they have been removed is more than we can undertake to say. The royal chancery is a large and fine structure in the Tuscan order. The old palace, in which Philip II. and III. were born, is an utter ruin: when Twiss visited the city its bare walls only were standing.

Valladolid has a university founded by Alonzo XI. in 1346: it was formerly distinguished for its school of jurisprudence, and continued to flourish till the end of last century; and it appears, even now, to be more frequented than Salamanca. Among the colleges of the city were one for the Scotch and one for the English, both of which were well endowed. The school of the fine arts is privileged as an academy, and has a superior collection of models in sculpture, architecture, and painting.

Valladolid was formerly an opulent commercial city; but its manufactures of woollen stuffs, hats, silk ribands, linen and cotton yarn, paper, liquorice, perfumery, earthenware, leather, &c., are now little more than sufficient for the supply of the town: its trade, however, would most probably increase were the Pisuerga made navigable to the Douro, only 16 m. distant. The country round produces white wine, of good quality, madder, silk, olives, &c., and coal is said to abound in the neighbourhood, though, if the statement be correct, little, if any, use is made of this valuable material.

Valladolid is the see of a bishop, under the archbishop of Burgos, the residence of a captain-general, military intendant, *corregimento*, &c. It was incorporated as a city, and made a bishop's see by Philip II., and was the residence of the court from his time till that of Philip IV., who removed to Madrid. Columbus, the discoverer of the New World, expired in this city on the 20th May, 1506. (*Miñano; Swinburne; Townsend; Twiss; Mod. Trav.*, xix., &c.)

VALLADOLID, or MECHOACAN, a city of Mexico, cap. intend. of same name, in a fine valley, 120 m. W.N.W. Mexico. Its population early in the present century was estimated at 18,000, and is probably about the same at present. Mr. Ward says, "I know few places the approach to which (from the N.) is so tedious as that to Valladolid. For more than two hours you see the city apparently below you, while the road continues to wind among the surrounding hills. At length a rapid descent conducts you to the plain, where a long causeway, built across a marsh, forms the entrance to the town. The suburbs are poor and insignificant, but the high street is fine, and the cathedral, standing alone and open, has a very imposing effect. The view of the town from the Mexico side is beautiful: gardens and orchards form the foreground; while the lofty aqueduct, erected toward the end of the last century, the gorgeous churches, and a bold range of mountains behind, fill up the remaining space. Nearly all the public edifices, not immediately connected with the government, are due to the munificence of the bishops, most of whom have contributed to enrich or adorn the town. The cathedral, hospitals, and aqueduct are all the works of the church. The first is a magnificent building, and wealthy, though despoiled of much of its treasures during the revolution." (*Ward's Mexico*, ii. 374-377.) Valladolid has a handsome public promenade; and its climate is temperate, as it stands nearly 6400 ft. above the level of the sea. Iturbide, the short-lived emperor of Mexico, was a native of this city.

Valladolid is the name of another Mexican town in Yucatan: but it is of no great importance.

VALOGNES, a town of France, dep. Manche, cap. arrond., on the Merderet, 12 m. S.E. Cherbourg. Pop., in 1836, ex. comm. 6034. It is a well built town, and has a handsome communal college, and public library, with 15,000 printed vols.; manufactures, on a small scale, of earthenware, lace, gloves, and leather; and some trade in provisions and shell fish for the Paris market. The ancient Ronuna town *Alauna* was situated in its vicinity. In the middle ages it had a strong castle, which, however, has been totally destroyed. Tourneur, the translator of Shakspere, and the celebrated anatomist Vicq-d'Azyr, were natives of Valognes. (*Hugo*, art. *Manche, &c.*)

VALPARAISO, the principal seaport town of Chili, prov. Santiago, on the Pacific, 60 m. N.N.W. Santiago. Lat. 33° 11′ 59″ S., long. 71° 31′ 8″ W. Pop. probably about 10,000. It is inconvenient and ill-built, but its appearance from the sea is imposing, being built, somewhat like Hastings, at the foot of a precipitous range of hills. It consists chiefly of a straggling, long street, or rather terrace, for it is built only on one side, with some narrow and inconvenient thoroughfares leading out of it up the several ravines. In one of these is the *plaza*, a small triangular space, where the market is held; and near it are the principal church, the Dominican and Franciscan chapels, &c. A little to the N. is the castle of Antonio, mounting about a dozen guns; and between it and the plaza are a number of low buildings and sheds, termed the arsenal. In the N. quarter of the town, in a recess larger than the other ravines, is a collection of narrow lanes and mean houses; and many isolated dwellings are scattered about among the hills, the only access to which is by winding footpaths. The suburb, Almendral, on the sandy shore to the S., is more regularly laid out: the houses here, where there has been more room to build, consist mostly of a ground floor only; but in the town of Valparaiso all have stories above the ground floor. They are in general painted of lively colours. About the middle of the Almendral are the ruins of the church and convent of La Merced, which, like many other buildings in Valparaiso, was destroyed by the earthquake of 1822.

The bay of Valparaiso is open to the N., but sheltered by heights on all other sides; and the holding ground, being mostly a stiff clay, offers a secure anchorage, except during N. gales. Opposite the custom house, 100 yards from the shore, there are 5 fathoms water, which suddenly deepens to 10 and 20 fathoms, at the distance of 300 yards, and in the centre of the bay are from 26 to 30 fathoms. (*Miers*, i. 444.) The harbour is defended by the castle, and two forts at the N. end of the Almendral, and another fort inland. There is no mole, nor any facility for landing goods, except by launches, which are moored to the shore, and across which all packages are brought on men's shoulders; or by boats, which, however, can land in all weathers, in the Fisherman's bay, between the castle and fort St. Antonio. There are generally a considerable number of vessels in the bay, the greater part British and American; and Valparaiso continues to be the central depôt for the trade of Chili. Large quantities of corn are shipped here for Callao and Panama, especially the former. Wheat, tallow, hides, copper, the precious metals, indigo, wool, and sarsaparilla, are among the principal exports. The market of Valparaiso is well supplied with meat, poultry, fish, bread, fruit, and vegetables, at very moderate prices and of good quality; and its climate is generally agreeable. (*Stevenson's S. America*, iii. 161.) Unless, however, it have materially altered of late years, it would appear, from Mr. Miers' report, to be subject to many drawbacks as a place of residence. "Independent of the want of society, there exists no public amusement, no theatre, commercial reading or news-room, no parade, not even a single spot to walk upon, nor any retirement or exit from the town, but over the barren steep hills, which renders the exercise more a toil than a pleasure. In short, in spite of its matchless and beautiful climate, I do not know in all Chili a spot presenting a more uncomfortable and cheerless place of residence than Valparaiso." (*Miers' Travels in Chili*, i. 449; *Commercial Diet. &c.*)

VAN, a city and considerable lake of Asiatic Turkey, pach. of same name; the city being on the E. bank of the lake, 140 m. N. by E. Mosul, and 145 m. S.E. Erzeroum. Pop., according to Col. Shiel, 12,000; but this is most probably much under the mark, seeing that it had been previously estimated by Kinneir at 50,000, and has since been estimated by Mr. Brant at 7000 families, which would amount to at least 40,000. The inhabitants consist of Turks, Kurds, and Armenians. It is situated in a fine plain, covered with gardens, nearly 5500 ft. above the sea, at the S.W. foot of an isolated rock, on which is its citadel. It is inclosed with double walls of mud and stone, having large round and small square bastions, and is farther defended, though not on all sides, by a ditch. "The streets are narrow, dirty, and ill paved; the exterior appearance of the houses in general mean; there was, however, to be seen occasionally a residence, which showed that it had once belonged to a man of consequence; but the general aspect of the city indicated decay. The bazaars were confined, and the shops ill-furnished; and I scarcely saw an article of European manufacture: there was, however, an abundance of Venetian glass beads, with which the Kurd females ornament their persons. The supply of fruit was superabundant." (*Brant in Geog. Journ.* x. 384.) The account of the city given by Kinneir is less unfavourable; but Brant's, being the later, is perhaps also the more accurate.

It has two large Armenian churches, four large mosques, several baths, caravanserais, &c. The great boast and dependence of Van is its gardens, which cover an extensive area between the city and the mountains. They comprise vineyards, orchards, melon-grounds, fields, &c. In summer the inhabitants of the town mostly reside in the gardens, the roads in which being lined with houses, the whole appears like an extensive village. Streams, bordered with willows, run through the main avenues. (*Brant, in Geog. Journ.* x. 391–393.)

The castle on the N.E. side of the town is built on a high and abrupt limestone rock, and, if the works were in proper repair and efficiently manned, would be all but impregnable. There are some very extensive excavations in this rock.

The trade of Van is at present inconsiderable. About 500 looms are employed in manufacturing coarse calicoes from cotton imported from Persia, mostly for home consumption: though some are sent to Bitlis to be dyed and exported. Almost the only other articles of export from Van are fruits, wine, and grain, the produce of the surrounding gardens and fields. Every person of respectability has a house in town, with a country house, an orchard and vineyard, and perhaps a few fields. Most of his wants are supplied from his garden, or from the profits of a petty trade carried on with a capital of from £20 to £100.

There can be no doubt that Van is very ancient. The walls of the castle are in part Cyclopean, and many inscriptions in the cuneiform character have been discovered in the town and its environs. It is even affirmed that the castle was originally found by Semiramis; but it is almost needless to say that there is no evidence by which to support this very improbable statement. Timour Bec, or Tamerlane, who took Van towards the close of the 14th century, is said, by the Persian writers, to have endeavoured, though ineffectually, to destroy its ancient monuments.

The lake Van, the *Arsisa* of Ptolemy, is of a very irregular shape; greatest length N.E. to S.W. 70 m.; greatest breadth about 28 m. Area estimated at 1000 geog. sq. m. (*Geog. Journ.* x.) It seldom freezes to any distance from the shore, except at its N.E. end, where, being shallow, in severe winters, it may be crossed on the ice. It has several islands, on one of which is an Armenian monastery. It is navigated by five or six crazy boats, which are sometimes employed to convey cotton cloths to Tadvan on their way to Bitlis, bringing back grain and timber. From the many wild fowl that frequent it, there is no doubt that fish abound in the lake; the fishery, however, occupies very few hands. (*Geog. Journ.; Kinneir's Pers. Emp.; Smith and Dwight's Missionary Researches, Introd., &c.*)

VAN BUREN, county, Mich. Situated towards the S.W. part of the state, and contains 633 sq. m. Bounded W. by

lake Michigan. Drained by Papaw r. and the S. branch of Black r. and their branches. Papaw river is boatable 70 m. from its mouth in Berrien county. It contained in 1840, 2125 neat cattle, 528 sheep, 3422 swine; and produced 15,640 bushels of wheat, 32,587 of Indian corn, 16,176 of oats, 90,532 of potatoes, 25,365 pounds of sugar. It had four stores, two grist-mills, eight saw-mills; 23 schools, 515 scholars. Pop. 1910. Capital, Papaw.

VAN BUREN, county, Iowa. Situated in the S.W. part of the territory, and contains 504 sq. m. Drained by Des Moines r. and its branches. Copper, iron, and tin ores are found, and fine marble and bituminous coal. It contained in 1840, 5364 neat cattle, 2667 sheep, 15,821 swine; and produced 8756 bushels of wheat, 151,481 of Indian corn. 1946 of buckwheat, 12,174 of oats, 18,150 of potatoes, 6369 pounds of sugar. It had 16 stores, 10 grist-mills, 12 saw-mills, one tannery, two potteries; six schools, 123 scholars. Pop. 6146. Capital, Keosauke.

VAN BUREN, county, Mo. Situated in the W. part of the state, and contains 648 sq. m. Bounded S. by Grand r., drained by Big cr. It contained in 1840, 1726 neat cattle, 3232 sheep, 19,573 swine; and produced 1644 bushels of wheat, 232,541 of Indian corn, 11,000 of oats, 9112 of potatoes, 8875 pounds of tobacco, 2890 of cotton. It had 11 stores, one flouring-mill, six grist-mills, two saw-mills, four tanneries; 13 schools, 298 scholars. Pop.: whites 4448; slaves, 214; free coloured, 31; total, 4693. Capital, Harrisonville.

VAN BUREN, county, Ark. Situated a little N. of the centre of the state, and contains 1230 sq. m. Drained by little Red r. and its branches. It contained in 1840, 2402 neat cattle, 520 sheep, 9213 swine; and produced 1545 bushels of wheat, 79,800 of Indian corn, 3237 of oats, 2815 of potatoes, 1220 pounds of tobacco, 8051 of cotton. It had five stores, seven grist-mills, three distilleries; four schools, 95 scholars. Pop.: whites, 1450; slaves, 59; total, 1512. Capital, Clinton.

VAN BUREN, p. t., Onondaga co., N. Y., 12 m. N.W. Syracuse, 141 m. W. by N. Albany, 208 W. Bounded N. and W. by Seneca r. Drained by Camp brook. It has eight saws, one fulling-mill, one flouring-mill, two grist-mills, eight saw-mills, one tannery, one distillery; one academy, 61 students; nine schools, 436 scholars. Pop. 3021.

VANDALIA, p. v., capital of Fayette co., Ill., until 1840, the capital of the state, 89 m. N.E. St. Louis, 782 W. It is regularly laid out with streets crossing each other at right angles, which are 80 ft. wide. It has a public square in the centre, and contains a courthouse, jail, a United States land office, two churches, a Presbyterian and Methodist, 11 stores, two printing-offices, each issuing a weekly newspaper, one steam and one water saw-mill, and about 900 inhabitants. The national road extends to this place.

VANDERBURG, county, Ia. Situated in the S.W. part of the state, and contains 235 sq. m. Bounded S. by Ohio r., along which the land is fertile. Drained by Big Pigeon cr. It contained in 1840, 3021 neat cattle, 713 sheep, 709 swine; and produced 4225 bushels of wheat, 165,739 of Indian corn, 4315 of oats, 3841 of potatoes, 335 pounds of sugar. It had one commission house in foreign trade, 49 stores, two flouring-mills, seven saw-mills, one tannery, one distillery, one brewery, two printing-offices, two weekly newspapers, seven schools, 212 scholars. Pop. 6250. Capital, Evansville.

VAN-DIEMEN'S LAND, or Tasmania, an island belonging to Great Britain, in the S. Pacific and eastern oceans, off the S. extremity of Australia, from which it is separated by Bass's straits, between lat. 41° 20' and 43° 40' S. and long. 144° 40' and 148° 20' E. It is shaped like a heart. Area, estimated at about 27,000 sq. m. Pop., in 1838, 43,545; of whom 26,055 were free, and 18,133 convicts; the number of aboriginal inhabitants is very trifling. The country is in general hilly or mountainous. Captain Fitzroy says, "The change of scene from Sidney to Hobart-town, was as striking as a view of Gibraltar or Madeira after leaving the Downs. Comparatively speaking, near Sydney all was light-coloured and level; while in Van-Diemen's Land we almost thought ourselves in another Tierra del Fuego." (*Voyage of the Adventure, &c.*) The mountains rise to 3500 or 4000 feet, not forming continuous ranges, as in the adjacent continent, but generally isolated peaks. The surface consists alternately of elevated table lands and fertile valleys, neat part of it being fit either for cultivation or pasturage. Several considerable rivers water the colony. The principal are the Derwent, and Tamar formed by the union of the N. and S. Esk, all which rise toward the centre of the island, in about lat. 42°; the first running to the S.E., and the rest northward. There are several large lakes in the interior, one of which is said to be 60 m. in circ. The coasts are very much indented, and abound with excellent harbours. The climate is comparatively healthy, being appropriate to European constitutions, than that of Sydney. The winters are colder; but the country seldom suffers from those

VAN-DIEMEN'S LAND.

ing continued droughts which are the case of New South Wales, nor from too much rain. Sandstone, limestone, and basalt, are among the principal geological rocks. Indications of coal have been met with, and iron ore has been dug up. some of which is said to yield as much as 50 per cent. of metal. Copper, lead, zinc, and manganese exist, but no mines ave been hitherto wrought. The upper soil is usually sandy r argillaceous; or else consists of a rich vegetable mould 'here appears to be comparatively a much larger proportion of good soil than in Australia: fine tracts of land are und quite down to the borders of the sea, a circumstance usual in the last-named continent: and extensive tracts vered with luxuriant herbage, and free from timber or underwood, and which, consequently, require no clearing on e part of settlers previously to being ploughed, are met ith in all parts of the interior. Timber, however, is by no ieans scarce; and in 1838, 435,632 cubic ft. were cut at ort Arthur, of which nearly three fourths were exported. 'he Huon and Adventure bay pines, and the black wood, re peculiar to the country; in most other respects the vegetable products, as well as the animals, are similar to those f Australia. A species of panther, which commits much avoc among the flocks, and kangaroos are found in great umbers; there is, however, no native dog. Poisonous reptiles are less numerous than in the adjacent continent. The boriginal inhabitants appear to belong to the negro race of he E. archipelago: they seem to be sunk in the lowest epths of barbarism; and are said to be ignorant of the ost useful and obvious arts, as fishing, and the construction of the rudest canoes. Their numbers have rapidly decreased since the establishment of the whites, and a few ears since the greater part of those remaining in the colony vere removed to Flinder's island in Bass's straits.

The only settled portions of Van Diemen's Land stretch cross the interior from the N. to the S.E. coast along the courses of the Derwent, Tamar, and other rivers. All the N. and the N.E. parts of the island are yet not merely unsettled, but even unexplored.

From 1824 to 1838 inclusive, the grants of land to settlers n the colony appear to have amounted to 1,128,000 acres, y far the greater proportion of which was in grants of 500 cres and upwards. The average price per acre in the last mentioned year was 5s. 10¼d.; and the sum total that accrued to the government from the sale of land during the iecennial period ending with 1838 amounted to £147,370. The plan of selling the unappropriated land by auction is now adopted in this colony, as in most others belonging to England. In 1839 the sales in this way amounted to 49,386 acres, at 10s. 1¾d., exclusive of certain town lots, the sum eceived for the whole being £23,256; and during the same ear 15,836 acres were granted by the crown in reward of military and naval services. In 1840 the sales of land amounted to 88,996 acres, at 11s. 4d., exclusive of town lots, he aggregate price of the whole being £55,305; the grants o military and naval men during the same year exceeded 4500 acres. It is estimated that there are still in the colony above 12,000,000 acres ungranted. In 1838 about 108,000 cres were said to be under crop, including about 41,760 in wheat, 13,500 do. in barley, 21,600 in oats, 3500 in potatoes, 1100 in turnips, and 17,000 or upwards in grasses. The total produce of corn during the same year was estimated at about 969,000 bushels; of which, 550,000 were wheat. If hese returns may be depended on, it would seem that at an average the produce of corn per acre amounts to little nore than 12 bushels! or to only about one third part of the produce of an acre of land in England! This lamentable deficiency of produce may be ascribed partly, we believe, to he backward state of agriculture, and the want of care in he preparing of the land, and partly to the inferior fertility f the latter. In fact, Van-Diemen's Land, though superior is a corn-growing country to New S. Wales, is notwithstanding better fitted for grazing than cropping. Wool, indeed, is here, as well as in Australia, the staple produce of he colony; and the increase in the breed of sheep has been o very great that the stock, which in 1829 amounted to 553,698 head, had in 1838 increased to 1,214,000 head; and while the imports of wool from Van-Diemen's Land into the United Kingdom in 1830, did not exceed 993,979 lbs., they amounted, in 1841, to 3,507,531 lbs. Maize is not raised in he colony, the climate being too cold. Apples, currants, gooseberries, &c., attain perfection, but the orange, citron, and pomegranate are not raised, and the grape and peach attain only an inferior degree of maturity.

In 1838, the stock of cattle was estimated at 75,000 head; of horses, at 9650; and goats, at 2400. All kinds of stock attain a much greater size than in the neighbourhood of Sydney. Between 1824 and 1836, there was, owing to a variety of causes, an unprecedented rise in the value of all sorts of property in the colony, particularly of land, sheep, and cattle. Land, which in 1824 would not have fetched more than from 5s. to 10s. an acre, was, in 1836, sold at prices ranging from £2 to £4; and sheep and cattle, which

in the former year might have been purchased for 5s. and 40s. respectively, could not, in the latter, be procured for less than £3 and £12. (Statist. Rep.) But since 1836, a fall has taken place in prices, wages, &c., which, though at first productive of inconvenience, has, by putting an end to much extravagant speculation, been of decided advantage; and the colony is steadily advancing in population and prosperity. In 1838, 101 vessels of the aggregate burden of 8382 tons, belonged to Van-Diemen's Land; of these 19, of about 2000 tons burden, were employed in the whale fishery, which imported oil, &c., to the gross value of £98,660. We subjoin an

ACCOUNT of the Import and Export Trade of the Colony in 1839 and 1840.

Countries from and to which Imports and Exports were made.	Value of Imports.		Value of Exports.	
	1839.	1840.	1839.	1840
	£.	£.	£.	£.
Great Britain	573,491	737,351	326,369	334,166
British Colonies	150,061	217,033	545,156	531,321
United States	6,013	6,996	3,640	1,620
Foreign States	17,323	27,176		
	746,887	988,356	875,165	867,007
	Men.	Tons.	Men.	Tons.
Ships:—Inwards	5,444	79,383	5,895	850,081
Outwards	5,481	77,556	6,173	86,701

The great articles of export are wool, whale oil, and whale-bone, and bark. The imports comprise every description of manufactured goods, colonial products, wines, farming utensils, &c.

The government is subordinate to that of New South Wales; but the lieutenant-governor, with the aid of the executive and legislative council, administers the local government independent of the parent colony. The executive council consists of five members, including, with the lieutenant-governor, the chief justice, colonial secretary, and treasurer, and the officer in command of the troops. The legislative council is composed of from 10 to 15 members, appointed by warrant of the sovereign; and all the above functionaries are ex officio members. The laws of England, and acts of the British parliament are generally administered in the colony, but special acts may be passed by the governor and council. The initiatory proceedings in the passing of all acts belong to the governor; and two thirds of the council must be present at their passing, and the majority vote with the governor, or no proposition can become law. Civil causes are tried before a judge and two assessors, and criminal cases by a jury of seven naval or military officers. In most other respects this colony is administered in a manner similar to New South Wales. Public revenue in 1838, £138,591; expenditure, £133,681. Of this expenditure £10,271 was for the public schools, in which 1380 pupils were instructed.

The settled part of Van-Diemen's Land is divided into 15 districts. Hobart town, or Hobarton, the cap., is in the district of same name, on the Derwent, about 20 m. from its mouth, lat. 42° 54' S., long. 147° 34' E. Pop. of the town and district in 1838, 14,382; of whom 3553 were convicts. The town covers about 1 sq. m.; it stands upon the declivities of two gentle hills, and is intersected by a fine stream from the heights of the Table mountain, which towers above the town on the W. to the height of 3936 ft. The streets are wide, and intersect each other at right angles; and, having been from the first laid out on a uniform plan, it is much more regularly built than Sydney. Its houses are substantial, and of two stories in height: it has some good public buildings, including a church constructed of brick, a jail, and a quay, close to which vessels of the largest burden load and unload. Hobarton possesses one of the finest harbours in the world. The Derwent, which here forms a fine sheet of water, is navigable for ships for 3 m. above the town, and continues to afford a safe passage for vessels of 50 tons as far as New Norfolk, 20 m. higher up, where a ridge of rocks abruptly puts an end to the navigation. In 1838, 79 ships, of the aggregate burden of 6079 tons, belonged to this port. The suburbs of Hobart town are increasing and receiving much embellishment; villas, enclosures, &c., are said to be springing up in every direction. The principal settlement on the N. side of the colony is Launceston, on the Tamar, about 40 m. from its mouth, and 105 m. N by W. Hobart town. The river is navigable to it for vessels of 300 tons. The other towns are insignificant.

This island was originally discovered by the navigator Tasman in 1642, and subsequently received its name in honour of a governor of the Dutch E. Indies. It was afterwards visited and partially explored by Cook, Furneaux, D'Entrecasteaux, &c.; but it was not ascertained to be an island till Bass sailed through the straits which bear his name in 1798. The first English penal settlement was established here in 1803, and down to 1813 it continued to be merely a place of transportation from New South Wales.

1607

The settlement continued to suffer from the depredations of escaped convicts, termed bush-rangers, till 1817, when these were finally put down; voluntary emigration began to take place to a considerable extent in 1831, and has since progressively increased. (*Statist. Rep. on Van-Diemen's Land, &c.*)

VANNES, a town and seaport of France, dep. Morbihan, of which it is the cap., at the bottom of the gulf of Morbihan; 63 m. W.N.W. Nantes. Lat. 47° 39' 28'' N., long., 2° 45' 4'' W. Pop., in 1836, ex. comm., 9398. It was formerly fortified, and entered by six gates, five of which, with some towers, still remain. It is clustered around its cathedral, and except one or two tolerable streets is irregularly and badly paved, and dirty. The cathedral, rebuilt in the 15th century, is a massive, but heavy edifice; in its interior, however, are some good monuments and paintings. It has another church, and several chapels, one of which, attached to the communal college, is of good architecture. The prefecture is an ancient castle; and a convent has been converted into the residence of the bishop. There were formerly many monastic establishments at Vannes; but their buildings have now mostly received other destinations, and one is appropriated to the *Institution du Pere Eternel*, which, besides boarding and educating 60 poor girls, has a great many out-scholars receiving primary instruction. The edifice, now the theatre, has served at different periods for the meetings of the states of Brittany, and the parliament of Rennes. Vannes has three hospitals, a communal college, school of navigation, a Polymathic society, established in 1826, for the culture of the arts and sciences, a public library of 8000 vols., and manufactures of coarse cloths, linen and cotton yarn, and lace. Its port is small, and the entrance being shallow, it is fit only for vessels of small burden; on one of its sides is a planted promenade, and on the other are slips for building boats and small craft. It has been supposed that Vannes occupies the site of *Dariorigum*, the cap. of the Veneti; but the better opinion seems to be that the latter was about 5 m. N. by W. Vannes, at Locmariaker, where the remains of a circus and Roman road are still extant. (*Hugo*, art. *Morbihan*; *D'Anville, Notice de L'Ancienne Gaule*, p. 262; *Martiniere, &c.*)

VAN WERT, county, O. Situated in the W. toward the N. part of the state, and contains 432 sq. m. Drained by St. Mary's, and Little Auglaize rivers. It contained in 1840, 1724 neat cattle, 335 sheep, 3800 swine; and produced 3090 bushels of wheat, 33,306 of Indian corn, 3957 of oats, 5671 of potatoes, 3070 pounds of sugar. It had four stores, two grist-mills, two saw-mills. Pop. 1577. Capital, Van Wert.

VAR, a maritime dep. of France, reg. S.E., between lat. 43° and 44° N., and long. 5° 40' and 7° 15' E.; having W. the dep. Bouches-du-Rhone, N. the Basses-Alpes, N.E. the Sardinian states (co. Nice), from which it is separated by the river Var, and E. and S. the Mediterranean. Area, 726,866 hectares. Pop., in 1836, 321,686. The surface, particularly in the N. and E., is mostly covered with ramifications of the Alps, consisting of primitive or calcareous formations covered with a gravelly or arid soil. The dep. is, however, well watered, and on the banks of some of the rivers, as the Var and Argens, are some very fertile tracts. The Var, whence the dep. derives its name, rises in the dep. of the Basses-Alpes, about 6 m. S.E. Colmar, and, after running generally southward, falls into the Mediterranean at St. Laurent, 3½ m. S.W. Nice, after a course of 68 m., for about half of which it is navigable for rafts. The Esteron and Vesoubia are its chief tributaries; no towns of much consideration are on its banks. Numerous lagoons and marshes border the coast of this dep., which is generally abrupt, rocky, and indented with numerous inlets. In 1834, the arable lands were estimated to comprise 118,052 hectares; vineyards, 67,657 do.; woods, 230,700 do.; and wastes, about 187,800 do. Agriculture is very backward, and the supply of corn is inadequate for the consumption of half the pop.: the produce in wine, olives, and fruits of various kinds is, however, considerable; and 115,494 kilogs. of silk were gathered in 1835. Capers, oranges, lemons, chestnuts, and perfumery are among the principal exports. Grasse is especially famous for its essences and liqueurs; and roses, jasmines, &c., are grown there and elsewhere in large quantities for their manufacture. The forests comprise many cork-trees, and bottle corks are made in several places Kermes and a little tobacco are among the other products. Few horses or cattle, but a good many mules, are bred. In 1830, there were estimated to be upwards of 370,000 sheep, and 18,000 goats in the dep., and the woods afford nourishment for numerous hogs. In 1835, of 109,560 properties subject to the *contrib. fonciere*, 42,012 were assessed at less than 5 fr., and 21,790 at from 5 to 10 fr., only 37 were assessed at 1000 fr. and upwards. Lead, coal, copper, iron, manganese, &c., are met with, though very few mines are wrought. The tunny and anchovy fisheries are of considerable importance, and coral of good quality is

1008

fished up on the coast. Manufactures of no great consequence; but silk twist, perfumery, soap, paper, and leather are produced to some extent: the dep. has, however, a considerable trade with the Sardinian states, &c. It is divided into four arronds.; chief towns, Draguignan, the cap. Toulon, Brignolles, and Grasse. It sends five members to the chamber of deputies. Registered electors in 1838-9, 1721. Total public revenue, in 1831, 9,210,927 fr. (*Hugo*, art. *Var; Dict. Geog.; French Official Tables.*)

VARICK, p. t., Seneca co., N. Y., 8 m. S. Waterloo, 196 W. Albany, 332 W. It extends from Cayuga lake on the E. to Seneca lake on the W. It has three stores; 11 schools, 615 scholars. Pop. 1971.

VARINAS, a town of Colombia, repub. Venezuela, cap. prov., on a tributary of the Apure, 360 m. S.W. Caracas Pop. uncertain, being variously estimated at from 8000 to 12,000. It is rather handsome, but has no public buildings worth notice. It is the principal mart for the excellent tobacco grown in its prov.; but has also a considerable trade in sugar, coffee, cotton, indigo, and cattle, which are mostly exported by way of the Apure and Orinoco. (*Codazzi; Geog. de Venezuela, &c.*)

VARNA, a fortified town and seaport of European Turkey, prov. Bulgaria, on the Black sea, at the mouth of the Pravadi, 47 m. E. Schumla; lat. 43° 12' N., long. 27° 54' E. Pop. estimated at about 16,000. Though the residence of a pacha, it is a poor town, and has only an open roadstead; but the latter, being sheltered from N.W. winds, which are the most to be feared in this sea, and the holding ground being good, it presents but little danger; and Varna is a principal *depot* for the export of Bulgarian produce to Constantinople. The town fell into the hands of the Russians in 1828, in consequence, as is supposed, of their having bribed the governor. (See *Keppel's Journey across the Balkan*, i. 358, 359.) Agreeably to the treaty of Adrianople, its fortifications were dismantled, but they are said to have been since replaced by new defensive works. (*Dict. Geog., Stein; &c.*)

A great battle was fought near Varna on the 10th of Nov. 1444, between the Hungarians and their allies, under their king Ladislaus, assisted by the famous John Huniades, and the Turks, under their sultan, Amurath II. The Christians, who had broken a truce which they had a short while before solemnly sworn to observe, suffered severely for their perfidy; their king having been killed in the battle, Huniades taken prisoner, and their army totally defeated with a prodigious slaughter. (*L'Art de Verifier les Dates*, 2e partie, v. 250, 8vo ed.)

VASSALBOROUGH, p. t., Kennebec co., Me., 11 N E Augusta, 606 W. Bounded W. by Kennebec river. By means of a dam across Kennebec river, at Augusta, and a lock at that place, vessels of considerable burthen pass from this place to the ocean. Incorporated in 1771. It has nine stores, two fulling-mills, one woollen factory, seven grist-mills, seven saw-mills, one paper-mill, six tanneries; one academy, 100 students; 22 schools, 1164 scholars. Pop. 2952.

VASTO (an. *Histonium*) a town and seaport of the Neapolitan dom., prov. Abruzzo-Citra, on the Adriatic, 21 m. S.E. Chieti. Pop., in 1830, 9016. It is walled, and has two collegiate churches, one of which is said to occupy the site of a temple of Ceres, several convents, hospitals, &c., with many good private buildings. Its inhabitants are mostly occupied in the manufacture of earthenware, in fishing, and the cultivation of olives and vines. On the 1st of April, 1816, this town was the theatre of an extraordinary catastrophe, by which it was partially destroyed by a sudden falling in and sliding of the soil, supposed to have been occasioned by the escape of subterranean water. (*Del Re, Descrizione delle Due Sicilie*, ii. 410.)

VAUCLUSE, a dep. of France, reg. S.E., between lat. 43° 40' and 44° 25' N., and long. 4° 40' and 5° 45' E. having N. the dep. Drome, E. Basses-Alpes, S. Bouches-du-Rhone, from which it is separated by the Durance, and W. the Rhone, separating it from Gard. Area, 347,377 hectares. Pop., in 1836, 239,113. The general slope of this dep. is to the W., in which direction it is traversed by many small tributaries of the Rhone. Soil, mostly calcareous, but it is only in the neighbourhood of the larger rivers that it possesses much fertility. In 1834, the arable lands were estimated to comprise 157,738 hectares; vineyards, 32,384 do.; woods, 62,141 do.; willow plantations, &c., 977 do.; and heaths, wastes, &c., 67,760 do. In 1835, of 81,149 properties subject to the *contrib. fonciere*, 38,304 were assessed at less than 5 fr., and only 22 at 1000 fr. and upwards. Agriculture, though still very backward, is said, of late years, to have made considerable progress. Oxen, horses, or mules are used indifferently for the plough; but the last are principally employed. The scarcity of other manure obliges the farmers to cut the wild box on the hills, which being macerated and suffered to rot, is used for the purpose, and found peculiarly suitable for manuring vines. Wine and silk are among the most important products of the dep.; the produce of wine may, perhaps, be estimated at about 500,000 hectolitres, of which a considerable portion is exported. The best wines are those of *Coteau-brulé*, near Avignon. and of *Châteauneuf*, 4½ leag. from Orange. Speaking generally, however, too little attention is paid to the culture and care of the vine. *On en compterait,* says Jullien. *beaucoup plus de bons vins, si les proprietaires apportaient plus de soin dans le choix des plantes et dans la culture.* (p. 255.) In 1835, upwards of 1,600,000 kilogr. of raw silk were gathered. The produce of wheat is insufficient for home consumption, but a good deal of rye, and some maize, barley, buckwheat, potatoes, &c., are grown, besides saffron, madder, and coriander and anise seeds. Almonds and other fruits, and essence of lavender are among the exports. Artificial grasses are sown; but the pasture lands are not extensive, and the quantity of stock is less than in any other of the S. deps. Bees are pretty numerous; the honey is of good quality; and about 60,000 kilogr. of wax are supposed to be exported annually. This dep. does not appear to be rich in metals; but building stone of good quality, lime, potters' clay, &c., are plentiful. Silk fabrics are made at Avignon and Orange; and the town of Apt is famous, in the S. of France, for its earthenware and confectionery, as is Avignon for its printing establishments, and manufactures of printing types, bells, and other metallic goods. Copper, lead, and iron plates, prepared madder, woollen stuffs, leather, paper, cordage, linen thread, glass wares, and tiles are made in this dep., which has also many distilleries and dyeing-houses. Vaucluse is subdivided into four arronds.: chief towns, Avignon, the cap., Apt, Carpentras and Orange. It sends four members to the chamber of deputies. Registered electors, in 1838–39, 1232. Total public revenue (1831), 5,482,148 fr. (*Hugo*, art. *Vaucluse; Dict. Géog.; French Official Tables.*)

VAUCLUSE, a famous fountain in the above dep., close by the small village of the same name, 4 m. W. Isle, and 15 m. W. by N. Apt. This fountain has its source in a vast cavern at the foot of a rock 300 feet in height at the bottom of a narrow gorge in the mountains. Within this cavern is a deep basin of the purest water, the surface of which appears to be perfectly smooth and placid, but which, notwithstanding, emits so great a quantity of water as to give birth to the small river Sorgues, an affluent of the Rhone. After the melting of snows, or the occurrence of long-continued rains, the flow of water is greatly increased; but even in the driest reasons it is astonishingly copious. In ordinary states of the fountain the water escapes by percolating through the rocks, stones, and gravel, that form the outward side of the abyss; whereas during floods it overflows its banks. It is possible, taking proper precautions, to descend, when the water is low, to its edge. Owing to its great depth it appears as if it were jet black, though, as already stated, it is limpid in the extreme.

This fountain is celebrated in the history of Petrarch. An old castle near the village, which belonged to his friend the bishop of Cavaillon, was frequently visited by Petrarch, and is thence called his castle. The poet, however, lived in the village. He here frequently saw Laura, who is generally supposed to have been the wife of the Count de la Sade, the lord of the village. But this is doubtful, as well as the Platonism of her lover. (*Hugo*, art. *Vaucluse, &c.*)

VAUD (Fr. *Pays de Vaud*, Germ. *Waadt*), a canton of Switzerland, holding the 19th rank in the confed., between lat. 49° and 47° N., and long. 6° and 7° 12′ E., having N. the cant. and lake of Neufchatel, E. Friburg and the Bernese Oberland, S. the Valais, Savoy, and the lake and territory of Geneva, and W. the dep. of Jura in France. Area, about 180 sq. m. Pop., in 1837, 180,582; all Protestants, except about 3000 Rom. Catholics. Both the S.E. and N.W. extremities of this canton are mountainous; the former quarter is covered by ramifications of the Alps, one summit of which, the Diablerets, rises to 11,190 ft. above the level of the sea; the latter region is traversed by ranges of the Jura. The middle of this canton, between the Jura mountains and the lakes of Geneva and Neufchatel, is a rich undulating country, and so celebrated for its agreeable character and picturesque beauty, that it has long been resorted to by visiters from many parts of Europe; in 1837, 14,500 of the inhabitants were citizens of other cantons, and nearly 4000 foreigners! The raising of corn and wine is the occupation of the population. The vineyards, reckoned the best in Switzerland, are supposed to comprise about 13,000 acres, and to employ 20,000 vine-dressers, independently of women and children. The wines of La Vaux and La Cote, both grown on the shores of lake Leman, are the best: the first is produced near Vevay, where the Romans, who erected a temple to Bacchus at Culy, are supposed to have originally planted the vine. A society of high antiquity in that town exercises a survey of the vineyards in the district, and celebrates a remarkable fête, attended by a great concourse of strangers, every 15th or 20th year. The num-

ber of cattle in Vaud has of late increased very much: in 1830, it was supposed to have about 75,000 head of cattle, 23,500 horses, and 77,000 sheep; and the improvement in the breeds is said to have kept pace with the increase of numbers. Almost the only salt springs in Switzerland exist at Bex in this cant. They were discovered in the 16th, and bought by the government of Bern in the succeeding century; in 1834, they produced a revenue of 18,439 Sw. francs. Marble, coal, crystal, sulphur, petroleum, and a few metallic products, are met with, though not raised in any large quantities. Manufactures are very unimportant, and mostly confined to those of leather and yarn. The transit trade from France into Switzerland and Germany is of more consequence to the inhabitants.

Vaud is subdivided into 19 districts and 60 circles: all the towns of more than 3000 inhabitants forming a separate circle. Its constitution is more aristocratic than that of most of the other Swiss cantons.

The legislative power is vested in a *grand council* of 184 members elected for five years, which meets yearly in May at Lausanne. The electors include a certain portion of those among the citizens who are taxed to the highest amount (*du nombre des trois quarts des citoyens le plus imposés a l' impôt foncier*), and must be neither domestic servants, recipients of public relief, bankrupt, nor under penal condemnation. Each circle sends one deputy to the grand council, who must be an inhabitant of the circle, 30 years of age, and possess property in the canton of the value of 2500 fr., or some equivalent thereto; the town of Lausanne sends four deputies; four candidates are also nominated by each of the circles, and from among the general list of the members already chosen elect 63 other members. Finally, an electoral commission, composed of the members of the executive body, of the court of appeal, and 40 members of the legislative council, chooses the remaining members, electing 36 from among the cantonal citizens 40 years of age and possessors of landed property to the value of 10,000 fr., and 18 above 25 years of age without property qualification. (*Picot.*) The legislature chooses the executive council, which consists of nine members, and has the initiative in all propositions of laws and taxation. Each district has a court of primary jurisdiction, and each circle a justice of the peace. The inhabitants understand and speak French; but their common *patois* is a dialect somewhat similar to the Romansche. Schools are pretty general, and in 1834 it was estimated that one in six of the population was receiving public instruction. Public revenue, in 1834, 1,276,977 fr.; expenditure, 1,115,665 do.

This territory nearly corresponds to the *Pagus Urbigenus* of Cæsar. It successively belonged to the Burgundian and Frankish kingdoms, the Germanic empire, and the dukes of Zahringen. From 1273 to 1536 it was possessed by Savoy; and by the government of Bern from the latter year till 1798, when it was erected into the canton of Leman. It reassumed its present name in 1803. It furnishes 2964 men to the army, and 59,273 fr. a year to the treasury of the Swiss confederacy. (*Picot, Statist. de la Suisse; Ebel; Bowring's Rep. on Switzerland.*)

VELEZ-MALAGA (an. *Menoba*), a town of Spain, in Andalusia, prov. Malaga, cap. distr. on the Velez, near its mouth, and 14 m. E.N.E. Malaga. Pop., with its vicinity, according to Miñano, about 14,000. "The town is slightly elevated above and on the left bank of the stream, and is commanded by the neighbouring hills. The streets are wide, clean, and well paved; but the thriving commerce, and abundant market naturally looked for in a place once so noted for the productiveness of its orchards and extent of its export trade, are no longer to be seen." (*Scott's Ronda and Granada,* i. 219.) Under the Moors, Velez was a place of considerable strength, and had a castle, now in ruins. It has two parish churches, six convents, several workhouses, a prison, public granary, &c., and some fine public promenades. It is said to be peculiarly well situated; its climate is not oppressively hot, the town being sheltered by the neighbouring heights; and its neighbourhood is very fertile, producing sugar, coffee, cotton, cochineal, large quantities of wine, silk, various fruits, the sweet potatoe, &c. It has some sugar-mills, and manufactures of hats, soap, brandy, &c. It was taken from the Moors in 1487, the siege having been conducted by Ferdinand in person. (*Miñano; Townsend; Scott; &c.*)

VELLETRI (an. *Velitræ*), a town of S. Italy, Papal States, Comarca di Roma, near the Appian Way, 20 m. S.E. Rome. Pop. about 10,000. It stands on a commanding eminence at the foot of mount Artemisio, and enjoys an extensive view of the Pontine Marshes, as far as Monte Circello, and of the sea, with the range of the mountains of Norba, Cora, and Segni, and even those beyond Palestrina. It is surrounded by ruined walls, with decayed towers, and ruinous curtains, and is but indifferently built. It has a town-hall by Bramante, and some fine palaces. The principal square has a fine fountain, and a bronze statue of

Pope Clemen VIII. The Borgian museum, now in Naples, was originally established in the Borgian palace, in this town. The inhabitants are mostly peasants, who work in the neighbouring fields and vineyards, and at night retire to the town, the air of which is extremely good. According to Simond, the inhabitants are "not quite so ragged as their neighbours, although the place is full as dirty, and the lazzaroni as miserable." *Velitra* appears to have been one of the most considerable cities of the Volsci, and is said by Dionysius to have been fortified by Coriolanus. It was the residence of the Octavian family before they settled in Rome; and is celebrated as the birth-place of Augustus, who was born here on the 22d September, anno 63 B.C., in the consulship of Cicero. (*Rampoldi; Gell's Topography of Rome, &c.,* ii. 340, &c.)

VELLORE, a town and fortress of British India, presid. Madras. distr. Arcot, 80 m. W. by S. Madras. The fortress, which is of considerable extent, comprises spacious barracks and other necessary accommodations for a garrison, with various handsome buildings round a square, on one side of which is a curious pagoda, now used as a magazine. The native town, which is also large and populous, is situated to the S. of the fort, with which it is connected by additional battlements. In 1677 it was taken by Sevajee; and, during the war of 1782, was relieved by Sir Eyre Coote, in the face of Hyder's army. After the capture of Seringapatam. Tippoo's family was removed thither. In 1806 a formidable revolt of the native troops, followed by a massacre of the Europeans, took place here, of which Tippoo's family were supposed to have been the instigators. This occasioned the removal of the latter to Calcutta. (*Hamilton's E. I. Gazetteer.*)

VENANGO, Pa. Situated toward the N.W. part of the state, and contains 1190 sq. m. Drained by Alleghany river and its branches. Organized in 1805. It contained in 1840, 16,070 neat cattle, 28,775 sheep, 24,512 swine; and produced 198,486 bushels of wheat, 70,068 of rye, 59,730 of Indian corn, 52,114 of buckwheat, 214,157 of oats, 118,719 of potatoes, 17,561 pounds of sugar. It had 43 stores, seven fulling-mills, three woollen-factories, 16 furnaces, one forge, 49 grist-mills, 59 saw-mills, two oil-mills, 17 tanneries, four distilleries, three potteries, one printing-office, one weekly newspaper; one academy, 20 students; 35 schools, 949 scholars. Pop. 17,900. Capital, Franklin.

VENANGO, t., Crawford co., Pa. It has three stores, one flouring-mill, six-saw-mills, one tannery, one pottery. Pop. 1900.

VENDEE (LA), a maritime dep. of France, reg. W., formerly included in the prov. of Poitou, and, exclusive of the islands Dieu and Noirmoutiers, extending between the 46th and 47th degrees of N. lat., long. 0° 30' and 2° 10' W.; having N., Loire-Inférieure and Maine-et-Loire; E., Deux-Sèvres; S., Charente-Inférieure; and S.W. and W. the Atlantic. Area, 781,700 hectares. Pop., in 1836, 330,350. No portion of the surface is much elevated; the E. part of the dep. is undulating, though nowhere rising to the height of 500 ft. The principal rivers are the Sevre-Nantaise and Niortaise, Autise, Vendée, Lay, Yon, and Vie. The Vendée rises in the department Deux-Sèvres, runs generally in a S.W. direction, and joins the Sèvres-Niortaise, after a course of about 16 leagues, only a small portion of which is navigable. Fontenay is the only town of consequence on its banks. The department is subdivided into three districts: the marshes, the plain country, and the *bocage*. The marshes extend principally along the coast; the *bocage*, so called from the wood sprinkled over it, occupies the centre and upper parts of La Vendée; and the plain country, a great part of which is very fertile, comprises the rest of the surface. In 1834, the arable lands were estimated to comprise 408,563 hectares; pastures, 109,766 do.; vineyards, 17,700 do.; and woods, 29,600 do. Agriculture is conducted in much the same way as in the adjacent department of Loire-Inférieure (which *see*). Leases run from three to seven years. Few of the farms are let for money; but the rent, paid in produce, may be equivalent for corn land to 26 fr.; for vineyards, 24 fr.; and for meadow lands, 40 fr. an acre. In the plains the ground is left idle the second year, as in Loire-Inférieure.

"The whole of this department, as well as that of the Loire-Inférieure is, with a very few exceptions, most miserably farmed. The farmers are without capital, and badly provided with implements, and have not half the number of labourers required for the land; the fields are always full of weeds, for the roots are left entire by the bad ploughs in use, and from the want of hands to pluck them out." (*Parl. Rep. on Agriculture,* 1834.) A greater quantity of wheat is grown than in any other of the W. departments; and, next to it, barley is the grain principally cultivated. Flax and hemp are grown to a considerable extent in the marshy tracts. A large quantity of wine, principally white, is produced; but, according to Jullien, it is *d' une qualité très-médiocre, et en général vert, plat et sujet à tourner à la*

1010

graisse dès la première année. (p. 147.) This infancy is ascribed to the circumstance of the cultivators being now anxious to increase the quantity than to improve the quality of their wines. Throughout a great part of the dry ment, estates are usually divided into farms of from 6 to 90 acres; and in the plain country few farms of this size have fewer than 60 or 80 sheep. The annual produce of wool is estimated by Hugo at 600,000 kilogr. Oxen are sold from the plain to the bocage farmers, who fatten a good deal of stock. In some places mules are used for ploughing; they are brought from Deux-Sèvres when young, and after having been worked lightly for four or five years are sold to merchants for the Spanish market. (*Rep. on Agriculture.*) In 1835, of 124,113 properties subject to the *contri. foncière,* 69,644 were assessed at less than 5 fr., and 17,825 at from 5 to 10 fr. (*Official Tables.*) A little lead and antimony, with some iron and coal are obtained; but the department is not rich in mineral products. Pilchards are taken on the coast, and the inhabitants of Sables d'Olonne are interested in the Newfoundland fishery. Manufactures few; being principally of hats and woollen stuffs, for home consumption. La Vendée is divided into three arrondissements; chief towns, Bourbon, Vendée, Fontenay, and Sables d'Olonne. It sends five members to the chamber of deputies. Registered electors in 1836–39, 1677. Total public revenue, in 1831, 6,671,303 fr. This department is distinguished for the chivalrous and obstinate stand made by its inhabitants during the progress of the French revolution, in favour of the rights or pretensions of the Bourbons. In this, no doubt they were a good deal animated by the nature of the country; but their gallantry and their sacrifices were alike remarkable, and were worthy of a better cause. Hugo, art. *La Vendée; Dict. Géog.; French Official Tables.*

VENDOME, a town of France, dep. Loir-et-Cher. cap. arrond.; on the Loir, 30 m. N.W. Blois. Pop., in 1836, etc. comm., 7030. It is well built, clean, and well-paved. The remains of an ancient castle, the communal college, cavalry barracks, theatre and public library, are the objects most worthy of notice. It has manufactures of kid gloves, cotton cloths, hosiery, yarn, paper, and leather.

VENEZUELA. *see* COLOMBIA.

VENICE (Ital. *Venezia*), a famous maritime city of Austrian Italy, formerly the cap. of the republic of the same name, and now of E. Lombardy, and a cluster of numerous small islands, in a shallow, but extensive lagoon, about 4 m. from the mainland, in the N.W. portion of the Adriatic, 75 m. W. by S. Trieste; lat. 45° 25' 57" N. long. 12° 20' 31" E. Pop., in 1837, 97,156. The appearance of Venice, from whatever side she may be approached, is striking and singular in the extreme. Owing to the lowness of the islands on which she is built, she seems at first upon the sea,

"From out the wave her structures rise,
As from the stroke of the enchanter's wand."

She is divided into two principal portions, of nearly equal size, by the Grand canal (*Il Canale Maggiore*), a serpentine channel, varying from about 100 to 100 ft. in width, crossed by the principal bridge of the city, the celebrated Rialto. The various islands, which form the foundations of these two grand divisions, are connected by numerous bridges, which, being very steep, and intended only for foot passengers, are cut into steps on either side. The canals, or *rii,* crossed by these bridges, intersect every part of the town, and form the "water-streets" of Venice, by far the greater part of the intercourse of the city being carried on by their means in gondolas or barges.

But, besides the canals, Venice is everywhere traversed by streets, or rather passages (*calle*), bearing a striking resemblance to Cranbourn-alley; and so very narrow, as to be in general only four, and seldom more than five or six in width! The only exception is the *Mercería,* a street in the centre of the city, lined on each side with handsome shops; but even this, which may be regarded as the Paternoster-row, or Regent-street of Venice, is only from 16 to 20 ft. across! To ride in a carriage, or on horseback, is here wholly out of the question. The streets, or lanes, are consequently not paved with round stones, or like it, but with flags, or marble slabs, having small sewers for carrying off the filth. Almost all the principal houses have on one side a canal, and on the other a lane, or calle. The former, however, is the grand thoroughfare; and gondolas, or canal-boats, are here the universal substitute for carriages and horses. They are generally long, narrow, light and, and, though rowed only by a single gondolier with one oar, cut their way through the water with extraordinary velocity. A sumptuary law of the old régime decreed that the gondolas should all be painted black. In the cabin is an apartment fitted with glass windows, blinds, doors, &c., for the accommodation of four persons. There are of gondolas, belonging to private families, are magnificently fitted up. The charge for a gondola is about 3d. an hour, and with it you may soon visit every part of the city.

VENICE.

In many parts there are small squares, or campi, in which are usually cisterns, for the careful preservation of rain water; but the only open space of any magnitude is the piazza of St. Mark, with the piazzette leading to it, and forming the state entrance to Venice from the sea. "The piazzetta is at right angles with the great square, branching off in a line with the church of St. Mark. On one side, and turning a side front to the port, is the old palace of the doges; on the other side are the zecca or mint, and the library of St. Mark, the regular architecture, and fresh and modern appearance of which seems to mock the fallen majesty of their antique neighbours. On the sea shore, which forms the fourth side of the piazzetta, stand two magnificent granite columns, each of a single block; one crowned with the winged lion of St. Mark, in bronze, and the other bearing the statue of St. Theodore. Between these two columns, in former times, public executions took place." (Condor's Italy.) The piazza of St. Mark is an oblong area, about 800 ft. by 350, flagged over. Two of its sides consist of regular buildings with deep arcades. Each side is itself uniform, though not similar to the other. On the N. is the Procuratorie Vecchie; on the S., the Procuratorie Nuove. The W. side was formerly occupied by the church of St. Geminiani; but this was taken down by the French, who erected in its stead the staircase of the new Imperial palace. At the opposite end are the cathedral of St. Mark, the Orologio, and the Campanile; and in front of the cathedral are three tall poles, supported on handsome bases of bronze, whence the flags of the Morea, Crete, and Cyprus were formerly displayed. From being the only piece of open ground in Venice of any consequence, this square is almost constantly thronged with company, and it is the scene of all the public masquerades, festivals, &c., that take place in the city.

Venice has a vast number of fine private palaces by Sansovino, Palladio, &c.; but many of its public buildings are more remarkable for gorgeousness and display, than for purity and taste. They present generally a heterogeneous intermixture of Byzantine, or other eastern, with Greek, Roman, and Gothic architecture. The celebrated church of St. Mark is not Gothic, Saracenic, or Roman, but a mixture of all those styles: neither a church nor a mosque, but something between the two; too low for grandeur, too heavy for beauty, no just proportion being preserved among the different parts. Yet it has the effect of grandeur, and a sort of beauty, from the richness of the materials, and the profusion of ornament. The original church was founded in 829; but the present edifice was founded in 977, under the direction of architects from Constantinople. The nave is 345 feet in length, the transept 201 feet; the middle dome is internally 90 feet in height, and the four other domes 80 feet each. The front is 170 feet in width and 72 feet in height, without its surmounting figures. In its lower part are five recessed doorways, each adorned with two stories of little columns, though these are mostly ill proportioned, and their capitals nearly all different. Over these arches is a gallery or balcony of marble, in the centre of which are the famous bronze horses, most probably of Chian origin, and carried to Constantinople by Theodosius, whence they were conveyed away by the Venetians, when they took and plundered the capital of the eastern empire, in 1206. For 18 years, or from 1797 to 1815, they crowned the triumphal arch in the Place du Carousel in Paris; but, though restored, we may well inquire,

> "Is not Doria's menace come to pass?
> Are they not bridled?"
>
> (Childe Harold, cant. iv., st. 13, and note.)

Immediately behind the horses is a large circular window, on either side of which an arched doorway opens upon the balcony. The front terminates in pointed arches, surmounted by a crowd of spires, pinnacles, statues, crosses, &c. The finishings are in the style of the Italian Gothic of the 15th century, but overcharged and heavy. Forsyth, speaking of this edifice, says, "Nowhere have I seen so many columns crowded into so small a space. Nearly 300 are stuck on the pillars of the front, and 300 more on the balustrade above. A like profusion prevails in the interior, which is heavy, dark, and barbarous." But, from Mr. Wood's account, it would appear that the same barbarous taste that prevails in the exterior, is not so obtrusive within. "The vaulting and great part of the walls are covered with mosaics, and the rest with rich marbles. The columns of porphyry, verd-antique, &c.; the pavement of minute pieces of white and coloured marbles, jasper, agate, lapis lazuli, &c., variously, and, for the most part, beautifully, disposed; the inlaid ornaments and gilded capitals produce a degree of astonishment and admiration in the mind of the spectator." (Letters of an Architect, i., 260.) The Orologia, or clock-tower, on the N. side of the cathedral, has little to recommend it; and the campanile is merely a large square tower, upwards of 300 feet in height, terminated by a pyramid. In it, however, Galileo made many of his astronomi-

cal observations. The loggia around its base, now converted into a lottery-office, is a beautiful building of the Corinthian order, from the designs of Sansovino. The Procuratorie Nuove, now the royal palace, is a rich line of building, fronted with all the different Greek orders.

The ducal palace was originally founded in the 9th century, but the present edifice dates only from the middle of the 14th, when it was erected by the doge Marino Faliero. Externally it presents a double range of arches, supporting a great wall of brickwork, pierced with a few windows. The corners are cut to admit thin spiral columns. Notwithstanding its many defects, this structure derives an imposing effect from its grandeur of dimensions, and unity of design. The palace is entered by eight gates, the principal leading into the cortile, which is surrounded on three sides by two stories of arcades. One side is richly ornamented, though the whole be in bad taste. A noble flight of steps, called, from the colossal statues of Mars and Neptune, by Sansovino, the Giants' Staircase, leads up from this cortile to the open arcade, where, under the republic, the lions' mouths gaped to receive communications or surmises of plots against the state! From this exterior corridor the state apartments are entered. The walls of the Sala di Quattro Porte are covered with paintings by Tintoretto, Vicentino, and Titian. The hall of the Great Council, Sala del Gran Consiglio, 153 feet in length, is now principally appropriated to the library, of which Petrarch was one of the founders; but it is also rich in frescoes, by Bassano, &c., and contains a fine collection of ancient sculpture, the portraits of the Venetian doges, &c. The Sala dei Pregadi, and numerous other apartments, are richly gilt, and exhibit all the glories of the Venetian school of painting, "which spreads over the walls and covers the ceilings, as if it had only cost a few shillings the square yard." In the hall of the Council of Ten; converted by Napoleon into the chamber of a court of cessation, the ceilings have been painted by Paul Veronese; and on every side the eye rests on pictorial representations of the achievements and glories of the republic. In the lower parts of the palace are the former tribunals and dungeons of the state inquisition, from which a passage leads across the Ponte de Sospiri, or Bridge of Sighs, to a door now walled up, but which formerly opened into a chamber where prisoners were despatched.

Next to the buildings in the neighbourhood of St. Mark's, those bordering the harbour and the canal of Giudecca appear to be the finest, including the Dogana or custom-house, the church of La Salute, and those of San Giorgio and Il Redemptore, both designed by Palladio. These last were on the point of being pulled down by the French, and only saved by being redeemed for a large sum of money. The Redemptore, Forsyth says, is admirable both in plan and elevation, and its interior is perhaps perfect in its proportions, simple, grand, and harmonious. It is, in fact, one of Palladio's chef d'œuvres, and is, perhaps, the finest church in Italy, though inferior to a great many in costliness and magnitude. Besides a Greek church and seven synagogues, Venice has altogether about 100 Roman Catholic churches, which are, on the whole, among its best buildings. There were formerly many more; but the French pulled them down, with a number of convents, in pursuance of their plans for the improvement of the city. Several of the churches date from the middle ages, though few of them are worth especial notice. One, however, is interesting, from its containing the remains of one of the greatest painters and of one of the greatest sculptors that Italy has produced—Titian and Canova: the grave of the former is marked by a short inscription on a plain stone let into the pavement. Canova has a fine pyramidical monument, from one of his own designs, the expense of which was defrayed by contributions collected in all parts of Europe. The famous painter, Paul Veronese, is buried in the church of St. Sebastian, where he has a monument. The belfries of all the churches are detached, and appear to be built on the model of St. Mark's.

The general cemetery or burying place for the city is on the island of San Cristoforo di Murano. Here, rich and poor, the noble and the beggar, are all interred, the expenses of the burial of the latter being defrayed by government. This cemetery was formed, and the plan of conveying all dead bodies to it from the city enforced, by the French; and, happily, it has since been continued. A gondola, moored to the island, is appropriated to the transmission of corpses. The Jews have their burying-ground at Malomocco.

VENICE.

responds with its fame and the attention it has excited. Its arch is 89 feet in span. The roadway of the bridge is divided into three parts, viz., a narrow street in the middle, with shops on each side, and two still narrower streets between the shops and the balustrades. The shops disfigure the bridge, and make it look heavy. It is lofty in the middle, and is ascended, like the other bridges, by long flights of steps at either end. The view from the summit, along the grand canal, frequently presents a very animated scene, and is one of the finest in Venice.

The palaces of Venice, built, like those of Amsterdam, on piles, are massive structures; but, except such as have been built by Palladio, Sansovino, Sanmichele, Longhena, Scamozzi, and a few other architects of eminence, they are mostly deficient in good taste. They in general exhibit too many orders in front. Venice, in truth, is more attractive from its singularities than its architectural elegance. Yet it is still highly interesting to the student of architecture, who may here " trace the gradation from the solid masses and round arches. the only remains of the ancient grand style in the 6th, 7th, 8th, and 9th centuries, through the fanciful forms and grotesque embellishments of the middle ages, to its revival and re-establishment in later times." (*Eustace, Classical Tour*, i., 167.) Many, however, of the old patrician mansions are deserted, and not a few of them have been pulled down. Necessity, too, has, in many instances, obliged their owners to part with the fine works of art, with which they were formerly embellished. Still, however, some of the palaces have yet to boast of good collections of pictures, statues, &c. The Manfrini palace has a splendid gallery of pictures; and the Palazzo Barberigo has some fine works by Titian. The Grimani palace contains the only extant statue of Marcus Agrippa, a fine bust of Caracalla, &c.; and in the Pisani palace is Dædalus fixing wings on his son, the first group executed by Canova.

But, notwithstanding their magnitude and imposing external appearance, the rooms inhabited by the family, in the greater number of the palaces which are still occupied, are often small, ill furnished, and uncomfortable. Personal accommodation and the enjoyment of good air have been sacrificed, that space might be found for the exhibition of statues, pictures, and other works of art. All the larger houses, or palazzi. are from three to four stories in height, being generally of a square form, with an inside court containing a cistern, in which the rainwater is carefully collected. As already stated, they have, for the most part, two entrances—the principal opening on a canal, and the other on a street or alley. Some of the finest palaces are built wholly of marble. The grand canal has on each side many such buildings.

The houses occupied by the middle and lower classes are built of brick, and are in general covered with wood. Few of them have arcades, but they are mostly provided with balconies. From the extreme narrowness of the streets, the houses are usually gloomy; and are miserably deficient in the appropriate distribution of their different parts, and in all those conveniences and adaptation to comfort that distinguish houses in this country.

The arsenal, which opens upon the port not far from St. Mark's, together with the dockyard, occupies an island between 2 and 3 m. in circuit, and is defended by lofty turreted walls. The entrance is guarded by two towers flanking a gateway, over which the winged lion still frowns defiance; and in front of the entrance are four lions, brought from the Piræus: two being of very fine proportions, and probably of Pentelic marble. The magazines and docks are kept in good order, and ship building is one of the chief branches of industry at Venice. Besides the armory, magazines, forges, foundries, and other necessary establishments, here is a rope-house, 1000 feet in length. One of the walls of the armory has a statue of Pisani, famous for his contests with the Genoese, and a beautiful monument by Canova, representing Fame crowning the Venetian admiral, Angelo Emo.

During the times of the republic the Bucentaur was the great lion of the arsenal. This was the state barge, in which the doge, accompanied by a splendid cortège, proceeded to espouse the Adriatic. The ceremony was performed by the doge dropping a ring of no great value into the sea, pronouncing at the same time the words *Desponsamus te, Mare, in signum veri perpetuique dominii*. In these days, however,

> " The spouseless Adriatic mourns her lord,
> And annual marriage now no more renewed."

Byron adds that the Bucentaur " lies rotting, unrestored ;" but, in fact, she was burned by the French soon after the downfall of the republic.

Venice has six theatres, the largest of which may contain 2500 spectators : the Venetian drama is, however, in a very low state. The Dogana, the old exchange, is a fantastic edifice of the 17th century; and the new prison, built by

Antonio da Ponte in 1589, is much too handsome for its purpose, being an elegant Doric edifice. The bishop's palace and seminary, the various hospitals and barracks, are among the other principal edifices. A flourishing academy of the fine arts, four schools of music, and a public school for each corporation of tradesmen, are among the principal institutions for education. " Venice holds a prominent place in Italy for its charitable institutions. There is one house within the city in which 700 poor people are lodged, and many more have free lodgings and receive pecuniary assistance out of the establishment. There is an orphan house for about 335 children; an infirmary for 36 women : a wealthy institution for the reception of penitent women ; a hospital, capable of receiving 1000 patients ; a house of education for 90 young girls ; a foundling hospital, &c.; and the yearly revenues, chiefly arising from endowments, amount to about 580,000 florins." (*Von Raumer's Italy*, i., 86.)

Venice has been represented as a delightful residence : but though it may be, and perhaps is, an aquatic paradise to the amphibious bipeds born within the sound of St. Mark's bells, it is very different to a stranger. At first, no doubt, it surprises and gratifies by its novelty ; but it soon becomes tiresome from its sameness, the incessant recourse to boats, the narrowness of the streets, the want of room, the absence of all rural beauty, and the constant sense of imprisonment. It would not, in fact, be habitable were the water fresh ; but the saltness of the water, and the flux and reflux of the tide, make it tolerably salubrious. The latter phenomena, however, which are at all times much less sensible in the Mediterranean than on the British shores, are in summer so inconsiderable, that the canals become stagnant, offensive, and unhealthy. The characteristics of the climate are, a summer much hotter than in England, accompanied with occasional visits of the sirocco ; a winter, not of great length, but sharp, particularly during the prevalence of the N.W. wind, which blows across the interior of Switzerland and the Alps. Rains are frequent, particularly in spring ; and, there being no springs or wells, the inhabitants, as already stated, supply themselves with water collected in cisterns, from the tops of the houses.

It should, however, be observed, that the Venetians are no longer wholly without trees and flowers ; very extensive gardens, with a fine street leading to them, were constructed by the French, and are a noble monument of their taste and munificence. " These gardens," says Mr. Pennington, " excite interest from the mode in which they were formed, more than from their beauty ; not that they are deficient in taste or variety. They were formed with immense labour by the introduction of artificial earth, brought at an immense expense from *terra-firma*, and no expense was spared in their completion. There are several serpentine walks over mounts, many trees and shrubs thriving very fast ; and all this, with the different views of the lagoon, the many islands interspersed, and Venice, make this promenade both agreeable and interesting. The gardens are nearly 2 m. round, and are connected by a handsome bridge." (*Tour*, ii., 225.)

Port.—The islands on which Venice is built lie within a line of long, low, narrow islands, running N. and S., and enclosing what is termed the lagoon, or shallows, that surround the city, and separate it from the main land. The principal entrance from the sea to the lagoon is at Malamocco, about 1½ leagues S. from the city ; but there are other, though less frequented, entrances, both to the S. and the N. of the latter. There is a bar outside Malamocco. on which there are not more than 10 feet at high water at spring tides; but there is a channel between the western point of the bar and the village of San Pietro, which has 16 feet water at springs, and 14 at neaps. Merchant vessels usually moor off the ducal palace ; but sometimes they come into the grand canal which intersects the city, and sometimes they moor in the wider channel of the Giudecca. Vessels coming from the S. for the most part make Piræus or Rovigno on the coast of Istria, where they take on board pilots, who carry them to the bar opposite to Malamocco. On arriving at the bar, ships are conducted across it and into port by pilots, whose duty it is to meet them outside, or on the bar, and of whose services they must avail themselves.

The chain of low, narrow islands which bounds the lagoon on the side next the sea being, in part, broken, the republic constructed, during last century, a mole several miles in length, to fill up the gaps in question, and protect the city and port from the storms and swells of the Adriatic. This vast work, formed of blocks of Istrian stone resembling marble, connects various little islands and towns, and is admirable alike for its magnitude, solidity, and utility. It bears the following inscription :

" *Ut sacra æstuaria, urbis et libertatis sedes, perpetuum conserventur, colossees moles ex solido marmore contra mare*

1030

VENICE.

posuere curatores aquarum. Anno salutis, 1751: ab urbe condita, 1330."

Money.—Formerly there were various methods of accounting here; but now accounts are kept in Austrian lire, divided into centisimi, or 100th parts. The lira is worth about 8d.

Weights and Measures.—The commercial weights are here, as at Genoa, of two sorts; the *peso sottile* and the *peso grosso.* The French kilogramme, called the libra Italiana, is also sometimes introduced. 100 lbs. peso grosso = 105·186 lbs. avoirdupois, and 100 lbs. peso sottile = 66·428 lbs. avoirdupois. The moggio, or measure for corn = 9 Winch. bushels. The measure for wine, anfora, contains 137 English wine gallons. The foot of Venice = 13·68 English inches.

The Old Bank of Venice was founded so far back as 1171, being the most ancient establishment of the kind in Europe. It was a bank of deposite: and such was the estimation in which it was held, that its paper continued to bear an agio as compared with coin down to 1797, when the bank fell with the government by which it had been guaranteed. At present there are no corporate banking establishments in the city; and no bank notes are in circulation. There are, however, several private banking houses, which buy, sell, and discount bills; and make advances on land and other securities. They are under no legal regulations of any sort, except formally declaring the amount of their capital to the authorities when they commence business. The legal and usual rate of interest and discount is 6 per cent. It is not the practice to allow interest on deposites. Bills on London are usually drawn at three months, and on Trieste at one month.

Morals and Manners.—Most travellers have been accustomed to represent Venice as distinguished by a peculiar profligacy of morals. It may be doubted, however, whether she be entitled to any peculiar pre-eminence in this respect over most of the other great cities of Italy; and the loss of her commerce and of that wealth which the expenditure of government brought into the city, has reduced alike the means of, and incentives to, corruption. It is now, we believe, pretty generally acknowledged that the impressions made on foreigners during the carnival season were in a great degree exaggerated; and that much of what they took for intrigue and profligacy, was no more than what the license of the period, and the universal use of masks, allowed even the most scrupulous persons to indulge in without any violation of propriety. Undoubtedly, however, the conduct of the government, the nature of her religion, and the vast wealth that formerly centred in Venice, all tended to corrupt the morals of the people, and to immerse them in sensual pleasures. We hardly think it was ever, as Addison has stated, a part of the policy of government " to encourage idleness and luxury in the nobility, to cherish ignorance and licentiousness in the clergy, to keep alive a continual faction in the common people, to connive at the viciousness and debauchery of convents, to breed dissensions among the nobles of the terra-firma, and to treat a brave man with scorn and infamy." (*Travels, art. Venice.*) But, whether intended or not, this, no doubt, was the effect of their jealous despotism, which, by its intolerance of all that was truly great, generous, and noble, shut up, in as far as possible, all the avenues to distinction in politics, literature, and even war, leaving little, save intrigue and licentiousness, to occupy the public mind. But, as already stated, society in Venice has been materially changed since the revolution of 1797. Lord Byron says, that " of the *gentiluomo Veneto,* the name is still known, and that is all. He is but the shadow of his former self, but he is noble and kind. It may surely be pardoned to him if he is querulous." But, notwithstanding the changes to which they have been subjected, and which have reduced them from haughty lords, but " one degree below kings," to abject subjects, the Venetians are now, as of old, most agreeable companions, and the Paphian queen still holds her court in the sea-girt city.

> "In Venice Tasso's echoes are no more,
> And silent rows the songless gondolier;
> Her palaces are crumbling to the shore,
> And music meets not always now the ear;
> Those days are gone—but Beauty still is here.
> States fall—arts fade—but Nature doth not die;
> Nor yet forget how Venice once was dear,
> The pleasant place of all festivity,
> The revel of the earth, the masque of Italy!"

Foreigners, especially, are extremely well received, and society is on a very easy footing in Venice. Owing to the facility with which the city is supplied with provisions from the interior of Lombardy and elsewhere, and the lowness of rents, living is here unusually cheap; and were it not for its disadvantages in other respects, it would be a very desirable residence.

The Government of Venice was one of the most singular that has ever existed. In her earlier period she appears to have been governed by doges, or princes, who were elected by the popular voice; but who, on being elected, became the absolute rulers of the state. The doge enjoyed, however, only a precarious dignity; for, in the event of any disaster occurring to the arms of the republic, or of his becoming unpopular, he was not unfrequently deposed, and sometimes assassinated. (*Daru,* i. 186, &c.) To obviate the disorders that grew out of this state of things, it was resolved, in the 12th century, that each of the six districts into which the city was then divided should nominate two individuals as electors, and that the twelve electors so chosen should nominate a grand council of 470 individuals which should represent the public, the general assemblies of which were henceforth discontinued. A senate was at the same time created, and six councillors were appointed to assist, or rather control, the proceedings of the doge. (*Daru,* i. 193, &c. But notwithstanding the influence of the popular voice was greatly lessened by the establishment of the grand council, which included all the most distinguished citizens, it was still very considerable; and on several occasions the people endeavoured by violence to recover the power they had lost. In this, however, they were wholly unsuccessful; and at length, after various struggles, it was resolved in 1319, that the grand council should no longer be elected, but that the dignity should be hereditary in its members! (*Daru,* i. 518.) The aristocracy was thus established on a solid foundation; but no sooner had this been done than the dignified families became jealous of each other; and to avert the chance of any individual acquiring a preponderating influence in the state, a carefully devised scheme of indirect election to all the higher offices was established, at the same time that the nobles subjected themselves, the doge, and every one else, to a system of despotism which not only determined the public and private conduct, but, in some measure, even the very thoughts of individuals! This was accomplished, partly by the institution of the council of ten, a committee chosen from the grand council, to which all the powers of the state were entrusted, and partly by the institution, in 1454, of three state inquisitors, selected from the council of ten, and invested with all but unlimited authority. The proceedings of this most formidable tribunal were shrouded in the most impenetrable secrecy; but it was believed at the time, and is now certain, that it did not wait for overt acts, but proceeded on suspicion and presumption; that it had secret prisons; and that it made free use of the agency of spies, torture, and even of assassins. An individual disappeared, by what means no one knew; but if it were supposed that he had fallen a victim to the fears or suspicions of the inquisitors, his relatives prudently abstained from all complaint, and even from making any inquiries respecting him. An unguarded expression, if reported, as was frequently the case, to the inquisitors, was sure to draw their attention to the offender, so that not merely the freedom of the press, but even of speech, at least on political matters, was completely annihilated. Although, however, this jealous tyranny did not fail to repress, or rather extinguish some of the nobler energies of the mind, it must be admitted, that it preserved for a lengthened period the peace of the republic. It is true, also, that its despotism pressed equally on all classes and all individuals; the doge was as liable, and as likely, if occasion required, to be called to account by the inquisitors as the humblest gondolier. Nothing, in fact, but implicit obedience to established authority, and a perfect abstinence from every sort of political preference and remark, could enable any individual, however high or low, to sleep soundly in Venice.

Historical Notice.—Venice was the earliest, and for a lengthened period the most considerable, commercial city of modern Europe. Her origin dates from the invasion of Italy by Attila, in 452. A number of the inhabitants of Aquileia, and the neighbouring territory, flying from the ravages of the barbarians, found a poor but secure asylum in the cluster of small islands opposite the mouths of the Brenta, on which the city is built. In this situation they were forced to cultivate commerce and its subsidiary arts, as the only means by which they could maintain themselves. At a very early period they began to trade with Constantinople and the Levant; and notwithstanding the competition of the Genoese and Pisans, they continued to engross the principal trade in eastern products, till the discovery of a route to India by the cape of Good Hope turned this traffic into a totally new channel. The crusades contributed to increase the wealth, and to extend the commerce and the possessions of Venice. Towards the middle of the 15th century, when the Turkish sultan, Mahomet II., entered Constantinople sword in hand, and placed himself on the throne of Constantine and Justinian, the power of the Venetians had attained its maximum. At that period, besides several extensive, populous, and well cultivated provs. in Lombardy, the republic was mistress of Crete and Cyprus, of the greater part of the Morea, and most of the isles in the Egean sea. She had secured a chain of forts and factories that extended along the coasts of Greece from the

1013

Morea to Dalmatia; while she monopolized almost the whole foreign trade of Egypt. The preservation of this monopoly, of the absolute dominion she had early usurped over the Adriatic, and of the dependence of her colonies and distant establishments, were among the principal objects of the Venetian government; and the measures it adopted in that view were at once skilfully devised, and prosecuted with inflexible constancy. With the single exception of Rome, Venice, in the 15th century, was by far the richest and most magnificent of European cities; and her singular situation in the midst of the sea contributed to impress those by whom she was visited with still higher notions of her wealth and grandeur. Sannazarius is not the only one who has preferred Venice to the ancient capital of the world; but none have so beautifully expressed their preference.

> Viderat Adriacis Venetam Neptunus in undis,
> Stare urbem, et toto ponere jura mari.
> Nunc mihi Tarpeias quantumvis, Jupiter, arces
> Objice, et illa tua mœnia Martis, ait,
> Si Tiberim pelago præfers, urbem aspice utramque,
> Illam homines dicas, hanc posuisse Deos.

Though justly regarded as one of the principal bulwarks of Christendom against the Turks, Venice had to contend, in the early part of the 16th century, against a combination of the European powers. The famous league of Cambray, of which Pope Julius II. was the real author, was formed for the avowed purpose of effecting the entire subjugation of the Venetians, and the partition of their territories. The emperor and the kings of France and Spain joined this powerful confederacy. But, owing less to the valour of the Venetians than to dissensions among their enemies, the league was speedily dissolved without materially weakening the power of the republic. From that period the policy of Venice was comparatively pacific and cautious. But notwithstanding her efforts to keep on good terms, with the Turks, the latter invaded Cyprus in 1570; and conquered it after a gallant resistance continued for eleven years. The Venetians had the principal share in the decisive victory gained over the Turks at Lepanto in 1571: but owing to the discordant views of the confederates, it was not properly followed up, and could not prevent the fall of Cyprus.

The war with the Turks in Candia commenced in 1645, and continued till 1670. The Venetians exerted all their energies in defence of this valuable island; and its acquisition cost the Turks above 200,000 men. The loss of Candia, and the rapid decline of the commerce of the republic, now almost wholly turned into other channels, reduced Venice, at the close of the 17th century, to a state of great exhaustion. She may be said, indeed, to have owed the last 100 years of her existence more to the forbearance and jealousies of others than to any strength of her own. Nothing, however, could avert that fate she had seen overwhelm so many once powerful states. In 1797, the "maiden city" submitted to the yoke of the conqueror: and the surviving witness of antiquity—the link that united the ancient to the modern world—stripped of independence and of wealth, now enjoys only a precarious existence, and is slowly sinking into the waves whence she arose.

The foundation of Venice is described by Gibbon, c. 25; and in his 60th chapter he has eloquently depicted her prosperity in the year 1500. Mr. Hallam, in his work on the *Middle Ages* (i. 470–487), has given a brief account of the changes of the Venetian government. Her history occupies a considerable space in the voluminous work of M. Sismondi on the *Italian Republics*; but his details as to her trade and commercial policy are singularly meagre and uninteresting. All previous histories of Venice have, however, been thrown into the shade by the admirable work of M. Daru (*Histoire de la Republique de Venise*, 2d ed. 8 vols. 8vo, Paris, 1821). Having had access to genuine sources of information, inaccessible to all his predecessors, M. Daru's work is as superior to theirs in accuracy, as it is in most other qualities required in history.

Trade, Navigation, and Manufactures of the Venetians in the 15th Century.—The Venetian ships of the largest class were denominated galeasses, and were fitted up for the double purpose of war and commerce. Some of them carried 50 pieces of cannon, and crews of 600 men. These vessels were sometimes, also, called *argosies* or argosies. They had early an intercourse with England; and argosies used to be common in our ports. In 1385, Edward II. entered into a commercial treaty with Venice, in which full liberty is given to them, for 10 years, to sell their merchandize in England, and to return home in safety, without being made answerable, as was the practice in those days, for the crimes or debts of other strangers. (*Anderson's Chron. Deduction, Anno* 1385.) Sir William Monson mentions, that the last argosie that sailed from Venice for England was lost, with a rich cargo and many passengers, on the coast of the Isle of Wight, in 1587.

In the beginning of the 15th century, the annual value of the goods exported from Venice by sea, exclusive of those

exported to the states adjoining her provinces in Lombardy, was estimated, by contemporary writers, at 10,000,000 ducats; the profits of the out and home voyage, including freight, being estimated at 4,000,000 ducats. At the period in question, the Venetian shipping consisted of 3000 vessels of from 100 to 200 tons burden, carrying 17,000 sailors; 36 ships with 8000 sailors; and 45 galleys of various sizes, kept afloat by the republic for the protection of her trade, &c., having 11,000 men on board. In the dock-yard, many labourers were usually employed.* The trade to Syria and Egypt seems to have been conducted principally by ready money; for 500,000 ducats are said to have been annually exported to these countries; 100,000 were sent to England (*Daru,* tome ii. p. 180, &c.) The vessels of Venice visited every port of the Mediterranean, and every coast of Europe; and her maritime commerce was, probably, not much inferior to that of all the rest of Christendom. So late as 1518, five Venetian galeasses arrived at Antwerp, laden with spices, drugs, silks, &c., for the fair at that city.

The Venetians did not, however, confine themselves to the supply of Europe with the commodities of the east, and to the extension and improvement of navigation. They attempted new arts, and prosecuted them with vigour and success, at a period when they were entirely unknown in other European countries. The glass manufacture of Venice was the first, and for a long time the most celebrated, of any in Europe; and her manufactures of silk, cloth of gold, leather, refined sugar, &c., were deservedly esteemed. The jealousy of the government, and their intolerance of any thing like free discussion, was unfavourable to the production of great literary works. Every scholar is, however, aware of the fame which Venice early acquired by the perfection to which she carried the art of printing. The classics that issued from the Aldine presses are still universally and justly admired for their beauty and correctness.

But the policy of government, though favourable to the introduction and establishment of manufactures, was fatal to their progressive advancement. The importation of foreign manufactured commodities into the territories of the republic for domestic consumption was forbidden under the severest penalties. The processes to be followed in the manufacture of most articles were regulated by law.—"*Dès l'année* 1172, *un tribunal avoit été créé pour la police des arts et métiers, la qualité et la quantité des matières furent soigneusement déterminées.*" (*Daru,* iii. 153.) Having, in this way, little to fear from foreign competition, and being tied down to a system of routine, there was nothing left to stimulate invention and discovery; and during the last century the manufactures of Venice were chiefly remarkable for evincing the extraordinary perfection to which they had early arrived, and the absence of all recent improvements. An unexceptionable judge, M. Berthollet, employed by the French government to report on the state of the arts of Venice observed, "*Que l'industrie des Vénitiens, comme celle des Chinois, avoit été précoce, mais étoit restée stationnaire.*" (*Daru,* iii. p. 161.)

M. Daru has given the following extract from an article in the statutes of the state inquisition, which strikingly displays the real character of the Venetian government and their jealousy of foreigners:—"If any workman or artisan carry his art to a foreign country, to the prejudice of the republic, he shall be ordered to return; if he do not obey, his nearest relations shall be imprisoned, that his regard for them may induce him to come back. If he return, the past shall be forgiven, and employment shall be provided for him at Venice. If, in despite of the imprisonment of his relations, he persevere in his absence, an *emissary shall be employed to despatch him*; and after his death his relations shall be set at liberty!" (Tome iii. p. 152.)

The 19th book of M. Daru's history contains a comprehensive and well-digested account of the commerce, manufactures, and navigation of Venice. But it was not possible, in a work on the general history of the republic, to enter so fully into the details as to these subjects as their importance would have justified. The *Storia Civile e Politica del Commercio de' Veneziani*, di Carlo Antonio Marin, 8 vols. 8vo, published at Venice at different periods, from 1798 to 1808, is unworthy of the title. It contains, indeed, a great many curious statements; but it is exceedingly prolix; and while the most unimportant and trivial subjects are frequently discussed at extreme length, many of great interest are either entirely omitted, or are treated in a very brief and unsatisfactory manner. The commercial history of Venice remains to be written; and were it executed by a person of competent attainments, it would be a most valuable acquisition.

Present Trade and Manufactures of Venice.—From the period when Venice came into the possession of Austria,

---

* The native authorities say 16,000; but there can be no doubt but this is a grossly exaggerated statement, and that 1600 would be much nearer the mark.

down to 1830, it seems to have been the policy of the government to encourage Trieste in preference to Venice; and the circumstance of the former being a free port, gave her a very decided advantage over the latter. Latterly, however, a more equitable policy has prevailed. In 1830, Venice was made a free port, and has since fully participated in every privilege conferred on Trieste. But, notwithstanding this circumstance, the latter still continues to preserve the ascendancy; and the revival of trade that has taken place at Venice has not been so great as might have been anticipated. The truth is, that except in so far as she is the *entrepôt* of the adjoining provs. of Lombardy, Venice has no considerable natural advantage as a trading city; and her extraordinary prosperity during the middle ages is more to be ascribed to the comparative security enjoyed by the inhabitants, and to their success in engrossing the principal share of the commerce of the Levant, than to any other circumstance. Still, however, her trade is far from inconsiderable. But, unfortunately, there are no means by which to ascertain its precise amount. The great articles of import are sugar, coffee, and other colonial products; indigo and other dye stuffs, olive oil, salted fish, various descriptions of cotton, woollen, and other manufactured goods; wheat and other grain, from the Black sea; tin plates and hardware, raw cotton, &c.; amounting, in all, to the value probably of £1,500,000 or £1,600,000 a year. The exports principally consist of silk and silk goods, wheat and other grain, paper, jewellery, glass, and glass wares, Venetian treacle, books, with a great variety of other articles, including portions of most of those that are imported. It should, however, be observed, that by far the greater part both of the import and export trade of the city is carried on through Trieste by coasting vessels, that are every day passing between the two cities. The smuggling of prohibited and over-taxed articles into Austrian Lombardy is practised to a great extent. It is believed that fully two thirds of the coffee made use of in Lombardy is clandestinely introduced; and sugar, British cottons and hardware, with a variety of other articles, are supplied through illegitimate channels. The facilities for smuggling, owing to the nature of the frontier, and the ease with which the officers are corrupted, are such, that the articles passing through the hands of the fair trader afford no test of the real extent of the business done. It is to be hoped that the Austrian government will take an enlightened view of this important matter. It cannot but be anxious for the suppression of smuggling; and it may be assured that this is not practicable otherwise than by a reduction of duties. The regulations as to the payment of the duties on goods destined for the interior, the clearing of ships, &c., are the same at Venice as at Trieste; which see.

The manufacture of Venice are very various, and more extensive than is generally supposed. The glass-works, which produce magnificent mirrors, with every variety of artificial pearls and gems, coloured beads, &c., situated on the island of Murano, employ, in all, about 4000 hands, including the women and children employed in arranging the beads. (*Bowring.*) Jewellery, including gold chains, is also extensively produced; as are gold and silver stuffs, velvets, silks, laces, and other expensive goods; and treacle, soap, earthenware, wax-lights, &c., to a greater or less extent. Printing is more extensively carried on in this than in any other city of Italy, and books form a considerable article of export. Ship-building is also carried on to some extent, both here and at Chiozza. In 1836, the first steam-engine seen in Venice was set up for a sugar-refinery.

From the circumstance of Venice being situated nearly opposite the mouths of the Brenta, which bring down large quantities of mud, the probability is that the lagoon by which she is surrounded will ultimately be filled up. Under the republic this was a subject of great apprehension, and every device was resorted to that seemed likely to avert a result so pregnant with danger to the independence of the city. But now that there is no particular motive for hindering the mud from accumulating in the lagoon, it is probable that, in the course of time, the shallows will be converted into terra firma, and Venice lose her insular position.

There belong to the city, exclusive of fishing-boats, about 30,000 tons of shipping, of which a large proportion is employed in the coasting trade. Many of the inhabitants depend for their subsistence on fishing in the lagoon, and the contiguous portion of the Adriatic. (Exclusive of the authorities already referred to, see *Bowring's Report on the Statistics of Italy; Commercial Dictionary; Commercial Circulars, &c.*)

VENLOO, a fortified town of Holland, prov. Limbourg, -cap. cant., on the Meuse, 40 m. N.E. Maestricht. Pop. in 1836, 6585. It is surrounded by a marshy tract of country, but is the centre of an active transit trade, and has manufactures of pins, wafers, tobacco-pipes, and various other small articles, with tanneries, sugar-refineries, and vinegar

distilleries. It was formerly one of the Hanse towns: in 1702 it was taken by the troops under Marlborough.

VENOSA, on the frontier of Lucania and Apulia), a town of the Neapolitan dom., prov. Basilicata, 24 m. N. Potenza. Pop. 6000. It stands on a perfectly flat, but not very extensive, plain. It is reached by a long, winding ascent, when it breaks on the sight under a favourable point of view, chiefly due to the venerable aspect of its castle, an edifice of the 15th century, which, though a complete ruin, exhibits such magnitude of dimensions and regularity of construction as to form a very striking feature in the landscape. The walls of Venosa have long since been levelled with the ground, but the gateways still exist. It is well built, and has numerous public edifices, including a large cathedral, five parish churches, an abbey church, a church erected in the 10th century from the materials of a Roman amphitheatre, in which are the tombs of Robert de Guiscard and other Norman chieftains; a hospital, two workhouses, a museum of antiquities, &c.

Venosa is celebrated as the birthplace of Horace, the prince of Latin lyric poets and satirists, born on the 8th of December, *anno* 66 B.C. (A.U.C. 688), in the consulship of L. Manlius Torquatus.

" O nata mecum consule Manlio."—III. Od. 21.

A bust of the poet, on a column of rough stone, has been set up in the city. In the vicinity are many places which have acquired interest from the references made to them by Horace. (*Craven's Excurs. in the Abruzzi*, 270, 290; *Tirabosschi, Della Letteratura Italiana*, i., 177, ed. Modena, 1787.)

VERA CRUZ, a town and the principal seaport of Mexico, on the S.W. side of the gulf of Mexico, cap. of the state of its own name, 195 m. E. by S. Mexico, and 235 m. S.E. Tampico. Lat. 19° 11′ 52″ N., long. 96° 8′ 45″ W. Pop. uncertain, but previously to the revolution it was estimated at 16,000. It is well built and clean, and its towers, cupolas, and battlements give it an imposing appearance from the sea. It is, however, surrounded by barren sand-hills and ponds of stagnant water, and is excessively unhealthy, being, in fact, the principal seat of the yellow fever. The older inhabitants and those accustomed to the climate are not so subject to this formidable visitation as strangers, all of whom, even if coming from the Havana and the W. India islands, are liable to the infection. No precautions prevent its attack, and numerous individuals have died at Xalapa, on the road to Mexico, who merely passed through this pestilential focus. The badness of the water at Vera Cruz is supposed to have some share in producing the complaint. The houses of Vera Cruz are mostly large, some of them being three stories high, built in the old Spanish or Moorish style, and generally enclosing a square court, with covered galleries. They have flat roofs, glass windows, and generally wooden balconies in front, their interior arrangement being the same as in Old Spain. The town and castle are built of madrepore, the lime that forms the cement being of the same material. There is one tolerably good square, of which the government house forms one side, and the principal church the other. The footpaths are frequently under arcades. No fewer than 16 cupolas or domes used to be counted from the sea, but only six churches are now in use; and most of the religious buildings have been neglected or abandoned since the Spaniards were expelled from the town. Rain-water is carefully preserved in tanks; and most sorts of provisions, excepting fish, are dear. Crowds of vultures and buzzards perform the office of scavengers. (*Bullock in Mod. Trav.*, xxv.)

The castle of St. Juan de Ulloa, which commands the town, is built on the small island of the same name, about 400 fathoms from the shore. It is a strong citadel, and its N.W. angle supports a lighthouse, with a brilliant revolving light, 79 ft. above the sea. The harbour of Vera Cruz is a mere roadstead between the town and castle, and is exceedingly insecure, the anchorage being so very bad that no vessel is considered safe unless made fast to brass rings fixed for the purpose, in the castle wall; nor are these always a sufficient protection during strong N. winds. But notwithstanding its numerous disadvantages, Vera Cruz maintains its commercial importance; though latterly Tampico, in a healthier situation, with a better port, has been growing into consequence. The precious metals, cochineal, sugar, flour, indigo, provisions, sarsaparilla, leather, vanilla, jalap, soap, logwood, and pimento, are the principal articles exported; and linen, cotton, woollen, and silk goods, paper, brandy, cocoa, quicksilver, iron, steel, wine, wax, &c. (See *ante*, 352.) During the period that the foreign trade of Mexico was carried on exclusively by the *flota*, which sailed periodically from Cadiz, Vera Cruz was celebrated for its fair, held at the arrival of the ships. It was then crowded with dealers from Mexico, and most parts of Spanish America; but the abolition of the system of regular fleets, in 1778, proved fatal to this fair, as well as to the still more celebrated fair of Portobello. We have already noticed (*Ante*, p. 352.) the wretched state of disrepair into which the great road from Vera Cruz to Mexico, formed by the Spaniards, has latterly been allowed to fall. This inflicts a great injury on the trade of the port.

Vera Cruz was founded towards the end of the 16th century, on the spot where Cortez first landed: it received the title and privileges of a city from Philip III. in 1615. The castle was taken by a French squadron, after a vigorous bombardment, in 1839, but was soon after restored to the Mexicans. (*Humboldt; Ward; Poinsett; Mod. Trav.,* xxv.; *Blunt's Coast Pilot; Comm. Dict.*)

VERCELLI (an. *Vercellæ*), a town of N. Italy, Sardinian dom., div. Novara, cap. prov. on the Sesia, and on the high road between Turin and Milan, 39 m. N.E. by E. Turin. Pop. in 1838, 18,353. It has a large market-place, one of the best cathedrals in Piedmont, several other churches, and good private buildings, a large and well kept hospital, museum, &c., with (in its environs) some fine promenades. Its fortifications were destroyed by the French in 1704. It is the see of an archbishop, and has some silk manufactures: but its chief trade is in rice, raised in the neighbourhood. A canal connects Vercelli with Ivrea. The date of its foundation is uncertain, but it was a town of some note in the time of the Romans. After suffering severely from the northern invaders, it revived under the Lombards, and took the lead of Turin till the latter became the residence of the court. (*Dict. Geog., &c.*)

VERDUN, (an. *Verodunum*), a town of France, dep. Meuse, cap. arrond., on the Meuse, where it begins to be navigable, and by which it is divided into five separate parts, 30 m. N.W. by N. Bar-le-Duc. Pop., 1836, ex comm., 9151. It has a citadel; and its defences were improved by Vauban. It is tolerably well built, but several of its streets are badly paved and steep. The bishop's palace, new cavalry barracks, military magazines, and theatre, are among the most remarkable buildings. It has six churches, including the cathedral; a Protestant church, a synagogue, a communal college, library with 14,000 volumes, &c. A planted esplanade separates the town from its citadel. Verdun has manufactures of fine striped serges, flannels, cotton yarn, liqueurs, &c., and several large tanneries. It was a station of importance under the Romans; and in the middle ages, under the Germanic emperors, it enjoyed the privileges of a free imperial city. It was definitively annexed to France in 1648; and is best known in modern times from its having been selected by Napoleon for the residence of the English prisoners detained in France after the rupture of 1803. (*Hugo,* art. *Meuse, &c.*)

VERGENNES, city, Addison co., Vt., 66 m. W. by S. Montpelier, 21 m. S. Burlington, 489 W. This is the only city in Vermont, chartered as such in 1783, having a territory 480 rods long and 400 wide. It is situated on both sides of Otter creek, over which is a bridge, 7 m. from its mouth. Vessels of 300 tons can come to its wharves, and here M'Donough's fleet, which gained the victory over the British in the battle of Plattsburgh, was fitted out. The falls in Otter creek, 37 ft. high, afford a very superior water-power, unsurpassed in the state; and extensive iron-works have here been erected, which are now in part suspended. It contains three churches, a Congregational, Episcopal, and Methodist; a bank, 13 stores, two fulling-mills, one woollen factory, two grist-mills, two saw-mills, three tanneries, one printing-office, one weekly newspaper; three schools, 230 scholars. Pop. 1017. Canal-boats ply regularly between this city and New-York.

VERMILLION, county, Ia. Situated in the W. part of the state, and contains 280 sq. m. Bounded E. by Wabash river. Watered by Vermillion river. It contained in 1840, 7632 neat cattle, 6408 sheep, 23,085 swine; and produced 51,185 bushels of wheat, 508,297 of Indian corn, 83,027 of oats, 18,690 of potatoes, 29,051 pounds of rice, 800 of tobacco. It had 22 stores, four flouring-mills, three grist-mills, six saw-mills, two tanneries, five distilleries, one printing-office, one weekly newspaper; 13 schools, 426 scholars. Pop. 8274. Capital, Newport.

VERMILLION, county, Ill. Situated in the E. part of the state, and contains 1000 sq. m. Drained by Big and Little Vermillion rivers and their branches. It contained in 1840, 13,725 neat cattle, 16,966 sheep, 35,984 swine; and produced 69,708 bushels of wheat, 6399 of rye, 941,810 of Indian corn, 180,353 of oats, 20,609 of potatoes, 5481 pounds of tobacco, 101,807 of sugar. It had 24 stores, four flouring-mills, 15 grist-mills, 27 saw-mills, three oil-mill, three tanneries, three distilleries, three potteries; 96 schools, 691 scholars. Pop. 9303. Capital, Danville.

VERMILLION, p. t., Erie co., O., 119 m. N. Columbus, 389 W. Bounded N. by lake Erie. Watered by Vermillion river. It has 12 schools, 575 scholars. Pop. 1334.

VERMONT, one of the northern United States, in the N.W. corner of New-England, and is between 42° 44' and 45° N. lat., and between 71° 33' and 73° 25' W. long., and

1016·

between 3° 35' and 5° 20' E. long. from Washington. The line surveyed and marked by Valentine and Collins previously to the year 1774 as the 45th degree of N. lat., and which has been known and understood as the line of division between Vermont and Lower Canada, is settled by the late treaty of compromise between the United States and Great Britain as the true line, though by the observations made in 1818 by Messrs. Hassler and Tiarks, surveyors under the treaty of Ghent, the 45th deg. of N. lat. was found to be a little S. of the line previously settled. The old line is now definitively settled as the true boundary, which gives Vermont a strip of land, not far from half a mile wide, more than she would have had by the true line of 45° N. lat. The state is bounded N. by Lower Canada, E. by the western bank of Connecticut river, which separates it from New-Hampshire, S. by Massachusetts, and W. by New-York, from which it is chiefly separated by lake Champlain. It is 157¼ m. long from N. to S., and 57½ m. wide from E. to W., and contains 9056½ sq. m., or 5,795,960 acres. The population in 1790 was 85,589; in 1800, 154,465, in 1810, 217,895; in 1820, 235,764; in 1830, 280,679; in 1840, 291,948. Of these, 146,378 were white males, 144,848 were white females; 364 were coloured males, 365 were coloured females. Employed in agriculture, 73,150; in commerce, 1303; in manufactures and trades, 13,174; in mining, 77; in navigating the ocean, 41; ditto lakes, rivers, and canals, 146; in the learned professions, 1563.

The state is divided into 14 counties, which, with their population in 1840, were as follow:

County	Population	County	Population
Addison	23,583	Lamoille	10,475
Bennington	16,872	Orange	27,873
Caledonian	21,891	Orleans	13,634
Chittenden	22,977	Rutland	30,699
Essex	4,226	Washington	23,506
Franklin	24,531	Windham	27,442
Grand Isle	3,883	Windsor	40,356

Montpelier, on Winooski or Onion river, 38 m. E.S.E. from Burlington, and about 10 m. N.E. of the geographical centre of the state, is the seat of government or political capital.

The surface of Vermont is hilly or mountainous. A few townships along the margin of lake Champlain may be regarded as level, extending from 5 to 10 m. from it; but otherwise the surface is generally uneven, consisting of hills and valleys, alluvial flats, gentle acclivities, elevated plains, and lofty mountains. The range of the Green mountains, so named by the French from the evergreens which cover them, and which have given name to the state, extend quite through it from N. to S. From the line of Massachusetts to the southern part of Washington county, it constitutes a lofty and unbroken range, keeping nearly a middle course between Connecticut river on the E. and lake Champlain on the W., and dividing the waters which fall into each. Though the passage across the mountains in this part is arduous, yet, by the construction of good roads, and a judicious location of them, it is much less so than formerly. In the southern part of Washington county the Green mountains are separated into two ranges. The highest of these ranges passes W. of the middle of the state to Canada line. The highest peaks lie in this range, which are Camel's Hump, generally called Camel's Rump, 4188 ft. high, and the Chin, in Mansfield mountain, 4279 ft. high; and it is remarkable that the whole is cloven down to its base, admitting a passage for Winooski or Onion river through it, the mountain approaching oftentimes so near the river as scarcely to admit a road along its banks, and affording much sublime and romantic scenery. The other range strikes off much more to the E., extending also near the Canada line. It is called the height of land, dividing the waters which fall into Connecticut river from those which, in the N. part of the state, flow into lake Champlain and lake Memphremagog. Though having no peaks so high as those of the western range, it is more uniformly elevated, yet with so gentle an acclivity as to admit of easy roads over it in various places; while the western range admits of roads across it chiefly where it is penetrated by the Winooski, Lamoille, and Missisque rivers, though in its northern parts it becomes less formidable. It is an interesting fact that the road from Burlington on lake Champlain, along Winooski river, through Montpelier, and thence down White river to Hartford, Vermont, on Connecticut river, passes over no considerable hills, and may hereafter be occupied by a line of railroads or canals from Connecticut river to lake Champlain.

The soil of Vermont is generally fertile, but better adapted to grazing than to grain. Throughout the western border of the state, near lake Champlain, the soil is strong, and adapted to wheat, and summer wheat succeeds well in most parts of the state. Indian corn is produced most extensively on the margins of the streams, but does well in other parts. Even on the Green mountains are fine grazing farms. The productions of the state are wheat, rye, Indian corn, barley, oats, buckwheat, potatoes, peas, and flax. The natural growth of the soil E. of the mountains is birch, beech, maple, ash, elm, and butternut; and W. of the mountains the growth of hardwood is intermixed with pine and other evergreens. The state is capable of raising sufficient grain for home consumption, but is particularly distinguished for its excellent pasturage, giving subsistence to numerous flocks of sheep and herds of cattle.

In 1840 there were in the state 82,402 horses and mules, 384,341 neat cattle, 1,681,819 sheep, 203,800 swine; poultry was raised to the value of $131,578. There were produced 495,800 bushels of wheat, 1,119,678 of Indian corn, 230,993 of rye, 54,781 of barley, 288,416 of buckwheat, 2,222,548 of oats, 8,869,751 of potatoes, 3,699,235 pounds of wool, 4,647,934 of sugar, 48,137 of hops, 4060 of wax, 4286 of silk cocoons, 836,739 tons of hay, 29 of hemp and flax. The products of the dairy amounted to $2,008,737; of the orchard to $213,944; of lumber to $346,939; 718 tons of pot and pearl ashes were produced.

The exports consist of beef, pork, butter, cheese, live cattle, and pot and pearl ashes. The trade E. of the mountains is chiefly with Boston and Hartford; and W. of the mountains to New-York and Montreal and Quebec, to which it has a ready access through lake Champlain, the Champlain canal, and the Hudson and St. Laurence rivers.

The climate is healthy, though the winters are severe. The ordinary extreme range of the thermometer is from 99° above to 29° below zero of Fahrenheit, though in some instances it has risen above and fallen below this. Snow generally lies from December to March, and is often 4 ft. deep upon the mountains. It is, on an average, several degrees colder on the eastern than on the western side of the mountains. Lake Champlain is not generally wholly frozen over, until about the first of February.

The most considerable rivers are on the W. side of the mountains, and flow into lake Champlain. They are Otter creek, 85 m. long, and navigable for sloops 6 m. to Vergennes; Onion or Winooski river, which is 80 m. long, and enters the lake 4 m. N. of the village of Burlington; Lamoille, which is 70 m. long; and Missisque, which is about the same length. Small boats may penetrate these rivers to their lower falls, and they have all fine sets of falls, which afford extensive water-power. The principal rivers on the E. side of the mountains flow into Connecticut river; they are Deerfield, West, White, Black, and Pasumic.

Lake Champlain lies partly in New-York, but more than half of it within the limits of Vermont. It extends in a straight line 102 m. from Whitehall to the 45th deg. of N. lat., and thence about 24 m. to St. John's in Canada. This lake is connected by a canal, 64 m. long, to Hudson river near Albany, which also forms, a junction with the Erie canal. This lake contains about 567 sq. m., two thirds of which lie within the limits of Vermont. Lake Memphremagog is 40 m. long, and 7 or 8 m. wide, lies on the N. line of the state, one third being in Vermont and two thirds in Canada. Lake Bombazine, on the W. border of Castleton, and lake Dunmore, in Salisbury, are considerable bodies of water. The islands of lake Champlain are considerably numerous, and some of them are large, fertile, and populous. The principal of these islands, with the harbours of Alburgh, constitute Grand Isle county. They are N. Hero, S. Hero, and La Motte. There are various harbours in Vermont on lake Champlain, the principal of which are St. Alban's, Burlington, Vergennes, and some others.

Burlington is the largest and most commercial town in the state. The other principal towns are Middlebury, St. Alban's, Rutland, and Bennington in the W., Montpelier in the centre, and Windsor, Woodstock, Danville, and Newbury on the E. side of the mountains.

The exports of this state for the year ending Sept. 30th, 1841, were $277,987; and the imports were $346,739: the tonnage entered was 13,560, and the tonnage cleared of the same amount.

Vermont is an agricultural, rather than a commercial and manufacturing state. There were in 1840, 747 retail stores, with a capital of $2,964,060; the lumber trade employed a capital of $43,506; home-made or family goods were produced to the amount of $674,548; 95 woollen factories and 239 fulling-mills, employing 1450 persons, produced articles to the amount of $1,331,953, with a capital of $1,406,950; seven cotton factories, with 7254 spindles, produced articles to the amount of $113,000, with a capital of $118,100; 26 furnaces produced 5743 tons of cast iron, and 14 forges produced 655 tons of bar iron, the whole employing 788 persons and a capital of $664,150; hats and caps were manufactured to the amount of $62,432, and straw bonnets to the amount of $2819, the whole employing 198 persons and a capital of $12,875; 17 paper-mills produced articles to the amount of $179,720, and all other manufactures of paper to the amount of $15,000, the whole employing 195 persons and a capital of $216,500; two glass-houses employed 70 persons, producing articles to the amount of $55,000, with a capital of $15,000; eight potteries produced articles to the amount of $23,000, with a capital of $10,350; 261 tanneries employed 500 persons and a capital of $403,093; 399 other manufactories of leather, as saddleries, &c., produced articles to the amount of $361,468, with a capital of $168,090; granite and marble was produced to the amount of $92,515; bricks and lime were made to the amount of $402,218; two distilleries and one brewery employed five persons and a capital of $8850; 87 persons produced machinery to the amount of $101,354; 33 persons produced hardware and cutlery to the amount of $16,650; 437 persons manufactured carriages and wagons to the amount of $162,097, with a capital of $101,570; 190 persons manufactured furniture to the amount of $83,275, with a capital of 49,859; 42 persons manufactured 1158 small arms; vessels were built to the amount of $73,000; 29 printing-offices, 14 binderies, two daily, 96 weekly, and two semi-weekly newspapers, and three periodicals, employed 156 persons and a capital of $194,900; 72 brick or stone houses and 468 wooden houses were built, employing 912 persons, and cost $344,896. The total capital employed in manufactures in the state was $4,396,640.

There are three colleges in Vermont. The university of Vermont, at Burlington, was founded in 1791; Middlebury college, at Middlebury, was founded in 1800; Norwich university was founded at Norwich in 1834. In these institutions there were in 1840, 328 students. There were in the state 48 academies with 4113 students, and 9402 common and primary schools with 82,117 scholars. There were in the state 2270 white persons over 20 years of age who could neither read nor write.

The principal religious denominations are the Congregationalists, Baptists, and Methodists. In 1836 the Congregationalists had 186 churches, 114 ministers, and 20,575 communicants; the Baptists had 125 churches, 78 ministers, and 10,525 communicants; the Methodists had 75 itinerant preachers; the Episcopalians had one bishop and 18 ministers. Besides these, there is a considerable number of Universalists and Christians, and a few Unitarians and Roman Catholics.

In September, 1839, there were 19 banks in the state, with an aggregate capital of $1,325,520, and a circulation of $1,966,812. Vermont has a state debt of about $250,000, about one half of which was contracted in building the new state house.

There is a state penitentiary located at Windsor.

The first constitution of this state was formed in 1777. The present constitution was formed July 4th, 1793, and has since been amended. The governor is elected annually by the people. He must be a citizen of the United States, and have resided in the state for four years next preceding his election. The lieutenant-governor is elected in the same manner, and must have the same qualifications; and in case of the absence of the governor, or his inability to serve, succeeds to the office. As lieutenant-governor, he is president of the senate. The supreme executive council consist of the governor, lieutenant-governor, and 12 persons chosen by the people. The senate consists of 30 members, each county choosing at least one. Some are entitled to more, according to their population. The house of representatives consists of 231 members, elected annually by the people. Every representative must be a citizen of the United States, must have resided in the state for two years, the latter of which must be in the town in which he is elected. The supreme court consists of five judges, elected annually by the legislature. The supreme court sits once a year, and the county courts twice, in each county. There is a court of chancery, which holds a session in each county, each judge of the supreme court being chancellor of a circuit. The secretary of state is elected annually by the joint vote of both houses of the legislature, and the treasurer is chosen annually by the people. The right of suffrage is enjoyed by every person who is full 21 years of age, a citizen of the United States, of quiet and peaceable behaviour, and who has resided in the state for one year next preceding an election. A council of censors is appointed once in seven years, whose duty it is to inquire whether the constitution has been preserved inviolate, and whether the legislature and executive branches have performed their duty as guardians of the commonwealth; whether the taxes have been justly laid and collected, and the public moneys have been properly disposed of, and the laws have been duly executed.

Excepting a few short canals for the improvement of the navigation of Connecticut river, no work of internal improvement has been undertaken, though several have been projected. The Champlain canal, extending from the lake to Hudson river, is a work in which this state has a deep interest, and from which it derives important advantages.

The first settlement of the state was at fort Dummer, in the S.E. part of the state, by emigrants from Massachusetts.

VERNON.

New-Hampshire claimed the territory from 1741 to 1764, and granted many townships in the state to proprietors, which were thence called the "New-Hampshire grants," which comprise many of the best towns in the state. New-York also claimed the territory, and obtained a grant of it from the British parliament in 1764. These conflicting claims exceedingly harassed the inhabitants. At the commencement of the revolutionary war, Congress dared not admit Vermont to the Union, though the state proclaimed itself independent, for fear of offending New-Hampshire and New-York, especially the latter. Vermont had a difficult part to act, and it conducted itself with great wisdom and patriotism. The British hoped to be able to detach it from the American confederacy ; and its leaders, without committing themselves, flattered these hopes, and saved its exposed frontier from attacks, while no portion of the Union showed a more devoted patriotism, or contributed more, according to its means, to the common cause. The "Green mountain boys" were characterized by daring bravery in the revolutionary struggle. In 1790 New-York was induced, by the payment of $30,000, to withdraw her claims to the territory ; and in 1791 Vermont was admitted to the Union, whose independence she had extensively contributed to acquire.

VERNON, p. t., Tolland co., Ct., 12 m. E.N.E. Hartford, 348 W. Watered by Hockanum river. It contains two churches, a Congregational and a Baptist ; four stores, 10 fulling-mills, 10 woollen factories, four cotton factories with 5104 spindles, two grist-mills, three saw-mills, two oil-mills, one paper-mill ; eight schools, 322 scholars. Pop. 1430.

VERNON, p. t., Oneida co., N. Y., 16 m. W. Utica, 108 m. W.N.W. Albany, 371 W. Drained by Oneida and Skenandoa creeks. It contains three churches, a Congregational, Baptist, and a Unitarian ; 13 stores, three fulling-mills, three woollen factories, two glass-houses, four grist-mills, 13 saw-mills, one printing-office, one weekly newspaper, two academies, 17 schools. Pop. 3043.

VERNON, p. t., Sussex co., N. J., 92 m. N. by E. Trenton, 260 W. Drained by tributaries of Wallkill river. It contains two stores, two forges, three grist-mills, four saw-mills, four distilleries ; 12 schools, 371 scholars. Pop. 3395.

VERNON, p. v., Jennings co., Ia., 65 m. S.E. Indianapolis, 560 W. Situated on the E. side of Vernon fork, Muscatatak river. It contains an elegant brick courthouse, a jail, a county clerk's office, a church common to Presbyterians, Methodists, and Baptists, two stores, and about 350 inhabitants.

VERNON, t., Washington co., Ia. It has six stores, one cotton factory ; one school, 20 scholars. Pop. 1936.

VERONA, a celebrated city of Austrian Italy, cap. deleg. of same name, on the Adige, at the point where the last declivities of the Alps sink into the great plain of Lombardy, 64 m. W. Venice ; lat. 45° 26' 9'' N., long. 11° 0' 54'' E. Pop. in 1837, 48,486. (*Berghaus.*) It is divided into two unequal parts by the Adige, which sweeps through it in a bold curve, and forms a peninsula, within which the whole of the ancient, and the greater part of the modern city is enclosed. The river, which is wide and rapid, is here crossed by four noble stone bridges.

Verona is an extremely well situated, well built, and most interesting city. "You enter it," says Mr. Rose, "by a magnificent approach, and a street probably the widest in Europe. This street is indeed short, and single in its breadth, but the city in general pleases by its picturesque appearance, to which an abundance of marble quarries has not a little contributed, 35 varieties of this species of stone being found in its neighbourhood." (*Letters*, i., 41.) The houses frequently present, in their form and ornaments, fine proportions, and beautiful workmanship. The old walls and towers still remain, and the city has five gates, two of which are fine structures by Sanmichele. Its former military defences were destroyed by the French, after the revolt of the inhabitants in 1797, but extensive fortifications are again in the course of being constructed.

But the great glory of Verona is its amphitheatre, one of the noblest existing monuments of the ancient Romans. Excepting the Colosseum at Rome, it is the largest extant edifice of its class. Like all other structures of the same kind, it is elliptical, the extreme length of its transverse and conjugate diameters to the outside of the outer walls being respectively about 510 and 410 ft. ; while those of the arena are 249 and 145 ft. (*Woods*, i., 226.) Its outer wall or clincture, which had 72 arches in every story, has been mostly destroyed, with the exception of one fragment containing three stories of four arches each, rising to the height of about 100 ft. Over this, however, there was a fourth story, so that the entire height of the building, when perfect, must have exceeded 120 ft. Internally it has suffered comparatively little ; and its concentric rows of benches or seats, of which 43 still remain, exclusive of two sunk below ground, with its corridors, stairs, &c., are wonderfully well preserved. Each row of seats is 1½ ft. in

VERONA.

height, and as much in breadth ; and allowing 1½ ft. of space to each individual, the Marquis Maffei supposes that the amphitheatre might have accommodated 22,000 spectators ! The interior of the amphitheatre having been in parts a good deal dilapidated, it was repaired at different periods in the 16th century, when the broken and wasted seats were replaced by others. But these repairs, though, on the whole, highly creditable to the Veronese, do not appear to have been well executed. The ancient benches were formed of vast blocks of marble, admirably cut and jointed ; whereas the modern benches consist, according to Maffei, of a soft flaky stone, which has, in parts, yielded to the weather ; and the same distinguished authority adds that but little care has been taken in laying the stones, and that in parts the elliptical curvature has not even been observed. These defects, however, are not visible on a cursory inspection of the building, which astonishes alike by its mass, its antiquity, and its preservation.

Owing to the want of inscriptions, and of all reference to its origin in the classical writers, we are without any authentic information either as to the founders or the era of this great work. Most probably, however, it was built somewhere between the reigns of Titus and Trajan, or in the early part of that of the latter. In the middle ages it was sometimes used for the exhibition of shows and sports, and sometimes as an arena for judicial combats. In more modern times, a bull fight was exhibited here in honour of the emperor Joseph II., then at Verona ; and at a still later period, the pope, in passing through the city, gave his benediction to a vast multitude collected in the amphitheatre. The French, when masters of Verona, had the bad taste to erect in the arena a wooden theatre, in which plays, farces, equestrian feats, &c., were performed for the amusement of the troops. This barbarous novelty is still kept up. (Full particulars respecting the amphitheatre are given in the *Verona Illustrata* of the learned Marquis Maffei. That part of the work which relates to amphitheatres having been printed separately, was translated into English, and published in 8vo. London, 1730.)

But the amphitheatre is not the only monument of antiquity that distinguishes Verona. In the middle of a street called the *Corso* is an ancient double gateway, which, on the strength of an inscription importing that the adjacent walls were built by Gallienus, has been named after that emperor, but though loaded with supernumerary ornaments, the Veronese antiquaries affirm that its style is too good for his age. Each gateway is ornamented with Corinthian pilasters supporting a light pediment ; and above are two arches with six small arched windows in each. The whole is of marble. The remains of another gateway, of a similar but chaster form, probably the entrance to the ancient forum, are to be seen in another street, and near the old Gothic castle, is the arch of the Gavii, perhaps part of a sepulchral edifice, but, at any rate, of very remote antiquity. Two arches in the purest style of Roman architecture make a part of one of the bridges ; and the remains of another bridge, and the traces of a very large ancient theatre and amusements, are still extant. Addison and Evelyn speak of a triumphal arch of Flaminius, as one of the noblest remains of antiquity in Europe, and of an arch commemorating the victory of Marius, with various temples and aqueducts ; but as few or no remains of these exist at present, it would seem that the antiquities of Verona had suffered greatly since the beginning of last century.

The ecclesiastical buildings comprise interesting specimens of middle-age architecture. The cathedral, an edifice of the 12th century, has nothing particularly remarkable except the Assumption by Titian, and the tomb of pope Lucius III., who, when driven from Rome in 1185, found a secure asylum in this city. In respect of architecture, merit the cathedral is very inferior to the church of St. Anastasia, built by the Dominicans at the beginning of the 13th century. Mr. Woods says that if the front were finished this edifice would probably be the most perfect specimen of the style of architecture to which it belongs. The church of St. Zeno, a curious structure, with a remarkable crypt, is said to have been founded by Pepin, but it was not completed till 1178. Its front is covered with bas-reliefs in one, its doors with sculpture in bronze of a very early date ; and near it are the remains of a palace in which the German emperors occasionally resided during the 12th and 13th centuries. Several other churches are worthy of notice : is that of San Giorgio is a fine picture of Paul Veronese. The tombs of the Scala family (*scaligeri*), once lords of Verona, stand in an inclosure in one of the thoroughfares. According to Forsyth, they are " models of the most of their Gothic, light, open, spiry, full of statues caged in their fretted niches ; yet slender as they seem, these tombs have stood entire for 500 years in a public street the frequent theatre of sedition." The Porto del Castel Vecchio, built in 1354, is remarkable for an arch 161 feet in span, forming part of a circle. (*Woods*.) The town-hall is ornamented externally with busts of the most celebrated natives of Verona, and has within it some fine paintings. The exchange; the *Musso Lapidario*, and *Philoti*, both having extensive collections of ancient monuments: the opera-house, the fine Ionic portico of which forms, with the arcades of the museum, three sides of a handsome square; the episcopal and new vice-regal palaces; the *Palazzo Bevilacqua*, an edifice by Sanmichele: the *Canossa* palace and several other noble residences; the lyceum, philharmonic academy founded by the Marquis Maffei, and arsenal, are among the most conspicuous edifices in the city. Verona is the seat of the high court of justice and of the superior military authorities for the Lombardo-Venetian kingdom, and of a court of primary jurisdiction for its deleg. It is a bishop's see, and has several gymnasia, a theological seminary, and numerous royal and other schools, learned societies, public and private libraries, galleries, &c. English travellers are shown what is called "Juliet's tomb," which is merely an old sarcophagus without a cover, lying in a garden where it has been made use of as a cistern.

Verona, according to Berghaus, is distinguished as the most industrious of Italian towns, "It has 60 silk twist factories, 9 establishments for weaving silks, large leather, earthenware, and soap factories, and numerous factories for the weaving of linen and woollen fabrics. Its trade is chiefly in the product of these, and in raw silk, grain, oil, sumach, and other agricultural produce." (*Sllg. Leander, &c.*) It has two weekly markets, and two considerable annual fairs, each lasting 15 days.

We have no certain details as to the origin of Verona. Under the Romans, however, she became a flourishing city; and in the time of Strabo was superior to Brixia, Mantua, Regium, Comum, &c. She was the capital of the kingdom of Italy from the time of Odoacer to that of Berengarius; and from the 12th to the 15th century she was the capital of a considerable territory, governed successively by the Scaligers, Visconti, &c. Under the former, in the 13th and 14th centuries, occurred the feuds between the Cappelletti and Montecchi, immortalized by Shakspere. In 1405 Verona submitted to Venice, of whose dominions it continued to form an important portion till the overthrow of the Venetian republic in 1797. In 1822 it was the seat of a congress. Perhaps no city of Italy has given birth to a greater number of distinguished men than Verona. Among these may be mentioned Catullus,

"Tantum magna suo debet Verona Catullo,
Quantum parva suo Mantua Virgilio."
Martial, lib. xiv. epig. 195.

Macer, Cornelius Nepos, Pliny the Elder, &c., who shed in antiquity an imperishable lustre over the place of their birth. At a later period Guarini, Calderini, Panvinius, and Fracastorius contributed to the revival of literature and of the ancient fame of their native city; which in more modern times has been still further extended by the labours of the famous painter Paolo Cagliari, surnamed Veronese, born here in 1530; Bianchini, distinguished alike as a mathematician, a historian, and a critic; Maffei, whose *Verona Illustrata*, already referred to (1731–1732, in folio, and 4 vols. small 4to), is a work of the greatest research and value; the poet Pindemonte, &c.

Vitruvius, in antiquity, and the famous Julius Cesar Scaliger, have also been included, though on no good grounds, among the illustrious natives of Verona. The latter, indeed, represented himself as the eldest son of one of the Scaligers, lords of Verona, and as entitled to that seigniory! But Maffei and Tiraboschi have shown that there is not so much as the shadow of a foundation for this statement; that it is a pure fabrication; that Scaliger was, in fact, the son of a miniature-painter of the name of Bordoni; and was most probably born at Padus. (*Verona Illustrata*, p. 300; *Tiraboschi*, vii. 1431.)

VERONA, t. Oneida co., N. Y., 20 m. W. Utica, 116 m. W.N.W. Albany. Bounded N. by Wood cr., W. by Oneida lake. The Erie canal passes through it, from which the Oneida lake canal branches off. It has 17 stores, one fulling mill, one woollen factory, one glass house, two grist-mills, 20 saw-mills, four tanneries; 94 schools, 1218 scholars. Pop. 4504.

VERSAILLES, a town of France, formerly the chief residence of the French court, dep. Seine-et-Oise, of which it is the cap., in a undulating plain, 9 m. S.W. Paris. Pop. in 1836, ex. comm., 28,776. It is one of the handsomest towns in the kingdom; it consists principally of three wide streets, lined with trees, diverging from the *Place d'Armes*, an open space in front of the palace: the central and widest of these streets is called the avenue de Paris: and those on the N. and S., the avenues of St. Cloud and Sceaux. The other streets, though of less width, are equally regular, cross each other at right angles, and are lined with handsome residences. The cathedral of St. Louis, founded by Louis XV in 1743, that of Notre Dame, built after the design of Mansard, in the previous reign; the church of St. Sympho-

rien, the town-hall, prefecture, theatre, royal college, public library with 48,000 volumes, civil and military hospital, barracks, *dépôt* of naval and colonial archives, and hall of the *jeu de paume*, in which the deputies of the national assembly made their famous declaration (see below), are among the principal public buildings. In one of the open spaces is a marble statue of General Hoche, a native of Versailles. The town is ornamented by many handsome fountains, &c.; but it wears a dull and deserted appearance, being no longer resorted to by the *beau monde*.

Versailles is wholly indebted for its celebrity, and, indeed, for its existence, to the royal palace in its immediate vicinity. Louis XIII. had a hunting-seat here; but the present edifice, which is of prodigious size and magnificence, was erected by Louis XIV., who expended immense sums on its construction and embellishment. On the E. side where it faces the place d'armes, it consists of only an irregular succession of buildings, inclosing a few small courts. But on the opposite side, facing the gardens, it presents a noble facade, 645 yards in length, three stories in elevation, ornamented with Ionic pilasters, and with 80 statues 16 feet in height, allegorically representing the months, seasons, arts, and sciences, &c., and crowned by a balustrade. Its galleries and saloons, enriched with every variety of coloured marbles, and splendidly gilt, are alike vast and magnificent. The Salon d'Hercule, and the Salles des Maréchaux, de Venus, Diane, Mercure, Mars, Apollon, l'Abondance, de la Guerre, &c., so named from the paintings on their ceilings, walls, or other appropriate devices, are all noble apartments. The Grande Galerie is 228 feet in length, by 32 feet in breadth, and 42 feet in height: the ceiling, painted by Le Brun, represents some of the most striking events in the early part of the reign of Louis XIV. At one of its extremities is the Salon de la Paix, corresponding with the Salon de la Guerre. Besides its innumerable apartments, the palace has an elegant chapel, in which the unfortunate Marie Antoinette was united to Louis XVI. on the 16th of May, 1770, an opera-house, or theatre, capable of accommodating 3000 spectators, a magnificent staircase, &c.

The palace has not been occupied by the court since 1789, and was getting into a state of disrepair, when it was entirely renovated and transformed by his present majesty, Louis Philippe, into what may be called a national museum, intended to illustrate the history, and to exhibit the progress of arts, arms, and civilization in France. In pursuance of this design, many small apartments, formerly appropriated to the lodging of the various functionaries attached to the court, have been converted into noble saloons. Of these, the Salle des Maréchaux, noticed above, containing portraits of all the marshals of France, the Galerie des Batailles, the Salle de 1830, the Galerie de Sculpture, &c., are among the most striking. The palace is, in fact, filled with an immense collection of statues and paintings, exhibiting all the principal personages and events in the history of the monarchy, from the reign of Clovis down to that of Louis Philippe. Of course, many of these statues and pictures must be very inferior; but these will be gradually replaced by others, and the collection is exceedingly interesting. The library is intended to comprise copies of all works having reference to the history and state of France. The fitting up of the apartments is, also, superb: and the whole reflects the highest credit on the taste and liberality of the king.

Immediately adjoining the palace on the W., is the little park, comprising the gardens, numerous reservoirs, fountains, and public walks; the orangery; Bains d'Apollon, and Bassin de Neptune, both having sculptured groups of much merit, &c. In this park are also the Great and Little Trianon, two royal palaces on a minor scale, and the grand water-works, of unrivalled magnitude, which, however, play only on great occasions. They are supplied from the Seine, by the aqueduct of Marly. The great park comprises a large tract of country, including several villages. (*Hugo*, art. *Versailles; Guide du Voyageur en France, &c.*)

Versailles has some manufactures of cotton yarn, and wax lights, but only on a limited scale; it had formerly an extensive factory of fire-arms, but this appears to be given up. It is the see of a bishop, and the seat of tribunals of primary jurisdiction and commerce, and of societies of literature and agriculture, &c. Philip V. of Spain, Louis XV., Louis XVI., Louis XVIII., Charles X., and several other eminent personages, were born at Versailles. It is also intimately connected with the history of the revolution. The states general met in the great hall of that palace on the 5th of May, 1789. And here, on the 17th of June, the *tiers état*, having been joined by the whole body of inferior clergy, and some of the nobles, constituted themselves the national assembly, and the sole representatives of the people. On the 20th of the same month the deputies, finding the doors of the hall in which they had been accustomed to meet shut against them, retired to the Tennis court, and took the famous oath, by which they bound themselves to continue their sittings till the constitution of the kingdom had been fixed on a solid basis! The revolution, thus fairly begun, set in with a violence which all the talent and firmness of purpose of Napoleon would, perhaps, have been unable to control, and which the vacillation of the good-natured imbecile king served only to aggravate. And so rapid was the progress of events, that on the 5th and 6th of October the palace of Versailles was forced by a mob, consisting of the lowest scum of Paris: who, in the end, succeeded in forcing off the king and his family to the capital, to encounter, at the hands of the vilest ruffians, imprisonment, every species of indignity, and an ignominious death !

VERSAILLES, L., Alleghany co., Pa., 12 m. S.E. Pittsburg. Bounded S. by Monongahela r., W. by Youghiogheny r. It has two schools, 113 scholars. Pop. 1456.

VERSAILLES, p. v., capital of Woodford co., Ky., 12 m. S.S.E. Frankfort, 534 W. Situated a little S. of the rail-road from Lexington to Frankfort, to which it is connected by a short branch. It contains a courthouse, jail, several stores, and 1044 inhabitants.

VERSAILLES, p. v., capital of Ripley co., Ia., 69 m. S.E. Indianapolis, 545 W. Situated on the W. side of Laughery cr., and contains a courthouse, jail, and about 400 inhabitants. It is surrounded by a fertile country.

VERSHIRE, p. t., Orange co., Vt., 31 m. S.E. Montpelier, 510 W. Chartered in 1781. Watered by Ompompanoosuc r. It has three churches, a Congregational, Baptist, and Freewill Baptist, two stores, one fulling-mill, one grist-mill, five saw-mills; 16 schools, 436 scholars. Pop. 1198.

VESTAL, p. t., Broome co., N. Y., 174 m. W.S.W. Albany, 289 W. Drained by tributaries of Susquehanna r. which bounds it on the N. It contains two grist-mills, 21 saw-mills; eight schools, 371 scholars. Pop. 1253. The inhabitants are extensively employed in getting out lumber.

VERVIERS, a town of Belgium, prov. Liege, cap. arrond., on the Vesdre, by which it is intersected, and which is here crossed by two stone bridges, 14 m. E. Liege. Pop., in 1836, 19,079. It is divided into the upper and lower town; some of its streets are wide and well built, but many others are quite the contrary. When erected into a town in 1651, it was surrounded with walls, but these were afterwards demolished by the French. A new church, the town-hall, and a little theatre with a front of the Ionic order, are handsome buildings. It has a tribunal and chamber of commerce, a college, hospital, and several asylums, nearly thirty public schools, a philanthropic society, bath establishment, &c. The Vesdre is divided into numerous canals, for the use of the various manufactories, which have increased rapidly since the peace. These comprise above 50 woollen cloth factories, and more than 30 dyeing-houses, with fulling and other mills, soap-works, breweries, iron and lead foundries, &c. It has two markets weekly, and four annual fairs. Great quantities of fuller's earth are dug up in the vicinity. (Vandermaelin, Dict. de Liège; &c.)

VESOUL, a town of France, dep. Haute-Saône, of which it is the cap.; in the valley of the Durgeon, at the foot of the Motte de Vesoul, a height covered with vineyards and meadows, 56 m. E.N.E. Dijon. Pop. in 1836, ex. comm. 5792. It is well built and clean; most of its streets are wide and straight, and it has several good public buildings, including cavalry barracks, civil and military, hospital, theatre, prefecture, parish church, public baths, courthouse, town-hall, &c. It has, also, a public library, said to comprise 21,000 vols., a museum, and a departmental nursery-ground; with manufactures of calico and gold lace, and some trade in corn, wine, salt, nails, and hardware. Near it are mineral waters, but they are turned to little account. (Hugo, art. Haute-Saône, &c.)

VESUVIUS (MOUNT), a celebrated mountain of S. Italy, Ætnei ignis imitator, being the only active volcano, of any consequence, at present existing on the European continent; on the E. shore of the bay of Naples, and 10 m. E. by S. from the city, the crater being in lat. 40° 48' N., long. 14° 27' E. Vesuvius does not belong to the Apennine system, but rises, altogether unconnected with any of its ramifications, out of the great plain of Campania. Including Somma on its inland side, it consists of a circular mass, the extreme height of which, about 3890 ft., is to its diameter, 8 m., nearly as 1 to 1½; it is somewhat less elevated than mount Hecla, and only two fifths the height, with considerably less than one third the circuit of Etna. Mr. Maclaren, by whom it has been carefully examined and elaborately described, gives the following account of its external appearance;—" To gain a distinct conception of the aspect of the hill, shape out for yourself, by a mental effort, the following objects. First, a sloping plain, 3 m. long, and 3 m. broad, stretching up, with a pretty rapid ascent, to an elevation of more than 3000 ft., very rugged in the surface, and covered everywhere with black burnt stones, like the scoriæ of an iron furnace; second, at the head of this plain, and towering over it, a cone of the same black burnt stones, with sides remarkably straight and uniform, shooting up in

the blue sky to a further elevation of 1300 feet; third, behind this cone a lofty circular precipice (the front of Monte Somma), 1400 feet high, and 3 m. long, standing like a vast wall, and of the same burnt appearance; fourth, at the lower side of the plain, between the burnt ground and the sea, a belt of land, 2 m. broad, laid out in vineyards, but intersected every one or two furlongs by terraces of the same black calcined matter, projecting like offshoots from the central mass, and now and then unveiling old currents of lava from beneath them. Very little lava is visible, but the course of the different currents is traced by the long terraces of scoriæ which cover and flank them.

"The top of the cone, which is about 3000 feet in diameter, presented the aspect of an uneven plane in the end of 1826; but when visited about five weeks after the eruption (1839) it had a regularly formed crater, shaped exactly like a tea-cup. I estimated its width at 1500 feet, and its depth at 500. The rim or crest of loose and solid matter which surrounded it was of very unequal breadth, 400 or 500 feet on the W. side, and apparently not 50 at some other parts of the circumference. Snow having fallen some days before, clouds of steam rose from the cavity, which, however, were neither so dense nor so constant as to prevent us from occasionally seeing the bottom of the crater very distinctly. It was nearly level, without crevices or openings, and covered with loose blocks of lava of no great size." (Notes on France and Italy, 134, 135.)

Geologically considered, Vesuvius is but the representative of a more ancient and much larger volcano, of which monte Somma is a remnant, and in the centre of which the modern vent has been upheaved. Monte Somma, on the N.E. side of Vesuvius, is a ridge extending 3 m. in length, forming about one third part of a circle, and rather less lofty than the present cone of Vesuvius. The average distance of the escarpment of Somma from the centre of the cone is about 1 m.; the back of the ridge dips outward at an angle of 25°, while the front towards Vesuvius is nearly vertical, rising 1377 ft. (420 metres) above the level space which divides it from Vesuvius, and which is called the Atrio dei Cavalli, or " vestibule of horses," because persons to the crater are obliged to leave their horses, and perform the rest of the journey on foot. The Atrio dei Cavalli forms a segment of a circular ring, about ¼ m. in breadth, at the base of the cone, dividing it from Somma, and having a continuation, in the shape of a depression, on the other side, where a slight projection, called the Pedimentis, is supposed to indicate the place of the rest of the ancient escarpment, which, when complete, must have formed a ring 6 m. in circumference (Maclaren, p. 140); being of greater extent than any crater with which we are acquainted. Hence, Mr. Lyell considers it probable that the ancient volcano was higher than Vesuvius, and that the first recorded explosion of the latter blew up a great part of the cone itself " so that the wall of Somma, and the ridge or terrace of the Pedimentis were never the margin of a crater of eruption, but are the relics of a ruined and truncated cone." This species of phenomenon has not been without its examples in modern times. During the eruption of Oct. 1822, more than 300 feet of the cone were carried away by explosions, reducing the height of the mountain from about 4200 to 3400 feet. (Forbe's Account of Mount Vesuvius; Lyell's Geology, ii. 80, 88, &c.)

According to Maclaren, the rocks of Somma and Vesuvius are mineralogically distinct. Somma, like Vesuvius, is composed of strata of fragmentary and stony matter intermixed; but the stony matter of Vesuvius consists of lava, more or less cellular, scoriaceous on the surface, and forming long narrow bands on the surface of the hill. That of Somma is a lenitic porphyry, containing steatite, and continuous with the beds under the tufa which forms the rest of the plain of Naples. Mr. Lyell says, " It is an extraordinary fact, that, in an area of 3 sq. m. round Vesuvius, a greater number of simple minerals have been found than at any other spot of the same dimensions on the surface of the globe. Haüy only enumerated 380 species of simple minerals as known to him; and no less than 82 had been found on Vesuvius before the end of the year 1828." Many of these are peculiar to that locality. (ii., 92.) The flora of Vesuvius is also peculiar to Italy, embracing several Dianthaceous and other plants not found elsewhere in the peninsula. The greater part of the mountain has, indeed, a bare and rugged aspect; but around its base, as above stated, is an extremely fertile and picturesque region, teeming with plantations, villages, and white country houses. The land here is divided into small farms of 5 or 4 acres, supporting each a whole family, and the population is ex-

¹ Lyell says :— The escarpment of Somma exhibits a structure precisely similar to that of the cone of Vesuvius. The principal part of Somma consists in the greater abundance in the older cones of fragments of stony sedimentary rocks ejected during eruptions. (Principles of Geology, ii. 80, Sect.) But it appears from Mr. Maclaren that latter discoveries have been made. (Notes, &c., p. 139.)

mated at not less than 5000 persons to a sq. league. The land is cultivated, like a garden, with the spade; and yields three crops a year, without fallows or manure. The proprietor of the soil usually receives two thirds of the gross produce in kind for his rent. The leases are long, and the intercourse between farmer and tenant is generally mild and liberal. It is on the slope of Vesuvius that the *Lacryma Christi* is grown. This, which is a red luscious wine, is better known by name than in reality, very little of it being produced, and that little being principally reserved for the royal cellars. The *vino Greco* and the muscadine wines of Vesuvius, are also deservedly celebrated. (*Simond's Italy*, p. 421, 422; *Henderson on Wine*, p. 240.)

Vesuvius, being so near Naples, is usually visited by strangers resorting to that city. M. Simond gives the following notice of his ascent to the mountain in 1818: "We left Portici, ascending gradually among cultivated fields and vineyards occasionally traversed by streams of old lava, black, rough, and sterile; and in 1½ hour reached the Hermitage, a convent where a few monks keep a sort of inn for the visiters of Vesuvius. Further up, we traversed large fields of lava, extremely rough; and at the base of the cone prepared for the ascent over a heap of crumbling ashes and cinders, extremely steep, of course, as it formed an angle of nearly 45°. In about one hour, stoppages included, we found ourselves on extremely hot ground, intolerable to the hand, and fatal to the soles of our shoes; it teemed with hot vapours, and was covered with beautiful efflorescences of sulphur. Smoke issued from numerous crevices; at the entrance of which a piece of paper or a stick took fire in a few seconds; and, what seems strange, a stone thrown into one of these openings increased the smoke at all the others. Stooping low, we could hear a noise very like that of a liquid boiling. The hard but thin crust upon which we stood appeared to have settled down in some places; a woeful indication of its hollow state. After a few steps more we came to the edge of a prodigious hole, on the very summit of the cone, being the crater formed by the last eruption, four months previously. This hole was not by any means the tremendous thing we expected—a fathomless abyss, fiery and black, with lava boiling at the bottom—but a slope of grey ashes and cinders, much like that by which we had ascended, or scarcely more precipitous, and ending, at the depth of 400 or 500 ft., in a level place, with grey ashes like the rest." The view from the summit is far inferior in extent and magnificence to that from Etna, but is, notwithstanding, extremely various, rich, and beautiful. The whole ascent and descent to and from Naples may be readily accomplished in seven or eight hours. (*Simond's Trav.*, p. 421-423; *Eustace, Class. Tour*, iii., 22-25; *Starke's Guide, &c.*)

From the period of the earliest records down to the days of Titus Vespasian, the volcano seems to have been inactive; the appearance of its crater and its cavernous structure being the only indications by which Strabo conjectured that it might at some distant period have been on fire! But in the first year of the reign of Titus (A.D. 79), the volcano that had so long been dormant burst forth with renewed and tremendous energy, in one of the most destructive eruptions of which history has preserved any account. The large and flourishing cities of Herculaneum and Pompeii, near the sea, were both overwhelmed by its lavas and ashes! Even the figure of the coast was materially changed; and for the space of more than 1600 years all trace of the buried cities was completely lost, and they were only accidentally discovered in the course of last century. The elder Pliny lost his life during this dreadful eruption, which has been described by the younger Pliny, by whom it was witnessed (*Epist.*, lib. vi., 16 and 20), and by Tacitus. *Luctum*, says the latter, *attulit atrox et continuus tremor terræ, quem secuta est horrenda Vesuvii montis conflagratio. Pulcherrima Campaniæ ora miseri fedata: obruta duæ urbes Herculanium et Pompeii: vasta hominum strages, quos inter periere Agrippa ejusque mater Drusilla. At studiorum famâ mors C. Plinii fuit insignior.* (App. Chron.)

Since the destruction of Herculaneum and Pompeii there have been 45 authenticated eruptions; but, luckily, none of them have been equal to it in destructive power. Of those which happened down to the 12th century, we have few accounts; and from 1138 to 1631 there were but two slight eruptions: during this interval, however, Etna was in a state of great activity, and the formation of monte Nuovo, &c., in the Phlegræan Fields took place. In 1631 a violent eruption occurred, during which seven streams of lava poured from the crater; and from 1666 to the present time, there has been a series of eruptions, the longest intervals between them having rarely exceeded ten years: the last was in January, 1834. The energy of Vesuvius, when in action, is extremely great, and the spectacle highly magnificent and sublime. In the eruption of 1779, Sir W. Hamilton says that jets of liquid lava were thrown up to the height of at least 10,000 ft.! having the appearance of a

column of fire; and in that of 1793, according to Dr. Clarke, millions of red-hot stones were shot into the air to full half the height of the cone itself, and then bending, fell all round in a fine arch, covering nearly half the cone with fire. The lava, however, does not always issue from the crater at the summit, but, as in Etna, sometimes from small cones raised in various parts of the declivity; and occasionally three or four of these cones are in a line, which generally points towards the great crater. The eruptions of 1760, 1794, and 1834, were of this description. (For further details, we beg to refer the reader to *Sir W. Hamilton's Observations on Vesuvius*, and his *Campi Phlegræi*, and to the works of *Clarke, Scrope, Dufrenoy, &c.*, with those of *Maclaren, Simond, Eustace, Rampoldi, &c.*)

VETERAN, p. t., Chemung co., N. Y., 11 m. N. Elmira, 190 m. W.S.W. Albany. Drained by the inlet of Seneca lake, which affords water-power. It has five stores, five tanneries; seven schools, 468 scholars. Pop. 2279.

VEVAY, p. t., capital of Switzerland co., Ia., 94 m. S.E. Indianapolis, 544 W. Pleasantly situated on the N.E. bank of Ohio. It was early settled by emigrants from Switzerland, who successfully attempted the cultivation of the vine, and it presents a pleasing assemblage of gardens, orchards, and vineyards, where taste and elegance are combined with use and comfort. It contains a courthouse of brick, a jail, seven stores, a printing-office, 200 dwellings, many of them brick, and about 1200 inhabitants. It presents a very pleasant appearance to those navigating the Ohio, and is regarded as one of the most interesting sites on the river.

VIATKA, a government of European Russia, chiefly between the 56th and 60th degs. of N. lat., and the 46th and 54th of E. long., having N. Vologda, E. Perm, S. Orenburg and Kasan, and W. Nijni-Novgorod, and Kostroma. Area estimated at about 52,900 sq. m. Pop., in 1838, 1,511,600. The slope of the country is towards the W. and S, in which directions the Viatka, a tributary of the Wolga, flows, traversing the government nearly in its centre. The Kama, which forms part of its E. and S. boundaries, also rises in this government. Surface generally undulating, and even mountainous towards the E., where it consists of the lower Ouralian ranges. The soil is mostly good, though encumbered in parts with extensive marshes. Climate severe in winter, but not usually unhealthy. Agriculture is the principal occupation of the inhabitants, particularly along the banks of the large rivers; and in ordinary years an excess of corn is grown. Rye, barley, and oats, are the principal grains; very little wheat is raised, but pease, lentils, and buckwheat, are grown, with large quantities of hemp and flax. The surplus produce goes chiefly to the N. Russian provinces. Potatoes are not much cultivated. Fruit is not plentiful; apples scarcely ripen. The forests are very extensive, they consist mostly of firs, intermixed with oak, elm, alder, lime, birch, and other trees. Cattle breeding, though a secondary branch of industry, is still of importance; and a good many small but robust horses are reared. Sheep are few. Furs, tar, iron, and copper, are among the chief products. Manufactures, though not extensive, appear to be on the increase: in 1832, there were 62 factories for woollen cloths, linen and cotton stuffs, paper, soap, potash, copper, and iron wares, &c., employing between 6000 and 7000 hands. About two million archines of woollen, and perhaps nearly double that quantity of linen cloth, are supposed to be annually made in the houses of the peasantry; and large quantities of spirits are distilled. Near Sarapoul is an extensive manufactory of arms, and at Votka anchors, gun carriages, and iron machinery of various kinds are made on a large scale. The government exports corn, flax, linseed, honey, tallow, leather, firs, silk goods, iron, and copper, to Archangel, and corn and timber to Sarasof and Astrakhan. It receives manufactured goods from Moscow and Nijni-Novgorod, tea from Irbit, and salt from Perm. Viatka, the capital, is the great emporium of the trade. It is subdivided into 11 districts; Viatka, Slobodskoi, and Sarapoul, being the chief towns. The inhabitants consist of various races; Russians, Votiaks (of a Finnish stock, and from whom the prov. has its name), Tartars, Baschkirs, Teptiars, &c., professing many different religions. The Mohammedans are estimated at nearly 50,000; and the Shamanists and idolators at 3500. In 1831, there were only nine public schools, in which, 1153 pupils were receiving instruction; but the number has since materially increased. This government is united under the same governor general with Kasan: but the Tartars and Finns are subordinate to the jurisdiction of their own chiefs.

VIATKA, a town of European Russia, cap. of the above gov.; on the Viatka, near the confluence of the Teheptsa, 230 m. W. by N. Perm, and 250 m. N.E. Nijni-Novgorod. Pop. 6890. (*Possart.*) It has several stone churches, one of which, the cathedral, with a silver altar ornamented with bas-reliefs, cost 130,000 roubles. Here are numerous convents, with an episcopal seminary, gymnasium, and high school, founded in 1899. It was annexed to Russia by Ivan Vasiliewitch, about the middle of the 16th century. (*Schnitzler, La Russie; Possart, Das Kaiserthum Russland, &c.*)

VICENZA (an. *Vicentia*, or *Vicetia*), a city of Austrian Italy, cap. deleg. of its own name, on the Bacchiglione, where it receives the Retrone, 26 m. E.N.E. Verona, and 37 m. W. by N. Venice. Pop., in 1837, 20,688. (*Berghaus*.) Though surrounded by dry moats and dilapidated walls, it is one of the best built cities of Italy. It has an astonishing number of well-designed houses, many of which are of very fine architecture; and even those which are less deserving of praise would, from their number and the richness of their ornaments, produce a great appearance of magnificence in the city, if they were well kept up; but they appear forlorn, neglected, and half uninhabited. (*Wood's Letters, &c.*, i., 238.) Vicenza, says Forsyth, "is full of Palladio," the modern Vitruvius, born here in 1518, who has lavished all his skill on his native place. Besides about 20 palaces, the townhouse, or *basilica*, the church of Sta. Maria dei Monte, the Rotunda, the Olympic triumphal arch leading to the Campo Marzo, the theatre of the Olympic academy, &c., are the works of this architect. The most celebrated of these is the Olympic theatre, a noble edifice, constructed upon the plan of the ancient theatres, and bearing a great resemblance in all essential particulars to those of Herculaneum and Pompeii. It is fully described by Eustace, who states that the palatial edifices of Vicenza, though inferior in materials and size to those of Genoa, are much superior in external appearance. (*Class. Tour.*, i., 132-5.) The Vicentine villas, which have been often imitated in England, are as beautiful as the larger buildings within the city. Many of them are on the Monte, a pleasant hill adjacent to the town, the favourite resort of the Vicentine gentry, and whence an extensive and rich view is obtained of the great plain of Lombardy. Vicenza has few Roman antiquities, and not many interesting specimens of middle-age architecture. The cathedral has a front exhibiting a mixture of different styles, and its interior presents a nave only, which is of great width; neither the length nor height being in proportion. The church of St. Corona has a fine "Adoration of the Magi," by Paul Veronese; the church of St. Lorenzo is now converted into a barn. All these buildings are in the pointed style, which prevailed in this part of Italy during the 13th century, and of which the church of St. Anastasia at Verona is one of the finest examples. (*Woods*.) Nine bridges cross the different rivers at Vicenza, one of which, the *Ponte de Sammichele*, is by Palladio, and, according to Rampoldi, may be compared with the Rialto at Venice.

Vicenza is a bishop's see, the seat of the council, and of the superior courts for the deleg., and has a lyceum, two gymnasiums, an ecclesiastical seminary, and many inferior schools, eleven hospitals and orphan houses, a government pawn-bank, public library, societies of agriculture, &c. The Olympic academy was founded in 1555, for the encouragement of polite literature, and still, as formerly, includes the most respectable citizens. (*Eustace*.) Some of the palaces have a few fine paintings; but they were mostly carried off by the French.

The Vicentines are said to manifest an aptitude for manufactures; and are, perhaps, inferior in industry only to the inhabitants of Verona. They weave silk and woollen fabrics, and make leather, whalebone articles, earthenware, hats, gold and silver articles, fire-engines, &c., and have a considerable trade in agricultural produce. "As you enter the Vicentine territory," says Mr. Rose, "you may observe a visible improvement in the mode of cultivation. The fields are kept cleaner, and everything indicates superior industry and exactness. If we except the resemblance of dialect, and some community of trifling customs, Calais and Dover are not more unlike than Padua and Vicenza, long subjected to the same government, and connected by facilities of communication both by land and water. To say nothing of the outward appearances of the two cities, which present a most remarkable contrast, it would seem as if the inhabitants were of different blood—as if a colony of Venetians, making a knight's move, had leaped over Padua and established themselves at Vicenza." (*Letters from the N. of Italy*, i., 154, 155.)

Vicenza was anciently a Roman *municipium*, but one of little consideration. (*Tacit., Hist.*, iii., 8.) It was sacked by Alaric in 401, and successively pillaged by Attila, the Lombards, and the emperor Frederick II. Early in the 15th century it came into the possession of the Venetians, who held it till the downfall of the republic in 1796.

Napoleon conferred the title of Duke of Vicenza on Caulaincourt. (*Rampoldi; Austr. Nat. Encyc.; Woods; Forsyth; Eustace, &c.*)

VICH (an. *Ausona*), a town of Spain, in Catalonia, prov. Barcelona, in an undulating plain, 36 m. N. Barcelona. Pop. 12,500. (*Miñano*.) It is of a very irregular figure;

some parts of it are well built, and two of its squares t e handsome. The cathedral is very mediocre, and is, indeed, inferior in many respects to the other churches. It has numerous convents, a seminary, college, and several hospitals, with manufactures of linen and hempen fabrics, printed cottons, woollen cloths, hats, and leather. (*Miñano, Dict. Geog.*)

VICKSBURG, city, capital of Warren co., Miss. 41 m. W. by N. Jackson, 1051 W. Situated on the E. side of Mississippi river, 400 m. above New-Orleans, and immediately below the Walnut hills. The city is situated on the shelving declivity of high hills, and the houses are scattered in groups on the terraces. It is surrounded by a fertile country, and, though recent in its origin, it has become a large and flourishing place. A number of boats are always lying in the harbour, and a great quantity of cotton is shipped here. It contains a courthouse, jail, four churches, a Presbyterian, Episcopal, Methodist, and Roman Catholic three academies, two male and one female; 50 wholesale grocery and commission stores, 50 retail dry goods stores, a printing-office, and about 3104 inhabitants. Steamboats regularly ply between this place and New-Orleans, and a railroad extends from Vicksburg, through Jackson, to Brandon.

VICTOR, p. t., Ontario co., N. Y., 12 m. N.W. Canandaigua, 351 m. W. by N. Albany, 285 W. Drained by Mud creek. It contains three churches, a Presbyterian, Methodist, and Universalist; three stores, two fulling-mills, two woollen factories, three flouring-mills, four grist-mills, six saw-mills, one tannery, one distillery; one academy, 25 students; 12 schools, 671 scholars. Pop. 2283.

VICTORY, p. t., Cayuga co., N. Y., 20 m. N. Auburn. 169 m. W.N.W. Albany, 259 W. Drained by Sodus creek. It has four stores, one fulling-mill, two grist-mills, six sawmills, two tanneries; one academy, 20 students, 13 schools. 856 scholars. Pop. 2271.

VIENNA (Germ. *Wien*, Lat. *Vindobona*), a city of Germany, cap. of the Austrian empire, prov. Lower Austria, on an arm of the Danube, where it is joined by the small rivers Wien and Alster, 190 m. E. Munich. 339 m. S.S.E. Berlin, and 800 m. N.W. Constantinople. Lat. of the observatory, in the centre of the city, 48° 12' 33" N.; long. 16° 22' 45" E. The pop., which in 1815 was about 250,000, had increased, in 1840, including the garrison of about 15,000 men, to 357,927, and may now (1843) be estimated at above 360,000.

The Danube, opposite Vienna, is divided into three or four separate arms, the most southerly of which washes the walls of the city. Between the third and fourth of these arms, however, is the important suburb of Leopoldstadt, with the Prater, the Augarten, and several other favourite promenades. This part of Vienna communicates with the city and the suburbs on the S. side of the Danube, by five bridges, of which the Ferdinands Brücke, in the centre, is the chief. Vienna stands in a plain, elevated about 529 feet above the level of the sea; but so little above that of the Danube in this part of its course, that, with the exception of its S. extremity, which is on the gradual ascent to the heights of Kahlenberg, most part of the city is liable to inundations. Vienna is of a nearly circular form, being about 10 m. in circumference. The city-proper, in the centre, is, however, scarcely 3 m. round. It is enclosed by ramparts of brickwork, and a beautiful glacis from one to three furlongs broad, planted with trees, laid out in public walks, forming, like the parks in London, the lungs of the metropolis; these separate the city from its numerous suburbs, which, on the S. side of the Danube are again enclosed by a line of ramparts, originally thrown up in 1703, when the passports of travellers are demanded.

Vienna, from its size, wealth, population, and activity, deserves to be compared with London and Paris better than any other European capital. Its chief points of external difference from these cities are that it preserves about it more antique grandeur, and that it is the old and not the new parts of the town which form the fashionable quarters. Most part of the principal edifices are within the city, where the houses are usually four or five stories high, and the streets irregular, narrow, and dark; but where the emperors family and most of the nobility reside. Nearly all the best shops, principal hotels, &c., are also in this quarter. In the suburbs, however, are several of the palaces and garden villas of the higher nobility, including those of Prince Liechtenstein, Esterhazy, Schwartzenberg, Auersberg, Metternich, &c.; the Belvidere palace, built by Prince Eugene, but appropriated by Joseph II. to the imperial picture gallery, and other public institutions; with immense barracks, magazines, and other military establishments, hospitals, &c. The streets in the suburbs are generally broad and straight; but some of them, being unpaved, are in wet weather dirty and muddy. The thoroughfares in the city proper are, on the contrary, uniformly clean, and well paved; but no part of the capital has as yet the advantage of footpaths.

VIENNA.

The houses, both in the city and suburbs, are in general huge edifices, and, as in Paris, are built around court-yards, and occupied by many different families. Some of these dwellings are of enormous extent, and quite towns in themselves. Prince Esterhazy has one comprising 150 different sets of apartments, and yielding a revenue of £1600 to £3000 a year: and one belonging to the Stahremberg family is said to be inhabited by 3000 persons, and to produce a rental of £4000 a year! (*Russell, Tour in Germany.*) The Burgher Spital, formerly a hospital for citizens, was converted by Joseph II. into a dwelling-house: it is six stories high, has 10 courts and 90 staircases; and several other houses are of equally colossal dimensions. No city in Europe has so large a number of resident nobility as Vienna: 94 families of princes, 70 of counts, and 60 of barons make it their home for the greater part of the year. These nobles may perhaps have fortunes of from 100,000 to 500,000 florins a year, and several, as Prince Esterhazy, Liechtenstein, &c., considerably more. Here, also, many private gentlemen spend 50,000 florins a year; and, with the exception of those of London, the citizens of Vienna are the richest in the world. Berlin and Dresden may, perhaps, have more cornices, pillars, and handsome public buildings, and in Munich and Paris these may have a more imposing effect, but in none of these capitals are there so many noble and extensive private edifices. The Herrengasse and other streets near the imperial residence are full of palaces of the higher nobility. These, as in London, frequently extend along narrow thoroughfares, and are not distinguished from humbler residences except by their greater size and elevation; their interiors are sumptuous.

Nearly all the so-called squares of Vienna are within the city. They are irregular, and comparatively small open spaces, none being so large as Waterloo-place: the cathedral stands in the centre of St. Stephen's platz, and the Graben is an incessant thoroughfare, and may be looked upon as the Charing cross or Mansion-house place of Vienna. Most of these open spaces are ornamented with one or more monuments, or fountains. These, however, are not always in good taste. In the Joseph platz is a fine equestrian statue of Joseph II., by Zanner. "The emperor, the resemblance to whom is said to be very striking, is attired in the Roman costume, and crowned with laurel; with one hand he curbs the impetuosity of his steed, and the other he extends to his people. The statue reposes on an elevated pedestal of granite, bearing the inscription *Sa luti publico vixit non diu sed totus*; and which, with its attendant pilasters are adorned with medallions, representing, not the remarkable events of his life, but his travels; it was erected by the late emperor, Francis II." (*Spencer's Germany, &c.*, ii., 134.) But, except this, Vienna has no other statue of her great men or benefactors; there is none of Montecuculi, Prince Eugene, Lacy, Laudohn, Louis of Baden, or John Sobieski; none of Daun or Kaunitz, Von Swinten, Mozart, or Haydn. In respect of such memorials, Berlin, and even London, are far before Vienna. In her bridges Vienna is also immeasurably behind London and Paris, having none worth notice. The Danube is here nowhere much more than 60 yards across; being, also, a sluggish and muddy, though a navigable stream. The Wien is little better than a mere filthy ditch. The drainage of the town is effected by good underground sewers.

Public Buildings, &c.—The chief of these is the cathedral of St. Stephen, almost in the centre of the city, and from which the principal thoroughfares diverge. It is an elegant Gothic building, ranking in elevation and richness of architecture with the cathedrals of Strasburg and Antwerp. Its length is 350 ft., and its greatest breadth 230 ft. Flanking its great W. doorway are two towers, the remains of the original church, consecrated in 1163 and at the angles of this front are two magnificent piles of a similar kind, though only the most southerly has been finished. This tower and spire is 450 ft. in height, or barely 16 ft. lower than that of Strasburg; it has a bell weighing 357½ cwt., cast from cannon taken from the Turks, and declines towards the N. about 3 ft. from the perpendicular. The exterior of the cathedral has a good deal of rich tracery. Within are some good wooden carving, a few good pictures, the monuments of Prince Eugene, the emperor Frederick III., &c., and a gorgeous chapel of the Liechtenstein family; but, on the whole, its interior is but little decorated. A crypt beneath it served for three centuries as the burial-place of the imperial family: at present, however, only parts of their viscera are preserved here; their hearts being deposited in the Augustine church, and the rest of their bodies in that of the Capuchins. The church of the Augustines is one of the handsomest in Vienna, and contains the monument of the archduchess Christine, one of the finest works of Canova; besides those of Leopold II., Daun, Von Swieten, &c. The church of St. Charles Borromeo is an imposing edifice, in the Byzantine style; Metastasio is buried in that of St. Michael, and the Carmelite

church has some fine stained glass. Vienna has, in all, nearly 60 churches, a third part of which are in the city; 17 conventual establishments, a Scotch Lutheran, three Greek churches, and two synagogues.

The *Burg*, or imperial palace, occupies a large extent of ground in the S.W. quarter of the city. It is externally a gloomy and shapeless congeries of buildings, erected from the 14th to the 17th century, on a par, Mr. Spencer says, in point of architecture, with St. James's. It comprises extensive suites of rooms; though these, in the simplicity of their furniture and decorations, show the unostentatious habits of the Austrian princes. The state-apartments, with their ancient gilding, and faded velvet hangings, remain in the same condition as in the time of Maria Theresa. But the palace has some fine collections in art and science. The imperial library, which comprised, according to Balbi, in 1835, 284,000 printed volumes, and 16,000 MSS., is placed in a handsome edifice built for the emperor Charles VI., whose statue, with that of many other Austrian monarchs, is placed in the centre of the grand hall, an apartment 240 ft. in length, by 45 in width and 62 in height, with a fine dome rising 30 ft. above the ceiling. The library increases by about 3500 volumes a year, a copy of every work published in the empire being deposited here; besides which, a fund of nearly £9000 a year is spent in the purchase of new works. This library is open to the public without introduction for five hours a day. It has, among other curiosities, an act of the Roman senate prohibiting the *bacchanalia*, engraved on bronze, and bearing date A.U.C. 567 (or B.C. 186); the 5th decade of Livy, a unique MS.; the Peutingerian Table, a military map of the Roman empire in the 4th century; several MSS. of succeeding centuries, the earliest book printed with a date; the MSS. of the *Gerusalemme Liberata, &c.* Here, also, are collections of music and engravings, the last comprising about 300,000 pieces. The museum of antiquities comprises a cabinet of medals, second only to that of Paris, and an unrivalled collection of intaglios and cameos. One of the latter, representing the apotheosis of Augustus on an enormous sardonyx, is supposed to be the finest existing: and the coins and medals amount to 80,000, including 18,000 Greek, and 23,000 Roman. (*Turner; Murray's Handbook.*) The collection of ancient sculpture is far inferior to the collections of either Dresden or Munich; but there are excellent museums of natural history and botany, and the cabinet of minerals surpasses every other in Europe. Here are also Egyptian and Brazilian museums, a good collection of Greek vases, and the imperial jewel-office; in which, including a number of relics, are the Austrian and Hungarian regalia, the Florentine diamond, the iron crown sceptre. &c., of Charlemagne, the sword of Tamerlane, &c.

The Belvidere palace is appropriated to the Ambras museum, and to one of the finest collections of paintings in Europe, being especially rich in works of the Flemish and German schools. The Ambras museum, formed late in the 16th century, includes, besides other curiosities, a most interesting historical collection of armour. The paintings in the imperial gallery are classed in separate rooms, according to schools. In those of the Italian schools are the famous *Ecce Homo* of Titian, formerly belonging to Charles I. of England; a superb Holy Family by Raphael; many other pictures by these artists, and by P. Veronese, the Caracci, S. Rosa, &c. In those of the Flemish school are three masterpieces by Rubens: St. Ignatius driving out evil spirits; St. Ildefonso; St. Ambrosius closing the church door at Milan against the Emperor Theodosius; some of the best works of Rembrandt and Vandyck; and pictures by Teniers, Cuyp, G. Dow, &c. Other rooms are appropriated to the German, Austrian, and Spanish schools, works of the middle ages, a comparative series of Italian paintings from the 14th to the 9th century. In the Belvidere gallery is the mosaic copy of Da Vinci's Last Supper, for which Napoleon engaged to pay 15,000 zecchinos, and which was afterwards bought for the same sum by the late Emperor Francis II. At Dresden," says Mr. Turnbull, "the gallery comprises perhaps the grandest *ensemble* in Europe, but is so neglected, so involved in gloom and dirt, as to afford too often a feeling more akin to pain than to pleasure. At Berlin, the condition, care, and arrangement, are perfect, but the works themselves are rarely first-rate specimens. The gallery of Vienna is good alike in intrinsic excellence, in order, and in condition. Of the museums generally, as, indeed, of most of the institutions under the Austrian government, the high and eminent excellence is their admirable adaptation to practical utility. In those of other countries we had seen articles of greater individual rarity; entire assemblages of certain branches, more copious and complete; but in no one were the various objects, to our apprehension, so ably and lucidly arranged, labelled, described, and exhibited, as at Vienna; and this, too, in a city where space and light are so defective. They are fully exhibited to the public, during a convenient number of hours; and the student has ample opportunity of following up his researches therein, in connexion with lectures gratuitously afforded on the principal branches of science." (*Turnbull's Austria*, i., 236–238.)

The imperial arsenal has one of the richest armouries in Europe. In the upper rooms 150,000 stand of arms are kept; and, besides a large store of weapons and armour of different dates, we have here the buff coat worn by Gustavus Adolphus at the battle of Lutzen, the arms of Marlborough, Eugene, Stahremberg, and Montecuculi, numerous standards, the enormous chain thrown across the Danube by the Turks in 1529, &c. The city arsenal is a fine building, constructed by the citizens at their own expense, and has, with many curiosities similar to the above, arms sufficient for 25,000 civic guards. The imperial riding-school is also a handsome edifice by Fischer of Erlach, but lost among the buildings of the palace. The Royal Stallung, in the suburbs facing the Burg-thor, is a noble palace appropriated to the Royal Hungarian guard.

Vienna has five theatres; the principal are, the Hof-theatre attached to the palace, and that at the Kärnthner-thor (*Carinthian-gate*). The first is devoted solely to the performance of the regular German drama; and, though not the largest, is by far the finest theatre in Vienna. It is both clean and well lighted, and is said somewhat to resemble Drury Lane. The acting here is at least equal to that at Berlin; and the performers have, after ten years' service, a handsome pension settled on them for life by the government, with an annuity after their death for their widows. (*Strang's Germany*, ii., 253.) The Kärnthner-thor is the opera-house of Vienna, and the singers and orchestra are unsurpassed in Germany. This house is very large, having six complete rows of boxes and a half circle next the pit: but the largest theatre is one on the Wien, appropriated to equestrian pieces. The really national theatre of the Viennese is the *Beyn Caspert*, in the Leopoldstadt. This theatre, the Adelphi or *Ambigu Comique* of Vienna, is appropriated to farces, and is the arena on which the national character is painted in the most lively colours and broadest manner. Here, says Mr. Strang, one circumstance is noticeable, as indicative of the power of "the million," even in Austria. The police, though exceedingly strict in the regular theatres, are said to wink hard at the political jokes that are frequently cracked on this stage; while the pulse of the public is not unfrequently felt here, by somewhat the same means as the old council of ten used to adopt at Venice, through the tricks and colloquies of Punchinello. (*Germany in* 1831, ii., 255, 256.)

Vienna has several handsome gates, the chief of which is the Burg-thor, near the palace; but none is comparable for magnificence to the Brandenburg gate, Berlin. Indeed, every object brought under public observation appears externally more splendid, elegant, and attractive in the Prussian capital; "yet it is merely as shadow to substance, for in Vienna, behind dark walls, there is far more sterling value than in the finest palaces of its rival." (*Spencer's Germany*, &c., ii., 138.)

Schools, Libraries, Galleries, &c.—Vienna has a university, founded in 1237, but which was wholly remodelled by Von Swieten in the time of Maria Theresa. It is celebrated on the continent as a school of medicine, and is probably attended by a greater number of students than any other German university, except that of Berlin: in 1832 it had 1619 students, of whom 309 studied divinity, 332 law, 519 medicine and surgery, and 459 philosophy, &c. (*Journal of Education*, vol. ix.) There are between 70 and 80 professors, all of whom are paid by government, and are neither permitted to receive fees on their own account, nor to give private lessons. The theological, surgical, and veterinary courses are delivered gratuitously; but the student has to pay a fee of 18 florins (about £1 11s. 6d.) for attendance on the lectures in philosophy, and of 30 florins (£3 12s. 6d.) for attending those in medicine and jurisprudence. The whole amount of the moneys thus paid for tuition during the session is expended in stipends to indigent students, and divided among them, without reference to their religious creeds, in allowances varying from 50 to 150 florins (£4 10s. to £13 10s.) Nearly all the lectures are delivered in the German language. The university has a library of above 100,000 volumes, and £150 a year is expended in the purchase of new works, and it receives, gratis, a copy of all works printed in Lower Austria. (*Journ. of Education*, 1834; *Statistical Journal*, 1841.) An observatory, and a botanic garden, are attached to this establishment.

The Polytechnic Institute, a handsome structure facing the glacis, was founded by the Emperor Francis in 1816, to afford instruction in the practical sciences, arts and commerce; and a few years since had about 750 pupils and 35 masters. Besides the ordinary branches of knowledge, the pupils are taught the history of commerce, the knowl-

edge of merchandise, mercantile law, and correspondence, natural history and chemistry as applied to commerce, the art of drawing, mathematics, &c.; for which instruction the pupils pay only three florins a month, and, for a small extra sum, are taught Latin, English, French, and Italian. Among other collections, this school has a museum of the products of arts and manufactures, both Austrian and foreign, and a valuable library. The *Theresianum*, for the sons of the aristocracy, and the normal school of St. Anne, were both established by Maria Theresa. The former was suppressed by Joseph II., but restored by Francis; and it has now a library of 30,000 printed volumes, besides MSS. and pamphlets. Joseph II., established both the Oriental academy and the *Josephinum*; the latter, an institution for the education of army surgeons, which has attached to it a hospital capable of receiving 1200 patients, a collection of anatomical figures in wax, by Fontana, &c. Besides these establishments Vienna has a special seminary for the education of the secular clergy, a Protestant seminary, founded in 1821; six military colleges, with nearly 1000, and 49 minor establishments for military education, with nearly 3000 pupils; an academy of the fine arts for about 1300, and a musical academy for 200 students; besides about 60 inferior public schools. (*Journ. of Educ., &c.*)

In addition to the libraries already mentioned, the Archduke Charles has one of 25,000 volumes, Prince Liechtenstein of 40,000, Prince Metternich of 23,000 do., Prince Esterhazy of 20,000, exclusive of many inferior collections. But with all these appliances for knowledge, Vienna cannot be considered so much a resort of learning as of the fine arts. In paintings she is, perhaps, the richest capital of Europe. The gallery of Prince Liechtenstein consists of 25 splendid apartments, filled with exquisite originals of the Italian, Flemish, French, and Dutch masters, including *chefs d'œuvre* of Raphael, Guido, Rubens, Vandyck, Domenichino, Guercino, Claude, S. Rosa, Carlo Dolce, &c. The Esterhazy gallery contains upwards of 600 pictures, of which 54 are by Spanish masters, whose works are rarely found out of Spain; with a collection of sculptures, including works by Canova and Thorwaldsen; 50,000 engravings, &c. Counts Czernin, Schönborn, Harrach, Lemberg, and many other noblemen, have collections of choice paintings; and in the palace of the Archduke Charles is a collection of 160,000 engravings. The foregoing galleries are all open to the public at stated times. (*Balbi, Essai sur les Bibliothèques de Vienne*, p. 84–113.)

In statuary, also (though not in public statues of celebrated men), Vienna is very rich. Canova's group of Theseus killing the Centaur deserves especial mention. It was originally intended by Napoleon to surmount the grand arch at Milan, but is now placed in the Theseum, a Doric temple, on the Volks-garten, in imitation of the temple of Theseus at Athens. "This group is of Carrara marble. The hero is in the act of grasping with his left hand the throat of the Centaur; while his right arm, raised behind his helmeted head, clenches the club with which he prepares to inflict the fatal blow. The whole character of the group is in Canova's most effective style." (*Turnbull*, i. 242.)

Hospitals and other Charities.—Few capitals are so abundantly furnished with charitable institutions as Vienna. Many of the principal, as the general hospital, house of invalids, deaf and dumb asylum, &c., were founded by Joseph II. The general hospital is a vast building, ranged around seven quadrangles, having 2000 beds. It is said to receive annually from 10,000 to 16,000 patients. It partly answers the purpose of a sanatorium, there being separate bed-rooms, which, with medical attendance, and every comfort necessary for an invalid, are within the reach of persons of limited income, on the payment of a small sum daily. The hospital of the charitable brethren, supported partly by voluntary contributions, in a monastic establishment, but open equally to Jews, Turks, and Christians of all persuasions.* The house of invalids is similar in its kind to Chelsea hospital, having been founded for 600 old soldiers. In its great hall there are two large pictures of the battles of Leipsic and Asperne. The deaf and dumb asylum is well conducted; and those among the pupils who evince intelligence are often afterwards employed in state affairs requiring secrecy. There are schools for the blind, &c., and a lunatic asylum, which is, however, said, to be not so well conducted as most of the other public establishments. Attached to the general hospital are the *maison d'accouchement* and foundling hospital. In the former of these "not even the name of the applicant is demanded; she may enter veiled or masked, and remain incognito the whole time she continues in the house; she

---

* It may be remarked here, that there is an increase among modern travels religious toleration in Vienna; at least in the Austrian government. Mr. Strang believes that there is a real inclination on the part of those in power in Austria to tolerate every kind of religious belief, provided it be not accompanied with offensive material, &c.

has merely to deliver a sealed paper to the superintendent, containing her name and real address, that, in the event of death ensuing, her relations may be apprised of her fate." (*Spencer*, 163.) The person who brings a child to the foundling hospital receives a ticket, by presenting which, the child may at any time be reclaimed: if it be not taken away, it is, at the proper age, brought up to some employment. It is probable that an institution of this kind may prevent a few cases of infanticide; but the mortality in this, as in all similar institutions, is quite excessive, and there can be no doubt that it acts as a powerful incentive to vice and immorality.

Commerce, Hotels, Shops, &c.—Vienna is the great emporium of the Austrian provinces N. of the Alps, and an important *dépôt* for the interchange of goods between E. and W. Europe. According to Becker, there were in Vienna, in 1834, 547 silk and velvet, and 340 cotton cloth manufacturers, 24 cotton-printing establishments, and many cotton mills. The porcelain manufacture of Vienna is among the most celebrated on the continent, and it has as imperial cannon foundry, and a manufacture of small arms, said to employ 500 workmen. Cutlery, watches, and jewellery, bronze and other metallic goods, meerschaum pipes, musical instruments, paper, chemical products, gloves, leather, hosiery, chocolate, and liqueurs, are among the other principal products: it has several large printers and music engravers. Many of the most wealthy mercantile houses belong to Greeks. The national bank of Vienna, established during the seven years' war, is divided into 51,820 shares. In 1837 it made a nett profit of 3,943,303 florins, being at the rate of 77 florins a share. It discounts, advances money on pledges, negotiates bills of exchange, &c., and is said to be very well managed.

The hotels are of two classes : living in those of first-rate excellence costs about one third more than in Paris; but those of the second class are very good of their kind. Lodgings are twice as dear in the city as in the suburbs, where a room tolerably furnished may be had for 5 florins a month. The *cafés* of this city are not decorated with the same splendour as those of Paris, but they are quite as much frequented, being resorted to in the evenings by both sexes of the middle classes, and at other times by gentlemen to play at billiards, or to smoke, which is not permitted in the streets. (*Murray's Hand Book, &c.*)

From their number being limited by government, the profits realised in the coffee-houses are great, and they frequently bear a value so high, that Russell mentions one on the Graben, for the privilege belonging to which a purchaser paid upwards of £3000 in addition to an extravagant price for the house itself. (ii., 945.) Vienna is well supplied with provisions of all kinds, which are generally cheap. House rent is said to be lower than in Paris; servants' wages are much less; furniture is still cheaper; and a pair of good Hungarian carriage horses, the keep of which will cost about £30 a year, may be bought for £40. "No town exhibits such an appearance of people living amidst plenty, such an absence of *uneasy classes*, and of anything that can represent poverty. The hackney coaches are as neat, clean, and showy as private carriages; the horses are generally in excellent condition. The shops, though in such narrow streets, are as *dashing* as those of London or Paris; and most of them have signs, with paintings almost worthy of museums. The booksellers' and picture shops are numerous and large ; and besides the literature of every state in Germany, you may find many popular books and the principal engravings published in England and France." (*Austria and the Austrians,* i., 49, 50.)

Parks, Amusements, &c.—The principal amusements of the Viennese are music, dancing, the theatres, and frequenting the Prater and other fine promenades which encircle the city. The Prater, Hyde Park or Champs Elysées of this capital, is handsomer than either, and may in fact be considered the finest public park in Europe. It is nearly 4 m. in length by half as much in breadth, being enclosed between two arms of the Danube. Besides the fashionable drives, the Prater contains a great number of coffee and ice houses, pavilions, shows, &c., and is generally filled with a throng of people, on Sundays and holidays. The glacis is studded in a similar manner with places of entertainment, and the Augarten and Brigitten-au, both N. of the Danube, and the Volksgarten, within the city, are promenades in much the same style. The dancing saloons, or public ball rooms, are not in general what can be called *fashionable* places of amusement, though the imperial family and higher nobility attend the balls in the *Redoutensaal* at the carnival and other times. They are, however, resorted to by great numbers of the middle and also of the upper classes, and one of the principal, the *Apollo Saal*, can accommodate with ease 10,000 persons. The music here is of a superior order, the celebrated bands of Strauss and Lanner, and others little inferior, being constantly engaged. [Mr. Russell says, "The Viennese take to themselves the reputation of being the most musical public in Europe, and this is the only part of their character about which they display much jealousy or anxiety. So long as it is granted that they can produce among their citizens a greater number of decent performers on the violin or piano than any other capital, they have no earthly objection to have it said that they can likewise produce a greater number of blockheads and debauchees." (*Tour, &c.,* ii., 271.) Nearly all the greatest modern composers, Mozart, Haydn, Beethoven, Gluck, &c., have composed their best works in or near Vienna.

Morals, &c.—Vienna has acquired the character of being the most dissolute capital in Europe. But without stopping to inquire whether it be entitled to this distinction, it is, at all events, a most agreeable place for a stranger. A liveliness and *bonhommie* pervades society ; in bustle and activity Vienna rivals London and Paris; and the pursuit of pleasure appears one of the main occupations of the great mass of the inhabitants. The peace of the city is preserved with the utmost care. The arrivals, departures, residences, &c., of strangers are carefully noted ; passports are strictly examined, and great care is taken that visiters shall show that they have the means of paying their way. With residents, however, the police interfere but little, and never obtrusively. Among the drawbacks on a residence here are, the furious driving in the crowded thoroughfares, through which pedestrians have to wind their way among heaps of fuel, the hewing of which is incessantly carried on before the doors of the houses ; the great variability of the climate, and the indifference of the water.

Vienna is an archbishop's see, the residence of the Protestant superintendent for all the S.W. provinces of the empire, the seat of the high judicial tribunals, and central bureaux of the Austrian dominion, of the court of appeal for the archduchy of Austria, and the provincial government of the province below the Enns. Though not in general famous as a seat of literature, it has, among many other associations, a literary society ; of which Von Hammer, the Orientalist, the poet Grillparzer, the historian Mailath, the novelist Caroline Pichler, the mineralogist Mohs, Balbi, &c., are members. Such is the influence of the censorship, that the city produces only two newspapers, and those worth little or nothing, with a few weekly scientific and fashionable papers, a monthly and a quarterly journal. The upper classes speak English, French, and Italian almost as well as their native language; and the *Times, Galignani's Messenger,* the *Edinburgh* and *Quarterly Reviews, Athenæum,* and other British and foreign journals, may be had with but little difficulty.

The *environs* are picturesque, but the roads around are very bad. About 2 m. from the city is Schönbrunn, the favourite summer residence of the emperor. It stands in a large park, stocked with deer and game of all kinds. The palace, built by Maria Theresa, is a vast monotonous pile, but richly furnished, and possesses many interesting portraits of the Imperial family. It was twice occupied by Napoleon : the vicinity of Schönbrunn was signed in it in 1809, and here the duke of Reichstadt, son of Napoleon, died in 1832. In the grounds are the *Gloriette,* a large columnar temple, from which a fine view is obtained ; a menagerie, a splendid botanic conservatory and gardens, with eating-houses, music and dancing-rooms, &c., for the public. Not far from Schönbrunn are Lacksenburg, Brühl, Baden, &c., frequented by pleasure parties from the metropolis, in much the same way as Richmond, Greenwich, or St. Cloud.

History.—Vindabona was remarkable in antiquity as the place where Marcus Aurelius expired. It was successively taken by the Goths and the Huns, and subsequently by Charlemagne, who placed it under the government of the margraves of the E. part of his dominion, thence called *Oester-reich,* and Austria. The margraves, afterwards dukes, held Vienna till the middle of the 13th century, soon after which it came into the possession of the house of Hapsburg. In 1484, it was taken by the Hungarians, whose king, Mathias, made it the seat of his court. Since the time of Maximilian I., it has been the usual residence of the arch-

it was relieved by Charles V., and on the second by John Sobieski of Poland, who totally defeated the enemy beneath its walls. In 1619 it was unsuccessfully blockaded by the Bohemian Protestants. In 1797, it was threatened by the French, but its siege was averted by the peace of Leoben. The French took it, however, in 1805 and 1809. The congress which parcelled out Europe into its modern subdivisions sat here from the 3d Nov. 1814, to the 9th June 1815. It suffered greatly from the cholera in 1831 and 1832. (*Penal.*

and the Germans, ii., 124-190; *Strang's Germany in 1831,* ii., 225-353, &c.)

VIENNA, p. t., Oneida co., N. Y., 3 m. N. Utica, 120 m. W. by S. Albany, 398 W. Bounded S. by Oneida lake and Fish creek. It has seven stores, one fulling-mill, one woollen factory, one grist-mill, 12 saw-mills; 22 schools, 822 scholars. Pop. 2530.

VIENNE, a dep. of France, reg. W., principally between the 46th and 47th degs. of N. lat., and long. 0° and 1° E., having N. Indre-et-Loire, E. Indre, S.E. Haute-Vienne, S. Charente, and W. Deux Sevres. Area, 676,000 hectares. Pop., in 1836, 292,731. It derives its name from the river Vienne, an. *Vigenna,* which rises in the dep. Creuse, and after traversing Haute-Vienne, a part of Charente, Vienne, and Indre-et-Loire, at first in a W., and afterward in a N. direction, enters the Loire after a lengthened course. Its principal affluents are the Thorison, Issoire, and Creuse from the E., and the Briance, Vaire, and Clain from the S. and W. Limoges, Confolens, Chatelherault, Chinon, &c., are on its banks. Nearly all the other rivers of the department are tributaries of the Vienne or of its affluents. Surface, mostly level; but in the S., a chain of heights separates the basin of the Loire from that of the Charente. The soil in the level ground is moderately good, but in the S. it is thin and chalky. In 1834, the arable lands were estimated to comprise 413,131 hectares; pastures, 42,732 do.; vineyards, 23,744 do.; woods, 80,372 do.; and health, wastes, &c., 75,167 do. Wheat and oats are the grains principally cultivated: rye and millet are raised for home consumption; but in years of scarcity, chestnuts are a principal resource of the population. From 500,000 to 700,000 hectol. wine are annually produced; but, on the whole, its quality is inferior, and large quantities are converted into *eau-de-vie,* frequently of great excellence. The white wines are the most extensively produced; the red wines are generally *très-colorés, durs, et âpres; ils se gardent long-temps, et s'améliorent en vieillissant: on en a vu qui conservaient encore leur qualité au bout de 40 ans.* (*Julien,* p. 150.) A good many cattle are reared, and the sheep in some of the cantons are said to be of a superior kind: the produce of wool is estimated at 400,000 kilogr. a year. About 45,000 hogs are said to be annually exported from this department, by way of the Atlantic quality. Bees and poultry are extensively reared. In 1835, of 190,518 properties, subject to the *contrib. foncière,* 72,589 were assessed at less than 5 francs, and 16,675 at from 5 to 10 francs: the number of those assessed at 1000 francs, or upward, was only 46. The vicinity of Chatelherault produces very superior lithographic stone; and marble, whetstone, millstone, &c., are found in other parts of the department. Some iron mines are wrought, and there are numerous iron forges. Chatelherault has rather extensive manufactures of fire-arms; and cutlery, lace, coarse woollen cloths and woollen yarn, paper, furs and skins, biscuits, beer, and vinegar are among the other goods made in Vienne. This department is divided into five arronds.; chief towns, Poitiers, the cap., Chatelherault, Civray, Loudon, and Montmorillon. It sends five members to the chamber of deputies. Registered electors, in 1838-39, 1799. Total public revenue, in 1831, 6,012,262 francs. According to the *Dict. Géog.,* " *L'instruction est arriérée dans ce pays. Le caractère dominant des habitans paraît être l'insouciance et l'inertie: ils sont invinciblement attachés aux usages, et aux préjugés de leurs pères.*" (*Hugo,* art. *Vienne; Dict. Géographique,* &c.)

VIENNE (an. *Vienna*), a town of France, dep. Isère, cap. arrond., on the Rhone, where it is joined by the Gere, the former being here crossed by a suspension bridge, 16 m. S. by E. Lyons. Pop., in 1836, 14,000, or, including its comm., 16,484. "Vienne, situated under a high cliff, with the castle upon its summit, is a striking and beautiful object in descending the river; and after passing it, there is a perfect union of the beautiful and the picturesque in its scenery. It is here also, about 1½ league after passing Vienne, that the vineyards lie so celebrated for their produce of *Cote-rotie.*" (*Inglis's Switzerland, &c.,* 179.) A handsome quay stretches along the Rhone; and the lower part of the town, on the high road between Lyons and Marseilles, has broad and well built streets; but the rest of the thoroughfares, along the narrow valley of the Gere, and up steep declivities, are ill laid out, and lined generally with mean houses. Vienne has several remains of Roman buildings, and other antiquities. In the centre of the town is a temple anciently dedicated to Augustus and Livia, having a good deal of resemblance to the *Maison carrée* at Nismes, though not in such good preservation. It has been used for a church, a club-house, and a tribunal of commerce, and is now appropriated to a museum of antiquities. Outside the town, and in much better preservation, is a pyramidal monument nearly 45 feet in height, and apparently a tomb. The traces of a bridge across the Rhone, an amphitheatre, a naumachia, theatre, &c., also exist. Here, also, are several middle-age antiquities, among which is the cathedral,

considered one of the best Gothic edifices in France. It stands in an elevated position; its grand entrance is ornamented with sculptures, and flanked by two high towers; the roof is supported by 48 lofty columns in the interior; the galleries have Gothic balustrades; and it has a fine monument of one of the archbishops of Vienne. The church of an ancient abbey is also worth notice. The other principal buildings are the cavalry barracks, corps-hospital, work-house, corn exchange, abattoir, and public library with 14,000 volumes.

Vienne has manufactures of woollen cloths, pasteboard, iron and copper plates, &c.; and near it are some argentiferous lead mines producing about 1500 quintals a year of metal. If was anciently a city of consequence, having been successively the capital of the Allobroges; of its province in Narbonnese Gaul, under the Romans; and of the first and second kingdoms of Burgundy: and in the early ages of Christianity, it was the see of the archbishop, primate of Gaul. It was united with Dauphiny to the French dominion by Louis XI. The famous council held in 1311, which abolished the order of the Templars, met in this town. (*Hugo,* art. *Isère; Dict. Géog.,* &c.)

VIENNE (HAUTE), a dep. of France, reg. W., between lat. 45° 22' and 46° 25' N., and long. 0° 35' and 1° 43' E., having N.W. and N. Vienne and Indre; E. Creuse; S.E. Correze; S.W. Dordogne; and W. Charente. Area, 554,956 hectares. Pop., in 1836, 303,192. The surface is hilly, particularly in the E., and the mean elevation of the dep. is estimated at between 1800 and 1900 feet above the level of the sea. The rivers, the principal of which are the Vienne (*see* previous art.) and the Gartempe, with their tributaries, have generally a W. direction. The soil, being mostly composed of the debris of granite, and other primary rocks, is, in general, of inferior fertility. In 1834, the arable lands were estimated to comprise 213,354 hectares; pastures, 129,899 do.; woods, 36,852 do.; and heaths, wastes, &c., 93,244 do. Wheat is but little grown; its place being supplied by rye, buckwheat, chestnuts, and potatoes. Very little wine is grown. The pasture lands are comparatively good; and, in 1830, the sheep in the department were estimated at nearly 610,000 head, and the cattle at 145,000 do. The wool produced in the department is, however, of inferior quality. Rural industry of all kinds is very backward. In 1835, of 56,733 properties, subject to the *contrib. foncière,* 25,483 were assessed at less than 5 francs and only 18 at 1000 francs, and upward. The fine *potter's clay* at St. Yriex is, perhaps, the most valuable of the mineral products; there is in the mine at Vaustry, the only one in France at which that metal is found; and copper, iron, lead, antimony, manganese, coal, &c., are met with in greater or less abundance. The manufacture of earthenware is the most important: and among its other products are iron and copper plates, cutlery, and other steel articles, nails, linens, woollen and cotton yarn, coarse woollen cloths, paper, leather hats, wooden shoes, &c. It is estimated, that 15,000 inhabitants of this department migrate annually as masons, sawyers, carpenters, &c., into the departments of Seine and Rhone, and the arsenals on the W. coast. Haute Vienne is divided into four arronds.; chief towns, Limoges, the cap., Bellac, Rochechouart, and St. Yriex. It sends five members to the chamber of deputies. Registered electors, in 1838-39, 1666. Total public revenue, in 1831, 5,162,677 francs. (*Hugo,* arts. *Vienne and Haute Vienne; French Official Tables,* &c.)

VIERZON-VILLE, a town of France, dep. Cher, cap. cant., on the Evre, near its junction with the Cher, in a fertile plain, 19 m. N.W. Bourges. Pop., in 1836, 4889. It consists principally of one street, which would be among the best in France, if furnished with footways. (*Guide du Voyageur,* &c.) Its houses are mostly elated. It has manufactures of woollen cloths, earthenware, and iron ware. Its castle was destroyed by Richard I. of England in 1196; and in 1356, it was pillaged by the army of the Black Prince. (*Dict. Géog.; Hugo;* &c.)

VIGAN (LE), a town of France, dep. Gard, cap. arrond., on the Arre, a tributary of the Herault, 40 m. W.N.W. Nismes. Pop., in 1836, 4686. The *Dict. Géog.* says that it is old and ill built; but according to the *Guide du Voyageur,* it is the pleasantest and most healthy of all the small towns in the Cevennes, and one to which the opulent inhabitants of Nismes and Montpelier resort during the heats of summer. In one of its squares has been erected a fine bronze statue of the Chevalier d'Assas, a native of the town. It has manufactures of cotton and silk hosiery, cotton yarn, leather, and paper.

VIGEVANO, a town of N. Italy, dom. of Sardinia, div. Novara, prov. Lomellina, cap. mand., on the Mora, near the Ticino, and 14 m. S.S.E. Novara. Pop., in 1838, 11,531. Its site is elevated, and it enjoys a salubrious climate. It is enclosed by walls, has an old castle, a cathedral which stands in a square surrounded on three sides by arcades, one of the best cavalry barracks in Piedmont, numerous convents, a hospital, government pawn-bank, a communal college, and a *sanatorium*, established in 1832. Near it is a large and handsome Dominican convent. The town has manufactures of silk stuffs, hats, soap, macaroni, &c.; two annual fairs of eight days each, and markets twice a week. This town gave birth to Francis Sforza II., duke of Milan, and is much indebted to the munificence of the Sforza family. Under the French it was the capital of an arrond., in the department of Agogna. (*Rampoldi, &c.*)

VIGO, co., Ia. Situated in the W. toward the S. part of the state, and contains 400 sq. m. Watered by Wabash river and its tributaries. It contained in 1840, 12,096 neat cattle, 8541 sheep, 31,730 swine; and produced 17,654 bushels of wheat, 718,485 of Indian corn, 2268 of buckwheat, 104,683 of oats, 33,706 of potatoes, 10,117 pounds of sugar. It had five commission houses in foreign trade, 54 retail stores, one furnace, two flouring-mills, nine grist-mills, 18 saw-mills, one oil-mill, four tanneries, four distilleries, one brewery, two printing-offices, two weekly newspapers; 29 schools, 1307 scholars. Pop. 12,076. Capital, Terra Haute.

VILLANOVA, p. t., Chautauque co., N. Y., 25 N.E. Mayville, 323 W. by S. Albany, 346 W. Drained by Conewango creek and its tributaries. It has two stores, two fulling-mills, two grist-mills, six saw-mills; 13 schools, 585 scholars. Pop. 1655.

VILLA-REAL, a town of Spain, in Valencia, prov. Castellon de la Plana, on the Mijares, here crossed by a fine bridge of 13 arches, within about 4 m. of the sea, and 33 m. N.N.E. Valencia. Pop., according to Miñano, about 8000. It originated in a country palace of James I., king of Aragon. It has one regular and well-built street; several religious edifices, a prison, a large suburb, and some silk and woollen manufactures, distilleries, &c. It was formerly fortified, and in the war of the succession was garrisoned for the archduke Charles; but, having been taken by the troops of Philip V., in 1706, its defensive works and most of its buildings were destroyed, and great part of its inhabitants put to the sword.

The town of the same name in Portugal, prov. Trasos-Montes, cap. Comarca, is said by Miñano to have 4100 inhabitants. (*Dict. d'Esp., &c.; Mod. Trav.*)

VILLA-RICA, a town of Brazil, cap. of the prov. Minas-Geraes, on the Ouro-preto, by which it is intersected, and which is here crossed by four stone bridges, 190 m. N.N.W. Rio Janeiro; lat. 20° 25′ 30″ S., long. 43° 19′ 23″ W. Pop. uncertain, fluctuating with the state of the mines; in the early part of the present century it was estimated at 90,000, principally whites, but is now, probably, a good deal less. It occupies an elevated site, but it has no very striking approach; nor, on a nearer view, does it present to the eye of a traveller any object corresponding with the grandeur of its name. It is situated on the declivity of a high mountain, forming part of an immense chain. Most of the streets range in parallel rows along the side of the mountain, being crossed by others leading up the acclivity. These have numerous public fountains, and the town generally is admirably supplied with water, which is conveniently conducted into almost every house. (*Mawe's Brazil, p. 238, &c.*) The streets are ill-paved; but there are many good houses two stories in height, built of stone, tiled, and whitewashed. The governor's residence, the town-hall, two parish churches, numerous chapels, the mint, college, and theatre, were, a few years ago, the principal edifices. Some of these are superior to the public buildings in most other parts of Brazil. The governor's house commands a view of nearly the whole town; and from it there is an open space surrounded by a sort of parapet, on which a few brass swivels are mounted. Several of the houses, &c., are richly ornamented. The mint is in the lower part of the town, attached to the treasury and custom-house. The climate of Villa Rica, owing to its elevated situation, is very agreeable; the usual range of the thermometer is from 64° to 80° Fah. in summer, and from 48° to 70° in winter. Thunder-storms, though common, are not violent. The gardens here, which extend in raised terraces along the side of the mountain, produce excellent kitchen vegetables; but beyond these, the vicinity of the town, notwithstanding its fertility, is wholly uncultivated, and the cattle and other stock are allowed to pasture at random. The markets are accordingly ill supplied; and when Mawe visited the country, most sorts of provisions and vegetables brought a very high price. The inhabitants, in fact, are chiefly interested in mining speculations; Villa Rica being, or, at all events, having been, the headquarters of the gold-mining district of Brazil. The metal, found in the mountain on which the town is built, is imbedded in a matrix of slaty clay whilst resting on granite, gneiss, or sandstone. Bars of this valuable product, with precious stones, cotton, hides, marmalade, cheese, &c., are sent to Rio, where they are exchanged for slaves, manufactured goods, wines, hams, &c. Owing, however, to the falling off in the productiveness of the mines, this trade is now

much less considerable than formerly. The goldsmith trade is prohibited in Villa Rica; but almost all other handicrafts are carried on. There are also manufactures of gunpowder, hats, pottery, &c. The inhabitants generally depend on mining; and in consequence of the uncertain, hazardous nature of their employment, which has greatly declined, they are very generally idle, poor, and dissolute. (*Mod. Trav. xxix.; Mawe's Brazil; Dict. Geog. &c.*)

VILLEFRANCHE, a town of France, dép. Aveyron, cap. arrond., on the Aveyron, 26 m. W. Rhodez. Pop. in 1836, ex. com., 8147. It is well built: four parallel thoroughfares divide the town into nine parts, besides which it has several suburbs interspersed with plantations. The old collegiate church, and the hospital, formerly a conventual building, are remarkable specimens of Gothic architecture. The public establishments comprise a college, a public library, museum, and club, or subscription rooms. The principal manufactures consist of linens and copper wares; it has, also, a considerable trade in corn, cattle, and other rural produce, and 12 annual fairs.

Another town of the same name is the cap. of an arrond. in the dep. of the Rhone, on the Saone, 17 m. N.N.W. Lyons. Pop. in 1836, 7553. It consists chiefly of one very long and wide street, is well built, and has agreeable environs. Its manufactures consist principally of linen fabrics, cotton, thread, and leather, in which articles, with the addition of wine, cattle, hemp, flax, hempen cloths, &c., it has a brisk trade with other towns in the S. of France. Near it are some lead-mines, which were wrought under the Romans. (*Hugo, &c.*)

VILLENA (an. *Turbula* or *Arbacula*), a town of Spain, prov. Murcia, cap. distr., in a fine plain, 32 m. N.W. Alicante. Pop. about 10,000. Inglis says, "it has its rock, castle, and huerta, and is a place of some size, with several convents and churches. The vine is extensively grown upon the lower acclivities of the neighbouring sierra, and is almost all converted into brandy. The pop. of this town appeared to me to present a singularly disreputable appearance; beggarly, idle, ragged, and ruffian-like." (*Spain in 1830, ii. 239.*) Villena gives title to a marquis, whose palace, a town-hall, two churches, many chapels and convents, a hospital, and some barracks are its principal edifices. It has some soap factories; and in the neighbouring marshes a good deal of salt is made. (*Miñano, &c.*)

VINALHAVEN, p. t., Waldo co., Me., 12 m. S.E. Camden. It consists of Fox islands at the mouth of Penobscot bay, of which there are three large, and several small ones. They contain good harbours, well situated for navigation, and especially for the fisheries. It has seven stores, two grist-mills, two saw-mills; 13 schools, 766 scholars. Pop. 1930.

VINCENNES, a town of France, Seine, cap. cant. on the road to Coulommiers, within a short distance of Paris. Pop. in 1836, 2895. Vincennes owes its origin to Philip Augustus, who surrounded the wood of its name with walls, and built at one of its extremities a royal residence, on the site of which, in 1339, the present castle was erected. This castle continued to be a place of resort for the French kings till the time of Louis XI., when it was made a state prison, a destination which it retained. With little intermission, till 1784; the great Condé, Diderot, and Mirabeau, having been among the number of those confined within its walls. Under Napoleon it again served the same purpose, and here, on the 21st of March, 1804, the Duke d'Enghien was shot. The castle of Vincennes is of an oblong form, about 360 yards in length by 210 in breadth, surrounded by dry ditches, and entered by two drawbridges. The keep is a square tower, five stories in height, with four turrets, and a balcony outside the fourth story. The chapel, founded by Charles V., in 1379, but mostly rebuilt under his successors, is a rich Gothic edifice, with some fine stained glass. The *cour royale* is surrounded by modern buildings, in which are some well furnished apartments, and a large collection of arms. In the fosse, a plain column of red granite, on a foot of black marble, and bearing the inscription, "*Hic cecidit*," points out the spot where the Duke d'Enghien met his fate. The wood of Vincennes, comprising about 1500 acres, is, with the town, a good deal resorted to by the Parisians on holidays, particularly the *fête patronale*, on the 15th of August. (*Guides du Voyageur en France: Hugo; &c.*)

VINCENNES, p. v., capital of Knox co., Ia., 118 m. S.W. Indianapolis, 662 W. Situated on the E. bank of Wabash river, 100 m. above its entrance into Ohio river. It has a fertile prairie of several thousand acres on the north, east and south. It is the oldest town in the state, having been established as a trading post by the French in 1730. It contains a brick courthouse, a jail, a county seminary, St. Clare's female school, conducted by the sisters of charity, three churches of brick, 22 stores, three steam-mills, one windmill, two cotton factories, two printing-offices, two weekly newspapers, a public library of over 1500 vols., 380 dwellings, and 2000 inhabitants, one fifth of whom are

French. The river is navigable to this place for steamboats, excepting at very low water.

VINCENT (ST.), one of the W. India islands, belonging to Great Britain, in the centre of the Windward group, about lat. 13° 10′ N., and long. 60° 37′ W., 91 m. S.S.W. St. Lucia, and 108 m. W. Barbadoes. It is of an elliptical shape, 17 m. in length, and from 7 to 8 m. in mean breadth. Area, about 85,000 acres. Pop., by last census, 96,533; of whom 22,997 were blacks. The centre of the island is occupied by a lofty range of mountains, which in some parts attain the height of 4000 ft.; but the mountains decline rapidly towards the sea; and there are some considerable and well-watered valleys, the soil of which, consisting of a fine black mould of sand and clay, is especially adapted for the culture of sugar. In the upper grounds the soil is light and sandy. St. Vincent is of volcanic origin, and a tremendous eruption of one of its mountains in 1812, occasioned great mischief. The mountains are clothed from their base to their summits with immense forest trees; but the ground having every where the advantage of a gradual slope, and there being little jungle or brush-wood, ventilation is not impeded. The valleys also are sufficiently wide, and free from excessive vegetation, to give a healthy character even to the uncultivated portion of the island; and there is little swampy ground, except in a few places near the sea. Only about one third part of its surface is under cultivation. The atmosphere is generally humid, and the dews heavy; but notwithstanding, St. Vincent is justly considered one of the most healthy of the W. India islands. (*Tulloch's Rep.*) We subjoin

An Account of the Principal Articles of Produce imported into the United Kingdom from St. Vincent during each of the Three Years ending with 1841.

Articles.	1839.	1840.	1841.
Sugar	cwts.	151,399	101,920	110,306
Molasses	cwts.	34,051	16,529	31,267
Rum	galls.	139,697	145,909	88,999
Coffee	lbs.	35	699	—
Cocoa	lbs.	760	6,422	1,755

The government is vested in a governor, a council of 19, and an assembly of 19 members. Representatives of the house of assembly must have an income of £300 a year, or, if representing the town of Kingston, a house in that town of the yearly value of £100. Electors must possess a freehold of 10 acres, worth £20 a year in Kingston, or £10 a year elsewhere. St. Vincent, with its dependency, the Grenadines, is divided into six parts. Kingston, the cap. lies at the bottom of a bay, near the S.W. extremity of the island, with an amphitheatre of wooded hills in its rear. The troops, amounting to nearly 960 men, are principally quartered at fort Charlotte, on a very steep hill, about 1¼ m. N.W. the town, and 600 ft. above the level of the sea.

St. Vincent was discovered by Columbus, but was inhabited only by Caribs till the latter part of the 17th century, when a slave ship from Guinea having run ashore on the island the blacks mostly escaped, and settling here became in the sequel the most formidable enemies of the Caribs. It subsequently fell into the hands of the French, who ceded it to the English in 1763. In 1779, it was re-captured by the French; but it reverted, in 1783, to Great Britain. The sum awarded, in 1835, for the manumission of the slaves in St. Vincent amounted to £592,509. (*Edward's West Indies; Tulloch's Rep. on the Health of the Troops in the W. Indies; Parl. Papers, &c.*)

VINCENT (CAPE ST.), the *Sacrum Promontorium* of the ancients, a promontory forming the S.W. extremity of Portugal, prov. Algarve, 110 m. S. Lisbon, lat. 37° 2′ 54″ N., long. 8° 59′ 39″ W. This cape is celebrated in naval history for the great victory gained in its vicinity on the 14th of February, 1797, by the British fleet, under Sir John Jervis, over a Spanish fleet. The British fleet comprised only 15, whereas that of the Spaniards amounted to 27 sail of the line. But notwithstanding this disparity, the latter were completely defeated, with the loss of two ships of 119, one of 84, and one of 74 guns. The victorious admiral, in acknowledgment of his gallantry and success, was elevated to the peerage by the title of Earl St. Vincent.

VINEYARD, p. t., Grand Isle co., Vt., 78 m. N.W. Montpelier. 551 W. It consists of Isle La Motte or Vineyard in Lake Champlain, containing 4690 acres. Pop. 435.

VIRE, a town of France, dép. Calvados, cap. arrond. near the source of the river of its own name, 35 m. S.W. Caen. Pop. in 1836, ex. comm., 7900. It is well-built, principally on the declivity of a hill, on the summit of which is the Foundling Asylum, and at the base the general hospital; on the ascent, among other buildings, are the courthouse, subprefecture, town-hall, and new prison, with a handsome square. In the middle ages, Vire had a castle, of which some remains still exist; but the greater part of its site is occupied by the town-hall and a planted promenade. The

principal church is a fine Gothic building. A great deal of activity prevails in Vire, which has manufactures of coarse and fine woollens, woollen yarn, paper of all kinds, needles, and other steel articles, horn articles, &c., with tanneries and fulling-mills. It has tribunals of primary jurisdiction, and commerce, a chamber of manufactures, council of *prud'-hommes*, communal college, and public library. De Harnel, and some other eminent personages, were natives of Vire. (*Hugo*, art. *Calvados; Dict. Ency.*)

VIRGIL, p. t., Cortland co., N. Y., 149 m. W. by S. Albany, 308 W. Drained by Toughalogs river, and East Owego creek. It contains three churches, a Presbyterian, Methodist and Baptist, four stores, one forge, one fulling-mill, two grist-mills, 13 saw-mills, two tanneries; 35 schools, 1315 scholars. Pop. 4509.

VIRGINIA, the northernmost of the southern United States is between 36° 33′ and 40° 43′ N. lat. and between 75° 25′ and 83° 40′ W. long. and between 1° 35′ E. and 6° 39′ W. long. from W. It is 370 m. long, and 200 broad at its greatest width, containing 64,000 sq. m., or 40,960,000 acres. It is bounded N. by Pennsylvania and Maryland, from which latter it is separated by Potomac river; E. by the Atlantic; S. by North Carolina and Tennessee; and W. by Kentucky; N.W. by Ohio. The population in 1790, was 747,610; in 1800, 886,149; in 1810, 974,622; in 1820, 1,065,366; in 1830, 1,211,272; in 1840, 1,239,797, of whom 448,987 were slaves. Of the free population, 371,223 were white males; 382,745 were white females; 22,814 were coloured males; 26,220 were coloured females. Employed in agriculture, 318,771; in commerce, 6301; in manufactures and trades, 54,147; in navigating the ocean, 399; do. canals and rivers, &c., 2952; in the learned professions, 3098.

The state is divided into 119 counties, which with their population in 1840, were as follows:

Counties.	Total Pop.	Counties.	Total Pop.
Eastern District.		Surry	6,480
Accomac	17,096	Sussex	11,229
Albemarle	22,924	Warwick	1,456
Amelia	10,320	Westmoreland	8,019
Amherst	12,576	York	4,720
Bedford	20,203		
Brunswick	14,346	Total	469,861
Buckingham	18,786		
Campbell	21,030	*Western District.*	
Caroline	17,813	Alleghany	2,749
Charles City	4,774	Augusta	19,628
Charlotte	14,595	Bath	4,300
Chesterfield	17,148	Berkeley	10,972
Culpeper	11,393	Botetourt	11,679
Cumberland	10,399	Braxton	2,575
Dinwiddie	22,559	Brooke	7,948
Elizabeth City	3,706	Cabell	8,163
Essex	11,309	Clarke	6,353
Fairfax	9,370	Fayette	3,924
Fauquier	21,897	Floyd	4,443
Fluvanna	8,812	Frederick	14,242
Franklin	15,832	Giles	5,307
Gloucester	10,715	Grayson	9,087
Goochland	9,760	Greenbrier	8,695
Greensville	6,360	Hampshire	12,295
Gregg	4,235	Hardy	7,458
Halifax	25,936	Harrison	17,669
Hanover	14,968	Jackson	4,890
Henrico	33,076	Jefferson	14,082
Henry	7,335	Kanawha	13,567
Isle of Wight	9,972	Lee	8,441
James City	3,879	Logan	4,309
King George	5,927	Marshall	6,937
King William	9,258	Mason	6,777
King and Queen	10,862	Mercer	2,233
Lancaster	4,628	Monongalia	17,368
Loudoun	20,431	Monroe	8,422
Louisa	15,433	Montgomery	7,405
Lunenburg	11,055	Morgan	4,253
Madison	8,107	Nicholas	2,515
Mathews	7,442	Ohio	13,357
Mecklenburg	20,724	Page	6,194
Middlesex	4,392	Pendleton	6,940
Nansemond	10,795	Pocahontas	2,922
Nelson	12,287	Preston	6,866
New Kent	6,230	Randolph	6,208
Norfolk	27,569	Rockbridge	14,284
Northampton	7,715	Rockingham	17,344
Northumberland	7,924	Russell	7,878
Nottoway	9,719	Scott	7,303
Orange	9,125	Shenandoah	11,618
Patrick	8,032	Smythe	6,522
Pittsylvania	26,398	Tazewell	6,290
Powhatan	7,924	Tyler	6,954
Princess Anne	7,285	Washington	13,001
Prince Edward	14,069	Wood	7,923
Prince George	7,175	Wythe	9,375
Prince William	8,144		
Rappahannock	9,257	Total	445,939
Richmond	6,055		
Southampton	14,525	Total of State	1,239,797
Spotsylvania	15,161		
Stafford	8,454		

Richmond, on James river, at the head of tide water, and just below the lower falls, is the capital.

The state has a great variety of surface and of soil. From the sea coast to the termination of tide water, which excludes a tract from 110 to 190 m. in width, the country is low and flat, in some places marshy, the soil is sandy, covered with pitch pine, with a light soil of little fertility, except on the margins of the rivers, where it is often productive. This is denominated the low country, and is unhealthy from August to October. Between the head of tide water to the Blue ridge, the country gradually rises and becomes uneven, and near the mountains, often abrupt and broken, though the soil is frequently fertile. The first ridge of mountains in this state, is generally about 150 m. from the ocean, beyond which the country is mountainous, traversed by various ridges of the Alleghany, which occupy a greater breadth of country in Virginia than in any other state. Between these ridges are extensive valleys of table land, composed of mould resting on a bed of limestone, and is often possesses great fertility, and is some of the most pleasant land in the state. The farms are here smaller than in the eastern parts of the state, better cultivated, and there are fewer slaves. The country beyond the Alleghany mountains is generally wild and broken, in some places fertile, but generally barren. Indian corn, wheat and tobacco are the staple productions of the state; cotton is raised for home consumption. But the agricultural statistics for 1840 will exhibit most correctly the productions of the state. There were 326,438 horses and mules; 1,034,148 neat cattle, 1,293,772 sheep, 1,992,155 swine; poultry was raised to the value of $754,698. There were produced 10,109,716 bushels of wheat, 1,482,799 of rye, 87,480 of barley, 943,892 of buckwheat, 34,577,501 of Indian corn, 13,451,062 of oats, 2,944,860 of potatoes, 2,538,374 pounds of wool, 75,347,106 of tobacco, 3656 of rice, 3,494,483 of cotton, 1,541,833 of sugar, 10,597 of hops, 65,020 of wax, 3191 of silk cocoons, 364,708 tons of hay, 25,594 of hemp and flax. The products of the dairy were valued at $1,480,488, of the orchard at $705,765, of lumber at $538,092; and 13,911 gallons of wine were made.

The mineral wealth of Virginia is very great. Gold, copper, lead, iron, coal, salt, limestone, and marble are found; and it has a number of very celebrated mineral springs, particularly those impregnated with sulphur. Mining has recently received much attention; and in 1840, 9000 persons were engaged in it. The belt of country in which gold is found is in Spottsylvania county and the adjacent country, and extending in a S.W. direction, it passes into North Carolina, South Carolina, Georgia and Alabama. But the iron and coal are much more valuable. The coal fields are very extensive, both anthracite and bituminous, and large quantities have been exported, particularly from the neighbourhood of Richmond. Salt springs are found, and large quantities of salt are exported from the banks of the Great Kanawha.

The Natural bridge in Rockbridge county (see NATURAL BRIDGE), and Weyer's cave in Augusta county, are great curiosities.

James river is the largest in the state, being 500 m. long, flowing from the region behind the Blue ridge. It is navigable for sloops of 125 tons 190 m. to Richmond, and for boats much above that, and it enters Chesapeake bay. Appomattox is 130 m. long, enters James river 100 m. above Hampton roads, and is navigable 12 m. to Petersburg. The Rappahannock rises in the Blue ridge, and flowing 130 m., it enters Chesapeake bay. It is navigable 110 m. for sloops. York river enters Chesapeake bay, 40 m. below the Rappahannock, and is navigable 40 m. for ships. Potomac river separates the state from Maryland, and just before its passage through the Blue ridge at Harper's ferry, it is joined by the Shenandoah from the S. The Potomac is navigable for ships of the line to Washington, and enters Chesapeake bay by a broad estuary. Great and little Kanawha are principal rivers west of the mountains, and enter Ohio river. The Monoogahela rises in this state, though it runs principally in Pennsylvania. Staunton and Dan rivers in the S. part of the state, unite to form Roanoke river, flowing into North Carolina.

The lower part of Chesapeake bay lies wholly in this state, is 15 m. wide at its mouth, and enters the Atlantic between cape Charles and cape Henry.

Norfolk, 8 m. from Hampton roads, near the mouth of James river, has much the best harbour in the state; spacious, safe, and well defended, with a sufficient depth of water for the largest ships. It is the most commercial place in the state; but Richmond and Petersburg are more populous, and better situated for trade with the interior. Lynchburg, Fredericksburg, Winchester and Wheeling are also principal places.

The exports of the state for the year ending Sept. 30th, 1841, were $5,630,286, and the imports were $337,237. The tonnage entered was 34,975, and the tonnage cleared was 63,243. There were in the state in 1840, 31 commercial and 64 commission houses in foreign trade, with a capital of $4,299,500; 9736 retail stores, with a capital of $16,684,413; 1454 persons employed in the lumber trade, with a capital of $113,210; 931 persons engaged in internal transportation,

who with 103 butchers, packers, &c., employed a capital of $100,680; 556 persons engaged in the fisheries, with a capital of $28,383.

The manufactures of Virginia are not so extensive as those of many states inferior to it in territory and population. Domestic or family manufactures amounted in 1840, to $2,441,672; 47 fulling-mills and 41 woollen manufactories, employed 922 persons, produced articles to the amount of $147,792, with a capital of $112,350; 23 cotton manufactories with 42,362 spindles, employed 1816 persons, producing articles to the amount of $446,063, with a capital of $1,299,090; 42 furnaces produced 18,810 tons of cast iron, and 59 forges produced 5686 tons of bar iron, the whole employing 1742 persons, and a capital of $1,246,650; 11 smelting houses employed 131 persons, and produced gold to the amount of $51,758, with a capital of $103,650; 5 smelting houses employed 73 persons, and produced 878,648 pounds of lead, with a capital of $21,500; 12 paper-mills produced articles to the amount of $216,845, and other manufactories of paper produced $1960, the whole employing 181 persons, and a capital of $287,750; 3349 persons manufactured tobacco to the amount of $2,406,671, employing a capital of $1,595,080; hats and caps were manufactured to the amount of $155,778, and straw bonnets to the amount of $14,700, the whole employing 340 persons, and a capital of $45,640; 650 tanneries employed 1422 persons, and a capital of $353,141; 962 other manufactories of leather, as saddleries, &c., produced articles to the amount of $220,507, and employed a capital of $341,957; four glass-houses, and two glass cutting works, employed 164 persons, producing articles to the amount of $146,500, with a capital of $132,000; 33 potteries employed 64 persons, producing articles to the amount of $31,380, with a capital of $10,225; 36 persons produced drugs, paints, &c., to the amount of $66,633, with a capital of $61,727; 445 persons produced machinery to the amount of $429,858; 150 persons produced hardware and cutlery to the amount of $50,501; 262 persons manufactured 6339 small arms; 40 persons manufactured granite and marble to the amount of $16,652; 1004 persons produced bricks and lime to the amount of $203,251; 1592 persons manufactured carriages and wagons to the amount of $617,815, with a capital of $311,625; 1454 distilleries produced 865,725 gallons of distilled spirits, and five breweries produced 32,960 gallons of beer, the whole employing 1631 persons, and a capital of $187,212; 764 flouring-mills produced 1,041,535 barrels of flour, and with other mills employed 3961 persons, producing articles to the amount of $7,855,499, with a capital of $5,184,669; vessels were built to the amount of $130,607; 675 persons manufactured furniture to the amount of $229,391; there were 402 brick or stone houses, and 3604 wooden houses erected, employing 4604 persons, and cost $1,367,393; 50 printing-offices, 13 binderies, four daily, 12 semi-weekly, and 35 weekly newspapers, and five periodicals, employed 310 persons, and a capital of $168,830. The total amount of capital employed in manufactures in the state was $11,360,861.

William and Mary college at Williamsburg, is the oldest in the state, and one of the oldest in the country, and was founded in 1691; Hampden Sydney college, in Prince Edward county, was founded in 1783, and is flourishing; Washington college, at Lexington, was founded in 1812; Randolph Macon college, at Boydton, was founded in 1832. There are theological schools at Richmond, in Prince Edward county, and in Fairfax county. But the most important literary institution in the state, is the university of Virginia, at Charlottesville, founded in 1819. Its plan is extensive, its endowment has been munificent, and it is a prosperous institution. Mr. Jefferson valued himself on having exerted his influence to found this institution, and left an inscription for his monument, since inscribed on it, written by himself; "Thomas Jefferson, author of the declaration of independence, and founder of the university of Virginia." In the above colleges, and a few others, there were in 1840, 1097 students; there were in the state 382 academies, with 11,083 students; and 1561 common and primary schools, with 35,331 scholars. There were in the state 58,787 white persons over 90 years of age who could neither read nor write.

The Baptists, the most numerous religious denomination, had in 1836, 435 churches, 261 ministers, and 54,302 communicants; the Methodists had 188 ministers and 41,763 communicants; the Presbyterians had 117 churches, 90 ministers, and 11,413 communicants; the Episcopalians had one bishop, one assistant bishop, 65 ministers, and about 3000 communicants; the Lutherans had 24 congregations and seven ministers; the Reformed Baptists (Campbellites) had about 10,000 communicants; the Roman Catholics had 10 congregations and five ministers; the Unitarians had one minister; and there were some Friends, and a few Tunkers and Jews.

In January 1841, there were in the state six banks and 21 branches, with an aggregate capital of $10,283,633, and a

1029

VIRGINIA

circulation of $6,858,485. The state debt amounted to $7,650,000. The property of the state, invested in bank and other stocks, was $12,500,000.

There is a state penitentiary located at Richmond.

Virginia has undertaken several important works of internal improvement, by chartering private companies, which have been aided liberally by the state. The Dismal Swamp canal connects Chesapeake bay with Albemarle sound, extending from Deep creek, a tributary of the former, to Joice's creek, a branch of Pasquotank river of Albemarle sound, 23 m. long, and cost $879,864. It has branches of 11 m. A canal extends along James river, from Richmond to Lynchburg; and this communication is designed to be extended by canal and railroad to the Ohio river by the great Kanawha. No other spot between New-York and Georgia presents an equally favourable country for a line of communication across the Alleghany mountains. The whole length would be 425 m. A railroad extends from the Potomac river, at the mouth of Aquia creek to Fredericksburg, and thence to Richmond, in the whole 75 m., and cost $1,370,000. It proceeds from Richmond S. to Petersburg 23 m., which cost $750,000; and from Petersburg to Weldon on Roanoke river, 59 m., which cost $600,000, where it unites with other southern railroads. A railroad extends from Petersburg to City point, 12 m. A railroad extends from Winchester 32 m. to Harper's Ferry, where it unites with the Baltimore and Ohio railroad. Greensville railroad connects the Petersburg and Roanoke railroad with the Raleigh and Gaston railroad, N. C., extending from Hicksford 18 m. to Gaston, and cost $250,000. The Chesterfield railroad, 13½ m. long, extends from the Chesterfield coal basin to tide water in James river, at Manchester, opposite Richmond. Other railroads have been projected.

The first constitution of Virginia was formed in 1776. This was amended by a convention assembled for that purpose, in 1830. The governor is elected for three years by the joint vote of the two houses of assembly, but is ineligible for the next succeeding three years. He must be at least 30 years of age, must be a native born citizen of the United States, and have resided in the state for five years next preceding his election. The council of state consists of three members, elected for three years, one of the number retiring annually. They are the advisers of the governor. The senior councillor is lieutenant governor, and acts as governor in case of the death, resignation, absence, or inability of the governor. The senate consists of 32 members, elected by the people for four years, one fourth of the number retiring annually. A member must be at least 30 years of age, and have a residence and a freehold in the district for which he is elected. The house of delegates consists of 134 members, elected annually by the people. A member must be at least 25 years of age, and have a residence and a freehold in the district for which he is elected. No person holding a lucrative office, no ministers of the gospel, or priests of any denomination, can be elected a member of the assembly. The judges of the supreme court of appeals, and the superior courts, are elected by the joint vote of both houses of assembly, and hold their offices during good behaviour. The attorney-general is appointed in the same manner, and holds his office during the pleasure of the general assembly. Every white male citizen of the state, of 21 years of age or upwards, who owns an interest in a freehold estate of the value of $25, or who is a housekeeper or head of a family, and has paid taxes, excepting paupers, non-commissioned officers, soldiers, seamen or marines in the service of the United States, and persons convicted of infamous crimes, is entitled to the right of suffrage. The general assembly meet annually at Richmond on the first Monday of December. At all elections, votes are given openly, or viva voce, and not by ballot, as in most of the other states.

Virginia has obtained the name of the ancient dominion, having been settled in April, 1607, at Jamestown on James river, and was the first white settlement in the United States, being over four years older than the settlement at Plymouth. It was named in honour of queen Elizabeth by Sir Walter Raleigh, to whom she granted the country; but the settlement of it which he attempted failed. The grant was vacated by the execution and attainder of that nobleman, under James I. The territory was then granted to two companies, the London company and the Plymouth company, and called North and South Virginia. By the former the country was settled, and Jamestown was named in honour of their royal patron. The first settlers were a different set of men from the settlers at Plymouth, and the country suffered many disasters from the contentions among the leaders, the turbulence of the citizens, the wars with the Indians, and the tyranny of the royal governors. Virginia was a very loyal province, and was attached to the royal party during the protectorate of Cromwell; and was among the first to proclaim Charles II. at the restoration. The church of England was established by law in 1662.

VITERBO.

Virginia had the high honour in 1732, of being the birthplace of George Washington, and it was as an officer of her colonial militia, when a very young man, during the French war, that he commenced his career of military glory. The state was among the first to resist the aggressions of the mother country, and her sages and yeomen bore a distinguished part in the Revolutionary struggle. Her Patrick Henry as a Revolutionary orator, her Washington as the greatest and the best in the field and in the cabinet, and her Jefferson, Madison, and Monroe, as presidents of the United States, and her Marshall, as chief justice, have conferred distinguished honour on their country, and given her a high place in the American confederacy. In convention, June 25th, 1788, the constitution of the United States was adopted,—yeas 89, nays 79; majority 10. It will be found that the constitution was adopted with the greatest struggle and the most powerful opposition, where the spirit of liberty was the most active and jealous. If after its excellency has been fully tested, conventions could now be called in the several states, to pass upon it, it would be adopted, probably, without a dissenting voice. Its excellency is becoming daily more apparent, and its authority more commanding and undisputed.

VISTULA (Germ. Weichsel), one of the great rivers of central Europe, flowing from S. to N. through Poland. The basin of the Vistula is situated between those of the Elbe to the W., the Niemen and Dnieper to the N.E. and N., and the Dniepr to the S.E. It rises in Moravia, in a branch of the Carpathians, close on the frontier of Galicia, and about 20 m. S.E. Teschen; and at a short distance from its source is precipitated over a fall 180 feet in height. It proceeds at first N. for about 40 m., and then turns to the E., separating Silesia, the territory of Cracow, and the kingdom of Poland on the N., from Galicia on the S. Shortly after passing Sandomir, it again flows northward, which course it retains through the centre of Poland to beyond Warsaw. It now turns W.N.W., and pursues generally the same direction to the influx of the Bran, 90 m. from Thorn; after which its course varies little from N.N.E. to its mouth in the Baltic. Its entire length is estimated at 550 m. It receives a vast number of tributaries, the principal of which are the Nida, Kamiena, Pilica, and Bran, from the W.; and the San, Wiepra, and Bug,* with its tributaries, from the E. At Cracow it is only about 150 feet in width; at Warsaw it is crossed by a bridge of boats 1600 feet in length. After receiving the Bug, a stream nearly equal in size to itself, at Modlin, it proceeds generally in a very wide channel past Plock. Thorn, Culm, and Marienwerder, about 15 m. below which last, and about 30 m. from the Baltic, it divides into two great arms, the most easterly of which, called the Nogat, flows past Marienburg and Elbing into the Frische Haf. The W. arm, or main stream, subdivides again at about 16 m. from the sea, the E. branch falling into the Frische Haff, and the W. making a long detour round by Dantzic. It is said, however, that during some heavy floods in the autumn of 1840 the eastern arm of the river opened for itself a direct course to the Baltic, a result which it is said the Prussian government had contemplated effecting by artificial means. This river is of very considerable commercial importance, being, as it were, the great highway of the extensive countries through which it flows; the channel by which the wheat, timber, and other products are conveyed to Dantzic and Elbing for exportation, and by which they receive supplies of colonial and other foreign produce. As it flows for the most part through a level country, it is navigable throughout the greater part of its extent. Large flat-bottomed boats convey the produce of the countries through which it passes to the port of Dantzic, and Warsaw is thus placed in direct communication with the Baltic; while by means of the navigation of the Nogat, the colonial produce imported into Konigsberg finds a ready access to Poland, Moravia, and Hungary." (Granville's Travels, i., 345.) The Vistula is connected with the Elbe by a canal from the Bran to the Netz, between Bromberg and Nakel; and with the Niemen by means of the canal of Augustowo. (Dict. Geog., &c.)

VITERBO, a city of central Italy, Papal states, cap. deleg. of same name, in the Campagna, and on the road between Rome and Sienna, 37½ m. N.N.E. the former. Population stated by Rampoldi to be little short of 13,000. It is well built, with volcanic tufa, and well paved, having 1 large and handsome square, 16 parish churches, and numerous noble residences, and public fountains. Wraxall says it is a curious looking city, with an abundance of caverns in the perpendicular faces of the rocks, bordering a little valley passing through it. The cathedral has a range of columns on each side, with grotesque capitals supporting semicircular arches. The Trinita is a handsome modern church, in the form of a Latin cross, with a dome in the

-centre. The church of St. Francis is a large building; the transept has pointed vaulting, and there are two fine archways of the pointed style, leading into chapels, and some Gothic tombs. It boasts also a painting by Sebastian del Piombo, from designs by Michael Angelo. (*Letters of an Architect*, i., 323.) The pontifical palace is a fine building. Viterbo is the seat of a cardinal delegate, and a court of primary jurisdiction. It has no manufactures worthy of notice; though alum, vitriol, sulphur, and other volcanic products are obtained in its neighbourhood, which abounds with mineral springs.

Viterbo is supposed to occupy the site of the *Fanum Voltumnæ*, the place where the general assembly of the Etruscan nations was held on solemn occasions. The modern town was encircled with turreted walls by Desiderius, the last king of the Lombards. It has been the residence of numerous popes, several of whom are buried in its churches. In its vicinity are many villas belonging to some of the more opulent Roman families. (*Rampoldi; Woods; Cramer's Anc. Italy.*)

VITRÉ, a town of France, dep. Vilaine, cap. arrond., on the Vilaine, 23 m. W. Rennes. Pop. in 1836, ex. comm., 7899. It is enclosed by walls of Gothic character, and flanked by round towers. Vitré is ill built, dirty, *triste*, and destitute of any public promenade; though the environs are agreeable, and in the vicinity are two parks open to the public. About 1½ m. S. from the town is the Chateau des Rochers, the seat of Madame de Sevigné, the most accomplished of letter writers, who sometimes also occupied a house in the town. Near the town are also the ruins of the castle, formerly belonging to the dukes De la Trimouille. The peasantry of the neighbourhood wear winter cloaks of goat skins, which, with cotton hosiery, sailcloth, flannels, leather, and barrels, are among the principal articles manufactured in Vitré. Wax, honey, and cantharides are here considerable articles of trade, and the town has no fewer than 22 annual fairs. Savary, the traveller, was a native of Vitré, where he first saw the light in 1750. (*Hugo*, art. *Ille-et-Vilaine.*)

VITRY-LE-FRANÇOIS, a town of France, dep. Marne, cap. arrond., on the Marne, 20 m. S.S.E. Chalons. Pop. in 1836, ex. comm., 6616. It is square shaped, and is enclosed by earth ramparts, and bastions, outside which is a deep moat. It is tolerably well laid out; and, though most of its houses are old and unprepossessing, it has a good many new buildings erected since the peace. Its church, an edifice in the Corinthian and Composite styles, was the earliest of any consequence built after the restoration of the arts, in the time of Francis I., under whom this town was founded. Vitry has a good public hall and theatre, and agreeable public walks, with some manufactures of cotton yarn and hosiery, hats, leather, &c. (*Hugo; Dict. Géog., &c.*)

VITTORIA (Span. *Vitoria*), a town of Spain in Biscay, cap. prov., on the high road between Burgos and Bayonne, 60 m. N.W. the former. Population estimated by Miñano at 12,000. It consists of an old and a new town, very different in appearance; the latter being clean and handsome, while the former is quite the contrary. The *Plaza Nueva*, a square, which, according to Inglis, is little inferior to the *Place Vendome* in Paris, has arcades at its sides, under which are very good shops: the S. side is occupied by the townhall, and the area serves for a market-place. The hall of the Biscayan society, orphan asylum, and general hospital are among the principal edifices. Vittoria has a collegiate and four parish churches, six conventual establishments, a school of design, public library, cabinet of coins and Roman antiquities, post-house, &c. Its manufactures comprise chairs and cabinet furniture, copper utensils, earthenware, cutlery, linens, &c., and, being one of the principal entrepôts for the trade between Navarre and Old Castile, and the ports of St. Sebastian and Bilbao, it has a considerable traffic in iron, wool, woollen and silk fabrics, articles of clothing, and colonial produce. In modern times Vittoria is famous for the decisive victory gained in its vicinity, on the 21st of June, 1813, by the Anglo-Spanish army, under the Duke of Wellington, over a French army commanded by King Joseph and Marshal Jourdan. Though the loss of men in the battle was nearly equal, the French were totally defeated, with the loss of all their artillery, baggage, ammunition, and treasure, and obliged to make a rapid retreat across the Pyrenees; this battle having all but annihilated their power in Spain. (*Napier's Peninsular War*, v., 555-580.)

VIZAGAPATAM, a seaport town of British India, presid. Madras, coast of Coromandel, cap. of a district of same name, in the N. Circars, at the mouth of a small river, lat. 17° 42' 30" N., long. 83° 24' E. It is not a place of any strength, its only defensive works being a thick wall enclosing the Zillah courthouse, hospital, other European buildings, and a bazaar in the centre of the town. The barracks and other public edifices are outside this wall. A good many well-built houses stretch along the shore; but the great insalu-

brity of the town has driven most of the former European residents to Waltair, a village at some little distance. (*Hamilton's E. I. Gazetteer.*)

VLADIMIR, a government of European Russia, between the 55th and 57th degs. of N. lat., and the 38th and 43d of E. long., having N. Jaroslavl and Kostroma, E. Nijni Novgorod, W. Tver, and S. Moscow, Riazan, and Tambof. Area estimated at 17,600 sq. m. Pop. in 1838, 1,133,900. Surface almost a level plain, watered by numerous rivers, the principal being the Oka in the E., the Wolga in the W., and the Kliazma, a tributary of the Oka, in the centre; all of which have, more or less, a N.E. course. The soil is not generally fertile, and a large part of the government is covered with forests, marshes, pools, and heaths. Rye, barley, oats, summer and winter wheat, millet, pease, hemp, and flax are grown; but the crops of corn are insufficient for the consumption. The gardens and orchards are pretty numerous and well attended to; and Vladimir is famous for its cherries and apples. A good many cucumbers and some hops are raised. Cattle rearing is a secondary business, and is far behind. The forests are of vast extent, those belonging to the crown alone covering about one ninth part of the entire surface. Extensive and valuable beds of iron ore have been found in the forest of Mourom; and at Vixa, on the Oka, are some of the most extensive iron-works in Russia. (*Lyall's Russia*, ii., 321-323.) The poverty of the soil, and other concurring circumstances, have turned the attention of the inhabitants toward manufactures, which appear to have succeeded better in this than in most other Russian governments. In 1830 the manufacturing establishments in the government employed 48,176 workmen. The cotton manufacture, which is by far the most extensive, is principally carried on at Choula and Ivanova, where it employed, in 1828, 15,612 looms and 24,257 work-people. It would seem, however, that this and other branches of industry have rapidly increased in the interval; for the official returns show that, in 1839, 315 factories afforded employment to 83,655 work-people, being little short of double the number employed in the government in 1830! The manufacture of woollen and linen is of less importance; but about 4000 hands are employed in iron foundries; and about 1300 in glass and crystal works, exclusive of those employed in the production of leather, earthenware, &c. The various products of the government are sent down the Kliazma and Oka, or else to Moscow, by means of land-carriages. Corn, cotton-twist, and flax from the neighbouring governments of Kostroma, Jaroslavl, and Nijni Novgorod, are the chief articles of import. Vladimir is divided into 13 districts; chief towns, Vladimir, the capital, Choula, and Mourom. Total public revenue about 4,000,000 roubles.

VLADIMIR, a town of European Russia, cap. of the above government, near the Kliazma, and on one of its small tributaries, 110 m. E. by N. Moscow; lat. 56° 7' 38" N., long. 40° 21' 45" E. Population estimated at 7000. (*Possart.*) Vladimir occupies a site rather more elevated than the rest of its government. It is surrounded by a ditch and earth rampart, and, like almost all the ancient towns of Russia, is divided into three portions. Its principal street is long, wide, and lined with houses of wood and stone intermixed. The cross streets are mostly mean. The principal structure is the cathedral of the Assumption, a square edifice, surmounted by five domes, and richly ornamented inside, though much less magnificent than formerly. There are about a dozen other churches. The former palace of the archbishop now serves for a seminary. The governor's house, courthouse, gymnasium, a nunnery, &c., are brick edifices. Vladimir is not considered a wealthy town or a principal emporium, owing partly to its distance from any large navigable river, and partly to the proximity of Moscow. Being, however, on the great road to the fairs of Nijni Novgorod and Irbit, and on the grand line of communication between Russia and Siberia, it often presents a busy and cheerful aspect. Some of its inhabitants are occupied in making linen cloths and leather; and many others in the cultivation of fruit, particularly cherries, which are grown in great quantities in the neighbourhood. The æra of its foundation is uncertain; some authors place it in the 10th, and others in the 12th century. Vladimir was, however, the capital of the grand duchy of Russia from 1157 till 1328, when that distinction was transferred to Moscow. (*Lyall's Trav. in Russia; Possart; Schnitzler, &c.*)

VOGHERA, a town of N. Italy, Sardinian dom., div. Alessandria, cap. prov., on the Staffora, 21 m. E. by N. Alessandria. Pop. in 1838, 10,706. It is well built, is surrounded by walls, has a good market-place, a magnificent collegiate church, a Jesuits' college, several monasteries, large barracks, and a good hospital. It is the residence of a governor, and the seat of a provincial court of justice: it has an active trade in corn, wine, and silk. (*Rampoldi, &c.*)

VOLCANO. *See* LIPARI ISLANDS.

VOLHYNIA, a gov. of European Russia, formerly comprised in the kingd. of Poland; principally between the 50th and 52d degs. of N. lat., and the 24th and 29th of E. long., having 'N.W. and N. the govs. of Grodno and Minsk, E. and S.E. Kief. S. Podolia, S.W. Austrian Poland. and W. the palatinate of Lublin. Area estimated at 22,740 sq. m. Pop. in 1838, 1,394,100. It is in general an undulating plain; and the hills, which are the last ramifications of the Carpathians, though they nowhere rise to 300 feet above the sea, give an agreeable variety to the scenery. The Bug rises in this province: the other principal rivers are the Styr, Goryne, &c., tributaries of the Pripats. Along some of these are extensive marshes and beds of turf; but in general the land is very fertile, producing, at an average, a considerable surplus of corn above the consumption. A good deal of flax and hemp is also grown. Agriculture is, however, not more advanced than in the rest of Russian Poland, and the gardens and orchards, particularly the former, are much neglected. The climate, though comparatively mild, is not warm enough for the vine. The forests comprise oak, beech, lindens, firs, pines, &c., and are very extensive, though only about 44,754 deciatines of forest land belong to the crown. The pastures are excellent, and well adapted for the fattening of cattle; a good many sheep, hogs, and poultry are kept. Volhynia has a breed of horses smaller than the generality of those of Poland. Fishing is an occupation of some importance; bog-iron, mill-stones, potter's clay, nitre, and flint are among the mineral products. Though agriculture is the chief occupation of the inhabitants, the manufacturing industry of Volhynia is greater than that of most other parts of Russian Poland. The women, almost everywhere, spin and weave different fabrics; and leather, glass, and earthenware, paper, potash, tar, charcoal, &c., are generally made. The principal exports are, however, corn, cattle, hides, flour, wool, wax, honey, and other rural produce. In 1835 the value of the exports was estimated at about 19,165,000 roubles, and that of the imports at 15,073,000 roubles. The trade is principally in the hands of the Jews, of whom there are about 40,000 in the government. The rest of the population consists of Russniaks, with Poles in the towns, and some Great Russians, gipsies, Tartars, Moldavians, and Germans. The inhabitants are mostly of the Greek, or united church. Volhynia is divided into 12 districts; principal town, Zytomir or Jitomir, the capital. A large annual fair is held at Easter at Berditchef. Public education appears to be less backward in this than in most of the Russian governments; and in 1832, besides the government printing press, there were six others, and a lithographic press. Volhynia, like Podolia, is subordinate to the military governor of Kief, but is one of the Polish provinces, which preserves, in some degree, its ancient constitution and laws. (Schnitzler, La Russie; Possart, Das Kaiserth; Russland, &c.)

VOLNEY, p. t., Oswego co., N. Y., 353 W.N.W. Albany, 375 W. Bounded S.W. by Oswego r. Drained by Black creek. It contains two churches, a Methodist and Presbyterian, seven stores, one grist-mill, five saw-mills; one academy, 147 students; 18 schools, 939 scholars. Pop. 3155.

VOLOGDA, the largest government of European Russia, after that of Archangel, between the 58th and 64th degs. of N. lat., and the 38th and 60th of E. long., having N. Archangel, W. Olonetz and Novgorod, S. Jaroslavl, Kostroma, and Viatka, and E. the Ouralian mountains, separating it from Tobolsk. Area estimated at upwards of 145,800 sq. m. Pop. in 1838, 747,500. Except in the E., where it is covered with the Ouralian mountains, the surface generally is an undulating plain, comprised in the basin of the N. Dwina, which is its largest river. The general slope is accordingly to the N.W. In the S. and S.W. the soil is fertile, but elsewhere it is sandy or thin, and the greater part of the surface is covered with marshes and forests of pine, birch, oak, &c. Though the climate varies with the situation, it is, speaking generally, very severe; it is far, however, from being unhealthy, and instances of longevity are frequent. The grains principally cultivated are rye and barley; but the produce of corn is insufficient for the consumption. Hemp, flax, and hops succeed, as do beans and peas. Cattle and horses are numerous and good; but a large part of the government being unoccupied and in a state of nature, the chase necessarily occupies much attention. The forests, which are its principal source of wealth, are of great extent, those of the crown only covering 29,558,000 deciatines of land. Granite, marble, salt, flints, copper, and iron are all obtained in Vologda. In 1834, there were 114 manufacturing establishments, principally for woollen and linen fabrics, soap, leather, potash, glass wares, and paper. Distillation is also very extensively carried on. Furs, tallow, pitch, wooden articles, masts and timber, turpentine, and other raw products are the great articles of export; being sent, for the most part, into the governments of Archangel and Tobolsk. The population is principally Russian, but include some Zyrians or Surjans of Finnish stock; and in the N. are some wandering Samoy-

ede tribes. Public instruction, owing to the the thinness of the population, is necessarily very limited; but it has been materially increased of late years. This territory is divided into 10 districts: the chief towns are Vologda, the capital, and Ustiug-Velild.

VOLOGDA, a town of European Russia, cap. of the above government, near its S.W. extremity, 353 m. E. by S. Petersburg; lat. 59° 13' 30" N., long. 40° 21' 15" E. Pop. in 1834 estimated at 14,000. It is built on both sides the river Vologda, and is supposed to be one of the most ancient towns in Russia. Most part of its houses are still of wood, but the buildings in stone are increasing, and several of the churches are of that material. It has two cathedrals, one of which was rebuilt in 1832. The palaces of the archbishop and governor, the prison, gymnasium, hospital, various asylums, and an episcopal seminary, are conspicuous edifices. Near the town is a famous convent, founded in 1371. Vologda has manufactures of soap, potash, cordage, bells, and tallow candles; for which last it is famous over all the N. of Russia. Its trade is considerable, which is principally with the Baltic, Germany, and England; also with Siberia, to the boundaries of the Chinese empire. (Possart; Schnitzler, &c.)

VOLTERRA (an. Volaterræ), a town of central Italy, in Tuscany, prov. Pisa, on a steep hill near a small tributary of the Cecina, 33 m. S.W. Florence. Pop. including its suburbs, about 6000. It has a citadel, now used as a state prison; a hospital, a cathedral, and many other churches; a very tolerable inn, a large and fine town hall, a theatre, a Piarist college, and a seminary. Its inhabitants are principally agriculturists, but a few of them are engaged in the manufacture of earthenware vases and plaster figures. Even if we had not the express authority of Dion. Halicarnassus, (iii., 51,) for assigning to Volterra a place among the 12 principal cities of ancient Etruria, the extent of its remains, its massive walls, vast sepulchral chambers, and numerous objects of Etruscan art, would alone suffice to show its antique splendour and importance, and claim for it that rank. Its walls were formed, as may yet be seen, of huge massive stones piled on each other without cement; and their circuit which is still distinctly marked, embraced a circumference of between 3 and 4 m. (Cramer's Anc. Italy, i. 185.) Two of its original gates are still in existence; one, called the Gate of Hercules, consisting of two arches, is a very perfect state, and the other leads to an ancient Etruscan burial ground, in which are some remarkable tombs. Under the Romans it was a colony and municipium, and the walls of the modern town, 2 m. in circuit, are said to have been built by the emperor Otho, and are still in good preservation. There are several other Roman antiquities, including a piscina and what are called the baths of Otho. Volterra has also a public museum, containing numerous remains of antiquity discovered in the neighbourhood. Persius, the satirist, is generally supposed to have been a native of Volterra, where he is said to have been born A.D. 34. (Rampoldi; Cramer's Anc. Italy, &c.)

VOLUNTOWN, p. t., Windham co., Ct. 35 E. by S. Hartford, 374 W. Drained by Patchaug r. and a branch of Wood r. It contains two churches, a Congregational and Baptist, two stores, two cotton factories with 3760 spindles; nine schools, 378 scholars. Pop. 1185.

VORONEJE, or WORONETZ, a gov. of European Russia, between lat. 48° 40' and 53° N., and 38° and 43° E. long.; having N. the governments of Riazan and Tambof, E. Saratof and the territory of the Don Cossacks; S. the latter and the government of Ekaterinoslaf; and W. Kharkof, Koursk, and Orlof. Area estimated at 28,760 sq. m. Pop. in 1838, 1,507,200. Surface undulating, and soil in general good; this being, in fact, one of the most productive governments in the empire. Principal rivers, the Don and some of its tributaries. Climate comparatively mild, the rivers being covered with ice for only two or three months of the year, and the government producing most of the products of temperate climates. Of 6,876,609 deciatines (1 deciat. = 2·7 acres) comprised in the government, the arable lands have been estimated to include 2,711,000, pasture lands, 2,818,000 do., and forests, 690,735 do. In good years a surplus is raised of about 1,500,000 chetwerts of corn beyond the home consumption. Besides wheat, peas, and beans, poppies, tobacco, hemp, and flax are grown; and in the gardens melons, cucumbers, onions, &c., in large quantities. Water-melons, indeed, are cultivated for the markets of Moscow and Petersburg, being planted in open fields covering whole acres of land. In some parts canes and reeds are used for fuel, but in general the forests furnish a sufficient supply of fire-wood. Oaks are numerous and luxuriant; pine woods are few. In 1832 the cattle in the government were estimated at 550,000 head; the sheep at nearly a million, and 11,000 horses were kept in the studs. Honey is an important product. Iron, limestone, and saltpetre are among the minerals. Manufactures of coarse woollens and other fabrics are rapidly increasing, having more than doubled between 1821 and 1835. The number of distilleries decreased during the same period; but we are not aware whether the production of spirits has undergone any corresponding decrease. The exports from the government consist principally of corn, cattle, skins, honey and wax, fruits, &c.

This government is divided into 12 districts: chief town, Voroneje, the capital. Except a colony of Germans near Ostrogojok and some gipsies, the population consists in the S. of Cossacks and White Russians, in the N. of Great Russians. Voroneje is under the same governor-general with Riazan, Orel, Tambof, and Saratof. The total revenue of the crown from this government amounts to about 15,000,000 roubles.

VORONEJE, a town of Russia, cap. of the above gov., on the river of the same name, near its confluence with the Don, and 290 m. S.S.E. Moscow. Pop. 18,600. (*Possart*.) It stands on a steep height, and might easily be rendered a fortress of some strength, as it is not commanded by any other hill, and is partly surrounded by a marsh for several months of the year. It consists of three portions, the upper town, lower town, and suburbs. " It has some spacious streets, but a great many which are very mean : the suburbs are as black and gloomy as a country village. The principal street has a noble appearance, its sides being lined with massy and handsome edifices, many of them the property of the crown, as the governor's and vice-governor's houses, the tribunals, postoffice, commissariat academy, &c. The Moscow (Moskovkaya) street is also very fine, and in it are the archbishop's palace, with an adjoining cathedral. The shops, or bazaars, are very respectable." (*Lyall's Trav.*, ii., 303.)

Voroneje has 18 stone churches, two convents, an exchange or *gostinof-door*, for the warehousing, exhibition, and sale of merchandise, an Episcopal seminary, schools for the children of the clergy, military, civil employés, and citizens, a hospital for 310 sick persons, military orphan asylum, &c. It is one of the most flourishing towns in the S. of Russia ; and its merchants carry on a lucrative trade with the Black sea, Crimea, and Turkey, and travel annually to Tobolsk, to buy furs, which they afterward take to the great German fairs. The town has also some soap, tallow, leather, and woollen cloth factories. It is supposed to be among the oldest Russian towns, and is spoken of as existing in the 12th century. Here Peter the Great built a palace, and established a dock-yard, arsenal, &c.; but the latter establishments were afterward removed successively to Ustea, Tavrof, and Rostof; and nearly all traces of the palace and magazines have been obliterated by the frequent fires which the town has since suffered. (*Schnitzler, La Russie ; Possart ; Lyall's Trav. in Russia, &c.*)

VOSGES, a dep. of France, reg. N.E., principally between the 48° and 49° of N. lat., and the 5° and 7½° of E. long., having N. the deps. Meurthe and Meuse, E. the dep. of the Rhine, S. Haute-Saône, and W. Haute-Marne. Area, 585,963 hectares. Pop. in 1836, 397,967. This department derives its name from the Vosges (Germ. *Wasgau*) mountains, a chain which extends parallel with the Rhine, separating the departments of Haute and Bas-Rhin on the E. from those of Haute-Saône, Vosges, and Meurthe on the W., stretching also into Rhenish Bavaria, and terminating to the N.E. in mont Tonnerre. These mountains usually rise between 4000 and 5000 feet above the sea, and their summits are covered with snow for most part of the year. They send off a remarkable continuation, the Faucilles mountains, E. and W. through this department, by the ramifications of which nearly its whole surface is covered. The Moselle, Meuse, Meurthe, Madon, Saone, &c., rise in this department, all of which, except the Saone, have a N. course. Small lakes are numerous. The arable land, which is said to comprise 244,745 hectares, is not generally fertile ; the meadows comprise 76,330 hectares ; woods, 129,474 hectares ; and heaths, wastes, &c., 36,550 hectares. In a portion of the department called " the plain," to the W. of Epinal, agriculture is said by Hugo to be pretty well advanced. The land is divided into very small properties ; so much so, that in 1835, of 148,699 properties subject to the *contributions foncière*, 87,600 were assessed at less than five francs, and only 43 at 1000 francs or upward. In 1835 the produce of corn, principally oats, wheat, and rye, was estimated at less than 3,000,000 hectolitres, and the potato crop was about as much. The rearing of stock is the most important branch of husbandry, and a greater number of cattle are kept in this than in any other of the N.E. departments ; in 1830 they were reckoned at about 140,000 head. Sheep are much less numerous, and the annual produce of wool is said by Hugo not to exceed 45,000 kilogr. The annual produce of cheese may be about 20,000 kilogr.; and that of wine (of very indifferent quality) about 150,000 hectolitres ; about 120,000 kilogr. of hops are annually sent to Paris. Cherries are grown in large quantities, and the department is famous for its *kirschenwasser*. A good many hogs are

fattened in the mountains. The forests abound in good fir timber, great quantities of which are floated down the rivers, as deals and rough timber. Iron is the chief mineral product; but it also produces coal, argentiferous lead, copper, manganese, granite, marble, porphyry, &c., though many of these resources are much neglected. The manufacture of steel and iron goods hold the first rank. Knives and forks are made at Bruyères; bayonets, &c., at Sionne, and nails at Neufchateau ; plate iron is made in large quantities at various places ; and Plombières is famous for its cutlery. Cotton stuffs are made in the arrondissements of Remiremont and St. Dié. Lace, musical instruments, barrels, and wooden shoes are considerable articles of manufacture ; and there are various glass and marble works, tanneries, breweries, &c. Vosges is divided into five arrondissements ; chief towns, Epinal, the capital ; Mirecourt, Neufchateau, Remiremont, and St. Dié. It sends five members to the chamber of deputies. Number of electors, in 1838-39, 999. Total public revenue, 1831, 7,165,857 francs. (*Hugo, art. Vosges ; French Official Tables, &c.*)

W.

WAAL, a river of the Netherlands, *see* RHINE.

WABASH r., Ia., rises in Darke co., O., and, flowing N.W. to Huntington co., Ia., it approaches the junction of St. Joseph's river with St. Mary's river, which form Maumee river. It then crosses the state in a W. and S.W. direction, and a little below Terre Haute it becomes the boundary between Indiana and Illinois, until it enters Ohio river, 128 m. above the Mississippi, and 68 m. above the mouth of Cumberland river. It is the largest tributary of the Ohio on the N. side, and probably the second in the whole course of that river. It is navigable by steamboats to Lafayette, and, by means of the Wabash and Erie canal, it connects lake Erie with the Ohio river. This canal, which runs along its border, and is now completed, is one of the most important works of the kind in the western states. It receives several important tributaries, the principal of which are Salamanie, Missisinewa, Tippecanoe, Vermillion, Embarrass, and, toward its mouth, White, and Little Wabash rivers.

WABASH, county, Ia. Situated centrally in the N. part of the state, and contains 415 sq. m. Watered by Wabash, Eel, and Salamanie rivers. It contained in 1840, 9775 neat cattle, 1065 sheep, 9468 swine ; and produced 8705 bushels of wheat, 75,644 of Indian corn, 9556 of oats, 14,397 of potatoes, 90,190 pounds of sugar. It had 11 stores, six grist-mills, nine saw-mills, four tanneries ; two schools, 45 scholars. Pop. 2756. Capital, Wabash.

WABASH, county, Ill. Situated toward the S.E. part of the state, and contains 180 sq. m. Bounded S.E. by Wabash river, W. by Bon Pas river. It contained in 1840, 5570 neat cattle, 3602 sheep, 10,392 swine ; and produced 19,135 bushels of wheat, 160,659 of Indian corn, 1394 of barley, 51,002 of oats, 12,533 of potatoes, 9078 pounds of tobacco, 2295 of sugar. It had 11 stores, one flouring-mill, five grist-mills, eight saw-mills, one oil-mill, six tanneries, four distilleries, two printing-offices, two weekly newspapers ; two academies, 32 students ; two schools, 80 scholars. Pop. 4240. Capital, Mount Carmel.

WACHUSETT, mt., an isolated peak in the N. part of Princeton t., Worcester co., Mass., 3000 feet above the level of the sea, and is the highest land in the state, E. of Connecticut river. It is often visited for the grand and beautiful prospect presented from its summit.

WAGRAM, a village of the archduchy of Austria, country below the Enns, on the left bank of the Rossbach, 11 m. N.E. Vienna. This village is celebrated in military history for the great battle fought in its vicinity, on the 6th of July, 1809, by the grand French army under Napoleon, and the Austrians under the archduke Charles. The former gained a complete victory ; the Austrians lost above 20,000 men taken prisoners, besides a vast number killed and wounded. This great victory led to an armistice, followed by the treaty of Schoenbrun.

WAITSFIELD, p. t., Washington co., Vt., 19 m. S.W. Montpelier, 506 W. Watered by Mad river and its tributaries. It contains three churches, a Congregational, Methodist, and Union, two stores, two fulling-mills, three grist-mills, five saw-mills, two tanneries; nine schools, 396 scholars. Pop. 1048.

WAKE, county, N. C. Situated in the centre of the state, and contains 1140 sq. m. Watered by Neuse river and its branches. It contained in 1840, 15,948 neat cattle, 11,574 sheep, 38,883 swine ; and produced 38,379 bushels of wheat, 79,011 of rye, 535,274 of Indian corn, 79,011 of oats, 55,965 of potatoes, 550 pounds of rice, 54,247 of tobacco, 2,391,996 of cotton. It had 53 stores, two flouring-mills, 32 grist-mills, 16 saw-mills one paper-mill, 67 distilleries, four printing-offices, five weekly and one semi-weekly newspapers ; one college, 77 students ; 13 academies. 283 students ; 10 schools, 169 scholars. Pop. : whites. 12,113 ; slaves, 7996 ; free coloured, 1009 ; total, 21,118. Capital, Raleigh.

WAKEFIELD, a parl. bor., market town, and par. of England, W. Riding, co. York, lower div. of Agbrigg, weapont. Agbrigg and Morley, on the Calder, 30 m. S.W. York. and 9 m. S. Leeds. Area of parish, comprising the townships of Wakefield, Stanley-cum-Wrenthorpe, Alverthorpe-with-Thornes, and the chapelry of Horbury, 9390 acres. Pop. in 1831, 24,538 ; in 1841, 98,321. The parliamentary borough, however, includes only the township of Wakefield (population in 1841, 13,777), with small portions of Alverthorpe and Stanley. The town is situated on the declivity of a hill sloping to the river, which is here crossed by a handsome stone bridge of nine arches. It is well built, the houses being mostly of brick ; streets spacious and regular, paved, and lighted with gas ; and since 1839 the town has been plentifully supplied with pure water by the W. Waterworks company. The market-place is small, but is well supplied with butchers' meat, fruit, vegetables, and other articles.

Wakefield is one of the principal country corn markets in England ; and the new corn exchange at the head of Westgate is in all respects suitable for the despatch of the important business of which it is the centre. On the S.W. side of Wakefield township the buildings advance in a continuous street into that of Alverthorpe, now embodied in the parliamentary borough ; and at the W. end of the town, and in Stanley township, are a great many buildings known by the name of East Moor, which also form part of the borough, which further comprises the small village of Thornes on the S. The latter, in fact, is connected with the town by an almost continuous line of houses and warehouses. (*Boundary Report.*) The parish church of All Saints is a handsome edifice of English architecture, 156 feet in length and 69 feet in width, founded in the reign of Henry III., but retaining few of its ancient features. It has a square tower, with battlements and pinnacles, surmounted by a spire, 237 feet in height, said to be the highest in the county. The living, a vicarage in the gift of the crown, is worth £547 nett. The church of St. John, in the district of the same name, erected in 1795, was, in 1815, rendered parochial jointly with All Saints : the living, a perpetual curacy, in the gift of the vicar of Wakefield, is worth £116 a year. In 1840 Trinity church, in George-street, built by subscription, and vested in trustees, under the Church Building Acts, was licensed for public worship. Here is also a Roman Catholic chapel, two places of worship for Independents, two for Wesleyans, and others for Friends, Primitive Methodists, Unitarians, and Baptists. In the centre of the bridge, projecting from its E. side, is a richly ornamented Gothic chapel, 30 feet in length by 24 in breadth, believed to have been founded by Edward III. ; but rebuilt and decorated by Edward IV., to commemorate the death of his father, Richard, duke of York, and his partisans at the battle of Wakefield. But (*quantum mutatus !*) this fine old building has latterly been degraded into a counting-house. In the market-place is a Doric cross with an open colonnade supporting a dome, and containing a small room in which the street commissioners transact business. The music saloon, subscription library and news room in Wood-street, is a handsome building. Here, also, is a literary and philosophical society, a mechanics' institute, a masonic lodge, a theatre, &c. The new and commodious corn exchange, at the top of Westgate, contains, exclusive of the exchange and several offices and shops, a very large assembly-room, with ante-rooms. The building called the Tammy-hall, for the exhibition and sale of woollens, has long been occupied as a worsted manufactory. The free grammar school, founded by Queen Elizabeth in 1592, and since enriched by various private benefactions, has a considerable income, and has long enjoyed a high reputation. It is open, free of expense, to the sons of the inhabitants desirous of a classical education, and has at present (1842) upward of 90 scholars. It has an attached writing school, and four exhibitions to Cambridge, and one to Oxford. Some very distinguished personages have been educated in this school, among whom may be mentioned Dr. John Potter, archbishop of Canterbury, author of the popular and excellent work on Grecian Antiquities, a native of the town ; Dr. Radcliffe, founder of the library at Oxford which bears his name, also a native of the town ; and Dr. Bentley, the eminent critic and scholar, a native of Oulton, in the vicinity. The green-coat school, founded in 1707, with an income of above £500 a year, clothes and instructs about 75 boys and 50 girls ; and among other schools is a charity school for 106 poor boys and 50 girls, two national schools established in 1839, a school of industry, a Lancastrian and several Sunday schools, in all affording instruction to nearly 2000 children. The proprietary school, opened in 1834, is a fine, collegiate-looking building, in which about 200 pupils receive a classical and com-

mercial education. The West Riding Pauper Lunatic asylum, erected in 1817, 1 m. N.E. of the town, is a noble building capable of accommodating above 400 patients. A dispensary and fever ward was established a few years since ; and there are well endowed almshouses for both sexes. Wakefield had formerly an extensive manufacture of woollens and worsted yarn ; but this, owing to the superior facilities for carrying on the manufacture enjoyed by other places, or their greater attention to the business, has greatly declined : still, however, a considerable business is carried on in the manufacture of spinning worsted and in dyeing ; and it is an important mart, not merely for corn, but also for wool and cattle. Great quantities of wool are sent from all parts of the surrounding country to be disposed of by the wool factors ; the cattle fairs, held every fortnight, are very extensive ; malting is also carried on to a considerable extent ; and there is a soap-work at Walton, in the vicinity, which produced, in 1839, 1,127,227 lbs. hard soap. The coal mines in the parish employ a great many hands.

Wakefield, though an inland situation, communicates by the Aire and Calder Navigation and various canals with Leeds, Hull, Manchester, and Liverpool, and several branch railways lead from the town to the different collieries in the vicinity. The W. Riding bank of Leatham, Tew and co. and the Wakefield and Barnsley banking company, have establishments here. The North Midland railway from Leeds to Derby passes by Oakenshaw (Wakefield station). 1½ m. S.E. from the town ; and the Leeds and Manchester railway passes the S. end of the town. The station near Kirkgate is within a short distance on the town side of the bridge over the Calder ; but the most important station is at Normanton, 5 m. N.E. from the town, at the junction of the North Midland, Manchester and Leeds, and York and North Midland railways. A very handsome hotel has been erected at this central station for the accommodation of passengers. The town is under the jurisdiction of a constable elected by the inhabitants. Quarter sessions are held in the courthouse, a handsome edifice in Wood-street, and petty sessions for the district in the courthouse every Monday, by the county magistrates. A court for petty causes, and the recovery of debts under £5, is held every three weeks by the steward of the manor. Here also is the W. Riding register office ; the office of the clerk of the peace, and the rolls office for the extensive manor of Wakefield. The house of correction for the W. Riding of Yorkshire at Wakefield ; it is built on an improved plan, and comprises a tread-mill, 307 cells, separate yards, a chapel, &c., and is now (1842) about to be considerably improved and enlarged. The prisoners are employed in weaving coarse cloth, calicoes, &c. The Reform Act conferred on Wakefield, for the first time, the privilege of sending a member to the House of Commons. Reg. electors in 1841–42. 689.

At the era of Domesday Survey, Wakefield, with its dependencies, was in the hands of the crown. The battle of Wakefield fought in the vicinity of this town on the 24th of December 1460, was one of the most important gained by the Lancastrians during the civil wars : the latter under Queen Margaret, having totally defeated the Yorkists under the duke of York, who fell in the battle, and whose son, the earl of Rutland, was assassinated immediately thereafter. In 1554, Wakefield was united to the Duchy of Lancaster. The manor which extends for more than 30 m. W. of the town, including above 150 towns. villages, and hamlets, and about one eighth part of the entire population of Yorkshire has belonged to the family of the duke of Leeds since 1700, and was devised by the late duke to his nephew, Walter Sackville Lane Fox, Esq., M.P., the present lord of the manor. Market days, Fridays, and every alternate Wednesday, for cattle and sheep. Fairs, July 4th and 5th and November 11th and 12th, for horses, cattle, hardware, &c. (*Allen's Yorkshire ; Baines' Gaz. and Direct. of York ; Parl. Reps., and private Information.*)

WAKEFIELD, p. t., Carroll co., N. H., 46 N.E. Concord, 527 W. Watered by ponds which give rise to Salmon Falls r. Lovewell's pond in the S. part is noted for a celebrated battle with the Indians in 1725. It contains two churches, a Congregational and Free-will Baptist, five stores, two fulling-mills, one cotton factory with 500 spindles, four grist mills, five saw-mills ; nine schools. 361 scholars. Pop. 1385.

WALCHEREN, the most W. of the islands of Holland, prov. Zealand, between the E. and W. Scheldt, having to the W. the N. sea, or Atlantic, and on the E. the Narrow Channel, by which it is separated from the island of Beveland. Middelburg, the cap., in the centre of the island is in lat. 51° 30' 6" N., long. 3° 37' 30" E. It is of a nearly circular shape ; length, E. and W., about 12 m. ; greatest breadth, 10 m. Pop. about 45,000. The surface is quite level, and below high water mark. Its W. side, or that facing the N. sea, is defended against its encroachments partly and principally by a line of sand-hills or *dunes*, and partly (at W. Capelle) by a very strong dyke. in other.

washed by the E. and W. Scheldt, are also defended by prodigious dykes. This is the most fertile, most populous, and best cultivated of all the Dutch islands : the inhabitants are mostly in easy circumstances; and besides Middleburg, the capital, it has the towns of Flushing and Vere, and several flourishing villages. It produces excellent crops of wheat and madder, considerable quantities of the last being sent to England. The climate, though not injurious to natives, is apt to exercise an exceedingly unfavourable influence over strangers. This was strikingly exemplified in the result of the ill-fated expedition of the British troops to Walcheren under the earl of Chatham in 1809 ; a great proportion of the force died on the spot from the attacks of a malignant marsh fever ; while many of those who survived had their constitutions shattered for ever. (*Geographie de Busching*, xiv. 140, Fr. ed. ; *Dict. Geog., &c.*)

WALDECK-PYRMONT, a principality of W. Germany, consisting of two separate portions, the most southerly and principal of which has Prussian Westphalia on the N. and W., while the most northerly and smaller, including the town of Pyrmont and adjacent territory, is almost surrounded by Lippe Detmold and Hanover. Aggregate area, 466 sq. m. Pop., in 1838, 56,480 ; all Lutherans, except about 500 Roman Catholics and 500 Jews. Surface hilly, having a mean elevation of 1000 ft. above the sea : its mean annual temp. is about 45½° Fah. Both Waldeck and Pyrmont belong to the basin of the Weser, their principal rivers being its tributaries, the Eder, Diemel, and Emmer. About 152,300 morgen of land, or nearly one third part of the surface is covered with forests; the trees being principally beech and oak. It produces an adequate supply of corn for home consumption, with potatoes, fruit of various kinds, and flax. Cattle-breeding is an important branch of industry, and within the last 20 years the cattle have been greatly improved by crossing with the breeds of Switzerland and the Tyrol. Copper, iron, salt, alabaster, marble, slates, &c. are raised ; and a large proportion of the prince's revenue is derived from the mineral waters of Pyrmont, which is one of the principal spas of Germany. Manufacture unimportant ; those of iron goods, which were formerly considerable, having, of late years, greatly declined : at present the principal are those of linen and woollen stuffs, paper, leather, and cotton hosiery. The chief exports are fine wool, corn, cattle, iron, mineral waters, and a few manufactured articles. The constitution is a limited monarchy, the diet consisting of 18 members of the nobility, 13 representatives of towns, and 10 deputies from the rural districts. The diet has the voting of the supplies, &c., but most part of the public business is carried on by a committee consisting of three members from each of the three estates composing the diet. The latter and the committee meet once a year. The territory is divided into five districts. Arolsen, a town of 2050 inhabitants, on the Aar, a tributary of the Diemel, is the cap. and seat of government ; Pyrmont, on the Emmer, one of the oldest watering-places in Europe, with about 3000 resident inhabitants. is the other principal town. Total public revenue estimated at 230,000 rix dollars a year ; public debt 680,000 do. Waldeck-Pyrmont holds the 29th place in the German confederacy, having one vote in the full diet, and one in the committee along with the principalities of Hohenzollern, Reuss, Lippe, and Liechtenstein. It contributes 519 men to the confederate army. (*Berghaus ; Allg. Lander u. Völkerkunde*, iv. &c.)

WALDO, county, Me. Centrally situated toward the S. part of the state, and contains 812 sq. m. Bounded E. by Penobscot river and bay. Watered by St. George and Marsh rivers. Penobscot bay and river afford great facilities for navigation. It contained, in 1840, 27,826 neat cattle, 55,318 sheep, 10,451 swine ; and produced 78,304 bushels of wheat. 6749 of rye, 52,784 of Indian corn, 30,103 of barley, 84,068 of oats, 1,066,424 of potatoes, 4675 pounds of sugar. It had 196 stores, capital $368,895 ; 18 lumber yards, capital $54,400 ; capital invested in the fisheries $79,373 ; 13 fulling-mills, 41 grist-mills, 117 saw-mills, 26 tanneries ; four academies, 1946 students ; 267 schools, 14,740 scholars. Pop. 41,509. Capital, Belfast.

WALDOBOROUGH, p. t. and port of entry, Lincoln co., Me. Bounded S.W. by Muscongus bay. Watered by Muscongus river. It has considerable navigation and ship-building. and contains 23 stores, three fulling-mills, four grist-mills, 10 saw-mills, three tanneries, one printing-office, one weekly newspaper ; one academy, 45 students ; 29 schools, 1168 scholars. Pop. 3661.

WALES. *See* ENGLAND AND WALES.

WALES, p. t., Erie county, N. Y., 18 m. S.E. Buffalo, 272 m W. Albany, 369 W. Drained by Seneca creek. It has three stores, one fulling-mill, one woollen factory, three grist-mills, nine saw-mills, one tannery, one pottery ; 16 schools, 678 scholars. Pop. 3661.

WALES (NEW SOUTH). *See* AUSTRALIA and SYDNEY. It appears from the official returns, that the pop. of New South Wales amounted, in 1841, to 128,726, having

more than doubled within seven years ! This extraordinary increase is principally to be ascribed to the stimulus given of late years to immigration.

WALKER, county, Ga. Situated in the N.W. part of the state, and contains 700 sq. m. Drained by Chatooga river and Chickamauga river. It contained, in 1840, 6306 neat cattle, 2087 sheep. 15,012 swine ; and produced 17,363 bushels of wheat, 1014 of rye, 236,494 of Indian corn, 16,650 of oats, 8985 of potatoes, 14,468 pounds of tobacco, 76,307 of cotton. It had seven stores, eight grist-mills, 14 saw-mills, two tanneries, eight distilleries ; one college, 15 students; nine schools, 17½ scholars. Pop. : whites, 5583 ; slaves, 946 ; free coloured, 43 ; total, 6572. Capital, La Fayette.

WALKER, county, Ala. Situated toward the N.W. part of the state, and contains 1500 sq. m. Drained by branches of Black Warrior river. It contained in 1840, 5043 neat cattle, 1493 sheep, 12,360 swine ; and produced 5587 bushels of wheat, 128,030 of Indian corn, 7584 of oats, 4767 of potatoes, 1210 pounds of rice, 4588 of tobacco, 110,917 of cotton. It had one store, six grist-mills, one saw-mill, one tannery, one distillery ; 111 scholars in schools. Pop.: whites, 3920 ; slaves, 941 ; free coloured, 1 ; total, 4032. Capital, Jasper.

WALKER, p. t., Centre co., Pa., 94 m. N.W. Harrisburg, 186 W Drained by Little Fishing creek. It has four stores, one furnace, two grist-mills, two saw-mills, two tanneries ; four schools, 160 scholars. Pop. 1180.

WALLACHIA and MOLDAVIA (an. *Dacia Transalpina*), two contiguous principalities of S.E. Europe, nominally included in European Turkey, but in reality under the protection of Russia. They lie principally between the 44° and 48° N. lat., and the 22° and 28° E. long., and are together of a crescent shape, enclosing Transylvania on the W. and N.W. Wallachia comprises the S., and Moldavia the E. and N. parts of the united territory ; the former, from the W. round to the S.E., is divided from Persia and Bulgaria by the Danube, and the latter on the E. and N. from Bessarabia by the Pruth, and on the N.W. from the Bukowine by some branches of the Carpathians. The Sereth forms the principal line of separation between the two provs., the area, pop., &c. of which are as follow :—

	Area in sq. m.	Population in 1837-39.	Chief Towns.
Wallachia . .	27,500	1,747,815	Bucharest.
Moldavia . .	17,020	1,419,105	Jassy.
Total . .	44,520	3,166,920	

The two principalities have each their own peculiar government ; but their history is so intimately connected, and the forms of their respective governments, the language, manners, and customs of their inhabitants have always been so much alike that both may be best noticed together.

Physical Geography and Products.—The Carpathians, where they separate these provs. from Transylvania, usually vary in height between 3000 and 5000 ft., though some summits rise to 7000 or 8000 ft. in elevation. (*Boué Turquie.*) From these mountains the surface gradually declines to the S. and E. through regions of a most picturesque character, and hill ranges and valleys of great fertility, till it terminates in a level and marshy plain from 12 to 20 leagues in breadth, which, with parts of Bulgaria and Bessarabia, constitutes what may be considered the second in point of size and importance of the great European plains. The whole country is thoroughly well watered, being intersected by the Olt, or Aluta, Jalomnitza, Argish, Sereth, and other large affluents of the Danube, most of which are navigable for a considerable distance, and which annually inundate the surrounding country. The winter is very severe, particularly in Moldavia. which prov. is open to the full force of the N.E. wind : and the Danube, with its various tributaries, is generally frozen over for six weeks, during which period the ice is often strong enough to bear the passage of the heaviest artillery. In the first two months of the year, the snow is so very thick, that the communication is every where carried on by means of sledges. A damp spring succeeds ; in May the summer bursts in on a sudden, during which season, though the heat during the day be excessive, the nights are cool, or even cold. The pleasantest season is the autumn, from September to the middle of November. The climate, generally speaking, cannot be said to be unhealthy ; but in the plains along the Danube endemic fevers occasionally prevail, and in the hill region goitres are extremely common. Earthquakes sometimes occur, but happily they are rarely very violent. Most part of the country towards the Danube consists of a rich alluvial soil ; elsewhere tertiary and calcareous, and in the Carpathians primary formations are prevalent. In the latter, ores of gold, silver, mercury, iron, copper, and other metals are found, and several mines were opened during the Russian occupation of these provinces. At present, except salt mines, few others are wrought, and the gold obtained is chiefly by washing the river deposits, an occupation almost solely confined to the gipsies, who pay their tribute partly in gold dust. Petro teum, sulphur, nitre, and coal are met with, but not much sought after. The salt of Wallachia, which is of the purest kind, forms an article of sale in all the bazaars of the country. Oak, pine, fir, beech, maple, elm, ash, walnut, white mulberry, &c., are the chief forest trees. The climate is unsuitable for the fig and olive; but apples, pears, plums, cherries, apricots, &c., come to perfection with little culture. Asparagus is indigenous; cabbages and artichokes grow to a great size, and cucumbers and melons are among the principal articles of food. Deer, wild goats, and hares are very numerous. Wolves, when pressed by hunger, come down from the mountains, and commit much devastation among the flocks and herds; but at other times, like the bears and other wild animals in these provs., they rarely attack a man.

Land and Agriculture.—The land principally belongs to the nobles, or *boyars*, though it is seldom cultivated by them on their own account. No regular system seems to be pursued as respects the arrangements between the landholders and cultivators; but, for the most part, the cultivators pay to their landlords a tithe of their whole produce of corn, and, in addition to this, they are bound to pay the land-tax and other burdens, and to work 30 days in the year for their landlords. M. Hagemeister says that, though nominally free, the cultivators are virtually enslaved, and that they are still subject to great oppression. (*Commerce of the Black Sea*, 112.) Owing to their subdivision among the children, on the death of a parent, there are now but few large properties. The mode of tillage does not much differ from that in other parts of eastern Europe. Oxen are usually employed for field labour. Manure is never used; but after a crop of corn, the land is left fallow for a season, and then sown with wheat, barley, or maize, which are the principal crops. Oats and rye are but little cultivated, maize constituting the principal part of the food of the people, and barley being used in distillation, and in the feeding of horses and poultry. No probable estimate can be formed of the ordinary produce of the wheat crops; though there can be no doubt that if the cultivators could calculate on a ready market, it might be very greatly increased. Speaking generally, the wheat of Moldavia, though inferior to that of England or Poland, is from 15 to 90 per cent. superior to that of Wallachia. In the latter prov. the wheat is mostly soft, whereas in Moldavia it is mostly hard. (*Hagemeister*, 104.)

It is extremely difficult to form any fair estimate of the price at which Moldavian and Wallachian wheat might be shipped from the Danube. The kilo, or principal corn measure, differs in the two provinces, and the value of the piastre is also constantly fluctuating; and in addition to these causes of uncertainty, the duties laid on corn when exported are liable to vary, and every body knows how much the price is affected by variations of demand. On the whole, however, we are inclined to think that the cost of corn at Galacz (which see), the principal port of the principalities, has been materially underrated. In some years it has been as low as 15s. a quarter; but, in 1838, when 171,813 quarters were exported, the price varied from 23s. to 28s. 7d. a quarter. Perhaps, under a fair average demand, it might vary from 20s. to 22s. a quarter.

Immediately after having been gathered in, the corn is trodden out by horses and cattle, and laid up in pits, in consequence of which it acquires an earthy flavour, unfavourable to its sale in foreign markets. (*Hagemeister, ubi suprà.*) A good deal of wine is made on the hill slopes, particularly in Moldavia. It is mostly of very indifferent quality; but Mr. Thornton says that some of the wines are pleasant and wholesome, resembling the light wines of Provence, and that they are largely exported to Russia and Transylvania. The strength and spirit of the wine, he further states, are increased by a process common among the rich proprietors, and practised also in Russia. "At the first approach of severe cold, the wine tubs are exposed to the severity of the weather in the open air: in a few nights the body of wine is encircled with a thick crust of ice; this is perforated by means of a hot iron, and the wine, thus deprived of its aqueous parts, is drawn off clear, strong, and capable of being preserved for a long time." (ii. 320.) Ovid may have had a similar practice in view when he says, of the wine at the place of his banishment,

"Udaque consistunt formam servantia testæ
Vina; nec hausta meri, sed data frusta bibunt."
De Trist, l. iii. el. x. 23, 24.

The rearing of cattle, however, rather than agriculture, has always been the principal employment of the Wallachians; and the Illyrian word *vlack*, signifying a herdsman, appears to have been the original root whence their name, and that of the country they inhabit, has been derived. The chief sources of wealth in both principalities are, in fact, their flocks and herds, which find abundant and nutritive pastures in winter in the plains, and in summer on the Carpathians. The number of sheep and goats in Wallachia has been estimated at 4,000,000, and the annual produce of wool in both provs. at 40,000 quintals. The latter, which is divided into three sorts, *cigeus*, *stroose*, and canary, is sold at from 3 to 3½, from 2 to 2½, and from 1 to 1½ piastres the oke. There is no public establishment for the washing of wool, but private individuals sometimes wash it at home, which commonly increases the price about 40 per cent.; but the greater part is sold in the grease. Manures have not yet been introduced into the country. About one sixth part of the whole clip is sent to France.

"The cattle of Moldavia are larger and better than those of Wallachia, for the simple reason that all branches are more backward in the latter country. Taking their size into account, Moldavia is richer in horned cattle than Wallachia, which, on the contrary, takes the lead in the number of sheep. In both the principalities the cattle and sheep are bought up by the dealers, who pasture them throughout the summer, in the view of selling or slaughtering them in August and September. A pair of good oxen commonly costs from 70 to 83 roubles; a cow is worth from 25 to 30 roubles; a sheep 3 to 4 roubles. Pasturage for a herd of 150 head is hired at a price of 150 to 250 roubles. The number of horned cattle fattened in this manner for sale may amount to 60,000 or 80,000 head. The buyers commonly advance one third or a half of the price for some months. The high price of cattle proceeds from the great consumption in Austria, into which numerous herds are annually sent, particularly from Moldavia. Bessarabia and Wallachia also furnish some, and these are always the largest and finest animals. Many of the inhabitants on the Austrian frontier are engaged in this trade. A part of the cattle, and especially the sheep, traverse the Danube for sale in Bulgaria; and previously to 1826 the number of cattle and sheep thus exported was estimated at 150,000 for Wallachia, and at 100,000 for Moldavia." (*Hagemeister*, p. 149.) There are several breeds of horses; and the best, which are those of Moldavia, are bought up in large numbers for the Austrian and Prussian cavalry. (*Thornton*, ii. 322-3.) The buffalo thrives in Wallachia, and poultry and game of all kinds are in great plenty; but the flesh of the latter, as well as beef, pork, and mutton, are said to be insipid and inferior. The honey, wax, and hare-skins are, however, of the best quality; of the last, about 500,000 are annually exported. Timber, yellow berries, butter, cheese, hides, staves, and masts of various sizes and descriptions, are the other chief articles of export: the Moldavian oak timber, which is much finer than the Wallachian, is well calculated for the construction of vessels, and many of the Turkish ships are built of it, and fitted out with masts and ropes of Moldavian growth. The yellow berries are inferior to those of Smyrna, and only in demand when the crop of Asia Minor is deficient: the export is, however, estimated at 600,000 okes a year. (*Wilkinson*, p. 75-78. &c.

Manufactures and Trade.—Coarse woollen cloth, hardware, common linen fabrics, glass, jewellery, saddlery, vehicles, &c., are made, and there are estimated to be altogether 3000 factories of different kinds in the two principalities, inclusive of distilleries. (*Colson, Moldavie*, &c.) But, for the most part, manufactured goods are imported from other parts of Europe, in return for the raw produce of the principalities. Galacz is the grand centre for the foreign trade of the provs., and the point whence the imported goods are sent to Bucharest and Jassy. Coffee, sugar, pepper, rum, lemons, oranges, and foreign wines are the principal imports. The local consumption of the first in both provs. is estimated at 600,000 okes, of the second, 900,000 do., and of the third, 30,000 okes yearly. Cotton and woollen goods, earthenware, &c., are brought chiefly from Germany and England, and Russia supplies the principalities with large quantities of furs, taking in return spirits, wines, and specie. (*Wilkinson, &c.*)

People and Condition.—Notwithstanding the various irruptions of the Goths, Gepidæ, Lombards, Huns, Turks, and Turks into these provs., their inhabs. at the present day appear to be, with comparatively little intermixture, the descendants of the ancient Dacians, to whom, as represented on Trajan's column at Rome, both in features and costume, Mr. Paget says, the modern Wallachs bear a remarkable resemblance. (*Paget's Hungary*, &c., ii. 169.) They still call themselves *Roumeni*, or *Romans*, and their country *Zara Rouman-Eska*; and it is a curious fact, that in a province which was among the last annexed to the Roman empire, and in a situation more exposed than any other to the irruptions of invaders from the E., the common dialect now spoken combines, together with many Sclavonic and Greek terms, a very large infusion of purely Latin words; so much so that a stranger speaking in Latin is generally understood by the natives. "In appearance the common Wallach present a decided difference from either Bulgar, Slave, or German. In height he is below the median, and generally rather slightly built and thin. His features are often fine, the nose arched, the eyes dark, the hair long, black, and wavy; but the expression is too often one of fear and cunning to be agreeable. The dull, heavy look of the Slowak is seldom seen among them, but still more rarely the proud self-respecting carriage of the Magyar." (*Paget.*) The long continued misgovernment to which the Wallachians and Moldavians have been subject, has corrupted their morals, weakened their energies, and given them most of the vices of slaves. All the worst features of Turkish despotism were exhibited in these provinces in their most revolting and odious forms. And hence we need not be surprised, that though not without hospitality, and some other redeeming qualities, the inhabitants are treacherous, revengeful, indolent, besotted, and often cowardly. The women, indeed, on whom most of the labour devolves, do not share in the idleness of the men, but their industry exhibits much want of method and thriftiness, and "to be as busy as a Wallach woman, and do as little," is a proverbial comparison among the German settlers in Transylvania, &c. (*Paget's Hungary*, ii. c. vi. vii., to which, and the art. *Hungary*, vol. i. p. 1139, in this Dict. we beg to refer the reader.)

Mr. Paget says, "I had pitied the Wallachs of Transylvania, till I saw their brethren of the principalities of Wallachia and Moldavia. I never saw two countries of their extent so fruitful in resources. Yet, with all their advantages, I never saw a country so thinly populated, nor a population so excessively poor and miserable." (ii. 497.) The peasants' dwellings throughout the country are all built in the same style and of the same size. The walls are of clay, and the roofs thatched with straw, neither of which is calculated to protect the inmates from the inclemency of the weather. In winter the people retire to caves under ground, kept warm by fires made of dried dung and branches of trees; and which, at the same time, serve for cooking their scanty food. Each family, however numerous, sleeps in one of these subterraneous habitations, their beds consisting of a piece of coarse woollen cloth, which serves in the double capacity of mattress and covering. These underground dwellings, have, in fact, been the winter residence of the inhabitants of Scythia from the remotest antiquity, and have been admirably described by Virgil:

> " Ipsi in defossis specubus secura sub alta
> Otia agunt terra, congestaque robora totasque
> Advolvere focie ulmos, ignique dedere.
> Hic noctam ludo ducunt, et poculá laeti
> Fermento atque acidis imitantur vitea sorbia."
>
> *Georg.* lib. iii. line 376, &c.

The ordinary food of the peasants consists of the flour of Indian corn, mixed into a dough with milk. For the first few days after Lent some indulge themselves in meat, but the greater part cannot afford this, and content themselves with eggs fried in butter. (*Wilkinson*, 157-8.) In their holidays they spend most of their time in the village winehouses, where they amuse themselves with dancing, witnessing the vagaries of gipsies, &c. They are no longer *adscripti glebæ*; and if dissatisfied with their masters, may, on giving due notice, quit their habitations and pass over to the estate of another, with their families and movables: this, however, is more an apparent than a real advantage, and, as already stated, the peasants are still in a very oppressed condition. The gipsies continue in a state of regular slavery, belonging either to the government or to private individuals. Their entire number in both principalities is estimated at 150,000, about 80,000 of whom belong to the government. Some are employed as domestic servants; the rest are suffered to stroll about the country, breeding cattle or horses, manufacturing wooden and iron utensils, or employing themselves as showmen, musicians, &c. For this liberty they bind themselves not to quit the country, and pay an annual tribute of 30 piastres per man, if belonging to the government; it is said that desertions by gipsies are rare.

The nobility and clergy are in general exempted from taxes for the service of the state, and from the demands of private creditors! They are in consequence overbearing, extravagant, and dissolute. Their education has hitherto been little superior to that of the common people; and though ostentatious in their dress and equipage, their manners present little refinement. In Moldavia, which is the most civilized of these provinces, the great landed proprietors bestow considerable attention to the management of their estates; but in Wallachia these are mostly left to the care of agents. The boyars who hold no place under government, spend their leisure in absolute idleness, or in visiting each other, to kill time. "They have," says Thornton, " adopted indiscriminately the vices, without inheriting the vivacity of the Greeks." This statement applies, however, rather to their past than to their present state. Of late years some considerable improvements have been introduced; and though society be still very backward, it is, at all events, much superior to what it was under the Turkish regime.

Government, &c.—For a lengthened period these provinces were governed by *vaivodes*, or princes appointed by the Sultan from among the Greeks of Constantinople, and, during the continuance of this system, the country was a prey to every species of abuse. Since 1829, however, they have been placed under the sole protection of Russia; and whatever may be the defects of Russian policy, there can be no question that they have gained immensely by the change. The prince, or hospodar, both in Wallachia and Moldavia, is elected for life from among the boyars of the first rank, by an assembly composed of those boyars, and of deputies of the inferior boyars, the academic bodies, and merchants; but his election must be approved by Russia, and the investiture is then given by Turkey. A diet of the clergy and boyars (the class which contribute nothing to the state) meets to vote the supplies, and to discuss other propositions of the prince; but no organic changes can be made without the sanction of Russia. Wallachia is divided into 18, and Moldavia into 13 districts, each of which has a prefect governor, a receiver-general of taxes, a civil tribunal, consisting of a president and two other judges; and Moldavia has a director of police and a town council in each municipality. Judges are removable at the pleasure of the superior authorities. The legal codes are founded upon the civil law and the customs of the principalities; but though the system of jurisprudence has been much amended, many reforms remain to be effected, especially in the administration of the laws, which is said to be most corrupt. Nearly all the population belongs to the Greek church, and every village has a small church or chapel, with one or more priests, who act as curates. The ecclesiastics of this order are chosen from among the people, from whom they are little distinguished in appearance, and whose avocations they follow when not engaged in their clerical functions. The generality of them can neither read nor write, and merely recite the formula of their service from memory: they have, however, an unbounded influence over the ignorant population of these countries. There are many large and rich monasteries, and four or five seminaries for the education of the superior clergy.

Public instruction, though still backward, appears to have advanced since 1832. Colleges and Lancastrian schools are established in the principal towns; and the latter have by this time probably spread into the rural districts. According to the *Journ. of Educ.*, the higher classes in these provinces have of late set about improving their national dialect with remarkable vigour; and it appears probable that their language will ultimately be rendered much nearer akin to the ancient Roman than even the Italian. There is a printing-press at Bucharest, which is in active employment; and translations of foreign as well as original works are continually being produced by native authors. (Vol. vii. p. 173.) This is a consequence, and by no means the only salutary one, that is likely to follow the establishment of the principalities under Russian protection, or at least their enfranchisement from the debasing rules of the Turks.

The *military force* is organized on the plan of the Russian army, and the staff officers are principally Russians. The militia is formed by the peasantry, in the proportion of two members for every 100 families; but along the banks of the Danube all the inhabitants capable of bearing arms are organized into a military force, auxiliary to the quarantine service. The total effective force of the provinces amounts to about 53,000 men, including nearly 5000 irregular cavalry. There is no artillery, nor are there any fortresses in either province.

The public revenues are derived from the capitation tax of 30 piastres (the piastre is about 3¼d. sterling) per head on the rural population; from 30 to 120 do. a year on the manufacturing classes, and 60 to 240 do. on merchants, &c.; from the government lands and rights of pasturage, fees, fines, customs, salt monopoly, and in Moldavia a tax on the incomes of the clergy. These sources produced in 1839, in Moldavia, 10,467,209 piastres, and in Wallachia, 16,293,299 do. In the latter province the expenditure nearly balanced this sum; but in the former there was a surplus of 2,797,800 piastres. Both provinces pay an annual but not very heavy tribute to the Porte. (*Colson, Etat présent de Moldavie, &c.*)

History.—Since the conquest of this country by Trajan, it has never ceased to be under foreign dominion. It was alternately in the power of the barbarians and the Greek emperors till the 13th century, when it appears to have been occupied by the Hungarians. Early in the 15th century it was conquered by the Turks, to whose empire it has since been generally tributary, but the last war between Turkey and Russia entirely destroyed the influence of the former, and rendered the latter paramount. (*Colson, Etat présent de Moldavie et de Valachie; Wilkinson's Wallachia and Moldavia; Thornton's Turkey; Boué, La Turquie; Walsh; Paget, &c.*, passim.)

WALLINGFORD, a parl. and mun. bor., market town and par. of England, co. Berks., hund. Moreton, on the Thames, 12½ m. N.W. Reading. The old parl. and mun. bor. were co-extensive with four small pars., having an area of 370 acres and a pop., in 1831, of 2545. But the modern parl. bor. includes several additional pars., partly in Berks and partly in Oxfordshire, having an aggregate area of about 18,000 acres, and a pop., in 1841, of about 6000. The modern mun. bor. is about twice as extensive as formerly. The town 3 m. from the main road, between London and Oxford, is pretty well built, paved, and lighted with portable gas, and, for its size, contains a considerable number of neat private dwellings, and a few of a superior character. (*Bound Rep.*) The river is here crossed by a handsome stone bridge, 300 yds. in length, with 19 arches, and four draw bridges, built in 1809, upon the site of a former structure of the same description. St. Mary's church has a fine tower, crowned with pinnacles, which appears to have been erected in 1658: the living, a rectory, worth £148 a year, is in the gift of the crown. St. Peter's church, a handsome edifice, rebuilt about 50 years since, is surmounted by a spire: the living, a rectory in the gift of W. S. Blackstone, Esq., is worth £100 a year. The living of St. Leonard's, worth £103 a year, is in the gift of the crown. There are also places of worship for Friends, Independents, Wesleyans, and Baptists. The market-house and town-hall are respectable buildings. Among numerous charitable institutions are the free grammar school, founded in 1659, and several almshouses. No manufacture of any consequence is now carried on: the chief business is that of malting; and, in general terms, Wallingford "may be described as a neat country town, respectably inhabited, and a place of no other importance than as the market-town for the surrounding country." (*Bound. Rep.*) It is a borough by prescription, its earliest existing charter being a copy of one dating from the reign of Henry I. It has returned two members to the House of Commons since the 23d of Edward I.; the right of voting, previously to the Reform Act, having been vested in individuals paying church and poor rates. Registered electors, in 1839-40, 366, under the Municipal Reform Act; it is governed by a mayor, four aldermen, and 12 councillors, and other officers. It has a commission of the peace, court of record, court leet, and a jail for the confinement of offenders before committal.

WALSALL, a parl. and munic. bor. and par. of England, co. Stafford, hund. Offlow, on a small tributary of the Tame, in the centre of one of the principal coal and iron districts 7 m. N.W. Birmingham. Area of par. 7990 acres, all of which is included in the parl. bor., except an outlying portion called Walsall Wood. Pop., in 1831, 15,066. It is situated on the declivity and summit of a low hill, and is pretty well built, having "the appearance of a compact and flourishing town: several houses of the value of from £90 to £40 a year are now building, and every indication of a thriving community is here exhibited." (*Municipal Bound. Report,* 1837.) The main streets are broad, well paved, lighted with gas, and well supplied with water; though there be but little that is prepossessing in their general appearance. In the environs, however, there are many handsome villas, with some picturesque scenery. Bloxwich, about 2 m. N. from the town, but included in the parl. bor., is a populous village, the inhabitants of which, like those of Walsall, are mostly occupied in the hardware manufactures. The parish church occupies a commanding situation on the top of the hill: it is an ancient, spacious, cruciform structure, with a tower surmounted by a lofty spire; and was thoroughly repaired in 1821. The living, a vicarage worth £368 a year, is in the gift of the earl of Bradford, lord of the manor. St. Paul's chapel, a handsome Grecian edifice, completed in 1826, is a perpetual curacy worth £50 a year. There are places of worship for Independents, Wesleyans, and Unitarians, and two Roman Catholic chapels, one of which is a handsome Greek building. The subscription library, established in 1800, has recently been enlarged: it contains reading and news rooms, and has a Doric colonnade, 30 ft. in height. The grammar school, founded and endowed by Queen Mary, in 1557, open to all the boys of the parish, has an annual income of £780; and subsidiary schools, dependant on the principal, have been established in different parts of the parish. It has also an English school in which 120 boys are instructed; a bluecoat charity; a national and several Sunday schools; and numerous charitable benefactions for the relief of the poor. In the time of Henry VI. an endowment was left for the annual distribution of 1d. to every person in the parish; but, in 1825, this useless endowment was judiciously appropriated to the erection and maintenance of 11 almshouses. Walsall is said to have been a borough by prescription: its earliest existing charter was granted by Henry VI. Under the Municipal Reform Act, it is divided into three wards, and is governed by a mayor, five other aldermen, and 18 councillors. The Reform Act conferred on it, for the first time, the important privilege of sending one member to the House of Commons. Registered electors, in 1839-40, 837.

It has a commission of the peace, a weekly court of petty sessions, and a court of record for the recovery of debts between 60s. and £20. The jail is of a very indifferent description. Its command of coal and iron has made Walsall a considerable seat of the hardware business: the manufacture of saddlers' ironmongery, that is, the making and plating of bridles, spurs, and stirrups, the mountings for coach and carriage harness, &c., being the staple employment of the town. It has, also, some brass and iron foundries, and a considerable trade in malt. A branch of the old Birmingham canal comes up to the W. end of the town, which is 3½ m. from Willenhall station on the Grand Junction railway. The bank of Walsall and S. Staffordshire, established in 1835, has its principal office in this town. Walsall, though not mentioned in Domesday book, is known to be a town of considerable antiquity; but it is not connected with any historical event of importance. Market days Tuesdays. Fairs, 24th February, Whit Tuesday, and the Tuesday before Michaelmas; the two latter are mostly for cattle and cheese. (*Parl. Rep. and Priv. Inf.*)

WALSHAM (NORTH) a market town and par. of England, co. Norfolk, hund. Tunstead, on a level, about 6 m. from the sea, and 13 m. N.N.E. Norwich. Area of par., 4018 acres. Pop. of do., in 1831, 2615. The town consists of three streets which meet so as to form an irregular triangle. The parish church is a large venerable old edifice. Its lower fell down in 1724; but it has a fine south porch of flint and stone, and a font with a very rich wooden cover, of tabernacle work. The vicarage is annexed to the rectory of Antingham; the livings being together worth £328 a year; patron, the crown. Here are several dissenting chapels, a free grammar school, with an income of nearly £250 a year, at which Lord Kelson was partly educated; a Sunday school, with a small endowment, and several minor charities. The market-cross, originally erected in the reign of Edward III., was rebuilt in 1600. Two annual courts-baron are held here, and petty sessions for the hund. by the co. magistrates. Market-day, Tuesday. Fairs, three times a year, chiefly for cattle and the hiring of servants. In 1600, this town was visited by a most destructive fire, which consumed 118 houses, besides barns, stables, &c. On Walsham heath, near the town, is a stone cross, erected to commemorate a victory of Henry Spencer, bishop of Norwich, over a band of rebels, in 1382. (*Beauties of England and Wales,* vol. xi.; *Parl. Reports, &c.*)

WALTHAM ABBEY. or HOLY CROSS, a market town and par. of England, on the W. border of the co. of Essex, hund. Waltham, on the Lea, 11½ m. N.N.E. London. Area of par., 11,870 acres. Pop. of do., in 1841, 4388. The town originally founded in the time of Canute, consists chiefly of one spacious and irregular street. The church was formerly the nave of the church of an opulent and famous monastery, founded here by Harold II., of which there are now but few remains. This venerable relic, though much disfigured and mutilated, contains some most interesting specimens of the ornamented columns, semicircular arches, and other characteristics of the Norman style of architecture. It is about 90 ft. in length, by about 48 ft. in breadth. At the W. end is a heavy square embattled tower, 86 ft. in height; but this is comparatively modern, and bears the date of 1558. The inside of the church bears witness to the iconoclastic zeal of the Reformers, and to the bad taste and miserable parsimony of those by whom they have been followed; the ornamental parts having been defaced and whitewashed, and the brasses torn from the gravestones! Harold, and his two brothers slain with him at the battle of Hastings, were interred in this church. The living, a curacy worth £337 a year, is in the gift of the trustees of the earl of Norwich. The Baptist and Wesleyans have, also, places of worship. The educational and charitable institutions comprise a free school for 20 boys and 90 girls, with an endowment producing about £138 a year; another endowed school, for the education of five boys, and several bequests for the support of Sunday schools, almshouses, and the general relief of the poor. The revenues of the monastery amounted, at its dissolution in 1538, to £900 according to Dugdale, and to £1080 according to Speed. Market-day, Tuesday; fairs twice a year. At present, however, the town derives its entire importance from the gunpowder mills established here on account of government. These, which were originally acquired from private parties in 1787, consist (in 1842) of four mills each having two pair of stones. The moving power is water; the establishment is in all respects, in the most efficient state, and the powder produced of the very best quality. During peace the consumption of powder by government amounts to about 10,000 barrels a year, of which about 600 are supplied by the works now under consideration. The latter, however, were not erected in the view of providing the entire supply of gunpowder, which, in periods of war sometimes exceeds 80,000 or 100,000 barrels a year, but partly as a check on the combinations of the manufacturers, and partly as affording the means of readily trying and fairly appreciating each new discoveries and experiments as may from time to time be made or suggested in the proper

ation of so important an article, and in these respects they have been completely successful. At Enfield Lock, about 2 m. below Waltham, a manufactory of small arms is also carried on upon account of government. At present (1842) it is almost wholly occupied in the making of new percussion muskets, of which it supplies about 10,000 a year. (*Private Information*.)

In the hamlet of West Waltham, or Waltham cross, about 1 m. W. from Waltham Abbey, in Hertfordshire, is one of the stone crosses erected by Edward I., at the different places where the corpse of his beloved wife, Queen Eleanor, rested on its way from Hareby, near Grantham, where she died, to Westminster Abbey. Only three of these crosses now remain. It had originally been a very fine structure; but the ornaments are now much defaced. (*Beauties of England*, vols. vii. and x.; *Farmer's History of Waltham Abbey; Parl. Papers, &c.*)

WALTHAMSTOW, a village and par. of England, co. Essex, hundred Precunctee, on the Lea, a tributary of the Thames, 5 m. N.E. London. Area of parish, 3690 acres. Pop. of do., in 1831, 4258. The village, on the borders of Epping Forest, is formed by the union of several hamlets; the houses, among which are many villas and country seats, being generally detached, and interspersed with trees and gardens. The church, built on an eminence, is a spacious structure in a mixed style of architecture, with a tower at its W. end: it was repaired and enlarged in 1817, and has several ancient monuments. The living is a vicarage, to which is attached the curacy of St. John's being worth, together, £772 a year: patron, W. Wilson, Esq. The Unitarians and Independents have each chapels; and to the latter a school is attached, in which 30 girls are educated, 90 of whom are clothed at the expense of the foundation. The free school, founded in 1542, has an endowment yielding £85 a year. Although established for the education of more than 30 pupils, it was lately attended by only five. It has also national and infant schools, with numerous well endowed almshouses, and benefactions to a considerable amount, for the relief of the poor. Some copper-mills and other works are established in this parish, on the banks of the Lea. The parish has an exclusive local jurisdiction; and is governed by a council of 17 members, presided over by the vicar and churchwardens. Courts leet and baron are held here when required. (*Guide to London and its Environs, &c.*)

WALTON, county, Ga. Situated N. of the centre of the state, and contains 390 sq. m. Bounded N.E. by Appalachee r. Drained by branches of Oxmulgee r. It contained in 1840, 3590 neat cattle, 1570 sheep, 11,703 swine; and produced 18,658 bushels of wheat, 171,495 of Indian corn, 18,560 of oats, 2800 of potatoes, 627,750 pounds of cotton. It had eight stores, 12 grist-mills, 10 saw-mills, three tanneries, seven distilleries. Pop.; whites, 6583; slaves, 3825; free coloured, 1; total, 10,909. Capital, Monroe.

WALTON, county, Flor. Situated in the W. part of the ter., and contains 1584 sq. m. Bounded E. by Choctawhatchee river, S. by Choctawhatchee bay. Drained by branches of Yellow r., the principal of which is Shoal r. Alaqua river is navigable 15 m. from the bay. It contained in 1840, 10,959 neat cattle, 366 sheep, 3950 swine: and produced 25,046 bushels of Indian corn, 6902 of potatoes, 52,192 pounds of cotton. It had four stores, six grist-mills; seven schools, 131 scholars. Pop.; whites, 1189; slaves, 231; free coloured, 41; total, 1401. Capital, Euchee Anna.

WALTON, p. t., Delaware co., N. Y., 97 m. S.W. Albany, 321 W. Drained by the W. branch of Delaware r. and its tributaries. It contains two churches, a Presbyterian and an Episcopal, four stores, one fulling-mill, two grist-mills, 13 saw-mills; 17 schools, 603 scholars. Pop. 1846.

WALWORTH, county, Wis. Situated in the S.E. part of the ter., and contains 675 sq. m. Drained by Turtle river and its branches, and by branches of Fox r., one of which issues from Geneva lake in its S. part. It contained in 1840, 2861 neat cattle, 410 sheep, 6380 swine; and produced 59,560 bushels of wheat, 1499 of barley, 40,837 of Indian corn, 35,155 of oats, 48,455 of potatoes, 13,050 pounds of sugar. It had one store, three grist-mills, seven saw-mills, one distillery; four schools, 107 scholars. Pop. 2611. Capital, Delavan.

WALWORTH, p. t., Wayne co., N. Y., 18 m. N.W. Lyons, 205 W. Albany, 362 W. Drained by branches of Mud or., and small streams flowing into lake Ontario. It contains a Baptist and Methodist church, three stores, two saw-mills, two tanneries; nine schools, 236 scholars. Pop. 1734.

WANDSWORTH, a large village and parish of England, co. Surrey, hund. Brixton; on the Wandle, near its confluence with the Thames, 5 m. S.W. London. Area of parish 1890 acres. Pop. of do., in 1831, 5879.

Wandsworth consists principally of one broad thoroughfare between two eminences called the E. and W. hills. "It is in parts noisy and bustling, in other parts rural and quiet; in parts clean and pleasant, in others low and dirty;

and the residents are for the most part in the extremes of rich and poor." (*Guide to London and its Environs*.) The old church, which was mostly rebuilt in 1780, is a plain brick edifice with a heavy square tower at its W. extremity; the living, a valuable vicarage, worth £840 a year, is in the gift of W. Borrodaile, Esq. The new church of St. Anne, erected by act of parliament in 1824, is an elegant edifice of Grecian architecture, with a handsome portico and a steeple of two circular arches; the living, a curacy worth £162 a year, is in the gift of the vicar of Wandsworth. Here also are meeting-houses for Friends, (to which two schools are attached,) Baptists, Independents, and Wesleyans. The free endowed school, founded in 1710, has been incorporated with the national school, and affords instruction to above 200 boys and 100 girls, to some of whom clothing is supplied. It has also a Lancastrian school, in which more than 200 children are educated. A school of industry, attended by 40 girls, and various other charities, among which those of Alderman Smith, a native of the town, who died in 1627, are the most valuable. The manufactures of Wandsworth are more considerable than might have been expected; that of hats was introduced by the French refugees towards the end of the 17th century; and there are works for bolting cloth, printing calicoes and kerseymeres, making coach and livery lace, dyeing (especially in scarlet), with corn, oil, iron, and white lead mills, vinegar works, and distilleries. The Southampton railway passes a little to the E. of the village. Petty sessions for the hundred are held weekly, and there is a court of record for the recovery of debts under £5. Fairs on the first three days of Whitsun week, for horses, cattle, pigs, and toys. (*Beauties of England and Wales, Surrey; Guide to Env. of London, &c.*)

WANTAGE, a market town and parish of England, co. Bucks., hund. Wantage, on a small tributary of the Thames, at the E. extremity of the vale of the White Horse, 23½ m. W.N.W. Reading. Area of parish, which includes the hamlets of Chariton and Grove, 7530 acres. Pop., in 1831, 3223. The town is irregularly built at the intersection of the high roads from Hungerford to Oxford, and from Farringdon to Wallingford, which form its principal streets. The church, a handsome cruciform structure, has a square embattled tower rising from its centre, and some fine monuments. The living, a vicarage, worth £503 a year, is in the gift of the dean and canons of Windsor. It has also places of worship for Independents and Wesleyans; a free grammar school, with an income of about £300 a year; some almshouses, founded in 1650, with an endowment of £100 a year, Sunday schools, &c. The town lands produce an income of about £450 a year, which is spent on the relief of the poor, the repair of highways, and the support of a school. Sacking, twine, and tarpaulins, are manufactured on a small scale. The market is celebrated for its fine corn, a great deal of the best seed-wheat being brought thither by the vale farmers. Its trade is facilitated by a branch of the Wilts and Berks canal, which comes up to the town. Wantage was made a borough after the conquest; but it no longer retains that distinction. A manorial court is, however, held in it once a year, and petty sessions for the hundred every Saturday.

WANTAGE t., Sussex county, N. J. Bounded E. by Wallkill r., along which is a swamp known by the name of the drowned lands. It has 11 stores, one fulling-mill, 10 grist-mills, eight saw-mills, two tanneries, five distilleries; one academy, 26 students, 22 schools, 573 scholars. Pop. 3008.

WARD, t., Randolph co., Ia. It has three stores, two grist-mills, two saw-mills; four schools, 116 scholars. Pop. 1138.

WARDEIN (GROSS or GREAT; Hungar, *Nagy-Varad*), a fortified town of Hungary, co. Bihar, of which it is the cap.; on the Körös, towards the borders of Transylvania, 30 m. S.W. Debreczin. Pop. estimated in the Austr. Nat. Encyc. at 10,000; but, according to Berghaus, it amounts, including its suburbs, to upwards of 16,000. It is the residence of a Roman Catholic and a united Greek bishop, a Greek protopapas, &c., and the seat of the county assembly, council, a commissariat department, &c. It has a royal academy, many other superior schools, an abbey, and various religious establishments, with manufactures of silk stuffs and earthenware. Mr. Paget says, "Gross Wardein is really one of the prettiest towns I have seen for a long time. Its wide well-built streets of one-storied houses, and extensive market-places, are quite to the taste of the Magyar, who loves not the narrow lanes, and high houses of his German neighbours. But the glory of Gross Wardein is in its gilded steeples, its episcopal palace, its convents, and its churches; and, although of the latter, the 70 which it formerly boasted are now reduced to 22, they are quite sufficient for the inhabitants. Prince Hohenlohe, of miracle-working memory, is now the occupant of this see. His elevation to the bishopric has, however, completely extinguished the light of miracle!" (*Paget's Hungary, &c.* ii. 518.)

WARDSBOROUGH, p. t., Windham co., Vt., 15 m. N.W.

Brattleboro', 190 m. S. by W. Montpelier, 447 W. Drained by a branch of West r., which affords water-power. It contains two churches, a Congregational and Baptist, three stores, one fulling-mill, one cotton factory with 60 spindles, three grist-mills, six saw-mills, three tanneries; seven schools, 396 scholars. Pop. 1102.

WARE, a market town and parish of England, co. Herts, hund. Braughin, on the great N. road, and on the Lea, 18½ miles N. London. Area of parish, 4430 acres. Pop., in 1831, 4214. It consists chiefly of one thoroughfare, nearly a mile in length, and lined in general with substantial and well built houses. The church is a large cruciform structure, mostly in the decorated and perpendicular styles; it has an embattled tower at the W. end, and within are several fine monuments, and a handsome font. The living, a vicarage, annexed to that of Thundridge, worth £333 a year, is in the patronage of Trinity college, Cambridge. Here are also chapels for Independents, Wesleyans, and Friends. The market-house, supported on arches, comprises a good assembly-room. The educational and charitable institutions include a free grammar-school, of very old foundation, attended by about 30 pupils; another free school, with a small endowment, established in 1834: a national, and two other schools, chiefly supported by subscription; numerous almshouses, a lying-in charity, and funds for distribution among the poor, yielding an income of £326 a year. (Analytical Digest of Charity Reps.) Ware has a considerable trade in corn, and malting is extensively carried on. It is governed by three constables and four head-boroughs: petty sessions for the division are held weekly, and a court baron once a year. Market-day, Tuesday. Fairs twice a year for horses and cattle.

Chadwell spring, near the town, assisted by a cut from the Lea, gives rise to the New river, an artificial stream brought from Hertfordshire, for the supply of water to the metropolis. Though the source of the New river, in a direct line, be not more than 20 m. from London, its course, including its windings, is nearly 40 m. This important work was completed in 1613, principally by the exertions of the famous Sir Hugh Middleton. Though very unproductive at first, it has since been a source of vast wealth to its proprietors, as well as of advantage to the city.

WARE, county, Ga. Situated in the S. towards the E. part of the state, and contains 3440 sq. m. Drained by Santilla and Suwannee rivers. Okefinoke swamp occupies its S.E. part. This great marshy lake is 280 miles in circumference, and extends into Florida. In wet seasons it appears like an inland sea, with several fertile islands. St. Mary's river rises in this swamp. The county contained in 1840, 20,918 neat cattle, 302 sheep, 12,269 swine; and produced 893 bushels of wheat, 26,746 of Indian corn, 18,658 of oats, 2900 of potatoes, 627,750 pounds of cotton, 11,935 of sugar. It had nine grist-mills; one college, 13 schools, 478 scholars. Pop.: whites, 2189; slaves, 132; free coloured, 2; total, 2323. Capital, Waresboro'.

WARE, p. t., Hampshire co., Mass., 22 m. E. by S. Northampton, 385 W. Bounded W. by Swift r. Watered by Ware r., which affords good water-power. Incorporated in 1761. The village in the E. part of the township, on Ware river at the falls, has become a large manufacturing place, with upwards of 1500 inhabitants. The township contains four churches, two Congregational, a Methodist, and a Baptist, a bank, six stores, one cotton factory with 350 spindles, two grist-mills, four saw-mills; 14 schools, 611 scholars. Pop. 1890.

WAREHAM, a parl. and munic. bor., market town, river-port, and par. of England, co. Dorset, hund. Winfrith, in Blandford div., on a peninsula between the rivers Frome and Piddle, about 1 m. above their confluence with Wareham harbour, the most westerly arm of Poole harbour, 30 m. S.S.W. Salisbury, and 102 m. S.W. London. The modern borough includes the whole of the three parishes, portions only of which were comprised in the ancient borough, together with those of Corfe Castle and Bere-Regis, and parts of two other adjacent parishes; having a total area of 23,890 acres, and a population, in 1831, of 5774.

The site of the town shelves gradually towards the S., and is mostly surrounded by flat marshy land. Having been nearly destroyed by fire on the 25th July, 1762, it has been built on a regular plan, and consists chiefly of two wide streets, intersecting each other at right angles. The houses, built of brick, and tiled or slated, are generally in good condition. It is surrounded by a remarkable ancient mound, the space between which and the town is now laid out in market gardens. Each of its rivers is here crossed by a bridge, that over the Frome being a handsome structure, erected in 1779. Down to a late period it was neither watched nor lighted. (Bound and Municipal Reports.) Of eight churches which formerly existed here, only one, St. Mary's, is now used for public worship, though two more, Trinity and St. Martin's, are made use of for other purposes; the former being converted into a national school, and the

1040

latter being merely used for reading the funeral service. St. Mary's, a spacious and ancient edifice, originally attached to a priory, is built in a mixed style, though principally of the decorated character. (Rickman.) It has a handsome tower, and contains some ancient monuments. All the livings of Wareham are now united in one rectory. Two more ancient churches that had fallen to decay, were taken down within the last century, and on the site of one of them the present town hall has been erected. The Independents, Wesleyans, and Unitarians have each places of worship. The educational and charitable institutions comprise a national school, held in the old church of the Holy Trinity; a small endowment for the education of 30 poor children; alms-houses for six men and four women, and some minor charities.

The trade of Wareham consists chiefly in the export of the fine clay found in the neighbourhood to the Staffordshire and other potteries, and in shipping of vegetables from the market gardens round the town for Poole and Portsmouth. A good many of the inhabitants are also employed in knitting stockings, and in the manufacture of shirt buttons. The port, which was formerly considerable, is now nearly choked up, being only accessible to vessels of from 25 to 30 tons; but vessels of 60 tons ascend to within about 1 m. of the town, and those of 200 tons may anchor at Russell's quay, about 3 m. from the town. "The inhabitants may be said to consist of persons of middling circumstances, and a few retired officers and independent persons, retail tradesmen, and those deriving a subsistence from the small craft. There is no poverty in the place, but its population and the number of its houses are probably less now than formerly." (Parl. Bound. Report.) We doubt, however, whether there be any good foundation for this surmise. The borough returned two members to the House of Commons from the 13th of Edward I. down to the Reform Act, the right of voting being exercised, since 1747, by the inhabitants paying scot and lot; but under the act now referred to, it returns only one member along with Corfe-Castle, Bere-Regis, &c. Registered electors in 1830–40, 496. The borough has a commission of the peace, a court leet, held annually; and a court of record, opened monthly, but now of little utility. Market day, Saturday. Fairs nine times a year, mostly for cattle, cheese, and hogs. (Bound. Rep.; Munic. Corp. Append., &c.)

WARHAM, p. t., Plymouth co., Mass., 52 m. S.S.E. Boston, 446 W. Bounded S. by the head of Buzzard's bay. Watered by Agawon r., a fine mill stream which flows into the head of the bay. It contains a Congregational church, a bank, eight stores, one cotton factory with 660 spindles, four grist-mills, three saw-mills, one paper-mill; five schools, 139 scholars. Pop. 2009.

WARMINSTER, a market town and parish of England, co. Wilts, hund. Warminster, on the Willey, at the W. extremity of Salisbury plain, 20 m. N.W. Salisbury, and 15 m. S.E. Bath. Area of parish, 5450 acres. Population of do. in 1831, 6106. The town consists chiefly of one spacious, clean, and well-paved thoroughfare, nearly 1 m. in length, the houses being mostly of freestone. The church of St. Denis is a spacious structure, in the perpendicular style, with a square central tower; the living, a vicarage, worth £394 a year, is in the gift of the bishop of Salisbury. The new church, erected in 1830, partly by subscription, and partly by a grant from the parliamentary commissioners, is a perpetual curacy, worth £100 a year, in the gift of the vicar of Warminster. Besides a chapel of ease, there are several dissenting places of worship; a free-endowed grammar-school, affording instruction to 30 boys, with national and Lancastrian schools, supported by subscription. Warminster had formerly the most extensive trade in malt of any town in the W. of England: and this branch of industry, though it has declined, is still largely carried on. The manufacture of broad cloths and kerseymeres has been, in a great measure, superseded by that of silk, in which many women and children are employed. The trade in corn is considerable; the market being one of the most extensive in this part of the country. The town is under the jurisdiction of a high constable, deputy constables, and tything-men chosen at the annual manorial court. The quarter sessions for the county are held here in July; petty sessions, monthly, by the county magistrates: and a court of requests for the recovery of debts under £5 is held alternately in this town and Westbury. Warminster is supposed to have been a Roman station, from the discovery of coins, weapons, a tessellated pavement, and other antiquities in the vicinity. Market-day, Saturday; fairs three times a year, for corn, sheep, hogs, and cheese.

The manor and lordship of Warminster is the property of the marquis of Bath; and about 4½ m. W. from the town, on the confines of Somersetshire, is Longleat house, the only magnificent seat of that nobleman. The park in which it is situated is of great extent, and is finely laid out. (Lewis's of England and Wales, &c.)

WARNER. p. t., Merrimac co., N. H., 17 m. W.N.W. Concord, 481 W. Drained by Warner r. a tributary of Contoocok r. Incorporated in 1774. It contains two churches, a Congregational and Baptist, five stores, three grist-mills, six saw-mills, one paper-mill, two tanneries; 23 schools, 361 scholars. Pop. 2139.

WARREN, county, N. Y., Situated in the E. towards the N. part of the state, and contains 912 sq. m. Drained by Hudson r. and Schroon branch, which here unite. Iron ore is abundant. It contained in 1840, 9696 neat cattle, 22,775 sheep, 8053 swine; and produced 12,961 bushels of wheat, 17,567 of rye, 63,476 of Indian corn. 24,647 of buckwheat, 103,733 of oats, 291,134 of potatoes, 43,821 pounds of sugar. It had 48 stores, seven fulling-mills, one woollen-factory, 14 grist mills, 99 saw-mills, two powder-mills, nine tanneries; 104 schools, 4119 scholars. Pop. 13,422. Capital, Caldwell.

WARREN, county, N. J. Situated in the W. towards the N. part of the state, and contains 350 sq. m. Drained by Paulinskill, Pequest, and Pohatcong rivers. Organized in 1824. Musconetcong or Schooley's mountain bounds it on the S.E. It contained in 1840, 12,718 neat cattle, 19,599 sheep, 22,517 swine; and produced 194,796 bushels of wheat, 184,877 of rye, 311,955 of Indian corn, 106,300 of buckwheat, 173,902 of oats, 142,662 of potatoes. It had 53 stores, one furnace, one cotton factory with 216 spindles, 21 flouring-mills, 35 grist-mills, 94 saw-mills, four oil-mills, nine tanneries, two distilleries, one brewery; 92 schools, 5533 scholars. Pop. 90,366. Capital, Belvidere.

WARREN, county, Pa. Situated in the N.W. part of the state, and contains 833 sq. m. Drained by Alleghany river and its tributaries, Conewango, Broken Straw, and other creeks. It contained in 1840, 8340 neat cattle, 13,081 sheep, 6334 swine; and produced 30,980 bushels of wheat, 9247 of rye, 37,298 of Indian corn, 13,570 of buckwheat, 64,890 of oats, 147,274 of potatoes, 91,318 pounds of sugar. It had 98 stores, five fulling mills, three furnaces, two flouring-mills, 16 grist-mills, 123 saw-mills, six tanneries, two printing-offices, two weekly newspapers; one academy, 50 students; 69 schools, 2874 scholars. Pop. 9278. Capital, Warren.

WARREN, county, Va. Situated in the N. part of the state, and contains 200 sq. m. Watered by Shenandoah r. It contained in 1840, 5424 neat cattle, 7229 sheep, 6790 swine; and produced 148,379 bushels of wheat, 17,300 of rye, 219,390 of Indian corn, 4881 of buckwheat, 57,844 of oats, 16,325 of potatoes. It had 11 stores, one fulling-mill, 13 flouring-mills, 20 grist-mills, 12 saw-mills, three tanneries, eight distilleries; one academy, 14 students; six schools, 290 scholars. Pop.: whites, 3851; slaves, 1434; free coloured. 342; total, 5627. Capital. Front Royal.

WARREN, county, N.C. Situated in the N. part of the state and contains 391 sq. m. Drained by Roanoke r. and its tributaries, and Fishing creek and its branches. It contained in 1840, 11,100 neat cattle, 7860 sheep, 30,856 swine; and produced 25,851 bushels of wheat, 395,351 of Indian corn, 80,113 of oats, 24,730 of potatoes 2,574,296 pounds of tobacco, 380,954 of cotton. It had 10 stores, four flouring-mills, 26 grist-mills, seven saw-mills, one printing-office, one weekly newspaper; three academies, 71 students; 10 schools, 192 scholars. Pop.: whites, 4390; slaves, 8200; free coloured 329; total, 12,919. Capital, Warrenton.

WARREN, county, Ga. Situated towards the E. part of the state, and contains 560 sq. m. Bounded S.W. by Ogeechee r., N. by Little r. The Georgia railroad passes through it. It contained in 1840, 6349 neat cattle, 4505 sheep. 17,934 swine; and produced 13,351 bushels of wheat, 268,320 of Indian corn, 6860 of oats, 5892 of potatoes, 262,555 pounds of cotton. It had 16 stores, two cotton factories with 1123 spindles, four flouring-mills, five grist-mills, six saw-mills; one academy, 75 students; 5 schools, 147 scholars. Pop.; whites, 5176; slaves, 4536; free coloured, 77; total, 9879. Capital, Warrenton.

WARREN, county, Miss. Situated in the W. part of the state, and contains 600 sq. m. Bounded W. by Mississippi r., S.E. by Big Black r. Watered by Yazoo r. The Vicksburg and Brandon railroad passes through it. It contained in 1840, 22,648 neat cattle, 3812 sheep, 25,890 swine; and produced 543,140 bushels of Indian corn, 32,119 of oats, 62,390 of potatoes, 16,049,200 pounds of cotton. It had 26 commission houses in foreign trade, 112 retail stores, 10 saw-mills; two academies, 184 students; seven schools, 176 scholars. Pop.: whites, 5223; slaves, 10,494; free coloured, 104; total, 15,820. Capital, Vicksburg.

WARREN, county, Tenn. Situated near the middle of the state, and contains 960 sq. m. Bounded N.E. by the Caney fork of Cumberland r. Drained by its tributary, Collins r. and its branches. It contained in 1840, 11,107 neat cattle, 9722 sheep. 10,141 swine; and produced 53.347 bushels of wheat, 3835 of rye, 462,085 of Indian corn, 108,117 of oats, 12,012 of potatoes, 4384 pounds of cotton. It had five stores, 10 grist-mills, four saw-mills, one powder-mill, five tanneries, six distilleries, one printing-office, one weekly newspa-

per; 29 schools, 1161 scholars. Pop.: whites, 9366; slaves, 1349; free coloured, 88; total, 10,803. Capital, McMinnville.

WARREN, county, Ky. Situated S.W. of the centre of the state. and contains 612 sq. m. Bounded N.W. by Green r. Drained by Big Barren r. and its branches. It contained in 1840, 34,523 neat cattle, 21,729 sheep, 52,329 swine; and produced 80,180 bushels of wheat, 1976 of rye, 715,566 of Indian corn, 183,210 of oats, 18,576 of potatoes, 1,039,890 pounds of tobacco. It had one commission house in foreign trade, 20 retail stores, seven grist-mills, three saw-mills, 12 tanneries, 23 distilleries, one printing-office, one weekly newspaper; one academy, 30 students; four schools, 125 scholars. Pop.; whites, 11,078; slaves, 4307; free coloured, 161; total, 15,446. Capital, Bowling Green.

WARREN, county, O. Situated toward the S.W. part of the state, and contains 400 sq. m. Bounded N.W. by great Miami r. Watered by little Miami r. and its branches. It has some remarkable ancient mounds; and contained in 1840, 21,308 neat cattle, 33,031 sheep, 56,847 swine; and produced 248,665 bushels of wheat, 7750 of rye, 1,231,321 of Indian corn, 327,314 of oats, 44,912 of potatoes, 187,295 pounds of sugar. It had 80 stores, four fulling-mills, three woollen factories, five flouring-mills, 22 grist-mills, 19 saw-mills, two paper-mills, 13 tanneries, one distillery, four potteries; 60 schools, 5356 scholars. Pop. 23,141. Capital, Lebanon.

WARREN, county, Ia. Situated in the W. part of the state, and contains 350 sq. m. Organized in 1832. Watered by Wabash r. and its branches. It contained in 1840, 7963 neat cattle, 9515 sheep, 15,851 swine; and produced 32,192 bushels of wheat, 1447 of rye, 414,046 of Indian corn, 80,955 of oats, 15,543 of potatoes, 8900 pounds of sugar. It had eight stores, one flouring-mill, four grist-mills, 19 saw-mills, six tanneries, one distillery; eight schools, 185 scholars. Pop. 5656. Capital, Williamsport.

WARREN, county, Ill. Situated in the W. towards the N. part of the state, and contains 900 sq. m. Formed in 1825, organized in 1830. Bounded W. by Mississippi river. Watered by Henderson r. and its branches, and by head branches of Spoon river. It contained in 1840, 9760 neat cattle, 7141 sheep, 23,082 swine; and produced 75,243 bushels of wheat, 4120 of rye, 293,843 of Indian corn, 1394 of barley, 97,400 of oats, 21,884 of potatoes, 22,619 of sugar. It had two commission houses in foreign trade, 10 retail stores, three flouring-mills, four grist-mills, 20 saw-mills, one tannery, one distillery; 19 schools, 569 scholars. Pop. 6720. Capital, Monmouth.

WARREN, county, Mo. Situated towards the E. part of state, and contains 350 sq. m. Bounded S.W. by Missouri river. It contained in 1840, 4686 neat cattle, 3683 sheep, 10,950 swine; and produced 13,693 bushels of wheat, 132,345 of Indian corn, 21,185 of oats, 6596 of potatoes, 338,400 pounds of tobacco, 4582 of sugar. It had 27 stores, five grist-mills, six saw-mills, one oil-mill, one tannery, two distilleries; five schools, 104 scholars. Pop.: whites, 3555; slaves, 696; free coloured, 2; total, 4253. Capital, Warrenton.

WARREN, p. t., one of the capitals of Lincoln co., Me.. 37 miles S.E. Augusta, 613 W. Incorporated in 1776. Watered by St. George r., which affords water-power, and is navigable for sloops 15 m. from the sea to Warren. It contains a courthouse, 14 stores, two grist-mills, five saw-mills, one academy, 80 students; 19 schools, 853 scholars. Pop. 2228.

WARREN, p. t., Worcester co., Mass., 66 m. W.S.W. Boston, 385 W. Incorporated by the name of Western, in 1741, received its present name in 1834. Chickapee river affords good water-power. It contains two churches, a Congregational and Universalist, six stores, two fulling-mills, one woollen factory, three cotton factories with 5590 spindles, one forge, three grist-mills, four saw-mills; 10 schools, 373 scholars. Pop. 1290.

WARREN, p. t., Bristol co., R. I., 14 m. S.E. Providence, 414 W. Bounded W. by Narragansett bay. The village is situated on the E. side of Warren r., 2 m. N. of the bay, and has an elevated and healthy site, 7 m. S.E. Providence in a direct line. It is well situated for navigation. There are engaged in the whaling business, 15 ships, four barques, and one brig, comprising 6022 tons, and in other branches of the foreign and coasting trade one ship, eight brigs, three schooners, and six sloops, with a tonnage of 2304, making the total tonnage of the port 8326. There are 14 wharves, with commodious store-houses, &c. It contains three churches, a Baptist, Methodist, and Episcopal, two banks, an academy, a female seminary, 11 private and four public schools, and about 200 dwellings. The Warren asylum for the poor has commodious buildings, and a farm of 75 acres under good cultivation. There are in the township seven commercial houses in foreign trade, 22 retail stores; four academies, 270 students; seven schools, 195 scholars. Pop. 2437. Rhode Island college was founded here in 1766, and removed to Providence in 1770.

WARREN, p. t., Herkimer co., N. Y., 64 m. N.W. Albany,

375 W. It has seven stores, three fulling-mills, two woollen factories, three grist-mills, six saw-mills, two tanneries; 11 schools, 495 scholars. Pop. 3003.

WARREN, t., Somerset co., N. J., 6 m. N.E. Somerville. Watered by Middle Brook. It contains five stores, three grist-mills, three saw-mills, four distilleries; four schools, 85 scholars. Pop. 1601.

WARREN, p. b., capital of Warren co., Pa., 205 m. N.W. Harrisburg. 297 W. Situated on the N. side of Alleghany r., at the junction of Conewango cr. The streets cross each other at right angles, dividing the whole into 16 blocks, with 500 lots, one third of an acre each. Near the centre, at the intersection of two streets, 100 ft. wide, is a public ground containing four acres for public buildings. It contains a courthouse, jail, county offices, an academy, two churches, a Methodist and Presbyterian, a bank, 14 stores, one fulling-mill, one flouring-mill, two grist mills, four saw-mills, two printing-offices, two weekly newspapers; one academy, 50 students; three schools, 90 scholars. Pop. 737.

WARREN, p. t., capital of Trumbull co., O., 163 N.E. Columbus, 303 W. Situated on Mahoning r., on the Pennsylvania and Ohio canal. The village contains one mile square, with streets crossing each other at right angles, and has three churches, 16 stores, various mechanic shops, two printing-offices, 200 dwellings, and about 1500 inhabitants. There are in the township 12 schools, 640 scholars. Pop. 1996.

WARREN, t., Belmont co., O. It has one academy, 14 students; four schools, 112 scholars. Pop. 2408.

WARREN, t., Jefferson co., O. It has five stores, four flouring-mills, one grist-mill, four saw-mills, one oil-mill; five schools, 250 scholars. Pop. 1945.

WARREN, t., Tuscarawas co., O. It has five stores, two grist-mills, four saw-mills; 10 schools, 402 scholars. Pop. 1173.

WARREN, t., Marion co., Ia. It has nine stores, one cotton factory with 500 spindles, one grist-mill, three-saw-mills, two tanneries, one brewery; two schools, 69 scholars Pop. 1374.

WARREN, t., Putnam co., Ia. It has nine stores, one fulling-mill, one flouring-mill, five grist-mills, three saw-mills, two tanneries, one distillery, one pottery; six schools, 486 scholars. Pop. 2901.

WARRENSBURG, p. t., Warren co., N. Y., 6 m. N.W. Caldwell, 69 m. N. Albany, 444 W. Hudson river here unites with Schroon branch. It contains two churches, a Methodist and Presbyterian, five stores, one fulling-mill, one grist-mill, nine saw-mills, two tanneries; 10 schools, 405 scholars. Pop. 1466.

WARRICK, county, Ia. Situated in the S.W. part of the state, and contains 360 sq. m. Bounded S. by Ohio river. Drained by Big and Little Pigeon creeks. It had in 1840, 5400 neat cattle. 5400 sheep, 1890 swine; and produced 20,500 bushels of wheat, 250,800 of Indian corn, 32,500 of oats, 20,000 of potatoes, 516,000 pounds of tobacco. It had 90 stores, two flouring-mills, 10 grist-mills, five saw-mills, six tanneries. six distilleries; 22 schools, 710 scholars. Pop. 6321. Capital, Booneville.

WARRINGTON, a parl. bor., market-town, and parish of England; county Lancaster, hund. W. Derby; in a low situation on the Mersey, here crossed by a stone bridge, 17 m. E. by S. Liverpool. Area of parish, 12.260 acres. Pop., in 1831, 19,155, in 1841, 21,837. The parliamentary borough, however, comprises only the townships of Warrington and Lachford, with portions of that of Thelwall, and may have at present (1842) a population of about 20,000. The town principally consists of four main streets, one or two of which are spacious and contain some handsome buildings; but the other streets are for the most part narrow and badly drained, and very little appears to have been done for the improvement of the town, except its being paved and lighted with gas in 1821. (Bound. Rep., &c.) The most important public buildings are the sessions-house erected in 1830; the market-hall over which are the assembly rooms; three cloth halls, theatre, &c. The parish church, which is of Saxon origin, and existed at the time of the conquest, is a large massive cruciform structure. The tower, which rises from the intersections of the transepts, was rebuilt in 1696; the interior of the church, which is lofty and handsome, contains two chapels, and some fine ancient monuments. Beneath the chancel has lately been discovered an ancient crypt, now converted into a vestry. At the entrance of the churchyard are two handsome gates. The living, a rectory, has under it the perpetual curacies of St. Paul and the Holy Trinity. There are two Roman Catholic chapels, founded severally in 1823 and 1836, and 10 other places of worship for different sects. The free grammar school, founded by Sir T. Boteler in 1526, has an annual income of between £700 and £800; but at present (1842) it is attended by only about 29 day scholars and 12 boarders. The blue-coat school, established in 1677, has since received legacies and benefactions amounting to up-
1042

wards of £2080, and has now an annual income of £691; 30 boys and 20 girls; children of settled inhabitants of the town, are lodged, maintained, and clothed in the building, and 170 boys and 40 girls admitted as day scholars. A general subscription library was established in 1758. A society was formed here early in the last century for the purpose of affording assistance to widows and orphans of clergymen in the archdeaconry of Chester; and the relief dispensed by it amounts at present to about £1000 a year. A handsome building, in the Elizabethan style of architecture, is in progress for the education of the orphan daughters of clergymen. A dispensary and branch of the royal humane society are among the other numerous charities, and there is a small but commodious infirmary, to which two wards for cases of fever are attached.

The appearance of Warrington is far less bustling and animated than formerly. Until the opening of the railway it was the great thoroughfare between Manchester and Liverpool; 70 public carriages daily passing through it between these great emporiums, whereas at present (1842) there is not one! Its traffic with the above towns is, however, very considerable; for though not strictly speaking a port, it possesses, by means of the Mersey and Irwell navigation, many of the advantages of a port. At spring-tides the Mersey rises from 10 to 12 ft. at Warrington bridge, and vessels of from 70 to 100 tons burden navigate the river up to this point. The Warrington and Newton railway, a branch of that between Manchester and Liverpool, 4½ m. in length, was completed in 1831, at a cost of £65,968. It now forms a portion of the Grand Junction line.

Warrington is distinguished by the number and variety of its manufactures. The making of sailcloth and sacking was formerly carried on here upon a very large scale, but it is dwindled to insignificance. At present, among the many that are carried on, cotton spinning and power loom weaving occupy a prominent place. In 1832, there were six cotton-mills at work in the parish, employing in all 1126 hands. (Rep. on Mills and Factories.) The refining of sugar, though not entirely relinquished, is not a leading branch of industry; but the soap manufacture continues to be of considerable importance. 365,011 lbs. of soft soap having been made in the town in 1839. The manufacture of flint and plate glass is carried on upon a large scale, and has long ranked among the principal businesses of the town. Warrington is also the principal seat of the manufacture of "Lancashire tools," under which designation are comprised files of the very best quality, chisels, graver's tools, watch and clock maker's tools, &c., and in some of its factories may be seen collections of the articles in quantities of unrivalled excellence. Pin-making is also carried on; and Warrington has long been celebrated for its malt and ale. The soil, too, in the neighbourhood being especially suitable for horticultural purposes, gardening is here well understood and successfully practised. The government of the town is vested in police commissioners and constables chosen annually in October at the court leet of the lord of the manor, acting under the superintendence of the magistrates of the hundred. Petty sessions are held every fortnight. The Reform Act conferred on Warrington, for the first time, the privilege of sending one member to the House of Commons. Registered electors, in 1839–1840, 632. Market days, Wednesday and Saturday. Fairs, 18th July and 30th November, each lasting 10 days, for horses, cattle, and cloth, and every other Wednesday for cattle. (Baines's Hist. of Lancaster; Parl. Papers; Priv. Infor.)

WARRINGTON, p. v., capital of Fauquier county, Va. 102 m. N. by W. Richmond, 51 W. It contains a courthouse, jail, a town-hall, three churches, an Episcopal, a Methodist and Presbyterian, several stores, a printing-office, issuing a weekly newspaper, 200 dwellings, and about 1300 inhabitants.

WARSAW (Pol. Warszawa, Fr. Varsovie), the capital city of Poland, palat. Masovia, on the Vistula, 658 m. S.E. Petersburg, lon. 52° 14' 28" N., long. 21° 2' 47" E. Pop. in 1839, including its suburbs, 139,671. The city, which with its gardens and suburbs, covers a great extent of ground, is on the left or W. bank of the river, which is here about as broad as the Thames at Westminster bridge, being connected with the suburb of Praga on the right, by a bridge of boats. A suspension bridge was some years since projected instead of the latter, but the project has not hitherto been carried into effect.

Warsaw, being situated partly in a plain and partly on an ascent gradually rising to the river's bank, has a magnificent appearance from the Petersburg road. But though the contrary has been affirmed by some travellers, the impression of grandeur is not supported on entering the town. It has, indeed, many fine palaces, public buildings and noble mansions, and, latterly, its private houses have been improved, by prohibiting the construction of new buildings of wood. But its streets, though spacious, are badly paved, badly lighted, and dirty; the greater part of the houses in the city, and sh.. more in the suburbs, are mean and ill constructed, above one fourth part of their number being at this moment of wood; and the whole town exhibits a painful contrast of wealth and poverty, civilization and barbarism, luxury and misery. The suburb of Praga, on the E. bank of the river, once strongly fortified and extensive, is now all but deserted. There are still, however, several other suburbs of large extent; and those adjacent to the city proper are included within its rampart and ditch.

The principal public building is the *Zamek*, a huge edifice, formerly the palace of the kings of Poland, and that in which the emperor still resides when he visits Warsaw. The hall of the Polish diet, a splendid gilt ball-room, and the national archives of Poland, are in this building; but the fine paintings of Canaletti, Bacciarelli, &c., with the library and other treasures, have been removed since 1831 to the Russian capital. There are several other royal palaces. That called the palace of Casimir, which was appropriated to the university, has in its square a statue of Copernicus. The Palais de Saxe is a large building in one of the finest squares. "At the back of this palace are the principal public gardens in the interior of Warsaw, which resemble in some respects the park at Brussels, though considerably larger. Another handsome public garden, much frequented at the fashionable hour of 12, belongs to what is called the government palace. This latter is, perhaps, one of the most chaste and really beautiful architectural elevations in the Polish capital. It is strictly in the Italian style, and contains the national theatre, custom-house, high tribunals, and offices of the minister of the interior. The palace of the minister of finance, which is quite modern, forms, with the new exchange, a very imposing object at the end of the street leading to the Breslau gate. The Marieville bazaar is a large square, the four sides of which consist of covered arcades, with dwellings for the merchants above, and shops for the merchandise under them; the latter amount to about 300, besides several warehouses. A great number of churches are to be found in the city; some of which are of really colossal dimensions, as the cathedral of St. John, and the church of the Holy Cross. In the former are an altarpiece of great merit by Palma Nova, and a large standard wrested from the Turks by Sobieski at the siege of Vienna. The Lutherans have also a magnificent church, erected at an expense of £25,000, and superior in beauty and boldness of design to all the Catholic churches in the place, having a dome and tower of prodigious elevation. Which way soever a traveller turns, he cannot fail to pass some one of the monuments which stand in the squares to commemorate the reign of a sovereign, or the achievements of a Polish warrior. The colossal statue of Sigismund III., cast in bronze, gilt, and placed on a lofty pillar of marble of the country, produces a very good effect; and the equestrian group in bronze of Poniatowski, &c., by Thorwaldsen, is another monument worthy of admiration.

"Independently of the public gardens, Warsaw may be said to have in its vicinity some of the finest drives and promenades in Europe for width and extent. The numerous avenues of the Ujasdow, planted with lofty lime and chestnut trees, are the rendezvous of nearly the entire population of Warsaw on Sundays and other holidays, and are admirably calculated for horse and sledge races, both of which take place here. In the immediate vicinity is the royal villa, formerly the country residence of Stanislaus Augustus. The palace is built in the Italian style: Bacciarelli's paintings decorate one of the principal rooms; and it has a ball-room, ornamented with colossal statues in white marble; a chapel, with some curious works in mosaic, &c. In the park is a stone bridge, on which is erected the equestrian statue of John Sobieski. The view of the Vistula from the park is very fine; and a large island lying in the middle stream is much frequented in summer by the amateurs of aquatic expeditions." (*Granville*, ii., 541-547.)

Among the other public buildings, may be specified the Radzivil and Krasinski palaces, the barracks, mint, six hospitals, five theatres, and several good inns. Since the late insurrection, a strong citadel has been erected partly in the view of protecting, but more of overawing the town. This citadel was built from the produce of a loan raised in Poland; and, in 1835, when the emperor Nicholas visited Warsaw in his way from the congress at Töplitz, he distinctly informed the civic authorities that on the first disturbance breaking out in the city, the guns of the citadel should level it with the ground! A cast-iron obelisk has been erected in the citadel in honour of the late emperor, inscribed "To Alexander, the conqueror and benefactor of Poland!"

The university of Warsaw, established in 1816, had faculties of theology, jurisprudence, medicine, philosophy, belles-lettres, and fine arts, and a library containing, it is said, 150,000 volumes of printed books, exclusive of rare MSS., with an observatory and botanic garden, cabinets of

natural philosophy, zoology, mineral, models, and coins, and printing and lithographic presses. Unfortunately, however, the university no longer exists, having been suppressed subsequently to the late ill-fated insurrection, its fine library being then, also, removed to Petersburg. Of late years there has been a Roman Catholic college at Warsaw, with twelve professors; but the adherents of the Russo-Greek church are rapidly increasing here, as in all other countries subject to Russia, and have now a cathedral and other churches in the city. The Jews, of whom there are about 25,000, have several synagogues; the Armenians, too, have their places of worship, and the English have a chapel. Among the educational establishments, are numerous superior, special, and elementary schools; all of them being modelled on the new system, and having attached to each a native Russian, as a teacher of his own language, a considerable proficiency in which is now an indispensable qualification for holding any public office, how trifling soever.

Warsaw has, also, a deaf and dumb asylum, a musical conservatory, societies of friends of literature and natural science, a bible society, &c., and some newspapers and other periodical publications. These, however, are subjected to a rigorous censorship, and are, consequently, worth little or nothing. Its manufactures comprise woollen and linen cloths, saddlery, leather, carriages of different kinds, ironmongery, paper, and tobacco, with chemical and cotton printing works, and numerous breweries. Warsaw is the great commercial entrepôt for Poland; and has two large annual fairs, in May and September, attended by traders from many parts of Europe and Asia, five banks, an assurance society, &c.

In comparing this city with Petersburg, Dr. Granville says, "There is a notable difference between the general aspect of the inhabitants of Warsaw and those of the capital he had just left. The women here are handsomer than the men: at Petersburg the impression I received was of an opposite nature. The absence of those semi-Asiatic costumes, which are so prevalent in all the streets of the Russian capital, tends, in a great measure, to give to the capital of Poland a more European aspect; but there is something else that contributes to produce that effect. The Poles are uniformly merry; they are loud chatterers, fond of amusement, and as partial to living in the open air, doing nothing, as the Parisian *fainéants* and the *habitués* of the Palais Royal, the Tuileries, the Boulevards, or the Luxembourg; to which class of people I should be tempted to compare them in many respects. They also do business differently: their shops and public places of amusement are more like those of any other European city farther S.; and their *ménage* appears to be much nearer to that of the French than of the Russians." (*Granville*, ii., 527, 528.)

Warsaw, though a very ancient town, was not the capital of Poland till 1566, after the union with Lithuania; when the Polish diet was transferred to it from Cracow. The city was occupied by the Swedes in the middle of the 17th century, and surrendered, without opposition, to Charles XII. in 1703. In 1793, the inhabitants expelled the Russian garrison previously in occupation; and the town was successfully defended against the Prussians in the succeeding year, by Kosciusko. But the suburb of Praga, being soon after taken and sacked by the Russians under Suwarrow, by whom a large portion of the inhabitants were put to the sword, the city, threatened with a similar fate, submitted to the conquerors. In 1795, Warsaw was assigned to Prussia: in 1806, she was made the capital of the grand duchy of Poland; and in 1815, she became the capital of the new kingdom of Poland. She was the principal seat of the ill-fated insurrection of 1831, which has entailed so much mischief on her and the rest of Poland. (*Malte Brun, Tableau de Pologne; Dict. Géog.; Granville, &c.*)

WARSAW, p. t., capital of Wyoming co., N. Y., 251 m. W. Albany, 363 W. Watered by Allen's creek. It contains a courthouse, jail, county clerk's office, four churches, two Presbyterian, a Methodist and Baptist, nine stores, three fulling-mills, one woollen factory, one furnace, two grist mills, four saw-mills, two tanneries; two academies, 80 students; 25 schools, 1068 scholars. Pop. 2841.

WARSAW, p. v., Hancock county, Ill., 114 m. W.N.W. Springfield, 892 W. Situated on the E. side of Mississippi river, at the foot of Des Moines rapids. It contains several stores, a steam-mill and about 800 inhabitants. It is near the site of old fort Edwards.

WARRIOR MARK, p. t., Huntingdon co., Pa., 108 m. W. Harrisburg, 171 W. It contains three stores, one furnace, two grist-mills, five saw-mills, two tanneries, two distilleries; nine schools, 500 scholars. Pop. 1689.

WARWICK, a co. of England, situated nearly in the centre of the kingd., having N.E. the co. of Leicester, E. Northampton, S. Oxford and Gloucester, W. Worcester, and N.W. Stafford. It contains 574,080 acres; of which above 500,000 are arable, meadow, and pasture. The northern and largest part of Warwickshire was formerly an extensive forest, and still retains something of its former character, being interspersed with heaths and moors, and sprinkled with woods; but the former have greatly diminished within the present century. The S. portion is in general very fertile. Both the dairy and grazing systems are successfully practised, but the former has been gaining on the latter. The long-horned breed of cattle is preferred for the dairies; the average produce of a cow being about 2½ cwt. of cheese. Short-woolled sheep have been almost entirely banished from this county. The standing sheep stock is supposed to amount to about 350,000 head, and the produce of wool to be between 8000 and 9000 packs. Arable husbandry is not so well understood as grazing; and in some districts it is far behind. Wheat, barley, oats, and beans, are extensively cultivated. The first is generally drilled; and when such is the case, it is not a little singular that turnips should be almost everywhere sown broadcast, and beans generally dibbled. The system of top-dressing is more commonly followed in this than in any other county. Estates of various sizes; some very large, and others small, Farms vary from 80 to 500 acres; but the smaller class predominate so much, that the average is not supposed to exceed 150 acres. Old enclosures average about 10 acres, new about 15. Leases getting more uncommon, and farms mostly held from year to year. Tenants bound not to exceed three crops to a fallow; but there is no restriction as to the quantity of wheat sown. Average rent of land in 1810, £1 2s. 5½d. an acre. Little can be said in favour of the farm buildings. The old houses and offices were sometimes built of timber; sometimes the walls were of stone, and sometimes of mud or clay, or thatched; they are in general injudiciously placed, ill-planned, and inconvenient. The new farm-houses and offices are of brick, covered with tile, and are very substantial; but convenience is said not to be much studied. There are no open sheds for wintering cattle, nor feeding-sheds for soiling with turnips, and other green food. (Survey, p. 30.) Coal is wrought to a considerable extent at various places; but Birmingham is supplied with coal brought by canal from Staffordshire. Warwick ranks high as a manufacturing county. Birmingham is the principal seat of the hardware manufacture; and nowhere, perhaps, has the combined influence of ingenuity, skill, and capital been more astonishingly displayed than in the immense variety, beauty, utility, and cheapness of the articles produced in this great workshop. Coventry has been long distinguished for its proficiency in the silk trade, particularly in the manufacture of ribands. Needles and fish-hooks are made at Alcester; hats at Atherstone; and flax-mills have been erected at Tamworth, and in other places. Principal rivers Avon, Tame, Alne, and Leam. The Birmingham and Fazely canal runs along the N.W. side of the county; and it is intersected by the Warwick and Birmingham canal, the Warwick and Napton canal, the Oxford canal, &c.; the county is also intersected by the railway from London to Birmingham, and thence to Manchester. It is divided into four hundreds, and four subsidiary districts, and contains 205 parishes. It sends 10 members to the House of Commons; viz. four for the county, and two each for the city of Coventry and boroughs of Birmingham and Warwick. Registe'ed electors for the county, in 1839-40, 11,039, whereof 6786 belong to the E., and 4253 to the W. division. In 1841 Warwick had 81,445 inhabited houses, 402,121 inhabitants, of whom 195,967 were males, and 206,154 females. Sum expended for the relief of the poor in 1838-39, £108,711. Total annual value of real property assessed to the poor rates, in 1841, £1,609,747.

WARWICK, a part. and mun. bor. and par. of England, near the centre of the co. Warwick, of which it is the cap., hund. Knightlow, on the Avon, 2½ m. W. Leamington, and 89 m. N.W. London. Area of parl. bor., the limits of which were not altered by the Boundary Act, and which is co-extensive with the two parishes of St. Mary and St. Nicholas, 5360 acres. Pop., in 1831, 9109. It stands on an abrupt acclivity on the N. bank of the river, which is here crossed by a handsome stone bridge of a single arch; and is regularly laid out, consisting of two principal thoroughfares crossing each other towards the centre of the town, with a number of smaller cross streets. The principal streets are well built, paved, lighted with gas, kept remarkably clean, and ornamented with several handsome public buildings. The most conspicuous of these is St. Mary's church, which, having been nearly burnt down in 1694, was rebuilt in 1704. It exhibits a singular union of various styles: the square tower, which was designed by Sir Christopher Wren, and is finely proportioned, rises to a height of 130 ft.; it is supported on four pointed arches, with a thoroughfare underneath, and crowned with pinnacles. Rickman says, that "The whole of this church, except the chancel and its adjuncts, is a composition of the greatest barbarity, but the chancel is an uncommonly beautiful specimen of

perpendicular work, and the east front is remarkably fine, simple in its arrangement, yet rich from the elegance of its parts and the excellent execution of its details. The interior is equally beautiful, and there are, on the N. side, a monumental chapel and vestry of very good character, but the great feature of the church is the Beauchamp chapel erected in 1464. It is completely enriched both within and without; its details of the most elegant character and excellent execution, and in very good preservation. It consists of a chapel, of several arches, and a small aisle, or rather passage, on the N. side, between the chapel and the church. In the centre of the chapel stands a very rich altar tomb, with the effigies of Richard Beauchamp, earl of Warwick, whose executors commenced the erection of this chapel, which, however, was not completed until the 3d Edw. IV. There are some other monuments, (including a fine one of Dudley, earl of Leicester, favourite of Elizabeth), but others are of much later date, and rather disfigure the chapel than add to its beauty." The living of St. Mary's is a vicarage, in the gift of the corporation of Warwick, worth £300 a year. The church of St. Nicholas is small and plain; the living, also in the gift of the corporation, is worth £218 per annum. Other churches formerly existed, of which there are now no remains. The Independents, Friends, Wesleyans, Unitarians, and Baptists, have their respective meeting-houses.

The courthouse in the High-street is a respectable stone building erected in 1730. The county hall, a spacious and handsome edifice 94 ft. in length, and 36 ft. in width, has an elegant stone front supported by a range of Corinthian pillars. In this building the courts of justice are held; and attached to it on the N. side is the county jail, a large and well designed building, surrounded by a strong wall 23 ft. in height, enclosing nearly an acre of ground. The county bridewell in which is a corn-mill, (worked by the male prisoners,) &c., and the market-house are large and substantial modern structures. It has also a public subscription library and news-room, and a small theatre.

But the great glory of Warwick is its castle, the seat of the earl of Warwick, and the most magnificent of the ancient feudal mansions of the English nobility, still used as a residence. It stands on a rock overhanging the Avon, a little to the S.E. of the town. It retains much of its ancient grandeur of appearance, and, uninjured by time, presents an interesting memorial of by-gone ages. Its foundation is attributed to Ethelfleda, daughter of Alfred, in 915; but no authentic trace now remains of the original building. Cæsar's tower, 147 ft. in height, supposed to have been built at least 700 years ago, is in a perfect state of preservation. Guy's tower, 128 ft. in height, and built in 1394, is, also, nearly perfect; it appears to be of a decorated character; and though very plain, is perhaps the most perfect remain of its kind in existence, and comes alike as to construction and construction. The principal entrance faces the E. side of the town, and the approach to it is a broad winding road cut in the solid rock. Before the front is a disused moat, a stone arch over which has replaced the ancient drawbridge. On passing the double gateway, the visiter finds himself in the inner court of the castle, surrounded on all sides by lofty embattled walls and ramparts. This castle was formerly a strong fortress and by means of open flights of stone steps and passages on the tops of the walls there is a line of communication all round the building. The parts of this vast and venerable pile that are occupied by the family are magnificently fitted up, but so as to harmonize in all respects with the style and character of the building. The collection of paintings is at once extensive and valuable.

In a greenhouse attached to the castle is the WARWICK VASE, one of the noblest remains of ancient art. It is of white marble, and of large dimensions, being capable of holding 136 gallons. Its handles are exquisitely formed of interwoven vine branches. On the body of the vase are the heads of satyrs, bound with wreaths of ivy, the skin of the panther, with the head and claws beautifully sculptured, and other appropriate ornaments. This most splendid relic was found at the bottom of a lake, at Adrian's Villa, at Tivoli (which see), of which, no doubt, it had formed a principal ornament; and having been purchased by Sir William Hamilton, at whose expense it was brought to England, and by whose liberality it has been placed in a situation where it may at all times be seen by the public.

The remains of several monastic establishments exist in and near Warwick; and at the E. and W. extremities of the town are gates, each containing some ancient work with modern additions. Leicester's hospital, an ancient building at the W. extremity of High-street, was originally a hall belonging to two guilds, and was converted to its present use by Robert Dudley, earl of Leicester, for the reception of 12 poor men, chiefly disabled veterans, and for a professor of divinity as master. In 1811, the clear value of the estates with which it is endowed amounted to £2000 per annum. In 1813, the master's salary was raised from £50 to £400 a year, and the number of inmates increased to 22. The college school, originally founded by Henry VIII. as a free grammar school, and endowed out of the revenues of the dissolved monasteries, is open to all the boys of the town. It has two exhibitions of £70 to each of the universities. Of late years it has fallen into a state of decay; but recently the number of youths educated here has increased, and at present (1842) the school is comparatively prosperous. Here is, also, a charity school founded and endowed for the instruction of 39 boys and 36 girls; a national school; school of industry, &c., and not less than 40 almshouses. Large funds are vested in the hands of the corporation for distribution among the poor.

The manufactures, which are unimportant, comprise a few descriptions of cotton and woollen goods, a worsted mill, with a ropery, and a brass and iron foundry. There are several large malting houses, and lime, timber, and coal-wharves, on the banks of the Stratford canal, which comes up to the N. part of the town, and by which it communicates with Oxford, Birmingham, and the Severn. It is within about 10 m. of Coventry, and consequently, also, of the London and Birmingham railway. The *Bound. Rep.* says, "The town is thriving, and distinguished by an appearance of respectability and neatness. Trade seems to be rapidly increasing, which may be accounted for in great measure by its proximity to and connexion with Leamington, where most of the Warwick tradespeople have shops, and where the chief speculators from this town invest their capital." But circumstances have changed materially in the interval. The great increase of Leamington, and its superior advantages, have taken much of the capital and business from Warwick, which it formerly enjoyed, and proportionally depressed the latter. Very few Warwick tradesmen have also shops in Leamington. Warwick is a borough by prescription: its earliest charter dates from the 45th of Henry III., but it was not regularly incorporated till 1553. Under the Municipal Reform Act, it is divided into two wards, and is governed by a mayor, five aldermen, and 18 councillors. Quarterly courts of session are held for all offences not capital; a court-leet annually; and there is occasionally a court of record for the recovery of debts under £40. The borough has returned two members to the House of Commons since the reign of Edward I., the franchise having been vested, previously to the Reform Act, in the payers of church and poor-rates. Registered electors, in 1839-40, 977. Corporation revenue, in 1840, £1053. It has a weekly newspaper; and the Warwick and Leamington Banking company, established in 1834, has its head office in the town. The Leamington Priory Banking company has also a branch here.

Warwick is conjectured by Dugdale and other writers to have been a Roman station; but there are no proofs of its having existed before the Saxon times. It was in great part destroyed by fire in 1694. Market-day, Saturday. Fairs, twelve times a year; mostly for horses, cattle, and cheese. (*Hist. of Warwick; Bound. and Munic. Corp. Rep.; Cooke's Warwick Castle; Private Inform.*)

WARWICK, county, Va. Situated in the S.E. part of the state, and contains 95 sq. m. Bounded S.W. by James river. It occupies the S.W. part of the peninsula between James and York rivers. It contained in 1840, 1573 neat cattle, 973 sheep, 3096 swine; and produced 10,901 bushels of wheat, 45,975 of Indian corn, 8698 of oats, 1705 of potatoes. It had 25 stores, one cotton factory with 1064 spindles, two flouring-mills, 11 grist-mills, six saw-mills; two schools, 32 scholars. Pop.: whites, 604; slaves, 831; free coloured, 21; total, 1456. Capital, Warwick C.H.

WARWICK, p. t. Franklin co., Mass., 79 m. W.N.W. Boston, 420 W. Drained by a small branch of Miller's river. Incorporated in 1763. It contains two churches, a Congregational and Unitarian; two stores, one fulling-mill, two grist-mills, 12 saw-mills, three tanneries; one academy, 42 students; 11 schools, 371 scholars. Pop. 1071.

WARWICK, p. t., Kent co., R. I., 11 m. S.W. Providence, 401 W. Watered by Pawtuxet river, which affords extensive water-power. The village of Apponaug in its S. part is on a branch of Narragansett bay, and has a good harbour, about 1 m. from it, for vessels of any size, and those of from 20 to 50 tons come up to the village. A mile from the village is "Drum Rock," a huge rock so exactly balanced upon another rock, that a boy of 14 years of age can move it, causing a noise more sonorous than that of a drum, whence it derives its name. Pawtuxet village, at the mouth of Pawtuxet river, is a large manufacturing village. It contains two churches, a Congregational and Baptist; a number of stores, and an extensive cotton factory. There are in the t. 68 stores, two woollen factories, 28 cotton factories with 73,041 spindles, eight grist-mills, two saw-mills, two tanneries; five academies, 186 students; 36 schools, 1390 scholars. Pop. 6726.

WASHINGTON, p. t., Orange co., N. Y., 24 m. S.W. Newburg, 111 m. S.S.W. Albany, 208 W. Drained by Wallkill river. It contains two churches, a Dutch Reformed and Baptist; 16 stores, one fulling-mill, one woollen factory, eight grist-mills, 14 saw-mills, four tanneries, 15 distilleries; one academy, 147 students; 20 schools, 1034 scholars. Pop. 5113.

WASHINGTON, t., Bucks co., Pa., 5 m. S.E. Doylestown, 21 m. N. Philadelphia. Drained by Nishaminy creek. It contains a church, three stores, one flouring-mill, four grist-mills, three saw-mills, two oil-mills; two schools, 56 scholars. Pop. 1959.

WASHINGTON, t., Lancaster co., Pa., 9 m. N. Lancaster. Drained by Great Chiques, Hanmer, Cocalico, and Moravia creeks. It has 12 stores, two flouring-mills, 10 grist-mills, seven saw-mills, one oil-mill, four tanneries, six distilleries, one brewery, one pottery; one academy, 104 students; one school, 35 scholars. Pop. 3725.

WASHINGTON, county, Me. Situated in the S.E. part of the state, and contains 3500 sq. m. Bounded E. by St. Croix river, and Passamaquody bay, which separates it from New-Brunswick. Drained by Machias, Pleasant, and Narraguagus river. The Atlantic. which bounds it on the S., is indented with numerous bays, that afford fine harbours. It contained in 1840, 12,031 neat cattle, 20,561 sheep, 4297 swine; and produced 20,871 bushels of wheat, 2094 of rye, 394 of Indian corn, 3218 of buckwheat, 5914 of barley, 22,111 of oats, 410,868 of potatoes, 1338 pounds of sugar. It had three commercial houses in foreign trade, 166 retail stores, 16 lumber-yards, with a capital of $208,300; 50,000 bushels of domestic salt were produced; three fulling-mills, one furnace, 36 grist-mills, 159 saw-mills, 14 tanneries; two academies, 191 students; 211 schools, 9273 scholars. Pop. 28,327. Capital, Machias.

WASHINGTON, county, R. I. Situated in the S. part of the state, and contains 367 sq. m. Bounded S. by the Atlantic, E. by Narragansett bay. Drained by Pawcatuck river and its branches, the principal of which are Wood and Charles rivers. It has several good harbours, and contained in 1840, 10,741 neat cattle, 35,126 sheep, 8599 swine; and produced 1426 bushels of wheat, 8967 of rye, 113,501 of Indian corn, 12,500 of barley, 60,317 of oats, 257,731 of potatoes. It had 80 stores, six lumber-yards, 16 fulling-mills, 25 woollen factories, 23 cotton factories with 23,660 spindles, 29 grist-mills, 21 saw-mills, 10 tanneries, one printing-office; six academies, 270 students; 19 schools, 742 scholars. Pop. 14,324. Capital, North Kingston.

WASHINGTON, county, Vt. Situated near the centre of the state, and contains 625 sq. m. Organized in 1810 by the name of Jefferson, which was changed to Washington county in 1814. Watered by Onion or Winooski river, which flows through the Green mountains, opening a passage for a road with no heavy hills, though the mountains tower on its sides. It contained in 1840, 25,415 neat cattle, 110,872 sheep, 12,150 swine; and produced 44,110 bushels of wheat, 5736 of rye, 63,108 of Indian corn, 23,066 of buckwheat, 4038 of barley, 200,294 of oats, 698,745 of potatoes. It had 45 stores, one furnace, 32 fulling-mills, three woollen factories, six flouring-mills, 36 grist-mills, 85 saw-mills, five oil-mills, one paper-mill, 10 tanneries, six printing-offices, two daily and six weekly newspapers, and one periodical; two academies, 231 students; 196 schools, 6866 scholars. Pop. 23,506. Capital, Montpelier.

WASHINGTON, county, N. Y. Situated in the E. towards the N. part of the state, and contains 807 sq. m. Bounded W. by Hudson river, N.W. by lake George. N.E. by part of lake Champlain, N. by Poultney river. Watered by Battenkill and Hoosick rivers and their tributaries, and Wood creek. It contained in 1840, 39,150 neat cattle, 210,610 sheep, 27,688 swine; and produced 49,189 bushels of wheat, 136,510 of rye, 297,465 of Indian corn, 32,642 of buckwheat, 9569 of barley, 448,064 of oats, 851,545 of potatoes, 4246 pounds of sugar. It had 140 stores, one furnace, three forges, 15 fulling-mills, 13 woollen factories, two cotton factories with 3144 spindles, two flouring-mills, 31 grist-mills, 93 saw-mills, one powder-mill, 19 tanneries, two potteries, four printing-offices, four weekly newspapers; 10 academies, 482 students; 205 schools, 7526 scholars. Pop. 41,080. Capital, Sandy Hill and Salem.

WASHINGTON, county, Pa. Situated in the S.W. part of the state, and contains 1000 sq. m. Bounded E. by Monongahela river, and drained by its tributaries, and by Raccoon creek. which flows into Ohio river. Bituminous coal is abundant. It contained in 1840, 35,344 neat cattle, 222,631 sheep, 54,245 swine; and produced 666,200 bushels of wheat, 32,157 of rye, 653,292 of Indian corn, 29,368 of buckwheat, 11,913 of barley, 1,862,528 of oats, 581,589 of potatoes, 56,032 pounds of sugar. It had 120 stores, three fulling-mills, three woollen-factories, two glass houses, 23 flouring-mills, 66 grist-mills, 72 saw-mills, 30 tanneries, 19 distilleries, two potteries; two colleges, 349 students; six academies, 175 students; 170 schools, 6490 scholars. Pop. 41,279.

WASHINGTON.

WASHINGTON, county, Md. Situated towards the W. part of the state, and contains 440 sq. m. Bounded S. by Potomac river. Drained by Conecocheague, Antietam, and other creeks flowing into Potomac river. It had in 1840, 15,841 neat cattle, 15,798 sheep, 36,418 swine; and produced 668,787 bushels of wheat, 111,829 of rye, 635,041 of Indian corn, 1864 of buckwheat, 436,864 of oats, 75,783 of potatoes. It had two commission houses in foreign trade, 96 stores, one furnace, two forges, seven fulling-mills, two woollen factories, 52 flouring-mills, nine grist-mills, 43 saw-mills, one oil-mill, 20 tanneries, 16 distilleries, three breweries, three potteries, one ropewalk, six printing-offices, five weekly newspapers; two academies, 95 students; 53 schools, 1823 scholars. Pop.: whites, 24,724; slaves, 2546; free coloured, 1580; total, 28,850. Capital, Hagerstown.

WASHINGTON, county, Va. Situated in the S.W. part of the state, and contains 764 sq. m. Drained by New river, and by branches of Great Kanawha and Clinch rivers. It contained in 1840, 13,810 neat cattle, 18,530 sheep, 38,450 swine; and produced 106,750 bushels of wheat, 8116 of rye, 397,476 of Indian corn, 4226 of buckwheat, 1168 of barley, 295,770 of oats, 60,180 of potatoes, 62,740 pounds of sugar. It had nine stores, 40 grist-mills, 36 saw-mills, two powder-mills, nine tanneries, 103 distilleries; one printing-office, one weekly newspaper; one college, 150 students; 18 schools, 551 scholars. Pop.: whites, 10,731; slaves, 2058; free coloured, 212; total, 13,001. Capital, Abingdon.

WASHINGTON, county, N. C. Situated on the E. part of the state, and contains 360 sq. m. Bounded N. by Albemarle sound, N.W. by Roanoke river. It contained in 1840, 2928 neat cattle, 2125 sheep, 9319 swine; and produced 6707 bushels of wheat, 135,023 of Indian corn, 4349 of oats, 19,672 of potatoes, 6050 pounds of rice, 79,925 of cotton. It had one commercial and two commission houses in foreign trade, 13 retail stores, three flouring-mills, four grist-mills, four saw-mills, 13 distilleries; two schools, 38 scholars. Pop.: whites, 2639; slaves, 1727; free coloured, 159; total, 4325. Capital, Plymouth.

WASHINGTON, county, Ga. Situated E. of the centre of the state, and contains 760 sq. m. Bounded N.E. by Ogeechee river, S.W. by Oconee river, by branches of which it is drained. It contained in 1840, 10,806 neat cattle, 4439 sheep, 26,426 swine; and produced 20,066 bushels of wheat, 2274 of rye, 298,747 of Indian corn, 2639 of oats, 15,877 of potatoes, 1,190,770 pounds of cotton. It had 17 stores, two flouring-mills, 18 grist-mills, 11 saw-mills. 29 distilleries; five academies, 133 students; eight schools, 108 scholars. Pop.: whites, 3982; slaves, 4566; free coloured, 37; total, 10,585. Capital, Sandersville.

WASHINGTON, county, Flor. Situated towards the W. part of the territory, and contains 1500 sq. m. Bounded S. by the gulf of Mexico, W. by Choctawhatchee river. Drained by Econfind river. The fine bay of St. Andrews contains a large quantity of live oak on its borders. It contained in 1840, 4601 neat cattle, 52 sheep, 2637 swine; and produced 24,965 bushels of Indian corn, 5365 of potatoes, 2100 pounds of rice, 176,120 of cotton, 4170 of sugar. It had five grist-mills; three academies, 52 students; five schools, 106 scholars. Pop.: whites, 504; slaves, 353; free coloured, 2; total, 859. Capital, Roche's Bluff.

WASHINGTON, county, Ala. Situated towards the S.W. part of the state, and contains 840 sq. m. Bounded E. by Tombigbee r. Drained by Escatawpa river. It contained in 1840, 20,361 neat cattle, 2221 sheep, 14,083 swine; and produced 326 bushels of wheat, 152,049 of Indian corn, 2383 of oats, 23,934 of potatoes, 355,748 pounds of cotton. It had eight stores, 10 grist-mills, four saw-mills, 14 oil mills, two tanneries, two potteries; 12 schools, 201 scholars. Pop.: whites, 2843; slaves, 2434; free coloured, 23; total, 5300. Capital, Barryton.

WASHINGTON, county, Miss. Situated in the W. part of the state, and contains 2420 sq. m. Bounded W. by Mississippi river, S.E. by Yazoo river. Watered by Sunflower river and its branches. It produced in 1840, 15,100,400 pounds of cotton. Pop.: whites, 654; slaves, 6637; free coloured, 6; total, 7287. Capital, Princeton.

WASHINGTON, parish, La. Situated in the E. part of the state, and contains 792 sq. m. Bounded E. by Pearl river. Drained by Bocuechitto and Chifungte rivers. It contained in 1840, 9000 neat cattle, 1050 sheep, 14250 swine; and produced 98,790 bushels of Indian corn, 59,000 of oats, 9850 of potatoes, 251,250 pounds of rice, 375,300 of cotton. It had eight stores, one cotton factory with six spindles, 11 grist-mills, 13 saw-mills; two academies, 70 students; two schools, 50 scholars. Pop.: whites, 1856; slaves, 791; free coloured, 2; total, 2649. Capital, Franklinton.

WASHINGTON, county, Tenn. Situated in the E. part of the state, and contains 590 sq. m. Drained by Nolachucky river and its branches. It contained in 1840, 9663 neat cattle, 19,032 sheep, 70,326 swine; and produced 121,424 bushels of wheat, 3069 of rye, 330,509 of Indian corn, 165,758 of oats, 13,936 of potatoes, 15,102 pounds of tobacco. It had

1046

18 stores, two furnaces, eight forges, six flouring-mills. 17 grist-mills, 28 saw-mills, three oil-mills, 11 tanneries, 36 distilleries, two printing-offices, two weekly newspapers. one college, 10 students; one academy, 34 students; 18 schools, 3d0 scholars. Pop.: whites, 10,600; slaves, 915; free coloured, 236; total, 11,751. Capital, Jonesboro'.

WASHINGTON, county, Ky. Situated a little N. of the centre of the state, and contains 475 sq. m. Bounded N. by Chaplin's fork of Salt river, by branches of which it is drained. It contained in 1840, 5955 neat cattle, 9966 sheep. 28,200 swine; and produced 27,350 bushels of wheat, 21,633 of rye, 281,730 of Indian corn, 28,739 of oats, 7088 of potatoes, 72,000 pounds of tobacco. It had 16 stores, two flouring-mills, 30 grist-mills, seven saw-mills, six tanneries. 5 distilleries, 11 breweries; one academy, 80 students; 22 schools, 636 scholars. Pop.: whites, 7900; slaves, 3636; free coloured, 38; total, 10,506. Capital, Springfield.

WASHINGTON, county, O. Situated in the S.E. part of the state, and contains 713 sq. m. Bounded S.E. by Ohio river. Watered by Muskingum and Little Muskingum rivers, on which bituminous coal abounds. It was the first organized county in the state, and contained in 1840, 12,122 neat cattle, 34,790 sheep, 17,399 swine; and produced 166,650 bushels of wheat, 390,471 of Indian corn, 1454 of buckwheat, 147,210 of oats, 96,362 of potatoes, 11,390 pounds of tobacco, 11,996 of sugar. It had 38 stores, two fulling-mills, one flouring-mill, 41 grist-mills, 60 saw-mills. one oil-mill, 10 tanneries, four distilleries; one college, 100 students; one academy, 53 students; 107 schools, 4476 scholars. Pop.: 20,823. Capital, Marietta.

WASHINGTON, county, Ia. Situated towards the S. part of the state, and contains 540 sq. m. Bounded N. by Muscatatuck river. Drained by Great Blue river. Organized in 1813. It contained in 1840, 10,434 neat cattle, 17,247 sheep, 21,621 swine; and produced 52,508 bushels of wheat, 1500 of rye, 164,401 of Indian corn, 50,047 of oats, 10,731 of potatoes, 71,161 pounds of tobacco. It had 30 stores, two cotton factories with 564 spindles, eight flouring-mills, 32 grist-mills, 37 saw-mills, one oil-mill, 22 tanneries, two printing-offices, two weekly newspapers; two academies, 96 students; 19 schools, 519 scholars. Pop. 15,269. Capital, Salem.

WASHINGTON, county, Ill. Situated towards the S. part of the state, and contains 656 sq. m. Bounded N.W. by Kaskaskia river, and drained by its branches. It contained in 1840, 10,360 neat cattle, 5300 sheep, 18,989 swine, and produced 28,540 bushels of wheat, 272,980 of Indian corn, 81,089 of oats, 7175 of potatoes. It had five stores, 14 schools, 290 scholars. Pop. 4610. Capital, Nashville.

WASHINGTON, county, Wis. Situated in the E. part of the territory, and contains 675 sq. m. Bounded E. by lake Michigan. Watered by Milwaukee and Menomonee rivers. It had in 1840, 277 neat cattle, 288 swine; and produced 122 bushels of wheat, 568 of Indian corn, 165 of oats, 650 of potatoes, 4650 pounds of sugar. It had three saw-mills, one school, 7 scholars. Pop. 343. Capital, Washington.

WASHINGTON, county, Iowa. Situated in the W. towards the S. part of the territory, and contains 648 sq. m. Watered by Iowa river, and Checauque or Skunk river, and their tributaries. It contained in 1840, 600 neat cattle, 30 sheep, 864 swine; and produced 2210 bushels of wheat, 100 055 of Indian corn, 2155 of oats, 1357 of potatoes. It had five stores, two saw-mills. Pop. 1594. Capital, Washington.

WASHINGTON, county, Mo. Situated towards the S.E. part of the state, and contains 829 sq. m. Watered by head branches of Big, Big Black, and St. Francis rivers. It contains the celebrated "Iron Mountain," which consists of a mass of magnetic iron ore, so pure as to yield from 70 to 80 per cent of malleable iron under the ordinary process of preparing it. It contained in 1840, 9573 neat cattle, 5610 sheep, 19,430 swine; and produced 34,933 bushels of wheat, 268,285 of Indian corn, 3600 of buckwheat, 50,344 of oats, 4748 of potatoes, 5448 pounds of tobacco, 10,227 of sugar. It had 16 stores, one furnace, one forge, five grist-mills, four saw-mills, three tanneries, one distillery, one pottery; one academy, 83 students; 15 schools, 399 scholars. Pop. whites, 6848, slaves, 923, free coloured, 61; total, 7692. Capital, Potosi.

WASHINGTON, county, Ark. Situated in the N.W. part of the state, and contains 900 sq. m. Drained by a head branch of White river. It contained in 1840, 11,734 neat cattle, 7947 sheep, 33,899 swine; and produced 16,364 bushels of wheat, 380,490 of Indian corn, 47,368 of oats, 94,364 of potatoes, 6070 pounds of tobacco, 18,685 of cotton. It had 32 stores, one cotton factory with 84 spindles, four flouring mills, 19 grist-mills, eight saw-mills, nine tanneries, even distilleries, one printing-office; two academies, 75 students; 13 schools, 358 scholars. Pop.: whites, 6946; slaves, 823; free coloured, 19; total, 7148. Capital, Fayetteville.

WASHINGTON, p. t., Lincoln co. Me., 31 m. E. Augusta, 619 W. Drained by head branches of Damariscotta and Muscongus rivers. Incorporated in 1811 by the name of

Putnam. It has two stores, two grist-mills, five saw-mills; 12 schools, 645 scholars. Pop. 1600.

WASHINGTON, p. t., Sullivan co., N. H., 33 m. W. Concord, 473 W. Drained by head branches of Ashuelot and Contoocook rivers, issuing from ponds. Incorporated in 1768. It contains three churches, a Congregational, Baptist, and Universalist; three stores, one fulling-mill, two grist-mills, six saw-mills; four schools, 363 scholars. Pop. 1103.

WASHINGTON, p. t., Orange co., Vt., 15 m. S.E. Montpelier, 512 W. Watered by Jail branch of Winooski or Onion river, and by branches of Wait's and White rivers. Chartered in 1780. It contains two churches, three stores, one fulling-mill, five saw-mills; 20 schools, 517 scholars. Pop. 1359.

WASHINGTON, p. t., Litchfield co., Ct., 47 m. W.S.W. Hartford, 390 W. Watered by Shepaug river, a tributary of Housatonic river. It contains two Congregational churches, six stores, two fulling-mills, one cotton factory with 1064 spindles, one furnace, three forges, three grist-mills, seven saw-mills, one tannery, two distilleries; one academy, 20 students; 10 schools, 418 scholars. Pop. 1692.

WASHINGTON, p. t., Dutchess co., N. Y., 12 m. N.E. Poughkeepsie, 79 m. S. by E. Albany, 314 W. Drained by tributaries of Wappinger's and Ten Mile creeks. It contains three churches, a Presbyterian and two Friends, six stores, three cotton factories with 1848 spindles, two flouring-mills, three grist-mills, seven saw-mills; eight schools, 326 scholars. Pop. 2833.

WASHINGTON, t., Bergen co., N. J. Bounded E. by Hackensack river, W. by Saddle river. It has six stores, six grist-mills, 14 saw-mills, three tanneries, one distillery; four schools, 135 scholars. Pop. 1833.

WASHINGTON, t., Burlington co., N. J., 22 m. S. Woodbury. Drained by Little Egg Harbour river and its branches. It has four stores, two furnaces, five grist-mills, eight saw-mills, one paper-mill; eight schools, 470 scholars. Pop. 1630.

WASHINGTON, t., Gloucester co., N. J., 12 m. W. Woodbury. Drained by Pensauken and Cooper's creeks. It contains an Episcopal church, six stores, one woollen factory, one glass house, one grist-mill, eight saw-mills; four schools, 331 scholars. Pop. 1545.

WASHINGTON, t., Morris co., N. J., 18 m. W. Morristown. Drained by the S. branch of Raritan river. Schooley's mountain has a mineral spring and hotels which are a celebrated place of summer resort. It contains a Presbyterian and a Lutheran church, 17 stores, one forge, four flouring-mills, six grist-mills, eight saw-mills, four oil-mills, one tannery, one brewery; 15 schools, 753 scholars. Pop. 2451.

WASHINGTON, p. b., capital of Washington co., Pa., 209 m. W. Harrisburg, 236 W. It is situated on the National road, and contains a courthouse, which cost $20,000, a jail, four churches, a Presbyterian, Associate Reformed, Methodist Episcopal, and a Cumberland Presbyterian; a female seminary, 23 stores, a steam woollen factory, one saw-mill, one tannery, one pottery, two printing-offices, two weekly newspapers; one academy, 100 students; seven schools, 300 scholars, and 2082 inhabitants. It is the seat of Washington college, founded as an academy in 1787, chartered as a college in 1806, which has a president and five professors, 524 alumni, 59 students in the college proper, and including the preparatory department, 150 students, and 3300 volumes in its libraries. The commencement is on the last Wednesday in September.

WASHINGTON, t., Luzerne co., Pa. It has two stores, one fulling-mill, two grist-mills, five saw-mills; four schools, 128 scholars. Pop. 1255.

WASHINGTON, t., Berks co., Pa. It has four stores, one forge, four grist-mills, three oil-mills, one tannery. Pop. 1210.

WASHINGTON, t., Franklin co., Pa., 11 m. S.W. Chambersburg. Drained by two branches of Antietam creek. It has seven stores, two fulling-mills, two woollen factories, six flouring-mills, one grist-mill, 14 saw-mills, three tanneries, seven distilleries, one brewery, two potteries; 10 schools, 310 scholars. Pop. 2404.

WASHINGTON, t., York co., Pa. It has three stores, four grist-mills, four saw-mills, two tanneries, one distillery, one pottery. Pop. 1226.

WASHINGTON, t., Fayette co., Pa. It has nine stores, one woollen factory, one grist-mill, five saw-mills, one paper-mill, two tanneries; four schools, 160 scholars. Pop. 1515.

WASHINGTON, t., Cambria co., Pa. Drained by Kiskiminitas river. It has five stores, one grist-mill, eight saw-mills; six schools, 210 scholars. Pop. 1266.

WASHINGTON, t., Erie co., Pa. Drained by head branches of French creek. It has two saw-mills, 18 schools, 700 scholars. Pop. 1851.

WASHINGTON, t., Westmoreland co., Pa., 16 m. N. Greensburg. Bounded N.E. by Kiskiminitas river. It has four stores, five grist-mills, seven saw-mills; 10 schools, 300 scholars. Pop. 2004.

WASHINGTON, city, District of Columbia, capital of the United States, is in 38° 53′ 23″ N. lat., and 77° 1′ 24″ W. long. It is 38 m. S.W. Baltimore; 136 m. S.S.W. Philadelphia; 225 m. S.W. New-York; 439 m. S.W. Boston; 856 m. E. by S. St. Louis; 544 m. N.N.E. Charleston; 662 m. N.N.E. Savannah, Ga.; 1203 m. N.E. New-Orleans. The population in 1800, was 8902; in 1820, 13,247; in 1830, 18,827; in 1840, 23,364, of whom 1713 were slaves. Employed in commerce, 103; in manufactures and trades, 886; in navigating the ocean, 45; do. rivers and canals, 95; in the learned professions, 83. It is situated on the E. side of Potomac river, at the junction of the Anacostia or eastern branch, 295 m. from the ocean, by the course of the river and bay. The city contains a little over 8 sq. m., or 5120 acres. The two rivers at the junction of which it stands, add to its natural beauty, and afford great facilities for commerce; but the large city of Baltimore in the vicinity will be likely long to eclipse it in commerce and manufactures, and prevent it from becoming a very large city. In addition to commerce and manufactures, large European capitals are extensively aided by luxurious and expensive courts, which last source of business and prosperity, it is to be hoped, Washington will forever want. Although it derives extensive advantages from the officers of the government, the American congress, and the foreign ministers which the capital collects around it; yet such is the economy of the government of the United States, that this source of business is but a fraction of that afforded by a European capital. The annual allowance of the princess of Wales, while travelling for the pleasure of her husband, would have paid the salary of the president of the United States for eight years, the longest time which that office has been held, by any one incumbent. The city is encompassed by a fine range of hills, forming a natural amphitheatre, and covered in part with trees and underwood, and presenting to the eye verdant and cultivated slopes. These elevations form beautiful sites for villas and private residences, and command an extensive and varied prospect of the city, of the surrounding country, and of the meanderings of the noble Potomac, as far as the eye can reach. These things doubtless attracted the eye of the father of his country, at whose suggestion this spot was early fixed on as the future capital of the nation; and though its growth has been less than was once anticipated, it has been considerable and is increasing. Probably the bustle of a large city would not much improve it as the seat of the national government. It enjoys two important requisites for health, pure air and good water; and there is sufficient refined and elegant society, particularly during the session of congress in the winter season, to render it a pleasant place of residence. The ground on which the city stands originally belonged to the state of Maryland, and was ceded to the United States on the 23d of December, 1788; and the original owners of the soil surrendered to the government the ground to be laid out as a city, and gave one half of the lots for the purpose of raising funds for the erection of the necessary public buildings. Three commissioners were appointed in January, 1791, who proceeded to plant the corner stone at Jones's point. The city was laid out under the direction of Andrew Ellicott, who was appointed in 1792. For so much of the land as might be appropriated to the use of the United States, the commissioners were to pay at the rate of $66⅔ the acre, not including streets; and this was to be paid out of the proceeds of the sale of lots given by the proprietors. The city is laid out on a magnificent plan, but it is but very partially built on; but posterity may reap the advantages of it. Millions of dollars would have been saved, if the same forecast had been manifested in the early laying out of the city of New-York. The general elevation of the ground on which the city is built is about 40 feet above the level of the river, and there are some moderate elevations, on two of which the capitol and the president's house are built. It is regularly laid out with streets running N. and S., and crossed at right angles by others running E. and W. But the different parts of the city are connected by broad avenues, which traverse the rectangular divisions diagonally. This would be very inconvenient, if ground were scarce. But where the intersection of these avenues with each other and with the streets would form many acute angles, considerable rectangular or circular open grounds are left, which, when the city shall be built up, will give it an open appearance. The avenues and streets leading to important public points, are from 120 to 160 feet wide, and the other streets are from 70 to 110 feet wide. The avenues are named after the states of the union, and the other streets, beginning at the capitol, are denoted by the letters of the alphabet, as A north and A south, B north and B south, &c.; and east and west, they are designated by numbers, as 1st east, 1st west, &c. Pennsylvania avenue, from the capitol to the president's house, contains the most dense population, is Macadamized, has a fine flagged side walk on one side, and is much the handsomest street in the city. Five avenues radiate from the capitol and five others from the president's house, giving these prominent places the most ready communication with all parts of the city. The buildings of Washington consist mostly of scattered clusters; nor is it probable that the magnificent plan of the city will soon be built up.

The public buildings of Washington have a magnificence becoming a great nation. The capitol is the finest building in the United States, and not inferior to any senate-house in the world. It is every way suitable that the representatives of the sovereign people should be accommodated in a building which would do honour to royalty, and be worthy of the most august legislative assembly in the world. The capitol is universally regarded as an honour to the nation. It is elevated 73 ft. above tide water, and affords a commanding view of the different parts of the city, and of the surrounding country. The building is of freestone, and covers an area of more than an acre and an half; the length of the front is 352 feet, including the wings; the depth of the wings is 121 feet. The projection or. the east or main front, including the steps, is 65 feet wide; and another on the west front, with the steps, is 83 feet wide. In the projection on the east front, there is a splendid portico of 24 lofty Corinthian columns, 38 feet high; and in the west front there is a portico of 10 Corinthian columns. The height of the building to the top of the dome is 190 feet. Under the dome in the middle of the building is the rotundo, a circular room, 95 feet in diameter, and of the same height, adorned with sculptures in stone pannels in bold relief, containing groups of figures representing Smith delivered by the interposition of Pocahontas; the landing of the pilgrims on Plymouth rock; the conflict of Boone with the Indians; and Penn treating with the Indians; and four magnificent paintings by Trumbull, with figures as large as life, representing the presentation to congress of the Declaration of Independence, in which all the figures, 47 in number, in that august assembly, which William Pitt in the British parliament pronounced superior in wisdom to any body of men whom he had ever heard or read of, are correct likenesses; the surrender of Burgoyne to General Gates; the surrender of Cornwallis at Yorktown; and Washington resigning his commission to congress at Annapolis. To these have recently been added the baptism of Pocahontas by Chapman, and the embarcation of the pilgrims by Weir. These paintings possess great merit as works of art, in addition to their commemoration of important events in American history. The rotunado has recently received a splendid additional ornament in Greenough's statue of Washington, a colossal figure in a sitting posture, twice as large as life. On the west of the rotundo is the library room of congress, 92 by 34 feet, and is 36 feet in height, containing in arched alcoves, over 23,000 volumes. The foundation of this library, after the burning of the capitol and its library by the British in the last war, was laid by the purchase of the entire private library of Mr. Jefferson, consisting of about 7000 volumes, many of them rare and valuable, for $23,000. This library has been enlarged from time to time by an annual appropriation by congress of $5000 for the purpose. In the second story of the south wing of the capitol is the hall of the house of representatives, of a semi-circular form, 96 feet long and 60 feet high, with a dome supported by 24 beautiful columns of variegated marble from the Potomac, with capitals of Italian marble, of the Corinthian order. The circular wall opposite the speaker, is surrounded by a gallery for men, and the chord of the arc, back of the speaker's chair, has a gallery for ladies. The room is ornamented with some fine statuary and paintings, and the whole furniture is elegant. The senate chamber is in the second story of the north wing of the capitol, and is semicircular, like that of the representatives, but smaller, being 78 feet long and 45 feet high. The vice president's chair is canopied with a rich crimson drapery, held by the talons of a hovering eagle. A light gallery of bronze running round the arc in front of the vice president's chair, is mainly appropriated to ladies. There is another gallery above and behind the chair, supported by fine Ionic columns of variegated marble. A magnificent chandelier hangs in the centre of the room, and the whole appearance of the furniture of the room is splendid. Below the senate chamber, and nearly of the same form and dimensions, though much less elegant, is the room of the supreme court of the United States, and there are in the building 70 rooms for the accommodation of committees, and officers of congress. The grounds around the capitol are spacious, containing 22 acres, highly ornamented with gravelled walks, shrubbery and trees, a naval monument ornamented with statuary, erected in honour of the youthful officers who fell in the battle of Tripoli, and fountains; and the whole is enclosed by a handsome iron fence. The whole cost of the building and its accompaniments has exceeded $2,000,000.

The president's house is an elegant edifice of freestone, two lofty stories high, at the intersection of Pennsylvania,

Virginia, New-York, Connecticut, and Vermont avenues, and stands near the centre of a plat of ground of 20 acres, at an elevation of 44 feet above high tide water. The entrance front, faces north upon La Fayette square, and the southern front toward the garden, presents a fine view of the improved part of the city, of the Potomac river and bridge, and of the opposite shores of Maryland and Virginia. The building is 170 feet front, and 86 feet deep. The north front is ornamented with a fine Ionic portico of four lofty columns, projecting with three columns. The outer intercolumniation is for carriages to drive into to place the company under shelter. The middle space is the entrance for those visiters who come on foot; the steps from both lead to a broad platform in front of the door of entrance. The garden front is varied by having a rusticated basement, and by a semicircular projecting colonnade of six Ionic columns, with two flights of steps leading from the ground to the level of the principal story. The apartments within are admirably fitted to their purpose, and splendidly furnished, and every way fitted for the residence of the chief of a great nation, and for the reception of his various company. On the E. side of the president's house are two large buildings, and on the W. side two large buildings for the departments of state, of the treasury, of war and of the navy. The general postoffice and the patent office are also extensive buildings. These, with the new treasury building, have been recently erected, to supply the place of those which were burned a few years since. The new treasury building contains 150 rooms, and when completed will contain $250. It has a splendid colonnade, 457 feet in length. The general postoffice contains about 80 rooms, and is of the Corinthian order, with columns and pilasters, on a rustic base. The patent office, in addition to other specimen apartments, has one room in the upper story, 275 feet long and 65 feet wide, and when completed by wings according to the original design, will be upward of 400 feet in length. It is regarded as one of the most splendid rooms in the United States, and is devoted to the grand and increasing collections of the national institution. The portico of this building is of the same extent as that of the Parthenon at Athens, consisting of 16 columns, in double rows, 50 feet high. In the war-office was formerly kept the fine collection of Indian portraits, painted from the original heads by King. These valuable pictures are now in the custody, and adorn the collections of the national institution in the building of the patent office.

The navy yard is on the eastern branch, about three fourths of a mile S.E. of the capitol, and contains 27 acres of ground. It has houses for the officers, and shops and warehouses, two large ship houses, a neat armoury, and every kind of naval stores. One 74 gun ship, and four frigates of 44 guns each, have been built at this yard. The navy magazine is a substantial brick edifice on 70 acres of ground belonging to the United States, and a wharf for the convenience of receiving and carrying off the powder extends from the shore a short distance into the river. There are also in the city, a city hall, a hospital, a penitentiary, a theatre, &c.

The city contains 21 places of worship, four Presbyterian, three Episcopal, three Methodist, one Protestant Methodist, three Baptist, three Roman Catholic, one Unitarian, one Friends, and two African. There are three orphan asylums. There are three banks, with an aggregate capital of $1,500,000; and two insurance companies, with a total capital of $450,000.

Washington is separated from Georgetown by Rock creek, over which are two bridges. A substantial bridge on piles, over a mile in length, crosses the Potomac, leading to Alexandria. There is a bridge also over the Anacostia or eastern branch. This river has a sufficient depth of water for frigates to ascend to the navy yard, without being lightened. Vessels requiring 14 feet of water come up to the Potomac bridge. By means of the Chesapeake and Ohio canal, now completed to Hancock, 136 m., a communication is opened to a rich back country, and it may be expected that the commerce of Washington will increase. The Washington canal is a continuation of this canal through the city to the eastern branch. This canal has several basins, which, with the canal, are walled in with stone on both sides. It cost $930,000.

There were in the city in 1840, 106 retail stores, with a capital of $926,040; six lumber-yards, capital $57,800; the precious metals were manufactured to the amount of $13,000; various metals to the amount of $37,300; hardware and cutlery was manufactured to the amount of $990; bricks and lime were produced to the amount of $88,80 with a capital of $25,300; 48 persons manufactured hats and caps to the amount of $39,500, with a capital of $88,80; two tanneries produced 500 sides of sole leather, and 380 sides of upper leather, employing five persons, and a capital of $2000; other manufactories of leather, as saddlers, &c., produced articles to the amount of $43,950, with a capital of $90,300; six persons produced 100,000 pounds of soap, and 25,000 pounds of tallow candles, with a capital of $7000; one brewery produced 165,000 gallons of beer, employing 20 persons, and a capital of $63,000; two potteries, employing five persons, produced articles to the amount of $3500, with a capital of $3250; one rope-walk, employing four persons, produced cordage to the amount of $1000, with a capital of $425; 68 persons produced carriages and wagons to the amount of $44,700, with a capital of $29,800; 65 persons manufactured furniture to the amount of $14,750, with a capital of $7050; 11 printing-offices, nine binderies, three daily, five weekly, and five semi-weekly or tri-weekly newspapers, and three periodicals, employed 271 persons, and a capital of $149,500; 30 brick or stone, and 23 wooden houses were erected, employing 54 persons, and cost $36,910. The total amount of capital employed in manufactures was $336,275.

The Columbian college was incorporated by an act of congress, in 1821. It is pleasantly situated on elevated ground N. of the president's house, about 3¼ m. from the capitol. The buildings are the college edifice, five stories high, including the basement and the attic, having 48 rooms for students, with two dormitories attached to each; two dwelling houses for professors, and a philosophical hall, all of brick. It has a medical department attached, with an edifice, at the corner of tenth and E streets, at equal distances from the capitol and president's house. In the several departments are a president, 10 professors, and in the college proper 25 students, with 4200 volumes in its libraries. The commencement is on the first Wednesday of October. The whole number of alumni is 97. It is under the direction of the Baptists.

There were in the city in 1840, 12 academies, with 809 students; nine common and primary schools, with 380 scholars.

The national institution for the promotion of science was organized in 1840. The president of the United States is patron; the heads of departments constitute six directors on the part of the government, and six literary and scientific gentlemen are directors on the part of the institution. Its stated monthly meetings are held in the patent office building. Its collections are deposited in the grand hall of this building, 275 ft. long and 65 ft. wide, and constitute a rapidly increasing scientific museum. The United States exploring expedition has added largely to its curiosities. The historical society, and the Columbian institute, have united with it, with their libraries and collections. They have a valuable mineralogical cabinet. It is proposed to bring out regularly, volumes of transactions. If properly patronized, it may rise to eminence, and become an honour to the nation. The Union literary society has been in existence for many years, and holds a weekly discussion in the lecture room of the medical college, and is well attended. Sectarian religion and party politics are excluded from its discussions. The city library contains about 6000 volumes.

The congressional burying ground is in the eastern section of the city, about a mile and a half from the capitol, containing 10 acres of ground near the Anacostia or eastern branch. The grounds are neatly laid out and tastefully kept. It has already become the final resting place of a considerable number of eminent men, has some fine monuments; and contains a vault in which bodies are placed that are awaiting a removal.

One of the most interesting public works in Washington is the electrical telegraph of Professor S. F. B. Morse, of the New-York university. The United States congress appropriated $30,000 to test the value of Professor Morse's invention, by means of a telegraphic communication to Baltimore, 38 m. distant. The thing has been done with the most complete success, and for a few thousand dollars less than the appropriation. This telegraph is undoubtedly superior to anything of the kind in the world, and Professor Morse claims that the invention is original with him, and is prepared to establish the fact; and all other electrical telegraphs must, probably, either adopt his arrangements, or have something less perfect. Nothing can well be conceived of more simple and efficient than this machine. By touching the keys of the machine at Washington, a corresponding machine at Baltimore impresses on strips of paper a set of signs, consisting of dots and dashes answering to the letters of the alphabet, in characters raised like the letters in the printing for the blind. Every letter is impressed in triplicate. It makes no difference whether the person is attending to the operations of the machine at Baltimore or not. If the machine at Washington were to write all night, whatever was written would be found in the morning impressed on rolls of paper. In characters as intelligible as the plainest print; and there being three copies of each letter impressed, they could be cut into three strips, and different parts given, if necessary, into the hands of different printers. At the late meeting of the National Democratic convention at Baltimore, for the nomination of

president and vice-president for the approaching election, intelligence was conveyed by the telegraph to the Hon. Silas Wright, then at Washington, that he had been nominated as the candidate for vice president. He answered by the same means, that he could not consent to be a candidate. His friends at Baltimore requested him by the telegraph to reconsider the matter. He replied that he had done so, and could in no case consent to be a candidate. These several communications, back and forth, were made in a few minutes, which could not have been done in as many hours, even by the railroad; and the convention proceeded to select another candidate before they adjourned. Professor Morse operates with a single circuit of wires. The first plan was to coat them with cotton twine, covered with gum shellac, and bury them in the earth in leaden tubes; but this was found to be troublesome, and needlessly expensive; and they are now coated with tar or other substance in addition to the shellac, and placed on the top of posts, 20 feet high. Professor Morse has even discovered that he can operate with a single wire, employing the earth to complete the circuit. It is not improbable that this invention may yet bring to light some new principles of electro-magnetism.

The seat of the federal government was removed to this city in the year 1800, and congress assumed the jurisdiction of the district on the 27th of February, 1801. The laws of Maryland are in force on the E. side of the Potomac, and of Virginia on the W. side, in the district, unless modified or repealed by congress. The north wing of the capitol was commenced in 1793, and finished in 1800, at an expense of $480,302; the south wing was commenced in 1803, and finished in 1808, at an expense of $303,808. The centre building was commenced and completed since the war, and the whole building, excepting the walls, rebuilt. In August, 1814, Washington was captured by the British, under General Ross, who burned the capitol, the president's house, the public offices, with the exception of the patent office, which was saved by the solicitation of its superintendent. The library of congress was burned. In the sanguinary war which grew out of the French revolution, the capitals of Europe were successively in the hands of conquerors, who waged no Vandal war against public buildings, not connected with warlike operations, and with public libraries and public papers. Whatever disgrace was incurred, on the part of Americans, by the capture of Washington, was more than wiped off by the signal defences and victories of Plattsburg, Baltimore, and New-Orleans, soon after accruing. General Ross, who lead the expedition against Washington, was killed in the attack on Baltimore. It is not improbable that these victories resulted from the fact, that the spirit of the nation was roused, and its differences were harmonized, by the conflagration of Washington; and the truth was evinced that the Americans united, are invincible when defending their own soil. Had the war continued, more evidences of this truth would probably have been given.

WASHINGTON, p. v., capital of Beaufort co., N. C., 127 m. E. by S. Raleigh, 313 W. Situated on the N. side of Tar river, at its entrance into Pamlico river, at the head of ship navigation. It has a courthouse, jail, a church, several stores, and about 1200 inhabitants. Tonnage of the port in 1840 was 5401 tons.

WASHINGTON, p. v., capital of Wilkes co., Ga., 50 m. W.N.W. Augusta, 78 m. N.E. Milledgeville, 570 W. Situated on Kettle creek, a branch of Little river. It contains a courthouse, jail, a bank, an academy, all of brick, three churches, a Presbyterian, Methodist, and Baptist; 33 stores, a printing-office, and about 800 inhabitants, about one half of whom are coloured.

WASHINGTON, p. v., Adams co., Miss., 6 m. N.E. Natchez, 84 m. S.W. Jackson, 1104 W. It was incorporated to contain 1 m. square, and contains a hospital and poorhouse of brick, two churches, a Methodist and Baptist, a female seminary, a United States' land office, and about 400 inhabitants. It is the seat of Jefferson college, founded in 1802. It has a large brick edifice, capable of accommodating 100 students, has a fund of $200,000, and an annual income of $8000. The library contains 1522 volumes.

WASHINGTON, p. v., capital of Mason co., Ky., 77 m. E.N.E. Frankfort, 465 W. It contains a courthouse, jail, an academy, four churches, and about 550 inhabitants.

WASHINGTON, t., Clermont co., O. Bounded S. by Ohio river. It has six stores, one grist-mill, two saw-mills, two tanneries, one distillery, 10 schools, 510 scholars. Pop. 2100.

WASHINGTON, t., Guernsey, co., O. It has three schools, 105 scholars. Pop. 1353.

WASHINGTON, t., Holmes co., O. It contains four stores, three flouring-mills, three grist-mills, two saw-mills; four schools, 215 scholars. Pop. 1461.

WASHINGTON, t., Montgomery co., O. It has 12 schools, 915 scholars. Pop. 2910.

WASHINGTON, t., Shelby co., O. Watered by a branch

of Miami river. It has three stores, two flouring-mills, one grist-mill, three saw-mills; five schools, 118 scholars. Pop. 1668.

WASHINGTON, t. Preble co., O. Watered by branches of Great Miami river. It contains four flouring-mills, four saw-mills. Pop. 2450.

WASHINGTON, t, Miami co., O. It contains Piqua village, and has three commission houses in foreign trade, 22 retail stores. two fulling-mills, two grist-mills, three saw-mills, one oil-mill. Pop. 1161.

WASHINGTON, t., Starke co., O. It has 600 scholars in schools. Pop. 1389.

WASHINGTON, t., Marion co., Ia. It has four stores, five grist-mills, 12 saw-mills, two tanneries, four distilleries; eight schools, 335 scholars. Pop. 1859.

WASHINGTON, t., Putnam co., Ia. It has five stores, one flouring-mill, five grist-mills, four saw-mills, one tannery, one distillery, one pottery; five schools, 385 scholars. Pop. 1873.

WASHINGTON, t., Washington co., Ia. It has 11 stores, two cotton factories with 504 spindles, four grist-mills, four saw-mills. one oil-mill, five tanneries, two printing-offices, two weekly newspapers. Pop. 1992. . Washington is, as it deserves to be, a favourite name; and there are many other smaller towns of this name.

WASHITA. parish. La. Situated towards the N.E. part of the state, and contains 2090 sq. m. Drained by Washita river and its branches. It contained in 1840, 2642 neat cattle, 350 sheep, 4194 swine ; and produced 32,894 bushels of Indian corn, 2240 of oats, 5300 of potatoes, 1,724,658 pounds of cotton. It had three grist-mills, two saw-mills ; one academy, 30 students ; five schools, 115 scholars. Pop.: whites, 2188; slaves, 2438; free coloured, 14; total, 4640. Capital, Monroe.

WASHTENAW, county, Mich. Situated towards the S.E. part of the state, and contains 720 sq. m. Watered by Huron river and its branches. It contained in 1840, 22,208 neat cattle, 19,273 sheep, 30,141 swine ; and produced 216,597 bushels of wheat, 2941 of rye, 220,0.6 of Indian corn, 22,600 of buckwheat, 31,050 of barley, 284,181 of oats, 201,224 of potatoes, 108,047 pounds of sugar. It had 65 stores. two furnaces, three fulling-mills, two woollen factories, 11 flouring-mills. seven grist-mills, 41 saw-mills, four tanneries, four distilleries. Pop. 23,571. Capital, Ann Arbor.

WATERBOROUGH, p. t., York co., Me., 23 m. N. York, 76 m. S.W. Augusta, 521 W. Bounded N. by Little Ossipee river. Drained by head branches of Moresam river. Incorporated in 1787. It has one fulling-mill, three grist-mills, three saw-mills ; 16 schools, 691 scholars. Pop. 1944.

WATERBURY, p. t., Washington co., Vt., 12 m. N.W. Montpelier, 525 W. Bounded S. by Onion or Winooski river. Watered by Waterbury river, which affords water-power. In the S.W. corner, the rocks have fallen down into Onion river, forming a curious natural bridge. It contains two churches, a Congregational and Methodist ; two stores, two fulling-mills. one grist-mill; three saw-mills ; 19 schools, 595 scholars. Pop. 1992.

WATERBURY, p. t., New-Haven co., Ct., 52 m. S.S.W. Hartford, 310 W. Naugatuck river affords good water-power. It contains four churches, a Congregational, Episcopal, Methodist, and Baptist; 17 stores, five fulling mills, three woollen factories, three cotton factories with 570 spindles, five grist-mills, 16 saw-mills, two tanneries, three distilleries ; two academies, 75 students ; 23 schools, 735 scholars. Pop. 3668.

WATEREE, river, rises in N. Carolina, which is called Catawba river, and flows into S. C., and unites with Congaree river to form Santee river.

WATERFORD, a marit. co. of Ireland, prov. Munster, having S. St. George's channel ; E. Waterford harbour, by which it is separated from Wexford and Kilkenny ; N. Tipperary ; and W. Cork. Area 471,281 acres, of which 118,034 are unimproved mountain, with but little bog. Though generally coarse, there is a considerable extent of fine land in this county, particularly in its S.E. quarter, and the mountains afford good pasturage for cattle. Estates, for the most part, very large ; the largest, which belongs to the duke of Devonshire, is managed on the most liberal principles, and greatly improved. Here, indeed, and generally throughout Ireland (and, we believe, that the same thing may be truly affirmed of England), tenants and occupiers on large estates are decidedly better off than those on the smaller class of properties. This is the principal dairy county of Ireland. When it was visited by Mr. Young, not 1-30th part was under the plough. (Tour in Ireland, 4to ed., p. 329.) This proportion in tillage is now, however, much larger. This has principally arisen from the vicious custom of dividing farms. "In this county," says Mr. Wakefield, "when the eldest daughter of a farmer marries, the father, instead of giving her a portion, divides his farm between himself and his son-in-law ; the next daughter gets a half of the remainder, and this division and subdivision is continued as long

as there are daughters to be disposed of. The sons are left to shift for themselves the best way they can." (I. 29. Some of the dairy farmers are in easy circumstances ; but the condition of the tillage farmers and cottiers is much the same as in other parts of Munster. Some very material improvements have, however, been introduced since 1829, into this and the contiguous counties. Improved swing-ploughs, made of iron, drawn by two horses driven by the ploughman, are now become very general. Land is kept cleaner ; there has been a very great increase in the quantity of lime, used as manure ; green crops are more attended to : and the quantity of wheat raised within these few years has been more than doubled ; while there has been a decided falling off in the production of oats. There has also been a large increase of the exports of butter and bacon. Average rent of land, 12s. 6d. an acre. The minerals, which comprise copper, iron, &c., are but little wrought, and are unimportant ; which, also, is the case with manufactures : a considerable cotton manufacture has, however, been established at Portlaw, and some glass is made in Waterford. Principal rivers, Blackwater, Suir, and Bride. Waterford is divided into seven baronies, and 74 parishes ; and returns five members to the House of Commons ; being two for the county, two for the borough of Waterford, and one for Dungarvan. Registered electors for the county, in 1-39-40, 1675. In 1831, Waterford had 34,848 inhabited houses, 36,191 families, and 177,054 inhabitants ; of whom 85,217 were males, and 91,837 females.

WATERFORD, a city, parl. bor., and seaport of Ireland, prov. Munster, on the estuary of the river Suir, about 10 m. from the sea, and 82 m. S.S.W. Dublin, lat. 52° 17′ N., long. 7° 10′ W. It is a county of itself, comprising 9885 acres, but it is locally situated near the W. extremity of the county of Waterford, of which it is the capital.

In 1831, the population of the county of the city amounted to 28,821, and that of the city itself to 26,377. The Boundary Act made no change in the limits of the part bor., which embrace the whole county of the city. The city is situated on the S., or S.W. side of the river ; but a considerable portion of the county of the city is situated on its opposite side, the communication between them being maintained by a fine wooden bridge, 832 feet in length by 40 feet in width, constructed by an American artist. The quay, fronting the river, about 1 m. in length, is one of the finest in Europe, and is bounded on the land side by a range of well built houses. The other principal streets are the Mall, Beresford-street, Broad-street, &c. ; but the city is very irregularly laid out ; and in the older parts the streets are in many narrow and dirty, with mean thatched houses, or rather hovels, occupied by a very poor and wretched population. In the more modern parts, however, the streets are comparatively broad, and the houses well built and substantial. The county of the city is divided into 12 parishes, of which three are entirely rural. The cathedral of the see of Waterford (now merged in that of Cashel), is a fine modern building, with an ornamented spire: near it is the bishops' palace, also, a handsome modern structure. Here are three parish churches, and four Roman Catholic chapels, the largest of which is the cathedral. The Presbyterians, Baptists, Methodists, Independents, and Quakers have, also, their respective places of worship. Among the public buildings, exclusive of the churches, may be noticed the town-hall, chamber of commerce, county and city prisons and court-houses, artillery barracks, penitentiary, custom-house. St. Reginald's tower, on the quay, an ancient fortress, now a police barrack, &c.

The educational establishments comprise an endowed classical school, under the patronage of the corporation, which provides a residence and salary for the head master ; the college of St. John, a plain but spacious building, provides instruction for candidates for the Roman Catholic ministry, preparatory to their entrance into the college of Maynooth ; and there are various other public and private schools. Mr. Inglis states that he visited a Catholic school at which upwards of 700 children were educated by an association of young men, called the "Brothers of the Christian Schools." Here, also, is the Waterford institution for the diffusion of knowledge, with a library and a small museum, a literary and scientific society, in which lectures are delivered, and agricultural and horticultural societies. In 1841 it had no fewer than five weekly newspapers.

The charitable institutions comprise a Blue-coat school for Protestant boys, founded about 1700, and possessing an estate of 1400 acres ; a Blue-coat hospital for Protestant girls, founded in 1740 ; Widows' apartments, erected in 1702, for the maintenance of 10 poor clergymen's widows ; the Leper hospital, founded by King John, and now used as an infirmary, is capable of accommodating 400 patients ; the Holy Ghost hospital, founded in 1546, and now appropriated to the reception of females ; the fever hospital, the first of the kind in Ireland, opened in 1799, and capable of receiving 150 patients ; the lunatic asylum, for the county and city, a large modern building, has accommodation for 100 patients; the house of industry, opened in 1779, for the relief of the poor, and the punishment of vagrants and sturdy beggars; a mendicity institution, with similar objects, was opened in 1820. There is also a lying-in hospital, a charitable loan fund, and several orphan societies.

The manufactures of Waterford are unimportant, comprising only a glass-work and several breweries. It is, however, better situated for trade than any other town of Ireland. The harbour is excellent, vessels of 800 tons burden coming up to the quays. The Suir, which is navigable for barges as far as Clonmell, gives it a considerable command of inland navigation, and it is also the principal *entrepot* for the produce brought down by the Barrow and its important tributary the Nore, as well as for the produce which is to be conveyed inland by these channels; its trade is in consequence great, and rapidly increasing. Its exports of raw produce, including corn and flour, butter, beef, pork, and bacon, hides, tallow, &c., exceed those from any other Irish port, and amount, at present, to above £2,000,000 a year. The opening of a steam communication between Waterford and Bristol, and other towns, has been of peculiar advantage to the first. Subjoined is a

STATEMENT of the quantity and value of the principal articles exported from Waterford during the year ending the 30th April, 1835:

Principal Articles.		Quantity.	Value.
			£ s. d
Beef	tierces	33 at 130s.	214 10 0
Pork	do.	672 — 65s.	3,024 0 0
Pork	barrels	2,219 — 46s.	5,103 14 0
Bacon	fliches	399,133 — 26s.	518,879 8 0
Butter	cwts.	118,471 — 80s.	473,884 0 0
Lard	do.	18,397 — 50s.	45,992 10 0
Wheat	barrels	63,773 — 21s.	66,963 15 0
Oats	do.	203,167 — 12s.	121,900 4 0
Barley	do.	57,731 — 14s.	40,411 14 0
Flour	cwts.	489,×34 — 16s.	397,170 8 0
Oatmeal	do.	11,321 — 12s.	6,792 12 0
Bread	do.	1,241 — 12s. 6d.	£00 12 8
Cattle	head	4,410 — 16s.	30,570 0 0
Live Sheep		1,855 — 45s	4,189 10 0
Do. Pigs		58,313 — 35s.	102 017 15 0
Total			1,818,333 12 6

There belonged to the port in 1841, 115 vessels of the aggregate burden of 19,309 tons; and the gross customs duty received at the port in 1840, amounted to £196,389. The management of the port is vested in 24 harbour commissioners, nominated partly by the chamber of commerce, incorporated in 1815, and partly by the corporation. Postoffice revenue, in 1830, £4727; in 1836, £5536. Branches of the Bank of Ireland, and the Provincial bank, and the Agricultural bank, have been opened in the town.

Being the place at which Henry II. landed, in 1172, to take possession of his conquests in Ireland, Waterford was early distinguished by marks of royal favour. It received its first charter from King John, and had no fewer than 12 additional charters from succeeding monarchs. It appears, however, that the right to send two representatives to the Irish House of Commons was not conferred by charter, but rested only on prescription, the practice having commenced in 1374. At the union, Waterford was authorized to send one member to the imperial House of Commons, and under the Reform Act she sends two members. Registered electors, in 1839-40, 1963. The limits of the municipal borough are much less extensive than those of the parliamentary borough. Under the act 3 and 4 Victoria, cap. 108, the city is divided into five wards, and is governed by a mayor, five other aldermen, and 30 councillors. It has a court of record, which decides pleas to any amount; a civil bill court for debts from £2 to £10; and a mayor's court for the decision of claims as to wages. Assizes for the county and city are held here twice a year, and general sessions of the peace fifteen times.

It is rather singular, that notwithstanding the increase of its trade, there is a great deal of abject poverty and misery in Waterford. Some improvements have, however, taken place, both in respect of cleanliness and of the dress of the lower orders. When Mr. Inglis was here, whiskey drinking prevailed to a frightful extent; but, thanks to the exertions of Father Mathew, this has been materially diminished. *(Parl. Reports and Private Information.)*

WATERFORD, p. t., Oxford co., Me., 53 m. W. Augusta, 588 W. Incorporated in 1797. It has four stores, two fulling-mills, three grist-mills, eight saw-mills, one oil-mill, two tanneries; 12 schools, 527 scholars. Pop. 1381.

WATERFORD, p. t., Caledonia co., Vt., 49 m. E.N.E. Montpelier, 547 W. Bounded S.E. by Connecticut r., in which, opposite to the t., is a portion of the 15 m. falls. Watered by Passumsic river. It has three churches, a Congregational, Methodist, and Free-will Baptist, three stores, eight grist-mills, one saw-mill; one academy, 41 students, 15 schools, 501 scholars. Pop. 1388.

WATERTOWN, p. t., New-London co., Ct., 66 m. S.E. Hartford, 451 W. Bounded E. by Thames river, S. by Long Island sound. Watered by Niantic river, which is navigable 3 or 4 m. from the sound, by a bay of the same name into which it enters, for sloops of 20 tons. It contains three Baptist churches, four stores, three grist-mills, one oil-mill; 10 schools, 532 scholars. Pop. 2399.

WATERFORD, p. t., Saratoga co., N. Y., 10 m. N. Albany. 380 W. The Mohawk here enters Hudson river, which bounds it on the E. There is an outlet of the canal by three locks of 11 ft. fall each. Sloops navigating Hudson river ascend by the dam and lock at Troy to this place. The village has four churches, a Presbyterian, Methodist, Dutch Reformed, and an Episcopal, an academy, a female seminary, a bank, various mechanic establishments, 200 dwellings, and about 1400 inhabitants. A bridge connects it with Lansingburg, 800 ft. long across the Hudson, which cost $70,000. The Champlain canal and the Troy and Saratoga railroad pass through it. The town has 28 stores, one grist-mill, one saw-mill, one tannery; five schools, 375 scholars. Pop. 1824.

WATERFORD, t., Gloucester co., N. J., 12 m. W. Woodbury, 51 S.S.W. Trenton, 161 W. Drained by Pensauken and Cooper's creeks. It contains an Episcopal church, 10 stores, two glass-houses, two grist-mills, 10 saw-mills, one oil-mill, four distilleries; eight schools, 425 scholars. Pop. 3467.

WATERLOO, a hamlet of Belgium, prov. Brabant, on the verge of the forest of Soignies, and on the road from Brussels to Charleroi, 9 m. S. by E. the former. This village will be forever memorable in military history for the great battle fought in its vicinity on the 18th of June, 1815, between the allied army under the duke of Wellington, and the French, under Napoleon. There is some discrepancy in the statements on the subject, but each army probably consisted of about 70,000 men. The French began the attack between 11 and 12 o'clock forenoon. The object of Napoleon was to defeat the British, or force them to retreat, before the Prussians, who he knew were coming up, could arrive on the field; while that of the duke of Wellington was to maintain his ground till he could be joined by his allies, when it might be in his power to become the assailant. The attacks of the French were repeated with the greatest fury; but they made no serious impression on the British, by whom they were sustained and repelled with invincible courage and resolution. At length, about ½ past 6 o'clock, the Prussians came into the field, with a strong force of from 15,000 to 20,000 men. The English then became the assailants; and though Napoleon brought forward his guard, which had not previously been engaged, it could not stem the torrent, and, having been forced to give way, the whole army got into inextricable confusion, and the rout became universal. The slaughter was enormous. The British lost, besides officers, about 15,000 men killed and wounded. The French loss is not exactly known; but it was not, perhaps, less, in the battle and pursuit, than 30,000 men. All their cannon and baggage also fell into the hands of the conquerors; and it may, indeed, be said that the French army was entirely destroyed.

Such was the battle of Waterloo, in which the star of Napoleon set never to shine again! It is not, however, to be denied that he did all that was possible in the desperate circumstances under which he was placed. He had already defeated and beat back the Prussians; and his only chance of being able to make head against the forces that were marching against him, of awakening the enthusiasm of the French, and paralysing his enemies, depended on his being able to defeat the army under the duke of Wellington before it could form a junction with the Prussians. The skill of the British general, and the invincible courage and resolution of his troops, defeated this project; but it was worthy the genius of Napoleon, whose efforts in this "death struggle" were well seconded by his troops, who, though unsuccessful, did all that brave men could do.

WATERLOO, p. t., semi-capital of Seneca co., N. Y., 173 m. W. by N. Albany, 346 W. Bounded S. by Seneca river, along which passes the Seneca and Cayuga canal. It contains four churches, a Presbyterian, Episcopal, Methodist, and Baptist, 10 stores, one fulling-mill, one woollen-factory, one furnace, one flouring-mill, two saw-mills, one oil-mill, one tannery, two distilleries, one printing-office, one semi-weekly newspaper; one academy, 66 students, six schools, 278 scholars. Pop. 3036.

WATERTOWN, p. t., Middlesex co., Mass., 7 m. W. Boston, 433 W. Bounded S. by Charles river, which is navigable to this place for vessels requiring 6 or 7 ft. water. It contains three churches, a Unitarian, Baptist, and a Universalist, a United States arsenal, containing 40 acres of ground, with several large brick buildings enclosed by a high fence, on the N. bank of Charles river. The Provincial congress held their sessions here in 1775. It contains 11 stores, seven tanneries, one paper-mill; one academy, 30 students, seven schools, 402 scholars. Pop. 1810.

WATERTOWN, p. t., Litchfield co., Ct., 42 m. S.S.W. Hartford, 290 W. It contains two churches, a Congregational and an Episcopal, four stores, two fulling-mills, two grist-mills, six saw-mills, two tanneries; nine schools, 354 scholars. Pop. 1442. It was the birth-place of John Trumbull, the celebrated author of McFingal.

WATERTOWN, p. t., capital of Jefferson co., N. Y., 164 m. N.W. Albany, 416 W. Bounded N. by Black river, which affords immense water-power. It contains a courthouse, jail, seven churches, two Presbyterian, an Episcopal, Methodist, Baptist, Universalist, and Roman Catholic, a state arsenal, two banks, an insurance company, 45 stores, five woollen-factories, one cotton-factory with 1000 spindles, one paper-mill, four grist-mills, four tanneries, one brewery, three printing-offices, three weekly newspapers; 19 schools, 3192 scholars. Pop. 5027.

WATERVILLE, p. t., Kennebec co., Me., 18 m. N. Augusta, 613 W. Bounded E. by Kennebec river, which has here a fall of 18 ft., affording an immense water-power. Emerson's stream, which flows into Kennebec river, has the highest fall in the state, and affords good water-power. The great water-power here concentrated is but partially improved. The Kennebec is navigable from this place to Augusta for boats of 40 tons. It has four churches, 39 stores, two fulling-mills; one college, 94 students; two academies, 392 students, 19 schools, 1274 scholars. Pop. 2971. It is the seat of Waterville college, under the direction of the Baptists, founded in 1820, which has a president and six professors or other instructors, 210 alumni, of whom 70 have been ministers of the gospel, 70 students, and 7000 vols. in its libraries. The commencement is on the second Wednesday in August. It has two edifices containing rooms for students, a chapel, and a commons hall.

WATERVLIET, p. t., Albany co., N. Y., 5 m. N. Albany, 373 W. Bounded N. by Mohawk river, E. by Hudson river. It contains several islands formed by the mouths of the Mohawk. The Erie and Champlain canals here separate, and in the former is a succession of locks to overcome the falls of the Cohoes. At West Troy village is the United States arsenal. (See TROY.) There are in the town 94 stores, four lumber yards, three fulling-mills, three woollen-factories, two cotton-factories, with 3160 spindles, two flouring-mills, 10 grist-mills, 12 saw-mills, one paper-mill, one brewery, four potteries, one printing-office, one weekly newspaper; 13 schools, 1600 scholars. Pop. 10,141.

WATFORD, a market town and par. of England, co. Herts, hund. Cashio; on the Colne, here crossed by a bridge, 10 m. N.W. London. Area of par., including besides Watford four adjacent hamlets, 10,980 acres. Pop. of do., in 1831, 5693; and of the town 2960. The latter, which is well built, consists principally of a main street, about 1 m. in length, on the high road from London to Birmingham. The church, in the centre of the town, is a large edifice, consisting of a nave, three aisles, and a chancel: it has, at the W. end, a massive embattled tower, 80 feet in height, surmounted by a small spire rising about 20 feet higher. It has some fine monuments, especially two by Nicholas Stone. The living, a valuable vicarage, worth £730 a year, is in the gift of the earl of Essex. Here, also, are chapels for Baptists and Wesleyans. The educational establishments comprise a free school, founded in 1704, for 40 boys and 14 girls, with an income of about £180 a year, which, in 1832, afforded instruction to 60 scholars; another free school, with a small endowment, established in 1641; a national school supported by subscription, &c.: the funds for the general charities yielded, at the date of last inquiry, an income of nearly £850 a year. Next to agriculture, the chief branches of industry pursued here are the spinning and winding of silk, straw plaiting, and malting. In 1838, there were two silk mills at work in the parish, which furnished employment for 290 hands. There are some very extensive paper-mills on the Colne, in the vicinity; and Watford is a considerable market for corn, sheep, cattle, and hogs. Its trade in these is facilitated by the Grand Junction canal, which passes about 2 m. W. of the town, where it is joined by the Colne, which has been rendered navigable to St. Alban's. The London and Birmingham railway has a station immediately to the E. of Watford, near which the line passes through a tunnel 1930 yards in length. A council of magistrates, and a court of requests for the recovery of small debts, are held in the town weekly. Markets on Tuesdays; and fairs four times a year for cattle, horses, pedlery, and the hiring of servants. Adjoining Watford on the W. is Cashiobury park, the seat of the earl of Essex, lord of the manor. The house has a good deal of the appearance of a monastery; it has some good pictures, and a valuable collection of books. (*Private Information.*)

WATSON, p. t., Lewis co., N. Y., 10 m. N.E. Martinsburg, 140 m. N.W. Albany, 439 W. Bounded W. by Black river. It has one store, 12 saw-mills; 17 schools, 354 scholars. Pop. 1707.

WAWARSING, p. t., Ulster co., N. Y., 22 m. S.W. Kingston, 82 m. S.S.W. Albany, 297 W. Drained by Rondout creek and its tributaries. Iron ore, plumbago, and some lead are found here. It has 17 stores, eight fulling-mills, 11 woollen-factories, one cotton-factory with 2136 spindles, one glass-house, three grist-mills, 43 saw-mills, six tanneries, one pottery; 20 schools, 924 scholars. Pop. 4044.

WAYNE, county, N. Y. Situated in the N. part of the state, and contains 572 sq. m. Bounded N. by lake Ontario. Canandaigua outlet, after receiving Mud creek at Lyons, becomes Clyde river. It contains gypsum, marl, water limestone, and several sulphur springs. It had, in 1840, 33,298 neat cattle, 100,986 sheep, 44,130 swine; and produced 571,083 bushels of wheat, 4460 of rye, 338,356 of Indian corn, 39,002 of buckwheat, 25,067 of barley, 4,52,969 of oats, 512,710 of potatoes, 159,554 pounds of sugar. It had 115 stores, four furnaces, one forge, 11 fulling-mills, 12 flouring-mills, 25 grist-mills, 118 saw-mills, two glass-houses, one rope-walk, 25 tanneries, six distilleries, one brewery, four printing-offices, four weekly newspapers; eight academies, 475 students; 182 schools, 9637 scholars. Pop. 42,057. Capital, Lyons.

WAYNE, county, Pa. Situated in the N.E. part of the state, and contains 648 sq. m. Bounded N.E. by Delaware river. Drained by Lackawaxen creek and its branches. It contained in 1840, 17,412 neat cattle, 34,371 sheep, 3122 swine; and produced in 1840, 461,155 of Indian corn, 8855 of oats, 20,566 of swine; and produced 15,210 bushels of wheat, 11,041 of rye, 10,393 of Indian corn, 36,555 of buckwheat, 3749 of barley, 102,140 of oats, 156,702 of potatoes, 2940 pounds of sugar. It had 26 stores, three fulling-mills, one glass-house, one flouring-mill, 14 grist-mills, 66 saw-mills, one oil-mill, two printing-offices, two weekly newspapers; three academies, 96 students; 97 schools, 9494 scholars. Pop. 11,848. Capital, Honesdale.

WAYNE, county, N. C. Situated S.E. of the centre of the state, and contains 790 sq. m. Watered by Neuse river and its branches. It contained in 1840, 8492 neat cattle, 6980 sheep, 46,594 swine; and produced 12,744 bushels of wheat, 1443 of rye, 461,155 of Indian corn, 8855 of oats, 20,566 of potatoes, 402,175 pounds of cotton. It had 19 stores, 49 grist-mills, 25 saw-mills; four academies, 129 students; six schools, 213 scholars. Pop.: whites, 6734; slaves, 367; free coloured, 464; total, 10,891. Capital, Waynesboro'.

WAYNE, county, Ga. Situated in the S.E. part of the state, and contains 900 sq. m. Bounded N.E. by Altamaha river. Watered by Santilla river. It contained in 1840, 7854 neat cattle, 404 sheep, 3933 swine; and produced 16,372 bushels of corn, 15,125 of potatoes, 19,183 pounds of cotton, 3465 of sugar. It had three stores, one grist mill; one academy, 12 students. Pop.: whites, 877; slaves, 367; free coloured, 14; total, 1258. Capital, Wayne, C. H.

WAYNE, county, Miss. Situated in the E., toward the S., part of the state, and contains 790 sq. m. Drained by Chickasawha river and its branches. It contained in 1840, 14,297 neat cattle, 937 sheep, 7671 swine; and produced 5327 bushels of Indian corn, 9154 of potatoes, 354,699 pounds of cotton. It had two stores, one tannery. Pop.: whites, 1211; slaves, 979; total, 2190. Capital, Winchester.

WAYNE, county, Tenn. Situated in the S. toward the W., part of the state, and contains 504 sq. m. Bounded N.W. by Tennessee river. Drained by Buffalo creek and its branches, and branches of Shoal creek. It contained in 1840, 8490 neat cattle, 5004 sheep, 34,351 swine; and produced 27,406 bushels of wheat, 1722 of rye, 428,569 of Indian corn, 1722 of buckwheat, 3,718 of oats, 6942 of potatoes, 17,516 pounds of tobacco, 39,382 of cotton, 6754 of sugar. It had nine stores, one forge, 17 grist-mills, two saw-mills, one tannery, 19 distilleries, one pottery, one school, 29 scholars. Pop.: whites, 7151; slaves, 329, free coloured, 25; total, 7705. Capital, Waynesboro'.

WAYNE, county, Ky. Situated in the S., toward the E., part of the state, and contains 570 sq. m. Bounded S. by Cumberland river, by the S. fork of which, and by other tributaries, it is drained. It contained in 1840, 12,022 neat cattle, 15,322 sheep, 46,016 swine; and produced 1011 bushels of wheat, 7408 of rye, 732,904 of Indian corn, 5515 of oats, 12,270 of potatoes, 5060 pounds of tobacco, 2868 of cotton, 3550 of sugar. It had 18 stores, three flouring-mills, 16 grist-mills, three saw-mills, one oil-mill, two tanneries, 11 distilleries; seven schools, 173 scholars. Pop.: whites, 6754; slaves, 630; free coloured, 15; total, 7399. Capital, Monticello.

WAYNE, county, O. Situated N.E. of the centre of the state, and contains 660 sq. m. Watered by Killbuck river, and by lake fork of Mohican creek, Sugar and Chippewa creeks. It contained in 1840, 90,614 neat cattle, 6,349 sheep, 55,899 swine; and produced 753,621 bushels of wheat, 49,666 of rye, 305,030 of Indian corn, 549,885 of oats, 132,003 of potatoes, 177,092 pounds of sugar. It had 66 stores, four fulling-mills, two woollen factories, 22 flouring-mills, 17 grist-mills, 81 saw-mills, six oil-mills, 27 tanneries, seven distilleries, three breweries, one pottery, two printing-offices, two weekly newspapers; one academy, 25 students; 10 schools, 278 scholars. Pop.: 35,808. Capital, Wooster.

WAYNE, county, Mich. Situated in the S.E. part of the state, and contains 600 sq. m. Drained by Rouge and Huron rivers and their branches, flowing into Detroit river, which bounds it on the W. Limestone, iron ore, and sulphur springs are found. It contained in 1840, 14,574 neat cattle, 10,181 sheep, 17,062 swine; and produced 89,769 bushels of wheat, 5575 of rye, 138,739 of Indian corn, 4975 of barley, 94,981 of oats, 204,313 of potatoes, 122,667 pounds of sugar. It had 11 commission houses in foreign trade, 141 retail stores, five furnaces, eight flouring-mills, three grist-mills, 28 saw-mills, three tanneries, one distillery, two breweries, three potteries, three printing-offices, three daily and four weekly newspapers; one college, 30 students; one academy, 40 students; 72 schools, 3097 scholars. Pop. 24,173. Capital, Detroit.

WAYNE, county, Ia. Situated in the E. part of the state, and contains 420 sq. m. Drained by branches of Whitewater river, which afford good water-power. It contained in 1840, 13,768 neat cattle, 28,173 sheep, 35,413 swine; and produced 124,705 bushels of wheat, 2938 of rye, 882,477 of Indian corn, 284,437 of oats, 21,412 of potatoes, 66,116 pounds of sugar. It had 63 stores, three furnaces, four fulling-mills, 10 flouring-mills, 37 grist-mills, 41 saw-mills, three oil-mills, one paper-mill, 14 tanneries, 14 distilleries, one brewery, three potteries, three printing-offices, three weekly newspapers; six academies, 315 students; 51 schools, 5079 scholars. Pop. 23,222. Capital, Centreville.

WAYNE, county, Ill. Situated toward the S.E. part of the state, and contains 720 sq. m. Drained by Elm river and by Skillet fork of Little Wabash river. It contained in 1840, 7801 neat cattle, 5183 sheep, 19,713 swine; and produced 17,515 bushels of wheat, 1443 of rye, 947,880 of Indian corn, 60,471 of oats, 10,987 of potatoes, 68,510 pounds of tobacco, 7784 of cotton, 1345 of sugar. It had 10 stores, five flouring-mills, seven grist-mills, four saw-mills, one oil-mill, four tanneries; 15 schools, 373 scholars. Pop. 5133. Capital, Fairfield.

WAYNE, county, Mo. Situated in the S.E. part of the state, and contains 1200 sq. m. Watered by St. Francis and Big Black rivers. It contained in 1840, 5849 neat cattle, 3084 sheep, 25,871 swine; and produced 12,949 bushels of wheat, 160,165 of Indian corn, 20,980 of oats, 3478 of potatoes, 22,434 pounds of tobacco, 6840 of cotton, 23,294 of sugar. It had 17 stores, 15 grist-mills, one saw-mill, one tannery, eight distilleries; 13 schools, 234 scholars. Pop.: whites, 3069; slaves, 392; free coloured, 12; total, 3403. Capital, Greenville.

WAYNE, p. t., Kennebec co., Me., 15 m. W. Augusta, 595 W. Watered by part of Androscoggin Great pond. Incorporated in 1798. It contains five stores, one fulling-mill, one furnace, two grist-mills, six saw-mills; one academy, 13 students; nine schools, 445 scholars. Pop. 1201.

WAYNE, p. t., Steuben co., N. Y., 198 m. W. by S. Albany, 316 W. Bounded W. by Crooked lake. It contains three stores, one oil-mill, one tannery; 10 schools, 522 scholars. Pop. 1377.

WAYNE, t., Mifflin co., Pa. Watered by Juniata river. It has four stores, one fulling-mill, one furnace, two grist-mills, eight saw-mills; eight schools, 297 scholars. Pop. 1350.

WAYNE, t., Schuylkill co., Pa. It has two grist-mills, 24 saw-mills, two tanneries. Pop. 1691.

WAYNE, t., Belmont co., O. Captina creek affords water-power. Pop. 1734.

WAYNE, t., Champaign co., O. Drained by branches of Little Darby creek and Mad river. It has one store, one fulling-mill, one woollen factory, two saw-mills; four schools, 98 scholars. Pop. 1300.

WAYNE, t., La Fayette co., O. Drained by Paint creek. It has one store, two grist-mills, two saw-mills; 14 schools, 350 scholars. Pop. 1540.

WAYNE, t., Jefferson co., O. Drained by Cross creek. It has two stores, two flouring-mills, one grist-mill, three saw-mills, two tanneries; six schools, 390 scholars. Pop. 1746.

WAYNE, t., Warren co., O. Watered by Little Miami river and its branches. It has eight schools, 636 scholars. Pop. 3392.

WAYNE, t., Tuscarawas co., O. It has one fulling-mill, four grist-mills, eight saw-mills, two distilleries; 13 schools, 844 scholars. Pop. 2142.

WEAKLY, county, Tenn. Situated in the N.W. part of the state, and contains 660 sq. m. Drained by Obion river and its branches. It contained in 1840, 9031 neat cattle, 5882 sheep, 24,085 swine; and produced 44,929 bushels of wheat, 1240 of rye, 454,248 of Indian corn, 64,421 of oats, 22,849 of potatoes, 493,986 pounds of tobacco, 24,167 of cotton. It had 11 stores, two flouring-mills, 10 grist-mills, nine saw-mills, two tanneries, six distilleries, one printing-office, one weekly newspaper; four academies, 117 students; 14

schools, 295 scholars. Pop.: whites, 8072; slaves, 1796 free coloured, 2; total, 9870. Capital, Dresden.

WEARE, p. t., Hillsborough co., N. H., 14 m. S.W. Concord, 475 W. Drained by a branch of Piscataquog river. Incorporated in 1764. It has three churches, a Baptist, Freewill Baptist, and Universalist, five stores, four fulling-mills, one woollen factory, one cotton factory with 764 spindles, seven grist-mills, 12 saw-mills, four tanneries; one academy, 50 students; 26 schools, 734 scholars. Pop. 2375.

WEARMOUTH. See SUNDERLAND.

WEATHERSFIELD, p. t., Windsor co., Vt., 74 m. S. by E. Montpelier, 464 W. Bounded E. by Connecticut river. Watered by Black river. It contains four churches, two Congregational, a Methodist, and a Union, six stores, three fulling-mills, three woollen factories, three grist-mills, nine saw-mills, two tanneries; 12 schools, 703 scholars. Pop. 2081.

WEBSTER, p. t., Worcester co., Mass., 56 m. W.S.W. Boston, 394 W. Watered by French river, which affords good water-power. It contains two churches, a Methodist and Baptist, four stores, one woollen factory, five cotton factories, with 9100 spindles, two grist-mills, two saw-mills; four schools, 278 scholars. Pop. 1403. A remnant of the Nipmuck Indians reside here.

WEBSTER, p. t., Monroe co., N. Y., 214 m. W. by N. Albany, 271 W. Bounded N. by lake Ontario. It has two stores; 15 schools, 747 scholars. Pop. 2235.

WEDNESBURY, a market town and par. of England, co. Stafford, hund. Offlow, near the source of the Tame, in the great coal and iron district of which Birmingham is the centre, 7 m. N.W. Birmingham. Area of par., 2190 acres. Pop. of do., in 1831, 8437. The church, which stands on a hill, and is supposed to occupy the site of an ancient castle, repaired within these few years, is a fine structure, with a tower surmounted by a lofty spire.. It has an octagonal E. end, and other portions in the perpendicular style, and within are some exquisitely carved prebendal stalls, and a curious moveable reading-desk. (Rickman.) The living, a vicarage worth £300 a year, is in the gift of the crown. Here are chapels for Independents and Wesleyans; a Lancastrian school, supported by subscription: a small endowed school for poor children, an almshouse, and some minor charities. The inhabitants are mostly employed in various branches of the hardware manufacture, especially in the production of the numerous articles included under the term saddlers' ironmongery, with nails, hinges, edge-tools, and cast-iron works of almost every description. Enamel painting is also extensively carried on; and it has a soap manufactory, which produced, in 1839, 2,971,221 lbs. hard soap. A valuable potter's earth is obtained in the vicinity; in which are, also, several corn mills. A branch of the Birmingham canal approaches within a short distance of the town on the one hand, and the Grand Junction railway on the other. The local authority is vested in a constable chosen at the manorial court, held here annually; a court of requests is opened occasionally, for the recovery of debts under £3. Market day, Wednesday; fairs, twice a year, for cattle and pedlery.

WEEDSPORT, p. v., Brutus t., Cayuga co., N. Y., 7 m N. Auburn, 155 m. W. Albany, 341 W. It is a flourishing village on the Erie canal.

WEIMAR-EISENACH (GRAND DUCHY OF SAXE), a state of central Germany, the most important of the minor Saxon states, consisting of several detached portions of territory, inclosed on different sides by the dominions of Prussia, Hesse-Cassel, Bavaria, the kingdom of Saxony, the duchies of Coburg, Meiningen, &c., the capital being in lat. 50° 59' 19" N., long. 11° 21' E. Total area estimated at 1416 sq. m. Pop., in 1839, 247,603, all Protestants, except about 10,300 Roman Catholics, and 1400 Jews. The principality of the Saxe-Weimar is subdivided into the circles of Weimar-Jena, and Neustadt; that of Eisenach in the W. forms a circle of itself. The greater part of the country belongs to what is called the Thüringerwald, or Thuringian forest, and to the basins of the Elbe and Weser; its principal rivers being the Ilm, Saale, White Elster, and Unstrut. Agriculture is the principal occupation of the inhabitants; the soil, consisting of a clayey loam upon a calcareous basis, is moderately fertile. Owing to the minute subdivision of the land the occupiers are for the most part poor. Mr. Jacob says that they live harder than day labourers, and that, despite their industry and economy, they are unable to increase their resources. (View of Germany, 328.) The country had not then, however, fully recovered from the devastation produced by the late war, and it is now a good deal improved. In the vicinity near Weimar the soil is a rich black mould, producing, notwithstanding its defective culture, very superior crops. The villages in this part of the duchy are thickly placed and populous, but, in consequence of the smallness of the farms, there is a great scarcity of cattle. Of late years, however, the rearing of cattle has been a good deal more attended to; and the stock of sheep has been greatly increased in consequence of the ready and advantageous market for wool afforded in England. Game is extremely plentiful; and the woods comprise about one million Prussian *morgen* of land. Most of the peasants' houses are built of timber.

Coal and salt are both raised, the former in no great quantities, but the production of the latter, at Kreuzburg, may amount to 1,100,000 lbs. a year. Iron and manganese are the chief metallic products. Manufactures are not very important; woollen cloths, carpets, hosiery, linen stuffs, iron, hardware, and tobacco-pipes, are the principal products. There are a good many breweries and distilleries. Manufacturing industry is most active in the circle of Eisenach.

The government is a limited monarchy; and the royal family of Weimar took the lead in Germany after the peace, in introducing a free representative system into their dominion. The constitution dates from 5th May, 1816, and is certainly one of the most liberal in Germany. "The miniature parliament forms only one house, for it consists of only 31 members; 10 are chosen by the proprietors of estates-noble, 10 by the citizens of the towns, 10 by the peasantry, and one by the university of Jena. The last is elected by the Senatus Academicus; and, besides being a professor, must have taken a regular degree in the juridical faculty. At the general election, which occurs every 7th year, not only the representatives themselves are chosen, but likewise a substitute for every member, that the representation may be always full. The 10 members for the nobility are chosen directly by all the possessors of patents of nobility, or estates-noble. Even ladies in possession of such estates have a vote; but if unmarried, they must vote by proxy.

"In the representation of the towns and peasantry, the election is indirect. The towns are distributed into 10 districts, each of which sends one member. In these, every resident citizen has a vote, without distinction of religion: even Jews possess the franchise, though they cannot be elected. The whole body of voters in a town choose a certain number of delegates, in the proportion of 1 for every 50 houses the town contains, and these deputies elect the member for the district. The member for a district of towns must have a certain independent income of about £75 sterling, if he be elected for Weimar or Eisenach, and £45 if chosen to represent the towns of any other district. The election of the 10 representatives of the peasantry proceeds in a similar way. Their representatives must belong *to themselves*; they are not allowed to take them from the higher classes of landed proprietors, which they certainly would have been easily brought to do, had it not been expressly prohibited. Neither brothers, nor father and son, are capable of sitting in the chamber at the same time. The parliament elects its own president, whose election is confirmed by the grand duke, and who holds office during two parliaments. Regularly the chamber meets only once in three years, and the budget is voted for the whole of that period; but a standing committee of nine members continues during the adjournment. During the session the members have an allowance of about 10s. a day, besides a certain sum per mile for travelling expenses. The powers of the chamber extend to all the branches of legislation, and its consent is indispensable to the validity of all legislative measures. The majority of voices determines every question. The members have full privilege of parliament; their persons are inviolable from the commencement till eight days after the close of the session: they are secured in liberty of speech, and legal proceedings cannot be instituted against them without the consent of the chamber." (*Russell*, i., 111-117.)

The ministry is in three departments, those of justice, finance, and public instruction. There are courts of primary jurisdiction in the principal towns, and courts of appeal in Weimar and Eisenach, in which, with Weida and Dermbach, are also criminal courts; all having appeal to the supreme court of Jena, which is also the supreme tribunal for the states of Saxe-Coburg, Meiningen, and Reuss. According to Berghaus, public education is nowhere in Germany so widely diffused, and so well attended to, as in Saxe-Weimar. In 1830, 35,365 children were receiving public instruction. According to the budget for 1839-41, the revenue of the Grand Duchy amounts to 773,093 *thalers*, and the expenditure to 664,748 do.: the public debt amounts to 3,500,000 thalers. The contingent to the army of the confederation consists of 2100 men, all persons being liable to service from their 20th to 25th year. Saxe-Weimar holds the 15th place in the confederacy; having one vote in the full diet, and with Saxe-Coburg, Meiningen and Altenburg, one in the committee.

WEIMAR, a city of central Germany, cap. of the above Grand Duchy, and the usual residence of the grand duke, on the Ilm, an affluent of the Saale, 104 m. W. by S. Dresden, 116 m. S.E. Hanover, and 136 m. S.W. Berlin. Pop., 10054

in 1838, 11,212. The city, which is partially surrounded with walls, though irregularly laid out, has several good and clean streets, and handsome houses; and deserves to rank with German towns of the second order. The Ilm, which flows along its E. side, is crossed by three bridges; it traverses the centre of the ducal park, the chief promenade of the inhabitants; and has on or near its W. bank the ducal palace and mews, the riding-house, *reths schloss* or red castle, public library, and several other public buildings. The ducal residence is a good building, and is tastefully furnished; but comfort rather than magnificence was the object of the late grand duke, by whom it was built. The town church has a large organ; an altar-piece of the Crucifixion, by Luke Cranach, in which are introduced portraits of his friends Luther and Melancthon, and of himself; and monuments to Herder, and numerous members of the ducal family interred here. In the park is a handsome temple containing some beautiful arabesques, and a portrait by Angelica Kauffmann. An avenue from this promenade conducts to the Belvidere, a summer palace of the grand duke, about 2 m. distant.

If Dresden be the Florence, Weimar was once fairly entitled to be called the Athens of Germany; having been the residence of Wieland, Schiller, Göethe, Herder, &c., invited thither by the late grand duke. Göethe and Schiller are buried in the new cemetery, one on each side their patron. During their superintendence, the theatre at Weimar was among the most celebrated in Germany; and no opera is still very well conducted and supported, the inhabitants of the city being great lovers of music. The grand ducal library holds a high rank, and has 98,000 printed volumes, besides MSS. (*Statist. Journ.* 1841.) It is open to the public, who are even allowed the use of the books at their own houses. In this library are some fine busts, and some paintings by A. Durer. One of the principal objects of notice in Weimar is the *Landes-Industrie-Comptoir*, a vast printing and publishing establishment, in which a great number of persons are employed in translating such foreign works as are likely to be read in Germany; and such is the rapidity with which this office is performed, that frequently the translation of a book published in London at the beginning of one month is in full circulation by the end of the same month throughout Saxony, and the independent states of Germany, from the press of the *Industrie-Comptoir*." (*Granville's Trav.*, i., 222.) From this press issues the *Weimar Almanac*, frequently quoted in this work. Weimar is the seat of nearly all the superior educational institutions of the grand duchy. It has a famous academy attended by young English gentlemen, several hospitals, an orphan asylum, central prison, &c. The manufactures of Weimar are inconsiderable: a few woollen and linen cloths, iron-wares, paper, and beer are the principal products. The town has some trade in corn and wool. Weimar was the birth-place of Kotzebue. (*Berghaus, Allg. Länder, &c.*, iv.; *Stein's Handbook; Russell, Jacob, and Granville's Trav., &c.*)

WEISENBURG, t., Lehigh co., Pa., 72 m. E. by N. Harrisburg, 180 W. Watered by Jordan creek and its tributaries. It contains two churches, two stores, six grist-mills, three saw-mills, four tanneries, 15 distilleries; four schools, 111 scholars. Pop. 1427.

WELD, p. t., Franklin co., Me., 47 m. W.N.W. Augusta, 624 W. Watered by a large pond and its inlets; and an outlet, which flows S. into Androscoggin river. It has one store, three grist mills, four saw-mills; 11 schools, 425 scholars. Pop. 1045.

WELDON, p. v. Halifax co., N. C., 93 m. N.E. Raleigh, 204 W. Situated on the S.W. side of Roanoke river, at the great falls, 12 m. above Halifax. Below the falls the river is navigable to Albemarle sound. There is a canal around the falls, 12 m. long, with a lockage of 100 ft. Above the falls the Roanoke is made boatable to Staunton river, over 200 m. A railroad extends from this place to Petersburg, Richmond, and Fredericksburg, Va. Another proceeds 161 m. to Wilmington, whence steamboats ply 180 m. to Charleston, S. C.

WELLFLEET, p.t., Barnstable co., Mass., 105 S.E. Boston, 506 W. Situated on cape Cod, and extending across it, about half way between the ocean and the N. point. Bounded E. by the Atlantic, W. by Cape Cod bay. Wellfleet bay sets up from Cape Cod bay in the S.E. part of the town, and forms a good harbour, protected by several islands at its mouth. It has considerable shipping employed in the coasting trade, and the cod and mackarel fisheries. It has two Congregational churches, 15 stores, two grist mills, one tannery; 10 schools, 640 scholars. Salt is extensively manufactured. Pop. 2377.

WELLINGBOROUGH, a market town and par. of England, co. Northampton, hund. Hamfordshoe, on the slope of a hill, 10 m. E.N.E. Northampton. Area of par. 4880 acres. Pop. of do., in 1831, 4698. The town, which was a place of some consequence in the time of the Saxons, consists principally of four streets, meeting in a market-place. The houses are built of red sandstone, and the town having been almost wholly destroyed by a tremendous fire in 1738, has now a comparatively modern appearance. The church, a large edifice, with a tower and spire, is, like most churches in this county, of a mixed style. It is, however, richly decorated with carved work; in its E. window is some stained glass, and on each side of the chancel are three stalls like those in cathedral choirs. The living, a vicarage worth £400 a year, is in the gift of — Vivian, Esq. Here, also, are places of worship for Baptists, Friends, Wesleyans, &c. The free school, founded by Edward VI., has an income of £112 a year, and is open to all boys belonging to the parish. The number varies from 12 to 20, who are taught Latin gratis, but who pay £1 1s. a quarter for English, writing, and arithmetic. The governors are the trustees of the town estate; the right of appointing the master and usher is vested in the inhabitants paying taxes. (Digest of Charity Reps.) The town estate, yielding an income of £350 a year, partly supports the free school; and the usher's salary, with a charity school for the primary instruction of 50 children, is partly dependent on Fisher's endowment of £137 10s. a year. There are several charities for supplying bread to the poor, &c. The manufacture of boots and shoes was carried on very extensively in this town during the war, and, though fallen off, still continues to be its staple business. The corn market, on Wednesdays, is considerable.

Wellingborough derived its name from the wells or mineral springs around it, which formerly enjoyed such celebrity that, in 1626, Charles I. and his queen resided here in tents for a considerable period that they might drink the waters pure from their source. The county magistrates hold petty sessions for the division weekly in the townhall. Fairs, Easter and Whit Wednesday, and Oct. 29, for live stock and cheese.

WELLINGTON, a market town and par. of England, co. Salop, head of a div. of Bradford hundred, on the ancient Watling-street, 10 m. E. Shrewsbury. Area of par., which, besides the town, includes six townships, 7950 acres. Pop. of do., in 1831, 9671. The town consists mostly of narrow streets; but, of late years, these have been lighted with gas, and otherwise much improved, and are mostly lined with well built houses. The parish church is a handsome structure, of freestone with cast-iron pillars, the window frames being also of iron. The living, a vicarage conjoined with the rectory of Eyton, worth £842 a year, is in the gift of — Eyton, Esq. (Eccl. Rev. Reps.) Here are several dissenting places of worship, free and national schools, and some almshouses and several minor charities. Near the town are some chalybeate and sulphureous springs, frequented by visiters. Most of the inhabitants of the parish are employed in working coal and iron mines, and limestone quarries; and there are in the parish many smelting furnaces, wrought by machinery, with nail-works, glass-works, &c. Malting and some trade in timber are also carried on. The town is governed by a mayor and two constables, appointed annually at a manorial court, who hold petty sessions weekly, and a court of record for debts under £90 at specified times. Market-day, Thursday. Fairs four times yearly, for cattle and stock.

WELLINGTON, a market town and par. of England, co. Somerset, hund. Kingsbury West, on the high road from Bath to Exeter, 46 m. S.W. the former. Area of par. 4830 acres. Pop. of ditto in 1831, 4762. The town is regularly laid out, and has been mostly rebuilt during last century: it principally consists of two spacious thoroughfares, crossing each other at right angles, the main street being about ½ m. in length. The church at the N. entrance is a handsome structure of mixed architecture, 110 ft. in length, and 50 ft. in breadth, comprising a nave, chancel, two aisles, and two small chapels. At its W. end is a fine embattled tower, 100 ft. in height, crowned with a turret and pinnacles. Within are several monuments, including a magnificent tomb in honour of Sir John Popham, chief justice of England in the reign of Elizabeth, and a great benefactor of the town, and a new altar-piece, ranking among the finest in the W. of England. The living, a very valuable vicarage, worth £894 a year, is in the gift of W. P. Thomas, Esq. A very elegant chapel of ease, of Grecian architecture, has been erected at the S. extremity of the town by the Rev. N. P. Thomas; and there are chapels for Baptists, Independents, Wesleyans, and Friends, the latter being here a very numerous sect. Several schools are attached to the various meeting-houses, and there are endowed almshouses for both sexes, founded by Sir John Popham, with some minor charities. A new market-house, over which is the town-hall, was erected in the centre of the town in 1832.

Wellington had formerly a flourishing manufacture of woollen goods, but it is now much fallen off. It still, however, produces druggets and serges, and has a small manufacture of earthenware. In 1838 there were in the parish three woollen-mills at work, furnishing employment to 298 hands. The corn market on Thursday is large and well attended. The Bath and Exeter railway will, when completed, pass close to the E. of the town. Wellington is governed by a bailiff and subordinate officers chosen at the annual manorial court.

This town enjoys the distinction of having successively conferred on Arthur Wellesley (third surviving son of the second earl of Mornington), the greatest of English generals, the titles of viscount, earl, marquis, and duke. An obelisk upwards of 120 ft. in height has also been erected, in honour of the illustrious duke, on a lofty hill about 3 m. S.E. from the town. (Beauties of England and Wales, art. Somerset; Priv. Inf.)

WELLS, a city, and a parl. and mus. bor. of England, co. Somerset, hund. Wells Forum, at the S. foot of the Mendip hills, 17 m. S.W. Bath. It is situated in the centre of the large parish of St. Cuthbert, which contains numerous hamlets, and has an area of about 14,000 acres, with, in 1831, a pop. of 6649. The parliamentary and municipal borough, which are now co-extensive, do not however, include the whole parish of St. Cuthbert, but comprise only the old city, the liberty of St. Andrew, and some additional portions, having, in 1841, a population of 4603. The city consists mostly of four principal streets, named from the four verderies into which it is divided. They are well paved and lighted, and have many good houses. The market-place, an extensive area, communicating by an ancient gateway with the cathedral-close, has in it the town-hall and a handsome public conduit, by which the city is well supplied with water. The great objects of interest in Wells are its fine ecclesiastical edifices. The cathedral, at the E. extremity of the city, is not only one of the most perfect in its original plan, but is more complete as respects its appendages, than any other in the kingdom. It was principally designed in the early part of the reign of Henry III. It is built in the form of a cross, from the intersection of which rises a large quadrangular tower 178 ft. in height. The W. front, 150 ft. in breadth, is flanked by two smaller towers, each 130 ft. in height: the total length of the church, from E. to W., is about 380 ft.; its breadth, 131 ft. "The situation of this edifice and the adjoining palace is beautiful; and though no whole side, except the W. front is visible in any one view, the cathedral is well displayed from several points, particularly the N.W. As at Peterborough, the palace and several other buildings adjoin the cathedral, and add much to its general appearance. The character of a large portion of the building is early English, with portions of the two later styles, which are very beautifully accommodated in their forms to the older parts. The nave and transepts, and part of the towers, are early English; the W. front is remarkably rich in niches and statues, and not less so in shafts and other small ornaments appropriate to the style. The lower parts of the sides of the western towers are similarly enriched, but the whole of the remaining exterior of the building is rather plain than otherwise; the upper parts of all the towers are much later than the lower, and much accommodated to the earlier portions as to lines and forms. The eastern part of the cross and the chapter-house are of decorated character, and remarkably elegant. The cloisters are perpendicular: the nave and transepts, and a N. porch out of the nave, present an early English arrangement very remarkable for simplicity and elegance. There are various excellent portions of stone screen-work, chapels, and monuments, and some stained glass, the effect of which is peculiarly good. This cathedral is very rich in details of the best as well as the most singular kinds, and, in point of composition, some of its best parts yield to no edifice in the kingdom. The bishop's palace, though it has been altered, and in some parts much modernized, contains some fine portions, an early decorated chapel, and some parts of earlier date. Taken altogether, the palace is one of the most valuable remains in the kingdom. The gates and other buildings in the precincts of the cathedral deserve careful examination." (Rickman.) Wells was first erected into a bishop's see in 905. In the 12th century it was united to the abbey-church at Bath; but the writ of congé d'élire for the election of the bishop is still addressed to the dean and chapter of Wells. The chapter consists of a dean and six other canons, four priest-vicars, and 42 prebendaries. Wolsey and Laud were bishops of this see. St. Cuthbert's church is a handsome perpendicular edifice: but its principal feature is its tower, one of the finest of the kind. The living, a vicarage in the gift of the dean and chapter of Wells, is worth £564 a year. There are, also, places of worship for Independents, Baptists, and Wesleyans. Here is a collegiate school, under the patronage of the dean and chapter, and a united charity school, founded in 1654, which affords instruction, clothing, &c., to 34 boys and 20 girls, and has an income of £314 a year. Under the town-hall is a hospital, founded and endowed in the 15th century, for aged men and women, which has now an income of above £350 a year, and 30 inmates. There are numerous other charities, including several well-endowed almshouses. Several manufactures that were formerly carried on in the town have either ceased altogether or have greatly declined; and that of silk has been wholly given up. It is said in the *Parliamentary Boundary Report*, that "no trade or manufactures are carried on in the town. It probably has been for a great number of years in the same state as it is now. There is no reason to believe that it will increase. Much of the property of the city belongs to the ecclesiastical or the city corporations, which cannot grant long leases, and give no encouragement to building." The corn-market, which used to be very considerable, has materially declined; but the market for cheese is still one of the most considerable in the W. of England. The trade of the place is mostly, however, confined to the retail supply of the inhabitants. The West of England and South Wales District bank has an office in the city. The earliest charter of Wells dates from the 3d of John, but the governing charter, previously to the Municipal Reform Act, was granted by Elizabeth. By the last mentioned statute, the town is governed by a mayor, three other aldermen, and 12 councillors. A court of quarter sessions, the jurisdiction of which is confined to cases of misdemeanour, is held four times a year; but it is merely a matter of form, all trials being referred to the county assize-court, and the court of record has also fallen into disuse. The county assizes are held alternately here and at Taunton. Corporation revenue in 1840, £698. Wells has returned two members to the House of Commons since the reign of Edward I.; the right of voting, down to the Reform Act, having been vested in the mayor, masters, burgesses, and persons admitted to the freedom of the city, which was obtainable by birth, marriage, or apprenticeship. Registered electors in 1839–40, 414. It is also a polling-place for the E. division of the county. Markets, Wednesday and Saturday; and every fourth Saturday a large market for corn, cattle, and cheese. Fairs five times a year, mostly for cattle horses, and pedlery. (*Parl. and Mun. Bound. Reps.*; *Mun. Corp. Append.*)

WELLS, a seaport town and par. of England, co. Norfolk, hund. N. Greenhoe, on a small creek, about 1 m. from the sea, 28 m. N.W. Norwich. Area of par., 2250 acres. Pop. of dtto in 1831, 3694. The town consists principally of two streets, only partially paved. The parish church, a spacious edifice built mostly of flint, with a lofty embattled tower, has some curious sculpture and paintings. The living, a valuable rectory, being worth £785 a year, is in the gift of — Hopper, Esq. Here are several dissenting chapels: a free school for 60 poor children, supported by a part of Ringar's endowment in 1678 of £120 a year, and other endowed charities to the amount of £66 a year for the general relief of the poor. (*New Digest of Charity Reports.*) It has a neat theatre, a subscription library, &c. The harbour of Wells is indifferent, and apt to be choked up with shifting sands; but it has been considerably improved of late years, through the exertions of the harbour commissioners. The principal trade consists in the shipment of corn and malt, and in the import of coals, timber, &c. There belonged to the port in 1841, 38 vessels of less than 50 tons burden, and 27 of more than that amount, the aggregate burden of the latter being 2213 tons. Gross customs duty received at the port in 1840, only £596. The oyster-fishing gives employment to a considerable number of persons. Petty sessions for the hundred are held once a fortnight, and courts leet and baron once a year. Fair, Shrove-Tuesday. The races formerly held at Wells are now discontinued. (*Parl. Rep., &c.*)

About 3 m. W. from the town is Holkham, the magnificent seat of the Earl of Leicester; and the country in the vicinity affords some of the best specimens of what is called the Norfolk system of farming.

WELLS, county, Ia. Situated in the E. towards the N. part of the state, and contains 373 sq. m. Organized in 1837. Wabash river, running centrally through it, affords great water-power. It contained in 1840, 1554 neat cattle, 531 sheep, 3466 swine; and produced 4746 bushels of wheat, 48,381 of Indian corn, 3911 of oats, 6697 of potatoes, 7519 pounds of sugar. It had four stores, one saw-mill, one tannery; one school, 25 scholars. Pop. 1822. Cap., Bluffton.

WELLS. p. t., York co., Me., 30 m. S.W. by S. Portland, 79 m. S.W. by S. Augusta, 513 W. Chartered in 1643. Bounded S.E. by the Atlantic. It has a harbour which admits small vessels, of which it has a number employed in the coasting trade and the fisheries. Some vessels are built. It contains 13 stores, one fulling-mill, four grist-mills, 16 saw-mills, three tanneries; 21 schools, 1915 scholars. Pop. 2978.

WELLSBURG, p. v., cap. of Brooke co., Va., 16 m. N. by E. Wheeling. 344 m. N.W. Richmond, 905 W. Situated on the E. side of Ohio river, at the mouth of Buffalo creek.

It contains a courthouse, jail, county offices, a markethouse, a bank, an academy, the Jefferson seminary, a female seminary, four churches, a Presbyterian, Episcopal, Methodist, and Disciples'; various mills and manufactories, 250 dwellings, and about 2000 inhabitants. It is surrounded by inexhaustible beds of bituminous coal.

WELSHPOOL, a parl. and mun. bor., market town, and par. of Wales, co. Montgomery, hunds. Pool and Caurse, on a branch of the Ellesmere canal, about ½ m. W. from the Severn, and 16 m. W. Shrewsbury. Pop. of par. in 1831, 4533. The parliamentary borough, which was formerly much larger, may now be considered as nearly co-extensive with the parish, but the municipal borough is of much smaller extent. It is principally in a hollow, but partly also on the acclivity of an eminence leading towards Powys park and castle, a little S. from the town, but included within the parliamentary borough; it is well lighted with gas, and consists of one long and wide street, intersected by others of smaller dimensions, all well paved, and well supplied with water. The houses, which are of brick, have an unusual degree of regularity for this part of the country, the town being, on the whole, neat, cheerful, and English looking. The church of St. Mary, rebuilt in 1774, is a spacious edifice, in the pointed style, with a lofty square tower: it is situated at the base of an eminence, on which is the churchyard, which in some parts overtops the church, and commands a fine view of the town and adjacent country. The living, a vicarage in the gift of the bishop of St. Asaph, is worth £273 a year nett. A new and handsome church on the W. side of the town has recently been erected on a site given by Lord Powys. Here, also, are places of worship for Baptists, Independents, Calvinists, and Wesleyans. The county-hall, in the centre of the main street, is a handsome brick building with a colonnade: the second floor is 64 ft. in length by 55 ft. in breadth, and is 18 ft. in height. Underneath is a spacious corn market and a court room for the county assizes. A national school for both sexes, in which 250 children are instructed, was opened in 1821; and it has, also, a free school with a small endowment, almshouses for eight females, a dispensary, and several charitable bequests for the education of children, and the distribution of charity among the poor.

From 1782 to 1834 Welshpool was the chief market in North Wales for the sale of Welsh flannels; but in the latter year the greater part of the trade was transferred to Newtown. The flannel manufacture carried on here is not of much importance: in 1838 two woollen-mills were at work in the parish, employing together about 30 hands. Flannel markets are still held once a fortnight: but the business is mostly conducted by private sales. (*Hand-loom Weavers' Rep.; Factory Returns.*) Malting is carried on to a considerable extent, and there are several rather large tanneries. The Severn is navigable to within a short distance of the town; and by means of the Ellesmere canal it communicates with the Birmingham and Chester canal lines. Under the Municipal Reform Act the town is governed by four aldermen and 12 councillors; it has a commission of the peace, petty sessions for the hundred of Caurse, a court leet, &c., and is, twice a year, the seat of the county assizes.

Welshpool was formerly joined with Montgomery in the exercise of the elective franchise, but was disfranchised in 1728. Under the Reform Act, however, it has been again reinvested with the franchise, and is united with Llanidloes, Llanfyllin, Machynlleth, Montgomery, and Newtown, in returning one member to the House of Commons. Registered electors in this borough in 1839–40, 295; in the united boroughs, 1021. About 1 m. to the S. is Powys castle, the magnificent seat of Earl Powys. It stands on an elevated site, in the centre of an extensive and finely wooded park. In 1823 the entire building underwent a thorough repair under the direction of its present proprietor. The principal entrance is a gateway between two massive round towers. It has in front two immense terraces rising one above another, the ascent to the castle being by a vast flight of steps. It is superbly fitted up, and has many fine pictures and works of art, including several pieces of sculpture from Herculaneum. In the vicinity are the Freiddyn hills, on the loftiest of which an obelisk has been erected in honour of Lord Rodney. Market-day, Monday; fairs six times a year. (*Parl. Reports; Nicholson's Cambrian Guide; Panorama of North Wales.*)

WEM, a market town, parish, and township of England, co. Salop, hund. Bradford, on the Roden, a tributary of the Severn, 11 m. N. by E. Shrewsbury. Area of par. 11,339 acres. Pop. of ditto in 1831, 3973. The town consists principally of one spacious street, from which several smaller streets branch off. The parish church, a handsome edifice, has a lofty tower and a fine chancel: the living, a very valuable rectory, with a curacy, worth £1767 a year, is in the gift of the duke of Cleveland. The market-house is a neat brick and stone building; and in one of its rooms courts leet are held, at which the two bailiffs governing the town are chosen. The free school, founded in 1651, by Sir T. Adams, a native of the town, who became lord mayor of London, and who also founded an Arabic professorship in Cambridge, has a total income of about £340 a year, and two of Careswell's exhibitions in Bridgenorth school. There are charities making provisions for the poor, but to no considerable amount. The inhabitants of the town are mostly engaged in tanning and malting. Horsley supposes that Wem occupies the site of the ancient *Rutunium :* it formerly had a castle, but of this edifice nothing remains. The manor of Wem having come into possession of the crown by the attainder of Philip, earl of Arundel, in the reign of Elizabeth, it was conferred by James II. on his fitting tool, chancellor Jeffries of bloody memory, who had the estate, and who was also created Baron Wem. Wycherley, the dramatist, was born near this town in 1640. Fairs, six times a year, for cattle and stock, linen, and flax seed. (*Parl. Reps., &c.*)

WENDOVER, a market town and par. of England, co. Buckingham, hund. Aylesbury, in the vale of that name; 32 m. N.E. London. Area of par. 5250 acres. Pop. of ditto in 1831, 2008. The town is inconsiderable, and meanly built; and would not, indeed, have been worth noticing in a work like this but for the circumstance of its having enjoyed, from the 21st James I. down to the passing of the Reform Act, by which it was disfranchised, the privilege of sending two members to the House of Commons. The right of election was nominally in the housekeepers not receiving alms, but really in the lord of the manor. The famous John Hampden (to whose family the manor belonged) represented Wendover in five successive parliaments. The living, a vicarage worth £300 a year, is in the gift of the crown. Petty sessions are held once a fortnight, and courts leet and baron arealso held in the town.

WENER and WETTER LAKES, two large lakes of Sweden, which see (*ante,* 874).

WENLOCK (MUCH or GREAT), a parl. and mun. bor., market town, and par. of England, co. Salop, hund. Wenlock, on a small tributary of the Severn, 11 m. S.E. Shrewsbury. Area of par., 8420 acres. Pop. of ditto in 1831, 2424. The parliamentary borough is, however, co-extensive with the large district called "Wenlock franchise," consisting of 16 parishes, in addition to that of Much Wenlock, the whole having, in 1831, a population of 19,435. The limits of the old municipal borough were formerly identical with those of the parliamentary borough; but the modern municipal borough is of much less extent, comprising only the three parishes of Dawley, Madely, and Broseley; the town of Wenlock itself being altogether excluded. The latter, though an inconsiderable place, and indifferently built, has some handsome residences. It consists chiefly of two streets, the houses being mostly of brick. The church, a spacious edifice in the Norman and early English style, has a tower surmounted by a spire at its W. extremity. The living, a vicarage worth £180 a year, is in the gift of Sir W. W. Wyan. It has, also, a Wesleyan chapel, a free endowed school for 12 boys, almshouses for four women, and several minor charities. There are extensive limestone quarries in the vicinity, and copper mines, now abandoned, were formerly wrought to a considerable extent. The municipal borough is divided into three wards, and is governed by six aldermen and 12 councillors. It has a commission of the peace, which is opened twice a year; petty sessions once a fortnight, or oftener if required; and a court of record, also once a fortnight: the last, however, has latterly fallen nearly into disuse. Corporation revenue in 1841, £98.

Wenlock received its first charter from Edward IV., under which it sent, in 1478, one member to the House of Commons, but Broseley and Little Wenlock being afterward added to the borough, it was empowered to return two members, a privilege it has since continued to enjoy. Previously to the Reform Act, the franchise was vested in the freemen of the borough, such freedom being acquired by birth or election. Registered electors in 1839-40, 949. Wenlock probably owed its origin to the foundation of a famous abbey, of which the ruins still exist, a little S. from the town. This edifice, founded towards the end of the 7th century, was mostly rebuilt, soon after the Conquest, in the Norman and early English styles. The entrance from Wenlock was by a strong gateway, one massive tower of which is still standing. Of the church, which dates from 1080, a large portion of the S. side of the nave, the whole S. wing of the transept, several arches on the N., and the foundations of the choir and Lady chapel remain. The ruins sufficiently attest the former magnificence and splendour of the structure, the precincts of which included an area of 30 acres. Its revenues amounted, at the dissolution, to £401 a year. Markets on Mondays; fairs, five times a year, principally for horses, cattle, and sheep. (*Parl, and Munic. Bound. Rep.; Munic. Corp. Append.*)

WEOBLY, a market town and par. of England, county Hereford hund. Stretford. 10 m. N.W. Hereford. Area of par., 3160 acres. Pop. of do., in 1831, 819. The town consists principally of one street, having several modern and well built houses. The church is a spacious edifice, to which are attached two or three ancient burial-chapels. The living, a vicarage worth £236 a year, is in the gift of the bishop of Hereford. A free grammar school, founded in 1655, affords instruction to 15 boys. Here is also a national school for both sexes, supported by subscription. No particular branch of industry is carried on in the town, which, indeed, would not have been worth notice in a work of this kind, but for the circumstance of its having sent two members to the House of Commons, from the reign of Edward III. down to the passing of the Reform Act, by which it was disfranchised. It has a market on Thursdays, but this is little more than nominal.

WESEL (Germ. *Niederwesel*), a frontier and strongly fortified town of Rhenish Prussia. reg. Dusseldorf, circ. Rees, of which it is the cap., on the Rhine. where it receives the Lippe, 20 m. S.E. Cleves. Pop., in 1838, 10,634. (*Berghaus.*) It is of high antiquity, and was formerly one of the Hanse towns. It has some manufactures of cotton and woollen stuffs, leather, and tobacco, with distilleries, &c.: its port is convenient, and packets ply regularly between it and Amsterdam. Its defences have been a good deal strengthened by the erection of fort Blucher on the opposite or W. bank of the Rhine. (*Schreiber, Guide du Rhin, &c.*)

WESER (an. *Visurgis*), a river of N.W. Germany, its embouchure being in the North sea, and its basin having that of the Elbe to the E., the Ems to the W., and the Rhine and Mayn to the S.W. and S. It is formed by the union of the Fulda and Werra. The former of these rivers has its source in the Rhongebirge, about lat. 50° 27' N., and long. 10° E. : and traversing the electorate of Hesse-Cassel, it passes the cities of Fulda and Cassel. The Werra rises in the Thuringian forest, in about the same lat., and long. 11° ; and intersects several of the smaller Saxon territories, with parts of Prussian Saxony, Hesse-Cassel, and Hanover. Both have a general northerly direction, and unite at Minden, near the southern extremity of the Hanoverian dominion. The combined river, or Weser, flows in a N. course, though with numerous windings, through the territories of Hesse-Cassel, Prussian Westphalia, Brunswick, Lippe-Detmold, Hanover, Bremen, and Oldenburg; falling into the North sea by a wide estuary, about lat. 53° 30' N., and long. 8° 30' E. Its entire length is estimated at nearly 200 m. Its chief affluents are the Aller and Wumme. The Weser is of considerable commercial importance, Bremen being on its banks. Vessels drawing 7 ft. water navigate it up to that city; and it is navigable for boats nearly to its commencement. Vessels drawing from 13 to 14 ft. water ascend the river to Vegesack, 13 m. below Bremen. Ships of large size stop at Bremerhafen, where a new harbour has been formed. Besides the foregoing towns, Minden, Nienburg, Rinteln, Hameln, &c., are on the Weser; and Hanover, Brunswick, Oldenburg, &c. are on its tributaries. (*Berghaus ; Von Reden ; Dict. Geog., &c.*)

WEST, t., Huntingdon co., Pa., 8 m. N. Huntingdon bor. Watered by Frankstown branch of Juniata river. It has three stores, one fulling-mill, one woollen factory, two forges, one flouring-mill, three grist-mills, seven saw-mills, one distillery ; four schools, 144 scholars. Pop. 1629.

WEST, t., Columbiana co., O. Drained by branches of Sandy creek. The line of the Beaver and Sandy canal passes through it. It has seven schools, 205 scholars. Pop. 1915.

WEST BETHLEHEM, t., Washington co., Pa., 12 m. S.E. Washington bor. It abounds with bituminous coal. The National road passes through it. It has two stores, three grist-mills, two saw-mills. Pop. 1743.

WEST BLOOMFIELD, p. t., Ontario co., N. Y., 219 m. W. Albany, 355 W. Drained by Honeoye creek. It contains three churches, two Presbyterian and a Methodist, nine stores, two fulling-mills, one woollen factory, four grist-mills, six saw-mills, one tannery, one distillery, three potteries ; one academy, 95 students ; 17 schools, 646 schol ars. Pop. 2094.

WESTBOROUGH, p. t., Worcester co., Mass., 30 m. W Boston, 412 W. Watered by branches of Concord and Sudbury rivers, which afford water-power. It contains two churches, a Congregational and Baptist, three stores, two grist-mills, three saw-mills ; eight schools, 453 scholars. Pop. 1658.

WEST BOYLESTON, p. t., Worcester co., Mass., 42 m. W. Boston, 405 W. It has two cotton factories with 3500 spindles ; five schools, 309 scholars. Pop. 1187.

WEST BRADFORD, t., Chester county, Pa., 11 m. S.W. Chester, 33 m. W. Philadelphia. Bounded S.W. by Brandywine river, by branches of which it is drained, and which afford water-power. It contains five stores, one fulling-mill, one woollen factory, two grist-mills, two saw-mills ; six schools. 230 scholars. Pop. 1562.

WEST BRIDGEWATER, p. t., Plymouth co., Mass., 25 m. S. Boston. Chartered in 1822, first settled in 1641. Drained by a branch of Taunton river, which affords water-power. It contains three churches, a Unitarian, Baptist, and New Jerusalem, four stores, three furnaces, one grist-mill, three saw-mills ; seven schools, 285 scholars. Pop. 1901.

WEST BROMWICH, a market town and par. of England, co. Stafford, hund. Offlow. on the high-road from Birmingham to Liverpool, 5 m. N.N.W. Birmingham. Area of par., 5385 acres. Pop., in 1831, 15,327 ; in 1841, 26,121. This, which was formerly an inconsiderable village, has increased rapidly in size and population, in consequence of its situation in the centre of one of the principal coal and iron districts of the empire, and of the grand seat of the hardware manufacture. The main street, nearly 1 m. in length, contains some good houses ; but the town is, for the most part, very irregularly laid out, and its proximity to coal-pits, gas, and iron-works, &c., gives it a black and very unprepossessing appearance. The old church of All Saints, on an eminence, in the N.E. part of the town, is a mixed style of architecture, and surmounted by a tower. The living, a perpetual curacy, worth £566 a year, is in the gift of the earl of Dartmouth. Christchurch, a handsome Gothic edifice, erected in 1822, is a curacy worth £330 a year. Besides a very fine Roman Catholic chapel, there are various places of worship for Protestant Dissenters, and a national, and some other schools. The gas-works in this town, belonging to the Staffordshire and Birmingham gas company, are probably the most extensive of any in existence. They supply Birmingham, Bilston, Wednesbury, and a vast number of other towns and villages within a radius of 16 m. Here are also some extensive crown-glass works, which, with the gas manufacture, and the extensive collieries in the vicinity, furnish the chief occupation of the labouring classes. The Birmingham and Dudley canals, in the immediate vicinity of the town, give it the benefit of very extensive water communications. Market day, Saturday. E. of the town is Sandwell park, the seat of the earl of Dartmouth. The house is built on the site of a priory of Benedictine monks, founded in the reign of Henry II., of which some trifling remains may still be seen. (*Parl. Reps.* ; *Railway Guide* ; *Beauties of England and Wales, &c.*)

WESTBROOK, t., Cumberland co., Me., 52 m. S.S.W. Augusta. Presumpscot river affords good water-power. The Cumberland and Oxford canal passes through it. It contains the villages of Sacarappa and Stroudwater, and has 24 stores, two furnaces, two fulling-mills, 10 grist-mills, 13 saw-mills, eight tanneries ; two academies, 80 students ; 15 schools, 1439 scholars. Pop. 4116.

WESTBROOK, p. t., Middlesex county, Ct., 47 m. S. by E. Hartford, 289 W. Watered by Pochaug river. Bounded S. by Long Island sound. Ship building is carried on, and 15 vessels are owned here, employed chiefly in the coasting trade. It contains two churches, a Congregational and Methodist, three stores, two grist-mills, two saw-mills ; six schools, 190 scholars. Pop. 1182.

WEST BUFFALO, t., Union co., Pa., 8 m. N.W. New-Berlin. Drained by White Deer, Buffalo, and Penn's creek. It has four grist-mills, six saw-mills, two tanneries, two distilleries ; seven schools, 380 scholars. Pop. 1460.

WESTBURY, a pari. and munic. bor., market town, and par. of England, co. Wilts, hund. Westbury : at the N.W. extremity of Salisbury plain, 22 m. N.W. Salisbury. The ancient bor. comprised only some portions of the town, but the modern bor. is co-extensive with the par. and hund. of Westbury, having an area of 11,340 acres, with a pop., in 1831, of 7324, and in 1841, of 7596. The town, which is insignificant and irregularly built, is scattered over a considerable surface, the principal street running nearly N. and S. The church, an old cruciform structure, has a tower rising from its centre, and some fine monuments. The living, a vicarage, to which are annexed the curacies of Bratton and Dilton, worth £838 a year, is in gift of the precentor of Salisbury cathedral. Here are several Dissenting chapels, a national school for 40 boys, endowed with £1000 by a benevolent burgess, who also bequeathed a like sum for the annual clothing of 20 poor women. The town hall, a handsome building, was erected in 1815. Westbury and its vicinity had formerly an extensive woollen manufacture, and though much fallen off, this branch of industry is still carried on, there being at work in the parish, in 1838, eight woollen-mills. employing altogether 461 hands. (*Mills and Factories' Rep.*) Some malting is also carried on : and, upon the whole, the trade of the town, such as it is, may be considered in a thriving state. (*Munic. Corp. Appendix.*)

The charter by which Westbury was incorporated is not extant ; the corporation, by which it has till lately been 1076

governed, consisted of a mayor, recorder, and 13 capital burgesses. Westbury returned two members to the House of Commons from the 27th of Henry VI. down to the passing of the Reform Act, which deprived it of one member. Previously to the act now referred to, the franchise was vested in the occupiers of 61 burgage tenements comprised in the old borough. Registered electors in 1839–1840, 291. The borough court, held annually on the 2d of November, is the only court held within and for the borough. The election of the borough officers appears to be its principal duty. (*Munic. Corp. Append.*)

Westbury, though a place of considerable antiquity, is not connected with any historical event of importance. It has two annual fairs, and a nominal market every Tuesday. (*Bound. and Munic. Rep., &c.*)

WEST CALN, t., Chester co., Pa., 44 m. W. by N. Philadelphia. Bounded E. by the West branch of Brandywine river which affords water-power. It has five stores, two woollen factories, two cotton factories with 1094 spindles, two flouring-mills, three grist mills, five saw-mills ; six schools, 230 scholars. Pop. 1363.

WEST CAMBRIDGE, p. t., Middlesex co., Mass., 6 m. N.W. Boston, 446 W. Incorporated in 1807. Alewife brook separates it from Cambridge, and affords water-power. It contains three churches, a Congregational, Baptist, and Universalist, 16 stores, two grist-mills, one saw-mill ; one academy, nine students ; three schools, 277 scholars. Pop. 1363.

WESTCHESTER, county, N. Y. Situated in the S.E. part of the state, and contains 470 sq. m. Bounded W. by Hudson river, S.E. by Long Island sound. Drained by Bronx, Saw-mill, and Croton rivers. Marble abounds at Sing Sing and the vicinity. It contained in 1840, 26,121 neat cattle, 20,043 sheep, 153,187 swine ; and produced 25,267 bushels of wheat, 907,574 of rye, 315,692 of Indian corn, 57,296 of buckwheat, 449,090 of oats, 620,920 of potatoes. It had 196 stores, six furnaces, seven fulling-mills, five woollen factories, five flouring-mills, 40 grist-mills, 35 saw-mills, one paper-mill, 11 tanneries, one distillery, two potteries, five printing-offices, five weekly newspapers ; 24 academies, 703 students ; 121 schools, 3082 scholars. Pop. 48,686. Capital, White Plains.

WESTCHESTER, p. t., Westchester co., N. Y., 146 m. S. Albany, 237 W. Bounded S. by Harlem and East river. drained by Bronx river. It contains the villages of Morsiana and Fordham. The Harlem rail cars pass several times daily from the city-hall, N. Y., through the latter, which is the seat of a Roman Catholic college. It contains three churches, an Episcopal, Methodist and French, several stores ; four schools, 142 scholars. Pop. 4134.

WEST CHESTER, p. b., capital of Chester co., Pa., 27 m. W. Philadelphia. 73 m. E.S.E. Harrisburg, 118 W. Situated 2 m. W. of Brandywine river. Incorporated as a borough in 1799. It is regularly laid out with streets crossing each other at right angles, on elevated ground, and contains a courthouse, a jail on the principle of solitary confinement, county offices, an academy, a market house, a bank, five churches, a Presbyterian, Methodist, two Friends, and a Roman Catholic, an athenæum with a reading room, a cabinet of natural science, which has connected lectures on scientific subjects, a public library, 33 stores, one brewery, one pottery, four printing-offices, four weekly newspapers ; five academies, 364 students ; six schools, 113 scholars. Pop. 2152.

WEST EARL, t., Lancaster co., Pa., 13 m. N.W. Lancaster. Drained by Conestoga creek. It has four stores, two fulling-mills, two woollen factories, five flouring-mills, two grist-mills, six saw-mills, two tanneries, three distilleries ; two schools, 89 scholars. Pop. 1793.

WESTERLOO, p. t., Albany co., N. Y., 22 m. S.W. Albany, 358 W. Drained by Provost creek and other tributaries of Catskill creek. It contains eight stores, three grist-mills, nine saw-mills ; 24 schools, 1142 scholars. Pop. 3026.

WESTERLY, p. t., Washington co., R. I., 43 m. S.S.W. Providence, 373 W. Bounded S. by the Atlantic. W by Pawcatuck river which separates it from Connecticut. Pawcatuck village is situated on Pawcatuck river, 6 m. from the ocean, to which vessels of 40 tons come, and of 90 tons 3 m. below. A bridge crosses the river to Stonington, where is a small connected village. It contains three churches, an Episcopal, Baptist, and a Free church. A railroad passes through this place to Providence. In the town there are two banks, two academies, 12 stores, two fulling-mills, four woollen factories, two cotton factories with 2536 spindles, three grist mills, two saw-mills, two tanneries ; 11 schools, 574 scholars. Pop. 1912.

WESTERN, t., Oneida co., N. Y., 18 m. N. Utica, 19 m. W.N.W. Albany. Drained by Mohawk river and its tributaries. It has two stores, one fulling-mill, one woollen factory, three grist-mills, 13 saw-mills, three tanneries, 20 schools. 218 scholars. Pop. 3448.

WEST FALLOWFIELD, t., Chester co., Pa., 37 m. W.

Philadelphia. It has four stores, two woollen factories, two grist-mills ; seven schools, 375 scholars. Pop. 1717.

WESTFIELD, p. t., Hampden co., Mass., 100 m. W. by S. Boston, 364 W. Drained by Westfield river and its tributary, Little river, which afford water-power. The Hampshire and Hampden canal passes through it. Incorporated in 1669. It has 25 stores, three grist-mills, six saw-mills, four powder-mills, two paper-mills, one printing-office, one weekly newspaper ; one academy, 219 students ; 19 schools, 667 scholars. Pop. 3526. The village is pleasantly situated on a plain, with a handsome public ground in the centre, and has two churches, a Congregational and Baptist, a bank, several whip factories, several of the above stores, and about 150 dwellings.

WESTFIELD, t., Richmond co., N. Y., 158 m. S. Albany. Situated on the S.W. part of Staten Island. Bounded E. by New-York bay, W. by Staten Island sound. It contains Richmond village, the capital of the county. It has six schools, 222 scholars. Pop. 2396.

WESTFIELD, p. t., Chautauque. co., N. Y., 342 m. W. by S. Albany, 345 W. Bounded N.W. by lake Erie. Drained by Chautauque creek. It contains three churches, a Presbyterian, Methodist and an Episcopal, three commission houses in foreign trade, 22 retail stores, three fulling-mills, one flouring-mill, four grist-mills, 13 saw-mills, one oil-mill, two tanneries, two distilleries ; one academy, 145 students ; 21 schools, 1316 scholars. Pop. 3190.

WESTFIELD, p. t., Essex co., N. J., 45 m. N.E. Trenton, 212 W. Bounded E. by Rahway river, W. by Green brook. It contains a Presbyterian church. 12 stores, five flouring-mills, four grist-mills, one saw-mill, one paper-mill ; two academies, 70 students ; 14 schools, 144 scholars. Pop. 3150.

WESTFORD, p. t., Chittenden co., Vt., 13 m. N.E. Burlington, 56 m. N.W. Montpelier, 531 W. Watered by Brown's river. Chartered in 1763. It contains three churches, a Congregational, Methodist and Baptist, four stores, one fulling-mill, one grist-mill, six saw-mills ; 11 schools, 413 scholars. Pop. 1352.

WESTFORD, p. t., Middlesex co., Mass., 8 m. W. by S. Lowell, 32 m. N.W. Boston, 436 W. Watered by Stony river. It contains two churches, a Congregational and a Unitarian, five stores, one fulling-mill, one woollen factory, two forges, four grist-mills, three saw-mills, two tanneries ; one academy, 84 students ; nine schools, 558 scholars. Pop. 1436.

WESTFORD, p. t., Otsego co. N. Y., 65 m. W. Albany, 379 W. Drained by Cherryvalley and Elk creeks. It has four stores, two fulling-mills, three grist-mills, four saw-mills ; nine schools, 474 scholars. Pop. 1478.

WEST GREENWICH, p. t., Kent co., R. I., 18 m. S.W. Providence, 390 W. Watered by Wood river and its branches. Chartered in 1741. It contains a Baptist church, eight stores, three cotton factories with 2204 spindles, seven grist-mills, 10 saw-mills ; 11 schools, 291 scholars. Pop. 1415.

WEST HEMPFIELD, t., Lancaster co., Pa., 8 m. W. Lancaster. Bounded W. by Susquehanna river, N. by Chiques creek. It contains Columbia bay, and has two commission houses in foreign trade, 28 retail stores, one furnace, five flouring-mills, six grist-mills, four saw-mills, four tanneries, four distilleries, one brewery, two printing-offices, two weekly newspapers. Pop. 1913.

WEST MANCHESTER, t., York county, Pa. Bounded N.W. by Conewago creek. York borough lies partly in this town. It has two stores, one flouring-mill, four grist-mills, one saw-mill, seven distilleries ; one school, 44 scholars. Pop. 1282.

WESTMEATH, an inland co. of Ireland, prov. Leinster, having N. Cavan and Meath, E. the latter, S. King's county, and W. Roscommon. (from which it is separated by the Shannon) and Longford. Area, 386,251 acres, of which 55,982 are unimproved bog and mountain, and 16,334 lakes. Surface agreeably diversified with woods, lakes, streams, hills, and bogs. The substratum being limestone, the verdure of the fields is remarkably fine, and the soil generally excellent. Property in modern-sized estates. Leases commonly granted for 21 years, and a life. Grazing-grounds extensive. Tillage farms much subdivided, and husbandry, in most respects, similar to that of Meath, which see. Average rent of land, 13s. 7d. an acre. Principal rivers, Shannon, Inny, and Breena. Westmeath is divided into 12 baronies and 62 parishes ; and returns three members to the House of Commons, viz. two for the county, and one for the borough of Athlone. Registered electors for the county, in 1839-40, 2497. In 1831, Westmeath had 23,803 inhabited houses, 23,531 families, and 136,872 inhabitants, of whom 67,700 were males, and 69,172 females.

WEST MILFORD, p. t., Passaic county, N. J., 109 m. M.N.E. Trenton, 266 W. It has three stores, 10 forges, two grist-mills, two saw-mills, two tanneries ; 11 schools, 400 scholars. Pop. 2108.

WESTMINSTER. See LONDON.

WESTMINSTER, p. t., Windham co., Vt., 103 m. S. Montpelier, 440 W. Bounded E. by Connecticut river. The principal village is beautifully situated on a plain, on the W. bank of Connecticut river, enclosed by a semi-circle of hills. The town contains two Congregational churches, one store, two grist-mills, eight saw-mills, one paper-mill ; 12 schools, 479 scholars. Pop. 1556.

WESTMINSTER, p. v., capital of Carroll co., Md., 58 m. N.W. by N. Annapolis, 68 W. It contains a courthouse, jail, a church, several stores, and about 350 inhabitants.

WESTMORELAND, a marit. co. of the N. of England, having N. Cumberland, Durham, and Yorkshire, S. Lancashire, and the extremity of Morecumbe bay. and W. Lancashire. Area, 487,680 acres, of which only 180,000 are said to be arable, meadow, and pasture. It is what its name (West-moor-land) imports, a region of lofty mountains, naked hills, and black barren moors ; but the valleys, particularly those of the Eden in the N., and of Kendal in the S., are fertile and well cultivated. The agriculture, state of property, character, and condition of the occupiers of Westmoreland, are so similar to those of Cumberland, that the statements as to the latter may be applied, with very little variation, to the former. Average rent of land, in 1810, 9s. 1d. an acre, being the lowest of any in England. Westmoreland abounds in slate of the finest quality ; is lead ; and, in some few places on its S. and W. borders, there are coal mines. Principal rivers, Eden, Lune, and Kent. The lakes are even more celebrated than those of Cumberland : Windermere, on its W. border, is the most extensive lake in England. The county is divided into four wards and 32 parishes. It returns four members to the House of Commons ; viz. two for the county and two for the borough of Kendal and some adjoining districts. In 1841, Westmoreland had 10,848 inhabited houses, and also 56,469 inhabitants, of whom 28,234 were males and 28,235 females. Sum expended on the relief of the poor, in 1836-39, £14,739. Total annual value of real property assessed to the poor rates in 1841, £266,335.

WESTMORELAND, county, Pa. Situated toward the S.W. part of the state, and contains 1050 sq. m. Bounded N.E. by Conemaugh or Keskiminetas river, on the N.W, for the distance of 12 m., by the Alleghany river, W. by Youghiogheny river. Drained by Loyalhanna river, and Big Sewickly creek. It contained, in 1840, 30,003 neat cattle, 43,632 sheep, 57,069 swine ; and produced 466,362 bushels of wheat. 103,884 of rye, 462,512 of Indian corn, 56,141 of buckwheat, 757,941 of oats, 116,052 of potatoes, 18,400 pounds of tobacco, 39,536 of sugar. It had 74 stores, six fulling-mills, two woollen factories, 70 flouring-mills, six grist-mills, 86 saw-mills, two paper-mills, three oil-mills, 36 tanneries, 53 distilleries, three potteries, four printing-offices, three weekly newspapers ; two academies, 80 students ; 135 schools, 4151 scholars. Pop. 42,699. Capital, Greensburg.

WESTMORELAND, county, Va. Situated in the E. part of the state, and contains 316 sq. m. Bounded N.E. by Potomac river, S.W. by Rappahannock river. It contained, in 1840, 5474 neat cattle, 4317 sheep, 6150 swine ; and produced 59,958 bushels of wheat, 1450 of rye, 943,670 of Indian corn, 27,751 of oats, 6899 of potatoes, 1040 pounds of tobacco, 5065 of cotton. It had 15 stores, 13 grist-mills, two saw-mills, three tanneries, one distillery ; one academy, 36 students ; five schools, 127 scholars. Pop.: whites, 3466 ; slaves, 3390 ; free coloured, 963 ; total, 8019. The true footing of the census makes 7992. Capital, Westmoreland C.H.

WESTMORELAND, p. t., Cheshire co., N. H., 65 m. S.W. Concord, 436 W. Bounded W. by Connecticut river. Chartered in 1752. It contains three churches, a Congregational, Universalist, and Christian, two stores, one fulling-mill, four grist-mills, seven saw-mills ; 13 schools, 494 scholars. Pop. 1546.

WESTMORELAND. p. t., Oneida co., N. Y., 103 m. W.N.W. Albany, 392 W. Watered by Oriskany creek. It contains four churches, a Presbyterian, two Methodist, and a Union, five stores, one fulling-mill, one woollen factory, 10 saw-mills, one paper-mill, four tanneries ; 18 schools, 1080 scholars. Pop. 3105.

WEST NANTMEAL, t., Chester co., Pa., 34 m. N.W. Philadelphia. Drained by the E. branch of Brandywine river and its tributaries. It has eight stores, one furnace, one forge, two flouring-mills, two grist-mills, five saw-mills, one oil-mill ; six schools, 376 scholars. Pop. 1731.

WEST NEWBURY, p. t., Essex co., Mass., 39 m. N. by E. boston, 470 W. Bounded N.W. and N. by Merrimac river. It contains three churches, a Congregational, Methodist, and Universalist, four stores, one grist-mill, one tannery ; six schools, 375 scholars. Pop. 1560.

WESTON, p. t., Windsor. co., Vt., 91 m. S. by W. Montpelier, 456 W. West river affords water-power. It contains three churches, two stores, one fulling-mill, two grist-mills, eight saw-mills ; 12 schools, 368 scholars. Pop. 1038.

WESTON, p. t., Middlesex co., Mass., 14 m. W. Boston, 436 W. Watered by a branch of Concord river. It contains three churches, a Congregational, Methodist, and Baptist, three stores, one cotton factory with 512 spindles, two grist-mills, one saw-mill, one tannery, one pottery; six schools, 287 scholars. Pop. 1092.

WESTON, p. t., Fairfield co., Conn., 63 m. S.W. Hartford, 281 W. Drained by Saugatuck and Mill rivers. It contains five churches, two Congregational, an Episcopal, Methodist, and Baptist, eight stores, one flouring-mill, eight grist-mills, 13 saw-mills; two academies, 26 students, 17 schools, 459 scholars. Pop. 2651.

WEST PENN, p. t., Schuylkill co., Pa., 79 m. N.E. Harrisburg, 189 W. Watered by Little Schuylkill river. It abounds with anthracite coal, and contains a Lutheran church. Pop. 1362.

WEST PENNSBORO', t. Cumberland co., Pa., 8 m. W. Carlisle. Bounded N. by Conedogwinit creek. It has one commission house in foreign trade, three retail stores, eight flouring-mills, two saw-mills, two tanneries, three distilleries; nine schools, 270 scholars. Pop. 1867.

WESTPHALIA, prov. containing all the N. portion of the Prussian dominions to the W. of the Weser, having N., Hanover; E., the latter, and some of the smaller German states; S., the latter, and the Prussian prov. of the Rhine; and W., Holland. Area, 7801 m. Pop., in 1837, 1,317,541; of whom 749,782 are Catholics, 554,649 Protestants, and 13,016 Jews. Principal towns, Munster, Minden, Paderborn, Arnsberg, Hamm, &c. It is divided into three regencies, and these again into 37 circles. Principal rivers, Ems, Weser, Lippe, &c. Surface in the E., N.E., and S., hilly or mountainous; but it is level in the middle of the province, and in the N.W. adjoining Holland. In some places the soil is very fertile, but there are some pretty extensive marshes and heaths. Most part of this, as well as of the adjoining province of the Rhine, is divided into small farms, the occupiers of which live together in villages. The rent is paid sometimes in money, but frequently in produce or services, or both. The occupiers are a kind of copy-holders, their lands descending from father to son. (Jacob's Tour in Germany, p. 168.) The Rhine provinces being in possession of the French, when the famous edict of 1811 was published, making copy-holders freeholders in the old states of the Prussian monarchy, it did not apply to them. (See PRUSSIA.) Principal products, corn, flax, and potatoes. It is also productive of horses, cattle, sheep, and hogs. The latter furnish the Westphalian hams, so abundant in our markets and so excellent. There is also great plenty of game and honey. There are vast beds of coal, with mines of iron, lead, copper, rock salt, &c., with salt springs. Manufactures, principally linen, in the production of which 32,331 looms were wholly and occasionally employed in 1837; cottons, hardware, and cutlery, the latter being largely produced at Iserlohn, Dortmund, Hagen, and other places; with paper, spirits, tobacco, and various other articles.

WEST POINT, p. v., Cornwall t., Orange co., N.Y., 92 m. S. by W. Albany, 278 W. Pleasantly situated on the W. bank of the Hudson, 52 m. N. of the city of New-York. It derives its importance from the United States military academy, located here. In 1824 congress appropriated $10,000 for the purchase of a farm which included the site of the academy, containing the ruins of the revolutionary fort Putnam; and in 1826 the state of New-York ceded to the United States the jurisdiction of 250 acres of land, which includes only a part of the land purchased by the United States. The remainder is under the jurisdiction of the state. There is a number of buildings, well adapted to the purpose of the institution. There is a stone building of the Gothic architecture containing the philosophical and astronomical rooms and the library, 150 feet long and 60 feet wide, two stories high, with three towers. The library room, which is ornamented by paintings and by architectural designs, contains an appropriate library of 14,554 vols., of which 3654 are duplicates. The philosophical room is furnished with a fine apparatus, and the astronomical part has an observatory covered by a dome in the central tower, weighing 4 tons, and revolving on four cannon balls. It has some very complete instruments. The chapel is devoted to religious purposes, is 50 feet in front by 70 feet deep and 24 feet high, and stands between the library and the academy. The academy is a fine stone building 275 feet long by 75 feet wide, and three stories high, containing a riding hall, a number of recitation rooms, various offices, and the engineering room, furnished with beautiful models of fortification and civil engineering, as well as of architecture, &c. There are in the building two spacious galleries for paintings and sculpture. In the basement is a chemical laboratory. The hospital is a stone edifice near the bank of Hudson river commanding a fine prospect, and admirably fitted up. The military laboratory has towers designed as workshops, and enclosed within the walls are various kinds of ordnance, some of revolutionary memory, and among them the remnants of the immense chain which was stretched

across Hudson r. There are also a mess hall, and two barracks for the accommodation of the cadets, the number of which is limited to 360. In the N. part is a village called Camp Town, comprising barracks for the soldiers, and some other buildings for some persons employed about the establishment; and farther N. near the Hudson, is the cemetery for the officers and cadets. Houses for the accommodation of the officers are located S.W. of the encampment ground. There is also a fine hotel for the accommodation of visitors. About the grounds are several monuments, the most conspicuous of which is that of the Polish general Koskiusko, erected by the cadets, at an expense of $5000. There have been from April 20th, 1812, to January 29th, 1844, 111 graduates at West Point; of whom 147 died in the service in the army, and 537 are now in the service, including five professors. The education given at West Point, both scientific and military, is of a high order. Near the military academy is the private academy of Mr. J. D. Kinsley, formerly instructor in the military academy, which is in high repute.

West Point was the strong hold which Arnold by his treason stipulated to surrender to the British. But the treason was detected; Arnold escaped, while Andre, his British coadjutor, paid the forfeiture with his life.

WESTPORT, a seaport town of Ireland, co. Mayo, prov. Galway, on a small river near the S.W. angle of Clew bay, 42 m. N. by W. Galway, and 135 m. W.N.W. Dublin. Pop., in 1831, 4428. This is a modern, neat, and rather nice-looking town. It occupies a narrow valley, and on each side and parallel to the stream by which it is intersected is a street planted with trees: the other streets branch from these on either side, and are for the most part inconveniently steep. The parish church is situated within Lord Sligo's park, immediately adjoining the town, and it has also a large Roman Catholic chapel, with places of worship for Presbyterians and Methodists, several schools, a barrack, a market-house, a linen-hall, and courthouse. A manor court, with jurisdiction to the amount of £10 sterling, is held once a month; general sessions in April and October, and petty sessions on Thursdays. It is also a constabulary and coast-guard station. The linen trade was, at no very distant period, carried on extensively in this town and its vicinity. Latterly, however, this branch of industry has greatly declined, and the inhabitants are now principally dependent on the fishery, and on the export of corn, provisions, &c. The gross value of the exports of these, in 1835, was estimated at £97,943; of which corn, meal, and flour made no less than £35,547, and linen only £560! Several large flour-mills have been opened in the neighbourhood; and there are also distilleries and breweries. Branches of the bank of Ireland and of the National bank have been opened here. Post-office revenue, in 1830, £811; do., in 1836, £949.

The port and corn warehouses are situated a little below the town, on the bay, vessels drawing 13 ft. water coming close to the quays. Clew bay has at its mouth Clare island, on the most northerly point of which is a lighthouse; there are many small islands within the bay, which, in many places, affords convenient and secure anchorage. Gross amount of customs duty collected here in 1844, £7555.

Croagh Patrick, or the Reek, rising 2499 ft. above the sea, is situated immediately on the S. side of the bay, 4 or 5 m. S.W. from the town. This is not only one of the highest, but also one of the most celebrated mountains in Ireland, being the spot where St. Patrick is said to have collected the snakes and other venomous reptiles from all parts of the island, and from which he precipitated them headlong into the sea! An altar or cairn is erected on the summit of the mountain in memory of this grand achievement, and it continues to be a frequent place of pilgrimage and devotion. The view from the summit is very magnificent. The land in the vicinity of the town is divided into very small portions, and the occupiers are, for the most part, miserably poor. Lord Sligo's park or demesne, to which strangers have access, is very fine; but the rest of his immense estate is but little removed from a state of nature. (Official Returns; Inglis, ii., 98, &c.)

WEST PORT, p. t., Bristol co., Mass., 59 m. S. Boston, 426 W. Bounded S. by the Atlantic, at the mouth of Buzzard's bay. Drained by Westport or Nochachuck river, which flows by a wide mouth into Buzzard's bay. It contains three churches, a Congregational, Methodist, and Baptist, six stores, one falling-mill, one cotton factory with 900 spindles, one furnace, three grist-mills, five saw-mills, 22 schools, 462 scholars. Pop. 2659.

WEST PORT, p. t., Fairfield co., Ct., 64 m. S.W. Boston, 273 W. Watered by Saugatuck r. It contains a range on both sides of the river, connected by a bridge; and has in the town three churches, a Congregational, a Methodist, and an Episcopal, 14 stores, one cotton factory with 1900 spindles, four grist-mills, one saw-mill; three academies, 57 students; six schools, 244 scholars. Pop. 1662.

WEST PORT, t., Essex co., N. Y., 125 m. N. Albany. Bounded E. by lake Champlain, into which flows Black cr. It contains nine stores, one fulling mill, two woollen factories, two forges, one grist-mill, 10 saw-mills; one academy, 100 students; six schools, 307 scholars. Pop. 1932.

WEST ROCKHILL, t., Bucks co., Pa., 33 m. N.N.W. Philadelphia, 99 m. E.N.E. Harrisburg, 171 W. Drained by Perkiomen creek and a branch of Tohickon cr. It contains four stores, three grist mills, three saw-mills, two oil-mills, four tanneries. Pop. 1631.

WEST SALEM, t., Mercer co., Pa., 17 m. N.W. Mercer. Drained by Shenango cr. It has one store, one fulling-mill, two grist-mills, six saw-mills, two tanneries; 16 schools, 716 scholars. Pop. 2528.

WEST SPRINGFIELD, p. t., Hampden co., Mass., 93 m. W. Boston, 363 W. Bounded E. by Connecticut river, over which is a fine bridge connecting it with Springfield. Westfield river affords good water power. It contains seven churches, four Congregational, a Methodist, and two Baptist, 10 stores, one fulling-mill, one cotton factory with 3400 spindles; 27 schools, 791 scholars. Pop. 3626. The Boston and Albany railroad passes through it.

WEST STOCKBRIDGE, p. t., Berkshire co., Mass., 138 m. W. Boston, 305 W. Drained by Williams r. It contains an abundance of marble, variously coloured, from a pure white to black. The Boston and Albany railroad here meets the Housatonic railroad, and the Hudson and Berkshire railroad. It contains two churches, a Congregational and a Methodist, six stores, one fulling-mill, one grist-mill, 12 saw-mills; six schools, 281 scholars. Pop. 1448.

WEST TURIN, t., Lewis co., N. Y., 196 m. N.W. Albany. Bounded E. by Black river, in whichcare here the high falls, 63 feet high, and immediately bel w the Black river canal enters it. Drained by Salmon river and Fish cr. It possesses great water-power, and has four stores, one flouring-mill, eight saw-mills; 16 schools, 472 scholars. Pop. 2042.

WEST UNION, p. v., Tiffin t., capital of Adams co., O., 106 m. S. Columbus, 443 W. It contains a courthouse of stone, a jail, county offices, a market-house, eight stores, various mechanic shops, a printing-office, issuing a weekly newspaper, and about 500 inhabitants.

WEST VINCENT, t., Chester co., Pa., 30 m. N.W. Philadelphia. Watered by Stony, French, and Ring creeks. It has five stores, four grist-mills, four saw-mills; five schools, 175 scholars. Pop. 1232.

WETHERSFIELD, p. t., Hartford co., Ct., 4 m. S. Hartford, 336 W. Bounded E. by Connecticut r. The village is beautifully situated on the W. side of the river, handsomely laid out with broad and finely shaded streets, and is well built, containing three churches, a Congregational, Methodist and Baptist, an academy, a female seminary, the Connecticut state prison, which contains apartments for the warden and 200 cells for convicts, and two ranges of workshops, the whole enclosed by a wall 20 feet high. The town has besides a Methodist and two Congregational churches; and nine stores, one fulling-mill, three grist-mills two saw-mills, two tanneries; one academy, 56 students; 13 schools, 632 scholars. Pop. 3924.

WETHERSFIELD, p. t., Wyoming co., N. Y., 258 m. W. Albany, 361 W. Drained by Tonawanda and Wiskoy creeks. It has four stores, three fulling-mills, one woollen factory, two grist-mills, five saw-mills, two tanneries; nine schools, 560 scholars. Pop. 1788.

WETHERSFIELD, p. t., Trumbull co., O., 169 m. N.E. Columbus, 298 W. Drained by Mahoning r. The Pennsylvania and Ohio canal passes through it. It has nine schools, 500 scholars. Pop. 1447.

WETUMPKA, city, Montgomery co., Ala., 119 m. S.E. Tuscaloosa. Situated on the E. side of Coosa r. It contains four churches, a Baptist, Methodist, Episcopal and Presbyterian, an academy, the Alabama state prison, 18 stores, 275 dwellings, 2000 inhabitants. The Harrowgate springs, containing valuable mineral properties, are situated on the S. side of the city, and are much resorted to during the summer months.

WEXFORD, a marit. county of Ireland, prov. Leinster, having on the S. and E., St. George's channel; on the N., the co. of Wicklow; and on the W., Carlow, Kilkenny, and Waterford harbour, by which it is separated from Waterford. Area, 564,479 acres, of which 18,500 are unimproved mountain and bog. Surface, pleasantly diversified; climate mild. Soils either light or stiff clays. Property in pretty considerable estates : farms of various sizes; but there is less of the extreme subdivision of land in this, than in most other Irish counties. Dairies numerous, but badly managed; some districts have been long noted for their great crops of barley. Average rent of land 14s. an acre; but, in general, it is too high, and the competition for small patches is carried beyond all reasonable bounds. The barony of Forth, occupying the S.W. angle of Wexford, differs widely from the rest of the county, and, in-

deed, from every other district of Ireland. It was settled at a distant period by a colony from South Wales; and, till very recently, the Welsh language was spoken by every one, and is still understood by the older inhabitants. The people are industrious, provident, peaceable, and cleanly. The farms are small, running from 10 up to 50 or 60 acres, but those from 30 to 40 are most common. Mr. Inglis says, that the land is clean, and well cultivated ; that the crops of wheat and beans, both of which are extensively grown, are excellent ; and that the improved Scotch plough with two horses is in universal use. The farm-houses are substantial, and the cottages clean and comfortable, forming, in this respect, a striking contrast with those in most other parts of the county. Persons of different religious creeds live in this barony harmoniously together. Land here, as in the rest of the county, very high-rented. (I., 49.) Wexford has neither minerals nor manufactures of any importance. Principal rivers, Slaney and Barrow. Wexford is divided into eight baronies, and 142 parishes; and returns four members to the House of Commons, viz. two for the county, and one each for the boroughs of Wexford and New Ross. Registered electors for the county, in 1839-40, 3604. In 1831, Wexford had 29,923 inhabited houses, 32,856 families, and 182,713 inhabitants, of whom 87,995 were males, and 94,718 females.

WEXFORD, a parl. bor. and seaport town of Ireland, prov. Leinster, co. Wexford, of which it is the cap., at the mouth of the estuary of the Slaney, on the W. or inner side of Wexford Haven, 67 m. S. by W. Dublin; lat. 52° 22' N., long. 6° 24' W. Area of modern parl. bor. 450 acres ; pop. of ditto, in 1831, 10,670. It is built wholly on the S. side of the river, and consists of a row of houses along the quay fronting the harbour, of a street parallel to the latter, and of numerous cross streets ; but, excepting the quay, they are mostly narrow, irregular, ill-paved, and dirty. A long and poor suburb, principally occupied by fishermen, extends to a considerable distance S. from the town. Of 1723 houses in the parliamentary borough in 1831, 1113 were slated and 670 thatched. The communication with the country on the opposite side of the harbour is maintained by means of a wooden bridge, which, with its embankments, is nearly 1700 ft. in length (measured on the plan in the Boundary Report), having a draw-bridge in the centre, and the river being navigable to Enniscorthy. The expense of keeping it up is defrayed by a toll. The public buildings and establishments comprise two Protestant churches, several Roman Catholic chapels, two Methodist meeting-houses, the county courthouse, a large and handsome jail, a fever hospital, house of industry, dispensary, lunatic asylum, a diocesan school for the see of Ferns, and various other schools ; a Roman Catholic college, supported by private endowment, a priory, a nunnery, a public library, chamber of commerce, assembly rooms, club-house, barracks, &c. The old walls, by which the town was formerly surrounded, were repaired in 1804, but they have since been allowed to go to ruin, and the suburbs extend considerably beyond them. Malting is carried on to a very considerable extent, and, in 1835, duty was paid on 202,449 gallons of spirits distilled in the town. The exertions of Father Mathew, however, lessened this business.

Wexford Haven is of great extent, and has a fine appearance on a map; but it is shallow, and owing to a bar at its mouth between the two low, long, sandy peninsulas which form its external boundary, it cannot be entered by vessels drawing more than 9 or 10 ft. water, and even those should, with neap tides, have four hours flood to enter. But notwithstanding these drawbacks, Wexford, from her situation in a fertile county, and on a river navigable to a considerable distance by barges, and other circumstances, has a considerable trade, and is, in fact, one of the principal secondary ports of Ireland for the exportation of corn, meal, and flour, butter of superior quality, provisions, and cattle. This is evident from the subjoined table.

STATEMENT of the Quantity and Value of the principal Articles exported from Wexford in 1835.

Articles Exported.		Quantity.	Estimated Value. L.
Corn, meal, and flour	cwts.	491,485	175,180
Butter	—	18,000	54,000
Feathers	—	400	400
Wool	lbs.	60,000	3,000
Eggs	crates	90	180
Spirits	galls.	490	96
Cows and oxen	head	3,500	36,666
Horses	—	20	300
Sheep	—	7,500	12,000
Swine	—	2,300	13,000
Other Articles	value	—	16,000
Total			314 126

Steamers ply once a week between Liverpool and Wexford. Excellent oysters are found in the bay, and the fishing business is carried on to a considerable extent. There belonged to the port, in 1841, 69 vessels of the aggregate burden of 7114 tons, exclusive of those below 50 tons. Gross customs revenue in same year, £9357. Branches of the Provincial bank, the bank of Ireland, and the National bank, have been opened here. Postoffice revenue in 1830, £1971, ditto in 1836, £9134. Two weekly newspapers are published in the town. A patent slip and a building yard are attached to the harbour, and some little business is done in the building and repairing of vessels and boats.

Wexford has several charters, the first having been granted by the earl of Pembroke, in 1318. Under the Irish Municipal Reform Act, 3 and 4 Victoria, cap. 109, the corporate body consists of a mayor, bailiffs, free burgesses, and commonalty. The borough returned two members to the Irish House of Commons from 1374 down to the union; and it has since returned one member to the imperial House of Commons. Registered electors in 1839–40, 405. The assizes for the county are held here, with general sessions in January and at midsummer, special road-sessions twice a year, and petty sessions and a borough court for debts not exceeding £2, once a week. Markets on Wednesdays and Fridays. Fairs six times a year.

In 1649 Wexford was taken by Cromwell, when a considerable number of the inhabitants were put to the sword. During the rebellion of 1798 it was, for awhile, the head quarters of the insurgents, by whom it was evacuated after the battle of Vinegar Hill. The town is possessed of considerable property: but it is let on long leases, at a low rate (Boundary Reports, and other official Returns.)

Mr. Inglis says, "There are many good shops in Wexford, and I heard no complaint of the want of trade; and the best illustration I can give of the comfortable condition of the people is, that during the two days I spent in the town, I was not once asked for charity. I do not mean to say that there is not a pauper or a person out of employment in Wexford, but it may be said that it is a flourishing town." (I. 45.)

WEYHILL, a village of England, co. Hants, hund. Andover, within a short distance of the W. verge of Salisbury plain, 15 m. N.W. Winchester. It is celebrated for its great annual fair, held for six or seven days from the 9th of October. This, perhaps, is the largest fair in the S. of England for sheep, and is, also, a considerable mart for horses, cheese, and hops. At the fair which began on the 10th of October, 1840, it was estimated that above 150,000 sheep were exhibited for sale. A row of booths, called Farnham-row, is assigned exclusively to the dealers in Farnham hops, but hops from Kent, Sussex, &c., are also brought thither in considerable quantities. The horses exposed for sale are principally cart colts, nags, and hunters; the cheese consists mostly of N. Wilts, Somerset, and Gloucester. The booths are formed into regular streets, and exhibit all the features of a large town, every part of which presents a scene of bustle and activity. On old Michaelmas day many farm servants are hired.

WEYMOUTH and MELCOMBE REGIS, two contiguous towns of England, forming together a parl. munic. bor., and seaport, co. Dorset, hund. Culliford Tree; on the English channel, at the mouth of the small river Wey, 3 m. N. from the headland called the Isle of Portland, and 120 m. W.S.W. London. The modern parl. bor., comprising the chapelry of Weymouth and the par. of Melcombe Regis, with portions of the adjacent pars. of Wyke Regis and Radipole, had, in 1831, a pop. of 8095. Weymouth and Melcombe lie on opposite sides of the harbour, that is, of the narrow outlet of an arm of the sea, called the Backwater, nearly 2 m. in length, being connected by a handsome stone bridge, erected over this outlet in 1770. The most considerable part of Melcombe consists of ranges of buildings situated on the N. side of the harbour, extending for nearly 1 m. in front of the esplanade along the sea, which here forms a fine semicircular bay. The houses here, which are large and handsome, are chiefly occupied by visiters, attracted to the town by its unequalled advantages for sea-bathing, and other conveniences. But, with one or two exceptions, the greater number of the other streets are mostly narrow and dirty, with mean houses. Much of the ground on which Melcombe stands is low, and has been reclaimed, at different times, from the Backwater. On the N., about 99 acres of this land have been recently enclosed for the purpose of forming a park, and between this and the sea there is a narrow space, which will probably be built over with good houses.

Weymouth, on the S. side of the harbour, not being resorted to by visiters, retains, in great measure, its original character of a fishing town, and is irregularly and ill built. It is chiefly, in fact, inhabited by the labouring classes; in the outskirts, however, there are some good ranges of houses, and handsome detached dwellings. The parish church of Melcombe, rebuilt in 1817, though a spacious edifice, is not remarkable for its architecture: the living, a rectory in the gift of W. Wyndham, Esq., is worth £698 a

1080

year nett. The chapel of ease in Weymouth, opposite the foot of the bridge, is a tasteful modern building in the Gothic style; the living, a perpetual curacy, is attached to the rectory of Wyke Regis, to which parish Weymouth belongs. There are several dissenting chapels; and in Melcombe are spacious assembly-rooms, a small town-hall, meat theatre, good libraries, baths, and other establishments usual at a watering place. There is a jail, but of very inferior description. Two national schools have been founded, and among other charitable institutions for the education and relief of the poor is a bequest of £75 a year for the apprenticing of poor children. The town is lighted with gas; but the inferior streets are badly paved. Facing the sea is the lodge built for the residence of the royal family, by whom the town was frequently visited during the reign of George III. An equestrian statue of that monarch has, also, been erected at the N. extremity of the main street. On the Dorchester road, near the town, are barracks, now occupied as private houses. The harbour which, as already stated, consists of the inlet between Weymouth and Melcombe as far as the bridge is narrow, and the bar at its mouth has only 6 feet water at low ebb; and, as spring tides have run only 6 or 7 feet, the port cannot be entered by large vessels. There is, however, good anchorage in the bay, in 7 or 8 fathoms water. A battery, and several small forts have been erected at the entrance to the harbour.

Freestone from the isle of Portland, Roman cement, bricks, tiles, &c., are among the principal exports. Shipbuilding, and rope and sail making are carried on to a small extent. In 1842, there belonged to the port 56 vessels, of the aggregate burden of 6037 tons. Gross customs revenue, in 1840, £14,798. Weymouth was of but little consequence till George III. made it his summer residence. Since then the town has continued to increase and may be considered as flourishing; it is frequented by numerous visiters during the summer season, and a great number of respectable families have made it their permanent residence. The fine sands along the shore, and the gradually increasing depth and purity of the water, render it highly suitable for a bathing place. A raised terrace or esplanade has been constructed round a great portion of its picturesque bay, which constitutes the fashionable promenade of the inhabitants. Races, and a regatta, take place annually in August.

Weymouth and Melcombe were originally distinct boroughs. In consequence of the continued dissensions between the municipal authorities of the two towns, a union of the boroughs took place in the 13th of Elizabeth; and from that period down to the passing of the Reform Act, the united borough possessed the privilege of returning four members to the House of Commons; but the above act reduced the number of members to two: while, at the same time, some additions were made to the boundaries of the old borough. Registered electors, in 1839–40, 680. Under the Municipal Reform Act, the town is divided into two wards; and is governed by a mayor, five other aldermen, and 18 councillors. It has a commission of the peace, generally held once a year, with jurisdiction over all the capital offences; a weekly court of record, and two courts leet, held annually at the same time and place as the sessions. The petty customs, harbour dues, &c., belonging to the corporation, are let by auction. Corp. revenue, in 1840, £629 4s. Markets, Tuesdays and Fridays; fairs, three times yearly. (Parl. Reps.; Priv. Information.)

WEYMOUTH, p. t., Norfolk co., Mass., 12 m. S.S.E. Boston, 446 W. Watered by Fore and Back rivers which are branches of Boston harbour, and navigable for large vessels, and into one of them a considerable mill stream enters. It contains three churches, two Congregational, and a Methodist, 23 stores, four grist-mills, four saw-mills, one tannery, four academies, 172 students, eight schools, 1005 scholars. Pop. 3738.

WEYMOUTH, t., Atlantic co., N. J. Bounded E. by Great Egg Harbour r. It has eight stores, one forge, one grist-mill, six saw-mills; six schools, 336 scholars. Pop. 1159.

WHARTON, t., Fayette co., Pa., 11 m. S.E. Connellsville. Bounded E. and N. by Youghiogheny r. It has two cxnns, one furnace, two grist-mills, 11 saw-mills; four schools, 193 scholars. Pop. 1385.

WHATELY, p. t., Franklin co., Mass., 9 m. N. Northampton, 94 m. W. Boston, 369 W. Bounded E. by Connecticut r. Drained by Mill r. Incorporated in 1771. It contains two churches, a Congregational and a Baptist, two stores, one fulling-mill, one woollen factory, one cotton factory with 200 spindles, three grist-mills, four saw-mills, two tanneries; six schools, 380 scholars. Pop. 1072.

WHEATLAND, p. t., Monroe co., N. Y., 14 m. S.W. Rochester, 273 m. W. by N. Albany, 369 W. Bounded E. by Genesee r. Drained by Allen's cr. It contains 13 stores, one furnace, one fulling-mill, one woollen factory, two grist-mills, four saw-mills, three tanneries; 13 schools, 695 scholars. Pop. 2671.

WHEELER, p. t., Steuben co., N. Y., 216 m. W. by S

Albany, 306 W. Drained by Five Mile cr., a branch of Conhocton r. It has one fulling mill, also saw mills; nine schools, 381 scholars. Pop. 1294.

WHEELING, city, port of entry, and capital of Ohio co. Va., 351 m. N.W. Richmond, 92 m. below Pittsburg, 837 m. above the mouth of the Ohio, 296 W. Pop. in 1810, 914; in 1820, 1567; in 1830, 5221; in 1840, 7885. Situated on the E. bank of Ohio r., at the mouth of Wheeling cr. It is surrounded by bold and precipitous hills, which contain an inexhaustible quantity of bituminous coal, and which approach the river so nearly as to leave rather a small area for the city. It is the largest place in regard to commerce, manufactures, and population, in western Virginia. The National Road passes through it, constituting it a great thoroughfare for persons travelling to the west. The city is situated on a high bank of the river, and extends along it for a mile and a half. It was laid out in 1793, and has received ten additions, and now contains 1270 lots, 856 on the N., and 414 on the S. side of Wheeling cr. Lane's Island lies in Ohio r., immediately in front of the city, one half of which is laid out in 923 lots, and has received the name of Columbia city. Wheeling contains a handsome courthouse, a jail, county offices, an academy, the Wheeling Institute, a theatre, a masonic hall, eight churches, two Presbyterian, an Episcopal, Methodist, Baptist, a Campbellite, a Friends, and a Roman Catholic, seven commission and forwarding houses, 97 retail stores, four iron foundries, four steam engine factories, eight glass houses, in several of which cut glass is manufactured, four woollen and cotton factories with carding machines; four saw-mills, two paper-mills, three manufactories of white lead, sheet lead and copperas; and in the vicinity, 134 flouring mills. The city contains about 1000 dwellings, many of them elegant. There are exported 1,500,000 bushels of bituminous coal, annually. Over 30 steamboats are owned here, and many others passing up and down the Ohio stop at its wharves. The tonnage, September 30th, 1842, was 1212. In 1840, it was 2490.

WHITBY, a parl. bor. and seaport town of England, N. riding co. York, liberty of Whitby Strand, at the mouth of the Esk, here crossed by a swing iron bridge, erected about six years ago, instead of an old drawbridge, 42 m. N.N.E. York. Lat. of lighthouse, 54° 30′ N.; long. 0° 37′ W. Pop. of parl. bor., which includes the townships of Whitby, Ruswarp, Hawkser-cum-Stainsacre, in 1831, 10,399, in 1841, 9975. It is built along both banks of the Esk, the direction of which, from S. to N., determines that of the town; but, as the level ground by the river is of very limited dimensions, the buildings on both sides are carried up its banks, which, on the E. side especially, are precipitous. The houses, partly of stone and partly of brick, in the lower part of the town, are closely packed together, and the streets are for the most part narrow, while those on the banks are inconveniently steep; they are, however, well paved and lighted with gas, and some new streets on the W. side of the town are comparatively handsome; and since the new bridge was erected, the streets leading to it and to the pier have been widened, and the lower parts of the town much improved. The more opulent inhabitants have residences in the environs, which are beautiful and romantic. The church is inconveniently situated on the top of a cliff nearly 200 ft. above the sea, on the E. side of the town, the ascent to it being by a flight of 190 stone steps; but a spacious proprietary chapel was, about 60 years ago, constructed in the lower part of the town. The living, a curacy in the gift of the archbishop of York, is worth £396 a year nett. It has also places of worship for Methodists, Quakers, Presbyterians, independents, Roman Catholics, &c. Among the educational and charitable institutions may be mentioned, Lancastrian schools for children of both sexes; two infant schools; a seaman's hospital, for 42 widows and their children; a dispensary, established in 1786; a large work-house, and several minor charities. The public buildings include the town hall, erected in 1788; a building with public baths, and apartments for the subscription library, and the literary and philosophical society, erected in 1826; a handsome news room, custom-house, &c. The Union mill, for supplying the members with flour at a reduced price, on the W. side of the town, is a conspicuous object. On the whole, "the appearance of the town is that of a substantial and wealthy seaport." (Bound. Report.)

The harbour is formed by two piers; that on the W. side, 640 yards in length, terminates in a circular head, on which a lighthouse, having the lantern elevated 89 ft. above the sea, has been constructed. A noble quay, extending from the bridge to the west pier, is now being built, and, when completed, will form one of the finest promenades in the kingdom. The opposite pier extends about 215 yards from the cliff on the E. side of the river. The channel between the piers forms the harbour, which, however, labours under the drawback of nearly drying at low water; and in rough weather, when the wind is from the N. or N.E., a heavy sea is thrown in. To obviate the latter inconvenience, an internal pier has been built, and the inner harbour, or that above the drawbridge, is but little affected by the weather. There are dry docks and slips for the construction and repair of ships, the building of which, though less now than formerly, is carried on to a considerable extent; ships of the burden of 3913 tons having been built here in 1841. The manufacture of sail-cloth, cordage, &c., is also carried on.

Whitby is principally indebted for its rise to the alum works in its vicinity, commenced in 1595; the exports of the alum, and the import of the coal required in its manufacture, giving birth to a considerable trade. This, however, is by no means so extensive as formerly. There are no alum works now very near to Whitby. Sandsend and Kitleness works, belonging to the marquis of Normandy, are from 3 to 5, and the Peak works 8 m. off. The shipments of alum amounted, in 1841, to 3937 tons; and in 1840, to 2747 do. There belonged to the port in 1842, 291 ships of the aggregate burden of 47,537 tons. Several of these ships were formerly employed in the N. whale fishery; but this having declined, the shipping belonging to the port is now principally employed in the Baltic, American, and East Indian trades. Most of the large ships sail from London, Bristol, Liverpool, and Hull. Gross customs duty received at the port in 1841, £6968, having increased considerably since the establishment, about that time, of bonding warehouses. The town is under the superintendence of the magistrates of the N. riding; and a manorial court of pleas for the recovery of debts is held every third Monday.

The Reform Act conferred on Whitby, for the first time, the privilege of sending one member to the House of Commons. Registered electors, in 1839–40, 412.

Whitby abbey, which, having been destroyed by the Danes, was rebuilt after the conquest, appears to have been a magnificent edifice. The ruins, in a commanding situation on the cliff near the church, are of considerable extent, and exhibit different styles of architecture. The neighbourhood of Whitby abounds with natural curiosities. In the alum rocks are found an immense variety of petrified shells, trunks of trees, pieces of wood, bones of fishes, &c., and several highly interesting specimens of the large marine animals called ichthyosaurus and plesiosaurus, with fossil crocodiles, of which the finest hitherto discovered adorns the Whitby museum, which is particularly rich in specimens of the various organic remains found in the vicinity.

The country about Whitby is highly picturesque, and the beautiful valley of the Esk is also rich in ironstone of superior quality, with an admixture of lime. This has become of late an article of export to the iron works on the Tyne to the extent of from 20,000 to 30,000 tons a year. Freestone of several varieties, for building and engineering purposes, is also extensively exported to the London and other markets on the E. coast, as is also the whinstone (found in the same vicinity), for the repairing of roads when broken, and paving of streets, when shaped into blocks, for which purposes it is considered superior to granite. It is only since the formation of a railway from Whitby to Pickering, opened in 1836, that these mineral stores have been developed and made available. This railway passes for 24 m. through a succession of varied and highly picturesque scenery, affording, perhaps, the most beautiful ride of the kind in the kingdom. It is worked by horse power, and, through its medium there is a daily communication (except on Sundays) with York. There are also daily coaches to and from Scarborough and the North. Since the opening of the railway, the fishery at Whitby has very materially increased. There are several mineral springs in the neighbourhood. Good lodgings may be had in the summer and autumn, rendering Whitby a desirable place of resort, where galety is not the only object. Market-day, Saturday. Fairs, August 25 and Martinmas day. (Private Information.)

WHITCHURCH, a market town and par. of England, co. Hants., hund. Evingar, in div. Kingsclere, on the road from London to Salisbury, 12 m. N. Winchester. Area of par., 7690 acres. Pop. of do. in 1831, 1673. The town is quite insignificant, and would be wholly unworthy of notice in a work of this kind, were it not for the fact of its having sent two members to the House of Commons, from the 27th year of Elizabeth down to the passing of the Reform Act, by which it was disfranchised. The living is a vicarage, worth £140 a year, in the gift of the Bishop of Winchester. An endowment to the amount of £102 a year is applied to the purchase of bedding and clothing for the poor.

A little to the E. of the church, near the London road, are the paper mills, at which the paper for the notes of the Bank of England has been manufactured since the reign of George I. down to the present time.

WHITCHURCH, or BLANCMINSTER, a market town and par. of England, co. Salop, hund. E. Bradford, on the borders of Wales and Cheshire, 18 m. N. by W. Shrewsbury. Area of par., 15,380 acres. Pop. of do., in 1831, 5819. The town is built on an acclivity, the summit of which is crowned by the church, a freestone edifice, erected in 1722 in the Tuscan order, with an embattled square tower 108 feet in height. The interior is handsome, and it has a good altarpiece, and two recumbent stone figures brought from the ruins of the old church, one of which represents Talbot, earl of Shrewsbury, celebrated for his exploits in the wars with France under Henry V., and immortalized by Shakspere in the 1st part of Henry VI. The living, a rectory, united with the rectory of Marbury, in Cheshire, is in the gift of the Countess of Bridgwater, and is one of the most valuable in the county, being worth £1458 a year nett. Here also are chapels for Unitarians, Baptists, Independents, Wesleyans, &c., and public schools and charities having an aggregate income of above £900 a year. The grammar-school, which has an income of above £450 a year, instructs a certain number of boys, free of expense. In classics and mathematics, writing and accounts being paid for separately. The master and usher are allowed £312 12s. a year. (Digest, &c. of Charity Reps.) Courts leet and baron are held in the town hall by a high steward appointed by the lord of the manor. The inhab. of Whitchurch are principally engaged in the malt and hop trade, in the manufacture of shoes, and lime and brick making. The Ellesmere canal comes up to the town. Markets on Fridays; fairs, four times a year, for farming stock, linen, and hempen and some woollen cloths. (Parl. Rep, &c.)

WHITE r., Vt. rises by one of its branches in Williamstown, in the same swamp with Steven's branch of Onion or Winooski r. It has several other branches, all of which afford water-power. It is about 55 m. long and enters Connecticut r. in Hartford, Vt. A road passes along White r., and Winooski r., from Connecticut r. to lake Champlain, with no high hills, though it passes through the Green mountains. The country affords great facilities for a railroad. A canal also might probably be practicable.

WHITE r., Ia. is composed of the East and West Forks, which can be made navigable 150 m. above the junction, which is 30 m. from its mouth. The West Fork is navigable for steamboats, in high water, as far as Indianapolis, and for flat boats much farther.

WHITE r., Ark., rises by the Buffalo fork in Washington co. Ark., and flows N.E. into Missouri, and by a broad sweep to the N. and E. it turns southwardly and again enters Ark. Its course is then S.E. to the junction with Big Black river, which proceeds from the N. in Missouri. The junction takes place on the W. border of Jackson co. It then flows S. by E., receiving Little Red r. on the W. side and Cache r. on the E. side, and enters Mississippi r. about 90 m. above the mouth of the Arkansas. Five miles from the mouth of White r. there is a bayou or gut which connects it with the Arkansas r., 25 m. from its mouth, with the current setting alternately from the one river to the other, as the flood in either may predominate. White r. has a very winding course, and is boatable 500 miles, a considerable part of the year.

WHITE, county, Tenn. Situated in the central part of the state, and contains 672 sq. m. Drained by Caney fork of Cumberland river, and its branches. It contained in 1840, 9709 neat cattle, 8970 sheep, 34,277 swine; and produced 51,062 bushels of wheat, 6691 of rye, 405,149 of Indian corn, 85,899 of oats, 15,563 of potatoes, 23,014 pounds of tobacco, 8193 of cotton, 1096 of sugar. It had eight stores, two forges, 10 flouring-mills, 21 grist-mills, 13 saw-mills, 68 tanneries, 32 distilleries, three potteries, one printing-office, one weekly newspaper. Pop. : whites, 9640; slaves, 933; free colored, 174; total, 10,747. Capital, Sparta.

WHITE, county, Ia. Situated toward the N.W. part of the state, and contains 530 sq. m. Drained by Tippecanoe river and its tributaries. It contained in 1840, 3617 neat cattle, 2254 sheep, 6071 swine; and produced 17,981 bushels of wheat, 124,782 of Indian corn, 36,528 of oats, 7849 of potatoes. It had two flouring-mills, 53 saw-mills; five schools, 95 scholars. Pop. 1832. Capital, Monticello.

WHITE, county, Ill. Situated toward the S.E. part of the state, and contains 480 sq. m. Bounded E. by Wabash river. Drained by Little Wabash river and its branches. It contained in 1840, 6912 neat cattle, 6602 sheep, 20,987 swine; and produced 38,853 bushels of wheat, 382,710 of Indian corn, 42,982 of oats, 6232 of potatoes, 68,061 pounds of tobacco, 2908 of cotton. It had one commission house in foreign trade, 20 retail stores, two flouring-mills, eight grist-mills, four saw-mills, seven tanneries, three distilleries; 14 schools, 352 scholars. Pop. 7919. Capital, Carmi.

WHITE, county, Ark. Situated a little N.E. the centre of the state, and contains 1000 sq. m. Bounded E. by White river. Drained by Little Red river. It contained in 1840, 3155 neat cattle, 367 sheep, 3927 swine; and produced 257 bushels of wheat, 37,910 of Indian corn, 1745 of oats, 3879 of potatoes. It had three stores, five grist-mills, one saw-mill. Pop.: whites, 841; slaves, 98; total, 939. Capital, Searcy.

WHITE CREEK, p. t., Washington co., N. Y., 12 m. S. Salem, 43 N.E. Albany, 413 W. Bounded W. by Owl creek, a tributary of Hoosic river. It contains four churches, a Presbyterian, two Baptist, and a Friends, eight stores, one fulling-mill, two woollen-factories, two grist-mills, four saw-mills; eight schools, 930 scholars. Pop. 2195.

WHITE DEER, t., Union co., Pa. Bounded E. by the W. branch of Susquehanna river. Drained by White Deer creek. It has two stores, two fulling-mills, two grist-mills, six saw-mills, one oil-mill, two tanneries, two distilleries; five schools, 500 scholars. Pop. 1922.

WHITEFIELD, p. t., Lincoln co., Me., 14 m. N. Wiscasset, 16 S.E. Augusta, 602 W. Drained by Sheepscot river. It has seven stores, two fulling mills, six tanneries; 16 schools, 917 scholars. Pop. 2159.

WHITEHALL, p. t., Washington co., N. Y., 21 m. N. Sandy hill, 73 N. Albany, 443 W. Bounded N. by Poultney river, lake Champlain, and South bay. At the entrance of Wood creek into lake Champlain, and the N. termination of the Champlain canal, is the village, finely situated for commerce by the lake and canal. Two large and elegant steamboats ply to St. John's, Canada, stopping at Burlington and other intermediate places. It contains two churches, a Presbyterian and Methodist, and societies of Baptists, Universalists and Roman Catholics. A considerable number of sloops, schooners and canal boats belong to the place. The town contains 27 stores, one grist-mill, two saw-mills, one tannery, one printing-office, one weekly newspaper; three academies, 164 students; 16 schools, 838 scholars. Pop. 3813.

WHITEHAVEN, a parl. bor. and seaport town of Eng-co. Cumberland, about 3. m. N.E. from St. Bees head, and 35 m. S.W. Carlisle; lat. 54° 32' 50" N., long. 3° 34' 36" W. Pop. of the parl. bor., which includes the village of Preston, to the S. of the town, and a small rural district, in 1834, 16,694. The creek, on which the town is situated, is surrounded on the land side by heights which approach close to the buildings. It is regularly laid out; the streets, which are of considerable width, cross each other at right angles, but they are, at the same time, ill-paved and dirty; and though there are many good houses and shops, a considerable proportion of the labouring population live in cellars. Among the public buildings are the three churches or chapels of St. Nicholas, St. James, and Trinity; the first erected in 1693, the second in 1752, and the third in 1715; the livings, which are perpetual curacies, worth respectively £185, £300, and £255, are in the gift of the earl of Lonsdale, on whose estate the town is built. There also are chapels or meeting-houses for Methodists, Presbyterians, Anabaptists, Roman Catholics, &c. The educational establishments comprise a marine school for 60 boys, erected on ground given by Lord Lonsdale, and endowed by a citizen of the town; a national and an infant school, &c. It has, also, a theatre, erected in 1769, a subscription library and news-room, a custom-house, market-house, an infirmary, a dispensary, cold and hot baths, &c. There are dry docks and patent slips for the building of ships, which is carried on to a considerable extent; and there are considerable manufactures of sail-cloth and cordage, copperas, tobacco-pipes, &c.; with a small soap-work, and iron and brass foundries. It has also two weekly newspapers, and two joint stock banks.

The harbour, formed by piers, constructed at different periods, dried till recently at low water; but it has been so much improved by the construction of a new pier on its S. side, projecting N., that a portion of it has now 9 feet water at low ebb, and above 20 feet at springs. Harbour lighthouses have been erected on the outer and inner pier-head. Whitehaven, which, in the beginning of the 17th century, was a miserable fishing village, is wholly indebted for its rise and importance to the working of the coal mines in its vicinity, belonging to the earl of Lonsdale. Some of these mines extend below the sea; and in the largest of them all, the William Pit, about 500 acres are excavated under the sea, the distance being about 2½ m. from the shaft, 130 fathoms deep, close to the shore, to the remotest part of the workings. There is, in this immense pit, a stable under the sea for 45 horses. (Appendix, 1., 239, to Report on Employment of Children in Mines.) A new pit is now being sunk which, when completed, will run some miles under the sea, and will be the deepest in the country. The coal, which is of excellent quality, is principally shipped coastwise for Dublin, the Isle of Man, and the S. of Scotland. Thus, in 1841, the exports of coal coastwise from Whitehaven amounted to 451,370 tons; and those to foreign parts to 23,478 tons. Exclusive of its coal, Whitehaven exports considerable quantities of lime, freestone, iron ore, &c.; and carries on a considerable trade with the W. Indies, N. America, and other foreign countries. Gross customs' revenue in 1840, £92,932. The town, however, is not flourishing, and both its trade and population have latterly declined; a consequence, most probably, of the superior facilities enjoyed by Liverpool, both as respects the trade with Ireland, and that with the West Indies, America, &c. The shipping of the port has fallen off materially during the last dozen years; still, however, it had, in 1842, 387 ships, of the burden of 62,115 tons. A piece of ground in the neighbourhood of the town, the gift of Lord Lonsdale, has been fitted up for bowls and other games.

The government of the town, and the care of the docks and harbour, is vested, under an act of Queen Anne, in 21 trustees, 14 of whom are chosen every three years by the inhabitants carrying on trade and paying harbour dues; and six are nominated by the lord of the manor (earl of Lonsdale), who, being himself added to the list, completes the number. Courts-leet are held annually, and courts-baron monthly: the latter decide in cases of 40s. and under; petty sessions are also held by the county magistrates. Markets, Tuesdays, Thursdays, and Saturdays.

The Reform Act conferred on this town, for the first time, the important privilege of sending one member to the House of Commons. Registered electors in 1839-40, 508.

Adjoining the town on the E. side is Whitehaven castle, a seat of the earl of Lonsdale. The town is in the parish of St. Bees, about 4 m. from the village of that name, in which are two valuable institutions; the one being a college where persons of limited means are prepared for the church, and the other a richly endowed foundation school, affording great advantages to the youth of the neighbourhood.

A lighthouse of the first class has been erected on St. Bees-head. It exhibits a fixed light elevated 333 feet above the level of the sea. (*Parl. Papers; Private Information.*)

WHITELEY, p. t., Greene co., Pa. Drained by Whiteley creek and branches of Dunkard's creek. It has 10 stores, one woollen factory, three grist-mills, three saw-mills, three tanneries, three distilleries; seven schools, 114 scholars. Pop. 3043.

WHITELEY county, Ky. Situated in the S.E. part of the state, and contains 800 sq. m. Watered by Cumberland river and its branches. It contained in 1840, 6174 neat cattle, 5050 sheep, 15,991 swine; and produced 4949 bushels of wheat, 191,170 of Indian corn, 35,099 of oats, 9459 of potatoes, 8297 pounds of tobacco, 3074 of cotton, 2418 of sugar. It had 10 stores, 44 grist-mills, five saw-mills, eight distilleries; two schools, 44 scholars. Pop.: whites, 4508; slaves, 146; free coloured, 19; total, 4673. Capital, Williamsburg.

WHITELEY county, Ia. Situated toward the N.E. part of the state, and contains 324 sq. m. Drained by Eel river and its branches, which afford water-power. It contained in 1840, 9207 neat cattle, 407 sheep, 3067 swine; and produced 5688 bushels of wheat, 27,135 of Indian corn, 3074 of oats, 90,091 of potatoes, 22,593 pounds of sugar. It had two stores, one grist-mill, three saw-mills; one school, 30 scholars. Pop. 1227. Capital, Columbia.

WHITE MARSH, p. t., Montgomery co., Pa., 11 m. N.W. Philadelphia, 104 E. Harrisburg, 150 W. Drained by Wissahickon creek. It contains an Episcopal church, 12 stores, one fulling-mill, two cotton-factories with 72 spindles, five flouring-mills, one grist-mill, two saw-mills; six schools, 246 scholars. Pop. 2079.

WHITE MOUNTAINS, N. H. Situated in the southern part of Coos co. These mountains extend about 20 m. from S.W. to N.E., being the more elevated parts of a range extending many miles in that direction. Their base is 8 or 10 m. broad, and they are 25 m. S.E. from Lancaster, 70 N. Concord, 82 N. by W. Portsmouth, in 42° 15' N. lat., and 71° 20' W. long. Until recently they were regarded as the highest land in the United States E. of the Rocky mountains; but it has been ascertained that there are higher peaks in Yancey co., N. C. Although distant more than 60 m. from the nearest part of the Atlantic coast, their snow-white summits are distinctly visible many leagues at sea, and along the coast of Maine. Mount Washington, the highest peak, has been found by barometrical measurement to be 6234 ft. above the level of the sea, and its base to be 1770 ft. Black mountain, in Yancey co., N. C., is 6476 ft. and therefore a little higher. Mount Washington is easily known by its being the southern of the three highest peaks. The other peaks are Adams, 5388 ft.; Jefferson 5058; Madison, 4866; Monroe, 4711; Franklin, 4356; and Pleasant, 4339 ft. Mount Washington is frequently visited by travellers, who commonly ascend by the S.E. side, commencing at the town of Conway, and following the course of Saco river, which rises high in the mountain. After climbing up the mountain for a considerable distance, the trees begin to diminish in height, till at the elevation of about 4000 ft., you come to a region of dwarfish evergreens, about the height of a man's head, putting forth numerous horizontal branches, closely interwoven with each other, and surrounding the mountain with a formidable edge. On emerging from this thicket you are above all woods, and what is called the bald part of the mountain, which consists of a huge pile of naked rocks of granite and gneiss.

After attaining its summit, the traveller is recompensed for his toil, if the sky be clear, by a prospect wonderfully grand and picturesque. Innumerable mountains, lakes, ponds, rivers, towns, and villages, meet the delighted eye, and the dim Atlantic stretches along the eastern horizon. To the N. are seen the lofty summits of Adams and Jefferson, and to the E., a little detached from the range, stands mount Madison; far to the N.E. the Katadin mountains are visible near the sources of Penobscot river; on the W. the Green mountains in Vermont; and 120 m. to the S.W. the lofty Monadnock in N. H. Among the near picturesque objects are Umbagog and Winnipiseogee lakes, Androscoggin and Connecticut rivers, and numerous smaller lakes and streams. Beneath and around, the mountains extend many miles from the summit, in every direction, resembling in their shape the waves of the sea in a storm. The elevated summits of the White mountains are covered with snow 9 or 10 months in the year. The *Notch* or Gap, on the W. side of the mountain, is a deep and narrow defile, extending two miles in length, between two huge cliffs. The entrance of the chasm is formed by two rocks standing perpendicular, at the distance of 22 ft. from each other; the one about 20 ft. high, and the other 12 ft. The road from Lancaster, N. H. to Portland, Me. passes through this notch, following the course of the head stream of the Saco. The mountain, otherwise a continued range, is here eleven down to its base, opening a passage for Saco river. The gap is so narrow, that a space has with difficulty been found for a road. Several brooks, the tributaries of the Saco, fall down the sides of the mountains, forming a succession of beautiful cascades, some of them within sight of the road, presenting the wildest and most romantic scenery. The White mountains belong to a range which extends southwardly to Belchertown, Mass., where it divides into two branches, called the Lyme range and mount Tom range.

WHITE PLAINS, p. t., semi-capital of Westchester co., N. Y., 129 m. S. Albany, 254 W. Drained by Bronx river and Mamaroneck creek. The village contains a courthouse, jail, county clerk's office, four churches, two Methodist, an Episcopal, and Presbyterian, an academy, a female seminary, three stores, a printing-office, issuing a weekly newspaper, 80 dwellings, and 650 inhabitants. It is built chiefly on one wide street. In the town are four academies, 100 students, four schools, 102 scholars. Pop. 1087.

WHITESIDES, county, Ill. Situated in the N.W. part of the state, and contains 770 sq. m. Watered by Rock river and its branches. It contained in 1840, 2882 neat cattle, 765 sheep, 6751 swine; and produced 37,206 bushels of wheat, 80,680 of Indian corn, 26,886 of oats, 33,918 of potatoes, 1020 pounds of sugar. It had 12 stores, four grist-mills, 13 saw-mills; eight schools, 194 scholars. Pop. 2514. Capital, Lyndon.

WHITESTOWN, p. t., semi-capital of Oneida co., N. Y., 96 m. W.N.W. Albany, 392 W. Drained by Mohawk river and Sanquoit creek. The village, situated at the confluence of the above streams, contains a courthouse of brick, a jail, an academy, the Oneida institute, a manual labour seminary, which has a president and several professors, and a farm of 114 acres, on which is a number of commodious buildings. The whole expense of the student, excepting clothing, is $82 60 cents per year. There are in the village three churches, a Presbyterian, Baptist, and Methodist, 200 dwellings, and about 1600 inhabitants. In the town are 24 stores, 10 fulling-mills, five woollen-factories, two cotton-factories with 15,100 spindles, two flouring-mills, two grist-mills, five saw-mills, one oil-mill, one paper-mill, two tanneries, one printing-office, one weekly newspaper; two academies, 284 students, 15 schools, 851 scholars. Pop. 5156.

WHITE SULPHUR SPRINGS, p. v., Greenbrier co., Va., 205 m. W. by N. Richmond, 242 W. This is the most celebrated watering-place of the whole state. The spring discharges 18 gallons of water per minute, at a uniform temperature of 60 degrees of Farenheit. It contains sulphate of lime, sulphate of soda, carbonate of magnesia, chloride of calcium, sulpho-hydrate of sodium, carbonate of lime, chloride of sodium, per-oxide of iron, organic matter, iodine, sulphate of magnesia, phosphate of lime, and precipitated sulphur. The gaseous contents are carbonic acid, sulphuretted hydrogen, oxygen and nitrogen. It is found to be very efficacious in dyspepsia, liver complaints, jaundice, gout, rheumatism, diseases of the skin, and various other complaints. The accommodations are elegant and extensive, sufficient for 1500 persons. From 4000 to 6000 persons visit it annually; the water is extensively exported, and retains its most valuable properties when removed from the spring. The Sharon springs in Schoharie co., N. Y., are said strongly to resemble the White Sulphur springs in Greenbrier co., Va., and are also much and increasingly frequented.

WHITHORN, a royal and parl. bor. of Scotland, the Wig town, the Burrow Head, the S.E. extremity of the co., and one of the principal headlands on the S. Scottish coast, being within the parish. Pop. of the bor., in 1841, 1513. The latter, about 4 m. N.N.W. from the Burrow Head, and 3 m. N.W. from its harbour at Isle of Whithorn, built partly on level ground and partly on a gentle acclivity, consists principally of one long and generally wide and well kept street, intersected in the middle by a rivulet concealed by a bridge the entire width of the street. The houses, which are all of stone, and mostly covered with slate, have been greatly improved of late years, many of those that were old and inferior having been pulled down, and replaced by others of a superior quality. The present church, on a rising ground, a little W. from the main street, was built in 1822. It is a substantial and commodious edifice; but is totally devoid of architectural beauty, forming, in this respect, a striking contrast to most of the churches lately erected in this part of the country. Here also are places of worship for members belonging to the Associate Synod, and the Reformed Presbyterian Synod, and for Roman Catholics. The last two, however, have no settled pastors; and the individuals attached to their communion are widely dispersed over the country. The parish school is in a very efficient state, and there are several other schools, and a subscription library. The only public building is the town-house and jail, surmounted by a tower and spire without any pretensions to elegance: the jail is, also, illcontrived and defective. Except the tanning of leather, which is carried on to some extent, the town has no manufactures: the inhabitants being either retail dealers, tailors, shoemakers, and other tradesmen, required for the accommodation of the surrounding district, or agricultural labourers. Some of them have small patches of land close to the town, for which they pay high rents. Two branch banks have, of late years, been opened in the bor. The inferior houses are mostly occupied by Irish immigrants, who are very numerous in this part of the country.

Whithorn was made a royal bor., by James IV., in 1511. It is governed, under the Municipal Reform Act, by a provost, two baillies, and 15 councillors. Corporation revenue, in 1839-40, £230 11s. 4½d. It unites with Wigtown, Stranraer, and New Gallaway, in sending one member to the House of Commons. Registered electors in the bor., in 1841-42, 54.

Whithorn, which is of high antiquity, is supposed to be identical with the Leucophibia of Ptolemy, and is certainly the Candida Casa of the venerable Bede. It was early distinguished in ecclesiastical annals, from the circumstance of a church being founded here late in the 4th or early in the 5th century, by St. Ninian, who was buried within its walls. At a later period, or in the 12th century, a magnificent priory for monks of the Premonstratensian order was erected here by Fergus, lord of Galloway, of which there still remains a fine Saxon arch, embodied in the wall of the old par. church, and some extensive vaults. The real, or supposed, relics of St. Ninian having been collected in this building, it was regarded with feelings of extraordinary veneration, and was, for a lengthened period, a place of pilgrimage and adoration. Several of the kings and queens of Scotland were among the number of its visitors. On one occasion, James IV. made a pilgrimage thither, on foot, to secure the good offices of the saint in behalf of his queen, then dangerously ill! The bishopric of Galloway, or Whithorn, was one of the oldest in Scotland; and this was a principal residence of the bishops till the abolition of the see. (*Murray's Literary Hist. of Galloway*, Append. p. 339.)

The *Isle of Whithorn*, about 2 m. N.N.E. from the Burrow Head, and 3 m. S.E. from the bor., may be regarded as the seaport of the latter. The isle, now united to the mainland by a causeway, is of very limited dimensions, not probably exceeding 30 or 40 acres. The village, which is built partly on the mainland and partly on the isle, has about 450 inhabitants. The par. church is at Whithorn, and there is not, at present, any place of worship in the village. This, however, had not been the case formerly, there being on the isle the ruins of a small church said, though probably on no good grounds, to be one of the oldest in Scotland. A little ship-building is carried on; and there is some trade in the shipping of corn and other products to Liverpool and Whitehaven, and in the importation of coal, slates, freestone, timber, &c. In the angle between the W. side of the island and the mainland is the harbour. The only danger in entering is the *screens*, a ledge of rock, overflowed at high water, bounding the harbour on the S.W. But there is a broad channel between the extremity of the screens and the isle. Small vessels may run in at any time, and anchor within the screens till the tide makes, when there is water to bring large ships to the pier. The latter, constructed about 40 years ago, by means of a grant made by the convention of royal bors., is a very considerable and useful work. At high springs there are from 18 to 20 ft. water at the pier, and at neps from 12 to 14 ft. The ground at the pier-head, and in the harbour is soft, so that vessels are but

little injured by grounding. The steamers from Liverpool and Whitehaven seldom touch at the isle of Whithorn, which is comparatively secluded; but in purity of air and water, mildness of climate, dryness of soil, cheerfulness, and salubrity, it is superior to most bathing-places on the Scottish coast.

It may probably be thought, seeing their limited pop. and importance, that this notice of Whithorn and its port has been extended to an unnecessary length. But not being of the number of those who care nothing for the place to which they belong, we may, perhaps be excused, if, towards the close of this lengthened and laborious survey of so many countries and places, we have lingered for a moment over scenes once familiar, and still well remembered. The associations which the mention of this locality calls up are all "redolent of joy and youth," and are too soothing and pleasing to be instantly dismissed.

WHITINGHAM, p. t., Windham co., Vt., 141 m. S. by W. Montpelier, 422 W. Drained by Deerfield river and its branches. It contains an abundance of limestone, and has a valuable mineral spring. It contains a church and several religious denominations, five stores, two grist-mills, nine saw-mills, two tanneries; 12 schools, 450 scholars. Pop. 1301.

WICK, a royal and parl. bor. and seaport town of Scotland, E. coast of the co. of Caithness, of which it is the cap., on the river Wick, at the bottom of a deep bay, 15 m. S. by W. Duncansby-head. Pop. of parl. bor., which includes Wick, Pulteney-town, and Louisburg, in 1841, 4599. Wick, properly so called, or the old town, is on the N. side of the river, and is irregularly and meanly built, and dirty: it has to the N. the suburb of Louisburg, and is connected by a bridge with Pulteney-town, on the S. side of the river. The latter, built on rising ground, feued from Lord Duffus by the British Fishery society, is laid out on a uniform plan, and is one of the handsomest fishing villages that is any where to be met with. Both towns are lighted with gas, supplied by a company recently established. The par. church, at the W. end of the town, with 1835 sittings, was erected, in 1830, at a cost of £4781; but a chapel of ease is now (1842) about being finished in Pulteney-town; and the United Associate Seceders, Independents, Reformed Presbyterians, Roman Catholics, &c., have their respective places of worship. The educational and literary establishments comprise a parochial school, an excellent academy, several unendowed schools, two Sunday schools, a subscription library, two reading-rooms, and a weekly newspaper. Exclusive of the churches, the town-hall, county buildings and prison, and commercial hall, are the principal public edifices.

Wick has been for upwards of half a century the principal seat of the herring fishery of Scotland; and, besides its own boats, its harbour is frequented, in the fishing season, by great numbers of boats from other parts of Scotland, and from Holland. Its port at the mouth of the Wick being small, inconvenient, and unsafe, a new harbour was commenced, in 1810, by the British Fishery society, which they completed at a cost of £14,000, of which £9,500 was defrayed by government. But the accommodation being still insufficient, a new or outer harbour, of larger dimensions, and having deeper water, was completed, in 1831, at an expense of not less than £40,000. In 1840, the town had 765 boats at the fishery, of which 498 belonged to the town, and 337 to other places. The crews of these boats amounted to 3893 individuals; and they are supposed to have given employment to about 4000 other individuals, as gutters, coopers, sailors in coasters for carrying away the fish, &c. There were cured, at Wick, during the year ending the 5th of April, 1840, 91,465 barrels of herrings, being about one sixth part of the total quantity taken and cured in Scotland during the same year, including Sunderland and Whitby. Unluckily, however, the fishery is subject to great fluctuations, above 150,000 barrels having been taken in some years, and in others not more than from 30,000 to 40,000 barrels. In consequence, the business partakes considerably of the nature of a gambling pursuit, and has not the beneficial influence that might otherwise have been expected.

The building of ships and boats, especially the latter, is carried on to a considerable extent; and the town has, besides, four rope-walks, a brewery and a distillery, an iron foundry, &c. The principal, or rather sole occupation of the females in and round the town consists in the spinning of yarn for, and in the making and mending of herring nets. The trade of the port is limited to the export of herrings, and of corn, wool, cattle, and other farm produce, and in the importation of coals, timber, groceries, &c. It has an intercourse by steam with Leith, Aberdeen, Kirkwall, and Lerwick. There belonged to the port, in 1841, 34 vessels, of the aggregate burden of 1753 tons. Gross customs revenue collected at the port, in 1840, £1141. The Commercial bank, and the Aberdeen Town and Country bank have each branches in the town.

Wick was made a royal bor. by James VI. in 1589. It unites with Kirkwall, Dornoch, Cromarty, and Dingwall, in sending one member to the House of Commons. Registered electors in Wick in 1840-41, 308. It is governed by a provost, two baillies, and nine councillors. Corporation revenue, in 1840-41, £214 4s.

The country in the vicinity of Wick has been very greatly improved within the last half century; even so late as in 1790, there was not a cart in the county, nor potatoes, turnip, nor rye grass; and such a thing as a rotation of crops had not then been heard of. The land was split into minute portions, and held under a system subversive of all industry and improvement. The advance in the interval has been astonishingly great; and though there are still an excess of small occupiers, their numbers are gradually decreasing; and all sorts of improvements have been introduced, and are extending themselves on all sides. To show the increase in the value of land, it may suffice to mention that the estate of Hemprigs, in the parish of Wick, let in 1753 for £642, was let in 1830 (exclusive of Pulteney-town) for £5608! (*New Statistical Account of Scotland,* art. *Wick; Official Returns, &c.*)

WICKFORD, p. v., North Kingston t., Washington co., R. I., 22 m. S. Providence, 401 W. Situated on a peninsula or point of land on the W. side of Narragansett bay, and has a good harbour. It has considerable trade, and a number of vessels employed in the coasting and West India trade and the fisheries. It contains three churches, a Baptist, Episcopal, and Friends, a town-house, a bank, 30 stores and 100 dwellings. The Washington academy, here located, has a substantial and elegant edifice, 60 by 30 ft., on elevated ground, a little out of the village, encircled with trees, giving it a romantic and rural appearance. It contains a valuable library.

WICKLOW, a marit. cb. of Ireland, prov. Leinster, having N. the co. of Dublin, E. St. George's Channel, S. Wexford, and W. Carlow and Kildare. Area, 494,704 acres, of which 94,000 are unimproved mountain and bog. This is a very mountainous county. In some places it is well wooded, and extremely picturesque and beautiful. Estates mostly large; the most extensive, valuable, and best cultivated belongs to Earl Fitzwilliam. Farms of various sizes; many small. Average rent of land, 12s. an acre. Speaking generally, rents are much too high; the labouring population not half employed; and their condition, and that of the small farmers, as bad as possible. But little wheat is raised, and that principally in the E. parts of the county. Wicklow has to boast of considerable mineral treasures; and some gold has been found in stream-works in different parts of the county. These, however, have been wholly abandoned, the produce of metal being insufficient to repay the expenses. From 600 to 900 persons may be employed in the copper mines of Cronebane, Ballymurtagh, Conorree, &c. The ores are shipped at Wicklow, and are smelted in Wales. Bismuth, manganese, zinc, &c., have also been met with, but in inconsiderable quantities. Marl is very abundant in parts of the county, and is said to have wonderfully increased the fertility of some districts. Principal rivers, Slaney, Fustia, and Ovoca. Wicklow contains six baronies, and 58 pars.; and returns two members to the House of Commons, both being for the county. Registered electors for the latter, in 1839-40, 2340. In 1841, Wicklow had 18,412 inhabited houses, 19,970 families, and 121,557 inhabitants, of whom 61,052 were males, and 60,505 females.

WICKLOW, a marit. town of Ireland, prov. Leinster, co. Wicklow, of which it is the cap., at the mouth of the Vartry, 25 m. S. by E. Dublin, and 24 m. W. by N. Wicklowhead, on which there are two lighthouses with fixed lights. Pop. in 1831, 2472. This, the poorest of assize towns, is irregularly built, and principally derives its means of support from the concourse of persons on county business, and for bathing during the summer months. Its public buildings comprise the par. church, a Roman Catholic chapel, meeting-houses for Methodists and Quakers, the courthouse and prison for the county, diocesan school, market-house, county infirmary, fever hospital, &c. Races are held annually on the Morrough, a flat sandy tract, extending several miles along the shore. The corporation, under a charter of James I. in 1613, consisted of a portreeve, 12 burgesses, and a commonalty, which returned two members to the Irish House of Commons till the union, when the bor. was disfranchised. The assizes for the county, general sessions and petty sessions, and a weekly court for the adjudication of small debts, are held here. The town is a constabulary and coast guard station. Markets on Saturdays: fairs March 28, May 24, August 12, and November 25. The harbour is fit only for small craft, having a bar at the entrance which has only 9 ft. at high water springs, and 6 ft. at neaps. It has, however, some trade in the shipping of copper ore, corn, &c.; the value of its exports, in 1835, being £36,566. Post-office revenue in 1830, £342, in 1838, £356.

WIDIN or VIDIN, a fortified town of Bulgaria, cap. san-

jack, on the Danube, 130 m. S.E. Belgrade. Pop. estimated at from 20,000 to 25,000, being one of the most populous towns in this part of Turkey. It presents an imposing appearance from a distance, having numerous mosques and minarets; and its streets, though equally dirty, are said to be broader than in most other Turkish towns. It is the residence of a pacha of three tails, and a Greek archbishop; its trade is principally in rock salt, corn, wine, and other products of the surrounding territory. Its fortifications are in a pretty good state of repair, and it is one of the strongest towns in Turkey. (*Bout,* ii. 336; *Dict. Geog., &c.*)

WIESBADEN, a town of W. Germany, cap. of the duchy of Nassau, and one of the most frequented watering places in the confederation; on an affluent of the Rhine, 4 m. N. from the latter, and 6 m. N.N.W. Mentz. Its resident pop. was, in 1831, under 10,000; but during the height of the season, the total number of its inhabitants has sometimes exceeded 25,000. The interior of Wiesbaden is badly laid out; but in its outskirts are many good streets and terraces. Being the seat of the government, superior courts, &c., of the duchy, it has some showy public edifices, including the ducal residence, the infantry and new artillery barracks, mint, Roman Catholic church, theatre, &c. But most of its buildings consist of hotels and lodging-houses for the accommodation of visiters; its chief prosperity and consequence being derived from its baths and mineral waters. These, the *Aquæ Mattiaci* of Pliny, are hot saline springs, containing muriate and sulphate of soda, muriate and carbonate of lime, muriate of magnesia and potassa, with some silica, oxide of iron, and free carbonic acid. There are fifteen different springs; the principal of which is the *Kochbrunnen* or "boiling spring," though its temperature is not boiling, but only about 158° Fahr. The other springs are not so hot, but all have the same general character, and are efficacious in cases of gout, rheumatism, paralysis, rigidity of the skin, &c. The usual time for drinking the waters, and taking baths, is from the middle of June to the end of August, during which all the usual attractions of a watering place are to be met with. The chief scene of gaiety is the *Kursaal,* a large edifice, with a central Ionic portico, which encloses, with its two wings, three sides of a spacious lawn, and comprises many magnificent apartments, including a noble saloon about 140 ft. in length, and 50 in height. A band of music attends here every day during the season, and the *table d'hôte* is well and respectably attended. Besides this establishment, Wiesbaden has a public library with 45,000 vols., museums of antiquities, and paintings, a well managed hospital, &c.; with manufactures of chocolate, sealing wax, and glue. The climate is very hot and oppressive in the height of the summer, and there is a deficiency of good ordinary drinking water; but the neighbourhood is very pleasant, and abounds with fine views and vineyards. Numerous Roman antiquities have been discovered in and about the town. (*Granville's Spas of Germany; Schreiber, Guide du Rhin; Chambers' Tour; Murray's Handbook for N. Germany; Berghaus, &c.*)

WIGAN, a parl. and mun. bor., market town, and par. of England, co. Lancaster, hund. W. Derby, on the Douglas, in the centre of an extensive coal-field, 17 m. N.E. by E. Liverpool, and 17 m. N.W. Manchester. The limits of the old parliamentary and municipal bor. were not affected by the Boundary and Municipal Reform acts, and are co-extensive with the township of Wigan, which has an area of 2170 acres, and had, in 1831, a pop. of 20,774, and in 1841, of 25,517. The town, situated on a hill, is spread over a large extent of ground, and though irregular, is pretty well built; its appearance, however, is not prepossessing, as the employments carried on in it, and the abundance of coal, give it a dirty, blackened look. Of late years its manufactures, buildings, and population have rapidly increased; several new streets, containing many good houses, have been erected; the approaches have been improved; and the whole town, which has been widely extended, especially towards the E., has every indication of augmenting wealth and prosperity. The charge for lighting, watching, and police, is defrayed by private subscription; and an ample supply of water is furnished by an incorporated company. The par. church is a stately old edifice, in the perpendicular style, with a handsome square tower; its interior is spacious and lofty, and it has several fine ancient monuments. Except on the S.W., it is closely surrounded by buildings. The living, one of the most valuable rectories in the county, being worth £2230 a year, is in the gift of the earl of Bradford. Besides a chapel-of-ease and eight places of worship for Protestant Dissenters, there are two large and handsome Roman Catholic chapels. The town-hall, in a small market-place, nearly in the centre of the town, a large brick edifice, was rebuilt in 1780, and has a colonnade, added in 1828; the moot or sessions-hall was rebuilt in 1829. Here, also, is a large commercial-hall, for the sale of woollen and cotton goods and hardware, erected in 1816; a small bor.

jail, a subscription library, established in 1787; a dispensary, founded in 1798, with a mechanics' institute, savings bank, &c. Towards the N. extremity of the town is a pillar erected in 1679, in memory of Sir T. Tildesly, killed in the engagement at Wigan on the 25th of August, 1650, between the royalists under the earl of Derby, and the parliamentary troops under Colonel Lilburn. The free grammar-school, founded in the reign of James I., has since received various endowments, and is now in the possession of a considerable income. It is conducted under an act passed in 1812, and affords instruction to numerous scholars in classical learning, mathematics, the modern languages, &c. Here, also, is a blue-coat school, established in 1773; a school of industry, established in 1823, for the education of girls for domestic servants and housewives; and numerous Sunday schools. The income arising from private legacies for the education and apprenticing of children, and general relief of the poor, amounts to between £350 and £450 a year.

The principal branches of industry carried on in the town comprise the carding and spinning of cotton, the weaving of muslins, calicoes, fustains, &c., by power and hand-looms; the manufacture of coarse linens; and more recently of silks. There are, also, in the parish, bleach works, brass and powder, nail and machine factories, iron foundries, &c. In 1836, there were in the town only 23 cotton factories, which employed 4202 hands; and in 1838, there were at work in the parish 37 cotton factories, which employed 6137 hands, and two flax-mills employing 400 hands. (*Baines' Lancashire; Rep. on Mills and Factories.*)

So early as 1720, an act of parliament was obtained for making the Douglas navigable from Wigan to the Ribble, at the point where the latter empties itself into the sea. This navigation was subsequently purchased by the Leeds and Liverpool canal company, and now forms an important branch of their works, conveying vast quantities of coal from Wigan to Liverpool and the N. of Lancashire. The North Union railway, a branch of the Liverpool and Manchester line, passes through the town to Preston and Lancaster.

Wigan is a bor. by prescription; its earliest extant charter dates from the 3d of Henry III.; but numerous others have been granted by subsequent sovereigns. It sent two members to the House of Commons in the 23d of Edward I.; and made another return 12 years afterwards; but from that epoch till the 16th century, a period of more than 200 years, the privilege remained dormant. At its revival the right of election was vested in the free burgesses by custom, of whom, previously to the Reform Act, there were generally about 100. Registered electors in 1839–40, 565.

Under the Mun. Reform Act, the bor. is divided into five wards, and is governed by 10 aldermen, and 18 councillors. It has a commission of the peace, two courts leet annually, petty sessions three times a week; and in 1841, it was purposed to establish a court of record for the recovery of small debts. Corp. rev. in 1840, £2584. A newspaper, called the *Wigan Gazette*, was first issued in 1836.

Wigan, anciently called Wibiggin, though not mentioned in Domesday-book, is certainly of considerable antiquity. A patent for paving the town, and erecting a bridge over the Douglas, was granted in the 7th Edward III. During the civil wars it was zealously attached to the royalists. Dr. Leland, author of a "View of Deistical Writers," and of several other publications, was a native of this town, having been born here in 1691. Market days, Monday and Friday; fairs three times a year, chiefly for cattle, pediery, and toys. (*Parl. Reps.; Baines' Lancashire, &c.*)

WIGHT (ISLE OF), the *Vectis* of the Romans, an island off the S. coast of England, opposite to, and included in, the co. of Hants, being separated from it by the road of Spithead on the E., and by the Solent, or W. channel, on the W. The E. channel, from below Gosport across to Ryde, is about 3 m. in width; but from Hurst Castle across to the island, the W. channel is little more than 1 m. in width. The figure of the island is rhomboidal, having its shores parallel to the opposite shores of the mainland. From its E. to its W. angle the distance is about 22 m.; and from the N. to the S., about 13 m. Area, 86,810 acres. Pop. of the island, in 1841, 35,431. This is one of the most beautiful districts in the kingdom, being finely diversified with hills, dales, woods, towns, villages, and gentlemen's seats. A range of chalk hills extends lengthwise across the island, affording excellent pasture for sheep, and some very commanding views. The cliffs on the S. coast are bold and precipitous, and hollowed into chasms, the resort of vast numbers of sea-fowl. The cliffs, called the Needles, on the W. angle, are lofty, almost perpendicular, and strikingly picturesque. One of the tallest of these cliffs, being undermined by the action of the waves, was overthrown in 1772, and totally submerged. Climate extremely mild, and, perhaps, the most salubrious of any in England. Soil dry, loamy, and mostly very fertile; being well adapted for all

1086

sorts of agricultural purposes. The husbandry is similar to that followed on the good soils of the mainland (*see HAMPSHIRE*); and the island has large quantities of agricultural produce to dispose of after supplying its own inhabitants. It is divided into two nearly equal parts by the river Medina. Large quantities of fine sand are shipped from Freshwater bay, for the glass and china manufactures in different parts of the country; but it has no other minerals of any importance. Since the opening of the railway from London to Southampton, the isle of Wight has been a great object of attraction, and is visited by crowds of tourists from the metropolis. A constant intercourse is kept up by steam between Southampton and Cowes; and individuals pressed for time may now leave London in the morning, and, after seeing a good deal of the island, return to town in the evening! Under the Reform Act, the isle of Wight sends one member to the House of Commons. Registered electors, in 1839–40, 1167.

WIGTON, a market town and par, of England, co. Cumberland, on the Wiza, 10 m. W.S.W. Carlisle. Area of par. 11,800 acres. Pop. of do. in 1831, 6501. The town, which is commodious and well built, consists principally of a main and transverse street, and has several superior inns and dwelling-houses. The par. church, a handsome building, was erected, instead of an older church, which had become ruinous, in 1788: it has attached to it a spacious Sunday school, built by voluntary subscription in 1830. The living, a vicarage worth £120 a year, is in the gift of the bishop of Carlisle. The grammar-school, established by subscription in 1714, but afterwards endowed, has now an income of above £70 a year: at the time of the late charity inquiry 35 children were taught on the foundation. A hospital for six widows of Protestant clergymen, founded in 1723, has an income of £58 a year; and there are some minor charities. The Wesleyans, Roman Catholics and Friends, have meeting houses at Wigton, and the Friends have, near the town, a school for 60 boys, founded in 1825. Wigton is a place of some manufacturing activity: checks, ginghams, muslins, fustains, &c., being made in the town and par., in which about 430 persons were employed as weavers in 1839. The gross average weekly earnings of the weavers at the same period was stated to be 5s. 10½d., or nett 4s. 3d. per loom. (*Handloom Rep.*, part v.) Several breweries and tanneries are established here, iron and coal are brought to the town from within a distance of 5 m. Petty sessions are held monthly, and an annual court leet and baron in September. Markets, Tuesday and Friday; the former a considerable corn mart. Large fairs for horses, cattle, Yorkshire cloth, and hardware, Feb. 20 and April 5; and one on Dec. 21, for butchers' meat, apples and honey. About 1 m. S. Wigton is Old Carlisle, probably a Roman station, of the ruins of which Wigton and church was built. Ewan Clarke, the Cumberland poet, and Sir R. Smirke, were natives of Wigton. (*Parl. Reps., &c.*)

WIGTOWN, a marit. co. of Scotland, occupying the S.W. extremity of that kingdom, and forming the part of the district known by the name of Galloway, has on the S. and W. the Irish sea, N. Ayrshire, and E. the Stewartry of Kirkcudbright or E. division of Galloway. It contains 293,760 acres, of which about a third part may be arable. Surface hilly, but the hills do not rise to any considerable height. It is divided into three districts, viz. the Moors, extending from Wigtown and Portwilliam to the Bay of Head; the Rhynns, comprising the peninsula formed by loch Ryan and the bay of Luce, terminating in the Mull of Galloway on the S., and Corsewall Point on the N.; and the Moors, or upper district. The soil of the first two districts is, for the most part, a hazelly loam, dry, and well adapted for the turnip husbandry; but near the town of Wigtown there is a considerable extent of rich alluvial land. The moors, which are bleak and barren, comprise more than a third part of the county. Climate mild but rather moist. Property has, for a long series of years, been gradually accumulating in fewer hands, and is now, for the most part, distributed in large estates held generally under entail; farms middle sized, and uniformly almost let on leases for 19 years. Agriculture in this, as in most other Scotch cos., was formerly in the most barbarous and wretched state imaginable. There was no rotation of crops; the process and implements were alike execrable; the pasture land was overstocked; and the occupiers steeped in poverty. Marl, of which Galloway contained immense quantities, began to be discovered and applied to the land about 1730; and for a while it caused an astonishing improvement in the corn crops. But their unceasing repetition reduced the soil to its former sterility, and convinced the landlords that marling, which promised so much, and by which so much had been realised, could be of no permanent utility to their estates, unless the tenants were restrained from over-cropping. In consequence, principally of this feeling, but partly also to the diffusion of intelligence to so much subjects, it was the usual practice, previously to the American war, to prohibit tenants from taking more than three white crops in succession; and it was also usual to prohibit them from breaking up pasture land until it had been at least six or nine years in grass. This practice, barbarous as it is, was a vast improvement on that by which it had been preceded; and it prevailed generally throughout Galloway and Dumfriesshire till the beginning of the present century; and in some backward parts lingers even to this day. But in all the best parts of the district two white crops are now rarely seen in succession; and every department of husbandry has been signally improved. Generally, however, the county is more suitable for pasture than for tillage; and it, as well as Kirkcudbright, suffered a good deal from over-cropping between 1809 and 1815. Oats and barley principal crops: wheat, however, is now raised in considerable quantities. Potatoes largely cultivated. Turnips have been long introduced; but it is only since 1825 that their culture has become an object of general and profitable attention; it is now rapidly extending, and large quantities of bone dust are imported as manure for the turnip lands. Farm houses and offices mostly new, substantial, and commodious. Roads new, and for the most part excellent. Breed of cattle polled, and one of the best in the empire. Breed of sheep in the low grounds, various; in the moors, principally the black faced, or Linton, variety. Average rent of land in 1810, 8s. 6d. an acre. Minerals and manufactures, quite unimportant. The condition and habits of the people have been materially improved since the commencement of the last war with France. "They are now more cleanly, more comfortably lodged, and both their diet and their dress are of a better description. They are generally, also, intelligent and well informed. Their morality has also kept pace with the progress of society, and the more extensive diffusion of knowledge." (*New Statistical Account of Scotland*, art. *Wigtown*.) Principal rivers, Cree, Bladnoch, and Luce. It is divided into 17 pars., and returns two members to the House of Commons, viz. one for the county, and one for the boro. of Wigtown, Whithorn, and Stranraer, in this county, with which the inconsiderable bor. of New Galloway, in Kirkcudbright, is associated. Reg. electors for the co., in 1841-42, 993. In 1841, Wigtown had 7440 inhabited houses, and 39,179 inhabitants, of whom 18,258 were males, and 20,921 females.

WIGTOWN, a royal and parl. bor. and seaport of Scotland, cap. of the above co., on rising ground near the mouth of the Bladnoch in Wigtown bay, 13 m. N.N.W. the Burrow head, and 37 m. W.S.W. Dumfries. The parl. bor., which includes a small village at the bridge of Bladnoch, about 1 m. S. from the town, had, in 1841. a pop. of 1870. This, which is a well situated and well built country town, has been much improved of late years. It consists principally of a main street of great width, the centre of which has been enclosed, and is now occupied with a shrubbery, bowling-green, &c. A new and handsome granite cross has, also, been erected within the last few years. The only public building is the town-house and prison, with a low spire at one end. The prison, however, is both insecure and unhealthy; and will most probably be renovated at no distant period. The church, which is old and mean-looking, is situated in a retired churchyard, in which are monuments to various individuals put to death during the persecutions under Charles II., for their adherence to the Covenant. The members of the United Associate Synod and the Relief have, also, meeting-houses: a considerable number of the Irish settlers in the town and parish are Roman Catholics. The educational establishments comprise a well conducted parish school, two unendowed schools, and two Sunday schools. A subscription library, founded in 1794, has a good collection of books. If we except a distillery, established at Bladnoch bridge, no manufacture of any kind is carried on in the bor. The harbour, on the Bladnoch, about ½ m. from the town, included in the port of Dumfries, nearly dries at low water, and the navigation is rather difficult. The only trade is in the shipping of corn, cattle, and other farm produce, constwise for Liverpool and other ports, and in the importation of coal, timber, freestone, &c. The port is visited about once a fortnight by a steamer from Liverpool. A branch bank of the British linen company is established in the town. Wigtown was made a royal bor. by James III. It unites with Stranraer, Whithorn, and New Galloway, in sending one member to the House of Commons. Registered electors, in 1840-41, 98. Under the Municipal Reform Act, it has a provost, two baillies, and 15 councillors. Corporation revenue, in 1840-41, £367 12s. 7½d., principally arising from the rent of land.

A castle and a monastery, both of which were founded in the 13th century, formerly existed here; but few or no traces of either the one or the other are now discoverable. (Art. *Wigtown* in *New Statistical Account of Scotland*; *Private Information*.)

WILBRAHAM, Hampden co., Mass., 83 m. W.S.W. Boston, 368 W. Bounded N. by Chickapee river. The Boston and Springfield railroad passes through its N. part. It contains the Wesleyan seminary, under the direction of the Methodists, founded in 1824. It is an academy of a high order, and generally has over 200 students of both sexes. It contains four churches, two Congregational and two Methodist, four stores, three fulling-mills, three grist-mills, eight saw-mills, two tanneries; one academy, 225 students; 11 schools, 304 scholars. Pop. 1864.

WILCOX, county, Ala. Situated in the S.W. part of the state, and contains 1200 sq. m. Watered by Alabama river and its tributaries. It contained in 1840, 15,800 neat cattle, 3630 sheep, 38,000 swine; and produced 9450 bushels of wheat, 1050 of rye, 650,000 of Indian corn, 256,510 of oats, 667,417 of potatoes, 28,825 pounds of rice, 1,729,030 of cotton. It had 21 stores, two flouring-mills, 16 grist-mills, 10 saw-mills, three tanneries, one printing-office, one weekly newspaper; five academies, 205 students; 14 schools, 425 scholars. Pop.: whites, 5960; slaves, 9294; free coloured, 24; total, 15,278. Capital, Barboursville.

WILKES, county, N. C. Situated in the N.W. part of the state, and contains 864 sq. m. Drained by Yadkin river and its numerous branches. It has the Blue ridge on its N.W. border, which has some very elevated summits. It contained in 1840, 13,977 neat cattle, 13,359 sheep, 40,965 swine; and produced 30,268 bushels of wheat, 16,685 of rye, 463,793 of Indian corn, 64,210 of oats, 30,920 of potatoes, 24,567 pounds of tobacco, 12,468 of cotton. It had eight stores, one flouring-mill, 100 grist-mills, 36 saw-mills, one oil-mill, four tanneries, 209 distilleries, two potteries; one academy, 30 students; one school, 20 scholars. Pop.: whites, 10,976; slaves, 1430; free coloured, 171; total, 12,577. Capital, Wilkesboro'.

WILKES, county, Ga. Situated in the E., toward the N., part of the state, and contains 550 sq. m. Bounded N. by Broad river, S. by Little river. Drained by Fishing creek and its branches. Its inhabitants were singularly united in their opposition to British oppression in the Revolutionary war, which caused the county to be denominated by the Tories "the Hornet's nest." It contained in 1840, 9391 neat cattle, 4481 sheep, 15,939 swine; and produced 38,906 bushels of wheat, 2c8,259 of Indian corn, 64,180 of oats, 6394 of potatoes, 2,345,087 pounds of cotton. It had nine stores, two flouring-mills, 15 grist-mills, four saw-mills, one distillery, two printing-offices, two weekly newspapers and one periodical; one college, 116 students; two academies, 86 students; 14 schools, 300 scholars. Pop.: whites, 3630; slaves, 6501; free coloured, 17; total, 10,148. Capital, Washington.

WILKESBARRE, p. t., capital of Luzerne co., Pa., 127 m. N.E. Harrisburg, 231 W. Wyoming mountain crosses the township centrally, in a N.E. direction, and the country between the mountain and the Susquehanna is a rich alluvian. It is drained by Solomon's and Mill creeks and Laurel run, flowing into Susquehanna river, and Bear creek and its tributaries, flowing into Lehigh river. Solomon's creek has a beautiful cascade of 30 feet fall, 3 m. from the borough. The t. abounds with anthracite coal. *Wilkesbarre borough* is situated on the E. side of Susquehanna river. It is regularly laid out, with streets crossing each other at right angles, with a square of four acres at the centre, the sides of which form an angle of 45 degrees with the street, so that the four principal streets enter the square at its corners. It contains a courthouse, jail, county offices, a bank, three churches, a Methodist, Episcopal, and Presbyterian, an academy, a seminary for young ladies, 22 stores, two furnaces, one flouring-mill, two tanneries, one brewery, one pottery, two printing-offices, one weekly newspaper; two academies, 65 students; seven schools, 330 scholars. Pop. 1718. It was laid out in 1773, under the Susquehanna Land company of Connecticut, and named in honour of Wilkes and Barre, two distinguished friends of the American cause in the British parliament. It is in the beautiful valley of Wyoming, on a plain elevated 18 or 20 feet above the ordinary level of Susquehanna river. It was burned by the British and Indians, under Colonel Butler, in the Revolutionary war. There are in the t., exclusive of the borough, one store, three flouring-mills, two saw-mills, one tannery, one distillery; one academy, 15 students; six schools, 280 scholars. Pop. 1513.

WILKINSON, county, Ga. Situated a little S. of the centre of the state, and contains 430 sq. m. Bounded E. by Oconee river, by branches of which it is drained. It contained in 1840, 6981 neat cattle, 1859 sheep, 16,047 swine; and produced 12,767 bushels of wheat, 149,643 of Indian corn, 5117 of oats, 10 828 of potatoes, 1780 pounds of rice, 1,809,612 of cotton. It had 12 stores, 16 grist-mills, 19 saw-mills, four distilleries; four academies, 159 students; 10 schools, 153 scholars. Pop.: whites, 4987; slaves, 1866; free coloured, 19; total, 6482. Capital, Irwinton.

WILKINSON, county, Miss. Situated in the S.W. corner of the state, and contains 580 sq. m. Bounded W. by Mississippi river, N. and N.W. by Homochitto river. Drained by Buffalo creek. It contained in 1840, 2219 neat cattle, 9729 sheep, 26,598 swine; and produced 473,603 bushels of Indian corn, 19,635 of oats, 99,565 of potatoes, 4555 pounds of rice, 15,250,907 of cotton. It had 32 stores, nine gristmills, four saw-mills, three tanneries, one printing-office, one weekly newspaper; six academies, 264 students; five schools, 87 scholars. Pop.: whites, 3309; slaves, 10,864; free coloured, 30; total, 14,193. Capital, Woodville.

WILL, county, Ill. Situated toward the N.E. part of the state, and contains 504 sq. m. Drained by Kankakee river and its branches, and by Desplaines river. It contained in 1840, 9278 neat cattle, 2584 sheep, 15,890 swine; and produced 110,464 bushels of wheat, 173,383 of Indian corn, 1561 of buckwheat, 5799 of barley, 271,587 of oats, 112,462 of potatoes, 8100 pounds of sugar. It had 41 stores, two flouring-mills, five grist-mills, 14 saw-mills, one tannery, one distillery, one printing-office, one weekly newspaper; three academies, 68 students; 35 schools, 1089 scholars. Pop. 10,167. Capital, Juliet.

WILLIAMS, county, O. Situated in the N.W. part of the state, and contains 600 sq. m. Watered by Maumee, Tiffin, Auglaize, and Little St. Joseph rivers. The Wabash and Erie canal passes through its S.E. corner. It contained in 1840, 3602 neat cattle, 1082 sheep, 8283 swine; and produced 29,288 bushels of wheat, 846 of rye, 86,403 of Indian corn, 94,805 of oats, 35,156 of potatoes, 97,567 pounds of sugar. It had 17 stores, 10 grist-mills, 13 saw-mills, three printing-offices, three weekly newspapers; 24 schools, 570 scholars. Pop. 4465. Capital, Defiance.

WILLIAMS, t., Northampton co., Pa. It has four stores, eight fulling-mills, one woollen factory, one grist-mill, three saw-mills; one school, 65 scholars. Pop. 1937.

WILLIAMSBURG, district, S. C. Situated toward the E. part of the state, and contains 1110 sq. m. Bounded S.W. by Santee river, N.E. by Lynches creek, and drained by its branches, by Black Mingo creek, and by Black river and its tributaries. It contained in 1840, 17,638 neat cattle, 3183 sheep, 29,035 swine; and produced 4460 bushels of wheat, 200,319 of Indian corn, 5294 of oats, 60,849 of potatoes, 95,500 pounds of rice, 515,036 of cotton. It had seven stores, nine flouring-mills, nine saw-mills; four academies 70 students; 12 schools, 164 scholars. Pop.: whites, 3387; slaves, 6908; free coloured, 32; total, 10,327. Capital, Kingstree.

WILLIAMSBURG, p. t., Hampshire co., Mass., 8 m. N.W. Northampton, 100 m. W. Boston, 386 W. Mill river affords water-power. It contains three churches, two Congregational and a Methodist, four stores, two fulling-mills, two woollen factories, one furnace, one grist-mill, seven saw-mills; one academy; nine schools, 442 scholars. Pop. 1309.

WILLIAMSBURG, p. t., King's co., N. Y., 147 m. S. Albany, 227 W. Bounded W. by East river or strait, which separates it from the N.E. part of the city of New-York, with which it is connected by three steam-ferries. It contains seven churches, a Presbyterian, Dutch Reformed, Episcopal, two Methodist, and two Roman Catholic. 23 stores, capital $65,000; two lumber-yards, capital $37,000; 25 persons manufactured hats and caps to the amount of $38,000, with a capital of $15,000; four distilleries. employing 65 persons, produced 1,050,000 gallons of distilled spirits, with a capital of $200,000; six rope-walks employed 123 persons, and produced cordage to the amount of $203,000, with a capital of $40,000; two printing-offices and two weekly newspapers employed eight persons, and a capital of $3000. The total amount of capital employed in manufactures was $243,000. There was one academy, 35 students; two schools, 107 scholars. Pop. 5094. The village is pleasantly situated at the W. end of Long Island, and contains most of the above churches, a village hall, including a jail, and has a considerable number of pleasant streets and handsome dwellings. It is closely allied to the city of New-York, where many of its inhabitants do business, and must grow with the growth of the city. It was incorporated in 1827, and its powers were enlarged in 1835.

WILLIAMSBURG, p. v., capital of James City co., Va., 58 m. E.S.E. Richmond, 175 W. Situated on elevated ground, between James and York rivers. It is the oldest incorporated town in Virginia, first settled in 1632, and in 1698 became the seat of the colonial government. It is regularly laid out, with streets intersecting each other at right angles, and contains a courthouse, jail, and county offices, and a public square; three churches, an Episcopal, Methodist, and Baptist, a lunatic asylum, a number of stores, various mechanic shops, 200 dwellings, some of which bear the marks of decay, and about 1600 inhabitants. It is the seat of William and Mary college, founded in 1693, has a president and three professors or other instructers, 96 students, and 5000 volumes in its libraries. The commencement is on July 4th. It is under the direction of the Episcopalians.

WILLIAMSBURG, p. t., Clermont co., O., 29 m. E. Cincinnati, 105 m. S.W. Columbus, 472 W. Watered by the E branch of Little Miami river. The village is 14 m. from Ohio river, and was formerly the capital of the county The township has six stores, one grist-mill, five saw-mills three schools, 85 scholars. Pop. 1458.

WILLIAMSON, county, Tenn. Situated near the centre of the state, and contains 476 sq. m. Drained by Harpeth river and its branches, flowing into Tennessee river. It contained in 1840, 23,417 neat cattle, 32,123 sheep, 54,919 swine; and produced 145,062 bushels of wheat, 22,871 of rye, 942,035 of Indian corn. 1225 of barley, 184,496 of oats, 30,390 of potatoes, 1,198,989 pounds of tobacco, 1,218,176 of cotton. It had 25 stores, one furnace, two forges, one cotton factory with 1052 spindles, 23 grist-mills, 11 saw-mills, nine tanneries, nine distilleries, one printing-office, one weekly newspaper; four academies, 143 students; 35 schools, 870 scholars. Pop.: whites, 15,641; slaves, 11,251, free coloured, 114; total, 27,006. Capital, Franklin.

WILLIAMSON, county, Ill. Situated in the S. part of the state, and contains 432 sq. m. Drained by branches of Muddy creek, which flows into Mississippi river, and of Saline creek, which flows into Ohio river. It contained in 1840, 4736 neat cattle, 3198 sheep, 12,902 swine; and produced 11,932 bushels of wheat, 172,890 of Indian corn, 29,379 of oats, 3881 of potatoes, 115,419 pounds of tobacco, 45,423 of cotton. It had seven stores, seven grist-mills, two saw-mills, two tanneries, two distilleries two printing-offices, two weekly newspapers; 10 schools, 365 scholars. Pop. 4457. Capital, Bainbridge.

WILLIAMSON, p. t., Wayne co., N. Y., 200 m. W. by N. Albany; 364 W. Bounded N. by lake Ontario, into which its streams flow. It contains four churches, two Presbyterian, a Methodist, and a Baptist; four stores, one grist-mill, nine saw-mills, two tanneries; 11 schools, 373 scholars. Pop. 2147.

WILLIAMSPORT, p. b., cap. of Lycoming co., Pa., 120 m. N.W. Philadelphia, 93 m. N. Harrisburg, 203 W. Situated on the N. bank of the W. branch of Susquehanna river, and contains a courthouse of brick, having a yard enclosed with an iron railing, and planted with trees; a jail of stone; five churches, two Presbyterian. an Episcopal a Methodist, and a German Reformed; a large academy of brick, 15 stores, three tanneries, two printing-offices, two weekly newspapers; two academies, 34 students; three schools, 150 scholars. It has a number of handsome dwellings and 1253 inhabitants. The Pennsylvania canal passes through the place. A railroad extends from this place 73½ m. to Elmira, New-York.

WILLIAMSPORT, p. v., Washington co., Md., 167 m. S.W. Annapolis, 80 W. Situated on the E. side of Potomac river, at the mouth of Conecocheaque creek, and contains about 500 inhabitants. The Chesapeake and Ohio canal passes through it.

WILLIAMSTOWN, p. t., Orange co., Vt. 11 m. S.E. Montpelier, 510 W. It occupies the height of land between lake Champlain and Connecticut river; a stream from the mountains naturally divides, a part flowing to White river, and a part to Onion or Winooski river. The Gulf road passes through a chasm in the mountains, and constitutes a remarkable passage through the Green mountain range. Chartered in 1781. It contains five churches, a Congregational, Methodist, Baptist, Free-will Baptist, and Universalist; two stores, one fulling-mill, one grist-mill, six saw-mills, two tanneries; 17 schools, 562 scholars. Pop. 1682.

WILLIAMSTOWN, p. t., Berkshire co., Mass., 99 m. N. Pittsfield, 131 m. W. by N. Boston, 383 W. Watered by Hoosic and Green rivers, which afford water-power. The village is situated on uneven ground, the main street passing over several ridges, and contains a Congregational church, several stores, an academy, the buildings of Williams college, and about 50 dwellings, some of them neat and handsome. The township contains also a Baptist church, and one common to Baptists and Congregationalists; seven stores, one fulling-mill, two cotton factories with 1788 spindles, two grist-mills, five saw-mills, two tanneries; one college, 130 students; one academy, 25 students; 13 schools, 340 scholars. Pop. 2153. Williams college, located here, was founded in 1793; has now (1844) a president and seven professors or other instructers, 982 alumni, 331 of whom have been ministers of the gospel, 146 students, and 7500 volumes in its libraries. The commencement is on the third Wednesday of August.

WILLINGTON, p. t., Tolland co., Ct., 28 m. W. Hartford, 364 W. Bounded W. by Willimantic river, by branches of which it is drained, affording water-power. It contains iron ore, and has two churches, a Congregational and Baptist; three stores, two grist-mills, two saw-mills; 10 schools, 437 scholars. Pop. 1288.

WILLISTON, p. t., Chittenden co., Vt., 32 m. N.N.W. Montpelier, 518 W. Chartered in 1763. First settled in 1774 by Thomas Chittenden, one of the fathers of Vermont, and the first governor of the state, who died here in 1797.

WILLOUGHBY.

It contains a handsome Congregational church of brick, 40 by 60 ft.; a town-house, 30 by 42 ft.; an academy, 26 by 36 ft., two stories high; four stores, six saw-mills, two tanneries: 10 schools, 350 scholars. Pop. 1554.

WILLISTON, t., Chester co., Pa., 20 m. N.W. Philadelphia. Drained by Ridley and Crum creeks. It has three stores, one flouring-mill, one grist-mill, four saw-mills; five schools, 206 scholars. Pop. 1460.

WILLOUGHBY, p. t., Lake co., O., 164 m. N.N.E. Columbus, 358 W. Watered by Chagrine river. It contains two churches, nine stores, one flouring-mill, two tanneries, and Willoughby university, the medical department of which only is in operation. It has five professors, 57 graduates, and 40 students. The lectures commence on the last Monday in October annually. It has been in contemplation to remove it to Cleveland. The township contains 13 schools, 800 scholars. Pop. 1943.

WILLSBOROUGH, p. t., Essex co., N. Y., 15 m. N.E. Elizabeth, 145 m. N. Albany, 516 W. Bounded E. by lake Champlain. Drained by Boquet river, flowing into the lake. It contains a Presbyterian church, one furnace, four forges, one fulling mill, one grist-mill, eight saw-mills; 11 schools, 424 scholars. Pop. 1658.

WILLS, t., Guernsey co., O. Drained by branches of Wills creek. It contains the villages of Elizabethtown and Washington on the National road, and has one college, three schools, 125 scholars. Pop. 1916.

WILMINGTON, p. t., Windham co., Vt., 17 m. E. Bennington, 124 m. S. by W. Montpelier, 429 W. Branches of Deerfield river afford water-power. Chartered in 1763. It contains two churches, a Congregational and Baptist; four stores, one fulling-mill, two grist mills, eight saw-mills; 13 schools, 436 scholars. Pop. 1295.

WILMINGTON, city, port of entry, and semi-cap. of New-castle co., Del. Situated between Brandywine river and Christiana creek, 1 m. above their junction, and 2 m. from Delaware river. 47 m. N. Dover, 28 m. S.W. Philadelphia, 70 m. N.E. Baltimore, 110 W. It is in 39° 41' N. lat., and 75° 28' W. long. Pop. in 1836, 6628; in 1840, 8367. It is the largest place in the state, and is considerably engaged in foreign commerce and the whale-fishery. Built on ground gradually rising to the height of 112 ft. above tidewater, and is regularly laid out with broad streets, crossing each other at right angles. The houses are generally built of brick, and some of them are elegant. It contains a city hall, two market-houses, three banks, an almshouse, an arsenal, a Friends' boarding-school for young ladies, a public library, and 16 churches, three Episcopal, three Presbyterian, three Methodist, two Friends', a Baptist, Roman Catholic, and some others. It has a daily communication with Baltimore and Philadelphia by railroad. Christiana creek is navigable to this place for vessels requiring 14 ft. of water. On Brandywine river are some of the finest flouring-mills in the United States, to which vessels can come requiring 8 ft. of water. It had in 1840, 95 stores, capital $344,850; three lumber-yards, capital $60,000; capital employed in the fisheries, $170,000; machinery was produced to the amount of $252,500; it had one cotton-factory with 1140 spindles, six flouring-mills, one grist-mill, two tanneries, three breweries, two potteries, one rope-walk, three printing-offices, two binderies, three weekly and three semi-weekly newspapers, and one periodical; total capital employed in manufactures, $459,600; nine academies, 437 students; 14 schools, 991 scholars. In the vicinity are many mills and manufactories besides the above. Tonnage of the port in 1842, 7068 tons.

WILMINGTON, p. v., port of New-Hanover co., N. C., 90 m. S.E. Fayetteville, 148 m. S.S.E. Raleigh, 365 W. Situated on the E. side of Cape Fear river, just below the entrance of the N.E. branch, 35 m. from the ocean. It is the most populous town in the state, but the situation is unhealthy. It contains a courthouse, jail, two banks, an academy, two churches, a Presbyterian and Episcopal; numerous stores, about 600 dwellings, and 4744 inhabitants. The harbour admits vessels of 300 tons, but there is a dangerous shoal at its entrance. Opposite the town are two islands, dividing the river into three channels. The islands afford the finest rice fields in the state. The tonnage in 1840 was 10,193. In 1819, 300 buildings were destroyed by fire, at a loss of $1,000,000.

WILMINGTON, p. v., Union t., cap. of Clinton co., O., 72 m. S W. Columbus, 445 W. Situated on Todd's fork of Little Miami river. It contains a brick courthouse. 40 ft. square; a jail, four churches, a Methodist, Baptist, Presbyterian, and Friends'; 16 stores, numerous mechanic shops; a printing-office, issuing a weekly newspaper; 125 dwellings, and about 750 inhabitants. It was laid out in 1810.

WILMOT, p. t., Merrimac co., N. H., 27 m. N.W. Concord, 302 W. Drained by small tributaries of Blackwater river. It contains two churches, a Methodist and Free-will Baptist; one store, one fulling-mill, two saw mills; nine schools, 350 scholars. Pop. 1212.

WILSON.

WILNA, or VILNA, a government of European Russia, comprising a large proportion of the ancient Lithuania and Samogitia; principally between the 54th and 56th degrees of N. lat., and the 21st and 27th of E. long.; having N. Courland, E. Minsk, S. Grodno, and S.W. Poland and Prussia. Area, estimated at 24,600 sq. m. Pop., in 1818, 1,513,800. It is a vast plain; there being only, in different parts, a few sandhills, reaching sometimes to the height of 208 ft., and abounding with fossil, shells, &c. Principal rivers, the Wilna, a tributary of the Niemen, and the Wilna, which forms its S.W. boundary. Lakes are numerous, particularly in the E. and N.E. The soil is partly sandy, and partly marshy; but in many places it consists of a fertile alluvial deposit. The climate, though severe, is not so cold as in some of the adjacent governments; the mean temp. of the year is about 45° Fahr. Agriculture is almost the sole occupation of the inhabs.; and rather more corn is grown than is required for home consumption. Rye is the grain principally cultivated. Hemp and flax are rarely grown; and hops and pulse are raised in gardens: fruits are neglected. The forests are very extensive, the crown possessing above 400,000 deciatines of forest land: and there is a considerable trade in deals, timber, tar, potash, and other woodland products. Lime trees are very abundant; and to this cause is attributed the excellence of the honey, for which this government is famous.

The breeding of stock is neglected; the horses are, however, strong and active, though of small size. Game is very plentiful: elks, wild boars, bears, wolves, &c., are numerous; occasionally the urus is met with; and fox, martin, and squirrel skins are articles of trade. Mineral products unimportant. Manufactures have increased a little of late; but they are still quite inconsiderable. Dr. Granville says of Shavel, a town of 2000 inhabitants in this government. "It consists of a long street of low gable-roofed huts of wood plastered over, and presenting a general appearance of the most squalid misery. This may be considered as a fair specimen of the second-rate towns in the government of Wilna, and indeed all over Russia and Poland." (Travels, ii. 516.) The trade, which is almost entirely in the hands of the Jews, is principally in timber and agricultural produce, sent down the Dwina to Riga, or by land into Prussia. Wilna is divided into 11 districts; chief towns Wilna, the capital, and Kovno. It is not subject to the government monopoly of ardent spirits; and preserves several of its old forms of administration. As respects education it is, though far behind, in advance of many of the governments.

WILNA, a town of the Russian empire, cap. of the above gov. and formerly the cap. of Lithuania, at the confluence of the Wilenka and Wilna, 90 m. N.E. Grodno. Pop., in 1834, 35,697, of whom 20,000 were Jews, they being more numerous here than in any other town of Polish Russia. It is surrounded by undulating hills, and enclosed by a wall. Its streets are narrow and crooked, and its houses mostly of timber, though it has several hundred dwellings built of brick or stone. Formerly a royal castle of the Jagellons existed here, but nothing is left of it except its ruins. The cathedral, founded in 1387, has some good paintings, and many chapels, one of which, appropriated to St. Casimir, and built wholly of marble, is very handsome. The body of the saint is preserved here in a silver coffin, made by order of Sigismund III., and weighing, it is said, 30 cwt! The church of St. John is surrounded by the buildings of the university, founded in 1578, and suppressed in 1832. Here are in all about 40 churches, numerous convents, a mosque, and 4 synagogues, a magnificent town-hall, an arsenal, exchange, theatre, two hospitals, barracks, magazines, &c. The governor's palace and some residences of the nobility are fine buildings. Previously to its dissolution, the university of Wilna was in a flourishing state, and possessed an observatory, collections in mineralogy and anatomy, and a library of 52,000 volumes. A medico-chirurgical school, to which are attached the botanic garden and some of the university collections, an ecclesiastical seminary, and two gymnasia, are the principal public schools: most part of the university establishment has been removed to Kief. Wilna has deaf and dumb and founding asylums, various other charitable institutions, a few manufactures, and a considerable trade. It was founded in 1322, and is reported to have had, in the middle of the 16th century, 100.000 inhabitants, though this, no doubt, is a gross exaggeration. It has often suffered severely from fire. (Schnitzler; Possart; Kaiserth, Russland.)

WILNA, p. t., Jefferson co., N. Y., 157 m. N.W Albany, 302 W. Drained by Indian r. Bounded S W. by Black river, which here has the long falls, above which the river is navigable for steamboats. 40 miles to the high falls. It contains two communication houses in foreign trade, six retail stores, one furnace, one forge, two fulling-mills, one grist-mill, six saw-mills; 20 schools, 810 scholars. Pop. 2391.

WILSON, county, Tenn. Situated a little N. of the centre of the state and contains 625 sq. m. Bounded N. by

1071

Cumberland river, by small tributaries of which it is drained. It contained in 1840, 20,109 neat cattle, 19,309 sheep, 48,115 swine; and produced 106,413 bushels of wheat, 13,554 of rye, 2,601,652 of Indian corn, 164,334 of oats, 153,990 of potatoes, 2,313,000 pounds of tobacco, 295,373 of cotton. It had 19 stores, six woollen factories, two cotton factories, with 1556 spindles, 94 grist-mills, 47 saw-mills, 81 tanneries, 23 distilleries, one printing-office, one weekly newspaper; 36 schools, 887 scholars. Pop.: whites, 169,003; slaves, 5088; free coloured, 969; total, 24,460. Capital Lebanon.

WILSON, p. t., Niagara co., N. Y., 10 m. N.W. Lockport, 290 m. W. by N. Albany, 420 W. Drained by Howell's and Tuscarora creeks, flowing N. into lake Ontario, which bounds it on the N. It has one store, one grist-mill, three saw-mills; 12 schools, 476 scholars. Pop. 1783.

WILTON, a parl. and munic. bor., and par. of England, co. Wilts, hund. Branch-and-Dole; on the Willy, a tributary stream of the Avon, 4 m. W. by N. Salisbury. The old borough comprised only the greater portion of the town; but the modern borough includes, besides the whole parish of Wilton, 11 adjacent parishes, and parts of five others, with an extra-parochial district; it has a total area of about 38,000 acres, and had, in 1831, a population of 8315.

It is a neat country town; the main street is partially paved and lighted, and is crossed by a smaller street nearly in its centre. The public buildings, the principal of which are the church and town-hall, are not remarkable, nor do either these or the private houses appear to be increasing. (Bound. Rep, &c.)

The church is in a mixed style of architecture; the living, a rectory annexed to the livings of Netherhampton, Ditchhampton, and Builbridge, worth £450 a year, is in the gift of the earl of Pembroke. It has, also, chapels for Independents, Wesleyans, an endowed free-school, established early in the 18th century, for the education and apprenticeship of 20 boys; a bequest of £1000, the interest of which is annually distributed in marriage portions to four young women belonging to the town, and several minor charities.

The hospital of St. John is the only one remaining of the numerous monastic establishments formerly existing here. The hospital itself is an old priory building, consisting of four distinct apartments under one roof, with a garden for the use of the inmates, who consist of two brethren and two sisters, presided over by a prior nominated by the dean of Salisbury cathedral. The rents reserved for the maintenance of the inmates amount to about £30 a year. (Charity Commissioners' Rep.)

Wilton had, for a lengthened period, a flourishing manufacture of woollen goods, especially of carpets, and it was here, indeed, that the first carpet made in England was manufactured. This business, however, gradually declined; and though it has somewhat revived of late years, it is still but inconsiderable.

The earliest existing charter of the borough dates from the 1st of Henry I., but from a very early period it has been governed by a mayor and an unlimited number of burgesses, including a recorder and five aldermen. This borough sent two members to the House of Commons from the 23d of Edward I. down to the passing of the Reform Act, which deprived it of one of its members. Previously to the last-mentioned act the franchise was vested " in the mayor and burgesses, who are to do all corporate acts and receive the sacrament." In point of fact, however, it was a nomination borough belonging to the earl of Pembroke. Registered electors, in 1839-40, 205. The only court that ever appears to have been held in the borough, except the court leet is a court of record, which has not been opened since 1781. There is a small prison, now used only as a lock-up-house.

Wilton is very ancient. It had a famous abbey, originally founded in 773, and greatly improved and enlarged after the conquest. Wilton-house, the magnificent seat of the earls of Pembroke, occupies the site of this abbey. It is built in a fine park, watered by the Willy; its garden front was rebuilt from designs by Inigo Jones, and more recently it was enlarged and considerably altered by Wyatt, especially with a view to the better display of its superb collection of ancient statues and other works of art, paintings, &c. Wilton was most probably the birth-place of the dramatic poet Massinger; and archdeacon Coxe, author of " Travels in Switzerland and the north of Europe," and of various valuable historical works, was, for a lengthened period, rector of Bemerton, in its immediate vicinity. The town has no market, but four annual fairs, that on the 19th of September being one of the largest sheep fairs in the W. of England. (Bound. and Munic. Reps.; Beauties of Wiltshire, &c.)

WILTON, p. t., Franklin co., Me., 38 m. W.N.W. Augusta, 475 W. Incorporated in 1803. Drained by branches of Sandy river, issuing from two ponds. It has seven stores, three fulling mills, one woollen factory, one furnace, three grist mills, four saw-mills, three tanneries; 14 schools, 785 scholars. Pop. 2198.

WILTON. p. t., Hillsborough co., N. H., 40 m. S. by W. Concord, 454 W. Drained by Souhegan river and its branches. It contains three churches, a Congregational Baptist and a Universalist, two stores, two fulling mills, five grist-mills, five saw-mills, two tanneries; nine schools, 330 scholars. Pop. 1033.

WILTON, p. t., Fairfield co., Ct., 74 m. S. W. Hartford, 275 W. Incorporated in 1802. Drained by Norwalk river, which affords water-power. It contains two churches, a Congregational and an Episcopal, seven stores, two grist-mills, two saw-mills; two academies, 63 students, 11 schools, 631 scholars. Pop. 2053.

WILTON, p. t., Saratoga co., N. Y., 12 m. N. by E. Ballston Spa, 43 m. N. Albany, 413 W. Drained by a tributary of Hudson r. It contains a Baptist church, two stores, one fulling-mill, one grist-mill, six saw-mills, two tanneries, seven schools, 289 scholars. Pop. 1438.

WILTSHIRE, an inland co. of England, in the S. part of the kingdom, having N. the co. of Gloucester, E. Berks and Hants, S. the latter and Dorset, and W. Somerset and Gloucester. Area, 874,880 acres, of which about 808,000 are arable, meadow, and pasture. It is divided by the rivers Kennet and Avon, and the canal by which they are united, into two grand divisions, popularly termed, from their situation, North and South Wiltshire. The latter consists, in great part, of Salisbury plain, extending from Westbury and Warminster, on the W., across the county to Hampshire, and from Lavington on the N. to near the city of Salisbury on the S. It consists principally of chalky downs, intermixed however with some fertile, well watered, and beautiful valleys. Though called a plain, the surface, as in all chalk land, is undulating; the most level part lies round Stonehenge. There is a good deal of rich land in the S. division, between Trowbridge and Pewsey, and between the Willy and the Dorsetshire border, E. to Wilton and Salisbury. Marlborough Downs, which bear in most respects a close resemblance to Salisbury plain, lie in the N. division of the county, between Marlborough and Swindon; but, with this exception, this division consists principally of rich vale land considerably exceeding in extent and importance the cultivated grounds of the S. division. There are some large estates; but property is, notwithstanding, a good deal subdivided. Farms of all sizes; and generally let on leases for 7, 14, and 21 years, with unobjectionable conditions as to entry. Farmhouses, in the S. division, were formerly built together for the convenience of water, but the more modern ones are generally detached. Agriculture in Wilts. is in an advanced state, and reflects great credit on the skill and enterprise of the farmers. The land under the plough is remarkably clean and in good order. It is believed, however, that tillage on the down lands has been carried to a much greater extent. When once broken up, it is extremely difficult to get them again into good condition as pasture; while, unless corn fetches a pretty high price, they are most productive in the latter. Principal corn crops, wheat and barley, the latter being, however, confined, in a great measure, to the light chalky soils. Turnips, rape or cole seed, and potatoes, largely cultivated. Much of the rich land in the S. division is appropriated to the dairy husbandry and increase of cattle. The cheese, which, excepting the butter made from the whey, is the only product of the dairies, was formerly sold as Gloucestershire cheese; but is now well known and much valued in London and elsewhere in its own proper name of North Wiltshire cheese. Breed of cattle various; they are partly slaughtered in Bath, Salisbury, &c., but the greater number are sold to the London butchers. In despite, however, of the encroachments made by the plough on the downs, sheep continue to be regarded, in the greater part of the county, as the principal support of the farms. They afford the chief article of manure used on the land, while the sale of lambs and wool furnishes the principal means of paying the rent. In consequence, as it would seem of this dependence, and of the high price of wool during the last 10 or 12 years, there have been fewer complaints among the Wiltshire farmers than among those of most southern counties. The sheep stock, consisting partly of the native horned breed, but in a far greater degree of South Downs, and crosses between the two, is estimated at about 700,000; of which about 585,000 are depastured on the downs, and the rest on the cultivated land; the fleeces of the former are supposed to weigh at an average of 3 lbs., and those of the latter 4 lbs.; producing together about 430 packs of wool. The irrigation of water meadows is to be seen in the greatest perfection in S. Wiltshire, and is practised on a large scale. Many hogs are kept; and Wiltshire bacon is highly esteemed.

In the vicinity of some of the towns of S. Wilts. a good deal of garden husbandry is carried on. Average rent of land, in 1810, 18s. 6½d. an acre. Stonehenge stands in rude magnificence, in the middle of Salisbury plain. (See STONEHENGE.) The manufactures of Wiltshire are considerable; they consist principally of various descriptions of superfine woollen goods, made at Bradford, Trowbridge, Westbury, &c.; thicksets, and other sorts of cotton goods, are also prepared, though in small quantities. Wilton was long celebrated for a carpet manufactory, established by one of the earls of Pembroke; but this, though it has latterly increased, is not nearly so considerable as formerly. Speaking generally, manufactures of all sorts are here, as in other southern counties, on the decline. Principal rivers, Thames, Upper and Lower Avon, and Kennet. Exclusive of some local jurisdiction, Wilts contains 28 hundreds, and 300 parishes. It returns 18 members to the House of Commons, viz., four for the county; two for the city of Salisbury; two each for the boroughs of Chippenham, Cricklade, Devizes, and Marlborough; and one each for Calne, Malmsbury, Westbury, and Wilton. Registered electors for the county, in 1839–40, 8172, whereof 5259 were for the N., and 2913 for the S. division of the county. In 1841, Wilts had 50,966 inhabited houses, and 260,007 inhabitants, of whom 128,904 were males, 131,103 females. Sum paid for the relief of the poor in 1838–39, £120,595. Annual value of real property assessed to the poor in 1840–41, £1,175,616.

WIMBORNE MINSTER, a market town and par. of England, co. Dorset, hund. Bradbury; in a valley between the rivers Stour and Allen, each of which is here crossed by a bridge, 20 m. S.S.W. Salisbury. Area of par. 11,880 acres: pop. of ditto. in 1831, 4009. The town is pleasantly situated; but the streets, though clean, are irregular, and the houses have but little uniformity: it is well supplied with water, and has been considerably improved of late years. It is principally remarkable for its magnificent minster or church, which was formerly collegiate. The date of its original foundation is uncertain; but it has been usually referred to the 8th century, when a monastery was established here by a sister of Ina, king of the W. Saxons. Dr. Stukely, however, and some other antiquaries are of opinion, that the E. tower, and most part of the church, are posterior to, but soon after the conquest. It is a large cruciform structure, 180 ft. in length from E. to W., with two towers: one, a short, massive, Norman tower, rising from the middle of the roof, formerly surmounted by a lofty spire destroyed by lightning early in the 17th century: the other tower, in the perpendicular style, at the W. end of the building, has a fine window, which has, however, been closed up. The interior is divided after the manner of a cathedral; and till within the last few years the cathedral service was performed here. In the chancel are 16 stalls, with canopies of carved oak. It has some monuments of distinguished personages; but time, and the hand of violence appear to have destroyed a great many more. This edifice underwent extensive repairs and improvements from 1836 to 1840, at the joint expense of Mr. Banks, the earl of Devon, and the duke of Beaufort. (Rickman's Gothic Architecture.) The living is a rectory in the patronage of the earl of Shaftesbury. Here also are chapels for Independents, Wesleyans, and Baptists. The free grammar-school, originally founded in 1497, and re-established by Queen Elizabeth, has an income of about £100 a year. Its freedom is unlimited, but in 1836 there were only 25 pupils on the foundation. Another free-school, founded by the endowment of a private individual in 1695, has an income of £20 a year, and in 1836 was attended by 16 pupils. St. Margaret's almshouse, which is of very ancient foundation, has an income of about £120 a year, and the aggregate produce of the funds for charitable and religious uses in that quarter amounted, at the period of the late inquiry by the charity commissioners, to nearly £1000 a year. The trade of Wimborne is limited to that arising from a small manufacture of woollen goods, and stocking-knitting. Petty sessions are held here for the Wimborne division of the hundred, and an annual court at Michaelmas in a tything of the town, at which two bailiffs are appointed by 12 jurymen; but these have no authority over other parts of the town. Wimborne is supposed, from the various coins and antiquities found there, to have been a Roman station. Market day, Friday; fairs frequently, for cattle and cheese. (Beauties of England and Wales, art. Dorset, &c.)

WINCANTON, a market town and par. of England, co. Somerset, hund. Norton Ferris; on a declivity beside the small river Cale, here crossed by a stone bridge, 23 m. S. Bath. Area of par. 3890 acres. Pop. of do., 1831, 2193. Wincanton, having been destroyed by fire in 1747, has been since regularly laid out in four principal streets. The church, a spacious edifice, partially rebuilt in 1748, has a square embattled tower. The living, a perpetual curacy worth £123 a year, is in the gift of U. and G. Messiter. Esqs. It has, also, a chapel for Independents, a neat market-house, and several small charitable institutions. The manufacture of serges, bed ticking, and dowlas, though much fallen off, is still carried on, and the silk manufacture has been introduced on a small scale. Wincanton is an important mart for the cheese made in the surrounding country. The

town, divided into a borough and tything, is under the jurisdiction of separate officers; two constables for the former, and a tything-man for the latter, being chosen annually at the manorial court: besides which last, an annual court-leet is held here. Market day, Wednesday; fairs, twice a year. Wincanton is a place of remote antiquity, and is mentioned in Domesday Book.

WINCHCOMBE, a market town and par. of England, co. Gloucester, hund. Kiftsgate, amidst the Cotswold hills, 6 m. N.E. Cheltenham. Area of par., 5700 acres. Pop. of do., in 1831, 2514. The town consists mostly of three streets: the houses are in general of stone. The church, the erection of which commenced in the reign of Henry VI., is a noble Gothic structure, with a lofty square tower embattled and pinnacled; the nave is separated from the aisle by octagonal pillars and compressed arches, and from the chancel by a screen of carved oak. The living, a vicarage, worth £134 a year, is in the gift of — Tracy, Esqr. The free-school, founded in the 19th of James I., has a yearly income of £49 4s., but has long declined from the condition of a grammar school; and at the last inquiry, it afforded only rudimentary instruction to 34 boys. An endowment of £20 a year supplies clothing to the poor. (Charity Reps.) The inhabitants are principally occupied in the manufacture of silk goods, paper, leather, cotton stockings, &c.: in 1838, one silk mill in the parish employed 76 hands. (Mills and Factories' Rep.) Winchcombe is of great antiquity, and before the time of Canute formed a county of itself, being then surrounded with walls, and having a famous abbey, founded during the Heptarchy, but of which, as of its ancient castle, there are now few or no traces. The town was made a borough in the time of Edward the Confessor, but its charter has long been obsolete. About 1½ m. distant are the remains of Sudeley castle, now the property of the duke of Buckingham. (Parslms. Reps. &c.)

WINCHELSEA, a bor. cinque port, and market town of England, co. Sussex, E. div. hund. Stapie, on the small river Breed, about 1½ m. from the English channel, 2 m. S. by W. Rye. Area of par. 1190 acres. Pop. in 1831, 772. Old Winchelsea, a town of importance under the Romans, situated on the coast at the mouth of the Rother, was destroyed by the encroachments of the sea, between 1280 and 1287. Before its destruction was completed, the inhabitants removed to New Winchelsea, as it was called, on a slight eminence about 2 m. S.W. from the ruined town. The new town, which was surrounded with walls by Edward I., covered a space of about 2 m. in circumference, had three churches, was regularly laid out, and enjoyed a considerable share of the trade with France, especially of that with Bordeaux. But, by a singular fatality, it was ruined by a cause precisely the opposite of that which had destroyed the old borough. Instead of encroaching, the sea began in the 16th century to recede from this part of the coast, leaving, in the end, the town without a harbour, 1½ m. from the sea, and in part surrounded with a salt marsh! since this change was effected, it has progressively declined, and would most probably have been wholly deserted, but for the circumstance of its having enjoyed, from the 49d of Edward III. down to the passing of the Reform Act, by which it was disfranchised, the important privilege of sending two members to the House of Commons. It is now included in the borough of Rye. Of its three churches, only a portion of that of St. Thomas now remains. It has several old monuments, one of which is said by Rickman to be hardly exceeded by any in the kingdom for beauty of composition. The living, a rectory in private patronage, is worth £878 a year. The courthouse and jail underneath are of Saxon architecture. About a mile N.E. of the town are the ruins of Winchelsea castle, built in the reign of Henry VIII. An inconsiderable market is held on Saturday, and a fair on the 14th of May, for cattle and pedlery. (Bound. and Mun. Reps.)

WINCHENDON, p. t. Worcester co., Mass., 30 m. N. by W. Worcester, 60 W.N.W. Boston, 419 W. Drained by branches of Miller's river, which afford water-power. It contains three churches, a Congregational, Methodist, and Baptist, three stores, one cotton-factory with 3500 spindles, one grist-mill, seven saw-mills; 13 schools, 345 scholars. Pop. 1754.

WINCHESTER, a city, parl. and mun. bor. of England, co. Hants, of which it is the cap., hund. Buddlesgate, on the Itchin, here crossed by a modern stone bridge, 12 m. N. by E. Southampton, and 63 m. S.W. London; lat. 51° 3' 40'' N., long. 10 19' 28'' W. The ancient parl. bor. was of very limited extent, and did not even include the whole city; but it is now wholly comprised in the modern parl. bor. (identical with the municipal bor.) with the adjacent hamlets of Winnal and St. Cross. Pop., in 1831, 9292. The town, standing on the declivity of a hill gently rising from the river, is regularly laid out, clean, well paved, and lighted with gas. In its centre is the High street, a spacious thoroughfare, running from E. to W., with parallel streets on either side, crossed by others of nearly an equal length. Most of the S.E. part of the town, but without the limits of the city-proper, is occupied by the cathedral and some other ecclesiastical edifices and their precincts. The houses are mostly substantial, and well built, many of them being in an antique style, and having a venerable appearance. It was formerly surrounded with walls; but these no longer exist; and of four ancient gates only the W. now remains.

Of the public edifices the cathedral is by far the most interesting, partly from its vast size and antiquity, partly from the variety of its architecture, and partly from its ancient importance. It was founded in 1079 by Bishop Walkelyn, a relative of William the Conqueror, who constructed the crypts, transepts, and tower; the work was continued under succeeding prelates, and was nearly completed by the famous William of Wykeham, between 1370 and 1400. It is of a cruciform shape, with a low tower rising from the centre; and, though rather heavy, has a grand and imposing appearance. Its extreme length, from E. to W. is 545 ft.; length of the nave from the W. porch to the iron doors at the entrance of the choir, 351 ft.; length of the choir, 136 ft.; breadth of the cathedral, 87 ft., and of the choir, 40 ft.; length of the transept, 186 ft.; height of the central tower, 150 ft. (Milner.) The character of the building was originally pure Norman, and the transepts and centre tower, built by Bishop Walkelyn, are admirable specimens of that style. The W. front, in the decorated Gothic, built by William of Wykeham, is singularly beautiful: it has a large and two smaller doorways, and a magnificent window, with two slender lantern turrets. The Norman parts of the building are bold, simple, and well executed; the tower massive and finely ornamented. The groining is varied in different parts of the church, and that of the nave is remarkable for its intricacy and richness. In its smaller structures, screens, monumental chapels, and stall-work, this cathedral is very rich. The altar-piece is of late perpendicular; and over it is a painting, by West, of the resurrection of Lazarus." (Rickman, 174.)

The coup-d'œil, on entering the cathedral by the W. door, is grand and imposing: the vast size of the building; the loftiness and long line of its vaulted roof; the lancet pointed windows shedding on the different objects a " dim religious light;" the lines of clustered pillars and branching aisles; the numerous chantries and monuments of eminent men; and the silence that prevails within its walls; conspire to impress the mind with a deep sense of awe and sublimity. In the middle of the presbytery, between the choir and the altar, is a coffin tomb said to inclose the remains of William Rufus, killed while hunting in the New forest, and buried here in 1100. Several Saxon monarchs are also interred in this cathedral. Among the episcopal monuments, the most interesting are those of William of Wykeham and Wayn- flete, two of the most illustrious prelates of whom England has to boast. The first, who was bishop of this see from 1366 till his death in 1404, besides completing the cathedral, founded and endowed a college, or school, in the city, the scholars educated in which were afterwards to be sent to finish their university education in New college, in the university of Oxford, of which Wykeham was also the munificent founder. Waynflete, who was bishop of Winchester from 1447 to 1486, founded Magdalen college, Oxford, one of the wealthiest foundations in that university. Here also are monuments in honour of the celebrated Bishop Hoadley, and of old Isaac Walton, the prince of anglers.

The bishopric of Winchester has long been one of the most valuable in the kingdom, its nett revenue having amounted, at an average of three years ending with 1831, to £11,151 a year: but in pursuance of the recommendations of the commissioners of ecclesiastical inquiry, its revenue will be reduced on the decease of the present incumbent.

The diocese includes 384 pars., comprising, together with Hants, the greater part of Surrey and the Channel islands. The cathedral establishment consists of a dean, 12 canons, and eight minor canons, who enjoy among them a gross annual income of about £12,000 a year. Winchester is said to have had at one time no fewer than 90 churches and chapels; but of these many were attached to monasteries and other religious establishments destroyed at the Reformation. There are still, however, as many as nine churches in the city and suburbs. Of these the small church of St. Lawrence, scarcely visible for the buildings by which it is surrounded, is supposed to be the mother church of the city, and the bishop takes possession of the diocese by making a solemn entry into it. St. Maurice, the principal parochial church, was pulled down in 1840, and an elegant and commodious structure has been erected on the site of the ancient edifice, which had become inconvenient and ruinous; the expense of its construction was defrayed by subscription. Among the other churches are St. Swithun's, built over a postern gate; St. Michael's, a handsome modern edifice in the pointed style, &c. The livings, except St. Bartholomew-

Hyde, a vicarage, and St. John's, a perpetual curacy, are all rectories, and are in the patronage either of the crown or the bishop of Winchester. But notwithstanding the number of its established churches, dissent is here pretty prevalent; and the Independents, Baptists, Wesleyans, and Rom. Catholics, have all places of worship. The chapel of the latter, a handsome edifice in the pointed style of architecture, was constructed in 1792.

The college, or school, founded, as already stated, by William of Wykeham, and completed in 1393, stands upon the site of a more ancient scholastic establishment. Its buildings enclose two large quadrangular courts, entered by spacious gateways; and besides apartments for the accommodation of the warden, scholars, &c., it has a noble hall and chapel. The whole structure is richly ornamented with pinnacles, buttresses, statues, &c. It is principally in the perpendicular style. Over the door of the school, a noble hall, constructed in 1687, at the expense of gentlemen educated in the college, is a fine bronze statue of the founder by Cibber, the sculptor, father of the hero of the " Dunciad." A building contiguous to the college is appropriated to the residence of the boys attending the school, but not on the foundation, where they are placed under the special inspection of the head-master. The building formerly used for this purpose being found to be inconvenient was pulled down, as well as the house of the head-master, in 1838, and a new and splendid edifice has been since erected in their stead, at an expense of about £25,000, defrayed by subscription.

This magnificent institution is the most ancient of the existing public schools of England, and formed the model for those of Eton, Westminster, &c. The establishment consists of a warden, a schoolmaster, and usher, 10 fellows, 3 chaplains, 3 clerks, 16 choristers, and 70 scholars; but there are in general about 200 boys in the school, including those not on the foundation. Boys on the foundation are provided with board and lodging within the walls of the college; the only payments to which they are subject, exclusive of travelling expenses, amounting to about £30 a year. Boys not on the foundation, lodge, as already stated, in an adjoining pile of buildings, under the superintendence of the head-master; and subject to college discipline. Scholars are sent, as vacancies occur, from this school to New college, Oxford. (See Oxford.) Among the distinguished individuals educated in this school may be specified Bishop Lowth, Sir Thomas Brown, Sir Henry Wotton, Otway, the tragedian, Young, author of "Night Thoughts," Collins, the two Wartons, &c.

Among the public buildings in the High street is St. John's house, an ancient structure, formerly the property of the knights templars, and a hospital. The great room in this building, 68 ft. in length and finely proportioned, was fitted up, in its present elegant style, by Geo. Brydges, Esq., a connection of the Chandos family, and a liberal benefactor of this city, of which he was long a parliamentary representative. It has a fine whole length portrait of Charles II., by Lely. In the rear of the building are neat edifices occupied by six poor widows, who, exclusive of their lodging, receive a weekly allowance of 10s. and other advantages. In 1833, an important addition was made to this charity, by the erection of a building, in the Elizabethan style of architecture, with a spacious quadrangle, intended to serve as an asylum for 12 aged females and six old men. Christ's hospital, founded in 1586, supports six old men, three boys, and a woman. An almshouse, founded by Bishop Morley in 1672, supports 10 clergymen's widows. Here, also, are charity schools for boys and girls, a national school for children of both sexes, and a mechanics' institute.

Among the public buildings may be specified the cross in the High street, a fine structure of the age of Henry VI. in the later pointed style, above 40 ft. in height. The guildhall, rebuilt in 1713, has in front a statue of queen Anne. The original Winchester bushel of king Edgar, and other ancient standards of length and capacity, formerly preserved in this building, have been removed to an apartment over the W. gate of the city. A bridewell and house of correction has been erected on the site of a magnificent monastery, in which the remains of the great Alfred are said to have been interred. On the N. side of the city is the county jail. The county-hospital, in Parchment-street, is a fine structure, which has recently been considerably enlarged; a new corn exchange was erected in 1838. Beyond the W. gate is an obelisk, erected in 1759, to commemorate a dreadful visitation of the plague to which the city was subjected in 1666. The places of amusement include a theatre, assembly-rooms, &c.; races take place in July, about 4 m. from the city. It has also a public library and reading-rooms, and a savings bank.

Winchester has no manufactures, but a very considerable retail trade, and all the public business for the county is transacted within its limits. " It may be considered, without hesitation, as gradually increasing in houses, population, and wealth. A large extent of the surrounding district belongs to ecclesiastical and other corporate bodies, which, not being empowered to grant long leases, give no encouragement to building; but wherever land held in fee-simple, or freehold land, can be procured, it was obvious, from our own inspection, that houses calculated for the habitations of the industrious classes were augmenting." (*Parl. Bound. Rep.*) The assize courts for the county are held, and other public business transacted in what was once the chapel of the castle, built by the Conqueror. (See *post.*) At the E. end of the hall is suspended a large round wooden table, 18 ft. in diameter, popularly called "Arthur's Round Table," but which is, no doubt, of a much less remote antiquity. It was painted in the time of Henry VIII.

The circumstance of its being the capital of the county makes Winchester the residence of a great number of gentlemen connected with the law; and being also the residence of a number of superior clergymen, and of gentlemen attracted to the city by the beauty of the situation, and the facilities for education and amusement, the society is very superior. There are, indeed, but few places in England that seem so well fitted for the residence of people of slender fortune. Being within a very short distance of the Southampton railway, the access to London on the one hand, and to the S. coast on the other, is as easy and expeditious as can possibly be desired.

Winchester was first incorporated in the reign of Henry II. Under the Mun. Reform Act, it is divided into three wards, and is governed by a mayor, five other aldermen, and 18 councillors. It has a recorder, who holds courts, and a commission of the peace. Petty sessions are held twice a week, and also quarter-sessions. Corp. rev., in 1840, £2975. Winchester has sent two members to the House of Commons since the 23d of Edward I. Previously to the Reform Act, the right of election was vested in the members of the corporation, who had power to augment their number. Registered electors, in 1839–40, 618.

This is certainly one of the most ancient of the English towns. Under the Romans it was a place of considerable importance, and it subsequently became the capital of the West Saxons. William the Conqueror erected a castle here, under the pretence of protecting the city, which had suffered much from the incursions of the Danes, but really, perhaps, in the view of overawing the inhabitants. From this period, however, London became the capital of the kingdom, and Winchester gradually declined in importance. But its castle was repeatedly occupied by the Norman monarchs; Henry III., hence called Henry of Winchester, was born here in 1207, and various parliaments were held in the city in the 14th and 15th centuries. Here, also, in 1552, Henry VIII. entertained his illustrious guest, the emperor Charles V; and here the marriage of Mary, daughter of Henry, with Philip II., eldest son of Charles, was solemnised July 25, 1554. In the reign of Charles II. it again became, though for a short period only, a royal residence. Charles, indeed, was so much attached to Winchester, that in 1683 he employed the famous architect, Sir Christopher Wren, to erect a new and magnificent palace on the site of the old castle. The death of the king put a stop to the progress of the building, before it was finished; and, after various mutations, it is now used as barracks.

About 1 m. S. from the city is the ancient hospital of St. Cross, founded in 1132 by Henry of Blois, bishop of Winchester, the brother to king Stephen. The present establishment, which approaches nearer to a monastery than any other in England, consists of a master, a chaplain, a steward, and 12 resident brethren. The buildings once composed two courts; but they have been partly pulled down. The chapel, in the interior court, is built in the cathedral form, with a nave and transepts, and a low, massive tower at their intersection, and affords a fine specimen of the transition of the Norman into the early English style of architecture. The entrance gateway has a handsome tower, with a statue of the founder, Cardinal Beaufort.

No traces now remain of the monastery previously referred to as containing the remains of the great Alfred, rebuilt in the reign of Henry II., the revenues of which amounted at the dissolution to £985 a year. Neither are there any remains of a university founded in the reign of Alfred, and other similar establishments.

About 4 m. E. by N. from Winchester is Avington, a fine seat belonging to the duke of Buckingham. (*Parl. Reports; Camden's Britannia, Gibson's ed., i. 138, &c.; Guide to the City of Winchester; and Private Information.*)

WINCHESTER, p. t., Cheshire co., N. H., 64 m. S.W. Concord. 418 W. Drained by Ashuelot river and its branches, which afford water-power. It contains three churches, a Congregational, Methodist, and Universalist; six stores, three fulling-mills, three woollen factories, one cotton factory with 400 spindles, one furnace, four grist-mills. 15 saw-mills, one oil-mill, two tanneries; 14 schools, 570 scholars. Pop. 3065.

WINCHESTER, p. t., Litchfield co., Ct., 96 m. N.W. Hartford, 350 W. Watered by Mad river, a branch of Farmington river. It contains four churches, two Congregational, a Methodist, and a Universalist; eight stores, three fulling mills, two woollen factories, one furnace, nine forges, one grist-mill, five saw-mills, three tanneries; one academy, 100 students; 13 schools, 420 scholars. Pop. 1667.

WINCHESTER, p. v., capital of Frederick co., Va., 146 m. N.N.W. Richmond, 74 W. Founded in 1752. The streets cross each other at right angles, and are generally paved, and the houses are mostly built of brick or stone. It contains a courthouse, jail, a market-house, a lyceum, and 13 churches, two Presbyterian, an Episcopal, two Methodist, three Baptist, two Lutheran, a German Reformed, Friends, and Roman Catholic; an academy, two banks, besides a bank for savings, numerous stores, two furnaces, one carpet factory, one cotton factory, three carriage factories, seven flouring-mills, two breweries, two printing-offices, two weekly newspapers, and 3454 inhabitants. It is supplied with water brought from a spring, half a mile distant, through cast iron pipes. A railroad extends to Harper's Ferry, 32 miles, where it connects with the Baltimore and Ohio railroad. It is well situated for trade, and has great advantages for intercourse. The remains of old fort London, famous in the French war, are here seen; and the inhabitants afforded important aid to the youthful Washington, after the defeat of Braddock, and in the final capture of fort Du Quesne.

WINCHESTER, p. v., capital of Scott co., Ill., 15 m. S.W. Jacksonville, 50 m. W.S.W. Springfield, 830 W. It has a courthouse, jail, three churches, a Congregational, Methodist, and Baptist; 11 stores, one flouring mill, one grist-mill, one saw-mill, two tanneries, one pottery; one academy, 60 students; two schools, 60 scholars. Pop. 720.

WINDHAM, county, Vt., situated in the S.E. part of the state, and contains 780 sq. m. Bounded E. by Connecticut river. Drained by Williams, West, and Saxton's rivers. It contained in 1840, 42,661 neat cattle, 114,336 sheep, 29,425 swine; and produced 23,796 bushels of wheat, 33,502 of rye, 139,923 of Indian corn, 13,387 of buckwheat, 8129 of barley, 178,761 of oats, 743,366 of potatoes, 423,400 pounds of sugar. It had 80 stores, two furnaces, 19 fulling-mills, eight woollen factories, one cotton factory with 60 spindles, 49 grist-mills, 134 saw-mills, one oil-mill, seven paper-mills, 23 tanneries, four printing-offices, three weekly newspapers, and one periodical; four academies, 501 students; 248 schools, 9138 scholars. Pop. 27,442. Capital, Fayetteville village, in Newfane township.

WINDHAM, county, Ct., situated in the N.E. part of the state, and contains 620 sq. m. Organized in 1726. Drained by Quinnebaug and Shetucket rivers, and their branches. The Norwich and Worcester railroad passes through it, from S. to N. It contained in 1840, 23,507 neat cattle, 39,504 sheep, 12,562 swine; and produced 8999 bushels of wheat, 48,378 of rye, 173,003 of Indian corn, 30,653 of buckwheat, 9404 of barley, 179,097 of oats, 497,991 of potatoes, 3001 pounds of silk cocoons, 2549 of sugar. It had 115 stores, one furnace, eight fulling-mills, 15 woollen factories, 54 cotton factories with 105,016 spindles, 57 grist-mills, 100 saw-mills, 50 oil mills, three paper-mills, 19 tanneries, one distillery; eight academies, 317 students; 166 schools, 7749 scholars. Pop. 28,080. Capital, Brooklyn.

WINDHAM, p. t., Cumberland co., Me., 14 m. N.N.W. Portland, 56 m. S.S.E. Augusta, 553 W. Bounded N.W. by Sebago pond, S.W. by Presumpscot river. Incorporated in 1762. It has five stores, 18 schools, 620 scholars. Pop.

WINDHAM, p. t. Windham co. Ct., 31 m. E. Hartford, 358 W. It was formerly the capital of the county, which was removed to Brooklyn, as more central, in 1820. Chartered in 1692, first settled in 1686. Watered by Shetucket river. The village near the centre contains two churches, a Congregational and Episcopal, a bank, an academy, six stores, and about 60 dwellings. It has not increased since the American revolution. Willimantic village is a flourishing manufacturing place, 3 m. from the principal village, and in the N.W. part of the township, on Willimantic river, one of the constituents of Shetucket river. It is built chiefly on one street, and contains three churches, a Congregational, Methodist, and Baptist, and the principal manufactories of the township. In the E. part is the parish of Scotland, with a Congregational church, and a small v. There are in the township 11 stores, two fulling-mills, three woollen factories, five cotton factories with 11,950 spindles, three grist-mills, seven saw-mills, two paper-mills; 18 schools, 928 scholars. Pop. 3382.

WINDHAM, p. t., Greene co., N. Y., 18 m. W. Catskill, 45 S.W. Albany, 363 W. Watered by Batviakill creek. It has 16 stores, three grist-mills, 17 saw-mills, five tanneries; 10 schools, 203 scholars. Pop. 2417.

WINDHAM, t., Luzerne co., Pa., 25 m. N.W. Wilkesbarre. Drained by Big and Little Mahoopeny creeks. It has two stores, one fulling-mill, two flouring-mills, one saw-mill; nine schools, 304 scholars. Pop. 1647.

WINDHAM, p. t., Portage co., O., 150 m. N.E. Columbus, 315 W. Called Sharon until 1820. First settled in 1811, by 14 emigrants from Berkshire co., Mass., who purchased the township, and organized a church before their departure. It has much of the industrious and steady habits of New-England, and is a fine farming township. It has one academy, 76 students; eight schools, 217 scholars. Pop. 907.

WINDSOR, a town, parl. and mun. bor., par. and royal residence of England, co. Berks, hund. Ripplesmere on the Thames, 20 m. W. by S. London. Previously to the Reform Act, the parl. bor. was nearly co-extensive with the par.; the modern parl. and mun. bor. comprises about half the par., with the lower ward of the castle, and a small portion of the adjacent par. of Clewer; having an area of about 4500 acres, with, in 1831, a pop. of 7071.[*] The town, partly situated on low ground, along the river, and partly on the declivity of the ridge occupied by the castle, the W. end of which is surrounded by its buildings, consists of six principal and several smaller streets, and is well paved, and lighted with gas. It communicates with Eton, on the opposite bank of the Thames, by a handsome iron bridge of three arches, raised on granite piers. Of late years, many buildings of a superior kind have been erected in the W. part of the town, in that portion of Clewer par. included in the modern parl. bor. The par. church is a handsome new Gothic structure: the living, a vicarage worth £400 a year, is in the gift of the crown. The guildhall, a neat edifice supported on columns and arches of Portland stone, occupies a conspicuous site in the High-street. On its N. side is a statue of Queen Anne, and on its S. one of Prince George of Denmark: in the interior are numerous portraits of royal and other distinguished persons. The cavalry and infantry barracks, the new royal stables, a neat theatre built in 1815, and several dissenting chapels, are among the other principal buildings. The charity school, founded in 1705, had at the date of the late enquiry an income of £167 a year, and was attended by 55 children; it has, also, a ladies' charity school for girls, national and Sunday schools, &c. George III. established a hospital for invalid soldiers in 1784; a lying in charity was founded in 1801, and the royal general dispensary in 1818, and there are numerous minor charities having an aggregate income of about £800 a year. Windsor was first chartered in 1276, by Edward I., in the 30th of whose reign it began to send members to the House of Commons, though returns have been regularly made only since the 25th Henry VI. The right of voting was formerly in householders paying scot and lot, who had resided for six months within the bor. Under the Boundary Act the alterations previously alluded to were made in the limits of the bor. Reg. electors, in 1839–40, 667. Under the Mun. Reform Act the bor. is divided into two wards, and is governed by a mayor, five other aldermen, and 18 councillors. It has a commission of the peace, and a court for the recovery of debts under 40s. Corp. revenue, in 1840, £1653. Windsor has no manufacture of importance; and being out of any principal line of road, its trade is merely one of retail, being confined to the supply of goods to the inhabitants and visiters. It has numerous inns and lodging houses, though, considering the resort of company to the town, the former are certainly of a very inferior description to what might have been expected. The ale of Windsor enjoys a considerable reputation, and is sent to London and other places. The town supports two weekly newspapers, Markets, Wednesday and Saturday, the latter principally for corn. Fairs, Easter Tuesday, July 5 and October 4, for horses, cattle, sheep, and wool. (*Bound. and Munic. Corp. Rep.*; *Priv. Inf., &c.*)

WINDSOR, county, Vt. Situated toward the S.E. part of the state, and contains 900 sq. m. Bounded E. by Connecticut river. Watered by White, Queeche, Black and Williams rivers. Soap stone is found in Plymouth, Bridgewater, and Bethel, and extensively wrought and exported. Limestone is abundant. It had, in 1840, 51,863 neat cattle, 234,896 sheep, 22,834 swine; and produced 56,659 bushels of wheat, 46,196 of rye, 168,897 of Indian corn, 49,380 of buckwheat, 5164 of barley, 301,026 of oats, 1,072,733 of potatoes, 1146 pounds of silk cocoons, 2100 of hops, 462,444 of sugar. It had 100 stores, four furnaces, 57 fulling-mills, 27 woollen factories, one cotton factory, with 650 spindles, 46 grist-mills, 154 saw-mills, two paper-mills, one oil-mill, one rope-walk, 45 tanneries, four printing-offices, two binderies, five weekly newspapers, and two periodicals; one college, 70 stu

dents; seven academies, 258 students; 257 schools, 13,824 scholars. Pop. 40,356. Capital, Woodstock.

WINDSOR, p. t., Kennebec co., Me., 10 m. E. Augusta, 685 W. Drained by Sheepscot river, and by several ponds which discharge their waters into it. Chartered in 1809 by the name of Malta. It has two stores, one fulling-mill, one grist-mill, seven saw-mills; 15 schools, 756 scholars. Pop 1788.

WINDSOR, p. t., Windsor co., Vt., 65 m. S.S.E. Montpelier, 471 W. Bounded E. by Connecticut river. Mill brook affords good water-power, having a fall of 60 ft. in one third of a mile. The t. contains six churches, two Congregational, a Methodist, Episcopal, Baptist, and one common to Freewill Baptists and Universalists, nine stores, three fulling-mills, two woollen factories, one furnace, five grist-mills, eight saw-mills, three tanneries, one printing-office, two weekly newspapers, and two periodicals; 14 schools, 848 scholars. Pop. 2744. The village is one of the pleasantest in this part of the state, situated on the W. bank of Connecticut river, one fourth of a mile distant, with a beautiful meadow in front. The main street has several turns at very obtuse angles, and is well built, having some elegant dwellings, finely ornamented with trees and shrubbery. It contains a courthouse for the United States circuit and district courts, three churches, the Vermont state prison, nine stores, various mechanic shops and mills and manufactories, a printing-office, 135 dwellings, and about 625 inhabitants. Ascutney mountain, on the S. line of the t. is an isolated peak, 3320 ft. above tidewater, and 3116 ft. above Connecticut river at Windsor. It is often ascended, and presents an extensive and beautiful prospect from its summit. It forms an interesting object as seen from the v.

WINDSOR, p. t., Hartford co. Ct., seven m. N. Hartford, 343 W. It is one of the oldest towns in Connecticut, having been settled in 1636. Bounded E. by Connecticut river. Watered by Farmington or Tunxis river. It contains three churches, two Congregational and a Methodist, six stores, one fulling-mill, one woollen factory, three cotton factories with 570 spindles, five grist-mills, two saw-mills, three paper-mills; one academy, 30 students, nine schools, 237 scholars. Pop. 2283. The principal village is built on a single street, from 2 to 3 m. long, a little back from the river, and contains a Congregational church, several stores, and many elegant houses, beautifully shaded with trees, and ornamented with shrubbery. There are two other villages in the t.

WINDSOR, p. t., Broome co., N. Y., 12 m. S.E. Binghamton, 196 m. S.W. Albany, 300 W. Watered by Susquehanna river and its tributaries. It contains two churches, a Presbyterian and Methodist, three stores, one fulling-mill, three grist-mills, 11 saw-mills, two tanneries; 20 schools, 685 scholars. Pop. 2368.

WINDSOR, t., Berks co., Pa. Bounded W. by Schuylkill river, E. by Maiden creek. It contains two churches, one of them common to Lutherans and Presbyterians, one store, one grist-mill, one oil-mill, two tanneries, six distilleries. Pop. 2662.

WINDSOR, p. v., capital of Bertie co., N. C. 157 m. E by N. Raleigh, 360 W. Situated on the W. side of Cashie river, a branch of Roanoke river. The river is navigable to this place for vessels of from 80 to 100 tons. The country around is fertile, and the place is well situated for trade, but the location is unhealthy.

WINDSOR CASTLE is the principal country seat of the sovereigns of England, and one of the most magnificent royal residences in Europe. It appears to have been founded by William I. soon after the conquest, and it has been enlarged or embellished by the greater number of his successors, particularly by Edward III, George III., and George IV. Under the latter it was, indeed, in great part rebuilt, and throughout renovated by Sir Jeffrey Wyattville, and has been fitted up in the most splendid style. Being placed on the summit of a lofty eminence rising abruptly on the S. side of the river, it commands very extensive views, and is, at the same time, a most conspicuous and interesting object from all the surrounding country. It is of an oblong form, and is divided into an upper, a middle, and a lower ward, the entire area comprised within its outer wall being about 12 acres.

The upper or E. ward consists of a quadrangle, having on the N. the state apartments shown to the public; on the S. the apartments appropriated to the use of visiters; and on the E. the private apartments of the sovereign: on the W. the upper ward communicates, by the Norman and Saint George's gateways, with the middle ward, a narrow inclosure round the base of the round tower, which crowns the summit of an artificial mound in the centre of the castle. The lower ward, which is considerably smaller than the upper, has on its S. and W. sides the houses of the military knights and the Salisbury, Garter, and Bell towers; and on the N. St. George's chapel, and Wolsey's tomb-house, behind which are other buildings inclosing several smaller

---

quadrangle: it is entered from the town of Windsor at the S.W. corner by Henry VIII.'s gateway. On the N. side of the castle, outside the state apartments and middle ward, is the North terrace, originally constructed by Queen Elizabeth, and afterward enlarged and improved by Charles II. This noble walk, resting partly on precipitous ledges of rock, and partly on masonry, rises about 70 ft. over the meadows at its base, and is at once the finest terrace of its kind in the kingdom, and a distinguishing feature of the castle. On the E. side of the castle, opposite her majesty's private apartments, are the sunk or royal gardens, comprising about two acres.

The principal and most magnificent entrance to the castle is on the S., by the gateway of George IV., between the York and Lancaster towers. The York tower, on the right hand, formed part of the ancient edifice; but the Lancaster tower is wholly new, its foundation having been laid on the 12th of August, 1824. The towers are symmetrical, being about 100 ft. in height, with machicolated battlements. Immediately opposite to this gateway is the principal entrance to the state apartments. The grand staircase, with the guard-room at its top, is, perhaps, among the happiest efforts of Wyatville's genius. The staircase is lighted by an octagonal lantern 100 ft. above the pavement, and has a marble statue of George IV. by Chantrey. In the vestibule is the collection of paintings by West, representing the exploits of Edward III.; and in the guard chamber are the coats of mail worn by John king of France, and David king of Scotland, while prisoners in the castle, with busts of Marlborough, Wellington, and Nelson, the latter on a pedestal formed of a portion of one of the masts of the Victory.

The decorations of the king's drawing-room are very superb: the ceiling is painted in compartments, representing the restoration of Charles II., the labours of Hercules, and other subjects; and on each side the room are numerous paintings by Rubens, and the arms of several of the English kings. The ceiling of the audience-chamber has an allegorical representation of the church of England; and in the same apartment are West's Installation of the Knights of the Garter, and several portraits. The ballroom, 96 ft. in length, 32 ft. in width, 31 ft. in height, is finished in the gorgeous style of Louis XIV. It is hung in part with Gobelin tapestry, representing the story of Jason and the Golden Fleece, said to have belonged to the unfortunate Marie Antoinette. St. George's hall, the banqueting room of the knights of the garter, is 200 ft. in length, with an arched ceiling divided into compartments and panels, in which are nearly 700 shields, emblazoned with the arms of the knights down to the present time. At the E. end is the throne, under a rich canopy: and on the S. side of the hall are the portraits of the different sovereigns, from James I. to George IV., by Vandyck, Lely, Kneller, Lawrence, &c. The Waterloo chamber, 100 ft. in length by 46 in width, has portraits, principally painted by Sir Thomas Lawrence, of most of the sovereigns, warriors, and statesmen who took a prominent part in the contest with France terminated by the battle of Waterloo. The other state apartments do not require any particular notice; they have the inconvenience of all entering from each other, so that to get to the last in the range all the others must be gone through.

The entrance to her majesty's private apartments is at the S.E. corner of the upper ward, through a handsome hall, from which a double staircase leads to a magnificent corridor 500 ft. in length. The private apartments consist of a dining-room, 50 ft. in length by 37 in width; a drawing-room, 66 ft. in length by 30 in width; a smaller drawing-room, 40 ft. in length by 25 in width; library, 50 ft. in length by 40 in width; with bedrooms, dressing-rooms, boudoirs, &c. These rooms are, as might be expected, most sumptuously furnished. The apartments for servants occupy the lower and higher stories of the palace.

The round tower was originally built by the celebrated William of Wykeham, the architect employed by Edward III. It stands on an artificial mound, and is approached by a covered flight of 100 steps. From a court in the interior, another flight of steps leads to the battlements, whence, in a clear day, portions may be seen of no fewer than 12 counties. This tower, which has been much modernised, is 32 ft. higher than formerly, and is surmounted by a turret 20 ft. in height, whence the royal standard is displayed: it is the residence of the governor of the castle. At the base of the tower is a bronze equestrian statue of Charles II., erected by one of his pages.

The great object of interest in the lower ward is St. George's chapel. "This is one of the finest perpendicular buildings in the kingdom: it is regular in its plan, and nearly all in one style. It is a cross church, with the transepts ending in octagonal projections which have two heights of windows. At each end of the aisles are also small octagonal projections sideways; all these are separated by screens, and form monumental chapels. In the E. wall of

the chapel is a doorway of early English date; and perhaps other portions of a date prior to the present chapel may remain; but the whole of the chapel is a specimen of the perpendicular style in its advanced, but not latest period. The roof of the nave is painted with armorial bearings, and the whole highly enriched, so that it now presents one of the best examples of the capability of English architecture for the reception of splendid colouring and gilding. (Rickman's Gothic Architecture, 134.)

The interior is divided by the screen and organ gallery into two parts, the body of the chapel and the choir. The W. end of the former is wholly occupied by an enormous window, fitted with painted glass, which, however, is deficient in brilliancy and richness of colouring. The fittings of the choir are mostly modern. St. George's chapel was built between 1474 and 1516, chiefly under the direction of Beauchamp, bishop of Salisbury, and Sir Reginald Bray, minister of Henry VII. It has served as the burial place of Henry VI. (removed hither by Richard III. from Chertsey), Edward IV. and his queen, Henry VIII. and Jane Seymour, and Charles I. It has a few old monuments, including that of Edward IV., of hammered steel. Here also is a monument in honour of the late Princess Charlotte, but it is generally admitted to be in bad taste, stiff, and unnatural. Adjoining the chapel on the E. is Wolsey's tombhouse, built by Henry VII., but which afterward came into possession of the cardinal. James II. fitted it up as a Roman Catholic chapel. It remained unoccupied from this down to that of George III., by whom it was repaired, and a vault beneath it fitted up as a mausoleum for the royal family; and in it are now deposited the bodies of George III. and his queen, George IV., William IV., the dukes of York and Kent, the Princess Charlotte, the Princess Augusta, &c.

But, despite its magnificence, we confess that Windsor castle appears to us to be extremely deficient in many things that one should expect to meet with in an ancient and favourite seat of the kings of England. Except the associations connected with the building, and the names of some of its towers and apartments, it has but little to connect it with the nation, or to make it an object of interest. In its interior, everything has been sacrificed to gratify the taste of George IV. for ostentation and vulgar finery. Not a single apartment has been allowed to continue in its ancient state, to carry us back to the days of the Edwards, the Henrys, Elizabeth, or even the Stuarts. Everything that was venerable for its antiquity, or interesting from its history or associations, has been demolished or changed; so that one might suppose it had been wholly constructed within the last 20 years. Nor is there anything in the fittings-up and embellishments of the apartments to atone for the destruction or metamorphosis of all that was old and interesting in the building. It has nothing to mark it out as the chosen seat of the constitutional sovereign of the British empire. Excepting the busts of Marlborough, Wellington, and Nelson, there is hardly, in the state apartments, any memorial of any one of the many great men whose exertions have contributed to increase the power and glory of the British nation. We look in vain for either busts or portraits of Shakspeare, Bacon, Milton, Newton, Locke, Dryden, and Pope. Much as the English nation owes to Watt and Arkwright, Windsor castle has no monument of either the one or the other. And the same may be said of most of our great parliamentary leaders, and even of the men who brought about the Revolution, and placed the Hanoverian family on the throne. The library is wretchedly deficient in books connected with the history and state of the country and its colonies; and, vast as is the building, it has neither a theatre nor an opera-house. One, in fact, might suppose that it had been fitted up for the residence of some opulent upholsterer; and, except in its fine situation, size, and external appearance, it has but little appropriate to or worthy of its destination.

The Little park is a fine expanse of lawn, comprising nearly 500 acres round the E. and N. sides of the castle. In it is the tree usually supposed to be identical with the Herne's oak of Shakspeare. Windsor great park comprises about 3800 acres on the S. side of the castle, being well wooded, and exhibiting a great variety of ground. Here is the long walk, a noble avenue, nearly 3 m. in length, extending in a straight line from the grand entrance to the castle to the top of a hill, on which a colossal bronze equestrian statue of George III., by Westmacott, has been erected. On the S. side of the hill is Virginia water, an artificial lake, with a fishing temple in the Chinese style. Windsor forest, the theme of Pope's fine poem, is a tract 56 m. in circumference, laid out by William the Conqueror for the purposes of hunting, and kept up by the succeeding sovereigns. Latterly, however, it has been mostly enclosed. Its limits embrace one market town, Wokingham, and numerous villages. Old Windsor, where the Saxon monarchs are said to have had a residence, is on the Thames, about

1 m. S.E. Windsor. (For accounts of the castle, the reader may consult *Jesse's Windsor and Eton*, and the different guide-books.)

WINFIELD, p. t., Herkimer co., N. Y., 76 m. W. Albany, 375 W. Drained by branches of Unadilla river. It contains three churches, a Presbyterian, Baptist, and one other; five stores, one fulling-mill, one woollen factory, one furnace, three grist-mills, 11 saw-mills, one tannery, one distillery; 11 schools, 490 scholars. Pop. 1632.

WINNEBAGO, lake, Wis. ter., 24 m. long and 10 broad. It receives Fox river in its W. part, and discharges it from its N. part, whence it flows into Green bay, lake Michigan.

WINNEBAGO, county, Wis. ter. Situated in the N. part of the settled portion of the territory, and contains 500 sq. m. Watered by Neenah creek, Fox river, and Panwaygun lake, through which passes a considerable river. Bounded E. by Winnebago lake. It contained in 1840, 184 neat cattle, 147 swine; and produced 1690 bushels of Indian corn, 1980 of potatoes, 4400 pounds of sugar. It had one grist-mill, one saw-mill. Pop. 125. Capital, Oshkosh.

WINNEBAGO, county, Ill. Situated in the N. part of the state, and contains 504 sq. m. Watered by Rock river and its tributary, Pekatonokee river. Formed in 1836, It contained in 1840, 4711 neat cattle, 894 sheep, 12,374 swine; and produced 68,315 bushels of wheat, 197,377 of Indian corn, 1399 of buckwheat, 50,117 of oats, 79,136 of potatoes, 115,419 pounds of tobacco, 13,893 of sugar. It had 18 stores, one fulling-mill, one flouring-mill, one grist-mill, 16 saw-mills, two distilleries, 23 potteries, two printing-offices, two weekly newspapers; 26 schools, 606 scholars. Pop. 4609. Capital, Rockford.

WINNIPISEOGEE, lake and river, N. H. Situated between Belknap and Carroll counties, and is 22 m. long and from 1 to 10 broad. The waters are remarkably pure, it being fed extensively by springs at its bottom, and it is very deep. In picturesque beauty, and fine surrounding mountain scenery it may vie with the lakes of Scotland or Switzerland. Its outlet, proceeding from Great bay on the S.W., and through two small lakes, forms Winnipiseogee river, which, uniting with Pemigewasset river, forms Merrimac river. Winnipiseogee river, in its short course, has a fall of 232 ft., affording a great water-power. The lake is 479 ft. above tidewater. It has a large number of islands, said to be not less than 365, and some of them are sufficiently extensive for farms, one of them containing 500 acres.

WINNSBOROUGH, p. v., cap. of Fairfield distr., S. C., 29 m. N.N.W. Columbia, 150 m. N.N.W. Charleston, 480 W. Situated on the dividing ridge between Wateree and Broad rivers. It contains a handsome courthouse, a jail, market-house, a male and a female academy; four churches, two Presbyterian, an Episcopal, and a Methodist; 12 stores, 100 dwellings, some of them handsome, and about 700 inhabitants. It is celebrated for the manufacture of cotton gins. In the vicinity is "the Furnace institution," a literary and theological seminary, under the direction of the Baptists, the object of which is to educate young men for the ministry.

WINSLOW, p. t., Kennebec co., Me., 20 m. N. by E. Augusta, 615 W. Bonded W. by Kennebec river. Watered by Sebasticook river and its tributaries, one of them issuing from a pond near its centre. It has two stores, one fulling-mill, four saw-mills; 14 schools, 733 scholars. Pop. 1792.

WINSTON, county, Miss. Situated a little N.E. of the centre of the state, and contains 720 sq. m. Drained by branches of Pearl and Tombigbee rivers. It contained in 1840, 8729 neat cattle, 1224 sheep, 12,896 swine; and produced 8843 bushels of wheat, 203,165 of Indian corn, 8495 of oats, 22,971 of potatoes, 4005 pounds of rice, 2475 of tobacco, 1,106,807 of cotton. It had three stores, nine grist-mills, five saw-mills, two tanneries; one academy, 18 students; one school, 29 scholars. Pop.: whites, 3061; slaves, 1589; total, 4650. Capital, Louisville.

WINTHROP, p. t., Kennebec co., Me., 10 m. W. Augusta, 598 W. It has a large pond or small lake in its E. part, 10 m. long, and from 1 to 3 broad; and two others, one of which extends N. into Readfield. Cobbesseconte river flows through these ponds to Kennebec river. There is a village at the union of the N. lake, the outlet of which affords water-power. There is another village at the N. end of the large lake or pond. The township contains five stores, one fulling-mill, one grist-mill, three saw-mills, two tanneries, one printing-office, one weekly newspaper; five schools, 601 scholars. Pop. 1915.

WINTON, p. v., cap. of Hertford co., N. C., 35 m. N.W. Edenton, 155 m. N.E. by E. Raleigh, 227 W. Situated on the S.W. side of Chowan river, 2 m. below the junction of Nottaway and Meherrin rivers, which form it. It contains a courthouse, jail, and about 100 inhabitants.

WIRKSWORTH, a market town and par. of England, hund. of same name, co. Derby, at the S. extremity of the lead mining district, 12 m. N.N.W. Derby. Area of parish,

14,646 acres. Pop. of ditto in 1831, 7754; of the town, 3787. The latter, in a valley nearly encircled by hills, consists principally of two streets formed by the intersection of two turnpike roads at right angles. The church of St. Mary, a spacious edifice in a mixed style of architecture, consists of a nave and side aisles, a N. and a S. transept, a chancel, and a square tower supported on four massive pillars. The living, a vicarage worth £164 a year, is in the gift of the dean of Lincoln. There are chapels for Baptists, Independents, and Wesleyans. A free grammar school, founded and endowed in 1579 by Anthony Gell, Esq., for an unlimited number of scholars, has an income of above £380 a year. There are several almshouses, and the funds for general charities yield an income of above £130 a year. The moot and sessions hall, erected in 1772, is a handsome stone building, with shambles underneath. The lead mines in the vicinity, though now comparatively neglected, still furnish employment for a considerable number of the inhabitants. In 1836, there were three cotton-mills at work in the parish, employing 619 hands. The other branches of industry consist of silk weaving, wool combing, and the making of hats, tapes, and hosiery. The Cromford canal passes about 1½ m. to the N. of the town, crossing the Derwent by an aqueduct of one arch 80 ft. in span; and the High Peak railway has also its terminus a little to the N. of Wirksworth. The town is under the jurisdiction of a constable and headborough. Petty sessions for the hundred are held weekly, and four manorial courts every year in the moot hall, in which all business relating to the mines is decided. The manor of Wirksworth forms a part of the duchy of Lancaster. It was acquired by Sir Richard Arkwright, the great founder of the cotton manufacture, who died at his house at Cromford in this parish, in 1792, and is now held by his son, Richard Arkwright, Esq. Market day, Tuesday; fairs four times a year for horned cattle. (*Beauties of England and Wales*, art. Derby; *Report on Mills and Factories*; *Charity Reports, &c.*)

WIRT, t., Alleghany co., N. Y., 15 m. S. Angelica. 270 m. W.S.W. Albany. Drained by Little Genesee creek. It has one grist-mill, seven saw-mills; one school, 36 scholars. Pop. 7207.

WIRTEMBERG (Germ. *Würtemberg*), KINGDOM OF. one of the secondary states of the German confederation, in the S. part of which it is situated, between lat. 47° 3' N., and long. 8° 15' and 10° 30' E. having N., W., and S.W. the territory of Baden; E. and S.E. Bavaria; and S., the lake of Constance, and the Hohenzollern principalities, which last it nearly encloses. Its area, population, subdivisions, &c., are given as follows in the *Almanach de Gotha* for 1849:

Circles.	Area in English sq. m.	Pop. in 1846.	Pop. to sq. m.	Chief Town.
Neckar . .	1,306	458,143	351	Stuttgard
Black Forest	1,961	415,197	236	Reutlingen
Danube . .	2,549	360,892	142	Ulm
Jaxt . . .	2,124	369,479	172	Ellwangen
Total .	7,940	1,643,330	224	

Physical Geography, &c.—The surface is in great part mountainous, being covered by ranges of the Black forest (*Schwarzwalde*), Suabian, and Raabe mountains. The Oberhohenberg, near Deilingen, rises to about 3276, and the Kniebis to 3100 ft. above the sea; but, in general, the various ranges are less than 3000 ft. in height. Würtemberg belongs partly to the basin of the Danube, and partly to that of the Rhine; besides which the principal rivers are the Neckar, with its tributaries, and the Iller, a tributary of the Danube. Except that of Constance, it has no lake of any importance.

The climate is mild in the sheltered valleys: at Stuttgard the mean temperature of the year is about 51° Fahr. The quantity of rain that falls varies, in different places, from 20 to 46 inches a year; but at Stuttgard is about 22.9 inches. As respects its productions, the country may be divided into three zones: the region of the vine, which extends to about 1000 ft. above the sea; that of fruit and corn, to 2000 ft.; and that of corn and forests, comprising all above the latter elevation.

Agriculture is the principal occupation of the mass of the population. The arable lands have been estimated to comprise about 2,440,000 morgen; vineyards, 64,778 ditto; gardens and orchards, 150,650 ditto; pasture lands, 728,000 ditto; and forests nearly 2,000,000 ditto. Spelt, oats, barley, rye, and wheat, are the grains principally cultivated; and a larger quantity of corn is usually produced than is required for home consumption: in 1834, a surplus of 149,000 scheffels was exported. Potatoes are raised in large quantities; and Berghaus has estimated the annual crop at 3,000,000 scheffels. Peas, beans, turnips, hops, and tobacco, are only partially cultivated. The wines grown on the Neckar are tolerably good; and altogether Würtemberg produces annually about 155,800 eimers of wine, of the or-timated value of 3,100,000 florins; but a great deal of this is grown at a considerable elevation on the banks of the lake of Constance, and elsewhere, and is of a thin and indifferent quality. Apples, pears, apricots, and other fruits of temperate climates, including even figs and melons, come to perfection, and small quantities of cider and perry are made. The forests are an important source of wealth. Pine, fir, and cedar are the principal kinds of trees, but oaks, beeches, &c., are also numerous. The timber cut in the Black forest is estimated to produce upwards of 400,000 florins a year. According to Berghaus, there are upwards of 390,000 head of cattle, 93,000 horses, 586,000 sheep, and 122,000 hogs in the kingdom; and the value of the wool produced annually may be about 1,713,000 florins. (*Allg. Länder, &c.*, iv., 254.) There are numerous associations for improving the different branches of rural industry; and government spends considerable sums in the encouragement of agriculture.

Salt is one of the principal mineral products, and between 400,000 and 500,000 centners a year are obtained from salt springs; of which quantity, about 230,000 centners are consumed in the country, the rest being exported, principally to Switzerland. Coal and lignite are found, but in no great quantities; iron ore, slate, building and mill stone, alabaster, gypsum, nitre, and potters' clay are, however, more or less abundant; and in the Black forest are several mineral springs frequented by visiters.

Manufactures of linen and linsey-wolsey fabrics, hosiery, and woollen cloths are carried on in most of the peasants' houses; and in some places to such an extent that, in the little village of Laichingen, there are 400 hand-looms, which annually produce 400,000 ells of linen. The manufacture of wooden clocks, toys, &c., exported to all parts of Europe and America, is extensively carried on in the Black forest and other parts. Cotton yarn is spun, and cotton cloth woven by machinery, in Stuttgard and Obendorf; and woollen yarn, stockings, leather, paper, glass, and tobacco, and tobacco-pipes are manufactured in the principal towns: and there are also a good many dyeing-houses, glue factories, breweries, distilleries, &c. The principal exports consist, however, of cattle, wool, corn, timber, fruit, wine, seeds, hemp, iron, salt, pitch, tar, oil, and other raw products, which are sent down the Neckar, Rhine, and Danube. The total value of the exports and imports may be respectively estimated at from 15,000,000 to 18,000,000 florins a year. Wirtemberg, at one period, proposed with Bavaria to join Austria in a commercial union, for the S. of Germany, in opposition to that established by Prussia in the N.; but the proposal being rejected, Wirtemberg, like Bavaria and Baden, is now a member of the Prussian commercial league.

Accounts are kept in gulden or florins, worth about 1s. 8¼d., divided into 15 *batzen*, or 60 *kreutzers* of 6 *hellers* each. The ell, about 2 ft. English; the *morgen*=nearly two thirds of an acre; the *scheffel*=7538 cubic inches.

The Government is an hereditary monarchy, limited by the constitution of 1819. The parliament consists of two chambers, called together every three years, or oftener, if necessary. The first chamber is composed of the subordinate members of the royal family, the mediatized princes, and the heads of the principal noble families: the second chamber, or house of representatives, consists of 94 members, including 13 of the inferior nobility, six Protestant superintendents, the Roman Catholic bishop, and other dignitaries, the university-chancellor, deputies from the towns of Stuttgard, Tubingen, Ludwigsburg, Ellwangen, Ulm, Heilbronn, and Reutlingen; and a representative from each rural district, chosen every six years. Members of the second chamber must be 30 years of age. The administration is conducted by five ministers of state, who, with a president, form the privy council. The supreme judicial court in Stuttgard is divided into civil, criminal, educational, and matrimonial tribunals; and in each of the circles, districts, and communes there are courts of justice. The reigning house, which had been previously Protestant, became Roman Catholic in 1772, without, however, the change having any material influence over the religious persuasion of the people. Of the population in 1836, 1,194,929 were 'Lutherans; 418,290 Roman Catholics; and 11,968 Jews. In respect of education, Wirtemberg ranks very high. In 1830 it was estimated that one in seven of the population was receiving public instruction; and, according to Berghaus, every individual in the kingdom is able to read and write, except a few in that part of the country called the Suabian terrace, where the Neckar rises. It has a university (at Tubingen), a lyceum, and gymnasiums, in all the principal towns, with ecclesiastical, citizens', and other schools; and a primary school in every village. The total armed force consists of 19,500 men, including eight regiments of infantry, four of cavalry, and three companies of artillery, with train, &c. The public revenue, at an average of the four years ending with 1841, amounted to 9,667,525 Rhenish florins a year; the public debt in 1839 was 24,024,179 florins.

Mr. Loudon, who travelled over most part of Wirtemberg, in 1828, says, "From what I have seen of this country (Wirtemberg) I am inclined to regard it as one of the most highly civilized in Europe. I am convinced that the great object of government is more perfectly attained here, than even in Great Britain: because, with an almost equal degree of individual liberty, there are incomparably fewer crimes, as well as far less poverty and misery. Every individual in Wirtemburg reads and thinks, and to satisfy oneself that such is the case, he has only to enter into conversation with the first peasant he meets: and observe the number and style of the journals that are everywhere circulated; and the multitude of libraries in the towns and villages. I did not meet with a single beggar in Wirtemberg, and with only one or two in Bavaria and Baden. The dress of the inhabitants of Wirtemberg, as well as those of a great part of Bavaria and Baden, appeared to me to indicate a greater degree of comfort than I had ever observed in any other country, with the exception, perhaps, of Sweden and of the Lowlands of Scotland." (*Loudon's Letter to Count Lasteyrie.*)

History.—Wirtemberg derives its name from a castle near Stuttgard, the principal seat of the reigning family. It was formerly a dukedom. The French overran the country in 1796; but the sovereign having made his peace with the conquerors, important additions were made to his territories in 1800; and soon after the battle of Austerlitz, Napoleon raised the duke to the rank of king. Under the congress of Vienna, Wirtemberg holds the sixth rank in the German confederation, with four votes in the full diet and one in committee; and contributes 13,955 men to the confederated army. (*Berghaus; Allg. Länder, &c.; Stein's Handbuk der Geog.; Dict. Geog.; Menninger; Beschreiben. Wirtemberg.*)

WISBEACH, a mun. bor., market town, river-port, and par. of England, co. Cambridge, hund. Wisbeach, in the Isle of Ely, close to the border of Norfolk, on the Nene, here crossed by a fine stone bridge of one arch, 11 m. S. from the mouth of the Nene, in the Wash, and 39 m. N. Cambridge. The old borough of very irregular shape, was co-extensive with the parish of St. Peter's: the modern borough is much more compact, and of less extent, but comprises a suburb of Wisbeach called New-Walsoken, on the E. side of the river, excluded from the old borough, and has an area of about 1900 acres, with a population, in 1831, of 8100. The central and main portion of the town lies in an angle between the Nene and Wisbeach canal: other streets extend for some distance N. and S. along both banks of the river, and along the canal, by which the town communicates with the Ouse in a S.E. direction. Most part of the houses on the E. side of the canal belong to the parish of Walsoken, in the county of Norfolk: this suburb has been built within the last thirty years, and appears to be still extending. The inhabitants consist chiefly of the labouring classes employed in or connected with the interests of Wisbeach. (*Mun. Corp. and Bound. Reps.*) The town is irregularly laid out, but has, of late, been much improved: in its centre is a handsome crescent, erected in 1816 on ground formerly occupied by a castle founded soon after the conquest. Most parts of the thoroughfares are well paved and lighted with gas. The parish church of St. Peter is a spacious but singular edifice, having two naves and two aisles. It exhibits a mixture of the Norman, decorated, and perpendicular styles. It has a fine tower, and has within several monuments of distinguished families, but is at the same time much encumbered with modern galleries. The living, a vicarage, to which the curacy of St. Mary's is attached, is one of the most valuable in this part of the country, being worth £1779 a year: it is in the gift of the bishop of Ely. A chapel-of-ease has an endowment yielding £900 a year; and the Baptists, Friends, Unitarians, Independents, and Wesleyans have their respective places of worship. The other principal buildings include a town-hall and custom-house, comprised in one building erected in 1804; a corn-exchange, assembly-rooms, and a theatre. Here, also, are some good libraries, and literary and other societies. The free grammar school, of ancient foundation, affords instruction to 20 boys, and has two exhibitions of about £70 a year to Magdalen college, Cambridge. Among the other educational establishments, are two charity schools, partly supported by voluntary contributions, but chiefly by bequests; one for boys having an income of above £250 a year, and one for girls, of above £220 a year. There are 18 well-endowed almshouses, and the funds in the hands of the corporation for the support of schools, almshouses, apprenticing of children, loans, and the general relief of the poor, are said by the commissioners of charity inquiry to have amounted to above £1130 a year.

Wisbeach has no staple manufacture; but it has an iron
1079

foundry, several yards for building and repairing vessels and boats, rope-walks, an extensive brewery, and several large malting establishments. The trade of the town is considerable, from its being the emporium of a considerable tract of country. The exports principally consist of corn, wool, rape-seed, and other products of the fens; and the imports of coal, timber, groceries, &c.

The trade of the port has been largely benefited by the great improvements that have been made, under acts passed in 1827 and 1829, in the course of the Nene from Wisbeach to its outfall in the Wash. These consisted principally in deepening and straightening the bed of the river, and in the drainage and embankment of the adjacent fens. This important work cost about £300,000, of which £30,000 was contributed by the corporation of Wisbeach, who were at the same time authorized to levy increased port-dues (now 3d. per ton) on vessels frequenting the port. The latter may now be reached by a comparatively safe and speedy navigation, at spring tides, by vessels of 190 tons burden, and, at other times, by vessels of 60 tons. There belonged to the port, in 1841, 56 vessels of 50 tons and upwards, having an aggregate burden of 3600 tons. The gross customs duty collected at this port, in 1840, amounted to £8501. The increase in the amount of shipping belonging to the port, which has nearly doubled since 1835 when the Nene Outfall was finished, and of the customs duties, shows the substantial advantages it has derived from that improvement.

"The town is apparently prosperous; its trade is extending, the value of property in the neighbourhood has increased, local taxation is stationary, and labour well paid. The inhabitants are said to be characterized by industry and prudence. Education among them is general, and few serious crimes are committed." (*Appendix to Munic. Corp. Report.*)

Wisbeach received its first charter of incorporation from Edw. VI., others being granted to it by James I. and Charles II. Under the Municipal Reform Act, it is divided into two wards, and governed by six aldermen and 18 councillors. The corporation revenue, which principally arises from harbour dues, exceeds £2000 a year. Wisbeach has a commission of the peace, petty sessions, and a court of record for the recovery of small debts; and assizes are held here annually. The ancient castle of Wisbeach was long the episcopal palace of the bishops of Ely, but no traces of it now exist. A cattle market was established here in 1810. Wisbeach has frequently suffered from inundations. Market-day, Saturday. Fairs five times yearly, chiefly for cattle, horses, hemp, and flax. (*Parl. Reps.*, art. BEDFORD LEVEL, in this work, and *Private Information.*)

WISBY, a town of the island of Gottland, which see.

WISCASSET, p. t., port of entry, and one of the capitals of Lincoln co., Me., 27 m. S.S.E. Augusta, 588 W. It has a spacious harbour at the mouth of Sheepscot river, which enters an arm of the ocean. It is open and safe, has a depth of water sufficient for the largest vessels and is not often obstructed by ice. It has considerable shipping, employed in the foreign and coasting trade and the fisheries. Ship building is a considerable business. During the prosperous days of American commerce, previously to 1807, it advanced rapidly in wealth; but it received a check from the non-intercourse and war, from which it is gradually recovering. Its harbour is one of the finest in the state, but its back country is limited. The village is beautifully located on rising ground, overlooking the harbour, and contains a courthouse, several churches, many stores, and some large and handsome dwellings. In the town, chiefly in the village, are five commercial houses in foreign trade, with a capital of $103,600; 21 retail stores, capital $27,800; one furnace produced 25 tons of cast iron; bricks and lime were produced to the amount of $6000; two tanneries produced 75 sides of sole leather and 340 sides of upper leather; two manufactories of leather, as saddleries, &c., produced to the amount of $700; one pottery produced $300; carriages and wagons were produced to the amount of $550; two grist-mills and four saw-mills produced articles to the amount of $35,500, with a capital of $46,500; vessels were built to the amount of $25,500; furniture was manufactured by 20 persons, with a capital of $21,200. Total capital employed in manufactures, $71,150. There were eight schools, 983 scholars. Pop., in 1830, 2255; in 1840, 2314. Tonnage of the port in 1842, was 12,989 tons.

WISCONSIN, territory, is bounded N. by the British possessions; N.E. by Montreal and Menomonee rivers, and a line connecting their sources, separating it from Michigan; E. by lake Michigan, separating it from Michigan proper; S. by Illinois; and W. by Mississippi river, and a line due N. from its source, to the British possessions, separating it from Iowa territory. It is between 42° 30' and 49° N. lat. and between 87° and 93° 54' W. long. It is 600 m. long and 150 broad, containing about 90,009 sq m., or 57,600,000 acres.

It contained in 1840, 30,945 inhabitants. Of these ?, ? were white males; 11,992 were white females; 10? ?? coloured males; 84 were coloured females. Employed in agriculture, 7947; in commerce, 479; in manufactures and trades, 1814; in mining, 479; in navigating lakes, rivers, canals, &c., 229; in the learned professions, 250.

It is divided into 22 counties, which, with their population in 1840, were as follows:

Counties.	Total Pop.	Counties.	Total Pop.
Brown	4,107	Milwaukee	5,808
Calumet	275	Portage	1,623
Crawford	1,502	Racine	3,475
Dane	3,114	Rock	1,701
Dodge	67	St Croix	891
Fond du Lac	139	Sauk	102
Grant	3,926	Sheboygan	133
Green	933	Walworth	2,611
Iowa	3,974	Washington	343
Jefferson	914	Winnebago	135
Manitowoc	235		
Marquette	18	Total	30,945

Madison, between the third and fourth lakes, of the chain of the Four lakes, in Dane county, is the capital. The U. States' government have appropriated $40,000 for the erection of the public buildings, and $5000 for the public library.

The surveyed part of the territory, south of Green bay, Fox and Wisconsin rivers, is composed of timbered land and prairie, with some swamps or wet prairies, having generally a soil from one foot to several feet deep. The crops generally raised in the same latitude may be cultivated with success. They are wheat, rye, Indian corn, buckwheat, barley, oats and potatoes. The abundant pasturage on the prairies render it peculiarly favourable to the raising of cattle. North of the Wisconsin river, the country becomes hilly, and thence northward it swells into a rugged and broken country, producing rapids and water-falls in the streams, and affording much wild and picturesque scenery. Near the sources of the Mississippi is an elevated table land, abounding with lakes and swamps, covered with wild rice, and affording abundance of fish. Bordering on the Mississippi and Wisconsin rivers, the soil is rich, and the surface is generally covered with a heavy growth of timber. White pine is found on the upper Mississippi. About Rock river, and thence to Mississippi river, there is much excellent land, but generally a deficiency of timber. This territory has not been long settled, and its productions do not as present show what the country is capable of producing.

There were in the territory in 1840, 5735 horses and mules; 30,289 neat cattle; 3468 sheep; 51,3?0 swine; poultry was raised to the value of $16,167. There were produced 212,116 bushels of wheat, 1965 of rye, 379,338 of Indian corn; 10,654 of buckwheat; 11,068 of barley; 498,514 of oats; 419,608 of potatoes; 6777 pounds of wool 1474 of wax; 135,289 of sugar. The products of the dairy were valued at $33,677,

The southern part of Wisconsin is an exceedingly rich mineral region, which extends into Illinois and Iowa. Lead ore, yielding 75 per cent of metal, is abundant. This has been, for a considerable time, extensively wrought. Copper ore is also extensively found, and is beginning to be wrought. Iron ore also exists.

Mississippi river washes its western border. (See Mississippi river.) The Wisconsin, a large tributary of Mississippi river, is 500 m. long; Chippewey river is a large stream, and enters the Mississippi further to the N.W. Rock river, another tributary, runs partly in this territory, but principally in Illinois. Neenah or Fox river, which passes so near the Wisconsin that in time of high water the country between them is often overflowed, and can be passed in boats, flows through lake Winnebago, and enters Green bay. Though obstructed by rapids, boats at times pass up it for 180 m. In future time, it may be an important medium of communication between lake Michigan and Mississippi river.

There are many lakes and ponds, particularly in its northern parts. The four lakes, between two of which Madison, the capital is situated, are strung along a stream called Catfish river, which enters Rock river, 25 or 30 m. above the boundary of Illinois, and are beautiful sheets of water. Winnebago lake is 24 m. long and ten broad, and has its outlet, by Fox river, into Green bay. This bay is a branch of lake Michigan which bounds the territory on the E. furnishing facilities for commerce.

The most important place in the territory is Milwaukee, on lake Michigan. It has had a rapid growth, is extensively visited by steamboats from Buffalo and different ports on lake Erie, and is an important commercial centre. It has the only good harbour between Chicago and Green Bay. Green Bay is near the mouth of Neenah or Fox river, at the head of Green bay, and has a good harbour and an extensive trade. Racine and Sheboygan on lake Michigan, and Prairie du Chien on the Mississippi river, just above the mouth of Wisconsin river, are important stations, and growing places.

There were in this territory in 1840, one commercial and seven commission houses in foreign trade, with a capital of $53,000; 178 retail dry goods and other stores, with a capital of $661,550; 153 persons employed in the lumber trade, with a capital of $21,180; 62 persons engaged in internal transportation, who, with three butchers, packers, &c., employed a capital of $14,100; the capital employed in the fisheries was $61,300; the amount of lumber produced, was $302,293; of skins and furs, $124,776.

No great extent of manufactures can be expected in this recently settled territory. Home-made or family manufactures amounted to $12,567. There was one furnace, employing a capital of $4000; 49 smelting houses produced 15,129,350 pounds of lead, employing 520 persons, and a capital of $664,600; one person manufactured 12 small arms; five persons manufactured various metals to the amount of $3500; 43 persons made bricks and lime to the amount of $6527; one tannery produced 150 sides of sole leather, and 150 sides of upper leather, employing three persons, and a capital of $2000; 13 other manufactures of leather, as saddleries, &c., employed 13 persons, producing articles to the amount of $11,800, with a capital of $7002; five persons produced 64,317 pounds of soap, 12,909 pounds of tallow candles, and 48 pounds of wax or spermaceti candles, with a capital of $3432; three distilleries produced $300 gallons of distilled spirits, and three breweries, 14,290 gallons of beer, the whole employing 11 persons, and a capital of $14,400; eight persons produced carriages and wagons to the amount of $2000, with a capital of $325; four flouring-mills produced 900 barrels of flour, and with 29 grist-mills and 124 saw-mills, employed 850 persons, producing articles to the amount of $350,903, with a capital of $561,650; vessels were built to the amount of $7150; 29 persons manufactured furniture to the amount of $6945, with a capital of $5740; six printing offices and six weekly newspapers employed 24 persons, and a capital of $10,300; seven brick or stone, and 509 wooden houses were built, employing 644 persons, and cost $212,085. The total amount of capital employed in manufactures was $635,926.

No college has been established in this territory; but 23,040 acres of land have been granted for a university. The land has been advantageously located. There were in 1840, two academies, with 65 students, and 77 common and primary schools, with 1937 scholars. The 16th section in every surveyed township has been reserved from sale, for the use of schools, but yet belongs to the general government. Schools are supported in most of the neighbourhoods by a public tax.

The government of the territory was organized in 1836. The governor is appointed by the president of the United States, with the consent of the senate: and he is, ex-officio, the superintendent of Indian affairs. The legislative assembly consists of a council of 13 members elected for two years, and the house of representatives of 26 members, elected annually. Their pay is three dollars a day, and three dollars for every 20 m. travel. All the town and county officers are elected by the people.

WISMAR, a town and seaport of N. Germany, in Mecklenburg-Schwerin, cap. lordship of its own name, at the bottom of a deep bay at the confluence of the Stor with the sea, 18 m. N. by E. Schwerin; lat. 53° 49′ 25″ N., long. 11° 36′ 15″ E. Pop. 11,000. The harbour, which is very extensive, is commodious and safe, being nearly land-locked by the islands of Poel and Wallfisch. Close to the town there is from 8 to 8½ ft. water; in the inner roads there is from 12 to 13 ft., and in the outer from 16 to 20 ft. water. The town is fortified, and has six churches, three hospitals, several schools, &c.; with manufactures of tobacco, playing cards, sail-cloth, and other fabrics, breweries and distilleries. Ship-building is also carried on to some extent; and Wismar is second in commercial importance to no town in the grand duchy but Rostock. The articles of import and export are the same as at Rostock (which see), but the trade of the town is more limited. It appears to have been founded in 1229, and afterwards became one of the Hanse towns. It has, however, generally belonged to Mecklenburg since 1648. (*Dict. Géog. and Com. Dict.*)

WISSEMBOURG, a town of France, dep. Bas-Rhin, cap. arrond., on the Lauter, on the Bavarian frontier, 33 m. N. by E. Strasbourg. Pop. in 1836, 5575. It is a fortified town, and of some importance, as it commands a defile leading from the plain of the Rhine into the Vosges mountains, and is connected with a system of military works stretching along the course of the Lauter for some distance, called the lines of Wissembourg. It has an old collegiate church, built in 1288; a Protestant church, in which is a bust of Luther; barracks, &c.; and manufactures of hosiery, straw hats, soap and earthenware. It originated in an abbey founded here by Dagobert II. in the 7th century, and became a free town of the empire in 1297; it was annexed to France by the treaty of Ryswick. (*Hugo; Dict. Géog., &c.*)

WITNEY, a market town and par. of England, co. Oxford, hund. Bampton; on the Windrush, a tributary of the Thames, 10 m. W. by N. Oxford. Area of par. 7450 acres: pop. of ditto, in 1831, 5336; of the town, 3190. It is well-built and cheerful; the main street being on the high road between Burford and Woodstock. The town-hall, a stone building, has beneath it an area used for a market-place. Near it is the market cross, erected in 1683, and repaired by subscription in 1811; and in the High-street is the staple or blanket hall, a handsome stone edifice, built in 1721. The church, at the S. extremity of the principal street, is one of the handsomest in the county, being a large cruciform structure in the early English, decorated, and perpendicular styles, with a tower and lofty spire, ornamented with minarets. In the N. transept is a fine window; and within the building are several ancient monuments, and a handsomely carved and gilded burial chapel. The living, a very valuable rectory, being worth £1290 a year, is in the gift of the bishop of Winchester. Here, also, are places of worship for Wesleyans and Independents. The free grammar school, founded in 1660, for 30 boys, comprising a spacious school-room, library, and apartments for the master. Another free school, with a small endowment, was established in 1663. A school for the education, clothing, and apprenticeship of weavers' sons, was founded in 1723, and it has besides a national school and several almshouses.

Witney was long celebrated as a principal seat of the blanket manufacture; and, in the reign of Queen Anne, the weavers of the town and adjacent district were incorporated into a company. But the trade has long been of very inferior importance, and the weavers' corporation has fallen into disuse. Since the peace especially, and the introduction of machinery into the business, blanket weaving has rapidly declined at Witney, and most part of the fabrics now sold as Witney blankets are, in fact, made in Glamorganshire, and elsewhere. Some rough coatings, tidings for barges and wagons, and felting for paper-makers, are, however, made here; and, in 1838, there were in the parish five woollen-mills, employing 238 hands. (*Mills and Factories' Report, &c.*) The glove manufacture also employs a few hands; wool stapling is carried on to some extent, and the town has considerable trade in malt.

Witney was made a parliamentary borough in the time of Edward II., but its privilege was withdrawn on the petition of the inhabitants in the succeeding reign. It is governed by two bailiffs and two constables, chosen at the annual court leet; and a court baron, presided over by the duke of Marlborough, is opened twice a year. Witney is of considerable antiquity, and its manor is stated to have been one of those given to the monastery of St. Swithin, Winchester, in the reign of Edward the Confessor. Market day, Thursday; fairs five times a year, for cattle and cheese. (*Beauties of England*, art. *Oxford; Lewis's Topog. Dict., &c.*)

WITEPSK, or VITEBSK, a gov. of European Russia, principally between the 55th and 57th degs. of N. lat., and the 26th and 32d of E. long.; having N.E. the gov. of Pskof, S.E. Smolensko and Moghilef, S.W. Minsk and Courland, and N.W. Riga. Area, 18,500 sq. m. Pop., in 1838, 717,708. Surface generally level, though on the banks of the rivers there are occasionally some low hills. Rivers and small lakes are numerous: of the former, which all flow towards the Baltic, the Dwina is the principal. Notwithstanding the soil is but of medium fertility, and agriculture is in a very backward state, more corn is produced than is required to supply the wants of the inhabitants. Hemp and flax are grown on a large scale, peas, beans, hops, fruits, &c. in the smaller inclosures. The forests are very extensive, 121,600 deciatines of forest land belonging to the crown. The grass lands are also extensive, and a good many horses and cattle are reared, though of inferior breeds. The sheep yield only coarse wool; and honey is, also, of inferior quality. The mineral products and manufactures are insignificant; the last being, with the exception of a few cloth factories, almost wholly restricted to distilleries and tanneries. The trade of the government is facilitated by the Dwina and the canal of Berezina: it is principally in the hands of the merchants of the principal towns, many of whom are Jews. This government is divided into 12 circles; chief towns, Witepsk, the capital, Wieliz, Dunaburg, and Polotsk. In 1832 it had 24 public schools, attended by about 1100 pupils, besides three lyceums, and seven private schools.

WITEPSK, a town of Russia, cap. of the above gov., on both banks of the Dwina, where it receives the Viteba, 230 m. S. by W. Petersburg. Pop. about 14,600. It is irregularly built, and is surrounded by old walls: it has numerous Greek and some Roman Catholic churches, convents, and synagogues. Though by far the greater number of its thouses be of wood, it has some dwellings of stone, a high school, a bazaar, an old castle, several hospitals. &c.; with manufactures of woollen cloths and tanneries. The Grand Duke Constantine, brother of the late and present emperor of Russia, died at Witepsk on the 27th June, 1839. (*Schnitzler; Possart, &c.*)

WITTENBERG, a town of Prussian Saxony, formerly the cap. of the Electorate Saxony, now the cap. of a circ. of the reg. of Merseburg, on the Elbe, here crossed by a long wooden bridge, and on the road between Potsdam and Leipsic, 39 m. S.W. the former. Pop., 8400. (*Berghaus.*) Though metamorphosed from the quiet seat of a university into a garrison and fortified town, Wittenberg has a peculiarly dull and melancholy aspect. It is, however, highly interesting, as the cradle of the Reformation; Luther and Melancthon having been professors in its university, and their remains being deposited in its cathedral. A statue of the great reformer in bronze, by Schadow, of Berlin, was erected in the market place, in 1821. "It represents, in colossal proportions, the full length figure of Luther, supporting on his left hand the Bible, kept open by the right, pointing to a passage in the inspired volume. The pedestal on which the statue stands is formed of a solid block of red polished granite, 20 ft. in height, 10 ft. in width, and 8 ft. in depth. On each of its sides is a central tablet, bearing a poetical inscription, the import of the principal being that 'if the Reformation be God's work, it is imperishable; if the work of man, it will fail.' Over the figure is a very handsome light Gothic canopy, supported by four corner pillars, and surmounted by eight filigree-pointed pinnacles. This canopy is beautifully cast in iron. Taken altogether, the monument is a most creditable piece of workmanship, and does honour to the present state of the arts in Prussia." (*Granville's Tour.*, i., 947-8.) The graves of Luther and Melancthon in the cathedral are marked by two plain tablets. The altarpiece is by Lucas Cranach, a burgomaster of Wittenberg, the town-hall of which he has embellished with pictures of Luther and of the subjects of the ten commandments. It was against the walls of this church that Luther suspended his 95 theses against papal indulgences; and outside the E. gate of the town he publicly burned the bull for his excommunication. Luther's apartment in the old Augustine convent remains in much the same state as in his time; and the autograph of Peter the Great on the wall is preserved by a glass covering. Wittenberg, having ceased to be a capital, was found inadequate to the support of its university, which was accordingly removed to Halle. It still has, however, a gymnasium, an ecclesiastical seminary, &c., and is the seat of a board of taxation and of the usual circle courts. From its situation on the Elbe, in a fertile country, with both iron and coal in its neighbourhood, it possesses great commercial advantages; but its trade is insignificant, and it has only a few manufactures of linen and woollen goods. It has frequently suffered from sieges, particularly in 1756; and in 1814, when it was taken by storm from the French. (*Dict. Geog., Hodgskin, Trav. in the N. of Germany; Streng, &c.*)

WOBURN, a market town, and par. of England, county Bedford, hund. Manshead, on the great N. road, 38 m. N.N. W. London. Area of par., 3900 acres. Pop. in 1830, 1827. The town consists of a main street, about 1-3d m. in length, with the market-cross nearly in its centre; and having been nearly burnt down in 1724, it is comparatively well laid out and well built. The market-house, a handsome edifice, originally erected by the Bedford family, was rebuilt at their expense, by Blore, in 1830. The same artist has, also, restored the church, an edifice in the perpendicular style built by the last abbot of Woburn, having a tower detached from its main body. In the chancel is, among others, a curious monument to Sir F. Staunton and his family. The living, a curacy, in the gift of the duke of Bedford, is worth £251 a year. The free school, founded by the earl of Bedford in 1582, has an income of £50 a year, and furnishes instruction to 150 boys on the Lancastrian plan. Almshouses, founded in 1672 for 24 widows, have an income of £30 a year, and there are several minor charities. Petty sessions monthly, and manorial courts occasionally, are held in Woburn. The inhabitants are either occupied in lace-making and straw plaiting or are employed by the Bedford family. Markets on Fridays: fairs four times yearly for farm stock.

Immediately on the E. of the town is Woburn park, with Woburn Abbey, the principal seat of the Duke of Bedford. It derives its name from its occupying the site of a Cistercian abbey, founded here in 1145, and granted to the Russell family in the time of Edward VI. The present mansion, which was built about 1745, has since been greatly improved and enlarged. It is a quadrangular edifice, its principal or W. front being of the Ionic order with a rustic basement. The interior of this noble pile is splendidly fitted up, and many of the apartments are enriched with val-

uable paintings, both by the old masters and British artists. The drawing room, thence called Venetian, has a fine series of 24 views in Venice, by Canaletti. In the hall is an ancient mosaic pavement, brought from Rome. A sculpture gallery, 138 ft. in length by 25 ft. in breadth, with a flat dome in its centre supported by 8 antique marble columns has a fine collection of antique marbles, including the famous Lanti Vase, of Parian marble, 6 feet 3 inches in diameter, and 6 ft. in height, exclusive of the plinth on which it stands. It is of the lotus form, has two magnificent handles, and is beautifully sculptured. This admirable specimen of ancient art was found among the ruins of Hadrian's Villa at Tivoli, of which it had no doubt formed a principal ornament. Here, also, is a fine cast of the Apollo Belvidere, Westmacott's statue of Psyche, &c. In the W. wing of the edifice is the temple of the Graces, erected in 1818, to receive Canova's magnificent group of the Graces, placed on a circular pedestal in the centre. The library is both extensive and valuable; and at one of its extremities is a room appropriated to Etruscan antiquities. The stables, riding-house, tennis-court, &c., are in a detached building, connected with the mansion by a colonnade, ⅓ m. in length. The greenhouse, designed by Sir J. Wyatville, is a handsome building, 160 ft. in length, and in which, with a great variety of valuable plants, are some fine statues. The park, 12 m. in circuit, surrounded by a wall 8 ft. in height, is beautifully diversified, abounding in wood and water, and well stocked with deer. (*Parl. Rep.; Railway Handbook. &c.*)

WOBURN, p. t., Middlesex co., Mass., 10 m. N.W. Boston, 450 W. Incorporated in 1642. The Middlesex canal and the Lowell railroad pass through it. Horn pond, near its centre, is a beautiful sheet of water, furnished with a pleasure sail-boat, and has its outlet into Mystic river. It contains four churches, a Congregational, two Baptist, and a Universalist; twelve stores, 14 forges, two grist-mills, 23 saw-mills, seven tanneries; one academy, 170 students; nine schools, 120 scholars. Pop. 2993.

WOKINGHAM, or OAKINGHAM, a market town and par. of England, co. Berks, hund. Sonning, on the confines of Windsor forest, 6½ m. S.E. Reading. Area of par., 5430 acres. Pop., in 1831, 3139. The town consists of several streets, meeting in a central market-place, in which is the market-house and town-hall. The church, in the perpendicular style, is a fine old edifice. The living, a perpetual curacy, worth £126 a year, is in the gift of — Jacob, Esq. The Wesleyans, Baptists, &c., have meeting-houses here; and, besides Sunday schools, here is a free school, with an income of above £45 a year, at which between 30 and 40 boys are instructed on the Madras system. The aggregate income of the various charities in Wokingham amounts to nearly £530 a year. (*Digest of Charity Rep.*)

About 1 m. from the town is a hospital, under the direction of the Drapers' company, London, founded in 1663, for 16 poor men and a master. The inhabitants of Wokingham are employed principally in weaving silk stuffs and gauze, and in making shoes. The town was incorporated by James I., and is governed by an alderman, recorder, two burgesses, and other officers, chosen on Easter Wednesday, who hold petty sessions and some other courts. Formerly all the courts for Windsor forest were held at Wokingham. Markets, Tuesdays; fairs, April 23, June 11, October 11, and November 2, for horses and cattle. (*Parl. Rep., &c.*)

WOLCOTT, p. t., Lamoille co., Vt., 23 m. N. Montpelier, 530 W. Watered by Lamoille river and its branches which afford water-power. Chartered in 1780. It has one store, one fulling-mill, one grist-mill, five saw-mills; 11 schools, 299 scholars. Pop. 910.

WOLCOTT, p. t., Wayne co., N. Y., 180 m. W. by N. Albany, 360 W. Bounded N. by lake Ontario, into which its streams flow. It contains three churches, a Presbyterian, Methodist, and Baptist; eight stores, one furnace, one fulling-mill, two grist-mills, 12 saw-mills, two tanneries; two academies, 97 students; 13 schools, 730 scholars. Pop. 2481.

WOLFBOROUGH, p. t., Carroll co., N. H., 39 m. N.N.W. Concord, 529 W. Bounded W. by Winnipiseogee lake, into which Smith's pond has its outlet. It contains six churches, a Congregational, two Methodist, two Free-will Baptist, and a Christian; seven stores, two fulling-mills, one woollen factory, three grist-mills, 16 saw-mills, five tanneries; one academy, 95 students; 15 schools, 730 scholars. Pop. 1918.

WOLFENBUTTEL, a town of Germany, duchy of Brunswick, cap. circ., on the Ocker, 8 m. S. Brunswick. Pop. 6500. It was formerly fortified, but its defences are now in a ruinous state. It consists of the citadel, the town proper, called Heinrichstadt, and the quarters or suburbs of Auguststadt and Gottesiager. It is well built, and is clean, which are broad and regular, are paved, and watered by branches of the Ocker. It has several churches worthy of notice, and a magnificent ducal library, comprising not less

1100

than 190,000 vols. It includes a large collection of bibles, among which is the one that belonged to Luther, with autograph notes: his marriage ring, doctor's ring, spoon, drinking-glass, and one of his many portraits by L. Cranach are also preserved here. (*Murray's Handbook.*) The old castle of the lords of Wolfenbuttel has been converted into a prison; and the ducal castle now serves for a factory! It has a large workhouse, hospital, orphan asylum, gymnasium, and several city schools; and is the seat of the superior court of appeal for the states of Brunswick, Waldeck, and Lippe, and of several subordinate courts. It has manufactures of lacquered and japanned wares, paper hangings, leather, tobacco, &c.; with some trade in corn and linen yarn, and five annual fairs. Its neighbourhood is fertile, but marshy and unhealthy. (*Berghaus; Dict. Géog.*)

WOLGA, or VOLGA (an. *Rha*), the largest river of Europe, through the E. part of which it flows; its basin comprising the central part of European Russia, has the basin of the Dwina to the N., of the Don and Dniepr to the S., and of the Oural to the E. The Wolga was formerly considered as constituting a part of the boundary line between Europe and Asia; but since the limits of these continents have been removed to the Caucasus and the Caspian, its basin, with those of its tributaries, lie wholly within Europe. From its source to its mouth its length is estimated at about 2000 m., being about 200 m. longer than the Danube. The area of its basin has been supposed to include upwards of 636,000 sq. m., or considerably more than twice as much as the basin of the Danube, and eight times as much as that of the Rhine. (*Müller, Stromsystem der Wolga.* 79.)

The Wolga has its source in a small lake at the W. extremity of the gov. Tver, in lat. 57° 10' N., long. 32° 20' E., 220 m. S.S.E. Petersburg; on the E. declivity of the Valdai plateau, near the source of the S. Dwina, the Dniepr, and other large rivers, at an elevation of 895 ft. above the level of the sea. (*Müller*, 113.) It flows at first S.E., and afterward N.E. through the govs. of Tver and Jaroslavl; at Molega it turns to the E.S.E., which direction it generally pursues through Jaroslavl, Kostroma, Nijni Novgorod, and Kazan to the confluence of the Kama, about lat. 55° 8' and long. 49° 20'. Thenceforward it runs generally S.S.W. through the govs. of Simbirsk and Saratof to Tzaritzin, where it approaches within 32 m. of the main stream of the Don, their confluence being prevented by an intervening chain of hills.

It then turns again to the S.E. through the gov. Astrakhan, and pours itself into the Caspian, on its N.W. side, through an extensive delta, by more than 70 mouths, the W. and largest of these being in lat. 46° N., and long. 49° E. Throughout its long course it waters, with its tributaries, some of the most productive portions of European Russia and the region which was anciently the nucleus of the Russian monarchy. Tver, Jaroslavl, Kostroma, Nijni Novgorod, Simbirsk, Saratof, Astrakhan, and several other towns are situated on the Wolga; and Kazan is on one of its tributaries, within a short distance of the main stream.

The principal affluents of the Wolga are the Tvertsa, Molega, Shexna, Unja, Vetluga, and Kama, from the N. and E., and the Oka and Sura from the W. and S. The Kama, which is by far the largest, is, also, the last tributary of any consequence which it receives. It rises in the gov. of Viatka, about lat. 58° N., long. 53½° E., and flows with a very tortuous course, at first N.E. to about lat. 58° 20', but afterward in general S. or S.W. through the govs. of Perm and Kazan, and between those of Viatka and Orenburg. After a course of nearly 1000 m. it joins the Wolga, bringing with it a volume of water nearly equal to that of the latter. Its basin is supposed to comprise about one third part of that of the Wolga. Perm is among the towns on its banks.

The Oka rises in lat. 52° 10' N., long. 36° E., in the gov. of Orel; through which, and the govs. of Tula, Kaluga, Moscow, Riazan, Tambof, Vladimir, and Nijni Novgorod, it flows in a very tortuous, but mostly N.E. direction, joining the Wolga at Nijni Novgorod after a course of nearly 700 m. Its basin is supposed to comprise 127,000 sq. m. (*Müller*.) It has several important affluents. Though rapid, it is navigable to Orel not far from its source. The waters of the Kama and Oka are, like those of the Wolga, remarkable for their purity; and all of them are famous for their fish. The Wolga is, in fact, believed to be more prolific of fish than any other European river; and its fisheries are an abundant source of employment and of food. The fish usually taken comprise sturgeon, the roes of which furnish the caviar, of which vast quantities are sent from Astrakhan to all parts of Russia, with salmon, sterlet, tench, pike, perch, belugas, &c. The sterlet (*acipenser stellatus?*), a small kind of sturgeon, supposed to be peculiar to the Russian and Siberian rivers, is much prized by the Russian epicures. Exclusive of caviar, the exports from Astrakhan include large quantities of cured fish. (*Müller*, p. 627; *Med. Trav.*, xvii.)

From its abounding with islands, particularly in the lower part of its course, the breadth of the Wolga is very variable. At Tver, however, it is nearly 600 ft in breadth; at Nijni Novgorod, after it has received the Oka, about a verst, or 1200 ft.; and at Astrakhan it is usually 1½ m. across. But this is not the case during the entire year, for, on the melting of the ice and snow in spring, it is subject to great risings, and inundates large tracts of the surrounding country. The rise begins in April; its height varies greatly in different places, but is greatest in the middle portion of the river's course. At Tver the total rise is about 12 ft. above its summer level; at Jaroslavl and Nijni Novgorod 18 or 20; at Kasan 25 or 30; and at Saratof from 30 to 40 ft.! But downwards beyond this point, after which the Wolga receives no affluent of any consequence, and its bed becomes more capacious; the height of its rise gradually diminishes, being at Tsaritzin from 25 to 30 ft., and at Astrakhan only from 6 to 8, or seldom so high as 12 ft. The time of subsidence also varies considerably in different parts: at Nijni Novgorod the river is commonly confined again within its bed by the beginning of June; at Kasan not till the middle of the same month; and at Astrakhan it does not diminish to its ordinary height till after the summer solstice. According to recent discoveries, the surface of the Caspian is 101 ft. below the level of the Black sea, which would give to the Wolga (estimating its course at 2000 m.) an average descent of about 5·4 inches per mile; from the junction of the Kasan with the Wolga, the fall of the latter, Humboldt says, is greater than that of either the Amazon or the Nile, and almost as great as that of the Oder. (See *Geog. Journ.*, viii., 135.) Though rather a rapid river (*Pallas*, i., 25), yet, as it runs through a flat country, with an immense volume of water, in a bed unbroken by cataracts, though not free from sandbanks, it is navigable for flat-bottomed boats nearly to its source. Not far below this point it is connected by a canal with the S. Dwina, establishing a direct water communication between the Caspian and the Baltic. The Ivanofska canal, in the government of Tula (which unites the Upa, a tributary of the Oka, with the Don), opens a communication between the Caspian and the Black sea; and, by means of the Vischnej Volotchok canal, between the Msta and Tvertsa rivers, and the canal between the Sestra and Istra, in the gov. of Moscow, Petersburg and Moscow are directly connected. Other canals connect the basin of the Wolga with that of the N. Dwina, the lake Onega, &c.: and nowhere else has so extensive a system of inland navigation been effected by artificial means, with so little labour. This navigation is, however, suspended by the frost for at least 160 days each year.

Though the situation of the Wolga, remote from the great marts of Europe, Asia, and Africa, with its embouchure in the Caspian, renders it of much less commercial importance than it would be under other circumstances, it is still the main artery of Russia, and the grand route of the internal traffic of that empire. It has been estimated that in the first 20 years of the present century, from 600 to 700 vessels a year came down the Wolga to Astrakhan, while from 300 to 400 sailed from that port to others on the upper course of the river. Unfortunately, it would seem as if the Wolga had been for some considerable period decreasing in depth; and it is said that of late years sandbanks have accumulated so much, particularly between Nijni Novgorod and Kasan, that the vessels laden with salt from Perm, which in the early part of last century used to bring cargoes of from 130,000 to 150,000 pounds, can now only convey cargoes of about 90,000 pounds; and in the portion of its course now referred to, it is navigated with difficulty even by the two-masted vessels of Astrakhan. (*Müller, Das Stromsystem der Wolga; Berghaus; Stein's Handbook der Geog.; Dict. Géog., &c.*)

WOLSINGHAM, or WALSINGHAM, a market town and par. of England, co. Durham, Darlington ward, on the Wear, 19¼ m. W.S.W. Durham. Area of par., 24,780 acres. Pop., in 1831, 2239. The town is pleasantly situated, but irregularly built. The church is an ancient structure, with a beautiful font of Weardale marble. The living, a valuable rectory worth £791 a year nett, is in the gift of the bishop of Durham. The remains of an old manor-house, belonging to the former bishops, and enclosed by a moat, are near the church. The grammar-school, founded in 1612, with an income of above £65 a year, besides 30 pay scholars, supplies gratis instruction to 26 boys, in reading, writing, arithmetic, and the classics. Here, also, are several other

par. of England, co. Stafford, hund. Seisdon, in one of the principal iron manufacturing districts, and at the junction of six principal roads, 13 m. N.W. Birmingham. The par., which is of great extent, comprises five townships, four of which, including the towns of Bilston and Willenhall, with the adjoining par. of Sedgeley, are included in the parl. bor.; the area of which amounts to 16,630 acres. Pop. of do., in 1831, 67,514, of whom 24,732 belonged to Wolverhampton township. Pop. of the latter, in 1841, 36,189. Wolverhampton stands on an eminence commanding fine views of the surrounding country, and though irregularly laid out, is not ill built; but, from the many furnaces and forges in the town and neighbourhood it has a blackened appearance. There are some good modern residences in the suburbs. Four of the principal streets diverge from the market-place. The town is well lighted, partially paved, and supplied with water from wells sunk in the rock on which it is built. The collegiate church of St. Peter, on the most elevated position in the town, is a large cruciform structure, chiefly in the perpendicular, but partly, also, in the decorated and early English styles. It has a tower, the upper part of which is late perpendicular, and a much finer composition than the other portions of the church. The chancel is modern; the nave has a rich stone pulpit, and in the church yard is a rudely sculptured cross, much defaced by time. (Rickman.) This church was formerly considered one of the king's free chapels, and was attached, by Edward IV., to the deanery of Windsor. The living, a vicarage worth £193 a year, is in the gift of the dean of Windsor. St. John's church, a handsome stone edifice, on the S. side of the town, erected by subscription in 1761, is a curacy worth £300 a year, in the gift of the earl of Stamford. St. George's, a building of Grecian architecture, and St. Paul's, a Gothic structure, are comparatively recent, and two other churches are now (1842) in the course of being erected. Besides a Roman Catholic chapel, there are several Dissenting meeting houses, to all of which, as well as to the churches, well attended Sunday schools are attached. The free grammar-school, founded by Sir Stephen Jennings, a native of the town, who was lord mayor of London, in 1668, has an endowment yielding about £1200 a year. It is managed by 40 trustees, who allow the head master £500 a year. It is open to all boys of the par., and is, at present (1842), attended by about 140 pupils. Sir W. Congreve and Abernethy, the surgeon, were educated at this school. Besides a blue coat charity, for 36 boys and 30 girls, of very ancient foundation, with an income of £940 a year, there are national and British schools; and considerable funds exist for distribution among the poor. A dispensary was established, in 1821, for which an eligible building was erected in 1826, and in 1813, a union-mill, for grinding corn for the poor at a cheap rate, was built by shares at a cost of £14,000. Wolverhampton has also a public news-room with an extensive library, assembly and concert rooms, and mechanics' institute. W. of the town is a fine race course, with a grand stand, where races take place annually in August. Two weekly newspapers are published in the town. Most departments of the hardware manufacture, excepting cutlery, are carried on here and in the adjacent town of Bilston and the district. Wolverhampton has been long celebrated for her locks, of which she almost exclusively engrossed the manufacture down to a comparatively late period. At present, however, the most expensive and best locks are made in the metropolis, and the lock trade is also carried on in Birmingham; but this town still enjoys the largest share of the business. Probably, however, the manufacture of japanned ware and tinned plates may be regarded as her staple business; and in addition she furnishes carpenters' tools, files, screws, hinges, steel mills, machinery, &c. Immense quantities of nails are made in the surrounding villages.

Wolverhampton, Bilston, and the other places within the limits of the parl. borough, are wholly indebted for their rapid rise and large population to the facilities they enjoy for carrying on the iron trade. In the vicinity are all but inexhaustible mines of coal and ironstone, the main bed of coal being 30 ft. thick, with strata of ironstone above and below! The district has also the further advantage of being connected by numerous canals with all the great shipping ports of the empire. Under such circumstances, we need not be surprised at the rapid progress it has made since 1779, when there was only one blast furnace at Bilston! whereas in 1842 there was in the parl. borough no fewer than 55 blast furnaces, capable of producing more than 4000 tons of iron a week! Indeed, the whole country to the S. and E. of the town is covered with furnaces, forges, rolling mills, foundries, ironstone and coal pits; and though the trade be at present (1842) much depressed, there being about 24 furnaces out of blast, the advantages enjoyed by the district are such as can hardly fail to ensure its prosperity.

The Grand Junction railway has a station 1 m. E. of the town, and here the line attains its greatest elevation of 440 ft. above the sea level. The Birmingham, Staffordshire

and Worcestershire, and Birmingham and Liverpool canals unite about 1 m. N. from the town, affording, as already stated, a ready communication with all parts of the kingdom. The Wolverhampton and Staffordshire Banking company, established in 1831, has its office in this town.

The town is under the jurisdiction of the county magistrates, but is locally governed by two constables and other officers, chosen at an annual manorial court. Petty sessions for the hund. are held here, and a court of requests monthly for debts under £5.

The Reform Act conferred on Wolverhampton, Bilston, and the district included within the parl. borough, the important privilege of sending 2 members to the House of Commons. Reg. electors in 1839-40, 2643.

The new poor law was introduced here in 1836, when the townships of W. Hampton, Bilston, Willenhall, and Wednesfield were formed into the "W. Hampton Union." The population of these townships amounted, in 1831, to 46,931; and in 1841, to 66,185. A superior and spacious "Union-house" has been lately erected.

Though of great antiquity, the earliest records of Wolverhampton date only from the end of the 10th century, when Wülfruna, duchess of Northampton, founded a monastery here, of which, however, there are now no remains. A fire, which continued for five days, destroyed the greater part of the town in 1590.

Market days, Wednesday and Saturday; fairs, July 10th, 11th, and 19th, for cattle and various merchandise. (Parl. Rep.; Priv. Inf.)

WOOD, county, Va. Situated in the N.W. part of the state, and contains 1923 sq. m. Bounded N.W. by Ohio river. Drained by Little Kanawha river, and its branches. It contained, in 1840, 7691 neat cattle, 13,885 sheep, 19,482 swine; and produced 71,339 bushels of wheat, 263,637 of Indian corn, 85,199 of oats, 21,491 of potatoes, 87,891 pounds of tobacco, 5844 of sugar. It had 25 stores, four flouring-mills, 10 grist-mills, 21 saw-mills, one oil-mill, four tanneries, one printing-office, one weekly newspaper: four academies, 119 students; 13 schools, 307 scholars. Pop.: whites, 7061; slaves, 624; free coloured, 56; total, 7923. Cap. Parkersburg.

WOOD, county, O. Situated toward the N.W. part of the state, and contains 590 sq. m. Bounded N.W. by Maumee river, which is navigable for vessels requiring 6 ft. of water to Perrysburg, 18 m. from Lake Erie. Drained by Portage river, and its branches. It contains a large portion of the "Black Swamp," an unhealthy spot, but capable of being drained. It contained, in 1840, 6497 neat cattle, 2799 sheep, 10,694 swine; and produced 29,004 bushels of wheat, 1382 of rye, 119,508 of Indian corn, 4150 of buckwheat, 24,49 of oats, 54,250 of potatoes, 2949 pounds of tobacco, 43,175 of sugar. It had 12 stores, four flouring-mills, one grist-mill, 10 saw-mills; 97 schools, 717 scholars. Pop. 5357. Capital, Perrysburg.

WOODBRIDGE, a market town, par., and river-port of England, co. Suffolk, on the Deben, 8 m. from the sea, and 7½ m. E.N.E. Ipswich. Area of par. 1630 acres. Pop. of do., in 1831, 4769. The town is on the slope of a hill, and consists of two principal streets, an open space called Market-hill, and some narrow thoroughfares; it has many good houses, and is tolerably well paved and lighted. In the centre of the Market-hill is the sessions house, a brick edifice, in the lower part of which the corn market is held. The church, a noble edifice, said to date from the time of Edward III., consists of a nave, chancel, and two side aisles, the roof of which are supported by 14 fine slender pillars; its outer walls are constructed of black flints, and it has a square tower 108 ft. in height. On its S. side was anciently an Augustine priory, founded by one of the Ross family in the 13th century. The living, a perpetual curacy, worth £430 a year, is in the gift of the Rev. T. Salmon. Here, also, is a free grammar school for 10 boys, some of the poorer inhabitants of the town, who are to be instructed in Latin and Greek, and fitted for the university; with national, Lancastrian, and Sunday schools. Almshouses for 12 poor men and 3 women, founded and endowed in 1587, had, in 1825, an income of £570 a year; but as this income was derived from a lease of property in Clerkenwell, in the city of London, entered into about 60 years previously, and that about to expire, it is now probably much greater. The income of the town lands is chiefly applied to parochial repairs, &c. A small theatre was erected in 1843.

Woodbridge is a member of the port of Yarmouth, and the Deben being navigable thus far, for vessels of 120 tons, it has a considerable trade with London, Hull, Newcastle, &c., exporting corn, flour, and malt, and importing coal, timber, wines, spirits, groceries, &c. It has several docks for the building of vessels, with convenient wharfs and quays. Woodbridge is governed by a visitor and two constables, chosen by the parishioners. Quarter sessions for the liberty of St. Etheired and for six adjacent hundreds are held here; and petty sessions weekly. Market day, Wednesday; fairs, April 5 and Oct. 31, for cattle.

WOODBRIDGE, p. t., Middlesex co., N. J., 43 m. N.E. Trenton, 209 W. Drained by Rahway river. It contains a part of Rahway v. It was incorporated in 1680, and in 1668 there were 120 families in the township. It has a church, 22 stores, one cotton factory; one academy, 43 students; 16 schools, 491 scholars. Pop. 4291.

WOODBURY, p. t., Litchfield co., Ct., 50 m. W. Hartford, 308 W. Drained by Pomperaug river, and its branches. It contains four churches, two Congregational, a Methodist, and an Episcopal, eight stores, three fulling-mills, two woollen factories, two grist-mills, four saw-mills, three tanneries; 12 schools, 327 scholars. Pop. 1948. The village is built chiefly on one extended street, and contains three of the churches, and about 80 dwellings, some of them neat and handsome.

WOODBURY, p. v., capital of Gloucester co., N. J., 37 m. S.S.W. Trenton, 148 W. Situated on Woodbury creek, which flows into Delaware river. It contains a spacious courthouse of brick, fire proof county offices, a jail of stone, three churches, a Presbyterian, Methodist and Friends, an academy, two public libraries, one of which belongs to the ladies, 10 stores, over 700 dwellings, and about 800 inhabitants.

WOODBURY, t., Huntingdon co., Pa. It has three stores, three furnaces, two forges, one flouring-mill, five grist-mills, 13 saw-mills, two tanneries, three distilleries; five schools, 146 scholars. Pop. 3108.

WOODBURY, p. t., Bedford co., Pa., 113 m. W. Harrisburg, 129 W. Drained by Frankstown branch of Juniata river, and by Yellow creek. It has 11 stores, four furnaces, eight flouring-mills, 11 saw-mills, one tannery, four potteries; one academy, 33 students; eight schools, 947 scholars. Pop. 3044.

WOODCOCK, p. t., Crawford co., Pa., 213 m. N.W. Harrisburg, 315 W. Drained by Woodcock creek. It has two stores, two fulling-mills, one flouring-mill, five grist-mills, 10 saw-mills, one oil-mill, one paper-mill, three tanneries, two distilleries, one brewery. Pop. 1991.

WOODFORD, county, Ky. Situated N.E. of the centre of the state, and contains 154 sq. m. Bounded W. by Kentucky river, by branches of which it is drained. It contained in 1840, 9633 neat cattle, 16,229 sheep, 21,852 swine; and produced 93,591 bushels of wheat, 35,681 of rye, 643,735 of Indian corn. 109,502 of oats, 15,845 of potatoes, 13,860 pounds of tobacco, 1410 of sugar. It had three commission houses in foreign trade, 27 retail stores, one cotton factory with 250 spindles, 30 flouring-mills, 29 grist-mills, 19 saw-mills, three tanneries, 12 distilleries; four academies, 104 students; four schools, 94 scholars. Pop.: whites, 5816; slaves, 5752; free coloured, 172; total, 11,740. Capital, Versailles.

WOODFORD, county, Ill. Situated N. of the centre of the state, and contains 470 sq. m. Drained by Mackinaw river, a tributary of Illinois river. Bounded W. by Peoria lake. Formed since the census of 1840. Capital, Woodford.

WOODSTOCK (NEW), a parl. and music. bor., and market town of England, co. Oxford, hund. Wootton, par. of Bladon; on the small river Glyme, which supplies the magnificent piece of water in Blenheim park, 8 m. N.N.W. Oxford. The old parl. bor. included only a portion of the town; but the modern parl. bor. includes the whole of the latter, with a district extending about 4 m. on every side, comprising several adjacent villages and hamlets, and having an area of 21,640 acres, and a pop., in 1831, of 7855. This is a clean, well built, country town; the streets are well paved, and well kept, and many of the houses, which are mostly of stone, are of a superior class; but it has, notwithstanding, a dull and inanimate appearance, and is not prosperous. The church is a handsome structure, partly rebuilt in 1785, when a tower was added to its W. extremity. The living is a curacy annexed to the rectory of Bladon, worth £399 a year, and in the gift of the duke of Marlborough. Here, also, are places of worship for Baptists and Wesleyans. The town hall, erected in 1766, at the expense of the then duke of Marlborough, has under it the market place. A free grammar school, founded in 1585, affords instruction to about 30 boys; but the endowment for its support is small, and the master's salary of £30 a year is partly made up by the corporation: it has, besides, an endowed school, with an income of £75 a year, in which 24 children are educated, partially clothed, and apprenticed, with almshouses for widows, and several minor charities.

Woodstock had formerly a considerable manufacture of polished steel articles, much esteemed for their delicate workmanship; but this business is now nearly or wholly extinct. The manufacture of doe-skin gloves, which was introduced at a later date, is now almost the only branch of industry carried on in the town. In 1835, upwards of 1200 hands mostly women and girls, were employed in this manufacture in Woodstock and its vicinity, who were supposed to produce about 500 pairs of gloves per week. But this branch has, also, declined in the interval, principally,

as is understood, in consequence of the substitution of thread and cotton for leather gloves; and in consequence of this decline, and of the embarrassed situation of the Marlborough family, the prosperity and population of the town have both fallen off.

The borough received its present charter of incorporation from Henry VI.; but that by which it has been latterly governed dates from the 16th of Charles II., the corporate body consisting of a mayor, four other aldermen, a high steward, recorder, and other officers. The annual court leet or sessions, granted to the borough by charter, has been discontinued since 1829; but petty sessions, and a court of record are opened monthly. The borough has no jail, but a small lock-up house.

Woodstock was a borough by prescription previously to its incorporation, and returned two members to the House of Commons from the 13th Elizabeth down to the passing of the Reform Act, which deprived it of one member, at the same time that the boundary of the parl. borough was altered, as already stated. Previously to the Reform Act the franchise was vested in the mayor, aldermen, and freemen of the borough; but substantially, and in fact, it was a nomination borough, belonging to the duke of Marlborough.. Registered electors, in 1839-40, 369. Market day, Tuesday; fairs, seven times a year, chiefly for cattle, horses, cheese, and hardware.

Old Woodstock stood in a low situation a little N. of the town, on the Glyme, and has now only a few houses and one ancient mansion. Woodstock was long a royal residence. A palace, or manor-house, on the N. bank of the Glyme, was the residence of Henry II., and the scene of some of the adventures of the fair Rosamond; but all traces of this building have long since disappeared. Edward I. held, in 1275, a parliament at Woodstock: and it was also the birth place of his second son Edmund, and of the eldest son of Edward III., the illustrious Black Prince. It was subsequently inhabited by several of our monarchs; and Elizabeth was for awhile imprisoned here. But every part of this more recent palace has also been pulled down.

Chaucer, the great improver of the English language and versification, is supposed by many to have been a native of Woodstock; where it is alleged, he first saw the light in 1328. At all events he frequently resided in the town: and some traces still exist of the house which he occupied.

At present Woodstock derives its whole importance from its being in the immediate vicinity of Blenheim palace and park, the seat of the duke of Marlborough. In previously noticing this noble seat (see BLENHEIM PARK, vol. i.) we omitted to notice the library, originally intended for a picture gallery, and one of the finest apartments in England. It extends along the whole of the W. front, being 183 ft. in length, and beautifully proportioned. It contains the Sunderland collection of books, comprising 17,000 vols.; and a fine statue of Queen Anne by Rysbrack. In the chapel is a splendid monument by the same sculptor, in honour of the great duke of Marlborough and his duchess. (*Bound. Rep.*; *Munic. Corp. Append.*; *Priv. Information.*)

WOODSTOCK, p. t., capital of Windsor co., Vt., 51 m. S. Montpelier, 485 W. Drained by Otta Quechee river and its branches, which afford water-power. The principal village is one of the largest in the county, and among the principal in the state, built on both sides of Otta Quechee river. The houses are mostly situated around a beautiful park, and it contains a handsome courthouse, a jail, five churches, a Congregational, Methodist, Episcopal, Christian, and a Universalist, the Vermont Medical college, 350 dwellings, and about 1400 inhabitants. The town contains one church, besides the above, 12 stores, one fulling-mill, two woollen factories, three grist-mills, five saw-mills, three tanneries, two printing-offices, two weekly newspapers; one academy, 25 students; 16 schools, 1042 scholars. Pop. 3315.

WOODSTOCK, p. t., Windham county, Ct., 43 m. E.N.E. Hartford, 379 W. Watered by Muddy brook, a tributary of Quinnebaug river, and by a branch of Nachaug river. Muddy brook affords good water-power. It contains six churches, four Congregational and two Baptist, 16 stores, three woollen factories, three cotton factories with 3392 spindles; three academies, 95 students; 90 schools, 909 scholars. Pop. 3053.

WOODSTOCK, p. t., Ulster co., N. Y., 69 m. S. Albany, 398 W. Drained by tributaries of Esopus creek. It contains two churches, two stores, one glass house, one glass-cutting works, four grist-mills, 12 saw-mills; seven schools, 372 scholars. Pop. 1691.

WOODSTOCK, p. v., capital of Shenandoah co., Va., 150 m. N.N.W. Richmond, 104 W. Situated on the W. side of the N. fork of Shenandoah river, about 1 m. from its bank. It contains a courthouse, jail, county offices, a masonic hall, an academy, three churches, a Lutheran, Methodist, and German Reformed, a number of stores, 195 dwellings, and about 1000 inhabitants.

WOODVILLE, p. v., capital of Wilkinson co., Miss., 135 m. S.W. Jackson, 1145 W. It contains a courthouse, jail, a market-house, a bank, two academies, three churches, a Methodist, Episcopal and Baptist, a number of stores, and about 800 inhabitants. A railroad 27½ m. long, connects with St. Francisville, La.

WOOLER, a market town and par. of England, co. Northumberland, Glendale ward, on the E. declivity of the Cheviot hills, 42 m. N.N.W. Newcastle. Area of par. 4690 acres. Pop. in 1831, 1996. The town is of high antiquity, and at one period was a good deal resorted to by invalids. It consists of several streets branching from the market-place, and has a public library, mechanics' institute, dispensary, and many dissenting places of worship. The church is a neat but plain building, erected about the middle of last century; the living a vicarage worth £478 a year, is in the gift of the bishop of Durham. A free school, with a small endowment, is the only charity mentioned as existing here. Courts leet and baron are held annually by the lord of the manor. In the neighbourhood are the vestiges of ancient encampments. At Homildon, about 2 m. from the town, is a pillar, erected to commemorate the total defeat in 1402, of an army of 12,000 Scotchmen, under Earl Douglas, by the forces of the Earls Percy and March. Markets on Thursdays; fairs, May 4th, October 27th, and Whit-Tuesday.

WOOLWICH, a parl. bor., market, and seaport town of England, co. Kent, on the S. Bank of the Thames, 7 m. E. London, and 2½ m. E. Greenwich. Pop. in 1831, 17,661. The older parts of the town, near the river, have narrow streets, and are mean, dirty, and badly built, with but little prospect of being much improved; but in the more modern portions, and especially between the old town and the barracks, the streets and houses are of an improved and far more respectable description. Woolwich, however, derives its entire importance from its dockyard, arsenal, and other great naval and military establishments; and is principally inhabited by individuals dependent upon or connected with them. The church in a conspicuous situation, with a tower at the W. end, was rebuilt towards the middle of the last century. The living a valuable rectory, worth £740 a year nott, is in the gift of the bishop of Rochester. Besides the church, Woolwich has an ordnance chapel on the road to Plumstead, a chapel in the barracks, a proprietary episcopal church near the arsenal, a Scotch church, and various meeting-houses for different classes of dissenters. The charitable institutions comprise an almshouse for 5 poor widows, endowed in 1560, by Sir Martin Bowes, lord mayor of London, and two charity schools, one for 30 girls, and one for educating, clothing, and apprenticing poor orphan boys, sons of such shipwrights as have served their apprenticeship in the dock-yard; and national and other public schools have been established, in which about 900 children are daily instructed, exclusive of the Sunday-schools attached to the parish church and the dissenting chapels, where upwards of 2000 children are taught. Among the places of amusement is a small theatre. A mineral spring on the N. of the common possesses valuable medicinal properties.

Dock yard.—This, though not the most extensive, is the most ancient royal dock-yard in the kingdom. Some uncertainty exists as to the precise date, but it is believed to have been established as early as 1512; and it is certain it was placed upon a permanent footing in the latter part of the reign of Henry VIII. It presents a frontage to the river of nearly 4000 ft., but is of irregular breadth. It has been of late years, greatly improved and extended, and contains at present about 55 acres; but it is in contemplation to increase it to about 62 acres. It has six building slips; two for first-rates, two for line-of-battle ships of inferior size, and two for smaller ships.

Although some very large ships have been built in this dock-yard at different periods, such as the Royal George (lost at Spithead), the Nelson in 1814, and the Trafalgar in 1841, yet considerable disadvantages are felt in the constructing of such ships from the shallowness of the water and the accumulation of mud in the river; and it has, in consequence, been determined to build the largest class of men-of-war in the other dock-yards, and to make this the principal yard for steam-ships belonging to the royal navy. With this intention a factory was built in 1839, consisting of two ranges of handsome and substantial buildings, each 440 feet in length by 95 in breadth, for the manufacture of steam-engines and boilers. The factory contains a foundry and forges, with fan-blowing machines for the furnaces, and a variety of machines for punching, shearing, plate-bending, drilling, boring, planing, turning, shaping, bolt-screwing, &c., &c. The machinery has been made by the most eminent manufacturers in the kingdom, and embodies the results of all the improvements of the last twenty years; its skilful selection and adaption reflecting infinite credit on the chief engineer.

The moving power of the factory is a steam-engine of 30

horse-power. Another building is to be immediately erected, at an expense of £14,000, 222 ft. in length by 55 ft. in breadth, with a chimney 190 ft. in height, into which all the flues of the factory will be led. This addition will give the means of doubling the work performed, and will, it is expected, enable the admiralty to furnish the whole of the steam ships with the necessary machinery.

There is at present an outer basin, comprising an area of 120,000 sq. ft., in which ten or a dozen steam-ships of different sizes may conveniently lie. An inner basin of 100,000 sq. ft. area, on the site of the mast pond, is also in a forward state of construction, and when completed will allow two first class steam-ships to lie alongside the factory and be simultaneously fitted with their engines, boilers, and other machinery. A magnificent dock, entirely composed of massive blocks of granite, has lately been built, capable of receiving a 120 gun-ship: a second is nearly completed, and it is proposed to construct two similar docks for steam-ships, making four in all.

The smithery, constructed by the late Mr. Rennie, is on a very grand scale, and is suitable for the forging of the largest anchors, and other heavy articles. It contains 67 forges, with 8 lift hammers weighing 4½ tons each, and 3 tilt hammers of 18 cwt. each; these are also 3 air and 2 blast furnaces, with a blasting apparatus of a very scientific description. These are two steam engines in the smithery, one of 20 and another of 14 horse-power. The anchors are tested by a powerful hydraulic machine, made by Bramah, capable of applying a pressure of 100 tons. The pitch and tar vault is constructed so as to admit of its easy inundation in case of fire. The ships now (1842) building are the Royal Albert, 120; Boscawen, 70; Colchester, 50; Amphion, 36 guns; with two steam-ships. Frames are in preparation for the Hannibal of 90 guns, a brig and 2 steam-ships.

The number of workmen of all kinds employed amounts at present (1842) to 1306, exclusive of about 700 convicts, landed daily from the Warrior hulk, moored off the yard who are generally employed in the most laborious occupations.

The mast houses, mast slip, mast ponds, warehouses, &c., are all on a very extensive scale, and the whole is thoroughly organised and in the most efficient state. A large ropery was formerly attached to the dock-yard, but, its situation not admitting of the introduction of machinery, it has been abandoned, the buildings removed, and the site built upon.

Arsenal.—Woolwich is the head quarters of the ordnance military corps, viz., the royal horse and foot artillery, the royal sappers and miners, &c.; and it is also the principal establishment in the kingdom for the provision of warlike materiel for the navy and army. The royal arsenal, formerly called the Warren, admirably situated on the E. side of the town, appears to have been used as a gun-wharf and powder magazine from about the middle of the 17th century; but it was not until the removal of the foundry from Moorfields, in 1720, that it began to assume the importance it has since attained. It comprises within its boundaries about 110 acres, or including a part of the Plumstead marshes, used as a review and practising ground, from which it is separated by a canal communicating with the Thames, it may extend over nearly 300 acres. The following are the principal departments into which it is divided: 1st. The royal brass foundry, erected by Sir John Vanburgh, in 1719, on a site chosen by a young Swiss named Schalch, under whose direction the foundry had been placed on its removal from London. In it are now cast all the best ordnance used in the service. It contains 3 furnaces the largest capable of melting 16 tons of metal, sufficient for about twelve 24-pounders; with accommodation for making moulds, boring, turning, and polishing the guns. Iron guns are also bored up from smaller to larger calibres; these, however, are not made by the board of ordnance, but are supplied by private manufacturers. In 1840, 355 guns were cast in the foundry, which appears to be the highest number cast in one year. The statue for the duke of York's monument was cast here. Horse power has hitherto been used, but a steam-engine of 12 horse-power has recently been erected. In this department, which is under the inspector of artillery, all guns are proved before being used.

2nd. The royal laboratory, in which every kind of ammunition, viz. ball and blank cartridge, &c., is made up; Congreve and other rockets, grenades and fireworks manufactured; shells and spherical case-shot, or Shrapnell shells, filled, &c. A machine, invented by Napier, for making musket and pistol balls by compression, has been amply introduced, which acts with great ease and rapidity, and seems to be far preferable to the old mode of casting. Percussion caps for the service have also, within the last few years, been entirely made in the laboratory; the whole of the machinery designed for filling and finishing the caps is exceedingly ingenious, and performs its work with great rapidity and safety, one man and four boys being able to ma-

distance, able to fill and complete 150,000 caps a day! All this machinery is, however, at present worked by hand. Two model rooms are attached to this department, containing specimens of every firework used in war, and many other interesting objects.

3rd. The *royal carriage department* in which every kind of ship and land gun carriage is made and repaired; with traversing platforms, ammunition wagons, carts, &c. Copper-lined powder-cases are also made. The sheets of copper are tinned instantly, on both sides, by a process displaying much ingenuity, the invention of one of the working tinmen employed here, superseding a tedious and unwholesome operation before in use, and effecting a considerable annual saving. This establishment contains a planing machine, invented by Bramah, set up in 1806, with apparatus for various operations in turning of wood and metal, worked by a steam-engine of 12 horse-power. Here is also a saw-mill, by Sir J. Brunel, set up in 1812, containing six frames of saws, wrought by a steam-engine of 20 horse-power. The smithy has about 90 forges.

The works in the manufacturing departments of the arsenal are in a perfectly efficient state; but, generally speaking, the machinery, except for percussion caps and musket balls, does not seem to be of the most modern or scientific kind. The separation of the departments under different authorities appears also to be in some degree adverse to the more general employment of steam power, except at an increased expense, thus leaving more to be effected by manual labour than would be requisite if a system of greater concentration and combination were adopted.

4th. The guns, which are arranged in the open air, comprise complete field and battering trains, mortars, howitzers, carronades, &c., with the guns belonging to many of the ships of war out of commission, numbering, in the whole, about 3000 pieces of brass, and 19,000 pieces of iron ordnance, of 200 different varieties. The shot and shells, arranged in pyramidal piles, amount to nearly 2,000,000 in number. These, as well as every other description of store required for naval or military equipment, are kept in constant readiness, under the charge of the *storekeeper's department*. The various piles of brick buildings containing these articles, or appertaining to the departments before enumerated, are constructed on the grandest scale: and are as complete and efficient as can well be imagined. There are two *butts*, one within the arsenal and the other in the marshes, for the practice and proof of ordnance; but, as the shot frequently cross the river, the service is liable to constant and serious interruptions from the passing of ships. A new practice range of nearly 3 m. extent, towards Erith, is therefore in course of formation, which will be free from the inconveniences now experienced; and will besides allow of the practice, at long ranges, of the new heavy guns (42 to 84 pounders), on Monk's and Millar's principles. The number of artificers, labourers, and boys employed in the various departments of the arsenal may amount to about 700, exclusive of the convicts belonging to the Justitia hulk, stationed on the river, opposite to the arsenal, amounting to about 300.

The royal military academy, an institution that has considerably raised the professional character of the corps of royal engineers and the royal regiment of artillery was founded in 1719, but was not finally arranged until 1741. The establishment, which has varied at different periods, has at present 165 pupils, termed gentlemen cadets. It is under the direction of the master-general and board of ordnance for the time being, and has a lieutenant governor, inspector, and other officers. For the literary department, there is a professor of mathematics, and 23 other professors and masters for mathematics, fortification, plan-drawing, surveying, chemistry, landscape-drawing, German, French, history, geography, &c. The academy has numbered among its professors several eminent men, among whom may be specified Mr. Thomas Simpson, Dr. Hutton, author of the *Mathematical Dictionary* and other valuable works, Mr. Bonnycastle, Dr. Olynthus Gregory, &c. The establishment formerly cost the public £6000 or £10,000 per annum, for half the number of cadets now educated; but for several years past it has been conducted upon a self-supporting principle: and with a much enlarged and a more efficient establishment, it now nearly maintains itself. The scale of payments to be made by the friends of cadets is as follows:

	Per ann.
Sons of noblemen and private gentlemen, not being officers of the army or navy	£195
Sons of admirals, and generals with regiments	80
Sons of generals without regiments	70
Sons of captains and commanders in the navy, and colonels and regimental field officers of the army	60
Sons of all officers of the navy and army under the above ranks	40
Sons of officers who have died in the service, and whose families are in pecuniary distress	20

The sons of civil officers of the army and ordnance are admitted upon payment of the sums required from military officers of corresponding ranks.

The cadets receive an annual allowance of £45 12s. 6d. which is considered sufficient to supply every necessary article except linen. The education is excellent; the standard having been much raised of late: when the course is completed, the cadets, if found duly qualified, receive commissions in the royal engineers or royal artillery, according to their merit.

The academy, formerly within the arsenal, was removed in 1806 to a fine building on the upper and of the common, about 1 m. S. from the town. This edifice, which was built by Wyatt, consists of a centre and two wings, united by corridors, with a range of building behind, containing the hall, servants' offices, &c. The centre forms a quadrangle, with octagonal towers at the angles; and, besides a variety of other apartments, contains the four teaching rooms or academies, as they are termed. The wings contain the apartments for the cadets and chief officers. The building, which is about 200 yards in length, is of brick, stuccoed over.

The barracks, erected for the accommodation of the royal artillery at different periods from 1783 to 1810, are situated on the N. brow of the common, and form a most extensive pile of building, calculated to accommodate 3000 officers and men, and 1900 horses. The principal front, 340 yards in length, consists of six ranges of brick building. The entrance consists of a handsome gateway, with Doric columns and military trophies. Exclusive of other apartments, this noble building contains a library and reading-room for the officers, a mess room, a guard room, a chapel capable of accommodating 1000 persons, &c. At a little distance from the back of the chapel is a riding-school, on the model of an ancient temple.

The barracks for the corps of royal sappers and miners are a short distance to the N.E. of the artillery barracks: they are constructed for 230 men.

The parade is in front of the barracks; and the open space on the common affords sufficient room for exercising the soldiers in the throwing of shells, and ball-firing. On the E. side of the barracks, on the descent leading to the arsenal, is the ordnance hospital, an extensive edifice, calculated to accommodate about 500 patients. It has a valuable library and museum. Several detached buildings, for the use of the artillery, have also been raised on different parts of the common; among which is a veterinary hospital for the horse brigade, with stalls for 66 sick horses. A division of the royal marines have also barracks and a hospital at Woolwich.

Perhaps, however, the most interesting establishment at Woolwich is the repository, on the S.W. side of the barrack-field, for the reception of models of different fortified places, ships, warlike instruments and machines of all kinds, trophies taken in war, &c. The collection, which is alike extensive and valuable, is partly contained in the rotunda, a circular apartment 115 feet in diameter, originally erected by George IV., in Carlton gardens, for the entertainment of the allied sovereigns when on a visit to this country in 1814. Near the repository has lately been erected an observatory for the use of the officers, containing a telescope and other philosophical instruments, and a museum.

The parish of Woolwich is governed, under a local act, by 30 commissioners, chosen by the parishioners, besides the rector and churchwardens. The reform act constituted Woolwich a portion of the parliamentary borough of Greenwich (which *see*). Petty sessions are held here by the county magistrates on Mondays and Fridays, and a court of requests for the adjudication of claims under £5 every alternate Friday. Market days, Wednesdays, Fridays, and Saturdays. *(Parl. Rep. and Private Information, obtained from the best sources.)*

WOONSOCKET FALLS, p. v., Cumberland t., Providence co., R. I., 15 m. N.N.W. Providence, 414 W. Situated on the falls of Blackstone river where it descends over a precipice of rocks by two falls, one of 16 and the other of 14 feet, making 30 feet in the whole. The village is partly in Smithfield, where it is called *Bernon*. On the Cumberland side it contains four churches, a Baptist, Methodist, Congregational, and Universalist, a bank, 100 stores, 16 cotton and two satinet factories with 50,826 spindles, 1988 looms, the whole employing 1162 hands, producing 9,739,717 yards of cloth, of which 5,645,000 were cotton prints, 3,117,000 cotton sheeting, 156,000 satinet, 364,000 jean, 62,400 negro cloth, 548,000 flannel. It has about 2500 inhabitants. *Bernon* has an Episcopal church, two cotton factories with 8000 spindles, and about 500 inhabitants. The Blackstone canal passes through the village.

WOOSTER, p. t., capital of Wayne co., O., 51 m. S. by W. Cleveland, 93 m. N.E. by N. Columbus, 344 W. The village is situated near the junction of Killbuck and Apple creeks, and contains a courthouse, jail, county offices, a bank, a U. States land office, five churches, a Presbyterian, Covenanters, Methodist, Baptist, and Dutch Reformed, 25 stores, two printing-offices; one academy, 25 students; two schools, 64 scholars; 150 dwellings, and 1913 inhabitants. The t. contains 29 stores, eight flouring-mills, one grist-mill, 11 saw-mills, one oil-mill, seven tanneries, one distillery, two printing-offices, two weekly newspapers; and, exclusive of the village, 1907 inhabitants.

WOOTTON BASSETT, a mun. bor., market-town, and par. of England, co. Wilts, hund. Kingsbridge, 78 m. W. London. Area of the par., 4380 acres. Pop. of do., in 1831, 1896. The town, which stands on a hill, consists almost wholly of one street, about ½ m. in length, and tolerably well built, in the centre of which is the town-hall. The parish church is an old building, in the mixed style; the living, a vicarage worth £461 a year, is in the gift of the earl of Clarendon. It has, also, a chapel for Independents; a free school, founded in 1688, affording instruction to about 90 children; with charity and Sunday schools. It has no manufactures of any kind; and would hardly, indeed, have been worth notice, but for the circumstance of its having returned two members to the House of Commons from the 25th of Henry VI., down to the passing of the Reform Act, by which it was disfranchised. It was reckoned too inconsiderable to be included in the provisions of the Municipal Reform Act. (*Mun. Corp. Appendix; Private Information.*)

WOOTTON-UNDER-EDGE, a market town and par. of England, co. Gloucester, hund. Berkeley, on the declivity of a hill, as its name implies. 17 m. S. by W. Gloucester. Area of par., 4890 acres. Pop. of do., in 1831, 5482; of the town only, 5004. The latter, traversed by a small stream, on which are several cloth mills, consists principally of two well built parallel streets. The church, which is large and handsome, has some curious old monuments. The living, a vicarage, worth £112 a year, is in the gift of the dean and chapter of Oxford. The grammar school, founded in the 8th of Richard II., and revived in the time of James I., had, at the date of last inquiry, an income of £376 12s. a year: it supports 10 foundation scholars and is free to all other boys born in, or inhabiting Woton and North Nibley. The boys are instructed in writing and accounts, and are " to use no language in the school but Latin." The Bluecoat school, established in 1693, has an income of £94 a year; the general hospital, for 12 almspeople, has a clear income of £346 a year; and Perry's hospital, also for 12 persons, an income of nearly £170 a year. The aggregate amount of the public endowments in the parish reached, at the period referred to above, £1130 a year. (*Digest of Charity Reps.*) Wotton-under-Edge is a borough by prescription, but has no extant charter: its corporation, consisting of a mayor and 12 aldermen, elected at an annual court-leet, has no municipal functions, revenues, or emoluments. The inhabitants of the town and surrounding district are chiefly occupied in the weaving of woollen cloth. In 1838, there were in the parish five woollen-mills, wrought principally by steam, employing, in all, 189 hands (*Mills and Factories' Report*); besides which, there were many looms wrought by the weavers in their own houses. Petty sessions for the hund. are held in Wotton. Markets on Fridays; fairs, Sept. 25th, for cattle and cheese. (*Parl. Reps., &c.*)

WORCESTER, an inland co. of England, having a very irregular outline and several detached portions, is bounded on the N. by the cos. of Salop and Stafford, W. by Hereford, S. by Gloucester, and E. by Warwick. Area, 462,790 acres, of which about 400,000 are supposed to be arable, meadow, and pasture. This is an extremely beautiful and well-watered county. It is traversed from N. to S. by the Severn, and in part, also, by its important tributaries the Avon from the E., and the Teme from the W. Surface finely diversified: the Malvern hills divide the S.W. part of the county from Herefordshire; the Bredon hills, to the S. of Pershore, have an elevation of nearly 900 feet; and there are some considerable hills on its N. frontier between Hales-Owen and Bromsgrove. The vales of Worcester and Evesham, or rather of the Severn and Avon, are alike beautiful and fertile; but the soil in other parts, especially on the E. side of the county, is cold and poor. Besides corn, cattle, and dairy produce, Worcester produces large quantities of fine wool, apples, hops, and excellent cider. We are sorry, however, to have to state that agriculture is by no means in an advanced state. "The system followed is, in itself, a bad one, and is carelessly and negligently conducted. There is no rotation as to cropping; nor are any pains taken to relieve the ground from water, though it be in many places very wet. Ploughing is badly performed, and the whole management of a slovenly description." (*Kennedy and Grainger, on the Tenancy of Land.* §, 358.) In 1838 there were in this county, 1392 acres under hops. Estates of all sizes; farms for the most part
1086

small. Average rent of land, in 1810, 22s. 4d. an acre. Coal is found in the N. parts of the county, and the brine springs of Droitwich furnish immense quantities of salt. The city of Worcester is the principal seat of the leather glove manufacture; the iron, hardware, and glass manufactures are carried on with spirit and success at Dudley; Kidderminster is famous for its carpets; and needles and fish-hooks are made to a greater extent at Redditch and Feckenham, in this county, than anywhere else in England or, indeed, in the world.

Worcestershire is divided into five hundreds, and 171 parishes. It sends 12 members to the House of Commons, viz., four for the county, two each for the city of Worcester and the borough of Evesham, and one each for the borough of Kidderminster, Bewdley, Droitwich, and Dudley. Registered electors for the county, in 1839-40, 10,917, whereof 6328 were for the E., and 4589 for the W. division of the county. In 1841, Worcester had 46,962 inhabited houses, and 233,484 inhabitants, of whom 114,753 were males, and 118,731 females. Sum expended on the relief of the poor in 1838-39, £62,168. Total annual value of real property in the county assessed to poor-rates, in 1840-41, £363,942.

WORCESTER, a city, pari. and mun. bor. of England, locally situated in the county of Worcester, of which it is the cap., but forming a county of itself; hund. Oswaldslow, on the Severn, crossed here by a handsome stone bridge of five arches, 25 m. S.W. Birmingham, and 100 m. W.S.W. London; lat. 52° 9' 30" N., long. 2° 0' 15" W. The city and old municipal borough, comprising 250 acres, constitute about 1-4th part of the modern parliamentary and municipal borough, which had, in 1831, a population of 26,225; but the population of the modern parliamentary borough in 1841, has not yet been stated.

Worcester is finely situated on the E. bank of the river, in a fertile and beautiful country; and is one of the best built, handsomest towns in the kingdom, having every appearance of wealth and respectability. "The main streets are wide, well paved, and lighted with gas; the central street, which traverses the city from N. to S., is of considerable length, and kept particularly clean and neat. A considerable extent of ground comprised in the suburbs is unoccupied by houses, and consists principally of gardens and meadows. It is probable, however, that a great portion of it will, in process of time, form the site of new buildings. On the E. side of the canal the number of houses is rapidly increasing; and it appears probable, that the area included within the new boundary to the N. will be gradually filled up by villa residences. To the N. both within and beyond the boundary of the Tything, the building of streets, and rows of dwelling houses, as well as of detached residences of a very superior class, is going on rapidly." (*Parliamentary Boundary Report.*) Of the public buildings, the principal is the cathedral, originally founded in 1084, but not completed till 1374. It stands towards the S. extremity of the town, and its appearance has recently been greatly improved by taking down the ancient church of St. Michael, a most ungainly edifice, which stood close to the N.E. extremity of the cathedral, and greatly marred the prospect. Its exterior is extremely plain, and its architecture consist principally in its size and the lightness of its architecture. The central tower, though the stone be much decayed, is extremely fine. It is built in the form of a double cross; its external length, including buttresses, is 426 feet, and the internal 394 feet; the nave, from the front to the W. transept is 180 feet in length: the tower, which is about 200 feet in height, is ornamented with light and elegant pinnacles. The general character of the building is early English; there are, however, some earlier parts. A crypt, part of the nave, the chapter-house, &c. are Norman; parts of the nave and aisles are decorated: the cloisters, and a fine S. porch are perpendicular. The interior is very spacious, mostly in the early English style, with elegant details, and good groining. Across the S.E. transept is the monumental chapel of Prince Arthur, son of Henry VII., and in the middle of the choir is the tomb of King John, the top stage of which, with the effigy, is evidence of a date soon after his decease. (*Rickman, 388.*) There are several fine ancient monuments in different parts of the church, including that of the celebrated Judge Littleton, one of the founders of the English law, a justice of the common pleas, under Edward IV., who died in 1481. Among the more modern monuments is that of the celebrated scholar Stillingfleet, author of the *Origines Sacra*, and other valuable works, bishop of the see from 1689 till his death in 1699, with an extravagantly eulogistic inscription written by the famous Dr. Bentley. Adjoining the S. side of the cloister is the ancient refectory of the monastery attached to the cathedral, a lofty and spacious hall, 120 feet in length, by 38 feet in width, now occupied as the king's school. On the E. side of the cloisters is the chapter house, the library belonging to which comprises a valuable collection of books and manuscripts, chiefly relating to theology and common law. The nett revenue of the bishopric amounted, at an average of the three years ending with 1831, to £6569 a year. The chapter, previously to 1836, consisted of a dean, 10 prebendaries, and eight minor canons. The canons or prebendaries were then reduced to eight, including the dean, and their emoluments were also considerably reduced. The income of the establishment averages about £12,000 a year. The bishop's palace, on the bank of the river, is an incongruous but commodious edifice formerly surrounded with walls. It has lately been arranged that the episcopal palace shall in future be the deanery; and the bishop is to reside at his house at Hartlebury, 10 m. distant. An ancient gateway, called Edgar's Tower, leads into the precincts of the cathedral. There are remains of several monastic establishments, including a commandery of the hospital of St. John, in different parts of the city; and it had formerly a castle, every trace of which has been obliterated, excepting the mound on which the keep stood. Several of the parish churches deserve notice: that of St. Andrews has a square tower, 90 feet in height, surmounted by an octagonal spire 155 feet 6 inches in height, making the total elevation of the latter 245 feet 6 inches. The church and tower are very ancient; but the spire is comparatively modern, having been added in 1751. All the livings in the city, except that of St. Peter, are rectories; and, except All Saints (which is in the gift of the crown), they are all in the patronage of the dean and chapter. The most valuable are St. Martin's, worth £378; St. Nicholas, worth £260; and St. Peter's, worth £233 a year. Exclusive of its numerous churches, Worcester has chapels for Roman Catholics, Independents, Calvinists, Wesleyans, Friends, &c. The guild-hall, a large brick edifice, constructed in 1723, the front of which is ornamented with columns, statues, &c., has a hall for the accommodation of the courts of assize, a council chamber, &c. The old county jail and house of correction was defective in its plan and accommodation; but a new jail at the top of Foregate-street was built on Howard's plan, in 1824. The old city jail, in Friar-street, was formerly a Franciscan convent. The market-house, in the High-street, is a spacious and convenient building; and the public subscription library, in Foregate-street, contains reading and news rooms, and a considerable collection of books. Here is also a small theatre, built in 1780.

The royal grammar school attached to the cathedral was founded in the reign of Henry VIII., for 40 scholars, who are prepared for the universities, and instructed besides in various subordinate departments of knowledge. It has two exhibitions to Balliol college, Oxford. The free grammar school, founded by Queen Elizabeth, in 1561, for 12 boys, has 14 exhibitions to the universities, of £30 each, and scholarships at Worcester college, Magdalen-hall, Oxford. The great Lord Somers and Samuel Butler, author of *Hudibras*, were educated in this school. Here also are subscription schools on the Madras and Lancasterian plans, with several other schools for children of both sexes. The whole number of children at school may be estimated at about 3000. A diocesan board of education for superintending all the schools in connexion with the church in the diocese has recently been established. It has various almshouses, the oldest of which appears to be those of St. Oswald's hospital, founded in 1268: a city and county infirmary, erected in 1770; a lying-in institution, house of industry, female penitentiary, dispensary, &c. Several medical and other societies have been formed: a music meeting is held every third year in the hall of the king's school, the meetings in the intervening years being at Gloucester and Hereford; the proceeds are appropriated to the relief of widows and orphans of the clergy. Races take place in July and November.

Worcester had formerly a considerable manufacture of woollen goods; but this has been discontinued, and the chief business of the city consists at present (1842) of the manufacture of gloves and china and fine earthenware. The produce of the glove manufacture was 10 years ago estimated at about 50,000 dozen pairs oil-leather or beaver gloves, and from 450,000 to 500,000 dozen pairs kid and lamb skin gloves. This trade has, however, greatly fallen off. In May, 1842, there were only 98 glove manufactures, whereas, in 1826, there were 140; and the total annual produce does not at present (1842) exceed 5000 dozen pairs beaver or oil-leather gloves, and 250,000 dozen pairs kid and all other kinds, including 80,000 dozen pairs thread gloves. China ware is produced here on a very extensive scale; and it is not too much to affirm that it is equal, both as respects quality and beauty of design, to any made anywhere else in the kingdom. The materials used in the manufacture are mostly brought from Cornwall. The other principal products are lace, spirits, tanned leather, nails, and turnery ware, &c. There are some large iron foundries on the canal and river banks. The inland trade is carried on by means of the Worcester and Birmingham canal and the Birmingham and Gloucester railway, the nearest station of which is at Spitchley, 3 m. from the city. The canal communicates with the Severn, which is here navigable for large barges; and its banks have been furnished with good quays and spacious warehouses. After several frustrated attempts on the part of joint stock companies to procure an act for improving the navigation of the river, a bill for that purpose received the royal assent in May, 1842. It places the management of the contemplated improvement in the hands of public commissioners, who are to be elected annually by the cities, towns, and land owners along the banks of the river, within the distance which is intended to be improved. A continuous depth of 8 feet water is expected to be obtained from the entrance of the Berkeley canal at Gloucester to Worcester; and from thence a depth of 6 feet to the entrance of the Staffordshire and Worcestershire canal at Stourport. The whole extent of the river to be improved is nearly 40 m. Mr. Cubitt, the engineer, has drawn up the plan adopted for the improvement. Its leading features are a succession of solid weirs built obliquely, instead of in the usual way, at right angles, across the stream, with side cuts or locks. Ascending the river, the first weir is placed a little below the city of Worcester. The channel is left clear to that point. The fall in this portion of the river being very gradual, it is proposed to cut out the shallows, and keep them clear by dredging. The greatly higher scale of gradients in the inclination of the river above Worcester rendered the employment of weirs and locks inevitable. The hop plantations of Worcestershire extend over about 1832 acres, and most part of the produce is brought hither for sale. The society of Worcester is extremely good, it being the residence of many respectable families, attracted hither by the beauty of the situation, cheapness of living, facilities of education, and the variety of amusements furnished by its theatre, assemblies, concerts, clubs, races, &c. It publishes four weekly newspapers, and has a savings' bank and four other banks, three of which are native establishments, and the fourth a branch of a metropolitan joint stock bank. In the year ending the 19th of June, 1841, there were sold in the Worcester corn-market 49,562 qrs. wheat, 13,968 do. barley, 1274 do. oats, 4912 do. beans, 912 do. peas, and 25 do. rye.

Worcester was chartered in the 1st of Richard I.; but the charter was not confirmed until the 2d of Henry III. Various other charters, &c., were afterward granted by different sovereigns; but that by which the city was governed previously to the Reform Act dated from the 19th of James I. It erected the city and liberties of Worcester into a county separate from, and independent of, the county of Worcester. Under the Municipal Reform Act, the city is governed by a mayor, 11 other aldermen, and 36 councillors. Corporation revenue in 1840, £3300. It has a commission of the peace, with jurisdiction in nearly all felonies, excepting such as affect life and limb. A sheriffs' court is held once a month by the under-sheriff; and a court of common pleas, and petty sessions weekly, besides several inferior courts. The assizes for the county are also held here. A police force has been organized, and the peace of the town is well maintained. Worcester has returned two members to the House of Commons since the 23d of Edward I.: the right of voting previously to the Reform Act having been vested in the freemen. Registered electors in 1840-41, 3037. It is also the seat of election and principal polling-place for the W. division of the county.

Worcester is of great, but uncertain, antiquity. It is principally celebrated in history from its giving name to the decisive victory obtained here by Cromwell over the forces of Charles II., on the 3d of September, 1651. Among other eminent individuals, Worcester gave birth to the distinguished statesman, Lord Somers, born here in 1652. Markets, Wednesday and Saturday. Fairs, eleven times a year, mostly for cattle, lambs, horses, linen, hops, cheese, &c. (*Parl. Round. and Mun. Reps.; Rees' Cyclopædia; Priv. Information.*)

WORCESTER, county, Mass. Situated in the central part of the state, and contains 1360 sq. m., being the largest county in the state extending quite across it, from N. to S. Watered by Blackstone, Quinnebaug, Nashua, Ware, Miller's, and Mill rivers. The Blackstone canal extends through its S. part, from Worcester to Providence; the railroad from Boston through Springfield to Albany passes through it from E. to W., and at Worcester is met by the Norwich and Worcester railroad. It contained in 1840, 67,566 neat cattle, 26,128 sheep, 26,272 swine; and produced 45,799 bushels of wheat, 84,914 of rye, 372,591 of Indian corn, 13,871 of buckwheat, 54,254 of barley, 375,471 of oats, 1,146,092 of potatoes, 15,000 pounds of tobacco. It had 289 stores, four furnaces, one forge, 74 fulling mills, 42 woollen factories, 71 cotton factories with 137,358 spindles, six flouring mills, 144 grist mills, 320 saw-mills, 15 paper-mills, two powder mills, 58 tanneries, one distillery, one brewery, 10 printing-offices, six weekly newspapers and one periodical;

23 academies, 1949 students; 549 schools, 24,406 scholars. Pop. 95,313. Capital, Worcester.

WORCESTER, county, Md. Situated in the S.E. part of the state, and contains 700 sq. m. Bounded S.E. by the Atlantic, W. by Chesapeake bay. Drained by Pocomoke river and its branches. Several islands extend along its coast, and enclose Sinepuxent bay. It contained in 1840, 14,765 neat cattle, 14,994 sheep, 19,900 swine; and produced 30,679 bushels of wheat, 487,462 of Indian corn, 161,759 of oats, 33,441 of potatoes. It had 45 stores, 24 grist-mills, 15 saw-mills, five tanneries; four academies, 212 students; 24 schools, 600 scholars. Pop.: whites, 11,765; slaves, 3539; free coloured, 3073; total, 18,377. Capital, Snow Hill.

WORCESTER, p. t., capital of Worcester co., Mass., 39 m. N.N.W. Providence, 42 m. W. by S. Boston, 394 W. It is in 46° 16′ 13″ N. lat., and 71° 48′ 10″ W. long. Pop. in 1830, 4172; in 1840, 7497. Incorporated in 1684, but not organized until 1722. Watered by Blackstone river and its branches, which afford water-power. The village is one of the finest in New-England, and among the finest inland places in the United States. The houses, which are mostly of brick, and many of them elegant, are built on many streets. Main-street is a broad and handsome street, over a mile in length, crossed by other streets, generally at right angles. It contains a courthouse, jail, four banks, the hall of the American Antiquarian society, the state lunatic asylum, and 11 churches, three Congregational, a Unitarian, a Methodist, two Baptist, Episcopal, Friends, a Universalist, and a Roman Catholic. There is a manual labour high school, and the Roman Catholic college of the holy cross. A new and splendid courthouse is in the process of erection. The hall of the Antiquarian society has a central building 46 feet long and 36 feet wide, with a neat Doric portico; and has two wings 28 feet long and 21 feet wide. It has a library of 12,000 volumes, containing many rare and valuable works relating to American history, and interesting specimens of early printing, nearly half of them a donation from Isaiah Thomas, Esq., its first president, and author of the "History of Printing." The lunatic asylum consists of a centre building 76 feet long and 40 feet wide, four stories high, with two wings each 90 feet long and 36 feet wide, and three stories high. At each end of the wings are two other buildings, 134 feet long and 34 feet wide, forming, with the main building, three sides of a spacious square, all built of brick. The interior arrangements are admirably fitted for the accommodation of the different classes of patients; and, on the whole, it is one of the best institutions of the kind in the country. Worcester enjoys great facilities for trade. It is surrounded by a fertile and highly cultivated country. The Blackstone canal extends alternately on both sides of Blackstone river, 45 m. to Providence, and cost $750,000, more than half of which was contributed by citizens of Rhode Island. The Boston and Worcester railroad extends 44 m. to Boston, and cost $1,500,000. It is extended westward through Springfield, and is continued to Albany. The part in Massachusetts cost $3,000,000. The Norwich and Worcester railroad extends 58 m. to Norwich, Conn., and has been continued down Thames river to Gale's ferry, in Ledyard, whence the line is continued by steamboats to New-York city, and by a steam ferry-boat to the Long Island railroad at Greenport, and thence 96 m. to New-York. These several modes of communication render Worcester one of the greatest thoroughfares in the country, facilitate its trade, and cannot fail to add to its growth and prosperity. In the t. there were, in 1840, 99 stores, with a capital of 243,000; machinery was produced to the value of $90,000; one furnace employed a capital of $20,000; granite and marble were produced to the amount of $12,000; lumber was produced to the value of $3000; bricks and lime to the amount of $3600; one cotton factory, with 1672 spindles, employing 30 persons, produced to the amount of $95,000, with a capital of $20,000; one woollen factory produced to the amount of $75,000, with a capital of $40,000; 28 persons produced hats and caps to the amount of $28,000, and straw bonnets to the amount of $11,000, with a capital of $18,000; manufactories of leather, as saddleries, &c., produced to the amount of $35,800, with a capital of $18,000; two persons produced 8000 pounds of soap and 1000 pounds of tallow candles, with a capital of $3000; two paper-mills, employing 14 persons, produced to the amount of $20,000, with a capital of $14,000; 30 persons produced carriages and wagons to the amount of $20,000, with a capital of $6000; 12 persons manufactured furniture to the amount of $30,000, with a capital of $15,000. There were two grist-mills, two saw-mills, four printing-offices, two binderies, four weekly newspapers and one periodical, 15 brick or stone and 10 wooden houses were built. The total amount of capital employed in manufactures was $600,000. There were three academies, 190 students; 30 schools, 1468 scholars.

WORCESTER, p. t., Otsego co., N. Y., 57 m. S W. Albany, 371 W. Drained by Charlotte river and it tribut r e-. It

contains two churches, six stores, two fulling-mills, four grist-mills, 16 saw-mills; 10 schools, 580 scholars. Pop. 2300.

WORKINGTON, a market town, par., and seaport of England, co. Cumberland, ward Allerdale, on the Derwent, near its mouth, 7 m. N. by E. Whitehaven. Area of parish, 7730 acres. Pop. of parish in 1831, 7196, of whom about 6400 inhabited Workington township. The streets are mostly narrow and inconvenient; but of late years many good houses have been erected. In the upper town a new square has been built, in which the corn-market is held. It has a small neat theatre, assembly and news rooms, and various other public buildings. The Derwent is crossed here by a stone bridge, of three arches. The parish church, rebuilt in 1760, has a fine altar-piece: the living is in the gift of —— Curwen, Esq., whose mansion, Workington-hall, a fine castellated structure, on a richly-wooded height, overlooks the town. A chapel-of-ease, in the Tuscan style, was built in 1823; and here, also, are chapels for Independents, Methodists, Presbyterians, Roman Catholics, &c. A grammar school, founded in 1664 by Sir P. Curwen, has since ceased to exist; the founder having had only a life interest in the property with which it was endowed. There are, however, Lancastrian and female schools, a dispensary, and various institutions for the benefit of the poor, supported by subscription. Workington has manufactures of sailcloth and cordage, and a valuable salmon fishery on the Derwent, the property of the earl of Lonsdale; but it derives its principal importance from the extensive collieries in its vicinity, belonging to the Curwen family, which furnish considerable quantities for shipment to Ireland, the Isle of Man, &c. Workington harbour is protected by a breakwater, and has good quays; but it nearly dries at low water. It is a creek of the port of Whitehaven; but in 1842 it had 46 ships of the burden of 6715 tons. A little business is carried on here in ship-building, and in the importation of timber and other articles. Markets, Wednesday and Saturday; fairs, May 18th and October 18th. Races are held here annually in August.

WORKSOP, a market town and par. of England, co. Nottingham, wapent. Bassetlaw, on the Ryton, a tributary of the Idle, 23 m. N. Nottingham. Area of par., 18,290 acres. Pop. in 1831, 5566. The town, consisting chiefly of one street, crossed by two others, is well built, paved, and lighted. Its church, which formerly belonged to an Augustine priory, is a fine old edifice, with two lofty towers, and has within several ancient monuments. The living, a vicarage worth £388 a year clear, is in the gift of the duke of Norfolk. Here, also, are places of worship for various sects, a national school, and some small endowments for parochial and charitable purposes. The town is celebrated for its malt, and was formerly, also, celebrated for its liquorice, of which large quantities were raised in the adjoining district; latterly, however, its culture has been wholly abandoned. Worksop may be regarded as the capital of the district popularly called the "dukery," from its containing Worksop-manor, formerly a seat of the duke of Norfolk, Clumber-park, the seat of the duke of Newcastle, and Welbeck abbey, the seat of the duke of Portland. There are also magnificent residences in fine parks. Recently, however, the duke of Newcastle has purchased Worksop-manor and the house, which had been rebuilt in 1763, came to been, or is to be immediately, pulled down. Clumber park, now about 11 m. in circumference, and finely laid out and wooded, was, so late as the era of the American war, into better than a black heath, interspersed with bogs and morasses; in this district also, is Thoresby, the seat of Earl Manvers. Markets on Wednesday. Fairs, March 31st and June 21st, for cattle and sheep; October 14th, for horses and cattle; and, about three weeks after the last, a statute fair. (Beauties of England and Wales, art. Nottingham, and Private Information.)

WORMS (an. Borbetomagus), a city of W. Germany, grand duchy of Hesse Darmstadt, prov. Lower Hesse, cantant, on the W. bank of the Rhine, near the border of Rhenish Bavaria, and 26 m. S.S.E. Mentz. Pop. about 8500. (Berghaus.) It was formerly an important free city of the empire, but is now much decayed, and is surrounded by dismantled and ruined walls. "Its interior," says Mr. Chambers, "consists of a single good street, lined with its mansions, inhabited by persons of an inferior order, and a number of back lanes and detached buildings, many of them vacant and desolate. In a piece of open ground behind the main street, stands the cathedral, a building of red sandstone, its foundation dating as far back as the beginning of the 11th century (Schreiber, Guide du Rhin, says for 966). The original part of this edifice is Gothic, but the larger portion is in the Byzantine style; the interior arch being all rounded, and the pinnacles and domes tinted in the Moorish taste. The building contains a number of excellent pieces of sculpture, and the high altar at the E. end is environed with ancient carvings, in oak." (*Tour on the Rhine*, 60.) This cathedral has two choirs, each surmounted by a cupola, one of which rises 137 feet (*Schreiber*) above the pavement. The W. choir is a good specimen of the architecture of the 12th century, and has a magnificent rose-window of that period. The Lutheran church in the market-place, in which is a painting of Luther before the diet of Worms, in 1521, occupies the site of the council-hall, in which that event took place. This hall was destroyed when the city was bombarded by the French, in 1689; at which time, also, a vast number of houses were destroyed. From this period, in fact, the decay of Worms may be dated; many of the inhabitants having afterward settled in other German towns, and in Holland. Some of the other churches deserve notice; and there are two synagogues. Charlemagne was married at Worms; and it was frequently inhabited, both before and after his time, by the Frankish sovereigns: but no remains of the imperial palace exist, except a few fragments of a wall, forming part of the *Bürgerhop*, a prison, and police-office. Worms is the seat of a consistory, about half its inhabitants being Protestants; and it has several convents and hospitals, a gymnasium, and elementary schools, supported by different religious sects. It has manufactures of tobacco, sealing-wax, hats, &c.; but its principal trade is in wine, and other agricultural produce. The vicinity of Worms, celebrated by the ancient Minnesingers as the *Wonnegau*, or "land of joy," is in great measure covered with vineyards, producing some of the best growths of the Rhine. The famous *Liebfrauenmilch* is grown around the church of Notre Dame, close to the city. Worms is supposed to owe its origin to a fort erected here by Drusus: many Roman antiquities have been discovered in and near it. Among the councils held at Worms, that in 1122 was the most famous. Diets were also held here in 1492, 1517, and 1521. The latter is famous from the fact of Luther having, as already stated, appeared before it, to explain and answer for his opinions. On appearing before the diet, he displayed equal firmness and moderation. An edict was, however, issued against him on the 26th of April, by which he was excommunicated as an obstinate heretic. But previously to this, in consequence of the determination of the emperor and the other princes who had given him a safe conduct, not to forfeit their word, he was allowed to withdraw from the city in safety. (*Schreiber*; *Berghaus*; *Robertson's Charles V.*, book ii.)

WORTHING, a maritime town and fashionable watering-place of England, on the English channel, co. Sussex, rape Bramber, hund. Brightford, in the par. of Broadwater, a village about a mile to the N., 49 m. S. by W. London, and 10 m. W. Brighton. Area of par., 2240 acres. Pop. of do. in 1831, 4576, of whom nearly 4000 were supposed to belong to the town. Until within the last 50 or 60 years, Worthing was only an inconsiderable fishing village; and is much indebted for its increase and celebrity to the visits of the royal family during the latter part of last century. The buildings of the town extend along the coast for about three quarters of a mile, and the main street runs for somewhat more than half a mile in a N. direction. Excepting in its centre, the houses are mostly arranged in regular terraces, though many are isolated and interspersed with fields and gardens. The ranges of building fronting the sea are generally faced with cream-coloured brick made from a peculiar clay found in the vicinity, and the town is well paved, lighted, and has an ample supply of water. The chapel-of-ease, erected in 1812, at a cost of £12,000, is a very neat edifice, with a Doric portico. The living, a curacy worth £150 a year, is in the gift of the rector of Broadwater. Here, also, are chapels for Independents and Wesleyans, to which Sunday schools are attached; with well attended national schools for both sexes, supported by subscription; a savings' bank, and a small, but elegant, theatre, opened in 1807. The market place consists of ranges of covered stalls built around a square area. The Esplanade, a raised causeway, extends along the shore for the whole length of the town: near its W. extremity are the Royal baths, comprising two complete suites of apartments. The New Parisian baths, recently built, adjoin the Sea House hotel. It is almost superfluous to add, that it has numerous hotels, with assembly-rooms; libraries, reading and news rooms, convenient bathing machines; and the other accommodations incident to a well-attended watering-place. Fine sands extend along the coast for 7 m. to the W., and 3 m. to the E., of the town. These, with the gradually increasing depth of the water, which gives the opportunity of bathing at any time of the tide, added to the mildness of the climate, in consequence of the shelter afforded on the N. and E. by the South Downs, render Worthing especially suitable as a place of resort for invalids. No manufacture of any kind is carried on; but the mackerel and herring fisheries are usually very productive, and contribute largely to the supply of the London markets. An annual fair is held on the 20th of July; market day, Saturday and every alternate Wed-

nesday, for corn. (*Guide to Watering-places of England; Priv. Inf.*)

WORTHINGTON, p. t., Hampshire co., Mass., 17 m. W.N.W. Northampton, 112 m. W. Boston, 395 W. Drained by Westfield river. Incorporated in 1762. It contains two churches, a Congregational and a Methodist, seven stores, one fulling-mill, three grist-mills, seven saw-mills, two tanneries; one academy, 155 students; 11 schools, 429 scholars. Pop. 1197.

WRENTHAM, p. t., Norfolk co., Mass., 22 m. S.S.W. Boston, 418 W. Drained by branches of Charles, Taunton, and Neponset rivers. It contains three churches, a Congregational, Baptist, and Universalist, six stores, four cotton factories with 3500 spindles, three grist-mills, 10 saw-mills; one academy, 150 students; 18 schools, 767 scholars. Pop. 2915.

WREXHAM, a parl. bor., market town, and par. of Wales, co. Denbigh, hund. Bromfield, on the high road between Shrewsbury and Chester, 11 m. S. by W. the latter. The parish of Wrexham includes no fewer than 12 townships, two only of which, and a small detached portion of a third, are included in the parliamentary borough, which has an area of about 1145 acres, and had, in 1831, a population of 5484.

Wrexham is a handsome and lively town, with spacious streets crossing each other at right angles, and neatly and substantially built houses; it is also well paved, lighted with gas, and plentifully supplied with water. The church, dedicated to St. Giles, a large and venerable structure, is deservedly regarded as one of the principal ecclesiastical edifices in the principality. It was erected about 1472, on the site of a more ancient structure: it is in the perpendicular style, and is covered with grotesque sculpture: but in correctness of design and proportion it is surpassed by few buildings of the same date. It consists of a nave, with side aisles and a chancel, the whole length being 178 feet, and the breadth 73 feet; the aisles are separated from the nave by clustered columns supporting pointed arches; and the ceiling is of oak, in imitation of groined stone. The tower, which was not completed till about 1506, is 135 feet in height: it consists of several successive stages pannelled throughout, and decorated with numerous statues of saints placed in niches of the buttresses; which latter are surmounted by four light open-work turrets rising 94 feet above the balustrade that surrounds the summit of the tower. It has a fine altar-piece, and some interesting monuments; among which are two to members of the Middleton family, admirably sculptured by Roubiliac. The living, a valuable rectory, worth £746 a year, is in the gift of the bishop of St. Asaph. It has also a Roman Catholic chapel, and four places of worship for Protestant dissenters; a house of correction for the county, with seven wards; a free endowed grammar school; two parochial national schools; a public library; reading, news, and lecture rooms; agricultural and horticultural societies; a neat theatre; and a property, yielding £230 a year, for distribution among the poor, and other charitable purposes. The town-hall, at the head of High-street, has a large room used for public meetings. Annual races take place in October. The town is under the jurisdiction of the county magistrates, who here hold monthly petty sessions for the hundreds of Bromfield and Gale. The Reform Act conferred on Wrexham the privilege of voting in the return of a member to the House of Commons, along with the boroughs of Denbigh, Holt, and Ruthin. Registered electors for Wrexham in 1839–40, 259; do. for the entire boroughs, 941. It is one of the polling-places for the county. No particular branch of trade or manufacture is now carried on here; though Leland describes it, some centuries since, as containing "cum mercatoribus and good brokeler (buckler) makers." It owes its present degree of activity principally to its situation on the main road from North Wales through Chester to Liverpool. Coal, iron, and lead mines are extensively wrought in the parish, which has also some large iron-works. Exclusive of several of minor importance, a large fair, which continues for 14 days from the 23d of March, and is attended by traders from a great way round, is held here annually, for the sale of horses, cattle, Manchester, Birmingham, and Sheffield goods, Irish linens, Welsh flannels, Yorkshire and other woollen cloths, &c. Five large areas in the town are fitted up with booths and temporary shops, for the accommodation of the dealers in the fair. Market-days, Monday and Thursday. (*Nicholson's Cambrian Guide; Panorama of N. Wales; Bound. Rep., &c.*)

WURZBURG, a city of Bavaria, circ. Lower Franconia, of which it is the cap., on the Mayn, by which it is divided into two parts, 62 m. S.E. by E. Frankfort, lat. 49° 47' 48" N., long. 7° 35' 9" E. Pop., exclusive of students and troops, about 22,500. Würzburg is finely situated, in a hollow surrounded by vine-covered hills, and traversed by the Mayn, here a large and fine stream, covered with boats and barges. The greater part of the city is on the right or N bank of the river, the communication with the citadel, and a suburb on the opposite bank, being kept up by means of a handsome bridge. Würzburg is inclosed by walls, and, being one of the oldest towns of Germany, is irregularly laid out, its streets being generally narrow and angular; it has, however, some venerable edifices. The cathedral was originally founded in the 8th century, but the earliest portions of the present building appear to date from the 11th or 12th. The interior has been modernized with little taste; but it has some monuments worth notice, including those of a long series of the prince-bishops of Würsburg, the sovereigns of the city and adjacent territory, for upwards of 1000 years. There are 32 other churches, the finest of which is the Marienkirche, in the pointed Gothic style. The royal, formerly the episcopal, residence, in a small square, was erected early in the last century; it is of an oblong form, on the plan of the palace at Versailles, and is of great extent, including, besides a magnificent staircase, upwards of 280 apartments, mostly fitted up in the style of Louis XIV. The gardens attached to it form a favourite promenade. The great hospital is an extensive and well conducted establishment, partly subsidiary to the school of medicine, for which the university of Würzburg is famous. This university was founded in 1403, and revived in 1582: at different periods it has been in a very flourishing state. It has some good scientific collections, and a library of 190,000 volumes. In 1832, it had 521 students, of whom 244 attended the medical classes, 109 the law, 118 the divinity, and 50 the philosophical, &c. (Journ. of Education); but the number of students has since declined, and probably does not at present (1842) amount to 400. It has, also, a gymnasium, a teachers' seminary, musical and polytechnic institutions, a society of arts and sciences, and an infirmary for the cure of deformities. One three or four of the numerous monastic institutions formerly established in the city now exist. Würzburg is the seat of the court of appeal for the circle, and a bishop's see. Its manufactures consist principally of woollen stuffs, hats, leather, sealing-wax, and surgical instruments. It is the principal depôt for Franconian wines, which are mostly sent down the Mayn to Frankfort; and its trade will most probably be materially increased by the opening of the canal from Bamberg on the Mayn to Dietfurth on the Altmuhl, an affluent of the Danube. Würzburg was secularized and given to the Archduke Ferdinand of Austria in 1803, and was ceded to Bavaria in 1815. (Berghaus; Dict. Geog., &c.)

WYALUSING, p. t., Bradford co., Pa., 149 m. N. Harrisburg, 250 W. Drained by Wyalusing creek and its tributaries, flowing into Susquehanna river. It has seven stores, one fulling mill, one woollen factory, four grist-mills, 12 saw-mills; six schools, 318 scholars. Pop. 1400.

WYCOMBE (CHIPPING, or HIGH), a parl. and mun. bor., market town, and par. of England, co. Buckingham, hund. Desborough, on the Wick, a small tributary of the Thames, and on the high road from London to Oxford, 27 m. W.N.W. the former. The old parl. and mun. bor., which were co-extensive, did not include the whole of the town; but the modern mun. bor. is rather more than three times the size of the former, and the modern parl. bor. is identical with the par. The latter has an area of 6360 acres; with, in 1831, a population of 6299. Wycombe extends for 1½ m. along the valley in which it is situated; and, though it has increased but little of late years, its general appearance is that of a well-built, prosperous market town. The principal roads communicating with the country to the N.W. and S.E., diverge from the market place in the centre of the High-street. The church, a large and venerable structure in the perpendicular and early-decorated styles, has a tower at its W. end, 108 feet in height, erected in 1592; but the rest of the church dates chiefly from the latter part of the 13th century. The interior has a fine altar-piece, and several monuments, among which is one by Scheemakers, to the earl of Shelburne, father of the first marquis of Lansdowne, who died in 1761, and another by Carlini to a countess of Shelburne. The living, a vicarage worth £140 a year, is in the gift of the marquis of Lansdowne. The Independents, Baptists, and Friends, have places of worship here. The town-hall, erected in 1775, is a large and respectable brick building, supported on stone pillars. The free grammar school, founded by Queen Elizabeth, has attached to it some almshouses, and an income of about £390 year. Though established for an unlimited number of scholars, only 27 were receiving instruction at the date of the late inquiry. Here, also, is a girls' Lancastrian school, with numerous bequests for the general relief of the poor. The manufacture of chairs is the only one of any importance carried on in the town. Some years since a considerable quantity of pillow lace was produced here; but this branch of industry has been nearly superseded by the machine-made lace of Nottingham and other places. There are several con-

siderable paper-mills near the town, on the Wick, and others in different parts of the parish. But the prosperity of High Wycombe is mainly owing to its being a place of considerable thoroughfare, though, since the opening of the Great Western railway, that has materially diminished; and to its being the market town for a district of 10 m. round. It has an extensive corn-market.

The earliest extant charter dates from 1558. Under the Municipal Reform Act it is governed by four aldermen, and 12 councillors.

The borough has returned two members to the House of Commons since the 28th of Edward I. Previously to the Reform Act the right of election was vested in the mayor, aldermen, bailiffs, and burgesses, of whom there were usually about 180. As already seen, the limits of the modern parliamentary borough have been considerably enlarged. Registered electors, in 1839–40, 389. Waller, the poet, was member for this borough in 1625. A little S.W. from the town is Wycombe abbey, the seat of Lord Carrington, by whose ancestors it was purchased from the Lansdowne family. Market day, Friday; fair, Monday before Michaelmas. (Mun. Corp. and Bound. Rep., &c.)

WYMONDHAM, or WYNDHAM, a market town and par. of England, co. Norfolk, hund. Forehoe, on a hill, 9 m. S.W. by W. Norwich. Area of par. 11,940 acres. Pop. of ditto, in 1831, 5465. The town, on the high-road from London to Norwich, has a market place with an ancient cross. The church, a venerable structure, in a mixed style, consists of a nave with aisles, a large W. tower, and another at the intersection of the nave with the transepts. Originally it formed a part of a monastery founded in the time of Henry I., to which the town appears to have owed its earliest importance. Within are many curious monuments, including that of the founder, William de Albini; a large carved font, &c. The living, a vicarage worth £535 a year, is in the gift of the bishop of Ely. Here, also, are chapels for Independents, Wesleyans, Baptists, Friends, &c. The grammar school, founded by Edward VI., has a total income of about £180 a year. It is governed by 12 trustees and 15 governors, and is free to all boys of the parish; but, in consequence of a pending chancery suit, it had no pupils at the date of last inquiry. It has two exhibitions at Cambridge, and a share of an exhibition for scholarships. A fuel allotment charity has an income of £96 15s. a year; and an endowment of £65 yearly is expended in payments to the vicar, to old unmarried women of the parish, and in the partial support of five schools. (Digest of Charity Reps.)

A national school affords instruction to about 260 boys and 70 girls: 600 children attend a Sunday school attached to the church, and about the same number frequent one attached to a dissenting chapel. It was estimated, in 1839, that about one sixth part of the population was supported by weaving, principally bombazines, crapes, and other Norwich goods. There were about 300 looms at work in the town; about one third for a resident manufacturer, one third for Norwich manufacturers, and the rest for various employers. The average wages of weavers at that period amounted to only about 7s. a week per loom, many of the weavers having only half-work; and few members of the weavers' families were brought up to manufacture, but sought rather for agricultural or other employment. Handloom Weavers' Reports, ii., 388, 329.) A court leet is held annually, and a manorial court occasionally. A little to the N. of the town is Kimberley-hall, the seat of Lord Wodehouse, in an extensive and finely-wooded park. The Wyndham family, one of the most illustrious in the county of Norfolk, which derived its name from this town, has produced, among other eminent individuals, the distinguished parliamentary leaders, Sir William Wyndham of the reigns of George I. and II., and Mr. Wyndham, of that of George III. Markets at Wymondham, on Fridays; fairs for cattle, &c., Feb. 12th, May 16th, and Sept. 22nd. (Parl. Reps. &c.)

WYOMING, county, N. Y. Situated toward the W part of the state, and contains 500 sq. m. Taken from Genesee county in 1841. Drained by Susquehanna river and its branches. Capital, Warsaw.

WYOMING, county, Pa. Situated in the N.E. part of the state, and contains 480 sq. m. Drained by Susquehanna river and its branches. Taken from Luzerne county in 1841. Capital, Tunkhannock.

WYSOX, p. t., Bradford co., Pa., 136 m. N. Harrisburg, 246 W. Drained by Rumfield and Wysox creeks. It has seven stores, one fulling-mill, one woollen factory, three grist-mills, 11 saw mills. Pop. 1871.

WYTHE, county, Va. Situated toward the S.W. part of the state, and contains 700 sq. m. Drained by New river and its branches, and by branches of Holston river. It contained in 1840, 13,530 neat cattle, 19,782 sheep, 32,736 swine; and produced 65,602 bushels of wheat, 6 000 of rye, 233,793 of Indian corn, 12,689 of buckwheat, 152,446 of oats, 36,307 of potatoes, 13,343 pounds of sugar. It had 15 stores, one fulling-mill, two flouring-mills, 36 grist-mills, 25 saw-mills, one oil-mill, two powder-mills, 13 tanneries, 28 distilleries, one printing-office, one weekly newspaper; one academy, 40 students; 12 schools, 969 scholars. Pop.: whites, 7632 ; slaves, 1618 ; free coloured, 125; total, 9375 Capital, Wytheville.

X.

XALAPA, or JALAPA, a town of Mexico, state Vera Cruz, on the high-road from Vera Cruz to Mexico, 55 m. N.W. the former. Pop. estimated at 13,000. It stands on a platform, about 4300 ft. above the level of the sea, surrounded by fine mountain scenery, and sometimes subject to heavy fogs. Its climate is generally mild and salubrious, though it is said to be neither so clean nor so well built as Vera Cruz; but it has numerous houses of two stories, built after the old Spanish manner, in a square, enclosing a court planted with trees and flowers, with a fountain in the centre. The cathedral and other churches, though in an indifferent style of architecture, are very gorgeous. This was formerly a great entrepôt for the European trade with Mexico, and large fairs were held here ; but its trade has greatly diminished, and its shops and warehouses do not now make much show: at present, indeed, Xalapa is chiefly celebrated for its washing; and many of the inhabitants of Vera Cruz send their clothes hither to have them cleaned. (*Bullock.*) The more wealthy inhabitants of Vera Cruz, and, indeed, of all the adjacent coast district, or *tierras calientes*, resort to Xalapa in the summer to avoid the heat, insects, and fevers of the low country, from all which it is free. The well known medicinal herb *jalap*, grows abundantly in the vicinity of this town, to which it is indebted for its name. (*Ward's Mexico*, ii., 19, 20 ; *Poinsett's Mexico*, 35-37.)

XENIA, p. t., capital of Greene co., O., 61 m. W.S.W. Columbus, 454 W. Drained by Little Miami river and its branches. It has 18 stores, one fulling-mill, one woollen factory, five flouring-mills, eight saw-mills, three printing-offices, three weekly newspapers ; 11 schools, 297 scholars. Pop. 4913. The village is situated on Shawnee creek, 3 m. from its entrance into Little Miami river, and is regularly laid out with streets crossing each other at right angles. It contains a courthouse and county offices of brick, a jail of stone, four churches in the immediate vicinity, 10 stores, and about 1900 inhabitants. It has many handsome dwellings, and is surrounded by a highly cultivated country.

XERES DE BADAJOS, or DE LOS CABALLEROS, a town of Spain, in Estremadura, prov. Badajos, near the Ardilla, a tributary of the Guadiana, 40 m. S. Badajos. Pop. estimated at about 9000. It is walled, and, like most Spanish towns, had numerous monastic institutions ; but from its being out of any great route, it is rarely visited by travellers. It has manufactures of linen fabrics, leather, hats, soap, &c., and a large trade in cattle, which are extensively reared in its neighbourhood.

XERES DE LA FRONTERA, a city of Spain, in Andalusia, prov. Cadiz, on the road from Cadiz to Seville, near the Gaudalets, 17 m. N.N.E. Cadiz. Pop., according to Miñano, 31,000, which, though perhaps somewhat underrated, is, we suspect, much nearer the mark than Captain Scott's estimate of 50,000. "Xeres is situated in the lap of two rounded hillocks, which shelter it to the E. and W ; and it covers a considerable extent of ground. The city, properly so called, is embraced by an old crenated Moorish wall, enclosing a labyrinth of narrow, ill-built and worse drained streets; but this wall is of no great circuit, and is so intermixed with the houses of the suburbs as to be visible only here and there. The limits of the ancient town are, however, well defined by the numerous gateways still standing. Some of the old buildings and narrow streets are striking in appearance, and the number of gables and chimneys cannot fail to strike one who has been long accustomed to the flat-roofed cities of Andalusia." (*Scott's Ronda, &c.*, ii., 79.) It has eight parish churches, among which is one that is collegiate, with a library and a collection of coins ; a town-hall, numerous convents ; a foundling, an orphan, and other hospitals ; several schools, a college, a public granary, infantry barracks, and an old fortress, are the principal public edifices. The streets, even in the best parts of the city, are disgustingly filthy ; and the want of cleanliness is no doubt the main source of the destructive epidemics with which the town is frequently visited.

Xeres derives its principal or rather its sole importance from its being the great emporium of the well-known wine, called sherry, grown in its vicinity. The principal wine merchants reside mostly in the suburbs, where are, also, the largest warehouses. These are all above ground, and

are immense buildings, with lofty roofs supported on arches, springing from rows of slender columns, having their walls pierced with numerous windows to admit of the thorough circulation of air. (*Scott*, ii., 79, 80.)

The vineyards, mostly situated on slopes, are scattered at considerable distances ; in 1818 they were estimated to comprise an area of about 8000 acres; and at present, perhaps, they may extend over 12,000 acres. It is not easy to form any very accurate estimate of the produce of the sherry vineyards, partly because there is no accurate account of the exports and of the stocks on hand, and partly because a considerable quantity of the light wine, called *suguer*, grown on the right bank of the Guadalquivir, is mixed up with the inferior sherries. Probably, however, the average annual export of sherries may amount to about 20,000 butts (the butt contains about 105 wine gallons), worth from £12 to £65 a butt! It is a mistake to suppose that good sherry is a cheap wine. "It may," says Inglis, "be laid down as a fact, that genuine sherry, one year old, cannot be imported under 30s. a dozen ; and if to this be added the profit of the merchant, and the accumulation of interest, it is obvious that genuine sherry, eight years old, cannot be purchased in England under 45s. a dozen." (*Inglis's Spain in 1830*, ii., 82.) When, therefore, any one sees fine old sherry advertised at 30s. or 35s. a dozen, he may be pretty well satisfied that it is a hoax, and that Marsala or cape Madeira, and not sherry, is the staple of the wine.

The finer sherries are all made from the Xeres grape, with the addition of only about two bottles of brandy to a butt, and sometimes of a little Paxarete, or sweet sherry, and of Amontillado. The wines are mostly all kept in very large casks, approaching in some degree to the Heidelberg tun, and when any wine is drawn off from one of these *madre* butts, it is replaced by an equal quantity taken from the next oldest butt, so that it is idle to talk of the sherry found in the market belonging to any particular vintage. The dark or deep brown sherries are occasionally produced by boiling a quantity of pale sherry to one fifth part its bulk, and mixing up this residuum with paler sherries, in quantities proportioned to the shade required. Amontillado, made in imitation of the wine of Montilla, near Cordova, the driest of sherries, is made from a variety of grapes plucked before they are quite ripe. It is the purest of the sherries, and will bear no admixture of either brandy or boiled wine.

England is and has long been the principal market for sherries. They used originally to be introduced and sold under the name of *sack*; but it is only of late years, and especially since the decline in the taste for Madeira, that they have come into all but universal use among all classes as a dinner wine. It is not easy, indeed, to account for their extraordinary popularity ; for, though sherry of good quality, and kept to a proper age, be a very superior wine, the finer varieties bear no proportion to those that are inferior ; and it is, besides, too powerful to be used with any degree of freedom.

To show its popularity it is sufficient to mention that, in 1841, of 6,184,900 gallons wine entered for consumption in the United Kingdom, no fewer than 2,412,821 were sherry! The entries of port during the same year amounted to 2,387,017 gallons, making together 4,799,838 gallons, leaving only 1,385,142 gallons for all the other sorts of wine! (*Scott's Ronda and Granada*, ii., 78, &c. ; *Inglis*, ii., 78, &c. ; *Henderson on Wine*, 190, &c.)

Of late years port St. Mary, on the N. side of the bay of Cadiz, 10 m. S.W. Xeres, has absorbed a considerable part of the trade of the latter, the wine-merchants who have settled there having the additional advantage of being able to superintend the shipping of their wines. Xeres has a few manufactures of serges, leather, soap, &c., but only for the consumption of its own inhabitants. On the plain outside its walls was fought, A.D. 714, the battle which finally overturned the Visigothic monarchy of Spain, and gave a great part of that country to the Moors. On the Guadalete, near the scene of this battle, is a Carthusian monastery, founded in 1571, once the most celebrated in Spain, but now in decay.

Y.

YADKIN, river, N. C., rises on the E. side of the Alleghany mountains, and flowing across the state, enters S. C. In Montgomery county, N. C., it passes through the Narrows, occasioned by mountains contracting it from 200 to 30 feet wide. In the lower part of its course it becomes Great Pedee river, and enters Winyaw bay near Georgetown.

YAKUTSK, a town of E. Siberia, cap. of the immense prov. of its own name; on the Lena, about 1150 m. N.E.

Irkutsk, lat. 62° 1' 50" N., long. 147° 44' E. Pop. about 4060. According to Wrangell, "Yakutsk has all the character of the cold and gloomy north. It is situated on a barren flat, near the river. The streets are wide, but the houses and cottages are poor in appearance, and are surrounded by tall wooden fences. Here are five churches, a convent, a stone building for commercial purposes, and an olden wooden fortress with its ruined tower, built in 1647, by the Cossack conquerors of Siberia. The town has, however, undergone great improvements in the last thirty years. The Yakut huts have been replaced by substantial houses; the windows of ice or talc, have given way to glass in the better class of houses, and the more wealthy inhabitants begin to have higher rooms, larger windows, double doors, &c.

"Yakutsk is the centre of the interior trade of E. Siberia. All the most costly furs, as well as the more common kinds, walrus teeth, and fossil remains, are brought here for sale, or barter, during the ten weeks of summer, from Anabor and Behring's straits, the coasts of the Polar sea, and even from Okhotsk and Kamtschatka. It is not easy to imagine the mountain-like piles of furs of all kinds seen here; their value often exceeds 2,500,000 roubles. Almost all the Russian settlers in Yakutsk employ their little capital in purchasing furs from the Yakuti during the winter; on which they realise a good profit at the time of the fair, when they sell them to the Irkutsk merchants.

"As soon as the Lena is clear of ice, the merchants begin to arrive from Irkutsk, bringing with them for barter, corn, meal, the pungent Circassian tobacco, tea, sugar, brandy, rum, Chinese cotton, and silk stuffs, yarn, cloth of inferior quality, hardware, glass, &c. But at the annual fair there is not the appearance of animation and bustle, which might naturally be expected. The goods are not exposed for sale, and most of the purchases are effected in the houses or enclosures of the citizens." (*Wrangell's Siberia,* 11–13.) Dobell says, that the inhabitants are hospitable and gay. Several balls were given during his stay, and the dress, manners, and appearance of the people far surpassed what he expected in so remote a situation. (*Trav.* ii. 12.) The variations of climate are extraordinary; for, though, on the whole, cold predominates to a very great extent, the thermometer in winter often falling to 40° R., or 56° below the zero of Fahr! The heat in summer is sometimes not inferior to that of the torrid zone!

The vast province of Yakutsk comprises, at least, three fifths of E. Siberia, and is watered by the great rivers Lena, Yana, Indigirka, and Kolyma, which supply vast quantities of fish. Iron, salt, and excellent talc, are the chief mineral products: game, of many kinds, abounds. Large herds of cattle, &c., are reared near Yakutsk, and notwithstanding the severity of the winters, rye, barley, and even wheat, are said to succeed well throughout the province, except in those parts which are so far N. as to render the summer too short for ripening grain. (See *Dobell's Travels in Siberia,* ii. 21, &c.: *Wrangell's Siberia and the Polar Sea, &c.*)

YALABUSHA, county, Miss. Situated centrally toward the N. part of the state, and contains 720 sq. m. Drained by Yalabusha river and its branches. It contained in 1840, 14,976 neat cattle, 4298 sheep, 33,519 swine; and produced 14,958 bushels of wheat, 543,685 of Indian corn, 32,680 of oats, 27,739 of potatoes, 4110 pounds of tobacco, 4,030,644 of cotton. It had one commission house in foreign trade, 33 retail stores, one flouring-mill, 44 grist-mills, 13 saw-mills, five tanneries, two distilleries, one printing-office, one weekly newspaper; two academies, 65 students, 31 schools, 558 scholars. Pop.: whites, 6640; slaves, 5601; free coloured, 7; total, 12,948. Capital, Coffeeville.

YANCEY, county, N. C. Situated in the W. part of the state, and contains 1760 sq. m. It contains, in its E. part, the highest land in the United States, E. of the Rocky mountains. The three highest peaks are Black mountain, 6476 ft. high, Roane, 6038, Grandfather, 5556 ft. above the level of the ocean, which are higher than the three highest peaks of the White mountains in N. H. Drained by Nolachucky river and its branches. It contained in 1840, 5665 neat cattle, 5041 sheep, 18,718 swine; and produced 6390 bushels of wheat, 2646 of rye, 405,390 of Indian corn, 33,670 of oats, 89,731 of potatoes, 4830 pounds of tobacco. It had two forges, 42 grist-mills, three saw-mills. Pop.: whites, 5681; slaves, 254; free coloured, 27; total, 5962. Capital, Burnsville.

YANINA, improperly JOANNINA, (probably the an. *Eurœa,*) a city of European Turkey prov. Albania, of which it is the cap.; on the W. bank of the lake of its own name, 60 m. W. by N. Larissa. Lat. 39° 47' N., long. 21° 1' E. Population said to be at present about 19,000. It occupies a small peninsula, extending into the lake, and part of the adjacent shore, its site, being tolerably level. Twenty five years ago Yanina was a town of 30,000 inhabitants, with numerous mosques, many large and well-built houses, and several palaces. It had then a considerable trade with

the rest of Epirus, Roumelia, Wallachia, &c.; and a large annual fair, to which a good deal of Italian produce, with French and German manufactures, &c., were brought. It was, however, set on fire by order of its barbarian and bloodthirsty tyrant, Ali Pacha, in 1820, and was almost wholly ruined. The streets are narrow and crooked, and the houses now are mostly of mud. Numerous vacant spaces especially about the citadel, are covered with ruins, and all its animation is confined to the bazaar. (*Burgess's Greece,* i. 61–63.)

The lake of Yanina is about 6 m. in length, and nearly 3 in its greatest breadth: it is narrowest at the N., and gradually expands towards the S. Mr. Burgess says that the scenery around it would be fine if its banks were wooded; but as it is, the lake is far inferior in respect of beauty to those of Italy or Switzerland, and is excelled by some of the Scottish lakes. An island opposite the peninsula has a church and monastery. The description given by modern travellers of the site of Yanina and its lake, answers perfectly to that of the city and lake of Emma by Procopius. Justinian built a fortress at Eurœa, apparently on the identical site now occupied by the citadel of Yanina. (*Cramer's Ancient Greece,* i. 141, 142; *Burgess's Greece, &c.; Holhouse; Hughes; Cramer's Greece, &c.*)

YANTIC, river, Ct., rises in Lebanon, and enters Thames river at Norwich city, through a cove which sets up a mile from the river. At its entrance into the cove, it has falls which afford great water-power, and where is a flourishing manufacturing village.

YARKUND, the chief city of Chinese Turkestan, in a fertile plain, on the river of its own name, lat. 38° 19' N., long. 76° 17' 45" E. Its population has been variously estimated, but may probably amount to 50,000, exclusive of the Chinese garrison. (*Burnes's Bokhara, &c.,* iii. 193.) It is enclosed by an earth rampart, pierced with five gateways, outside which are extensive suburbs: there are two citadels, one in the suburbs, and the other in the town. The houses, built of stone and clay, are mostly only one story in height, the streets are intersected by numerous canals and aqueducts which bring water from the river for the use of the inhabitants. Yarkund has two large bazaars, numerous mosques, and 10 or 12 Mohammedan colleges, most of the native inhabitants being Mussulmen, though much more lax in their religious prejudices than their neighbours to the W. About 200 Chinese merchants reside in the place, and some Cashmerians and Persians, but only a few Hindoos, and neither Jews nor Nogai Tartars. When Mons Polo visited this city, he found some Nestorian Christians among the inhabitants. (*Ritter, Erdkunde von Asien,* ii. 39–68.)

"The productions of China," says Sir A. Burnes, "are transmitted to this province, and sold to the natives of Bokhara and Thibet, who are permitted to frequent certain fixed markets. No Chinese crosses the frontiers; the trade to Bokhara being carried on by Mohammedans, who visit Yarkund for that purpose. The same vigilance to prevent the ingress of foreigners is here exhibited as upon the sea-coast." Horses are a great article of trade; and it was chiefly to open a traffic to these animals, that Moorcroft desired to reach Yarkund, in which attempt, however, he was frustrated by the jealousy of the Chinese authorities. Yarkund with the adjacent province of Cashgar, formed the principality of a Mohammedan Khojir. Dissensions arose in the reigning family about 80 years since (in the time of Kienlong), and they called on the Chinese government as a mediator, which, as frequently happens, acted the part of a conqueror. The period which has elapsed since the capture of Yarkund has in no way diminished the precautions of the Chinese government. Yarkund is still considered but an outpost, and the communication between it and Pekin is maintained in a most characteristic manner." (*Burnes,* iii. 193–196.) The Mohammedan designs, indeed, fill the subordinate offices of state, but under the strict superintendence of the Chinese authorities. The garrison, consisting of from 5000 to 7000 soldiers, are recruited from boys of 14 and 15 years old, who are sent back, after about as long a period of service. (*Burnes.*) According to some Chinese documents, the annual tribute of the Yarkund people to the Chinese comprises 30 oz. of gold, 33,069 oz. of silver, 30,000 sacks of corn, 69 lbs. of oil, 57,699 pieces of linen, 15,000 lbs. of cotton, 3,000 lbs. of copper, &c. (*Ritter, Erdkunde,* v. 409; *Klaproth; Marco Polo; Watham, &c., in Cale. Asiat. Journ., &c.*)

YARMOUTH, a parl. and munic. bor. and seaport of England, partly and principally in the co. Norfolk, but partly, also, in that of Suffolk, on the Yare, at its mouth, in the North sea, 19 m. E. by S. Norwich, and 168 m. N.E. London. Lat. 52° 36' 40" N., long. 1° 44' 22" E. The old parl. borough, which included the hamlet of Southtown, in Suffolk, in the parish of Gorlestone, on the S. side of the river, had an area of 2110 acres, with, in 1831, a population of 23,326; but the modern parl. borough includes the whole parish of Gorlestone, comprising the villages of that name, and has an area of 3940 acres, with, in 1831, a population of 25,448, and, in 1841, of about 28,000. The part of the town on the E. side of the Yare, or Yarmouth properly so called, occupies a narrow peninsula, between the sea on the one hand, and the river on the other. "It consists of four principal lines of streets running nearly parallel with the river, and of an immense number of narrow lanes, or rows, that form the lateral communications between these streets. Very few of the rows exceed from 5 to 8 ft., and only two of them at the opposite ends of the town were passable for common wheel-carriages until the widening of some others in the centre of the town, not long since formed the street called Regent-street. The principal streets are well built, and wide, opening in some places into a spacious quay, market-place, and squares; and the town presents, on the whole, a thriving appearance. The best dwelling-houses are situated along the quay; many of these are substantial and handsome, as are, also, many of those in the other principal streets. Most of the shops are situated in the market-place, King-street, Broad-row, and Market-row; and the warehouses, granaries, malt-houses, and fish-offices, together with the inferior dwelling-houses, are in the different rows. On the N., E., and S., the town is enclosed by old walls, beyond which is an intermixture of every description of buildings; but principally of extensive warehousing premises, and residences of an inferior class. This part is not paved, and only partially lighted, and principally with oil lamps; but the whole of the town within the walls is well paved, and lighted with gas. A considerable extension beyond the old walls has taken place on the E. side; and long lines of streets, besides many detached dwellings and extensive fish-offices, and other premises connected with the trade of the place, now occupy a great part of the space between the town and the sea shore, particularly towards the new jetty. New houses are in progress also on the N. side of the town, along the road leading toward Caistor; and along the road toward Nelson's monument, S. of the town, the whole of which neighbourhood, and more especially the line fronting the sea, it is contemplated to cover with a new town.

"Yarmouth is connected by a bridge over the Yare with Southtown or Little Yarmouth. This suburb, forming the N. part of Gorlestone parish, consists principally of neat and substantial private residences; with docks, timber wharfs, building yards, &c., on the river, and in which much of the business of this port is carried on. The other distinct group of buildings in Gorlestone, which forms the village or town of that name, lies considerably more to the S., nearer to the entrance of the harbour. South town is very imperfectly lighted; and Gorlestone is neither paved nor lighted." (*Mun. Corp. Rep.*)

Yarmouth quay is one of the most extensive and finest in England: it is upwards of 1 m. in length, and in some places 150 yards in breadth, being in its centre a planted promenade. Here is the town-hall, a handsome edifice with a Tuscan portico; the council chamber, which is highly decorated, has a full-length portrait of George II. The Star inn, near the town-hall, was once the residence of Bradshaw, president of the high court of justice which condemned Charles I.; and some of its apartments still remain apparently as he left them, or even as they were at an earlier period, for the house is of the Elizabethan age. Yarmouth parish church is one of the largest ecclesiastical edifices in the kingdom. It was originally founded in the time of William II.; but the most ancient parts of the present edifice date no further back than about 1250; and, according to Rickman, only a portion of the building is early English; other parts, particularly the windows, being of the decorated and perpendicular styles. (*Rickman's Gothic Architecture*, p. 208.) It is 230 feet in its greatest length, by 108 feet in breadth. At the W. end are four octangular towers, the outermost of which are surmounted with plain pinnacles, as are the octangular towers at each angle of the S. transept. The tower, at the intersection of the transepts with the nave, formerly decorated with pinnacles, is now embattled, and supports a tall turned spire, erected in 1807, a conspicuous mark from the sea. The part of the interior W. of the tower forms a spacious choir, the ceiling of which is panelled into compartments, having coats of arms of different branches of the royal family of England, and of the Fastolfs, Gournays, Bardolfs, and other proprietors of the neighbouring castle of Caistor. The organ in this church is one of the finest in England; and it has many interesting monuments. The churchyard, comprising about 6 acres, is entered by handsome iron gates: on its W. side was standing, till the present summer, the leaf gate, a curious brick archway of the time of Charles I. The living, a perpetual curacy, worth £243 a year, is in the gift of the dean and chapter of Norwich. A chapel-of-ease was built in 1715; and the living, a curacy worth £100 a year, is in the gift of the corporation of Yarmouth, the minister's salary being derived from a local duty on coal. The living of St. Peter's

church, an edifice in the Tudor style, built under a recent act, is a perpetual curacy, worth £160 a year, in the gift of the incumbent of St. Nicholas' (the parish) church. Attached to the living of Gorlestone and Southtown is a curacy, with a separate church, a modern erection, called St. Mary's chapel, worth £300 a year, lately in the gift of trustees. (*Ecclesiast. Rev. Rep.*) Here, also, are chapels for Roman Catholics, Independents, Unitarians, Friends, Baptists, & Jews' synagogue, &c.; and some remains exist of various convents suppressed at the Reformation. A free grammar-school was formerly supported by the corporation, but it ceased to exist about the end of last century. The children's hospital, founded in the reign of Edward I., is under the government of the corporation, and has an income of £857 a year: it serves as a workhouse, and, also, maintains and clothes 30 boys and 90 girls, and 100 children are taught as day scholars. All children of parents belonging to the town are eligible to be admitted as day scholars, and, as vacancies occur, to become boarders by rotation, according to seniority. The charity school, founded in 1713, clothes and educates a considerable number of children; here, also, is a Lancastrian school, which educates about 150 boys; a girls' school, founded in 1810, for educating and clothing 80 poor girls; and a proprietary grammar school, founded in Southtown in 1833. The Fishermen's hospital, established and built in 1763, was chiefly supported by an annual government grant of £160 a year, deducted from the beer duties, till 1832, when it was discontinued, in consequence of the repeal of the beer duty. The hospital is, however, otherwise endowed, and, at the date of the late inquiry, had an income of £56 10s. a year. Warren's charity, established in 1694 for the general relief of the poor, sick, orphans, widows, &c., has an income of about £375 a year; and there are several other endowments for schools and other charitable purposes. (*Charities' Rep. XXVI.; and Digest, &c.*)

There is in the town a very extensive silk factory for the winding and throwing of silk, and the weaving of crapes and other silk goods. The municipal corporation report of 1836 states that "Ship-building, and the various trades with it, are also carried on to some extent in Yarmouth; nevertheless, it cannot be considered as a manufacturing town, but derives its importance and prosperity from the trade and commerce which it owes to its situation and port. The rivers, Yare, Waveney, and Bure, which unite in Braydon water, adjoining the town, are navigable; the first to Norwich, the second to Bungay, and the Bure to Aylsham; and they secure to Yarmouth an extensive trade in the exportation of the agricultural produce of the districts traversed by these rivers, and in supplying them with coals and other heavy goods. The export of grain and malt from this port is considerable, of barley greater than from any other part in England; but the principal business of Yarmouth is the herring and mackerel fisheries, and the curing and exportation of the herrings to foreign countries, particularly the states bordering on the Mediterranean. An extensive timber trade with the Baltic is also carried on; and a considerable number of square-rigged vessels belong to the port. Yarmouth roads have long been the principal rendezvous of the vessels in the collier trade; and the town derives some advantages from the supply of fresh provisions to them.

"The harbour of Yarmouth is formed by the river Yare: it has an awkward entrance obstructed by a bar. Great attention, however, appears to be bestowed on remedying this defect, and on the improvement of the port generally. Vessels drawing about 12 ft. water, or of about 200 tons burden, can cross the bar, and proceed up to the town at spring tides." The chief improvements of the harbour were effected by a Dutchman, named Johnson, employed for the purpose, who first erected piers at the mouth of the river.

Yarmouth roads, between the town and a line of outer

the aggregate burden of the latter being 34,320 tons.

Yarmouth is the principal seat of the herring English fishery. The herrings usually make their appearance in the roads about the middle of September, when the fishery begins, and continues till towards the end of November. They are partly cured, and partly sent fresh into the metropolis. The fishery of cod, mackerel, skate, soles, red-mullet, whitings, &c., is also extensively carried on. In 1840, the gross customs duties received at Yarmouth, amounted to £46,782.

Yarmouth has been, for a long time, more or less frequented as a bathing place, for which, indeed, it is well fitted by its salubrity and its firm shelving sea-beach. It has, also, a pier, projecting 450 ft. into the sea, with public baths, assembly rooms, a neat theatre, a public library, public gardens, and all the establishments usual at a watering-place. To the N. and S. of the town, facing the sea, are open and level pieces of ground covered with verdure, called the Denes; and on the most southerly of these is a beautiful fluted column designed by Wilkins, erected in 1817 in honour of Nelson: it is 144 ft. in height, and is surmounted by a statue of Britannia. On other parts of the Denes, are various batteries, the barracks, a fine edifice, formerly a naval hospital, built in 1809, at an expense of £190,000; a new workhouse, erected in 1839, at an expense of £8000; numerous windmills, a race-course, &c. On other sides the environs of Yarmouth have no particular beauty; but the country is well cultivated, and the markets of the town are well supplied. Within a few miles, on the Suffolk side, are extensive remains of the Roman station *Garianonum*, so called from its situation at the mouth of the *Garienis* or Yare; and within a similar distance, on the Norfolk side, are the ruins of Caistor castle, formerly a sumptuous mansion, erected by Sir J. Fastolfe soon after the battle of Agincourt.

The first charter of incorporation possessed by Yarmouth appears to have been granted by John in 1208; but the governing charter previously to the late acts was that granted by Queen Anne, in 1702. Under the Municipal Reform Act the borough is divided into six wards, and is governed by a mayor, 11 other aldermen, and 36 councillors. Corporation revenue, about £6000 a year.

Yarmouth has sent two members to the House of Commons, with little intermission, since the time of Edward I., the right of voting, down to the Reform Act, having been in the sons of freemen, and in apprentices serving a seven years' apprenticeship to freemen within the borough. Registered electors, in 1839–40, 1904. It has two banks. The borough has a commission of the peace, and a jail, an admiralty court held weekly by the mayor, a 40s. court of requests, courts leet, piepoudre, &c. The maritime jurisdiction of the corporation extends for 10 m. up the rivers Bure, Yare, and Waveney. Two markets are held weekly; that on Saturday is extensive, that on Wednesday of minor importance. A fair is held on Friday and Saturday in Easter week; a regatta annually, &c.

Yarmouth communicates with Norwich daily by three coaches and two steamboats, London, Colchester, Ipswich, Bury, &c., by daily coaches, and with London and Hull by steam packets; the Leith, Dundee, Aberdeen, and other steamers, usually call here in fine weather. (*Bound., Mun., Corp., and other Parl. Reps., and Local Information.*)

YARMOUTH, a market town and par. of England, on the N.W. shore of the Isle of Wight, at the mouth of the little river Yar, immediately opposite Lymington, and 9 m. W. Newport. Area of par. 50 acres. Pop. of ditto, in 1831, 588. This town, which has long been in a stationary state, would have been wholly unworthy of notice in a work of this kind but for the circumstance of its having enjoyed the important privilege of sending two members to the House of Commons from the æra of Edward I. down to the passing of the Reform Act, by which it was most properly disfranchised. It was, in fact, one of the most perfect specimens of a proprietary borough. (*Parl. Rep., &c.*)

YARMOUTH, p. t. Barnstable co., Mass., 78 m. S.E. Boston, 479 W. It extends across Cape Cod. Bounded S. by the Atlantic, N. by Cape Cod bay. It contains four churches, a Congregational, Methodist, Baptist, and Universalist, 13 stores, six grist-mills, one tannery, one printing-office, one weekly newspaper; one academy, 45 students; 12 schools, 627 scholars. Pop. 2554. It has extensive salt works, and a number of vessels employed in the cod and mackerel fisheries.

YATES, county, N. Y. Situated W. of the centre of the state, and contains 398 sq. m. Bounded E. by Seneca lake and N.W. by Canandaigua lake. Crooked lake enters its S. part, and has its outlet through it into Seneca lake, affording good water-power. Crooked Lake canal connects that lake with Seneca lake, and thus with the Erie canal. Organized in 1823. It contained, in 1840, 16,989 neat cattle, 36,876 sheep, 18,473 swine; and produced 352,814 bushels of wheat, 2102 of rye, 104,066 of Indian corn, 50,591 of buckwheat, 30,994 of barley, 162,463 of oats, 170,318 of potatoes, 2200 pounds of hops, 39,394 of sugar. It had 51 stores, four furnaces, 14 fulling-mills, three woollen factories, seven flouring-mills, 13 grist-mills, 62 saw-mills, one oil-mill, eight tanneries, two distilleries, two printing-offices, two weekly newspapers, and one periodical; 119 schools, 6907 scholars. Pop. 20,444. Capital, Penn Yan.

YATES, p. t., Orleans co., N. Y., 12 m. N. Albion, 266 m. W. by N. Albany, 409 W. Bounded N. by lake Ontario. Drained by Johnson's creek. It has five stores, one furnace, one fulling-mill, one grist-mill, five saw-mills; 14 schools, 663 scholars. Pop. 2230.

YAZOO, river, Miss. is formed by the junction of Tallahatchee and Yalabusha rivers in Carroll co., and flows S.S.W. into Mississippi river, 12 m. above the Walnut Hills. It is 100 yards wide at its mouth, and is navigable for boats for 50 miles, in high stages of the water.

YAZOO, county, Miss. Situated W. of the centre of the

state, and contains 630 sq. m. Bounded N.W. by Yazoo, S.E. by Big Black river, and drained by their tributaries. It contained, in 1840, 19032 neat cattle, 2636 sheep, 3172 swine; and produced 834 bushels of wheat, 1746 of rye, 536,340 of Indian corn, 81,005 of oats, 129,964 of potatoes, 7350 pounds of rice, 2705 of tobacco, 12,615,802 of cotton. It had nine commission houses in foreign trade, 14 retail stores, two grist-mills, three saw-mills, two printing-offices, two weekly newspapers; 11 schools, 256 scholars. Pop. whites, 3116; slaves, 7339; free coloured, 25; total, 10,480. Capital, Benton.

YECLA, a town of Spain, prov. Murcia, at the foot and on the declivity of a hill, 43 m. N. by E. Murcia. Pop., according to Miñano, 11,600. It was formerly walled and had a fortress, but of these there are now no remains. Its principal buildings comprise two parish churches, some convents, a hospital, an ecclesiastical tribunal, and a prison. Its neighbourhood is very fertile; and its inhabitants are mostly occupied in the production of corn, flour, wine, brandy, and leather. Near Yecla are the traces of a more ancient town, where various Roman antiquities have been discovered. (*Miñano; Dict. Géog.*)

YEDDO, or JEDDO, the chief city of Japan, and the residence of the *siógun* or military emperor; on the S.E. shore of the island of Niphon, prov. Musasaa, at the bottom of the bay of Yeddo; lat. 39° 37′ N., long. 140° E. Its population has been variously estimated at from 700,000 to 1,500,000; but the probability is that the first of these numbers is beyond the mark. Yeddo is said to be surrounded by a ditch, and intersected by numerous canals and branches of the river Toniak, which are navigable for vessels of moderate burden. It has two large suburbs. Its usual plan would appear to be less regular than that of most other Japanese cities; but its streets and squares are clean, and some of the former are of prodigious length. Each street is appropriated to persons of one trade only, lined with covered arcades, and closed at night by gates at each extremity. The houses are mostly two stories in height; but being built almost wholly of wood, destructive fires are very frequent. Yeddo has many temples, Buddhic convents, and other large public buildings: the emperor's palace occupies a large extent of ground. This city has a considerable trade; but their are no materials for forming any estimate of its amount. For further information as to Yeddo, we must refer the reader to the works of Kämpfer, Thunberg, Siebold, &c.

YELL, county, Ark. Situated W. of the centre of the state, and contains 936 sq. m. Bounded N.E. by Arkansas river, and drained by its tributaries. Formed since the census of 1840. Capital, Danville.

YEMEN, a district of Arabia, which see.

YENISEI, a great river of N. Asia, in Siberia, through the central part of which it flows. Its basin lying between those of the Lena to the E., and the Obi to the W., is supposed to comprise an area of near 1,000,000 sq. m., being about the same as that of the province of Yenisesk. The Yenisei rises within the Chinese empire, not far from lat. 51° N., long. 98° E., and proceeds at first W. for about 5° of long., to near the point where it leaves the Chinese frontier. It then turns northward, and pursues generally a northerly course to the Arctic ocean, which it enters by a wide estuary called the bay of the 72 islands, the mouth of which is in about lat. 72½° N., long. 85° E., about 290 m. E. of the gulf of Obi. The entire course of the Yenisei has been estimated at 2600 m. Its chief affluents join it from the E., its tributaries from the W. being of much less importance. Various towns in the upper, with Krasnoyarsk, Yenisesk, &c., in the middle and lower part of its course, are on its banks; Irkutsk is on its great tributary the Yercheni-Tungooska, which flows out of lake Baikal. As far as Krasnoyarsk it runs through a mountainous country, and thenceforward to Yenisesk, where its width, when highest, is about 1 m.; its banks are elevated and precipitous. A survey of the river was completed last century by the Russian government, up to this town; and from this it appears that its channel varies from 2 to 8 fathoms in depth. This noble stream, however, like the other large rivers of Siberia, is but of little use, inasmuch as it flows, for the most part, through desolate wastes; its embouchure being also in a frozen sea, and the river itself being frozen over for the greater part of the year. The Russian surveyors were stopped in their progress upwards, by the ice at Turruchansk on the 1st of October, and by the 16th the river was completely frozen over; and it was not till the succeeding 6th of June that they were enabled to proceed with their survey. (See *Wrangell's Siberia, &c. Introd.; Dict. Géog.*)

YEOVIL, a market town and par. of England, county Somerset, hund. Stone, on the border of Dorset, on the Yeo or Ivel, a tributary of the Parrott, here crossed by a stone bridge, 18 m. S. Wells. Area of parish, 3220 acres. Pop. of do., in 1831, 5921, of whom about 3500 inhabited the town. Yeovil comprises about twenty streets and lanes, some of which are wide and open thoroughfares, the houses being generally good and built of freestone. The parish church, a light Gothic structure, with a large plain tower at the W. end, is supposed to date from the time of Henry VI. An ancient crypt, an adjoining chapel, and a handsome altar in the church, are worthy of notice. The living, a vicarage, with the curacy of Preston, of the clear annual value of £391, is in the gift of —— Philips, Esq. Here, also, are places of worship for Unitarians, Baptists, Wesleyans, Independents, &c. The free-school, endowed in 1707 and subsequently, had recently an income of £114 a year, and at the date of the last inquiry, 30 boys were taught reading, writing, and arithmetic, 14 of whom were clothed and apprenticed, and one taught Latin. An almshouse for a master, two wardens, and 12 poor men and women, had an income of £391 a year; and, exclusive of these, there is an almshouse for four poor women, and several minor charities. (*Digest of Charity Reports.*)

Yeovil was at one period celebrated for its woollen manufactures. But these appear to have been early superseded by the glove trade, the latter having attained to considerable importance in the town so far back as the middle of the 16th century. "At present the manufacturers are employed in making men and women's fine gloves; which pass in the retail shops as kid gloves, but are, in reality, made from lamb skins imported from Italy, Spain, and Germany. These skins are mostly dressed into leather in Yeovil, in which place the manufacturers are leather dressers and large dealers in wool, as well as glovers. The quantity of gloves made in Yeovil, of all sorts, may be estimated at 300,000 dozens annually; and the number of men, women, and children employed in the place and the adjoining districts, (spreading over 20 m.) amounts, perhaps, to 20,000. (*Hull's History of the Glove Trade,* 70, 71.)

The use of cotton and woollen gloves, and the importation of French and other foreign leather gloves, has of late years seriously depressed the trade of Yeovil; though, according to the statement in the appendix to the Municipal Corporation Report, about £70,000 a year are said to have been paid as wages (in 1836) to persons engaged in the glove trade of the town and its vicinity, and that the town was increasing. Yeovil claims to be a borough by prescription, its government having been till lately vested in a portreeve and 11 burgesses; but their authority was very circumscribed, and all the courts formerly held by the corporation have fallen into disuse. Market-day, Friday, when a good deal of butter cheese, corn, &c., is sent into the town; and large quantities of the butter made in the surrounding district are purchased, and sent to London, to be sold as Dorset butter. Fairs, June 28th and Nov. 17, chiefly for farm stock. (*Parliament Reports; Hull's History of the Glove Trade. &c.*)

YEZD, a considerable city of Persia, in the E. part of which it is situated, about 250 m. E. by S. Ispahan. Its population has been variously estimated; but if the statement of Kipneir, who assigns to it 24,000 houses, be near the mark, it must certainly be one of the most populous cities in the empire. It is situated in a sandy desert, near a range of high mountains, and has a fort, but no other defensive works. Being at the point of union of the principal roads connecting Ispahan, Kirman, Meshed, and Herat, it is consequently a considerable emporium. Its bazaar is said to be well supplied with provisions; though, from the sterility of the adjacent country, its supplies of corn have, for the most part, to be brought from Ispahah, and cattle are both scarce and dear.

The manufacture of silk stuffs in this city was, some years ago, superior to any other in Persia, and the village of Tuft, about 8 m. S.W., was equally famous for its sumuds. Here, also, are some fabrics of arms and sugar refineries. Formerly a good many Hindoos were settled in the town, but these were driven away by the exactions of a late Persian governor: it has still, however, numerous families of Parsees (Guebres, or fire-worshippers) among the population, this being almost the only town in the Persian dominions where they are now met with. Though an oppressed, they are an exceedingly industrious people. (*Kinneir's Pers. Emp.,* 113, 114; *Dict. Geog., &c.*)

YONKERS, p. t., Westchester co., N. Y., 132 m. S. Albany, 242 W. Watered by Bronx and Saw-mill rivers. It has three academies, 75 students; five schools, 216 scholars. Pop. 2958. The village is on the E. side of Hudson river, at the mouth of Saw-mill river, where is a steamboat landing.

YONNE, a dep. of France, neg. centre, formerly comprised for the most part, in Champagne and Burgundy, principally between lat. 47° 30' and 48° 29' N., and the 3rd and 4th degrees of E. long., having N.W. the dep. Seine-et-Marne, N.E. Aube, E. Cote d'Or, S. Nievre, and W. Loiret. Area, 726,747 hectares. Pop., in 1836, 352,487. Surface undulating; the hills scarcely anywhere rising to more than 600 ft. in height: the most elevated are in the S.W.,

separating the basin of the Seine from that of the Loire. The Yonne, whence the dep. takes its name, rises in Nievre, near Château-Chinon, and runs generally northward to the Seine, which it enters at Montereau, in the dep. Seine-et-Marne, after a course of about 177 m.; for 70 of which, or as high as Auxerre, it is navigable. It traverses the dep. of Yonne nearly in its centre, receiving within its limits the Cure, Serein, and Armancon from the E., its tributaries from the opposite side being inconsiderable. It communicates with the Loire by the canal of Nivernais; and with the Saone by that of Burgundy. A great part of the soil is calcareous, or gravelly, but about 300,000 hectares are said, in the *Official Tables,* to consist of rich land; and more corn is grown than is required for home consumption. In 1841, the arable lands were estimated to comprise 453,100 hectares; meadows, 31,265 do.; vineyards, 37,543 do.; and woods, 146,570 do. Hugo estimates the annual produce of corn, mostly wheat and oats, at 2,060,000 hectolitres, and that of wine at 1,110,000 do. The growths of this dep. are known as those of Lower Burgundy; the red wines of Tonnerre and Auxerre are especially esteemed; and the secondary growths of Epineuil, Irancy, &c., are also in high estimation. Chablis, the best of the white wines, is served up by the French epicures with oysters. Wines of this class *sont spiritueux sans être trop fumeux, ont du corps, de la finesse, et un parfum tres agréable.* The consumption of the department does not exceed 250,000 hectolitres, the rest being mostly sent to Paris, the N. of France, and foreign countries, little brandy being made. The orchards, which comprise nearly 6000 hectares, are of importance; and Yonne, along with Loiret, supplies Paris with all the *raisinné* consumed by its inhabitants. Fewer cattle and sheep are reared in this than any other part of the central departments. The forests abound with game, and produce great quantities of charcoal, the trade in which is extensive. Iron, marble, lithographic, and many other kinds of stone, gun flints at Cerilly, lime, and clay, are the principal minerals. Bricks and tiles are made in large quantities, and glass, earthenware, &c., in various places. The manufactured products include woollen stuffs and yarn, blankets, serges, &c.; beet-root sugar, paper, glue, &c.; hydraulic clocks made at Sens, barrels at Avallon, &c. The chief trade of the department consists in the export of its wines, corn, timber, and other agricultural produce. In 1835, of 190,786 properties subject to the *contribution foncière,* 108,342 were assessed at less than 5 fr., and 23,578 at from 5 to 10 fr.; the number of those assessed at 1,000 fr. and upwards, amounted to only 161. Yonne is divided into five arronds.; chief towns, Auxerre the capital, Avallon, Joigny, Sens, and Tonnerre. It sends five members to the chamber of deputies. Number of electors, in 1838-39, 1839. Total public revenue, in 1831, 11,344,964 francs. (*Hugo, art. Yonne; Dict. Géog.; French Offic. Tables.*)

YORK, a marit. co. of England, being by far the largest and most important in that part of the United Kingdom, is bounded on the N. by the co. Durham, E. by the North sea, S. by the counties of Lincoln (from which it is separated by the Humber), Nottingham, and Derby, and W. by Lancaster and Westmoreland, and a small part of Chester. Area, 3,669,510 acres, of which about 2,500,000 are supposed to be arable, meadow, and pasture. It is divided into the districts of the North, East, and West Ridings, being respectively as large as counties, and each of them having its particular lord lieutenant: there is besides a separate smaller district called the city of York and Ainsty; but the latter, except in so far as the city is concerned, has been united to the W. Riding. The extent, population, &c., of these different divisions are as follows:

	Area, Acres.	Inhabited Houses in 1841.	Population in 1841.
North Riding	1,275,520	61,509	204,502
East Riding	711,980	29,990	198,874
West Riding	1,668,560	296,473	1,154,904
City and Ainsty	33,440	7,710	26,323
Totals	3,669,510	31 5,048	1,581,584

Owing to its extent and various capacities, Yorkshire presents an epitome of the whole kingdom with respect to surface, soil, products, and industry. Some of the mountains on its W. border, are among the highest in the island: central ridge extending from Scotland S. to the middle of Derbyshire; and both there and in its N. division are very extensive tracts of high, sterile, moor ground. In the E. Riding a large tract of wolds extend from Flamborough head to Filey head, on the coast, to Pocklington and Market Wighton: but, notwithstanding these deductions, Yorkshire contains a great extent of most excellent land. The vale of York, the district of Cleveland in the N; and that of Holderness in the S.E., besides various other extensive tracts in different parts of the county, are exceedingly fertile, possessing soils suitable for every purpose, either of arable or stock husbandry. The climate is as various as the soil and elevation; but, except on the high grounds, it is mild and early, and is everywhere salubrious, except on the low, marshy grounds along the Humber. Agriculture in a medium state of improvement; not so far advanced as in Northumberland or Lincoln, but not so backward as in several other counties. There is, in this respect, however, a great difference in the different ridings, agriculture being in a much more advanced state in the W. riding than in either of the others. The general rotation is there—1st, turnips or fallow; 2nd, barley; 3rd, seeds; 4th, wheat. Bone manure is much used, but not to so great an extent as rape-dust; the latter, however, is principally used for wheat, the bone manure being decidedly superior for turnips. Drainage is too much neglected in the N. and E. ridings. In the latter no system is acted upon, except in the wolds where the rotation is—1st, turnips; 2nd, barley; 3rd, seeds, 4th, wheat. In other parts of this riding, and in the N. riding, two corn crops not unfrequently follow in succession, and but few operations are performed as they ought to be. (*Kennedy and Grainger*, i. 367.) York is more of a grazing than of an agricultural county. Vast numbers of horses are bred in most parts. Those in the highest estimation are called Cleveland bays, partly from the district in which they were originally found in the greatest perfection, and partly from their colour; but they are now very widely diffused. They are in extensive demand as carriage-horses. Cattle very various: they consist mostly of the long-horned breed; but there are considerable numbers of short horns, with endless varieties produced by crosses between these and other breeds. At present, the Teeswater and Holderness breeds are the greatest favourites with the graziers; but the long horns, or a cross between them and the short horns, are preferred by the dairy farmers. Yorkshire supplies most of the cows used in the London dairies. Their average yield of milk may be estimated at from 22 to 24 quarts a day, but it does not yield a proportionate quantity of butter. Sheep of all varieties, and stock very large, supposed to amount to about 1,300,000 head, producing annually about 28,000 packs of wool. Many hogs are kept, and Yorkshire hams are celebrated in all parts of the country. Property in the N. and W. ridings very subdivided; but in the E. riding it is less subdivided than in most parts of England, and many families in this riding have held their estates for centuries. Farms of all sizes; but the majority seem to be unusually small. Most part of these farms are held from year to year, or by tenants at will; and, notwithstanding the statements that have been made to the contrary, we have no doubt that this species of tenure, by diminishing the security of the farmer, has operated in no ordinary degree to retard the progress of improvement. Farm-houses and buildings for the most part rather indifferent. Average rent of land, in 1810, 16s. 7¼d. an acre.

The W. riding of this county stands in the very first rank as a manufacturing district. Leeds, Bradford, Huddersfield, Halifax, and Wakefield are the great seats of the woollen manufacture; flax-spinning is extensively carried on at Leeds; and the hardware manufacturers of Sheffield rival, and in some departments, as that of cutlery, far surpass, those of Birmingham. There are extensive iron-works at Rotherham; and latterly the iron-works of Yorkshire have made considerable progress. Their total produce in 1839 was estimated by Mr. Scrivenor at about 90,000 tons, though we incline to think that this is beyond the mark. Cotton manufactures have been established at Easingwold, and in some other parts of the W. Riding. The manufactures in the other ridings are but of trivial importance. The valuable beds of coals found in the vicinity of Leeds, Sheffield, Bradford, Wakefield, &c., have, no doubt, been the principal source of their prosperity. Besides coal and iron, Yorkshire has mines of lead, and veins of copper; alum works were established near Whitby in the reign of Elizabeth, and are still worked (*see* WHITBY); and there are in various places excellent limestone and freestone quarries. Principal rivers Ouse, Swale, Ure, Wharfe, Aire, Calder, Don, Derwent, Hull, and Esk; the waters of all these, except the last, being poured into the great estuary of the Humber. The canals, particularly in the W. Riding, are numerous, and some of them are of great importance; and the principal towns are now, also, connected with railways. The county is divided into wapentakes and liberties, and contains 613 parishes. It sends 39 members to the House of Commons, viz. six for the county, being two for each riding; two each for the city of York, and the boroughs of Leeds, Sheffield, Hull, Beverley, Bradford, Halifax, Doncaster, Pontefract, Ripon, Knaresborough, Malton, Richmond, and Scarborough; and one each for the boroughs of Huddersfield, Whitby, Wakefield, Northallerton, and Thirsk. Registered electors for the N. Riding in 1839–40, 11,911; for the E. Riding, 7496; and for the W. Riding, 30,129! Sum expended for the relief of the poor, in 1839, N. Riding, £53,642, E.

Riding, £87,539, W. Riding, £390,994. Total value of real property assessed to the county, in 1843-41, in the N. Riding, £1,011,885, E. Riding, £1,111,807, and W. Riding, £1,824,222 Total amount of money levied for poor-rates for the county during the year ended Lady-day, 1814, £468,947.

YORK (an. *Eboracum*), an ancient and celebrated city of England, being, under the Romans, the capital of Britain, and at present the second city of the kingdom in respect of rank, though not of importance. It is a county of itself, and a parliamentary and municipal borough, locally situated near the centre of the county York, of which it is the capital, at the junction of the N., E., and W. Ridings, on the Ouse, at the confluence of the Foss, 22 m. N.E. Leeds, 23 m. N.W. Hull, 170 m. N.N.W. London, and 199 m. S.S.E. Edinburgh. Lat. 53° 57′ 45″ N., long. 1° 4′ 31″ W. The area of the old parliamentary borough and city comprised 1779 acres, and had in 1831 a population of 26,260. But by the late act, some additions were made of parts of the townships of Clifton and Heworth on the N., and of Fulford on the S.E., to the parliamentary borough, which is nearly co-extensive with the municipal borough, comprising, in part or wholly, 36 parishes, with some extra-parochial districts.

The city is enclosed by its ancient walls, supposed to have been erected by Edw. I., about 1990; they are flanked with numerous towers, and having been repaired and renovated in 1831, form a most delightful promenade, extending round the greater part of the city. They are pierced by five principal gates, termed bars, and by five smaller gates, or posterns; some of the former being remarkable structures. The Ouse and the Foss traverse the interior of the city, meeting at its S. extremity. The Foss is crossed by four bridges, and the Ouse by a single bridge, a handsome structure of three arches, constructed between 1810 and 1820, at a cost of £80,000. The span of the central arch is 75 ft., that of the other arches 65 ft. each; the total width of the bridge within the parapet is 40 ft. Handsome flights of steps at each end conduct to spacious quays on both sides the river, called the King's and Queen's staiths, to which vessels of 90 tons may be moored.

York consists of several parallel lines of thoroughfares, running N. and S., crossed by others, which are generally shorter and more irregular, in an opposite direction. The principal of the former, nearly 2 m. in length, consists of Bootham, Petergate, Collergate, Walmgate, &c., with their continuations. The line crossing it, and composed of Micklegate, Ousegate, Pavement, St. Saviour's-gate, &c. is almost as long. In the centre of the city is a fine broad open space called Parliament-street, terminating at one end in Sampson-square; and at the other end in the Pavement, the site of the corn, poultry, and other markets; and meat and leather fairs are held in Peasholme Green, an open space in the E. part of the city. There are a few other open spaces in the heart of the city, as St. Helen's square, &c.; but none of them deserves any particular notice. York has been much improved and modernised of late years, but it still preserves an air of antiquity in its narrow streets and old-fashioned houses. Many of the latter French overhung the streets, the upper stories projecting beyond the lower; but a good many of them have been taken down, and buildings in a modern style have been erected in their stead. Some of the streets also have been widened, and the city generally is well paved, and lighted with gas. In consequence of the rise of Liverpool, Manchester, &c., the increasing importance of many of the large towns of Yorkshire, and the greater facilities of communication between the different parts of the kingdom, York no longer enjoys that pre-eminence in the N. she possessed in the earlier part of last century. Still, however, she is not declining in any respect; but is, on the contrary, increasing in size. In the outskirts many substantial and even superior buildings have been recently erected; and the city is extending itself in an equal degree in almost all directions. To the N.E. of the town was formerly an open space known by the name of Heworth moor. In 1817 this was enclosed, and in this neighbourhood a great number of substantial and excellent houses have been built since the period of the enclosure. Here also many market gardens are cultivated, and altogether the district is thriving and populous, and presents undoubted testimony of progressive and prosperous industry. (*Hound. Report.*) On the W. of the Ouse also the road from Leeds to the "Micklegate" are several good houses, many of which have been recently built, and are occupied chiefly by persons who have either retired from business, or are engaged in business in the other part of the town; the number of these houses is constantly increasing. The parishes beyond the Foss, through which the road passes, contain for the most part a population of poor description." (*Mun. Corp. Rep.*)

York minster, or cathedral, is the finest edifice of its kind in the kingdom. It stands in the N. part of the city, and except on its N. side, where a considerable space of ground has been cleared, is closely hemmed in by these small buildings. The present edifice, said to have been raised on the site of a church originally founded by Edwin, king of Northumberland, in the 7th century, was principally erected during the 13th and 14th centuries. It is without cloisters, and built in the form of a cross; consisting internally of a nave and two aisles; a transept, with aisles and a lantern in the centre; a choir, with aisles, and vestries or chapels on the S. side; and a chapter-house, with a vestibule, on the N. side. Its principal measurements are as follow:— length, internally, 524½ ft. (being greater than that of any other cathedral in England, except Winchester); internal length of transept, 222 ft.; length of nave, 264 ft.; do. of choir, 131 ft.; height of both, 99 ft.; breadth of nave, 109 ft.; height of great tower, 234 ft.; height of W. towers, each 196 ft. (*Stranger's Guide to York*, p. 103.)

This magnificent structure has a portion of all the styles of English architecture; but the Norman only appears in a fine crypt, under a part of the choir, which reduces the general appearance to the three later styles; of these, the transepts are early English; the nave and arches supporting the great tower are decorated; and the choir and upper part of the great tower are perpendicular. (*Rickman's Gothic Architecture.*) The W. front has been compared to that of the cathedral of Rheims for richness, sublimity, and beauty of architectural design. It is divided into three compartments, by two massive graduated buttresses enriched on every face with tabernacle-work, and the elevated battlemented gable is covered with ornamental tracery of the most florid kind. There are three entrances in this front; over the central of which is the unrivalled W. window, divided into eight portions by upright mullions, which in the upper part beautifully diverge into the leafy tracery peculiar to the 14th century. The magnificent towers which flank this side exactly correspond; they are supported by buttresses, and have at their summits eight crocketed pinnacles connected by a battlement. Almost the whole of the W. front is filled with niches, but these, with few exceptions, are empty. The S. side, though finished less elaborately than the W. front, is very imposing. The porch in the S. transept is the most usual entrance to the church, and is deeply recessed by numerous mouldings; over it is a beautiful marygold window, and the gable is surmounted by an enriched pinnacle. The N. side is in a similar style to the S., though finished in a plainer manner; and in its transept is the remarkable stained glass window termed the "five sisters." The E., like the W. front, is in three grand divisions separated by buttresses, the central of which is wholly occupied by a magnificent window. Like the W. front also, it is covered with niches, though only a very few of the statues formerly occupying them now exist. On this front the influence of time is very perceptible. The central tower, 234 ft. in height, is probably unfinished. It has two large windows, with two tiers of mullions, in each of its four sides. But it wants a spire; and when contrasted with the W. towers, has a heavy appearance.

The interior of the minster corresponds in beauty and grandeur with the exterior. A careful restoration of the cathedral in most of its parts had been completed, when, on 2d February, 1829, it was set on fire by a lunatic; the conflagration thence ensuing destroyed the fine organ, and all the woodwork and roof of the choir. Another destructive fire broke out on the 20th of May, 1840, in the S.W. tower, by which its fine ring of 10 bells and the clock, with part of the roof of the nave, were burnt. These injuries are now (1842) in the course of being completely repaired: the choir was renovated after the fire of 1829, under the superintendence of Sir R. Smirke. The new roof is wholly constructed of teak, presented by government; and is covered with lead procured from the mines of the Greenwich hospital estates. The remarkable stone screen, which separates the choir from the nave, stands in its original position, and is of a most gorgeous and florid style, ornamented with 15 statues of the kings of England, from William I. to Henry VI., all of which, except the last, are of ancient sculpture. The new organ placed above this screen, and presented by the late earl of Scarborough, is of the most superb description, and has some pipes 32 ft. in length. A great deal of fine stained glass, many sculptured coats of arms, and the tombs of many of the archbishops of York, attract the visiter's notice in the interior; though, on the whole, this cathedral is less rich in monuments than many others in the kingdom. From the N. transept, a vestibule leads to the chapter-house; this is an octagonal building, 63 ft. in diameter, and 67 ft. 10 in. in height, supported on the outside by eight massive buttresses. "The more minutely," says Rickman, "this magnificent edifice is examined, the more will its great value appear. The simplicity and boldness, and at the same time the great richness of the nave, and the very great chastity of design and harmony of composition of the choir and great tower, render the building more completely one whole than any of our mixed cathedrals; while the exquisite beauty of the early character of the chapter-house,

and its approach, forms a valuable link to unite the early English transepts and the decorated nave. This chapter-house is by far the finest polygonal room without a central pier in the kingdom, and the delicacy and variety of its details are nearly unequalled. Too much praise cannot be given to the dean and chapter for their careful restoration of every decayed portion. By this restoration the whole of the W. front may be considered in as good a state as when first erected; a considerable portion of the S. side is also restored." (*Gothic Architect*, p. 265.) The vestries on the S. side of the church contain, among many other antiquities, a chair in which several of the Saxon kings were crowned, and which is said to be older than the cathedral itself; and the drinking horn of Ulphus, a Saxon prince of Deira, presented to the cathedral in 1036, with a large extent of country to the E. of York, still in the possession of the see. The library is a short distance from the cathedral on the N. side. (*Winkle's Cathedrals; Guide to York, &c.*)

The gross annual income of the see of York amounted, at an average of the three years ending with 1831, to about £13,800, and the nett income to £12,629. The chapter consists of a dean and four canons residentiary, sharing an income of £1352 a year, and 26 prebendaries having separate revenues. (*Eccl. Rev. Report.*) The archbishop of York has the title of primate of England, with the privilege of crowning the queen-consort, and ecclesiastical authority over the province of York, comprising the sees of York, Durham, Carlisle, Chester, Ripon, and Sodor and Man.

Previously to the dissolution of the religious houses by Henry VIII., besides 17 chapels, 16 hospitals, and nine religious houses, there were in the city 41 parish churches, but of these last only 23 now remain. Many of these would be worthy of notice elsewhere; but they sink into insignificance after the cathedral. St. Michael-le-Belfrey, in the minster-yard, is the largest and most elegant, and with St. Martin's in Coney-street is in the late perpendicular style. All Saints, North-street, and St. Mary's, Castle-gate, have towers and lofty spires, and are mostly perpendicular with some earlier portions; St. Denis, St. Lawrence, and St. Margaret, have good Norman doors, with portions of later date; and St. Mary Bishophill, the Elder, has portions of good early English and decorated work, amidst various alterations and insertions. In many of the churches are considerable quantities of old stained glass. (*Rickman*, p. 226.) All Saints in the Pavement is of very ancient foundation; and Drake says that its N. side is almost wholly built out of the ruins of *Eboracum*, though other parts of the edifice are quite modern. A large lamp still preserved here used to be hung at the summit of this building, as a beacon for travellers at night through the forest of Galtres, which extended from Bootham-bar a considerable distance N. of the city. Most of the livings of these churches are rectories or vicarages in the gift of the crown or the dean and chapter of York; the most valuable are these of St. Cuthbert, worth £433 a year nett, and St. Mary Bishophill, the Elder, worth £396 a year, both in the gift of the crown: St. Martin's, Micklegate, worth £343 a year, is in the gift of a private family. (*Eccl. Rev. Rep., &c.*)

The remains of St. Mary's abbey, originally founded by William Rufus in 1088, and refounded in 1270 for black monks of the Benedictine order, are very interesting. The abbot was mitred, and had a seat in parliament; and at the time of the dissolution the revenues of the abbey amounted to £2085 1s. 3d. a year. The buildings were for the most part destroyed between 1701 and 1717, and their materials used for rebuilding the castle of York and St. Olave's church, and repairing Beverley minster. Almost the only parts remaining are a gateway, and the N. wall of the abbey church, 371 ft. in length, having fine light Gothic window-arches, with highly finished carved capitals. Mr. Rickman says that the remaining part of the church furnishes the richest and most beautiful specimens of transitions from early English to decorated that remain for examination; but being entirely exposed, it is fast decaying. The abbey had an extensive and strongly fortified precinct without the ancient walls of the city; and some of its walls and towers, forming an extraneous portion of the old city defences, may still be seen between Bootham-bar and the Ouse. The remains of St. William's college, founded by Henry VI., exist in a street near the cathedral. St. William's chapel stood on the old bridge over the Ouse, and was consequently taken down with that structure. The cloisters of St. Leonard's and St. Peter's hospitals, curious remains of the architecture of the time of William I. and II., are now used as wine vaults.

The dissenters, who form a numerous and respectable body in York, have at least a dozen places of worship; the oldest of which is the Presbyterian (Unitarian) chapel, in St. Saviour-gate. The Wesleyans have an elegant new chapel in the same street, with a massive Ionic portico, &c., besides three other chapels. The Independents have two chapels, one of which (Salem chapel), erected at the end of St. Saviour-gate, is a large and handsome edifice. There are also meeting houses for Primitive and other Methodists, and Friends; a fine Roman Catholic chapel, a nunnery, and chapel outside Micklegate-bar, &c.

York castle, towards the S. extremity of the city, between the Ouse and Foss, near their confluence, occupies a space of nearly four acres. It was originally built by William the Conqueror, who also erected another fortress at York, on the other side of the Ouse. But only a small portion of the original structure of the castle remains, except Clifford's tower, a keep added by the Conqueror to the rest of the edifice, and erected upon an artificial mound, which had probably served for the site of a Roman fortress. York castle, which was long garrisoned for the king in the civil wars, is not now a defensive military post, but has been converted into the county prison and hall. The basilica, or county hall, on the W. side of the great area, is entered by a portico, supported by Ionic columns, and internally divided into civil and criminal courts, with handsome rooms, for the use of the grand and petit juries, counsel, &c. The building, on the E. side of the area, which is uniform in design with the courthouse, is chiefly appropriated to female prisoners. Between 1821 and 1836, a new prison was built here, at an expense of £203,530, on the panopticon principle, with eight airing courts, the whole being surrounded by a lofty stone wall, 35 ft. high. This building is said to be the most excellent of its kind. In 1838-39, the committals to York castle amounted to 632. (*Jail Returns*, 1840.) The city jail and house of correction is on the W. side of the Ouse; its outer wall encloses an area of nearly three fourths of a mile in circuit: it is appropriated partly to prisoners before trial. Near it is the *vetus ballium*, or old baile, a mound corresponding with that on which Clifford's tower is built, having probably had the same origin and purpose.

Most of the other edifices of public interest, are in the N. part of the city. The mansion-house, a large and handsome edifice, erected in 1725, has in front a rustic basement supporting an Ionic colonnade, with a pediment on which are the arms of the city. The state-room, 49½ ft. in length by 27⅓ ft. in breadth, has paintings of William III., George II. and IV., and of several noblemen and gentlemen. The guildhall, behind this edifice, built in 1446, comprises one of the finest Gothic halls in the kingdom, 96 ft. in length, 43 ft. in width, and 29½ ft. in height, the roof being supported by 10 octagon pillars on stone bases. In the windows are some fine specimens of stained glass, and over the entrance is a full-length statue of George II. In this hall, the lords-president of the north formerly held their court; and here also the Scotch received the £200,000 paid them by parliament for the assistance they afforded against Charles I. At the end of the court, and house of assize and sessions court; and adjoining, are the council chambers of the corporation. The assembly-rooms were erected, by subscription, in 1730, and are entered under a portico, resting upon light stone columns, supporting a balustrade. The walls are supported by 44 light and elegant Corinthian columns, with a beautiful cornice, the upper part of the building being of the composite order, and richly adorned. The rooms are lighted by 44 windows. The grand assembly-room is constructed from a design by Palladio, and measures 112 ft. by 40 ft., and 40 ft. in height. It was used for concerts till about 1825, when, being found too small, a magnificent concert-hall adjoining was built, 95 ft. in length, 60 ft. in breadth, and 45 ft. in height, capable of containing 1700 persons, 400 being accommodated in a gallery supported by cast-iron pillars. It is lighted with gas, and fitted up in a style of much elegance: its cost, including the purchase of the ground, amounted to £9400. The theatre, built by Tate Wilkinson, in 1765, and recently altered externally in the Elizabethan style, is extremely commodious. The Yorkshire Philosophical society, founded in 1822, obtained, in 1826, a grant of three acres of land, part of the site of St. Mary's abbey, from government, for a suitable building, botanic garden, &c. The museum, built between 1827 and 1830, is an elegant edifice, with a front towards the Ouse, 200 ft. in length, and has a spacious hall, a library, a theatre for lectures, with large collections in geology, mineralogy, zoology, comparative anatomy, a chemical laboratory, &c. The affairs of the society are conducted by a council of 12 members, and officers elected once a year. Annual subscription of members £2 (ladies, £1), with £5 entrance. The York Subscription library, with 17,000 vols., occupies a spacious suite of rooms in St. Leonard's-place: it is supported by about 400 members, who pay 10 guineas each on entrance, and 25s. a year afterwards. There are other subscription, and several good private circulating libraries, with three news-rooms, at which the London, Edinburgh, and many provincial newspapers are taken in. Four weekly newspapers are published in York.

Outside Monk-bar is the county hospital, founded in 1740 by Lady Hastings, with (in 1842) an income of about £1400

a year. The building has a front of 75 ft., by a depth of 49 ft., and encloses a court measuring 36 ft. by 35 ft. It is remarkably well kept, and capable of accommodating 100 patients. Without Bootham-bar is a lunatic asylum, built by subscription in 1777, three stories in height, having a front 132 ft. in length, with extensive grounds, &c.; and about 1 m. from the city is the Retreat, an establishment of a similar nature under the careful management of the society of Quakers. Here, also, is a dispensary, founded in 1788; an eye infirmary, established in 1831, and various medical and other charities for the benefit of the poor. The educational establishments are on a very extensive scale, at least in so far as elementary and the more ordinary branches of instruction are concerned. A manners' school, in connection with the York Diocesan society, occupies the extensive premises in Monkgate, formerly used as the Unitarian college. This last, the chief seminary of the Unitarians in England, was removed from Manchester to York, in 1803, but has lately been again removed to Manchester. Here also are national schools, established in 1812, in which above 700 children of both sexes are educated; British schools for about 200 boys and 300 girls; the blue-coat boys' and grey coat girls' schools, established in 1705, having an income of about £1500 a year; Haughton's charity school, for the education of 20 poor children of the parish of St. Crux; the spinning school, established by two ladies in 1782, where about 60 girls are instructed in reading, knitting, and sewing, and principally clothed; with Sunday schools, &c.

The cavalry barracks on the Fulford road, the manor-house at different times a royal palace and mint, and Bishopsthorpe palace, the seat of the archbishop, about 3 m. S.E. from the city, are the principal edifices which remain to be noticed. The grounds of the last are frequently resorted to in summer by the inhabitants, whose principal public promenade in the city is the New walk, a gravelled terrace planted with elms, &c., extending from the neighbourhood of the castle for nearly 1 m. along the Ouse. Near the New walk is a cold bath; and hot, cold, and vapour-baths are established in Lendal, for the accommodation of the public at large. Swimming baths have also been recently erected.

The city of York claims to be a corporation by prescription. Its earliest extant charter is one of Henry II., without date; but its governing charters before the Municipal Reform Act were of the 16th Charles II. and the 10th George IV. By the latter, the corporation officers were the mayor, 12 aldermen, and the two acting and the former sheriffs, the recorder, city counsel, town clerk, coroners, 72 common councilmen, &c., who sat, as in London, in two separate courts. All the corporate officers were freemen, the freedom of the city being acquired by birth or apprenticeship to a freeman within the city liberty, and by gift or purchase from the upper house, the price of purchase varying from £35 to £150. Under the Municipal Reform Act, the borough is divided into six wards, and is governed by a mayor, recorder, 12 aldermen, and 36 councillors, six from each ward. The chief magistrate has the title of lord mayor, conferred by Richard II., in 1389, which title he consequently enjoyed before the . chief magistrate of the metropolis. York sent two members to the House of Commons in the 49th of Henry III., and has continued to do so regularly from the time of Edward I., the right of election having been formerly vested in the corporation and freemen. The greatest number of voters within the 30 years previously to the parl. Reform Act, was 3715 polled in 1830. Registered electors in 1839-40, 3396. The election for the N. riding of the county of York is held here. Courts of assizes for the county and the city are also held here twice a year, besides quarter sessions, a court of pleas, and petty sessions twice a week; there were formerly several other courts, now obsolete. (*Mun. Rep., Appendix*, III.) Corporation revenue, in 1840, about £3000. The corporation of York had exclusive jurisdiction over the Ainsty, a large district comprising about 35 towns and villages, from the time of Henry VI. till a late act annexed the Ainsty to the W. riding of the county.

Under the Romans, York was, no doubt, the commercial emporium of the N. part of the island, and it appears to have been a city of some commercial importance in the time of Edward III., who established a woollen manufacture in the city, which continued to flourish for a lengthened period. At present its trade is comparatively small; for its size; and the largest amount of capital now employed in any one branch by the citizens is supposed to be in the drug trade. Considerable business has, however, been done latterly in the iron trade, and there are several large foundries; printing, brewing, and comb making are also extensively carried on. The glass manufacture was established at York at a somewhat early period; and phials and flint glass wares are still made here. Linen cloth, sacking, twine, leather, gloves, jewellery, paper-hangings, fringe, musical instruments, brass wares, tobacco pipes, &c., are among the other goods made at York. Many guilds or trading companies formerly existed, but all of them except three appear to be de-solved. The company of merchant adventurers of York is an ancient corporation by prescription, now consisting of about 120 members, under a governor, deputy governor, &c., having property yielding £300 a year, with a chapel and hall, and a hospital, in Foxgate. The merchant tailors' company, incorporated by charter 14 Charles II., consists of from 20 to 25 members, with exclusive privileges in the city and an income of £136 a year. The other company is the goldsmiths', authorized by act of parliament. The Ouse trustees have lately spent large sums on the improvement of the river navigation; and steamers now ply to and from Hull at all times of the tide. Coals are brought to the town by water and by railway. A decided increase of trade has been experienced since the completion of the railways, by which York communicates with Newcastle, Durham, Carlisle, &c., northward, and with Leeds, Hull, the Liverpool lines, and other parts to the S. The York station of these railways is an elegant building, immediately within the walls near Micklegate. Large sales of cattle and horses take place at fairs held here once a fortnight, besides which there are monthly fairs for leather; many others in the year for flax, wool, &c. Markets, Tuesday and Saturday, the latter chiefly for corn. A new cattle-market was opened in 1828 outside Fishergate. Races, which are extremely well attended, are held three times a year on Knavesmire, a large plain about 1 m. S. from the city, where is a spacious grand stand. Four joint-stock banking companies and two private banks are established in the city.

The society of York is superior to that of most provincial towns. From its being the capital of the most extensive county in the kingdom, it is the residence of a greater number of gentlemen connected with the law and the administration of public affairs. It is necessarily, also, the residence of a number of gentlemen connected with its cathedral and different ecclesiastical establishments. And in addition to its fixed inhabitants, it is the winter residence of many of the provincial gentry attracted hither by its superior society, amusements, facilities for education, &c.

Antiquities and History.—York, though successively the residence of Hadrian, Severus, Geta and Caracalla, Constantius Chlorus, Constantine the Great, &c., has few striking Roman antiquities. Such as do exist comprise a remarkable multangular tower, a long wall, with altars, *paters*, tombs, monuments, and the foundations of ancient buildings. The *palatium* of the Roman emperors is supposed by Drake to have occupied several acres near the cathedral, extending from Christchurch through all the space between Goodramgate and St. Andrewgate to Aldwark. Not far from this, in St. Cuthbert's cemetery, many Roman sepulchral remains have been found. Outside Micklegate-bar, a Roman vault, with a perfect skeleton, was opened in 1807; and a tessellated pavement was discovered within the same bar in 1814. Severus died at York, A.D. 212; and his funeral obsequies would appear to have been performed on some heights a little W. of the city, still called Severus' hills. Constantius, who died in 307, is traditionally said to have been buried in the parish church of St. Helen's. Under the Saxons, York was successively the capital of the kingdoms of Northumberland and Deira. It was taken, and its neighbourhood devastated by William the Conqueror in 1069. Several parliaments have been held in York, the first being that summoned by Henry II. in 1180. In 1540, Henry VIII. established in this city an officer called the lord president of the north, and a council with very extensive powers which existed till the civil wars, when York was frequently a principal station and residence of Charles I.; it, however, surrendered to the parliament in 1644. (*Bound., Munic., and other Parl. Reps.; Bellerby's Guide to York; Rickman's Archit. in England; Private Information, &c.*)

YORK, river, Va., is formed by the junction of Pamunkey and Mattapony rivers, and enters Chesapeake bay between York and Gloucester counties. The lower part forms a bay, generally 2 or 3 m. wide, but at Yorktown it is contracted to the width of 1 m., where it forms one of the best harbours in the state. It is navigable 20 m. above this for large vessels. It is about 120 m. long to its remote sources.

YORK, county, Me. Situated in the S.W. part of the state, and contains 818 sq. m. Bounded S.E. by the Atlantic, along which is a number of good harbours, and several lighthouses. It contained in 1840, 5010 neat cattle, 60,477 sheep, 14,391 swine; and produced 55,866 bushels of wheat, 23,140 of rye, 230,650 of Indian corn, 6046 of buckwheat, 40,670 of barley, 109,413 of oats, 1,123,441 of potatoes, 26,084 of sugar. It had 945 stores, 22 fulling-mills, seven woollen factories, five cotton factories with 25,736 spindles, 90 grist-mills, 157 saw-mills, two oil-mills 101 tanneries, six potteries, four printing-offices, four weekly newspapers; total capital in manufactures $1,604,425. Pop. 54,034. Capital, Alfred.

YORK, county, Pa. Situated in the S., toward the E.,

part of the state, and contains 864 sq. m. Bounded N.E. by Susquehanna river, N. by Yellow Breeches creek. Drained by Codorus and Conewago creeks. It contained in 1840, 34,495 neat cattle, 36,347 sheep, 56,297 swine; and produced 357,515 bushels of wheat, 363,886 of rye, 600,692 of Indian corn, 12,999 of buckwheat, 1714 of barley, 597,044 of oats, 172,946 of potatoes, 162,748 pounds of tobacco. It had 159 stores, 17 lumber-yards, seven fulling-mills, 10 woollen factories, seven flouring-mills, 132 grist-mills, 104 saw-mills, four paper-mills, nine oil-mills, 53 tanneries, 216 distilleries, four breweries, nine potteries, six printing-offices, seven weekly newspapers; two academies, 50 students; 130 schools, 3749 scholars. Pop. 47,010. Capital, York.

YORK, county, Va. Situated in the S.E. part of the state, and contains 150 sq. m. Bounded N.E. by York river, S.E. by Chesapeake bay. It has six schools, 170 scholars. Pop. 4720. Capital, Yorktown.

YORK, district, S. C. Situated in the N. part of the state, and contains 700 sq. m. Bounded S.W. by Broad river, N.E. by Catawba river, and drained by their tributaries. It contained in 1840, 13,695 neat cattle, 7049 sheep, 24,021 swine; and produced 64,021 bushels of wheat, 44,738 of rye, 478,833 of Indian corn, 44,148 of oats, 19,175 of potatoes, 5900 pounds of rice, 2942 of tobacco, 866,504 of cotton. It had 14 stores, six flouring-mills, 21 grist-mills, 15 saw-mills, three tanneries, 24 distilleries, one pottery, one printing-office, one weekly newspaper; one academy, 33 students; 29 schools, 679 scholars. Pop.: whites, 11,449; slaves, 6695; free coloured, 109; total, 18,383. Capital, Yorkville.

YORK, p. t., port of entry, York co., Me., 45 m. S.W. by S. Portland, 92 m. S.W. Augusta, 502 W. Bounded S.E. by the Atlantic, into which York river enters by a broad æstuary. Incorporated in 1653. Agamenticus mountain lies in its N.E. part, a noted landmark for seamen. The village is situated on York river, 1 m. from the ocean, and has a harbour which admits vessels of 250 tons. It was originally laid out for a large city, with streets crossing each other at right angles, but has not equalled the designs of its early founders. On cape Neddock, a rocky promontory on the S. side of a river of the same name, is a lighthouse, and on this river, a mile from the ocean, is a village, with a harbour, navigable at full tides only. The town has 11 stores, two fulling-mills, five grist-mills, five saw-mills; 886 scholars in schools. Pop. 3111. Tonnage of the district in 1843, 2072 tons.

YORK, p. t., Livingston co., N. Y., 238 m. W. Albany, 354 W. Bounded E. by Genesee river, and drained by its tributaries. It has 12 stores, one woollen factory, two furnaces, three grist-mills, five saw-mills, two distilleries; 17 schools, 1039 scholars. Pop. 3049.

YORK, t., York co., Pa., 5 m. S.W. York borough. It has two grist-mills, one saw-mill, two tanneries. Pop. 1294.

YORK, p. b., capital of York co., Pa., 24 m. S. Harrisburg, 90 W. Situated in Spring Garden and West Manchester townships, on Codorus creek, 11 m. from Susquehanna river. It is regularly laid out, with streets crossing each other at right angles, and contains a spacious and elegant new courthouse, a jail, a bank, an academy, a county lyceum, with a cabinet of mineralogy and Natural history, which sustains scientific lectures, and 10 churches belonging to Lutherans, German Reformed, Presbyterians, Moravians, Episcopalians, Methodists, Friends, Roman Catholics, and African Methodists, 38 stores, 11 lumber-yards, one woollen-factory, five tanneries, three breweries, four printing-offices, five weekly newspapers, three of which are in German;

cemetery of the German Reformed church.

YORK, t., Athens co., O. Drained by Hockhocking river, on which is Nelsonville village. It has 10 schools, 191 scholars. Pop. 1601.

YORK, t., Belmont co., O. Bounded E. by Ohio river. Captina creek affords water-power It has one school, 25 scholars. Pop. 1294.

YORK, p. t., Washtenaw co., Mich., 46 m. W Detroit, 511 W. It has one store, one flouring-mill, one grist-mill, four saw-mills. Pop. 1146.

YORKSHIRE, p. t., Cattaraugus co., N. Y., 278 m. W. Albany, 341 W. Drained by Cattaraugus creek and its tributaries. It has one church, two stores, one fulling-mill, two grist-mills, six saw-mills, three tanneries. Pop. 1992.

YORKTOWN, p. t., Westchester co., N. Y., 112 m. S. Albany, 275 W. Drained by Croton river. It has 10 schools, 201 scholars. Pop. 2819.

YORKTOWN, p. v., port of entry, and capital of York co. Va., 70 m. E.S.E. Richmond, 185 W. Situated on the S. side of York river, opposite to Gloucester. It contains a courthouse, jail, county offices, and 40 dwellings, some of them in a dilapidated condition, and about 300 inhabitants. It is memorable as the place where Lord Cornwallis surrendered the British army to Gen. Washington, Oct. 19th 1781, which event terminated the Revolutionary war.

YORKVILLE, v., New-York co., N. Y. Situated on Harlem railroad, 5 m. N. of the city hall, through which the cars ply many times daily. The railroad here passes through a deep cut, and immediately N. of the village is a tunnel, 595 ft. long, 24 wide and 21 high to the crown of the arch, cut through solid rock, which supersedes the necessity of masonry, and which cost $90,000. In the vicinity of the village is the receiving reservoir of the Croton water-works, containing 35 acres, enclosed by a high and substantial wall. It is a little remarkable that, in the hottest weather in the summer, the water in this reservoir is as cool as when it enters from the aqueduct.

YOUGHALL, a parl. bor. and seaport town of Ireland, prov. Munster, co. Cork, on the W. side of the estuary of the Blackwater, immediately within its mouth, 27 m. E. by N. Cork. Area of modern parl. bor., 212 acres. Pop. in 1831, 9600. It is built close to the water's edge, along the foot of a pretty steep hill, and consists principally of a main street, extending for about 1 m. parallel to the strand, and of various other smaller streets and lanes. It was formerly surrounded by walls; and these in part remain, and form, on the summit of the hill to the W., the boundary of the town. Of 1900 houses comprised within the parl. bor., in 1831, 1000 were slated, and 900 thatched; the greater proportion of the latter are in the suburbs, which are "large and bad," extending in every direction up the hill. The principal public building is the parish church, a large Gothic edifice; in its immediate vicinity are the ruins of an old abbey, one of the windows of which "is extremely beautiful, and quite entire. The churchyard, too, is one of the largest and finest, in point of situation, I have ever seen. It is interspersed with lime and other trees; and like everything else about Youghall, has many remnants of antiquity, old tombs, old ivied moss-grown stones, and luxuriant weeds." (*Inglis*, i. 178.) It has, also, a chapel of ease, several Roman Catholic chapels, and meeting-houses for various classes of dissenters, an infirmary, a dispensary, a barrack for infantry, numerous public schools, a convent, the college, now in a neglected state, the property of the duke of Devonshire, a courthouse, custom-house, fever and lying-in hospitals, &c. The house occupied by Sir Walter Raleigh is still preserved in good repair, and with but little change.

Youghall sent two members to the Irish House of Commons; and it has sent one member to the Imperial House of Commons from the era of the union downwards. Registered electors for 1839-40, 670. Under the Irish Municipal Reform Act, 3 and 4 Victoria. cap. 108, the corporate body is styled the mayor, bailiffs, burgesses, and commonalty of Youghall.

The manufactures of the town are inconsiderable, consisting only of two small potteries and brick-works. It is too near Cork to have much foreign trade; but owing to its situation on a fine navigable river, it is the emporium of a considerable tract of country. The great articles of export consist of grain and meal, provisions, cattle, and pigs, their aggregate value having amounted, in 1836, to £315,316. The principal articles of import are timber and coal. The bar at the river's mouth has only 4 ft. water at ebb tide, and it is inaccessible for vessels drawing more than 12 or 13 feet water, except at high springs. Youghall is included in the port of Cork, but its shipping is inconsiderable. Postage, in 1830, £1114; ditto, in 1836, £1299. Branches of the bank of Ireland and the Provincial banks are established in the town. The beach is fine and the town is well fitted for sea-bathing; though, in this respect, but little advantage has been taken of its capabilities.

It is believed, apparently on good grounds, that the introduction of the potato cultivation into Ireland dates f.om 1610, when Sir Walter Raleigh sent a few to be planted on his estate in the vicinity of this town! But such has been the progress of this exotic, that it now furnishes, and has for a lengthened period furnished more than three fourths of the food of the people of Ireland; and its astonishing increase has been at once a cause and a consequence of the equally astonishing increase of population in the island. (See *Boundary Report; Railway Report; Commercial Dict.*, art. *Potatoes*, &c.)

YOUGHIOGHENY, river, Va., Md. and Pa., rises in Preston co., Va., and flowing through Md. and Pa. enters

Monongahela river, 18 m. S.E. of Pittsburg. It runs through the Laurel hills, and Chestnut ridge. The Ohiopyle falls, about 30 m. from its entrance into Monongahela river, have a perpendicular descent of 20 ft., presenting, in high water, a grand and interesting appearance. It is navigable to these falls.

YPRES (Flem. *Ypern*,) a fortified town of Belgium, prov. W. Flanders. cap. arrond. and two cants., on the Yperlee, 29 m. S.W. Bruges, and 16 m. N.N.E. Lille; lat. 50° 51' 10" N., long. 2° 53' 4" E. In the 14th century it is said to have been nearly equal in population and importance to Bruges; whereas in 1836, it had only 15,964 inhabitants (*Hasselingh*.) It is well built, and, like most towns in Flanders, it has extensive water communications, being connected by canals with Nieuport, Bruges, &c.

The courthouse and cloth-hall occupy a vast Gothic building of the 14th century, surmounted by a fine tower. The cathedral, a Gothic edifice, has a painting attributed to Van Eyck; the tomb of Jansen, bishop of Ypres, and founder of the sect of Jansenists in the 17th century, &c. There are several other churches and chapels, two hospitals, an exchange, a royal college, &c. Ypres was formerly famous for its manufactures of woollen and linen cloths, and the fabric called *diaper* (originally *d'Ypres*) derives its name from having been originally made in this town. Linen yarn and lace are now the principal articles manufactured; but there are still some woollen and linen cloth factories at Ypres, with tanneries, bleaching and dyeing-houses, one or more salt refineries, &c. Ypres experienced many reverses in the wars of the 17th and 18th centuries. Under the French it was the capital of the dep. Lys. (*Vandermaelen, Dict. du Fland. Occident.; De Cloet, &c.*)

YPSILANTI, p. t., Washtenaw co., Mich., 30 m. W. by S. Detroit, 517 W. Watered by Huron river and Stony creek. The village is situated on the W. side of Huron river. It contains three churches, a Presbyterian, Methodist, and Episcopal, an academy, 16 stores, three flouring-mills, three saw-mills, one woollen factory, two carding machines, one iron foundry, one tannery, 380 dwellings and about 1500 inhabitants. The town contains 16 stores, one furnace, four flouring-mills, three grist-mills, nine saw-mills, one distillery. Pop. 2419.

YRIEX (ST.), a town of France, dep. Haute-Vienne, cap. arrond., on the Loue, a tributary of the Isle, 22 m. S.S.W. Limoges. Pop. in 1836, incl. comm., 6040. It owes its origin to a monastery founded here in the 6th century; and is old and ill built. It has a collegiate church, a recent Gothic edifice of the 12th century, four other par. churches, a hospital, a court of primary jurisdiction, a society of agriculture, &c., with manufactures of woollen stuffs, linen yarn, and porcelain, and 12 annual fairs. Here are some iron works, and works for the preparation of antimony: all the porcelain clay used in the china manufactory of Sèvres comes from St. Yriex. (*Hugo; Guide du Voyageur, &c.*)

YSSENGEAUX, a town of France, dep. Haute-Loire, cap. arrond., on a rocky and elevated site. 14 m. N.E. Le Puy. Pop., incl. comm., 7621. Though irregularly built, and *assez triste*, it has a good modern church, and is improving. It has no manufactures worthy of notice, its inhabitants being principally engaged in agriculture and cattle-dealing. (*Hugo, &c.*)

YUCATAN, the most E. state of the Mexican confederation, consisting of a peninsula, projecting northwards, between the Carribean sea on the E., and the gulf of Mexico on the W., and between the 18th and 21st degs. of N. lat. and the 87th and 91st of W. long., having S. the states of Tabasco, Chiapa, Vera Paz, and the British territ. of Honduras; length N. and S., about 250 m.; average breadth, 200 m. Area about 50,000 sq. m. The population has been estimated at about 500,000. The most striking accounts of the productiveness of this region have been frequently put forth in geographical works. But according to Mr. Ward, " Yucatan is one of the poorest states in the federation. On parts of it, maize, cotton rice, tobacco, pepper, and the sugar cane, are produced; with dye woods, hides, soap, &c. But the scarcity of water in the central parts of the peninsula, where not a stream of any kind is known to exist, and the uncertainty of the rainy season, render the crops very variable; and years frequently occur in which the poorer classes are driven to seek a subsistence by collecting roots in the woods, when a great mortality ensues in consequence of their exposure to a very detrimental climate. Yucatan has no mines. An active intercourse was formerly carried on with Havannah, which Yucatan supplied with Campeachy wood, salt, hides, deer skins, salted meat, and the *jenequen*, a plant from which a sort of coarse thread was made, and wrought up into sacking, cordage, and hammocks. This trade was cut short by the war; and as few foreigners have been induced to settle in Yucatan, the inhabitants have derived but little advantage from the late change of institutions. The receipts of the estates in 1626 amounted to 213,127 dolls., the expenditure was 207,199 do. ; so that a small surplus revenue remained." (*Ward's Mexico*, ii. 390, 391.)

This state is divided into 15 departments; its chief towns are Merida, the capital, Valladolid, Bacalar, Campeachy, and Vittoria; but none is of much importance. In 1829 it separated itself for a time from Mexico; and we incline to think that at present (1842) it is but little, if in any degree, dependent on the central government.

YUTHIA, the an. cap. of the kingdom of Siam, on the Menam, 40 m. N. Bangkok: it appears to have been formerly a place of much magnificence, but it has now fallen into decay.

YVERDUN (Germ. *Ifertos*, an. *Ebrodunum*), a town of Switzerland, cant. Vaud, cap. distr., on the Thiele, at its mouth in the S. extremity of the lake of Neufchatel, 17 m. N. by W. Lausanne. Population between 3000 and 4000. It is well built, consisting of three principal streets, with a handsome square, a new church, and town-hall, several bridges across the Thiele, &c. Its principal edifice is a castle, built in the 12th century, and which, from 1805 to 1825, served for Pestalozzi's central school, conducted by himself. Yverdun has a college, a public library, with a museum of antiquities, and a tolerable harbour on the Thiele. Its trade is brisk, it being the great depôt for the wine of the canton exported northward. (*Ebel; Dict. Geog., &c.*)

YVETOT, a town of France, dep. Seine-Inférieure, cap. arrond., on the road between Havre and Rouen, 20 m. N.W. the latter. Pop. in 1836, ex. comm., 7983. It is situated on a bare and arid hill, destitute of any running water, the inhabitants being supplied from wells. It consists chiefly of one long street; but this has few good houses, and the rest of the town is very meanly built. It has, however, a planted promenade; and the surrounding country is fertile and populous. Yvetot is the seat of courts of primary jurisdiction and commerce, a chamber of manufactures, &c., and has manufactures of linen and cotton cloths, cotton velvet, handkerchiefs, hosiery, cutlery, and hardware. It has also a considerable trade in corn and sheep. Toward the end of the 13th century the Spanish, Italian, and other merchants used to proceed from Harfleur to Yvetot, where they conducted their chief mercantile transactions with the French; and, perhaps in the view of encouraging commerce, the fief of Yvetot was declared, in 1370, free of all feudal service to the French crown. Its lords soon afterward coined their own money, and assumed the title of *king*. The exploits of one of these petty monarchs form the subject of one of Beranger's national songs. (*Hugo*, art. *Seine Inférieure; Dict. Géog., &c.*)

Z.

ZAANDAM, improperly SAARDAM, a town of N. Holland, on the Zaan, a tributary of the Y, by which it is divided into E. and W. Zaandam, 4½ m. N.W. Amsterdam. Pop. about 10,000. Mr. M'Gregor says, " We have only visited one place (Broek) so trim, quiet, and minutely clean. The streets are paved with clinkers, and daily washed; the houses are built of wood, and painted white and green, and their principal door, that of ceremony, is only opened at baptisms, marriages, and funerals. The dockyard, in which 200 vessels were built and repaired annually, has disappeared; its herring and whale fisheries have also vanished; but its vast number of windmills employed in sawing timber, &c., appear, with their dependent operations, to give full occupation to the inhabitants." (*Nets Book*, i., 197, 198.)

At one period Zaandam ranked among the greatest naval arsenals in Europe; but the principal celebrity of the arsenal, and, indeed, of the town, is derived from the circumstance of Peter the Great having wrought in it as an ordinary ship carpenter during his visit to Holland in 1697. The hut which he occupied is still kept up, and has been visited by numerous distinguished personages, including Napoleon, Alexander, emperor of Russia, &c. (*De Cloet; Voltaire, Hist. de Russie, partie i., cap. 9.*)

ZACATECAS, a city of Mexico, cap. of the state of its own name, in a narrow valley, 290 m. N.W. Mexico. Ward estimated its population at 22,000, and that of its suburb, Veta Grande, at 6000. (*Mexico*, ii., 342.) At a distance, its numerous churches and convents give it a fine appearance, and it has many excellent houses; but its streets are narrow and filthy. Its markets appear to be abundantly supplied with fish, fruits, vegetables, &c. Gunpowder and some cotton fabrics are manufactured here; and Zacatecas is next to Guanaxuato, the principal mining city, and one of the chief mints in Mexico. In the latter establishment, some years ago, 300 people were constantly employed. The machinery, of brass, and made in the town, was ponderous, and a great deal of labour was wasted; still, however, the coinage from January, 1821, to June, 1826, amounted to upwards of 17,570,000 dollars.

The state of Zacatecas, with an area of about 20,000 sq. m., and a population of 280,000, is one of the richest mining provinces in America. "As a mining district it differs materially from Guanaxuato, for, in lieu of one great mother vein, it has three lodes nearly equal in importance, with many inferior lodes; upon all of which nearly 3000 pits or shafts have been opened." (*Ward's Mexico*, ii., 333.) N. and E. of Zacatecas, the country is divided into vast breeding estates, and is very thinly peopled. The state has no manufactures, except those of the capital, and a few in Aguas Calientes; the population living by mining and rural industry. After the capital, the principal towns are Sombureti, Fresnillo, Jerez, Pinos, &c., which, according to Mr. Ward, have a population of from 14,000 to 18,000 each.

ZAFRA (an. *Segeda*), a town of Spain, in Estremadura, prov. Badajoz, 40 m. S.E. Badajoz, on the road between it and Seville. Pop. 7500. (*Miñano.*) It is regularly built, and has two squares surrounded with arcades, and many houses of a superior class. Among the latter is the magnificent residence of the dukes of Medina Celi. (See *Miñano*, x., 58.) The collegiate church is also a fine edifice, and several other churches are richly adorned. This town had formerly manufactures of gloves and jewellery; but these have decayed, and earthenware and leather are now the principal articles made at Zaffra. It was taken from the Moors by Ferdinand III. in 1240. (*Dict. Geog., &c.*)

ZAMORA, a city of Spain, in Leon, cap. prov. of its own name, near the confines of Portugal, on the Douro, here crossed, according to Twiss, by an ancient and clumsy stone bridge, with 16 unequal arches; 34 m. N.N.W. Salamanca. Pop. about 10,000. (*Miñano.*) Its fortifications are of considerable extent, and some years ago enclosed upwards of 20 churches, 16 convents, three hospitals, infantry and cavalry barracks, a courthouse, public granary, bishop's palace, &c. The cathedral was much admired by Townsend (ii., 71), for its variety of marbles and the beauty of its hangings. Without the walls are the remains of an ancient castle. The inhabitants manufacture hats, serges, leather, liqueurs, and gunpowder, and have several dyeing-houses. The city, which is supposed to have been the ancient *Seutica*, derives its modern name from the turquoises found in its vicinity, for which *Zamora* is the Moorish term. Alphonso, the Catholic, took it from the Moors in 748, but it was retaken by the latter in 985. Ferdinand the Great finally annexed it to Castile in 1063, and it was the seat of the Cortes in 1297 and 1309. (*Miñano; Townsend; Twiss; Dict. Geog., &c.*)

ZANESVILLE, p. v., capital of Muskingum co., O., 54 m. E. Columbus, 339 W. Beautifully located at the point where the National road crosses Muskingum river, opposite to the mouth of Licking river. Two bridges cross Muskingum river, of fine workmanship, connecting it with Putnam and West Zanesville. It contains an elegant courthouse and county offices, a jail, a market-house, two academies, an atheneum with an extensive reading room, a cabinet of minerals, and a library of 9000 volumes, a juvenile lyceum with a reading room and library, a bank, and another in Putnam, opposite; an insurance office, nine churches, 40 wholesale and retail stores, two flouring-mills, and six grist-mills in the immediate vicinity, three of which are moved by steam power; three woollen factories, one cotton factory, three saw-mills, one oil-mill, one paper-mill, three iron foundries, two brass foundries, two steam engine and machine shops, two glass houses, two rope-walks, two breweries, three coach factories, a steam hat-body factory, a shoe-last factory, five printing-offices, five weekly newspapers. The population of the village is 4766; and including West and South Zanesville and Putnam adjacent, 7000. The Ohio canal is connected with the Muskingum river by a side cut, 2¼ m. long at Dresden; and by means of the dam at Zanesville, and a dam and lock between this and Dresden, a communication is opened to the Ohio canal, to which steamboats are continually plying. By dams and locks, the Muskingum is made navigable to its mouth, in the Ohio river. There is a canal around the falls of the Muskingum at Zanesville, which affords great water-power. At his decease, the proprietor of Zanesville left his property, amounting to over $35,000, to found a free school. This place is on a tract granted by the United States congress to Ebenezer Zane in 1796, on condition that he should open a bridle track from Wheeling, Va., to Maysville, Ky. The place was laid out in 1799, and the first cabin was built. The population of the t., exclusive of the village, is 355.

ZANTE (an. *Zacynthus*), one of the Ionian Islands (which see), of which it is the third in point of magnitude and importance, about 10 m. off the W. coast of the Morea, its capital being in lat. 37° 47′ 17″ N., long. 20° 54′ 32″ E. It is of a somewhat oblong shape; greatest length, N.W. to S.E., about 20 m.; greatest breadth, 10 m. Area estimated at 156 sq. m. Pop., in 1836, 35,348. It is mostly mountainous, particularly its W. portion, where several summits rise to the height of 1300 ft.; but on the E. side, behind the town of Zante, is an extensive and fertile vale, so covered with currant bushes (*Vitis Corinthiaca*), olive-trees, cypresses, &c., as to entitle the island now, as of old, to the epithet of " woody."

 Jam medio apparet fluctu hemerum Zacynthus.*
           Æneid, iii., 270.

" About 9,000,000 lbs. currants are annually produced in this fertile vale. They are accounted better than those of Cephalonia, but inferior to those of the Morea. They are gathered in August, and spread out to dry for three weeks; and for this purpose a plot of ground is levelled and kept dry before every house in the valley. Much depends upon the process of drying : a shower of rain will sometimes diminish the value of the article by one third, and a second entirely ruin the crop." (*Burgess's Greece, &c.,* i. 120.) The learned traveller, Dr. Chandler, has given the following details with respect to the treatment of currants, which, perhaps, may be worth quoting ; " When dried by the sun and air, they are transported to the city on horses and mules, guarded by armed peasants; and poured down a hole into magazines, where they cake together. When about to be shipped, the fruit is dug up with iron crows, and stamped into casks by men with bare legs and feet. In the ships it sweats, and, as we experienced, often fills the vessel with a stench scarcely tolerable. The islanders believe it is purchased to be used in dyeing, and in general are ignorant of the many dishes of which currants are an ingredient." (*Travels in Greece,* cap. 79.) The honey, oil, and wine of the island are much esteemed ; of the latter no fewer than 40 different sorts are said to be made. Oranges, lemons, and citrons, are also exported, and about 40,000 barrels of salt are annually produced from the salt-works of the island. The pitch wells, visited and described by Herodotus (iv., 195), are situated towards the S. extremity of the island, in a small plain, open on one side to the sea, but elsewhere circumscribed by hill ranges. It is partly, at least, of volcanic formation, and occasionally suffers from earthquakes, one of which, in 1840 (October 30), committed the most extensive ravages. In the wells, a dark substance is continually forcing itself from the bottom through the water, boiling up in large globules, which burst when they come to the surface. The pitch is collected with large spoon-like implements : the average annual produce is about 100 barrels, used for smearing ships' bottoms, &c.

The town of Zante, on the E. shore of the island, is the largest in the Ionian islands, having about 20,000 inhabitants. (*Burgess.*) It stands partly on the level shore, and partly on some acclivities, one of which is crowned by its citadel, anciently called *Psophis,* founded by the Arcadian Zacynthus. The town, which is well kept and clean, is supplied with water by an aqueduct constructed by the British. The reflection of the sun renders it extremely hot in summer, though the heat be a good deal moderated by the action of the sea breeze, which blows during the day. The harbour is capacious, and protected from the N.E. winds by a mole, at the extremity of which a lighthouse is erected. Ships anchor opposite the town, at from 500 to 1000 yards distance, in from 12 to 15 fathoms water. Zante is the see of a Greek protopapas, and of a Roman Catholic bishop, and has numerous churches, two synagogues, a lazaretto, a lyceum, &c.; with some manufactures of linen, cotton, and woollen stuffs, liqueurs, soap, jewellery, &c. This town suffered severely from the earthquake already alluded to.

At the time of the Peloponnesian war, Zacynthus belonged to Athens; it was at an after period alternately a possession of the Macedonians and the Romans. Several curious antiquities have been discovered in the island, and it has been supposed that the remains of Cicero were deposited in a tomb discovered here in 1544. (See *Hughes' Travels,* i., 155, 156, &c.; *Burgess ; Chandler's Greece ; Cramer's Anc. Greece ; Commerc. Dict., &c.*)

ZARA (an. *Jadera*), the cap. of Dalmatia, circ. of same name on the Adriatic, opposite the island Ugliano, 150 m. S.E. Venice. Lat. 44° 2′ 25″ N., lon. 15° 9′ 32″ E. Pop. about 6500; principally of Italian descent. It stands on a small peninsula, and is fortified with bastioned walls and several outworks. It has many good private dwellings, but its streets are narrow and ill drained, and it suffers from a deficiency of water. It has a cathedral and several other churches, 10 convents, a naval and military arsenal, and a theatre, with a lyceum, gymnasium, Episcopal seminary, many inferior schools, a museum of antiquities, &c. Its harbour is spacious, but exposed to N. winds, which sometimes blow with tremendous violence. The coasting trade and fisheries employ most part of the inhabitants, and a great number of vessels are owned in the port. The manufacture of *rosoglio* is almost the only other branch of industry carried on, and that at present to a very limited extent. (*Oesterr. Nat. Encyc.*) Zara is an archbishop's see, the residence of a general commandant, and the seat of all the superior provincial courts of Dalmatia. Without its walls

1104

are the remains of an ancient aqueduct; but, with this exception, few other Roman antiquities exist in Zara, in consequence of their having been mostly employed in the building of the fortifications. (*Fortis's Dalmatia,* 115; *Berghaus, &c.*)

ZEALAND, the largest and most important of the Danish islands, being that on which Copenhagen is situated. It lies mostly between the 55th and 56th degs. of N. lat., and long. 11° and 12° 40′ E., at the entrance of the Baltic, being separated from Sweden by the Sound, and from Funen and Langeland by the Great Belt. Its area may be estimated at 2830 sq. m. Pop. in 1834, including that of the small and thinly peopled islands of Moen and Samsoe, 439,982. Like the rest of the Danish islands, it is flat, or, at most, gently undulating, and is in parts intersected by canals. The climate is mild, and similar to that of the E. of Scotland. It is well cultivated, and is exceedingly fertile, producing grain of all sorts, especially rye, barley, oats, and wheat. The pastures are excellent, and the island is celebrated for its breed of horses. It is, also, well stocked with cattle and sheep. Wood is plentiful, except in the middle of the island, where turf is used for fuel. It is studded with cottages, farms, and country-houses; bearing a greater resemblance to England than is exhibited by most continental districts. It is also the principal seat of the manufactures and trade of Denmark. It is subdivided into five bailiwicks, and is governed by a grand-bailiff: it forms, of itself, a separate ecclesiastical superintendency. (See DENMARK, in this work; *Inglis ; Bremner, &c.*)

ZEALAND (NEW), a group of two large and numerous small islands in the S. Pacific ocean, belonging to the Australian continent, and now forming a dependency of the British crown. The group extends between 35° and 48° of S. lat., and 166° and 179° of E. long., about 12° E. of S. Australia and Van Diemen's Land, being the land nearest to the antipodes of Great Britain. The two large islands, now called New Ulster and New Munster, stretch lengthways from N. to S., being separated by the narrow channel called Cook's strait : a strait of about the same width separates the most southerly of the above islands from New Leinster, formerly called Stewart's, or the S. island. The length of curved line extending through these three islands from the N. to the S. cape would be about 900 m. The middle island, or New Munster, is the largest ; but New Ulster, which alone has been colonised by the British, is the widest, being about 300 m. in its greatest breadth. Altogether, the area of New Zealand is estimated by Mr. Terry at about 86,000 sq. m., or, 55,000,000 acres ; and the total population may perhaps be 200,000, of whom nearly 10,000 are Europeans. (*Terry's New Zealand, &c.*)

"The mountains of New Zealand stretch along the centre of the middle island in its whole length and along the better half of the N. island, and sloping gradually down towards the sea level, leave an immense extent of forest, plain, and pasture, on both sides of the mountain range, between it and the sea. Here and there along the line of this cordillera, several huge mountains, overtopping the rest, rise into the region of perpetual snow. Some are more than 14,000 feet high ; an elevation nearly equal to that of Mont Blanc. There are likewise several subordinate ranges of hills, and a few detached outliers of vast dimensions. A few of the smaller mountains are barren, or clothed with fern; but by far the greater number are covered, up to the range of perpetual snow, by magnificent timber of enormous size and great variety." (*Present State of N. Zealand,* 73.) The country is well watered; an abundance of streams descend from the central chain on both sides. Few of the larger rivers have been surveyed to any great distance, but the Wai-kato, with its affluent the Wai-pa, the Wai-hou, or Thames, and others, are of considerable size and length. The shores are in parts iron-bound and dangerous ; but the N. island especially is indented with many excellent bays and harbours.

Our knowledge of the geology and mineralogy of New Zealand is very imperfect. It has, however, several active volcanoes; in the N. island, also, are various cavities, which appear to be extinct craters, in the vicinity of which numerous hot springs are met with ; some of these, as they rise to the boiling point, the natives use for cooking. (*N. Zealand,* 80.) Pumice-stone is abundant, and used by the natives for polishing their spears; coal is supposed to exist in the middle island, whence also comes the green talc of which the natives make some of their weapons. Iron-stone is plentiful ; some pigments used by the natives appear to consist of manganese; sulphur, whinstone, slate, granite, marble, &c., are found ; and in every part hitherto explored there is clay fitted for brick-burning.

The climate is temperate, being analogous to that of France and the S. part of England. The mean annual temperature at Auckland is about 59° Fahr., that of the summer months 67°, and of the winter months 52°. (*Terry, 62.* The country is free from the oppressive heats that prevail in the middle of the day at Sydney, and, what is of still greater importance, it is not subject to the long-continued droughts that prevail in the Australian continent. Strong winds, principally from the N.E. or S.W., always occur at the changes of the moon, frequently bringing rain, particularly in the winter months: the rains, however, do not last for days together, and the W. and more prevalent winds very seldom blow so hard as to interfere with navigation. In the interior the weather is colder, but more equable. The climate appears to be generally salubrious, and favourable to longevity; the prevalent diseases are mostly those which have been introduced by Europeans, though in some situations the natives suffer from scrofulous and glandular affections. (*Parl. Rep. on New-Zealand.*)

The country presents the aspect of perpetual vegetation, most of its indigenous vegetable products being evergreens; and the soil, which, in most parts yet explored, is a rich loam or vegetable mould of much fertility, is apparently well adapted to the culture of nearly all the useful vegetables of Europe. However, as Mr. Terry says, " the exaggerated statements circulated in England of the colony and its products, soil, and climate, have led generally to the very erroneous impression and opinion, that the necessaries of life, especially food, would be abundant and cheap. But New Zealand has neither a tropical climate, nor is it a country in which edible vegetables and fruits, indigenous to such regions, grow and flourish spontaneously and abundantly, nor is it a land inhabited by native animals adapted for the food of man, and easily obtained by the toils of the chase. The islands are, at present, uncultivated wastes, consisting either of mountains covered with dense forests, of plains and low lands covered with impenetrable high fern and shrubs, or of swamps and marshes covered with rushes and flax, without any open spots of grass land for pasturage, or of verdant downs and hills for sheep. In these vast tracts there is not to be seen a living animal, wild or domestic; and whatever is produced from the soil for the food of the population, either of grain from arable land, or of stock from pasturage, must be the work of time, of great labour, and of much expense. Small are and will be the resources of food from the actual produce of the islands, either animal or vegetable, for the subsistence of any great number of settlers until they themselves bring the land into cultivation and pasture, and import cattle and sheep from New South Wales, to stock it for the supply of their wants." (*Terry,* 57, 58.)

Timber and flax are at present the most valuable native products. But the former, owing to the extravagantly high rate of wages, the difficulty of the country, and the want of roads, is very high priced. even in the ports of the colony; while the freight to England, which is from £5 to £6 per load, amounts of itself to a prohibition of its importation into this country. No doubt, however, its price in the colony will be progressively reduced; and being admirably adapted for ship-building, this will probably become, at no very distant date, a profitable business. Timber will, also, there can be no question, become an important article of export to Sydney, Hobart Town, and other ports in Australia. The forest-trees are principally of the pine species. The kauri is of enormous size; one was cut and shipped in 1841, 150 feet in length and 25 feet in circumference at the base; and another now standing on the E. coast measures 75 feet round at its base; its height being unknown, owing to the thickness of the surrounding forest.

Flax, it has been supposed, will continue, for a lengthened period, to be the staple of the country. It is obtained from the leaves, and not from the stem, of the *Phormium tenax,* an indigenous plant. It is said to be distinguished by the length, strength, and flexibility of its fibres, and to be preferable to the flax of the N. of Europe. But, in point of fact, there is a great diversity of opinion as to its real merits, and it fetches at present but a low price. It is alleged that this is a consequence of its imperfect preparation, which is left solely to the native women: they separate the fibre from the external epidermis in a green state by means of a muscle-shell, and then expose it to the air for a few days, which bleaches the flax, and dries the thin inner epidermis. In heckling and properly freeing the flax afterward from this substance, there is a loss in quantity of 25 per cent., besides the trouble and expense. But, without presuming to say whether the defects with which it is charged are inherent in the flax itself, or depend on its preparation, it is abundantly obvious that, unless it be. furnished of a superior quality, it will never become a considerable article of export.

The principal food of the natives, when other kinds of vegetables have failed, has, until recently, been the roots of fern, which is found in inexhaustible abundance, and in the greatest variety throughout New Zealand. The natives had made some progress in agriculture even before their country was visited by Captain Cook, who saw along many parts of the coast small patches of ground turned up

and cultivated, each separate district being fenced in with reeds. In these places sweet potatoes, coccos or eddas, gourds, &c., were grown. Captain Cook planted and left with intelligent natives the seeds of wheat, peas, cabbage, onions, turnips; potatoes, &c.; though all these seem to have perished, through neglect or otherwise, except turnips and potatoes, which last are now the chief dependence of the natives during winter. Taro, water-melons, and pumpkins, are merely temporary food in their season. Maize is grown to some extent; but very little of any other kind of European grain. In general, the New Zealanders cultivate only virgin soils, raising two crops in succession, and then proceeding to break up new soil. Their native implements are of the rudest description; but latterly European spades, hoes, axes, &c., have been plentifully introduced among them. For these they show a great avidity; and, in fact, it is rather by payment of such articles and others of utility, than of finery and mere baubles, that purchases of land have been effected by Europeans.

Except a few cattle or sheep in the possession of the missionaries, and a small number of goats, no kind of live stock existed in New Zealand, down to a very late epoch, except pigs. These, however, from the great abundance of fern roots, their favourite food, multiply exceedingly, and have been allowed to run wild by the natives, who, when they require them, catch them by means of dogs. It is remarkable that when New Zealand was first discovered, it had no *indigenous mammalia whatever;* indeed its only quadrupeds were a few species of lizards, which the inhabitants held in veneration or terror. Even the rat and dog were introduced by Europeans; and the rat is at present the principal species of *game!* A good many parrots, parroquets, wild ducks, pigeons of a large size and fine flavour, &c., inhabit the forests; and poultry are found to thrive very well, though not yet reared to any great extent. Indeed, if we except their prisoners of war, almost the only animal food hitherto used by the New Zealanders has been fish, which abound round the coasts.

The New Zealanders probably belong to the Malay family, and if so, are by far its best specimens. In general, the men are tall; many individuals of the upper classes reaching the height of 6 feet and upwards. They are strong, active, and almost uniformly well shaped. Their hair is commonly straight, but sometimes curly, particularly the women, who are frequently handsome. Their colour resembles that of a European gypsy; but varies in individuals from a dark chesnut to the light tinge of an English brunette. Some individuals of the negro race of Australia and the E. archipelago are said to inhabit parts of the middle island; but the New Zealanders bear no sort of analogy to these, to whom they are greatly superior in every respect. They make excellent seamen, in which capacity they are extensively known. If we except occasional cannibalism and infanticide (both of which are said to have greatly decreased of late years), they manifest fewer of the vices of savages than almost any other savage people. Their manufactures, as may be supposed, are few, and mostly confined to the furniture of their huts, articles of dress, weapons, and other necessaries. But they prepare mats and other articles in flax of great beauty; evince much ingenuity in carving and building canoes; and, with cultivation, would, perhaps, attain to a considerable degree of civilization. They have an abundance of poetry of a lyrical kind, in a metre which appears to be regulated by a regard to quantity; and are passionately fond of music. They have also a kind of astronomy; and, according to Baron Hügel, there is not a single tree or even weed, a fish or a bird, in the N. island, for which the natives have not a name universally known. Unlike many other savages, they have evinced the greatest aptitude for sharing in the usages of civilized life; and this, probably, is a main cause of their ready adoption of the doctrines inculcated by the missionaries, and of their acquiescence in the rule of Great Britain. A considerable proportion of the natives are slaves to others, who are themselves dependent, to some extent, on certain *arekees* or head chiefs; but the holders of slaves appear, notwithstanding, to have independent control over their own lands, and are able to dispose of them at will, without the consent of the *arekee.* (See *Parl. Rep.*) Polygamy is practised by such of the New Zealanders as continue attached to their ancient superstition: but the missionaries, who have establishments in various parts of the N. island, have, according to their own account, been eminently successful in the conversion of many of the natives to Christianity; and these have given up the practice of polygamy, cannibalism, &c.

New Zealand was discovered by Tasman in 1642; but its extent and character were not ascertained till the voyages of Cook in 1769 and 1774. From that period, the coasts were occasionally visited by whalers, and some communication was held with the natives; but no permanent settlement appears to have been made by any people till about

1815, when a missionary station was established in the Bay of Islands, towards the N. extremity of the N. island. Though the right of Great Britain to these islands was recognized at the general peace, no constituted authority was placed over New Zealand till 1833, when a resident, subordinate to the government of New South Wales, was sent hither, but with very limited powers. Meantime the shores had become infested by marauding traders, runaway convicts, and other unscrupulous characters, who introduced a taste for ardent spirits, various diseases, and much demoralization. These persons also swindled, or attempted to swindle, the natives out of large tracts of land, by getting them to subscribe contracts, of the real import of which they certainly knew little or nothing by which entire districts were conveyed away for the merest trifle. Under these circumstances, it became necessary to establish a government sufficiently strong to protect the aborigines and the real interests of the colonists. Accordingly, in Jan. 1840, New Zealand was constituted a colony dependent on New South Wales; and a lieut. governor appointed, who immediately proclaimed, among other announcements, that *all purchases of land would, in future, be void unless conducted through the British local government*. But shortly before the formal occupation of the islands, the mania for speculating in land attained to an enormous extent; and vast tracts, equal, in fact, to provinces, were acquired by a few individuals, belonging to the islands, to Sydney, and other parts. It was not, therefore, enough to prevent such wholesale acquisitions in future. Justice to the natives, on the one hand, and the best interests of the colony on the other, made it imperatively necessary that the grounds on which the claims to land were made should be carefully inquired into; that in all cases in which the natives had been swindled the grants should be cancelled; and that, when confirmed, their extent should be limited. And, in consequence of these considerations, a commission has been appointed to inquire into the validity of all claims to land; and the commissioners have been instructed to recognize those only which are founded on fair and equitable considerations, with the important proviso, that no claim for land, when affirmed, shall be allowed to a greater extent than 2560 acres. Lands acquired by the government will be sold by auction, as in the other colonies.

In April, 1841, New Zealand was separated from New South Wales, and is now placed under a governor, with whom the colonial secretary and treasurer, the attorney-general, and three senior justices of the peace, compose the legislative council. Besides a bishop and 12 clergymen of the church of England, 63 other ministers of the Church Missionary, Wesleyan, Scotch church, and Roman Catholic associations, are established in the country. Auckland, the capital, on the Waitemata, in lat. 36° 51′ 27″ S., long. 174° 45′ 30″ E., is rapidly rising into a town, and has a spacious harbour. Russell, on the Bay of Islands, and Port Nicholson, are the other principal stations.

New Zealand has, till lately, been much frequented by whaling ships, not only British, but American, French, &c.; but the establishment of custom houses and a regular government has driven many of these to other islands of the Pacific; and a long time will probably elapse before the colonists be much benefited by whale fisheries of their own. Neither are New Zealand colonists likely to be speedily enriched by farming on a large scale, on account of the great expense necessary in the clearing the lands of fern. According to Mr. Terry, "the class of emigrants to which New Zealand at present offers the most certain advantages and success are those who have been accustomed to husbandry. If such persons, having families, would be content with small farms of from 20 to 50 acres, according to their means for outlay in stock and buildings, and then by their own personal industry and labour, cultivate the land, and rear cattle, poultry, &c., with moderate views and expectations, looking to frugality, perseverance, and time, to acquire competence and independence, instead of resorting to land-jobbing and speculation for sudden wealth, they would be certain of success in their undertakings, and of realising property in a few years.

"If New Zealand, by the power and means of abundance and judicious emigration, becomes extensively cultivated and plentifully stocked with cattle, so as to render the necessaries of life cheap, and consequently diminish, proportionally, the rate of wages, it will most probably become in the end, the seat of extensive manufactures. In addition to moderate wages and cheap food, there would be the further important auxiliaries of coal, timber, and clay, with endless excellent localities, having water communication. New Zealand would then bring into profitable production her timber, for ship-building; flax, for canvass, ropes, &c.; copper, for sheathing her ships, and all other purposes; sulphur, for brimstone, &c.; alum and dye woods, in manufacturing the wool of Australia or the cotton of India; tan, for leather from the hides of her own cattle, or from Australia and S. America; tobacco which could be manufactured; breweries and distilleries, for barley and hops of native growth, &c. But it is far more rational to conceive that, instead of attempting fruitlessly to compete in the exports of raw produce, the colonists in the first instance will endeavour to render themselves independent of any other colony for the supply of food; and when food and labour are cheap, they will direct their capital and energies to being into play the other national products, in manufactures for their own wants, as well as to supply Australia, India, China, and Spanish America, all of which are not far distant." (*Terry*, p. 250-261, &c.; *Parl. Rep. on New Zealand*; *Pres. State of New Zealand*; *Nicholas*; *Polack*, &c.)

ZEITZ, a town of Prussian Saxony, reg. Merseburg, cap. circ., on the White Elster, here crossed by a stone bridge, 22 m. W.S.W. Leipsic. Pop. 10,600. (*Berghaus*.) It is walled, is divided into an upper and lower town, and has a cathedral, and several other churches, various hospitals, two castles, one of which was formerly the residence of its princes, but now serves for a house of correction, a gymnasium, with a public library of 14,000 vols., and manufactures of cotton goods, earthenware, leather, shoes, &c.; with cotton-printing establishments, breweries, and distilleries. It is the seat of the ordinary circle courts, of an ecclesiastical board, and a Calvinist college. The gardens and grounds in its vicinity are celebrated for their seasoness, and the attention bestowed on them. (*Berghaus*; *Dict. Geog.*, &c.)

ZELL, or CELLE, a town of the Hanoverian dom., distr. Lüneburg, on the Aller, where it receives the Fuse, on the road between Hanover and Lüneburg, 22 m. N.E. the former. Pop. 11,200. It is well built and paved, and has Lutheran, Calvinist and Roman Catholic churches, an old castle once the residence of the dukes of Lüneburg, a large penitentiary, a medical college, Latin school, society of agriculture, and a famous royal breeding stud. Its inhabitants manufacture linen cloths, hosiery, flannel, hats, tobacco, &c., and have a brisk transit trade both by the Aller and by land. Celle is the seat of the high court of appeal for the kingdom of Hanover, the decisions of which were final, till recently they were interfered with by the government. (See HANOVER, l. 1062.)

It was the residence, during the latter years of her life, of the unfortunate Matilda, queen of Denmark, and sister of George I. of England; and a monument to her memory stands in the palace garden. Zell has also been for a lengthened period the favourite abode of such of the nobility of Lüneburg as do not live in Hanover. (*Hodgskin's, Travels in the N. of Germany*, l. 155-163.)

ZERBST, a town of N. Germany, territory of Anhalt-Dessau, on a small tributary of the Elbe, 17¼ m. S.E. Magdeburg. Pop. 9200. It is walled and entered by six gates; has an old castle, several churches, one of which is among the finest structures of its class in Germany, two well endowed charitable institutions, an orphan asylum, a house of correction, and a large school termed the Francisceum. It is the seat of the high court of appeal for the Anhalt and Schwartzburg principalities; and till near the end of last century it was the residence of the ducal family of Anhalt. It has manufactures of jewellery and earthenware. The empress Catherine II. of Russia was a princess of Zerbst; but she was not born here, as is stated in the Dict. Geographique, but at Stettin (in 1729, of which her father was governor. (*Berghaus*; *Dict. Geog.*, &c.)

ZITTAU, a town of the kingdom of Saxony, circ. Bautzen, on the Mandau, a tributary of the Neisse, 30 m. E.S.E. Dresden. Pop. in 1836, 9874. It is tolerably well built, and has numerous churches, a gymnasium, public library of 13,000 vols., house of correction, and various charitable institutions. It is the centre of the linen manufactures of Lusatia; and most of its inhabitants are occupied in the weaving of damasks, ticks, and other linen fabrics, or of cotton and woollen cloths; and in bleaching, printing, carding and other auxiliary occupations. Zittau has also porcelain factories, paper-mills, and breweries, and a large trade in flax. It was the birth place of the great orientalist, B. Michaelis. (*B. Ritter*; *Berghaus*, &c.)

ZOAR, p. v., Lawrence t., Tuscarawas co., O., 120 m. N.E. Columbus, 319 W. Situated on the E. side of Tuscarawas river, a short distance from the Ohio canal. The singular community was settled from Germany, by a colony called Separatists, from their secession from the Lutheran and other churches. It is under the government of a patriarch, and chooses its own officers. They were at first poor, and purchased their lands on credit, but they have paid for them, and added 1000 acres to their possessions, which are in a high state of cultivation. Their cattle are of a superior breed, and carefully selected; and their sheep, though of the finest wool, are bred with a view to their productiveness, and are conducted to the hills in the morning by the herdsmen, and are returned by them to their fold in the evening. The village contains one store, one grist-mill, two saw mills one oil-mill, one woollen factory, one linen factory, upwards of 60 dwellings, and over 300 inhabitants. On the canal they have several warehouses and two blast furnaces. They are tenants in common, and each member seeks to advance his own interest by promoting that of the community. They have an extensive pleasure garden with a hot-house, which contains rare plants and exotic fruits, including lemons, oranges, figs, pomegranates, &c. The inhabitants of large places on the canal send their rare plants here for preservation in the winter. Their village is kept with the most prefect neatness, and is often visited as a curiosity.

ZOMBOR, a royal free town of Hungary, co. Bacs, of which it is the cap., in an extensive plain near the Francis canal, uniting the Danube and the Theiss, 118 m. S. by E. Pesth. Pop. about 21,000, mostly of the Greek church. It has several fine buildings, including a county hall, townhouse, several churches, barracks, and the government offices. Here, also, is a Greek ecclesiastical seminary, and a Roman Catholic high school, with some silk manufactures, and a considerable trade in corn, wine, and cattle. (*Oest., Nat. Encyc.*)

ZUG or ZONG, a canton, lake, and town of Switzerland, in the central part of the confederation. The canton, which is the smallest in Switzerland, is enclosed between the territory of Zurich on the N., Schwytz on the E. and S., and a small part of Lucerne and Aargau on the W.; from which last it is separated by the Reuss. Area, about 85 sq. m.; a considerable part of which is occupied by the lakes of Zug and Egeri. Pop. in 1837, 15,322. Except a small plain to the N. of Zug, the surface is wholly mountainous, but the mountains do not rise to any great elevation; the highest the Rossberg, on the S. border, being little more than 5000 ft. above the sea. Principal rivers, Reuss, Sihl, which forms the N.E. boundary, and Lortz, which brings the waters of the Egeri lake into that of Zug, and forms also the outlet of the latter towards the Reuss. The lake of Zug, principally comprised in this canton, but partly in that of Schwytz, and intermediate in situation, as in character, between the lakes of Zurich and Lucerne, is about 8¼ m. in length, N. to S., by 3 m. in its greatest breadth. Its area has been estimated at about 10 sq. m., and the height of its surface above the level of the sea, at 1385 English ft. Its waters are of a very dark blue colour; and though near the town of Zug, its depth appears to be only about 900 ft.; at its S. extremity it is said to exceed 1900 ft. ! (*Ebel ; Picot, &c.*)

The banks of the lake of Zug are cultivated, richly wooded, and in general gradually sloping, except on the S. and S.W. sides, where the Righi and Rossberg rise abruptly from the water's edge. The lake abounds with fish, the taking of which forms an important occupation of the inhabitants of its vicinity. Some indifferent wine, with cider, kirschwasser, &c., are made, and considerable quantities of apples and other fruits are grown for exportation; but the principal employment of the population is cattle breeding. A few silk and cotton fabrics are woven, cotton yarn is spun, and at Zug, Cham, and Baar are some tanneries and paper-mills; but the manufactures of the canton are comparatively insignificant. The government is strictly democratic. The cantonal council is composed of 54 deputies, elected for two years by all the male citizens of the canton above the age of 19 years, who are not bankrupt, pauper, or under penal condemnation. This council exercises all the ordinary administrative functions. The legislative power is exercised by the *triple council*, composed of the cantonal council and two additional members for each deputy, chosen like the deputies triennially by the communes. The general assembly meets annually in May, its *landesmann* or president being taken alternately from the two circles into which the canton is divided. The deputies are paid for their services, at such rates as can be afforded by the communes which send them. The sum paid by the town of Zug to its representatives is about £4 each per annum ! and, besides this, every councillor entering Zug to attend a council, which takes place perhaps once a month, receives about 9d. English. And some of the communes are so poor as not to be able to pay even this pittance to their representatives. (*Inglis.*) The chief criminal tribunal consists of 25 members, and the ordinary civil tribunal of 6 assessors and the *statshelts* : the latter becomes a final court of appeal by the addition of 6 members chosen annually by the cantonal council. Civil causes below the amount of 12 francs, misdemeanours, and other matters of minor importance, are decided by the communal assemblies and tribunals. There is no tax of any kind in the canton. The whole state expenses, amounting to about £160 a year, are defrayed from the general Swiss fund, drawn from the entry of foreign merchandise, and from a monopoly of salt, which is farmed by the government, and brings in about £80 a year. Zug furnishes a contingent of 250 men to the army, and 2497 francs a year to the treasury of the Swiss confederation.

Zug, the capital on the N.E. side of the lake of the same name, 15 m. S. Zurich, with about 2500 inhabitants, is the only town worth notice. It is pleasantly situated, and has several good churches, to one of which is attached a curious *goigetha*, containing many hundreds of skulls, each labelled with the name of its original possessor. Provisions are cheap at Zug ; and though without the pretension to rank with Zurich in importance, a residence here would seem to be the more agreeable of the two from the greater cordiality and gaiety of the people.

The people of this canton are of a German stock, and for the most part similar to those of Schwytz, though less ignorant and superstitious. They are all Roman Catholics ; and, small as is the extent of the canton, Ebel says, " *Le pays de Zug fournit des prêtres à une grande partie de la Suisse Catholique.*" (*Picot, Statistique de la Suisse ; Ebel's Manual, &c. ; Inglis's Switzerland ; Dict. Géog,. &c.*)

ZURICH (CANTON OF), a canton of Switzerland, ranking first in the confederation, and being superior also, in population and importance, to most of the other cantons. It extends between lat. 47° 10′ and 47° 40′ N., and long. 8° 20′ and 9′ E ; having E. Thurgau and St. Gall, S. the lake of Zurich and the canton Zug, W. Aargau, and N. Schaffhausen and Baden, from which it is partly separated by the Rhine. Length, N. and S., 30 m. ; greatest breadth, 25 m. Area, 685 sq. m. Pop., in 1837, 251,576 ; nearly all Protestants. Surface generally undulating ; and though picturesque, it presents none of those grand natural features which arrest the traveller's attention in the cantons further S. Several mountain, or rather hill ranges, enter Zurich, but the highest summit, the Hörnli, near the E. border, scarcely rises to 3800 ft. above the sea.

After the Rhine, the principal rivers are its tributaries, the Limmat, which drains the lake of Zurich, Thur, Toss, Sihl, &c., with the Reuss forming a part of the W. border. Of these, however, only the Limmat is navigable. The Greiffen, famous for its fine eels, and several smaller lakes, are in this canton. Climate mild ; the mean annual temperature at Zurich is about 48½° Fahr. Nowhere in the canton is the ground perpetually covered with snow; and the soil is in general productive. Agriculture is perhaps better conducted in this than in most other parts of Switzerland; manuring is well understood ; and irrigation is successfully practised. Inglis says, " Anywhere in the neighborhood of Zurich, one is struck with the extraordinary industry of the inhabitants; and if we learn that a proprietor here has a return of 10 per cent., we are inclined to say, ' He deserves it.' It is impossible to look at a field, a garden, a hedge, scarcely even a tree, a flower, or a vegetable, without perceiving proofs of the extreme care and industry that are bestowed upon the cultivation of the soil. If, for example, a path leads through or by the side of a field of grain, the corn is not, as in England, permitted to hang over the path, but is everywhere bounded by a fence. If you look into a field towards evening, where there are large beds of cauliflower or cabbage, you will find that every single plant has been watered. In the gardens, which, around Zurich, are extremely large, the most punctitious care is evinced in regard to the culture of every product." (*Inglis's Switzerland, &c., 23.*)

The labouring classes in this canton are almost universally proprietors of the small farms and cottages which they cultivate and inhabit. The corn grown is insufficient for the population, but great quantities of fruit and garden vegetables are raised. The vine is pretty generally cultivated. But, though improved, the wine is still very inferior. The pasture lands are not extensive, and no great quantities of farm stock are reared : a very large breed of cattle is, however, produced by a cross between those of this canton and those of Schwytz. Some iron, coal, salt, &c., are met with, but mining industry is not of much consequence. (*Meyer, Der Kant. Zurich.*)

Zurich is one of the principal manufacturing cantons of Switzerland; its inhabitants generally dividing their attention between the labours of agriculture and those of the loom. " I have seldom entered," says Dr. Bowring, " a rural dwelling without finding one or more looms in it, employed in the weaving of silk or cotton. If the labours of the field demand the hands of the peasant, his wife or children are occupied in manufacturing industry. When lighter toils suffice for the agricultural part of the family exertions, the females and the young people resign the loom to the father or the brothers. The interstices of agricultural labour are filled up by manufacturing employment: and in more than half of the operations of Zurich the farmer and the weaver are united." (*Rep., p. 69, 70.*) Cotton and silk fabrics are those principally produced. The silk fabrics consist of Florentines, gros de Naples, marcelines, taffetas, levantines, handkerchiefs, crapes, shawls, velvets, &c. Early in the present century about 5000 looms were employed upon these goods, but since the peace they have rapidly increased. The disturbances at Lyons, also, in 1834, were the cause of many Lyons workmen settling in Zurich. In 1840, Villermé estimated the number of silk-looms in the canton at 11,000, and the weavers of all ages at 16,000. (*Tableau des Ouvriers*, i., 490.) The annual value of the total produce of the silk-loom has been estimated at £600,000 sterling. The male weavers of Florentines and serges get from 3½ to 4½ francs a week wages, and of gros d'Orleans and marcelines, at an average from 6½ to 7½ francs a week. (*Hand-loom Weavers' Rep.*) The cotton manufactures of Zurich had their origin in the 5th century, their two principal seats being then, as now, Zurich and Winterthur. There are said to be about 12,000 cotton weavers in the canton, and 4000 persons engaged in other trades connected with the cotton manufacture, producing annually 800,000 pieces of cotton. In 1836 there were 15 cotton printing establishments in the canton, employing about 1000 workmen, and printing 100,000 pieces a year of cloth. (*Bowring's Rep.*, p. 76.) Cotton spinning is, also, extensively carried on, there being, in 1836, 292,960 spindles in operation. (*Hand-loom Weavers' Rep.*, p. 196.) At Winterthur and elsewhere numbers 120 and 130 are made; but the yarns spun are mostly from 20 to 40, the higher numbers being imported from England. Forty thousand hundred weight of raw cotton are supposed to be annually consumed in the canton. The general average rate of wages in the Zurich mills is, for a man, about 7½ francs, girls 4½ francs, and children 3 francs a week. (See also a table of the spinners' and weavers' gains in *Villermé's Tableau des Ouvriers*, i., 429.)

The other manufactures are not of any great importance. The woollen trade does not employ 300 hands, and the linen manufacture is now almost wholly extinguished. The imports of Zurich mainly consist of cotton and cotton yarn (Nos. 80 to 150), woollen cloths, colonial products, bark, straw hats, linens, furs, glass, stationery; wheat, principally from Swabia; wine, brandy, fruits, tobacco, fir-wood, raw silk, butter and cheese, minerals, &c. The exports are cotton cloths, particularly Turkey reds; silk goods, chiefly plain; machinery, tanned leather, kirschwasser, and sometimes an excess of agricultural produce to the neighbouring districts. (*Bowring's Rep.*, p. 77.)

"Most of the families in Zurich canton, consisting of father and mother and two or three children, earn among them, or possess in the produce of their land, an income fully equal to 20s. a week in England. The working classes are, compared with those of England, more moral, and better educated. With regard to education, the law compels it, and consequently there are scarcely any persons to be found who cannot read, and very few who cannot write. Music is much cultivated in this canton; and the whole demeanour and appearance of the working classes present a most gratifying picture of high prosperity, contentment, morality, and intelligence. Few cantons are really more flourishing: the entire poor rates a few years since was only 2½d. per head per annum." (*Symons in Hand-loom Rep.*) In point of fact, however, the state of things is mainly to be ascribed to the extreme economy of the people, a consequence, in part, of severe sumptuary laws, and to their avoiding all superfluous expenditure.

The constitution of Zurich underwent a great change in 1831. The cantonal assembly, or greater council, still consists, as formerly, of 212 members; but instead of 130 being elected by the grand council itself, only 32 are now so nominated, the remainder being chosen by the different guilds, and the population at large. Every male above the age of 19, not a domestic, a bankrupt, a recipient of public relief, or under penal condemnation, has a right to vote in the election of representatives: citizens must, however, be 30 years of age to sit in the chamber. The members of the greater council are elected for four years; but half their number goes out biennially. By the new constitution, the executive and judicial powers, formerly united in the same individuals, are separated; the functions of the former are exercised by a body of 19 members chosen by the greater council, and those of the latter by a high court of appeal composed of 11 members, a criminal court of primary jurisdiction in Zurich, district courts, &c. The proceedings of the council and of the law courts are public; freedom of trade and of the press is guaranteed; and each individual contributes to the exigencies of the state in proportion to his income. (*Meyer, Kant. Zurich.*)

The cantonal government compels a general system of insurance against fire, being itself the insurer. Public revenue in 1834, 1,333,380 Swiss francs; expenditure, 1,291,483 francs. Zurich has no public debt. This canton contributes 3858 men to the army, and 77,153 francs a year to the treasury of the Swiss confederation. (*Picot, Statist. de la Suisse.*) It is divided into 11 districts, but it has no town, except its capital, deserving of notice.

ZURICH (an. *Turicum*), a town of Switzerland, cap. of the above canton, and alternately with Bern and Lucerne the seat of the confederation government, on the Limmat, at

its efflux from the N.W. extremity of the lake of Zurich, 58½ m. N.E. Bern. Lat. of the observatory, 47° 22′ 21″ N., long. 26° 31′ 30″ E. Pop. in 1833, 11,326. It is beautifully situated, the river dividing it into two parts, which are connected by three bridges; and considerable improvements are going on in the town. It has some fine public walks, but few public buildings are worth notice. The principal are the cathedral, a massive edifice of the 10th or 11th century, in which Zwinglius denounced, though in comparatively mild and measured terms, the errors of the church of Rome, and enforced the principles of the Reformation; St. Peter's church, of which Lavater was the minister; the town hall, a square edifice, in which the diet meets; the old arsenal; the town library, a spacious edifice, containing about 60,000 volumes, with portraits of Zwinglius and many of the burgomasters of Zurich, a bust of Lavater by Danecker, a bas-relief model of a great part of Switzerland, a collection of fossils, &c.; and, in the middle of the Limmat, the town of Wollenberg, formerly a state prison.

The principal manufactures are those of silk and cotton goods; and numerous factories and country houses stud the banks of the lake in the environs. "In Zurich," says Inglis, "it is all work and no play; there are no amusements of any kind, nor probably do the inhabitants feel the want of them. There is no theatre; there are no public concerts: balls, in a canton where leave to dance must be asked, are out of the question. The great object of the Zurickers is to get money; distinction in wealth is the chief distinction of rank known in Zurich. Learning, however, has kept its place here; and teachers, perhaps, in Europe is the study of the classics more general than in this city. Here are an academy for theology and various other branches of philosophy; another academy preparatory to the former; an institution for medicine and surgery; another for the education of merchants; an institution for the instruction of the deaf and dumb, and for the blind, the model of which was considered so excellent that upon it Napoleon formed that of Paris; academies of sculpture and music; a society of public utility; and many schools for instruction in languages and for the education of the poor. Two newspapers are published in Zurich, one appearing weekly, the other twice a week; and there is also a monthly literary journal. If houses and ways out of the question, one might live cheaply enough at Zurich or in its neighbourhood. Beef usually sells at about 3d. per lb., mutton at 2½d., and veal 1d. higher: fowls average 1s. 6d. a pair, butter 7d. per lb., and eggs two dozen for 6d. But the reasonable price at which most necessaries may be obtained in Zurich is more than counterbalanced by the high rent of houses, for which indeed at least three times the rent asked that would command the same accommodation in England! and to those desirous of selecting a remotely agreeable residence I dare not recommend Zurich. A visitor's residence could not be otherwise than dull in a city where amusement is confounded with crime, and where men and women do not meet each other in society." (*Inglis's Switzerland.*) There is, however, a museum club, with a good reading-room. Where the leading English newspapers and periodical publications are taken in, a perpetual communication is kept up by diligences with Basle, Bern, Constance, and the other chief Swiss towns; and by steamboat twice a day with places on the bank of the lake. Zurich was one of the earliest cities that joined the Swiss confederation; and here the Reformation in Switzerland commenced, under Zwinglius, in 1519. Among its distinguished natives have been the two Gesners, Zimmermann, Fuseli, Lavater, Bodmer, and Pestalozzi. (*Ehel; Inglis; Diet. Geog. &c.*)

ZURICH (LAKE OF), one of the principal lakes of Switzerland, in the E. part of which it is situated, being bounded by the cantons of Zurich, Schwytz, and St. Gall. It curves in a semicircular manner, from S.E. bend to N.W. Length, about 24 m.; breadth, varying to about 3 m.; but at Rapperschwyl it is contracted to less than ½ m., and is crossed there by a wooden bridge. Area estimated at about 23 sq. m.; height above the sea, 1326 English feet. Its depth in some places exceeds 600 ft.; but for several hundred yards from its banks it is (near Zurich at least) seldom more than from 5 to 19 ft. in depth. At an alt.

---

* Zwinglius, or rather Zuingle, was born January 1, 1484, at Wildhaus, a small village in the Tockenburg. "Of all the Reformers," says Coxe, "the mild and elegant Melancthon alone excepted, Zuingle seems to me the most amiable. He possessed to a great degree the spirit of moderation, toleration, and charity which are the characteristics of true Christianity, and, amid all the disputes between the Lutherans and Calvinists, he was a constant advocate for peace and reconciliation. Even purged from narrow bigotry, which makes so illustrious between them of different religious persuasions, his religious opinions were the greatest importance, and free from that bigoted pride, which, while it violently condemns the opinions of all others, holds infallibility with respect to its own. In a word, it was his decided opinion that no disagreement on subjects less inconvertible, and which do not affect morals." (*Coxe's Switzerland*, i., 83, 8vo ed.)

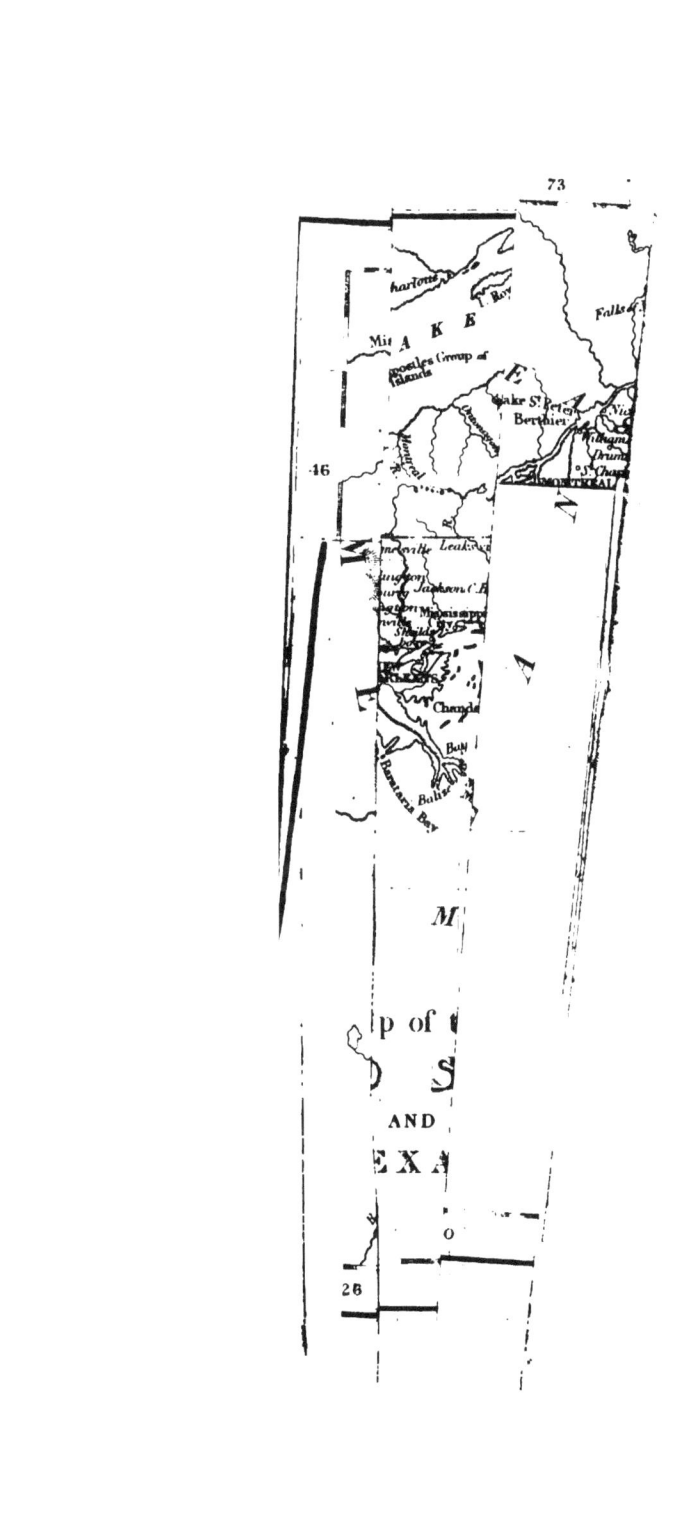

extremity it receives the Linth canal, which brings to it the superfluous waters of the lake Wallenstadt; at its N.W. extremity it discharges itself by the Limmat. Zurich, Meilen, Rapperschwyl, and Richtenschwyl, are on its banks. This lake has none of that savage sublimity which characterizes most of the Swiss lakes: its scenery is, in fact, comparatively tame. Inglis calls it "the Winandermere of Switzerland." The hills around it scarcely rise to 3000 ft. above the sea, and they descend in gentle and cultivated slopes to the water's edge; where the banks, from one end of the lake to the other, are studded with villages, country houses, and other habitations. Good carriage roads run along both sides of the lake; and it is daily traversed by steamers between Zurich and Rapperschwyl. (*Ebel; Picot; Inglis's Switzerland; Stein's Handb.; Mayer, Du Canton Zurich; Murray's Handb. for Switzerland, &c.*)

ZUTPHEN, a fortified town of the Netherlands, prov. Guelderland, cap. arrond., on the Yssel, crossed here by a stone bridge, where it is joined by the Brickel, 15 m. N.E. Arnhem. Pop., according to Stein, about 10,000. It is strong by its situation, and, though in the midst of fens, is not considered unhealthy. It is divided by the Birckel into an old and a new town. The principal church is an old and stately edifice: the town-hall, the college of deputies, and the palace of the former counts of Zutphen, are the other most conspicuous buildings. Here, also, is a Latin school, a society of physical science, a court of primary jurisdiction, manufactures of cotton fabrics, with tanneries, paper and glue factories, oil and flour mills, &c.

Zutphen was one of the Hanse towns. It was taken and pillaged by the Spaniards in 1572 and 1583, but was retaken by the troops under Prince Maurice in 1591. In this siege the famous Sir Philip Sidney, the flower of the chivalry of Elizabeth's reign, received a wound of which he died on the 17th of October, at the early age of 32. (*De Cloet; Stein; Dict. Géog.*)

ZVORNIK, or ISVORNIK, a fortified town of Bosnia, cap. sandjak, on the Drin, 72 m. W.S.W. Belgrade. It is situated on a rocky height, and has two castles, and a large collection of mud houses, with several mosques, and Greek and Roman Catholic churches. Its pop. has been estimated at 15,000. It is said to have a considerable trade in timber and fuel with Belgrade, Semlin, &c.; but from its

lying out of any great road, it is very seldom visited by travellers from W. Europe; and our information respecting it (as, indeed, of the whole of Bosnia) is very imperfect.

ZWICKAU, a town of the kingdom of Saxony, cap. circ. of its own name, on the Mulda, 58 m. S.W. Dresden. Pop. in 1837, 7239. It was formerly strongly fortified, and suffered repeatedly in the wars of last century between Austria and Prussia. Its principal buildings are St. Mary's church, with some fine paintings by Wohlgemuth, and a lofty tower, which was often ascended by Luther; and an old castle, now used for a house of correction. The gymnasium has a library of 18,000 volumes (*B. Ritter*); and there are also some military storehouses, a hospital, &c. Zwickau has manufactures of woollen cloths, hosiery, cotton goods, and hardware; which she owes to the coal fields on both sides the Mulda, in her vicinity.

ZWOLLE, a fortified town of the Netherlands, province Overyssel, of which it is the cap., on the Zwarte-water, about 10 m. from the Zuyder-zee, and 50 m. E.N.E. Amsterdam. Pop. between 15,000 and 16,000. It is well built, in the style of most other Dutch towns, and has several suburbs, eight churches, including a fine old cathedral, a house of correction, tribunals of primary jurisdiction and commerce, and some agreeable promenades in the vicinity. It was formerly one of the Hanse towns, and its trade is still considerable in cattle and other live stock, dried fish, corn, wool, hides, honey, leather, &c. It has some salt and sugar refineries, tanneries, &c. It was taken by the Dutch in 1580. The famous Thomas-à-Kempis was, for 64 years, a monk of an Augustine priory in this town, where he died in 1471. (*De Cloet; Dict. Géog.; Murray's Handbook, &c.*)

ZYTOMIERS, or *Jitomir*, a town of Russian Poland, gov. Volhynia, of which it is the cap.; on a tributary of the Dniepr, 75 m. W.S.W. Kief. Pop. in 1838, 17,434. (*Berghaus.*) It has three Russo-Greek, a Lutheran, and two Roman Catholic churches, various government buildings, a gymnasium, seminary, public library, &c. It has increased greatly in importance since it came into the possession of the Russians; it has manufacture of hats, leather, &c., and a considerable trade in woollen, silk, and linen fabrics, honey, wax, salt, and wines, chiefly with Galicia, Hungary, and Wallachia. (*Schnitz.; Poss.; Berghaus, &c.*)

1109

THE END.

Lightning Source UK Ltd.
Milton Keynes UK
UKHW012227210119
335963UK00009B/222/P